SECOND EDITION

COMPUTER SCIENCE

HANDBOOK

EDITOR-IN-CHIEF
ALLEN B. TUCKER

CHAPMAN & HALL/CRC

Published in Cooperation with ACM, The Association for Computing Machinery

Library of Congress Cataloging-in-Publication Data

Computer science handbook / editor-in-chief, Allen B. Tucker—2nd ed.
    p. cm.
Includes bibliographical references and index.
ISBN 1-58488-360-X (alk. paper)
    1.Computer science-Handbooks, manuals, etc. 2. Engineering—Hanbooks, manuals, etc.
    I. Tucker, Allen B.

QA76.C54755 2004
004—dc22   2003068758

Visit the CRC Press Web site at www.crcpress.com

Preface to the Second Edition

Purpose

The purpose of *The Computer Science Handbook* is to provide a single comprehensive reference for computer scientists, software engineers, and IT professionals who wish to broaden or deepen their understanding in a particular subfield of computer science. Our goal is to provide the most current information in each of the following eleven subfields in a form that is accessible to students, faculty, and professionals in computer science:

> algorithms, architecture, computational science, graphics, human-computer interaction, information management, intelligent systems, net-centric computing, operating systems, programming languages, and software engineering.

Each of the eleven sections of the *Handbook* is dedicated to one of these subfields. In addition, the appendices provide useful information about professional organizations in computer science, standards, and languages. Different points of access to this rich collection of theory and practice are provided through the table of contents, two introductory chapters, a comprehensive subject index, and additional indexes.

A more complete overview of this *Handbook* can be found in Chapter 1, which summarizes the contents of each of the eleven sections. This chapter also provides a history of the evolution of computer science during the last 50 years, as well as its current status, and future prospects.

New Features

Since the first edition of the *Handbook* was published in 1997, enormous changes have taken place in the discipline of computer science. The goals of the second edition of the *Handbook* are to incorporate these changes by:

1. Broadening its reach across all 11 subject areas of the discipline, as they are defined in *Computing Curricula 2001* (the new standard taxonomy)
2. Including a heavier proportion of applied computing subject matter
3. Bringing up to date all the topical discussions that appeared in the first edition

This new edition was developed by the editor-in-chief and three editorial advisors, whereas the first edition was developed by the editor and ten advisors. Each edition represents the work of over 150 contributing authors who are recognized as experts in their various subfields of computer science.

Readers who are familiar with the first edition will notice the addition of many new chapters, reflecting the rapid emergence of new areas of research and applications since the first edition was published. Especially exciting are the addition of new chapters in the areas of computational science, information management, intelligent systems, net-centric computing, and software engineering. These chapters explore topics like cryptography, computational chemistry, computational astrophysics, human-centered software development, cognitive modeling, transaction processing, data compression, scripting languages, multimedia databases, event-driven programming, and software architecture.

Acknowledgments

A work of this magnitude cannot be completed without the efforts of many individuals. During the 2-year process that led to the first edition, I had the pleasure of knowing and working with ten very distinguished, talented, and dedicated editorial advisors:

Harold Abelson (MIT), Mikhail Atallah (Purdue), Keith Barker (Uconn), Kim Bruce (Williams), John Carroll (VPI), Steve Demurjian (Uconn), Donald House (Texas A&M), Raghu Ramakrishnan (Wisconsin), Eugene Spafford (Purdue), Joe Thompson (Mississippi State), and Peter Wegner (Brown).

For this edition, a new team of trusted and talented editorial advisors helped to reshape and revitalize the *Handbook* in valuable ways:

Robert Cupper (Allegheny), Fadi Deek (NJIT), Robert Noonan (William and Mary)

All of these persons provided valuable insights into the substantial design, authoring, reviewing, and production processes throughout the first eight years of this *Handbook*'s life, and I appreciate their work very much.

Of course, it is the chapter authors who have shared in these pages their enormous expertise across the wide range of subjects in computer science. Their hard work in preparing and updating their chapters is evident in the very high quality of the final product. The names of all chapter authors and their current professional affiliations are listed in the contributor list.

I want also to thank Bowdoin College for providing institutional support for this work. Personal thanks go especially to Craig McEwen, Sue Theberge, Matthew Jacobson-Carroll, Alice Morrow, and Aaron Olmstead at Bowdoin, for their various kinds of support as this project has evolved over the last eight years. Bob Stern, Helena Redshaw, Joette Lynch, and Robert Sims at CRC Press also deserve thanks for their vision, perseverance and support throughout this period.

Finally, the greatest thanks is always reserved for my wife Meg – my best friend and my love – for her eternal influence on my life and work.

Allen B. Tucker
Brunswick, Maine

Editor-in-Chief

 **Allen B. Tucker** is the Anne T. and Robert M. Bass Professor of Natural Sciences in the Department of Computer Science at Bowdoin College, where he has taught since 1988. Prior to that, he held similar positions at Colgate and Georgetown Universities. Overall, he has served eighteen years as a department chair and two years as an associate dean. At Colgate, he held the John D. and Catherine T. MacArthur Chair in Computer Science.

Professor Tucker earned a B.A. in mathematics from Wesleyan University in 1963 and an M.S. and Ph.D. in computer science from Northwestern University in 1970. He is the author or coauthor of several books and articles in the areas of programming languages, natural language processing, and computer science education. He has given many talks, panel discussions, and workshop presentations in these areas, and has served as a reviewer for various journals, NSF programs, and curriculum projects. He has also served as a consultant to colleges, universities, and other institutions in the areas of computer science curriculum, software design, programming languages, and natural language processing applications.

A Fellow of the ACM, Professor Tucker co-authored the 1986 Liberal Arts Model Curriculum in Computer Science and co-chaired the ACM/IEEE-CS Joint Curriculum Task Force that developed Computing Curricula 1991. For these and other related efforts, he received the ACM's 1991 Outstanding Contribution Award, shared the IEEE's 1991 Meritorious Service Award, and received the ACM SIGCSE's 2001 Award for Outstanding Contributions to Computer Science Education. In Spring 2001, he was a Fulbright Lecturer at the Ternopil Academy of National Economy (TANE) in Ukraine. Professor Tucker has been a member of the ACM, the NSF CISE Advisory Committee, the IEEE Computer Society, Computer Professionals for Social Responsibility, and the Liberal Arts Computer Science (LACS) Consortium.

Contributors

Eric W. Allender
Rutgers University

James L. Alty
Loughborough University

Thomas E. Anderson
University of Washington

M. Pauline Baker
National Center for
   Supercomputing
   Applications

Steven Bellovin
AT&T Research Labs

Andrew P. Bernat
Computer Research
   Association

Brian N. Bershad
University of Washington

Christopher M. Bishop
Microsoft Research

Guy E. Blelloch
Carnegie Mellon University

Philippe Bonnet
University of Copenhagen

Jonathan P. Bowen
London South Bank University

Kim Bruce
Williams College

Steve Bryson
NASA Ames Research Center

Douglas C. Burger
University of Wisconsin
   at Madison

Colleen Bushell
National Center for
   Supercomputing
   Applications

Derek Buzasi
U.S. Air Force Academy

William L. Bynum
College of William and Mary

Bryan M. Cantrill
Sun Microsystems, Inc.

Luca Cardelli
Microsoft Research

David A. Caughy
Cornell University

Vijay Chandru
Indian Institute of Science

Steve J. Chapin
Syracuse University

Eric Chown
Bowdoin College

Jacques Cohen
Brandeis University

J.L. Cox
Brooklyn College, CUNY

Alan B. Craig
National Center for
   Supercomputing
   Applications

Maxime Crochemore
University of Marne-la-Vallée
   and King's College London

Robert D. Cupper
Allegheny College

Thomas Dean
Brown Univeristy

Fadi P. Deek
New Jersey Institute
   of Technology

Gerald DeJong
University of Illinois at
   Urbana-Champaign

Steven A. Demurjian Sr.
University of Connecticut

Peter J. Denning
Naval Postgraduate School

Angel Diaz
IBM Research

T.W. Doeppner Jr.
Brown University

Henry Donato
College of Charleston

Chitra Dorai
IBM T.J. Watson
Research Center

Wolfgang Dzida
Pro Context GmbH

David S. Ebert
Purdue University

Raimund Ege
Florida International
University

Osama Eljabiri
New Jersey Institute
of Technology

David Ferbrache
U.K. Ministry of Defence

Raphael Finkel
University of Kentucky

John M. Fitzgerald
Adept Technology

Michael J. Flynn
Stanford University

Kenneth D. Forbus
Northwestern University

Stephanie Forrest
University of New Mexico

Michael J. Franklin
University of California
at Berkeley

John D. Gannon
University of Maryland

Carlo Ghezzi
Politecnico di Milano

Benjamin Goldberg
New York University

James R. Goodman
University of Wisconsin
at Madison

Jonathan Grudin
Microsoft Research

Gamil A. Guirgis
College of Charleston

Jon Hakkila
College of Charleston

Sandra Harper
College of Charleston

Frederick J. Heldrich
College of Charleston

Katherine G. Herbert
New Jersey Institute
of Technology

Michael G. Hinchey
NASA Goddard Space
Flight Center

Ken Hinckley
Microsoft Research

Donald H. House
Texas A&M University

Windsor W. Hsu
IBM Research

Daniel Huttenlocher
Cornell University

Yannis E. Ioannidis
University of Wisconsin

Robert J.K. Jacob
Tufts University

Sushil Jajodia
George Mason University

Mehdi Jazayeri
Technical University of Vienna

Tao Jiang
University of California

Michael J. Jipping
Hope College

Deborah G. Johnson
University of Virginia

Michael I. Jordan
University of California
at Berkeley

David R. Kaeli
Northeastern University

Erich Kaltofen
North Carolina State University

Subbarao Kambhampati
Arizona State University

Lakshmi Kantha
University of Colorado

Gregory M. Kapfhammer
Allegheny College

Jonathan Katz
University of Maryland

Arie Kaufman
State University of New York
at Stony Brook

Samir Khuller
University of Maryland

David Kieras
University of Michigan

David T. Kingsbury
Gordon and Betty Moore
Foundation

Danny Kopec
Brooklyn College, CUNY

Henry F. Korth
Lehigh University

Kristin D. Krantzman
College of Charleston

Edward D. Lazowska
University of Washington

Thierry Lecroq
University of Rouen

D.T. Lee
Northwestern University

Miriam Leeser
Northeastern University

Henry M. Levy
University of Washington

Frank L. Lewis
University of Texas at Arlington

Ming Li
University of Waterloo

Ying Li
IBM T.J. Watson
    Research Center

Jianghui Liu
New Jersey Institute
    of Technology

Kai Liu
Alcatel Telecom

Kenneth C. Louden
San Jose State University

Michael C. Loui
University of Illinois at
    Urbana-Champaign

James J. Lu
Emory University

Abby Mackness
Booz Allen Hamilton

Steve Maddock
University of Sheffield

Bruce M. Maggs
Carnegie Mellon University

Dino Mandrioli
Politecnico di Milano

M. Lynne Markus
Bentley College

Tony A. Marsland
University of Alberta

Edward J. McCluskey
Stanford University

James A. McHugh
New Jersey Institute
    of Technology

Marshall Kirk McKusick
Consultant

Clyde R. Metz
College of Charleston

Keith W. Miller
University of Illinois

Subhasish Mitra
Stanford University

Stuart Mort
U.K. Defence and Evaluation
    Research Agency

Rajeev Motwani
Stanford University

Klaus Mueller
State University of New York
    at Stony Brook

Sape J. Mullender
Lucent Technologies

Brad A. Myers
Carnegie Mellon University

Peter G. Neumann
SRI International

Jakob Nielsen
Nielsen Norman Group

Robert E. Noonan
College of William and Mary

Ahmed K. Noor
Old Dominion University

Vincent Oria
New Jersey Institute
    of Technology

Jason S. Overby
College of Charleston

M. Tamer Özsu
University of Waterloo

Victor Y. Pan
Lehman College, CUNY

Judea Pearl
University of California
    at Los Angeles

Jih-Kwon Peir
University of Florida

Radia Perlman
Sun Microsystems Laboratories

Patricia Pia
University of Connecticut

Steve Piacsek
Naval Research Laboratory

Roger S. Pressman
R.S. Pressman & Associates,
    Inc.

J. Ross Quinlan
University of New South Wales

Balaji Raghavachari
University of Texas at Dallas

Prabhakar Raghavan
Verity, Inc.

Z. Rahman
College of William and Mary

M.R. Rao
Indian Institute of
Management

Bala Ravikumar
University of Rhode Island

Kenneth W. Regan
State University of New York
at Buffalo

Edward M. Reingold
Illinois Institute of Technology

Alyn P. Rockwood
Colorado School of Mines

Robert S. Roos
Allegheny College

Erik Rosenthal
University of New Haven

Kevin W. Rudd
Intel, Inc.

Betty Salzberg
Northeastern University

Pierangela Samarati
Universitá degli Studi di
Milano

Ravi S. Sandhu
George Mason University

David A. Schmidt
Kansas State University

Stephen B. Seidman
New Jersey Institute
of Technology

Stephanie Seneff
Massachusetts Institute
of Technology

J.S. Shang
Air Force Research

Dennis Shasha
Courant Institute
New York University

William R. Sherman
National Center for
Supercomputing
Applications

Avi Silberschatz
Yale University

Gurindar S. Sohi
University of Wisconsin
at Madison

Ian Sommerville
Lancaster University

Bharat K. Soni
Mississippi State University

William Stallings
Consultant and Writer

John A. Stankovic
University of Virginia

S. Sudarshan
IIT Bombay

Earl E. Swartzlander Jr.
University of Texas at Austin

Roberto Tamassia
Brown University

Patricia J. Teller
University of Texas at ElPaso

Robert J. Thacker
McMaster University

Nadia Magnenat Thalmann
University of Geneva

Daniel Thalmann
Swiss Federal Institute of
Technology (EPFL)

Alexander Thomasian
New Jersey Institute of
Technology

Allen B. Tucker
Bowdoin College

Jennifer Tucker
Booz Allen Hamilton

Patrick Valduriez
INRIA and IRIN

Jason T.L. Wang
New Jersey Institute
of Technology

Colin Ware
University of New Hampshire

Alan Watt
University of Sheffield

Nigel P. Weatherill
University of Wales Swansea

Peter Wegner
Brown University

Jon B. Weissman
University of Minnesota-Twin
Cities

Craig E. Wills
Worcester Polytechnic
Institute

George Wolberg
City College of New York

Donghui Zhang
Northeastern University

Victor Zue
Massachusetts Institute
of Technology

Contents

1 Computer Science: The Discipline and its Impact
Allen B. Tucker and Peter Wegner 1-1

2 Ethical Issues for Computer Scientists
Deborah G. Johnson and Keith W. Miller 2-1

Section I: Algorithms and Complexity

3 Basic Techniques for Design and Analysis of Algorithms
Edward M. Reingold 3-1

4 Data Structures
Roberto Tamassia and Bryan M. Cantrill 4-1

5 Complexity Theory
Eric W. Allender, Michael C. Loui, and Kenneth W. Regan 5-1

6 Formal Models and Computability
Tao Jiang, Ming Li, and Bala Ravikumar 6-1

7 Graph and Network Algorithms
Samir Khuller and Balaji Raghavachari 7-1

8 Algebraic Algorithms
Angel Diaz, Erich Kaltófen, and Victor Y. Pan 8-1

9 Cryptography
Jonathan Katz 9-1

10 Parallel Algorithms
Guy E. Blelloch and Bruce M. Maggs 10-1

11 Computational Geometry
D. T. Lee 11-1

12 Randomized Algorithms
Rajeev Motwani and Prabhakar Raghavan **12**-1

13 Pattern Matching and Text Compression Algorithms
Maxime Crochemore and Thierry Lecroq **13**-1

14 Genetic Algorithms
Stephanie Forrest **14**-1

15 Combinatorial Optimization
Vijay Chandru and M. R. Rao **15**-1

Section II: Architecture and Organization

16 Digital Logic
Miriam Leeser **16**-1

17 Digital Computer Architecture
David R. Kaeli **17**-1

18 Memory Systems
Douglas C. Burger, James R. Goodman, and Gurindar S. Sohi **18**-1

19 Buses
Windsor W. Hsu and Jih-Kwon Peir **19**-1

20 Input/Output Devices and Interaction Techniques
Ken Hinckley, Robert J. K. Jacob, and Colin Ware **20**-1

21 Secondary Storage Systems
Alexander Thomasian **21**-1

22 High-Speed Computer Arithmetic
Earl E. Swartzlander Jr. **22**-1

23 Parallel Architectures
Michael J. Flynn and Kevin W. Rudd **23**-1

24 Architecture and Networks
Robert S. Roos **24**-1

25 Fault Tolerance
Edward J. McCluskey and Subhasish Mitra **25**-1

Section III: Computational Science

26 Geometry-Grid Generation
    Bharat K. Soni and Nigel P. Weatherill **26**-1

27 Scientific Visualization
    William R. Sherman, Alan B. Craig, M. Pauline Baker, and Colleen Bushell **27**-1

28 Computational Structural Mechanics
    Ahmed K. Noor **28**-1

29 Computational Electromagnetics
    J. S. Shang **29**-1

30 Computational Fluid Dynamics
    David A. Caughey **30**-1

31 Computational Ocean Modeling
    Lakshmi Kantha and Steve Piacsek **31**-1

32 Computational Chemistry
    Frederick J. Heldrich, Clyde R. Metz, Henry Donato, Kristin D. Krantzman,
    Sandra Harper, Jason S. Overby, and Gamil A. Guirgis **32**-1

33 Computational Astrophysics
    Jon Hakkila, Derek Buzasi, and Robert J. Thacker **33**-1

34 Computational Biology
    David T. Kingsbury **34**-1

Section IV: Graphics and Visual Computing

35 Overview of Three-Dimensional Computer Graphics
    Donald H. House **35**-1

36 Geometric Primitives
    Alyn P. Rockwood **36**-1

37 Advanced Geometric Modeling
    David S. Ebert **37**-1

38 Mainstream Rendering Techniques
    Alan Watt and Steve Maddock **38**-1

39 Sampling, Reconstruction, and Antialiasing
    George Wolberg **39**-1

40 Computer Animation
Nadia Magnenat Thalmann and Daniel Thalmann **40**-1

41 Volume Visualization
Arie Kaufman and Klaus Mueller **41**-1

42 Virtual Reality
Steve Bryson **42**-1

43 Computer Vision
Daniel Huttenlocher **43**-1

Section V: Human-Computer Interaction

44 The Organizational Contexts of Development and Use
Jonathan Grudin and M. Lynne Markus **44**-1

45 Usability Engineering
Jakob Nielsen **45**-1

46 Task Analysis and the Design of Functionality
David Kieras **46**-1

47 Human-Centered System Development
Jennifer Tucker and Abby Mackness **47**-1

48 Graphical User Interface Programming
Brad A. Myers **48**-1

49 Multimedia
James L. Alty **49**-1

50 Computer-Supported Collaborative Work
Fadi P. Deek and James A. McHugh **50**-1

51 Applying International Usability Standards
Wolfgang Dzida **51**-1

Section VI: Information Management

52 Data Models
Avi Silberschatz, Henry F. Korth, and S. Sudarshan **52**-1

53 Tuning Database Design for High Performance
Dennis Shasha and Philippe Bonnet **53**-1

54 Access Methods
Betty Salzberg and Donghui Zhang..54-1

55 Query Optimization
Yannis E. Ioannidis..55-1

56 Concurrency Control and Recovery
Michael J. Franklin..56-1

57 Transaction Processing
Alexander Thomasian..57-1

58 Distributed and Parallel Database Systems
M. Tamer Özsu and Patrick Valduriez..58-1

59 Multimedia Databases: Analysis, Modeling, Querying, and Indexing
Vincent Oria, Ying Li, and Chitra Dorai..59-1

60 Database Security and Privacy
Sushil Jajodia..60-1

Section VII: Intelligent Systems

61 Logic-Based Reasoning for Intelligent Systems
James J. Lu and Erik Rosenthal..61-1

62 Qualitative Reasoning
Kenneth D. Forbus..62-1

63 Search
D. Kopec, T.A. Marsland, and J.L. Cox..63-1

64 Understanding Spoken Language
Stephanie Seneff and Victor Zue..64-1

65 Decision Trees and Instance-Based Classifiers
J. Ross Quinlan..65-1

66 Neural Networks
Michael I. Jordan and Christopher M. Bishop..66-1

67 Planning and Scheduling
Thomas Dean and Subbarao Kambhampati..67-1

68 Explanation-Based Learning
Gerald DeJong..68-1

69 Cognitive Modeling
Eric Chown..69-1

70 Graphical Models for Probabilistic and Causal Reasoning
    Judea Pearl **70**-1

71 Robotics
    Frank L. Lewis, John M. Fitzgerald, and Kai Liu **71**-1

Section VIII: Net-Centric Computing

72 Network Organization and Topologies
    William Stallings **72**-1

73 Routing Protocols
    Radia Perlman **73**-1

74 Network and Internet Security
    Steven Bellovin **74**-1

75 Information Retrieval and Data Mining
    Katherine G. Herbert, Jason T.L. Wang, and Jianghui Liu **75**-1

76 Data Compression
    Z. Rahman **76**-1

77 Security and Privacy
    Peter G. Neumann **77**-1

78 Malicious Software and Hacking
    David Ferbrache and Stuart Mort **78**-1

79 Authentication, Access Control, and Intrusion Detection
    Ravi S. Sandhu and Pierangela Samarati **79**-1

Section IX: Operating Systems

80 What Is an Operating System?
    Raphael Finkel **80**-1

81 Thread Management for Shared-Memory Multiprocessors
    Thomas E. Anderson, Brian N. Bershad, Edward D. Lazowska,
    and Henry M. Levy **81**-1

82 Process and Device Scheduling
    Robert D. Cupper **82**-1

83 Real-Time and Embedded Systems
    John A. Stankovic **83**-1

84 Process Synchronization and Interprocess Communication
  Craig E. Wills **84**-1

85 Virtual Memory
  Peter J. Denning **85**-1

86 Secondary Storage and Filesystems
  Marshall Kirk McKusick **86**-1

87 Overview of Distributed Operating Systems
  Sape J. Mullender **87**-1

88 Distributed and Multiprocessor Scheduling
  Steve J. Chapin and Jon B. Weissman **88**-1

89 Distributed File Systems and Distributed Memory
  T. W. Doeppner Jr. **89**-1

Section X: Programming Languages

90 Imperative Language Paradigm
  Michael J. Jipping and Kim Bruce **90**-1

91 The Object-Oriented Language Paradigm
  Raimund Ege **91**-1

92 Functional Programming Languages
  Benjamin Goldberg **92**-1

93 Logic Programming and Constraint Logic Programming
  Jacques Cohen **93**-1

94 Scripting Languages
  Robert E. Noonan and William L. Bynum **94**-1

95 Event-Driven Programming
  Allen B. Tucker and Robert E. Noonan **95**-1

96 Concurrent/Distributed Computing Paradigm
  Andrew P. Bernat and Patricia Teller **96**-1

97 Type Systems
  Luca Cardelli **97**-1

98 Programming Language Semantics
  David A. Schmidt **98**-1

99 Compilers and Interpreters
Kenneth C. Louden 99-1

100 Runtime Environments and Memory Management
Robert E. Noonan and William L. Bynum 100-1

Section XI: Software Engineering

101 Software Qualities and Principles
Carlo Ghezzi, Mehdi Jazayeri, and Dino Mandrioli 101-1

102 Software Process Models
Ian Sommerville 102-1

103 Traditional Software Design
Steven A. Demurjian Sr. 103-1

104 Object-Oriented Software Design
Steven A. Demurjian Sr. and Patricia J. Pia 104-1

105 Software Testing
Gregory M. Kapfhammer 105-1

106 Formal Methods
Jonathan P. Bowen and Michael G. Hinchey 106-1

107 Verification and Validation
John D. Gannon 107-1

108 Development Strategies and Project Management
Roger S. Pressman 108-1

109 Software Architecture
Stephen B. Seidman 109-1

110 Specialized System Development
Osama Eljabiri and Fadi P. Deek 110-1

Appendix A: Professional Societies in Computing A-1

Appendix B: The ACM Code of Ethics and Professional Conduct B-1

Appendix C: Standards-Making Bodies and Standards C-1

Appendix D: Common Languages and Conventions D-1

Index Index-1

SECOND EDITION

COMPUTER SCIENCE

HANDBOOK

1

Computer Science: The Discipline and its Impact

1.1   Introduction **1**-1
1.2   Growth of the Discipline and the Profession **1**-2
      Curriculum Development • Growth of Academic Programs
      • Academic R&D and Industry Growth
1.3   Perspectives in Computer Science **1**-6
1.4   Broader Horizons: From HPCC
      to Cyberinfrastructure **1**-7
1.5   Organization and Content 1-10
      Algorithms and Complexity • Architecture • Computational
      Science • Graphics and Visual Computing • Human–Computer
      Interaction • Information Management • Intelligent Systems
      • Net-Centric Computing • Operating Systems • Programming
      Languages • Software Engineering
1.6   Conclusion 1-15

Allen B. Tucker
Bowdoin College

Peter Wegner
Brown University

1.1 Introduction

The field of computer science has undergone a dramatic evolution in its short 70-year life. As the field has matured, new areas of research and applications have emerged and joined with classical discoveries in a continuous cycle of revitalization and growth.

In the 1930s, fundamental mathematical principles of computing were developed by Turing and Church. Early computers implemented by von Neumann, Wilkes, Eckert, Atanasoff, and others in the 1940s led to the birth of scientific and commercial computing in the 1950s, and to mathematical programming languages like Fortran, commercial languages like COBOL, and artificial-intelligence languages like LISP. In the 1960s the rapid development and consolidation of the subjects of algorithms, data structures, databases, and operating systems formed the core of what we now call traditional computer science; the 1970s saw the emergence of software engineering, structured programming, and object-oriented programming. The emergence of personal computing and networks in the 1980s set the stage for dramatic advances in computer graphics, software technology, and parallelism. The 1990s saw the worldwide emergence of the Internet, both as a medium for academic and scientific exchange and as a vehicle for international commerce and communication.

This Handbook aims to characterize computer science in the new millenium, incorporating the explosive growth of the Internet and the increasing importance of subject areas like human–computer interaction, massively parallel scientific computation, ubiquitous information technology, and other subfields that would not have appeared in such an encyclopedia even ten years ago. We begin with the following short definition, a variant of the one offered in [Gibbs 1986], which we believe captures the essential nature of "computer science" as we know it today.

> *Computer science* is the study of computational processes and information structures, including their hardware realizations, their linguistic models, and their applications.

The Handbook is organized into eleven sections which correspond to the eleven major subject areas that characterize computer science [ACM/IEEE 2001], and thus provide a useful modern taxonomy for the discipline. The next section presents a brief history of the computing industry and the parallel development of the computer science curriculum. Section 1.3 frames the practice of computer science in terms of four major conceptual paradigms: theory, abstraction, design, and the social context. Section 1.4 identifies the "grand challenges" of computer science research and the subsequent emergence of information technology and cyber-infrastructure that may provide a foundation for addressing these challenges during the next decade and beyond. Section 1.5 summarizes the subject matter in each of the Handbook's eleven sections in some detail.

This Handbook is designed as a professional reference for researchers and practitioners in computer science. Readers interested in exploring specific subject topics may prefer to move directly to the appropriate section of the Handbook — the chapters are organized with minimal interdependence, so that they can be read in any order. To facilitate rapid inquiry, the Handbook contains a Table of Contents and three indexes (Subject, Who's Who, and Key Algorithms and Formulas), providing access to specific topics at various levels of detail.

1.2   Growth of the Discipline and the Profession

The computer industry has experienced tremendous growth and change over the past several decades. The transition that began in the 1980s, from centralized mainframes to a decentralized networked microcomputer–server technology, was accompanied by the rise and decline of major corporations. The old monopolistic, vertically integrated industry epitomized by IBM's comprehensive client services gave way to a highly competitive industry in which the major players changed almost overnight. In 1992 alone, emergent companies like Dell and Microsoft had spectacular profit gains of 77% and 53%. In contrast, traditional companies like IBM and Digital suffered combined record losses of $7.1 billion in the same year [Economist 1993] (although IBM has since recovered significantly). As the 1990s came to an end, this euphoria was replaced by concerns about new monopolistic behaviors, expressed in the form of a massive antitrust lawsuit by the federal government against Microsoft. The rapid decline of the "dot.com" industry at the end of the decade brought what many believe a long-overdue rationality to the technology sector of the economy. However, the exponential decrease in computer cost and increase in power by a factor of two every 18 months, known as Moore's law, shows no signs of abating in the near future, although underlying physical limits will eventually be reached.

Overall, the rapid 18% annual growth rate that the computer industry had enjoyed in earlier decades gave way in the early 1990s to a 6% growth rate, caused in part by a saturation of the personal computer market. Another reason for this slowing of growth is that the performance of computers (speed, storage capacity) has improved at a rate of 30% per year in relation to their cost. Today, it is not unusual for a laptop or hand-held computer to run at hundreds of times the speed and capacity of a typical computer of the early 1990s, and at a fraction of its cost. However, it is not clear whether this slowdown represents a temporary plateau or whether a new round of fundamental technical innovations in areas such as parallel architectures, nanotechnology, or human–computer interaction might generate new spectacular rates of growth in the future.

1.2.1 Curriculum Development

The computer industry's evolution has always been affected by advances in both the theory and the practice of computer science. Changes in theory and practice are simultaneously intertwined with the evolution of the field's undergraduate and graduate curricula, which have served to define the intellectual and methodological framework for the discipline of computer science itself.

The first coherent and widely cited curriculum for computer science was developed in 1968 by the ACM Curriculum Committee on Computer Science [ACM 1968] in response to widespread demand for systematic undergraduate and graduate programs [Rosser 1966]. "Curriculum 68" defined computer science as comprising three main areas: information structures and processes, information processing systems, and methodologies. Curriculum 68 defined computer science as a discipline and provided concrete recommendations and guidance to colleges and universities in developing undergraduate, master's, and doctorate programs to meet the widespread demand for computer scientists in research, education, and industry. Curriculum 68 stood as a robust and exemplary model for degree programs at all levels for the next decade.

In 1978, a new ACM Curriculum Committee on Computer Science developed a revised and updated undergraduate curriculum [ACM 1978]. The "Curriculum 78" report responded to the rapid evolution of the discipline and the practice of computing, and to a demand for a more detailed elaboration of the computer science (as distinguished from the mathematical) elements of the courses that would comprise the core curriculum.

During the next few years, the IEEE Computer Society developed a model curriculum for engineering-oriented undergraduate programs [IEEE-CS 1976], updated and published it in 1983 as a "Model Program in Computer Science and Engineering" [IEEE-CS 1983], and later used it as a foundation for developing a new set of accreditation criteria for undergraduate programs. A simultaneous effort by a different group resulted in the design of a model curriculum for computer science in liberal arts colleges [Gibbs 1986]. This model emphasized science and theory over design and applications, and it was widely adopted by colleges of liberal arts and sciences in the late 1980s and the 1990s.

In 1988, the ACM Task Force on the Core of Computer Science and the IEEE Computer Society [ACM 1988] cooperated in developing a fundamental redefinition of the discipline. Called "Computing as a Discipline," this report aimed to provide a contemporary foundation for undergraduate curriculum design by responding to the changes in computing research, development, and industrial applications in the previous decade. This report also acknowledged some fundamental methodological changes in the field. The notion that "computer science = programming" had become wholly inadequate to encompass the richness of the field. Instead, three different paradigms—called *theory, abstraction,* and *design*—were used to characterize how various groups of computer scientists did their work. These three points of view — those of the theoretical mathematician or scientist (theory), the experimental or applied scientist (abstraction, or modeling), and the engineer (design) — were identified as essential components of research and development across all nine subject areas into which the field was then divided.

"Computing as a Discipline" led to the formation of a joint ACM/IEEE-CS Curriculum Task Force, which developed a more comprehensive model for undergraduate curricula called "Computing Curricula 91" [ACM/IEEE 1991]. Acknowledging that computer science programs had become widely supported in colleges of engineering, arts and sciences, and liberal arts, Curricula 91 proposed a core body of knowledge that undergraduate majors in all of these programs should cover. This core contained sufficient theory, abstraction, and design content that students would become familiar with the three complementary ways of "doing" computer science. It also ensured that students would gain a broad exposure to the nine major subject areas of the discipline, including their social context. A significant laboratory component ensured that students gained significant abstraction and design experience.

In 2001, in response to dramatic changes that had occurred in the discipline during the 1990s, a new ACM/IEEE-CS Task Force developed a revised model curriculum for computer science [ACM/IEEE 2001]. This model updated the list of major subject areas, and we use this updated list to form the organizational basis for this Handbook (see below). This model also acknowledged that the enormous growth of the computing field had spawned four distinct but overlapping subfields — "computer science," "computer engineering," "software engineering," and "information systems." While these four subfields share significant knowledge in common, each one also underlies a distinctive academic and professional field. While the computer science dimension is directly addressed by this Handbook, the other three dimensions are addressed to the extent that their subject matter overlaps that of computer science.

1.2.2 Growth of Academic Programs

Fueling the rapid evolution of curricula in computer science during the last three decades was an enormous growth in demand, by industry and academia, for computer science professionals, researchers, and educators at all levels. In response, the number of computer science Ph.D.-granting programs in the U.S. grew from 12 in 1964 to 164 in 2001. During the period 1966 to 2001, the annual number of Bachelor's degrees awarded in the U.S. grew from 89 to 46,543; Master's degrees grew from 238 to 19,577; and Ph.D. degrees grew from 19 to 830 [ACM 1968, Bryant 2001].

Figure 1.1 shows the number of bachelor's and master's degrees awarded by U.S. colleges and universities in computer science and engineering (CS&E) from 1966 to 2001. The number of Bachelor's degrees peaked at about 42,000 in 1986, declined to about 24,500 in 1995, and then grew steadily toward its current peak during the past several years. Master's degree production in computer science has grown steadily without decline throughout this period.

The dramatic growth of BS and MS degrees in the five-year period between 1996 and 2001 parallels the growth and globalization of the economy itself. The more recent falloff in the economy, especially the collapse of the "dot.com" industry, may dampen this growth in the near future. In the long run, future increases in Bachelor's and Master's degree production will continue to be linked to expansion of the technology industry, both in the U.S and throughout the world.

Figure 1.2 shows the number of U.S. Ph.D. degrees in computer science during the same 1966 to 2001 period [Bryant 2001]. Production of Ph.D. degrees in computer science grew throughout the early 1990s, fueled by continuing demand from industry for graduate-level talent and from academia to staff growing undergraduate and graduate research programs. However, in recent years, Ph.D. production has fallen off slightly and approached a steady state. Interestingly, this last five years of non-growth at the Ph.D. level is coupled with five years of dramatic growth at the BS and MS levels. This may be partially explained by the unusually high salaries offered in a booming technology sector of the economy, which may have lured some

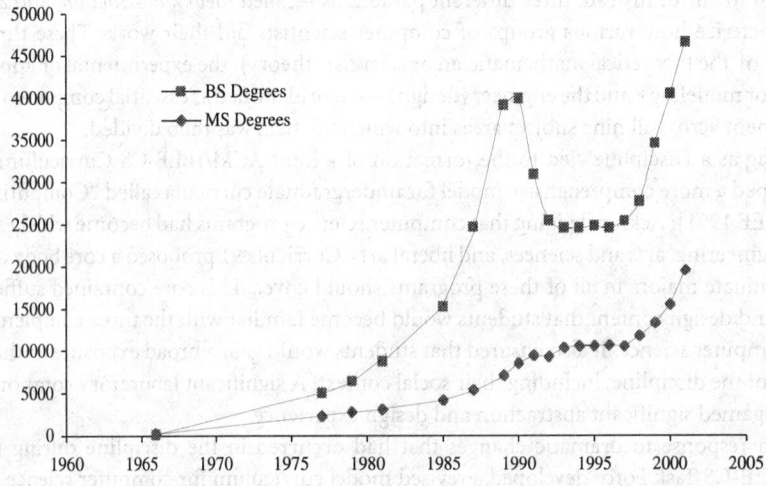

FIGURE 1.1   U.S. bachelor's and master's degrees in CS&E.

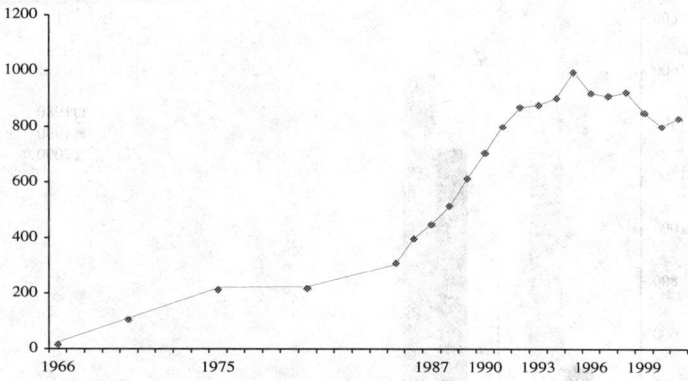

FIGURE 1.2   U.S. Ph.D. degrees in computer science.

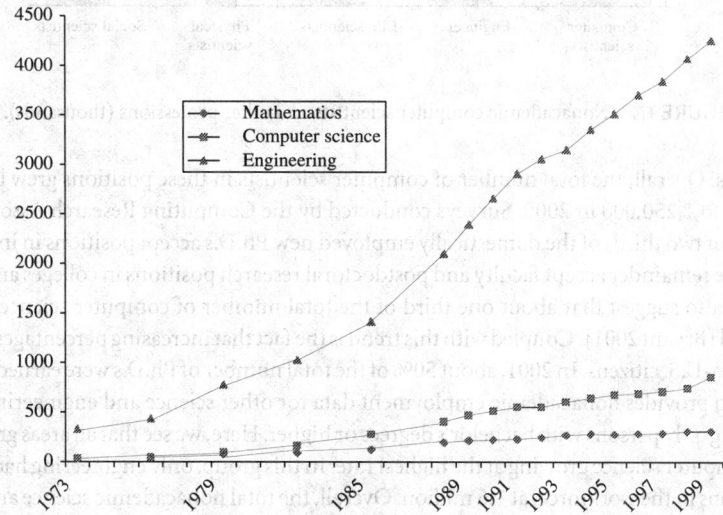

FIGURE 1.3   Academic R&D in computer science and related fields (in millions of dollars).

undergraduates away from immediate pursuit of a Ph.D. The more recent economic slowdown, especially in the technology industry, may help to normalize these trends in the future.

1.2.3   Academic R&D and Industry Growth

University and industrial research and development (R&D) investments in computer science grew rapidly in the period between 1986 and 1999. Figure 1.3 shows that academic research and development in computer science nearly tripled, from $321 million to $860 million, during this time period. This growth rate was significantly higher than that of academic R&D in the related fields of engineering and mathematics. During this same period, the overall growth of academic R&D in engineering doubled, while that in mathematics grew by about 50%. About two thirds of the total support for academic R&D comes from federal and state sources, while about 7% comes from industry and the rest comes from the academic institutions themselves [NSF 2002].

Using 1980, 1990, and 2000 U.S. Census data, Figure 1.4 shows recent growth in the number of persons with at least a bachelor's degree who were employed in nonacademic (industry and government) computer

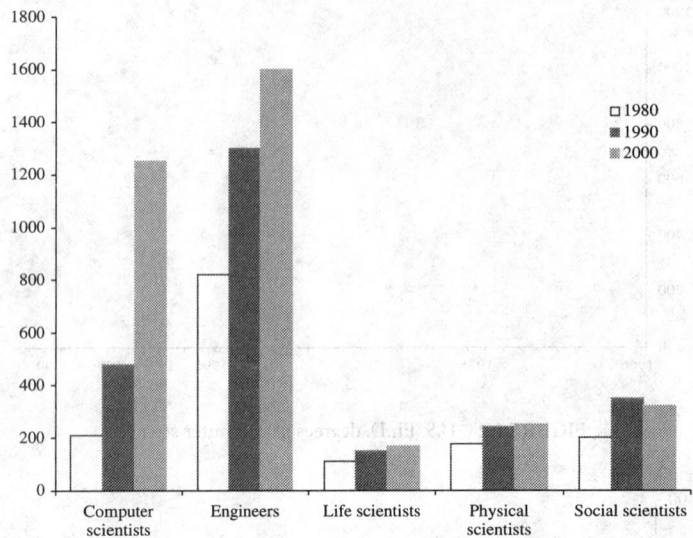

FIGURE 1.4   Nonacademic computer scientists and other professions (thousands).

science positions. Overall, the total number of computer scientists in these positions grew by 600%, from 210,000 in 1980 to 1,250,000 in 2000. Surveys conducted by the Computing Research Association (CRA) suggest that about two thirds of the domestically employed new Ph.D.s accept positions in industry or government, and the remainder accept faculty and postdoctoral research positions in colleges and universities.

CRA surveys also suggest that about one third of the total number of computer science Ph.D.s accept positions abroad [Bryant 2001]. Coupled with this trend is the fact that increasing percentages of U.S. Ph.D.s are earned by non-U.S. citizens. In 2001, about 50% of the total number of Ph.D.s were earned by this group.

Figure 1.4 also provides nonacademic employment data for other science and engineering professions, again considering only persons with bachelor's degrees or higher. Here, we see that all areas grew during this period, with computer science growing at the highest rate. In this group, only engineering had a higher total number of persons in the workforce, at 1.6 million. Overall, the total nonacademic science and engineering workforce grew from 2,136,200 in 1980 to 3,664,000 in 2000, an increase of about 70% [NSF 2001].

1.3   Perspectives in Computer Science

By its very nature, computer science is a multifaceted discipline that can be viewed from at least four different perspectives. Three of the perspectives — *theory, abstraction,* and *design* — underscore the idea that computer scientists in all subject areas can approach their work from different intellectual viewpoints and goals. A fourth perspective — *the social and professional context* — acknowledges that computer science applications directly affect the quality of people's lives, so that computer scientists must understand and confront the social issues that their work uniquely and regularly encounters.

The *theory* of computer science draws from principles of mathematics as well as from the formal methods of the physical, biological, behavioral, and social sciences. It normally includes the use of abstract ideas and methods taken from subfields of mathematics such as logic, algebra, analysis, and statistics. Theory includes the use of various proof and argumentation techniques, like induction and contradiction, to establish properties of formal systems that justify and explain underlying the basic algorithms and data structures used in computational models. Examples include the study of algorithmically unsolvable problems and the study of upper and lower bounds on the complexity of various classes of algorithmic problems. Fields like algorithms and complexity, intelligent systems, computational science, and programming languages have different theoretical models than human–computer interaction or net-centric computing; indeed, all 11 areas covered in this Handbook have underlying theories to a greater or lesser extent.

Abstraction in computer science includes the use of scientific inquiry, modeling, and experimentation to test the validity of hypotheses about computational phenomena. Computer professionals in all 11 areas of the discipline use abstraction as a fundamental tool of inquiry — many would argue that computer science is itself the science of building and examining abstract computational models of reality. Abstraction arises in computer architecture, where the Turing machine serves as an abstract model for complex real computers, and in programming languages, where simple semantic models such as lambda calculus are used as a framework for studying complex languages. Abstraction appears in the design of heuristic and approximation algorithms for problems whose optimal solutions are computationally intractable. It is surely used in graphics and visual computing, where models of three-dimensional objects are constructed mathematically; given properties of lighting, color, and surface texture; and projected in a realistic way on a two-dimensional video screen.

Design is a process that models the essential structure of complex systems as a prelude to their practical implementation. It also encompasses the use of traditional engineering methods, including the classical life-cycle model, to implement efficient and useful computational systems in hardware and software. It includes the use of tools like cost/benefit analysis of alternatives, risk analysis, and fault tolerance that ensure that computing applications are implemented effectively. Design is a central preoccupation of computer architects and software engineers who develop hardware systems and software applications. Design is an especially important activity in computational science, information management, human–computer interaction, operating systems, and net-centric computing.

The *social and professional context* includes many concerns that arise at the computer–human interface, such as liability for hardware and software errors, security and privacy of information in databases and networks (e.g., implications of the Patriot Act), intellectual property issues (e.g., patent and copyright), and equity issues (e.g., universal access to technology and to the profession). All computer scientists must consider the ethical context in which their work occurs and the special responsibilities that attend their work. Chapter 2 discusses these issues, and Appendix B presents the ACM Code of Ethics and Professional Conduct. Several other chapters address topics in which specific social and professional issues come into play. For example, security and privacy issues in databases, operating systems, and networks are discussed in Chapter 60 and Chapter 77. Risks in software are discussed in several chapters of Section XI.

1.4   Broader Horizons: From HPCC to Cyberinfrastructure

In 1989, the Federal Office of Science and Technology announced the "High Performance Computing and Communications Program," or HPCC [OST 1989]. HPCC was designed to encourage universities, research programs, and industry to develop specific capabilities to address the "grand challenges" of the future. To realize these grand challenges would require both fundamental and applied research, including the development of high-performance computing systems with speeds two to three orders of magnitude greater than those of current systems, advanced software technology and algorithms that enable scientists and mathematicians to effectively address these grand challenges, networking to support R&D for a gigabit National Research and Educational Network (NREN), and human resources that expand basic research in all areas relevant to high-performance computing.

The grand challenges themselves were identified in HPCC as those fundamental problems in science and engineering with potentially broad economic, political, or scientific impact that can be advanced by applying high-performance computing technology and that can be solved only by high-level collaboration among computer professionals, scientists, and engineers. A list of grand challenges developed by agencies such as the NSF, DoD, DoE, and NASA in 1989 included:

- Prediction of weather, climate, and global change
- Challenges in materials sciences
- Semiconductor design
- Superconductivity
- Structural biology

- Design of drugs
- Human genome
- Quantum chromodynamics
- Astronomy
- Transportation
- Vehicle dynamics and signature
- Turbulence
- Nuclear fusion
- Combustion systems
- Oil and gas recovery
- Ocean science
- Speech
- Vision
- Undersea surveillance for anti-submarine warfare

The 1992 report entitled "Computing the Future" (CTF) [CSNRCTB 1992], written by a group of leading computer professionals in response to a request by the Computer Science and Technology Board (CSTB), identified the need for computer science to broaden its research agenda and its educational horizons, in part to respond effectively to the grand challenges identified above. The view that the research agenda should be broadened caused concerns among some researchers that this funding and other incentives might overemphasize short-term at the expense of long-term goals. This Handbook reflects the broader view of the discipline in its inclusion of computational science, information management, and human–computer interaction among the major subfields of computer science.

CTF aimed to bridge the gap between suppliers of research in computer science and consumers of research such as industry, the federal government, and funding agencies such as the NSF, DARPA, and DoE. It addressed fundamental challenges to the field and suggested responses that encourage greater interaction between research and computing practice. Its overall recommendations focused on three priorities:

1. To sustain the core effort that creates the theoretical and experimental science base on which applications build
2. To broaden the field to reflect the centrality of computing in science and society
3. To improve education at both the undergraduate and graduate levels

CTF included recommendations to federal policy makers and universities regarding research and education:

- *Recommendations to federal policy makers regarding research:*
  - The High-Performance Computing and Communication (HPCC) program passed by Congress in 1989 [OST 1989] should be fully supported.
  - Application-oriented computer science and engineering research should be strongly encouraged through special funding programs.
- *Recommendations to universities regarding research:*
  - Academic research should broaden its horizons, embracing application-oriented and technology-transfer research as well as core applications.
  - Laboratory research with experimental as well as theoretical content should be supported.
- *Recommendation to federal policy makers regarding education:*
  - Basic and human resources research of HPCC and other areas should be expanded to address educational needs.

- *Recommendations to universities regarding education:*
  - Broaden graduate education to include requirements and incentives to study application areas.
  - Reach out to women and minorities to broaden the talent pool.

Although this report was motivated by the desire to provide a rationale for the HPCC program, its message that computer science must be responsive to the needs of society is much broader. The years since publication of CTF have seen a swing away from pure research toward application-oriented research that is reflected in this edition of the Handbook. However, it remains important to maintain a balance between short-term applications and long-term research in traditional subject areas.

More recently, increased attention has been paid to the emergence of information technology (IT) research as an academic subject area having significant overlap with computer science itself. This development is motivated by several factors, including mainly the emergence of electronic commerce, the shortage of trained IT professionals to fill new jobs in IT, and the continuing need for computing to expand its capability to manage the enormous worldwide growth of electronic information. Several colleges and universities have established new IT degree programs that complement their computer science programs, offering mainly BS and MS degrees in information technology. The National Science Foundation is a strong supporter of IT research, earmarking $190 million in this priority area for FY 2003. This amounts to about 35% of the entire NSF computer science and engineering research budget [NSF 2003a].

The most recent initiative, dubbed "Cyberinfrastructure" [NSF 2003b], provides a comprehensive vision for harnessing the fast-growing technological base to better meet the new challenges and complexities that are shared by a widening community of researchers, professionals, organizations, and citizens who use computers and networks every day. Here are some excerpts from the executive summary for this initiative:

> . . . a new age has dawned in scientific and engineering research, pushed by continuing progress in computing, information, and communication technology, and pulled by the expanding complexity, scope, and scale of today's challenges. The capacity of this technology has crossed thresholds that now make possible a comprehensive "cyberinfrastructure" on which to build new types of scientific and engineering knowledge environments and organizations and to pursue research in new ways and with increased efficacy.
>
> Such environments . . . are required to address national and global priorities, such as understanding global climate change, protecting our natural environment, applying genomics-proteomics to human health, maintaining national security, mastering the world of nanotechnology, and predicting and protecting against natural and human disasters, as well as to address some of our most fundamental intellectual questions such as the formation of the universe and the fundamental character of matter.
>
> This panel's overarching recommendation is that the NSF should establish and lead a large-scale, interagency, and internationally coordinated Advanced Cyberinfrastructure Program (ACP) to create, deploy, and apply cyberinfrastructure in ways that radically empower all scientific and engineering research and allied education. We estimate that sustained new NSF funding of $1 billion per year is needed to achieve critical mass and to leverage the coordinated co-investment from other federal agencies, universities, industry, and international sources necessary to empower a revolution.

It is too early to tell whether the ambitions expressed in this report will provide a new rallying call for science and technology research in the next decade. Achieving them will surely require unprecedented levels of collaboration and funding.

Nevertheless, in response to HPCC and successive initiatives, the two newer subject areas of "computational science" [Stevenson 1994] and "net-centric computing" [ACM/IEEE 2001] have established themselves among the 11 that characterize computer science at this early moment in the 21st century. This Handbook views "computational science" as the application of computational and mathematical models and methods to science, having as a driving force the fundamental interaction between computation and scientific research. For instance, fields like computational astrophysics, computational biology, and computational chemistry all unify the application of computing in science and engineering with underlying mathematical concepts, algorithms, graphics, and computer architecture. Much of the research and accomplishments of the computational science field is presented in Section III.

Net-centric computing, on the other hand, emphasizes the interactions among people, computers, and the Internet. It affects information technology systems in professional and personal spheres, including the implementation and use of search engines, commercial databases, and digital libraries, along with their risks and human factors. Some of these topics intersect in major ways with those of human–computer interaction, while others fall more directly in the realm of management information systems (MIS). Because MIS is widely viewed as a separate discipline from computer science, this Handbook does not attempt to cover all of MIS. However, it does address many MIS concerns in Section V (human–computer interaction) Section VI (information management), and Section VIII (net-centric computing).

The remaining sections of this Handbook cover relatively traditional areas of computer science — algorithms and complexity, computer architecture, operating systems, programming languages, artificial intelligence, software engineering, and computer graphics. A more careful summary of these sections appears below.

1.5 Organization and Content

In the 1940s, computer science was identified with number crunching, and numerical analysis was considered a central tool. Hardware, logical design, and information theory emerged as important subfields in the early 1950s. Software and programming emerged as important subfields in the mid-1950s and soon dominated hardware as topics of study in computer science. In the 1960s, computer science could be comfortably classified into theory, systems (including hardware and software), and applications. Software engineering emerged as an important subdiscipline in the late 1960s. The 1980 Computer Science and Engineering Research Study (COSERS) [Arden 1980] classified the discipline into nine subfields:

1. Numerical computation
2. Theory of computation
3. Hardware systems
4. Artificial intelligence
5. Programming languages
6. Operating systems
7. Database management systems
8. Software methodology
9. Applications

This Handbook's organization presents computer science in the following 11 sections, which are the subfields defined in [ACM/IEEE 2001].

1. Algorithms and complexity
2. Architecture and organization
3. Computational science
4. Graphics and visual computing
5. Human–computer interaction
6. Information management
7. Intelligent systems
8. Net-centric computing
9. Operating systems
10. Programming languages
11. Software engineering

This overall organization shares much in common with that of the 1980 COSERS study. That is, except for some minor renaming, we can read this list as a broadening of numerical analysis into computational science, and an addition of the new areas of human–computer interaction and graphics. The other areas appear in both classifications with some name changes (theory of computation has become algorithms and complexity, artificial intelligence has become intelligent systems, applications has become net-centric computing, hardware systems has evolved into architecture and networks, and database has evolved into information management). The overall similarity between the two lists suggests that the discipline of computer science has stabilized in the past 25 years.

However, although this high-level classification has remained stable, the content of each area has evolved dramatically. We examine below the scope of each area individually, along with the topics in each area that are emphasized in this Handbook.

1.5.1 Algorithms and Complexity

The subfield of algorithms and complexity is interpreted broadly to include core topics in the theory of computation as well as data structures and practical algorithm techniques. Its chapters provide a comprehensive overview that spans both theoretical and applied topics in the analysis of algorithms. Chapter 3 provides an overview of techniques of algorithm design like divide and conquer, dynamic programming, recurrence relations, and greedy heuristics, while Chapter 4 covers data structures both descriptively and in terms of their space–time complexity.

Chapter 5 examines topics in complexity like P vs. NP and NP-completeness, while Chapter 6 introduces the fundamental concepts of computability and undecidability and formal models such as Turing machines. Graph and network algorithms are treated in Chapter 7, and algebraic algorithms are the subject of Chapter 8.

The wide range of algorithm applications is presented in Chapter 9 through Chapter 15. Chapter 9 covers cryptographic algorithms, which have recently become very important in operating systems and network security applications. Chapter 10 covers algorithms for parallel computer architectures, Chapter 11 discusses algorithms for computational geometry, while Chapter 12 introduces the rich subject of randomized algorithms. Pattern matching and text compression algorithms are examined in Chapter 13, and genetic algorithms and their use in the biological sciences are introduced in Chapter 14. Chapter 15 concludes this section with a treatment of combinatorial optimization.

1.5.2 Architecture

Computer architecture is the design of efficient and effective computer hardware at all levels, from the most fundamental concerns of logic and circuit design to the broadest concerns of parallelism and high-performance computing. The chapters in Section II span these levels, providing a sampling of the principles, accomplishments, and challenges faced by modern computer architects.

Chapter 16 introduces the fundamentals of logic design components, including elementary circuits, Karnaugh maps, programmable array logic, circuit complexity and minimization issues, arithmetic processes, and speedup techniques. Chapter 17 focuses on processor design, including the fetch/execute instruction cycle, stack machines, CISC vs. RISC, and pipelining. The principles of memory design are covered in Chapter 18, while the architecture of buses and other interfaces is addressed in Chapter 19. Chapter 20 discusses the characteristics of input and output devices like the keyboard, display screens, and multimedia audio devices. Chapter 21 focuses on the architecture of secondary storage devices, especially disks.

Chapter 22 concerns the design of effective and efficient computer arithmetic units, while Chapter 23 extends the design horizon by considering various models of parallel architectures that enhance the performance of traditional serial architectures. Chapter 24 focuses on the relationship between computer architecture and networks, while Chapter 25 covers the strategies employed in the design of fault-tolerant and reliable computers.

1.5.3   Computational Science

The area of computational science unites computation, experimentation, and theory as three fundamental modes of scientific discovery. It uses scientific visualization, made possible by simulation and modeling, as a window into the analysis of physical, chemical, and biological phenomena and processes, providing a virtual microscope for inquiry at an unprecedented level of detail.

This section focuses on the challenges and opportunities offered by very high-speed clusters of computers and sophisticated graphical interfaces that aid scientific research and engineering design. Chapter 26 introduces the section by presenting the fundamental subjects of computational geometry and grid generation. The design of graphical models for scientific visualization of complex physical and biological phenomena is the subject of Chapter 27.

Each of the remaining chapters in this section covers the computational challenges and discoveries in a specific scientific or engineering field. Chapter 28 presents the computational aspects of structural mechanics, Chapter 29 summarizes progress in the area of computational electromagnetics, and Chapter 30 addresses computational modeling in the field of fluid dynamics. Chapter 31 addresses the grand challenge of computational ocean modeling. Computational chemistry is the subject of Chapter 32, while Chapter 33 addresses the computational dimensions of astrophysics. Chapter 34 closes this section with a discussion of the dramatic recent progress in computational biology.

1.5.4   Graphics and Visual Computing

Computer graphics is the study and realization of complex processes for representing physical and conceptual objects visually on a computer screen. These processes include the internal modeling of objects, rendering, projection, and motion. An overview of these processes and their interaction is presented in Chapter 35.

Fundamental to all graphics applications are the processes of modeling and rendering. Modeling is the design of an effective and efficient internal representation for geometric objects, which is the subject of Chapter 36 and Chapter 37. Rendering, the process of representing the objects in a three-dimensional scene on a two-dimensional screen, is discussed in Chapter 38. Among its special challenges are the elimination of hidden surfaces and the modeling of color, illumination, and shading.

The reconstruction of scanned and digitally photographed images is another important area of computer graphics. Sampling, filtering, reconstruction, and anti-aliasing are the focus of Chapter 39. The representation and control of motion, or animation, is another complex and important area of computer graphics. Its special challenges are presented in Chapter 40.

Chapter 41 discusses volume datasets, and Chapter 42 looks at the emerging field of virtual reality and its particular challenges for computer graphics. Chapter 43 concludes this section with a discussion of progress in the computer simulation of vision.

1.5.5   Human–Computer Interaction

This area, the study of how humans and computers interact, has the goal of improving the quality of such interaction and the effectiveness of those who use technology in the workplace. This includes the conception, design, implementation, risk analysis, and effects of user interfaces and tools on the people who use them.

Modeling the organizational environments in which technology users work is the subject of Chapter 44. Usability engineering is the focus of Chapter 45, while Chapter 46 covers task analysis and the design of functionality at the user interface. The influence of psychological preferences of users and programmers and the integration of these preferences into the design process is the subject of Chapter 47.

Specific devices, tools, and techniques for effective user-interface design form the basis for the next few chapters in this section. Lower-level concerns for the design of interface software technology are addressed in Chapter 48. The special challenges of integrating multimedia with user interaction are presented in Chapter 49. Computer-supported collaboration is the subject of Chapter 50, and the impact of international standards on the user interface design process is the main concern of Chapter 51.

1.5.6 Information Management

The subject area of information management addresses the general problem of storing large amounts of data in such a way that they are reliable, up-to-date, accessible, and efficiently retrieved. This problem is prominent in a wide range of applications in industry, government, and academic research. Availability of such data on the Internet and in forms other than text (e.g., CD, audio, and video) makes this problem increasingly complex.

At the foundation are the fundamental data models (relational, hierarchical, and object-oriented) discussed in Chapter 52. The conceptual, logical, and physical levels of designing a database for high performance in a particular application domain are discussed in Chapter 53.

A number of basic issues surround the effective design of database models and systems. These include choosing appropriate access methods (Chapter 54), optimizing database queries (Chapter 55), controlling concurrency (Chapter 56), and processing transactions (Chapter 57).

The design of databases for distributed and parallel systems is discussed in Chapter 58, while the design of hypertext and multimedia databases is the subject of Chapter 59. The contemporary issue of database security and privacy protection, in both stand-alone and networked environments, is the subject of Chapter 60.

1.5.7 Intelligent Systems

The field of intelligent systems, often called artificial intelligence (AI), studies systems that simulate human rational behavior in all its forms. Current efforts are aimed at constructing computational mechanisms that process visual data, understand speech and written language, control robot motion, and model physical and cognitive processes. Robotics is a complex field, drawing heavily from AI as well as other areas of science and engineering.

Artificial intelligence research uses a variety of distinct algorithms and models. These include fuzzy, temporal, and other logics, as described in Chapter 61. The related idea of qualitative modeling is discussed in Chapter 62, while the use of complex specialized search techniques that address the combinatorial explosion of alternatives in AI problems is the subject of Chapter 63. Chapter 64 addresses issues related to the mechanical understanding of spoken language.

Intelligent systems also include techniques for automated learning and planning. The use of decision trees and neural networks in learning and other areas is the subject of Chapter 65 and Chapter 66. Chapter 67 presents the rationale and uses of planning and scheduling models, while Chapter 68 contains a discussion of deductive learning. Chapter 69 addresses the challenges of modeling from the viewpoint of cognitive science, while Chapter 70 treats the challenges of decision making under uncertainty.

Chapter 71 concludes this section with a discussion of the principles and major results in the field of robotics: the design of effective devices that simulate mechanical, sensory, and intellectual functions of humans in specific task domains such as navigation and planning.

1.5.8 Net-Centric Computing

Extending system functionality across a networked environment has added an entirely new dimension to the traditional study and practice of computer science. Chapter 72 presents an overview of network organization and topologies, while Chapter 73 describes network routing protocols. Basic issues in network management are addressed in Chapter 74.

The special challenges of information retrieval and data mining from large databases and the Internet are addressed in Chapter 75. The important topic of data compression for internetwork transmission and archiving is covered in Chapter 76.

Modern computer networks, especially the Internet, must ensure system integrity in the event of inappropriate access, unexpected malfunction and breakdown, and violations of data and system security or individual privacy. Chapter 77 addresses the principles surrounding these security and privacy issues. A discussion of some specific malicious software and hacking events appears in Chapter 78. This section concludes with Chapter 79, which discusses protocols for user authentication, access control, and intrusion detection.

1.5.9   Operating Systems

An operating system is the software interface between the computer and its applications. This section covers operating system analysis, design, and performance, along with the special challenges for operating systems in a networked environment. Chapter 80 briefly traces the historical development of operating systems and introduces the fundamental terminology, including process scheduling, memory management, synchronization, I/O management, and distributed systems.

The "process" is a key unit of abstraction in operating system design. Chapter 81 discusses the dynamics of processes and threads. Strategies for process and device scheduling are presented in Chapter 82. The special requirements for operating systems in real-time and embedded system environments are treated in Chapter 83. Algorithms and techniques for process synchronization and interprocess communication are the subject of Chapter 84.

Memory and input/output device management is also a central concern of operating systems. Chapter 85 discusses the concept of virtual memory, from its early incarnations to its uses in present-day systems and networks. The different models and access methods for secondary storage and filesystems are covered in Chapter 86.

The influence of networked environments on the design of distributed operating systems is considered in Chapter 87. Distributed and multiprocessor scheduling are the focus in Chapter 88, while distributed file and memory systems are discussed in Chapter 89.

1.5.10   Programming Languages

This section examines the design of programming languages, including their paradigms, mechanisms for compiling and runtime management, and theoretical models, type systems, and semantics. Overall, this section provides a good balance between considerations of programming paradigms, implementation issues, and theoretical models.

Chapter 90 considers traditional language and implementation questions for imperative programming languages such as Fortran, C, and Ada. Chapter 91 examines object-oriented concepts such as classes, inheritance, encapsulation, and polymorphism, while Chapter 92 presents the view of functional programming, including lazy and eager evaluation. Chapter 93 considers declarative programming in the logic/constraint programming paradigm, while Chapter 94 covers the design and use of special purpose scripting languages. Chapter 95 considers the emergent paradigm of event-driven programming, while Chapter 96 covers issues regarding concurrent, distributed, and parallel programming models.

Type systems are the subject of Chapter 97, while Chapter 98 covers programming language semantics. Compilers and interpreters for sequential languages are considered in Chapter 99, while the issues surrounding runtime environments and memory management for compilers and interpreters are addressed in Chapter 100.

Brief summaries of the main features and applications of several contemporary languages appear in Appendix D, along with links to Web sites for more detailed information on these languages.

1.5.11   Software Engineering

The section on software engineering examines formal specification, design, verification and testing, project management, and other aspects of the software process. Chapter 101 introduces general software qualities such as maintainability, portability, and reuse that are needed for high-quality software systems, while Chapter 109 covers the general topic of software architecture.

Chapter 102 reviews specific models of the software life cycle such as the waterfall and spiral models. Chapter 106 considers a more formal treatment of software models, including formal specification languages.

Chapter 103 deals with the traditional design process, featuring a case study in top-down functional design. Chapter 104 considers the complementary strategy of object-oriented software design. Chapter 105 treats the subject of validation and testing, including risk and reliability issues. Chapter 107 deals with the use of rigorous techniques such as formal verification for quality assurance.

Chapter 108 considers techniques of software project management, including team formation, project scheduling, and evaluation, while Chapter 110 concludes this section with a treatment of specialized system development.

1.6   Conclusion

In 2002, the ACM celebrated its 55th anniversary. These five decades of computer science are characterized by dramatic growth and evolution. While it is safe to reaffirm that the field has attained a certain level of maturity, we surely cannot assume that it will remain unchanged for very long. Already, conferences are calling for new visions that will enable the discipline to continue its rapid evolution in response to the world's continuing demand for new technology and innovation.

This Handbook is designed to convey the modern spirit, accomplishments, and direction of computer science as we see it in 2003. It interweaves theory with practice, highlighting "best practices" in the field as well as emerging research directions. It provides today's answers to computational questions posed by professionals and researchers working in all 11 subject areas. Finally, it identifies key professional and social issues that lie at the intersection of the technical aspects of computer science and the people whose lives are impacted by such technology.

The future holds great promise for the next generations of computer scientists. These people will solve problems that have only recently been conceived, such as those suggested by the HPCC as "grand challenges." To address these problems in a way that benefits the world's citizenry will require substantial energy, commitment, and real investment on the part of institutions and professionals throughout the field. The challenges are great, and the solutions are not likely to be obvious.

References

ACM Curriculum Committee on Computer Science 1968. Curriculum 68: recommendations for the undergraduate program in computer science. *Commun. ACM*, 11(3):151–197, March.

ACM Curriculum Committee on Computer Science 1978. Curriculum 78: recommendations for the undergraduate program in computer science. *Commun. ACM*, 22(3):147–166, March.

ACM Task Force on the Core of Computer Science: Denning, P., Comer, D., Gries, D., Mulder, M., Tucker, A., and Young, P., 1988. *Computing as a Discipline*. Abridged version, *Commun. ACM*, Jan. 1989.

ACM/IEEE-CS Joint Curriculum Task Force. Computing Curricula 1991. ACM Press. Abridged version, *Commun. ACM*, June 1991, and *IEEE Comput.* Nov. 1991.

ACM/IEEE-CS Joint Task Force. Computing Curricula 2001: Computer Science Volume. ACM and IEEE Computer Society, December 2001, (http://www.acm.org/sigcse/cc2001).

Arden, B., Ed., 1980. *What Can be Automated?* Computer Science and Engineering Research (COSERS) Study. MIT Press, Boston, MA.

Bryant, R.E. and M.Y. Vardi, 2001. 2000–2001 Taulbee Survey: Hope for More Balance in Supply and Demand. Computing Research Assoc (http://www.cra.org).

CSNRCTB 1992. Computer Science and National Research Council Telecommunications Board. *Computing the Future: A Broader Agenda for Computer Science and Engineering*. National Academy Press, Washington, D.C.

Economist 1993. The computer industry: reboot system and start again. *Economist*, Feb. 27.

Gibbs, N. and A. Tucker 1986. A Model Curriculum for a Liberal Arts Degree in Computer Science. *Communications of the ACM*, March.

IEEE-CS 1976. Education Committee of the IEEE Computer Society. *A Curriculum in Computer Science and Engineering*. IEEE Pub. EH0119-8, Jan. 1977.

IEEE-CS 1983. Educational Activities Board. *The 1983 Model Program in Computer Science and Engineering*. Tech. Rep. 932. Computer Society of the IEEE, December.

NSF 2002. National Science Foundation. *Science and Engineering Indicators* (Vol. I and II), National Science Board, Arlington, VA.

NSF 2003a. National Science Foundation. *Budget Overview FY 2003* (http://www.nsf.gov/bfa/bud/fy2003/overview.htm).

NSF 2003b. National Science Foundation. *Revolutionizing Science and Engineering through Cyberinfrastructure*, report of the NSF Blue-Ribbon Advisory Panel on Cyberinfrastructure, January.

OST 1989. Office of Science and Technology. *The Federal High Performance Computing and Communication Program*. Executive Office of the President, Washington, D.C.

Rosser, J.B. et al. 1966. *Digital Computer Needs in Universities and Colleges*. Publ. 1233, National Academy of Sciences, National Research Council, Washington, D.C.

Stevenson, D.E. 1994. Science, computational science, and computer science. *Commun. ACM*, December.

2

Ethical Issues for Computer Scientists

2.1 Introduction: Why a Chapter on Ethical Issues? 2-1
2.2 Ethics in General 2-3
  Utilitarianism • Deontological Theories • Social Contract
  Theories • A Paramedic Method for Computer Ethics • Easy
  and Hard Ethical Decision Making
2.3 Professional Ethics 2-7
2.4 Ethical Issues That Arise from Computer Technology ... 2-9
  Privacy • Property Rights and Computing • Risk, Reliability,
  and Accountability • Rapidly Evolving Globally Networked
  Telecommunications
2.5 Final Thoughts 2-12

Deborah G. Johnson
University of Virginia

Keith W. Miller
University of Illinois

2.1 Introduction: Why a Chapter on Ethical Issues?

Computers have had a powerful impact on our world and are destined to shape our future. This observation, now commonplace, is the starting point for any discussion of professionalism and ethics in computing. The work of computer scientists and engineers is part of the social, political, economic, and cultural world in which we live, and it affects many aspects of that world. Professionals who work with computers have special knowledge. That knowledge, when combined with computers, has significant power to change people's lives — by changing socio-technical systems; social, political and economic institutions; and social relationships.

In this chapter, we provide a perspective on the role of computer and engineering professionals and we examine the relationships and responsibilities that go with having and using computing expertise. In addition to the topic of professional ethics, we briefly discuss several of the social–ethical issues created or exacerbated by the increasing power of computers and information technology: privacy, property, risk and reliability, and globalization.

Computers, digital data, and telecommunications have changed work, travel, education, business, entertainment, government, and manufacturing. For example, work now increasingly involves sitting in front of a computer screen and using a keyboard to make things happen in a manufacturing process or to keep track of records. In the past, these same tasks would have involved physically lifting, pushing, and twisting or using pens, paper, and file cabinets. Changes such as these in the way we do things have, in turn, fundamentally changed who we are as individuals, communities, and nations. Some would argue, for example, that new kinds of communities (e.g., cyberspace on the Internet) are forming, individuals are developing new types of personal identities, and new forms of authority and control are taking hold as a result of this evolving technology.

1-58488-360-X/$0.00+$1.50
© 2004 by CRC Press, LLC

Computer technology is shaped by social–cultural concepts, laws, the economy, and politics. These same concepts, laws, and institutions have been pressured, challenged, and modified by computer technology. Technological advances can antiquate laws, concepts, and traditions, compelling us to reinterpret and create new laws, concepts, and moral notions. Our attitudes about work and play, our values, and our laws and customs are deeply involved in technological change.

When it comes to the social–ethical issues surrounding computers, some have argued that the issues are not unique. All of the ethical issues raised by computer technology can, it is said, be classified and worked out using traditional *moral* concepts, distinctions, and theories. There is nothing new here in the sense that we can understand the new issues using traditional moral concepts, such as privacy, property, and responsibility, and traditional moral values, such as individual freedom, autonomy, accountability, and community. These concepts and values predate computers; hence, it would seem there is nothing unique about *computer ethics.*

On the other hand, those who argue for the uniqueness of the issues point to the fundamental ways in which computers have changed so many human activities, such as manufacturing, record keeping, banking, international trade, education, and communication. Taken together, these changes are so radical, it is claimed, that traditional moral concepts, distinctions, and theories, if not abandoned, must be significantly reinterpreted and extended. For example, they must be extended to computer-mediated relationships, computer software, computer art, datamining, virtual systems, and so on.

The uniqueness of the ethical issues surrounding computers can be argued in a variety of ways. Computer technology makes possible a scale of activities not possible before. This includes a larger scale of record keeping of personal information, as well as larger-scale calculations which, in turn, allow us to build and do things not possible before, such as undertaking space travel and operating a global communication system. Among other things, the increased scale means finer-grained personal information collection and more precise data matching and datamining. In addition to scale, computer technology has involved the creation of new kinds of entities for which no rules initially existed: entities such as computer files, computer programs, the Internet, Web browsers, cookies, and so on. The uniqueness argument can also be made in terms of the power and pervasiveness of computer technology. Computers and information technology seem to be bringing about a magnitude of change comparable to that which took place during the Industrial Revolution, transforming our social, economic, and political institutions; our understanding of what it means to be human; and the distribution of power in the world. Hence, it would seem that the issues are at least special, if not unique.

In this chapter, we will take an approach that synthesizes these two views of computer ethics by assuming that the analysis of computer ethical issues involves both working on something new and drawing on something old. We will view issues in computer ethics as new species of older ethical problems [Johnson 1994], such that the issues can be understood using traditional moral concepts such as autonomy, privacy, property, and responsibility, while at the same time recognizing that these concepts may have to be extended to what is new and special about computers and the situations they create.

Most ethical issues arising around computers occur in contexts in which there are already social, ethical, and legal norms. In these contexts, often there are implicit, if not formal (legal), rules about how individuals are to behave; there are familiar practices, social meanings, interdependencies, and so on. In this respect, the issues are not new or unique, or at least cannot be resolved without understanding the prevailing context, meanings, and values. At the same time, the situation may have special features because of the involvement of computers — features that have not yet been addressed by prevailing norms. These features can make a moral difference. For example, although property rights and even intellectual property rights had been worked out long before the creation of software, when software first appeared, it raised a new form of property issue. Should the arrangement of icons appearing on the screen of a user interface be ownable? Is there anything intrinsically wrong in copying software? Software has features that make the distinction between idea and expression (a distinction at the core of copyright law) almost incoherent. As well, software has features that make standard intellectual property laws difficult to enforce. Hence, questions about what should be owned when it comes to software and how to evaluate violations of software ownership rights are not new in the sense that they are property rights issues, but they are new in the sense that nothing with the characteristics of software had been addressed before. We have, then, a new species of traditional property rights.

Similarly, although our understanding of rights and responsibilities in the employer–employee relationship has been evolving for centuries, never before have employers had the capacity to monitor their workers electronically, keeping track of every keystroke, and recording and reviewing all work done by an employee (covertly or with prior consent). When we evaluate this new monitoring capability and ask whether employers should use it, we are working on an issue that has never arisen before, although many other issues involving employer–employee rights have. We must address a new species of the tension between employer–employee rights and interests.

The social–ethical issues posed by computer technology are significant in their own right, but they are of special interest here because computer and engineering professionals bear responsibility for this technology. It is of critical importance that they understand the social change brought about by their work and the difficult social–ethical issues posed. Just as some have argued that the social–ethical issues posed by computer technology are not unique, some have argued that the issues of professional ethics surrounding computers are not unique. We propose, in parallel with our previous genus–species account, that the professional ethics issues arising for computer scientists and engineers are species of generic issues of professional ethics. All professionals have responsibilities to their employers, clients, co-professionals, and the public. Managing these types of responsibilities poses a challenge in all professions. Moreover, all professionals bear some responsibility for the impact of their work. In this sense, the professional ethics issues arising for computer scientists and engineers are generally similar to those in other professions. Nevertheless, it is also true to say that the issues arise in unique ways for computer scientists and engineers because of the special features of computer technology.

In what follows, we discuss ethics in general, professional ethics, and finally, the ethical issues surrounding computer and information technology.

2.2 Ethics in General

Rigorous study of ethics has traditionally been the purview of philosophers and scholars of religious studies. Scholars of ethics have developed a variety of ethical theories with several tasks in mind:

  To explain and justify the idea of morality and prevailing moral notions
  To critique ordinary moral beliefs
  To assist in rational, ethical decision making

Our aim in this chapter is not to propose, defend, or attack any particular ethical theory. Rather, we offer brief descriptions of three major and influential ethical theories to illustrate the nature of ethical analysis. We also include a decision-making method that combines elements of each theory.

Ethical analysis involves giving reasons for moral claims and commitments. It is not just a matter of articulating intuitions. When the reasons given for a claim are developed into a moral theory, the theory can be incorporated into techniques for improved technical decision making. The three ethical theories described in this section represent three traditions in ethical analysis and problem solving. The account we give is not exhaustive, nor is our description of the three theories any more than a brief introduction. The three traditions are utilitarianism, deontology, and social contract theory.

2.2.1 Utilitarianism

Utilitarianism has greatly influenced 20th-century thinking, especially insofar as it influenced the development of cost–benefit analysis. According to utilitarianism, we should make decisions about what to do by focusing on the consequences of actions and policies; we should choose actions and policies that bring about the best consequences. Ethical rules are derived from their usefulness (their utility) in bringing about happiness. In this way, utilitarianism offers a seemingly simple moral principle to determine what to do in a given situation: everyone ought to act so as to bring about the greatest amount of happiness for the greatest number of people.

According to utilitarianism, happiness is the only value that can serve as a foundational base for ethics. Because happiness is the ultimate good, morality must be based on creating as much of this good as possible. The utilitarian principle provides a decision procedure. When you want to know what to do, the right action is the alternative that produces the most overall net happiness (happiness-producing consequences minus unhappiness-producing consequences). The right action may be one that brings about some unhappiness, but that is justified if the action also brings about enough happiness to counterbalance the unhappiness or if the action brings about the least unhappiness of all possible alternatives.

Utilitarianism should not be confused with egoism. Egoism is a theory claiming that one should act so as to bring about the most good consequences *for oneself*. Utilitarianism does not say that you should maximize your own good. Rather, total happiness in the world is what is at issue; when you evaluate your alternatives, you must ask about their effects on the happiness of everyone. It may turn out to be right for you to do something that will diminish your own happiness because it will bring about an increase in overall happiness.

The emphasis on consequences found in utilitarianism is very much a part of personal and policy decision making in our society, in particular as a framework for law and public policy. Cost–benefit and risk–benefit analysis are, for example, consequentialist in character.

Utilitarians do not all agree on the details of utilitarianism; there are different kinds of utilitarianism. One issue is whether the focus should be on *rules* of behavior or individual *acts*. Utilitarians have recognized that it would be counter to overall happiness if each one of us had to calculate at every moment what the consequences of every one of our actions would be. Sometimes we must act quickly, and often the consequences are difficult or impossible to foresee. Thus, there is a need for general rules to guide our actions in ordinary situations. Hence, *rule-utilitarians* argue that we ought to adopt rules that, if followed by everyone, would, in general and in the long run, maximize happiness. *Act-utilitarians*, on the other hand, put the emphasis on judging individual actions rather than creating rules.

Both rule-utilitarians and act-utilitarians, nevertheless, share an emphasis on consequences; deontological theories do not share this emphasis.

2.2.2 Deontological Theories

Deontological theories can be understood as a response to important criticisms of utilitarian theories. A standard criticism is that utilitarianism seems to lead to conclusions that are incompatible with our most strongly held moral intuitions. Utilitarianism seems, for example, open to the possibility of justifying enormous burdens on some individuals for the sake of others. To be sure, every person counts equally; no one person's happiness or unhappiness is more important than any other person's. However, because utilitarians are concerned with the total amount of happiness, we can imagine situations where great overall happiness would result from sacrificing the happiness of a few. Suppose, for example, that having a small number of slaves would create great happiness for large numbers of people; or suppose we kill one healthy person and use his or her body parts to save ten people in need of transplants.

Critics of utilitarianism say that if utilitarianism justifies such practices, then the theory must be wrong. Utilitarians have a defense, arguing that such practices could not be justified in utilitarianism because of the long-term consequences. Such practices would produce so much fear that the happiness temporarily created would never counterbalance the unhappiness of everyone living in fear that they might be sacrificed for the sake of overall happiness.

We need not debate utilitarianism here. The point is that deontologists find utilitarianism problematic because it puts the emphasis on the consequences of an act rather than on the quality of the act itself. Deontological theories claim that the internal character of the act is what is important. The rightness or wrongness of an action depends on the principles inherent in the action. If an action is done from a sense of duty, and if the principle of the action can be universalized, then the action is right. For example, if I tell the truth because it is convenient for me to do so or because I fear the consequences of getting caught in a lie, my action is not worthy. A worthy action is an action that is done from duty, which involves respecting other people and recognizing them as ends in themselves, not as means to some good effect.

According to deontologists, utilitarianism is wrong because it treats individuals as means to an end (maximum happiness). For deontologists, what grounds morality is not happiness, but human beings as rational agents. Human beings are capable of reasoning about what they want to do. The laws of nature determine most activities: plants grow toward the sun, water boils at a certain temperature, and objects accelerate at a constant rate in a vacuum. Human action is different in that it is self-determining; humans initiate action after thinking, reasoning, and deciding. The human capacity for rational decisions makes morality possible, and it grounds deontological theory. Because each human being has this capacity, each human being must be treated accordingly — with respect. No one else can make our moral choices for us, and each of us must recognize this capacity in others.

Although deontological theories can be formulated in a number of ways, one formulation is particularly important: Immanuel Kant's categorical imperative [Kant 1785]. There are three versions of it, and the second version goes as follows: *Never treat another human being merely as a means but always as an end*. It is important to note the *merely* in the categorical imperative. Deontologists do not insist that we never use another person; only that we never *merely* use them. For example, if I own a company and hire employees to work in my company, I might be thought of as using those employees as a means to my end (i.e., the success of my business). This, however, is not wrong if the employees agree to work for me and if I pay them a fair wage. I thereby respect their ability to choose for themselves, and I respect the value of their labor. What would be wrong would be to take them as slaves and make them work for me, or to pay them so little that they must borrow from me and remain always in my debt. This would show disregard for the value of each person as a freely choosing, rationally valuing, efficacious person.

2.2.3 Social Contract Theories

A third tradition in ethics thinks of ethics on the model of a social contract. There are many different social contract theories, and some, at least, are based on a deontological principle. Individuals are rational free agents; hence, it is immoral to exert undue power over them, that is, to coerce them. Government and society are problematic insofar as they seem to force individuals to obey rules, apparently treating individuals as means to social good. Social contract theories get around this problem by claiming that morality (and government policy) is, in effect, the outcome of rational agents agreeing to social rules. In agreeing to live by certain rules, we make a contract. Morality and government are not, then, systems imposed on individuals; they do not exactly involve coercion. Rather, they are systems created by freely choosing individuals (or they are institutions that rational individuals would choose if given the opportunity).

Philosophers such as Rousseau, Locke, Hobbes, and more recently Rawls [1971] are generally considered social contract theorists. They differ in how they get to the social contract and what it implies. For our purposes, however, the key idea is that principles and rules guiding behavior may be derived from identifying what it is that rational (even self-interested) individuals would agree to in making a social contract. Such principles and rules are the basis of a shared morality. For example, it would be rational for me to agree to live by rules that forbid killing and lying. Even though such rules constrain me, they also give me some degree of protection: if they are followed, I will not be killed or lied to.

It is important to note, however, that social contract theory cannot be used simply by asking what rules you would agree to *now*. Most theorists recognize that what you would agree to now is influenced by your present position in society. Most individuals would opt for rules that would benefit their particular situation and characteristics. Hence, most social contract theorists insist that the principles or rules of the social contract must be derived by assuming certain things about human nature or the human condition. Rawls, for example, insists that we imagine ourselves behind a *veil of ignorance*. We are not allowed to know important features about ourselves (e.g., what talents we have, what race or gender we are), for if we know these things, we will not agree to just rules, but only to rules that will maximize our self-interest. Justice consists of the rules we would agree to when we do not know who we are, for we would want rules that would give us a fair situation no matter where we ended up in the society.

2.2.4   A Paramedic Method for Computer Ethics

Drawing on elements of the three theories described, Collins and Miller [1992] have proposed a decision-assisting method, called the *paramedic method for computer ethics*. This is *not* an algorithm for solving ethical problems; it is not nearly detailed or objective enough for that designation. It is merely a guideline for an organized approach to ethical problem solving.

Assume that a computer professional is faced with a decision that involves human values in a significant way. There may already be some obvious alternatives, and there also may be creative solutions not yet discovered. The paramedic method is designed to help the professional to analyze alternative actions and to encourage the development of creative solutions. To illustrate the method, suppose you are in a tight spot and do not know exactly what the right thing to do is. The method proceeds as follows:

1. Identify alternative actions; list the few alternatives that seem most promising. If an action requires a long description, summarize it as a title with just a few words. Call the alternative actions A_1, $A_2, \ldots, A_a$. No more than five actions should be analyzed at a time.

2. Identify people, groups of people, or organizations that will be affected by each of the alternative decision-actions. Again, hold down the number of entities to the five or six that are affected most. Label the people $P_1, P_2, \ldots, P_p$.

3. Make a table with the horizontal rows labeled by the identified people and the vertical columns labeled with the identified actions. We call such a table a $P \times A$ *table*. Make two copies of the P × A table; label one the *opportunities* table and the other the *vulnerabilities* table. In the opportunities table, list in each interior cell of the table at entry $[x, y]$ the possible good that is likely to happen to person x if action y is taken. Similarly, in the vulnerability table, at position $[x, y]$ list all of the things that are likely to happen badly for x if the action y is taken. These two graphs represent benefit–cost calculations for a consequentialist, utilitarian analysis.

4. Make a new table with the set of persons marking both the columns and the rows (a P × P table). In each cell $[x, y]$ name any responsibilities or duties that x owes y in this situation. (The cells on the diagonal $[x, x]$ are important; they list things one owes oneself.) Now, make copies of this table, labeling one copy for each of the alternative actions being considered. Work through each cell $[x, y]$ of each table and place a + next to a duty if the action for that table is likely to fulfill the duty x owes y; mark the duty with a − if the action is unlikely to fulfill that duty; mark the duty with a +/− if the action partially fulfills it and partially does not; and mark the duty with a ? if the action is irrelevant to the duty or if it is impossible to predict whether or not the duty will be fulfilled. (Few cells generally fall into this last category.)

5. Review the tables from steps 3 and 4. Envision a meeting of all of the parties (or one representative from each of the groups) in which no one knows which role they will take or when they will leave the negotiation. Which alternative do you think such a group would adopt, if any? Do you think such a group could discover a new alternative, perhaps combining the best elements of the previously listed actions? If this thought experiment produces a new alternative, expand the P × A tables from step 3 to include the new alternative action, make a new copy of the P × P table in step 4, and do the + and − marking for the new table.

6. If any one of the alternatives seems to be clearly preferred (i.e., it has high opportunity and low vulnerability for all parties and tends to fulfill all the duties in the P × P table), then that becomes the recommended decision. If no one alternative action stands out, the professionals can examine trade-offs using the charts or can iteratively attempt step 5 (perhaps with outside consultations) until an acceptable alternative is generated.

Using the paramedic method can be time consuming, and it does not eliminate the need for judgment. But it can help organize and focus analysis as an individual or a group works through the details of a situation to arrive at a decision.

2.2.5 Easy and Hard Ethical Decision Making

Sometimes ethical decision making is easy; for example, when it is clear that an action will prevent a serious harm and has no drawbacks, then that action is the right thing to do. Sometimes, however, ethical decision making is more complicated and challenging. Take the following case: your job is to make decisions about which parts to buy for a computer manufacturing company. A person who sells parts to the company offers you tickets to an expensive Broadway show. Should you accept the tickets? In this case, the right thing to do is more complicated because you may be able to accept the tickets and not have this affect your decision about parts. You owe your employer a decision on parts that is in the best interests of the company, but will accepting the tickets influence future decisions?

Other times, you know what the right thing to do is, but doing it will have such great personal costs that you cannot bring yourself to do it. For example, you might be considering blowing the whistle on your employer, who has been extremely kind and generous to you, but who now has asked you to cheat on the testing results on a life-critical software system designed for a client.

To make good decisions, professionals must be aware of potential issues and must have a fairly clear sense of their responsibilities in various kinds of situations. This often requires sorting out complex relationships and obligations, anticipating the effects of various actions, and balancing responsibilities to multiple parties. This activity is part of professional ethics.

2.3 Professional Ethics

Ethics is not just a matter for individuals as individuals. We all occupy a variety of social roles that involve special responsibilities and privileges. As parents, we have special responsibilities for children. As citizens, members of churches, officials in clubs, and so on, we have special rights and duties — and so it is with professional roles. Being a professional is often distinguished from merely having an occupation, because a professional makes a different sort of commitment. Being a professional means more than just having a job. The difference is commitment to doing the right thing because you are a member of a group that has taken on responsibility for a domain of activity. The group is accountable to society for this domain, and for this reason, professionals must behave in ways that are worthy of public trust.

Some theorists explain this commitment in terms of a social contract between a profession and the society in which it functions. Society grants special rights and privileges to the professional group, such as control of admission to the group, access to educational institutions, and confidentiality in professional–client relationships. Society, in turn, may even grant the group a monopoly over a domain of activity (e.g., only licensed engineers can sign off on construction designs, and only doctors can prescribe drugs). In exchange, the professional group promises to self-regulate and practice its profession in ways that are beneficial to society, that is, to promote safety, health, and welfare. The social contract idea is a way of illustrating the importance of the trust that clients and the public put in professionals; it shows the importance of professionals acting so as to be worthy of that trust.

The special responsibilities of professionals have been accounted for in other theoretical frameworks, as well. For example, Davis [1995] argues that members of professions implicitly, if not explicitly, agree among themselves to adhere to certain standards because this elevates the level of activity. If all computer scientists and engineers, for example, agreed never to release software that has not met certain testing standards, this would prevent market pressures from driving down the quality of software being produced. Davis's point is that the special responsibilities of professionals are grounded in what members of a professional group owe to one another: they owe it to one another to live up to agreed-upon rules and standards. Other theorists have tried to ground the special responsibilities of professionals in ordinary morality. Alpern [1991] argues, for example, that the engineer's responsibility for safety derives from the ordinary moral edict *do no harm*. Because engineers are in a position to do greater harm than others, engineers have a special responsibility in their work to take greater care.

In the case of computing professionals, responsibilities are not always well articulated because of several factors. Computing is a relatively new field. There is no single unifying professional association that controls membership, specifies standards of practice, and defines what it means to be a member of the profession. Moreover, many computer scientists and engineers are employees of companies or government agencies, and their role as computer professional may be somewhat in tension with their role as an employee of the company or agency. This can blur an individual's understanding of his or her professional responsibilities. Being a professional means having the independence to make decisions on the basis of special expertise, but being an employee often means acting in the best interests of the company, i.e., being loyal to the organization. Another difficulty in the role of computing professional is the diversity of the field. Computing professionals are employed in a wide variety of contexts, have a wide variety of kinds of expertise, and come from diverse educational backgrounds. As mentioned before, there is no single unifying organization, no uniform admission standard, and no single identifiable professional role.

To be sure, there are pressures on the field to move more in the direction of professionalization, but this seems to be happening to factions of the group rather than to the field as a whole. An important event moving the field in the direction of professionalization was the decision of the state of Texas to provide a licensing system for software engineers. The system specifies a set of requirements and offers an exam that must be passed in order for a computer professional to receive a software engineering license.

At the moment, Texas is the only state that offers such a license, so the field of computing remains loosely organized. It is not a strongly differentiated profession in the sense that there is no single characteristic (or set of characteristics) possessed by all computer professionals, no characteristic that distinguishes members of the group from anyone who possesses knowledge of computing. At this point, the field of computing is best described as a large group of individuals, all of whom work with computers, many of whom have expertise in subfields; they have diverse educational backgrounds, follow diverse career paths, and engage in a wide variety of job activities.

Despite the lack of unity in the field, there are many professional organizations, several professional codes of conduct, and expectations for professional practice. The codes of conduct, in particular, form the basis of an emerging professional ethic that may, in the future, be refined to the point where there will be a strongly differentiated role for computer professionals.

Professional codes play an important role in articulating a collective sense of both the ideal of the profession and the minimum standards required. Codes of conduct state the consensus views of members while shaping behavior.

A number of professional organizations have codes of ethics that are of interest here. The best known include the following:

The Association for Computing Machinery (ACM) Code of Ethics and Professional Conduct (see Appendix B)

The Institute of Electrical and Electronic Engineers (IEEE) Code of Ethics

The Joint ACM/IEEE Software Engineering Code of Ethics and Professional Practice

The Data Processing Managers Association (DPMA, now the Association of Information Technology Professionals [AITP]) Code of Ethics and Standards of Conduct

The Institute for Certification of Computer Professionals (ICCP) Code of Ethics

The Canadian Information Processing Society Code of Ethics

The British Computer Society Code of Conduct

Each of these codes has different emphases and goals. Each in its own way, however, deals with issues that arise in the context in which computer scientists and engineers typically practice.

The codes are relatively consistent in identifying computer professionals as having responsibilities to be faithful to their employers and clients, and to protect public safety and welfare. The most salient ethical issues that arise in professional practice have to do with balancing these responsibilities with personal (or nonprofessional) responsibilities. Two common areas of tension are worth mentioning here, albeit briefly.

As previously mentioned, computer scientists may find themselves in situations in which their responsibility as professionals to protect the public comes into conflict with loyalty to their employer. Such situations sometimes escalate to the point where the computer professional must decide whether to blow the whistle. Such a situation might arise, for example, when the computer professional believes that a piece of software has not been tested enough but her employer wants to deliver the software on time and within the allocated budget (which means immediate release and no more resources being spent on the project). Whether to blow the whistle is one of the most difficult decisions computer engineers and scientists may have to face. Whistle blowing has received a good deal of attention in the popular press and in the literature on professional ethics, because this tension seems to be built into the role of engineers and scientists, that is, the combination of being a professional with highly technical knowledge and being an employee of a company or agency.

Of course, much of the literature on whistle blowing emphasizes strategies that avoid the need for it. Whistle blowing can be avoided when companies adopt mechanisms that give employees the opportunity to express their concerns without fear of repercussions, for example, through ombudspersons to whom engineers and scientists can report their concerns anonymously. The need to blow the whistle can also be diminished when professional societies maintain hotlines that professionals can call for advice on how to get their concerns addressed.

Another important professional ethics issue that often arises is directly tied to the importance of being worthy of client (and, indirectly, public) trust. Professionals can find themselves in situations in which they have (or are likely to have) a conflict of interest. A conflict-of-interest situation is one in which the professional is hired to perform work for a client and the professional has some personal or professional interest that may (or may appear to) interfere with his or her judgment on behalf of the client. For example, suppose a computer professional is hired by a company to evaluate its needs and recommend hardware and software that will best suit the company. The computer professional does precisely what is requested, but fails to mention being a silent partner in a company that manufactures the hardware and software that has been recommended. In other words, the professional has a personal interest — financial benefit — in the company's buying certain equipment. If the company were told this upfront, it might expect the computer professional to favor his own company's equipment; however, if the company finds out about the affiliation later on, it might rightly think that it had been deceived. The professional was hired to evaluate the needs of the company and to determine how best to meet those needs, and in so doing to have the best interests of the company fully in mind. Now, the company suspects that the professional's judgment was biased. The professional had an interest that might have interfered with his judgment on behalf of the company.

There are a number of strategies that professions use to avoid these situations. A code of conduct may, for example, specify that professionals reveal all relevant interests to their clients before they accept a job. Or the code might specify that members never work in a situation where there is even the appearance of a conflict of interest.

This brings us to the special character of computer technology and the effects that the work of computer professionals can have on the shape of the world. Some may argue that computer professionals have very little say in what technologies get designed and built. This seems to be mistaken on at least two counts. First, we can distinguish between computer professionals as individuals and computer professionals as a group. Even if individuals have little power in the jobs they hold, they can exert power collectively. Second, individuals can have an effect if they think of themselves as professionals and consider it their responsibility to anticipate the impact of their work.

2.4 Ethical Issues That Arise from Computer Technology

The effects of a new technology on society can draw attention to an old issue and can change our understanding of that issue. The issues listed in this section — privacy, property rights, risk and reliability, and global communication — were of concern, even problematic, before computers were an important technology. But computing and, more generally, electronic telecommunications, have added new twists and new intensity to each of these issues. Although computer professionals cannot be expected to be experts on all of these issues, it is important for them to understand that computer technology is shaping the world. And it is important for them to keep these impacts in mind as they work with computer technology. Those who are aware of privacy issues, for example, are more likely to take those issues into account when they design database management systems; those who are aware of risk and reliability issues are more likely to articulate these issues to clients and attend to them in design and documentation.

2.4.1 Privacy

Privacy is a central topic in computer ethics. Some have even suggested that privacy is a notion that has been antiquated by technology and that it should be replaced by a new openness. Others think that computers must be harnessed to help restore as much privacy as possible to our society. Although they may not like it, computer professionals are at the center of this controversy. Some are designers of the systems that facilitate information gathering and manipulation; others maintain and protect the information. As the saying goes, *information is power* — but power can be used or abused.

Computer technology creates wide-ranging possibilities for tracking and monitoring of human behavior. Consider just two ways in which personal privacy may be affected by computer technology. First, because of the capacity of computers, massive amounts of information can be gathered by record-keeping organizations such as banks, insurance companies, government agencies, and educational institutions. The information gathered can be kept and used indefinitely, and shared with other organizations rapidly and frequently. A second way in which computers have enhanced the possibilities for monitoring and tracking of individuals is by making possible new kinds of information. When activities are done using a computer, transactional information is created. When individuals use automated bank teller machines, records are created; when certain software is operating, keystrokes on a computer keyboard are recorded; the content and destination of electronic mail can be tracked, and so on. With the assistance of newer technologies, much more of this transactional information is likely to be created. For example, television advertisers may be able to monitor television watchers with scanning devices that record who is sitting in a room facing the television. Highway systems allow drivers to pass through toll booths without stopping as a beam reading a bar code on the automobile charges the toll, simultaneously creating a record of individual travel patterns. All of this information (transactional and otherwise) can be brought together to create a detailed portrait of a person's life, a portrait that the individual may never see, although it is used by others to make decisions about the individual.

This picture suggests that computer technology poses a serious threat to personal privacy. However, one can counter this picture in a number of ways. Is it computer technology *per se* that poses the threat or is it just the way the technology has been used (and is likely to be used in the future)? Computer professionals might argue that they create the technology but are not responsible for how it is used. This argument is, however, problematic for a number of reasons and perhaps foremost because it fails to recognize the potential for solving some of the problems of abuse in the design of the technology. Computer professionals are in the ideal position to think about the potential problems with computers and to design so as to avoid these problems. When, instead of deflecting concerns about privacy as out of their purview, computer professionals set their minds to solve privacy and security problems, the systems they design can improve.

At the same time we think about changing computer technology, we also must ask deeper questions about privacy itself and what it is that individuals need, want, or are entitled to when they express concerns about the loss of privacy. In this sense, computers and privacy issues are ethical issues. They compel us to ask deep questions about what makes for a good and just society. Should individuals have more choice about who has what information about them? What is the proper relationship between citizens and government, between individuals and private corporations? How are we to negotiate the tension between the competing needs for privacy and security? As previously suggested, the questions are not completely new, but some of the possibilities created by computers are new, and these possibilities do not readily fit the concepts and frameworks used in the past. Although we cannot expect computer professionals to be experts on the philosophical and political analysis of privacy, it seems clear that the more they know, the better the computer technology they produce is likely to be.

2.4.2 Property Rights and Computing

The protection of intellectual property rights has become an active legal and ethical debate, involving national and international players. Should software be copyrighted, patented, or free? Is computer software a process, a creative work, a mathematical formalism, an idea, or some combination of these? What is society's stake in protecting software rights? What is society's stake in widely disseminating software? How do corporations and other institutions protect their rights to ideas developed by individuals? And what are the individuals' rights? Such questions must be answered publicly through legislation, through corporate policies, and with the advice of computing professionals. Some of the answers will involve technical details, and all should be informed by ethical analysis and debate.

An issue that has received a great deal of legal and public attention is the ownership of software. In the course of history, software is a relatively new entity. Whereas Western legal systems have developed property laws that encourage invention by granting certain rights to inventors, there are provisions against ownership of things that might interfere with the development of the technological arts and sciences. For this reason, copyrights protect only the expression of ideas, not the ideas themselves, and we do not grant patents on laws of nature, mathematical formulas, and abstract ideas. The problem with computer software is that it has not been clear that we could grant ownership of it without, in effect, granting ownership of numerical sequences or mental steps. Software can be copyrighted, because a copyright gives the holder ownership of the *expression* of the idea (not the idea itself), but this does not give software inventors as much protection as they need to compete *fairly*. Competitors may see the software, grasp the idea, and write a somewhat different program to do the same thing. The competitor can sell the software at less cost because the cost of developing the first software does not have to be paid. Patenting would provide stronger protection, but until quite recently the courts have been reluctant to grant this protection because of the problem previously mentioned: patents on software would appear to give the holder control of the building blocks of the technology, an ownership comparable to owning ideas themselves. In other words, too many patents may interfere with technological development.

Like the questions surrounding privacy, property rights in computer software also lead back to broader ethical and philosophical questions about what constitutes a just society. In computing, as in other areas of technology, we want a system of property rights that promotes invention (creativity, progress), but at the same time, we want a system that is fair in the sense that it rewards those who make significant contributions but does not give anyone so much control that others are prevented from creating. Policies with regard to property rights in computer software cannot be made without an understanding of the technology. This is why it is so important for computer professionals to be involved in public discussion and policy setting on this topic.

2.4.3 Risk, Reliability, and Accountability

As computer technology becomes more important to the way we live, its risks become more worrisome. System errors can lead to physical danger, sometimes catastrophic in scale. There are security risks due to hackers and crackers. Unreliable data and intentional misinformation are risks that are increased because of the technical and economic characteristics of digital data. Furthermore, the use of computer programs is, in a practical sense, inherently unreliable.

Each of these issues (and many more) requires computer professionals to face the linked problems of risk, reliability, and accountability. Professionals must be candid about the risks of a particular application or system. Computing professionals should take the lead in educating customers and the public about what predictions we can and cannot make about software and hardware reliability. Computer professionals should make realistic assessments about costs and benefits, and be willing to take on both for projects in which they are involved.

There are also issues of sharing risks as well as resources. Should liability fall to the individual who buys software or to the corporation that developed it? Should society acknowledge the inherent risks in using software in life-critical situations and shoulder some of the responsibility when something goes wrong? Or should software providers (both individuals and institutions) be exclusively responsible for software safety? All of these issues require us to look at the interaction of technical decisions, human consequences, rights, and responsibilities. They call not just for technical solutions but for solutions that recognize the kind of society we want to have and the values we want to preserve.

2.4.4 Rapidly Evolving Globally Networked Telecommunications

The system of computers and connections known as the Internet provides the infrastructure for new kinds of communities — electronic communities. Questions of individual accountability and social control, as well as matters of etiquette, arise in electronic communities, as in all societies. It is not just that we have societies forming in a new physical environment; it is also that ongoing electronic communication changes the way individuals understand their identity, their values, and their plans for their lives. The changes that are taking place must be examined and understood, especially the changes affecting fundamental social values such as democracy, community, freedom, and peace.

Of course, speculating about the Internet is now a popular pastime, and it is important to separate the hype from the reality. The reality is generally much more complex and much more subtle. We will not engage in speculation and prediction about the future. Rather, we want to emphasize how much better off the world would be if (instead of watching social impacts of computer technology after the fact) computer engineers and scientists were thinking about the potential effects early in the design process. Of course, this can only happen if computer scientists and engineers are encouraged to see the social–ethical issues as a component of their professional responsibility. This chapter has been written with that end in mind.

2.5 Final Thoughts

Computer technology will, no doubt, continue to evolve and will continue to affect the character of the world we live in. Computer scientists and engineers will play an important role in shaping the technology. The technologies we use shape how we live and who we are. They make every difference in the moral environment in which we live. Hence, it seems of utmost importance that computer scientists and engineers understand just how their work affects humans and human values.

References

Alpern, K.D. 1991. Moral responsibility for engineers. In *Ethical Issues in Engineering*, D.G. Johnson, Ed., pp. 187–195. Prentice Hall, Englewood Cliffs, NJ.

Collins, W.R., and Miller, K. 1992. A paramedic method for computing professionals. *J. Syst. Software.* 17(1): 47–84.

Davis, M. 1995. Thinking like an engineer: the place of a code of ethics in the practice of a profession. In *Computers, Ethics, and Social Values*, D.G. Johnson and H. Nissenbaum, Eds., pp. 586–597. Prentice Hall, Englewood Cliffs, NJ.

Johnson, D.G. 2001. *Computer Ethics, 3rd edition.* Prentice Hall, Englewood Cliffs, NJ.

Kant, I. 1785. *Foundations of the Metaphysics of Morals.* L. Beck, trans., 1959. Library of Liberal Arts, 1959.

Rawls, J. 1971. *A Theory of Justice.* Harvard Univ. Press, Cambridge, MA.

I

Algorithms and Complexity

This section addresses the challenges of solving hard problems algorithmically and efficiently. These chapters cover basic methodologies (divide and conquer), data structures, complexity theory (space and time measures), parallel algorithms, and strategies for solving hard problems and identifying unsolvable problems. They also cover some exciting contemporary applications of algorithms, including cryptography, genetics, graphs and networks, pattern matching and text compression, and geometric and algebraic algorithms.

3 Basic Techniques for Design and Analysis of Algorithms
   Edward M. Reingold **3**-1
   Introduction · Analyzing Algorithms · Some Examples of the Analysis
   of Algorithms · Divide-and-Conquer Algorithms · Dynamic Programming
   · Greedy Heuristics

4 Data Structures   *Roberto Tamassia and Bryan M. Cantrill* **4**-1
   Introduction · Sequence · Priority Queue · Dictionary

5 Complexity Theory   *Eric W. Allender, Michael C. Loui, and Kenneth W. Regan* ... **5**-1
   Introduction · Models of Computation · Resources and Complexity
   Classes · Relationships between Complexity Classes · Reducibility and
   Completeness · Relativization of the P vs. NP Problem · The Polynomial
   Hierarchy · Alternating Complexity Classes · Circuit Complexity · Probabilistic
   Complexity Classes · Interactive Models and Complexity Classes · Kolmogorov
   Complexity · Research Issues and Summary

6 Formal Models and Computability   *Tao Jiang, Ming Li, and Bala Ravikumar* **6**-1
   Introduction · Computability and a Universal Algorithm · Undecidability
   · Formal Languages and Grammars · Computational Models

7 Graph and Network Algorithms   *Samir Khuller and Balaji Raghavachari* **7**-1
   Introduction · Tree Traversals · Depth-First Search · Breadth-First Search
   · Single-Source Shortest Paths · Minimum Spanning Trees · Matchings and Network
   Flows · Tour and Traversal Problems

8 Algebraic Algorithms   *Angel Diaz, Erich Kaltófen, and Victor Y. Pan* **8**-1
   Introduction · Matrix Computations and Approximation of Polynomial Zeros
   · Systems of Nonlinear Equations and Other Applications · Polynomial Factorization

9 Cryptography   *Jonathan Katz* **9**-1
   Introduction · Cryptographic Notions of Security · Building Blocks
   · Cryptographic Primitives · Private-Key Encryption · Message
   Authentication · Public-Key Encryption · Digital Signature Schemes

10 Parallel Algorithms *Guy E. Blelloch and Bruce M. Maggs* **10**-1
Introduction • Modeling Parallel Computations • Parallel Algorithmic Techniques
• Basic Operations on Sequences, Lists, and Trees • Graphs • Sorting
• Computational Geometry • Numerical Algorithms
• Parallel Complexity Theory

11 Computational Geometry *D. T. Lee* **11**-1
Introduction • Problem Solving Techniques • Classes of Problems • Conclusion

12 Randomized Algorithms *Rajeev Motwani and Prabhakar Raghavan* **12**-1
Introduction • Sorting and Selection by Random Sampling • A Simple Min-Cut
Algorithm • Foiling an Adversary • The Minimax Principle and Lower
Bounds • Randomized Data Structures • Random Reordering and Linear
Programming • Algebraic Methods and Randomized Fingerprints

13 Pattern Matching and Text Compression Algorithms *Maxime Crochemore
and Thierry Lecroq* **13**-1
Processing Texts Efficiently • String-Matching Algorithms • Two-Dimensional Pattern
Matching Algorithms • Suffix Trees • Alignment • Approximate String Matching
• Text Compression • Research Issues and Summary

14 Genetic Algorithms *Stephanie Forrest* **14**-1
Introduction • Underlying Principles • Best Practices • Mathematical Analysis
of Genetic Algorithms • Research Issues and Summary

15 Combinatorial Optimization *Vijay Chandru and M. R. Rao* **15**-1
Introduction • A Primer on Linear Programming • Large-Scale Linear Programming
in Combinatorial Optimization • Integer Linear Programs • Polyhedral
Combinatorics • Partial Enumeration Methods • Approximation in Combinatorial
Optimization • Prospects in Integer Programming

3

Basic Techniques for Design and Analysis of Algorithms

3.1 Introduction 3-1
3.2 Analyzing Algorithms 3-1
     Linear Recurrences • Divide-and-Conquer Recurrences
3.3 Some Examples of the Analysis of Algorithms 3-5
     Sorting • Priority Queues
3.4 Divide-and-Conquer Algorithms 3-10
3.5 Dynamic Programming 3-12
3.6 Greedy Heuristics 3-17

Edward M. Reingold
Illinois Institute of Technology

3.1 Introduction

We outline the basic methods of algorithm design and analysis that have found application in the manipulation of discrete objects such as lists, arrays, sets, graphs, and geometric objects such as points, lines, and polygons. We begin by discussing recurrence relations and their use in the analysis of algorithms. Then we discuss some specific examples in algorithm analysis, sorting, and priority queues. In the final three sections, we explore three important techniques of algorithm design: divide-and-conquer, dynamic programming, and greedy heuristics.

3.2 Analyzing Algorithms

It is convenient to classify algorithms based on the relative amount of time they require: how fast does the time required grow as the size of the problem increases? For example, in the case of arrays, the "size of the problem" is ordinarily the number of elements in the array. If the size of the problem is measured by a variable n, we can express the time required as a function of n, $T(n)$. When this function $T(n)$ grows rapidly, the algorithm becomes unusable for large n; conversely, when $T(n)$ grows slowly, the algorithm remains useful even when n becomes large.

We say an algorithm is $\Theta(n^2)$ if the time it takes quadruples when n doubles; an algorithm is $\Theta(n)$ if the time it takes doubles when n doubles; an algorithm is $\Theta(\log n)$ if the time it takes increases by a constant, independent of n, when n doubles; an algorithm is $\Theta(1)$ if its time does not increase at all when n increases. In general, an algorithm is $\Theta(T(n))$ if the time it requires on problems of size n grows proportionally to $T(n)$ as n increases. Table 3.1 summarizes the common growth rates encountered in the analysis of algorithms.

TABLE 3.1   Common Growth Rates of Times of Algorithms

Rate of Growth	Comment	Examples
$\Theta(1)$	Time required is constant, independent of problem size	Expected time for hash searching
$\Theta(\log \log n)$	Very slow growth of time required	Expected time of interpolation search
$\Theta(\log n)$	Logarithmic growth of time required — doubling the problem size increases the time by only a constant amount	Computing x^n; binary search of an array
$\Theta(n)$	Time grows linearly with problem size — doubling the problem size doubles the time required	Adding/subtracting n-digit numbers; linear search of an n-element array
$\Theta(n \log n)$	Time grows worse than linearly, but not much worse — doubling the problem size more than doubles the time required	Merge sort; heapsort; lower bound on comparison-based sorting
$\Theta(n^2)$	Time grows quadratically — doubling the problem size quadruples the time required	Simple-minded sorting algorithms
$\Theta(n^3)$	Time grows cubically — doubling the problem size results in an eight fold increase in the time required	Ordinary matrix multiplication
$\Theta(c^n)$	Time grows exponentially — increasing the problem size by 1 results in a c-fold increase in the time required; doubling the problem size *squares* the time required	Traveling salesman problem

The analysis of an algorithm is often accomplished by finding and solving a recurrence relation that describes the time required by the algorithm. The most commonly occurring families of recurrences in the analysis of algorithms are linear recurrences and divide-and-conquer recurrences. In the following subsection we describe the "method of operators" for solving linear recurrences; in the next subsection we describe how to transform divide-and-conquer recurrences into linear recurrences by substitution to obtain an asymptotic solution.

3.2.1   Linear Recurrences

A *linear recurrence with constant coefficients* has the form

$$c_0 a_n + c_1 a_{n-1} + c_2 a_{n-2} + \cdots + c_k a_{n-k} = f(n), \tag{3.1}$$

for some constant k, where each c_i is constant. To solve such a recurrence for a broad class of functions f (that is, to express a_n in closed form as a function of n) by the *method of operators*, we consider two basic operators on sequences: $\mathcal{S}$, which shifts the sequence left,

$$\mathcal{S}\langle a_0, a_1, a_2, \ldots \rangle = \langle a_1, a_2, a_3, \ldots \rangle,$$

and C, which, for any constant C, multiplies each term of the sequence by C:

$$C\langle a_0, a_1, a_2, \ldots \rangle = \langle Ca_0, Ca_1, Ca_2, \ldots \rangle.$$

Then, given operators A and B, we define the sum and product

$$(A + B)\langle a_0, a_1, a_2, \ldots \rangle = A\langle a_0, a_1, a_2, \ldots \rangle + B\langle a_0, a_1, a_2, \ldots \rangle,$$

$$(AB)\langle a_0, a_1, a_2, \ldots \rangle = A(B\langle a_0, a_1, a_2, \ldots \rangle).$$

Thus, for example,

$$(\mathcal{S}^2 - 4)\langle a_0, a_1, a_2, \ldots \rangle = \langle a_2 - 4a_0, a_3 - 4a_1, a_4 - 4a_2, \ldots \rangle,$$

which we write more briefly as

$$(\mathcal{S}^2 - 4)\langle a_i \rangle = \langle a_{i+2} - 4a_i \rangle.$$

With the operator notation, we can rewrite Equation (3.1) as

$$P(S)\langle a_i \rangle = \langle f(i) \rangle,$$

where

$$P(S) = c_0 S^k + c_1 S^{k-1} + c_2 S^{k-2} + \cdots + c_k$$

is a polynomial in S.

Given a sequence $\langle a_i \rangle$, we say that the operator $P(S)$ *annihilates* $\langle a_i \rangle$ if $P(S)\langle a_i \rangle = \langle 0 \rangle$. For example, $S^2 - 4$ annihilates any sequence of the form $\langle u2^i + v(-2)^i \rangle$, with constants u and v. In general,

The operator $S^{k+1} - c$ annihilates $\langle c^i \times$ a polynomial in i of degree $k \rangle$.

The *product* of two annihilators annihilates the *sum* of the sequences annihilated by each of the operators, that is, if A annihilates $\langle a_i \rangle$ and B annihilates $\langle b_i \rangle$, then AB annihilates $\langle a_i + b_i \rangle$. Thus, determining the annihilator of a sequence is tantamount to determining the sequence; moreover, it is straightforward to determine the annihilator from a recurrence relation.

For example, consider the Fibonacci recurrence

$$F_0 = 0$$

$$F_1 = 1$$

$$F_{i+2} = F_{i+1} + F_i.$$

The last line of this definition can be rewritten as $F_{i+2} - F_{i+1} - F_i = 0$, which tells us that $\langle F_i \rangle$ is annihilated by the operator

$$S^2 - S - 1 = (S - \phi)(S + 1/\phi),$$

where $\phi = (1 + \sqrt{5})/2$. Thus we conclude that

$$F_i = u\phi^i + v(-\phi)^{-i}$$

for some constants u and v. We can now use the initial conditions $F_0 = 0$ and $F_1 = 1$ to determine u and v: These initial conditions mean that

$$u\phi^0 + v(-\phi)^{-0} = 0$$

$$u\phi^1 + v(-\phi)^{-1} = 1$$

and these linear equations have the solution

$$u = v = 1/\sqrt{5},$$

and hence

$$F_i = \phi^i/\sqrt{5} + (-\phi)^{-i}/\sqrt{5}.$$

In the case of the similar recurrence,

$$G_0 = 0$$

$$G_1 = 1$$

$$G_{i+2} = G_{i+1} + G_i + i,$$

TABLE 3.2   Rate of Growth of the Solution to the
Recurrence $T(n) = g(n) + uT(n/v)$: The
Divide-and-Conquer Recurrence Relations

$g(n)$	u, v	Growth Rate of $T(n)$
$\Theta(1)$	$u = 1$	$\Theta(\log n)$
	$u \neq 1$	$\Theta(n^{\log_v u})$
$\Theta(\log n)$	$u = 1$	$\Theta[(\log n)^2]$
	$u \neq 1$	$\Theta(n^{\log_v u})$
$\Theta(n)$	$u < v$	$\Theta(n)$
	$u = v$	$\Theta(n \log n)$
	$u > v$	$\Theta(n^{\log_v u})$
$\Theta(n^2)$	$u < v^2$	$\Theta(n^2)$
	$u = v^2$	$\Theta(n^2 \log n)$
	$u > v^2$	$\Theta(n^{\log_v u})$

u and v are positive constants, independent of n, and $v > 1$.

the last equation tells us that

$$(\mathcal{S}^2 - \mathcal{S} - 1)\langle G_i \rangle = \langle i \rangle,$$

so the annihilator for $\langle G_i \rangle$ is $(\mathcal{S}^2 - \mathcal{S} - 1)(\mathcal{S} - 1)^2$ since $(\mathcal{S} - 1)^2$ annihilates $\langle i \rangle$ (a polynomial of degree 1 in i) and hence the solution is

$$G_i = u\phi^i + v(-\phi)^{-i} + \text{(a polynomial of degree 1 in i)};$$

that is,

$$G_i = u\phi^i + v(-\phi)^{-i} + wi + z.$$

Again, we use the initial conditions to determine the constants u, v, w, and x.

In general, then, to solve the recurrence in Equation 3.1, we factor the annihilator

$$P(\mathcal{S}) = c_0\mathcal{S}^k + c_1\mathcal{S}^{k-1} + c_2\mathcal{S}^{k-2} + \cdots + c_k,$$

multiply it by the annihilator for $\langle f(i) \rangle$, write the form of the solution from this product (which is the annihilator for the sequence $\langle a_i \rangle$), and the use the initial conditions for the recurrence to determine the coefficients in the solution.

3.2.2 Divide-and-Conquer Recurrences

The divide-and-conquer paradigm of algorithm construction that we discuss in Section 4 leads naturally to divide-and-conquer recurrences of the type

$$T(n) = g(n) + uT(n/v),$$

for constants u and v, $v > 1$, and sufficient initial values to define the sequence $\langle T(0), T(1), T(2), \ldots \rangle$. The growth rates of $T(n)$ for various values of u and v are given in Table 3.2. The growth rates in this table are derived by transforming the divide-and-conquer recurrence into a linear recurrence for a subsequence of $\langle T(0), T(1), T(2), \ldots \rangle$.

To illustrate this method, we derive the penultimate line in Table 3.2. We want to solve

$$T(n) = n^2 + v^2 T(n/v).$$

So, we want to find a subsequence of $\langle T(0), T(1), T(2), \ldots \rangle$ that will be easy to handle. Let $n_k = v^k$; then,

$$T(n_k) = n_k^2 + v^2 T(n_k/v),$$

or

$$T(v^k) = v^{2k} + v^2 T(v^{k-1}).$$

Defining $t_k = T(v^k)$,

$$t_k = v^{2k} + v^2 t_{k-1}.$$

The annihilator for t_k is then $(\mathcal{S} - v^2)^2$ and thus

$$t_k = v^{2k}(ak + b),$$

for constants a and b. Expressing this in terms of $T(n)$,

$$T(n) \approx t_{\log_v n} = v^{2 \log_v n}(a \log_v n + b) = an^2 \log_v n + bn^2,$$

or,

$$T(n) = \Theta(n^2 \log n).$$

3.3 Some Examples of the Analysis of Algorithms

In this section we introduce the basic ideas of analyzing algorithms by looking at some data structure problems that commonly occur in practice, problems relating to maintaining a collection of n objects and retrieving objects based on their relative size. For example, how can we determine the smallest of the elements? Or, more generally, how can we determine the kth largest of the elements? What is the running time of such algorithms in the worst case? Or, on average, if all $n!$ permutations of the input are equally likely? What if the set of items is dynamic — that is, the set changes through insertions and deletions — how efficiently can we keep track of, say, the largest element?

3.3.1 Sorting

The most demanding request that we can make of an array of n values $\texttt{x[1]}, \texttt{x[2]}, \ldots, \texttt{x[n]}$ is that they be kept in perfect order so that $\texttt{x[1]} \leq \texttt{x[2]} \leq \cdots \leq \texttt{x[n]}$. The simplest way to put the values in order is to mimic what we might do by hand: take item after item and insert each one into the proper place among those items already inserted:

```
 1  void insert (float x[], int i, float a) {
 2    // Insert a into x[1] ... x[i]
 3    // x[1] ... x[i-1] are sorted;  x[i] is unoccupied
 4    if (i == 1 || x[i-1] <= a)
 5      x[i] = a;
 6    else {
 7      x[i] = x[i-1];
 8      insert(x, i-1, a);
 9    }
10  }
11
12  void insertionSort (int n, float x[]) {
13    // Sort  x[1] ... x[n]
```

```
14  if (n > 1) {
15      insertionSort(n-1, x);
16      insert(x, n, x[n]);
17  }
18  }
```

To determine the time required in the worst case to sort n elements with `insertionSort`, we let t_n be the time to sort n elements and derive and solve a recurrence relation for t_n. We have,

$$t_n \begin{cases} \Theta(1) & \text{if } n = 1, \\ t_{n-1} + s_{n-1} + \Theta(1) & \text{otherwise,} \end{cases}$$

where s_m is the time required to insert an element in place among m elements using `insert`. The value of s_m is also given by a recurrence relation:

$$s_m \begin{cases} \Theta(1) & \text{if } m = 1, \\ s_{m-1} + \Theta(1) & \text{otherwise.} \end{cases}$$

The annihilator for $\langle s_i \rangle$ is $(S-1)^2$, so $s_m = \Theta(m)$. Thus, the annihilator for $\langle t_i \rangle$ is $(S-1)^3$, so $t_n = \Theta(n^2)$. The analysis of the average behavior is nearly identical; only the constants hidden in the Θ-notation change.

We can design better sorting methods using the divide-and-conquer idea of the next section. These algorithms avoid $\Theta(n^2)$ worst-case behavior, working in time $\Theta(n \log n)$. We can also achieve time $\Theta(n \log n)$ using a clever way of viewing the array of elements to be sorted as a tree: consider `x[1]` as the root of the tree and, in general, `x[2*i]` is the root of the left subtree of `x[i]` and `x[2*i+1]` is the root of the right subtree of `x[i]`. If we further insist that parents be greater than or equal to children, we have a *heap*; Figure 3.1 shows a small example.

A heap can be used for sorting by observing that the largest element is at the root, that is, `x[1]`; thus, to put the largest element in place, we swap `x[1]` and `x[n]`. To continue, we must restore the heap property, which may now be violated at the root. Such restoration is accomplished by swapping `x[1]` with its larger child, if that child is larger than `x[1]`, and the continuing to swap it downward until either it reaches the bottom or a spot where it is greater or equal to its children. Because the tree-cum-array has height $\Theta(\log n)$, this restoration process takes time $\Theta(\log n)$. Now, with the heap in `x[1]` to `x[n-1]` and `x[n]` the largest value in the array, we can put the second largest element in place by swapping `x[1]` and `x[n-1]`; then we restore the heap property in `x[1]` to `x[n-2]` by propagating `x[1]` downward; this takes time $\Theta(\log(n-1))$. Continuing in this fashion, we find we can sort the entire array in time

$$\Theta(\log n + \log(n-1) + \cdots + \log 1) = \Theta(n \log n).$$

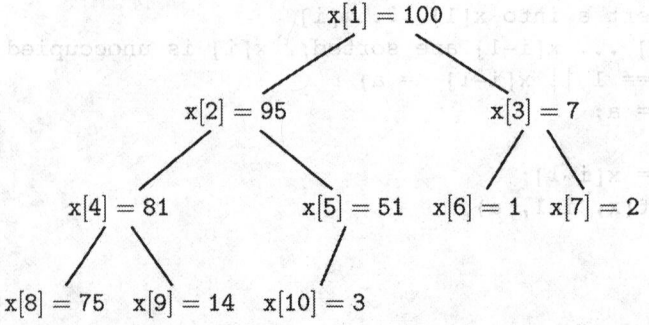

FIGURE 3.1 A heap — that is, an array, interpreted as a binary tree.

The initial creation of the heap from an unordered array is done by applying the restoration process successively to $x[n/2], x[n/2-1], \ldots, x[1]$, which takes time $\Theta(n)$.

Hence, we have the following $\Theta(n \log n)$ sorting algorithm:

```
1   void heapify (int n, float x[], int i) {
2     // Repair heap property below x[i] in x[1] ... x[n]
3     int largest = i;   // largest of x[i], x[2*i], x[2*i+1]
4     if (2*i <= n && x[2*i] > x[i])
5       largest = 2*i;
6     if (2*i+1 <= n && x[2*i+1] > x[largest])
7       largest = 2*i+1;
8     if (largest != i) {
9       // swap x[i] with larger child and repair heap below
10       float t = x[largest]; x[largest] = x[i]; x[i] = t;
11       heapify(n, x, largest);
12     }
13  }
14
15  void makeheap (int n, float x[]) {
16    // Make x[1] ... x[n] into a heap
17    for (int i=n/2; i>0; i--)
18      heapify(n, x, i);
19  }
20
21  void heapsort (int n, float x[]) {
22    // Sort  x[1] ... x[n]
23    float t;
24    makeheap(n, x);
25    for (int i=n; i>1; i--) {
26      // put x[1] in place and repair heap
27      t = x[1]; x[1] = x[i]; x[i] = t;
28      heapify(i-1, x, 1);
29    }
30  }
```

Can we find sorting algorithms that take less time than $\Theta(n \log n)$? The answer is no if we are restricted to sorting algorithms that derive their information from comparisons between the values of elements. The flow of control in such sorting algorithms can be viewed as binary trees in which there are $n!$ leaves, one for every possible sorted output arrangement. Because a binary tree with height h can have at most 2^h leaves, it follows that the height of a tree with $n!$ leaves must be at least $\log_2 n! = \Theta(n \log n)$. Because the height of this tree corresponds to the longest sequence of element comparisons possible in the flow of control, any such sorting algorithm must, in its worst case, use time proportional to $n \log n$.

3.3.2 Priority Queues

Aside from its application to sorting, the heap is an interesting data structure in its own right. In particular, heaps provide a simple way to implement a *priority queue*; a priority queue is an abstract data structure that keeps track of a dynamically changing set of values allowing the operations

create: Create an empty priority queue.
insert: Insert a new element into a priority queue.
decrease: Decrease an element in a priority queue.
minimum: Report the minimum element in a priority queue.

`deleteMinimum:` Delete the minimum element in a priority queue.
`delete:` Delete an element in a priority queue.
`merge:` Merge two priority queues.

A heap can implement a priority queue by altering the heap property to insist that parents are less than or equal to their children, so that that smallest value in the heap is at the root, that is, in the first array position. Creation of an empty heap requires just the allocation of an array, an $\Theta(1)$ operation; we assume that once created, the array containing the heap can be extended arbitrarily at the right end. Inserting a new element means putting that element in the $(n+1)$st location and "bubbling it up" by swapping it with its parent until it reaches either the root or a parent with a smaller value. Because a heap has logarithmic height, insertion to a heap of n elements thus requires worst-case time $O(\log n)$. Decreasing a value in a heap requires only a similar $O(\log n)$ "bubbling up." The minimum element of such a heap is always at the root, so reporting it takes $\Theta(1)$ time. Deleting the minimum is done by swapping the first and last array positions, bubbling the new root value downward until it reaches its proper location, and truncating the array to eliminate the last position. `Delete` is handled by decreasing the value so that it is the least in the heap and then applying the `deleteMinimum` operation; this takes a total of $O(\log n)$ time.

The `merge` operation, unfortunately, is not so economically accomplished; there is little choice but to create a new heap out of the two heaps in a manner similar to the `makeheap` function in heapsort. If there are a total of n elements in the two heaps to be merged, this re-creation will require time $O(n)$.

There are better data structures than a heap for implementing priority queues, however. In particular, the *Fibonacci heap* provides an implementation of priority queues in which the `delete` and `deleteMinimum` operations take $O(\log n)$ time and the remaining operations take $\Theta(1)$ time, *provided we consider the times required for a sequence of priority queue operations, rather than individual times.* That is, we must consider the cost of the individual operations *amortized over the sequence of operations*: Given a sequence of n priority queue operations, we will compute the total time $T(n)$ for all n operations. In doing this computation, however, we do not simply add the costs of the individual operations; rather, we subdivide the cost of each operation into two parts: the *immediate cost* of doing the operation and the *long-term savings* that result from doing the operation. The long-term savings represent costs *not* incurred by later operations as a result of the present operation. The immediate cost minus the long-term savings give the amortized cost of the operation.

It is easy to calculate the immediate cost (time required) of an operation, but how can we measure the long-term savings that result? We imagine that the data structure has associated with it a bank account; at any given moment, the bank account must have a non-negative balance. When we do an operation that will save future effort, we are making a deposit to the savings account; and when, later on, we derive the benefits of that earlier operation, we are making a withdrawal from the savings account. Let $\mathcal{B}(i)$ denote the balance in the account after the ith operation, $\mathcal{B}(0) = 0$. We define the amortized cost of the ith operation to be

Amortized cost of ith operation = (Immediate cost of ith operation) + (Change in bank account)

$$= \text{(Immediate cost of } i\text{th operation)} + (\mathcal{B}(i) - \mathcal{B}(i-1)).$$

Because the bank account $\mathcal{B}$ can go up or down as a result of the ith operation, the amortized cost may be less than or more than the immediate cost. By summing the previous equation, we get

$$\sum_{i=1}^{n}(\text{Amortized cost of } i\text{th operation}) = \sum_{i=1}^{n}(\text{Immediate cost of } i\text{th operation}) + (\mathcal{B}(n) - \mathcal{B}(0))$$

$$= \text{(Total cost of all } n \text{ operations)} + \mathcal{B}(n)$$

$$\geq \text{Total cost of all } n \text{ operations}$$

$$= T(n)$$

because $B(i)$ is non-negative. Thus defined, the sum of the amortized costs of the operations gives us an upper bound on the total time $T(n)$ for all n operations.

It is important to note that the function $B(i)$ is not part of the data structure, but is just our way to measure how much time is used by the sequence of operations. As such, we can choose *any rules* for B, provided $B(0) = 0$ and $B(i) \geq 0$ for $i \geq 1$. Then the sum of the amortized costs defined by

Amortized cost of ith operation = (Immediate cost of ith operation) + $(B(i) - B(i - 1))$

bounds the overall cost of the operation of the data structure.

Now to apply this method to priority queues. A *Fibonacci heap* is a list of heap-ordered trees (not necessarily binary); because the trees are heap ordered, the minimum element must be one of the roots and we keep track of which root is the overall minimum. Some of the tree nodes are *marked*. We define

$$B(i) = (\text{Number of trees after the } i\text{th operation})$$
$$+ 2 \times (\text{Number of marked nodes after the } i\text{th operation}).$$

The clever rules by which nodes are marked and unmarked, and the intricate algorithms that manipulate the set of trees, are too complex to present here in their complete form, so we just briefly describe the simpler operations and show the calculation of their amortized costs:

Create: To create an empty Fibonacci heap we create an empty list of heap-ordered trees. The immediate cost is $\Theta(1)$; because the numbers of trees and marked nodes are zero before and after this operation, $B(i) - B(i - 1)$ is zero and the amortized time is $\Theta(1)$.

Insert: To insert a new element into a Fibonacci heap we add a new one-element tree to the list of trees constituting the heap and update the record of what root is the overall minimum. The immediate cost is $\Theta(1)$. $B(i) - B(i - 1)$ is also 1 because the number of trees has increased by 1, while the number of marked nodes is unchanged. The amortized time is thus $\Theta(1)$.

Decrease: Decreasing an element in a Fibonacci heap is done by cutting the link to its parent, if any, adding the item as a root in the list of trees, and decreasing its value. Furthermore, the marked parent of a cut element is itself cut, propagating upward in the tree. Cut nodes become unmarked, and the unmarked parent of a cut element becomes marked. The immediate cost of this operation is $\Theta(c)$, where c is the number of cut nodes. If there were t trees and m marked elements before this operation, the value of B before the operation was $t + 2m$. After the operation, the value of B is $(t + c) + 2(m - c + 2)$, so $B(i) - B(i - 1) = 4 - c$. The amortized time is thus $\Theta(c) + 4 - c = \Theta(1)$ *by changing the definition of B by a multiplicative constant large enough to dominate the constant hidden in $\Theta(c)$.*

Minimum: Reporting the minimum element in a Fibonacci heap takes time $\Theta(1)$ and does not change the numbers of trees and marked nodes; the amortized time is thus $\Theta(1)$.

DeleteMinimum: Deleting the minimum element in a Fibonacci heap is done by deleting that tree root, making its children roots in the list of trees. Then, the list of tree roots is "consolidated" in a complicated $O(\log n)$ operation that we do not describe. The result takes amortized time $O(\log n)$.

Delete: Deleting an element in a Fibonacci heap is done by decreasing its value to $-\infty$ and then doing a `deleteMinimum`. The amortized cost is the sum of the amortized cost of the two operations, $O(\log n)$.

Merge: Merging two Fibonacci heaps is done by concatenating their lists of trees and updating the record of which root is the minimum. The amortized time is thus $\Theta(1)$.

Notice that the amortized cost of each operation is $\Theta(1)$ except `deleteMinimum` and `delete`, both of which are $O(\log n)$.

3.4   Divide-and-Conquer Algorithms

One approach to the design of algorithms is to decompose a problem into subproblems that resemble the original problem, but on a reduced scale. Suppose, for example, that we want to compute x^n. We reason that the value we want can be computed from $x^{\lfloor n/2 \rfloor}$ because

$$x^n = \begin{cases} 1 & \text{if } n = 0, \\ (x^{\lfloor n/2 \rfloor})2 & \text{if } n \text{ is even}, \\ x \times (x^{\lfloor n/2 \rfloor})2 & \text{if } n \text{ is odd}. \end{cases}$$

This recursive definition can be translated directly into

```
 1   int power (float x, int n) {
 2    // Compute the n-th power of x
 3    if (n == 0)
 4       return 1;
 5    else {
 6       int t = power(x, floor(n/2));
 7       if ((n % 2) == 0)
 8          return t*t;
 9       else
10          return x*t*t;
11    }
12  }
```

To analyze the time required by this algorithm, we notice that the time will be proportional to the number of multiplication operations performed in lines 8 and 10, so the divide-and-conquer recurrence

$$T(n) = 2 + T(\lfloor n/2 \rfloor),$$

with $T(0) = 0$, describes the rate of growth of the time required by this algorithm. By considering the subsequence $n_k = 2^k$, we find, using the methods of the previous section, that $T(n) = \Theta(\log n)$. Thus, the above algorithm is considerably more efficient than the more obvious

```
 1   int power (int k, int n) {
 2    // Compute the n-th power of k
 3    int product = 1;
 4    for (int i = 1; i <= n; i++)
 5       // at this point power is k*k*k*...*k (i times)
 6       product = product * k;
 7    return product;
 8  }
```

which requires time $\Theta(n)$.

An extremely well-known instance of divide-and-conquer algorithms is *binary search* of an ordered array of n elements for a given element; we "probe" the middle element of the array, continuing in either the lower or upper segment of the array, depending on the outcome of the probe:

```
 1   int binarySearch (int x, int w[], int low, int high) {
 2    // Search for x among sorted array w[low..high]. The integer returned
 3    // is either the location of x in w, or the location where x belongs.
 4    if (low > high) // Not found
 5       return low;
```

```
 6    else {
 7       int middle := (low+high)/2;
 8       if (w[middle] < x)
 9          return binarySearch(x, w, middle+1, high);
10       else if (w[middle] == x)
11          return middle;
12       else
13          return binarySearch(x, w, low, middle-1);
14    }
15 }
```

The analysis of binary search in an array of n elements is based on counting the number of probes used in the search, because all remaining work is proportional to the number of probes. But, the number of probes needed is described by the divide-and-conquer recurrence

$$T(n) = 1 + T(n/2),$$

with $T(0) = 0$, $T(1) = 1$. We find from Table 3.2 (the top line) that $T(n) = \Theta(\log n)$. Hence, binary search is much more efficient than a simple linear scan of the array.

To multiply two very large integers x and y, assume that x has exactly $l \geq 2$ digits and y has at most l digits. Let $x_0, x_1, x_2, \ldots, x_{l-1}$ be the digits of x and let $y_0, y_1, \ldots, y_{l-1}$ be the digits of y (some of the significant digits at the end of y may be zeros, if y is shorter than x), so that

$$x = x_0 + 10x_1 + 10^2 x_2 + \cdots + 10^{l-1} x_{l-1},$$

and

$$y = y_0 + 10y_1 + 10^2 y_2 + \cdots + 10^{l-1} y_{l-1},$$

We apply the divide-and-conquer idea to multiplication by chopping x into two pieces — the leftmost n digits and the remaining digits:

$$x = x_{\text{left}} + 10^n x_{\text{right}},$$

where $n = l/2$. Similarly, chop y into two corresponding pieces:

$$y = y_{\text{left}} + 10^n y_{\text{right}},$$

because y has at most the number of digits that x does, y_{right} might be 0. The product $x \times y$ can be now written

$$x \times y = (x_{\text{left}} + 10^n x_{\text{right}}) \times (y_{\text{left}} + 10^n y_{\text{right}}),$$

$$= x_{\text{left}} \times y_{\text{left}}$$

$$+ 10^n (x_{\text{right}} \times y_{\text{left}} + x_{\text{left}} \times y_{\text{right}})$$

$$+ 10^{2n} x_{\text{right}} \times y_{\text{right}}.$$

If $T(n)$ is the time to multiply two n-digit numbers with this method, then

$$T(n) = kn + 4T(n/2);$$

the kn part is the time to chop up x and y and to do the needed additions and shifts; each of these tasks involves n-digit numbers and hence $\Theta(n)$ time. The $4T(n/2)$ part is the time to form the four needed subproducts, each of which is a product of about $n/2$ digits.

The line for $g(n) = \Theta(n)$, $u = 4 > v = 2$ in Table 3.2 tells us that $T(n) = \Theta(n^{\log_2 4}) = \Theta(n^2)$, so the divide-and-conquer algorithm is no more efficient than the elementary-school method of multiplication. However, we can be more economical in our formation of subproducts:

$$x \times y = (x_{\text{left}} + 10^n x_{\text{right}}) \times (y_{\text{left}} + 10^n y_{\text{right}}),$$
$$= B + 10^n C + 10^{2n} A,$$

where

$$A = x_{\text{right}} \times y_{\text{right}}$$
$$B = x_{\text{left}} \times y_{\text{left}}$$
$$C = (x_{\text{left}} + x_{\text{right}}) \times (y_{\text{left}} + y_{\text{right}}) - A - B.$$

The recurrence for the time required changes to

$$T(n) = kn + 3T(n/2).$$

The kn part is the time to do the two additions that form $x \times y$ from A, B, and C and the two additions and the two subtractions in the formula for C; each of these six additions/subtractions involves n-digit numbers. The $3T(n/2)$ part is the time to (recursively) form the three needed products, each of which is a product of about $n/2$ digits. The line for $g(n) = \Theta(n)$, $u = 3 > v = 2$ in Table 3.2 now tells us that

$$T(n) = \Theta(n^{\log_2 3}).$$

Now,

$$\log_2 3 = \frac{\log_{10} 3}{\log_{10} 2} \approx 1.5849625\cdots,$$

which means that this divide-and-conquer multiplication technique will be faster than the straightforward $\Theta(n^2)$ method for large numbers of digits.

Sorting a sequence of n values efficiently can be done using the divide-and-conquer idea. Split the n values arbitrarily into two piles of $n/2$ values each, sort each of the piles separately, and then merge the two piles into a single sorted pile. This sorting technique, pictured in Figure 3.2, is called *merge sort*. Let $T(n)$ be the time required by merge sort for sorting n values. The time needed to do the merging is proportional to the number of elements being merged, so that

$$T(n) = cn + 2T(n/2),$$

because we must sort the two halves (time $T(n/2)$ each) and then merge (time proportional to n). We see by Table 3.2 that the growth rate of $T(n)$ is $\Theta(n \log n)$, since $u = v = 2$ and $g(n) = \Theta(n)$.

3.5 Dynamic Programming

In the design of algorithms to solve optimization problems, we need to make the optimal (lowest cost, highest value, shortest distance, etc.) choice from among a large number of alternative solutions. *Dynamic programming* is an organized way to find an optimal solution by systematically exploring all possibilities without unnecessary repetition. Often, dynamic programming leads to efficient, polynomial-time algorithms for problems that appear to require searching through exponentially many possibilities.

Like the divide-and-conquer method, dynamic programming is based on the observation that many optimization problems can be solved by solving similar subproblems and the composing the solutions of those subproblems into a solution for the original problem. In addition, the problem is viewed as

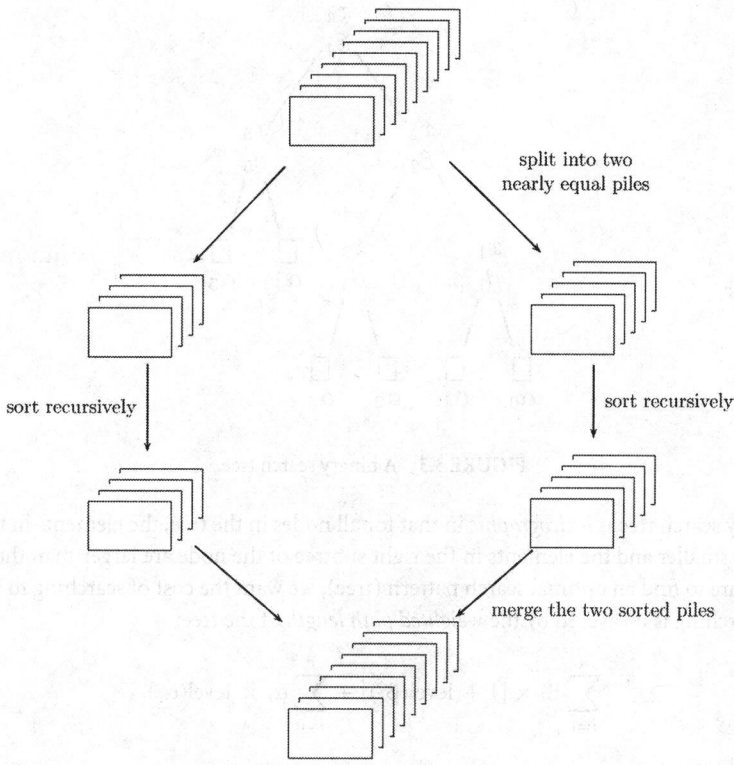

split into two
nearly equal piles

sort recursively

sort recursively

merge the two sorted piles

FIGURE 3.2 Schematic description of merge sort.

a sequence of decisions, each decision leading to different subproblems; if a wrong decision is made, a suboptimal solution results, so all possible decisions need to be accounted for.

As an example of dynamic programming, consider the problem of constructing an optimal search pattern for probing an ordered sequence of elements. The problem is similar to searching an array. In the previous section we described binary search, in which an interval in an array is repeatedly bisected until the search ends. Now, however, suppose we know the frequencies with which the search will seek various elements (both in the sequence and missing from it). For example, if we know that the last few elements in the sequence are frequently sought — binary search does not make use of this information — it might be more efficient to begin the search at the right end of the array, not in the middle. Specifically, we are given an ordered sequence $x_1 < x_2 < \cdots < x_n$ and associated frequencies of access $\beta_1, \beta_2, \ldots, \beta_n$, respectively; furthermore, we are given $\alpha_0, \alpha_1, \ldots, \alpha_n$ where α_i is the frequency with which the search will fail because the object sought, z, was missing from the sequence, $x_i < z < x_{i+1}$ (with the obvious meaning when $i = 0$ or $i = n$). What is the optimal order to search for an unknown element z? In fact, how should we describe the optimal search order?

We express a search order as a *binary search tree*, a diagram showing the sequence of probes made in every possible search. We place at the root of the tree the sequence element at which the first probe is made, for example, x_i; the left subtree of x_i is constructed recursively for the probes made when $z < x_i$, and the right subtree of x_i is constructed recursively for the probes made when $z > x_i$. We label each item in the tree with the frequency that the search ends at that item. Figure 3.3 shows a simple example. The search of sequence $x_1 < x_2 < x_3 < x_4 < x_5$ according the tree of Figure 3.3 is done by comparing the unknown element z with x_4 (the root); if $z = x_4$, the search ends. If $z < x_2$, z is compared with x_2 (the root of the left subtree); if $z = x_2$, the search ends. Otherwise, if $z < x_2$, z is compared with x_1 (the root of the left subtree of x_2); if $z = x_1$, the search ends. Otherwise, if $z < x_1$, the search ends unsuccessfully at the leaf labeled α_0. Other results of comparisons lead along other paths in the tree from the root downward. By its

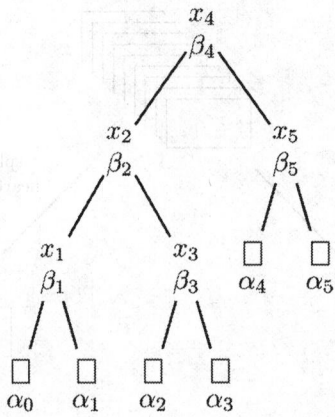

FIGURE 3.3   A binary search tree.

nature, a binary search tree is *lexicographic* in that for all nodes in the tree, the elements in the left subtree of the node are smaller and the elements in the right subtree of the node are larger than the node.

Because we are to find an optimal search pattern (tree), we want the cost of searching to be minimized. The cost of searching is measured by the *weighted path length* of the tree:

$$\sum_{i=1}^{n} \beta_i \times [1 + \text{level}(\beta_i)] + \sum_{i=0}^{n} \alpha_i \times \text{level}(\alpha_i),$$

defined formally as

$$W(\square) = 0,$$

$$W\left(T = \bigwedge_{T_l \ \ T_r} \right) = W(T_l) + W(T_r) + \sum \alpha_i + \sum \beta_i,$$

where the summations $\sum \alpha_i$ and $\sum \beta_i$ are over all α_i and β_i in T. Because there are exponentially many possible binary trees, finding the one with minimum weighted path length could, if done naïvely, take exponentially long.

The key observation we make is that a *principle of optimality* holds for the cost of binary search trees: subtrees of an optimal search tree must themselves be optimal. This observation means, for example, that if the tree shown in Figure 3.3 is optimal, then its left subtree must be the optimal tree for the problem of searching the sequence $x_1 < x_2 < x_3$ with frequencies $\beta_1, \beta_2, \beta_3$ and $\alpha_0, \alpha_1, \alpha_2, \alpha_3$. (If a subtree in Figure 3.3 were *not* optimal, we could replace it with a better one, reducing the weighted path length of the entire tree because of the recursive definition of weighted path length.) In general terms, the principle of optimality states that subsolutions of an optimal solution must themselves be optimal.

The optimality principle, together with the recursive definition of weighted path length, means that we can express the construction of an optimal tree recursively. Let $C_{i,j}$, $0 \le i \le j \le n$, be the cost of an optimal tree over $x_{i+1} < x_{i+2} < \cdots < x_j$ with the associated frequencies $\beta_{i+1}, \beta_{i+2}, \ldots, \beta_j$ and $\alpha_i, \alpha_{i+1}, \ldots, \alpha_j$. Then,

$$C_{i,i} = 0,$$

$$C_{i,j} = \min_{i < k \le j}(C_{i,k-1} + C_{k,j}) + W_{i,j},$$

where

$$W_{i,i} = \alpha_i,$$

$$W_{i,j} = W_{i,j-1} + \beta_j + \alpha_j.$$

These two recurrence relations can be implemented directly as recursive functions to compute $C_{0,n}$, the cost of the optimal tree, leading to the following two functions:

```
1   int W (int i, int j) {
2     if (i == j)
3       return alpha[j];
4     else
5       return W(i,j-1) + beta[j] + alpha[j];
6   }
7
8   int C (int i, int j) {
9     if (i == j)
10      return 0;
11    else {
12      int minCost = MAXINT;
13      int cost;
14      for (int k = i+1; k <= j; k++) {
15        cost = C(i,k-1) + C(k,j) + W(i,j);
16        if (cost < minCost)
17          minCost = cost;
18      }
19      return minCost;
20    }
21  }
```

These two functions correctly compute the cost of an optimal tree; the tree itself can be obtained by storing the values of k when cost < minCost in line 16.

However, the above functions are unnecessarily time consuming (requiring exponential time) because the same subproblems are solved repeatedly. For example, each call W(i,j) uses time $\Theta(j-i)$ and such calls are made repeatedly for the same values of i and j. We can make the process more efficient by caching the values of W(i,j) in an array as they are computed and using the cached values when possible:

```
1   int W[n][n];
2   for (int i = 0; i < n; i++)
3     for (int j = 0; j < n; j++)
4       W[i][j] = MAXINT;
5
6   int W (int i, int j) {
7     if (W[i][j] = MAXINT)
8       if (i == j)
9         W[i][j] = alpha[j];
10      else
11        W[i][j] = W(i,j-1) + beta[j] + alpha[j];
12    return W[i][j];
13  }
```

In the same way, we should cache the values of C(i,j) in an array as they are computed:

```
1   int C[n][n];
2   for (int i = 0; i < n; i++)
3     for (int j = 0; j < n; j++)
4       C[i][j] = MAXINT;
5
```

```
 6   int C (int i, int j) {
 7     if (C[i][j] == MAXINT)
 8       if (i == j)
 9         C[i][j] = 0;
10       else {
11         int minCost = MAXINT;
12         int cost;
13         for (int k = i+1; k <= j; k++) {
14           cost = C(i,k-1) + C(k,j) + W(i,j);
15           if (cost < minCost)
16             minCost = cost;
17         }
18         C[i][j] = minCost;
19       }
20     return C[i][j];
21   }
```

The idea of caching the solutions to subproblems is crucial to making the algorithm efficient. In this case, the resulting computation requires time $\Theta(n^3)$; this is surprisingly efficient, considering that an optimal tree is being found from among exponentially many possible trees.

By studying the pattern in which the arrays C and W are filled in, we see that the main diagonal C[i][i] is filled in first, then the first upper super-diagonal C[i][i+1], then the second upper super-diagonal C[i][i+2], and so on until the upper-right corner of the array is reached. Rewriting the code to do this directly, and adding an array R[][] to keep track of the roots of subtrees, we obtain:

```
 1   int W[n][n];
 2   int R[n][n];
 3   int C[n][n];
 4
 5   // Fill in main diagonal
 6   for (int i = 0; i < n; i++) {
 7     W[i][i] = alpha[i];
 8     R[i][i] = 0;
 9     C[i][i] = 0;
10   }
11
12   int minCost, cost;
13   for (int d = 1; d < n; d++)
14     // Fill in d-th upper super-diagonal
15     for (i = 0; i < n-d; i++) {
16       W[i][i+d] = W[i][i+d-1] + beta[i+d] + alpha[i+d];
17       R[i][i+d] = i+1;
18       C[i][i+d] = C[i][i] + C[i+1][i+d] + W[i][i+d];
19       for (int k = i+2; k <= i+d; k++) {
20         cost = C[i][k-1] + C[k][i+d] + W[i][i+d];
21         if (cost < C[i][i+d]) {
22           R[i][i+d] = k;
23           C[i][i+d] = cost;
24         }
25       }
26   }
```

which more clearly shows the $\Theta(n^3)$ behavior.

As a second example of dynamic programming, consider the *traveling salesman problem* in which a salesman must visit n cities, returning to his starting point, and is required to minimize the cost of the trip. The cost of going from city i to city j is $C_{i,j}$. To use dynamic programming we must specify an optimal tour in a recursive framework, with subproblems resembling the overall problem. Thus we define

$$T(i; j_1, j_2, \ldots, j_k) = \begin{cases} \text{cost of an optimal tour from city } i \text{ to city} \\ 1 \text{ that goes through each of the cities } j_1, \\ j_2, \ldots, j_k \text{ exactly once, in any order, and} \\ \text{through no other cities.} \end{cases}$$

The principle of optimality tells us that

$$T(i; j_1, j_2, \ldots, j_k) = \min_{1 \le m \le k} \{C_{i,j_m} + T(j_m; j_1, j_2, \ldots, j_{m-1}, j_{m+1}, \ldots, j_k)\},$$

where, by definition,

$$T(i; j) = C_{i,j} + C_{j,1}.$$

We can write a function T that directly implements the above recursive definition, but as in the optimal search tree problem, many subproblems would be solved repeatedly, leading to an algorithm requiring time $\Theta(n!)$. By caching the values $T(i; j_1, j_2, \ldots, j_k)$, we reduce the time required to $\Theta(n^2 2^n)$, still exponential, but considerably less than without caching.

3.6 Greedy Heuristics

Optimization problems always have an objective function to be minimized or maximized, but it is not often clear what steps to take to reach the optimum value. For example, in the optimum binary search tree problem of the previous section, we used dynamic programming to systematically examine all possible trees. But perhaps there is a simple rule that leads directly to the best tree; say, by choosing the largest β_i to be the root and then continuing recursively. Such an approach would be less time-consuming than the $\Theta(n^3)$ algorithm we gave, but it does not necessarily give an optimum tree (if we follow the rule of choosing the largest β_i to be the root, we get trees that are no better, on the average, than a randomly chosen trees). The problem with such an approach is that it makes decisions that are *locally optimum*, although perhaps not *globally optimum*. But such a "greedy" sequence of locally optimum choices does lead to a globally optimum solution in some circumstances.

Suppose, for example, $\beta_i = 0$ for $1 \le i \le n$, and we remove the lexicographic requirement of the tree; the resulting problem is the determination of an optimal prefix code for $n + 1$ letters with frequencies $\alpha_0, \alpha_1, \ldots, \alpha_n$. Because we have removed the lexicographic restriction, the dynamic programming solution of the previous section no longer works, but the following simple greedy strategy yields an optimum tree: repeatedly combine the two lowest-frequency items as the left and right subtrees of a newly created item whose frequency is the sum of the two frequencies combined. Here is an example of this construction; we start with five leaves with weights

$$\alpha_0 = 25 \qquad \alpha_1 = 34 \qquad \alpha_2 = 38 \qquad \alpha_3 = 58 \qquad \alpha_4 = 95 \qquad \alpha_5 = 20$$

First, combine leaves $\alpha_0 = 25$ and $\alpha_5 = 20$ into a subtree of frequency $25 + 20 = 45$:

$$25 + 20 = 45 \qquad\qquad \alpha_1 = 34 \qquad \alpha_2 = 38 \qquad \alpha_3 = 58 \qquad \alpha_5 = 95$$

$$\alpha_0 = 25 \quad \alpha_5 = 20$$

Then combine leaves $\alpha_1 = 34$ and $\alpha_2 = 38$ into a subtree of frequency $34 + 38 = 72$:

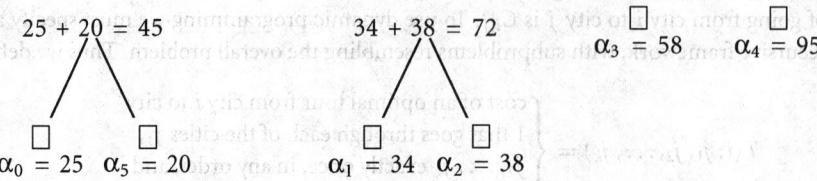

Next, combine the subtree of frequency $\alpha_0 + \alpha_5 = 45$ with $\alpha_3 = 58$:

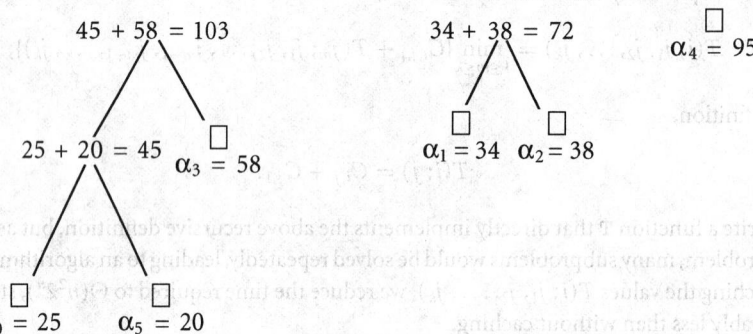

Then combine the subtree of frequency $\alpha_1 + \alpha_2 = 72$ with $\alpha_4 = 95$:

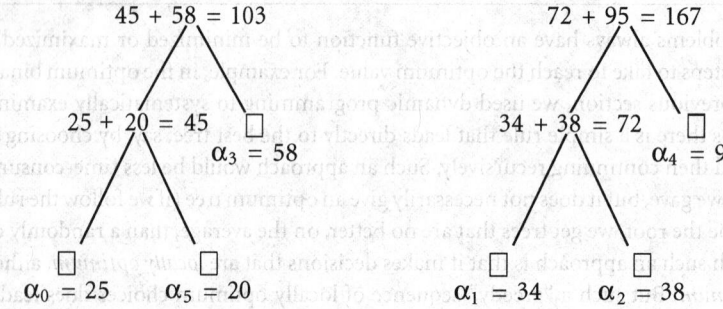

Finally, combine the only two remaining subtrees:

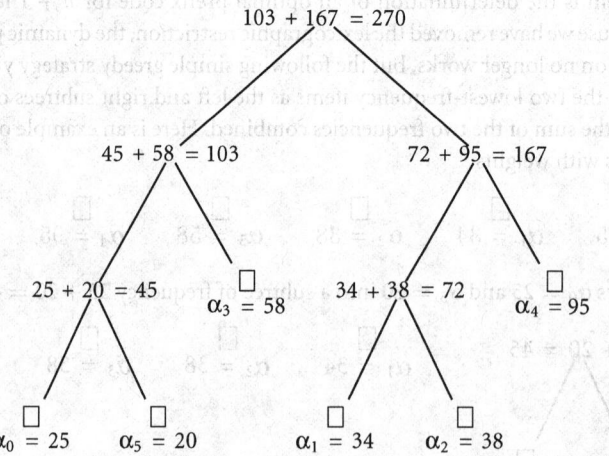

How do we know that the above-outlined process leads to an optimum tree? The key to proving that the tree is optimum is to assume, by way of contradiction, that it is not optimum. In this case, the greedy strategy must have erred in one of its choices, so let's look at the *first* error this strategy made. Because all previous greedy choices were not errors, and hence lead to an optimum tree, we can assume that we have a sequence of frequencies $\alpha_0, \alpha_1, \ldots, \alpha_n$ such that the first greedy choice is erroneous — without loss of generality assume that α_0 and α_1 are two smallest frequencies, those combined erroneously by the greedy strategy. For this combination to be erroneous, there must be no optimum tree in which these two leaves are siblings, so consider an optimum tree, the locations of α_0 and α_1, and the location of the two deepest leaves in the tree, α_i and α_j:

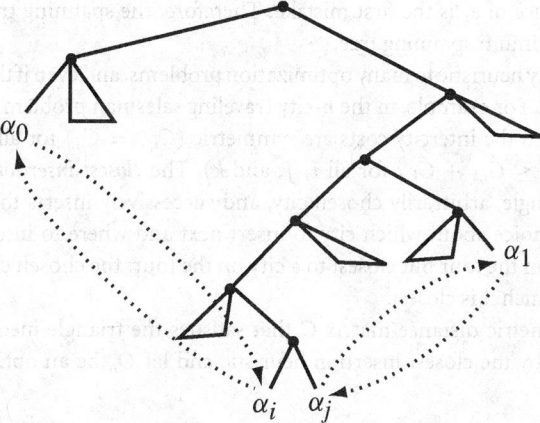

By interchanging the positions of α_0 and α_i and α_1 and α_j (as shown), we obtain a tree in which α_0 and α_1 are siblings. Because α_0 and α_1 are the two lowest frequencies (because they were the greedy algorithm's choice) $\alpha_0 \leq \alpha_i$ and $\alpha_1 \leq \alpha_j$, the weighted path length of the modified tree is no larger than before the modification since $\mathrm{level}(\alpha_0) \geq \mathrm{level}(\alpha_i)$, $\mathrm{level}(\alpha_1) \geq \mathrm{level}(\alpha_j)$ and, hence,

$$\mathrm{level}(\alpha_i) \times \alpha_0 + \mathrm{level}(\alpha_j) \times \alpha_1 \leq \mathrm{level}(\alpha_0) \times \alpha_0 + \mathrm{level}(\alpha_1) \times \alpha_1.$$

In other words, the first so-called mistake of the greedy algorithm was in fact not a mistake because there is an optimum tree in which α_0 and α_1 are siblings. Thus we conclude that the greedy algorithm never makes a first mistake — that is, it never makes a mistake at all!

The greedy algorithm above is called *Huffman's algorithm*. If the subtrees are kept on a priority queue by cumulative frequency, the algorithm needs to insert the $n + 1$ leaf frequencies onto the queue, and then repeatedly remove the two least elements on the queue, unite those to elements into a single subtree, and put that subtree back on the queue. This process continues until the queue contains a single item, the optimum tree. Reasonable implementations of priority queues will yield $O(n \log n)$ implementations of Huffman's greedy algorithm.

The idea of making greedy choices, facilitated with a priority queue, works to find optimum solutions to other problems too. For example, a spanning tree of a weighted, connected, undirected graph $G = (V, E)$ is a subset of $|V| - 1$ edges from E connecting all the vertices in G; a spanning tree is minimum if the sum of the weights of its edges is as small as possible. *Prim's algorithm* uses a sequence of greedy choices to determine a minimum spanning tree: start with an arbitrary vertex $v \in V$ as the spanning-tree-to-be. Then, repeatedly add the cheapest edge connecting the spanning-tree-to-be to a vertex not yet in it. If the vertices not yet in the tree are stored in a priority queue implemented by a Fibonacci heap, the total time required by Prim's algorithm will be $O(|E| + |V| \log |V|)$. But why does the sequence of greedy choices lead to a minimum spanning tree?

Suppose Prim's algorithm does *not* result in a minimum spanning tree. As we did with Huffman's algorithm, we ask what the state of affairs must be when Prim's algorithm makes its first mistake; we will see that the assumption of a first mistake leads to a contradiction, thus proving the correctness of Prim's algorithm. Let the edges added to the spanning tree be, in the order added, $e_1, e_2, e_3, \ldots$, and let e_i be the first mistake. In other words, there is a minimum spanning tree T_{min} containing $e_1, e_2, \ldots, e_{i-1}$, but no minimum spanning tree contains $e_1, e_2, \ldots, e_i$. Imagine what happens if we add the edge e_i to T_{min}: because T_{min} is a spanning tree, the addition of e_i causes a cycle containing e_i. Let e_{max} be the highest-cost edge on that cycle. Because Prim's algorithm makes a greedy choice — that is, chooses the lowest cost available edge — the cost of e_{max} is at least that of e_i, so the cost of the spanning $T_{min} - \{e_{max}\} \cup \{e_i\}$ is at most that of T_{min}; in other words, $T_{min} - \{e_{max}\} \cup \{e_i\}$ is also a minimum spanning tree, contradicting our assumption that the choice of e_i is the first mistake. Therefore, the spanning tree constructed by Prim's algorithm must be a minimum spanning tree.

We can apply the greedy heuristic to many optimization problems, and even if the results are not optimal, they are often quite good. For example, in the n-city traveling salesman problem, we can get near-optimal tours in time $O(n^2)$ when the intercity costs are symmetric ($C_{i,j} = C_{j,i}$ for all i and j) and satisfy the triangle inequality ($C_{i,j} \leq C_{i,k} + C_{k,j}$ for all i, j, and k). The *closest insertion algorithm* starts with a "tour" consisting of a single, arbitrarily chosen city, and successively inserts the remaining cities to the tour, making a greedy choice about which city to insert next and where to insert it: the city chosen for insertion is the city not on the tour but closest to a city on the tour; the chosen city is inserted adjacent to the city on the tour to which it is closest.

Given an $n \times n$ symmetric distance matrix C that satisfies the triangle inequality, let I_n be the tour of length $|I_n|$ produced by the closest insertion heuristic and let O_n be an optimal tour of length $|O_n|$. Then,

$$\frac{|I_n|}{|O_n|} < 2.$$

This bound is proved by an incremental form of the optimality proofs for greedy heuristics we saw seen above: we ask not where the first error is, but by how much we are in error at each greedy insertion to the tour; we establish a correspondence between edges of the optimal tour and cities inserted on the closest insertion tour. We show that at each insertion of a new city to the closest insertion tour, the cost of that insertion is at most twice the cost of corresponding edge of the optimal tour.

To establish the correspondence, imagine the closest insertion algorithm keeping track not only of the current tour, but also of a spider-like configuration including the edges of the current tour (the body of the spider) and pieces of the optimal tour (the legs of the spider). We show the current tour in solid lines and the pieces of optimal tour as dotted lines:

Initially, the spider consists of the arbitrarily chosen city with which the closest insertion tour begins and the legs of the spider consist of all the edges of the optimal tour *except* for one edge eliminated arbitrarily. As each city is inserted into the closest insertion tour, the algorithm will delete from the spider-like configuration one of the dotted edges from the optimal tour. When city k is inserted between cities l and m, the edge deleted is the one attaching spider to the leg containing the city inserted (from city x to city y), shown here in bold:

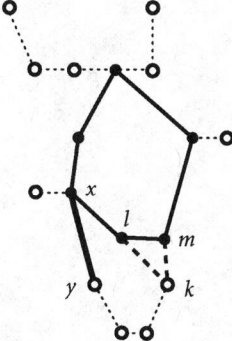

Now,

$$C_{k,m} \leq C_{x,y}$$

because of the greedy choice to add city k to the tour and not city y. By the triangle inequality,

$$C_{l,k} \leq C_{l,m} + C_{m,k},$$

and by symmetry, we can combine these two inequalities to get

$$C_{l,k} \leq C_{l,m} + C_{x,y}.$$

Adding this last inequality to the first one above,

$$C_{l,k} + C_{k,m} \leq C_{l,m} + 2C_{x,y},$$

that is,

$$C_{l,k} + C_{k,m} - C_{l,m} \leq 2C_{x,y}.$$

Thus, adding city k between cities l and m adds no more to I_n than $2C_{x,y}$. Summing these incremental amounts over the cost of the entire algorithm tells us that

$$I_n \leq 2O_n,$$

as we claimed.

References

Cormen, T. H., C. E. Leiserson, R. L. Rivest, and C. Stein, *Introduction to Algorithms*, McGraw-Hill, New York, 2nd ed., 2001.

Greene, D. H. and D. E. Knuth, *Mathematics for the Analysis of Algorithms*, 3rd ed., Birkhäuser, Boston, 1990.

Knuth, D. E., *The Art of Computer Programming, Volume 1: Fundamental Algorithms*, Addison-Wesley, Reading, MA, 3rd ed., 1997.

Knuth, D. E., *The Art of Computer Programming, Volume 3: Sorting and Searching*, Addison-Wesley, Reading, MA, 2nd ed., 1998.

Lueker, G. S., "Some techniques for solving recurrences," *Computing Surveys*, **12**, 419–436, 1980.

Reingold, E. M. and W. J. Hansen, *Data Structures in Pascal*, Little, Brown and Company, Boston, 1986.

Reingold, E. M., J. Nievergelt, and N. Deo, *Combinatorial Algorithms: Theory and Practice*, Prentice Hall, Englewood Cliffs, NJ, 1977.

4

Data Structures

4.1 Introduction 4-1
Containers, Elements, and Positions or Locators
• Abstract Data Types • Main Issues in the Study of Data
Structures • Fundamental Data Structures • Organization
of the Chapter

4.2 Sequence 4-4
Introduction • Operations • Implementation with an Array
• Implementation with a Singly Linked List • Implementation
with a Doubly Linked List

4.3 Priority Queue 4-6
Introduction • Operations • Realization with a Sequence
• Realization with a Heap • Realization with a Dictionary

4.4 Dictionary 4-12
Operations • Realization with a Sequence • Realization with a
Search Tree • Realization with an (a, b)-Tree • Realization with
an AVL-Tree • Realization with a Hash Table

Roberto Tamassia
Brown University

Bryan M. Cantrill
Sun Microsystems, Inc.

4.1 Introduction

The study of data structures — that is, methods for organizing data that are suitable for computer processing — is one of the classic topics of computer science. At the hardware level, a computer views storage devices such as internal memory and disk as holders of elementary data units (bytes), each accessible through its address (an integer). When writing programs, instead of manipulating the data at the byte level, it is convenient to organize them into higher-level entities called *data structures*.

4.1.1 Containers, Elements, and Positions or Locators

Most data structures can be viewed as **containers** that store a collection of objects of a given type, called the *elements* of the container. Often, a total order is defined among the elements (e.g., alphabetically ordered names, points in the plane ordered by x-coordinate). Following the approach of Goodrich and Tamassia [2001], we assume that the elements of a container can be accessed by means of variables called **positions** or **locators**. When an object is inserted into the container, a position or locator is returned, which can be later used to access or delete the object. A position represents a "place" where an element is stored, Examples of positions are array cells and list nodes. A locator "tracks" the position of an element in the data structure as it changes over time. A locator is typically implemented with an object that stores a pointer to a position.

A data structure has an associated repertory of operations, classified into *queries*, which retrieve information on the data structure (e.g., return the number of elements, or test the presence of a given element), and *updates*, which modify the data structure (e.g., insertion and deletion of elements). The performance of a data structure is characterized by the space requirement and the time complexity of the operations in its repertory. The *amortized* time complexity of an operation is the average time over a suitably defined sequence of operations.

However, efficiency is not the only quality measure of a data structure. Simplicity and ease of implementation should be taken into account when choosing a data structure for solving a practical problem.

4.1.2 Abstract Data Types

Data structures are concrete implementations of **abstract data types** (ADTs). A *data type* is a collection of objects. A data type can be mathematically specified (e.g., real number, directed graph) or concretely specified within a programming language (e.g., **int** in C, **set** in Pascal). An ADT is a mathematically specified data type equipped with operations that can be performed on the objects. Object-oriented programming languages, such as C++, provide support for expressing ADTs by means of *classes*. ADTs specify the data stored and the operations to be performed on them.

4.1.3 Main Issues in the Study of Data Structures

The following issues are of foremost importance in the study of data structures.

4.1.3.1 Static vs. Dynamic

A *static* data structure supports only queries, whereas a *dynamic* data structure also supports updates. A *dynamic* data structure is often more complicated than its static counterpart supporting the same repertory of queries. A *persistent* data structure (see, e.g., Driscoll et al. [1989]) is a dynamic data structure that supports operations on past versions. There are many problems for which no efficient dynamic data structures are known.

4.1.3.2 Implicit vs. Explicit

Two fundamental data organization mechanisms are used in data structures. In an *explicit* data structure, pointers (i.e., memory addresses) are used to link the elements and access them (e.g., a singly linked list, where each element has a pointer to the next one). In an *implicit* data structure (see, e.g., [Munro and Suwanda 1980]), mathematical relationships support the retrieval of elements (e.g., array representation of a heap, see Section 4.3). Explicit data structures must use additional space to store pointers. However, they are more flexible for complex problems. Most programming languages support pointers and basic implicit data structures, such as arrays.

4.1.3.3 Internal vs. External Memory

In a typical computer, there are two levels of memory: internal memory (also called random access memory, i.e., RAM) and external memory (disk). The internal memory is much faster than external memory but has much smaller capacity. Data structures designed to work for data that fit into internal memory may not perform well for large amounts of data that need to be stored in external memory. For large-scale problems, data structures need to be designed that take into account the two levels of memory [Aggarwal and Vitter 1988]. For example, two-level indices such as B-trees [Comer 1979] have been designed to efficiently search in large databases.

4.1.3.4 Space vs. Time

Data structures often exhibit a trade-off between space and time complexity. For example, suppose we want to represent a set of integers in the range $[0, N]$ (e.g., for a set of social security numbers $N = 10^{10} - 1$) such that we can efficiently query whether a given element is in the set, insert an element, or delete an element. Two possible data structures for this problem are an N-element bit array (where the bit in position i indicates the presence of integer i in the set), and a balanced search tree (such as a 2–3 tree or a red–black tree). The bit array has optimal time complexity because it supports queries, insertions, and deletions in constant time. However, it uses space proportional to the size N of the range, irrespective of the number of elements actually stored. The balanced **search tree** supports queries, insertions, and deletions in logarithmic time but uses optimal space proportional to the current number of elements stored.

4.1.3.5 Theory vs. Practice

A large and ever-growing body of theoretical research on data structures is available, where the performance is measured in asymptotic terms (big-Oh notation). Although asymptotic complexity analysis is an important mathematical subject, it does not completely capture the notion of efficiency of data structures in practical scenarios, where constant factors cannot be disregarded and the difficulty of implementation substantially affects design and maintenance costs. Experimental studies comparing the practical efficiency of data structures for specific classes of problems should be encouraged to bridge the gap between the theory and practice of data structures.

4.1.4 Fundamental Data Structures

The following data structures are ubiquitously used in the description of discrete algorithms, and serve as basic building blocks for realizing more complex data structures. They are covered in detail in the textbooks listed in the "Further Information" section and in the additional references provided.

4.1.4.1 Sequence

A **sequence** is a container that stores elements in a certain linear order, which is imposed by the operations performed. The basic operations supported are retrieving, inserting, and removing an element given its position. Special types of sequences include stacks and queues, where insertions and deletions can be done only at the head or tail of the sequence. The basic realization of sequences are by means of arrays and linked lists. Concatenable queues (see, e.g., Hoffman et al. [1986]) support additional operations such as splitting and splicing, and determining the sequence containing a given element. In external memory, a sequence is typically associated with a file.

4.1.4.2 Priority Queue

A **priority queue** is a container of elements from a totally ordered universe that supports the basic operations of inserting an element and retrieving/removing the largest element. A key application of priority queues is sorting algorithms. A **heap** is an efficient realization of a priority queue that embeds the elements into the ancestor/descendant partial order of a **binary tree**. A heap also admits an implicit realization where the nodes of the tree are mapped into the elements of an array (see Section 4.3). Sophisticated variations of priority queues include min–max heaps, pagodas, deaps, binomial heaps, and Fibonacci heaps. The buffer tree is an efficient external-memory realization of a priority queue.

4.1.4.3 Dictionary

A **dictionary** is a container of elements from a totally ordered universe that supports the basic operations of inserting/deleting elements and searching for a given element. **Hash tables** provide an efficient implicit realization of a dictionary. Efficient explicit implementations include skip lists [Pugh 1990], tries, and balanced search trees (e.g., **AVL-trees**, red–black trees, 2–3 trees, 2–3–4 trees, weight-balanced trees, biased search trees, splay trees). The technique of fractional cascading [Chazelle and Guibas 1986] speeds up searching for the same element in a collection of dictionaries. In external memory, dictionaries are typically implemented as B-trees and their variations.

The above data structures are widely used in the following application domains:

1. *Graphs and networks:* adjacency matrix, adjacency lists, link-cut tree [Sleator and Tarjan 1983], dynamic expression tree [Cohen and Tamassia 1995], topology tree [Frederickson 1997], SPQR-tree [Di Battista and Tamassia 1996], sparsification tree [Eppstein et al. 1997]. See also, for example, Di Battista et al. [1999], Even [1979], Mehlhorn [1984], and Tarjan [1983].

2. *Text processing:* string, suffix tree, Patricia tree. See, for example, Gonnet and Baeza-Yates [1991].
3. *Geometry and graphics:* binary space partition tree, chain tree, trapezoid tree, range tree, segment tree, interval tree, priority search tree, hull tree, quad tree, R-tree, grid file, metablock tree. For example, see Chiang and Tamassia [1992], Edelsbrunner [1987], Foley et al. [1990], Mehlhorn [1984], Nievergelt and Hinrichs [1993], O'Rourke [1994], and Preparata and Shamos [1985].

4.1.5 Organization of the Chapter

The remainder of this chapter focuses on three fundamental abstract data types: sequences, priority queues, and dictionaries. Examples of efficient data structures and algorithms for implementing them are presented in detail in Section 4.2 through Section 4.4, respectively. Namely, we cover arrays, singly and doubly linked lists, heaps, search trees, (a, b)-trees, AVL-trees, bucket arrays, and hash tables.

4.2 Sequence

4.2.1 Introduction

A *sequence* is a container that stores elements in linear order, which is imposed by the operations performed. The basic operations supported are:

- INSERTRANK: insert an element in a given position.
- REMOVE: remove an element.

Sequences are a basic form of data organization, and are typically used to realize and implement other data types and data structures.

4.2.2 Operations

Using positions (see Section 4.1.1), we can define a more complete repertory of operations for a sequence S:

SIZE(N): return the number of elements N of S.

HEAD(p): assign to p the position of the first element of S; if S is empty, then p is set to null.

TAIL(p): assign to p the position of the last element of S; if S is empty, then p is set to null.

POSITIONRANK(r, p): assign to p the position of the rth element of S; if $r < 1$ or $r > N$, where N is the size of S, then p is set to null.

PREV(p', p''): assign to p'' the position of the element of S preceding the element with position p'; if p' is the position of the first element of S, then p'' is set to null.

NEXT(p', p''): assign to p'' the position of the element of S following the element with position p'; if p' is the position of the last element of S, then p'' is set to null.

INSERTAFTER(e, p', p''): insert element e into S after the element with position p', and return the position p'' of e.

INSERTBEFORE(e, p', p''): insert element e into S before the element with position p', and return the position p'' of e.

INSERTHEAD(e, p): insert element e at the beginning of S, and return the position p of e.

INSERTTAIL(e, p): insert element e at the end of S, and return the position p of e.

INSERTRANK(e, r, p): insert element e in the rth position of S; if $r < 1$ or $r > N + 1$, where N is the current size of S, then p is set to null.

REMOVE(p, e): remove from S and return element e with position p.

MODIFY(p, e): replace with e the element with position p.

Some of the preceding operations can be easily expressed by means of other operations of the repertory. For example, operations HEAD and TAIL can be easily expressed by means of POSITIONRANK and SIZE.

TABLE 4.1 Performance of a Sequence Implemented with an Array

Operation	Time
SIZE	$O(1)$
HEAD	$O(1)$
TAIL	$O(1)$
POSITIONRANK	$O(1)$
PREV	$O(1)$
NEXT	$O(1)$
INSERTAFTER	$O(N)$
INSERTBEFORE	$O(N)$
INSERTHEAD	$O(N)$
INSERTTAIL	$O(1)$
INSERTRANK	$O(N)$
REMOVE	$O(N)$
MODIFY	$O(1)$

TABLE 4.2 Performance of a Sequence Implemented with a Singly Linked List

Operation	Time
SIZE	$O(1)$
HEAD	$O(1)$
TAIL	$O(1)$
POSITIONRANK	$O(N)$
PREV	$O(N)$
NEXT	$O(1)$
INSERTAFTER	$O(1)$
INSERTBEFORE	$O(N)$
INSERTHEAD	$O(1)$
INSERTTAIL	$O(1)$
INSERTRANK	$O(N)$
REMOVE	$O(N)$
MODIFY	$O(1)$

4.2.3 Implementation with an Array

The simplest way to implement a sequence is to use a (one-dimensional) array, where the ith element of the array stores the ith element of the list, and to keep a variable that stores the size N of the sequence. With this implementation, accessing elements takes $O(1)$ time, whereas insertions and deletions take $O(N)$ time. Table 4.1 shows the time complexity of the implementation of a sequence by means of an array.

4.2.4 Implementation with a Singly Linked List

A sequence can also be implemented with a singly linked list, where each position has a pointer to the next one. We also store the size of the sequence and pointers to the first and last position of the sequence.

With this implementation, accessing elements by rank takes $O(N)$ time because we need to traverse the list, whereas some insertions and deletions take $O(1)$ time. Table 4.2 shows the time complexity of the implementation of a sequence by means of a singly linked list.

4.2.5 Implementation with a Doubly Linked List

Better performance can be achieved, at the expense of using additional space, by implementing a sequence with a doubly linked list, where each position has pointers to the next and previous positions. We also

TABLE 4.3 Performance of a Sequence
Implemented with a Doubly Linked List

Operation	Time
SIZE	$O(1)$
HEAD	$O(1)$
TAIL	$O(1)$
POSITIONRANK	$O(N)$
PREV	$O(1)$
NEXT	$O(1)$
INSERTAFTER	$O(1)$
INSERTBEFORE	$O(1)$
INSERTHEAD	$O(1)$
INSERTTAIL	$O(1)$
INSERTRANK	$O(N)$
REMOVE	$O(1)$
MODIFY	$O(1)$

store the size of the sequence and pointers to the first and last positions of the sequence. Table 4.3 shows the time complexity of the implementation of sequence by means of a doubly linked list.

4.3 Priority Queue

4.3.1 Introduction

A priority queue is a container of elements from a totally ordered universe that supports the following two basic operations:

1. INSERT: insert an element into the priority queue.
2. REMOVEMAX: remove the largest element from the priority queue.

Here are some simple applications of a priority queue:

- *Scheduling.* A scheduling system can store the tasks to be performed into a priority queue, and select the task with highest priority to be executed next.
- *Sorting.* To sort a set of N elements, we can insert them one at a time into a priority queue by means of N INSERT operations, and then retrieve them in decreasing order by means of N REMOVEMAX operations. This two-phase method is the paradigm of several popular sorting algorithms, including *selection sort*, *insertion sort*, and *heap-sort*.

4.3.2 Operations

Using locators, we can define a more complete repertory of operations for a priority queue Q:

SIZE(N): return the current number of elements N in Q.

MAX(c): return a locator c to the maximum element of Q.

INSERT(e, c): insert element e into Q and return a locator c to e.

REMOVE(c, e): remove from Q and return element e with locator c.

REMOVEMAX(e): remove from Q and return the maximum element e from Q.

MODIFY(c, e): replace with e the element with locator c.

Note that operation REMOVEMAX(e) is equivalent to MAX(c) followed by REMOVE(c, e).

TABLE 4.4 Performance of a Priority Queue Realized by an Unsorted Sequence, Implemented with a Doubly Linked List

Operation	Time
SIZE	$O(1)$
MAX	$O(N)$
INSERT	$O(1)$
REMOVE	$O(1)$
REMOVEMAX	$O(N)$
MODIFY	$O(1)$

TABLE 4.5 Performance of a Priority Queue Realized by a Sorted Sequence, Implemented with a Doubly Linked List

Operation	Time
SIZE	$O(1)$
MAX	$O(1)$
INSERT	$O(N)$
REMOVE	$O(1)$
REMOVEMAX	$O(1)$
MODIFY	$O(N)$

4.3.3 Realization with a Sequence

We can realize a priority queue by reusing and extending the sequence abstract data type (see Section 4.2). Operations SIZE, MODIFY, and REMOVE correspond to the homonymous sequence operations.

4.3.3.1 Unsorted Sequence

We can realize INSERT by an INSERTHEAD or an INSERTTAIL, which means that the sequence is not kept sorted. Operation MAX can be performed by scanning the sequence with an iteration of NEXT operations, keeping track of the maximum element encountered. Finally, as observed earlier, operation REMOVEMAX is a combination of MAX and REMOVE. Table 4.4 shows the time complexity of this realization, assuming that the sequence is implemented with a doubly linked list. In the table we denote with N the number of elements in the priority queue at the time the operation is performed. The space complexity is $O(N)$.

4.3.3.2 Sorted Sequence

An alternative implementation uses a sequence that is kept sorted. In this case, operation MAX corresponds to simply accessing the last element of the sequence. However, operation INSERT now requires scanning the sequence to find the appropriate position to insert the new element. Table 4.5 shows the time complexity of this realization, assuming that the sequence is implemented with a doubly linked list. In the table we denote with N the number of elements in the priority queue at the time the operation is performed. The space complexity is $O(N)$.

Realizing a priority queue with a sequence, sorted or unsorted, has the drawback that some operations require linear time in the worst case. Hence, this realization is not suitable in many applications where fast running times are sought for all the priority queue operations.

4.3.3.3 Sorting

For example, consider the sorting application (see the first introduction to this section). We have a collection of N elements from a totally ordered universe, and we want to sort them using a priority queue Q. We

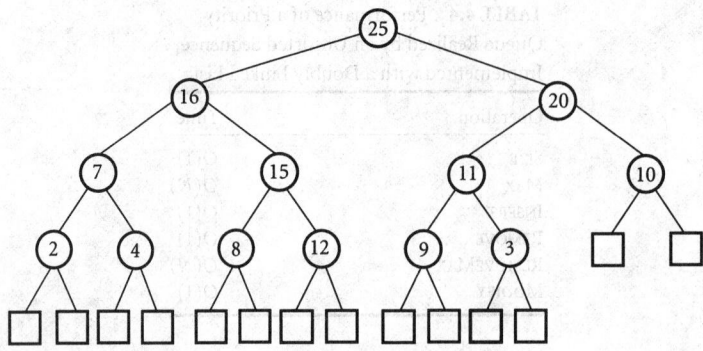

FIGURE 4.1   Example of a heap storing 13 elements.

assume that each element uses $O(1)$ space, and any two elements can be compared in $O(1)$ time. If we realize Q with an unsorted sequence, then the first phase (inserting the N elements into Q) takes $O(N)$ time. However, the second phase (removing N times the maximum element) takes time

$$O\left(\sum_{i=1}^{N} i\right) = O(N^2)$$

Hence, the overall time complexity is $O(N^2)$. This sorting method is known as *selection sort*.

However, if we realize the priority queue with a sorted sequence, then the first phase takes time

$$O\left(\sum_{i=1}^{N} i\right) = O(N^2)$$

while the second phase takes time $O(N)$. Again, the overall time complexity is $O(N^2)$. This sorting method is known as *insertion sort*.

4.3.4   Realization with a Heap

A more sophisticated realization of a priority queue uses a data structure called a *heap*. A heap is a binary tree T whose internal nodes each store one element from a totally ordered universe, with the following properties (see Figure 4.1):

Level property. All of the levels of T are full, except possibly for the bottommost level, which is left filled.
Partial order property. Let μ be a node of T distinct from the root, and let ν be the parent of μ; then the element stored at μ is less than or equal to the element stored at ν.

The leaves of a heap do not store data and serve only as placeholders. The level property implies that heap T is a minimum-height binary tree. More precisely, if T stores N elements and has height h, then each level i with $0 \leq i \leq h - 2$ stores exactly 2^i elements, whereas level $h - 1$ stores between 1 and 2^{h-1} elements. Note that level h contains only leaves. We have

$$2^{h-1} = 1 + \sum_{i=0}^{h-2} 2^i \leq N \leq \sum_{i=0}^{h-1} 2^i = 2^h - 1$$

from which we obtain:

$$\log_2(N + 1) \leq h \leq 1 + \log_2 N$$

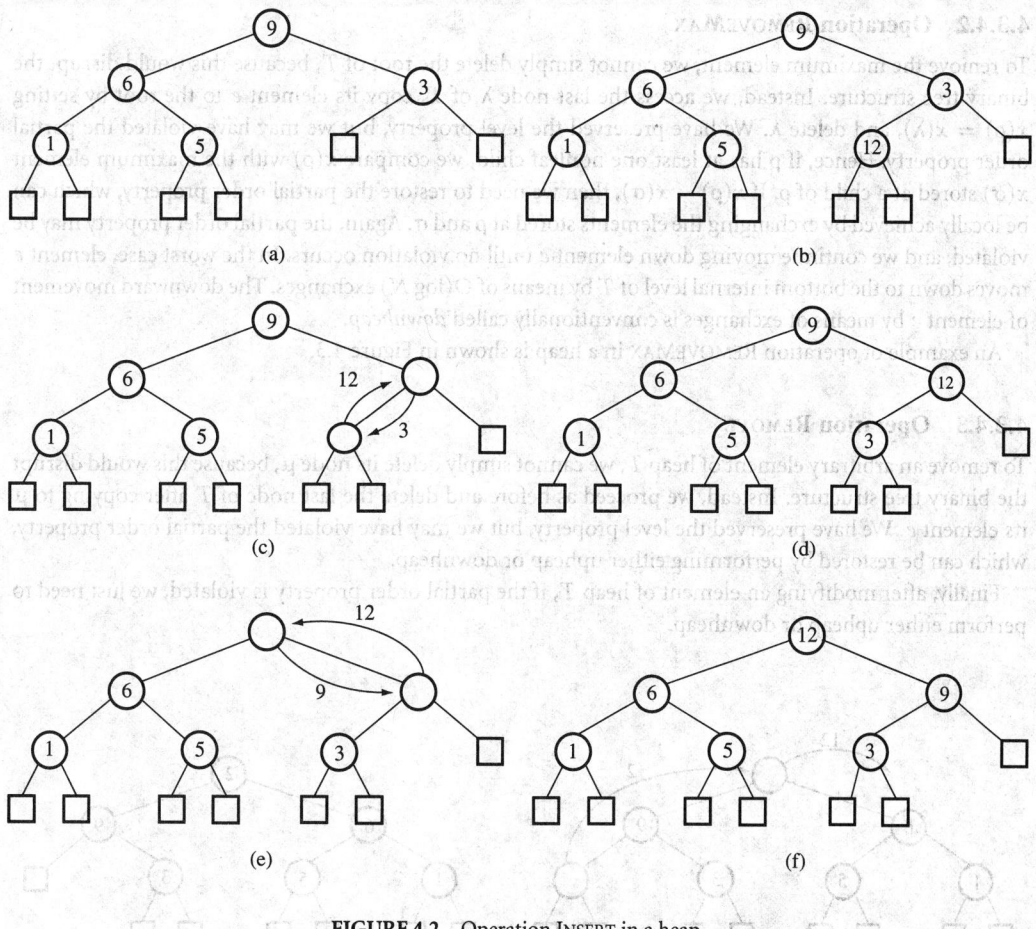

FIGURE 4.2   Operation INSERT in a heap.

Now we show how to perform the various priority queue operations by means of a heap T. We denote with $x(\mu)$ the element stored at an internal node μ of T. We denote with ρ the root of T. We call the *last node* of T the rightmost internal node of the bottommost internal level of T.

By storing a counter that keeps track of the current number of elements, SIZE consists of simply returning the value of the counter. By the partial order property, the maximum element is stored at the root and, hence, operation MAX can be performed by accessing node ρ.

4.3.4.1   Operation INSERT

To insert an element e into T, we add a new internal node μ to T such that μ becomes the new last node of T, and set $x(\mu) = e$. This action ensures that the level property is satisfied, but may violate the partial order property. Hence, if $\mu \neq \rho$, we compare $x(\mu)$ with $x(\nu)$, where ν is the parent of μ. If $x(\mu) > x(\nu)$, then we need to restore the partial order property, which can be locally achieved by exchanging the elements stored at μ and ν. This causes the new element e to move up one level. Again, the partial order property may be violated, and we may have to continue moving up the new element e until no violation occurs. In the worst case, the new element e moves up to the root ρ of T by means of $O(\log N)$ exchanges. The upward movement of element e by means of exchanges is conventionally called *upheap*.

An example of a sequence of insertions into a heap is shown in Figure 4.2.

4.3.4.2 Operation RemoveMax

To remove the maximum element, we cannot simply delete the root of T, because this would disrupt the binary tree structure. Instead, we access the last node λ of T, copy its element e to the root by setting $x(\rho) = x(\lambda)$, and delete λ. We have preserved the level property, but we may have violated the partial order property. Hence, if ρ has at least one nonleaf child, we compare $x(\rho)$ with the maximum element $x(\sigma)$ stored at a child of ρ. If $x(\rho) < x(\sigma)$, then we need to restore the partial order property, which can be locally achieved by exchanging the elements stored at ρ and σ. Again, the partial order property may be violated, and we continue moving down element e until no violation occurs. In the worst case, element e moves down to the bottom internal level of T by means of $O(\log N)$ exchanges. The downward movement of element e by means of exchanges is conventionally called *downheap*.

An example of operation RemoveMax in a heap is shown in Figure 4.3.

4.3.4.3 Operation Remove

To remove an arbitrary element of heap T, we cannot simply delete its node μ, because this would disrupt the binary tree structure. Instead, we proceed as before and delete the last node of T after copying to μ its element e. We have preserved the level property, but we may have violated the partial order property, which can be restored by performing either upheap or downheap.

Finally, after modifying an element of heap T, if the partial order property is violated, we just need to perform either upheap or downheap.

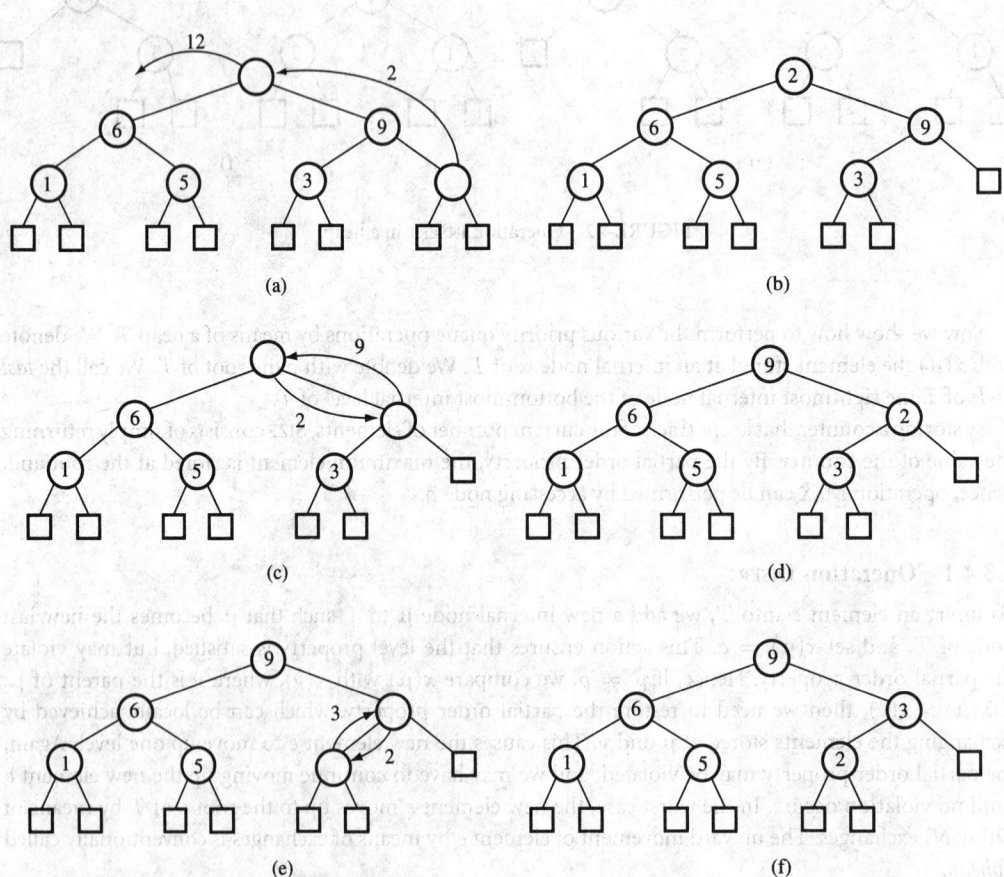

FIGURE 4.3 Operation RemoveMax in a heap.

TABLE 4.6 Performance of a Priority Queue Realized by a Heap, Implemented with a Suitable Binary Tree Data Structure

Operation	Time
SIZE	$O(1)$
MAX	$O(1)$
INSERT	$O(\log N)$
REMOVE	$O(\log N)$
REMOVEMAX	$O(\log N)$
MODIFY	$O(\log N)$

4.3.4.4 Time Complexity

Table 4.6 shows the time complexity of the realization of a priority queue by means of a heap. In the table we denote with N the number of elements in the priority queue at the time the operation is performed. The space complexity is $O(N)$. We assume that the heap is itself realized by a data structure for binary trees that supports $O(1)$-time access to the children and parent of a node. For instance, we can implement the heap explicitly with a linked structure (with pointers from a node to its parents and children), or implicitly with an array (where node i has children $2i$ and $2i + 1$). Let N be the number of elements in a priority queue Q realized with a heap T at the time an operation is performed. The time bounds of Table 4.6 are based on the following facts:

- In the worst case, the time complexity of upheap and downheap is proportional to the height of T.
- If we keep a pointer to the last node of T, we can update this pointer in time proportional to the height of T in operations INSERT, REMOVE, and REMOVEMAX, as illustrated in Figure 4.4.
- The height of heap T is $O(\log N)$.

The $O(N)$ space complexity bound for the heap is based on the following facts:

- The heap has $2N + 1$ nodes (N internal nodes and $N + 1$ leaves).
- Every node uses $O(1)$ space.
- In the array implementation, because of the level property, the array elements used to store heap nodes are in the contiguous locations 1 through $2N - 1$.

Note that we can reduce the space requirement by a constant factor implementing the leaves of the heap with null objects, such that only the internal nodes have space associated with them.

4.3.4.5 Sorting

Realizing a priority queue with a heap has the advantage that all of the operations take $O(\log N)$ time, where N is the number of elements in the priority queue at the time the operation is performed. For example, in the sorting application (see Section 4.3.1), both the first phase (inserting the N elements) and the second phase (removing N times the maximum element) take time

$$O\left(\sum_{i=1}^{N} \log i\right) = O(N \log N)$$

Hence, sorting with a priority queue realized with a heap takes $O(N \log N)$ time. This sorting method is known as *heap sort*, and its performance is considerably better than that of selection sort and insertion sort (see Section 4.3.3.3), where the priority queue is realized as a sequence.

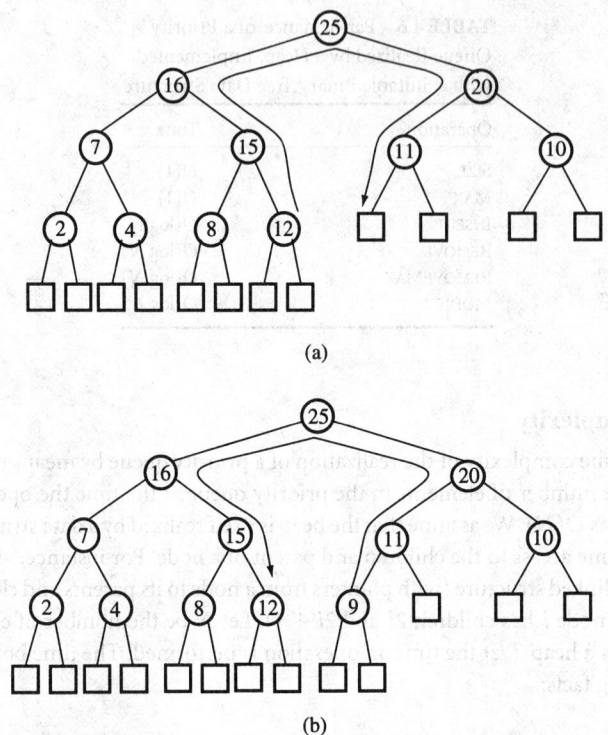

FIGURE 4.4   Update of the pointer to the last node: (a) INSERT and (b) REMOVE or REMOVEMAX.

4.3.5   Realization with a Dictionary

A priority queue can be easily realized with a dictionary (see Section 4.4). Indeed, all of the operations in the priority queue repertory are supported by a dictionary. To achieve $O(1)$ time for operation MAX, we can store the locator of the maximum element in a variable, and recompute it after an update operation. This realization of a priority queue with a dictionary has the same asymptotic complexity bounds as the realization with a heap, provided the dictionary is suitably implemented, for example, with an (a, b)-tree (see section "Realization with an (a, b)-tree") or an AVL-tree (see section "Realization with an AVL-tree"). However, a heap is simpler to program than an (a, b)-tree or an AVL-tree.

4.4   Dictionary

A dictionary is a container of elements from a totally ordered universe that supports the following basic operations:

- FIND: search for an element.
- INSERT: insert an element.
- REMOVE: delete an element.

A major application of dictionaries is database systems.

4.4.1   Operations

In the most general setting, the elements stored in a dictionary are pairs (x, y), where x is the *key* giving the ordering of the elements and y is the auxiliary information. For example, in a database storing student records, the key could be the student's last name, and the auxiliary information the student's transcript. It is convenient to augment the ordered universe of keys with two *special keys* ($+\infty$ and $-\infty$) and assume that each dictionary has, in addition to its *regular elements*, two *special elements*, with keys $+\infty$ and $-\infty$, respectively. For simplicity, we will also assume that no two elements of a dictionary have the same key. An insertion of an element with the same key as that of an existing element will be rejected by returning a null locator.

Using locators (see Section 4.1), we can define a more complete repertory of operations for a dictionary D:

SIZE(N): return the number of regular elements N of D.

FIND(x, c): if D contains an element with key x, assign to c a locator to such as an element; otherwise; set c equal to a null locator.

LOCATEPREV(x, c): assign to c a locator to the element of D with the largest key less than or equal to x; if x is smaller than all of the keys of the regular elements, then c is a locator to the special element with key $-\infty$; if $x = -\infty$, then c is a null locator.

LOCATENEXT(x, c): assign to c a locator to the element of D with the smallest key greater than or equal to x; if x is larger than all of the keys of the regular elements, then c is a locator to the special element with key $+\infty$; then, if $x = +\infty$, c is a null locator.

PREV(c', c''): assign to c'' a locator to the element of D with the largest key less than that of the element with locator c'; if the key of the element with locator c' is smaller than all of the keys of the regular elements, then this operation returns a locator to the special element with key $-\infty$.

NEXT(c', c''): assign to c'' a locator to the element of D with the smallest key larger than that of the element with locator c'; if the key of the element with locator c' is larger than all of the keys of the regular elements, then this operation returns a locator to the special element with key $+\infty$.

MIN(c): assign to c a locator to the regular element of D with minimum key; if D has no regular elements, then c is a null locator.

MAX(c): assign to c a locator to the regular element of D with maximum key; if D has no regular elements, then c is a null locator.

INSERT(e, c): insert element e into D, and return a locator c to e; if there is already an element with the same key as e, then this operation returns a null locator.

REMOVE(c, e): remove from D and return element e with locator c.

MODIFY(c, e): replace with e the element with locator c.

Some of these operations can be easily expressed by means of other operations of the repertory. For example, operation FIND is a simple variation of LOCATEPREV or LOCATENEXT.

4.4.2 Realization with a Sequence

We can realize a dictionary by reusing and extending the sequence abstract data type (see Section 4.2). Operations SIZE, INSERT, and REMOVE correspond to the homonymous sequence operations.

4.4.2.1 Unsorted Sequence

We can realize INSERT by an INSERTHEAD or an INSERTTAIL, which means that the sequence is not kept sorted. Operation FIND(x, c) can be performed by scanning the sequence with an iteration of NEXT operations, until we either find an element with key x, or we reach the end of the sequence. Table 4.7 shows the time complexity of this realization, assuming that the sequence is implemented with a doubly linked list. In the table we denote with N the number of elements in the dictionary at the time the operation is performed. The space complexity is $O(N)$.

4.4.2.2 Sorted Sequence

We can also use a sorted sequence to realize a dictionary. Operation INSERT now requires scanning the sequence to find the appropriate position to insert the new element. However, in a FIND operation, we can stop scanning the sequence as soon as we find an element with a key larger than the search key.

TABLE 4.7 Performance of a Dictionary
Realized by an Unsorted Sequence,
Implemented with a Doubly Linked List

Operation	Time
SIZE	$O(1)$
FIND	$O(N)$
LOCATEPREV	$O(N)$
LOCATENEXT	$O(N)$
NEXT	$O(N)$
PREV	$O(N)$
MIN	$O(N)$
MAX	$O(N)$
INSERT	$O(1)$
REMOVE	$O(1)$
MODIFY	$O(1)$

TABLE 4.8 Performance of a Dictionary
Realized by a Sorted Sequence,
Implemented with a Doubly Linked List

Operation	Time
SIZE	$O(1)$
FIND	$O(N)$
LOCATEPREV	$O(N)$
LOCATENEXT	$O(N)$
NEXT	$O(1)$
PREV	$O(1)$
MIN	$O(1)$
MAX	$O(1)$
INSERT	$O(N)$
REMOVE	$O(1)$
MODIFY	$O(N)$

Table 4.8 shows the time complexity of this realization by a sorted sequence, assuming that the sequence is implemented with a doubly linked list. In the table we denote with N the number of elements in the dictionary at the time the operation is performed. The space complexity is $O(N)$.

4.4.2.3 Sorted Array

We can obtain a different performance trade-off by implementing the sorted sequence by means of an array, which allows constant-time access to any element of the sequence given its position. Indeed, with this realization we can speed up operation FIND(x, c) using the *binary search* strategy, as follows. If the dictionary is empty, we are done. Otherwise, let N be the current number of elements in the dictionary. We compare the search key k with the key x_m of the middle element of the sequence, that is, the element at position $\lfloor N/2 \rfloor$. If $x = x_m$, we have found the element. Else, we recursively search in the subsequence of the elements preceding the middle element if $x < x_m$, or following the middle element if $x > x_m$. At each recursive call, the number of elements of the subsequence being searched halves. Hence, the number of sequence elements accessed and the number of comparisons performed by binary search is $O(\log N)$. While searching takes $O(\log N)$ time, inserting or deleting elements now takes $O(N)$ time.

TABLE 4.9 Performance of a Dictionary
Realized by a Sorted Sequence, Implemented
with an Array

Operation	Time
SIZE	$O(1)$
FIND	$O(\log N)$
LOCATEPREV	$O(\log N)$
LOCATENEXT	$O(\log N)$
NEXT	$O(1)$
PREV	$O(1)$
MIN	$O(1)$
MAX	$O(1)$
INSERT	$O(N)$
REMOVE	$O(N)$
MODIFY	$O(N)$

Table 4.9 shows the performance of a dictionary realized with a sorted sequence, implemented with an array. In the table we denote with N the number of elements in the dictionary at the time the operation is performed. The space complexity is $O(N)$.

4.4.3 Realization with a Search Tree

A *search tree* for elements of the type (x, y), where x is a key from a totally ordered universe, is a rooted ordered tree T such that:

- Each internal node of T has at least two children and stores a nonempty set of elements.
- A node μ of T with d children $\mu_1, \ldots, \mu_d$ stores $d - 1$ elements $(x_1, y_1) \cdots (x_{d-1}, y_{d-1})$, where $x_1 \leq \cdots \leq x_{d-1}$.
- For each element (x, y) stored at a node in the subtree of T rooted at μ_i, we have $x_{i-1} \leq x \leq x_i$, where $x_0 = -\infty$ and $x_d = +\infty$.

In a search tree, each internal node stores a nonempty collection of keys, whereas the leaves do not store any key and serve only as placeholders. An example search tree is shown in Figure 4.5a. A special type of search tree is a *binary search tree*, where each internal node stores one key and has two children.

We will recursively describe the realization of a dictionary D by means of a search tree T because we will use dictionaries to implement the nodes of T. Namely, an internal node μ of T with children $\mu_1, \ldots, \mu_d$ and elements $(x_1, y_1) \cdots (x_{d-1}, y_{d-1})$ is equipped with a dictionary $D(\mu)$ whose regular elements are the pairs $(x_i, (y_i, \mu_i))$, $i = 1, \ldots, d - 1$ and whose special element with key $+\infty$ is $(+\infty, (\cdot, \mu_d))$. A regular element (x, y) stored in D is associated with a regular element $(x, (y, \nu))$ stored in a dictionary $D(\mu)$, for some node μ of T. See the example in Figure 4.5b.

4.4.3.1 Operation FIND

Operation FIND(x, c) on dictionary D is performed by means of the following recursive method for a node μ of T, where μ is initially the root of T [see Figure 4.5b]. We execute LOCATENEXT(x, c') on dictionary $D(\mu)$ and let $(x', (y', \nu))$ be the element pointed by the returned locator c'. We have three cases:

1. Case $x = x'$: we have found x and return locator c to (x', y').
2. Case $x \neq x'$ and ν is a leaf: we have determined that x is not in D and return a null locator c.
3. Case $x \neq x'$ and ν is an internal node: we set $\mu = \nu$ and recursively execute the method.

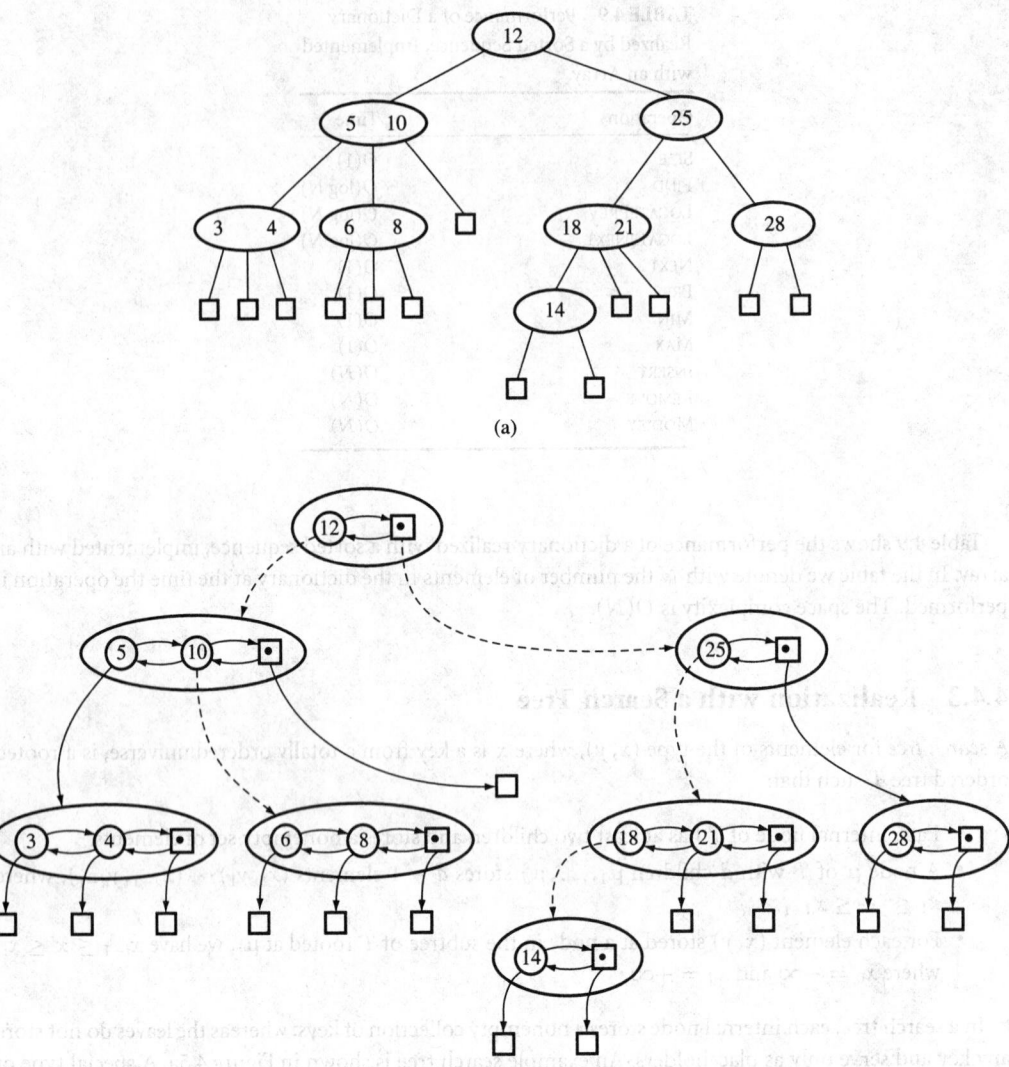

FIGURE 4.5 Realization of a dictionary by means of a search tree: (a) a search tree T, (b) realization of the dictionaries at the nodes of T by means of sorted sequences. The search paths for elements 9 (unsuccessful search) and 14 (successful search) are shown with dashed lines.

4.4.3.2 Operation INSERT

Operations LOCATEPREV, LOCATENEXT, and INSERT can be performed with small variations of the previously described method. For example, to perform operation INSERT(e, c), where $e = (x, y)$, we modify the previous cases as follows (see Figure 4.6):

1. Case $x = x'$: an element with key x already exists, and we return a null locator.
2. Case $x \neq x'$ and ν is a leaf: we create a new leaf node λ, insert a new element $(x, (y, \lambda))$ into $D(\mu)$, and return a locator c to (x, y).
3. Case $x \neq x'$ and ν is an internal node: we set $\mu = \nu$ and recursively execute the method.

Note that new elements are inserted at the bottom of the search tree.

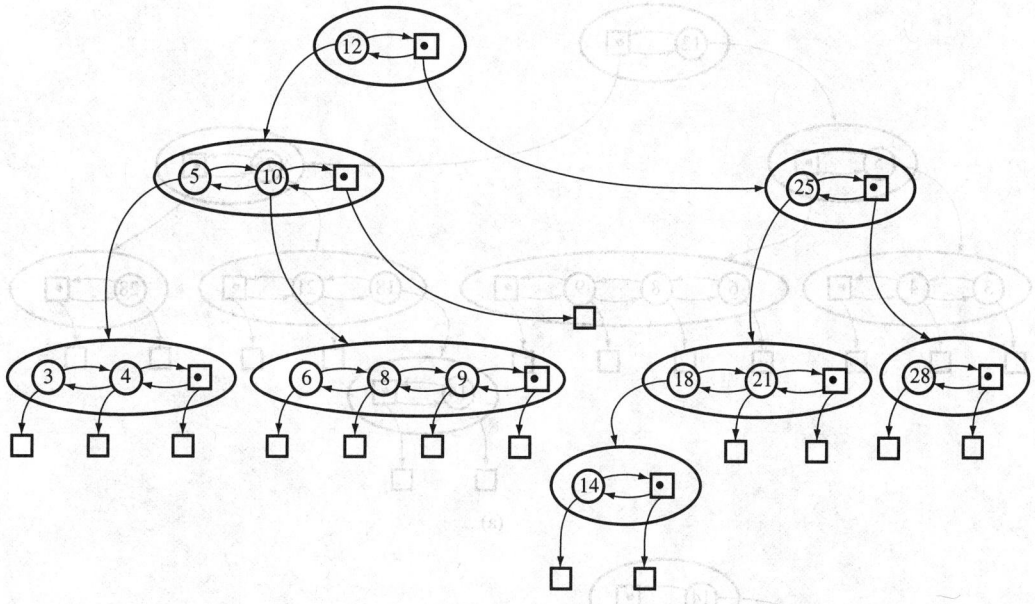

FIGURE 4.6 Insertion of element 9 into the search tree of Figure 4.5.

4.4.3.3 Operation REMOVE

Operation REMOVE(e, c) is more complex (see Figure 4.7). Let the associated element of $e = (x, y)$ in T be $(x, (y, v))$, stored in dictionary $D(\mu)$ of node μ:

- If node v is a leaf, we simply delete element $(x, (y, v))$ from $D(\mu)$.
- Else (v is an internal node), we find the successor element $(x', (y', v'))$ of $(x, (y, v))$ in $D(\mu)$ with a NEXT operation in $D(\mu)$. (1) If v' is a leaf, we replace v' with v, that is, change element $(x', (y', v'))$ to $(x', (y', v))$, and delete element $(x, (y, v))$ from $D(\mu)$. (2) Else (v' is an internal node), while the leftmost child v'' of v' is not a leaf, we set $v' = v''$. Let $(x'', (y'', v''))$ be the first element of $D(v')$ (node v'' is a leaf). We replace $(x, (y, v))$ with $(x'', (y'', v))$ in $D(\mu)$ and delete $(x'', (y'', v''))$ from $D(v')$.

The listed actions may cause dictionary $D(\mu)$ or $D(v')$ to become empty. If this happens, say for $D(\mu)$ and μ is not the root of T, we need to remove node μ. Let $(+\infty, (\cdot, \kappa))$ be the special element of $D(\mu)$ with key $+\infty$, and let $(z, (w, \mu))$ be the element pointing to μ in the parent node π of μ. We delete node μ and replace $(z, (w, \mu))$ with $(z, (w, \kappa))$ in $D(\pi)$.

Note that if we start with an initially empty dictionary, a sequence of insertions and deletions performed with the described methods yields a search tree with a single node. In the next sections, we show how to avoid this behavior by imposing additional conditions on the structure of a search tree.

4.4.4 Realization with an (a, b)-Tree

An (a, b)-*tree*, where a and b are integer constants such that $2 \leq a \leq (b + 1)/2$, is a a search tree T with the following additional restrictions:

Level property. All of the levels of T are full, that is, all of the leaves are at the same depth.

Size property. Let μ be an internal node of T, and d be the number of children of μ; if μ is the root of T, then $d \geq 2$, else $a \leq d \leq b$.

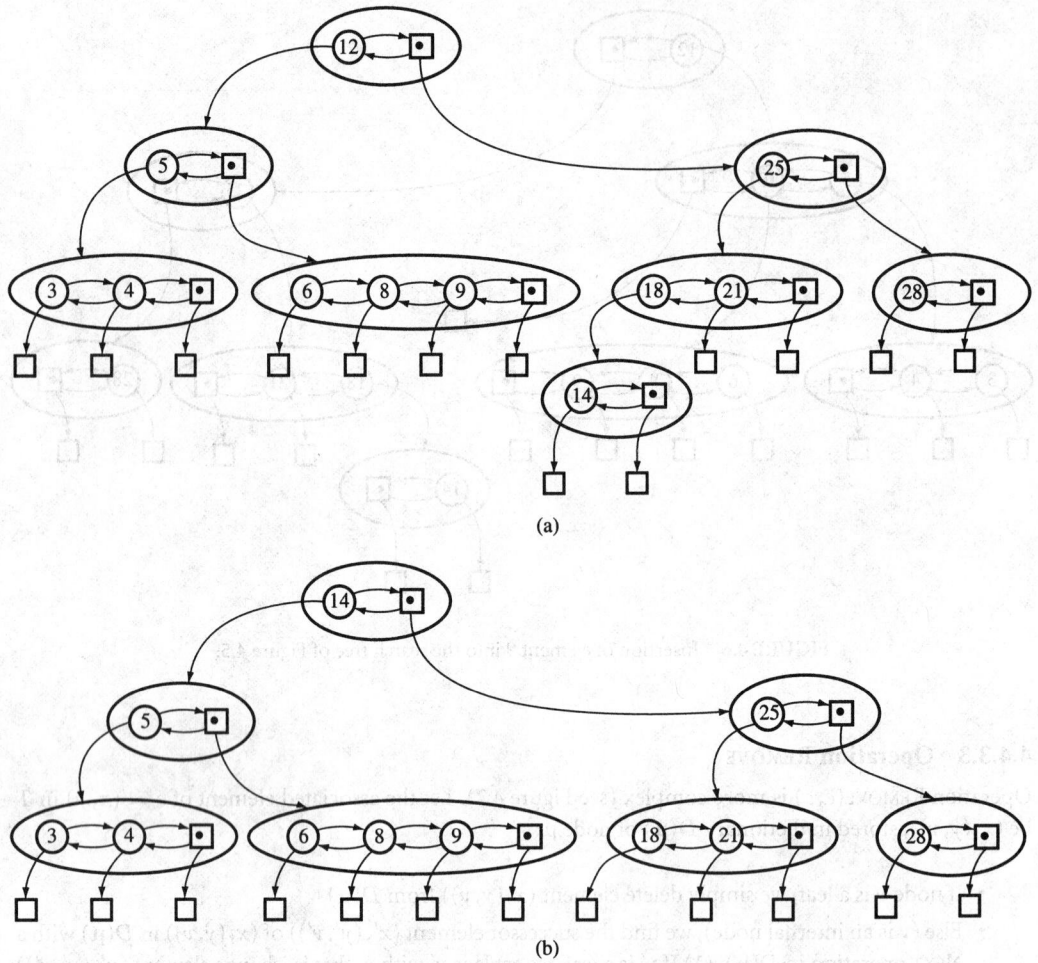

(a)

(b)

FIGURE 4.7 (a) Deletion of element 10 from the search tree of Figure 4.6. (b) Deletion of element 12 from the search tree of part a.

The height of an (a, b)-tree storing N elements is $O(\log_a N) = O(\log N)$. Indeed, in the worst case, the root has two children and all of the other internal nodes have a children.

The realization of a dictionary with an (a, b)-tree extends that with a search tree. Namely, the implementation of operations INSERT and REMOVE need to be modified in order to preserve the level and size properties. Also, we maintain the current size of the dictionary, and pointers to the minimum and maximum regular elements of the dictionary.

4.4.4.1 Insertion

The implementation of operation INSERT for search trees given earlier in this section adds a new element to the dictionary $D(\mu)$ of an existing node μ of T. Because the structure of the tree is not changed, the level property is satisfied. However, if $D(\mu)$ had the maximum allowed size $b - 1$ before insertion (recall that the size of $D(\mu)$ is one less than the number of children of μ), then the size property is violated at μ because $D(\mu)$ has now size b. To remedy this *overflow* situation, we perform the following *node split* (see Figure 4.8):

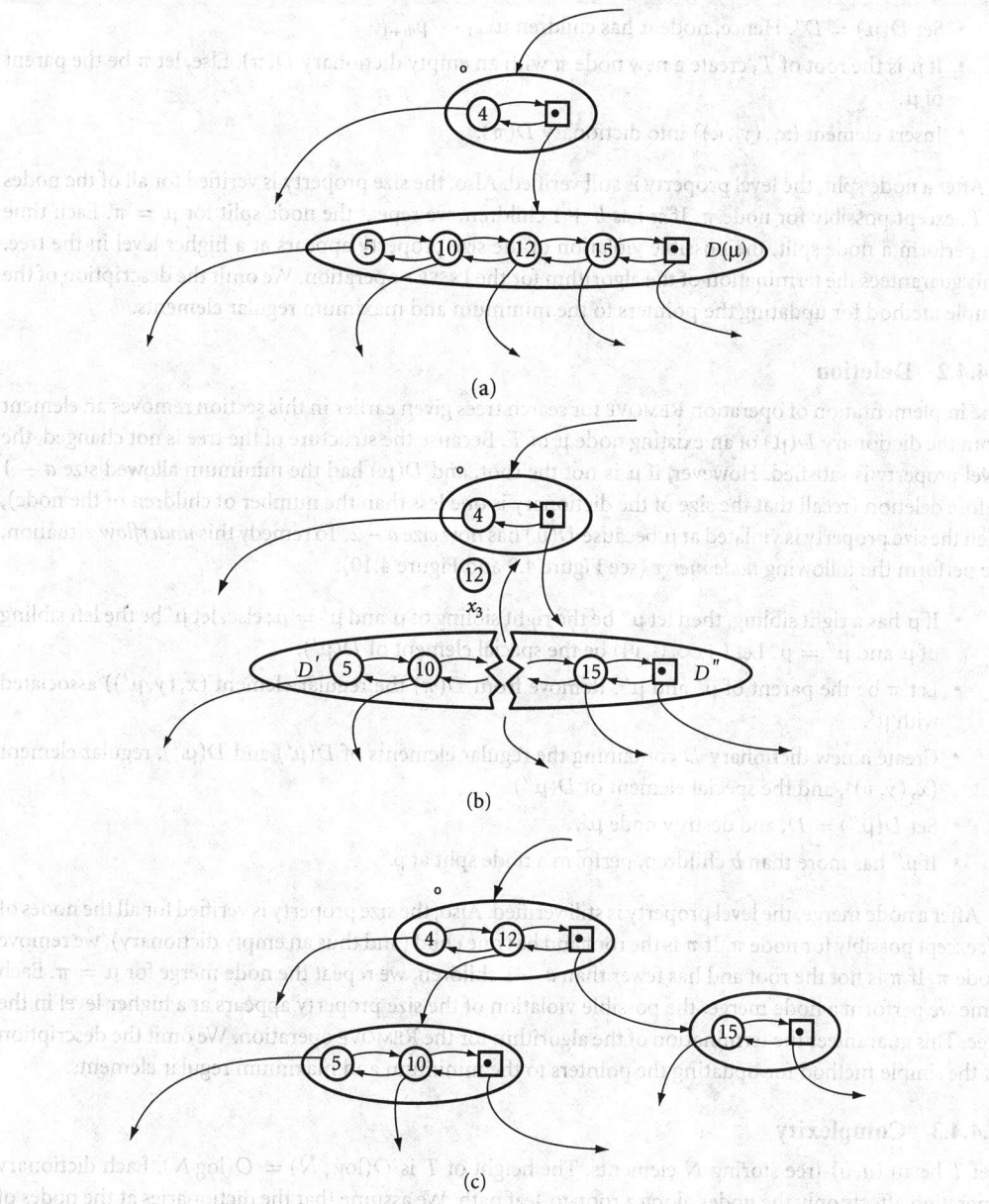

FIGURE 4.8 Example of node split in a 2–4 tree: (a) initial configuration with an overflow at node μ, (b) split of the node μ into μ' and μ'' and insertion of the median element into the parent node π, and (c) final configuration.

- Let the special element of $D(\mu)$ be $(+\infty, (\cdot, \mu_{b+1}))$. Find the median element of $D(\mu)$, that is, the element $e_i = (x_i, (y_i, \mu_i))$ such that $i = \lceil (b+1)/2 \rceil$.
- Split $D(\mu)$ into: (1) dictionary D' containing the $\lceil (b-1)/2 \rceil$ regular elements $e_j = (x_j, (y_j, \mu_j))$, $j = 1 \cdots i - 1$ and the special element $(+\infty, (\cdot, \mu_i))$; (2) element e; and (3) dictionary D'', containing the $\lfloor (b-1)/2 \rfloor$ regular elements $e_j = (x_j, (y_j, \mu_j))$, $j = i + 1 \cdots b$ and the special element $(+\infty, (\cdot, \mu_{b+1}))$.
- Create a new tree node κ, and set $D(\kappa) = D'$. Hence, node κ has children $\mu_1 \cdots \mu_i$.

- Set $D(\mu) = D''$. Hence, node μ has children $\mu_{i+1} \cdots \mu_{b+1}$.
- If μ is the root of T, create a new node π with an empty dictionary $D(\pi)$. Else, let π be the parent of μ.
- Insert element $(x_i, (y_i, \kappa))$ into dictionary $D(\pi)$.

After a node split, the level property is still verified. Also, the size property is verified for all of the nodes of T, except possibly for node π. If π has $b + 1$ children, we repeat the node split for $\mu = \pi$. Each time we perform a node split, the possible violation of the size property appears at a higher level in the tree. This guarantees the termination of the algorithm for the INSERT operation. We omit the description of the simple method for updating the pointers to the minimum and maximum regular elements.

4.4.4.2 Deletion

The implementation of operation REMOVE for search trees given earlier in this section removes an element from the dictionary $D(\mu)$ of an existing node μ of T. Because the structure of the tree is not changed, the level property is satisfied. However, if μ is not the root, and $D(\mu)$ had the minimum allowed size $a - 1$ before deletion (recall that the size of the dictionary is one less than the number of children of the node), then the size property is violated at μ because $D(\mu)$ has now size $a - 2$. To remedy this *underflow* situation, we perform the following *node merge* (see Figure 4.9 and Figure 4.10):

- If μ has a right sibling, then let μ'' be the right sibling of μ and $\mu' = \mu$; else, let μ' be the left sibling of μ and $\mu'' = \mu$. Let $(+\infty, (\cdot, \nu))$ be the special element of $D(\mu')$.
- Let π be the parent of μ' and μ''. Remove from $D(\pi)$ the regular element $(x, (y, \mu'))$ associated with μ'.
- Create a new dictionary D containing the regular elements of $D(\mu')$ and $D(\mu'')$, regular element $(x, (y, \nu))$, and the special element of $D(\mu'')$.
- Set $D(\mu'') = D$, and destroy node μ'.
- If μ'' has more than b children, perform a node split at μ''.

After a node merge, the level property is still verified. Also, the size property is verified for all the nodes of T, except possibly for node π. If π is the root and has one child (and thus an empty dictionary), we remove node π. If π is not the root and has fewer than $a - 1$ children, we repeat the node merge for $\mu = \pi$. Each time we perform a node merge, the possible violation of the size property appears at a higher level in the tree. This guarantees the termination of the algorithm for the REMOVE operation. We omit the description of the simple method for updating the pointers to the minimum and maximum regular elements.

4.4.4.3 Complexity

Let T be an (a, b)-tree storing N elements. The height of T is $O(\log_a N) = O(\log N)$. Each dictionary operation affects only the nodes along a root-to-leaf path. We assume that the dictionaries at the nodes of T are realized with sequences. Hence, processing a node takes $O(b) = O(1)$ time. We conclude that each operation takes $O(\log N)$ time.

Table 4.10 shows the performance of a dictionary realized with an (a, b)-tree. In the table we denote with N the number of elements in the dictionary at the time the operation is performed. The space complexity is $O(N)$.

4.4.5 Realization with an AVL-Tree

An *AVL-tree* is a search tree T with the following additional restrictions:

  Binary property. T is a binary tree, that is, every internal node has two children (left and right child), and stores one key.

  Balance property. For every internal node μ, the heights of the subtrees rooted at the children of μ differ at most by one.

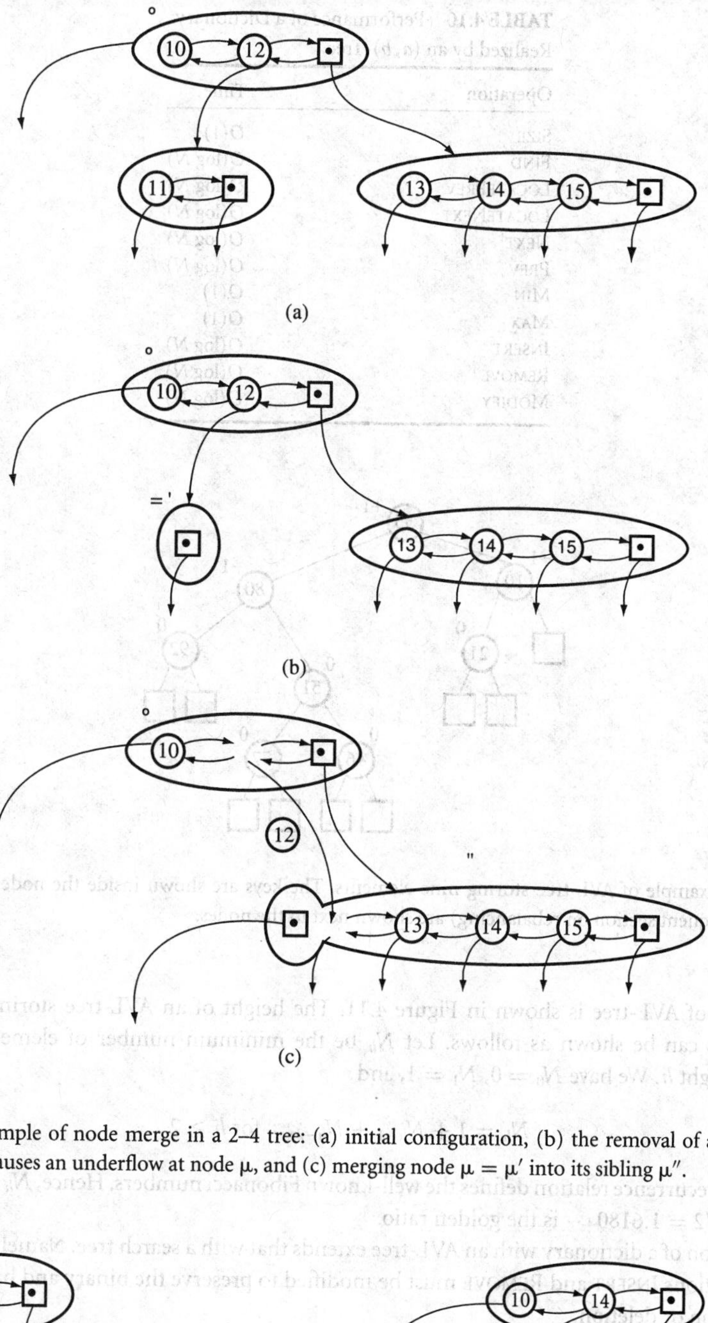

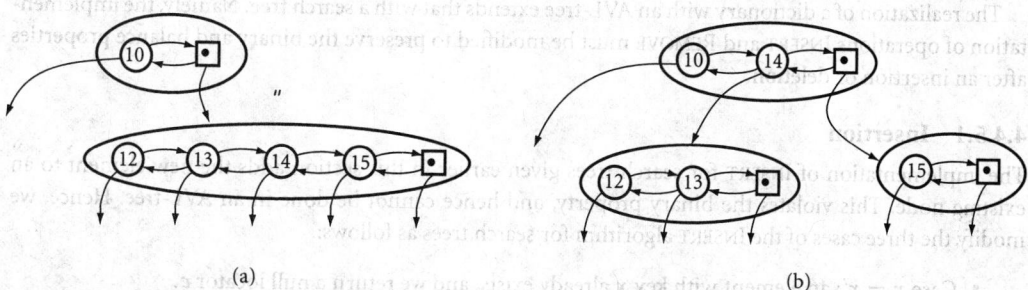

FIGURE 4.9 Example of node merge in a 2–4 tree: (a) initial configuration, (b) the removal of an element from dictionary $D(\mu)$ causes an underflow at node μ, and (c) merging node $\mu = \mu'$ into its sibling μ''.

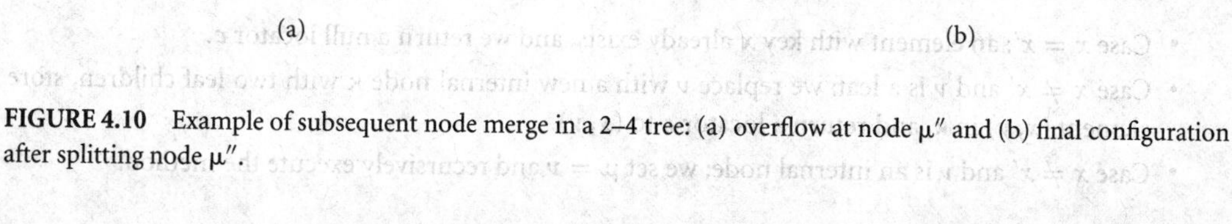

FIGURE 4.10 Example of subsequent node merge in a 2–4 tree: (a) overflow at node μ'' and (b) final configuration after splitting node μ''.

TABLE 4.10   Performance of a Dictionary
Realized by an (a, b)-Tree

Operation	Time
SIZE	$O(1)$
FIND	$O(\log N)$
LOCATEPREV	$O(\log N)$
LOCATENEXT	$O(\log N)$
NEXT	$O(\log N)$
PREV	$O(\log N)$
MIN	$O(1)$
MAX	$O(1)$
INSERT	$O(\log N)$
REMOVE	$O(\log N)$
MODIFY	$O(\log N)$

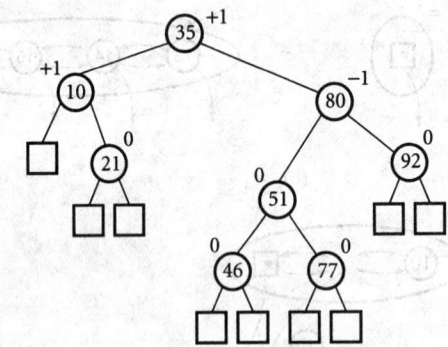

FIGURE 4.11   Example of AVL-tree storing nine elements. The keys are shown inside the nodes, and the balance factors (see subsequent section on rebalancing) are shown next to the nodes.

An example of AVL-tree is shown in Figure 4.11. The height of an AVL-tree storing N elements is $O(\log N)$. This can be shown as follows. Let N_h be the minimum number of elements stored in an AVL-tree of height h. We have $N_0 = 0$, $N_1 = 1$, and

$$N_h = 1 + N_{h-1} + N_{h-2}, \quad \text{for } h \geq 2$$

The preceding recurrence relation defines the well-known Fibonacci numbers. Hence, $N_h = \Omega(\phi^N)$, where $\phi = (1 + \sqrt{5})/2 = 1.6180 \cdots$ is the golden ratio.

The realization of a dictionary with an AVL-tree extends that with a search tree. Namely, the implementation of operations INSERT and REMOVE must be modified to preserve the binary and balance properties after an insertion or deletion.

4.4.5.1 Insertion

The implementation of INSERT for search trees given earlier in this section adds the new element to an existing node. This violates the binary property, and hence cannot be done in an AVL-tree. Hence, we modify the three cases of the INSERT algorithm for search trees as follows:

- Case $x = x'$: an element with key x already exists, and we return a null locator c.
- Case $x \neq x'$ and v is a leaf: we replace v with a new internal node κ with two leaf children, store element (x, y) in κ, and return a locator c to (x, y).
- Case $x \neq x'$ and v is an internal node: we set $\mu = v$ and recursively execute the method.

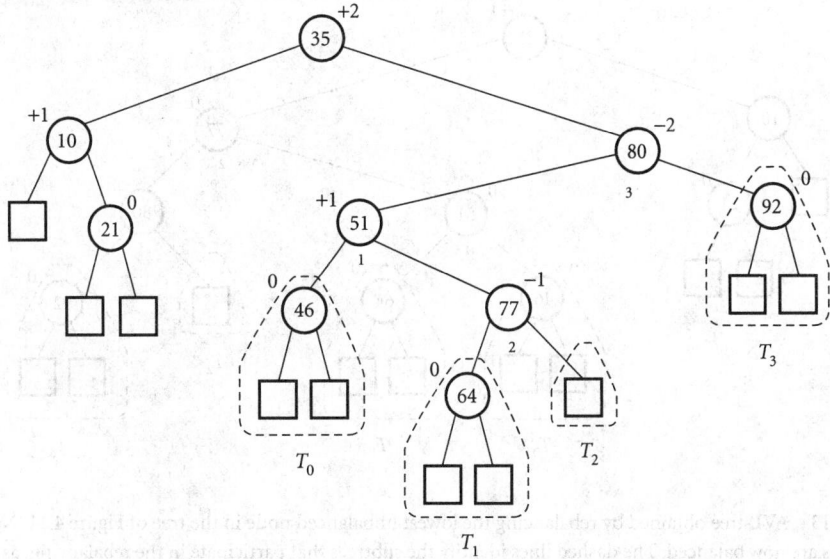

FIGURE 4.12 Insertion of an element with key 64 into the AVL-tree of Figure 4.11. Note that two nodes (with balance factors +2 and −2) have become unbalanced. The dashed lines identify the subtrees that participate in the rebalancing, as illustrated in Figure 4.14.

We have preserved the binary property. However, we may have violated the balance property because the heights of some subtrees of T have increased by one. We say that a node is balanced if the difference between the heights of its subtrees is −1, 0, or 1, and is unbalanced otherwise. The unbalanced nodes form a (possibly empty) subpath of the path from the new internal node κ to the root of T. See the example of Figure 4.12.

4.4.5.2 Rebalancing

To restore the balance property, we *rebalance* the lowest node μ that is unbalanced, as follows:

- Let μ' be the child of μ whose subtree has maximum height, and μ'' be the child of μ' whose subtree has maximum height.
- Let (μ_1, μ_2, μ_3) be the left-to-right ordering of nodes $\{\mu, \mu', \mu''\}$, and (T_0, T_1, T_2, T_3) be the left-to-right ordering of the four subtrees of $\{\mu, \mu', \mu''\}$ not rooted at a node in $\{\mu, \mu', \mu''\}$.
- Replace the subtree rooted at μ with a new subtree rooted at μ_2, where μ_1 is the left child of μ_2 and has subtrees T_0 and T_1, and μ_3 is the right child of μ_2 and has subtrees T_2 and T_3.

Two examples of rebalancing are schematically shown in Figure 4.14. Other symmetric configurations are possible. In Figure 4.13 we show the rebalancing for the tree of Figure 4.12.

Note that the rebalancing causes all the nodes in the subtree of μ_2 to become balanced. Also, the subtree rooted at μ_2 now has the same height as the subtree rooted at node μ before insertion. This causes all of the previously unbalanced nodes to become balanced. To keep track of the nodes that become unbalanced, we can store at each node a *balance factor*, which is the difference of the heights of the left and right subtrees. A node becomes unbalanced when its balance factor becomes +2 or −2. It is easy to modify the algorithm for operation INSERT such that it maintains the balance factors of the nodes.

4.4.5.3 Deletion

The implementation of REMOVE for search trees given earlier in this section preserves the binary property, but may cause the balance property to be violated. After deleting a node, there can be only one unbalanced node, on the path from the deleted node to the root of T.

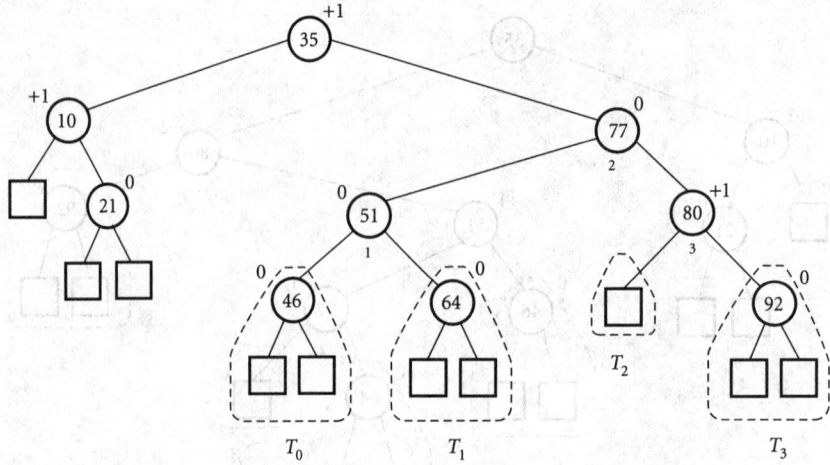

FIGURE 4.13   AVL-tree obtained by rebalancing the lowest unbalanced node in the tree of Figure 4.11. Note that all of the nodes are now balanced. The dashed lines identify the subtrees that participate in the rebalancing, as illustrated in Figure 4.14.

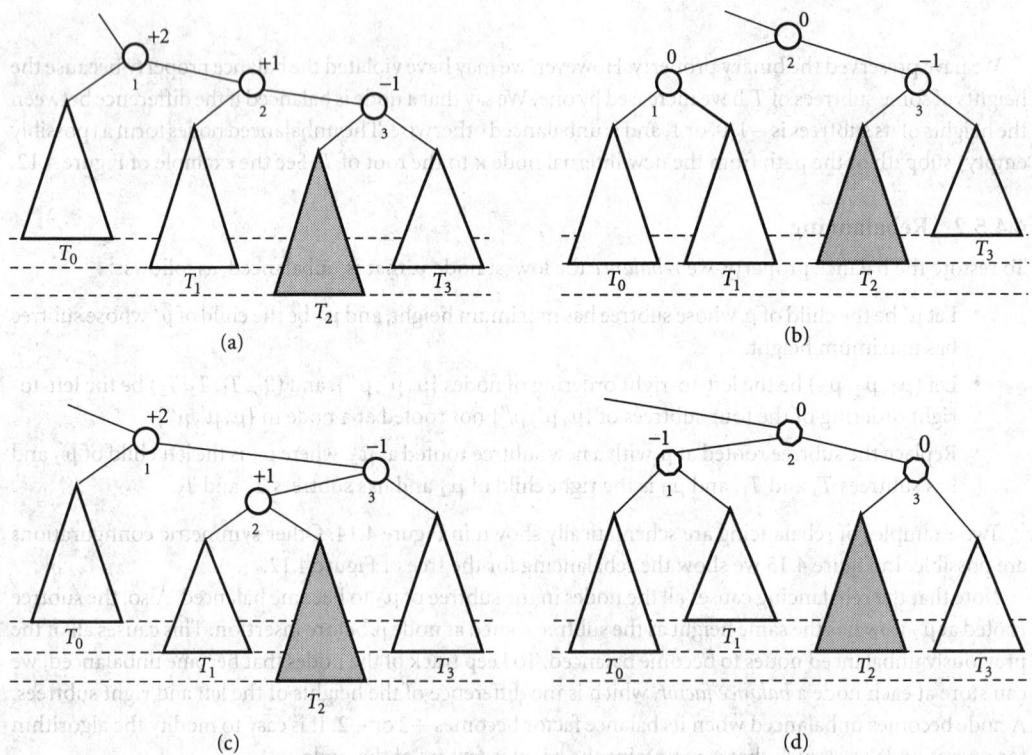

FIGURE 4.14   Schematic illustration of rebalancing a node in the INSERT algorithm for AVL-trees. The shaded subtree is the one where the new element was inserted. (a) and (b) Rebalancing by means of a single rotation. (c) and (d) Rebalancing by means of a double rotation.

TABLE 4.11 Performance of a Dictionary
Realized by an AVL-Tree

Operation	Time
SIZE	$O(1)$
FIND	$O(\log N)$
LOCATEPREV	$O(\log N)$
LOCATENEXT	$O(\log N)$
NEXT	$O(\log N)$
PREV	$O(\log N)$
MIN	$O(1)$
MAX	$O(1)$
INSERT	$O(\log N)$
REMOVE	$O(\log N)$
MODIFY	$O(\log N)$

To restore the balance property, we *rebalance* the unbalanced node using the previous algorithm, with minor modifications. If the subtrees of μ' have the same height, the height of the subtree rooted at μ_2 is the same as the height of the subtree rooted at μ before rebalancing, and we are done. If, instead, the subtrees of μ' do not have the same height, then the height of the subtree rooted at μ_2 is one less than the height of the subtree rooted at μ before rebalancing. This may cause an ancestor of μ_2 to become unbalanced, and we repeat the above computation. Balance factors are used to keep track of the nodes that become unbalanced, and can be easily maintained by the REMOVE algorithm.

4.4.5.4 Complexity

Let T be an AVL-tree storing N elements. The height of T is $O(\log N)$. Each dictionary operation affects only the nodes along a root-to-leaf path. Rebalancing a node takes $O(1)$ time. We conclude that each operation takes $O(\log N)$ time.

Table 4.11 shows the performance of a dictionary realized with an AVL-tree. In this table we denote with N the number of elements in the dictionary at the time the operation is performed. The space complexity is $O(N)$.

4.4.6 Realization with a Hash Table

The previous realizations of a dictionary make no assumptions on the structure of the keys and use comparisons between keys to guide the execution of the various operations.

4.4.6.1 Bucket Array

If the keys of a dictionary D are integers in the range $[1, M]$, we can implement D with a *bucket array B*. An element (x, y) of D is represented by setting $B[x] = y$. If an integer x is not in D, the location $B[x]$ stores a null value. In this implementation, we allocate a bucket for every possible element of D.

Table 4.12 shows the performance of a dictionary realized with a bucket array. In this table the keys in the dictionary are integers in the range $[1, M]$. The space complexity is $O(M)$.

The bucket array method can be extended to keys that are easily mapped to integers. For example, three-letter airport codes can be mapped to the integers in the range $[1, 26^3]$.

4.4.6.2 Hashing

The bucket array method works well when the range of keys is small. However, it is inefficient when the range of keys is large. To overcome this problem, we can use a *hash function h* that maps the keys of the original dictionary D into integers in the range $[1, M]$, where M is a parameter of the hash function. Now, we can apply the bucket array method using the *hashed value* $h(x)$ of the keys. In general, a *collision* may

TABLE 4.12 Performance of a Dictionary
Realized by Bucket Array

Operation	Time
SIZE	$O(1)$
FIND	$O(1)$
LOCATEPREV	$O(M)$
LOCATENEXT	$O(M)$
NEXT	$O(M)$
PREV	$O(M)$
MIN	$O(M)$
MAX	$O(M)$
INSERT	$O(1)$
REMOVE	$O(1)$
MODIFY	$O(1)$

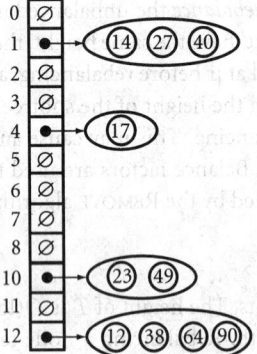

FIGURE 4.15 Example of a hash table of size 13 storing 10 elements. The hash function is $h(x) = x$ mod 13.

happen, where two distinct keys x_1 and x_2 have the same hashed value, that is, $x_1 \neq x_2$ and $h(x_1) = h(x_2)$. Hence, each bucket must be able to accommodate a collection of elements.

A hash table of size M for a function $h(x)$ is a bucket array B of size M (primary structure) whose entries are dictionaries (secondary structures), such that element (x, y) is stored in the dictionary $B[h(x)]$. For simplicity of programming, the dictionaries used as secondary structures are typically realized with sequences. An example of a hash table is shown in Figure 4.15.

If all of the elements in the dictionary D collide, they are all stored in the same dictionary of the bucket array, and the performance of the hash table is the same as that of the kind of dictionary used for the secondary structures. At the other end of the spectrum, if no two elements of the dictionary D collide, they are stored in distinct one-element dictionaries of the bucket array, and the performance of the hash table is the same as that of a bucket array.

A typical hash function for integer keys is $h(x) = x$ mod M (here, the range is $[0, M-1]$). The size M of the hash table is usually chosen as a prime number. An example of a hash table is shown in Figure 4.15. It is interesting to analyze the performance of a hash table from a probabilistic viewpoint. If we assume that the hashed values of the keys are uniformly distributed in the range $[0, M-1]$, then each bucket holds on average N/M keys, where N is the size of the dictionary. Hence, when $N = O(M)$, the average size of the secondary data structures is $O(1)$.

Table 4.13 shows the performance of a dictionary realized with a hash table. Both the worst-case and average time complexity in the preceding probabilistic model are indicated. In this table we denote with N the number of elements in the dictionary at the time the operation is performed. The space complexity

TABLE 4.13 Performance of a Dictionary Realized by a Hash Table of Size M

	Time	
Operation	Worst Case	Average
SIZE	$O(1)$	$O(1)$
FIND	$O(N)$	$O(N/M)$
LOCATEPREV	$O(N+M)$	$O(N+M)$
LOCATENEXT	$O(N+M)$	$O(N+M)$
NEXT	$O(N+M)$	$O(N+M)$
PREV	$O(N+M)$	$O(N+M)$
MIN	$O(N+M)$	$O(N+M)$
MAX	$O(N+M)$	$O(N+M)$
INSERT	$O(1)$	$O(1)$
REMOVE	$O(1)$	$O(1)$
MODIFY	$O(1)$	$O(1)$

is $O(N+M)$. The average time complexity refers to a probabilistic model where the hashed values of the keys are uniformly distributed in the range $[1, M]$.

Acknowledgments

Work supported in part by the National Science Foundation under grant DUE–0231202. Bryan Cantrill contributed to this work while at Brown University.

Defining Terms

(a, b)-Tree: Search tree with additional properties (each node has between a and b children, and all the levels are full).

Abstract data type: Mathematically specified data type equipped with operations that can be performed on the objects.

AVL-tree: Binary search tree such that the subtrees of each node have heights that differ by at most one.

Binary search tree: Search tree such that each internal node has two children.

Bucket array: Implementation of a dictionary by means of an array indexed by the keys of the dictionary elements.

Container: Abstract data type storing a collection of objects (elements).

Dictionary: Container storing elements from a sorted universe supporting searches, insertions, and deletions.

Hash table: Implementation of a dictionary by means of a bucket array storing secondary dictionaries.

Heap: Binary tree with additional properties storing the elements of a priority queue.

Position: Object representing the place of an element stored in a container.

Locator: Mechanism for tracking an element stored in a container.

Priority queue: Container storing elements from a sorted universe that supports finding the maximum element, insertions, and deletions.

Search tree: Rooted ordered tree with additional properties storing the elements of a dictionary.

Sequence: Container storing objects in a linear order, supporting insertions (in a given position) and deletions.

References

Aggarwal, A. and Vitter, J.S. 1988. The input/output complexity of sorting and related problems. *Commun. ACM*, 31:1116–1127.

Aho, A.V., Hopcroft, J.E., and Ullman, J.D. 1983. *Data Structures and Algorithms*. Addison-Wesley, Reading, MA.

Chazelle, B. and Guibas, L.J. 1986. Fractional cascading. I. A data structuring technique. *Algorithmica*, 1:133–162.

Chiang, Y.-J. and Tamassia, R. 1992. Dynamic algorithms in computational geometry. *Proc. IEEE*, 80(9):1412–1434.

Cohen, R.F. and Tamassia, R. 1995. Dynamic expression trees. *Algorithmica*, 13:245–265.

Comer, D. 1979. The ubiquitous B-tree. *ACM Comput. Surv.*, 11:121–137.

Cormen, T.H., Leiserson, C.E., Rivest, R.L., and Stein, C. 2001. *Introduction to Algorithms*. MIT Press, Cambridge, MA.

Di Battista, G. and Tamassia, R. 1996. On-line maintenance of triconnected components with SPQR-trees. *Algorithmica*, 15:302–318.

Di Battista, G., Eades, P., Tamassia, R., and Tollis, I.G. 1999. *Graph Drawing: Algorithms for the Visualization of Graphs*. Prentice Hall, Upper Saddle River, NJ.

Driscoll, J.R., Sarnak, N., Sleator, D.D., and Tarjan, R.E. 1989. Making data structures persistent. *J. Comput. Syst. Sci.* 38:86–124.

Edelsbrunner, H. 1987. *Algorithms in Combinatorial Geometry*, Vol. 10, *EATCS Monographs on Theoretical Computer Science*. Springer–Verlag, Heidelberg, Germany.

Eppstein, D., Galil, Z., Italiano, G.F., and Nissenzweig, A. 1997. Sparsification: a technique for speeding up dynamic graph algorithms. *J. ACM*, 44:669–696.

Even, S. 1979. *Graph Algorithms*. Computer Science Press, Potomac, MD.

Foley, J.D., van Dam, A., Feiner, S.K., and Hughes, J.F. 1990. *Computer Graphics: Principles and Practice*. Addison-Wesley, Reading, MA.

Frederickson, G.N. 1997. A data structure for dynamically maintaining rooted trees. *J. Algorithms*, 24:37–65.

Galil, Z. and Italiano, G.F. 1991. Data structures and algorithms for disjoint set union problems. *ACM Comput. Surv.*, 23(3):319–344.

Gonnet, G.H. and Baeza-Yates, R. 1991. *Handbook of Algorithms and Data Structures*. Addison-Wesley, Reading, MA.

Goodrich, M.T. and Tamassia, R. 2001. *Data Structures and Algorithms in Java*. Wiley, New York.

Hoffmann, K., Mehlhorn, K., Rosenstiehl, P., and Tarjan, R.E. 1986. Sorting Jordan sequences in linear time using level-linked search trees. *Inf. Control*, 68:170–184.

Horowitz, E., Sahni, S., and Metha, D. 1995. *Fundamentals of Data Structures in C++*. Computer Science Press, Potomac, MD.

Knuth, D.E. 1968. *Fundamental Algorithms*. Vol. I. In *The Art of Computer Programming*. Addison-Wesley, Reading, MA.

Knuth, D.E. 1973. *Sorting and Searching*, Vol. 3. In *The Art of Computer Programming*. Addison-Wesley, Reading, MA.

Mehlhorn, K. 1984. *Data Structures and Algorithms*. Vol. 1–3. Springer–Verlag.

Mehlhorn, K. and Näher, S. 1999. *LEDA: a Platform for Combinatorial and Geometric Computing*. Cambridge University Press.

Mehlhorn, K. and Tsakalidis, A. 1990. Data structures. In *Algorithms and Complexity*. J. van Leeuwen, Ed., Vol. A, *Handbook of Theoretical Computer Science*. Elsevier, Amsterdam.

Munro, J.I. and Suwanda, H. 1980. Implicit Data Structures for Fast Search and Update. *J. Comput. Syst. Sci.*, 21:236–250.

Nievergelt, J. and Hinrichs, K.H. 1993. *Algorithms and Data Structures: With Applications to Graphics and Geometry*. Prentice Hall, Englewood Cliffs, NJ.

O'Rourke, J. 1994. *Computational Geometry in C*. Cambridge University Press,

Overmars, M.H. 1983. *The Design of Dynamic Data Structures*, Vol. 156, *Lecture Notes in Computer Science*. Springer-Verlag.

Preparata, F.P. and Shamos, M.I. 1985. *Computational Geometry: An Introduction*. Springer-Verlag, New York.

Pugh, W. 1990. Skip lists: a probabilistic alternative to balanced trees. *Commun. ACM*, 35:668–676.

Sedgewick, R. 1992. *Algorithms in C++*. Addison-Wesley, Reading, MA.

Sleator, D.D. and Tarjan, R.E. 1993. A data structure for dynamic trees. *J. Comput. Syst. Sci.*, 26(3):362–381.

Tamassia, R., Goodrich, M.T., Vismara, L., Handy, M., Shubina, G., Cohen R., Hudson, B., Baker, R.S., Gelfand, N., and Brandes, U. 2001. JDSL: the data structures library in Java. *Dr. Dobb's Journal*, 323:21–31.

Tarjan, R.E. 1983. *Data Structures and Network Algorithms, Vol. 44, CBMS-NSF Regional Conference Series in Applied Mathematics*. Society for Industrial Applied Mathematics.

Vitter, J.S. and Flajolet, P. 1990. Average-case analysis of algorithms and data structures. In *Algorithms and Complexity*, J. van Leeuwen, Ed., Vol. A, *Handbook of Theoretical Computer Science*, pp. 431–524. Elsevier, Amsterdam.

Wood, D. 1993. *Data Structures, Algorithms, and Performance*. Addison-Wesley, Reading, MA.

Further Information

Many textbooks and monographs have been written on data structures, for example, Aho et al. [1983], Cormen et al. [2001], Gonnet and Baeza-Yates [1990], Goodrich and Tamassia [2001], Horowitz et al. [1995], Knuth [1968, 1973], Mehlhorn [1984], Nievergelt and Hinrichs [1993], Overmars [1983], Preparata and Shamos [1995], Sedgewick [1992], Tarjan [1983], and Wood [1993].

Papers surveying the state-of-the art in data structures include Chiang and Tamassia [1992], Galil and Italiano [1991], Mehlhorn and Tsakalidis [1990], and Vitter and Flajolet [1990].

JDSL is a library of fundamental data structures in Java [Tamassia et al. 2000]. LEDA is a library of advanced data structures in C++ [Mehlhorn and Näher 1999].

5

Complexity Theory

5.1	Introduction	5-1
5.2	Models of Computation	5-2
	Computational Problems and Languages • Turing Machines • Universal Turing Machines • Alternating Turing Machines • Oracle Turing Machines	
5.3	Resources and Complexity Classes	5-5
	Time and Space • Complexity Classes	
5.4	Relationships between Complexity Classes	5-8
	Constructibility • Basic Relationships • Complementation • Hierarchy Theorems and Diagonalization • Padding Arguments	
5.5	Reducibility and Completeness	5-12
	Resource-Bounded Reducibilities • Complete Languages • Cook-Levin Theorem • Proving NP-Completeness • Complete Problems for Other Classes	
5.6	Relativization of the P vs. NP Problem	5-17
5.7	The Polynomial Hierarchy	5-18
5.8	Alternating Complexity Classes	5-19
5.9	Circuit Complexity	5-20
5.10	Probabilistic Complexity Classes	5-22
5.11	Interactive Models and Complexity Classes	5-23
	Interactive Proofs • Probabilistically Checkable Proofs	
5.12	Kolmogorov Complexity	5-25
5.13	Research Issues and Summary	5-26

Eric W. Allender
Rutgers University

Michael C. Loui
*University of Illinois
at Urbana-Champaign*

Kenneth W. Regan
State University of New York at Buffalo

5.1 Introduction

Computational complexity is the study of the difficulty of solving computational problems, in terms of the required computational resources, such as time and space (memory). Whereas the analysis of algorithms focuses on the time or space of an *individual* algorithm for a *specific* problem (such as sorting), complexity theory focuses on the **complexity class** of problems solvable in the same amount of time or space. Most common computational problems fall into a small number of complexity classes. Two important complexity classes are P, the set of problems that can be solved in polynomial time, and NP, the set of problems whose solutions can be verified in polynomial time.

By quantifying the resources required to solve a problem, complexity theory has profoundly affected our thinking about computation. Computability theory establishes the existence of undecidable problems, which cannot be solved in principle regardless of the amount of time invested. However, computability theory fails to find meaningful distinctions among decidable problems. In contrast, complexity theory establishes the existence of decidable problems that, although solvable in principle, cannot be solved in practice because the time and space required would be larger than the age and size of the known universe [Stockmeyer and Chandra, 1979]. Thus, complexity theory characterizes the computationally feasible problems.

The quest for the boundaries of the set of feasible problems has led to the most important unsolved question in all of computer science: is P different from NP? Hundreds of fundamental problems, including many ubiquitous optimization problems of operations research, are **NP-complete**; they are the hardest problems in NP. If someone could find a polynomial-time algorithm for any one NP-complete problem, then there would be polynomial-time algorithms for all of them. Despite the concerted efforts of many scientists over several decades, no polynomial-time algorithm has been found for any NP-complete problem. Although we do not yet know whether P is different from NP, showing that a problem is NP-complete provides strong evidence that the problem is computationally infeasible and justifies the use of heuristics for solving the problem.

In this chapter, we define P, NP, and related complexity classes. We illustrate the use of **diagonalization** and **padding** techniques to prove relationships between classes. Next, we define NP-completeness, and we show how to prove that a problem is NP-complete. Finally, we define complexity classes for probabilistic and interactive computations.

Throughout this chapter, all numeric functions take integer arguments and produce integer values. All logarithms are taken to base 2. In particular, $\log n$ means $\lceil \log_2 n \rceil$.

5.2   Models of Computation

To develop a theory of the difficulty of computational problems, we need to specify precisely what a problem is, what an algorithm is, and what a measure of difficulty is. For simplicity, complexity theorists have chosen to represent problems as languages, to model algorithms by off-line multitape **Turing machines**, and to measure computational difficulty by the time and space required by a Turing machine. To justify these choices, some theorems of complexity theory show how to translate statements about, say, the time complexity of language recognition by Turing machines into statements about computational problems on more realistic models of computation. These theorems imply that the principles of complexity theory are not artifacts of Turing machines, but intrinsic properties of computation.

This section defines different kinds of Turing machines. The deterministic Turing machine models actual computers. The nondeterministic Turing machine is not a realistic model, but it helps classify the complexity of important computational problems. The alternating Turing machine models a form of parallel computation, and it helps elucidate the relationship between time and space.

5.2.1   Computational Problems and Languages

Computer scientists have invented many elegant formalisms for representing data and control structures. Fundamentally, all representations are patterns of symbols. Therefore, we represent an instance of a computational problem as a sequence of symbols.

Let Σ be a finite set, called the *alphabet*. A *word* over Σ is a finite sequence of symbols from Σ. Sometimes a word is called a *string*. Let Σ^* denote the set of all words over Σ. For example, if $\Sigma = \{0, 1\}$, then

$$\Sigma^* = \{\lambda, 0, 1, 00, 01, 10, 11, 000, \ldots\}$$

is the set of all binary words, including the empty word λ. The *length* of a word w, denoted by $|w|$, is the number of symbols in w. A *language* over Σ is a subset of Σ^*.

A *decision problem* is a computational problem whose answer is simply yes or no. For example, is the input graph connected, or is the input a sorted list of integers? A decision problem can be expressed as a membership problem for a language A: for an input x, does x belong to A? For a language A that represents connected graphs, the input word x might represent an input graph G, and $x \in A$ if and only if G is connected.

For every decision problem, the representation should allow for easy parsing, to determine whether a word represents a legitimate instance of the problem. Furthermore, the representation should be concise. In particular, it would be unfair to encode the answer to the problem into the representation of an instance of the problem; for example, for the problem of deciding whether an input graph is connected, the representation should not have an extra bit that tells whether the graph is connected. A set of integers $S = \{x_1, \ldots, x_m\}$ is represented by listing the binary representation of each x_i, with the representations of consecutive integers in S separated by a nonbinary symbol. A graph is naturally represented by giving either its adjacency matrix or a set of adjacency lists, where the list for each vertex v specifies the vertices adjacent to v.

Whereas the solution to a decision problem is yes or no, the solution to an optimization problem is more complicated; for example, determine the shortest path from vertex u to vertex v in an input graph G. Nevertheless, for every optimization (minimization) problem, with objective function g, there is a corresponding decision problem that asks whether there exists a feasible solution z such that $g(z) \leq k$, where k is a given target value. Clearly, if there is an algorithm that solves an optimization problem, then that algorithm can be used to solve the corresponding decision problem. Conversely, if an algorithm solves the decision problem, then with a binary search on the range of values of g, we can determine the optimal value. Moreover, using a decision problem as a subroutine often enables us to construct an optimal solution; for example, if we are trying to find a shortest path, we can use a decision problem that determines if a shortest path starting from a given vertex uses a given edge. Therefore, there is little loss of generality in considering only decision problems, represented as language membership problems.

5.2.2 Turing Machines

This subsection and the next three give precise, formal definitions of Turing machines and their variants. These subsections are intended for reference. For the rest of this chapter, the reader need not understand these definitions in detail, but may generally substitute "program" or "computer" for each reference to "Turing machine."

A k-worktape **Turing machine** M consists of the following:

- A finite set of states Q, with special states q_0 (initial state), q_A (accept state), and q_R (reject state).
- A finite alphabet Σ, and a special blank symbol $\square \notin \Sigma$.
- The $k + 1$ linear tapes, each divided into cells. Tape 0 is the *input tape*, and tapes $1, \ldots, k$ are the *worktapes*. Each tape is infinite to the left and to the right. Each cell holds a single symbol from $\Sigma \cup \{\square\}$. By convention, the input tape is read only. Each tape has an access head, and at every instant, each access head scans one cell (see Figure 5.1).

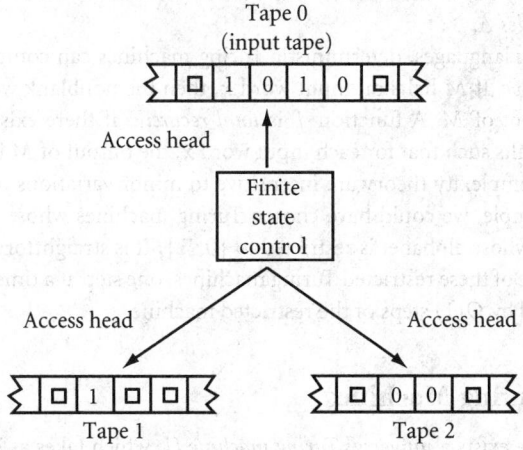

FIGURE 5.1 A two-tape Turing machine.

- A finite transition table δ, which comprises tuples of the form

$$(q, s_0, s_1, \ldots, s_k, q', s_1', \ldots, s_k', d_0, d_1, \ldots, d_k)$$

  where $q, q' \in Q$, each $s_i, s_i' \in \Sigma \cup \{\square\}$, and each $d_i \in \{-1, 0, +1\}$.

  A tuple specifies a step of M: if the current state is q, and $s_0, s_1, \ldots, s_k$ are the symbols in the cells scanned by the access heads, then M replaces s_i by s_i' for $i = 1, \ldots, k$ simultaneously, changes state to q', and moves the head on tape i one cell to the left ($d_i = -1$) or right ($d_i = +1$) or not at all ($d_i = 0$) for $i = 0, \ldots, k$. Note that M cannot write on tape 0, that is, M can write only on the worktapes, not on the input tape.

- In a tuple, no s_i' can be the blank symbol $\square$. Because M may not write a blank, the worktape cells that its access heads previously visited are nonblank.

- No tuple contains q_A or q_R as its first component. Thus, once M enters state q_A or state q_R, it stops.

- Initially, M is in state q_0, an input word in Σ^* is inscribed on contiguous cells of the input tape, the access head on the input tape is on the leftmost symbol of the input word, and all other cells of all tapes contain the blank symbol $\square$.

The Turing machine M that we have defined is *nondeterministic*: δ may have several tuples with the same combination of state q and symbols $s_0, s_1, \ldots, s_k$ as the first $k + 2$ components, so that M may have several possible next steps. A machine M is *deterministic* if for every combination of state q and symbols $s_0, s_1, \ldots, s_k$, at most one tuple in δ contains the combination as its first $k + 2$ components. A deterministic machine always has at most one possible next step.

A *configuration* of a Turing machine M specifies the current state, the contents of all tapes, and the positions of all access heads.

A *computation path* is a sequence of configurations $C_0, C_1, \ldots, C_t, \ldots$, where C_0 is the initial configuration of M, and each C_{j+1} follows from C_j in one step by applying the changes specified by a tuple in δ. If no tuple is applicable to C_t, then C_t is *terminal*, and the computation path is *halting*. If M has no infinite computation paths, then M *always halts*.

A halting computation path is *accepting* if the state in the last configuration C_t is q_A; otherwise it is *rejecting*. By adding tuples to the program if needed, we can ensure that every rejecting computation ends in state q_R. This leaves the question of computation paths that do not halt. In complexity theory, we rule this out by considering only machines whose computation paths *always halt*. M accepts an input word x if there exists an accepting computation path that starts from the initial configuration in which x is on the input tape. For nondeterministic M, it does not matter if some other computation paths end at q_R. If M is deterministic, then there is at most one halting computation path, hence at most one accepting path.

The *language accepted by* M, written $L(M)$, is the set of words accepted by M. If $A = L(M)$, and M always halts, then M *decides* A.

In addition to deciding languages, deterministic Turing machines can compute functions. Designate tape 1 to be the *output tape*. If M halts on input word x, then the nonblank word on tape 1 in the final configuration is the output of M. A function f is *total recursive* if there exists a deterministic Turing machine M that always halts such that for each input word x, the output of M is the value of $f(x)$.

Almost all results in complexity theory are insensitive to minor variations in the underlying computational models. For example, we could have chosen Turing machines whose tapes are restricted to be only one-way infinite or whose alphabet is restricted to $\{0, 1\}$. It is straightforward to simulate a Turing machine as defined by one of these restricted Turing machines, one step at a time: each step of the original machine can be simulated by $O(1)$ steps of the restricted machine.

5.2.3 Universal Turing Machines

Chapter 6 states that there exists a *universal Turing machine* U, which takes as input a string $\langle M, x \rangle$ that encodes a Turing machine M and a word x, and simulates the operation of M on x, and U accepts $\langle M, x \rangle$ if and only if M accepts x. A theorem of Hennie and Stearns [1966] implies that the machine U can be constructed to have only two worktapes, such that U can simulate any t steps of M in only $O(t \log t)$ steps of its own, using only $O(1)$ times the worktape cells used by M. The constants implicit in these big-O bounds may depend on M.

We can think of U with a fixed M as a machine U_M and define $L(U_M) = \{x : U \text{ accepts } \langle M, x \rangle\}$. Then $L(U_M) = L(M)$. If M always halts, then U_M always halts; and if M is deterministic, then U_M is deterministic.

5.2.4   Alternating Turing Machines

By definition, a nondeterministic Turing machine M accepts its input word x if there exists an accepting computation path, starting from the initial configuration with x on the input tape. Let us call a *configuration* C accepting if there is a computation path of M that starts in C and ends in a configuration whose state is q_A. Equivalently, a configuration C is accepting if either the state in C is q_A or there exists an accepting configuration C' reachable from C by one step of M. Then M accepts x if the initial configuration with input word x is accepting.

The *alternating Turing machine* generalizes this notion of acceptance. In an alternating Turing machine M, each state is labeled either existential or universal. (Do not confuse the universal state in an alternating Turing machine with the universal Turing machine.) A nonterminal configuration C is existential (respectively, universal) if the state in C is labeled existential (universal). A terminal configuration is accepting if its state is q_A. A nonterminal existential configuration C is accepting if there *exists* an accepting configuration C' reachable from C by one step of M. A nonterminal universal configuration C is accepting if for *every* configuration C' reachable from C by one step of M, the configuration C' is accepting. Finally, M accepts x if the initial configuration with input word x is an accepting configuration.

A nondeterministic Turing machine is thus a special case of an alternating Turing machine in which every state is existential.

The computation of an alternating Turing machine M alternates between existential states and universal states. Intuitively, from an existential configuration, M guesses a step that leads toward acceptance; from a universal configuration, M checks whether each possible next step leads toward acceptance — in a sense, M checks all possible choices in parallel. An alternating computation captures the essence of a two-player game: player 1 has a winning strategy if there exists a move for player 1 such that for every move by player 2, there exists a subsequent move by player 1, etc., such that player 1 eventually wins.

5.2.5   Oracle Turing Machines

Some computational problems remain difficult even when solutions to instances of a particular, different decision problem are available for free. When we study the complexity of a problem *relative* to a language A, we assume that answers about membership in A have been precomputed and stored in a (possibly infinite) table and that there is no cost to obtain an answer to a membership query: Is w in A? The language A is called an **oracle**. Conceptually, an algorithm queries the oracle whether a word w is in A, and it receives the correct answer in one step.

An *oracle Turing machine* is a Turing machine M with a special *oracle tape* and special states QUERY, YES, and NO. The computation of the oracle Turing machine M^A, with oracle language A, is the same as that of an ordinary Turing machine, except that when M enters the QUERY state with a word w on the oracle tape, in one step, M enters either the YES state if $w \in A$ or the NO state if $w \notin A$. Furthermore, during this step, the oracle tape is erased, so that the time for setting up each query is accounted for separately.

5.3   Resources and Complexity Classes

In this section, we define the measures of difficulty of solving computational problems. We introduce complexity classes, which enable us to classify problems according to the difficulty of their solution.

5.3.1 Time and Space

We measure the difficulty of a computational problem by the running time and the space (memory) requirements of an algorithm that solves the problem. Clearly, in general, a finite algorithm cannot have a table of all answers to infinitely many instances of the problem, although an algorithm could look up precomputed answers to a finite number of instances; in terms of Turing machines, the finite answer table is built into the set of states and the transition table. For these instances, the running time is negligible — just the time needed to read the input word. Consequently, our complexity measure should consider a whole problem, not only specific instances.

We express the complexity of a problem, in terms of the growth of the required time or space, as a function of the length n of the input word that encodes a problem instance. We consider the worst-case complexity, that is, for each n, the maximum time or space required among all inputs of length n.

Let M be a Turing machine that always halts. The *time* taken by M on input word x, denoted by $\text{Time}_M(x)$, is defined as follows:

- If M accepts x, then $\text{Time}_M(x)$ is the number of steps in the shortest accepting computation path for x.
- If M rejects x, then $\text{Time}_M(x)$ is the number of steps in the longest computation path for x.

For a deterministic machine M, for every input x, there is at most one halting computation path, and its length is $\text{Time}_M(x)$. For a nondeterministic machine M, if $x \in L(M)$, then M can guess the correct steps to take toward an accepting configuration, and $\text{Time}_M(x)$ measures the length of the path on which M always makes the best guess.

The *space* used by a Turing machine M on input x, denoted by $\text{Space}_M(x)$, is defined as follows. The space used by a halting computation path is the number of nonblank worktape cells in the last configuration; this is the number of different cells ever written by the worktape heads of M during the computation path, since M never writes the blank symbol. Because the space occupied by the input word is not counted, a machine can use a sublinear ($o(n)$) amount of space.

- If M accepts x, then $\text{Space}_M(x)$ is the minimum space used among all accepting computation paths for x.
- If M rejects x, then $\text{Space}_M(x)$ is the maximum space used among all computation paths for x.

The **time complexity** of a machine M is the function

$$t(n) = \max\{\text{Time}_M(x) : |x| = n\}$$

We assume that M reads all of its input word, and the blank symbol after the right end of the input word, so $t(n) \geq n + 1$. The **space complexity** of M is the function

$$s(n) = \max\{\text{Space}_M(x) : |x| = n\}$$

Because few interesting languages can be decided by machines of sublogarithmic space complexity, we henceforth assume that $s(n) \geq \log n$.

A function $f(x)$ is *computable in polynomial time* if there exists a deterministic Turing machine M of polynomial time complexity such that for each input word x, the output of M is $f(x)$.

5.3.2 Complexity Classes

Having defined the time complexity and space complexity of individual Turing machines, we now define classes of languages with particular complexity bounds. These definitions will lead to definitions of P and NP.

Let $t(n)$ and $s(n)$ be numeric functions. Define the following classes of languages:

- DTIME$[t(n)]$ is the class of languages decided by deterministic Turing machines of time complexity $O(t(n))$.
- NTIME$[t(n)]$ is the class of languages decided by nondeterministic Turing machines of time complexity $O(t(n))$.
- DSPACE$[s(n)]$ is the class of languages decided by deterministic Turing machines of space complexity $O(s(n))$.
- NSPACE$[s(n)]$ is the class of languages decided by nondeterministic Turing machines of space complexity $O(s(n))$.

We sometimes abbreviate DTIME$[t(n)]$ to DTIME$[t]$ (and so on) when t is understood to be a function, and when no reference is made to the input length n.

The following are the **canonical complexity classes**:

- L $=$ DSPACE$[\log n]$ (deterministic log space)
- NL $=$ NSPACE$[\log n]$ (nondeterministic log space)
- P $=$ DTIME$[n^{O(1)}] = \bigcup_{k \geq 1}$ DTIME$[n^k]$ (polynomial time)
- NP $=$ NTIME$[n^{O(1)}] = \bigcup_{k \geq 1}$ NTIME$[n^k]$ (nondeterministic polynomial time)
- PSPACE $=$ DSPACE$[n^{O(1)}] = \bigcup_{k \geq 1}$ DSPACE$[n^k]$ (polynomial space)
- E $=$ DTIME$[2^{O(n)}] = \bigcup_{k \geq 1}$ DTIME$[k^n]$
- NE $=$ NTIME$[2^{O(n)}] = \bigcup_{k \geq 1}$ NTIME$[k^n]$
- EXP $=$ DTIME$[2^{n^{O(1)}}] = \bigcup_{k \geq 1}$ DTIME$[2^{n^k}]$ (deterministic exponential time)
- NEXP $=$ NTIME$[2^{n^{O(1)}}] = \bigcup_{k \geq 1}$ NTIME$[2^{n^k}]$ (nondeterministic exponential time)

The space classes L and PSPACE are defined in terms of the DSPACE complexity measure. By Savitch's Theorem (see Theorem 5.2), the NSPACE measure with polynomial bounds also yields PSPACE.

The class P contains many familiar problems that can be solved efficiently, such as (decision problem versions of) finding shortest paths in networks, parsing for context-free languages, sorting, matrix multiplication, and linear programming. Consequently, P has become accepted as representing the set of computationally feasible problems. Although one could legitimately argue that a problem whose best algorithm has time complexity $\Theta(n^{99})$ is really infeasible, in practice, the time complexities of the vast majority of known polynomial-time algorithms have low degrees: they run in $O(n^4)$ time or less. Moreover, P is a robust class: although defined by Turing machines, P remains the same when defined by other models of sequential computation. For example, random access machines (RAMs) (a more realistic model of computation defined in Chapter 6) can be used to define P because Turing machines and RAMs can simulate each other with polynomial-time overhead.

The class NP can also be defined by means other than nondeterministic Turing machines. NP equals the class of problems whose solutions can be *verified* quickly, by deterministic machines in polynomial time. Equivalently, NP comprises those languages whose membership proofs can be checked quickly.

For example, one language in NP is the set of satisfiable Boolean formulas, called SAT. A Boolean formula ϕ is satisfiable if there exists a way of assigning true or false to each variable such that under this truth assignment, the value of ϕ is true. For example, the formula $x \wedge (\overline{x} \vee y)$ is satisfiable, but $x \wedge \overline{y} \wedge (\overline{x} \vee y)$ is not satisfiable. A nondeterministic Turing machine M, after checking the syntax of ϕ and counting the number n of variables, can nondeterministically write down an n-bit 0-1 string a on its tape, and then deterministically (and easily) evaluate ϕ for the truth assignment denoted by a. The computation path corresponding to each individual a accepts if and only if $\phi(a) =$ true, and so M itself accepts ϕ if and only if ϕ is satisfiable; that is, $L(M) =$ SAT. Again, this checking of given assignments differs significantly from trying to *find* an accepting assignment.

Another language in NP is the set of undirected graphs with a *Hamiltonian circuit*, that is, a path of edges that visits each vertex exactly once and returns to the starting point. If a solution exists and is given, its correctness can be verified quickly. Finding such a circuit, however, or proving one does not exist, appears to be computationally difficult.

The characterization of NP as the set of problems with easily verified solutions is formalized as follows: $A \in$ NP if and only if there exist a language $A' \in$ P and a polynomial p such that for every x, $x \in A$ if and only if there exists a y such that $|y| \leq p(|x|)$ and $(x, y) \in A'$. Here, whenever x belongs to A, y is interpreted as a positive solution to the problem represented by x, or equivalently, as a proof that x belongs to A. The difference between P and NP is that between solving and checking, or between finding a proof of a mathematical theorem and testing whether a candidate proof is correct. In essence, NP represents all sets of theorems with proofs that are short (i.e., of polynomial length) and checkable quickly (i.e., in polynomial time), while P represents those statements that can proved or refuted quickly from scratch.

Further motivation for studying L, NL, and PSPACE comes from their relationships to P and NP. Namely, L and NL are the largest space-bounded classes known to be contained in P, and PSPACE is the smallest space-bounded class known to contain NP. (It is worth mentioning here that NP does not stand for "non-polynomial time"; the class P is a subclass of NP.) Similarly, EXP is of interest primarily because it is the smallest deterministic time class known to contain NP. The closely related class E is not known to contain NP.

5.4 Relationships between Complexity Classes

The P versus NP question asks about the relationship between these complexity classes: Is P a proper subset of NP, or does P = NP? Much of complexity theory focuses on the relationships between complexity classes because these relationships have implications for the difficulty of solving computational problems. In this section, we summarize important known relationships. We demonstrate two techniques for proving relationships between classes: diagonalization and padding.

5.4.1 Constructibility

The most basic theorem that one should expect from complexity theory would say, "If you have more resources, you can do more." Unfortunately, if we are not careful with our definitions, then this claim is false:

Theorem 5.1 (Gap Theorem)   *There is a computable, strictly increasing time bound $t(n)$ such that* DTIME[$t(n)$] = DTIME[$2^{2^{t(n)}}$] [Borodin, 1972].

That is, there is an empty gap between time $t(n)$ and time doubly-exponentially greater than $t(n)$, in the sense that anything that can be computed in the larger time bound can already be computed in the smaller time bound. That is, even with much more time, you can not compute more. This gap can be made much larger than doubly-exponential; for any computable r, there is a computable time bound t such that DTIME[$t(n)$] = DTIME[$r(t(n))$]. Exactly analogous statements hold for the NTIME, DSPACE, and NSPACE measures.

Fortunately, the gap phenomenon cannot happen for time bounds t that anyone would ever be interested in. Indeed, the proof of the Gap Theorem proceeds by showing that one can define a time bound t such that no machine has a running time that is between $t(n)$ and $2^{2^{t(n)}}$. This theorem indicates the need for formulating only those time bounds that actually describe the complexity of some machine.

A function $t(n)$ is **time-constructible** if there exists a deterministic Turing machine that halts after exactly $t(n)$ steps for every input of length n. A function $s(n)$ is **space-constructible** if there exists a deterministic Turing machine that uses exactly $s(n)$ worktape cells for every input of length n. (Most authors consider only functions $t(n) \geq n + 1$ to be time-constructible, and many limit attention to $s(n) \geq \log n$ for space bounds. There do exist sub-logarithmic space-constructible functions, but we prefer to avoid the tricky theory of $o(\log n)$ space bounds.)

For example, $t(n) = n + 1$ is time-constructible. Furthermore, if $t_1(n)$ and $t_2(n)$ are time-constructible, then so are the functions $t_1 + t_2$, $t_1 t_2$, $t_1^{t_2}$, and c^{t_1} for every integer $c > 1$. Consequently, if $p(n)$ is a polynomial, then $p(n) = \Theta(t(n))$ for some time-constructible polynomial function $t(n)$. Similarly, $s(n) = \log n$ is space-constructible, and if $s_1(n)$ and $s_2(n)$ are space-constructible, then so are the functions $s_1 + s_2$, $s_1 s_2$, $s_1^{s_2}$, and c^{s_1} for every integer $c > 1$. Many common functions are space-constructible: for example, $n \log n$, n^3, 2^n, $n!$.

Constructibility helps eliminate an arbitrary choice in the definition of the basic time and space classes. For general time functions t, the classes DTIME$[t]$ and NTIME$[t]$ may vary depending on whether machines are required to halt within t steps on all computation paths, or just on those paths that accept. If t is time-constructible and s is space-constructible, however, then DTIME$[t]$, NTIME$[t]$, DSPACE$[s]$, and NSPACE$[s]$ can be defined without loss of generality in terms of Turing machines that always halt.

As a general rule, any function $t(n) \geq n + 1$ and any function $s(n) \geq \log n$ that one is interested in as a time or space bound, is time- or space-constructible, respectively. As we have seen, little of interest can be proved without restricting attention to constructible functions. This restriction still leaves a rich class of resource bounds.

5.4.2 Basic Relationships

Clearly, for all time functions $t(n)$ and space functions $s(n)$, DTIME$[t(n)] \subseteq$ NTIME$[t(n)]$ and DSPACE $[s(n)] \subseteq$ NSPACE$[s(n)]$ because a deterministic machine is a special case of a nondeterministic machine. Furthermore, DTIME$[t(n)] \subseteq$ DSPACE$[t(n)]$ and NTIME$[t(n)] \subseteq$ NSPACE$[t(n)]$ because at each step, a k-tape Turing machine can write on at most $k = O(1)$ previously unwritten cells. The next theorem presents additional important relationships between classes.

Theorem 5.2 *Let $t(n)$ be a time-constructible function, and let $s(n)$ be a space-constructible function, $s(n) \geq \log n$.*

(a) NTIME$[t(n)] \subseteq$ DTIME$[2^{O(t(n))}]$
(b) NSPACE$[s(n)] \subseteq$ DTIME$[2^{O(s(n))}]$
(c) NTIME$[t(n)] \subseteq$ DSPACE$[t(n)]$
(d) (**Savitch's Theorem**) NSPACE$[s(n)] \subseteq$ DSPACE$[s(n)^2]$ [Savitch, 1970]

As a consequence of the first part of this theorem, NP $\subseteq$ EXP. No better general upper bound on deterministic time is known for languages in NP, however. See Figure 5.2 for other known inclusion relationships between canonical complexity classes.

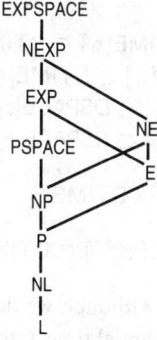

FIGURE 5.2 Inclusion relationships between the canonical complexity classes.

Although we do not know whether allowing nondeterminism strictly increases the class of languages decided in polynomial time, Savitch's Theorem says that for space classes, nondeterminism does not help by more than a polynomial amount.

5.4.3 Complementation

For a language A over an alphabet Σ, define $\overline{A}$ to be the complement of A in the set of words over Σ: that is, $\overline{A} = \Sigma^* - A$. For a class of languages $\mathcal{C}$, define co-$\mathcal{C} = \{\overline{A} : A \in \mathcal{C}\}$. If $\mathcal{C} = $ co-$\mathcal{C}$, then $\mathcal{C}$ is **closed under complementation.**

In particular, co-NP is the class of languages that are complements of languages in NP. For the language SAT of satisfiable Boolean formulas, $\overline{\text{SAT}}$ is essentially the set of unsatisfiable formulas, whose value is `false` for every truth assignment, together with the syntactically incorrect formulas. A closely related language in co-NP is the set of Boolean tautologies, namely, those formulas whose value is `true` for every truth assignment. The question of whether NP equals co-NP comes down to whether every tautology has a short (i.e., polynomial-sized) proof. The only obvious general way to prove a tautology ϕ in m variables is to verify all 2^m rows of the truth table for ϕ, taking exponential time. Most complexity theorists believe that there is no general way to reduce this time to polynomial, hence that NP $\neq$ co-NP.

Questions about complementation bear directly on the P vs. NP question. It is easy to show that P is closed under complementation (see the next theorem). Consequently, if NP $\neq$ co-NP, then P $\neq$ NP.

Theorem 5.3 (Complementation Theorems)   *Let t be a time-constructible function, and let s be a space-constructible function, with $s(n) \geq \log n$ for all n. Then,*

1. DTIME$[t]$ *is closed under complementation.*
2. DSPACE$[s]$ *is closed under complementation.*
3. NSPACE$[s]$ *is closed under complementation* [Immerman, 1988; Szelepcsényi, 1988].

The Complementation Theorems are used to prove the Hierarchy Theorems in the next section.

5.4.4 Hierarchy Theorems and Diagonalization

A hierarchy theorem is a theorem that says, "If you have more resources, you can compute more." As we saw in Chapter 5.4.1, this theorem is possible only if we restrict attention to constructible time and space bounds. Next, we state hierarchy theorems for deterministic and nondeterministic time and space classes. In the following, $\subset$ denotes *strict* inclusion between complexity classes.

Theorem 5.4 (Hierarchy Theorems)   *Let t_1 and t_2 be time-constructible functions, and let s_1 and s_2 be space-constructible functions, with $s_1(n), s_2(n) \geq \log n$ for all n.*

(a) *If $t_1(n) \log t_1(n) = o(t_2(n))$, then* DTIME$[t_1] \subset$ DTIME$[t_2]$.
(b) *If $t_1(n + 1) = o(t_2(n))$, then* NTIME$[t_1] \subset$ NTIME$[t_2]$ [Seiferas et al., 1978].
(c) *If $s_1(n) = o(s_2(n))$, then* DSPACE$[s_1] \subset$ DSPACE$[s_2]$.
(d) *If $s_1(n) = o(s_2(n))$, then* NSPACE$[s_1] \subset$ NSPACE$[s_2]$.

As a corollary of the Hierarchy Theorem for DTIME,

$$P \subseteq \text{DTIME}[n^{\log n}] \subset \text{DTIME}[2^n] \subseteq E;$$

hence, we have the strict inclusion P $\subset$ E. Although we do not know whether P $\subset$ NP, there exists a problem in E that cannot be solved in polynomial time. Other consequences of the Hierarchy Theorems are NE $\subset$ NEXP and NL $\subset$ PSPACE.

In the Hierarchy Theorem for DTIME, the hypothesis on t_1 and t_2 is $t_1(n) \log t_1(n) = o(t_2(n))$, instead of $t_1(n) = o(t_2(n))$, for technical reasons related to the simulation of machines with multiple worktapes by a single universal Turing machine with a fixed number of worktapes. Other computational models, such as random access machines, enjoy tighter time hierarchy theorems.

All proofs of the Hierarchy Theorems use the technique of **diagonalization**. For example, the proof for DTIME constructs a Turing machine M of time complexity t_2 that considers all machines $M_1, M_2, \ldots$ whose time complexity is t_1; for each i, the proof finds a word x_i that is accepted by M if and only if $x_i \notin L(M_i)$, the language decided by M_i. Consequently, $L(M)$, the language decided by M, differs from each $L(M_i)$, hence $L(M) \notin \text{DTIME}[t_1]$. The diagonalization technique resembles the classic method used to prove that the real numbers are uncountable, by constructing a number whose j^{th} digit differs from the jth digit of the jth number on the list. To illustrate the diagonalization technique, we outline the proof of the Hierarchy Theorem for DSPACE. In this subsection, $\langle i, x \rangle$ stands for the string $0^i 1x$, and *zeroes*(y) stands for the number of 0's that a given string y starts with. Note that *zeroes*$(\langle i, x \rangle) = i$.

Proof (of the DSPACE Hierarchy Theorem)

We construct a deterministic Turing machine M that decides a language A such that $A \in \text{DSPACE}[s_2] - \text{DSPACE}[s_1]$.

Let U be a deterministic universal Turing machine, as described in Section 5.2.3. On input x of length n, machine M performs the following:

1. Lay out $s_2(n)$ cells on a worktape.
2. Let $i = zeroes(x)$.
3. Simulate the universal machine U on input $\langle i, x \rangle$. Accept x if U tries to use more than s_2 worktape cells. (We omit some technical details, and the way in which the constructibility of s_2 is used to ensure that this process halts.)
4. If U accepts $\langle i, x \rangle$, then reject; if U rejects $\langle i, x \rangle$, then accept.

Clearly, M always halts and uses space $O(s_2(n))$. Let $A = L(M)$.

Suppose $A \in \text{DSPACE}[s_1(n)]$. Then there is some Turing machine M_j accepting A using space at most $s_1(n)$. Since the space used by U is $O(1)$ times the space used by M_j, there is a constant k depending only on j (in fact, we can take $k = |j|$), such that U, on inputs z of the form $z = \langle j, x \rangle$, uses at most $ks_1(|x|)$ space.

Since $s_1(n) = o(s_2(n))$, there is an n_0 such that $ks_1(n) \leq s_2(n)$ for all $n \geq n_0$. Let x be a string of length greater than n_0 such that the first $j + 1$ symbols of x are $0^j 1$. Note that the universal Turing machine U, on input $\langle j, x \rangle$, simulates M_j on input x and uses space at most $ks_1(n) \leq s_2(n)$. Thus, when we consider the machine M defining A, we see that on input x the simulation does not stop in step 3, but continues on to step 4, and thus $x \in A$ if and only if U rejects $\langle j, x \rangle$. Consequently, M_j does not accept A, contrary to our assumption. Thus, $A \notin \text{DSPACE}[s_1(n)]$. □

Although the diagonalization technique successfully separates some pairs of complexity classes, diagonalization does not seem strong enough to separate P from NP. (See Theorem 5.10 below.)

5.4.5 Padding Arguments

A useful technique for establishing relationships between complexity classes is the **padding argument**. Let A be a language over alphabet Σ, and let # be a symbol not in Σ. Let f be a numeric function. The f-**padded version of** L is the language

$$A' = \{x\#^{f(n)} : x \in A \text{ and } n = |x|\}.$$

That is, each word of A' is a word in A concatenated with $f(n)$ consecutive # symbols. The padded version A' has the same information content as A, but because each word is longer, the computational complexity of A' is smaller.

The proof of the next theorem illustrates the use of a padding argument.

Theorem 5.5   *If* P = NP, *then* E = NE [Book, 1974].

Proof   Since E $\subseteq$ NE, we prove that NE $\subseteq$ E.

Let $A \in$ NE be decided by a nondeterministic Turing machine M in at most $t(n) = k^n$ time for some constant integer k. Let A' be the $t(n)$-padded version of A. From M, we construct a nondeterministic Turing machine M' that decides A' in linear time: M' checks that its input has the correct format, using the time-constructibility of t; then M' runs M on the prefix of the input preceding the first # symbol. Thus, $A' \in$ NP.

If P = NP, then there is a deterministic Turing machine D' that decides A' in at most $p'(n)$ time for some polynomial p'. From D', we construct a deterministic Turing machine D that decides A, as follows. On input x of length n, since $t(n)$ is time-constructible, machine D constructs $x\#^{t(n)}$, whose length is $n + t(n)$, in $O(t(n))$ time. Then D runs D' on this input word. The time complexity of D is at most $O(t(n)) + p'(n + t(n)) = 2^{O(n)}$. Therefore, NE $\subseteq$ E.   □

A similar argument shows that the E = NE question is equivalent to the question of whether NP − P contains a subset of 1*, that is, a language over a single-letter alphabet.

5.5   Reducibility and Completeness

In this section, we discuss relationships between problems: informally, if one problem reduces to another problem, then in a sense, the second problem is harder than the first. The hardest problems in NP are the NP-complete problems. We define NP-completeness precisely, and we show how to prove that a problem is NP-complete. The theory of NP-completeness, together with the many known NP-complete problems, is perhaps the best justification for interest in the classes P and NP. All of the other canonical complexity classes listed above have natural and important problems that are complete for them; we give some of these as well.

5.5.1   Resource-Bounded Reducibilities

In mathematics, as in everyday life, a typical way to solve a new problem is to reduce it to a previously solved problem. Frequently, an instance of the new problem is expressed completely in terms of an instance of the prior problem, and the solution is then interpreted in the terms of the new problem. For example, the maximum weighted matching problem for bipartite graphs (also called the assignment problem) reduces to the network flow problem (see Chapter 7). This kind of reduction is called **many-one reducibility**, and is defined below.

A different way to solve the new problem is to use a subroutine that solves the prior problem. For example, we can solve an optimization problem whose solution is feasible and maximizes the value of an objective function g by repeatedly calling a subroutine that solves the corresponding decision problem of whether there exists a feasible solution x whose value $g(x)$ satisfies $g(x) \geq k$. This kind of reduction is called **Turing reducibility**, and is also defined below.

Let A_1 and A_2 be languages. A_1 is many-one reducible to A_2, written $A_1 \leq_m A_2$, if there exists a total recursive function f such that for all x, $x \in A_1$ if and only if $f(x) \in A_2$. The function f is called the **transformation function.** A_1 is Turing reducible to A_2, written $A_1 \leq_T A_2$, if A_1 can be decided by a deterministic oracle Turing machine M using A_2 as its oracle, that is, $A_1 = L(M^{A_2})$. (Total recursive functions and oracle Turing machines are defined in Section 5.2). The oracle for A_2 models a hypothetical efficient subroutine for A_2.

If f or M above consumes too much time or space, the reductions they compute are not helpful. To study complexity classes defined by bounds on time and space resources, it is natural to consider resource-bounded reducibilities. Let A_1 and A_2 be languages.

- A_1 is **Karp reducible** to A_2, written $A_1 \leq_m^p A_2$, if A_1 is many-one reducible to A_2 via a transformation function that is computable deterministically in polynomial time.
- A_1 is **log-space reducible** to A_2, written $A_1 \leq_m^{\log} A_2$, if A_1 is many-one reducible to A_2 via a transformation function that is computable deterministically in $O(\log n)$ space.
- A_1 is **Cook reducible** to A_2, written $A_1 \leq_T^p A_2$, if A_1 is Turing reducible to A_2 via a deterministic oracle Turing machine of polynomial time complexity.

The term "polynomial-time reducibility" usually refers to Karp reducibility. If $A_1 \leq_m^p A_2$ and $A_2 \leq_m^p A_1$, then A_1 and A_2 are **equivalent** under Karp reducibility. Equivalence under Cook reducibility is defined similarly.

Karp and Cook reductions are useful for finding relationships between languages of high complexity, but they are not at all useful for distinguishing between problems in P, because all problems in P are equivalent under Karp (and hence Cook) reductions. (Here and later we ignore the special cases $A_1 = \emptyset$ and $A_1 = \Sigma^*$, and consider them to reduce to any language.)

Log-space reducibility [Jones, 1975] is useful for complexity classes within P, such as NL, for which Karp reducibility allows too many reductions. By definition, for every nontrivial language A_0 (i.e., $A_0 \neq \emptyset$ and $A_0 \neq \Sigma^*$) and for every A in P, necessarily $A \leq_m^p A_0$ via a transformation that simply runs a deterministic Turing machine that decides A in polynomial time. It is not known whether log-space reducibility is different from Karp reducibility, however; all transformations for known Karp reductions can be computed in $O(\log n)$ space. Even for decision problems, L is not known to be a proper subset of P.

Theorem 5.6 *Log-space reducibility implies Karp reducibility, which implies Cook reducibility:*

1. *If $A_1 \leq_m^{\log} A_2$, then $A_1 \leq_m^p A_2$.*
2. *If $A_1 \leq_m^p A_2$, then $A_1 \leq_T^p A_2$.*

Theorem 5.7 *Log-space reducibility, Karp reducibility, and Cook reducibility are transitive:*

1. *If $A_1 \leq_m^{\log} A_2$ and $A_2 \leq_m^{\log} A_3$, then $A_1 \leq_m^{\log} A_3$.*
2. *If $A_1 \leq_m^p A_2$ and $A_2 \leq_m^p A_3$, then $A_1 \leq_m^p A_3$.*
3. *If $A_1 \leq_T^p A_2$ and $A_2 \leq_T^p A_3$, then $A_1 \leq_T^p A_3$.*

The key property of Cook and Karp reductions is that they preserve polynomial-time feasibility. Suppose $A_1 \leq_m^p A_2$ via a transformation f. If M_2 decides A_2, and M_f computes f, then to decide whether an input word x is in A_1, we can use M_f to compute $f(x)$, and then run M_2 on input $f(x)$. If the time complexities of M_2 and M_f are bounded by polynomials t_2 and t_f, respectively, then on each input x of length $n = |x|$, the time taken by this method of deciding A_1 is at most $t_f(n) + t_2(t_f(n))$, which is also a polynomial in n. In summary, if A_2 is feasible, and there is an efficient reduction from A_1 to A_2, then A_1 is feasible. Although this is a simple observation, this fact is important enough to state as a theorem (Theorem 5.8). First, however, we need the concept of "closure."

A class of languages $\mathcal{C}$ is **closed under a reducibility** $\leq_r$ if for all languages A_1 and A_2, whenever $A_1 \leq_r A_2$ and $A_2 \in \mathcal{C}$, necessarily $A_1 \in \mathcal{C}$.

Theorem 5.8

1. P *is closed under log-space reducibility, Karp reducibility, and Cook reducibility.*
2. NP *is closed under log-space reducibility and Karp reducibility.*
3. L *and* NL *are closed under log-space reducibility.*

We shall see the importance of closure under a reducibility in conjunction with the concept of completeness, which we define in the next section.

5.5.2 Complete Languages

Let C be a class of languages that represent computational problems. A language A_0 is C-**hard** under a reducibility $\leq_r$ if for all A in C, $A \leq_r A_0$. A language A_0 is C-**complete** under $\leq_r$ if A_0 is C-hard and $A_0 \in C$. Informally, if A_0 is C-hard, then A_0 represents a problem that is at least as difficult to solve as any problem in C. If A_0 is C-complete, then in a sense, A_0 is one of the most difficult problems in C.

There is another way to view completeness. Completeness provides us with tight lower bounds on the complexity of problems. If a language A is complete for complexity class C, then we have a lower bound on its complexity. Namely, A is as hard as the most difficult problem in C, assuming that the complexity of the reduction itself is small enough not to matter. The lower bound is tight because A is in C; that is, the upper bound matches the lower bound.

In the case $C = $ NP, the reducibility $\leq_r$ is usually taken to be Karp reducibility unless otherwise stated. Thus, we say

- A language A_0 is **NP-hard** if A_0 is NP-hard under Karp reducibility.
- A_0 is **NP-complete** if A_0 is NP-complete under Karp reducibility.

However, many sources take the term "NP-hard" to refer to Cook reducibility.

Many important languages are now known to be NP-complete. Before we get to them, let us discuss some implications of the statement "A_0 is NP-complete," and also some things this statement does not mean.

The first implication is that *if* there exists a deterministic Turing machine that decides A_0 in polynomial time — that is, if $A_0 \in $ P — then because P is closed under Karp reducibility (Theorem 5.8 in Section 5.5.1), it would follow that NP $\subseteq$ P, hence P $=$ NP. In essence, the question of whether P is the same as NP comes down to the question of whether any particular NP-complete language is in P. Put another way, *all* of the NP-complete languages stand or fall together: if one is in P, then all are in P; if one is not, then all are not. Another implication, which follows by a similar closure argument applied to co-NP, is that if $A_0 \in $ co-NP, then NP $=$ co-NP. It is also believed unlikely that NP $=$ co-NP, as was noted in connection with whether all tautologies have short proofs in Section 5.4.3.

A common misconception is that the above property of NP-complete languages is actually their definition, namely: if $A \in $ NP and $A \in $ P implies P $=$ NP, then A is NP-complete. This "definition" is wrong if P $\neq$ NP. A theorem due to Ladner [1975] shows that P $\neq$ NP if and only if there exists a language A' in NP $-$ P such that A' is not NP-complete. Thus, if P $\neq$ NP, then A' is a counterexample to the "definition."

Another common misconception arises from a misunderstanding of the statement "If A_0 is NP-complete, then A_0 is one of the most difficult problems in NP." This statement is true on one level: if there is any problem at all in NP that is not in P, then the NP-complete language A_0 is one such problem. However, note that there are NP-complete problems in NTIME$[n]$ — and these problems are, in some sense, much *simpler* than many problems in NTIME$[n^{10^{500}}]$.

5.5.3 Cook-Levin Theorem

Interest in NP-complete problems started with a theorem of Cook [1971] that was proved independently by Levin [1973]. Recall that SAT is the language of Boolean formulas $\phi(z_1, \ldots, z_r)$ such that there exists a truth assignment to the variables $z_1, \ldots, z_r$ that makes ϕ true.

Theorem 5.9 (Cook-Levin Theorem) SAT *is NP-complete.*

Proof We know already that SAT is in NP, so to prove that SAT is NP-complete, we need to take an arbitrary given language A in NP and show that $A \leq_m^p$ SAT. Take N to be a nondeterministic Turing machine that decides A in polynomial time. Then the relation $R(x, y) =$ "y is a computation path of N that leads it to accept x" is decidable in deterministic polynomial time depending only on $n = |x|$. We can assume that the length m of possible y's encoded as binary strings depends only on n and not on a particular x.

It is straightforward to show that there is a polynomial p and for each n a Boolean circuit C_n^R with $p(n)$ wires, with $n + m$ input wires labeled $x_1, \ldots, x_n, y_1, \ldots, y_m$ and one output wire w_0, such that $C_n^R(x, y)$ outputs 1 if and only if $R(x, y)$ holds. (We describe circuits in more detail below, and state a theorem for this principle as part 1. of Theorem 5.14.) Importantly, C_n^R itself can be designed in time polynomial in n, and by the universality of NAND, may be composed entirely of binary NAND gates. Label the wires by variables $x_1, \ldots, x_n, y_1, \ldots, y_m, w_0, w_1, \ldots, w_{p(n)-n-m-1}$. These become the variables of our Boolean formulas. For each NAND gate g with input wires u and v, and for each output wire w of g, write down the subformula

$$\phi_{g,w} = (u \vee w) \wedge (v \vee w) \wedge (\bar{u} \vee \bar{v} \vee \bar{w})$$

This subformula is satisfied by precisely those assignments to u, v, w that give $w = u \text{ NAND } v$. The conjunction ϕ_0 of $\phi_{g,w}$ over the polynomially many gates g and their output wires w thus is satisfied only by assignments that set every gate's output correctly given its inputs. Thus, for any binary strings x and y of lengths n, m, respectively, the formula $\phi_1 = \phi_0 \wedge w_0$ is satisfiable by a setting of the wire variables $w_0, w_1, \ldots, w_{p(n)-n-m-1}$ if and only if $C_n^R(x, y) = 1$ — that is, if and only if $R(x, y)$ holds.

Now given any fixed x and taking $n = |x|$, the Karp reduction computes ϕ_1 via C_n^R and ϕ_0 as above, and finally outputs the Boolean formula ϕ obtained by substituting the bit-values of x into ϕ_1. This ϕ has variables $y_1, \ldots, y_m, w_0, w_1, \ldots, w_{p(n)-m-m-1}$, and the computation of ϕ from x runs in deterministic polynomial time. Then $x \in A$ if and only if N accepts x, if and only if there exists y such that $R(x, y)$ holds, if and only if there exists an assignment to the variables $w_0, w_1, \ldots, w_{p(n)-n-m-1}$ *and* $y_1, \ldots, y_m$ that satisfies ϕ, if and only if $\phi \in \text{SAT}$. This shows $A \leq_m^P \text{SAT}$. $\square$

We have actually proved that SAT remains NP-complete even when the given instances ϕ are *restricted* to Boolean formulas that are a conjunction of *clauses*, where each clause consists of (here, at most three) disjuncted literals. Such formulas are said to be in *conjunctive normal form*. Theorem 5.9 is also commonly known as Cook's Theorem.

5.5.4 Proving NP-Completeness

After one language has been proved complete for a class, others can be proved complete by constructing transformations. For NP, if A_0 is NP-complete, then to prove that another language A_1 is NP-complete, it suffices to prove that $A_1 \in \text{NP}$, and to construct a polynomial-time transformation that establishes $A_0 \leq_m^P A_1$. Since A_0 is NP-complete, for every language A in NP, $A \leq_m^P A_0$, hence, by transitivity (Theorem 5.7), $A \leq_m^P A_1$.

Beginning with Cook [1971] and Karp [1972], hundreds of computational problems in many fields of science and engineering have been proved to be NP-complete, almost always by reduction from a problem that was previously known to be NP-complete. The following NP-complete decision problems are frequently used in these reductions — the language corresponding to each problem is the set of instances whose answers are yes.

- 3-SATISFIABILITY (3SAT)

    Instance: A Boolean expression ϕ in conjunctive normal form with three literals per clause [e.g., $(w \vee x \vee \bar{y}) \wedge (\bar{x} \vee y \vee z)$].

    Question: Is ϕ satisfiable?

- VERTEX COVER

  Instance:　A graph G and an integer k.

  Question:　Does G have a set W of k vertices such that every edge in G is incident on a vertex of W?

- CLIQUE

  Instance:　A graph G and an integer k.

  Question:　Does G have a set K of k vertices such that every two vertices in K are adjacent in G?

- HAMILTONIAN CIRCUIT

  Instance:　A graph G.

  Question:　Does G have a circuit that includes every vertex exactly once?

- THREE-DIMENSIONAL MATCHING

  Instance:　Sets W, X, Y with $|W| = |X| = |Y| = q$ and a subset $S \subseteq W \times X \times Y$.

  Question:　Is there a subset $S' \subseteq S$ of size q such that no two triples in S' agree in any coordinate?

- PARTITION

  Instance:　A set S of positive integers.

  Question:　Is there a subset $S' \subseteq S$ such that the sum of the elements of S' equals the sum of the elements of $S - S'$?

Note that our ϕ in the above proof of the Cook-Levin Theorem already meets a form of the definition of 3SAT relaxed to allow "at most 3 literals per clause." Padding ϕ with some extra variables to bring up the number in each clause to exactly three, while preserving whether the formula is satisfiable or not, is not difficult, and establishes the NP-completeness of 3SAT. Here is another example of an NP-completeness proof, for the following decision problem:

- TRAVELING SALESMAN PROBLEM (TSP)

  Instance:　A set of m "cities" $C_1, \ldots, C_m$, with an integer distance $d(i, j)$ between every pair of cities C_i and C_j, and an integer D.

  Question:　Is there a tour of the cities whose total length is at most D, that is, a permutation $c_1, \ldots, c_m$ of $\{1, \ldots, m\}$, such that

  $$d(c_1, c_2) + \cdots + d(c_{m-1}, c_m) + d(c_m, c_1) \leq D?$$

First, it is easy to see that TSP is in NP: a nondeterministic Turing machine simply guesses a tour and checks that the total length is at most D.

Next, we construct a reduction from Hamiltonian Circuit to TSP. (The reduction goes from the known NP-complete problem, Hamiltonian Circuit, to the new problem, TSP, not vice versa.)

From a graph G on m vertices $v_1, \ldots, v_m$, define the distance function d as follows:

$$d(i, j) = \begin{cases} 1 & \text{if } (v_i, v_j) \text{ is an edge in } G \\ m + 1 & \text{otherwise.} \end{cases}$$

Set $D = m$. Clearly, d and D can be computed in polynomial time from G. Each vertex of G corresponds to a city in the constructed instance of TSP.

If G has a Hamiltonian circuit, then the length of the tour that corresponds to this circuit is exactly m. Conversely, if there is a tour whose length is at most m, then each step of the tour must have distance 1, not $m + 1$. Thus, each step corresponds to an edge of G, and the corresponding sequence of vertices in G is a Hamiltonian circuit.

5.5.5 Complete Problems for Other Classes

Besides NP, the following canonical complexity classes have natural complete problems. The three problems now listed are complete for their respective classes under log-space reducibility.

- NL: GRAPH ACCESSIBILITY PROBLEM
  Instance: A directed graph G with nodes $1, \ldots, N$.
  Question: Does G have a directed path from node 1 to node N?
- P: CIRCUIT VALUE PROBLEM
  Instance: A Boolean circuit (see Section 5.9) with output node u, and an assignment I of $\{0, 1\}$ to each input node.
  Question: Is **1** the value of u under I?
- PSPACE: QUANTIFIED BOOLEAN FORMULAS
  Instance: A Boolean expression with all variables quantified with either $\forall$ or $\exists$ [e.g., $\forall x \forall y \exists z(x \wedge (\overline{y} \vee z))$].
  Question: Is the expression `true`?

These problems can be used to prove other problems are NL-complete, P-complete, and PSPACE-complete, respectively.

Stockmeyer and Meyer [1973] defined a natural decision problem that they proved to be complete for NE. If this problem were in P, then by closure under Karp reducibility (Theorem 5.8), we would have NE $\subseteq$ P, a contradiction of the hierarchy theorems (Theorem 5.4). Therefore, this decision problem is infeasible: it has no polynomial-time algorithm. In contrast, decision problems in NEXP $-$ P that have been constructed by diagonalization are artificial problems that nobody would want to solve anyway. Although diagonalization produces unnatural problems by itself, the combination of diagonalization and completeness shows that *natural* problems are intractable.

The next section points out some limitations of current diagonalization techniques.

5.6 Relativization of the P vs. NP Problem

Let A be a language. Define P^A (respectively, NP^A) to be the class of languages accepted in polynomial time by deterministic (nondeterministic) oracle Turing machines with oracle A.

Proofs that use the diagonalization technique on Turing machines without oracles generally carry over to oracle Turing machines. Thus, for instance, the proof of the DTIME hierarchy theorem also shows that, for *any* oracle A, $\text{DTIME}^A[n^2]$ is properly contained in $\text{DTIME}^A[n^3]$. This can be seen as a *strength* of the diagonalization technique because it allows an argument to "relativize" to computation carried out relative to an oracle. In fact, there are examples of lower bounds (for deterministic, "unrelativized" circuit models) that make crucial use of the fact that the time hierarchies relativize in this sense.

But it can also be seen as a weakness of the diagonalization technique. The following important theorem demonstrates why.

Theorem 5.10 *There exist languages A and B such that $P^A = NP^A$, and $P^B \neq NP^B$* [Baker et al., 1975].

This shows that resolving the P vs. NP question requires techniques that do not relativize, that is, that do not apply to oracle Turing machines too. Thus, diagonalization as we currently know it is unlikely to succeed in separating P from NP because the diagonalization arguments we know (and in fact *most* of the arguments we know) relativize. Important non-relativizing proof techniques have appeared only recently, in connection with interactive proof systems (Section 5.11.1).

5.7   The Polynomial Hierarchy

Let $\mathcal{C}$ be a class of languages. Define:

- $NP^{\mathcal{C}} = \bigcup_{A \in \mathcal{C}} NP^{A}$
- $\Sigma_0^P = \Pi_0^P = P$

and for $k \geq 0$, define:

- $\Sigma_{k+1}^P = NP^{\Sigma_k^P}$
- $\Pi_{k+1}^P = \text{co-}\Sigma_{k+1}^P$.

Observe that $\Sigma_1^P = NP^P = NP$ because each of polynomially many queries to an oracle language in P can be answered directly by a (nondeterministic) Turing machine in polynomial time. Consequently, $\Pi_1^P = \text{co-}NP$. For each k, $\Sigma_k^P \cup \Pi_k^P \subseteq \Sigma_{k+1}^P \cap \Pi_{k+1}^P$, but this inclusion is not known to be strict. See Figure 5.3.

The classes Σ_k^P and Π_k^P constitute the **polynomial hierarchy**. Define:

$$PH = \bigcup_{k \geq 0} \Sigma_k^P.$$

It is straightforward to prove that $PH \subseteq PSPACE$, but it is not known whether the inclusion is strict. In fact, if $PH = PSPACE$, then the polynomial hierarchy collapses to some level, that is, $PH = \Sigma_m^P$ for some m. In the next section, we define the polynomial hierarchy in two other ways, one of which is in terms of alternating Turing machines.

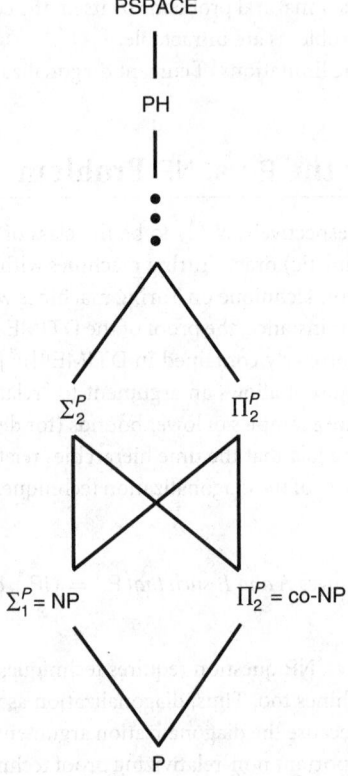

FIGURE 5.3   The polynomial hierarchy.

5.8   Alternating Complexity Classes

In this section, we define time and space complexity classes for alternating Turing machines, and we show how these classes are related to the classes introduced already. The possible computations of an alternating Turing machine M on an input word x can be represented by a tree T_x in which the root is the initial configuration, and the children of a nonterminal node C are the configurations reachable from C by one step of M. For a word x in $L(M)$, define an **accepting subtree** S of T_x to be a subtree of T_x with the following properties:

- S is finite.
- The root of S is the initial configuration with input word x.
- If S has an existential configuration C, then S has exactly one child of C in T_x; if S has a universal configuration C, then S has all children of C in T_x.
- Every leaf is a configuration whose state is the accepting state q_A.

Observe that each node in S is an accepting configuration.

We consider only alternating Turing machines that always halt. For $x \in L(M)$, define the time taken by M to be the height of the shortest accepting tree for x, and the space to be the maximum number of non-blank worktape cells among configurations in the accepting tree that minimizes this number. For $x \notin L(M)$, define the time to be the height of T_x, and the space to be the maximum number of non-blank worktape cells among configurations in T_x.

Let $t(n)$ be a time-constructible function, and let $s(n)$ be a space-constructible function. Define the following complexity classes:

- $\text{ATIME}[t(n)]$ is the class of languages decided by alternating Turing machines of time complexity $O(t(n))$.
- $\text{ASPACE}[s(n)]$ is the class of languages decided by alternating Turing machines of space complexity $O(s(n))$.

Because a nondeterministic Turing machine is a special case of an alternating Turing machine, for every $t(n)$ and $s(n)$, $\text{NTIME}[t] \subseteq \text{ATIME}[t]$ and $\text{NSPACE}[s] \subseteq \text{ASPACE}[s]$. The next theorem states further relationships between computational resources used by alternating Turing machines, and resources used by deterministic and nondeterministic Turing machines.

Theorem 5.11 (Alternation Theorems)   [Chandra et al., 1981]. *Let $t(n)$ be a time-constructible function, and let $s(n)$ be a space-constructible function, $s(n) \geq \log n$.*

   (a) $\text{NSPACE}[s(n)] \subseteq \text{ATIME}[s(n)^2]$
   (b) $\text{ATIME}[t(n)] \subseteq \text{DSPACE}[t(n)]$
   (c) $\text{ASPACE}[s(n)] \subseteq \text{DTIME}[2^{O(s(n))}]$
   (d) $\text{DTIME}[t(n)] \subseteq \text{ASPACE}[\log t(n)]$

In other words, space on deterministic and nondeterministic Turing machines is polynomially related to time on alternating Turing machines. Space on alternating Turing machines is exponentially related to time on deterministic Turing machines. The following corollary is immediate.

Theorem 5.12

   (a) $\text{ASPACE}[O(\log n)] = \text{P}$
   (b) $\text{ATIME}[n^{O(1)}] = \text{PSPACE}$
   (c) $\text{ASPACE}[n^{O(1)}] = \text{EXP}$

In Chapter 5.7, we defined the classes of the polynomial hierarchy in terms of oracles, but we can also define them in terms of alternating Turing machines with restrictions on the number of alternations between existential and universal states. Define a *k-alternating Turing machine* to be a machine such that on every computation path, the number of changes from an existential state to universal state, or from a universal state to an existential state, is at most $k - 1$. Thus, a nondeterministic Turing machine, which stays in existential states, is a 1-alternating Turing machine.

Theorem 5.13   [Stockmeyer, 1976; Wrathall, 1976]. *For any language A, the following are equivalent:*

1. $A \in \Sigma_k^P$.
2. *A is decided in polynomial time by a k-alternating Turing machine that starts in an existential state.*
3. *There exists a language B in* P *and a polynomial p such that for all x, x $\in$ A if and only if*

$$(\exists y_1 : |y_1| \le p(|x|))(\forall y_2 : |y_2| \le p(|x|)) \cdots (Q y_k : |y_k| \le p(|x|))[(x, y_1, \ldots, y_k) \in B]$$

where the quantifier Q is $\exists$ if k is odd, $\forall$ if k is even.

Alternating Turing machines are closely related to Boolean circuits, which are defined in the next section.

5.9 Circuit Complexity

The hardware of electronic digital computers is based on digital logic gates, connected into combinational circuits (see Chapter 16). Here, we specify a model of computation that formalizes the combinational circuit.

A *Boolean circuit* on n input variables $x_1, \ldots, x_n$ is a directed acyclic graph with exactly n input nodes of indegree 0 labeled $x_1, \ldots, x_n$, and other nodes of indegree 1 or 2, called *gates*, labeled with the Boolean operators in $\{\wedge, \vee, \neg\}$. One node is designated as the output of the circuit. See Figure 5.4. Without loss of generality, we assume that there are no extraneous nodes; there is a directed path from each node to the output node. The indegree of a gate is also called its *fan-in*.

An *input assignment* is a function I that maps each variable x_i to either 0 or 1. The value of each gate g under I is obtained by applying the Boolean operation that labels g to the values of the immediate predecessors of g. The function computed by the circuit is the value of the output node for each input assignment.

A Boolean circuit computes a finite function: a function of only n binary input variables. To decide membership in a language, we need a circuit for each input length n.

A *circuit family* is an infinite set of circuits $C = \{c_1, c_2, \ldots\}$ in which each c_n is a Boolean circuit on n inputs. C *decides* a language $A \subseteq \{0,1\}^*$ if for every n and every assignment $a_1, \ldots, a_n$ of $\{0,1\}$ to the n inputs, the value of the output node of c_n is **1** if and only if the word $a_1 \cdots a_n \in A$. The *size complexity* of C is the function $z(n)$ that specifies the number of nodes in each c_n. The *depth complexity* of C is the function $d(n)$ that specifies the length of the longest directed path in c_n. Clearly, since the fan-in of each

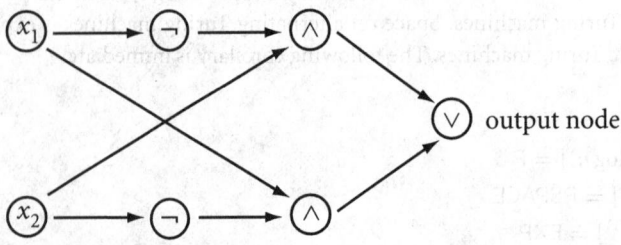

FIGURE 5.4   A Boolean circuit.

gate is at most 2, $d(n) \geq \log z(n) \geq \log n$. The class of languages decided by polynomial-size circuits is denoted by P/poly.

With a different circuit for each input length, a circuit family could solve an undecidable problem such as the halting problem (see Chapter 6). For each input length, a table of all answers for machine descriptions of that length could be encoded into the circuit. Thus, we need to restrict our circuit families. The most natural restriction is that all circuits in a family should have a concise, uniform description, to disallow a different answer table for each input length. Several uniformity conditions have been studied, and the following is the most convenient.

A circuit family $\{c_1, c_2, \ldots\}$ of size complexity $z(n)$ is *log-space uniform* if there exists a deterministic Turing machine M such that on each input of length n, machine M produces a description of c_n, using space $O(\log z(n))$.

Now we define complexity classes for uniform circuit families and relate these classes to previously defined classes. Define the following complexity classes:

- SIZE[$z(n)$] is the class of languages decided by log-space uniform circuit families of size complexity $O(z(n))$.

- DEPTH[$d(n)$] is the class of languages decided by log-space uniform circuit families of depth complexity $O(d(n))$.

In our notation, SIZE[$n^{O(1)}$] equals P, which is a proper subclass of P/poly.

Theorem 5.14

1. If $t(n)$ is a time-constructible function, then DTIME[$t(n)$] $\subseteq$ SIZE[$t(n) \log t(n)$] [Pippenger and Fischer, 1979].
2. SIZE[$z(n)$] $\subseteq$ DTIME[$z(n)^{O(1)}$].
3. If $s(n)$ is a space-constructible function and $s(n) \geq \log n$, then NSPACE[$s(n)$] $\subseteq$ DEPTH[$s(n)^2$] [Borodin, 1977].
4. If $d(n) \geq \log n$, then DEPTH[$d(n)$] $\subseteq$ DSPACE[$d(n)$] [Borodin, 1977].

The next theorem shows that size and depth on Boolean circuits are closely related to space and time on alternating Turing machines, provided that we permit sublinear running times for alternating Turing machines, as follows. We augment alternating Turing machines with a random-access input capability. To access the cell at position j on the input tape, M writes the binary representation of j on a special tape, in $\log j$ steps, and enters a special reading state to obtain the symbol in cell j.

Theorem 5.15 [Ruzzo, 1979].

Let $t(n) \geq \log n$ and $s(n) \geq \log n$ be such that the mapping $n \mapsto (t(n), s(n))$ (in binary) is computable in time $s(n)$.

1. Every language decided by an alternating Turing machine of simultaneous space complexity $s(n)$ and time complexity $t(n)$ can be decided by a log-space uniform circuit family of simultaneous size complexity $2^{O(s(n))}$ and depth complexity $O(t(n))$.
2. If $d(n) \geq (\log z(n))^2$, then every language decided by a log-space uniform circuit family of simultaneous size complexity $z(n)$ and depth complexity $d(n)$ can be decided by an alternating Turing machine of simultaneous space complexity $O(\log z(n))$ and time complexity $O(d(n))$.

In a sense, the Boolean circuit family is a model of parallel computation, because all gates compute independently, in parallel. For each $k \geq 0$, NC^k denotes the class of languages decided by log-space uniform bounded fan-in circuits of polynomial size and depth $O((\log n)^k)$, and AC^k is defined analogously for unbounded fan-in circuits. In particular, AC^k is the same as the class of languages decided by a parallel machine model called the CRCW PRAM with polynomially many processors in parallel time $O((\log n)^k)$ [Stockmeyer and Vishkin, 1984].

5.10   Probabilistic Complexity Classes

Since the 1970s, with the development of randomized algorithms for computational problems (see Chapter 12). Complexity theorists have placed randomized algorithms on a firm intellectual foundation. In this section, we outline some basic concepts in this area.

A **probabilistic Turing machine** M can be formalized as a nondeterministic Turing machine with exactly two choices at each step. During a computation, M chooses each possible next step with independent probability $1/2$. Intuitively, at each step, M flips a fair coin to decide what to do next. The probability of a computation path of t steps is $1/2^t$. The probability that M accepts an input string x, denoted by $p_M(x)$, is the sum of the probabilities of the accepting computation paths.

Throughout this section, we consider only machines whose time complexity $t(n)$ is time-constructible. Without loss of generality, we can assume that every computation path of such a machine halts in exactly t steps.

Let A be a language. A probabilistic Turing machine M decides A with

		for all $x \in A$	for all $x \notin A$
unbounded two-sided error	if	$p_M(x) > 1/2$	$p_M(x) \leq 1/2$
bounded two-sided error	if	$p_M(x) > 1/2 + \epsilon$	$p_M(x) < 1/2 - \epsilon$
		for some positive constant ϵ	
one-sided error	if	$p_M(x) > 1/2$	$p_M(x) = 0$

Many practical and important probabilistic algorithms make one-sided errors. For example, in the primality testing algorithm of Solovay and Strassen [1977], when the input x is a prime number, the algorithm *always* says "prime"; when x is composite, the algorithm *usually* says "composite," but may occasionally say "prime." Using the definitions above, this means that the Solovay-Strassen algorithm is a one-sided error algorithm for the set A of composite numbers. It also is a bounded two-sided error algorithm for $\overline{A}$, the set of prime numbers.

These three kinds of errors suggest three complexity classes:

1. PP is the class of languages decided by probabilistic Turing machines of polynomial time complexity with unbounded two-sided error.
2. BPP is the class of languages decided by probabilistic Turing machines of polynomial time complexity with bounded two-sided error.
3. RP is the class of languages decided by probabilistic Turing machines of polynomial time complexity with one-sided error.

In the literature, RP is also called R.

A probabilistic Turing machine M is a PP-**machine** (respectively, a BPP-**machine**, an RP-**machine**) if M has polynomial time complexity, and M decides with two-sided error (bounded two-sided error, one-sided error).

Through repeated Bernoulli trials, we can make the error probabilities of BPP-machines and RP-machines arbitrarily small, as stated in the following theorem. (Among other things, this theorem implies that RP $\subseteq$ BPP.)

Theorem 5.16   *If $A \in$ BPP, then for every polynomial $q(n)$, there exists a BPP-machine M such that $p_M(x) > 1 - 1/2^{q(n)}$ for every $x \in A$, and $p_M(x) < 1/2^{q(n)}$ for every $x \notin A$.*

If $L \in$ RP, then for every polynomial $q(n)$, there exists an RP-machine M such that $p_M(x) > 1 - 1/2^{q(n)}$ for every x in L.

It is important to note just how minuscule the probability of error is (provided that the coin flips are truly random). If the probability of error is less than $1/2^{5000}$, then it is less likely that the algorithm produces an incorrect answer than that the computer will be struck by a meteor. An algorithm whose probability of

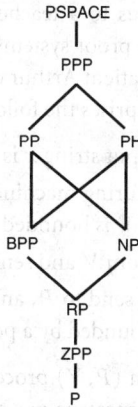

FIGURE 5.5 Probabilistic complexity classes.

error is $1/2^{5000}$ is essentially as good as an algorithm that makes no errors. For this reason, many computer scientists consider BPP to be the class of practically feasible computational problems.

Next, we define a class of problems that have probabilistic algorithms that make no errors. Define:

$$ZPP = RP \cap co\text{-}RP$$

The letter Z in ZPP is for zero probability of error, as we now demonstrate. Suppose $A \in ZPP$. Here is an algorithm that checks membership in A. Let M be an RP-machine that decides A, and let M' be an RP-machine that decides $\overline{A}$. For an input string x, alternately run M and M' on x, repeatedly, until a computation path of one machine accepts x. If M accepts x, then accept x; if M' accepts x, then reject x. This algorithm works correctly because when an RP-machine accepts its input, it does not make a mistake. This algorithm might not terminate, but with very high probability, the algorithm terminates after a few iterations.

The next theorem expresses some known relationships between probabilistic complexity classes and other complexity classes, such as classes in the polynomial hierarchy. See Section 5.7 and Figure 5.5.

Theorem 5.17

(a) $P \subseteq ZPP \subseteq RP \subseteq BPP \subseteq PP \subseteq PSPACE$ [Gill, 1977]

(b) $RP \subseteq NP \subseteq PP$ [Gill, 1977]

(c) $BPP \subseteq \Sigma_2^P \cap \Pi_2^P$ [Lautemann, 1983; Sipser, 1983]

(d) $BPP \subset P/poly$

(e) $PH \subseteq P^{PP}$ [Toda, 1991]

An important recent research area called **de-randomization** studies whether randomized algorithms can be converted to deterministic ones of the same or comparable efficiency. For example, if there is a language in E that requires Boolean circuits of size $2^{\Omega(n)}$ to decide it, then $BPP = P$ [Impagliazzo and Wigderson, 1997].

5.11 Interactive Models and Complexity Classes

5.11.1 Interactive Proofs

In Section 5.3.2, we characterized NP as the set of languages whose membership proofs can be checked quickly, by a deterministic Turing machine M of polynomial time complexity. A different notion of proof involves interaction between two parties, a prover P and a verifier V, who exchange messages. In an **interactive proof system** [Goldwasser et al., 1989], the prover is an all-powerful machine, with unlimited computational resources, analogous to a teacher. The verifier is a computationally limited machine, analogous to a student. Interactive proof systems are also called "Arthur-Merlin games": the wizard Merlin corresponds to P, and the impatient Arthur corresponds to V [Babai and Moran, 1988].

Formally, an **interactive proof system** comprises the following:

- A read-only input tape on which an input string x is written.
- A *verifier* V, which is a probabilistic Turing machine augmented with the capability to send and receive messages. The running time of V is bounded by a polynomial in $|x|$.
- A *prover* P, which receives messages from V and sends messages to V.
- A tape on which V writes messages to send to P, and a tape on which P writes messages to send to V. The length of every message is bounded by a polynomial in $|x|$.

A computation of an interactive proof system (P, V) proceeds in rounds, as follows. For $j = 1, 2, \ldots$, in round j, V performs some steps, writes a message m_j, and temporarily stops. Then P reads m_j and responds with a message m'_j, which V reads in round $j + 1$. An interactive proof system (P, V) **accepts** an input string x if the probability of acceptance by V satisfies $p_V(x) > 1/2$.

In an interactive proof system, a prover can convince the verifier about the truth of a statement without exhibiting an entire proof, as the following example illustrates.

Consider the graph non-isomorphism problem: the input consists of two graphs G and H, and the decision is yes if and only if G is not isomorphic to H. Although there is a short proof that two graphs *are* isomorphic (namely: the proof consists of the isomorphism mapping G onto H), nobody has found a general way of proving that two graphs are *not* isomorphic that is significantly shorter than listing all $n!$ permutations and showing that each fails to be an isomorphism. (That is, the graph non-isomorphism problem is in co-NP, but is not known to be in NP.) In contrast, the verifier V in an interactive proof system is able to take statistical evidence into account, and determine "beyond all reasonable doubt" that two graphs are non-isomorphic, using the following protocol.

In each round, V randomly chooses either G or H with equal probability; if V chooses G, then V computes a random permutation G' of G, presents G' to P, and asks P whether G' came from G or from H (and similarly if V chooses H). If P gave an erroneous answer on the first round, and G is isomorphic to H, then after k subsequent rounds, the probability that P answers all the subsequent queries correctly is $1/2^k$. (To see this, it is important to understand that the prover P does not see the coins that V flips in making its random choices; P sees only the graphs G' and H' that V sends as messages.) V accepts the interaction with P as "proof" that G and H are non-isomorphic if P is able to pick the correct graph for 100 consecutive rounds. Note that V has ample grounds to accept this as a convincing demonstration: if the graphs are indeed isomorphic, the prover P would have to have an incredible streak of luck to fool V.

It is important to comment that de-randomization techniques applied to these proof systems have shown that under plausible hardness assumptions, proofs of non-isomorphism of sub-exponential length (or even polynomial length) do exist [Klivans and van Melkebeek, 2002]. Thus, many complexity theoreticians now conjecture that the graph isomorphism problem lies in NP ∩ co-NP.

The complexity class IP comprises the languages A for which there exists a verifier V and a positive ϵ such that

- There exists a prover $\hat{P}$ such that for all x in A, the interactive proof system $(\hat{P}, V)$ accepts x with probability greater than $1/2 + \epsilon$; and
- For every prover P and every $x \notin A$, the interactive proof system (P, V) rejects x with probability greater than $1/2 + \epsilon$.

By substituting random choices for existential choices in the proof that $\mathrm{ATIME}(t) \subseteq \mathrm{DSPACE}(t)$ (Theorem 5.11), it is straightforward to show that IP ⊆ PSPACE. It was originally believed likely that IP was a small subclass of PSPACE. Evidence supporting this belief was the construction of an oracle language B for which co-$\mathrm{NP}^B - \mathrm{IP}^B \neq \emptyset$ [Fortnow and Sipser, 1988], so that IP^B is strictly included in PSPACE^B. Using a proof technique that does not relativize, however, Shamir [1992] proved that, in fact, IP and PSPACE are the same class.

Theorem 5.18 IP = PSPACE. [Shamir, 1992].

If NP is a proper subset of PSPACE, as is widely believed, then Theorem 5.18 says that interactive proof systems can decide a larger class of languages than NP.

5.11.2 Probabilistically Checkable Proofs

In an interactive proof system, the verifier does not need a complete conventional proof to become convinced about the membership of a word in a language, but uses random choices to query parts of a proof that the prover may know. This interpretation inspired another notion of "proof": a proof consists of a (potentially) large amount of information that the verifier need only inspect in a few places in order to become convinced. The following definition makes this idea more precise.

A language A has a **probabilistically checkable proof** if there exists an oracle BPP-machine M such that:

- For all $x \in A$, there exists an oracle language B_x such that M^{B_x} accepts x with probability 1.
- For all $x \notin A$, and for every language B, machine M^B accepts x with probability strictly less than $1/2$.

Intuitively, the oracle language B_x represents a proof of membership of x in A. Notice that B_x can be finite since the length of each possible query during a computation of M^{B_x} on x is bounded by the running time of M. The oracle language takes the role of the prover in an interactive proof system — but in contrast to an interactive proof system, the prover cannot change strategy adaptively in response to the questions that the verifier poses. This change results in a potentially stronger system, since a machine M that has bounded error probability relative to all languages B might not have bounded error probability relative to some adaptive prover. Although this change to the proof system framework may seem modest, it leads to a characterization of a class that seems to be much larger than PSPACE.

Theorem 5.19 *A has a probabilistically checkable proof if and only if $A \in$ NEXP* [Babai et al., 1991].

Although the notion of probabilistically checkable proofs seems to lead us away from feasible complexity classes, by considering natural restrictions on how the proof is accessed, we can obtain important insights into familiar complexity classes.

Let $\mathrm{PCP}[r(n), q(n)]$ denote the class of languages with probabilistically checkable proofs in which the probabilistic oracle Turing machine M makes $O[r(n)]$ random binary choices, and queries its oracle $O[q(n)]$ times. (For this definition, we assume that M has either one or two choices for each step.) It follows from the definitions that $\mathrm{BPP} = \mathrm{PCP}(n^{O(1)}, 0)$, and $\mathrm{NP} = \mathrm{PCP}(0, n^{O(1)})$.

Theorem 5.20 (The PCP Theorem) NP = $\mathrm{PCP}[\emptyset \log n, \emptyset(1)]$ [Arora et al., 1998].

Theorem 5.20 asserts that for every language A in NP, a proof that $x \in A$ can be encoded so that the verifier can be convinced of the correctness of the proof (or detect an incorrect proof) by using only $O(\log n)$ random choices, and inspecting only a *constant* number of bits of the proof.

5.12 Kolmogorov Complexity

Until now, we have considered only dynamic complexity measures, namely, the time and space used by Turing machines. Kolmogorov complexity is a static complexity measure that captures the difficulty of describing a string. For example, the string consisting of three million zeroes can be described with fewer than three million symbols (as in this sentence). In contrast, for a string consisting of three million randomly generated bits, with high probability there is no shorter description than the string itself.

Let U be a universal Turing machine (see Section 5.2.3). Let λ denote the empty string. The **Kolmogorov complexity** of a binary string y with respect to U, denoted by $K_U(y)$, is the length of the shortest binary string i such that on input $\langle i, \lambda \rangle$, machine U outputs y. In essence, i is a description of y, for it tells U how to generate y.

The next theorem states that different choices for the universal Turing machine affect the definition of Kolmogorov complexity in only a small way.

Theorem 5.21 (Invariance Theorem)　*There exists a universal Turing machine U such that for every universal Turing machine U', there is a constant c such that for all y, $K_U(y) \leq K_{U'}(y) + c$.*

Henceforth, let K be defined by the universal Turing machine of Theorem 5.21. For every integer n and every binary string y of length n, because y can be described by giving itself explicitly, $K(y) \leq n + c'$ for a constant c'. Call y **incompressible** if $K(y) \geq n$. Since there are 2^n binary strings of length n and only $2^n - 1$ possible shorter descriptions, there exists an incompressible string for every length n.

Kolmogorov complexity gives a precise mathematical meaning to the intuitive notion of "randomness." If someone flips a coin 50 times and it comes up "heads" each time, then intuitively, the sequence of flips is not random — although from the standpoint of probability theory, the all-heads sequence is precisely as likely as any other sequence. Probability theory does not provide the tools for calling one sequence "more random" than another; Kolmogorov complexity theory does.

Kolmogorov complexity provides a useful framework for presenting combinatorial arguments. For example, when one wants to prove that an object with some property P exists, then it is sufficient to show that any object that does *not* have property P has a short description; thus, any incompressible (or "random") object must have property P. This sort of argument has been useful in proving lower bounds in complexity theory.

5.13　Research Issues and Summary

The core research questions in complexity theory are expressed in terms of separating complexity classes:

- Is L different from NL?
- Is P different from RP or BPP?
- Is P different from NP?
- Is NP different from PSPACE?

Motivated by these questions, much current research is devoted to efforts to understand the power of nondeterminism, randomization, and interaction. In these studies, researchers have gone well beyond the theory presented in this chapter:

- Beyond Turing machines and Boolean circuits, to restricted and specialized models in which non-trivial lower bounds on complexity can be proved
- Beyond deterministic reducibilities, to nondeterministic and probabilistic reducibilities, and refined versions of the reducibilities considered here
- Beyond worst-case complexity, to average-case complexity

Recent research in complexity theory has had direct applications to other areas of computer science and mathematics. Probabilistically checkable proofs were used to show that obtaining approximate solutions to some optimization problems is as difficult as solving them exactly. Complexity theory has provided new tools for studying questions in finite model theory, a branch of mathematical logic. Fundamental questions in complexity theory are intimately linked to practical questions about the use of cryptography for computer security, such as the existence of one-way functions and the strength of public key cryptosystems.

This last point illustrates the urgent practical need for progress in computational complexity theory. Many popular cryptographic systems in current use are based on unproven assumptions about the difficulty of computing certain functions (such as the factoring and discrete logarithm problems). All of these systems are thus based on wishful thinking and conjecture. Research is needed to resolve these open questions and replace conjecture with mathematical certainty.

Acknowledgments

Donna Brown, Bevan Das, Raymond Greenlaw, Lane Hemaspaandra, John Jozwiak, Sung-il Pae, Leonard Pitt, Michael Roman, and Martin Tompa read earlier versions of this chapter and suggested numerous helpful improvements. Karen Walny checked the references.

Eric W. Allender was supported by the National Science Foundation under Grant CCR-0104823. Michael C. Loui was supported by the National Science Foundation under Grant SES-0138309. Kenneth W. Regan was supported by the National Science Foundation under Grant CCR-9821040.

Defining Terms

Complexity class: A set of languages that are decided within a particular resource bound. For example, $\text{NTIME}(n^2 \log n)$ is the set of languages decided by nondeterministic Turing machines within $O(n^2 \log n)$ time.

Constructibility: A function $f(n)$ is time (respectively, space) constructible if there exists a deterministic Turing machine that halts after exactly $f(n)$ steps (after using exactly $f(n)$ worktape cells) for every input of length n.

Diagonalization: A technique for constructing a language A that differs from every $L(M_i)$ for a list of machines $M_1, M_2, \ldots$.

NP-complete: A language A_0 is NP-complete if $A_0 \in \text{NP}$ and $A \leq_m^p A_0$ for every A in NP; that is, for every A in NP, there exists a function f computable in polynomial time such that for every x, $x \in A$ if and only if $f(x) \in A_0$.

Oracle: An oracle is a language A to which a machine presents queries of the form "Is w in A" and receives each correct answer in one step.

Padding: A technique for establishing relationships between complexity classes that uses padded versions of languages, in which each word is padded out with multiple occurrences of a new symbol — the word x is replaced by the word $x\#^{f(|x|)}$ for a numeric function f — in order to artificially reduce the complexity of the language.

Reduction: A language A reduces to a language B if a machine that decides B can be used to decide A efficiently.

Time and space complexity: The time (respectively, space) complexity of a deterministic Turing machine M is the maximum number of steps taken (nonblank cells used) by M among all input words of length n.

Turing machine: A Turing machine M is a model of computation with a read-only input tape and multiple worktapes. At each step, M reads the tape cells on which its access heads are located, and depending on its current state and the symbols in those cells, M changes state, writes new symbols on the worktape cells, and moves each access head one cell left or right or not at all.

References

Allender, E., Loui, M.C., and Regan, K.W. 1999. Chapter 27: Complexity classes, Chapter 28: Reducibility and completeness, Chapter 29: Other complexity classes and measures. In *Algorithms and Theory of Computation Handbook*, Ed. M. J. Atallah, CRC Press, Boca Raton, FL.

Arora, S., Lund, C., Motwani, R., Sudan, M., and Szegedy, M. 1998. Proof verification and hardness of approximation problems. *J. ACM*, 45(3):501–555.

Babai, L. and Moran, S. 1988. Arthur-Merlin games: a randomized proof system, and a hierarchy of complexity classes. *J. Comput. Sys. Sci.*, 36(2):254–276.

Babai, L., Fortnow, L., and Lund, C. 1991. Nondeterministic exponential time has two-prover interactive protocols. *Computational Complexity*, 1:3–40.

Baker, T., Gill, J., and Solovay, R. 1975. Relativizations of the P = NP? question. *SIAM J. Comput.*, 4(4): 431–442.

Balcázar, J.L., Díaz, J., and Gabarró, J. 1990. *Structural Complexity II*. Springer-Verlag, Berlin.

Balcázar, J.L., Díaz, J., and Gabarró, J. 1995. *Structural Complexity I*. 2nd ed. Springer-Verlag, Berlin.

Book, R.V. 1974. Comparing complexity classes. *J. Comp. Sys. Sci.*, 9(2):213–229.

Borodin, A. 1972. Computational complexity and the existence of complexity gaps. *J. Assn. Comp. Mach.*, 19(1):158–174.

Borodin, A. 1977. On relating time and space to size and depth. *SIAM J. Comput.*, 6(4):733–744.

Bovet, D.P. and Crescenzi, P. 1994. *Introduction to the Theory of Complexity*. Prentice Hall International Ltd; Hertfordshire, U.K.

Chandra, A.K., Kozen, D.C., and Stockmeyer, L.J. 1981. Alternation. *J. Assn. Comp. Mach.*, 28(1):114–133.

Cook, S.A. 1971. The complexity of theorem-proving procedures. In *Proc. 3rd Annu. ACM Symp. Theory Comput.*, pp. 151–158. Shaker Heights, OH.

Du, D-Z. and Ko, K.-I. 2000. *Theory of Computational Complexity*. Wiley, New York.

Fortnow, L. and Sipser, M. 1988. Are there interactive protocols for co-NP languages? *Inform. Process. Lett.*, 28(5):249–251.

Garey, M.R. and Johnson, D.S. 1979. *Computers and Intractability: A Guide to the Theory of NP-Completeness*. W.H. Freeman, San Francisco.

Gill, J. 1977. Computational complexity of probabilistic Turing machines. *SIAM J. Comput.*, 6(4):675–695.

Goldwasser, S., Micali, S., and Rackoff, C. 1989. The knowledge complexity of interactive proof systems. *SIAM J. Comput.*, 18(1):186–208.

Hartmanis, J., Ed. 1989. *Computational Complexity Theory*. American Mathematical Society, Providence, RI.

Hartmanis, J. 1994. On computational complexity and the nature of computer science. *Commun. ACM*, 37(10):37–43.

Hartmanis, J. and Stearns, R.E. 1965. On the computational complexity of algorithms. *Trans. Amer. Math. Soc.*, 117:285–306.

Hemaspaandra, L.A. and Ogihara, M. 2002. *The Complexity Theory Companion*. Springer, Berlin.

Hemaspaandra, L.A. and Selman, A.L., Eds. 1997. *Complexity Theory Retrospective II*. Springer, New York.

Hennie, F. and Stearns, R.A. 1966. Two–way simulation of multitape Turing machines. *J. Assn. Comp. Mach.*, 13(4):533–546.

Immerman, N. 1988. Nondeterministic space is closed under complementation. *SIAM J. Comput.*, 17(5):935–938.

Impagliazzo, R. and Wigderson, A. 1997. P = BPP if E requires exponential circuits: Derandomizing the XOR lemma. *Proc. 29th Annu. ACM Symp. Theory Comput.*, ACM Press, pp. 220–229. El Paso, TX.

Jones, N.D. 1975. Space-bounded reducibility among combinatorial problems. *J. Comp. Sys. Sci.*, 11(1):68–85. Corrigendum *J. Comp. Sys. Sci.*, 15(2):241, 1977.

Karp, R.M. 1972. Reducibility among combinatorial problems. In *Complexity of Computer Computations*. R.E. Miller and J.W. Thatcher, Eds., pp. 85–103. Plenum Press, New York.

Klivans, A.R. and van Melkebeek, D. 2002. Graph nonisomorphism has subexponential size proofs unless the polynomial-time hierarchy collapses. *SIAM J. Comput.*, 31(5):1501–1526.

Ladner, R.E. 1975. On the structure of polynomial-time reducibility. *J. Assn. Comp. Mach.*, 22(1):155–171.

Lautemann, C. 1983. BPP and the polynomial hierarchy. *Inf. Proc. Lett.*, 17(4):215–217.

Levin, L. 1973. Universal search problems. *Problems of Information Transmission*, 9(3):265–266 (in Russian).

Li, M. and Vitányi, P.M.B. 1997. *An Introduction to Kolmogorov Complexity and Its Applications*. 2nd ed. Springer-Verlag, New York.

Papadimitriou, C.H. 1994. *Computational Complexity*. Addison-Wesley, Reading, MA.

Pippenger, N. and Fischer, M. 1979. Relations among complexity measures. *J. Assn. Comp. Mach.*, 26(2):361–381.

Ruzzo, W.L. 1981. On uniform circuit complexity. *J. Comp. Sys. Sci.*, 22(3):365–383.

Savitch, W.J. 1970. Relationship between nondeterministic and deterministic tape complexities. *J. Comp. Sys. Sci.*, 4(2):177–192.

Seiferas, J.I., Fischer, M.J., and Meyer, A.R. 1978. Separating nondeterministic time complexity classes. *J. Assn. Comp. Mach.*, 25(1):146–167.

Shamir, A. 1992. IP = PSPACE. *J. ACM* 39(4):869–877.

Sipser, M. 1983. Borel sets and circuit complexity. In *Proc. 15th Annual ACM Symposium on the Theory of Computing*, pp. 61–69.

Sipser, M. 1992. The history and status of the P versus NP question. In *Proc. 24th Annu. ACM Symp. Theory Comput.*, ACM Press, pp. 603–618. Victoria, B.C., Canada.

Solovay, R. and Strassen, V. 1977. A fast Monte-Carlo test for primality. *SIAM J. Comput.*, 6(1):84–85.

Stearns, R.E. 1990. Juris Hartmanis: the beginnings of computational complexity. In *Complexity Theory Retrospective*. A.L. Selman, Ed., pp. 5–18, Springer-Verlag, New York.

Stockmeyer, L.J. 1976. The polynomial time hierarchy. *Theor. Comp. Sci.*, 3(1):1–22.

Stockmeyer, L.J. 1987. Classifying the computational complexity of problems. *J. Symb. Logic*, 52:1–43.

Stockmeyer, L.J. and Chandra, A.K. 1979. Intrinsically difficult problems. *Sci. Am.*, 240(5):140–159.

Stockmeyer, L.J. and Meyer, A.R. 1973. Word problems requiring exponential time: preliminary report. In *Proc. 5th Annu. ACM Symp. Theory Comput.*, ACM Press, pp. 1–9. Austin, TX.

Stockmeyer, L.J. and Vishkin, U. 1984. Simulation of parallel random access machines by circuits. *SIAM J. Comput.*, 13(2):409–422.

Szelepcsényi, R. 1988. The method of forced enumeration for nondeterministic automata. *Acta Informatica*, 26(3):279–284.

Toda, S. 1991. PP is as hard as the polynomial-time hierarchy. *SIAM J. Comput.*, 20(5):865–877.

van Leeuwen, J. 1990. *Handbook of Theoretical Computer Science, Volume A: Algorithms and Complexity*. Elsevier Science, Amsterdam, and M.I.T. Press, Cambridge, MA.

Wagner, K. and Wechsung, G. 1986. *Computational Complexity*. D. Reidel, Dordrecht, The Netherlands.

Wrathall, C. 1976. Complete sets and the polynomial-time hierarchy. *Theor. Comp. Sci.*, 3(1):23–33.

Further Information

This chapter is a short version of three chapters written by the same authors for the *Algorithms and Theory of Computation Handbook* [Allender et al., 1999].

The formal theoretical study of computational complexity began with the paper of Hartmanis and Stearns [1965], who introduced the basic concepts and proved the first results. For historical perspectives on complexity theory, see Hartmanis [1994], Sipser [1992], and Stearns [1990].

Contemporary textbooks on complexity theory are by Balcázar et al. [1990, 1995], Bovet and Crescenzi [1994], Du and Ko [2000], Hemaspaandra and Ogihara [2002], and Papadimitriou [1994]. Wagner and Wechsung [1986] is an exhaustive survey of complexity theory that covers work published before 1986. Another perspective of some of the issues covered in this chapter can be found in the survey by Stockmeyer [1987].

A good general reference is the *Handbook of Theoretical Computer Science* [van Leeuwen, 1990], Volume A. The following chapters in that *Handbook* are particularly relevant: "Machine Models and Simulations," by P. van Emde Boas, pp. 1–66; "A Catalog of Complexity Classes," by D.S. Johnson, pp. 67–161; "Machine-Independent Complexity Theory," by J.I. Seiferas, pp. 163–186; "Kolmogorov Complexity and its Applications," by M. Li and P.M.B. Vitányi, pp. 187–254; and "The Complexity of Finite Functions," by R.B. Boppana and M. Sipser, pp. 757–804, which covers circuit complexity.

A collection of articles edited by Hartmanis [1989] includes an overview of complexity theory, and chapters on sparse complete languages, on relativizations, on interactive proof systems, and on applications of complexity theory to cryptography. A collection edited by Hemaspaandra and Selman [1997] includes chapters on quantum and biological computing, on proof systems, and on average case complexity.

For specific topics in complexity theory, the following references are helpful. Garey and Johnson [1979] explain NP-completeness thoroughly, with examples of NP-completeness proofs, and a collection of hundreds of NP-complete problems. Li and Vitányi [1997] provide a comprehensive, scholarly treatment of Kolmogorov complexity, with many applications.

Surveys and lecture notes on complexity theory that can be obtained via the Web are maintained by A. Czumaj and M. Kutylowski at:

```
http://www.uni-paderborn.de/fachbereich/AG/agmadh/WWW/english/scripts.html
```

As usual with the Web, such links are subject to change. Two good stem pages to begin searches are the site for SIGACT (the ACM Special Interest Group on Algorithms and Computation Theory) and the site for the annual IEEE Conference on Computational Complexity:

```
http://sigact.acm.org/
http://www.computationalcomplexity.org/
```

The former site has a pointer to a "Virtual Address Book" that indexes the personal Web pages of over 1000 computer scientists, including all three authors of this chapter. Many of these pages have downloadable papers and links to further research resources. The latter site includes a pointer to the *Electronic Colloquium on Computational Complexity* maintained at the University of Trier, Germany, which includes downloadable prominent research papers in the field, often with updates and revisions.

Research papers on complexity theory are presented at several annual conferences, including the annual ACM Symposium on Theory of Computing; the annual International Colloquium on Automata, Languages, and Programming, sponsored by the European Association for Theoretical Computer Science (EATCS); and the annual Symposium on Foundations of Computer Science, sponsored by the IEEE. The annual Conference on Computational Complexity (formerly Structure in Complexity Theory), also sponsored by the IEEE, is entirely devoted to complexity theory. Research articles on complexity theory regularly appear in the following journals, among others: *Chicago Journal on Theoretical Computer Science, Computational Complexity, Information and Computation, Journal of the ACM, Journal of Computer and System Sciences, SIAM Journal on Computing, Theoretical Computer Science,* and *Theory of Computing Systems* (formerly *Mathematical Systems Theory*). Each issue of *ACM SIGACT News* and *Bulletin of the EATCS* contains a column on complexity theory.

6

Formal Models and Computability

Tao Jiang
University of California

Ming Li
University of Waterloo

Bala Ravikumar
University of Rhode Island

6.1 Introduction **6-1**
6.2 Computability and a Universal Algorithm **6-2**
   Some Computational Problems • A Universal Algorithm
6.3 Undecidability **6-7**
   Diagonalization and Self-Reference • Reductions and More
   Undecidable Problems
6.4 Formal Languages and Grammars **6-11**
   Representation of Languages • Hierarchy of Grammars
   • Context-Free Grammars and Parsing
6.5 Computational Models **6-22**
   Finite Automata • Turing Machines

6.1 Introduction

The concept of **algorithms** is perhaps almost as old as human civilization. The famous Euclid's algorithm is more than 2000 years old. Angle trisection, solving diophantine equations, and finding polynomial roots in terms of radicals of coefficients are some well-known examples of algorithmic questions. However, until the 1930s the notion of algorithms was used informally (or rigorously but in a limited context). It was a major triumph of logicians and mathematicians of this century to offer a rigorous definition of this fundamental concept. The revolution that resulted in this triumph was a collective achievement of many mathematicians, notably Church, Gödel, Kleene, Post, and Turing. Of particular interest is a machine model proposed by Turing in 1936, which has come to be known as a **Turing machine** [Turing 1936].

This particular achievement had numerous significant consequences. It led to the concept of a general-purpose computer or universal computation, a revolutionary idea originally anticipated by Babbage in the 1800s. It is widely acknowledged that the development of a universal Turing machine was prophetic of the modern all-purpose digital computer and played a key role in the thinking of pioneers in the development of modern computers such as von Neumann [Davis 1980]. From a mathematical point of view, however, a more interesting consequence was that it was now possible to show the *nonexistence* of algorithms, hitherto impossible due to their elusive nature. In addition, many apparently different definitions of an algorithm proposed by different researchers in different continents turned out to be equivalent (in a precise technical sense, explained later). This equivalence led to the widely held hypothesis known as the *Church–Turing thesis* that mechanical solvability is the same as solvability on a Turing machine.

Formal languages are closely related to algorithms. They were introduced as a way to convey mathematical proofs without errors. Although the concept of a formal language dates back at least to the time of Leibniz, a systematic study of them did not begin until the beginning of this century. It became a vigorous field of study when Chomsky formulated simple grammatical rules to describe the syntax of a language

[Chomsky 1956]. **Grammars** and **formal languages** entered into computability theory when Chomsky and others found ways to use them to classify algorithms.

The main theme of this chapter is about formal models, which include Turing machines (and their variants) as well as grammars. In fact, the two concepts are intimately related. Formal computational models are aimed at providing a framework for computational problem solving, much as electromagnetic theory provides a framework for problems in electrical engineering. Thus, formal models guide the way to build computers and the way to program them. At the same time, new models are motivated by advances in the technology of computing machines. In this chapter, we will discuss only the most basic computational models and use these models to classify problems into some fundamental classes. In doing so, we hope to provide the reader with a conceptual basis with which to read other chapters in this Handbook.

6.2   Computability and a Universal Algorithm

Turing's notion of mechanical computation was based on identifying the basic steps of such computations. He reasoned that an operation such as multiplication is not primitive because it can be divided into more basic steps such as digit-by-digit multiplication, shifting, and adding. Addition itself can be expressed in terms of more basic steps such as add the lowest digits, compute, carry, and move to the next digit, etc. Turing thus reasoned that the most basic features of mechanical computation are the abilities to read and write on a storage medium (which he chose to be a linear tape divided into cells or squares) and to make some simple logical decisions. He also restricted each tape cell to hold only one among a finite number of symbols (which we call the *tape alphabet*).* The decision step enables the computer to control the sequence of actions. To make things simple, Turing restricted the next action to be performed on a cell neighboring the one on which the current action occurred. He also introduced an instruction that told the computer to stop. In summary, Turing proposed a model to characterize mechanical computation as being carried out as a sequence of instructions of the form: write a symbol (such as 0 or 1) on the tape cell, move to the next cell, observe the symbol currently scanned and choose the next step accordingly, or stop.

These operations define a language we call the GOTO language.** Its instructions are

> PRINT i (i is a tape symbol)
> GO RIGHT
> GO LEFT
> GO TO STEP j IF i IS SCANNED
> STOP

A **program** in this language is a sequence of instructions (written one per line) numbered $1 - k$. To run a program written in this language, we should provide the *input*. We will assume that the input is a string of symbols from a finite input alphabet (which is a subset of the tape alphabet), which is stored on the tape before the computation begins. How much memory should we allow the computer to use? Although we do not want to place any bounds on it, allowing an infinite tape is not realistic. This problem is circumvented by allowing *expandable memory*. In the beginning, the tape containing the input defines its boundary. When the machine moves beyond the current boundary, a new memory cell will be attached with a special symbol B (blank) written on it. Finally, we define the result of computation as the contents of the tape when the computer reaches the STOP instruction.

We will present an example program written in the GOTO language. This program accomplishes the simple task of doubling the number of 1s (Figure 6.1). More precisely, on the input containing k 1s, the

---

*This bold step of using a discrete model was perhaps the harbinger of the digital revolution that was soon to follow.

**Turing's original formulation is closer to our presentation in Section 6.5. But the GOTO language presents an equivalent model.

```
1   PRINT 0
2   GO LEFT
3   GO TO STEP 2 IF 1 IS SCANNED
4   PRINT 1
5   GO RIGHT
6   GO TO STEP 5 IF 1 IS SCANNED
7   PRINT 1
8   GO RIGHT
9   GO TO STEP 1 IF 1 IS SCANNED
10  STOP
```

FIGURE 6.1   The doubling program in the GOTO language.

program produces $2k$ 1s. Informally, the program achieves its goal as follows. When it reads a 1, it changes the 1 to 0, moves left looking for a new cell, writes a 1 in the cell, returns to the starting cell and rewrites as 1, and repeats this step for each 1. Note the way the GOTO instructions are used for repetition. This feature is the most important aspect of programming and can be found in all of the imperative style programming languages.

The simplicity of the GOTO language is rather deceptive. There is strong reason to believe that it is powerful enough that any mechanical computation can be expressed by a suitable program in the GOTO language. Note also that the programs written in the GOTO language may not always halt, that is, on certain inputs, the program may never reach the STOP instruction. In this case, we say that the output is undefined.

We can now give a precise definition of what an algorithm is. An algorithm is any program written in the GOTO language with the additional property that it halts on all inputs. Such programs will be called *halting programs*. Throughout this chapter, we will be interested mainly in computational problems of a special kind called *decision problems* that have a yes/no answer. We will modify our language slightly when dealing with decision problems. We will augment our instruction set to include ACCEPT and REJECT (and omit STOP). When the ACCEPT (REJECT) instruction is reached, the machine will output yes or 1 (no or 0) and halt.

6.2.1 Some Computational Problems

We will temporarily shift our focus from the tool for problem solving (the computer) to the problems themselves. Throughout this chapter, a computational problem refers to an input/output relationship. For example, consider the problem of squaring an integer input. This problem assigns to each integer (such as 22) its square (in this case 484). In technical terms, this input/output relationship defines a function. Therefore, solving a computational problem is the same as computing the function defined by the problem. When we say that an algorithm (or a program) solves a problem, what we mean is that, for all inputs, the program halts and produces the correct output. We will allow inputs of arbitrary size and place no restrictions. A reader with primary interest in software applications is apt to question the validity (or even the meaningfulness) of allowing inputs of arbitrary size because it makes the set of all *possible* inputs infinite, and thus unrealistic, in real-world programming. But there are no really good alternatives. Any finite bound is artificial and is likely to become obsolete as the technology and our requirements change. Also, in practice, we do not know how to take advantage of restrictions on the size of the inputs. (See the discussion about nonuniform models in Section 6.5.) Problems (functions) that can be solved by an algorithm (or a halting GOTO program) are called *computable*.

As already remarked, we are interested mainly in decision problems. A decision problem is said to be decidable if there is a halting GOTO program that solves it correctly on all inputs. An important class of problems called **partially decidable decision problems** can be defined by relaxing our requirement a little bit; a decision problem is partially decidable if there is a GOTO program that halts and outputs 1 on all inputs for which the output should be 1 and either halts and outputs 0 or loops forever on the other inputs.

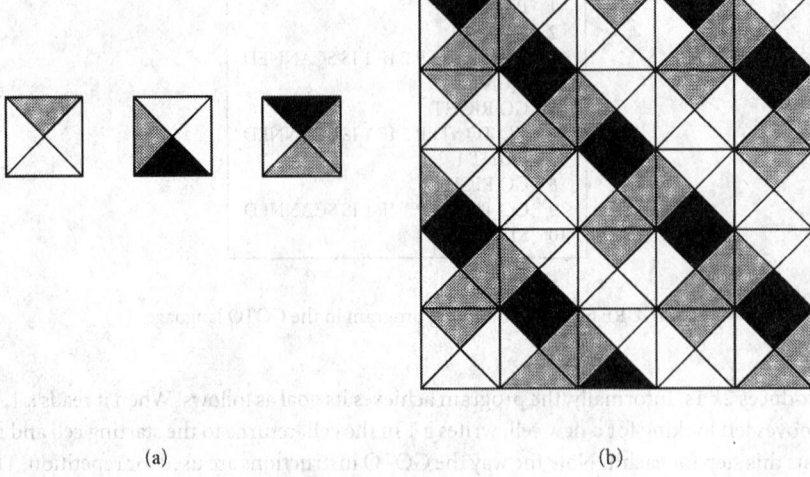

(a)   (b)

FIGURE 6.2   An example of tiling.

This means that the program may never give a wrong answer but is not required to halt on negative inputs (i.e., inputs with 0 as output).

We now list some problems that are fundamental either because of their inherent importance or because of their historical roles in the development of computation theory:

Problem 1 (halting problem). The input to this problem is a program P in the GOTO language and a binary string x. The expected output is 1 (or yes) if the program P halts when run on the input x, 0 (or no) otherwise.

Problem 2 (universal computation problem). A related problem takes as input a program P and an input x and produces as output what (if any) P would produce on input x. (Note that this is a decision problem if P is restricted to a yes/no program.)

Problem 3 (string compression). For a string x, we want to find the shortest program in the GOTO language that when started with the empty tape (i.e., tape containing one B symbol) halts and prints x. Here shortest means the total number of symbols in the program is as small as possible.

Problem 4 (tiling). A tile* is a square card of unit size (i.e., 1×1) divided into four quarters by two diagonals, each quarter colored with some color (selected from a finite set of colors). The tiles have fixed orientation and cannot be rotated. Given some finite set T of such tiles as input, the program is to determine if finite rectangular areas of all sizes (i.e., $k \times m$ for all positive integers k and m) can be tiled using only the given tiles such that the colors on any two touching edges are the same. It is assumed that an unlimited number of cards of each type is available. Figure 6.2(b) shows how the base set of tiles given in Figure 6.2(a) can be used to tile a 5×5 square area.

Problem 5 (linear programming). Given a system of linear inequalities (called constraints), such as $3x - 4y \leq 13$ with integer coefficients, the goal is to find if the system has a solution satisfying all of the constraints.

Some remarks must be made about the preceding problems. The problems in our list include nonnumerical problems and *meta problems*, which are problems about other problems. The first two problems are motivated by a quest for reliable program design. An algorithm for problem 1 (if it exists) can be used to test if a program contains an infinite loop. Problem 2 is motivated by an attempt to design a **universal**

---

*More precisely, a Wang tile, after Hao Wang, who wrote the first research paper on it.

algorithm, which can simulate any other. This problem was first attempted by Babbage, whose analytical engine had many ingredients of a modern electronic computer (although it was based on mechanical devices). Problem 3 is an important problem in information theory and arises in the following setting. Physical theories are aimed at creating simple laws to explain large volumes of experimental data. A famous example is Kepler's laws, which explained Tycho Brahe's huge and meticulous observational data. Problem 3 asks if this compression process can be automated. When we allow the inference rules to be sufficiently strong, this problem becomes **undecidable**. We will not discuss this problem further in this section but will refer the reader to some related formal systems discussed in Li and Vitányi [1993]. The tiling problem is not merely an interesting puzzle. It is an art form of great interest to architects and painters. Tiling has recently found applications in crystallography. Linear programming is a problem of central importance in economics, game theory, and operations research.

In the remainder of the section, we will present some basic algorithm design techniques and sketch how these techniques can be used to solve some of the problems listed (or their special cases). The main purpose of this discussion is to present techniques for showing the decidability (or partial decidability) of these problems. The reader can learn more advanced techniques of algorithm design in some later sections of this chapter as well as in many later chapters of this volume.

6.2.1.1 Table Lookup

The basic idea is to create a table for a function f, which needs to be computed by tabulating in one column an input x and the corresponding $f(x)$ in a second column. Then the table itself can be used as an algorithm. This method cannot be used directly because the set of all inputs is infinite. Therefore, it is not very useful, although it can be made to work in conjunction with the technique described subsequently.

6.2.1.2 Bounding the Search Domain

The difficulty of establishing the decidability of a problem is usually caused by the fact that the object we are searching for may have no known upper limit. Thus, if we can place such an upper bound (based on the structure of the problem), then we can reduce the search to a finite domain. Then table lookup can be used to complete the search (although there may be better methods in practice). For example, consider the following special case of the tiling problem: Let k be a fixed integer, say 1000. Given a set of tiles, we want to determine whether all rectangular rooms of shape $k \times n$ can be tiled for all n. (Note the difference between this special case and the general problem. The general one allows k and n both to have unbounded value. But here we allow only n to be unbounded.) It can be shown (see Section 6.5 for details) that there are two bounds n_0 and n_1 (they depend on k) such that if there is at least one tile of size $k \times t$ that can be tiled for some $n_0 \le t \le n_1$ then every tile of size $k \times n$ can be tiled. If no $k \times t$ tile can be tiled for any t between n_0 and n_1, then obviously the answer is no. Thus, we have reduced an infinite search domain to a finite one.

As another example, consider the linear programming problem. The set of possible solutions to this problem is infinite, and thus a table search cannot be used. But it is possible to reduce the search domain to a finite set using the geometric properties of the set of solutions of the linear programming problem. The fact that the set of solutions is convex makes the search especially easy.

6.2.1.3 Use of Subroutines

This is more of a program design tool than a tool for algorithm design. A central concept of programming is repetitive (or iterative) computation. We already observed how GOTO statements can be used to perform a sequence of steps repetitively. The idea of a subroutine is another central concept of programming. The idea is to make use of a program P itself as a single step in another program Q. Building programs from simpler programs is a natural way to deal with the complexity of programming tasks. We will illustrate the idea with a simple example. Consider the problem of multiplying two positive integers i and j. The input to the problem will be the form $11 \ldots 1011 \ldots 1$ (i 1s followed by a 0, followed by j 1s) and the output will be $i * j$ 1s (with possibly some 0s on either end). We will use the notation $1^i 0 1^j$ to denote the starting configuration of the tape. This just means that the tape contains i 1s followed by a 0 followed by j 1s.

TABLE 6.1   Coding the GOTO Instructions

Instruction	Code
PRINT i	0001^{i+1}
GO LEFT	001
GO RIGHT	010
GO TO j IF i IS SCANNED	$0111^{j}01^{i+1}$
STOP	100

The basic idea behind a GOTO program for this problem is simple; add j 1s on the right end of tape exactly $i - 1$ times and then erase the original sequence of i 1s on the left. A little thought reveals that the subroutine we need here is to duplicate a string of 1s so that if we start with $x02^{k}1^{j}$ a call to the subroutine will produce $x02^{k+j}1^{j}$. Here x is just any sequence of symbols. Note the role played by the symbol 2. As new 1s are created on the right, the old 1s change to 2s. This will ensure that there are exactly j 1s on the right end of the tape all of the time. This duplication subroutine is very similar to the doubling program, and the reader should have very little difficulty writing this program. Finally, the multiplication program can be done using the copy subroutine $(i - 1)$ times.

6.2.2   A Universal Algorithm

We will now present in some detail a (partial) solution to problem 2 by arguing that there is a program U written in the GOTO language, which takes as input a program P (also written using the GOTO language) and an input x and produces as output $P(x)$, the output of P on input x. For convenience, we will assume that all programs written in the GOTO language use a fixed alphabet containing just 0, 1, and B. Because we have assumed this for all programs in the GOTO language, we should first address the issue of how an input to program U will look. We cannot directly place a program P on the tape because the alphabet used to write the program P uses letters G, O, T, O, etc. This minor problem can be easily circumvented by coding. The idea is to represent each instruction using only 0 and 1. One such coding scheme is shown in Table 6.1.

To encode an entire program, we simply write down in order (without the line numbers) the code for each instruction as given in the table. For example, here is the code for the doubling program shown in Figure 6.1:

$$00010010111101100011010011111110110001101001110111100$$

Note that the encoded string contains all of the information about the program so that the encoding is completely reversible. From now on, if P is a program in the GOTO language, then code(P) will denote its binary code as just described. When there is no confusion, we will identify P and code(P). Before proceeding further, the reader may want to test his/her understanding of the encoding/decoding process by decoding the following string: 010011101100.

The basic idea behind the construction of a universal algorithm is simple, although the details involved in actually constructing one are enormous. We will present the central ideas and leave out the actual construction. Such a construction was carried out in complete detail by Turing himself and was simplified by others.* U has as its input code(P) followed by the string x. U simulates the computational steps of P on input x. It divides the input tape into three segments, one containing the program P, the second one essentially containing the contents of the tape of P as it changes with successive moves, and the third one containing the line number in program P of the instruction being currently simulated (similar to a *program counter* in an actual computer).

---

*A particularly simple exposition can be found in Robinson [1991].

We now describe a *cycle* of computation by U, which is similar to a central processing unit (CPU) cycle in a real computer. A single instruction of P is implemented by U in one cycle. First, U should know which location on the tape that P is currently reading. A simple artifact can handle this as follows: U uses in its tape alphabet two special symbols $0'$ and $1'$. U stores the tape of P in the tape segment alluded to in the previous paragraph exactly as it would appear when the program P is run on the input x with one minor modification. The symbol currently being read by program P is stored as the *primed version* ($0'$ is the primed version of 0, etc.). As an example, suppose after completing 12 instructions, P is reading the fourth symbol (from left) on its tape containing 01001001. Then the tape region of U after 12 cycles looks like $0100'1001$. At the beginning of a new cycle, U uses a subroutine to move to the region of the tape that contains the ith instruction of program P where i is the value of the program counter. It then decodes the ith instruction. Based on what type it is, U proceeds as follows: If it is a PRINT i instruction, then U scans the tape until the unique primed symbol in the *tape region* is reached and rewrites it as instructed. If it is a GO LEFT or GO RIGHT symbol, U locates the primed symbol, unprimes it, and primes its left or right neighbor, as instructed. In both cases, U returns to the program counter and increments it. If the instruction is GO TO i IF j IS SCANNED, U reads the primed symbol, and if it is j', U changes the program counter to i. This completes a cycle. Note that the three regions may grow and contract while U executes the cycles of computation just described. This may result in one of them running into another. U must then shift one of them to the left or right and make room as needed.

It is not too difficult to see that all of the steps described can be done using the instructions of the GOTO language. The main point to remember is that these actions will have to be coded as a single program, which has nothing whatsoever to do with program P. In fact, the program U is totally independent of P. If we replace P with some other program Q, it should simulate Q as well. The preceding argument shows that problem 2 is partially decidable. But it does not show that this problem is decidable. Why? It is because U may not halt on all inputs; specifically, consider an input consisting of a program P and a string x such that P does not halt on x. Then U will also keep executing cycle after cycle the moves of P and will never halt. In fact, in Section 6.3, we will show that problem 2 is not decidable.

6.3 Undecidability

Recall the definition of an undecidable problem. In this section, we will establish the undecidability of Problem 2, Section 6.2. The simplest way to establish the existence of undecidable problems is as follows: There are more problems than there are programs, the former set being uncountable, whereas the latter is countably infinite.* But this argument is purely existential and does not identify any specific problem as undecidable. In what follows, we will show that Problem 2 introduced in Section 6.2 is one such problem.

6.3.1 Diagonalization and Self-Reference

Undecidability is inextricably tied to the concept of self-reference, and so we begin by looking at this rather perplexing and sometimes paradoxical concept. The idea of self-reference seems to be many centuries old and may have originated with a barber in ancient Greece who had a sign board that read: "I shave all those who do not shave themselves." When the statement is applied to the barber himself, we get a self-contradictory statement. Does he shave himself? If the answer is yes, then he is one of those who shaves himself, and so the barber should not shave him. The contrary answer no is equally untenable. So neither yes nor no seems to be the correct answer to the question; this is the essence of the paradox. The barber's

---

*The reader who does not know what countable and uncountable infinities are can safely ignore this statement; the rest of the section does not depend on it.

paradox has made entry into modern mathematics in various forms. We will present some of them in the next few paragraphs.*

The first version, called Berry's paradox, concerns English descriptions of natural numbers. For example, the number 7 can be described by many different phrases: seven, six plus one, the fourth smallest prime, etc. We are interested in the *shortest* of such descriptions, namely, the one with the fewest letters in it. Clearly there are (infinitely) many positive integers whose shortest descriptions exceed 100 letters. (A simple counting argument can be used to show this. The set of positive integers is infinite, but the set of positive integers with English descriptions in fewer than or equal to 100 letters is finite.) Let D denote the set of positive integers that do not have English descriptions with fewer than 100 letters. Thus, D is not empty. It is a well-known fact in set theory that any nonempty subset of positive integers has a smallest integer. Let x be the smallest integer in D. Does x have an English description with fewer than or equal to 100 letters? By the definition of the set D and x, we have: x is "the smallest positive integer that cannot be described in English in fewer than 100 letters." This is clearly absurd because part of the last sentence in quotes is a description of x and it contains fewer than 100 letters in it. A similar paradox was found by the British mathematician Bertrand Russell when he considered the set of all sets that do not include themselves as elements, that is, $S = \{x \mid x \notin x\}$. The question "Is $S \in S$?" leads to a similar paradox.

As a last example, we will consider a charming self-referential paradox due to mathematician William Zwicker. Consider the collection of all two-person games (such as chess, tic-tac-toe, etc.) in which players make alternate moves until one of them loses. Call such a game *normal* if it has to end in a finite number of moves, no matter what strategies the two players use. For example, tic-tac-toe must end in at most nine moves and so it is normal. Chess is also normal because the 50-move rule ensures that the game cannot go forever. Now here is *hypergame*. In the first move of the hypergame, the first player calls out a normal game, and then the two players go on to play the game, with the second player making the first move. The question is: "Is hypergame normal?" Suppose it is normal. Imagine two players playing hypergame. The first player can call out hypergame (since it is a normal game). This makes the second player call out the name of a normal game, hypergame can be called out again and they can keep saying hypergame without end, and this contradicts the definition of a normal game. On the other hand, suppose it is not a normal game. But now in the first move, player 1 cannot call out hypergame and would call a normal game instead, and so the infinite move sequence just given is not possible, and so hypergame is normal after all!

In the rest of the section, we will show how these paradoxes can be modified to give nonparadoxical but surprising conclusions about the decidability of certain problems. Recall the encoding we presented in Section 6.2 that encodes any program written in the GOTO language as a binary string. Clearly this encoding is reversible in the sense that if we start with a program and encode it, it is possible to decode it back to the program. However, not every binary string corresponds to a program because there are many strings that cannot be decoded in a meaningful way, for example, 11010011000110. For the purposes of this section, however, it would be convenient if we can treat *every* binary string as a program. Thus, we will simply stipulate that any undecodable string be decoded to the program containing the single statement

1. REJECT

In the following discussion, we will identify a string x with a GOTO program to which it decodes. Now define a function f_D as follows: $f_D(x) = 1$ if x, decoded into a GOTO program, does not halt when started with x itself as the input. Note the self-reference in this definition. Although the definition of f_D seems artificial, its importance will become clear in the next section when we use it to show the undecidability of Problem 2. First we will prove that f_D is not computable. Actually, we will prove a stronger statement, namely, that f_D is not even partially decidable. [Recall that a function is partially decidable if there is a GOTO

---

*The most enchanting discussions of self-reference are due to the great puzzlist and mathematician R. Smullyan who brings out the breadth and depth of this concept in such delightful books as *What is the name of this book?* published by Prentice–Hall in 1978 and *Satan, Cantor, and Infinity* published by Alfred A. Knopf in 1992. We heartily recommend them to anyone who wants to be amused, entertained, and, more importantly, educated on the intricacies of mathematical logic and computability.

program (not necessarily halting) that computes it. An important distinction between computable and semicomputable functions is that a GOTO program for the latter need not halt on inputs with output = 0.]

Theorem 6.1 *Function f_D is not partially decidable.*

The proof is by contradiction. Suppose a GOTO program P' computes the function f_D. We will modify P' into another program P in the GOTO language such that P computes the same function as P' but has the additional property that it will never terminate its computation by ending up in a REJECT statement.* Thus, P is a program with the property that it computes f_D and halts on an input y if and only if $f_D(y) = 1$. We will complete the proof by showing that there is at least one input in which the program produces a wrong output, that is, there is an x such that $f_D(x) \neq P(x)$.

Let x be the encoding of program P. Now consider the question: Does P halt when given x as input? Suppose the answer is yes. Then, by the way we constructed P, here $P(x) = 1$. On the other hand, the definition of f_D implies that $f_D(x) = 0$. (This is the punch line in this proof. We urge the reader to take a few moments and read the definition of f_D a few times and make sure that he or she is convinced about this fact!) Similarly, if we start with the assumption that $P(x) = 0$, we are led to the conclusion that $f_D(x) = 1$. *In both cases,* $f_D(x) \neq P(x)$ and thus P is not the correct program for f_D. Therefore, P' is not the correct program for f_D either because P and P' compute the same function. This contradicts the hypothesis that such a program exists, and the proof is complete.

Note the crucial difference between the paradoxes we presented earlier and the proof of this theorem. Here we do not have a paradox because our conclusion is of the form $f_D(x) = 0$ if and only if $P(x) = 1$ and not $f_D(x) = 1$ if and only if $f_D(x) = 0$. But in some sense, the function f_D was motivated by Russell's paradox. We can similarly create another function f_Z (based on Zwicker's paradox of hypergame). Let f be any function that maps binary strings to $\{0, 1\}$. We will describe a method to generate successive functions f_1, f_2, etc., as follows: Suppose $f(x) = 0$ for all x. Then we cannot create any more functions, and the sequence stops with f. On the other hand, if $f(x) = 1$ for some x, then choose one such x and decode it as a GOTO program. This defines another function; call it f_1 and repeat the same process with f_1 in the place of f. We call f a normal function if no matter how x is selected at each step, the process terminates after a finite number of steps. A simple example of a nonnormal function is as follows: Suppose $P(Q) = 1$ for some program P and input Q and at the same time $Q(P) = 1$ (note that we are using a program and its code interchangeably), then it is easy to see that the functions defined by both P and Q are not normal. Finally, define $f_Z(X) = 1$ if X is a normal program, 0 if it is not. We leave it as an instructive exercise to the reader to show that f_Z is not semicomputable. A perceptive reader will note the connection between Berry's paradox and problem 3 in our list (string compression problem) just as f_Z is related to Zwicker's paradox. Such a reader should be able to show the undecidability of problem 3 by imitating Berry's paradox.

6.3.2 Reductions and More Undecidable Problems

Theory of computation deals not only with the behavior of individual problems but also with relations among them. A **reduction** is a simple way to relate two problems so that we can deduce the (un)decidability of one from the (un)decidability of the other. Reduction is similar to using a subroutine. Consider two problems A and B. We say that problem A can be reduced to problem B if there is an algorithm for B provided that A has one. To define the reduction (also called a *Turing reduction*) precisely, it is convenient to augment the instruction set of the GOTO programming language to include a new instruction CALL X, i, j where X is a (different) GOTO program, and i and j are line numbers. In detail, the execution of such augmented programs is carried out as follows: When the computer reaches the instruction CALL X,

---

*The modification needed to produce P from P' is straightforward. If P' did not have any REJECT statements at all, then no modification would be needed. If it had, then we would have to replace each one by a looping statement, which keeps repeating the same instruction forever.

i, j, the program will simply start executing the instructions of the program from line 1, treating whatever is on the tape currently as the input to the program X. When (if at all) X finishes the computation by reaching the ACCEPT statement, the execution of the original program continues at line number i and, if it finishes with REJECT, the original program continues from line number j.

We can now give a more precise definition of a reduction between two problems. Let A and B be two computational problems. We say that A is reducible to B if there is a halting program Y in the GOTO language for problem A in which calls can be made to a halting program X for problem B. The algorithm for problem A described in the preceding reduction does not assume the availability of program X and cannot use the details behind the design of this algorithm. The right way to think about a reduction is as follows: Algorithm Y, from time to time, needs to know the solutions to different instances of problem B. It can query an algorithm for problem B (as a black box) and use the answer to the query for making further decisions. An important point to be noted is that the program Y actually can be implemented even if program X was never built as long as someone can correctly answer some questions asked by program Y about the output of problem B for certain inputs. Programs with such calls are sometimes called *oracle programs*. Reduction is rather difficult to assimilate at the first attempt, and so we will try to explain it using a puzzle. How do you play two chess games, one each with Kasparov and Anand (perhaps currently the world's two best players) and ensure that you get at least one point? (You earn one point for a win, 0 for a loss, and 1/2 for a draw.) Because you are a novice and are pitted against two Goliaths, you are allowed a concession. You can choose to play white or black on either board. The well-known answer is the following: Take white against one player, say, Anand, and black against the other, namely, Kasparov. Watch the first move of Kasparov (as he plays white) and make the same move against Anand, get his reply and play it back to Kasparov and keep playing back and forth like this. It takes only a moment's thought that you are guaranteed to win (exactly) 1 point. The point is that your game involves taking the position of one game, applying the algorithm of one player, getting the result and applying it to the other board, etc., and you do not even have to know the rules of chess to do this. This is exactly how algorithm Y is required to use algorithm X.

We will use reductions to show the undecidability as follows: Suppose A can be reduced to B as in the preceding definition. If there is an algorithm for problem B, it can be used to design a program for A by essentially imitating the execution of the augmented program for A (with calls to the oracle for B) as just described. But we will turn it into a negative argument as follows: If A is undecidable, then so is B. Thus, a reduction from a problem known to be undecidable to problem B will prove B's undecidability.

First we define a new problem, Problem 2′, which is a special case of Problem 2. Recall that in Problem 2 the input is (the code of) a program P in GOTO language and a string x. The output required is $P(x)$. In Problem 2′, the input is (only) the code of a program P and the output required is $P(P)$, that is, instead of requiring P to run on a given input, this problem requires that it be run on its own code. This is clearly a special case of problem 2. The reader may readily see the self-reference in Problem 2′ and suspect that it may be undecidable; therefore, the more general Problem 2 may be undecidable as well. We will establish these claims more rigorously as follows.

We first observe a general statement about the decidability of a function f (or problem) and its *complement*. The complement function is defined to take value 1 on all inputs for which the original function value is 0 and vice versa. The statement is that a function f is decidable if and only if the complement $\bar{f}$ is decidable. This can be easily proved as follows. Consider a program P that computes f. Change P into $\bar{P}$ by interchanging all of the ACCEPT and REJECT statements. It is easy to see that $\bar{P}$ actually computes $\bar{f}$. The converse also is easily seen to hold. It readily follows that the function defined by problem 2′ is undecidable because it is, in fact, the complement of f_D.

Finally, we will show that problem 2 is uncomputable. The idea is to use a reduction from problem 2′ to problem 2. (Note the direction of reduction. This always confuses a beginner.) Suppose there is an algorithm for problem 2. Let X be the GOTO language program that implements this algorithm. X takes as input code(P) (for any program P) followed by x, produces the result $P(x)$, and halts. We want to design a program Y that takes as input code(P) and produce the output $P(P)$ using calls to program X. It is clear what needs to be done. We just create the input in proper form code(P) followed by code(P) and call X. This requires first duplicating the input, but this is a simple programming task similar to the one we demonstrated in our first program in Section 6.2. Then a call to X completes the task. This shows that Problem 2′ reduces to Problem 2, and thus the latter is undecidable as well.

By a more elaborate reduction (from f_D), it can be shown that tiling is not partially decidable. We will not do it here and refer the interested reader to Harel [1992]. But we would like to point out how the undecidability result can be used to infer a result about tiling. This deduction is of interest because the result is an important one and is hard to derive directly. We need the following definition before we can state the result. A different way to pose the tiling problem is whether a given set of tiles can tile *an entire plane* in such a way that all of the adjacent tiles have the same color on the meeting quarter. (Note that this question is different from the way we originally posed it: Can a given set of tiles tile any *finite* rectangular region? Interestingly, the two problems are identical in the sense that the answer to one version is yes if and only if it is yes for the other version.) Call a tiling of the plane periodic if one can identify a $k \times k$ square such that the entire tiling is made by repeating this $k \times k$ square tile. Otherwise, call it *aperiodic*. Consider the question: Is there a (finite) set of unit tiles that can tile the plane, but only aperiodically? The answer is yes and it can be shown from the total undecidability of the tiling problem. Suppose the answer is no. Then, for any given set of tiles, the entire plane can be tiled if and only if the plane can be tiled periodically. But a periodic tiling can be found, if one exists, by trying to tile a $k \times k$ region for successively increasing values of k. This process will eventually succeed (in a finite number of steps) if the tiling exists. This will make the tiling problem partially decidable, which contradicts the total undecidability of the problem. This means that the assumption that the entire plane can be tiled if and only if some $k \times k$ region can be tiled is wrong. Thus, there exists a (finite) set of tiles that can tile the entire plane, but only aperiodically.

6.4 Formal Languages and Grammars

The universe of strings is probably the most general medium for the representation of information. This section is concerned with sets of strings called *languages* and certain systems generating these languages such as *grammars*. Every programming language including Pascal, C, or Fortran can be precisely described by a grammar. Moreover, the grammar allows us to write a computer program (called the lexical analyzer in a compiler) to determine if a piece of code is syntactically correct in the programming language. Would not it be nice to also have such a grammar for English and a corresponding computer program which can tell us what English sentences are grammatically correct?* The focus of this brief exposition is the formalism and mathematical properties of various languages and grammars. Many of the concepts have applications in domains including natural language and computer language processing, string matching, etc. We begin with some standard definitions about languages.

Definition 6.1 An *alphabet* is a finite nonempty set of *symbols*, which are assumed to be *indivisible*.

For example, the alphabet for English consists of 26 uppercase letters $A, B, \ldots, Z$ and 26 lowercase letters $a, b, \ldots, z$. We usually use the symbol Σ to denote an alphabet.

Definition 6.2 A *string* over an alphabet Σ is a finite sequence of symbols of Σ.

The number of symbols in a string x is called its *length*, denoted $|x|$. It is convenient to introduce an empty string, denoted ϵ, which contains no symbols at all. The length of ϵ is 0.

Definition 6.3 Let $x = a_1 a_2 \cdots a_n$ and $y = b_1 b_2 \cdots b_m$ be two strings. The *concatenation* of x and y, denoted xy, is the string $a_1 a_2 \cdots a_n b_1 b_2 \cdots b_m$.

---

*Actually, English and the other natural languages have grammars; but these grammars are not precise enough to tell apart the correct and incorrect sentences with 100% accuracy. The main problem is that *there is no universal agreement* on what are grammatically correct English sentences.

Thus, for any string x, $\epsilon x = x\epsilon = x$. For any string x and integer $n \geq 0$, we use x^n to denote the string formed by sequentially concatenating n copies of x.

Definition 6.4 The set of all strings over an alphabet Σ is denoted Σ^* and the set of all nonempty strings over Σ is denoted Σ^+. The empty set of strings is denoted $\emptyset$.

Definition 6.5 For any alphabet Σ, a *language* over Σ is a set of strings over Σ. The members of a language are also called the *words* of the language.

Example 6.1

The sets $L_1 = \{01, 11, 0110\}$ and $L_2 = \{0^n 1^n \mid n \geq 0\}$ are two languages over the binary alphabet $\{0, 1\}$. The string 01 is in both languages, whereas 11 is in L_1 but not in L_2.

Because languages are just sets, standard set operations such as union, intersection, and complementation apply to languages. It is useful to introduce two more operations for languages: *concatenation* and *Kleene closure*.

Definition 6.6 Let L_1 and L_2 be two languages over Σ. The concatenation of L_1 and L_2, denoted $L_1 L_2$, is the language $\{xy \mid x \in L_1, y \in L_2\}$.

Definition 6.7 Let L be a language over Σ. Define $L^0 = \{\epsilon\}$ and $L^i = LL^{i-1}$ for $i \geq 1$. The Kleene closure of L, denoted L^*, is the language

$$L^* = \bigcup_{i \geq 0} L^i$$

and the *positive closure* of L, denoted L^+, is the language

$$L^+ = \bigcup_{i \geq 1} L^i$$

In other words, the Kleene closure of language L consists of all strings that can be formed by concatenating some words from L. For example, if $L = \{0, 01\}$, then $LL = \{00, 001, 010, 0101\}$ and L^* includes all binary strings in which every 1 is preceded by a 0. L^+ is the same as L^* except it excludes ϵ in this case. Note that, for any language L, L^* always contains ϵ and L^+ contains ϵ if and only if L does. Also note that Σ^* is in fact the Kleene closure of the alphabet Σ when viewed as a language of words of length 1, and Σ^+ is just the positive closure of Σ.

6.4.1 Representation of Languages

In general, a language over an alphabet Σ is a subset of Σ^*. How can we describe a language rigorously so that we know if a given string belongs to the language or not? As shown in the preceding paragraphs, a finite language such as L_1 in Example 6.1 can be explicitly defined by enumerating its elements, and a simple infinite language such as L_2 in the same example can be described using a rule characterizing all members of L_2. It is possible to define some more systematic methods to represent a wide class of languages. In the following, we will introduce three such methods: regular expressions, pattern systems, and grammars. The languages that can be described by this kind of system are often referred to as *formal languages*.

Definition 6.8 Let Σ be an alphabet. The *regular expressions* over Σ and the languages they represent are defined inductively as follows.

1. The symbol $\emptyset$ is a regular expression, denoting the empty set.
2. The symbol ϵ is a regular expression, denoting the set $\{\epsilon\}$.

3. For each $a \in \Sigma$, a is a regular expression, denoting the set $\{a\}$.
4. If r and s are regular expressions denoting the languages R and S, then $(r + s)$, (rs), and (r^*) are regular expressions that denote the sets $R \cup S$, RS, and R^*, respectively.

For example, $((0(0 + 1)^*) + ((0 + 1)^*0))$ is a regular expression over $\{0, 1\}$, and it represents the language consisting of all binary strings that begin or end with a 0. Because the set operations union and concatenation are both associative, many parentheses can be omitted from regular expressions if we assume that Kleene closure has higher precedence than concatenation and concatenation has higher precedence than union. For example, the preceding regular expression can be abbreviated as $0(0 + 1)^* + (0 + 1)^*0$. We will also abbreviate the expression rr^* as r^+. Let us look at a few more examples of regular expressions and the languages they represent.

Example 6.2

The expression $0(0 + 1)^*1$ represents the set of all strings that begin with a 0 and end with a 1.

Example 6.3

The expression $0 + 1 + 0(0 + 1)^*0 + 1(0 + 1)^*1$ represents the set of all nonempty binary strings that begin and end with the same bit.

Example 6.4

The expressions 0^*, 0^*10^*, and $0^*10^*10^*$ represent the languages consisting of strings that contain no 1, exactly one 1, and exactly two 1s, respectively.

Example 6.5

The expressions $(0 + 1)^*1(0 + 1)^*1(0 + 1)^*$, $(0 + 1)^*10^*1(0 + 1)^*$, $0^*10^*1(0 + 1)^*$, and $(0 + 1)^*10^*10^*$ all represent the same set of strings that contain at least two 1s.

For any regular expression r, the language represented by r is denoted as $L(r)$. Two regular expressions representing the same language are called *equivalent*. It is possible to introduce some identities to algebraically manipulate regular expressions to construct equivalent expressions, by tailoring the set identities for the operations union, concatenation, and Kleene closure to regular expressions. For more details, see Salomaa [1966]. For example, it is easy to prove that the expressions $r(s + t)$ and $rs + rt$ are equivalent and $(r^*)^*$ is equivalent to r^*.

Example 6.6

Let us construct a regular expression for the set of all strings that contain no consecutive 0s. A string in this set may begin and end with a sequence of 1s. Because there are no consecutive 0s, every 0 that is not the last symbol of the string must be followed by at least a 1. This gives us the expression $1^*(01^+)^*1^*(\epsilon + 0)$. It is not hard to see that the second 1^* is redundant, and thus the expression can in fact be simplified to $1^*(01^+)^*(\epsilon + 0)$.

Regular expressions were first introduced in Kleene [1956] for studying the properties of neural nets. The preceding examples illustrate that regular expressions often give very clear and concise representations of languages. Unfortunately, not every language can be represented by regular expressions. For example, it will become clear that there is no regular expression for the language $\{0^n1^n \mid n \geq 1\}$. The languages represented by regular expressions are called the **regular languages**. Later, we will see that regular languages are exactly the class of languages generated by the so-called **right-linear grammars**. This connection allows one to prove some interesting mathematical properties about regular languages as well as to design an efficient algorithm to determine whether a given string belongs to the language represented by a given **regular expression**.

Another way of representing languages is to use *pattern systems* [Angluin 1980, Jiang et al. 1995].

Definition 6.9   A *pattern system* is a triple (Σ, V, p), where Σ is the alphabet, V is the set of *variables* with $\Sigma \cap V = \emptyset$, and p is a string over $\Sigma \cup V$ called the *pattern*.

An example pattern system is $(\{0, 1\}, \{v_1, v_2\}, v_1 v_1 0 v_2)$.

Definition 6.10   The language generated by a pattern system (Σ, V, p) consists of all strings over Σ that can be obtained from p by replacing each variable in p with a string over Σ.

For example, the language generated by $(\{0, 1\}, \{v_1, v_2\}, v_1 v_1 0 v_2)$ contains words $0, 00, 01, 000, 001,$ $010, 011, 110,$ etc., but does not contain strings, $1, 10, 11, 100, 101,$ etc. The pattern system $(\{0, 1\}, \{v_1\},$ $v_1 v_1)$ generates the set of all strings, which is the concatenation of two equal substrings, that is, the set $\{xx \mid x \in \{0, 1\}^*\}$. The languages generated by pattern systems are called the *pattern languages*.

Regular languages and pattern languages are really different. One can prove that the pattern language $\{xx \mid x \in \{0, 1\}^*\}$ is not a regular language and the set represented by the regular expression 0^*1^* is not a pattern language. Although it is easy to write an algorithm to decide if a string is in the language generated by a given pattern system, such an algorithm most likely would have to be very inefficient [Angluin 1980].

Perhaps the most useful and general system for representing languages is based on grammars, which are extensions of the pattern systems.

Definition 6.11   A grammar is a quadruple (Σ, N, S, P), where:

1. Σ is a finite nonempty set called the alphabet. The elements of Σ are called the *terminals*.
2. N is a finite nonempty set disjoint from Σ. The elements of N are called the *nonterminals* or *variables*.
3. $S \in N$ is a distinguished nonterminal called the *start symbol*.
4. P is a finite set of *productions* (or *rules*) of the form

$$\alpha \rightarrow \beta$$

where $\alpha \in (\Sigma \cup N)^* N (\Sigma \cup N)^*$ and $\beta \in (\Sigma \cup N)^*$, that is, α is a string of terminals and nonterminals containing at least one nonterminal and β is a string of terminals and nonterminals.

Example 6.7

Let $G_1 = (\{0, 1\}, \{S, T, O, I\}, S, P)$, where P contains the following productions:

$$S \rightarrow OT$$
$$S \rightarrow OI$$
$$T \rightarrow SI$$
$$O \rightarrow 0$$
$$I \rightarrow 1$$

As we shall see, the grammar G_1 can be used to describe the set $\{0^n 1^n \mid n \geq 1\}$.

Example 6.8

Let $G_2 = (\{0, 1, 2\}, \{S, A\}, S, P)$, where P contains the following productions.

$$S \rightarrow 0SA2$$
$$S \rightarrow \epsilon$$
$$2A \rightarrow A2$$
$$0A \rightarrow 01$$
$$1A \rightarrow 11$$

This grammar G_2 can be used to describe the set $\{0^n 1^n 2^n \geq n \geq 0\}$.

Example 6.9

To construct a grammar G_3 to describe English sentences, the alphabet Σ contains all words in English. N would contain nonterminals, which correspond to the structural components in an English sentence, for example, ⟨sentence⟩, ⟨subject⟩, ⟨predicate⟩, ⟨noun⟩, ⟨verb⟩, ⟨article⟩, etc. The start symbol would be ⟨sentence⟩. Some typical productions are

$$\langle \text{sentence} \rangle \rightarrow \langle \text{subject} \rangle \langle \text{predicate} \rangle$$

$$\langle \text{subject} \rangle \rightarrow \langle \text{noun} \rangle$$

$$\langle \text{predicate} \rangle \rightarrow \langle \text{verb} \rangle \langle \text{article} \rangle \langle \text{noun} \rangle$$

$$\langle \text{noun} \rangle \rightarrow \text{mary}$$

$$\langle \text{noun} \rangle \rightarrow \text{algorithm}$$

$$\langle \text{verb} \rangle \rightarrow \text{wrote}$$

$$\langle \text{article} \rangle \rightarrow \text{an}$$

The rule ⟨sentence⟩ → ⟨subject⟩⟨predicate⟩ follows from the fact that a sentence consists of a subject phrase and a predicate phrase. The rules ⟨noun⟩ → mary and ⟨noun⟩ → algorithm mean that both mary and algorithms are possible nouns.

To explain how a grammar represents a language, we need the following concepts.

Definition 6.12 Let (Σ, N, S, P) be a grammar. A *sentential form* of G is any string of terminals and nonterminals, that is, a string over $\Sigma \cup N$.

Definition 6.13 Let (Σ, N, S, P) be a grammar and γ_1 and γ_2 two sentential forms of G. We say that γ_1 *directly derives* γ_2, denoted $\gamma_1 \Rightarrow \gamma_2$, if $\gamma_1 = \sigma \alpha \tau$, $\gamma_2 = \sigma \beta \tau$, and $\alpha \rightarrow \beta$ is a production in P.

For example, the sentential form $00S11$ directly derives the sentential form $00OT11$ in grammar G_1, and $A2A2$ directly derives $AA22$ in grammar G_2.

Definition 6.14 Let γ_1 and γ_2 be two sentential forms of a grammar G. We say that γ_1 *derives* γ_2, denoted $\gamma_1 \Rightarrow^* \gamma_2$, if there exists a sequence of (zero or more) sentential forms $\sigma_1, \ldots, \sigma_n$ such that

$$\gamma_1 \Rightarrow \sigma_1 \Rightarrow \cdots \Rightarrow \sigma_n \Rightarrow \gamma_2$$

The sequence $\gamma_1 \Rightarrow \sigma_1 \Rightarrow \cdots \Rightarrow \sigma_n \Rightarrow \gamma_2$ is called a derivation from γ_1 to γ_2.

For example, in grammar G_1, $S \Rightarrow^* 0011$ because

$$S \Rightarrow \underline{O}T \Rightarrow 0\underline{T} \Rightarrow 0S\underline{I} \Rightarrow 0\underline{S}1 \Rightarrow 0\underline{O}I1 \Rightarrow 00\underline{I}1 \Rightarrow 0011$$

and in grammar G_2, $S \Rightarrow^* 001122$ because

$$S \Rightarrow 0\underline{S}A2 \Rightarrow 00\underline{S}A2A2 \Rightarrow 00\underline{A}2A2 \Rightarrow 0012\underline{A}2 \Rightarrow 0011\underline{A}22 \Rightarrow 001122$$

Here the left-hand side of the relevant production in each derivation step is underlined for clarity.

Definition 6.15 Let (Σ, N, S, P) be a grammar. The language generated by G, denoted $L(G)$, is defined as

$$L(G) = \{x \mid x \in \Sigma^*, S \Rightarrow^* x\}$$

The words in $L(G)$ are also called the *sentences* of $L(G)$.

Clearly, $L(G_1)$ contains all strings of the form $0^n 1^n$, $n \geq 1$, and $L(G_2)$ contains all strings of the form $0^n 1^n 2^n$, $n \geq 0$. Although only a partial definition of G_3 is given, we know that $L(G_3)$ contains sentences such as "mary wrote an algorithm" and "algorithm wrote an algorithm" but does not contain sentences such as "an wrote algorithm."

The introduction of formal grammars dates back to the 1940s [Post 1943], although the study of rigorous description of languages by grammars did not begin until the 1950s [Chomsky 1956]. In the next subsection, we consider various restrictions on the form of productions in a grammar and see how these restrictions can affect the power of a grammar in representing languages. In particular, we will know that regular languages and pattern languages can all be generated by grammars under different restrictions.

6.4.2 Hierarchy of Grammars

Grammars can be divided into four classes by gradually increasing the restrictions on the form of the productions. Such a classification is due to Chomsky [1956, 1963] and is called the *Chomsky hierarchy*.

Definition 6.16 Let $G = (\Sigma, N, S, P)$ be a grammar.

1. G is also called a *type-0 grammar* or an *unrestricted grammar*.
2. G is *type-1* or **context sensitive** if each production $\alpha \to \beta$ in P either has the form $S \to \epsilon$ or satisfies $|\alpha| \leq |\beta|$.
3. G is *type-2* or **context free** if each production $\alpha \to \beta$ in P satisfies $|\alpha| = 1$, that is, α is a nonterminal.
4. G is *type-3* or right linear or regular if each production has one of the following three forms:

$$A \to aB, \qquad A \to a, \qquad A \to \epsilon$$

where A and B are nonterminals and a is a terminal.

The language generated by a type-i is called a type-i language, $i = 0, 1, 2, 3$. A type-1 language is also called a **context-sensitive language** and a type-2 language is also called a **context-free language**. It turns out that every type-3 language is in fact a regular language, that is, it is represented by some regular expression, and vice versa. See the next section for the proof of the equivalence of type-3 (right-linear) grammars and regular expressions.

The grammars G_1 and G_3 given in the last subsection are context free and the grammar G_2 is context sensitive. Now we give some examples of unrestricted and right-linear grammars.

Example 6.10

Let $G_4 = (\{0, 1\}, \{S, A, O, I, T\}, S, P)$, where P contains

$$S \to AT$$
$$A \to 0AO \qquad A \to 1AI$$
$$0O \to 0O \qquad O1 \to 1O$$
$$I0 \to 0I \qquad I1 \to 1I$$
$$OT \to 0T \qquad IT \to 1T$$
$$A \to \epsilon \qquad T \to \epsilon$$

Then G_4 generates the set $\{xx \mid x \in \{0, 1\}^*\}$. For example, we can derive the word 0101 from S as follows:

$$S \Rightarrow \underline{A}T \Rightarrow 0\underline{A}OT \Rightarrow 01\underline{A}IOT \Rightarrow 01\underline{I}\,\underline{OT} \Rightarrow 01\underline{I0}T \Rightarrow 010\underline{I}T \Rightarrow 0101\underline{T} \Rightarrow 0101$$

Example 6.11

We give a right-linear grammar G_5 to generate the language represented by the regular expression in Example 6.3, that is, the set of all nonempty binary strings beginning and ending with the same bit. Let $G_5 = (\{0, 1\}, \{S, O, I\}, S, P)$, where P contains

$$
\begin{aligned}
S &\to 0O & S &\to 1I \\
S &\to 0 & S &\to 1 \\
O &\to 0O & O &\to 1O \\
I &\to 0I & I &\to 1I \\
O &\to 0 & I &\to 1
\end{aligned}
$$

The following theorem is due to Chomsky [1956, 1963].

Theorem 6.2 *For each $i = 0, 1, 2$, the class of type-i languages properly contains the class of type-$(i + 1)$ languages.*

For example, one can prove by using a technique called *pumping* that the set $\{0^n 1^n \mid n \geq 1\}$ is context free but not regular, and the sets $\{0^n 1^n 2^n \mid n \geq 0\}$ and $\{xx \mid x \in \{0, 1\}^*\}$ are context sensitive but not context free [Hopcroft and Ullman 1979]. It is, however, a bit involved to construct a language that is of type-0 but not context sensitive. See, for example, Hopcroft and Ullman [1979] for such a language.

The four classes of languages in the Chomsky hierarchy also have been completely characterized in terms of Turing machines and their restricted versions. We have already defined a Turing machine in Section 6.2. Many restricted versions of it will be defined in the next section. It is known that type-0 languages are exactly those recognized by Turing machines, context-sensitive languages are those recognized by Turing machines running in linear space, context-free languages are those recognized by Turing machines whose worktapes operate as pushdown stacks [called **pushdown automata** (PDA)], and regular languages are those recognized by Turing machines without any worktapes (called **finite-state machine** or **finite automata**) [Hopcroft and Ullman 1979].

Remark 6.1 Recall our definition of a Turing machine and the function it computes from Section 6.2. In the preceding paragraph, we refer to *a language recognized* by a Turing machine. These are two seemingly different ideas, but they are essentially the same. The reason is that the function f, which maps the set of strings over a finite alphabet to $\{0, 1\}$, corresponds in a natural way to the language L_f over Σ defined as: $L_f = \{x \mid f(x) = 1\}$. Instead of saying that a Turing machine computes the function f, we say equivalently that it recognizes L_f.

Because $\{xx \mid x \in \{0, 1\}^*\}$ is a pattern language, the preceding discussion implies that the class of pattern languages is not contained in the class of context-free languages. The next theorem shows that the class of pattern languages is contained in the class of context-sensitive languages.

Theorem 6.3 *Every pattern language is context sensitive.*

The theorem follows from the fact that every pattern language is recognized by a Turing machine in linear space [Angluin 1980] and linear space-bounded Turing machines recognize exactly context-sensitive languages. To show the basic idea involved, let us construct a context-sensitive grammar for the pattern language $\{xx \mid x \in \{0, 1\}^*\}$. The grammar G_4 given in Example 6.10 for this language is almost context-sensitive. We just have to get rid of the two ϵ-productions: $A \to \epsilon$ and $T \to \epsilon$. A careful modification of G_4 results in the following grammar $G_6 = (\{0, 1\}, \{S, A_0, A_1, O, I, T_0, T_1\}, S, P)$, where P contains

$$S \to \epsilon$$

$$S \to A_0 T_0 \qquad S \to A_1 T_1$$

$$A_0 \to 0 A_0 O \qquad A_0 \to 1 A_0 I$$

$$A_1 \to 0 A_1 O \qquad A_1 \to 1 A_1 I$$

$$A_0 \to 0 \qquad A_1 \to 1$$

$$O0 \to 0O \qquad O1 \to 1O$$

$$I0 \to 0I \qquad I1 \to 1I$$

$$O T_0 \to 0 T_0 \qquad I T_0 \to 1 T_0$$

$$O T_1 \to 0 T_1 \qquad I T_1 \to 1 T_1$$

$$T_0 \to O \qquad T_1 \to 1,$$

which is context sensitive and generates $\{xx \mid x \in \{0, 1\}^*\}$. For example, we can derive 011011 as

$$\Rightarrow \underline{A_1} T_1 \Rightarrow 0 \underline{A_1} O T_1 \Rightarrow 01 \underline{A_1} I O T_1$$

$$\Rightarrow 011 \underline{I O} T_1 \Rightarrow 011 \underline{I 0} T_1 \Rightarrow 0110 \underline{I} T_1 \Rightarrow 01101 \underline{T_1} \Rightarrow 011011$$

For a class of languages, we are often interested in the so-called *closure properties* of the class.

Definition 6.17   A class of languages (e.g., regular languages) is said to be *closed* under a particular operation (e.g., union, intersection, complementation, concatenation, Kleene closure) if each application of the operation on language(s) of the class results in a language of the class.

These properties are often useful in constructing new languages from existing languages as well as proving many theoretical properties of languages and grammars. The closure properties of the four types of languages in the Chomsky hierarchy are now summarized [Harrison 1978, Hopcroft and Ullman 1979, Gurari 1989].

Theorem 6.4

1. *The class of type-0 languages is closed under union, intersection, concatenation, and Kleene closure but not under complementation.*
2. *The class of context-free languages is closed under union, concatenation, and Kleene closure but not under intersection or complementation.*
3. *The classes of context-sensitive and regular languages are closed under all five of the operations.*

For example, let $L_1 = \{0^m 1^n 2^p \mid m = n \text{ or } n = p\}$, $L_2 = \{0^m 1^n 2^p \mid m = n\}$, and $L_3 = \{0^m 1^n 2^p \mid n = p\}$. It is easy to see that all three are context-free languages. (In fact, $L_1 = L_2 \cup L_3$.) However, intersecting L_2 with L_3 gives the set $\{0^m 1^n 2^p \mid m = n = p\}$, which is not context free.

We will look at context-free grammars more closely in the next subsection and introduce the concept of **parsing** and ambiguity.

6.4.3   Context-Free Grammars and Parsing

From a practical point of view, for each grammar $G = (\Sigma, N, S, P)$ representing some language, the following two problems are important:

1. (Membership) Given a string over Σ, does it belong to $L(G)$?
2. (Parsing) Given a string in $L(G)$, how can it be derived from S?

The importance of the membership problem is quite obvious: given an English sentence or computer program we wish to know if it is grammatically correct or has the right format. Parsing is important because a derivation usually allows us to interpret the meaning of the string. For example, in the case of a Pascal program, a derivation of the program in Pascal grammar tells the compiler how the program should be executed. The following theorem illustrates the decidability of the membership problem for the four classes of grammars in the Chomsky hierarchy. The proofs can be found in Chomsky [1963], Harrison [1978], and Hopcroft and Ullman [1979].

Theorem 6.5 *The membership problem for type-0 grammars is undecidable in general and is decidable for any context-sensitive grammar (and thus for any context-free or right-linear grammars).*

Because context-free grammars play a very important role in describing computer programming languages, we discuss the membership and parsing problems for context-free grammars in more detail. First, let us look at another example of context-free grammar. For convenience, let us abbreviate a set of productions with the same left-hand side nonterminal

$$A \to \alpha_1, \ldots, A \to \alpha_n$$

as

$$A \to \alpha_1 \mid \cdots \mid \alpha_n$$

Example 6.12

We construct a context-free grammar for the set of all valid Pascal real values. In general, a real constant in Pascal has one of the following forms:

$$m.n, \qquad m\mathbf{e}q, \qquad m.n\mathbf{e}q,$$

where m and q are signed or unsigned integers and n is an unsigned integer. Let $\Sigma = \{0, 1, 2, 3, 4, 5, 6, 7, 8, 9, \mathbf{e}, +, -, .\}$, $N = \{S, M, N, D\}$, and the set P of the productions contain

$$S \to M.N \mid M\mathbf{e}M \mid M.N\mathbf{e}M$$
$$M \to N \mid +N \mid -N$$
$$N \to DN \mid D$$
$$D \to 0 \mid 1 \mid 2 \mid 3 \mid 4 \mid 5 \mid 7 \mid 8 \mid 9$$

Then the grammar generates all valid Pascal real values (including some absurd ones like $001.200\mathbf{e}000$). The value $12.3\mathbf{e} - 4$ can be derived as

$$S \Rightarrow \underline{M}.NeM \Rightarrow \underline{N}.NeM \Rightarrow \underline{D}N.NeM \Rightarrow 1\underline{N}.NeM \Rightarrow 1\underline{D}.NeM$$

$$\Rightarrow 12.\underline{N}eM \Rightarrow 12.\underline{D}eM \Rightarrow 12.3\mathbf{e}\underline{M} \Rightarrow 12.3\mathbf{e} - \underline{N} \Rightarrow 12.3\mathbf{e} - \underline{D} \Rightarrow 12.3\mathbf{e} - 4$$

Perhaps the most natural representation of derivations for a context-free grammar is *a derivation tree* or *a parse tree*. Each *internal node* of such a tree corresponds to a nonterminal and each *leaf* corresponds to a terminal. If A is an internal node with children $B_1, \ldots, B_n$ ordered from left to right, then $A \to B_1 \cdots B_n$ must be a production. The concatenation of all leaves from left to right yields the string being derived. For example, the derivation tree corresponding to the preceding derivation of $12.3\mathbf{e} - 4$ is given in Figure 6.3. Such a tree also makes possible the extraction of the parts 12, 3, and -4, which are useful in the storage of the real value in a computer memory.

Definition 6.18 A context-free grammar G is **ambiguous** if there is a string $x \in L(G)$, which has two distinct derivation trees. Otherwise G is *unambiguous*.

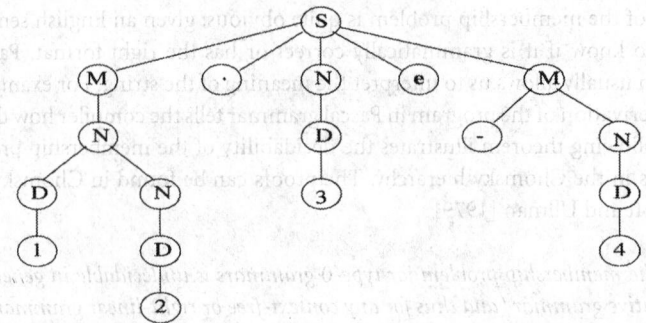

FIGURE 6.3 The derivation tree for 12.3e − 4.

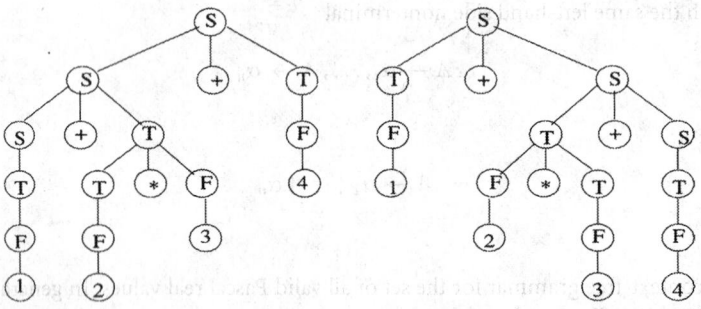

FIGURE 6.4 Different derivation trees for the expression $1 + 2 * 3 + 4$.

Unambiguity is a very desirable property to have as it allows a unique interpretation of each sentence in the language. It is not hard to see that the preceding grammar for Pascal real values and the grammar G_1 defined in Example 6.7 are all unambiguous. The following example shows an ambiguous grammar.

Example 6.13

Consider a grammar G_7 for all valid arithmetic expressions that are composed of unsigned positive integers and symbols $+, *, (,)$. For convenience, let us use the symbol n to denote any unsigned positive integer. This grammar has the productions

$$S \rightarrow T + S \mid S + T \mid T$$
$$T \rightarrow F * T \mid T * F \mid F$$
$$F \rightarrow n \mid (S)$$

Two possible different derivation trees for the expression $1 + 2 * 3 + 4$ are shown in Figure 6.4. Thus, G_7 is ambiguous. The left tree means that the first addition should be done before the second addition and the right tree says the opposite.

Although in the preceding example different derivations/interpretations of any expression always result in the same value because the operations addition and multiplication are associative, there are situations where the difference in the derivation can affect the final outcome. Actually, the grammar G_7 can be made unambiguous by removing some (redundant) productions, for example, $S \rightarrow T + S$ and $T \rightarrow F * T$. This corresponds to the convention that a sequence of consecutive additions (or multiplications) is always evaluated from left to right and will not change the language generated by G_7. It is worth noting that there are context-free languages which cannot be generated by any unambiguous context-free grammar [Hopcroft and Ullman 1979]. Such languages are said to be *inherently ambiguous*. An example of inherently ambiguous languages is the set

$$\{0^m 1^m 2^n 3^n \mid m, n > 0\} \cup \{0^m 1^n 2^m 3^n \mid m, n > 0\}$$

We end this section by presenting an efficient algorithm for the membership problem for context-free grammars. The algorithm is due to Cocke, Younger, and Kasami [Hopcroft and Ullman 1979] and is often called the CYK algorithm. Let $G = (\Sigma, N, S, P)$ be a context-free grammar. For simplicity, let us assume that G does not generate the empty string ϵ and that G is in the so-called **Chomsky normal form** [Chomsky 1963], that is, every production of G is either in the form $A \rightarrow BC$ where B and C are nonterminals, or in the form $A \rightarrow a$ where a is a terminal. An example of such a grammar is G_1 given in Example 6.7. This is not a restrictive assumption because there is a simple algorithm which can convert every context-free grammar that does not generate ϵ into one in the Chomsky normal form.

Suppose that $x = a_1 \cdots a_n$ is a string of n terminals. The basic idea of the CYK algorithm, which decides if $x \in L(G)$, is *dynamic programming*. For each pair i, j, where $1 \le i \le j \le n$, define a set $X_{i,j} \subseteq N$ as

$$X_{i,j} = \{A \mid A \Rightarrow^* a_i \cdots a_j\}$$

Thus, $x \in L(G)$ if and only if $S \in X_{1,n}$. The sets $X_{i,j}$ can be computed inductively in the ascending order of $j - i$. It is easy to figure out $X_{i,i}$ for each i because $X_{i,i} = \{A \mid A \rightarrow a_i \in P\}$. Suppose that we have computed all $X_{i,j}$ where $j - i < d$ for some $d > 0$. To compute a set $X_{i,j}$, where $j - i = d$, we just have to find all of the nonterminals A such that there exist some nonterminals B and C satisfying $A \rightarrow BC \in P$ and for some k, $i \le k < j$, $B \in X_{i,k}$, and $C \in X_{k+1,j}$. A rigorous description of the algorithm in a Pascal style pseudocode is given as follows.

Algorithm CYK($x = a_1 \cdots a_n$):

1.   for $i \leftarrow 1$ to n do
2. $X_{i,i} \leftarrow \{A \mid A \rightarrow a_i \in P\}$
3.   for $d \leftarrow 1$ to $n - 1$ do
4.   for $i \leftarrow 1$ to $n - d$ do
5. $X_{i,i+d} \leftarrow \emptyset$
6.   for $t \leftarrow 0$ to $d - 1$ do
7. $X_{i,i+d} \leftarrow X_{i,i+d} \cup \{A \mid A \rightarrow BC \in P$ for some $B \in X_{i,i+t}$ and $C \in X_{i+t+1,i+d}\}$

Table 6.2 shows the sets $X_{i,j}$ for the grammar G_1 and the string $x = 000111$. It just so happens that every $X_{i,j}$ is either empty or a singleton. The computation proceeds from the main diagonal toward the upper-right corner.

TABLE 6.2 An Example Execution of the CYK Algorithm

		0	0	0	1	1	1
		\multicolumn{6}{c}{$j \rightarrow$}					
		1	2	3	4	5	6
	1	O					S
	2		O			S	T
i	3			O	S	T	
$\downarrow$	4				I		
	5					I	
	6						I

6.5   Computational Models

In this section, we will present many restricted versions of Turing machines and address the question of what kinds of problems they can solve. Such a classification is a central goal of computation theory. We have already classified problems broadly into (totally) decidable, partially decidable, and totally undecidable. Because the decidable problems are the ones of most practical interest, we can consider further classification of decidable problems by placing two types of restrictions on a Turing machine. The first one is to restrict its structure. This way we obtain many machines of which a finite automaton and a pushdown automaton are the most important. The other way to restrict a Turing machine is to bound the amount of resources it uses, such as the number of time steps or the number of tape cells it can use. The resulting machines form the basis for *complexity theory*.

6.5.1   Finite Automata

The finite automaton (in its deterministic version) was first introduced by McCulloch and Pitts [1943] as a logical model for the behavior of neural systems. Rabin and Scott [1959] introduced the nondeterministic version of the finite automaton and showed the equivalence of the nondeterministic and deterministic versions. Chomsky and Miller [1958] proved that the set of languages that can be recognized by a finite automaton is precisely the regular languages introduced in Section 6.4. Kleene [1956] showed that the languages accepted by finite automata are characterized by regular expressions as defined in Section 6.4.

In addition to their original role in the study of neural nets, finite automata have enjoyed great success in many fields such as sequential circuit analysis in circuit design [Kohavi 1978], asynchronous circuits [Brzozowski and Seger 1994], lexical analysis in text processing [Lesk 1975], and compiler design. They also led to the design of more efficient algorithms. One excellent example is the development of linear-time string-matching algorithms, as described in Knuth et al. [1977]. Other applications of finite automata can be found in computational biology [Searls 1993], natural language processing, and distributed computing.

A finite automaton, as in Figure 6.5, consists of an input tape which contains a (finite) sequence of input symbols such as $aabab \cdots$, as shown in the figure, and a finite-state control. The tape is read by the one-way *read-only* input head from left to right, one symbol at a time. Each time the input head reads an input symbol, the finite control changes its state according to the symbol and the current state of the machine. When the input head reaches the right end of the input tape, if the machine is in a final state, we say that the input is accepted; if the machine is not in a final state, we say that the input is rejected. The following is the formal definition.

Definition 6.19   A *nondeterministic finite automaton* (NFA) is a quintuple $(Q, \Sigma, \delta, q_0, F)$, where:

- Q is a finite set of *states*.
- Σ is a finite set of *input symbols*.
- δ, the *state transition function*, is a mapping from $Q \times \Sigma$ to subsets of Q.
- $q_0 \in Q$ is the *initial state* of the NFA.
- $F \subseteq Q$ is the set of *final states*.

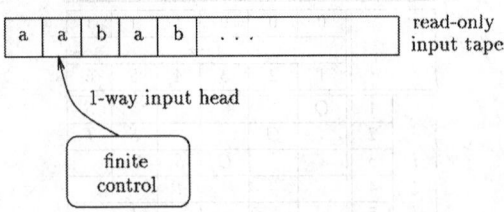

FIGURE 6.5   A finite automaton.

If δ maps $| Q | \times \Sigma$ to singleton subsets of Q, then we call such a machine a *deterministic finite automaton* (DFA).

When an automaton, M, is nondeterministic, then from the current state and input symbol, it may go to one of several different states. One may imagine that the device goes to all such states in parallel. The DFA is just a special case of the NFA; it always follows a single deterministic path. The device M *accepts* an input string x if, starting with q_0 and the read head at the first symbol of x, one of these parallel paths reaches an accepting state when the read head reaches the end of x. Otherwise, we say M *rejects* x. A language, L, is accepted by M if M accepts all of the strings in L and nothing else, and we write $L = L(M)$. We will also allow the machine to make ϵ-*transitions*, that is, changing state without advancing the read head. This allows transition functions such as $\delta(s, \epsilon) = \{s'\}$. It is easy to show that such a generalization does not add more power.

Remark 6.2 The concept of a nondeterministic automaton is rather confusing for a beginner. But there is a simple way to relate it to a concept which must be familiar to all of the readers. It is that of a solitaire game. Imagine a game like *Klondike*. The game starts with a certain arrangement of cards (the input) and there is a well-defined final position that results in success; there are also dead ends where a further move is not possible; you lose if you reach any of them. At each step, the precise rules of the game dictate how a new arrangement of cards can be reached from the current one. But the most important point is that there are many possible moves at each step. (Otherwise, the game would be no fun!) Now consider the following question: What starting positions are *winnable*? These are the starting positions for which *there is a winning move sequence*; of course, in a typical play a player may not achieve it. But that is beside the point in the definition of what starting positions are winnable. The connection between such games and a nondeterministic automaton should be clear. The multiple choices at each step are what make it *nondeterministic*. Our definition of winnable positions is similar to the concept of acceptance of a string by a nondeterministic automaton. Thus, an NFA may be viewed as a formal model to define solitaire games.

Example 6.14

We design a DFA to accept the language represented by the regular expression $0(0 + 1)^*1$ as in Example 6.2, that is, the set of all strings in $\{0, 1\}$ which begin with a 0 and end with a 1. It is usually convenient to draw our solution as in Figure 6.6. As a convention, each circle represents a state; the state a, pointed at by the initial arrow, is the initial state. The darker circle represents the final states (state c). The transition function δ is represented by the labeled edges. For example, $\delta(a, 0) = \{b\}$. When a transition is missing, for example on input 1 from a and on inputs 0 and 1 from c, it is assumed that all of these lead to an implicit nonaccepting trap state, which has transitions to itself on all inputs.

The machine in Figure 6.6 is nondeterministic because from b on input 1 the machine has two choices: stay at b or go to c.

Figure 6.7 gives an equivalent DFA, accepting the same language.

Example 6.15

The DFA in Figure 6.8 accepts the set of all strings in $\{0, 1\}^*$ with an even number of 1s. The corresponding regular expression is $(0^*10^*1)^*0^*$.

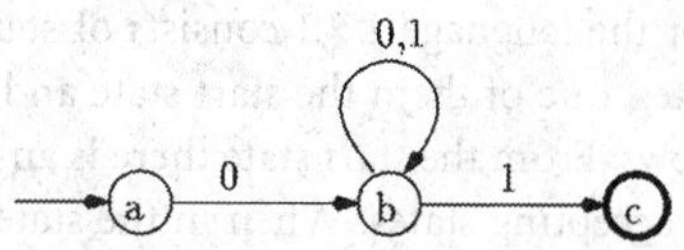

FIGURE 6.6 An NFA accepting $0(0 + 1)^*1$.

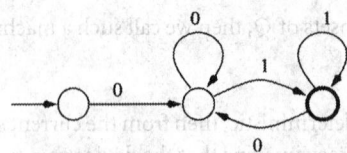

FIGURE 6.7 A DFA accepting $0(0+1)^*1$.

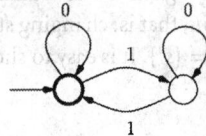

FIGURE 6.8 A DFA accepting $(0^*10^*1)^*0^*$.

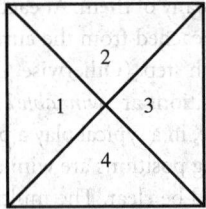

FIGURE 6.9 Numbering the quarters of a tile.

Example 6.16

As a final example, consider the special case of the tiling problem that we discussed in Section 6.2. This version of the problem is as follows: Let k be a fixed positive integer. Given a set of unit tiles, we want to know if they can tile any $k \times n$ area for all n. We show how to deal with the case $k = 1$ and leave it as an exercise to generalize our method for larger values of k. Number the quarters of each tile as in Figure 6.9. The given set of tiles will tile the area if we can find a sequence of the given tiles $T_1, T_2, \ldots, T_m$ such that (1) the third quarter of T_1 has the same color as the first quarter of T_2, and the third quarter of T_2 has the same color as the first quarter of T_3, etc., and (2) the third quarter of T_m has the same color as T_1. These conditions can be easily understood as follows. The first condition states that the tiles T_1, T_2, etc., can be placed adjacent to each other along a row in that order. The second condition implies that the whole sequence $T_1 T_2 \cdots T_m$ can be replicated any number of times. And a little thought reveals that this is all we need to answer yes on the input. But if we cannot find such a sequence, then the answer must be no. Also note that in the sequence no tile needs to be repeated and so the value of m is bounded by the number of tiles in the input. Thus, we have reduced the problem to searching a finite number of possibilities and we are done.

How is the preceding discussion related to finite automata? To see the connection, define an alphabet consisting of the unit tiles and define a language $L = \{T_1 T_2 \cdots T_m \mid T_1 T_2 \cdots T_m$ is a valid tiling, $m \geq 0\}$. We will now construct an NFA for the language L. It consists of states corresponding to *distinct* colors contained in the tiles plus two states, one of them the start state and another state called the dead state. The NFA makes transitions as follows: From the start state there is an ϵ-transition to each color state, and all states except the dead state are accepting states. When in the state corresponding to color i, suppose it receives input tile T. If the first quarter of this tile has color i, then it moves to the color of the third quarter of T; otherwise, it enters the dead state. The basic idea is to remember the only relevant piece

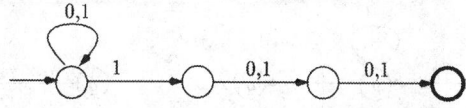

FIGURE 6.10 An NFA accepting L_3.

of information after processing some input. In this case, it is the third quarter color of the last tile seen. Having constructed this NFA, the question we are asking is if the language accepted by this NFA is infinite. There is a simple algorithm for this problem [Hopcroft and Ullman 1979].

The next three theorems show a satisfying result that all the following language classes are identical:

- The class of languages accepted by DFAs
- The class of languages accepted by NFAs
- The class of languages generated by regular expressions, as in Definition 6.8
- The class of languages generated by the right-linear, or type-3, grammars, as in Definition 6.16

Recall that this class of languages is called the *regular languages* (see Section 6.4).

Theorem 6.6 *For each NFA, there is an equivalent DFA.*

Proof   An NFA might look more powerful because it can carry out its computation in parallel with its nondeterministic branches. But because we are working with a *finite number* of states, we can simulate an NFA $M = (Q, \Sigma, \delta, q_0, F)$ by a DFA $M' = (Q', \Sigma, \delta', q_0', F')$, where

- $Q' = \{[S] : S \subseteq Q\}$.
- $q_0' = [\{q_0\}]$.
- $\delta'([S], a) = [S'] = [\cup_{q_l \in S} \delta(q_l, a)]$.
- F' is the set of all subsets of Q containing a state in F.

It can now be verified that $L(M) = L(M')$.   □

Example 6.17

Example 6.1 contains an NFA and an equivalent DFA accepting the same language. In fact, the proof provides an effective procedure for converting an NFA to a DFA. Although each NFA can be converted to an equivalent DFA, the resulting DFA might be exponentially large in terms of the number of states, as we can see from the previous procedure. This turns out to be the best thing one can do in the worst case. Consider the language: $L_k = \{x : x \in \{0, 1\}^*$ and the kth letter from the right of x is a $1\}$. An NFA of $k + 1$ states (for $k = 3$) accepting L_k is given in Figure 6.10. A counting argument shows that any DFA accepting L_k must have at least 2^k states.

Theorem 6.7 *L is generated by a right-linear grammar if it is accepted by an* NFA.

Proof   Let L be accepted by a right-linear grammar $G = (\Sigma, N, S, P)$. We design an NFA $M = (Q, \Sigma, \delta, q_0, F)$ where $Q = N \cup \{f\}$, $q_0 = S$, $F = \{f\}$. To define the δ function, we have $C \in \delta(A, b)$ if $A \to bC$. For rules $A \to b$, $\delta(A, b) = \{f\}$. Obviously, $L(M) = L(G)$.

Conversely, if L is accepted by an NFA $M = (Q, \Sigma, \delta, q_0, F)$, we define an equivalent right-linear grammar $G = (\Sigma, N, S, P)$, where $N = Q$, $S = q_0$, $q_i \to aq_j \in N$ if $q_j \in \delta(q_i, a)$, and $q_j \to \epsilon$ if $q_j \in F$. Again it is easily seen that $L(M) = L(G)$.   □

Theorem 6.8 *L is generated by a regular expression if it is accepted by an* NFA.

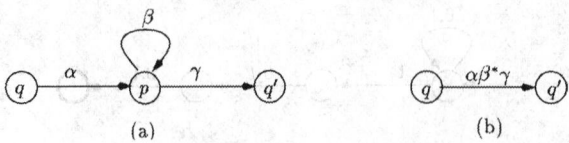

FIGURE 6.11   Converting an NFA to a regular expression.

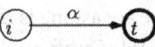

FIGURE 6.12   The reduced NFA.

Proof (Idea)   **Part 1.** We inductively convert a regular expression to an NFA which accepts the language generated by the regular expression as follows.

- Regular expression ϵ converts to $(\{q\}, \Sigma, \emptyset, q, \{q\})$.
- Regular expression $\emptyset$ converts to $(\{q\}, \Sigma, \emptyset, q, \emptyset)$.
- Regular expression a, for each $a \in \Sigma$ converts to $(\{q, f\}, \Sigma, \delta(q, a) = \{f\}, q, \{f\})$.
- If α and β are regular expressions, converting to NFAs M_α and M_β, respectively, then the regular expression $\alpha \cup \beta$ converts to an NFA M, which connects M_α and M_β in parallel: M has an initial state q_0 and all of the states and transitions of M_α and M_β; by ϵ-transitions, M goes from q_0 to the initial states of M_α and M_β.
- If α and β are regular expressions, converting to NFAs M_α and M_β, respectively, then the regular expression $\alpha\beta$ converts to NFA M, which connects M_α and M_β sequentially: M has all of the states and transitions of M_α and M_β, with M_α's initial state as M's initial state, ϵ-transition from the final states of M_α to the initial state of M_β, and M_β's final states as M's final states.
- If α is a regular expression, converting to NFA M_α, then connecting all of the final states of M_α to its initial state with ϵ-transitions gives α^+. Union of this with the NFA for ϵ gives the NFA for α^*.

Part 2. We now show how to convert an NFA to an equivalent regular expression. The idea used here is based on Brzozowski and McCluskey [1963]; see also Brzozowski and Seger [1994] and Wood [1987].

   Given an NFA M, expand it to M' by adding two extra states i, the initial state of M', and t, the only final state of M', with ϵ transitions from i to the initial state of M and from all final states of M to t. Clearly, $L(M) = L(M')$. In M', remove states other than i and t one by one as follows. To remove state p, for each triple of states q, p, q' as shown in Figure 6.11a, add the transition as shown in Figure 6.11(b). $\square$

   If p does not have a transition leading back to itself, then $\beta = \epsilon$. After we have considered all such triples, delete state p and transitions related to p. Finally, we obtain Figure 6.12 and $L(\alpha) = L(M)$.

   Apparently, DFAs cannot serve as our model for a modern computer. Many extremely simple languages cannot be accepted by DFAs. For example, $L = \{xx : x \in \{0, 1\}^*\}$ cannot be accepted by a DFA. One can prove this by counting, or using the so-called pumping lemmas; one can also prove this by arguing that x contains more information than a *finite* state machine can *remember*. We refer the interested readers to textbooks such as Hopcroft and Ullmann [1979], Gurari [1989], Wood [1987], and Floyd and Beigel [1994] for traditional approaches and to Li and Vitányi [1993] for a nontraditional approach. One can try to generalize the DFA to allow the input head to be *two way* but still read only. But such machines are not more powerful, they can be simulated by normal DFAs. The next step is apparently to add *storage* space such that our machines can *write* information in.

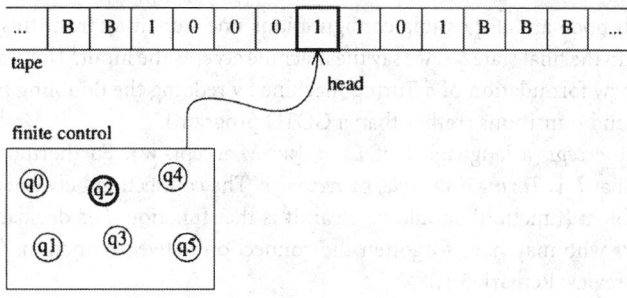

FIGURE 6.13 A Turing machine.

6.5.2 Turing Machines

In this section we will provide an alternative definition of a Turing machine to make it compatible with our definitions of a DFA, PDA, etc. This also makes it easier to define a nondeterministic Turing machine. But this formulation (at least the deterministic version) is essentially the same as the one presented in Section 6.2.

A Turing machine (TM), as in Figure 6.13, consists of a *finite control*, an infinite *tape* divided into cells, and a read/write *head* on the tape. We refer to the two directions on the tape as *left* and *right*. The finite control can be in any one of a finite set Q of states, and each tape cell can contain a 0, a 1, or a *blank B*. Time is discrete and the time instants are ordered $0, 1, 2, \ldots$ with 0 the time at which the machine starts its computation. At any time, the head is positioned over a particular cell, which it is said to *scan*. At time 0 the head is situated on a distinguished cell on the tape called the *start cell*, and the finite control is in the initial state q_0. At time 0 all cells contain Bs, except a contiguous finite sequence of cells, extending from the start cell to the right, which contain 0s and 1s. This binary sequence is called the *input*.

The device can perform the following basic operations:

1. It can write an element from the tape alphabet $\Sigma = \{0, 1, B\}$ in the cell it scans.
2. It can shift the head one cell left or right.

Also, the device executes these operations at the rate of one operation per time unit (a *step*). At the conclusion of each step, the finite control takes on a state in Q. The device operates according to a finite set P of *rules*.

The rules have format (p, s, a, q) with the meaning that if the device is in state p and s is the symbol under scan then write a if $a \in \{0, 1, B\}$ or move the head according to a if $a \in \{L, R\}$ and the finite control changes to state q. At some point, if the device gets into a special *final* state q_f, the device stops and accepts the input.

If every pair of distinct quadruples differs in the first two elements, then the device is *deterministic*. Otherwise, the device is *nondeterministic*. Not every possible combination of the first two elements has to be in the set; in this way we permit the device to perform *no* operation. In this case, we say the device *halts*. In this case, if the machine is not in a final state, we say that the machine *rejects* the input.

Definition 6.20 A Turing machine is a quintuple $M = (Q, \Sigma, P, q_0, q_f)$ where each of the components has been described previously.

Given an input, a deterministic Turing machine carries out a uniquely determined succession of operations, which may or may not terminate in a finite number of steps. If it terminates, then the nonblank symbols left on the tape are the output. Given an input, a **nondeterministic Turing machine** behaves much like an NFA. One may imagine that it carries out its computation in parallel. Such a computation may be viewed as a (possibly infinite) tree. The root of the tree is the starting configuration of the machine.

The children of each node are all possible configurations one step away from this node. If any of the branches terminates in the final state q_f, we say the machine accepts the input. The reader may want to test understanding this new formulation of a Turing machine by redoing the doubling program on a Turing machine with states and transitions (rather than a GOTO program).

A Turing machine *accepts* a language L if $L = \{w : M \text{ accepts } w\}$. Furthermore, if M halts on all inputs, then we say that L is *Turing decidable*, or *recursive*. The connection between a recursive language and a decidable problem (function) should be clear. It is that function f is decidable if and only if L_f is recursive. (Readers who may have forgotten the connection between function f and the associated language L_f should review Remark 6.1.)

Theorem 6.9   *All of the following generalizations of Turing machines can be simulated by a one-tape deterministic Turing machine defined in Definition 6.20.*

- *Larger tape alphabet Σ*
- *More work tapes*
- *More access points, read/write heads, on each tape*
- *Two- or more dimensional tapes*
- *Nondeterminism*

Although these generalizations do not make a Turing machine compute more, they do make a Turing machine more efficient and easier to program. Many more variants of Turing machines are studied and used in the literature. Of all simulations in Theorem 6.9, the last one needs some comments. A nondeterministic computation branches like a tree. When simulating such a computation for n steps, the obvious thing for a deterministic Turing machine to do is to try all possibilities; thus, this requires up to c^n steps, where c is the maximum number of nondeterministic choices at each step.

Example 6.18

A DFA is an extremely simple Turing machine. It just reads the input symbols from left to right. Turing machines naturally accept more languages than DFAs can. For example, a Turing machine can accept $L = \{xx : x \in \{0, 1\}^*\}$ as follows:

- Find the middle point first: it is trivial by using two heads; with one head, one can mark one symbol at the left and then mark another on the right, and go back and forth to eventually find the middle point.
- Match the two parts: with two heads, this is again trivial; with one head, one can again use the marking method matching a pair of symbols each round; if the two parts match, accept the input by entering q_f.

There are types of storage media other than a tape:

- A *pushdown store* is a semi-infinite work tape with one head such that each time the head moves to the left, it erases the symbol scanned previously; this is a last-in first-out storage.
- A *queue* is a semi-infinite work tape with two heads that move only to the right, the leading head is write-only and the trailing head is read-only; this is a first-in first-out storage.
- A *counter* is a pushdown store with a single-letter alphabet (except its one end, which holds a special marker symbol). Thus, a counter can store a nonnegative integer and can perform three operations.

A queue machine can simulate a normal Turing machine, but the other two types of machines are not powerful enough to simulate a Turing machine.

Example 6.19

When the Turing machine tape is replaced by a pushdown store, the machine is called a *pushdown automaton*. Pushdown automata have been thoroughly studied because they accept the class of context-free languages defined in Section 6.4. More precisely, it can be shown that if L is a context-free language, then it is accepted by a PDA, and if L is accepted by a PDA, then there is a CFG generating L. Various types of PDAs have fundamental applications in compiler design.

The PDA is more restricted than a Turing machine. For example, $L = \{xx : x \in \{0, 1\}^*\}$ cannot be accepted by a PDA, but it can be accepted by a Turing machine as in Example 6.18. But a PDA is more powerful than a DFA. For example, a PDA can accept the language $L' = \{0^k 1^k : k \geq 0\}$ easily. It can read the 0s and push them into the pushdown store; then, after it finishes the 0s, each time the PDA reads a 1, it removes a 0 from the pushdown store; at the end, it accepts if the pushdown store is empty (the number of 0s matches that of 1s). But a DFA cannot accept L', because after it has read all of the 0s, it cannot remember k when k has higher information content than the DFA's finite control.

Two pushdown stores can be used to simulate a tape easily. For comparisons of powers of pushdown stores, queues, counters, and tapes, see van Emde Boas [1990] and Li and Vitányi [1993].

The idea of the universal algorithm was introduced in Section 6.2. Formally, a *universal Turing machine*, U, takes an encoding of a pair of parameters (M, x) as input and simulates M on input x. U accepts (M, x) iff M accepts x. The universal Turing machines have many applications. For example, the definition of Kolmogorov complexity [Li and Vitányi 1993] fundamentally relies on them.

Example 6.20

Let $L_u = \{\langle M, w \rangle : M \text{ accepts } w\}$. Then L_u can be accepted by a Turing machine, but it is not Turing decidable. The proof is omitted.

If a language is Turing acceptable but not Turing decidable, we call such a language *recursively enumerable* (r.e.). Thus, L_u is r.e. but not recursive. It is easily seen that if both a language and its complement are r.e., then both of them are recursive. Thus, $\bar{L}_u$ is not r.e.

6.5.2.1 Time and Space Complexity

With Turing machines, we can now formally define what we mean by **time and space complexities**. Such a formal investigation by Hartmanis and Stearns [1965] marked the beginning of the field of *computational complexity*. We refer the readers to Hartmanis' Turing Award lecture [Hartmanis 1994] for an interesting account of the history and the future of this field.

To define the space complexity properly (in the sublinear case), we need to slightly modify the Turing machine of Figure 6.13. We will replace the tape containing the input by a read-only input tape and give the Turing machine some extra work tapes.

Definition 6.21   Let M be a Turing machine. If for each n, for each input of length n, and for each sequence of choices of moves when M is nondeterministic, M makes at most $T(n)$ moves we say that M is of *time complexity* $T(n)$; similarly, if M uses at most $S(n)$ tape cells of the work tape, we say that M is of space complexity $S(n)$.

Theorem 6.10   *Any Turing machine using $s(n)$ space can be simulated by a Turing machine, with just one work tape, using $s(n)$ space. If a language is accepted by a k-tape Turing machine running in time $t(n)$ [space $s(n)$], then it also can be accepted by another k-tape Turing machine running in time $ct(n)$ [space $cs(n)$], for any constant $c > 0$.*

To avoid writing the constant c everywhere, we use the standard big-O notation: we say $f(n)$ is $O(g(n))$ if there is a constant c such that $f(n) \leq cg(n)$ for all but finitely many n. The preceding theorem is called the linear speedup theorem; it can be proved easily by using a larger tape alphabet to encode several cells into one and hence compress several steps into one. It leads to the following definitions.

Definition 6.22

DTIME[$t(n)$] is the set of languages accepted by multitape deterministic TMs in time $O(t(n))$.

NTIME[$t(n)$] is the set of languages accepted by multitape nondeterministic TMs in time $O(t(n))$.

DSPACE[$s(n)$] is the set of languages accepted by multitape deterministic TMs in space $O(s(n))$.
NSPACE[$s(n)$] is the set of languages accepted by multitape nondeterministic TMs in space $O(s(n))$.
P is the complexity class $\bigcup_{c \in \mathcal{N}} \text{DTIME}[n^c]$.
NP is the complexity class $\bigcup_{c \in \mathcal{N}} \text{NTIME}[n^c]$.
PSPACE is the complexity class $\bigcup_{c \in \mathcal{N}} \text{DSPACE}[n^c]$.

Example 6.21

We mentioned in Example 6.18 that $L = \{xx : x \in \{0, 1\}^*\}$ can be accepted by a Turing machine. The procedure we have presented in Example 6.18 for a one-head one-tape Turing machine takes $O(n^2)$ time because the single head must go back and forth marking and matching. With two heads, or two tapes, L can be easily accepted in $O(n)$ time.

It should be clear that any language that can be accepted by a DFA, an NFA, or a PDA can be accepted by a Turing machine in $O(n)$ time. The type-1 grammar in Definition 6.16 can be accepted by a Turing machine in $O(n)$ space. Languages in P, that is, languages acceptable by Turing machines in *polynomial* time, are considered as *feasibly* computable. It is important to point out that all generalizations of the Turing machine, except the nondeterministic version, can all be simulated by the basic one-tape deterministic Turing machine with at most polynomial slowdown. The class NP represents the class of languages accepted in polynomial time by a nondeterministic Turing machine. The nondeterministic version of PSPACE turns out to be identical to PSPACE [Savitch 1970]. The following relationships are true:

$$P \subseteq NP \subseteq PSPACE$$

Whether or not either of the inclusions is proper is one of the most fundamental open questions in computer science and mathematics. Research in computational complexity theory centers around these questions. To solve these problems, one can identify the hardest problems in NP or PSPACE. These topics will be discussed in Chapter 8. We refer the interested reader to Gurari [1989], Hopcroft and Ullman [1979], Wood [1987], and Floyd and Beigel [1994].

6.5.2.2 Other Computing Models

Over the years, many alternative computing models have been proposed. With reasonable complexity measures, they can all be simulated by Turing machines with at most a polynomial slowdown. The reference van Emde Boas [1990] provides a nice survey of various computing models other than Turing machines. Because of limited space, we will discuss a few such alternatives very briefly and refer our readers to van Emde Boas [1990] for details and references.

Random Access Machines. The *random access machine* (RAM) [Cook and Reckhow 1973] consists of a finite control where a program is stored, with several arithmetic registers and an infinite collection of memory registers $R[1], R[2], \ldots$. All registers have an unbounded word length. The basic instructions for the program are LOAD, ADD, MULT, STORE, GOTO, ACCEPT, REJECT, etc. Indirect addressing is also used. Apparently, compared to Turing machines, this is a closer but more complicated approximation of modern computers. There are two standard ways for measuring time complexity of the model:

- The *unit-cost RAM:* in this case, each instruction takes one unit of time, no matter how big the operands are. This measure is convenient for analyzing some algorithms such as sorting. But it is unrealistic or even meaningless for analyzing some other algorithms, such as integer multiplication.

- The *log-cost RAM:* each instruction is charged for the sum of the lengths of all data manipulated implicitly or explicitly by the instruction. This is a more realistic model but sometimes less convenient to use.

Log-cost RAMs and Turing machines can simulate each other with polynomial overheads. The unit-cost RAM might be exponentially (but unrealistically) faster when, for example, it uses its power of multiplying two large numbers in one step.

Pointer Machines. The pointer machines were introduced by Kolmogorov and Uspenskii [1958] (also known as the Kolmogorov–Uspenskii machine) and by Schönhage in 1980 (also known as the storage modification machine, see Schönhage [1980]). We informally describe the pointer machine here. A pointer machine is similar to a RAM but differs in its memory structure. A pointer machine operates on a storage structure called a Δ structure, where Δ is a finite alphabet of size greater than one. A Δ-structure S is a finite directed graph (the Kolmogorov–Uspenskii version is an undirected graph) in which each node has $k = |\Delta|$ outgoing edges, which are labeled by the k symbols in Δ. S has a distinguished node called the *center*, which acts as a starting point for addressing, with words over Δ, other nodes in the structure. The pointer machine has various instructions to redirect the pointers or edges and thus modify the storage structure. It should be clear that Turing machines and pointer machines can simulate each other with at most polynomial delay if we use the log-cost model as with the RAMs. There are many interesting studies on the efficiency of the preceding simulations. We refer the reader to van Emde Boas [1990] for more pointers on the pointer machines.

Circuits and Nonuniform Models. A *Boolean circuit* is a finite, labeled, directed acyclic graph. Input nodes are nodes without ancestors; they are labeled with input variables $x_1, \ldots, x_n$. The internal nodes are labeled with functions from a finite set of Boolean operations, for example, {and, or, not} or {$\oplus$}. The number of ancestors of an internal node is precisely the number of arguments of the Boolean function that the node is labeled with. A node without successors is an output node. The circuit is naturally evaluated from input to output: at each node the function labeling the node is evaluated using the results of its ancestors as arguments. Two cost measures for the circuit model are:

- *Depth:* the length of a longest path from an input node to an output node
- *Size:* the number of nodes in the circuit

These measures are applied to a family of circuits $\{C_n : n \geq 1\}$ for a particular problem, where C_n solves the problem of size n. If C_n can be computed from n (in polynomial time), then this is a *uniform measure*. Such circuit families are equivalent to Turing machines. If C_n cannot be computed from n, then such measures are *nonuniform* measures, and such classes of circuits are more powerful than Turing machines because they simply can compute any function by encoding the solutions of all inputs for each n. See van Emde Boas [1990] for more details and pointers to the literature.

Acknowledgment

We would like to thank John Tromp and the reviewers for reading the initial drafts and helping us to improve the presentation.

Defining Terms

Algorithm A finite sequence of instructions that is supposed to solve a particular problem.

Ambiguous context-free grammar For some string of terminals the grammar has two distinct derivation trees.

Chomsky normal form: Every rule of the context-free grammar has the form $A \to BC$ or $A \to a$, where A, B, and C are nonterminals and a is a terminal.

Computable or decidable function/problem: A function/problem that can be solved by an algorithm (or equivalently, a Turing machine).

Context-free grammar: A grammar whose rules have the form $A \to \beta$, where A is a nonterminal and β is a string of nonterminals and terminals.

Context-free language: A language that can be described by some context-free grammar.

Context-sensitive grammar: A grammar whose rules have the form $\alpha \to \beta$, where α and β are strings of nonterminals and terminals and $|\alpha| \leq |\beta|$.

Context-sensitive language: A language that can be described by some context-sensitive grammar.

Derivation or parsing: An illustration of how a string of terminals is obtained from the start symbol by successively applying the rules of the grammar.

Finite automaton or finite-state machine: A restricted Turing machine where the head is read only and shifts only from left to right.

(Formal) grammar: A description of some language typically consisting of a set of terminals, a set of nonterminals with a distinguished one called the start symbol, and a set of rules (or productions) of the form $\alpha \rightarrow \beta$, depicting what string α of terminals and nonterminals can be rewritten as another string β of terminals and nonterminals.

(Formal) language: A set of strings over some fixed alphabet.

Halting problem: The problem of deciding if a given program (or Turing machine) halts on a given input.

Nondeterministic Turing machine: A Turing machine that can make any one of a prescribed set of moves on a given state and symbol read on the tape.

Partially decidable decision problem: There exists a program that always halts and outputs 1 for every input expecting a positive answer and either halts and outputs 0 or loops forever for every input expecting a negative answer.

Program: A sequence of instructions that is not required to terminate on every input.

Pushdown automaton: A restricted Turing machine where the tape acts as a pushdown store (or a stack).

Reduction: A computable transformation of one problem into another.

Regular expression: A description of some language using operators union, concatenation, and Kleene closure.

Regular language: A language that can be described by some right-linear/regular grammar (or equivalently by some regular expression).

Right-linear or regular grammar: A grammar whose rules have the form $A \rightarrow aB$ or $A \rightarrow a$, where A, B are nonterminals and a is either a terminal or the null string.

Time/space complexity: A function describing the maximum time/space required by the machine on any input of length n.

Turing machine: A simplest formal model of computation consisting of a finite-state control and a semi-infinite sequential tape with a read–write head. Depending on the current state and symbol read on the tape, the machine can change its state and move the head to the left or right.

Uncomputable or undecidable function/problem: A function/problem that cannot be solved by any algorithm (or equivalently, any Turing machine).

Universal algorithm: An algorithm that is capable of simulating any other algorithms if properly encoded.

References

Angluin, D. 1980. Finding patterns common to a set of strings. *J. Comput. Syst. Sci.* 21:46–62.

Brzozowski, J. and McCluskey, E., Jr. 1963. Signal flow graph techniques for sequential circuit state diagram. *IEEE Trans. Electron. Comput.* EC-12(2):67–76.

Brzozowski, J. A. and Seger, C.-J. H. 1994. *Asynchronous Circuits*. Springer–Verlag, New York.

Chomsky, N. 1956. Three models for the description of language. *IRE Trans. Inf. Theory* 2(2):113–124.

Chomsky, N. 1963. Formal properties of grammars. In *Handbook of Mathematical Psychology*, Vol. 2, pp. 323–418. John Wiley and Sons, New York.

Chomsky, N. and Miller, G. 1958. Finite-state languages. *Information and Control* 1:91–112.

Cook, S. and Reckhow, R. 1973. Time bounded random access machines. *J. Comput. Syst. Sci.* 7:354–375.

Davis, M. 1980. What is computation? In *Mathematics Today–Twelve Informal Essays*. L. Steen, ed., pp. 241–259. Vintage Books, New York.

Floyd, R. W. and Beigel, R. 1994. *The Language of Machines: An Introduction to Computability and Formal Languages*. Computer Science Press, New York.

Gurari, E. 1989. *An Introduction to the Theory of Computation*. Computer Science Press, Rockville, MD.

Harel, D. 1992. *Algorithmics: The Spirit of Computing*. Addison–Wesley, Reading, MA.

Harrison, M. 1978. *Introduction to Formal Language Theory*. Addison–Wesley, Reading, MA.

Hartmanis, J. 1994. On computational complexity and the nature of computer science. *Commun. ACM* 37(10):37–43.

Hartmanis, J. and Stearns, R. 1965. On the computational complexity of algorithms. *Trans. Amer. Math. Soc.* 117:285–306.

Hopcroft, J. and Ullman, J. 1979. *Introduction to Automata Theory, Languages and Computation.* Addison–Wesley, Reading, MA.

Jiang, T., Salomaa, A., Salomaa, K., and Yu, S. 1995. Decision problems for patterns. *J. Comput. Syst. Sci.* 50(1):53–63.

Kleene, S. 1956. Representation of events in nerve nets and finite automata. In *Automata Studies*, pp. 3–41. Princeton University Press, Princeton, NJ.

Knuth, D., Morris, J., and Pratt, V. 1977. Fast pattern matching in strings. *SIAM J. Comput.* 6:323–350.

Kohavi, Z. 1978. *Switching and Finite Automata Theory.* McGraw–Hill, New York.

Kolmogorov, A. and Uspenskii, V. 1958. On the definition of an algorithm. *Usp. Mat. Nauk.* 13:3–28.

Lesk, M. 1975. LEX–a lexical analyzer generator. *Tech. Rep. 39. Bell Labs.* Murray Hill, NJ.

Li, M. and Vitányi, P. 1993. *An Introduction to Kolmogorov Complexity and Its Applications.* Springer–Verlag, Berlin.

McCulloch, W. and Pitts, W. 1943. A logical calculus of ideas immanent in nervous activity. *Bull. Math. Biophys.* 5:115–133.

Post, E. 1943. Formal reductions of the general combinatorial decision problems. *Am. J. Math.* 65:197–215.

Rabin, M. and Scott, D. 1959. Finite automata and their decision problems. *IBM J. Res. Dev.* 3:114–125.

Robinson, R. 1991. Minsky's small universal Turing machine. *Int. J. Math.* 2(5):551–562.

Salomaa, A. 1966. Two complete axiom systems for the algebra of regular events. *J. ACM* 13(1):158–169.

Savitch, J. 1970. Relationships between nondeterministic and deterministic tape complexities. *J. Comput. Syst. Sci.* 4(2)177–192.

Schönhage, A. 1980. Storage modification machines. *SIAM J. Comput.* 9:490–508.

Searls, D. 1993. The computational linguistics of biological sequences. In *Artificial Intelligence and Molecular Biology.* L. Hunter, ed., pp. 47–120. MIT Press, Cambridge, MA.

Turing, A. 1936. On computable numbers with an application to the Entscheidungs problem. *Proc. London Math. Soc., Ser. 2* 42:230–265.

van Emde Boas, P. 1990. Machine models and simulations. In *Handbook of Theoretical Computer Science.* J. van Leeuwen, ed., pp. 1–66. Elsevier/MIT Press.

Wood, D. 1987. *Theory of Computation.* Harper and Row.

Further Information

The fundamentals of the theory of computation, automata theory, and formal languages can be found in many text books including Floyd and Beigel [1994], Gurari [1989], Harel [1992], Harrison [1978], Hopcroft and Ullman [1979], and Wood [1987]. The central focus of research in this area is to understand the relationships between the different resource complexity classes. This work is motivated in part by some major open questions about the relationships between resources (such as time and space) and the role of control mechanisms (nondeterminism/randomness). At the same time, new computational models are being introduced and studied. One such recent model that has led to the resolution of a number of interesting problems is the interactive proof systems. They exploit the power of randomness and interaction. Among their applications are new ways to encrypt information as well as some unexpected results about the difficulty of solving some difficult problems even approximately. Another new model is the quantum computational model that incorporates quantum-mechanical effects into the basic move of a Turing machine. There are also attempts to use molecular or cell-level interactions as the basic operations of a computer. Yet another research direction motivated in part by the advances in hardware technology is the study of neural networks, which model (albeit in a simplistic manner) the brain structure of mammals. The following chapters of this volume will present state-of-the-art information about many of these developments. The following annual conferences present the leading research work in computation theory: Association of Computer Machinery (ACM) Annual Symposium on Theory of Computing; Institute of Electrical and Electronics Engineers (IEEE) Symposium on the Foundations of Computer Science; IEEE Conference on Structure in Complexity Theory; International Colloquium on Automata,

Languages and Programming; Symposium on Theoretical Aspects of Computer Science; Mathematical Foundations of Computer Science; and Fundamentals of Computation Theory. There are many related conferences such as Computational Learning Theory, ACM Symposium on Principles of Distributed Computing, etc., where specialized computational models are studied for a specific application area. Concrete algorithms is another closely related area in which the focus is to develop algorithms for specific problems. A number of annual conferences are devoted to this field. We conclude with a list of major journals whose primary focus is in theory of computation: *The Journal of the Association of Computer Machinery, SIAM Journal on Computing, Journal of Computer and System Sciences, Information and Computation, Mathematical Systems Theory, Theoretical Computer Science, Computational Complexity, Journal of Complexity, Information Processing Letters, International Journal of Foundations of Computer Science,* and *ACTA Informatica.*

7

Graph and Network Algorithms

7.1	Introduction	7-1
7.2	Tree Traversals	7-2
7.3	Depth-First Search	7-3
	The Depth-First Search Algorithm • Sample Execution • Analysis • Directed Depth-First Search • Sample Execution • Applications of Depth-First Search	
7.4	Breadth-First Search	7-6
	Sample Execution • Analysis	
7.5	Single-Source Shortest Paths........................	7-7
	Dijkstra's Algorithm • Bellman–Ford Algorithm	
7.6	Minimum Spanning Trees	7-9
	Prim's Algorithm • Kruskal's Algorithm	
7.7	Matchings and Network Flows........................	7-11
	Matching Problem Definitions • Applications of Matching • Matchings and Augmenting Paths • Bipartite Matching Algorithm • Assignment Problem • B-Matching Problem • Network Flows • Network Flow Problem Definitions • Blocking Flows • Applications of Network Flow	
7.8	Tour and Traversal Problems	7-20

Samir Khuller
University of Maryland

Balaji Raghavachari
University of Texas at Dallas

7.1 Introduction

Graphs are useful in modeling many problems from different scientific disciplines because they capture the basic concept of objects (vertices) and relationships between objects (edges). Indeed, many optimization problems can be formulated in graph theoretic terms. Hence, algorithms on graphs have been widely studied. In this chapter, a few fundamental graph algorithms are described. For a more detailed treatment of graph algorithms, the reader is referred to textbooks on graph algorithms [Cormen et al. 2001, Even 1979, Gibbons 1985, Tarjan 1983].

An undirected *graph* $G = (V, E)$ is defined as a set V of *vertices* and a set E of *edges*. An edge $e = (u, v)$ is an unordered pair of vertices. A *directed graph* is defined similarly, except that its edges are ordered pairs of vertices; that is, for a directed graph, $E \subseteq V \times V$. The terms *nodes* and vertices are used interchangeably. In this chapter, it is assumed that the graph has neither self-loops, edges of the form (v, v), nor multiple edges connecting two given vertices. A graph is a **sparse graph** if $|E| \ll |V|^2$.

Bipartite graphs form a subclass of graphs and are defined as follows. A graph $G = (V, E)$ is bipartite if the vertex set V can be partitioned into two sets X and Y such that $E \subseteq X \times Y$. In other words, each edge of G connects a vertex in X with a vertex in Y. Such a graph is denoted by $G = (X, Y, E)$. Because bipartite graphs occur commonly in practice, algorithms are often specially designed for them.

A vertex w is *adjacent* to another vertex v if $(v, w) \in E$. An edge (v, w) is said to be *incident* on vertices v and w. The *neighbors* of a vertex v are all vertices $w \in V$ such that $(v, w) \in E$. The number of edges incident to a vertex v is called the **degree** of vertex v. For a directed graph, if (v, w) is an edge, then we say that the edge goes from v to w. The *out-degree* of a vertex v is the number of edges from v to other vertices. The *in-degree* of v is the number of edges from other vertices to v.

A **path** $p = [v_0, v_1, \ldots, v_k]$ from v_0 to v_k is a sequence of vertices such that (v_i, v_{i+1}) is an edge in the graph for $0 \leq i < k$. Any edge may be used only once in a path. A **cycle** is a path whose end vertices are the same, that is, $v_0 = v_k$. A path is *simple* if all its internal vertices are distinct. A cycle is *simple* if every node has exactly two edges incident to it in the cycle. A **walk** $w = [v_0, v_1, \ldots, v_k]$ from v_0 to v_k is a sequence of vertices such that (v_i, v_{i+1}) is an edge in the graph for $0 \leq i < k$, in which edges and vertices may be repeated. A walk is *closed* if $v_0 = v_k$. A graph is **connected** if there is a path between every pair of vertices. A directed graph is **strongly connected** if there is a path between every pair of vertices in each direction. An acyclic, undirected graph is a **forest**, and a **tree** is a connected forest. A directed graph without cycles is known as a **directed acyclic graph** (DAG). Consider a binary relation C between the vertices of an undirected graph G such that for any two vertices u and v, uCv if and only if there is a path in G between u and v. It can be shown that C is an equivalence relation, partitioning the vertices of G into equivalence classes, known as the connected components of G.

There are two convenient ways of representing graphs on computers. We first discuss the *adjacency list* representation. Each vertex has a linked list: there is one entry in the list for each of its adjacent vertices. The graph is thus represented as an array of linked lists, one list for each vertex. This representation uses $O(|V| + |E|)$ storage, which is good for sparse graphs. Such a storage scheme allows one to scan all vertices adjacent to a given vertex in time proportional to its degree. The second representation, the *adjacency matrix*, is as follows. In this scheme, an $n \times n$ array is used to represent the graph. The $[i, j]$ entry of this array is 1 if the graph has an edge between vertices i and j, and 0 otherwise. This representation permits one to test if there is an edge between any pair of vertices in constant time. Both these representation schemes can be used in a natural way to represent directed graphs. For all algorithms in this chapter, it is assumed that the given graph is represented by an adjacency list.

Section 7.2 discusses various types of tree traversal algorithms. Sections 7.3 and 7.4 discuss depth-first and breadth-first search techniques. Section 7.5 discusses the single source shortest path problem. Section 7.6 discusses minimum spanning trees. Section 7.7 discusses the bipartite matching problem and the single commodity maximum flow problem. Section 7.8 discusses some traversal problems in graphs, and the Further Information section concludes with some pointers to current research on graph algorithms.

7.2   Tree Traversals

A tree is *rooted* if one of its vertices is designated as the root vertex and all edges of the tree are oriented (directed) to point away from the root. In a rooted tree, there is a directed path from the root to any vertex in the tree. For any directed edge (u, v) in a rooted tree, u is v's *parent* and v is u's *child*. The *descendants* of a vertex w are all vertices in the tree (including w) that are reachable by directed paths starting at w. The *ancestors* of a vertex w are those vertices for which w is a descendant. Vertices that have no children are called **leaves**. A *binary tree* is a special case of a rooted tree in which each node has at most two children, namely, the left child and the right child. The trees rooted at the two children of a node are called the *left subtree* and *right subtree*.

In this section we study techniques for processing the vertices of a given binary tree in various orders. We assume that each vertex of the binary tree is represented by a record that contains fields to hold attributes of that vertex and two special fields *left* and *right* that point to its left and right subtree, respectively.

The three major tree traversal techniques are *preorder*, *inorder*, and *postorder*. These techniques are used as procedures in many tree algorithms where the vertices of the tree have to be processed in a specific order. In a preorder traversal, the root of any subtree has to be processed *before* any of its descendants. In a postorder traversal, the root of any subtree has to be processed *after* all of its descendants. In an inorder traversal, the root of a subtree is processed after all vertices in its left subtree have been processed, but before any of the vertices in its right subtree are processed. Preorder and postorder traversals generalize to arbitrary rooted trees. In the example to follow, we show how postorder can be used to count the number of descendants of each node and store the value in that node. The algorithm runs in linear time in the size of the tree:

Postorder Algorithm. *PostOrder* (*T*):

```
1 if T ≠ nil then
2     lc ← PostOrder (left[T]) .
3     rc ← PostOrder (right[T]) .
4     desc[T] ← lc + rc + 1.
5     return desc[T] .
6 else
7     return 0.
8 end-if
end-proc
```

7.3 Depth-First Search

Depth-first search (DFS) is a fundamental graph searching technique [Tarjan 1972, Hopcroft and Tarjan 1973]. Similar graph searching techniques were given earlier by Tremaux (see Fraenkel [1970] and Lucas [1882]). The structure of DFS enables efficient algorithms for many other graph problems such as biconnectivity, triconnectivity, and planarity [Even 1979].

The algorithm first initializes all vertices of the graph as being unvisited. Processing of the graph starts from an arbitrary vertex, known as the root vertex. Each vertex is processed when it is first discovered (also referred to as *visiting* a vertex). It is first marked as visited, and its adjacency list is then scanned for unvisited vertices. Each time an unvisited vertex is discovered, it is processed recursively by DFS. After a node's entire adjacency list has been explored, that invocation of the DFS procedure returns. This procedure eventually visits all vertices that are in the same connected component of the root vertex. Once DFS terminates, if there are still any unvisited vertices left in the graph, one of them is chosen as the root and the same procedure is repeated.

The set of edges such that each one led to the discovery of a new vertex form a maximal forest of the graph, known as the **DFS forest**; a *maximal forest* of a graph *G* is an acyclic subgraph of *G* such that the addition of any other edge of *G* to the subgraph introduces a cycle. The algorithm keeps track of this forest using parent pointers. In each connected component, only the root vertex has a *nil* parent in the DFS tree.

7.3.1 The Depth-First Search Algorithm

DFS is illustrated using an algorithm that labels vertices with numbers $1, 2, \ldots$ in such a way that vertices in the same component receive the same label. This labeling scheme is a useful preprocessing step in many problems. Each time the algorithm processes a new component, it numbers its vertices with a new label.

Depth-First Search Algorithm. *DFS-Connected-Component* (*G*):

```
1   c ← 0.
2   for all vertices v in G do
3       visited[v] ← false.
4       finished[v] ← false.
5       parent[v] ← nil.
6   end-for
7   for all vertices v in G do
8       if not visited [v] then
```

9        $c \leftarrow c + 1$.
10   DFS (v, c).
11   **end-if**
12 **end-for**
end-proc

DFS (v, c):
1   *visited*$[v] \leftarrow$ *true*.
2   *component*$[v] \leftarrow c$.
3   **for** all vertices w in *adj*$[v]$ **do**
4   **if not** *visited*$[w]$ **then**
5   *parent*$[w] \leftarrow v$.
6   DFS (w, c).
7   **end-if**
8   **end-for**
9   *finished*$[v] \leftarrow$ *true*.
end-proc

7.3.2 Sample Execution

Figure 7.1 shows a graph having two connected components. DFS was started at vertex a, and the DFS forest is shown on the right. DFS visits the vertices b, d, c, e, and f, in that order. DFS then continues with vertices g, h, and i. In each case, the recursive call returns when the vertex has no more unvisited neighbors. Edges $(d, a), (c, a), (f, d)$, and (i, g) are called *back edges* (these do not belong to the DFS forest).

7.3.3 Analysis

A vertex v is processed as soon as it is encountered, and therefore at the start of DFS (v), *visited*$[v]$ is *false*. Since *visited*$[v]$ is set to true as soon as DFS starts execution, each vertex is visited exactly once. Depth-first search processes each edge of the graph exactly twice, once from each of its incident vertices. Since the algorithm spends constant time processing each edge of G, it runs in $O(|V| + |E|)$ time.

Remark 7.1   In the following discussion, there is no loss of generality in assuming that the input graph is connected. For a rooted DFS tree, vertices u and v are said to be *related*, if either u is an ancestor of v, or vice versa.

   DFS is useful due to the special way in which the edges of the graph may be classified with respect to a DFS tree. Notice that the DFS tree is not unique, and which edges are added to the tree depends on the

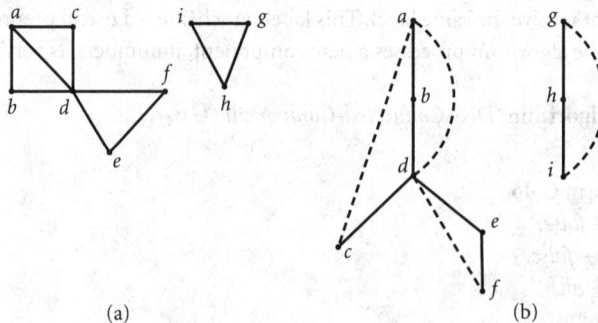

(a)   (b)

FIGURE 7.1   Sample execution of DFS on a graph having two connected components: (a) graph, (b) DFS forest.

order in which edges are explored while executing DFS. Edges of the DFS tree are known as *tree* edges. All other edges of the graph are known as *back* edges, and it can be shown that for any edge (u, v), u and v must be related. The graph does not have any *cross* edges, edges that connect two vertices that are unrelated. This property is utilized by a DFS-based algorithm that classifies the edges of a graph into **biconnected** components, maximal subgraphs that cannot be disconnected by the removal of any single vertex [Even 1979].

7.3.4 Directed Depth-First Search

The DFS algorithm extends naturally to directed graphs. Each vertex stores an adjacency list of its outgoing edges. During the processing of a vertex, first mark it as visited, and then scan its adjacency list for unvisited neighbors. Each time an unvisited vertex is discovered, it is processed recursively. Apart from tree edges and back edges (from vertices to their ancestors in the tree), directed graphs may also have *forward* edges (from vertices to their descendants) and *cross* edges (between unrelated vertices). There may be a cross edge (u, v) in the graph only if u is visited after the procedure call DFS (v) has completed execution.

7.3.5 Sample Execution

A sample execution of the directed DFS algorithm is shown in Figure 7.2. DFS was started at vertex a, and the DFS forest is shown on the right. DFS visits vertices b, d, f, and c in that order. DFS then returns and continues with e, and then g. From g, vertices h and i are visited in that order. Observe that (d, a) and (i, g) are back edges. Edges $(c, d), (e, d)$, and (e, f) are cross edges. There is a single forward edge (g, i).

7.3.6 Applications of Depth-First Search

Directed DFS can be used to design a linear-time algorithm that classifies the edges of a given directed graph into *strongly connected* components: maximal subgraphs that have directed paths connecting any pair of vertices in them, in each direction. The algorithm itself involves running DFS twice, once on the original graph, and then a second time on G^R, which is the graph obtained by reversing the direction of all edges in G. During the second DFS, we are able to obtain all of the strongly connected components. The proof of this algorithm is somewhat subtle, and the reader is referred to Cormen et al. [2001] for details.

Checking if a graph has a cycle can be done in linear time using DFS. A graph has a cycle if and only if there exists a back edge relative to any of its depth-first search trees. A directed graph that does not have any cycles is known as a directed acyclic graph. DAGs are useful in modeling precedence constraints in scheduling problems, where nodes denote jobs/tasks, and a directed edge from u to v denotes the constraint that job u must be completed before job v can begin execution. Many problems on DAGs can be solved efficiently using dynamic programming.

A useful concept in DAGs is that of a **topological order**: a linear ordering of the vertices that is consistent with the partial order defined by the edges of the DAG. In other words, the vertices can be labeled with

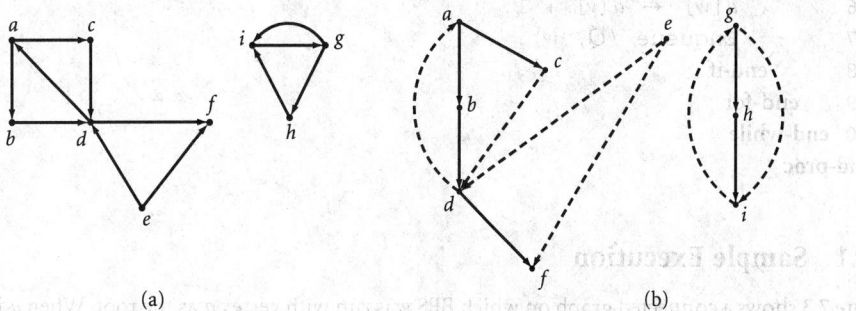

(a)   (b)

FIGURE 7.2 Sample execution of DFS on a directed graph: (a) graph, (b) DFS forest.

distinct integers in the range $[1 \ldots |V|]$ such that if there is a directed edge from a vertex labeled i to a vertex labeled j, then $i < j$. The vertices of a given DAG can be ordered topologically in linear time by a suitable modification of the DFS algorithm. We keep a counter whose initial value is $|V|$. As each vertex is marked finished, we assign the counter value as its topological number and decrement the counter. Observe that there will be no back edges; and that for all edges (u, v), v will be marked finished before u. Thus, the topological number of v will be higher than that of u. Topological sort has applications in diverse areas such as project management, scheduling, and circuit evaluation.

7.4 Breadth-First Search

Breadth-first search (BFS) is another natural way of searching a graph. The search starts at a root vertex r. Vertices are added to a queue as they are discovered, and processed in (first-in–first-out) (FIFO) order.

Initially, all vertices are marked as unvisited, and the queue consists of only the root vertex. The algorithm repeatedly removes the vertex at the front of the queue, and scans its neighbors in the graph. Any neighbor not visited is added to the end of the queue. This process is repeated until the queue is empty. All vertices in the same connected component as the root are scanned and the algorithm outputs a spanning tree of this component. This tree, known as a **breadth-first tree**, is made up of the edges that led to the discovery of new vertices. The algorithm labels each vertex v by $d[v]$, the distance (length of a shortest path) of v from the root vertex, and stores the BFS tree in the array p, using parent pointers. Vertices can be partitioned into levels based on their distance from the root. Observe that edges not in the BFS tree always go either between vertices in the same level, or between vertices in adjacent levels. This property is often useful.

Breadth-First Search Algorithm. *BFS-Distance* (G, r):

```
 1  MakeEmptyQueue (Q).
 2  for all vertices v in G do
 3      visited[v] ← false.
 4      d[v] ← ∞.
 5      p[v] ← nil.
 6  end-for
 7  visited[r] ← true.
 8  d[r] ← 0.
 9  Enqueue (Q, r).
10  while not Empty (Q) do
11      v ← Dequeue (Q).
12      for all vertices w in adj[v] do
13          if not visited[w] then
14              visited[w] ← true.
15              p[w] ← v.
16              d[w] ← d[v] + 1.
17              Enqueue (Q, w).
18          end-if
19      end-for
20  end-while
end-proc
```

7.4.1 Sample Execution

Figure 7.3 shows a connected graph on which BFS was run with vertex a as the root. When a is processed, vertices b, d, and c are added to the queue. When b is processed, nothing is done since all its neighbors have been visited. When d is processed, e and f are added to the queue. Finally c, e, and f are processed.

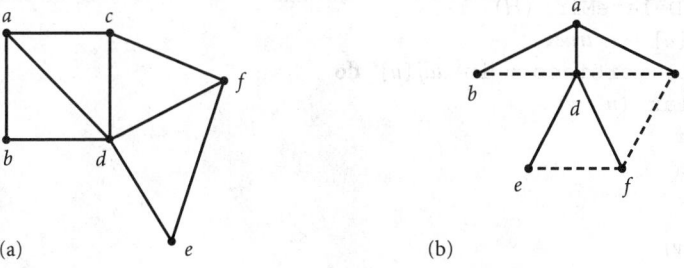

FIGURE 7.3 Sample execution of BFS on a graph: (a) graph, (b) BFS tree.

7.4.2 Analysis

There is no loss of generality in assuming that the graph G is connected, since the algorithm can be repeated in each connected component, similar to the DFS algorithm. The algorithm processes each vertex exactly once, and each edge exactly twice. It spends a constant amount of time in processing each edge. Hence, the algorithm runs in $O(|V| + |E|)$ time.

7.5 Single-Source Shortest Paths

A natural problem that often arises in practice is to compute the shortest paths from a specified node to all other nodes in an undirected graph. BFS solves this problem if all edges in the graph have the same length. Consider the more general case when each edge is given an arbitrary, non-negative length, and one needs to calculate a shortest length path from the root vertex to all other nodes of the graph, where the length of a path is defined to be the sum of the lengths of its edges. The distance between two nodes is the length of a shortest path between them.

7.5.1 Dijkstra's Algorithm

Dijkstra's algorithm [Dijkstra 1959] provides an efficient solution to this problem. For each vertex v, the algorithm maintains an upper bound to the distance from the root to vertex v in $d[v]$; initially $d[v]$ is set to infinity for all vertices except the root. The algorithm maintains a set S of vertices with the property that for each vertex $v \in S, d[v]$ is the length of a shortest path from the root to v. For each vertex u in $V - S$, the algorithm maintains $d[u]$, the shortest known distance from the root to u that goes entirely within S, except for the last edge. It selects a vertex u in $V - S$ of minimum $d[u]$, adds it to S, and updates the distance estimates to the other vertices in $V - S$. In this update step, it checks to see if there is a shorter path to any vertex in $V - S$ from the root that goes through u. Only the distance estimates of vertices that are adjacent to u are updated in this step. Because the primary operation is the selection of a vertex with minimum distance estimate, a priority queue is used to maintain the d-values of vertices. The priority queue should be able to handle a DecreaseKey operation to update the d-value in each iteration. The next algorithm implements Dijkstra's algorithm.

Dijkstra's Algorithm. *Dijkstra-Shortest Paths* (G, r):

```
1 for all vertices v in G do
2     visited[v] ← false.
3     d[v] ← ∞.
4     p[v] ← nil.
5 end-for
6 d[r] ← 0.
7 BuildPQ (H, d).
8 while not Empty (H) do
```

```
 9    u ← DeleteMin (H).
10    visited[u] ← true.
11    for all vertices v in adj[u] do
12        Relax (u, v).
13    end-for
14 end-while
end-proc

Relax (u, v)
1 if not visited[v] and d[v] > d[u] + w(u, v) then
2    d[v] ← d[u] + w(u, v).
3    p[v] ← u.
4    DecreaseKey (H, v, d[v]).
5 end-if
end-proc
```

7.5.1.1 Sample Execution

Figure 7.4 shows a sample execution of the algorithm. The column titled Iter specifies the number of iterations that the algorithm has executed through the **while** loop in step 8. In iteration 0, the initial values of the distance estimates are ∞. In each subsequent line of the table, the column marked u shows the vertex that was chosen in step 9 of the algorithm, and the change to the distance estimates at the end of that iteration of the **while** loop. In the first iteration, vertex r was chosen, after that a was chosen because it had the minimum distance label among the unvisited vertices, and so on. The distance labels of the unvisited neighbors of the visited vertex are updated in each iteration.

7.5.1.2 Analysis

The running time of the algorithm depends on the data structure that is used to implement the priority queue H. The algorithm performs $|V|$ DELETEMIN operations and, at most, $|E|$ DECREASEKEY operations. If a binary heap is used to update the records of any given vertex, each of these operations runs in $O(\log |V|)$ time. There is no loss of generality in assuming that the graph is connected. Hence, the algorithm runs in $O(|E| \log |V|)$. If a Fibonacci heap is used to implement the priority queue, the running time of the algorithm is $O(|E| + |V| \log |V|)$. Although the Fibonacci heap gives the best asymptotic running time, the binary heap implementation is likely to give better running times for most practical instances.

7.5.2 Bellman–Ford Algorithm

The shortest path algorithm described earlier directly generalizes to directed graphs, but it does not work correctly if the graph has edges of negative length. For graphs that have edges of negative length, but no

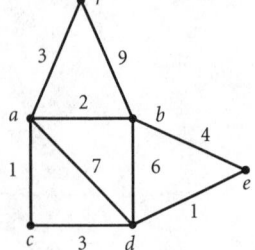

Iter	u	$d[a]$	$d[b]$	$d[c]$	$d[d]$	$d[e]$
0	—	∞	∞	∞	∞	∞
1	r	3	9	∞	∞	∞
2	a	3	5	4	10	∞
3	c	3	5	4	7	∞
4	b	3	5	4	7	9
5	d	3	5	4	7	8
6	e	3	5	4	7	8

FIGURE 7.4   Dijkstra's shortest path algorithm.

cycles of negative length, there is a different algorithm due to Bellman [1958] and Ford and Fulkerson [1962] that solves the single source shortest paths problem in $O(|V||E|)$ time.

The key to understanding this algorithm is the RELAX operation applied to an edge. In a single scan of the edges, we execute the RELAX operation on each edge. We then repeat the step $|V| - 1$ times. No special data structures are required to implement this algorithm, and the proof relies on the fact that a shortest path is simple and contains at most $|V| - 1$ edges (see Cormen et al. [2001] for a proof).

This problem also finds applications in finding a feasible solution to a system of linear equations, where each equation specifies a bound on the difference of two variables. Each constraint is modeled by an edge in a suitably defined directed graph. Such systems of equations arise in real-time applications.

7.6 Minimum Spanning Trees

The following fundamental problem arises in network design. A set of sites needs to be connected by a network. This problem has a natural formulation in graph-theoretic terms. Each site is represented by a vertex. Edges between vertices represent a potential link connecting the corresponding nodes. Each edge is given a nonnegative cost corresponding to the cost of constructing that link. A tree is a minimal network that connects a set of nodes. The cost of a tree is the sum of the costs of its edges. A minimum-cost tree connecting the nodes of a given graph is called a minimum-cost spanning tree, or simply a **minimum spanning tree**.

The problem of computing a minimum spanning tree (MST) arises in many areas, and as a subproblem in combinatorial and geometric problems. MSTs can be computed efficiently using algorithms that are greedy in nature, and there are several different algorithms for finding an MST. One of the first algorithms was due to Boruvka [1926]. The two algorithms that are popularly known as Prim's algorithm and Kruskal's algorithm are described here. (Prim's algorithm was first discovered by Jarnik [1930].)

7.6.1 Prim's Algorithm

Prim's [1957] algorithm for finding an MST of a given graph is one of the oldest algorithms to solve the problem. The basic idea is to start from a single vertex and gradually grow a tree, which eventually spans the entire graph. At each step, the algorithm has a tree that covers a set S of vertices, and looks for a *good* edge that may be used to extend the tree to include a vertex that is currently not in the tree. All edges that go from a vertex in S to a vertex in $V - S$ are candidate edges. The algorithm selects a minimum-cost edge from these candidate edges and adds it to the current tree, thereby adding another vertex to S.

As in the case of Dijkstra's algorithm, each vertex $u \in V - S$ can attach itself to only one vertex in the tree (so that cycles are not generated in the solution). Because the algorithm always chooses a minimum-cost edge, it needs to maintain a minimum-cost edge that connects u to some vertex in S as the candidate edge for including u in the tree. A priority queue of vertices is used to select a vertex in $V - S$ that is incident to a minimum-cost candidate edge.

Prim's Algorithm. *Prim-MST* (G, r):

```
 1  for all vertices v in G do
 2      visited[v] ← false.
 3      d[v] ← ∞.
 4      p[v] ← nil.
 5  end-for
 6  d[r] ← 0.
 7  BuildPQ (H, d).
 8  while not Empty (H) do
 9      u ← DeleteMin (H).
10      visited[u] ← true.
11      for all vertices v in adj[u] do
12          if not visited[v] and d[v] > w(u,v) then
```

```
13              d[v] ← w(u,v).
14              p[v] ← u.
15              DecreaseKey (H, v, d[v]).
16          end-if
17       end-for
18   end-while
end-proc
```

7.6.1.1 Analysis

First observe the similarity between Prim's and Dijkstra's algorithms. Both algorithms start building the tree from a single vertex and grow it by adding one vertex at a time. The only difference is the rule for deciding when the current label is updated for vertices outside the tree. Both algorithms have the same structure and therefore have similar running times. Prim's algorithm runs in $O(|E| \log |V|)$ time if the priority queue is implemented using binary heaps, and it runs in $O(|E| + |V| \log |V|)$ if the priority queue is implemented using Fibonacci heaps.

7.6.2 Kruskal's Algorithm

Kruskal's [1956] algorithm for finding an MST of a given graph is another classical algorithm for the problem, and is also greedy in nature. Unlike Prim's algorithm, which grows a single tree, Kruskal's algorithm grows a forest. First, the edges of the graph are sorted in nondecreasing order of their costs. The algorithm starts with the empty spanning forest (no edges). The edges of the graph are scanned in sorted order, and if the addition of the current edge does not generate a cycle in the current forest, it is added to the forest. The main test at each step is: does the current edge connect two vertices in the same connected component? Eventually, the algorithm adds $|V| - 1$ edges to make a spanning tree in the graph.

The main data structure needed to implement the algorithm is for the maintenance of connected components, to ensure that the algorithm does not add an edge between two nodes in the same connected component. An abstract version of this problem is known as the Union-Find problem for a collection of disjoint sets. Efficient algorithms are known for this problem, where an arbitrary sequence of UNION and FIND operations can be implemented to run in almost linear time [Cormen et al. 2001, Tarjan 1983].

Kruskal's Algorithm. *Kruskal-MST* (G):

```
1  T ← φ.
2  for all vertices v in G do
3    Makeset(v).
4  Sort the edges of G by nondecreasing order of costs.
5  for all edges e = (u,v) in G in sorted order do
6    if Find (u) ≠ Find (v) then
7        T ← T ∪ (u,v).
8        Union (u, v).
9  end-proc
```

7.6.2.1 Analysis

The running time of the algorithm is dominated by step 4 of the algorithm in which the edges of the graph are sorted by nondecreasing order of their costs. This takes $O(|E| \log |E|)$ [which is also $O(|E| \log |V|)$] time using an efficient sorting algorithm such as Heap-sort. Kruskal's algorithm runs faster in the following special cases: if the edges are presorted, if the edge costs are within a small range, or if the number of different edge costs is bounded by a constant. In all of these cases, the edges can be sorted in linear time, and the algorithm runs in near-linear time, $O(|E| \alpha (|E|, |V|))$, where $\alpha(m, n)$ is the inverse Ackermann function [Tarjan 1983].

Remark 7.2 The MST problem can be generalized to directed graphs. The equivalent of trees in directed graphs are called **arborescences** or **branchings**; and because edges have directions, they are rooted spanning trees. An incoming branching has the property that every vertex has a unique path to the root. An outgoing branching has the property that there is a unique path from the root to each vertex in the graph. The input is a directed graph with arbitrary costs on the edges and a root vertex r. The output is a minimum-cost branching rooted at r. The algorithms discussed in this section for finding minimum spanning trees do not directly extend to the problem of finding optimal branchings. There are efficient algorithms that run in $O(|E| + |V| \log |V|)$ time using Fibonacci heaps for finding minimum-cost branchings [Gibbons 1985, Gabow et al. 1986]. These algorithms are based on techniques for weighted matroid intersection [Lawler 1976]. Almost linear-time deterministic algorithms for the MST problem in undirected graphs are also known [Fredman and Tarjan 1987].

7.7 Matchings and Network Flows

Networks are important both for electronic communication and for transporting goods. The problem of efficiently moving entities (such as bits, people, or products) from one place to another in an underlying network is modeled by the **network flow** problem. The problem plays a central role in the fields of operations research and computer science, and much emphasis has been placed on the design of efficient algorithms for solving it. Many of the basic algorithms studied earlier in this chapter play an important role in developing various implementations for network flow algorithms.

First the **matching** problem, which is a special case of the flow problem, is introduced. Then the **assignment problem**, which is a generalization of the matching problem to the weighted case, is studied. Finally, the network flow problem is introduced and algorithms for solving it are outlined.

The maximum matching problem is studied here in detail only for bipartite graphs. Although this restricts the class of graphs, the same principles are used to design polynomial time algorithms for graphs that are not necessarily bipartite. The algorithms for general graphs are complex due to the presence of structures called *blossoms*, and the reader is referred to Papadimitriou and Steiglitz [1982, Chapter 10], or Tarjan [1983, Chapter 9] for a detailed treatment of how blossoms are handled. Edmonds (see Even [1979]) gave the first algorithm to solve the matching problem in polynomial time. Micali and Vazirani [1980] obtained an $O(\sqrt{|V|}|E|)$ algorithm for nonbipartite matching by extending the algorithm by Hopcroft and Karp [1973] for the bipartite case.

7.7.1 Matching Problem Definitions

Given a graph $G = (V, E)$, a matching M is a subset of the edges such that no two edges in M share a common vertex. In other words, the problem is that of finding a set of independent edges that have no incident vertices in common. The cardinality of M is usually referred to as its *size*.

The following terms are defined with respect to a matching M. The edges in M are called *matched edges* and edges not in M are called *free edges*. Likewise, a vertex is a *matched vertex* if it is incident to a matched edge. A *free vertex* is one that is not matched. The *mate* of a matched vertex v is its neighbor w that is at the other end of the matched edge incident to v. A matching is called *perfect* if all vertices of the graph are matched in it. The objective of the maximum matching problem is to maximize $|M|$, the size of the matching. If the edges of the graph have weights, then the *weight* of a matching is defined to be the sum of the weights of the edges in the matching. A path $p = [v_1, v_2, \ldots, v_k]$ is called an *alternating path* if the edges (v_{2j-1}, v_{2j}), $j = 1, 2, \ldots$, are free and the edges (v_{2j}, v_{2j+1}), $j = 1, 2, \ldots$, are matched. An **augmenting path** $p = [v_1, v_2, \ldots, v_k]$ is an alternating path in which both v_1 and v_k are free vertices. Observe that an augmenting path is defined with respect to a specific matching. The symmetric difference of a matching M and an augmenting path P, $M \oplus P$, is defined to be $(M - P) \cup (P - M)$. The operation can be generalized to the case when P is any subset of the edges.

7.7.2 Applications of Matching

Matchings are the underlying basis for many optimization problems. Problems of assigning workers to jobs can be naturally modeled as a bipartite matching problem. Other applications include assigning a collection of jobs with precedence constraints to two processors, such that the total execution time is minimized [Lawler 1976]. Other applications arise in chemistry, in determining structure of chemical bonds, matching moving objects based on a sequence of photographs, and localization of objects in space after obtaining information from multiple sensors [Ahuja et al. 1993].

7.7.3 Matchings and Augmenting Paths

The following theorem gives necessary and sufficient conditions for the existence of a perfect matching in a bipartite graph.

Theorem 7.1 (Hall's Theorem.) *A bipartite graph $G = (X, Y, E)$ with $|X| = |Y|$ has a perfect matching if and only if $\forall S \subseteq X, |N(S)| \geq |S|$, where $N(S) \subseteq Y$ is the set of vertices that are neighbors of some vertex in S.*

Although Theorem 7.1 captures exactly the conditions under which a given bipartite graph has a perfect matching, it does not lead directly to an algorithm for finding maximum matchings. The following lemma shows how an augmenting path with respect to a given matching can be used to increase the size of a matching. An efficient algorithm that uses augmenting paths to construct a maximum matching incrementally is described later.

Lemma 7.1 *Let P be the edges on an augmenting path $p = [v_1, \ldots, v_k]$ with respect to a matching M. Then $M' = M \oplus P$ is a matching of cardinality $|M| + 1$.*

Proof 7.1 Since P is an augmenting path, both v_1 and v_k are free vertices in M. The number of free edges in P is one more than the number of matched edges. The symmetric difference operator replaces the matched edges of M in P by the free edges in P. Hence, the size of the resulting matching, $|M'|$, is one more than $|M|$. □

The following theorem provides a necessary and sufficient condition for a given matching M to be a maximum matching.

Theorem 7.2 *A matching M in a graph G is a maximum matching if and only if there is no augmenting path in G with respect to M.*

Proof 7.2 If there is an augmenting path with respect to M, then M cannot be a maximum matching, since by Lemma 7.1 there is a matching whose size is larger than that of M. To prove the converse we show that if there is no augmenting path with respect to M, then M is a maximum matching. Suppose that there is a matching M' such that $|M'| > |M|$. Consider the set of edges $M \oplus M'$. These edges form a subgraph in G. Each vertex in this subgraph has degree at most two, since each node has at most one edge from each matching incident to it. Hence, each connected component of this subgraph is either a path or a simple cycle. For each cycle, the number of edges of M is the same as the number of edges of M'. Since $|M'| > |M|$, one of the paths must have more edges from M' than from M. This path is an augmenting path in G with respect to the matching M, contradicting the assumption that there were no augmenting paths with respect to M. □

7.7.4 Bipartite Matching Algorithm

7.7.4.1 High-Level Description

The algorithm starts with the empty matching $M = \emptyset$, and augments the matching in phases. In each phase, an augmenting path with respect to the current matching M is found, and it is used to increase the size of the matching. An augmenting path, if one exists, can be found in $O(|E|)$ time, using a procedure similar to breadth-first search described in Section 7.4.

The search for an augmenting path proceeds from the free vertices. At each step when a vertex in X is processed, all its unvisited neighbors are also searched. When a matched vertex in Y is considered, only its matched neighbor is searched. This search proceeds along a subgraph referred to as the *Hungarian tree*.

Initially, all free vertices in X are placed in a queue that holds vertices that are yet to be processed. The vertices are removed one by one from the queue and processed as follows. In turn, when vertex v is removed from the queue, the edges incident to it are scanned. If it has a neighbor in the vertex set Y that is free, then the search for an augmenting path is successful; procedure AUGMENT is called to update the matching, and the algorithm proceeds to its next phase. Otherwise, add the mates of all of the matched neighbors of v to the queue if they have never been added to the queue, and continue the search for an augmenting path. If the algorithm empties the queue without finding an augmenting path, its current matching is a maximum matching and it terminates.

The main data structure that the algorithm uses consists of the arrays *mate* and *free*. The array *mate* is used to represent the current matching. For a matched vertex $v \in G$, *mate*[v] denotes the matched neighbor of vertex v. For $v \in X$, *free*[v] is a vertex in Y that is adjacent to v and is free. If no such vertex exists, then *free*[v] = 0.

Bipartite Matching Algorithm. *Bipartite Matching* $(G = (X, Y, E))$:

```
 1  for all vertices v in G do
 2      mate[v] ← 0.
 3  end-for
 4  found ← false.
 5  while not found do
 6      Initialize.
 7      MakeEmptyQueue (Q).
 8      for all vertices x ∈ X do
 9          if mate[x] = 0 then
10              Enqueue (Q,x).
11              label[x] ← 0.
12          endif
13      end-for
14      done ← false.
15      while not done and not Empty (Q) do
16          x ← Dequeue (Q).
17          if free[x] ≠ 0 then
18              Augment (x).
19              done ← true.
20          else
21              for all edges (x,x') ∈ A do
22                  if label[x'] = 0 then
23                      label[x'] ← x.
24                      Enqueue (Q,x').
25                  end-if
26              end-for
```

```
27              end-if
28              if Empty (Q) then
29                  found ← true.
30              end-if
31          end-while
32  end-while
end-proc
```

Initialize :
```
1  for all vertices x ∈ X do
2      free[x] ← 0.
3  end-for
4  A ← ∅.
5  for all edges (x, y) ∈ E do
6      if mate[y] = 0 then free[x] ← y
7      else if mate[y] ≠ x then A ← A ∪ (x, mate[y]) .
8      end-if
9  end-for
end-proc
```

Augment(x):
```
1  if label[x] = 0 then
2      mate[x] ← free[x] .
3      mate[free[x]] ← x
4  else
5      free[label[x]] ← mate[x]
6      mate[x] ← free[x]
7      mate[free[x]] ← x
8      Augment (label[x])
9  end-if
end-proc
```

7.7.4.2 Sample Execution

Figure 7.5 shows a sample execution of the matching algorithm. We start with a partial matching and show the structure of the resulting Hungarian tree. An augmenting path from vertex b to vertex u is found by the algorithm.

7.7.4.3 Analysis

If there are augmenting paths with respect to the current matching, the algorithm will find at least one of them. Hence, when the algorithm terminates, the graph has no augmenting paths with respect to the current matching and the current matching is optimal. Each iteration of the main **while** loop of the algorithm runs in $O(|E|)$ time. The construction of the auxiliary graph A and computation of the array *free* also take $O(|E|)$ time. In each iteration, the size of the matching increases by one and thus there are, at most, $\min(|X|, |Y|)$ iterations of the **while** loop. Therefore, the algorithm solves the matching problem for bipartite graphs in time $O(\min(|X|, |Y|)|E|)$. Hopcroft and Karp [1973] showed how to improve the running time by finding a maximal set of shortest disjoint augmenting paths in a single phase in $O(|E|)$ time. They also proved that the algorithm runs in only $O(\sqrt{|V|})$ phases.

7.7.5 Assignment Problem

We now introduce the assignment problem, which is that of finding a maximum-weight matching in a given bipartite graph in which edges are given nonnegative weights. There is no loss of generality in

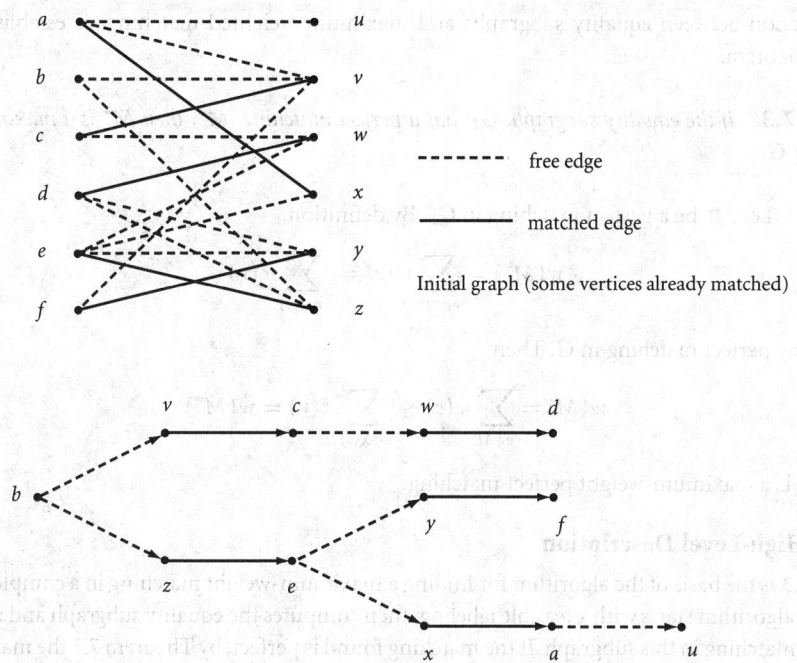

FIGURE 7.5 Sample execution of matching algorithm.

assuming that the graph is complete, since zero-weight edges may be added between pairs of vertices that are nonadjacent in the original graph without affecting the weight of a maximum-weight matching. The minimum-weight perfect matching can be reduced to the maximum-weight matching problem as follows: choose a constant M that is larger than the weight of any edge. Assign each edge a new weight of $w'(e) = M - w(e)$. Observe that maximum-weight matchings with the new weight function are minimum-weight perfect matchings with the original weights. We restrict our attention to the study of the maximum-weight matching problem for bipartite graphs. Similar techniques have been used to solve the maximum-weight matching problem in arbitrary graphs (see Lawler [1976] and Papadimitriou and Steiglitz [1982]).

The input is a complete bipartite graph $G = (X, Y, X \times Y)$ and each edge e has a nonnegative weight of $w(e)$. The following algorithm, known as the Hungarian method, was first given by Kuhn [1955]. The method can be viewed as a primal-dual algorithm in the linear programming framework [Papadimitriou and Steiglitz 1982]. No knowledge of linear programming is assumed here.

A *feasible vertex-labeling* ℓ is defined to be a mapping from the set of vertices in G to the real numbers such that for each edge (x_i, y_j) the following condition holds:

$$\ell(x_i) + \ell(y_j) \geq w(x_i, y_j)$$

The following can be verified to be a feasible vertex labeling. For each vertex $y_j \in Y$, set $\ell(y_j)$ to be 0; and for each vertex $x_i \in X$, set $\ell(x_i)$ to be the maximum weight of an edge incident to x_i,

$$\ell(y_j) = 0,$$
$$\ell(x_i) = \max_j w(x_i, y_j)$$

The *equality subgraph*, G_ℓ, is defined to be the subgraph of G, which includes all vertices of G but only those edges (x_i, y_j) that have weights such that

$$\ell(x_i) + \ell(y_j) = w(x_i, y_j)$$

The connection between equality subgraphs and maximum-weighted matchings is established by the following theorem.

Theorem 7.3   *If the equality subgraph, G_ℓ, has a perfect matching, M^*, then M^* is a maximum-weight matching in G.*

Proof 7.3   Let M^* be a perfect matching in G_ℓ. By definition,

$$w(M^*) = \sum_{e \in M^*} w(e) = \sum_{v \in X \cup Y} \ell(v)$$

Let M be any perfect matching in G. Then,

$$w(M) = \sum_{e \in M} w(e) \leq \sum_{v \in X \cup Y} \ell(v) = w(M^*)$$

Hence, M^* is a maximum-weight perfect matching.   □

7.7.5.1   High-Level Description

Theorem 7.3 is the basis of the algorithm for finding a maximum-weight matching in a complete bipartite graph. The algorithm starts with a feasible labeling, then computes the equality subgraph and a maximum cardinality matching in this subgraph. If the matching found is perfect, by Theorem 7.3 the matching must be a maximum-weight matching and the algorithm returns it as its output. Otherwise, more edges need to be added to the equality subgraph by *revising* the vertex labels. The revision keeps edges from the current matching in the equality subgraph. After more edges are added to the equality subgraph, the algorithm grows the Hungarian trees further. Either the size of the matching increases because an augmenting path is found, or a new vertex is added to the Hungarian tree. In the former case, the current phase terminates and the algorithm starts a new phase, because the matching size has increased. In the latter, new nodes are added to the Hungarian tree. In n phases, the tree includes all of the nodes, and therefore there are at most n phases before the size of the matching increases.

It is now described in more detail how the labels are updated and which edges are added to the equality subgraph G_ℓ. Suppose M is a maximum matching in G_ℓ found by the algorithm. Hungarian trees are grown from all the free vertices in X. Vertices of X (including the free vertices) that are encountered in the search are added to a set S, and vertices of Y that are encountered in the search are added to a set T. Let $\overline{S} = X - S$ and $\overline{T} = Y - T$. Figure 7.6 illustrates the structure of the sets S and T. Matched edges are shown in bold; the other edges are the edges in G_ℓ. Observe that there are no edges in the equality subgraph from S to $\overline{T}$, although there may be edges from T to $\overline{S}$. Let us choose δ to be the smallest value such that some edge of $G - G_\ell$ enters the equality subgraph. The algorithm now revises the labels as follows. Decrease all of the labels of vertices in S by δ and increase the labels of the vertices in T by δ. This ensures that edges in the matching continue to stay in the equality subgraph. Edges in G (not in G_ℓ) that go from vertices in S to vertices in $\overline{T}$ are candidate edges to enter the equality subgraph, since one label is decreasing and the other is unchanged. Suppose this edge goes from $x \in S$ to $y \in \overline{T}$. If y is free, then an augmenting path has been found. On the other hand, if y is matched, the Hungarian tree is grown by moving y to T and its matched neighbor to S, and the process of revising labels continues.

7.7.6   B-Matching Problem

The B-Matching problem is a generalization of the matching problem. In its simplest form, given an integer $b \geq 1$, the problem is to find a subgraph H of a given graph G such that the degree of each vertex is exactly equal to b in H (such a subgraph is called a *b-regular subgraph*). The problem can also be formulated as an optimization problem by seeking a subgraph H with most edges, with the degree of each vertex to

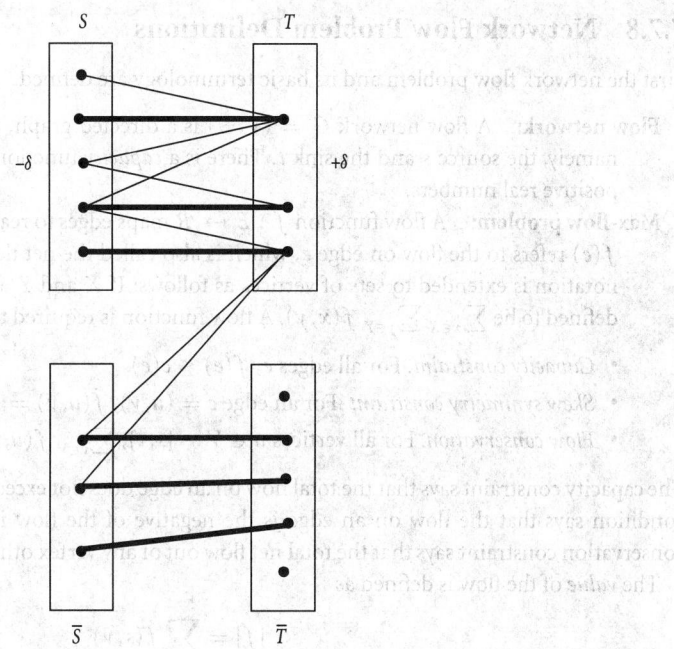

FIGURE 7.6 Sets S and T as maintained by the algorithm. Only edges in G_ℓ are shown.

be at most b in H. Several generalizations are possible, including different degree bounds at each vertex, degrees of some vertices unspecified, and edges with weights. All variations of the B-Matching problem can be solved using the techniques for solving the Matching problem.

In this section, we show how the problem can be solved for the unweighted B-Matching problem in which each vertex v is given a degree bound of $b[v]$, and the objective is to find a subgraph H in which the degree of each vertex v is exactly equal to $b[v]$. From the given graph G, construct a new graph G_b as follows. For each vertex $v \in G$, introduce $b[v]$ vertices in G_b, namely $v_1, v_2, \ldots, v_{b[v]}$. For each edge $e = (u, v)$ in G, add two new vertices e_u and e_v to G_b, along with the edge (e_u, e_v). In addition, add edges between v_i and e_v, for $1 \le i \le b[v]$ (and between u_j and e_u, for $1 \le j \le b[u]$). We now show that there is a natural one-to-one correspondence between B-Matchings in G and perfect matchings in G_b.

Given a B-Matching H in G, we show how to construct a perfect matching in G_b. For each edge $(u, v) \in H$, match e_u to the next available u_j, and e_v to the next available v_i. Since u is incident to exactly $b[u]$ edges in H, there are exactly enough nodes $u_1, u_2 \ldots u_{b[v]}$ in the previous step. For all edges $e = (u, v) \in G - H$, we match e_u and e_v. It can be verified that this yields a perfect matching in G_b.

We now show how to construct a B-Matching in G, given a perfect matching in G_b. Let M be a perfect matching in G_b. For each edge $e = (u, v) \in G$, if $(e_u, e_b) \in M$, then do not include the edge e in the B-Matching. Otherwise, e_u is matched to some u_j and e_v is matched to some v_i in M. In this case, we include e in our B-Matching. Since there are exactly $b[u]$ vertices $u_1, u_2, \ldots u_{b[u]}$, each such vertex introduces an edge into the B-Matching, and therefore the degree of u is exactly $b[u]$. Therefore, we get a B-Matching in G.

7.7.7 Network Flows

A number of polynomial time flow algorithms have been developed over the past two decades. The reader is referred to Ahuja et al. [1993] for a detailed account of the historical development of the various flow methods. Cormen et al. [2001] review the preflow push method in detail; and to complement their coverage, an implementation of the blocking flow technique of Malhotra et al. [1978] is discussed here.

7.7.8 Network Flow Problem Definitions

First the network flow problem and its basic terminology are defined.

Flow network: A flow network $G = (V, E)$ is a directed graph, with two specially marked nodes, namely, the source s and the sink t. There is a *capacity* function $c : E \mapsto R^+$ that maps edges to positive real numbers.

Max-flow problem: A flow function $f : E \mapsto R$ maps edges to real numbers. For an edge $e = (u, v)$, $f(e)$ refers to the flow on edge e, which is also called the net flow from vertex u to vertex v. This notation is extended to sets of vertices as follows: If X and Y are sets of vertices then $f(X, Y)$ is defined to be $\sum_{x \in X} \sum_{y \in Y} f(x, y)$. A flow function is required to satisfy the following constraints:

- *Capacity constraint*. For all edges e, $f(e) \leq c(e)$.
- *Skew symmetry constraint*. For an edge $e = (u, v)$, $f(u, v) = -f(v, u)$.
- *Flow conservation*. For all vertices $u \in V - \{s, t\}$, $\sum_{v \in V} f(u, v) = 0$.

The capacity constraint says that the total flow on an edge does not exceed its capacity. The skew symmetry condition says that the flow on an edge is the negative of the flow in the reverse direction. The flow conservation constraint says that the total net flow out of any vertex other than the source and sink is zero.

The *value* of the flow is defined as

$$|f| = \sum_{v \in V} f(s, v)$$

In other words, it is the net flow out of the source. In the *maximum-flow problem*, the objective is to find a flow function that satisfies the three constraints, and also maximizes the total flow value $|f|$.

Remark 7.3 This formulation of the network flow problem is powerful enough to capture generalizations where there are many sources and sinks (single commodity flow), and where both vertices and edges have capacity constraints, etc.

First, the notion of cuts is defined, and the max-flow min-cut theorem is introduced. Then, residual networks, layered networks, and the concept of blocking flows are introduced. Finally, an efficient algorithm for finding a blocking flow is described.

An *s–t cut* of the graph is a partitioning of the vertex set V into two sets S and $T = V - S$ such that $s \in S$ and $t \in T$. If f is a flow, then the net flow across the cut is defined as $f(S, T)$. The capacity of the cut is similarly defined as $c(S, T) = \sum_{x \in X} \sum_{y \in Y} c(x, y)$. The net flow across a cut may include negative net flows between vertices, but the capacity of the cut includes only nonnegative values, that is, only the capacities of edges from S to T.

Using the flow conservation principle, it can be shown that the net flow across an *s–t* cut is exactly the flow value $|f|$. By the capacity constraint, the flow across the cut cannot exceed the capacity of the cut. Thus, the value of the maximum flow is no greater than the capacity of a minimum *s–t* cut. The well-known *max-flow min-cut theorem* [Elias et al. 1956, Ford and Fulkerson 1962] proves that the two numbers are actually equal. In other words, if f^* is a maximum flow, then there is some cut $(X, \overline{X})$ such that $|f^*| = c(X, \overline{X})$. The reader is referred to Cormen et al. [2001] and Tarjan [1983] for further details.

The *residual capacity* of a flow f is defined to be a function on vertex pairs given by $c'(v, w) = c(v, w) - f(v, w)$. The residual capacity of an edge (v, w), $c'(v, w)$, is the number of additional units of flow that can be pushed from v to w without violating the capacity constraints. An edge e is *saturated* if $c(e) = f(e)$, that is, if its residual capacity, $c'(e)$, is zero. The residual graph $G_R(f)$ for a flow f is the graph with vertex set V, source and sink s and t, respectively, and those edges (v, w) for which $c'(v, w) > 0$.

An augmenting path for f is a path P from s to t in $G_R(f)$. The residual capacity of P, denoted by $c'(P)$, is the minimum value of $c'(v, w)$ over all edges (v, w) in the path P. The flow can be increased by $c'(P)$, by increasing the flow on each edge of P by this amount. Whenever $f(v, w)$ is changed, $f(w, v)$ is also correspondingly changed to maintain skew symmetry.

Most flow algorithms are based on the concept of augmenting paths pioneered by Ford and Fulkerson [1956]. They start with an initial zero flow and augment the flow in stages. In each stage, a residual graph $G_R(f)$ with respect to the current flow function f is constructed and an augmenting path in $G_R(f)$ is found to increase the value of the flow. Flow is increased along this path until an edge in this path is saturated. The algorithms iteratively keep increasing the flow until there are no more augmenting paths in $G_R(f)$, and return the final flow f as their output.

The following lemma is fundamental in understanding the basic strategy behind these algorithms.

Lemma 7.2 *Let f be any flow and f^* a maximum flow in G, and let $G_R(f)$ be the residual graph for f. The value of a maximum flow in $G_R(f)$ is $|f^*| - |f|$.*

Proof 7.4 Let f' be any flow in $G_R(f)$. Define $f + f'$ to be the flow defined by the flow function $f(v, w) + f'(v, w)$ for each edge (v, w). Observe that $f + f'$ is a feasible flow in G of value $|f| + |f'|$. Since f^* is the maximum flow possible in G, $|f'| \leq |f^*| - |f|$. Similarly define $f^* - f$ to be a flow in $G_R(f)$ defined by $f^*(v, w) - f(v, w)$ in each edge (v, w), and this is a feasible flow in $G_R(f)$ of value $|f^*| - |f|$, and it is a maximum flow in $G_R(f)$. $\square$

Blocking flow: A flow f is a **blocking flow** if every path in G from s to t contains a saturated edge. It is important to note that a blocking flow is not necessarily a maximum flow. There may be augmenting paths that increase the flow on some edges and decrease the flow on other edges (by increasing the flow in the reverse direction).

Layered networks: Let $G_R(f)$ be the residual graph with respect to a flow f. The level of a vertex v is the length of a shortest path (using the least number of edges) from s to v in $G_R(f)$. The level graph L for f is the subgraph of $G_R(f)$ containing vertices reachable from s and only the edges (v, w) such that $\text{dist}(s, w) = 1 + \text{dist}(s, v)$. L contains all shortest-length augmenting paths and can be constructed in $O(|E|)$ time.

The Maximum Flow algorithm proposed by Dinitz [1970] starts with the zero flow, and iteratively increases the flow by augmenting it with a blocking flow in $G_R(f)$ until t is not reachable from s in $G_R(f)$. At each step the current flow is replaced by the sum of the current flow and the blocking flow. Since in each iteration the shortest distance from s to t in the residual graph increases, and the shortest path from s to t is at most $|V| - 1$, this gives an upper bound on the number of iterations of the algorithm.

An algorithm to find a blocking flow that runs in $O(|V|^2)$ time is described here, and this yields an $O(|V|^3)$ max-flow algorithm. There are a number of $O(|V|^2)$ blocking flow algorithms available [Karzanov 1974, Malhotra et al. 1978, Tarjan 1983], some of which are described in detail in Tarjan [1983].

7.7.9 Blocking Flows

Dinitz's algorithm to find a blocking flow runs in $O(|V||E|)$ time [Dinitz 1970]. The main step is to find paths from the source to the sink and saturate them by pushing as much flow as possible on these paths. Every time the flow is increased by pushing more flow along an augmenting path, one of the edges on this path becomes saturated. It takes $O(|V|)$ time to compute the amount of flow that can be pushed on the path. Since there are $|E|$ edges, this yields an upper bound of $O(|V||E|)$ steps on the running time of the algorithm.

Malhotra–Kumar–Maheshwari Blocking Flow Algorithm. The algorithm has a current flow function f and its corresponding residual graph $G_R(f)$. Define for each node $v \in G_R(f)$, a quantity $tp[v]$ that specifies its maximum throughput, that is, either the sum of the capacities of the incoming arcs or the sum of the capacities of the outgoing arcs, whichever is smaller. $tp[v]$ represents the maximum flow that could pass through v in any feasible blocking flow in the residual graph. Vertices for which the throughput is zero are deleted from $G_R(f)$.

The algorithm selects a vertex u for which its throughput is a minimum among all vertices with nonzero throughput. It then greedily pushes a flow of $tp[u]$ from u toward t, level by level in the layered residual graph. This can be done by creating a queue, which initially contains u and which is assigned the task of pushing $tp[u]$ out of it. In each step, the vertex v at the front of the queue is removed, and the arcs going out of v are scanned one at a time, and as much flow as possible is pushed out of them until v's allocated flow has been pushed out. For each arc (v, w) that the algorithm pushed flow through, it updates the residual capacity of the arc (v, w) and places w on a queue (if it is not already there) and increments the net incoming flow into w. Also, $tp[v]$ is reduced by the amount of flow that was sent through it now. The flow finally reaches t, and the algorithm never comes across a vertex that has incoming flow that exceeds its outgoing capacity since u was chosen as a vertex with the smallest throughput. The preceding idea is again repeated to pull a flow of $tp[u]$ from the source s to u. Combining the two steps yields a flow of $tp[u]$ from s to t in the residual network that goes through u. The flow f is augmented by this amount. Vertex u is deleted from the residual graph, along with any other vertices that have zero throughput.

This procedure is repeated until all vertices are deleted from the residual graph. The algorithm has a blocking flow at this stage since at least one vertex is saturated in every path from s to t. In the algorithm, whenever an edge is saturated, it may be deleted from the residual graph. Since the algorithm uses a greedy strategy to send flows, at most $O(|E|)$ time is spent when an edge is saturated. When finding flow paths to push $tp[u]$, there are at most n times, one each per vertex, when the algorithm pushes a flow that does not saturate the corresponding edge. After this step, u is deleted from the residual graph. Hence, in $O(|E| + |V|^2) = O(|V|^2)$ steps, the algorithm to compute blocking flows terminates.

Goldberg and Tarjan [1988] proposed a preflow push method that runs in $O(|V||E| \log |V|^2/|E|)$ time without explicitly finding a blocking flow at each step.

7.7.10 Applications of Network Flow

There are numerous applications of the Maximum Flow algorithm in scheduling problems of various kinds. See Ahuja et al. [1993] for further details.

7.8   Tour and Traversal Problems

There are many applications for finding certain kinds of paths and tours in graphs. We briefly discuss some of the basic problems.

The **traveling salesman problem (TSP)** is that of finding a shortest tour that visits all of the vertices in a given graph with weights on the edges. It has received considerable attention in the literature [Lawler et al. 1985]. The problem is known to be computationally intractable (NP-hard). Several heuristics are known to solve practical instances. Considerable progress has also been made for finding optimal solutions for graphs with a few thousand vertices.

One of the first graph-theoretic problems to be studied, the **Euler tour problem** asks for the existence of a closed walk in a given connected graph that traverses each edge exactly once. Euler proved that such a closed walk exists if and only if each vertex has even degree [Gibbons 1985]. Such a graph is known as an **Eulerian graph**. Given an Eulerian graph, a Euler tour in it can be computed using DFS in linear time.

Given an edge-weighted graph, the **Chinese postman problem** is that of finding a shortest closed walk that traverses each edge at least once. Although the problem sounds very similar to the TSP problem, it can be solved optimally in polynomial time by reducing it to the matching problem [Ahuja et al. 1993].

Acknowledgments

Samir Khuller's research is supported by National Science Foundation (NSF) Awards CCR-9820965 and CCR-0113192.

Balaji Raghavachari's research is supported by the National Science Foundation under Grant CCR-9820902.

Defining Terms

Assignment problem: That of finding a perfect matching of maximum (or minimum) total weight.

Augmenting path: An alternating path that can be used to augment (increase) the size of a matching.

Biconnected graph: A graph that cannot be disconnected by the removal of any single vertex.

Bipartite graph: A graph in which the vertex set can be partitioned into two sets X and Y, such that each edge connects a node in X with a node in Y.

Blocking flow: A flow function in which any directed path from s to t contains a saturated edge.

Branching: A spanning tree in a rooted graph, such that the root has a path to each vertex.

Chinese postman problem: Asks for a minimum length tour that traverses each edge at least once.

Connected: A graph in which there is a path between each pair of vertices.

Cycle: A path in which the start and end vertices of the path are identical.

Degree: The number of edges incident to a vertex in a graph.

DFS forest: A rooted forest formed by depth-first search.

Directed acyclic graph: A directed graph with no cycles.

Eulerian graph: A graph that has an Euler tour.

Euler tour problem: Asks for a traversal of the edges that visits each edge exactly once.

Forest: An acyclic graph.

Leaves: Vertices of degree one in a tree.

Matching: A subset of edges that do not share a common vertex.

Minimum spanning tree: A spanning tree of minimum total weight.

Network flow: An assignment of flow values to the edges of a graph that satisfies flow conservation, skew symmetry, and capacity constraints.

Path: An ordered list of edges such that any two consecutive edges are incident to a common vertex.

Perfect matching: A matching in which every node is matched by an edge to another node.

Sparse graph: A graph in which $|E| \ll |V|^2$.

s–t cut: A partitioning of the vertex set into S and T such that $s \in S$ and $t \in T$.

Strongly connected: A directed graph in which there is a directed path in each direction between each pair of vertices.

Topological order: A linear ordering of the edges of a DAG such that every edge in the graph goes from left to right.

Traveling salesman problem: Asks for a minimum length tour of a graph that visits all of the vertices exactly once.

Tree: An acyclic graph with $|V| - 1$ edges.

Walk: An ordered sequence of edges (in which edges could repeat) such that any two consecutive edges are incident to a common vertex.

References

Ahuja, R.K., Magnanti, T., and Orlin, J. 1993. *Network Flows*. Prentice Hall, Upper Saddle River, NJ.

Bellman, R. 1958. On a routing problem. *Q. App. Math.*, 16(1):87–90.

Boruvka, O. 1926. O jistem problemu minimalnim. *Praca Moravske Prirodovedecke Spolecnosti*, 3:37–58 (in Czech).

Cormen, T.H., Leiserson, C.E., Rivest, R.L., and Stein, C. 2001. *Introduction to Algorithms, second edition*. The MIT Press.

DiBattista, G., Eades, P., Tamassia, R., and Tollis, I. 1994. Annotated bibliography on graph drawing algorithms. *Comput. Geom.: Theory Applic.*, 4:235–282.

Dijkstra, E.W. 1959. A note on two problems in connexion with graphs. *Numerische Mathematik*, 1:269–271.

Dinitz, E.A. 1970. Algorithm for solution of a problem of maximum flow in a network with power estimation. *Soviet Math. Dokl.*, 11:1277–1280.

Elias, P., Feinstein, A., and Shannon, C.E. 1956. Note on maximum flow through a network. *IRE Trans. Inf. Theory*, IT-2:117–119.

Even, S. 1979. *Graph Algorithms*. Computer Science Press, Potomac, MD.

Ford, L.R., Jr. and Fulkerson, D.R. 1956. Maximal flow through a network. *Can. J. Math.*, 8:399–404.

Ford, L.R., Jr. and Fulkerson, D.R. 1962. *Flows in Networks*. Princeton University Press.

Fraenkel, A.S. 1970. Economic traversal of labyrinths. *Math. Mag.*, 43:125–130.

Fredman, M. and Tarjan, R.E. 1987. Fibonacci heaps and their uses in improved network optimization algorithms. *J. ACM*, 34(3):596–615.

Gabow, H.N., Galil, Z., Spencer, T., and Tarjan, R.E. 1986. Efficient algorithms for finding minimum spanning trees in undirected and directed graphs. *Combinatorica*, 6(2):109–122.

Gibbons, A.M. 1985. *Algorithmic Graph Theory*. Cambridge University Press, New York.

Goldberg, A.V. and Tarjan, R.E. 1988. A new approach to the maximum-flow problem. *J. ACM*, 35:921–940.

Hochbaum, D.S., Ed. 1996. *Approximation Algorithms for NP-Hand Problems*. PWS Publishing.

Hopcroft, J.E. and Karp, R.M. 1973. An $n^{2.5}$ algorithm for maximum matching in bipartite graphs. *SIAM J. Comput.*, 2(4):225–231.

Hopcroft, J.E. and Tarjan, R.E. 1973. Efficient algorithms for graph manipulation. *Commun. ACM*, 16:372–378.

Jarnik, V. 1930. O jistem problemu minimalnim. *Praca Moravske Prirodovedecke Spolecnosti*, 6:57–63 (in Czech).

Karzanov, A.V. 1974. Determining the maximal flow in a network by the method of preflows. *Soviet Math. Dokl.*, 15:434–437.

Kruskal, J.B., 1956. On the shortest spanning subtree of a graph and the traveling salesman problem. *Proc. Am. Math. Soc.*, 7:48–50.

Kuhn, H.W. 1955. The Hungarian method for the assignment problem. *Nav. Res. Logistics Q.*, 2:83–98.

Lawler, E.L. 1976. *Combinatorial Optimization: Networks and Matroids*. Holt, Rinehart and Winston.

Lawler, E.L., Lenstra, J.K., Rinnooy Kan, A.H.G., and Shmoys, D.B. 1985. *The Traveling Salesman Problem: A Guided Tour of Combinatorial Optimization*. Wiley, New York.

Lucas, E. 1882. *Recreations Mathematiques*. Paris.

Malhotra, V.M., Kumar, M.P., and Maheshwari, S.N. 1978. An $O(|V|^3)$ algorithm for finding maximum flows in networks. *Inf. Process. Lett.*, 7:277–278.

Micali, S. and Vazirani, V.V. 1980. An $O(\sqrt{|V|}|E|)$ algorithm for finding maximum matching in general graphs, pp. 17–27. In *Proc. 21st Annu. Symp. Found. Comput. Sci.*

Papadimitriou, C.H. and Steiglitz, K. 1982. *Combinatorial Optimization: Algorithms and Complexity*. Prentice Hall, Upper Saddle River, NJ.

Prim, R.C. 1957. Shortest connection networks and some generalizations. *Bell Sys. Tech. J.*, 36:1389–1401.

Tarjan, R.E. 1972. Depth first search and linear graph algorithms. *SIAM J. Comput.*, 1:146–160.

Tarjan, R.E. 1983. *Data Structures and Network Algorithms*. SIAM.

Further Information

The area of graph algorithms continues to be a very active field of research. There are several journals and conferences that discuss advances in the field. Here we name a partial list of some of the important meetings: ACM Symposium on Theory of Computing, IEEE Conference on Foundations of Computer Science, ACM–SIAM Symposium on Discrete Algorithms, the International Colloquium on Automata, Languages and Programming, and the European Symposium on Algorithms. There are many other regional algorithms/theory conferences that carry research papers on graph algorithms. The journals that carry articles on current research in graph algorithms are *Journal of the ACM, SIAM Journal on Computing, SIAM Journal on Discrete Mathematics, Journal of Algorithms, Algorithmica, Journal of Computer and System Sciences, Information and Computation, Information Processing Letters,* and *Theoretical Computer Science.*

To find more details about some of the graph algorithms described in this chapter we refer the reader to the books by Cormen et al. [2001], Even [1979], and Tarjan [1983]. For network flows and matching, a more detailed survey regarding various approaches can be found in Tarjan [1983]. Papadimitriou and Steiglitz [1982] discuss the solution of many combinatorial optimization problems using a primal–dual framework.

Current research on graph algorithms focuses on approximation algorithms [Hochbaum 1996], dynamic algorithms, and in the area of graph layout and drawing [DiBattista et al. 1994].

8

Algebraic Algorithms

8.1 Introduction 8-1
8.2 Matrix Computations and Approximation of
    Polynomial Zeros 8-1
    Products of Vectors and Matrices, Convolution of Vectors
    • Some Computations Related to Matrix Multiplication
    • Gaussian Elimination Algorithm • Singular Linear Systems of
    Equations • Sparse Linear Systems (Including Banded
    Systems), Direct and Iterative Solution Algorithms • Dense and
    Structured Matrices and Linear Systems • Parallel Matrix
    Computations • Rational Matrix Computations, Computations
    in Finite Fields and Semirings • Matrix Eigenvalues and
    Singular Values Problems • Approximating Polynomial Zeros
    • Fast Fourier Transform and Fast Polynomial Arithmetic

Angel Diaz
IBM Research

8.3 Systems of Nonlinear Equations and Other
    Applications 8-10
    Resultant Methods • Gröbner Bases

Erich Kaltófen
North Carolina State University

8.4 Polynomial Factorization 8-14
    Polynomials in a Single Variable over a Finite Field
    • Polynomials in a Single Variable over Fields
    of Characteristic Zero • Polynomials in Two Variables
    • Polynomials in Many Variables

Victor Y. Pan
Lehman College, CUNY

8.1 Introduction

The title's subject is the algorithmic approach to algebra: arithmetic with numbers, polynomials, matrices, differential polynomials, such as $y'' + (1/2 + x^4/4)y$, truncated series, and algebraic sets, i.e., quantified expressions such as $\exists x \in \mathbb{R} : x^4 + p \cdot x + q = 0$, which describes a subset of the two-dimensional space with coordinates p and q for which the given quartic equation has a real root. Algorithms that manipulate such objects are the backbone of modern symbolic mathematics software such as the Maple and Mathematica systems, to name but two among many useful systems. This chapter restricts itself to algorithms in four areas: linear matrix algebra, root finding of univariate polynomials, solution of systems of nonlinear algebraic equations, and polynomial factorization.

8.2 Matrix Computations and Approximation of Polynomial Zeros

This section covers several major algebraic and numerical problems of scientific and engineering computing that are usually solved numerically, with rounding off or chopping the input and computed values to a fixed number of bits that fit the computer precision (Sections 8.2 and 8.3 are devoted to some fundamental infinite precision symbolic computations, and within Section 8.2 we comment on the infinite precision techniques for some matrix computations). We also study approximation of polynomial zeros, which is an important, fundamental, as well as very popular subject. In our presentation, we will very briefly list the major subtopics of our huge subject and will give some pointers to the references. We will include brief coverage of the topics of the algorithm design and analysis, regarding the complexity of matrix computation and of approximating polynomial zeros. The reader may find further material on these subjects in the survey articles by Pan [1984a, 1991, 1992a, 1995b] and in the books by Bini and Pan [1994, 1996].

8.2.1 Products of Vectors and Matrices, Convolution of Vectors

An $m \times n$ matrix $A = (a_{i,j}, i = 0, 1, \ldots, m - 1; j = 0, 1, \ldots, n - 1)$ is a two-dimensional array, whose (i, j) entry is $(A)_{i,j} = a_{i,j}$. A is a column vector of dimension m if $n = 1$ and is a row vector of dimension n if $m = 1$. Transposition, hereafter, indicated by the superscript T, transforms a row vector $v^T = [v_0, \ldots, v_{n-1}]$ into a column vector $v = [v_0, \ldots, v_{n-1}]^T$.

For two vectors, $u^T = (u_0, \ldots, u_{m-1})$ and $v^T = (v_0, \ldots, v_{n-1})^T$, their *outer product* is an $m \times n$ matrix,

$$W = uv^T = [w_{i,j}, i = 0, \ldots, m - 1; j = 0, \ldots, n - 1]$$

where $w_{i,j} = u_i v_j$, for all i and j, and their *convolution* vector is said to equal

$$w = u \circ v = (w_0, \ldots, w_{m+n-2})^T, \qquad w_k = \sum_{i=0}^{k} u_i v_{k-i}$$

where $u_i = v_j = 0$, for $i \geq m$, $j \geq n$; in fact, w is the coefficient vector of the product of two polynomials,

$$u(x) = \sum_{i=0}^{m-1} u_i x^i \qquad \text{and} \qquad v(x) = \sum_{i=0}^{n-1} v_i x^i$$

having coefficient vectors u and v, respectively.

If $m = n$, the scalar value

$$v^T u = u^T v = u_0 v_0 + u_1 v_1 + \cdots + u_{n-1} v_{n-1} = \sum_{i=0}^{n-1} u_i v_i$$

is called the *inner* (*dot*, or *scalar*) *product* of u and v.

The straightforward algorithms compute the inner and outer products of u and v and their convolution vector by using $2n - 1$, mn, and $mn + (m - 1)(n - 1) = 2mn - m - n + 1$ arithmetic operations (hereafter, referred to as **ops**), respectively.

These upper bounds on the numbers of ops for computing the inner and outer products are sharp, that is, cannot be decreased, for the general pair of the input vectors u and v, whereas (see, e.g., Bini and Pan [1994]) one may apply the *fast fourier transform* (FFT) in order to compute the convolution vector $u \circ v$ much faster, for larger m and n; namely, it suffices to use $4.5K \log K + 2K$ ops, for $K = 2^k$, $k = \lceil \log(m + n + 1) \rceil$. (Here and hereafter, all logarithms are binary unless specified otherwise.)

If $A = (a_{i,j})$ and $B = (b_{j,k})$ are $m \times n$ and $n \times p$ matrices, respectively, and $v = (v_k)$ is a p-dimensional vector, then the straightforward algorithms compute the vector

$$w = Bv = (w_0, \ldots, w_{n-1})^T, \qquad w_i = \sum_{j=0}^{p-1} b_{i,j} v_j, \qquad i = 0, \ldots, n - 1$$

by using $(2p - 1)n$ ops (sharp bound), and compute the *matrix product*

$$AB = (w_{i,k}, i = 0, \ldots, m - 1; k = 0, \ldots, p - 1)$$

by using $2mnp - mp$ ops, which is $2n^3 - n^2$ if $m = n = p$. The latter upper bound is not sharp: the subroutines for $n \times n$ matrix multiplication on some modern computers, such as CRAY and Connection

Machines, rely on algorithms using $O(n^{2.81})$ ops, and some nonpractical algorithms involve $O(n^{2.376})$ ops [Bini and Pan 1994, Golub and Van Loan 1989].

In the special case, where all of the input entries and components are bounded integers having short binary representation, each of the preceding operations with vectors and matrices can be reduced to a single multiplication of 2 longer integers, by means of the techniques of *binary segmentation* (cf. Pan [1984b, Section 40], Pan [1991], Pan [1992b], or Bini and Pan [1994, Examples 36.1–36.3]).

For an $n \times n$ matrix B and an n-dimensional vector v, one may compute the vectors $B^i v$, $i = 1, 2, \ldots, k - 1$, which define *Krylov sequence* or *Krylov matrix*

$$[B^i v, \; i = 0, 1, \ldots, k - 1]$$

used as a basis of several computations. The straightforward algorithm takes on $(2n - 1)nk$ ops, which is order n^3 if k is of order n. An alternative algorithm first computes the matrix powers

$$B^2, B^4, B^8, \ldots, B^{2^s}, \quad s = \lceil \log k \rceil - 1$$

and then the products of $n \times n$ matrices B^{2^i} by $n \times 2^i$ matrices, for $i = 0, 1, \ldots, s$,

$$
\begin{array}{ll}
B & v \\
B^2 & (v, \; Bv) = (B^2 v, \; B^3 v) \\
B^4 & (v, \; Bv, \; B^2 v, \; B^3 v) = (B^4 v, \; B^5 v, \; B^6 v, \; B^7 v) \\
\vdots &
\end{array}
$$

The last step completes the evaluation of the Krylov sequence, which amounts to $2s$ matrix multiplications, for $k = n$, and, therefore, can be performed (in theory) in $O(n^{2.376} \log k)$ ops.

8.2.2 Some Computations Related to Matrix Multiplication

Several fundamental matrix computations can be ultimately reduced to relatively few [that is, to a constant number, or, say, to $O(\log n)$] $n \times n$ matrix multiplications. These computations include the evaluation of det A, the **determinant** of an $n \times n$ matrix A; of its *inverse* A^{-1} (where A is nonsingular, that is, where det $A \neq 0$); of the coefficients of its **characteristic polynomial**, $c_A(x) = \det(xI - A)$, x denoting a scalar variable and I being the $n \times n$ identity matrix, which has ones on its diagonal and zeros elsewhere; of its *minimal polynomial*, $m_A(x)$; of its *rank*, rank A; of the solution vector $x = A^{-1} v$ to a nonsingular *linear system of equations*, $Ax = v$; of various *orthogonal* and *triangular factorizations* of A; and of a submatrix of A having the maximal rank, as well as some fundamental computations with singular matrices. Consequently, all of these operations can be performed by using (theoretically) $O(n^{2.376})$ ops (cf. Bini and Pan [1994, Chap. 2]). The idea is to represent the input matrix A as a block matrix and, operating with its blocks (rather than with its entries), to apply fast matrix multiplication algorithms. In practice, due to various other considerations (accounting, in particular, for the overhead constants hidden in the O notation, for the memory space requirements, and particularly, for numerical stability problems), these computations are based either on the straightforward algorithm for matrix multiplication or on other methods allowing order n^3 arithmetic operations (cf. Golub and Van Loan [1989]). Many block matrix algorithms supporting the (nonpractical) estimate $O(n^{2.376})$, however, become practically important for parallel computations (see Section 8.2.7).

In the next six sections, we will more closely consider the solution of a linear system of equations, $Av = b$, which is the most frequent operation in practice of scientific and engineering computing and is highly important theoretically. We will partition the known solution methods depending on whether the coefficient matrix A is *dense and unstructured*, **sparse**, or *dense and* **structured**.

8.2.3   Gaussian Elimination Algorithm

The solution of a nonsingular linear system $Ax = v$ uses only about n^2 ops if the system is lower (or upper) triangular, that is, if all subdiagonal (or superdiagonal) entries of A vanish. For example (cf. Pan [1992b]), let $n = 3$,

$$x_1 + 2x_2 - x_3 = 3$$
$$-2x_2 - 2x_3 = -10$$
$$-6x_3 = -18$$

Compute $x_3 = 3$ from the last equation, substitute into the previous ones, and arrive at a triangular system of $n - 1 = 2$ equations. In $n - 1$ (in our case, 2) such recursive substitution steps, we compute the solution.

The triangular case is itself important; furthermore, every nonsingular linear system is reduced to two triangular ones by means of *forward elimination* of the variables, which essentially amounts to computing the PLU factorization of the input matrix A, that is, to computing two lower triangular matrices L and U^T (where L has unit values on its diagonal) and a permutation matrix P such that $A = PLU$. [A permutation matrix P is filled with zeros and ones and has exactly one nonzero entry in each row and in each column; in particular, this implies that $P^T = P^{-1}$. Pu has the same components as u but written in a distinct (fixed) order, for any vector u]. As soon as the latter factorization is available, we may compute $x = A^{-1} v$ by solving two triangular systems, that is, at first, $Ly = P^T v$, in y, and then $Ux = y$, in x. Computing the factorization (elimination stage) is more costly than the subsequent *back substitution stage*, the latter involving about $2n^2$ ops. The Gaussian classical algorithm for elimination requires about $2n^3/3$ ops, not counting some comparisons, generally required in order to ensure appropriate *pivoting*, also called *elimination ordering*. Pivoting enables us to avoid divisions by small values, which could have caused numerical stability problems. Theoretically, one may employ fast matrix multiplication and compute the matrices P, L, and U in $O(n^{2.376})$ ops [Aho et al. 1974] [and then compute the vectors y and x in $O(n^2)$ ops]. Pivoting can be dropped for some important classes of linear systems, notably, for *positive definite* and for *diagonally dominant* systems [Golub and Van Loan 1989, Pan 1991, 1992b, Bini and Pan 1994].

We refer the reader to Golub and Van Loan [1989, pp. 82–83], or Pan [1992b, p. 794], on sensitivity of the solution to the input and roundoff errors in numerical computing. The output errors grow with the **condition number** of A, represented by $\|A\| \|A^{-1}\|$ for an appropriate matrix norm or by the ratio of maximum and minimum singular values of A. Except for ill-conditioned linear systems $Ax = v$, for which the condition number of A is very large, a rough initial approximation to the solution can be rapidly refined (cf. Golub and Van Loan [1989]) via the *iterative improvement algorithm*, as soon as we know P and rough approximations to the matrices L and U of the PLU factorization of A. Then b correct bits of each output value can be computed in $(b + n)n^2$ ops as $b \to \infty$.

8.2.4   Singular Linear Systems of Equations

If the matrix A is **singular** (in particular, if A is rectangular), then the linear system $Ax = v$ is either overdetermined, that is, has no solution, or underdetermined, that is, has infinitely many solution vectors. All of them can be represented as $\{x_0 + y\}$, where x_0 is a fixed solution vector and y is a vector from the *null space* of A, $\{y : Ay = 0\}$, that is, y is a solution of the homogeneous linear system $Ay = 0$. (The null space of an $n \times n$ matrix A is a linear space of the dimension n–rank A.) A vector x_0 and a basis for the null-space of A can be computed by using $O(n^{2.376})$ ops if A is an $n \times n$ matrix or by using $O(mn^{1.736})$ ops if A is an $m \times n$ or $n \times m$ matrix and if $m \geq n$ (cf. Bini and Pan [1994]).

For an overdetermined linear system $Ax = v$, having no solution, one may compute a vector x minimizing the norm of the residual vector, $\|v - Ax\|$. It is most customary to minimize the Euclidean norm,

$$\|u\| = \left(\sum_i |u_i|^2 \right)^{1/2}, \quad u = v - Ax = (u_i)$$

This defines a least-squares solution, which is relatively easy to compute both practically and theoretically ($O(n^{2.376})$ ops suffice in theory) (cf. Bini and Pan [1994] and Golub and Van Loan [1989]).

8.2.5 Sparse Linear Systems (Including Banded Systems), Direct and Iterative Solution Algorithms

A matrix is sparse if it is filled mostly with zeros, say, if its all nonzero entries lie on 3 or 5 of its diagonals. In many important applications, in particular, solving partial and ordinary differential equations (PDEs and ODEs), one has to solve linear systems whose matrix is sparse and where, moreover, the disposition of its nonzero entries has a certain structure. Then, memory space and computation time can be dramatically decreased (say, from order n^2 to order $n \log n$ words of memory and from n^3 to $n^{3/2}$ or $n \log n$ ops) by using some special data structures and special solution methods. The methods are either direct, that is, are modifications of Gaussian elimination with some special policies of elimination ordering that preserve sparsity during the computation (notably, *Markowitz rule* and *nested dissection* [George and Liu 1981, Gilbert and Tarjan 1987, Lipton et al. 1979, Pan 1993]), or various iterative algorithms. The latter algorithms rely either on computing Krylov sequences [Saad 1995] or on multilevel or multigrid techniques [McCormick 1987, Pan and Reif 1992], specialized for solving linear systems that arise from discretization of PDEs. An important particular class of sparse linear systems is formed by *banded linear systems* with $n \times n$ coefficient matrices $A = (a_{i,j})$ where $a_{i,j} = 0$ if $i - j > g$ or $j - i > h$, for $g + h$ being much less than n. For banded linear systems, the nested dissection methods are known under the name of *block cyclic reduction* methods and are highly effective, but Pan et al. [1995] give some alternative algorithms, too. Some special techniques for computation of Krylov sequences for sparse and other special matrices A can be found in Pan [1995a]; according to these techniques, Krylov sequence is recovered from the solution of the associated linear system $(I - A) \, x = v$, which is solved fast in the case of a special matrix A.

8.2.6 Dense and Structured Matrices and Linear Systems

Many dense $n \times n$ matrices are defined by $O(n)$, say, by less than $2n$, parameters and can be multiplied by a vector by using $O(n \log n)$ or $O(n \log^2 n)$ ops. Such matrices arise in numerous applications (to signal and image processing, coding, algebraic computation, PDEs, integral equations, particle simulation, Markov chains, and many others). An important example is given by $n \times n$ *Toeplitz matrices* $T = (t_{i,j})$, $t_{i,j} = t_{i+1,j+1}$ for $i, j = 0, 1, \ldots, n-1$. Such a matrix can be represented by $2n - 1$ entries of its first row and first column or by $2n - 1$ entries of its first and last columns. The product Tv is defined by vector convolution, and its computation uses $O(n \log n)$ ops. Other major examples are given by *Hankel matrices* (obtained by reflecting the row or column sets of Toeplitz matrices), *circulant* (which are a subclass of Toeplitz matrices), and *Bezout, Sylvester, Vandermonde,* and *Cauchy* matrices. The known solution algorithms for linear systems with such dense structured coefficient matrices use from order $n \log n$ to order $n \log^2 n$ ops. These properties and algorithms are extended via associating some linear operators of displacement and scaling to some more general classes of matrices and linear systems. We refer the reader to Bini and Pan [1994] for many details and further bibliography.

8.2.7 Parallel Matrix Computations

Algorithms for matrix multiplication are particularly suitable for parallel implementation; one may exploit natural association of processors to rows and/or columns of matrices or to their blocks, particularly, in the implementation of matrix multiplication on loosely coupled multiprocessors (cf. Golub and Van Loan [1989] and Quinn [1994]). This motivated particular attention to and rapid progress in devising effective parallel algorithms for block matrix computations. The complexity of parallel computations is usually represented by the computational and communication time and the number of processors involved; decreasing all of these parameters, we face a tradeoff; the product of time and processor bounds (called potential work of parallel algorithms) cannot usually be made substantially smaller than the sequential time bound for the solution. This follows because, according to a variant of *Brent's scheduling principle*, a single processor can simulate the work of s processors in time $O(s)$. The usual goal of designing a parallel algorithm is in decreasing its parallel time bound (ideally, to a constant, logarithmic or polylogarithmic level, relative to n) and keeping its work bound at the level of the record sequential time bound for the same computational problem (within constant, logarithmic, or at worst polylog factors). This goal has been easily achieved for matrix and vector multiplications, but turned out to be nontrivial for linear system solving, inversion, and some other related computational problems. The recent solution for general matrices [Kaltofen and Pan 1991, 1992] relies on computation of a Krylov sequence and the coefficients of the minimum polynomial of a matrix, by using randomization and auxiliary computations with structured matrices (see the details in Bini and Pan [1994]).

8.2.8 Rational Matrix Computations, Computations in Finite Fields and Semirings

Rational algebraic computations with matrices are performed for a rational input given with no errors, and the computations are also performed with no errors. The precision of computing can be bounded by reducing the computations modulo one or several fixed primes or prime powers. At the end, the exact output values $z = p/q$ are recovered from $z \mod M$ (if M is sufficiently large relative to p and q) by using the continued fraction approximation algorithm, which is the Euclidean algorithm applied to integers (cf. Pan [1991, 1992a], and Bini and Pan [1994, Section 3 of Chap. 3]). If the output z is known to be an integer lying between $-m$ and m and if $M > 2m$, then z is recovered from $z \mod M$ as follows:

$$z = \begin{cases} z \mod M & \text{if } z \mod M < m \\ -M + z \mod M & \text{otherwise} \end{cases}$$

The reduction modulo a prime p may turn a nonsingular matrix A and a nonsingular linear system $Ax = v$ into singular ones, but this is proved to occur only with a low probability for a random choice of the prime p in a fixed sufficiently large interval (see Bini and Pan [1994, Section 3 of Chap. 4]). To compute the output values z modulo M for a large M, one may first compute them modulo several relatively prime integers $m_1, m_2, \ldots, m_k$ having no common divisors and such that $m_1, m_2, \ldots, m_k > M$ and then easily recover $z \mod M$ by means of the Chinese remainder algorithm. For matrix and polynomial computations, there is an effective alternative technique of p-adic (*Newton–Hensel*) *lifting* (cf. Bini and Pan [1994, Section 3 of Chap. 3]), which is particularly powerful for computations with dense structured matrices, since it preserves the structure of a matrix. We refer the reader to Bareiss [1968] and Geddes et al. [1992] for some special techniques, which enable one to control the growth of all intermediate values computed in the process of performing rational Gaussian elimination, with no roundoff and no reduction modulo an integer.

Gondran and Minoux [1984] and Pan [1993] describe some applications of matrix computations on semirings (with no divisions and subtractions allowed) to graph and combinatorial computations.

8.2.9 Matrix Eigenvalues and Singular Values Problems

The matrix eigenvalue problem is one of the major problems of matrix computation: given an $n \times n$ matrix A, one seeks a $k \times k$ diagonal matrix Λ and an $n \times k$ matrix V of full rank k such that

$$AV = \Lambda V \tag{8.1}$$

The diagonal entries of Λ are called the *eigenvalues* of A; the entry (i, i) of Λ is associated with the ith column of V, called an *eigenvector* of A. The eigenvalues of an $n \times n$ matrix A coincide with the zeros of the characteristic polynomial

$$c_A(x) = \det(xI - A)$$

If this polynomial has n distinct zeros, then $k = n$, and V of Equation 8.1 is a nonsingular $n \times n$ matrix. The matrix $A = I + Z$, where $Z = (z_{i,j})$, $z_{i,j} = 0$ unless $j = i + 1$, $z_{i,i+1} = 1$, is an example of a matrix for which $k = 1$, so that the matrix V degenerates to a vector.

In principle, one may compute the coefficients of $c_A(x)$, the characteristic polynomial of A, and then approximate its zeros (see Section 8.3) in order to approximate the eigenvalues of A. Given the eigenvalues, the corresponding eigenvectors can be recovered by means of the inverse power iteration [Golub and Van Loan 1989, Wilkinson 1965]. Practically, the computation of the eigenvalues via the computation of the coefficients of $c_A(x)$ is not recommended, due to arising numerical stability problems [Wilkinson 1965], and most frequently, the eigenvalues and eigenvectors of a general (unsymmetric) matrix are approximated by means of the QR *algorithm* [Wilkinson 1965, Watkins 1982, Golub and Van Loan 1989]. Before application of this algorithm, the matrix A is simplified by transforming it into the more special (*Hessenberg*) *form H*, by a *similarity transformation*,

$$H = UAU^H \tag{8.2}$$

where $U = (u_{i,j})$ is a unitary matrix, where $U^H U = I$, where $U^H = (\overline{u}_{j,i})$ is the Hermitian transpose of U, with $\overline{z}$ denoting the complex conjugate of z; $U^H = U^T$ if U is a real matrix [Golub and Van Loan 1989]. Similarity transformation into Hessenberg form is one of examples of *rational transformations* of a matrix into special *canonical forms*, of which transformations into *Smith* and *Hermite forms* are two other most important representatives [Kaltofen et al. 1990, Geddes et al. 1992, Giesbrecht 1995].

In practice, the eigenvalue problem is very frequently symmetric, that is, arises for a real symmetric matrix A, for which

$$A^T = (a_{j,i}) = A = (a_{i,j})$$

or for complex Hermitian matrices A, for which

$$A^H = (\overline{a}_{j,i}) = A = (a_{i,j})$$

For real symmetric or Hermitian matrices A, the eigenvalue problem (called symmetric) is treated much more easily than in the unsymmetric case. In particular, in the symmetric case, we have $k = n$, that is, the matrix V of Equation 8.1 is a nonsingular $n \times n$ matrix, and moreover, all of the eigenvalues of A are real and little sensitive to small input perturbations of A (according to the Courant–Fisher minimization criterion [Parlett 1980, Golub and Van Loan 1989]).

Furthermore, similarity transformation of A to the Hessenberg form gives much stronger results in the symmetric case: the original problem is reduced to one for a symmetric tridiagonal matrix H of Equation 8.2 (this can be achieved via the Lanczos algorithm, cf. Golub and Van Loan [1989] or Bini and Pan [1994, Section 3 of Chap. 2]). For such a matrix H, application of the QR algorithm is dramatically simplified; moreover, two competitive algorithms are also widely used, that is, the *bisection* [Parlett 1980] (a slightly slower but very robust algorithm) and the *divide-and-conquer* method [Cuppen 1981, Golub and Van Loan 1989]. The latter method has a modification [Bini and Pan 1991] that only uses $O(n \log^2 n (\log n + \log^2 b))$ arithmetic operations in order to compute all of the eigenvalues of an $n \times n$ symmetric tridiagonal matrix A within the output error bound $2^{-b} \|A\|$, where $\|A\| \le n \max |a_{i,j}|$.

The eigenvalue problem has a generalization, where generalized eigenvalues and eigenvectors for a pair A, B of matrices are sought, such that

$$AV = B\Lambda V$$

(the solution algorithm should proceed without computing the matrix $B^{-1}A$, so as to avoid numerical stability problems).

In another highly important extension of the symmetric eigenvalue problem, one seeks a singular value decomposition (SVD) of a (generally unsymmetric and, possibly, rectangular) matrix A: $A = U\Sigma V^T$, where U and V are unitary matrices, $U^H U = V^H V = I$, and Σ is a diagonal (generally rectangular) matrix, filled with zeros, except for its diagonal, filled with (positive) singular values of A and possibly, with zeros. The SVD is widely used in the study of numerical stability of matrix computations and in numerical treatment of singular and ill-conditioned (close to singular) matrices. An alternative tool is orthogonal (QR) factorization of a matrix, which is not as refined as SVD but is a little easier to compute [Golub and Van Loan 1989]. The squares of the singular values of A equal the eigenvalues of the Hermitian (or real symmetric) matrix $A^H A$, and the SVD of A can be also easily recovered from the eigenvalue decomposition of the Hermitian matrix

$$\begin{bmatrix} 0 & A^H \\ A & 0 \end{bmatrix}$$

but more popular are some effective direct methods for the computation of the SVD [Golub and Van Loan 1989].

8.2.10 Approximating Polynomial Zeros

Solution of an nth degree polynomial equation,

$$p(x) = \sum_{i=0}^{n} p_i \, x^i = 0, \quad p_n \neq 0$$

(where one may assume that $p_{n-1} = 0$; this can be ensured via shifting the variable x) is a classical problem that has greatly influenced the development of mathematics throughout the centuries [Pan 1995b]. The problem remains highly important for the theory and practice of present day computing, and dozens of new algorithms for its approximate solution appear every year. Among the existent implementations of such algorithms, the practical heuristic champions in efficiency (in terms of computer time and memory space used, according to the results of many experiments) are various modifications of *Newton's iteration*, $z(i + 1) = z(i) - a(i)p(z(i))/p'(z(i))$, $a(i)$ being the step-size parameter [Madsen 1973], *Laguerre's method* [Hansen et al. 1977, Foster 1981], and the randomized *Jenkins–Traub algorithm* [1970] [all three for approximating a single zero z of $p(x)$], which can be extended to approximating other zeros by means of deflation of the input polynomial via its numerical division by $x - z$. For simultaneous approximation of all of the zeros of $p(x)$ one may apply the Durand–Kerner algorithm, which is defined by the following recurrence:

$$z_j(i+1) = \frac{z_j(i) - p((z_j(i)))}{z_j(i) - z_k(i)}, \quad j = 1, \ldots, n, \quad i = 1, 2, \ldots \tag{8.3}$$

Here, the customary choice for the n initial approximations $z_j(0)$ to the n zeros of

$$p(x) = p_n \prod_{j=1}^{n} (x - z_j)$$

is given by $z_j(0) = Z \exp(2\pi\sqrt{-1}/n), j = 1, \ldots, n$, with Z exceeding (by some fixed factor $t > 1$) $\max_j |z_j|$; for instance, one may set

$$Z = 2t \max_{i<n} (p_i/p_n) \tag{8.4}$$

For a fixed i and for all j, the computation according to Equation 8.3 is simple, only involving order n^2 ops, and according to the results of many experiments, the iteration Equation 8.3 rapidly converges to the solution, though no theory confirms or explains these results. Similar is the situation with various modifications of this algorithm, which are now even more popular than the original algorithms and many of which are listed in Pan [1992a, 1992b] (also cf. Bini and Pan [1996] and McNamee [1993]).

On the other hand, there are two groups of algorithms that, when implemented, promise to be competitive or even substantially superior to Newton's and Laguerre's iteration, the algorithm by Jenkins and Traub, and all of the algorithms of the Durand–Kerner type. One such group is given by the modern modifications and improvements (due to Pan [1987, 1994a, 1994b] and Renegar [1989]) of *Weyl's quadtree construction* of 1924. In this approach, an initial square S, containing all the zeros of $p(x)$ [say, $S = \{x, |Im\ x| < Z, |Re\ x| < Z\}$ for Z of Eq. (8.4)], is recursively partitioned into four congruent subsquares. In the center of each of them, a proximity test is applied that estimates the distance from this center to the closest zero of $p(x)$. If such a distance exceeds one-half of the diagonal length, then the subsquare contains no zeros of $p(x)$ and is discarded. When this process ensures a strong isolation from each other for the components formed by the remaining squares, then certain extensions of Newton's iteration [Renegar 1989, Pan 1994a, 1994b], or some iterative techniques based on numerical integration [Pan 1987] are applied and very rapidly converge to the desired approximations to the zeros of $p(x)$, within the error bound $2^{-b}Z$ for Z of Equation 8.4. As a result, the algorithms of Pan [1987, 1994a, 1994b] solve the entire problem of approximating (within $2^{-b}Z$) all of the zeros of $p(x)$ at the overall cost of performing $O((n^2 \log n) \log(bn))$ ops (cf. Bini and Pan [1996]), versus order n^2 operations at each iteration of Durand–Kerner type.

The second group is given by the divide-and-conquer algorithms. They first compute a sufficiently wide annulus A, which is free of the zeros of $p(x)$ and contains comparable numbers of such zeros (that is, the same numbers up to a fixed constant factor) in its exterior and its interior. Then the two factors of $p(x)$ are numerically computed, that is, $F(x)$ having all its zeros in the interior of the annulus, and $G(x) = p(x)/F(x)$ having no zeros there. The same process is recursively repeated for $F(x)$ and $G(x)$ until factorization of $p(x)$ into the product of linear factors is computed numerically. From this factorization, approximations to all of the zeros of $p(x)$ are obtained. The algorithms of Pan [1995a, 1996] based on this approach only require $O(n \log(bn) (\log n)^2)$ ops in order to approximate all of the n zeros of $p(x)$ within $2^{-b}Z$ for Z of Eq. (8.4). (Note that this is a quite sharp bound: at least n ops are necessary in order to output n distinct values.)

The computations for the polynomial zero problem are ill conditioned, that is, they generally require a high precision for the worst-case input polynomials in order to ensure a required output precision, no matter which algorithm is applied for the solution. Consider, for instance, the polynomial $(x - \frac{6}{7})^n$ and perturb its x-free coefficient by 2^{-bn}. Observe the resulting jumps of the zero $x = 6/7$ by 2^{-b}, and observe similar jumps if the coefficients p_i are perturbed by $2^{(i-n)b}$ for $i = 1, 2, \ldots, n - 1$. Therefore, to ensure the output precision of b bits, we need an input precision of at least $(n - i)b$ bits for each coefficient $p_i, i = 0, 1, \ldots, n - 1$. Consequently, for the worst-case input polynomial $p(x)$, any solution algorithm needs at least about a factor n increase of the precision of the input and of computing versus the output precision.

Numerically unstable algorithms may require even a higher input and computation precision, but inspection shows that this is not the case for the algorithms of Pan [1987, 1994a, 1994b, 1995a, 1996] and Renegar [1989] (cf. Bini and Pan [1996]).

8.2.11 Fast Fourier Transform and Fast Polynomial Arithmetic

To yield the record complexity bounds for approximating polynomial zeros, one should exploit fast algorithms for basic operations with polynomials (their multiplication, division, and transformation under the shift of the variable), as well as FFT, both directly and for supporting the fast polynomial arithmetic. The FFT and fast basic polynomial algorithms (including those for multipoint polynomial evaluation and interpolation) are the basis for many other fast polynomial computations, performed both numerically and symbolically (compare the next sections). These basic algorithms, their impact on the field of algebraic computation, and their complexity estimates have been extensively studied in Aho et al. [1974], Borodin and Munro [1975], and Bini and Pan [1994].

8.3 Systems of Nonlinear Equations
and Other Applications

Given a system $\{p_1(x_1,\ldots,x_n), p_2(x_1,\ldots,x_n),\ldots, p_r(x_1,\ldots,x_n)\}$ of nonlinear polynomials with rational coefficients [each $p_i(x_1,\ldots,x_n)$ is said to be an element of $\mathbb{Q}[x_1,\ldots,x_n]$, the ring of polynomials in $x_1,\ldots,x_n$ over the field $\mathbb{Q}$ of rational numbers], the n-tuple of complex numbers $(a_1,\ldots,a_n)$ is a common solution of the system, if $f_i(a_1,\ldots,a_n) = 0$ for each i with $1 \le i \le r$. In this section, we explore the problem of exactly solving a system of nonlinear equations over the field $\mathbb{Q}$. We provide an overview and cite references to different symbolic techniques used for solving systems of algebraic (polynomial) equations. In particular, we describe methods involving *resultant* and *Gröbner basis* computations.

The *Sylvester resultant method* is the technique most frequently utilized for determining a common zero of two polynomial equations in one variable [Knuth 1981]. However, using the Sylvester method successively to solve a system of multivariate polynomials proves to be inefficient. Successive resultant techniques, in general, lack efficiency as a result of their sensitivity to the ordering of the variables [Kapur and Lakshman 1992]. It is more efficient to eliminate all variables together from a set of polynomials, thus leading to the notion of the *multivariate resultant*. The three most commonly used multivariate resultant formulations are the *Dixon* [Dixon 1908, Kapur and Saxena 1995], *Macaulay* [Macaulay 1916, Canny 1990, Kaltofen and Lakshman 1988], and *sparse resultant formulations* [Canny and Emiris 1993a, Sturmfels 1991].

The theory of Gröbner bases provides powerful tools for performing computations in multivariate polynomial rings. Formulating the problem of solving systems of polynomial equations in terms of polynomial ideals, we will see that a Gröbner basis can be computed from the input polynomial set, thus allowing for a form of back substitution (cf. Section 8.2) in order to compute the common roots.

Although not discussed, it should be noted that the *characteristic set algorithm* can be utilized for polynomial system solving. Ritt [1950] introduced the concept of a characteristic set as a tool for studying solutions of algebraic differential equations. Wu [1984, 1986], in search of an effective method for automatic theorem proving, converted Ritt's method to ordinary polynomial rings. Given the before mentioned system P, the characteristic set algorithm transforms P into a triangular form, such that the set of common zeros of P is equivalent to the set of roots of the triangular system [Kapur and Lakshman 1992].

Throughout this exposition we will also see that these techniques used to solve nonlinear equations can be applied to other problems as well, such as computer-aided design and automatic geometric theorem proving.

8.3.1 Resultant Methods

The question of whether two polynomials $f(x), g(x) \in \mathbb{Q}[x]$,

$$f(x) = f_n x^n + f_{n-1} x^{n-1} + \cdots + f_1 x + f_0$$
$$g(x) = g_m x^m + g_{m-1} x^{m-1} + \cdots + g_1 x + g_0$$

have a common root leads to a condition that has to be satisfied by the coefficients of both f and g. Using a derivation of this condition due to Euler, the *Sylvester matrix* of f and g (which is of order $m + n$) can be formulated. The vanishing of the determinant of the Sylvester matrix, known as the *Sylvester resultant*, is a necessary and sufficient condition for f and g to have common roots [Knuth 1981].

As a running example let us consider the following system in two variables provided by Lazard [1981]:

$$f = x^2 + xy + 2x + y - 1 = 0$$
$$g = x^2 + 3x - y^2 + 2y - 1 = 0$$

The Sylvester resultant can be used as a tool for eliminating several variables from a set of equations [Kapur and Lakshman 1992]. Without loss of generality, the roots of the Sylvester resultant of f and g treated as polynomials in y, whose coefficients are polynomials in x, are the x-coordinates of the common zeros of

f and g. More specifically, the Sylvester resultant of the Lazard system with respect to y is given by the following determinant:

$$\det\left(\begin{bmatrix} x+1 & x^2+2x-1 & 0 \\ 0 & x+1 & x^2+2x-1 \\ -1 & 2 & x^2+3x-1 \end{bmatrix}\right) = -x^3 - 2x^2 + 3x$$

The roots of the Sylvester resultant of f and g are $\{-3, 0, 1\}$. For each x value, one can substitute the x value back into the original polynomials yielding the solutions $(-3, 1), (0, 1), (1, -1)$.

The method just outlined can be extended recursively, using *polynomial GCD computations*, to a larger set of multivariate polynomials in $\mathbb{Q}[x_1, \ldots, x_n]$. This technique, however, is impractical for eliminating many variables, due to an explosive growth of the degrees of the polynomials generated in each elimination step.

The Sylvester formulations have led to a *subresultant theory*, developed simultaneously by G. E. Collins and W. S. Brown and J. Traub. The subresultant theory produced an efficient algorithm for computing polynomial GCDs and their resultants, while controlling intermediate expression swell [Brown 1971, Brown and Traub 1971, Collins 1967, 1971, Knuth 1981].

It should be noted that by adopting an implicit representation for symbolic objects, the intermediate expression swell introduced in many symbolic computations can be palliated. Recently, polynomial GCD algorithms have been developed that use implicit representations and thus avoid the computationally costly content and primitive part computations needed in those GCD algorithms for polynomials in explicit representation [Díaz and Kaltofen 1995, Kaltofen 1988, Kaltofen and Trager 1990].

The solvability of a set of nonlinear multivariate polynomials over the field $\mathbb{Q}$ can be determined by the vanishing of a generalization of the Sylvester resultant of two polynomials in a single variable.

Due to the special structure of the Sylvester matrix, Bézout developed a method for computing the resultant as a determinant of order $\max(m, n)$ during the 18th century. Cayley [1865] reformulated Bézout's method leading to Dixon's [1908] extension to the bivariate case. Dixon's method can be generalized to a set

$$\{p_1(x_1, \ldots, x_n), p_2(x_1, \ldots, x_n), \ldots, p_{n+1}(x_1, \ldots, x_n)\}$$

of $n + 1$ generic n-degree polynomials in n variables [Kapur et al. 1994]. The vanishing of the Dixon resultant is a necessary and sufficient condition for the polynomials to have a nontrivial projective common zero, and also a necessary condition for the existence of an affine common zero. The Dixon formulation gives the resultant up to a multiple, and hence in the affine case it may happen that the vanishing of the Dixon resultant does not necessarily indicate that the equations in question have a common root. A nontrivial multiple, known as the *projection operator*, can be extracted via a method based on so-called *rank subdeterminant computation* (RSC) [Kapur et al. 1994]. It should be noted that the RSC method can also be applied to the Macaulay and sparse resultant formulations as is detailed here.

In 1916, Macaulay constructed a resultant for n homogeneous polynomials in n variables, which simultaneously generalizes the Sylvester resultant and the determinant of a system of linear equations [Canny et al. 1989, Kapur and Lakshman 1992]. Like the Dixon formulation, the Macaulay resultant is a multiple of the resultant (except in the case of generic homogeneous polynomials, where it produces the exact resultant). For the Macaulay formulation, Canny [1990] has invented a general method that perturbs any polynomial system and extracts a nontrivial projection operator.

Using recent results pertaining to sparse polynomial systems [Gelfand et al. 1994, Sturmfels 1991, Sturmfels and Zelevinsky 1992], the mixed sparse resultant of a system of $n + 1$ sparse polynomials in n variables in its matrix form was given by Canny and Emiris [1993a] and consequently improved in Canny and Emiris [1993b, 1994]. Here, sparsity denotes that only certain monomials in each of the $n + 1$ polynomials have nonzero coefficients. The determinant of the sparse resultant matrix, such as the Macaulay and Dixon matrices, only yields a projection operation, not the exact resultant.

Suppose we are asked to find the common zeros of a set of n polynomials in n variables $\{p_1(x_1, \ldots, x_n), p_2(x_1, \ldots, x_n), \ldots, p_n(x_1, \ldots, x_n)\}$. By augmenting the polynomial set by a generic linear form [Canny 1990, Canny and Manocha 1991, Kapur and Lakshman 1992], one can construct the *u-resultant* of a given system of polynomials. The u-resultant factors into linear factors over the complex numbers, providing the common zeros of the given polynomials equations. The u-resultant method takes advantage of the properties of the multivariate resultant, and hence can be constructed using either Dixon's, Macaulay's, or sparse formulations.

Consider the previous example augmented by a generic linear form

$$f_1 = x^2 + xy + 2x + y - 1 = 0$$
$$f_2 = x^2 + 3x - y^2 + 2y - 1 = 0$$
$$f_1 = ux + vy + w = 0$$

As described in Canny et al. [1989], the following matrix M corresponds to the Macaulay u-resultant of the preceding system of polynomials, with z being the homogenizing variable:

$$M = \begin{bmatrix}
1 & 0 & 0 & 1 & 0 & 0 & 0 & 0 & 0 & 0 \\
1 & 1 & 0 & 0 & 1 & 0 & u & 0 & 0 & 0 \\
2 & 0 & 1 & 3 & 0 & 1 & 0 & u & 0 & 0 \\
0 & 1 & 0 & -1 & 0 & 0 & v & 0 & 0 & 0 \\
1 & 2 & 1 & 2 & 3 & 0 & w & v & u & 0 \\
-1 & 0 & 2 & -1 & 0 & 3 & 0 & w & 0 & u \\
0 & 0 & 0 & 0 & -1 & 0 & 0 & 0 & 0 & 0 \\
0 & 1 & 0 & 0 & 2 & -1 & 0 & 0 & v & 0 \\
0 & -1 & 1 & 0 & -1 & 2 & 0 & 0 & w & v \\
0 & 0 & -1 & 0 & 0 & -1 & 0 & 0 & 0 & w
\end{bmatrix}$$

It should be noted that

$$\det(M) = (u - v + w)(-3u + v + w)(v + w)(u - v)$$

corresponds to the affine solutions $(1, -1)$, $(-3, 1)$, $(0, 1)$, and one solution at infinity. An empirical comparison of the detailed resultant formulations can be found in Kapur and Saxena [1995]. Recently, the multivariate resultant formulations are being used for other applications such as *algebraic and geometric reasoning* [Kapur et al. 1994], *computer-aided design* [Stederberg and Goldman 1986], and for *implicitization and finding base points* [Chionh 1990].

8.3.2 Gröbner Bases

Solving systems of nonlinear equations can be formulated in terms of polynomial ideals [Becker and Weispfenning 1993, Geddes et al. 1992, Winkler 1996]. Let us first establish some terminology.

The ideal generated by a system of polynomial equations $p_1, \ldots, p_r$ over $\mathbb{Q}[x_1, \ldots, x_n]$ is the set of all linear combinations

$$(p_1, \ldots, p_r) = \{h_1 p_1 + \cdots + h_r p_r \mid h_1, \ldots, h_r \in \mathbb{Q}[x_1, \ldots, x_n]\}$$

The algebraic variety of $p_1, \ldots, p_r \in \mathbb{Q}[x_1, \ldots, x_n]$ is the set of their common zeros,

$$V(p_1, \ldots, p_r) = \{(a_1, \ldots, a_n) \in \mathbb{C}^n \mid f_1(a_1, \ldots, a_n) = \cdots = f_r(a_1, \ldots, a_n) = 0\}$$

A version of the *Hilbert Nullstellensatz* states that

$$V(p_1, \ldots, p_r) = \text{the empty set } \emptyset \iff 1 \in (p_1, \ldots, p_r) \text{ over } \mathbb{Q}[x_1, \ldots, x_n]$$

which relates the solvability of polynomial systems to the ideal membership problem.

A term $t = x_1^{e_1} x_2^{e_2} \ldots x_n^{e_n}$ of a polynomial is a product of powers with $\deg(t) = e_1 + e_2 + \cdots + e_n$. In order to add needed structure to the polynomial ring we will require that the terms in a polynomial be ordered in an admissible fashion [Geddes et al. 1992, Kapur and Lakshman 1992]. Two of the most common admissible orderings are the **lexicographic order** ($\prec_l$), where terms are ordered as in a dictionary, and the **degree order**

($\prec_d$), where terms are first compared by their degrees with equal degree terms compared lexicographically. A variation to the lexicographic order is the *reverse lexicographic order*, where the lexicographic order is reversed [Davenport et al. 1988, p. 96].

It is this previously mentioned structure that permits a type of simplification known as polynomial reduction. Much like a polynomial remainder process, the process of polynomial reduction involves subtracting a multiple of one polynomial from another to obtain a smaller degree result [Becker and Weispfenning 1993, Geddes et al. 1992, Kapur and Lakshman 1992, Winkler 1996].

A polynomial g is said to be reducible with respect to a set $P = \{p_1, \ldots, p_r\}$ of polynomials if it can be reduced by one or more polynomials in P. When g is no longer reducible by the polynomials in P, we say that g is *reduced* or is *a normal form* with respect to P.

For an arbitrary set of basis polynomials, it is possible that different reduction sequences applied to a given polynomial g could reduce to different normal forms. A basis $G \subseteq \mathbb{Q}[x_1, \ldots, x_n]$ is a *Gröbner basis* if and only if every polynomial in $\mathbb{Q}[x_1, \ldots, x_n]$ has a unique normal form with respect to G. Buchberger [1965, 1976, 1983, 1985] showed that every basis for an ideal $(p_1, \ldots, p_r)$ in $\mathbb{Q}[x_1, \ldots, x_n]$ can be converted into a Gröbner basis $\{p_1^*, \ldots, p_s^*\} = GB(p_1, \ldots, p_r)$, concomitantly designing an algorithm that transforms an arbitrary ideal basis into a Gröbner basis. Another characteristic of Gröbner bases is that by using the previously mentioned reduction process we have

$$g \in (p_1, \ldots, p_r) \quad \Longleftrightarrow \quad (g \bmod p_1^*, \ldots, p_s^*) = 0$$

Further, by using the Nullstellensatz it can be shown that $p_1, \ldots, p_r$ viewed as a system of algebraic equations is solvable if and only if $1 \notin GB(p_1, \ldots, p_r)$.

Depending on which admissible term ordering is used in the Gröbner bases construction, an ideal can have different Gröbner bases. However, an ideal cannot have different (reduced) Gröbner bases for the same term ordering.

Any system of polynomial equations can be solved using a lexicographic Gröbner basis for the ideal generated by the given polynomials. It has been observed, however, that Gröbner bases, more specifically lexicographic Gröbner bases, are hard to compute [Becker and Weispfenning 1993, Geddes et al. 1992, Lakshman 1990, Winkler 1996]. In the case of zero-dimensional ideals, those whose varieties have only isolated points, Faugère, et al. [1993] outlined a change of basis algorithm which can be utilized for solving zero-dimensional systems of equations. In the zero-dimensional case, one computes a Gröbner basis for the ideal generated by a system of polynomials under a degree ordering. The so-called *change of basis algorithm* can then be applied to the degree ordered Gröbner basis to obtain a Gröbner basis under a lexicographic ordering.

Turning to Lazard's example in the form of a polynomial basis,

$$f_1 = x^2 + xy + 2x + y - 1$$
$$f_2 = x^2 + 3x - y^2 + 2y - 1$$

one obtains (under lexicographical ordering with $x \prec_l y$) a Gröbner basis in which the variables are triangularized such that the finitely many solutions can be computed via back substitution:

$$f_1^* = x^2 + 3x + 2y - 2$$
$$f_2^* = xy - x - y + 1$$
$$f_3^* = y^2 - 1$$

It should be noted that the final univariate polynomial is of minimal degree and the polynomials used in the back substitution will have degree no larger than the number of roots.

As an example of the process of polynomial reduction with respect to a Gröbner basis, the following demonstrates two possible reduction sequences to the same normal form. The polynomial $x^2 y^2$ is reduced with respect to the previously computed Gröbner basis $\{f_1^*, f_2^*, f_3^*\} = GB(f_1, f_2)$ along the following two distinct reduction paths, both yielding $-3x - 2y + 2$ as the normal form.

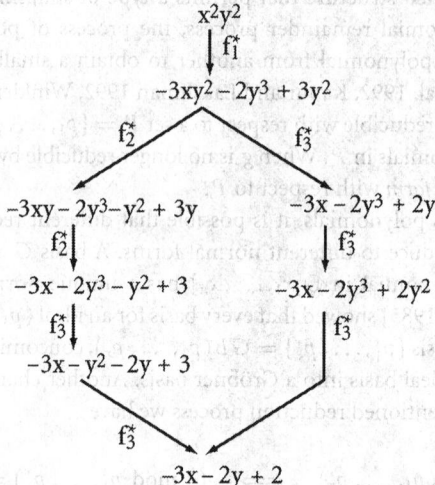

There is a strong connection between lexicographic Gröbner bases and the previously mentioned resultant techniques. For some types of input polynomials, the computation of a reduced system via resultants might be much faster than the computation of a lexicographic Gröbner basis. A good comparison between the Gröbner computations and the different resultant formulations can be found in Kapur and Saxena [1995].

In a survey article, Buchberger [1985] detailed how Gröbner bases can be used as a tool for many polynomial ideal theoretic operations. Other applications of Gröbner basis computations include automatic geometric theorem proving [Kapur 1986, Wu 1984, 1986], multivariate polynomial factorization and GCD computations [Gianni and Trager 1985], and polynomial interpolation [Lakshman and Saunders 1994, 1995].

8.4 Polynomial Factorization

The problem of factoring polynomials is a fundamental task in symbolic algebra. An example in one's early mathematical education is the factorization $x^2 - y^2 = (x + y) \cdot (x - y)$, which in algebraic terms is a factorization of a polynomial in two variables with integer coefficients. Technology has advanced to a state where most polynomial factorization problems are doable on a computer, in particular, with any of the popular mathematical software, such as the Mathematica or Maple systems. For instance, the factorization of the determinant of a 6×6 symmetric Toeplitz matrix over the integers is computed in Maple as

```
> readlib(showtime):
> showtime():
O1 := T := linalg[toeplitz]([a, b, c, d, e, f]);
```

$$T := \begin{bmatrix} a & b & c & d & e & f \\ b & a & b & c & d & e \\ c & b & a & b & c & d \\ d & c & b & a & b & c \\ e & d & c & b & a & b \\ f & e & d & c & b & a \end{bmatrix}$$

```
time 0.03 words7701
O2 := factor(linalg[det](T));
```

$$-(2dca - 2bce + 2c^2a - a^3 - da^2 + 2d^2c + d^2a + b^3 + 2abc - 2c^2b$$
$$+ d^3 + 2ab^2 - 2dcb - 2cb^2 - 2ec^2 + 2eb^2 + 2fcb + 2bae$$
$$+ b^2f + c^2f + be^2 - ba^2 - fdb - fda - fa^2 - fba + e^2a - 2db^2$$
$$+ dc^2 - 2deb - 2dec - dba)(2dca - 2bce - 2c^2a + a^3$$
$$- da^2 - 2d^2c - d^2a + b^3 + 2abc - 2c^2b + d^3 - 2ab^2 + 2dcb$$
$$+ 2cb^2 + 2ec^2 - 2eb^2 - 2fcb + 2bae + b^2f + c^2f + be^2 - ba^2$$
$$- fdb + fda - fa^2 + fba - e^2a - 2db^2 + dc^2 + 2deb - 2dec$$
$$+ dba)$$

```
time 27.30 words 857700
```

Clearly, the Toeplitz determinant factorization requires more than tricks from high school algebra. Indeed, the development of modern algorithms for the polynomial factorization problem is one of the great successes of the discipline of symbolic mathematical computation. Kaltofen [1982, 1990, 1992] has surveyed the algorithms until 1992, mostly from a computer science perspective. In this chapter we shall focus on the applications of the known fast methods to problems in science and engineering. For a more extensive set of references, please refer to Kaltofen's survey articles.

8.4.1 Polynomials in a Single Variable over a Finite Field

At first glance, the problem of factoring an integer polynomial modulo a prime number appears to be very similar to the problem of factoring an integer represented in a prime radix. That is simply not so. The factorization of the polynomial $x^{511} - 1$ can be done modulo 2 on a computer in a matter of milliseconds, whereas the factorization of the integer $2^{511} - 1$ into its integer factors is a computational challenge. For those interested: the largest prime factors of $2^{511} - 1$ have 57 and 67 decimals digits, respectively, which makes a tough but not undoable 123 digit product for the number field sieve factorizer [Leyland 1995]. Irreducible factors of polynomials modulo 2 are needed to construct finite fields. For example, the factor $x^9 + x^4 + 1$ of $x^{511} - 1$ leads to a model of the finite field with 2^9 elements, GF(2^9), by simply computing with the polynomial remainders modulo $x^9 + x^4 + 1$ as the elements. Such irreducible polynomials are used for setting up error-correcting codes, such as the BCH codes [MacWilliams and Sloan 1977]. Berlekamp's [1967, 1970] pioneering work on factoring polynomials over a finite field by linear algebra is done with this motivation. The linear algebra tools that Berlekamp used seem to have been introduced to the subject as early as in 1937 by Petr (cf. Št. Schwarz [1956]).

Today, factoring algorithms for univariate polynomials over finite fields form the innermost subalgorithm to lifting-based algorithms for factoring polynomials in one [Zassenhaus 1969] and many [Musser 1975] variables over the integers. When Maple computed the factorization of the previous Toeplitz determinant, it began with factoring a univariate polynomial modulo a prime integer. The case when the prime integer is very large has led to a significant development in computer science itself. As it turns out, by selecting random residues the expected performance of the algorithms can be speeded up exponentially [Berlekamp 1970, Rabin 1980]. Randomization is now an important tool for designing efficient algorithms and has proliferated to many fields of computer science. Paradoxically, the random elements are produced by a congruential random number generator, and the actual computer implementations are quite deterministic, which leads some computer scientists to believe that random bits can be eliminated in general at no exponential slow down. Nonetheless, for the polynomial factoring problem modulo a large prime, no fast methods are known to date that would work without this *probabilistic* approach.

One can measure the computing time of selected algorithms in terms of n, the degree of the input polynomial, and p, the cardinality of the field. When counting arithmetic operations modulo p (including reciprocals), the best known algorithms are quite recent. Berlekamp's 1970 method performs

$O(n^\omega + n^{1+o(1)} \log p)$ residue operations. Here and subsequently, ω denotes the exponent implied by the used linear system solver, i.e., $\omega = 3$ when classical methods are used, and $\omega = 2.376$ when asymptotically fast (though impractical) matrix multiplication is assumed. The correction term $o(1)$ accounts for the $\log n$ factors derived from the FFT-based fast polynomial multiplication and remaindering algorithms. An approach in the spirit of Berlekamp's but possibly more practical for $p = 2$ has recently been discovered by Niederreiter [1994]. A very different technique by Cantor and Zassenhaus [1981] first separates factors of different degrees and then splits the resulting polynomials of equal degree factors. It has $O(n^{2+o(1)} \log p)$ complexity and is the basis for the following two methods. Algorithms by von zur Gathen and Shoup [1992] have running time $O(n^{2+o(1)} + n^{1+o(1)} \log p)$ and those by Kaltofen and Shoup [1995] have running time $O(n^{1.815} \log p)$, the latter with fast matrix multiplication.

For n and p simultaneously large, a variant of the method by Kaltofen and Shoup [1995] that uses classical linear algebra and runs in $O(n^{2.5} + n^{1+o(1)} \log p)$ residue operations is the current champion among the practical algorithms. With it Shoup [1996], using his own fast polynomial arithmetic package, has factored a randomlike polynomial of degree 2048 modulo a 2048-bit prime number in about 12 days on a Sparc-10 computer using 68 megabyte of main memory. For even larger n, but smaller p, parallelization helps, and Kaltofen and Lobo [1994] could factor a polynomial of degree $n = 15\,001$ modulo $p = 127$ in about 6 days on 8 computers that are rated at 86.1 MIPS. At the time of this writing, the largest polynomial factored modulo 2 is $X^{216\,091} + X + 1$; this was accomplished by Peter Montgomery in 1991 by using Cantor's fast polynomial multiplication algorithm based on additive transforms [Cantor 1989].

8.4.2 Polynomials in a Single Variable over Fields of Characteristic Zero

As mentioned before, generally usable methods for factoring univariate polynomials over the rational numbers begin with the Hensel lifting techniques introduced by Zassenhaus [1969]. The input polynomial is first factored modulo a suitable prime integer p, and then the factorization is lifted to one modulo p^k for an exponent k of sufficient size to accommodate all possible integer coefficients that any factors of the polynomial might have. The lifting approach is fast in practice, but there are hard-to-factor polynomials on which it runs an exponential time in the degree of the input. This slowdown is due to so-called parasitic modular factors. The polynomial $x^4 + 1$, for example, factors modulo all prime integers but is irreducible over the integers: it is the cyclotomic equation for eighth roots of unity. The products of all subsets of modular factors are candidates for integer factors, and irreducible integer polynomials with exponentially many such subsets exist [Kaltofen et al. 1983].

The elimination of the exponential bottleneck by giving a polynomial-time solution to the integer polynomial factoring problem, due to Lenstra et al. [1982] is considered a major result in computer science algorithm design. The key ingredient to their solution is the construction of integer relations to real or complex numbers. For the simple demonstration of this idea, consider the polynomial

$$x^4 + 2x^3 - 6x^2 - 4x + 8$$

A root of this polynomial is $\alpha \approx 1.236067977$, and $\alpha^2 \approx 1.527864045$. We note that $2\alpha + \alpha^2 \approx 4.000000000$, hence $x^2 + 2x - 4$ is a factor. The main difficulty is to efficiently compute the integer linear relation with relatively small coefficients for the high-precision big-float approximations of the powers of a root. Lenstra et al. [1982] solve this diophantine optimization problem by means of their now famous lattice reduction procedure, which is somewhat reminiscent of the ellipsoid method for linear programming.

The determination of linear integer relations among a set of real or complex numbers is a useful task in science in general. Very recently, some stunning identities could be produced by this method, including the following formula for π [Finch 1995]:

$$\pi = \sum_{n=0}^{\infty} \frac{1}{16^n} \left(\frac{4}{8n+1} - \frac{2}{8n+4} - \frac{1}{8n+5} - \frac{1}{8n+6} \right)$$

Even more surprising, the lattice reduction algorithm can prove that no linear integer relation with integers smaller than a chosen parameter exists among the real or complex numbers. There is an efficient alternative to the lattice reduction algorithm, originally due to Ferguson and Forcade [1982] and recently improved by Ferguson and Bailey.

The complexity of factoring an integer polynomial of degree n with coefficients of no more than l bits is thus a polynomial in n and l. From a theoretical point of view, an algorithm with a low estimate is by Miller [1992] and has a running time of $O(n^{5+o(1)}l^{1+o(1)} + n^{4+o(1)}l^{2+o(1)})$ bit operations. It is expected that the relation-finding methods will become usable in practice on hard-to-factor polynomials in the near future. If the hard-to-factor input polynomial is irreducible, an alternate approach can be used to prove its irreducibility. One finds an integer evaluation point at which the integral value of the polynomial has a large prime factor, and the irreducibility follows by mathematical theorems. Monagan [1992] has proven large hard-to-factor polynomials irreducible in this way, which would be hopeless by the lifting algorithm.

Coefficient fields other than finite fields and the rational numbers are of interest. Computing the factorizations of univariate polynomials over the complex numbers is the root finding problem described in the earlier section Approximating Polynomial Zeros. When the coefficient field has an extra variable, such as the field of fractions of polynomials (rational functions) the problem reduces, by an old theorem of Gauss, to factoring multivariate polynomials, which we discuss subsequently. When the coefficient field is the field of Laurent series in t with a finite segment of negative powers,

$$\frac{c_{-k}}{t^k} + \frac{c_{-k+1}}{t^{k-1}} + \cdots + \frac{c_{-1}}{t} + c_0 + c_1 t + c_2 t^2 + \cdots, \quad \text{where } k \geq 0$$

fast methods appeal to the theory of Puiseux series, which constitute the domain of algebraic functions [Walsh 1993].

8.4.3 Polynomials in Two Variables

Factoring bivariate polynomials by reduction to univariate factorization via homomorphic projection and subsequent lifting can be done similarly to the univariate algorithm [Musser 1975]. The second variable y takes the role of the prime integer p and $f(x, y) \bmod y = f(x, 0)$. Lifting is possible only if $f(x, 0)$ had no multiple root. Provided that $f(x, y)$ has no multiple factor, which can be ensured by a simple GCD computation, the *squarefreeness* of $f(x, 0)$ can be obtained by variable translation $\hat{y} = y + a$, where a is an easy-to find constant in the coefficient field. For certain domains, such as the rational numbers, any irreducible multivariate polynomial $h(x, y)$ can be mapped to an irreducible univariate polynomial $h(x, b)$ for some constant b. This is the important *Hilbert irreducibility theorem*, whose consequence is that the combinatorial explosion observed in the univariate lifting algorithm is, in practice, unlikely. However, the magnitude and probabilistic distribution of good points b is not completely analyzed.

For so-called non-Hilbertian coefficient fields good reduction is not possible. An important such field is the complex number. Clearly, all $f(x, b)$ completely split into linear factors, while $f(x, y)$ may be irreducible over the complex numbers. An example of an irreducible polynomial is $f(x, y) = x^2 - y^3$. Polynomials that remain irreducible over the complex numbers are called absolutely irreducible. An additional problem is the determination of the algebraic extension of the ground field in which the absolutely irreducible factors can be expressed. In the example

$$x^6 - 2x^3 y^2 + y^4 - 2x^3 = (x^3 - \sqrt{2}x - y^2) \cdot (x^3 + \sqrt{2}x - y^2)$$

the needed extension field is $\mathbb{Q}(\sqrt{2})$. The relation-finding approach proves successful for this problem. The root is computed as a Taylor series in y, and the integrality of the linear relation for the powers of the series means that the multipliers are polynomials in y of bounded degree. Several algorithms of polynomial-time complexity and pointers to the literature are found in Kaltofen [1995].

Bivariate polynomials constitute implicit representations of algebraic curves. It is an important operation in geometric modeling to convert from implicit to parametric representation. For example, the circle

$$x^2 + y^2 - 1 = 0$$

has the rational parameterization

$$x = \frac{2t}{1 + t^2}, \qquad y = \frac{1 - t^2}{1 + t^2}, \quad \text{where } -\infty \le t \le \infty$$

Algorithms are known that can find such rational parameterizations provided that they exist [Sendra and Winkler 1991]. It is crucial that the inputs to these algorithms are absolutely irreducible polynomials.

8.4.4 Polynomials in Many Variables

Polynomials in many variables, such as the symmetric Toeplitz determinant previously exhibited, are rarely given explicitly, due to the fact that the number of possible terms grows exponentially in the number of variables: there can be as many as $\binom{n+v}{n} \ge 2^{\min\{n,v\}}$ terms in a polynomial of degree n with v variables. Even the factors may be dense in canonical representation, but could be sparse in another basis: for instance, the polynomial

$$(x_1 - 1)(x_2 - 2) \cdots (x_v - v) + 1$$

has only two terms in the shifted basis, whereas it has 2^v terms in the power basis, i.e., in expanded format.

Randomized algorithms are available that can efficiently compute a factor of an implicitly given polynomial, say, a matrix determinant, and even can find a shifted basis with respect to which a factor would be sparse, provided, of course, that such a shift exists. The approach is by manipulating polynomials in so-called black box representations [Kaltofen and Trager 1990]: a black box is an object that takes as input a value for each variable, and then produces the value of the polynomial it represents at the specified point. In the Toeplitz example the representation of the determinant could be the Gaussian elimination program which computes it. We note that the size of the polynomial in this case would be nearly constant, only the variable names and the dimension need to be stored. The factorization algorithm then outputs procedures which will evaluate all irreducible factors at an arbitrary point (supplied as the input). These procedures make calls to the black box given as input to the factorization algorithm in order to evaluate them at certain points, which are derived from the point at which the procedures computing the values of the factors are probed. It is, of course, assumed that subsequent calls evaluate one and the same factor and not associates that are scalar multiples of one another. The algorithm by Kaltofen and Trager [1990] finds procedures that with a controllably high probability evaluate the factors correctly. Randomization is needed to avoid parasitic factorizations of homomorphic images which provide some static data for the factor boxes and cannot be avoided without mathematical conjecture. The procedures that evaluate the individual factors are deterministic.

Factors constructed as black box programs are much more space efficient than those represented in other formats, for example, the straight-line program format [Kaltofen 1989]. More importantly, once the black box representation for the factors is found, sparse representations can be rapidly computed by any of the new sparse interpolation algorithms. See Grigoriev and Lakshman [1995] for the latest method allowing shifted bases and pointers to the literature of other methods, including those for the standard power bases.

The black box representation of polynomials is normally not supported by commercial computer algebra systems such as Axiom, Maple, or Mathematica. Díaz is currently developing the FOXBOX system in C++ that makes black box methodology available to users of such systems. It is anticipated that factorizations as those of large symmetric Toeplitz determinants will be possible on computers. Earlier implementations based on the straight-line program model [Freeman et al. 1988] could factor 16×16 group determinants, which represent polynomials of over 300 million terms.

Acknowledgment

This material is based on work supported in part by the National Science Foundation under Grants CCR-9319776 (first and second author) and CCR-9020690 (third author), by GTE under a Graduate Computer Science Fellowship (first author), and by PSC CUNY Awards 665301 and 666327 (third author). Part of this work was done while the second author was at the Department of Computer Science at Rensselaer Polytechnic Insititute in Troy, New York.

Defining Terms

Characteristic polynomial: A polynomial associated with a square matrix, the determinant of the matrix when a single variable is subtracted to its diagonal entries. The roots of the characteristic polynomial are the eigenvalues of the matrix.

Condition number: A scalar derived from a matrix that measures its relative nearness to a singular matrix. Very close to singular means a large condition number, in which case numeric inversion becomes an unstable process.

Degree order: An order of the terms in a multivariate polynomial; for two variables x and y with $x \prec y$ the ascending chain of terms is $1 \prec x \prec y \prec x^2 \prec xy \prec y^2 \cdots$.

Determinant: A polynomial in the entries of a square matrix with the property that its value is nonzero if and only if the matrix is invertible.

Lexicographic order: An order of the terms in a multivariate polynomial; for two variables x and y with $x \prec y$ the ascending chain of terms is $1 \prec x \prec x^2 \prec \cdots \prec y \prec xy \prec x^2 y \cdots \prec y^2 \prec xy^2 \cdots$.

Ops: Arithmetic operations, i.e., additions, subtractions, multiplications, or divisions; as in floating point operations (*flops*).

Singularity: A square matrix is singular if there is a nonzero second matrix such that the product of the two is the zero matrix. Singular matrices do not have inverses.

Sparse matrix: A matrix where many of the entries are zero.

Structured matrix: A matrix where each entry can be derived by a formula depending on few parameters. For instance, the Hilbert matrix has $1/(i + j - 1)$ as the entry in row i and column j.

References

Anderson, E. et al. 1992. *LAPACK Users' Guide*. SIAM Pub., Philadelphia, PA.

Aho, A., Hopcroft, J., and Ullman, J. 1974. *The Design and Analysis of Algorithms*. Addison–Wesley, Reading, MA.

Bareiss, E. H. 1968. Sylvester's identity and multistep integers preserving Gaussian elimination. *Math. Comp.* 22:565–578.

Becker, T. and Weispfenning, V. 1993. *Gröbner Bases: A Computational Approach to Commutative Algebra*. Springer–Verlag, New York.

Berlekamp, E. R. 1967. Factoring polynomials over finite fields. *Bell Systems Tech. J.* 46:1853–1859; rev. 1968. *Algebraic Coding Theory*. Chap. 6, McGraw–Hill, New York.

Berlekamp, E. R. 1970. Factoring polynomials over large finite fields. *Math. Comp.* 24:713–735.

Bini, D. and Pan, V. Y. 1991. Parallel complexity of tridiagonal symmetric eigenvalue problem. In *Proc. 2nd Annu. ACM-SIAM Symp. on Discrete Algorithms*, pp. 384–393. ACM Press, New York, SIAM Pub., 1994. Philadelphia, PA.

Bini, D. and Pan, V. Y. 1994. *Polynomial and Matrix Computations Vol. 1, Fundamental Algorithms*. Birkhäuser, Boston, MA.

Bini, D. and Pan, V. Y. 1996. *Polynomial and Matrix Computations, Vol. 2*. Birkhäuser, Boston, MA.

Borodin, A. and Munro, I. 1975. *Computational Complexity of Algebraic and Numeric Problems*. American Elsevier, New York.

Brown, W. S. 1971. On Euclid's algorithm and the computation of polynomial greatest common divisors. *J. ACM* 18:478–504.

Brown, W. S. and Traub, J. F. 1971. On Euclid's algorithm and the theory of subresultants. *J. ACM* 18:505–514.

Buchberger, B. 1965. *Ein Algorithmus zum Auffinden der Basiselemente des Restklassenringes nach einem nulldimensionalen Polynomideal.* Ph.D. dissertation. University of Innsbruck, Austria.

Buchberger, B. 1976. A theoretical basis for the reduction of polynomials to canonical form. *ACM SIGSAM Bull.* 10(3):19–29.

Buchberger, B. 1983. A note on the complexity of constructing Gröbner-bases. In *Proc. EUROCAL '83*, J. A. van Hulzen, ed. *Lecture Notes in Computer Science*, pp. 137–145. Springer.

Buchberger, B. 1985. Gröbner bases: an algorithmic method in polynomial ideal theory. In *Recent Trends in Multidimensional Systems Theory*, N. K. Bose, ed., pp. 184–232. D. Reidel, Dordrecht, Holland.

Cantor, D. G. 1989. On arithmetical algorithms over finite fields. *J. Combinatorial Theory, Serol. A* 50:285–300.

Canny, J. 1990. Generalized characteristic polynomials. *J. Symbolic Comput.* 9(3):241–250.

Canny, J. and Emiris, I. 1993a. An efficient algorithm for the sparse mixed resultant. In *Proc. AAECC-10*, G. Cohen, T. Mora, and O. Moreno, ed. Vol. 673, *Lecture Notes in Computer Science*, pp. 89–104. Springer.

Canny, J. and Emiris, I. 1993b. A practical method for the sparse resultant. In *ISSAC '93, Proc. Internat. Symp. Symbolic Algebraic Comput.*, M. Bronstein, ed., pp. 183–192. ACM Press, New York.

Canny, J. and Emiris, I. 1994. Efficient incremental algorithms for the sparse resultant and the mixed volume. Tech. Rep., Univ. California-Berkeley, CA.

Canny, J., Kaltofen, E., and Lakshman, Y. 1989. Solving systems of non-linear polynomial equations faster. In *Proc. ACM-SIGSAM Internat. Symp. Symbolic Algebraic Comput.*, pp. 121–128.

Canny, J. and Manocha, D. 1991. Efficient techniques for multipolynomial resultant algorithms. In *ISSAC '91, Proc. Internat. Symp. Symbolic Algebraic Comput.*, S. M. Watt, ed., pp. 85–95, ACM Press, New York.

Cantor, D. G. and Zassenhaus, H. 1981. A new algorithm for factoring polynomials over finite fields. *Math. Comp.* 36:587–592.

Cayley, A. 1865. On the theory of eliminaton. *Cambridge and Dublin Math. J.* 3:210–270.

Chionh, E. 1990. *Base Points, Resultants and Implicit Representation of Rational Surfaces.* Ph.D. dissertation. Department of Computer Science, University of Waterloo, Waterloo, Canada.

Collins, G. E. 1967. Subresultants and reduced polynomial remainder sequences. *J. ACM* 14:128–142.

Collins, G. E. 1971. The calculation of multivariate polynomial resultants. *J. ACM* 18:515–532.

Cuppen, J. J. M. 1981. A divide and conquer method for the symmetric tridiagonal eigenproblem. *Numer. Math.* 36:177–195.

Davenport, J. H., Tournier, E., and Siret, Y. 1988. *Computer Algebra Systems and Algorithms for Algebraic Computation.* Academic Press, London.

Díaz, A. and Kaltofen, E. 1995. On computing greatest common divisors with polynomials given by black boxes for their evaluation. In *ISSAC '95 Proc. 1995 Internat. Symp. Symbolic Algebraic Comput.*, A. H. M. Levelt, ed., pp. 232–239, ACM Press, New York.

Dixon, A. L. 1908. The elimination of three quantics in two independent variables. In *Proc. London Math. Soc.* Vol. 6, pp. 468–478.

Dongarra, J. et al. 1978. *LAPACK Users' Guide.* SIAM Pub., Philadelphia, PA.

Faugère, J. C., Gianni, P., Lazard, D., and Mora, T. 1993. Efficient computation of zero-dimensional Gröbner bases by change of ordering. *J. Symbolic Comput.* 16(4):329–344.

Ferguson, H. R. P. and Forcade, R. W. 1982. Multidimensional Euclidean algorithms. *J. Reine Angew. Math.* 334:171–181.

Finch, S. 1995. The miraculous Bailey–Borwein–Plouffe pi algorithm. Internet document, Mathsoft Inc., http://www.mathsoft.com/asolve/plouffe/plouffe.html, Oct.

Foster, L. V. 1981. Generalizations of Laguerre's method: higher order methods. *SIAM J. Numer. Anal.* 18:1004–1018.

Freeman, T. S., Imirzian, G., Kaltofen, E., and Lakshman, Y. 1988. Dagwood: a system for manipulating polynomials given by straight-line programs. *ACM Trans. Math. Software* 14(3):218–240.

Garbow, B. S. et al. 1972. *Matrix Eigensystem Routines: EISPACK Guide Extension*. Springer, New York.

Geddes, K. O., Czapor, S. R., and Labahn, G. 1992. *Algorithms for Computer Algebra*. Kluwer Academic.

Gelfand, I. M., Kapranov, M. M., and Zelevinsky, A. V. 1994. *Discriminants, Resultants and Multidimensional Determinants*. Birkhäuser Verlag, Boston, MA.

George, A. and Liu, J. W.-H. 1981. *Computer Solution of Large Sparse Positive Definite Linear Systems*. Prentice–Hall, Englewood Cliffs, NJ.

Gianni, P. and Trager, B. 1985. GCD's and factoring polynomials using Gröbner bases. *Proc. EUROCAL '85*, Vol. 2, *Lecture Notes in Computer Science*, 204, pp. 409–410.

Giesbrecht, M. 1995. Nearly optimal algorithms for canonical matrix forms. *SIAM J. Comput.* 24(5):948–969.

Gilbert, J. R. and Tarjan, R. E. 1987. The analysis of a nested dissection algorithm. *Numer. Math.* 50:377–404.

Golub, G. H. and Van Loan, C. F. 1989. *Matrix Computations*. Johns Hopkins Univ. Press, Baltimore, MD.

Gondran, M. and Minoux, M. 1984. *Graphs and Algorithms*. Wiley–Interscience, New York.

Grigoriev, D. Y. and Lakshman, Y. N. 1995. Algorithms for computing sparse shifts for multivariate polynomials. In *ISSAC '95 Proc. 1995 Internat. Symp. Symbolic Algebraic Comput.*, A. H. M. Levelt, ed., pp. 96–103, ACM Press, New York.

Hansen, E., Patrick, M., and Rusnack, J. 1977. Some modifications of Laguerre's method. *BIT* 17:409–417.

Heath, M. T., Ng, E., and Peyton, B. W. 1991. Parallel algorithms for sparse linear systems. *SIAM Rev.* 33:420–460.

Jenkins, M. A., and Traub, J. F. 1970. A three-stage variable-shift iteration for polynomial zeros and its relation to generalized Rayleigh iteration. *Numer. Math.* 14:252–263.

Kaltofen, E. 1982. Polynomial factorization. In 2nd ed. *Computer Algebra*, B. Buchberger, G. Collins, and R. Loos, eds., pp. 95–113. Springer–Verlag, Vienna.

Kaltofen, E. 1988. Greatest common divisors of polynomials given by straight-line programs. *J. ACM* 35(1):231–264.

Kaltofen, E. 1989. Factorization of polynomials given by straight-line programs. In *Randomness and Computation*, S. Micali, ed. Vol. 5 of Advances in computing research, pp. 375–412. JAI Press, Greenwich, CT.

Kaltofen, E. 1990. Polynomial factorization 1982–1986. 1990. In *Computers in Mathematics*, D. V. Chudnovsky and R. D. Jenks, eds. Vol. 125, *Lecture Notes in Pure and Applied Mathematics*, pp. 285–309. Marcel Dekker, New York.

Kaltofen, E. 1992. Polynomial factorization 1987–1991. In *Proc. LATIN '92*, I. Simon, ed. Vol. 583, *Lecture Notes in Computer Science*, pp. 294–313.

Kaltofen, E. 1995. Effective Noether irreducibility forms and applications. *J. Comput. Syst. Sci.* 50(2):274–295.

Kaltofen, E., Krishnamoorthy, M. S., and Saunders, B. D. 1990. Parallel algorithms for matrix normal forms. *Linear Algebra Appl.* 136:189–208.

Kaltofen, E. and Lakshman, Y. 1988. Improved sparse multivariate polynomial interpolation algorithms. *Proc. ISSAC '88*, Vol. 358, *Lecture Notes in Computer Science*, pp. 467–474.

Kaltofen, E. and Lobo, A. 1994. Factoring high-degree polynomials by the black box Berlekamp algorithm. In *ISSAC '94, Proc. Internat. Symp. Symbolic Algebraic Comput.*, J. von zur Gathen and M. Giesbrecht, eds., pp. 90–98, ACM Press, New York.

Kaltofen, E., Musser, D. R., and Saunders, B. D. 1983. A generalized class of polynomials that are hard to factor. *SIAM J. Comp.* 12(3):473–485.

Kaltofen, E. and Pan, V. 1991. Processor efficient parallel solution of linear systems over an abstract field. In *Proc. 3rd Ann. ACM Symp. Parallel Algor. Architecture*, pp. 180–191, ACM Press, New York.

Kaltofen, E. and Pan, V. 1992. Processor-efficient parallel solution of linear systems II: the positive characteristic and singular cases. In *Proc. 33rd Annual Symp. Foundations of Comp. Sci.*, pp. 714–723, Los Alamitos, CA. IEEE Computer Society Press.

Kaltofen, E. and Shoup, V. 1995. Subquadratic-time factoring of polynomials over finite fields. In *Proc. 27th Annual ACM Symp. Theory Comp.*, pp. 398–406, ACM Press, New York.

Kaltofen, E. and Trager, B. 1990. Computing with polynomials given by black boxes for their evaluations: greatest common divisors, factorization, separation of numerators and denominators. *J. Symbolic Comput.* 9(3):301–320.

Kapur, D. 1986. Geometry theorem proving using Hilbert's nullstellensatz. *J. Symbolic Comp.* 2:399–408.

Kapur, D. and Lakshman, Y. N. 1992. Elimination methods: an introduction. In *Symbolic and Numerical Computation for Artificial Intelligence*. B. Donald, D. Kapur, and J. Mundy, eds. Academic Press.

Kapur, D. and Saxena, T. 1995. Comparison of various multivariate resultant formulations. In *Proc. Internat. Symp. Symbolic Algebraic Comput. ISSAC '95*, A. H. M. Levelt, ed., pp. 187–195, ACM Press, New York.

Kapur, D., Saxena, T., and Yang, L. 1994. Algebraic and geometric reasoning using Dixon resultants. In *ISSAC '94, Proc. Internat. Symp. Symbolic Algebraic Comput.* J. von zur Gathen and M. Giesbrecht, ed., pp. 99–107, ACM Press, New York.

Knuth, D. E. 1981. *The Art of Computer Programming, Vol. 2, Seminumerical Algorithms*, 2nd ed. Addison–Wesley, Reading, MA.

Lakshman, Y. N. 1990. *On the complexity of computing Gröbner bases for zero dimensional polynomia.* Ph.D. thesis, Dept. Comput. Sci., Rensselaer Polytechnic Inst. Troy, NY, Dec.

Lakshman, Y. N. and Saunders, B. D. 1994. On computing sparse shifts for univariate polynomials. In *ISSAC '94, Proc. Internat. Symp. Symbolic Algebraic Comput.*, J. von zur Gathen and M. Giesbrecht, eds., pp. 108–113, ACM Press, New York.

Lakshman, Y. N. and Saunders, B. D. 1995. Sparse polynomial interpolation in non-standard bases. *SIAM J. Comput.* 24(2):387–397.

Lazard, D. 1981. Résolution des systèmes d'équation algébriques. *Theoretical Comput. Sci.* 15:77–110. (In French).

Lenstra, A. K., Lenstra, H. W., and Lovász, L. 1982. Factoring polynomials with rational coefficients. *Math. Ann.* 261:515–534.

Leyland, P. 1995. Cunningham project data. Internet document, Oxford Univ., ftp://sable.ox.ac.uk/pub/math/cunningham/, November.

Lipton, R. J., Rose, D., and Tarjan, R. E. 1979. Generalized nested dissection. *SIAM J. on Numer. Analysis* 16(2):346–358.

Macaulay, F. S. 1916. Algebraic theory of modular systems. *Cambridge Tracts* 19, Cambridge.

MacWilliams, F. J. and Sloan, N. J. A. 1977. *The Theory of Error-Correcting Codes*. North–Holland, New York.

Madsen, K. 1973. A root-finding algorithm based on Newton's method. *BIT* 13:71–75.

McCormick, S., ed. 1987. *Multigrid Methods*. SIAM Pub., Philadelphia, PA.

McNamee, J. M. 1993. A bibliography on roots of polynomials. *J. Comput. Appl. Math.* 47(3):391–394.

Miller, V. 1992. Factoring polynomials via relation-finding. In *Proc. ISTCS '92*, D. Dolev, Z. Galil, and M. Rodeh, eds. Vol. 601, *Lecture Notes in Computer Science*, pp. 115–121.

Monagan, M. B. 1992. A heuristic irreducibility test for univariate polynomials. *J. Symbolic Comput.* 13(1):47–57.

Musser, D. R. 1975. Multivariate polynomial factorization. *J. ACM* 22:291–308.

Niederreiter, H. 1994. New deterministic factorization algorithms for polynomials over finite fields. In *Finite Fields: Theory, Applications and Algorithms*, L. Mullen and P. J.-S. Shiue, eds. Vol. 168, Contemporary mathematics, pp. 251–268, Amer. Math. Soc., Providence, RI.

Ortega, J. M., and Voight, R. G. 1985. Solution of partial differential equations on vector and parallel computers. *SIAM Rev.* 27(2):149–240.

Pan, V. Y. 1984a. How can we speed up matrix multiplication? *SIAM Rev.* 26(3):393–415.

Pan, V. Y. 1984b. How to multiply matrices faster. *Lecture Notes in Computer Science*, 179.

Pan, V. Y. 1987. Sequential and parallel complexity of approximate evaluation of polynomial zeros. *Comput. Math. (with Appls.)*, 14(8):591–622.

Pan, V. Y. 1991. Complexity of algorithms for linear systems of equations. In *Computer Algorithms for Solving Linear Algebraic Equations (State of the Art)*, E. Spedicato, ed. Vol. 77 of *NATO ASI Series, Series F: computer and systems sciences*, pp. 27–56, Springer–Verlag, Berlin.

Pan, V. Y. 1992a. Complexity of computations with matrices and polynomials. *SIAM Rev.* 34(2):225–262.

Pan, V. Y. 1992b. Linear systems of algebraic equations. In *Encyclopedia of Physical Sciences and Technology*, 2nd ed. Marvin Yelles, ed. Vol. 8, pp. 779–804, 1987. 1st ed. Vol. 7, pp. 304–329.

Pan, V. Y. 1993. Parallel solution of sparse linear and path systems. In *Synthesis of Parallel Algorithms*, J. H. Reif, ed. Ch. 14, pp. 621–678. Morgan Kaufmann, San Mateo, CA.

Pan, V. Y. 1994a. Improved parallel solution of a triangular linear system. *Comput. Math. (with Appl.)*, 27(11):41–43.

Pan, V. Y. 1994b. *On approximating polynomial zeros: modified quadtree construction and improved Newton's iteration*. Manuscript, Lehman College, CUNY, Bronx, New York.

Pan, V. Y. 1995a. Parallel computation of a Krylov matrix for a sparse and structured input. *Math. Comput. Modelling* 21(11):97–99.

Pan, V. Y. 1995b. *Solving a polynomial equation: some history and recent progress*. Manuscript, Lehman College, CUNY, Bronx, New York.

Pan, V. Y. 1996. Optimal and nearly optimal algorithms for approximating polynomial zeros. *Comput. Math. (with Appl.)*.

Pan, V. Y. and Preparata, F. P. 1995. Work-preserving speed-up of parallel matrix computations. *SIAM J. Comput.* 24(4):811–821.

Pan, V. Y. and Reif, J. H. 1992. Compact multigrid. *SIAM J. Sci. Stat. Comput.* 13(1):119–127.

Pan, V. Y. and Reif, J. H. 1993. Fast and efficient parallel solution of sparse linear systems. *SIAM J. Comp.*, 22(6):1227–1250.

Pan, V. Y., Sobze, I. and Atinkpahoun, A. 1995. On parallel computations with band matrices. *Inf. and Comput.* 120(2):227–250.

Parlett, B. 1980. *Symmetric Eigenvalue Problem*. Prentice–Hall, Englewood Cliffs, NJ.

Quinn, M. J. 1994. *Parallel Computing: Theory and Practice*. McGraw–Hill, New York.

Rabin, M. O. 1980. Probabilistic algorithms in finite fields. *SIAM J. Comp.* 9:273–280.

Renegar, J. 1989. On the worst case arithmetic complexity of approximating zeros of systems of polynomials. *SIAM J. Comput.* 18(2):350–370.

Ritt, J. F. 1950. *Differential Algebra*. AMS, New York.

Saad, Y. 1992. *Numerical Methods for Large Eigenvalue Problems: Theory and Algorithms*. Manchester Univ. Press, U.K., Wiley, New York. 1992.

Saad, Y. 1995. *Iterative Methods for Sparse Linear Systems*. PWS Kent, Boston, MA.

Sendra, J. R. and Winkler, F. 1991. Symbolic parameterization of curves. *J. Symbolic Comput.* 12(6):607–631.

Shoup, V. 1996. A new polynomial factorization algorithm and its implementation. *J. Symbolic Comput.*

Smith, B. T. et al. 1970. *Matrix Eigensystem Routines: EISPACK Guide*, 2nd ed. Springer, New York.

St. Schwarz, 1956. On the reducibility of polynomials over a finite field. *Quart. J. Math. Oxford Ser. (2)*, 7:110–124.

Stederberg, T. and Goldman, R. 1986. Algebraic geometry for computer-aided design. *IEEE Comput. Graphics Appl.* 6(6):52–59.

Sturmfels, B. 1991. Sparse elimination theory. In *Proc. Computat. Algebraic Geom. and Commut. Algebra*, D. Eisenbud and L. Robbiano, eds. Cortona, Italy, June.

Sturmfels, B. and Zelevinsky, A. 1992. Multigraded resultants of the Sylvester type. *J. Algebra*.

von zur Gathen, J. and Shoup, V. 1992. Computing Frobenius maps and factoring polynomials. *Comput. Complexity* 2:187–224.

Walsh, P. G. 1993. *The computation of Puiseux expansions and a quantitative version of Runge's theorem on diophantine equations*. Ph.D. dissertation. University of Waterloo, Waterloo, Canada.

Watkins, D. S. 1982. Understanding the QR algorithm. *SIAM Rev.* 24:427–440.

Watkins, D. S. 1991. Some perspectives on the eigenvalue problem. *SIAM Rev.* 35(3):430–471.

Wilkinson, J. H. 1965. *The Algebraic Eigenvalue Problem*. Clarendon Press, Oxford, England.

Winkler, F. 1996. *Introduction to Computer Algebra*. Springer–Verlag, Heidelberg, Germany.

Wu, W. 1984. Basis principles of mechanical theorem proving in elementary geometries. *J. Syst. Sci. Math Sci.* 4(3):207–235.

Wu, W. 1986. Basis principles of mechanical theorem proving in elementary geometries. *J. Automated Reasoning* 2:219–252.

Zassenhaus, H. 1969. On Hensel factorization I. *J. Number Theory* 1:291–311.

Zippel, R. 1993. *Effective Polynomial Computations*, p. 384. Kluwer Academic, Boston, MA.

Further Information

The books by Knuth [1981], Davenport et al. [1988], Geddes et al. [1992], and Zippel [1993] provide a much broader introduction to the general subject. There are well-known libraries and packages of subroutines for the most popular numerical matrix computations, in particular, Dongarra et al. [1978] for solving linear systems of equations, Smith et al. [1970] and Garbow et al. [1972] approximating matrix eigenvalues, and Anderson et al. [1992] for both of the two latter computational problems. There is a comprehensive treatment of numerical matrix computations [Golub and Van Loan 1989], with extensive bibliography, and there are several more specialized books on them [George and Liu 1981, Wilkinson 1965, Parlett 1980, Saad 1992, 1995], as well as many survey articles [Heath et al. 1991, Watkins 1991, Ortega and Voight 1985, Pan 1992b] and thousands of research articles.

Special (more efficient) parallel algorithms have been devised for special classes of matrices, such as sparse [Pan and Reif 1993, Pan 1993], banded [Pan et al. 1995], and dense structured [Bini and Pan (cf. [1994])]. We also refer to Pan and Preparata [1995] on a simple but surprisingly effective extension of Brent's principle for improving the processor and work efficiency of parallel matrix algorithms and to Golub and Van Loan [1989], Ortega and Voight [1985], and Heath et al. [1991] on practical parallel algorithms for matrix computations.

9
Cryptography

9.1   Introduction 9-1
9.2   Cryptographic Notions of Security..................... 9-2
       Information-Theoretic Notions of Security • Toward a
       Computational Notion of Security • Notation
9.3   Building Blocks 9-4
       One-Way Functions • Trapdoor Permutations
9.4   Cryptographic Primitives.............................. 9-7
       Pseudorandom Generators • Pseudorandom Functions
       and Block Ciphers • Cryptographic Hash Functions
9.5   Private-Key Encryption 9-11
9.6   Message Authentication 9-14
9.7   Public-Key Encryption 9-15
9.8   Digital Signature Schemes 9-17

Jonathan Katz

University of Maryland

9.1 Introduction

Cryptography is a vast subject, and we cannot hope to give a comprehensive account of the field here. Instead, we have chosen to narrow our focus to those areas of cryptography having the most practical relevance to the problem of *secure communication*. Broadly speaking, secure communication encompasses two complementary goals: the **secrecy** (sometimes called "privacy") and **integrity** of communicated data. These terms can be illustrated using the simple example of a user A sending a message m to a user B over a public channel. In the simplest sense, techniques for data secrecy ensure that an eavesdropping adversary (i.e., an adversary who sees all communication occurring on the channel) cannot get any information about m and, in particular, cannot determine m. Viewed in this way, such techniques protect against a *passive* adversary who listens to — but does not otherwise interfere with — the parties' communication. Techniques for data integrity, on the other hand, protect against an *active* adversary who may arbitrarily modify the data sent over the channel or may interject messages of his own. Here, secrecy is not necessarily an issue; instead, security in this setting requires only that any modifications performed by the adversary to the transmitted data will be detected by the receiving party.

In the cases of both secrecy and integrity, two different assumptions regarding the initial setup of the communicating parties can be considered. In the **private-key** setting (also known as the "shared-key," "secret-key," or "symmetric-key" setting), the assumption is that parties A and B have securely shared a random key s in advance. This key, which is completely hidden from the adversary, is used to secure their future communication. (We do not comment further on how such a key might be securely generated and shared; for our purposes, it is simply an assumption of the model.) Techniques for secrecy in this setting are called **private-key encryption** schemes, and those for data integrity are termed **message authentication codes (MACs)**.

In the **public-key** setting, the assumption is that one (or both) of the parties has generated a pair of keys: a *public key* that is widely disseminated throughout the network and an associated *secret key* that is kept private. The parties generating these keys may now use them to ensure secret communication using a **public-key encryption** scheme; they can also use these keys to provide data integrity (for messages they send) using a **digital signature scheme**.

We stress that, in the public-key setting, widespread distribution of the public key is assumed to occur before any communication over the public channel and without any interference from the adversary. In particular, if A generates a public/secret key, then B (for example) knows the correct public key and can use this key when communicating with A. On the flip side, the fact that the public key is widely disseminated implies that the adversary also knows the public key, and can attempt to use this knowledge when attacking the parties' communication.

We examine each of the above topics in turn. In Section 9.2 we introduce the information-theoretic approach to cryptography, describe some information-theoretic solutions for the above tasks, and discuss the severe limitations of this approach. We then describe the modern, computational (or complexity-theoretic) approach to cryptography that will be used in the remaining sections. This approach requires computational "hardness" assumptions of some sort; we formalize these assumptions in Section 9.3 and thus provide cryptographic building blocks for subsequent constructions. These building blocks are used to construct some basic cryptographic primitives in Section 9.4.

With these primitives in place, we proceed in the remainder of the chapter to give solutions for the tasks previously mentioned. Sections 9.5 and 9.6 discuss private-key encryption and message authentication, respectively, thereby completing our discussion of the private-key setting. Public-key encryption and digital signature schemes are described in Sections 9.7 and 9.8. We conclude with some suggestions for further reading.

9.2 Cryptographic Notions of Security

Two central features distinguish modern cryptography from "classical" (i.e., pre-1970s) cryptography: precise definitions and rigorous proofs of security. Without a precise definition of security for a stated goal, it is meaningless to call a particular protocol "secure." The importance of rigorous proofs of security (based on a set of well-defined assumptions) should also be clear: if a given protocol is not proven secure, there is always the risk that the protocol can be "broken." That protocol designers have not been able to find an attack does not preclude a more clever adversary from doing so. A proof that a given protocol is secure (with respect to some precise definition and using clearly stated assumptions) provides much more confidence in the protocol.

9.2.1 Information-Theoretic Notions of Security

With this in mind, we present one possible definition of security for private-key encryption and explore what can be achieved with respect to this definition. Recall the setting: two parties A and B share a random secret key s; this key will be used to secure their future communication and is completely hidden from the adversary. The data that A wants to communicate to B is called the **plaintext**, or simply the *message*. To transmit this message, A will **encrypt** the message using s and an encryption algorithm $\mathcal{E}$, resulting in **ciphertext** C. We write this as $C = \mathcal{E}_s(m)$. This ciphertext is sent over the public channel to B. Upon receiving the ciphertext, B recovers the original message by **decrypting** it using s and decryption algorithm $\mathcal{D}$; we write this as $m = \mathcal{D}_s(C)$.

We stress that the adversary is assumed to know the encryption and decryption algorithms; the only information hidden from the adversary is the secret key s. It is a mistake to require that the details of the encryption scheme be hidden in order for it to be secure, and modern cryptosystems are designed to be secure even when the full details of all algorithms are publicly available.

A plausible definition of security is to require that an adversary who sees ciphertext C (recall that C is sent over a public channel) — but does not know s — learns no information about the message m. In particular, even if the message m is known to be one of two possible messages m_1, m_2 (each being chosen with probability $1/2$), the adversary should not learn which of these two messages was actually sent. If we abstract this by requiring the adversary to, say, output "1" when he believes that m_1 was sent, this requirement can be formalized as:

> For all possible m_1, m_2 and for any adversary A, the probability that A guesses "1" when C is an encryption of m_1 is equal to the probability that A guesses "1" when C is an encryption of m_2.

That is, the adversary is no more likely to guess that m_1 was sent when m_1 is the actual message than when m_2 is the actual message. An encryption scheme satisfying this definition is said to be *information-theoretically secure* or to achieve *perfect secrecy*.

Perfect secrecy can be achieved by the **one-time pad** encryption scheme, which works as follows. Let ℓ be the length of the message m, where m is viewed as a binary string. The parties share in advance a secret key s that is uniformly distributed over strings of length ℓ (i.e., s is an ℓ-bit string chosen uniformly at random). To encrypt message m, the sender computes $C = m \oplus s$ where $\oplus$ represents binary exclusive-or and is computed bit-by-bit. Decryption is performed by setting $m = C \oplus s$. Clearly, decryption always recovers the original message. To see that the scheme is perfectly secret, let M, C, K be random variables denoting the message, ciphertext, and key, respectively, and note that for *any* message m and observed ciphertext c, we have:

$$\Pr[M = m | C = c] = \frac{\Pr[C = c | M = m] \Pr[M = m]}{\Pr[C = c]}$$
$$= \frac{\Pr[K = c \oplus m] \Pr[M = m]}{\Pr[C = c]} = \frac{2^{-\ell} \Pr[M = m]}{\Pr[C = c]}$$

Thus, if m_1, m_2 have equal *a priori* probability, then $\Pr[M = m_1 | C = c] = \Pr[M = m_2 | C = c]$ and the ciphertext gives no further information about the actual message sent.

While this scheme is provably secure, it has limited value for most common applications. For one, *the length of the shared key is equal to the length of the message*. Thus, the scheme is simply impractical when long messages are sent. Second, it is easy to see that the scheme is secure *only when it is used to send a single message* (hence the name "one-time pad"). This will not do for applications is which multiple messages must be sent. Unfortunately, it can be shown that the one-time pad is optimal if perfect secrecy is desired. More formally, any scheme achieving perfect secrecy requires the key to be at least as long as the (total) length of all messages sent.

Can information-theoretic security be obtained for other cryptographic goals? It is known that perfectly-secure message authentication is possible (see, e.g., [51, Section 4.5]), although constructions achieving perfect security are similarly inefficient and require impractically long keys to authenticate multiple messages. In the public-key setting, the situation is even worse: perfectly secure public-key encryption or digital signature schemes are simply unachievable.

In summary, it is impossible to design perfectly secure yet practical protocols achieving the basic goals outlined in Section 9.1. To obtain reasonable solutions for our original goals, it will be necessary to (slightly) relax our definition of security.

9.2.2 Toward a Computational Notion of Security

The observation noted at the end of the previous section has motivated a shift in modern cryptography toward *computational* notions of security. Informally, whereas information-theoretic security guarantees that a scheme is absolutely secure against all (even arbitrarily powerful) adversaries, computational security ensures that a scheme is secure except with "negligible" probability against all "efficient" adversaries (we formally define these terms below). Although information-theoretic security is a strictly stronger notion, computational security suffices in practice and allows the possibility of more efficient schemes. However, it should be noted that computational security ultimately relies on currently unproven assumptions regarding the computational "hardness" of certain problems; that is, the security guarantee provided in the computational setting is not as iron-clad as the guarantee given by information-theoretic security.

In moving to the computational setting, we introduce a *security parameter* $k \in \mathbb{N}$ that will be used to precisely define the terms "efficient" and "negligible." An *efficient* algorithm is defined as a probabilistic algorithm that runs in time polynomial in k; we also call such an algorithm "probabilistic, polynomial-time (PPT)." A *negligible* function is defined as one asymptotically smaller than any inverse polynomial; that is, a function $\varepsilon : \mathbb{N} \to \mathbb{R}^+$ is negligible if, for all $c \geq 0$ and for all n large enough, $\varepsilon(n) < 1/n^c$.

A cryptographic construction will be indexed by the security parameter k, where this value is given as input (in unary) to the relevant algorithms. Of course, we will require that these algorithms are all efficient and run in time polynomial in k. A typical definition of security in the computational setting requires that some condition hold for all PPT adversaries with all but negligible probability or, equivalently, that a PPT adversary will succeed in "breaking" the scheme with at most negligible probability. Note that the security parameter can be viewed as corresponding to a higher level of security (in some sense) because, as the security parameter increases, the adversary may run for a longer amount of time but has even lower probability of success.

Computational definitions of this sort will be used throughout the remainder of this chapter, and we explicitly contrast this type of definition with an information-theoretic one in Section 9.5 (for the case of private-key encryption).

9.2.3 Notation

Before continuing, we introduce some mathematical notation (following [30]) that will provide some useful shorthand. If A is a deterministic algorithm, then $y = A(x)$ means that we set y equal to the output of A on input x. If A is a probabilistic algorithm, the notation $y \leftarrow A(x_1, x_2, \ldots)$ denotes running A on inputs $x_1, x_2, \ldots$ and setting y equal to the output of A. Here, the "$\leftarrow$" is an explicit reminder that the process is probabilistic, and thus running A twice on the same inputs, for example, may not necessarily give the same value for y. If S represents a finite set, then $b \leftarrow S$ denotes assigning b an element chosen uniformly at random from S. If $p(x_1, x_2, \ldots)$ is a predicate that is either true or false, the notation

$$\Pr[x_1 \leftarrow S; x_2 \leftarrow A(x_1, y_2, \ldots); \cdots : p(x_1, x_2, \ldots)]$$

denotes the probability that $p(x_1, x_2, \ldots)$ is true after ordered execution of the listed experiment. The key features of this notation are that everything to the left of the colon represents the experiment itself (whose components are executed in order, from left to right, and are separated by semicolons) and the predicate is written to the right of the colon. To give a concrete example: $\Pr[b \leftarrow \{0, 1, 2\} : b = 2]$ denotes the probability that b is equal to 2 following the experiment in which b is chosen at random from $\{0, 1, 2\}$; this probability is, of course, $1/3$.

The notation $\{0, 1\}^\ell$ denotes the set of binary strings of length ℓ, while $\{0, 1\}^{\leq \ell}$ denotes the set of binary strings of length at most ℓ. We let $\{0, 1\}^*$ denote the set of finite-length binary strings. 1^k represents k repetitions of the digit "1", and has the value k in unary notation.

We assume familiarity with basic algebra and number theory on the level of [11]. We let $\mathbb{Z}_N = \{0, \ldots, N-1\}$ denote the set of integers modulo N; also, $\mathbb{Z}_N^* \subset \mathbb{Z}_N$ is the set of integers between 0 and N that are relatively prime to N. The Euler totient function is defined as $\varphi(N) \overset{\text{def}}{=} |\mathbb{Z}_N^*|$; of importance here is that $\varphi(p) = p - 1$ for p prime, and $\varphi(pq) = (p-1)(q-1)$ if p, q are distinct primes. For any N, the set $\mathbb{Z}_N^*$ forms a group under multiplication modulo N [11].

9.3 Building Blocks

As hinted at previously, cryptography seeks to exploit the presumed existence of computationally "hard" problems. Unfortunately, the mere existence of computationally hard problems does not appear to be sufficient for modern cryptography as we know it. Indeed, it is not currently known whether it is possible to have, say, secure private-key encryption (in the sense defined in Section 9.5) based only on the conjecture that $P \neq NP$ (where P refers to those problems solvable in polynomial time and NP [informally] refers to those problems whose solutions can be verified in polynomial time; cf. [50] and Chapter 6). Seemingly stronger assumptions are currently necessary in order for cryptosystems to be built. On the other hand — fortunately for cryptographers — such assumptions currently seem very reasonable.

9.3.1 One-Way Functions

The most basic building block in cryptography is a **one-way function**. Informally, a one-way function f is a function that is "easy" to compute but "hard" to invert. Care must be taken, however, in interpreting this informal characterization. In particular, the formal definition of one-wayness requires that f be hard to invert *on average* and not merely hard to invert *in the worst case*. This is in direct contrast to the situation in complexity theory, where a problem falls in a particular class based on the worst-case complexity of solving it (and this is one reason why $P \neq NP$ does not seem to be sufficient for much of modern cryptography).

A number of equivalent definitions of one-way functions are possible; we present one such definition here. Note that the security parameter is explicitly given as input (in unary) to all algorithms.

Definition 9.1 Let $F = \{f_k : \mathcal{D}_k \to \mathcal{R}_k\}_{k \geq 1}$ be an infinite collection of functions where $\mathcal{D}_k \subseteq \{0,1\}^{\leq \ell(k)}$ for some fixed polynomial $\ell(\cdot)$. Then F is one-way (more formally, F is a one-way function family) if the following conditions hold:

"Easy" to compute There is a deterministic, polynomial-time algorithm A such that for all k and for all $x \in \mathcal{D}_k$ we have $A(1^k, x) = f_k(x)$.

"Hard" to invert For all PPT algorithms B, the following is negligible (in k):

$$\Pr[x \leftarrow \mathcal{D}_k; y = f_k(x); x' \leftarrow B(1^k, y) : f_k(x') = y].$$

Efficiently sampleable There is a PPT algorithm S such that $S(1^k)$ outputs a uniformly distributed element of $\mathcal{D}_k$.

It is not hard to see that the existence of a one-way function family implies $P \neq NP$. Thus, we have no hope of proving the unequivocal existence of a one-way function family given our current knowledge of complexity theory. Yet, certain number-theoretic problems appear to be one-way (and have thus far resisted all attempts at proving otherwise); we mention three popular candidates:

1. **Factoring.** Let $\mathcal{D}_k$ consist of pairs of k-bit primes, and define f_k such that $f_k(p, q) = pq$. Clearly, this function is easy to compute. It is also true that the domain $\mathcal{D}_k$ is efficiently sampleable because efficient algorithms for generating random primes are known (see, e.g., Appendix A.7 in [14]). Finally, f_k is hard to invert — and thus the above construction is a one-way function family — under the conjecture that factoring is hard (we refer to this simply as "the factoring assumption"). Of course, we have no proof for this conjecture; rather, evidence favoring the conjecture comes from the fact that no polynomial-time algorithm for factoring has been discovered in roughly 300 years of research related to this problem.

2. **Computing discrete logarithms.** Let $\mathcal{D}_k$ consist of tuples (p, g, x) in which p is a k-bit prime, g is a generator of the multiplicative group $\mathbb{Z}_p^*$, and $x \in \mathbb{Z}_{p-1}$. Furthermore, define f_k such that $f_k(p, g, x) = (p, g, g^x \bmod p)$. Given p, g as above and for any $y \in \mathbb{Z}_p^*$, define $\log_g y$ as the unique value $x \in \mathbb{Z}_{p-1}$ such that $g^x = y \bmod p$ (that a unique such x exists follows from the fact that $\mathbb{Z}_p^*$ is a cyclic group for p prime). Although exponentiation modulo p can be done in time polynomial in the lengths of p and the exponent x, it is not known how to efficiently compute $\log_g y$ given p, g, y. This suggests that this function family is indeed one-way (we note that there exist algorithms to efficiently sample from $\mathcal{D}_k$; see e.g., Chapter 6 in [14]).

    It should be clear that the above construction generalizes to other collections of finite, cyclic groups in which exponentiation can be done in polynomial time. Of course, the function family thus defined is one-way only if the discrete logarithm problem in the relevant group is hard. Other popular examples in which this is believed to be the case include the group of points on certain elliptic curves (see Chapter 6 in [34]) and the subgroup of quadratic residues in $\mathbb{Z}_p^*$ when p and $\frac{p-1}{2}$ are both prime.

3. **RSA [45].** Let $\mathcal{D}_k$ consist of tuples (N, e, x), where N is a product of two distinct k-bit primes, $e < N$ is relatively prime to $\varphi(N)$, and $x \in \mathbb{Z}_N^*$. Furthermore, define f_k such that $f_k(N, e, x) = (N, e, x^e \bmod N)$. Following the previous examples, it should be clear that this function is easy to compute and has an efficiently sampleable domain (note that $\varphi(N)$ can be efficiently computed if p, q are known), It is conjectured that this function is hard to invert [45] and thus constitutes a one-way function family; we refer to this assumption simply as "the RSA assumption." For reasons of efficiency, the RSA function family is sometimes restricted by considering only $e = 3$ (and choosing N such that $\varphi(N)$ is not divisible by 3), and this is also believed to give a one-way function family.

It is known that if RSA is a one-way function family, then factoring is hard (see the discussion of RSA as a trapdoor permutation, below). The converse is not believed to hold, and thus the RSA assumption appears to be strictly stronger than the factoring assumption (of course, all other things being equal, the weaker assumption is preferable).

9.3.2 Trapdoor Permutations

One-way functions are sufficient for many cryptographic applications. Sometimes, however, an "asymmetry" of sorts — whereby one party can efficiently accomplish some task which is infeasible for anyone else — must be introduced. **Trapdoor permutations** represent one way of formalizing this asymmetry. Recall that a one-way function has the property (informally) that it is "easy" to compute but "hard" to invert. Trapdoor permutations are also "easy" to compute and "hard" to invert *in general*; however, there is some trapdoor information that makes the permutation "easy" to invert. We give a formal definition now, and follow with some examples.

Definition 9.2 Let $\mathcal{K}$ be a PPT algorithm which, on input 1^k (for any $k \geq 1$), outputs a pair (key, td) such that key defines a permutation f_{key} over some domain $\mathcal{D}_{\text{key}}$. We say $\mathcal{K}$ is a trapdoor permutation generator if the following conditions hold:

"Easy" to compute There is a deterministic, polynomial-time algorithm A such that for all k, all (key, td) output by $\mathcal{K}(1^k)$, and all $x \in \mathcal{D}_{\text{key}}$ we have $A(1^k, \text{key}, x) = f_{\text{key}}(x)$.

"Hard" to invert For all PPT algorithms B, the following is negligible (in k):

$$\Pr[(\text{key}, \text{td}) \leftarrow \mathcal{K}(1^k); x \leftarrow \mathcal{D}_{\text{key}}; y = f_{\text{key}}(x); x' \leftarrow B(1^k, \text{key}, y) : f_{\text{key}}(x') = y].$$

Efficiently sampleable There is a PPT algorithm S such that for all (key, td) output by $\mathcal{K}(1^k)$, $S(1^k, \text{key})$ outputs a uniformly distributed element of $\mathcal{D}_{\text{key}}$.

"Easy" to invert with trapdoor There is a deterministic, polynomial-time algorithm I such that for all (key, td) output by $\mathcal{K}(1^k)$ and all $y \in \mathcal{D}_{\text{key}}$ we have $I(1^k, \text{td}, y) = f_{\text{key}}^{-1}(y)$.

It should be clear that the existence of a trapdoor permutation generator immediately implies the existence of a one-way function family. Note that one could also define the completely analogous notion of trapdoor *function* generators; however, these have (thus far) had much more limited applications to cryptography.

It seems that the existence of a trapdoor permutation generator is a strictly stronger assumption than the existence of a one-way function family. Yet, number theory again provides examples of (conjectured) candidates:

9.3.2.1 RSA

We have seen in the previous section that RSA gives a one-way function family. It can also be used to give a trapdoor permutation generator. Here, we let $\mathcal{K}$ be an algorithm which, on input 1^k, chooses two distinct

k-bit primes p, q at random, sets $N = pq$, and chooses $e < N$ such that e and $\varphi(N)$ are relatively prime (note that $\varphi(N) = (p-1)(q-1)$ is efficiently computable because the factorization of N is known to $\mathcal{K}$). Then, $\mathcal{K}$ computes d such that $ed = 1 \mod \varphi(N)$. The output is $((N, e), d)$, where (N, e) defines the permutation $f_{N,e} : \mathbb{Z}_N^* \to \mathbb{Z}_N^*$ given by $f_{N,e}(x) = x^e \mod N$. It is not hard to verify that this is indeed a permutation. That this permutation satisfies the first three requirements of the definition above follows from the fact that RSA is a one-way function family. To verify the last condition ("easiness" of inversion given the trapdoor d), note that

$$f_{N,d}(x^e \mod N) = (x^e)^d \mod N = x^{ed \mod \varphi(N)} \mod N = x,$$

and thus $f_{N,d} = f_{N,e}^{-1}$. So, the permutation $f_{N,e}$ can be efficiently inverted given d.

9.3.2.2 A Trapdoor Permutation Based on Factoring [42]

Let $\mathcal{K}$ be an algorithm which, on input 1^k, chooses two distinct k-bit primes p, q at random such that $p = q = 3 \mod 4$, and sets $N = pq$. The output is $(N, (p, q))$, where N defines the permutation $f_N : \mathcal{QR}_N \to \mathcal{QR}_N$ given by $f_N(x) = x^2 \mod N$; here, $\mathcal{QR}_N$ denotes the set of quadratic residues modulo N (i.e., the set of $x \in \mathbb{Z}_N^*$ such that x is a square modulo N). It can be shown that f_N is a permutation, and it is immediate that f_N is easy to compute. $\mathcal{QR}_N$ is also efficiently sampleable: to choose a random element in $\mathcal{QR}_N$, simply pick a random $x \in \mathbb{Z}_N^*$ and square it. It can also be shown that the trapdoor information p, q (i.e., the factorization of N) is sufficient to enable efficient inversion of f_N (see Section 3.6 in [14]). We now prove that this permutation is hard to invert as long as factoring is hard.

Lemma 9.1 *Assuming the hardness of factoring N of the form generated by $\mathcal{K}$, algorithm $\mathcal{K}$ described above is a trapdoor permutation family.*

Proof The lemma follows by showing that the squaring permutation described above is hard to invert (without the trapdoor). For any PPT algorithm B, define

$$\delta(k) \stackrel{\text{def}}{=} \Pr[(N, (p, q)) \leftarrow \mathcal{K}(1^k); y \leftarrow \mathcal{QR}_N; z \leftarrow B(1^k, N, y) : z^2 = y \mod N]$$

(this is exactly the probability that B inverts a randomly-generated f_N). We use B to construct another PPT algorithm B' which factors the N output by $\mathcal{K}$. Algorithm B' operates as follows: on input $(1^k, N)$, it chooses a random $\tilde{x} \in \mathbb{Z}_N^*$ and sets $y = \tilde{x}^2 \mod N$. It then runs $B(1^k, N, y)$ to obtain output z. If $z^2 = y \mod N$ and $z \neq \pm \tilde{x}$, we claim that $\gcd(z - \tilde{x}, N)$ is a nontrivial factor of N. Indeed, $z^2 - \tilde{x}^2 = 0 \mod N$, and thus

$$(z - \tilde{x})(z + \tilde{x}) = 0 \mod N.$$

Since $z \neq \pm \tilde{x}$, it must be the case that $\gcd(z - \tilde{x}, N)$ gives a nontrivial factor of N, as claimed.

Now, conditioned on the fact that $z^2 = y \mod N$ (which is true with probability $\delta(k)$), the probability that $z \neq \pm \tilde{x}$ is exactly $1/2$; this follows from the fact that y has exactly four square roots, two of which are $\tilde{x}$ and $-\tilde{x}$. Thus, the probability that B' factors N is exactly $\delta(k)/2$. Because this quantity is negligible under the factoring assumption, $\delta(k)$ must be negligible as well. □

9.4 Cryptographic Primitives

The building blocks of the previous section can be used to construct a variety of primitives, which in turn have a wide range of applications. We explore some of these primitives here.

9.4.1 Pseudorandom Generators

Informally, a **pseudorandom generator (PRG)** is a deterministic function that takes a short, random string as input and returns a longer, "random-looking" (i.e., pseudorandom) string as output. But to properly understand this, we must first ask: what does it mean for a string to "look random"? Of course, it is meaningless (in the present context) to talk about the "randomness" of any particular string — once a string is fixed, it is no longer random! Instead, we must talk about the randomness — or pseudorandomness — of a *distribution* of strings. Thus, to evaluate $G : \{0,1\}^k \to \{0,1\}^{k+1}$ as a PRG, we must compare the uniform distribution on strings of length $k + 1$ with the distribution $\{G(x)\}$ for x chosen uniformly at random from $\{0,1\}^k$.

It is rather interesting that although the design and analysis of PRGs has a long history [33], it was not until the work of [9, 54] that a definition of PRGs appeared which was satisfactory for cryptographic applications. Prior to this work, the quality of a PRG was determined largely by ad hoc techniques; in particular, a PRG was deemed "good" if it passed a specific battery of statistical tests (for example, the probability of a "1" in the final bit of the output should be roughly $1/2$). In contrast, the approach advocated by [9, 54] is that a PRG is good if it passes *all possible* (efficient) statistical tests! We give essentially this definition here.

Definition 9.3   Let $G : \{0,1\}^* \to \{0,1\}^*$ be an efficiently computable function for which $|G(x)| = \ell(|x|)$ for some fixed polynomial $\ell(k) > k$ (i.e., fixed-length inputs to G result in fixed-length outputs, and the output of G is always longer than its input). We say G is a pseudorandom generator (PRG) with expansion factor $\ell(k)$ if the following is negligible (in k) for all PPT statistical tests T:

$$\left| \Pr[x \leftarrow \{0,1\}^k : T(G(x)) = 1] - \Pr[y \leftarrow \{0,1\}^{\ell(k)} : T(y) = 1] \right|.$$

Namely, no PPT algorithm can distinguish between the output of G (on uniformly selected input) and the uniform distribution on strings of the appropriate length.

Given this strong definition, it is somewhat surprising that PRGs can be constructed at all; yet, they can be constructed from any one-way function (see below). As a step toward the construction of PRGs based on general assumptions, we first define and state the existence of a *hard-core bit* for any one-way function. Next, we show how this hard-core bit can be used to construct a PRG from any one-way *permutation*. (The construction of a PRG from arbitrary one-way *functions* is more complicated and is not given here.) This immediately extends to give explicit constructions of PRGs based on some specific assumptions.

Definition 9.4   Let $F = \{f_k : \mathcal{D}_k \to \mathcal{R}_k\}_{k \geq 1}$ be a one-way function family, and let $H = \{h_k : \mathcal{D}_k \to \{0,1\}\}_{k \geq 1}$ be an efficiently computable function family. We say that H is a hard-core bit for F if $h_k(x)$ is hard to predict with probability significantly better than $1/2$ given $f_k(x)$. More formally, H is a hard-core bit for F the following is negligible (in k) for all PPT algorithms A:

$$\left| \Pr[x \leftarrow \mathcal{D}_k; y = f_k(x) : A(1^k, y) = h_k(x)] - 1/2 \right|.$$

(Note that this is the "best" one could hope for in a definition of this sort, since an algorithm that simply outputs a random bit will guess $h_k(x)$ correctly half the time.)

We stress that *not* every H is a hard-core bit for a given one-way function family F. To give a trivial example: for the one-way function family based on factoring (in which $f_k(p,q) = pq$), it is easy to predict the last bit of p (and also q), which is always 1! On the other hand, a one-way function family with a hard-core bit can be constructed from any one-way function family; we state the following result to that effect without proof.

Theorem 9.2 ([27])   *If there exists a one-way function family F, then there exists (constructively) a one-way function family F' and an H which is a hard-core bit for F'.*

Hard-core bits for specific functions are known without recourse to the general theorem above [1, 9, 21, 32, 36]. We discuss a representative result for the case of RSA (this function family was introduced in Section 9.3, and we assume the reader is familiar with the notation used there). Let $H = \{h_k\}$ be a function family such that $h_k(N, e, x)$ returns the least significant bit of x mod N. Then H is a hard-core bit for RSA [1, 21]. Reiterating the definition above and assuming that RSA is a one-way function family, this means that given N, e, and x^e mod N (for randomly chosen N, e, and x from the appropriate domains), it is hard for any PPT algorithm to compute the least significant bit of x mod N with probability better than $1/2$.

We show now a construction of a PRG with expansion factor $k + 1$ based on any one-way *permutation* family $F = \{f_k\}$ with hard-core bit $H = \{h_k\}$. For simplicity, assume that the domain of f_k is $\{0, 1\}^k$; furthermore, for convenience, let $f(x), h(x)$ denote $f_{|x|}(x), h_{|x|}(x)$, respectively. Define:

$$G(x) = f(x) \circ h(x).$$

We claim that G is a PRG. As some intuition toward this claim, let $|x| = k$ and note that the first k bits of $G(x)$ are indeed uniformly distributed if x is uniformly distributed; this follows from the fact that f is a permutation over $\{0, 1\}^k$. Now, because H is a hard-core bit of F, $h(x)$ cannot be predicted by any efficient algorithm with probability better than $1/2$ even when the algorithm is given $f(x)$. Informally, then, $h(x)$ "looks random" to a PPT algorithm even conditioned on the observation of $f(x)$; hence, the entire string $f(x) \circ h(x)$ is pseudorandom.

It is known that given any PRG with expansion factor $k + 1$, it is possible to construct a PRG with expansion factor $\ell(k)$ for any polynomial $\ell(\cdot)$. The above construction, then, may be extended to yield a PRG that expands its input by an essentially arbitrary amount. Finally, although the preceding discussion focused only on the case of one-way *permutations*, it can be generalized (with much difficulty!) for the more general case of one-way *functions*. Putting these known results together, we obtain:

Theorem 9.3 ([31]) *If there exists a one-way function family, then for any polynomial $\ell(\cdot)$, there exists a PRG with stretching factor $\ell(k)$.*

9.4.2 Pseudorandom Functions and Block Ciphers

A pseudorandom generator G takes a short random string x and yields a polynomially-longer pseudo-random string $G(x)$. This in turn is useful in many contexts; see Section 9.5 for an example. However, a PRG has the following "limitations." First, for $G(x)$ to be pseudorandom, it is necessary that (1) x be chosen uniformly at random and also that (2) x be unknown to the distinguishing algorithm (clearly, once x is known, $G(x)$ is determined and hence no longer looks random). Furthermore, a PRG generates pseudorandom output whose length must be polynomially related to that of the input string x. For some applications, it would be nice to circumvent these limitations in some way.

These considerations have led to the definition and development of a more powerful primitive: a (family of) **pseudorandom functions (PRFs)**. Informally, a PRF $F : \{0, 1\}^k \times \{0, 1\}^m \to \{0, 1\}^n$ is a keyed function, so that fixing a particular key $s \in \{0, 1\}^k$ may be viewed as defining a function $F_s : \{0, 1\}^m \to \{0, 1\}^n$. (For simplicity in the rest of this and the following paragraph, we let $m = n = k$ although in general $m, n = \text{poly}(k)$.) Informally, a PRF F acts like a random function in the following sense: no efficient algorithm can distinguish the input/output behavior of F (with a randomly chosen key which is fixed for the duration of the experiment) from the input/output behavior of a truly random function. We stress that this holds even when the algorithm is allowed to interact with the function in an arbitrary way. It may be helpful to picture the following imaginary experiment: an algorithm is given access to a box that implements a function over $\{0, 1\}^k$. The algorithm can send inputs of its choice to the box and observe the corresponding outputs, but may not experiment with the box in any other way. Then F is a PRF if no efficient algorithm can distinguish whether the box implements a truly random function over $\{0, 1\}^k$ (i.e., a function chosen uniformly at random from the space of all 2^{k2^k} functions over $\{0, 1\}^k$) or whether it implements an instance of F_s (for uniformly chosen key $s \in \{0, 1\}^k$).

Note that this primitive is much stronger than a PRG. For one, the key s can be viewed as encoding an *exponential* amount of pseudorandomness because, roughly speaking, $F_s(x)$ is an independent pseudo-random value for each $x \in \{0,1\}^k$. Second, note that $F_s(x)$ is pseudorandom even if x is known, and even if x was not chosen at random. Of course, it must be the case that the key s is unknown and is chosen uniformly at random. We now give a formal definition of a PRF.

Definition 9.5 Let $\mathcal{F} = \{F_s : \{0,1\}^{m(k)} \rightarrow \{0,1\}^{n(k)}\}_{k \geq 1; s \in \{0,1\}^k}$ be an efficiently computable function family where $m, n = \mathrm{poly}(k)$, and let $\mathrm{Rand}_{m(k)}^{n(k)}$ denote the set of all functions from $\{0,1\}^{m(k)}$ to $\{0,1\}^{n(k)}$. We say $\mathcal{F}$ is a pseudorandom function family (PRF) if the following is negligible in k for all PPT algorithms A:

$$\left| \Pr[s \leftarrow \{0,1\}^k : A^{F_s(\cdot)}(1^k) = 1] - \Pr[f \leftarrow \mathrm{Rand}_{m(k)}^{n(k)} : A^{f(\cdot)}(1^k) = 1] \right|,$$

where the notation $A^{f(\cdot)}$ denotes that A has oracle access to function f; that is, A can send (as often as it likes) inputs of its choice to f and receive the corresponding outputs.

We do not present any details about the construction of a PRF based on general assumptions, beyond noting that they can be constructed from any one-way function family.

Theorem 9.4 ([25]) *If there exists a one-way function family F, then there exists (constructively) a PRF $\mathcal{F}$.*

An efficiently computable permutation family $\mathcal{P} = \{P_s : \{0,1\}^{m(k)} \rightarrow \{0,1\}^{m(k)}\}_{k \geq 1; s \in \{0,1\}^k}$ is an efficiently computable function family for which P_s is a permutation over $\{0,1\}^{m(k)}$ for each $s \in \{0,1\}^k$; and furthermore P_s^{-1} is efficiently computable (given s). By analogy with the case of a PRF, we say that $\mathcal{P}$ is a pseudorandom permutation (PRP) if P_s (with s randomly chosen in $\{0,1\}^k$) is indistinguishable from a truly random *permutation* over $\{0,1\}^{m(k)}$. A pseudorandom permutation can be constructed from any pseudorandom function [37].

What makes PRFs and PRPs especially useful in practice (especially as compared to PRGs) is that very efficient implementations of (conjectured) PRFs are available in the form of **block ciphers**. A block cipher is an efficiently computable permutation family $\mathcal{P} = \{P_s : \{0,1\}^m \rightarrow \{0,1\}^m\}_{s \in \{0,1\}^k}$ *for which keys have a fixed length* k. Because keys have a fixed length, we can no longer speak of a "negligible function" or a "polynomial-time algorithm" and consequently there is no notion of asymptotic security for block ciphers; instead, concrete security definitions are used. For example, a block cipher is said to be a (t, ε)-secure PRP, say, if no adversary running in time t can distinguish P_s (for randomly chosen s) from a random permutation over $\{0,1\}^m$ with probability better than ε. See [3] for further details.

Block ciphers are particularly efficient because they are *not* based on number-theoretic or algebraic one-way function families but are instead constructed directly, with efficiency in mind from the outset. One popular block cipher is DES (the Data Encryption Standard) [17, 38], which has 56-bit keys and is a permutation on $\{0,1\}^{64}$. DES dates to the mid-1970s, and recent concerns about its security — particularly its relatively short key length — have prompted the development* of a new block cipher termed AES (the Advanced Encryption Standard). This cipher supports 128-, 192-, and 256-bit keys, and is a permutation over $\{0,1\}^{128}$. Details of the AES cipher and the rationale for its construction are available [13].

9.4.3 Cryptographic Hash Functions

Although hash functions play an important role in cryptography, our discussion will be brief and informal because they are used sparingly in the remainder of this survey.

Hash functions — functions that compress long, often variable-length strings to much shorter strings — are widely used in many areas of computer science. For many applications, constructions of hash functions

---

*See http://csrc.nist.gov/CryptoToolkit/aes/ for a history and discussion of the design competition resulting in the selection of a cipher for AES.

with the necessary properties are known to exist without any computational assumptions. For cryptography, however, hash functions with very strong properties are often needed; furthermore, it can be shown that the existence of a hash function with these properties would imply the existence of a one-way function family (and therefore any such construction must be based on a computational assumption of some sort). We discuss one such property here.

The security property that arises most often in practice is that of collision resistance. Informally, H is said to be a **collision-resistant hash function** if an adversary is unable to find a "collision" in H; namely, two inputs x, x' with $x \neq x'$ but $H(x) = H(x')$. As in the case of PRFs and block ciphers (see the previous section), we can look at either the asymptotic security of a function family $\mathcal{H} = \{H_s : \{0,1\}^* \rightarrow \{0,1\}^k\}_{k \geq 1; s \in \{0,1\}^k}$ or the concrete security of a fixed hash function $H : \{0,1\}^* \rightarrow \{0,1\}^m$. The former are constructed based on specific computational assumptions, while the latter (as in the case of block ciphers) are constructed directly and are therefore much more efficient.

It is not hard to show that a collision-resistant hash function family mapping arbitrary-length inputs to fixed-length outputs is itself a one-way function family. Interestingly, however, collision-resistant hash function families are believed to be impossible to construct based on (general) one-way function families or trapdoor permutation generators [49]. On the other hand, constructions of collision-resistant hash function families based on specific computational assumptions (e.g., the hardness of factoring) are known; see Section 10.2 in [14].

In practice, customized hash functions — designed with efficiency in mind and not derived from number-theoretic problems — are used. One well-known example is MD5 [44], which hashes arbitrary-length inputs to 128-bit outputs. Because collisions in *any* hash function with output length k can be found in expected time (roughly) $2^{k/2}$ via a "birthday attack" (see, for example, Section 3.4.2 in [14]) and because computations on the order of 2^{64} are currently considered just barely outside the range of feasibility, hash functions with output lengths longer than 128 bits are frequently used. A popular example is SHA-1 [19], which hashes arbitrary-length inputs to 160-bit outputs. SHA-1 is considered collision-resistant for practical purposes, given current techniques and computational ability.

Hash functions used in cryptographic protocols sometimes require properties stronger than collision resistance in order for the resulting protocol to be provably secure [5]. It is fair to say that, in many cases, the exact properties needed by the hash function are not yet fully understood.

9.5 Private-Key Encryption

As discussed in Section 9.2.1, perfectly secret private-key encryption is achievable using the one-time pad encryption scheme; however, perfectly secret encryption requires that the shared key be at least as long as the communicated message. Our goal was to beat this bound by considering computational notions of security instead. We show here that this is indeed possible.

Let us first see what a definition of computational secrecy might involve. In the case of perfect secrecy, we required that for all messages m_0, m_1 of the same length ℓ, *no possible* algorithm could distinguish *at all* whether a given ciphertext is an encryption of m_0 or m_1. In the notation we have been using, this is equivalent to requiring that for all adversaries A,

$$\left| \Pr[s \leftarrow \{0,1\}^\ell : A(\mathcal{E}_s(m_0)) = 1] - \Pr[s \leftarrow \{0,1\}^\ell : A(\mathcal{E}_s(m_1)) = 1] \right| = 0.$$

To obtain a computational definition of security, we make two modifications: (1) we require the above to hold only for *efficient* (i.e., PPT) algorithms A; and (2) we only require the "distinguishing advantage" of the algorithm to be *negligible*, and not necessarily 0. The resulting definition of computational secrecy is that for all PPT adversaries A, the following is negligible:

$$\left| \Pr[s \leftarrow \{0,1\}^k : A(1^k, \mathcal{E}_s(m_0)) = 1] - \Pr[s \leftarrow \{0,1\}^k : A(1^k, \mathcal{E}_s(m_1)) = 1] \right|. \tag{9.1}$$

The one-time pad encryption scheme, together with the notion of a PRG as defined in Section 9.4.1, suggest a computationally secret encryption scheme in which the shared key is shorter than the message (we reiterate that this is simply not possible if perfect secrecy is required). Specifically, let G be a PRG with expansion factor $\ell(k)$ (recall $\ell(k)$ is a polynomial with $\ell(k) > k$). To encrypt a message of length $\ell(k)$, the parties share a key s of length k; message m is then encrypted by computing $C = m \oplus G(s)$. Decryption is done by simply computing $m = C \oplus G(s)$.

For some intuition as to why this is secure, note that the scheme can be viewed as implementing a "pseudo"-one-time pad in which the parties share the pseudorandom string $G(s)$ instead of a uniformly random string of the same length. (Of course, to minimize the secret key length, the parties actually share s and regenerate $G(s)$ when needed.) But because the pseudorandom string $G(s)$ "looks random" to a PPT algorithm, the pseudo-one-time pad scheme "looks like" the one-time pad scheme to any PPT adversary. Because the one-time pad scheme is secure, so is the pseudo-one-time pad. (This is not meant to serve as a rigorous proof, but can easily be adapted to give one.)

We re-cap the discussion thus far in the following lemma.

Lemma 9.5 *Perfectly secret encryption is possible if and only if the shared key is at least as long as the message. However, if there exists a PRG, then there exists a computationally secret encryption scheme in which the message is (polynomially) longer than the shared key.*

Let us examine the pseudo-one-time pad encryption scheme a little more critically. Although the scheme allows encrypting messages longer than the secret key, the scheme is secure only when it is used *once* (as in the case of the one-time pad). Indeed, if an adversary views ciphertexts $C_1 = m_1 \oplus G(s)$ and $C_2 = m_2 \oplus G(s)$ (where m_1 and m_2 are unknown), the adversary can compute $m_1 \oplus m_2 = C_1 \oplus C_2$ and hence learn something about the relation between the two messages. Even worse, if the adversary somehow learns (or later determines), say, m_1, then the adversary can compute $G(s) = C_1 \oplus m_1$ and can thus decrypt any ciphertexts subsequently transmitted. We stress that such attacks (called *known-plaintext attacks*) are not merely of academic concern, because there are often messages sent whose values are uniquely determined, or known to lie in a small range. Can we obtain secure encryption even in the face of such attacks?

Before giving a scheme that prevents such attacks, let us precisely formulate a definition of security. First, the scheme should be "secure" even when used to encrypt multiple messages; in particular, an adversary who views the ciphertexts corresponding to multiple messages should not learn any information about the relationships among these messages. Second, the secrecy of the scheme should remain intact if some encrypted messages are known by the adversary. In fact, we can go beyond this last requirement and mandate that the scheme remain "secure" even if the adversary can request the encryption of messages of his choice (a *chosen-plaintext attack* of this sort arises when an adversary can influence the messages sent).

We model chosen-plaintext attacks by giving the adversary unlimited and unrestricted access to an *encryption oracle* denoted $\mathcal{E}_s(\cdot)$. This is simply a "black-box" that, on inputting a message m, returns an encryption of m using key s (in case $\mathcal{E}$ is randomized, the oracle chooses fresh randomness each time). Note that the resulting attack is perhaps stronger than what a real-world adversary can do (a real-world adversary likely cannot request as many encryptions — of arbitrary messages — as he likes); by the same token, if we can construct a scheme secure against this attack, then certainly the scheme will be secure in the real world. A formal definition of security follows.

Definition 9.6 A private-key encryption scheme $(\mathcal{E}, \mathcal{D})$ is said to be secure against chosen-plaintext attacks if, for all messages m_1, m_2 and all PPT adversaries A, the following is negligible:

$$\left| \Pr[s \leftarrow \{0,1\}^k : A^{\mathcal{E}_s(\cdot)}(1^k, \mathcal{E}_s(m_1)) = 1] - \Pr[s \leftarrow \{0,1\}^k : A^{\mathcal{E}_s(\cdot)}(1^k, \mathcal{E}_s(m_2)) = 1] \right|.$$

Namely, a PPT adversary cannot distinguish between the encryption of m_1 and m_2 *even if the adversary is given unlimited access to an encryption oracle.*

We stress one important corollary of the above definition: an encryption scheme secure against chosen-plaintext attacks *must be randomized* (in particular, the one-time pad does not satisfy the above definition).

This is so for the following reason: if the scheme were deterministic, an adversary could obtain $C_1 = \mathcal{E}_s(m_1)$ and $C_2 = \mathcal{E}_s(m_2)$ from its encryption oracle and then compare the given ciphertext to each of these values; thus, the adversary could immediately tell which message was encrypted. Our strong definition of security forces us to consider more complex encryption schemes.

Fortunately, many encryption schemes satisfying the above definition are known. We present two examples here; the first is mainly of theoretical interest (but is also practical for short messages), and its simplicity is illuminating. The second is more frequently used in practice.

Our first encryption scheme uses a key of length k to encrypt messages of length k (we remind the reader, however, that this scheme will be a tremendous improvement over the one-time pad because the present scheme can be used to encrypt polynomially-many messages). Let $\mathcal{F} = \{F_s : \{0,1\}^k \to \{0,1\}^k\}_{k \geq 1; s \in \{0,1\}^k}$ be a PRF (cf. Section 9.4.2); alternatively, one can think of k as being fixed and using a block cipher for $\mathcal{F}$ instead. We define encryption using key s as follows [26]: on input a message $m \in \{0,1\}^k$, choose a random $r \in \{0,1\}^k$ and output $\langle r, F_s(r) \oplus m \rangle$. To decrypt ciphertext $\langle r, c \rangle$ using key s, simply compute $m = c \oplus F_s(r)$.

We give some intuition for the security of this scheme against chosen-plaintext attacks. Assume the adversary queries the encryption oracle n times, receiving in return the ciphertexts $\langle r_1, c_1 \rangle, \ldots, \langle r_n, c_n \rangle$ (the messages to which these ciphertexts correspond are unimportant). Let the ciphertext given to the adversary — corresponding to the encryption of either m_1 or m_2 — be $\langle r, c \rangle$. By the definition of a PRF, the value $F_s(r)$ "looks random" to the PPT adversary A unless $F_s(\cdot)$ was previously computed on input r; in other words, $F_s(r)$ "looks random" to A unless $r \in \{r_1, \ldots, r_n\}$ (we call this occurrence a *collision*). Security of the scheme is now evident from the following: (1) assuming a collision does *not* occur, $F_s(r)$ is pseudorandom as discussed and hence the adversary cannot determine whether m_1 or m_2 was encrypted (as in the one-time pad scheme); furthermore, (2) the probability that a collision occurs is $\frac{n}{2^k}$, which is negligible (because n is polynomial in k). We thus have Theorem 9.6.

Theorem 9.6 ([26]) *If there exists a PRF $\mathcal{F}$, then there exists an encryption scheme secure against chosen-plaintext attacks.*

The previous construction applies to small messages whose length is equal to the output length of the PRF. From a theoretical point of view, an encryption scheme (secure against chosen-plaintext attacks) for longer messages follows immediately from the construction given previously; namely, to encrypt message $M = m_1, \ldots, m_\ell$ (where $m_i \in \{0,1\}^k$), simply encrypt each block of the message using the previous scheme, giving ciphertext:

$$\langle r_1, F_s(r_1) \oplus m_1, \ldots, r_\ell, F_s(r_\ell) \oplus m_\ell \rangle.$$

This approach gives a ciphertext twice as long as the original message and is therefore not very practical.

A better idea is to use a **mode of encryption**, which is a method for encrypting long messages using a block cipher with fixed input/output length. Four modes of encryption were introduced along with DES [18], and we discuss one such mode here (not all of the DES modes of encryption are secure). In cipher block chaining (CBC) mode, a message $M = m_1, \ldots, m_\ell$ is encrypted using key s as follows:

$$\text{Choose } C_0 \in \{0,1\}^k \text{ at random}$$
$$\text{For } i = 1 \text{ to } \ell:$$
$$C_i = F_s(m_i \oplus C_{i-1})$$
$$\text{Output } \langle C_0, C_1, \ldots, C_\ell \rangle$$

Decryption of a ciphertext $\langle C_0, \ldots, C_\ell \rangle$ is done by reversing the above steps:

$$\text{For } i = 1 \text{ to } \ell:$$
$$m_i = F_s^{-1}(C_i) \oplus C_{i-1}$$
$$\text{Output } m_1, \ldots, m_\ell$$

It is known that CBC mode is secure against chosen-plaintext attacks [3].

9.6 Message Authentication

The preceding section discussed how to achieve message *secrecy*; we now discuss techniques for message *integrity*. In the private-key setting, this is accomplished using message authentication codes (MACs). We stress that secrecy and authenticity are two incomparable goals, and it is certainly possible to achieve either one without the other. As an example, the one-time pad — which achieves perfect secrecy — provides no message integrity whatsoever because *any* ciphertext C of the appropriate length decrypts to some valid message. Even worse, if C represents the encryption of a particular message m (so that $C = m \oplus s$ where s is the shared key), then flipping the first bit of C has the effect of flipping the first bit of the resulting decrypted message.

Before continuing, let us first define the semantics of a MAC.

Definition 9.7   A message authentication code consists of a pair of PPT algorithms $(\mathcal{T}, \mathsf{Vrfy})$ such that (here, the length of the key is taken to be the security parameter):

- The tagging algorithm $\mathcal{T}$ takes as input a key s and a message m and outputs a tag $t = \mathcal{T}_s(m)$.
- The verification algorithm Vrfy takes as input a key s, a message m, and a (purported) tag t and outputs a bit signifying acceptance (1) or rejection (0).

We require that for all m and all t output by $\mathcal{T}_s(m)$ we have $\mathsf{Vrfy}_s(m, t) = 1$.

Actually, a MAC should also be defined over a particular message space and this must either be specified or else clear from the context.

Schemes designed to detect "random" modifications of a message (e.g., error-correcting codes) do not constitute secure MACs because they are not designed to provide message authenticity in an *adversarial* setting. Thus, it is worth considering carefully the exact security goal we desire. Ideally, even if an adversary can request tags for multiple messages $m_1, \dots$ of his choice, it should be impossible for the adversary to "forge" a valid-looking tag t on a new message m. (As in the case of encryption, this adversary is likely stronger than what is encountered in practice; however, if we can achieve security against even this strong attack so much the better!) To formally model this, we give the adversary access to an oracle $\mathcal{T}_s(\cdot)$, which returns a tag t for any message m of the adversary's choice. Let $m_1, \dots, m_\ell$ denote the messages submitted by the adversary to this oracle. We say a forgery occurs if the adversary outputs (m, t) such that $m \notin \{m_1, \dots, m_\ell\}$ and $\mathsf{Vrfy}_s(m, t) = 1$. Finally, we say a MAC is secure if the probability of a forgery is negligible for all PPT adversaries A. For completeness, we give a formal definition following [4].

Definition 9.8   MAC $(\mathcal{T}, \mathsf{Vrfy})$ is said to be secure against adaptive chosen-message attacks if, for all PPT adversaries A, the following is negligible:

$$\Pr[s \leftarrow \{0,1\}^k; (m, t) \leftarrow A^{\mathcal{T}_s(\cdot)}(1^k) : \mathsf{Vrfy}_s(m, t) = 1 \wedge m \notin \{m_1, \dots, m_\ell\}],$$

where $m_1, \dots, m_\ell$ are the messages that A submitted to $\mathcal{T}_s(\cdot)$.

We now give two constructions of a secure MAC. For the first, let $\mathcal{F} = \{F_s : \{0,1\}^k \to \{0,1\}^k\}_{k \geq 1; s \in \{0,1\}^k}$ be a PRF (we can also let $\mathcal{F}$ be a block cipher for some fixed value k). The discussion of PRFs in Section 9.4.2 should motivate the following construction of a MAC for messages of length k [26]: the tagging algorithm $\mathcal{T}_s(m)$ (where $|s| = |m| = k$) returns $t = F_s(m)$, and the verification algorithm $\mathsf{Vrfy}_s(m, t)$ outputs 1 if and only if $F_s(m) = t$. A proof of security for this construction is immediate: Let $m_1, \dots, m_\ell$ denote those messages for which adversary A has requested a tag from $\mathcal{T}_s(\cdot)$. Because $\mathcal{F}$ is a PRF, $\mathcal{T}_s(m) = F_s(m)$ "looks random" for any $m \notin \{m_1, \dots, m_\ell\}$ (call m of this sort *new*). Thus, the adversary's probability of outputting (m, t) such that $t = F_s(m)$ and m is new is (roughly) 2^{-k}; that is, the probability of guessing the output of a random function with output length k at a particular point m. This is negligible, as desired.

Because PRFs exist for any (polynomial-size) input length, the above construction can be extended to achieve secure message authentication for polynomially-long messages. We summarize the theoretical implications of this result in Theorem 9.7.

Theorem 9.7 ([26]) *If there exists a PRF $\mathcal{F}$, then there exists a MAC secure against adaptive chosen-message attack.*

Although the above result gives a theoretical solution to the problem of message authentication (and can be made practical for short messages by using a block cipher to instantiate the PRF), it does not give a practical solution for authenticating long messages. So, we conclude this section by showing a practical and widely used MAC construction for long messages. Let $\mathcal{F} = \{F_s : \{0,1\}^n \rightarrow \{0,1\}^n\}_{s \in \{0,1\}^k}$ denote a block cipher. For fixed ℓ, define the CBC-MAC for messages of length $(\{0,1\}^n)^{\ell}$ as follows (note the similarity with the CBC mode of encryption from Section 9.5): the tag of a message $m_1, \ldots, m_{\ell}$ with $m_i \in \{0,1\}^n$ is computed as:

$$C_0 = 0^n$$
$$\text{For } i = 1 \text{ to } \ell:$$
$$C_i = F_s(m_i \oplus C_{i-1})$$
$$\text{Output } C_{\ell}$$

Verification of a tag t on a message $m_1, \ldots, m_{\ell}$ is done by re-computing C_{ℓ} as above and outputting 1 if and only if $t = C_{\ell}$. It is known that the CBC-MAC is secure against adaptive chosen-message attacks [4] for n sufficiently large. We stress that this is true only when fixed-length messages are authenticated (this was why ℓ was fixed, above). Subsequent work has focused on extending CBC-MAC to allow authentication of arbitrary-length messages [8, 41].

9.7 Public-Key Encryption

The advent of public-key encryption [15, 39, 45] marked a revolution in the field of cryptography. For hundreds of years, cryptographers had relied exclusively on shared, secret keys to achieve secure communication. Public-key cryptography, however, enables two parties to secretly communicate without having arranged for any *a priori* shared information. We first describe the semantics of a public-key encryption scheme, and then discuss two general ways such a scheme can be used.

Definition 9.9 A public-key encryption scheme is a triple of PPT algorithms $(\mathcal{K}, \mathcal{E}, \mathcal{D})$ such that:

- The key generation algorithm $\mathcal{K}$ takes as input a security parameter 1^k and outputs a public key PK and a secret key SK.
- The encryption algorithm $\mathcal{E}$ takes as input a public key PK and a message m and outputs a ciphertext C. We write this as $C \leftarrow \mathcal{E}_{PK}(m)$.
- The deterministic decryption algorithm $\mathcal{D}$ takes as input a secret key SK and a ciphertext C and outputs a message m. We write this as $m = \mathcal{D}_{SK}(C)$.

We require that for all k, all (PK, SK) output by $\mathcal{K}(1^k)$, for all m, and for all C output by $\mathcal{E}_{PK}(m)$, we have $\mathcal{D}_{SK}(C) = m$.

For completeness, a message space must be specified; however, the message space is generally taken to be $\{0,1\}^*$.

There are a number of ways in which a public-key encryption scheme can be used to enable communication between a sender $\mathcal{S}$ and a receiver $\mathcal{R}$. First, we can imagine that when $\mathcal{S}$ and $\mathcal{R}$ wish to communicate, $\mathcal{R}$ executes algorithm $\mathcal{K}$ to generate the pair of keys (PK, SK). The public key PK is sent (in the clear) to $\mathcal{S}$, and the secret key SK is (of course) kept secret by $\mathcal{R}$. To send a message m, $\mathcal{S}$ computes $C \leftarrow \mathcal{E}_{PK}(m)$ and transmits C to $\mathcal{R}$. The receiver $\mathcal{R}$ can now recover the original message by computing $m = \mathcal{D}_{SK}(C)$. Note that to fully ensure secrecy against an eavesdropping adversary, it must be the case that m remains hidden even if the adversary sees both PK and C (i.e., the adversary eavesdrops on the *entire* communication between $\mathcal{S}$ and $\mathcal{R}$).

A second way to picture the situation is to imagine that $\mathcal{R}$ runs $\mathcal{K}$ to generate keys (PK, SK) *independent of any particular sender S* (indeed, the identity of S need not be known at the time the keys are generated). The public key PK of $\mathcal{R}$ is then widely distributed — for example, published on $\mathcal{R}$'s personal homepage — and may be used by *anyone* wishing to securely communicate with $\mathcal{R}$. Thus, when a sender S wishes to confidentially send a message m to $\mathcal{R}$, the sender simply looks up $\mathcal{R}$'s public key PK, computes $C \leftarrow \mathcal{E}_{PK}(m)$, and sends C to $\mathcal{R}$; decryption by $\mathcal{R}$ is done as before. In this way, multiple senders can communicate multiple times with $\mathcal{R}$ using the same public key PK for all communication.

Note that, as was the case above, secrecy must be guaranteed even when an adversary knows PK. This is so because, by necessity, $\mathcal{R}$'s public key is widely distributed so that anyone can communicate with $\mathcal{R}$. Thus, it is only natural to assume that the adversary also knows PK. The following definition of security extends the definition given in the case of private-key encryption.

Definition 9.10 A public-key encryption scheme $(\mathcal{K}, \mathcal{E}, \mathcal{D})$ is said to be secure against chosen-plaintext attacks if, for all messages m_1, m_2 and all PPT adversaries A, the following is negligible:

$$\left| \Pr[(PK, SK) \leftarrow \mathcal{K}(1^k) : A(PK, \mathcal{E}_{PK}(m_0)) = 1] - \Pr[(PK, SK) \leftarrow \mathcal{K}(1^k) : A(PK, \mathcal{E}_{PK}(m_1)) = 1] \right|.$$

The astute reader will notice that this definition is analogous to the definition of one-time security for private-key encryption (with the exception that the adversary is now given the public key as input), but seems inherently different from the definition of security against chosen-plaintext attacks (cf. Definition 9.6). Indeed, the above definition makes no mention of any "encryption oracle" as does Definition 9.6. However, it is known for the case of public-key encryption that the definition above implies security against chosen-plaintext attacks (of course, we have seen already that the definitions are *not* equivalent in the private-key setting).

Definition 9.10 has the following immediate and important consequence, first noted by Goldwasser and Micali [29]: for a public-key encryption scheme to be secure, encryption *must* be probabilistic. To see this, note that if encryption were deterministic, an adversary could always tell whether a given ciphertext C corresponds to an encryption of m_1 or m_2 by simply computing $\mathcal{E}_{PK}(m_1)$ and $\mathcal{E}_{PK}(m_2)$ himself (recall the adversary knows PK) and comparing the results to C.

The definition of public-key encryption — in which determining the message corresponding to a ciphertext is "hard" in general, but becomes "easy" with the secret key — is reminiscent of the definition of trapdoor permutations. Indeed, the following feasibility result is known.

Theorem 9.8 ([54]) *If there exists a trapdoor permutation (generator), there exists a public-key encryption scheme secure against chosen-plaintext attacks.*

Unfortunately, public-key encryption schemes constructed via this generic result are generally quite inefficient, and it is difficult to construct *practical* encryption schemes secure in the sense of Definition 9.10. At this point, some remarks about the practical efficiency of public-key encryption are in order. Currently known public-key encryption schemes are roughly three orders of magnitude slower (per bit of plaintext) than private-key encryption schemes with comparable security. For encrypting long messages, however, all is not lost: in practice, a long message m is encrypted by first choosing at random a "short" (i.e., 128-bit) key s, encrypting this key using a public-key encryption scheme, and then encrypting m using a private-key scheme with key s. So, the public-key encryption of m under public key PK is given by:

$$\langle \mathcal{E}_{PK}(s) \circ \mathcal{E}'_s(m) \rangle,$$

where $\mathcal{E}$ is the public-key encryption algorithm and $\mathcal{E}'$ represents a private-key encryption algorithm. If both the public-key and private-key components are secure against chosen-plaintext attacks, so is the scheme above. Thus, the problem of designing efficient public-key encryption schemes for long messages is reduced to the problem of designing efficient public-key encryption for short messages.

We discuss the well-known El Gamal encryption scheme [16] here. Let G be a cyclic (multiplicative) group of order q with generator $g \in G$. Key generation consists of choosing a random $x \in \mathbb{Z}_q$ and setting $y = g^x$. The public key is (G, q, g, y) and the secret key is x. To encrypt a message $m \in G$, the sender chooses a random $r \in \mathbb{Z}_q$ and sends:

$$\langle g^r, y^r m \rangle.$$

To decrypt a ciphertext $\langle A, B \rangle$ using secret key x, the receiver computes $m = B/A^x$. It is easy to see that decryption correctly recovers the intended message.

Clearly, security of the scheme requires the discrete logarithm problem in G to be hard; if the discrete logarithm problem were easy, then the secret key x could be recovered from the information contained in the public key. Hardness of the discrete logarithm problem is not, however, sufficient for the scheme to be secure in the sense of Definition 9.10; a stronger assumption (first introduced by Diffie and Hellman [15] and hence called the *decisional Diffie-Hellman (DDH) assumption*) is, in fact, needed. (See [52] or [7] for further details.)

We have thus far *not* mentioned the "textbook RSA" encryption scheme. Here, key generation results in public key (N, e) and secret key d such that $ed = 1 \bmod \varphi(N)$ (see Section 9.3.2 for further details) and encryption of message $m \in \mathbb{Z}_N^*$ is done by computing $C = m^e \bmod N$. The reason for its omission is that this scheme is simply *not secure* in the sense of Definition 9.10; for one thing, encryption in this scheme is deterministic and therefore cannot possibly be secure.

Of course — and as discussed in Section 9.3.2 — the RSA assumption gives a trapdoor permutation generator, which in turn can be used to construct a secure encryption scheme (cf. Theorem 9.8). Such an approach, however, is inefficient and not used in practice. The public-key encryption schemes used in practice that are based on the RSA problem seem to require additional assumptions regarding certain hash functions; we refer to [5] for details that are beyond our present scope.

We close this section by noting that current, widely used encryption schemes in fact satisfy stronger definitions of security than that of Definition 9.10; in particular, encryption schemes are typically designed to be secure against chosen-ciphertext attacks (see [7] for a definition). Two efficient examples of encryption schemes meeting this stronger notion of security include the Cramer-Shoup encryption scheme [12] (based on the DDH assumption) and OAEP-RSA [6, 10, 22, 48] (based on the RSA assumption and an assumption regarding certain hash functions [5]).

9.8 Digital Signature Schemes

As public-key encryption is to private-key encryption, so are digital signature schemes to message authentication codes. Digital signature schemes are the public-key analog of MACs; they allow a *signer* who has established a public key to "sign" messages in a way that is verifiable to anyone who knows the signer's public key. Furthermore (by analogy with MACs), no adversary can forge valid-looking signatures on messages that were not explicitly authenticated (i.e., signed) by the legitimate signer.

In more detail, to use a signature scheme, a user first runs a key generation algorithm to generate a public-key/private-key pair (PK, SK); the user then publishes and widely distributes PK (as in the case of public-key encryption). When the user wants to authenticate a message m, she may do so using the signing algorithm along with her secret key SK; this results in a signature σ. Now, *anyone* who knows PK can verify correctness of the signature by running the public verification algorithm using the known public key PK, message m, and (purported) signature σ. We formalize the semantics of digital signature schemes in the following definition.

Definition 9.11 A signature scheme consists of a triple of PPT algorithms $(\mathcal{K}, \mathsf{Sign}, \mathsf{Vrfy})$ such that:

- The key generation algorithm $\mathcal{K}$ takes as input a security parameter 1^k and outputs a public key PK and a secret key SK.

- The signing algorithm Sign takes as input a secret key SK and a message m and outputs a signature $\sigma = \mathsf{Sign}_{SK}(m)$.
- The verification algorithm Vrfy takes as input a public key PK, a message m, and a (purported) signature σ and outputs a bit signifying acceptance (1) or rejection (0).

We require that for all (PK, SK) output by $\mathcal{K}$, for all m, and for all σ output by $\mathsf{Sign}_{SK}(m)$, we have $\mathsf{Vrfy}_{PK}(m, \sigma) = 1$.

As in the case of MACs, the message space for a signature scheme should be specified. This is also crucial when discussing the security of a scheme.

A definition of security for signature schemes is obtainable by a suitable modification of the definition of security for MACs* (cf. Definition 9.8) with oracle $\mathsf{Sign}_{SK}(\cdot)$ replacing oracle $\mathcal{T}_s(\cdot)$, and the adversary now having as additional input the signer's public key. For reference, the definition (originating in [30]) is included here.

Definition 9.12   Signature scheme $(\mathcal{K}, \mathsf{Sign}, \mathsf{Vrfy})$ is said to be secure against adaptive chosen-message attacks if, for all PPT adversaries A, the following is negligible:

$$\Pr\left[(PK, SK) \leftarrow \mathcal{K}(1^k); (m, \sigma) \leftarrow A^{\mathsf{Sign}_{SK}(\cdot)}(1^k, PK) : \right.$$
$$\left. \mathsf{Vrfy}_{PK}(m, \sigma) = 1 \wedge m \notin \{m_1, \ldots, m_\ell\}\right],$$

where $m_1, \ldots, m_\ell$ are the messages that A submitted to $\mathsf{Sign}_{SK}(\cdot)$.

Under this definition of security, a digital signature emulates (the ideal qualities of) a handwritten signature. The definition shows that a digital signature on a message or document is easily verifiable by any recipient who knows the signer's public key; furthermore, a secure signature scheme is unforgeable in the sense that a third party cannot affix someone else's signature to a document without the signer's agreement.

Signature schemes also possess the important quality of *non-repudiation*; namely, a signer who has digitally signed a message cannot later deny doing so (of course, he can claim that his secret key was stolen or otherwise illegally obtained). Note that this property is not shared by MACs, because a tag on a given message could have been generated by either of the parties who share the secret key. Signatures, on the other hand, uniquely bind one party to the signed document.

It will be instructive to first look at a simple proposal of a signature scheme based on the RSA assumption, which is *not secure*. Unfortunately, this scheme is presented in many textbooks as a secure implementation of a signature scheme; hence, we refer to the scheme as the "textbook RSA scheme." Here, key generation involves choosing two large primes p, q of equal length and computing $N = pq$. Next, choose $e < N$ which is relatively prime to $\varphi(N)$ and compute d such that $ed = 1 \bmod \varphi(N)$. The public key is (N, e) and the secret key is (N, d). To sign a message $m \in \mathbb{Z}_N^*$, the signer computes

$$\sigma = m^d \bmod N;$$

verification of signature σ on message m is performed by checking that

$$\sigma^e \stackrel{?}{=} m \bmod N.$$

That this is indeed a signature scheme follows from the fact that $(m^d)^e = m^{de} = m \bmod N$ (see Section 9.3.2). What can we say about the security of the scheme?

---

*Historically, the definition of security for MACs was based on the earlier definition of security for signatures.

It is not hard to see that the textbook RSA scheme is completely insecure! An adversary can forge a valid message/signature pair as follows: choose arbitrary $\sigma \in \mathbb{Z}_N^*$ and set $m = \sigma^e \bmod N$. It is clear that the verification algorithm accepts σ as a valid signature on m.

In the previous attack, the adversary generates a signature on an essentially random message m. Here, we show how an adversary can forge a signature on a *particular* message m. First, the adversary finds arbitrary m_1, m_2 such that $m_1 m_2 = m \bmod N$; the adversary then requests and obtains signatures σ_1, σ_2 on m_1, m_2, respectively (recall that this is allowed by Definition 9.12). Now we claim that the verification algorithm accepts $\sigma = \sigma_1 \sigma_2 \bmod N$ as a valid signature on m. Indeed:

$$(\sigma_1 \sigma_2)^e = \sigma_1^e \sigma_2^e = m_1 m_2 = m \bmod N.$$

The two preceding examples illustrate that textbook RSA is *not* secure. The general approach, however, may be secure if the message is hashed (using a cryptographic hash function) before signing; this approach yields the *full-domain hash* (FDH) signature scheme [5]. In more detail, let $H : \{0,1\}^* \to \mathbb{Z}_N^*$ be a cryptographic hash function that might be included as part of the signer's public key. Now, message m is signed by computing $\sigma = H(m)^d \bmod N$; a signature σ on message m is verified by checking that $\sigma^e \stackrel{?}{=} H(m) \bmod N$. The presence of the hash (assuming a "good" hash function) prevents the two attacks mentioned above: for example, an adversary will still be able to generate σ, m' with $\sigma^e = m' \bmod N$ as before, but now the adversary will *not* be able to find a message m for which $H(m) = m'$. Similarly, the second attack is foiled because it is is difficult for an adversary to find m_1, m_2, m with $H(m_1)H(m_2) = H(m) \bmod N$. The use of the hash H has the additional advantage that messages of arbitrary length can now be signed.

It is, in fact, possible to prove the security of the FDH signature scheme based on the assumption that RSA is a trapdoor permutation and a (somewhat non-standard) assumption about the hash function H; however, it is beyond the scope of this work to discuss the necessary assumptions on H in order to enable a proof of security. We refer the interested reader to [5] for further details.

The Digital Signature Algorithm (DSA) (also known as the Digital Signature Standard [DSS]) [2, 20] is another widely used and standardized signature scheme whose security is related to the hardness of computing discrete logarithms (and which therefore offers an alternative to schemes whose security is based on, e.g., the RSA problem). Let p, q be primes such that $|q| = 160$ and q divides $p - 1$; typically, we might have $|p| = 512$. Let g be an element of order q in the multiplicative group $\mathbb{Z}_p^*$, and let $\langle g \rangle$ denote the subgroup of $\mathbb{Z}_p^*$ generated by g. Finally, let $H : \{0,1\}^* \to \{0,1\}^{160}$ be a cryptographic hash function. Parameters (p, q, g, H) are public, and can be shared by multiple signers. A signer's personal key is computed by choosing a random $x \in \mathbb{Z}_q$ and setting $y = g^x \bmod p$; the signer's public key is y and their private key is x. (Note that if computing discrete logarithms in $\langle g \rangle$ were easy, then it would be possible to compute a signer's secret key from their public key and the scheme would immediately be insecure.)

To sign a message $m \in \{0,1\}^*$ using secret key x, the signer generates a random $k \in \mathbb{Z}_q$ and computes

$$r = (g^k \bmod p) \bmod q$$
$$s = (H(m) + xr)k^{-1} \bmod q$$

The signature is (r, s). Verification of signature (r, s) on message m with respect to public key y is done by checking that $r, s \in \mathbb{Z}_q^*$ and

$$r \stackrel{?}{=} (g^{H(m)s^{-1}} y^{rs^{-1}} \bmod p) \bmod q.$$

It can be easily verified that signatures produced by the legitimate signer are accepted (with all but negligible probability) by the verification algorithm.

It is beyond the scope of this work to discuss the security of DSA; we refer the reader to a recent survey article [53] for further discussion and details.

Finally, we state the following result, which is of great theoretical importance but (unfortunately) of limited practical value.

Theorem 9.9 ([35, 40, 46]) *If there exists a one-way function family $\mathcal{F}$, then there exists a digital signature scheme secure against adaptive chosen-message attack.*

Defining Terms

Block cipher: An efficient instantiation of a pseudorandom function.

Ciphertext: The result of encrypting a message.

Collision-resistant hash function: Hash function for which it is infeasible to find two different inputs mapping to the same output.

Data integrity: Ensuring that modifications to a communicated message are detected.

Data secrecy: Hiding the contents of a communicated message.

Decrypt: To recover the original message from the transmitted ciphertext.

Digital signature scheme: Method for protecting data integrity in the public-key setting.

Encrypt: To apply an encryption scheme to a plaintext message.

Message-authentication code: Algorithm preserving data integrity in the private-key setting.

Mode of encryption: A method for using a block cipher to encrypt arbitrary-length messages.

One-time pad: A private-key encryption scheme achieving perfect secrecy.

One-way function: A function that is "easy" to compute but "hard" to invert.

Plaintext: The communicated data, or message.

Private-key encryption: Technique for ensuring data secrecy in the private-key setting.

Private-key setting: Setting in which communicating parties secretly share keys in advance of their communication.

Pseudorandom function: A keyed function that is indistinguishable from a truly random function.

Pseudorandom generator: A deterministic function that converts a short, random string to a longer, pseudorandom string.

Public-key encryption: Technique for ensuring data secrecy in the public-key setting.

Public-key setting: Setting in which parties generate public/private keys and widely disseminate their public keys.

Trapdoor permutation: A one-way permutation that is "easy" to invert if some trapdoor information is known.

References

[1] Alexi, W.B., Chor, B., Goldreich, O., and Schnorr, C.P. 1988. RSA/Rabin functions: certain parts are as hard as the whole. *SIAM J. Computing*, 17(2):194–209.

[2] ANSI X9.30. 1997. Public key cryptography for the financial services industry. Part 1: The digital signature algorithm (DSA). American National Standards Institute. American Bankers Association.

[3] Bellare, M., Desai, A., Jokipii, E., and Rogaway, P. 1997. A concrete security treatment of symmetric encryption. *Proceedings of the 38th Annual Symposium on Foundations of Computer Science*, IEEE, pp. 394–403.

[4] Bellare, M., Kilian, J., and Rogaway, P. 2000. The security of the cipher block chaining message authentication code. *J. of Computer and System Sciences*, 61(3):362–399.

[5] Bellare, M. and Rogaway, P. 1993. Random oracles are practical: a paradigm for designing efficient protocols. *First ACM Conference on Computer and Communications Security*, ACM, pp. 62–73.

[6] Bellare, M. and Rogaway, P. 1995. Optimal asymmetric encryption. *Advances in Cryptology — Eurocrypt '94*, Lecture Notes in Computer Science, Vol. 950, A. De Santis, Ed., Springer-Verlag, pp. 92–111.

[7] Bellare, M. and Rogaway, P. January 2003. Introduction to modern cryptography. Available at http://www.cs.ucsd.edu/users/mihir/cse207/classnotes.html.

[8] Black, J. and Rogaway, P. 2000. CBC MACs for arbitrary-length messages: the three-key constructions. *Advances in Cryptology — Crypto 2000*, Lecture Notes in Computer Science, Vol. 1880, M. Bellare, Ed., Springer-Verlag, pp. 197–215.

[9] Blum, M. and Micali, S. 1984. How to generate cryptographically strong sequences of pseudorandom bits. *SIAM J. Computing*, 13(4):850–864.

[10] Boneh, D. 2001. Simplified OAEP for the RSA and Rabin functions. *Advances in Cryptology — Crypto 2001*, Lecture Notes in Computer Science, Vol. 2139, J. Kilian, Ed., Springer-Verlag, pp. 275–291.

[11] Childs, L.N. 2000. *A Concrete Introduction to Higher Algebra*. Springer-Verlag, Berlin.

[12] Cramer, R. and Shoup, V. 1998. A practical public-key cryptosystem provably secure against adaptive chosen ciphertext attack. *Advances in Cryptology — Crypto '98*, Lecture Notes in Computer Science, Vol. 1462, H. Krawczyk, Ed., Springer-Verlag, pp. 13–25.

[13] Daemen, J. and Rijmen, V. 2002. *The Design of Rijndael: AES — The Advanced Encryption Standard*. Springer-Verlag, Berlin.

[14] Delfs, H. and Knebl, H. 2002. *Introduction to Cryptography: Principles and Applications*. Springer-Verlag, Berlin.

[15] Diffie, W. and Hellman, M. 1976. New directions in cryptography. *IEEE Transactions on Information Theory*, 22(6): 644–654.

[16] El Gamal, T. 1985. A public-key cryptosystem and a signature scheme based on discrete logarithms. *IEEE Transactions on Information Theory*, 31(4):469–472.

[17] *Federal Information Processing Standards* publication #46. 1977. Data encryption standard. U.S. Department of Commerce/National Bureau of Standards.

[18] *Federal Information Processing Standards* publication #81. 1980. DES modes of operation. U.S. Department of Commerce/National Bureau of Standards.

[19] *Federal Information Processing Standards* Publication #180-1. 1995. Secure hash standard. U.S. Department of Commerce/National Institute of Standards and Technology.

[20] *Federal Information Processing Standards* Publication #186-2. 2000. Digital signature standard (DSS). U.S. Department of Commerce/National Institute of Standards and Technology.

[21] Fischlin, R. and Schnorr, C.P. 2000. Stronger security proofs for RSA and Rabin bits. *J. Cryptology*, 13(2):221–244.

[22] Fujisaki, E., Okamoto, T., Pointcheval, D., and Stern, J. 2001. RSA-OAEP is secure under the RSA assumption. *Advances in Cryptology — Crypto 2001*, Lecture Notes in Computer Science, Vol. 2139, J. Kilian, Ed., Springer-Verlag, pp. 260–274.

[23] Goldreich, O. 2001. *Foundations of Cryptography: Basic Tools*. Cambridge University Press.

[24] Goldreich, O. Foundations of cryptography, Vol. 2: basic applications. Available at http://www.wisdom.weizmann.ac.il/~oded/foc-vol2.html.

[25] Goldreich, O., Goldwasser, S., and Micali, S. 1986. How to construct random functions. *Journal of the ACM*, 33(4):792–807.

[26] Goldreich, O., Goldwasser, S., and Micali, S. 1985. On the cryptographic applications of random functions. *Advances in Cryptology — Crypto '84*, Lecture Notes in Computer Science, Vol. 196, G.R. Blakley and D. Chaum, Eds., Springer-Verlag, pp. 276–288.

[27] Goldreich, O. and Levin, L. 1989. Hard-core predicates for any one-way function. *Proceedings of the 21st Annual ACM Symposium on Theory of Computing*, ACM, pp. 25–32.

[28] Goldwasser, S. and Bellare, M. August, 2001. Lecture notes on cryptography. Available at http://www.cs.ucsd.edu/users/mihir/papers/gb.html.

[29] Goldwasser, S. and Micali, S. 1984. Probabilistic encryption. *J. Computer and System Sciences*, 28(2):270–299.

[30] Goldwasser, S., Micali, S., and Rivest, R. 1988. A digital signature scheme secure against adaptive chosen-message attacks. *SIAM J. Computing*, 17(2):281–308.

[31] Håstad, J., Impagliazzo, R., Levin, L., and Luby, M. 1999. A pseudorandom generator from any one-way function. *SIAM J. Computing*, 28(4):1364–1396.

[32] Håstad, J., Schrift, A.W., and Shamir, A. 1993. The discrete logarithm modulo a composite hides $O(n)$ bits. *J. Computer and System Sciences*, 47(3):376–404.

[33] Knuth, D.E. 1997. *The Art of Computer Programming, Vol. 2: Seminumerical Algorithms (third edition)*. Addison-Wesley Publishing Company.

[34] Koblitz, N. 1999. *Algebraic Aspects of Cryptography*. Springer-Verlag, Berlin.

[35] Lamport, L. 1979. Constructing digital signatures from any one-way function. Technical Report CSL-98, SRI International, Palo Alto.

[36] Long, D.L. and Wigderson, A. 1988. The discrete logarithm problem hides $O(\log n)$ bits. *SIAM J. Computing*, 17(2):363–372.

[37] Luby, M. and Rackoff, C. 1988. How to construct pseudorandom permutations from pseudorandom functions. *SIAM J. Computing*, 17(2):412–426.

[38] Menezes, A.J., van Oorschot, P.C., and Vanstone, S.A. 2001. *Handbook of Applied Cryptography*. CRC Press.

[39] Merkle, R. and Hellman, M. 1978. Hiding information and signatures in trapdoor knapsacks. *IEEE Transactions on Information Theory*, 24:525–530.

[40] Naor, M. and Yung, M. 1989. Universal one-way hash functions and their cryptographic applications. *Proceedings of the 21st Annual ACM Symposium on Theory of Computing*, ACM, pp. 33–43.

[41] Petrank, E. and Rackoff, C. 2000. CBC MAC for real-time data sources. *J. of Cryptology*, 13(3): 315–338.

[42] Rabin, M.O. 1979. Digitalized signatures and public key functions as intractable as factoring. MIT/LCS/TR-212, MIT Laboratory for Computer Science.

[43] Rivest, R. 1990. Cryptography. Chapter 13 of *Handbook of Theoretical Computer Science, Vol. A: Algorithms and Complexity*, J. van Leeuwen, Ed., MIT Press.

[44] Rivest, R. 1992. The MD5 message-digest algorithm. RFC 1321, available at ftp://ftp.rfc-editor.org/in-notes/rfc1321.txt.

[45] Rivest, R., Shamir, A., and Adleman, L.M. 1978. A method for obtaining digital signatures and public-key cryptosystems. *Communications of the ACM*, 21(2):120–126.

[46] Rompel, J. 1990. One-way functions are necessary and sufficient for secure signatures. *Proceedings of the 22nd Annual ACM Symposium on Theory of Computing*, ACM, pp. 387–394.

[47] Schneier, B. 1995. *Applied Cryptography: Protocols, Algorithms, and Source Code in C (second edition)*. John Wiley & Sons.

[48] Shoup, V. 2001. OAEP reconsidered. *Advances in Cryptology — Crypto 2001*, Lecture Notes in Computer Science, Vol. 2139, J. Kilian, Ed., Springer-Verlag, pp. 239–259.

[49] Simon, D. 1998. Finding collisions on a one-way street: can secure hash functions be based on general assumptions? *Advances in Cryptology — Eurocrypt '98*, Lecture Notes in Computer Science, Vol. 1403, K. Nyberg, Ed., Springer-Verlag, pp. 334–345.

[50] Sipser, M. 1996. *Introduction to the Theory of Computation*. Brooks/Cole Publishing Company.

[51] Stinson, D.R. 2002. *Cryptography: Theory and Practice (second edition)*. Chapman & Hall.

[52] Tsiounis, Y. and Yung, M. 1998. On the security of El Gamal based encryption. *Public Key Cryptography — PKC '98*, Lecture Notes in Computer Science, Vol. 1431, H. Imai and Y. Zheng, Eds., Springer-Verlag, pp. 117–134.

[53] Vaudenay, S. 2003. The security of DSA and ECDSA. *Public-Key Cryptography — PKC 2003*, Lecture Notes in Computer Science, Vol. 2567, Y. Desmedt, Ed., Springer-Verlag, pp. 309–323.

[54] Yao, A.C. 1982. Theory and application of trapdoor functions. *Proceedings of the 23rd Annual Symposium on Foundations of Computer Science*, IEEE, pp. 80–91.

Further Information

A number of excellent sources are available for the reader interested in more information about modern cryptography. An excellent and enjoyable review of the field up to 1990 is given by Rivest [43]. Details on the more practical aspects of cryptography appear in the approachable textbooks of Stinson [51] and Schneier [47]; the latter also includes detail on implementing many popular cryptographic algorithms.

More formal and mathematical approaches to the subject (of which the present treatment is an example) are available in a number of well-written textbooks and online texts, including those by Goldwasser and Bellare [28], Goldreich [23, 24], Delfs and Knebl [14], and Bellare and Rogaway [7]. We also mention the comprehensive reference book by Menezes, van Oorschot, and Vanstone [38].

The International Association for Cryptologic Research (IACR) sponsors a number of conferences covering all areas of cryptography, with Crypto and Eurocrypt being perhaps the best known. Proceedings of these conferences (dating, in some cases, to the early 1980s) are published as part of Springer-Verlag's *Lecture Notes in Computer Science*. Research in theoretical cryptography often appears at the ACM Symposium on Theory of Computing, the Annual Symposium on Foundations of Computer Science (sponsored by IEEE), and elsewhere; more practice-oriented aspects of cryptography are covered in many security conferences, including the ACM Conference on Computer and Communications Security.

The IACR publishes the *Journal of Cryptology*, which is devoted exclusively to cryptography. Articles on cryptography frequently appear in the *Journal of Computer and System Sciences*, the *Journal of the ACM*, and the *SIAM Journal of Computing*.

10

Parallel Algorithms

10.1 Introduction **10**-1
10.2 Modeling Parallel Computations **10**-2
      Multiprocessor Models • Work-Depth Model
      • Assigning Costs to Algorithms • Emulations Among Models
      • Model Used in This Chapter
10.3 Parallel Algorithmic Techniques **10**-11
      Divide-and-Conquer • Randomization
      • Parallel Pointer Techniques • Other Techniques
10.4 Basic Operations on Sequences, Lists, and Trees **10**-14
      Sums • Scans • Multiprefix and Fetch-and-Add
      • Pointer Jumping • List Ranking • Removing Duplicates
10.5 Graphs **10**-19
      Graphs and Their Representation • Breadth-First Search
      • Connected Components
10.6 Sorting **10**-28
      QuickSort • Radix Sort
10.7 Computational Geometry **10**-29
      Closest Pair • Planar Convex Hull
10.8 Numerical Algorithms **10**-35
      Matrix Operations • Fourier Transform
10.9 Parallel Complexity Theory **10**-37

Guy E. Blelloch
Carnegie Mellon University

Bruce M. Maggs
Carnegie Mellon University

10.1 Introduction

The subject of this chapter is the design and analysis of parallel algorithms. Most of today's computer algorithms are sequential, that is, they specify a sequence of steps in which each step consists of a single operation. As it has become more difficult to improve the performance of sequential computers, however, researchers have sought performance improvements in another place: parallelism. In contrast to a sequential algorithm, a parallel algorithm may perform multiple operations in a single step. For example, consider the problem of computing the sum of a sequence, A, of n numbers. The standard sequential algorithm computes the sum by making a single pass through the sequence, keeping a running sum of the numbers seen so far. It is not difficult, however, to devise an algorithm for computing the sum that performs many operations in parallel. For example, suppose that, in parallel, each element of A with an even index is paired and summed with the next element of A, which has an odd index, i.e., $A[0]$ is paired with $A[1]$, $A[2]$ with $A[3]$, and so on. The result is a new sequence of $\lceil n/2 \rceil$ numbers whose sum is identical to the sum that we wish to compute. This pairing and summing step can be repeated, and after $\lceil \log_2 n \rceil$ steps, only the final sum remains.

The parallelism in an algorithm can yield improved performance on many different kinds of computers. For example, on a parallel computer, the operations in a parallel algorithm can be performed simultaneously by different processors. Furthermore, even on a single-processor computer it is possible to exploit the parallelism in an algorithm by using multiple functional units, pipelined functional units, or pipelined memory systems. As these examples show, it is important to make a distinction between the parallelism in an algorithm and the ability of any particular computer to perform multiple operations in parallel. Typically, a parallel algorithm will run efficiently on a computer if the algorithm contains at least as much parallelism as the computer. Thus, good parallel algorithms generally can be expected to run efficiently on sequential computers as well as on parallel computers.

The remainder of this chapter consists of eight sections. Section 10.2 begins with a discussion of how to model parallel computers. Next, in Section 10.3 we cover some general techniques that have proven useful in the design of parallel algorithms. Section 10.4 to Section 10.8 present algorithms for solving problems from different domains. We conclude in Section 10.9 with a brief discussion of parallel complexity theory. Throughout this chapter, we assume that the reader has some familiarity with sequential algorithms and asymptotic analysis.

10.2 Modeling Parallel Computations

To analyze parallel algorithms it is necessary to have a formal model in which to account for costs. The designer of a sequential algorithm typically formulates the algorithm using an abstract model of computation called a *random-access machine* (RAM) [Aho et al. 1974, ch. 1]. In this model, the machine consists of a single processor connected to a memory system. Each basic central processing unit (CPU) operation, including arithmetic operations, logical operations, and memory accesses, requires one time step. The designer's goal is to develop an algorithm with modest time and memory requirements. The random-access machine model allows the algorithm designer to ignore many of the details of the computer on which the algorithm ultimately will be executed, but it captures enough detail that the designer can predict with reasonable accuracy how the algorithm will perform.

Modeling parallel computations is more complicated than modeling sequential computations because in practice parallel computers tend to vary more in their organizations than do sequential computers. As a consequence, a large proportion of the research on parallel algorithms has gone into the question of modeling, and many debates have raged over what the *right* model is, or about how practical various models are. Although there has been no consensus on the right model, this research has yielded a better understanding of the relationships among the models. Any discussion of parallel algorithms requires some understanding of the various models and the relationships among them.

Parallel models can be broken into two main classes: **multiprocessor models** and **work-depth models**. In this section we discuss each and then discuss how they are related.

10.2.1 Multiprocessor Models

A multiprocessor model is a generalization of the sequential RAM model in which there is more than one processor. Multiprocessor models can be classified into three basic types: local memory machines, modular memory machines, and **parallel random-access machines** (PRAMs). Figure 10.1 illustrates the structures of these machines. A local memory machine consists of a set of n processors, each with its own local memory. These processors are attached to a common communication network. A modular memory machine consists of m memory modules and n processors all attached to a common network. A PRAM consists of a set of n processors all connected to a common shared memory [Fortune and Wyllie 1978, Goldshlager 1978, Savitch and Stimson 1979].

The three types of multiprocessors differ in the way memory can be accessed. In a local memory machine, each processor can access its own local memory directly, but it can access the memory in another processor only by sending a memory request through the network. As in the RAM model, all local operations, including local memory accesses, take unit time. The time taken to access the memory in another processor,

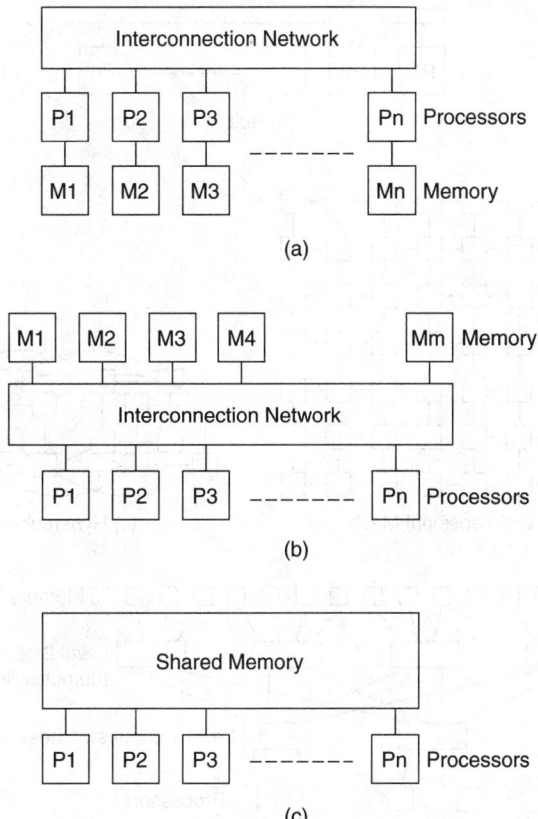

FIGURE 10.1 The three classes of multiprocessor machine models: (a) a local memory machine, (b) a modular memory machine, and (c) a parallel random-access machine (PRAM).

however, will depend on both the capabilities of the communication network and the pattern of memory accesses made by other processors, since these other accesses could congest the network. In a modular memory machine, a processor accesses the memory in a memory module by sending a memory request through the network. Typically, the processors and memory modules are arranged so that the time for any processor to access any memory module is roughly uniform. As in a local memory machine, the exact amount of time depends on the communication network and the memory access pattern. In a PRAM, in a single step each processor can simultaneously access any word of the memory by issuing a memory request directly to the shared memory.

The PRAM model is controversial because no real machine lives up to its ideal of unit-time access to shared memory. It is worth noting, however, that the ultimate purpose of an abstract model is not to directly model a real machine but to help the algorithm designer produce efficient algorithms. Thus, if an algorithm designed for a PRAM (or any other model) can be translated to an algorithm that runs efficiently on a real computer, then the model has succeeded. Later in this section, we show how algorithms designed for one parallel machine model can be translated so that they execute efficiently on another model.

The three types of multiprocessor models that we have defined are very broad, and these models further differ in network topology, network functionality, control, synchronization, and cache coherence. Many of these issues are discussed elsewhere in this volume. Here we will briefly discuss some of them.

10.2.1.1 Network Topology

A network is a collection of switches connected by communication channels. A processor or memory module has one or more communication ports that are connected to these switches by communication

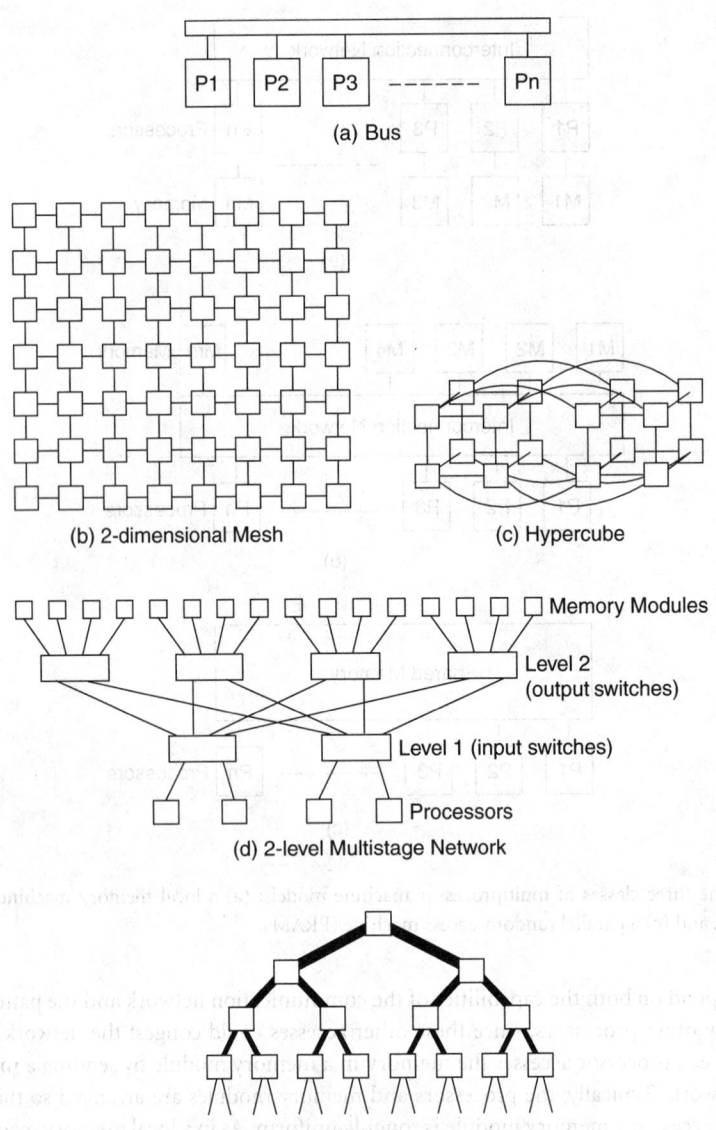

FIGURE 10.2 Various network topologies: (a) bus, (b) two-dimensional mesh, (c) hypercube, (d) two-level multistage network, and (e) fat-tree.

channels. The pattern of interconnection of the switches is called the network topology. The topology of a network has a large influence on the performance and also on the cost and difficulty of constructing the network. Figure 10.2 illustrates several different topologies.

The simplest network topology is a bus. This network can be used in both local memory machines and modular memory machines. In either case, all processors and memory modules are typically connected to a single bus. In each step, at most one piece of data can be written onto the bus. This datum might be a request from a processor to read or write a memory value, or it might be the response from the processor or memory module that holds the value. In practice, the advantages of using buses are that they are simple to build, and, because all processors and memory modules can observe the traffic on a bus, it is relatively easy to develop protocols that allow processors to cache memory values locally.

The disadvantage of using a bus is that the processors have to take turns accessing the bus. Hence, as more processors are added to a bus, the average time to perform a memory access grows proportionately.

A two-dimensional *mesh* is a network that can be laid out in a rectangular fashion. Each switch in a mesh has a distinct label (x, y) where $0 \le x \le X - 1$ and $0 \le y \le Y - 1$. The values X and Y determine the length of the sides of the mesh. The number of switches in a mesh is thus $X \cdot Y$. Every switch, except those on the sides of the mesh, is connected to four neighbors: one to the north, one to the south, one to the east, and one to the west. Thus, a switch labeled (x, y), where $0 < x < X - 1$ and $0 < y < Y - 1$ is connected to switches $(x, y + 1)$, $(x, y - 1)$, $(x + 1, y)$, and $(x - 1, y)$. This network typically appears in a local memory machine, i.e., a processor along with its local memory is connected to each switch, and remote memory accesses are made by routing messages through the mesh. Figure 10.2b shows an example of an 8×8 mesh.

Several variations on meshes are also popular, including three-dimensional meshes, toruses, and hypercubes. A *torus* is a mesh in which the switches on the sides have connections to the switches on the opposite sides. Thus, every switch (x, y) is connected to four other switches: $(x, y + 1 \bmod Y)$, $(x, y - 1 \bmod Y)$, $(x + 1 \bmod X, y)$, and $(x - 1 \bmod X, y)$. A hypercube is a network with 2^n switches in which each switch has a distinct n-bit label. Two switches are connected by a communication channel in a hypercube if their labels differ in precisely one-bit position.

A *multistage network* is used to connect one set of switches called the *input switches* to another set called the *output switches* through a sequence of stages of switches. Such networks were originally designed for telephone networks [Beneš 1965]. The stages of a multistage network are numbered 1 through L, where L is the **depth** of the network. The input switches form stage 1 and the output switches form stage L. In most multistage networks, it is possible to send a message from any input switch to any output switch along a path that traverses the stages of the network in order from 1 to L. Multistage networks are frequently used in modular memory computers; typically, processors are attached to input switches, and memory modules to output switches. There are many different multistage network topologies. Figure 10.2d, for example, shows a 2-stage network that connects 4 processors to 16 memory modules. Each switch in this network has two channels at the bottom and four channels at the top. The ratio of processors to memory modules in this example is chosen to reflect the fact that, in practice, a processor is capable of generating memory access requests faster than a memory module is capable of servicing them.

A *fat-tree* is a network whose overall structure is that of a tree [Leiserson 1985]. Each edge of the tree, however, may represent many communication channels, and each node may represent many network switches (hence the name fat). Figure 10.2e shows a fat-tree whose overall structure is that of a binary tree. Typically the capacities of the edges near the root of the tree are much larger than the capacities near the leaves. For example, in this tree the two edges incident on the root represent 8 channels each, whereas the edges incident on the leaves represent only 1 channel each. One way to construct a local memory machine is to connect a processor along with its local memory to each leaf of the fat-tree. In this scheme, a message from one processor to another first travels up the tree to the least common ancestor of the two processors and then down the tree.

Many algorithms have been designed to run efficiently on particular network topologies such as the mesh or the hypercube. For an extensive treatment such algorithms, see Leighton [1992]. Although this approach can lead to very fine-tuned algorithms, it has some disadvantages. First, algorithms designed for one network may not perform well on other networks. Hence, in order to solve a problem on a new machine, it may be necessary to design a new algorithm from scratch. Second, algorithms that take advantage of a particular network tend to be more complicated than algorithms designed for more abstract models such as the PRAM because they must incorporate some of the details of the network. Nevertheless, there are some operations that are performed so frequently by a parallel machine that it makes sense to design a fine-tuned network-specific algorithm. For example, the algorithm that routes messages or memory access requests through the network should exploit the network topology. Other examples include algorithms for broadcasting a message from one processor to many other processors, for collecting the results computed in many processors in a single processor, and for synchronizing processors.

An alternative to modeling the topology of a network is to summarize its routing capabilities in terms of two parameters, its latency and bandwidth. The latency L of a network is the time it takes for a message to traverse the network. In actual networks this will depend on the topology of the network, which particular ports the message is passing between, and the congestion of messages in the network. The latency, however, often can be usefully modeled by considering the worst-case time assuming that the network is not heavily congested. The bandwidth at each port of the network is the rate at which a processor can inject data into the network. In actual networks this will depend on the topology of the network, the bandwidths of the network's individual communication channels, and, again, the congestion of messages in the network. The bandwidth often can be usefully modeled as the maximum rate at which processors can inject messages into the network without causing it to become heavily congested, assuming a uniform distribution of message destinations. In this case, the bandwidth can be expressed as the minimum *gap g* between successive injections of messages into the network.

Three models that characterize a network in terms of its latency and bandwidth are the postal model [Bar-Noy and Kipnis 1992], the bulk-synchronous parallel (BSP) model [Valiant 1990a], and the LogP model [Culler et al. 1993]. In the postal model, a network is described by a single parameter, L, its latency. The bulk-synchronous parallel model adds a second parameter, g, the minimum ratio of computation steps to communication steps, i.e., the gap. The LogP model includes both of these parameters and adds a third parameter, o, the overhead, or wasted time, incurred by a processor upon sending or receiving a message.

10.2.1.2 Primitive Operations

As well as specifying the general form of a machine and the network topology, we need to define what operations the machine supports. We assume that all processors can perform the same instructions as a typical processor in a sequential machine. In addition, processors may have special instructions for issuing nonlocal memory requests, for sending messages to other processors, and for executing various global operations, such as synchronization. There can also be restrictions on when processors can simultaneously issue instructions involving nonlocal operations. For example a machine might not allow two processors to write to the same memory location at the same time. The particular set of instructions that the processors can execute may have a large impact on the performance of a machine on any given algorithm. It is therefore important to understand what instructions are supported before one can design or analyze a parallel algorithm. In this section we consider three classes of nonlocal instructions: (1) how global memory requests interact, (2) synchronization, and (3) global operations on data.

When multiple processors simultaneously make a request to read or write to the same resource — such as a processor, memory module, or memory location — there are several possible outcomes. Some machine models simply forbid such operations, declaring that it is an error if more than one processor tries to access a resource simultaneously. In this case we say that the machine allows only *exclusive* access to the resource. For example, a PRAM might allow only exclusive read or write access to each memory location. A PRAM of this type is called an **exclusive-read exclusive-write (EREW)** PRAM. Other machine models may allow unlimited access to a shared resource. In this case we say that the machine allows *concurrent* access to the resource. For example, a **concurrent-read concurrent-write (CRCW)** PRAM allows both concurrent read and write access to memory locations, and a **CREW** PRAM allows **concurrent reads but only exclusive writes**. When making a concurrent write to a resource such as a memory location there are many ways to resolve the conflict. Some possibilities are to choose an arbitrary value from those written, to choose the value from the processor with the lowest index, or to take the *logical or* of the values written. A final choice is to allow for *queued* access, in which case concurrent access is permitted but the time for a step is proportional to the maximum number of accesses to any resource. A queue-read queue-write (QRQW) PRAM allows for such accesses [Gibbons et al. 1994].

In addition to reads and writes to nonlocal memory or other processors, there are other important primitives that a machine may supply. One class of such primitives supports synchronization. There are a variety of different types of synchronization operations and their costs vary from model to model. In the PRAM model, for example, it is assumed that all processors operate in lock step, which provides implicit synchronization. In a local-memory machine the cost of synchronization may be a function of the particular network topology. Some machine models supply more powerful primitives that combine arithmetic operations with communication. Such operations include the prefix and **multiprefix** operations, which are defined in the subsections on scans and multiprefix and fetch-and-add.

10.2.2 Work-Depth Models

Because there are so many different ways to organize parallel computers, and hence to model them, it is difficult to select one multiprocessor model that is appropriate for all machines. The alternative to focusing on the machine is to focus on the algorithm. In this section we present a class of models called work-depth models. In a work-depth model, the cost of an algorithm is determined by examining the total number of operations that it performs and the dependencies among those operations. An algorithm's **work** W is the total number of operations that it performs; its *depth* D is the longest chain of dependencies among its operations. We call the ratio $\mathcal{P} = W/D$ the *parallelism* of the algorithm. We say that a parallel algorithm is work-efficient relative to a sequential algorithm if it does at most a constant factor more work.

The work-depth models are more abstract than the multiprocessor models. As we shall see, however, algorithms that are efficient in the work-depth models often can be translated to algorithms that are efficient in the multiprocessor models and from there to real parallel computers. The advantage of a work-depth model is that there are no machine-dependent details to complicate the design and analysis of algorithms. Here we consider three classes of work-depth models: circuit models, vector machine models, and language-based models. We will be using a language-based model in this chapter, and so we will return to these models later in this section.

The most abstract work-depth model is the *circuit model*. In this model, an algorithm is modeled as a family of directed acyclic circuits. There is a circuit for each possible size of the input. A circuit consists of nodes and arcs. A node represents a basic operation, such as adding two values. For each input to an operation (i.e., node), there is an incoming arc from another node or from an input to the circuit. Similarly, there are one or more outgoing arcs from each node representing the result of the operation. The work of a circuit is the total number of nodes. (The work is also called the *size*.) The depth of a circuit is the length of the longest directed path between any pair of nodes. Figure 10.3 shows a circuit in which the inputs are at the top, each + is an adder circuit, and each of the arcs carries the result of an adder circuit. The final

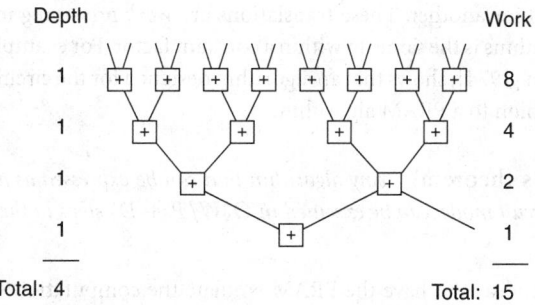

FIGURE 10.3   Summing 16 numbers on a tree. The total depth (longest chain of dependencies) is 4 and the total work (number of operations) is 15.

sum is returned at the bottom. Circuit models have been used for many years to study various theoretical aspects of parallelism, for example, to prove that certain problems are hard to solve in parallel (see Karp and Ramachandran [1990] for an overview).

In a *vector model*, an algorithm is expressed as a sequence of steps, each of which performs an operation on a vector (i.e., sequence) of input values, and produces a vector result [Pratt and Stockmeyer 1976, Blelloch 1990]. The work of each step is equal to the length of its input (or output) vector. The work of an algorithm is the sum of the work of its steps. The depth of an algorithm is the number of vector steps.

In a *language* model, a work-depth cost is associated with each programming language construct [Blelloch and Greiner 1995, Blelloch 1996]. For example, the work for calling two functions in parallel is equal to the sum of the work of the two calls. The depth, in this case, is equal to the maximum of the depth of the two calls.

10.2.3 Assigning Costs to Algorithms

In the work-depth models, the cost of an algorithm is determined by its work and by its depth. The notions of work and depth also can be defined for the multiprocessor models. The work W performed by an algorithm is equal to the number of processors times the time required for the algorithm to complete execution. The depth D is equal to the total time required to execute the algorithm.

The depth of an algorithm is important because there are some applications for which the time to perform a computation is crucial. For example, the results of a weather-forecasting program are useful only if the program completes execution before the weather does!

Generally, however, the most important measure of the cost of an algorithm is the work. This can be justified as follows. The cost of a computer is roughly proportional to the number of processors in the computer. The cost for purchasing time on a computer is proportional to the cost of the computer times the amount of time used. The total cost of performing a computation, therefore, is roughly proportional to the number of processors in the computer times the amount of time used, i.e., the work.

In many instances, the cost of running a computation on a parallel computer may be slightly larger than the cost of running the same computation on a sequential computer. If the time to completion is sufficiently improved, however, this extra cost often can be justified. As we shall see, in general there is a tradeoff between work and time to completion. It is rarely the case, however, that a user is willing to give up any more than a small constant factor in cost for an improvement in time.

10.2.4 Emulations Among Models

Although it may appear that a different algorithm must be designed for each of the many parallel models, there are often automatic and efficient techniques for translating algorithms designed for one model into algorithms designed for another. These translations are *work preserving* in the sense that the work performed by both algorithms is the same, to within a constant factor. For example, the following theorem, known as Brent's theorem [1974], shows that an algorithm designed for the circuit model can be translated in a work-preserving fashion to a PRAM algorithm.

Theorem 10.1 (Brent's theorem)   *Any algorithm that can be expressed as a circuit of size (i.e., work) W and depth D in the circuit model can be executed in $O(W/P + D)$ steps in the PRAM model.*

Proof 10.1   The basic idea is to have the PRAM emulate the computation specified by the circuit in a level-by-level fashion. The level of a node is defined as follows. A node is on level 1 if all of its inputs are also inputs to the circuit. Inductively, the level of any other node is one greater than the maximum of the level of the nodes with arcs into it. Let l_i denote the number of nodes on level i. Then, by assigning $\lceil l_i/P \rceil$ operations to each of the P processors in the PRAM, the operations for level i can be performed in $O(\lceil l_i/P \rceil)$ steps. Summing the time over all D levels, we have

$$T_{\text{PRAM}}(W, D, P) = O\left(\sum_{i=1}^{D}\left\lceil\frac{l_i}{P}\right\rceil\right)$$

$$= O\left(\sum_{i=1}^{D}\left(\frac{l_i}{P}+1\right)\right)$$

$$= O\left(\frac{1}{P}\left(\sum_{i=1}^{D}l_i\right)+D\right)$$

$$= O\left(W/P + D\right) \qquad \square$$

The total work performed by the PRAM, i.e., the processor-time product, is $O(W+PD)$. This emulation is work preserving to within a constant factor when the parallelism ($\mathcal{P} = W/D$) is at least as large as the number of processors P, in this case the work is $O(W)$. The requirement that the parallelism exceed the number of processors is typical of work-preserving emulations.

Brent's theorem shows that an algorithm designed for one of the work-depth models can be translated in a work-preserving fashion on to a multiprocessor model. Another important class of work-preserving translations is those that translate between different multiprocessor models. The translation we consider here is the work-preserving translation of algorithms written for the PRAM model to algorithms for a more realistic machine model. In particular, we consider a *butterfly machine* in which P processors are attached through a butterfly network of depth $\log P$ to P memory banks. We assume that, in constant time, a processor can hash a virtual memory address to a physical memory bank and an address within that bank using a sufficiently powerful hash function. This scheme was first proposed by Karlin and Upfal [1988] for the EREW PRAM model. Ranade [1991] later presented a more general approach that allowed the butterfly to efficiently emulate CRCW algorithms.

Theorem 10.2 *Any algorithm that takes time T on a P-processor PRAM can be translated into an algorithm that takes time $O(T(P/P' + \log P'))$, with high probability, on a P'-processor butterfly machine.*

Sketch of proof Each of the P' processors in the butterfly machine emulates a set of P/P' PRAM processors. The butterfly machine emulates the PRAM in a step-by-step fashion. First, each butterfly processor emulates one step of each of its P/P' PRAM processors. Some of the PRAM processors may wish to perform memory accesses. For each memory access, the butterfly processor hashes the memory address to a physical memory bank and an address within the bank and then routes a message through the network to that bank. These messages are pipelined so that a processor can have multiple outstanding requests. Ranade proved that if each processor in a P-processor butterfly machine sends at most P/P' messages whose destinations are determined by a sufficiently powerful hash function, then the network can deliver all of the messages, along with responses, in $O(P/P' + \log P')$ time. The $\log P'$ term accounts for the latency of the network and for the fact that there will be some congestion at memory banks, even if each processor sends only a single message.

This theorem implies that, as long as $P \geq P' \log P'$, i.e., if the number of processors employed by the PRAM algorithm exceeds the number of processors in the butterfly machine by a factor of at least $\log P'$, then the emulation is work preserving. When translating algorithms from a guest multiprocessor model (e.g., the PRAM) to a host multiprocessor model (e.g., the butterfly machine), it is not uncommon to require that the number of guest processors exceed the number of host processors by a factor proportional to the latency of the host. Indeed, the latency of the host often can be hidden by giving it a larger guest to emulate. If the bandwidth of the host is smaller than the bandwidth of a comparably sized guest, however, it usually is much more difficult for the host to perform a work-preserving emulation of the guest.

For more information on PRAM emulations, the reader is referred to Harris [1994] and Valiant [1990].

10.2.5   Model Used in This Chapter

Because there are so many work-preserving translations between different parallel models of computation, we have the luxury of choosing the model that we feel most clearly illustrates the basic ideas behind the algorithms, a work-depth language model. Here we define the model we will use in this chapter in terms of a set of language constructs and a set of rules for assigning costs to the constructs. The description we give here is somewhat informal, but it should suffice for the purpose of this chapter. The language and costs can be properly formalized using a profiling semantics [Blelloch and Greiner 1995].

Most of the syntax that we use should be familiar to readers who have programmed in Algol-like languages, such as Pascal and C. The constructs for expressing parallelism, however, may be unfamiliar. We will be using two parallel constructs — a parallel *apply-to-each* construct and a *parallel-do* construct — and a small set of parallel primitives on sequences (one-dimensional arrays). Our language constructs, syntax, and cost rules are based on the NESL language [Blelloch 1996].

The apply-to-each construct is used to apply an expression over a sequence of values in parallel. It uses a setlike notation. For example, the expression

$$\{a * a : a \in [3, -4, -9, 5]\}$$

squares each element of the sequence $[3, -4, -9, 5]$ returning the sequence $[9, 16, 81, 25]$. This can be read: "in parallel, for each a in the sequence $[3, -4, -9, 5]$, square a." The apply-to-each construct also provides the ability to subselect elements of a sequence based on a filter. For example,

$$\{a * a : a \in [3, -4, -9, 5] \mid a > 0\}$$

can be read: "in parallel, for each a in the sequence $[3, -4, -9, 5]$ such that a is greater than 0, square a." It returns the sequence $[9, 25]$. The elements that remain maintain their relative order.

The parallel-do construct is used to evaluate multiple statements in parallel. It is expressed by listing the set of statements after an **in parallel do**. For example, the following fragment of code calls FUN1(X) and assigns the result to A and in parallel calls FUN2(Y) and assigns the result to B:

in parallel do

$A := \text{FUN1}(X)$

$B := \text{FUN2}(Y)$

The parallel-do completes when all the parallel subcalls complete.

Work and depth are assigned to our language constructs as follows. The work and depth of a scalar primitive operation is one. For example, the work and depth for evaluating an expression such as $3 + 4$ is one. The work for applying a function to every element in a sequence is equal to the sum of the work for each of the individual applications of the function. For example, the work for evaluating the expression

$$\{a * a : a \in [0..n]\}$$

which creates an n-element sequence consisting of the squares of 0 through $n - 1$, is n. The depth for applying a function to every element in a sequence is equal to the maximum of the depths of the individual applications of the function. Hence, the depth of the previous example is one. The work for a parallel-do construct is equal to the sum of the work for each of its statements. The depth is equal to the maximum depth of its statements. In all other cases, the work and depth for a sequence of operations is the sum of the work and depth for the individual operations.

In addition to the parallelism supplied by apply-to-each, we will use four built-in functions on sequences, *dist*, ++ (append), *flatten*, and ← (write), each of which can be implemented in parallel. The function *dist* creates a sequence of identical elements. For example, the expression *dist* (3, 5) creates the sequence

$$[3, 3, 3, 3, 3]$$

The ++ function appends two sequences. For example, $[2, 1] + + [5, 0, 3]$ create the sequence $[2, 1, 5, 0, 3]$. The *flatten* function converts a nested sequence (a sequence for which each element is itself a sequence) into a flat sequence. For example,

$$flatten([[3, 5], [3, 2], [1, 5], [4, 6]])$$

creates the sequence

$$[3, 5, 3, 2, 1, 5, 4, 6]$$

The $\leftarrow$ function is used to write multiple elements into a sequence in parallel. It takes two arguments. The first argument is the sequence to modify and the second is a sequence of integer-value pairs that specify what to modify. For each pair (i, v), the value v is inserted into position i of the destination sequence. For example,

$$[0, 0, 0, 0, 0, 0, 0, 0] \leftarrow [(4, -2), (2, 5), (5, 9)]$$

inserts the -2, 5, and 9 into the sequence at locations 4, 2, and 5, respectively, returning

$$[0, 0, 5, 0, -2, 9, 0, 0]$$

As in the PRAM model, the issue of concurrent writes arises if an index is repeated. Rather than choosing a single policy for resolving concurrent writes, we will explain the policy used for the individual algorithms. All of these functions have depth one and work n, where n is the size of the sequence(s) involved. In the case of the $\leftarrow$, the work is proportional to the length of the sequence of integer-value pairs, not the modified sequence, which might be much longer. In the case of ++, the work is proportional to the length of the second sequence.

We will use a few shorthand notations for specifying sequences. The expression $[-2..1]$ specifies the same sequence as the expression $[-2, -1, 0, 1]$. Changing the left or right brackets surrounding a sequence omits the first or last elements, i.e., $[-2..1)$ denotes the sequence $[-2, -1, 0]$. The notation $A[i..j]$ denotes the subsequence consisting of elements $A[i]$ through $A[j]$. Similarly, $A[i, j)$ denotes the subsequence $A[i]$ through $A[j - 1]$. We will assume that sequence indices are zero based, i.e., $A[0]$ extracts the first element of the sequence A.

Throughout this chapter, our algorithms make use of random numbers. These numbers are generated using the functions *rand_bit()*, which returns a random bit, and *rand_int(h)*, which returns a random integer in the range $[0, h - 1]$.

10.3 Parallel Algorithmic Techniques

As with sequential algorithms, in parallel algorithm design there are many general techniques that can be used across a variety of problem areas. Some of these are variants of standard sequential techniques, whereas others are new to parallel algorithms. In this section we introduce some of these techniques, including parallel divide-and-conquer, randomization, and parallel pointer manipulation. In later sections on algorithms we will make use of them.

10.3.1 Divide-and-Conquer

A divide-and-conquer algorithm first splits the problem to be solved into subproblems that are easier to solve than the original problem either because they are smaller instances of the original problem, or because they are different but easier problems. Next, the algorithm solves the subproblems, possibly recursively. Typically, the subproblems can be solved independently. Finally, the algorithm merges the solutions to the subproblems to construct a solution to the original problem.

The divide-and-conquer paradigm improves program modularity and often leads to simple and efficient algorithms. It has, therefore, proven to be a powerful tool for sequential algorithm designers. Divide-and-conquer plays an even more prominent role in parallel algorithm design. Because the subproblems created in the first step are typically independent, they can be solved in parallel. Often the subproblems are solved recursively and thus the next divide step yields even more subproblems to be solved in parallel. As a consequence, even divide-and-conquer algorithms that were designed for sequential machines typically have some inherent parallelism. Note, however, that in order for divide-and-conquer to yield a highly parallel algorithm, it often is necessary to parallelize the divide step and the merge step. It is also common in parallel algorithms to divide the original problem into as many subproblems as possible, so that they all can be solved in parallel.

As an example of parallel divide-and-conquer, consider the sequential mergesort algorithm. Mergesort takes a set of n keys as input and returns the keys in sorted order. It works by splitting the keys into two sets of $n/2$ keys, recursively sorting each set, and then merging the two sorted sequences of $n/2$ keys into a sorted sequence of n keys. To analyze the sequential running time of mergesort we note that two sorted sequences of $n/2$ keys can be merged in $O(n)$ time. Hence, the running time can be specified by the recurrence

$$T(n) = \begin{cases} 2T(n/2) + O(n) & n > 1 \\ O(1) & n = 1 \end{cases}$$

which has the solution $T(n) = O(n \log n)$. Although not designed as a parallel algorithm, mergesort has some inherent parallelism since the two recursive calls can be made in parallel. This can be expressed as:

Algorithm: MERGESORT(A).

1 **if** ($|A| = 1$) **then return** A
2 **else**
3   **in parallel do**
4        $L :=$ MERGESORT($A[0..|A|/2]$)
5        $R :=$ MERGESORT($A[|A|/2..|A|]$)
6   **return** MERGE(L, R)

Recall that in our work-depth model we can analyze the depth of an algorithm that makes parallel calls by taking the maximum depth of the two calls, and the work by taking the sum. We assume that the merging remains sequential so that the work and depth to merge two sorted sequences of $n/2$ keys is $O(n)$. Thus, for mergesort the work and depth are given by the recurrences:

$$W(n) = 2W(n/2) + O(n)$$
$$D(n) = \max(D(n/2), D(n/2)) + O(n)$$
$$= D(n/2) + O(n)$$

As expected, the solution for the work is $W(n) = O(n \log n)$, i.e., the same as the time for the sequential algorithm. For the depth, however, the solution is $D(n) = O(n)$, which is smaller than the work. Recall that we defined the parallelism of an algorithm as the ratio of the work to the depth. Hence, the parallelism of this algorithm is $O(\log n)$ (not very much). The problem here is that the merge step remains sequential, and this is the bottleneck.

As mentioned earlier, the parallelism in a divide-and-conquer algorithm often can be enhanced by parallelizing the divide step and/or the merge step. Using a parallel merge [Shiloach and Vishkin 1982], two sorted sequences of $n/2$ keys can be merged with work $O(n)$ and depth $O(\log n)$. Using this merge algorithm, the recurrence for the depth of mergesort becomes

$$D(n) = D(n/2) + O(\log n)$$

which has solution $D(n) = O(\log^2 n)$. Using a technique called **pipelined divide-and-conquer**, the depth of mergesort can be further reduced to $O(\log n)$ [Cole 1988]. The idea is to start the merge at the top level before the recursive calls complete.

Divide-and-conquer has proven to be one of the most powerful techniques for solving problems in parallel. In this chapter we will use it to solve problems from computational geometry, sorting, and performing fast Fourier transforms. Other applications range from linear systems to factoring large numbers to n-body simulations.

10.3.2 Randomization

The use of random numbers is ubiquitous in parallel algorithms. Intuitively, randomness is helpful because it allows processors to make local decisions which, with high probability, add up to good global decisions. Here we consider three uses of randomness.

10.3.2.1 Sampling

One use of randomness is to select a representative sample from a set of elements. Often, a problem can be solved by selecting a sample, solving the problem on that sample, and then using the solution for the sample to guide the solution for the original set. For example, suppose we want to sort a collection of integer keys. This can be accomplished by partitioning the keys into buckets and then sorting within each bucket. For this to work well, the buckets must represent nonoverlapping intervals of integer values and contain approximately the same number of keys. **Random sampling** is used to determine the boundaries of the intervals. First, each processor selects a random sample of its keys. Next, all of the selected keys are sorted together. Finally, these keys are used as the boundaries. Such random sampling also is used in many parallel computational geometry, graph, and string matching algorithms.

10.3.2.2 Symmetry Breaking

Another use of randomness is in **symmetry breaking**. For example, consider the problem of selecting a large independent set of vertices in a graph in parallel. (A set of vertices is *independent* if no two are neighbors.) Imagine that each vertex must decide, in parallel with all other vertices, whether to join the set or not. Hence, if one vertex chooses to join the set, then all of its neighbors must choose not to join the set. The choice is difficult to make simultaneously by each vertex if the local structure at each vertex is the same, for example, if each vertex has the same number of neighbors. As it turns out, the impasse can be resolved by using randomness to break the symmetry between the vertices [Luby 1985].

10.3.2.3 Load Balancing

A third use is load balancing. One way to quickly partition a large number of data items into a collection of approximately evenly sized subsets is to randomly assign each element to a subset. This technique works best when the average size of a subset is at least logarithmic in the size of the original set.

10.3.3 Parallel Pointer Techniques

Many of the traditional sequential techniques for manipulating lists, trees, and graphs do not translate easily into parallel techniques. For example, techniques such as traversing the elements of a linked list, visiting the nodes of a tree in postorder, or performing a depth-first traversal of a graph appear to be inherently sequential. Fortunately, these techniques often can be replaced by parallel techniques with roughly the same power.

10.3.3.1 Pointer Jumping

One of the earliest parallel pointer techniques is **pointer jumping** [Wyllie 1979]. This technique can be applied to either lists or trees. In each pointer jumping step, each node in parallel replaces its pointer with that of its successor (or parent). For example, one way to label each node of an n-node list (or tree) with the label of the last node (or root) is to use pointer jumping. After at most $\lceil \log n \rceil$ steps, every node points to the same node, the end of the list (or root of the tree). This is described in more detail in the subsection on pointer jumping.

10.3.3.2 Euler Tour

An Euler tour of a directed graph is a path through the graph in which every edge is traversed exactly once. In an undirected graph each edge is typically replaced with two oppositely directed edges. The Euler tour of an undirected tree follows the perimeter of the tree visiting each edge twice, once on the way down and once on the way up. By keeping a linked structure that represents the Euler tour of a tree, it is possible to compute many functions on the tree, such as the size of each subtree [Tarjan and Vishkin 1985]. This technique uses linear work and parallel depth that is independent of the depth of the tree. The Euler tour often can be used to replace standard traversals of a tree, such as a depth-first traversal.

10.3.3.3 Graph Contraction

Graph contraction is an operation in which a graph is reduced in size while maintaining some of its original structure. Typically, after performing a graph contraction operation, the problem is solved recursively on the contracted graph. The solution to the problem on the contracted graph is then used to form the final solution. For example, one way to partition a graph into its connected components is to first contract the graph by merging some of the vertices into their neighbors, then find the connected components of the contracted graph, and finally undo the contraction operation. Many problems can be solved by contracting trees [Miller and Reif 1989, 1991], in which case the technique is called **tree contraction**. More examples of graph contraction can be found in Section 10.5.

10.3.3.4 Ear Decomposition

An ear decomposition of a graph is a partition of its edges into an ordered collection of paths. The first path is a cycle, and the others are called ears. The endpoints of each ear are anchored on previous paths. Once an ear decomposition of a graph is found, it is not difficult to determine if two edges lie on a common cycle. This information can be used in algorithms for determining biconnectivity, triconnectivity, 4-connectivity, and planarity [Maon et al. 1986, Miller and Ramachandran 1992]. An ear decomposition can be found in parallel using linear work and logarithmic depth, independent of the structure of the graph. Hence, this technique can be used to replace the standard sequential technique for solving these problems, depth-first search.

10.3.4 Other Techniques

Many other techniques have proven to be useful in the design of parallel algorithms. Finding small graph separators is useful for partitioning data among processors to reduce communication [Reif 1993, ch. 14]. Hashing is useful for load balancing and mapping addresses to memory [Vishkin 1984, Karlin and Upfal 1988]. Iterative techniques are useful as a replacement for direct methods for solving linear systems [Bertsekas and Tsitsiklis 1989].

10.4 Basic Operations on Sequences, Lists, and Trees

We begin our presentation of parallel algorithms with a collection of algorithms for performing basic operations on sequences, lists, and trees. These operations will be used as subroutines in the algorithms that follow in later sections.

10.4.1 Sums

As explained at the opening of this chapter, there is a simple recursive algorithm for computing the sum of the elements in an array:

Algorithm: SUM(A).

1 **if** $|A| = 1$ **then return** $A[0]$
2 **else return** SUM($\{A[2i] + A[2i + 1] : i \in [0..|A|/2)\}$)

The work and depth for this algorithm are given by the recurrences

$$W(n) = W(n/2) + O(n) = O(n)$$
$$D(n) = D(n/2) + O(1) = O(\log n)$$

which have solutions $W(n) = O(n)$ and $D(n) = O(\log n)$. This algorithm also can be expressed without recursion (using a **while** loop), but the recursive version forshadows the recursive algorithm for implementing the **scan** function.

As written, the algorithm works only on sequences that have lengths equal to powers of 2. Removing this restriction is not difficult by checking if the sequence is of odd length and separately adding the last element in if it is. This algorithm also can easily be modified to compute the sum relative to any associative operator in place of $+$. For example, the use of max would return the maximum value of a sequence.

10.4.2 Scans

The *plus-scan* operation (also called **all-prefix-sums**) takes a sequence of values and returns a sequence of equal length for which each element is the sum of all previous elements in the original sequence. For example, executing a plus-scan on the sequence $[3, 5, 3, 1, 6]$ returns $[0, 3, 8, 11, 12]$. The scan operation can be implemented by the following algorithm [Stone 1975]:

Algorithm: SCAN(A).

1 **if** $|A| = 1$ **then return** $[0]$
2 **else**
3    $S = $ SCAN($\{A[2i] + A[2i + 1] : i \in [0..|A|/2)\}$)
4    $R = \{$**if** $(i \bmod 2) = 0$ **then** $S[i/2]$ **else** $S[(i - 1)/2] + A[i - 1] : i \in [0..|A|)\}$
5 **return** R

The algorithm works by elementwise adding the even indexed elements of A to the odd indexed elements of A and then recursively solving the problem on the resulting sequence (line 3). The result S of the recursive call gives the plus-scan values for the even positions in the output sequence R. The value for each of the odd positions in R is simply the value for the preceding even position in R plus the value of the preceding position from A.

The asymptotic work and depth costs of this algorithm are the same as for the SUM operation, $W(n) = O(n)$ and $D(n) = O(\log n)$. Also, as with the SUM operation, any associative function can be used in place of the $+$. In fact, the algorithm described can be used more generally to solve various recurrences, such as the first-order linear recurrences $x_i = (x_{i-1} \otimes a_i) \oplus b_i, 0 \leq i \leq n$, where $\otimes$ and $\oplus$ are both associative [Kogge and Stone 1973].

Scans have proven so useful in the implementation of parallel algorithms that some parallel machines provide support for scan operations in hardware.

10.4.3 Multiprefix and Fetch-and-Add

The multiprefix operation is a generalization of the scan operation in which multiple independent scans are performed. The input to the multiprefix operation is a sequence A of n pairs (k, a), where k specifies a key and a specifies an integer data value. For each key value, the multiprefix operation performs an independent scan. The output is a sequence B of n integers containing the results of each of the scans such that if $A[i] = (k, a)$ then

$$B[i] = \text{sum}(\{b : (t, b) \in A[0..i] | t = k\})$$

In other words, each position receives the sum of all previous elements that have the same key. As an example,

$$\text{MULTIPREFIX}([(1, 5), (0, 2), (0, 3), (1, 4), (0, 1), (2, 2)])$$

returns the sequence

$$[0, 0, 2, 5, 5, 0]$$

The *fetch-and-add* operation is a weaker version of the multiprefix operation, in which the order of the input elements for each scan is not necessarily the same as their order in the input sequence A. In this chapter we omit the implementation of the multiprefix operation, but it can be solved by a function that requires work $O(n)$ and depth $O(\log n)$ using concurrent writes [Matias and Vishkin 1991].

10.4.4 Pointer Jumping

Pointer jumping is a technique that can be applied to both linked lists and trees [Wyllie 1979]. The basic pointer jumping operation is simple. Each node i replaces its pointer $P[i]$ with the pointer of the node that it points to, $P[P[i]]$. By repeating this operation, it is possible to compute, for each node in a list or tree, a pointer to the end of the list or root of the tree. Given set P of pointers that represent a tree (i.e., pointers from children to their parents), the following code will generate a pointer from each node to the root of the tree. We assume that the root points to itself.

Algorithm: POINT_TO_ROOT(P).

```
1  for j from 1 to ⌈log|P|⌉
2      P := {P[P[i]] : i ∈ [0..|P|)}
```

The idea behind this algorithm is that in each loop iteration the distance spanned by each pointer, with respect to the original tree, will double, until it points to the root. Since a tree constructed from $n = |P|$ pointers has depth at most $n - 1$, after $\lceil \log n \rceil$ iterations each pointer will point to the root. Because each iteration has constant depth and performs $\Theta(n)$ work, the algorithm has depth $\Theta(\log n)$ and work $\Theta(n \log n)$.

10.4.5 List Ranking

The problem of computing the distance from each node to the end of a linked list is called *list ranking*. Algorithm POINT_TO_ROOT can be easily modified to compute these distances, as follows.

Algorithm: LIST_RANK(P).

```
1  V = {if P[i] = i then 0 else 1 : i ∈ [0..|P|)}
2  for j from 1 to ⌈log|P|⌉
3      V := {V[i] + V[P[i]] : i ∈ [0..|P|)}
4      P := {P[P[i]] : i ∈ [0..|P|)}
5  return V
```

In this function, $V[i]$ can be thought of as the distance spanned by pointer $P[i]$ with respect to the original list. Line 1 initializes V by setting $V[i]$ to 0 if i is the last node (i.e., points to itself), and 1 otherwise. In each iteration, line 3 calculates the new length of $P[i]$. The function has depth $\Theta(\log n)$ and work $\Theta(n \log n)$.

It is worth noting that there is a simple sequential solution to the list-ranking problem that performs only $O(n)$ work: you just walk down the list, incrementing a counter at each step. The preceding parallel algorithm, which performs $\Theta(n \log n)$ work, is not **work efficient**. There are, however, a variety of work-efficient parallel solutions to this problem.

The following parallel algorithm uses the technique of random sampling to construct a pointer from each node to the end of a list of n nodes in a work-efficient fashion [Reid-Miller 1994]. The algorithm is easily generalized to solve the list-ranking problem:

1. Pick m list nodes at random and call them the *start* nodes.
2. From each start node u, follow the list until reaching the next start node v. Call the list nodes between u and v the *sublist* of u.
3. Form a shorter list consisting only of the start nodes and the final node on the list by making each start node point to the next start node on the list.
4. Using pointer jumping on the shorter list, for each start node create a pointer to the last node in the list.
5. For each start node u, distribute the pointer to the end of the list to all of the nodes in the sublist of u.

The key to analyzing the work and depth of this algorithm is to bound the length of the longest sublist. Using elementary probability theory, it is not difficult to prove that the expected length of the longest sublist is at most $O((n \log m)/m)$. The work and depth for each step of the algorithm are thus computed as follows:

1. $W(n, m) = O(m)$ and $D(n, m) = O(1)$.
2. $W(n, m) = O(n)$ and $D(n, m) = O((n \log m)/m)$.
3. $W(n, m) = O(m)$ and $D(n, m) = O(1)$.
4. $W(n, m) = O(m \log m)$ and $D(n, m) = O(\log m)$.
5. $W(n, m) = O(n)$ and $D(n, m) = O((n \log m)/m)$.

Thus, the work for the entire algorithm is $W(m, n) = O(n + m \log m)$, and the depth is $O((n \log m)/m)$. If we set $m = n/\log n$, these reduce to $W(n) = O(n)$ and $D(n) = O(\log^2 n)$.

Using a technique called **contraction**, it is possible to design a list ranking algorithm that runs in $O(n)$ work and $O(\log n)$ depth [Anderson and Miller 1988, 1990]. This technique also can be applied to trees [Miller and Reif 1989, 1991].

10.4.6 Removing Duplicates

Given a sequence of items, the remove-duplicates algorithm removes all duplicates, returning the resulting sequence. The order of the resulting sequence does not matter.

10.4.6.1 Approach 1: Using an Array of Flags

If the items are all nonnegative integers drawn from a small range, we can use a technique similar to bucket sort to remove the duplicates. We begin by creating an array equal in size to the range and initializing all of its elements to 0. Next, using concurrent writes we set a flag in the array for each number that appears in the input list. Finally, we extract those numbers whose flags are set. This algorithm is expressed as follows.

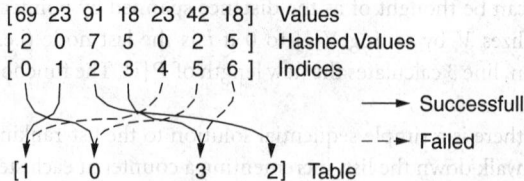

FIGURE 10.4   Each key attempts to write its index into a hash table entry.

Algorithm: REM_DUPLICATES (V).

```
1   RANGE := 1 + MAX(V)
2   FLAGS := dist(0, RANGE) ← {(i, 1) : i ∈ V}
3   return {j : j ∈ [0..RANGE] | FLAGS[j] = 1}
```

This algorithm has depth $O(1)$ and performs work $O(\text{MAX}(V))$. Its obvious disadvantage is that it explodes when given a large range of numbers, both in memory and in work.

10.4.6.2   Approach 2: Hashing

A more general approach is to use a hash table. The algorithm has the following outline. First, we create a hash table whose size is prime and approximately two times as large as the number of items in the set V. A prime size is best, because it makes designing a good hash function easier. The size also must be large enough that the chances of collisions in the hash table are not too great. Let m denote the size of the hash table. Next, we compute a hash value, $hash(V[j], m)$, for each item $V[j] \in V$ and attempt to write the index j into the hash table entry $hash(V[j], m)$. For example, Figure 10.4 describes a particular hash function applied to the sequence $[69, 23, 91, 18, 23, 42, 18]$. We assume that if multiple values are simultaneously written into the same memory location, one of the values will be correctly written. We call the values $V[j]$ whose indices j are successfully written into the hash table *winners*. In our example, the winners are $V[0]$, $V[1]$, $V[2]$, and $V[3]$, that is, 69, 23, 91, and 18. The winners are added to the duplicate-free set that we are constructing, and then set aside. Among the losers, we must distinguish between two types of items: those that were defeated by an item with the same value, and those that were defeated by an item with a different value. In our example, $V[5]$ and $V[6]$ (23 and 18) were defeated by items with the same value, and $V[4]$ (42) was defeated by an item with a different value. Items of the first type are set aside because they are duplicates. Items of the second type are retained, and we repeat the entire process on them using a different hash function. In general, it may take several iterations before all of the items have been set aside, and in each iteration we must use a different hash function.

Removing duplicates using hashing can be implemented as follows:

Algorithm: REMOVE_DUPLICATES (V).

```
1    m := NEXT_PRIME (2 * |V|)
2    TABLE := dist(−1, m)
3    i := 0
4    R := {}
5    while |V| > 0
6        TABLE := TABLE ← {(hash(V[j], m, i), j) : j ∈ [0..|V|)}
7        W := {V[j] : j ∈ [0..|V|)| TABLE [hash(V[j], m, i)] = j}
8        R := R ++W
9        TABLE := TABLE ← {(hash(k, m, i), k) : k ∈ W}
10       V := {k ∈ V| TABLE [hash(k, m, i)] ≠ k}
11       i := i + 1
12   return R
```

The first four lines of function REMOVE_DUPLICATES initialize several variables. Line 1 finds the first prime number larger than $2 * |V|$ using the built-in function NEXT_PRIME. Line 2 creates the hash table and initializes its entries with an arbitrary value (-1). Line 3 initializes i, a variable that simply counts iterations of the **while** loop. Line 4 initializes the sequence R, the result, to be empty. Ultimately, R will contain a single copy of each distinct item in the sequence V.

The bulk of the work in function REMOVE_DUPLICATES is performed by the **while** loop. Although there are items remaining to be processed, we perform the following steps. In line 6, each item $V[j]$ attempts to write its index j into the table entry given by the hash function $hash(V[j], m, i)$. Note that the hash function takes the iteration i as an argument, so that a different hash function is used in each iteration. Concurrent writes are used so that if several items attempt to write to the same entry, precisely one will win. Line 7 determines which items successfully wrote their indices in line 6 and stores their values in an array called W (for *winners*). The winners are added to the result array R in line 8. The purpose of lines 9 and 10 is to remove all of the items that are either winners or duplicates of winners. These lines reuse the hash table. In line 9, each winner writes its value, rather than its index, into the hash table. In this step there are no concurrent writes. Finally, in line 10, an item is retained only if it is not a winner, and the item that defeated it has a different value.

It is not difficult to prove that, with high probability, each iteration reduces the number of items remaining by some constant fraction until the number of items remaining is small. As a consequence, $D(n) = O(\log n)$ and $W(n) = O(n)$.

The remove-duplicates algorithm is frequently used for set operations; for instance, there is a trivial implementation of the set union operation given the code for REMOVE_DUPLICATES.

10.5 Graphs

Graphs present some of the most challenging problems to parallelize since many standard sequential graph techniques, such as depth-first or priority-first search, do not parallelize well. For some problems, such as minimum spanning tree and biconnected components, new techniques have been developed to generate efficient parallel algorithms. For other problems, such as single-source shortest paths, there are no known efficient parallel algorithms, at least not for the general case.

We have already outlined some of the parallel graph techniques in Section 10.3. In this section we describe algorithms for breadth-first search, connected components, and minimum spanning trees. These algorithms use some of the general techniques. In particular, randomization and graph contraction will play an important role in the algorithms. In this chapter we will limit ourselves to algorithms on sparse undirected graphs. We suggest the following sources for further information on parallel graph algorithms Reif [1993, Chap. 2 to 8], JáJá [1992, Chap. 5], and Gibbons and Ritter [1990, Chap. 2].

10.5.1 Graphs and Their Representation

A *graph* $G = (V, E)$ consists of a set of *vertices* V and a set of *edges* E in which each edge connects two vertices. In a *directed graph* each edge is directed from one vertex to another, whereas in an *undirected graph* each edge is symmetric, i.e., goes in both directions. A *weighted graph* is a graph in which each edge $e \in E$ has a weight $w(e)$ associated with it. In this chapter we will use the convention that $n = |V|$ and $m = |E|$. Qualitatively, a graph is considered sparse if $m \ll n^2$ and dense otherwise. The *diameter* of a graph, denoted $D(G)$, is the maximum, over all pairs of vertices (u, v), of the minimum number of edges that must be traversed to get from u to v.

There are three standard representations of graphs used in sequential algorithms: edge lists, adjacency lists, and adjacency matrices. An *edge list* consists of a list of edges, each of which is a pair of vertices. The list directly represents the set E. An *adjacency list* is an array of lists. Each array element corresponds to one vertex and contains a linked list of the neighboring vertices, i.e., the linked list for a vertex v would contain pointers to the vertices $\{u \mid (v, u) \in E\}$. An *adjacency matrix* is an $n \times n$ array A such that A_{ij} is 1 if $(i, j) \in E$ and 0 otherwise. The adjacency matrix representation is typically used only when the graph

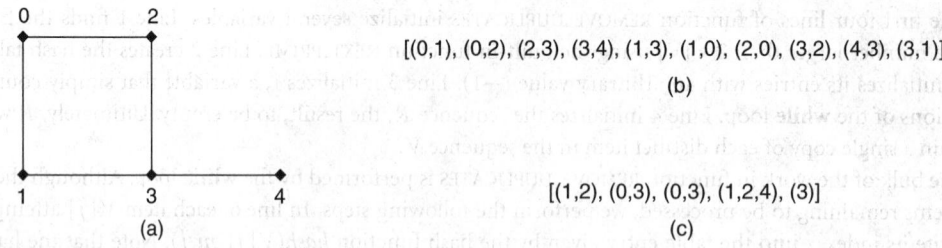

FIGURE 10.5 Representations of an undirected graph: (a) a graph, G, with 5 vertices and 5 edges, (b) the edge-list representation of G, and (c) the adjacency-list representation of G. Values between square brackets are elements of an array, and values between parentheses are elements of a pair.

is dense since it requires $\Theta(n^2)$ space, as opposed to $\Theta(m)$ space for the other two representations. Each of these representations can be used to represent either directed or undirected graphs.

For parallel algorithms we use similar representations for graphs. The main change we make is to replace the linked lists with arrays. In particular, the edge list is represented as an array of edges and the adjacency list is represented as an array of arrays. Using arrays instead of lists makes it easier to process the graph in parallel. In particular, they make it easy to grab a set of elements in parallel, rather than having to follow a list. Figure 10.5 shows an example of our representations for an undirected graph. Note that for the edge-list representation of the undirected graph each edge appears twice, once in each direction. We assume these double edges for the algorithms we describe in this chapter.* To represent a directed graph we simply store the edge only once in the desired direction. In the text we will refer to the left element of an edge pair as the *source vertex* and the right element as the *destination vertex*.

In algorithms it is sometimes more efficient to use the edge list and sometimes more efficient to use an adjacency list. It is, therefore, important to be able to convert between the two representations. To convert from an adjacency list to an edge list (representation c to representation b in Fig. 10.5) is straightforward. The following code will do it with linear work and constant depth:

$$flatten(\{\{(i, j) : j \in G[i]\} : i \in [0 \cdots |G|])$$

where G is the graph in the adjacency list representation. For each vertex i this code pairs up each of i's neighbors with i and then flattens the results.

To convert from an edge list to an adjacency list is somewhat more involved but still requires only linear work. The basic idea is to sort the edges based on the source vertex. This places edges from a particular vertex in consecutive positions in the resulting array. This array can then be partitioned into blocks based on the source vertices. It turns out that since the sorting is on integers in the range $[0 \ldots |V|]$, a radix sort can be used (see radix sort subsection in Section 10.6), which can be implemented in linear work. The depth of the radix sort depends on the depth of the multiprefix operation. (See previous subsection on multiprefix.)

10.5.2 Breadth-First Search

The first algorithm we consider is parallel breadth-first search (BFS). BFS can be used to solve various problems such as finding if a graph is connected or generating a spanning tree of a graph. Parallel BFS is similar to the sequential version, which starts with a source vertex s and visits levels of the graph one after the other using a queue. The main difference is that each level is going to be visited in parallel and no queue is required. As with the sequential algorithm, each vertex will be visited only once and each edge, at most twice, once in each direction. The work is therefore linear in the size of the graph $O(n + m)$. For a graph with diameter D, the number of levels processed by the algorithm will be at least $D/2$ and at most

---

*If space is of serious concern, the algorithms can be easily modified to work with edges stored in just one direction.

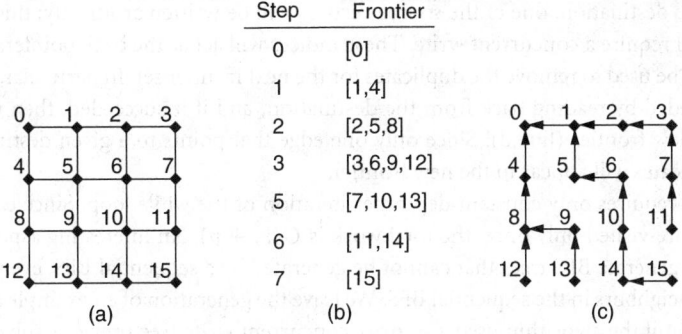

Step	Frontier
0	[0]
1	[1,4]
2	[2,5,8]
3	[3,6,9,12]
5	[7,10,13]
6	[11,14]
7	[15]

(a)   (b)   (c)

FIGURE 10.6 Example of parallel breadth-first search: (a) a graph, G, (b) the frontier at each step of the BFS of G with $s = 0$, and (c) a BFS tree.

D, depending on where the search is initiated. We will show that each level can be processed in constant depth assuming a concurrent-write model, so that the total depth of parallel BFS is $O(D)$.

The main idea of parallel BFS is to maintain a set of frontier vertices, which represent the current level being visited, and to produce a new frontier on each step. The set of frontier vertices is initialized with the singleton s (the source vertex) and during the execution of the algorithm each vertex will be visited only once. A new frontier is generated by collecting all of the neighbors of the current frontier vertices in parallel and removing any that have already been visited. This is not sufficient on its own, however, since multiple vertices might collect the same unvisited vertex. For example, consider the graph in Figure 10.6. On step 2 vertices 5 and 8 will both collect vertex 9. The vertex will therefore appear twice in the new frontier. If the duplicate vertices are not removed, the algorithm can generate an exponential number of vertices in the frontier. This problem does not occur in the sequential BFS because vertices are visited one at a time. The parallel version therefore requires an extra step to remove duplicates.

The following algorithm implements the parallel BFS. It takes as input a source vertex s and a graph G represented as an adjacency array and returns as its result a breadth-first search tree of G. In a BFS tree each vertex processed at level i points to one of its neighbors processed at level $i - 1$ [see Figure 10.6c]. The source s is the root of the tree.

Algorithm: BFS (s, G).

1    $Fr := [s]$
2    $Tr := dist(-1, |G|)$
3    $Tr[s] := s$
4   **while** $(|Fr| \neq 0)$
5        $E := flatten(\{\{(u, v) : u \in G[v]\} : v \in Fr\})$
6        $E' := \{(u, v) \in E \mid Tr[u] = -1\}$
7        $Tr := Tr \leftarrow E'$
8        $Fr := \{u : (u, v) \in E' \mid v = Tr[u]\}$
9   **return** Tr

In this code Fr is the set of frontier vertices, and Tr is the current BFS tree, represented as an array of indices (pointers). The pointers in Tr are all initialized to -1, except for the source s, which is initialized to point to itself. The algorithm assumes the arbitrary concurrent-write model.

We now consider each iteration of the algorithm. The iterations terminate when there are no more vertices in the frontier (line 4). The new frontier is generated by first collecting together the set of edges from the current frontier vertices to their neighbors into an edge array (line 5). An edge from v to u is represented as the pair (u, v). We then remove any edges whose destination has already been visited (line 6). Now each edge writes its source index into the destination vertex (line 7). In the case that more than one edge has the same destination, one of the source vertices will be written arbitrarily; this is the only place the algorithm will require a concurrent write. These indices will act as the back pointers for the BFS tree, and they also will be used to remove the duplicates for the next frontier set. In particular, each edge checks whether it succeeded by reading back from the destination, and if it succeeded, then the destination is included in the new frontier (line 8). Since only one edge that points to a given destination vertex will succeed, no duplicates will appear in the new frontier.

The algorithm requires only constant depth per iteration of the while loop. Since each vertex and its associated edges are visited only once, the total work is $O(m + n)$. An interesting aspect of this parallel BFS is that it can generate BFS trees that cannot be generated by a sequential BFS, even allowing for any order of visiting neighbors in the sequential BFS. We leave the generation of an example as an exercise. We note, however, that if the algorithm used a priority concurrent write (see previous subsection describing the model used in this chapter) on line 7, then it would generate the same tree as a sequential BFS.

10.5.3 Connected Components

We now consider the problem of labeling the connected components of an undirected graph. The problem is to label all of the vertices in a graph G such that two vertices u and v have the same label if and only if there is a path between the two vertices. Sequentially, the connected components of a graph can easily be labeled using either depth-first or breadth-first search. We have seen how to implement breadth-first search, but the technique requires a depth proportional to the diameter of a graph. This is fine for graphs with a small diameter, but it does not work well in the general case. Unfortunately, in terms of work, even the most efficient polylogarithmic depth parallel algorithms for depth-first search and breadth-first search are very inefficient. Hence, the efficient algorithms for solving the connected components problem use different techniques.

The two algorithms we consider are based on graph contraction. Graph contraction proceeds by contracting the vertices of a connected subgraph into a single vertex to form a new smaller graph. The techniques we use allow the algorithms to make many such contractions in parallel across the graph. The algorithms, therefore, proceed in a sequence of steps, each of which contracts a set of subgraphs, and forms a smaller graph in which each subgraph has been converted into a vertex. If each such step of the algorithm contracts the size of the graph by a constant fraction, then each component will contract down to a single vertex in $O(\log n)$ steps. By running the contraction in reverse, the algorithms can label all of the vertices in the components. The two algorithms we consider differ in how they select subgraphs for contraction. The first uses randomization and the second is deterministic. Neither algorithm is work efficient because they require $O((n + m) \log n)$ work for worst-case graphs, but we briefly discuss how they can be made to be work efficient in the subsequent improved version subsection. Both algorithms require the concurrent-write model.

10.5.3.1 Random Mate Graph Contraction

The random mate technique for graph contraction is based on forming a set of star subgraphs and contracting the stars. A *star* is a tree of depth one; it consists of a root and an arbitrary number of children. The random mate algorithm finds a set of nonoverlapping stars in a graph and then contracts each star into a single vertex by merging the children into their parents. The technique used to form the stars uses randomization. It works by having each vertex flip a coin and then identify itself as either a parent or a child based on the outcome. We assume the coin is unbiased so that every vertex has a 50% probability of being a parent. Now every vertex that has come up a child looks at its neighbors to see if any are parents. If at least one is a parent, then the child picks one of the neighboring parents as its parent. This process has selected a set of stars, which can be contracted. When contracting, we relabel all of the edges that were incident on a contracting child to its parent's label. Figure 10.7 illustrates a full contraction step. This contraction step is repeated until all components are of size 1.

To analyze the costs of the algorithm we need to know how many vertices are expected to be removed on each contraction step. First, we note that the step is going to remove only children and only if they have

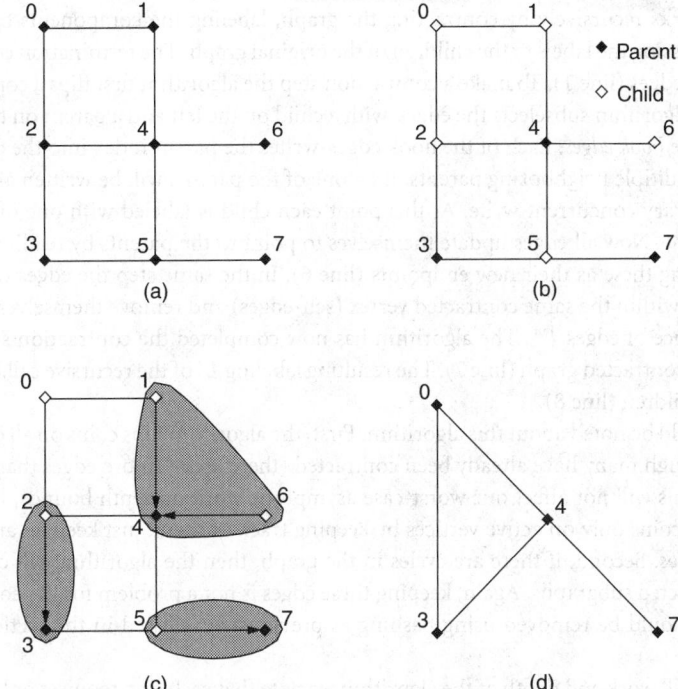

FIGURE 10.7 Example of one step of random mate graph contraction: (a) the original graph G, (b) G after selecting the parents randomly, (c) contracting the children into the parents (the shaded regions show the subgraphs), and (d) the contracted graph G'.

a neighboring parent. The probability that a vertex will be deleted is therefore the probability that it is a child multiplied by the probability that at least one of its neighbors is a parent. The probability that it is a child is 1/2 and the probability that at least one neighbor is a parent is at least 1/2 (every vertex has one or more neighbors, otherwise it would be completed). We, therefore, expect to remove at least 1/4 of the remaining vertices at each step and expect the algorithm to complete in no more than $\log_{4/3} n$ steps. The full probabilistic analysis is somewhat more involved since we could have a streak of bad flips, but it is not too hard to show that the algorithm is very unlikely to require more than $O(\log n)$ steps.

The following algorithm implements the random mate technique. The input is a graph G in the edge list representation (note that this is a different representation than used in BFS), along with the labels L of the vertices. We assume the labels are initialized to the index of the vertex. The output of the algorithm is a label for each vertex, such that all vertices in a component will be labeled with one of the original labels of a vertex in the component.

Algorithm: CC_RANDOM_MATE (L, E).

```
1  if (|E| = 0) then return L
2  else
3      CHILD := {rand_bit() : v ∈ [1..n]}
4      H := {(u, v) ∈ E | CHILD[u] ∧ ¬CHILD[v]}
5      L := L ← H
6      E' := {(L[u], L[v]) : (u, v) ∈ E | L[u] ≠ L[v]}
7      L' := CC_RANDOM_MATE(L, E')
8      L' := L' ← {(u, L'[v]) : (u, v) ∈ H}
9      return L'
```

The algorithm works recursively by contracting the graph, labeling the components of the contracted graph, and then passing the labels to the children of the original graph. The termination condition is when there are no more edges (line 1). To make a contraction step the algorithm first flips a coin on each vertex (line 3). Now the algorithm subselects the edges-with a child on the left and a parent on the right (line 4). These are called the *hook edges*. Each of the hook edges-writes the parent index into the child's label (line 5). If a child has multiple neighboring parents, then one of the parents will be written arbitrarily; we are assuming an arbitrary concurrent write. At this point each child is labeled with one of its neighboring parents, if it has one. Now all edges update themselves to point to the parents by reading from their two endpoints and using these as their new endpoints (line 6). In the same step the edges can check if their two endpoints are within the same contracted vertex (self-edges) and remove themselves if they are. This gives a new sequence of edges E^1. The algorithm has now completed the contraction step and is called recursively on the contracted graph (line 7). The resulting labeling L' of the recursive call is used to update the labels of the children (line 8).

Two things should be noted about this algorithm. First, the algorithm flips coins on all of the vertices on each step even though many have already been contracted (there are no more edges that point to them). It turns out that this will not affect our worst-case asymptotic work or depth bounds, but in practice it is not hard to flip coins only on active vertices by keeping track of them: just keep an array of the labels of the active vertices. Second, if there are cycles in the graph, then the algorithm will create redundant edges in the contracted subgraphs. Again, keeping these edges is not a problem for the correctness or cost bounds, but they could be removed using hashing as previously discussed in the section on removing duplicates.

To analyze the full work and depth of the algorithm we note that each step requires only constant depth and $O(n+m)$ work. Since the number of steps is $O(\log n)$ with high probability, as mentioned earlier, the total depth is $O(\log n)$ and the work is $O((n+m)\log n)$, both with high probability. One might expect that the work would be linear since the algorithm reduces the number of vertices on each step by a constant fraction. We have no guarantee, however, that the number of edges also is going to contract geometrically, and in fact for certain graphs they will not. Subsequently, in this section we will discuss how this can be improved to lead to a work-efficient algorithm.

10.5.3.2 Deterministic Graph Contraction

Our second algorithm for graph contraction is deterministic [Greiner 1994]. It is based on forming trees as subgraphs and contracting these trees into a single vertex using pointer jumping. To understand the algorithm, consider the graph in Figure 10.8a. The overall goal is to contract all of the vertices of the

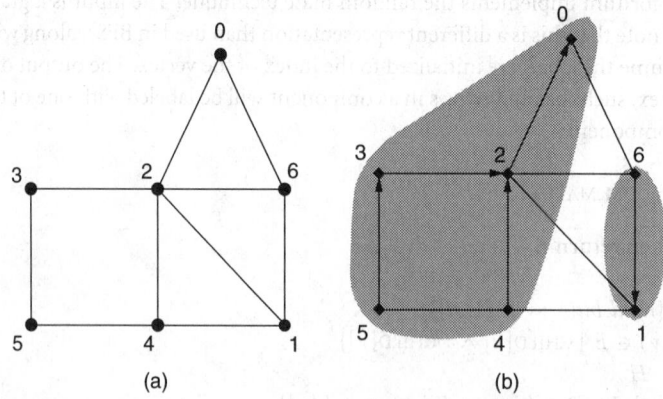

         (a)   (b)

FIGURE 10.8 Tree-based graph contraction: (a) a graph, G, and (b) the hook edges induced by hooking larger to smaller vertices and the subgraphs induced by the trees.

graph into a single vertex. If we had a spanning tree that was imposed on the graph, we could contract the graph by contracting the tree using pointer jumping as discussed previously. Unfortunately, finding a spanning tree turns out to be as hard as finding the connected components of the graph. Instead, we will settle for finding a number of trees that cover the graph, contract each of these as our subgraphs using pointer jumping, and then recurse on the smaller graph. To generate the trees, the algorithm hooks each vertex into a neighbor with a smaller label. This guarantees that there are no cycles since we are only generating pointers from larger to smaller numbered vertices. This hooking will impose a set of disjoint trees on the graph. Figure 10.8b shows an example of such a hooking step. Since a vertex can have more than one neighbor with a smaller label, there can be many possible hookings for a given graph. For example, in Figure 10.8, vertex 2 could have hooked into vertex 1.

The following algorithm implements the tree-based graph contraction. We assume that the labels L are initialized to the index of the vertex.

Algorithm: CC_TREE_CONTRACT(L, E).

1   **if**($|E| = 0$)
2   **then return** L
3   **else**
4      $H := \{(u, v) \in E \mid u < v\}$
5      $L := L \leftarrow H$
6      $L := \text{POINT\_TO\_ROOT}(L)$
7      $E' := \{(L[u], L[v]) : (u, v) \in E \mid L[u] \neq L[v]\}$
8   **return** CC_TREE_CONTRACT(L, E')

The structure of the algorithm is similar to the random mate graph contraction algorithm. The main differences are in how the hooks are selected (line 4), the pointer jumping step to contract the trees (line 6), and the fact that no relabeling is required when returning from the recursive call. The hooking step simply selects edges that point from smaller numbered vertices to larger numbered vertices. This is called a *conditional hook*. The pointer jumping step uses the algorithm given earlier in Section 10.4. This labels every vertex in the tree with the root of the tree. The edge relabeling is the same as in a random mate algorithm. The reason we do not need to relabel the vertices after the recursive call is that the pointer jumping will do the relabeling.

Although the basic algorithm we have described so far works well in practice, in the worst case it can take $n - 1$ steps. Consider the graph in Figure 10.9a. After hooking and contracting, only one vertex has been removed. This could be repeated up to $n - 1$ times. This worst-case behavior can be avoided by trying to hook in both directions (from larger to smaller and from smaller to larger) and picking the hooking that hooks more vertices. We will make use of the following lemma.

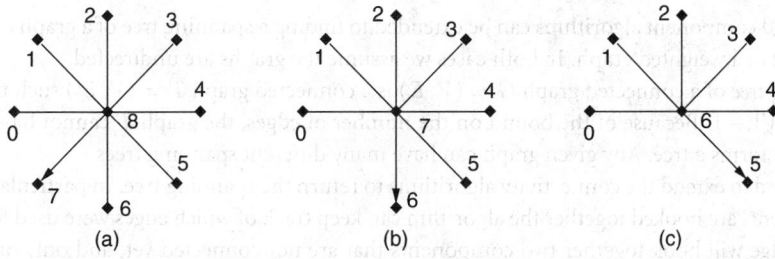

FIGURE 10.9 A worst-case graph: (a) a star graph, G, with the maximum index at the root of the star, (b) G after one step of contraction, and (c) G after two steps of contraction.

Lemma 10.1 *Let $G = (V, E)$ be an undirected graph in which each vertex has at least one neighbor, then either $|\{u|(u, v) \in E, u < v\}| \geq |V|/2$ or $|\{u|(u, v) \in E, u > v\}| > |V|/2$.*

Proof 10.2 Every vertex must have either a neighbor with a lesser index or a neighbor with a greater index. This means that if we consider the set of vertices with a lesser neighbor and the set of vertices with a greater neighbor, then one of those sets must consist of at least one-half the vertices. □

This lemma will guarantee that if we try hooking in both directions and pick the better one we will remove at least one-half of the vertices on each step, so that the number of steps is bounded by $\log n$.

We now consider the total cost of the algorithm. The hooking and relabeling of edges on each step takes $O(m)$ work and constant depth. The tree contraction using pointer jumping on each step requires $O(n \log n)$ work and $O(\log n)$ depth, in the worst case. Since there are $O(\log n)$ steps, in the worst case, the total work is $O((m + n \log n) \log n)$ and depth $O(\log^2 n)$. However, if we keep track of the active vertices (the roots) and only pointer jump on active vertices, then the work is reduced to $O((m + n) \log n)$ since the number of vertices geometrically decreases. This requires that the algorithm relabels on the way back up the recursion as done for the random mate algorithm. The total work with this modification is the same work as the randomized technique, although the depth has increased.

10.5.3.3 Improved Versions of Connected Components

There are many improvements to the two basic connected component algorithms we described. Here we mention some of them.

The deterministic algorithm can be improved to run in $O(\log n)$ depth with the same work bounds [Awerbuch and Shiloach 1987, Shiloach and Vishkin 1982]. The basic idea is to interleave the hooking steps with the **shortcutting** steps. The one tricky aspect is that we must always hook in the same direction (i.e., from smaller to larger), so as not to create cycles. Our previous technique to solve the star-graph problem, therefore, does not work. Instead, each vertex checks if it belongs to any tree after hooking. If it does not, then it can hook to any neighbor, even if it has a larger index. This is called an *unconditional hook*.

The randomized algorithm can be improved to run in optimal work $O(n + m)$ [Gazit 1991]. The basic idea is to not use all of the edges for hooking on each step and instead use a sample of the edges. This basic technique developed for parallel algorithms has since been used to improve some sequential algorithms, such as deriving the first linear work algorithm for minimum spanning trees [Klein and Tarjan 1994].

Another improvement is to use the EREW model instead of requiring concurrent reads and writes [Halperin and Zwick 1994]. However, this comes at the cost of greatly complicating the algorithm. The basic idea is to keep circular linked lists of the neighbors of each vertex and then to splice these lists when merging vertices.

10.5.3.4 Extensions to Spanning Trees and Minimum Spanning Trees

The connected component algorithms can be extended to finding a spanning tree of a graph or minimum spanning tree of a weighted graph. In both cases we assume the graphs are undirected.

A *spanning tree* of a connected graph $G = (V, E)$ is a connected graph $T = (V, E')$ such that $E' \subseteq E$ and $|E'| = |V| - 1$. Because of the bound on the number of edges, the graph T cannot have any cycles and therefore forms a tree. Any given graph can have many different spanning trees.

It is not hard to extend the connectivity algorithms to return the spanning tree. In particular, whenever two components are hooked together the algorithm can keep track of which edges were used for hooking. Since each edge will hook together two components that are not connected yet, and only one edge will succeed in hooking the components, the collection of these edges across all steps will form a spanning tree (they will connect all vertices and have no cycles). To determine which edges were used for contraction, each edge checks if it successfully hooked after the attempted hook.

A minimum spanning tree of a connected weighted graph $G = (V, E)$ with weights $w(e)$ for $e \in E$ is a spanning tree $T = (V, E')$ of G such that

$$w(T) = \sum_{e \in E'} w(e)$$

is minimized. The connected component algorithms also can be extended to determine the minimum spanning tree. Here we will briefly consider an extension of the random mate technique. The algorithm will take advantage of the property that, given any $W \subset V$, the minimum edge from W to $V - W$ must be in some minimum spanning tree. This implies that the minimum edge incident on a vertex will be on a minimum spanning tree. This will be true even after we contract subgraphs into vertices since each subgraph is a subset of V.

To implement the minimum spanning tree algorithm we therefore modify the random mate technique so that each child u, instead of picking an arbitrary parent to hook into, finds the incident edge (u, v) with minimum weight and hooks into v if it is a parent. If v is not a parent, then the child u does nothing (it is left as an orphan). Figure 10.10 illustrates the algorithm. As with the spanning tree algorithm, we keep track of the edges we use for hooks and add them to a set E'. This new rule will still remove 1/4 of the vertices on each step on average since a vertex has 1/2 probability of being a child, and there is 1/2 probability that the vertex at the other end of the minimum edge is a parent. The one complication in this minimum spanning tree algorithm is finding for each child the incident edge with minimum weight. Since we are keeping an edge list, this is not trivial to compute. If we had an adjacency list, then it would be easy, but since we are updating the endpoints of the edges, it is not easy to maintain the adjacency list. One way to solve this problem is to use a priority concurrent write. In such a write, if multiple values are written to the same location, the one coming from the leftmost position will be written. With such a scheme the minimum edge can be found by presorting the edges by their weight so that the lowest weighted edge will always win when executing a concurrent write. Assuming a priority write, this minimum spanning tree algorithm has the same work and depth as the random mate connected components algorithm.

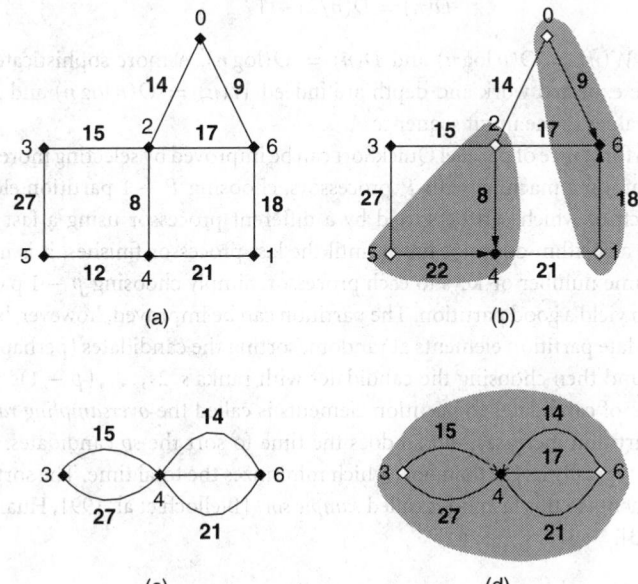

(a)

(b)

(c)

(d)

FIGURE 10.10 Example of the minimum spanning tree algorithm. (a) The original weighted graph G. (b) Each child (light) hooks across its minimum weighted edge to a parent (dark), if the edge is incident on a parent. (c) The graph after one step of contraction. (d) The second step in which children hook across minimum weighted edges to parents.

10.6   Sorting

Sorting is a problem that admits a variety of parallel solutions. In this section we limit our discussion to two parallel sorting algorithms, QuickSort and radix sort. Both of these algorithms are easy to program, and both work well in practice. Many more sorting algorithms can be found in the literature. The interested reader is referred to Akl [1985], JáJá [1992], and Leighton [1992] for more complete coverage.

10.6.1   QuickSort

We begin our discussion of sorting with a parallel version of QuickSort. This algorithm is one of the simplest to code.

Algorithm: QUICKSORT(A).

1   **if** $|A| = 1$ **then return** A
2   $i := rand\_int(|A|)$
3   $p := A[i]$
4   **in parallel do**
5       $L := $ QUICKSORT($\{a : a \in A \mid a < p\}$)
6       $E := \{a : a \in A \mid a = p\}$
7       $G := $ QUICKSORT($\{a : a \in A \mid a > p\}$)
8   **return** $L ++ E ++ G$

We can make an optimistic estimate of the work and depth of this algorithm by assuming that each time a partition element, p, is selected, it divides the set A so that neither L nor H has more than half of the elements. In this case, the work and depth are given by the recurrences

$$W(n) = 2W(n/2) + O(n)$$
$$D(n) = D(n/2) + 1$$

whose solutions are $W(n) = O(n \log n)$ and $D(n) = O(\log n)$. A more sophisticated analysis [Knuth 1973] shows that the expected work and depth are indeed $W(n) = O(n \log n)$ and $D(n) = O(\log n)$, independent of the values in the input sequence A.

In practice, the performance of parallel QuickSort can be improved by selecting more than one partition element. In particular, on a machine with P processors, choosing $P - 1$ partition elements divides the keys into P sets, each of which can be sorted by a different processor using a fast sequential sorting algorithm. Since the algorithm does not finish until the last processor finishes, it is important to assign approximately the same number of keys to each processor. Simply choosing $p - 1$ partition elements at random is unlikely to yield a good partition. The partition can be improved, however, by choosing a larger number, sp, of candidate partition elements at random, sorting the candidates (perhaps using some other sorting algorithm), and then choosing the candidates with ranks $s, 2s, \ldots, (p - 1)s$ to be the partition elements. The ratio s of candidates to partition elements is called the *oversampling ratio*. As s increases, the quality of the partition increases, but so does the time to sort the sp candidates. Hence, there is an optimum value of s, typically larger than one, which minimizes the total time. The sorting algorithm that selects partition elements in this fashion is called *sample sort* [Blelloch et al. 1991, Huang and Chow 1983, Reif and Valiant 1983].

10.6.2   Radix Sort

Our next sorting algorithm is radix sort, an algorithm that performs well in practice. Unlike QuickSort, radix sort is not a *comparison sort*, meaning that it does not compare keys directly in order to determine the relative ordering of keys. Instead, it relies on the representation of keys as b-bit integers.

The basic radix sort algorithm (whether serial or parallel) examines the keys to be sorted one *digit* at a time, starting with the least significant digit in each key. Of fundamental importance is that this intermediate sort on digits be *stable*: the output ordering must preserve the input order of any two keys whose bits are the same.

The most common implementation of the intermediate sort is as a counting sort. A counting sort first counts to determine the *rank* of each key — its position in the output order — and then we permute the keys to their respective locations. The following algorithm implements radix sort assuming one-bit digits.

Algorithm: RADIX_SORT(A, b)

```
1  for i from 0 to b − 1
2      B := {(a ≫ i) mod 2 : a ∈ A}
3      NB := {1 − b : b ∈ B}
4      R_0 := SCAN(NB)
5      s_0 := SUM(NB)
6      R_1 := SCAN(B)
7      R := {if B[j] = 0 then R_0[j] else R_1[j] + s_0 : j ∈ [0..|A|)}
8      A := A ← {(R[j], A[j]) : j ∈ [0..|A|)}
9  return A
```

For keys with b bits, the algorithm consists of b sequential iterations of a **for** loop, each iteration sorting according to one of the bits. Lines 2 and 3 compute the value and inverse value of the bit in the current position for each key. The notation $a \gg i$ denotes the operation of shifting a i bit positions to the right. Line 4 computes the rank of each key whose bit value is 0. Computing the ranks of the keys with bit value 1 is a little more complicated, since these keys follow the keys with bit value 0. Line 5 computes the number of keys with bit value 0, which serves as the rank of the first key whose bit value is 1. Line 6 computes the relative order of the keys with bit value 1. Line 7 merges the ranks of the even keys with those of the odd keys. Finally, line 8 permutes the keys according to their ranks.

The work and depth of RADIX_SORT are computed as follows. There are b iterations of the **for** loop. In each iteration, the depths of lines 2, 3, 7, 8, and 9 are constant, and the depths of lines 4, 5, and 6 are $O(\log n)$. Hence, the depth of the algorithm is $O(b \log n)$. The work performed by each of lines 2–9 is $O(n)$. Hence, the work of the algorithm is $O(bn)$.

The radix sort algorithm can be generalized so that each b-bit key is viewed as b/r blocks of r bits each, rather than as b individual bits. In the generalized algorithm, there are b/r iterations of the **for** loop, each of which invokes the SCAN function 2^r times. When r is large, a multiprefix operation can be used for generating the ranks instead of executing a SCAN for each possible value [Blelloch et al. 1991]. In this case, and assuming the multiprefix runs in linear work, it is not hard to show that as long as $b = O(\log n)$, the total work for the radix sort is $O(n)$, and the depth is the same order as the depth of the multiprefix.

Floating-point numbers also can be sorted using radix sort. With a few simple bit manipulations, floating-point keys can be converted to integer keys with the same ordering and key size. For example, IEEE double-precision floating-point numbers can be sorted by inverting the mantissa and exponent bits if the sign bit is 1 and then inverting the sign bit. The keys are then sorted as if they were integers.

10.7 Computational Geometry

Problems in computational geometry involve determining various properties about sets of objects in a k-dimensional space. Some standard problems include finding the closest distance between a pair of points (closest pair), finding the smallest convex region that encloses a set of points (convex hull), and finding line or polygon intersections. Efficient parallel algorithms have been developed for most standard problems in computational geometry. Many of the sequential algorithms are based on divide-and-conquer and lead in a relatively straightforward manner to efficient parallel algorithms. Some others are based on a technique called plane sweeping, which does not parallelize well, but for which an analogous parallel technique, the *plane sweep tree* has been developed [Aggarwal et al. 1988, Atallah et al. 1989]. In this section we describe parallel algorithms for two problems in two dimensions — closest pair and convex hull. For the convex hull we describe two algorithms. These algorithms are good examples of how sequential algorithms can be parallelized in a straightforward manner.

We suggest the following sources for further information on parallel algorithms for computational geometry: Reif [1993, Chap. 9 and Chap. 11], JáJá [1992, Chap. 6], and Goodrich [1996].

10.7.1   Closest Pair

The *closest pair problem* takes a set of points in k dimensions and returns the two points that are closest to each other. The distance is usually defined as Euclidean distance. Here we describe a closest pair algorithm for two-dimensional space, also called the planar closest pair problem. The algorithm is a parallel version of a standard sequential algorithm [Bentley and Shamos 1976], and, for n points, it requires the same work as the sequential versions $O(n \log n)$ and has depth $O(\log^2 n)$. The work is optimal.

The algorithm uses divide-and-conquer based on splitting the points along lines parallel to the y axis and is implemented as follows.

Algorithm: CLOSEST_PAIR(P).

```
1   if (|P| < 2) then return (P, ∞)
2   x_m := MEDIAN ({x : (x, y) ∈ P})
3   L := {(x, y) ∈ P | x < x_m}
4   R := {(x, y) ∈ P | x ≥ x_m}
5   in parallel do
6       (L', δ_L) := CLOSEST_PAIR(L)
7       (R', δ_R) := CLOSEST_PAIR(R)
8   P' := MERGE_BY_Y(L', R')
9   δ_P := BOUNDARY_MERGE(P', δ_L, δ_R, x_m)
10  return (P', δ_P)
```

This function takes a set of points P in the plane and returns both the original points sorted along the y axis and the distance between the closest two points. The sorted points are needed to help merge the results from recursive calls and can be thrown away at the end. It would be easy to modify the routine to return the closest pair of points in addition to the distance between them. The function works by dividing the points in half based on the median x value, recursively solving the problem on each half, and then merging the results. The MERGE_BY_Y function merges L' and R' along the y axis and can use a standard parallel merge routine. The interesting aspect of the code is the BOUNDARY_MERGE routine, which works on the same principle as described by Bentley and Shamos [1976] and can be computed with $O(\log n)$ depth and $O(n)$ work. We first review the principle and then show how it is implemented in parallel.

The inputs to BOUNDARY_MERGE are the original points P sorted along the y axis, the closest distance within L and R, and the median point x_m. The closest distance in P must be either the distance δ_L, the distance δ_R, or the distance between a point in L and a point in R. For this distance to be less than δ_L or δ_R, the two points must lie within $\delta = \min(\delta_L, \delta_R)$ of the line $x = x_m$. Thus, the two vertical lines at $x_r = x_m + \delta$ and $x_l = x_m - \delta$ define the borders of a region M in which the points must lie (see Figure 10.11). If we could find the closest distance in M, call it δ_M, then the closest overall distance is $\delta_P = \min(\delta_L, \delta_R, \delta_M)$.

To find δ_M, we take advantage of the fact that not many points can be packed closely together within M since all points within L or R must be separated by at least δ. Figure 10.11 shows the tightest possible packing of points in a $2\delta \times \delta$ rectangle within M. This packing implies that if the points in M are sorted along the y axis, each point can determine the minimum distance to another point in M by looking at a fixed number of neighbors in the sorted order, at most seven in each direction. To see this, consider one of the points along the top of the $2\delta \times \delta$ rectangle. To find if there are any points below it that are closer

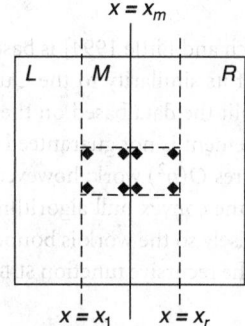

FIGURE 10.11 Merging two rectangles to determine the closest pair. Only 8 points can fit in the $2\delta \times \delta$ dashed rectangle.

than δ, it needs only to consider the points within the rectangle (points below the rectangle must be farther than δ away). As the figure illustrates, there can be at most seven other points within the rectangle. Given this property, the following function implements the border merge.

Algorithm: BOUNDARY_MERGE($P, \delta_L, \delta_R, x_m$).

1   $\delta := \min(\delta_L, \delta_R)$
2   $M := \{(x, y) \in P \mid (x \geq x_m - \delta) \wedge (x \leq x_m + \delta)\}$
3   $\delta_M := \min(\{= \min(\{distance(M[i], M[i + j]) : j \in [1..7]\})$
4        $: i \in [0..|P - 7)\}$
5   **return** $\min(\delta, \delta_M)$

In this function each point in M looks at seven points in front of it in the sorted order and determines the distance to each of these points. The minimum over all distances is taken. Since the distance relationship is symmetric, there is no need for each point to consider points behind it in the sorted order.

The work of BOUNDARY_MERGE is $O(n)$ and the depth is dominated by taking the minimum, which has $O(\log n)$ depth.* The work of the merge and median steps in CLOSEST_PAIR is also $O(n)$, and the depth of both is bounded by $O(\log n)$. The total work and depth of the algorithm therefore can be solved with the recurrences

$$W(n) = 2W(n/2) + O(n) \quad = O(n \log n)$$
$$D(n) = D(n/2) + O(\log n) = O(\log^2 n)$$

10.7.2   Planar Convex Hull

The convex hull problem takes a set of points in k dimensions and returns the smallest convex region that contains all of the points. In two dimensions, the problem is called the planar convex hull problem and it returns the set of points that form the corners of the region. These points are a subset of the original points. We will describe two parallel algorithms for the planar convex hull problem. They are both based on divide-and-conquer, but one does most of the work before the divide step, and the other does most of the work after.

---

*The depth of finding the minimum or maximum of a set of numbers actually can be improved to $O(\log \log n)$ with concurrent reads [Shiloach and Vishkin 1981].

10.7.2.1 QuickHull

The parallel *QuickHull* algorithm [Blelloch and Little 1994] is based on the sequential version [Preparata and Shamos 1985], so named because of its similarity to the QuickSort algorithm. As with QuickSort, the strategy is to pick a *pivot* element, split the data based on the pivot, and recurse on each of the split sets. Also as with QuickSort, the pivot element is not guaranteed to split the data into equally sized sets, and in the worst case the algorithm requires $O(n^2)$ work; however, in practice the algorithm is often very efficient, probably the most practical of the convex hull algorithms. At the end of the section we briefly describe how the splits can be made precisely so the work is bounded by $O(n \log n)$.

The QuickHull algorithm is based on the recursive function SUBHULL, which is implemented as follows.

Algorithm: SUBHULL(P, p_1, p_2).

```
1   P' := {p ∈ P | RIGHT_OF ?(p, (p₁, p₂))}
2   if (|P'| < 2)
3   then return [p₁] ++ P'
4   else
5       i := MAX_INDEX({DISTANCE(p, (p₁, p₂)) : p ∈ P'})
6       pₘ := P'[i]
7   in parallel do
8       Hₗ := SUBHULL(P', p₁, pₘ)
9       Hᵣ := SUBHULL(P', pₘ, p₂)
10      return Hₗ ++ Hᵣ
```

This function takes a set of points P in the plane and two points p_1 and p_2 that are known to lie on the convex hull and returns all of the points that lie on the hull clockwise from p_1 to p_2, inclusive of p_1, but not of p_2. For example, in Figure 10.12 SUBHULL($[A, B, C, \ldots, P], A, P$) would return the sequence $[A, B, J, O]$.

The function SUBHULL works as follows. Line 1 removes all of the elements that cannot be on the hull because they lie to the right of the line from p_1 to p_2. This can easily be calculated using a cross product. If the remaining set P' is either empty or has just one element, the algorithm is done. Otherwise, the algorithm finds the point p_m farthest from the line (p_1, p_2). The point p_m must be on the hull since as a line at infinity parallel to (p_1, p_2) moves toward (p_1, p_2), it must first hit p_m. In line 5, the function MAX_INDEX returns the index of the maximum value of a sequence, using $O(n)$ work $O(\log n)$ depth, which is then used to extract the point p_m. Once p_m is found, SUBHULL is called twice recursively to find

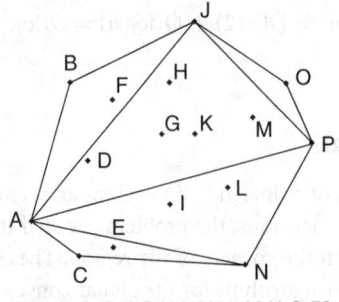

[A B C D E F G H I J K L M N O P]

A [B D F G H J K M O] P [C E I L N]

A [B F] J [O] P N [C E]

A B J O P N C

FIGURE 10.12 An example of the QuickHull algorithm.

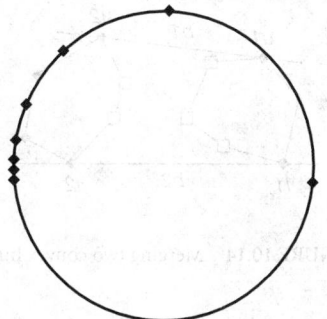

FIGURE 10.13 Contrived set of points for worst-case QuickHull.

the hulls from p_1 to p_m and from p_m to p_2. When the recursive calls return, the results are appended. The algorithm function uses SUBHULL to find the full convex hull.

Algorithm: QUICK_HULL(P).

1 $X := \{x : (x, y) \in P\}$
2 $x_{\min} := P[min\_index(X)]$
3 $x_{\max} := P[max\_index(X)]$
4 **return** SUBHULL($P, x_{\min}, x_{\max}$) ++ SUBHULL($P, x_{\max}, x_{\min}$)

We now consider the costs of the parallel QuickHull. The cost of everything other than the recursive calls is $O(n)$ work and $O(\log n)$ depth. If the recursive calls are balanced so that neither recursive call gets much more than half the data, then the number of levels of recursion will be $O(\log n)$. This will lead to the algorithm running in $O(\log^2 n)$ depth. Since the sum of the sizes of the recursive calls can be less than n (e.g., the points within the triangle AJP will be thrown out when making the recursive calls to find the hulls between A and J and between J and P), the work can be as little as $O(n)$ and often is in practice. As with QuickSort, however, when the recursive calls are badly partitioned, the number of levels of recursion can be as bad as $O(n)$ with work $O(n^2)$. For example, consider the case when all of the points lie on a circle and have the following unlikely distribution: $x_{\min}$ and $x_{\max}$ appear on opposite sides of the circle. There is one point that appears halfway between $x_{\min}$ and $x_{\max}$ on the sphere and this point becomes the new $x_{\max}$. The remaining points are defined recursively. That is, the points become arbitrarily close to $x_{\min}$ (see Figure 10.13). Kirkpatrick and Seidel [1986] have shown that it is possible to modify QuickHull so that it makes provably good partitions. Although the technique is shown for a sequential algorithm, it is easy to parallelize. A simplification of the technique is given by Chan et al. [1995]. This parallelizes even better and leads to an $O(\log^2 n)$ depth algorithm with $O(n \log h)$ work where h is the number of points on the convex hull.

10.7.2.2 MergeHull

The MergeHull algorithm [Overmars and Van Leeuwen 1981] is another divide-and-conquer algorithm for solving the planar convex hull problem. Unlike QuickHull, however, it does most of its work after returning from the recursive calls. The algorithm is implemented as follows.

Algorithm: MERGEHULL(P).

1 **if** ($|P| < 3$) **then return** P
2 **else**
3   **in parallel do**
4      $H_1 = $ MERGEHULL ($P[0..|P|/2]$)
5      $H_2 = $ MERGEHULL ($P[|P|/2..|P|]$)
6   **return** JOIN_HULLS(H_1, H_2)

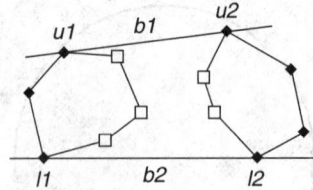

FIGURE 10.14   Merging two convex hulls.

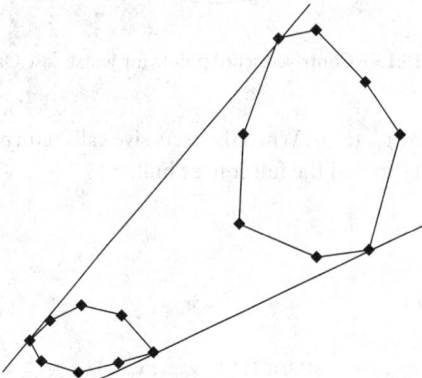

FIGURE 10.15   A bridge that is far from the top of the convex hull.

This function assumes the input P is presorted according to the x coordinates of the points. Since the points are presorted, H_1 is a convex hull on the left and H_2 is a convex hull on the right. The JOIN_HULLS routine is the interesting part of the algorithm. It takes the two hulls and merges them into one. To do this, it needs to find upper and lower points u_1 and l_1 on H_1 and u_2 and l_2 on H_2 such that u_1, u_2 and l_1, l_2 are successive points on H (see Figure 10.14). The lines b_1 and b_2 joining these upper and lower points are called the upper and lower bridges, respectively. All of the points between u_1 and l_1 and between u_2 and l_2 on the *outer* sides of H_1 and H_2 are on the final convex hull, whereas the points on the *inner* sides are not on the convex hull. Without loss of generality we consider only how to find the upper bridge b_1. Finding the lower bridge b_2 is analogous.

To find the upper bridge, one might consider taking the points with the maximum y. However, this does not work in general; u_1 can lie as far down as the point with the minimum x or maximum x value (see Figure 10.15). Instead, there is a nice solution based on binary search. Assume that the points on the convex hulls are given in order (e.g., clockwise). At each step the search algorithm will eliminate half the remaining points from consideration in either H_1 or H_2 or both. After at most $\log|H_1| + \log|H_2|$ steps the search will be left with only one point in each hull, and these will be the desired points u_1 and u_2. Figure 10.16 illustrates the rules for eliminating part of H_1 or H_2 on each step.

We now consider the cost of the algorithm. Each step of the binary search requires only constant work and depth since we only need to consider the middle two points M_1 and M_2, which can be found in constant time if the hull is kept sorted. The cost of the full binary search to find the upper bridge is therefore bounded by $D(n) = W(n) = O(\log n)$. Once we have found the upper and lower bridges, we need to remove the points on H_1 and H_2 that are not on H and append the remaining convex hull points. This requires linear work and constant depth. The overall costs of MERGEHULL are, therefore,

$$D(n) = D(n/2) + \log n = O(\log^2 n)$$
$$W(n) = 2W(n/2) + \log n + n = O(n \log n)$$

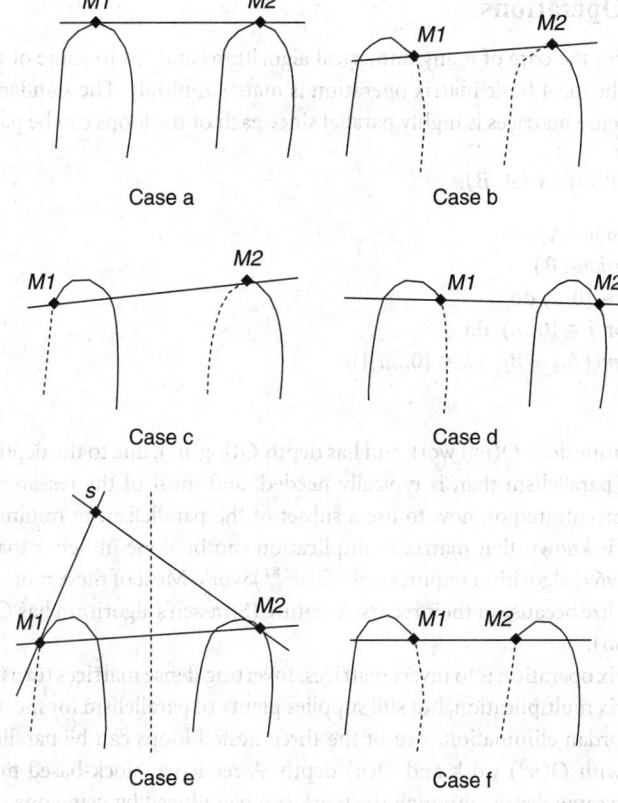

FIGURE 10.16 Cases used in the binary search for finding the upper bridge for the MergeHull. The points *M*1 and *M*2 mark the middle of the remaining hulls. The dotted lines represent the part of the hull that can be eliminated from consideration. The mirror images of cases b–e are also used. In case e, the region to eliminate depends on which side of the separating line the intersection of the tangents appears.

This algorithm can be improved to run in $O(\log n)$ depth using one of two techniques. The first involves implementing the search for the bridge points such that it runs in constant depth with linear work [Atallah and Goodrich 1988]. This involves sampling every $\sqrt{n}$th point on each hull and comparing all pairs of these two samples to narrow the search region down to regions of size $\sqrt{n}$ in constant depth. The patches then can be finished in constant depth by comparing all pairs between the two patches. The second technique [Aggarwal et al. 1988, Atallah and Goodrich 1986] uses a divide-and-conquer to separate the point set into $\sqrt{n}$ regions, solves the convex hull on each region recursively, and then merges all pairs of these regions using the binary search method. Since there are $\sqrt{n}$ regions and each of the searches takes $O(\log n)$ work, the total work for merging is $O((\sqrt{n})^2 \log n) = O(n \log n)$ and the depth is $O(\log n)$. This leads to an overall algorithm that runs in $O(n \log n)$ work and $O(\log n)$ depth.

10.8　Numerical Algorithms

There has been an immense amount of work on parallel algorithms for numerical problems. Here we briefly discuss some of the problems and results. We suggest the following sources for further information on parallel numerical algorithms: Reif [1993, Chap. 12 and Chapter 14], JáJá [1992, Chap. 8], Kumar et al. [1994, Chap. 5, Chapter 10 and Chapter 11], and Bertsekas and Tsitsiklis [1989].

10.8.1 Matrix Operations

Matrix operations form the core of many numerical algorithms and led to some of the earliest work on parallel algorithms. The most basic matrix operation is matrix multiply. The standard triply nested loop for multiplying two dense matrices is highly parallel since each of the loops can be parallelized:

Algorithm: MATRIX_MULTIPLY (A, B).

```
1   (l, m) := dimensions(A)
2   (m, n) := dimensions(B)
3   in parallel for i ∈ [0..l] do
4       in parallel for j ∈ [0..n] do
5           R_ij := sum({A_ik * B_kj : k ∈ [0..m]})
6   return R
```

If $l = m = n$, this routine does $O(n^3)$ work and has depth $O(\log(n))$, due to the depth of the summation. This has much more parallelism than is typically needed, and most of the research on parallel matrix multiplication has concentrated on how to use a subset of the parallelism to minimize communication costs. Sequentially, it is known that matrix multiplication can be done in better than $O(n^3)$ work. For example, Strassen's [1969] algorithm requires only $O(n^{2.81})$ work. Most of these more efficient algorithms are also easy to parallelize because of their recursive nature (Strassen's algorithm has $O(\log n)$ depth using a simple parallelization).

Another basic matrix operation is to invert matrices. Inverting dense matrices turns out to be somewhat less parallel than matrix multiplication, but still supplies plenty of parallelism for most practical purposes. When using Gauss–Jordan elimination, two of the three nested loops can be parallelized leading to an algorithm that runs with $O(n^3)$ work and $O(n)$ depth. A recursive block-based method using matrix multiplies leads to the same depth, although the work can be reduced by using one of the more efficient matrix multiplies.

Parallel algorithms for many other matrix operations have been studied, and there has also been significant work on algorithms for various special forms of matrices, such as tridiagonal, triangular, and general sparse matrices. Iterative methods for solving sparse linear systems have been an area of significant activity.

10.8.2 Fourier Transform

Another problem for which there has been a long history of parallel algorithms is the discrete Fourier transform (DFT). The fast Fourier transform (FFT) algorithm for solving the DFT is quite easy to parallelize and, as with matrix multiplication, much of the research has gone into reducing communication costs. In fact, the butterfly network topology is sometimes called the FFT network since the FFT has the same communication pattern as the network [Leighton 1992, Section 3.7]. A parallel FFT over complex numbers can be expressed as follows.

Algorithm: FFT(A).

```
1   n := |A|
2   if (n = 1) then return A
3   else
4       in parallel do
5           E := FFT({A[2i] : i ∈ [0..n/2)})
6           O := FFT({A[2i + 1] : i ∈ [0..n/2)})
7       return {E[j] + O[j]e^{2πij/n} : j ∈ [0..n/2)} ++ {E[j] − O[j]e^{2πij/n} : j ∈ [0..n/2)}
```

It simply calls itself recursively on the odd and even elements and then puts the results together. This algorithm does $O(n \log n)$ work, as does the sequential version, and has a depth of $O(\log n)$.

10.9   Parallel Complexity Theory

Researchers have developed a complexity theory for parallel computation that is in some ways analogous to the theory of NP-completeness. A problem is said to belong to the class NC (Nick's class) if it can be solved in depth polylogarithmic in the size of the problem using work that is polynomial in the size of the problem [Cook 1981, Pippenger 1979]. The class NC in parallel complexity theory plays the role of P in sequential complexity, i.e., the problems in NC are thought to be tractable in parallel. Examples of problems in NC include sorting, finding minimum cost spanning trees, and finding convex hulls. A problem is said to be P-complete if it can be solved in polynomial time and if its inclusion in NC would imply that $NC = P$. Hence, the notion of P-completeness plays the role of NP-completeness in sequential complexity. (And few believe that $NC = P$.)

Although much early work in parallel algorithms aimed at showing that certain problems belong to the class NC (without considering the issue of efficiency), this work tapered off as the importance of work efficiency became evident. Also, even if a problem is P-complete, there may be efficient (but not necessarily polylogarithmic time) parallel algorithms for solving it. For example, several efficient and highly parallel algorithms are known for solving the maximum flow problem, which is P-complete.

We conclude with a short list of P-complete problems. Full definitions of these problems and proofs that they are P-complete can be found in textbooks and surveys such as Gibbons and Rytter [1990], JáJá [1992], and Karp and Ramachandran [1990]. P-complete problems are:

1. **Lexicographically first maximal independent set and clique.** Given a graph G with vertices $V = 1, 2, \ldots, n$, and a subset $S \subseteq V$, determine if S is the lexicographically first maximal independent set (or maximal clique) of G.
2. **Ordered depth-first search.** Given a graph $G = (V, E)$, an ordering of the edges at each vertex, and a subset $T \subset E$, determine if T is the depth-first search tree that the sequential depth-first algorithm would construct using this ordering of the edges.
3. **Maximum flow.**
4. **Linear programming.**
5. **The circuit value problem.** Given a Boolean circuit, and a set of inputs to the circuit, determine if the output value of the circuit is one.
6. **The binary operator generability problem.** Given a set S, an element e not in S, and a binary operator·, determine if e can be generated from S using·.
7. **The context-free grammar emptiness problem.** Given a context-free grammar, determine if it can generate the empty string.

Defining Terms

CRCW: This refers to a shared memory model that allows for concurrent reads (CR) and concurrent writes (CW) to the memory.

CREW: This refers to a shared memory model that allows for concurrent reads (CR) but only exclusive writes (EW) to the memory.

Depth: The longest chain of sequential dependences in a computation.

EREW: This refers to a shared memory model that allows for only exclusive reads (ER) and exclusive writes (EW) to the memory.

Graph contraction: Contracting a graph by removing a subset of the vertices.

List contraction: Contracting a list by removing a subset of the nodes.

Multiprefix: A generalization of the scan (prefix sums) operation in which the partial sums are grouped by keys.

Multiprocessor model: A model of parallel computation based on a set of communicating sequential processors.

Pipelined divide-and-conquer: A divide-and-conquer paradigm in which partial results from recursive calls can be used before the calls complete. The technique is often useful for reducing the depth of various algorithms.

Pointer jumping: In a linked structure replacing a pointer with the pointer it points to. Used for various algorithms on lists and trees. Also called recursive doubling.

PRAM model: A multiprocessor model in which all of the processors can access a shared memory for reading or writing with uniform cost.

Prefix sums: A parallel operation in which each element in an array or linked list receives the sum of all of the previous elements.

Random sampling: Using a randomly selected sample of the data to help solve a problem on the whole data.

Recursive doubling: Same as pointer jumping.

Scan: A parallel operation in which each element in an array receives the sum of all of the previous elements.

Shortcutting: Same as pointer jumping.

Symmetry breaking: A technique to break the symmetry in a structure such as a graph which can locally look the same to all of the vertices. Usually implemented with randomization.

Tree contraction: Contracting a tree by removing a subset of the nodes.

Work: The total number of operations taken by a computation.

Work-depth model: A model of parallel computation in which one keeps track of the total work and depth of a computation without worrying about how it maps onto a machine.

Work efficient: When an algorithm does no more work than some other algorithm or model. Often used when relating a parallel algorithm to the best known sequential algorithm but also used when discussing emulations of one model on another.

References

Aggarwal, A., Chazelle, B., Guibas, L., Ò'Dùnlaing, C., and Yap, C. 1988. Parallel computational geometry. *Algorithmica* 3(3):293–327.

Aho, A. V., Hopcroft, J. E., and Ullman, J. D. 1974. *The Design and Analysis of Computer Algorithms*. Addison–Wesley, Reading, MA.

Akl, S. G. 1985. *Parallel Sorting Algorithms*. Academic Press, Toronto, Canada.

Anderson, R. J. and Miller, G. L. 1988. Deterministic parallel list ranking. In *Aegean Workshop on Computing: VLSI Algorithms and Architectures*. J. Reif, ed. Vol. 319, Lecture notes in computer science, pp. 81–90. Springer–Verlag, New York.

Anderson, G. L. and Miller, G. L. 1990. A simple randomized parallel algorithm for list-ranking. *Inf. Process. Lett.* 33(5):269–273.

Atallah, M. J., Cole, R., and Goodrich, M. T. 1989. Cascading divide-and-conquer: a technique for designing parallel algorithms. *SIAM J. Comput.* 18(3):499–532.

Atallah, M. J. and Goodrich, M. T. 1986. Efficient parallel solutions to some geometric problems. *J. Parallel Distrib. Comput.* 3(4):492–507.

Atallah, M. J. and Goodrich, M. T. 1988. Parallel algorithms for some functions of two convex polygons. *Algorithmica* 3(4):535–548.

Awerbuch, B. and Shiloach, Y. 1987. New connectivity and MSF algorithms for shuffle-exchange network and PRAM. *IEEE Trans. Comput.* C-36(10):1258–1263.

Bar-Noy, A. and Kipnis, S. 1992. Designing broadcasting algorithms in the postal model for message-passing systems, pp. 13–22. In *Proc. 4th Annu. ACM Symp. Parallel Algorithms Architectures*. ACM Press, New York.

Beneš, V. E. 1965. *Mathematical Theory of Connecting Networks and Telephone Traffic*. Academic Press, New York.

Bentley, J. L. and Shamos, M. 1976. Divide-and-conquer in multidimensional space, pp. 220–230. In *Proc. ACM Symp. Theory Comput.* ACM Press, New York.

Bertsekas, D. P. and Tsitsiklis, J. N. 1989. *Parallel and Distributed Computation: Numerical Methods.* Prentice–Hall, Englewood Cliffs, NJ.

Blelloch, G. E. 1990. *Vector Models for Data-Parallel Computing.* MIT Press, Cambridge, MA.

Blelloch, G. E. 1996. Programming parallel algorithms. *Commun. ACM* 39(3):85–97.

Blelloch, G. E., Chandy, K. M., and Jagannathan, S., eds. 1994. *Specification of Parallel Algorithms.* Vol. 18, DIMACS series in discrete mathematics and theoretical computer science. American Math. Soc. Providence, RI.

Blelloch, G. E. and Greiner, J. 1995. Parallelism in sequential functional languages, pp. 226–237. In *Proc. ACM Symp. Functional Programming Comput. Architecture.* ACM Press, New York.

Blelloch, G. E., Leiserson, C. E., Maggs, B. M., Plaxton, C. G., Smith, S. J., and Zagha, M. 1991. A comparison of sorting algorithms for the connection machine CM-2, pp. 3–16. In *Proc. ACM Symp. Parallel Algorithms Architectures.* Hilton Head, SC, July. ACM Press, New York.

Blelloch, G. E. and Little, J. J. 1994. Parallel solutions to geometric problems in the scan model of computation. *J. Comput. Syst. Sci.* 48(1):90–115.

Brent, R. P. 1974. The parallel evaluation of general arithmetic expressions. *J. Assoc. Comput. Mach.* 21(2):201–206.

Chan, T. M. Y., Snoeyink, J., and Yap, C. K. 1995. Output-sensitive construction of polytopes in four dimensions and clipped Voronoi diagrams in three, pp. 282–291. In *Proc. 6th Annu. ACM–SIAM Symp. Discrete Algorithms.* ACM–SIAM, ACM Press, New York.

Cole, R. 1988. Parallel merge sort. *SIAM J. Comput.* 17(4):770–785.

Cook, S. A. 1981. Towards a complexity theory of synchronous parallel computation. *Enseignement Mathematique* 27:99–124.

Culler, D., Karp, R., Patterson, D., Sahay, A., Schauser, K. E., Santos, E., Subramonian, R., and von Eicken, T. 1993. LogP: towards a realistic model of parallel computation, pp. 1–12. In *Proc. 4th ACM SIGPLAN Symp. Principles Pract. Parallel Programming.* ACM Press, New York.

Cypher, R. and Sanz, J. L. C. 1994. *The SIMD Model of Parallel Computation.* Springer–Verlag, New York.

Fortune, S. and Wyllie, J. 1978. Parallelism in random access machines, pp. 114–118. In *Proc. 10th Annu. ACM Symp. Theory Comput.* ACM Press, New York.

Gazit, H. 1991. An optimal randomized parallel algorithm for finding connected components in a graph. *SIAM J. Comput.* 20(6):1046–1067.

Gibbons, P. B., Matias, Y., and Ramachandran, V. 1994. The QRQW PRAM: accounting for contention in parallel algorithms, pp. 638–648. In *Proc. 5th Annu. ACM–SIAM Symp. Discrete Algorithms.* Jan. ACM Press, New York.

Gibbons, A. and Ritter, W. 1990. *Efficient Parallel Algorithms.* Cambridge University Press, Cambridge, England.

Goldshlager, L. M. 1978. A unified approach to models of synchronous parallel machines, pp. 89–94. In *Proc. 10th Annu. ACM Symp. Theory Comput.* ACM Press, New York.

Goodrich, M. T. 1996. Parallel algorithms in geometry. In *CRC Handbook of Discrete and Computational Geometry.* CRC Press, Boca Raton, FL.

Greiner, J. 1994. A comparison of data-parallel algorithms for connected components, pp. 16–25. In *Proc. 6th Annu. ACM Symp. Parallel Algorithms Architectures.* June. ACM Press, New York.

Halperin, S. and Zwick, U. 1994. An optimal randomized logarithmic time connectivity algorithm for the EREW PRAM, pp. 1–10. In *Proc. ACM Symp. Parallel Algorithms Architectures.* June. ACM Press, New York.

Harris, T. J. 1994. A survey of pram simulation techniques. *ACM Comput. Surv.* 26(2):187–206.

Huang, J. S. and Chow, Y. C. 1983. Parallel sorting and data partitioning by sampling, pp. 627–631. In *Proc. IEEE Comput. Soc. 7th Int. Comput. Software Appl. Conf.* Nov.

JáJá, J. 1992. *An Introduction to Parallel Algorithms.* Addison–Wesley, Reading, MA.

Karlin, A. R. and Upfal, E. 1988. Parallel hashing: an efficient implementation of shared memory. *J. Assoc. Comput. Mach.* 35:876–892.

Karp, R. M. and Ramachandran, V. 1990. Parallel algorithms for shared memory machines. In *Handbook of Theoretical Computer Science — Volume A: Algorithms and Complexity*. J. Van Leeuwen, ed. MIT Press, Cambridge, MA.

Kirkpatrick, D. G. and Seidel, R. 1986. The ultimate planar convex hull algorithm? *SIAM J. Comput.* 15:287–299.

Klein, P. N. and Tarjan, R. E. 1994. A randomized linear-time algorithm for finding minimum spanning trees. In *Proc. ACM Symp. Theory Comput.* May. ACM Press, New York.

Knuth, D. E. 1973. *Sorting and Searching. Vol. 3. The Art of Computer Programming.* Addison–Wesley, Reading, MA.

Kogge, P. M. and Stone, H. S. 1973. A parallel algorithm for the efficient solution of a general class of recurrence equations. *IEEE Trans. Comput.* C-22(8):786–793.

Kumar, V., Grama, A., Gupta, A., and Karypis, G. 1994. *Introduction to Parallel Computing: Design and Analysis of Algorithms.* Benjamin Cummings, Redwood City, CA.

Leighton, F. T. 1992. *Introduction to Parallel Algorithms and Architectures: Arrays, Trees, and Hypercubes.* Morgan Kaufmann, San Mateo, CA.

Leiserson, C. E. 1985. Fat-trees: universal networks for hardware-efficient supercomputing. *IEEE Trans. Comput.* C-34(10):892–901.

Luby, M. 1985. A simple parallel algorithm for the maximal independent set problem, pp. 1–10. In *Proc. ACM Symp. Theory Comput.* May. ACM Press, New York.

Maon, Y., Schieber, B., and Vishkin, U. 1986. Parallel ear decomposition search (eds) and st-numbering in graphs. *Theor. Comput. Sci.* 47:277–298.

Matias, Y. and Vishkin, U. 1991. On parallel hashing and integer sorting. *J. Algorithms* 12(4):573–606.

Miller, G. L. and Ramachandran, V. 1992. A new graph triconnectivity algorithm and its parallelization. *Combinatorica* 12(1):53–76.

Miller, G. and Reif, J. 1989. Parallel tree contraction part 1: fundamentals. In *Randomness and Computation. Vol. 5. Advances in Computing Research*, pp. 47–72. JAI Press, Greenwich, CT.

Miller, G. L. and Reif, J. H. 1991. Parallel tree contraction part 2: further applications. *SIAM J. Comput.* 20(6):1128–1147.

Overmars, M. H. and Van Leeuwen, J. 1981. Maintenance of configurations in the plane. *J. Comput. Syst. Sci.* 23:166–204.

Padua, D., Gelernter, D., and Nicolau, A., eds. 1990. *Languages and Compilers for Parallel Computing Research Monographs in Parallel and Distributed.* MIT Press, Cambridge, MA.

Pippenger, N. 1979. On simultaneous resource bounds, pp. 307–311. In *Proc. 20th Annu. Symp. Found. Comput. Sci.*

Pratt, V. R. and Stockmeyer, L. J. 1976. A characterization of the power of vector machines. *J. Comput. Syst. Sci.* 12:198–221.

Preparata, F. P. and Shamos, M. I. 1985. *Computational Geometry — An Introduction.* Springer–Verlag, New York.

Ranade, A. G. 1991. How to emulate shared memory. *J. Comput. Syst. Sci.* 42(3):307–326.

Reid-Miller, M. 1994. List ranking and list scan on the Cray C-90, pp. 104–113. In *Proc. 6th Annu. ACM Symp. Parallel Algorithms Architectures.* June. ACM Press, New York.

Reif, J. H., ed. 1993. *Synthesis of Parallel Algorithms.* Morgan Kaufmann, San Mateo, CA.

Reif, J. H. and Valiant, L. G. 1983. A logarithmic time sort for linear size networks, pp. 10–16. In *Proc. 15th Annu. ACM Symp. Theory Comput.* April. ACM Press, New York.

Savitch, W. J. and Stimson, M. 1979. Time bounded random access machines with parallel processing. *J. Assoc. Comput. Mach.* 26:103–118.

Shiloach, Y. and Vishkin, U. 1981. Finding the maximum, merging and sorting in a parallel computation model. *J. Algorithms* 2(1):88–102.

Shiloach, Y. and Vishkin, U. 1982. An $O(\log n)$ parallel connectivity algorithm. *J. Algorithms* 3:57–67.

Stone, H. S. 1975. Parallel tridiagonal equation solvers. *ACM Trans. Math. Software* 1(4):289–307.

Strassen, V. 1969. Gaussian elimination is not optimal. *Numerische Mathematik* 14(3):354–356.

Tarjan, R. E. and Vishkin, U. 1985. An efficient parallel biconnectivity algorithm. *SIAM J. Comput.* 14(4):862–874.

Valiant, L. G. 1990a. A bridging model for parallel computation. *Commun. ACM* 33(8):103–111.

Valiant, L. G. 1990b. General purpose parallel architectures, pp. 943–971. In *Handbook of Theoretical Computer Science.* J. van Leeuwen, ed. Elsevier Science, B. V., Amsterdam, The Netherlands.

Vishkin, U. 1984. Parallel-design distributed-implementation (PDDI) general purpose computer. *Theor. Comp. Sci.* 32:157–172.

Wyllie, J. C. 1979. The Complexity of Parallel Computations. *Department of Computer Science, Tech. Rep.* TR-79-387, Cornell University, Ithaca, NY. Aug.

11

Computational Geometry

11.1 Introduction 11-1
11.2 Problem Solving Techniques 11-2
 Incremental Construction • Plane Sweep
 • Geometric Duality • Locus • Divide-and-Conquer
 • Prune-and-Search • Dynamization • Random Sampling
11.3 Classes of Problems 11-10
 Convex Hull • Proximity • Point Location • Motion
 Planning: Path Finding Problems • Geometric Optimization
 • Decomposition • Intersection • Geometric Searching
11.4 Conclusion 11-26

D. T. Lee
Northwestern University

11.1 Introduction

Computational geometry evolves from the classical discipline of design and analysis of algorithms, and has received a great deal of attention in the past two decades since its identification in 1975 by Shamos. It is concerned with the computational complexity of geometric problems that arise in various disciplines such as pattern recognition, computer graphics, computer vision, robotics, very large-scale integrated (VLSI) layout, operations research, statistics, etc. In contrast with the classical approach to proving mathematical theorems about geometry-related problems, this discipline emphasizes the computational aspect of these problems and attempts to exploit the underlying geometric properties possible, e.g., the metric space, to derive efficient algorithmic solutions.

The classical theorem, for instance, that a set S is convex if and only if for any $0 \leq \alpha \leq 1$ the convex combination $\alpha p + (1 - \alpha)q = r$ is in S for any pair of elements $p, q \in S$, is very fundamental in establishing convexity of a set. In geometric terms, a body S in the Euclidean space is convex if and only if the line segment joining any two points in S lies totally in S. But this theorem per se is not suitable for computational purposes as there are infinitely many possible pairs of points to be considered. However, other properties of convexity can be utilized to yield an algorithm. Consider the following problem. Given a simple closed Jordan polygonal curve, determine if the interior region enclosed by the curve is convex. This problem can be readily solved by observing that if the line segments defined by all pairs of vertices of the polygonal curve, $\overline{v_i, v_j}, i \neq j, 1 \leq i, j \leq n$, where n denotes the total number of vertices, lie totally inside the region, then the region is convex. This would yield a straightforward algorithm with time complexity $O(n^3)$, as there are $O(n^2)$ line segments, and to test if each line segment lies totally in the region takes $O(n)$ time by comparing it against every polygonal segment. As we shall show, this problem can be solved in $O(n)$ time by utilizing other geometric properties.

At this point, an astute reader might have come up with an $O(n)$ algorithm by making the observation: Because the interior angle of each vertex must be strictly less than π in order for the region to be convex, we just have to check for every consecutive three vertices v_{i-1}, v_i, v_{i+1} that the angle at vertex v_i is less than π. (A vertex whose internal angle has a measure less than π is said to be *convex*; otherwise, it is said to be *reflex*.) One may just be content with this solution. Mathematically speaking, this solution is fine and indeed runs in $O(n)$ time. The problem is that the algorithm implemented in this straightforward manner without care may produce an incorrect answer when the input polygonal curve is ill formed. That is, if the input polygonal curve is not simple, i.e., it self-intersects, then the *enclosed* region by this closed curve is not well defined. The algorithm, without checking this simplicity condition, may produce a wrong answer. Note that the preceding observation that all of the vertices must be convex in order to have a convex region is only a necessary condition. Only when the input polygonal curve is *verified* to be simple will the algorithm produce a correct answer. But to verify whether the input polygonal curve self-intersects or not is no longer as straightforward. The fact that we are dealing with computer solutions to geometric problems may make the task of designing an algorithm and proving its correctness nontrivial.

An objective of this discipline in the theoretical context is to prove lower bounds of the complexity of geometric problems and to devise algorithms (giving upper bounds) whose complexity *matches* the lower bounds. That is, we are interested in the *intrinsic* difficulty of geometric computational problems under a certain computation model and at the same time are concerned with the algorithmic solutions that are provably optimal in the worst or average case. In this regard, the asymptotic time (or space) complexity of an algorithm is of interest. Because of its applications to various science and engineering related disciplines, researchers in this field have begun to address the efficacy of the algorithms, the issues concerning robustness and numerical stability [Fortune 1993], and the actual running times of their implementions.

In this chapter, we concentrate mostly on the theoretical development of this field in the context of sequential computation. Parallel computation geometry is beyond the scope of this chapter. We will adopt the *real* random access machine (RAM) model of computation in which all arithmetic operations, comparisons, kth-root, exponential or logarithmic functions take unit time. For more details refer to Edelsbrunner [1987], Mulmuley [1994], and Preparata and Shamos [1985]. We begin with a summary of problem solving techniques that have been developed [Lee and Preparata 1982, O'Rourke 1994, Yao 1994] and then discuss a number of topics that are central to this field, along with additional references for further reading about these topics.

11.2 Problem Solving Techniques

We give an example for each of the eight major problem-solving paradigms that are prevalent in this field. In subsequent sections we make reference to these techniques whenever appropriate.

11.2.1 Incremental Construction

This is the simplest and most intuitive method, also known as *iterative method*. That is, we compute the solution in an iterative manner by considering the input incrementally.

Consider the problem of computing the line arrangements in the plane. Given is a set $\mathcal{L}$ of n straight lines in the plane, and we want to compute the partition of the plane induced by $\mathcal{L}$. One obvious approach is to compute the partition iteratively by considering one line at a time [Chazelle et al. 1985]. As shown in Figure 11.1, when line i is inserted, we need to traverse the regions that are intersected by the line and construct the new partition at the same time. One can show that the traversal and repartitioning of the intersected regions can be done in $O(n)$ time per insertion, resulting in a total of $O(n^2)$ time. This algorithm is asymptotically optimal because the running time is proportional to the amount of space required to represent the partition. This incremental approach also generalizes to higher dimensions. We conclude with the theorem [Edelsbrunner et al. 1986].

Theorem 11.1 *The problem of computing the arrangement $\mathcal{A}(H)$ of a set H of n hyperplanes in $\Re^k$ can be solved iteratively in $O(n^k)$ time and space, which is optimal.*

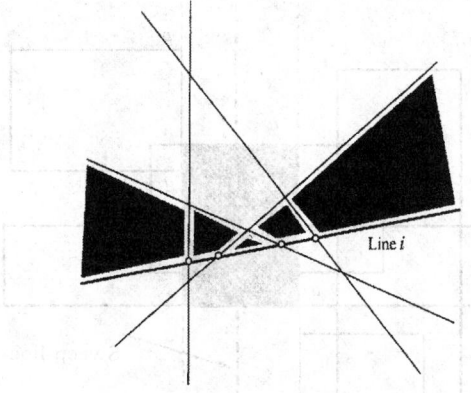

FIGURE 11.1 Incremental construction of line arrangement: phase i.

11.2.2 Plane Sweep

This approach works most effectively for two-dimensional problems for which the solution can be computed incrementally as the entire input is scanned in a certain order. The concept can be easily generalized to higher dimensions [Bieri and Nef 1982]. This is also known as the *scan-line* method in computer graphics and is used for a variety of applications such as shading and polygon filling, among others.

Consider the problem of computing the *measure* of the union of n isothetic rectangles, i.e., whose sides are parallel to the coordinate axes. We would proceed with a *vertical* sweep line, sweeping across the plane from left to right. As we sweep the plane, we need to keep track of the rectangles that intersect the current sweep line and those that are yet to be visited. In the meantime we compute the area covered by the union of the rectangles seen so far. More formally, associated with this approach there are two basic data structures containing all *relevant* information that should be maintained.

1. *Event schedule* defines a sequence of *event points* that the sweep-line status will change. In this example, the sweep-line status will change only at the left and right boundary edges of each rectangle.
2. *Sweep-line status* records the information of the geometric structure that is being swept. In this example the sweep-line status keeps track of the set of rectangles intersecting the current sweep line.

The event schedule is normally represented by a *priority queue*, and the list of events may change dynamically. In this case, the events are static; they are the x-coordinates of the left and right boundary edges of each rectangle. The sweep-line status is represented by a suitable data structure that supports insertions, deletions, and computation of the partial solution at each event point. In this example a *segment tree* attributed to Bentley is sufficient [Preparata and Shamos 1985]. Because we are computing the area of the rectangles, we need to be able to know the *new* area covered by the current sweep line between two adjacent event points. Suppose at event point x_{i-1} we maintain a partial solution $\mathcal{A}_{i-1}$. In Figure 11.2 the shaded area S needs to be added to the partial solution, that is, $\mathcal{A}_i = \mathcal{A}_{i-1} + S$. The shaded area is equal to the total measure, denoted sum_ℓ, of the union of vertical line segments representing the intersection of the rectangles and the current sweep line times the distance between the two event points x_i and x_{i-1}. If the next event corresponds to the left boundary of a rectangle, the corresponding vertical segment, $\overline{p,q}$ in Figure 11.2, needs to be inserted to the segment tree. If the next event corresponds to a right boundary edge, the segment, $\overline{u,v}$ needs to be deleted from the segment tree. In either case, the total measure sum_ℓ should be updated accordingly. The correctness of this algorithm can be established by observing that the partial solution obtained for the rectangles to the *left* of the sweep line is maintained correctly. In fact, this property is typical of any algorithm based on the plane-sweep technique.

Because the segment tree structure supports segment insertions and deletions and the update (of sum_ℓ) operation in $O(\log n)$ time per event point, the total amount of time needed is $O(n \log n)$.

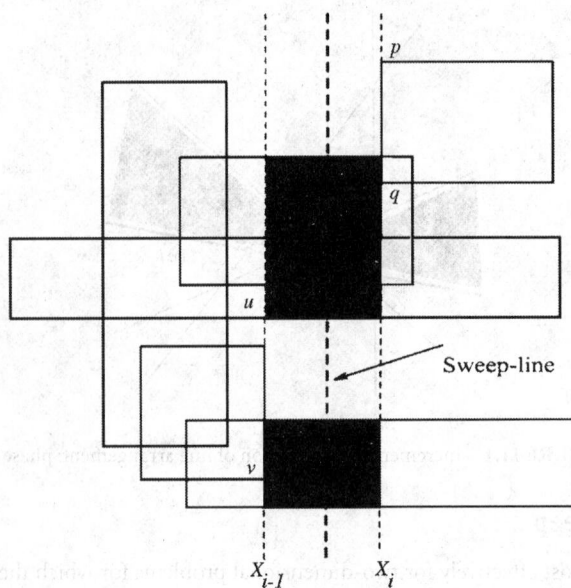

FIGURE 11.2   The plane-sweep approach to the measure problem in two dimensions.

The measure of the union of rectangles in higher dimensions also can be solved by the plane-sweep technique with quad trees, a generalization of segment trees.

Theorem 11.2   *The problem of computing the measure of n isothetic rectangles in k dimensions can be solved in $O(n \log n)$ time, for $k \leq 2$ and in $O(n^{k-1})$ time for $k \geq 3$.*

The time bound is asymptotically optimal. Even in one dimension, i.e., computing the total length of the union of n intervals requires $\Omega(n \log n)$ time (see Preparata and Shamos [1985]).

We remark that the sweep line used in this approach is not necessarily a straight line. It can be a topological line as long as the objects stored in the sweep line status are ordered, and the method is called *topological sweep* [Asano et al. 1994, Edelsbrunner and Guibas 1989]. Note that the measure of isothetic rectangles can also be solved using the divide-and-conquer paradigm to be discussed.

11.2.3   Geometric Duality

This is a geometric transformation that maps a given problem into its equivalent form, preserving certain geometric properties so as to manipulate the objects in a more convenient manner. We will see its usefulness for a number of problems to be discussed. Here let us describe a transformation in k-dimensions, known as *polarity* or *duality*, denoted $\mathcal{D}$, that maps d-dimensional varieties to $(k - 1 - d)$-dimensional varieties, $0 \leq d < k$.

Consider any point $p = (\pi_1, \pi_2, \ldots, \pi_k) \in \Re^k$ other than the origin. The dual of p, denoted $\mathcal{D}(p)$, is the hyperplane $\pi_1 x_1 + \pi_2 x_2 + \cdots + \pi_k x_k = 1$. Similarly, a hyperplane that does not contain the origin is mapped to a point such that $\mathcal{D}(\mathcal{D}(p)) = p$. Geometrically speaking, point p is mapped to a hyperplane whose normal is the vector determined by p and the origin and whose distance to the origin is the reciprocal of that between p and the origin. Let S denote the unit sphere $S : x_1^2 + x_2^2 + \cdots + x_k^2 = 1$. If point p is external to S, then it is mapped to a hyperplane $\mathcal{D}(p)$ that intersects S at those points q that admit supporting hyperplanes h such that $h \cap S = q$ and $p \in h$. In two dimensions a point p outside of the unit disk will be mapped to a line intersecting the disk at two points, q_1 and q_2, such that line segments $\overline{p, q_1}$ and $\overline{p, q_2}$ are tangent to the disk. Note that the distances from p to the origin and from the line $\mathcal{D}(p)$ to the origin are reciprocal to each other. Figure 11.3a shows the duality transformation in two dimensions. In

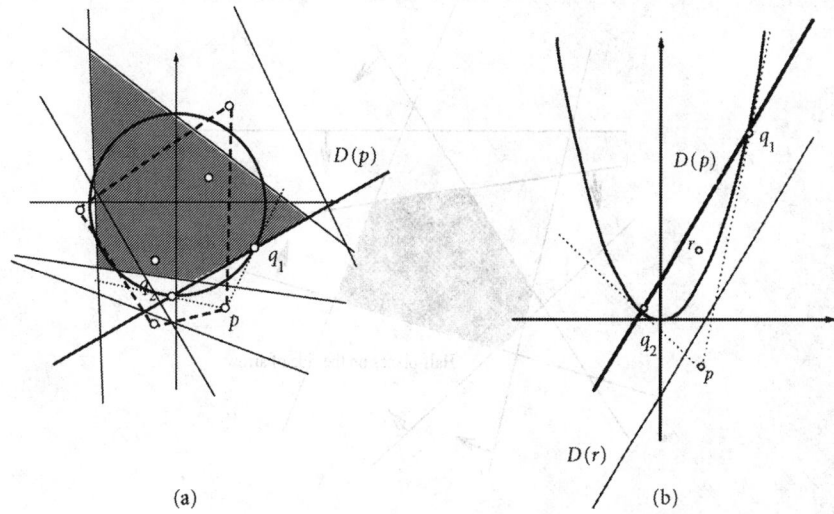

(a)   (b)

FIGURE 11.3   Geometric duality transformation in two dimensions.

particular, point p is mapped to the line shown in boldface. For each hyperplane $\mathcal{D}(p)$, let $\mathcal{D}(p)^+$ denote the half-space that contains the origin and let $\mathcal{D}(p)^-$ denote the other half-space.

The duality transformation not only leads to dual arrangements of hyperplanes and configurations of points and vice versa, but also preserves the following properties.

Incidence: Point p belongs to hyperplane h if and only if point $\mathcal{D}(\langle\rangle)$ belongs to hyperplane $\mathcal{D}(p)$.
Order: Point p lies in half-space h^+ (respectively, h^-) if and only if point $\mathcal{D}(\langle\rangle)$ lies in half-space $\mathcal{D}(p)^+$ (respectively, $\mathcal{D}(p)^-$).

Figure 11.3a shows the convex hull of a set of points that are mapped by the duality transformation to the shaded region, which is the common intersection of the half-planes $\mathcal{D}(p)^+$ for all points p.

Another transformation using the unit paraboloid U, represented as $U : x_k = x_1^2 + x_2^2 + \cdots + x_{k-1}^2$, can also be similarly defined. That is, point $p = (\pi_1, \pi_2, \ldots, \pi_k) \in R^k$ is mapped to a hyperplane $\mathcal{D}_{\sqcap}(\sqrt{})$ represented by the equation $x_k = 2\pi_1 x_1 + 2\pi_2 x_2 + \cdots + 2\pi_{k-1} x_{k-1} - \pi_k$. And each nonvertical hyperplane is mapped to a point in a similar manner such that $\mathcal{D}_u(\mathcal{D}_u(p)) = p$. Figure 11.3b illustrates the two-dimensional case, in which point p is mapped to a line shown in boldface. For more details see, e.g., Edelsbrunner [1987] and Preparata and Shamos [1985].

11.2.4   Locus

This approach is often used as a preprocessing step for a geometric *searching* problem to achieve faster query-answering response time. For instance, given a *fixed* database consisting of geographical locations of post offices, each represented by a point in the plane, one would like to be able to efficiently answer queries of the form: "what is the nearest post office to location q?" for some query point q. The locus approach to this problem is to partition the plane into n regions, each of which consists of the locus of *query* points for which the *answer* is the same. The partition of the plane is the so-called *Voronoi* diagram discussed subsequently. In Figure 11.7, the post office closest to query point q is site s_i. Once the Voronoi diagram is available, the query problem reduces to that of locating the region that contains the query, an instance of the point-location problem discussed in Section 11.3.

11.2.5   Divide-and-Conquer

This is a classic problem-solving technique and has proven to be very powerful for geometric problems as well. This technique normally involves partitioning of the given problem into several subproblems,

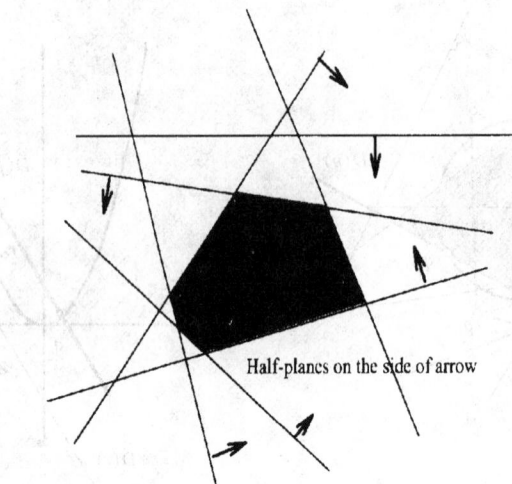

Half-planes on the side of arrow

FIGURE 11.4 The common intersection of half-planes.

recursively solving each subproblem, and then combining the solutions to each of the subproblems to obtain the final solution to the original problem. We illustrate this paradigm by considering the problem of computing the common intersection of n half-planes in the plane. Given is a set S of n half-planes, h_i, represented by $a_i x + b_i y \leq c_i, i = 1, 2, \ldots, n$. It is well known that the common intersection of half-planes, denoted $CI(S) = \bigcap_{i=1}^{n} h_i$, is a convex set, which may or may not be bounded. If it is bounded, it is a convex polygon. See Figure 11.4, in which the shaded area is the common intersection.

The divide-and-conquer paradigm consists of the following steps.

Algorithm Common_Intersection_D&C (S)

1. *If* $|S| \leq 3$, *compute the intersection* $CI(S)$ *explicitly.* **Return** $(CI(S))$.
2. *Divide S into two approximately equal subsets S_1 and S_2.*
3. $CI(S_1) = $ **Common_Intersection_D&C**(S_1).
4. $CI(S_2) = $ **Common_Intersection_D&C**(S_2).
5. $CI(S) = $ **Merge**$(CI(S_1), CI(S_2))$.
6. **Return** $(CI(S))$.

The key step is the *merge* of two common intersections. Because $CI(S_1)$ and $CI(S_2)$ are convex, the merge step basically calls for the computation of the intersection of two convex polygons, which can be solved in time proportional to the size of the polygons (cf. subsequent section on intersection). The running time of the divide-and-conquer algorithm is easily shown to be $O(n \log n)$, as given by the following recurrence formula, where $n = |S|$:

$$T(3) = O(1)$$
$$T(n) = 2T\left(\frac{n}{2}\right) + O(n) + M\left(\frac{n}{2}, \frac{n}{2}\right)$$

where $M(n/2, n/2) = O(n)$ denotes the merge time (step 5).

Theorem 11.3 *The common intersection of n half-planes can be solved in $O(n \log n)$ time by the divide-and-conquer method.*

The time complexity of the algorithm is asymptotically optimal, as the problem of sorting can be reduced to it [Preparata and Shamos 1985].

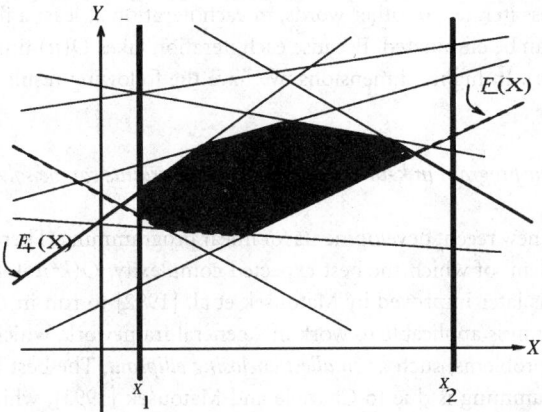

FIGURE 11.5 Feasible region defined by upward- and downward-convex piecewise linear functions.

11.2.6 Prune-and-Search

This approach, developed by Dyer [1986] and Megiddo [1983a, 1983b, 1984], is a very powerful method for solving a number of geometric optimization problems, one of which is the well-known linear programming problem. Using this approach, they obtained an algorithm whose running time is linear in the number of constraints. For more development of linear programming problems, see Megiddo [1983c, 1986]. The main idea is to prune away a fraction of *redundant* input constraints in each iteration while searching for the solution. We use a two-dimensional linear programming problem to illustrate this approach. Without loss of generality, we consider the following linear programming problem:

$$\text{Minimize} \quad Y$$
$$\text{subject to} \quad \alpha_i X + \beta_i Y + \gamma_i \leq 0, \quad i = 1, 2, \ldots, n$$

These n constraints are partitioned into three classes, C_0, C_+, C_-, depending on whether β_i is zero, positive, or negative, respectively. The constraints in class C_0 define an X-interval $[x_1, x_2]$, which constrains the solution, if any. The constraints in classes C_+ and C_- define, however, upward- and downward-convex piecewise linear functions $F_+(X)$ and $F_-(X)$ delimiting the feasible region* (Figure 11.5). The problem now becomes

$$\text{Minimize} \quad F_-(X)$$
$$\text{subject to} \quad F_-(X) \leq F_+(X)$$
$$x_1 \leq X \leq x_2$$

Let λ^* denote the optimal solution, if it exists. The values of $F_-(\lambda)$ and $F_+(\lambda)$ for any λ can be computed in $O(n)$ time, based on the slopes $-\alpha_i/\beta_i$. Thus, in $O(n)$ time one can determine for any $\lambda' \in [x_1, x_2]$ if (1) λ' is infeasible, and there is no solution, (2) λ' is infeasible, and we know a feasible solution is less or greater than λ', (3) $\lambda' = \lambda^*$, or (4) λ' is feasible, and whether λ^* is less or greater than λ'.

To choose λ' we partition constraints in classes C_- and C_+ into pairs and find the abscissa $\lambda_{i,j}$ of their intersection. If $\lambda_{i,j} \notin [x_1, x_2]$ then one of the constraints can be eliminated as redundant. For those $\lambda_{i,j}$ that are in $[x_1, x_2]$ we find in $O(n)$ time [Dobkin and Munro 1981] the median $\lambda'_{i,j}$ and compute $F_-(\lambda'_{i,j})$ and $F_+(\lambda'_{i,j})$. By the preceding arguments that we can determine where λ^* should lie, we know one-half of the $\lambda_{i,j}$ do not lie in the region containing λ^*. Therefore, one constraint of the corresponding pair can

---

*These upward- and downward-convex functions are also known as the upper and lower *envelopes* of the line arrangements for lines belonging to classes C_- and C_+, respectively.

be eliminated. The process iterates. In other words, in each iteration at least a fixed fraction $\delta = 1/4$ of the current constraints can be eliminated. Because each iteration takes $O(n)$ time, the total time spent is $Cn + C\delta n + \cdots = O(n)$. In higher dimensions, we have the following result due to Dyer [1986] and Clarkson [1986].

Theorem 11.4 *A linear program in k-dimensions with n constraints can be solved in $O(3^{k^2}n)$ time.*

We note here some of the new recent developments for linear programming. There are several randomized algorithms for this problem, of which the best expected complexity, $O(k^2n + k^{k/2+O(1)} \log n)$ is due to Clarkson [1988], which is later improved by Matoušek et al. [1992] to run in $O(k^2n + e^{O(\sqrt{k \ln k})} \log n)$. Clarkson's [1988] algorithm is applicable to work in a general framework, which includes various other geometric optimization problems, such as *smallest enclosing ellipsoid*. The best known deterministic algorithm for linear programming is due to Chazelle and Matoušek [1993], which runs in $O(k^{7k+o(k)}n)$ time.

11.2.7 Dynamization

Techniques have been developed for query-answering problems, classified as *geometric searching* problems, in which the underlying database is changing over (discrete) time. A typical geometric searching problem is the *membership* problem, i.e., given a set $\mathcal{D}$ of objects, determine if x is a member of $\mathcal{D}$, or the *nearest neighbor searching* problem, i.e., given a set $\mathcal{D}$ of objects, find an object that is closest to x according to some distance measure. In the database area, these two problems are referred to as the *exact match* and *best match* queries. The idea is to make use of good data structures for a static database and enhance them with dynamization mechanisms so that updates of the database can be accommodated on line and yet queries to the database can be answered efficiently.

A general query Q contains a variable of type $T1$ and is asked of a set of objects of type $T2$. The answer to the query is of type $T3$. More formally, Q can be considered as a mapping from $T1$ and subsets of $T2$ to $T3$, that is, $Q: T1 \times 2^{T2} \rightarrow T3$. The class of geometric searching problems to which the dynamization techniques are applicable is the class of *decomposable searching problems* [Bentley and Saxe 1980].

Definition 11.1 A searching problem with query Q is decomposable if there exists an efficiently computable associative, and communtative binary operator @ satisfying the condition

$$Q(x, A \cup B) = @(Q(x, A), Q(x, B))$$

In other words, the answer to a query Q in $\mathcal{D}$ can be computed by the answers to two subsets $\mathcal{D}_\infty$ and $\mathcal{D}_\epsilon$ of $\mathcal{D}$. The membership problem and the nearest-neighbor searching problem previously mentioned are decomposable.

To answer queries efficiently, we have a data structure to support various update operations. There are typically three measures to evaluate a static data structure $\mathcal{A}$. They are:

1. $P_{\mathcal{A}}(N)$, the preprocessing time required to build $\mathcal{A}$
2. $S_{\mathcal{A}}(N)$, the storage required to represent $\mathcal{A}$
3. $Q_{\mathcal{A}}(N)$, the query response time required to search in $\mathcal{A}$

where N denotes the number of elements represented in $\mathcal{A}$. One would add another measure $U_{\mathcal{A}}(N)$ to represent the *update* time.

Consider the nearest-neighbor searching problem in the Euclidean plane. Given a set of n points in the plane, we want to find the nearest neighbor of a query point x. One can use the Voronoi diagram data structure $\mathcal{A}$ (cf. subsequent section on Voronoi diagrams) and point location scheme (cf. subsequent section on point location) to achieve the following: $P_{\mathcal{A}}(n) = O(n \log n)$, $S_{\mathcal{A}}(n) = O(n)$, and $Q_{\mathcal{A}}(n) = O(\log n)$. We now convert the static data structure $\mathcal{A}$ to a dynamic one, denoted $\mathcal{D}$, to support insertions and deletions as well. There are a number of dynamization techniques, but we describe the technique developed by van Leeuwan and Wood [1980] that provides the general flavor of the approach.

The general principle is to decompose $\mathcal{A}$ into a collection of separate data structures so that each update can be confined to one or a small, fixed number of them; however, to avoid degrading the query response time we cannot afford to have excessive fragmentation because queries involve the entire collection.

Let $\{x_k\}_{k \geq 1}$ be a sequence of increasing integers, called *switch points*, where x_k is divisible by k and $x_{k+1}/(k+1) > x_k/k$. Let $x_0 = 0$, $y_k = x_k/k$, and n denote the current size of the point set. For a given *level* k, $\mathcal{D}$ consists of $(k+1)$ static structures of the same type, one of which, called *dump* is designated to allow for insertions. Each substructure $\mathcal{B}$ has size $y_k \leq s(\mathcal{B}) \leq y_{k+1}$, and the dump has size $0 \leq s(dump) < y_{k+1}$. A block $\mathcal{B}$ is called *low* or *full* depending on whether $s(\mathcal{B}) = y_k$ or $s(\mathcal{B}) = y_{k+1}$, respectively, and is called *partial* otherwise. When an insertion to the dump makes its size equal to y_{k+1}, it becomes a full block and any nonfull block can be used as the dump. If all blocks are full, we switch to the next level. Note that at this point the total size is $y_{k+1} * (k+1) = x_{k+1}$. That is, at the beginning of level $k+1$, we have $k+1$ low blocks and we create a new dump, which has size 0. When a deletion from a low block occurs, we need to borrow an element either from the dump, if it is not empty, or from a partial block. When all blocks are low and $s(dump) = 0$, we switch to level $k-1$, making the low block from which the latest deletion occurs the *dump*. The level switching can be performed in $O(1)$ time. We have the following:

Theorem 11.5 *Any static data structure $\mathcal{A}$ used for a decomposable searching problem can be transformed into a dynamic data structure $\mathcal{D}$ for the same problem with the following performance. For $x_k \leq n < x_{k+1}$, $Q_{\mathcal{D}}(n) = O(k Q_{\mathcal{A}}(y_{k+1}))$, $U_{\mathcal{D}}(n) = O(C(n) + U_{\mathcal{A}}(y_{k+1}))$, and $S_{\mathcal{D}}(n) = O(k S_{\mathcal{A}}(y_{k+1}))$, where $C(n)$ denotes the time needed to look up the block which contains the data when a deletion occurs.*

If we choose, for example, x_k to be the first multiple of k greater than or equal to 2^k, that is, $k = \log_2 n$, then y_k is about $n/\log_2 n$. Because we know there exists an $\mathcal{A}$ with $Q_{\mathcal{A}}(n) = O(\log n)$ and $U_{\mathcal{A}}(n) = P_{\mathcal{A}}(n) = O(n \log n)$, we have the following corollary.

Corollary 11.1 The nearest-neighbor searching problem in the plane can be solved in $O(\log^2 n)$ query time and $O(n)$ update time. [Note that $C(n)$ in this case is $O(\log n)$.]

There are other dynamization schemes that exhibit various query-time/space and query-time/update-time tradeoffs. The interested reader is referred to Chiang and Tamassia [1992], Edelsbrunner [1987], Mehlhorn [1984], Overmars [1983], and Preparata and Shamos [1985] for more information.

11.2.8 Random Sampling

Randomized algorithms have received a great deal of attention recently because of their potential applications. See Chapter 4 for more information. For a variety of geometric problems, randomization techniques help in building geometric subdivisions and data structures to quickly answer queries about such subdivisions. The resulting randomized algorithms are simpler to implement and/or asymptotically faster than those previously known. It is important to note that the focus of randomization is *not* on random input, such as a collection of points randomly chosen uniformly and independently from a region. We are concerned with algorithms that use a source of random numbers and analyze their performance for an arbitrary input. Unlike *Monte Carlo* algorithms, whose output may be incorrect (with very low probability), the randomized algorithms, known as *Las Vegas* algorithms, considered here are guaranteed to produce a correct output.

There are a good deal of newly developed randomized algorithms for geometric problems. See Du and Hwang [1992] for more details. Randomization gives a general way to divide and conquer geometric problems and can be used for both parallel and serial computation. We will use a familiar example to illustrate this approach.

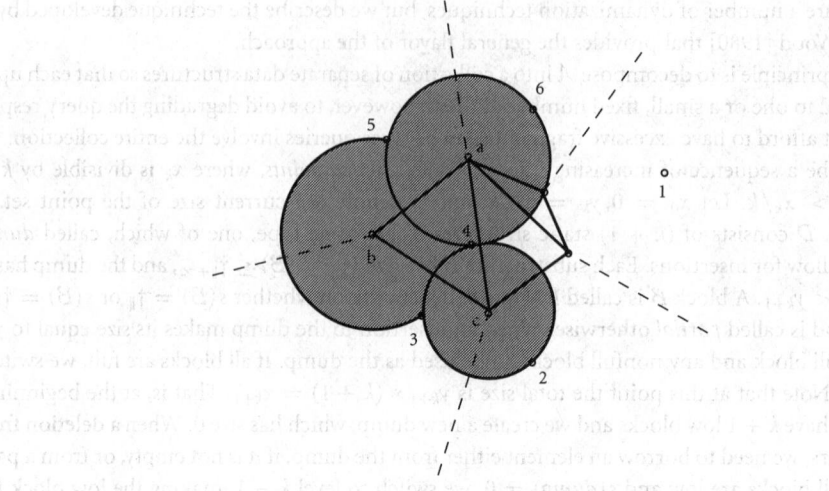

FIGURE 11.6   A triangulation of the Voronoi diagram of six sites and $K_{\mathcal{R}}(T)$, $T = \Delta(a, b, c)$.

Let us consider the problem of nearest-neighbor searching discussed in the preceding subsection. Let $\mathcal{D}$ be a set of n points in the plane and q be the query point. A simple approach to this problem is:

Algorithm S

- *Compute the distance to q for each point $p \in \mathcal{D}$.*
- *Return the point p whose distance is the smallest.*

It is clear that Algorithm S, requiring $O(n)$ time, is not suitable if we need to answer many queries of this type. To obtain faster query response time one can use the technique discussed in the preceding subsection. An alternative is to use the *random sampling* technique as follows. We pick a random sample, a subset $\mathcal{R} \subset \mathcal{D}$ of size r. Let point $p \in \mathcal{R}$ be the nearest neighbor of q in $\mathcal{R}$. The open disk $K_{\mathcal{R}}(q)$ centered at q and passing through p does not contain any other point in $\mathcal{R}$. The answer to the query is either p or some point of $\mathcal{D}$ that lies in $K_{\mathcal{R}}(q)$.

We now extend the above observation to a finite region G in the plane. Let $K_{\mathcal{R}}(G)$ be the union of disks $K_{\mathcal{R}}(r)$ for all $r \in G$. If a query q lies in G, the nearest neighbor of q must be in $K_{\mathcal{R}}(G)$ or in $\mathcal{R}$. Let us consider the Voronoi diagram, $\mathcal{V}(\mathcal{R})$ of $\mathcal{R}$ and a triangulation, $\Delta(\mathcal{V}(\mathcal{R}))$. For each triangle T with vertices a, b, c of $\Delta(\mathcal{V}(\mathcal{R}))$ we have $K_{\mathcal{R}}(T) = K_{\mathcal{R}}(a) \cup K_{\mathcal{R}}(b) \cup K_{\mathcal{R}}(c)$, shown as the shaded area in Figure 11.6. A probability lemma [Clarkson 1988] shows that with probability at least $1 - O(1/n^2)$ the candidate set $\mathcal{D} \cap K_{\mathcal{R}}(T)$ for all $T \in \Delta(\mathcal{V}(\mathcal{R}))$ contains $O(\log n)n/r$ points. More precisely, if $r > 5$ then with probability at least $1 - e^{-C/2 + 36nr}$ each open disk $K_{\mathcal{R}}(r)$ for $r \in \mathcal{R}$ contains no more than Cn/r points of $\mathcal{D}$. If we choose r to be $\sqrt{n}$, the query time becomes $O(\sqrt{n} \log n)$, a speedup from Algorithm S. If we apply this scheme recursively to the candidate sets of $\Delta(\mathcal{V}(\mathcal{R}))$, we can get a query time $O(\log n)$ [Clarkson 1988].

There are many applications of these random sampling techniques. Derandomized algorithms were also developed. See, e.g., Chazelle and Friedman [1990] for a deterministic view of random sampling and its use in geometry.

11.3   Classes of Problems

In this section we aim to touch upon classes of problems that are fundamental in this field and describe solutions to them, some of which may be nontrivial. The reader who needs further information about these problems is strongly encouraged to refer to the original articles cited in the references.

11.3.1 Convex Hull

The convex hull of a set of points in $\Re^k$ is the most fundamental problem in computational geometry. Given is a set of points, and we are interested in computing its convex hull, which is defined to be the smallest convex body containing these points. Of course, the first question one has to answer is how to represent the convex hull. An implicit representation is just to list all of the extreme points,[*] whereas an explicit representation is to list all of the extreme d-faces of dimensions $d = 0, 1, \ldots, k - 1$. Thus, the complexity of any convex hull algorithm would have two parts, computation part and the output part. An algorithm is said to be *output sensitive* if its complexity depends on the size of output.

Definition 11.2   The convex hull of a set S of points in $\Re^k$ is the smallest convex set containing S. In two dimensions, the convex hull is a convex polygon containing S; in three dimensions it is a convex polyhedron.

11.3.1.1   Convex Hulls in Two and Three Dimensions

For an arbitrary set of n points in two and three dimensions, we can compute the convex hull using the *Graham scan*, *gift-wrapping*, or *divide-and-conquer* paradigm, which are briefly described next.

Recall that the convex hull of an arbitrary set of points in two dimensions is a convex polygon. The Graham scan computes the convex hull by (1) sorting the input set of points with respect to an interior point, say, O, which is the centroid of the first three noncollinear points, (2) connecting these points into a star-shaped polygon P centered at O, and (3) performing a linear scan to compute the convex hull of the polygon [Preparata and Shamos 1985]. Because step 1 is the dominating step, the Graham scan takes $O(n \log n)$ time.

One can also use the gift-wrapping technique to compute the convex polygon. Starting with a vertex that is known to be on the convex hull, say, the point O, with the smallest y-coordinate, we sweep a half-line emanating from O counterclockwise. The first point v_1 we hit will be the next point on the convex polygon. We then march to v_1, repeat the same process, and find the next vertex v_2. This process terminates when we reach O again. This is similar to wrapping an object with a *rope*. Finding the next vertex takes time proportional to the number of points remaining. Thus, the total time spent is $O(n\mathcal{H})$, where $\mathcal{H}$ denotes the number of points on the convex polygon. The gift-wrapping algorithm is output sensitive and is more efficient than Graham scan if the number of points on the convex polygon is small, that is, $o(\log n)$.

One can also use the divide-and-conquer paradigm. As mentioned previously, the key step is the merge of two convex hulls, each of which is the solution to a subproblem derived from the recursive step. In the division step, we can recursively separate the set into two subsets by a vertical line L. Then the merge step basically calls for computation of two common tangents of these two convex polygons. The computation of the common tangents, also known as *bridges* over line L, begins with a segment connecting the rightmost point l of the left convex polygon to the leftmost point r of the right convex polygon. Advancing the endpoints of this segment in a *zigzag* manner we can reach the top (or the bottom) common tangent such that the entire set of points lies on one side of the line containing the tangent. The running time of the divide-and-conquer algorithm is easily shown to be $O(n \log n)$.

A more sophisticated output-sensitive and optimal algorithm, which runs in $O(n \log \mathcal{H})$ time, has been developed by Kirkpatrick and Seidel [1986]. It is based on a variation of the divide-and-conquer paradigm. The main idea in achieving the optimal result is that of eliminating *redundant* computations. Observe that in the divide-and-conquer approach after the common tangents are obtained, some vertices that used to belong to the left and right convex polygons must be deleted. Had we known these vertices were not on the final convex hull, we could have saved time by not computing them. Kirkpatrick and Seidel capitalized on this concept and introduced the *marriage-before-conquest* principle. They construct the convex hull by

---

[*]A point in S is an extreme point if it cannot be expressed as a convex combination of other points in S. In other words, the convex hull of S would change when an extreme point is removed from S.

computing the upper and lower hulls of the set; the computations of these two hulls are symmetric. It performs the *divide* step as usual that decomposes the problem into two subproblems of approximately equal size. Instead of computing the upper hulls recursively for each subproblem, it finds the common tangent segment of the two yet-to-be-computed upper hulls and proceeds recursively. One thing that is worth noting is that the points known not to be on the (convex) upper hull are discarded before the algorithm is invoked recursively. This is the key to obtaining a time bound that is both output sensitive and asymptotically optimal.

The divide-and-conquer scheme can be easily generalized to three dimensions. The merge step in this case calls for computing common supporting faces that *wrap* two recursively computed convex polyhedra. It is observed by Preparata and Hong that the common supporting faces are computed from connecting two *cyclic* sequences of edges, one on each polyhedron [Preparata and Shamos 1985]. The computation of these supporting faces can be accomplished in linear time, giving rise to an $O(n \log n)$ time algorithm. By applying the marriage-before-conquest principle Edelsbrunner and Shi [1991] obtained an $O(n \log^2 \mathcal{H})$ algorithm.

The gift-wrapping approach for computing the convex hull in three dimensions would mimic the process of wrapping a gift with a piece of paper and has a running time of $O(n\mathcal{H})$.

11.3.1.2 Convex Hulls in k-Dimensions, $k > 3$

For convex hulls of higher dimensions, a recent result by Chazelle [1993] showed that the convex hull can be computed in time $O(n \log n + n^{\lfloor k/2 \rfloor})$, which is optimal in all dimensions $k \geq 2$ in the worst case. But this result is insensitive to the output size. The gift-wrapping approach generalizes to higher dimensions and yields an output-sensitive solution with running time $O(n\mathcal{H})$, where $\mathcal{H}$ is the total number of i-faces, $i = 0, 1, \ldots, k - 1$, and $\mathcal{H} = O(n^{\lfloor k/2 \rfloor})$ [Edelsbrunner 1987]. One can also use the *beneath-beyond* method of adding points one at a time in ascending order along one of the coordinate axes.* We compute the convex hull $CH(S_{i-1})$ for points $S_{i-1} = \{p_1, p_2, \ldots, p_{i-1}\}$. For each added point p_i, we update $CH(S_{i-1})$ to get $CH(S_i)$, for $i = 2, 3, \ldots, n$, by deleting those t-faces, $t = 0, 1, \ldots, k-1$, that are internal to $CH(S_{i-1} \cup \{p_i\})$. It is shown by Seidel (see Edelsbrunner [1987]) that $O(n^2 + \mathcal{H} \log n)$ time is sufficient. Most recently Chan [1995] obtained an algorithm based on the gift-wrapping method that runs in $O(n \log \mathcal{H} + (n\mathcal{H})^{1-1/(\lfloor k/2 \rfloor + 1)} \log^{O(1)} n)$ time. Note that the algorithm is optimal when $k = 2, 3$. In particular, it is optimal when $\mathcal{H} = o(n^{1-\epsilon})$ for some $0 < \epsilon < 1$.

We conclude this subsection with the following theorem [Chan 1995].

Theorem 11.6 *The convex hull of a set S of n points in $\Re^k$ can be computed in $O(n \log \mathcal{H})$ time for $k = 2$ or $k = 3$, and in $O(n \log \mathcal{H} + (n\mathcal{H})^{1-1/(\lfloor k/2 \rfloor + 1)} \log^{O(1)} n)$ time for $k > 3$, where $\mathcal{H}$ is the number of i-faces, $i = 0, 1, \ldots, k - 1$.*

11.3.2 Proximity

In this subsection we address proximity related problems.

11.3.2.1 Closest Pair

Consider a set S of n points in $\Re^k$. The closest pair problem is to find in S a pair of points whose distance is the minimum, i.e., find p_i and p_j, such that $d(p_i, p_j) = \min_{k \neq l} \{d(p_k, p_l)$, for all points $p_k, p_l \in S\}$, where $d(a, b)$ denotes the Euclidean distance between a and b. (The subsequent result holds for any distance metric in Minkowski's norm.) The brute force method takes $O(d \cdot n^2)$ time by computing all $O(n^2)$ interpoint distances and taking the minimum; the pair that gives the minimum distance is the closest pair.

---

*If the points of S are not given a priori, the algorithm can be made *on line* by adding an extra step of checking if the newly added point is internal or external to the current convex hull. If internal, just discard it.

In one dimension, the problem can be solved by sorting these points and then scanning them in order, as the two closest points must occur consecutively. And this problem has a lower bound of $\Omega(n \log n)$ even in one dimension following from a linear time transformation from the *element uniqueness problem*. See Preparata and Shamos [1985].

But sorting is not applicable for dimension $k > 1$. Indeed this problem can be solved in optimal time $O(n \log n)$ by using the divide-and-conquer approach as follows. Let us first consider the case when $k = 2$. Consider a vertical cutting line λ that divides S into S_1 and S_2 such that $|S_1| = |S_2| = n/2$. Let δ_i be the minimum distance defined by points in $S_i, i = 1, 2$. Observe that the minimum distance defined by points in S can be either δ_1, δ_2, or defined by two points, one in each set. In the former case, we are done. In the latter, these two points must lie in the vertical strip of width $\delta = \min\{\delta_1, \delta_2\}$ on each side of the cutting line λ. The problem now reduces to that of finding the closest pair between points in S_1 and S_2 that lie inside the strip of width 2δ. This subproblem has a special property, known as the *sparsity* condition, i.e., the number of points in a box* of length 2δ is bounded by a constant $c = 4 \cdot 3^{k-1}$, because in each set S_i, there exists no point that lies in the interior of the δ-ball centered at each point in $S_i, i = 1, 2$ [Preparata and Shamos 1985]. It is this sparsity condition that enables us to solve the bichromatic closest pair problem (cf. the following subsection for more information) in $O(n)$ time. Let $\overline{S}_i \subseteq S_i$ denote the set of points that lies in the vertical strip. In two dimensions, the sparsity condition ensures that for each point in $\overline{S}_1$ the number of candidate points in $\overline{S}_2$ for the closest pair is at most 6. We therefore can scan these points $\overline{S}_1 \cup \overline{S}_2$ in order along the cutting line λ and compute the distance between each point scanned and its six candidate points. The pair that gives the minimum distance δ_3 is the bichromatic closest pair. The minimum distance of all pairs of points in S is then equal to $\delta_S = \min\{\delta_1, \delta_2, \delta_3\}$.

Since the merge step takes linear time, the entire algorithm takes $O(n \log n)$ time. This idea generalizes to higher dimensions, except that to ensure the sparsity condition the cutting hyperplane should be appropriately chosen to obtain an $O(n \log n)$ algorithm [Preparata and Shamos 1985].

11.3.2.2 Bichromatic Closest Pair

Given two sets of *red* and *blue* points, denoted R and B, respectively, find two points, one in R and the other in B, that are closest among all such mutual pairs.

The special case when the two sets satisfy the sparsity condition defined previously can be solved in $O(n \log n)$ time, where $n = |R| + |B|$. In fact a more general problem, known as *fixed radius all nearest-neighbor problem in a sparse set* [Bentley 1980, Preparata and Shamos 1985], i.e., given a set M of points in $\Re^k$ that satisfies the sparsity condition, find all pairs of points whose distance is less than a given parameter δ, can be solved in $O(|M| \log |M|)$ time [Preparata and Shamos 1985]. The bichromatic closest pair problem in general, however, seems quite difficult. Agarwal et al. [1991] gave an $O(n^{2(1-1/(\lceil k/2 \rceil+1))+\epsilon})$ time algorithm and a randomized algorithm with an expected running time of $O(n^{4/3} \log^c n)$ for some constant c. Chazelle et al. [1993] gave an $O(n^{2(1-1/(\lfloor k/2 \rfloor+1))+\epsilon})$ time algorithm for the bichromatic farthest pair problem, which can be used to find the diameter of a set S of points by setting $R = B = S$.

A lower bound of $\Omega(n \log n)$ for the bichromatic closest pair problem can be established. (See e.g., Preparata and Shamos [1985].) However, when the two sets are given as two simple polygons, the bichromatic closest pair problem can be solved relatively easily. Two problems can be defined. One is the *closest visible vertex pair* problem, and the other is the *separation problem*. In the former, one looks for a red–blue pair of vertices that are visible to each other and are the closest; in the latter, one looks for two boundary points that have the shortest distance. Both the closest visible vertex pair problem and the separation problem can be solved in linear time [Amato 1994, 1995]. But if both polygons are convex, the separation problem can be solved in $O(\log n)$ time [Chazelle and Dobkin 1987, Edelsbrunner 1985].

Additional references about different variations of closest pair problems can be found in Bespamyatnikh [1995], Callahan and Kosaraju [1995], Kapoor and Smid [1996], Schwartz et al. [1994], and Smid [1992].

---

*A box is also known as a hypercube.

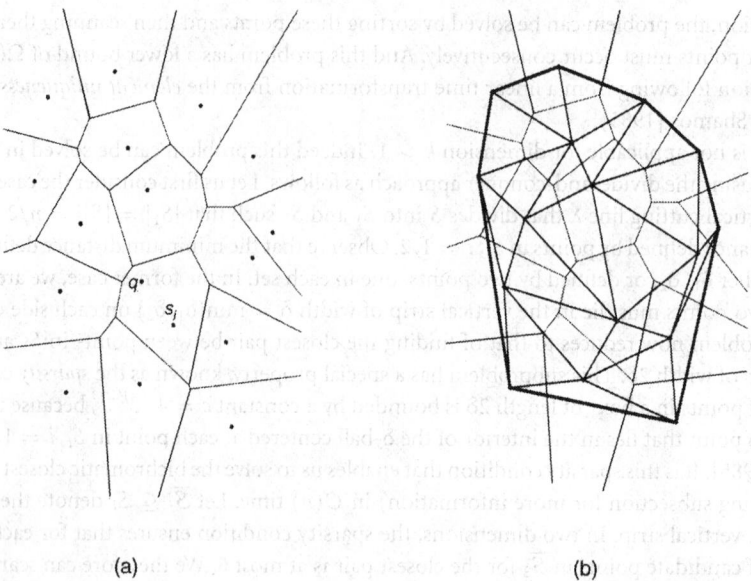

(a)   (b)

FIGURE 11.7   The Voronoi diagram of a set of 16 points in the plane.

11.3.2.3   Voronoi Diagrams

The Voronoi diagram $\mathcal{V}(S)$ of a set S of points, called *sites*, $S = \{s_1, s_2, \ldots, s_n\}$ in $\mathfrak{R}^k$ is a partition of $\mathfrak{R}^k$ into Voronoi cells $V(s_i)$, $i = 1, 2, \ldots, n$, such that each cell contains points that are closer to site s_i than to any other site s_j, $j \neq i$, i.e.,

$$V(s_i) = \{x \in \mathfrak{R}^k \mid d(x, s_i) \leq d(x, s_j) \forall s_j \in \mathfrak{R}^k, j \neq i\}$$

Figure 11.7a shows the Voronoi diagram of 16 point sites in two dimensions. Figure 11.7b shows the straight-line dual graph of the Voronoi diagram, which is called the Delaunay triangulation.

In two dimensions, $\mathcal{V}(S)$ is a planar graph and is of size linear in $|S|$. In dimensions $k \geq 2$, the total number of d-faces of dimensions $d = 0, 1, \ldots, k-1$, in $\mathcal{V}(S)$ is $O(n^{\lceil d/2 \rceil})$.

11.3.2.3.1   Construction of Voronoi Diagram in Two Dimensions

The Voronoi diagram possesses many properties that are proximity related. For instance, the closest pair problem for S can be solved in linear time after the Voronoi diagram has been computed. Because this pair of points must be adjacent in the Delaunay triangulation, all one has to do is examine all adjacent pairs of points and report the pair with the smallest distance. A divide-and-conquer algorithm to compute the Voronoi diagram of a set of points in the Euclidean plane was first given by Shamos and Hoey and generalized by Lee to L_p-metric for all $1 \leq p \leq \infty$ [Preparata and Shamos 1985]. A *plane-sweep* technique for constructing the diagram is proposed by Fortune [1987] that runs in $O(n \log n)$ time. There is a rich body of literature concerning the Voronoi diagram. The interested reader is referred to a recent survey by Fortune in Du and Hwang [1992, pp. 192–234].

Although $\Omega(n \log n)$ is the lower bound for computing the Voronoi diagram for an arbitrary set of n sites, this lower bound does not apply to special cases, e.g., when the sites are on the vertices of a convex polygon. In fact the Voronoi diagram of a convex polygon can be computed in linear time [Aggarwal et al. 1989]. This demonstrates further that an additional property of the input is to help reduce the complexity of the problem.

11.3.2.3.2 Construction of Voronoi Diagrams in Higher Dimensions

The Voronoi diagrams in $\Re^k$ are related to the convex hulls $\Re^{k+1}$ via a geometric transformation similar to duality discussed earlier in the subsection on geometric duality. Consider a set of n sites in $\Re^k$, which is the hyperplane $\mathcal{H}^0$ in $\Re^{k+1}$ such that $x_{k+1} = 0$, and a paraboloid $\mathcal{P}$ in $\Re^{k+1}$ represented as $x_{k+1} = x_1^2 + x_2^2 + \cdots + x_k^2$. Each site $s_i = (\mu_1, \mu_2, \ldots, \mu_k)$ is transformed into a hyperplane $\mathcal{H}(s_i)$ in $\Re^{k+1}$ denoted as

$$x_{k+1} = 2\sum_{j=1}^{k} \mu_j x_j - \left(\sum_{j=1}^{k} \mu_j^2\right)$$

That is, $\mathcal{H}(s_i)$ is tangent to the paraboloid $\mathcal{P}$ at a point $\mathcal{P}(s_i) = (\mu_1, \mu_2, \ldots, \mu_k, \mu_1^2 + \mu_2^2 + \cdots + \mu_k^2)$, which is just the vertical projection of site s_i onto the paraboloid $\mathcal{P}$. The half-space defined by $\mathcal{H}(s_i)$ and containing the paraboloid $\mathcal{P}$ is denoted as $\mathcal{H}^+(s_i)$. The intersection of all half-spaces, $\bigcap_{i=1}^{n} \mathcal{H}^+(s_i)$ is a convex body, and the boundary of the convex body is denoted $CH(\mathcal{H}(S))$. Any point $p \in \Re^k$ lies in the Voronoi cell $V(s_i)$ if the vertical projection of p onto $CH(\mathcal{H}(S))$ is contained in $\mathcal{H}(s_i)$. In other words, every κ-face of $CH(\mathcal{H}(S))$ has a vertical projection on the hyperplane $\mathcal{H}^0$ equal to the κ-face of the Voronoi diagram of S in $\mathcal{H}^0$.

We thus obtain the result which follows from Theorem 11.6 [Edelsbrunner 1987].

Theorem 11.7 *The Voronoi diagram of a set S of n points in $\Re^k$, $k \geq 3$, can be computed in $O(CH_{RH}(n))$ time and $O(n^{\lceil k/2 \rceil})$ space, where $CH_\ell(n)$ denotes the time for constructing the convex hull of n points in $\Re^\ell$.*

For more results concerning the Voronoi diagrams in higher dimensions and duality transformation see Aurenhammer [1990].

11.3.2.4 Farthest-Neighbor Voronoi Diagram

The Voronoi diagram defined in the preceding subsection is also known as the nearest-neighbor Voronoi diagram. A variation of this partitioning concept is a partition of the space into cells, each of which is associated with a site, which contains all points that are farther from the site than from any other site. This diagram is called the *farthest-neighbor* Voronoi diagram. Unlike the nearest-neighbor Voronoi diagram, only a subset of sites have a Voronoi cell associated with them. Those sites that have a nonempty Voronoi cell are those that lie on the convex hull of S. A similar partitioning of the space is known as the order κ-nearest-neighbor Voronoi diagram, in which each Voronoi cell is associated with a subset of κ sites in S for some fixed integer κ such that these κ sites are the closest among all other sites. For $\kappa = 1$ we have the nearest-neighbor Voronoi diagram, and for $\kappa = n - 1$ we have the farthest-neighbor Voronoi diagram. The higher order Voronoi diagrams in k-dimensions are related to the levels of hyperplane arrangements in $k + 1$ dimensions using the paraboloid transformation [Edelsbrunner 1987].

Because the farthest-neighbor Voronoi diagram is related to the convex hull of the set of sites, one can use the marriage-before-conquest paradigm of Kirkpatrick and Seidel [1986] to compute the farthest-neighbor Voronoi diagram of S in two dimensions in time $O(n \log \mathcal{H})$, where $\mathcal{H}$ is the number of sites on the convex hull.

11.3.2.5 Weighted Voronoi Diagrams

When the sites are associated with weights such that the distance function from a point to the sites is weighted, the structure of the Voronoi diagram can be drastically different than the unweighted case.

11.3.2.5.1 Power Diagrams

Suppose each site s in $\Re^k$ is associated with a nonnegative weight, w_s. For an arbitrary point p in $\Re^k$ the weighted distance from p to s is defined as

$$\delta(s, p) = d(s, p)^2 - w_s^2$$

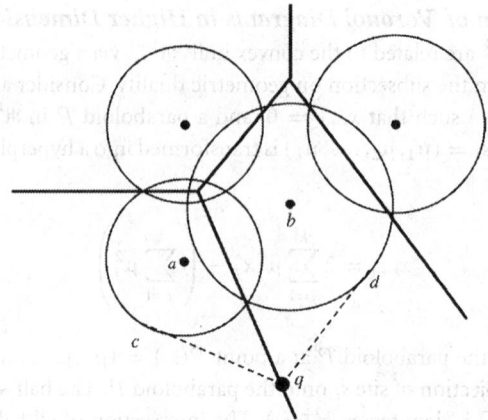

FIGURE 11.8 The power diagram in two dimensions; solid lines are equidistant to two sites.

If w_s is positive, and if $d(s, p) \geq w_s$, then $\sqrt{\delta(s, p)}$ is the length of the tangent of p to the ball $b(s)$ of radius w_s and centered at s. Here $\delta(s, p)$ is also called the *power of* p with respect to the ball $b(s)$. The locus of points p equidistant from two sites $s \neq t$ of equal weight will be a hyperplane called the *chordale* of s and t. See Figure 11.8. Point q is equidistant to sites a and b, and the distance is the length of the tangent line $\overline{q,c} = \overline{q,d}$.

The power diagram of two dimensions can be used to compute the contour of the union of n disks and the connected components of n disks in $O(n \log n)$ time, and in higher dimensions it can be used to compute the union or intersection of n axis-parallel cones in $\Re^k$ with apices in a common hyperplane in time $O(CH_{k+1}(n))$, the multiplicative weighted nearest-neighbor Voronoi diagram (defined subsequently) for n points in $\Re^k$ in time $O(CH_{k+2}(n))$, and the Voronoi diagrams for n spheres in $\Re^k$ in time $O(CH_{k+2}(n))$, where $CH_\ell(n)$ denotes the time for constructing the convex hull of n points in $\Re^\ell$ [Aurenhammer 1987]. For the best time bound for $CH_\ell(n)$ consult the subsection on convex hulls.

11.3.2.5.2 Multiplicative-Weighted Voronoi Diagrams

Each site $s \in \Re^k$ has a positive weight w_s, and the distance from a point p to s is defined as

$$\delta_{\text{multi}-w}(s, p) = d(p, s)/w_s$$

In two dimensions, the locus of points equidistant to two sites $s \neq t$ is a circle, if $w_s \neq w_t$, and a perpendicular bisector of line segment $\overline{s,t}$, if $w_s = w_t$. Each cell associated with a site s consists of all points closer to s than to any other site and may be disconnected. In the worst case the nearest-neighbor Voronoi diagram of a set S of n points in two dimensions can have an $O(n^2)$ regions and can be found in $O(n^2)$ time. In one dimension, the diagram can be computed optimally in $O(n \log n)$ time. However, the farthest-neighbor multiplicative-weighted Voronoi diagram has a very different characteristic. Each Voronoi cell associated with a site remains connected, and the size of the diagram is still linear in the number of sites. An $O(n \log^2 n)$ time algorithm for constructing such a diagram is given in Lee and Wu [1993]. See Schaudt and Drysdale [1991] for more applications of the diagram.

11.3.2.5.3 Additive-Weighted Voronoi Diagrams

The distance of a point p to a site s of a weight w_s is defined as

$$\delta_{\text{add}-w}(s, p) = d(p, s) - w_s$$

In two dimensions, the locus of points equidistant to two sites $s \neq t$ is a branch of a hyperbola, if $w_s \neq w_t$, and a perpendicular bisector of line segment $\overline{s,t}$ if $w_s = w_t$. The Voronoi diagram has properties similar to the ordinary unweighted diagram. For example, each cell is still connected and the size of the diagram is linear. If the weights are positive, the diagram is the same as the Voronoi diagram of a set of spheres centered at site s and of radius w_s, in two dimensions this diagram for n disks can be computed in $O(n \log^2 n)$ time [Lee and Drysdale 1981, Sharir 1985], and in $k \geq 3$ one can use the notion of power diagram to compute the diagram [Aurenhammer 1987].

11.3.2.6 Other Generalizations

The sites mentioned so far are point sites. They can be of different shapes. For instance, they can be line segments, disks, or polygonal objects. The metric used can also be a convex distance function or other norms. See Alt and Schwarzkopf [1995], Boissonnat et al. [1995], Klein [1989], and Yap [1987a] for more information.

11.3.3 Point Location

Point location is yet another fundamental problem in computational geometry. Given a planar subdivision and a query point, one would like to find which region contains the point in question.

In this context, we are mostly interested in fast response time to answer repeated queries to a fixed database. An earlier approach is based on the *slab method* [Preparata and Shamos 1985], in which parallel lines are drawn through each vertex, thus partitioning the plane into parallel slabs. Each parallel slab is further divided into subregions by the edges of the subdivision that can be ordered. Any given point can thus be located by two binary searches: one to locate the slab containing the point among the $n + 1$ horizontal slabs, followed by another to locate the region defined by a pair of consecutive edges that are ordered from left to right. This requires preprocessing of the planar subdivision, and setting up suitable search tree structures for the slabs and the edges crossing each slab. We use a three-tuple, $(P(n), S(n), Q(n)) =$ (preprocessing time, space requirement, query time) to denote the performance of the search strategy (cf. section on dynamization). The slab method gives an $(O(n^2), O(n^2), O(\log n))$ algorithm. Because preprocessing time is only performed once, the time requirement is not as critical as the space requirement. The primary goal of the query processing problems is to minimize the query time and the space required.

Lee and Preparata first proposed a *chain decomposition* method to decompose a monotone planar subdivision with n points into a collection of $m \leq n$ monotone chains organized in a complete binary tree [Preparata and Shamos 1985]. Each node in the binary tree is associated with a monotone chain of at most n edges, ordered in the y-coordinate. Between two adjacent chains, there are a number of disjoint regions. Each query point is compared with the node, hence the associated chain, to decide on which side of the chain the query point lies. Each chain comparison takes $O(\log n)$ time, and the total number of nodes visited is $O(\log m)$. The search on the binary tree will lead to two adjacent chains and hence identify a region that contains the point. Thus, the query time is $O(\log m \log n) = O(\log^2 n)$. Unlike the slab method in which each edge may be stored as many as $O(n)$ times, resulting in $O(n^2)$ space, it can be shown that each edge in the planar subdivision, with an appropriate chain assignment scheme, is stored only once. Thus, the space requirement is $O(n)$. The chain decomposition scheme gives rise to an $(O(n \log n), O(n), O(\log^2 n))$ algorithm. The binary search on the chains is not efficient enough. Recall that after each *chain comparison*, we will move down the binary search tree to perform the next chain comparison and start over another binary search on the y-coordinate to find an edge of the chain, against which a comparison is made to decide if the point lies to the left or right of the chain. A more efficient scheme is to perform a binary search of the y-coordinate at the root node and to spend only $O(1)$ time per node as we go down the chain tree, shaving off an $O(\log n)$ factor from the query time [Edelsbrunner et al. 1986]. This scheme is similar to the ones adopted by Chazelle and Guibas [1986] in a fractional cascading search paradigm and by Willard [1985] in his range tree search method. With the linear time algorithm for triangulating a simple polygon due to Chazelle [1991] (cf. subsequent subsection on triangulation) we conclude with the following optimal search structure for planar point location.

Theorem 11.8   *Given a planar subdivision of n vertices, one can preprocess the subdivision in linear time and space such that each point location query can be answered in $O(\log n)$ time.*

The point location problem in arrangements of hyperplanes is also of significant interest. See, e.g., Chazelle and Friedman [1990]. Dynamic versions of the point location problem have also been investigated. See Chiang and Tamassia [1992] for a survey of dynamic computational geometry.

11.3.4   Motion Planning: Path Finding Problems

The problem is mostly cast in the following setting. Given are a set of obstacles O, an object, called *robot*, and an initial and final position, called source and destination, respectively. We wish to find a path for the robot to move from the source to the destination, avoiding all of the obstacles. This problem arises in several contexts. For instance, in robotics this is referred to as the *piano movers' problem* [Yap 1987b] or *collision avoidance* problem, and in VLSI routing this is the *wiring* problem for 2-terminal nets. In most applications we are searching for a collision avoidance path that has a shortest length, where the distance measure is based on the Euclidean or L_1-metric. For more information regarding motion planning see, e.g., Alt and Yap [1990] and Yap [1987b].

11.3.4.1   Path Finding in Two Dimensions

In two dimensions, the Euclidean shortest path problem in which the robot is a point and the obstacles are simple polygons, is well studied. A most fundamental approach is by using the notion of *visibility graph*. Because the shortest path must make turns at polygonal vertices, it is sufficient to construct a graph whose vertices are the vertices of the polygonal obstacles and the source and destination and whose edges are determined by vertices that are mutually *visible*, i.e., the segment connecting the two vertices does not intersect the interior of any obstacle. Once the visibility graph is constructed with edge weight equal to the Euclidean distance between the two vertices, one can then apply Dijkstra's shortest path algorithms [Preparata and Shamos 1985] to find a shortest path between the source and destination. The Euclidean shortest path between two points is referred to as the *geodesic path* and the distance as the *geodesic distance*. The computation of the visibility graph is the dominating factor for the complexity of any visibility graph-based shortest path algorithm. Research results aiming at more efficient algorithms for computing the visibility graph and for computing the geodesic path in time proportional to the size of the graph have been obtained. Ghosh and Mount [1991] gave an output-sensitive algorithm that runs in $O(E + n \log n)$ time for computing the visibility graph, where E denotes the number of edges in the graph.

Mitchell [1993] used the so-called *continuous Dijkstra* wave front approach to the problem for the general polygonal domain of n obstacle vertices and obtained an $O(n^{5/3+\epsilon})$ time algorithm. He constructed a *shortest path map* that partitions the plane into regions such that all points q that lie in the same region have the same vertex sequence in the shortest path from the given source to q. The shortest path map takes $O(n)$ space and enables us to perform shortest path queries, i.e., find a shortest path from the given source to any query points, in $O(\log n)$ time. Hershberger and Suri [1993] on the other hand, used a plane subdivision approach and presented an $O(n \log^2 n)$-time and $O(n \log n)$-space algorithm to compute the shortest path map of a given source point. They later improved the time bound to $O(n \log n)$. If the source-destination path is confined in a simple polygon with n vertices, the shortest path can be found in $O(n)$ time [Preparata and Shamos 1985].

In the context of VLSI routing one is mostly interested in rectilinear paths (L_1-metric) whose edges are either horizontal or vertical. As the paths are restricted to be rectilinear, the shortest path problem can be solved more easily. Lee et al. [1996] gave a survey on this topic.

In a two-layer VLSI routing model, the number of segments in a rectilinear path reflects the number of *vias*, where the wire segments change layers, which is a factor that governs the fabrication cost. In robotics, a straight-line motion is not as costly as making turns. Thus, the number of segments (or *turns*) has also become an objective function. This motivates the study of the problem of finding a path with the smallest number of segments, called the *minimum link path problem* [Mitchell et al. 1992, Suri 1990].

These two cost measures, length and number of links, are in conflict with each other. That is, a shortest path may have far too many links, whereas a minimum link path may be arbitrarily long compared with a shortest path. Instead of optimizing both measures *simultaneously*, one can seek a path that either optimizes a linear function of both length and the number of links or optimizes them in a lexicographical order. For example, we optimize the length first, and then the number of links, i.e., among those paths that have the same shortest length, find one whose number of links is the smallest, and vice versa.

A generalization of the collision-avoidance problem is to allow collision with a cost. Suppose each obstacle has a weight, which represents the cost if the obstacle is *penetrated*. Mitchell and Papadimitriou [1991] first studied the weighted region shortest path problem. Lee et al. [1991] studied a similar problem in the rectilinear case. Another generalization is to include in the set of obstacles some subset $F \subset O$ of obstacles, whose vertices are *forbidden* for the solution path to make turns. Of course, when the weight of obstacles is set to be ∞, or the forbidden set $F = \emptyset$, these generalizations reduce to the ordinary collision-avoidance problem.

11.3.4.2 Path Finding in Three Dimensions

The Euclidean shortest path problem between two points in a three-dimensional polyhedral environment turns out to be much harder than its two-dimensional counterpart. Consider a convex polyhedron P with n vertices in three dimensions and two points s, d on the surface of P. A shortest path from s to d on the surface will cross a sequence of edges, denoted $\xi(s, d)$. Here $\xi(s, d)$ is called the *shortest path edge sequence* induced by s and d and consists of distinct edges. If the edge sequence is known, the shortest path between s and d can be computed by a planar unfolding procedure so that these faces crossed by the path lie in a common plane and the path becomes a straight-line segment.

Mitchell et al. [1987] gave an $O(n^2 \log n)$ algorithm for finding a shortest path between s and d even if the polyhedron may not be convex. If s and d lie on the surface of two different polyhedra, Sharir [1987] gave an $O(N^{O(k)})$ algorithm, where N denotes the total number of vertices of k obstacles. In general, the problem of determining the shortest path edge sequence of a path between two points among k polyhedra is NP-hard [Canny and Reif 1987].

11.3.4.3 Motion Planning of Objects

In the previous sections, we discussed path planning for moving a point from the source to a destination in the presence of polygonal or polyhedral obstacles. We now briefly describe the problem of moving a polygonal or polyhedral object from an initial position to a final position subject to translational and/or rotational motions.

Consider a set of k convex polyhedral obstacles, $O_1, O_2, \ldots, O_k$, and a convex polyhedral robot, R in three dimensions. The motion planning problem is often solved by using the so-called *configuration space*, denoted $\mathcal{C}$, which is the space of parametric representations of possible robot placements [Lozano-Pérez 1983]. The free placement (FP) is the subspace of $\mathcal{C}$ of points at which the robot does not intersect the interior of any obstacle. For instance, if only translations of R are allowed, the free configuration space will be the union of the Minkowski sums $M_i = O_i \oplus (-R) = \{a - b \mid a \in O_i, b \in R\}$ for $i = 1, 2, \ldots, k$. A *feasible* path exists if the initial placement of R and final placement belong to the same connected component of FP. The problem is to find a continuous curve connecting the initial and final positions in FP. The combinatorial complexity, i.e., the number of vertices, edges, and faces on the boundary of FP, largely influences the efficiency of any $\mathcal{C}$-based algorithm. For translational motion planning, Aronov and Sharir [1994] showed that the combinatorial complexity of FP is $O(nk \log^2 k)$, where k is the number of obstacles defined above and n is the total complexity of the Minkowski sums $M_i, 1 \leq i \leq k$.

Moving a ladder (represented as a line segment) among a set of polygonal obstacles of size n can be done in $O(K \log n)$ time, where K denotes the number of pairs of obstacle vertices whose distance is less than the length of the ladder and is $O(n^2)$ in general [Sifrony and Sharir 1987]. If the moving robot is also a polygonal object, Avnaim et al. [1988] showed that $O(n^3 \log n)$ time suffices. When the obstacles are *fat** Van der Stappen and Overmars [1994] showed that the two preceding two-dimensional motion planning problems can be solved in $O(n \log n)$ time, and in three dimensions the problem can be solved in $O(n^2 \log n)$ time, if the obstacles are ℓ-fat for some positive constant ℓ.

11.3.5 Geometric Optimization

The geometric optimization problems arise in operations research, pattern recognition, and other engineering disciplines. We list some representative problems.

11.3.5.1 Minimum Cost Spanning Trees

The minimum (cost) spanning tree MST of an undirected, weighted graph $G(V, E)$, in which each edge has a nonnegative weight, is a well-studied problem in graph theory and can be solved in $O(|E| \log |V|)$ time [Preparata and Shamos 1985]. When cast in the Euclidean or other L_p-metric plane in which the input consists of a set S of n points, the complexity of this problem becomes different. Instead of constructing a *complete* graph whose edge weight is defined by the distance between its two endpoints, from which to extract an MST, a sparse graph, known as the *Delaunay triangulation* of the point set, is computed. It can be shown that the MST of S is a subgraph of the Delaunay triangulation. Because the MST of a planar graph can be found in linear time [Preparata and Shamos 1985], the problem can be solved in $O(n \log n)$ time. In fact, this is asymptotically optimal, as the closest pair of the set of points must define an edge in the MST, and the closest pair problem is known to have an $\Omega(n \log n)$ lower bound, as mentioned previously.

This problem in three or more dimensions can be solved in subquadratic time. For instance, in three dimensions $O((n \log n)^{1.5})$ time is sufficient [Chazelle 1985] and in $k \geq 3$ dimensions $O(n^{2(1-1/(\lceil k/2 \rceil+1))+\epsilon})$ time suffices [Agarwal et al. 1991].

11.3.5.2 Minimum Diameter Spanning Tree

The minimum *diameter* spanning tree (MDST) of an undirected, weighted graph $G(V, E)$ is a spanning tree such that the total weight of the longest path in the tree is minimum. This arises in applications to communication networks where a tree is sought such that the maximum delay, instead of the total cost, is to be minimized. A graph-theoretic approach yields a solution in $O(|E||V| \log |V|)$ time [Handler and Mirchandani 1979]. Ho et al. [1991] showed that by the triangle inequality there exists an MDST such that the longest path in the tree consists of no more than *three* segments. Based on this an $O(n^3)$ time algorithm was obtained.

Theorem 11.9 *Given a set S of n points, the minimum diameter spanning tree for S can be found in $\theta(n^3)$ time and $O(n)$ space.*

We remark that the problem of finding a spanning tree whose total cost and the diameter are both bounded is NP-complete [Ho et al. 1991]. A similar problem that arises in VLSI clock tree routing is to find a tree from a source to multiple sinks such that every source-to-sink path is the shortest and the total wire length is to be minimized. This problem still is not known to be solvable in polynomial time or NP-hard. Recently, we have shown that the problem of finding a minimum spanning tree such that the longest source-to-sink path is bounded by a given parameter is NP-complete [Seo and Lee 1995].

11.3.5.3 Minimum Enclosing Circle Problem

Given a set S of points, the problem is to find the smallest disk enclosing the set. This problem is also known as the (unweighted) one-center problem. That is, find a center such that the maximum distance

---

*An object $O \subseteq R^k$ is said to be ℓ-fat if for all hyperspheres S centered inside O and not fully containing O we have $\ell \cdot volume\,(O \cap S) \geq volume(S)$.

from the center to the points in S is minimized. More formally, we need to find the center $c \in \Re^2$ such that $\max_{p_j \in S} d(c, p_j)$ is minimized. The weighted one-center problem, in which the distance function $d(c, p_j)$ is multiplied by the weight w_j, is a well-known minimax problem, also known as the *emergency center problem* in operations research. In two dimensions, the one-center problem can be solved in $O(n)$ time [Dyer 1986, Megiddo 1983b]. The minimum enclosing ball problem in higher dimensions is also solved by using a linear programming technique [Megiddo 1983b, 1984].

11.3.5.4 Largest Empty Circle Problem

This problem, in contrast to the minimum enclosing circle problem, is to find a circle centered in the interior of the convex hull of the set S of points that does not contain any given point and the radius of the circle is to be maximized. This is mathematically formalized as a maximin problem; the minimum distance from the center to the set is maximized. The weighted version is also known as the *obnoxious center* problem in facility location. An $O(n \log n)$ time solution for the unweighted version can be found in [Preparata and Shamos 1985].

11.3.5.5 Minimum Annulus Covering Problem

The *minimum annulus covering problem* is defined as follows. Given a set of S of n points find an annulus (defined by two concentric circles) whose center lies internal to the convex hull of S such that the *width* of the annulus is minimized. The problem arises in mechanical part design. To measure whether a circular part is *round*, an American National Standards Institute (ANSI) standard is to use the width of an annulus covering the set of points obtained from a number of measurements. This is known as the *roundness* problem [Le and Lee 1991]. It can be shown that the center of the annulus is either at a vertex of the nearest-neighbor Voronoi diagram, a vertex of the farthest-neighbor Voronoi diagram, or at the intersection of these two diagrams [Le and Lee 1991]. If the input is defined by a simple polygon P with n vertices, and the problem is to find a minimum-width annulus that contains the boundary of P, the problem can be solved in $O(n \log n + k)$, where k denotes the number of intersection points of the *medial axis* of the simple polygon and the boundary of P [Le and Lee 1991]. When the polygon is known to be convex, a linear time is sufficient [Swanson et al. 1995]. If the center of the smallest annulus of a point set can be arbitrarily placed, the center may lie at infinity and the annulus degenerates to a pair of parallel lines enclosing the set of points. This problem is different from the problem of finding the width of a set, which is to find a pair of parallel lines enclosing the set such that the distance between them is minimized. The width of a set of n points can be found in $O(n \log n)$ time, which is optimal [Lee and Wu 1986]. In three dimensions the *width* of a set is also used as a measure for flatness of a *plate—flatness* problem. Houle and Toussaint [1988] gave an $O(n^2)$ time algorithm, and Chazelle et al. [1993] improved it to $O(n^{8/5+\epsilon})$.

11.3.6 Decomposition

Polygon decomposition arises in pattern recognition in which recognition of a shape is facilitated by first decomposing it into simpler parts, called *primitives*, and comparing them to templates previously stored in a library via some similarity measure. The primitives are often convex, with the simplest being the shape of a triangle.

We consider two types of decomposition, *partition* and *covering*. In the former type, the components are pairwise disjoint except they may have some boundary edges in common. In the latter type, the components may overlap. A minimum decomposition is one such that the number of components is minimized. Sometimes additional points, called *Steiner points*, may be introduced to obtain a minimum decomposition. Unless otherwise specified, we assume that no Steiner points are used.

11.3.6.1 Triangulation

Triangulating a simple polygon or, in general, triangulating a planar straight-line graph, is a process of introducing noncrossing edges so that each face is a triangle. It is also a fundamental problem in computer graphics, geographical information systems, and finite-element methods.

Let us begin with the problem of triangulating a simple polygon with n vertices. It is obvious that for a simple polygon with n edges, one needs to introduce at most $n - 3$ diagonals to triangulate the interior into $n - 2$ triangles. This problem has been studied very extensively. A pioneering work is due to Garey et al., which gave an $O(n \log n)$ algorithm and a linear algorithm if the polygon is monotone [O'Rourke 1994, Preparata and Shamos 1985]. A breakthrough linear time triangulation result of Chazelle [1991] settled the long-standing open problem. As a result of this linear triangulation algorithm, a number of problems can be solved in linear time, for example, the simplicity test, defined subsequently, and many other shortest path problems inside a simple polygon [Guibas and Hershberger 1989]. Note that if the polygons have holes, the problem of triangulating the interior requires $\Omega(n \log n)$ time [Asano et al. 1986].

Sometimes we want to look for *quality* triangulation instead of just an arbitrary one. For instance, triangles with large or small angles are not desirable. It is well known that the Delaunay triangulation of points in general position is unique, and it will maximize the minimum angle. In fact, the characteristic angle vector* of the Delaunay triangulation of a set of points is *lexicographically maximum* [Lee 1978]. The notion of Delaunay triangulation of a set of points can be generalized to a planar straight-line graph $G(V, E)$. That is, we would like to have G as a subgraph of a triangulation $G'(V, E')$, $E \subseteq E'$, such that each triangle satisfies the *empty circumcircle* property; no vertex visible from the vertices of a triangle is contained in the interior of the circle. This *generalized* Delaunay triangulation was first introduced by Lee [1978] and an $O(n^2)$ (respectively, $O(n \log n)$) algorithm for constructing the generalized triangulation of a planar graph (respectively, a simple polygon) with n vertices was given in Lee and Lin [1986b]. Chew [1989] later improved the result and gave an $O(n \log n)$ time algorithm using divide-and-conquer. Triangulations that minimize the maximum angle or maximum edge length were also studied. But if constraints on the measure of the triangles, for instance, each triangle in the triangulation must be nonobtuse, then Steiner points must be introduced. See Bern and Eppstein (in Du and Hwang [1992, pp. 23–90]) for a survey of different criteria of triangulations and discussions of triangulations in two and three dimensions.

The problem of triangulating a set P of points in $\Re^k$, $k \geq 3$, is less studied. In this case, the convex hull of P is to be partitioned into $\mathcal{F}$ nonoverlapping simplices, the vertices of which are points in P. A simplex in k-dimensions consists of exactly $k + 1$ points, all of which are extreme points. Avis and ElGindy [1987] gave an $O(k^4 n \log_{1+1/k} n)$ time algorithm for triangulating a simplicial set of n points in $\Re^k$. In $\Re^3$ an $O(n \log n + \mathcal{F})$ time algorithm was presented and $\mathcal{F}$ is shown to be linear if no three points are collinear and at most $O(n^2)$ otherwise. See Du and Hwang [1992] for more references on three-dimensional triangulations and Delaunay triangulations in higher dimensions.

11.3.6.2 Other Decompositions

Partitioning a simple polygon into shapes such as convex polygons, star-shaped polygons, spiral polygons, monotone polygons, etc., has also been investigated [Toussaint 1985]. A linear time algorithm for partitioning a polygon into star-shaped polygons was given by Avis and Toussaint [1981] after the polygon has been triangulated. This algorithm provided a very simple proof of the traditional art gallery problem originally posed by Klee, i.e., $\lfloor n/3 \rfloor$ vertex guards are always sufficient to see the entire region of a simple polygon with n vertices. But if a minimum partition is desired, Keil [1985] gave an $O(n^5 N^2 \log n)$ time, where N denotes the number of reflex vertices. However, the problem of *covering* a simple polygon with a minimum number of star-shaped parts is NP-hard [Lee and Lin 1986a]. The problem of partitioning a polygon into a minimum number of convex parts can be solved in $O(N^2 n \log n)$ time [Keil 1985]. The minimum covering problem by star-shaped polygons for rectilinear polygons is still open. For variations and results of art gallery problems the reader is referred to O'Rourke [1987] and Shermer [1992]. Polynomial time algorithms for computing the minimum partition of a simple polygon into simpler parts while allowing Steiner points can be found in Asano et al. [1986] and Toussaint [1985].

---

*The characteristic angle vector of a triangulation is a vector of minimum angles of each triangle arranged in nondescending order. For a given point set, the number of triangles is the same for all triangulations, and therefore each of them is associated with a characteristic angle vector.

The minimum partition or covering problem for simple polygons becomes NP-hard when the polygons are allowed to have *holes* [Keil 1985, O'Rourke and Supowit 1983]. Asano et al. [1986] showed that the problem of partitioning a simple polygon with h holes into a minimum number of trapezoids with two horizontal sides can be solved in $O(n^{h+2})$ time and that the problem is NP-complete if h is part of the input. An $O(n \log n)$ time 3-approximation algorithm was presented. Imai and Asano [1986] gave an $O(n^{3/2} \log n)$ time and $O(n \log n)$ space algorithm for partitioning a rectilinear polygon with holes into a minimum number of rectangles (allowing Steiner points). The problem of covering a rectilinear polygon (without holes) with a minimum number of rectangles, however, is also NP-hard [Culberson and Reckhow 1988].

The problem of minimum partition into convex parts and the problem of determining if a nonconvex polyhedron can be partitioned into tetrahedra without introducing Steiner points are NP-hard [O'Rourke and Supowit 1983, Ruppert and Seidel 1992].

11.3.7 Intersection

This class of problems arises in architectural design, computer graphics [Dorward 1994], etc., and encompasses two types of problems, *intersection detection* and *intersection computation*.

11.3.7.1 Intersection Detection Problems

The intersection detection problem is of the form: Given a set of objects, do any two intersect? The intersection detection problem has a lower bound of $\Omega(n \log n)$ [Preparata and Shamos 1985]. The pairwise intersection detection problem is a precursor to the general intersection detection problem.

In two dimensions the problem of detecting if two polygons of r and b vertices intersect was easily solved in $O(n \log n)$ time, where $n = r + b$ using the red–blue segment intersection algorithm [Mairson and Stolfi 1988]. However, this problem can be reduced in linear time to the problem of detecting the self-intersection of a polygonal curve. The latter problem is known as the *simplicity* test and can be solved optimally in linear time by Chazelle's [1991] linear time triangulation algorithm. If the two polygons are convex, then $O(\log n)$ suffices [Chazelle and Dobkin 1987, Edelsbrunner 1985]. We remark here that, although detecting whether two convex polygons intersect can be done in logarithmic time, detecting whether the boundary of the two convex polygons intersects requires $\Omega(n)$ time [Chazelle and Dobkin 1987].

In three dimensions, detecting if two convex polyhedra intersect can be solved in linear time by using a hierarchical representation of the convex polyhedron, or by formulating it as a linear programming problem in three variables [Chazelle and Dobkin 1987, Dobkin and Kirkpatrick 1985, Dyer 1984, Megiddo 1983b].

For some applications, we would not only detect intersection but also *report* all such intersecting pairs of objects or *count* the number of intersections, which is discussed next.

11.3.7.2 Intersection Reporting/Counting Problems

One of the simplest of such intersecting reporting problems is that of *reporting* all intersecting pairs of line segments in the plane. Using the plane sweep technique, one can obtain an $O((n + \mathcal{F}) \log n)$ time, where $\mathcal{F}$ is the output size. It is not difficult to see that the lower bound for this problem is $\Omega(n \log n + \mathcal{F})$; thus the preceding algorithm is $O(\log n)$ factor from the optimal. Recently, this segment intersection reporting problem was solved optimally by Chazelle and Edelsbrunner [1992], who used several important algorithm design and data structuring techniques as well as some crucial combinatorial analysis. In contrast to this asymptotically optimal *deterministic* algorithm, a simpler randomized algorithm for this problem that takes $O(n \log n + \mathcal{F})$ time but requires only $O(n)$ space (instead of $O(n + \mathcal{F})$) was obtained [Du and Hwang 1992]. Balaban [1995] recently reported a deterministic algorithm that solves this problem optimally both in time and space.

On a separate front, the problem of finding intersecting pairs of segments from different sets was considered. This is called the *bichromatic line segment* intersection problem. Nievergelt and Preparata [1982] considered the problem of merging two planar convex subdivisions of total size n and showed that the resulting subdivision can be computed in $O(n \log n + \mathcal{F})$ time. This result [Nievergelt and Preparata 1982] was extended in two ways. Mairson and Stolfi [1988] showed that the bichromatic line segment intersection reporting problem can be solved in $O(n \log n + \mathcal{F})$ time. Guibas and Seidel [1987] showed that merging two convex subdivisions can actually be solved in $O(n + \mathcal{F})$ time using topological plane sweep.

Most recently, Chazelle et al. [1994] used *hereditary segment trees* structure and *fractional cascading* [Chazelle and Guibas 1986] and solved both segment intersection reporting and counting problems optimally in $O(n \log n)$ time and $O(n)$ space. (The term $\mathcal{F}$ should be included for reporting.)

The *rectangle intersection reporting* problem arises in the design of VLSI circuitry, in which each rectangle is used to model a certain circuitry component. This is a well-studied classic problem and optimal algorithms ($O(n \log n + \mathcal{F})$ time) have been reported (see Lee and Preparata [1984] for references). The k-dimensional hyperrectangle intersection reporting (respectively, counting) problem can be solved in $O(n^{k-2} \log n + \mathcal{F})$ time and $O(n)$ space [respectively, in time $O(n^{k-1} \log n)$ and space $O(n^{k-2} \log n)$].

11.3.7.3 Intersection Computation

Computing the actual intersection is a basic problem, whose efficient solutions often lead to better algorithms for many other problems.

Consider the problem of computing the common intersection of half-planes discussed previously. Efficient computation of the intersection of two convex polygons is required. The intersection of two convex polygons can be solved very efficiently by plane sweep in linear time, taking advantage of the fact that the edges of the input polygons are ordered. Observe that in each vertical strip defined by two consecutive sweep lines, we only need to compute the intersection of two trapezoids, one derived from each polygon [Preparata and Shamos 1985].

The problem of intersecting two convex polyhedra was first studied by Muller and Preparata [Preparata and Shamos 1985], who gave an $O(n \log n)$ algorithm by reducing the problem to the problems of intersection detection and convex hull computation. From this one can easily derive an $O(n \log^2 n)$ algorithm for computing the common intersection of n half-spaces in three dimensions by the divide-and-conquer method. However, using geometric duality and the concept of separating plane, Preparata and Muller [Preparata and Shamos 1985] obtained an $O(n \log n)$ algorithm for this problem, which is asymptotically optimal. There appears to be a difference in the approach to solving the common intersection problem of half-spaces in two and three dimensions. In the latter, we resorted to geometric duality instead of divide-and-conquer. This *inconsistency* was later resolved. Chazelle [1992] combined the hierarchical representation of convex polyhedra, geometric duality, and other ingenious techniques to obtain a linear time algorithm for computing the intersection of two convex polyhedra. From this result several problems can be solved optimally: (1) the common intersection of half-spaces in three dimensions can now be solved by divide-and-conquer optimally, (2) the merging of two Voronoi diagrams in the plane can be done in linear time by observing the relationship between the Voronoi diagram in two dimensions and the convex hull in three dimensions (cf. subsection on Voronoi diagrams), and (3) the medial axis of a simple polygon or the Voronoi diagram of vertices of a convex polygon can be solved in linear time.

11.3.8 Geometric Searching

This class of problems is cast in the form of query answering as discussed in the subsection on dynamization. Given a collection of objects, with preprocessing allowed, one is to find objects that satisfy the queries. The problem can be static or dynamic, depending on whether the database is allowed to change over the course of query-answering sessions, and it is studied mostly in modes, *count-mode* and *report-mode*. In the former case only the number of objects satisfying the query is to be answered, whereas in the latter the actual identity of the objects is to be reported. In the report mode the query time of the algorithm consists of two components, *search time* and *output*, and expressed as $Q_A(n) = O(f(n) + \mathcal{F})$, where n denotes the size of the database, $f(n)$ a function of n, and $\mathcal{F}$ the size of output. It is obvious that algorithms that handle the report-mode queries can also handle the count-mode queries ($\mathcal{F}$ is the answer). It seems natural to expect that the algorithms for count-mode queries would be more efficient (in terms of the order of magnitude of the space required and query time), as they need not search for the objects. However, it was argued that in the report-mode range searching, one could take advantage of the fact that since reporting takes time, the more there is to report, the *sloppier* the search can be. For example, if we were to know that the ratio $n/\mathcal{F}$ is $O(1)$, we could use a sequential search on a linear list. Chazelle in his seminal paper on filtering search capitalizes on this observation and improves the time complexity for searching for several problems [Chazelle 1986]. As indicated subsequently, the count-mode range searching problem is harder than the report-mode counterpart.

11.3.8.1 Range Searching Problems

This is a fundamental problem in database applications. We will discuss this problem and the algorithm in two dimensions. The generalization to higher dimensions is straightforward using a known technique [Bentley 1980]. Given is a set of n points in the plane, and the ranges are specified by a product $(l_1, u_1) \times (l_2, u_2)$. We would like to find points $p = (x, y)$ such that $l_1 \leq x \leq u_1$ and $l_2 \leq y \leq u_2$. Intuitively we want to find those points that lie inside a query rectangle specified by the range. This is called *orthogonal range searching*, as opposed to other kinds of range searching problems discussed subsequently. Unless otherwise specified, a range refers to an orthogonal range. We discuss the static case; as this belongs to the class of decomposable searching problems, the dynamization transformation techniques can be applied. We note that the range tree structure mentioned later can be made dynamic by using a weight-balanced tree, called a $BB(\alpha)$ tree [Mehlhorn 1984, Willard and Luecker 1985].

For count-mode queries this problem can be solved by using the locus method as follows. Divide the plane into $O(n^2)$ cells by drawing horizontal and vertical lines through each point. The answer to the query q, i.e., find the number of points dominated by q (those points whose x- and y-coordinates are both no greater than those of q) can be found by locating the cell containing q. Let it be denoted by $Dom(q)$. Thus, the answer to the count-mode range queries can be obtained by some simple arithmetic operations of $Dom(q_i)$ for the four corners of the query rectangle. We have $Q(k, n) = O(k \log n)$, $S(k, n) = P(k, n) = O(n^k)$. To reduce the space requirement at the expense of query time has been a goal of further research on this topic. Bentley [1980] introduced a data structure, called *range trees*. Using this structure the following results were obtained: for $k \geq 2$, $Q(k, n) = O(\log^{k-1} n)$, $S(k, n) = P(k, n) = O(n \log^{k-1} n)$. (See Lee and Preparata [1984] and Willard [1985] for more references.)

For report-mode queries, Chazelle [1986] showed that by using a filtering search technique the space requirement can be further reduced by a $\log \log n$ factor. In essence we use less space to allow for more objects than necessary to be found by the search mechanism, followed by a filtering process leaving out unwanted objects for output. If the range satisfies additional conditions, e.g., *grounded* in one of the coordinates, say, $l_1 = 0$, or the aspect ratio of the intervals specifying the range is fixed, then less space is needed. For instance, in two dimensions, the space required is linear (a saving of $\log n / \log \log n$ factor) for these two cases. By using the so-called functional approach to data structures Chazelle [1988] developed a *compression* scheme to encode the *downpointers* used by Willard [1985] to reduce further the space requirement. Thus in k-dimensions, $k \geq 2$, for the count-mode range queries we have $Q(k, n) = O(\log^{k-1} n)$ and $S(k, n) = O(n \log^{k-2} n)$ and for report-mode range queries $Q(k, n) = O(\log^{k-1} n + \mathcal{F})$, and $S(k, n) = O(n \log^{k-2+\epsilon} n)$ for some $0 < \epsilon < 1$.

11.3.8.2 Other Range Searching Problems

There are other range searching problems, called the simplex range searching problem and the half-space range searching problem that have been well studied. A simplex range in $\Re^k$ is a range whose boundary is specifed by $k + 1$ hyperplanes. In two dimensions it is a triangle.

The report-mode half-space range searching problem in the plane is optimally solved by Chazelle et al. [1985] in $Q(n) = O(\log n + \mathcal{F})$ time and $S(n) = O(n)$ space, using geometric duality transform. But this method does not generalize to higher dimensions. For $k = 3$, Chazelle and Preparata [1986] obtained an optimal $O(\log n + \mathcal{F})$ time algorithm using $O(n \log n)$ space. Agarwal and Matoušek [1995] obtained a more general result for this problem: for $n \leq m \leq n^{\lfloor k/2 \rfloor}$, with $O(m^{1+\epsilon})$ space and preprocessing,

$Q(k, n) = O((n/m^{1/\lfloor k/2 \rfloor}) \log n + \mathcal{F})$. As the half-space range searching problem is also decomposable (cf. earlier subsection on dynamization) standard dynamization techniques can be applied.

A general method for simplex range searching is to use the notion of the *partition tree*. The search space is partitioned in a hierarchical manner using cutting hyperplanes, and a search structure is built in a tree structure. Willard [1982] gave a sublinear time algorithm for count-mode half-space query in $O(n^\alpha)$ time using linear space, where $\alpha \approx 0.774$, for $k = 2$. Using Chazelle's cutting theorem Matoušek showed that for k-dimensions there is a linear space search structure for the simplex range searching problem with query time $O(n^{1-1/k})$, which is optimal in two dimensions and within $O(\log n)$ factor of being optimal for $k > 2$. For more detailed information regarding geometric range searching see Matoušek [1994].

The preceding discussion is restricted to the case in which the database is a collection of points. One may consider other kinds of objects, such as line segments, rectangles, triangles, etc., depending on the needs of the application. The inverse of the orthogonal range searching problem is that of the *point enclosure searching problem*. Consider a collection of isothetic rectangles. The point enclosure searching problem is to find all rectangles that contain the given query point q. We can cast these problems as the *intersection searching* problems, i.e., given a set S of objects and a query object q, find a subset $\mathcal{F}$ of S such that for any $f \in \mathcal{F}$, $f \cap q \neq \emptyset$. We then have the rectangle enclosure searching problem, rectangle containment problem, segment intersection searching problem, etc. We list only a few references about these problems [Bistiolas et al. 1993, Imai and Asano 1987, Lee and Preparata 1982]. Janardan and Lopez [1993] generalized intersection searching in the following manner. The database is a collection of *groups* of objects, and the problem is to find all groups of objects intersecting a query object. A group is considered to be intersecting the query object if any object in the group intersects the query object. When each group has only one object, this reduces to the ordinary searching problems.

11.4 Conclusion

We have covered in this chapter a wide spectrum of topics in computational geometry, including several major problem solving paradigms developed to date and a variety of geometric problems. These paradigms include incremental construction, plane sweep, geometric duality, locus, divide-and-conquer, prune-and-search, dynamization, and random sampling. The topics included here, i.e., convex hull, proximity, point location, motion planning, optimization, decomposition, intersection, and searching, are not meant to be exhaustive. Some of the results presented are classic, and some of them represent the state of the art of this field. But they may also become classic in months to come. The reader is encouraged to look up the literature in major computational geometry journals and conference proceedings given in the references. We have not discussed parallel computational geometry, which has an enormous amount of research findings. Atallah [1992] gave a survey on this topic.

We hope that this treatment will provide sufficient background information about this field and that researchers in other science and engineering disciplines may find it helpful and apply some of the results to their own problem domains.

Acknowledgment

This material is based on work supported in part by the National Science Foundation under Grant CCR-9309743 and by the Office of Naval Research under Grants N00014-93-1-0272 and N00014-95-1-1007.

References

Agarwal, P., Edelsbrunner, H., Schwarzkopf, O., and Welzl, E. 1991. Euclidean minimum spanning trees and bichromatic closest pairs. *Discrete Comput. Geom.* 6(5):407–422.

Agarwal, P. and Matoušek, J. 1995. Dynamic half-space range reporting and its applications. *Algorithmica* 13(4):325–345.

Aggarwal, A., Guibas, L. J., Saxe, J., and Shor, P. W. 1989. A linear-time algorithm for computing the Voronoi diagram of a convex polygon. *Discrete Comput. Geom.* 4(6):591–604.

Alt, H. and Schwarzkopf, O. 1995. The Voronoi diagram of curved objects, pp. 89–97. In *Proc. 11th Ann. ACM Symp. Comput. Geom.*, June.

Alt, H. and Yap, C. K. 1990. Algorithmic aspect of motion planning: a tutorial, part 1 & 2. *Algorithms Rev.* 1(1, 2):43–77.

Amato, N. 1994. Determining the separation of simple polygons. *Int. J. Comput. Geom. Appl.* 4(4):457–474.

Amato, N. 1995. Finding a closest visible vertex pair between two polygons. *Algorithmica* 14(2):183–201.

Aronov, B. and Sharir, M. 1994. On translational motion planning in 3-space, pp. 21–30. In *Proc. 10th Ann. ACM Comput. Geom.*, June.

Asano, T., Asano, T., and Imai, H. 1986. Partitioning a polygonal region into trapezoids. *J. ACM* 33(2):290–312.

Asano, T., Asano, T., and Pinter, R. Y. 1986. Polygon triangulation: efficiency and minimality. *J. Algorithms* 7:221–231.

Asano, T., Guibas, L. J., and Tokuyama, T. 1994. Walking on an arrangement topologically. *Int. J. Comput. Geom. Appl.* 4(2):123–151.

Atallah, M. J. 1992. Parallel techniques for computational geometry. *Proc. of IEEE* 80(9):1435–1448.

Aurenhammer, F. 1987. Power diagrams: properties, algorithms and applications. *SIAM J. Comput.* 16(1):78–96.

Aurenhammer, F. 1990. A new duality result concerning Voronoi diagrams. *Discrete Comput. Geom.* 5(3):243–254.

Avis, D. and ElGindy, H. 1987. Triangulating point sets in space. *Discrete Comput. Geom.* 2(2):99–111.

Avis, D. and Toussaint, G. T. 1981. An efficient algorithm for decomposing a polygon into star-shaped polygons. *Pattern Recog.* 13:395–398.

Avnaim, F., Boissonnat, J. D., and Faverjon, B. 1988. A practical exact motion planning algorithm for polygonal objects amidst polygonal obstacles, pp. 67–86. In *Proc. Geom. Robotics Workshop*. J. D. Boissonnat and J. P. Laumond, eds. LNCS Vol. 391.

Balaban, I. J. 1995. An optimal algorithm for finding segments intersections, pp. 211–219. In *Proc. 11th Ann. Symp. Comput. Geom.*, June.

Bentley, J. L. 1980. Multidimensional divide-and-conquer. *Comm. ACM* 23(4):214–229.

Bentley, J. L. and Saxe, J. B. 1980. Decomposable searching problems I: static-to-dynamic transformation. *J. Algorithms* 1:301–358.

Bespamyatnikh, S. N. 1995. An optimal algorithm for closest pair maintenance, pp. 152–166. In *Proc. 11th Ann. Symp. Comput. Geom.*, June.

Bieri, H. and Nef, W. 1982. A recursive plane-sweep algorithm, determining all cells of a finite division of R^d. *Computing* 28:189–198.

Bistiolas, V., Sofotassios, D., and Tsakalidis, A. 1993. Computing rectangle enclosures. *Comput. Geom. Theory Appl.* 2(6):303–308.

Boissonnat, J.-D., Sharir, M., Tagansky, B., and Yvinec, M. 1995. Voronoi diagrams in higher dimensions under certain polyhedra distance functions, pp. 79–88. In *Proc. 11th Ann. ACM Symp. Comput. Geom.*, June.

Callahan, P. and Kosaraju, S. R. 1995. Algorithms for dynamic closests pair and n-body potential fields, pp. 263–272. In *Proc. 6th ACM–SIAM Symp. Discrete Algorithms*.

Canny, J. and Reif, J. R. 1987. New lower bound techniques for robot motion planning problems, pp. 49–60. In *Proc. 28th Annual Symp. Found. Comput. Sci.*, Oct.

Chan, T. M. 1995. Output-sensitive results on convex hulls, extreme points, and related problems, pp. 10–19. In *Proc. 11th ACM Ann. Symp. Comput. Geom.*, June.

Chazelle, B. 1985. How to search in history. *Inf. Control* 64:77–99.

Chazelle, B. 1986. Filtering search: a new approach to query-answering, *SIAM J. Comput.* 15(3):703–724.

Chazelle, B. 1988. A functional approach to data structures and its use in multidimensional searching. *SIAM J. Comput.* 17(3):427–462.

Chazelle, B. 1991. Triangulating a simple polygon in linear time. *Discrete Comput. Geom.* 6:485–524.

Chazelle, B. 1992. An optimal algorithm for intersecting three-dimensional convex polyhedra. *SIAM J. Comput.* 21(4):671–696.

Chazelle, B. 1993. An optimal convex hull algorithm for point sets in any fixed dimension. *Discrete Comput. Geom.* 8(2):145–158.

Chazelle, B. and Dobkin, D. P. 1987. Intersection of convex objects in two and three dimensions. *J. ACM* 34(1):1–27.

Chazelle, B. and Edelsbrunner, H. 1992. An optimal algorithm for intersecting line segments in the plane. *J. ACM* 39(1):1–54.

Chazelle, B., Edelsbrunner, H., Guibas, L. J., and Sharir, M. 1993. Diameter, width, closest line pair, and parametric searching. *Discrete Comput. Geom.* 8(2):183–196.

Chazelle, B., Edelsbrunner, H., Guibas, L. J., and Sharir, M. 1994. Algorithms for bichromatic line-segment problems and polyhedral terrains. *Algorithmica* 11(2):116–132.

Chazelle, B. and Friedman, J. 1990. A deterministic view of random sampling and its use in geometry. *Combinatorica* 10(3):229–249.

Chazelle, B. and Friedman, J. 1994. Point location among hyperplanes and unidirectional ray-shooting. *Comput. Geom. Theory Appl.* 4(2):53–62.

Chazelle, B. and Guibas, L. J. 1986. Fractional cascading: I. a data structuring technique. *Algorithmica* 1(2):133–186.

Chazelle, B., Guibas, L. J., and Lee, D. T. 1985. The power of geometric duality. *BIT* 25:76–90.

Chazelle, B. and Matoušek, J. 1993. On linear-time deterministic algorithms for optimization problems in fixed dimension, pp. 281–290. In *Proc. 4th ACM–SIAM Symp. Discrete Algorithms.*

Chazelle, B. and Preparata, F. P. 1986. Halfspace range search: an algorithmic application of k-sets. *Discrete Comput. Geom.* 1(1):83–93.

Chew, L. P. 1989. Constrained Delaunay triangulations. *Algorithmica* 4(1):97–108.

Chiang, Y.-J. and Tamassia, R. 1992. Dynamic algorithms in computational geometry. *Proc. IEEE* 80(9):1412–1434.

Clarkson, K. L. 1986. Linear programming in $O(n3^{d^2})$ time. *Inf. Proc. Lett.* 22:21–24.

Clarkson, K. L. 1988. A randomized algorithm for closest-point queries. *SIAM J. Comput.* 17(4): 830–847.

Culberson, J. C. and Reckhow, R. A. 1988. Covering polygons is hard, pp. 601–611. In *Proc. 29th Ann. IEEE Symp. Found. Comput. Sci.*

Dobkin, D. P. and Kirkpatrick, D. G. 1985. A linear algorithm for determining the separation of convex polyhedra. *J. Algorithms* 6:381–392.

Dobkin, D. P. and Munro, J. I. 1981. Optimal time minimal space selection algorithms. *J. ACM* 28(3):454–461.

Dorward, S. E. 1994. A survey of object-space hidden surface removal. *Int. J. Comput. Geom. Appl.* 4(3):325–362.

Du, D. Z. and Hwang, F. K., eds. 1992. *Computing in Euclidean Geometry.* World Scientific, Singapore.

Dyer, M. E. 1984. Linear programs for two and three variables. *SIAM J. Comput.* 13(1):31–45.

Dyer, M. E. 1986. On a multidimensional search technique and its applications to the Euclidean one-center problem. *SIAM J. Comput.* 15(3):725–738.

Edelsbrunner, H. 1985. Computing the extreme distances between two convex polygons. *J. Algorithms* 6:213–224.

Edelsbrunner, H. 1987. *Algorithms in Combinatorial Geometry.* Springer–Verlag.

Edelsbrunner, H. and Guibas, L. J. 1989. Topologically sweeping an arrangement. *J. Comput. Syst. Sci.* 38:165–194; (1991) *Corrigendum* 42:249–251.

Edelsbrunner, H., Guibas, L. J., and Stolfi, J. 1986. Optimal point location in a monotone subdivision. *SIAM J. Comput.* 15(2):317–340.

Edelsbrunner, H., O'Rourke, J., and Seidel, R. 1986. Constructing arrangements of lines and hyperplanes with applications. *SIAM J. Comput.* 15(2):341–363.

Edelsbrunner, H. and Shi, W. 1991. An $O(n \log^2 h)$ time algorithm for the three-dimensional convex hull problem. *SIAM J. Comput.* 20(2):259–269.

Fortune, S. 1987. A sweepline algorithm for Voronoi diagrams. *Algorithmica* 2(2):153–174.

Fortune, S. 1993. Progress in computational geometry. In *Directions in Geom. Comput.*, pp. 81–128. R. Martin, ed. Information Geometers Ltd.

Ghosh, S. K. and Mount, D. M. 1991. An output-sensitive algorithm for computing visibility graphs. *SIAM J. Comput.* 20(5):888–910.

Guibas, L. J. and Hershberger, J. 1989. Optimal shortest path queries in a simple polygon. *J. Comput. Syst. Sci.* 39:126–152.

Guibas, L. J. and Seidel, R. 1987. Computing convolutions by reciprocal search. *Discrete Comput. Geom.* 2(2):175–193.

Handler, G. Y. and Mirchandani, P. B. 1979. *Location on Networks: Theory and Algorithm*. MIT Press, Cambridge, MA.

Hershberger, J. and Suri, S. 1993. Efficient computation of Euclidean shortest paths in the plane, pp. 508–517. In *Proc. 34th Ann. IEEE Symp. Found. Comput. Sci.*

Ho, J. M., Chang, C. H., Lee, D. T., and Wong, C. K. 1991. Minimum diameter spanning tree and related problems. *SIAM J. Comput.* 20(5):987–997.

Houle, M. E. and Toussaint, G. T. 1988. Computing the width of a set. *IEEE Trans. Pattern Anal. Machine Intelligence* PAMI-10(5):761–765.

Imai, H. and Asano, T. 1986. Efficient algorithms for geometric graph search problems. *SIAM J. Comput.* 15(2):478–494.

Imai, H. and Asano, T. 1987. Dynamic orthogonal segment intersection search. *J. Algorithms* 8(1):1–18.

Janardan, R. and Lopez, M. 1993. Generalized intersection searching problems. *Int. J. Comput. Geom. Appl.* 3(1):39–69.

Kapoor, S. and Smid, M. 1996. New techniques for exact and approximate dynamic closest-point problems. *SIAM J. Comput.* 25(4):775–796.

Keil, J. M. 1985. Decomposing a polygon into simpler components. *SIAM J. Comput.* 14(4):799–817.

Kirkpatrick, D. G. and Seidel, R. 1986. The ultimate planar convex hull algorithm? *SIAM J. Comput.*, 15(1):287–299.

Klein, R. 1989. *Concrete and Abstract Voronoi Diagrams*. LNCS Vol. 400, Springer–Verlag.

Le, V. B. and Lee, D. T. 1991. Out-of-roundness problem revisited. *IEEE Trans. Pattern Anal. Machine Intelligence* 13(3):217–223.

Lee, D. T. 1978. *Proximity and Reachability in the Plan*. Ph.D. Thesis, *Tech. Rep.* R-831, Coordinated Science Lab., University of Illinois, Urbana.

Lee, D. T. and Drysdale, R. L., III 1981. Generalization of Voronoi diagrams in the plane. *SIAM J. Comput.* 10(1):73–87.

Lee, D. T. and Lin, A. K. 1986a. Computational complexity of art gallery problems. *IEEE Trans. Inf. Theory* 32(2):276–282.

Lee, D. T. and Lin, A. K. 1986b. Generalized Delaunay triangulation for planar graphs. *Discrete Comput. Geom.* 1(3):201–217.

Lee, D. T. and Preparata, F. P. 1982. An improved algorithm for the rectangle enclosure problem. *J. Algorithms* 3(3):218–224.

Lee, D. T. and Preparata, F. P. 1984. Computational geometry: a survey. *IEEE Trans. Comput.* C-33(12):1072–1101.

Lee, D. T. and Wu, V. B. 1993. Multiplicative weighted farthest neighbor Voronoi diagrams in the plane, pp. 154–168. In *Proc. Int. Workshop Discrete Math. and Algorithms*. Hong Kong, Dec.

Lee, D. T. and Wu, Y. F. 1986. Geometric complexity of some location problems. *Algorithmica* 1(2):193–211.

Lee, D. T., Yang, C. D., and Chen, T. H. 1991. Shortest rectilinear paths among weighted obstacles. *Int. J. Comput. Geom. Appl.* 1(2):109–124.

Lee, D. T., Yang, C. D., and Wong, C. K. 1996. Rectilinear paths among rectilinear obstacles. In *Perspectives in Discrete Applied Math.* K. Bogart, ed.

Lozano-Pérez, T. 1983. Spatial planning: a configuration space approach. *IEEE Trans. Comput.* C-32(2):108–120.

Mairson, H. G. and Stolfi, J. 1988. Reporting and counting intersections between two sets of line segments, pp. 307–325. In *Proc. Theor. Found. Comput. Graphics CAD.* Vol. F40, Springer–Verlag.

Matoušek, J. 1994. Geometric range searching. *ACM Computing Sur.* 26:421–461.

Matoušek, J., Sharir, M., and Welzl, E. 1992. A subexponential bound for linear programming, pp. 1–8. In *Proc. 8th Ann. ACM Symp. Comput. Geom.*

Megiddo, N. 1983a. Applying parallel computation algorithms in the design of serial algorithms. *J. ACM* 30(4):852–865.

Megiddo, N. 1983b. Linear time algorithm for linear programming in R^3 and related problems. *SIAM J. Comput.* 12(4):759–776.

Megiddo, N. 1983c. Towards a genuinely polynomial algorithm for linear programming. *SIAM J. Comput.* 12(2):347–353.

Megiddo, N. 1984. Linear programming in linear time when the dimension is fixed. *J. ACM* 31(1):114–127.

Megiddo, N. 1986. New approaches to linear programming. *Algorithmica* 1(4):387–394.

Mehlhorn, K. 1984. *Data Structures and Algorithms*, Vol. 3, Multi-dimensional searching and computational geometry. Springer–Verlag.

Mitchell, J. S. B. 1993. Shortest paths among obstacles in the plane, pp. 308–317. In *Proc. 9th ACM Symp. Comput. Geom.*, May.

Mitchell, J. S. B., Mount, D. M., and Papadimitriou, C. H. 1987. The discrete geodesic problem. *SIAM J. Comput.* 16(4):647–668.

Mitchell, J. S. B. and Papadimitriou, C. H. 1991. The weighted region problem: finding shortest paths through a weighted planar subdivision. *J. ACM* 38(1):18–73.

Mitchell, J. S. B., Rote, G., and Wöginger, G. 1992. Minimum link path among obstacles in the planes. *Algorithmica* 8(5/6):431–459.

Mulmuley, K. 1994. *Computational Geometry: An Introduction through Randomized Algorithms.* Prentice–Hall, Englewood Cliffs, NJ.

Nievergelt, J. and Preparata, F. P. 1982. Plane-sweep algorithms for intersecting geometric figures. *Commun. ACM* 25(10):739–747.

O'Rourke, J. 1987. *Art Gallery Theorems and Algorithms.* Oxford University Press, New York.

O'Rourke, J. 1994. *Computational Geometry in C.* Cambridge University Press, New York.

O'Rourke, J. and Supowit, K. J. 1983. Some NP-hard polygon decomposition problems. *IEEE Trans. Inform. Theory* IT-30(2):181–190,

Overmars, M. H. 1983. *The Design of Dynamic Data Structures.* LNCS Vol. 156, Springer–Verlag.

Preparata, F. P. and Shamos, M. I. 1985. In *Computational Geometry: An Introduction.* Springer–Verlag.

Ruppert, J. and Seidel, R. 1992. On the difficulty of triangulating three-dimensional non-convex polyhedra. *Discrete Comput. Geom.* 7(3):227–253.

Schaudt, B. F. and Drysdale, R. L. 1991. Multiplicatively weighted crystal growth Voronoi diagrams, pp. 214–223. In *Proc. 7th Ann. ACM Symp. Comput. Geom.*

Schwartz, C., Smid, M., and Snoeyink, J. 1994. An optimal algorithm for the on-line closest-pair problem. *Algorithmica* 12(1):18–29.

Seo, D. Y. and Lee, D. T. 1995. On the complexity of bicriteria spanning tree problems for a set of points in the plane. *Tech. Rep. Dept.* EE/CS, Northwestern University, June.

Sharir, M. 1985. Intersection and closest-pair problems for a set of planar discs. *SIAM J. Comput.* 14(2):448–468.

Sharir, M. 1987. On shortest paths amidst convex polyhedra. *SIAM J. Comput.* 16(3):561–572.

Shermer, T. C. 1992. Recent results in art galleries. *Proc. IEEE* 80(9):1384–1399.

Sifrony, S. and Sharir, M. 1987. A new efficient motion planning algorithm for a rod in two-dimensional polygonal space. *Algorithmica* 2(4):367–402.

Smid, M. 1992. Maintaining the minimal distance of a point set in polylogarithmic time. *Discrete Comput. Geom.* 7:415–431.

Suri, S. 1990. On some link distance problems in a simple polygon. *IEEE Trans. Robotics Automation* 6(1):108–113.

Swanson, K., Lee, D. T., and Wu, V. L. 1995. An optimal algorithm for roundness determination on convex polygons. *Comput. Geom. Theory Appl.* 5(4):225–235.

Toussaint, G. T., ed. 1985. *Computational Geometry.* North–Holland.

Van der Stappen, A. F. and Overmars, M. H. 1994. Motion planning amidst fat obstacle, pp. 31–40. In *Proc. 10th Ann. ACM Comput. Geom.*, June.

van Leeuwen, J. and Wood, D. 1980. Dynamization of decomposable searching problems. *Inf. Proc. Lett.* 10:51–56.

Willard, D. E. 1982. Polygon retrieval. *SIAM J. Comput.* 11(1):149–165.

Willard, D. E. 1985. New data structures for orthogonal range queries. *SIAM J. Comput.* 14(1):232–253.

Willard, D. E. and Luecker, G. S. 1985. Adding range restriction capability to dynamic data structures. *J. ACM* 32(3):597–617.

Yao, F. F. 1994. Computational geometry. In *Handbook of Theoretical Computer Science*, Vol. A: Algorithms and Complexity, J. van Leeuwen, ed., pp. 343–389.

Yap, C. K. 1987a. An $O(n \log n)$ algorithm for the Voronoi diagram of a set of simple curve segments. *Discrete Comput. Geom.* 2(4):365–393.

Yap, C. K. 1987b. Algorithmic motion planning. In *Advances in Robotics, Vol I: Algorithmic and Geometric Aspects of Robotics*. J. T. Schwartz and C. K. Yap, eds., pp. 95–143. Lawrence Erlbaum, London.

Further Information

We remark that there are new efforts being made in the applied side of algorithm development. A library of geometric software including visualization tools and applications programs is under development at the Geometry Center, University of Minnesota, and a concerted effort is being put together by researchers in Europe and in the United States to organize a system library containing primitive geometric abstract data types useful for geometric algorithm developers and practitioners.

Those who are interested in the implementations or would like to have more information about available software may consult the Proceedings of the Annual ACM Symposium on Computational Geometry, which has a video session, or the WWW page on *Geometry in Action* by David Eppstein (http://www.ics.uci.edu/~eppstein/geom.html).

12

Randomized Algorithms

Rajeev Motwani*
Stanford University

Prabhakar Raghavan
Verity, Inc.

12.1 Introduction **12-1**
12.2 Sorting and Selection by Random Sampling **12-2**
    Randomized Selection
12.3 A Simple Min-Cut Algorithm **12-4**
    Classification of Randomized Algorithms
12.4 Foiling an Adversary **12-6**
12.5 The Minimax Principle and Lower Bounds **12-7**
    Lower Bound for Game Tree Evaluation
12.6 Randomized Data Structures **12-9**
12.7 Random Reordering and Linear Programming **12-13**
12.8 Algebraic Methods and Randomized Fingerprints **12-15**
    Freivalds' Technique and Matrix Product Verification
    • Extension to Identities of Polynomials • Detecting Perfect
    Matchings in Graphs

12.1 Introduction

A **randomized algorithm** is one that makes random choices during its execution. The behavior of such an algorithm may thus be random even on a fixed input. The design and analysis of a randomized algorithm focus on establishing that it is likely to behave well on *every* input; the likelihood in such a statement depends only on the probabilistic choices made by the algorithm during execution and not on any assumptions about the input. It is especially important to distinguish a randomized algorithm from the *average-case analysis* of algorithms, where one analyzes an algorithm assuming that its input is drawn from a fixed probability distribution. With a randomized algorithm, in contrast, no assumption is made about the input.

Two benefits of randomized algorithms have made them popular: simplicity and efficiency. For many applications, a randomized algorithm is the simplest algorithm available, or the fastest, or both. In the following, we make these notions concrete through a number of illustrative examples. We assume that the reader has had undergraduate courses in algorithms and complexity, and in probability theory. A comprehensive source for randomized algorithms is the book by Motwani and Raghavan [1995]. The articles

*Supported by an Alfred P. Sloan Research Fellowship, an IBM Faculty Partnership Award, an ARO MURI Grant DAAH04-96-1-0007, and NSF Young Investigator Award CCR-9357849, with matching funds from IBM, Schlumberger Foundation, Shell Foundation, and Xerox Corporation.

by Karp [1991], Maffioli et al. [1985], and Welsh [1983] are good surveys of randomized algorithms. The book by Mulmuley [1993] focuses on randomized geometric algorithms.

Throughout this chapter, we assume the random access memory (RAM) model of computation, in which we have a machine that can perform the following operations involving registers and main memory: input–output operations, memory–register transfers, indirect addressing, branching, and arithmetic operations. Each register or memory location may hold an integer that can be accessed as a unit, but an algorithm has no access to the representation of the number. The arithmetic instructions permitted are $+$, $-$, $\times$, and $/$. In addition, an algorithm can compare two numbers and evaluate the square root of a positive number. In this chapter, $\mathbf{E}[X]$ will denote the expectation of random variable X, and $\mathbf{Pr}[A]$ will denote the probability of event A.

12.2   Sorting and Selection by Random Sampling

Some of the earliest randomized algorithms included algorithms for sorting the set S of numbers and the related problem of finding the kth smallest element in S. The main idea behind these algorithms is the use of *random sampling*: a randomly chosen member of S is unlikely to be one of its largest or smallest elements; rather, it is likely to be near the middle. Extending this intuition suggests that a random sample of elements from S is likely to be spread roughly uniformly in S. We now describe randomized algorithms for sorting and selection based on these ideas.

Algorithm RQS

Input: A set of numbers, S.
Output: The elements of S sorted in increasing order.

1. *Choose element y uniformly at random from S: every element in S has equal probability of being chosen.*
2. *By comparing each element of S with y, determine the set S_1 of elements smaller than y and the set S_2 of elements larger than y.*
3. *Recursively sort S_1 and S_2. Output the sorted version of S_1, followed by y, and then the sorted version of S_2.*

Algorithm RQS is an example of a *randomized algorithm* — an algorithm that makes random choices during execution. It is inspired by the Quicksort algorithm due to Hoare [1962], and described in Motwani and Raghavan [1995]. We assume that the random choice in Step 1 can be made in unit time. What can we prove about the running time of RQS?

We now analyze the *expected* number of comparisons in an execution of RQS. Comparisons are performed in Step 2, in which we compare a randomly chosen element to the remaining elements. For $1 \le i \le n$, let $S_{(i)}$ denote the element of *rank i* (the ith smallest element) in the set S. Define the random variable X_{ij} to assume the value 1 if $S_{(i)}$ and $S_{(j)}$ are compared in an execution and the value 0 otherwise. Thus, the total number of comparisons is $\sum_{i=1}^{n} \sum_{j>i} X_{ij}$. By linearity of expectation, the expected number of comparisons is

$$\mathbf{E}\left[\sum_{i=1}^{n} \sum_{j>i} X_{ij} \right] = \sum_{i=1}^{n} \sum_{j>i} \mathbf{E}[X_{ij}] \tag{12.1}$$

Let p_{ij} denote the probability that $S_{(i)}$ and $S_{(j)}$ are compared during an execution. Then,

$$\mathbf{E}[X_{ij}] = p_{ij} \times 1 + (1 - p_{ij}) \times 0 = p_{ij} \tag{12.2}$$

To compute p_{ij}, we view the execution of RQS as binary tree T, each node of which is labeled with a distinct element of S. The root of the tree is labeled with the element y chosen in Step 1; the left subtree of

y contains the elements in S_1 and the right subtree of y contains the elements in S_2. The structures of the two subtrees are determined recursively by the executions of RQS on S_1 and S_2. The root y is compared to the elements in the two subtrees, but no comparison is performed between an element of the left subtree and an element of the right subtree. Thus, there is a comparison between $S_{(i)}$ and $S_{(j)}$ if and only if one of these elements is an ancestor of the other.

Consider the permutation π obtained by visiting the nodes of T in increasing order of the level numbers and in a left-to-right order within each level; recall that the ith level of the tree is the set of all nodes at a distance exactly i from the root. The following two observations lead to the determination of p_{ij}:

1. There is a comparison between $S_{(i)}$ and $S_{(j)}$ if and only if $S_{(i)}$ or $S_{(j)}$ occurs earlier in the permutation π than any element $S_{(\ell)}$ such that $i < \ell < j$. To see this, let $S_{(k)}$ be the earliest in π from among all elements of rank between i and j. If $k \notin \{i, j\}$, then $S_{(i)}$ will belong to the left subtree of $S_{(k)}$ and $S_{(j)}$ will belong to the right subtree of $S_{(k)}$, implying that there is no comparison between $S_{(i)}$ and $S_{(j)}$. Conversely, when $k \in \{i, j\}$, there is an ancestor–descendant relationship between $S_{(i)}$ and $S_{(j)}$, implying that the two elements are compared by RQS.
2. Any of the elements $S_{(i)}, S_{(i+1)}, \ldots, S_{(j)}$ is equally likely to be the first of these elements to be chosen as a partitioning element and hence to appear first in π. Thus, the probability that this first element is either $S_{(i)}$ or $S_{(j)}$ is exactly $2/(j - i + 1)$.

It follows that $p_{ij} = 2/(j - i + 1)$. By Eqs. (12.1) and (12.2), the expected number of comparisons is given by:

$$\sum_{i=1}^{n} \sum_{j>i} p_{ij} = \sum_{i=1}^{n} \sum_{j>i} \frac{2}{j - i + 1}$$

$$\leq \sum_{i=1}^{n-1} \sum_{k=1}^{n-i} \frac{2}{k + 1}$$

$$\leq 2 \sum_{i=1}^{n} \sum_{k=1}^{n} \frac{1}{k}$$

It follows that the expected number of comparisons is bounded above by $2nH_n$, where H_n is the nth *harmonic number*, defined by $H_n = \sum_{k=1}^{n} 1/k$.

Theorem 12.1 *The expected number of comparisons in an execution of RQS is at most $2nH_n$.*

Now $H_n = \ell n\, n + \Theta(1)$, so that the expected running time of RQS is $O(n \log n)$. Note that this expected running time *holds for every input*. It is an expectation that depends only on the random choices made by the algorithm and *not* on any assumptions about the distribution of the input.

12.2.1 Randomized Selection

We now consider the use of random sampling for the problem of selecting the kth smallest element in set S of n elements drawn from a totally ordered universe. We assume that the elements of S are all distinct, although it is not very hard to modify the following analysis to allow for multisets. Let $r_S(t)$ denote the rank of element t (the kth smallest element has rank k) and recall that $S_{(i)}$ denotes the ith smallest element of S. Thus, we seek to identify $S_{(k)}$. We extend the use of this notation to subsets of S as well. The following algorithm is adapted from one due to Floyd and Rivest [1975].

Algorithm LazySelect

Input: A set, S, of n elements from a totally ordered universe and an integer, k, in $[1, n]$.
Output: The kth smallest element of S, $S_{(k)}$.

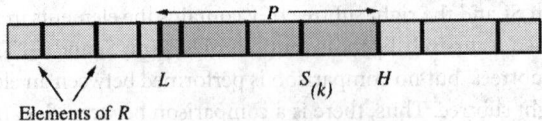

FIGURE 12.1   The LazySelect algorithm.

1. *Pick $n^{3/4}$ elements from S, chosen independently and uniformly at random with replacement; call this multiset of elements R.*
2. *Sort R in $O(n^{3/4} \log n)$ steps using any optimal sorting algorithm.*
3. *Let $x = kn^{-1/4}$. For $\ell = \max\{\lfloor x - \sqrt{n} \rfloor, 1\}$ and $h = \min\{\lceil x + \sqrt{n} \rceil, n^{3/4}\}$, let $a = R_{(\ell)}$ and $b = R_{(h)}$. By comparing a and b to every element of S, determine $r_S(a)$ and $r_S(b)$.*
4. *if $k < n^{1/4}$, let $P = \{y \in S \mid y \leq b\}$ and $r = k$;*
   else if $k > n - n^{1/4}$, let $P = \{y \in S \mid y \geq a\}$ and $r = k - r_S(a) + 1$;
   else if $k \in [n^{1/4}, n - n^{1/4}]$, let $P = \{y \in S \mid a \leq y \leq b\}$ and $r = k - r_S(a) + 1$;
   Check whether $S_{(k)} \in P$ and $|P| \leq 4n^{3/4} + 2$. If not, repeat Steps 1–3 until such a set, P, is found.
5. *By sorting P in $O(|P| \log |P|)$ steps, identify P_r, which is $S_{(k)}$.*

Figure 12.1 illustrates Step 3, where small elements are at the left end of the picture and large ones are to the right. Determining (in Step 4) whether $S_{(k)} \in P$ is easy because we know the ranks $r_S(a)$ and $r_S(b)$ and we compare either or both of these to k, depending on which of the three *if* statements in Step 4 we execute. The sorting in Step 5 can be performed in $O(n^{3/4} \log n)$ steps.

Thus, the idea of the algorithm is to identify two elements a and b in S such that both of the following statements hold with high probability:

1. The element $S_{(k)}$ that we seek is in P, the set of elements between a and b.
2. The set P of elements is not very large, so that we can sort P inexpensively in Step 5.

As in the analysis of RQS, we measure the running time of LazySelect in terms of the number of comparisons performed by it. The following theorem is established using the *Chebyshev bound* from elementary probability theory; a full proof can be found in Motwani and Raghavan [1995].

Theorem 12.2   *With probability $1 - O(n^{-1/4})$, LazySelect finds $S_{(k)}$ on the first pass through Steps 1–5 and thus performs only $2n + o(n)$ comparisons.*

This adds to the significance of LazySelect — the best-known deterministic selection algorithms use $3n$ comparisons in the worst case and are quite complicated to implement.

12.3   A Simple Min-Cut Algorithm

Two events $\mathcal{E}_1$ and $\mathcal{E}_2$ are said to be *independent* if the probability that they both occur is given by

$$\Pr[\mathcal{E}_1 \cap \mathcal{E}_2] = \Pr[\mathcal{E}_1] \times \Pr[\mathcal{E}_2] \tag{12.3}$$

More generally, when $\mathcal{E}_1$ and $\mathcal{E}_2$ are not necessarily independent,

$$\Pr[\mathcal{E}_1 \cap \mathcal{E}_2] = \Pr[\mathcal{E}_1 \mid \mathcal{E}_2] \times \Pr[\mathcal{E}_2] = \Pr[\mathcal{E}_2 \mid \mathcal{E}_1] \times \Pr[\mathcal{E}_1] \tag{12.4}$$

where $\Pr[\mathcal{E}_1 \mid \mathcal{E}_2]$ denotes the *conditional probability* of $\mathcal{E}_1$ given $\mathcal{E}_2$. When a collection of events is not independent, the probability of their intersection is given by the following generalization of Eq. (12.4):

$$\Pr\left[\bigcap_{i=1}^{k} \mathcal{E}_i\right] = \Pr[\mathcal{E}_1] \times \Pr[\mathcal{E}_2 \mid \mathcal{E}_1] \times \Pr[\mathcal{E}_3 \mid \mathcal{E}_1 \cap \mathcal{E}_2] \cdots \Pr\left[\mathcal{E}_k \,\middle|\, \bigcap_{i=1}^{k-1} \mathcal{E}_i\right] \tag{12.5}$$

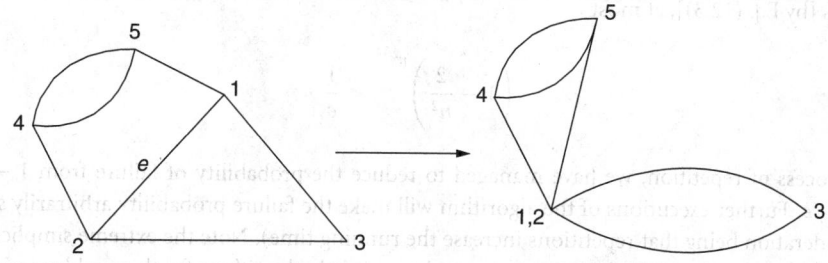

FIGURE 12.2   A step in the min-cut algorithm; the effect of contracting edge $e = (1, 2)$ is shown.

Let G be a connected, undirected multigraph with n vertices. A *multigraph* may contain multiple edges between any pair of vertices. A *cut* in G is a set of edges whose removal results in G being broken into two or more components. A *min-cut* is a cut of minimum cardinality. We now study a simple algorithm due to Karger [1993] for finding a min-cut of a graph.

We repeat the following step: Pick an edge uniformly at random and merge the two vertices at its end points. If as a result there are several edges between some pairs of (newly formed) vertices, retain them all. Remove edges between vertices that are merged, so that there are never any self-loops. This process of merging the two endpoints of an edge into a single vertex is called the *contraction* of that edge. See Figure 12.2. With each contraction, the number of vertices of G decreases by one. Note that as long as at least two vertices remain, an edge contraction does not reduce the min-cut size in G. The algorithm continues the contraction process until only two vertices remain; at this point, the set of edges between these two vertices is a cut in G and is output as a candidate min-cut. What is the probability that this algorithm finds a min-cut?

Definition 12.1   For any vertex v in the multigraph G, the *neighborhood* of G, denoted $\Gamma(v)$, is the set of vertices of G that are adjacent to v. The *degree* of v, denoted $d(v)$, is the number of edges incident on v. For the set S of vertices of G, the neighborhood of S, denoted $\Gamma(S)$, is the union of the neighborhoods of the constituent vertices.

Note that $d(v)$ is the same as the cardinality of $\Gamma(v)$ when there are no self-loops or multiple edges between v and any of its neighbors.

Let k be the min-cut size and let C be a particular min-cut with k edges. Clearly, G has at least $kn/2$ edges (otherwise there would be a vertex of degree less than k, and its incident edges would be a min-cut of size less than k). We bound from below the probability that no edge of C is ever contracted during an execution of the algorithm, so that the edges surviving until the end are exactly the edges in C.

For $1 \leq i \leq n - 2$, let $\mathcal{E}_i$ denote the event of *not* picking an edge of C at the ith step. The probability that the edge randomly chosen in the first step is in C is at most $k/(nk/2) = 2/n$, so that $\Pr[\mathcal{E}_1] \geq 1 - 2/n$. Conditioned on the occurrence of $\mathcal{E}_1$, there are at least $k(n-1)/2$ edges during the second step so that $\Pr[\mathcal{E}_2 \mid \mathcal{E}_1] \geq 1 - 2/(n-1)$. Extending this calculation, $\Pr[\mathcal{E}_i \mid \cap_{j=1}^{i-1} \mathcal{E}_j] \geq 1 - 2/(n - i + 1)$. We now invoke Eq. (12.5) to obtain

$$\Pr\left[\bigcap_{i=1}^{n-2} \mathcal{E}_i \right] \geq \prod_{i=1}^{n-2} \left(1 - \frac{2}{n - i + 1} \right) = \frac{2}{n(n-1)}$$

Our algorithm may err in declaring the cut it outputs to be a min-cut. But the probability of discovering a particular min-cut (which may in fact be the unique min-cut in G) is larger than $2/n^2$, so that the probability of error is at most $1 - 2/n^2$. Repeating the preceding algorithm $n^2/2$ times and making independent random choices each time, the probability that a min-cut is not found in any of the $n^2/2$

attempts is [by Eq. (12.3)], at most

$$\left(1 - \frac{2}{n^2}\right)^{n^2/2} < \frac{1}{e}$$

By this process of repetition, we have managed to reduce the probability of failure from $1 - 2/n^2$ to less than $1/e$. Further executions of the algorithm will make the failure probability arbitrarily small (the only consideration being that repetitions increase the running time). Note the extreme simplicity of this randomized min-cut algorithm. In contrast, most deterministic algorithms for this problem are based on network flow and are considerably more complicated.

12.3.1 Classification of Randomized Algorithms

The randomized sorting algorithm and the min-cut algorithm exemplify two different types of randomized algorithms. The sorting algorithm *always* gives the correct solution. The only variation from one run to another is its running time, whose distribution we study. Such an algorithm is called a **Las Vegas algorithm**.

In contrast, the min-cut algorithm may sometimes produce a solution that is incorrect. However, we prove that the probability of such an error is bounded. Such an algorithm is called a **Monte Carlo algorithm**. We observe a useful property of a Monte Carlo algorithm: If the algorithm is run repeatedly with independent random choices each time, the failure probability can be made arbitrarily small, at the expense of running time. In some randomized algorithms, both the running time and the quality of the solution are random variables; sometimes these are also referred to as Monte Carlo algorithms. The reader is referred to Motwani and Raghavan [1995] for a detailed discussion of these issues.

12.4 Foiling an Adversary

A common paradigm in the design of randomized algorithms is that of *foiling an adversary*. Whereas an adversary might succeed in defeating a **deterministic algorithm** with a carefully constructed *bad* input, it is difficult for an adversary to defeat a randomized algorithm in this fashion. Due to the random choices made by the randomized algorithm, the adversary cannot, while constructing the input, predict the precise behavior of the algorithm. An alternative view of this process is to think of the randomized algorithm as first picking a series of random numbers, which it then uses in the course of execution as needed. In this view, we can think of the random numbers chosen at the start as *selecting* one of a family of deterministic algorithms. In other words, a randomized algorithm can be thought of as a probability distribution on deterministic algorithms. We illustrate these ideas in the setting of AND–OR *tree evaluation*; the following algorithm is due to Snir [1985].

For our purposes, an AND–OR tree is a rooted complete binary tree in which internal nodes at even distance from the root are labeled AND and internal nodes at odd distance are labeled OR. Associated with each leaf is a Boolean *value*. The *evaluation* of the game tree is the following process. Each leaf *returns* the value associated with it. Each OR node returns the Boolean OR of the values returned by its children, and each AND node returns the Boolean AND of the values returned by its children. At each step, an evaluation algorithm chooses a leaf and reads its value. We do not charge the algorithm for any other computation. We study the number of such steps taken by an algorithm for evaluating an AND–OR tree, the worst case being taken over all assignments of Boolean values of the leaves.

Let T_k denote an AND-OR tree in which every leaf is at distance $2k$ from the root. Thus, any root-to-leaf path passes through k AND nodes (including the root itself) and k OR nodes, and there are 2^{2k} leaves. An algorithm begins by specifying a leaf whose value is to be read at the first step. Thereafter, it specifies such a leaf at each step based on the values it has read on previous steps. In a deterministic algorithm, the choice of the next leaf to be read is a deterministic function of the values at the leaves read thus far. For a randomized algorithm, this choice may be randomized. It is not difficult to show that for any deterministic evaluation algorithm, there is an instance of T_k that forces the algorithm to read the values on all 2^{2k} leaves.

We now give a simple randomized algorithm and study the expected number of leaves it reads on any instance of T_k. The algorithm is motivated by the following simple observation. Consider a single AND node with two leaves. If the node were to return 0, at least one of the leaves must contain 0. A deterministic algorithm inspects the leaves in a fixed order, and an adversary can therefore always *hide* the 0 at the second of the two leaves inspected by the algorithm. Reading the leaves in a random order foils this strategy. With probability 1/2, the algorithm chooses the hidden 0 on the first step, so that its expected number of steps is 3/2, which is better than the worst case for any deterministic algorithm. Similarly, in the case of an OR node, if it were to return a 1, then a randomized order of examining the leaves will reduce the expected number of steps to 3/2. We now extend this intuition and specify the complete algorithm.

To evaluate an AND node, v, the algorithm chooses one of its children (a subtree rooted at an OR node) at random and evaluates it by recursively invoking the algorithm. If 1 is returned by the subtree, the algorithm proceeds to evaluate the other child (again by recursive application). If 0 is returned, the algorithm returns 0 for v. To evaluate an OR node, the procedure is the same with the roles of 0 and 1 interchanged. We establish by induction on k that the expected cost of evaluating any instance of T_k is at most 3^k.

The basis ($k = 0$) is trivial. Assume now that the expected cost of evaluating any instance of T_{k-1} is at most 3^{k-1}. Consider first tree T whose root is an OR node, each of whose children is the root of a copy of T_{k-1}. If the root of T were to evaluate to 1, at least one of its children returns 1. With probability 1/2, this child is chosen first, incurring (by the inductive hypothesis) an expected cost of at most 3^{k-1} in evaluating T. With probability 1/2 both subtrees are evaluated, incurring a net cost of at most $2 \times 3^{k-1}$. Thus, the expected cost of determining the value of T is

$$\leq \frac{1}{2} \times 3^{k-1} + \frac{1}{2} \times 2 \times 3^{k-1} = \frac{3}{2} \times 3^{k-1} \qquad (12.6)$$

If, on the other hand, the OR were to evaluate to 0 both children must be evaluated, incurring a cost of at most $2 \times 3^{k-1}$.

Consider next the root of the tree T_k, an AND node. If it evaluates to 1, then both its subtrees rooted at OR nodes return 1. By the discussion in the previous paragraph and by linearity of expectation, the expected cost of evaluating T_k to 1 is at most $2 \times (3/2) \times 3^{k-1} = 3^k$. On the other hand, if the instance of T_k evaluates to 0, at least one of its subtrees rooted at OR nodes returns 0. With probability 1/2 it is chosen first, and so the expected cost of evaluating T_k is at most

$$2 \times 3^{k-1} + \frac{1}{2} \times \frac{3}{2} \times 3^{k-1} \leq 3^k$$

Theorem 12.3 *Given any instance of T_k, the expected number of steps for the preceding randomized algorithm is at most 3^k.*

Because $n = 4^k$, the expected running time of our randomized algorithm is $n^{\log_4 3}$, which we bound by $n^{0.793}$. Thus, the expected number of steps is smaller than the worst case for any deterministic algorithm. Note that this is a Las Vegas algorithm and always produces the correct answer.

12.5 The Minimax Principle and Lower Bounds

The randomized algorithm of the preceding section has an expected running time of $n^{0.793}$ on any uniform binary AND–OR tree with n leaves. Can we establish that *no randomized algorithm* can have a lower expected running time? We first introduce a standard technique due to Yao [1977] for proving such lower bounds. This technique applies only to algorithms that terminate in finite time on all inputs and sequences of random choices.

The crux of the technique is to relate the running times of randomized algorithms for a problem to the running times of deterministic algorithms for the problem *when faced with randomly chosen inputs*. Consider a problem where the number of distinct inputs of a fixed size is finite, as is the number of distinct (deterministic, terminating, and always correct) algorithms for solving that problem. Let us define the **distributional complexity** of the problem at hand as the expected running time of the best deterministic algorithm for the worst distribution on the inputs. Thus, we envision an adversary choosing a probability distribution on the set of possible inputs and study the best deterministic algorithm for this distribution. Let p denote a probability distribution on the set $\mathcal{I}$ of inputs. Let the random variable $C(I_p, A)$ denote the running time of deterministic algorithm $A \in \mathcal{A}$ on an input chosen according to p. Viewing a randomized algorithm as a probability distribution q on the set $\mathcal{A}$ of deterministic algorithms, we let the random variable $C(I, A_q)$ denote the running time of this randomized algorithm on the worst-case input.

Proposition 12.1 (Yao's Minimax Principle)   For all distributions p over $\mathcal{I}$ and q over $\mathcal{A}$,

$$\min_{A \in \mathcal{A}} \mathrm{E}[C(I_p, A)] \le \max_{I \in \mathcal{I}} \mathrm{E}[C(I, A_q)]$$

In other words, the expected running time of the optimal deterministic algorithm for an arbitrarily chosen input distribution p is a lower bound on the expected running time of the optimal (Las Vegas) randomized algorithm for Π. Thus, to prove a lower bound on the randomized complexity, it suffices to choose any distribution p on the input and prove a lower bound on the expected running time of deterministic algorithms for that distribution. The power of this technique lies in the flexibility in the choice of p and, more importantly, the reduction to a lower bound on deterministic algorithms. It is important to remember that the deterministic algorithm "knows" the chosen distribution p.

The preceding discussion dealt only with lower bounds on the performance of Las Vegas algorithms. We briefly discuss Monte Carlo algorithms with error probability $\epsilon \in [0, 1/2]$. Let us define the distributional complexity with error ϵ, denoted $\min_{A \in \mathcal{A}} \mathrm{E}[C_\epsilon(I_p, A)]$, to be the minimum expected running time of any deterministic algorithm that errs with probability at most ϵ under the input distribution p. Similarly, we denote by $\max_{I \in \mathcal{I}} \mathrm{E}[C_\epsilon(I, A_q)]$ the expected running time (under the worst input) of any randomized algorithm that errs with probability at most ϵ (again, the randomized algorithm is viewed as probability distribution q on deterministic algorithms). Analogous to Proposition 12.1, we then have:

Proposition 12.2   For all distributions p over $\mathcal{I}$ and q over $\mathcal{A}$ and any $\epsilon \in [0, 1/2]$,

$$\frac{1}{2} \left(\min_{A \in \mathcal{A}} \mathrm{E}[C_{2\epsilon}(I_p, A)] \right) \le \max_{I \in \mathcal{I}} \mathrm{E}[C_\epsilon(I, A_q)]$$

12.5.1   Lower Bound for Game Tree Evaluation

We now apply Yao's minimax principle to the AND-OR tree evaluation problem. A randomized algorithm for AND-OR tree evaluation can be viewed as a probability distribution over deterministic algorithms, because the length of the computation as well as the number of choices at each step are both finite. We can imagine that all of these coins are tossed before the beginning of the execution.

The tree T_k is equivalent to a balanced binary tree, all of whose leaves are at distance $2k$ from the root and all of whose internal nodes compute the NOR function; a node returns the value 1 if both inputs are 0, and 0 otherwise. We proceed with the analysis of this tree of NORs of depth $2k$.

Let $p = (3 - \sqrt{5})/2$; each leaf of the tree is independently set to 1 with probability p. If each input to a NOR node is independently 1 with probability p, its output is 1 with probability

$$\left(\frac{\sqrt{5} - 1}{2} \right)^2 = \frac{3 - \sqrt{5}}{2} = p$$

Thus, the value of every node of NOR tree is 1 with probability p, and the value of a node is independent of the values of all of the other nodes on the same level. Consider a deterministic algorithm that is evaluating a tree furnished with such random inputs, and let v be a node of the tree whose value the algorithm is trying to determine. Intuitively, the algorithm should determine the value of one child of v before inspecting any leaf of the other subtree. An alternative view of this process is that the deterministic algorithm should inspect leaves visited in a depth-first search of the tree, except of course that it ceases to visit subtrees of node v when the value of v has been determined. Let us call such an algorithm a *depth-first pruning* algorithm, referring to the order of traversal and the fact that subtrees that supply no additional information are pruned away without being inspected. The following result is due to Tarsi [1983]:

Proposition 12.3   Let T be a NOR tree each of whose leaves is independently set to 1 with probability q for a fixed value $q \in [0,1]$. Let $W(T)$ denote the minimum, over all deterministic algorithms, of the expected number of steps to evaluate T. Then, there is a depth-first pruning algorithm whose expected number of steps to evaluate T is $W(T)$.

Proposition 12.3 tells us that for the purposes of our lower bound, we can restrict our attention to depth-first pruning algorithms. Let $W(h)$ be the expected number of leaves inspected by a depth-first pruning algorithm in determining the value of a node at distance h from the leaves, when each leaf is independently set to 1 with probability $(3 - \sqrt{5})/2$. Clearly,

$$W(h) = W(h-1) + (1-p) \times W(h-1)$$

where the first term represents the work done in evaluating one of the subtrees of the node, and the second term represents the work done in evaluating the other subtree (which will be necessary if the first subtree returns the value 0, an event occurring with probability $1 - p$). Letting h be $\log_2 n$ and solving, we get $W(h) \geq n^{0.694}$.

Theorem 12.4   *The expected running time of any randomized algorithm that always evaluates an instance of T_k correctly is at least $n^{0.694}$, where $n = 2^{2k}$ is the number of leaves.*

Why is our lower bound of $n^{0.694}$ less than the upper bound of $n^{0.793}$ that follows from Theorem 12.3? The reason is that we have not chosen the best possible probability distribution for the values of the leaves. Indeed, in the NOR tree if both inputs to a node are 1, no reasonable algorithm will read leaves of both subtrees of that node. Thus, to prove the best lower bound we have to choose a distribution on the inputs that precludes the event that both inputs to a node will be 1; in other words, the values of the inputs are chosen at random but not independently. This stronger (and considerably harder) analysis can in fact be used to show that the algorithm of section 12.4 is optimal; the reader is referred to the paper of Saks and Wigderson [1986] for details.

12.6   Randomized Data Structures

Recent research into data structures has strongly emphasized the use of randomized techniques to achieve increased efficiency without sacrificing simplicity of implementation. An illustrative example is the randomized data structure for dynamic dictionaries called *skip list* that is due to Pugh [1990].

The dynamic dictionary problem is that of maintaining the set of keys X drawn from a totally ordered universe so as to provide efficient support of the following operations: find(q, X) — decide whether the query key q belongs to X and return the information associated with this key if it does indeed belong to X; insert(q, X) — insert the key q into the set X, unless it is already present in X; delete(q, X) — delete the key q from X, unless it is absent from X. The standard approach for solving this problem involves the use of a binary search tree and gives worst-case time per operation that is $O(\log n)$, where n is the size of X at the time the operation is performed. Unfortunately, achieving this time bound requires the use of

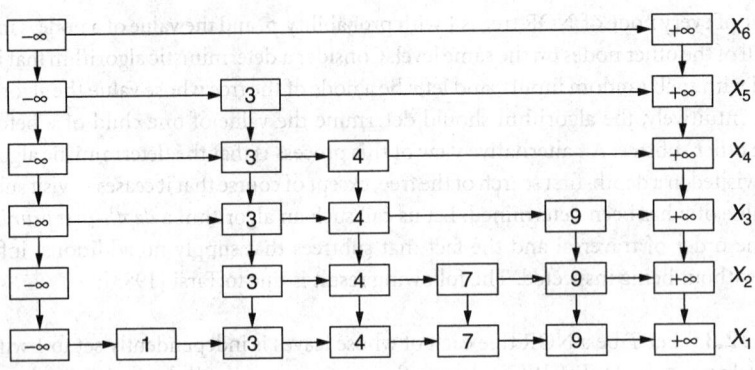

FIGURE 12.3   A skip list.

complex rebalancing strategies to ensure that the search tree remains balanced, that is, has depth $O(\log n)$. Not only does rebalancing require more effort in terms of implementation, but it also leads to significant overheads in the running time (at least in terms of the constant factors subsumed by the big-O notation). The skip list data structure is a rather pleasant alternative that overcomes both of these shortcomings.

Before getting into the details of randomized skip lists, we will develop some of the key ideas without the use of randomization. Suppose we have a totally ordered data set, $X = \{x_1 < x_2 < \cdots < x_n\}$. A *gradation of X* is a sequence of nested subsets (called *levels*)

$$X_r \subseteq X_{r-1} \subseteq \cdots \subseteq X_2 \subseteq X_1$$

such that $X_r = \emptyset$ and $X_1 = X$. Given an ordered set, X, and a gradation for it, the level of any element $x \in X$ is defined as

$$L(x) = \max\{i \mid x \in X_i\}$$

that is, $L(x)$ is the largest index i such that x belongs to the ith level of the gradation. In what follows, we will assume that two special elements $-\infty$ and $+\infty$ belong to each of the levels, where $-\infty$ is smaller than all elements in X and $+\infty$ is larger than all elements in X.

We now define an ordered list data structure with respect to a gradation of the set X. The first level, X_1, is represented as an ordered linked list, and each node x in this list has a stack of $L(x) - 1$ additional nodes directly above it. Finally, we obtain the skip list with respect to the gradation of X by introducing horizontal and vertical pointers between these nodes as illustrated in Figure 12.3. The skip list in Figure 12.3 corresponds to a gradation of the data set $X = \{1, 3, 4, 7, 9\}$ consisting of the following six levels:

$$X_6 = \emptyset$$
$$X_5 = \{3\}$$
$$X_4 = \{3, 4\}$$
$$X_3 = \{3, 4, 9\}$$
$$X_2 = \{3, 4, 7, 9\}$$
$$X_1 = \{1, 3, 4, 7, 9\}$$

Observe that starting at the ith node from the bottom in the leftmost column of nodes and traversing the horizontal pointers in order yields a set of nodes corresponding to the elements of the ith level X_i.

Additionally, we will view each level i as defining a set of *intervals*, each of which is defined as the set of elements of X spanned by a horizontal pointer at level i. The sequence of levels X_i can be viewed as successively coarser partitions of X. In Figure 12.3, the levels determine the following

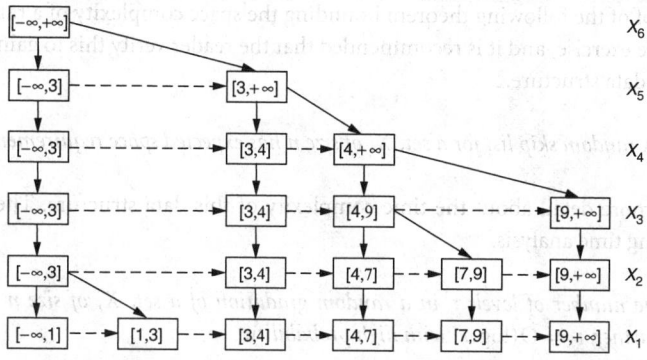

FIGURE 12.4 Tree representation of a skip list.

partitions of X into intervals:

$$X_6 = [-\infty, +\infty]$$
$$X_5 = [-\infty, 3] \cup [3, +\infty]$$
$$X_4 = [-\infty, 3] \cup [3, 4] \cup [4, +\infty]$$
$$X_3 = [-\infty, 3] \cup [3, 4] \cup [4, 9] \cup [9, +\infty]$$
$$X_2 = [-\infty, 3] \cup [3, 4] \cup [4, 7] \cup [7, 9] \cup [9, +\infty]$$
$$X_1 = [-\infty, 1] \cup [1, 3] \cup [3, 4] \cup [4, 7] \cup [7, 9] \cup [9, +\infty]$$

An alternative view of the skip list is in terms of a tree defined by the interval partition structure, as illustrated in Figure 12.4 for the preceding example. In this tree, each node corresponds to an interval, and the intervals at a given level are represented by nodes at the corresponding level of the tree. When the interval J at level $i+1$ is a superset of the interval I at level i, then the corresponding node J has the node I as a child in this tree. Let $C(I)$ denote the number of children in the tree of a node corresponding to the interval I; that is, it is the number of intervals from the previous level that are subintervals of I. Note that the tree is not necessarily binary because the value of $C(I)$ is arbitrary. We can view the skip list as a threaded version of this tree, where each thread is a sequence of (horizontal) pointers linking together the nodes at a level into an ordered list. In Figure 12.4, the broken lines indicate the threads, and the full lines are the actual tree pointers.

Finally, we need some notation concerning the membership of element x in the intervals already defined, where x is not necessarily a member of X. For each possible x, let $I_j(x)$ be the interval at level j containing x. In the degenerate case where x lies on the boundary between two intervals, we assign it to the leftmost such interval. Observe that the nested sequence of intervals containing y,

$$I_r(y) \subseteq I_{r-1}(y) \subseteq \cdots \subseteq I_1(y),$$

corresponds to a root-leaf path in the tree corresponding to the skip list.

It remains to specify the choice of the gradation that determines the structure of a skip list. This is precisely where we introduce randomization into the structure of a skip list. The idea is to define a random gradation. Our analysis will show that, with high probability, the search tree corresponding to a random skip list is balanced, and then the dictionary operations can be efficiently implemented.

We define the *random gradation* for X as follows. Given level X_i, the next level X_{i+1} is determined by independently choosing to retain each element $x \in X_i$ with probability 1/2. The random selection process begins with $X_1 = X$ and terminates when for the first time the resulting level is empty. Alternatively, we may view the choice of the gradation as follows. For each $x \in X$, choose the level $L(x)$ independently from the geometric distribution with parameter $p = 1/2$ and place x in the levels $X_1, \ldots, X_{L(x)}$. We define r to be one more than the maximum of these level numbers. Such a random level is chosen for every element of X upon its insertion and remains fixed unit until its deletion.

We omit the proof of the following theorem bounding the space complexity of a randomized skip list. The proof is a simple exercise, and it is recommended that the reader verify this to gain some insight into the behavior of this data structure.

Theorem 12.5   *A random skip list for a set, X, of size n has expected space requirement $O(n)$.*

We will go into more detail about the time complexity of this data structure. The following lemma underlies the running time analysis.

Lemma 12.1   *The number of levels r in a random gradation of a set, X, of size n has expected value $E[r] = O(\log n)$. Further, $r = O(\log n)$ with high probability.*

***Proof* 12.1**   We will prove the high probability result; the bound on the expected value follows immediately from this. Recall that the level numbers $L(x)$ for $x \in X$ are independent and identically distributed (i.i.d.) random variables distributed geometrically with parameter $p = 1/2$; notationally, we will denote these random variables by $Z_1, \ldots, Z_n$. Now, the total number of levels in the skip list can be determined as

$$r = 1 + \max_{x \in X} L(x) = 1 + \max_{1 \le i \le n} Z_i$$

that is, as one more than the maximum of n i.i.d. geometric random variables.

For such geometric random variables with parameter p, it is easy to verify that for any positive real t, $\Pr[Z_i > t] \le (1 - p)^t$. It follows that

$$\Pr[\max_i Z_i > t] \le n(1 - p)^t = \frac{n}{2^t}$$

because $p = 1/2$ in this case. For any $\alpha > 1$, setting $t = \alpha \log n$, we obtain

$$\Pr[r > \alpha \log n] \le \frac{1}{n^{\alpha - 1}} \qquad\qquad \Box$$

We can now infer that the tree representing the skip list has height $O(\log n)$ with high probability. To show that the overall search time in a skip list is similarly bounded, we must first specify an efficient implementation of the find operation. We present the implementation of the dictionary operations in terms of the tree representation; it is fairly easy to translate this back into the skip list representation.

To implement find (y, X), we must walk down the path

$$I_r(y) \subseteq I_{r-1}(y) \subseteq \cdots \subseteq I_1(y)$$

For this, at level j, starting at the node $I_j(y)$, we use the vertical pointer to descend to the leftmost child of the current interval; then, via the horizontal pointers, we move rightward until the node $I_j(y)$ is reached. Note that it is easily determined whether y belongs to a given interval or to an interval to its right. Further, in the skip list, the vertical pointers allow access only to the leftmost child of an interval, and therefore we must use the horizontal pointers to scan its children.

To determine the expected cost of find(y, X) operation, we must take into account both the number of levels and the number of intervals/nodes scanned at each level. Clearly, at level j, the number of nodes visited is no more than the number of children of $I_{j+1}(y)$. It follows that the cost of find can be bounded by

$$O\left(\sum_{j=1}^{r} (1 + C(I_j(y))) \right)$$

The following lemma shows that this quantity has expectation bounded by $O(\log n)$.

Lemma 12.2 *For any y, let* $I_r(y), \ldots, I_1(y)$ *be the search path followed by* find(y, X) *in a random skip list for a set, X, of size n. Then,*

$$\mathrm{E}\left[\sum_{j=1}^{r}(1 + C(I_j(y)))\right] = O(\log n)$$

Proof 12.2 We begin by showing that for any interval I in a random skip list, $\mathrm{E}[C(I)] = O(1)$. By Lemma 12.1, we are guaranteed that $r = O(\log n)$ with his probability, and so we will obtain the desired bound. It is important to note that we really do need the high-probability bound on Lemma 12.1 because it is incorrect to multiply the expectation of r with that of $1 + C(I)$ (the two random variables need not be independent). However, in the approach we will use, because $r > \alpha \log n$ with probability at most $1/n^{\alpha-1}$ and $\sum_j (1 + C(I_j(y))) = O(n)$, it can be argued that the case $r > \alpha \log n$ does not contribute significantly to the expectation of $\sum_j C(I_j(y))$.

To show that the expected number of children of interval J at level i is bounded by a constant, we will show that the expected number of siblings of J (children of its parent) is bounded by a constant; in fact, we will bound only the number of right siblings because the argument for the number of left siblings is identical. Let the intervals to the right of J be the following:

$$J_1 = [x_1, x_2]; J_2 = [x_2, x_3]; \ldots; J_k = [x_k, +\infty]$$

Because these intervals exist at level i, each of the elements $x_1, \ldots, x_k$ belongs to X_i. If J has s right siblings, then it must be the case that $x_1, \ldots, x_s \notin X_{i+1}$, and $x_{s+1} \in X_{i+1}$. The latter event occurs with probability $1/2^{s+1}$ because each element of X_i is independently chosen to be in X_{i+1} with probability 1/2. Clearly, the number of right siblings of J can be viewed as a random variable that is geometrically distributed with parameter 1/2. It follows that the expected number of right siblings of J is at most 2. $\square$

Consider now the implementation of the insert and delete operations. In implementing the operation insert(y, X), we assume that a random level, $L(y)$, is chosen for y as described earlier. If $L(y) > r$, then we start by creating new levels from $r + 1$ to $L(y)$ and then redefine r to be $L(y)$. This requires $O(1)$ time per level because the new levels are all empty prior to the insertion of y. Next we perform find(y, X) and determine the search path $I_r(y), \ldots, I_1(y)$, where r is updated to its new value if necessary. Given this search path, the insertion can be accomplished in time $O(L(y))$ by splitting around y the intervals $I_1(y), \ldots, I_{L(y)}(y)$ and updating the pointers as appropriate. The delete operation is the converse of the insert operation; it involves performing find(y, X) followed by collapsing the intervals that have y as an endpoint. Both operations incur costs that are the cost of a find operation and additional cost proportional to $L(y)$. By Lemmas 12.1 and 12.2, we obtain the following theorem.

Theorem 12.6 *In a random skip list for a set, X, of size n, the operations* find, insert, *and* delete *can be performed in expected time* $O(\log n)$.

12.7 Random Reordering and Linear Programming

The *linear programming problem* is a particularly notable example of the two main benefits of randomization: simplicity and speed. We now describe a simple algorithm for linear programming based on a paradigm for randomized algorithms known as *random reordering*. For many problems, it is possible to design natural algorithms based on the following idea. Suppose that the input consists of n elements. Given any subset of these n elements, there is a solution to the partial problem defined by these elements. If we start with the empty set and add the n elements of the input one at a time, maintaining a partial solution after each addition, we will obtain a solution to the entire problem when all of the elements have been added. The usual difficulty with this approach is that the running time of the algorithm depends heavily on the order in which the input elements are added; for any fixed ordering, it is generally possible to force this algorithm to behave badly. The key idea behind random reordering is to *add the elements in a random order*. This simple device often avoids the pathological behavior that results from using a fixed order.

The linear programming problem is to find the extremum of a linear objective function of d real variables subject to set H of n constraints that are linear functions of these variables. The intersection of the n half-spaces defined by the constraints is a polyhedron in d-dimensional space (which may be empty, or possibly unbounded). We refer to this polyhedron as the *feasible region*. Without loss of generality [Schrijver 1986] we assume that the feasible region is nonempty and bounded. (Note that we are not assuming that we can *test* an arbitrary polyhedron for nonemptiness or boundedness; this is known to be equivalent to solving a linear program.) For a set of constraints, S, let $B(S)$ denote the optimum of the linear program defined by S; we seek $B(S)$.

Consider the following algorithm due to Seidel [1991]: Add the n constraints in random order, one at a time. After adding each constraint, determine the optimum subject to the constraints added so far. This algorithm also may be viewed in the following backwards manner, which will prove useful in the sequel.

Algorithm SLP

Input: A set of constraints H, and the dimension d.
Output: The optimum $B(H)$.

    0. *If there are only d constraints, output $B(H) = H$.*
    1. *Pick a random constraint $h \in H$;*
        Recursively find $B(H \setminus \{h\})$.
   2.1. *If $B(H \setminus \{h\})$ does not violate h, output $B(H \setminus \{h\})$ to be the optimum $B(H)$.*
   2.2. *Else project all of the constraints of $H \setminus \{h\}$ onto h and recursively solve this new linear programming problem of one lower dimension.*

The idea of the algorithm is simple. Either h (the constraint chosen randomly in Step 1) is redundant (in which case we execute Step 2.1), or it is not. In the latter case, we know that the vertex formed by $B(H)$ must lie on the hyperplane bounding h. In this case, we project all of the constraints of $H \setminus \{h\}$ onto h and solve this new linear programming problem (which has dimension $d - 1$).

The optimum $B(H)$ is defined by d constraints. At the top level of recursion, the probability that random constraint h violates $B(H \setminus \{h\})$ is at most d/n. Let $T(n, d)$ denote an upper bound on the expected running time of the algorithm for any problem with n constraints in d dimensions. Then, we may write

$$T(n, d) \leq T(n - 1, d) + O(d) + \frac{d}{n}[O(dn) + T(n - 1, d - 1)] \qquad (12.7)$$

In Equation (12.7), the first term on the right denotes the cost of recursively solving the linear program defined by the constraints in $H \setminus \{h\}$. The second accounts for the cost of checking whether h violates $B(H \setminus \{h\})$. With probability d/n it does, and this is captured by the bracketed expression, whose first term counts the cost of projecting all of the constraints onto h. The second counts the cost of (recursively) solving the projected problem, which has one fewer constraint and dimension. The following theorem may be verified by substitution and proved by induction.

Theorem 12.7   *There is a constant b such that the recurrence (12.7) satisfies the solution $T(n, d) \leq bnd!$.*

In contrast, if the choice in Step 1 of SLP were not random, the recurrence (12.7) would be

$$T(n, d) \leq T(n - 1, d) + O(d) + O(dn) + T(n - 1, d - 1) \qquad (12.8)$$

whose solution contains a term that grows quadratically in n.

12.8   Algebraic Methods and Randomized Fingerprints

Some of the most notable randomized results in theoretical computer science, particularly in complexity theory, have involved a nontrivial combination of randomization and algebraic methods. In this section, we describe a fundamental randomization technique based on algebraic ideas. This is the randomized fingerprinting technique, originally due to Freivalds [1977], for the verification of identities involving matrices, polynomials, and integers. We also describe how this generalizes to the so-called Schwartz–Zippel technique for identities involving multivariate polynomials (independently due to Schwartz [1987] and Zippel [1979]; see also DeMillo and Lipton [1978]. Finally, following Lovász [1979], we apply the technique to the problem of detecting the existence of perfect matchings in graphs.

The *fingerprinting* technique has the following general form. Suppose we wish to decide the equality of two elements x and y drawn from some large universe U. Assuming any reasonable model of computation, this problem has a deterministic complexity $\Omega(\log|U|)$. Allowing randomization, an alternative approach is to choose a random function from U into a smaller space V such that with high probability x and y are identical if and only if their images in V are identical. These images of x and y are said to be their *fingerprints*, and the equality of fingerprints can be verified in time $O(\log|V|)$. Of course, for any fingerprint function the average number of elements of U mapped to an element of V is $|U|/|V|$; thus, it would appear impossible to find good fingerprint functions that work for arbitrary or worst-case choices of x and y. However, as we will show subsequently, when the identity checking is required to be correct only for x and y chosen from the small subspace S of U, particularly a subspace with some algebraic structure, it is possible to choose good fingerprint functions without any a priori knowledge of the subspace, provided the size of V is chosen to be comparable to the size of S.

Throughout this section, we will be working over some unspecified field $\mathcal{F}$. Because the randomization will involve uniform sampling from a finite subset of the field, we do not even need to specify whether the field is finite. The reader may find it helpful in the infinite case to assume that $\mathcal{F}$ is the field $\mathcal{Q}$ of rational numbers and in the finite case to assume that $\mathcal{F}$ is $\mathcal{Z}_p$, the field of integers modulo some prime number p.

12.8.1   Freivalds' Technique and Matrix Product Verification

We begin by describing a fingerprinting technique for verifying matrix product identities. Currently, the fastest algorithm for matrix multiplication (due to Coppersmith and Winograd [1990]) has running time $O(n^{2.376})$, improving significantly on the obvious $O(n^3)$ time algorithm; however, the fast matrix multiplication algorithm has the disadvantage of being extremely complicated. Suppose we have an implementation of the fast matrix multiplication algorithm and, given its complex nature, are unsure of its correctness. Because program verification appears to be an intractable problem, we consider the more reasonable goal of verifying the correctness of the output produced by executing the algorithm on specific inputs. (This notion of verifying programs on specific inputs is the basic tenet of the theory of *program checking* recently formulated by Blum and Kannan [1989].) More concretely, suppose we are given three $n \times n$ matrices X, Y, and Z over field $\mathcal{F}$, and would like to verify that $XY = Z$. Clearly, it does not make sense to use a simpler but slower matrix multiplication algorithm for the verification, as that would defeat the whole purpose of using the fast algorithm in the first place. Observe that, in fact, there is no need to recompute Z; rather, we are merely required to verify that the product of X and Y is indeed equal to Z. Freivalds' technique gives an elegant solution that leads to an $O(n^2)$ time randomized algorithm with bounded error probability.

The idea is to first pick the random vector $r \in \{0, 1\}^n$, that is, each component of r is chosen independently and uniformly at random from the set $\{0, 1\}$ consisting of the additive and multiplicative identities of the field $\mathcal{F}$. Then, in $O(n^2)$ time, we can compute $y = Yr$, $x = Xy = XYr$, and $z = Zr$. We would like to claim that the identity $XY = Z$ can be verified merely by checking that $x = z$. Quite clearly, if $XY = Z$, then $x = z$; unfortunately, the converse is not true in general. However, given the random choice of r, we can show that for $XY \neq Z$, the probability that $x \neq z$ is at least $1/2$. Observe that the fingerprinting algorithm errs only if $XY \neq Z$ but x and z turn out to be equal, and this has a bounded probability.

Theorem 12.8 *Let X, Y, and Z be $n \times n$ matrices over some field $\mathcal{F}$ such that $XY \neq Z$; further, let r be chosen uniformly at random from $\{0, 1\}^n$ and define $x = XYr$ and $z = Zr$. Then,*

$$\Pr[x = z] \leq 1/2$$

Proof 12.3 Define $W = XY - Z$ and observe that W is not the all-zeroes matrix. Because $Wr = XYr - Zr = x - z$, the event $x = z$ is equivalent to the event that $Wr = 0$. Assume, without loss of generality, that the first row of W has a nonzero entry and that the nonzero entries in that row precede all of the zero entries. Define the vector w as the first row of W, and assume that the first $k > 0$ entries in w are nonzero. Because the first component of Wr is w^Tr, giving an upper bound on the probability that the inner product of w and r is zero will give an upper bound on the probability that $x = z$.

Observe that $w^Tr = 0$ if and only if

$$r_1 = \frac{-\sum_{i=2}^{k} w_i r_i}{w_1} \tag{12.9}$$

Suppose that while choosing the random vector r, we choose $r_2, \ldots, r_n$ before choosing r_1. After the values for $r_2, \ldots, r_n$ have been chosen, the right-hand side of Equation (12.9) is fixed at some value $v \in \mathcal{F}$. If $v \notin \{0, 1\}$, then r_1 will never equal v; conversely, if $v \in \{0, 1\}$, then the probability that $r_1 = v$ is 1/2. Thus, the probability that $w^Tr = 0$ is at most 1/2, implying the desired result. $\qquad \square$

We have reduced the matrix multiplication verification problem to that of verifying the equality of two vectors. The reduction itself can be performed in $O(n^2)$ time and the vector equality can be checked in $O(n)$ time, giving an overall running time of $O(n^2)$ for this Monte Carlo procedure. The error probability can be reduced to $1/2^k$ via k independent iterations of the Monte Carlo algorithm. Note that there was nothing magical about choosing the components of the random vector r from $\{0, 1\}$, because any two distinct elements of $\mathcal{F}$ would have done equally well. This suggests an alternative approach toward reducing the error probability, as follows: Each component of r is chosen independently and uniformly at random from some subset $\mathcal{S}$ of the field $\mathcal{F}$; then, it is easily verified that the error probability is no more than $1/|\mathcal{S}|$.

Finally, note that Freivalds' technique can be applied to the verification of any matrix identity $A = B$. Of course, given A and B, just comparing their entries takes only $O(n^2)$ time. But there are many situations where, just as in the case of matrix product verification, computing A explicitly is either too expensive or possibly even impossible, whereas computing Ar is easy. The random fingerprint technique is an elegant solution in such settings.

12.8.2 Extension to Identities of Polynomials

The fingerprinting technique due to Freivalds is fairly general and can be applied to many different versions of the identity verification problem. We now show that it can be easily extended to identity verification for symbolic polynomials, where two polynomials $P_1(x)$ and $P_2(x)$ are deemed identical if they have identical coefficients for corresponding powers of x. Verifying integer or string equality is a special case because we can represent any string of length n as a polynomial of degree n by using the kth element in the string to determine the coefficient of the kth power of a symbolic variable.

Consider first the polynomial product verification problem: Given three polynomials $P_1(x)$, $P_2(x)$, $P_3(x) \in \mathcal{F}[x]$, we are required to verify that $P_1(x) \times P_2(x) = P_3(x)$. We will assume that $P_1(x)$ and $P_2(x)$ are of degree at most n, implying that $P_3(x)$ has degree at most $2n$. Note that degree n polynomials can be multiplied in $O(n \log n)$ time via fast Fourier transforms and that the evaluation of a polynomial can be done in $O(n)$ time.

The randomized algorithm we present for polynomial product verification is similar to the algorithm for matrix product verification. It first fixes set $\mathcal{S} \subseteq \mathcal{F}$ of size at least $2n + 1$ and chooses $r \in \mathcal{S}$ uniformly at random. Then, after evaluating $P_1(r)$, $P_2(r)$, and $P_3(r)$ in $O(n)$ time, the algorithm declares the identity $P_1(x)P_2(x) = P_3(x)$ to be correct if and only if $P_1(r)P_2(r) = P_3(r)$. The algorithm makes an error only in the case where the polynomial identity is false but the value of the three polynomials at r indicates otherwise. We will show that the error event has a bounded probability.

Consider the degree $2n$ polynomial $Q(x) = P_1(x)P_2(x) - P_3(x)$. The polynomial $Q(x)$ is said to be *identically zero*, denoted by $Q(x) \equiv 0$, if each of its coefficients equals zero. Clearly, the polynomial identity $P_1(x)P_2(x) = P_3(x)$ holds if and only if $Q(x) \equiv 0$. We need to establish that if $Q(x) \not\equiv 0$, then with high probability $Q(r) = P_1(r)P_2(r) - P_3(r) \neq 0$. By elementary algebra we know that $Q(x)$ has at most $2n$ distinct roots. It follows that unless $Q(x) \equiv 0$, not more that $2n$ different choices of $r \in S$ will cause $Q(r)$ to evaluate to 0. Therefore, the error probability is at most $2n/|S|$. The probability of error can be reduced either by using independent iterations of this algorithm or by choosing a larger set S. Of course, when $\mathcal{F}$ is an infinite field (e.g., the reals), the error probability can be made 0 by choosing r uniformly from the entire field $\mathcal{F}$; however, that requires an infinite number of random bits!

Note that we could also use a deterministic version of this algorithm where each choice of $r \in S$ is tried once. But this involves $2n + 1$ different evaluations of each polynomial, and the best known algorithm for multiple evaluations needs $\Theta(n \log^2 n)$ time, which is more than the $O(n \log n)$ time requirement for actually performing a multiplication of the polynomials $P_1(x)$ and $P_2(x)$.

This verification technique is easily extended to a generic procedure for testing any polynomial identity of the form $P_1(x) = P_2(x)$ by converting it into the identity $Q(x) = P_1(x) - P_2(x) \equiv 0$. Of course, when P_1 and P_2 are explicitly provided, the identity can be deterministically verified in $O(n)$ time by comparing corresponding coefficients. Our randomized technique will take just as long to merely evaluate $P_1(x)$ and $P_2(x)$ at a random value. However, as in the case of verifying matrix identities, the randomized algorithm is quite useful in situations where the polynomials are implicitly specified, for example, when we have only a *black box* for computing the polynomials with no information about their coefficients, or when they are provided in a form where computing the actual coefficients is expensive. An example of the latter situation is provided by the following problem concerning the determinant of a symbolic matrix. In fact, the determinant problem will require a technique for the verification of polynomial identities of *multivariate* polynomials that we will discuss shortly.

Consider the $n \times n$ matrix M. Recall that the determinant of the matrix M is defined as follows:

$$\det(M) = \sum_{\pi \in S_n} \text{sgn}(\pi) \prod_{i=1}^{n} M_{i,\pi(i)} \tag{12.10}$$

where S_n is the symmetric group of permutations of order n, and $\text{sgn}(\pi)$ is the sign of a permutation π. [The sign function is defined to be $\text{sgn}(\pi) = (-1)^t$, where t is the number of pairwise exchanges required to convert the identity permutation into π.] Although the determinant is defined as a summation with $n!$ terms, it is easily evaluated in polynomial time provided that the matrix entries M_{ij} are explicitly specified. Consider the Vandermonde matrix $M(x_1, \ldots, x_n)$, which is defined in terms of the indeterminates $x_1, \ldots, x_n$ such that $M_{ij} = x_i^{j-1}$, that is,

$$M = \begin{pmatrix} 1 & x_1 & x_1^2 & \cdots & x_1^{n-1} \\ 1 & x_2 & x_2^2 & \cdots & x_2^{n-1} \\ & & \cdot & & \\ & & \cdot & & \\ & & \cdot & & \\ 1 & x_n & x_n^2 & \cdots & x_n^{n-1} \end{pmatrix}$$

It is known that for the Vandermonde matrix, $\det(M) = \prod_{i < j}(x_i - x_j)$. Consider the problem of verifying this identity without actually devising a formal proof. Computing the determinant of a symbolic matrix is infeasible as it requires dealing with a summation over $n!$ terms. However, we can formulate the identity verification problem as the problem of verifying that the polynomial $Q(x_1, \ldots, x_n) = \det(M) - \prod_{i < j}(x_i - x_j)$ is identically zero. Based on our discussion of Freivalds' technique, it is natural to consider the substitution of random values for each x_i. Because the determinant can be computed in polynomial time for any specific assignment of values to the symbolic variables $x_1, \ldots, x_n$, it is easy to evaluate the polynomial Q for random values of the variables. The only issue is that of bounding the error probability for this randomized test.

We now extend the analysis of Freivalds' technique for univariate polynomials to the multivariate case. But first, note that in a multivariate polynomial $Q(x_1, \ldots, x_n)$, the degree of a term is the sum of the exponents of the variable powers that define it, and the total degree of Q is the maximum over all terms of the degrees of the terms.

Theorem 12.9 *Let $Q(x_1, \ldots, x_n) \in \mathcal{F}[x_1, \ldots, x_n]$ be a multivariate polynomial of total degree m. Let S be a finite subset of the field $\mathcal{F}$, and let $r_1, \ldots, r_n$ be chosen uniformly and independently from S. Then*

$$\Pr[Q(r_1 \ldots, r_n) = 0 \mid Q(x_1, \ldots, x_n) \not\equiv 0] \le \frac{m}{|S|}$$

Proof 12.4 We will proceed by induction on the number of variables n. The basis of the induction is the case $n = 1$, which reduces to verifying the theorem for a univariate polynomial $Q(x_1)$ of degree m. But we have already seen for $Q(x_1) \not\equiv 0$ the probability that $Q(r_1) = 0$ is at most $m/|S|$, taking care of the basis.

We now assume that the induction hypothesis holds for multivariate polynomials with at most $n - 1$ variables, where $n > 1$. In the polynomial $Q(x_1, \ldots, x_n)$ we can factor out the variable x_1 and thereby express Q as

$$Q(x_1, \ldots, x_n) = \sum_{i=0}^{k} x_1^i P_i(x_2, \ldots, x_n)$$

where $k \le m$ is the largest exponent of x_1 in Q. Given our choice of k, the coefficient $P_k(x_2, \ldots, x_n)$ of x_1^k cannot be identically zero. Note that the total degree of P_k is at most $m - k$. Thus, by the induction hypothesis, we conclude that the probability that $P_k(r_2, \ldots, r_n) = 0$ is at most $(m - k)/|S|$.

Consider now the case where $P_k(r_2, \ldots, r_n)$ is indeed not equal to 0. We define the following univariate polynomial over x_1 by substituting the random values for the other variables in Q:

$$q(x_1) = Q(x_1, r_2, r_3, \ldots, r_n) = \sum_{i=0}^{k} x_1^i P_i(r_2, \ldots, r_n)$$

Quite clearly, the resulting polynomial $q(x_1)$ has degree k and is not identically zero (because the coefficient of x_i^k is assumed to be nonzero). As in the basis case, we conclude that the probability that $q(r_1) = Q(r_1, r_2, \ldots, r_n)$ evaluates to 0 is bounded by $k/|S|$.

By the preceding arguments, we have established the following two inequalities:

$$\Pr[P_k(r_2, \ldots, r_n) = 0] \le \frac{m - k}{|S|}$$

$$\Pr[Q(r_1, r_2, \ldots, r_n) = 0 \mid P_k(r_2, \ldots, r_n) \ne 0] \le \frac{k}{|S|}$$

Using the elementary observation that for any two events $\mathcal{E}_1$ and $\mathcal{E}_2$, $\Pr[\mathcal{E}_1] \le \Pr[\mathcal{E}_1 \mid \bar{\mathcal{E}_2}] + \Pr[\mathcal{E}_2]$, we obtain that the probability that $Q(r_1, r_2, \ldots, r_n) = 0$ is no more than the sum of the two probabilities on the right-hand side of the two obtained inequalities, which is $m/|S|$. This implies the desired results. □

This randomized verification procedure has one serious drawback: when working over large (or possibly infinite) fields, the evaluation of the polynomials could involve large intermediate values, leading to inefficient implementation. One approach to dealing with this problem in the case of integers is to perform all computations modulo some small random prime number; it can be shown that this does not have any adverse effect on the error probability.

12.8.3 Detecting Perfect Matchings in Graphs

We close by giving a surprising application of the techniques from the preceding section. Let $G(U, V, E)$ be a bipartite graph with two independent sets of vertices $U = \{u_1, \ldots, u_n\}$ and $V = \{v_1, \ldots, v_n\}$ and edges E that have one endpoint in each of U and V. We define a matching in G as a collection of edges $M \subseteq E$ such that each vertex is an endpoint of at most one edge in M; further, a perfect matching is defined to be a matching of size n, that is, where each vertex occurs as an endpoint of exactly one edge in M. Any perfect matching M may be put into a one-to-one correspondence with the permutations in S_n, where the matching corresponding to a permutation $\pi \in S_n$ is given by the collection of edges $\{(u_i, v_{\pi(i)}) \mid 1 \leq i \leq n\}$. We now relate the matchings of the graph to the determinant of a matrix obtained from the graph.

Theorem 12.10 *For any bipartite graph $G(U, V, E)$, define a corresponding $n \times n$ matrix A as follows:*

$$A_{ij} = \begin{cases} x_{ij} & (u_i, v_j) \in E \\ 0 & (u_i, v_j) \notin E \end{cases}$$

Let the multivariate polynomial $Q(x_{11}, x_{12}, \ldots, x_{nn})$ denote the determinant $\det(A)$. Then G has a perfect matching if and only if $Q \not\equiv 0$.

Proof 12.5 We can express the determinant of A as follows:

$$\det(A) = \sum_{\pi \in S_n} \text{sgn}(\pi) A_{1,\pi(1)} A_{2,\pi(2)} \cdots A_{n,\pi(n)}$$

Note that there cannot be any cancellation of the terms in the summation because each indeterminate x_{ij} occurs at most once in A. Thus, the determinant is not identically zero if and only if there exists some permutation π for which the corresponding term in the summation is nonzero. Clearly, the term corresponding to a permutation π is nonzero if and only if $A_{i,\pi(i)} \neq 0$ for each $i, 1 \leq i \leq n$; this is equivalent to the presence in G of the perfect matching corresponding to π. □

The matrix of indeterminates is sometimes referred to as the *Edmonds matrix* of a bipartite graph. The preceding result can be extended to the case of nonbipartite graphs, and the corresponding matrix of indeterminates is called the Tutte matrix. Tutte [1947] first pointed out the close connection between matchings in graphs and matrix determinants; the simpler relation between bipartite matchings and matrix determinants was given by Edmonds [1967].

We can turn the preceding result into a simple randomized procedure for testing the existence of perfect matchings in a bipartite graph (due to Lovász [1979]) — using the algorithm from the preceding subsection, determine whether the determinant is identically zero. The running time of this procedure is dominated by the cost of computing a determinant, which is essentially the same as the time required to multiply two matrices. Of course, there are algorithms for *constructing* a maximum matching in a graph with m edges and n vertices in time $O(m\sqrt{n})$ (see Hopcroft and Karp [1973], Micali and Vazirani [1980], Vazirani [1994], and Feder and Motwani [1991]). Unfortunately, the time required to compute the determinant exceeds $m\sqrt{n}$ for small m, and so the benefit in using this randomized *decision* procedure appears marginal at best. However, this technique was extended by Rabin and Vazirani [1984, 1989] to obtain simple algorithms for the actual *construction* of maximum matchings; although their randomized algorithms for matchings are simple and elegant, they are still slower than the deterministic $O(m\sqrt{n})$ time algorithms known earlier. Perhaps more significantly, this randomized decision procedure proved to be an essential ingredient in devising fast *parallel* algorithms for computing maximum matchings [Karp et al. 1988, Mulmuley et al. 1987].

Defining Terms

Deterministic algorithm: An algorithm whose execution is completely determined by its input.

Distributional complexity: The expected running time of the best possible deterministic algorithm over the worst possible probability distribution of the inputs.

Las Vegas algorithm: A randomized algorithm that always produces correct results, with the only variation from one run to another being in its running time.

Monte Carlo algorithm: A randomized algorithm that may produce incorrect results but with bounded error probability.

Randomized algorithm: An algorithm that makes random choices during the course of its execution.

Randomized complexity: The expected running time of the best possible randomized algorithm over the worst input.

References

Aleliunas, R., Karp, R. M., Lipton, R. J., Lovász, L., and Rackoff, C. 1979. Random walks, universal traversal sequences, and the complexity of maze problems. In *Proc. 20th Ann. Symp. Found. Comput. Sci.*, pp. 218–223. San Juan, Puerto Rico, Oct.

Aragon, C. R. and Seidel, R. G. 1989. Randomized search trees. In *Proc. 30th Ann. IEEE Symp. Found. Comput. Sci.*, pp. 540–545.

Ben-David, S., Borodin, A., Karp, R. M., Tardos, G., and Wigderson, A. 1994. On the power of randomization in on-line algorithms. *Algorithmica* 11(1):2–14.

Blum, M. and Kannan, S. 1989. Designing programs that check their work. In *Proc. 21st Annu. ACM Symp. Theory Comput.*, pp. 86–97. ACM.

Coppersmith, D. and Winograd, S. 1990. Matrix multiplication via arithmetic progressions. *J. Symbolic Comput.* 9:251–280.

DeMillo, R. A. and Lipton, R. J. 1978. A probabilistic remark on algebraic program testing. *Inf. Process. Lett.* 7:193–195.

Edmonds, J. 1967. Systems of distinct representatives and linear algebra. *J. Res. Nat. Bur. Stand.* 71B, 4:241–245.

Feder, T. and Motwani, R. 1991. Clique partitions, graph compression and speeding-up algorithms. In *Proc. 25th Annu. ACM Symp. Theory Comput.*, pp. 123–133.

Floyd, R. W. and Rivest, R. L. 1975. Expected time bounds for selection. *Commun. ACM* 18:165–172.

Freivalds, R. 1977. Probabilistic machines can use less running time. In *Inf. Process. 77, Proc. IFIP Congress 77*, B. Gilchrist, Ed., pp. 839–842, North-Holland, Amsterdam, Aug.

Goemans, M. X. and Williamson, D. P. 1994. 0.878-approximation algorithms for MAX-CUT and MAX-2SAT. In *Proc. 26th Annu. ACM Symp. Theory Comput.*, pp. 422–431.

Hoare, C. A. R. 1962. Quicksort. *Comput. J.* 5:10–15.

Hopcroft, J. E. and Karp, R. M. 1973. An $n^{5/2}$ algorithm for maximum matching in bipartite graphs. *SIAM J. Comput.* 2:225–231.

Karger, D. R. 1993. Global min-cuts in $\mathcal{RNC}$, and other ramifications of a simple min-cut algorithm. In *Proc. 4th Annu. ACM–SIAM Symp. Discrete Algorithms.*

Karger, D. R., Klein, P. N., and Tarjan, R. E. 1995. A randomized linear-time algorithm for finding minimum spanning trees. *J. ACM* 42:321–328.

Karger, D., Motwani, R., and Sudan, M. 1994. Approximate graph coloring by semidefinite programming. In *Proc. 35th Annu. IEEE Symp. Found. Comput. Sci.*, pp. 2–13.

Karp, R. M. 1991. An introduction to randomized algorithms. *Discrete Appl. Math.* 34:165–201.

Karp, R. M., Upfal, E., and Wigderson, A. 1986. Constructing a perfect matching is in random $\mathcal{NC}$. *Combinatorica* 6:35–48.

Karp, R. M., Upfal, E., and Wigderson, A. 1988. The complexity of parallel search. *J. Comput. Sys. Sci.* 36:225–253.

Lovász, L. 1979. On determinants, matchings and random algorithms. In *Fundamentals of Computing Theory*. L. Budach, Ed. Akademia-Verlag, Berlin.

Maffioli, F., Speranza, M. G., and Vercellis, C. 1985. Randomized algorithms. In *Combinatorial Optimization: Annotated Bibliographies*, M. O'Eigertaigh, J. K. Lenstra, and A. H. G. Rinooy Kan, Eds., pp. 89–105. Wiley, New York.

Micali, S. and Vazirani, V. V. 1980. An $O(\sqrt{|V|}|e|)$ algorithm for finding maximum matching in general graphs. In *Proc. 21st Annu. IEEE Symp. Found. Comput. Sci.*, pp. 17–27.

Motwani, R., Naor, J., and Raghavan, P. 1996. Randomization in approximation algorithms. In *Approximation Algorithms*, D. Hochbaum, Ed. PWS.

Motwani, R. and Raghavan, P. 1995. *Randomized Algorithms*. Cambridge University Press, New York.

Mulmuley, K. 1993. *Computational Geometry: An Introduction through Randomized Algorithms*. Prentice Hall, New York.

Mulmuley, K., Vazirani, U. V., and Vazirani, V. V. 1987. Matching is as easy as matrix inversion. *Combinatorica* 7:105–113.

Pugh, W. 1990. Skip lists: a probabilistic alternative to balanced trees. *Commun. ACM* 33(6):668–676.

Rabin, M. O. 1980. Probabilistic algorithm for testing primality. *J. Number Theory* 12:128–138.

Rabin, M. O. 1983. Randomized Byzantine generals. In *Proc. 24th Annu. Symp. Found. Comput. Sci.*, pp. 403–409.

Rabin, M. O. and Vazirani, V. V. 1984. Maximum matchings in general graphs through randomization. *Aiken Computation Lab. Tech. Rep.* TR-15-84, Harvard University, Cambridge, MA.

Rabin, M. O. and Vazirani, V. V. 1989. Maximum matchings in general graphs through randomization. *J. Algorithms* 10:557–567.

Raghavan, P. and Snir, M. 1994. Memory versus randomization in on-line algorithms. *IBM J. Res. Dev.* 38:683–707.

Saks, M. and Wigderson, A. 1986. Probabilistic Boolean decision trees and the complexity of evaluating game trees. In *Proc. 27th Annu. IEEE Symp. Found. Comput. Sci.*, pp. 29–38. Toronto, Ontario.

Schrijver, A. 1986. *Theory of Linear and Integer Programming*. Wiley, New York.

Schwartz, J. T. 1987. Fast probabilistic algorithms for verification of polynomial identities. *J. ACM* 27(4):701–717.

Seidel, R. G. 1991. Small-dimensional linear programming and convex hulls made easy. *Discrete Comput. Geom.* 6:423–434.

Sinclair, A. 1992. *Algorithms for Random Generation and Counting: A Markov Chain Approach, Progress in Theoretical Computer Science*. Birkhauser, Boston, MA.

Snir, M. 1985. Lower bounds on probabilistic linear decision trees. *Theor. Comput. Sci.* 38:69–82.

Solovay, R. and Strassen, V. 1977. A fast Monte-Carlo test for primality. *SIAM J. Comput.* 6(1):84–85. *See also* 1978. *SIAM J. Comput.* 7(Feb.):118.

Tarsi, M. 1983. Optimal search on some game trees. *J. ACM* 30:389–396.

Tutte, W. T. 1947. The factorization of linear graphs. *J. London Math. Soc.* 22:107–111.

Valiant, L. G. 1982. A scheme for fast parallel communication. *SIAM J. Comput.* 11:350–361.

Vazirani, V. V. 1994. A theory of alternating paths and blossoms for proving correctness of $O(\sqrt{VE})$ graph maximum matching algorithms. *Combinatorica* 14(1):71–109.

Welsh, D. J. A. 1983. Randomised algorithms. *Discrete Appl. Math.* 5:133–145.

Yao, A. C.-C. 1977. Probabilistic computations: towards a unified measure of complexity. In *Proc. 17th Annu. Symp. Found. Comput. Sci.*, pp. 222–227.

Zippel, R. E. 1979. Probabilistic algorithms for sparse polynomials. In *Proc. EUROSAM 79*, Vol. 72, Lecture Notes in Computer Science., pp. 216–226. Marseille, France.

Further Information

In this section we give pointers to a plethora of randomized algorithms not covered in this chapter. The reader should also note that the examples in the text are but a (random!) sample of the many randomized algorithms for each of the problems considered. These algorithms have been chosen to illustrate the main ideas behind randomized algorithms rather than to represent the state of the art for these problems. The reader interested in other algorithms for these problems is referred to Motwani and Raghavan [1995].

Randomized algorithms also find application in a number of other areas: in load balancing [Valiant 1982], approximation algorithms and combinatorial optimization [Goemans and Williamson 1994, Karger et al. 1994, Motwani et al. 1996], graph algorithms [Aleliunas et al. 1979, Karger et al. 1995], data structures [Aragon and Seidel 1989], counting and enumeration [Sinclair 1992], parallel algorithms [Karp et al. 1986, 1988], distributed algorithms [Rabin 1983], geometric algorithms [Mulmuley 1993], on-line algorithms [Ben-David et al. 1994, Raghavan and Snir 1994], and number-theoretic algorithms [Rabin 1983, Solovay and Strassen 1977]. The reader interested in these applications may consult these articles or Motwani and Raghavan [1995].

13

Pattern Matching and Text Compression Algorithms

13.1 Processing Texts Efficiently **13-1**

13.2 String-Matching Algorithms **13-2**
Karp–Rabin Algorithm • Knuth–Morris–Pratt Algorithm
• Boyer–Moore Algorithm • Quick Search Algorithm
• Experimental Results • Aho–Corasick Algorithm

13.3 Two-Dimensional Pattern Matching Algorithms **13-14**
Zhu–Takaoka Algorithm • Bird/Baker Algorithm

13.4 Suffix Trees **13-17**
McCreight Algorithm

13.5 Alignment **13-20**
Global alignment • Local Alignment • Longest Common
Subsequence of Two Strings • Reducing the Space: Hirschberg
Algorithm

13.6 Approximate String Matching **13-29**
Shift-Or Algorithm • String Matching with k Mismatches
• String Matching with k Differences • Wu–Manber
Algorithm

13.7 Text Compression **13-36**
Huffman Coding • Lempel–Ziv–Welsh (LZW) Compression
• Mixing Several Methods

13.8 Research Issues and Summary **13-45**

Maxime Crochemore
University of Marne-la-Vallée
and King's College London

Thierry Lecroq
University of Rouen

13.1 Processing Texts Efficiently

The present chapter describes a few standard algorithms used for processing texts. They apply, for example, to the manipulation of texts (text editors), to the storage of textual data (text compression), and to data retrieval systems. The algorithms of this chapter are interesting in different respects. First, they are basic components used in the implementations of practical software. Second, they introduce programming methods that serve as paradigms in other fields of computer science (system or software design). Third, they play an important role in theoretical computer science by providing challenging problems.

Although data is stored in various ways, text remains the main form of exchanging information. This is particularly evident in literature or linguistics where data is composed of huge corpora and dictionaries. This applies as well to computer science, where a large amount of data is stored in linear files. And this is also the case in molecular biology where biological molecules can often be approximated as sequences of nucleotides or amino acids. Moreover, the quantity of available data in these fields tends to double every 18 months. This is the reason why algorithms should be efficient even if the speed of computers increases at a steady pace.

Pattern matching is the problem of locating a specific pattern inside raw data. The pattern is usually a collection of strings described in some formal language. Two kinds of textual patterns are presented: single strings and approximated strings. We also present two algorithms for matching patterns in images that are extensions of string-matching algorithms.

In several applications, texts need to be structured before being searched. Even if no further information is known about their syntactic structure, it is possible and indeed extremely efficient to build a data structure that supports searches. From among several existing data structures equivalent to represent indexes, we present the suffix tree, along with its construction.

The comparison of strings is implicit in the approximate pattern searching problem. Because it is sometimes required to compare just two strings (files or molecular sequences), we introduce the basic method based on longest common subsequences.

Finally, the chapter contains two classical text compression algorithms. Variants of these algorithms are implemented in practical compression software, in which they are often combined together or with other elementary methods. An example of mixing different methods is presented there.

The efficiency of algorithms is evaluated by their running times, and sometimes by the amount of memory space they require at runtime as well.

13.2   String-Matching Algorithms

String matching is the problem of finding one or, more generally, all the **occurrences** of a pattern in a text. The pattern and the text are both strings built over a finite alphabet (a finite set of symbols). Each algorithm of this section outputs all occurrences of the pattern in the text. The pattern is denoted by $x = x[0 .. m - 1]$; its length is equal to m. The text is denoted by $y = y[0 .. n - 1]$; its length is equal to n. The alphabet is denoted by Σ and its size is equal to σ.

String-matching algorithms of the present section work as follows: they first align the left ends of the pattern and the text, then compare the aligned symbols of the text and the pattern — this specific work is called an attempt or a scan, and after a whole match of the pattern or after a mismatch, they shift the pattern to the right. They repeat the same procedure again until the right end of the pattern goes beyond the right end of the text. This is called the scan and shift mechanism. We associate each attempt with the position j in the text, when the pattern is aligned with $y[j .. j + m - 1]$.

The brute-force algorithm consists of checking, at all positions in the text between 0 and $n - m$, whether an occurrence of the pattern starts there or not. Then, after each attempt, it shifts the pattern exactly one position to the right. This is the simplest algorithm, which is described in Figure 13.1.

The time complexity of the brute-force algorithm is $O(mn)$ in the worst case but its behavior in practice is often linear on specific data.

```
BF(x, m, y, n)
1   ▷ Searching
2   for j ← 0 to n − m
3       do i ← 0
4           while i < m and x[i] = y[i + j]
5               do i ← i + 1
6           if i ≥ m
7               then OUTPUT(j)
```

FIGURE 13.1   The brute-force string-matching algorithm.

13.2.1 Karp–Rabin Algorithm

Hashing provides a simple method for avoiding a quadratic number of symbol comparisons in most practical situations. Instead of checking at each position of the text whether the pattern occurs, it seems to be more efficient to check only if the portion of the text aligned with the pattern "looks like" the pattern. To check the resemblance between these portions, a hashing function is used. To be helpful for the string-matching problem, the hashing function should have the following properties:

- Efficiently computable
- Highly discriminating for strings
- $hash(y[j + 1 .. j + m])$ must be easily computable from $hash(y[j .. j + m - 1])$;
  $hash(y[j + 1 .. j + m]) = \text{ReHash}(y[j], y[j + m], hash(y[j .. j + m - 1]))$

For a word w of length k, its symbols can be considered as digits, and we define $hash(w)$ by:

$$hash(w[0 .. k - 1]) = (w[0] \times 2^{k-1} + w[1] \times 2^{k-2} + \cdots + w[k - 1]) \bmod q$$

where q is a large number. Then, ReHash has a simple expression

$$\text{ReHash}(a, b, h) = ((h - a \times d) \times 2 + b) \bmod q$$

where $d = 2^{k-1}$ and q is the computer word-size (see Figure 13.2).

During the search for the pattern x, $hash(x)$ is compared with $hash(y[j - m + 1 .. j])$ for $m - 1 \le j \le n - 1$. If an equality is found, it is still necessary to check the equality $x = y[j - m + 1 .. j]$ symbol by symbol.

In the algorithms of Figures 13.2 and 13.3, all multiplications by 2 are implemented by shifts (operator $<<$). Furthermore, the computation of the modulus function is avoided by using the implicit modular

```
ReHash(a, b, h)
1    return ((h - a × d) << 1) + b
```

FIGURE 13.2   Function ReHash

```
KR(x, m, y, n)
 1    ▷ Preprocessing
 2    d ← 1
 3    for i ← 1 to m − 1
 4        do d ← d << 1
 5    h_x ← 0
 6    h_y ← 0
 7    for i ← 0 to m − 1
 8        do h_x ← (h_x << 1) + x[i]
 9           h_y ← (h_y << 1) + y[i]
10    ▷ Searching
11    if h_x = h_y and x = y[0 .. m − 1]
12       then Output(0)
13    j ← m
14    while j < n
15        do h_y ← ReHash(y[j − m], y[j], h_y)
16           if h_x = h_y and x = y[j − m + 1 .. j]
17              then Output(j − m + 1)
18           j ← j + 1
```

FIGURE 13.3   The Karp–Rabin string-matching algorithm.

arithmetic given by the hardware that forgets carries in integer operations. Thus, q is chosen as the maximum value of an integer of the system.

The worst-case time complexity of the Karp–Rabin algorithm (Figure 13.3) is quadratic (as it is for the brute-force algorithm), but its expected running time is $O(m + n)$.

Example 13.1

Let $x = \mathbf{ing}$. Then, $hash(x) = 105 \times 2^2 + 110 \times 2 + 103 = 743$ (symbols are assimilated with their ASCII codes):

$$
\begin{array}{lccccccccccccccc}
y = & s & t & r & i & n & g & & m & a & t & c & h & i & n & g \\
hash = & & 806 & 797 & 776 & 743 & 678 & 585 & 443 & 746 & 719 & 766 & 709 & 736 & 743
\end{array}
$$

13.2.2 Knuth–Morris–Pratt Algorithm

This section presents the first discovered linear-time string-matching algorithm. Its design follows a tight analysis of the brute-force algorithm, and especially the way this latter algorithm wastes the information gathered during the scan of the text.

Let us look more closely at the brute-force algorithm. It is possible to improve the length of shifts and simultaneously remember some portions of the text that match the pattern. This saves comparisons between characters of the text and of the pattern, and consequently increases the speed of the search.

Consider an attempt at position j, that is, when the pattern $x[0 \, .. \, m - 1]$ is aligned with the segment $y[j \, .. \, j + m - 1]$ of the text. Assume that the first mismatch (during a left-to-right scan) occurs between symbols $x[i]$ and $y[i + j]$ for $0 \leq i < m$. Then, $x[0 \, .. \, i - 1] = y[j \, .. \, i + j - 1] = u$ and $a = x[i] \neq y[i + j] = b$. When shifting, it is reasonable to expect that a **prefix** v of the pattern matches some **suffix** of the portion u of the text. Moreover, if we want to avoid another immediate mismatch, the letter following the prefix v in the pattern must be different from a. (Indeed, it should be expected that v matches a suffix of ub, but elaborating along this idea goes beyond the scope of the chapter.) The longest such prefix v is called the **border** of u (it occurs at both ends of u). This introduces the notation: let $next[i]$ be the length of the longest (proper) border of $x[0 \, .. \, i - 1]$, followed by a character c different from $x[i]$. Then, after a shift, the comparisons can resume between characters $x[next[i]]$ and $y[i + j]$ without missing any occurrence of x in y and having to backtrack on the text (see Figure 13.4).

Example 13.2

Here,

$$
\begin{array}{lccccccccccccccc}
y = & . & . & . & a & b & a & b & a & a & b & . & . & . & . & . \\
x = & & & & \underline{a} & \underline{b} & \underline{a} & \underline{b} & \underline{a} & \underline{b} & a \\
x = & & & & & & & \underline{a} & \underline{b} & a & b & a & b & a
\end{array}
$$

Compared symbols are <u>underlined</u>. Note that the empty string is the suitable border of **ababa**. Other borders of **ababa** are **aba** and **a**.

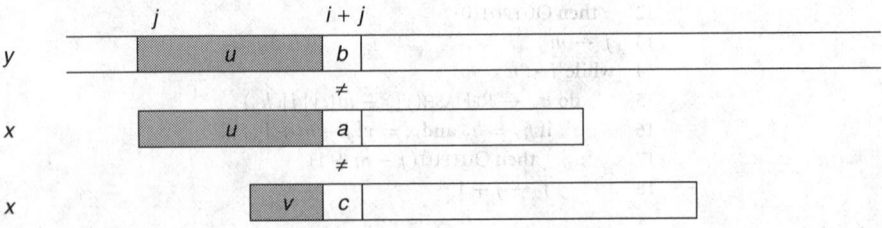

FIGURE 13.4 Shift in the Knuth–Morris–Pratt algorithm (v suffix of u).

```
KMP(x, m, y, n)
 1  ▷ Preprocessing
 2  next ← PREKMP(x, m)
 3  ▷ Searching
 4  i ← 0
 5  j ← 0
 6  while j < n
 7      do while i > −1 and x[i] ≠ y[j]
 8          do i ← next[i]
 9         i ← i + 1
10         j ← j + 1
11         if i ≥ m
12             then OUTPUT(j − i)
13                 i ← next[i]
```

FIGURE 13.5   The Knuth–Morris–Pratt string-matching algorithm.

```
PREKMP(x, m)
 1  i ← −1
 2  j ← 0
 3  next[0] ← −1
 4  while j < m
 5      do while i > −1 and x[i] ≠ x[j]
 6          do i ← next[i]
 7         i ← i + 1
 8         j ← j + 1
 9         if x[i] = x[j]
10             then next[j] ← next[i]
11             else next[j] ← i
12  return next
```

FIGURE 13.6   Preprocessing phase of the Knuth–Morris–Pratt algorithm: computing *next*.

The Knuth–Morris–Pratt algorithm is displayed in Figure 13.5. The table *next* it uses is computed in $O(m)$ time before the search phase, applying the same searching algorithm to the pattern itself, as if $y = x$ (see Figure 13.6). The worst-case running time of the algorithm is $O(m + n)$ and it requires $O(m)$ extra space. These quantities are independent of the size of the underlying alphabet.

13.2.3   Boyer–Moore Algorithm

The Boyer–Moore algorithm is considered the most efficient string-matching algorithm in usual applications. A simplified version of it, or the entire algorithm, is often implemented in text editors for the search and substitute commands.

The algorithm scans the characters of the pattern from right to left, beginning with the rightmost symbol. In case of a mismatch (or a complete match of the whole pattern), it uses two precomputed functions to shift the pattern to the right. These two shift functions are called the *bad-character shift* and the *good-suffix shift*. They are based on the following observations.

Assume that a mismatch occurs between the character $x[i] = a$ of the pattern and the character $y[i+j] = b$ of the text during an attempt at position j. Then, $x[i+1 .. m−1] = y[i+j+1 .. j+m−1] = u$ and $x[i] \neq y[i + j]$. The good-suffix shift consists in aligning the **segment** $y[i + j + 1 .. j + m − 1]$ with its rightmost occurrence in x that is preceded by a character different from $x[i]$ (see Figure 13.7). If there exists no such segment, the shift consists in aligning the longest suffix v of $y[i + j + 1 .. j + m − 1]$ with a matching prefix of x (see Figure 13.8).

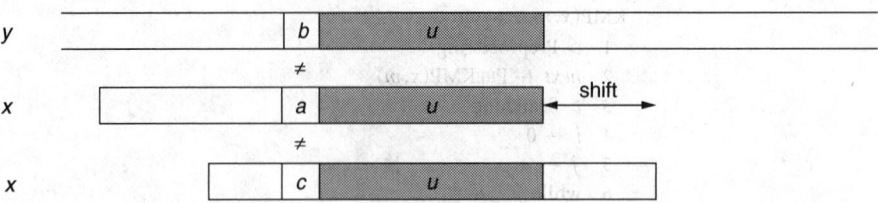

FIGURE 13.7   The good-suffix shift, when *u* reappears, preceded by a character different from *a*.

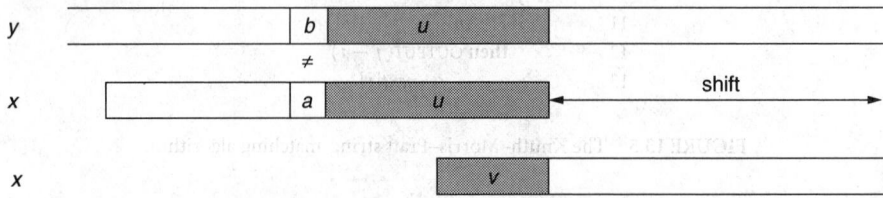

FIGURE 13.8   The good-suffix shift, when the situation of Figure 13.7 does not happen, only a suffix of *u* reappears as a prefix of *x*.

Example 13.3

Here,

$$
\begin{array}{cccccccccccccccccc}
y = & . & . & . & a & b & b & a & a & b & b & a & b & b & a & . & . & . \\
x = & a & b & b & a & a & b & b & a & b & b & a & & & & & & \\
x = & & & & a & b & b & a & a & b & b & a & b & b & a & & & \\
\end{array}
$$

The shift is driven by the suffix **abba** of *x* found in the text. After the shift, the segment **abba** in the middle of *y* matches a segment of *x* as in Figure 13.7. The same mismatch does not recur.

Example 13.4

Here,

$$
\begin{array}{cccccccccccccccc}
y = & . & . & . & a & b & b & a & a & b & b & a & b & b & a & b & b & a & . & . \\
x = & & b & b & a & b & b & a & b & b & a & & & & & & \\
x = & & & & & & & & b & b & a & b & b & a & b & b & a & \\
\end{array}
$$

The segment **abba** found in *y* partially matches a prefix of *x* after the shift, as in Figure 13.8.

The bad-character shift consists in aligning the text character $y[i + j]$ with its rightmost occurrence in $x[0 .. m - 2]$ (see Figure 13.9). If $y[i + j]$ does not appear in the pattern *x*, no occurrence of *x* in *y* can overlap the symbol $y[i + j]$, then the left end of the pattern is aligned with the character at position $i + j + 1$ (see Figure 13.10).

Example 13.5

Here,

$$
\begin{array}{cccccccccccc}
y = & . & . & . & . & . & a & b & c & d & . & . & . \\
x = & c & d & a & h & g & f & e & b & c & d & \\
x = & & & & c & d & a & h & g & f & e & b & c & d \\
\end{array}
$$

The shift aligns the symbol **a** in *x* with the mismatch symbol **a** in the text *y* (Figure 13.9).

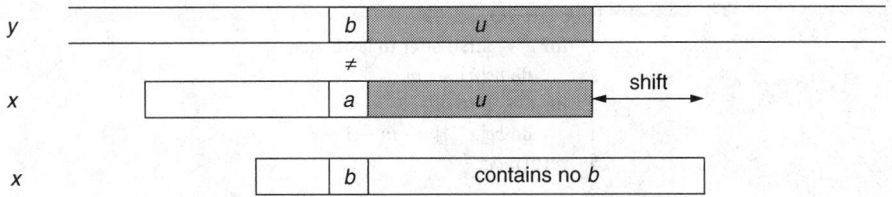

FIGURE 13.9 The bad-character shift, b appears in x.

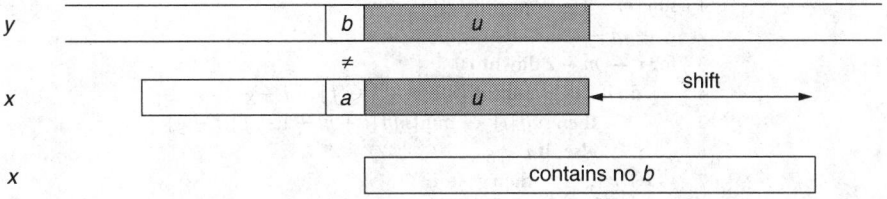

FIGURE 13.10 The bad-character shift, b does not appear in x (except possibly at $m - 1$).

```
BM(x, m, y, n)
 1   ▷ Preprocessing
 2   gs ← PREGS(x, m)
 3   bc ← PREBC(x, m)
 4   ▷ Preprocessing
 5   j ← 0
 6   while j ≤ n − m
 7       do i ← m − 1
 8           while i ≥ 0 and x[i] = y[i + j]
 9               do i ← i − 1
10           if i < 0
11               then OUTPUT(j)
12           j ← max{gs[i + 1], bc[y[i + j] − m + i + 1]}
```

FIGURE 13.11 The Boyer–Moore string-matching algorithm.

Example 13.6
Here,

$$
\begin{array}{llllllllllll}
y = & . & . & . & . & . & \mathbf{a} & \mathbf{b} & \mathbf{c} & \mathbf{d} & . & . & . & . & . & . & . \\
x = & \mathbf{c} & \mathbf{d} & \mathbf{h} & \mathbf{g} & \mathbf{f} & \mathbf{e} & \underline{\mathbf{b}} & \underline{\mathbf{c}} & \underline{\mathbf{d}} \\
x = & & & & & & \mathbf{c} & \mathbf{d} & \mathbf{h} & \mathbf{g} & \mathbf{f} & \mathbf{e} & \mathbf{b} & \mathbf{c} & \underline{\mathbf{d}}
\end{array}
$$

The shift positions the left end of x right after the symbol **a** of y (Figure 13.10).

The Boyer–Moore algorithm is shown in Figure 13.11. For shifting the pattern, it applies the maximum between the bad-character shift and the good-suffix shift. More formally, the two shift functions are defined as follows. The bad-character shift is stored in a table bc of size σ and the good-suffix shift is stored in a table gs of size $m + 1$. For $a \in \Sigma$

$$
bc[a] = \begin{cases} \min\{i \mid 1 \le i < m \text{ and } x[m - 1 - i] = a\} & \text{if } a \text{ appears in } x, \\ m & \text{otherwise.} \end{cases}
$$

PreBC(x, m)

1   **for** $a \leftarrow$ firstLetter **to** lastLetter
2   **do** $bc[a] \leftarrow m$
3   **for** $i \leftarrow 0$ **to** $m - 2$
4   **do** $bc[x[i]] \leftarrow m - 1 - i$
5   **return** bc

FIGURE 13.12   Computation of the bad-character shift.

Suffixes(x, m)

1   $suff[m - 1] \leftarrow m$
2   $g \leftarrow m - 1$
3   **for** $i \leftarrow m - 2$ **downto** 0
4   **do if** $i > g$ **and** $suff[i + m - 1 - f] \neq i - g$
5   **then** $suff[i] \leftarrow \min\{suff[i + m - 1 - f], i - g\}$
6   **else if** $i < g$
7   **then** $g \leftarrow i$
8                        $f \leftarrow i$
9   **while** $g \geq 0$ **and** $x[g] = x[g + m - 1 - f]$
10   **do** $g \leftarrow g - 1$
11                       $suff[i] \leftarrow f - g$
12 **return** $suff$

FIGURE 13.13   Computation of the table *suff*.

Let us define two conditions,

$$\begin{cases} cond_1(i, s): \text{ for each } k \text{ such that } i < k < m, s \geq k \text{ or } x[k - s] = x[k], \\ cond_2(i, s): \text{ if} s < i \text{ then } x[i - s] \neq x[i]. \end{cases}$$

Then, for $0 \leq i < m$,

$$gs[i + 1] = \min\{s > 0 \mid cond_1(i, s) \text{ and } cond_2(i, s) \text{ hold}\}$$

and we define $gs[0]$ as the length of the smallest period of x.

To compute the table gs, a table *suff* is used. This table can be defined as follows: for $i = 0, 1, \ldots, m - 1$,

$$suff[i] = \text{longest common suffix between } x[0 .. i] \text{ and } x.$$

It is computed in linear time and space by the function Suffixes (see Figure 13.13).

Tables bc and gs can be precomputed in time $O(m + \sigma)$ before the search phase and require an extra space in $O(m + \sigma)$ (see Figure 13.12 and Figure 13.14). The worst-case running time of the algorithm is quadratic. However, on large alphabets (relative to the length of the pattern), the algorithm is extremely fast. Slight modifications of the strategy yield linear-time algorithms (see the bibliographic notes). When searching for a^m in $(a^{m-1}b)^{\lfloor n/m \rfloor}$, the algorithm makes only $O(n/m)$ comparisons, which is the absolute minimum for any string-matching algorithm in the model where the pattern only is preprocessed.

13.2.4   Quick Search Algorithm

The bad-character shift used in the Boyer–Moore algorithm is not very efficient for small alphabets; but when the alphabet is large compared with the length of the pattern, as is often the case with the ASCII table and ordinary searches made under a text editor, it becomes very useful. Using it alone produces a practically very efficient algorithm that is described now.

After an attempt where x is aligned with $y[j .. j + m - 1]$, the length of the shift is at least equal to one. Thus, the character $y[j + m]$ is necessarily involved in the next attempt, and thus can be used for

```
PREGS(x, m)
 1   gs ← SUFFIXES(x, m)
 2   for i ← 0 to m − 1
 3       do gs[i] ← m
 4   j ← 0
 5   for i ← m − 1 downto −1
 6       do if i = −1 or suff[i] = i + 1
 7           then while j < m − 1 − i
 8               do if gs[j] = m
 9                   then gs[j] ← m − 1 − i
10                   j ← j + 1
11   for i ← 0 to m − 2
12       do gs[m − 1 − suff[i]] ← m − 1 − i
13   return gs
```

FIGURE 13.14   Computation of the good-suffix shift.

```
QS(x, m, y, n)
 1   ▷ Preprocessing
 2   for a ← firstLetter to lastLetter
 3       do bc[a] ← m + 1
 4   for i ← 0 to m − 1
 5       do bc[x[i]] ← m − i
 6   ▷ Searching
 7   j ← 0
 8   while j ≤ n − m
 9       do i ← 0
10           while i ≥ 0 and x[i] = y[i + j]
11               do i ← i + 1
12           if i ≥ m
13               then OUTPUT(j)
14           j ← bc[y[j + m]]
```

FIGURE 13.15   The Quick Search string-matching algorithm.

the bad-character shift of the current attempt. In the present algorithm, the bad-character shift is slightly modified to take into account the observation as follows ($a \in \Sigma$):

$$bc[a] = 1 + \begin{cases} \min\{i \mid 0 \le i < m \text{ and } x[m − 1 − i] = a\} & \text{if } a \text{ appears in } x, \\ m & \text{otherwise.} \end{cases}$$

Indeed, the comparisons between text and pattern characters during each attempt can be done in any order. The algorithm of Figure 13.15 performs the comparisons from left to right. It is called Quick Search after its inventor and has a quadratic worst-case time complexity but good practical behavior.

Example 13.7

Here,

```
y = s t r i n g - m a t c h i n g
x = i n g
x =       i n g
x =                 i n g
x =                   i n g
x =                         i n g
```

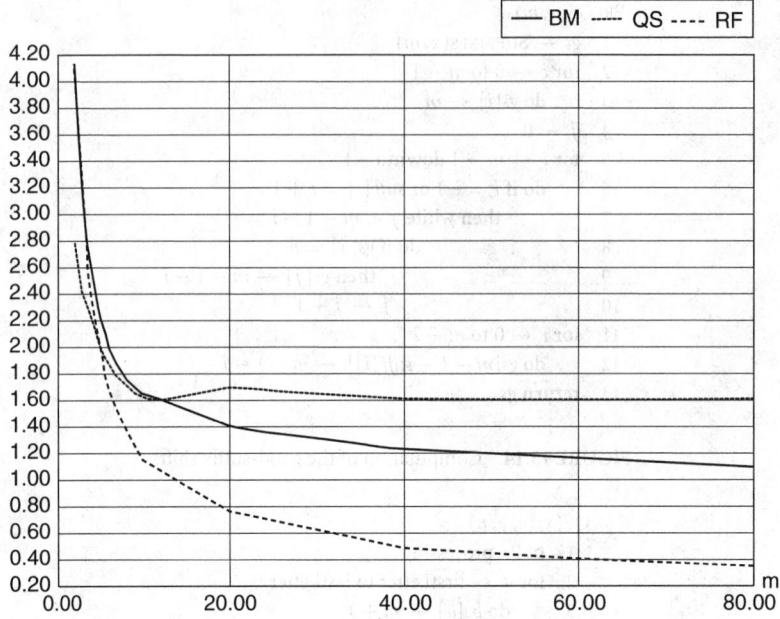

FIGURE 13.16   Running times for a DNA sequence.

The Quick Search algorithm makes only nine comparisons to find the two occurrences of **ing** inside the text of length 15.

13.2.5 Experimental Results

In Figure 13.16 and Figure 13.17, we present the running times of three string-matching algorithms: the Boyer–Moore algorithm (BM), the Quick Search algorithm (QS), and the Reverse-Factor algorithm (RF). The Reverse–Factor algorithm can be viewed as a variation of the Boyer–Moore algorithm where factors (segments) rather than suffixes of the pattern are recognized. The RF algorithm uses a data structure to store all the factors of the reversed pattern: a suffix automaton or a **suffix tree**.

Tests have been performed on various types of texts. In Figure 13.16 we show the results when the text is a DNA sequence on the four-letter alphabet of nucleotides **A**, **C**, **G**, **T**. In Figure 13.17 English text is considered.

For each pattern length, we ran a large number of searches with random patterns. The average time according to the length is shown in the two figures. The running times of both preprocessing and searching phases are added. The three algorithms are implemented in a homogeneous way in order to keep the comparison significant.

For the genome, as expected, the QS algorithm is the best for short patterns. But for long patterns it is less efficient than the BM algorithm. In this latter case, the RF algorithm achieves the best results. For rather large alphabets, as is the case for an English text, the QS algorithm remains better than the BM algorithm whatever the pattern length is. In this case, the three algorithms have similar behaviors; however, the QS is better for short patterns (which is typical of search under a text editor) and the RF is better for large patterns.

13.2.6 Aho–Corasick Algorithm

The Unix operating system provides standard text (or file) facilities. Among them is the series of **grep** commands that locate patterns in files. We describe in this section the algorithm underlying the **fgrep**

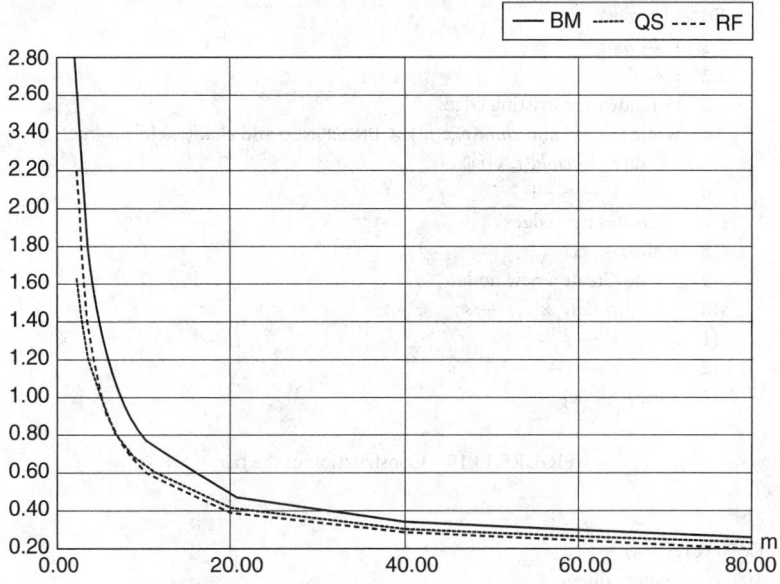

FIGURE 13.17 Running times for an English text.

```
PREAC(X, k)
 1    Create a new node root
 2    ▷ creates a loop on the root of the trie
 3    for a ∈ Σ
 4        do child(root, a) ← root
 5    ▷ enters each pattern in the trie
 6    for i ← 0 to k − 1
 7        do ENTER(X[i], root)
 8    ▷ completes the trie with failure links
 9    COMPLETE(root)
10    return root
```

FIGURE 13.18 Preprocessing phase of the Aho–Corasick algorithm.

command of Unix. It searches files for a finite set of strings, and can, for instance, output lines containing at least one of the strings.

If we are interested in searching for all occurrences of all patterns taken from a finite set of patterns, a first solution consists in repeating some string-matching algorithm for each pattern. If the set contains k patterns, this search runs in time $O(kn)$. The solution described in the present section and designed by Aho and Corasick runs in time $O(n \log \sigma)$. The algorithm is a direct extension of the Knuth–Morris–Pratt algorithm, and the running time is independent of the number of patterns.

Let $X = \{x_0, x_1, \ldots, x_{k-1}\}$ be the set of patterns, and let $|X| = |x_0| + |x_1| + \cdots + |x_{k-1}|$ be the total size of the set X. The Aho–Corasick algorithm first builds a **trie** $T(X)$, a digital tree recognizing the patterns of X. The trie $T(X)$ is a tree in which edges are labeled by letters and in which branches spell the patterns of X. We identify a node p in the trie $T(X)$ with the unique word w spelled by the path of $T(X)$ from its root to p. The root itself is identified with the empty word ε. Notice that if w is a node in $T(X)$ then w is a prefix of some $x_i \in X$. If w is a node in $T(X)$ and $a \in \Sigma$ then $child(w, a)$ is equal to wa if wa is a node in $T(X)$; it is equal to UNDEFINED otherwise.

The function PREAC in Figure 13.18 returns the trie of all patterns. During the second phase, where patterns are entered in the trie, the algorithm initializes an output function *out*. It associates the singleton

```
ENTER(x, root)
 1   r ← root
 2   i ← 0
 3   ▷ follows the existing edges
 4   while i < |x| and child(r, x[i]) ≠ UNDEFINED and child(r, x[i]) ≠ root
 5       do r ← child(r, x[i])
 6          i ← i + 1
 7   ▷ creates new edges
 8   while i < |x|
 9       do Create a new node s
10          child(r, x[i]) ← s
11          r ← s
12          i ← i + 1
13   out(r) ← {x}
```

FIGURE 13.19   Construction of the trie.

```
COMPLETE(root)
 1   q ← empty queue
 2   ℓ ← list of the edges (root, a, p) for any character a ∈ Σ and any node p ≠ root
 3   while the list ℓ is not empty
 4       do (r, a, p) ← FIRST(ℓ)
 5          ℓ ← NEXT(ℓ)
 6          ENQUEUE(q, p)
 7          fail(p) ← root
 8   while the queue q is not empty
 9       do r ← DEQUEUE(q)
10          ℓ ← list of the edges (r, a, p) for any character a ∈ Σ and any node p
11          while the list ℓ is not empty
12              do (r, a, p) ← FIRST(ℓ)
13                 ℓ ← NEXT(ℓ)
14                 ENQUEUE(q, p)
15                 s ← fail(r)
16                 while child(s, a) = UNDEFINED
17                     do s ← fail(s)
18                 fail(p) ← child(s, a)
19                 out(p) ← out(p) ∪ out(child(s, a))
```

FIGURE 13.20   Completion of the output function and construction of failure links.

$\{x_i\}$ with the nodes x_i $(0 \leq i < k)$, and associates the empty set with all other nodes of $T(X)$ (see Figure 13.19).

Finally, the last phase of function PREAC (Figure 13.18) consists in building the failure link of each node of the trie, and simultaneously completing the output function. This is done by the function COMPLETE in Figure 13.20. The failure function *fail* is defined on nodes as follows (w is a node):

$$fail(w) = u \quad \text{where } u \text{ is the longest proper suffix of } w \text{ that belongs to } T(X).$$

Computation of failure links is done during a breadth-first traversal of $T(X)$. Completion of the output function is done while computing the failure function *fail* using the following rule:

$$\text{if } fail(w) = u \text{ then } out(w) = out(w) \cup out(u).$$

Example 13.8

Here, $X = \{\text{search}, \text{ear}, \text{arch}, \text{chart}\}$

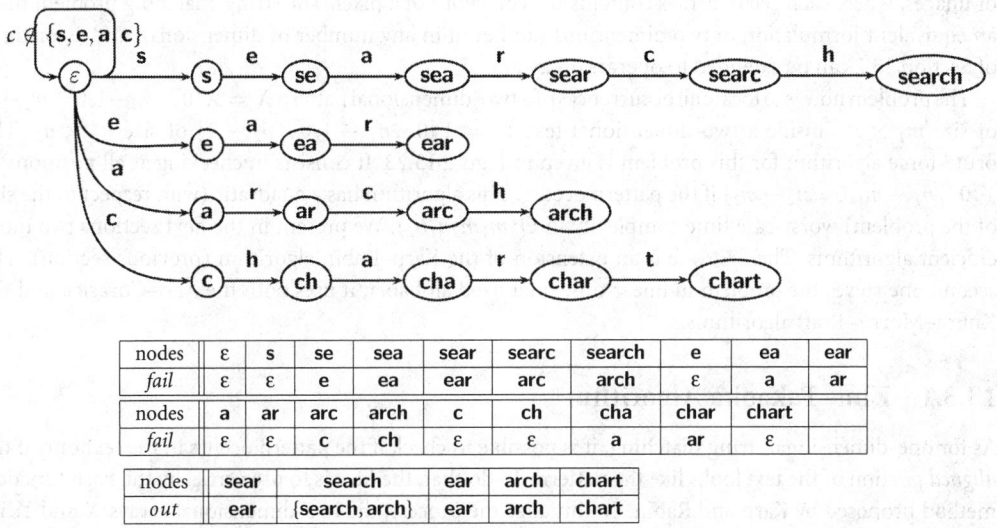

nodes	ε	s	se	sea	sear	searc	search	e	ea	ear
fail	ε	ε	e	ea	ear	arc	arch	ε	a	ar

nodes	a	ar	arc	arch	c	ch	cha	char	chart	
fail	ε	ε	c	ch	ε	ε	a	ar	ε	

nodes	sear	search	ear	arch	chart
out	ear	{search, arch}	ear	arch	chart

To stop going back with failure links during the computation of the failure links, and also to overpass text characters for which no transition is defined from the root, a loop is added on the root of the trie for these symbols. This is done at the first phase of function PREAC.

After the preprocessing phase is completed, the searching phase consists in parsing all the characters of the text y with $T(X)$. This starts at the root of $T(X)$ and uses failure links whenever a character in y does not match any label of outgoing edges of the current node. Each time a node with a nonempty output is encountered, this means that the patterns of the output have been discovered in the text, ending at the current position. Then, the position is output.

An implementation of the Aho–Corasick algorithm from the previous discussion is shown in Figure 13.21. Note that the algorithm processes the text in an on-line way, so that the buffer on the text can be limited to only one symbol. Also note that the instruction $r \leftarrow fail(r)$ in Figure 13.21 is the exact analogue of instruction $i \leftarrow next[i]$ in Figure 13.5. A unified view of both algorithms exists but is beyond the scope of the chapter.

The entire algorithm runs in time $O(|X| + n)$ if the *child* function is implemented to run in constant time. This is the case for any fixed alphabet. Otherwise, a $\log \sigma$ multiplicative factor comes from access to the children nodes.

```
AC(X, k, y, n)
1   ▷ Preprocessing
2   r ← PREAC(X, k)
3   ▷ Searching
4   for j ← 0 to n − 1
5       do while child(r, y[j]) = UNDEFINED
6              do r ← fail(r)
7           r ← child(r, y[j])
8           if out(r) ≠ ∅
9              then OUTPUT((out(r), j))
```

FIGURE 13.21 The complete Aho–Corasick algorithm.

13.3 Two-Dimensional Pattern Matching Algorithms

In this section we consider only two-dimensional arrays. Arrays can be thought of as bit map representations of images, where each cell of arrays contains the codeword of a pixel. The string-matching problem finds an equivalent formulation in two dimensions (and even in any number of dimensions), and algorithms of Section 13.2 can be extended to operate on arrays.

The problem now is to locate all occurrences of a two-dimensional pattern $X = X[0 .. m_1 - 1, 0 .. m_2 - 1]$ of size $m_1 \times m_2$ inside a two-dimensional text $Y = Y[0 .. n_1 - 1, 0 .. n_2 - 1]$ of size $n_1 \times n_2$. The brute-force algorithm for this problem is given in Figure 13.22. It consists in checking at all positions of $Y[0 .. n_1 - m_1, 0 .. n_2 - m_2]$ if the pattern occurs. This algorithm has a quadratic (with respect to the size of the problem) worst-case time complexity in $O(m_1 m_2 n_1 n_2)$. We present in the next sections two more efficient algorithms. The first one is an extension of the Karp–Rabin algorithm (previous section). The second one solves the problem in linear time on a fixed alphabet; it uses both the Aho–Corasick and the Knuth–Morris–Pratt algorithms.

13.3.1 Zhu–Takaoka Algorithm

As for one-dimensional string matching, it is possible to check if the pattern occurs in the text only if the *aligned* portion of the text looks like the pattern. To do that, the idea is to use vertically the hash function method proposed by Karp and Rabin. To initialize the process, the two-dimensional arrays X and Y are translated into one-dimensional arrays of numbers x and y. The translation from X to x is done as follows $(0 \le i < m_2)$:

$$x[i] = hash(X[0,i]X[1,i] \cdots X[m_1 - 1, i])$$

and the translation from Y to y is done by $(0 \le i < m_2)$:

$$y[i] = hash(Y[0,i]Y[1,i] \cdots Y[m_1 - 1, i]).$$

The fingerprint y helps to find occurrences of X starting at row $j = 0$ in Y. It is then updated for each new row in the following way $(0 \le i < m_2)$:

$$hash(Y[j + 1, i]Y[j + 2, i] \cdots Y[j + m_1, i])$$
$$= \text{REHASH}(Y[j,i], Y[j + m_1, i], hash(Y[j,i]Y[j + 1, i] \cdots Y[j + m_1 - 1, i]))$$

(functions *hash* and REHASH are described in the section on the Karp–Rabin algorithm).

```
BF2D(X, m₁, m₂, Y, n₁, n₂)
 1   ▷ Searching
 2   for j₁ ← 0 to n₁ − m₁
 3       do for j₂ ← 0 to n₂ − m₂
 4           do i ← 0
 5               while i < m₁ and x[i, 0 .. m₂ − 1] = y[j₁ + i, j₂ .. j₂ + m₂ − 1]
 6                   do i ← i + 1
 7               if i ≥ m₁
 8                   then OUTPUT(j₁, j₂)
```

FIGURE 13.22   The brute-force two-dimensional pattern matching algorithm.

```
KMP-IN-LINE(X, m₁, m₂, Y, n₁, n₂, x, y, next, j₁)
1   i₂ ← 0
2   j₂ ← 0
3   while j₂ < n₂
4       do while i₂ > −1 and x[i₂] ≠ y[j₂]
5           do i₂ ← next[i₂]
6       i₂ ← i₂ + 1
7       j₂ ← j₂ + 1
8       if i₂ ≥ m₂
9           then DIRECT-COMPARE(X, m₁, m₂, Y, n₁, n₂, j₁, j₂ − 1)
10              i₂ ← next[m₂]
```

FIGURE 13.23   Search for x in y using KMP algorithm.

```
DIRECT-COMPARE(X, m₁, m₂, Y, row, column)
1   j₁ ← row − m₁ + 1
2   j₂ ← column − m₂ + 1
3   for i₁ ← 0 to m₁ − 1
4       do for i₂ ← 0 to m₂ − 1
5           do if X[i₁, i₂] ≠ Y[i₁ + j₁, i₂ + j₂]
6               then return
7   OUTPUT(j₁, j₂)
```

FIGURE 13.24   Naive check of an occurrence of x in y at position (*row, column*).

Example 13.9

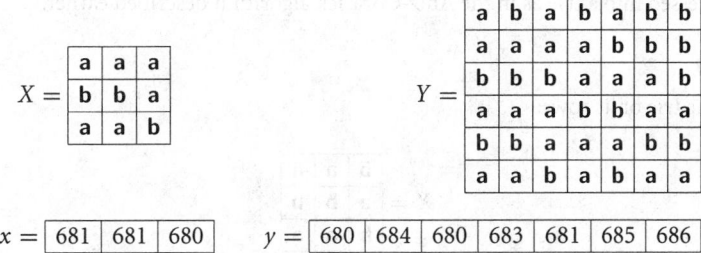

Next value of y is $\boxed{681 \quad 681 \quad 681 \quad 680 \quad 684 \quad 683 \quad 685}$. The occurrence of x at position 1 on y corresponds to an occurrence of X at position $(1, 1)$ on Y.

Since the alphabet of x and y is large, searching for x in y must be done by a string-matching algorithm for which the running time is independent of the size of the alphabet: the Knuth–Morris–Pratt suits this application perfectly. Its adaptation is shown in Figure 13.23.

When an occurrence of x is found in y, then we still have to check if an occurrence of X starts in Y at the corresponding position. This is done naively by the procedure of Figure 13.24.

The Zhu–Takaoka algorithm as explained above is displayed in Figure 13.25. The search for the pattern is performed row by row starting at row 0 and ending at row $n_1 − m_1$.

13.3.2   Bird/Baker Algorithm

The algorithm designed independently by Bird and Baker for the two-dimensional pattern matching problem combines the use of the Aho–Corasick algorithm and the Knuth–Morris–Pratt (KMP) algorithm.

ZT(X, m_1, m_2, Y, n_1, n_2)

```
 1   ▷ Preprocessing
 2   ▷ Computes x
 3   for i₂ ← 0 to m₂ − 1
 4       do x[i₂] ← 0
 5           for i₁ ← 0 to m₁ − 1
 6               do x[i₂] ← (x[i₂] << 1) + X[i₁, i₂]
 7   ▷ Computes the first value of y
 8   for j₂ ← 0 to n₂ − 1
 9       do y[j₂] ← 0
10           for j₁ ← 0 to m₁ − 1
11               do y[j₂] ← (y[j₂] << 1) + Y[j₁, j₂]
12   d ← 1
13   for i ← 1 to m₁ − 1
14       do d ← d << 1
15   next ← PREKMP(X′, m2)
16   ▷ Searching
17   j₁ ← m₁ − 1
18   while j₁ < n₁
19       do KMP-IN-LINE(X, m₁, m₂, Y, n₁, n₂, x, y, next, j₂)
20           if j₁ < n₁ − 1
21               then for j₂ ← 0 to n₂ − 1
22                   do y[j₂] ← REHASH(Y[j₁ − m₁ + 1, j₂], Y[j₁ + 1, j₂], y[j₂])
23           j₁ ← j₁ + 1
```

FIGURE 13.25   The Zhu-Takaoka two-dimensional pattern matching algorithm.

The pattern X is divided into its m_1 rows $R_0 = X[0, 0 .. m_2 − 1]$ to $R_{m_1−1} = x[m_1 − 1, 0 .. m_2 − 1]$. The rows are preprocessed into a trie as in the Aho–Corasick algorithm described earlier.

Example 13.10

Pattern X and the trie of its rows:

$$X = \begin{array}{|c|c|c|} \hline b & a & a \\ \hline a & b & b \\ \hline b & a & a \\ \hline \end{array}$$

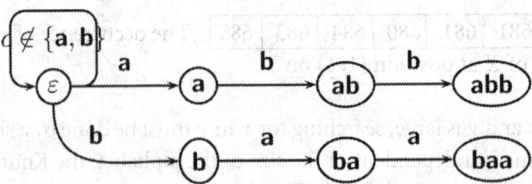

The search proceeds as follows. The text is read from the upper left corner to the bottom right corner, row by row. When reading the character $Y[j_1, j_2]$, the algorithm checks whether the portion $Y[j_1, j_2 − m_2 + 1 .. j_2] = R$ matches any of $R_0, \ldots, R_{m_1−1}$ using the Aho–Corasick machine. An additional one-dimensional array a of size n_1 is used as follows: $a[j_2] = k$ means that the $k − 1$ first rows $R_0, \ldots, R_{k−2}$ of the pattern match, respectively, the portions of the text: $Y[j_1 − k + 1, j_2 − m_2 + 1 .. j_2], \ldots, Y[j_1 − 1,$ $j_2 − m_2 + 1 .. j_2]$. Then, if $R = R_{k−1}, a[j_2]$ is incremented to $k + 1$. If not, $a[j_2]$ is set to $s + 1$ where s is the maximum i such that

$$R_0 \cdots R_i = R_{k−s+1} \cdots R_{k−2} R.$$

PRE-KMP-FOR-B(X, m_1, m_2)

```
1   i ← 0
2   next[0] ← -1
3   j ← -1
4   while i < m₁
5       do while j > -1 and X[i, 0 .. m₂ - 1] ≠ X[j, 0 .. m₂ - 1]
6           do j ← next[j]
7       i ← i + 1
8       j ← j + 1
9       if X[i, 0 .. m₂ - 1] ≠ X[j, 0 .. m₂ - 1]
10          then next[i] ← next[j]
11          else next[i] ← j
12  return next
```

FIGURE 13.26   Computes the function *next* for rows of X.

B$(X, m_1, m_2, Y, n_1, n_2)$

```
1   ▷ Preprocessing
2   for i ← 0 to m₂ - 1
3       do a[i] ← 0
4   root ← PREAC(m₁)
5   next ← PRE-KMP-FOR-B(X, m₁, m₂)
6   for j₁ ← 0 to n₁ - 1
7       do r ← root
8           for j₂ ← 0 to n₂ - 1
9               do while child(r, Y[j₁, j₂]) = UNDEFINED
10                  do r ← fail(r)
11                  r ← child(r, Y[j₁, j₂])
12                  if out(r) ≠ ∅
13                      then k ← a[j₂]
14                          while k > 0 and X[k, 0 .. m₂ - 1] = out(r)
15                              do k ← next[k]
16                          a[j₂] ← k + 1
17                          if k ≥ m₁ - 1
18                              then OUTPUT(j₁ - m₁ + 1, j₂ - m₂ + 1)
19                      else a[j₂] ← 0
```

FIGURE 13.27   The Bird/Baker two-dimensional pattern matching algorithm.

The value s is computed using the KMP algorithm vertically (in columns). If there exists no such s, $a[j_2]$ is set to 0. Finally, if at some point $a[j_2] = m_1$, an occurrence of the pattern appears at position $(j_1 - m_1 + 1, j_2 - m_2 + 1)$ in the text.

The Bird/Baker algorithm is presented in Figure 13.26 and Figure 13.27. It runs in time $O((n_1 n_2 + m_1 m_2) \log \sigma)$.

13.4   Suffix Trees

The suffix tree $S(y)$ of a string y is a trie (as described earlier) containing all the suffixes of the string, and having the properties described subsequently. This data structure serves as an index on the string: it provides a direct access to all segments of the string, and gives the positions of all their occurrences in the string.

Once the suffix tree of a text y is built, searching for x in y remains to spell x along a branch of the tree. If this walk is successful, the positions of the pattern can be output. Otherwise, x does not occur in y.

SUFFIX-TREE(y, n)
1   $T_{-1} \leftarrow$ one node tree
2   **for** $j \leftarrow 0$ **to** $n - 1$
3   **do** $T_j \leftarrow$ INSERT($T_{j-1}, y[j .. n - 1]$)
4   **return** T_{n-1}

FIGURE 13.28   Construction of a suffix tree for y.

INSERT($T_{j-1}, y[j .. n - 1]$)
1   locate the node h associated with $head_j$ in T_{j-1}, possibly breaking an edge
2   add a new edge labeled $tail_j$ from h to a new leaf representing $y[j .. n - 1]$
3   **return** the modified tree

FIGURE 13.29   Insertion of a new suffix in the tree.

Any kind of trie that represents the suffixes of a string can be used to search it. But the suffix tree has additional features which imply that its size is linear. The suffix tree of y is defined by the following properties:

- All branches of $S(y)$ are labeled by all suffixes of y.
- Edges of $S(y)$ are labeled by strings.
- Internal nodes of $S(y)$ have at least two children (when y is not empty).
- Edges outgoing an internal node are labeled by segments starting with different letters.
- The preceding segments are represented by their starting positions on y and their lengths.

Moreover, it is assumed that y ends with a symbol occurring nowhere else in it (the dollar sign is used in examples). This avoids marking nodes, and implies that $S(y)$ has exactly n leaves (number of nonempty suffixes). The other properties then imply that the total size of $S(y)$ is $O(n)$, which makes it possible to design a linear-time construction of the trie. The algorithm described in the present section has this time complexity provided the alphabet is fixed, or with an additional multiplicative factor $\log \sigma$ otherwise.

The algorithm inserts all nonempty suffixes of y in the data structure from the longest to the shortest suffix, as shown in Figure 13.28. We introduce two definitions to explain how the algorithm works:

- $head_j$ is the longest prefix of $y[j .. n - 1]$ which is also a prefix of $y[i .. n - 1]$ for some $i < j$.
- $tail_j$ is the word such that $y[j .. n - 1] = head_j tail_j$.

The strategy to insert the ith suffix in the tree is based on these definitions and described in Figure 13.29.

The second step of the insertion (Figure 13.29) is clearly performed in constant time. Thus, finding the node h is critical for the overall performance of the algorithm. A brute-force method to find it consists in spelling the current suffix $y[j .. n - 1]$ from the root of the tree, giving an $O(|head_j|)$ time complexity for the insertion at step j, and an $O(n^2)$ running time to build $S(y)$. Adding short-cut links leads to an overall $O(n)$ time complexity, although there is no guarantee that insertion at step j is realized in constant time.

Example 13.11

The different tries during the construction of the suffix tree of $y =$ **CAGATAGAG**. Leaves are black and labeled by the position of the suffix they represent. Plain arrows are labeled by pairs: the pair (j, ℓ) stands for the segment $y[j .. j + \ell - 1]$. Dashed arrows represent the nontrivial suffix links.

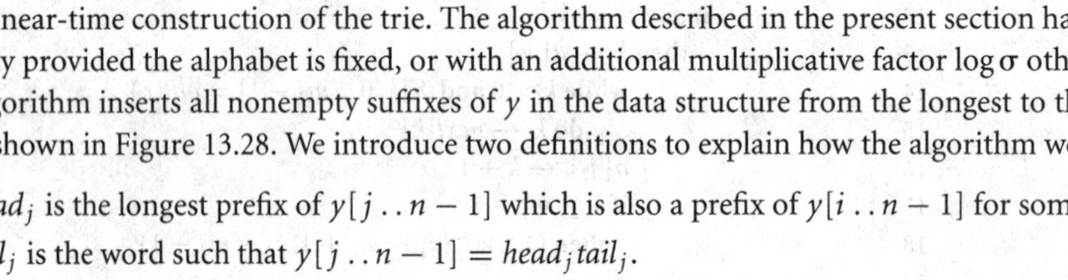

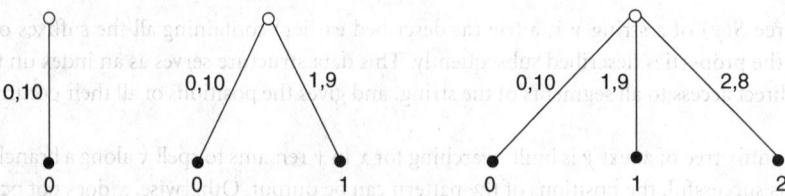

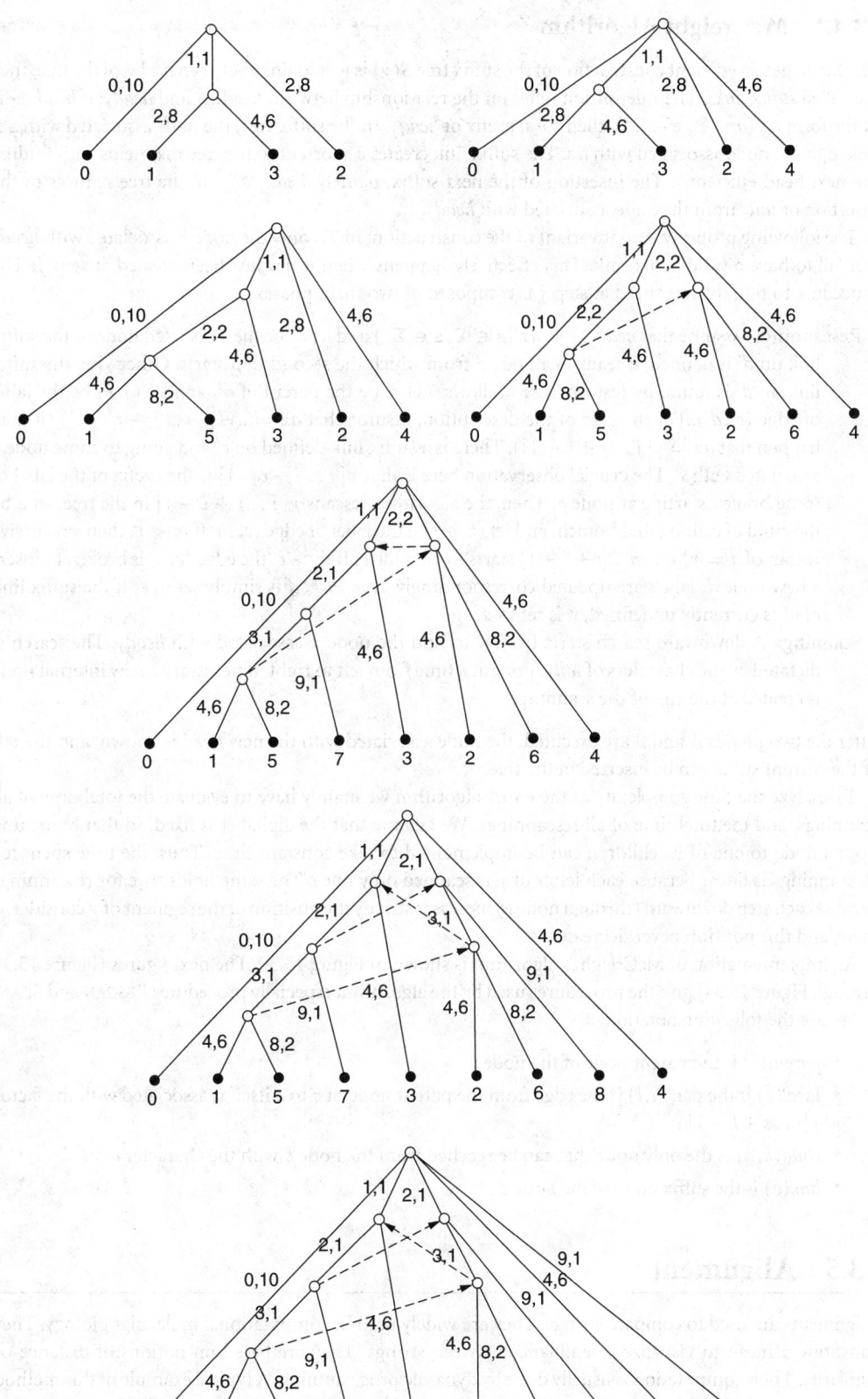

13.4.1 McCreight Algorithm

The key to get an efficient construction of the suffix tree $S(y)$ is to add links between nodes of the tree: they are called *suffix links*. Their definition relies on the relationship between $head_{j-1}$ and $head_j$: if $head_{j-1}$ is of the form az ($a \in \Sigma, z \in \Sigma^*$), then z is a prefix of $head_j$. In the suffix tree, the node associated with z is linked to the node associated with az. The suffix link creates a shortcut in the tree that helps with finding the next head efficiently. The insertion of the next suffix, namely, $head_j tail_j$, in the tree reduces to the insertion of $tail_j$ from the node associated with $head_j$.

The following property is an invariant of the construction: in T_j, only the node h associated with $head_j$ can fail to have a valid suffix link. This effectively happens when h has just been created at step j. The procedure to find the next head at step j is composed of two main phases:

A Rescanning: Assume that $head_{j-1} = az$ ($a \in \Sigma, z \in \Sigma^*$) and let d' be the associated node. If the suffix link on d' is defined, it leads to a node d from which the second step starts. Otherwise, the suffix link on d' is found by rescanning as follows. Let c' be the parent of d', and let (j, ℓ) be the label of edge (c', d'). For the ease of the description, assume that $az = av(y[j .. j + \ell - 1])$ (it may happen that $az = y[j .. j + \ell - 1]$). There is a suffix link defined on c' and going to some node c associated with v. The crucial observation here is that $y[j .. j + \ell - 1]$ is the prefix of the label of some branch starting at node c. Then, the algorithm rescans $y[j .. j + \ell - 1]$ in the tree: let e be the child of c along that branch, and let (k, m) be the label of edge (c, e). If $m < \ell$, then a recursive rescan of $q = y[j + m .. j + \ell - 1]$ starts from node e. If $m > \ell$, the edge (c, e) is broken to insert a new node d; labels are updated correspondingly. If $m = \ell$, d is simply set to e. If the suffix link of d' is currently undefined, it is set to d.

B Scanning: A downward search starts from d to find the node h associated with $head_j$. The search is dictated by the characters of $tail_{j-1}$ one at a time from left to right. If necessary a new internal node is created at the end of the scanning.

After the two phases A and B are executed, the node associated with the new head is known, and the tail of the current suffix can be inserted in the tree.

To analyze the time complexity of the entire algorithm we mainly have to evaluate the total time of all scannings, and the total time of all rescannings. We assume that the alphabet is fixed, so that branching from a node to one of its children can be implemented to take constant time. Thus, the time spent for all scannings is linear because each letter of y is scanned only once. The same holds true for rescannings because each step downward (through node e) increases strictly the position of the segment of y considered there, and this position never decreases.

An implementation of McCreight's algorithm is shown in Figure 13.30. The next figures (Figure 13.31 through Figure 13.34) give the procedures used by the algorithm, especially procedures RESCAN and SCAN.

We use the following notation:

- *parent*(c) is the parent node of the node c
- *label*(c) is the pair (i, l) if the edge from the parent node of c to c itself is associated with the factor $y[i .. i + l - 1]$
- *child*(c, a) is the only node that can be reached from the node c with the character a
- *link*(c) is the suffix node of the node c

13.5 Alignment

Alignments are used to compare strings. They are widely used in computational molecular biology. They constitute a mean to visualize resemblance between strings. They are based on notions of distance or similarity. Their computation is usually done by dynamic programming. A typical example of this method is the computation of the longest common subsequence of two strings. The reduction of the memory space presented on it can be applied to similar problems. We consider three different kinds of alignment of two

M(y, n)
1   $root \leftarrow$ INIT(y, n)
2   $head \leftarrow root$
3   $tail \leftarrow child(root, y[0])$
4   $n \leftarrow n - 1$
5   **while** $n > 0$
6   **do if** $head = root$ ▷ Phase A (rescanning)
7   **then** $d \leftarrow root$
8               $(j, \ell) \leftarrow label(tail)$
9               $\gamma \leftarrow (j + 1, \ell - 1)$
10   **else** $\gamma \leftarrow label(tail)$
11   **if** $link(head) \neq$ UNDEFINED
12   **then** $d \leftarrow link(head)$
13   **else** $(j, \ell) \leftarrow label(head)$
14   **if** $parent(head) = root$
15   **then** $d \leftarrow$ RESCAN($root, j + 1, \ell - 1$))
16   **else** $d \leftarrow$ RESCAN($link(parent(head)), j, \ell$))
17                      $link(head) \leftarrow d$
18          $(head, \gamma) \leftarrow$ SCAN(d, γ)   ▷ Phase B (scanning)
19   create a new node $tail$
20          $parent(tail) \leftarrow head$
21          $label(tail) \leftarrow \gamma$
22          $(j, \ell) \leftarrow \gamma$
23          $child(head, y[j]) \leftarrow tail$
24          $n \leftarrow n - 1$
25 **return** $root$

FIGURE 13.30   Suffix tree construction.

INIT(y, n)
1   create a new node $root$
2   create a new node c
3   $parent(root) \leftarrow$ UNDEFINED
4   $parent(c) \leftarrow root$
5   $child(root, y[0]) \leftarrow c$
6   $label(root) \leftarrow$ UNDEFINED
7   $label(c) \leftarrow (0, n)$
8   **return** $root$

FIGURE 13.31   Initialization procedure.

RESCAN(c, j, ℓ)
1   $(k, m) \leftarrow label(child(c, y[j]))$
2   **while** $\ell > 0$ **and** $\ell \geq m$
3   **do** $c \leftarrow child(c, y[j])$
4           $\ell \leftarrow \ell - m$
5           $j \leftarrow j + m$
6           $(k, m) \leftarrow label(child(c, y[j]))$
7   **if** $\ell > 0$
8   **then return** BREAK-EDGE($child(c, y[j]), \ell$)
9   **else return** c

FIGURE 13.32   The crucial rescan operation.

BREAK -EDGE(c, k)

```
 1   create a new node g
 2   parent(g) ← parent(c)
 3   (j, ℓ) ← label(c)
 4   child(parent(c), y[j]) ← g
 5   label(g) ← (j, k)
 6   parent(c) ← g
 7   label(c) ← (j + k, ℓ − k)
 8   child(g, y[j + k]) ← c
 9   link(g) ← UNDEFINED
10   return g
```

FIGURE 13.33   Breaking an edge.

SCAN(d, γ)

```
 1   (j, ℓ) ← γ
 2   while child(d, y[j]) ≠ UNDEFINED
 3        do g ← child(d, y[j])
 4           k ← 1
 5           (s, lg) ← label(g)
 6           s ← s + 1
 7           ℓ ← ℓ − 1
 8           j ← j + 1
 9           while k < lg and y[j] = y[s]
10                do j ← j + 1
11                   s ← s + 1
12                   k ← k + 1
13                   ℓ ← ℓ − 1
14           if k < lg
15                then return (BREAK -EDGE(g, k), (j, ℓ))
16           d ← g
17   return (d, (j, ℓ))
```

FIGURE 13.34   The scan operation.

strings x and y: global alignment (that consider the whole strings x and y), local alignment (that enable to find the segment of x that is closer to a segment of y), and the longest common subsequence of x and y.

An **alignment** of two strings x and y of length m and n, respectively, consists in aligning their symbols on vertical lines. Formally, an alignment of two strings $x, y \in \Sigma$ is a word w on the alphabet $(\Sigma \cup \{\varepsilon\}) \times (\Sigma \cup \{\varepsilon\}) \setminus (\{(\varepsilon, \varepsilon)\}$ (ε is the empty word) whose projection on the first component is x and whose projection of the second component is y.

Thus, an alignment $w = (\overline{x}_0, \overline{y}_0)(\overline{x}_1, \overline{y}_1) \cdots (\overline{x}_{p-1}, \overline{y}_{p-1})$ of length p is such that $x = \overline{x}_0 \overline{x}_1 \cdots \overline{x}_{p-1}$ and $y = \overline{y}_0 \overline{y}_1 \cdots \overline{y}_{p-1}$ with $\overline{x}_i \in \Sigma \cup \{\varepsilon\}$ and $\overline{y}_i \in \Sigma \cup \{\varepsilon\}$ for $0 \leq i \leq p - 1$. The alignment is represented as follows

$$\overline{x}_0 \quad \overline{x}_1 \quad \cdots \quad \overline{x}_{p-1}$$

$$\overline{y}_0 \quad \overline{y}_1 \quad \cdots \quad \overline{y}_{p-1}$$

with the symbol $-$ instead of the symbol ε.

Example 13.12

$$\begin{array}{cccccc} A & C & G & - & - & A \\ A & T & G & C & T & A \end{array}$$

is an alignment of **ACGA** and **ATGCTA**.

13.5.1 Global alignment

A global alignment of two strings x and y can be obtained by computing the distance between x and y. The notion of distance between two strings is widely used to compare files. The **diff** command of Unix operating system implements an algorithm based on this notion, in which lines of the files are treated as symbols. The output of a comparison made by **diff** gives the minimum number of operations (substitute a symbol, insert a symbol, or delete a symbol) to transform one file into the other.

Let us define the edit distance between two strings x and y as follows: it is the minimum number of elementary edit operations that enable to transform x into y. The elementary edit operations are:

- The substitution of a character of x at a given position by a character of y
- The deletion of a character of x at a given position
- The insertion of a character of y in x at a given position

A cost is associated to each elementary edit operation. For $a, b \in \Sigma$:

- $Sub(a, b)$ denotes the cost of the substitution of the character a by the character b,
- $Del(a)$ denotes the cost of the deletion of the character a, and
- $Ins(a)$ denotes the cost of the insertion of the character a.

This means that the costs of the edit operations are independent of the positions where the operations occur. We can now define the edit distance of two strings x and y by

$$edit(x, y) = \min\{\text{cost of } \gamma \mid \gamma \in \Gamma_{x,y}\}$$

where $\Gamma_{x,y}$ is the set of all the sequences of edit operations that transform x into y, and the cost of an element $\gamma \in \Gamma_{x,y}$ is the sum of the costs of its elementary edit operations.

To compute $edit(x, y)$ for two strings x and y of length m and n, respectively, we make use of a two-dimensional table T of $m + 1$ rows and $n + 1$ columns such that

$$T[i, j] = edit(x[i], y[j])$$

for $i = 0, \ldots, m - 1$ and $j = 0, \ldots, n - 1$. It follows $edit(x, y) = T[m - 1, n - 1]$.

The values of the table T can be computed by the following recurrence formula:

$$T[-1, -1] = 0$$
$$T[i, -1] = T[i - 1, -1] + Del(x[i])$$
$$T[-1, j] = T[-1, j - 1] + Ins(y[j])$$
$$T[i, j] = \min \begin{cases} T[i - 1, j - 1] + Sub(x[i], y[j]) \\ T[i - 1, j] + Del(x[i]) \\ T[i, j - 1] + Ins(y[j]) \end{cases}$$

for $i = 0, 1, \ldots, m - 1$ and $j = 0, 1, \ldots, n - 1$.

GENERIC-DP(x, m, y, n, MARGIN, FORMULA)

1   MARGIN(T, x, m, y, n)

2   **for** $j \leftarrow 0$ **to** $n - 1$

3   **do for** $i \leftarrow 0$ **to** $m - 1$

4   **do** $T[i, j] \leftarrow$ FORMULA(T, x, i, y, j)

5   **return** T

FIGURE 13.35   Computation of the table T by dynamic programming.

MARGIN-GLOBAL(T, x, m, y, n)

1    $T[-1, -1] \leftarrow 0$

2   **for** $i \leftarrow 0$ **to** $m - 1$

3   **do** $T[i, -1] \leftarrow T[i - 1, -1] + Del(x[i])$

4   **for** $j \leftarrow 0$ **to** $n - 1$

5   **do** $T[-1, j] \leftarrow T[-1, j - 1] + Ins(y[j])$

FIGURE 13.36   Margin initialization for the computation of a global alignment.

FORMULA-GLOBAL(T, x, i, y, j)

1   **return** min $\begin{cases} T[i - 1, j - 1] + Sub(x[i], y[j]) \\ T[i - 1, j] + Del(x[i]) \\ T[i, j - 1] + Ins(y[j]) \end{cases}$

FIGURE 13.37   Computation of $T[i, j]$ for a global alignment.

The value at position (i, j) in the table T only depends on the values at the three neighbor positions $(i - 1, j - 1)$, $(i - 1, j)$, and $(i, j - 1)$.

The direct application of the above recurrence formula gives an exponential time algorithm to compute $T[m - 1, n - 1]$. However, the whole table T can be computed in quadratic time technique known as "dynamic programming." This is a general technique that is used to solve the different kinds of alignments.

The computation of the table T proceeds in two steps. First it initializes the first column and first row of T; this is done by a call to a generic function MARGIN, which is a parameter of the algorithm and that depends on the kind of alignment considered. Second, it computes the remaining values of T, which is done by a call to a generic function FORMULA, which is a parameter of the algorithm and that depends on the kind of alignment considered. Computing a global alignment of x and y can be done by a call to GENERIC-DP with the following parameters $(x, m, y, n, \text{MARGIN-GLOBAL}, \text{FORMULA-GLOBAL})$ (see Figure 13.35, Figure 13.36, and Figure 13.37). The computation of all the values of the table T can thus be done in quadratic space and time: $O(m \times n)$.

An optimal alignment (with minimal cost) can then be produced by a call to the function ONE-ALIGNMENT($T, x, m - 1, y, n - 1$) (see Figure 13.38). It consists in tracing back the computation of the values of the table T from position $[m - 1, n - 1]$ to position $[-1, -1]$. At each cell $[i, j]$, the algorithm determines among the three values $T[i - 1, j - 1] + Sub(x[i], y[j])$, $T[i - 1, j] + Del(x[i])$, and $T[i, j - 1] + Ins(y[j])$ which has been used to produce the value of $T[i, j]$. If $T[i - 1, j - 1] + Sub(x[i], y[j])$ has been used it adds $(x[i], y[j])$ to the optimal alignment and proceeds recursively with the cell at $[i - 1, j - 1]$. If $T[i - 1, j] + Del(x[i])$ has been used, it adds $(x[i], -)$ to the optimal alignment and proceeds recursively with cell at $[i - 1, j]$. If $T[i, j - 1] + Ins(y[j])$ has been used, it adds $(-, y[j])$ to the optimal alignment

ONE-ALIGNMENT(T, x, i, y, j)

```
1   if i = -1 and j = -1
2       then return (ε, ε)
3       else  if i = -1
4           then return ONE-ALIGNMENT(T, x, -1, y, j - 1) · (ε, y[j])
5           elseif j = -1
6           then return ONE-ALIGNMENT(T, x, i - 1, y, -1) · (x[i], ε)
7           else   if T[i, j] = T[i - 1, j - 1] + Sub(x[i], y[j])
8               then return ONE-ALIGNMENT(T, x, i - 1, y, j - 1) · (x[i], y[j])
9               elseif T[i, j] = T[i - 1, j] + Del(x[i])
10                  then return ONE-ALIGNMENT(T, x, i - 1, y, j) · (x[i], ε)
11                  else   return ONE-ALIGNMENT(T, x, i, y, j - 1) · (ε, y[j])
```

FIGURE 13.38 Recovering an optimal alignment.

and proceeds recursively with cell at $[i, j - 1]$. Recovering all the optimal alignments can be done by a similar technique.

Example 13.13

T	j	-1	0	1	2	3	4	5
i	$y[j]$		A	T	G	C	T	A
-1	$x[i]$	0	1	2	3	4	5	6
0	A	1	0	1	2	3	4	5
1	C	2	1	1	2	2	3	4
2	G	3	2	2	1	2	3	4
3	A	4	3	3	2	2	3	3

The values of the above table have been obtained with the following unitary costs: $Sub(a, b) = 1$ if $a \neq b$ and $Sub(a, a) = 0$, $Del(a) = Ins(a) = 1$ for $a, b \in \Sigma$.

13.5.2 Local Alignment

A local alignment of two strings x and y consists in finding the segment of x that is closer to a segment of y. The notion of distance used to compute global alignments cannot be used in that case because the segments of x closer to segments of y would only be the empty segment or individual characters. This is why a notion of similarity is used based on a scoring scheme for edit operations.

A score (instead of a cost) is associated to each elementary edit operation. For $a, b \in \Sigma$:

- $Sub_S(a, b)$ denotes the score of substituting the character b for the character a.
- $Del_S(a)$ denotes the score of deleting the character a.
- $Ins_S(a)$ denotes the score of inserting the character a.

This means that the scores of the edit operations are independent of the positions where the operations occur. For two characters a and b, a positive value of $Sub_S(a, b)$ means that the two characters are close to each other, and a negative value of $Sub_S(a, b)$ means that the two characters are far apart.

We can now define the edit score of two strings x and y by

$$sco(x, y) = \max\{\text{score of } \gamma \mid \gamma \in \Gamma_{x,y}\}$$

where $\Gamma_{x,y}$ is the set of all the sequences of edit operations that transform x into y and the score of an element $\sigma \in \Gamma_{x,y}$ is the sum of the scores of its elementary edit operations.

To compute $sco(x, y)$ for two strings x and y of length m and n, respectively, we make use of a two-dimensional table T of $m + 1$ rows and $n + 1$ columns such that

$$T[i, j] = sco(x[i], y[j])$$

for $i = 0, \ldots, m - 1$ and $j = 0, \ldots, n - 1$. Therefore, $sco(x, y) = T[m - 1, n - 1]$.

The values of the table T can be computed by the following recurrence formula:

$$T[-1, -1] = 0,$$
$$T[i, -1] = 0,$$
$$T[-1, j] = 0,$$
$$T[i, j] = \max \begin{cases} T[i - 1, j - 1] + Sub_S(x[i], y[j]), \\ T[i - 1, j] + Del_S(x[i]), \\ T[i, j - 1] + Ins_S(y[j]), \\ 0, \end{cases}$$

for $i = 0, 1, \ldots, m - 1$ and $j = 0, 1, \ldots, n - 1$.

Computing the values of T for a local alignment of x and y can be done by a call to GENERIC-DP with the following parameters $(x, m, y, n, \text{MARGIN-LOCAL}, \text{FORMULA-LOCAL})$ in $O(mn)$ time and space complexity (see Figure 13.35, Figure 13.39, and Figure 13.40). Recovering a local alignment can be done in a way similar to what is done in the case of a global alignment (see Figure 13.38) but the trace back procedure must start at a position of a maximal value in T rather than at position $[m - 1, n - 1]$.

MARGIN-LOCAL(T, x, m, y, n)

1    $T[-1, -1] \leftarrow 0$
2   **for** $i \leftarrow 0$ **to** $m - 1$
3   **do** $T[i, -1] \leftarrow 0$
4   **for** $j \leftarrow 0$ **to** $n - 1$
5   **do** $T[-1, j] \leftarrow 0$

FIGURE 13.39   Margin initialization for computing a local alignment.

FORMULA-LOCAL(T, x, i, y, j)

1   **return** $\max \begin{cases} T[i - 1, j - 1] + Sub_S(x[i], y[j]) \\ T[i - 1, j] + Del_S(x[i]) \\ T[i, j - 1] + Ins_S(y[j]) \\ 0 \end{cases}$

FIGURE 13.40   Recurrence formula for computing a local alignment.

Example 13.14

Computation of an optimal local alignment of $x =$ **EAWACQGKL** and $y =$ **ERDAWCQPGKWY** with scores:

$Sub_S(a,a) = 1$, $Sub_S(a,b) = -3$ and $Del_S(a) = Ins_S(a) = -1$ for $a, b \in \Sigma, a \neq b$.

T	j	−1	0	1	2	3	4	5	6	7	8	9	10	11
i		y[j]	E	R	D	A	W	C	Q	P	G	K	W	Y
−1	x[i]	0	0	0	0	0	0	0	0	0	0	0	0	0
0	E	0	1	0	0	0	0	0	0	0	0	0	0	0
1	A	0	0	0	0	1	0	0	0	0	0	0	0	0
2	W	0	0	0	0	0	2	1	0	0	0	0	1	0
3	A	0	0	0	0	1	1	0	0	0	0	0	0	0
4	C	0	0	0	0	0	0	2	1	0	0	0	0	0
5	Q	0	0	0	0	0	0	1	3	2	1	0	0	0
6	G	0	0	0	0	0	0	0	2	1	3	2	1	0
7	K	0	0	0	0	0	0	0	1	0	2	4	3	2
8	L	0	0	0	0	0	0	0	0	0	1	3	2	1

The corresponding optimal local alignment is:

```
A  W  A  C  Q  -  G  K
A  W  -  C  Q  P  G  K
```

13.5.3 Longest Common Subsequence of Two Strings

A subsequence of a word x is obtained by deleting zero or more characters from x. More formally, $w[0 .. i - 1]$ is a subsequence of $x[0 .. m - 1]$ if there exists an increasing sequence of integers $(k_j \mid j = 0, \ldots, i - 1)$ such that for $0 \leq j \leq i - 1$, $w[j] = x[k_j]$. We say that a word is an lcs(x, y) if it is a **longest common subsequence** of the two words x and y. Note that two strings can have several longest common subsequences. Their common length is denoted by llcs(x, y).

A brute-force method to compute an lcs(x, y) would consist in computing all the subsequences of x, checking if they are subsequences of y, and keeping the longest one. The word x of length m has 2^m subsequences, and so this method could take $O(2^m)$ time, which is impractical even for fairly small values of m.

However, llcs(x, y) can be computed with a two-dimensional table T by the following recurrence formula:

$$T[-1, -1] = 0,$$
$$T[i, -1] = 0,$$
$$T[-1, j] = 0,$$
$$T[i, j] = \begin{cases} T[i - 1, j - 1] + 1 & \text{if } x[i] = y[j], \\ \max(T[i - 1, j], T[i, j - 1]) & \text{otherwise,} \end{cases}$$

for $i = 0, 1, \ldots, m - 1$ and $j = 0, 1, \ldots, n - 1$. Then, $T[i, j] = $ llcs$(x[0 .. i], y[0 .. j])$ and llcs$(x, y) = T[m - 1, n - 1]$.

Computing $T[m - 1, n - 1]$ can be done by a call to GENERIC-DP with the following parameters $(x, m, y, n, \text{MARGIN-LOCAL}, \text{FORMULA-LCS})$ in $O(mn)$ time and space complexity (see Figure 13.35, Figure 13.39, and Figure 13.41).

FORMULA-LCS(T, x, i, y, j)

1   **if** $x[i] = y[j]$
2   **then return** $T[i-1, j-1] + 1$
3   **else return** $\max\{T[i-1, j], T[i, j-1]\}$

FIGURE 13.41   Recurrence formula for computing an *lcs*.

It is possible afterward to trace back a path from position $[m-1, n-1]$ in order to exhibit an lcs(x, y) in a similar way as for producing a global alignment (see Figure 13.38).

Example 13.15

The value $T[4, 8] = 4$ is llcs(x, y) for $x =$ **AGCGA** and $y =$ **CAGATAGAG**. String **AGGA** is an lcs of x and y.

T	j	-1	0	1	2	3	4	5	6	7	8
i		$y[j]$	**C**	**A**	**G**	**A**	**T**	**A**	**G**	**A**	**G**
-1	$x[i]$	0	0	0	0	0	0	0	0	0	0
0	**A**	0	0	1	1	1	1	1	1	1	1
1	**G**	0	0	1	2	2	2	2	2	2	2
2	**C**	0	1	1	2	2	2	2	2	2	2
3	**G**	0	1	1	2	2	2	2	3	3	3
4	**A**	0	1	2	2	3	3	3	3	4	4

13.5.4 Reducing the Space: Hirschberg Algorithm

If only the length of an lcs(x, y) is required, it is easy to see that only one row (or one column) of the table T needs to be stored during the computation. The space complexity becomes $O(\min(m, n))$, as can be checked on the algorithm of Figure 13.42. Indeed, the Hirschberg algorithm computes an lcs(x, y) in linear space and not only the value llcs(x, y). The computation uses the algorithm of Figure 13.43.

Let us define

$$T^*[i, n] = T^*[m, j] = 0, \quad \text{for } 0 \le i \le m \quad \text{and} \quad 0 \le j \le n$$

$$T^*[m-i, n-j] = \text{llcs}((x[i \mathinner{..} m-1])^R, (y[j \mathinner{..} n-1])^R)$$

$$\text{for } 0 \le i \le m-1 \quad \text{and} \quad 0 \le j \le n-1$$

and

$$M(i) = \max_{0 \le j < n} \{T[i, j] + T^*[m-i, n-j]\}$$

where the word w^R is the reverse (or mirror image) of the word w. The following property is the key observation to compute an lcs(x, y) in linear space:

$$M(i) = T[m-1, n-1], \quad \text{for } 0 \le i < m.$$

In the algorithm shown in Figure 13.43, the integer j is chosen as $n/2$. After $T[i, j-1]$ and $T^*[m-i, n-j]$ $(0 \le i < m)$ are computed, the algorithm finds an integer k such that $T[i, k] + T^*[m-i, n-k] = T[m-1, n-1]$. Then, recursively, it computes an lcs$(x[0 \mathinner{..} k-1], y[0 \mathinner{..} j-1])$ and an lcs$(x[k \mathinner{..} m-1], y[j \mathinner{..} n-1])$, and concatenates them to get an lcs(x, y).

```
                        LLCS(x, m, y, n)
          1   for i ← −1 to m − 1
          2       do C[i] ← 0
          3   for j ← 0 to n − 1
          4       do last ← 0
          5           for i ← −1 to m − 1
          6               do if last > C[i]
          7                   then C[i] ← last
          8                   elseif last < C[i]
          9                       then last ← C[i]
         10                   elseif x[i] = y[j]
         11                       then C[i] ← C[i] + 1
         12                           last ← last + 1
         13   return C
```

FIGURE 13.42 $O(m)$-space algorithm to compute llcs(x, y).

```
      HIRSCHBERG(x, m, y, n)
       1   if m = 0
       2       then return ε
       3       else if m = 1
       4           then if x[0] ∈ y
       5               then return x[0]
       6               else return ε
       7           else  j ← ⌊n/2⌋
       8                 C ← LLCS(x, m, y[0 . . j − 1], j)
       9                 C* ← LLCS(x^R, m, y[j . . n − 1]^R, n − j)
      10                 k ← m − 1
      11                 M ← C[m − 1] + C*[m − 1]
      12                 for j ← −1 to m − 2
      13                     do if C[j] + C*[j] > M
      14                         then M ← C[j] + C*[j]
      15                             k ← j
      16                 return HIRSCHBERG(x[0 . . k − 1], k, y[0 . . j − 1], j)·
                                 HIRSCHBERG(x[k . . m − 1], m − k, y[j . . n − 1], n − j)
```

FIGURE 13.43 $O(\min(m, n))$-space computation of lcs(x, y).

The running time of the Hirschberg algorithm is still $O(mn)$ but the amount of space required for the computation becomes $O(\min(m, n))$, instead of being quadratic when computed by dynamic programming.

13.6 Approximate String Matching

Approximate string matching is the problem of finding all approximate occurrences of a pattern x of length m in a text y of length n. Approximate occurrences of x are segments of y that are close to x according to a specific distance: the distance between segments and x must be not greater than a given integer k. We consider two distances in this section: the **Hamming distance** and the **Levenshtein distance**.

With the Hamming distance, the problem is also known as approximate string matching with k mismatches. With the Levenshtein distance (or edit distance), the problem is known as approximate string matching with k differences.

The Hamming distance between two words w_1 and w_2 of the same length is the number of positions with different characters. The Levenshtein distance between two words w_1 and w_2 (not necessarily of the same length) is the minimal number of differences between the two words. A difference is one of the following operations:

- A substitution: a character of w_1 corresponds to a different character in w_2.
- An insertion: a character of w_1 corresponds to no character in w_2.
- A deletion: a character of w_2 corresponds to no character in w_1.

The *Shift-Or algorithm* of the next section is a method that is both very fast in practice and very easy to implement. It solves the Hamming distance and the Levenshtein distance problems. We initially describe the method for the exact string-matching problem and then show how it can handle the cases of k mismatches and k differences. The method is flexible enough to be adapted to a wide range of similar approximate matching problems.

13.6.1   Shift-Or Algorithm

We first present an algorithm to solve the exact string-matching problem using a technique different from those developed previously, but which extends readily to the approximate string-matching problem.

Let $\mathbf{R}^0$ be a bit array of size m. Vector $\mathbf{R}^0_j$ is the value of the entire array $\mathbf{R}^0$ after text character $y[j]$ has been processed (see Figure 13.44). It contains information about all matches of prefixes of x that end at position j in the text. It is defined, for $0 \leq i \leq m - 1$, by

$$\mathbf{R}^0_j[i] = \begin{cases} 0 & \text{if } x[0 .. i] = y[j - i .. j] \\ 1 & \text{otherwise.} \end{cases}$$

Therefore, $\mathbf{R}^0_j[m - 1] = 0$ is equivalent to saying that an (exact) occurrence of the pattern x ends at position j in y.

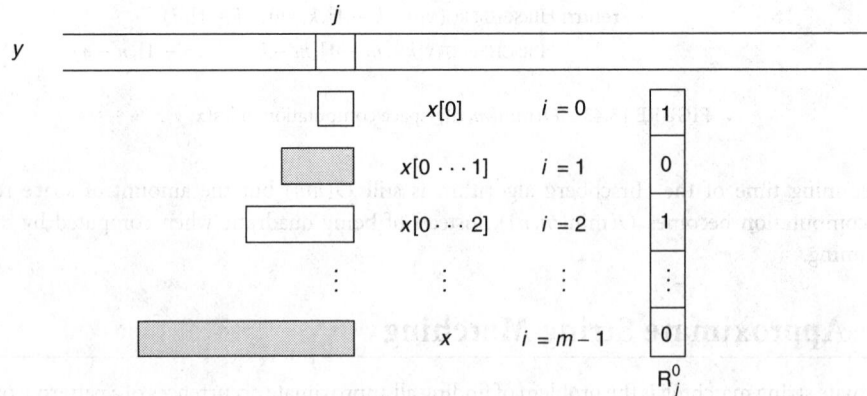

FIGURE 13.44   Meaning of vector $\mathbf{R}^0_j$.

The vector $\mathbf{R}_j^0$ can be computed after $\mathbf{R}_{j-1}^0$ by the following recurrence relation:

$$\mathbf{R}_j^0[i] = \begin{cases} 0 & \text{if } \mathbf{R}_{j-1}^0[i-1] = 0 \text{ and } x[i] = y[j], \\ 1 & \text{otherwise,} \end{cases}$$

and

$$\mathbf{R}_j^0[0] = \begin{cases} 0 & \text{if } x[0] = y[j], \\ 1 & \text{otherwise.} \end{cases}$$

The transition from $\mathbf{R}_{j-1}^0$ to $\mathbf{R}_j^0$ can be computed very fast as follows. For each $a \in \Sigma$, let S_a be a bit array of size m defined, for $0 \le i \le m-1$, by

$$S_a[i] = 0 \quad \text{if} \quad x[i] = a.$$

The array S_a denotes the positions of the character a in the pattern x. All arrays S_a are preprocessed before the search starts. And the computation of $\mathbf{R}_j^0$ reduces to two operations, SHIFT and OR:

$$\mathbf{R}_j^0 = \text{SHIFT}(\mathbf{R}_{j-1}^0) \quad \text{OR} \quad S_{y[j]}.$$

Example 13.16

String $x = $ **GATAA** occurs at position 2 in $y = $ **CAGATAAGAGAA**.

$S_\mathbf{A}$	$S_\mathbf{C}$	$S_\mathbf{G}$	$S_\mathbf{T}$
1	1	0	1
0	1	1	1
1	1	1	0
0	1	1	1
0	1	1	1

	C	A	G	A	T	A	A	G	A	G	A	A
G	1	1	0	1	1	1	1	0	1	0	1	1
A	1	1	1	0	1	1	1	1	0	1	0	1
T	1	1	1	1	0	1	1	1	1	1	1	1
A	1	1	1	1	1	0	1	1	1	1	1	1
A	1	1	1	1	1	1	0	1	1	1	1	1

13.6.2 String Matching with k Mismatches

The Shift-Or algorithm easily adapts to support approximate string matching with k mismatches. To simplify the description, we shall present the case where at most one substitution is allowed.

We use arrays $\mathbf{R}^0$ and S as before, and an additional bit array $\mathbf{R}^1$ of size m. Vector $\mathbf{R}_{j-1}^1$ indicates all matches with at most one substitution up to the text character $y[j-1]$. The recurrence on which the computation is based splits into two cases.

1. There is an exact match on the first i characters of x up to $y[j-1]$ (i.e., $\mathbf{R}_{j-1}^0[i-1] = 0$). Then, substituting $x[i]$ to $y[j]$ creates a match with one substitution (see Figure 13.45). Thus,

$$\mathbf{R}_j^1[i] = \mathbf{R}_{j-1}^0[i-1].$$

2. There is a match with one substitution on the first i characters of x up to $y[j-1]$ and $x[i] = y[j]$. Then, there is a match with one substitution of the first $i+1$ characters of x up to $y[j]$

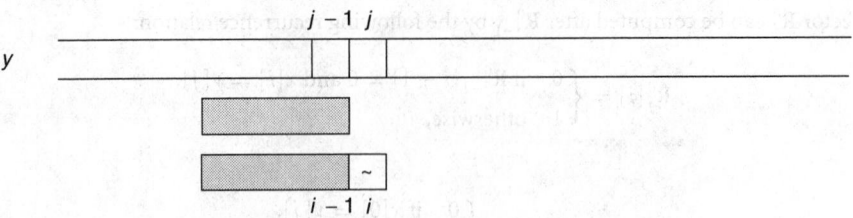

FIGURE 13.45   If $\mathbf{R}^0_{j-1}[i-1] = 0$, then $\mathbf{R}^1_j[i] = 0$.

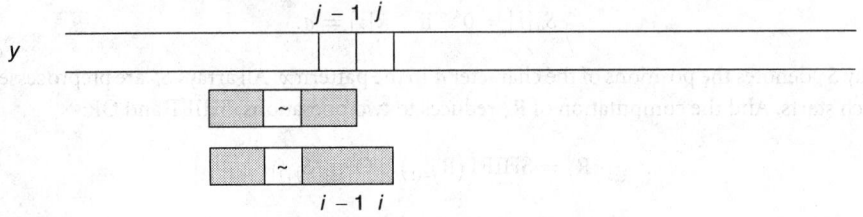

FIGURE 13.46 $\mathbf{R}^1_j[i] = \mathbf{R}^1_{j-1}[i-1]$ if $x[i] = y[j]$.

(see Figure 13.46). Thus,

$$\mathbf{R}^1_j[i] = \begin{cases} \mathbf{R}^1_{j-1}[i-1] & \text{if } x[i] = y[j], \\ 1 & \text{otherwise.} \end{cases}$$

This implies that $\mathbf{R}^1_j$ can be updated from $\mathbf{R}^1_{j-1}$ by the relation:

$$\mathbf{R}^1_j = (\text{SHIFT}(\mathbf{R}^1_{j-1}) \quad \text{OR} \quad S_{y[j]}) \quad \text{AND} \quad \text{SHIFT}(\mathbf{R}^0_{j-1}).$$

Example 13.17

String $x = $ **GATAA** occurs at positions 2 and 7 in $y = $ **CAGATAAGAGAA** with no more than one mismatch.

	C	A	G	A	T	A	A	G	A	G	A	A
G	0	0	0	0	0	0	0	0	0	0	0	0
A	1	0	1	0	1	0	0	1	0	1	0	0
T	1	1	1	1	0	1	1	1	1	0	1	0
A	1	1	1	1	1	0	1	1	1	1	0	1
A	1	1	1	1	1	1	0	1	1	1	1	0

13.6.3   String Matching with k Differences

We show in this section how to adapt the Shift-Or algorithm to the case of only one insertion, and then dually to the case of only one deletion. The method is based on the following elements.

One insertion is allowed: here, vector $\mathbf{R}^1_{j-1}$ indicates all matches with at most one insertion up to text character $y[j-1]$. $\mathbf{R}^1_{j-1}[i-1] = 0$ if the first i characters of x ($x[0..i-1]$) match i symbols of the last $i+1$ text characters up to $y[j-1]$. Array $\mathbf{R}^0$ is maintained as before, and we show how to maintain array $\mathbf{R}^1$. Two cases arise.

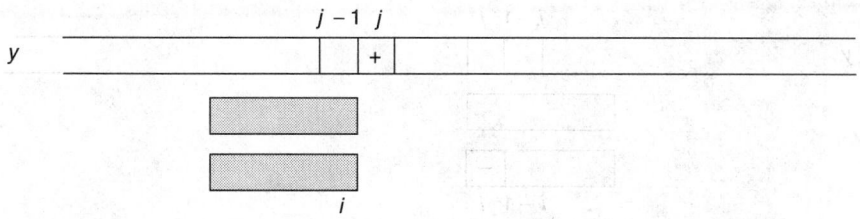

FIGURE 13.47 If $\mathbf{R}^0_{j-1}[i] = 0$, then $\mathbf{R}^1_j[i] = 0$.

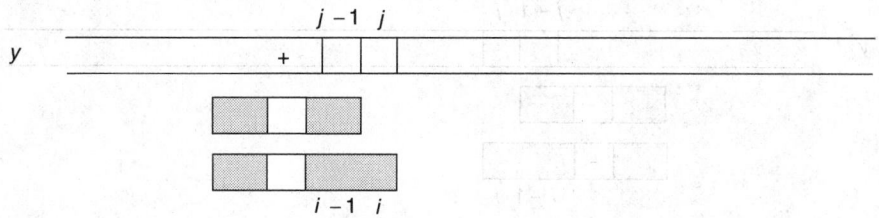

FIGURE 13.48 $\mathbf{R}^1_j[i] = \mathbf{R}^1_{j-1}[i-1]$ if $x[i] = y[j]$.

1. There is an exact match on the first $i + 1$ characters of x ($x[0..i]$) up to $y[j-1]$. Then inserting $y[j]$ creates a match with one insertion up to $y[j]$ (see Figure 13.47). Thus,

$$\mathbf{R}^1_j[i] = \mathbf{R}^0_{j-1}[i].$$

2. There is a match with one insertion on the i first characters of x up to $y[j-1]$. Then if $x[i] = y[j]$, there is a match with one insertion on the first $i + 1$ characters of x up to $y[j]$ (see Figure 13.48). Thus,

$$\mathbf{R}^1_j[i] = \begin{cases} \mathbf{R}^1_{j-1}[i-1] & \text{if } x[i] = y[j], \\ 1 & \text{otherwise.} \end{cases}$$

This shows that $\mathbf{R}^1_j$ can be updated from $\mathbf{R}^1_{j-1}$ with the formula

$$\mathbf{R}^1_j = (\text{SHIFT}(\mathbf{R}^1_{j-1}) \quad \text{OR} \quad S_{y[j]}) \quad \text{AND} \quad \mathbf{R}^0_{j-1}.$$

Example 13.18

Here, **GATAAG** is an occurrence of $x =$ **GATAA** with exactly one insertion in $y =$ **CAGATAAGAGAA**

	C	A	G	A	T	A	A	G	A	G	A	A
G	1	1	1	0	1	1	1	1	0	1	0	1
A	1	1	1	1	0	1	1	1	1	0	1	0
T	1	1	1	1	1	0	1	1	1	1	1	1
A	1	1	1	1	1	1	0	1	1	1	1	1
A	1	1	1	1	1	1	1	0	1	1	1	1

One deletion is allowed: we assume here that $\mathbf{R}^1_{j-1}$ indicates all possible matches with at most one deletion up to $y[j-1]$. As in the previous solution, two cases arise.

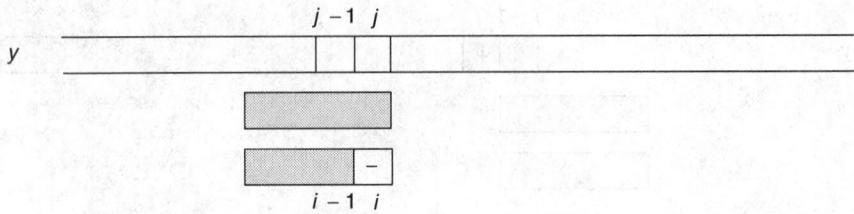

FIGURE 13.49 If $\mathbf{R}_j^0[i] = 0$, then $\mathbf{R}_j^1[i] = 0$.

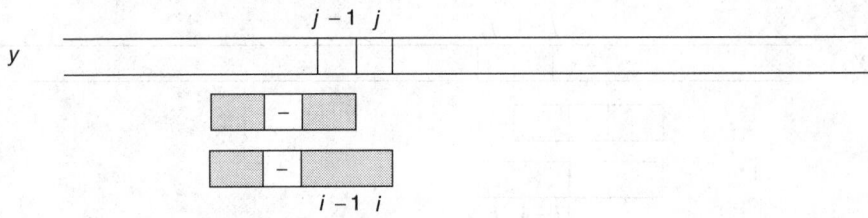

FIGURE 13.50 $\mathbf{R}_j^1[i] = \mathbf{R}_{j-1}^1[i-1]$ if $x[i] = y[j]$.

1. There is an exact match on the first $i + 1$ characters of x ($x[0..i]$) up to $y[j]$ (i.e., $\mathbf{R}_j^0[i] = 0$). Then, deleting $x[i]$ creates a match with one deletion (see Figure 13.49). Thus,

$$\mathbf{R}_j^1[i] = \mathbf{R}_j^0[i].$$

2. There is a match with one deletion on the first i characters of x up to $y[j-1]$ and $x[i] = y[j]$. Then, there is a match with one deletion on the first $i + 1$ characters of x up to $y[j]$ (see Figure 13.50). Thus,

$$\mathbf{R}_j^1[i] = \begin{cases} \mathbf{R}_{j-1}^1[i-1] & \text{if } x[i] = y[j], \\ 1 & \text{otherwise.} \end{cases}$$

The discussion provides the following formula used to update $\mathbf{R}_j^1$ from $\mathbf{R}_{j-1}^1$:

$$\mathbf{R}_j^1 = (\text{SHIFT}(\mathbf{R}_{j-1}^1) \quad \text{OR} \quad S_{y[j]}) \quad \text{AND} \quad \text{SHIFT}(\mathbf{R}_j^0).$$

Example 13.19

GATA and **ATAA** are two occurrences with one deletion of $x = $ **GATAA** in $y = $ **CAGATAAGAGAA**

	C	A	G	A	T	A	A	G	A	G	A	A
G	0	0	0	0	0	0	0	0	0	0	0	0
A	1	0	0	0	1	0	0	0	0	0	0	0
T	1	1	1	0	0	1	1	1	0	1	0	1
A	1	1	1	1	0	0	1	1	1	1	1	0
A	1	1	1	1	1	0	0	1	1	1	1	1

13.6.4 Wu–Manber Algorithm

We present in this section a general solution for the approximate string-matching problem with at most k differences of the types: insertion, deletion, and substitution. It is an extension of the problems presented above. The following algorithm maintains $k+1$ bit arrays $\mathbf{R}^0, \mathbf{R}^1, \ldots, \mathbf{R}^k$ that are described now. The vector $\mathbf{R}^0$ is maintained similarly as in the exact matching case (Section 13.6.1). The other vectors are computed with the formula ($1 \leq \ell \leq k$)

$$\mathbf{R}_j^\ell = (\text{SHIFT}(\mathbf{R}_{j-1}^\ell) \quad \text{OR} \quad S_{y[j]})$$
$$\text{AND} \quad \text{SHIFT}(\mathbf{R}_j^{\ell-1})$$
$$\text{AND} \quad \text{SHIFT}(\mathbf{R}_{j-1}^{\ell-1})$$
$$\text{AND} \quad \mathbf{R}_{j-1}^{\ell-1}$$

which can be rewritten into

$$\mathbf{R}_j^\ell = (\text{SHIFT}(\mathbf{R}_{j-1}^\ell) \quad \text{OR} \quad S_{y[j]})$$
$$\text{AND} \quad \text{SHIFT}(\mathbf{R}_j^{\ell-1} \quad \text{AND} \quad \mathbf{R}_{j-1}^{\ell-1})$$
$$\text{AND} \quad \mathbf{R}_{j-1}^{\ell-1}.$$

Example 13.20

Here, $x = \mathbf{GATAA}$ and $y = \mathbf{CAGATAAGAGAA}$ and $k = 1$. The output 5, 6, 7, and 11 corresponds to the segments **GATA**, **GATAA**, **GATAAG**, and **GAGAA**, which approximate the pattern **GATAA** with no more than one difference.

	C	A	G	A	T	A	A	G	A	G	A	A
G	0	0	0	0	0	0	0	0	0	0	0	0
A	1	0	0	0	0	0	0	0	0	0	0	0
T	1	1	1	0	0	0	1	1	0	0	0	0
A	1	1	1	1	0	0	0	1	1	1	0	0
A	1	1	1	1	1	0	0	0	1	1	1	0

The method, called the Wu–Manber algorithm, is implemented in Figure 13.51. It assumes that the length of the pattern is no more than the size of the memory word of the machine, which is often the case in applications.

```
WM(x, m, y, n, k)
 1    for each character a ∈ Σ
 2       do Sₐ ← 1ᵐ
 3    for i ← 0 to m − 1
 4       do S_{x[i]}[i] ← 0
 5    R⁰ ← 1ᵐ
 6    for ℓ ← 1 to k
 7       do Rℓ ← SHIFT(Rℓ⁻¹)
 8    for j ← 0 to n − 1
 9       do T ← R⁰
10          R⁰ ← SHIFT(R⁰)  OR  S_{y[j]}
11          for ℓ ← 1 to k
12             do T′ ← Rℓ
13                Rℓ ← (SHIFT(Rℓ) OR S_{y[j]}) AND (SHIFT((T AND Rℓ⁻¹)) AND T
14                T ← T′
15          if Rᵏ[m − 1] = 0
16             then OUTPUT(j)
```

FIGURE 13.51 Wu–Manber approximate string-matching algorithm.

The preprocessing phase of the algorithm takes $O(\sigma m + km)$ memory space, and runs in time $O(\sigma m + k)$. The time complexity of its searching phase is $O(kn)$.

13.7 Text Compression

In this section we are interested in algorithms that compress texts. Compression serves both to save storage space and to save transmission time. We shall assume that the uncompressed text is stored in a file. The aim of compression algorithms is to produce another file containing the compressed version of the same text. Methods in this section work with no loss of information, so that decompressing the compressed text restores exactly the original text.

We apply two main strategies to design the algorithms. The first strategy is a statistical method that takes into account the frequencies of symbols to build a uniquely decipherable code optimal with respect to the compression. The code contains new codewords for the symbols occurring in the text. In this method, fixed-length blocks of bits are encoded by different codewords. *A contrario*, the second strategy encodes variable-length segments of the text. To put it simply, the algorithm, while scanning the text, replaces some already read segments just by a pointer to their first occurrences.

Text compression software often use a mixture of several methods. An example of that is given in Section 13.7.3, which contains in particular two classical simple compression algorithms. They compress efficiently only a small variety of texts when used alone, but they become more powerful with the special preprocessing presented there.

13.7.1 Huffman Coding

The Huffman method is an optimal statistical coding. It transforms the original code used for characters of the text (ASCII code on 8 b, for instance). Coding the text is just replacing each symbol (more exactly, each occurrence of it) by its new codeword. The method works for any length of blocks (not only 8 b), but the running time grows exponentially with the length. In the following, we assume that symbols are originally encoded on 8 b to simplify the description.

The Huffman algorithm uses the notion of **prefix code**. A prefix code is a set of words containing no word that is a prefix of another word of the set. The advantage of such a code is that decoding is immediate. Moreover, it can be proved that this type of code does not weaken the compression.

A prefix code on the binary alphabet $\{0, 1\}$ can be represented by a trie (see section on the Aho–Corasick algorithm) that is a binary tree. In the present method codes are complete: they correspond to complete tries (internal nodes have exactly two children). The leaves are labeled by the original characters, edges are labeled by 0 or 1, and labels of branches are the words of the code. The condition on the code implies that codewords are identified with leaves only. We adopt the convention that, from an internal node, the edge to its left child is labeled by 0, and the edge to its right child is labeled by 1.

In the model where characters of the text are given new codewords, the Huffman algorithm builds a code that is optimal in the sense that the compression is the best possible (the length of the compressed text is minimum). The code depends on the text, and more precisely on the frequencies of each character in the uncompressed text. The more frequent characters are given short codewords, whereas the less frequent symbols have longer codewords.

13.7.1.1 Encoding

The coding algorithm is composed of three steps: count of character frequencies, construction of the prefix code, and encoding of the text.

The first step consists in counting the number of occurrences of each character in the original text (see Figure 13.52). We use a special end marker (denoted by END), which (virtually) appears only once at the end of the text. It is possible to skip this first step if fixed statistics on the alphabet are used. In this case, the method is optimal according to the statistics, but not necessarily for the specific text.

COUNT(*fin*)
1   **for** each character $a \in \Sigma$
2   **do** *freq*(*a*) ← 0
3   **while** not end of file *fin* and *a* is the next symbol
4   **do** *freq*(*a*) ← *freq*(*a*) + 1
5   *freq*(END) ← 1

FIGURE 13.52   Counts the character frequencies.

BUILD-TREE()
1   **for** each character $a \in \Sigma \cup \{\text{END}\}$
2   **do if** *freq*(*a*) ≠ 0
3   **then** create a new node *t*
4   *weight*(*t*) ← *freq*(*a*)
5   *label*(*t*) ← *a*
6   *lleaves* ← list of all the nodes in increasing order of weight
7   *ltrees* ← empty list
8   **while** LENGTH(*lleaves*) + LENGTH(*ltrees*) > 1
9   **do** (ℓ, r) ← extract the two nodes of smallest weight (among the two nodes at the
        beginning of *lleaves* and the two nodes at the beginning of *ltrees*)
10   create a new node *t*
11   *weight*(*t*) ← *weight*(ℓ) + *weight*(*r*)
12   *left*(*t*) ← ℓ
13   *right*(*t*) ← *r*
14   insert *t* at the end of *ltrees*
15   **return** *t*

FIGURE 13.53   Builds the coding tree.

The second step of the algorithm builds the tree of a prefix code using the character frequency *freq*(*a*) of each character *a* in the following way:

- Create a one-node tree *t* for each character *a*, setting *weight*(*t*) = *freq*(*a*) and *label*(*t*) = *a*,
- Repeat (1), extract the two least weighted trees t_1 and t_2, and (2) create a new tree t_3 having left subtree t_1, right subtree t_2, and weight *weight*(t_3) = *weight*(t_1) + *weight*(t_2),
- Until only one tree remains.

The tree is constructed by the algorithm BUILD-TREE in Figure 13.53. The implementation uses two linear lists. The first list contains the leaves of the future tree, each associated with a symbol. The list is sorted in the increasing order of the weight of the leaves (frequency of symbols). The second list contains the newly created trees. Extracting the two least weighted trees consists in extracting the two least weighted trees among the two first trees of the list of leaves and the two first trees of the list of created trees. Each new tree is inserted at the end of the list of the trees. The only tree remaining at the end of the procedure is the coding tree.

After the coding tree is built, it is possible to recover the codewords associated with characters by a simple depth-first search of the tree (see Figure 13.54); *codeword*(*a*) is then the binary code associated with the character *a*.

BUILD-CODE(*t*, *length*)

1   **if** *t* is not a leaf
2   **then** *temp*[*length*] ← 0
3   BUILD-CODE(*left*(*t*), *length* + 1)
4   *temp*[*length*] ← 1
5   BUILD-CODE(*right*(*t*), *length* + 1)
6   **else** *codeword*(*label*(*t*)) ← *temp*[0 .. *length* − 1]

FIGURE 13.54   Builds the character codes from the coding tree.

CODE-TREE(*fout*, *t*)

1   **if** *t* is not a leaf
2   **then** write a 0 in the file *fout*
3   CODE-TREE(*fout*, *left*(*t*))
4   CODE-TREE(*fout*, *right*(*t*))
5   **else** write a 1 in the file *fout*
6   write the original code of *label*(*t*) in the file *fout*

FIGURE 13.55   Memorizes the coding tree in the compressed file.

CODE-TEXT(*fin*, *fout*)

1   **while** not end of file *fin* and *a* is the next symbol
2   **do** write *codeword*(*a*) in the file *fout*
3   write *codeword*(END) in the file *fout*

FIGURE 13.56   Encodes the characters in the compressed file.

CODING(*fin*, *fout*)

1   COUNT(*fin*)
2   *t* ← BUILD-TREE()
3   BUILD-CODE(*t*, 0)
4   CODE-TREE(*fout*, *t*)
5   CODE-TEXT(*fin*, *fout*)

FIGURE 13.57   Complete function for Huffman coding.

    In the third step, the original text is encoded. Since the code depends on the original text, in order to be able to decode the compressed text, the coding tree and the original codewords of symbols must be stored with the compressed text. This information is placed in a header of the compressed file, to be read at decoding time just before the compressed text. The header is made via a depth-first traversal of the tree. Each time an internal node is encountered, a 0 is produced. When a leaf is encountered, a 1 is produced, followed by the original code of the corresponding character on 9 b (so that the end marker can be equal to 256 if all the characters appear in the original text). This part of the encoding algorithm is shown in Figure 13.55. After the header of the compressed file is computed, the encoding of the original text is realized by the algorithm of Figure 13.56.

    A complete implementation of the Huffman algorithm, composed of the three steps just described, is given in Figure 13.57.

Example 13.21

Here, $y =$ **CAGATAAGAGAA**. The length of $y = 12 \times 8 = 96$ b (assuming an 8-b code). The character frequencies are

A	C	G	T	END
7	1	3	1	1

The different steps during the construction of the coding tree are

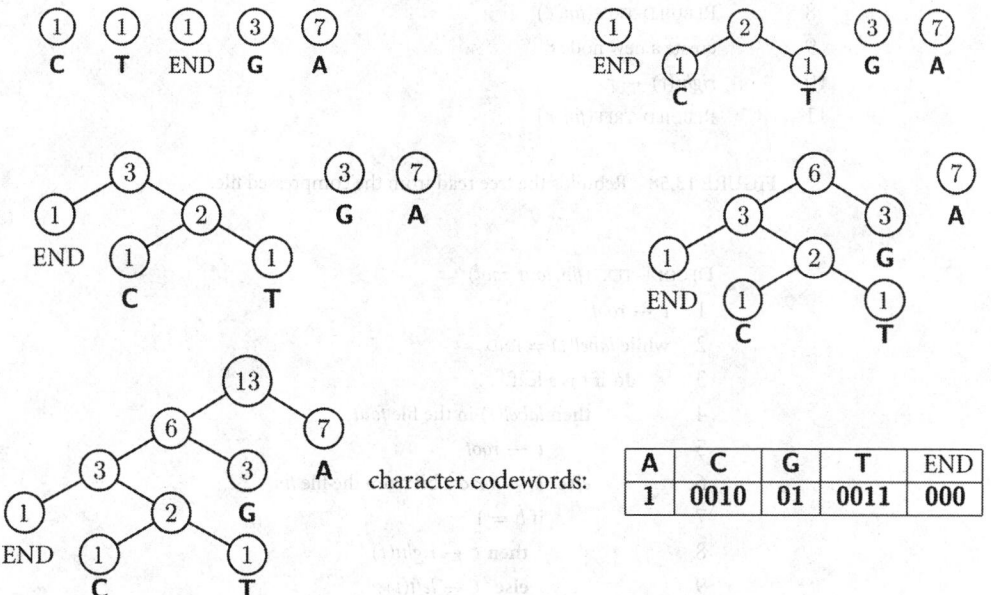

character codewords:

A	C	G	T	END
1	0010	01	0011	000

The encoded tree is **0001** binary (END, 9)**01**binary (**C**, 9)**1**binary(**T**, 9) **1**binary (**G**, 9)**1**binary (**A**, 9), which produces a header of length 54 b,

$$0001\ 100000000\ 01\ 001000011\ 1\ 001010100\ 1\ 001000111\ 1\ 001000001$$

The encoded text

$$0010\ 1\ 01\ 1\ 0011\ 1\ 1\ 01\ 1\ 01\ 1\ 1\ 000$$

is of length 24 b. The total length of the compressed file is 78 b.

The construction of the tree takes $O(\sigma \log \sigma)$ time if the sorting of the list of the leaves is implemented efficiently. The rest of the encoding process runs in linear time in the sum of the sizes of the original and compressed texts.

13.7.1.2 Decoding

Decoding a file containing a text compressed by the Huffman algorithm is a mere programming exercise. First, the coding tree is rebuilt by the algorithm of Figure 13.58. Then, the uncompressed text is recovered by parsing the compressed text with the coding tree. The process begins at the root of the coding tree and follows a left edge when a 0 is read or a right edge when a 1 is read. When a leaf is encountered, the corresponding character (in fact the original codeword of it) is produced and the parsing phase resumes at the root of the tree. The parsing ends when the codeword of the end marker is read. An implementation of the decoding of the text is presented in Figure 13.59.

REBUILD-TREE(*fin, t*)

```
 1   b ← read a bit from the file fin
 2   if b = 1                          ▷ leaf
 3     then left(t) ← NIL
 4          right(t) ← NIL
 5          label(t) ← symbol corresponding to the 9 next bits in the file fin
 6     else create a new node ℓ
 7          left(t) ← ℓ
 8          REBUILD-TREE(fin, ℓ)
 9          create a new node r
10          right(t) ← r
11          REBUILD-TREE(fin, r)
```

FIGURE 13.58   Rebuilds the tree read from the compressed file.

DECODE-TEXT(*fin, fout, root*)

```
 1   t ← root
 2   while label(t) ≠ END
 3     do if t is a leaf
 4          then label(t) in the file fout
 5               t ← root
 6          else b ← read a bit from the file fin
 7               if b = 1
 8                 then t ← right(t)
 9                 else t ← left(t)
```

FIGURE 13.59   Reads the compressed text and produces the uncompressed text.

DECODING(*fin, fout*)

```
 1   create a new node root
 2   REBUILD-TREE(fin, root)
 3   DECODE-TEXT(fin, fout, root)
```

FIGURE 13.60   Complete function for Huffman decoding.

The complete decoding program is given in Figure 13.60. It calls the preceding functions. The running time of the decoding program is linear in the sum of the sizes of the texts it manipulates.

13.7.2   Lempel–Ziv–Welsh (LZW) Compression

Ziv and Lempel designed a compression method using encoding segments. These segments are stored in a dictionary that is built during the compression process. When a segment of the dictionary is encountered later while scanning the original text, it is substituted by its index in the dictionary. In the model where portions of the text are replaced by pointers on previous occurrences, the Ziv–Lempel compression scheme can be proved to be asymptotically optimal (on large enough texts satisfying good conditions on the probability distribution of symbols).

The dictionary is the central point of the algorithm. It has the property of being prefix closed (every prefix of a word of the dictionary is in the dictionary), so that it can be implemented as a tree. Furthermore, a hashing technique makes its implementation efficient. The version described in this section is called the Lempel–Ziv–Welsh method after several improvements introduced by Welsh. The algorithm is implemented by the **compress** command existing under the Unix operating system.

13.7.2.1 Compression Method

We describe the scheme of the compression method. The dictionary is initialized with all the characters of the alphabet. The current situation is when we have just read a segment w in the text. Let a be the next symbol (just following w). Then we proceed as follows:

- If wa is not in the dictionary, we write the index of w to the output file, and add wa to the dictionary. We then reset w to a and process the next symbol (following a).
- If wa is in the dictionary, we process the next symbol, with segment wa instead of w.

Initially, the segment w is set to the first symbol of the source text.

Example 13.22

Here $y = $ **CAGTAAGAGAA**

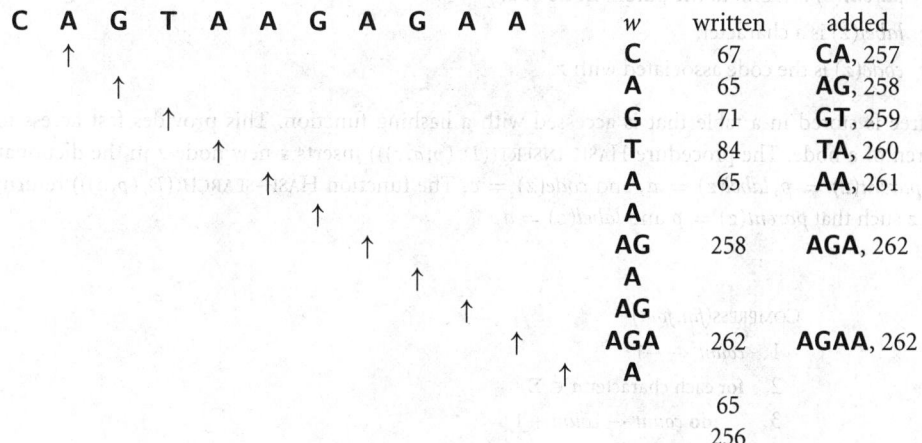

	w	written	added
	C	67	**CA**, 257
	A	65	**AG**, 258
	G	71	**GT**, 259
	T	84	**TA**, 260
	A	65	**AA**, 261
	A		
	AG	258	**AGA**, 262
	A		
	AG		
	AGA	262	**AGAA**, 262
	A		
		65	
		256	

13.7.2.2 Decompression Method

The decompression method is symmetrical to the compression algorithm. The dictionary is recovered while the decompression process runs. It is basically done in this way:

- Read a code c in the compressed file.
- Write in the output file the segment w that has index c in the dictionary.
- Add to the dictionary the word wa where a is the first letter of the next segment.

In this scheme, a problem occurs if the next segment is the word that is being built. This arises only if the text contains a segment $azazax$ for which az belongs to the dictionary but aza does not. During the compression process, the index of az is written into the compressed file, and aza is added to the dictionary. Next, aza is read and its index is written into the file. During the decompression process, the index of aza is read while the word az has not been completed yet: the segment aza is not already in the dictionary. However, because this is the unique case where the situation arises, the segment aza is recovered, taking the last segment az added to the dictionary concatenated with its first letter a.

Example 13.23

Here, the decoding is 67, 65, 71, 84, 65, 258, 262, 65, 256

read	written	added
67	**C**	
65	**A**	**CA**, 257
71	**G**	**AG**, 258
84	**T**	**GT**, 259
65	**A**	**TA**, 260
258	**AG**	**AA**, 261
262	**AGA**	**AGA**, 262
65	**A**	**AGAA**, 263
256		

13.7.2.3 Implementation

For the compression algorithm shown in Figure 13.61, the dictionary is stored in a table D. The dictionary is implemented as a tree; each node z of the tree has the three following components:

- $parent(z)$ is a link to the parent node of z.
- $label(z)$ is a character.
- $code(z)$ is the code associated with z.

The tree is stored in a table that is accessed with a hashing function. This provides fast access to the children of a node. The procedure HASH-INSERT$((D, (p, a, c)))$ inserts a new node z in the dictionary D with $parent(z) = p, label(z) = a$, and $code(z) = c$. The function HASH-SEARCH$((D, (p, a)))$ returns the node z such that $parent(z) = p$ and $label(z) = a$.

COMPRESS(*fin, fout*)

```
 1   count ← −1
 2   for each character a ∈ Σ
 3       do count ← count + 1
 4           HASH-INSERT(D, (−1, a, count))
 5   count ← count + 1
 6   HASH-INSERT(D, (−1, END, count))
 7   p ← −1
 8   while not end of file fin
 9       do a ← next character of fin
10           q ← HASH-SEARCH(D, (p, a))
11           if q = NIL
12               then write code(p) on 1 + log(count) bits in fout
13                   count ← count + 1
14                   HASH-INSERT(D, (p, a, count))
15                   p ← HASH-SEARCH(D, (−1, a))
16               else p ← q
17   write code(p) on 1 + log(count) bits in fout
18   write code(HASH-SEARCH(D, (−1, END))) on 1 + log(count) bits in fout
```

FIGURE 13.61   LZW compression algorithm.

UNCOMPRESS(*fin*, *fout*)
1   *count* ← −1
2   **for** each character *a* ∈ Σ
3   **do** *count* ← *count* + 1
4   HASH-INSERT(*D*, (−1, *a*, *count*))
5   *count* ← *count* + 1
6   HASH-INSERT(*D*, (−1, END, *count*))
7   *c* ← first code on 1 + log(*count*) bits in *fin*
8   write *string*(*c*) in *fout*
9   *a* ← *first*(*string*(*c*))
10   **while** TRUE
11   **do** *d* ← next code on 1 + log(*count*) bits in *fin*
12   **if** *d* > *count*
13   **then** *count* ← *count* + 1
14   *parent*(*count*) ← *c*
15   *label*(*count*) ← *a*
16   write *string*(*c*)*a* in *fout*
17   *c* ← *d*
18   **else** *a* ← *first*(*string*(*d*))
19   **if** *a* ≠ END
20   **then** *count* ← *count* + 1
21   *parent*(*count*) ← *c*
22   *label*(*count*) ← *a*
23   write *string*(*d*) in *fout*
24   *c* ← *d*
25   **else break**

FIGURE 13.62   LZW decompression algorithm.

For the decompression algorithm, no hashing technique is necessary. Having the index of the next segment, a bottom-up walk in the trie implementing the dictionary produces the mirror image of the segment. A stack is used to reverse it. We assume that the function *string*(*c*) performs this specific work for a code *c*. The bottom-up walk follows the parent links of the data structure. The function *first*(*w*) gives the first character of the word *w*. These features are part of the decompression algorithm displayed in Figure 13.62.

The Ziv–Lempel compression and decompression algorithms run both in linear time in the sizes of the files provided a good hashing technique is chosen. Indeed, it is very fast in practice. Its main advantage compared to Huffman coding is that it captures long repeated segments in the source file.

13.7.3 Mixing Several Methods

We describe simple compression methods and then an example of a combination of several of them, the basis of the popular **bzip** software.

13.7.3.1 Run Length Encoding

The aim of Run Length Encoding (RLE) is to efficiently encode repetitions occurring in the input data. Let us assume that it contains a good quantity of repetitions of the form *aa* . . . *a* for some character *a* (*a* ∈ Σ). A repetition of *k* consecutive occurrences of letter *a* is replaced by &*ak*, where the symbol & is a new character (& ∉ Σ).

The string &ak that encodes a repetition of k consecutive occurrences of a is itself encoded on the binary alphabet $\{0, 1\}$. In practice, letters are often represented by their ASCII code. Therefore, the codeword of a letter belongs to $\{0, 1\}^k$ with $k = 7$ or 8. Generally, there is no problem in choosing or encoding the special character &. The integer k of the string &ak is also encoded on the binary alphabet, but it is not sufficient to translate it by its binary representation, because we would be unable to recover it at decoding time inside the stream of bits. A simple way to cope with this is to encode k by the string $0^\ell \text{bin}(k)$, where $\text{bin}(k)$ is the binary representation of k, and ℓ is the length. This works well because the binary representation of k starts with a 1 so there is no ambiguity to recover ℓ by counting during the decoding phase. The size of the encoding of k is thus roughly $2 \log k$. More sophisticated integer representations are possible, but none is really suitable for the present situation. Simpler solution consists in encoding k on the same number of bits as other symbols, but this bounds values of ℓ and decreases the power of the method.

13.7.3.2 Move To Front

The Move To Front (MTF) method can be regarded as an extension of Run Length Encoding or a simplification of Ziv–Lempel compression. It is efficient when the occurrences of letters in the input text are localized into a relatively short segment of it. The technique is able to capture the proximity between occurrences of symbols and to turn it into a short encoded text.

Letters of the alphabet Σ of the input text are initially stored in a list that is managed dynamically. Letters are represented by their rank in the list, starting from 1, rank that is itself encoded as described above for RLE.

Letters of the input text are processed in an on-line manner. The clue of the method is that each letter is moved to the beginning of the list just after it is translated by the encoding of its rank.

The effect of MTF is to reduce the size of the encoding of a letter that reappears soon after its preceding occurrence.

13.7.3.3 Integrated Example

Most compression software combines several methods to be able to efficiently compress a large range of input data. We present an example of this strategy, implemented by the UNIX command **bzip**.

Let $y = y[0]y[1] \cdots y[n - 1]$ be the input text. The k-th rotation (or conjugate) of y, $0 \le k \le n - 1$, is the string $y_k = y[k]y[k + 1] \cdots y[n - 1]y[0]y[1] \cdots y[k - 1]$.

We define the BW transformation as $BW(y) = y[p_0]y[p_1] \cdots y[p_{n-1}]$, where $p_i + 1$ is such that y_{p_i+1} has rank i in the sorted list of all rotations of y.

It is remarkable that y can be recovered from both $BW(y)$ and a position on it, starting position of the inverse transformation (see Figure 13.63). This is possible due to the following property of the transformation. Assume that $i < j$ and $y[p_i] = y[p_j] = a$. Since $i < j$, the definition implies $y_{p_i+1} < y_{p_j+1}$. Since $y[p_i] = y[p_j]$, transferring the last letters of y_{p_i+1} and y_{p_j+1} to the beginning of these words does not change the inequality. This proves that the two occurrences of a in $BW(y)$ are in the same relative order as in the sorted list of letters of y. Figure 13.63 illustrates the inverse transformation.

Transformation BW obviously does not compress the input text y. But $BW(y)$ is compressed more efficiently with simple methods. This is the strategy applied for the command **bzip**. It is a combination of the BW transformation followed by MTF encoding and RLE encoding. Arithmetic coding, a method providing compression ratios slightly better than Huffman coding, can also be used.

Table 13.1 contains a sample of experimental results showing the behavior of compression algorithms on different types of texts from the Calgary Corpus: bib (bibliography), book1 (fiction book), news (USENET batch file), pic (black and white fax picture), progc (source code in C), and trans (transcript of terminal session).

The compression algorithms reported in the table are the Huffman coding algorithm implemented by **pack**, the Ziv–Lempel algorithm implemented by **gzip-b**, and the compression based on the BW transform implemented by **bzip2-1**.

Additional compression results can be found at http://corpus.canterbury.ac.nz.

TABLE 13.1 Compression Results with Three Algorithms. Huffman coding (**pack**), Ziv–Lempel coding (**gzip-b**) and Burrows-Wheeler coding (**bzip2-1**). Figures give the number of bits used per character (letter). They show that **pack** is the less efficient method and that **bzip2-1** compresses a bit more than **gzip-b**.

| Sizes in bytes | 111,261 | 768,771 | 377,109 | 513,216 | 39,611 | 93,695 | |
Source Texts	bib	book1	news	pic	progc	trans	Average
pack	5.24	4.56	5.23	1.66	5.26	5.58	4.99
gzip-b	2.51	3.25	3.06	0.82	2.68	1.61	2.69
bzip2-1	2.10	2.81	2.85	0.78	2.53	1.53	2.46

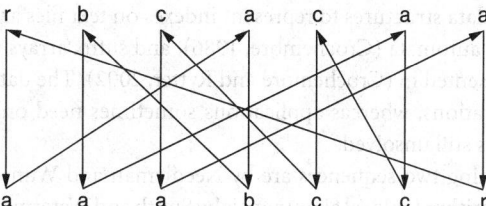

FIGURE 13.63 Example of text $y = $ **baccara**. Top line is $BW(y)$ and bottom line the sorted list of letters of it. Top-down arrows correspond to succession of occurrences in y. Each bottom-up arrow links the same occurrence of a letter in y. Arrows starting from equal letters do not cross. The circular path is associated with rotations of the string y. If the starting point is known, the only occurrence of letter **b** here, following the path produces the initial string y.

13.8 Research Issues and Summary

The algorithm for string searching by hashing was introduced by Harrison in 1971, and later fully analyzed by Karp and Rabin (1987).

The linear-time string-matching algorithm of Knuth, Morris, and Pratt is from 1976. It can be proved that, during the search, a character of the text is compared to a character of the pattern no more than $\log_\Phi(|x| + 1)$ (where Φ is the golden ratio $(1 + \sqrt{5})/2$). Simon (1993) gives an algorithm similar to the previous one but with a delay bounded by the size of the alphabet (of the pattern x). Hancart (1993) proves that the delay of Simon's algorithm is, indeed, no more than $1 + \log_2 |x|$. He also proves that this is optimal among algorithms searching the text through a window of size 1.

Galil (1981) gives a general criterion to transform searching algorithms of that type into real-time algorithms.

The Boyer–Moore algorithm was designed by Boyer and Moore (1977). The first proof on the linearity of the algorithm when restricted to the search of the first occurrence of the pattern is in Knuth et al. (1977). Cole (1994) proves that the maximum number of symbol comparisons is bounded by $3n$, and that this bound is tight.

Knuth et al. (1977) consider a variant of the Boyer–Moore algorithm in which all previous matches inside the current window are memorized. Each window configuration becomes the state of what is called the Boyer–Moore automaton. It is still unknown whether the maximum number of states of the automaton is polynomial or not.

Several variants of the Boyer–Moore algorithm avoid the quadratic behavior when searching for all occurrences of the pattern. Among the more efficient in terms of the number of symbol comparisons are the algorithm of Apostolico and Giancarlo (1986), Turbo–BM algorithm by Crochemore et al. (1992) (the two algorithms are analyzed in Lecroq (1995)), and the algorithm of Colussi (1994).

The general bound on the expected time complexity of string matching is $O(|y| \log |x|/|x|)$. The probabilistic analysis of a simplified version of the Boyer–Moore algorithm, similar to the Quick Search algorithm of Sunday (1990) described in the chapter, was studied by several authors.

String searching can be solved by a linear-time algorithm requiring only a constant amount of memory in addition to the pattern and the (window on the) text. This can be proved by different techniques presented in Crochemore and Rytter (2002).

The Aho–Corasick algorithm is from Aho and Corasick (1975). It is implemented by the **fgrep** command under the UNIX operating system. Commentz-Walter (1979) has designed an extension of the Boyer-Moore algorithm to several patterns. It is fully described in Aho (1990).

On general alphabets the two-dimensional pattern matching can be solved in linear time, whereas the running time of the Bird/Baker algorithm has an additional $\log \sigma$ factor. It is still unknown whether the problem can be solved by an algorithm working simultaneously in linear time and using only a constant amount of memory space (see Crochemore and Rytter 2002).

The suffix tree construction of Section 13.2 is by McCreight (1976). An on-line construction is given by Ukkonen (1995). Other data structures to represent indexes on text files are: direct acyclic word graph (Blumer et al., 1985), suffix automata (Crochemore, 1986), and suffix arrays (Manber and Myers, 1993). All these techniques are presented in (Crochemore and Rytter, 2002). The data structures implement full indexes with standard operations, whereas applications sometimes need only incomplete indexes. The design of compact indexes is still unsolved.

First algorithms for aligning two sequences are by Needleman and Wunsch (1970) and Wagner and Fischer (1974). Idea and algorithm for local alignment is by Smith and Waterman (1981). Hirschberg (1975) presents the computation of the lcs in linear space. This is an important result because the algorithm is classically run on large sequences. Another implementation is given in Durbin et al. (1998). The quadratic time complexity of the algorithm to compute the Levenshtein distance is a bottleneck in practical string comparison for the same reason.

Approximate string searching is a lively domain of research. It includes, for instance, the notion of regular expressions to represent sets of strings. Algorithms based on regular expression are commonly found in books related to compiling techniques. The algorithms of Section 13.6 are by Baeza-Yates and Gonnet (1992) and Wu and Manber (1992).

The statistical compression algorithm of Huffman (1951) has a dynamic version where symbol counting is done at coding time. The current coding tree is used to encode the next character and then updated. At decoding time, a symmetrical process reconstructs the same tree, so the tree does not need to be stored with the compressed text; see Knuth (1985). The command **compact** of UNIX implements this version.

Several variants of the Ziv and Lempel algorithm exist. The reader can refer to Bell et al. (1990) for further discussion. Nelson (1992) presents practical implementations of various compression algorithms. The *BW* transform is from Burrows and Wheeler (1994).

Defining Terms

Alignment: An alignment of two strings x and y is a word of the form $(\overline{x}_0, \overline{y}_0)(\overline{x}_1, \overline{y}_1) \cdots (\overline{x}_{p-1}, \overline{y}_{p-1})$
where each $(\overline{x}_i, \overline{y}_i) \in (\Sigma \cup \{\varepsilon\}) \times (\Sigma \cup \{\varepsilon\}) \setminus (\{(\varepsilon, \varepsilon)\})$ for $0 \leq i \leq p-1$ and both $x = \overline{x}_0 \overline{x}_1 \cdots \overline{x}_{p-1}$
and $y = \overline{y}_0 \overline{y}_1 \cdots \overline{y}_{p-1}$.

Border: A word $u \in \Sigma^*$ is a border of a word $w \in \Sigma^*$ if u is both a prefix and a suffix of w (there exist
two words $v, z \in \Sigma^*$ such that $w = vu = uz$). The common length of v and z is a period of w.

Edit distance: The metric distance between two strings that counts the minimum number of insertions
and deletions of symbols to transform one string into the other.

Hamming distance: The metric distance between two strings of same length that counts the number of
mismatches.

Levenshtein distance: The metric distance between two strings that counts the minimum number of
insertions, deletions, and substitutions of symbols to transform one string into the other.

Occurrence: An occurrence of a word $u \in \Sigma^*$, of length m, appears in a word $w \in \Sigma^*$, of length n, at
position i if for $0 \leq k \leq m-1$, $u[k] = w[i+k]$.

Prefix: A word $u \in \Sigma^*$ is a prefix of a word $w \in \Sigma^*$ if $w = uz$ for some $z \in \Sigma^*$.

Prefix code: Set of words such that no word of the set is a prefix of another word contained in the set. A prefix code is represented by a coding tree.

Segment: A word $u \in \Sigma^*$ is a segment of a word $w \in \Sigma^*$ if u occurs in w (see occurrence); that is, $w = vuz$ for two words $v, z \in \Sigma^*$ (u is also referred to as a factor or a subword of w).

Subsequence: A word $u \in \Sigma^*$ is a subsequence of a word $w \in \Sigma^*$ if it is obtained from w by deleting zero or more symbols that need not be consecutive (u is sometimes referred to as a subword of w, with a possible confusion with the notion of segment).

Suffix: A word $u \in \Sigma^*$ is a suffix of a word $w \in \Sigma^*$ if $w = vu$ for some $v \in \Sigma^*$.

Suffix tree: Trie containing all the suffixes of a word.

Trie: Tree in which edges are labeled by letters or words.

References

Aho, A.V. 1990. Algorithms for finding patterns in strings. In *Handbook of Theoretical Computer Science*, Vol. A. *Algorithms and Complexity*, J. van Leeuwen, Ed., pp. 255–300. Elsevier, Amsterdam.

Aho, A.V. and Corasick, M.J. 1975. Efficient string matching: an aid to bibliographic search. *Comm. ACM*, 18(6):333–340.

Baeza-Yates, R.A. and Gonnet, G.H. 1992. A new approach to text searching. *Comm. ACM*, 35(10):74–82.

Baker, T.P. 1978. A technique for extending rapid exact-match string matching to arrays of more than one dimension. *SIAM J. Comput.*, 7(4):533–541.

Bell, T.C., Cleary, J.G., and Witten, I.H. 1990. *Text Compression*. Prentice Hall, Englewood Cliffs, NJ.

Bird, R.S. 1977. Two-dimensional pattern matching. *Inf. Process. Lett.*, 6(5):168–170.

Blumer, A., Blumer, J., Ehrenfeucht, A., Haussler, D., Chen, M.T., and Seiferas, J. 1985. The smallest automaton recognizing the subwords of a text. *Theor. Comput. Sci.*, 40:31–55.

Boyer, R.S. and Moore, J.S. 1977. A fast string searching algorithm. *Comm. ACM*, 20(10):762–772.

Breslauer, D., Colussi, L., and Toniolo, L. 1993. Tight comparison bounds for the string prefix matching problem. *Inf. Process. Lett.*, 47(1):51–57.

Burrows, M. and Wheeler, D. 1994. A block sorting lossless data compression algorithm. Technical Report 124, Digital Equipment Corporation.

Cole, R. 1994. Tight bounds on the complexity of the Boyer-Moore pattern matching algorithm. *SIAM J. Comput.*, 23(5):1075–1091.

Colussi, L. 1994. Fastest pattern matching in strings. *J. Algorithms*, 16(2):163–189.

Crochemore, M. 1986. Transducers and repetitions. *Theor. Comput. Sci.*, 45(1):63–86.

Crochemore, M. and Rytter, W. 2002. *Jewels of Stringology*. World Scientific.

Durbin, R., Eddy, S., and Krogh, A., and Mitchison G. 1998. *Biological Sequence Analysis Probabilistic Models of Proteins and Nucleic Acids*. Cambridge University Press.

Galil, Z. 1981. String matching in real time. *J. ACM*, 28(1):134–149.

Hancart, C. 1993. On Simon's string searching algorithm. *Inf. Process. Lett.*, 47(2):95–99.

Hirschberg, D.S. 1975. A linear space algorithm for computing maximal common subsequences. *Comm. ACM*, 18(6):341–343.

Hume, A. and Sunday, D.M. 1991. Fast string searching. *Software — Practice Exp.*, 21(11):1221–1248.

Karp, R.M. and Rabin, M.O. 1987. Efficient randomized pattern-matching algorithms. *IBM J. Res. Dev.*, 31(2):249–260.

Knuth, D.E. 1985. Dynamic Huffman coding. *J. Algorithms*, 6(2):163–180.

Knuth, D.E., Morris, J.H., Jr, and Pratt, V.R. 1977. Fast pattern matching in strings. *SIAM J. Comput.*, 6(1):323–350.

Lecroq, T. 1995. Experimental results on string-matching algorithms. *Software — Practice Exp.* 25(7): 727–765.

McCreight, E.M. 1976. A space-economical suffix tree construction algorithm. *J. Algorithms*, 23(2): 262–272.

Manber, U. and Myers, G. 1993. Suffix arrays: a new method for on-line string searches. *SIAM J. Comput.*, 22(5):935–948.

Needleman, S.B. and Wunsch, C.D. 1970. A general method applicable to the search for similarities in the amino acid sequence of two proteins. *J. Mol. Biol.*, 48:443–453.

Nelson, M. 1992. *The Data Compression Book*. M&T Books.

Simon, I. 1993. String matching algorithms and automata. In *First American Workshop on String Processing*, Baeza-Yates and Ziviani, Eds., pp. 151–157. Universidade Federal de Minas Gerais.

Smith, T.F. and Waterman, M.S. 1981. Identification of common molecular sequences. *J. Mol. Biol.*, 147:195–197.

Stephen, G.A. 1994. *String Searching Algorithms*. World Scientific Press.

Sunday, D.M. 1990. A very fast substring search algorithm. *Commun. ACM* 33(8):132–142.

Ukkonen, E. 1995. On-line construction of suffix trees. *Algorithmica*, 14(3):249–260.

Wagner, R.A. and Fischer, M. 1974. The string-to-string correction problem. *J. ACM*, 21(1):168–173.

Welch, T. 1984. A technique for high-performance data compression. *IEEE Comput.* 17(6):8–19.

Wu, S. and Manber, U. 1992. Fast text searching allowing errors. *Commun. ACM*, 35(10):83–91.

Zhu, R.F. and Takaoka, T. 1989. A technique for two-dimensional pattern matching. *Commun. ACM*, 32(9):1110–1120.

Further Information

Problems and algorithms presented in the chapter are just a sample of questions related to pattern matching. They share the formal methods used to design solutions and efficient algorithms. A wider panorama of algorithms on texts can be found in books, other including:

Apostolico, A. and Galil, Z., Editors. 1997. *Pattern Matching Algorithms*. Oxford University Press.

Bell, T.C., Cleary, J.G., and Witten, I.H. 1990. *Text Compression*. Prentice Hall, Englewood Cliffs, NJ.

Crochemore, M. and Rytter, W. 2002. *Jewels of Stringology*. World Scientific.

Gusfield D. 1997. *Algorithms on Strings, Trees and Sequences: Computer Science and Computational Biology*. Cambridge University Press.

Navarro, G. and Raffinot M. 2002. *Flexible Pattern Matching in Strings: Practical On-line Search Algorithms for Texts and Biological Sequences*. Cambridge University Press.

Nelson, M. 1992. *The Data Compression Book*. M&T Books.

Salomon, D. 2000. *Data Compression: the Complete Reference*. Springer-Verlag.

Stephen, G.A. 1994. *String Searching Algorithms*. World Scientific Press.

Research papers in pattern matching are disseminated in a few journals, among which are: *Communications of the ACM, Journal of the ACM, Theoretical Computer Science, Algorithmica, Journal of Algorithms, SIAM Journal on Computing,* and *Journal of Discrete Algorithms*.

Finally, three main annual conferences present the latest advances of this field of research and Combinatorial Pattern Matching, which started in 1990. Data Compression Conference, which is regularly held at Snowbird. The scope of SPIRE (String Processing and Information Retrieval) includes the domain of data retrieval.

General conferences in computer science often have sessions devoted to pattern matching algorithms.

Several books on the design and analysis of general algorithms contain chapters devoted to algorithms on texts. Here is a sample of these books:

Cormen, T.H., Leiserson, C.E., and Rivest, R.L. 1990. *Introduction to Algorithms*. MIT Press.

Gonnet, G.H. and Baeza-Yates, R.A. 1991. *Handbook of Algorithms and Data Structures*. Addison-Wesley.

Animations of selected algorithms can be found at:

`http://www-igm.univ-mlv.fr/~lecroq/string/` (Exact String Matching Algorithms),

`http://www-igm.univ-mlv.fr/~lecroq/seqcomp/` (Alignments).

14

Genetic Algorithms

14.1   Introduction **14**-1
14.2   Underlying Principles **14**-1
14.3   Best Practices **14**-4
       Function Optimization • Ordering Problems • Automatic
       Programming • Genetic Algorithms for Making Models
14.4   Mathematical Analysis of Genetic Algorithms **14**-9
14.5   Research Issues and Summary **14**-12

Stephanie Forrest
University of New Mexico

14.1 Introduction

A genetic algorithm is a form of evolution that occurs in a computer. Genetic algorithms are useful, both as search methods for solving problems and for modeling evolutionary systems. This chapter describes how genetic algorithms work, gives several examples of genetic algorithm applications, and reviews some mathematical analysis of genetic algorithm behavior.

In genetic algorithms, strings of binary digits are stored in a computer's memory, and over time the properties of these strings evolve in much the same way that populations of individuals evolve under natural selection. Although the computational setting is highly simplified when compared with the natural world, genetic algorithms are capable of evolving surprisingly complex and interesting structures. These structures, called **individuals**, can represent solutions to problems, strategies for playing games, visual images, or computer programs. Thus, genetic algorithms allow engineers to use a computer to evolve problem solutions over time, instead of designing them by hand. Although genetic algorithms are known primarily as a problem-solving method, they can also be used to study and model evolution in various settings, including biological (such as ecologies, immunology, and population genetics), social (such as economies and political systems), and cognitive systems.

14.2 Underlying Principles

The basic idea of a genetic algorithm is quite simple. First, a population of individuals is created in a computer, and then the population is evolved using the principles of variation, selection, and inheritance. Random variations in the population result in some individuals being more fit than others (better suited to their environment). These individuals have more offspring, passing on successful variations to their children, and the cycle is repeated. Over time, the individuals in the population become better adapted to their environment. There are many ways of implementing this simple idea. Here I describe the one invented by Holland [1975, Goldberg 1989].

The idea of using selection and variation to evolve solutions to problems goes back at least to Box [1957], although his work did not use a computer. In the late 1950s and early 1960s there were several independent

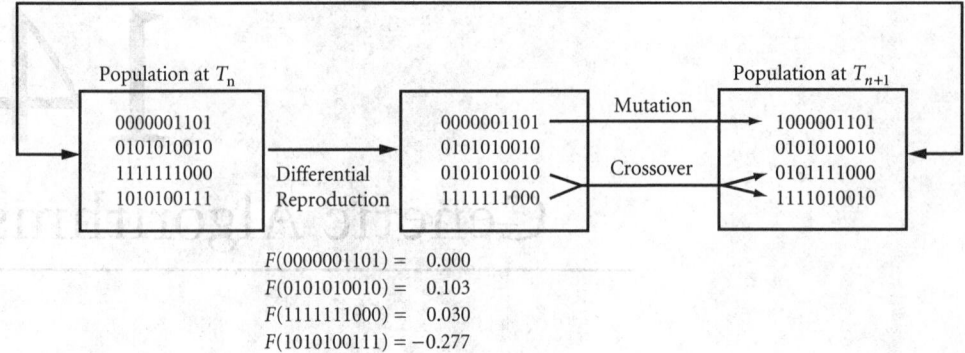

$$F(0000001101) = \quad 0.000$$
$$F(0101010010) = \quad 0.103$$
$$F(1111111000) = \quad 0.030$$
$$F(1010100111) = -0.277$$

FIGURE 14.1 (See Plate 14.1 in the color insert following page **29**-22.) Genetic algorithm overview: A population of four individuals is shown. Each is assigned a fitness value by the function $F(x, y) = yx^2 - x^4$. (See Figure 14.3.) On the basis of these fitnesses, the selection phase assigns the first individual (0000001101) one copy, the second (0101010010) two copies, the third (1111111000) one copy, and the fourth (1010100111) zero copies. After selection, the genetic operators are applied probabilistically; the first individual has its first bit mutated from a 0 to a 1, and crossover combines the last two individuals into two new ones. The resulting population is shown in the box labeled $T_{(N+1)}$.

efforts to incorporate ideas from evolution in computation. Of these, the best known are genetic algorithms [Holland 1962], evolutionary programming [Fogel et al. 1966], and evolutionary strategies [Back and Schwefel 1993]. Rechenberg [Back and Schwefel 1993] emphasized the importance of selection and mutation as mechanisms for solving difficult real-valued optimization problems. Fogel et al. [1966] developed similar ideas for evolving intelligent agents in the form of finite state machines. Holland [1962, 1975] emphasized the adaptive properties of entire populations and the importance of recombination mechanisms such as **crossover**. In recent years, genetic algorithms have taken many forms, and in some cases bear little resemblance to Holland's original formulation. Researchers have experimented with different types of representations, crossover and mutation operators, special-purpose operators, and different approaches to reproduction and selection. However, all of these methods have a family resemblance in that they take some inspiration from biological evolution and from Holland's original genetic algorithm. A new term, *evolutionary computation*, has been introduced to cover these various members of the genetic algorithm family, evolutionary programming, and evolution strategies.

Figure 14.1 gives an overview of a simple genetic algorithm. In its simplest form, each individual in the population is a bit string. Genetic algorithms often use more complex representations, including richer alphabets, diploidy, redundant encodings, and multiple **chromosomes**. However, the binary case is both the simplest and the most general. By analogy with genetics, the string of bits is referred to as the **genotype**. Each individual consists only of its genetic material, and it is organized into one (haploid) chromosome. Each bit position (set to 1 or 0) represents one gene. I will use the term bit string to refer both to genotypes and the individuals that they define. A natural question is how genotypes built from simple strings of bits can specify a solution to a specific problem. In other words, how are the binary genes expressed? There are many techniques for mapping bit strings to different problem domains, some of which are described in the following subsections.

The initial population of individuals is usually generated randomly, although it need not be. For example, prior knowledge about the problem solution can be encoded directly into the initial population, as in Hillis [1990]. Each individual is tested empirically in an environment, receiving a numerical evaluation of its merit, assigned by a **fitness function** F. The environment can be almost anything: another computer simulation, interactions with other individuals in the population, actions in the physical world (by a robot for example), or a human's subjective judgment. The fitness function's evaluation typically returns a single number (usually, higher numbers are assigned to fitter individuals). This constraint is sometimes relaxed so that the fitness function returns a vector of numbers [Fonseca and Fleming 1995], which can be appropriate for problems with multiple objectives. The fitness function determines how each gene (bit)

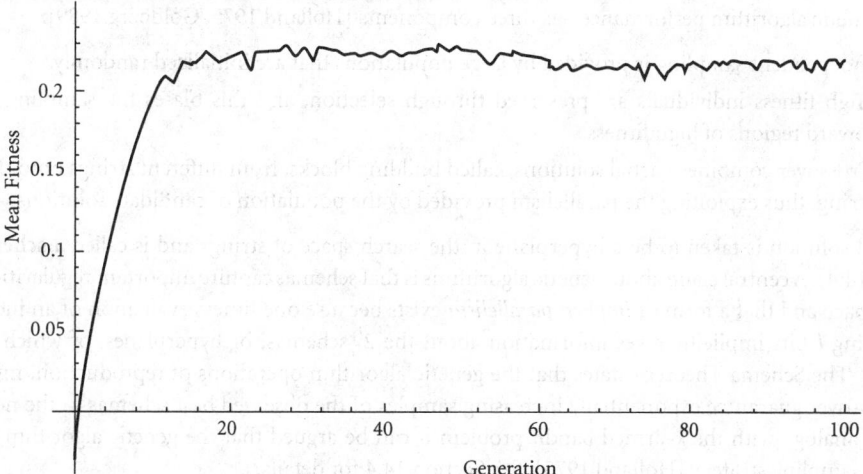

FIGURE 14.2 Mean fitness of a population evolving under the genetic algorithm. The population size is 100 individuals, each of which is 10 bits long (5 bits for x, 5 bits for y, as described in Figure 14.3), mutation probability is 0.0026/bit, crossover probability is 0.6 per pair of individuals, and the fitness function is $F = yx^2 - x^4$. Population mean is shown every generation for 100 generations.

of an individual will be interpreted and thus what specific problem the population will evolve to solve. The fitness function is the primary place where the traditional genetic algorithm is tailored to a specific problem.

Once all individuals in the population have been evaluated, their fitnesses form the basis for selection. **Selection** is implemented by eliminating low-fitness individuals from the population, and inheritance is implemented by making multiple copies of high-fitness individuals. Genetic operators such as **mutation** (flipping individual bits) and crossover (exchanging substrings of two individuals to obtain new offspring) are then applied probabilistically to the selected individuals to produce a new population (or **generation**) of individuals. The term crossover is used here to refer to the exchange of homologous substrings between individuals, although the biological term crossing over generally implies exchange within an individual. New generations can be produced either synchronously, so that the old generation is completely replaced, or asynchronously, so that generations overlap.

By transforming the previous set of good individuals to a new one, the operators generate a new set of individuals that ideally have a better than average chance of also being good. When this cycle of evaluation, selection, and genetic operations is iterated for many generations, the overall fitness of the population generally improves, as shown in Figure 14.2, and the individuals in the population represent improved solutions to whatever problem was posed in the fitness function.

There are many details left unspecified by this description. For example, selection can be performed in any of several ways — it could arbitrarily eliminate the least fit 50% of the population and make one copy of all of the remaining individuals, it could replicate individuals in direct proportion to their fitness (fitness-proportionate selection), or it could scale the fitnesses in any of several ways and replicate individuals in direct proportion to their scaled values (a more typical method). Similarly, the crossover operator can pass on both offspring to the new generation, or it can arbitrarily choose one to be passed on; the number of crossover points can be restricted to one per pair, two per pair, or N per pair. These and other variations of the basic algorithm have been discussed extensively in Goldberg [1989], in Davis [1991], and in the Proceedings of the International Conference on Genetic Algorithms. (See Further Information section.)

The genetic algorithm is interesting from a computational standpoint, at least in part, because of the claims that have been made about its effectiveness as a biased sampling algorithm. The classical argument about genetic algorithm performance has three components [Holland 1975, Goldberg 1989]:

- Independent sampling is provided by large populations that are initialized randomly.
- High-fitness individuals are preserved through selection, and this biases the sampling process toward regions of high fitness.
- Crossover combines partial solutions, called building blocks, from different strings onto the same string, thus exploiting the parallelism provided by the population of candidate solutions.

A partial solution is taken to be a hyperplane in the search space of strings and is called a **schema** (see Section 14.4). A central claim about genetic algorithms is that schemas capture important regularities in the search space and that a form of *implicit parallelism* exists because one fitness evaluation of an individual comprising l bits implicitly gives information about the 2^l schemas, or hyperplanes, of which it is an instance. The Schema Theorem states that the genetic algorithm operations of reproduction, mutation, and crossover guarantee exponentially increasing samples of the observed best schemas in the next time step. By analogy with the k-armed bandit problem it can be argued that the genetic algorithm uses an optimal sampling strategy [Holland 1975]. See Section 14.4 for details.

14.3 Best Practices

The simple computational procedure just described can be applied in many different ways to solve a wide range of problems. In designing a genetic algorithm to solve a specific problem there are two major design decisions: (1) specifying the mapping between binary strings and candidate solutions (this is commonly referred to as the representation problem) and (2) defining a concrete measure of fitness. In some cases the best representation and fitness function are obvious, but in many cases they are not, and in all cases, the particular representation and fitness function that are selected will determine the ultimate success of the genetic algorithm on the chosen problem. Possibly the simplest representation is a *feature list* in which each bit, or gene, represents the presence or absence of a single feature. This representation is useful for learning pattern classes defined by a critical set of features. For example, in spectroscopic applications, an important problem is selecting a small number of spectral frequencies that predict the concentration of some substance (e.g., concentration of glucose in human blood). The feature list approach to this problem assigns 1 bit to represent the presence or absence of each different observable frequency, and high fitness is assigned to those individuals whose feature settings correspond to good predictors for high (or low) glucose levels [Thomas 1993].

Genetic algorithms in various forms have been applied to many scientific and engineering problems, including optimization, automatic programming, machine and robot learning, modeling natural systems, and artificial life. They have been used in a wide variety of optimization tasks, including numerical optimization (see section on function optimization) and combinatorial optimization problems such as circuit design and job shop scheduling (see section on ordering problems). Genetic algorithms have also been used to evolve computer programs for specific tasks (see section on automatic programming) and to design other computational structures, e.g., cellular automata rules and sorting networks. In machine learning, they have been used to design neural networks, to evolve rules for rule-based systems, and to design and control robots. For an overview of genetic algorithms in machine learning, see DeJong [1990a, 1990b] and Schaffer et al. [1992].

Genetic algorithms have been used to model processes of innovation, the development of bidding strategies, the emergence of economic markets, the natural immune system, and ecological phenomena such as biological arms races, host–parasite coevolution, symbiosis, and resource flow. They have been used to study evolutionary aspects of social systems, such as the evolution of cooperation, the evolution of communication, and trail-following behavior in ants. They have been used to study questions in population genetics, such as "under what conditions will a gene for recombination be evolutionarily viable?" Finally, genetic algorithms are an important component in many artificial-life models, including systems that model interactions between species evolution and individual learning. See Further Information section and Mitchell and Forrest [1994] for details about genetic algorithms in modeling and artificial life.

The remainder of this section describes four illustrative examples of how genetic algorithms are used: numerical encodings for function optimization, permutation representations and special operators for sequencing problems, computer programs for automated programming, and endogenous fitness and other extensions for ecological modeling. The first two cover the most common classes of engineering applications. They are well understood and noncontroversial. The third example illustrates one of the most promising recent advances in genetic algorithms, but it was developed more recently and is less mature than the first two. The final example shows how genetic algorithms can be modified to more closely approximate natural evolutionary processes.

14.3.1 Function Optimization

Perhaps the most common application of genetic algorithms, pioneered by DeJong [1975], is multiparameter function optimization. Many problems can be formulated as a search for an optimal value, where the value is a complicated function of some input parameters. In some cases, the parameter settings that lead to the exact greatest (or least) value of the function are of interest. In other cases, the exact optimum is not required, just a near optimum, or even a value that represents a slight improvement over the previously best-known value. In these latter cases, genetic algorithms are often an appropriate method for finding good values.

As a simple example, consider the function $f(x, y) = yx^2 - x^4$. This function is solvable analytically, but if it were not, a genetic algorithm could be used to search for values of x and y that produce high values of $f(x, y)$ in a particular region of $\Re^2$. The most straightforward representation (Figure 14.3) is to assign regions of the bit string to represent each parameter (variable). Once the order in which the parameters are to appear is determined (in the figure x appears first and y appears second), the next step is to specify the domain for x and y (that is, the set of values for x and y that are candidate solutions). In our example, x and y will be real values in the interval $[0, 1)$. Because x and y are real valued in this example, and we are using a bit representation, the parameters need to be discretized. The precision of the solution is determined by how many bits are used to represent each parameter. In the example, 5 bits are assigned for x and 5 for y, although 10 is a more typical number. There are different ways of mapping between bits and decimal numbers, and so an encoding must also be chosen, and here we use gray coding.

Once a representation has been chosen, the genetic algorithm generates a random population of bit strings, decodes each bit string into the corresponding decimal values for x and y, applies the fitness function ($f(x, y) = yx^2 - x^4$) to the decoded values, selects the most fit individuals [those with the highest $f(x, y)$] for copying and variation, and then repeats the process. The population will tend to converge on a set of bit strings that represents an optimal or near optimal solution. However, there will always be some variation in the population due to mutation (Figure 14.2).

The standard binary encoding of decimal values has the drawback that in some cases all of the bits must be changed in order to increase a number by one. For example, the bit pattern 011 translates to 3 in decimal,

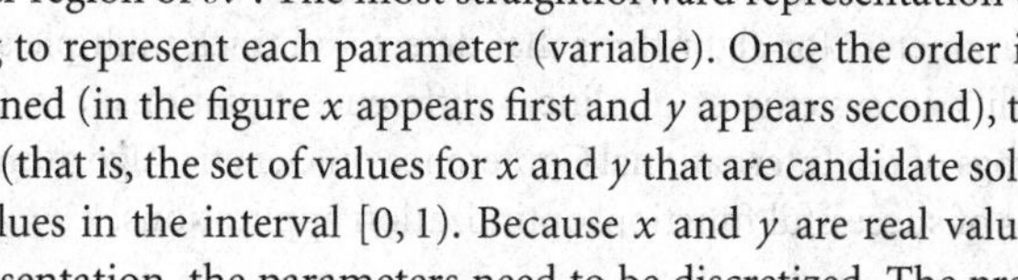

FIGURE 14.3 Bit-string encoding of multiple real-valued parameters. An arbitrary string of 10 bits is interpreted in the following steps: (1) segment the string into two regions with the first 5 bits reserved for x and the second 5 bits for y; (2) interpret each 5-bit substring as a Gray code and map back to the corresponding binary code; (3) map each 5-bit substring to its decimal equivalent; (4) scale to the interval $[0, 1)$; (5) substitute the two scaled values for x and y in the fitness function F; (6) return $F(x, y)$ as the fitness of the original string.

but 4 is represented by 100. This can make it difficult for an individual that is close to an optimum to move even closer by mutation. Also, mutations in high-order bits (the leftmost bits) are more significant than mutations in low-order bits. This can violate the idea that bit strings in successive generations will have a better than average chance of having high fitness, because mutations may often be disruptive. Gray codes address the first of these problems. Gray codes have the property that incrementing or decrementing any number by one is always 1 bit change. In practice, Gray-coded representations are often more successful for multiparameter function optimization applications of genetic algorithms.

Many genetic algorithm practitioners encode real-valued parameters directly without converting to a bit-based representation. In this approach, each parameter can be thought of as a gene on the chromosome. Crossover is defined as before, except that crosses take place only between genes (between real numbers). Mutation is typically redefined so that it chooses a random value that is close to the current value. This representation strategy is often more effective in practice, but it requires some modification of the operators [Back and Schwefel 1993, Davis 1991]. There are a number of other representation tricks that are commonly employed for function optimization, including logarithmic scaling (interpreting bit strings as the logarithm of the true parameter value), dynamic encoding (a technique that allows the number and interpretation of bits allocated to a particular parameter to vary throughout a run), variable-length representations, delta coding (the bit strings express a distance away from some previous partial solution), and a multitude of nonbinary encodings.

This completes our description of a simple method for encoding parameters onto a bit string. Although a function of two variables was used as an example, the strength of the genetic algorithm lies in its ability to manipulate many parameters, and this method has been used for hundreds of applications, including aircraft design, tuning parameters for algorithms that detect and track multiple signals in an image, and locating regions of stability in systems of nonlinear difference equations. See Goldberg [1989], Davis [1991], and the Proceedings of the International Conference on Genetic Algorithms for more detail about these and other examples of successful function-optimization applications.

14.3.2 Ordering Problems

A common problem involves finding an optimal ordering for a sequence of N items. Examples include various NP-complete problems such as finding a tour of cities that minimizes the distance traveled (the traveling salesman problem), packing boxes into a bin to minimize wasted space (the bin packing problem), and graph coloring problems.

For example, in the traveling salesman problem, suppose there are four cities: 1, 2, 3, and 4 and that each city is labeled by a unique bit string.* A common fitness function for this problem is the length of the candidate tour. A natural way to represent a tour is as a permutation, so that 3 2 1 4 is one candidate tour and 4 1 2 3 is another. This representation is problematic for the genetic algorithm because mutation and crossover do not necessarily produce legal tours. For example, a crossover between positions two and three in the example produces the individuals 3 2 2 3 and 4 1 1 4, both of which are illegal tours — not all of the cities are visited and some are visited more than once.

Three general methods have been proposed to address this representation problem: (1) adopting a different representation, (2) designing specialized crossover operators that produce only legal tours, and (3) penalizing illegal solutions through the fitness function. Of these, the use of specialized operators has been the most successful method for applications of genetic algorithms to ordering problems such as the traveling salesman problem (for example, see Mühlenbein et al. [1988]), although a number of generic representations have been proposed and used successfully on other sequencing problems. Specialized crossover operators tend to be less general, and I will describe one such method, **edge recombination**, as an example of a special-purpose operator that can be used with the permutation representation already described.

---

*For simplicity, we will use integers in the following explanation rather than the bit strings to which they correspond.

3 6 2 1 4 5
5 2 1 3 6 4   ⟶   3 6 4 1 2 5

Original Individuals   New Individual

Adjacency List

Key	Adjacent Keys
1	2, 2, 3, 4
2	1, 1, 3, 6
3	1, 6, 6
4	1, 5, 6
5	2, 4
6	2, 3, 3, 4

FIGURE 14.4   Example of edge-recombination operator. The adjacency list is constructed by examining each element in the parent permutations (labeled Key) and recording its adjacent elements. The new individual is constructed by selecting one parent arbitrarily (the top parent) and assigning its first element (3) to be the first element in the new permutation. The adjacencies of 3 are examined, and 6 is chosen to be the second element because it is a shared adjacency. The adjacencies of 6 are then examined, and of the unused ones, 4 is chosen randomly. Similarly, 1 is assigned to be the fourth element in the new permutation by random choice from {1, 5}. Then 2 is placed as the fifth element because it is a shared adjacency, and then the one remaining element, 5, is placed in the last position.

When designing special-purpose operators it is important to consider what information from the parents is being transmitted to the offspring, that is, what information is correlated with high-fitness individuals. In the case of traditional bitwise crossover, the answer is generally short, low-order schemas. (See Section 14.4.) But in the case of sequences, it is not immediately obvious what this means. Starkweather et al. [1991] identified three potential kinds of information that might be important for solving an ordering problem and therefore important to preserve through recombination: absolute position in the order, relative ordering (e.g., precedence relations might be important for a scheduling application), and adjacency information (as in the traveling salesman problem). They designed the edge-recombination operator to emphasize adjacency information. The operator is rather complicated, and there are many variants of the originally published operator. A simplified description follows (for details, see Starkweather et al. [1991]). For each pair of individuals to be crossed: (1) construct a table of adjacencies in the parents (see Figure 14.4) and (2) construct one new permutation (offspring) by combining information from the two parents:

- Select one parent at random and assign the first element in its permutation to be the first one in the child.
- Select the second element for the child, as follows: If there is an adjacency common to both parents, then choose that element to be the next one in the child's permutation; if there is an unused adjacency available from one parent, choose it; or if (1) and (2) fail, make a random selection.
- Select the remaining elements in order by repeating step 2.

An example of the edge-recombination operator is shown in Figure 14.4. Although this method has proved effective, it should be noted that it is more expensive to build the adjacency list for each parent and to perform edge recombination operation than it is to use a more standard crossover operator.

A final consideration in the choice of special-purpose operators is the amount of random information that is introduced when the operator is applied. This can be difficult to assess, but it can have a large effect (positive or negative) on the performance of the operator.

14.3.3   Automatic Programming

Genetic algorithms have been used to evolve a special kind of computer program [Koza 1992]. These programs are written in a subset of the programming language Lisp and more recently other languages.

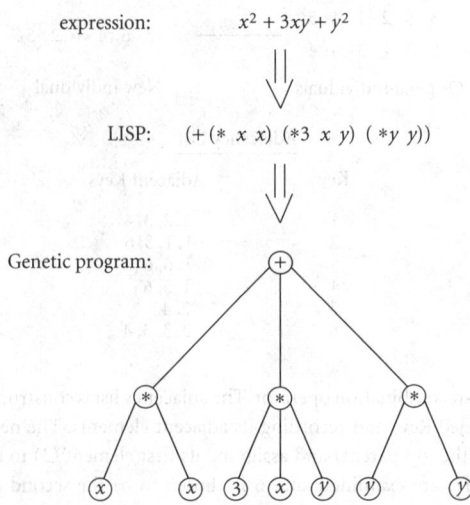

expression: $x^2 + 3xy + y^2$

LISP: $(+ (* x x) (*3 x y) (*y y))$

Genetic program:

FIGURE 14.5 Tree representation of computer programs: The displayed tree corresponds to the expression $x^2 + 3xy + y^2$. Operators for each expression are displayed as a root, and the operands for each expression are displayed as children. (From Forrest, S. 1993a. *Science* 261:872–878. With permission.)

Lisp programs can naturally be represented as trees (Figure 14.5). Populations of random program trees are generated and evaluated as in the standard genetic algorithm. All other details are similar to those described for binary genetic algorithms with the exception of crossover. Instead of exchanging substrings, **genetic programs** exchange subtrees between individual program trees. This modified form of crossover appears to have many of the same advantages as traditional crossover (such as preserving partial solutions).

Genetic programming has the potential to be extremely powerful, because Lisp is a general-purpose programming language and genetic programming eliminates the need to devise an explicit chromosomal representation. In practice, however, genetic programs are built from subsets of Lisp tailored to particular problem domains, and at this point considerable skill is required to select just the right set of primitives for a particular problem. Although the method has been tested on a wide variety of problems, it has not yet been used extensively in real applications.

The genetic programming method is intriguing because its solutions are so different from human-designed programs for the same problem. Humans try to design elegant and general computer programs, whereas genetic programs are often needlessly complicated, not revealing the underlying algorithm. For example, a human-designed program for computing $\cos 2x$ might be $1 - 2 \sin^2 x$, expressed in Lisp as $(-1(*2(*(\sin x)(\sin x))))$, whereas genetic programming discovered the following program (Koza 1992, p. 241):

$$(\sin(-(-2(*x2)))(\sin(\sin(\sin(\sin(\sin(\sin(*(\sin(\sin 1))(\sin(\sin 1))))))))))$$

For anyone who has studied computer programming this is apparently a major drawback because the evolved programs are inelegant, redundant, inefficient, difficult for a human to read, and do not reveal the underlying structure of the algorithm. However, genetic programs do resemble the kinds of ad hoc solutions that evolve in nature through gene duplication, mutation, and modifying structures from one purpose to another. There is some evidence that the junk components of a genetic program sometimes turn out to be useful components in other contexts. Thus, if the genetic programming endeavor is successful, it could revolutionize software design.

14.3.4   Genetic Algorithms for Making Models

The past three examples concentrated on understanding how genetic algorithms can be applied to solve problems. This subsection discusses how the genetic algorithm can be used to model other systems. Genetic algorithms have been employed as models of a wide variety of dynamical processes, including induction in psychology, natural evolution in ecosystems, evolution in immune systems, and imitation in social systems. Making computer models of evolution is somewhat different from many conventional models because the models are highly abstract. The data produced by these models are unlikely to make exact numerical predictions. Rather, they can reveal the conditions under which certain qualitative behaviors are likely to arise — diversity of phenotypes in resource-rich (or poor) environments, cooperation in competitive nonzero-sum games, and so forth. Thus, the models described here are being used to discover qualitative patterns of behavior and, in some cases, critical parameters in which small changes have drastic effects on the outcomes. Such modeling is common in nonlinear dynamics and in artificial intelligence, but it is much less accepted in other disciplines. Here we describe one of these examples: ecological modeling. This exploratory research project is still in an early stage of development. For examples of more mature modeling projects, see Holland et al. [1986] and Axelrod [1986].

The Echo system [Holland 1995] shows how genetic algorithms can be used to model ecosystems. The major differences between Echo and standard genetic algorithms are: (1) there is no explicit fitness function, (2) individuals have local storage (i.e., they consist of more than their genome), (3) the genetic representation is based on a larger alphabet than binary strings, and (4) individuals always have a spatial location. In Echo, fitness evaluation takes place implicitly. That is, individuals in the population (called *agents*) are allowed to make copies of themselves anytime they acquire enough *resources* to replicate their genome. Different resources are modeled by different letters of the alphabet (say, A, B, C, D), and genomes are constructed out of those same letters. These resources can exist independently of the agent's genome, either free in the environment or stored internally by the agent. Agents acquire resources by interacting with other agents through trading relationships and combat. Echo thus relaxes the constraint that an explicit fitness function must return a numerical evaluation of each agent. This **endogenous fitness function** is much closer to the way fitness is assessed in natural settings. In addition to trade and combat, a third form of interaction between agents is mating. Mating provides opportunities for agents to exchange genetic material through crossover, thus creating hybrids. Mating, together with mutation, provides the mechanism for new types of agents to evolve.

Populations in Echo exist on a two-dimensional grid of sites, although other connection topologies are possible. Many agents can cohabit one site, and agents can migrate between sites. Each site is the source of certain renewable resources. On each time step of the simulation, a fixed amount of resources at a site becomes available to the agents located at that site. Different sites can produce different amounts of different resources. For example, one site might produce 10 As and 5 Bs each time step, and its neighbor might produce 5 As, 0 Bs, and 5 Cs. The idea is that an agent will do well (reproduce often) if it is located at a site whose renewable resources match well with its genomic makeup or if it can acquire the relevant resources from other agents at its site.

In preliminary simulations, the Echo system has demonstrated surprisingly complex behaviors, including something resembling a biological arms race (in which two competing species develop progressively more complex offensive and defensive strategies), functional dependencies among different species, trophic cascades, and sensitivity (in terms of the number of different phenotypes) to differing levels of renewable resources. Although the Echo system is still largely untested, it illustrates how the fundamental ideas of genetic algorithms can be incorporated into a system that captures important features of natural ecological systems.

14.4   Mathematical Analysis of Genetic Algorithms

Although there are many problems for which the genetic algorithm can evolve a good solution in reasonable time, there are also problems for which it is inappropriate (such as problems in which it is important to find the exact global optimum). It would be useful to have a mathematical characterization of how the genetic algorithm works that is predictive. Research on this aspect of genetic algorithms has not produced definitive answers. The domains for which one is likely to choose an adaptive method such as the genetic algorithm are precisely those about which we typically have little analytical knowledge — they are complex, noisy, or dynamic (changing over time). These characteristics make it virtually impossible to predict with certainty how well a particular algorithm will perform on a particular problem instance, especially if the algorithm is stochastic, as is the case with the genetic algorithm. In spite of this difficulty, there are fairly extensive theories about how and why genetic algorithms work in idealized settings.

Analysis of genetic algorithms begins with the concept of a search space. The genetic algorithm can be viewed as a procedure for searching the space of all possible binary strings of fixed length l. Under this interpretation, the algorithm is searching for points in the l-dimensional space $\{0, 1\}^l$ that have high fitness. The search space is identical for all problems of the same size (same l), but the locations of good points will generally differ. The surface defined by the fitness of each point, together with the neighborhood relation imposed by the operators, is sometimes referred to as the **fitness landscape**. The longer the bit strings, corresponding to higher values of l, the larger the search space is, growing exponentially with the length of l. For problems with a sufficiently large l, only a small fraction of this size search space can be examined, and thus it is unreasonable to expect an algorithm to locate the global optimum in the space. A more reasonable goal is to search for good regions of the search space corresponding to regularities in the problem domain. Holland [1975] introduced the notion of a *schema* to explain how genetic algorithms search for regions of high fitness. Schemas are theoretical constructs used to explain the behavior of genetic algorithms, and are not processed directly by the algorithm. The following description of schema processing is excerpted from Forrest and Mitchell [1993b].

A schema is a template, defined over the alphabet $\{0, 1, *\}$, which describes a pattern of bit strings in the search space $\{0, 1\}^l$ (the set of bit strings of length l). For each of the l bit positions, the template either specifies the value at that position (1 or 0), or indicates by the symbol $*$ (referred to as don't care) that either value is allowed.

For example, the two strings A and B have several bits in common. We can use schemas to describe the patterns these two strings share:

A = 100111
B = 010011
**0*11
****11
0*
01

A bit string x that matches a schema s's pattern is said to be an *instance* of s; for example, A and B are both instances of the schemas just shown. In schemas, 1s and 0s are referred to as *defined bits*; the *order* of a schema is the number of defined bits in that schema, and the *defining length* of a schema is the distance between the leftmost and rightmost defined bits in the string. For example, the defining length of $**0**1$ is 3.

Schemas define hyperplanes in the search space $\{0, 1\}^l$. Figure 14.6 shows four hyperplanes, corresponding to the schemas $0****$, $1****$, $*0***$, and $*1***$. Any point in the space is simultaneously an instance

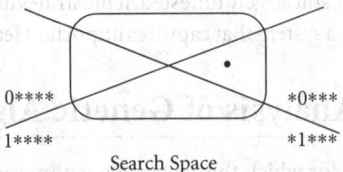

Search Space

FIGURE 14.6   Schemas define hyperplanes in the search space. (From Forrest, S. and Mitchell, M. 1993b. *Machine Learning* 13:285–319. With permission.)

of two of these schemas. For example, the point shown in Figure 14.6 is an instance of both 1**** and
*0*** (and also of 10***).

The fitness of any bit string in the population gives some information about the average fitness of the 2^l
different schemas of which it is an instance, and so an explicit evaluation of a population of M individual
strings is also an implicit evaluation of a much larger number of schemas. This is referred to as implicit
parallelism. At the explicit level the genetic algorithm searches through populations of bit strings, but the
genetic algorithm's search can also be interpreted as an implicit schema sampling process. Feedback from
the fitness function, combined with selection and recombination, biases the sampling procedure over time
away from those schemas that give negative feedback (low average fitness) and toward those that give
positive feedback (high average fitness). Ultimately, the search procedure should identify regularities, or
patterns, in the environment that lead to high fitness. Because the space of possible patterns is larger than
the space of possible individuals (3^l vs. 2^l), implicit parallelism is potentially advantageous.

An important theoretical result about genetic algorithms is the Schema Theorem [Holland 1975,
Goldberg 1989], which states that the observed best schemas will on average be allocated an exponentially
increasing number of samples in the next generation. Figure 14.7 illustrates the rapid convergence on fit
schemas by the genetic algorithm. This strong convergence property of the genetic algorithm is a two-edged

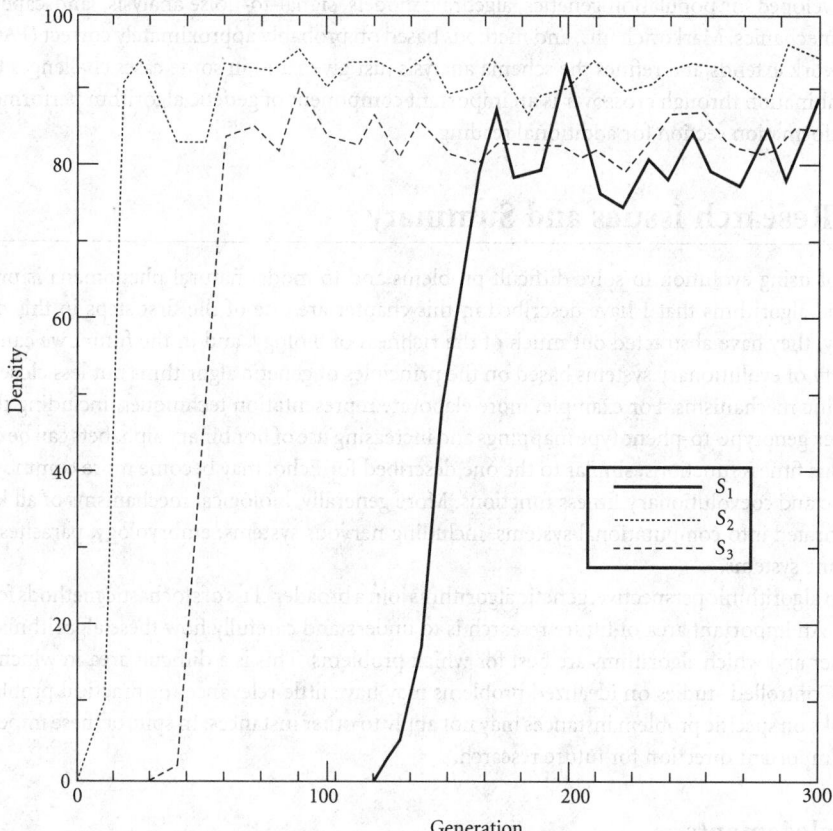

FIGURE 14.7 Schema frequencies over time. The graph plots schema frequencies in the population over time for
three schemas:

$$s_1 = 1111111111111111***;$$

$$s_2 = ****************1111111111111111*****************************;$$

$$s_3 = ****************************1111111111111111111111111111111.$$

The function plotted was a royal road function [Forrest and Mitchell 1993a] in which the optimum value is the string
of all 1s. (From Forrest, S. 1993a. *Science* 261:872–878. With permission.)

sword. On the one hand, the fact that the genetic algorithm can close in on a fit part of the space very quickly is a powerful property; on the other hand, because the genetic algorithm always operates on finite-size populations, there is inherently some sampling error in the search, and in some cases the genetic algorithm can magnify a small sampling error, causing premature convergence on local optima.

According to the **building blocks hypothesis** [Holland 1975, Goldberg 1989], the genetic algorithm initially detects biases toward higher fitness in some low-order schemas (those with a small number of defined bits), and converges on this part of the search space. Over time, it detects biases in higher-order schemas by combining information from low-order schemas via crossover, and eventually it converges on a small region of the search space that has high fitness. The building blocks hypothesis states that this process is the source of the genetic algorithm's power as a search and optimization method. If this hypothesis about how genetic algorithms work is correct, then crossover is of primary importance, and it distinguishes genetic algorithms from other similar methods, such as simulated annealing and greedy algorithms. A number of authors have questioned the adequacy of the building blocks hypothesis as an explanation for how genetic algorithms work and there are several active research efforts studying schema processing in genetic algorithms. Nevertheless, the explanation of schemas and recombination that I have just described stands as the most common account of why genetic algorithms perform as they do.

There are several other approaches to analyzing mathematically the behavior of genetic algorithms: models developed for population genetics, algebraic models, signal-to-noise analysis, landscape analysis, statistical mechanics, Markov chains, and methods based on probably approximately correct (PAC) learning. This work extends and refines the schema analysis just given and in some cases challenges the claim that recombination through crossover is an important component of genetic algorithm performance. See Further Information section for additional reading.

14.5   Research Issues and Summary

The idea of using evolution to solve difficult problems and to model natural phenomena is promising. The genetic algorithms that I have described in this chapter are one of the first steps in this direction. Necessarily, they have abstracted out much of the richness of biology, and in the future we can expect a wide variety of evolutionary systems based on the principles of genetic algorithms but less closely tied to these specific mechanisms. For example, more elaborate representation techniques, including those that use complex genotype-to-phenotype mappings and increasing use of nonbinary alphabets can be expected. Endogenous fitness functions, similar to the one described for Echo, may become more common, as well as dynamic and coevolutionary fitness functions. More generally, biological mechanisms of all kinds will be incorporated into computational systems, including nervous systems, embryology, parasites, viruses, and immune systems.

From an algorithmic perspective, genetic algorithms join a broader class of stochastic methods for solving problems. An important area of future research is to understand carefully how these algorithms relate to one another and which algorithms are best for which problems. This is a difficult area in which to make progress. Controlled studies on idealized problems may have little relevance for practical problems, and benchmarks on specific problem instances may not apply to other instances. In spite of these impediments, this is an important direction for future research.

Acknowledgments

The author gratefully acknowledges support from the National Science Foundation (Grant IRI-9157644), the Office of Naval Research (Grant N00014-95-1-0364), ATR Human Information Processing Research Laboratories, and the Santa Fe Institute. Ron Hightower prepared Figure 14.2.

Significant portions of this chapter are excerpted with permission from Forrest, S. 1993. Genetic algorithms: principles of adaption applied to computation. *Science* 261 (Aug. 13):872–878. © 1993 American Association for the Advancement of Science.

Defining Terms

Building blocks hypothesis: The hypothesis that the genetic algorithm searches by first detecting biases toward higher fitness in some low-order schemas (those with a small number of defined bits) and converging on this part of the search space. Over time, it then detects biases in higher-order schemas by combining information from low-order schemas via crossover and eventually converges on a small region of the search space that has high fitness. The building blocks hypothesis states that this process is the source of the genetic algorithm's power as a search and optimization method [Holland 1975, Goldberg 1989].

Chromosome: A string of symbols (usually in bits) that contains the genetic information about an individual. The chromosome is interpreted by the fitness function to produce an evaluation of the individual's fitness.

Crossover: An operator for producing new individuals from two parent individuals. The operator works by exchanging substrings between the two individuals to obtain new offspring. In some cases, both offspring are passed to the new generation; in others, one is arbitrarily chosen to be passed on; the number of crossover points can be restricted to one per pair, two per pair, or N per pair.

Edge recombination: A special-purpose crossover operator designed to be used with permutation representations for sequencing problems. The edge-recombination operator attempts to preserve adjacencies between neighboring elements in the parent permutations [Starkweather et al. 1991].

Endogenous fitness function: Fitness is not assessed explicitly using a fitness function. Some other criterion for reproduction is adopted. For example, individuals might be required to accumulate enough internal resources to copy themselves before they can reproduce. Individuals who can gather resources efficiently would then reproduce frequently and their traits would become more prevalent in the population.

Fitness function: Each individual is tested empirically in an environment, receiving a numerical evaluation of its merit, assigned by a fitness function F. The environment can be almost anything — another computer simulation, interactions with other individuals in the population, actions in the physical world (by a robot for example), or a human's subjective judgment.

Fitness landscape: The surface defined by the fitness of each point in the search space, together with the neighborhood relation imposed by the operators.

Generation: One iteration, or time step, of the genetic algorithm. New generations can be produced either synchronously, so that the old generation is completely replaced (the time step model), or asynchronously, so that generations overlap. In the asynchronous case, generations are defined in terms of some fixed number of fitness-function evaluations.

Genetic programs: A form of genetic algorithm that uses a tree-based representation. The tree represents a program that can be evaluated, for example, an S-expression.

Genotype: The string of symbols, usually bits, used to represent an individual. Each bit position (set to 1 or 0) represents one gene. The term bit string in this context refers both to genotypes and to the individuals that they define.

Individuals: The structures that are evolved by the genetic algorithm. They can represent solutions to problems, strategies for playing games, visual images, or computer programs. Typically, each individual consists only of its genetic material, which is organized into one (haploid) chromosome.

Mutation: An operator for varying an individual. In mutation, individual bits are flipped probabilistically in individuals selected for reproduction. In representations other than bit strings, mutation is redefined to an appropriate smallest unit of change. For example, in permutation representations, mutation is often defined to be the swap of two neighboring elements in the permutation; in real-valued representations, mutation can be a creep operator that perturbs the real number up or down some small increment.

Schema: A theoretical construct used to explain the behavior of genetic algorithms. Schemas are not processed directly by the algorithm. Schemas are coordinate hyperplanes in the search space of strings.

Selection: Some individuals are more fit than others (better suited to their environment). These individuals have more offspring, that is, they are selected for reproduction. Selection is implemented by eliminating low-fitness individuals from the population, and inheritance is implemented by making multiple copies of high-fitness individuals.

References

Axelrod, R. 1986. An evolutionary approach to norms. *Am. Political Sci. Rev.* 80 (Dec).

Back, T. and Schwefel, H. P. 1993. An overview of evolutionary algorithms. *Evolutionary Comput.* 1:1–23.

Belew, R. K. and Booker, L. B., eds. 1991. *Proc. 4th Int. Conf. Genet. Algorithms.* July. Morgan Kaufmann, San Mateo, CA.

Booker, L. B., Riolo, R. L., and Holland, J. H. 1989. Learning and representation in classifier systems. *Art. Intelligence* 40:235–282.

Box, G. E. P. 1957. Evolutionary operation: a method for increasing industrial productivity. *J. R. Stat. Soc.* 6(2):81–101.

Davis, L., ed. 1991. *The Genetic Algorithms Handbook.* Van Nostrand Reinhold, New York.

DeJong, K. A. 1975. *An analysis of the behavior of a class of genetic adaptive systems.* Ph.D. thesis, University of Michigan, Ann Arbor.

DeJong, K. A. 1990a. Genetic-algorithm-based learning. *Machine Learning* 3:611–638.

DeJong, K. A. 1990b. Introduction to second special issue on genetic algorithms. *Machine Learning.* 5(4):351–353.

Eshelman, L. J., ed. 1995. *Proc. 6th Int. Conf. Genet. Algorithms.* Morgan Kaufmann, San Francisco.

Filho, J. L. R., Treleaven, P. C., and Alippi, C. 1994. Genetic-algorithm programming environments. *Computer* 27(6):28–45.

Fogel, L. J., Owens, A. J., and Walsh, M. J. 1966. *Artificial Intelligence Through Simulated Evolution.* Wiley, New York.

Fonseca, C. M. and Fleming, P. J. 1995. An overview of evolutionary algorithms in multiobjective optimization. *Evolutionary Comput.* 3(1):1–16.

Forrest, S. 1993a. Genetic algorithms: principles of adaptation applied to computation. *Science* 261:872–878.

Forrest, S., ed. 1993b. *Proc. Fifth Int. Conf. Genet. Algorithms.* Morgan Kaufmann, San Mateo, CA.

Forrest, S. and Mitchell, M. 1993a. Towards a stronger building-blocks hypothesis: effects of relative building-block fitness on ga performance. In *Foundations of Genetic Algorithms,* Vol. 2, L. D. Whitley, ed., pp. 109–126. Morgan Kaufmann, San Mateo, CA.

Forrest, S. and Mitchell, M. 1993b. What makes a problem hard for a genetic algorithm? Some anomalous results and their explanation. *Machine Learning* 13(2/3).

Goldberg, D. E. 1989. *Genetic Algorithms in Search, Optimization, and Machine Learning.* Addison Wesley, Reading, MA.

Grefenstette, J. J. 1985. *Proc. Int. Conf. Genet. Algorithms Appl.* NCARAI and Texas Instruments.

Grefenstette, J. J. 1987. *Proc. 2nd Int. Conf. Genet. Algorithms.* Lawrence Erlbaum, Hillsdale, NJ.

Hillis, W. D. 1990. Co-evolving parasites improve simulated evolution as an optimization procedure. *Physica D* 42:228–234.

Holland, J. H. 1962. Outline for a logical theory of adaptive systems. *J. ACM* 3:297–314.

Holland, J. H. 1975. *Adaptation in Natural and Artificial Systems.* University of Michigan Press, Ann Arbor, MI; 1992. 2nd ed. MIT Press, Cambridge, MA.

Holland, J. H. 1992. Genetic algorithms. *Sci. Am.,* pp. 114–116.

Holland, J. H. 1995. *Hidden Order: How Adaptation Builds Complexity.* Addison–Wesley, Reading, MA.

Holland, J. H., Holyoak, K. J., Nisbett, R. E., and Thagard, P. 1986. *Induction: Processes of Inference, Learning, and Discovery.* MIT Press, Cambridge, MA.

Koza, J. R. 1992. *Genetic Programming.* MIT Press, Cambridge, MA.

Männer, R. and Manderick, B., eds. 1992. *Parallel Problem Solving From Nature 2*. North Holland, Amsterdam.

Mitchell, M. 1996. *An Introduction to Genetic Algorithms*. MIT Press, Cambridge, MA.

Mitchell, M. and Forrest, S. 1994. Genetic algorithms and artificial life. *Artif. Life* 1(3):267–289; reprinted 1995. In *Artificial Life: An Overview*, C. G. Langton, ed. MIT Press, Cambridge, MA.

Mühlenbein, H., Gorges-Schleuter, M., and Kramer, O. 1988. *Parallel Comput.* 6:65–88.

Rawlins, G., ed. 1991. *Foundations of Genetic Algorithms*. Morgan Kaufmann, San Mateo, CA.

Schaffer, J. D., ed. 1989. *Proc. 3rd Int. Conf. Genet. Algorithms*. Morgan Kaufmann, San Mateo, CA.

Schaffer, J. D., Whitley, D., and Eshelman, L. J. 1992. Combinations of genetic algorithms and neural networks: a survey of the state of the art. In *Int. Workshop Combinations Genet. Algorithms Neural Networks*, L. D. Whitley and J. D. Schaffer, eds., pp. 1–37. IEEE Computer Society Press, Los Alamitos, CA.

Schwefel, H. P. and Männer, R., eds. 1990. Parallel problem solving from nature. *Lecture Notes in Computer Science*. Springer–Verlag, Berlin.

Srinivas, M. and Patnaik, L. M. 1994. Genetic algorithms: a survey. *Computer* 27(6):17–27.

Starkweather, T., McDaniel, S., Mathias, K., Whitley, D., and Whitley, C. 1991. A comparison of genetic sequencing operators. In *4th Int. Conf. Genet. Algorithms*, R. K. Belew and L. B. Booker, eds., pp. 69–76. Morgan Kaufmann, Los Altos, CA.

Thomas, E. V. 1993. Frequency Selection Using Genetic Algorithms. *Sandia National Lab. Tech. Rep.* SAND93-0010, Albuquerque, NM.

Whitley, L. D., ed. 1993. *Foundations of Genetic Algorithms 2*. Morgan Kaufmann, San Mateo, CA.

Whitley, L. D. and Vose, M., eds. 1995. *Foundations of Genetic Algorithms 3*. Morgan Kaufmann, San Francisco.

Further Information

Review articles on genetic algorithms include Booker et al. [1989], Holland [1992], Forrest [1993a], Mitchell and Forrest [1994], Srinivas and Patnaik [1994] and Filho et al. [1994]. Books that describe the theory and practice of genetic algorithms in greater detail include Holland [1975], Goldberg [1989], Davis [1991], Koza [1992], Holland et al. [1986], and Mitchell [1996]. Holland [1975] was the first book-length description of genetic algorithms, and it contains much of the original insight about the power and breadth of adaptive algorithms. The 1992 reprinting contains interesting updates by Holland. However, Goldberg [1989], Davis [1991], and Mitchell [1996] are more accessible introductions to the basic concepts and implementation issues. Koza [1992] describes genetic programming and Holland et al. [1986] discuss the relevance of genetic algorithms to cognitive modeling.

Current research on genetic algorithms is reported many places, including the Proceedings of the International Conference on Genetic Algorithms [Grefenstette 1985, 1987, Schaffer 1989, Belew and Booker 1991, Forrest 1993b, Eshelman 1995], the proceedings of conferences on Parallel Problem Solving from Nature [Schwefel and Männer 1990, Männer and Manderick 1992], and the workshops on Foundations of Genetic Algorithms [Rawlins 1991, Whitley 1993, Whitley and Vose 1995]. Finally, the artificial-life literature contains many interesting papers about genetic algorithms.

There are several archival journals that publish articles about genetic algorithms. These include *Evolutionary Computation* (a journal devoted to GAs), *Complex Systems*, *Machine Learning*, *Adaptive Behavior*, and *Artificial Life*.

Information about genetic algorithms activities, public domain packages, etc., is maintained through the WWW at URL http://www.aic.nrl.navy.mil/galist/ or through anonymous ftp at ftp.aic.nrl.navy.mil [192.26.18.68] in/pub/galist.

15

Combinatorial Optimization

15.1   Introduction **15**-1

15.2   A Primer on Linear Programming **15**-3
     Algorithms for Linear Programming

15.3   Large-Scale Linear Programming in
     Combinatorial Optimization **15**-13
     Cutting Stock Problem • Decomposition and
     Compact Representations

15.4   Integer Linear Programs...........................**15**-15
     Example Formulations • Jeroslow's Representability
     Theorem • Benders's Representation

15.5   Polyhedral Combinatorics **15**-20
     Special Structures and Integral Polyhedra • Matroids
     • Valid Inequalities, Facets, and Cutting Plane Methods

15.6   Partial Enumeration Methods **15**-27
     Branch and Bound • Branch and Cut

15.7   Approximation in Combinatorial Optimization **15**-30
     LP Relaxation and Randomized Rounding • Primal–Dual
     Approximation • Semidefinite Relaxation and Rounding
     • Neighborhood Search • Lagrangian Relaxation

15.8   Prospects in Integer Programming...................**15**-35

Vijay Chandru
Indian Institute of Science

M. R. Rao
Indian Institute of Management

15.1 Introduction

Bin packing, routing, scheduling, layout, and network design are generic examples of combinatorial optimization problems that often arise in computer engineering and decision support. Unfortunately, almost all interesting generic classes of combinatorial optimization problems are $\mathcal{N}P$-hard. The scale at which these problems arise in applications and the explosive exponential complexity of the search spaces preclude the use of simplistic enumeration and search techniques. Despite the worst-case intractability of combinatorial optimization, in practice we are able to solve many large problems and often with off-the-shelf software. Effective software for combinatorial optimization is usually problem specific and based on sophisticated algorithms that combine approximation methods with search schemes and that exploit mathematical (and not just syntactic) structure in the problem at hand.

Multidisciplinary interests in combinatorial optimization have led to several fairly distinct paradigms in the development of this subject. Each paradigm may be thought of as a particular combination of a *representation scheme* and a *methodology* (see Table 15.1). The most established of these, the **integer programming** paradigm, uses implicit algebraic forms (linear constraints) to represent combinatorial

TABLE 15.1   Paradigms in Combinatorial Optimization

Paradigm	Representation	Methodology
Integer programming	Linear constraints, Linear objective, Integer variables	Linear programming and extensions
Search	State space, Discrete control	Dynamic programming, $\mathcal{A}^*$
Local improvement	Neighborhoods Fitness functions	Hill climbing, Simulated annealing, Tabu search, Genetic algorithms
Constraint logic programming	Horn rules	Resolution, constraint solvers

optimization and **linear programming** and its extensions as the workhorses in the design of the solution algorithms. It is this paradigm that forms the central theme of this chapter.

Other well known paradigms in combinatorial optimization are **search, local improvement,** and **constraint logic programming.** Search uses state-space representations and partial enumeration techniques such as $\mathcal{A}^*$ and dynamic programming. Local improvement requires only a representation of neighborhood in the solution space, and methodologies vary from simple hill climbing to the more sophisticated techniques of simulated annealing, tabu search, and genetic algorithms. Constraint logic programming uses the syntax of Horn rules to represent combinatorial optimization problems and uses resolution to orchestrate the solution of these problems with the use of domain-specific constraint solvers. Whereas integer programming was developed and nurtured by the mathematical programming community, these other paradigms have been popularized by the artificial intelligence community.

An abstract formulation of combinatorial optimization is

$$\text{(CO)} \quad \min\{f(I) : I \in \mathcal{I}\}$$

where $\mathcal{I}$ is a collection of subsets of a finite ground set $E = \{e_1, e_2, \ldots, e_n\}$ and f is a criterion (objective) function that maps 2^E (the power set of E) to the reals. A **mixed integer linear program** (MILP) is of the form

$$\text{(MILP)} \quad \min_{x \in \Re^n}\{\mathbf{cx} : A\mathbf{x} \geq \mathbf{b}, \, \mathbf{x}_j \text{ integer } \forall \, j \in J\}$$

which seeks to minimize a linear function of the decision vector $\mathbf{x}$ subject to linear inequality constraints and the requirement that a subset of the decision variables is integer valued. This model captures many variants. If $J = \{1, 2, \ldots, n\}$, we say that the integer program is *pure*, and *mixed* otherwise. Linear equations and bounds on the variables can be easily accommodated in the inequality constraints. Notice that by adding in inequalities of the form $0 \leq \mathbf{x}_j \leq 1$ for a $j \in J$ we have forced $\mathbf{x}_j$ to take value 0 or 1. It is such Boolean variables that help capture combinatorial optimization problems as special cases of MILP.

Pure integer programming with variables that take arbitrary integer values is a class which has strong connections to number theory and particularly the geometry of numbers and Presburgher arithmetic. Although this is a fascinating subject with important applications in cryptography, in the interests of brevity we shall largely restrict our attention to MILP where the integer variables are Boolean.

The fact that mixed integer linear programs subsume combinatorial optimization problems follows from two simple observations. The first is that a collection $\mathcal{I}$ of subsets of a finite ground set E can always be represented by a corresponding collection of incidence vectors, which are $\{0, 1\}$-vectors in $\Re^E$. Further, arbitrary nonlinear functions can be represented via piecewise linear approximations by using linear constraints and mixed variables (continuous and Boolean).

The next section contains a primer on linear inequalities, polyhedra, and linear programming. These are the tools we will need to analyze and solve integer programs. Section 15.4, is a testimony to the earlier cryptic comments on how integer programs model combinatorial optimization problems. In addition to working a number of examples of such integer programming formulations, we shall also review a formal representation theory of (Boolean) mixed integer linear programs.

With any mixed integer program we associate a **linear programming relaxation** obtained by simply ignoring the integrality restrictions on the variables. The point being, of course, that we have polynomial-time (and practical) algorithms for solving linear programs. Thus, the linear programming relaxation of (MILP) is given by

$$(LP) \quad \min_{x \in \Re^n} \{ \mathbf{cx} : A\mathbf{x} \geq \mathbf{b} \}$$

The thesis underlying the integer linear programming approach to combinatorial optimization is that this linear programming relaxation retains enough of the structure of the combinatorial optimization problem to be a useful weak representation. In Section 15.5 we shall take a closer look at this thesis in that we shall encounter special structures for which this relaxation is *tight*. For general integer programs, there are several alternative schemes for generating linear programming relaxations with varying qualities of approximation. A general principle is that we often need to disaggregate integer formulations to obtain higher quality linear programming relaxations. To solve such huge linear programs we need specialized techniques of large-scale linear programming. These aspects will be the content of Section 15.3.

The reader should note that the focus in this chapter is on solving hard combinatorial optimization problems. We catalog the special structures in integer programs that lead to tight linear programming relaxations (Section 15.5) and hence to polynomial-time algorithms. These include structures such as network flows, matching, and matroid optimization problems. Many hard problems actually have pieces of these nice structures embedded in them. Practitioners of combinatorial optimization have always used insights from special structures to devise strategies for hard problems.

The computational art of integer programming rests on useful interplays between search methodologies and linear programming relaxations. The paradigms of branch and bound and branch and cut are the two enormously effective partial enumeration schemes that have evolved at this interface. These will be discussed in Section 15.6. It may be noted that all general purpose integer programming software available today uses one or both of these paradigms.

The inherent complexity of integer linear programming has led to a long-standing research program in approximation methods for these problems. Linear programming relaxation and Lagrangian relaxation are two general approximation schemes that have been the real workhorses of computational practice. Primal–dual strategies and semidefinite relaxations are two recent entrants that appear to be very promising. Section 15.7 of this chapter reviews these developments in the approximation of combinatorial optimization problems.

We conclude the chapter with brief comments on future prospects in combinatorial optimization from the algebraic modeling perspective.

15.2 A Primer on Linear Programming

Polyhedral combinatorics is the study of embeddings of combinatorial structures in Euclidean space and their algebraic representations. We will make extensive use of some standard terminology from polyhedral theory. Definitions of terms not given in the brief review below can be found in Nemhauser and Wolsey [1988].

A (convex) **polyhedron** in $\Re^n$ can be algebraically defined in two ways. The first and more straightforward definition is the *implicit* representation of a polyhedron in $\Re^n$ as the solution set to a finite system of linear inequalities in n variables. A single linear inequality $\mathbf{ax} \leq a_0$; $\mathbf{a} \neq \mathbf{0}$ defines a *half-space* of $\Re^n$. Therefore, geometrically a polyhedron is the intersection set of a finite number of half-spaces.

A *polytope* is a bounded polyhedron. Every polytope is the convex closure of a finite set of points. Given a set of points whose convex combinations generate a polytope, we have an explicit or *parametric* algebraic representation of it. A *polyhedral cone* is the solution set of a system of homogeneous linear inequalities.

Every (polyhedral) cone is the conical or positive closure of a finite set of vectors. These generators of the cone provide a parametric representation of the cone. And finally, a polyhedron can be alternatively defined as the Minkowski sum of a polytope and a cone. Moving from one representation of any of these polyhedral objects to another defines the essence of the computational burden of polyhedral combinatorics. This is particularly true if we are interested in *minimal* representations.

A set of points $\mathbf{x}^1, \ldots, \mathbf{x}^m$ is *affinely independent* if the unique solution of $\sum_{i=1}^{m} \lambda_i \mathbf{x}^i = 0$, $\sum_{i=1}^{m} \lambda_i = 0$ is $\lambda_i = 0$ for $i = 1, \ldots, m$. Note that the maximum number of affinely independent points in $\Re^n$ is $n + 1$. A polyhedron P is of *dimension* k, dim $P = k$, if the maximum number of affinely independent points in P is $k + 1$. A polyhedron $P \subseteq \Re^n$ of dimension n is called *full dimensional*. An inequality $\mathbf{a}\mathbf{x} \leq a_0$ is called *valid* for a polyhedron P if it is satisfied by all $\mathbf{x}$ in P. It is called *supporting* if in addition there is an $\bar{\mathbf{x}}$ in P that satisfies $\mathbf{a}\bar{\mathbf{x}} = a_0$. A *face* of the polyhedron is the set of all $\mathbf{x}$ in P that also satisfies a valid inequality as an equality. In general, many valid inequalities might represent the same face. Faces other than P itself are called *proper*. A *facet* of P is a maximal nonempty and proper face. A facet is then a face of P with a dimension of dim $P - 1$. A face of dimension zero, i.e., a point v in P that is a face by itself, is called an **extreme point** of P. The extreme points are the elements of P that cannot be expressed as a strict convex combination of two distinct points in P. For a full-dimensional polyhedron, the valid inequality representing a facet is unique up to multiplication by a positive scalar, and facet-inducing inequalities give a minimal implicit representation of the polyhedron. Extreme points, on the other hand, give rise to minimal parametric representations of polytopes.

The two fundamental problems of linear programming (which are polynomially equivalent) follow:

- *Solvability.* This is the problem of checking if a system of linear constraints on real (rational) variables is solvable or not. Geometrically, we have to check if a polyhedron, defined by such constraints, is nonempty.

- *Optimization.* This is the problem (LP) of optimizing a linear objective function over a polyhedron described by a system of linear constraints.

Building on polarity in cones and polyhedra, duality in linear programming is a fundamental concept which is related to both the complexity of linear programming and to the design of algorithms for solvability and optimization. We will encounter the solvability version of duality (called Farkas Lemma) while discussing the Fourier elimination technique subsequently. Here we will state the main duality results for optimization. If we take the *primal* linear program to be

$$(P) \quad \min_{x \in \Re^n}\{\mathbf{cx} : Ax \geq \mathbf{b}\}$$

there is an associated *dual* linear program

$$(D) \quad \max_{y \in \Re^m}\{\mathbf{b}^T\mathbf{y} : A^T\mathbf{y} = \mathbf{c}^T, \mathbf{y} \geq \mathbf{0}\}$$

and the two problems satisfy the following:

1. For any $\hat{\mathbf{x}}$ and $\hat{\mathbf{y}}$ feasible in (P) and (D) (i.e., they satisfy the respective constraints), we have $\mathbf{c}\hat{\mathbf{x}} \geq \mathbf{b}^T\hat{\mathbf{y}}$ (weak duality). Consequently, (P) has a finite optimal solution if and only if (D) does.
2. The pair $\mathbf{x}^*$ and $\mathbf{y}^*$ are optimal solutions for (P) and (D), respectively, if and only if $\mathbf{x}^*$ and $\mathbf{y}^*$ are feasible in (P) and (D) (i.e., they satisfy the respective constraints) and $\mathbf{c}\mathbf{x}^* = \mathbf{b}^T\mathbf{y}^*$ (strong duality).
3. The pair $\mathbf{x}^*$ and $\mathbf{y}^*$ are optimal solutions for (P) and (D), respectively, if and only if $\mathbf{x}^*$ and $\mathbf{y}^*$ are feasible in (P) and (D) (i.e., they satisfy the respective constraints) and $(Ax^* - \mathbf{b})^T\mathbf{y}^* = 0$ (complementary slackness).

The strong duality condition gives us a good stopping criterion for optimization algorithms. The complementary slackness condition, on the other hand, gives us a constructive tool for moving from dual to primal solutions and vice versa. The weak duality condition gives us a technique for obtaining lower bounds for minimization problems and upper bounds for maximization problems.

Note that the properties just given have been stated for linear programs in a particular form. The reader should be able to check that if, for example, the primal is of the form

$$(P') \quad \min_{\mathbf{x}\in\Re^n}\{\mathbf{cx} : A\mathbf{x} = \mathbf{b}, \mathbf{x} \geq \mathbf{0}\}$$

then the corresponding dual will have the form

$$(D') \quad \max_{\mathbf{y}\in\Re^m}\{\mathbf{b}^T\mathbf{y} : A^T\mathbf{y} \leq \mathbf{c}^T\}$$

The tricks needed for seeing this are that any equation can be written as two inequalities, an unrestricted variable can be substituted by the difference of two nonnegatively constrained variables, and an inequality can be treated as an equality by adding a nonnegatively constrained variable to the lesser side. Using these tricks, the reader could also check that duality in linear programming is involutory (i.e., the dual of the dual is the primal).

15.2.1 Algorithms for Linear Programming

We will now take a quick tour of some algorithms for linear programming. We start with the classical technique of Fourier, which is interesting because of its really simple syntactic specification. It leads to simple proofs of the duality principle of linear programming (solvability) that has been alluded to. We will then review the simplex method of linear programming, a method that has been finely honed over almost five decades. We will spend some time with the ellipsoid method and, in particular, with the polynomial equivalence of solvability (optimization) and separation problems, for this aspect of the ellipsoid method has had a major impact on the identification of many tractable classes of combinatorial optimization problems. We conclude the primer with a description of Karmarkar's [1984] breakthrough, which was an important landmark in the brief history of linear programming. A noteworthy role of interior point methods has been to make practical the theoretical demonstrations of tractability of various aspects of linear programming, including solvability and optimization, that were provided via the ellipsoid method.

15.2.1.1 Fourier's Scheme for Linear Inequalities

Constraint systems of linear *inequalities* of the form $A\mathbf{x} \leq \mathbf{b}$, where A is an $m \times n$ matrix of real numbers, are widely used in mathematical models. Testing the solvability of such a system is equivalent to linear programming.

Suppose we wish to eliminate the first variable $\mathbf{x}_1$ from the system $A\mathbf{x} \leq \mathbf{b}$. Let us denote

$$I^+ = \{i : A_{i1} > 0\} \qquad I^- = \{i : A_{i1} < 0\} \qquad I^0 = \{i : A_{i1} = 0\}$$

Our goal is to create an equivalent system of linear inequalities $\bar{A}\bar{\mathbf{x}} \leq \bar{\mathbf{b}}$ defined on the variables $\bar{\mathbf{x}} = (\mathbf{x}_2, \mathbf{x}_3, \ldots, \mathbf{x}_n)$:

- If I^+ is empty then we can simply delete all the inequalities with indices in I^- since they can be trivially satisfied by choosing a large enough value for $\mathbf{x}_1$. Similarly, if I^- is empty we can discard all inequalities in I^+.
- For each $k \in I^+, l \in I^-$ we add $-A_{l1}$ times the inequality $A_k\mathbf{x} \leq \mathbf{b}_k$ to A_{k1} times $A_l\mathbf{x} \leq \mathbf{b}_l$. In these new inequalities the coefficient of $\mathbf{x}_1$ is wiped out, that is, $\mathbf{x}_1$ is eliminated. Add these new inequalities to those already in I^0.
- The inequalities $\{\bar{A}_{i1}\bar{\mathbf{x}} \leq \bar{\mathbf{b}}_i\}$ for all $i \in I^0$ represent the equivalent system on the variables $\bar{\mathbf{x}} = (\mathbf{x}_2, \mathbf{x}_3, \ldots, \mathbf{x}_n)$.

Repeat this construction with $\tilde{A}\tilde{\mathbf{x}} \leq \tilde{\mathbf{b}}$ to eliminate $\mathbf{x}_2$ and so on until all variables are eliminated. If the resulting $\tilde{\mathbf{b}}$ (after eliminating $\mathbf{x}_n$) is nonnegative, we declare the original (and intermediate) inequality systems as being consistent. Otherwise,* $\tilde{\mathbf{b}} \not\geq 0$ and we declare the system inconsistent.

As an illustration of the power of elimination as a tool for theorem proving, we show now that Farkas Lemma is a simple consequence of the correctness of Fourier elimination. The lemma gives a direct proof that solvability of linear inequalities is in $\mathcal{NP} \cap co\mathcal{NP}$.

FARKAS LEMMA 15.1 (Duality in Linear Programming: Solvability). *Exactly one of the alternatives*

$$I. \quad \exists\, \mathbf{x} \in \Re^n : A\mathbf{x} \leq \mathbf{b}$$

$$II. \quad \exists\, \mathbf{y} \in \Re_+^m : \mathbf{y}^t A = \mathbf{0}, \mathbf{y}^t \mathbf{b} < 0$$

is true for any given real matrices $A, \mathbf{b}$.

Proof 15.1   Let us analyze the case when Fourier elimination provides a proof of the inconsistency of a given linear inequality system $A\mathbf{x} \leq \mathbf{b}$. The method clearly converts the given system into $RA\mathbf{x} \leq R\mathbf{b}$ where RA is zero and $R\mathbf{b}$ has at least one negative component. Therefore, there is some row of R, say, $\mathbf{r}$, such that $\mathbf{r}A = \mathbf{0}$ and $\mathbf{r}\mathbf{b} < 0$. Thus $\neg I$ implies II. It is easy to see that I and II cannot both be true for fixed $A, \mathbf{b}$.   □

In general, the Fourier elimination method is quite inefficient. Let k be any positive integer and n the number of variables be $2^k + k + 2$. If the input inequalities have left-hand sides of the form $\pm\mathbf{x}_r \pm \mathbf{x}_s \pm \mathbf{x}_t$ for all possible $1 \leq r < s < t \leq n$, it is easy to prove by induction that after k variables are eliminated, by Fourier's method, we would have at least $2^{n/2}$ inequalities. The method is therefore exponential in the worst case, and the explosion in the number of inequalities has been noted, in practice as well, on a wide variety of problems. We will discuss the central idea of minimal generators of the projection cone that results in a much improved elimination method.

First, let us identify the set of variables to be eliminated. Let the input system be of the form

$$P = \{(\mathbf{x}, \mathbf{u}) \in \Re^{n_1 + n_2} \mid A\mathbf{x} + B\mathbf{u} \leq \mathbf{b}\}$$

where $\mathbf{u}$ is the set to be eliminated. The projection of P onto $\mathbf{x}$ or equivalently the effect of eliminating the $\mathbf{u}$ variables is

$$P_{\mathbf{x}} = \{\mathbf{x} \in \Re^{n_1} \mid \exists\, \mathbf{u} \in \Re^{n_2} \text{ such that } A\mathbf{x} + B\mathbf{u} \leq \mathbf{b}\}$$

Now W, the *projection cone* of P, is given by

$$W = \{\mathbf{w} \in \Re^m \mid \mathbf{w}B = \mathbf{0}, \mathbf{w} \geq \mathbf{0}\}$$

A simple application of Farkas Lemma yields a description of $P_{\mathbf{x}}$ in terms of W.

PROJECTION LEMMA 15.2   *Let G be any set of generators (e.g., the set of extreme rays) of the cone W. Then* $P_{\mathbf{x}} = \{\mathbf{x} \in \Re^{n_1} \mid (\mathbf{g}A)\mathbf{x} \leq \mathbf{g}\mathbf{b}\, \forall\, \mathbf{g} \in G\}$.

The lemma, sometimes attributed to Černikov [1961], reduces the computation of $P_{\mathbf{x}}$ to enumerating the extreme rays of the cone W or equivalently the extreme points of the polytope $W \cap \{\mathbf{w} \in \Re^m \mid \sum_{i=1}^m \mathbf{w}_i = 1\}$.

---

*Note that the final $\tilde{\mathbf{b}}$ may not be defined if all of the inequalities are deleted by the monotone sign condition of the first step of the construction described. In such a situation, we declare the system $A\mathbf{x} \leq \mathbf{b}$ *strongly consistent* since it is consistent for any choice of $\mathbf{b}$ in $\Re^m$. To avoid making repeated references to this exceptional situation, let us simply assume that it does not occur. The reader is urged to verify that this assumption is indeed benign.

15.2.1.2 Simplex Method

Consider a polyhedron $\mathcal{K} = \{\mathbf{x} \in \mathfrak{R}^n : A\mathbf{x} = \mathbf{b}, \mathbf{x} \geq \mathbf{0}\}$. Now $\mathcal{K}$ cannot contain an infinite (in both directions) line since it is lying within the nonnegative orthant of $\mathfrak{R}^n$. Such a polyhedron is called a *pointed* polyhedron. Given a pointed polyhedron $\mathcal{K}$ we observe the following:

- If $\mathcal{K} \neq \emptyset$, then $\mathcal{K}$ has at least one extreme point.
- If $\min\{\mathbf{c}\mathbf{x} : A\mathbf{x} = \mathbf{b}, \mathbf{x} \geq \mathbf{0}\}$ has an optimal solution, then it has an optimal extreme point solution.

These observations together are sometimes called the fundamental theorem of linear programming since they suggest simple finite tests for both solvability and optimization. To generate all extreme points of $\mathcal{K}$, in order to find an optimal solution, is an impractical idea. However, we may try to run a partial search of the space of extreme points for an optimal solution. A simple local improvement search strategy of moving from extreme point to adjacent extreme point until we get to a local optimum is nothing but the simplex method of linear programming. The local optimum also turns out to be a global optimum because of the convexity of the polyhedron $\mathcal{K}$ and the linearity of the objective function $\mathbf{c}\mathbf{x}$.

The simplex method walks along edge paths on the combinatorial graph structure defined by the boundary of convex polyhedra. Since these graphs are quite dense (Balinski's theorem states that the graph of d-dimensional polyhedron must be d-connected [Ziegler 1995]) and possibly large (the Lower Bound Theorem states that the number of vertices can be exponential in the dimension [Ziegler 1995]), it is indeed somewhat of a miracle that it manages to get to an optimal extreme point as quickly as it does. Empirical and probabilistic analyses indicate that the number of iterations of the simplex method is just slightly more than linear in the dimension of the primal polyhedron. However, there is no known variant of the simplex method with a worst-case polynomial guarantee on the number of iterations. Even a polynomial bound on the diameter of polyhedral graphs is not known.

Procedure 15.1   Primal Simplex $(\mathcal{K}, c)$:

0. **Initialize:**

   $\mathbf{x}_0 :=$ an extreme point of $\mathcal{K}$
   $k := 0$

1. **Iterative step:**

   do
   If for all edge directions $\mathcal{D}_k$ at $\mathbf{x}_k$, the objective function is nondecreasing, i.e.,

$$\mathbf{c}\mathbf{d} \geq 0 \quad \forall \, \mathbf{d} \in \mathcal{D}_k$$

   then exit and return optimal $\mathbf{x}_k$.
   Else pick some $\mathbf{d}_k$ in $\mathcal{D}_k$ such that $\mathbf{c}\mathbf{d}_k < 0$.
   If $\mathbf{d}_k \geq \mathbf{0}$ **then** declare the linear program unbounded in objective value and exit.
   Else $\mathbf{x}_{k+1} := \mathbf{x}_k + \theta_k * \mathbf{d}_k$, where

$$\theta_k = \max\{\theta : \mathbf{x}_k + \theta * \mathbf{d}_k \geq \mathbf{0}\}$$

   $k := k + 1$
   od

2. **End**

Remark 15.1   In the initialization step, we assumed that an extreme point $\mathbf{x}_0$ of the polyhedron $\mathcal{K}$ is available. This also assumes that the solvability of the constraints defining $\mathcal{K}$ has been established. These assumptions are reasonable since we can formulate the solvability problem as an optimization problem, with a self-evident extreme point, whose optimal solution either establishes unsolvability of $Ax = b, x \geq 0$ or provides an extreme point of $\mathcal{K}$. Such an optimization problem is usually called a phase I model. The point being, of course, that the simplex method, as just described, can be invoked on the phase I model and, if successful, can be invoked once again to carry out the intended minimization of cx. There are several different formulations of the phase I model that have been advocated. Here is one:

$$\min\{v_0 : Ax + bv_0 = b, x \geq 0, v_0 \geq 0\}$$

The solution $(x, v_0)^T = (0, \ldots, 0, 1)$ is a self-evident extreme point and $v_0 = 0$ at an optimal solution of this model is a necessary and sufficient condition for the solvability of $Ax = b, x \geq 0$.

Remark 15.2   The scheme for generating improving edge directions uses an algebraic representation of the extreme points as certain bases, called feasible bases, of the vector space generated by the columns of the matrix A. It is possible to have linear programs for which an extreme point is geometrically overdetermined (degenerate), i.e., there are more than d facets of $\mathcal{K}$ that contain the extreme point, where d is the dimension of $\mathcal{K}$. In such a situation, there would be several feasible bases corresponding to the same extreme point. When this happens, the linear program is said to be *primal degenerate*.

Remark 15.3   There are two sources of nondeterminism in the primal simplex procedure. The first involves the choice of edge direction d_k made in step 1. At a typical iteration there may be many edge directions that are improving in the sense that $cd_k < 0$. Dantzig's rule, the maximum improvement rule, and steepest descent rule are some of the many rules that have been used to make the choice of edge direction in the simplex method. There is, unfortunately, no clearly dominant rule and successful codes exploit the empirical and analytic insights that have been gained over the years to resolve the edge selection nondeterminism in simplex methods. The second source of nondeterminism arises from degeneracy. When there are multiple feasible bases corresponding to an extreme point, the simplex method has to pivot from basis to adjacent basis by picking an entering basic variable (a pseudoEdge direction) and by dropping one of the old ones. A wrong choice of the leaving variables may lead to cycling in the sequence of feasible bases generated at this extreme point. Cycling is a serious problem when linear programs are highly degenerate as in the case of linear relaxations of many combinatorial optimization problems. The lexicographic rule (perturbation rule) for the choice of leaving variables in the simplex method is a provably finite method (i.e., all cycles are broken). A clever method proposed by Bland (cf. Schrijver [1986]) preorders the rows and columns of the matrix A. In the case of nondeterminism in either entering or leaving variable choices, Bland's rule just picks the lowest index candidate. All cycles are avoided by this rule also.

The simplex method has been the veritable workhorse of linear programming for four decades now. However, as already noted, we do not know of a simplex method that has worst-case bounds that are polynomial. In fact, Klee and Minty exploited the sensitivity of the original simplex method of Dantzig, to projective scaling of the data, and constructed exponential examples for it. Recently, Spielman and Tang [2001] introduced the concept of smoothed analysis and smoothed complexity of algorithms, which is a hybrid of worst-case and average-case analysis of algorithms. Essentially, this involves the study of performance of algorithms under small random Gaussian perturbations of the coefficients of the constraint matrix. The authors show that a variant of the simplex algorithm, known as the *shadow vertex simplex algorithm* (Gass and Saaty [1955]) has polynomial smoothed complexity.

The ellipsoid method of Shor [1970] was devised to overcome poor scaling in convex programming problems and, therefore, turned out to be the natural choice of an algorithm to first establish polynomial-time solvability of linear programming. Later Karmarkar [1984] took care of both projection and scaling simultaneously and arrived at a superior algorithm.

15.2.1.3 The Ellipsoid Algorithm

The ellipsoid algorithm of Shor [1970] gained prominence in the late 1970s when Hačijan [1979] (pronounced Khachiyan) showed that this convex programming method specializes to a polynomial-time algorithm for linear programming problems. This theoretical breakthrough naturally led to intense study of this method and its properties. The survey paper by Bland et al. [1981] and the monograph by Akgül [1984] attest to this fact. The direct theoretical consequences for combinatorial optimization problems was independently documented by Padberg and Rao [1981], Karp and Papadimitriou [1982], and Grötschel et al. [1988]. The ability of this method to implicitly handle linear programs with an exponential list of constraints and maintain polynomial-time convergence is a characteristic that is the key to its applications in combinatorial optimization. For an elegant treatment of the many deep theoretical consequences of the ellipsoid algorithm, the reader is directed to the monograph by Lovász [1986] and the book by Grötschel et al. [1988].

Computational experience with the ellipsoid algorithm, however, showed a disappointing gap between the theoretical promise and practical efficiency of this method in the solution of linear programming problems. Dense matrix computations as well as the slow average-case convergence properties are the reasons most often cited for this behavior of the ellipsoid algorithm. On the positive side though, it has been noted (cf. Ecker and Kupferschmid [1983]) that the ellipsoid method is competitive with the best known algorithms for (nonlinear) convex programming problems.

Let us consider the problem of testing if a polyhedron $Q \in \mathfrak{R}^d$, defined by linear inequalities, is nonempty. For technical reasons let us assume that Q is rational, i.e., all extreme points and rays of Q are rational vectors or, equivalently, that all inequalities in some description of Q involve only rational coefficients. The ellipsoid method does not require the linear inequalities describing Q to be explicitly specified. It suffices to have an oracle representation of Q. Several different types of oracles can be used in conjunction with the ellipsoid method (Karp and Papadimitriou [1982], Padberg and Rao [1981], Grötschel et al. [1988]). We will use the *strong separation oracle:*

Oracle: **Strong Separation**(Q, **y**)

> Given a vector $\mathbf{y} \in \mathfrak{R}^d$, decide whether $\mathbf{y} \in Q$, and if not find a hyperplane that separates $\mathbf{y}$ from Q; more precisely, find a vector $\mathbf{c} \in \mathfrak{R}^d$ such that $\mathbf{c}^T \mathbf{y} < \min\{\mathbf{c}^T\mathbf{x} \mid \mathbf{x} \in Q\}$.

The ellipsoid algorithm initially chooses an ellipsoid large enough to contain a part of the polyhedron Q if it is nonempty. This is easily accomplished because we know that if Q is nonempty then it has a rational solution whose (binary encoding) length is bounded by a polynomial function of the length of the largest coefficient in the linear program and the dimension of the space.

The center of the ellipsoid is a feasible point if the separation oracle tells us so. In this case, the algorithm terminates with the coordinates of the center as a solution. Otherwise, the separation oracle outputs an inequality that separates the center point of the ellipsoid from the polyhedron Q. We translate the hyperplane defined by this inequality to the center point. The hyperplane slices the ellipsoid into two halves, one of which can be discarded. The algorithm now creates a new ellipsoid that is the minimum volume ellipsoid containing the remaining half of the old one. The algorithm questions if the new center is feasible and so on. The key is that the new ellipsoid has substantially smaller volume than the previous one. When the volume of the current ellipsoid shrinks to a sufficiently small value, we are able to conclude that Q is empty. This fact is used to show the polynomial-time convergence of the algorithm.

The crux of the complexity analysis of the algorithm is on the a priori determination of the iteration bound. This in turn depends on three factors. The volume of the initial ellipsoid E_0, the rate of volume shrinkage ($vol(E_{k+1})/vol(E_k) < e^{-\frac{1}{(2d)}}$), and the volume threshold at which we can safely conclude that Q must be empty. The assumption of Q being a rational polyhedron is used to argue that Q can be modified into a full-dimensional polytope without affecting the decision question: "Is Q non-empty?" After careful accounting for all of these technical details and some others (e.g., compensating for the roundoff errors caused by the square root computation in the algorithm), it is possible to establish the following fundamental result.

Theorem 15.1   *There exists a polynomial $g(d, \phi)$ such that the **ellipsoid method** runs in time bounded by $T g(d, \phi)$ where ϕ is an upper bound on the size of linear inequalities in some description of Q and T is the maximum time required by the oracle* **Strong Separation**$(Q, \mathbf{y})$ *on inputs* $\mathbf{y}$ *of size at most* $g(d, \phi)$.

The size of a linear inequality is just the length of the encoding of all of the coefficients needed to describe the inequality. A direct implication of the theorem is that solvability of linear inequalities can be checked in polynomial time if strong separation can be solved in polynomial time. This implies that the standard linear programming solvability question has a polynomial-time algorithm (since separation can be effected by simply checking all of the constraints). Happily, this approach provides polynomial-time algorithms for much more than just the standard case of linear programming solvability. The theorem can be extended to show that the optimization of a linear objective function over Q also reduces to a polynomial number of calls to the strong separation oracle on Q. A converse to this theorem also holds, namely, separation can be solved by a polynomial number of calls to a solvability/optimization oracle (Grötschel et al. [1982]). Thus, optimization and separation are polynomially equivalent. This provides a very powerful technique for identifying tractable classes of optimization problems. Semidefinite programming and submodular function minimization are two important classes of optimization problems that can be solved in polynomial time using this property of the ellipsoid method.

15.2.1.4 Semidefinite Programming

The following optimization problem defined on symmetric $(n \times n)$ real matrices

$$\text{(SDP)} \quad \min_{X \in \mathfrak{R}^{n \times n}} \left\{ \sum_{ij} C \bullet X : A \bullet X = B, \ X \succeq 0 \right\}$$

is called a semidefinite program. Note that $X \succeq 0$ denotes the requirement that X is a positive semidefinite matrix, and $F \bullet G$ for $n \times n$ matrices F and G denotes the product matrix $(F_{ij} * G_{ij})$. From the definition of positive semidefinite matrices, $X \succeq 0$ is equivalent to

$$\mathbf{q}^T X \mathbf{q} \geq 0 \quad \text{for every } \mathbf{q} \in \mathfrak{R}^n$$

Thus semidefinite programming (SDP) is really a linear program on $O(n^2)$ variables with an (uncountably) infinite number of linear inequality constraints. Fortunately, the strong separation oracle is easily realized for these constraints. For a given symmetric X we use Cholesky factorization to identify the minimum eigenvalue $\lambda_{\min}$. If $\lambda_{\min}$ is nonnegative then $X \succeq 0$ and if, on the other hand, $\lambda_{\min}$ is negative we have a separating inequality

$$\gamma_{\min}^T X \gamma_{\min} \geq 0$$

where $\gamma_{\min}$ is the eigenvector corresponding to $\lambda_{\min}$. Since the Cholesky factorization can be computed by an $O(n^3)$ algorithm, we have a polynomial-time separation oracle and an efficient algorithm for SDP via the ellipsoid method. Alizadeh [1995] has shown that interior point methods can also be adapted to solving SDP to within an additive error ϵ in time polynomial in the size of the input and $\log 1/\epsilon$.

This result has been used to construct efficient approximation algorithms for maximum stable sets and cuts of graphs, Shannon capacity of graphs, and minimum colorings of graphs. It has been used to define hierarchies of relaxations for integer linear programs that strictly improve on known exponential-size linear programming relaxations. We shall encounter the use of SDP in the approximation of a maximum weight cut of a given vertex-weighted graph in Section 15.7.

15.2.1.5 Minimizing Submodular Set Functions

The minimization of submodular set functions is another important class of optimization problems for which ellipsoidal and projective scaling algorithms provide polynomial-time solution methods.

Definition 15.1 Let N be a finite set. A real valued set function f defined on the subsets of N is submodular if $f(X \cup Y) + f(X \cap Y) \leq f(X) + f(Y)$ for $X, Y \subseteq N$.

Example 15.1

Let $G = (V, E)$ be an undirected graph with V as the node set and E as the edge set. Let $c_{ij} \geq 0$ be the weight or capacity associated with edge $(ij) \in E$. For $S \subseteq V$, define the cut function $c(S) = \sum_{i \in S, j \in V \setminus S} c_{ij}$. The cut function defined on the subsets of V is submodular since $c(X) + c(Y) - c(X \cup Y) - c(X \cap Y) = \sum_{i \in X \setminus Y, j \in Y \setminus X} 2c_{ij} \geq 0$.

The optimization problem of interest is

$$\min\{f(X) : X \subseteq N\}$$

The following remarkable construction that connects submodular function minimization with convex function minimization is due to Lovász (see Grötschel et al. [1988]).

Definition 15.2 The Lovász extension $\hat{f}(.)$ of a submodular function $f(.)$ satisfies

- $\hat{f} : [0,1]^N \to \Re$.
- $\hat{f}(\mathbf{x}) = \sum_{I \in \mathcal{I}} \lambda_I f(\mathbf{x}_I)$ where $\mathbf{x} = \sum_{I \in \mathcal{I}} \lambda_I \mathbf{x}_I$, $\mathbf{x} \in [0,1]^N$, $\mathbf{x}_I$ is the incidence vector of I for each $I \in \mathcal{I}$, $\lambda_I > 0$ for each I in $\mathcal{I}$, and $\mathcal{I} = \{I_1, I_2, \ldots, I_k\}$ with $\emptyset \neq I_1 \subset I_2 \subset \cdots \subset I_k \subseteq N$. Note that the representation $\mathbf{x} = \sum_{I \in \mathcal{I}} \lambda_I \mathbf{x}_I$ is unique given that the $\lambda_I > 0$ and that the sets in $\mathcal{I}$ are nested.

It is easy to check that $\hat{f}(.)$ is a convex function. Lovász also showed that the minimization of the submodular function $f(.)$ is a special case of convex programming by proving

$$\min\{f(X) : X \subseteq N\} = \min\{\hat{f}(\mathbf{x}) : \mathbf{x} \in [0,1]^N\}$$

Further, if $\mathbf{x}^*$ is an optimal solution to the convex program and

$$\mathbf{x}^* = \sum_{I \in \mathcal{I}} \lambda_I \mathbf{x}_I$$

then for each $\lambda_I > 0$, it can be shown that $I \in \mathcal{I}$ minimizes f. The ellipsoid method can be used to solve this convex program (and hence submodular minimization) using a polynomial number of calls to an oracle for f [this oracle returns the value of $f(X)$ when input X].

15.2.1.6 Interior Point Methods

The announcement of the polynomial solvability of linear programming followed by the probabilistic analyses of the simplex method in the early 1980s left researchers in linear programming with a dilemma. We had one method that was good in a theoretical sense but poor in practice and another that was good in practice (and on average) but poor in a theoretical worst-case sense. This left the door wide open for a method that was good in both senses. Narendra Karmarkar closed this gap with a breathtaking new projective scaling algorithm. In retrospect, the new algorithm has been identified with a class of nonlinear programming methods known as logarithmic barrier methods. Implementations of a primal–dual variant of the logarithmic barrier method have proven to be the best approach at present. It is this variant that we describe.

It is well known that moving through the interior of the feasible region of a linear program using the negative of the gradient of the objective function, as the movement direction, runs into trouble because of getting *jammed* into corners (in high dimensions, corners make up most of the interior of a polyhedron). This jamming can be overcome if the negative gradient is balanced with a *centering* direction. The centering direction in Karmarkar's algorithm is based on the *analytic center* $\mathbf{y}_c$ of a full-dimensional polyhedron $\mathcal{D} = \{\mathbf{y} : A^T\mathbf{y} \leq c\}$ which is the unique optimal solution to

$$\max\left\{\sum_{j=1}^{n} \ell n\,(\mathbf{z}_j)\;:\; A^T\mathbf{y} + \mathbf{z} = \mathbf{c}\right\}$$

Recall the primal and dual forms of a linear program may be taken as

$$(P) \quad \min\{\mathbf{cx} : A\mathbf{x} = \mathbf{b}, \mathbf{x} \geq \mathbf{0}\}$$

$$(D) \quad \max\{\mathbf{b}^T\mathbf{y} : A^T\mathbf{y} \leq \mathbf{c}\}$$

The logarithmic barrier formulation of the dual (D) is

$$(D_\mu) \quad \max\left\{\mathbf{b}^T\mathbf{y} + \mu\sum_{j=1}^{n} \ell n\,(\mathbf{z}_j)\;:\; A^T\mathbf{y} + \mathbf{z} = \mathbf{c}\right\}$$

Notice that (D_μ) is equivalent to (D) as $\mu \to 0^+$. The optimality (Karush–Kuhn–Tucker) conditions for (D_μ) are given by

$$D_{\mathbf{x}}D_{\mathbf{z}}\mathbf{e} = \mu\mathbf{e}$$

$$A\mathbf{x} = \mathbf{b} \qquad\qquad (15.1)$$

$$A^T\mathbf{y} + \mathbf{z} = \mathbf{c}$$

where $D_{\mathbf{x}}$ and $D_{\mathbf{z}}$ denote $n \times n$ diagonal matrices whose diagonals are $\mathbf{x}$ and $\mathbf{z}$, respectively. Notice that if we set μ to 0, the above conditions are precisely the primal–dual optimality conditions: complementary slackness, primal and dual feasibility of a pair of optimal (P) and (D) solutions. The problem has been reduced to solving the equations in $\mathbf{x}, \mathbf{y}, \mathbf{z}$. The classical technique for solving equations is Newton's method, which prescribes the directions,

$$\Delta\mathbf{y} = -\left(AD_{\mathbf{x}}D_{\mathbf{z}}^{-1}A^T\right)^{-1}AD_{\mathbf{z}}^{-1}(\mu\mathbf{e} - D_{\mathbf{x}}D_{\mathbf{z}}\mathbf{e})\Delta\mathbf{z} = -A^T\Delta\mathbf{y}\Delta\mathbf{x}$$
$$= D_{\mathbf{z}}^{-1}(\mu\mathbf{e} - D_{\mathbf{x}}D_{\mathbf{z}}\mathbf{e}) - D_{\mathbf{x}}D_{\mathbf{z}}^{-1}\Delta\mathbf{z} \qquad\qquad (15.2)$$

The strategy is to take one Newton step, reduce μ, and iterate until the optimization is complete. The criterion for stopping can be determined by checking for feasibility $(\mathbf{x}, \mathbf{z} \geq \mathbf{0})$ and if the duality gap $(\mathbf{x}^t\mathbf{z})$ is close enough to 0. We are now ready to describe the algorithm.

Procedure 15.2   Primal-Dual Interior:

0. **Initialize:**

   $\mathbf{x}_0 > \mathbf{0}, \mathbf{y}_0 \in \mathfrak{R}^m, \mathbf{z}_0 > \mathbf{0}, \mu_0 > 0, \epsilon > 0, \rho > 0$
   $k := 0$

1. **Iterative step:**

   do
   Stop if $A\mathbf{x}_k = \mathbf{b}$, $A^T\mathbf{y}_k + \mathbf{z}_k = \mathbf{c}$ and $\mathbf{x}_k^T\mathbf{z}_k \leq \epsilon$.
   $\mathbf{x}_{k+1} \leftarrow \mathbf{x}_k + \alpha_k^P \Delta\mathbf{x}_k$
   $\mathbf{y}_{k+1} \leftarrow \mathbf{y}_k + \alpha_k^D \Delta\mathbf{y}_k$
   $\mathbf{z}_{k+1} \leftarrow \mathbf{z}_k + \alpha_k^D \Delta\mathbf{z}_k$
   /* $\Delta\mathbf{x}_k, \Delta\mathbf{y}_k, \Delta\mathbf{z}_k$ **are the Newton directions from (1)** */
   $\mu_{k+1} \leftarrow \rho\mu_k$
   $k := k + 1$
   od

2. **End**

Remark 15.4 The step sizes α_k^P and α_k^D are chosen to keep $\mathbf{x}_{k+1}$ and $\mathbf{z}_{k+1}$ strictly positive. The ability in the primal–dual scheme to choose separate step sizes for the primal and dual variables is a major advantage that this method has over the pure primal or dual methods. Empirically this advantage translates to a significant reduction in the number of iterations.

Remark 15.5 The stopping condition essentially checks for primal and dual feasibility and near complementary slackness. Exact complementary slackness is not possible with interior solutions. It is possible to maintain primal and dual feasibility through the algorithm, but this would require a phase I construction via artificial variables. Empirically, this feasible variant has not been found to be worthwhile. In any case, when the algorithm terminates with an interior solution, a post-processing step is usually invoked to obtain optimal extreme point solutions for the primal and dual. This is usually called the *purification* of solutions and is based on a clever scheme described by Megiddo [1991].

Remark 15.6 Instead of using Newton steps to drive the solutions to satisfy the optimality conditions of (D_μ), Mehrotra [1992] suggested a predictor–corrector approach based on power series approximations. This approach has the added advantage of providing a rational scheme for reducing the value of μ. It is the predictor–corrector based primal–dual interior method that is considered the current winner in interior point methods. The OB1 code of Lustig et al. [1994] is based on this scheme.

Remark 15.7 CPLEX 6.5 [1999], a general purpose linear (and integer) programming solver, contains implementations of interior point methods. A computational study of parallel implementations of simplex and interior point methods on the SGI power challenge (SGI R8000) platform indicates that on all but a few small linear programs in the NETLIB linear programming benchmark problem set, interior point methods dominate the simplex method in run times. New advances in handling Cholesky factorizations in parallel are apparently the reason for this exceptional performance of interior point methods. For the simplex method, CPLEX 6.5 incorporates efficient methods of solving triangular linear systems and faster updating of reduced costs for identifying improving edge directions. For the interior point method, the same code includes improvements in computing Cholesky factorizations and better use of level-two cache available in modern computing architectures. Using CPLEX 6.5 and CPLEX 5.0, Bixby et al. [2001] in a recent study have done extensive computational testing comparing the two codes with respect to the performance of the Primal simplex, Dual simplex and Interior Point methods as well as a comparison of the performance of these three methods. While CPLEX 6.5 considerably outperformed CPLEX 5.0 for all the three methods, the comparison among the three methods is inconclusive. However, as stated by Bixby et al. [2001], the computational testing was biased against interior point method because of the inferior floating point performance of the machine used and the nonimplementation of the parallel features on shared memory machines.

Remark 15.8 Karmarkar [1990] has proposed an interior-point approach for integer programming problems. The main idea is to reformulate an integer program as the minimization of a quadratic energy function over linear constraints on continuous variables. Interior-point methods are applied to this formulation to find local optima.

15.3 Large-Scale Linear Programming in Combinatorial Optimization

Linear programming problems with thousands of rows and columns are routinely solved either by variants of the simplex method or by interior point methods. However, for several linear programs that arise in combinatorial optimization, the number of columns (or rows in the dual) are too numerous to be enumerated explicitly. The columns, however, often have a structure which is exploited to generate the columns as and when required in the simplex method. Such an approach, which is referred to as **column**

generation, is illustrated next on the *cutting stock problem* (Gilmore and Gomory [1963]), which is also known as the *bin packing problem* in the computer science literature.

15.3.1 Cutting Stock Problem

Rolls of sheet metal of standard length L are used to cut required lengths $l_i, i = 1, 2, \ldots, m$. The jth cutting pattern should be such that a_{ij}, the number of sheets of length l_i cut from one roll of standard length L, must satisfy $\sum_{i=1}^{m} a_{ij} l_i \leq L$. Suppose $n_i, i = 1, 2, \ldots, m$ sheets of length l_i are required. The problem is to find cutting patterns so as to minimize the number of rolls of standard length L that are used to meet the requirements. A linear programming formulation of the problem is as follows.

Let $\mathbf{x}_j, j = 1, 2, \ldots, n$, denote the number of times the jth cutting pattern is used. In general, $\mathbf{x}_j, j = 1, 2, \ldots, n$ should be an integer but in the next formulation the variables are permitted to be fractional.

$$(P1) \quad \text{Min} \sum_{j=1}^{n} \mathbf{x}_j$$

$$\text{Subject to} \quad \sum_{j=1}^{n} a_{ij} \mathbf{x}_j \geq n_i \quad i = 1, 2, \ldots, m$$

$$\mathbf{x}_j \geq 0 \quad j = 1, 2, \ldots, n$$

$$\text{where} \quad \sum_{i=1}^{m} l_i a_{ij} \leq L \quad j = 1, 2, \ldots, n$$

The formulation can easily be extended to allow for the possibility of p standard lengths $L_k, k = 1, 2, \ldots, p$, from which the n_i units of length $l_i, i = 1, 2, \ldots, m$, are to be cut.

The cutting stock problem can also be viewed as a bin packing problem. Several bins, each of standard capacity L, are to be packed with n_i units of item i, each of which uses up capacity of l_i in a bin. The problem is to minimize the number of bins used.

15.3.1.1 Column Generation

In general, the number of columns in (P1) is too large to enumerate all of the columns explicitly. The simplex method, however, does not require all of the columns to be explicitly written down. Given a basic feasible solution and the corresponding simplex multipliers $\mathbf{w}_i, i = 1, 2, \ldots, m$, the column to enter the basis is determined by applying dynamic programming to solve the following knapsack problem:

$$(P2) \quad z = \text{Max} \sum_{i=1}^{m} \mathbf{w}_i a_i$$

$$\text{Subject to} \quad \sum_{i=1}^{m} l_i a_i \leq L$$

$$a_i \geq 0 \text{ and integer,} \quad \text{for } i = 1, 2, \ldots, m$$

Let $a_i^*, i = 1, 2, \ldots, m$, denote an optimal solution to (P2). If $z > 1$, the kth column to enter the basis has coefficients $a_{ik} = a_i^*, i = 1, 2, \ldots, m$.

Using the identified columns, a new improved (in terms of the objective function value) basis is obtained, and the column generation procedure is repeated. A major iteration is one in which (P2) is solved to identify, if there is one, a column to enter the basis. Between two major iterations, several minor iterations may be performed to optimize the linear program using only the available (generated) columns.

If $z \leq 1$, the current basic feasible solution is optimal to (P1). From a computational point of view, alternative strategies are possible. For instance, instead of solving (P2) to optimality, a column to enter the basis can be indentified as soon as a feasible solution to (P2) with an objective function value greater than 1 has been found. Such an approach would reduce the time required to solve (P2) but may increase the number of iterations required to solve (P1).

A column once generated may be retained, even if it comes out of the basis at a subsequent iteration, so as to avoid generating the same column again later on. However, at a particular iteration some columns, which appear unattractive in terms of their reduced costs, may be discarded in order to avoid having to store a large number of columns. Such columns can always be generated again subsequently, if necessary. The rationale for this approach is that such unattractive columns will rarely be required subsequently.

The dual of (P1) has a large number of rows. Hence column generation may be viewed as row generation in the dual. In other words, in the dual we start with only a few constraints explicitly written down. Given an optimal solution **w** to the current dual problem (i.e., with only a few constraints which have been explicitly written down) find a constraint that is violated by **w** or conclude that no such constraint exists. The problem to be solved for identifying a violated constraint, if any, is exactly the separation problem that we encountered in the section on algorithms for linear programming.

15.3.2 Decomposition and Compact Representations

Large-scale linear programs sometimes have a block diagonal structure with a few additional constraints linking the different blocks. The linking constraints are referred to as the master constraints and the various blocks of constraints are referred to as subproblem constraints. Using the representation theorem of polyhedra (see, for instance, Nemhauser and Wolsey [1988]), the decomposition approach of Dantzig and Wolfe [1961] is to convert the original problem to an equivalent linear program with a small number of constraints but with a large number of columns or variables. In the cutting stock problem described in the preceding section, the columns are generated, as and when required, by solving a knapsack problem via dynamic programming. In the Dantzig–Wolfe decomposition scheme, the columns are generated, as and when required, by solving appropriate linear programs on the subproblem constraints.

It is interesting to note that the reverse of decomposition is also possible. In other words, suppose we start with a statement of a problem and an associated linear programming formulation with a large number of columns (or rows in the dual). If the column generation (or row generation in the dual) can be accomplished by solving a linear program, then a *compact* formulation of the original problem can be obtained. Here compact refers to the number of rows and columns being bounded by a polynomial function of the input length of the original problem. This result due to Martin [1991] enables one to solve the problem in the polynomial time by solving the compact formulation using interior point methods.

15.4 Integer Linear Programs

Integer linear programming problems (ILPs) are linear programs in which all of the variables are restricted to be integers. If only some but not all variables are restricted to be integers, the problem is referred to as a mixed integer program. Many combinatorial problems can be formulated as integer linear programs in which all of the variables are restricted to be 0 or 1. We will first discuss several examples of combinatorial optimization problems and their formulation as integer programs. Then we will review a general representation theory for integer programs that gives a formal measure of the expressiveness of this algebraic approach. We conclude this section with a representation theorem due to Benders [1962], which has been very useful in solving certain large-scale combinatorial optimization problems in practice.

15.4.1 Example Formulations

15.4.1.1 Covering and Packing Problems

A wide variety of location and scheduling problems can be formulated as set covering or set packing or set partitioning problems. The three different types of **covering and packing** problems can be succinctly stated as follows: Given (1) a finite set of elements $\mathcal{M} = \{1, 2, \ldots, m\}$, and (2) a family F of subsets of $\mathcal{M}$ with each member $F_j, j = 1, 2, \ldots, n$ having a profit (or cost) $\mathbf{c}_j$ associated with it, find a collection, S, of the members of F that maximizes the profit (or minimizes the cost) while ensuring that every element of $\mathcal{M}$ is in one of the following:

(P3): at most one member of S (set packing problem)
(P4): at least one member of S (set covering problem)
(P5): exactly one member of S (set partitioning problem)

The three problems (P3), (P4), and (P5) can be formulated as ILPs as follows:
   Let A denote the $m \times n$ matrix where

$$A_{ij} = \begin{cases} 1 & \text{if element } i \in F_j \\ 0 & \text{otherwise} \end{cases}$$

The decision variables are $\mathbf{x}_j, j = 1, 2, \ldots, n$ where

$$\mathbf{x}_{ij} = \begin{cases} 1 & \text{if } F_j \text{ is chosen} \\ 0 & \text{otherwise} \end{cases}$$

The set packing problem is

$$\text{(P3)} \quad \text{Max } \mathbf{cx}$$
$$\text{Subject to} \quad A\mathbf{x} \leq \mathbf{e}_m$$
$$\mathbf{x}_j = 0 \text{ or } 1, \quad j = 1, 2, \ldots, n$$

where $\mathbf{e}_m$ is an m-dimensional column vector of ones.

The set covering problem (P4) is (P3) with less than or equal to constraints replaced by greater than or equal to constraints and the objective is to minimize rather than maximize. The set partitioning problem (P5) is (P3) with the constraints written as equalities. The set partitioning problem can be converted to a set packing problem or set covering problem (see Padberg [1995]) using standard transformations. If the right-hand side vector $\mathbf{e}_m$ is replaced by a nonnegative integer vector $\mathbf{b}$, (P3) is referred to as the generalized set packing problem.

The airline crew scheduling problem is a classic example of the set partitioning or the set covering problem. Each element of $\mathcal{M}$ corresponds to a flight segment. Each subset F_j corresponds to an acceptable set of flight segments of a crew. The problem is to cover, at minimum cost, each flight segment exactly once. This is a set partitioning problem. If *dead heading* of crew is permitted, we have the set covering problem.

15.4.1.2 Packing and Covering Problems in a Graph

Suppose A is the node-edge incidence matrix of a graph. Now, (P3) is a weighted matching problem. If in addition, the right-hand side vector $\mathbf{e}_m$ is replaced by a nonnegative integer vector $\mathbf{b}$, (P3) is referred to as a weighted $\mathbf{b}$-matching problem. In this case, each variable $\mathbf{x}_j$ which is restricted to be an integer may have a positive upper bound of u_j. Problem (P4) is now referred to as the weighted edge covering problem. Note that by substituting for $\mathbf{x}_j = 1 - \mathbf{y}_j$, where $\mathbf{y}_j = 0$ or 1, the weighted edge covering problem is transformed to a weighted $\mathbf{b}$-matching problem in which the variables are restricted to be 0 or 1.

Suppose A is the edge-node incidence matrix of a graph. Now, (P3) is referred to as the weighted vertex packing problem and (P4) is referred to as the weighted vertex covering problem. The *set packing* problem can be transformed to a weighted vertex packing problem in a graph G as follows:

G contains a node for each $\mathbf{x}_j$ and an edge between nodes j and k exists if and only if the columns $A_{.j}$ and $A_{.k}$ are not orthogonal. G is called the *intersection graph* of A. The set packing problem is equivalent to the weighted vertex packing problem on G. Given G, the complement graph $\overline{G}$ has the same node set as G and there is an edge between nodes j and k in $\overline{G}$ if and only if there is no such corresponding edge in G. A clique in a graph is a subset, k, of nodes of G such that the subgraph induced by k is complete. Clearly, the weighted vertex packing problem in G is equivalent to finding a maximum weighted clique in $\overline{G}$.

15.4.1.3 Plant Location Problems

Given a set of customer locations $N = \{1, 2, \ldots, n\}$ and a set of potential sites for plants $M = \{1, 2, \ldots, m\}$, the plant location problem is to identify the sites where the plants are to be located so that the customers are served at a minimum cost. There is a fixed cost $\mathbf{f}_i$ of locating the plant at site i and the cost of serving customer j from site i is $\mathbf{c}_{ij}$. The decision variables are: $\mathbf{y}_i$ is set to 1 if a plant is located at site i and to 0 otherwise; $\mathbf{x}_{ij}$ is set to 1 if site i serves customer j and to 0 otherwise.

A formulation of the problem is

$$(\text{P6}) \quad \text{Min} \sum_{i=1}^{m} \sum_{j=1}^{n} \mathbf{c}_{ij} \mathbf{x}_{ij} + \sum_{i=1}^{m} \mathbf{f}_i \mathbf{y}_i$$

$$\text{subject to} \quad \sum_{i=1}^{m} \mathbf{x}_{ij} = 1 \qquad j = 1, 2, \ldots, n$$

$$\mathbf{x}_{ij} - \mathbf{y}_i \leq 0 \qquad i = 1, 2, \ldots, m; \quad j = 1, 2, \ldots, n$$

$$\mathbf{y}_i = 0 \quad \text{or} \quad 1 \quad i = 1, 2, \ldots, m$$

$$\mathbf{x}_{ij} = 0 \quad \text{or} \quad 1 \quad i = 1, 2, \ldots, m; \quad j = 1, 2, \ldots, n$$

Note that the constraints $\mathbf{x}_{ij} - \mathbf{y}_i \leq 0$ are required to ensure that customer j may be served from site i only if a plant is located at site i. Note that the constraints $\mathbf{y}_i = 0$ or 1 force an optimal solution in which $\mathbf{x}_{ij} = 0$ or 1. Consequently, the $\mathbf{x}_{ij} = 0$ or 1 constraints may be replaced by nonnegativity constraints $\mathbf{x}_{ij} \geq 0$.

The linear programming relaxation associated with (P6) is obtained by replacing constraints $\mathbf{y}_i = 0$ or 1 and $\mathbf{x}_{ij} = 0$ or 1 by nonnegativity contraints on $\mathbf{x}_{ij}$ and $\mathbf{y}_i$. The upper bound constraints on $\mathbf{y}_i$ are not required provided $\mathbf{f}_i \geq 0, i = 1, 2, \ldots, m$. The upper bound constraints on $\mathbf{x}_{ij}$ are not required in view of constraints $\sum_{i=1}^{m} \mathbf{x}_{ij} = 1$.

Remark 15.9 It is frequently possible to formulate the same combinatorial problem as two or more different ILPs. Suppose we have two ILP formulations (F1) and (F2) of the given combinatorial problem with both (F1) and (F2) being minimizing problems. Formulation (F1) is said to be stronger than (F2) if (LP1), the the linear programming relaxation of (F1), always has an optimal objective function value which is greater than or equal to the optimal objective function value of (LP2), which is the linear programming relaxation of (F2).

It is possible to reduce the number of constraints in (P6) by replacing the constraints $\mathbf{x}_{ij} - \mathbf{y}_i \leq 0$ by an aggregate:

$$\sum_{j=1}^{n} \mathbf{x}_{ij} - n\mathbf{y}_i \leq 0 \quad i = 1, 2, \ldots, m$$

However, the disaggregated (P6) is a stronger formulation than the formulation obtained by aggregrating the constraints as previously. By using standard transformations, (P6) can also be converted into a set packing problem.

15.4.1.4 Satisfiability and Inference Problems:

In propositional logic, a truth assignment is an assignment of true or false to each atomic proposition $\mathbf{x}_1, \mathbf{x}_2, \ldots \mathbf{x}_n$. A literal is an atomic proposition $\mathbf{x}_j$ or its negation $\neg\mathbf{x}_j$. For propositions in conjunctive normal form, a clause is a disjunction of literals and the proposition is a conjunction of clauses. A clause is obviously satisfied by a given truth assignment if at least one of its literals is true. The satisfiability problem consists of determining whether there exists a truth assignment to atomic propositions such that a set S of clauses is satisfied.

Let T_i denote the set of atomic propositions such that if any one of them is assigned true, the clause $i \in S$ is satisfied. Similarly, let F_i denote the set of atomic propositions such that if any one of them is assigned false, the clause $i \in S$ is satisfied.

The decision variables are

$$x_j = \begin{cases} 1 & \text{if atomic proposition } j \text{ is assigned true} \\ 0 & \text{if atomic proposition } j \text{ is assigned false} \end{cases}$$

The satisfiability problem is to find a feasible solution to

$$\text{(P7)} \quad \sum_{j \in T_i} x_j - \sum_{j \in F_i} x_j \geq 1 - |F_i| \quad i \in S$$

$$x_j = 0 \quad \text{or} \quad 1 \quad \text{for } j = 1, 2, \ldots, n$$

By substituting $x_j = 1 - y_j$, where $y_j = 0$ or 1, for $j \in F_i$, (P7) is equivalent to the set covering problem

$$\text{(P8)} \quad \text{Min} \sum_{j=1}^{n} (x_j + y_j) \tag{15.3}$$

$$\text{subject to} \quad \sum_{j \in T_i} x_j + \sum_{j \in F_i} y_j \geq 1 \quad i \in S \tag{15.4}$$

$$x_j + y_j \geq 1 \quad j = 1, 2, \ldots, n \tag{15.5}$$

$$x_j, y_j = 0 \quad \text{or} \quad 1 \quad j = 1, 2, \ldots, n \tag{15.6}$$

Clearly (P7) is feasible if and only if (P8) has an optimal objective function value equal to n.

Given a set S of clauses and an additional clause $k \notin S$, the logical inference problem is to find out whether every truth assignment that satisfies all of the clauses in S also satisfies the clause k. The logical inference problem is

$$\text{(P9)} \quad \text{Min} \sum_{j \in T_k} x_j - \sum_{j \in F_k} x_j$$

$$\text{subject to} \quad \sum_{j \in T_i} x_j - \sum_{j \in F_i} x_j \geq 1 - |F_i| \quad i \in S$$

$$x_j = 0 \quad \text{or} \quad 1 \quad j = 1, 2, \ldots, n$$

The clause k is implied by the set of clauses S, if and only if (P9) has an optimal objective function value greater than $-|F_k|$. It is also straightforward to express the MAX-SAT problem (i.e., find a truth assignment that maximizes the number of satisfied clauses in a given set S) as an integer linear program.

15.4.1.5 Multiprocessor Scheduling

Given n jobs and m processors, the problem is to allocate each job to one and only one of the processors so as to minimize the make span time, i.e., minimize the completion time of all of the jobs. The processors may not be identical and, hence, job j if allocated to processor i requires p_{ij} units of time. The multiprocessor scheduling problem is

$$\text{(P10)} \quad \text{Min} \ T$$

$$\text{subject to} \quad \sum_{i=1}^{m} x_{ij} = 1 \quad j = 1, 2, \ldots, n$$

$$\sum_{j=1}^{n} p_{ij} x_{ij} - T \leq 0 \quad i = 1, 2, \ldots, m$$

$$x_{ij} = 0 \quad \text{or} \quad 1$$

Note that if all p_{ij} are integers, the optimal solution will be such that T is an integer.

15.4.2 Jeroslow's Representability Theorem

Jeroslow [1989], building on joint work with Lowe in 1984, characterized subsets of n-space that can be represented as the feasible region of a mixed integer (Boolean) program. They proved that a set is the feasible region of some mixed integer/linear programming problem (MILP) if and only if it is the union of finitely many polyhedra having the same recession cone (defined subsequently). Although this result is not widely known, it might well be regarded as the fundamental theorem of mixed integer modeling.

The basic idea of Jeroslow's results is that any set that can be represented in a mixed integer model can be represented in a disjunctive programming problem (i.e., a problem with either/or constraints). A *recession direction* for a set S in n-space is a vector $\mathbf{x}$ such that $s + \alpha \mathbf{x} \in S$ for all $s \in S$ and all $\alpha \geq 0$. The set of recession directions is denoted $rec(S)$. Consider the general mixed integer constraint set

$$\mathbf{f}(\mathbf{x}, \mathbf{y}, \boldsymbol{\lambda}) \leq \mathbf{b}$$
$$\mathbf{x} \in \Re^n, \qquad \mathbf{y} \in \Re^p \qquad\qquad (15.7)$$
$$\boldsymbol{\lambda} = (\lambda_1, \ldots, \lambda_k), \quad \text{with} \quad \lambda_j \in \{0, 1\} \quad \text{for } j = 1, \ldots, k$$

Here $\mathbf{f}$ is a vector-valued function, so that $\mathbf{f}(\mathbf{x}, \mathbf{y}, \boldsymbol{\lambda}) \leq \mathbf{b}$ represents a set of constraints. We say that a set $S \subset \Re^n$ is *represented* by Eq. (15.6) if,

$$\mathbf{x} \in S \quad \text{if and only if } (\mathbf{x}, \mathbf{y}, \boldsymbol{\lambda}) \text{ satisfies Eq. (15.6) for some } y, \lambda.$$

If $\mathbf{f}$ is a linear transformation, so that Equation 15.6 is a MILP constraint set, we will say that S is *MILP representable*. The main result can now be stated.

Theorem 15.2 [Jeroslow and Lowe 1984, Jeroslow 1989]. *A set in n-space is MILP representable if and only if it is the union of finitely many polyhedra having the same set of recession directions.*

15.4.3 Benders's Representation

Any mixed integer linear program can be reformulated so that there is only one continuous variable. This reformulation, due to Benders [1962], will in general have an exponential number of constraints. Analogous to column generation, discussed earlier, these rows (constraints) can be generated as and when required.

Consider the (MILP)

$$\max \{\mathbf{cx} + \mathbf{dy} : A\mathbf{x} + G\mathbf{y} \leq \mathbf{b}, \mathbf{x} \geq \mathbf{0}, \mathbf{y} \geq \mathbf{0} \text{ and integer}\}$$

Suppose the integer variables $\mathbf{y}$ are fixed at some values, then the associated linear program is

$$\text{(LP)} \quad \max \{\mathbf{cx} : \mathbf{x} \in \mathcal{P} = \{\mathbf{x} : A\mathbf{x} \leq \mathbf{b} - G\mathbf{y}, \mathbf{x} \geq \mathbf{0}\}\}$$

and its dual is

$$\text{(DLP)} \quad \min \{\mathbf{w}(\mathbf{b} - G\mathbf{y}) : \mathbf{w} \in \mathcal{Q} = \{\mathbf{w} : \mathbf{w}A \geq \mathbf{c}, \mathbf{w} \geq \mathbf{0}\}\}$$

Let $\{\mathbf{w}^k\}$, $k = 1, 2, \ldots, K$ be the extreme points of Q and $\{\mathbf{u}^j\}$, $j = 1, 2, \ldots, J$ be the extreme rays of the recession cone of Q, $\mathcal{C}_Q = \{\mathbf{u} : \mathbf{u}A \geq \mathbf{0}, \mathbf{u} \geq \mathbf{0}\}$. Note that if Q is nonempty, the $\{\mathbf{u}^j\}$ are all of the extreme rays of Q.

From linear programming duality, we know that if Q is empty and $\mathbf{u}^j(\mathbf{b} - G\mathbf{y}) \geq 0$, $j = 1, 2, \ldots, J$ for some $\mathbf{y} \geq \mathbf{0}$ and integer then (LP) and consequently (MILP) have an unbounded solution. If Q is nonempty and $\mathbf{u}^j(\mathbf{b} - G\mathbf{y}) \geq 0$, $j = 1, 2, \ldots, J$ for some $\mathbf{y} \geq \mathbf{0}$ and integer then (LP) has a finite optimum given by

$$\min_k \{\mathbf{w}^k(\mathbf{b} - G\mathbf{y})\}$$

Hence an equivalent formulation of (MILP) is

$$\text{Max } \alpha$$

$$\alpha \leq \mathbf{dy} + \mathbf{w}^k(\mathbf{b} - G\mathbf{y}), \quad k = 1, 2, \ldots, K$$

$$\mathbf{u}^j(\mathbf{b} - G\mathbf{y}) \geq 0, \quad j = 1, 2, \ldots, J$$

$$y \geq 0 \text{ and integer}$$

$$\alpha \quad \text{unrestricted}$$

which has only one continuous variable α as promised.

15.5 Polyhedral Combinatorics

One of the main purposes of writing down an algebraic formulation of a combinatorial optimization problem as an integer program is to then examine the linear programming relaxation and understand how well it represents the discrete integer program. There are somewhat special but rich classes of such formulations for which the linear programming relaxation is sharp or tight. These correspond to linear programs that have integer valued extreme points. Such polyhedra are called **integral polyhedra**.

15.5.1 Special Structures and Integral Polyhedra

A natural question of interest is whether the LP associated with an ILP has only integral extreme points. For instance, the linear programs associated with matching and edge covering polytopes in a bipartite graph have only integral vertices. Clearly, in such a situation, the ILP can be solved as LP. A polyhedron or a polytope is referred to as being integral if it is either empty or has only integral vertices.

Definition 15.3 A $0, \pm 1$ matrix is totally unimodular if the determinant of every square submatrix is 0 or ± 1.

Theorem 15.3 [Hoffman and Kruskal 1956]. *Let*

$$A = \begin{pmatrix} A_1 \\ A_2 \\ A_3 \end{pmatrix}$$

be a $0, \pm 1$ *matrix and*

$$\mathbf{b} = \begin{pmatrix} \mathbf{b}_1 \\ \mathbf{b}_2 \\ \mathbf{b}_3 \end{pmatrix}$$

be a vector of appropriate dimensions. Then A is totally unimodular if and only if the polyhedron

$$P(A, \mathbf{b}) = \{\mathbf{x} : A_1\mathbf{x} \leq \mathbf{b}_1; A_2\mathbf{x} \geq \mathbf{b}_2; A_3\mathbf{x} = \mathbf{b}_3; \mathbf{x} \geq 0\}$$

is integral for all integral vectors $\mathbf{b}$.

The constraint matrix associated with a network flow problem (see, for instance, Ahuja et al. [1993]) is totally unimodular. Note that for a given integral $\mathbf{b}$, $P(A, \mathbf{b})$ may be integral even if A is not totally unimodular.

Definition 15.4 A polyhedron defined by a system of linear constraints is totally dual integral (TDI) if for each objective function with integral coefficient the dual linear program has an integral optimal solution whenever an optimal solution exists.

Theorem 15.4 [Edmonds and Giles 1977]. *If $P(A) = \{\mathbf{x} : A\mathbf{x} \leq \mathbf{b}\}$ is TDI and $\mathbf{b}$ is integral, then $P(A)$ is integral.*

Hoffman and Kruskal [1956] have, in fact, shown that the polyhedron $P(A, \mathbf{b})$ defined in Theorem 15.3 is TDI. This follows from Theorem 15.3 and the fact that A is totally unimodular if and only if A^T is totally unimodular.

Balanced matrices, first introduced by Berge [1972] have important implications for packing and covering problems (see also Berge and Las Vergnas [1970]).

Definition 15.5 A $0, 1$ matrix is balanced if it does not contain a square submatrix of odd order with two ones per row and column.

Theorem 15.5 [Berge 1972, Fulkerson et al. 1974]. *Let A be a balanced $0, 1$ matrix. Then the set packing, set covering, and set partitioning polytopes associated with A are integral, i.e., the polytopes*

$$P(A) = \{\mathbf{x} : \mathbf{x} \geq \mathbf{0}; A\mathbf{x} \leq \mathbf{1}\}$$
$$Q(A) = \{\mathbf{x} : \mathbf{0} \leq \mathbf{x} \leq \mathbf{1}; A\mathbf{x} \geq \mathbf{1}\}$$
$$R(A) = \{\mathbf{x} : \mathbf{x} \geq \mathbf{0}; A\mathbf{x} = \mathbf{1}\}$$

are integral.

Let

$$A = \begin{pmatrix} A_1 \\ A_2 \\ A_3 \end{pmatrix}$$

be a balanced $0, 1$ matrix. Fulkerson et al. [1974] have shown that the polytope $P(A) = \{\mathbf{x} : A_1\mathbf{x} \leq \mathbf{1}; A_2\mathbf{x} \geq \mathbf{1}; A_3\mathbf{x} = \mathbf{1}; \mathbf{x} \geq \mathbf{0}\}$ is TDI and by the theorem of Edmonds and Giles [1977] it follows that $P(A)$ is integral.

Truemper [1992] has extended the definition of balanced matrices to include $0, \pm 1$ matrices.

Definition 15.6 A $0, \pm 1$ matrix is balanced if for every square submatrix with exactly two nonzero entries in each row and each column, the sum of the entries is a multiple of 4.

Theorem 15.6 [Conforti and Cornuejols 1992b]. *Suppose A is a balanced $0, \pm 1$ matrix. Let $\mathbf{n}(A)$ denote the column vector whose ith component is the number of -1s in the ith row of A. Then the polytopes*

$$P(A) = \{\mathbf{x} : A\mathbf{x} \leq \mathbf{1} - \mathbf{n}(A); \mathbf{0} \leq \mathbf{x} \leq \mathbf{1}\}$$
$$Q(A) = \{\mathbf{x} : A\mathbf{x} \geq \mathbf{1} - \mathbf{n}(A); \mathbf{0} \leq \mathbf{x} \leq \mathbf{1}\}$$
$$R(A) = \{\mathbf{x} : A\mathbf{x} = \mathbf{1} - \mathbf{n}(A); \mathbf{0} \leq \mathbf{x} \leq \mathbf{1}\}$$

are integral.

Note that a $0, \pm 1$ matrix A is balanced if and only if A^T is balanced. Moreover, A is balanced (totally unimodular) if and only if every submatrix of A is balanced (totally unimodular). Thus, if A is balanced (totally unimodular) it follows that Theorem 15.6 (Theorem 15.3) holds for every submatrix of A.

Totally unimodular matrices constitute a subclass of balanced matrices, i.e., a totally unimodular $0, \pm 1$ matrix is always balanced. This follows from a theorem of Camion [1965], which states that a $0, \pm 1$ is totally unimodular if and only if for every square submatrix with an even number of nonzero entries in each row and in each column, the sum of the entries equals a multiple of 4. The 4×4 matrix in Figure 15.1

$$A = \begin{bmatrix} 1 & 1 & 0 & 0 \\ 1 & 1 & 1 & 1 \\ 1 & 0 & 1 & 0 \\ 1 & 0 & 0 & 1 \end{bmatrix} \qquad A = \begin{bmatrix} 1 & 1 & 0 \\ 0 & 1 & 1 \\ 1 & 0 & 1 \\ 1 & 1 & 1 \end{bmatrix}$$

FIGURE 15.1   A balanced matrix and a perfect matrix. (From Chandru, V. and Rao, M. R. Combinatorial optimization: an integer programming perspective. ACM *Comput. Surveys*, 28, 1. March 1996.)

illustrates the fact that a balanced matrix is not necessarily totally unimodular. Balanced $0, \pm 1$ matrices have implications for solving the satisfiability problem. If the given set of clauses defines a balanced $0, \pm 1$ matrix, then as shown by Conforti and Cornuejols [1992b], the satisfiability problem is trivial to solve and the associated MAXSAT problem is solvable in polynomial time by linear programming. A survey of balanced matrices is in Conforti et al. [1994].

Definition 15.7   A $0, 1$ matrix A is perfect if the set packing polytope $P(A) = \{\mathbf{x} : A\mathbf{x} \leq \mathbf{1}; \mathbf{x} \geq \mathbf{0}\}$ is integral.

The chromatic number of a graph is the minimum number of colors required to color the vertices of the graph so that no two vertices with the same color have an edge incident between them. A graph G is perfect if for every node induced subgraph H, the chromatic number of H equals the number of nodes in the maximum clique of H. The connections between the integrality of the set packing polytope and the notion of a perfect graph, as defined by Berge [1961, 1970], are given in Fulkerson [1970], Lovasz [1972], Padberg [1974], and Chvátal [1975].

Theorem 15.7 [Fulkerson 1970, Lovasz 1972, Chvátal 1975]   *Let A be $0, 1$ matrix whose columns correspond to the nodes of a graph G and whose rows are the incidence vectors of the maximal cliques of G. The graph G is perfect if and only if A is perfect.*

Let G_A denote the intersection graph associated with a given $0, 1$ matrix A (see Section 15.4). Clearly, a row of A is the incidence vector of a clique in G_A. In order for A to be perfect, every maximal clique of G_A must be represented as a row of A because inequalities defined by maximal cliques are facet defining. Thus, by Theorem 15.7, it follows that a $0, 1$ matrix A is perfect if and only if the undominated (a row of A is dominated if its support is contained in the support of another row of A) rows of A form the clique-node incidence matrix of a perfect graph.

Balanced matrices with $0, 1$ entries, constitute a subclass of $0, 1$ perfect matrices, i.e., if a $0, 1$ matrix A is balanced, then A is perfect. The 4×3 matrix in Figure 15.1 is an example of a matrix that is perfect but not balanced.

Definition 15.8   A $0, 1$ matrix A is ideal if the set covering polytope

$$Q(A) = \{\mathbf{x} : A\mathbf{x} \geq \mathbf{1}; \mathbf{0} \leq \mathbf{x} \leq \mathbf{1}\}$$

is integral.

Properties of ideal matrices are described by Lehman [1979], Padberg [1993], and Cornuejols and Novick [1994]. The notion of a $0, 1$ perfect (ideal) matrix has a natural extension to a $0, \pm 1$ perfect (ideal) matrix. Some results pertaining to $0, \pm 1$ ideal matrices are contained in Hooker [1992], whereas some results pertaining to $0, \pm 1$ perfect matrices are given in Conforti et al. [1993].

An interesting combinatorial problem is to check whether a given $0, \pm 1$ matrix is totally unimodular, balanced, or perfect. Seymour's [1980] characterization of totally unimodular matrices provides a polynomial-time algorithm to test whether a given matrix $0, 1$ matrix is totally unimodular. Conforti et al. [1999] give a polynomial-time algorithm to check whether a 0, 1 matrix is balanced. This has been extended by Conforti et al. [1994] to check in polynomial time whether a $0, \pm1$ matrix is balanced. An open problem is that of checking in polynomial time whether a 0, 1 matrix is perfect. For linear matrices (a matrix is linear if it does not contain a 2×2 submatrix of all ones), this problem has been solved by Fonlupt and Zemirline [1981] and Conforti and Rao [1993].

15.5.2 Matroids

Matroids and submodular functions have been studied extensively, especially from the point of view of combinatorial optimization (see, for instance, Nemhauser and Wolsey [1988]). Matroids have nice properties that lead to efficient algorithms for the associated optimization problems. One of the interesting examples of a matroid is the problem of finding a maximum or minimum weight spanning tree in a graph. Two different but equivalent definitions of a matroid are given first. A greedy algorithm to solve a linear optimization problem over a matroid is presented. The matroid intersection problem is then discussed briefly.

Definition 15.9 Let $N = \{1, 2, \cdot, n\}$ be a finite set and let $\mathcal{F}$ be a set of subsets of N. Then $I = (N, \mathcal{F})$ is an independence system if $S_1 \in \mathcal{F}$ implies that $S_2 \in \mathcal{F}$ for all $S_2 \subseteq S_1$. Elements of $\mathcal{F}$ are called independent sets. A set $S \in \mathcal{F}$ is a maximal independent set if $S \cup \{j\} \notin \mathcal{F}$ for all $j \in N \backslash S$. A maximal independent set T is a maximum if $|T| \geq |S|$ for all $S \in \mathcal{F}$.

The rank $r(Y)$ of a subset $Y \subseteq N$ is the cardinality of the maximum independent subset $X \subseteq Y$. Note that $r(\phi) = 0$, $r(X) \leq |X|$ for $X \subseteq N$ and the rank function is nondecreasing, i.e., $r(X) \leq r(Y)$ for $X \subseteq Y \subseteq N$.

A matroid $M = (N, \mathcal{F})$ is an independence system in which every maximal independent set is a maximum.

Example 15.2

Let $G = (V, E)$ be an undirected connected graph with V as the node set and E as the edge set.

1. Let $I = (E, \mathcal{F})$ where $F \in \mathcal{F}$ if $F \subseteq E$ is such that at most one edge in F is incident to each node of V, that is, $F \in \mathcal{F}$ if F is a matching in G. Then $I = (E, \mathcal{F})$ is an independence system but not a matroid.
2. Let $M = (E, \mathcal{F})$ where $F \in \mathcal{F}$ if $F \subseteq E$ is such that $G_F = (V, F)$ is a forest, that is, G_F contains no cycles. Then $M = (E, \mathcal{F})$ is a matroid and maximal independent sets of M are spanning trees.

An alternative but equivalent definition of matroids is in terms of submodular functions.

Definition 15.10 A nondecreasing integer valued submodular function r defined on the subsets of N is called a matroid rank function if $r(\phi) = 0$ and $r(\{j\}) \leq 1$ for $j \in N$. The pair (N, r) is called a matroid.

A nondecreasing, integer-valued, submodular function f, defined on the subsets of N is called a polymatroid function if $f(\phi) = 0$. The pair (N, r) is called a polymatroid.

15.5.2.1 Matroid Optimization

To decide whether an optimization problem over a matroid is polynomially solvable or not, we need to first address the issue of representation of a matroid. If the matroid is given either by listing the independent sets or by its rank function, many of the associated linear optimization problems are trivial to solve. However, matroids associated with graphs are completely described by the graph and the condition for independence. For instance, the matroid in which the maximal independent sets are spanning forests, the graph $G = (V, E)$ and the independence condition of no cycles describes the matroid.

Most of the algorithms for matroid optimization problems require a test to determine whether a specified subset is independent. We assume the existence of an oracle or subroutine to do this checking in running time, which is a polynomial function of $|N| = n$.

Maximum Weight Independent Set. Given a matroid $M = (N, \mathcal{F})$ and weights w_j for $j \in N$, the problem of finding a maximum weight independent set is $\max_{F \in \mathcal{F}} \left\{ \sum_{j \in F} w_j \right\}$. The greedy algorithm to solve this problem is as follows:

Procedure 15.3   Greedy:

0. **Initialize:** Order the elements of N so that $w_i \geq w_{i+1}$, $i = 1, 2, \ldots, n - 1$. Let $T = \phi, i = 1$.
1. **If** $w_i \leq 0$ or $i > n$, **stop** T is optimal, i.e., $x_j = 1$ for $j \in T$ and $x_j = 0$ for $j \notin T$. If $w_i > 0$ and $T \cup \{i\} \in \mathcal{F}$, add element i to T.
2. **Increment** i by 1 and return to step 1.

Edmonds [1970, 1971] derived a complete description of the *matroid polytope*, the convex hull of the characteristic vectors of independent sets of a matroid. While this description has a large (exponential) number of constraints, it permits the treatment of linear optimization problems on independent sets of matroids as linear programs. Cunningham [1984] describes a polynomial algorithm to solve the separation problem for the matroid polytope. The matroid polytope and the associated greedy algorithm have been extended to polymatroids (Edmonds [1970], McDiarmid [1975]).

The separation problem for a polymatroid is equivalent to the problem of minimizing a submodular function defined over the subsets of N (see Nemhauser and Wolsey [1988]). A class of submodular functions that have some additional properties can be minimized in polynomial time by solving a maximum flow problem [Rhys 1970, Picard and Ratliff 1975]. The general submodular function can be minimized in polynomial time by the ellipsoid algorithm [Grötschel et al. 1988].

The uncapacitated plant location problem formulated in Section 15.4 can be reduced to maximizing a submodular function. Hence, it follows that maximizing a submodular function is $\mathcal{NP}$-hard.

15.5.2.2   Matroid Intersection

A matroid intersection problem involves finding an independent set contained in two or more matroids defined on the same set of elements.

Let $G = (V_1, V_2, E)$ be a bipartite graph. Let $M_i = (E, \mathcal{F}_i), i = 1, 2$, where $F \in \mathcal{F}_i$ if $F \subseteq E$ is such that no more than one edge of F is incident to each node in V_i. The set of matchings in G constitutes the intersection of the two matroids $M_i, i = 1, 2$. The problem of finding a maximum weight independent set in the intersection of two matroids can be solved in polynomial time [Lawler 1975, Edmonds 1970, 1979, Frank 1981]. The two (poly) matroid intersection polytope has been studied by Edmonds [1979].

The problem of testing whether a graph contains a Hamiltonian path is $\mathcal{NP}$-complete. Since this problem can be reduced to the problem of finding a maximum cordinality independent set in the intersection of three matroids, it follows that the matroid intersection problem involving three or more matroids is $\mathcal{NP}$-hard.

15.5.3   Valid Inequalities, Facets, and Cutting Plane Methods

Earlier in this section, we were concerned with conditions under which the packing and covering polytopes are integral. But, in general, these polytopes are not integral, and additional inequalities are required to have a complete linear description of the convex hull of integer solutions. The existence of finitely many such linear inequalities is guaranteed by Weyl's [1935] Theorem.

Consider the feasible region of an ILP given by

$$P_I = \{\mathbf{x} : A\mathbf{x} \leq \mathbf{b}; \mathbf{x} \geq \mathbf{0} \text{ and integer}\} \tag{15.8}$$

Recall that an inequality $\mathbf{f}\mathbf{x} \leq f_0$ is referred to as a valid inequality for P_I if $\mathbf{f}\mathbf{x}^* \leq f_0$ for all $\mathbf{x}^* \in P_I$. A valid linear inequality for $P_I(A, \mathbf{b})$ is said to be facet defining if it intersects $P_I(A, \mathbf{b})$ in a face of dimension one less than the dimension of $P_I(A, \mathbf{b})$. In the example shown in Figure 15.2, the inequality $\mathbf{x}_2 + \mathbf{x}_3 \leq 1$ is a facet defining inequality of the integer hull.

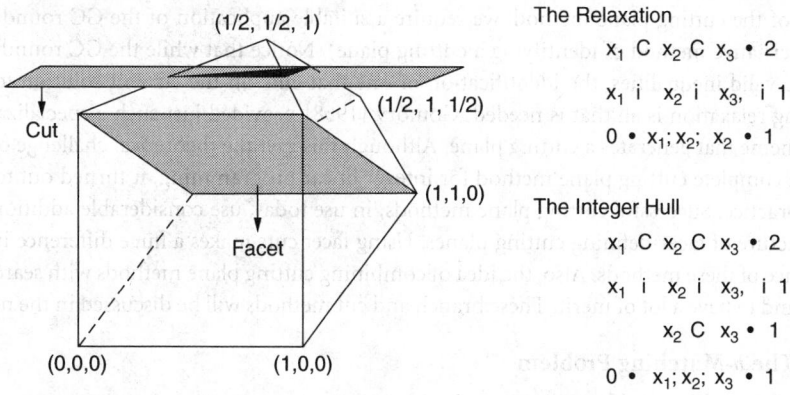

The Relaxation

$$x_1 \subseteq x_2 \subseteq x_3 \bullet 2$$
$$x_1 \; i \; x_2 \; i \; x_3, \; i \; 1$$
$$0 \bullet x_1; x_2; \; x_2 \bullet 1$$

The Integer Hull

$$x_1 \subseteq x_2 \subseteq x_3 \bullet 2$$
$$x_1 \; i \; x_2 \; i \; x_3, \; i \; 1$$
$$x_2 \subseteq x_3 \bullet 1$$
$$0 \bullet x_1; x_2; x_3 \bullet 1$$

FIGURE 15.2 Relaxation, cuts, and facets (From Chandru, V. and Rao, M. R. Combinatorial optimization: an integer programming perspective. *ACM Comput. Surveys,* 28, 1. March 1996.)

Let $\mathbf{u} \geq \mathbf{0}$ be a row vector of appropriate size. Clearly $\mathbf{u}A\mathbf{x} \leq \mathbf{u}\mathbf{b}$ holds for every $\mathbf{x}$ in P_I. Let $(\mathbf{u}A)_j$ denote the jth component of the row vector $\mathbf{u}A$ and $\lfloor(\mathbf{u}A)_j\rfloor$ denote the largest integer less than or equal to $(\mathbf{u}A)_j$. Now, since $\mathbf{x} \in P_I$ is a vector of nonnegative integers, it follows that $\sum_j \lfloor(\mathbf{u}A)_j\rfloor x_j \leq \lfloor\mathbf{u}\mathbf{b}\rfloor$ is a valid inequality for P_I. This scheme can be used to generate many valid inequalities by using different $\mathbf{u} \geq \mathbf{0}$. Any set of generated valid inequalities may be added to the constraints in Equation 15.7 and the process of generating them may be repeated with the enhanced set of inequalities. This iterative procedure of generating valid inequalities is called Gomory–Chvátal (GC) rounding. It is remarkable that this simple scheme is complete, i.e., every valid inequality of P_I can be generated by finite application of GC rounding (Chvátal [1973], Schrijver [1986]).

The number of inequalities needed to describe the convex hull of P_I is usually exponential in the size of A. But to solve an optimization problem on P_I, one is only interested in obtaining a partial description of P_I that facilitates the identification of an integer solution and prove its optimality. This is the underlying basis of any cutting plane approach to combinatorial problems.

15.5.3.1 The Cutting Plane Method

Consider the optimization problem

$$\max\{\mathbf{c}\mathbf{x} : \mathbf{x} \in P_I = \{\mathbf{x} : A\mathbf{x} \leq \mathbf{b}; \mathbf{x} \geq \mathbf{0} \text{ and integer}\}\}$$

The generic **cutting plane** method as applied to this formulation is given as follows.

Procedure 15.4   Cutting Plane:

1. Initialize $A' \leftarrow A$ and $\mathbf{b}' \leftarrow \mathbf{b}$.
2. Find an optimal solution $\bar{\mathbf{x}}$ to the linear program

$$\max\{\mathbf{c}\mathbf{x} : A'\mathbf{x} \leq \mathbf{b}'; \mathbf{x} \geq \mathbf{0}\}$$

   If $\bar{\mathbf{x}} \in P_I$, stop and return $\bar{\mathbf{x}}$.
3. Generate a valid inequality $\mathbf{f}\mathbf{x} \leq f_0$ for P_I such that $\mathbf{f}\bar{\mathbf{x}} > f_0$ (the inequality "cuts" $\bar{\mathbf{x}}$).
4. Add the inequality to the constraint system, update

$$A' \leftarrow \begin{pmatrix} A' \\ \mathbf{f} \end{pmatrix}, \qquad \mathbf{b}' \leftarrow \begin{pmatrix} \mathbf{b}' \\ f_0 \end{pmatrix}$$

   Go to step 2.

In step 3 of the cutting plane method, we require a suitable application of the GC rounding scheme (or some alternative method of identifying a cutting plane). Notice that while the GC rounding scheme will generate valid inequalities, the identification of one that cuts off the current solution to the linear programming relaxation is all that is needed. Gomory [1958] provided just such a specialization of the rounding scheme that generates a cutting plane. Although this met the theoretical challenge of designing a sound and complete cutting plane method for integer linear programming, it turned out to be a weak method in practice. Successful cutting plane methods, in use today, use considerable additional insights into the structure of facet-defining cutting planes. Using facet cuts makes a huge difference in the speed of convergence of these methods. Also, the idea of combining cutting plane methods with search methods has been found to have a lot of merit. These branch and cut methods will be discussed in the next section.

15.5.3.2 The *b*-Matching Problem

Consider the **b**-matching problem:

$$\max\{\mathbf{cx} : A\mathbf{x} \leq \mathbf{b}, \ \mathbf{x} \geq \mathbf{0} \text{ and integer}\} \tag{15.9}$$

where A is the node-edge incidence matrix of an undirected graph and $\mathbf{b}$ is a vector of positive integers. Let G be the undirected graph whose node-edge incidence matrix is given by A and let $W \subseteq V$ be any subset of nodes of G (i.e., subset of rows of A) such that

$$\mathbf{b}(W) = \sum_{i \in W} \mathbf{b}_i$$

is odd. Then the inequality

$$\mathbf{x}(W) = \sum_{e \in E(W)} \mathbf{x}_e \leq \frac{1}{2}(\mathbf{b}(W) - 1) \tag{15.10}$$

is a valid inequality for integer solutions to Equation 15.8 where $E(W) \subseteq E$ is the set of edges of G having both ends in W. Edmonds [1965] has shown that the inequalities Equation 15.8 and Equation 15.9 define the integral **b**-matching polytope. Note that the number of inequalities Equation 15.9 is exponential in the number of nodes of G. An instance of the successful application of the idea of using only a partial description of P_I is in the blossom algorithm for the matching problem, due to Edmonds [1965].

As we saw, an implication of the ellipsoid method for linear programming is that the linear program over P_I can be solved in polynomial time if and only if the associated separation problem (also referred to as the constraint identification problem, see Section 15.2) can be solved in polynomial time, see Grötschel et al. [1982], Karp and Papadimitriou [1982], and Padberg and Rao [1981]. The separation problem for the **b**-matching problem with or without upper bounds was shown by Padberg and Rao [1982], to be solvable in polynomial time. The procedure involves a minor modification of the algorithm of Gomory and Hu [1961] for multiterminal networks. However, no polynomial (in the number of nodes of the graph) linear programming formulation of this separation problem is known. A related unresolved issue is whether there exists a polynomial size (compact) formulation for the **b**-matching problem. Yannakakis [1988] has shown that, under a symmetry assumption, such a formulation is impossible.

15.5.3.3 Other Combinatorial Problems

Besides the matching problem, several other combinatorial problems and their associated polytopes have been well studied and some families of facet defining inequalities have been identified. For instance, the set packing, graph partitioning, plant location, max cut, traveling salesman, and Steiner tree problems have been extensively studied from a polyhedral point of view (see, for instance, Nemhauser and Wolsey [1988]).

These combinatorial problems belong to the class of $\mathcal{NP}$-complete problems. In terms of a worst-case analysis, no polynomial-time algorithms are known for these problems. Nevertheless, using a cutting plane approach with branch and bound or branch and cut (see Section 15.6), large instances of these problems have been successfully solved, see Crowder et al. [1983], for general $0 - 1$ problems, Barahona et al. [1989] for the max cut problem, Padberg and Rinaldi [1991] for the traveling salesman problem, and Chopra et al. [1992] for the Steiner tree problem.

15.6 Partial Enumeration Methods

In many instances, to find an optimal solution to integer linear programing problems (ILP), the structure of the problem is exploited together with some sort of partial enumeration. In this section, we review the branch and bound (B-and-B) and branch and cut (B-and-C) methods for solving an ILP.

15.6.1 Branch and Bound

The branch bound (B-and-B) method is a systematic scheme for implicitly enumerating the finitely many feasible solutions to an ILP. Although, theoretically the size of the enumeration tree is exponential in the problem parameters, in most cases, the method eliminates a large number of feasible solutions. The key features of branch and bound method are:

1. **Selection/removal** of one or more problems from a candidate list of problems
2. **Relaxation** of the selected problem so as to obtain a lower bound (on a minimization problem) on the optimal objective function value for the selected problem
3. **Fathoming**, if possible, of the selected problem
4. **Branching** strategy is needed if the selected problem is not fathomed. Branching creates subproblems, which are added to the candidate list of problems.

The four steps are repeated until the candidate list is empty. The B-and-B method sequentially examines problems that are added and removed from a candidate list of problems.

15.6.1.1 Initialization

Initially, the candidate list contains only the original ILP, which is denoted as

$$(P) \quad \min\{\mathbf{cx} : A\mathbf{x} \le \mathbf{b}, \mathbf{x} \ge \mathbf{0} \text{ and integer}\}$$

Let $F(P)$ denote the feasible region of (P) and $z(P)$ denote the optimal objective function value of (P). For any $\bar{\mathbf{x}}$ in $F(P)$, let $z_P(\bar{\mathbf{x}}) = \mathbf{c}\bar{\mathbf{x}}$.

Frequently, heuristic procedures are first applied to get a good feasible solution to (P). The best solution known for (P) is referred to as the current incumbent solution. The corresponding objective function value is denoted as z_I. In most instances, the initial heuristic solution is neither optimal nor at least immediately certified to be optimal. Thus, further analysis is required to ensure that an optimal solution to (P) is obtained. If no feasible solution to (P) is known, z_I is set to ∞.

15.6.1.2 Selection/Removal

In each iterative step of B-and-B, a problem is selected and removed from the candidate list for further analysis. The selected problem is henceforth referred to as the candidate problem (CP). The algorithm terminates if there is no problem to select from the candidate list. Initially, there is no issue of selection since the candidate list contains only the problem (P). However, as the algorithm proceeds, there would be many problems on the candidate list and a selection rule is required. Appropriate selection rules, also referred to as branching strategies, are discussed later. Conceptually, several problems may be simultaneously selected and removed from the candidate list. However, most sequential implementations of B-and-B select only one problem from the candidate list and this is assumed henceforth. Parallel aspects of B-and-B on $0 - 1$ integer linear programs are discussed in Cannon and Hoffman [1990] and for the case of traveling salesman problems in Applegate et al. [1994].

The computational time required for the B-and-B algorithm depends crucially on the order in which the problems in the candidate list are examined. A number of clever heuristic rules may be employed in devising such strategies. Two general purpose selection strategies that are commonly used are as follows:

1. Choose the problem that was added last to the candidate list. This last-in–first-out rule (LIFO) is also called depth first search (DFS) since the selected candidate problem increases the depth of the active enumeration tree.
2. Choose the problem on the candidate list that has the least lower bound. Ties may be broken by choosing the problem that was added last to the candidate list. This rule would require that a lower bound be obtained for each of the problems on the candidate list. In other words, when a problem is added to the candidate list, an associated lower bound should also be stored. This may be accomplished by using ad hoc rules or by solving a relaxation of each problem before it is added to the candidate list.

Rule 1 is known to empirically dominate rule 2 when storage requirements for candidate list and computation time to solve (P) are taken into account. However, some analysis indicates that rule 2 can be shown to be superior if minimizing the number of candidate problems to be solved is the criterion (see Parker and Rardin [1988]).

15.6.1.3 Relaxation

In order to analyze the selected candidate problem (CP), a **relaxation** (CP_R) of (CP) is solved to obtain a lower bound $z(CP_R) \leq z(CP)$. (CP_R) is a relaxation of (CP) if:

1. $F(CP) \subseteq F(CP_R)$
2. For $\bar{\mathbf{x}} \in F(CP)$, $z_{CP_R}(\bar{\mathbf{x}}) \leq z_{CP}(\bar{\mathbf{x}})$
3. For $\bar{\mathbf{x}}, \hat{\mathbf{x}} \in F(CP)$, $z_{CP_R}(\bar{\mathbf{x}}) \leq z_{CP_R}(\hat{\mathbf{x}})$ implies that $z_{CP}(\bar{\mathbf{x}}) \leq z_{CP}(\hat{\mathbf{x}})$

Relaxations are needed because the candidate problems are typically hard to solve. The relaxations used most often are either linear programming or Lagrangian relaxations of (CP), see Section 15.7 for details. Sometimes, instead of solving a relaxation of (CP), a lower bound is obtained by using some ad hoc rules such as penalty functions.

15.6.1.4 Fathoming

A candidate problem is fathomed if:

(FC1) analysis of (CP_R) reveals that (CP) is infeasible. For instance, if $F(CP_R) = \phi$, then $F(CP) = \phi$.
(FC2) analysis of (CP_R) reveals that (CP) has no feasible solution better than the current incumbent solution. For instance, if $z(CP_R) \geq z_I$, then $z(CP) \geq z(CP_R) \geq z_I$.
(FC3) analysis of (CP_R) reveals an optimal solution of (CP). For instance, if the optimal solution, $\mathbf{x}_R$, to (CP_R) is feasible in (CP), then $(\mathbf{x}_R)$ is an optimal solution to (CP) and $z(CP) = \mathbf{c}\mathbf{x}_R$.
(FC4) analysis of (CP_R) reveals that (CP) is dominated by some other problem, say, CP^*, in the candidate list. For instance, if it can shown that $z(CP^*) \leq z(CP)$, then there is no need to analyze (CP) further.

If a candidate problem (CP) is fathomed using any of the preceding criteria, then further examination of (CP) or its descendants (subproblems) obtained by separation is not required. If (FC3) holds, and $z(CP) < z_I$, the incumbent is updated as $\mathbf{x}_R$ and z_I is updated as $z(CP)$.

15.6.1.5 Separation/Branching

If the candidate problem (CP) is not fathomed, then CP is separated into several problems, say, (CP_1), $(CP_2), \ldots, (CP_q)$, where $\bigcup_{t=1}^{q} F(CP_t) = F(CP)$ and, typically,

$$F(CP_i) \cap F(CP_j) = \phi \; \forall \, i \neq j$$

For instance, a separation of (CP) into $(CP_i), i = 1, 2, \ldots, q$, is obtained by fixing a single variable, say, $\mathbf{x}_j$, to one of the q possible values of $\mathbf{x}_j$ in an optimal solution to (CP). The choice of the variable to fix depends on the separation strategy, which is also part of the branching strategy. After separation, the subproblems are added to the candidate list. Each subproblem (CP_t) is a restriction of (CP) since $F(CP_t) \subseteq F(CP)$. Consequently, $z(CP) \leq z(CP_t)$ and $z(CP) = \min_t z(CP_t)$.

The various steps in the B-and-B algorithm are outlined as follows.

Procedure 15.5 B-and-B:

0. **Initialize:** Given the problem (P), the incumbent value z_I is obtained by applying some heuristic (if a feasible solution to (P) is not available, set $z_I = +\infty$). Initialize the candidate list $C \leftarrow \{(P)\}$.
1. **Optimality:** If $C = \emptyset$ and $z_I = +\infty$, then (P) is infeasible, stop. Stop also if $C = \emptyset$ and $z_I < +\infty$, the incumbent is an optimal solution to (P).
2. **Selection:** Using some candidate selection rule, select and remove a problem $(CP) \in C$.
3. **Bound:** Obtain a lower bound for (CP) by either solving a relaxation (CP_R) of (CP) or by applying some ad-hoc rules. If (CP_R) is infeasible, return to Step 1. Else, let x_R be an optimal solution of (CP_R).
4. **Fathom:** If $z(CP_R) \geq z_I$, return to step 1. Else if x_R is feasible in (CP) and $z(CP) < z_I$, set $z_I \leftarrow z(CP)$, update the incumbent as x_R and return to step 1. Finally, if x_R is feasible in (CP) but $z(CP) \geq z_I$, return to step 1.
5. **Separation:** Using some separation or branching rule, separate (CP) into $(CP_i), i = 1, 2, \ldots, q$ and set $C \leftarrow C \cup \{(CP_1), (CP_2), \ldots, (CP_q)\}$ and return to step 1.
6. **End Procedure.**

Although the B-and-B method is easy to understand, the implementation of this scheme for a particular ILP is a nontrivial task requiring the following:

1. A relaxation strategy with efficient procedures for solving these relaxations
2. Efficient data-structures for handling the rather complicated bookkeeping of the candidate list
3. Clever strategies for selecting promising candidate problems
4. Separation or branching strategies that could effectively prune the enumeration tree

A key problem is that of devising a relaxation strategy, that is, to find *good relaxations*, which are significantly easier to solve than the original problems and tend to give sharp lower bounds. Since these two are conflicting, one has to find a reasonable tradeoff.

15.6.2 Branch and Cut

In the past few years, the branch and cut (B-and-C) method has become popular for solving NP-complete combinatorial optimization problems. As the name suggests, the B-and-C method incorporates the features of both the branch and bound method just presented and the cutting plane method presented previously. The main difference between the B-and-C method and the general B-and-B scheme is in the bound step (step 3).

A distinguishing feature of the B-and-C method is that the relaxation (CP_R) of the candidate problem (CP) is a linear programming problem, and, instead of merely solving (CP_R), an attempt is made to solve (CP) by using cutting planes to tighten the relaxation. If (CP_R) contains inequalities that are valid for (CP) but not for the given ILP, then the GC rounding procedure may generate inequalities that are valid for (CP) but not for the ILP. In the B-and-C method, the inequalities that are generated are always valid for the ILP and hence can be used globally in the enumeration tree.

Another feature of the B-and-C method is that often heuristic methods are used to convert some of the fractional solutions, encountered during the cutting plane phase, into feasible solutions of the (CP) or more generally of the given ILP. Such feasible solutions naturally provide upper bounds for the ILP. Some of these upper bounds may be better than the previously identified best upper bound and, if so, the current incumbent is updated accordingly.

We thus obtain the B-and-C method by replacing the bound step (step 3) of the B-and-B method by steps 3(a) and 3(b) and also by replacing the fathom step (step 4) by steps 4(a) and 4(b) given subsequently.

3(a) **Bound:** Let (CP_R) be the LP relaxation of (CP). Attempt to solve (CP) by a cutting plane method which generates valid inequalities for (P). Update the constraint system of (P) and the incumbent as appropriate.

Let $F\mathbf{x} \leq \mathbf{f}$ denote all of the valid inequalities generated during this phase. Update the constraint system of (P) to include all of the generated inequalities, i.e., set $A^T \leftarrow (A^T, F^T)$ and $\mathbf{b}^T \leftarrow (\mathbf{b}^T, \mathbf{f}^T)$. The constraints for all of the problems in the candidate list are also to be updated.

During the cutting plane phase, apply heuristic methods to convert some of the identified fractional solutions into feasible solutions to (P). If a feasible solution, $\bar{\mathbf{x}}$, to (P), is obtained such that $\mathbf{c}\bar{\mathbf{x}} < z_I$, update the incumbent to $\bar{\mathbf{x}}$ and z_I to $\mathbf{c}\bar{\mathbf{x}}$. Hence, the remaining changes to B-and-B are as follows:

3(b) **If** (CP) is solved go to step 4(a). **Else,** let $\hat{\mathbf{x}}$ be the solution obtained when the cutting plane phase is terminated, (we are unable to identify a valid inequality of (P) that is violated by $\hat{\mathbf{x}}$). Go to step 4(b).

4(a) **Fathom by Optimality:** Let $\mathbf{x}^*$ be an optimal solution to (CP). If $z(CP) < z_I$, set $\mathbf{x}_I \leftarrow z(CP)$ and update the incumbent as $\mathbf{x}^*$. Return to step 1.

4(b) **Fathom by Bound:** If $\mathbf{c}\hat{\mathbf{x}} \geq z_I$, return to Step 1.
Else go to step 5.

The incorporation of a cutting plane phase into the B-and-B scheme involves several technicalities which require careful design and implementation of the B-and-C algorithm. Details of the state of the art in cutting plane algorithms including the B-and-C algorithm are reviewed in Jünger et al. [1995].

15.7 Approximation in Combinatorial Optimization

The inherent complexity of integer linear programming has led to a long-standing research program in approximation methods for these problems. Linear programming relaxation and Lagrangian relaxation are two general approximation schemes that have been the real workhorses of computational practice. Semidefinite relaxation is a recent entrant that appears to be very promising. In this section, we present a brief review of these developments in the approximation of combinatorial optimization problems.

In the past few years, there has been significant progress in our understanding of performance guarantees for approximation of $\mathcal{NP}$-hard combinatorial optimization problems. A **ρ-approximate** algorithm for an optimization problem is an approximation algorithm that delivers a feasible solution with objective value within a factor of ρ of optimal (think of minimization problems and $\rho \geq 1$). For some combinatorial optimization problems, it is possible to *efficiently* find solutions that are arbitrarily close to optimal even though finding the true optimal is hard. If this were true of most of the problems of interest, we would be in good shape. However, the recent results of Arora et al. [1992] indicate exactly the opposite conclusion.

A polynomial-time approximation scheme (PTAS) for an optimization problem is a family of algorithms, A_ρ, such that for each $\rho > 1$, A_ρ is a polynomial-time ρ-approximate algorithm. Despite concentrated effort spanning about two decades, the situation in the early 1990s was that for many combinatorial optimization problems, we had no PTAS and no evidence to suggest the nonexistence of such schemes either. This led Papadimitriou and Yannakakis [1991] to define a new complexity class (using reductions that preserve approximate solutions) called MAXSNP, and they identified several complete languages in this class. The work of Arora et al. [1992] completed this agenda by showing that, assuming $\mathcal{P} \neq \mathcal{NP}$, there is no PTAS for a MAXSNP-complete problem.

An implication of these theoretical developments is that for most combinatorial optimization problems, we have to be quite satisfied with performance guarantee factors ρ that are of some small fixed value. (There are problems, like the general traveling salesman problem, for which there are no ρ-approximate algorithms for any finite value of ρ, assuming of course that $\mathcal{P} \neq \mathcal{NP}$.) Thus, one avenue of research is to go problem by problem and knock ρ down to its smallest possible value. A different approach would be to look for other notions of good approximations based on probabilistic guarantees or empirical validation. Let us see how the polyhedral combinatorics perspective helps in each of these directions.

15.7.1 LP Relaxation and Randomized Rounding

Consider the well-known problem of finding the *smallest weight vertex cover* in a graph. We are given a graph $G(V, E)$ and a nonnegative weight $\mathbf{w}(v)$ for each vertex $v \in V$. We want to find the smallest total weight subset of vertices S such that each edge of G has at least one end in S. (This problem is known to be MAXSNP-hard.) An integer programming formulation of this problem is given by

$$\min \left\{ \sum_{v \in V} \mathbf{w}(v)\mathbf{x}(v) : \mathbf{x}(u) + \mathbf{x}(v) \geq 1, \ \forall (u,v) \in E, \ \mathbf{x}(v) \in \{0,1\} \ \forall v \in V \right\}$$

To obtain the linear programming relaxation we substitute the $\mathbf{x}(v) \in \{0,1\}$ constraint with $\mathbf{x}(v) \geq 0$ for each $v \in V$. Let $\mathbf{x}^*$ denote an optimal solution to this relaxation. Now let us round the fractional parts of $\mathbf{x}^*$ in the usual way, that is, values of 0.5 and up are rounded to 1 and smaller values down to 0. Let $\hat{\mathbf{x}}$ be the 0–1 solution obtained. First note that $\hat{\mathbf{x}}(v) \leq 2\mathbf{x}^*(v)$ for each $v \in V$. Also, for each $(u,v) \in E$, since $\mathbf{x}^*(u) + \mathbf{x}^*(v) \geq 1$, at least one of $\hat{\mathbf{x}}(u)$ and $\hat{\mathbf{x}}(v)$ must be set to 1. Hence $\hat{\mathbf{x}}$ is the incidence vector of a vertex cover of G whose total weight is within twice the total weight of the linear programming relaxation (which is a lower bound on the weight of the optimal vertex cover). Thus, we have a 2-approximate algorithm for this problem, which solves a linear programming relaxation and uses rounding to obtain a feasible solution.

The deterministic rounding of the fractional solution worked quite well for the vertex cover problem. One gets a lot more power from this approach by adding in randomization to the rounding step. Raghavan and Thompson [1987] proposed the following obvious randomized rounding scheme. Given a $0 - 1$ integer program, solve its linear programming relaxation to obtain an optimal $\mathbf{x}^*$. Treat the $\mathbf{x}_j^* \in [0,1]$ as probabilities, i.e., let probability $\{\mathbf{x}_j = 1\} = \mathbf{x}_j^*$, to randomly round the fractional solution to a $0 - 1$ solution. Using Chernoff bounds on the tails of the binomial distribution, Raghavan and Thompson [1987] were able to show, for specific problems, that with high probability, this scheme produces integer solutions which are close to optimal. In certain problems, this rounding method may not always produce a feasible solution. In such cases, the expected values have to be computed as conditioned on feasible solutions produced by rounding. More complex (nonlinear) randomized rounding schemes have been recently studied and have been found to be extremely effective. We will see an example of nonlinear rounding in the context of semidefinite relaxations of the max-cut problem in the following.

15.7.2 Primal–Dual Approximation

The linear programming relaxation of the vertex cover problem, as we saw previously, is given by

$$(P_{VC}) \quad \min \left\{ \sum_{v \in V} \mathbf{w}(v)\mathbf{x}(v) : \mathbf{x}(u) + \mathbf{x}(v) \geq 1, \ \forall (u,v) \in E, \ \mathbf{x}(v) \geq 0 \ \forall v \in V \right\}$$

and its dual is

$$(D_{VC}) \quad \max \left\{ \sum_{(u,v) \in E} \mathbf{y}(u,v) : \sum_{(u,v) \in E} \mathbf{y}(u,v) \leq \mathbf{w}(v), \ \forall v \in V, \ \mathbf{y}(u,v) \geq 0 \ \forall (u,v) \in E \right\}$$

The primal–dual approximation approach would first obtain an optimal solution $\mathbf{y}^*$ to the dual problem (D_{VC}). Let $\hat{V} \subseteq V$ denote the set of vertices for which the dual constraints are tight, i.e.,

$$\hat{V} = \left\{ v \in V : \sum_{(u,v) \in E} \mathbf{y}^*(u,v) = \mathbf{w}(v) \right\}$$

The approximate vertex cover is taken to be $\hat{V}$. It follows from complementary slackness that $\hat{V}$ is a vertex cover. Using the fact that each edge (u,v) is in the star of at most two vertices (u and v), it also follows that $\hat{V}$ is a 2-approximate solution to the minimum weight vertex cover problem.

In general, the primal–dual approximation strategy is to use a dual solution to the linear programming relaxation, along with complementary slackness conditions as a heuristic to generate an integer (primal) feasible solution, which for many problems turns out to be a good approximation of the optimal solution to the original integer program.

Remark 15.10   A recent survey of primal-dual approximation algorithms and some related interesting results are presented in Williamson [2000].

15.7.3   Semidefinite Relaxation and Rounding

The idea of using semidefinite programming to solve combinatorial optimization problems appears to have originated in the work of Lovász [1979] on the Shannon capacity of graphs. Grötschel et al. [1988] later used the same technique to compute a maximum stable set of vertices in perfect graphs via the ellipsoid method. Lovasz and Schrijver [1991] resurrected the technique to present a fascinating theory of semidefinite relaxations for general 0–1 integer linear programs. We will not present the full-blown theory here but instead will present a lovely application of this methodology to the problem of finding the maximum weight cut of a graph. This application of semidefinite relaxation for approximating MAXCUT is due to Goemans and Williamson [1994].

We begin with a quadratic Boolean formulation of MAXCUT

$$\max \left\{ \frac{1}{2} \sum_{(u,v) \in E} \mathbf{w}(u,v)(1 - \mathbf{x}(u)\mathbf{x}(v)) : \mathbf{x}(v) \in \{-1, 1\} \; \forall \, v \in V \right\}$$

where $G(V, E)$ is the graph and $\mathbf{w}(u,v)$ is the nonnegative weight on edge (u,v). Any $\{-1, 1\}$ vector of $\mathbf{x}$ values provides a bipartition of the vertex set of G. The expression $(1 - \mathbf{x}(u)\mathbf{x}(v))$ evaluates to 0 if u and v are on the same side of the bipartition and to 2 otherwise. Thus, the optimization problem does indeed represent exactly the MAXCUT problem.

Next we reformulate the problem in the following way:

- We square the number of variables by substituting each $\mathbf{x}(v)$ with $\boldsymbol{\chi}(v)$ an n-vector of variables (where n is the number of vertices of the graph).
- The quadratic term $\mathbf{x}(u)\mathbf{x}(v)$ is replaced by $\boldsymbol{\chi}(u) \cdot \boldsymbol{\chi}(v)$, which is the inner product of the vectors.
- Instead of the $\{-1, 1\}$ restriction on the $\mathbf{x}(v)$, we use the Euclidean normalization $\|\boldsymbol{\chi}(v)\| = 1$ on the $\boldsymbol{\chi}(v)$.

Thus, we now have a problem

$$\max \left\{ \frac{1}{2} \sum_{(u,v) \in E} \mathbf{w}(u,v)(1 - \boldsymbol{\chi}(u) \cdot \boldsymbol{\chi}(v)) : \|\boldsymbol{\chi}(v)\| = 1 \; \forall \, v \in V \right\}$$

which is a relaxation of the MAXCUT problem (note that if we force only the first component of the $\boldsymbol{\chi}(v)$ to have nonzero value, we would just have the old formulation as a special case).

The final step is in noting that this reformulation is nothing but a semidefinite program. To see this we introduce $n \times n$ Gram matrix Y of the unit vectors $\chi(v)$. So $Y = X^T X$ where $X = (\chi(v) : v \in V)$. Thus, the relaxation of MAXCUT can now be stated as a semidefinite program,

$$\max\left\{\frac{1}{2}\sum_{(u,v)\in E}\mathbf{w}(u,v)(1 - Y_{(u,v)}) : Y \succeq 0, \; Y_{(u,v)} = 1 \; \forall \; v \in V\right\}$$

Recall from Section 15.2 that we are able to solve such semidefinite programs to an additive error ϵ in time polynomial in the input length and $\log 1/\epsilon$ by using either the ellipsoid method or interior point methods.

Let χ^* denote the near optimal solution to the semidefinite programming relaxation of MAXCUT (convince yourself that χ^* can be reconstructed from an optimal Y^* solution). Now we encounter the final trick of Goemans and Williamson. The approximate maximum weight cut is extracted from χ^* by randomized rounding. We simply pick a random hyperplane H passing through the origin. All of the $v \in V$ lying to one side of H get assigned to one side of the cut and the rest to the other. Goemans and Williamson observed the following inequality.

Lemma 15.1 *For χ_1 and χ_2, two random n-vectors of unit norm, let $\mathbf{x}(1)$ and $\mathbf{x}(2)$ be ± 1 values with opposing signs if H separates the two vectors and with same signs otherwise. Then $\bar{E}(1 - \chi_1 \cdot \chi_2) \leq 1.1393 \cdot \bar{E}(1 - \mathbf{x}(1)\mathbf{x}(2))$ where $\bar{E}$ denotes the expected value.*

By linearity of expectation, the lemma implies that the expected value of the cut produced by the rounding is at least 0.878 times the expected value of the semidefinite program. Using standard conditional probability techniques for derandomizing, Goemans and Williamson show that a deterministic polynomial-time approximation algorithm with the same margin of approximation can be realized. Hence we have a cut with value at least 0.878 of the maximum cut value.

Remark 15.11 For semidefinite relaxations of mixed integer programs in which the integer variables are restricted to be 0 or 1, Iyengar and Cezik [2002] develop methods for generating Gomory–Chavatal and disjunctive cutting planes that extends the work of Balas et al. [1993]. Ye [2000] shows that strengthened semidefinite relaxations and mixed rounding methods achieve superior performance guarantee for some discrete optimization problems. A recent survey of semidefinite programming and applications is in Wolkowicz et al. [2000].

15.7.4 Neighborhood Search

A combinatorial optimization problem may be written succinctly as

$$\min\{f(x) : x \in X\}$$

The traditional neighborhood method starts at a feasible point x_0 (in X), and iteratively proceeds to a neighborhood point that is better in terms of the objective function f until a specified termination condition is attained. While the concept of neighborhood $N(x)$ of a point x is well defined in calculus, the specification of $N(x)$ is itself a matter of consideration in combinatorial optimization. For instance, for the traveling salesman problem the so-called *k-opt heuristic* (see Lin and Kernighan [1973]) is a neighborhood search method which for a given tour considers "neighborhood tours" in which k variables (edges) in the given tour are replaced by k other variables such that a tour is maintained. This search technique has proved to be effective though it is quite complicated to implement when k is larger than 3.

A neighborhood search method leads to a local optimum in terms of the neighborhood chosen. Of course, the chosen neighborhood may be large enough to ensure a global optimum but such a procedure is typically not practical in terms of searching the neighborhood for a better solution. Recently Orlin [2000] has presented very large-scale neighborhood search algorithms in which the neighborhood is searched using network flow or dynamic programming methods. Another method advocated by Orlin [2000] is to define the neighborhood in such a manner that the search process becomes a polynomially solvable special case of a hard combinatorial problem.

To avoid getting trapped at a local optimum solution, different strategies such as tabu search (see, for instance, Glover and Laguna [1997]), simulated annealing (see, for instance, Aarts and Korst [1989]), genetic algorithms (see, for instance, Whitley [1993]), and neural networks have been developed. Essentially these methods allow for the possibility of sometimes moving to an inferior solution in terms of the objective function or even to an infeasible solution. While there is no guarantee of obtaining a global optimal solution, computational experience in solving several difficult combinatorial optimization problems has been very encouraging. However, a drawback of these methods is that performance guarantees are not typically available.

15.7.5 Lagrangian Relaxation

We end our discussion of approximation methods for combinatorial optimization with the description of Lagrangian relaxation. This approach has been widely used for about two decades now in many practical applications. Lagrangian relaxation, like linear programming relaxation, provides bounds on the combinatorial optimization problem being relaxed (i.e., lower bounds for minimization problems).

Lagrangian relaxation has been so successful because of a couple of distinctive features. As was noted earlier, in many hard combinatorial optimization problems, we usually have embedded some nice tractable subproblems which have efficient algorithms. Lagrangian relaxation gives us a framework to *jerry-rig* an approximation scheme that uses these efficient algorithms for the subproblems as subroutines. A second observation is that it has been empirically observed that well-chosen Lagrangian relaxation strategies usually provide very tight bounds on the optimal objective value of integer programs. This is often used to great advantage within partial enumeration schemes to get very effective pruning tests for the search trees.

Practitioners also have found considerable success with designing heuristics for combinatorial optimization by starting with solutions from Lagrangian relaxations and constructing good feasible solutions via so-called *dual ascent* strategies. This may be thought of as the analogue of rounding strategies for linear programming relaxations (but with no performance guarantees, other than empirical ones).

Consider a representation of our combinatorial optimization problem in the form

$$(P) \quad z = \min\{\mathbf{cx} : A\mathbf{x} \geq \mathbf{b}, \ \mathbf{x} \in X \subseteq \Re^n\}$$

Implicit in this representation is the assumption that the explicit constraints ($A\mathbf{x} \geq \mathbf{b}$) are *small* in number. For convenience, let us also assume that that X can be replaced by a finite list $\{\mathbf{x}^1, \mathbf{x}^2, \ldots, \mathbf{x}^T\}$.

The following definitions are with respect to (P):

- Lagrangian. $L(\mathbf{u}, \mathbf{x}) = \mathbf{u}(A\mathbf{x} - \mathbf{b}) + \mathbf{cx}$ where $\mathbf{u}$ are the Lagrange multipliers.
- Lagrangian-dual function. $\mathcal{L}(\mathbf{u}) = \min_{\mathbf{x} \in X}\{L(\mathbf{u}, \mathbf{x})\}$.
- Lagrangian-dual problem. (D) $d = \max_{\mathbf{u} \geq 0}\{\mathcal{L}(\mathbf{u})\}$.

It is easily shown that (D) satisfies a weak duality relationship with respect to (P), i.e., $z \geq d$. The discreteness of X also implies that $\mathcal{L}(\mathbf{u})$ is a piecewise linear and concave function (see Shapiro [1979]). In practice, the constraints X are chosen such that the evaluation of the Lagrangian dual function $\mathcal{L}(\mathbf{u})$ is easily made (i.e., the *Lagrangian subproblem* $\min_{\mathbf{x} \in X}\{L(\mathbf{u}, \mathbf{x})\}$ is easily solved for a fixed value of $\mathbf{u}$).

Example 15.3

Traveling salesman problem (TSP). For an undirected graph G, with costs on each edge, the TSP is to find a minimum cost set H of edges of G such that it forms a Hamiltonian cycle of the graph. H is a Hamiltonian cycle of G if it is a simple cycle that spans all the vertices of G. Alternatively, H must satisfy:

(1) exactly two edges of H are adjacent to each node, and (2) H forms a connected, spanning subgraph of G.

Held and Karp [1970] used these observations to formulate a Lagrangian relaxation approach for TSP that relaxes the degree constraints (1). Notice that the resulting subproblems are minimum spanning tree problems which can be easily solved.

The most commonly used general method of finding the optimal multipliers in Lagrangian relaxation is subgradient optimization (cf. Held et al. [1974]). Subgradient optimization is the non differentiable counterpart of steepest descent methods. Given a dual vector $\mathbf{u}^k$, the iterative rule for creating a sequence of solutions is given by:

$$\mathbf{u}^{k+1} = \mathbf{u}^k + t_k \gamma(\mathbf{u}^k)$$

where t_k is an appropriately chosen step size, and $\gamma(\mathbf{u}^k)$ is a subgradient of the dual function $\mathcal{L}$ at $\mathbf{u}^k$. Such a subgradient is easily generated by

$$\gamma(\mathbf{u}^k) = A\mathbf{x}^k - \mathbf{b}$$

where $\mathbf{x}^k$ is a maximizer of $\min_{\mathbf{x} \in X}\{L(\mathbf{u}^k, \mathbf{x})\}$.

Subgradient optimization has proven effective in practice for a variety of problems. It is possible to choose the step sizes $\{t_k\}$ to guarantee convergence to the optimal solution. Unfortunately, the method is not finite, in that the optimal solution is attained only in the limit. Further, it is not a pure descent method. In practice, the method is heuristically terminated and the best solution in the generated sequence is recorded. In the context of nondifferentiable optimization, the ellipsoid algorithm was devised by Shor [1970] to overcome precisely some of these difficulties with the subgradient method.

The ellipsoid algorithm may be viewed as a scaled subgradient method in much the same way as variable metric methods may be viewed as scaled steepest descent methods (cf. Akgül [1984]). And if we use the ellipsoid method to solve the Lagrangian dual problem, we obtain the following as a consequence of the polynomial-time equivalence of optimization and separation.

Theorem 15.8 *The Lagrangian dual problem is polynomial-time solvable if and only if the Lagrangian subproblem is. Consequently, the Lagrangian dual problem is $\mathcal{NP}$-hard if and only if the Lagrangian subproblem is.*

The theorem suggests that, in practice, if we set up the Lagrangian relaxation so that the subproblem is tractable, then the search for optimal Lagrangian multipliers is also tractable.

15.8 Prospects in Integer Programming

The current emphasis in software design for integer programming is in the development of shells (for example, CPLEX 6.5 [1999], MINTO (Savelsbergh et al. [1994]), and OSL [1991]) wherein a general purpose solver like branch and cut is the driving engine. Problem-specific codes for generation of cuts and facets can be easily interfaced with the engine. Recent computational results (Bixby et al. [2001]) suggests that it is now possible to solve relatively large size integer programming problems using general purpose codes. We believe that this trend will eventually lead to the creation of general purpose problem solving languages for combinatorial optimization akin to AMPL (Fourer et al. [1993]) for linear and nonlinear programming.

A promising line of research is the development of an empirical science of algorithms for combinatorial optimization (Hooker [1993]). Computational testing has always been an important aspect of research on the efficiency of algorithms for integer programming. However, the standards of test designs and empirical analysis have not been uniformly applied. We believe that there will be important strides in this aspect of integer programming and more generally of algorithms. J. N. Hooker argues that it may be useful to stop looking at algorithmics as purely a deductive science and start looking for advances through repeated application of "hypothesize and test" paradigms, i.e., through empirical science. Hooker and Vinay [1995] developed a science of selection rules for the Davis–Putnam–Loveland scheme of theorem proving in propositional logic by applying the empirical approach.

The integration of logic-based methodologies and mathematical programming approaches is evidenced in the recent emergence of constraint logic programming (CLP) systems (Saraswat and Van Hentenryck [1995], Borning [1994]) and logico-mathematical programming (Jeroslow [1989], Chandru and Hooker [1991]). In CLP, we see a structure of Prolog-like programming language in which some of the predicates are constraint predicates whose truth values are determined by the solvability of constraints in a wide range of algebraic and combinatorial settings. The solution scheme is simply a clever orchestration of constraint solvers in these various domains and the role of conductor is played by resolution. The clean semantics of logic programming is preserved in CLP. A bonus is that the output language is symbolic and expressive. An orthogonal approach to CLP is to use constraint methods to solve inference problems in logic. Imbeddings of logics in mixed integer programming sets were proposed by Williams [1987] and Jeroslow [1989]. Efficient algorithms have been developed for inference algorithms in many types and fragments of logic, ranging from Boolean to predicate to belief logics (Chandru and Hooker [1999]).

A persistent theme in the integer programming approach to combinatorial optimization, as we have seen, is that the representation (formulation) of the problem deeply affects the efficacy of the solution methodology. A proper choice of formulation can therefore make the difference between a successful solution of an optimization problem and the more common perception that the problem is insoluble and one must be satisfied with the best that heuristics can provide. Formulation of integer programs has been treated more as an art form than a science by the mathematical programming community. (See Jeroslow [1989] for a refreshingly different perspective on representation theories for mixed integer programming.) We believe that progress in representation theory can have an important influence on the future of integer programming as a broad-based problem solving methodology in combinatorial optimization.

Defining Terms

Column generation: A scheme for solving linear programs with a huge number of columns.

Cutting plane: A valid inequality for an integer polyhedron that separates the polyhedron from a given point outside it.

Extreme point: A corner point of a polyhedron.

Fathoming: Pruning a search tree.

Integer polyhedron: A polyhedron, all of whose extreme points are integer valued.

Linear program: Optimization of a linear function subject to linear equality and inequality constraints.

Mixed integer linear program: A linear program with the added constraint that some of the decision variables are integer valued.

Packing and covering: Given a finite collection of subsets of a finite ground set, to find an optimal subcollection that is pairwise disjoint (packing) or whose union covers the ground set (covering).

Polyhedron: The set of solutions to a finite system of linear inequalities on real-valued variables. Equivalently, the intersection of a finite number of linear half-spaces in $\Re^n$.

ρ-Approximation: An approximation method that delivers a feasible solution with an objective value within a factor ρ of the optimal value of a combinatorial optimization problem.

Relaxation: An enlargement of the feasible region of an optimization problem. Typically, the relaxation is considerably easier to solve than the original optimization problem.

References

Aarts, E.H.L. and Korst, J.H. 1989. *Simulated annealing and Boltzmann machines: A stochastic approach to Combinatorial Optimization and neural computing*, Wiley, New York.

Ahuja, R. K., Magnati, T. L., and Orlin, J. B. 1993. *Network Flows: Theory, Algorithms and Applications*. Prentice–Hall, Englewood Cliffs, NJ.

Akgül, M. 1984. *Topics in Relaxation and Ellipsoidal Methods, Research Notes in Mathematics,* Pitman.

Alizadeh, F. 1995. Interior point methods in semidefinite programming with applications to combinatorial optimization. *SIAM J. Optimization* 5(1):13–51.

Applegate, D., Bixby, R. E., Chvátal, V., and Cook, W. 1994. Finding cuts in large TSP's. *Tech. Rep.* Aug.

Arora, S., Lund, C., Motwani, R., Sudan, M., and Szegedy, M. 1992. Proof verification and hardness of approximation problems. In *Proc. 33rd IEEE Symp. Found. Comput. Sci.* pp. 14–23.

Balas, E., Ceria, S. and Cornuejols, G. 1993. A lift and project cutting plane algorithm for mixed 0-1 programs. *Mathematical Programming* 58: 295–324.

Barahona, F., Jünger, M., and Reinelt, G. 1989. Experiments in quadratic $0 - 1$ programming. *Math. Programming* 44:127–137.

Benders, J. F. 1962. Partitioning procedures for solving mixed-variables programming problems. *Numerische Mathematik* 4:238–252.

Berge, C. 1961. Farbung von Graphen deren samtliche bzw. deren ungerade Kreise starr sind (Zusammenfassung). Wissenschaftliche Zeitschrift, Martin Luther Universitat Halle-Wittenberg, Mathematisch-Naturwiseenschaftliche Reihe. pp. 114–115.

Berge, C. 1970. Sur certains hypergraphes generalisant les graphes bipartites. In *Combinatorial Theory and its Applications I.* P. Erdos, A. Renyi, and V. Sos eds., Colloq. Math. Soc. Janos Bolyai, 4, pp. 119–133. North Holland, Amsterdam.

Berge, C. 1972. Balanced matrices. *Math. Programming* 2:19–31.

Berge, C. and Las Vergnas, M. 1970. Sur un theoreme du type Konig pour hypergraphes. pp. 31–40. In *Int. Conf. Combinatorial Math.,* Ann. New York Acad. Sci. 175.

Bixby, R. E. 1994. Progress in linear programming. *ORSA J. Comput.* 6(1):15–22.

Bixby, R.E., Fenelon, M., Gu, Z., Rothberg, E., and Wunderling, R. 2001. M.I.P: Theory and practice: Closing the gap. Paper presented at Padberg-Festschrift, Berlin.

Bland, R., Goldfarb, D., and Todd, M. J. 1981. The ellipsoid method: a survey. *Operations Res.* 29:1039–1091.

Borning, A., ed. 1994. *Principles and Practice of Constraint Programming,* LNCS Vol. 874, Springer–Verlag.

Camion, P. 1965. Characterization of totally unimodular matrices. *Proc. Am. Math. Soc.* 16:1068–1073.

Cannon, T. L. and Hoffman, K. L. 1990. Large-scale zero-one linear programming on distributed workstations. *Ann. Operations Res.* 22:181–217.

Černikov, R. N. 1961. The solution of linear programming problems by elimination of unknowns. *Doklady Akademii Nauk* 139:1314–1317 (translation in 1961. *Soviet Mathematics Doklady* 2:1099–1103).

Chandru, V. and Hooker, J. N. 1991. Extended Horn sets in propositional logic. *JACM* 38:205–221.

Chandru, V. and Hooker, J. N. 1999. *Optimization Methods for Logical Inference,* Wiley Interscience.

Chopra, S., Gorres, E. R., and Rao, M. R. 1992. Solving Steiner tree problems by branch and cut. *ORSA J. Comput.* 3:149–156.

Chvátal, V. 1973. Edmonds polytopes and a hierarchy of combinatorial problems. *Discrete Math.* 4:305–337.

Chvátal, V. 1975. On certain polytopes associated with graphs. *J. Combinatorial Theory B* 18:138–154.

Conforti, M. and Cornuejols, G. 1992a. Balanced 0, ±1 matrices, bicoloring and total dual integrality. Preprint, Carnegie Mellon University.

Conforti, M. and Cornuejols, G. 1992b. A class of logical inference problems solvable by linear programming. *FOCS* 33:670–675.

Conforti, M., Cornuejols, G., and De Francesco, C. 1993. Perfect 0, ±1 matrices. Preprint, Carnegie Mellon University.

Conforti, M., Cornuejols, G., Kapoor, A., and Vuskovic, K. 1994. Balanced 0, ±1 matrices. Pts. I–II. Preprints, Carnegie Mellon University.

Conforti, M., Cornuejols, G., Kapoor, A. Vuskovic, K., and Rao, M. R. 1994. Balanced matrices. In *Mathematical Programming, State of the Art 1994.* J. R. Birge and K. G. Murty, Eds., University of Michigan.

Conforti, M., Cornuejols, G., and Rao, M. R. 1999. Decomposition of balanced 0, 1 matrices. *Journal of Combinatorial Theory* Series B 77:292–406.

Conforti, M. and Rao, M. R. 1993. Testing balancedness and perfection of linear matrices. *Math. Programming* 61:1–18.

Cook, W., Lovász, L., and Seymour, P., eds. 1995. *Combinatorial Optimization: Papers from the DIMACS Special Year*. Series in discrete mathematics and theoretical computer science, Vol. 20, AMS.

Cornuejols, G. and Novick, B. 1994. Ideal 0, 1 matrices. *J. Combinatorial Theory* 60:145–157.

CPLEX 6.5. 1999. Using the CPLEX Callable Library and CPLEX Mixed Integer Library, Ilog Inc.

Crowder, H., Johnson, E. L., and Padberg, M. W. 1983. Solving large scale 0–1 linear programming problems. *Operations Res.* 31:803–832.

Cunningham, W. H. 1984. Testing membership in matroid polyhedra. *J. Combinatorial Theory* 36B:161–188.

Dantzig, G. B. and Wolfe, P. 1961. The decomposition algorithm for linear programming. *Econometrica* 29:767–778.

Ecker, J. G. and Kupferschmid, M. 1983. An ellipsoid algorithm for nonlinear programming. *Math. Programming* 27.

Edmonds, J. 1965. Maximum matching and a polyhedron with 0–1 vertices. *J. Res. Nat. Bur. Stand.* 69B:125–130.

Edmonds, J. 1970. Submodular functions, matroids and certain polyhedra. In *Combinatorial Structures and their Applications*, R. Guy, Ed., pp. 69–87. Gordon Breach, New York.

Edmonds, J. 1971. Matroids and the greedy algorithm. *Math. Programming* 127–136.

Edmonds, J. 1979. Matroid intersection. *Ann. Discrete Math.* 4:39–49.

Edmonds, J. and Giles, R. 1977. A min-max relation for submodular functions on graphs. *Ann. Discrete Math.* 1:185–204.

Edmonds, J. and Johnson, E. L. 1970. Matching well solved class of integer linear polygons. In *Combinatorial Structure and Their Applications*. R. Guy, Ed., Gordon and Breach, New York.

Fonlupt, J. and Zemirline, A. 1981. A polynomial recognition algorithm for $K_4 \backslash e$-free perfect graphs. *Res. Rep.*, University of Grenoble.

Fourer, R., Gay, D. M., and Kernighian, B. W. 1993. *AMPL: A Modeling Language for Mathematical Programming*, Scientific Press.

Fourier, L. B. J. 1827. In: Analyse des travaux de l'Academie Royale des Sciences, pendant l'annee 1824, Partie mathematique, *Histoire de l'Academie Royale des Sciences de l'Institut de France 7* (1827) xlvii–lv. (Partial English translation Kohler, D. A. 1973. *Translation of a Report by Fourier on his Work on Linear Inequalities. Opsearch* 10:38–42.)

Frank, A. 1981. A weighted matroid intersection theorem. *J. Algorithms* 2:328–336.

Fulkerson, D. R. 1970. The perfect graph conjecture and the pluperfect graph theorem. pp. 171–175. In *Proc. 2nd Chapel Hill Conf. Combinatorial Math. Appl.* R. C. Bose et al., Eds.

Fulkerson, D. R., Hoffman, A., and Oppenheim, R. 1974. On balanced matrices. *Math. Programming Study* 1:120–132.

Gass S. and Saaty, T. 1955. The computational algorithm for the parametric objective function. *Naval Research Logistics Quarterly* 2:39–45.

Gilmore, P. and Gomory, R. E. 1963. A linear programming approach to the cutting stock problem. Pt. I. *Operations Res.* 9:849–854; Pt. II. *Operations Res.* 11:863–887.

Glover, F. and Laguna, M. 1997. *Tabu Search*, Kluwer Academic Publishers.

Goemans, M. X. and Williamson, D. P. 1994. .878 approximation algorithms MAX CUT and MAX 2SAT. pp. 422–431. In *Proc. ACM STOC*.

Gomory, R. E. 1958. Outline of an algorithm for integer solutions to linear programs. *Bull. Am. Math. Soc.* 64:275–278.

Gomory, R. E. and Hu, T. C. 1961. Multi-terminal network flows. *SIAM J. Appl. Math.* 9:551–556.

Grötschel, M., Lovasz, L., and Schrijver, A. 1982. The ellipsoid method and its consequences in combinatorial optimization. *Combinatorica* 1:169–197.

Grötschel, M., Lovász, L., and Schrijver, A. 1988. *Geometric Algorithms and Combinatorial Optimization*. Springer–Verlag.

Hačijan, L. G. 1979. A polynomial algorithm in linear programming. *Soviet Math. Dokl.* 20:191–194.

Held, M. and Karp, R. M. 1970. The travelling-salesman problem and minimum spanning trees. *Operations Res.* 18:1138–1162, Pt. II. 1971. *Math. Programming* 1:6–25.

Held, M., Wolfe, P., and Crowder, H. P. 1974. Validation of subgradient optimization. *Math. Programming* 6:62–88.

Hoffman, A. J. and Kruskal, J. K. 1956. Integral boundary points of convex polyhedra. In *Linear Inequalities and Related Systems*, H. W. Kuhn and A. W. Tucker, Eds., pp. 223–246. Princeton University Press, Princeton, NJ.

Hooker, J. N. 1988. Resolution vs cutting plane solution of inference problems: some computational experience. *Operations Res. Lett.* 7:1–7.

Hooker, J. N. 1992. Resolution and the integrality of satisfiability polytopes. Preprint, GSIA, Carnegie Mellon University.

Hooker, J. N. 1993. Towards and empirical science of algorithms. *Operations Res.* 42:201–212.

Hooker, J. N. and Vinay, V. 1995. Branching rules for satisfiability. In *Automated Reasoning* 15:359–383.

Huynh, T., Lassez C., and Lassez, J.-L. 1992. Practical issues on the projection of polyhedral sets. *Ann. Math. Artif. Intell.* 6:295–316.

IBM. 1991. *Optimization Subroutine Library—Guide and Reference (Release 2)*, 3rd ed.

Iyengar, G. and Cezik, M. T. 2002. Cutting planes for mixed 0-1 semidefinite programs. *Proceedings of the VIII IPCO conference.*

Jeroslow, R. E. 1987. Representability in mixed integer programming, I: characterization results. *Discrete Appl. Math.* 17:223–243.

Jeroslow, R. E. and Lowe, J. K. 1984. Modeling with integer variables. *Math. Programming Stud.* 22:167–184.

Jeroslow, R. G. 1989. *Logic-Based Decision Support: Mixed Integer Model Formulation.* Ann. discrete mathematics, Vol. 40, North–Holland.

Jünger, M., Reinelt, G., and Thienel, S. 1995. Practical problem solving with cutting plane algorithms. In *Combinatorial Optimization: Papers from the DIMACS Special Year.* Series in discrete mathematics and theoritical computer science, Vol. 20, pp. 111–152. AMS.

Karmarkar, N. K. 1984. A new polynomial-time algorithm for linear programming. *Combinatorica* 4:373–395.

Karmarkar, N. K. 1990. An interior-point approach to NP-complete problems—Part I. In *Contemporary Mathematics*, Vol. 114, pp. 297–308.

Karp, R. M. and Papadimitriou, C. H. 1982. On linear characterizations of combinatorial optimization problems. *SIAM J. Comput.* 11:620–632.

Lawler, E. L. 1975. Matroid intersection algorithms. *Math. Programming* 9:31–56.

Lehman, A. 1979. On the width-length inequality, mimeographic notes (1965). *Math. Programming* 17:403–417.

Lin, S. and Kernighan, B.W. 1973. An effective heuristic algorithm for the travelling salesman problem. *Operations Research* 21: 498-516.

Lovasz, L. 1972. Normal hypergraphs and the perfect graph conjecture. *Discrete Math.* 2:253–267.

Lovasz, L. 1979. On the Shannon capacity of a graph. *IEEE Trans. Inf. Theory* 25:1–7.

Lovász, L. 1986. *An Algorithmic Theory of Numbers, Graphs and Convexity*, SIAM Press.

Lovasz, L. and Schrijver, A. 1991. Cones of matrices and setfunctions. *SIAM J. Optimization* 1:166–190.

Lustig, I. J., Marsten, R. E., and Shanno, D. F. 1994. Interior point methods for linear programming: computational state of the art. *ORSA J. Comput.* 6(1):1–14.

Martin, R. K. 1991. Using separation algorithms to generate mixed integer model reformulations. *Operations Res. Lett.* 10:119–128.

McDiarmid, C. J. H. 1975. Rado's theorem for polymatroids. *Proc. Cambridge Philos. Soc.* 78:263–281.

Megiddo, N. 1991. On finding primal- and dual-optimal bases. *ORSA J. Comput.* 3:63–65.

Mehrotra, S. 1992. On the implementation of a primal-dual interior point method. *SIAM J. Optimization* 2(4):575–601.

Nemhauser, G. L. and Wolsey, L. A. 1988. *Integer and Combinatorial Optimization*. Wiley.

Orlin, J. B. 2000. Very large-scale neighborhood search techniques. Featured Lecture at the International Symposium on Mathematical Programming, Atlanta, Georgia.

Padberg, M. W. 1973. On the facial structure of set packing polyhedra. *Math. Programming* 5:199–215.

Padberg, M. W. 1974. Perfect zero-one matrices. *Math. Programming* 6:180–196.

Padberg, M. W. 1993. Lehman's forbidden minor characterization of ideal 0, 1 matrices. *Discrete Math.* 111:409–420.

Padberg, M. W. 1995. *Linear Optimization and Extensions*. Springer–Verlag.

Padberg, M. W. and Rao, M. R. 1981. The Russian method for linear inequalities. Part III, bounded integer programming. Preprint, New York University, New York.

Padberg, M. W. and Rao, M. R. 1982. Odd minimum cut-sets and b-matching. *Math. Operations Res.* 7:67–80.

Padberg, M. W. and Rinaldi, G. 1991. A branch and cut algorithm for the resolution of large scale symmetric travelling salesman problems. *SIAM Rev.* 33:60–100.

Papadimitriou, C. H. and Yannakakis, M. 1991. Optimization, approximation, and complexity classes. *J. Comput. Syst. Sci.* 43:425–440.

Parker, G. and Rardin, R. L. 1988. *Discrete Optimization*. Wiley.

Picard, J. C. and Ratliff, H. D. 1975. Minimum cuts and related problems. *Networks* 5:357–370.

Pulleyblank, W. R. 1989. Polyhedral combinatorics. In *Handbooks in Operations Research and Management Science*. Vol. 1, Optimization, G. L. Nemhauser, A. H. G. Rinooy Kan, and M. J. Todd, eds., pp. 371–446. North–Holland.

Raghavan, P. and Thompson, C. D. 1987. Randomized rounding: a technique for provably good algorithms and algorithmic proofs. *Combinatorica* 7:365–374.

Rhys, J. M. W. 1970. A selection problem of shared fixed costs and network flows. *Manage. Sci.* 17:200–207.

Saraswat, V. and Van Hentenryck, P., eds. 1995. *Principles and Practice of Constraint Programming*, MIT Press, Cambridge, MA.

Savelsbergh, M. W. P., Sigosmondi, G. S., and Nemhauser, G. L. 1994. MINTO, a mixed integer optimizer. *Operations Res. Lett.* 15:47–58.

Schrijver, A. 1986. *Theory of Linear and Integer Programming*. Wiley.

Seymour, P. 1980. Decompositions of regular matroids. *J. Combinatorial Theory* B 28:305–359.

Shapiro, J. F. 1979. A survey of lagrangian techniques for discrete optimization. *Ann. Discrete Math.* 5:113–138.

Shmoys, D. B. 1995. Computing near-optimal solutions to combinatorial optimization problems. In *Combinatorial Optimization: Papers from the DI'ACS special year*. Series in discrete mathematics and theoretical computer science, Vol. 20, pp. 355–398. AMS.

Shor, N. Z. 1970. Convergence rate of the gradient descent method with dilation of the space. *Cybernetics* 6.

Spielman, D. A. and Tang, S.-H. 2001. Smoothed analysis of algorithms: Why the simplex method usually takes polynomial time. *Proceedings of the The Thirty-Third Annual ACM Symposium on Theory of Computing*, 296–305.

Truemper, K. 1992. Alpha-balanced graphs and matrices and GF(3)-representability of matroids. *J. Combinatorial Theory* B 55:302–335.

Weyl, H. 1935. Elemetere Theorie der konvexen polyerer. *Comm. Math. Helv.* Vol. pp. 3–18 (English translation 1950. *Ann. Math. Stud.* 24, Princeton).

Whitley, D. 1993. *Foundations of Genetic Algorithms 2*, Morgan Kaufmann.

Williams, H. P. 1987. Linear and integer programming applied to the propositional calculus. *Int. J. Syst. Res. Inf. Sci.* 2:81–100.

Williamson, D. P. 2000. The primal-dual method for approximation algorithms. *Proceedings of the International Symposium on Mathematical Programming*, Atlanta, Georgia.

Wolkowicz, W., Saigal, R. and Vanderberghe, L. eds. 2000. *Handbook of semidefinite programming*. Kluwer Acad. Publ.

Yannakakis, M. 1988. Expressing combinatorial optimization problems by linear programs. pp. 223–228. In *Proc. ACM Symp. Theory Comput.*

Ye, Y. 2000. Semidefinite programming for discrete optimization: Approximation and Computation. *Proceedings of the International Symposium on Mathematical Programming*, Atlanta, Georgia.

Ziegler, M. 1995. *Convex Polytopes*. Springer–Verlag.

II

Architecture and Organization

Computer architecture is the design and organization of efficient and effective computer hardware at all levels — from the most fundamental aspects of logic and circuit design to the broadest concerns of RISC, parallel, and high-performance computing. Individual chapters cover the design of the CPU, memory systems, buses, disk storage, and computer arithmetic devices. Other chapters treat important subjects such as parallel architectures, the interaction between computers and networks, and the design of computers that tolerate unanticipated interruptions and failures.

16 **Digital Logic**   *Miriam Leeser* **16**-1
Introduction • Overview of Logic • Concept and Realization of a Digital Gate
• Rules and Objectives in Combinational Design • Frequently Used Digital Components
• Sequential Circuits • ASICs and FPGAs — Faster, Cheaper, More Reliable Logic

17 **Digital Computer Architecture**   *David R. Kaeli* **17**-1
Introduction • The Instruction Set • Memory • Addressing • Instruction Execution
• Execution Hazards • Superscalar Design • Very Long Instruction Word
Computers • Summary

18 **Memory Systems**   *Douglas C. Burger, James R. Goodman, and Gurindar S. Sohi* .. **18**-1
Introduction • Memory Hierarchies • Cache Memories • Parallel and Interleaved
Main Memories • Virtual Memory • Research Issues • Summary

19 **Buses**   *Windsor W. Hsu and Jih-Kwon Peir* **19**-1
Introduction • Bus Physics • Bus Arbitration • Bus Protocol • Issues in SMP System
Buses • Putting It All Together — CCL-XMP System Bus • Historical Perspective and
Research Issues

20 **Input/Output Devices and Interaction Techniques**   *Ken Hinckley,*
Robert J. K. Jacob, and Colin Ware **20**-1
Introduction • Interaction Tasks, Techniques, and Devices • The Composition of
Interaction Tasks • Properties of Input Devices • Discussion of Common Pointing
Devices • Feedback and Perception — Action Coupling • Keyboards, Text Entry, and
Command Input • Modalities of Interaction • Displays and Perception • Color Vision
and Color Displays • Luminance, Color Specification, and Color Gamut • Information
Visualization • Scale in Displays • Force and Tactile Displays • Auditory
Displays • Future Directions

21 **Secondary Storage Systems**   *Alexander Thomasian* **21**-1
Introduction • Single Disk Organization and Performance • RAID Disk Arrays
• RAID1 or Mirrored Disks • RAID5 Disk Arrays • Performance Evaluation
Studies • Data Allocation and Storage Management in Storage Networks
• Conclusions and Recent Developments

22 High-Speed Computer Arithmetic   *Earl E. Swartzlander Jr.* **22**-1
   Introduction • Fixed Point Number Systems • Fixed Point Arithmetic
   Algorithms • Floating Point Arithmetic • Conclusion

23 Parallel Architectures   *Michael J. Flynn and Kevin W. Rudd* **23**-1
   Introduction • The Stream Model • SISD • SIMD • MISD • MIMD
   • Network Interconnections • Afterword

24 Architecture and Networks   *Robert S. Roos* **24**-1
   Introduction • Underlying Principles • Best Practices: Physical Layer Examples
   • Best Practices: Data-Link Layer Examples • Best Practices: Network Layer
   Examples • Research Issues and Summary

25 Fault Tolerance   *Edward J. McCluskey and Subhasish Mitra* **25**-1
   Introduction • Failures, Errors, and Faults • Metrics and Evaluation • System Failure
   Response • System Recovery • Repair Techniques • Common-Mode Failures and Design
   Diversity • Fault Injection • Conclusion • Further Reading

16

Digital Logic

16.1 Introduction 16-1
16.2 Overview of Logic 16-2
16.3 Concept and Realization of a Digital Gate 16-2
CMOS Binary Logic Is Low Power • CMOS Switching Model for NOT, NAND, and NOR • Multiple Inputs and Our Basic Primitives • Doing It All with NAND
16.4 Rules and Objectives in Combinational Design 16-7
Boolean Realization: Half Adders, Full Adders, and Logic Minimization • Axioms and Theorems of Boolean Logic • Design, Gate-Count Reduction, and SOP/POS Conversions • Minimizing with Don't Cares • Adder/Subtractor • Representing Negative Binary Numbers
16.5 Frequently Used Digital Components 16-19
Elementary Digital Devices: ENC, DEC, MUX, DEMUX • The Calculator Arithmetic and Logical Unit
16.6 Sequential Circuits 16-23
Concept of a Sequential Device • The Data Flip-Flop and the Register • From DFF to Data Register, Shift Register, and Stack • Datapath for a 4-bit Calculator
16.7 ASICs and FPGAs — Faster, Cheaper, More Reliable Logic 16-30
FPGA Architecture • Higher Levels of Complexity

Miriam Leeser
Northeastern University

16.1 Introduction

This chapter explores combinational and sequential Boolean logic design as well as technologies for implementing efficient, high-speed digital circuits. Some of the most common devices used in computers and general logic circuits are described. Sections 16.2 through 16.4 introduce the fundamental concepts of logic circuits and in particular the rules and theorems upon which *combinational logic*, logic with no internal memory, is based. Section 16.5 describes in detail some frequently used combinational logic components, and shows how they can be combined to build the Arithmetic and Logical Unit (ALU) for a simple calculator. Section 16.6 introduces the subject of *sequential logic*, logic in which feedback and thus internal memory exist. Two of the most important elements of sequential logic design, the *data flip-flop* and the *register*, are introduced. Memory elements are combined with the ALU to complete the design of a simple calculator. The final section of the chapter examines field-programmable gate arrays that now provide fast, economical solutions for implementing large logic designs for solving diverse problems.

1-58488-360-X/$0.00+$1.50
© 2004 by CRC Press, LLC

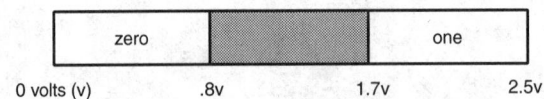

FIGURE 16.1   The states zero and one as defined in 2.5V CMOS logic.

16.2   Overview of Logic

Logic has been a favorite academic subject, certainly since the Middle Ages and arguably since the days of the greatness of Athens. That use of *logic* connoted the pursuit of orderly methods for defining theorems and proving their consistency with certain accepted propositions. In the middle of the 19th century, George Boole put the whole subject on a sound mathematical basis and spread "logic" from the Philosophy Department into Engineering and Mathematics. (Boole's original writings have recently been reissued [Boole 1998].) Specifically, what Boole did was to create an algebra of two-valued (*binary*) variables. Initially designated as *true* or *false*, these two values can represent any parameter that has two clearly defined states. Boolean algebras of more than two values have been explored, but the original binary variable of Boole dominates the design of circuitry for reasons that we will explore. This chapter presents some of the rules and methods of binary Boolean algebra and shows how it is used to design digital hardware to meet specific engineering applications.

One of the first things that must strike a reader who sees *true* or *false* proposed as the two identifiable, discrete states is that we live in a world with many half-truths, with hung juries that end somewhere between *guilty* and *not guilty*, and with "not bad" being a response that does not necessarily mean "good." The answer to the question: "Does a two-state variable really describe anything?" is properly: "Yes and no." This apparent conflict between the continuum that appears to represent life and the underlying reality of atomic physics, which is inherently and absolutely discrete, never quite goes away at any level. We use the words "quantum leap" to describe a transition between two states with no apparent state between them. Yet we know that the leaper spends some time between the two states.

A system that is well adapted to digital (discrete) representation is one that spends little time in a state of ambiguity. All digital systems spend some time in indeterminate states. One very common definition of the two states is made for systems operating between 2.5 volts (V) and ground. It is shown in Figure 16.1. One state, usually called *one*, is defined as any voltage greater than 1.7V. The other state, usually called *zero*, is defined as any voltage less than 0.8V.

The gray area in the middle is *ambiguous*. When an input signal is between 0.8 and 1.7V in a 2.5V CMOS (complementary metal–oxide–semiconductor) digital circuit, you cannot predict the output value. Most of what you will read in this chapter assumes that input variables are clearly assigned to the state *one* or the state *zero*. In real designs, there are always moments when the inputs are ambiguous. A good design is one in which the system never makes decisions based on ambiguous data. Such requirements limit the speed of response of real systems; they must wait for the ambiguities to settle out.

16.3   Concept and Realization of a Digital Gate

A *gate* is the basic building block of digital circuits. A gate is a circuit with one or more inputs and a single output. From a logical perspective in a binary system, any input or output can take on only the values *one* and *zero*. From an analog perspective (a perspective that will vanish for most of this chapter), the gates make transitions through the ambiguous region with great rapidity and quickly achieve an unambiguous state of *oneness* or *zeroness*.

In Boolean algebra, a good place to begin is with three operations: NOT, AND, and OR. These have similar meaning to their meaning in English. Given two input variables, called A and B, and an output variable X, $X = $ NOT A is true when A is false, and false when A is true. $X = A$ AND B is true when both inputs are true, and $X = A$ OR B is true when either A or B is true (or both are true). This is called an

TABLE 16.1   The Boolean Operators of Two Input Variables

Inputs A B	True	False	A	NOT(A)	AND	OR	XOR	NAND	NOR	XNOR
0 0	1	0	0	1	0	0	0	1	1	1
0 1	1	0	0	1	0	1	1	1	0	0
1 0	1	0	1	0	0	1	1	1	0	0
1 1	1	0	1	0	1	1	0	0	0	1

TABLE 16.2   The Boolean Operators Extended to More than Two Inputs

Operation	Input Variables	Operator Symbol	Output = 1 if
NOT	A	$\overline{A}$	$A = 0$
AND	$A, B, \ldots$	$A \cdot B \cdots$	All of the set $[A, B, \ldots]$ are 1.
OR	$A, B, \ldots$	$A + B + \cdots$	Any of the set $[A, B, \ldots]$ are 1.
NAND	$A, B, \ldots$	$\overline{(A \cdot B \cdots)}$	Any of the set $[A, B, \ldots]$ are 0.
NOR	$A, B, \ldots$	$\overline{(A + B + \cdots)}$	All of the set $[A, B, \ldots]$ are 0.
XOR	$A, B, \ldots$	$A \oplus B \oplus \cdots$	The set $[A, B, \ldots]$ contains an odd number of 1's.
XNOR	$A, B, \ldots$	$A \odot B \odot \cdots$	The set $[A, B, \ldots]$ contains an even number of 1's.

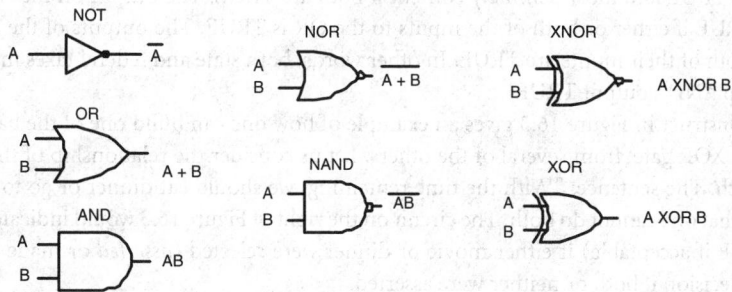

FIGURE 16.2   Commonly used graphical symbols for seven of the gates defined in Table 16.1.

"inclusive or" function because it includes the case where A and B are both true. There is another Boolean operator, *exclusive or*, that is true when either A or B, but not both, is true. In fact, there are 16 Boolean functions of two variables. The more useful functions are shown in truth table form in Table 16.1. These functions can be generalized to more than one variable, as is shown in Table 16.2.

AND, OR, and NOT are sufficient to describe all Boolean logic functions. Why do we need all these other operators?

Logic gates are themselves an abstraction. The actual physical realization of logic gates is with transistors. Most digital designs are implemented in CMOS technology. In CMOS and most other transistor technologies, logic gates are naturally inverting. In other words, it is very natural to build NOT, NAND, and NOR gates, even if it is more natural to think about positive logic: AND and OR. Neither XOR nor XNOR are natural building blocks of CMOS technology. They are included for completeness. AND and OR gates are implemented with NAND and NOR gates as we shall see.

One question that arises is: How many different gates do we really need? The answer is one. We normally admit three or four to our algebra, but one is sufficient. If we pick the right gate, we can build all the others. Later we will explore the minimal set.

There are widely used graphical symbols for these same operations. These are presented in Figure 16.2. The symbol for NOT includes both a buffer (the triangle) and the actual inversion operation (the open

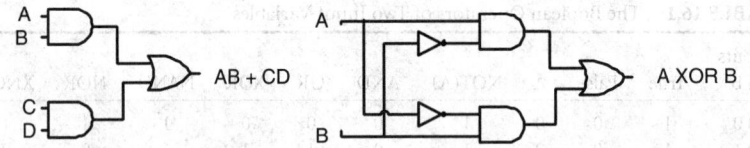

FIGURE 16.3 Two constructs built from the gates in column 1 of Figure 16.2. The first is a common construct in which if either of two paired propositions is TRUE, the output is TRUE. The second is XOR constructed from the more primitive gates, AND, OR, and NOT.

circle). Often, the inversion operation alone is used, as seen in the outputs of NAND, NOR, and XNOR. In writing Boolean operations we use the symbols $\overline{A}$ for NOT A, $A + B$ for A OR B, and $A \cdot B$ for A AND B. $A + B$ is called the sum of A and B and $A \cdot B$ is called the product. The operator for AND is often omitted, and the operation is implied by adjacency, just like in multiplication. To illustrate the use of these symbols and operators and to see how well these definitions fit common speech, Figure 16.3 shows two constructs made from the gates of Figure 16.2. These two examples show how to build the expression $AB + CD$ and how to construct an XOR from the basic gates AND, OR, and NOT.

The first construct of Figure 16.3 would fit the logic of the sentence: "I will be content if my federal and state taxes are lowered (A and B, respectively), or if the money that I send is spent on reasonable things and spent effectively (C and D, respectively)." You would certainly expect the speaker to be content if either pair is TRUE and most definitely content if both are TRUE. The output on the right side of the construct is TRUE if either or both of the inputs to the OR is TRUE. The outputs of the AND gates are TRUE when both of their inputs are TRUE. In other words, both state and federal taxes must be reduced to make the top AND's output TRUE.

The right construct in Figure 16.3 gives an example of how one can build one of the basic logic gates, in this case the XOR gate, from several of the others. Let us consider the relationship of this construct to common speech. The sentence: "With the time remaining, we should eat dinner or go to a movie." The implication is that one cannot do both. The circuit on the right of Figure 16.3 would indicate an acceptable decision (TRUE if acceptable) if either movie or dinner were selected (*asserted* or made TRUE) but an unacceptable decision if both or neither were asserted.

What makes logic gates so very useful is their speed and remarkably low cost. On-chip logic gates today can respond in less than a nanosecond and can cost less than 0.0001 cent each. Furthermore, a rather sophisticated decision-making apparatus can be designed by combining many simple-minded binary decisions. The fact that it takes many gates to build a useful apparatus leads us back directly to one of the reasons why binary logic is so popular. First we will look at the underlying technology of logic gates. Then we will use them to build some useful circuits.

16.3.1 CMOS Binary Logic Is Low Power

A modern microcomputer chip contains more than 10 million logic gates. If all of those gates were generating heat at all times, the chip would melt. Keeping them cool is one of the most critical issues in computer design. Good thermal designs were significant parts of the success of Cray, IBM, Intel, and Sun. One of the principal advantages of CMOS binary logic is that it can be made to expend much less energy to generate the same amount of calculation as other forms of circuitry.

Gates are classified as **active logic** or **saturated logic**, depending on whether they control the current continuously or simply switch it on or off. In active logic, the gate has a considerable voltage across it and conducts current in all of its states. The result is that power is continually being dissipated. In saturated logic, the TRUE–FALSE dichotomy has the gate striving to be perfectly connected to the power bus when the output voltage is high and perfectly connected to the ground bus when the voltage is low. These are zero-dissipation ideals that are not achieved in real gates, but the closer one gets to the ideal, the better the gate. When you start with more than 1 million gates per chip, small reductions in power dissipation make the difference between usable and unusable chips.

Saturated logic is *saturated* because it is driven hard enough to ensure that it is in a minimum-dissipation state. Because it takes some effort to bring such logic out of saturation, it is a little slower than active logic. Active logic, on the other hand, is always dissipative. It is very fast, but it is always getting hot. Although it has often been the choice for the most active circuits in the fastest computers, active logic has never been a major player, and it owns a diminishing role in today's designs. This chapter focuses on today's dominant family of binary, saturated logic, which is CMOS: Complementary Metal Oxide Semiconductor.

16.3.2 CMOS Switching Model for NOT, NAND, and NOR

The metal–oxide–semiconductor (MOS) transistor is the oldest transistor in concept and still the best in one particular aspect: its control electrode — also called a *gate* but in a different meaning of that word from *logic gate* — is a purely capacitive load. Holding it at constant voltage takes no energy whatsoever. These MOS transistors, like most transistors, come in two types. One turns on with a positive voltage; the other turns off with a positive voltage. This pairing allows one to build *complementary* gates, which have the property that they dissipate no energy except when switching. Given the large number of logic gates and the criticality of energy dissipation, zero dissipation in the static state is enormously compelling. It is small wonder that the complementary metal–oxide–semiconductor (CMOS) gate dominates today's digital technology.

Consider how we can construct a set of primitive gates in the CMOS family. The basic element is a pair of switches in series, the NOT gate. This basic building block is shown in Figure 16.4. The switching operation is shown in the two drawings to the right. If the input is low, the upper switch is closed and the lower one is open — complementary operation. This connects the output to the high side. Apart from voltage drops across the switch itself, the output voltage becomes the voltage of the high bus. If the input now goes high, both switches flip and the output is connected, through the resistance of the switch, to the ground bus. High–in, low–out, and vice versa. We have an inverter. Only while the switches are switching is there significant current flowing from one bus to the other. Furthermore, if the loads are other CMOS switches, only while the gates are charging is any current flowing from bus to load. Current flows when charging or discharging a load. Thus, in the static state, these devices dissipate almost no power at all. Once one has the CMOS switch concept, it is easy to show how to build NAND and NOR gates with multiple inputs.

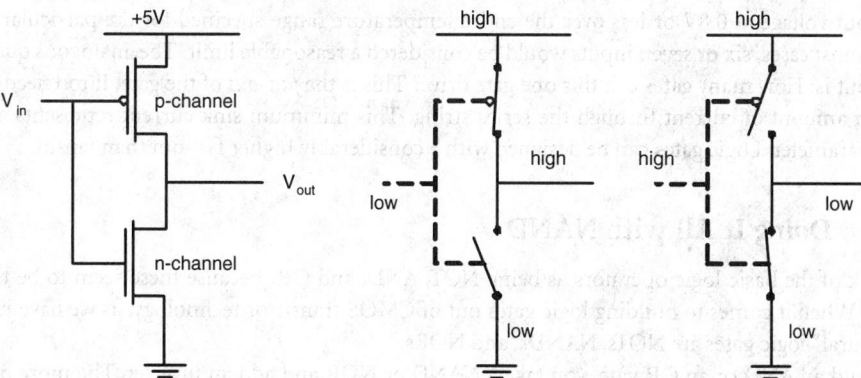

FIGURE 16.4   A CMOS inverter shown as a pair of transistors with voltage and ground and also as pairs of switches with logic levels. The open circle indicates logical negation (NOT).

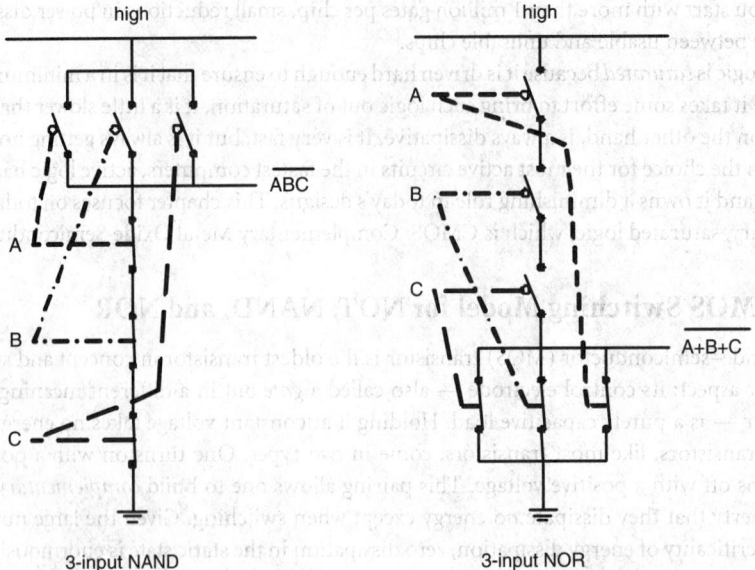

FIGURE 16.5 Three pairs of CMOS switches arranged on the left to execute the three-input NAND function and on the right the three-input NOR. The switches are shown with all the inputs high, putting the output in the low state.

16.3.3 Multiple Inputs and Our Basic Primitives

Let us look at the switching structure of a 3-input NAND and 3-input NOR, just to show how multiple-input gates are created. The basic inverter, or NOT gate of Figure 16.4 is our paradigm; if the lower switch is closed, the upper one is open, and vice versa. To go from NOT to an N-input NAND, make the single lower switch in the NOT a series of N switches, so only one of these need be open to open the circuit. Then change the upper complementary switch in the NOT into N parallel switches. With these, only one switch need be closed to connect the circuit. Such an arrangement with $N = 3$ is shown on the left in Figure 16.5. On the left, if any input is low, the output is high. On the right is the construction for NOR. All three inputs must be low to drive the output high.

An interesting question at this point is: How many inputs can such a circuit support? The answer is called the *fan-in* of the circuit. The fan-in depends mostly on the resistance of each switch in the series string. That series of switches must be able to *sink* a certain amount of current to ground and still hold the output voltage at 0.8V or less over the entire temperature range specified for the particular class of gate. In most cases, six or seven inputs would be considered a reasonable limit. The analogous question at the output is: How many gates can this one gate drive? This is the *fan-out* of the gate. It too needs to sink a certain amount of current through the series string. This minimum sink current represents a central design parameter. Logic gates can be designed with a considerably higher fan-out than fan-in.

16.3.4 Doing It All with NAND

We think of the basic logic operators as being NOT, AND, and OR, because these seem to be the most natural. When it comes to building logic gates out of CMOS transistor technology, as we have just seen, the "natural" logic gates are NOTs, NANDs, and NORs.

To build an AND or an OR gate, you take a NAND or NOR and add an inverter. The more primitive nature of NAND and NOR comes about because transistor switches are inherently inverting. Thus, a single-stage gate will be NAND or NOR; AND and OR gates require an extra stage. If this is the way one were to implement a design with a million gates, a million extra inverters would be required. Each extra stage requires extra area and introduces longer propagation delays. Simplifying logic to eliminate delay

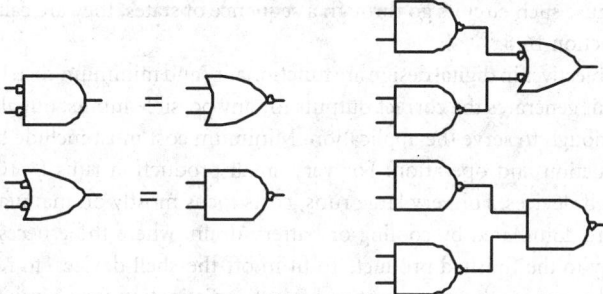

FIGURE 16.6 On the left, the two forms of De Morgan's theorem in logic gates. On the right, the two forms of the circuit on the left of Figure 16.3. In the upper form, we have replaced the lines between the ANDs and OR with two inverters in series. Then, we have used the lower form of De Morgan's theorem to replace the OR and its two inverters with a NAND. The resulting circuit is all-NAND and is simpler to implement than the construction from AND and OR in Figure 16.3.

and unnecessary heat are two of the most important objectives of logic design. Instead of using an inverter after each NAND or NOR gate, most designs use the inverting gates directly. We will see how Boolean logic helps us do this. Consider the declaration: "Fred and Jack will come over this afternoon." This is equivalent to saying: "Fred will stay away or Jack will stay away, NOT." This strange construct in English is an exact formulation of one of two relationships in Boolean logic known as *De Morgan's theorems*. More formally:

$$\overline{A \cdot B} = \overline{A} + \overline{B}$$

$$\overline{A + B} = \overline{A} \cdot \overline{B}$$

In other words, the NAND of A and B is equivalent to the OR of (not A) and (not B). Similarly, the NOR of A and B is equivalent to the AND of (not A) and (not B). These two statements can be represented at the gate level as shown in Figure 16.6.

De Morgan's theorems show that a NAND can be used to implement a NOR if we have inverters. It turns out that a NAND gate is the only gate required. Next we will show that a NOT gate (inverter) can be constructed from a NAND. Once we have shown that NORs and NOTs can be constructed out of NANDs, only NAND gates are required. An AND gate is a NAND followed by a NOT and an OR gate is a NOR followed by a NOT. Thus, all other logic gates can be implemented from NANDs. The same is true of NOR gates; all other logic gates can be implemented from NORs.

Take a NAND gate and connect both inputs to the same input A. The output is the function $\overline{(A \cdot A)}$. Since A AND A is TRUE only if A is TRUE ($AA = A$), we have just constructed our inverter. If we actually wanted an inverter, we would not use a two-input gate where a one-input gate would do. But we could. This exercise shows that the minimal number of logic gates required to implement all Boolean logic functions is one. In reality, we use AND, OR, and NOT when using positive logic, and NAND, NOR, and NOT when using negative logic or thinking about how logic gates are implemented with transistors.

16.4   Rules and Objectives in Combinational Design

Once the concept of a logic gate is established, the next natural question is: What useful devices can you build with them? We will look at a few basic building blocks and show how you can put them together to build a simple calculator.

The components of digital circuits can be divided into two classes. The first class of circuits has outputs that are simply some logical combination of their inputs. Such circuits are called **combinational**. Examples include the gates we have just looked at and those that we will examine in this section and in Section 16.5. The other class of circuits, constructed from combinational gates, but with the addition of internal feedback, have the property of *memory*. Thus, their output is a function not only of their inputs but also of their previous state(s). Because such circuits go through a sequence of states, they are called **sequential**. These will be discussed in Section 16.6.

The two principal objectives in digital design are functionality and minimum cost. Functionality requires not only that the circuit generates the correct outputs for any possible inputs, but also that those outputs be available quickly enough to serve the application. Minimum cost must include both the design effort and the cost of production and operation. For very small production runs ($<10,000$), one wants to "program" off-the-shelf devices. For very large runs, costs focus mostly on manufacture and operation. The *operation* costs are dominated by cooling or battery drain, where these necessary peripherals add weight and complexity to the finished product. To fit in off-the-shelf devices, to reduce delays between input and output, and to reduce the gate count and thus the dissipation for a given functionality, designs must be realized with the smallest number of gates possible. Many design tools have been developed for achieving designs with minimum gate count. In this section and the next, we will develop the basis for such minimization in a way that assures the design achieves logical functionality.

16.4.1 Boolean Realization: Half Adders, Full Adders, and Logic Minimization

One of the basic units central to a calculator or microprocessor is a binary adder. We will consider how an adder is implemented from logic gates. A straightforward way to specify a Boolean logic function is by using a truth table. This table enumerates the output for every possible combinations of inputs. Truth tables were used in Table 16.1 to specify different Boolean functions of two variables. Table 16.3 shows the truth table for a Boolean operation that adds two one bit numbers A and B and produces two outputs: the sum bit S and the carry-out C. Because binary numbers can only have the values 1 or 0, adding two binary numbers each of value 1 will result in there being a carry-out. This operation is called a *half adder*.

To implement the half adder with logic gates, we need to write Boolean logic equations that are equivalent to the truth table. A separate Boolean logic equation is required for each output. The most straightforward way to write an equation from the truth table is to use Sum of Products (SOP) form to specify the outputs as a function of the inputs. An SOP expression is a set of "products" (ANDs) that are "summed" (ORed) together. Note that any Boolean formula can be expressed in SOP or POS (Product of Sums) form.

Let's consider output S. Every line in the truth table that has a 1 value for an output corresponds to a term that is ORed with other terms in SOP form. This term is formed by ANDing together all of the input variables. If the input variable is a 1 to make the output 1, the variable appears as is in the AND term. If the input is a zero to make the output 1, the variable appears negated in the AND term. Let's apply these rules to the half adder. The S output has two combinations of inputs that result in its output being 1; therefore, its SOP form has two terms ORed together. The C output only has one AND or product term, because only one combination of inputs results in a 1 output. The entire truth table can be summarized as:

$$S = \overline{A} \cdot B + A \cdot \overline{B}$$

$$C = A \cdot B$$

Note that we are implicitly using the fact that A and B are Boolean inputs. The equation for C can be read "C is 1 when A and B are both 1." We are assuming that C is zero in all other cases. From the Boolean

TABLE 16.3 Truth Table for a Half Adder

Inputs		Outputs	
A	B	S	C
0	0	0	0
0	1	1	0
1	0	1	0
1	1	0	1

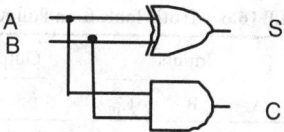

FIGURE 16.7   The gate level implementation of a half adder.

TABLE 16.4   Binary Representation
of Decimal Numbers

Decimal	Binary
0	000
1	001
2	010
3	011
4	100
5	101
6	110
7	111

TABLE 16.5   Adding 3-Bit Binary Numbers

1	0		carry bits
0	1	0	2
0	1	1	+3
1	0	1	= 5

logic equations, it is straightforward to implement S and C with logic gates, as shown as in Figure 16.7. The logical function for S is that of an XOR gate, so we show S as an XOR gate in the figure.

The half adder is a building block in an n-bit binary adder. An n-bit binary adder adds n-bit numbers represented in base 2. Table 16.4 shows the representation of 3-bit, unsigned binary numbers. The leftmost bit is the most significant bit. It is in the 4's place. The middle bit represents the 2's place and the rightmost bit represents the 1's place. The largest representable number, 111_2 represents $4 + 2 + 1$ or 7 in decimal.

Let's examine adding two binary numbers. Table 16.5 shows the operation $2 + 3 = 5$ in binary. The top row is the carry-out from the addition in the previous bit location. Notice that there is a carry-out bit with value 1 from the second position to the third (leftmost) bit position.

The half adder described above has two inputs: A and B. This can be used for the rightmost bit where there is no carry-in bit. For other bit positions we use a *full adder* with inputs A, B, and C_{in} and outputs S and C_{out}. The truth table for the full adder is given in Table 16.6. Note that a full adder adds one bit position; it is *not* an n-bit adder.

To realize the full adder as a circuit we need to design it using logic gates. We do this in the same manner as with the half adder, by writing a logic equation for each of the outputs separately. For each 1 in the truth table on the output of the function, there is an AND term in the Sum of Products representation. Thus, there are four AND terms for the S equation and four AND terms for the C_{out} equation. These equations are given below:

$$S = \overline{A} \cdot \overline{B} \cdot C_{in} + \overline{A} \cdot B \cdot \overline{C_{in}} + A \cdot \overline{B} \cdot \overline{C_{in}} + ABC_{in}$$

$$C_{out} = \overline{A} \cdot BC_{in} + A \cdot \overline{B} \cdot C_{in} + AB \cdot \overline{C_{in}} + ABC_{in}$$

These equations for S and C_{out} are logically correct, but we would also like to use the minimum number of logic gates to implement these functions. The fewer gates used, the fewer gates that need to switch, and

TABLE 16.6 Truth Table for a Full Adder

	Inputs			Outputs	
	A	B	C_{in}	S	C_{out}
0	0	0	0	0	0
1	0	0	1	1	0
2	0	1	0	1	0
3	0	1	1	0	1
4	1	0	0	1	0
5	1	0	1	0	1
6	1	1	0	0	1
7	1	1	1	1	1

hence the smaller amount of power that is dissipated. Next, we will look at applying the rules of Boolean logic to minimize our logic equations.

16.4.2 Axioms and Theorems of Boolean Logic

Our goal is to use the minimum number of logic gates to implement a design. We use logic rules or axioms. These were first described by George Boole, hence the term Boolean algebra. Many of the axioms and theorems of Boolean algebra will seem familiar because they are similar to the rules you learned in algebra in high school. Let us be formal here and state the axioms:

1. Variables are binary. This means that every variable in the algebra can take on one of two values and these two values are not the same. Usually, we will choose to call the two values 1 and 0, but other binary pairs, such as TRUE and FALSE, and HIGH and LOW, are widely used and often more descriptive. Two binary operators, AND ($\cdot$) and OR ($+$), and one unary operator, NOT, can transform variables into other variables. These operators were defined in Table 16.2.
2. Closure: The AND or OR of any two variables is also a binary variable.
3. Commutativity: $A \cdot B = B \cdot A$ and $A + B = B + A$.
4. Associativity: $(A \cdot B) \cdot C = A \cdot (B \cdot C)$ and $(A + B) + C = A + (B + C)$.
5. Identity elements: $A \cdot 1 = 1 \cdot A = A$ and $A + 0 = 0 + A = A$.
6. Distributivity: $A \cdot (B + C) = A \cdot B + A \cdot C$ and $A + (B \cdot C) = (A + B) \cdot (A + C)$. (The usual rules of algebraic hierarchy are used here where $\cdot$ is done before $+$.)
7. Complementary pairs: $A \cdot \overline{A} = 0$ and $A + \overline{A} = 1$.

These are the axioms of this algebra. They are used to prove further theorems. Each algebraic relationship in Boolean algebra has a *dual*. To get the dual of an axiom or a theorem, one simply interchanges AND and OR as well as 0 and 1. Because of this principle of *duality*, Boolean algebra axioms and theorems come in pairs. The principle of duality tells us that if a theorem is true, then its dual is also true.

In general, one can prove a Boolean theorem by exhaustion — that is, by listing all of the possible cases — although more abstract algebraic reasoning may be more efficient. Here is an example of a pair of theorems based on the axioms given above:

Theorem 16.1 (Idempotency). $A \cdot A = A$ *and* $A + A = A$.

Proof 16.1 The definition of AND in Table 16.1 can be used with exhaustion to complete the proof for the first form.

$$A \text{ is } 1: \quad 1 \cdot 1 = 1 = A$$
$$A \text{ is } 0: \quad 0 \cdot 0 = 0 = A$$

The second form follows as the dual of the first.

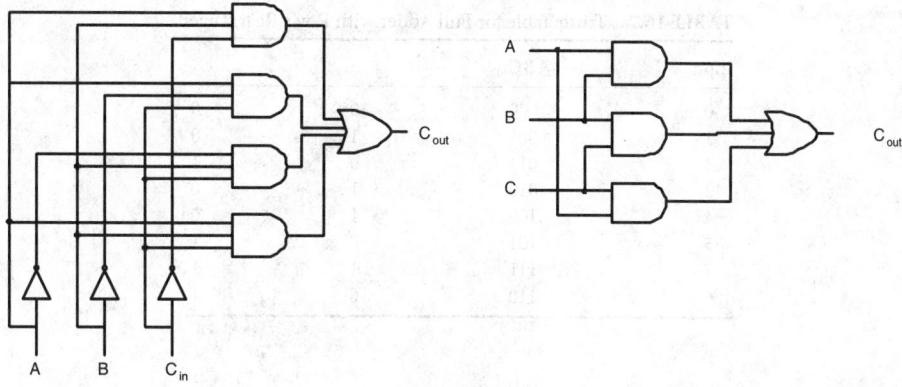

FIGURE 16.8 The direct and reduced circuits for computing the carry-out from the three inputs to the full adder.

Now let us consider reducing the expression from the previous section:

$$C_{out} = \overline{A} \cdot BC_{in} + A \cdot \overline{B} \cdot C_{in} + AB \cdot \overline{C_{in}} + ABC_{in}$$

First we apply idempotency twice to triplicate the last term on the right and put the extra terms after the first and second terms by repeated application of axiom 3:

$$C_{out} = \overline{A} \cdot BC_{in} + ABC_{in} + A \cdot \overline{B} \cdot C_{in} + ABC_{in} + AB \cdot \overline{C_{in}} + ABC_{in}$$

Now we apply axioms 4, 3, and 6 to obtain:

$$C_{out} = (\overline{A} + A)BC_{in} + A(\overline{B} + B)C_{in} + AB(\overline{C} + C)$$

And finally, we apply axioms 7 and 5 to obtain:

$$C_{out} = AB + AC_{in} + BC_{in}$$

The reduced equation certainly looks simpler; let's consider the gate representation of the two equations. This is shown in Figure 16.8. From four 3-input ANDs to three 2-input ANDs and from a 4-input OR to a 3-input OR is a major saving in a basically simple circuit.

The reduction is clear. The savings in a chip containing more than a million gates should build some enthusiasm for gate simplification. What is probably not so clear is how you could know that the key to all of this saving was knowing to make two extra copies of the fourth term in the direct expression. It turns out that there is a fairly direct way to see what you have to do, one that takes advantage of the eye's remarkable ability to see a pattern. This tool, the **Karnaugh map**, is the topic of the next section.   □

16.4.3   Design, Gate-Count Reduction, and SOP/POS Conversions

The *truth table* for the full adder was given is Table 16.6. All possible combinations of the three input bits appear in the second through fourth columns. Note that the first column is the *numerical value* if I interpret those three bits as an unsigned binary number. So, 000 is the value 0, 101 is the value 5, etc.

Let's rearrange the rows of the truth table so that rather than being in increasing numerical order, the truth table values are listed in a way that each row differs from its neighbors by only one bit value. (Note that the fourth and fifth rows (entries for 2 and 4) differ by more than one bit value.) It should become apparent soon why you would want to do this. The result will be Table 16.7.

Consider the last two lines in Table 16.7, corresponding to 6 and 7. Both have $C_{out} = 1$. On the input side, the pair is represented as $AB\overline{C_{in}} + ABC_{in} = AB(\overline{C_{in}} + C_{in})$.

TABLE 16.7 Truth Table for Full Adder with Rows Rearranged

Input	ABC_{in}	S	C_{out}
0	000	0	0
1	001	1	0
3	011	0	1
2	010	1	0
4	100	1	0
5	101	0	1
7	111	1	1
6	110	0	1

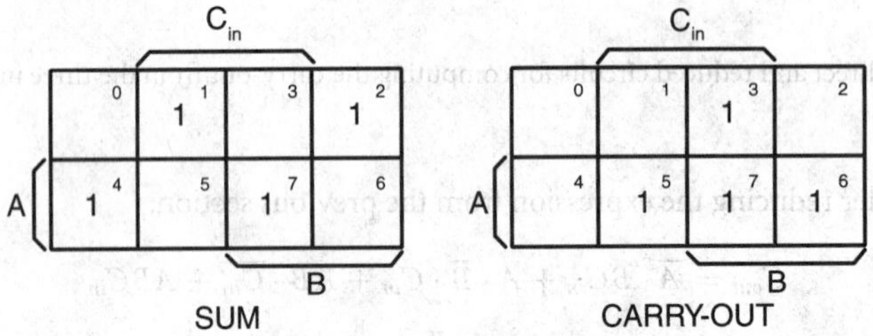

FIGURE 16.9 Karnaugh maps for SUM and CARRY-OUT. The numbers in the cell corners give the bit patterns of ABC_{in}. The cells whose outputs are 1 are marked; those whose outputs are 0 are left blank.

The algebraic reduction operation shows up as adjacency in the table. In the same way, the 5,7 pair can be reduced. The two are adjacent and both C_{out} outputs are 1. It is less obvious in the truth table, but notice that 3,7 also forms just such a pair. In other words, all of the steps proposed in algebra are "visible" in this truth table. To make adjacency even clearer, we arrange the groups of four, one above the other, in a table called a Karnaugh map after its inventor, M. Karnaugh [1953]. In this map, each possible combination of inputs is represented by a box. The contents of the box are the output for that combination of inputs. Adjacent boxes all have numerical values exactly one bit different from their neighbors on any side. It is customary to mark the asserted outputs (the 1's) but to leave the unasserted cells blank (for improved readability). The tables for S and C_{out} are shown in Figure 16.9. The two rows are just the first and second group of four from the truth table with the output values of the appropriate column. First convince yourself that each and every cell differs from any of its neighbors (no diagonals) by precisely one bit. The neighbors of an outside cell include the opposite outside cell. That is, they wrap around. Thus, 2 and 0 or 4 and 6 are neighbors. The Karnaugh map (or K-map) simply shows the relationships of the outputs of conjugate pairs, which are sets of inputs that differ in exactly one bit location. The item that most people find difficult about K-maps is the meaning and arrangement of the input variables around the map. If you think of these input variables as the bits in a binary number, the arrangement is more logical. The difference between the first four rows of the truth table and the second four is that A has the value 0 in the first four and the value 1 in the second four. In the map, this is shown by having A indicated as asserted in the second row. In other words, where the input parameter is placed, it is asserted. Where it is not placed, it is unasserted. Accordingly, the middle two columns are those cells that have C_{in} asserted. The right two columns have B asserted. Column 3 has both B and C_{in} asserted.

Let us look at how the carry-out map implies gate reduction while sum's K-map shows that no reduction is possible. Because we are looking for conjugate pairs of asserted cells, we simply look for adjacent pairs of 1's. The carry-out map has three such pairs; sum has none. We take pairs, pairs of pairs, or pairs of pairs of pairs — any rectangular grouping of 2^n cells with all 1's. With carry-out, this gives us the groupings shown in Figure 16.10.

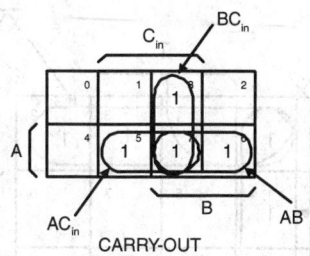

FIGURE 16.10   The groupings of conjugate pairs in CARRY-OUT.

The three groupings do the three things that we must always achieve:

1. The groups must cover all of the 1's (and none of the 0's).
2. Each group must include at least one cell not included in any other group.
3. Each group must be as large a rectangular box of 2^n cells as can be drawn.

The last rule says that in Figure 16.10 none of these groups can cover only one cell. Once we fulfill these three rules, we are assured of a minimal set, which is our goal. Although there is no ambiguity in the application of these rules in this example, there are other examples where more than one set of groups results in a correct, minimal set. K-maps can be used for functions of up to six input variables, and are useful aids for humans to minimize logic functions. Computer-aided design programs use different techniques to accomplish the same goal.

Writing down the solution once you have done the groupings is done by reading the specification of the groups. The vertical pair in Figure 16.10 is BC_{in}. In other words, that pair of cells is uniquely defined as having B and C_{in} both 1. The other two groups are indicated in the figure. The sum of those three (where "+" is OR) is the very function we derived algebraically in the last section. Notice how you could know to twice replicate cell 7. It occurs in three different groups. It is important to keep in mind that the Karnaugh map simply represents the algebraic steps in a highly visual way. It is not magical or intrinsically different from the algebra.

We have used the word "cell" to refer to a single box in the K-map. The formal name for a cell whose value is 1 is the *minterm* of the function. Its counterpart, the *maxterm*, comprises all the cells that represent an output value of 0. Note that all cells are both possible minterms and possible maxterms.

Two more examples will complete our coverage of K-maps. One way to specify a function is to list the minterms in the form of a summation. For example, $C_{out} = \sum(2, 5, 6, 7)$. Consider the arbitrary four-input function $F(X, Y, Z, T) = \sum(0, 1, 2, 3, 4, 8, 9, 12, 15)$. With four input variables, there are 16 possible input states, and every minterm must contact four neighbors. That can be accomplished in a 4×4 array of cells as shown in Figure 16.11. Convince yourself that each cell is properly adjacent to its neighbors. For example, 11_{10} (1011) is adjacent to 15 (1111), 9 (1001), 10_{10} (1010), and 3 (0011) with each neighbor differing by one bit. Now consider the groupings. Minterm 15 has no neighbors whose value is 1. Hence, it forms a group on its own, represented by the AND of all four inputs. The top row and first columns can each be grouped as a pair of pairs. It takes only two variables to specify such a group. For example, the top row includes all terms of the form $00xx$, and the first column includes all the terms of the form $xx00$. This leaves us but one uncovered cell, 9. You might be tempted to group it with its neighbor, 8, but rule 3 demands that we make as large a covering as possible. We can make a group of four by including the neighbors 0 and 1 on top. Had we not done that, the bottom pair would be $X \cdot \overline{Y} \cdot \overline{Z}$; but by increasing the coverage, we get that down to $\overline{Y} \cdot \overline{Z}$, a 2-input AND vs. a 3-input AND. The final expression is

$$F(X, Y, Z, T) = \overline{X} \cdot \overline{Y} + \overline{Y} \cdot \overline{Z} + \overline{Z} \cdot \overline{T} + XYZT$$

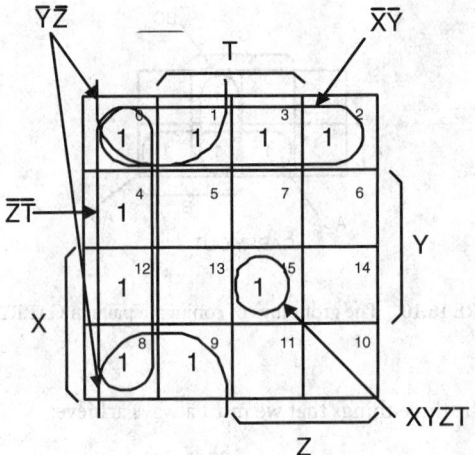

FIGURE 16.11   The K-map for $F(X, Y, Z, T) = \sum(0, 1, 2, 3, 4, 8, 9, 12, 15)$ with the minterm groupings shown.

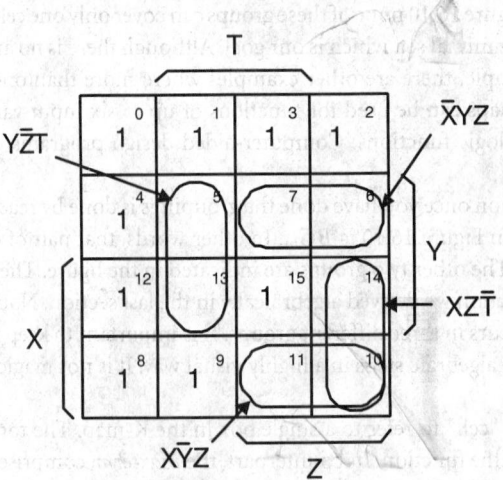

FIGURE 16.12   The K-map for the complement of F from Figure 16.11.

The above function is in Sum of Products (SOP) form because it is a set of products that are summed together. It is just as easy to generate the function with a Product of Sums (POS) where several OR gates are joined by a single AND. To get to that expression, we find $\overline{F}$, the complement of F, and then convert to F using De Morgan's theorem. $\overline{F}$ is obtained by grouping the cells where F is not asserted — the zero cells. This is shown in Figure 16.12, where we get the expression $\overline{F} = \overline{X}YZ + XZ\overline{T} + X\overline{Y}Z + Y\overline{Z}T$.

Let us convert from $\overline{F}$ to F using De Morgan's theorem to get the POS form:

$$F = \overline{(\overline{X}YZ + XZ\overline{T} + X\overline{Y}Z + Y\overline{Z}T)}$$
$$= (X + \overline{Y} + \overline{Z})(\overline{X} + \overline{Z} + T)(\overline{X} + Y + \overline{Z})(\overline{Y} + Z + \overline{T})$$

Why would one want to do this? Economy of gates. Sometimes the SOP form has fewer gates, sometimes the POS form does. In this example, the SOP form is somewhat more economical.

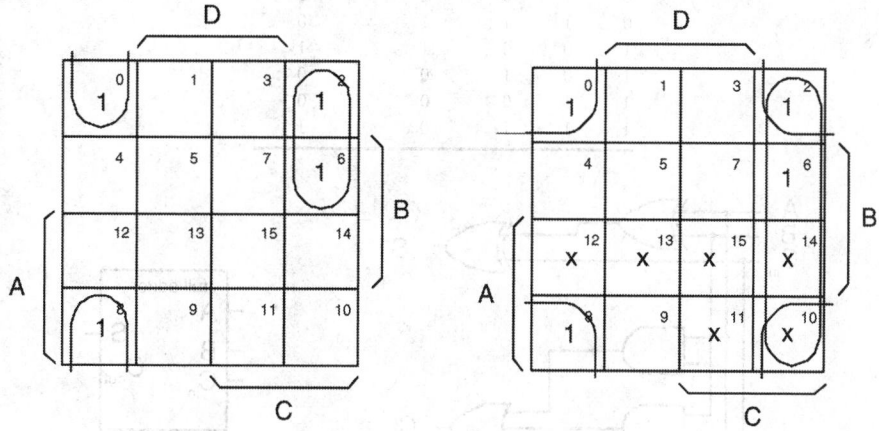

FIGURE 16.13 Segment *e* of the seven-segment display whose decoder we are going to minimize.

FIGURE 16.14 Minimization of Se without and with deliberate assignment of the *don't cares*.

16.4.4 Minimizing with Don't Cares

Sometimes, we can guarantee that some combination of inputs will never occur. I don't care what the output value is for that particular combination of inputs because I know that the output can never occur. This is known as an "output" don't care. I can set these outputs to any value I want. The best way to do this is to set these outputs to values that will minimize the gate count of the entire circuit.

An example is the classic seven-segment numerical display that is common in watches, calculators, and other digital displays. The input to a seven-segment display is a number coded in binary-coded-decimal (BCD), a 4-bit representation with 16 possible input combinations, but only the 10 numbers 0, ..., 9 ever occur. The states 10, ..., 15 are called **don't cares**. One can assign them to achieve minimum gate count. Consider the entire number set that one can display using seven line segments. We will consider the one line segment indicated by the arrows in Figure 16.13. It is generally referred to as "segment *e*," and it is asserted only for the numbers 0, 2, 6, and 8.

Now we will minimize Se(A, B, C, D) with and without the use of the *don't cares*. We put an "X" wherever the *don't cares* may lie in the K-map and then treat each one as either 0 or 1 in such a way as to minimize the gate count. This is shown in Figure 16.14.

We are not doing something intrinsically different on the right and left. On the left, all of the *don't cares* are assigned to 0. In other words, if someone enters a 14 into this 0 : 9 decoder, it will not light up segment *e*. But because this is a *don't care* event, we examine the map to see if letting it light up on 14 will help. The grouping with the choice of *don't care* values is decidedly better. We choose to assert *e* only for *don't cares* 10 and 14, but those assignments reduce the gates required from two 3-input ANDs to two 2-input ANDs. For this little circuit, that is a substantial reduction.

16.4.5 Adder/Subtractor

Let's return to the design of the full adder. A full (one-bit) adder can be implemented out of logic gates by implementing the equations for C and S. As we have seen, the simplified version for C is:

$$C_{out} = AB + AC_{in} + BC_{in}$$

TABLE 16.8   Truth Table for 3-Input XOR

Inputs			Outputs	
A	B	C	$A \oplus B$	$A \oplus B \oplus C$
0	0	0	0	0
0	0	1	0	1
0	1	0	1	1
0	1	1	1	0
1	0	0	1	1
1	0	1	0	0
1	1	0	0	0
1	1	1	0	1

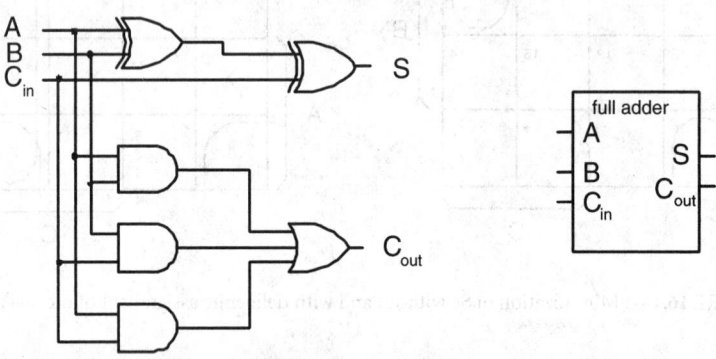

FIGURE 16.15   Implementation of full adder from logic gates on the left. Symbol of full adder on the right.

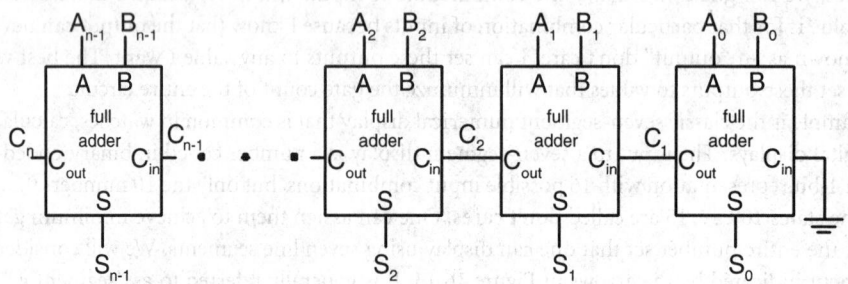

FIGURE 16.16   Implementation of an *n*-bit adder from full adders.

S cannot be simplified using K-maps. Instead, we will simplify S by inspection of the truth table for the full-adder given in Table 16.6. Note that S is high when exactly one of the three inputs is high, or when all the inputs are high. This is the same functionality as a 3-input XOR gate, as shown in Table 16.8. Thus we can implement S with the equation:

$$S = A \oplus B \oplus C_{in}$$

This completes our design of a full adder. Its implementation in logic gates is shown in Figure 16.15.

We would like to add n-bit binary numbers. To accomplish this we will connect N full adders as shown in Figure 16.16. This configuration connects the carry-out of one bit to the carry-in of the next bit, and is called a ripple carry adder. This design is small but slow because the carry must ripple through from the least significant bit to the most significant bit. For designs where speed is of the essence, such as modern high-speed microprocessors, various techniques are used to speed up the calculation of the carry chain. Such techniques are beyond the scope of this chapter.

TABLE 16.9   Binary Representation of Negative Numbers

Decimal	Sign-Magnitude	1's Complement	2's Complement
7	0111	0111	0111
6	0110	0110	0110
5	0101	0101	0101
4	0100	0100	0100
3	0011	0011	0011
2	0010	0010	0010
1	0001	0001	0001
0	0000	0000	0000
−0	1000	1111	
−1	1001	1110	1111
−2	1010	1101	1110
−3	1011	1100	1101
−4	1100	1011	1100
−5	1101	1010	1011
−6	1110	1001	1010
−7	1111	1000	1001
−8			1000

16.4.6   Representing Negative Binary Numbers

We would like to add negative numbers as well as positive numbers, and we would like to subtract numbers as well as add them. To do this we need a way of representing negative numbers in base 2. Two common methods are used: *sign-magnitude* and *2's complement*. A third method, *1's complement* will also be described to aid in the explanation of 2's complement.

In Table 16.9, four bits are used to represent the values 7 to 0. The three different methods, for representing negative numbers are shown. Note that for *all* three methods, the positive numbers have the same representation. Also, the leftmost bit is always the sign bit. It is 0 for a positive number and 1 for a negative number. It is important to note that all these representations are different ways that a *human* interprets the bit patterns. The bit patterns are not what is changing; the interpretations are. Given a bit pattern, you cannot tell which system is being used unless someone tells you. This discussion can be extended to numbers represented with any number of bits.

The sign-magnitude method is the closest to the method we use for representing positive and negative numbers in decimal. The sign bit indicates whether it is positive or negative, and the remaining bits represent the magnitude or value of the number. So, for example, to get the binary value of negative 3, you take the positive value 0011 and flip the sign bit to get 1011. While this is the most intuitive for humans to understand, it is not the easiest representation for computers to manipulate. This is true for several reasons. One is that there are two representations for 0: $+0$ and -0. A very common operation to perform is to compare a number to 0. With two representations of zero, this operation becomes more complex.

Instead, a representation called 2's complement is more frequently used. 2's complement has the feature that it has only one representation for zero. It has other advantages as well, including the fact that addition and subtraction are straightforward to implement; subtraction is the same as adding a negative number. If you add a number and its complement, the result is zero with no carry-out, as one would expect. To form the negative value of a positive number in 2's complement, simply invert all the bits and add 1. Inverting the bits results in the 1's complement, as shown in Table 16.7. 1's complement has many of the advantages of 2's complement, except that it still has two representations of zero. The 2's complement is formed by adding 1 to the 1's complement of a number. 3 is 0011. Its one's complement is 1100 and its 2's complement is 1101. Given a negative number, how can I tell its value? Due to properties of 2's complement numbers, if I take the 2's complement of a negative number, I get its positive value. Given 1101, I invert the bits to get 0010 and then add 1 to get 0011, its positive value 3. Note that the 2's complement representation has one representation for zero, and it is the value represented by all zeros. Because I can represent 16 values

TABLE 16.10 Choosing the B Input for an Adder/Subtractor

SB	B_i	Result
0	0	0
0	1	1
1	0	1
1	1	0

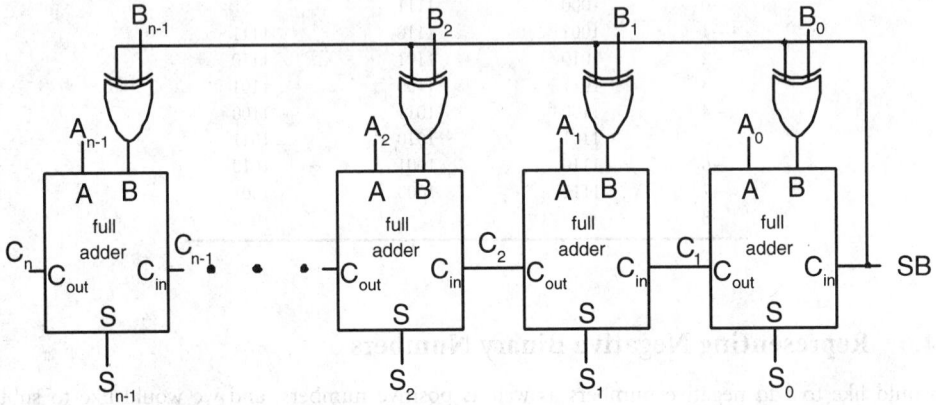

FIGURE 16.17 Connection of n full adders to form an N-bit ripple-carry adder/subtractor. At the rightmost adder, the subtract line (SB) is connected to Cin_0.

with 4 bits, this leaves me with a non-symmetric range. In other words, I can represent one more negative number 1000 than positive number. This number is -8. Its positive value cannot be represented in 4 bits. What happens if I take the 2's complement of -8? The 1's complement is 0111. When I add 1 to form the 2's complement, I get 1000. What is really happening is that the true value, $+8$, cannot be represented in the number of bits I have available. It overflows the range for the representation I am using.

Representing numbers in 2's complement makes it easy to do subtraction. To subtract two n-bit numbers $A - B$, you simply invert the bits of B and add 1 when you are summing $A + \overline{B}$.

We are now ready to expand our n-bit ripple carry adder to an n-bit adder/subtractor. Just as we do addition one digit at a time, the adder circuit handles two input bits, A_i and B_i, plus a carry-in C_{in_i}. We can arrange as many of these circuits in parallel as we have bits. The ith circuit gets the ith bits of two operands plus the carry-out of the previous stage. It is straightforward to modify the full-adder to be a one-bit adder/subtractor. A one-bit adder/subtractor performs the following tasks:

1. Choose B_i or the complement of B_i as the B input.
2. Form the sum of the three input bits, $S_i = A_i + B_i + C_{in_i}$.
3. Form the carry-out of the three bits, $C_{out_i} = f(A_i, B_i, C_{in_i})$.

On a bit-by-bit basis, the complement of B_i is just $\overline{B_i}$. For an n-bit number, the 2's complement is formed by taking the bit-by-bit complement and adding 1. I can add 1 to an n-bit subtraction by setting the carry-in bit C_{in_0} to 1. In other words, I want $C_{in_0} = 1$ when subtract is true, and $C_{in_0} = 0$ when subtract is false. This is accomplished by connecting the control signal for subtracting (SB) to C_{in_0}. Similarly when SB = 0, I want to use B_i as the input to my full adder. When SB = 1, I want to use $\overline{B_i}$ as input. This is summarized in Table 16.10. By inspection, the desired B input bit to the ith full adder is the XOR of SB and B_i. If I put all the components together — n full adders, the carry-in of the LSB set to SB, and the XOR of SB and B_i to form the complement of B — I get an n-bit adder/subtractor as shown in Figure 16.17.

TABLE 16.11   Truth Table for a 4-to-2 Encoder

D_0	D_1	D_2	D_3	Q_1	Q_0	V
0	0	0	0	0	0	0
1	0	0	0	0	0	1
0	1	0	0	0	1	1
0	0	1	0	1	0	1
0	0	0	1	1	1	1

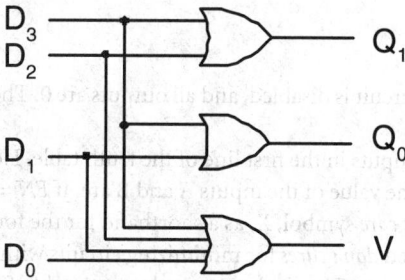

FIGURE 16.18   A 4-to-2 encoder with outputs Q_0 and Q_1 and valid signal.

16.5   Frequently Used Digital Components

Many components such as full adders, half adders, 4-bit adders, 8-bit adders, etc., are used over and over again in digital logic design. These are usually stored in a design library to be used by designers. In some libraries these components are parameterized. For example, a generator for creating an n-bit adder may be stored. When the designer wants a 6-bit adder, he or she must instantiate the specific bit width for the component.

Many other, more complex components are stored in these libraries as well. This allows components to be designed efficiently once and reused many times. These include encoders, multiplexers, demultiplexers, and decoders. Such designs are described in more detail below. Later, we will use them in the design of a calculator datapath.

16.5.1   Elementary Digital Devices: ENC, DEC, MUX, DEMUX

16.5.1.1   ENC

An ENCODER circuit has 2^n input lines and n output lines. The output is the number of the input line that is asserted. Such a circuit *encodes* the asserted line. The truth table of a 4-to-2 encoder is shown in Table 16.11. The inputs are D_0, D_1, D_2, and D_3, and the outputs are Q_0 and Q_1. Note that we assume at most one input can be high at any given time. More complicated encoders, such as priority encoders, that allow more than one input to be high at a time, are described below.

An encoder, like all other components, is built out of basic logic gates. The equations for the Q outputs can be determined by inspection. Q_0 is 1 if D_1 is 1 or D_3 is 1. Q_1 is 1 if D_2 is 1 or D_3 is 1.

Note that there is no difference in the outputs of this circuit if *no* input is asserted, or if input D_0 is asserted. To distinguish between these two cases, an output that indicates the Q outputs are valid, V is added. V is 1 if any of the inputs are 1, and 0 otherwise. V is considered a control output rather than a data output. A logic diagram of an encoder is shown in Figure 16.18.

16.5.1.2   DEC

A decoder performs the opposite function of an encoder. Exactly one of the outputs is true if the circuit is enabled. That output is the *decoded* value of the input, if I think of the input as a binary number. When

TABLE 16.12 Truth Table for a 2-to-4 Decoder with Enable

EN	A	B	Q_0	Q_1	Q_2	Q_3
0	X	X	0	0	0	0
1	0	0	1	0	0	0
1	1	0	0	1	0	0
1	0	1	0	0	1	0
1	1	1	0	0	0	1

the enable input (*EN*) is 0, the circuit is disabled, and all outputs are 0. The truth table of a 2-to-4 decoder is given in Table 16.12.

Note the use of X values for inputs in the first line of the truth table. Here, X stands for "don't care." In other words, I don't care what the value of the inputs A and B are. If *EN* = 0, the outputs will always have the value 0. I am using the *don't care* symbol, X, as a shorthand for the four combinations of input values for A and B. We have already used *don't cares* for minimizing circuits with K-maps above. In that case, the *don't cares* were "output" *don't cares*. The *don't cares* in the truth table for the encoder are "input" *don't cares*. They are shorthand for several combinations of inputs.

16.5.1.3   MUX

Many systems have multiple inputs which are handled one at a time. *Call waiting* is an example. You are talking on one connection when a clicking noise signals that someone else is calling. You switch to the other, talk briefly, and then switch back. You can toggle between calls as often as you like. Because you are using one phone to talk on two different circuits, you need a **multiplexer** or MUX to choose between the two. There is also an inverse MUX gate called either a DEMUX or a DECODER. Again, a telephone example is the selection of an available line among, say, eight lines between two exchanges. That is, you have one line in and eight possible out, but only one output line is connected at any time. An algorithm based on which lines are currently free determines the choice. Let us design these two devices, beginning with a 2-to-1 MUX.

What we want in a two-input MUX is a circuit with one output. The value of that output should be the same as the input that we select. We will call the two inputs A and B and the output Q. The select input S chooses which input to steer to the output.

Logically I can think of Q being equal to A when S = 0, and Q being equal to B when S = 1. I can write this as the Boolean equation: $Q = (A \cdot \overline{S}) + (B \cdot S)$. You can use a truth table to convince yourself that this equation captures the behavior of a two-input MUX. Note that we now have a new dichotomy of inputs. We call some of them *inputs* and the others *controls*. They are not inherently different, but from the human perspective, we would like to separate them. In logic circuit drawings, *inputs* come in from the left and outputs go out to the right. *Controls* are brought in from top or bottom. The select input for our multiplexer is a control input. Note that, although I talk about S being a control, in the logic equation it is treated the same as an input. An enable is another kind of control input. A *valid* signal can be viewed as a control output. A realization of a 2-to-1 multiplexer with enable is shown in Figure 16.19.

The 2-to-1 MUX circuit is quite useful and is found in many design libraries. Other similar circuits are the 4-to-1 MUX and the 8-to-1 MUX. In general, you can design n-to-1 MUX circuits, where n is a power of 2. The number of select bits needed for an n-to-1 MUX is $log_2(n)$.

Figure 16.20 shows a 4-to-1 multiplexer on the left, and a 1-to-4 demultiplexer on the right. The MUX chooses one of four inputs using the two select lines: S1 and S0. If we view the values on the select lines as the binary numbers $0, \ldots, 3$, we understand the selection process as enabling the top AND when the input is 00 and then progressively lower ANDs as the numbers become 01, 10, and 11. Essentially, there is a decoder circuit within the 4-to-1 multiplexer.

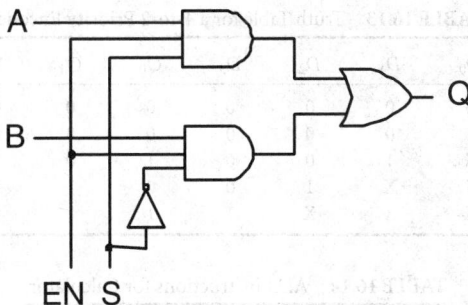

FIGURE 16.19 A 2-to-1 MUX with enable. If the enable is asserted, this circuit delivers at its output, Q, the value of A or the value of B, depending on the value of S. In this sense, the output is "connected" to one of the input lines. If the enable is not asserted, the output Q is low.

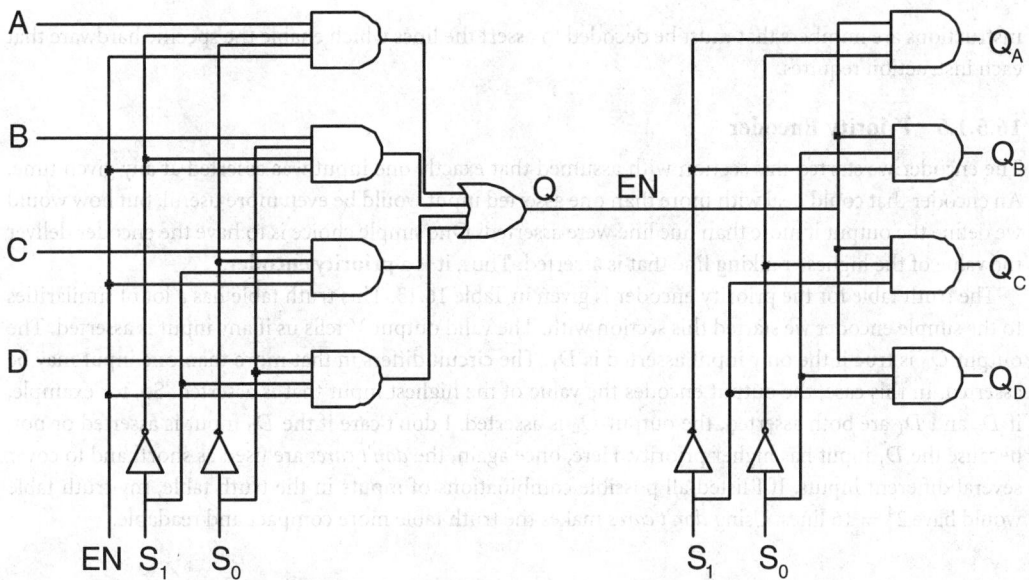

FIGURE 16.20 A 4-to-1 MUX feeding a 1-to-4 DEMUX. The value on MUX select lines S1:S0 determines the input connected to Q. EN, in turn, is connected to the output of choice by S1:S0 on the DEMUX.

16.5.1.4 DEMUX/DECODER

The inverse circuit to a multiplexer is a demultiplexer, or DEMUX. A DEMUX has one line in, which it switches to one of n possible output lines. A 1-to-4 DEMUX, used in conjunction with the 4-to-1 MUX, is shown on the right in Figure 16.20.

Note that the Q output line on the multiplexer is labeled as the *EN* input line on the DEMUX. If I treat this line as an *enable*, the DEMUX becomes a DECODER, in the sense that, when *EN* is asserted, one and only one of the four outputs is asserted, that being the output selected by the number on S1:S0. So a decoder and a demux are the same circuit. You usually do not find a demux in a design library. Rather, what you find is called a decoder, or sometimes a demux/decoder.

Decoding is an essential function in many places in computer design. For example, random-access memory (RAM) is fed an address — a number — and must return data based on that number. It does this by decoding the address to assert lines that enable the output of the selected data. Similarly, computer

TABLE 16.13   Truth Table for a 4-to-2 Priority Encoder

D_0	D_1	D_2	D_3	Q_0	Q_1	V
0	0	0	0	0	0	0
1	0	0	0	0	0	1
X	1	0	0	1	0	1
X	X	1	0	0	1	1
X	X	X	1	1	1	1

TABLE 16.14   ALU Instructions for Calculator

I_1	I_0	Result
0	0	A AND B
0	1	A OR B
1	0	$A + B$
1	1	$A - B$

instructions are numbers that must be decoded to assert the lines which enable the specific hardware that each instruction requires.

16.5.1.5   Priority Encoder

The encoder we started this section with assumed that exactly one input was asserted at any given time. An encoder that could deal with more than one asserted input would be even more useful, but how would we define the output if more than one line were asserted? One simple choice is to have the encoder deliver the value of the highest-ranking line that is asserted. Thus, it is a **priority encoder**.

The truth table for the priority encoder is given in Table 16.13. This truth table has a lot of similarities to the simple encoder we started this section with. The valid output V tells us if any input is asserted. The output Q_0 is true if the only input asserted is D_1. The circuit differs in that more than one input may be asserted. In this case, the output encodes the value of the highest input that is asserted. So, for example, if D_0 and D_1 are both asserted, the output Q_0 is asserted. I don't care if the D_0 input is asserted or not, because the D_1 input has higher priority. Here, once again, the *don't cares* are used as shorthand to cover several different inputs. If I listed all possible combinations of inputs in the truth table, my truth table would have $2^4 = 16$ lines. Using *don't cares* makes the truth table more compact and readable.

16.5.2   The Calculator Arithmetic and Logical Unit

Let's look at putting some of these components together to do useful work. The core of a calculator or microprocessor is its Arithmetic and Logic Unit (ALU). This is the part of the calculator that implements the arithmetic functions, such as addition, subtraction, and multiplication. In addition, logic functions are also implemented such as ANDing inputs and ORing inputs. A microprocessor may have several, sophisticated ALUs. We will examine the design of a very simple ALU for a 4-bit calculator. Our ALU will perform four different operations: AND, OR, addition, and subtraction on two 4-bit inputs, A and B. Two input control signals, I_1 and I_0, will be used to choose between the operations, as shown in Table 16.14. We will call the 4-bit result R.

A common way to implement such an ALU is to implement the various functions in parallel, then choose the requested result based on the setting of the control inputs. The 4-bit AND requires four AND gates, one for each bit position. Similarly, the 4-bit OR requires four OR gates. We will implement the adder and subtractor using a single adder/subtractor unit, since using two would waste area. The adder/subtractor unit in Figure 16.17 is perfect for our purposes here. Note, in Table 16.14, that I_0 is 0 for adding A and B and 1 for subtracting A and B, so we can use I_0 for the SB input to the adder/subtractor. These units will always operate on the A and B inputs; a multiplexer will select the correct output based on the values of I_1 and I_0. Four 4-to-1 multiplexers are used, one for each output bit. 4-to-1 multiplexers are used because

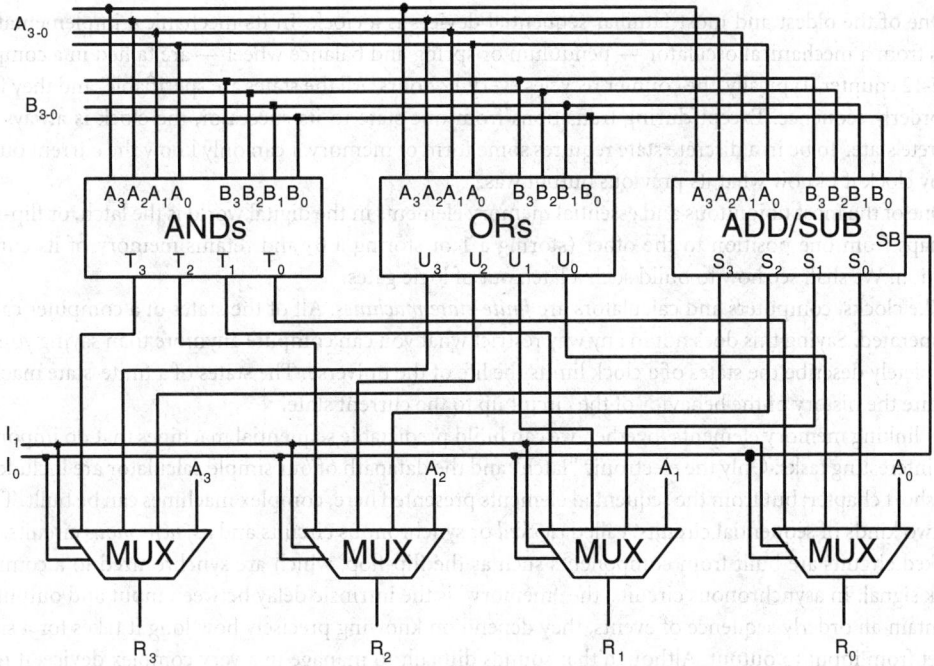

FIGURE 16.21   Implementation of an ALU from other components.

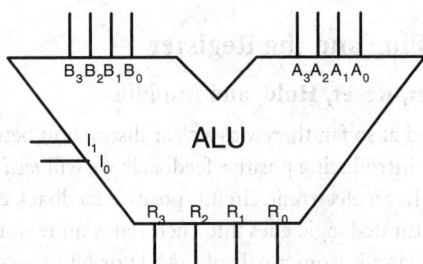

FIGURE 16.22   Symbol of an ALU component.

there are no 3-to-1 muxes. We will use the fourth input to pass the A input to the output R. The reason for doing this will become apparent when we use the ALU in a calculator datapath. To keep the diagram readable, we use the convention that signals with the same name are wired together. The resulting ALU implementation is shown in Figure 16.21. The symbol for this ALU is shown in Figure 16.22. We will use this symbol when we incorporate the ALU into a calculator datapath.

16.6   Sequential Circuits

16.6.1   Concept of a Sequential Device

So far, all the circuits we have discussed have been combinational. The current output can be determined by knowing the current inputs. Sequential circuits differ from combinational circuits because they have memory. For a circuit with memory, the current outputs depend on the current inputs *and* on the past history of the circuit. Memory elements dramatically change the way a circuit operates.

One of the oldest and most familiar sequential devices is a clock. In its mechanical implementation, ticks from a mechanical oscillator — pendulum or spring and balance wheel — are tallied in a complex, base-12 counter. Typically, the counter recycles every 12 hours. All the states are specifiable, and they form an orderly sequence. Except during transitions from one state to its successor, the clock is always in a discrete state. To be in a discrete state requires some form of memory. I can only know the current output of my clock if I know what its previous output was.

One of the most ubiquitous and essential memory elements in the digital world is the latch, or flip-flop. It snaps from one position to the other (storing a 1 or storing a 0) and retains memory of its current position. We shall see how to build such a latch out of logic gates.

Like clocks, computers and calculators are *finite-state machines*. All of the states of a computer can be enumerated. Saying this does not in any way restrict what you can compute anymore than saying you can completely describe the states of a clock limits the life of the universe. The states of a finite-state machine capture the history of the behavior of the circuit up to the current state.

By linking memory elements together, we can build predictable sequential machines that do important and interesting tasks. Only the electronic "latch" and the datapath of our simple calculator are included in this short chapter; but from the sequential elements presented here, complex machines can be built. There are two kinds of sequential circuits, called *clocked* or synchronous circuits and *asynchronous* circuits. The clocked circuits are built from components such as the flip-flop, which are synchronized to a common clock signal. In asynchronous circuits, the "memory" is the intrinsic delay between input and output. To maintain an orderly sequence of events, they depend on knowing precisely how long it takes for a signal to get from input to output. Although that sounds difficult to manage in a very complex device, it turns out that keeping a common clock synchronized over a large and complex circuit is nontrivial as well. We will limit our discussion to clocked sequential circuits. They are more common, but as computer speeds become faster, the asynchronous approach is receiving greater attention.

16.6.2 The Data Flip-Flop and the Register

16.6.2.1 The *SR* Latch: Set, Reset, Hold, and Muddle

In all the circuits we have looked at so far, there was a clear distinction between inputs and outputs. Now we will erase this distinction by introducing positive feedback; we will *feed back* the outputs of a circuit to the inputs of the same circuit. In an electronic circuit, positive feedback can be used to force the circuit into a "stable state." Because saturated logic goes into such states quite normally, it is a very small step to generate an electronic latching circuit from a pair of NAND or NOR gates. The simplest such circuit is shown in Figure 16.23.

Analyzing Figure 16.23 requires walking through the behavior of the circuit. Let's assume that Q has the value 1 and $\overline{Q}$ has the value 0. Start with both $\overline{S}$ and $\overline{R}$ deasserted. In other words, both have value 1 because they are active low signals. The inputs to B will be high, so $\overline{Q}$ will be low. This is a "steady state"

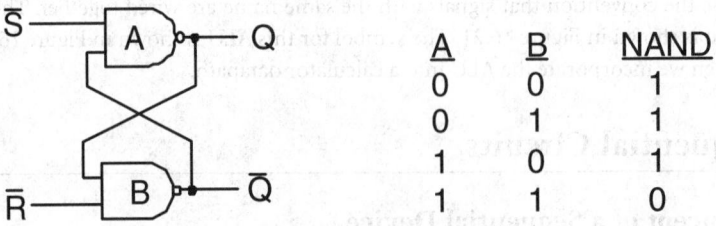

A	B	NAND
0	0	1
0	1	1
1	0	1
1	1	0

FIGURE 16.23   The basic set/reset (SR) latch is shown on the left. If $\overline{S}$ is asserted, Q is asserted (set). If $\overline{R}$ is asserted, Q is deasserted (reset). If neither $\overline{S}$ nor $\overline{R}$ is asserted (both high), the latch retains its current state. If both are asserted, the latch goes into a *muddle* state where Q is asserted and $\overline{Q}$ is deasserted (both high); but upon simultaneous release of the inputs, the next state is unpredictable. The truth table for an NAND gate is shown on the right.

of this circuit; the circuit will stay in this state for some time. This state is called "storing 1," or sometimes just "1" because Q has the value 1. You could toggle $\overline{S}$ (i.e., change its value to 0 and then back to 1) and no other change would take place in the circuit.

Now, with $\overline{S}$ high, let's assert $\overline{R}$ by setting it to 0. First, $\overline{Q}$ will go high because $\overline{R}$ is one of the inputs to B, and the NAND of 0 with anything is 1. This makes both of the inputs to A high, so Q goes low. Now the upper input to B is low, so deasserting $\overline{R}$ (setting it to 1) will have no effect. Thus, asserting $\overline{R}$ has reset the latch. The latch is in the other steady state, "storing 0" or "0."

At this point, asserting $\overline{S}$ will set the latch, or put it back into the state "1." For this reason, the S input is the "set" input to the latch, and the R input is the "reset" input.

What happens if both $\overline{S}$ and $\overline{R}$ are asserted at the same time? The initial result is to have both Q and $\overline{Q}$ go high simultaneously. Now, deassert both inputs simultaneously. What happens? You cannot tell. It may go into either the set or the reset state. Occasionally, the circuit may even oscillate, although this behavior is rare. For this reason, it is usually understood that the designer is *not allowed* to assert both $\overline{S}$ and $\overline{R}$ at the same time. This means that, if the designer asserts both $\overline{S}$ and $\overline{R}$ at the same time, the future behavior of the circuit cannot be guaranteed, until it is set or reset again into a known state.

There is another problem with this circuit. To hold its value, both $\overline{S}$ and $\overline{R}$ must be continuously deasserted. Glitches and other noise in a circuit might cause the state to flip when it should not.

With a little extra logic, we can improve upon this basic latch to build circuits less likely to go into an unknown state, oscillate, or switch inadvertently. These better designs eliminate the muddle state.

16.6.2.2 The Transparent *D*-Latch

A simple way to avoid having someone press two buttons at once is to provide them with a toggle switch. You can push it only one way at one time. We can also provide a single line to enable the latch. This enable control signal is usually called the **clock**. We will modify the SR latch above. First we will combine the S and R inputs into one input called the data, or D input. When the D line is a 1, we will set the latch. When the D line is a 0, we will reset the latch. Second, we will add a clock signal (CLK) to control when the latch updates. With the addition of two NANDs and an inverter, we can accomplish both purposes, as shown in Figure 16.24.

Note that we tie the data line, D, to the top NAND gate, and the inverse or $\overline{D}$ to the bottom NAND gate. This assures that only one of the two NAND outputs can be low at one time. The CLK signal allows us to open the latch (let data through) or latch the data at will. This device is called a transparent D-latch, and is found in many digital design libraries. This latch is called *transparent* because the current value of D appears at Q if the CLK signal is high. If CLK is low, then the latch retains the last value D had when CLK was high.

Has this device solved all the problems we described for the *SR* latch? No. Consider what might happen if D changes from low to high just as the clock changes from high to low. For the brief period before the change has propagated through the D-inverter, both NANDs see both inputs high. Thus, at least briefly,

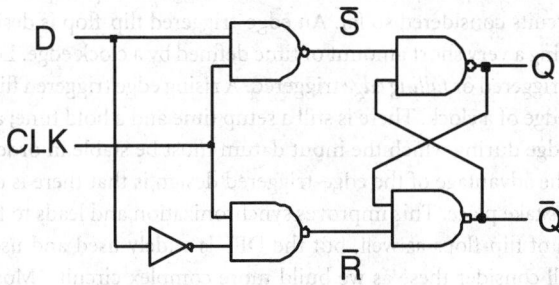

FIGURE 16.24 The transparent D-latch. The circuit is transparent when CLK is high (that is, the current value D appears at Q) and latched when CLK is low (the value of D when the clock went low is held at Q.)

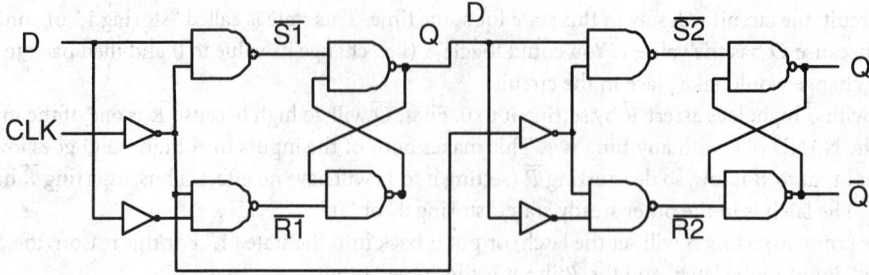

FIGURE 16.25   The master–slave data flip-flop constructed of two D-latches in series.

both $\overline{R}$ and $\overline{S}$ are asserted. This is the very situation we wanted to avoid. This muddle situation would last only for the propagation time of the inverter, but then the CLK signal arrives and drives both $\overline{S}$ and $\overline{R}$ high. The latch might oscillate or flip either way. In any case, it will be unpredictable.

There is another problem with this circuit. It is indeed transparent during the high clock signal. This means that Q will mirror D while CLK is high. If D changes rapidly, so will Q. Sometimes you may want transparency. However, frequently you want to be able to guarantee when the output will change and to only allow one transition on the output per clock cycle. In that case, you do not want transparency; what you really want is a different circuit: a flip-flop (FF).

16.6.2.3   Master–Slave DFF to Eliminate Transparency

The problem with transparent gates is not a new one. A solution that first appeared in King Solomon's time (9th century B.C.E.) will work here as well. The Solomonic gate was a series pair of two quite ordinary city gates. They were arranged so that both were never open at the same time. You entered the first and it was shut behind you. While you were stuck between the two gates, a well-armed, suspicious soldier asked your business. Only if you satisfied him was the second gate opened. The solution of putting out-of-phase transparent latches between input and output is certainly one obvious solution to generating a **data flip-flop** (DFF). Such an arrangement of two *D*-latches is shown in Figure 16.25.

The latch on the left is called the **master**; that on the right is called the **slave**. This master–slave (MS) DFF solves the transparency problem but does nothing to ameliorate the timing problems. While timing problems are not entirely solvable in any FF, accommodating the number of delays in this circuit tends to make the MSFF a slow device and thus a less attractive solution. Why should it be slow? The issue is that to be sure that you do not put either of these devices into a metastable or oscillatory state, you must hold D constant for a relatively long setup time (before the clock edge) and continue it past the clock transition for a sufficient hold time. This accommodation limits the speed with which the whole system can switch.

Can we do better? Yes, not perfect, but better. The device of choice is the edge-triggered DFF. We will not go into the details of the implementation of an edge-triggered flip-flop as it is considerably more complicated than the circuits considered so far. An edge-triggered flip-flop is designed to pass the input datum to the output during a very short amount of time defined by a clock edge. Edge-triggered flip-flops can either be *rising edge* triggered or *falling edge* triggered. A rising edge triggered flip-flop will only change its output on the rising edge of a clock. There is still a setup time and a hold time; a small amount of time right around the clock edge during which the input datum must be stable in order to avoid the flip-flop becoming metastable. The advantage of the edge-triggered design is that there is only one brief moment when any critical changes take place. This improves synchronization and leads to faster circuits.

There are other types of flip-flops as well, but the DFF is widely used and useful for building more complex circuits. We will consider these as we build more complex circuits. Most designs of a positive edge-triggered DFF include two additional **asynchronous** inputs, *preset* and *clear*. Asynchronous inputs cause the output to change independently of the clock input. The D input is a *synchronous* input; changes on the D input are only reflected at the output during a clock edge. An active signal on the preset input

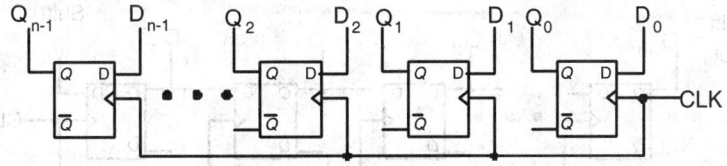

FIGURE 16.26 An *n*-bit data register built from *n* DFFs.

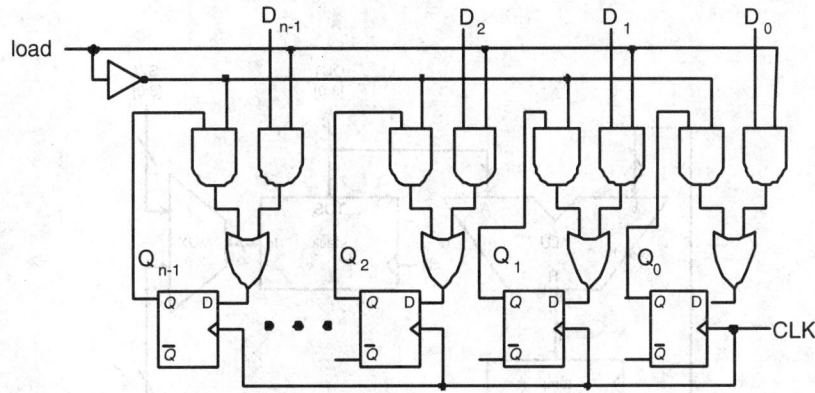

FIGURE 16.27 An *n*-bit shift register with load input. The upper layer is a set of *n* two-input MUXs. The bottom layer is a set of *n* positive-edge-triggered DFFs.

causes the output of the FF to be set to 1; an active signal on the clear input causes the output to be cleared, or set to 0. These inputs are useful for putting flip-flops in a design into a known state.

16.6.3 From DFF to Data Register, Shift Register, and Stack

The simplest and most useful device we can build with DFFs is a data register. A data register stores data. The simplest data register stores new data on every clock edge. Its design is shown in Figure 16.26. It is useful to control when the data register stores new data. We can extend the data register by adding a control signal called *load*. Now our register will load new data only when *load* is high. Otherwise it will store its old data. The new register design is shown in Figure 16.27.

We can also build a *shift* register, which shifts data, one bit at a time, into a parallel register. Shift registers are useful for converting serial data to parallel data. For example, you may receive data one bit at a time from a serial connection to your computer, but want to operate on the data in parallel. Shift registers can sometimes be loaded in parallel also, just like the data register. We will keep things simple; a serial-in parallel-out shift register is shown in Figure 16.28. This shift register also has a serial output.

16.6.3.1 A Stack for Holding Data

Our calculator requires memory to store variables that need to be manipulated. We could implement a random access memory with addressing, read, and write capabilities. Instead, to keep things simple, we will implement a *stack* for storing variables and results. A stack is a group of memory locations with limited access to its contents. At any given time, only the top of the stack can be read. This makes its implementation simple because general addressing does not need to be supported. Our stack supports three operations: push, pop, and hold. When a value is pushed onto the stack, it becomes the value at the top of the stack. All values already in the stack are pushed down by one. If the stack was full *before* the push operation, then the oldest (first in) value on the stack is lost. In a pop operation, the top of the stack is deleted, and all other values move up in the stack by one position. Note that this operation differs from a software *pop*

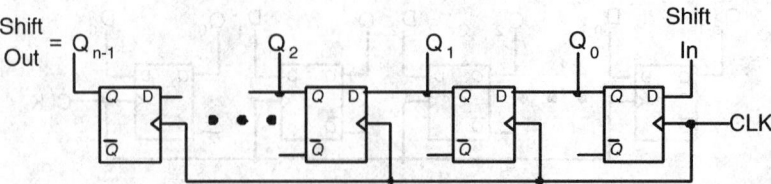

FIGURE 16.28   An *n*-bit shift register with serial input, parallel and serial output. This register shifts one bit to the left every clock cycle.

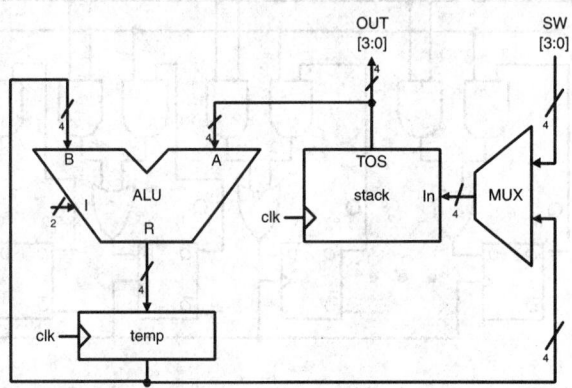

FIGURE 16.29   Calculator datapath. The temp register stores the ALU results every CLK edge.

operation. In software, the popped value is stored in a register. In our hardware implementation, there is no register storing the removed value, so this value is lost. A stack is sometimes called a Last In First Out (LIFO) because that is the order in which values are accessed. The hold operation ensures that the current contents of the stack are retained. It is important to explicitly support this so that the contents of the stack are not changed during operation. Push, pop, and hold all happen on a clock edge. By default, the stack holds its contents when there is no clock edge.

We will implement a stack to hold 4-bit variables. Our stack will contain four locations. One can implement this stack as four shift registers (one for each bit position) with each shift register containing four flip-flops. The total memory contents of our stack is held in 16 DFFs. In a real calculator implementation, the stack would be implemented with memory cells which use fewer transistors and consume less power than DFFs, but for our small design, DFFs will suffice.

16.6.4   Datapath for a 4-bit Calculator

Let's put the pieces together and implement the datapath of a 4-bit calculator. We will use the stack described above as our memory, and the ALU described in the previous section to implement logical operations. We need to add connections and a way to input data values. We also need an *instruction* to tell the calculator datapath what to do. We will use a 7-bit input that can be connected to toggle switches to provide the data and instructions. The datapath for the calculator is shown in Figure 16.29. The output is the top of stack (TOS). This is represented as the four bits OUT[3:0]. These could be hooked to a seven-segment display to display the results.

Our calculator will support the instructions: **push**, **pop**, **and**, **or**, **add**, and **sub**tract. Note that the last four are the operations that our ALU supports. These are operations that take two 4-bit input values and produce a 4-bit result. The operands will be found on the stack. The input values will be popped off the stack, and the output value will be pushed onto the top of the stack. For a push operation, the value to be

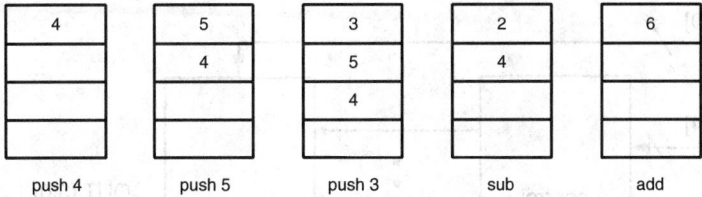

FIGURE 16.30   Stack contents for calculating $4 + (5 - 3)$.

TABLE 16.15   Switch Settings for Calculator Instructions

Instruction	$SW_6 \ldots SW_4$	$SW_3 \ldots SW_0$
AND	000	unused
OR	001	unused
ADD	010	unused
SUB	011	unused
PUSH	101	data
POP	110	unused

pushed onto the stack will be entered via the input switches. Suppose we want to calculate: $4 + (5 - 3)$. The operations required are:

```
PUSH 4
PUSH 5
PUSH 3
SUB
ADD
```

The contents of the stack for this sequence of operations is shown in Figure 16.30.

The format of the instructions and data on the input switches is shown in Table 16.15. Note that only the PUSH instruction uses external data.

What is missing from our calculator design? A controller to make sure that the right operations are performed on the datapath in the correct sequence for each instruction. While the design of such a controller is beyond the scope of this chapter, we will specify its behavior. For each instruction, the datapath goes through a sequence of operations or states. There are a finite number of such states to describe the behavior of the controller, hence the controller is a finite-state machine. Such a finite-state machine can be implemented with flip-flops to hold the state, and combinational logic to generate the next state and the outputs required. The inputs to the state machine are the current state and, for the calculator, the instruction that is entered via the switches. There is also a button the user needs to be able to press that indicates that the next instruction is ready to execute. The outputs are the control signals for the datapath. Our controller needs to output two bits for the ALU to tell it what to do; two bits for the stack to generate the push, pop, or hold inputs, and one bit for the MUX to tell it which input to use. The temporary register requires no control signals — it will be updated every clock cycle. The entire calculator design is shown in Figure 16.31.

Let's look at how the instructions are executed on our calculator. There are three *classes* of instruction: **push**, **pop**, and ALU (**add, sub, and, or**). The **pop** instruction is the simplest. All that is needed is one state to pop the stack. In that state, the control inputs to the stack should be "pop." We don't care what the ALU or MUX control inputs are set to **push** is also a one state instruction. In that state, the data on the switches is pushed onto the stack. So the control input for the MUX should be set to select the input switches and the control input for the stack should be set to "push."

The ALU instructions are more complicated; they require three control states. We assume that our calculator has the two operands on the stack before an ALU instruction is executed. In the first state, the data value

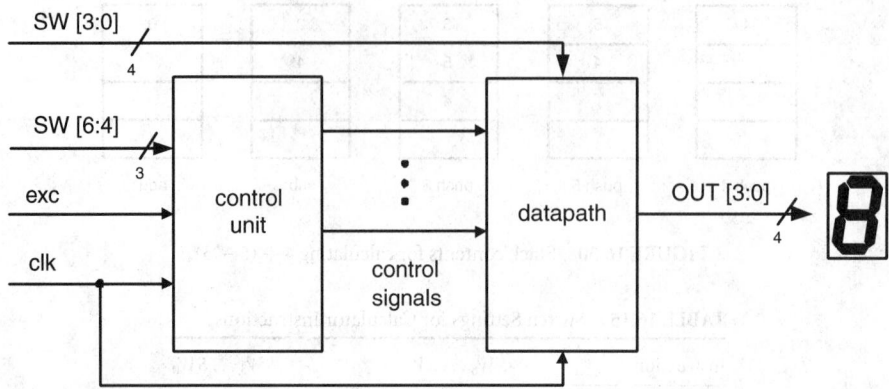

FIGURE 16.31   Connecting the controller, datapath, inputs, and outputs of the calculator.

on the top of the stack is stored in the temp register and the stack is popped. How can we accomplish this in one state? Note that the ALU is a combinational circuit and the TOS is already the A input to the ALU. If we select the ALU operation to pass A through to the output of the ALU, then on the clock edge, we will simultaneously store the current TOS value in the temp register, and pop the value off the stack. Now, at the start of the second state, the first operand is at input B of the ALU because the temp register output is connected to input B. The second operand is at input A of the ALU because the TOS is connected to input A. In this state we will execute the correct ALU instruction, store the result in the temp register, and pop the A operand off the stack. Finally, in the third state the contents of the temp register (i.e., the results of the ALU operation) are pushed onto the stack. At the end of an ALU operation, the two operands have been popped from the stack, and the result pushed on the stack as required. To summarize, the three states of an ALU operation are:

```
state 1:   temp <- operand A;  pop
state 2:   temp <- A op B  ;  pop
state 3:                      push temp
```

We have discussed what happens when an instruction is executing. What happens between instructions? The stack should "hold" its contents. It does not matter what the ALU is doing, or what input the MUX passes through. The memory of this design is in the stack. As long as the stack holds its contents, the memory is maintained. The temporary register is updated every clock cycle, but the results are not saved, so this does not affect the correct operation of the calculator datapath. Note that the output is always active, and it always shows what is currently on the top of the stack.

We have presented basic digital components, both combinational and sequential, and put them together to build a useful design: a simple calculator. Using essentially the same techniques, much more complicated devices can be constructed. Many of the digital devices we take for granted — digital watches, antilock braking systems, microwave oven controllers, etc. — are the implementations of designs using these techniques. In the final section of this chapter, we will consider how such devices are physically realized.

16.7   ASICs and FPGAs — Faster, Cheaper, More Reliable Logic

The circuits we have considered so far — encoders, decoders, muxes, flip-flops, etc. — can be bought in packages with several to a chip. For example, a 14-pin package containing four 2-input NANDs per package has been available for more than 30 years. These chips are called small-scale integrated circuits (SSI). No one who has watched the astonishing decline in the cost of digital electronics brought about by Very Large Scale Integrated circuits (VLSI) would expect to find engineers generating circuits by hooking up vast arrays of such packages. In a world where powerful computer chips roll off the line with more than 10 million transistors all properly connected and functioning at clock speeds in excess of 1 GHz, why would we be manually hooking up hundreds of these small packages with 16 transistors in a chip that takes 15 to 20 ns to get a signal through a single NAND? Today, the equivalent of many pages of random logic circuit diagrams can be implemented in a single chip. There are many different ways to specify such designs and many different ways to physically realize them. The dominant method of design entry is using a Hardware Description Language (HDL). HDLs resemble software programming languages with added features specifically for describing hardware such as bitwidth, I/O ports, and controller state specifications. Design tools translate HDL descriptions of a circuit to the final implementation. One of the goals of an HDL design is to separate the design from the implementation technology. The same HDL description, in theory at least, can be mapped to different target technologies.

There are also many technologies for physically realizing digital logic designs. Application Specific Integrated Circuits (ASICs) can be implemented as VLSI circuits where millions of transistors are realized on a single chip resulting in very high speeds and very low power dissipation. Such designs are manufactured at a foundry and cannot be changed after they have been implemented. Because high-performance VLSI designs are very expensive to manufacture, they are increasingly used only for very large volume designs and designs where low power is critical. For example, VLSI ASIC chips are found in mobile phones and other handheld devices that meet these criteria.

Designers are increasingly turning to *programmable* and *reconfigurable* devices for realizing their designs. These devices are manufactured in large quantities using VLSI techniques with the latest technology. They are specialized to a particular design after the fabrication process, and hence are programmable. Devices that can be reprogrammed, to fix errors or update functionality, are also called reconfigurable. One of the most popular of these devices is the Field Programmable Gate Array (FPGA). Modern FPGAs can implement designs with the equivalent of millions of transistors on a single chip and operate with clock speeds of several hundred MegaHertz. FPGAs are much more cost-effective than ASICs for many designs, and enjoy an increasing market share in digital hardware products.

For both FPGAs and ASICs, all of the steps that take the initial design to finished chip can be automated. The initial design can be described as a schematic, similar to the diagrams in this chapter, or using an HDL. In the case of FPGAs, design tools translate this specification into programming data that can be downloaded to the chip. As we shall see, FPGA chips are based on memory technology. Rather than downloading a data file to memory, you download a configuration file to an FPGA that changes the way the hardware functions. This *programming* of the chip is very rapid. One can make a change in a complex design and have a working realization in less than an hour. By comparison, ASIC fabrication can take several weeks. By tightening the design cycle, such rapid prototyping has dramatically reduced the cost of designing and producing complex circuits for specific applications.

The underlying technology of an FPGA, and what makes it programmable and reconfigurable, is memory. Writing HDL programs and programming the FPGA makes the design process sound more like software than hardware development. The major difference is that the underlying structures being programmed implement the hardware structures we have been discussing in this chapter. In this section we introduce FPGA technology and explain how digital designs are mapped onto the underlying structures. There are several companies that design and manufacture FPGAs. We will use the architecture of the Xilinx FPGA as an example.

16.7.1 FPGA Architecture

Let's consider the architecture of Xilinx FPGAs. Our objective is not to learn how to program them — a task normally accomplished by software — but rather to show the relationship of these sophisticated chips to the logic we have already developed. The Xilinx chip is made up of three basic building blocks:

1. **CLB.** The configurable logic blocks are where the computation of the user's circuit takes place.
2. **IOB.** The input/output blocks connect I/O pins to the circuitry on the chip.
3. **Interconnect.** Interconnect is essential for wiring between CLBs and from IOBs to CLBs.

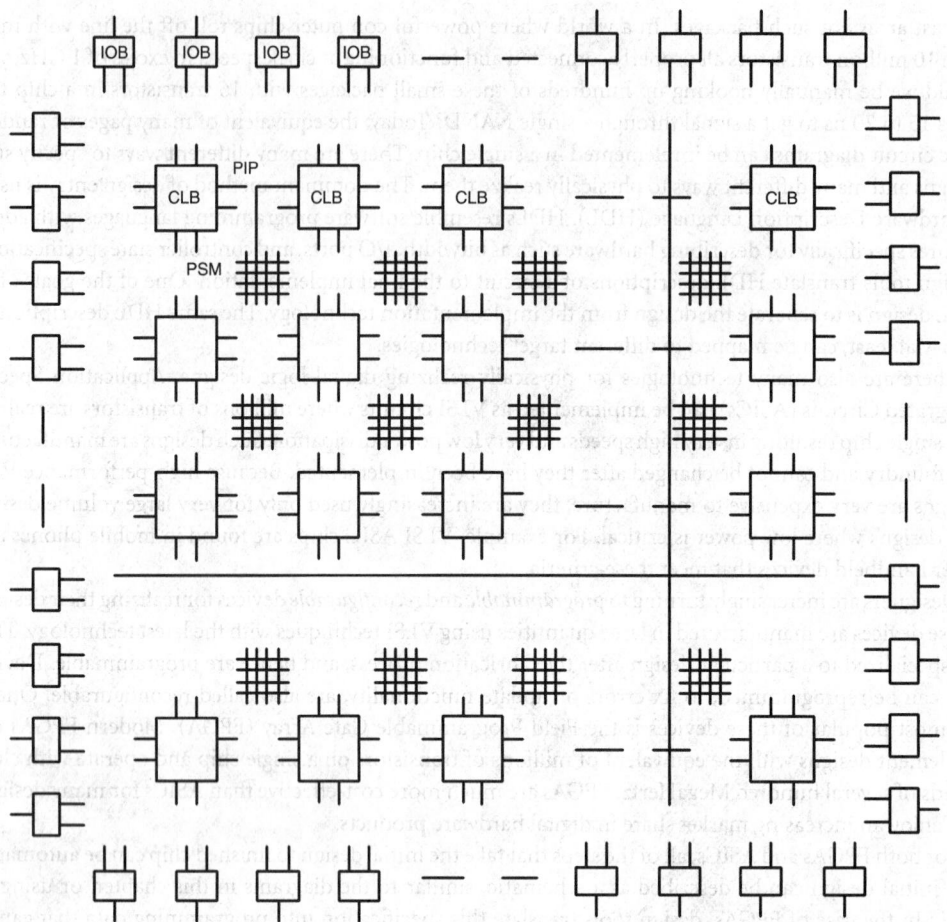

FIGURE 16.32 Overview of the Xilinx FPGA. I/O blocks (IOBs) are connected to pads on the chip, which are connected to the chip-carrier pins. Several different types of interconnect are shown, including Programmable Interconnect Points (PIPs), Programmable Switch Matrices (PSMs), and long line interconnect.

The Xilinx chip is organized with its configurable logic blocks (CLBs) in the middle, its I/O blocks (IOBs) on the periphery, and lots of different types of interconnect. Wiring is essential to support the versatility of the chips to implement different designs efficiently and to ensure that the resources on the FPGA can be utilized efficiently. The overview of a Xilinx chip is presented in Figure 16.32. Each CLB is programmable and can implement combinational or sequential logic or both. Data enter or exit the chip through the IOBs. The interconnect can be programmed so that the desired connections are made. Distributed configuration memory (not shown in the figure) controls the functionality of the CLBs and IOBs, as well as the wiring connections. Implementation of the CLBs, interconnect, and IOBs are described in more detail below.

16.7.1.1 The Configurable Logic Block

How do you use memory to implement different Boolean logic functions? You download the truth table of the function to the memory. Changing the contents of the memory cells changes the functionality of the hardware. One set of memory cells for a one-bit output is referred to as a lookup table (LUT) because you "look up" the result. The basic Xilinx logic slice contains a four-input LUT for realizing combinational logic. The result of this combinational function may or may not be stored in a D flip-flop. The implementation of a logic slice, with four-input LUT and optional flip-flop on the LUT output, is shown on the left in

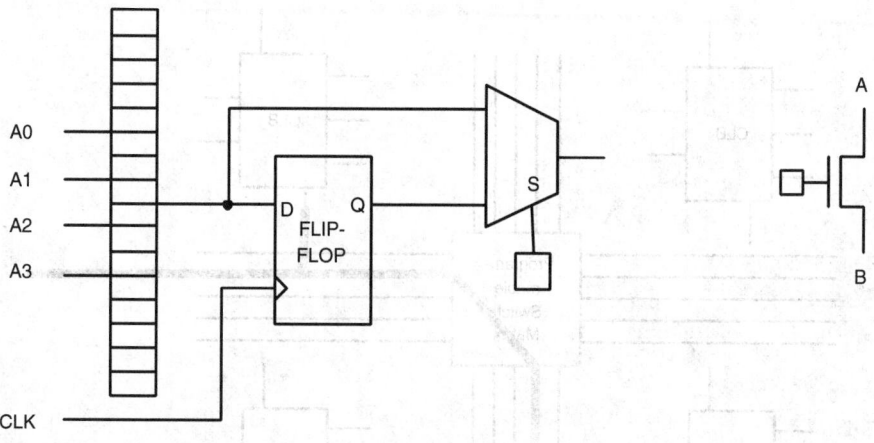

FIGURE 16.33 On the left, a simplified CLB logic slice containing one 4-input lookup table (LUT) and optional DFF. The 16 one-bit memory locations on the left implement the LUT. One additional bit of memory is used to configure the MUX so the output comes either directly from the LUT or from the DFF. On the right is a programmable interconnect point (PIP). LUTs, PIPs, and MUXes are three of the components that make FPGA hardware programmable.

Figure 16.33. Note that the multiplexer can be configured to output the combinational result of the LUT or the result of the LUT after it has been stored in the flip-flop by setting the memory bit attached to the MUX's select line. The logic is configured by downloading 17 bits of memory: the 16 bits in the lookup table and the one select bit for the multiplexer. Using this simple structure, all of the designs described so far in this chapter can be implemented.

The real Xilinx 4000 family CLB is considerably more complicated; however, the way it is programmed is the same. It essentially contains two of the logic slices shown in Figure 16.33. In addition, there is a third LUT whose three inputs can include the outputs of the two 4-input LUTs. The CLB is designed to be able to implement any Boolean function of five input variables, or two Boolean functions each of four input variables. With the third LUT, some functions of up to nine input variables can be realized. Extra logic and routing is also added to speed up the carry chain for an adder, because adders are such common digital components. In addition, additional routing and MUXes allow the flip-flops to be used independent of the LUTs as well as in conjunction with them.

16.7.1.2 Interconnect

Once the CLBs have been configured to implement combinational and sequential logic components, they need to be connected to implement larger circuits, just as we wired together the ALU, register, and MUXes to implement the calculator datapath. This requires programmable interconnect, so that an FPGA can be programmed with different connections depending on the circuit being implemented. The key is the programmable interconnect point (PIP) shown on the right in Figure 16.33. This simple device is a pass transistor with its gate connected to a memory bit. If that memory bit contains a 1, the two ends of the transistor are logically connected; if the memory bit contains a 0, no connection is made. By appropriately loading these memory bits, different wiring connections can be realized. Note that there is considerably more delay across the PIP than across a simple metal wire on a chip. This is the trade-off when using programmable interconnect.

Our FPGA architecture has CLBs arranged in a matrix over the surface of a chip, with routing channels for wiring between the CLBs. Programmable Switch Matrices (PSMs) are implemented at the intersection between a row and column of routing. These switch matrices support multiple connections, including signals on a row and a column, signals passing through on a row, and signals passing through on a column. Figure 16.34 shows a signal output from one CLB connecting to the inputs of two others. This signal passes through two PSMs and three PIPs, one for each CLB connection.

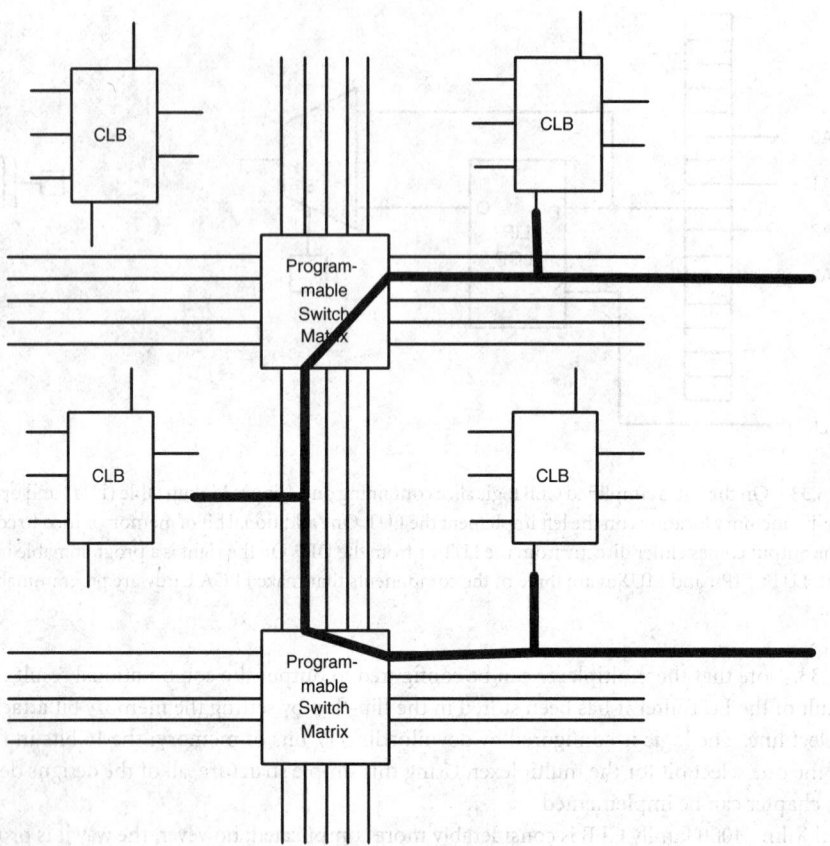

FIGURE 16.34 Programmable interconnect, including two programmable switch matrices (PSMs) for connecting the output of one CLB to the input of two other CLBs.

While programmable interconnect makes the FPGA versatile, each active device in the interconnection fabric slows the signal being routed. For this reason, early FPGA devices, where all the interconnect went through PIPs and PSMs, implemented designs that were considerably slower than their ASIC counterparts. More recent FPGA architectures have recognized the fact that high-speed interconnect is essential to high-performance designs. In addition to PIPs and PSMs, many other types of interconnect have been added. Many architectures have nearest neighbor connections, where wires connect from one CLB to its neighbors without going through a PIP. Lines that skip PSMs have been added. For example, double lines go through every other PSM in a row or a column. Long lines have been added to support signals that span the chip. Special channels for fast carry chains are available. Finally, global lines that transmit clock and reset signals are provided to ensure these signals are propagated with little delay. All of these types of interconnect are provided to support both versatility and performance.

16.7.1.3 The Xilinx Input/Output Block

Finally, we need a way to get signals into and out of the chip. This is done with I/O blocks (IOBs) that can be configured as input blocks, output blocks, or both (but not at the same time). The Output Enable (OE) signal enables the IOB as an output. If the OE signal is high, the output buffer drives its signal out to the I/O pad. If the OE signal is low, the output function is disabled, and the IOB does not interfere with reading the input from the pad. The OE signal can be produced from a CLB, thus allowing the IOB to sometimes be enabled as an output and sometimes not. In addition, IOBs contain DFFs for latching

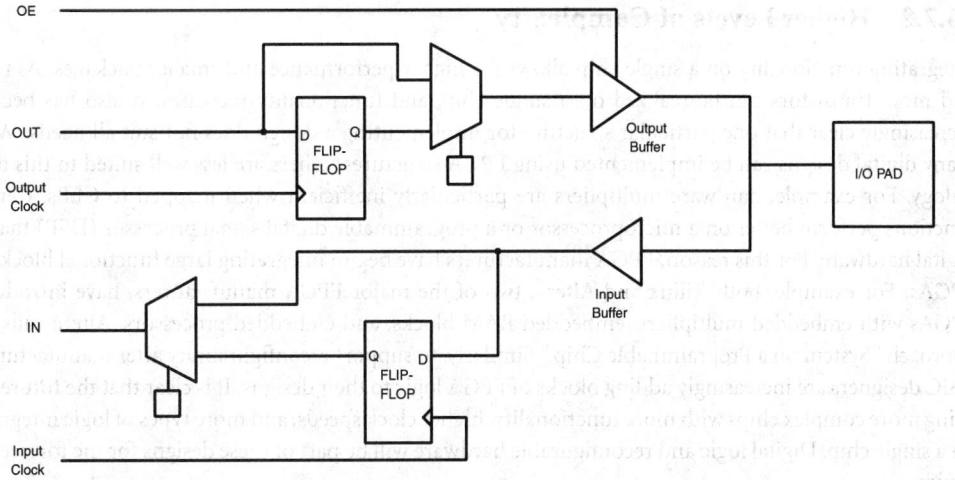

FIGURE 16.35 Simplified version of the IOB. IOBs can be configured to input or output signals to the FPGA. When OE is high, the output buffer is enabled so the output signal is driven on the I/O pad. When OE is low, the IOB functions as an input block. Buffers handle electrical issues with signals from the I/O pad.

the input and output signals. The latches can be bypassed by appropriately programming multiplexers. A simplified version of the IOB is shown in Figure 16.35. The actual IOB contains additional circuitry to properly deal with such electrical issues as voltage and current levels, ringing, and glitches, that are important when interfacing the chip to signals on a circuit board.

CLBs, IOBs, and interconnect form the basic architecture for implementing many different designs in a single FPGA. The configuration memory locations, distributed across the chip, need to be loaded to implement the appropriate design. For a Xilinx FPGA, these memory bits are SRAM, and are loaded on power-up. Special I/O pins that are not user configurable are provided to download the configuration bits that define the design to the FPGA. Other devices use different underlying technologies than SRAM to provide programmability and reconfigurability.

16.7.1.4 Mapping the Simple Calculator to an FPGA

Let's look at how the calculator design (Figure 16.31) is mapped onto a board containing an FPGA. The board used contains a Xilinx 4028E FPGA, switches, push-buttons, and seven-segment displays. The calculator was designed to map to this board, with switches used for entering instructions and data, a push-button for the EXC command, and a seven-segment display used to show the top of stack. The logic of the calculator is mapped to CLBs. The controller is made up of Boolean logic and DFFs to hold the state. The datapath is made up of the components developed in this chapter and mapped to LUTs and DFFs.

This calculator was developed as an undergraduate laboratory experiment. Students enter the design using a schematic capture tool, which involves drawing diagrams like the ones in this chapter. Synthesis tools translate the design to LUTs and flip-flops, breaking the logic up into four-input chunks of logic, each of which is implemented with one truth table. Alternatively, the design can be described using a hardware description language to specify behavior. A different synthesis tool is involved, but the end result is a set of LUTs and DFFs that implement the design. Placement tools map these components to CLBs on the chip, and routing tools route the connections, making use of the various kinds of interconnect available. The tools translate the logic design to a bitstream that is downloaded to the board from a PC through a download cable. The result is a functioning calculator on an FPGA board.

An advantage of this design flow is that designers can migrate their designs to the newest chip architecture without changing the specification. Only the tools need to change to target a faster or cheaper device.

16.7.2 Higher Levels of Complexity

Integrating functionality on a single chip allows for higher performance and smaller packages. As more and more transistors can be realized on a single chip, and functionality increases, it also has become increasingly clear that one particular structure for implementing a design does not suit all needs. While many digital designs can be implemented using FPGA structures, others are less well suited to this technology. For example, hardware multipliers are particularly inefficient when mapped to CLBs. Certain functions perform better on a microprocessor or a programmable digital signal processor (DSP) than in digital hardware. For this reason, FPGA manufacturers have begun integrating large functional blocks on FPGAs. For example, both Xilinx and Altera, two of the major FPGA manufacturers, have introduced FPGAs with embedded multipliers, embedded RAM blocks, and embedded processors. Altera calls this approach "System on a Programmable Chip." Similarly, to support reconfigurability after manufacturing, ASIC designers are increasingly adding blocks of FPGA logic to their designs. It is clear that the future will bring more complex chips with more functionality, higher clock speeds, and more types of logic integrated on a single chip. Digital logic and reconfigurable hardware will be part of these designs for the foreseeable future.

Acknowledgment

The author acknowledges the significant contribution of James Feldman, author of the original version of the chapter, to the current organization and content.

Defining Terms

Active logic: Digital logic that operates all of the time in the active, dissipative region of the electronic amplifiers from which it is constructed. Such logic is generally faster than saturated logic, but it dissipates much more energy. *See* **saturated logic**.

ASIC: Application-specific integrated circuit. Integrated circuits that are designed for a specific application. The term is used to describe VLSI circuits that can be configured before manufacturing to meet the specific needs of the application. High-performance ASICs are low power and high speed, but also expensive.

CLB: Configurable logic block in a Xilinx FPGA. This is where the computation occurs in an FPGA. CLBs implement truth tables and D flip-flops.

Clock: The input that provides the timing signal for a circuit. In general, the oscillator circuit that generates a synchronization signal.

CMOS: Complementary metal–oxide–semiconductor. The dominant family of binary, saturated logic. CMOS circuits are built from MOS transistors in complementary pairs. When one transistor is open, its complement is closed, ensuring that no current flows through the switch itself. Such a configuration has minimum power dissipation in any static state.

Combinational circuit: A logic circuit whose output is a function only of its inputs. Apart from propagation delays, the output always represents a logical combination of its present inputs. *See* **sequential circuit**.

Critical race: In a sequential circuit, a situation in which the "next state" is determined by which of two (or more) internal gates is first to reach saturated state. Such a race is dependent on minor variations in circuit parameters and on temperature, making the circuit unpredictable and possibly even unstable.

Decoder: A logic circuit with N inputs and 2^N outputs, one and only one of which is asserted to indicate the numerical value of the N input lines read as a binary number. A decoder and demultiplexer use the same internal circuitry.

Demultiplexer: (DEMUX) A logic circuit with K control inputs that steers the data input to the one of the 2^K outputs selected by the control inputs.

D flip-flop (DFF): Data flip-flop. A fundamental sequential circuit whose output changes only upon a clock signal and whose output represents the data on its input at the time of the last clock.

Don't care: In a truth table or Karnaugh-map, a state that is irrelevant to the correct functioning of the circuit (e.g., because it never occurs in the intended application). Thus, the designer "doesn't care" whether that state is asserted, and he or she may choose the output that best minimizes the number of gates.

Edge-triggered FF: A flip-flop that changes state on a clock transition from low to high or high to low rather than responding to the level of the clock signal. Contrast to **master–slave FF**.

Encoder: A logic circuit with 2^N inputs and N outputs, the outputs indicating the binary number of the one input line that is asserted. *See also* **priority encoder**.

Flip-flop: Any of several related bistable circuits that form the memory elements in clocked, sequential circuits.

FPGA: Field-programmable gate array. VLSI chips with a large number of reconfigurable gates that can be "programmed" to function as complex logic circuits. The programming can be done on site and in some cases may be dynamically (in circuit) reprogrammable.

Glitch: A transient transition between logic states caused by different delays through parallel paths in a logic circuit. They are unintentional transitions, so they do not correctly represent the logic of the intended design.

HDL: Hardware Description Language. A language that resembles a programming language with added features for specifying hardware designs.

IOB: I/O block in a Xilinx FPGA. Block on the periphery of an FPGA that supports the input and/or output of a signal from the FPGA to its external environment.

Karnaugh map: A mapping of a truth table into a rectangular array of cells in which the nearest neighbors of any cell differ from that cell by exactly one binary input variable. K-maps are useful for minimizing Boolean logic functions.

Master–slave FF: A flip-flop that changes state when the clock voltage reaches a threshold level. Contrast to **edge-triggered FF**.

Multiplexer: (MUX) A circuit with N control inputs to select among one of 2^N data inputs, and connect the appropriate data input to the single output line.

PIP: Programmable Interconnect Point on a Xilinx FPGA. A pass transistor with a memory bit connected to its gate terminal. If the memory is loaded with a "1" the two ends are connected; if loaded with a "0," the two ends are not connected. This is the basis of programmable interconnect.

Priority encoder: An encoder with the additional property that if several inputs are asserted simultaneously, the output number indicates the numerically highest input that is asserted. For example, if lines 1 and 3 were both asserted, the output value would be 3.

Saturated logic: Logic gates whose output is fully on or fully off. Saturated logic dissipates no power except while switching. The opposite of *saturated logic* is *active logic*.

Sequential circuit: A circuit that goes through a sequence of stable states, transitioning between such states at times determined by a clock signal. The output of a sequential circuit depends both on its current inputs and its history, which is captured in the states. Contrast with **combinational** circuit.

Transparent latch: Essentially, a flip-flop that continuously passes the input to the output (thus *transparent*) when the clock is high (low) but holds the last output during any interval when the clock is low (high). The circuit is said to have *latched* when it is holding its output constant regardless of the value of the input.

VLSI: Very Large Scale Integrated Circuit. A semiconductor device that integrates millions of transistors on a single chip. VLSI chips are typically very high speed and have very high power dissipation.

References

Ashenden, P.J. 2001. *The Designer's Guide to VHDL*, 2nd ed. Morgan Kaufmann.

Boole, G. 1998. *The Mathematical Analysis of Logic*. St. Augustine Press, Inc.

Karnaugh, M. 1953. A map method for synthesis of combinational circuits. *Trans. AIEE. Comm. and Electron.*, 72(1):593–599.

Katz, R.H. 2003. *Contemporary Logic Design*, 2nd ed. Addison-Wesley Publishing.

Mano, M.M. and Kime, C.R. 2000. *Logic and Computer Design Fundamentals*, 2nd ed. Prentice Hall.

Moorby, P.R. and Thomas, D.E. 2002. *The Verilog Hardware Description Language*, 5th ed. Kluwer Academic
    Publishers.

Salcic, Z. and Smailagic, A. 2000. *Digital Systems Design and Prototyping Using Field Programmable Logic
    and Hardware Description Language*, 2nd ed. Kluwer Academic Publishers.

Wakerly, J.F. 2000. *Digital Design: Principles and Practice*, 3rd ed. Prentice Hall.

Zeidman, R. 2002. *Designing with FPGAs and CPLDs*. CMP Books.

Further Information

This is a very quick pass through digital circuit design. What has been covered in this chapter provides
a good overview of the principles as well as information to help the reader understand the chapters on
computer architecture in this volume.

There are many textbooks devoted to the subject of digital logic design. Wakerly [2000] emphasizes
basic principles and the underlying technologies. Other digital logic texts emphasize logic design tools
[Katz 2003] and computer design fundamentals [Mano and Kime 2000].

There have also been many volumes published on design entry, tools for automating the digital desgin
process, and mapping designs onto Field Programmable Logic (FPL). The Hardware Description Languages
(HDLs) most widely used today are VHDL [Ashenden 2001] and Verilog [Moorby and Thomas 2002].
The interested reader may also wish to pursue the topic of design with field programmable logic [Zeidman
2002]. Salcic and Smailagic [2000] brings the subjects of logic design, FPL, and HDLs together in one
volume.

Ongoing research in this area is concerned with design entry, automation of the design process, and new
architectures and technologies for implementing digital designs. The research in design entry is focused on
raising the level of specification of digital logic designs. New HDLs based on high-level languages such as
Java and C are being developed. Another approach is design environments that incorporate sophisticated
libraries of very complex, parameterized components such as digital filters, ALUs, and Ethernet controllers.
The user can customize these blocks for a specific design.

Along with higher levels of design specification, researchers are investigating more sophisticated design
automation tools. The goal is to have designers specify the functionality of their designs, and to use synthesis
tools to automatically translate that functionality to efficient hardware implementations.

17

Digital Computer Architecture

17.1 Introduction 17-1
The Processor-Program Interface

17.2 The Instruction Set 17-2
ALU Instructions • Memory and Memory Referencing
Instructions • Control Transfer Instructions

17.3 Memory 17-5
Register File • Main Memory • Cache Memory
• Memory and Architecture • Secondary Storage

17.4 Addressing 17-7
Addressing Format • Physical and Virtual Memory

17.5 Instruction Execution 17-9
Instruction Fetch Unit • Instruction Decode Unit
• Execution Unit • Storeback Unit

17.6 Execution Hazards 17-12
Data Hazards • Control Hazards • Structural Hazards

17.7 Superscalar Design 17-15

17.8 Very Long Instruction Word Computers 17-15

17.9 Summary 17-16

David R. Kaeli
Northeastern University

17.1 Introduction

A computer architecture is a specification which defines the interface between the hardware and software. This specification is a contract, describing the features of the hardware which the software writer can depend upon and identifying design implementation issues which need to be supported in the hardware.

While the term computer architecture has commonly been used to define the interface between the instruction set of the processing element and the software that runs on on the processsing element, the term has also been applied more generally to define the overall structure of the computing system. This structure generally includes a processing element, memory subsystem, and input/output devices. We will first discuss the more traditional use of the term, focusing on how a program interacts with a processing element, and then discuss design issues associated with the broader definition of this term.

17.1.1 The Processor-Program Interface

Tasks are carried out on a computer by software specific to each program task. A simple program, written in the high-level language C, is shown in Figure 17.1. This program computes the difference of two integers. Even within a simple programming example we are able to identify some of the necessary elements in a computer architecture (e.g., arithmetic and assignment operations, integers).

```
x = 5;          /* Initialize x to 5 */
y = 3;          /* Initialize y to 3 */
z = x - y;      /* Compute the difference */
```

FIGURE 17.1   High-level language program.

```
x:      .int 5
y:      .int 3
z:      .
        load r1, x
        load r2, y
        sub r3, r1, r2
        store r3, z
```

FIGURE 17.2   Assembly-language version of HLL program.

High-level languages passes through a series of phases of transformation, including compilation, assembly, and linking. The result is a program that can run on an *execution processor*.

An assembly-code version of the subtraction machine code is shown in Figure 17.2. After the assembly code (stored in an object file) is then compiled, linking is performed. Linking merges all compiled elements of the program and constructs the final machine-language representation of the tasks to be performed.

The machine code of the computer system comprises a set of primitive operations which are performed with great rapidity. We refer to these operations as instructions. The set of instructions provided is defined in the specification of the architecture. An instruction set is just one aspect of defining an architecture.

Instructions are executed on the processor (commonly called the CPU or microprocessor), which may comprise a number of units, including (1) an *arithmetic logical unit* (ALU), (2) a *floating-point unit* (FPU), (3) local memory, and (4) external bus control. A particular computer architecture can be realized in hardware in a wide variety of hardware organizations. What remains constant across these different implementations of the architecture is a common software interface that programmers can depend upon. The Intel Pentium IV and SPARCV9 architectures are two good examples of well-defined computer architectures.

In this chapter various aspects of a computer architecture are presented. The design of a digital computer will also include the supporting memory system and input/output devices necessary. We will begin our discussion by describing the fundamentals of an instruction set.

17.2   The Instruction Set

An instruction set includes the machine-language primitives which can be directly executed on a processor. Instruction classes present in most instruction sets include:

1. ALU instructions (e.g., integer add/subtract, shift left/right, logical and/or/xor/inversion).
2. Memory accessing instructions (e.g., load and store).
3. Control transfer instructions (e.g., conditional/unconditional branch, call/return, and interrupt).

Two prevalent instruction set implementation paradigms are *reduced instruction set computers* (RISC) [Patterson 1985] and *complex instruction set computers* (CISC) [IBM 1981]. RISC microprocessors are designed around the concept that by keeping instructions simple, the compiler will have greater freedom to produce optimized code (i.e., code which will execute faster). CISC architectures use a very different principle, attempting to perform operations specified in the high-level language in a single instruction (commonly referred to as reducing the semantic gap). While RISC architectures have become the popular paradigm, the Intel *x*86 and Motorola 680*x*0 architectures continue to employ a CISC architecture (actually the more recent Intel $x86$ processors utilize a CISC instruction set, but are implemented as RISC machines at the hardware level).

While these CISC and RISC models are based on different principles, they both contain instructions from the three classes above. The underlying principles are that RISC instruction sets are simple (or reduced) and that CISC instruction sets are complicated (or complex).

Most architectures include floating-point instructions. Those implementations of the architecture which contain a floating-point unit (FPU) are able to execute floating-point operations at speeds approaching integer operations. If a floating-point unit is not provided, then floating-point instructions are emulated by the integer processor (using a software program). When the hardware encounters a floating-point instruction, and if no FPU is present, a message (i.e., a software interrupt) is presented to the operating system. In response to this message, the floating-point instruction is executed using a number of integer instructions (i.e., it is emulated). The performance of emulated floating-point instructions is typically 3–4 orders of magnitude slower than if an FPU were present. Specific instructions may also be provided to manipulate decimal or string data formats. Most modern architectures include instructions for graphics and multimedia (e.g., the Intel $x86$ provides an architectural extension for multimedia called MMX)[Peleg 1997].

17.2.1 ALU Instructions

An ALU, which builds on the adder/subtracter presented in the chapter on digital logic, is used to perform simple operations upon data values. The ALU performs a variety of operations on input data fed from its two input registers and stores the result in a third register. Figure 17.3 shows an example of an ALU which might be found in the CPU.

The ALU is supplied data from a pair of registers a and b (registers are typically designed using D-flip-flops). The resulting answer is placed into register c. The function performed is determined based on the value of the select lines, which is directly derived by the instruction currently being executed. Table 17.1 shows the possible values for the three control lines shown in Figure 17.3.

The select lines are decoded (e.g., with a 3-to-8 decoder) to specify the desired operation. For example, the machine code format for the subtract instruction's operation code would contain (or be decoded into) the bit values 001. The assignment of these values is defined by the architecture and generally appears in a programmer's reference manual for the particular instruction set.

17.2.2 Memory and Memory Referencing Instructions

To run our program, we must have a place to store the machine code, and also a place to act as a scratch pad for intermediate results. Main memory provides a place where we can temporarily store both the machine instructions and data values associated with our program.

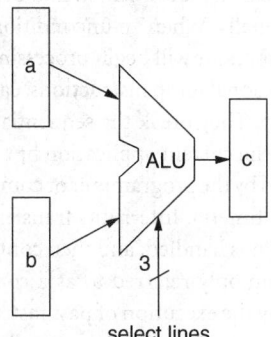

FIGURE 17.3 ALU with input and output registers.

TABLE 17.1   Sample ALU Operations

Operation Selected	Select-Line Value
$c = a + b$	000
$c = a - b$	001
$c = a$ shifted left b bits	010
$c = a$ shifted right b bits	011
$c = a$ OR b	100
$c = a$ AND b	101
$c = a$ XOR b	110
$c = $ NOT a	111

When a program begins execution, it is loaded into memory by the operating system. In our example, the values of the variables **x** and **y** in our high-level language program need to be initially stored in memory. The compiler will reserve memory locations to store these values and will provide instructions which will retrieve these values to initialize **x** and **y**. Then **x** and **y** are supplied to the input of the ALU via instructions that load them from registers **r1** and **r2**. The subtract instruction will tell the ALU to produce the difference (**r1**−**r2**) in register **r3**.

A computer architecture defines how data are stored in memory and how they are retrieved. To retrieve the values of **x** and **y** from memory, a load instruction is used. A load retrieves data values from memory and loads them into registers (e.g., data values **x** and **y**, and registers **r1** and **r2**). A store instruction takes the contents of a register (e.g., register **r3**) and stores it to the specified memory location (e.g., the memory location which the compiler assigned to **z**). These instructions are necessary for obtaining the data upon which the CPU will operate and to store away results produced by the CPU.

The CPU needs a way to differentiate between different data. Just as we are assigned Social Security numbers to differentiate one taxpayer from another, memory locations are assigned unique numbers called memory addresses. Using a memory address, the CPU can retrieve the desired datum. We will discuss addressing a little later and will focus on the aspects of addressing which are defined by the computer architecture.

RISC CPUs provide individual instructions for loading from and storing to memory. Pure RISC architectures provide only two memory-referencing instructions, load and store. All ALU instructions can have only input and output operands specified as registers or integer values. In contrast, CISC architectures provide a variety of instructions which both access memory and perform operations in a single instruction. A good discussion comparing the implications of these capabilities can be found in Colwell et al. [1985].

17.2.3   Control Transfer Instructions

Control transfer instructions cause a break in the sequential flow of program execution. The transfer can be performed unconditionally or conditionally. When an unconditional control transfer occurs (commonly referred to as a jump), the execution processor will begin processing instructions from a different portion of the program. Jumps include unconditional jump instructions, call instructions, and return instructions.

Interrupts are another form of jump. They break the sequential instruction flow unconditionally and transfer control not to another part of the current application but instead to an interrupt service routine. Interrupts can be programmed (inserted by the programmer or compiler) or they can occur due to hardware or software events (e.g., input/output, timers). Interrupts transfer control over to the operating system. The event which caused the interruption is handled, and then control is later passed back to the program.

A conditional control transfer (commonly referred to as a conditional branch) is dependent on the execution processor state determined by the execution of past instructions. Conditional branches are used for making decisions in control logic, checking for errors or overflows, or a variety of other situations. For instance, we may want to check if the result of adding our two integer numbers produced an overflow (a result which cannot be represented in the range of values provided for in the result).

Conditional branch instructions can perform both the decision making and the control transfer in a single branch instruction, or a separate comparison instruction (i.e., an ALU operation) can perform the comparison upon which the branch decision is dependent. Next, we explore the memory elements provided in the support of the execution processor.

17.3 Memory

The interface between the memory system and CPU is also defined by the architecture. This includes defining the amount of space addressable, the addressing format, and the management of memory at various levels of the memory hierarchy.

In traditional computer systems, memory is used for storing both the instructions and the data for the program to be executed. Instructions are fetched from memory by the CPU, and once decoded (instructions are typically in an encoded format, similar to that found in Table 17.1 for ALU instructions), the data operands to be operated upon by the instruction are retrieved from, or stored to, memory. Memory is typically organized in a hierarchy. The closer the memory is to the CPU, the faster (and more expensive) it is. The memory hierarchy shown in Figure 17.4 includes the following levels:

1. Register file.
2. Cache memory.
3. Main memory.
4. Secondary memory (disk, tapes, etc.).

Above each level in Figure 17.4 is a measure of its typical size and access speed in contemporary technology. Notice we include multiple levels of the cache, which are commonly found as on-chip elements in today's CPUs.

17.3.1 Register File

The register file is an integral part of the CPU and, as such, is clearly defined by the architecture. The register file can provide operands directly to the ALU. Memory referencing instructions either load to or

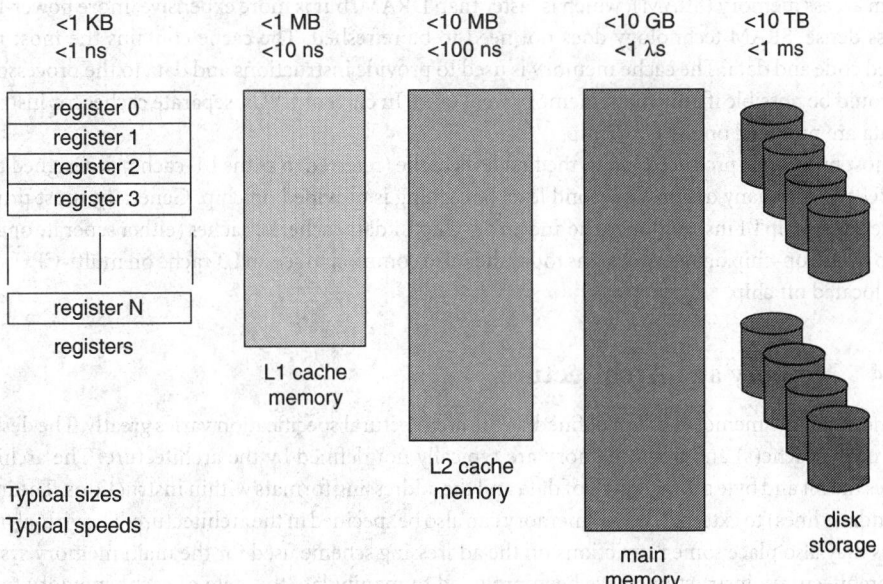

FIGURE 17.4   The memory hierarchy.

store from the registers contained in the register file. The register file can contain both general-purpose registers (GPRs) and floating-point registers (FPRs). Additional registers for managing the CPU state and addressing information are generally provided.

The register file represents the lowest level in the memory hierarchy, since it is closest to the processor. The registers are constructed out of fast flip-flops. The typical size of a GPR in current designs is 32 or 64 bits, as defined by the architecture. Registers are typically accessible on either a bit, byte, halfword, or fullword granularity (a word refers to either 32 or 64 bits). While a processor many have many hundred hardware registers, the architecture generally defines a smaller set of general purpose registers. In many instruction sets, use of particular registers is reserved for use by selected instructions (e.g., in the Intel $x86$ architecture, register CX holds the count value used by the *LOOP* instruction).

17.3.2 Main Memory

Main memory is physical (versus virtual) memory which typically resides off the CPU chip. The memory is usually organized in banks and supplies instructions and data to the processor. Main memory is typically byte-addressable (meaning that the smallest addressable quantity is 8 bits of instruction or data). Main memory is generally implemented in dynamic random-access memory (DRAM) to take advantage of DRAM's low cost, low power drain, and high storage density. The costs of using this memory technology include reduced storage response time and increased design complexity (DRAM needs to be periodically refreshed, since it is basically a tiny capacitor).

Main memory is typically organized to provide efficient access to sequential memory addresses. This technique of accessing many memory locations in parallel is called interleaving. Interleaved memory allows memory references to be multiplexed between different banks of memory. Since memory references tend to be sequential in nature, allowing the processor to obtain multiple addresses in a single access cycle can be advantageous. Multiplexing can also provide a substantial benefit when using cache memories.

17.3.3 Cache Memory

Cache memory is used to hold a small subset of the main memory. The cache is typically developed in static random access memory (SRAM), which is faster than DRAM, but is more expensive, more power-hungry, and less dense. SRAM technology does not need to be refreshed. The cache contains the most recently accessed code and data. The cache memory is used to provide instructions and data to the processor faster than would be possible if only main memory were used. In current CPUs, separate caches for instructions and data are provided on the CPU chip.

In most processors produced today, the first level cache (referred to as the L1-cache) is designed to be on the CPU chip. In many designs, a second level of caching is provided on chip. Generally, most processors provide an on-chip L1 instruction cache and an on-chip L1 data cache. L2 caches (either seperate or unified) are also found on-chip on many designs today. It is also common to see an L3 cache on multi-CPU systems, that is located off chip.

17.3.4 Memory and Architecture

The amount of the memory system defined by the architectural specification varies greatly. The design and layout of the cache(s) and main memory are typically not defined by the architecture. The architecture specifies the bit and byte arrangement of data and the addressing formats within instructions. The interface (e.g., address lines) to external (main) memory can also be specified in the architecture. The virtual memory systems may also place some restrictions on the addressing scheme used for the main memory system. In some architectures, instructions have been provided to manipulate the state of cache memory to ensure the memory coherency of the system (memory coherency refers to the issue of having only a single valid copy of a datum in the system).

Since the performance gap between accessing cache and main memory is so great, maximizing the probability of finding a memory request resident in cache when requested is of great interest. Some of the tradeoffs in the design of a cache include the block size (minimum unit of transfer between main memory and the cache), associativity (used to define the mapping between main memory and the cache), handling of cache writes (for data and mixed caches), number of entries, and mapping strategies. Handy provides a thorough discussion on a number of these topics [Handy 1993]. While many of these design parameters are important, they typically are not included in the architectural definition and are left as part of the design not specified.

17.3.5 Secondary Storage

When you first turn on your computer, the program code will reside in secondary storage (disk storage). When the program is run, it will be transferred to main memory, then to cache, and then to the processor (generally these last two transfers are performed simultaneously). Disk storage is commonly used for secondary storage, since it is nonvolatile storage (the contents are maintained even when power is turned off).

A disk is designed using magnetic media, and data are stored in the form of a magnetic polarization. The disk comprises one or more platters which are always spinning. When a request is made for instructions or data, the disk must rotate to the proper location in order for the read heads to be able to access the information at that location. Because disk rotation is a mechanical operation, disk accesses are many orders of magnitude slower than the access time of the DRAM used for main memory.

After a program has been run once, it will reside for a period of time in either the cache or main memory. Access to the program will be faster upon subsequent runs if the execution processor does not have to wait for the program to be reloaded from disk.

The architecture of a system does not typically impose any limitations on the organization of secondary storage besides defining the smallest addressable unit (typically called a block) and the total amount of addressable storage.

17.4 Addressing

All instructions and data in registers and memory are accessed by an address. Next we look at various aspects of addressing: addressing format, physical addressing, virtual addressing, and byte ordering.

17.4.1 Addressing Format

Defined within all architectures are the various addressing formats which are permissible with particular instructions. Addresses define where input and output operands will located in the system. Standard addressing formats provided with most architectures include:

1. *Register:* The register number is specified in the instruction. The register number is used directly to access the register file.
2. *Immediate:* A constant field is specified in the instruction. This field is used directly as an operand. While this format does not fit nicely into our concept of addressing, immediate addressing is a commonly used address format for specifying the operands.
3. *Direct:* The full address is provided in the instruction. No address calculation is necessary. The address field can be used directly (thus, direct addressing). Direct addressing is commonly used for accessing static data.
4. *Register indirect:* The number of the GPR which contains the memory address is specified. Register addressing is commonly used for accessing via a pointer value. The format is also referred to as *register deferred*.
5. *Memory indirect:* The number of the GPR is specified which contains the address where the address of the desired data is located. In this case, two memory references are performed to obtain the desired data. Indirect addressing is commonly used to dereference pointers. This format is also referred to as *memory deferred*.

6. *Base displacement:* The number of the GPR which contains the base address of some data structure is specified. The base value is added to a displacement field to obtain the final memory address. Base-displacement addressing is commonly used for addressing sequential data patterns (e.g., arrays, structures).

7. *Indexed:* A GPR is added to a base register (GPR) to obtain the memory address. Some architectures provide separate index registers for this purpose. Other architectures add an index to a base register and possibly even include a displacement field. Indexed addressing is commonly used for traversing complex data structures such as link lists of structures.

Register, immediate, and base displacement addressing are the most commonly used addressing formats. Some other addressing modes that are found in selected instruction sets include:

1. *Auto-increment/auto-decrement.* Similar to register indirect, with the addition that an index register is also incremented/decremented. This format is commonly used when accessing arrays within a loop.

2. *Scaled.* Similar to register indexed, except that a second index register is multiplied by a constant (typically the data type size) and this second index is added to the base and the index. This allows for efficiently traversing arrays with nonstandard data sizes.

17.4.2 Physical and Virtual Memory

Virtual memory refers to the ability of a computer system to address a larger address range than the amount of physically installed memory (DRAM). This is a very cost-effective approach to computing. We need to have only a small amount of DRAM memory installed on our system, compared to the actual addressable space. The range of the virtual address space places a limit on the amount of addressable secondary storage (including disks, printers, devices, etc.). Prior to the introduction of the concept of virtual memory, the programmer had to insert explicit commands to load programs in from disk storage. One segment supplanted another in memory (this was called and *overlay*). With the introduction of virtual memory, memory overlays were no longer necessary. Some motivating factors behind using virtual memory include: (1) providing efficient usage of physical memory without the need for explicit overlays, and (2) allowing full flexibility of placement of instructions and data.

A virtual memory address generally refers to any memory location in the available address range as defined by the architecture. A physical memory address refers to a memory location in main memory. The physical-memory-address range is defined by the amount of installed main memory and is bounded by the number of bits provided for in the address scheme of the architecture. The virtual-address range is defined by the largest permissible address value. Figure 17.5 provides an example of how a 32-bit virtual address can address over 4 billion different memory locations. Since the virtual-address range is generally 2 or more orders of magnitude greater than the installed physical memory, a mapping from virtual to physical memory is performed. See Figure 17.5 for an example of how different regions (i.e., pages) in the virtual address space map to the physical address space.

The operating system performs the mapping between the virtual and physical address spaces. The mapping is performed on either a fixed-size or variable-sized memory blocks (pages versus segments, respectively). In Figure 17.5 we see that a fixed-size page (4096 bytes) is being mapped to the physical memory address space. This mapping is stored in a table in memory (called a page table). The page table will hold virtual-to-physical mapping for all pages present in the physical memory. In Figure 17.5 we see that virtual page A' is mapped to physical memory location A and that virtual page M' is mapped to physical page M. Thus, sequentiality is limited to a page and generally does not extend past the page boundary, which is to say, two sequential pages in the virtual address space need not be mapped to two sequential physical pages.

Pages are brought from secondary storage into physical (main) memory when they are requested by the execution processor. This is referred to as *demand paging*. The size of a page is typically fixed (e.g., 4096 bytes in our example). Segmented virtual memory systems provide a variable-sized unit of

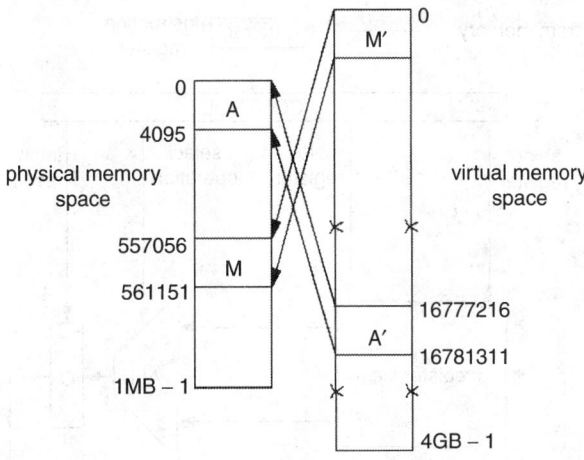

FIGURE 17.5 Mapping of physical to virtual memory addresses.

transfer between the virtual and physical memory spaces. A good discussion of the different types of virtual memory systems can be found in Feldman and Retter [1995] as well as in Chapter 85.

Since every instruction execution in the CPU would require that at least two memory accesses be performed to obtain instructions (one access to obtain the physical address stored in the page table, and a second access to obtain the instruction stored at that physical address), a hardware feature called a *translation lookaside buffer* (TLB) was proposed. The TLB caches the recently accessed portions of the page table and quickly provides a virtual-to-physical translation of the address. The TLB is generally located on the CPU to provide fast translation capability. Further discussion on TLBs can be found in Teller [1991].

Now that we know a little bit more about instructions, addressing, and memory, we can begin to understand how instructions are processed by the CPU.

17.5 Instruction Execution

To this point we have focused on *architected* features of a digital computer system. These will be used to document the interface to which compiler and application developers must program and the features which the hardware designer must provide. Next we will describe some *nonarchitected* features of a digital computer system which are typically left to the implementer to decide upon. The first feature looks at how we organize elements of the CPU in order to execute an instruction efficiently.

Instructions are requested from the memory system. This involves sending an address out to the memory and waiting until the memory system produces the requested address. Once retrieved, the instruction enters the processor. The execution of an instruction involves a number of steps. Figure 17.6 provides the buses and logic to be used to describe the steps for nonpipelined instruction execution.

In a nonpipelined system, a single instruction at a time enters the processor and waits the necessary amount of time for all of the steps associated with an instruction to be completed. For instance, in Figure 17.6 the subtract instruction is loaded into the instruction register and decoded by the control logic. Then the input registers to the ALU are enabled and the ALU is programmed to perform a subtract. After a sufficient delay to allow all of these operations to be performed and for the output of the ALU to reach a steady-state value, the output register is latched. This all happens during a single processor clock cycle (the CPU clock is typically used to synchronously capture the new state of memory devices).

If instructions were allowed to flow through the execution processor only one at a time, the throughput of the execution processor (the rate at which instructions exit the processor) would be low. This is because a majority of the elements of the CPU would remain idle while different phases of the instruction execution progressed.

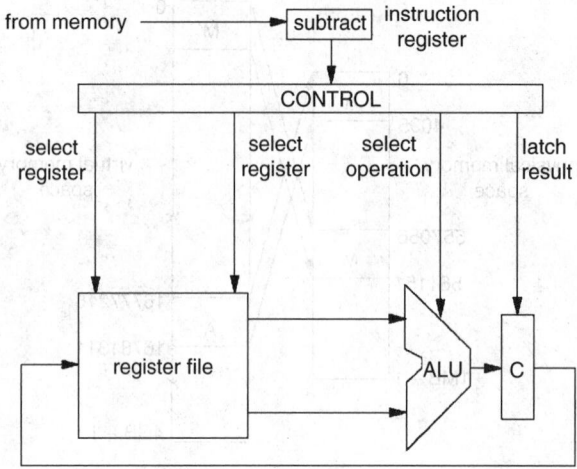

FIGURE 17.6 Nonpipelined instruction execution.

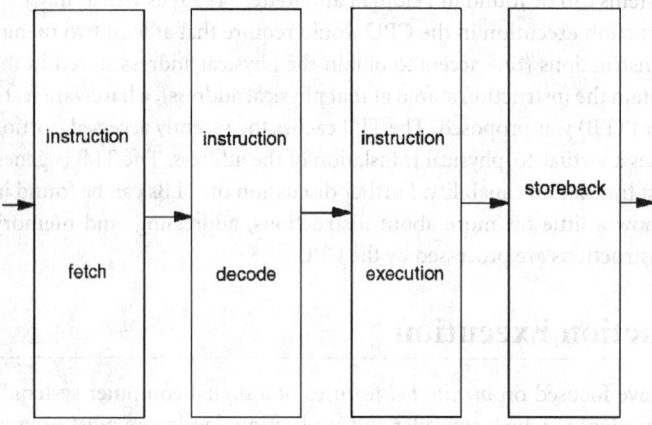

FIGURE 17.7 Pipelined execution of the subtraction example.

Instead, consider the execution of the same instruction broken up into a number of stages, as is shown in Figure 17.7. A single clock is used to latch results at the end of each stage (in practice, different clock edges may be used to latch particular elements in a stage). The key ideas here are that all stages work synchronously and that multiple instructions are being processed simultaneously (similar in concept to Henry Ford's assembly line). This is called *pipelining*. Given a pipeline of *n* stages, *n* instructions can be in process simultaneously (each at different stages of the pipeline). This can dramatically increase instruction throughput.

To complete the execution of an instruction, a series of tasks are performed by a number of functional units. To begin with, the instruction must be present in the processor. We will label this stage the instruction fetch stage of the pipeline, since it is fetching the next instruction to be executed.

The second stage of the pipeline is instruction decode. During this stage, the machine language is decoded and the operation to be performed is discovered. The decoding process generates the control bit values which will enable buses, latch registers, and program ALUs. These are represented in Figure 17.6 by the vertical lines produced by the control logic. Also during this stage, the necessary operands (e.g., **x** and **y** for the subtract) may also be discovered.

The third stage of the pipeline is the instruction execution stage. During this stage, the operation specified in the instruction operation code is actually performed (e.g., in our example program, the subtract will take place during this stage and the result will be latched into the ALU result register **c)**.

The final stage is the storeback stage. During this stage, the results of the execution stage are stored back to the specified register or storage location. While the discussion here has been for ALU-type instructions, it can be generalized for memory referencing and control-transfer instructions. Kogge describes a number of pipeline organizations [Kogge 1981].

Next we discuss the individual contents of each of these stages.

17.5.1 Instruction Fetch Unit

An executable computer program is a contiguous block of instructions which is stored in memory (copies may reside simultaneously in secondary, main, and cache memory). The *instruction fetch unit* (IFU) is responsible for retrieving the instructions which are next to be executed. To start (or restart) a program, the dispatch portion of the operating system will point the execution processor's *program counter* (PC) to the beginning (or current) address of the program which is to be executed. The IFU will begin to retrieve instructions from memory and feed them to the next stage of the pipeline, the *instruction decode unit* (IDU).

17.5.2 Instruction Decode Unit

Instructions are stored in memory in an encoded format. Encoding is done to reduce the length of instructions (common RISC architectures use a 32-bit instruction format). Shorter instructions will reduce the demands on the memory system, but then encoded instructions must be decoded to determine the desired control bit values and identify the accompanying operands. The instruction decode unit performs this decoding and will generate the necessary control signals, which will be fed to the *execution unit*.

17.5.3 Execution Unit

The execution unit will perform the operation specified in the instruction decoded operation code. The operands upon which the operation will be performed are present at the inputs of the ALU. If this is a memory referencing instruction (we will assume we are using a RISC processor for this discussion, so ALU operations are performed only on either immediate or register operands), address calculations will be performed during this stage. The execution unit will also perform any comparisons needed to execute conditional branch instructions. The result of the execution unit is then fed to the *storeback unit*.

17.5.4 Storeback Unit

Once the result of the requested operation is available, it must be stored away so that the next instruction can utilize the execution unit. The storeback unit is used to store the results of ALU operation to the register file (again, we are considering only RISC processors in this discussion), to update a register with a new value from memory (for LOAD instructions), and to update memory with a register value (for STORE instructions). The storeback unit is also used to update the program counter on branch instructions.

Figure 17.8 shows our subtraction code flowing through both a nonpipelined and a pipelined execution processor. The width of the boxes in the figures is meant to depict the length of a processor clock cycle. The nonpipelined clock cycle is longer (slower), since all of the work accomplished in the four separate stages of the instruction execution are completed in a single clock tick. The pipelined execution clock cycle is dominated by the time to stabilize and latch the result at the end of each stage. This is why a single pipelined instruction execution will typically take longer to execute than a nonpipelined instruction. The advantages of pipelining are reaped only when instructions are overlapped. As we can see, the time to execute the subtraction program is significantly smaller for the pipelined example (but not nearly four times smaller).

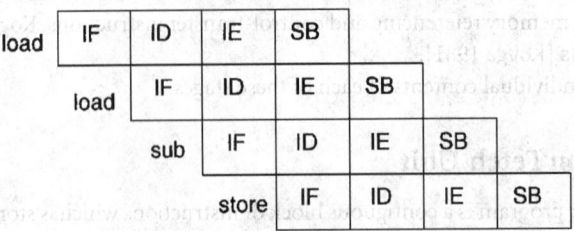

nonpipelined execution

pipelined execution

FIGURE 17.8   Comparison of nonpipelined and pipelined execution.

Also, other executions can be overlapped with this example code (i.e., instructions in the pipeline prior to the first load and after the store instruction). This is not the case for nonpipelined execution.

Note that in our examples in Figure 17.8 we are assuming that all nonpipelined instructions and pipelined stages take a single clock cycle. This is one of the underlying principles in RISC architectures. If instructions are kept simple enough, they can be executed in a single cycle.

Pipelining can provide an advantage only if the pipeline is supplied with instructions which can be issued without any delay or uncertainty. The benefits of pipelining can be greatly reduced if stalls occur due to different types of hazards. We will discuss this topic next.

17.6   Execution Hazards

Consider attempting to process the high-level language example in Figure 17.9, which builds upon our subtraction program. If we look at what the compiler would do with this code, it might look something like the instruction sequence shown in Figure 17.10.

A problem occurs if we try to execute this on the pipelined model. The cause of the problem is illustrated in Figure 17.11, which shows the instruction sequence given in Figure 17.10 flowing through the pipeline. The multiply instruction needs to get the most recent update of the variable **z** (which will reside in **r3**). Given the pipeline model described above, the multiply instruction will be attempting to direct the contents of **r3** to the inputs of the ALU during its instruction decode stage. In the same cycle the subtract instruction will be trying to store the results of the subtraction to **r3** during the storeback stage. This is just one example of a *hazard*, called a data hazard. There are three classes of hazards:

1. *data hazards*, which include read-after-write (RAW), write-after-read (WAR), and write-after-write hazards
2. *control hazards*, which include any instructions or interruptions which break sequential instruction execution
3. *structural hazards*, which occur when multiple instructions vie for a single functional element (e.g., an ALU)

17.6.1   Data Hazards

Data hazards occur when there exist dependencies between instructions. For the hazard to occur, the instructions need to be close enough in the execution stream to coexist in the pipeline. The example provided in Figure 17.10 is just one type of interinstruction dependence, called a *read-after-write* hazard.

```
x = 5;          /* Initialize x to 5 */
y = 3;          /* Initialize y to 3 */
z = x - y;      /* Compute the difference */
w = z * z;      /* compute the square */
```

FIGURE 17.9   Subtraction program which also computes the square of the difference.

```
w:      .
x:      .int 5
y:      .int 3
z:      .
        load r1, x
        load r2, y
        sub r3, r1, r2
        store r3, z
        mult r4, r3, r3
```

FIGURE 17.10   Assembly-language version of Figure 17.9.

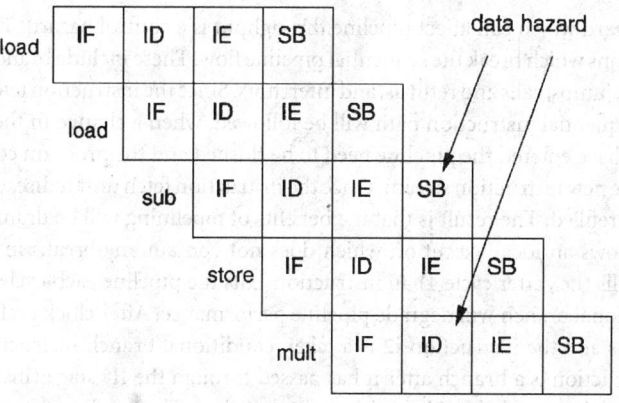

FIGURE 17.11   Example of a RAW hazard.

The multiply is attempting to read **r3** as the subtract is writing **r3**. This is the most commonly encountered type of data hazard.

A second type of data hazard occurs when a subsequent instruction modifies an operand of the current instruction before it is read by the current instruction. This is called a *write-after-read* hazard. This will occur only if writes are allowed to be performed early in the pipeline and reads are performed late in the pipeline.

A third type of data hazard occurs when a subsequent instruction modifies an operand of the current instruction before it is written by the current instruction. This is called a *write-after-write* hazard. This will occur only if writes are allowed to be performed both early and late in the pipeline.

A number of solutions to detecting data hazards have been proposed. Hardware techniques-of-forwarding data values can overcome a large number of these hazards which frequently occur. This technique is integrated into the decoding logic and keeps track of the source and destination of all instructions currently active in the pipeline. Hazard detection then becomes part of the instruction decoding stage. Compilers can also be tuned to eliminate the possibility of data hazards occurring.

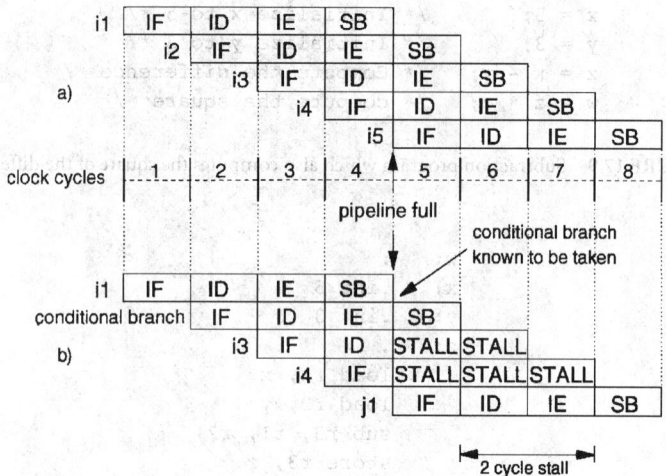

FIGURE 17.12   a) Ideal execution. b) A taken conditional branch.

17.6.2   Control Hazards

Another type of hazard which can affect pipeline throughput is a control hazard. This class of hazard includes any instructions which break the sequential pipeline flow. These include branches (both conditional and unconditional), jumps, calls and returns, and interrupts. Since the instruction fetch stage of the pipeline assumes that the sequential instruction path will be followed, when a change in the path takes place, the instructions which have entered the pipeline need to be flushed and the program counter needs to be adjusted to point to the new instruction stream. Once the instruction fetch unit redirects instruction fetching, the pipeline will be refilled. The result is that the benefits of pipelining will be dramatically reduced.

Figure 17.12a shows an ideal execution which does not contain any breaks in sequential execution. After the pipeline fills (i.e., after cycle 4) an instruction exits the pipeline each cycle. Figure 17.12b shows how a taken conditional branch will degrade pipeline performance. After clock cycle 4, the pipeline is full of valid instructions and the instruction **i2** is a taken conditional branch instruction. The pipeline will know that this instruction is a branch after it has passed through the ID stage (instruction decode). The pipeline forges ahead, hoping that the branch will not be taken. The pipeline detects that the conditional branch is taken when it completes the IE stage (at the end of clock cycle 4). Instructions **i3** and **i4** have already entered the pipeline and will need to be flushed. This introduces a 2-cycle stall into the pipeline. The branch is taken to instruction **j1**, and the pipeline is refilled. There is a 2-cycle delay in this example, since 2 cycles pass before a valid instruction (i.e., instruction **j1**) exits the pipeline. The average number of cycles per instruction (once the pipeline has filled) is one for the ideal execution and two for the code containing the conditional branch.

To handle control hazards many techniques have been proposed. History-based branch prediction is one mechanism which is now commonly used in most microprocessors to reduce the negative effects of control hazards. This type of mechanism attempts to predict the future behavior of branches by recording history of past execution. A table of entries, each containing information for a particular branch, is maintained. History for each branch is recorded (e.g., taken vs. not-taken, and branch destination). Then when the branch instruction is again instruction fetched, a lookup in the history table finds the outcome of the last execution(s) for this branch and then predicts that history will repeat itself. It has been found that past history is a very good predictor of future behavior. Using history-based branch prediction, a majority of the pipeline stalls encountered due to conditional branches can be eliminated. Cragon describes a number of these mechanisms [Cragon 1992].

Another method for hiding the effects of control hazards is called *branch predication*. Each predicated instruction is accompanied by a predication bit. Using predication, instructions from both paths of a conditional branch can be issued, though only those instructions on correct path are committed. The predicate bits indicate which instructions are committed, and which instructions are *squashed*. Interest in predicated execution has recently been renewed with the introduction of the Intel IA-64 architecture [Gwennap 1998].

17.6.3 Structural Hazards

Structural hazards are the third class of pipeline delays which can occur. They occur when multiple instructions active in the pipeline vie for shared resources. Some examples of structural hazards include: two instructions trying to compute a memory address in the same cycle when a single address-generation ALU is provided or two instructions both attempting to access the data cache in the same cycle when only a single data-cache access port is provided.

A number of approaches can be taken to alleviate the delays introduced by structural hazards. One approach is to further exploit the principle of pipelining by employing pipelined stages within each pipeline unit. This technique is called *superpipelining*. This will allow multiple instructions which are active in the pipeline to coexist in a single pipeline stage.

Another approach is to provide multiple functional units (e.g., two cache ports or multiple copies of the register file). This approach is commonly used when high performance is critical or when we want to be able to issue multiple instructions in a single clock cycle. Multiple issue processors, also called *superscalar* processors, are discussed next.

17.7 Superscalar Design

If we can solve all the problems associated with hazards, an instruction should be exiting the pipeline every processor clock cycle. While this level of performance is seldom achieved (mainly due to latencies in the memory system and the limitations of effectively handling control hazards), we would like to be able to see multiple instructions exit the pipeline in a single clock cycle if possible. This approach has been labeled *superscalar* design. The idea is that if the compiler can produce groups of instructions which can be issued in parallel (which do not contain any data or control dependencies), then we can attain our goal of having multiple instructions exit the pipeline in a single cycle.

Some of the initial ideas which have motivated this direction date back to the 1960s and were initially implemented in early IBM [Anderson et al. 1967] and CDC machines [Thornton 1964]. The problem with this approach is finding a large number of instructions which are independent of one another. The compiler cannot exploit the scheduling to perfection because some conflicts are data-dependent. We can instead design complex hazard detection logic in our execution processor. This has been the approach taken by most superscalar designers.

Two issues occur in superscalar execution. First, can we issue nonsequential instructions in parallel? This is referred to as *out-of-order issue*. A second question is whether we can allow instructions to exit the pipeline in nonsequential order. This is referred to as *out-of-order completion*. A thorough discussion of the tradeoffs associated with superscalar execution and issue/completion design can be found in Johnson [1990].

17.8 Very Long Instruction Word Computers

In contrast to superscalar execution, which using *dynamic scheduling* to select at runtime the sequence of instructions to issue to the functional unit, Very Long Instruction Word (VLIW) architectures employ *static scheduling*, relying on the compiler to produce an efficient sequence of instructions to execute. A VLIW processor packs multiple, RISC-like, instructions into a long instruction word. If the compiler can find multiple instructions that can be issued in the same cycle, we should be able to expose high instruction-level parallelism. A number of designs have been developed based on a VLIW architecture, including the Cydra 5 [Beck 1993] and the Intel Itanium processors [Gwennap 1998].

17.9 Summary

This chapter has introduced many of the features provided in a digital computer architecture. The instruction set, memory hierarchy, and memory addressing elements, which are central to the definition of a computer architecture, were covered. Then some optimization techniques, which attempt to improve the efficiency of instruction execution, were presented. The hope is that this introductory material provides enough background for the nonspecialist to gain an appreciation for the pipelining and superscalar techniques that are currently used in today's CPU designs.

Defining Terms

Branch prediction: A mechanism used to predict the outcome of branches prior to their execution.

Cache memory: Fast memory, located between the CPU and main storage, that stores the most recently accessed portions of memory for future use.

Control hazards: Breaks in sequential instruction execution flow.

Data hazards: Dependencies between instructions that coexist in the pipeline.

Memory coherency: Ensuring that there is only one valid copy of any memory address at any time.

Pipelining: Splitting the CPU into a number of stages, which allows multiple instructions to be executed concurrently.

Predication: Conditionally executing instructions and only committing results for those instructions with enable predicates.

Structural hazards: A situation where shared resources are simultaneously accessed by multiple instructions.

Superpipelining: Dividing each pipeline stage into substages, providing for further overlap of multiple instruction execution.

Superscalar: Having the ability to simultaneously issue multiple instructions to separate functional units in a CPU.

Very Long Instruction Word: Specifies a multiple (although fixed) number of primitive operations that are issued together and executed upon multiple functional units. VLIW relies upon effecive static (compile-time) scheduling.

References

Anderson, D. W., Sparacio, F. J., and Tomasulo, R. M. 1967. The IBM 360 Model 91: processor philosophy and instruction handling. *IBM J. Res. Dev.* 11(1):8–24, Jan.

Beck, G. R., Yen, D. W., and Anderson, T. L. 1993. The Cydra 5 Mini-Supercomputer: Architecture and Implemementation. *Journal of Supercomputing*, 7:140–180.

Colwell, R. P., Hitchcock, C. Y., Jensen, E. D., Sprunt, H. M. B., and Kollar, C. P. 1985. Computers, complexity, and controversy. *IEEE Comput. Mag.* Sept., pp. 8–19.

Cragon, H. C. 1992. *Branch Strategy Taxonomy and Performance Models.* IEEE Computer Society Press, Los Alamitos, CA.

Feldman, J. and Retter, C. 1995. *Computer Architecture: A Designer's Guide to a Generic RISC.* Prentice–Hall, Reading, MA.

Gwennap, L. 1998. Intel's Merced and IA-64: Technology and Market Forecast. MDR Technical Library Report.

Handy, J. 1993. *The Cache Memory Book.* Academic Press, Boston.

IBM Corp. 1981. *IBM System/370 Principles of Operation.* Document No. GA22-7000-7, IBM Corp. New York, Mar.

Intel Corporation. 2001. Desktop Performance and Optimization of the Intel Pentium 4 Processor. Intel Corporation, Order Number 249438-01, Feb.

Johnson, M. 1990. *Superscalar Microprocessor Design.* Prentice–Hall, Englewood Cliffs, NJ.

Kogge, P. M. 1981. *The Architecture of Pipelined Computers.* McGraw–Hill, New York.

Patterson, D. 1985. Reduced instruction set computers. *Commun. ACM* 28(1):8–21, Jan.

Peleg, A., Wilkie, S., and Weiser, U. 1997. Intel MMX for Multimedia PCs, 40(1):24–38.

Sun Microsystems. 2002. V9 (64-bit SPARC) Architecture Book. www.sparc.com/standards.html.

Teller, P. 1991. *Translation Lookaside Buffer Consistency in Highly-Parallel Shared Memory Multiprocessors*, Ph.D. dissertation. New York University, May. Also available as an *IBM Research Report*, RC 16858, #74685, May 14.

Thornton, J. E. 1964. Parallel operation in the Control Data 6600. In *AFIPS Proc. Fall Joint Computer Conf.* No. 27.

Wang, P. H., Wang, H., Collins, J. D., Grochowski, E., Kling, R. M., and Shen, J. P. 2002. Memory Latency-Tolerance Approaches for Itanium Processors: Out-of-Order Execution vs. Speculative Precomputation, *Proc. of HPCA-8*, Feb., pp. 187–196.

Further Information

To learn more about the recent advances in computer architecture, you will find articles on a variety of related subjects in the following list of IEEE and ACM publications:

IEEE Transactions on Computers.

IEEE Computer Architecture Letters.

ACM SIGARCH Newsletter.

IEEE TCCA Newsletter.

Proceedings of the International Symposium on Computer Architecture, IEEE Computer Society Press.

Proceedings of the International Conference on High-Performance Computer Architecture, IEEE Computer Society Press.

Proceedings of the Conference on Architectural Support for Programming Languages and Operating Systems, ACM.

18

Memory Systems

Douglas C. Burger
University of Wisconsin at Madison

James R. Goodman
University of Wisconsin at Madison

Gurindar S. Sohi
University of Wisconsin at Madison

18.1 Introduction 18-1
18.2 Memory Hierarchies 18-2
18.3 Cache Memories 18-4
18.4 Parallel and Interleaved Main Memories 18-7
18.5 Virtual Memory... 18-10
18.6 Research Issues.. 18-12
18.7 Summary 18-13

18.1 Introduction

The *memory system* serves as the repository of information (data) in a computer system. The processor (also called the central processing unit, or CPU) accesses (reads or loads) data from the memory system, performs computations on them, and stores (writes) them back to memory. The memory system is a collection of storage locations. Each storage location, or *memory word*, has a numerical *address*. A collection of storage locations form an *address space*. Figure 18.1 shows the essentials of how a processor is connected to a memory system via address, data, and control lines.

When a processor attempts to load the contents of a memory location, the request is very urgent. In virtually all computers, the work soon comes to a halt (in other words, the processor *stalls*) if the memory request does not return quickly. Modern computers are generally able to continue briefly by overlapping memory requests, but even the most sophisticated computers will frequently exhaust their ability to process data and stall momentarily in the face of long memory delays. Thus, a key performance parameter in the design of any computer, fast or slow, is the effective speed of its memory.

Ideally, the memory system must be both infinitely large, so that it can contain an arbitrarily large amount of information and infinitely fast, so that it does not limit the processing unit. Practically, however, this is not possible. There are three properties of memory that are inherently in conflict: speed, capacity, and cost. In general, technology tradeoffs can be employed to optimize any two of the three factors at the expense of the third. Thus it is possible to have memories that are (1) large and cheap, but not fast; (2) cheap and fast, but small; or (3) large and fast, but expensive. The last of the three is further limited by physical constraints. A large-capacity memory that is very fast is also physically large, and speed-of-light delays place a limit on the speed of such a memory system.

The **latency** (L) of the memory is the delay from when the processor first requests a word from memory until that word arrives and is available for use by the processor. The latency of a memory system is one attribute of performance. The other is **bandwidth** (BW), which is the rate at which information can be transferred from the memory system. The bandwidth and the latency are closely related. If R is the number of requests that the memory can service simultaneously, then

$$BW = \frac{R}{L} \qquad (18.1)$$

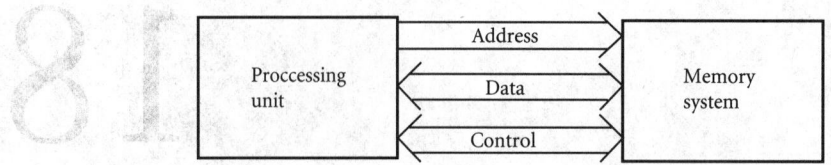

FIGURE 18.1   The memory interface. (*Source*: Dorf, R. C. 1992. *The Electrical Engineering Handbook*, 1st ed., p. 1928. CRC Press, Inc., Boca Raton, FL.)

From Equation (18.1) we see that a decrease in the latency will result in an increase in bandwidth, and vice versa, if R is unchanged. We can also see that the bandwidth can be increased by increasing R, if L does not increase proportionately. For example, we can build a memory system that takes 20 ns to service the access of a single 32-bit word. Its latency is 20 ns per 32-bit word, and its bandwidth is

$$\frac{32}{20 \times 10^{-9}} \text{ bits/s}$$

or 200 Mbyte/s. If the memory system is modified to accept a new (still 20-ns) request for a 32-bit word every 5 ns by overlapping requests, then its bandwidth is

$$\frac{32}{5 \times 10^{-9}} \text{ bits/s}$$

or 800 Mbyte/s. This memory system must be able to handle four requests at a given time.

Building an ideal memory system (infinite capacity, zero latency, and infinite bandwidth, with affordable cost) is not feasible. The challenge is, given the cost and technology constraints, to engineer a memory system whose abilities match the abilities that the processor demands of it. That is, engineering a memory system that performs as close to an ideal memory system (for the given processing unit) as is possible. For a processor that stalls when it makes a memory request (some current microprocessors are in this category), it is important to engineer a memory system with the lowest possible latency. For those processors that can handle multiple outstanding memory requests (vector processors and high-end CPUs), it is important not only to reduce latency but also to increase bandwidth (over what is possible by latency reduction alone) by designing a memory system that is capable of servicing multiple requests simultaneously.

Memory hierarchies provide decreased average latency and reduced bandwidth requirements, whereas parallel or **interleaved** memories provide higher bandwidth.

18.2   Memory Hierarchies

Technology does not permit memories that are cheap, large, and fast. By recognizing the nonrandom nature of memory requests, and emphasizing the *average* rather than worst-case latency, it is possible to implement a hierarchical memory system that performs well. A small amount of very fast memory, placed in front of a large, slow memory, can be designed to satisfy most requests at the speed of the small memory. This, in fact, is the primary motivation for the use of registers in the CPU: in this case, the programmer makes sure that the most commonly accessed variables are allocated to registers.

A variety of techniques, using hardware, software, or a combination of the two, can be employed to ensure that most memory references are satisfied by the faster memory. The foremost of these techniques is the exploitation of the *locality of reference* principle. This principle captures the fact that some memory locations are referenced much more frequently than others. *Spatial locality* is the property that an access to a given memory location greatly increases the probability that neighboring locations will be accessed immediately. This is largely, but not exclusively, a result of the tendency to access memory locations sequentially. *Temporal locality* is the property that an access to a given memory location greatly increases the probability that the same location will be accessed again soon. This is largely, but not exclusively, a result

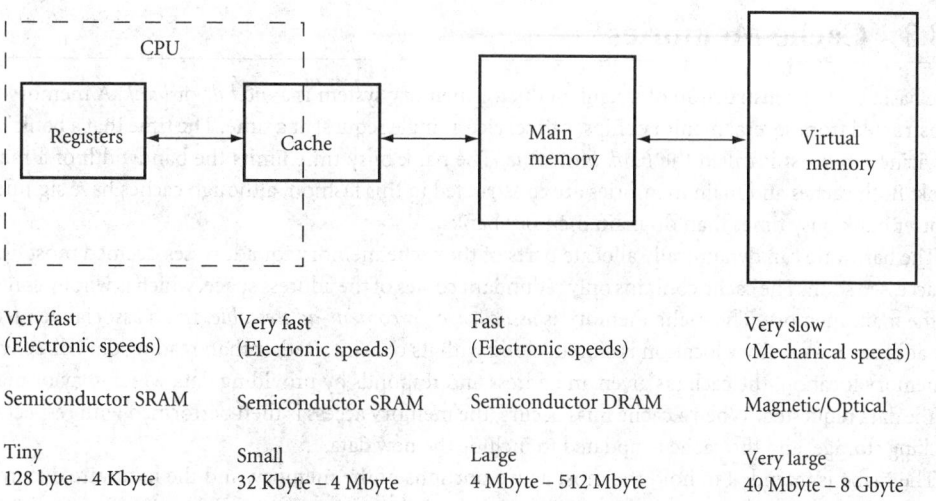

FIGURE 18.2 A memory hierarchy. (*Source:* Dorf, R. C. 1992. *The Electrical Engineering Handbook*, 1st ed., p. 1932. CRC Press, Inc., Boca Raton, FL.)

of the high frequency of programs' looping behavior. Particularly for temporal locality, a good predictor of the future is the past; the longer a variable has gone unreferenced, the less likely it is to be accessed soon.

Figure 18.2 depicts a common construction of a memory hierarchy. At the top of the hierarchy are the CPU registers, which are small and extremely fast. The next level down in the hierarchy is a special, high-speed semiconductor memory, known as a **cache memory**. The cache can actually be divided into multiple distinct levels; most current systems have between one and three levels of cache. Some of the levels of cache may be on the CPU chip itself, they may be on the same module as the CPU, or they may all be entirely distinct. Below the cache is the conventional memory, referred to as *main memory*, or *backing storage*. Like a cache, main memory is semiconductor memory, but it is slower, cheaper, and denser than a cache. Below the main memory is the virtual memory, which is generally stored on magnetic or optical disks. Accessing the virtual memory can be tens of thousands of times slower than accessing the main memory, since it involves moving, mechanical parts.

As requests go deeper into the memory hierarchy, they encounter levels that are larger (in terms of capacity) and slower than the higher levels (moving left to right in Figure 18.2). In addition to size and speed, the bandwidth between adjacent levels in the memory hierarchy is smaller for the lower levels. The bandwidth between the registers and top cache level, for example, is higher than that between cache and main memory or between main memory and virtual memory. Since each level presumably intercepts a fraction of the requests, the bandwidth to the level below need not be as great as that to the intercepting level.

A useful performance parameter is the *effective latency*. If the needed word is found in a level of the hierarchy, it is a *hit*; if a request must be sent to the next lower level, the request is said to *miss*. If the latency L_{HIT} is known in the case of a hit and the latency in the case of a miss is L_{MISS}, the effective latency for that level in the hierarchy can be determined from the *hit ratio* (H), the fraction of memory accesses that are hits:

$$L_{average} = L_{HIT} H + L_{MISS} (1 - H) \tag{18.2}$$

The portion of memory accesses that miss is called the *miss ratio* ($M = 1 - H$). The hit ratio is strongly influenced by the program being executed, but it is largely independent of the ratio of cache size to memory size. It is not uncommon for a cache with a capacity of a few thousand bytes to exhibit a hit ratio greater than 90%.

18.3   Cache Memories

The basic unit of construction of a semiconductor memory system is a *module* or *bank*. A memory bank, constructed from several memory chips, can service a single request at a time. The time that a bank is busy servicing a request is called the *bank busy time*. The bank busy time limits the bandwidth of a memory bank. Both caches and main memories are constructed in this fashion, although caches have significantly shorter bank busy times than do main memory banks.

The hardware can dynamically allocate parts of the cache memory for addresses deemed most likely to be accessed soon. The cache contains only redundant copies of the address space, which is wholly contained in the main memory. The cache memory is *associative*, or *content-addressable*. In an associative memory, the address of a memory location is stored, along with its content. Rather than reading data directly from a memory location, the cache is given an address and responds by providing data which may or may not be the data requested. When a cache miss occurs, the memory access is then performed with respect to the backing storage, and the cache is updated to include the new data.

The cache is intended to hold the most active portions of the memory, and the hardware dynamically selects portions of main memory to store in the cache. When the cache is full, bringing in new data must be matched by deleting old data. Thus a strategy for cache management is necessary. Cache management strategies exploit the principle of locality. Spatial locality is exploited by the choice of what is brought into the cache. Temporal locality is exploited by the choice of which block is removed. When a cache miss occurs, hardware copies a large, contiguous block of memory into the cache, which includes the requested word. This fixed-size region of memory, known as a cache *line* or *block*, may be as small as a single word, or up to several hundred bytes. A block is a set of contiguous memory locations, the number of which is usually a power of two. A block is said to be *aligned* if the lowest address in the block is exactly divisible by the block size. That is to say, for a block of size B beginning at location A, the block is aligned if

$$A \bmod B = 0 \tag{18.3}$$

Conventional caches require that all blocks be aligned.

When a block is brought into the cache, it is likely that another block must be evicted. The selection of the evicted block is based on an attempt to capture temporal locality. Since prescience is difficult to achieve, other methods are generally used to predict future memory accesses. A least-recently-used (LRU) policy is often the basis for the replacement choice. Other replacement policies are sometimes used, particularly because true LRU replacement requires extensive logic and hardware bookkeeping.

The cache often comprises two conventional memories: the data memory and the tag memory, shown in Figure 18.3. The address of each cache line contained in the data memory is stored in the tag memory, as well as other information (*state* information), particularly the fact that a valid cache line is present. The state also keeps track of which cache lines the processor has modified. Each line contained in the data memory is allocated a corresponding entry in the tag memory to indicate the full address of the cache line.

The requirement that the cache memory be associative (content-addressable) complicates the design. Addressing data by content is inherently more complicated than by its address. All the tags must be compared concurrently, of course, because the whole point of the cache is to achieve low latency. The cache can be made simpler, however, by introducing a mapping of memory locations to cache cells. This mapping limits the number of possible cells in which a particular line may reside. The extreme case is known as *direct mapping*, in which each memory location is mapped to a single location in the cache. Direct mapping makes many aspects of the design simpler, since there is no choice of where the line might reside and no choice as to which line must be replaced. Direct mapping, however, can result in poor utilization of the cache when two memory locations are alternately accessed and must share a single cache cell.

A hashing algorithm is used to determine the cache address from the memory address. The conventional mapping algorithm consists of a function with the form

$$A_{cache} = \frac{A_{memory} \bmod cache\_size}{cache\_line\_size} \tag{18.4}$$

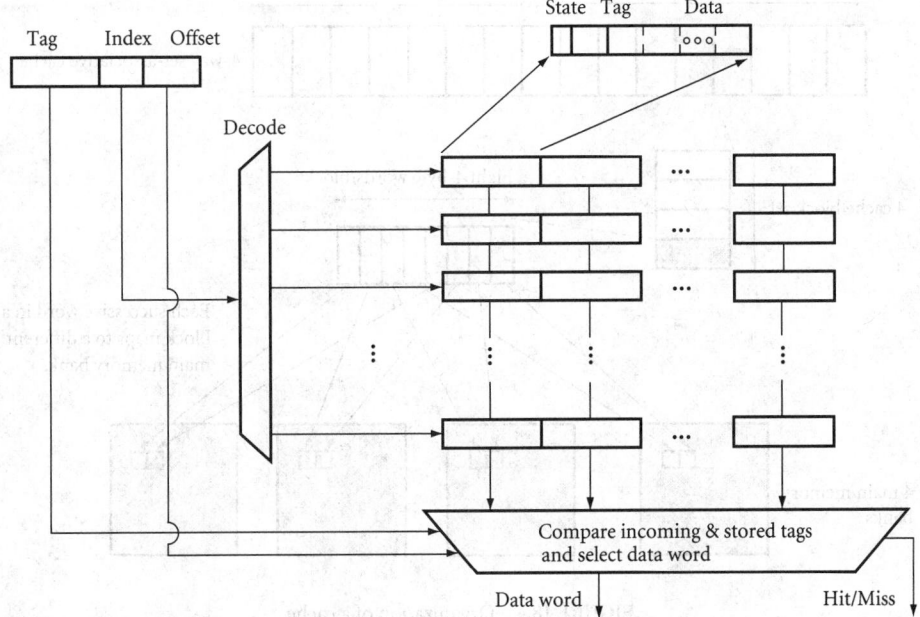

FIGURE 18.3 Components of a cache memory. (*Source:* Hill, M. D. 1988. A case for direct-mapped caches. *IEEE Comput.* 21(12):27. IEEE Computer Society, New York. With permission.)

where A_{cache} is the address within the cache for main memory location A_{memory}, cache_size is the capacity of the cache in addressable units (usually bytes), and cache_line_size is the size of the cache line in addressable units. Since the hashing function is simple bit selection, the tag memory need contain only the part of the address not implied by the result of the hashing function. That is,

$$A_{\text{tag}} = A_{\text{memory}} \text{ div size\_of\_cache} \qquad (18.5)$$

where A_{tag} is stored in the tag memory and *div* is the integer divide operation. In testing for a match, the complete address of a line stored in the cache can be inferred from the tag and its storage location within the cache.

A *two-way set-associative* cache maps each memory location into either of two locations in the cache, and can be constructed essentially as two identical direct-mapped caches. However, both caches must be searched at each memory access and the appropriate data selected and multiplexed on a tag match (hit). On a miss, a choice must be made between the two possible cache lines as to which is to be replaced. A single LRU bit can be saved for each such pair of lines to remember which line has been accessed more recently. This bit must be toggled to the current state each time either of the cache lines is accessed.

In the same way, an *M-way associative* cache maps each memory location into any of M memory locations in the cache and can be constructed from M identical direct-mapped caches. The problem of maintaining the LRU ordering of M cache lines quickly becomes hard, however, since there are $M!$ possible orderings, so it takes at least

$$\lceil \log_2(M!) \rceil \qquad (18.6)$$

bits to store the ordering. In practice, this requirement limits true LRU replacement to three- or four-way set associativity.

Figure 18.4 shows how a cache is organized into sets, blocks, and words. The cache shown is a 2-Kbyte, four-way set-associative cache, with 16 sets. Each set consists of four blocks. The cache block size in this example is 32 bytes, so each block contains eight 4-byte words. Also depicted at the bottom of Figure 18.4

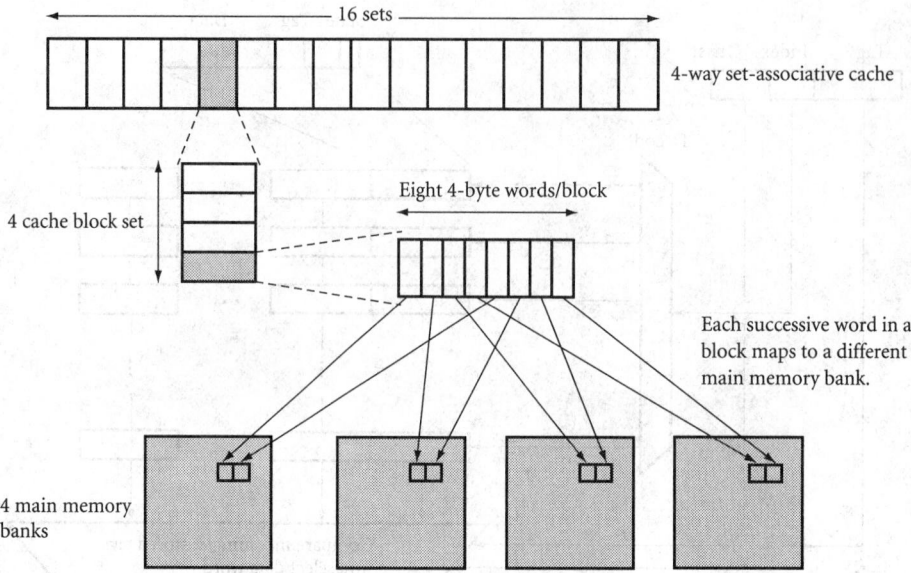

FIGURE 18.4   Organization of a cache.

is a four-way interleaved main memory system (see the next section for details). Each successive word in the cache block maps into a different main memory bank. Because of the cache's mapping restrictions, each cache block obtained from main memory will be loaded into its corresponding set, but it may appear anywhere within that set.

Write operations require special handling in the cache. If the main memory copy is updated with each write operation — a technique known as *write-through* or *store-through* — the writes may force operations to stall while the write operations are completing. This can happen after a series of write operations even if the processor is allowed to proceed before the write to the memory has completed. If the main memory copy is not updated with each write operation — a technique known as *write-back* or *copy-back* or *deferred writes* — the main memory locations become stale, that is, memory no longer contains the correct values and must not be relied upon to provide data. This is generally permissible, but care must be exercised to make sure that it is always updated before the line is purged from the cache and that the cache is never bypassed. Such a bypass could occur with *direct memory access* (DMA), in which the I/O system writes directly into main memory without the involvement of the processor.

Even for a system that implements a write-through policy, care must be exercised if memory requests may bypass the cache. While the main memory is never stale, a write that bypasses the cache, such as from I/O, could have the effect of making the cached copy stale. A later access by the CPU could then provide an incorrect value. This can be avoided only by making sure that cached entries are invalidated even if the cache is bypassed. The problem is relatively easy to solve for a single processor with I/O, but it becomes very difficult to solve for multiple processors, particularly so if multiple caches are involved as well. This is known in general as the cache *coherence* or *consistency* problem.

The cache exploits spatial locality by loading an entire cache line after a miss. This tends to result in bursty traffic to the main memory, since most accesses are filtered out by the cache. After a miss, however, the memory system must provide an entire line at once. Cache memory nicely complements an interleaved, high-bandwidth main memory (described in the next section), since a cache line can be interleaved across many banks in a regular manner — thus avoiding memory conflicts and being loaded rapidly into the cache. The example of main memory shown in Figure 18.4 can provide the entire cache line with two parallel memory accesses.

Conventional caches traditionally could not accept requests while they were servicing a miss request. In other words, they *locked-up* or *blocked* when servicing a miss. The growing penalty for cache misses has made it necessary for high-end commodity memory systems to continue to accept (and service) requests from the processor while a miss is being serviced. Some systems are able to service multiple miss requests simultaneously. To allow this mode of operation, the cache design is *lockup-free* or *nonblocking* [Kroft 1981]. Lockup-free caches have one structure for each simultaneous outstanding miss that they can service. This structure holds the information necessary to correctly return the loaded data to the processor, even if the misses come back in a different order than that in which they were sent.

Two factors drive the existence of multiple levels of cache memory in the memory hierarchy: access times and a limited number of transistors on the CPU chip. Larger banks with greater capacity are slower than smaller banks. If the time needed to access the cache limits the clock frequency of the CPU, then the first-level cache size may need to be constrained. Much of the benefit of a large cache may be obtained by placing a small first-level cache above a larger second-level cache; the first is accessed quickly, and the second holds more data close to the processor. Since many modern CPUs have caches on the CPU chip itself, the size of the cache is limited by the CPU silicon real estate. Some CPU designers have assumed that system designers will add large off-chip caches to the one or two levels of caches on the processor chip. The complexity of this part of the memory hierarchy may continue to grow as main memory access penalties continue to increase.

Caches that appear on the CPU chip are manufactured by the CPU vendor. Off-chip caches, however, are a commodity part sold in large volume. An incomplete list of major cache manufacturers includes Hitachi, IBM Micro, Micron, Motorola, NEC, Samsung, SGS-Thomson, Sony, and Toshiba. Although most personal computers and all major workstations now contain caches, very high-end machines (such as multimillion-dollar supercomputers) do not usually have caches. These ultraexpensive computers can afford to implement their main memory in a comparatively fast semiconductor technology such as static RAM (SRAM) and can afford so many banks that cacheless bandwidth out of the main memory system is sufficient. Massively parallel processors (MPPs), however, are often constructed out of workstation-like nodes to reduce cost. MPPs therefore contain cache hierarchies similar to those found in the workstations on which the nodes of the MPPs are based.

Cache sizes have been steadily increasing on personal computers and workstations. Intel Pentium-based personal computers come with 8 Kbyte each of instruction and data caches. Two of the Pentium chip sets, manufactured by Intel and OPTi, allow level-two caches ranging from 256 to 512 Kbyte and 64 Kbyte to 2 Mbyte, respectively. The newer Pentium Pro systems also have 8-Kbyte first-level instruction and data caches, but they also have either a 256-Kbyte or a 512-Kbyte second-level cache on the same module as the processor chip. Higher-end workstations — such as DEC Alpha 21164-based systems — are configured with substantially more cache. The 21164 also has 8-Kbyte first-level instruction and data caches. Its second-level cache is entirely on chip and is 96 Kbyte. The third-level cache is off chip, and can have a size ranging from 1 to 64 Mbyte.

For all desktop machines, cache sizes are likely to continue to grow — although the rate of growth compared to processor speed increases and main memory size increases is unclear.

18.4 Parallel and Interleaved Main Memories

Main memories are composed of a series of semiconductor memory chips. A number of these chips, like caches, form a *bank*. Multiple memory banks can be connected together to form an **interleaved** (or parallel) memory system. Since each bank can service a request, an interleaved memory system with K banks can service K requests simultaneously, increasing the peak bandwidth of the memory system to K times the bandwidth of a single bank. In most interleaved memory systems, the number of banks is a power of two, that is, $K = 2^k$. An n-bit memory word address is broken into two parts: a k-bit bank number and an m-bit address of a word within a bank. Though the k bits used to select a bank number could be any k bits of the n-bit word address, typical interleaved memory systems use the low-order k address bits to select the bank number; the higher order $m = n - k$ bits of the word address are used to access a word in the selected bank. The reason for using the low-order k bits will be discussed shortly. An interleaved memory system which uses the low-order k bits to select the bank is referred to as a *low-order* or a *standard* interleaved memory.

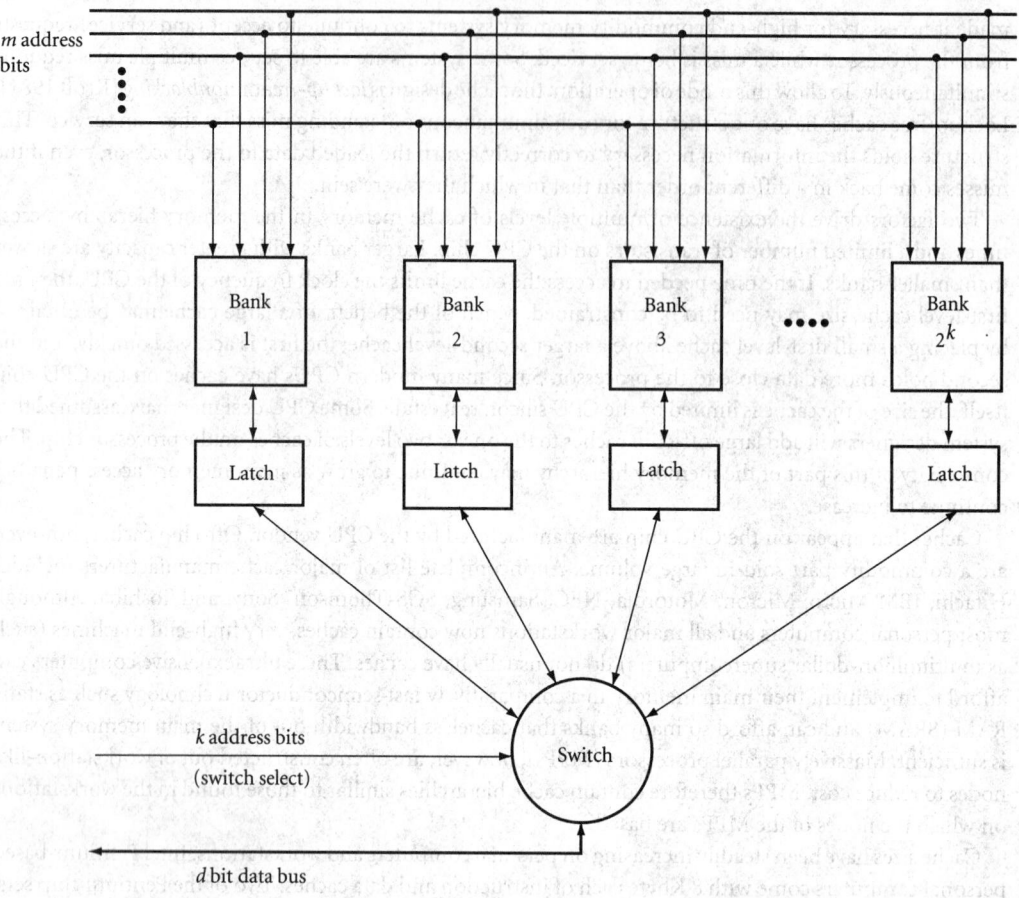

FIGURE 18.5 A simple interleaved memory system. (*Source:* Adapted from Kogge, P. M. 1981. *The Architecture of Pipelined Computers,* 1st ed., p. 41. McGraw–Hill, New York.)

There are two ways of connecting multiple memory banks: *simple interleaving* and *complex interleaving.* Sometimes simple interleaving is also referred to as *interleaving,* and complex interleaving is referred to as *banking.*

Figure 18.5 shows the structure of a simple interleaved memory system. m address bits are simultaneously supplied to every memory bank. All banks are also connected to the same read/write control line (not shown in Figure 18.5). For a read operation, the banks start the read operation and deposit the data in their latches. Data can then be read from the latches, one by one, by appropriately setting the switch. Meanwhile, the banks could be accessed again to carry out another read or write operation. For a write operation, the latches are loaded one by one. When all the latches have been written, their contents can be written into the memory banks by supplying m bits of address (they will be written into the same word in each of the different banks). In a simple interleaved memory, all banks are cycled at the same time; each bank starts and completes its individual operations at the same time as every other bank; a new memory cycle can start (for all banks) once the previous cycle is complete. Timing details of the accesses can be found in *The Architecture of Pipelined Computers* [Kogge 1981].

One use of a simple interleaved memory system is to back up a cache memory. To do so, the memory must be able to read blocks of contiguous words (a cache block) and supply them to the cache. If the low-order k bits of the address are used to select the bank number, then consecutive words of the block reside in different banks; they can all be read in parallel and supplied to the cache one by one. If some

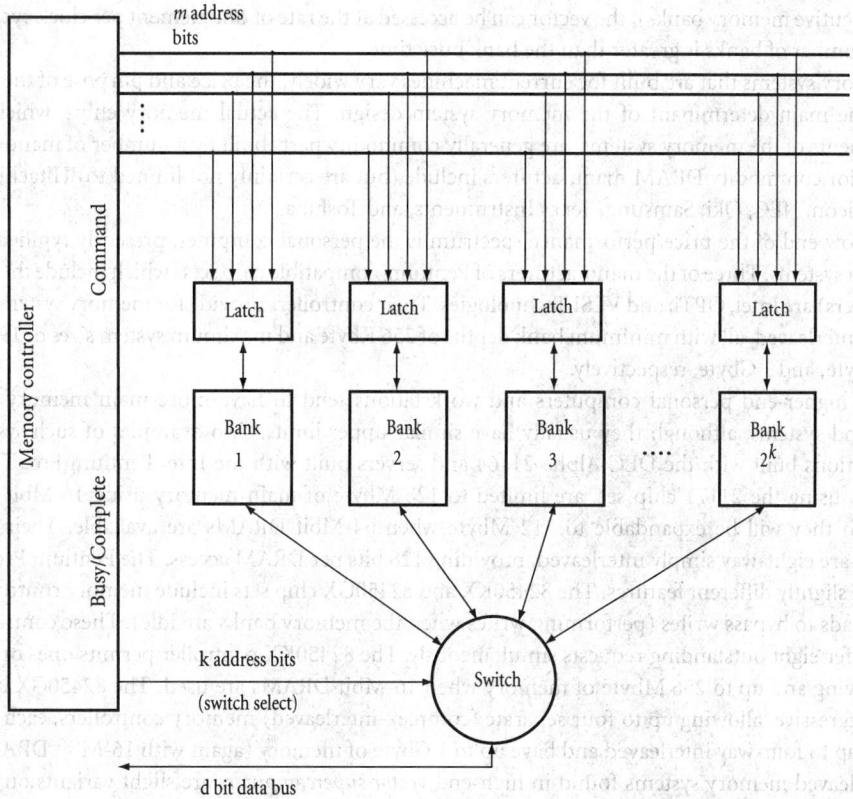

FIGURE 18.6 A complex interleaved memory system. (*Source:* Adapted from Kogge, P. M. 1981. *The Architecture of Pipelined Computers*, 1st ed., p. 42. McGraw–Hill, New York.)

other address bits are used for bank selection, then multiple words from the block might fall in the same memory bank, requiring multiple accesses to the same bank to fetch the block.

Figure 18.6 shows the structure of a complex interleaved memory system. In such a system, each bank is set up to operate on its own, independent of the other banks' operation. In this example, bank 1 could carry out a read operation on a particular memory address while bank 2 carries out a write operation on a completely unrelated memory address. (Contrast this with the operation in a simple interleaved memory where all banks are carrying out the same operation, read or write, and the locations accessed within each bank represent a contiguous block of memory.) Complex interleaving is accomplished by providing an address latch and a read/write command line for each bank. The *memory controller* handles the overall operation of the interleaved memory. The processing unit submits the memory request to the memory controller, which determines the bank that needs to be accessed. The controller then determines if the bank is busy (by monitoring a busy line for each bank). The controller holds the request if the bank is busy, submitting it later when the bank is available to accept the request. When the bank responds to a read request, the switch is set by the controller to accept the request from the bank and forward it to the processing unit. Timing details of the accesses can be found in *The Architecture of Pipelined Computers* [Kogge 1981].

A typical use of a complex interleaved memory system is in a *vector processor*. In a vector processor, the processing units operate on a vector, for example a portion of a row or a column of a matrix. If consecutive elements of a vector are present in different memory banks, then the memory system can sustain a bandwidth of one element per clock cycle. By arranging the data suitably in memory and using standard interleaving (for example, storing the matrix in row-major order will place consecutive elements in consecutive memory banks), the vector can be accessed at the rate of one element per clock cycle as long as the number of banks is greater than the bank busy time.

Memory systems that are built for current machines vary widely, the price and purpose of the machine being the main determinant of the memory system design. The actual memory chips, which are the components of the memory systems, are generally commodity parts built by a number of manufacturers. The major commodity DRAM manufacturers include (but are certainly not limited to) Hitachi, Fujitsu, LG Semicon, NEC, Oki, Samsung, Texas Instruments, and Toshiba.

The low end of the price/performance spectrum is the personal computer, presently typified by Intel Pentium systems. Three of the manufacturers of Pentium-compatible chip sets (which include the memory controllers) are Intel, OPTi, and VLSI Technologies. Their controllers provide for memory systems that are simply interleaved, all with minimum bank depths of 256 Kbyte and maximum system sizes of 192 Mbyte, 128 Mbyte, and 1 Gbyte, respectively.

Both higher-end personal computers and workstations tend to have more main memory than the lower-end systems, although they usually have similar upper limits. Two examples of such systems are workstations built with the DEC Alpha 21164 and servers built with the Intel Pentium Pro. The Alpha systems, using the 21171 chip set, are limited to 128 Mbyte of main memory using 16 Mbit DRAMs, although they will be expandable to 512 Mbyte when 64-Mbit DRAMs are available. Their memory systems are eight-way simply interleaved, providing 128 bits per DRAM access. The Pentium Pro systems support slightly different features. The 82450KX and 82450GX chip sets include memory controllers that allow reads to bypass writes (performing writes when the memory banks are idle). These controllers can also buffer eight outstanding requests simultaneously. The 82450KX controller permits one- or two-way interleaving and up to 256 Mbyte of memory when 16-Mbit DRAMs are used. The 82450GX chip set is more aggressive, allowing up to four separate (complex-interleaved) memory controllers, each of which can be up to four-way interleaved and have up to 1 Gbyte of memory (again with 16-Mbit DRAMs).

Interleaved memory systems found in high-end *vector supercomputers* are slight variants on the basic complex interleaved memory system of Figure 18.6. Such memory systems may have hundreds of banks, with multiple memory controllers that allow multiple independent memory requests to be made every clock cycle. Two examples of modern vector supercomputers are the Cray T-90 series and the NEC SX series. The Cray T-90 models come with varying numbers of processors — up to 32 in the largest configuration. Each of these processors is coupled with 256 Mbyte of memory, split into 16 banks of 16 Mbyte each. The T-90 has complex interleaving among banks. The largest configuration (the T-932) has 32 processors, for a total of 512 banks and 8 Gbyte of main memory. The T-932 can provide a peak of 800-GByte/s bandwidth out of its memory system. NEC's SX-4 product line, their most recent vector supercomputer series, has numerous models. Their largest single-node model (with one processor per node) contains 32 processors, with a maximum of 8 Gbyte of memory, and a peak bandwidth of 512 Gbyte/s out of main memory. Although the sizes of the memory systems are vastly different between workstations and vector machines, the techniques that both use to increase total bandwidth and minimize bank conflicts are similar.

18.5 Virtual Memory

Cache memory contains portions of the main memory in dynamically allocated cache lines. Since the data portion of the cache memory is itself a conventional memory, each line present in the cache has two addresses associated with it: its main memory address and its cache address. Thus, the main memory address of a word can be divorced from a particular storage location and abstractly thought of as an element in the address space. The use of a two-level hierarchy — consisting of main memory and a slower, larger disk storage device — evolved by making a clear distinction between the address space and the locations in memory. An address generated during the execution of a program is known as a *virtual address*, which must be translated to a *physical address* before it can be accessed in main memory. The total address space is only an abstraction.

A **virtual memory** address is mapped to a physical address, which indicates the location in main memory where the data actually reside [Denning 1970]. The mapping is maintained through a structure called the *page table*, which is maintained in software by the operating system. Like the tag memory of a cache memory, the page table is accessed through a virtual address to determine the physical (main memory) address of the entry. Unlike the tag memory, however, the table is usually sorted by virtual addresses, making the translation process a simple matter of an extra memory access to determine the physical address of the desired item. A system maintaining the page table in a way analogous to a cache tag memory is said to have *inverted page tables*. In addition to the physical address mapped to a virtual page, and an indication of whether the page is present at all, a page table entry often contains other information. For example, the page table may contain the location on the disk where each block of data is stored when not present in main memory.

The virtual memory can be thought of as a collection of blocks. These blocks are often aligned and of fixed size, in which case they are known as *pages*. Pages are the unit of transfer between the disk and main memory and are generally larger than a cache line — usually thousands of bytes. A typical page size for machines in 1995 is 4 Kbyte. A page's virtual address can be broken into two parts: a virtual page number and an offset. The page number specifies which page is to be accessed, and the page offset indicates the distance from the beginning of the page to the indicated address.

A physical address can also be broken into two parts: a physical page number (also called a *page frame* number) and an offset. This mapping is done at the level of pages, so the page table can be indexed by means of the virtual page number. The page frame number is contained in the page table and is read out during the translation along with other information about the page. In most implementations the page offset is the same for a virtual address and the physical address to which it is mapped.

The virtual memory hierarchy is different than the cache/main memory hierarchy in a number of respects, resulting primarily from the fact that there is a much greater difference in latency between accesses to the disk and to main memory. While a typical latency ratio for cache and main memory is one order of magnitude (main memory has a latency ten times larger than the cache), the latency ratio between disk and main memory is often four orders of magnitude or more. This large ratio exists because the disk is a mechanical device — with a latency partially determined by velocity and inertia — whereas main memory is limited only by electronic and energy constraints. Because of the much larger penalty for a page miss, many design decisions are affected by the need to minimize the frequency of misses. When a miss does occur, the processor could be idle for a period during which it could execute tens of thousands of instructions. Rather than stall during this time, as may occur upon a cache miss, the processor invokes the operating system and may switch to a different task. Because the operating system is being invoked anyway, it is convenient to rely on it to set up and maintain the page table, unlike cache memory, where it is done entirely in hardware. The fact that this accounting occurs in the operating system enables the system to use virtual memory to enforce protection on the memory. This ensures that no program can corrupt the data in memory that belong to any other program.

Hardware support provided for a virtual memory system generally includes the ability to translate the virtual addresses provided by the processor into the physical addresses needed to access main memory. Thus, only upon a virtual-address miss is the operating system invoked. An important aspect of a computer that implements virtual memory, however, is the necessity of freezing the processor at the point at which a miss occurs, servicing the page table fault, and later returning to continue the execution as if no page fault had occurred. This requirement means either that it must be possible to halt execution at any point — including possibly in the middle of a complex instruction — or that it must be possible to guarantee that all memory accesses will be to pages resident in main memory.

As described above, virtual memory requires two memory accesses to fetch a single entry from memory: one into the page table to map the virtual address into the physical address, and the second to fetch the actual data. This process can be sped up in a variety of ways. First, a special-purpose cache memory to store the active portion of the page table can be used to speed up the first access. This special-purpose cache is usually called a *translation lookaside buffer* (TLB). Second, if the system also employs a cache memory, it may be possible to overlap the access of the cache memory with the access to the TLB, ideally allowing the requested item to be accessed in a single cache access time. The two accesses can be fully overlapped if the virtual address supplies sufficient information to fetch the data from the cache before the virtual-to-physical address translation has been accomplished. This is true for an M-way set associative cache of capacity C if the following relationship holds:

$$\text{Page\_size} \geq \frac{C}{M} \qquad (18.7)$$

For such a cache, the index into the cache can be determined strictly from the page offset. Since the virtual page offset is identical to the physical page offset, no translation is necessary, and the cache can be accessed concurrently with the TLB. The physical address must be obtained before the tag can be compared, of course.

An alternative method applicable to a system containing both virtual memory and a cache is to store the virtual address in the tag memory instead of the physical address. This technique introduces consistency problems in virtual memory systems that either permit more than a single address space, or allow a single physical page to be mapped to more than one single virtual page. This problem is known as the *aliasing* problem.

Chapter 85 is devoted to virtual memory and contains significantly more material on this topic for the interested reader.

18.6 Research Issues

Research is occurring on all levels of the memory hierarchy. At the register level, researchers are exploring techniques to provide more registers than are architecturally visible to the compiler. A large volume of work exists (and is occurring) for cache optimizations and alternative cache organizations. For instance, modern processors now commonly split the top level of the cache into separate physical caches, one for instructions (code) and one for program data. Due to the increasing cost of cache misses (in terms of processor cycles), some research trades off increasing the complexity of the cache for reducing the miss rate. Two examples of cache research from opposite ends of the hardware/software spectrum are *blocking* [Lam et al. 1991] and *skewed-associative caches* [Seznec 1993]. Blocking is a software technique in which the programmer or compiler reorganizes algorithms to work on subsets of data that are smaller than the cache instead of streaming entire large data structures repeatedly through the cache. This reorganization greatly improves temporal locality. The skewed-associative cache is one example of a host of hardware techniques that map blocks into the cache differently, with the goal of reducing misses from set conflicts. In skewed-associative caches, either one of two hashing functions may determine where a block should be placed in the cache, as opposed to just the one hashing function (low-order index bits) that traditional caches use. An important cache-related research topic is *prefetching* [Mowry et al. 1992], in which the processor issues requests for data well before the data are actually needed. Speculative prefetching is also a current research topic. In speculative prefetching, prefetches are issued based on guesses as to which data will be needed soon. Other cache-related research examines placing special structures in parallel with the cache, trying to optimize for workloads that do not lend themselves well to caches. Stream buffers [Jouppi 1990] are one such example. A stream buffer automatically detects when a linear access through a data structure occurs. The stream buffer issues multiple sequential prefetches upon detection of a linear array access.

Much of the ongoing research on main memory involves improving the bandwidth from the memory system without greatly increasing the number of banks. Multiple banks are expensive, particularly with the large and growing capacity of modern DRAM chips. Rambus [Rambus 1992] and Ramlink [IEEE Computer Society 1993] are two such examples.

Research issues associated with improving the performance of the virtual memory system fall into the domain of operating system research. One proposed strategy for reducing page faults allows each running program to specify its own page replacement algorithm, enabling each program to optimize the choice of page replacements based on its reference pattern [Engler et al. 1995]. Other recent research focuses on improving the performance of the TLB. Two techniques for doing this are the use of a two-level TLB (the motivation is similar to that for a two-level cache) and the use of superpages [Talluri and Hill 1994]. With superpages, each TLB entry may represent a mapping for more than one consecutive page, thus increasing the total address range that a fixed number of TLB entries may cover.

18.7   Summary

A computer's memory system is the repository for all the information that the CPU uses and produces. A perfect memory system is one that can immediately supply any datum that the CPU requests. This ideal memory is not implementable, however, as the three factors of memory — capacity, speed, and cost — are directly in opposition.

By staging smaller, faster memories in front of larger, slower, and cheaper memories, the performance of the memory system may approach that of a perfect memory system — at a reasonable cost. The memory hierarchies of modern general-purpose computers generally contain registers at the top, followed by one or more levels of cache memory, main memory, and virtual memory on a magnetic or optical disk.

Performance of a memory system is measured in terms of latency and bandwidth. The latency of a memory request is how long it takes the memory system to produce the result of the request. The bandwidth of a memory system is the rate at which the memory system can accept requests and produce results. The memory hierarchy improves average latency by quickly returning results that are found in the higher levels of the hierarchy. The memory hierarchy generally reduces bandwidth requirements by intercepting a fraction of the memory requests at higher levels of the hierarchy. Some machines — such as high-performance vector machines — may have fewer levels in the hierarchy, increasing memory cost for better predictability and performance. Some of these machines contain no caches at all, relying on large arrays of main memory banks to supply very high bandwidth, with pipelined accesses of operands that mitigate the adverse performance impact of long latencies.

Cache memories are a general solution to improving the performance of a memory system. Although caches are smaller than typical main memory sizes, they ideally contain the most frequently accessed portions of main memory. By keeping the most heavily used data near the CPU, caches can service a large fraction of the requests without accessing main memory (the fraction serviced is called the hit rate). Caches assume locality of reference to work well transparently — they assume that accessed memory words will be accessed again quickly (temporal locality) and that memory words adjacent to an accessed word will be accessed soon after the access in question (spatial locality). When the CPU issues a request for a datum not in the cache (a cache miss), the cache loads that datum and some number of adjacent data (a cache block) into itself from main memory.

To reduce cache misses, some caches are associative — a cache may place a given block in one of several places, collectively called a set. This set is content-addressable; a block may or may not be accessed based on an address tag, one of which is coupled with each block. When a new block is brought into a set and the set is full, the cache's replacement policy dictates which of the old blocks should be removed from the cache to make room for the new block. Most caches use an approximation of least-recently-used (LRU) replacement, in which the block last accessed farthest in the past is the one that the cache replaces.

Main memory, or backing store, consists of banks of dense semiconductor memory. Since each memory chip has a small off-chip bandwidth, rows of these chips are placed together to form a bank, and multiple banks are used to increase the total bandwidth out of main memory. When a bank is accessed, it remains busy for a period of time, during which the processor may make no other accesses to that bank. By increasing the number of interleaved (parallel) banks, the chance that the processor issues two conflicting requests to the same bank is reduced.

Systems generally require a greater number of memory locations than are available in the main memory (i.e., a larger address space). The entire address space that the CPU uses is kept on large magnetic or optical disks; this is called the virtual address space, or virtual memory. The most frequently used sections of the virtual memory are kept in main memory (physical memory) and are moved back and forth in units

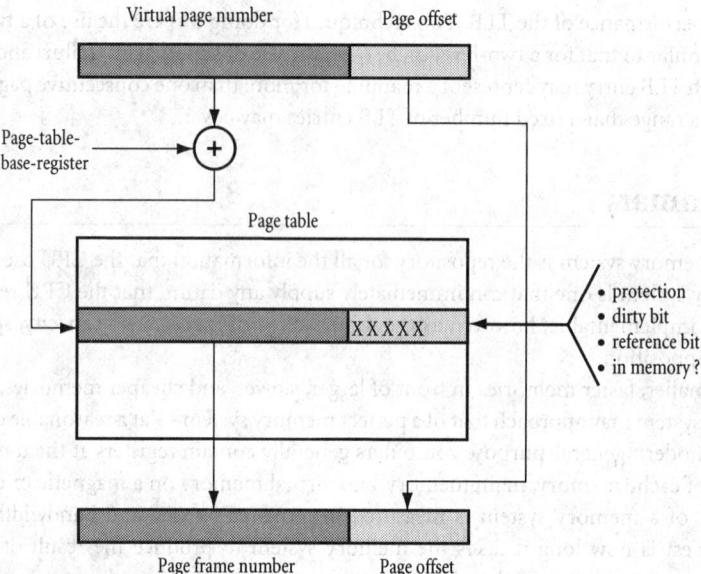

FIGURE 18.7   Virtual-to-physical address translation. (*Source:* Dorf, R. C. 1992. *The Electrical Engineering Handbook*, 1st ed., p. 1935. CRC Press, Inc., Boca Raton, FL.)

called pages. The place at which a virtual address lies in main memory is called its physical address. Since a much larger address space (virtual memory) is mapped onto a much smaller one (physical memory), the CPU must translate the memory addresses issued by a program (virtual addresses) into their corresponding locations in physical memory (physical addresses) (see Figure 18.7). This mapping is maintained in a memory structure called the page table. When the CPU attempts to access a virtual address that does not have a corresponding entry in physical memory, a page fault occurs. Since a page fault requires an access to a slow mechanical storage device (such as a disk), the CPU usually switches to a different task while the needed page is read from the disk.

Every memory request issued by the CPU requires an address translation, which in turn requires an access to the page table stored in memory. A translation lookaside buffer (TLB) is used to reduce the number of page table lookups. The most frequent virtual-to-physical mappings are kept in the TLB, which is a small associative memory tightly coupled with the CPU. If the needed mapping is found in the TLB, the translation is performed quickly and no access to the page table need be made. Virtual memory allows systems to run larger or more programs than are able to fit in main memory, enhancing the capabilities of the system.

Defining Terms

Bandwidth: The rate at which the memory system can service requests.

Cache memory: A small, fast, redundant memory used to store the most frequently accessed parts of the main memory.

Interleaving: Technique for connecting multiple memory modules together in order to improve the bandwidth of the memory system.

Latency: The time between the initiation of a memory request and its completion.

Memory hierarchy: Successive levels of different types of memory, which attempt to approximate a single large, fast, and cheap memory structure.

Virtual memory: A memory space implemented by storing the more frequently accessed data in main memory and less frequently accessed data on disk.

References

Denning, P. J. 1970. Virtual memory. *Comput. Surveys* 2(3):153–170.

Dorf, R. C. 1992. *The Electrical Engineering Handbook*, 1st ed. CRC Press, Inc., Boca Raton, FL.

Engler, D. R., Kaashoek, M. F., and O'Toole, J., Jr. 1995. Exokernel: an operating system architecture for application-level resource management, pp. 251–266. In *Proc. 15th Symp. on Operating Systems Principles*.

Hennessy, J. L. and Patterson, D. A. 1990. *Computer Architecture: A Quantitative Approach*, 1st ed. Morgan Kaufmann, San Mateo, CA.

Hill, M. D. 1988. A case for direct-mapped caches. *IEEE Comput.* 21(12):25–40.

IEEE Computer Society. 1993. *IEEE Standard for High-Bandwidth Memory Interface Based on SCI Signaling Technology (RamLink)*. Draft 1.00 IEEE P1596.4-199X.

Jouppi, N. 1990. Improving direct-mapped cache performance by the addition of a small fully-associative cache and prefetch buffers, pp. 364–373. In *Proc. 17th Annual Int. Symp. on Computer Architecture*.

Kogge, P. M. 1981. *The Architecture of Pipelined Computers*, 1st ed. McGraw–Hill, New York.

Kroft, D. 1981. Lockup-free instruction fetch/prefetch cache organization, pp. 81–87. In *Proc. 8th Annual Int. Symp. on Computer Architecture*.

Lam, M. S., Rothberg, E. E., and Wolf, M. E. 1991. The cache performance and optimizations of blocked algorithms, pp. 63–74. In *Proc. 4th Annual Symp. on Architectural Support for Programming Languages and Operating Systems*.

Mowry, T. C., Lam, M. S., and Gupta, A. 1992. Design and evaluation of a compiler algorithm for prefetching, pp. 62–73. In *Proc. 5th Annual Symp. on Architectural Support for Programming Languages and Operating Systems*.

Rambus 1992. *Rambus Architectural Overview*. Rambus Inc., Mountain View, CA.

Seznec, A. 1993. A case for two-way skewed-associative caches, pp. 169–178. In *Proc. 20th International Symposium on Computer Architecture*.

Smith, A. J. 1986. Bibliography and readings on CPU cache memories and related topics. *ACM SIGARCH Comput. Architecture News* 14(1):22–42.

Smith, A. J. 1991. Second bibliography on cache memories. *ACM SIGARCH Comput. Architecture News* 19(4):154–182.

Talluri, M. and Hill, M. D. 1994. Surpassing the TLB performance of superpages with less operating system support, pp. 171–182. In *Proc. 6th Int. Symp. on Architectural Support for Programming Languages and Operating Systems*.

Further Information

Some general information on the design of memory systems is available in *High-Speed Memory Systems* by A. V. Pohm and O. P. Agarwal. 1983. Reston Publishing, Reston, VA.

Computer Architecture: A Quantitative Approach by John Hennessy and David Patterson [Hennessy and Patterson 1990] contains a detailed discussion on the interaction between memory systems and computer architecture.

For information on memory system research, the recent proceedings of the *International Symposium on Computer Architecture* contain annual research papers in computer architecture, many of which focus on the memory system. To obtain copies, contact the IEEE Computer Society Press, 10662 Los Vaqueros Circle, P.O. Box 3014, Los Alamitos, CA 90720-1264.

19

Buses

19.1	Introduction	19-1
19.2	Bus Physics	19-2
	Transmission-Line Concepts • Signal Reflections • Wire-OR Glitches • Signal Skew • Cross-Coupling Effects	
19.3	Bus Arbitration	19-5
	Centralized Arbitration • Decentralized Arbitration	
19.4	Bus Protocol	19-7
	Asynchronous Protocol • Synchronous Protocol • Split-Transaction Protocol	
19.5	Issues in SMP System Buses	19-9
	Cache Coherence Protocols • Bus Arbitration • Bus Bandwidth • Memory Access Latency • Synchronization and Locking	
19.6	Putting It All Together — CCL-XMP System Bus	19-15
19.7	Historical Perspective and Research Issues	19-17

Windsor W. Hsu
IBM Research

Jih-Kwon Peir
University of Florida

19.1 Introduction

The *bus* is the most popular communication pathway among the various components of a computer system. The distinguishing feature of the bus is that it consists of a single set of shared communication links to which many components can be attached. The bus is not only a very cost-effective means of connecting various components together, but also is very versatile in that new components can be added easily. Furthermore, the bus has a broadcasting capability which can be extremely useful. The downside of the shared communication links is that they allow only one communication to occur at a time and the bandwidth does not scale with the number of components attached. Nevertheless, the bus is very popular because there are many situations where several components need to be connected together but they need not all transmit at the same time. This kind of requirement maps naturally onto a bus, allowing a very cost-effective solution. However, there are cases where the bus does become a communication bottleneck. In such cases, very aggressive bus designs have been attempted, but there comes a point where the fundamental characteristic of the bus cannot be overcome and more expensive solutions such as point-to-point links have to be used.

Buses are used at every level in the computer system. For instance, within the processor itself, the bus is often the means of communication between the register file and the various execution units. At a higher level, the processor is connected to the memory subsystem through the *system bus*. Today's computers typically have a fast peripheral bus called a *local bus* which directly interfaces onto the system bus to provide a high bandwidth for demanding devices such as the graphics adaptor. Other less demanding peripheral devices are attached to the *I/O bus*. In the old days, the processor, memory subsystem, and I/O devices were all plugged onto the *backplane bus,* which is so called because the bus runs physically

1-58488-360-X/$0.00+$1.50
© 2004 by CRC Press, LLC

along the backplane of the computer chassis. The various buses are each optimized for a particular set of performance requirements and cost constraints and may thus seem very different from one another. However, their underlying issues are fundamentally the same.

A major requirement for designing a bus or simply comprehending a bus design is a proper understanding of the electrical and mechanical behavior of the bus. As buses are pushed to provide higher data rates, physical phenomena such as signal reflection, crosstalk, skew, etc., are becoming more significant and have to be handled carefully. Because the communication medium of a bus is shared by multiple devices, at most one transmission can be initiated at any time by any device. The other devices can act only as receivers or **bus slaves** for the transmission. A device that is capable of initiating and controlling a communication is called a **bus master**. In order to ensure that only one bus master is talking at any one time, the bus masters have to go through a **bus arbitration** process before they can gain control of the bus. Once a bus master has been granted control of the bus, a **bus protocol** has to be followed by the master and the slave in order for them to understand one another. The specifics of the protocol can vary widely, depending on the functional and performance requirements. For instance, the protocol used in the system bus of a uniprocessor is dramatically different from that used in the system bus of a shared-memory **symmetric multiprocessor** (SMP).

In this chapter, an overview of the basic underlying physics of computer buses is presented first. This is followed by a discussion of important issues in bus designs, including bus arbitration and various communication protocols. Because of the increasing prevalence of SMP systems and the special challenges they pose for buses, a separate section is devoted to discussing special bus issues in such machines. The discussion is wrapped up with a case study of a modern SMP system bus design. Finally, a historical perspective of computer buses and the related research issues are given in the last section.

19.2 Bus Physics

Computer buses are becoming wider and are being run at higher frequencies in order to keep up with the phenomenal improvement in CPU performance. As the physical and temporal margins in bus designs are reduced, it is imperative that electrical phenomena such as signal reflections, skew, crosstalk, etc., be understood and properly handled. In this section, we introduce the basic ideas behind these phenomena. A more detailed discussion can be found in Giacomo [1990].

19.2.1 Transmission-Line Concepts

Electrical signals propagate with a finite speed that depends on the propagation medium. For instance, it takes approximately 5 ns for an electrical signal to travel 1 m along a typical copper wire. However, when analyzing electrical circuits, we often ignore their spatial properties. For example, we are seldom concerned with where each element of the circuit is located. We can do this because the circuits we usually encounter are lumped. In other words, their physical dimensions are small enough that for their particular applications, electromagnetic waves propagate across the circuits virtually instantaneously. However, this is not the case with high-speed buses. In this subsection, we introduce the basic concepts of the transmission-line model which are needed to understand the electrical behavior of today's buses.

In general, a connection is considered a transmission line if the propagation time of an electrical signal through it is a significant part of the rise time, fall time, or width of the signal. A good rule of thumb is to consider anything above $\frac{1}{4}$ as a "significant part." A transmission line has a characteristic impedance Z_0. If the line impedance Z_L changes as a signal propagates down the line, part of the power travels backwards as a reflected signal. The reflection coefficient is given by $\Gamma = (Z_L - Z_0)/(Z_L + Z_0)$. In other words, the strength of the reflected signal increases with the magnitude of the impedance mismatch.

Consider the circuit in Figure 19.1. It contains a voltage source V_s with internal resistance R_s connected by a pair of transmission lines to a load of resistance Z_L. We assume that it takes T units of time for an electrical signal to propagate one-way across the circuit. At time $t = 0$, the switch is closed and a voltage

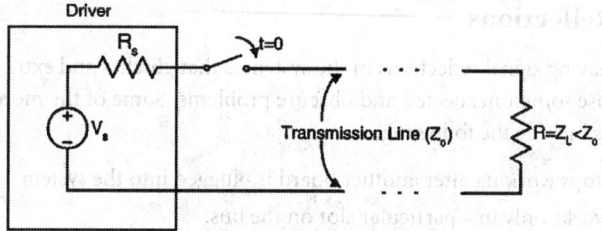

FIGURE 19.1   Circuit for transmission-line example.

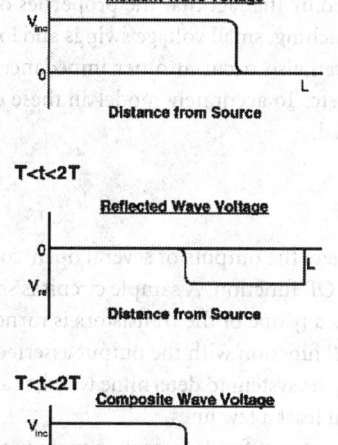

FIGURE 19.2   Transmission-line voltage waveform.

pulse is sent down the transmission lines toward the load. Figure 19.2 contains a short chronicle of the voltage waveform seen on the lines. For time $t < T$, the signal sees only Z_0, the impedance of the line. Thus, the voltage of the incident wave $V_{inc} = V_s Z_0/(R_s + Z_0)$. When the incident wave hits the load at time $t = T$, it sees an additional impedance Z_L which reflects the incident wave with a coefficient of $\Gamma = (Z_L - Z_0)/(Z_L + Z_0)$. In Figure 19.2, we assume that $Z_L < Z_0$ so that the reflected wave is negative with voltage $V_{re} = \Gamma V_{inc}$. If the transmission lines are not properly terminated at the driver, i.e., $R_s \neq Z_0$, there will be additional reflections before the system settles down. The composite signal at any point along the line is the instantaneous sum of all the incident and reflected signals. Note that all the signals are subject to line losses, especially in their high-frequency components. Thus, pulses eventually lose their shape.

When the circuit is in steady state, the above analysis should agree with the lumped circuit model. For simplicity, let us assume that the transmission lines are properly terminated at the driver, i.e., $R_s = Z_0$. In this case, there is only one reflection and the circuit settles down at time $t = 2T$. Thus, we would expect the steady-state voltage to be the sum of V_{inc} and V_{re}. This works out to $V_s Z_L/(Z_L + Z_0)$, a result which agrees with that predicted by the lumped-circuit model. An intuitive way to reason about transmission lines is to think in terms of feedback. In the beginning, the signal has no clear idea of how much impedance it will encounter in the circuit. Thus, it makes a guess based on the impedance it has already seen. As the signal propagates, it learns more about the circuit, and this information is fed back in the form of reflections, so that eventually Kirchhoff's circuit laws are satisfied.

19.2.2   Signal Reflections

The consequence of having signal reflections in the system is that glitches and extra pulses may appear on the bus. This may cause some unexpected and obscure problems. Some of the more common symptoms of reflection problems include the following:

- A board that stops working after another board is plugged into the system.
- A board that works only in a particular slot on the bus.
- A system that works only when the boards are arranged in a specific order.

Reflections are typically most significant at the various sources and loads on the bus. We can reduce the magnitude of these reflections by matching the impedance of the sources and loads to that of the lines.

This can be accomplished by adding a series or parallel resistance or by using a clamping diode. Impedance matching is complicated by the fact that the properties of bus drivers change as they switch on or off. For better impedance matching, small voltage swings and low-capacitance drivers and receivers are helpful. Note that reflections can also occur at other impedance discontinuities such as interboard connections, board layer changes, etc. To accurately model all these effects, computer simulations using tools such as SPICE are often needed.

19.2.3   Wire-OR Glitches

Wire-OR logic is a kind of logic where the outputs of several open-collector gates are connected together in such a way as to realize a logical *OR* function. A sample circuit is shown in Figure 19.3. Notice that the voltage on the line is low as long as any one of the transistors is turned on. Thus this circuit implements the logical *NOR* function or the *OR* function with the output asserted low. Wire-OR is very useful in bus arbitration. For instance, it enables the system to determine whether any bus master wishes to use the bus. Most buses use wire-OR logic for at least a few lines.

However, wire-OR lines are subject to a fundamental phenomenon known as *wire-OR glitch*. During an active to high-impedance transition, a glitch of up to one round-trip delay in width may appear on a wire-OR line. This phenomenon is a result of the finite propagation speed of electrical signals on a transmission line. Consider the case where only transistors 1 and *n* are initially turned on. Suppose that transistor *n* is now turned off. The current that it was previously sinking continues to flow, thus creating a signal which propagates along the line.

A more detailed explanation of the wire-OR glitch is given in Gustavson and Theus [1983]. Various ways of dealing with it are discussed in Gustavson and Theus [1983] and Taub [1983a, 1983b]. In the IEEE Futurebus, the wire-OR glitch problem is mitigated by the fact that the bus specification imposes constraints on when devices can switch on or off, effectively setting a limit on the maximum glitch duration [Taub 1984].

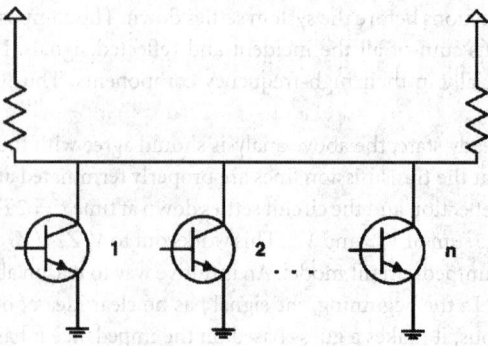

FIGURE 19.3   Wire-OR circuit.

19.2.4 Signal Skew

Another important electrical phenomenon in buses is *signal skew*. Because of differences in transmission lines, loading, etc., slight differences in the propagation delay of different bits in a word are inevitable. These differences are known as signal skew.

In a transmission, the receiver must be able to somehow determine when all the bits of a word have arrived and can be sampled. The effect of skew is to reduce the window during which the receiver can assume that the data are valid. This effectively limits the data rate of the bus. A wide bus consists of more parallel lines and is thus subject to more skew. In general, skew can be reduced by paying meticulous attention to the impedance of the bus lines. Synchronous buses have to deal with the additional problem of clock to data skew. An approach that has been taken to minimize this skew is to loop back the clock line at the end of the bus. When doing a data transfer, the clock signal that propagates in the same direction as the data transfer is used. Rambus uses this technique to minimize skew with respect to its aggressive 250-MHz clock [Rambus 1992].

19.2.5 Cross-Coupling Effects

A signal-carrying line sets up electrostatic and magnetic fields around it. In a bus, the lines run parallel and close to one another. Thus the fields from nearby lines intersect, causing a signal on one line to affect the signal on another. This is called crosstalk or coupling noise.

A simple way to reduce this effect is to spatially separate the bus lines so that the fields do not interfere with one another. However there is clearly a limit to how far we can carry this. Another way to reduce both the mutual capacitance and inductance of the lines is to introduce ground planes or wires near the bus lines. But this has undesirable side effects such as increasing the self-capacitance of the lines. An approach commonly taken to reduce coupling effects is to separate the lines with an insulator that has a low dielectric constant. Typically, combinations of these techniques are used in a bus design.

19.3 Bus Arbitration

Buses are ubiquitous in computer systems because they are a cost-effective and versatile means of connecting several devices together. The cost-effectiveness and versatility of buses stems from the fact that a bus has only one communication medium, which is shared by all the devices. In other words, at most one communication can occur on a bus at any one time. This implies that there must be some mechanism to decide which bus master has control of the bus at a given time. The process of arbitrating between requests for bus control is called bus arbitration. Bus arbitration can be handled in different ways depending on the performance requirements and cost constraints.

19.3.1 Centralized Arbitration

In *centralized arbitration*, there is a special device, the central arbiter, which is in charge of granting bus control to one of the bus masters. The fact that there is only one arbiter means that the centralized scheme has a single point of failure. In general, centralized arbitration can be further divided into two schemes, depending on how the bus masters are connected to the arbiter.

In the first scheme, the central arbiter is connected to each of the bus masters through private two-way connections. Because the bus requests can be made independently and in parallel by the bus masters, this is sometimes known as *centralized independent requests arbitration* or *centralized parallel arbitration*. Notice that the connections in this system form star networks emanating from the central arbiter. One such network carries the bus request signals from the bus masters to the arbiter. Another carries the bus grant signal from the arbiter back to one of the bus masters. Various arbitration policies can be implemented in the arbiter, making this a very flexible scheme. This scheme is fast because there are direct connections between any bus master and the arbiter. However, all these direct connections lead to a high implementation

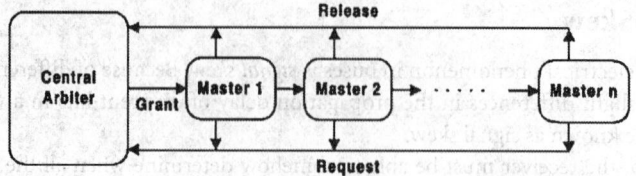

FIGURE 19.4   Daisy-chain bus arbitration.

cost. Furthermore, the use of direct connections means that the arbitration signals do not appear on the bus. This makes bus monitoring for debugging and diagnostic purposes difficult.

The second centralized arbitration scheme is known as *centralized serial priority arbitration*. In this scheme, there is a single bus grant signal line, which is routed through each of the bus masters as shown in Figure 19.4. This form of connection is known as a daisy chain. Hence, centralized serial priority arbitration is more commonly referred to as *daisy-chain arbitration*. In this scheme, there is a common wire-OR bus request line. A bus master may take control of the bus only if it has made a request and its incoming grant line is asserted. A bus master that does not wish to use the bus is required to forward the bus grant signal along the daisy chain. Notice that this implies an implicit priority assignment — the nearer a bus master is to the arbiter, the higher is its priority. The main advantage of daisy-chain arbitration is that it requires very few interconnections and the interface logic is simple. However, bus allocation in this scheme is slow because the grant signal has to travel along the daisy chain. Furthermore, the implicit priority scheduling may cause low-priority requests to be locked out indefinitely. Finally, as in centralized parallel arbitration, daisy-chain arbitration does not facilitate debugging and diagnosis. The VMEbus uses a variation of this scheme with four daisy-chained grant lines, which enable it to implement a variety of scheduling algorithms [Giacomo 1990].

19.3.2   Decentralized Arbitration

In *decentralized arbitration*, each bus master has its own arbitration and allocation logic. The responsibility of deciding who has control of the bus is distributed among the bus masters. Thus, this scheme is also known as *distributed arbitration*.

Typically, the bus contains n arbitration wire-OR lines and each bus master is assigned a unique n-bit arbitration number according to some priority scheme. During arbitration, if a master wishes to use the bus, it drives the arbitration lines with its arbitration number. If a master detects that the arbitration lines are carrying a higher priority number, it stops driving the less significant bits of its number. When the system settles down, the arbitration lines will indicate the winner, which is the participating master with the highest priority. The time for the system to settle down can be specified as a fixed time in the bus protocol. An alternative is to use an additional wire-OR line to indicate whether competing bus masters have completed arbitration. As with any priority scheme, the possibility of access starvation of the low-priority masters has to be considered. In the IEEE Futurebus, bus masters are divided into priority and fairness modules depending on whether they have any particularly urgent needs such as having to meet real-time constraints [Taub 1984]. A priority module can issue bus requests whenever it needs to. A fairness module, after winning an arbitration, will have to wait for all pending requests to be serviced before it can issue another request [Taub 1984]. Note that this does not guarantee but only helps to ensure that every module will be able to get a portion of the bus bandwidth.

The major disadvantage of the distributed scheme is that it requires relatively complex arbitration logic in each bus master and several arbitration lines on the bus. However, this scheme allows very fast bus allocation and a flexible assignment of priorities to the bus masters. Distributed arbitration is also more fault-tolerant than the centralized schemes in that the failure of a single bus master does not necessarily affect the operation of the bus. Variations of this scheme are widely used in buses such as the IEEE Futurebus, Nubus, Multibus II, Fastbus [Borrill 1985], and the Powerpath-2 System Bus of the SGI Challenge [Galles and Williams 1994].

The scheme described above is sometimes known as *distributed arbitration by self-selection* because each master decides whether it has won the race, effectively letting the winner select itself. Some computer networks, such as Ethernet, use another form of distributed arbitration which relies on collision detection [Metcalfe and Boggs 1976].

19.4 Bus Protocol

The bus is a shared resource. In order for it to function properly, all the devices on it must cooperate and adhere to a protocol or set of rules. This protocol defines precisely the bus signals that have to be asserted by the master and slave devices in each phase of the bus operation. In this section, we discuss some of the key options in designing bus protocols.

19.4.1 Asynchronous Protocol

In a communication, the sender and receiver must be coordinated so that the sender knows when to talk and the receiver knows when to listen. There are two basic ways to achieve proper coordination. This subsection discusses the asynchronous protocol. The next describes the synchronous design.

An asynchronous system does not have an explicit clock signal to coordinate the sender and receiver. Instead a *handshaking protocol* is used. In a handshaking protocol, the sender and receiver can proceed to the next step in the bus operation only if both of them are ready. In other words, both parties have to shake hands and agree to proceed before they can do so.

Figure 19.5 shows a basic handshaking protocol. Assume that *ReadReq* is used to request a read from memory, *DataReady* is used to indicate that data are ready on the data lines, and *Ack* is used to acknowledge the *ReadReq* or *DataReady* signals of the other party. Suppose a device wishes to initiate a memory read. When memory sees the *ReadReq*, it reads the address on the address bus and raises the *Ack* in acknowledgment of the request (step 1). When the device sees the *Ack*, it deasserts *ReadReq* and stops driving the address bus (step 2). Once memory sees that *ReadReq* has been deasserted, it drops the *Ack* line to acknowledge that it has seen the *ReadReq* signal (step 3). A similar exchange is carried out when memory is ready with the data that has been requested (steps 5–7). Notice that the two parties involved in the protocol take turns to respond to one another. They proceed in lockstep. This is how the handshaking protocol is able to coordinate the two devices.

The handshaking protocol is relatively insensitive to noise because of its self-timing nature. This self-timing nature also allows data to be transferred between devices of any speed, giving asynchronous designs the flexibility to handle new and faster devices as they appear. Thus, asynchronous designs are better able to scale with technology improvements. Furthermore, because clock skew is not an issue, asynchronous buses can be physically longer than their synchronous counterparts. The disadvantage of asynchronous designs lies in the fact that the handshaking protocol adds significant overhead to each data transfer and is thus slower than a synchronous protocol. As in any communication between parties with different clocks, there

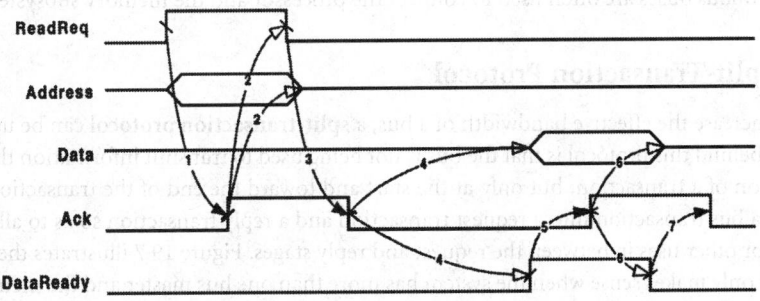

FIGURE 19.5 A basic handshaking protocol.

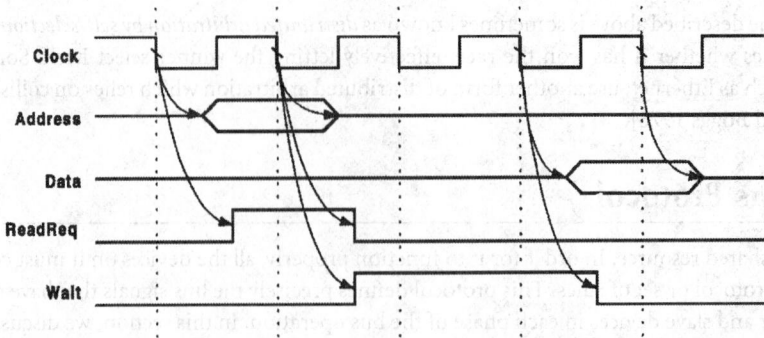

FIGURE 19.6   A basic synchronous protocol.

is an additional problem of synchronization failure when an asynchronous signal is sampled. Asynchronous designs are typically used when there is a need to accommodate many devices with a wide performance range and when the ability to incrementally upgrade the system is important. Thus, many of the asynchronous buses such as VMEbus, Futurebus, MCA, and IPI are backplane or I/O buses [Giacomo 1990].

19.4.2   Synchronous Protocol

In synchronous buses, the coordination of devices on the bus is achieved by distributing an explicit clock signal throughout the system. This clock signal is used as a reference to determine when the various bus signals can be assumed to be valid. Figure 19.6 shows a basic synchronous protocol coordinating a memory read transaction. Notice that all the signal changes happen with respect to the clock. In this particular example, the system is negative-edge triggered, which means that the signals are sampled on the falling edge of the clock.

An important design decision in synchronous buses is the choice of the clock frequency. Once a frequency is selected, it becomes locked into the protocol. The clock frequency must be chosen to allow sufficient time for the signals to propagate and settle throughout the system. Allowances must also be made for clock skew. Thus, the clock frequency is limited by the length of the bus and the speed of the interface logic. All things being equal, shorter buses can be designed to run at higher speeds.

The main advantage of the synchronous protocol is that it is fast. It also requires relatively few bus lines and simple interface logic, making it easy to implement and test. However, the synchronous protocol is less flexible than the asynchronous protocol in that it requires all the devices to support the same clock rate. Furthermore, this clock rate is fixed and cannot be raised compatibly to take advantage of technological advances. In addition, the length of synchronous buses is limited by the difficulty of distributing the clock signal to all the devices at the same time. Synchronous buses are typically used where there is a need to connect a small number of very tightly coupled devices and where speed is of paramount importance. Thus, synchronous buses are often used to connect the processor and the memory subsystem.

19.4.3   Split-Transaction Protocol

In order to increase the effective bandwidth of a bus, a **split-transaction protocol** can be used. The basic observation behind this protocol is that the bus is not being used to transmit information throughout the entire duration of a transaction, but only at the start and toward the end of the transaction. The idea is thus to split a bus transaction into a request transaction and a reply transaction so as to allow the bus to be released for other uses in between the request and reply stages. Figure 19.7 illustrates the idea. Clearly, this protocol only makes sense when the system has more than one bus master and the memory system is sophisticated enough to handle multiple overlapping transactions. This protocol is sometimes also known as a *connect/disconnect protocol*, *pipelined protocol*, or *packet-switched protocol*.

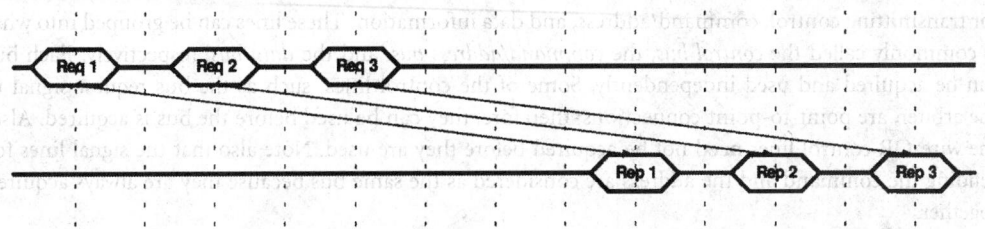

FIGURE 19.7 A split-transaction protocol.

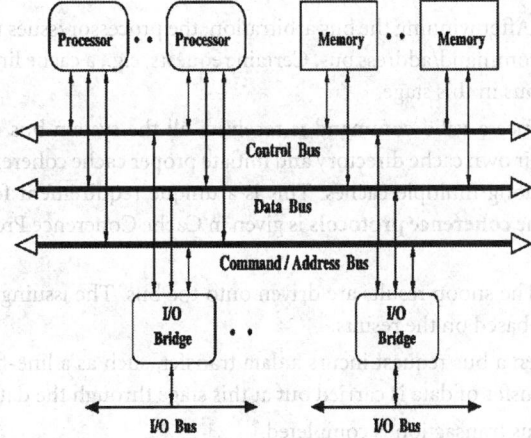

FIGURE 19.8 A typical multiprocessor system bus.

Although the split-transaction protocol allows more efficient utilization of bus bandwidth than a protocol that holds on to the bus for the whole transaction, it usually has a higher latency because the bus has to be acquired twice — once for the request and once for the reply. Furthermore, the split-transaction protocol is expensive to implement because it requires that the bus transactions be tagged and tracked by each device. Split-transaction protocols are widely used in the system buses of shared-memory SMPs, because bus bandwidth is a big issue in these machines.

Notice from Figure 19.7 that even with split transactions, the bus bandwidth is not totally utilized. This is because some bus cycles are needed to acquire the bus and to set up the transfer. Furthermore, some buses require a cycle of turnaround time between different masters driving the bus. Another way to increase effective bus bandwidth is thus to amortize this fixed cost over several words by allowing the bus to transfer multiple contiguous words back to back in one transaction. This is known as a *burst protocol* or *block transfer protocol* because a contiguous block of several words is transferred in each transaction.

19.5 Issues in SMP System Buses

Microprocessor-based symmetric multiprocessors (SMPs) with shared snooping buses have become an industry standard for building midrange departmental servers [Peir et al. 1993, Galles and Williams 1994]. In this section, we devote special attention to the bus issues that arise in SMP designs.

A typical SMP bus architecture (usually referred to as a *SMP system bus*, or simply *system bus*) is illustrated in Figure 19.8. SMP system buses require high bandwidth and low latency to connect multiple processors, memory modules, and I/O bridges. A system bus is composed of independent signal lines for transmitting control, command/address, and data information. These lines can be grouped into what is commonly called the *control bus*, the *command/address bus,* and the *data bus*, respectively. Each bus can be acquired and used independently. Some of the control lines, such as the bus request signal to the arbiter, are point-to-point connections; therefore, they can be used before the bus is acquired. Also, the wire-OR control lines need not be acquired before they are used. Note also that the signal lines for sending the command and the address are considered as the same bus because they are always acquired together.

In general, each system bus request traverses through a number of stages, each of which may take one or more bus cycles.

- *Bus arbitration:* The requesting processor needs to go through an arbitration process in order to gain access to the shared system bus.
- *Command issuing:* After winning the bus arbitration, the processor issues the command along with an address on the command/address bus. Certain requests, e.g., a cache line writeback, also require access to the data bus in this stage.
- *Cache snooping:* Once a valid command is received, all the system bus masters (processors, I/O bridges) search their own cache directory and initiate proper cache coherence activities to maintain data coherence among multiple caches. This is a unique requirement for SMP system buses. A description of **cache coherence protocols** is given in Cache Coherence Protocols subsection of this section.
- *Acknowledgment:* The snoop results are driven onto the bus. The issuing processor has to update its cache directory based on the results.
- *Data transfer:* When a bus request incurs a data transfer, such as a line-fill request issued upon a cache miss, the transfer of data is carried out at this stage through the data bus.
- *Completion:* The bus transaction is completed.

Several important and unique issues in SMP system bus design are outlined as follows.

1. *Bus physics:* Due to the high bandwidth requirement, SMP system buses face the challenge of a high speed and wide data bus, and have to overcome heavy signal loading to accommodate a reasonable number of ports for connecting multiple system devices.
2. *Cache coherence:* Cache memory is a critical component in SMP systems. In order to maintain data coherence among multiple caches, a cache coherence protocol must be incorporated into the bus protocol.
3. *Bus arbitration:* Each of the processors should be given an equal opportunity to gain access to the bus. Logic to prevent, detect, and resolve starvation and deadlock is typically needed.
4. *Bus bandwidth:* Besides faster and wider data buses, two other bandwidth-increasing features are important: pipelining bus requests using the split-transaction bus design, and transferring large data blocks in a *burst* transfer mode. As described before, the processing of a system bus request involves several distinct stages. Pipelining these requests can increase the effective bus bandwidth significantly.
5. *Memory access latency:* Memory access latency is a classic performance bottleneck. Techniques such as replicated arbiter and **bus parking** [Mahoney 1990] can reduce the latency in bus arbitration. Furthermore, early DRAM access before resolving cache coherence issues can reduce memory access latency. Such early DRAM accesses, however, may have to be canceled depending on the snoop results.
6. *Synchronization and locking:* SMP system buses need to provide an efficient way of implementing atomic read–modify–write instructions. Techniques such as address-based locking and nonblocking caches can reduce the interference between locking requests and other normal bus operations.
7. *Reliability and fault tolerance:* Parity or ECC can be used to detect and correct transmission errors. When an error is uncorrectable, the requester will receive a negative acknowledgment and will retry the request. The timeout scheme is commonly included to detect the loss of a command.

In the following, these important issues will be discussed in detail. Because the techniques for handling bus physics issues for the SMP system bus are basically the same as those in other buses, we will omit further discussion.

19.5.1 Cache Coherence Protocols

Cache memory is a critical component in SMP systems. It helps the processors to execute near their full speed by substantially reducing the average memory access time. This is achieved through fast cache hits for the majority of memory accesses and through reduced memory contention in the entire system. However, in designing a shared-memory SMP system where each processor is equipped with a cache memory, it is necessary to maintain coherence among the caches such that any memory access is guaranteed to return the latest version of the data in the system [Censier and Feautrier 1978]. Cache coherence can be enforced through a shared **snooping bus** [Goodman 1983, Sweazey and Smith 1986]. The basic idea is to rely on the broadcast nature of the bus to keep all the cache controllers informed of each other's activities so that they can perform the necessary operations to maintain coherency.

A number of snooping cache coherence protocols have been proposed [Archibald and Baer 1986, Sweazey and Smith 1986]. They can be broadly classified into the write-invalidate scheme and the write-broadcast scheme. In both schemes, read requests are carried out locally if a valid copy exists in the local cache. For write requests, these two schemes work differently. When a processor updates a cache line, all other copies of the same cache line must be invalidated according to the write-invalidate scheme to prevent other processors from accessing the stale data. Under the write-broadcast scheme, the new data of a write request will be broadcast to all the other caches to enable them to update any old copies. These two cache coherence schemes normally operate in conjunction with the *writeback* policy, because the *writethrough* policy generates memory traffic on every write request and is thus not suitable for a bus-based multiprocessor system.

The MESI (modified-exclusive-shared-invalid) write-invalidate cache coherence protocol with the writeback policy is considered in the following discussion. With minor variations, this protocol has been implemented in several commercial systems [Intel 1994, Greenley et al. 1995, Levitan et al. 1995]. In this protocol, each cache line has an associated MESI state recorded in the cache *directory* (also called cache tag array). The definitions of the four states are given in Table 19.1. When a memory request from either the processor or the snooping bus arrives at a cache controller, the cache directory is searched to determine cache hit/miss and the coherence action to be taken. The state transition diagram of the MESI protocol is illustrated in Figure 19.9, in which solid arrows represent state transitions due to requests issued by the local processor, and dashed arrows indicate state transitions due to requests from the snooping bus.

TABLE 19.1 Four States in MESI Coherence Protocol

State	Description
Modified (M)	The M-state indicates that the corresponding line is valid and is exclusive to the local cache. It also indicates that the line has been modified by the local processor. Therefore, the local cache has the latest copy of the line.
Exclusive (E)	The E-state indicates that the corresponding line is valid and is exclusive to the local cache. No modification has been made to the line. A write to an E-state line can be performed locally without producing any snooping bus traffic.
Shared (S)	The S-state indicates that the corresponding line is valid but may also exist in other caches in a multiprocessor system. Writing to an S-state line updates the local cache and generates a request to invalidate other shared copies.
Invalid (I)	The I-state indicates that the corresponding line is not available in the local cache. A cache miss occurs in accessing an I-state line. Typically, a line-fill request is issued to the memory to bring in the valid copy of the requested cache line.

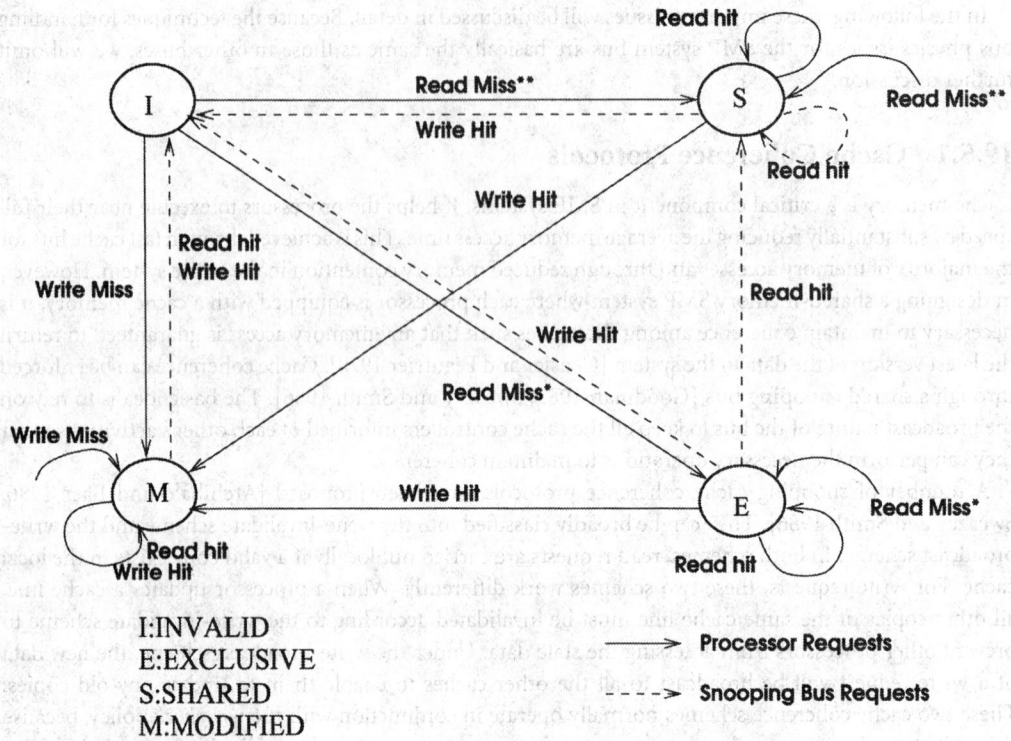

I:INVALID
E:EXCLUSIVE
S:SHARED
M:MODIFIED

⟶ Processor Requests
--→ Snooping Bus Requests

Read Miss*: No other cache has a copy of the requested line, or
the line is found in another cache in the M state.

Read Miss**: The requested line is found in other caches in the E or S state.

FIGURE 19.9 State transition diagram of MESI coherence protocol.

In general, when a read request from the processor hits a line in the local cache, the state of the cache line will not be altered. A write hit, on the other hand, will change the line state to M. In the meantime, if the original line state is S upon the write hit, an invalidation request will be issued to the snooping bus to invalidate the same line if it is present in any of the other caches. When a read miss occurs, the requested line will be brought into the cache in different states depending on whether the line is present in other caches and on its state in these caches. When a write miss occurs, the target line is fetched into the local cache; the new state becomes M and any copy of the cache line in any other cache is invalidated. For the requests from the snooping bus, a write hit always causes invalidation. A read hit to an M-state line will result in a writeback of the modified line and a transfer of ownership of the target line to the requesting processor. A read hit to an E- or S-state line for a snooping read request will cause a state transition to S.

19.5.2 Bus Arbitration

There are two issues in bus arbitration that are especially important for SMP system buses. The first is to ensure that the system is fair, deadlock-free, and starvation-free. In general, the fairness issue can be handled by using a first-come-first-served (FCFS) priority scheme, which will guarantee that requests are serviced in order of arrival. In fact, a simple random priority scheme may be adequate to provide a fair arbitration scheme for all the bus masters. The deadlock problem typically arises in a split-transaction bus where it is possible for a cycle of dependencies to form among the requests and replies. This can be handled by distinguishing between requests and replies and always ensuring that replies are able to make forward progress. At first sight, it might seem that a fair arbitration policy should be starvation-free. However, this is not the case, because when overlapping bus requests are allowed, a request that has been granted by the bus may have to be later rejected due to interference with other requests. On a subsequent retry, the request may again encounter another conflict. In the pathological case, such a request may retry indefinitely. This situation is worse if there are lock requests and other resource conflicts on the system bus.

There are two general solutions to this starvation problem. The first solution eliminates any request rejection to prevent the starvation from happening. Whenever a conflict occurs, the request is queued at the location where the conflict is encountered and the queued request will be processed once the conflict condition disappears. This solution implies a variable-length bus pipeline, which requires extra handshaking on the system bus. In addition, excessive queues must be implemented to handle the conflict situation. The second solution depends on the ability to detect the starvation condition and to resolve the condition once it occurs. This method, in general, counts the number of times each request has been rejected. When a certain threshold is exceeded, emergency logic is activated to ensure that the starved request is serviced quickly and successfully.

The second important issue in bus arbitration is to minimize the latency for acquiring the bus. There are two techniques for doing this. The first technique is to replicate the arbiter in each processor. Each arbiter receives the request signals from all the bus masters just like in the centralized arbiter design. However, after the arbitration is resolved, each replicated arbiter only needs to send the grant signal to the local processor. The delay through a long wire across multiple chips in the centralized implementation can thus be avoided. The second latency-reduction technique is bus parking, which essentially implements a round-robin priority among the bus masters with a token being passed around. Once a bus master owns the token, it can access the bus without arbitration. This scheme is effective, in general, with a small number of bus masters.

19.5.3 Bus Bandwidth

An SMP system bus provides the only pathway for multiple processors to access the memory and the I/O devices. Therefore, the bandwidth of the system bus effectively determines the number of processors that can be supported. The simplest system bus design is to allow only one request at a time. A second command cannot be issued before the current bus request completes. This so-called *simple bus* design has very limited bus performance. A split-transaction bus, on the other hand, allows the overlapping of multiple bus requests using the pipelining technique. When a bus transaction is split, it only occupies the bus when the bus is really needed. For example, when a request is at the cache snooping stage or at the stage of accessing memory, the bus is available for other bus transactions. Even when the request is at the data transfer stage, the command/address bus is free to accept another request which does not require the data bus in the same cycle. The split-transaction approach utilizes the critical system bus much better. Most of the modern SMP system buses employ split-transaction designs to maximize the bus bandwidth [Peir et al. 1993, Galles and Williams 1994].

Even with the aggressive split-transaction design, the system bus can support a very limited number of today's high-performance microprocessors. For instance, consider a 50-MHz system bus connecting multiple 100 MIPS processors together. Assume that each processor has on-chip instruction/data caches as well as an external second-level cache; together, the instruction-per-miss rate is 50. In this case, each processor will generate bus traffic for handling two million cache misses every second. As a result, 25 processors will produce enough traffic to use up 100% of the bus bandwidth. This calculation does not allow for the fact that about 30% of the cache lines being replaced are typically modified and have to be written back to memory. In addition, this calculation does not take into account bus traffic due to I/O instructions, cache coherence transactions, etc. All these factors will further limit the number of processors that can be effectively supported by the system bus.

In the above discussion, two ideal conditions are assumed. The first assumption is that the bus utilization can reach 100%. In reality, the rule-of-thumb is that heavy queuing will occur when the bus utilization reaches above 60%. The second assumption is that each bus request only occupies a single bus cycle. This is very difficult to achieve for a split-transaction bus running at 50 MHz. Typically, both the arbitration stage and the cache snooping stage may require more than one cycle. Even though these two stages do not occupy the bus directly, they need to access other related critical components such as the arbiter and the cache directory, respectively. Realistically, it takes a minimum of two cycles to initiate a new command on the system bus. This two-cycle-per-request design is already very aggressive, because the subsequent command is issued before the current command is acknowledged. In order to avoid potential interference between the active commands, protection hardware must be implemented. In addition, the data bus must be able to transfer a cache line in two cycles. For instance, a 128-bit data bus is required when the Intel Pentium processor is used in the above example, because the Pentium has 32-byte cache lines. Furthermore, any *idle* cycles during the data transfer must be eliminated. This typically requires a high-performance memory system with multiple modules each with independent memory banks.

Current projections suggest that processor performance will continue to improve at over 50% a year [Hennessy and Patterson 1996]. It is unlikely that improvements in the system bus will be able to keep up with the processor curve. Thus, the number of processors that the snooping bus can support is expected to decrease even further. There has been some work to extend the single bus architecture to multiple interleaved buses [Hopper et al. 1989] or hierarchical buses [Wilson 1987]. There have also been proposals to abandon the snooping bus approach and to use a directory method to enforce cache coherence [Lenoski et al. 1992, Gustavson 1992]. The detailed descriptions of these proposals are beyond the scope of this chapter.

19.5.4 Memory Access Latency

When a requested data item is not present in a processor's caches, the item must be fetched from memory. The delay in obtaining the data from memory may stall the requesting processor even when advanced techniques, such as out-of-order execution, nonblocking cache, relaxed consistency model, etc., are used to overlap processor execution with cache misses [Johnson 1990, Farkas and Jouppi 1992, Gharachorloo et al. 1990]. Therefore, it is important to design the system bus to minimize the number of cycles needed to return data upon a cache miss. Techniques such as replicated arbiter and bus parking to reduce arbitration cycles have been described in the subsection on Bus Arbitration above.

Another way to reduce the memory latency is to trigger DRAM access once the command arrives at the memory controller. This early DRAM access may have to be canceled if the requested line turns out to be present in a modified state in another processor's cache. Such a condition will only be known after the acknowledgment. Early DRAM access is very useful because the chance of hitting a modified line in another cache is not very high. In the unlikely case that the requested line is in fact modified in another cache, a cache-to-cache data and ownership transfer can be implemented to send the requested data directly to the requesting processor. The cache-to-cache transfer normally has higher priority and may bypass other requests in the write buffer of the processor that owns the modified line. In some implementations, the memory is also updated with the modified data during the cache-to-cache transfer.

19.5.5 Synchronization and Locking

The basic hardware primitive for implementing synchronization and locking is an atomic read–modify–write instruction such as Test&Set and Compare&Swap. These instructions must be guaranteed to read the contents of a memory location, test and modify the data, and write the result back to the same memory location in an indivisible sequence of operations. In a bus-based SMP system, the Test&Set instruction can be executed in two separate steps: a read-lock operation followed by a write-unlock operation. The read-lock operation reads the memory word and at the same time sets up certain hardware lock signals or registers so that no other requests for the same memory word will be permitted. After the memory word has been modified, the write-unlock operation returns the new result back to the memory location and releases the hardware lock.

There are two ways of implementing the hardware lock on the system bus. The first is to lock the bus completely during the period from the read-lock operation to the write-unlock operation; no other request is allowed in between. This approach has poor performance but may be suitable for a simple bus design which allows only one request at a time anyway. The second approach is to implement **address locking** on the system bus. When a read-lock operation is issued, an invalidation request is sent across the system bus to knock out any copy of the cache line from the other caches. In the meantime, the address of the cache line is recorded in a *lock register* to prevent any snooping on the same cache line until the lock is released by the write-unlock operation. Depending on the data alignment and the size of the synchronization variable, a read-lock operation may have to lock two cache lines at a time.

The address locking method minimizes the interference of a Test&Set instruction with other system bus requests, because only those requests that access the same cache line as the Test&Set will be rejected. Multiple lock requests, each from a different processor, are permitted as long as they target different cache lines. However, the lock request is still relatively expensive because the read-lock operation has to be broadcast to all the other processors and the issuing processor cannot proceed until confirmation is received from each processor.

19.6 Putting It All Together — CCL-XMP System Bus

CCL-XMP [Peir et al. 1993], an Intel Pentium-based SMP system, was designed and developed at the Computer and Communication Laboratories (CCL) of the Industrial Technology Research Institute (ITRI) of Taiwan under a collaborative effort with Acer, ICL, and Intel. The initial target system consists of eight 66-MHz Pentium processors, each with a 256-Kbyte second-level cache. The system bus, operating at 33 MHz, has independent control, command/address, and data buses. The MESI protocol is incorporated to enforce coherence among multiple caches. The system bus can arbitrate and accept one request every two cycles. A dual 64-bit data bus sustains a data transfer rate of over 400 Mbyte/s. Based on odd–even line interleaving, each data bus is connected to one memory module. In order to maximize the bus bandwidth, each memory module is divided into four independent banks. Furthermore, each bank is designed as a two-way interleaved DRAM array to provide zero-wait-state operation in the burst-mode data transfer at 33 MHz. CCL-XMP requires a powerful I/O subsystem. Both VESA local bus and PCI bus can be connected to the system bus through high-performance I/O bridges. The I/O bridges act as both a bus master and slave to transfer data between the system bus and the I/O buses.

The CCL-XMP system bus architecture is very similar to the general SMP bus architecture shown in Figure 19.8. Each request traverses through the same number of stages as described in Section 19.5. In this section, some of the relevant features of the CCL-XMP system bus design are described.

The CCL-XMP system bus is a synchronous bus. All the signals are driven at the rising edge of the clock and latched and sampled at the next rising edge. TTL voltage levels are used, and most signals are tristate with the exception of a few wire-OR control signals. The first version of the motherboard measures 32 cm × 32 cm. The system bus is 15 cm in length and supports eight bus slots. A single 64-bit data bus is included in the first prototype to support up to six Pentium processors. The clock tree is carefully laid out on the motherboard to limit the clock skew to within 2 ns. The total delay between the sender and the receiver through the system bus is less than 28 ns. This includes the delay of latches (flip-flops), the output TTL pad (with 120-pF bus loading), the input pad, boundary scan, clock skews, and other combinational logic. This is tolerable under the 33-MHz clock rate. In designing CCL-XMP, it is more challenging to manage the delay between the system bus interface chip and the second-level cache controller, which is operating at 66 MHz.

Figure 19.10 shows the detailed operation of a line-fill request on the CCL-XMP system bus. Arbitration of the command/address bus takes two cycles. Once the bus is granted, the memory read command and address are issued to the bus in the third cycle. It takes two cycles for all the bus masters (including I/O bridges) to search their cache directories and to update the respective line states upon a hit. At the same time, the memory controller initiates the DRAM access on the arrival of the read command. In the sixth cycle, an acknowledgment is sent back to the requester from each device based on the snooping result.

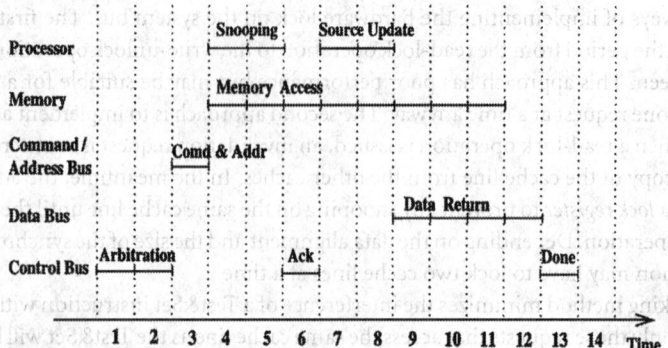

FIGURE 19.10   The pipeline cycles of a line-fill request on the CCL-XMP system bus.

The requester then updates its cache directory according to the acknowledgment and the cache coherence protocol. The memory controller begins arbitrating for the data bus two cycles before the data are fetched out of the DRAM array. A read request is given priority for using the data bus over a write request. After the data bus has been granted, the target 64-bit data is driven onto the bus in the ninth cycle. This is followed by three consecutive cycles to transfer the entire cache line to the requester. Finally, the command completes in the thirteenth cycle.

The CCL-XMP system bus uses replicated arbiters to achieve a two-cycle arbitration. Each arbiter receives *n* request signals, one from each bus master; but the grant signal is only sent to the local processor to avoid chip-crossing and long wiring delays. Basically, the arbiter latches the requests from all the bus masters at the end of the first cycle. In the second cycle, a resolution of the bus priority is carried out, and the winner is notified by the local arbiter. In addition, the CCL-XMP system bus uses the concept of *arbitration group* to achieve fairness without implementing FIFO queues. An arbitration group consists of the bus masters that are simultaneously requesting access to the system bus. The arbiter will serve all the members in the group according to some priority scheme. Any other master outside the current group is not allowed to join the group until all the requesters in the group have been granted the bus. After that, a new arbitration group can be formed.

Emergency logic is included in the arbiter design to resolve the starvation conditions. Each bus master has an associated *urgent counter*. The counter is incremented when a command receives a negative acknowledgment on the bus. A command can be rejected for several reasons, including resource busy, address locking, or some erroneous conditions. When the counter exceeds a certain threshold, the processor will raise its urgent request control line to the arbiter. The arbiter will then give the highest priority to the urgent requester and activate the emergency logic for resolving any blocking condition for the urgent request until the command is executed successfully and the urgent line is dropped.

The CCL-XMP system bus is a split-transaction bus; it allows the pipelining of multiple requests to sustain maximum bus bandwidth. Figure 19.11 illustrates two consecutive cache line-fill instructions on the command/address bus and the data transfer cycles for the two cache lines on the data buses. These two requests must have originated from different processors, because the Pentium processor only allows one outstanding memory request at a time. When the two requested lines are located in different memory modules, concurrent data transfers are possible. However, the data transfers have to be serialized when both accesses hit the same module. Note that there is a dead cycle on the data bus between the two cache line transfers on the same data bus. This is required to prevent a potential signal conflict. After the read command is issued, it takes six cycles including ECC check for the memory controller to return the target 64-bit data. Because the command bus can accept one command every two cycles, a dual 64-bit bus is designed to balance the performance. The dual data bus can transfer two 32-bytes of data every five bus cycles (including the dead cycle). This provides a sustained bus bandwidth of over 400 Mbyte/s.

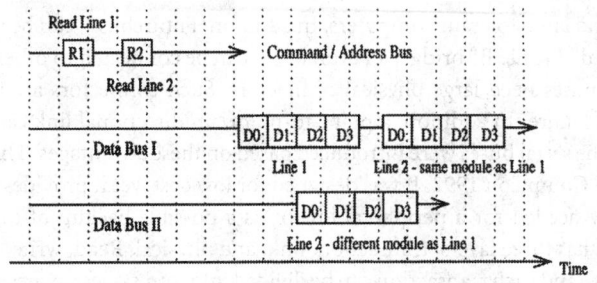

FIGURE 19.11 Split transactions on the CCL-XMP system bus.

CCL-XMP supports cache-to-cache data transfer with memory reflection when a request hits a modified line located in another cache. The memory copy is updated during the data transfer to the requester. The cache-to-cache transfer request has higher priority and may bypass other requests in the write buffer or the replacement writeback queues. This design reduces the penalty in accessing a modified line located in another processor's cache. The update of the memory copy provides the flexibility to transfer the ownership of the requested line to the requester and to switch the requested line to a shared state when there are subsequent read requests for the same line from other processors.

A pair of lock registers are implemented in each cache snooping controller to support address locking. At any given time, each processor can lock two cache lines to implement atomic read–modify–write memory instructions. A request is negative-acknowledged when a snoop hits an address in the lock register. The rejected request will be retried at a later time. Several protection registers are incorporated into the cache snooping controller to protect a cache line in a transient state from being accessed. For example, in addition to the lock registers, a current register (CR) is used to protect the active command until it completes. Also, a writeback register (WBR) records the modified line that is in the process of being written back to memory. As in the case of the lock register, a request will be rejected if it hits these protection registers. Parity bits are added to both the command/address bus and the data bus for detecting transmission errors.

19.7 Historical Perspective and Research Issues

Computer systems based on a shared backplane bus are very popular because they are both cost-effective and highly flexible. The early to mid 1980s saw the establishment of a number of standard buses such as the VME bus [Pri-Tal 1986], Fastbus [Gustavson 1986], Nubus [Taylor 1989], Multibus II [Mahoney 1990], and Futurebus [Borrill 1986]. All these bus architectures have been standardized by the IEEE to allow different vendors to design various boards that can be attached to each bus. Although the option of connecting multiple processors was considered in all these bus designs, only the Futurebus fully supports cache coherence protocols for both writethrough and writeback caches. Other important aspects, such as synchronous/asynchronous communication protocols, bus arbitration, bus bandwidth, locking mechanisms, TTL/BTL/ECL bus interfaces, connectors, pins, etc., are also significantly different among these open bus architectures. A comprehensive comparison can be found in [Borrill 1985].

The continuing search for higher-performance system interconnection has led to three more recent IEEE standards: the IEEE 896.x Futurebus+ [Aichinger 1992], the P1394 High-Performance Serial Bus [Teener 1992], and the P1596 Scalable Coherence Interface (SCI) [Gustavson 1992]. Futurebus+ is an expanded version of the original Futurebus (IEEE 896.1-1987). The work was initiated by the navy in 1988 and later gained widespread support from the working groups of the VMEbus and Multibus as well as many of the major computer companies. The Futurebus+ was designed to be a truly open standard. The standard defines in detail the architectural, electrical, and mechanical specifications. The data path of Futurebus+ varies from 32 to 256 bits wide to provide better performance scalability. The centralized arbitration mechanism is used for its simplicity and high performance. **Live insertion** is adopted to answer the needs of the market for high-availability and fault-tolerant computers. In addition, Futurebus+ can be reconfigured as an I/O bus. Using the standard "Profile B" bridge, a Futurebus+ can be connected to other open-system buses.

Parallel backplane buses need large physical connectors. Such connectors are usually costly and are the primary source of failure. In addition, a point-to-point, unidirectional link can provide much faster transmission speed. The serial buses were introduced based on these advantages. The P1394 serial bus was first presented at IEEE CompCon 1992. It was designed for low cost, yet it provides the data transmission speed and low latency needed for a peripheral bus or as a possible backup of the backplane bus. The transmission protocol has three layers: transaction, link, and physical. Read, write, and lock are the three transactions supported, and each transaction can be divided into four stages: request, indication, response, and confirmation. The link layer provides a half-duplex data-packet delivery service. Each packet delivery needs to go through three stages: an arbitration stage to gain the access of the physical bus, a packet transmission stage to deliver the packet to the physical layer, and an acknowledgment stage to confirm the transmission with the receiver. The last stage is only required for synchronous package delivery. The split-transaction protocol and a Test&Set instruction are implemented in the P1394 serial bus to enable higher transaction rates and to properly handle locking activities.

As microprocessor performance increased at the rate of over 50% per year [Hennessy and Patterson 1996], it soon became apparent that the use of shared buses as the fundamental interconnection in multi-processor systems creates a performance bottleneck. In July 1988, the P1596 SCI working group was formed with the aim of defining a scalable interconnect architecture for future shared-memory cache-coherent multiprocessor systems. There are a number of fundamental issues that must be solved in order to achieve this goal. First, signal speed has to be made independent of the size of the system. Second, multiple signal paths (links) have to be used so that multiple independent transfers can take place concurrently. Third, multiple cache coherence activities must be allowed to occur in parallel to overcome the bottleneck in the snooping-bus approach. To resolve the signal-speed problem, SCI uses point-to-point, unidirectional buses with low-voltage differential signals. *Ring* is the most common structure to connect multiple processing nodes (processors) through the fast SCI bus. *Distributed directory* is chosen to maintain cache coherence. The main memory and the directory which records the status of the associated lines in cache are distributed among all the nodes. Instead of maintaining full presence bits to indicate the caches where each line is located, a double link list is constructed to link all the copies of each cache line. Certain cache coherence actions, e.g., invalidation of shared copies of a cache line, need to go through the link list in a sequential fashion.

While all these standardization efforts were under way, a different approach was being taken by Rambus Inc. Rambus Inc. was set up in 1990 to develop technology that will enable systems to keep up with the increasing bandwidth requirements of processors [Rambus 1992]. The core of the Rambus technology is a proprietary chip-to-chip bus, the Rambus Channel. The Rambus Channel consists of a small number of very high-speed lines clocked at an aggressive 250 MHz. Data are transferred on both edges of the clock, allowing nine data lines to achieve a peak transfer rate of 500 Mbyte/s. A block-oriented protocol is used to allow effective utilization of this peak bandwidth. The key to achieving the high data rate lies in paying special attention to the physics of the bus. First, the Rambus interface consists of only 32 carefully terminated transmission lines. The small number of lines helps to reduce bus noise and power consumption, which is significant given the high clock rate. Second, clock-to-data skew is minimized by having a clock-to-master signal and a clock-from-master signal. The appropriate clock is chosen depending on the direction of the data transfer. Third, the channel is designed to operate with low-voltage swings, which further reduces power consumption. Finally, the maximum bus length is limited to about 10 cm and vertical surface-mount packages are used to enable devices to be densely packed onto the bus.

Although flexible and cost-effective, the bus is fundamentally not scalable in performance. This becomes increasingly apparent as processors become faster and SMP designs start pushing the performance limits of buses with fewer and fewer processors. Widening the data bus will not help, because of the bottleneck in the command/address bus. Using multiple buses [Hopper et al. 1989] to achieve more than one transfer at a time results in a complex bus interface design to maintain cache coherence with bus snooping mechanisms. The Scalable Coherence Interface uses a distributed-directory approach to build coherent shared-memory multiprocessors with high-performance SCI rings. Although the SCI approach allows concurrent coherence activities, the cost of the large directory and the latency of coherence transaction remain serious problems. The Stanford DASH project [Lenoski et al. 1992] advocates the same distributed directory method using mesh-connected networks. Again, the latency in accessing remote memory may cause severe performance degradation [Kuskin et al. 1994]. The "right" way to build a scalable cache-coherent shared-memory multiprocessor remains an active research area.

An issue intimately related to bus performance is the performance of the memory subsystem. The standard RAS/CAS DRAM interface is designed for low cost and high density. In order to achieve sufficient memory bandwidth for today's demanding processors, an interleaved memory subsystem is often required. For SMPs, the memory interleaving is typically very aggressive. Recently, a couple of improvements to the existing DRAM interface have been announced. These include the extended-data-out (EDO) mode and the pipelined burst mode (PBM) [Kumanoya et al. 1995]. In addition, several new DRAM interfaces promising higher bandwidth and lower latency have been proposed. These include the synchronous DRAM, cached DRAM, Rambus DRAM, and others designed specifically for graphics application. The interested reader is referred to [Przybylski 1993, Kumanoya et al. 1995] for more information on these novel DRAM interfaces.

Defining Terms

Address locking: A mechanism to protect a specific memory address so that it can be accessed exclusively by a single processor.

Bus arbitration: The process of determining which competing bus master should be granted control of the bus.

Bus master: A bus device that is capable of initiating and controlling a communication on the bus.

Bus parking: A priority scheme which allows a bus master to gain control of the bus without arbitration.

Bus protocol: The set of rules which define precisely the bus signals that have to be asserted by the master and slave devices in each phase of a bus operation.

Bus slave: A bus device that can only act as a receiver.

Cache coherence protocol: A mechanism to maintain data coherence among multiple caches so that every data access will always return the latest version of that datum in the system.

Cache line: A block of data associated with a cache tag.

Live insertion: The process of inserting devices into a system or removing them from the system while the system is up and running.

Snooping bus: A multiprocessor bus that is continually monitored by the cache controllers to maintain cache coherence.

Split-transaction bus: A bus that overlaps multiple bus transactions, in contrast to the *simple bus* that services one bus request at a time.

Symmetric multiprocessor (SMP): A multiprocessor system where all the processors, memories, and I/O devices are equally accessible without a master–slave relationship.

References

Aichinger, B. P. 1992. Futurebus+ as an I/O bus: profile B, pp. 300–307. In *Proc. 19th Int. Symp. on Computer Architecture*.

Archibald, J. and Baer, J. L. 1986. Cache-coherence protocols: evaluation using a multiprocessor simulation model. *ACM Trans. Comput. Systems* 4(4):273–298.

Agarwal, A., Simoni, R., Hennessy, J., and Horowitz, M. 1988. An evaluation of directory schemes for cache coherence, pp. 280–289. In *Proc. 15th Int. Symp. on Computer Architecture*.

Borrill, P. 1985. MicroStandards special feature: a comparison of 32-bit buses. *IEEE Micro* 5(6):71–79.

Borrill, P. 1986. Futurebus: the ultimate in advanced system buses, pp. 210–216. In *Proc. Buscon '86 West*.

Censier, L. and Feautrier, P. 1978. A new solution to coherence problems in multicache systems. *IEEE Trans. Comput.* C-27(12):1112–1118.

Chaiken, D., Fields, C., Kurihara, K., and Agarwal, A. 1990. Directory-based cache coherence in large-scale multiprocessors. *IEEE Comput.* 23(6):49–59.

Farkas, K. and Jouppi, N. 1992. Complexity/performance tradeoffs with non-blocking loads, pp. 211–222. In *Proc. 21st Int. Symp. on Computer Architecture.*

Galles, M. and Williams, E. 1994. Performance optimizations, implementation, and verification of the SGI challenge multiprocessor, pp. 134–143. In *Proc. 1994 Hawaii Int. Conf. on System Science, Architecture Track.*

Gharachorloo, K., et. al. 1990. Memory consistency and event ordering in scalable shared-memory multiprocessors, pp. 15–26. In *Proc. 17th Int. Symp. on Computer Architecture.*

Giacomo, J. D. 1990. *Digital Bus Handbook.* McGraw–Hill, New York.

Goodman, J. 1983. Using cache memory to reduce processor-memory traffic, pp. 124–131. In *Proc. 10th Int. Symp. on Computer Architecture.*

Greenley, D. et. al. 1995. Ultrasparc: the next generation superscalar 64-bit SPARC, pp. 442–451. In *Proc. COMPCON'95.*

Gustavson, D. B. 1984. Computer buses — a tutorial. *IEEE Micro* (4):7–22.

Gustavson, D. B. 1986. Introduction to the Fastbus. *Microprocessors and Microsystems* 10(2):77–85.

Gustavson, D. B. 1992. The scalable coherent interface and related standards projects. *IEEE Micro* 12(1): 10–22.

Gustavson, D. B. and Theus, J. 1983. Wire-OR logic on transmission lines. *IEEE Micro* 3(3):51–55.

Hennessy, J. and Patterson, D. 1996. *Computer Architecture, a Quantitative Approach*, 2nd ed. Morgan Kaufmann, San Francisco.

Hopper, A., Jones, A., and Lioupis, D. 1989. Multiple vs wide shared bus multiprocessors, pp. 300–306. In *Proc. 16th Int. Symp. on Computer Architecture.*

IBM Corp. 1982. IBM 3081 Functional Characteristics, GA22-7076. IBM Corp., Poughkeepsie, NY.

Intel Corp. 1994. *Pentium Processor User's Manual*, Vols. 1, 2. Intel Corp. Order nos. 241428, 241429.

Johnson, M. 1990. *Superscalar Microprocessor Design.* Prentice–Hall, Englewood Cliffs, NJ.

Kumanoya, M., Ogawa, T., and Inoue, K. 1995. Advances in DRAM interfaces. *IEEE Micro* 15(6):30–36.

Kuskin, J., et. al. The Stanford FLASH multiprocessor, pp. 302–313. In *Proc. 21st Int. Symp. on Computer Architecture.*

Lenoski, D. et. al. 1992. The Stanford Dash multiprocessor. *IEEE Comput.* 25(3):63–79.

Levitan, D., Thomas, T., and Tu, P. 1995. The PowerPC 620 microprocessor: a high performance superscalar RISC microprocessor, pp. 285–291. In *Proc. COMPCON '95.*

Mahoney, J. 1990. Overview of Multibus II architecture. *SuperMicro J.* No. 4, pp. 58–67.

Metcalfe, R. M., and Boggs, D. R. 1976. Ethernet: distributed packet switching for local computer networks. *Commun. ACM* 19(7):395–404.

Peir, J. K., et al. 1993. CCL-XMP: a Pentium-based symmetric multiprocessor system, pp. 545–550. In *Proc. 1993 Int. Conf. on Parallel and Distributed Systems.*

Pri-Tal, S. 1986. The VME subsystem bus. *IEEE Micro* 6(2):66–71.

Przybylski, S. 1993. DRAMs for new memory systems. parts 1, 2, 3. *Microprocessor Rep.*, Mar. 8, pp. 18–21.

Rambus. 1992. *Rambus Architectural Overview.* Rambus, Inc., Mountain View, CA.

Stenstorm, P. 1990. A survey of cache coherence schemes for multiprocessors. *IEEE Comput.* 23(6):12–25.

Sweazey, P. and Smith, A. J. 1986. A class of compatible cache consistency protocols and their support by IEEE Futurebus, pp. 414–423. In *Proc. 13th Int. Symp. on Computer Architecture.*

Taub, D. M. 1983a. Overcoming the effects of spurious pulses on wired-OR lines in computer bus systems. *Electron. Lett.* 19(9):340–341.

Taub, D. M. 1983b. Limitations of looped-line scheme for overcoming wired-OR glitch effects. *Electron. Lett.* 19(15):579–580.

Taub, D. M. 1984. Arbitration and control acquisition in the proposed IEEE 896 FutureBus. *IEEE Micro* 4(4):28–41.

Taylor, B. G. 1989. Developing for the Macintosh NuBus, pp. 143–175. In *Proc. Eurobus/UK Conference.*

Teener, M. 1992. A bus on a diet — the serial bus alternative: an introduction to the P1394 high performance serial bus, pp. 316–321. In *Compcon '92*.

Wilson, A. 1987. Hierarchical cache/bus architecture for shared memory multiprocessors, pp. 244–252. In *Proc. 14th Int. Symp. on Computer Architecture*.

Zalewski, J. 1995. *Advanced Multimicroprocessor Bus Architecture*. IEEE Computer Society Press.

Further Information

Advanced Multimicroprocessor Bus Architectures by J. Zalewski [Zalewski 1995] contains a comprehensive collection of papers covering bus basics, physics, arbitration and protocols, board and interface designs, cache coherence, various standard bus architectures, and their performance evaluations. The bibliography section at the end provides a complete list of references for each of the topics discussed in the book.

Digital Bus Handbook by J. D. Giacomo [Giacomo 1990] is a good source of information for the details of the various standard bus architectures. In addition, this book has several chapters devoted to the electrical and mechanical issues in bus design.

The bimonthly journal *IEEE Micro* and the *Proceedings of International Symposium on Computer Architecture* are good sources of the latest papers in computer architecture area including computer buses.

20

Input/Output Devices and Interaction Techniques

20.1 Introduction 20-1
20.2 Interaction Tasks, Techniques, and Devices 20-2
20.3 The Composition of Interaction Tasks 20-3
20.4 Properties of Input Devices 20-3
20.5 Discussion of Common Pointing Devices 20-6
20.6 Feedback and Perception — Action Coupling 20-7
20.7 Keyboards, Text Entry, and Command Input 20-8
20.8 Modalities of Interaction 20-10
    Voice and Speech • Pen-Based Gestures and Hand
    Gesture Input • Bimanual Input • Passive Measurement:
    Interaction in the Background
20.9 Displays and Perception 20-12
    Properties of Displays and Human Visual Perception
20.10 Color Vision and Color Displays 20-14
20.11 Luminance, Color Specification,
    and Color Gamut 20-14
20.12 Information Visualization 20-16
    General Issues in Information Coding • Color Information
    Coding • Integrated Control/Display Objects
    • Three-Dimensional Graphics and Virtual Environments
    • Augmented Reality
20.13 Scale in Displays 20-18
    Small Displays • Multiple Displays • Large-Format Displays
20.14 Force and Tactile Displays 20-20
20.15 Auditory Displays 20-21
    Nonspeech Audio • Speech Output
    • Spatialized Audio Displays
20.16 Future Directions 20-22

Ken Hinckley
Microsoft Research

Robert J. K. Jacob
Tufts University

Colin Ware
University of New Hampshire

20.1 Introduction

The computing literature often draws an artificial distinction between input and output; computer scientists are used to regarding a screen as a passive output device and a mouse as a pure input device. However, nearly all examples of human–computer interaction require *both* input and output to do anything useful.

For example, what good would a mouse be without the corresponding feedback embodied by the cursor on the screen, as well as the sound and feel of the buttons when they are clicked? The distinction between output devices and input devices becomes even more blurred in the real world. A sheet of paper can be used both to record ideas (input) and to display them (output). Clay reacts to the sculptor's fingers, yet it also provides feedback through the curvature and texture of its surface. Indeed, the complete and seamless integration of input and output is becoming a common research theme in advanced computer interfaces, such as ubiquitous computing [Weiser, 1991] and tangible interaction [Ishii and Ullmer, 1997].

Input and output bridge the chasm between a computer's inner world of bits and the real world perceptible to the human senses. *Input* to computers consists of sensed information about the physical environment. Familiar examples include the mouse, which senses movement across a planar surface, and the keyboard, which detects a contact closure when the user presses a key. However, any sensed information about physical properties of people, places, or things can serve as input to computer systems. *Output* from computers can comprise any emission or modification to the physical environment, such as a display (including the cathode ray tube [CRT], flat-panel displays, or even light-emitting diodes), speakers, or tactile and force feedback devices (sometimes referred to as *haptic* displays). An *interaction technique* is the fusion of input and output, consisting of all hardware and software elements, that provides a way for the user to accomplish a task. For example, in the traditional graphical user interface (GUI), users can scroll through a document by clicking or dragging the mouse (input) within a scroll bar displayed on the screen (output).

The fundamental task of human–computer interaction is to shuttle information between the brain of the user and the silicon world of the computer. Progress in this area attempts to increase the useful bandwidth across that interface by seeking faster, more natural, and more convenient means for users to transmit information to computers, as well as efficient, salient, and pleasant mechanisms to provide feedback to the user. On the user's side of the communication channel, interaction is constrained by the nature of human attention, cognition, and perceptual–motor skills and abilities; on the computer side, it is constrained only by the technologies and methods that we can invent.

Research in input and output focuses on the two ends of this channel: the devices and techniques computers can use for communicating with people, and the perceptual abilities, processes, and organs people can use for communicating with computers. It then attempts to find the common ground through which the two can be related by studying new modes of communication that could be used for human–computer interaction (HCI) and developing devices and techniques to use such modes. Basic research seeks theories and principles that inform us of the parameters of human cognitive and perceptual facilities, as well as models that can predict or interpret user performance in computing tasks. Advances can be driven by the need for new modalities to support the unique requirements of specific applications, by technological breakthroughs that HCI researchers attempt to apply to improving or extending the capabilities of interfaces, by theoretical insights suggested by studies of human abilities and behaviors, or even by problems uncovered during careful analyses of existing interfaces. These approaches complement one another; all have their value and contributions to the field, but the best research seems to have elements of all of these.

20.2 Interaction Tasks, Techniques, and Devices

A designer looks at the interaction tasks necessary for a particular application [Foley et al., 1984]. Interaction tasks are low-level primitive inputs required from the user, such as entering a text string or choosing a command. For each such task, the designer chooses an appropriate interaction technique. In selecting an interaction device and technique for each task in a human–computer interface, simply making an optimal choice for each task individually may lead to a poor overall design, with too many different or inconsistent types of devices or dialogues. Therefore, it is often desirable to compromise on individual choices to reach a better overall design.

There may be several different ways to accomplish the same task. For example, one could use a mouse to select a command by using a pop-up menu, a fixed menu (a palette or command bar), multiple clicking, circling the desired command, or even writing the name of the command with the mouse. Software might detect patterns of mouse use in the background, such as repeated surfing through menus, and automatically suggest commands or help topics [Horvitz et al., 1998]. The latter suggests a shift from the classical view of interaction as direct manipulation, in which the user is responsible for all actions and decisions, to one that uses background sensing techniques to allow technology to support the user with semiautomatic or implicit actions and services [Buxton, 1995a].

20.3 The Composition of Interaction Tasks

Early efforts in human–computer interaction sought to identify elemental tasks that appear repeatedly in human–computer dialogs. Foley et al. [1984] proposed that user interface transactions are composed of the following elemental tasks:

Selection — Choosing objects from a set of alternatives
Position — Specifying a position within a range, including picking a screen coordinate with a pointing device
Orientation — Specifying an angle or three-dimensional orientation
Path — Specifying a series of positions or orientations over time
Quantification — Specifying an exact numeric value
Text — Entering symbolic data

While these are commonly occurring tasks in many direct-manipulation interfaces, a problem with this approach is that the level of analysis at which one specifies elemental tasks is not well defined. For example, for position tasks, a screen coordinate could be selected using a pointing device such as a mouse, but it might be entered as a pair of numeric values (quantification) using a pair of knobs (like an Etch-a-Sketch) where precision is paramount. But if these represent elemental tasks, why do we find that we must subdivide position into a pair of quantification subtasks for some devices but not for others?

Treating all tasks as hierarchies of subtasks, known as *compound tasks*, is one way to address this. With appropriate design, and by using technologies and interaction metaphors that parallel as closely as possible the way the user thinks about a task, the designer can phrase together a series of elemental tasks into a single cognitive chunk. For example, if the user's task is to draw a rectangle, a device such as an Etch-a-Sketch is easier to use. For drawing a circle, a pen is far easier to use. Hence, the choice of device influences the level at which the user is required to think about the individual actions that must be performed to achieve a goal. See Buxton [1986] for further discussion of this important concept.

A problem with viewing tasks as assemblies of elemental tasks is that typically this view only considers explicit input in the classical direct-manipulation paradigm. Where do devices like cameras, microphones, and fingerprint scanners fit in? These support higher-level data types and concepts (e.g., images, audio, and identity). Advances in technology will continue to yield new "elemental" inputs. However, these new technologies also may make increasing demands on systems to move from individual samples to synthesis of meaningful structure from the resulting data [Fitzmaurice et al., 1999].

20.4 Properties of Input Devices

The breadth of input devices and displays on the market today can be completely bewildering. Fortunately, there are a number of organizing properties and principles that can help to make sense of the design space and performance issues. First, we consider continuous, manually operated pointing devices (as opposed to discrete input mechanisms such as buttons or keyboards, or other devices not operated with the hand, which we will discuss briefly later). For further insight, readers may also wish to consult complete taxonomies of devices [Buxton, 1983; Card et al., 1991]. As we shall see, however, it is nearly impossible to describe properties of input devices without reference to output — especially the resulting feedback on the screen — because, after all, input devices are only useful insofar as they support interaction techniques that allow the user to accomplish something.

Physical property sensed — Traditional pointing devices typically sense position, motion, or force. A tablet senses position, a mouse measures motion (i.e., change in position), and an isometric joystick senses force. An isometric joystick is a self-centering, force-sensing joystick such as the IBM TrackPoint ("eraser-head") found on many laptops. For a rotary device, the corresponding properties are angle, change in angle, and torque. Position-sensing devices are also known as *absolute* devices, whereas motion-sensing devices are *relative* devices. An absolute device can fully support relative motion, because it can calculate changes to position, but a relative device cannot fully support absolute positioning. In fact, it can only emulate position at all by introducing a cursor on the screen. Note that it is difficult to move the mouse cursor to a particular area of the screen (other than the edges) without looking at it, but with a tablet one can easily point to a region with the stylus using the kinesthetic sense [Balakrishnan and Hinckley, 1999]. This phenomenon is known informally as *muscle memory*.

Transfer function — In combination with the host operating system, a device typically modifies its signals using a mathematical transformation that scales the data to provide smooth, efficient, and intuitive operation. An *appropriate mapping* is a transfer function that matches the physical properties sensed by the input device. Appropriate mappings include force-to-velocity, position-to-position, and velocity-to-velocity functions. For example, an isometric joystick senses force; a nonlinear rate mapping transforms this into a velocity of movement [Rutledge and Selker, 1990; Zhai and Milgram, 1993; Zhai et al., 1997]. When using a rate mapping, the device ideally should also be *self-centering* (i.e., spring return to the zero input value), so that the user can stop quickly by releasing the device. A common inappropriate mapping is calculating a speed of scrolling based on the position of the mouse cursor, such as extending a selected region by dragging the mouse close to the edge of the screen. The user has no feedback about when or to what extent scrolling will accelerate, and the resulting interaction can be hard to learn how to use and difficult to control.

Gain — This is a simple multiplicative transfer function, which can also be described as a control–display (C:D) ratio: the ratio between the movement of the input device and the corresponding movement of the object it controls. For example, if a mouse (the control) must be moved 1 cm on the desk in order to move a cursor 2 cm on the screen (the display), the device has a 1:2 control–display ratio. However, on commercial pointing devices and operating systems, the gain is rarely constant*; an *acceleration function* is often used to modulate the gain depending on velocity. Acceleration function is simply another term for a transfer function that exhibits an exponential relationship between velocity and gain. Experts believe the primary benefit of acceleration is to reduce the footprint, or the physical movement space, required by an input device [Jellinek and Card, 1990; Hinckley et al., 2001]. One must also be very careful when studying the possible influence of gain settings on user performance: experts have criticized gain as a fundamental concept, since it confounds two separate concepts (device size and display size) in one arbitrary metric [Accot and Zhai, 2001]. Furthermore, user performance may exhibit speed–accuracy trade-offs, calling into question the assumption that there exists an optimal C:D ratio [MacKenzie, 1995].

Number of dimensions — Devices can measure one or more linear and angular dimensions. For example, a mouse measures two linear dimensions, a knob measures one angular dimension, and a six-degree-of-freedom magnetic tracker measures three linear dimensions and three angular. If the number of dimensions required by the user's interaction task does not match the number of dimensions provided by the input device, then special handling (e.g., interaction techniques that may require extra buttons, graphical widgets, mode switching, etc.) must be introduced.

---

*Direct input devices are an exception, since the C:D ratio is typically fixed at 1:1, but see also Sears and Shneiderman [1991].

This is a particular concern for three-dimensional user interfaces and interaction [Zhai, 1998; Hinckley et al., 1994b]. Numerous interaction techniques have been proposed to allow standard 2-D pointing devices to control 3-D positioning or orientation tasks (e.g., Conner et al., 1992; Chen et al., 1988, Bukowski and Sequin, 1995). Well designed interaction techniques using specialized multiple degree-of-freedom input devices can sometimes offer superior performance [Ware and Rose, 1999; Hinckley et al., 1997] but may be ineffective for standard desktop tasks, so overall performance must be considered [Balakrishnan et al., 1997; Hinckley et al., 1999].

Pointing speed and accuracy — The standard way to characterize pointing device performance employs the Fitts' law paradigm [Douglas et al., 1999]. Fitts' law relates the movement time to point at a target, the amplitude of the movement (the distance to the target), and the width of the target (i.e., the precision requirement of the pointing movement). While not emphasized in this chapter, Fitts' law is the single most important quantitative analysis, testing, and prediction tool available to input research and device evaluation. For an excellent overview of its application to the problems of HCI, including use of Fitts' law to characterize bandwidth (a composite measure of both speed and accuracy), see MacKenzie [1992]. For discussion of other accuracy metrics, see MacKenzie et al. [2001].

Recent years have seen a number of new insights and new applications for Fitts' law. Fitts' law was originally conceived in the context of rapid, aimed movements, but it can also be applied to tasks such as scrolling [Hinckley et al., 2001], multiscale navigation [Guiard et al., 2001], and crossing boundaries [Accot and Zhai, 2002]. Recently, researchers have also applied Fitts' law to expanding targets that double in width as the user approaches them. Even if the expansion begins after the user has already covered 90% of the distance from a starting point, the expanding target can be selected as easily as if it had been fully expanded since the movement began [McGuffin and Balakrishnan, 2002]. However, it remains unclear whether this can be successfully applied to improving pointing performance for multiple targets that are closely packed together, as typically found in menus and tool palettes. For tasks that exhibit continuous speed–accuracy requirements, such as moving through a hierarchical menu, Fitts' law cannot be applied, but researchers have recently formulated the *steering law*, which does addresses such tasks [Accot and Zhai, 1997; Accot and Zhai, 1999; Accot and Zhai 2001].

Input device states — To select a single point or region with an input device, users need a way to signal when they are selecting something, as opposed to when they are just moving over something to reach a desired target. The need for this fundamental signal of intention is often forgotten by researchers eager to explore new interaction modalities such as empty-handed pointing (e.g., using camera tracking or noncontact proximity sensing of hand position). The *three-state model* of input generalizes the states sensed by input devices [Buxton, 1990b] as *tracking*, which causes the cursor to move; *dragging*, which allows the selection of objects by clicking or the moving of objects by clicking and dragging them; and *out of range*, which occurs when the device moves out of its physical tracking range (e.g., a mouse is lifted from the desk, or a stylus is removed from a tablet). Most pointing devices sense only two of these three states: for example, a mouse senses tracking and dragging, but a touchpad senses tracking and the out-of-range state. Hence, to fully simulate the functionality offered by mice, touchpads need special procedures, such as tapping to click, which are prone to inadvertent activation (e.g., touching the pad by accident causes a click). For further discussion and examples, see Buxton, 1990b; Hinckley et al., 1998a; and MacKenzie and Oniszczak, 1998.

Direct vs. indirect control — A mouse is indirect (you move it on the table to point to a spot on the screen); a touchscreen is direct (the display surface is also the input surface). Direct devices raise several unique issues. Designers must consider the possibility of parallax error resulting from a gap between the input and display surfaces, reduced transmissivity of the screen introduced by a sensing layer, or occlusion of the display by the user's hands. Another issue is that touchscreens can support a cursor tracking state or a dragging state, but not both. Typically, touchscreens move directly from the out-of-range state to the dragging state when the user touches the screen, with no intermediate cursor feedback [Buxton, 1990b]. Techniques for touchscreen cursor feedback typically require that selection occurs on lift-off [Sears et al., 1991; Sears et al., 1992]. See also "Pen input" in Section 20.5 of this chapter.

Hardware criteria — Various other characteristics can distinguish input devices but are perhaps less important in distinguishing the fundamental types of interaction techniques that can be supported. Engineering parameters of a device's performance, such as sampling rate, resolution, accuracy, and linearity, can all influence performance, as well. *Latency* is the end-to-end delay between the user's physical movement, sensing this, and providing the ultimate system feedback to the user. Latency can be a devious problem because it is impossible to eliminate it completely from system performance. Latency of more than 75 to 100 milliseconds significantly impairs user performance for many interactive tasks [Robertson et al., 1989; MacKenzie and Ware, 1993]. For vibrotactile or haptic feedback, users may be sensitive to much smaller latencies of just a few milliseconds [Cholewiak and Collins, 1991].

Other user performance criteria — Devices can also be distinguished using various criteria: user performance, learning time, user preference, and so forth. Device *acquisition time*, which is the average time to pick up or put down a device, is often assumed to be a significant factor in user performance, but the Fitts' law bandwidth of a device tends to dominate this unless switching occurs frequently [Douglas and Mithal, 1994]. However, one exception is stylus- or pen-based input devices; pens are generally comparable to mice in general pointing performance [Accot and Zhai 1999] or even superior for some high-precision tasks [Guiard et al., 2001], but these benefits can easily be negated by the much greater time it takes to switch between using a pen and using a keyboard.

20.5   Discussion of Common Pointing Devices

Here, we briefly describe commonly available pointing devices and some issues that can arise with them in light of the properties discussed above.

Mouse — A mouse senses movement relative to a flat surface. Mice exhibit several properties that are well suited to the demands of desktop graphical interfaces [Balakrishnan et al., 1997]. The mouse is stable and does not fall over when released (unlike a stylus on a tablet). A mouse can also provide integrated buttons for selection, and because the force required to activate a mouse's buttons is orthogonal to the plane of movement, it helps to minimize accidental clicking or interference with motion. Another subtle benefit is the possibility for users to employ a combination of finger, hand, wrist, arm, and even shoulder muscles to span the range of tasks from short precise selections to large, ballistic movements [Zhai et al., 1996; Balakrishnan and MacKenzie, 1997]. Finally, Fitts' law studies show that users can point with the mouse about as well as with the hand itself [Card et al., 1978].

Trackball — A trackball is like a mouse that has been turned upside down, with a mechanical ball that rolls in place. The main advantage of trackballs is that they can be used on an inclined surface, and they often require a smaller footprint than mice. They also employ different muscle groups, which some users find more comfortable. However, trackballs cannot accommodate buttons as easily as mice, so tasks that require moving the trackball while holding down a button can be awkward [MacKenzie et al., 1991].

Tablets — Most tablets sense the absolute position of a mechanical intermediary such as a stylus or puck on the tablet surface. A *puck* is a mouse that is used on a tablet; the only difference is that it senses absolute position and it cannot be used on a surface other than the tablet. Absolute mode is generally preferable for tasks such as tracing, digitizing, drawing, free-hand inking, and signature capture. Tablets that sense contact with the bare finger are known as *touch tablets* [Buxton et al., 1985]. Touchpads are miniature touch tablets, as commonly found on portable computers [MacKenzie and Oniszczak 1998]. A touchscreen is a transparent, touch-sensitive tablet mounted on a display, but it demands different handling than a tablet (see "Direct vs. indirect control" in Section 20.4).

Pen input — Pen-based input for mobile devices is an area of increasing practical concern. Pens effectively support activities such as inking, marking, and gestural input (see Section 20.8.2), but pens raise a number of problems when supporting graphical interfaces originally designed for mouse input. Pen input raises the concerns about direct input devices described previously. There is no way to see exactly what position will be selected before selecting it: pen contact with the screen directly enters the dragging state of the three-state model [Buxton, 1990b]. There is neither a true equivalent of a hover state for tool tips nor an extra button for context menus. Pen dwell time on a target can be used to provide one of these two functions. For detecting double-tap, allow a longer interval between the taps (as compared to double-click on a mouse), and also allow a significant change to the screen position between taps. Finally, users often want to touch the screen of small devices using a bare finger, so applications should be designed to accommodate imprecise selections. Note that some pen-input devices, such as the Tablet PC, use an inductive sensing technology that can only sense contact from a specially instrumented stylus, and thus cannot be used as a touchscreen. However, this deficiency is made up for by the ability to track the pen when it is close to (but not touching) the screen, allowing support for a tracking state with cursor feedback (and hence tool tips).

Joysticks — There are many varieties of joystick. As mentioned earlier, an isometric joystick senses force and returns to center when released. Because isometric joysticks can have a tiny footprint, they are often used when space is at a premium, allowing integration with a keyboard and hence rapid switching between typing and pointing [Rutledge and Selker, 1990; Douglas and Mithal, 1994]. Isotonic joysticks sense the stick's angle of deflection, so they tend to move more than isometric joysticks, offering better feedback to the user. Such joysticks may or may not have a mechanical spring return to center. Some joysticks include force sensing, position sensing, and other special features. For a helpful organization of the complex design space of joysticks, see Lipscomb and Pique [1993].

Alternative devices — Researchers have explored using the feet [Pearson and Weiser, 1988], head tracking, and eye tracking as alternative approaches to pointing. Head tracking has much lower pointing bandwidth than the hands and may require the neck to be held in an awkward fixed position, but it has useful applications for intuitive coupling of head movements to virtual environments [Sutherland, 1968; Brooks, 1988] and interactive 3-D graphics [Hix et al., 1995; Ware et al., 1993]. Eye movement–based input, properly used, can provide an unusually fast and natural means of communication, because we move our eyes rapidly and almost unconsciously. The human eye fixates visual targets within the fovea, which fundamentally limits the accuracy of eye gaze tracking to 1 degree of the field of view [Zhai et al., 1999]. Eye movements are subconscious and must be interpreted carefully to avoid annoying the user with unwanted responses to his actions, known as the *Midas touch problem* [Jacob, 1991]. Eye-tracking technology is expensive and has numerous technical limitations, confining its use to research labs and disabled persons with few other options.

20.6 Feedback and Perception — Action Coupling

The ecological approach to human perception [Gibson, 1986] asserts that the organism, the environment, and the tasks the organism performs are inseparable and should not be studied in isolation. Hence, perception and action are intimately linked in a single motor–visual feedback loop, and any separation of the two is artificial. The lesson for interaction design is that techniques must consider both the motor control (input) and feedback (output) aspects of the design and how they interact with one another.

From the technology perspective, one can consider feedback passive or active. Active feedback is under computer control. This can be as simple as presenting a window on a display, or as sophisticated as simulating haptic contact forces with virtual objects when the user moves an input device. We will discuss active feedback techniques later in this chapter, when we discuss display technologies and techniques.

Passive feedback may come from sensations within the user's body, as influenced by physical properties of the device, such as the shape, color, and feel of buttons when they are depressed. The industrial design of a device suggests the purpose and use of a device even before a user touches it [Norman, 1990]. Mechanical sounds and vibrations that result from using the device provide confirming feedback of the user's action. The shape of the device and the presence of landmarks can help users orient a device without having to look at it [Hinckley et al., 1998b]. Proprioceptive and kinesthetic feedback are somewhat imprecise terms, often used interchangeably, that refer to sensations of body posture, motion, and muscle tension [Burdea, 1996]. These senses allow users to feel how they are moving an input device without looking at the device, and indeed without looking at the screen in some situations [Mine et al., 1997; Balakrishnan and Hinckley, 1999]. This may be important when the user's attention is divided between multiple tasks and devices [Fitzmaurice and Buxton, 1997]. Sellen et al. [1992] report that muscular tension from depressing a foot pedal makes modes more salient to the user than purely visual feedback. Although all of these sensations are passive and not under the direct control of the computer, these examples nonetheless demonstrate that they are relevant to the design of devices. Interaction techniques can consider these qualities and attempt to leverage them.

User performance may be influenced by correspondences between input and output. Some correspondences are obvious, such as the need to present confirming visual feedback in response to the user's actions. Ideally, feedback should indicate the results of an operation before the user commits to it (e.g., highlighting a button or menu item when the cursor moves over it). Kinesthetic correspondence and perceptual structure, described below, are less obvious.

Kinesthetic correspondence refers to the principle that graphical feedback on the screen should correspond to the direction in which the user moves the input device, particularly when 3-D rotation is involved [Britton et al., 1978]. Users can easily adapt to certain noncorrespondences: when the user moves a mouse forward and back, the cursor actually moves up and down on the screen; if the user drags a scrollbar downward, the text on the screen scrolls upward. With long periods of practice, users can adapt to almost anything. For example, for more than 100 years psychologists have known of the phenomenon of prism adaptation: people can eventually adapt to wearing prisms that cause everything to look upside down [Stratton, 1897]. However, one should not force users to adapt to a poor design.

Researchers also have found that the interaction of the input dimensions of a device with the control dimensions of a task can exhibit perceptual structure. Jacob et al. [1994] explored two input devices: a 3-D position tracker with integral (x, y, z) input dimensions and a standard 2-D mouse, with (x, y) input separated from (z) input by holding down a mouse button. For selecting the position and size of a rectangle, the position tracker is most effective. For selecting the position and grayscale color of a rectangle, the mouse is most effective. The best performance results when the integrality or separability of the input matches that of the output.

20.7 Keyboards, Text Entry, and Command Input

For over a century, keyboards and typewriters have endured as the mechanism of choice for text entry. The resiliency of the keyboard, in an era of unprecedented technological change, is the result of how keyboards complement human skills. This may make keyboards difficult to supplant with new input devices or technologies. We summarize some general issues surrounding text entry below, with a focus on mechanical keyboards; see also Lewis et al. [1997].

Skill acquisition and skill transfer — Procedural memory is a specific type of memory that encodes repetitive motor acts. Once an activity is encoded in procedural memory, it requires little conscious effort to perform [Anderson, 1980]. Because procedural memory automates the physical act of text entry, touch-typists can rapidly type words without interfering with the mental composition of text. The process of encoding an activity in procedural memory can be formalized as the *power law of practice*: $T = aP^b$, where T is the time to perform the task, P is the amount of practice, and a and b

are constants that fit the curve to observed data. This suggests that changing the keyboard can have a high relearning cost. However, a change to the keyboard can succeed if it does not interfere with existing skills or allows a significant transfer of skill. For example, some ergonomic keyboards have succeeded by preserving the basic key layout but altering the typing pose to help maintain neutral postures [Honan et al., 1995; Marklin and Simoneau, 1996], whereas the Dvorak key layout may have some small performance advantages but has not found wide adoption due to high retraining costs [Lewis et al., 1997].

Eyes-free operation — With practice, users can memorize the location of commonly used keys relative to the home position of the two hands, allowing typing with little or no visual attention [Lewis et al., 1997]. By contrast, soft keyboards (small on-screen virtual keyboards found on many handheld devices) require nearly constant visual monitoring, resulting in diversion of attention from one's work. Furthermore, with stylus-driven soft keyboards, the user can strike only one key at a time. Thus the design issues for soft keyboards differ tremendously from mechanical keyboards [Zhai et al., 2000].

Tactile feedback — On a mechanical keyboard, users can feel the edges and gaps between the keys, and the keys have an activation force profile that provides feedback of the key strike. In the absence of such feedback, as on touchscreen keyboards [Sears, 1993], performance may suffer and users may not be able to achieve eyes-free performance [Lewis et al., 1997].

Combined text, command, and navigation input — Finally, it is easy to forget that keyboards provide many secondary command and control actions in addition to pure text entry, such as power keys and navigation keys (Enter, Home/End, Delete, Backspace, Tab, Esc, Page Up/Down, arrow keys, etc.), chord key combinations (such as Ctrl+C for Copy) for frequently used commands, and function keys for miscellaneous functions defined by the current application. Without these keys, frequent interleaving of mouse and keyboard activity may be required to perform these secondary functions.

Ergonomic issues — Many modern information workers suffer from repetitive strain injury (RSI). Researchers have identified many risk factors for such injuries, such as working under stress or taking inadequate rest breaks. People often casually associate these problems with keyboards, but the potential for RSI is common to many manually operated tools and repetitive activities [Putz-Anderson, 1988]. Researchers have advocated themes for ergonomic design of keyboards and other devices [Pekelney and Chu, 1995], including reducing repetition, minimizing force required to hold and move the device or to press its buttons, avoiding sharp edges that put pressure on the soft tissues of the hand, and designing for natural and neutral postures of the user's hands and wrists [Honan et al., 1995; Marklin et al., 1997]. Communicating a clear orientation for gripping and moving the device through its industrial design also may help to discourage inappropriate, ergonomically unsound grips.

Other text entry mechanisms — One-handed keyboards can be implemented using simultaneous depression of multiple keys; such *chord keyboards* can sometimes allow one to achieve high peak performance (e.g., court stenographers) but take much longer to learn how to use [Noyes, 1983; Mathias et al., 1996; Buxton, 1990a]. They are often used in conjunction with wearable computers [Smailagic and Siewiorek, 1996] to keep the hands free as much as possible (but see also Section 20.8.1). With complex written languages, such as Chinese and Japanese, key chording and multiple stages of selection and disambiguation are currently necessary for keyboard-based text entry [Wang et al., 2001]. Handwriting and character recognition may ultimately provide a more natural solution, but for Roman languages, handwriting (even on paper, with no recognition involved) is much slower than skilled keyboard use. To provide reliable stylus-driven text input, some systems have adopted unistroke (single-stroke) gestural alphabets [Goldberg and Richardson, 1993] that reduce the demands on recognition technology while remaining relatively easy for users to learn [MacKenzie and Zhang, 1997]. However, small two-thumb keyboards [MacKenzie and Soukoreff, 2002] or fold-away peripheral keyboards are becoming increasingly popular for such devices.

20.8   Modalities of Interaction

Here, we briefly review a number of general strategies and input modalities that have been explored by researchers. These approaches generally transcend a specific type of input device, spanning a range of devices and applications.

20.8.1   Voice and Speech

Carrying on a full conversation with a computer as one might do with another person is well beyond the state of the art today and, even if possible, may be a naïve goal. Yet even without understanding the content of the speech, computers can digitize, store, edit, and replay segments of speech to augment human–human communication [Arons, 1993; Stifelman, 1996]. Conventional voice mail and the availability of MP3 music files on the Web are simple examples of this. Computers can also infer information about the user's activity from ambient audio, such as determining whether the user is present or perhaps engaging in a conversation with a colleague, allowing more timely delivery of information or suppression of notifications that may interrupt the user [Schmandt, 1993; Sawhney and Schmandt, 1999; Horvitz et al., 1999; Buxton, 1995b].

Understanding speech as input is a long-standing area of research. While progress is being made, it is slower than optimists originally predicted, and daunting unsolved problems remain. For limited vocabulary applications with native English speakers, speech recognition can excel at recognizing words that occur in the vocabulary. Error rates can increase substantially when users employ words that are out-of-vocabulary (i.e., words the computer is not "listening" for), when the grammatical complexity of possible phrases increases, or when the microphone is not a high-quality, close-talk headset. Dictation using continuous speech recognition is available on the market today, but the technology still has a long way to go; a recent study found that the corrected words-per-minute rate of text entry using a mouse and keyboard is about twice as fast as dictation input [Karat et al., 1999]. Even if the computer could recognize all of the user's words, the problem of understanding natural language is a significant and unsolved one. It can be avoided by using an artificial language of special commands or even a fairly restricted subset of natural language. Given the current state of the art, however, the closer the user moves toward full, unrestricted natural language, the more difficulties will be encountered.

20.8.2   Pen-Based Gestures and Hand Gesture Input

Pen-based gestures can indicate commands, such as crossing out a word to delete it or circling a paragraph and drawing an arrow to move it. Such gestures support cognitive chunking by integrating command selection with specification of the command's scope [Buxton et al., 1983; Kurtenbach and Buxton, 1991]. Marking menus use directional pen motion to provide extremely rapid menu selection [Kurtenbach and Buxton, 1993; Kurtenbach et al., 1993]. With pen-based interfaces, designers often face a difficult design trade-off between treating the user's marks as ink that is not subject to interpretation vs. providing pen-based input that treats the ink as a potential command [Kramer, 1994; Moran et al., 1997; Mynatt et al., 1999]. Pen input, via sketching, can be used to define 3-D objects [Zeleznik et al., 1996; Igarashi et al., 1999]. Researchers also have explored multimodal pen and voice input; this is a powerful combination because pen and voice have complementary strengths and weaknesses and can disambiguate one another [Cohen et al., 1997; Cohen and Sullivan, 1989; Oviatt, 1997].

Input using hand gestures represents another, less well understood area of inquiry. There are many human behaviors involving hand movements and gestures, but few have been thoroughly explored for human–computer interaction. Cadoz [1994] broadly categorizes hand gestures as semiotic, ergotic, or epistemic. *Semiotic* gestures are those used to communicate meaningful information, such as "thumbs up." *Ergotic* gestures are those used to manipulate physical objects. *Epistemic* gestures are exploratory movements to acquire haptic or tactile information; see also [Kirsh, 1995; Kirsh and Maglio, 1994]. Most research in hand gesture recognition focuses on empty-handed semiotic gestures [Cassell, 2003], which

Rime and Schiaratura further classify as follows [Rime and Schiaratura, 1991]:

Symbolic — Conventional symbolic gestures such as "OK"
Deictic — Pointing to fill in a semantic frame, analogous to deixis in natural language
Iconic — Illustrating a spatial relationship
Pantomimic — Mimicking an invisible tool, such as pretending to swing a golf club

Command input using recognition-based techniques raises a number of unique challenges [Bellotti et al., 2002]. In particular, with most forms of gestural input, errors of user intent and errors of computer interpretation seem inevitable. Deictic gesture in particular has received much attention, with several efforts using pointing (typically captured using instrumented gloves or camera-based recognition) to interact with "intelligent" environments [Baudel and Beaudouin-Lafon, 1993; Maes et al., 1996; Freeman and Weissman, 1995; Jojic et al., 2000]. Deictic gesture in combination with speech recognition has also been studied [Bolt, 1980; Hauptmann, 1989; Lucente et al., 1998; Wilson and Shafer, 2003]. However, there is more to the field than empty-handed semiotic gestures. Recent exploration of tangible interaction techniques [Ishii and Ullmer, 1997] and efforts to sense movements and handling of sensor-enhanced mobile devices perhaps fall more closely under ergotic (manipulative) gestures [Hinckley et al., 2003; Hinckley et al., 2000; Harrison et al., 1998].

20.8.3 Bimanual Input

Aside from touch typing, most of the devices and modes of operation discussed thus far and in use today involve only one hand at a time. But people use both hands in a wide variety of the activities associated with daily life. For example, when writing, a right-hander writes with the pen in the right hand, but the left hand also plays a crucial and distinct role. It holds the paper and orients it to a comfortable angle that suits the right hand. In fact, during many skilled manipulative tasks, Guiard observed that the hands take on asymmetric, complementary roles [Guiard, 1987]: for right-handers, the role of the left hand precedes the right (the left hand first positions the paper), the left hand sets the frame of reference for the action of the right hand (the left hand orients the paper), and the left hand performs infrequent, large-scale movements compared to the frequent, small-scale movements of the right hand (writing with the pen). Most applications for bimanual input to computers are characterized by asymmetric roles of the hands, including compound navigation/selection tasks such as scrolling a Web page and then clicking on a link [Buxton and Myers, 1986], command selection using the nonpreferred hand [Bier et al., 1993; Kabbash et al., 1994], as well as navigation, virtual camera control, and object manipulation in three-dimensional user interfaces [Kurtenbach et al., 1997; Balakrishnan and Kurtenbach, 1999; Hinckley et al., 1998b]. For some tasks, such as banging together a pair of cymbals, the hands may take on symmetric roles. For further discussion of bimanual symmetric tasks, see Guiard, 1987; Balakrishnan and Hinckley, 2000.

20.8.4 Passive Measurement: Interaction in the Background

Not all interaction with computers need consist of explicit, intentionally communicated commands. Think about walking into a modern grocery store. You approach the building; the doors sense this motion and open for you. No explicit communication has occurred, yet a computer has used your action of walking toward the store as an input to decide when to open the door. Intentional, explicit interaction takes place in the foreground, while implicitly sensed interaction takes place in the background, behind the fore of the user's attention [Buxton, 1995a]. Background interaction will be a major emphasis of future research in automation and sensing systems, as users become increasingly mobile and become saturated with information from many sources. Researchers are currently exploring ways to provide context awareness through location sensing, ambient sensing of light, temperature, and other environmental qualities; movement and handling of devices; detection of the user's the identity and physical objects in the environment; and possibly even physiological measures such as heart-rate variability. This type of information potentially can allow technology to interpret the context of a situation and respond more appropriately [Schilit et al., 1994; Schmidt et al., 1999; Dey et al., 2001; Hinckley et al., 2003].

Background interaction can also be applied to explicit input streams through passive behavioral measurements, such as observation of typing speed, manner of moving the cursor, sequence and timing of commands activated in a GUI [Horvitz et al., 1998], and other patterns of use. A carefully designed user interface could make intelligent use of such information to modify its dialogue with the user, based on, for example, inferences about the user's alertness or expertise. These measures do not require additional input devices, but rather gleaning of additional, typically neglected, information from the existing input stream. These are sometimes known as *intelligent* or *adaptive* user interfaces, but mundane examples also exist. For example, cursor control using the mouse or scrolling using a wheel can be optimized by modifying the device response depending on the velocity of movement [Jellinek and Card, 1990; Hinckley et al., 2001].

We must acknowledge the potential for misuse or abuse of information collected in the background. Users should always be made aware of what information is or may potentially be observed as part of a human–computer dialogue. Users should have control and the ability to block any information that they want to remain private [Nguyen and Mynatt, 2001].

20.9 Displays and Perception

We now turn our attention to the fundamental properties of displays and to techniques for effective use of displays. We focus on visual displays and visual human perception, because these represent the vast majority of displays, but we also discuss feedback through the haptic and audio channels.

20.9.1 Properties of Displays and Human Visual Perception

Display requirements, such as resolution in time and space, derive from the properties of human vision. Thus, we begin with the basic issues relating to display brightness, uniformity, and spatial and temporal resolution.

Dynamic range— The human eye has an enormous dynamic range. The amount of light reflected from surfaces on a bright day at the beach is about five orders of magnitude higher than the amount available under dim lamp lights. Yet the shapes, layouts, and colors of objects look nearly identical to the human eye across much of this range. Most displays in common use are self-luminous cathode ray tubes (CRTs) or back-lit liquid crystal displays (LCDs). The best of these devices have a dynamic range (the ratio between the maximum and minimum values produced) of a little more than two orders of magnitude. In practice, under typical room lighting conditions, 15 to 40% of the light reaching the user's eye is actually ambient room light reflected by the front surface of the phosphors, or off the screen surface. This means that the effective dynamic range of most devices, unless viewed in dark rooms, is no better than three or four to one. Fortunately, the human eye can tolerate extreme variation in the overall level of illumination, as well as in the amount of contrast produced by the display.

Spatial frequency — The ability of the human visual system to resolve fine targets is known as *visual acuity*. A standard way to measure visual acuity is to determine how fine a sinusoidal striped pattern can be discriminated from a uniform gray. Humans are capable of perceiving targets as fine as 50 to 60 cycles/degree of visual angle when the pattern is of very high contrast. Figure 20.1 illustrates the spatial sensitivity of the human eye as a function of spatial frequency. Specifically, it illustrates the degree of contrast required for sinusoidal gratings of different spatial frequencies to be perceived. The function has an inverted U shape with a peak at about 2 cycles/degree of visual angle. This means that 5-mm stripes at arm's length are optimally visible. The falloff at low spatial frequencies indicates that the human visual system is insensitive to gradual changes in overall screen luminance. Indeed, most CRTs have a brightness falloff toward the edges of as much as 20%, which we barely notice. This nonuniformity is even more pronounced with rear projection systems, due to the construction of screens that project light primarily in a forward direction. This is called the *screen gain;* a gain of 3.0 means that three times as much light is transmitted in the straight-through

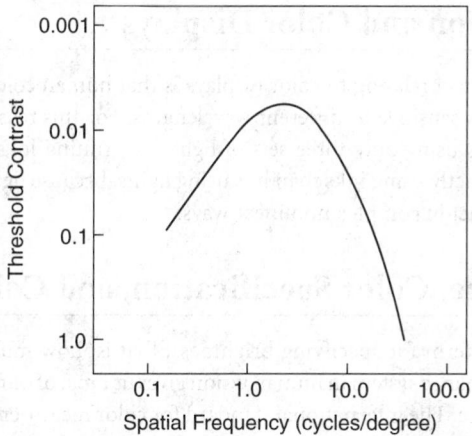

FIGURE 20.1 Spatial contrast sensitivity function of the human visual system. There is a falloff in sensitivity both to detailed patterns (high-spatial frequencies) and to gradually changing gray values (low-spatial frequencies).

direction, compared to a perfect Lambertian diffuser. At other angles, less light is transmitted, so that at a 45° off-axis viewing angle, only half as much light may be available, compared to a perfect diffuser. Screen gain is also available with front projection with similar nonuniformities as a consequence, although the use of curved screens can compensate to some extent.

Spatial resolution — The receptors in the human eye have a visual angle of about 0.8 seconds of arc. Modern displays provide approximately 40 pixels/cm. A simple calculation reveals that at about a 50-cm viewing distance, pixels will subtend about 1.5 seconds of arc, about twice the size of cone receptors in the center of vision. Viewed from 100 cm, such a screen has pixels that will be imaged on the retina at about the same size as the receptors. This might suggest that we are within reach of the perfect display in terms of spatial resolution; such a screen would require approximately 80 pixels/cm at normal viewing distances. However, under some conditions the human visual system is capable of producing superacuities that imply resolution better than the receptor size. For example, during fusion in stereo vision, disparities smaller than 5 seconds of arc can be detected [Westheimer, 1979] (see also Ware, 2000 for discussion of stereopsis and stereo displays aimed at the practitioner). Another example of superacuity is known as *aliasing*, resulting from the division of the screen into discrete pixels; for example, a line on the display that is almost (but not quite) horizontal may exhibit a jagged stairstep pattern that is very noticeable and unsatisfactory. This effect can be diminished by *antialiasing*, which computes pixel color values that are averages of all the different objects that contribute to the pixel, weighted by the percentage of the pixel they cover. Similar techniques can be applied to improve the appearance of text, particularly on LCD screens, where individual red, green, and blue display elements can be used for subpixel antialiasing [Betrisey et al., 2000; Platt, 2000].

Temporal resolution, refresh, and update rates — The *flicker fusion frequency* represents the least rapidly flickering light that the human eye does not perceive as steady. Flicker fusion frequency typically occurs around 50 Hz for a light that turns completely on and off [Wyszecki and Styles, 1982]. In discussing the performance of monitors, it is important to differentiate the refresh rate and the update rate. The *refresh rate* is the rate at which a screen is redrawn, and it is typically constant (values of 60 Hz up to 120 Hz are common). In contrast, the *update rate* is the rate at which the system software updates the output to be refreshed. Ideally, this should occur at or above the refresh rate, but with increasingly demanding applications and complex data sets, this may not be possible. A rule of thumb states that a 10 Hz update rate is a minimum for smooth animation [Robertson et al., 1989]. Motion blur (also known as *temporal antialiasing*) techniques can be applied to reduce the jerky effects resulting from low frame rates [Cook, 1986].

20.10   Color Vision and Color Displays

The single most important fact relating to color displays is that human color vision is trichromatic; our eyes contain three receptors sensitive to different wavelengths. For this reason, it is possible to generate nearly all perceptible colors using only three sets of lights or printing inks. However, it is much more difficult to specify colors exactly using inks than it is using lights, because lights can be treated as a simple vector space, but inks interact in complex nonlinear ways.

20.11   Luminance, Color Specification, and Color Gamut

Luminance is the standard term for specifying brightness, that is, how much light is emitted by a self-luminous display. The luminance system in human vision gives us most of our information about the shape and layout of objects in space. The international standard for color measurement is the CIE (Commission Internationale de l'Eclairage) standard. The central function in Figure 20.2 is the CIE $V(\lambda)$ function, which represents the amount that light of different wavelengths contributes to the overall sensation of brightness. As this curve demonstrates, short wavelengths (blue) and long wavelengths (red) contribute much less than green wavelengths to the sensation of brightness. The CIE tristimulus functions, also shown in Figure 20.2, are a set of color-matching functions that represent the color vision of a typical person. Humans are most sensitive to the green wavelengths around 560 nm. Specifying luminance and specifying a color in CIE tristimulus values are complex technical topics; for further discussion, see Ware, 2000; Wyszecki et al., 1982.

A chromaticity diagram can be used to map out all possible colors perceptible to the human eye, as illustrated in Figure 20.3. The pure spectral hues are given around the boundary of this diagram in nanometers (10^{-9} m). While the spacing of colors in tristimulus coordinates and on the chromaticity diagram is not perceptually uniform, uniform color spaces exist that produce a space in which equal metric distances are closer to matching equal perceptual differences [Wyszecki et al., 1982]. For example, this can be useful to produce color sequences in map displays [Robertson, 1988].

The *gamut* of all possible colors is the dark gray region of the chromaticity diagram, with pure hues at the edge and neutral tones in the center. The triangular region represents the gamut achievable by a particular color monitor, determined by the colors of the phosphors given at the corners of the triangle. Every color within this triangular region is achievable, and every color outside of the triangle is not. This

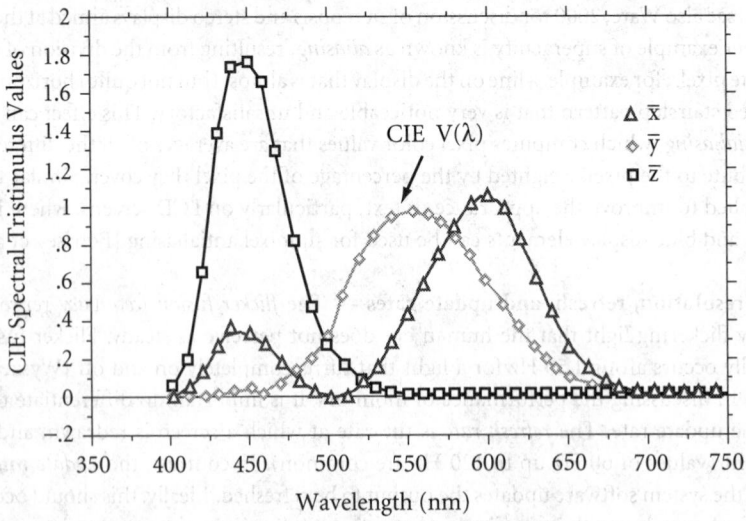

FIGURE 20.2 The CIE tristimulus functions. These are used to represent the standard observer in colorimetry. Short wavelengths at the left-hand side appear blue, in the middle they are green, and to the right they are red. Humans are most sensitive to the green wavelengths around 560 nm.

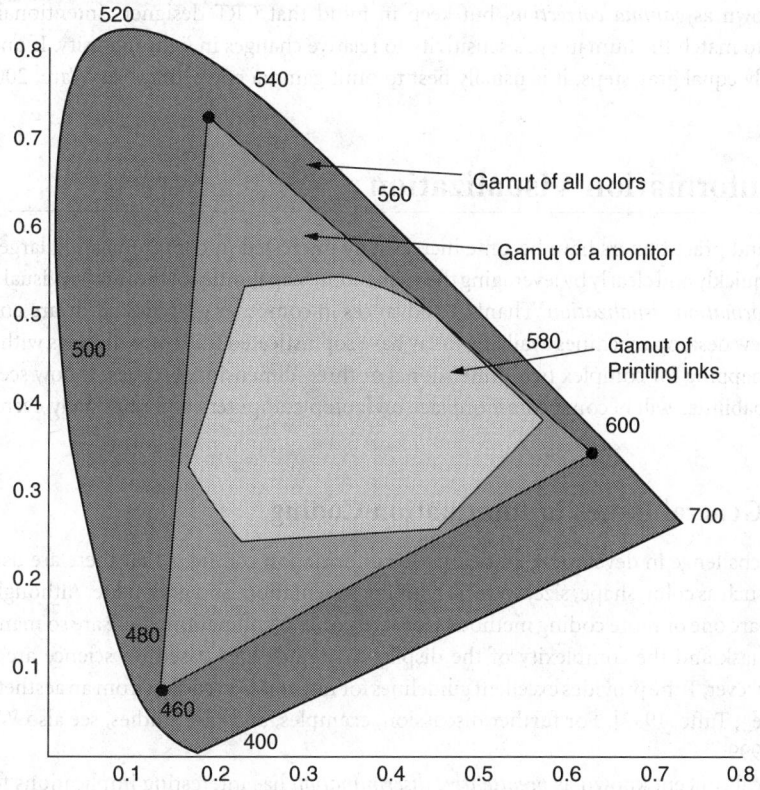

FIGURE 20.3 A CIE chromaticity diagram with a monitor gamut and a printing ink gamut superimposed. It can be seen that the range of available colors with a color printing is smaller than that available with a monitor, and both fall short of providing the full range of color that can be seen.

diagram nicely illustrates the trade-off faced by the designer of color displays. A phosphor that produces a very narrow wavelength band will have chromaticity coordinates close to the pure spectral colors, and this will produce more saturated colors (thus enlarging the triangle). However, this narrow band also means that little light is produced.

The irregular shape inside the triangle illustrates the gamut of colors obtainable using printing inks. Notice that this set of colors is still smaller, causing difficulties when we try to obtain a hard-copy reproduction of the colors on the monitor. Because the eye is relatively insensitive to overall color shifts and overall contrast changes, we can take the gamut from one device and map it into the gamut of the printing inks (or some other device) by compressing and translating it. This is known as *gamut mapping*, a process designed to preserve the overall color relationships while effectively using the range of a device [Stone et al., 1988]. However, it should be noted that the original colors will be lost in this process and that after a succession of gamut mappings, colors may become distorted from their original values.

A process known as *chromatic adaptation* occurs in the human eye receptors and in the early stages of visual processing. For example, we hardly notice that daylight is much bluer than the yellow cast of tungsten light produced from ordinary light bulbs. The CIE standard does not account for chromatic adaptation, nor does it account for color contrast (colors appear differently to the human eye depending on the surrounding visual field). The practical implication is that we can get by with color monitors and printers that are grossly out of calibration. However, accurate color is essential in some applications. It is possible to calibrate a color device precisely so that the particular inputs required to produce a color may be specified in CIE tristimulus values. For monitor calibration, see Cowan, 1983; for calibrating print devices, see Stone et al., 1988. It is also possible to correct for the nonlinear response of CRT displays, a process known as *gamma correction*, but keep in mind that CRT designers intentionally insert this nonlinearity to match the human eye's sensitivity to relative changes in light intensity. If one desires a set of perceptually equal gray steps, it is usually best to omit gamma correction. See Ware, 2000 for further discussion.

20.12 Information Visualization

Researchers and practitioners have become increasingly interested in communicating large quantities of information quickly and clearly by leveraging the tremendous capabilities of the human visual system, a field known as *information visualization*. Thanks to advances in computer graphics hardware and algorithms, virtually all new desktop machines available today have sophisticated full-color displays with transparency and texture mapping for complex two-dimensional or three-dimensional scenes. It now seems inevitable that these capabilities will become commonplace on laptop computers and ultimately even on handheld devices.

20.12.1 General Issues in Information Coding

The greatest challenge in developing guidelines for information coding is that there are usually effective alternatives, such as color, shape, size, texture, blinking, orientation, and gray value. Although a number of studies compare one or more coding methods separately or in combination, there are so many interactions between the task and the complexity of the display that guidelines based on science are not generally practical. However, Tufte provides excellent guidelines for information coding from an aesthetic perspective [Tufte, ; Tufte, ; Tufte, 1997]. For further discussion, examples, and case studies, see also Ware, 2000 and Card et al., 1999.

A theoretical concept known as *preattentive discrimination* has interesting implications for whether or not the coding used can be processed in parallel by the visual system. The fact that certain coding schemes are processed faster than others is called the *popout* phenomenon, and this is thought to be due to early preattentive processing by the visual system. Thus, for example, the shape of the word *bold* is not processed preattentively, and it will be necessary to scan this entire page to determine how many times the word appears. However, if all of the instances of the word **bold** are emphasized, they pop out at the viewer. This is true as long as there are not too many other emphasized words on the same page: if there are fewer than seven or so instances, they can be processed at a single glance. Preattentive processing is done for color, brightness, certain aspects of texture, stereo disparities, and object orientation and size. Codes that are preattentively discriminable are very useful if rapid search for information is desired [Triesman, 1985]. The following visual attributes are known to be preattentive codes and, therefore, useful in differentiating information belonging to different classes:

Color — Use no more than ten different colors for labeling purposes.
Orientation — Use no more than ten orientations.
Blink coding — Use no more than two blink rates.
Texture granularity — Use no more than five grain sizes.
Stereo depth — The number of depths that can be effectively coded is not known.
Motion — Objects moving out of phase with one another are perceptually grouped. The number of usable phases is not known.

Note that coding multiple dimensions by combining different popout cues is not necessarily effective [Ware, 2000].

20.12.2 Color Information Coding

When considering information display, one of the most important distinctions is between chromatic and luminance information, because these are treated quite differently in human perception. Gray scales are not perceived in the same way as rainbow-colored scales. A purely chromatic difference is one where two colors of identical luminance, such as red and green, are placed adjacent to one another. Research has shown that we are insensitive to a variety of information if it is presented through purely chromatic changes. This includes shape perception, stereo depth information, shape from shading, and motion. However, chromatic information helps us to classify the material properties of objects. A number of practical implications arise from the differences in the ways luminance information and chromatic information are processed in human vision:

  Our spatial sensitivity is lower for chromatic information, allowing image compression techniques to transmit less information about hue relative to luminance.

  To make text visible, it is important to make sure that there is a luminance difference between the color of the text and the color of the background. If the background may vary, it is a good idea to put a contrasting border around the letters (e.g., Harrison and Vicente, 1996).

  When spatial layout is shown either through a stereo display or through motion cues, ensure adequate luminance contrast.

  When fine detail must be shown, for example, with fine lines in a diagram, ensure that there is adequate luminance contrast with the background.

  Chromatic codes are useful for labeling objects belonging to similar classes.

  Color (both chromatic and gray scale) can be used as a quantitative code, such as on maps, where it commonly encodes height and depth. However, simultaneous contrast effects can change the appearance of a patch of color depending on the surrounding colors; careful selection of colors can minimize this [Ware, 1988].

A number of empirical studies have shown color coding to be an effective way to identify information. It is also effective if used in combination with other cues, such as shape. For example, users may respond to targets faster if the targets can be identified by both shape and color differences (for useful reviews, see Christ, 1975; Stokes et al., 1990; and Silverstein, 1977). Color codes are also useful in the perceptual grouping of objects. Thus, the relationship between a set of different screen objects can be made more apparent by giving them all the same color. However, it is also the case that only a limited number of color codes can be used effectively. The use of more than about ten will cause the color categories to become blurred. In general, there are complex relationships between the type of symbols displayed (e.g., point, line, area, or text), the luminance of the display, the luminance and color of the background, and the luminance and color of the symbol [Spiker et al., 1985].

20.12.3 Integrated Control/Display Objects

When the purpose of a display is to allow a user to integrate diverse pieces of information, it may make sense to integrate the information into a single visual object or *glyph* [Wickens, 1992]. For example, if the purpose is to represent a pump, the liquid temperature could be shown by changing the color of the pump, the capacity could be shown by the overall size of the pump, and the output pressure might be represented by the changing height of a bar attached to the output pipe, rather than a set of individual dials showing these attributes separately. However, perceptual distortions can result from an ill chosen display mapping, and the object display may introduce visual clutter: if there are 50 pumps to control, then the outlines of all the pumps may interfere with the data of interest [Tufte, 1983]. In object displays, input and output can be integrated in a manner analogous to widgets such as the scroll bar, or even more directly by having input devices that resemble the physical object being handled, known as a *prop* [Hinckley et al., 1994a]. For some good examples of the linking of output and input, see Ahlberg and Shneiderman, 1994; Ishii and Ullmer, 1997.

  This style of presentation and interaction can be especially relevant for telepresence or augmented reality applications, where the user interacts with actual physical objects with attributes that must be viewed and controlled [Tani et al., 1992; Feiner et al., 1993]. For more abstract data representation tasks, choosing the color, size, orientation, or texture to represent a particular data attribute may be difficult, and there seem to be practical limits on the number of attributes that one can encode simultaneously. Thus, object displays usually must be custom designed for each display problem. In general, this means that the display and controls should somehow match the user's cognitive model of the task [Norman, 1990; Cole, 1986; Hinckley et al., 1994a].

20.12.4 Three-Dimensional Graphics and Virtual Environments

Much research in three-dimensional information visualization and virtual environments is motivated by the observation that humans naturally operate in physical space and can intuitively move about and remember where things are (an ability known as *spatial memory*). However, translating these potential benefits into artificially generated graphical environments is difficult because of limitations in display and interaction technologies. Virtual environments research pushed this to the limit by totally immersing the user in an artificial world of graphics, but this comes at the cost of visibility and awareness of colleagues and objects in the real world. This has led to research in so-called *fish-tank virtual reality* displays, using a head tracking system in conjunction with a stereo display [Deering, 1992; Ware et al., 1993] or a mirrored setup, which allows superimposition of graphics onto the volume where the user's hands are located [Schmandt, 1983; Serra et al., 1997]. However, much of our ability to navigate without becoming lost depends on the vestibular system and spatial updating as we physically turn our bodies, neither of which is engaged with stationary displays [Loomis et al., 1999; Chance et al., 1998]. For further discussion of navigation in virtual environments, see Darken and Sibert, 1995; Darken and Sibert, 1993. For application of spatial memory to three-dimensional environments, see Robertson et al., 1998; Robertson et al., 1999.

20.12.5 Augmented Reality

Augmented reality superimposes information on the surrounding environment rather than blocking it out. For example, the user may wear a semitransparent display that has the effect of projecting labels and diagrams onto objects in the real world. It has been suggested that this may be useful for training people to use complex systems or for fault diagnosis. For example, a person repairing an aircraft engine could see the names and functions of parts, which could appear superimposed on the parts seen through the display, together with a maintenance record if desired [Caudell and Mizell, 1992; Feiner et al., 1993]. The computer must obtain a detailed model of the environment; otherwise it is not possible to match the synthetic objects with the real ones. Even with this information, correct registration of computer graphics with the physical environment is an extremely difficult technical problem due to measurement error and system latency. This technology has been applied to heads-up displays for fighter aircraft, with semitransparent information about flight paths and various threats in the environment projected on the screen in front of the pilot [Stokes et al., 1990]. It also has been applied to digitally augmented desk surfaces [Wellner, 1993].

20.13 Scale in Displays

It is important to consider the full range of scale for display devices and form-factors that may embody an interaction task. Computing devices increasingly span orders of magnitude in size and computational resources, from watches and handheld personal data assistants (PDAs) to tablet computers and desktop computers, all the way up to multiple-monitor and wall-size displays. A technique that works well on a desktop computer, such as a pull-down menu, may be awkward on a handheld device or even unusable on a wall-size display (where the top of the display may not be within the user's reach). Each class of device seems to raise unique challenges. The best approach may ultimately be to design special-purpose, appliancelike devices (see Want and Borriello, 2000 for a survey) that suit specific purposes.

20.13.1 Small Displays

Users increasingly want to do more and more on handheld devices, mobile phones, pagers, and watches that offer less and less screen real estate. Researchers have investigated various strategies for conserving screen real estate. Transparent overlays allow divided attention between foreground and background layers [Harrison et al., 1995a; Harrison et al., 1995b; Harrison and Vicente, 1996; Kamba et al., 1996], but some degree of interference seems inevitable. This can be combined with sensing which elements of an interface are being used, such as presenting widgets on the screen only when the user is touching a pointing device [Hinckley and Sinclair, 1999]. Researchers also have experimented with replacing graphical interfaces with graspable interfaces, which respond to tilting, movement, and physical gestures and do not need constant on-screen representations [Fitzmaurice et al., 1995; Rekimoto, 1996; Harrison et al., 1998; Hinckley et al., 2000]. Much research in focus plus context techniques, including fisheye magnification [Bederson, 2000] and zooming metaphors [Perlin and Fox, 1993; Bederson et al., 1996; Smith and Taivalsaari, 1999], has also been motivated by providing more space than the boundaries of the physical screen can provide. Researchers have started to identify principles and quantitative models to analyze the trade-offs between multiple views and zooming techniques [Baudisch et al., 2002; Plumlee and Ware, 2002]. There has been considerable effort devoted to supporting Web browsing in extremely limited screen space (e.g., Buyukkokten et al., 2001; Jones et al., 1999; Trevor et al., 2001).

20.13.2 Multiple Displays

Some very interesting issues arise, however, when multiple displays are considered. Having multiple monitors for a single computer is not like having one large display [Grudin, 2001]. Users employ the boundary between displays to partition their tasks: one monitor is reserved for a primary task, and other monitors are used for secondary tasks. Secondary tasks may support the primary task (e.g., reference material, help files, or floating tool palettes), may provide peripheral awareness of ongoing events (such as an e-mail client) or may provide other background information (to-do lists, calendars, etc.). Switching between applications has a small time penalty (incurred once to switch, and again to return). Perhaps more importantly, it may distract the user or force the user to remember information while switching between applications. Having additional screen space "with a dedicated purpose, always accessible with a glance" [Grudin, 2001] reduces these burdens, and studies suggest that providing multiple, distinct foci for interaction may aid users' memory and recall [Tan et al., 2001; Tan et al., 2002]. Finally, small displays can be used in conjunction with larger displays [Myers et al., 1998; Myers et al., 2000; Rekimoto, 1998], with controls and private information on the small device and shared public information on the larger display. Displays of different dimensions support completely different user activities and social conventions.

20.13.3 Large-Format Displays

Trends in display technology suggest that large-format displays will become increasingly affordable and common. A journal's recent special issue includes numerous articles on implementing large-format displays using projection, application design for large displays, and applications such as automotive design [Funkhouser and Li, 2000]. Large displays often implicitly suggest multiple simultaneous users, with many applications revolving around collaboration [Swaminathan and Sato, 1997; Funkhouser and Li, 2000] and giving a large-scale physical presence to virtual activities [Buxton et al., 2000]. To support input directly on whiteboard-size displays, researchers have explored gestural interaction techniques for pens or touchscreens [Guimbretiere et al., 2001; Moran et al., 1997]. Many technologies cannot handle more than one point of contact, so check this carefully if simultaneous use by multiple persons is desired.

Large displays also seem to lend themselves to interaction at a distance, although using laser pointers to support such interaction [Olsen and Nielsen, 2001] has met with mixed success due to the lack of separation between tracking and dragging states [Buxton, 1990b]. Using handheld devices to interact with the full area of a large display also is problematic, because the ratio of the display to the control surface may be very large [Myers et al., 1998]. Environmentally situated ambient displays share some properties of large displays but emphasize subtle presentation of information in the periphery of attention [Ishii and Ullmer, 1997; Wisneski et al., 1998]. A 3-D environment using multiple large-format displays that surround the user is known as a *cave* [Cruz-Neira et al., 1993; Buxton and Fitzmaurice, 1998].

Unless life-size viewing of large objects is necessary [Buxton et al., 2000], in general the performance benefits a single large display vs. multiple monitors with the same screen area partitioned by bezels is not yet clear. One recent study suggests that the increased field of view afforded by large-format displays can lead to improved 3-D navigation performance, especially for women [Czerwinski et al., 2002].

20.14   Force and Tactile Displays

Haptic feedback research has sought to provide an additional channel of sensory feedback by synthesizing forces on the skin of the operator. The touch sensation is extraordinarily complex. In fact, the sense of "touch" is a very imprecise term: it includes multiple sensory systems, including sensitivity to pressure, small shear forces in the skin, heat and cold, pain, kinesthesis and proprioception, and the vestibular system [Burdea, 1996].

There appears to be no physical means by which a complex tactile stimulus can be delivered except in a very localized way. As a result, most haptic feedback devices are limited to simulation of a single point of contact, analogous to feeling the world with the tip of a pencil, although a few examples of whole-hand force feedback devices exist [Iwata, 1990; Burdea, 1996]. Efforts in haptic feedback include *force* feedback (active presentation of forces to the user) and *tactile* feedback (active presentation of vibrotactile stimuli to the user). Haptic feedback is popular for gaming devices, such as force feedback steering wheels and joysticks, but general-purpose pointing devices with force or tactile feedback remain uncommon. For a comprehensive discussion of force and tactile feedback technologies and techniques, as well as perceptual properties of the skin and joints, see [Burdea, 1996].

Adding force feedback to a mouse or stylus may impose constraints on the mechanical design, since a physical linkage is typically needed to reflect true forces. This may prevent a force feedback mouse from functioning like a traditional mouse, as it may limit range of motion or preclude clutching by lifting the device. Some devices instead increase resistance between the mouse and the pad, but this prevents simulation of hard contact forces. One can also use a vibrotactile stimulus, such as a vibrating pin under the mouse button or a vibrating shaft in an isometric joystick [Campbell et al., 1999]. Combination devices also have been explored [Akamatsu and Mackenzie, 1996]. Vibrotactile feedback seems especially promising for small mobile devices, for example, to provide the user with feedback of command recognition when the user's attention may not be focused on the screen [Poupyrev et al., 2002]. Applications for remote controls and augmented handles also look promising [Snibbe and MacLean, 2001; MacLean et al., 2000].

Using force feedback to provide attractive forces that pull the user toward a target, or tactile feedback to provide additional feedback for the boundaries of the target, has been found to yield modest speed improvements in some target acquisition experiments, although error rates may also increase [Akamatsu and MacKenzie, 1996; MacKenzie, 1995]. However, there have been almost no published studies about tasks where multiple targets are present, as on a computer screen with many icons and menus. Haptic feedback for one target may interfere with the selection of another, unless one uses techniques such as reducing the haptic forces during rapid motion [Oakley et al., 2001]. Finally, one should also consider whether software constraints, such as snap-to grids, are sufficient to support the user's tasks.

The construction of force output devices is extremely technically demanding. They must be stiff in order to be able to create the sensation of solid contact, yet light so that they have little inertia themselves, and there must be a tight loop between input (position) and output (force). Sigoma [1993] has suggested that having this loop iterate at 5 kHz may be necessary for optimal fine motor control. It has been shown that force feedback improves performance in certain telerobotic applications, such as inserting a peg into a hole [Sheridan, 1992]. The most promising applications of force output seem to appear in applications in which simulation of force is essential, such as surgical simulation and telerobotics [Burdea, 1996].

Another fundamental challenge for haptic feedback techniques results from the interaction between the haptic and visual channels. Visual dominance deals with phenomena resulting from the tendency for vision to dominate other modalities [Wickens, 1992]. Campbell et al. [1999] show that tactile feedback improves steering through a narrow tunnel, but only if the visual texture matches the tactile texture; otherwise, tactile feedback harms performance.

20.15 Auditory Displays

Here, we consider computer-generated auditory feedback. Audio can consist of synthesized or recorded speech. All other audio feedback is known as *nonspeech audio*. With stereo speakers or a stereo headset, either type of audio can be presented so that it seems to come from a specific 3-D location around the user, known as *spatialized audio*. For speech input and technology-mediated human–human communication applications that treat stored voice as data, see Section 20.8.1.

20.15.1 Nonspeech Audio

Nonspeech auditory feedback is prevalent in video games but largely absent from interaction with computing devices. Providing an auditory echo of the visual interface has little or no practical utility and may annoy users. Audio should be reserved to communicate simple, short messages that complement visual feedback (if any) and will not be referred to later. Furthermore, one or more of the following conditions should hold [Deatherage, 1972; Buxton, 1995b]:

The message should deal with events in time.

The message should call for immediate action.

The message should take place when the user's visual attention may be overburdened or directed elsewhere.

For example, researchers have attempted to enhance scrollbars using audio feedback [Brewster et al., 1994]. However, the meaning of such sounds may not be clear. Gaver advocates ecological sounds that resemble real-world events with an analogous meaning. For example, an empty disc drive might sound like a hollow metal container [Gaver, 1989]. If a long or complex message must be delivered using audio, it will likely be quicker and clearer to deliver it using speech output. Audio feedback may be crucial to support tasks or functionality on mobile devices that must take place when the user is not looking at the display (for some examples, see Hinckley et al., 2000).

Nonspeech sounds can be especially useful for attracting the attention of the user. Auditory alerting cues have been shown to work well, but only in environments where there is low auditory clutter. However, the number of simple nonspeech alerting signals is limited, and this can easily result in misidentification or cause signals to mask one another. An analysis of sound signals in fighter aircraft [Doll and Folds, 1985] found that the ground proximity warning and the angle-of-attack warning on an F16 were both an 800-Hz tone — a dangerous confound because these conditions require opposite responses from the pilot. It can also be difficult to devise nonspeech audio events that convey information without provoking an alerting response that unnecessarily interrupts the user. For example, this design tension arises when considering nonspeech audio cues that convey various properties of an incoming e-mail message [Sawhney and Schmandt, 2000; Hudson and Smith, 1996].

20.15.2 Speech Output

Speech auditory output is generally delivered either through recorded speech segments or completely synthetic speech (also known as *text-to-speech* technology). There has been considerable interest, especially for military applications, in the use of speech in providing warnings to the operators of complex systems. Speech can provide information to direct the operator's attention in a way that alarms cannot; an unfamiliar alarm simply indicates a problem, without telling the user the nature or context of the problem. Synthetic speech is most useful when visual information is not available, for example, in touch-tone telephone menu systems or in screen reader software for blind or low-sighted users. Although progress is being made, synthetic voices still sound somewhat unnatural and may be more difficult for users to understand.

Recorded speech is often used to give applications, particularly games, a more personal feel, but it can only be used for a limited number of responses known in advance.

The rate at which words must be produced to sound natural is a narrow range. For warning messages, 178 words per minute is intelligible but hurried, 123 words per minute is distracting and irritatingly slow, and a more natural rate of 156 words per minute is preferred [Simpson and Marchionda-Frost, 1984]. The playback rate of speech can be increased by overlapping sample times so that one sample is presented to one ear and another sample to the other ear. Technologies to correct for pitch distortions and to remove pauses have also been developed [Arons, 1993; Stifelman, 1996; Sawhney and Schmandt, 2000]. It is recommended by the U.S. Air Force that synthetic speech be 10 dB above ambient noise levels [Stokes et al., 1990].

20.15.3 Spatialized Audio Displays

It is possible to synthesize spatially localized sounds with a quality such that spatial localization in the virtual space is almost as good as localization of sounds in the natural environment [Wenzel, 1992]. Auditory localization appears to be primarily a two-dimensional phenomenon. That is, observers can localize, to some degree of accuracy, in horizontal position (*azimuth*) and elevation angle. Azimuth and elevation accuracies are of the order of 15°. As a practical consequence, this means that sound localization is of little use in identifying sources in conventional screen displays. Where localized sounds are really useful is in providing an orienting cue or warning about events occurring behind the user, outside of the field of vision.

There is also a well known phenomenon called *visual capture of sound*. Given a sound and an apparent visual source for the sound, such as a talking face on a cinema screen, the sound is perceived to come from the apparent source despite the fact that the actual source may be off to one side. Thus, visual localization tends to dominate auditory localization when both kinds of cues are present.

20.16 Future Directions

The future of interaction with computers will be both very different and very much like it is today. Some of our current tools, such as mice and keyboards, have evolved to suit interaction with desktop GUIs and rapid text entry. As long as users' work continues to involve desktop computers and text entry, we will continue to see these devices in use — not only because they are familiar, but also because they closely match human skills and the requirements of the task. Thus, devising new techniques that provide more efficient pointing at a display than a mouse, for example, is difficult to achieve [Card et al., 1978]. Speech recognition will allow new types of interaction and may enable interaction where it previously has been difficult or infeasible. However, we will continue to interact with computers using our hands and physical intermediaries, not necessarily because our technology requires us to do so, but because touching, holding, and moving physical objects is the foundation of the long evolution of tool use in the human species [Wilson, 1998].

But our computers and the tasks they serve are rapidly evolving. Current handheld devices have the display and computational capabilities of common desktop machines from ten years ago. What is lacking is new methods of interacting with such devices that uniquely suit mobile interaction, rather than derivatives of the desktop interface. Researchers are still actively exploring and debating the best ways to achieve this. Meanwhile, technology advances and economic trends continue to drive the cost of commodity displays lower and lower, while the limits of the technology continue to increase. Thus, we will continue to see new innovations in both very small and very large displays. As these become commonplace, new forms of interaction will become prevalent. Very small displays invariably seem to be incorporated into input/output appliances such as watches, pagers, and handheld devices, so interaction techniques for very small form-factors will become increasingly important.

The Internet and wireless networking seem to be the main disruptive technologies of the current era. Indeed, it seems likely that 100 years from now the phrase "wireless network" will seem every bit as antiquated as the phrase "horseless carriage" does today. Nobody really understands yet what it will mean for everything and everyone to be connected, but many researchers are working to explore the vision of ubiquitous computing originally laid out by Mark Weiser [Weiser, 1991]. Techniques that allow users to communicate and share information will become increasingly important. Biometric sensors or other convenient means for establishing identity will make services such as personalization of the interface and data sharing much simpler [Rekimoto, 1997; Sugiura and Koseki, 1998]. Techniques that combine dissimilar input devices and displays in interesting ways also will be important to realize the full potential of these technologies (e.g., Myers et al., 2001; Streitz et al., 1999). Electronic tagging techniques for identifying objects [Want et al., 1999] may also become commonplace. Such diversity of locations, users, and task contexts points to the increasing importance of sensors to acquire contextual information, as well as machine learning techniques to interpret them and infer meaningful actions [Buxton, 1995a; Bellotti et al., 2002; Hinckley et al., 2003]. This may well lead to an age of ubiquitous sensors [Saffo, 1997] with devices that can see, feel, and hear through digital perceptual mechanisms.

References

Accot, J., and S. Zhai (1997). Beyond Fitts' law: Models for trajectory-Based HCI tasks. *Proc. CHI '97: ACM Conference on Human Factors in Computing Systems.* 295–302.

Accot, J., and S. Zhai (1999). Performance evaluation of input devices in trajectory-based tasks: An application of the steering law. *Proc. CHI '99*, Pittsburgh, PA. 466–472.

Accot, J., and S. Zhai (2001). Scale effects in steering law tasks. *Proc. CHI 2001 ACM Conference on Human Factors in Computing Systems.* 1–8.

Accot, J., and S. Zhai (2002). More than dotting the i's — foundations for crossing-based interfaces. *ACM CHI 2002 Conf. on Human Factors in Computing Systems.* 73–80.

Ahlberg, C., and B. Shneiderman (1994). The alphaslider: a compact and rapid selector. *CHI '94.* 365–371.

Akamatsu, M., and I.S. Mackenzie (1996). Movement characteristics using a mouse with tactile and force feedback. *International Journal of Human–Computer Studies* **45**: 483–493.

Anderson, J.R. (1980). Chapter 8: Cognitive Skills. *Cognitive Psychology and Its Implications.* San Francisco, W. H. Freeman: 222–254.

Arons, B. (1993). SpeechSkimmer: interactively skimming recorded speech. *UIST '93 Symp. on User Interface Software and Technology.* 187–195.

Balakrishnan, R., T. Baudel, G. Kurtenbach, and G. Fitzmaurice (1997). The rockin' mouse: integral 3-D manipulation on a plane. *CHI '97 Conf. on Human Factors in Computing Systems.* 311–318.

Balakrishnan, R., and K. Hinckley (1999). The role of kinesthetic reference frames in two-handed input performance. *Proc. ACM UIST '99 Symp. on User Interface Software and Technology.* 171–178.

Balakrishnan, R., and K. Hinckley (2000). Symmetric bimanual interaction. *Proc. ACM CHI 2000 Conference on Human Factors in Computing Systems.* 33–40.

Balakrishnan, R., and G. Kurtenbach (1999). Exploring bimanual camera control and object manipulation in 3-D graphics interfaces. *Proc. CHI '99 ACM Conf. on Human Factors in Computing Systems.* 56–63.

Balakrishnan, R., and I.S. MacKenzie (1997). Performance differences in the fingers, wrist, and forearm in computer input control. *Proc. CHI '97 ACM Conf. on Human Factors in Computing Systems.* 303–310.

Baudel, T., and M. Beaudouin-Lafon (1993). Charade: Remote control of objects using hand gestures. *Communications of the ACM* **36**(7): 28–35.

Baudisch, P., N. Good, V. Bellotti, and P. Schraedley (2002). Keeping things in context: a comparative evaluation of focus plus context screens, overviews, and zooming. *CHI 2002 Conf. on Human Factors in Computing Systems.* 259–266.

Bederson, B. (2000). Fisheye menus. *Proc. ACM UIST 2000 Symposium on User Interface Software and Technology.* 217–226.

Bederson, B., J. Hollan, K. Perlin, J. Meyer, D. Bacon and G. Furnas (1996). Pad++: a zoomable graphical sketchpad for exploring alternate interface physics. *Journal of Visual Languages and Computing* 7:3–31.

Bellotti, V., M. Back, W.K. Edwards, R. Grinter, C. Lopes, and A. Henderson (2002). Making sense of sensing systems: five questions for designers and researchers. *Proc. ACM CHI 2002 Conference on Human Factors in Computing Systems*. 415–422.

Betrisey, C., J. Blinn, B. Dresevic, B. Hill, G. Hitchcock, B. Keely, D. Mitchell, J. Platt, and T. Whitted (2000). Displaced filtering for patterned displays. *Proc. Society for Information Display Symposium*. 296–299.

Bier, E., M. Stone, K. Pier, W. Buxton, and T. DeRose (1993). Toolglass and magic lenses: the see-through interface. *Proceedings of SIGGRAPH 93*, Anaheim, CA. 73–80.

Bolt, R. (1980). Put-that-there: voice and gesture at the graphics interface. *Computer Graphics* (Aug.): 262–270.

Brewster, S.A., P.C. Wright, and A.D.N. Edwards (1994). The design and evaluation of an auditory-enhanced scrollbar. *Proc. ACM CHI '94 Conference Proceedings on Human Factors in Computing Systems*. 173–179.

Britton, E., J. Lipscomb, and M. Pique (1978). Making nested rotations convenient for the user. *Computer Graphics* 12(3): 222–227.

Brooks, J., F.P. (1988). Grasping reality through illusion: interactive graphics serving science. *Proceedings of CHI '88: ACM Conference on Human Factors in Computing Systems*, Washington, DC, ACM, New York. 1–11.

Bukowski, R., and C. Sequin (1995). Object associations: A simple and practical approach to virtual 3-D manipulation. *ACM 1995 Symposium on Interactive 3-D Graphics*. 131–138.

Burdea, G. (1996). *Force and Touch Feedback for Virtual Reality*. New York, John Wiley and Sons.

Buxton, B., and G. Fitzmaurice (1998). HMDs, caves and chameleon: a human-centric analysis of interaction in virtual space. *Computer Graphics* 32(8): 69–74.

Buxton, W. (1983). Lexical and pragmatic considerations of input structure. *Computer Graphics* 17(1): 31–37.

Buxton, W. (1986). Chunking and phrasing and the design of human–computer dialogues. *Information Processing '86, Proc. of the IFIP 10th World Computer Congress*, Amsterdam, North Holland Publishers. 475–480.

Buxton, W. (1990a). The pragmatics of haptic input. *Proceedings of CHI '90: ACM Conference on Human Factors in Computing Systems, Tutorial 26 Notes*, Seattle, WA, ACM, New York.

Buxton, W. (1990b). A three-state model of graphical input. *Proc. INTERACT '90*, Amsterdam, Elsevier Science. 449–456.

Buxton, W. (1995a). Integrating the periphery and context: a new taxonomy of telematics. *Proceedings of Graphics Interface '95*. 239–246.

Buxton, W. (1995b). Speech, language and audition. *Readings in Human–Computer Interaction: Toward the Year 2000*, R. Baecker, J. Grudin, W. Buxton, and S. Greenberg, Eds. San Francisco, Morgan Kaufmann Publishers. 525–537.

Buxton, W., G. Fitzmaurice, R. Balakrishnan, and G. Kurtenbach (2000). Large displays in automotive design. *IEEE Computer Graphics and Applications* (July/August): 68–75.

Buxton, W., E. Fiume, R. Hill, A. Lee, and C. Woo (1983). Continuous hand-gesture driven input. *Proceedings of Graphics Interface '83*. 191–195.

Buxton, W., R. Hill, and P. Rowley (1985). Issues and techniques in touch-sensitive tablet input. *Computer Graphics* 19(3): 215–224.

Buxton, W., and B. Myers (1986). A study in two-handed input. *Proceedings of CHI '86: ACM Conference on Human Factors in Computing Systems*, Boston, MA, ACM, New York. 321–326.

Buyukkokten, O., H. Garcia-Molina, and A. Paepcke (2001). Accordion summarization for end-game browsing on PDAs and cellular phones. *ACM CHI 2001 Conf. on Human Factors in Computing Systems*, Seattle, WA. 213–220.

Cadoz, C. (1994). *Les réalités virtuelles*. Dominos, Flammarion.

Campbell, C., S. Zhai, K. May, and P. Maglio (1999). What you feel must be what you see: adding tactile feedback to the trackpoint. *Proceedings of INTERACT '99: 7th IFIP conference on Human–Computer Interaction*. 383–390.

Card, S., W. English, and B. Burr (1978). Evaluation of mouse, rate-controlled isometric joystick, step keys, and text keys for text selection on a CRT. *Ergonomics* **21**: 601–613.

Card, S., J. Mackinlay, and G. Robertson (1991). A morphological analysis of the design space of input devices. *ACM Transactions on Information Systems* **9**(2): 99–122.

Card, S., J. Mackinlay, and B. Shneiderman (1999). *Readings in Information Visualization: Using Vision to Think*. San Francisco, Morgan Kaufmann.

Cassell, J. (2003). A framework for gesture generation and interpretation. In *Computer Vision in Human-Machine Interaction*, R. Cipolla and A. Pentland, Eds. Cambridge, Cambridge University Press. (In press.)

Caudell, T.P., and D.W. Mizell (1992). Augmented reality: an application of heads-up display technology to manual manufacturing processes. *Proc. HICSS '92*. 659–669.

Chance, S., F. Gaunet, A. Beall, and J. Loomis (1998). Locomotion mode affects the updating of objects encountered during travel: The contribution of vestibular and proprioceptive inputs to path integration. *Presence* **7**(2): 168–178.

Chen, M., S.J. Mountford, and A. Sellen (1988). A study in interactive 3-D rotation using 2-D control devices. *Computer Graphics* **22**(4): 121–129.

Cholewiak, R., and A. Collins (1991). Sensory and physiological bases of touch. *The Psychology of Touch*, M. Heller and W. Schiff, Eds. Hillsdale, NJ, Lawrence Erlbaum. 23–60.

Christ, R.E. (1975). Review and analysis of color coding research for visual displays. *Human Factors* **25**: 71–84.

Cohen, P., M. Johnston, D. McGee, S. Oviatt, J. Pittman, I. Smith, L. Chen, and J. Clow (1997). QuickSet: multimodal interaction for distributed applications. *ACM Multimedia 97*. 31–40.

Cohen, P.R., and J.W. Sullivan (1989). Synergistic use of direct manipulation and natural language. *Proc. ACM CHI '89 Conference on Human Factors in Computing Systems*. 227–233.

Cole, W.G. (1986). Medical cognitive graphics. *ACM CHI '86 Conf. on Human Factors in Computing Systems*. 91–95.

Conner, D., S. Snibbe, K. Herndon, D. Robbins, R. Zeleznik, and A. van Dam (1992). Three-dimensional widgets. *Computer Graphics (Proc. 1992 Symposium on Interactive 3-D Graphics)*. 183–188, 230–231.

Cook, R.L. (1986). Stochastic sampling in computer graphics. *ACM Trans. Graphics* **5**(1): 51–72.

Cowan, W.B. (1983). An inexpensive calibration scheme for calibrations of a color monitor in terms of CIE standard coordinates. *Computer Graphics* **17**(3): 315–321.

Cruz-Neira, C., D. Sandin, and T. DeFanti (1993). Surround-screen projection-based virtual reality: The design and implementation of the CAVE. *Computer Graphics (SIGGRAPH Proceedings)*. 135–142.

Czerwinski, M., D. S. Tan, and G. G. Robertson (2002). Women take a wider view. *Proc. ACM CHI 2002 Conference on Human Factors in Computing Systems*, Minneapolis, MN. 195–202.

Darken, R.P., and J.L. Sibert (1993). A toolset for navigation in virtual environments. *Proc. ACM UIST '93. Symposium on User Interface Software and Technology*. 157–165.

Darken, R.P., and J.L. Sibert (1995). Navigating large virtual spaces. *International Journal of Human–Computer Interaction* (Oct.).

Deatherage, B.H. (1972). Auditory and other sensory forms of information presentation. In *Human Engineering Guide to Equipment Design*, H. Van Cott and R. Kinkade, Eds. U.S. Government Printing Office.

Deering, M. (1992). High-resolution virtual reality. *Computer Graphics* **26**(2): 195–202.

Dey, A., G. Abowd, and D. Salber (2001). A conceptual framework and a toolkit for supporting the rapid prototyping of context-aware applications. *Journal of Human–Computer Interaction* **16**(2–4): 97–166.

Doll, T.J., and D.J. Folds (1985). Auditory signals in military aircraft: ergonomic principles versus practice. *Proc. 3rd Symp. Aviation Psych*, Ohio State University, Dept. of Aviation, Colombus, OH. 111–125.

Douglas, S., A. Kirkpatrick, and I. S. MacKenzie (1999). Testing pointing device performance and user assessment with the ISO 9241, part 9 standard. *Proc. ACM CHI '99 Conf. on Human Factors in Computing Systems*. 215–222.

Douglas, S., and A. Mithal (1994). The effect of reducing homing time on the speed of a finger-controlled isometric pointing device. *Proc. ACM CHI '94 Conf. on Human Factors in Computing Systems*. 411–416.

Feiner, S., B. Macintyre, and D. Seligmann (1993). Knowlege-based augmented reality. *Communications of the ACM* **36**(7): 53–61.

Fitzmaurice, G., and W. Buxton (1997). An empirical evaluation of graspable user interfaces: toward specialized, space-multiplexed input. *Proceedings of CHI '97: ACM Conference on Human Factors in Computing Systems*, Atlanta, GA, ACM, New York. 43–50.

Fitzmaurice, G., H. Ishii, and W. Buxton (1995). Bricks: laying the foundations for graspable user interfaces. *Proceedings of CHI '95: ACM Conference on Human Factors in Computing Systems*, Denver, CO, ACM, New York. 442–449.

Fitzmaurice, G.W., R. Balakrishnan, and G. Kurtenbach (1999). Sampling, synthesis, and input devices. *Commun. ACM* **42**(8): 54–63.

Foley, J.D., V. Wallace, and P. Chan (1984). The human factors of computer graphics interaction techniques. *IEEE Computer Graphics and Applications* (Nov.): 13–48.

Freeman, W.T., and C. Weissman (1995). Television control by hand gestures. *Intl. Workshop on Automatic Face and Gesture Recognition*, Zurich, Switzerland. 179–183.

Funkhouser, T., and K. Li (2000). Large format displays. *IEEE Comput. Graphics Appl.* (July–Aug. special issue): 20–75.

Gaver, W. (1989). The SonicFinder: an interface that uses auditory icons. *Human–Computer Interaction* **4**(1): 67–94.

Gibson, J. (1962). Observations on active touch. *Psychological Review* **69**(6): 477–491.

Gibson, J. (1986). *The Ecological Approach to Visual Perception*. Hillsdale, NJ, Lawrence Erlbaum Assoc.

Goldberg, D., and C. Richardson (1993). Touch-typing with a stylus. *Proc. INTERCHI '93 Conf. on Human Factors in Computing Systems*. 80–87.

Grudin, J. (2001). Partitioning digital worlds: focal and peripheral awareness in multiple monitor use. *Proc. ACM CHI 2001 Conference on Human Factors in Computing Systems*. 458–465.

Guiard, Y. (1987). Asymmetric division of labor in human skilled bimanual action: the kinematic chain as a model. *The Journal of Motor Behavior* **19**(4): 486–517.

Guiard, Y., F. Buourgeois, D. Mottet, and M. Beaudouin-Lafon (2001). Beyond the 10-bit barrier: Fitts' law in multi-scale electronic worlds. *People and Computers XV: Interaction with on Fractions. Joint Proc. IHM 2001 and HCI 2001*. Springs. 573–581.

Guimbretiere, F., M.C. Stone, and T. Winograd (2001). Fluid interaction with high-resolution wall-size displays. *Proc. UIST 2001 Symp. on User Interface Software and Technology*. 21–30.

Harrison, B., K. Fishkin, A. Gujar, C. Mochon, and R. Want (1998). Squeeze me, hold me, tilt me! An exploration of manipulative user interfaces. *Proc. ACM CHI '98 Conf. on Human Factors in Computing Systems*. 17–24.

Harrison, B., H. Ishii, K. Vicente, and W. Buxton (1995a). Transparent layered user interfaces: an evaluation of a display design to enhance focused and divided attention. *Proceedings of CHI '95: ACM Conference on Human Factors in Computing Systems*. 317–324.

Harrison, B., G. Kurtenbach, and K. Vicente (1995b). An experimental evaluation of transparent user interface tools and information content. *Proc. ACM UIST '95*. 81–90.

Harrison, B., and K. Vicente (1996). An experimental evaluation of transparent menu usage. *Proceedings of CHI '96: ACM Conference on Human Factors in Computing Systems*. 391–398.

Hauptmann, A. (1989). Speech and gestures for graphic image manipulation. *Proceedings of CHI '89: ACM Conference on Human Factors in Computing Systems*, Austin, TX, ACM, New York. 241–245.

Hinckley, K., E. Cutrell, S. Bathiche, and T. Muss (2001). Quantitative analysis of scrolling techniques. *Proc. ACM CHI 2002. Conf. on Human Factors in Computing Systems*. 65–72.

Hinckley, K., M. Czerwinski, and M. Sinclair (1998a). Interaction and modeling techniques for desktop two-handed input. *Proceedings of the ACM UIST '98 Symposium on User Interface Software and Technology*, San Francisco, CA, ACM, New York. 49–58.

Hinckley, K., R. Pausch, J. Goble, and N. Kassell (1994a). Passive real-world interface props for neurosurgical visualization. *Proceedings of CHI '94: ACM Conference on Human Factors in Computing Systems*, Boston, MA, ACM, New York. 452–458.

Hinckley, K., R. Pausch, J.C. Goble, and N.F. Kassell (1994b). A survey of design issues in spatial input. *Proceedings of the ACM UIST '94 Symposium on User Interface Software and Technology*, Marina del Rey, CA, ACM, New York. 213–222.

Hinckley, K., R. Pausch, D. Proffitt, and N. Kassell (1998b). Two-handed virtual manipulation. *ACM Transactions on Computer–Human Interaction* 5(3): 260–302.

Hinckley, K., J. Pierce, E. Horvitz, and M. Sinclair (2003). Foreground and background interaction with sensor-enhanced mobile devices. *ACM TOCHI*. Special issue on sensor-based interaction (to appear).

Hinckley, K., J. Pierce, M. Sinclair, and E. Horvitz (2000). Sensing techniques for mobile interaction. *ACM UIST 2000 Symp. on User Interface Software and Technology*. 91–100.

Hinckley, K., and M. Sinclair (1999). Touch-sensing input devices. *ACM CHI '99 Conf. on Human Factors in Computing Systems*. 223–230.

Hinckley, K., M. Sinclair, E. Hanson, R. Szeliski, and M. Conway (1999). The VideoMouse: A camera-based multi-degree-of-freedom input device. *ACM UIST '99 Symp. on User Interface Software and Technology*. 103–112.

Hinckley, K., J. Tullio, R. Pausch, D. Proffitt, and N. Kassell (1997). Usability analysis of 3-D rotation techniques. *Proc. ACM UIST '97 Symp. on User Interface Software and Technology*, Banff, Alberta, Canada, ACM, New York. 1–10.

Hix, D., J. Templeman, and R. Jacob (1995). Pre-screen projection: from concept to testing of a new interaction technique. *Proc. ACM CHI '95*. 226–233.

Honan, M., E. Serina, R. Tal, and D. Rempel (1995). Wrist postures while typing on a standard and split keyboard. *Proc. HFES Human Factors and Ergonomics Society 39th Annual Meeting*. 366–368.

Horvitz, E., J. Breese, D. Heckerman, D. Hovel, and K. Rommelse (1998). The Lumiere Project: Bayesian user modeling for inferring the goals and needs of software users. *Proceedings of the Fourteenth Conference on Uncertainty in Artificial Intelligence*, Madison, WI, San Francisco, Morgan Kaufmann. 256–265.

Horvitz, E., A. Jacobs, and D. Hovel (1999). Attention-sensitive alerting. *Proceedings of UAI '99, Conference on Uncertainty and Artificial Intelligence*. 305–313.

Hudson, S., and I. Smith (1996). Electronic mail previews using non-speech audio. *CHI '96 Companion Proceedings*. 237–238.

Igarashi, T., S. Matsuoka, and H. Tanaka (1999). Teddy: a sketching interface for 3-D freeform design. *ACM SIGGRAPH '99*, Los Angeles, CA. 409–416.

Ishii, H., and B. Ullmer (1997). Tangible bits: Toward seamless interfaces between people, bits, and atoms. *Proceedings of CHI '97: ACM Conference on Human Factors in Computing Systems*, Atlanta, GA, ACM, New York. 234–241.

Iwata, H. (1990). Artificial reality with force-feedback: development of desktop virtual space with compact master manipulator. *Computer Graphics* 24(4): 165–170.

Jacob, R. (1991). The use of eye movements in human–computer interaction techniques: what you look at is what you get. *ACM Transactions on Information Systems* 9(3): 152–169.

Jacob, R., L. Sibert, D. McFarlane, and M. Mullen, Jr. (1994). Integrality and separability of input devices. *ACM Transactions on Computer–Human Interaction* **1**(1): 3–26.

Jellinek, H., and S. Card (1990). Powermice and user performance. *Proc. ACM CHI '90 Conf. on Human Factors in Computing Systems*. 213–220.

Jojic, N., B. Brumitt, B. Meyers, and S. Harris (2000). Detecting and estimating pointing gestures in dense disparity maps. *Proceed. of IEEE Intl. Conf. on Automatic Face and Gesture Recognition*. 468.

Jones, M., G. Marsden, N. Mohd-Nasir, K. Boone, and G. Buchanan (1999). Improving Web interaction on small displays. *Computer Networks* **31**(11–16): 1129–1137.

Kabbash, P., W. Buxton, and A. Sellen (1994). Two-handed input in a compound task. *Proceedings of CHI '94: ACM Conference on Human Factors in Computing Systems*, Boston, MA, ACM, New York. 417–423.

Kamba, T., S.A. Elson, T. Harpold, T. Stamper, and P. Sukaviriya (1996). Using small screen space more efficiently. *Conference Proceedings on Human Factors in Computing Systems*. 383.

Karat, C., C. Halverson, D. Horn, and J. Karat (1999). Patterns of entry and correction in large vocabulary continuous speech recognition systems. *Proc. ACM CHI '99 Conf. on Human Factors in Computing Systems*. 568–575.

Kirsh, D. (1995). Complementary strategies: why we use our hands when we think. *Proceedings of 7th Annual Conference of the Cognitive Science Society*. Hillsdale, NJ, Lawrence Erlbaum. 212–217.

Kirsh, D., and P. Maglio (1994). On distinguishing epistemic from pragmatic action. *Cognitive Science* **18**(4): 513–549.

Kramer, A. (1994). Translucent patches — dissolving windows. *Proc. ACM UIST '94 Symp. on User Interface Software and Technology*. 121–130.

Kurtenbach, G., and W. Buxton (1991). Issues in combining marking and direct manipulation techniques. *Proc. UIST '91*. 137–144.

Kurtenbach, G., and W. Buxton (1993). The limits of expert performance using hierarchic marking menus. *Proc. INTERCHI '93*. 482–487.

Kurtenbach, G., G. Fitzmaurice, T. Baudel, and B. Buxton (1997). The design of a GUI paradigm based on tablets, two-hands, and transparency. *Proceedings of CHI '97: ACM Conference on Human Factors in Computing Systems*, Atlanta, GA, ACM, New York. 35–42.

Kurtenbach, G., A. Sellen, and W. Buxton (1993). An emprical evaluation of some articulatory and cognitive aspects of "marking menus." *Journal of Human–Computer Interaction* **8**(1).

Lewis, J., K. Potosnak, and R. Magyar (1997). Keys and keyboards. In *Handbook of Human–Computer Interaction*, M. Helander, T. Landauer, and P. Prabhu, Eds., Amsterdam, North-Holland. 1285–1316.

Lipscomb, J., and M. Pique (1993). Analog input device physical characteristics. *SIGCHI Bulletin* **25**(3): 40–45.

Loomis, J., R.L. Klatzky, R.G. Golledge, and J.W. Philbeck (1999). Human navigation by path integration. In *Wayfinding: Cognitive Mapping and Other Spatial Processes*, R.G. Golledge, Ed. Baltimore, Johns Hopkins. 125–151.

Lucente, M., G. Zwart, and A. George (1998). Visualization space: a testbed for deviceless multimodal user interface. *AAAI '98*.

MacKenzie, I.S. (1992). Fitts' law as a research and design tool in human-computer interaction. *Human–Computer Interaction* **7**: 91–139.

MacKenzie, I.S. (1995). Input devices and interaction techniques for advanced computing. In *Virtual Environments and Advanced Interface Design*, W. Barfield and T. Furness, Eds. Oxford, Oxford University Press. 437–470.

MacKenzie, I.S., T. Kauppinen, and M. Silfverberg (2001). Accuracy measures for evaluating computer pointing devices. *CHI 2001*. 9–16.

MacKenzie, I.S., and A. Oniszczak (1998). A comparison of three selection techniques for touchpads. *Proc. ACM CHI '98 Conf. on Human Factors in Computing Systems*. 336–343.

MacKenzie, I.S., A. Sellen, and W. Buxton (1991). A comparison of input devices in elemental pointing and dragging tasks. *Proc. ACM CHI '91 Conf. on Human Factors in Computing Systems*. 161–166.

MacKenzie, I.S., and R.W. Soukoreff (2002). A model of two-thumb text entry. *Proceedings of Graphics Interface*, Toronto, Canadian Information Processing Society. 117–124.

MacKenzie, I.S., and C. Ware (1993). Lag as a determinant of human performance in interactive systems. *Proc. ACM INTERCHI '93*. 488–493.

MacKenzie, I.S., and S. Zhang (1997). The immediate usability of graffiti. *Proc. Graphics Interface '97*. 129–137.

MacLean, K.E., S.S. Snibbe, and G. Levin (2000). Tagged handles: merging discrete and continuous control. *Proc. ACM CHI 2000 Conference on Human Factors in Computing Systems*, The Hague, Netherlands.

Maes, P., T. Darrell, B. Blumberg, and A. Pentland (1996). The ALIVE system: wireless, full-body interaction with autonomous agents. *ACM Multimedia Systems* (Special Issue on Multimedia and Multisensory Virtual Worlds).

Marklin, R., and G. Simoneau (1996). Upper extremity posture of typists using alternative keyboards. *ErgoCon '96*. 126–132.

Marklin, R., G. Simoneau, and J. Monroe (1997). The effect of split and vertically inclined computer keyboards on wrist and forearm posture. *Proc. HFES Human Factors and Ergonomics Society 41st Annual Meeting*. 642–646.

Mathias, E., I.S. MacKenzie, and W. Buxton (1996). One-handed touch typing on a QWERTY keyboard. *Human–Computer Interaction* 11(1): 1–27.

McGuffin, M., and R. Balakrishnan (2002). Acquisition of expanding targets. *CHI Letters* 4(1).

Mine, M., F. Brooks, and C. Sequin (1997). Moving objects in space: exploiting proprioception in virtual-environment interaction. *Computer Graphics* 31(Proc. SIGGRAPH '97): 19–26.

Moran, T., P. Chiu, and W. van Melle (1997). Pen-based interaction techniques for organizing material on an electronic whiteboard. *Proc. ACM UIST '97 Symp. on User Interface Software and Technology*. 45–54.

Myers, B., K. Lie, and B. Yang (2000). Two-handed input using a PDA and a mouse. *CHI 2000*. 41–48.

Myers, B., R. Miller, C. Evankovich, and B. Bostwick (2001). Individual use of hand-held and desktop computers simultaneously. (Submitted).

Myers, B., H. Stiel, and R. Gargiulo (1998). Collaboration using multiple PDAs connected to a PC. *Proc. ACM CSCW '98 Conf. on Computer Supported Cooperative Work*, Seattle, WA. 285–294.

Mynatt, E.D., T. Igarashi, W.K. Edwards, and A. LaMarca (1999). Flatland: new dimensions in office whiteboards. *ACM SIGCHI Conference on Human Factors in Computing Systems*, Pittsburgh, PA. 346–353.

Nguyen, D.H., and E. Mynatt (2001). Toward visibility of a ubicomp environment.

Norman, D. (1990). *The Design of Everyday Things*. New York, Doubleday.

Noyes, J. (1983). Chord keyboards. *Applied Ergonomics* 14: 55–59.

Oakley, I., S. Brewster, and P. Gray (2001). Solving multi-target haptic problems in menu interaction. *Proc. ACM CHI 2001 Conf. on Human Factors in Computing Systems: Extended Abstracts*. 357–358.

Olsen, D.R., and T. Nielsen (2001). Laser pointer interaction. *Proc. ACM CHI 2001 Conf. on Human Factors in Computing Systems*. 17–22.

Oviatt, S. (1997). Multimodal interactive maps: Designing for human performance. *Human–Computer Interaction* 12: 93–129.

Pearson, G., and M. Weiser (1988). Exploratory evaluation of a planar foot-operated cursor-positioning device. *Proc. ACM CHI '88 Conference on Human Factors in Computing Systems*. 13–18.

Pekelney, R., and R. Chu (1995). Design criteria of an ergonomic mouse computer input device. *Proc. HFES Human Factors and Ergonomics Society 39th Annual Meeting*. 369–373.

Perlin, K., and D. Fox (1993). Pad: an alternative approach to the computer interface. *SIGGRAPH '93*.

Platt, J. (2000). Optimal filtering for patterned displays. *IEEE Signal Processing Letters* 7(7): 179–83.

Plumlee, M., and C. Ware (2002). Modeling performance for zooming vs. multi-window interfaces based on visual working memory. *AVI '02: Advanced Visual Interfaces*, Trento Italy.

Poupyrev, I., S. Maruyama, and J. Rekimoto (2002). Ambient touch: designing tactile interfaces for hand-held devices. *UIST 2002 Symp. on User Interface Software and Technology*. 51–60.

Putz-Anderson, V. (1988). *Cumulative Trauma Disorders: a Manual for Musculoskeletal Diseases of the Upper Limbs*. Bristol, PA, Taylor and Francis.

Rekimoto, J. (1996). Tilting operations for small screen interfaces. *Proc. ACM UIST '96*. 167–168.

Rekimoto, J. (1997). Pick-and-drop: a direct manipulation technique for multiple computer environments. *Proc. ACM UIST '97*. 31–39.

Rekimoto, J. (1998). A multiple device approach for supporting whiteboard-based interactions. *CHI '98*. 344–351.

Rime, B., and L. Schiaratura (1991). Gesture and speech. In *Fundamentals of Nonverbal Behavior*, R.S. Feldman and B. Rimé, Eds., New York, Press Syndicate of the University of Cambridge. 239–281.

Robertson, G., M. Czerwinski, K. Larson, D. Robbins, D. Thiel, and M. van Dantzich (1998). Data Mountain: using spatial memory for document management. *UIST '98*.

Robertson, G., M. van Dantzich, D. Robbins, M. Czerwinski, K. Hinckley, K. Risden, V. Gorokhovsky, and D. Thiel (1999). The task gallery: a 3-D window manager. To appear in *ACM CHI 2000*.

Robertson, G.G., S.K. Card, and J.D. Mackinlay (1989). The cognitive coprocessor architecture for interactive user interfaces. *Proc. UIST '89 Symposium on User Interface Software and Technology*. 10–18.

Robertson, P.K. (1988). Perceptual color spaces. Visualizing color gamuts: a user interface for the effective use of perceptual color spaces in data display. *IEEE Comput. Graphics Appl.* **8**(5): 50–64.

Rutledge, J., and T. Selker (1990). Force-to-motion functions for pointing. *Proc. of Interact '90: The IFIP Conf. on Human–Computer Interaction*. 701–706.

Saffo, P. (1997). Sensors: The next wave of infotech innovation. *Institute for the Future: 1997 Ten-Year Forecast*. 115–122.

Sawhney, N., and C. Schmandt (1999). Nomadic radio: scaleable and contextual notification for wearable audio messaging. *CHI '99*. 96–103.

Sawhney, N., and C.M. Schmandt (2000). Nomadic radio: speech and audio interaction for contextual messaging in nomadic environments. *ACM Transactions on Computer–Human Interaction* **7**(3): 353–383.

Schilit, B.N., N.I. Adams, and R. Want (1994). Context-aware computing applications. *Proc. IEEE Workshop on Mobile Computing Systems and Applications*, Santa Cruz, CA, IEEE Computer Society. 85–90.

Schmandt, C.M. (1983). Spatial input/display correspondence in a stereoscopic computer graphic work station. *Computer Graphics (Proc. ACM SIGGRAPH '83)* **17**(3): 253–262.

Schmandt, C.M. (1993). From desktop audio to mobile access: opportunities for voice in computing. *Advances in Human–Computer Interaction*, H.R. Hartson and D. Hix, Eds. Norwood, NJ, Ablex. 251–283.

Schmidt, A., K. Aidoo, A. Takaluoma, U. Tuomela, K. Van Laerhove, and W. Van de Velde (1999). Advanced interaction in context. *Handheld and Ubiquitous Computing (HUC'99)*. Heidelberg, Springer-Verlag: 89–101.

Sears, A. (1993). Investigating touchscreen typing: the effect of keyboard size on typing speed. *Behaviour and Information Technology* **12**(1): 17–22.

Sears, A., C. Plaisant, and B. Shneiderman (1992). A new era for high precision touchscreens. *Advances in Human–Computer Interaction* **3**: 1–33.

Sears, A., and B. Shneiderman (1991). High precision touchscreens: design strategies and comparisons with a mouse. *International Journal of Man–Machine Studies* **34**(4): 593–613.

Sellen, A., G. Kurtenbach, and W. Buxton (1992). The prevention of mode errors through sensory feedback. *Human–Computer Interaction* **7**(2): 141–164.

Serra, L., N. Hern, C. Beng Choon, and T. Poston (1997). Interactive vessel tracing in volume data. *ACM/SIGGRAPH Symposium on Interactive 3-D Graphics*, Providence, RI, ACM, New York. 131–137.

Sheridan, T.B. (1992). *Telerobotics, Automation, and Human Supervisory Control*. Cambridge, MA, MIT Press.

Sigoma, K.B. (1993). A survey of perceptual feedback issues in dexterous telemanipulation: part I. Finger force feedback. *Proc. IEEE Virtual Reality Annu. Int. Symp.* 263–270.

Silverstein, D. (1977). Human factors for color display systems. *Color and the Computer: Concepts, Methods and Research*. New York, Academic Press. 27–61.

Simpson, C.A., and K. Marchionda-Frost (1984). Synthesized speech rate and pitch effects on intelligibility of warning messages for pilots. *Human Factors* 26: 509–517.

Smailagic, A., and D. Siewiorek (1996). Modalities of interaction with CMU wearable computers. *IEEE Personal Communications* (Feb.): 14–25.

Smith, R.B., and A. Taivalsaari (1999). Generalized and stationary scrolling. *Proc. UIST '99: CHI Letters* 1(1): 1–9.

Snibbe, S., and K. MacLean (2001). Haptic techniques for media control. *CHI Letters (Proc. UIST 2001)* 3(2): 199–208.

Spiker, A., S. Rogers, and J. Cicinelli (1985). Selecting color codes for a computer-generated topographic map based on perception experiments and functional requirements. *Proc. 3rd Symp. Aviation Psychology*, Ohio State University, Dept. of Aviation, Columbus, OH. 151–158.

Stifelman, L. (1996). Augmenting real-world objects: a paper-based audio notebook. *CHI '96 Conference Companion*. 199–200.

Stokes, A., C. Wickens, and K. Kite (1990). *Display Technology: Human Factors Concepts*. Warrendale, PA, SAE.

Stone, M.C., W.B. Cowan, and J.C. Beatty (1988). Color gamut mapping and the printing of digital color images. *ACM Trans. Graphics* 7(4): 249–292.

Stratton, G. (1897). Vision without inversion of the retinal image. *Psychological Review* 4: 360–361.

Streitz, N.A., J. Geißler, T. Holmer, S. Konomi, C. Müller-Tomfelde, W. Reischl, P. Rexroth, R.P. Seitz, and Steinmetz (1999). i-LAND: an interactive landscape for creativity and innovation. *ACM CHI '99 Conf. on Human Factors in Computing Systems*, Pittsburgh, PA. 120–127.

Sugiura, A., and Y. Koseki (1998). A user interface using fingerprint recognition — holding commands and data objects on fingers. *UIST '98 Symp. on User Interface Software and Technology*. 71–79.

Sutherland, I.E., (1968). A head-mounted three-dimensional display. *Proc. the Fall Joint Computer Conference*. 757–764.

Swaminathan, K., and S. Sato (1997). Interaction design for large displays. *Interactions* (Jan–Feb).

Tan, D.S., J.K. Stefanucci, D.R. Proffitt, and R. Pausch (2001). The infocockpit: providing location and place to aid human memory. *Workshop on Perceptive User Interfaces*, Orlando, FL.

Tan, D.S., J.K. Stefanucci, D.R. Proffitt, and R. Pausch (2002). Kinesthesis aids human memory. *CHI 2002 Extended Abstracts*, Minneapolis, MN.

Tani, M., K. Yamaashi, K. Tanikoshi, M. Futakawa, and S. Tanifuji (1992). Object-oriented video: interaction with real-world objects through live video. *Proceedings of ACM CHI '92 Conference on Human Factors in Computing Systems*. 593–598, 711–712.

Trevor, J., D.M. Hilbert, B.N. Schilit, and T.K. Koh (2001). From desktop to phonetop: a UI for Web interaction on very small devices. *Proc. UIST '01 Symp. on User Interface Software and Technology*. 121–130.

Triesman, A. (1985). Preattentive processing in vision. *Comput. Vision, Graphics and Image Process* 31: 156–177.

Tufte, E.R. (1983). *The Visual Display of Quantitative Information*. Cheshire, CT, Graphics Press.

Tufte, E.R. (1990). *Envisioning Information*. Cheshire, CT, Graphics Press.

Tufte, E.R. (1997). *Visual Explanations: Images and Quantities, Evidence and Narrative*. Cheshire, CT, Graphics Press.

Wang, J., S. Zhai, and H. Su (2001). Chinese input with keyboard and eye-tracking: an anatomical study. *Proceedings of the SIGCHI conference on Human Factors in Computing Systems*.

Want, R., and G. Borriello (2000). Survey on information appliances. *IEEE Personal Communications* (May/June): 24–31.

Want, R., K.P. Fishkin, A. Gujar, and B.L. Harrison (1999). Bridging physical and virtual worlds with electronic tags. *Proc. ACM CHI '99 Conf. on Human Factors in Computing Systems.* 370–377.

Ware, C. (1988). Color sequences for univariate maps: theory, experiments, and principles. *IEEE Comput. Graphics Appl.* **8**(5): 41–49.

Ware, C. (2000). *Information Visualization: Design for Perception.* San Francisco, Morgan Kaufmann.

Ware, C., K. Arthur, and K. S. Booth (1993). Fish tank virtual reality. *Proceedings of ACM INTERCHI '93 Conference on Human Factors in Computing Systems.* 37–41.

Ware, C., and J. Rose (1999). Rotating virtual objects with real handles. *ACM Transactions on CHI* **6**(2): 162–180.

Weiser, M. (1991). The computer for the 21st century. *Scientific American* (Sept.): 94–104.

Wellner, P. (1993). Interacting with paper on the DigitalDesk. *Communications of the ACM* **36**(7): 87–97.

Wenzel, E.M. (1992). Localization in virtual acoustic displays. *Presence* **1**(1): 80–107.

Westheimer, G. (1979). Cooperative nerual processes involved in stereoscopic acuity. *Exp. Brain Res.* **36**: 585–597.

Wickens, C. (1992). *Engineering Psychology and Human Performance.* New York, HarperCollins.

Wilson, A., and S. Shafer (2003). XWand: UI for intelligent spaces. *CHI 2003.* To appear.

Wilson, F. R. (1998). *The Hand: How Its Use Shapes the Brain, Language, and Human Culture.* New York, Pantheon Books.

Wisneski, C., H. Ishii, A. Dahley, M. Gorbet, S. Brave, B. Ullmer, and P. Yarin (1998). Ambient displays: turning architectural space into an interface between people and digital information. *Lecture Notes in Computer Science* **1370**: 22–32.

Wyszecki, G., and W.S. Styles (1982). *Color Science, 2nd Ed.* New York, Wiley.

Zeleznik, R., K. Herndon, and J. Hughes (1996). SKETCH: an interface for sketching 3-D scenes. *Proceedings of SIGGRAPH '96,* New Orleans, LA. 163–170.

Zhai, S. (1998). User performance in relation to 3-D input device design. *Computer Graphics* **32**(8): 50–54.

Zhai, S., M. Hunter, and B.A. Smith (2000). The Metropolis keyboard — an exploration of quantitative techniques for virtual keyboard design. *CHI Letters* **2**(2): 119–128.

Zhai, S., and P. Milgram (1993). Human performance evaluation of isometric and elastic rate controllers in a 6DoF tracking task. *Proc. SPIE Telemanipulator Technology.*

Zhai, S., P. Milgram, and W. Buxton (1996). The influence of muscle groups on performance of multiple degree-of-freedom input. *Proceedings of CHI '96: ACM Conference on Human Factors in Computing Systems,* Vancouver, British Columbia, Canada, ACM, New York. 308–315.

Zhai, S., C. Morimoto, and S. Ihde (1999). Manual and gaze input cascaded (MAGIC) pointing. *Proc. ACM CHI '99 Conf. on Human Factors in Computing Systems.* 246–253.

Zhai, S., B.A. Smith, and T. Selker (1997). Improving browsing performance: a study of four input devices for scrolling and pointing tasks. *Proc. INTERACT '97: The Sixth IFIP Conf. on Human–Computer Interaction.* 286–292.

21

Secondary Storage Systems

21.1 Introduction **21-1**
    Roadmap to the Chapter
21.2 Single Disk Organization and Performance **21-4**
    Disk Organization • Disk Arm Scheduling • Methods to
    Improve Disk Performance
21.3 RAID Disk Arrays **21-12**
    Motivation for RAID • RAID Concepts
    • RAID Fault-Tolerance and Classification • Caching and
    the Memory Hierarchy • RAID Reliability Modeling
21.4 RAID1 or Mirrored Disks **21-17**
    Request Scheduling with Mirrored Disks
    • Scheduling of Write Requests with an NVS Cache
    • Mirrored Disk Layouts
21.5 RAID5 Disk Arrays **21-19**
    RAID5 Operation in Normal Mode • RAID5 Operation in
    Degraded Mode • RAID5 Operation in Rebuild Mode
21.6 Performance Evaluation Studies **21-21**
    Single Disk Performance • RAID Performance
    • Analysis of RAID5 Systems
21.7 Data Allocation and Storage Management
    in Storage Networks **21-25**
    Requirements of a Storage Network • Storage Management
21.8 Conclusions and Recent Developments **21-26**

Alexander Thomasian
New Jersey Institute of Technology

21.1 Introduction

The magnetic disk technology was developed in the 1950s to provide a higher capacity than magnetic drums, which with one head per track constituted a *fixed head* system, while disks introduced a *movable head* system. The movable heads were shared by all tracks, resulting in a significant cost reduction. IBM's *RAMAC — Random Access Method for Accounting and Control* [Matick 1977] — announced in 1956 was the first magnetic disk storage system with a capacity of 5 Megabytes (MB) and a $10,000 price tag.

Magnetic disks have been a multibillion industry for many years. There has been a recent dramatic decrease in the cost per megabyte of disk capacity, with a resulting drop in disk prices. This is a result of a rapid increase in disk capacity: 62% *CAGR (compound annual growth rate)* [Wilkes 2003], which is due to increases in areal recording density.

The explosion in storage capacity in computer installations has led to a dramatic increase in the cost of storage management. In 1980, one data administrator was in charge of 10 GB (gigabytes) of data, while in

1-58488-360-X/$0.00+$1.50
© 2004 by CRC Press, LLC

2000 this number was 1 TB (terabyte) [Gray 2002]. Given that 5 TB costs $5000, the cost of administering it is an order of magnitude higher. Consequently, automating storage management and data allocation has emerged as a very important area [Wilkes 2003].

Despite the importance of disks and secondary storage, there has been relatively little discussion about them in the computer science and engineering literature. This is due to the fact that more attention has been paid traditionally to the **central processing unit (CPU)**, which determines the speed of the computation. This situation changed significantly with the publication of a paper on **RAID (Redundant Array of Independent Disks)** [Patterson et al. 1998]. More recently there has been an explosion in the level of research activity in storage systems.

In fact, there has been a constant stream of articles dealing with scheduling and performance analysis of disks and disk subsystems. Magnetic bubbles and charged couple devices (CCDs) were considered candidates for replacing disks, or at least serving as an intermediate level storage between disks and main memory, but neither of the two prophecies became a reality. Some other technologies, such as Flash memories, optical disks, and magnetic tapes, are reviewed in Hennessey and Patterson [2003].

DRAM memories, which are *volatile* because they lose their contents when power is removed, were projected to replace disks a decade ago (see Section 2.4 in Gibson [1992]). **Nonvolatile storage (NVS)** with DRAM (NVRAM) can be realized by means of *UPS (uninterruptible power supply)* and it has been argued that a duplexed NVS has about the same reliability as disk storage. This allows **fast writes**, that is, the writing of data to disk is considered completed as soon as the data is written onto duplexed NVS [Menon and Cortney 1993].

The longevity of disk storage with respect to DRAM is attributable to the lower cost of storage per megabyte by a factor of 1000 to one, the higher recording density, and also the nonvolatility.

The nonvolatility feature is quite important and *micro-electromechanical system-*(**MEMS) based storage**, which is nonvolatile and an order of magnitude faster than disks is a technology contending to replace disks, although it is currently ten times more expensive [Uysal et al. 2003]. MEMS-based storage differs from disks in that data can be accessed by moving the read/write heads in two dimensions, while magnetic disks are rotated (angularly) and the read-write heads are moved radially to access the data. Another interesting technology is **magnetic RAM (MRAM)**, which has approximately DRAM speeds but is less costly, stores 100s MB/chip, and is nonvolatile [Wilkes 2003]. It will take time for these technologies to replace disks, but fundamental principles of storage hierarchies are expected to prevail. Given the current dominance of magnetic disks and disk arrays, the current chapter solely concerns itself with this technology.

The impact of magnetic disks on computing is comparable to magnetic core memories, whose introduction resulted in a dramatic increase in feasible memory sizes. Programs that encountered fewer interruptions due to page faults, because their working sets could reside in main memory, could run much faster (see Chapter 85 and Chapter 86).

When the main memory sizes were small, only one program could be held in main memory and when that program stopped — due to (1) completion, (2) preemption because its time quantum expired, or (3) a page fault or an I/O request — switchover (context switching) to another program was slow. Switchover required the workspace of the current program to be swapped out onto a drum or a disk (this is not required in case (1)) before the next program can be loaded into main memory. In fact, this time-critical and frequent swapping activity was carried out onto drums, even after disks became available, because drums, unlike disks, do not incur a seek time and are hence faster. Swapping occurs less frequently in current systems, which utilize disks, whose performance (especially seek time) has improved considerably, so that disks have completely replaced drums.

The context switching overhead, which took tens of milliseconds, was eliminated when larger memories were introduced. Current DRAM memories are in gigabytes, while the largest and expensive core memories were in megabytes. Consequently, there is less need to swap out processes and, furthermore, several programs can be held in main memory, so that the CPU can switch from one program to another at a small cost. *Multiprogramming*, even when a single processor is available, results in an increased system throughput, which is due to the overlap in CPU and I/O processing.

Shared memory multiprocessors (SMPs) allow multiple programs residing in main memory to execute concurrently. This necessitates more disks and I/O paths to handle the higher rate of disk accesses.

The space held by user and system programs is quite small compared to the space held by data, so that it will not be considered further in our discussions. We are mainly concerned with accesses to datasets associated with *DBMSs (database management systems)*, such as the tables of an operational relational DBMS. In fact, *data warehouses*, which are utilized by *decision support systems (DSS)* [Ramakrishnan and Gehrke 2003] occupy more space than operational databases.

In this chapter we do not concern ourselves with the structure of file systems, and the metadata associated with them, because it is covered in Chapter 86 on *Secondary Storage and Filesystems*.

Magnetic disks made the transition from **batch processing** to **OLTP (online transaction processing)** possible. As an example of batch processing, consider a banking application. Debit and credit transactions originating from bank tellers were entered onto punched cards, which were then transferred to a magnetic tape (direct document entry onto tape or disk became possible later). Transactions were sorted by customer number before updating the master file (M/F), which was already in that order. A new M/F tape with an updated balance was created by means of a daily run of the batch update program. The batch update process was (1) error prone (due to manual data entry), (2) costly because it was labor intensive, (3) slow in that the updating was carried out once daily, and (4) risky in that overdrafts were a possibility.

To check the balance of a customer record in the M/F required accessing half of the file records on the average. It would have been impossible to implement relational DBMSs without the fast random access capability provided by disks.

Magnetic disks usually update data in place, while magnetic tapes write modified data sequentially onto another tape, which makes data compression/decompression possible. Data compression with in-place updating does not work with disks because modified data may have a lower compression ratio than the original data, so that the compressed data might not fit into its original space any longer. Compression in disks can be used in conjunction with the *log-structured file system (LFS)* paradigm, which is discussed in Section 21.3.3.

The time to access a record on a disk file depends on the file organization [Ramakrishnan and Gehrke 2003]. A *sequential scan* of an unordered files or a **heap file** is $O(N)$ and the cost is $N/2$, on average, if the record is found and N otherwise. A *binary search* on a sorted file with N records can be used to find the desired record in $log_2 N$ steps. **Hashed files** apply a hashing function to a *search field* to map a record into a bucket (e.g., a 4- or 8-KB disk block). The record can be retrieved with one disk access, provided that the bucket has not overflowed. *Linear hashing* and *extendible hashing* are two methods to deal with bucket overflows and underflows (empty buckets) when there are dynamic insertions and deletions [Ramakrishnan and Gehrke 2003].

B+ trees are the most popular indexing structure utilized in database management systems. The number of accesses for a file indexed with a B+ tree index is one access (for the page containing the desired record) plus the number of levels traversed in the index. The top two levels of the B+ tree, which typically has three or four levels, are usually cached, thus reducing the number of disk accesses by two. Given an indexed or hashed file organization, OLTP in the form of online updating of a customer record is possible, while the customer is at an *ATM (automatic teller machine)*. More importantly, balances can be checked for withdrawals and overdrafts can be prevented.

Despite this relatively short disk access time (less than 10 milliseconds), I/O time might be a significant contributor to application response time. With the advent of very high-speed CPUs, transaction response time for OLTP applications is determined by the number of disk accesses resulting from database buffer *misses*. A miss means that the requested page cannot be found in the database buffer, as opposed to a *hit*, when it is. The *miss ratio* is used to quantify this effect. The maximum throughput of an OLTP benchmark can be designated as the point where the 95th percentile of transaction response time exceeds 2 seconds.

Reducing disk access time is also important when users are involved in an online interaction with a computer system, such as program development, CAD/CAM, etc. Studies have shown that reducing computer response time for interactive applications results in a synergistic reduction in user think time. More specifically, for *subsecond computer response times*, user reaction time is faster because the user operates from short-term memory; consequently, there is a reduction in the overall duration of the activity [Hennessey and Patterson 2003]. More generally, disk access time affects the end-to-end response time and *Quality-of-Service (QoS)* for many applications.

21.1.1 Roadmap to the Chapter

In Section 21.2, we describe the organization of a single disk in some detail and then provide a rather comprehensive review of disk scheduling paradigms.

What makes an array of disks more interesting than a *just a bunch of disks (JBOD)* is that using them as an ensemble provides additional capabilities (e.g., for parallel access and/or to protect data against data failures. In Section 21.3 we discuss general concepts associated with RAID. In Section 21.4 and Section 21.5 we discuss the two most important RAID levels: RAID1 (mirrored disks) and RAID5 (rotated parity arrays). Performance evaluation studies of disks and disk arrays are discussed in Section 21.6.

Issues associated with data allocation and storage management in storage networks are discussed in Section 21.7. In Section 21.8, after summarizing the discussion in the chapter, we discuss topics that are not covered here, but are expected to gain importance in the future.

21.2   Single Disk Organization and Performance

Two disk performance metrics of interest to this discussion are the **data transfer rate** and the **mean access time**. The *sustained data transfer rate*, rather than the *instantaneous data transfer rate*, is of interest. The sustained rate is smaller than the instantaneous rate because of the inter-record gaps, the check block,* and disk addressing information separating data blocks (this is also to identify faulty sectors). The number of sectors per track and hence the disk transfer rate can be increased considerably by adopting the *noID sector format*, where the header information is stored in solid-state memory rather than on the disk surface. In modern disk drives, the data transfer rate from outer tracks is higher than inner tracks due to zoning, which is explained below.

The mean access time ($\bar{x}_{disk}$) is the metric of interest when disk accesses are to randomly placed blocks of data, as in OLTP applications. The maximum number of requests satisfied by the disk per second is the reciprocal of $\bar{x}_{disk}$, assuming a FCFS scheduling policy, although much higher throughputs can be obtained by the scheduling policies described later in this section.

The **disk positioning time**, or the time to place the R/W head at the beginning of the block being accessed, dominates the disk service time when small disk blocks are being accessed. The positioning time is the sum of seek time to move the arm and rotational latency until the desired data block reaches the R/W head. Early disk scheduling methods dealt with minimizing seek time, while it is minimizing the positioning time that provides the best performance.

In fact, the best way to reduce disk access time is to eliminate disk access altogether! This can be accomplished by caching and prefetching at any one of the levels of the memory hierarchy. The **five-minute rule** states: "Keep a data item in main memory if its access frequency is five minutes or higher, otherwise keep it in magnetic memory" [Gray and Reuter 1993]. Of course, the parameters of this rule change in time.

This section is organized as follows. In Section 21.2.1 we describe the organization of a modern disk drive. In Section 21.2.2 we review disk arm scheduling techniques for requests to discrete and continuous data (typically used in multimedia applications) and combinations of the two. In Section 21.2.3 we discuss various other methods to improve disk performance, which include reorganizing the data layout, etc.

---

*Things are more complicated than this. There is a first line of protection for a sector, a second line of protection for a group of sectors, and a third line of protection for blocks of blocks of sectors.

21.2.1 Disk Organization

A disk drive consists of one or more **platters** or disks on which data is recorded in concentric circular **tracks**. The platters are attached to a spindle, which rotates the platters at a **constant angular velocity (CAV)**, which means that the linear velocity is higher on outer tracks.*

There are a set of read/write **(R/W) heads** attached to an **actuator** or **disk arm**, where each head can read/write from a track on one side of a platter. At any time instant, the R/W heads can only access tracks located on the same disk **cylinder**, but only one R/W head is active at any time. The time to activate another R/W head is called the **track switching time.**

The arm can be moved to access any of the C disk cylinders and the time required to move the arm is called the **seek time.** The seek from the outermost to innermost cylinder, referred to as a **full-stroke** seek, has the maximum seek time, which is less than 10 milliseconds (ms) for modern disk drives. Cylinder-to-cylinder seeks tend to take less time than head switching time (both are under 1 ms). The **seek time characteritic** $t(d)$, $1 \leq d \leq C - 1$ is an increasing function of the seek distance d, but usually a few irregularities are observed in $t(d)$, which is measured by averaging repeated seeks with the same seek distances.

Modern disks accelerate the arm to some maximum speed, let it coast at the speed it has attained, and then decelerate the arm until it stops. Applying curve-fitting to measurement data yields the seek time $t(d)$ vs. the seek distance d. A representative function is $t(d) = a + b\sqrt{d}$ for $1 \leq d \leq d_0$ and $t(d) = e + fd$ for $d_0 \leq d \leq C - 1$, where d_0 designates the beginning of the linear portion of the seek.

When the blocks being accessed are uniformly distributed over all the blocks of a non-zoned disk (a disk with the same number of blocks per cylinder), the number of accesses over the C cylinders of the disk will be uniform. It follows from a geometrical probability argument that the average distance for a random seek is one third the disk cylinders: $\bar{d} \approx C/3$. Because the seek time characteristic ($t(d), 1 \leq d \leq C - 1$) of the disk is nonlinear, the mean seek time $\bar{x}_{seek}$ is not equal to $t(\bar{d})$, but rather $\bar{x}_{seek} = \sum_{d=1}^{C-1} P[d] t(d)$, where $P[d]$ is the probability that the seek distance is d.

The seek distance probability mass function, assuming that the probability of no seek is p and the other cylinders are accessed uniformly, is given by: $P[0] = p$ and $P[d] = 2(1 - p)(C - d)/[C(C - 1)], 1 \leq d \leq C - 1$. To derive the density function, we note that there are two seeks of distance $C - 1$ and $2(C - d)$ seeks of distance d. The normalization factor is $\sum_{d=1}^{C-1} 2(C - d) = C(C - 1)$. For uniform accesses to all cylinders, $P[1] = 1/C$ and $P[d] = 2(C - d)/C^2, 1 \leq d \leq C - 1$. Expressions for $P[d]$ in zoned disks are given in Thomasian and Menon [1997].

In addition to radial positioning of R/W heads to access the desired block, angular positioning is accomplished by rotating the disk at a fixed velocity, (e.g., 7200 rotations per minute [RPM]). The delay to access data, called **rotational latency**, is uniformly distributed over disk rotation time T_{rot}. The average latency is $T_{rot}/2$, that is, 4.17 ms for 7200 RPM. There have been suggestions to place duplicate data blocks [Zhang et al. 2002] or provide two access arms [Ng 1991], 180° apart in both cases.

It is best to write data files sequentially, that is, on consecutive sectors of a track, consecutive tracks of a cylinder, consecutive cylinders, so that sequential reads can be carried out efficiently. In fact, the layout of sectors has been optimized for this purpose by introducing **track skew** and **cylinder skew** to mask head and cylinder switching times. The first sector on a succeeding track (or cylinder) is skewed to match the head switching time (or a single cylinder seek), so that no additional latency is incurred in reading sequential data.

There has been a rapid increase in areal magnetic disk recording density due to an increase in the number of **tracks per inch** (TPI) and the number of **bits per inch** (BPI) on a track. A typical areal recording density of 35 Gigabits per square inch was possible at the end of the 20th century [Gray and Shenoy 2000] and a continued increase in this density has been projected, so that 100 Gigabits per square inch seems possible.

---

*CD-ROMs (compact disc read-only memory) adjust the speed of the motor so that the linear speed is always constant; hence, **constant linear velocity (CLV)**.

Obviously, the effective recording density is lower than the raw recording density. The increase in areal density has led to small form-factor (3.5-inch diameter) disks, whose capacity exceeds 100 Gigabytes (GB), while 500-GB disks have been projected to be available in a few years. Higher TPI has resulted in a reduction in the seek distance and seek time, but this is at the cost of head-settling time [Ng 1998]. Disk seek time remains a major contributor to positioning time.

Almost all disks follow the **Fixed Block Architecture (FBA)** format, as opposed to the variable block size **count-key-data (CKD)** format, where the (physical) block size is determined by application requirements. FBA organizes data as 512-byte blocks called **sectors**, which are addressable by their block number. Each sector is protected by an **ECC (error correcting code)** for error detection and correction, which can detect error bursts in tens of bits and correct several bytes in a sector. Such information is available in disk product manuals, which are retrievable online (see, e.g., the Maxtor Atlas 15K).

Roll-mode or **zero-latency** read (and write) capability allows the sectors of a block to be read out-of-order, that is, starting with any sector. When the data transfer starts from the middle of the block, the latency equals the time of a full rotation time (T_{rot}) minus the transfer time of a block ($\bar{x}_{block}$). The average latency is the weighted sum of two cases: (1) with probability $q = \#\_sector\_per\_block/\#\_sectors\_per\_track$, the R/W head lands inside the block to be read, in which case the average latency is $\bar{x}_{latency} = T_{rot} + \bar{x}_{sector}/2$; and (2) with probability $1 - q$, the head lands outside the block, so that $\bar{x}_{latency} = (T_{rot} - \bar{x}_{block})/2$. Given that q and $\bar{x}_{block}$ are negligibly small for smaller block sizes, it follows that $\bar{x}_{latency} \approx T_{rot}/2$. This capability reduces the latency considerably when reading larger blocks of data (e.g., reading a full track).

Until a decade ago, most disks stored a fixed number of bits on all disk tracks, so that all tracks had a fixed transfer rate. This resulted in a lower recording density than was technologically possible at outer disk tracks (with a longer circumference than inner tracks), thus resulting in a lower overall disk capacity.

Disk zoning or **zoned constant angular velocity (ZCAV)** or **zoned bit recording (ZBR)** or **notches** or **notched drives** (used in the SCSI standard) were introduced to deal with this inefficiency. The number of bits per track, and consequently the number of sectors per track in such disks, is increased from innermost to outermost tracks, keeping the linear recording density per track almost constant. The number of sectors on multiple neighboring cylinders, referred to as *zones*, is maintained at the same value to simplify bookkeeping. The number of disk zones may be higher than 20 and the number of tracks per zone may be in the thousands A consequence of zoning is that the data transfer rate from outer tracks is much higher than the inner tracks. If the outermost track has $1 + \alpha$ as many sectors as the innermost track, its transfer rate is higher by that factor, and the increase in disk capacity is roughly $1 + \alpha/2$, that is, 25% for $\alpha = 0.5$. The transfer time of frequently accessed large files can be reduced by placing them on the outermost zone (note that all tracks in a zone have the same transfer rate).

Higher recording densities have resulted in smaller diameter disks and fewer disk platters (even one), because the capacity is already met. This simplifies disk design and allows disks to rotate faster, that is, at higher RPMs, with less power and generating less heat. In fact, disk RPMs can be varied (reduced) to consume less power [Gurumurthi et al. 2003].

Higher RPMs combined with higher recording densities have resulted in higher disk transfer rates, which are of the order of tens of megabytes per second (MB/s). Higher RPMs also incur lower rotational latencies.

The disk transfer rate, which varies zone to zone, is lower than the bus data transfer rate. This mismatch is handled by the **onboard disk cache**, which buffers data that is read from disk before it is transmitted on the bus. The bus data transfer of an incompletely read block can be started as soon as enough data has been read from disk to ensure an uninterrupted transfer. This depends on the zone from which the data is being read.

Higher disk access bandwidths can be attained by accessing larger blocks, so as to amortize positioning overhead. *Track-aligned extents (traxtents)* use disk-specific knowledge to improve performance [Schindler et al. 2002]. Access time is reduced be eliminating head switching time by assigning blocks such that they do not span track or cylinder boundaries. Matching block sizes to track size and reading full tracks eliminates rotational latency, provided the disk has the zero-latency capability.

The onboard cache is also used for prefetching all or the remaining blocks of a track from which the data data was accessed. Prefetching is initiated after the disk controller detects sequentiality in the access pattern, but disk utilization due to prefetching may be wasteful because prefetching is speculative [Patterson et al. 1995]. Caching writes and deferring their processing for improving performance might be disabled because it affects data integrity. The performance degradation introduced by prefetching can be minimized by allowing preemption. Little has been written about onboard cache management policies because they are of proprietary nature.

Most modern disk drives, such as the Maxtor Atlas 15K, come in different sizes (18, 36, and 73 GB) which is achieved by increasing the number of disks from 1 to 2 to 4 and the number of R/W heads by twice these numbers. The RPM is 15K (surprised?); hence, the rotation time is $T_{rot} = 4$ ms, there are 24 zones per surface, 61 KTPI, 32,386 cylinders, the number of 512-byte sectors per track varies from 455 to 630. The maximum effective areal density is 18 Gbits^2. The head switching time (on the same track) is less than 0.3 ms for reads and 0.48 ms for writes, while sequential cylinder switch times are 0.25 ms and 0.40 ms, respectively. The random average seek time to read (resp. write) a random block is 3.2 (resp. 3.6) ms, while the full-stroke seek is less than 8 ms. The maximum transfer rate is 74.5 MB/s, and the onboard buffer size is 8 MB. A 45-byte Reed-Solomon code [MacWilliams and Sloane 1977] is used as an error correcting code (ECC), and uncorrectable read errors occur in one per 10^{15} bits read. The access time to a small randomly placed block approximately equals the positioning time, which is $2 + 3.2 = 5.2$ ms (half of the disk rotation time added to the mean seek time), so that the maximum access rate for read requests (ignoring controller overhead) is 192.3 requests per second.

We have discussed **fixed disks** as opposed to **removable disks**. Removable disks with large form-factors were popular in the days when mini- and mainframe computers dominated the market. The removable disk was inserted into a disk drive, which resembled a washing machine with a shaft that rotated the disk. The R/W heads were attached to an arm that was retracted when the disk was being loaded/unloaded.*

A good but dated text describing magnetic disk drives is Sierra [1990]. The organization and performance of modern disk drives are discussed in Ruemmler and Wilkes [1994] and Ng [1998]. In what follows we will discuss various techniques for reducing disk access time.

21.2.2 Disk Arm Scheduling

A disk is at its best when it is sequentially transferring large blocks of data, such that it is operating close to its maximum transfer rate. Large block sizes are usually associated with **synchronous requests**, which are based on processes that generate one request at a time, after a certain "think time." Prefetching, at least "double buffering," is desirable; that is, initiate the transfer of next block while the current block is being processed. Synchronous requests commonly occur in multimedia applications, such as video-on-demand (VoD). Requests are to successive blocks of a file and must be completed at regular time intervals to allow glitch-free video viewing. Disk scheduling is usually round-based, in that a set of streams are processed periodically in a fixed order. An **admission control policy** can be used to ensure that processing of a new stream is possible with satisfactory performance for all streams and that this is done by taking into account buffer requirements [Sitaram and Dan 2000].

In contrast, **discrete requests** originate from an *infinite number of sources*. The arrival process is usually assumed to be Poisson with parameter λ, which implies (1) the arrivals in a time interval are uniformly distributed over its duration, (2) exponentially distributed interarrival times with a mean equal to $1/\lambda$, and (3) the arrival process is *memoryless*, in that the time to the next arrival is not affected by the time that has already elapsed [Kleinrock 1975]. In an OLTP system, sources correspond to concurrent transactions that generate I/O requests. A disk access is required when the requested data block is not cached at a level of the memory hierarchy preceding the disk.

It is commonly known that OLTP applications generate accesses to small, randomly placed blocks of data. For example, the analysis of the I/O trace of an OLTP workload showed that 96% of disk accesses

---

*IBM developed an early disk drive with 30 MB of fixed and 30 MB of removable storage, which was called Winchester in honor of its 30/30 rifle. Winchester disks are now synonymous with hard disks.

are to 4-KB and 4% to 24-KB blocks of data [Ramakrishnan et al. 1992]. The discussion in this section is therefore based on accesses to small, random blocks. This is followed by a brief discussion of sequential and mixed requests in the next section.

The default FCFS scheduling policy provides rather poor performance in the case of randomly placed data. Excluding the time spent at the disk controller, the mean service time in this case is the sum of the mean seek time, the mean latency, and the transfer time. The transfer time of small (4 or 8 KB) blocks tends to be negligibly small with respect to positioning time, which is the sum of the seek time and rotational latency.

Disk scheduling methods require the availability of multiple enqueued requests to carry out their optimization. In fact, the improvement with respect to the FCFS policy increases with the number of requests that are available for scheduling. This optimization was done by the operating system; but with the advent of the SCSI protocol, request queueing occurs at the disk.

The observation that disk queue-lengths are short was used as an argument to discourage disk scheduling studies [Lynch 1972]. Short queue-lengths can be partially attributed to **data access skew**, which was prevalent in computer installations with a large number of disks. A few disks had a high utilization, but most disks had a low utilization.

Evidence to the contrary, that queue-lengths can be significant, is given in Figure 1 in Jacobson and Wilkes [1991]. Increasing disk capacities and the volume of data stored on them should lead to higher disk access rates, based on the fact that there is an inherent **access density** associated with each megabyte of data (stale data is deleted or migrated to tertiary storage). On the other hand, larger data buffers are possible due to increased memory sizes; that is, as the disk capacity increases, so does the capacity for caching, which limits the increase in disk access rate [Gray and Shenoy 2000]. The working set of disk data in main memory was shown to be a few percent of disk capacity in one study [Ruemmler and Wilkes 1993].

Most early disk scheduling policies concentrated on reducing seek time, because of its dominant effect on disk positioning time. This is best illustrated in Jacobson and Wilkes [1991], which gives the ratio of the maximum seek times and rotation times for various disks. This ratio is quite high for some early disk drives (IBM 2314). The **shortest seek time first (SSTF)** and **SCAN** policies address this issue [Denning 1967].

SSTF serves requests from the queue according to the seek distance (i.e., with the request with the smallest seek distance and seek time served first). SSTF is a greedy policy, so that requests in some disk areas (e.g., innermost and outermost disk cylinders) will be prone to starvation if there is a heavy load at other disk areas (e.g., the middle disk cylinders).

The SCAN scheduling policy moves the disk arm in alternate directions, making stops at cylinders to serve all pending requests, so that it is expected to be less prone to starvation and to produce a smaller variance in response time than SSTF. SCAN is also referred to as the **elevator algorithm**.

Cyclical SCAN (CSCAN) returns the disk arm to one end after each scan, thus alleviating the bias in serving requests on middle disk cylinders twice. A plot of the mean response time at cylinder c (R_c) vs. the cylinder number shows that R_c is lower at the middle disk cylinders and higher at outer cylinders for scan. LOOK and CLOOK are minor variations of SCAN and CSCAN that reverse the direction of the scan as soon as there are no more requests in that direction, rather than reaching the innermost and/or outermost cylinder if there are no requests in the direction of the scan.

The **shortest access time first (SATF)** or **shortest positioning time first (SPTF)**) gives priority to requests whose processing will minimize positioning time [Jacobson and Wilkes 1991]. This policy is desirable for modern disks with improved seek times. In effect, we have a *shortest job first (SJF)* policy with the desirable property that it minimizes the mean response time among all non-preemptive policies [Kleinrock 1976]. The difference between SATF and SJF is that the service time of individual disk requests depends on the order in which they are processed.

Several simulation experiments have shown that SPTF, which minimizes positioning time, outperforms SSTF and SCAN [Worthington et al. 1994, Thomasian and Liu 2002a, Thomasian and Liu 2002b]. SPTF is a greedy policy, and appropriate precautions are required to bound the waiting time of requests (e.g., by reducing the positioning time according to waiting time).

Prioritizing the processing of one category of requests with respect to another (e.g., reads vs. writes) is a simple way to improve the performance of reads in this case. The head-of-the-line priority queueing discipline serves requests (in FCFS order) from a lower priority queue, but only when all higher priority queues are empty [Kleinrock 1976].

The SATF policy can be modified to prioritize read requests with respect to write requests as follows Thomasian and Liu [2002a] and Thomasian and Liu [2002b]. An SATF winner read request is processed unconditionally, while the service time of an SATF winner write request is compared to that of the best read request and processed only if the ratio of its service time and that of the read request is below a threshold $0 \le t \le 1$. In effect, $t = 1$ corresponds to "pure SATF," while $t = 0$ prioritizes reads unconditionally. There are two considerations: (1) simulation results show that doing so results in a significant reduction in throughput with respect to SATF, and (2) such a scheme should take into account the space occupied by write requests at the onboard buffer. An intermediate value of t can be selected, which improves response time while achieving a small reduction in throughput.*

SATF performance can be improved by applying **lookahead**. For example, consider an SATF winner request A, whose processing will be followed by request X, also according to the SATF policy. There might be some other request B, which when followed by Y (according to SATF) yields a total processing time $T_B + T_Y < T_A + T_X$. With n requests in the queue, the cost of the algorithm increases from $O(n)$ to $O(n^2)$. In fact, the second requests (X or Y) may not be processed at all, so that in carrying out comparisons, their processing time is multiplied by a discount factor $0 \le \alpha \le 1$. The improvement in performance due to lookahead is insignificant for disk requests uniformly distributed over all disk blocks, but improves performance when requests are localized.

There have been many proposals for hybrid disk scheduling methods, which are combinations of other well-known methods. We discuss two variations of SCAN and two policies that combine SCAN with SSTF and SPTF policies in order to reduce the variance of response time.

In **N-step SCAN**, the request queue is segmented into subqueues of length N and requests are served in SCAN order from each subqueue. When $N = 1$, we have the FCFS policy; otherwise, when N is large, N-step SCAN is tantamount to the SCAN policy.

FSCAN is another variation of SCAN, which has $N = 2$ queues [Coffman et al. 1972]. Requests are served from one queue, while the other queue is being filled with requests. This allows the SCAN policy to serve requests that were there at the beginning of the SCAN.

The $V(R)$ disk scheduling algorithm ranges from $V(0) = $ SSTF to $V(1) = $ SCAN as its two extremes [Geist and Daniel 1987]. It provides a "continuum" of algorithms combining SCAN and SSTF for $0 \le R \le 1$. R is used to compute the bias $d_{bias} = R \times C$, which is subtracted from the seek distance in the forward direction (C is the number of disk cylinders). More precisely, the seek distance in the direction of the scan is given as $\max(0, d_{forward} - d_{bias})$, which is compared against $d_{backward}$. The value of R is varied in simulation results to determine the value that minimizes the sum of the mean response time and a constant k times its standard deviation, which is tantamount to a percentile of response time. It is shown that for lower arrival rates, SSTF is best (i.e., $R = 0$), while at higher arrival rates ($R = 0.2$) provide the best performance.

Grouped Shortest Time First (GSTF) is another hybrid policy combining SCAN and STF (same as SATF) [Seltzer et al. 1990]. A disk is divided into groups of consecutive cylinders and the disk arm completes the processing of requests in the current group according to SPTF, before proceeding to the next group. When there is only one group, we have SPTF and with as many groups as cylinders, and we effectively have SCAN (with SPTF at each cylinder). The **weighted shortest time first (WSTF)** also considered in this study multiplies the anticipated access time by $1 - w/W_M$, where w is the waiting time and W_M

---

*This technique can be utilized to make the mean response of requests to a failed disk equal to the mean response times at surviving disks. This is accomplished by prioritizing disk accesses on behalf of fork-join requests with respect to others (see Section 21.5).

is the maximum waiting time (the 99th percentile of waiting time is a more meaningful metric for this purpose).

21.2.2.1 Disk Scheduling for Continuous Data Requests

Continuous data requests have an implicit deadline associated with the delivery of the next data block (e.g., video segment in a video stream) for glitch-free viewing.

The **Earliest Deadline First (EDF)** scheduling policy is a natural choice because it attempts to minimize the number of missed deadlines. On the other hand, it incurs high positioning overhead, which is manifested by a reduction in the number of video streams that can be supported. SCAN-EDF improves performance by using SCAN while servicing requests with the same deadline.

Scheduling in "rounds" is a popular scheduling paradigm, so that the successive blocks of all active requests need to be completed by the end of the round. The size of the blocks being read and the duration of the rounds should be chosen carefully to allow for glitch-free viewing. Round-robin, SCAN, or **Group Sweeping Scheduling (GSS)** [Chen et al. 1993] policies have been proposed for this purpose.

In addition to requests to continuous or C-data, media servers also serve discrete or D-data. One method to serve requests is to divide a round into subrounds, which are used to serve requests of different types. More sophisticated scheduling methods are described in [Balafoutis et al. 2003], two of which are described here.

One scheduling method serves C-requests according to SCAN and intervening D-requests according to either SPTF or OPT(N). The latter determines the optimal schedule after enumerating all $N!$ schedules ($N = 6$ is used in the chapter).

The **FAair MIxed-scan ScHeduling (FAMISH)** method ensures that all C-requests are served in the current round and that D-requests are served in FCFS order. More specifically, this method constructs the SCAN schedule for C-requests but also incorporates the D-requests in FCFS order in the proper position in the SCAN queue, up to the point where no C-request misses its deadline. D-requests can also be selected according to SPTF. The relative performance of various methods as determined by simulation studies is reported in Balafoutis et al. [2003].

21.2.3 Methods to Improve Disk Performance

These methods include (1) disk arm prepositioning, (2) disk reorganization and defragmentation, (3) log-structured file systems, and (4) active disks and free-block disk scheduling.

21.2.3.1 Disk Arm Prepositioning

When disk utilization is low, the service time of future requests can be reduced by prepositioning the disk arm. By positioning the disk arm at the middle disk cylinder, the average seek distance for random requests will be reduced from $C/3$ to $C/4$ [King 1990].

With mirrored disks it is best to position one arm at $C/4$ and another arm at $3C/4$, when all requests are reads, so that the positioning time is reduced to $C/8$ [King 1990]. When a fraction w of requests are writes, a symmetry argument can be used to place the two arms at a distance s at the opposite sides of the middle disk cylinders; that is, $1/2 \pm s$ (the maximum seek distance is set to 1). Taking the derivative of the seek distance with respect to s yields $s_{opt} = 0.25(1 - 2w)/(1 - w)$ and the expected seek distance at s_{opt} is $E[d_{min}] = 0.25 - 0.125(1 - 2w)^2/(1 - w)$. Since $s_{opt} \leq 0$ for $w \geq 1/2$, it follows that s_{opt} should be set to zero, i.e., when write requests dominate, both disk arms should be placed at the middle disk cylinders.

The disadvantage of this method is that disk arm positioning can interfere with arriving requests and the chances for this increase with the arrival rate. Preemption of disk arm positioning requests is one method to deal with this problem.

Non-work-conserving schedulers, which introduce forced idleness [Kleinrock 1976], can be used to improve performance in some cases. In a general context, this is so if a current activity is not completed and switching to another activity will result in incurring the setup time twice, once to switch to the new activity and again to return to the current activity. Setup time in processing disk requests manifests itself as positioning time. It is best that the scheduler defers the processing of a new request until it has been ascertained that there are no further pending requests. The improvement in performance due to *anticipatory disk scheduling* is explored in the context of Apache file servers, Andrews filesystem, and the TPC-B benchmark in Iyer and Druschel [2001].

21.2.3.2 Disk Reorganization and Defragmentation

The access time to large files is improved by allocating them to contiguous disk blocks. This is not a problem if there is abundant disk space; but otherwise, a "flexible" file organization (the original UNIX file system) might allow blocks to be scattered all over the free disk space. The sequential reading of a file will then incur multiple positioning times, versus one in the best case. The Unix BSD FFS (Fast File System) achieved improved file performance by enforcing larger allocations and an improved directory organization [McKusick et al. 1984]. The performance of file systems allowing disk fragmentation can be improved by applying defragmentation programs, which even come packaged with some operating systems.

Seek distances, and hence overall performance, can be simply improved based on access frequencies, as in the well-known **organ pipe arrangement** [Wong 1980]; that is, the most popular file is placed on the middle disk cylinder and other files are placed next to it, alternating from one side to the other. File access frequencies, which vary over time, can be determined by monitoring file accesses. These frequencies are then used to adaptively reorganize the files to approximate an organ pipe organization [Akyurek and Salem 1995].

The average seek distance and time can be reduced by allocating disk files by taking into account the frequency with which files are accessed together. **Clustering algorithms** utilizing dynamic access patterns have been proposed for this purpose [Bennett and Franaszek 1977].

21.2.3.3 Log-Structured File Systems (LFS)

The **log-structured file systems (LFS)** paradigm is useful in an environment where read requests are not common, while there are many writes. This is so in a system with a large cache, where most read requests are satisfied by cache accesses and the disks are mainly used for writing (there is usually a timeout for modified data in the cache to be destaged). Caching can be at a client's workstation or PC, so that a user is involved in continuously modifying the locally cached file and saving it occasionally (or even frequently if the power supply is unreliable) at the disks of a centralized server.

LFS accumulates modified files in a large cache, which is destaged to disk in large chunks, called segments (e.g., the size of a disk cylinder), to prorate arm positioning overhead [Rosenblum and Ousterhout 1992]. Space previously allocated to altered files is designated as free space and a background garbage collection process is used to consolidate segments with free space to create almost full and absolutely empty segments for future destaging.

The write cost in LFS is a (steeply) increasing function of the utilization of disk capacity (u), which is the fraction of live data in segments (see Figure 3 in Rosenblum and Ousterhout [1992]) and a crossover point with respect to Unix's FFS occurs at $u = 0.5$. The following segment cleaning policies are investigated: (1) when should the segment cleaner be executed?, (2) how many segments should be cleaned?, (3) the choice of the segment to be cleaned, and (4) the grouping of live blocks into segments. The reader is referred to Rosenblum and Ousterhout [1992] for more details.

21.2.3.4 Active Disks and Free-Block Scheduling

Given the high value of disk arm utilization and the fact that each disk is equipped with a relatively powerful microprocessor with local memory, there have been recent proposals for **freeblock scheduling (FS)**, which utilizes opportunistic disk accesses as the arm moves in processing regular requests [Lumb et al. 2000]. FS can be used in conjunction with LFS to carry out *cleaning*, which is the defragmenting operation to prepare free segments. Other studies have considered nearest-neighbor search queries, data mining, etc. [Acharya et al. 1998, Riedel et al. 2000, Riedel et al. 2001].

The off-loading of CPU processing to the disk controller works well for simpler activities that can be localized to a disk; otherwise, activities at several disks must be coordinated at the server. In fact, despite their slower microprocessors, the overall processing capacity at (a large number of) disks might exceed the processing capacity of the server.

Another advantage of downloading database applications to disk controllers is a reduction in the volume of data to be transmitted. This is important when the I/O bus is a potential bottleneck. For example, computing the average salary ($\bar{S}$) of N employees, whose information is held in B byte records, will require the transmission of a few bytes vs. $N \times B$ bytes, allowing more disks to be connected to the bus. In case this data resides at K disks, then the kth disk sends the local average salary $\bar{S}_k$ and the number of employees N_k to the server, which then computes $\bar{S} = \sum_{k=1}^{K} \bar{S}_k N_k / N$, where $N = \sum_{k=1}^{K} N_k$.

21.3 RAID Disk Arrays

The main motivation for **RAID (Redundant Array of Inexpensive Disks)**, as reflected by its name, was to replace *Single Large Expensive Disks (SLEDs)* used in mainframe computers with an array of small form-factor, *inexpensive*, hard disk drives utilized in early PCs [Patterson et al. 1998] (see Section 21.3.1). In Section 21.3.2, common features of RAID levels are discussed under the title of RAID Concepts. Five RAID levels — RAID1–5 — were initially introduced, and two more levels (RAID0 and RAID6) were added later [Chen et al. 1994]. We proceed to discuss all RAID levels briefly, but dedicate separate sections to RAID1 (mirrored disks) and RAID5 (rotated parity arrays). The memory hierarchy and especially the disk controller cache are discussed next. We conclude with a brief discussion or reliability modeling in RAID.

21.3.1 Motivation for RAID

A design study reported in Gibson [1992] replaces one IBM 3390 with 12 useful actuators and 10.6-inch diameter disks with 70 IBM 0661 3.5-inch disks to maintain the same capacity. The increased number of disks and their low reliability (with respect to SLEDs) led to an inexpensive system with unacceptably low reliability. This issue was addressed by introducing fault-tolerance through redundancy (the R in RAID). An additional 14 inexpensive disks were added to the 70 disks, seven of which were for parity and seven were spares.

An added complication associated with RAID used in conjunction with mainframe computers available from IBM (and its compatibles), which ran the MVS/390 operating system (now renamed z/OS), was that MVS issued I/O commands to variable block size (count-key-data [CKD] and extended CKD [ECKD]) disks, which were unrecognizable by *fixed block architecture (FBA)* disks, for example, the IBM 0661 Lightning disk drives, with 512-byte sectors. Two solutions are:

- Rewrite the MVS file system software and I/O routines to access FBA disks. This solution, which is only available to IBM, was perhaps too costly to undertake because possibly most of the routines were written using the low-level assembler programming language.
- Provide a RAID controller that intercepts I/O requests originating from the mainframe to ECKD disks and emulates CKD/ECKD disks on FBA disks.

In effect, the RAID controller, in addition to providing access to the disks and controlling the cache, acts as a simulator of (no longer manufactured) SLEDs on FBA disks. Controllers' front-end processors receive I/O commands addressed to cylinder, track, and block numbers (on virtual 3380 or 3390 disks), which are translated to logical block numbers. The I/O commands to SCSI or IDE drives are issued by back-end processors.

All currently manufactured disks have a small form-factor, so that the I in RAID now stands for *independent*. The 3.5-inch disks are the most popular form-factor for servers, desktop, and notebook PCs, while smaller form-factor disks are used in mobile systems.

New capabilities associated with z/OS (a successor to MVS and OS/390) on IBM's mainframe computers for efficient access to disk arrays [Meritt et al. 2003]. It was possible for a request to be enqueued for an original disk drive, which had a request in progress. Given that the data on that drive was allocated over several disks made this enqueueing unnecessary when the request was for data on another physical drive. The unnecessary waiting is obviated by means of the *PAV (parallel access volume)* capability in z/OS.

21.3.2 RAID Concepts

We first introduce **striping**, which is utilized by most **RAID** levels. We then discuss fault-tolerance techniques used in RAID systems, followed by caching and prefetching in RAID systems. We finally discuss RAID reliability modeling.

21.3.2.1 Striping

Striping, later classified as RAID0 [Chen et al. 1994], is not a new concept and was used in an early, high-performance airline reservation systems. Data was stored on a few (middle) disk cylinders to reduce seek time. Striping was also used in Cray supercomputers for increased data rates, but such data rates can only be sustained by supercomputers. In this chapter we will deemphasize the topic of RAID parallelism because it is best discussed in conjunction with parallel file systems and associated applications [Jain et al. 1996, Jin et al. 2002].

Traditionally, files or datasets for commercial applications required a fraction of the capacity of a disk, so that multiple files with different access rates were usually assigned to the same disk. There was no guarantee that a random allocation of files would result in a balanced disk load and **data access skew** was observed in many systems. Several packages were developed to relocate datasets to balance the load. Local perturbations to the current allocation, rather than global reallocation, was generally considered to be sufficient [Wolf 1989].

Striping partitions (larger) files into **striping units**, which are allocated in a round-robin manner on all disks. The striping units in a row constitute a **stripe** (e.g., D0-D5, D6-D11, etc.) (see Figure 21.1).

Load balancing is attained because all disks are allocated striping units from all files, so that equal access rates to the blocks of a file result in uniform access to disks. Skew is possible for accesses to records of the same file; that is, some records can be accessed more frequently than others (e.g., according to a Zipfian distribution). The effect of highly skewed accesses to small files or nonuniform accesses to the blocks of a file is expected to be obviated by caching; that is, such small files and sets of records will tend to reside in main memory if the access rate is significant (e.g., once every "five minutes") [Gray and Reuter 1993].

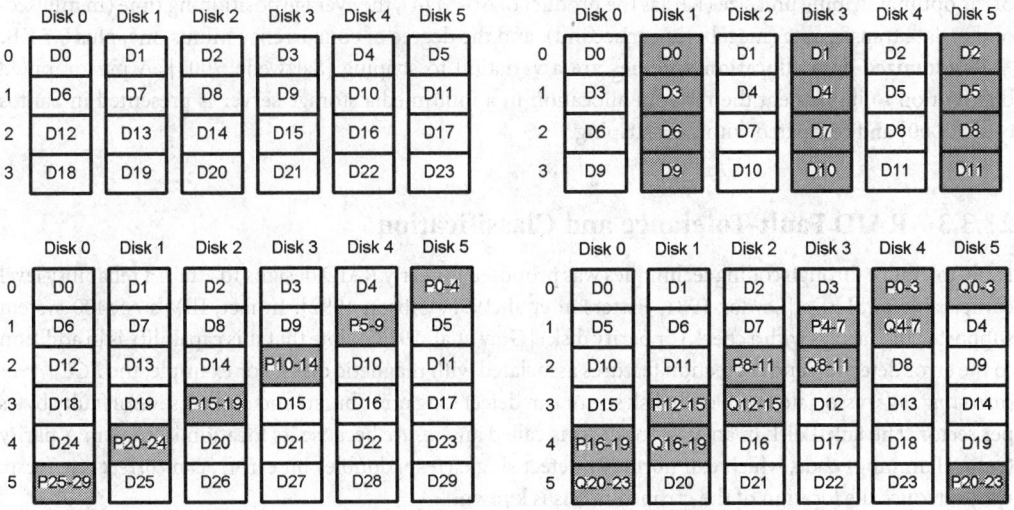

FIGURE 21.1   RAID0, RAID1, RAID5, and RAID6 layouts.

A possible disadvantage of striping is that distributing the segments of a file among several disks makes backup and recovery quite difficult. For example, if more than two disks fail in a RAID5 system, we need to reconstruct all disks, rather than just two. The **parity striping** data layout [Gray et al. 1990] described below could easily back up the data blocks at each disk separately, but then the parity blocks at all disks must be recomputed.

Striping over the disks of one array does not eliminate access skew across arrays, because each array holds a different set of files. Access skew can be eliminated by allocating data across disk array boundaries, but this requires coordination at a higher level.

The striping unit size has been an area of intense investigation. It is intuitively clear that it should be large enough to ensure that most commonly occurring request sizes can be satisfied by a single disk access. Access to small (4 or 8 KB) blocks are common in OLTP applications, while ad hoc queries and decisions support applications require accesses to much larger block sizes (e.g., 64 KB).

Data transfer time is a small fraction of arm positioning time for current disks, so that the stripe unit size should be selected to be large enough to accommodate 64-KB requests via a single disk access.

Servers run transactions at a high degree of concurrency and each transaction generates multiple I/O requests at short time intervals. Because of the high volume of I/O requests, the disks in the disk array are best utilized by not participating in parallel accesses. *Effective disk utilization* is the fraction of time that the disk is involved in "useful data transfers." Data transferred due to prefetching may not be useful because is it is not used by an application.

Accesses to very large blocks can be accommodated by large stripe units, but a very large stripe unit may introduce data access skew. In fact, very large blocks are best striped and accessed in parallel, so that the stripe unit size should be selected appropriately to optimize performance.

Parity striping maintains the regular data layout, but allows enough space on each disk for parity protection [Gray et al. 1990]. The parity blocks are placed in a separate area on each disk. Contrary to the reasoning given earlier regarding load balancing, it is argued in this paper that without striping, the user has direct control over data allocation and can ensure, for example, that two hot files are placed on two different disk drives, while this is not guaranteed in RAID5.

A performance comparison of parity striping with RAID5 is reported in Chen and Towsley [1993]. RAID5's stripe unit is assumed to be so small that there is no access skew; on the other hand, a logical request results in multiple disk accesses. The parity striped array is assumed to be susceptible to data skew, although each logical request is satisfied with one disk access. The relative performance of the two systems is compared using an analytic model based on M/G/1 queueing analysis [Kleinrock 1975].

A study for maximizing performance in striped disk arrays [Chen and Patterson 1990] estimates the size of the optimal striping unit size (KB) as the product of $S (\approx 1/4)$, the average positioning time (in milliseconds), data transfer rate (megabytes per second), and the degree of concurrency minus one, plus 0.5 KB.

Randomized data allocation schemes are a variation to striping [Salzwedel 2003]. A precomputed distribution to implement the random allocation in a multimedia storage server is presented in Santos et al. [2000] and compared with data striping.

21.3.3 RAID Fault-Tolerance and Classification

Fault-tolerance through coding techniques was proposed for early RAID designs to attain a reliability level comparable to SLEDs [Lawlor 1981, Patterson et al. 1998, Gibson 1992]. In fact, IBM's AS/400 system supported disk arrays with a check (or parity disk) [Gray et al. 1990]. Note that this capability is in addition to the error detection and correction features associated with magnetic disks. For example, the *ECC (error correcting code)* associated with each disk sector can detect long error bursts and correct several faulty bytes per sector. The failed disk in an array of disks is called an *erasure* because its location is known. A parity (resp. Hamming) code, which can normally detect single (resp. double) bit errors, can *correct* one (resp. two) bit, once the location of the erroneous bits is known.

Given that d_n denotes a single data bit on disk $1 \le n \le N$, then $p \leftarrow d_1 \oplus d_2 \cdots \oplus d_N$, where $\oplus$ is the *exclusive-OR (XOR)* operation. For example, given that d_1 (on disk one) is unavailable due to disk failure, then its value can be obtained by reading all the other disks in the parity group and computing:

$$d_1 \leftarrow p \oplus d_2 \oplus \cdots d_N$$

A single check (parity) disk is utilized in RAID3 and RAID4 disk arrays. RAID3 arrays have a small striping unit (e.g., a sector) and use parallel access to expedite large data transfers. Without synchronization, the time to complete parallel I/Os will be dominated by positioning times, so that larger block sizes should be transferred to prorate the positioning time overhead.

If disks are used exclusively for parallel processing, synchronization delay for seek times can be minimized by moving the disk arms together. *Spindle synchronization*, a functionality available to SCSI disk drives, can be used to minimize synchronization time due to latency.

In RAID3, single disk failures can be handled by reading the parity disk, in addition to the surviving data disks, to reconstruct missing data on demand. Similarly, when data is being written, the bits being written are XORed to generate and write the parity. Thus efficient **full stripe writes** can be easily achieved in RAID3.

RAID4 utilizes a larger stripe unit than RAID3, so that disks are usually accessed independently. A weakness of RAID4 is that the parity disk may become a bottleneck for write-intensive applications. This effect can be alleviated by rotating the parity at the stripe unit level, which is the only difference between RAID4 and RAID5. In fact, the **left symmetric** parity organization has been shown to have several desirable properties [Gibson 1992] (see Figure 21.1).

RAID1 or mirrored disks is a degenerate case of RAID4 with one disk protected by the other (see Figure 21.1). Several schemes to combine striping with RAID1 are discussed in Section 21.4.

RAID2 disk arrays utilize a *Hamming code*, which could be used to detect the location of up to two failed disks, but because the location of these disks is already known, it can correct two disk failures [Gibson 1992]. The number of check disks K required for N data disks satisfies $2^K \geq N + K + 1$; for example, $N = 4$ requires $K = 3$. The check disk overhead decreases as N increases. Other linear coding techniques to handle multiple disk failures are described in Hellerstein et al. [1994] but these techniques have not been adopted in practice.

Two-dimensional parity protection is based on an $N \times N$ array of data disks, with N disks providing horizontal and another N disks providing vertical parity [Lawlor 1981, Gibson 1992]. More elaborate methods for two-dimensional parity are described in Newberg and Wolfe [1994]. A clever scheme that deals with the unreliability of disks, as well as other components of a RAID5 disk array, is proposed in Ng [1994].

The RAID6 classification was introduced to classify disk arrays that have two check disks and can tolerate two disk failures, such as StorageTek's Iceberg disk array [Chen et al. 1994]. Iceberg uses $P + Q$ or more formally Reed-Solomon codes, which is a *non-binary symbol code* [MacWilliams and Sloane 1977]. Two different linear combinations of data bits, called P and Q, are computed and stored according to a left-symmetric distribution in parallel diagonals as in RAID5 (see Figure 21.1). In the case of one-disk failures, the parity P can be used to reconstruct the data, as in RAID5. In the case of two disk failures, where the data being referenced is on one of two unavailable data striping units, is the more complicated case. We need to access the corresponding $N - 2$ data blocks and the corresponding P and Q check blocks, from N out of $N + 2$ disks. To obtain the missing data block (and the one at the other failed disk), we solve two linear equations in two unknowns, except that these calculations are done over *finite fields*.

The EVENODD scheme has the same redundancy level as RAID6, which is the minimal level of redundancy, but only uses the parity operation [Blaum et al. 1995].

The previous dichotomy of order of magnitude differences in disk capacity has disappeared because SLEDs are no longer manufactured. The practice of using parity to protect against single disk failures has persisted because it has two advantages:

1. There is no disruption in data accesses. The system continues its operation by recreating requested blocks on demand, although performance is degraded.
2. The system can automatically rebuild the contents of the failed disk on a *hot spare*, if such a disk is available. A hot spare is required because some systems do not allow hot swapping (i.e., replacing a broken disk with a spare disk) while the system is running.

Self-Monitoring Analysis and Recording Technology (SMART) can be used to detect that a disk failure is imminent, so that the disk can be backed up before it fails. This is advantageous because it does not involve a complicated rebuild process, which is discussed in the context of RAID5 in Section 21.5.

21.3.4 Caching and the Memory Hierarchy

Disk blocks are cached in the database or file buffers in main memory, the cache associated with the RAID controller, and the onboard disk cache. The main advantage of the caches is in satisfying read requests, and the higher the level of the cache, the lower the latency. Data cached in main memory obviates the need for a request to the lower level of the memory hierarchy. Otherwise, the server issues an I/O request, which is intercepted by a processor at the disk array controller. An actual I/O request is issued to the disk, only if the data is not cached at this level.

Part of the disk array controller cache, which is nonvolatile storage — NVS or NVRAM — can be used to provide a *fast-write* capability; that is, there is an indication to the computer system as soon as a write to NVS is completed [Menon and Cortney 1993]. To attain a reliability comparable to disk reliability, the NVRAM might be duplexed.

Fast writes have the advantage that the destaging (writing out) of modified blocks to disk can be deferred, so that read requests that directly affect application response time can be processed at a higher priority than writes. Caching of disk blocks also occurs at the database buffer, so that when a data block is updated with a NO-FORCE transaction commit policy, only logging data has to be written to disk, while dirty blocks remain in the database buffer, until they are replaced by the cache replacement policy or by checkpointing [Ramakrishnan and Gehrke 2003].

Two additional advantages of caching in the NVS cache include: (1) dirty blocks in the cache may be modified several times, before the block is destaged to disk; and (2) multiple blocks in the cache can be processed at a much higher level of efficiency by taking advantage of disk geometry. We will discuss caching further in conjunction with RAID1 and RAID5 disk arrays.

An early study of miss ratios in disk controllers is Smith [1985]. Cache management in RAID caches is investigated via trace-driven simulations in Treiber and Menon [1995]. The destaging from the cache is started when its occupancy reaches a certain high mark and is stopped at a low mark. A more sophisticated destaging policy, which starts destaging based on the rate at which the cache is being filled, is given in Varma and Jacobson [1998].

21.3.5 RAID Reliability Modeling

Reliability and availability modeling of RAID systems was an early area of interest because the inexpensive disks were less reliable than the more expensive, large-form factor disks used in mainframes. There was also a significant increase in the number of disk drives to maintain the same capacity.

An investigation of the time-to-failure of disks showed that this time can be accurately approximated by an exponential distribution [Gibson 1992], which is a special case of the Weiball distribution [Trivedi 2002]. Denoting the parameter of the exponential distribution with λ, the **Mean Time To Failure (MTTF)** $= 1/\lambda$. To allow a simplified analysis with a Markov chain model, the repair time can be represented by an exponential distribution with rate μ, so that the **Mean Time to Repair (MTTR)** $= 1/\mu$.

Reliability modeling is concerned with determining the probability that the system will fail at a given time and the time to failure, starting with a system operating in normal mode [Trivedi 2002]. The duration of time the system has been operating in that mode is not relevant due to the *memoryless property* of the exponential distribution [Kleinrock 1975, Trivedi 2002], that the *residual lifetime* of a component with an exponential failure rate is independent of the age of the component. Given that disk failure rates and the repair rate are exponential, we can represent the behavior of the system with a Markov chain [Trivedi 2002].

A RAID5 disk array with $N + 1$ disks can be in one of three states: **normal**, **degraded**, and **failed**, which correspond to the three states of the Markov chain: $S_{N+1-i}, 0 \leq i \leq 2$, where i is the number of failed disks. At S_{N+1}, the system is in normal operating mode and the system failure rate is $(N + 1)\lambda$.

A disk failure leads to S_N, that is, operation in degraded mode. The system at S_N is repaired at rate μ, which will lead the system back to normal mode (S_{N+1}). The system at S_N fails with rate $N\lambda$, this leads to S_{N-1}, and this happens with probability $p_{fail} = N\lambda/(\mu + N\lambda)$, which is very small because μ is much larger than $N\lambda$. The transient solution to this birth-death system can be easily obtained using well-known techniques [Trivedi 2002]. The mean time to failure, which is called the **Mean Time to Data Loss (MTDL)** in this case, is the mean time to transition from S_{N+1} to S_{N-1} [Gibson 1992, Trivedi 2002]:

$$MTDL = \frac{(2N+1)\lambda + \mu}{N(N+1)\lambda^2} \approx \frac{MTTF^2}{N(N+1)MTTR}$$

More sophisticated techniques can be used to obtain the MTDL with a general repair time distribution, and it is observed that the MTDL is determined by mean repair time.

RAID6's MTDL is given in Chen et al. [1994]. Analytic solutions for the reliability and availability of various RAID models appear in Gibson [1992], which also reports experiences with an off-the-shelf reliability modeling package that was not able to solve all submitted problems. This was due to a state space explosion problem. Customized solutions that take advantage of symmetry, use state aggregation and hierarchical modeling techniques, or yield approximate solutions by ignoring highly improbable transitions are applicable. Rare event simulation is another technique to deal with this problem.

21.4 RAID1 or Mirrored Disks

This section is organized as follows. We first provide a categorization of routing and disk scheduling policies in mirrored disks. We consider two cases: when each disk has an independent queue and when the disks have a shared queue.

We next consider the scheduling of requests when an NVS cache is provided to hold write requests. This allows read requests to be processed at a higher priority than writes. Furthermore, the writes are processed in batches to take advantage of disk geometry to reduce their completion time.

Finally, we present several data layouts for mirrored disks and compare them from the viewpoint of reliability and balancedness of the load when operating with a failed disk.

21.4.1 Request Scheduling with Mirrored Disks

In addition to higher reliability, mirrored disks provide twice the access bandwidth of single disks for processing read requests, which tend to dominate write requests in many applications. This effect was first identified and quantified assuming that seeks are uniform over $(0, 1)$ in Ng [1987]. Similar results were obtained in Bitton and Gray [1988] by considering disks with C cylinders having equal capacities. Requests are unformly distributed over all disk blocks and hence disk cylinders. Sending a request to the disk providing the shorter seek distance results in a reduction in the mean seek distance from $C/3$ to $C/5$, but the mean seek distance for writes is $7C/15$, which is the expected value of the maximum of two seeks distances to update the data on both disks.

Mirrored disks can be classified into two configurations as far as their queueing structure is concerned [Thomasian et al. 2003]:

1. **Independent queues (IQ).** Read requests are immediately routed to one of the disks according to some routing policy, while write requests are sent to both disks.
2. **Shared queue (SQ).** Read requests are held in this queue, so that there are more opportunities for improving performance.
   Many variations are possible, for example, deferred forwarding of requests from the shared queue (at the router) to independent queues.

Request routing in IQ can be classified as *static* or *dynamic*. Uniform and round-robin routing are examples of static policies. Round-robin routing is simpler to implement than uniform routing and improves performance by making the arrival process more regular [Thomasian et al. 2003].

The router, in addition to checking whether a request is a read or a write, can determine other request attributes (e.g., the address of the data being accessed). Such affinity-based routing is beneficial for sequential accesses to the same file. Carrying out such requests on the same disk makes it possible to take advantage of onboard buffer hits due to prefetching.

A dynamic policy takes into account the number of requests at each disk or the composition of requests at a disk, etc. A *join the shortest queue (JSQ)* policy can be used in the first case, but this policy is known not to improve performance when requests have high variability. Simulation studies have shown that the routing policy has a negligible effect for random requests, so that performance is dominated by the local scheduling policy [Thomasian et al. 2003].

SQ provides more opportunities than IQ to improve performance because more requests are available to carry out optimization. For example, the SATF policy with SQ provides better performance than is possible with IQ, because the shared queue is twice the length of individual queues [Thomasian et al. 2003].

21.4.2 Scheduling of Write Requests with an NVS Cache

The performance of a mirrored disk system without an NVS cache can be improved by using a write-anywhere policy on one disk (to minimize disk arm utilization and susceptibility to data loss), while the data is written in place later on the primary disk. Writing in place allows efficient sequential accesses, while a special directory is required to keep track of blocks written. This is a brief description of the *distorted mirrors* method [Solworth and Orji 1991], which is one of several methods for improving mirrored disk performance.

Caching of write requests in NVS can be used to improve the performance of mirrored disks. Prioritizing the processing of read requests yields a significant improvement in response time, especially if the fraction of write requests is high. We can process write requests more efficiently by scheduling them in batches optimized with respect to the data layout on disk. The scheme proposed in Polyzois et al. [1993] runs mirrored disks in two phases; while one disk is processing read requests, the other disk is processing writes in a batch mode using CSCAN.

The performance of the above method can be improved as follows Thomasian and Liu [2003]: (1) eliminating the forced idleness in processing write requests individually; (2) using SATF or preferably an exhaustive enumeration, which is only possible for sufficiently small batch sizes, say 10, instead of CSCAN, to find an optimal destaging order; (3) introducing a threshold for the number of read requests, which when exceeded, defers the processing of write batches.

21.4.3 Mirrored Disk Layouts

With **mirrored pairs**, which is the configuration used in Tandem's NonSTOP SQL, the read load on one disk is doubled when the other disk fails.

The **interleaved declustering** method used in the Teradata/NCR DBC/1012 database machine (1) organizes disks as clusters of N disks, and (2) designates one half of each disk as the primary data area and the second half as secondary, which mirrors the primary data area. The primary area of each disk is partitioned conceptually into $N - 1$ partitions, which are allocated in round-robin manner over the secondary areas of the other $N - 1$ disks. If one disk fails, the read load on surviving disks will increase only by a factor of $1/(N - 1)$, as opposed to the doubling of load in standard mirrored disks. This method is less reliable than mirrored pairs in that the failure of any two disks in a cluster will result in data loss, while with mirrored pairs data loss occurs if two disks constituting a pair fail, which is less likely.

The **group rotate declustering** method is similar to interleaved declustering, but it adds striping. Copy 1 is a striped array and copy 2 stores the stripes of copy 1 in rotated manner [Chen and Towsley 1996]. Let's assume copy 1 and 2 both have four disks. The stripe units in the second row of copy 1 ($F5, F6, F7, F8$)

will appear as $(f8, f5, f6, f7)$ in copy 2, and this rotation continues. There is no rotation when (row_number) mod(4) = 0. This data layout has the advantage that the load of a failed disk is uniformly distributed over all disks, but has the drawback that more than two disk failures are more likely to result in data loss than standard mirrored disks or the following method.

The **chained declustering** method alleviates the aforementioned reduced reliability problem [Hsiao and DeWitt 1993]. All N disks constitute one cluster. Each disk has a primary and secondary area, so that the primary data on each disk is allocated on the secondary area of the following disk modulo(N). Consider four disks denoted by $D0$ through $D3$ whose contents are given as $[P0, S3]$, $[P1, S0]$, $[P2, S1]$, $[P3, S2]$, where P and S denote the primary and secondary data, respectively. Assume that originally all read requests are sent to the primary data and have an intensity of one request per time unit. When D1 is broken, requests can be routed to sustain the original load, while keeping disk loads balanced, as follows: $[P0, (1/3)S3]$, $[(1/3)P2, S1]$, $[(2/3)P3, (2/3)S2]$.

21.5 RAID5 Disk Arrays

In this section we describe the operation of a RAID5 system in normal, degraded, and rebuild modes. We also describe some interesting variations of RAID5.

21.5.1 RAID5 Operation in Normal Mode

Reads are not affected, but updating a single data block d_{old} to d_{new} requires the updating of the corresponding parity block p_{old}: $p_{new} \leftarrow d_{old} \oplus d_{new} \oplus p_{old}$. When d_{old} and p_{old} are not cached, the writing of a single block requires four disk accesses, which is referred to as the **small write penalty**. This penalty can be reduced by carrying out the reading and writing of data and parity blocks as **read-modify-write (RMW)** accesses, so that extra seeks are eliminated. On the other hand, it may be possible to process other requests opportunistically, before it is time to write after a disk rotation.

Simultaneously starting the RMW access for data and parity blocks may result in the parity disk being ready for writing *before* the data block has been read. One way to deal with this problem is to incur additional rotations at the parity disk until the necessary data becomes available. A more efficient way is to start the RMW for the parity block only after the data block has been read. Such precedence relationships are best represented by dags (directed acyclic graphs) [Courtright 1997].

Several techniques have been proposed to reduce the write overhead in RAID5. One method provides extra space for parity blocks, so that their writing incurs less rotational latency. These techniques are reviewed in Stodolski et al. [1994], which proposes another technique based on batch updating of parity blocks. The system logs the "difference blocks" (exclusive-OR of the old and new data blocks) on a dedicated disk. The blocks are then sorted in batches, according to their disk locations, so as to reduce the cost of updating the parity blocks.

Modified data blocks can be first written onto a duplexed NVS cache that has the same reliability level as disks. Destaging of modified blocks due to write requests and the updating of associated parity blocks can therefore be deferred and carried out at a lower priority than reads.

The small write penalty can be eliminated if we only have full stripe writes, so that the parity can be computed on-the-fly. Such writes are rare and can only occur when a batch application is updating all of the data blocks in the dataset sequentially. On the other hand, the aforementioned LFS paradigm can be used to store a stripe's worth of data in the cache to make such writes possible. The previous version of updated blocks is designated as free space, and garbage collection is carried out by a background process to convert two half-empty stripes to one full and one empty stripe. The **log-structured array (LSA)** proposed and analyzed in Menon [1995] accomplishes just that. While this analysis shows that LSA outperforms RAID5, it seems that updates of individual blocks will result in a data allocation that does not lend itself to efficient sequential data access, unless, of course, the block size is quite large, say 64 KB, so as to accommodate larger units of transfer for database applications.

21.5.2 RAID5 Operation in Degraded Mode

A RAID5 system can continue its operation in *degraded mode* with one failed disk. Blocks on the failed disks can be reconstructed on demand, by accessing all the corresponding blocks on surviving disks according to a *fork-join request* and XORing these blocks to recreate the missing block. A fork-join request takes more time, which is the maximum of the times at all disks.

Given that we have a balanced load due to striping and that each one of the surviving disks must process its own requests, in addition to the fork-join requests, results in a doubling of disk loads when all requests are reads.

In the case of write requests, there is no need to compute the parity block if the disk on which the parity resides is broken. In case d_{old} is not cached and the corresponding data disk, say the $N+1$st disk, is broken, then the parity block that resides on disk one can be computed as: $p_1 \leftarrow d_2 \oplus d_3 \ldots d_N$.

Given the increase in RAID5 disk utilizations in degraded mode, the disk utilizations should be below 50% in normal mode, when all requests are reads, although higher initial disk utilizations are possible otherwise. **Clustered RAID**, proposed in Muntz and Lui [1990], solves this problem using a *parity group size* that is smaller than the number of disks. This means that only a fraction of disks are involved in rebuilding a block [Muntz and Lui 1990], so that less intrusive reconstruction is possible. Complete or full block designs require infinite capacity, so that **Balanced Incomplete Block Designs (BIBDs)** [Hall 1986] have been proposed to deal with finite disk capacities [Ng and Mattson 1994, Holland et al. 1994]. **Nearly random permutations** is a different approach [Merchant and Yu 1996].

Six properties for ideal layouts are given in Holland et al. [1994]:

1. Single failure correcting: the stripe units of the same stripe are mapped to different disks.
2. Balanced load due to parity: all disks have the same number of parity stripes mapped onto them.
3. Balanced load in failed mode: the reconstruction workload should be balanced across all disks.
4. Large write optimization: each stripe should contain $N-1$ contiguous stripe units, where N is the parity group size.
5. Maximal read parallelism: reading $n \leq N$ disk blocks entails accessing different n disks.
6. Efficient mapping: the function that maps physical to logical addresses is easily computable.

The *Permutation Development Data Layout (PDDL)* is a mapping function described in Schwartz [1999] that has excellent properties and good performance in light loads (like the PRIME data layout [Alvarez et al. 1998]) and heavy loads (like the DATUM data layout [Alvarez et al. 1997]).

21.5.3 RAID5 Operation in Rebuild Mode

In addition to performance degradation, a RAID5 system operating in degraded mode is vulnerable to data loss if a second disk failure occurs. If a hot spare is provided, the rebuild process should be initiated as soon as possible when a disk fails. This is important because a failed disk may not be replaceable without bringing the system down (i.e., disrupting its operation).

Instead of wasting the bandwidth of the hot spare, the **distributed sparing** scheme distributes the spare areas over $N+2$ disks [Thomasian and Menon 1997]. Distributed sparing is also used in PDDL [Schwartz 1999]. The drawback of hot sparing is that a second round of data movement is required to return the system to its original configuration.

The **parity sparing** method utilizes two parity groups, one with $N+1$ and the other with $N'+1$ disks, which are combined to form a single group with $N+N'+1$ disks after one disk fails [Chandy and Reddy 1993].

Two variations of RAID5 rebuild have been proposed in the research literature. The **disk-oriented rebuild** scheme reads successive disk tracks from surviving disks, when the disk is idle. These tracks are XORed to obtain the track on the failed disk, which is then written onto a spare disk (or a spare area in distributed sparing). The performance of this method is studied via simulation in Holland et al. [1994] and analytically in Thomasian and Menon [1997]. The **stripe-oriented rebuild** scheme synchronizes at the stripe level, or more generally at the level of the *rebuild unit*. Its performance is evaluated via simulation and shown to be inferior compared to the former rebuild method in Holland et al. [1994]. Another technique is to allow one rebuild request at a time, which is evaluated in Merchant and Yu [1996] as a *queueing system with permanent customers* [Boxma and Cohen 1991].

A hybrid RAID1/5 disk array called AutoRAID is described in Wilkes et al. [1996], which provides RAID1 at the higher level (of disk) storage hierarchy and RAID5 at the lower level. Data is automatically transferred between the two levels to optimize performance.

21.6 Performance Evaluation Studies

In this section we provide an overview of analytic and simulation studies to evaluate the performance of disks and disk arrays. Attention is also paid to simulation tools available for this purpose.

A recent review article on performance analysis of storage systems is Shriver et al. [2000], which also pays attention to tertiary storage systems.

21.6.1 Single Disk Performance

Disk simulation studies can be classified into trace- and random-number-driven simulations. Trace-driven simulation is based on traces of I/O activity. A simplified entry in the trace consists of a timestamp, disk block (sector) numbers, number of blocks being accessed, and whether a request is a read or a write. The trace is run through the simulator of a particular disk drive to determine the performance characteristic of interest (e.g., the mean response time of disk accesses).

An influential study based on a trace-driven simulation showed that an extremely detailed model is required to achieve high accuracy in estimating disk performance [Ruemmler and Wilkes 1994]. A complementary simulation study that uses trace- and random-number-driven simulations is Worthington et al. [1994].

One of the problems with simulating disk drives is having access to their detailed characteristics. Although such information is usually published in a manual, finding this information is a time-consuming task. More importantly, some important information might be missing. The DIXTRAC tool automates the extraction of disk parameters [Schindler and Ganger 1999], which can be used by the DiskSim simulation tool [DiskSim]. Extracted disk parameters are available at [Diskspecs]. DiskSim is a successor to RAIDframe, which can be used for simulation of disk arrays, as well as fast prototyping [Courtright et al. 1996].

An I/O trace, in addition to being used in trace-driven simulations, can be analyzed to characterize the I/O request pattern, which can then be used in a random-number-driven simulation study. A positive aspect of the latter approach is that it is easy to vary the arrival rate of requests and especially the characteristics of disk requests, while basing these variations on a realistic workload.

Requests to a modern disk drive are specified by the address of the first (logical) block (sector) number and the number of blocks to be transferred. Logical block numbers are translated into physical block numbers by the disk controller, taking into account spare areas, as well as faulty sectors and tracks. It is a straightforward task to use the trace obtained on one disk model to estimate performance on another disk model, but there are some complications, such as track alignment, in case the first data allocation was optimized for the particular disk drive.

The analysis of an I/O trace for an OLTP environment showed that 96% of requests are to 4-KB blocks and 4% to 24-KB blocks [Ramakrishnan et al. 1992]. Other notable disk I/O characterization studies are Ruemmler and Wilkes [1992], Hsu et al. [2001], and Hsu and Smith [2003], where the second study is concerned with I/O requests at the logical level.

Referring back to the distinction between discrete and synchronous requests, it is meaningful to vary the arrival rate of discrete requests, although this may not be so for synchronous requests since a synchronous request is generated only after a previous one is completed.

In random-number-driven or analytic studies, the workload at a single disk is characterized by the arrival process, request sizes, and the fraction of requests of a certain size, the fraction of reads and writes, and the spatial distribution of requests (e.g., random or sequential).

The sum of the average positioning time with an FCFS policy and data transfer time yields the mean service of requests and hence the maximum throughput of the disk (λ_{max}). The disk controller overhead is known, but may be (partially) overlapped with disk accesses, so we will ignore it in this discussion. This is a lower bound to λ_{max} because positioning time can be reduced by judicious disk scheduling and we have ignored the possibility of hits at the onboard disk cache. Note that the hit ratio at the onboard cache is expected to be negligibly small for random I/O requests.

In dealing with discrete requests, most analytic and even simulation studies assume Poisson arrivals. With the further assumption that disk service times are independent, the mean disk response time with FCFS scheduling can be obtained using the *M/G/1 queueing model* [Kleinrock 1975, Trivedi 2002]. There is a dependence among successive disk requests as far as seek times are concerned, because an access to a middle disk cylinder is followed by a seek of distance $C/4$ on the average for random requests (C is the number of disk cylinders), while the distance of a random seek after an access to the innermost or outermost disk cylinders is $C/2$ on the average. Simulation studies have shown that this queueing model yields fairly accurate results.

The analysis is quite complicated for other scheduling policies [Coffman and Hofri 1990], even when simplifying assumptions are introduced to make the analysis tractable. For example, the analysis of the SCAN policy in Coffman and Hofri [1982] assumes: (1) Poisson arrivals to each cylinder; (2) the disk arm seeks cylinder-to-cylinder, even visiting cylinders not holding any requests; and (3) the processing at all cylinders takes the same amount of time. Clearly, this analysis cannot be used to predict the performance of the SCAN policy in a realistic environment.

Most early analytic and simulation studies were concerned with the relative performance of various disk scheduling methods; for example, SSTF has a better performance than FCFS, at high arrival rates, and SCAN outperforms SSTF [Coffman and Hofri 1990]. Other studies propose a new scheduling policy and carry out a simulation study to evaluate its performance with respect to standard policies [Geist and Daniel 1987, Seltzer et al. 1990]. Simulation studies of disk scheduling methods, which also review previous work, are Worthington et al. [1994], Thomasian and Liu [2002a], and Thomasian and Liu [2002b].

There have been numerous studies of multidisk configurations, where the delays associated with I/O bus contention are taken into account. **Rotational Positing Sensing (RPS)** is a technique to detect collisions when two or more disks connected to a single bus are ready to transmit at the same time, in which case only one disk is the winner and additional disk rotations are incurred at the other disks, which can result in a significant increase in disk response time. There is also a delay in initiating requests at a disk if the bus is busy. Reconnect delays are obviated by onboard caches, but a detailed analysis requires a thorough knowledge of the bus protocols Approximate analytic techniques to analyze the disk subsystem of mainframe computer systems given in Lazowska et al. [1984] are *not* applicable to current disk subsystems.

Bottleneck analysis is used in Section 7.11 in Hennessey and Patterson [2003] to determine the number of disks that can be supported on an I/O bus and an M/M/1 queueing analysis to estimate mean response time [Kleinrock 1975].

21.6.2 RAID Performance

There have been very few performance (measurement) studies of commercial RAID systems reported in the open literature. There have been several RAID prototypes — Hager at IBM, RAID prototypes at Berkeley [Chen et al. 1994], the TickerTAIP RAID system at HP [Cao et al. 1994], the Scotch prototype at CMU [Gibson et al. 1995], AutoRAID at HP Labs [Wilkes et al. 1996] — and performance results have been reported for most of them. Such studies are important in that they constitute the only way to identify bottlenecks in real systems and to develop detailed algorithms to ensure correct operation [Courtright 1997]. Some prototypes are developed for the purpose of estimating performance, in which case difficult-to-implement aspects, such as recovery, are usually omitted [Hartman and Ousterhout 1995].

One reason for the lack of performance results for RAID may be due to the unavailability of a common benchmark, but steps in this direction have been taken recently [Storage Performance Council 2003].

A performance evaluation tool for I/O systems is proposed in Chen and Patterson [1994] that is self-scaling and adjusts dynamically to the performance characteristics of the system being measured.

There have been numerous analytical and simulation studies of disk arrays. Markovian models (with Poisson arrivals and exponential service times) have been used successfully to investigate the relative performance of variations of RAID5 disk arrays [Menon 1994]. Several performance evaluation studies of RAID5 systems have been carried out based on an M/G/1 model. Most notably, RAID5 performance is compared with parity striping [Gray et al. 1990] in Chen and Towsley [1993] and with mirrored disks in Chen and Towsley [1996]. An analysis of clustered RAID in all three operating modes appears in Merchant and Yu [1996]. An analysis of RAID5 disk arrays in normal, degraded, and rebuild mode appears in Thomasian and Menon [1994] and Thomasian and Menon [1997], where the former (resp. latter) study deals with RAID5 systems with dedicated (resp. distributed) sparing. These analyses are presented in a unified manner in Thomasian [1998], which also reviews other analytical studies of disk arrays.

An analytic throughput model is reported in Uysal et al. [2001], which includes a cache model, a controller model, and a disk model. Validation against a state-of-the-art disk array shows an average 15% prediction error.

A simulation study of clustered RAID is reported in Holland et al. [1994], which compares the effect of disk-oriented and stripe-oriented rebuild. The effect of parity sparing on performance is investigated in Chandy and Reddy [1993]. The performance of a RAID system tolerating two-disk failure is investigated in Alvarez et al. [1997] via simulation, while an M/G/1-based model is used to evaluate and compare the performance of RAID0, RAID5, RAID6, RM2, and EVENODD organizations in Han and Thomasian [2003].

It is difficult to evaluate the performance of disk arrays via a trace-driven simulation because, in effect, we have a very large logical disk whose capacity equals the sum of the capacities of several disks. A simulation study investigating the performance of a heterogeneous RAID1 system (MEMS-based storage backed up by a magnetic disk) is reported in Uysal et al. [2003]. Specialized simulation and analytic tools have been developed to evaluate the performance of MEMS-based storage; see, for example, Griffin et al. [2000].

A comprehensive trace-driven simulation study of a cached RAID5 design is reported in Treiber and Menon [1995], while Varma and Jacobson [1998] consider an improved destaging policy based on the rate at which the cache is filled as well as its utilization.

21.6.3 Analysis of RAID5 Systems

As discussed, a tractable analysis of RAID5 systems can be obtained by postulating a Poisson arrival process. Two complications are estimating the mean response time of fork-join requests and the rebuild time. Analytical results pertinent to these analyses are presented here.

21.6.3.1 Fork-Join Synchronization

Analytic solutions of fork-join synchronization are required for the analysis of disk arrays in degraded mode (e.g., to estimate the mean response time of a read request to a failed disk). This results in read requests to all surviving disks, so that we need to determine the maximum of response times. In what follows, we give a brief overview of useful analytical results.

In a 2-way fork-join system with two servers, Poisson arrivals (with arrival rate λ), exponential service times (with rate μ), it has been shown that $R_2^{F/J} = (1.5 - \rho/8)R$, where ρ is the utilization factor $\rho = \lambda/\mu$ and $R = 1/(\mu - \lambda)$ is the mean response time of an M/M/1 queueing system.

Curve-fitting to simulation results was used in Nelson and Tantawi [1988] to extend this formula to $K > 2$ servers:

$$R_k^{F/J} = \left[\alpha(\rho) + (1 - \alpha(r)) \frac{H_k}{H_2} \right] R_2^{F/J}$$

where $\alpha(\rho) = 4\rho/11$ and $H_K = \sum_{k=1}^{K} 1/k$.

The assumption that arrivals are Poisson and that service times are exponential is relaxed in Varma and Makowski [1994], which combines heavy and light traffic limits to derive an interpolation approximation. Validation against simulation shows this approximation to be quite accurate.

The above fork-join synchronization delay formula was utilized in several analyses of RAID5 disk arrays with Markovian assumptions: Poisson arrivals and exponential service times (see, e.g., Menon [1994]). A more careful examination of RAID5 operation in degraded mode shows that each surviving disk, in addition to processing fork-join requests, also processes its regular workload, so that these requests interfere with fork-join requests. It is stated in Nelson and Tantawi [1988] that the above formula also holds when there is interfering traffic, but, in fact, $R_K^{F/J}$ is a lower bound and R_K^{max} (the expected value of the maximum of response times at individual nodes) tends to be an upper bound [Thomasian 1997]. This is because interfering requests make the components of a fork-join request less dependent. The two bounds are quite close when the service time has a small coefficient of variation, but interpolation is required otherwise.

Computing the expected value of the maximum is trivial when the response time distributions are exponentially distributed, which is the case with exponential arrivals (with rate λ) and service times (with rate μ). For an N-way fork-join queueing system with mean response time R at one of the servers we have: $R_N^{max} = R \sum_{n=1}^{N} 1/n$, where $R = 1/(\mu - \lambda)$ [Trivedi 2002]. Note that $R_2^{max} = 1.5R$, so that as $\rho \to 1$ $R_2^{max} - R_2^{F/J} \to R/8$.

The coefficient of variation of response time is $c_R = \sigma_R/R$, where σ_R is the standard deviation of response time, and tends to be smaller than one when requests are to small uniformly distributed disk blocks. In this case, the response time distribution can be approximated by an Erlang distribution [Kleinrock 1975, Trivedi 2002]. This distribution consists of k exponential stages, each with a rate $k\mu$, so that the mean is $\bar{x} = 1/\mu$ and the *coefficient of variation* $c_R^2 = 1/K$. Given R and c_R^2, the parameters of the Erlang distribution are $\mu = 1/R$ and $k = ceil(1/c_R^2)$. More generally, for any c_R^2, the response time can be approximated by a Coxian distribution [Kleinrock 1975, Trivedi 2002], but there are more parameters to be estimated in this case [Chen and Towsley 1993]. Given the mean and standard deviation of disk response time, R_N^{max} can be obtained approximately using the *extreme value distribution* $R_N^{max} = R + \sqrt{6}\sigma_R ln(N)/\pi$. More details are given in Thomasian [1997] and Thomasian [1998].

21.6.3.2 Vacationing Server Model for Rebuild

The vacationing server model for M/G/1 queueing systems is used to analyze disk-oriented rebuild in a RAID5 system with dedicated sparing in Thomasian and Menon [1994]. This analysis is extended to distributed sparing in Thomasian and Menon [1997] by introducing an additional iterative step that makes the analysis more complicated. Only the former analysis is described here for the sake of brevity.

Each disk behaves as an M/G/1 queue that undergoes busy and idle periods. After the disk has been idle, a busy period starts when a new request arrives. After completing the processing of (external) requests and the end of the *busy period* [Kleinrock 1975], a sequence of tracks are read by the rebuild process and each one of these reads is considered a *vacation* [Kleinrock 1975]. The durations of the vacations are different because the first track being read (in an idle period) incurs a seek, while the succeeding track reads do not incur a seek. This behavior can be modeled as a vacationing server model where successive vacations have different distribution times [Takagi 1991]. No new vacations are started after the arrival of the first external request.

The service time of the first request is elongated by the residual vacation time (with mean $\bar{r}$) [Thomasian and Menon 1994]. The busy period started by this first unusual request is different from an ordinary busy period and is called a *delay cycle* [Kleinrock 1976]. The mean waiting time for a regular disk request is $\bar{r} + W$, where W is given by the Pollaczek-Khinchine formula [Kleinrock 1975]; in other words, $\bar{r}$ is added to the mean waiting time of all requests, rather than just the first one.

The mean cycle time, defined as the sum of a delay cycle and the vacation time, can be easily computed. The mean rebuild time is equal to the number of tracks divided by the mean number of tracks read per cycle time. There is the additional complication that the rate of rebuild speeds up as more and more read requests are carried out via *read redirection*, by reading rebuilt data directly from the spare disk [Muntz and Lui 1990]. Zoning is a similar complication, and can be dealt with similarly by discretizing the solution space and solving a few cylinders at a time.

21.7 Data Allocation and Storage Management in Storage Networks

Two confusing new terms are **Storage Area Networks (SAN)** and **Network Attached Storage (NAS)** (as opposed to **Direct Attached Storage (DAS)**). NAS embeds the file system into storage, while a SAN does not [Gibson and Van Meter 2002]. A SAN is typically based on Ethernet while NAS is based on Fibre Channel (note spelling). Four typical NAS systems are described in Gibson and Van Meter [2002]:

1. Storage appliances such as products from Network Appliance and SNAP!
2. NASD (network attached secure devices), a project at CMU's Parallel Data Laboratory [Gibson et al. 1998]
3. The Petal project whose goal was to provide easily expandable storage with a block interface [Lee and Thekkath 1996]
4. **The Internet SCSI (iSCSI)** protocol which implements the *SCSI (Small Computer System Interface)* protocol on top of TCP/IP. Two recent publications on this topic are Sarkar et al. [2003] and Meth and Satran [2003].

This section is organized into two subsections. We first review the requirements of a storage network, and then proceed to review recent work at HP Labs in the area of storage allocation.

21.7.1 Requirements of a Storage Network

A storage network is defined as a set of disk drives connected by a network that can be a simple SCSI bus, an Ethernet switch, or a peer-to-peer network, where the disk drives are associated with cooperating servers. Requirements for a storage network can be summarized as follows [Salzwedel 2003]:

1. **Space and access balance.** The requirement to store additional data is growing rapidly. Given that disk capacity is cheap, the storage capacity of the network can be easily expanded. There is a need to allocate data carefully to utilize disk capacities and balance disk utilizations. We have already discussed striping and randomized methods for this purpose in our discussion of RAID5 systems.

    This problem is related to the *File Assignment Problem (FAP)* [Dowdy and Foster 1982]. Given the characteristics of the disk system and the access rates of files, FAP searches for a nearly optimal assignment of files to disks to optimize some performance measure such as disk response times [Wolf 1989]. Without such an optimized allocation, disks are susceptible to access skew. Recent work in this area is discussed in the following subsection.

2. **Availability.** High data availability can be attained utilizing various forms of redundancy, for example, mirroring, as well as parity. A distributed RAID system is described in Hartman and Ousterhout [1995]. It uses striping (to balance the load) and parity, but in fact it is a log-structured array (LSA) [Menon 1995], in that data from various clients is combined into a single stream, which is written as one stripe.

3. **Resource efficiency.** Higher availability implies wasted disk capacity in the form of check disks and even replicated data. It is important to quantify this overhead against higher performance and availability.

4. **Access efficiency.** This is determined by time and space requirements to access the data.

5. **Heterogeneity.** In a smaller computer installation, it is possible to upload disk data onto tape, replace old disks with new disks, and recreate the data on the new disks. Index building for relational tables may require a considerable amount of processing time, however, especially when there are several indices per table. With the advent of higher-capacity disks and the need for high availability, this is not possible in a large distributed system. The introduction of new disks, heterogeneous or not, should introduce as little disruption as possible. Furthermore, it might not be economically viable to replace all old disks with new disks, just for the sake of maintaining homogeneity.

There have been several studies of data layout with heterogeneous disks, and these are reviewed in Section 5 of Salzwedel [2003]. Consider the data layout for RAID0 in a heterogeneous disk array with N disks, N' of which are large. The stripe width can be set to N up to the point where the smaller disks are filled, and the striping is continued with a stripe width $N - N'$ thereafter. The data layout becomes more complicated with RAID5 because we will have variable stripe sizes and there is a need to balance disk loadings for updating parity. Solutions to these and other problems are given in Cortes and Labarta [2001].

6. **Adaptivity.** This is a measure of how fast can the system adapt to increases in the data volume and the addition of more disk capacity. More specifically, a fraction of the existing data must be redistributed, and the efficiency of this process is called *adaptivity*. The *faithfulness* property is to balance disk capacity utilizations, so that given m data units and that the ith disk has a fraction d_i of the total capacity, then this disk's share of the data is $(d_i + \epsilon)m$, where ϵ is a small value [Salzwedel 2003]. The issue of speed (access bandwidth) should also be taken into consideration because it is an important attribute of disk drives.

    Adaptivity can be measured by *competitive analysis*, which measures the efficiency of a scheme as multiples of an optimal scheme, which ensures faithfulness. For example, a placement scheme will be called c-competitive if for any sequence of changes it requires the replacement of, at most, c times the number of data units needed by an "optimal adaptive and perfectly faithful strategy" [Salzwedel 2003].

7. **Locality.** This is a measure of the degree of communication required for accessing data.

21.7.2 Storage Management

There is renewed interest in this area because of user demand for more storage and predictable performance. The former issue is easy to deal with because storage costs are dropping; however, the latter issue is a difficult problem because there are numerous schemes for storage structures with different performance characteristics, yet limited knowledge is available about the workload.

We briefly describe recent studies at HP Labs in this direction [E. Anderson et al. 2002]. Hyppodrome is conceptually an "iterative storage management loop" consisting of three steps: (1) design new system, (2) implement design, (3) analyze workload and go back to (1). The last step involves I/O tracing and its analysis to generate a workload summary, which includes *request rates and counts*; *run counts*, that is, mean number of accesses to contiguous locations, (device) *queue-length*, *on and off times* (mean period a stream is generating data); and *overlap fraction* of these periods. The *solver* module takes the workload description and outputs the design of the system that meets performance requirements.

Hyppodrome uses Ergastulum to generate a storage system design [E. Anderson et al. 2003]. The inputs to Ergastulum are a workload description, a set of potential devices, and user-specified constraints and goals [Wilkes 2001]. The workload description is in the form of *stores* (static aspect of datasets, such as size) and *streams* (dynamic aspects of workload: access rates, sequentiality, locality, etc.). There is an intermediate step for selecting RAID levels, configuring devices, and assigning stores onto devices. An *integer linear programming* formulation is easy, but it is very expensive to solve (takes several weeks to run) for larger problems. Ergastulum uses *generalized best-fit bin packing* with *randomization* to solve the problem quickly, thus benefiting from an earlier tool, Minerva, which was a solver for *attribute managed storage* [Alvarez et al. 2001].

21.8 Conclusions and Recent Developments

We have given a lengthy review of secondary storage systems, which are materialized as magnetic disks and disk arrays. We have emphasized methods that can be used to improve performance, especially in the form of disk scheduling.

While there are many varieties of disk arrays, only two — mirrored disks (RAID1) and rotated parity arrays (RAID5) — are popular. There are also commercial disk arrays with P + Q coding, one of which is also a log-structured array, and the AutoRaid dynamically adaptive RAID system. Much more work remains to be done to gain a better understanding of the performance of these systems. We have therefore covered analytic and simulation methods that can be used to evaluate the performance of disks and disk arrays.

While magnetic disks serve as a component, there is a significant amount of computing power associated with each disk that can be utilized for many useful functions. While there are many remaining problems to be solved, this is similar to a step taken by the Intelligent RAM (IRAM) project, which enhanced RAMs with processing power (or vice versa) [Kozyrakis and Patterson 2003].

The Storage Networking Industry Association (www.snia.org), among its other activities, has formed the Object-based Storage Devices (OSD) work group, which is concerned with the creation of self-managed, heterogeneous, shared storage. This entails moving low-level storage functions to the device itself and providing the appropriate interface. A recent paper on this topic is Mesnier et al. [2003].

The InfiniBand™ Architecture (IBA) was started because "processing power is substantially outstripping the capabilities of industry-standard I/O systems using busses" [Pfister 2002]. Commands and data are transferred among hosts as messages (not memory operations) over point-to-point connections.

Finally, there is increasing interest in constructing *information storage systems* that provide availability, confidentiality, and integrity policies against failures and even malicious attacks; see, for example, Wylie et al. [2000].

Acknowledgments

We acknowledge the support of NSF through Grant 0105485 in Computer System Architecture.

Defining Terms

Error correction code (ECC): ECC is computed for each sector and used to detect and even correct (within some limits) storage errors.

Floppy disks: A flexible plastic disk coated with magnetic material and enclosed in a protective envelope.

Gigabyte (GB): 2^{30} or approximately one thousand million bytes. Similarly, kilobyte (KB) is 2^{10} bytes, megabyte (MB) is 2^{20} or approximately one million bytes, and petabyte is $2^{40} \approx 10^{12}$ bytes. Powers of ten are used for disk sizes and powers of two for main memory sizes.

Hard disk drive (HDD): Consists of one or more rigid disk platters that are coated with magnetic material on which data is recorded. Also called Winchester disk drives for historical reasons.

Integrated device electronics (IDE): Disk drive interface, which includes drive controller and electronics. It is related to ATA (AT attachment) interface, which originated with IBM's PC AT (advanced technology) in mid-1980s.

Millisecond: One thousandth of a second. Similarly, a microsecond is one millionth (10^{-6}) of a second, and a nanosecond is 10^{-9} seconds.

Mirrored or Shadowed Disk: The same datasets are replicated on two disks. This corresponds to RAID level 1 or RAID1.

MTBF, MTTR, MTDL: Mean Time Between Failures, Mean Time To Repair, Mean Time to Data Loss.

Network Attached Storage (NAS): Storage, such as a disk array, that is directly attached to a network (see SAN) rather than a server.

Network File System (NFS): A file system originating from Sun Microsystems, but now accepted as an IETF (Internet Engineering Task Force [IETF]) standard for file services on TCP/IP networks.

Offline/nearline/online storage: Infrequently used data is stored in offline storage, which usually consists of a tape archive. Nearline storage refers to automated type library, while online storage for frequently accessed data usually resides on hard disk drives (as opposed to floppy disks).

Parity blocks: Blocks computed bit-by-bit using the exclusive-OR (XOR) operation from data blocks. It is used in RAID3, RAID4, and RAID5 to reconstruct data on a failed disk.

RAID (Redundant Array of Inexpensive Disks): An array of disks organized to provide higher availability, through fault-tolerance, and higher performance through parallel processing.

RAID1: RAID level one, which corresponds to mirrored disks.

RAID3: RAID3 associates a parity disk to N data disks. Because all disks are written together, the parity can be computed on-the-fly and all disks can be read together to provide for a higher level of data integrity than provided by ECC alone.

RAID4: RAID4 is similar to RAID3 but has a larger striping unit. Data accesses in RAID4, as well as RAID5, are usually satisfied by a single disk access.

RAID5: RAID5 is RAID4 with rotated parity.

RAID6: A disk array utilizing two check disks. Reed-Solomon codes or P+Q coding is used to protect against two-disk failures.

Read-modify-write (RMW): The data is read, modified with the new data block or by computing the new parity blocks, and written after one disk rotation. Read-after-write verifies that the data has been written correctly.

SAN (storage area network): A network (switch) to which servers and storage peripherals are attached.

SCSI (Small Computer System Interface): Hardware and software standards for connecting peripherals to a computer system. Fast SCSI has an 8-bit bus and 10-MB/second transfer rate, while ULTRA320 has a 16-bit bus and 320-MB/second transfer rate.

Sector: The unit of track formatting and data transmission to/from disk drives. Sectors are usually 512 bytes long and are protected by an ECC.

Striping: A method of data allocation that partitions a dataset into equal-sized striping units, which are allocated in a round-robin manner on the disks of a disk array. This corresponds to RAID level zero.

Zero latency read: Ability to read data out of order so that latency is reduced, especially for accesses to larger blocks.

Zoned disk: The cylinders on the disk are partitioned into groups, called zones, which have a fixed number of sectors.

References

[Acharya et al. 1998] A. Acharya, M. Uysal, and J.H. Saltz. "Active disks: programming model, algorithms, and evaluation," *Proc. ASPLOS VIII*, 1998, pp. 81–91.

[Akyurek and Salem 1995] S. Akyurek and K. Salem. "Adaptive block rearrangement," *ACM Trans. Computer Systems*, 13(2):89–121, May 1995.

[Alvarez et al. 1997] G.A. Alvarez, W.A. Burkhard, and F. Cristian. "Tolerating multiple failures in RAID architectures with optimal storage and uniform declustering," *Proc. 24th ISCA*, 1997, pp. 62–72.

[Alvarez et al. 1998] G.A. Alvarez, W.A. Burkhard, L.J. Stockmeyer, and F. Cristian. "Declustered disk array architectures with optimal and near optimal parallelism," *Proc. 25th ISCA*, 1998, pp. 109–120.

[Alvarez et al. 2001] G.A. Alvarez et al. "Minerva: an automated resource provisioning tool for large-scale storage systems," *ACM Trans. Computer Systems*, 19(4):483–518 (2001).

[D. Anderson et al. 2003] D. Anderson, J. Dykes, and E. Riedel. "More than an interface — SCSI vs. ATA," *Proc. 2nd USENIX Conf. File and Storage Technologies*, USENIX, 2003, pp. 245–257.

[E. Anderson et al. 2002] E. Anderson et al. "Hyppodrome: Running circles around storage administration," *Proc. 1st Conf. on File and Storage Technologies — FAST '02*, USENIX, 2002, pp. 175–188.

[E. Anderson et al. 2003] E. Anderson et al. "Ergastulum: quickly finding near-optimal storage system designs," http:// www.hpl.hp.com/research/ssp/papers/ergastulum-paper.pdf

[Balafoutis et al. 2003] E. Balafoutis et al. "Clustered scheduling algorithms for mixed media disk workloads in a multimedia server," *Cluster Computing*, 6(1):75–86 (2003).

[Bennett and Franaszek 1977] B.T. Bennett and P.A. Franaszek. "Permutation clustering: an approach to online storage reorganization," *IBM J. Research and Development*, 21(6):528–533 (1977).

[Bitton and Gray 1988] D. Bitton and J. Gray. "Disk shadowing," *Proc. 14th Int. VLDB Conf.*, 1988, pp. 331–338.

[Blaum et al. 1995] M. Blaum, J. Brady, J. Bruck, and J. Menon. "EVENODD: an optimal scheme for tolerating disk failure in RAID architectures," *IEEE Trans. Computers*, 44(2):192–202 (Feb. 1995).

[Boxma and Cohen 1991] O.J. Boxma and J.W. Cohen. "The M/G/1 queue with permanent customers," *IEEE J. Selected Topics in Communications*, 9(2):179–184 (1991).

[Cao et al. 1994] P. Cao, S.B. Lim, S. Venkataraman, and J. Wilkes. "The TickerTAIP parallel RAID architecture," *ACM Trans. Computer Systems*, 12(3):236–269 (Aug. 1994).

[Chandy and Reddy 1993] J. Chandy and A.L. Narasimha Reddy. "Failure evaluation of disk array organizations," *Proc. 13th Int. Conf. on Distributed Computing Systems — ICDCS*, 1993, pp. 319–326.

[Chen et al. 1993] M.S. Chen, D.D. Kandlur, and P.S. Yu. "Optimization of the grouped sweeping scheduling for heterogeneous disks," *Proc. 1st ACM Int. Conf. on Multimedia*, 1993, pp. 235–242.

[Chen and Patterson 1990] P.M. Chen and D.A. Patterson. "Maximizing performance on a striped disk array," *Proc. 17th ISCA*, 1990, pp. 322–331.

[Chen et al. 1994] P.M. Chen, E.K. Lee, G.A. Gibson, R.H. Katz, and D.A. Patterson. "RAID: High-performance, reliable secondary storage," *ACM Computing Surveys*, 26(2):145–185 (1994).

[Chen and Patterson 1994] P.M. Chen and D.A. Patterson. "A new approach to I/O performance evaluation: self-scaling I/O benchmarks, predicted I/O performance," *ACM Trans. Computer Systems*, 12(4):308–339 (Nov. 1994).

[Chen and Towsley 1993] S.-Z. Chen and D.F. Towsley. "The design and evaluation of RAID5 and parity striping disk array architectures," *J. Parallel and Distributed Computing*, 10(1/2):41–57 (1993).

[Chen and Towsley 1996] S.-Z. Chen and D.F. Towsley. "A performance evaluation of RAID architectures," *IEEE Trans. Computers*, 45(10):1116–1130 (1996).

[Coffman et al. 1972] E.G. Coffman Jr., E.G. Klimko, and B. Ryan. "Analyzing of scanning policies for reducing disk seek times," *SIAM J. Computing*, 1(3):269–279 (1972).

[Coffman and Hofri 1982] E.G. Coffman, Jr. and M. Hofri. "On the expected performance of scanning disks," *SIAM J. Computing*, 11(1):60–70 (1982).

[Coffman and Hofri 1990] E.G. Coffman, Jr. and M. Hofri. "Queueing models of secondary storage devices," in *Stochastic Analysis of Computer and Communication Systems*, H. Takagi (Ed.), North-Holland, 1990, pp. 549–588.

[Cortes and Labarta 2001] T. Cortes and J. Labarta. "Extending heterogeneity to RAID level 5," *Proc. USENIX Annual Technical Conf.*, 2001, pp. 119–132.

[Courtright et al. 1996] W.V. Courtright II et al. "RAIDframe: A Rapid Prototyping Tool for Raid Systems," http://www.pdl.cmu.edu/RAIDframe/raidframebook.pdf.

[Courtright 1997] W.V. Courtright II. "A Transactional Approach to Redundant Disk Array Implementation," Technical Report CMU-CS-97-141, 1997.

[Denning 1967] P.J. Denning. "Effects of scheduling in file memory operations," *Proc. AFIPS Spring Joint Computer Conf.*, 1967, pp. 9–21.

[Diskspecs] http://www.pdl.cmu.edu/DiskSim/diskspecs.html.

[DiskSim] http://www.pdl.cmu.edu/DiskSim/disksim2.0.html.

[Dowdy and Foster 1982] L.W. Dowdy and D.V. Foster. "Comparative models of the file assignment problem," *ACM Computing Surveys*, 14(2):287–313 (1982).

[Geist and Daniel 1987] R.M. Geist and S. Daniel. "A continuum of disk scheduling algorithm," *ACM Trans. Computer Systems*, 5(1):77–92 (1987).

[Gibson 1992] G.A. Gibson. *Redundant Disk Arrays: Reliable, Parallel Secondary Storage*, The MIT Press, 1992.

[Gibson et al. 1995] G.A. Gibson et al. "The Scotch parallel storage system," *Proc. IEEE CompCon Conf.*, 1995, pp. 403–410.

[Gibson et al. 1998] G.A. Gibson et al. "A cost-effective, high bandwidth storage architecture," *Proc. ASPLOS VIII*, 1998, 92–103.

[Gibson and Van Meter 2002] G.A. Gibson and R. Van Meter. "Network attached storage architecture," *Commun. ACM*, 43(11):37–45 (Nov. 2000).

[Gray and Reuter 1993] J. Gray and A. Reuter. *Transaction Processing: Concepts and Techniques*, Morgan-Kaufmann Publishers, 1993.

[Gray et al. 1990]   J. Gray, B. Horst, and M. Walker. "Parity striping of disk arrays: low-cost reliable storage with acceptable throughput," *Proc. 16th Int. VLDB Conf.*, 1990, 148–161.

[Gray and Shenoy 2000]   J. Gray and P. J. Shenoy. "Rules of thumb in data engineering," *Proc. 16th ICDE*, 2000, pp. 3–16.

[Gray 2002]   J. Gray. "Storage bricks have arrived" (Keynote Speech), *1st Conf. on File and Storage Technologies-FAST '02*, USENIX, 2002.

[Griffin et al. 2000]   J. L. Griffin, S. W. Shlosser, G. R. Ganger, and D. F. Nagle. "Modeling and performance of MEMS-based storage devices," *Proc. ACM SIGMETRICS 2000*, pp. 56–65.

[Gurumurthi et al. 2003]   S. Gurumurthi, A. Sivasubramaniam, M. Kandemir, and H. Franke. "DRPM: Dynamic speed control for power management in server-class disks," *Proc. 30th ISCA*, 2003, pp. 169–181.

[Hall 1986]   M. Hall. *Combinatorial Theory*, Wiley, 1986.

[Han and Thomasian 2003]   C. Han and A. Thomasian. "Performance of two-disk failure tolerant disk arrays," *Proc. Symp. Performance Evaluation of Computer and Telecomm. Systems — SPECTS '03*, 2003.

[Hartman and Ousterhout 1995]   J.H. Hartman and J.K. Ousterhout. "The Zebra striped network file system," *ACM Trans. Computer Systems*, 13(3):274–310 (1995).

[Hellerstein et al. 1994]   L. Hellerstein et al. "Coding techniques for handling failures in large disk arrays," *Algorithmica*, 12(2/3):182–208 (Aug./Sept. 1994).

[Hennessey and Patterson 2003]   J. Hennessey and D. Patterson. *Computer Organization: A Quantitative Approach, 3rd ed.* Morgan-Kauffman Publishers, 2003.

[Holland et al. 1994]   M.C. Holland, G.A. Gibson, and D.P. Siewiorek. "Architectures and algorithms for on-line failure recovery in redundant disk arrays," *Distributed and Parallel Databases*, 11(3):295–335 (1994).

[Hsiao and DeWitt 1993]   H.I. Hsiao and D.J. DeWitt. "A performance study of three high availability data replication strategies," *J. Distributed and Parallel Databases*, 1(1):53–80 (Jan. 1993).

[Hsu et al. 2001]   W.W. Hsu, A.J. Smith, and H.C. Young. "I/O reference behavior of production database workloads and the TPC benchmarks — an analysis at the logical level," *ACM Trans. Database Systems*, 26(1): 96–143 (2001).

[Hsu and Smith 2003]   W.W. Hsu and A.J. Smith. "Characteristics of I/O traffic in personal computer and server workloads," *IBM Systems J.*, 42(2):347–372 (2003).

[Iyer and Druschel 2001]   S. Iyer and P. Druschel. "Anticipatory scheduling: a disk scheduling framework to overcome deceptive idleness in synchronous I/O," *Proc. 17th SOSP*, 2001, pp. 117–130.

[Jacobson and Wilkes 1991]   D. Jacobson and J. Wilkes. "Disk scheduling algorithms based on rotational position," *HP Technical Report* HPL-CSP-91-7rev, 1991.

[Jain et al. 1996]   R. Jain, J. Werth, and J.C. Browne (Editors). *Input/Output in Parallel and Distributed Systems*, Kluwer Academic Publishers, 1996.

[Jin et al. 2002]   H. Jin, T. Cortes, and R. Buyya. *High Performance Mass Storage and Parallel I/O: Technologies and Applications*, Wiley Interscience, 2002.

[King 1990]   R.P. King. "Disk arm movement in anticipation of future requests," *ACM Trans. Computer Systems*, 8(3):214–229 (1990).

[Kleinrock 1975]   L. Kleinrock. *Queueing Systems, Vol. I: Theory*, Wiley Interscience, 1975.

[Kleinrock 1976]   L. Kleinrock. *Queueing Systems, Vol. II: Computer Applications*, Wiley Interscience, 1976.

[Kozyrakis and Patterson 2003]   C. Kozyrakis and D. Patterson. "Overcoming the limitations of current vector processors," *Proc. 30th ISCA*, 2003, 399–409.

[Lawlor 1981]   F.D. Lawlor. "Efficient mass storage parity recovery mechanism," *IBM Technical Disclosure Bulletin*, 24(2):986–987 (July 1981).

[Lazowska et al. 1984]   E.D. Lazowska, J. Zahorjan, G.S. Graham, and K.C. Sevcik. *Quantitative System Performance*, Prentice Hall, 1984. Also http://www.cs.washingtion.edu/homes/lazowska/qsp.

[Lee and Thekkath 1996]   E.K. Lee and C. Thekkath. "Petal: distributed virtual disks," *Proc. ASPLOS XII*, 1996, pp. 84–92.

[Lee and Katz 1993] E.K. Lee and R.H. Katz. "The performance of parity placements in disk arrays," *IEEE Trans. Computers*, 42(6):651–664 (June 1993).

[Lumb et al. 2000] C.R. Lumb, J. Schindler, G.R. Ganger, and D.F. Nagle. "Towards higher disk head utilization: extracting free bandwidth from busy disk drives," *Proc. 4th OSDI Symp.* USENIX, 2000, pp. 87–102.

[Lynch 1972] W.C. Lynch. "Do disk arms move?," *Performance Evaluation Review*, 1(4):3–16 (Dec. 1972).

[MacWilliams and Sloane 1977] F.J. MacWilliams and N.J.A. Sloane. *The Theory of Error Correcting Codes*, North-Holland, 1977.

[Matick 1977] R.E. Matick. *Computer Storage Systems and Technology*, Wiley, 1977.

[McKusick et al. 1984] M.K. McKusick, W.N. Joy, S.J. Leffler, and R.S. Fabry. "A fast file system for UNIX," *ACM Trans. Computer Systems*, 2(3):181–197 (1984).

[Menon and Cortney 1993] J. Menon and J. Cortney. "The architecture of a fault-tolerant cached RAID controller," *Proc. 20th ISCA*, 1993, pp. 76–86.

[Menon 1994] J. Menon. "Performance of RAID5 disk arrays with read and write caching," *Distributed and Parallel Databases*, 11(3):261–293 (1994).

[Menon 1995] J. Menon. "A performance comparison of RAID5 and log-structured arrays," *Proc. 4th IEEE HPDC*, 1995, pp. 167–178.

[Merchant and Yu 1996] A. Merchant and P.S. Yu. "Analytic modeling of clustered RAID with mapping based on nearly random permutation," *IEEE Trans. Computers*, 45(3):367–373 (1996).

[Meritt et al. 2003] A.S. Meritt et al. "z/OS support for IBM TotalStorage enterprise storage server," *IBM Systems J.*, 42(2):280–301 (2003).

[Mesnier et al. 2003] M. Mesnier, G.R. Ganger, and E. Riedel. "Object-based storage," *IEEE Communications Magazine*, 41(8):84–90 (2003).

[Meth and Satran 2003] K.Z. Meth and J. Satran. "Features of the iSCSI protocol," *IEEE Communications Magazine*, 41(8):72–75 (2003).

[Muntz and Lui 1990] R. Muntz and J.C.S. Lui. "Performance analysis of disk arrays under failure," *Proc. 16th Int. VLDB Conf.*, 1990, pp. 162–173.

[Nelson and Tantawi 1988] R. Nelson and A. Tantawi. "Approximate analysis of fork-join synchronization in parallel queues," *IEEE Trans. Computers*, 37(6):739–743 (1988).

[Newberg and Wolfe 1994] L. Newberg and D. Wolfe. "String layout for a redundant array of inexpensive disks," *Algorithmica*, 12(2/3):209–224 (Aug./Sept. 1994).

[Ng 1987] S.W. Ng. "Reliability, availability, and performance analysis of duplex disk systems," *Reliability and Quality Control*, M.H. Hamza (Ed.), Acta Press, 1987, pp. 5–9.

[Ng 1991] S.W. Ng. "Improving disk performance via latency reduction," *IEEE Trans. Computers*, 40(1):22–30 (Jan. 1991).

[Ng and Mattson 1994] S.W. Ng and R.L. Mattson. "Uniform parity distribution in disk arrays with multiple failures," *IEEE Trans. Computers*, 43(4):501–506 (1994).

[Ng 1994] S.W. Ng. "Crosshatch disk array for improved reliability and performance," *Proc. 21st ISCA*, 1994, pp. 255–264.

[Ng 1998] S.W. Ng. "Advances in disk technology: performance issues," *IEEE Computer*, 40(1):75–81 (1998).

[Patterson et al. 1998] D.A. Patterson, G.A. Gibson, and R.H. Katz. "A case study for redundant arrays of inexpensive disks," *Proc. ACM SIGMOD Int. Conf.*, 1998, pp. 109–116.

[Patterson et al. 1995] R.H. Patterson et al. "Informed prefetching and caching," *Proc. 15th SOSP*, 1995, pp. 79–95.

[Pfister 2002] G.F. Pfister. "An introduction to the InfiniBand™ architecture," in *High Performance Mass Storage and Parallel I/O*, H. Jin et al. (Eds.), Wiley, 2002, pp. 617–632.

[Polyzois et al. 1993] C. Polyzois, A. Bhide, and D.M. Dias. "Disk mirroring with alternating deferred updates," *Proc. 19th Int. VLDB Conf.*, 1993, 604–617.

[Ramakrishnan et al. 1992] K.K. Ramakrishnan, P. Biswas, and R. Karedla. "Analysis of file I/O traces in commercial computing environments," *Proc. Joint ACM SIGMETRICS/Performance '92 Conf.*, 1992, pp. 78–90.

[Ramakrishnan and Gehrke 2003] R. Ramakrishnan and J. Gehrke. *Database Management Systems, 3rd ed.*, McGraw-Hill, 2003.

[Riedel et al. 2000] E. Riedel, C. Faloutsos, G.R, Ganger, and D.F. Nagle. "Data mining in an OLTP system (nearly) for free," *Proc. ACM SIGMOD Int. Conf.*, 2000, pp. 13–21.

[Riedel et al. 2001] E. Riedel, C. Faloutsos, G.R. Ganger, and D.F. Nagle. "Active disks for large scale data processing," *IEEE Computer*, 34(6):68–74 (2001).

[Rosenblum and Ousterhout 1992] M. Rosenblum and J.K. Ousterhout. "The design and implementation of a log-structured file system," *ACM Trans. Computer Systems*, 10(1):26–52 (Feb. 1992).

[Ruemmler and Wilkes 1992] C. Ruemmler and J. Wilkes. "UNIX disk access patterns," *HP Labs Technical Report*, HPL-92-152, 1992.

[Ruemmler and Wilkes 1993] C. Ruemmler and J. Wilkes. "A trace-driven analysis of disk working set sizes," *HP Labs Technical Report*, HPL-93-23, 1993.

[Ruemmler and Wilkes 1994] C. Ruemmler and J. Wilkes. "An introduction to disk drive modeling," *IEEE Computer*, 27(3):17–28 (March 1994).

[Salzwedel 2003] K.A. Salzwedel. "Algorithmic approaches for storage networks," in *Algorithms for Memory Hierarchies, Advanced Lectures LNCS 2625*, Springer, 2003, pp. 251–272.

[Santos et al. 2000] J.R. Santos, R.R. Muntz, and B. Ribeiro-Neto. "Comparing random data allocation and data striping in multimedia servers," *Proc. ACM SIGMETRICS Conf.*, 2000, pp. 44–55.

[Sarkar et al. 2003] P. Sarkar, S. Uttamchandani, and K. Voruganti. "Storage over IP: when does hardware support help?," *Proc. 2nd Conf. on File and Storage Technologies — FAST '03*, USENIX, 2003.

[Schindler and Ganger 1999] J. Schindler and G. R. Ganger. "Automated disk drive characterization," *CMU SCS Technical Report*, CMU-CS-99-176, 1999.

[Schindler et al. 2002] J. Schindler, J.L. Griffin, C.R. Lumb, and G.R. Ganger. "Track-aligned extents: matching access patterns to disk drive characteristics," *Proc. Conf. on File and Storage Technologies — FAST '02*, USENIX, 2002.

[Schwartz 1999] T.J.E. Schwartz, J. Steinberg, and W.A. Burkhard. "Permutation development data layout (PDDL) disk array declustering," *Proc. 5th IEEE Symp. on High Performance Computer Architecture — HPCA*, 1999, pp. 214–217.

[Seltzer et al. 1990] M.I. Seltzer, P.M. Chen, and J.K. Ousterhout. "Disk scheduling revisited," *Proc. 1990 USENIX Summer Technical Conf.*, pp. 307–326.

[Shriver et al. 2000] E. Shriver, B.K. Hillyer, and A. Silberschatz. "Performance analysis of storage systems," in *Performance Evaluation: Origins and Directions — LNCS 1769*, G. Haring, C. Lindemann, and M. Reiser (Editors), Springer-Verlag, 2000.

[Sierra 1990] H.M. Sierra. *An Introduction to Direct Access Storage Devices*, Academic Press, 1990.

[Sitaram and Dan 2000] D. Sitaram and A. Dan. *Multimedia Servers: Applications, Environments, and Design*, Morgan-Kaufmann Publishers, 2000.

[Smith 1985] A.J. Smith. "Disk cache: miss ratio analysis and design considerations," *ACM Trans. Computer Systems*, 3(3):161–203 (Aug. 1985).

[Solworth and Orji 1991] J.A. Solworth and C.U. Orji. "Distorted mirrors," *Proc. 1st Int. Conf. on Parallel and Distributed Information Systems — PDIS*, 1991, pp. 10–17.

[Stodolsky et al. 1994] D. Stodolsky, M. Holland, W.C. Courtright II, and G.A. Gibson. "Parity logging disk arrays," *ACM Trans. Computer Systems*, 12(3):206–325 (1994).

[Storage Performance Council 2003] http://www.storageperformance.org.

[Takagi 1991] H. Takagi. *Queueing Analysis. Vol. 1: Vacation and Priority Systems*, North-Holland, 1991.

[Thomasian and Menon 1994] A. Thomasian and J. Menon. "Performance analysis of RAID5 disk arrays with a vacationing server model," *Proc. 10th ICDE Conf.*, 1994, pp. 111–119.

[Thomasian and Menon 1997] A. Thomasian and J. Menon. "RAID5 performance with distributed sparing," *IEEE Trans. Parallel and Distributed Systems*, 8(6):640–657 (June 1997).

[Thomasian 1997] A. Thomasian. "Approximate analysis for fork-join synchronization in RAID5," *Computer Systems: Science and Engineering*, 12(5):329–338 (1997).

[Thomasian 1998] A. Thomasian. "RAID5 disk arrays and their performance evaluation," in *Recovery Mechanisms in Database Systems*, V. Kumar and M. Hsu (Eds.), 1998 pp. 807–846.

[Thomasian and Liu 2002a] A. Thomasian and C. Liu. "Some new disk scheduling policies and their performance," *Proc. ACM SIGMETRICS Conf.*, 2002, pp. 266–267.

[Thomasian and Liu 2002b] A. Thomasian and C. Liu. "Disk scheduling policies with lookahead," *Performance Evaluation Review*, 30(2):31–40 (Sept. 2002).

[Thomasian et al. 2003] A. Thomasian et al. "Mirrored disk scheduling," *Proc. Symp. Performance Evaluation of Computer and Telecomm. Systems — SPECTS '03*, 2003.

[Thomasian and Liu 2003] A. Thomasian and C. Liu. "Performance of mirrored disks with a shared NVS cache," submitted for publication (available from authors).

[Toigo 2000] J.W. Toigo. *The Holy Grail of Data Storage Management*, Prentice-Hall, 2000.

[Treiber and Menon 1995] K. Treiber and J. Menon. "Simulation study of cached RAID5 designs," *Proc. 1st IEEE Symp. on High Performance Computer Architecture — HPCA*, 1995, pp. 186–197.

[Trivedi 2002] K.S. Trivedi. *Probability and Statistics with Reliability, Queueing and Computer Science Applications, 2nd edition*, Wiley Interscience, 2002.

[Uysal et al. 2001] M. Uysal, G.A. Alvarez, and A. Merchant. "A modular, analytical throughput model for modern disk drives," *Proc. MASCOTS-2001*.

[Uysal et al. 2003] M. Uysal, A. Merchant, and G.A. Alvarez. "Using MEMS-based storage in disk arrays," *Proc. 2nd Conf. on File and Storage Technologies — FAST '03*, USENIX, 2003.

[Varma and Jacobson 1998] A. Varma and Q. Jacobson. "Destage algorithms for disk arrays with non-volatile caches," *IEEE Trans. Computers*, 47(2):228–235 (1998).

[Varma and Makowski 1994] S. Varma and A. Makowski. "Interpolation approximations for symmetric fork-join queues," *Performance Evaluation*, 20(1–3):361–368 (1994).

[Wilkes et al. 1996] J. Wilkes, R.A. Golding, C. Staelin, and T. Sullivan. "The HP AutoRAID hierarchical storage system," *ACM Trans. Computer Systems*, 14(1):108–136 (1996).

[Wilkes 2001] J. Wilkes. "Traveling to Rome: QoS specifications for automated storage management," *Proc. Int. Workshop on Quality of Service — IWQOS*, Springer-Verlag, 2001.

[Wilkes 2003] J. Wilkes. "Data services — from data to containers," *2nd Conf. on File and Storage Technologies — FAST '03*, USENIX, (Keynote Speech). http://www.usenix.com/publications/library/proceedings/ fast03/tech.html

[Wolf 1989] J.L. Wolf. "The placement optimization program: a practical solution to the disk file assignment problem," *Proc. ACM SIGMETRICS Conf.*, 1989, pp. 1–10.

[Wong 1980] C.K. Wong. "Minimizing head movement in one dimensional and two dimensional mass storage systems," *ACM Computing Surveys*, 12(2):167–178 (1980).

[Wong 1983] C.K. Wong. *Algorithmic Studies in Mass Storage Systems*, Computer Science Press, 1983.

[Worthington et al. 1994] B.L. Worthington, G.R. Ganger, and Y.L. Patt. "Scheduling for modern disk drivers and non-random workloads," *Proc. ACM SIGMETRICS Conf. 1994*, pp. 241–251.

[Worthington et al. 1995] B.L. Worthington, G.R. Ganger, Y.N. Patt, and J. Wilkes. "On-line extraction of SCSI disk drive parameters," *Proc. 1995 ACM SIGMETRICS Conf.*, pp. 146–156 (also *HP Labs Technical Report* HPL-97-02, 1997).

[Wylie et al. 2000] J.J. Wylie et al. "Survivable information storage systems," *IEEE Computer*, 33(8):61–68 (Aug. 2000).

[Zhang et al. 2002] C. Zhang, X. Yu, A. Krishnamurthy, and R.Y. Wang. "Configuring and scheduling an eager-writing disk array for a transaction processing workload," *Proc. 1st Conf. on File and Storage Technologies — FAST '02*, USENIX, 2002, pp. 289–304.

Further Information

Chapters with overlapping content, whose topics should be of interest to the readers of this chapter, include the following:

[**Access Methods by B. Salzberg**]. Especially Section 1, Introduction: brief description of magnetic disks.

[**Process and Device Scheduling by R.D. Cupper**]. Especially Section 6, Device Scheduling. The discussion of Scheduling Shared Devices (disks) is not as complete as our discussion and SPTF is not mentioned at all.

[**Secondary Storage and Filesystems by M. K. McKusick**]. Especially Section 1 Introduction: Computer Storage Hierarchy. Section 2 Secondary Storage Devices: Magnetic disks, RAID. Section 3 Filesystems: Directory structure, file layout on disk, file transfers, disk space management, logging, including LFS, file allocation, and disk defragmentation.

In addition to Gibson [1992], two useful books are: J.W. Toigo, *The Holy Grail of Data Storage Management*, Addison-Wesley, 2000 (an overview of the field for a nontechnical reader) and H. Jin, T. Cortes, and R. Buyya, *High Performance Mass Storage and Parallel I/O: Technologies and Applications*, Wiley, 2002 (a collection of papers, with emphasis on parallel I/O).

Matick [1977] discusses early computer storage systems from hardware and software viewpoint. Sierra [1990] is a text dedicated to disk technology, but is somewhat dated.

The *IBM Systems J.*, Vol. 42, No. 2, 2003, is on *Storage Systems*. Similar information is available as "White Papers" from the Web pages of EMC, Hitachi, HP, etc.

Disk drive characteristics are available from the Web sites of disk drive manufacturers: Hitachi, Maxtor, Seagate, Western Digital, etc.

The *File and Storage Technologies Conf.* organized by USENIX and the *Mass Storage Systems Conf.* co-sponsored by NASA and IEEE are two specialized conferences.

Some web sites of interest are as follows:
HP Labs Storage Systems Program: http://www.hpl.hp.com/research/ssp
CMU's Parallel Data Laboratory: http://www.pdl.cmu.edu
UC Berkeley: http://roc.cs.berkeley.edu
UC Santa Cruz: http://csl.cse.ucsc.edu
U. Wisconsin: http://www.cs.wisc.edu/wind

22

High-Speed Computer Arithmetic

22.1   Introduction **22**-1
22.2   Fixed Point Number Systems **22**-1
      Two's Complement • Sign Magnitude • One's Complement
22.3   Fixed Point Arithmetic Algorithms **22**-4
      Fixed Point Addition • Fixed Point Subtraction • Fixed Point
      Multiplication • Fixed Point Division
22.4   Floating Point Arithmetic **22**-18
      Floating Point Number Systems
22.5   Conclusion **22**-20

Earl E. Swartzlander Jr.
University of Texas at Austin

22.1 Introduction

The speeds of memory and arithmetic units are the primary determinants of the speed of a computer. Whereas the speed of both units depends directly on the implementation technology, arithmetic unit speed also depends strongly on the logic design. Even for an integer adder, speed can easily vary by an order of magnitude, whereas the complexity varies by less than 50%.

This chapter begins with a discussion of binary fixed point number systems in Section 22.2. Section 22.3 provides examples of fixed point implementations of the four basic arithmetic operations (i.e., add, subtract, multiply, and divide). Finally, Section 22.4 describes algorithms that implement floating point arithmetic.

Regarding notation, capital letters represent digital numbers (i.e., words), whereas subscripted lowercase letters represent bits of the corresponding word. The subscripts range from $n - 1$ to 0, to indicate the bit position within the word (x_{n-1} is the most significant bit of X, x_0 is the least significant bit of X, etc.). The logic designs presented in this chapter are based on positive logic with AND, OR, and INVERT operations. Depending on the technology used for implementation, different operations (such as NAND and NOR) can be used, but the basic concepts are not likely to change significantly.

22.2 Fixed Point Number Systems

Most arithmetic is performed with fixed point numbers that have constant scaling (i.e., the position of the binary point is fixed). The numbers can be interpreted as fractions, integers, or mixed numbers, depending on the application. Pairs of fixed point numbers are used to create floating point numbers, as discussed in Section 22.4.

1-58488-360-X/$0.00+$1.50
© 2004 by CRC Press, LLC

At the present time, fixed point binary numbers are generally represented using the two's complement number system. This choice has prevailed over the sign magnitude and one's complement number systems, because the frequently performed operations of addition and subtraction are more easily performed on two's complement numbers. Sign magnitude numbers are more efficient for multiplication but the lower frequency of multiplication and the development of Booth's efficient two's complement multiplication algorithm have resulted in the nearly universal selection of the two's complement number system for most applications. The algorithms presented in this chapter assume the use of two's complement numbers.

Fixed point number systems represent numbers, for example, A, by n bits: a sign bit and $n - 1$ data bits. By convention, the most significant bit a_{n-1} is the sign bit, which is a 1 for negative numbers and a 0 for positive numbers. The $n - 1$ data bits are $a_{n-2}, a_{n-3}, \ldots, a_1, a_0$. In the following material, fixed point fractions will be described for each of the three systems.

22.2.1 Two's Complement

In the two's complement fractional number system, the value of a number is the sum of $n - 1$ positive binary fractional bits and a sign bit, which has a weight of -1:

$$A = -a_{n-1} + \sum_{i=0}^{n-2} a_i 2^{i-n+1} \tag{22.1}$$

Two's complement numbers are negated by complementing all bits and adding a 1 to the least significant bit (lsb) position. For example, to form $-3/8$,

$$+3/8 = 0011$$
$$\text{invert all bits} = 1100$$
$$\text{add 1 lsb} \quad 0001$$
$$1101 = -3/8$$

Check:

$$\text{invert all bits} = 0010$$
$$\text{add 1 lsb} \quad 0001$$
$$0011 = +3/8$$

22.2.2 Sign Magnitude

Sign magnitude numbers consist of a sign bit and $n - 1$ bits that express the magnitude of the number,

$$A = (1 - 2a_{n-1}) \sum_{i=0}^{n-2} a_i 2^{i-n+1} \tag{22.2}$$

Sign magnitude numbers are negated by complementing the sign bit. For example, to form $-3/8$,

$$+3/8 = 0011$$
$$\text{invert sign bit} = 1011 = -3/8$$

Check:

$$\text{invert sign bit} = 0011 = +3/8$$

22.2.3 One's Complement

One's complement numbers are negated by complementing all of the bits of the original number,

$$A = \sum_{i=0}^{n-2}(a_i - a_{n-1})2^{i-n+1} \tag{22.3}$$

In this equation, the subtraction $(a_i - a_{n-1})$ is an arithmetic operation (not a logical operation) that produces values of 1 or 0 (if $a_{n-1} = 0$) or values of 0 or -1 (if $a_{n-1} = 1$).

The negative of a one's complement number is formed by inverting all bits. For example, to form $-3/8$,

$$+3/8 = 0011$$
$$\text{invert all bits} = 1100 = -3/8$$

Check:

$$\text{invert all bits} = 0011 = +3/8$$

Table 22.1 compares 4-bit fractional fixed point numbers in the three number systems. Note that both the sign magnitude and one's complement number systems have two zeros (i.e., positive zero and negative zero) and that only two's complement is capable of representing -1. For positive numbers, all three number systems use identical representations.

A significant difference between the three number systems is their behavior under truncation. Figure 22.1 shows the effect of truncating high-precision fixed point fractions X, to form three bit fractions $T(X)$. Truncation of two's complement numbers never increases the value of the number (i.e., the truncated numbers have values that are unchanged or shift toward negative infinity), as can be seen from Equation 22.1 since any truncated bits have positive weight. This bias can cause an accumulation of errors for computations that involve summing many truncated numbers (which may occur in scientific, matrix, and signal processing applications). In both the sign magnitude and one's complement number systems, truncated numbers are unchanged or shifted toward zero, so that if approximately half of the numbers are positive and half are negative, the errors will tend to cancel.

TABLE 22.1 Example of 4-bit Fractional Fixed Point Numbers

Number	Two's Complement	Sign Magnitude	One's Complement
+7/8	0111	0111	0111
+3/4	0110	0110	0110
+5/8	0101	0101	0101
+1/2	0100	0100	0100
+3/8	0011	0011	0011
+1/4	0010	0010	0010
+1/8	0001	0001	0001
+0	0000	0000	0000
−0	N/A	1000	1111
−1/8	1111	1001	1110
−1/4	1110	1010	1101
−3/8	1101	1011	1100
−1/2	1100	1100	1011
−5/8	1011	1101	1010
−3/4	1010	1110	1001
−7/8	1001	1111	1000
−1	1000	N/A	N/A

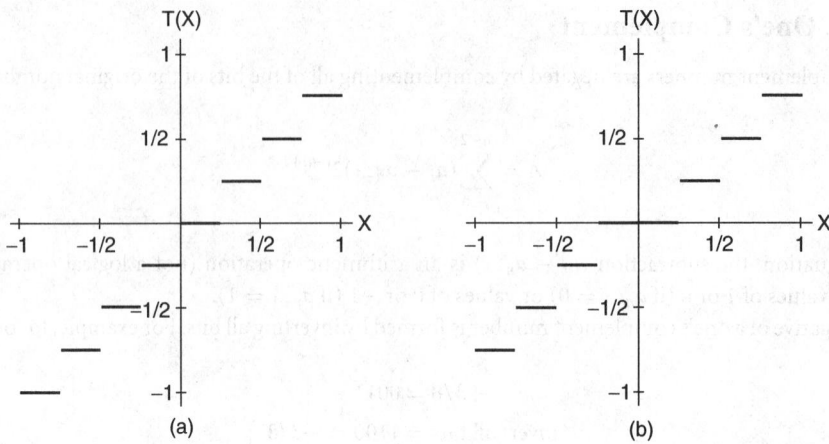

FIGURE 22.1 Behavior of fixed point fractions under truncation: (a) two's complement and (b) sign magnitude and one's complement.

TABLE 22.2 Full Adder Truth Table

Inputs			Outputs	
a_k	b_k	c_k	c_{k+1}	s_k
0	0	0	0	0
0	0	1	0	1
0	1	0	0	1
0	1	1	1	0
1	0	0	0	1
1	0	1	1	0
1	1	0	1	0
1	1	1	1	1

22.3   Fixed Point Arithmetic Algorithms

This section presents an assortment of typical fixed point algorithms for addition, subtraction, multiplication, and division.

22.3.1   Fixed Point Addition

Addition is performed by summing the corresponding bits of two n-bit numbers, including the sign bit. Subtraction is performed by summing the corresponding bits of the minuend and the two's complement of the subtrahend. Overflow is detected in a two's complement adder by comparing the carry signals into and out of the most significant adder stage (i.e., the stage which computes the sign bit). If the carries differ, an overflow has occurred and the result is invalid.

22.3.1.1   Full Adder

The full adder is the fundamental building block of most arithmetic circuits. Its operation is defined by the truth table shown in Table 22.2. The sum and the carry outputs are described by the following equations:

$$s_k = a_k\bar{b}_k\bar{c}_k + \bar{a}_k b_k\bar{c}_k + \bar{a}_k\bar{b}_k c_k + a_k b_k c_k = a_k \oplus b_k \oplus c_k \tag{22.4}$$

$$c_{k+1} = \bar{a}_k b_k c_k + a_k\bar{b}_k c_k + a_k b_k\bar{c}_k + a_k b_k c_k = a_k b_k + a_k c_k + b_k c_k \tag{22.5}$$

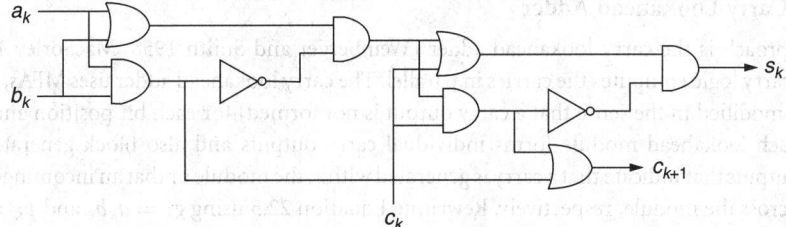

FIGURE 22.2 9-gate full adder.

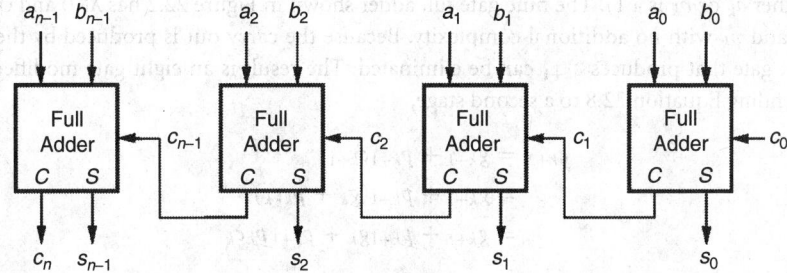

FIGURE 22.3 Ripple carry adder.

where a_k, b_k, and c_k are the inputs to the kth full adder stage, and s_k and c_{k+1} are the sum and carry outputs, respectively.

In evaluating the relative complexity of implementations, it is often convenient to assume a nine gate realization of the full adder, as shown in Figure 22.2. For this implementation, the delay from either a_k or b_k to s_k is six gate delays and the delay from c_k to c_{k+1} is two gate delays. Some technologies, such as CMOS, form inverting gates (e.g., NAND and NOR gates) more efficiently than the noninverting gates that are assumed in this chapter. Circuits with equivalent speed and complexity can be constructed with inverting gates.

22.3.1.2 Ripple Carry Adder

A ripple carry adder for n-bit numbers is implemented by concatenating n full adders as shown in Figure 22.3. At the kth-bit position, the kth bits of operands A and B and a carry signal from the preceding adder stage are used to form the kth bit of the sum, s_k, and the carry, c_{k+1}, to the next adder stage. This is called a ripple carry adder, since the carry signals *ripple* from the least significant bit position to the most significant bit position. If the ripple carry adder is implemented by concatenating n of the nine gate full adders, as shown in Figure 22.2 and Figure 22.3, an n-bit ripple carry adder requires $2n + 4$ gate delays to produce the most significant sum bit and $2n + 3$ gate delays to produce the carry output. A total of $9n$ logic gates are required to implement an n-bit ripple carry adder.

In comparing adders, the delay from data input to the slowest (usually the most significant sum) output and the complexity (i.e., the gate count) will be used. These will be denoted by DELAY and GATES (subscripted by RCA to indicate ripple carry adder), respectively. These simple metrics are suitable for first-order comparisons. More accurate comparisons require consideration of both the number and the types of gates (because gates with fewer inputs are generally faster and smaller than gates with more inputs).

$$\text{DELAY}_{\text{RCA}} = 2n + 4 \tag{22.6}$$

$$\text{GATES}_{\text{RCA}} = 9n \tag{22.7}$$

22.3.1.3 Carry Lookahead Adder

Another approach is the carry lookahead adder [Weinberger and Smith 1958, MacSorley 1961]. Here, specialized carry logic computes the carries in parallel. The carry lookahead adder uses MFAs, or modified full adders (modified in the sense that a carry output is not formed) for each bit position and lookahead modules. Each lookahead module forms individual carry outputs and also block generate and block propagate outputs that indicate that a carry is generated within the module or that an incoming carry would propagate across the module, respectively. Rewriting Equation 22.5 using $g_k = a_k b_k$ and $p_k = a_k + b_k$:

$$c_{k+1} = g_k + p_k c_k \tag{22.8}$$

This explains the concept of carry generation and carry propagation. At a given stage, a carry is generated if g_k is true (i.e., both a_k and b_k are 1), and a stage propagates a carry from its input to its output if p_k is true (i.e., either a_k or b_k is a 1). The nine gate full adder shown in Figure 22.2 has AND and OR gates that produce g_k and p_k with no additional complexity. Because the carry out is produced by the lookahead logic, the OR gate that produces c_{k+1} can be eliminated. The result is an eight gate modified full adder (MFA). Extending Equation 22.8 to a second stage,

$$\begin{aligned}
c_{k+2} &= g_{k+1} + p_{k+1} c_{k+1} \\
&= g_{k+1} + p_{k+1}(g_k + p_k c_k) \\
&= g_{k+1} + p_{k+1} g_k + p_{k+1} p_k c_k
\end{aligned} \tag{22.9}$$

This equation results from evaluating Equation 22.8 for the $k + 1$st stage and substituting c_{k+1} from Equation 22.8. Carry c_{k+2} exits from stage $k + 1$ if: (1) a carry is generated there; or (2) a carry is generated in stage k and propagates across stage $k + 1$; or (3) a carry enters stage k and propagates across both stages k and $k + 1$. Extending to a third stage,

$$\begin{aligned}
c_{k+3} &= g_{k+2} + p_{k+2} c_{k+2} \\
&= g_{k+2} + p_{k+2}(g_{k+1} + p_{k+1} g_k + p_{k+1} p_k c_k) \\
&= g_{k+2} + p_{k+2} g_{k+1} + p_{k+2} p_{k+1} g_k + p_{k+2} p_{k+1} p_k c_k
\end{aligned} \tag{22.10}$$

Although it would be possible to continue this process indefinitely, each additional stage increases the size (i.e., the number of inputs) of the logic gates. Four inputs (as required to implement Equation 22.10) is frequently the maximum number of inputs per gate for current technologies. To continue the process, generate and propagate signals are defined over four bit blocks (stages k to $k + 3$), $g_{k+3:k}$ and $p_{k+3:k}$, respectively.

$$g_{k+3:k} = g_{k+3} + p_{k+3} g_{k+2} + p_{k+3} p_{k+2} g_{k+1} + p_{k+3} p_{k+2} p_{k+1} g_k \tag{22.11}$$

and

$$p_{k+3:k} = p_{k+3} p_{k+2} p_{k+1} p_k \tag{22.12}$$

Equation 22.8 can be expressed in terms of the 4-bit block generate and propagate signals,

$$c_{k+4} = g_{k+3:k} + p_{k+3:k} c_k \tag{22.13}$$

Thus, the carryout from a 4-bit-wide block can be computed in only four gate delays (the first to compute p_i and g_i for $i = k - k + 3$, the second to evaluate $p_{k+3:k}$, the second and third to evaluate $g_{k+3:k}$, and the third and fourth to evaluate c_{k+4} using Equation 22.13.

An n-bit carry lookahead adder requires $\lceil (n - 1)/(r - 1) \rceil$ lookahead blocks, where r is the width of the block. A 4-bit lookahead block is a direct implementation of Equation 22.8 through Equation 22.12, requiring 14 logic gates. In general, an r-bit lookahead block requires $\frac{1}{2}(3r + r^2)$ logic gates. The Manchester carry chain [Kilburn et al. 1960] is an alternative switch-based technique for implementation of the lookahead block.

Figure 22.4 shows the interconnection of 16 modified full adders and 5 lookahead logic blocks to realize a 16-bit carry lookahead adder. The sequence of events that occurs during an add operation is as follows: (1) apply A, B, and carry in signals; (2) each adder computes P and G; (3) first-level lookahead logic computes the 4-bit propagate and generate signals; (4) second-level lookahead logic computes c_4, c_8, and c_{12}; (5) first-level lookahead logic computes the individual carries; and (6) each adder computes the sum outputs. This process may be extended to larger adders by subdividing the large adder into 16-bit blocks and using additional levels of carry lookahead (e.g., a 64-bit adder requires three levels).

The delay of carry lookahead adders is evaluated by recognizing that an adder with a single level of carry lookahead (for 4-bit words) has six gate delays and that each additional level of lookahead increases the maximum word size by a factor of four and adds four gate delays. More generally [Waser and Flynn 1982, pp. 83–88], the number of lookahead levels for an n-bit adder is $\lceil \log_r n \rceil$ where r is the maximum number of inputs per gate. Because an r-bit carry lookahead adder has six gate delays and there are four additional gate delays per carry lookahead level after the first,

$$\text{DELAY}_{\text{CLA}} = 2 + 4\lceil \log_r n \rceil \tag{22.14}$$

The complexity of an n-bit carry lookahead adder implemented with r-bit lookahead blocks is n modified full adders (each of which requires 8 gates) and $\lceil (n-1)/(r-1) \rceil$ lookahead logic blocks [each of which requires $\frac{1}{2}(3r + r^2)$ gates]:

$$\text{GATES}_{\text{CLA}} = 8n + \frac{1}{2}(3r + r^2)\left\lceil \frac{n-1}{r-1} \right\rceil \tag{22.15}$$

For the currently common case of $r = 4$,

$$\text{GATES}_{\text{CLA}} \approx 12\frac{2}{3}n - 4\frac{2}{3} \tag{22.16}$$

The carry lookahead approach reduces the delay of adders from increasing linearly with the word size (as for ripple carry adders) to increasing as the logarithm of the word size. As with ripple carry adders, the carry lookahead adder complexity grows linearly with the word size (for $r = 4$, this occurs at a 40% faster rate than for ripple carry adders).

22.3.1.4 Carry Skip Adder

The carry skip adder divides the words to be added into blocks (like the carry lookahead adder). The basic structure of a 16-bit carry skip adder implemented with five 3-bit blocks and one 1-bit-wide block is shown in Figure 22.5. Within each block, a ripple carry adder is used to produce the sum bits and the carry (which is used as a block generate). In addition, an AND gate is used to form the block propagate signal. These signals are combined using Equation 22.13 to produce a fast carry signal.

For example, with $k = 4$, Equation 22.13 yields $c_8 = g_{7:4} + p_{7:4}c_4$. The carry out of the second ripple carry adder is a block generate signal if it is evaluated when carries generated by the data inputs (i.e., $a_{7:4}$ and $b_{7:4}$ in Figure 22.5) are valid but before the carry that results from c_4. Normally, these two types of carries coincide in time, but in the carry skip adder the c_4 signal is produced by a 4-bit ripple carry adder, so the carry output is a block generate from 11 gate delays after application of A and B until it becomes c_8 at 19 gate delays after the application of A and B.

In the carry skip adder, the first and last blocks are simple ripple carry adders, whereas the $\lceil n/k \rceil - 2$ intermediate blocks are ripple carry adders augmented with three gates. The delay of a carry skip adder is the sum of $2k + 3$ gate delays to produce the carry in the first block, 2 gate delays through each of the intermediate blocks, and $2k + 1$ gate delays to produce the most significant sum bit in the last block.

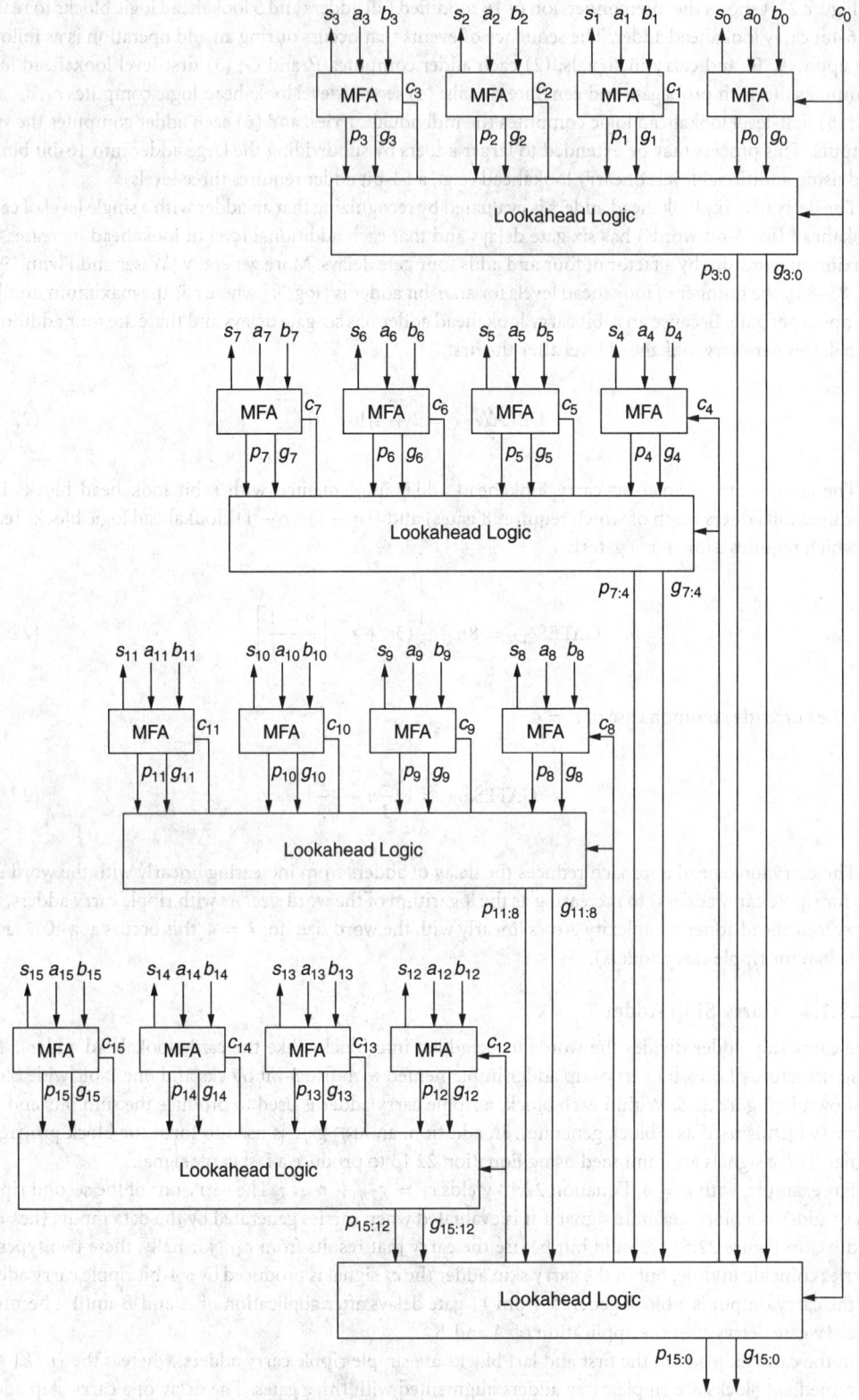

FIGURE 22.4   16-bit two level carry lookahead adder.

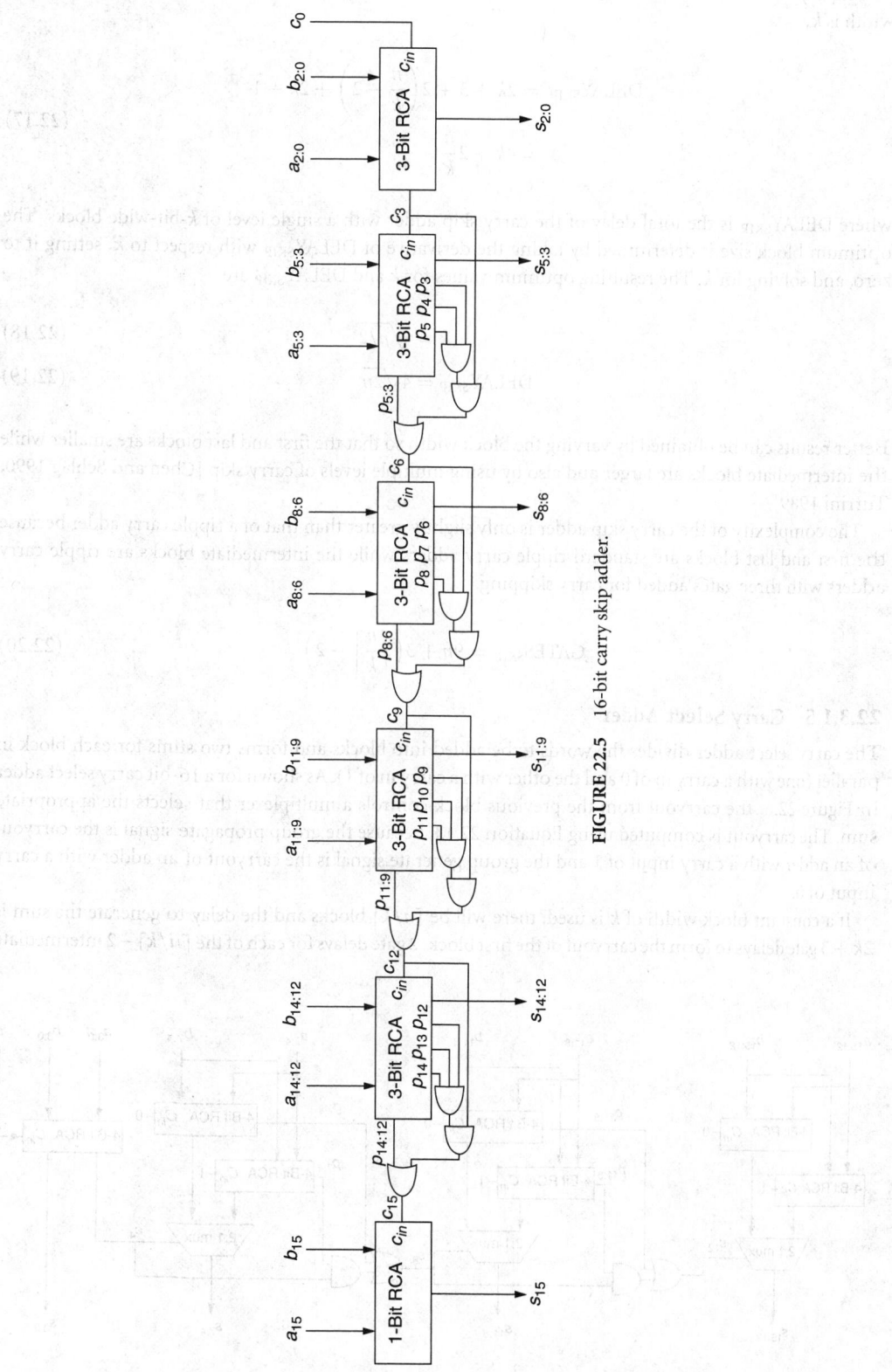

FIGURE 22.5 16-bit carry skip adder.

To simplify the analysis, the ceiling function in the count of intermediate blocks is ignored. If the block width is k,

$$
\begin{aligned}
\text{DELAY}_{\text{SKIP}} &= 2k + 3 + 2\left(\frac{n}{k} - 2\right) + 2k + 1 \\
&= 4k + 2\frac{n}{k}
\end{aligned}
\tag{22.17}
$$

where $\text{DELAY}_{\text{SKIP}}$ is the total delay of the carry skip adder with a single level of k-bit-wide blocks. The optimum block size is determined by taking the derivative of $\text{DELAY}_{\text{SKIP}}$ with respect to k, setting it to zero, and solving for k. The resulting optimum values for k and $\text{DELAY}_{\text{SKIP}}$ are

$$
k = \sqrt{n/2}
\tag{22.18}
$$

$$
\text{DELAY}_{\text{SKIP}} = 4\sqrt{2n}
\tag{22.19}
$$

Better results can be obtained by varying the block width so that the first and last blocks are smaller while the intermediate blocks are larger and also by using multiple levels of carry skip [Chen and Schlag 1990, Turrini 1989].

The complexity of the carry skip adder is only slightly greater than that of a ripple carry adder because the first and last blocks are standard ripple carry adders while the intermediate blocks are ripple carry adders with three gates added for carry skipping.

$$
\text{GATES}_{\text{SKIP}} = 9n + 3\left(\left\lceil\frac{n}{k}\right\rceil - 2\right)
\tag{22.20}
$$

22.3.1.5 Carry Select Adder

The carry select adder divides the words to be added into blocks and forms two sums for each block in parallel (one with a carry in of 0 and the other with a carry in of 1). As shown for a 16-bit carry select adder in Figure 22.6, the carryout from the previous block controls a multiplexer that selects the appropriate sum. The carryout is computed using Equation 22.13, because the group propagate signal is the carryout of an adder with a carry input of 1 and the group generate signal is the carryout of an adder with a carry input of 0.

If a constant block width of k is used, there will be $\lceil n/k \rceil$ blocks and the delay to generate the sum is $2k + 3$ gate delays to form the carryout of the first block, 2 gate delays for each of the $\lceil n/k \rceil - 2$ intermediate

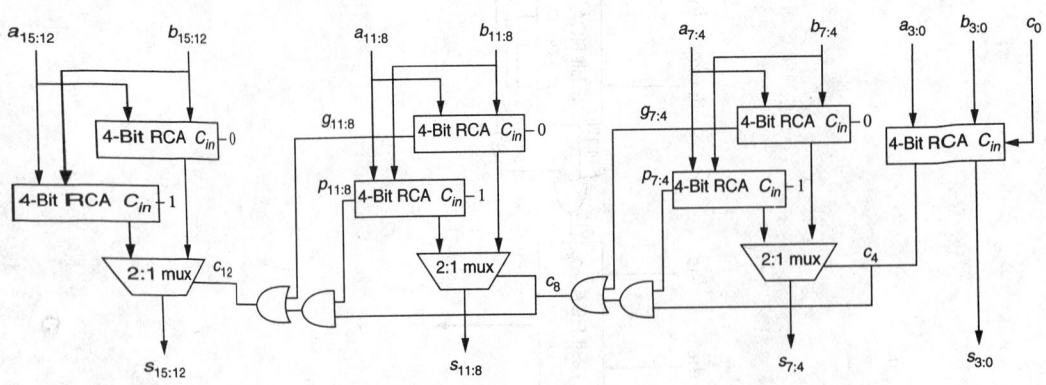

FIGURE 22.6   16-bit carry select adder.

blocks, and 3 gate delays (for the multiplexer) in the final block. To simplify the analysis, the ceiling function in the count of intermediate blocks is ignored. Thus, the total delay is

$$\mathrm{DELAY_{C-SEL}} = 2k + 2\frac{n}{k} + 2 \qquad (22.21)$$

The optimum block size is determined by taking the derivative of $\mathrm{DELAY_{SEL}}$ with respect to k, setting it to zero, and solving for k. The results are:

$$k = \sqrt{n} \qquad (22.22)$$

$$\mathrm{DELAY_{SEL}} = 2 + 4\sqrt{n} \qquad (22.23)$$

As for the carry skip adder, better results can be obtained by varying the width of the blocks. In this case, the optimum is to make the two least significant blocks the same size and each successively more significant block 1 bit larger. In this configuration, the delay for each block's most significant sum bit will equal the delay to the multiplexer control signal [Goldberg 1990, p. A-38].

The complexity of the carry select adder is $2n - k$ ripple carry adder stages, the intermediate carry logic and ($\lceil n/k \rceil - 1$) k-bit-wide 2:1 multiplexers:

$$\mathrm{GATES_{SEL}} = 9(2n - k) + 2\left(\left\lceil \frac{n}{k} \right\rceil - 2\right) + 3(n - k) + \left\lceil \frac{n}{k} \right\rceil - 1$$

$$= 21n - 12k + 3\left\lceil \frac{n}{k} \right\rceil - 5 \qquad (22.24)$$

This is slightly more than twice the complexity of a ripple carry adder of the same size.

22.3.2 Fixed Point Subtraction

To produce an adder/subtracter, the adder is modified as shown in Figure 22.7 by including EXCLUSIVE-OR gates to complement operand B when performing subtraction. It forms either $A + B$ or $A - B$. In the case of $A + B$, the mode selector is set to logic 0, which causes the EXCLUSIVE-OR gates to pass operand B through unchanged to the ripple carry adder. The carry into the least significant adder stage is also set to ZERO, so standard addition occurs. Subtraction is implemented by setting the mode selector to logic ONE, which causes the EXCLUSIVE-OR gates to complement the bits of B; formation of the two's complement of B is completed by setting the carry into the least significant adder stage to ONE.

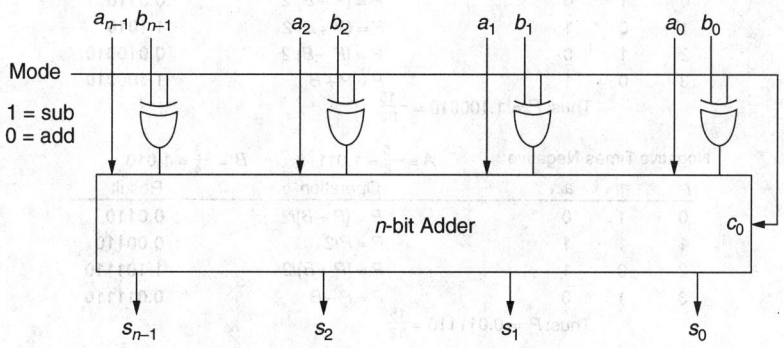

FIGURE 22.7   Adder/subtracter.

22.3.3 Fixed Point Multiplication

Multiplication is generally implemented either via a sequence of addition, subtraction, and shift operations or by direct logic implementations.

22.3.3.1 Sequential Booth Multiplier

The Booth algorithm [Booth 1951] is widely used for two's complement multiplication because it is easy to implement. Earlier two's complement multipliers (e.g., [Shaw 1950]) require data-dependent correction cycles if either operand is negative. In the Booth algorithm, to multiply A times B, the product P is initially set to 0. Then, the bits of the multiplier A are examined in pairs of adjacent bits, starting with the least significant bit (i.e., $a_0\ a_{-1}$) and assuming $a_{-1} = 0$:

- If $a_i = a_{i-1}, P = P/2$.
- If $a_i = 0$ and $a_{i-1} = 1, P = (P + B)/2$.
- If $a_i = 1$ and $a_{i-1} = 0, P = (P - B)/2$.

The division by 2 is not performed on the last stage (i.e., when $i = n - 1$). All of the divide by 2 operations are simple arithmetic right shifts (i.e., the word is shifted right one position and the old sign bit is repeated for the new sign bit), and overflows in the addition process are ignored. The algorithm is illustrated in Figure 22.8, in which products for all combinations of $\pm 5/8$ times $\pm 3/4$ are computed for 4-bit operands. The sequential Booth multiplier requires n cycles to form the product for a pair of n-bit

Positive Times Positive $\qquad A = \frac{5}{8} = 0.101 \qquad B = \frac{3}{4} = 0.110$

i	a_i	a_{i-1}	Operation	Result
0	1	0	$P = (P - B)/2$	1.1010
1	0	1	$P = (P + B)/2$	0.00110
2	1	0	$P = (P - B)/2$	1.101110
3	0	1	$P = P + B$	0.011110

Thus: $P = 0.011110 = \frac{15}{32}$

Negative Times Positive $\qquad A = -\frac{5}{8} = 1.011 \qquad B = \frac{3}{4} = 0.110$

i	a_i	a_{i-1}	Operation	Result
0	1	0	$P = (P - B)/2$	1.1010
1	1	1	$P = P/2$	1.11010
2	0	1	$P = (P + B)/2$	0.010010
3	1	0	$P = P - B$	1.100010

Thus: $P = 1.100010 = -\frac{15}{32}$

Positive Times Negative $\qquad A = \frac{5}{8} = 0.101 \qquad B = -\frac{3}{4} = 1.010$

i	a_i	a_{i-1}	Operation	Result
0	1	0	$P = (P - B)/2$	0.0110
1	0	1	$P = (P + B)/2$	1.1010
2	1	0	$P = (P - B)/2$	0.010010
3	0	1	$P = P + B$	1.100010

Thus: $P = 1.100010 = -\frac{15}{32}$

Negative Times Negative $\qquad A = -\frac{5}{8} = 1.011 \qquad B = -\frac{3}{4} = 1.010$

i	a_i	a_{i-1}	Operation	Result
0	1	0	$P = (P - B)/2$	0.0110
1	1	1	$P = P/2$	0.00110
2	0	1	$P = (P + B)/2$	1.101110
3	1	0	$P = P - B$	0.011110

Thus: $P = 0.011110 = \frac{15}{32}$

FIGURE 22.8   Example of sequential Booth multiplication.

TABLE 22.3 Operations for the Radix-4 Modified Booth Algorithm

a_{i+1}	a_i	a_{i-1}	Operation
0	0	0	$P = P/4$
0	0	1	$P = (P + B)/4$
0	1	0	$P = (P + B)/4$
0	1	1	$P = (P + 2B)/4$
1	0	0	$P = (P - 2B)/4$
1	0	1	$P = (P - B)/4$
1	1	0	$P = (P - B)/4$
1	1	1	$P = P/4$

numbers, where each cycle consists of either an n-bit addition and a shift, an n-bit subtraction and a shift, or a shift without any other arithmetic operation.

22.3.3.2 Sequential Modified Booth Multiplier

The radix-4 modified Booth multiplier [MacSorley 1961] uses $n/2$ cycles in which each cycle examines three adjacent bits; adds or subtracts 0, B, or $2B$; and shifts 2 bits to the right. Table 22.3 shows the operations as a function of the three bits: a_{i+1}, a_i, and a_{i-1}. The radix-4 modified Booth multiplier requires half as many cycles as the standard (radix-2) Booth multiplier although the operations performed during each cycle are slightly more complex (because it is necessary to select one of five possible addends instead of one of three). Extensions to higher radices that examine more than 3 bits are possible but generally are not attractive because the addition/subtraction operations involve multiplies of B that are not powers of two (such as 3, 5, etc.), which raises the complexity.

22.3.3.3 Array Multipliers

An alternative approach to multiplication involves the combinational generation of all bit products and their summation with an array of full adders. The block diagram for an 8-by-8 array multiplier is shown in Figure 22.9. It uses an 8-by-8 array of AND gates to form the bit products (some of the AND gates are denoted G in Figure 22.9 and the remainder are contained within the FA and HA blocks) and an 8-by-7 array of adders [48 full adders (denoted FA in Figure 22.9) and 8 half-adders (denoted HA in Figure 22.9)] to sum the 64-bit products. Summing the bits requires 14 adder delays.

Modification of the array multiplier to multiply two's complement numbers requires inverting most of the signals in the most significant row and column and also adding a few correction terms [Baugh and Wooley 1973, Blankenship 1974]. Array multipliers are easily laid out in a cellular fashion, making them suitable for very large scale integrated (VLSI) implementation, where minimizing the design effort may be more important than maximizing the speed of the multiplier.

22.3.3.4 Wallace Tree/Dadda Fast Multiplier

A method for fast multiplication was developed by Wallace [1964] and refined by Dadda [1965]. With this method, a three-step process is used to multiply two numbers: (1) the bit products are formed; (2) the bit product matrix is reduced to a two-row matrix in which the sum of the rows equals the sum of the bit products; and (3) the two numbers are summed using a fast adder to produce the product. Although this may seem to be a complex process, it yields multipliers with delay proportional to the logarithm of the operand word size, which is faster than the Booth multiplier, the modified Booth multiplier, or array multipliers, all of which have delays proportional to the word size.

The second step in the fast multiplication process is shown for an 8-by-8 Dadda multiplier in Figure 22.10. An input 8-by-8 matrix of dots (each dot represents a bit product) is shown as Matrix 0. Columns having more than six dots (or that will grow to more than six dots due to carries) are reduced by the use of half-adders (each half-adder takes in two dots and outputs one in the same column and one in the next

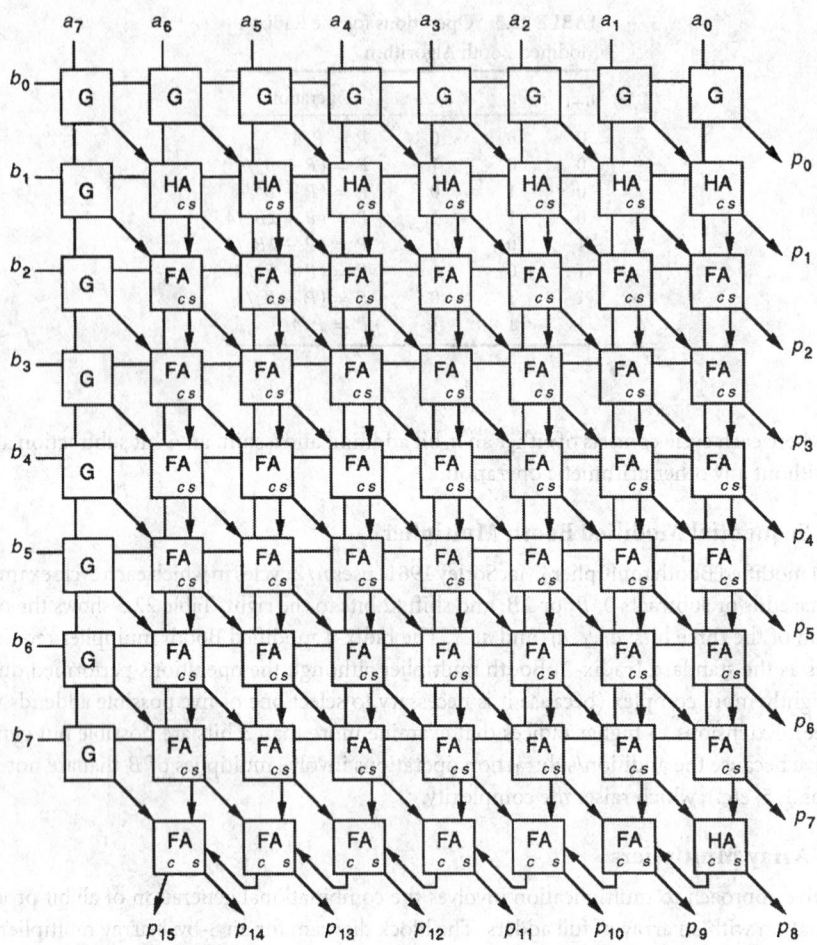

FIGURE 22.9 Unsigned 8-by-8 array multiplier.

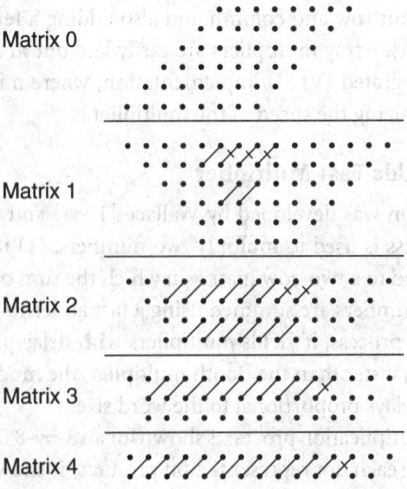

FIGURE 22.10 Unsigned 8-by-8 Dadda multiplier.

more significant column) and full adders (each full adder takes in three dots and outputs one in the same column and one in the next more significant column) so that no column in Matrix 1 will have more than six dots. Half-adders are shown by two dots connected with a crossed line in the succeeding matrix and full adders are shown by two dots connected with a line in the succeeding matrix. In each case, the rightmost dot of the pair connected by a line is in the column from which the inputs were taken for the adder. In the succeeding steps, reduction to Matrix 2 with no more than four dots per column, Matrix 3 with no more than three dots per column, and finally Matrix 4 with no more than two dots per column is performed. The height of the matrices is determined by working back from the final (two-row) matrix and limiting the height of each matrix to the largest integer that is no more than 1.5 times the height of its successor. Each matrix is produced from its predecessor in one adder delay. Because the number of matrices is logarithmically related to the number of bits in the words to be multiplied, the delay of the matrix reduction process is proportional to log n. Because the adder that reduces the final two-row matrix can be implemented as a carry lookahead adder (which also has logarithmic delay), the total delay for this multiplier is proportional to the logarithm of the word size.

22.3.4 Fixed Point Division

Division is traditionally implemented as a digit recurrence requiring a sequence of shift, subtract, and compare operations, in contrast to the shift and add approach employed for multiplication. The comparison operation is significant. It results in a serial process, which is not amenable to parallel implementation. There are several digit recurrence-based division algorithms, including binary restoring, nonperforming, nonrestoring, and Sweeney, Robertson, and Tocher (SRT) division algorithms for a variety of radices [Ercegovac and Lang 1994].

22.3.4.1 Nonrestoring Divider

Traditional nonrestoring division is based on selecting digits of the quotient Q (where $Q = N/D$) to satisfy the following equation:

$$P_{k+1} = r P_k - q_{n-k-1} D \quad \text{for } k = 1, 2, \ldots, n-1 \tag{22.25}$$

where P_k is the partial remainder after the selection of the kth quotient digit, $P_0 = N$ (subject to the constraint $|P_0| < |D|$), r is the radix ($r = 2$ for binary nonrestoring division), q_{n-k-1} is the kth quotient digit to the right of the binary point, and D is the divisor. In this section, it is assumed that both N and D are positive; see Ercegovac and Lang [1994] for details on handling the general case.

In nonrestoring division, the quotient digits are constrained to be ± 1 (i.e., q_k is selected from $\{1, \bar{1}\}$). The digit selection and resulting partial remainder are given for the kth iteration by the following relations:

$$\text{If } P_k \geq 0, \quad q_{n-k-1} = 1 \quad \text{and} \quad P_{k+1} = r P_k - D \tag{22.26}$$
$$\text{If } P_k < 0, \quad q_{n-k-1} = \bar{1} \quad \text{and} \quad P_{k+1} = r P_k + D \tag{22.27}$$

This process continues either for a set number of iterations or until the partial remainder is smaller than a specified error criterion. The kth most significant bit of the quotient is a 1 if P_k is 0 or positive and is a $\bar{1}(-1)$ if P_k is negative. The algorithm is illustrated in Figure 22.11, where $\frac{5}{16}$ is divided by $\frac{3}{8}$. The result ($\frac{13}{16}$) is the closest 4-bit fraction to the correct result of $\frac{5}{6}$.

The signed digit number (comprising ± 1 digits) can be converted into a conventional binary number by subtracting n, the number formed by the negative digits (with 0s where there are 1s in Q and 1s where

Nonrestoring division

$$N = \frac{5}{16} = 0.0101$$

$$D = \frac{3}{8} = 0.0110$$

$$P_{(0)} = N$$

Since $P_{(0)} \geq 0$, $q_3 = 1$ and $P_{(1)} = 2P_{(0)} - D$

$$
\begin{array}{ll}
2P_{(0)} = & 0.1010 \\
-D & 1.1010 \\
\hline
P_{(1)} = & 0.0100
\end{array}
$$

Since $P_{(1)} \geq 0$, $q_2 = 1$ and $P_{(2)} = 2P_{(1)} - D$

$$
\begin{array}{ll}
2P_{(1)} = & 0.1000 \\
-D & 1.1010 \\
\hline
P_{(2)} = & 0.0010
\end{array}
$$

Since $P_{(2)} \geq 0$, $q_1 = 1$ and $P_{(3)} = 2P_{(2)} - D$

$$
\begin{array}{ll}
2P_{(2)} = & 0.0100 \\
-D & 1.1010 \\
\hline
P_{(3)} = & 1.1110
\end{array}
$$

Since $P_{(3)} < 0$, $q_0 = \bar{1}$ and $P_{(4)} = 2P_{(3)} + D$

$$
\begin{array}{ll}
2P_{(3)} = & 1.1100 \\
+D & 0.0110 \\
\hline
P_{(4)} = & 0.0010
\end{array}
$$

$$Q = 0.111\bar{1} = 0.1101 = \frac{13}{16}$$

FIGURE 22.11 Example of nonrestoring division.

there are $\bar{1}$s in Q), from p, the number formed by the positive digits (with 0s where there are $\bar{1}$s in Q and 1s where there are 1s in Q). For the example of Figure 22.11:

$$Q = 0.111\bar{1}$$

$$n = 0.0001$$

$$p = 0.1110$$

$$p - n = 0.1110 - 0.0001$$

$$= 0.1110 + 1.1111$$

$$= 0.1101$$

Other digit recurrence division algorithms include the restoring algorithm and the SRT algorithms [Robertson 1958]. The restoring algorithm is similar to the nonrestoring algorithm presented here except that either subtract and shift or shift are permitted for each iteration instead of subtract and shift or add and shift as in Equations 22.26 and 22.27. Restoring division forms the quotient from the digits 0 and 1 so that no signed digit to binary conversion is required (as it is for nonrestoring division). For all of the digit recurrence algorithms, the number of steps is proportional to the number of quotient digits.

The more advanced SRT division process for radix-2 and higher radices is similar to nonrestoring division in that the recurrence of Equation 22.25 is used, but the set of allowable quotient digits is increased. The process of quotient digit selection is sufficiently complex that a research monograph has been devoted to SRT division [Ercegovac and Lang 1994].

22.3.4.2 Newton–Raphson Divider

A second division algorithm uses a form of Newton–Raphson iteration to derive a quadratically convergent approximation to the reciprocal of the divisor, which is then multiplied by the dividend to produce the quotient. In systems that include a fast multiplier, this process is often faster than conventional division [Ferrari 1967].

$$N = 0.625$$

$$D = 0.75$$

$$R_{(0)} = 2.915 - 2 \cdot D \qquad\qquad \text{1 Shift, 1 Subtract}$$
$$= 2.915 - 2 \cdot .75$$
$$R_{(0)} = 1.415$$

$$R_{(1)} = R_{(0)} (2 - D \cdot R_{(0)}) \qquad\qquad \text{2 Multiplys, 1 Subtract}$$
$$= 1.415 (2 - .75 \cdot 1.415)$$
$$= 1.415 \cdot .93875$$
$$R_{(1)} = 1.32833125$$

$$R_{(2)} = R_{(1)} (2 - D \cdot R_{(1)}) \qquad\qquad \text{2 Multiplys, 1 Subtract}$$
$$= 1.32833125 (2 - .75 \cdot 1.32833125)$$
$$= 1.32833125 \cdot 1.00375156$$
$$R_{(2)} = 1.3333145677$$

$$R_{(3)} = R_{(2)} (2 - D \cdot R_{(2)}) \qquad\qquad \text{2 Multiplys, 1 Subtract}$$
$$= 1.3333145677 (2 - .75 \cdot 1.3333145677)$$
$$= 1.3333145677 \cdot 1.00001407$$
$$R_{(3)} = 1.3333333331$$

$$Q = N \cdot R_{(3)} \qquad\qquad \text{1 Multiply}$$
$$= 0.625 \cdot 1.3333333331$$
$$Q = 0.83333333319$$

FIGURE 22.12 Example of Newton–Raphson division.

The Newton–Raphson division algorithm to compute $Q = N/D$ consists of three basic steps:

1. Calculate a starting estimate of the reciprocal of the divisor $R_{(0)}$. If the divisor D is normalized (i.e., $\frac{1}{2} \leq D < 1$), then $R_{(0)} = 3 - 2D$ exactly computes $1/D$ at $D = 0.5$ and $D = 1$ and exhibits maximum error (of approximately 0.17) at $D = \sqrt{\frac{1}{2}}$. Adjusting $R_{(0)}$ downward to by half the maximum error gives:

$$R_{(0)} = 2.915 - 2D \qquad\qquad (22.28)$$

This produces an initial estimate that is within about 0.087 of the correct value for all points in the interval $\frac{1}{2} \leq D < 1$.

2. Compute successively more accurate estimates of the reciprocal by the following iterative procedure:

$$R_{(i+1)} = R_{(i)}(2 - DR_{(i)}) \qquad \text{for } i = 0, 1, \ldots, k \qquad (22.29)$$

3. Compute the quotient by multiplying the dividend times the reciprocal of the divisor,

$$Q = NR_{(k)} \qquad\qquad (22.30)$$

where i is the iteration count and N is the numerator. Figure 22.12 illustrates the operation of the Newton–Raphson algorithm. For this example, three iterations (one shift, four subtractions, and seven multiplications) produce an answer accurate to nine decimal digits (approximately 30 bits).

With this algorithm, the error decreases quadratically, so that the number of correct bits in each approximation is roughly twice the number of correct bits on the previous iteration. Thus, from a $3\frac{1}{2}$-bit initial approximation, two iterations produce a reciprocal estimate accurate to 14 bits, four iterations produce a reciprocal estimate accurate to 56 bits, etc. The number of iterations is proportional to the logarithm of the number of accurate quotient digits.

The efficiency of this process is dependent on the availability of a fast multiplier, since each iteration of Equation 22.29 requires two multiplications and a subtraction. The complete process for the initial estimate, three iterations, and the final quotient determination requires a shift, four subtractions, and seven multiplications to produce a 16-bit quotient. This is faster than a conventional nonrestoring divider if multiplication is roughly as fast as addition, a condition that is usually satisfied for systems that include hardware multipliers.

22.4 Floating Point Arithmetic

Advances in VLSI have increased the feasibility of hardware implementations of floating point arithmetic units. The main advantage of floating point arithmetic is that its wide dynamic range virtually eliminates overflow for most applications.

22.4.1 Floating Point Number Systems

A floating point number, A, consists of a signed significand, S_a, and an exponent, E_a. The value of a number, A, is given by the equation

$$A = S_a r^{E_a} \tag{22.31}$$

where r is the radix (or base) of the number system. The significand is generally normalized by requiring that the most significant digit be nonzero. Use of the binary radix (i.e., $r = 2$) gives maximum accuracy but may require more frequent normalization than higher radices.

The IEEE Standard 754 single precision (32-bit) floating point format, which is widely implemented, has an 8-bit biased integer exponent, which ranges from 0 to 255 [IEEE 1985]. The exponent is expressed in excess 127 code, which means that its effective value is determined by subtracting 127 from the stored value. Thus, the range of effective values of the exponent is -126 to 127, corresponding to stored values of 1 to 254, respectively. A stored exponent value of 0 (E_{min}) (effective value $= -127$) serves as a flag for 0 (if the significand is 0) and for denormalized numbers (if the significand is nonzero). A stored exponent value of 255 (E_{max}) (effective value $= 128$) serves as a flag for infinity (if the significand is 0) and for "Not A Number"(NAN) (if the significand is nonzero). The significand is a 25-bit sign magnitude mixed number (the binary point is to the right of the most significant bit, which is always a 1 except for denormalized numbers). More details on floating point formats and on the various considerations that arise in the implementation of floating point arithmetic units are given in Gosling [1980], Goldberg [1990], Parhami [2000], Flynn and Oberman [2001], and Koren [2002].

22.4.1.1 Floating Point Addition

A flowchart for a floating point addition algorithm is shown in Figure 22.13. For this flowchart (and those that follow for multiplication and division), the significands are assumed to have magnitudes in the range $[1, 2)$.

On the flowchart, the operands are (E_a, S_a) and (E_b, S_b), and the result is (E_s, S_s). In step 1, the operand exponents are compared; if they are unequal, the significand of the number with the smaller exponent is shifted right in either step 3 or step 4 by the difference in the exponents to properly align the significands. For example, to add 0.867×10^5 and 0.512×10^4, the latter would be shifted right and 0.867 added to 0.0512 to give a sum of 0.9182×10^5. The addition of the significands is performed in step 5. Steps 6 through 8 test for overflow and correct if necessary by shifting the significand one position to the right and incrementing the exponent. Step 9 tests for a zero significand. The loop of steps 10 and 11 scales unnormalized (but non-0) significands upward to normalize the significand. Step 12 tests for underflow.

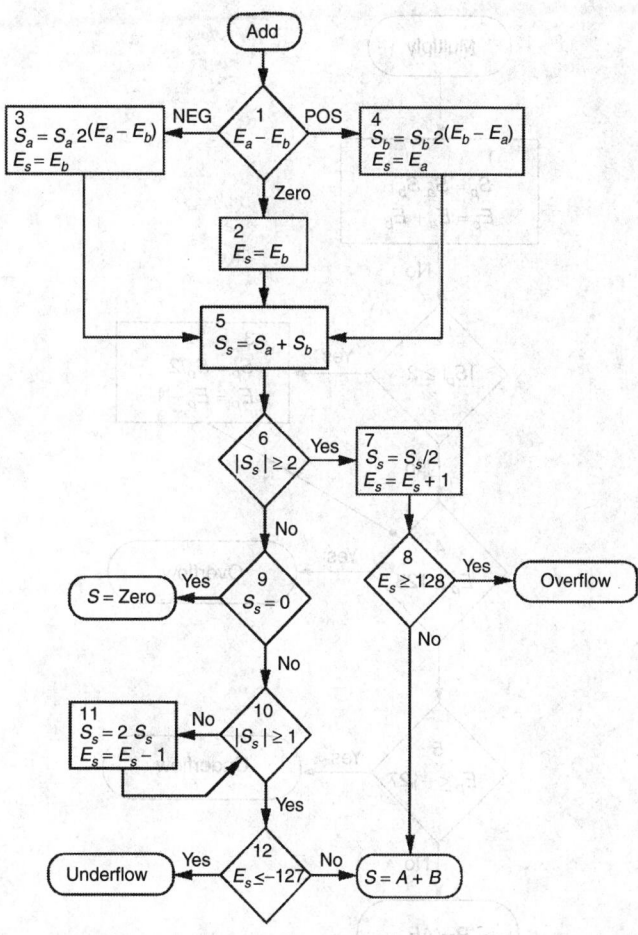

FIGURE 22.13 Floating point addition.

Floating point subtraction is implemented with a similar algorithm. Many refinements are possible to improve the speed of the addition and subtraction algorithms, but floating point addition will, in general, be much slower than fixed point addition as a result of the need for preaddition alignment and postaddition normalization.

22.4.1.2 Floating Point Multiplication

The algorithm for floating point multiplication is shown in the flowchart of Figure 22.14. In step 1, the product of the operand significands and the sum of the operand exponents are computed. Steps 2 and 3 normalize the significand if necessary. For radix-2 floating point numbers, if the operands are normalized, at most a single shift is required to normalize the product. Step 4 tests the exponent for overflow. Finally, step 5 tests for underflow.

22.4.1.3 Floating Point Division

The floating point division algorithm is shown in the flowchart of Figure 22.15. The quotient of the significands and the difference of the exponents are computed in step 1. The quotient is normalized (if necessary) in steps 2 and 3 by shifting the quotient significand while the quotient exponent is adjusted appropriately. For radix-2, if the operands are normalized, only a single shift is required to normalize the quotient. The computed exponent is tested for underflow in step 4. Finally, the fifth step tests for overflow.

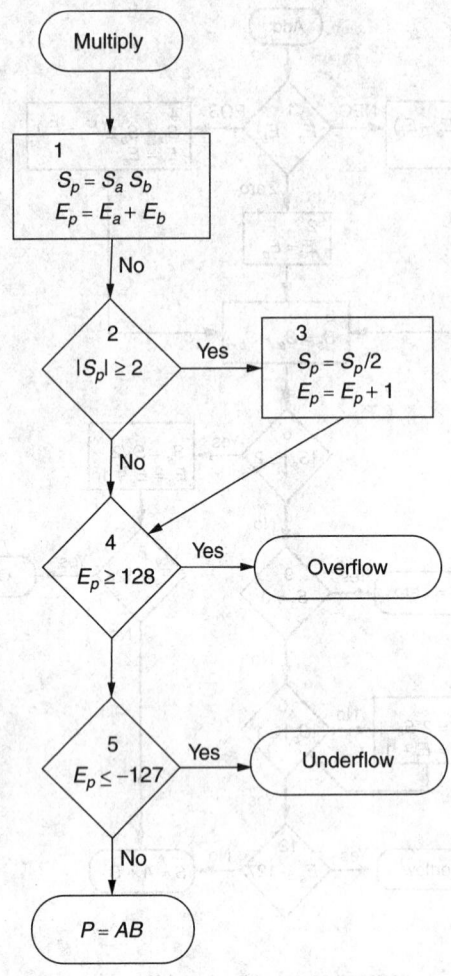

FIGURE 22.14   Floating point multiplication.

22.4.1.4 Floating Point Rounding

All floating point algorithms may require rounding to produce a result in the correct format. A variety of alternative rounding schemes have been developed for specific applications. Round to nearest, round toward ∞, round toward $-\infty$, and round toward 0 are all available in implementations of the IEEE floating point standard.

22.5   Conclusion

This chapter has presented an overview of binary number systems, algorithms for the basic integer arithmetic operations, and a brief introduction to floating point operations. When implementing arithmetic units there is often an opportunity to optimize both the performance and the area for the requirements of the specific application. In general, faster algorithms require either more area or more complex control: it is often useful to use the fastest algorithm that will fit the available area.

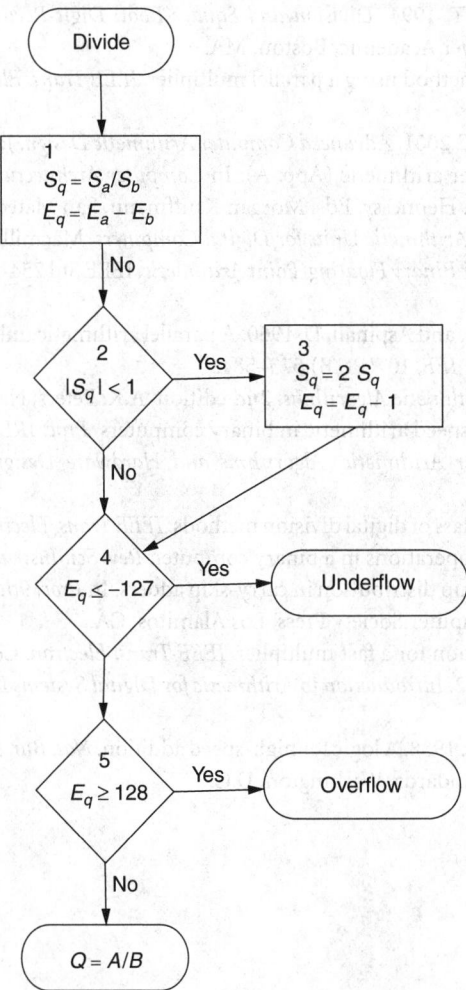

FIGURE 22.15 Floating point division.

Acknowledgments

Revision of chapter originally presented in Swartzlander, E.E., Jr. 1992. Computer arithmetic. In *Computer Engineering Handbook*. C.H. Chen, Ed., Ch. 4, pp. 4-1–4-20. McGraw–Hill, New York. With permission.

References

Baugh, C.R. and Wooley, B.A. 1973. A two's complement parallel array multiplication algorithm. *IEEE Trans. Comput.*, C-22:1045–1047.

Blankenship, P.E. 1974. Comments on "A two's complement parallel array multiplication algorithm." *IEEE Trans. Comput.*, C-23:1327.

Booth, A.D. 1951. A signed binary multiplication technique. *Q. J. Mech. Appl. Math.*, 4(Pt. 2):236–240.

Chen, P.K. and Schlag, M.D.F. 1990. Analysis and design of CMOS Manchester adders with variable carry skip. *IEEE Trans. Comput.*, 39:983–992.

Dadda, L. 1965. Some schemes for parallel multipliers. *Alta Frequenza*, 34:349–356.

Ercegovac, M.D. and Lang, T. 1994. *Division and Square Root: Digit-Recurrence Algorithms and Their Implementations*. Kluwer Academic, Boston, MA.

Ferrari, D. 1967. A division method using a parallel multiplier. *IEEE Trans. Electron. Comput.*, EC-16:224–226.

Flynn, M.J. and Oberman, S.F. 2001. *Advanced Computer Arithmetic Design*. John Wiley & Sons, New York.

Goldberg, D. 1990. Computer arithmetic (App. A). In *Computer Architecture: A Quantitative Approach*, D.A. Patterson and J.L. Hennessy, Eds. Morgan Kauffmann, San Mateo, CA.

Gosling, J.B. 1980. *Design of Arithmetic Units for Digital Computers*. Macmillan, New York.

IEEE. 1985. *IEEE Standard for Binary Floating-Point Arithmetic*. IEEE Std 754-1985, Reaffirmed 1990. IEEE Press, New York.

Kilburn, T. , Edwards, D.B.G., and Aspinall, D. 1960. A parallel arithmetic unit using a saturated transistor fast-carry circuit. *Proc. IEE*, 107(Pt. B):573–584.

Koren, I. 2002. *Computer Arithmetic Algorithms*. 2nd edition, A.K. Peters, Natick, MA.

MacSorley, O.L. 1961. High-speed arithmetic in binary computers. *Proc. IRE*, 49:67–91.

Parhami, B. 2000. *Computer Arithmetic: Algorithms and Hardware Design*. Oxford University Press, New York.

Robertson, J.E. 1958. A new class of digital division methods. *IEEE Trans. Electron. Comput.*, EC-7:218–222.

Shaw, R.F. 1950. Arithmetic operations in a binary computer. *Rev. Sci. Instrum.*, 21:687–693.

Turrini, S. 1989. Optimal group distribution in carry-skip adders. In *Proc. 9th Symp. Computer Arithmetic*, pp. 96–103. IEEE Computer Society Press, Los Alamitos, CA.

Wallace, C.S. 1964. A suggestion for a fast multiplier. *IEEE Trans. Electron. Comput.* EC-13:14–17.

Waser, S. and Flynn, M.J. 1982. *Introduction to Arithmetic for Digital Systems Designers*. Holt, Rinehart and Winston, New York.

Weinberger, A. and Smith, J.L. 1958. A logic for high-speed addition. *Nat. Bur. Stand. Circular 591*, pp. 3–12. National Bureau of Standards, Washington, D.C.

23

Parallel Architectures

23.1	Introduction	**23**-1
23.2	The Stream Model	**23**-1
23.3	SISD	**23**-2
23.4	SIMD	**23**-5
	Array Processors • Vector Processors	
23.5	MISD	**23**-8
23.6	MIMD	**23**-8
23.7	Network Interconnections	**23**-11
23.8	Afterword	**23**-12

Michael J. Flynn
Stanford University

Kevin W. Rudd
Intel, Inc.

23.1 Introduction

Parallel or concurrent operation has many different forms within a computer system. Multiple computers can be executing pieces of the same program in parallel, or a single computer can be executing multiple instructions in parallel, or some combination of the two. Parallelism can arise at a number of levels: task level, instruction level, or some lower machine level. The parallelism may be exhibited in space with multiple independently functioning units, or in time, where a single function unit is many times faster than several instruction-issuing units. This chapter attempts to remove some of the complexity regarding parallel architectures (unfortunately, there is no hope of removing the complexity of programming some of these architectures, but that is another matter).

With all the possible kinds of parallelism, a framework is needed to describe particular instances of parallel architectures. One of the oldest and simplest such structures is the stream approach [Flynn 1966] that is used here as a basis for describing developments in parallel architecture. Using the stream model, different architectures will be described and the defining characteristics for each architecture will be presented. These characteristics provide a qualitative feel for the architecture for high-level comparisons between different processors — they do not attempt to characterize subtle or quantitative differences that, while important, do not provide a significant benefit in a larger view of an architecture.

23.2 The Stream Model

A parallel architecture has, or at least appears to have, multiple interconnected **processor elements** (PE in Figure 23.1) that operate concurrently, solving a single overall problem. Initially, the various parallel architectures can be described using the *stream* concept. A stream is simply a sequence of objects or

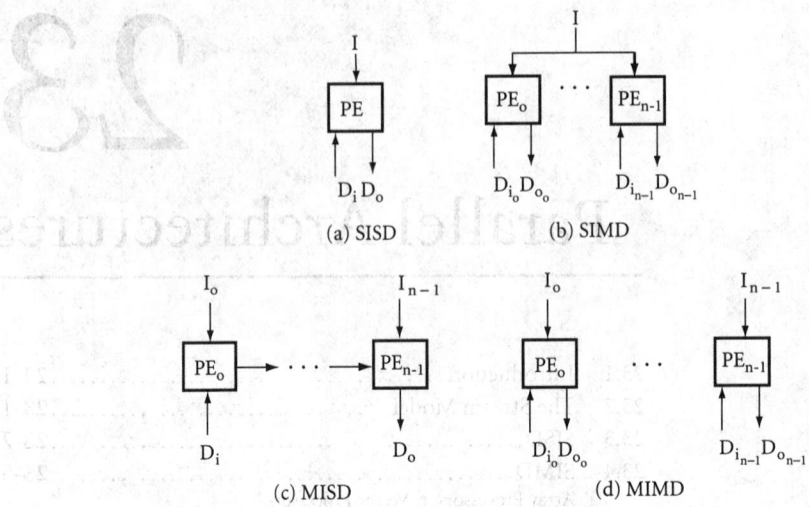

FIGURE 23.1   The stream model.

actions. Since there are both instruction streams and data streams (I and D in Figure 23.1), there are four combinations that describe most familiar parallel architectures:

1. SISD — single instruction, single data stream. This is the traditional uniprocessor [Figure 23.1a].
2. SIMD — single instruction, multiple data stream. This includes vector processors as well as massively parallel processors [Figure 23.1b].
3. MISD — multiple instruction, single data stream. These are typically systolic arrays [Figure 23.1c].
4. MIMD — multiple instruction, multiple data stream. This includes traditional multiprocessors as well as newer work in the area of networks of workstations [Figure 23.1d].

The stream description of architectures uses as its reference point the programmer's view of the machine. If the processor architecture allows for parallel processing of one sort or another, then this information must be visible to the programmer at some level for this reference point to be useful.

An additional limitation of the stream categorization is that, while it serves as a useful shorthand, it ignores many subtleties of an architecture or an implementation. Even an SISD processor can be highly parallel in its execution of operations. This parallelism is typically not visible to the programmer even at the assembly language level, but it becomes visible at execution time with improved performance.

There are many factors that determine the overall effectiveness of a parallel processor organization, and hence its eventual speedup when implemented. Some of these, including networks of interconnected streams, will be touched upon in the remainder of this chapter. The characterizations of both processors and networks are complementary to the stream model and, when coupled with the stream model, enhance the qualitative understanding of a given processor configuration.

23.3   SISD

The SISD class of processor architectures is the most familiar class — it can be found in video games, home computers, engineering workstations, and mainframe computers. From the reference point of the assembly language programmer there is no parallelism in an SISD organization, yet a good deal of concurrency can be present. **Pipelining** is an early technique that is used in almost all current processor implementations. Other techniques aggressively exploit parallelism in executing code whether it is declared statically or determined dynamically from an analysis of the code stream.

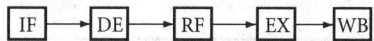

FIGURE 23.2   Canonical pipeline.

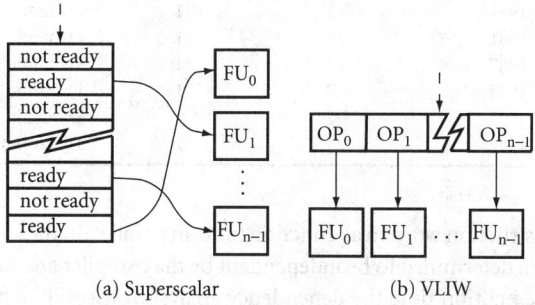

(a) Superscalar   (b) VLIW

FIGURE 23.3   Instruction issue for superscalar processors.

Pipelining is a straightforward approach to exploiting parallelism that is based on concurrently performing different phases of processing an instruction. These phases often include fetching an **instruction** from memory, decoding an instruction to determine its **operation** and **operands**, accessing its operands, performing the computation specified by the operation, and storing the computed value (IF, DE, RF, EX, and WB, respectively, in Figure 23.2). Pipelining assumes that these phases are independent between different operations and can be overlapped — when this condition does not hold, the processor stalls the downstream phases to enforce the dependency. Thus multiple operations can be processed simultaneously with each operation at a different phase of its processing. For a simple pipelined machine, only one operation is in each phase at any given time — thus one operation is being fetched, one operation is being decoded, one operation is accessing operands, one operation is executing, and one operation is storing results. With this scheme, assuming that each phase takes a single cycle to complete (which is not always the case), a pipelined processor could achieve a speedup of five over a traditional nonpipelined processor design.

Unfortunately, it can be argued that pipelining does not achieve true concurrency but that it only eliminates (in the limit) the overhead associated with instruction processing. With this viewpoint, the only phase of instruction processing that is not overhead is the evaluation of the result — everything else just supports this evaluation. Even so, there is still a speedup of five over a nonpipelined processor, but this speedup serves only to bring the maximum execution rate up from one operation every five cycles to one operation every cycle (for the pipeline just described). This is the best performance that can be achieved by a processor without true parallel execution.

While pipelining does not necessarily lead to achieving true concurrency, there are other techniques that do. These techniques use some combination of static scheduling and dynamic analysis to perform concurrently the actual evaluation phase of several different operations — potentially yielding an execution rate of greater than one operation every cycle. This kind of parallelism exploits concurrency at the computation level (in contrast to pipelining, which exploits concurrency at the overhead level). Since historically most instructions consist of only a single operation (and thus a single computation), this kind of parallelism has been named *instruction-level parallelism* (ILP).

Two architectures that exploit ILP — **superscalar** and **VLIW** (very long instruction word) — use radically different techniques to achieve more than one operation per cycle. A superscalar processor dynamically examines the instruction stream to determine which operations are independent and can be executed concurrently. Figure 23.3a shows the issue of ready instructions from a window of available instructions and the routing of these instructions to the appropriate function units (FU). A VLIW processor depends on the compiler to analyze the available operations (OP) and to schedule independent operations into wide instruction words; it then executes these operations in parallel with no further analysis. Figure 23.3b shows the issue of

TABLE 23.1 Typical Scalar Processors

Processor	Year of Introduction	Number of Function Units	Issue Width	Scheduling	Issue/Complete Order
Intel x86	1978	2	1	Dynamic	In-order/In-order
Stanford MIPS-X	1981	1	1	Dynamic	In-order/In-order
Berkeley RISC-I	1981	1	1	Dynamic	In-order/In-order
Sun SPARC	1987	2	1	Dynamic	In-order/In-order
MIPS R3000	1988	2	1	Dynamic	In-order/In-order
MIPS R4000	1992	2	1	Dynamic	In-order/In-order
HP PA-RISC 1.1	1992	1	1	Dynamic	In-order/In-order

operations from a wide instruction word to all function units in parallel. In the superscalar processor, even if two operations have been determined to be independent by the compiler and are scheduled properly for the current processor, at execution time the dependency analysis must still be performed to ensure that the proper ordering is maintained. In the VLIW processor, since the compiler is depended upon to ensure the proper scheduling of operations, any operations that are improperly scheduled result in indeterminate (and probably bad!) results. Considering the programmer's reference point, the kind of parallelism that superscalar processors exploit is invisible, while the kind of parallelism that VLIW processors exploit is visible only at the assembly level where the explicit packing of multiple operations into instructions is visible — the high-level language programmer in both cases is isolated from the machine-exploited ILP. Actually, these statements are true only in the general sense; for the superscalar processor, the assembly language programmer may be aware of the organization and characteristics of the machine and be able to schedule instructions so that they can be executed in parallel by the processor (this scheduling is usually performed by the compiler, although there are assemblers which perform minor scheduling transformations). For the VLIW processor, the assembly language programmer must be aware of the specific characteristics of the machine to ensure the proper scheduling of operations (the assembler could perform the analysis and scheduling although this is typically not desired by the programmer). For both processors, even at the high-level language level there are ways of writing programs that make it easy for the compiler to find the latent parallelism.

Both superscalar and VLIW use the same compiler techniques to achieve their superscalar performance. However, a superscalar processor is able to execute code scheduled for any instruction-compatible processor while a VLIW processor can only execute code that was specifically scheduled for execution on that particular processor.* This flexibility is not for free. While a superscalar processor does execute inappropriately scheduled code, the achieved performance can be significantly worse than if it were appropriately scheduled. Nevertheless, the flexibility is an important feature in a marketplace with a significant investment in software where binary compatibility is more important than raw performance.

A SISD processor has four defining characteristics. The first characteristic is whether or not the processor is capable of executing multiple operations concurrently. The second characteristic is the mechanism by which operations are scheduled for execution — statically at compile time, dynamically at execution, or possibly both. The third characteristic is the order in which operations are issued and retired relative to the original program order — these can be in order or out of order. The fourth characteristic is the manner in which exceptions are handled by the processor — precise, imprecise, or a combination. This last characteristic is not of immediate concern to the applications programmer, although it is certainly important to the compiler writer or operating system programmer, who must be able to properly handle exceptional conditions. Most processors implement precise exceptions, although a few high-performance architectures allow imprecise floating-point exceptions (with the ability to select between precise and imprecise exceptions).

Tables 23.1, 23.2, and 23.3 present some representative (pipelined) scalar and super-scalar (both superscalar and VLIW) processor families. As Table 23.1 and Table 23.2 show, the trend has been from a scalar

---

*This does not have to be true — although past VLIW processors have had this restriction, this has been due to engineering decisions in the implementation and not to anything inherent in the specification. For two variant approaches to providing support for dynamic scheduling in VLIW processors see Rau [1993] and Rudd [1994].

TABLE 23.2   Typical Superscalar Processors

Processor	Year of Introduction	Number of Function Units	Issue Width	Scheduling	Issue/Complete Order
DEC 21064	1992	4	2	Dynamic	In-order/In-order
Sun UltraSPARC	1992	9	4	Dynamic	In-order/Out-of-order
MIPS R8000	1994	6	4	Dynamic	In-order/In-order
DEC 21164	1994	4	2	Dynamic	In-order/In-order
Motorola PowerPC 620	1995	6	4	Dynamic	In-order/Out-of-order
HP PA-RISC 8000	1995	10	4	Dynamic	Out-of-order/Out-of-order
MIPS R10000	1995	5	5	Dynamic	Out-of-order/Out-of-order
Intel Pentium Pro	1996	5	3	Dynamic	In-order x86 Out-of-order uops/Out-of-order
AMD K5	1996	6	4	Dynamic	In-order x86 Out-of-order ROPs/Out-of-order

TABLE 23.3   Typical VLIW Processors

Processor	Year of Introduction	Number of Function Units	Issue Width	Scheduling	Issue/Complete Order
Multiflow Trace 7/200	1987	7	7	Static	In-order/In-order
Multiflow Trace 28/200	1987	28	28	Static	In-order/In-order
Cydrome Cydra 5	1987	7	7	Static	In-order/In-order
Philips TM-1	1996	27	5	Static	In-order/In-order

to a compatible superscalar processor (except for the DEC Alpha and the IBM/Motorola/Apple PowerPC processors, which were designed from the ground up to be capable of superscalar performance). There have been very few VLIW processors to date, although advances in compiler technology may cause this to change. Philips has explored VLIW processors internally for years, and the TM-1 is the first of a planned series of processors. After the demise of both Multiflow and Cydrome, HP acquired both the technology and some of the staff of these companies and has continued research in VLIW processors.

23.4   SIMD

The SIMD class of processor architectures includes both array and vector processors. The SIMD processor is a natural response to the use of certain regular data structures, such as vectors and matrices. From the reference point of an assembly language programmer, programming an SIMD architecture appears to be very similar to programming a simple SISD processor except that some operations perform computations on aggregate data. Since these regular structures are widely used in scientific programming, the SIMD processor has been very successful in these environments.

Two types of SIMD processor will be considered: the **array processor** and the **vector processor**. They differ both in their implementations and in their data organizations. An array processor consists of many interconnected processor elements that each have their own local memory space. A vector processor consists of a single processor that references a single global memory space and has special function units that operate specifically on vectors.

23.4.1   Array Processors

The array processor is a set of parallel processor elements (typically hundreds to tens of thousands) connected via one or more networks (possibly including local and global interelement data and control communications). Processor elements operate in lockstep in response to a single broadcast instruction

TABLE 23.4   Typical Array Processors

Processor	Year of Introduction	Memory Model	Processor Element	Number of Processors
Burroughs BSP	1979	Shared	General purpose	16
Thinking Machines CM-1	1985	Distributed	Bit-serial	Up to 65,536
Thinking Machines CM-2	1987	Distributed	Bit-serial	4,096–65,536
MasPar MP-1	1990	Distributed	Bit-serial	1,024–16,384

from a control processor. Each processor element has its own private memory, and data are distributed across the elements in a regular fashion that is dependent on both the actual structure of the data and also the computations to be performed on the data. Direct access to global memory or another processor element's local memory is expensive (although scalar values can be broadcast along with the instruction), so intermediate values are propagated through the array through local interelement connections. This requires that the data are distributed carefully so that the routing required to propagate these values is simple and regular. It is sometimes easier to duplicate data values and computations than it is to effect a complex or irregular routing of data between processor elements.

Since instructions are broadcast, there is no means local to a processor element of altering the flow of the instruction stream; however, individual processor elements can conditionally disable instructions based on local status information — these processor elements are idle when this condition occurs. The actual instruction stream consists of more than a fixed stream of operations — an array processor is typically coupled to a general-purpose control processor that provides both scalar operations (that operate locally within the control processor) as well as array operations (that are broadcast to all processor elements in the array). The control processor performs the scalar sections of the application, interfaces with the outside world, and controls the flow of execution; the array processor performs the array sections of the application as directed by the control processor.

A suitable application for use on an array processor has several key characteristics: a significant amount of data which has a regular structure; computations on the data which are uniformly applied to many or all elements of the data set; simple and regular patterns relating the computations and the data. An example of an application that has these characteristics is the solution of the Navier–Stokes equations, although any application that has significant matrix computations is likely to benefit from the concurrent capabilities of an array processor.

The programmer's reference point for an array processor is typically the high-level language level — the programmer is concerned with describing the relationships between the data and the computations, but is not directly concerned with the details of scalar and array instruction scheduling or the details of the interprocessor distribution of data within the processor. In fact, in many cases the programmer is not even concerned with size of the array processor. In general, the programmer specifies the size and any specific distribution information for the data, and the compiler maps the implied virtual processor array onto the physical processor elements that are available and generates code to perform the required computations. Thus, while the size of the processor is an important factor in determining the performance that the array processor can achieve, it is not a defining characteristic of an array processor.

The primary defining characteristic of a SIMD processor is whether the memory model is shared or distributed. In this chapter, only processors using a distributed memory model are described, since this is the configuration used by SIMD processors today and the cost of scaling a shared-memory SIMD processor to a large number of processor elements would be prohibitive. Processor element and network characteristics are also important in characterizing a SIMD processor, and these are described in Section 23.2 and Section 23.6.

There have not been a significant number of SIMD architectures developed, due to a limited application base and market requirement. Table 23.4 shows several representative architectures.

23.4.2   Vector Processors

A vector processor is a single processor that resembles a traditional SISD processor except that some of the function units (and registers) operate on vectors — sequences of data values that are seemingly operated on as a single entity. These function units are deeply pipelined and have a high clock rate; while the vector pipelines have as long or longer latency than a normal scalar function unit, their high clock rate and the rapid delivery of the input vector data elements results in a significant throughput that cannot be matched by scalar function units.

Early vector processors processed vectors directly from memory. The primary advantage of this approach was that the vectors could be of arbitrary lengths and were not limited by processor resources; however, the high startup cost, limited memory system bandwidth, and memory system contention proved to be significant limitations. Modern vector processors require that vectors be explicitly loaded into special vector registers and stored back into memory — the same course that modern scalar processors have taken for similar reasons. However, since vector registers can rapidly produce values for or collect results from the vector function units and have low startup costs, modern register-based vector processors achieve significantly higher performance than the earlier memory-based vector processors for the same implementation technology.

Modern processors have several features that enable them to achieve high performance. One feature is the ability to concurrently load and store values between the vector register file and main memory while performing computations on values in the vector register file. This is an important feature, since the limited length of vector registers requires that vectors that are longer than the registers be processed in segments — a technique called strip mining. Not being able to overlap memory accesses and computations would pose a significant performance bottleneck.

Just like their SISD cousins, vector processors support a form of result bypassing — in this case called chaining — which allows a follow-on computation to commence as soon as the first value is available from the preceding computation. Thus, instead of waiting for the entire vector to be processed, the follow-on computation can be significantly overlapped with the preceding computation that it is dependent on. Sequential computations can be efficiently compounded and behave as if they were a single operation with a total latency equal to the latency of the first operation with the pipeline and chaining latencies of the remaining operations but none of the startup overhead that would be incurred without chaining. For example, division could be synthesized by chaining a reciprocal with a multiply operation. Chaining typically works for the results of load operations as well as normal computations. Most vector processors implement some form of chaining.

A typical vector processor configuration might have a vector register file, one vector addition unit, one vector multiplication unit, and one vector reciprocal unit (used in conjunction with the vector multiplication unit to perform division); the vector register file contains multiple vector registers (eight registers with 64 double-precision floating-point values is typical). In addition to the vector registers there are also a number of auxiliary and control registers, the most important of which is the vector length register. The vector length register contains the length of the vector (or the loaded subvector if the full vector length is longer than the vector register itself) and is used to control the number of elements processed by vector operations — there is no reason to perform computations on nondata that are useless or could cause an exception.

As with the array processor, the programmer's reference point for a vector machine is the high-level language. In most cases, the programmer sees a traditional SISD machine; however, since vector machines excel on vectorizable loops, the programmer can often improve the performance of the application by carefully coding the application — in some cases explicitly writing the code to perform strip mining — and by providing hints to the compiler that help to locate the vectorizable sections of the code. This situation is purely an artifact of the fact that the programming languages are scalar-oriented and do not support the treatment of vectors as an aggregate data type but only as a collection of individual values. As languages are defined (such as Fortran 90 or High Performance Fortran) that make vectors a fundamental data type, then the programmer is exposed less to the details of the machine and to its SIMD nature.

The vector processor has one primary characteristic. This characteristic is the location of the vectors — vectors can be memory- or register-based. There are many features that vector processors have that are not included here due to their number and many variations. These include variations on chaining, masked vector operations based on a Boolean mask vector, indirectly addressed vector operations (scatter/gather), compressed/expanded vector operations, reconfigurable register files, multiprocessor support, etc.

TABLE 23.5 Typical Vector Processors

Processor	Year of Introduction	Memory-or Register-Based	Number of Processor Units	Maximum Vector Length	Number of Vector Units
Cray 1	1976	Register	1	64	7
CDC Cyber 205	1981	Memory	1	65,535	3–5
Cray X-MP	1982	Register	1–4	64	8
Hitachi HITAC S-810	1984	Register	1		4–8
Fujitsu FACOM VP-100/200[a]	1985	Register	3	32–1024	4
Cray 2	1985	Register	5	64	4
ETA ETA	1987	Memory	2–8	65,535	3–5
Cray C90		Register		64	
NEC SX-3	1990	Register	1–4		4–16
NEC SX-4	1994	Register	1–512		4–8
Cray T90	1995		1–32		2

[a]Sold as the Amdahl VP1100/VP1200 in the United States.

Vector processors have developed dramatically from simple memory-based vector processors to modern multiple-processor vector processors that exploit both SIMD vector and MIMD-style processing. Table 23.5 shows some representative vector processors.

23.5   MISD

While it is easy to both envision and design MISD processors, there has been little interest in this type of parallel architecture. The reason, so far anyway, is that there are no ready programming constructs that easily map programs into the MISD organization.

Abstractly, the MISD can be represented as multiple independently executing function units operating on a single stream of data, forwarding results from one function unit to the next. On the microarchitecture level, this is exactly what the vector processor does. However, in the vector pipeline the operations are simply fragments of an assembly-level operation, as distinct from being a complete operation in themselves. Surprisingly, some of the earliest attempts at computers in the 1940s could be seen as the MISD concept. They used plugboards for programs, where data in the form of a punched card were introduced into the first stage of a multistage processor. A sequential series of actions were taken where the intermediate results were forwarded from stage to stage until at the final stage a result was punched into a new card.

There are, however, more interesting uses of the MISD organization. Nakamura [1995] has pointed out the value of an MISD machine called the SHIFT machine. In the SHIFT machine, all data memory is decomposed into shift registers. Various function units are associated with each shift column. Data are initially introduced into the first column and are shifted across the shift-register memory. In the SHIFT-machine concept, data are regularly shifted from memory region to memory region (column to column) for processing by various function units. The purpose behind the SHIFT machine is to reduce memory latency. In a traditional organization, any function unit can access any region of memory and the worst-case delay path for accessing memory must be taken into account. In the SHIFT machine, we must allow for access time only to the worst element in a data column. The memory latency in modern machines is becoming a major problem — the SHIFT machine has a natural appeal for its ability to tolerate this latency.

23.6   MIMD

The MIMD class of parallel architecture brings together multiple processors with some form of interconnection. In this configuration, each processor executes completely independently, although most applications require some form of synchronization during execution to pass information and data between processors.

While there is no requirement that all processor elements be identical, most MIMD configurations are homogeneous with all processor elements being identical. There have been heterogeneous MIMD configurations that use different kinds of processor elements to perform different kinds of tasks, but (with the possible exception of recent work aimed at using networked workstations as a loosely coupled MIMD configuration) these configurations have not lent themselves to general-purpose applications. We limit ourselves to homogeneous MIMD organizations in the remainder of this section.

Up to this point, the MIMD processor with its multiple processor elements interconnected by a network appears to be very similar to a SIMD array processor. This similarity is deceptive, since there is a significant difference between these two configurations of processor elements — in the array processor the instruction stream delivered to each processor element is the same, while in the MIMD processor the instruction stream delivered to each processor element is independent and specific to each processor element. Recall that in the array processor, the instruction stream for each processor element is generated by the control processor and that the processor elements operate in lockstep. In the MIMD processor, the instruction stream for each processor element is generated independently by that processor element as it executes its program. While it is often the case that each processor element is running pieces of the same program, there is no reason that different processor elements could not run different programs.

The interconnection network in both the array processor and the MIMD processor passes data between processor elements; however, in the MIMD processor it is also used to synchronize the independent execution streams between processor elements. When the memory of the processor is distributed across all processors and only the local processor element has access to it, all data sharing is performed explicitly using messages and all synchronization is handled within the message system. When the memory of the processor is shared across all processor elements, synchronization is more of a problem — certainly messages can be used through the memory system to pass data and information between processor elements, but this is not necessarily the most effective use of the system.

When communications between processor elements is performed through a shared memory address space — either global or distributed between processor elements (called distributed shared memory to distinguish it from distributed memory) — there are two significant problems that arise. The first is maintaining **memory consistency** — the programmer-visible ordering effects of memory references both within a processor element and between different processor elements. The second is **cache coherency** — the programmer-invisible mechanism to ensure that all processor elements see the same value for a given memory location. Neither of these problems is significant in SISD or SIMD array processors. In a SISD processor, there is only one instruction stream and the amount of reordering is limited, so the hardware can easily guarantee the effects of perfect memory reference ordering and thus there is no consistency problem; since a SISD processor has only one processor element, cache coherency is not applicable. In a SIMD array processor (assuming distributed memory), there is still only one instruction stream and typically no instruction reordering; since all interprocessor element communication is via messages, there is neither a consistency problem nor a coherency problem.

The memory consistency problem is usually solved through a combination of hardware and software techniques. At the processor element level, the appearance of perfect memory consistency is usually guaranteed for local memory references only — this is usually a feature of the processor element itself. At the MIMD processor level, memory consistency is often guaranteed only through explicit synchronization between processors. In this case, all nonlocal references are only ordered relative to these synchronization points (such as fences or acquire/release points). While the programmer must be aware of the limitations imposed by the ordering scheme, the added performance achieved using nonsequential ordering can be significant. Table 23.6 shows the common memory consistency schemes and a brief description of their basic characteristics ($\sqrt{}$ indicates that the given feature exists, and $\sim \sqrt{}$ indicates that a restricted form of the feature exists).

The cache coherency problem is usually solved exclusively through hardware techniques. This problem is significant because of the possibility that multiple processor elements will have copies of data in their local caches and these copies could have differing values. There are two primary techniques to maintain cache coherency. The first technique is to ensure that all processor elements are informed of any change

TABLE 23.6   Simple Categorization of Consistency Models

Model[a]	Relaxation:[b] W → R Order	W → W Order	W → RW Order	Read Other's Write Early	Read Own Write Early	Explicit Synchronization[c]
SC					✓	
TSO, PC	✓			~✓	✓	RMW
PSO	✓	✓			✓	RMW, fence
WO	✓	✓	✓		✓	Fence
RC	✓	✓	✓	~✓	✓	Release, acquire, nsync, RMW

[a]SC = sequential consistency, TSO = total store order, PC = processor consistency, PSO = partial store order, WO = weak order, RC = release consistency.

[b]$x \to y$ represents the relaxation of the logical ordering between the reference x and a following reference y; R = read reference, W = write reference, RW = read or write reference.

[c]RMW is an atomic read–modify–write operation. Fetch-and-add is a common example of the general Fetch-and-Φ operation. (*Source:* Based on information in Adve and Gharachorloo [1995] and used with their permission.)

TABLE 23.7   Cache Coherency Summary

Coherency Model	Protocol Type	Modification Policy	Exclusive Use State	Exclusive Write State
Write once	Invalidate	Copyback on first write		
Synapse $N + 1$	Invalidate	Copyback	✓	
Berkeley	Invalidate	Copyback		✓
Illinois	Invalidate	Copyback	✓	
Firefly	Broadcast	Copyback private, writethrough shared	✓	

to the shared memory state — these changes are broadcast throughout the MIMD processor, and each processor element monitors these changes (commonly referred to as "snooping"). The second technique is to keep track of all users of a memory address or block in a directory structure and to specifically inform each user when there is a change made to the shared state. In either case the result of a change can be one of two things — either the new value is provided and the local value is updated, or all other copies of the value are invalidated.

As the number of processor elements in a system increases, a directory-based system becomes significantly better, since the amount of communications required to maintain coherency is limited to only those processors holding copies of the data. Snooping is frequently used within a small cluster of processor elements to track local changes — here the local interconnection can support the extra traffic used to maintain coherency since each cluster has only a few (typically two to eight) processor elements in it. Table 23.7 shows the common cache coherency schemes and a brief description of their basic characteristics (✓ indicates that the given state exists).

The primary characteristic of a MIMD processor is the nature of the memory address space — it is either separate or shared for all processor elements. The interconnection network is also important in characterizing a MIMD processor and is described in the next section. With a separate address space (distributed memory), the only means of communications between processor elements is through messages, and thus these processors force the programmer to use a message-passing paradigm. With a shared address space (shared memory), communication between processor elements is through the memory system — depending on the application needs or programmer preference, either a shared-memory or a message-passing paradigm can be used.

The implementation of a distributed-memory machine is far easier than the implementation of a shared-memory machine when memory consistency and cache coherency are taken into account. However, programming a distributed-memory processor can be much more difficult, since the applications must be written to exploit and not be limited by the use of message passing as the only form of communications between processor elements. On the other hand, despite the problems associated with maintaining

TABLE 23.8 Typical MIMD Systems

System	Year of Introduction	Processor Element	Number of Processors	Memory Distribution	Programming Paradigm	Interconnection Type
Alliant FX/2800	1990	Intel i860	4–28	Central	Shared memory	Bus + crossbar
Stanford DASH	1992	MIPS R3000	4–64	Distributed	Shared memory	Bus + mesh
CRAY T3D	1993	DEC Alpha 61064	128–2048	Distributed	Shared memory	3-D torus
MIT Alewife	1994	Sparcle	1–512	Distributed	Message passing	Mesh
Convex C4/XA	1994	Custom	1–4	Global	Shared memory	Crossbar
Thinking Machines CM-500	1995	SuperSPARC	16–2048	Distributed	Message passing	Fat tree
Tera Computers MTA	1995	Custom	16–256	Distributed	Shared memory	3-D torus
SGI Power Challenge XL	1995	MIPS R8000	2–18	Global	Shared memory	Bus
Convex SPP1200/XA	1995	HP PA-RISC 7200	8–128	Global	Shared memory	Crossbar + ring
CRAY T3E	1996	DEC Alpha 61164	16–2048	Distributed	Shared memory	3-D torus
Network of Workstations	1990s	Various	Any	Distributed	Message passing	Ethernet

consistency and coherency, programming a shared-memory processor can take advantage of whatever communications paradigm is appropriate for a given communications requirement and can be much easier to program. Both distributed and shared-memory processors can be extremely scalable and neither approach is significantly more difficult to scale than the other. Some typical MIMD systems are described in Table 23.8.

23.7   Network Interconnections

Both SIMD array processors and MIMD processors rely on networks for the transfer of data between processor elements or processors. A bus is a simple kind of network — it serves to interconnect all devices that are plugged into it — but is not commonly referred to as a network. We discuss here only the aspects of networks that are of interest in characterizing a processor — particularly the SIMD array processors and MIMD processors — and present some network characteristics that provide a qualitative sense that is useful for understanding the basic nature of a multiprocessor interconnect.

There are three primary characteristics of networks. The first is the method used to transfer the information through the network — either using packet routing or circuit switching. The second characteristic is the mechanism that connects source and destination nodes — either the connections are static and fixed or they are dynamic and reconfigurable. The third characteristic is whether the network is a single-level or a multiple-level network. While these characteristics leave out a significant amount of detail about the actual network, they qualitatively describe the network connections and how information is routed between processor elements.

Packet routing is efficient for small random packets, but it has the drawback that neither the latency nor the bandwidth is necessarily deterministic and thus packets may not be delivered in the same order that they were sent; circuit switching achieves high bandwidth for a given connection between processor nodes and guarantees uniform latency and proper receipt ordering, but it has the drawback that the latency for small packets becomes the latency for setting up and breaking down the connection.

Dynamic networks allow network reconfiguration so that there are essentially direct connections between nodes across the network, producing high bandwidth and low latency but limiting the scalability of the system; static networks improve the scalability, since connections are node to node and any two nodes can be connected either directly or through intermediate nodes, resulting in longer latency and lower-bandwidth connections. Use of multilevel networks, which use clusters of processor elements at each network node, increases the complexity of the system but reduces congestion on the global interconnect and leads to a more scalable system — intracluster communications are performed on a local interconnect that is much faster and does not leave the cluster. Single-level networks are more general but less scalable, since all communications must use the global interconnect, and traffic can be much higher for the same number of processor elements.

23.8 Afterword

In this chapter we have reviewed a number of different parallel architectures organized by the stream model. We have described some general characteristics that offer some insight into the qualitative differences between different parallel architectures but, in the general case, provide little quantitative information about the architectures themselves — this would be a much more significant task, although there is no reason why a significantly increased level of detail could not be described. Just as the stream model is incomplete and overlapping (consider that a vector processor can be considered to be a SIMD, MISD, or SISD processor depending on the particular categorization desired), so the characteristics for each class of architecture are also incomplete and overlapping. However, the general insight gained from considering these general characteristics leads to an understanding of the qualitative differences between gross classes of computer architectures, so the characteristics that we have described provide similar benefits and liabilities.

In a sense, the characterizations that we have provided for each architectural class of processor can be considered as specializations on the stream model. Thus a superscalar processor could be described as a SISD processor which supports concurrent execution of multiple operations that are scheduled dynamically, performs issue and retire out of order, and provides precise interrupts. A similar superscalar processor that does not provide precise interrupts would be described almost identically, but the description would provide an insight into one significant difference between these two processors — the first superscalar processor would be more complicated to design and might run more slowly than the second superscalar processor, but it would support more efficient exception recovery. While this comparison does not, in all likelihood, provide sufficient information to make a design choice in many cases, it does provide a basis for processor comparison.

For a MIMD or a SIMD processor, although the primary characteristic is the characterization of the memory address space, the system can be more completely described by including the description of both the processor element (most likely a SISD processor) along with a description of any networks that are included in the processor. This results in a description of the processor that provides a more complete understanding of the system as a whole but now including much more information about the remainder of the system.

This is not meant to imply that the aggregate of the stream model along with the relevant characteristics is a complete and formal extension to the original taxonomy — far from it. There are still a wide range of processors that are problematic to describe well in this (and likely in any) framework. The example was given earlier concerning the appropriate placement of a vector processor. Another example is the proper placement of an architecture that is designed to support multiple threads on a single processor element. These processor elements could be considered to be just SISD processors which have a specialized operating system that provides this support (albeit requiring hardware support as well), but there is some reason to believe that multiple threads are a significant feature, especially in the case of the Tera MTA architecture [Alverson et al. 1990], where the threads are interleaved through the execution units on a cycle-by-cycle basis — clearly a distinct difference beyond simply performing efficient task switches.

Whatever the problems with classifying and characterizing a given architecture, processor architectures, particularly multiprocessor architectures, are developing rapidly. Much of this growth is the result of significant improvements in compiler technology that allow the unique capabilities of an architecture to be efficiently exploited. In many cases, the design of a system is based on the ability of a compiler to produce code for it. It may be that a feature is unable to be utilized if a compiler cannot exploit it and thus the feature is wasted (although perhaps the inclusion of such a feature would spur compiler development). It may also be that an architectural feature is added specifically to support a capability that a compiler readily supports and thus performance is improved. Compiler development is clearly an integral part of system design and architectural effectiveness is no longer limited only to concerns for the processor itself.

Defining Terms

Array processor: An array of processor elements operating in lockstep in response to a single instruction and performing computations on data that are distributed accross the processor elements.

Cache coherency: The programmer-invisible mechanism that ensures that all caches within a computer system have the same value for the same shared-memory address.

Instruction: Specification of a collection of operations that may be treated as an atomic entity with a guarantee of no dependencies between these operations. A typical processor uses an instruction containing one operation.

Memory consistency: The programmer-visible mechanism that guarantees that multiple processor elements in a computer system receive the same value on a request to the same shared-memory address.

Operand: Specification of a storage location — typically either a register of a memory location — that provides data to or receives data from the results of an operation.

Operation: Specification of one or a set of computations on the specified source operands placing the results in the specified destination operands.

Pipelining: The technique used to overlap stages of instruction execution in a processor so that processor resources are more efficiently used.

Processor element: The element of a computer system that is able to process a data stream (sequence) based on the content of an instruction stream (sequence). A processor element may or may not be capable of operating as a stand-alone processor.

Superscalar processor: A popular term to describe a processor that dynamically analyzes the instruction stream and attempts to execute multiple ready operations independently of their ordering within the instruction stream.

Vector processor: A computer architecture with specialized function units designed to operate very efficiently on vectors represented as streams of data.

VLIW processor: A popular term to describe a processor that performs no dynamic analysis on the instruction stream and executes operations precisely as ordered in the instruction stream.

References

Adve, V. and Gharachorloo, K. 1995. Shared memory consistency models: a tutorial. Research Report 95/7. Digital Equipment Corp. Western Research Lab.

Alverson, R., Callahan, D., Cummings, D., Koblenz, B., Porterfield, A., and Smith, B. 1990. The Tera computer system. In *Int. Conf. on Supercomputing*, ACM, New York, June.

Flynn, M. J. 1966. Very high-speed computing systems. *Proc. IEEE* 54(12):1901–1909.

Flynn, M. J. 1995. *Computer Architecture: Pipelined and Parallel Processor Design.* Jones and Bartlett, Boston.

Hennessy, J. L. and Patterson, D. A. 1996. *Computer Architecture A Quantitative Approach*, 2nd ed. Morgan Kaufmann, San Francisco.

Hockney, R. W. and Jesshope, C. R. 1988. *Parallel Computers: Architecture, Programming and Algorithms.* Adam Hilger, Bristol.

Hwang, K. 1993. *Advanced Computer Architecture: Parallelism, Scalability, Programmability.* McGraw–Hill, New York.

Ibbett, R. N. and Topham, N. P. 1989a. *Architecture of High Performance Computers. Vol. I. Uniprocessors and Vector Processors.* Springer–Verlag, New York.

Ibbett, R. N. and Topham, N. P. 1989b. *Architecture of High Performance Computers, Vol. II. Array Processors and Multiprocessor Systems.* Springer–Verlag, New York.

Lilja, D. J. 1991. *Architectural Alternatives for Exploiting Parallelism.* IEEE Computer Society Press, Los Alamitos, CA.

Nakamura, T. 1995. The SHIFT machine. Personal correspondence.

Rau, B. R. 1993. Dynamically scheduled VLIW processors, pp. 80–92. In *26th Annual International Symp. on Microarchitecture.*

Rudd, K. W. 1994. Instruction level parallel processors — a new architectural model for simulation and
analysis. Technical Report CSL-TR-94-657. Stanford Univ.
Trew, A. and Wilson, G., eds. 1991. *Past, Present, Parallel: A Survey of Available Parallel Computing Systems*.
Springer–Verlag, London.

Further Information

There are many good sources of information on different aspects of parallel architectures. The references
for this chapter provide a selection of texts that cover a wide range of issues in this field. There are many
professional journals that cover different aspects of this area either specifically or as part of a wider coverage
of related areas. Some of these are:

IEEE Transactions on Computers, Transactions on Parallel and Distributed Systems, Computer, Micro.
ACM Transactions on Computer Systems, Computer Surveys.
Journal of Supercomputing.
Journal of Parallel and Distributed Computing.

There are also a number of conferences that deal with various aspects of parallel processing. The
proceedings from these conferences provide a current view of research on the topic. Some of these are:

International Symposium on Computer Architecture (ISCA).
Supercomputing.
International Symposium on Microarchitecture (MICRO).
International Conference on Parallel Processing (ICPP).
International Symposium on High Performance Computer Architecture (HPCA).
Symposium on the Frontiers of Massively Parallel Computation.

24

Architecture and Networks

24.1 Introduction 24-1
24.2 Underlying Principles 24-2
  LANs, WANs, MANs, and Topologies • Circuit Switching
  and Packet Switching • Protocol Stacks
  • Data Representation: Baseband Methods
  • Data Representation: Modulation
24.3 Best Practices: Physical Layer Examples 24-8
  Media • Repeaters and Regenerators • Voice Modems
  • Cable Modems • DSL • Hubs and Other LAN
  Switching Devices
24.4 Best Practices: Data-Link Layer Examples 24-13
  Network Interface Cards • An Example: Ethernet NIC
  • Bridges and Other Data-Link Layer Switches
24.5 Best Practices: Network Layer Examples 24-14
24.6 Research Issues and Summary 24-15

Robert S. Roos
Allegheny College

24.1 Introduction

The word "architecture," as it applies to computer networks, can be interpreted in at least two different ways. The overall design of a network (analogous to an architect's blueprint of a building) — including the interconnection pattern, communication media, choice of protocols, placement of network management resources, and other design decisions — is generally referred to as the network's architecture. However, we can also speak of the hardware and the hardware-specific protocols used to implement the various pieces of the network (much as we would speak of the beams, pipes, bricks, and mortar that comprise a building).

The goal of this overview is to concentrate on architectures of networks in both senses of the word, but with emphasis on the second, showing how established and relatively stable design technologies, such as multiplexing and signal modulation, are integrated with an array of hardware devices, such as routers and modems, into a physically realized network. More specifically, this chapter presents representative technologies whose general characteristics will remain applicable even as the hardware state of the art advances.

The organizing principle used in this study is the notion of a network protocol stack, defined in the next section, which provides a hierarchy of communication models that abstract away from the physical details of the network. Each of the network technologies considered in this survey is best explained in relation to a particular abstraction level in this hierarchy. At times, this classification scheme may be relaxed a bit for terminological reasons; for example, voice modems are most relevant to layer 1 of the protocol stack,

1-58488-360-X/$0.00+$1.50
© 2004 by CRC Press, LLC

whereas cable modems are really layer 2 devices, but these modems are all discussed in the same section of the chapter. On the other hand, in a few cases, the classification is difficult to apply. The word "switch" can be applied to network hubs and telephone switches (normally associated with layer 1) or to LAN (local area network) switches (normally thought of as layer 2 devices), or to routers (implementing layer 3 functionality). Unavoidably, the discussion of switching will overlap different layers.

Other chapters of this *Handbook* deal in greater depth with architecture in its first sense. Chapter 72 deals with topics such as network topologies and differences between local area and wide area networks; Chapter 73 discusses the important subject of routing in networks.

For the most part, the technologies presented in this chapter represent standards that are recognized worldwide. Several organizations provide standards for networking hardware and software. The International Telecommunication Union (which replaced the International Telegraph and Telephone Consultative Committee, or CCITT) has a sector (ITU-T) devoted to telecommunication standards. The ITU-T has recommended standards for modems (the "V" series of recommendations) and data networks and open system communication (the "X" series, including the well-known X.25 recommendation for virtual circuit connections in packet-switching networks), multiplexing standards, and many others, some of which are described in greater detail below. The International Organization for Standardization (ISO), formerly the International Standards Organization, has produced a huge number of standards documents in many areas, including information technology. (Their recommendation for the Open Systems Interconnection (OSI) reference model is described briefly in the next section.) The Institute for Electrical and Electronics Engineers (IEEE) is responsible for the "802" series of LAN standards, most of which have been adopted by the ISO. The American National Standards Institute (ANSI) is one of the founding members of the ISO and the originator of many networking standards. Gallo and Hancock [2002] provide a concise summary of most of the important regional, national, international, and industry, trade, and professional standards organizations.

24.2 Underlying Principles

We adopt Tanenbaum's definition of a computer network as an interconnected collection of autonomous computers [Tanenbaum, 1996]. We will refer to these as hosts or stations in the following discussion. Although this definition fails to take into account many network-like structures, from "networks-on-a-chip" [Benini and De Micheli, 2002] to interconnected "information appliances" such as Web-enabled cellular telephones and portable digital assistants, it is sufficient for an introductory look at the architectures of networks.

24.2.1 LANs, WANs, MANs, and Topologies

We will often make references to local area networks (LANs) and wide area networks (WANs). The phrases are fairly self-explanatory. LANs normally connect machines in a geographical area not much larger than a building or a small cluster of buildings, while WANs could connect machines separated by tens of thousands of kilometers. The higher-level architecture of LANs and WANs is described in more detail in Chapter 72. Another category, the metropolitan area network (MAN), is often mentioned in the literature. MANs lie in between LANs and WANs, perhaps joining machines distributed over the area of a large city. Classifications such as "personal area network" (PAN), "system area network" (SAN), "desk area network" (DAN), and others have been proposed, but their meanings are not yet well-established.

Many of the most popular LAN architectures represent broadcast networks: the host machines comprising the network share a common transmission medium (such as a wire in an Ethernet LAN or a block of the electromagnetic spectrum in a wireless LAN). Each host on the network sees every transmission (whether or not it is the intended recipient), and each must contend with the others for access to this medium when it wishes to transmit. The details of such multiple access methods can influence, for example, the manner in which the physical signal is encoded (for instance, code-division multiple access methods require each host to have a unique code word that is used to represent the bits transmitted by that host).

In some instances, particularly in the area of LAN networks, our discussion of networking devices will depend upon the network topology (the interconnection pattern of the hosts comprising the network). For example, a LAN may be organized using a bus topology (such as a single Ethernet cable), a star topology (several Ethernet lines joined at a central point), or a ring topology (such as an FDDI network). Repeaters may be needed to extend a bus further than the limits imposed by the physical properties of the medium. A hub might be used to join machines into a star configuration. Rings require some sort of switching mechanism to restore the ring topology when part of the network is damaged.

Topology is less of a concern with WANs, which present an entirely different set of equipment issues. Most wide area networks are comprised of a combination of media, ranging from copper twisted-pair wires to high-bandwidth fiber optic cables to terrestrial microwave or satellite transmission. Devices are needed to deal with the interfaces between these media and to perform routing and switching.

24.2.2 Circuit Switching and Packet Switching

Data transmission in networks can generally be classified into one of two categories. If data is conveyed as a bit stream by means of a dedicated "line" (in the form of optical fiber, copper wire, or reserved bandwidth along a transmission path), it is said to be **circuit-switched**. However, if it is subdivided into smaller units, each of which finds its way more or less independently from source to destination, it is said to be **packet-switched**. At the lowest layer of the protocol stack, where we deal only with streams of bits, we are usually not concerned with this distinction; however, at higher levels, devices such as switches must be designed according to which of the two methods is used. A technique known as **virtual circuits** combines features of both of these switching categories. In a virtual circuit, a path is set up in advance, after which data is relayed along the path using packet-switching methods. Virtual circuits can be either permanent or switched; in the latter case, each virtual circuit is preceded by a set-up phase that builds the virtual circuit and is followed by a tear-down phase.

In general, packets can vary in size (for example, IP packets can be anywhere from 20 to 65,536 bytes in length). A technology known as Asynchronous Transfer Mode (ATM) uses fixed-size packets of 53 bytes, which are usually referred to as cells. The term "packet-switching" is customarily used to describe packet-switching methods other than ATM. ATM is described in greater detail in Chapter 72; however, we will discuss some of the details of ATM switching in later sections.

24.2.3 Protocol Stacks

A **communication protocol** is a set of conventions and procedures agreed upon by two communicating parties. Protocols are usually grouped together into layers (a protocol stack); the actions required to implement a protocol at the $(i + 1)st$ layer of the stack require the services of the protocols at the ith layer.

This layered approach to network design has been used in many network implementations, dating back at least to the initial proposed three-layer protocol stack for the ARPA network [Meyer 1970]. In 1983, the International Standards Organization (ISO) proposed the seven-layer Open Systems Interconnection (OSI) Reference Model for network protocol stacks. Although there have been criticisms of this model, some of which are summarized by Tanenbaum [1996], it has served as a convenient framework for discussion and comparison of protocol architectures. A summary of the ISO/OSI Reference Model sufficient for the purposes of this discussion is given in Table 24.1.

Protocols in the upper layers are classified as "end-to-end" because they provide abstractions that deal only with the source and destination nodes of a network communication; they are usually more concerned with details that are global in nature, such as establishing a connection to a remote host or determining how to proceed when data is damaged or lost in transit. Details about routing, encoding, contention for network resources, etc., are reserved for layers 3 and lower, which are termed "point-to-point" because they mediate communication between adjacent nodes in the network.

TABLE 24.1 The ISO-OSI Reference Model

Layer	Name	Remarks
7	Application	
6	Presentation	
5	Session	End-to-end
4	Transport	
3	Network	
2	Data link	Point-to-point
1	Physical	

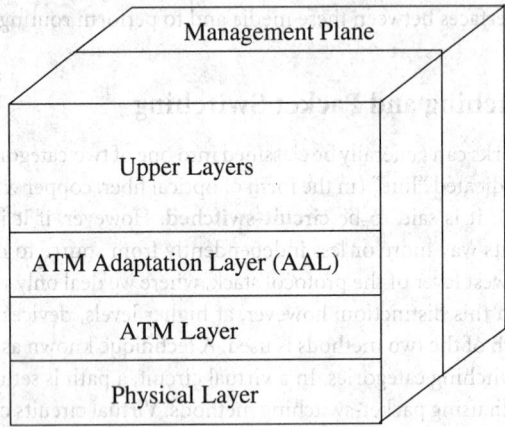

FIGURE 24.1 ATM reference model (simplified).

The protocol stack used in the Internet is often called the TCP/IP reference model, although technically it is not a model but an actual architecture. Our interest in this chapter is primarily with the lower levels of the protocol stack, and at these levels the distinctions between these two models are not critical to the discussion.

The ATM reference model is organized in a slightly different fashion, not completely analogous to the OSI stack. Figure 24.1 gives a simplified view. In addition to a vertical organization of protocols into layers, ATM specifies different *planes* that cut across the layers. The full specification includes a user plane and a control plane (not shown in the figure). The physical layer roughly corresponds to layer 1 in the OSI model, but the relationship between the ATM and AAL layers and the layers in OSI is not clear-cut. Certainly, layer 2 functionality is included in these two layers, and for the purposes of this chapter it is not necessary to make any finer distinctions.

The OSI seven-layer model does not represent the final level of granularity; most technology standards divide the OSI levels into two or more sublayers. To give just one example, the IEEE 802.3 standard for CSMA/CD networks [IEEE, 2002] subdivides the physical layer into four sublayers: PLS (physical layer signaling), AUI (attachment unit interface), MAU (medium attachment unit), and MDI (medium-dependent interface). Forouzan gives a very readable description of these sublayers [Forouzan, 2003]. In what follows, we limit ourselves to distinguishing between the logical link control and medium access control sublayers of the data-link layer and leave the remaining fine structure of this and other layers for further reading.

It is worth noting that alternatives to the layered network architecture approach have been proposed (see, for example, Braden et al. [2002]); however, this approach appears to be firmly established and unlikely to change.

24.2.4 Data Representation: Baseband Methods

There are many ways of physically representing, or keying, binary digits for transmission along some medium. To take the simplest example, consider a copper wire. We can use, for example, a voltage v_0 to represent a binary 0 and a different voltage v_1 to represent a binary 1. Although this is the most straightforward and obvious method, it suffers from several problems, most notably issues of synchronization and signal detection (a very long string of 0s or 1s, for example, would be represented as a steady voltage, making it difficult to count the bits and, if some drift in the signal value is possible, perhaps leading to the mistaken assumption that no signal is present at all). To correct this, we could use, not the voltage values themselves, but transitions between voltages (a transition from v_0 to v_1 or v_1 to v_0) to indicate bit values. This is the method used in Manchester encoding, which is the signaling method for Ethernet local area networks. Manchester encoding can be thought of as a phase modulation of a square-wave signal; phase modulation is discussed below. This encoding suffers from a different problem, namely that the transmitted signal must change values twice as fast as the bit rate, because two different voltages (low/high or high/low) are needed for each bit.

We need not restrict ourselves to just two values. If the sender and receiver are capable of distinguishing among a larger set of voltages, we could use, say, eight different voltages $v_0, \ldots, v_7$ to encode any of the eight 3-bit strings $000, 001, \ldots, 111$. The limits of this approach are dictated by the amount of noise in the transmission medium and the sensitivity of the receiver. If just three voltages are available, say $v_0 < 0, 0,$ and $v_1 > 0$, we can require the signal to return to the zero value after each bit (RZ schemes), or we can use a "non-return-to-zero" scheme (NRZ) that uses only the positive and negative values.

Some encoding schemes assume the existence of a clock that divides the signal into discrete time intervals; others, such as the Manchester scheme, are self-clocking because every time interval contains a transition. A compromise between these two is a block encoding scheme in which extra bits are inserted at regular intervals to provide a synchronization mechanism. One such method, called 4B5B, transmits 5 bits for every 4 bits of data; each combination of 4 bits has a corresponding 5-bit code. Fiber Distributed Data Interface (FDDI) uses a combination of 4B5B and NRZ.

Six examples of encoding schemes are shown in Figure 24.2. The straightforward binary encoding (a) would require a clocking mechanism for synchronization to deal with long constant bit strings. This can be avoided using three signal values and a return-to-zero encoding (b). The Manchester encoding (c) is an NRZ scheme because the zero is not used. A low-to-high transition signifies a binary zero; a high-to-low transition signifies a binary 1. The differential Manchester encoding (d) uses, not transitions of a particular type, but *changes* in the transition type to indicate bits — if a low-to-high transition is followed by a high-to-low transition (or vice versa), a 1 bit is indicated; but if two successive transitions are of the same type, a 0 bit is indicated. Bipolar alternate mark inversion, or AMI (e), uses alternating positive and negative values to represent the "1" bits in the signal. It, too, is subject to synchronization problems (long strings of zeros hold the signal constant, although long strings of ones are no longer a problem). Several variations of bipolar AMI counteract this effect. For example, in B8ZS, strings of eight consecutive zeros are replaced by the string "00011011," but in such a way that the encoding of the spurious "1"s violates the rule of sign alternation. NRZ-I (f) uses the occurrence of a signal change to indicate a 1 bit; it requires a clock or must be used in conjunction with a coding method such as 4B5B that guarantees regular changes in the signal value.

These are examples of baseband signaling — the digital signal (or an analog signal that closely approximates it) is the only thing sent. An alternative to baseband signaling is to use a carrier wave modulated by the digital signal. This process is explained in more detail below.

24.2.5 Data Representation: Modulation

In modulation, an underlying signal, for example, a sine wave, is used as a carrier and modifications to its amplitude, frequency, and phase are used to encode the digital information. Figure 24.3 gives a simplified view of this process, showing the same digital signal encoded three different ways: by amplitude modulation, by frequency modulation, and by phase modulation.

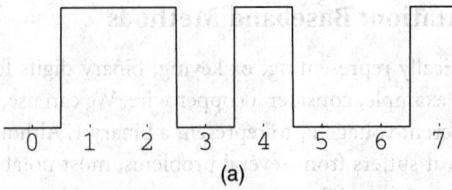

(a)

The 8-bit binary signal 01101001

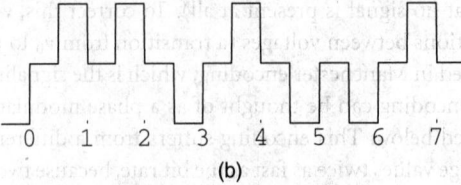

(b)

Return-to-zero (RZ) encoding

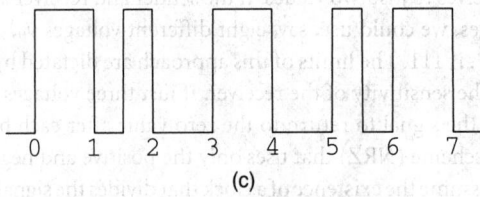

(c)

Manchester encoding — 0 is a
"low-to-high" transition, 1 is a "high-to-low"

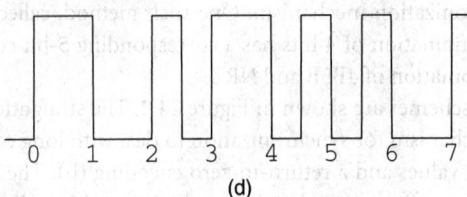

(d)

Differential Manchester encoding — 1 is represented
by a change in the transition direction

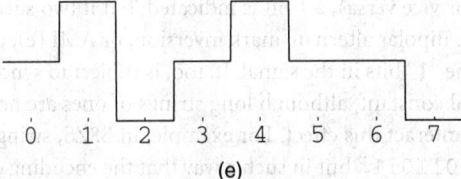

(e)

Bipolar AMI encoding — successive 1s
alternate between positive and negative

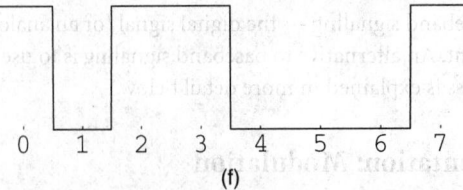

(f)

NRZ-I encoding — a 1
toggles the signal value

FIGURE 24.2   Six baseband signaling methods.

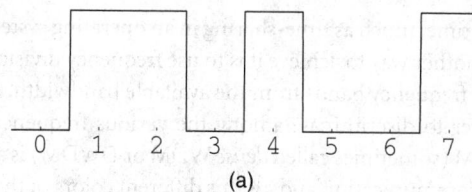

(a)

The 8-bit binary signal 01101001

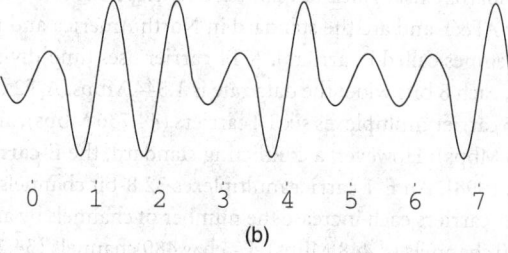

(b)

Amplitude modulation — 1 corresponds
to higher amplitude than 0

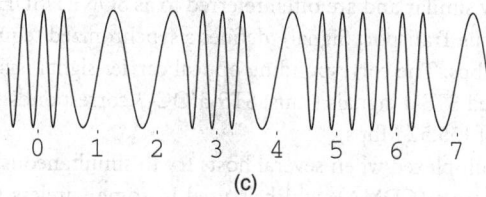

(c)

Frequency modulation — 1 corresponds
to lower frequency than 0

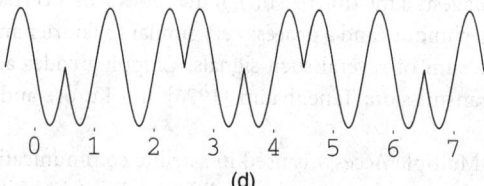

(d)

Phase modulation — 180° phase
difference between 0 and 1

FIGURE 24.3   Modulation of a signal by amplitude, frequency, and phase.

It is possible to combine several types of modulation in the same signal. Furthermore, there is no reason to limit the set of possible values to simple binary choices. Using a variety of amplitudes and phases, it is possible to encode several bits in one signaling time unit (e.g., with four phases and two amplitudes, we can encode eight distinct values, or 3 bits, in one signal). This technique is called quadrature amplitude modulation, or QAM.

24.2.5.1   Multiplexing

It is often the case that many signals must be sent over a single transmission path. For a variety of reasons it makes sense to **multiplex** these signals — to merge them together before placing them on the transmission medium and separate them at the other end (the separation process is called demultiplexing). One way to do this is called time division multiplexing. Each signal is divided into segments and each segment is allocated a certain amount of time, much as time-sharing in an operating system permits several processes to share a single processor. Another way to achieve it is to use frequency division multiplexing — different signals are allocated different frequency bands from the available bandwidth. (Standard signal processing techniques permit the receiver to discriminate among the various frequency bands used.) Wavelength division multiplexing or WDM (sometimes called *dense* WDM or DWDM) is a form of frequency division multiplexing used in optical communication and assigns different colors of the light spectrum to different signals.

Several multiplexing standards exist to facilitate connecting telephone and data networks to one another. **T-carriers** were defined by AT&T and are the standard in North America and (with slight discrepancies) Japan (where they are sometimes called J-carriers). A T1 carrier uses time-division methods to multiplex 24 voice channels together, each 8 bits wide; the data rate is 1.544 Mbps. A T2 carrier multiplexes four T1 carriers (6.312 Mbps), a T3 carrier multiplexes six T2 carriers (44.736 Mbps), and a T4 carrier multiplexes seven T3 carriers (274.176 Mbps). However, a conflicting standard, the **E-carrier**, is used in Europe and the rest of the world [ITU, 1998]. An E-1 carrier multiplexes 32 8-bit channels together and has a rate of 2.048 Mbps. E-2 and higher carriers each increase the number of channels by an additional factor of four; thus, an E-2 carrier has 120 channels (8.448 Mbps), E-3 has 480 channels (34.368 Mbps), etc.

With the increasing use of optical fiber, another set of multiplexing standards has been adopted. SONET (Synchronous Optical Network), ANSI standard T.105, and SDH (Synchronous Digital Hierarchy), developed by the ITU, are very similar and are often referred to as SONET/SDH. The basic building block, called STS-1 (for Synchronous Transport Signal), defines a synchronized sequence of 810-bit frames and a basic data rate of 51.84 Mbps. The corresponding optical carrier signal is called OC-1. STS-n (OC-n) corresponds to n multiplexed STS-1 signals (thus, STS-3/OC-3 corresponds to three multiplexed STS-1 signals and has a data rate of 155.52 Mbps).

Signals are sometimes multiplexed when several hosts try to simultaneously access a shared medium. In code division multiple access (CDMA), which is used in some wireless networks such as the IEEE 802.11b standard [IEEE, 1999b], each host is assigned a unique binary code word, also called a chip code or chipping code. Each 1 bit is transmitted using this code word, while a 0 bit uses the complementary bit sequence (the word "chip" suggests a fraction of a bit). If the codes satisfy certain orthogonality properties, a combination of statistical techniques and a process very similar to Fourier analysis can be used to extract a particular signal from the sum of several such signals. Chipping codes are used in Direct Sequence Spread Spectrum (DSSS) transmission; Tanenbaum [1996] and Kurose and Ross [2003] provide good introductions to CDMA.

SDMA (Spatial Division Multiple Access) is used in satellite communications. It allows two different signals to be broadcast along the same frequency but in different directions. If the recipients are sufficiently separated in space, each will receive only the intended signal.

Some wireless technologies, such as IEEE 802.11a [IEEE, 1999a] and the newly approved 54-Mbps 802.11g standard [Lawson, 2003], use a technique called ODFM (Orthogonal Frequency Division Multiplexing). Rather than combining signals from several transmitters, this variation of FDM is used for sending a single message by means of modulating multiple carriers to overcome effects such as multipath fading (in which a signal is canceled out by interference with delayed copies of itself).

24.3 Best Practices: Physical Layer Examples

24.3.1 Media

The lowest layer of the protocol stack concerns itself with simply delivering bits from one machine to another, regardless of any meaning that may be attached to the bits. Communicating a stream of bits between two computers requires a transmission medium and one or more protocols specifying the way in which the bits are represented and their rate of transmission. Typical media in modern networks include ordinary copper, usually in the form of "twisted pairs" of wires, fiber-optic cable, and wireless transmission via radio, microwave, or infrared waves. A detailed description of the physical properties of each of these would require a separate chapter; therefore, only brief descriptions of some of the more common media are given.

Ordinary copper wire remains one of the most common and most versatile media. Occurring in the form of twisted pairs of wires or coaxial cable, copper wiring occurs in telephone wiring (particular the "last mile" connection between homes and local telephone switching offices), many local area networks such as Ethernet and ATM LANs, and other places where an inexpensive and versatile short-haul transmission medium is needed.

Optical fiber is much more expensive but has the benefits of large bandwidth and low noise, permitting larger data transfer rates over much longer distances than copper. Fiber can be either glass or plastic and transmits data in the form of light beams, generally in the low infrared region. Multi-mode fiber permits simultaneous transmission of light beams of many different wavelengths; single-mode fiber allows transmission of just one wavelength.

Wireless transmission can be at any of a number of electromagnetic frequencies, such as radio, microwave, or infrared. This overview deals primarily with guided media.

At the physical layer, among the many challenges facing network designers are signal detection, attenuation, corruption due to noise, signal drift, synchronization, and multipath fading. Standard signal processing techniques handle many of these problems; Stein [2000] gives an overview of these techniques from a computer science point of view. Some of the more interesting problems deal with interfacing transmission media (particularly issues dealing with digital vs. analog representation) and maximizing the bandwidth usage of various media. Several representative technologies for dealing with these concerns are presented in this section.

24.3.2   Repeaters and Regenerators

A repeater is essentially just a signal amplifier; the term predates computer networks, originally referring to radio signal repeaters. The term "regenerator" is sometimes used to describe a device that amplifies a binary (as opposed to an analog) signal. Repeaters are used where the transmission medium (e.g., twisted pair copper wire or microwaves) has an inherent physical range limit beyond which the signal becomes too attenuated to use. Microwave antennas serve the function of repeaters in terrestrial microwave networks; communication satellites can be thought of as repeaters that amplify electromagnetic signals and retransmit them back to the ground. Optical repeaters, used with fiber-optic cable, sometimes have additional functionality, such as converting between single-mode and multimode fiber transmission. Because repeaters are relatively simple, they introduce very little delay in the signal. Repeaters are sometimes incorporated into more sophisticated devices, such as hubs connecting several local area networks; hubs are discussed in more detail in a later section.

An amplifier will amplify noise just as well as data. More sophisticated technologies are required to ensure the integrity of the bit stream. In the particular case of a digital signal regenerator, much of the noise can be eliminated using fairly standard techniques that involve synchronizing the data signal with a clock signal and using threshold methods; see, for example, Anttalainen's [1999] discussion of regenerative repeaters.

With the growth in popularity of wireless networks, new kinds of repeater mechanisms are coming into play; for example, "smart antennas" can selectively amplify some signals and inhibit others [Liberti and Rappaport, 1999].

24.3.3   Voice Modems

Modems are used when a digital signal must be encoded in analog fashion using modulation. This is often the case when home subscribers to an Internet service provider must use the analog lines of the public switched telephone network (PSTN) to go online. The word "modem" is a shortened form of modulator/demodulator.

Some manufacturers make a distinction between **hard modems**, which contain all of the necessary digital signal processing functions on the modem board, and the less-expensive **soft modems**, which off-load some of the signal processing responsibility to software on the host machine.

Modems designed for data transmission over telephone lines, also called voice-grade modems or simply telephone modems, use a combination of modulation techniques. QAM is used in all modems that adhere to ITU-T modem standards V.32 and higher.

The most recent ITU voice-grade modem recommendations as of this writing are V.90 and V.92, which specify an asymmetric data connection in which one direction (called "upstream") is analog and the other ("downstream") is digital. The maximum data rates for these are 33,000 to 48,000 bits per second for upstream and about 56,000 bits per second for downstream, respectively.

The difference in the upstream and downstream rates for V.90/V.92 modems is due to the fact that the digital-to-analog conversion used in the upstream connection incurs a penalty (the so-called "quantization noise"). The upstream data rate is nearly at the "Shannon limit" of roughly 38,000 bits per second, a theoretical upper bound on the achievable bit rate based on estimates of the best signal-to-noise ratio for analog telephone lines [Tanenbaum, 1996]. (Compression techniques used in V.92 permit some improvement here.) Lower signal-to-noise ratios result in slower data transfer rates. However, disregarding noise, both rates are limited to about 56,000 bits per second (56 Kbps) by the inherent properties of telephone voice transmission, which uses a very narrow bandwidth (about 3000 Hertz [3 kHz]), and a technique called pulse code modulation (PCM), which involves sampling the analog signal 8000 times per second. Each sample yields 8 bits, 7 of which contain data and 1 of which is used for control.

In the United States, the 56-Kbps downstream rate has, until recently, been unachievable owing to a Federal Communications Commission (FCC) restriction on power usage that effectively reduced the limit to 53K. In 2002, the FCC granted power to set such standards to a private industry committee, the Administrative Council for Terminal Attachments, which adopted new power limits permitting the 56-Kbps data rate for V.90 and V.92 modems [FCC, 2002; ACTA, 2002].

24.3.4 Cable Modems

The word "modem," once used almost exclusively to refer to voice-grade communication over telephone lines, has more recently been applied to other interfacing devices. A cable modem interfaces a computer to the cable television infrastructure, allowing digital signals to share bandwidth with the audio and visual broadcast content. Cable modem systems generally use coaxial cable or a combination of coaxial cable and fiber-optic cable (the latter are called hybrid fiber-coaxial, or HFC, networks). The much larger bandwidth of the cable TV connection permits faster data rates than telephone modems. As was the case with V.90 modems, the upstream and downstream data rates differ. With a cable modem, upstream data transfer rates can be millions of bits per second (Mbps), and downstream data rates can be in the tens of millions. The discrepancy between upstream and downstream rates is due, in part, to the fact that cable television systems were originally designed only for downstream data transfer from the cable provider to the television owner. A small portion of the bandwidth must be allocated for the upstream signal. (Cable companies offering cable modem service must upgrade their equipment by adding amplifiers for the upstream signal.)

Unlike telephone modem connections, which are dedicated point-to-point connections, cable modem connections are part of a broadcast network consisting of all the cable modem users in a particular neighborhood. When demand is high, the portion of bandwidth available to each user diminishes and data transfer rates decrease. In fact, a cable modem should be considered more as a layer 2, or data-link layer, device because it actually links a client computer to a local area network of cable modem users.

24.3.5 DSL

Sometimes, the word "modem" is used in a rather loose sense. For example, digital subscriber lines (DSL) use existing copper telephone wires and also require so-called digital modems. Technically, a digital modem is not a "modulator/demodulator," but the term has become standard. DSL exploits the fact that much more bandwidth is available than the narrow range (between about 300 and 3300 Hz) used for voice transmission. A portion of this bandwidth is reserved for ordinary voice transmission; the rest is divided into upstream and downstream bandwidth and uses digital rather than analog techniques. DSL comes in several varieties, the most common of which is asymmetric DSL, or ADSL. In ADSL, the upstream data rate is generally no more than about 600 Kbps, and the downstream rate is generally no more than about 8 Mbps.

There are several other varieties of DSL — rate-adaptive DSL (RADSL), high bit-rate DSL (HDSL), very high bit-rate DSL (VDSL), etc. Peden and Young [2001] describe the various flavors of DSL and provide an overview of how DSL modems have evolved from voice-grade modems.

24.3.6 Hubs and Other LAN Switching Devices

The commonplace meaning of switch, a device for making and breaking connections in a circuit, encompasses a great many components in networks at several levels of the protocol stack: hubs, telephone switches (part of the communication infrastructure supporting wide area networks), bridges (devices for connecting local area networks to other networks), routers (devices used to direct packets along an appropriate path), and others.

The word "switch" is sometimes specialized to a particular class of devices normally associated with layer 2 of the OSI stack, in order to distinguish them from less-intelligent devices such as hubs. Although they clearly must communicate via the physical layer, they possess additional capabilities to do data-link-layer processing such as interpreting physical device addresses. When the word "switch" is used in the sequel, it will always be more precisely qualified.

A *hub* is used to connect several data lines into a single local area network; it simply takes incoming bits and broadcasts them to all the other lines to which it is connected. Such a hub is sometimes called a repeating hub. Early hubs did not even perform the amplification needed to regenerate the signal. These "passive hubs" were equivalent to wiring concentrators that are used to gather wires together into a central location; hubs are still sometimes referred to as concentrators.

Most LANs are broadcast networks; that is, all messages sent across the LAN are seen by all hosts in the network. As the number of hosts in a network grows, it becomes inefficient to force every host machine to see every message (more formally, we say that these hosts all share the same **collision domain** because two messages are said to collide if they are broadcast at the same time). Better segmentation is achieved using the protocols of the data-link layer and more sophisticated devices called **bridges**, which are discussed in the next section. Commercial vendors are increasingly using phrases such as "intelligent hub" to describe more bridge-like, even router-like, devices.

Some physical layer switching devices are specialized to particular situations. FDDI (Fiber Distributed Data Interface) networks consist of a number of hosts connected to two counter-rotating rings of fiber-optic cable (i.e., one ring transmits data clockwise, the other transmits it counterclockwise). One of these rings is designated the primary ring and is normally used for all data transfer; the secondary ring is idle. However, from time to time, one of the network hosts will fail, or one of the fibers may become damaged. In this case, a switch called a dual attachment concentrator "wraps" the primary and secondary rings together to form a single ring. A second type of switch, called an optical bypass switch, can also be used to handle host outages. Unlike the concentrator, the bypass switch optically "splices" each of the rings across the broken section, retaining the dual ring structure. (It is worth mentioning here that physically splicing fiber is an extremely precise operation; the optical switch uses mirrors to achieve the effect of an actual splice operation.) Downes et al. [1998] provide a good introduction to FDDI and its associated protocols.

Telephone switching has, of course, resulted in the design of many kinds of switches. One of the earliest types of switch, the **crossbar switch** (so named because the original electromechanical implementation actually used two layers of bars with each bar in one layer crossing every bar in the other), contains connection points for every combination of communication endpoints. Thus, if n ports are involved, there are n^2 switching points, of which only n can be closed at any given time (or $n/2$ if we remove the symmetry and use only the upper triangle of the connection point grid). Despite this quadratic growth, the simplicity of the crossbar design has made it a popular choice for switching, even in high-performance networks. For example, Stanford University's Tiny Tera packet switch uses a 32-port crossbar [Gupta and

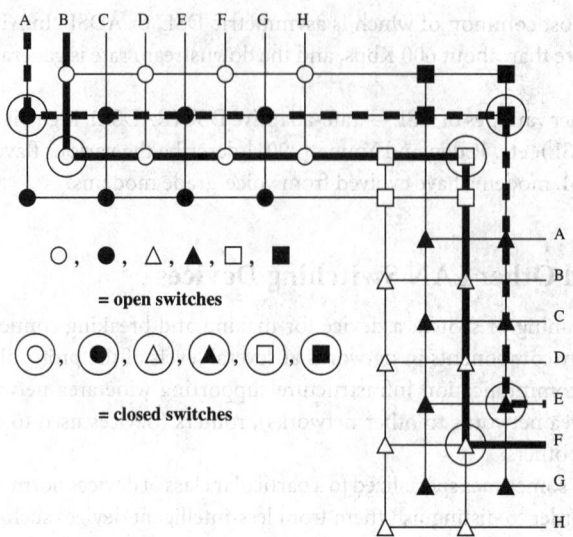

FIGURE 24.4   Space division switching.

McKeown, 1999]. Abel et al. [2003] describe a 256-port crossbar switch operating at data rates up to 4 terabits per second.

A necessary property of any switch is the ability to connect any pair of ports to one another. A desirable property is to permit a large number of simultaneous connections. A crossbar switch allows the maximum number of simultaneous connections, but in many cases it can be shown that fewer than the maximum will be required based on normal network traffic behavior. In such cases, it is possible to use small crossbar switches as modules for building larger multistage switches that allow any port pairs to be connected. This is called *space division switching*. Because there are fewer switches, some attempts to connect ports may "block" (blocking occurs when both endpoints are free but every path between them is unavailable due to conflicts with other connections). Figure 24.4 illustrates this. Any of the eight ports A,...,H can be connected to any other port. While a crossbar switch would require 64 connection points, this design uses a combination of four 4-by-2 crossbars (indicated by white and black circles and triangles) and two 2-by-2 crossbars (indicated by white and black rectangles) and has only 40 connection points. Connections are shown between A and E, and between B and F. Ports C and G are currently free, but it is not possible to connect them to each other because the wires needed to join them are already in use by other connections; an attempt to connect C to G would be blocked. In a full crossbar, this connection could be made.

One alternative to space division switching is time division switching. For this we use time division multiplexing to select one of the n input ports during each time slot, in round-robin fashion. Each input slice is placed into a buffer corresponding to its output port, so that after n time slices, all n buffers will be filled. On the output side, buffers are emptied in round-robin fashion and the contents of the ith buffer are sent to output port i. However, it requires n cycles to process one slice from each input port, so timing considerations come into play: the rate at which buffers are filled and emptied must be at least n times faster than the data rate for each port.

All the switching mechanisms described thus far have assumed a one-to-one pairing of input and output ports, an appropriate assumption when telephone conversations are being discussed. For data transfer, however, we often have multiple data streams directed to a single common destination. Switching now becomes just one component (the "switching fabric") of more complicated devices capable of dealing with such routing issues. These will be taken up once again when we describe switching functions in the higher layers of the protocol stack.

24.4   Best Practices: Data-Link Layer Examples

The data-link layer is sometimes divided into two sublayers. The higher of these is often called the logical link control layer (LLC); the lower sublayer is called the media access layer (MAC). The LLC is responsible for grouping bits together into units called frames and for handling things like flow and error control, while the MAC handles media-specific issues.

24.4.1   Network Interface Cards

Any networked device requires an interface to attach it to the network. A network interface card (NIC), also called an adapter, serves this function. It carries out the layer 2 protocols (framing, flow control, encoding/decoding, etc.) and the layer 1 protocols (transmitting a bit stream onto the network medium). There are two popular driver architectures for network adapters: 3Com/Microsoft's NDIS (Network Device Interface Specification) and Novell's ODI (Open Datalink Interface).

24.4.2   An Example: Ethernet NIC

Due to the many different data-link layer protocols in existence, it would be impractical to try to describe every variation in network adapter behavior. As an example of the type of processing required of an NIC, we describe the main features of the carrier-sense multiple-access protocol with collision detection (CSMA/CD) as described in IEEE [2002] and implemented by a traditional 10-Mbps Ethernet LAN. (Strictly speaking, Ethernet is not identical to IEEE 802.3; but we ignore the differences in what follows.)

Every Ethernet NIC has a 6-byte address, called its MAC address or Ethernet address, that uniquely distinguishes it from every other Ethernet adapter (IEEE controls the allocation of MAC addresses and enforces this uniqueness). Data packets received from the upper layers of the protocol stack are packaged by the data-link into frames of sizes ranging from 64 to 1518 bytes. These include the data (which might have been padded with extra bits in order to achieve minimum frame size), the MAC addresses of the sender and the intended recipient, and several other fields. The physical layer prepends an 8-byte preamble used for synchronization and timing purposes, and then transmits the resulting frame using the Manchester encoding described in a previous section.

When a host is ready to transmit, it sets a counter k to zero (this counter keeps track of the number of re-trys made by the sender). It then senses the wire to see if anyone else is transmitting. When the line is free, it waits an additional 96 bit-times (the time needed to transmit 1 bit) — about 9.6 microseconds for 10-Mbps Ethernet — and starts sending data. If, during the transmission, another host were also to begin transmitting, the resulting collision will be detectable by both senders. At this point, the CSMA/CD protocol requires both hosts to stop transmitting and to emit a 32-bit "jam signal." The sender then increments k and chooses a random number r in the range 0 through 2^{k-1} and waits $512 \times r$ bit-times before trying to retransmit. This is called *exponential backoff*.

The reason for requiring a minimum frame size has to do with collision detection. In the slowest Ethernet environment, 10 Mbps, a signal can travel no more than 2500 meters (the maximum allowed by the Ethernet standard). In this environment, conservatively assuming a propagation speed of one third the speed of light, or 10^8 meters per second (propagation in copper wire is usally twice this amount), a single bit could require as many as 25 microseconds (25 μs) to reach any host. If a signal were to travel this far just as a distant host began transmitting, it would take another 25 μs for news of the collision to return to the sender of the initial transmission. Therefore, a host might need to wait at least 50 μs to learn of a collision. It is important that the transmission last at least this long so that it can be terminated and restarted. About 500 bits (over 60 bytes) can be sent during this time at a speed of 10 Mbps, so 64 bytes is a conservative lower bound to enforce on the size of an Ethernet frame.

The IEEE 802 standards specify similarly precise behaviors for interfacing to Token Ring LANs, wireless LANs (in this case, the NIC is a wireless transceiver, often in the form of a PC card), and others. Most of these are described in nearly any published standard introduction to networks.

24.4.3 Bridges and Other Data-Link Layer Switches

A bridge connects several networks, forwarding data based upon the physical device addresses of the destination host as encoded in the headers added by the link layer protocol (unlike hubs, which indiscriminately forward bits to everyone, and unlike routers, which use logical addresses, such as IP addresses, rather than physical addresses). The word "switch" is sometimes used interchangeably with "bridge," but the primary difference between them is that LAN switches implement the switching function in hardware, while a bridge implements the switching function in software. For this reason, layer 2 switches can handle more traffic at higher data rates.

Some LAN switches have an additional capability; they are capable of switching a data stream immediately after determining the physical address of the destination (rather than waiting for the entire packet to arrive before forwarding it). These are called **cut-through switches**.

Bridges and switches are described in greater detail in Chapter 73; an excellent treatment of the subject can be found in Perlman's book [1999].

24.5 Best Practices: Network Layer Examples

Layer 3 of the OSI Reference Model deals mainly with routing. Protocols at this level are responsible for determining where to direct data based upon a *logical address*. In the current incarnation of the Internet protocol, IP version 4 (IPv4), the 32-bit IP address plays this role. However, the proposed IPv6 addressing scheme [Hinden and Deering, 2003] expands the size of IP addresses to 128 bits. For a description of the evolution of IP addressing and the impact of IPv6, see, for example, Kurose and Ross [2003].

The primary technology associated with layer 3 of the OSI stack is the router, which contains protocols from the lowest three layers of the OSI model. Unlike bridges, which use physical addresses to direct data, routers use logical addresses (such as IP addresses) to determine packet destinations. Routers fall into one of two categories: (1) edge routers, which are responsible for routing data between WANs and individual hosts or LANs; and (2) core routers, which perform routing between locations within the WAN.

Chapter 73 deals with routing algorithms (shortest path algorithms, multicast routing, distance vector vs. link state methods, etc.). The discussion here focuses more on the switching process used inside the router.

A router consists of a number of input ports, a number of output ports, and a switching fabric connecting them. We will assume that packets are all the same size (e.g., ATM cells). The small, fixed-size cells used in ATM make switching design much simpler, so a large amount of research has gone into designing high-speed ATM switches. In fact, this is one reason why ATM is a suitable technology for WANs — it lends itself to the design of high-performance switches. Several examples are described here.

Even assuming that all input ports and output ports have the same data rate and that the switching fabric is fast enough to handle all non-conflicting switching requests, it will eventually happen that data packets arriving at two different input ports will be destined for the same output port. The switch must place one or the other packet into a queue.

When queuing occurs at the output ports, there is always a danger of the queue becoming full, in which case routing software must make a decision about which packet to discard. A special kind of crossbar switch, called a *knockout switch* [Yeh, Hluchyj, and Acampora, 1987], uses a series of "competitions" between randomly chosen pairs of packets to eliminate some of them from entering the queue. For descriptions of more sophisticated switching fabrics, see Oie et al. [1990] and Robertazzi [1998].

An alternative to queueing the packet at the output port is to wait until the destination is free before forwarding the second data item. In this case, queueing occurs at the input ports. It is possible that a packet waiting in the input queue for its destination port, which is blocked, may prevent a packet further back in line, whose destination port is available, from being transmitted. This is called *head-of-the-line blocking*. Most switching fabric designs try to avoid input queueing.

A complete and thorough treatment of switching technologies and the associated scheduling and queueing algorithms is beyond the scope of this chapter. Although it has been in print for over a decade, the

IEEE Computer Society's collection of papers on switching is still a useful source of background information on switching technologies [Dhas, Konangi, and Sreetharan, 1991]. Turner and Yamanaka [1998] provide an overview of large-scale switching technologies (with particular reference to ATM networks). Li [2001] provides a modern mathematical treatment of switches with particular reference to networking applications.

24.6   Research Issues and Summary

The biggest challenges facing the next generation of networking devices appear to be in the areas of infrastructure, multimedia, and wireless networking. Most long-distance data transmission now takes place over fast media such as fiber-optic cable. However, replacing copper wire with fiber at the bottleneck points where most consumers access the Internet — the so-called "last mile" — is expensive and time-consuming. DSL and cable modem technologies provide some measure of broadband access, but they have reached the limits imposed by the copper wire medium. A number of alternatives to fiber have been proposed, most of them involving wireless technologies. In recent years, Multichannel Multipoint Distribution Service (MMDS, also sometimes called Multipoint Microwave Distribution System), originally a one-way microwave broadcast technology for subscription television, has been adopted for use in last-mile Internet connections [Ortiz, 2000]. Free-space optics involving infrared lasers could be used in a network of laser diodes to boost last-mile data rates into the gigabit per second range. Acampora and Krishnamurthy [1999] provide a description of a possible wireless broadband access network; and Acampora [2002] gives a nontechnical introduction to these technologies.

Infrastructure includes not only the hardware but also the protocols. TCP/IP has proven to be remarkably robust, to the embarrassment of experts who have occasionally predicted its collapse [Metcalfe, 1996], but address space considerations alone will soon force the issue of making the transition to IPv6.

Multimedia is one of the driving economic forces in the networking industry; services such as video-on-demand [Ma and Shin, 2002], Voice-over-IP [Rosenberg and Schulzrinne, 2000], and others require not only large amounts of bandwidth, but more attention to issues such as real-time networking, multicasting, and quality-of-service. These, in turn, will influence the design of not only the software protocols, but also the switches and routers that must implement them at the hardware level. *Convergence* is the word often used to describe the goal of integrating voice, data, audio, and video into a single network architecture. B-ISDN and ATM were designed with this in mind, but there are still many hurdles to overcome.

Defining Terms

Bridge: A data-link-layer device that connects LANs (usually of the same type) to one another and performs routing based upon physical addresses.

Circuit switching: A transmission method that uses a dedicated path over which the data bit stream is sent.

Cut-through switch: A switch that bypasses the store-and-forward approach of older packet switches and begins immediately to forward data as soon as the destination port has been determined.

IEEE 802: A collection of local area networking standards developed by the Institute of Electrical and Electronics Engineers, including 802.2 (Token Ring), 802.3 (CSMA/CD), and 802.11 (wireless OFDM and CSMA protocols).

Hub: A device used to join several LANs together; broadcasts all bits arriving from an attached network to every other attached network.

LAN switch: A device much like a bridge, but with switching functions implemented in hardware rather than software.

Modem (modulator/demodulator): A device for converting back and forth between digital data and an analog transmission medium.

Modulation: The use of a signal, usually digital, to modify an underlying carrier signal by varying the carrier's amplitude, frequency, phase, or some combination of these properties.

Multiplexing: The process of merging several signals into one; can be done in any of several domains: time, frequency, wavelength, etc. The inverse is called *demultiplexing*.

Packet switching: A transmission method involving the segmentation of a message into units called packets, each of which is separately and independently sent.

Protocol: A set of conventions and procedures agreed upon by two communicating parties.

Repeater (regenerator): A device used to simply retransmit the data that it receives; used to extend the range of transmission medium.

Router: A network layer device that forwards packets based upon their logical destination address.

Virtual circuit: A transmission method in which packets or cells are transmitted upon a path that has been reserved in advance; may be permanent or switched.

References

Abel, F., Minkenberg, C., Luijten, R., Gusat, P., and Iliadis, I. 2003. A four-terabit single-stage packet switch supporting long round-trip times. *IEEE Micro* 23(1): 10–24.

Acampora, A.S. 2002. Last mile by laser. *Scientific American*, 287(2):48–53.

Acampora, A.S. and Krishnamurthy, S.V. 1999. A broadband wireless access network based on mesh-connected free-space optical links. *IEEE Personal Communications*, 6(5):62–65.

ACTA (Administrative Council for Terminal Attachments), 2002. Telecommunications — telephone terminal equipment — technical requirements for connection of terminal equipment to the telephone network. Document TIA-968-A.

Anttalainen, Tarmo. 1999. *Introduction to Telecommunications Network Engineering*. Artech House, Inc., Boston, MA.

Benini, L. and De Micheli, G. 2002. Networks on chips: a new SoC paradigm. *Computer*. 35(1):70–78.

Braden, R., Faber, T., and Handley, M. 2002. From protocol stack to protocol heap — role-based architecture. *Comp. Commun. Rev.*, 33(1):17–22.

Dhas, C., Konangi, V.K., and Sreetharan, M. 1991. *Broadband Switching: Architecture, Protocols, Design, and Analysis*. IEEE Computer Society Press, Washington, D.C.

Downes, K., Ford, M., Lew, H.K., Spanier, S., and Stevenson, T. 1998. *Internetworking Technologies Handbook*, 2nd ed., Macmillan Technical Publishing, Indianapolis, IN.

FCC (Federal Communications Commission), 2002. Order on reconsideration in CC docket no. 99-216 and order terminating proceeding in CC docket no. 98-163. FCC document number 02-103, sec. III.G.1.

Forouzan, B.A. 2003. *Local Area Networks*. McGraw-Hill, Boston, MA.

Gallo, M.A. and Hancock, W.M. 2002. *Computer Communications and Networking Technologies*. Brooks/Cole, Pacific Grove, CA.

Gupta, P. and McKeown, N. 1999. Designing and implementing a fast crossbar scheduler. *IEEE Micro*, 20–28.

Hinden, R. and Deering, S. 2003. Internet Protocol Version 6 (IPv6) Addressing Architecture. Internet protocol version 6 (IPv6) addressing architecture (Network Working Group Request for Comments 3513). Internet Engineering Task Force (http://www.ietf.org/).

IEEE (Institute of Electrical and Electronics Engineers). 1999a. *IEEE Standard for Information technology — Telecommunications and information exchange between systems — Local and metropolitan area networks — Specific requirements — Part 11: Wireless LAN Medium Access Control (MAC) and Physical Layer (PHY) Specifications — Amendment 1: High-Speed Physical Layer in the 5 GHz Band.* IEEE Standard: IEEE 802.11a-1999 (8802-11:1999/Amd 1:2000(E)). (http://standards.ieee.org/getieee802/802.11.html).

IEEE. 1999b. *IEEE 802.11b-1999 Supplement to 802.11-1999, Wireless LAN MAC and PHY Specifications: Higher Speed Physical Layer (PHY) Extension in the 2.4 GHz Band.* IEEE Standard: IEEE 802.11b-1999. (http://standards.ieee.org/getieee802/802.11.html).

IEEE. 2002. *IEEE Standard for Information technology — Telecommunications and Information Exchange between Systems — Local and Metropolitan Area Networks — Specific Requirements — Part 3: Carrier Sense Multiple Access with Collision Detection (CSMA/CD) Access Method and Physical Layer Specifications.* IEEE Standard: IEEE 802.3-2002. (http://standards.ieee.org/getieee802/802.3.html).

IEEE. 2003. *IEEE Standard for IT — Telecommunications and Information Exchange between Systems LAN/MAN — Part II: Wireless LAN Medium Access Control (MAC) and Physical Layer (PHY) Specifications Amendment 4: Further Higher Data Rate Extension in the 2.4 GHz Band.* IEEE Standard: 802.11G-2003.

ITU (International Telecommunications Union). 1998. Synchronous frame structures used at 1544, 6312, 2048, 8448 and 44 736 kbit/s hierarchical levels. ITU-T Recommendation G.704 (10/98).

Klensin, J.C. 2002. A policy look at IPv6: a tutorial paper. International Telecommunications Union Tutorial Workshop on IPv6, Workshop Document WS IPv6-3. (http://www.itu.int/itudoc/itu-t/workshop/ipv6/003.html).

Krishnamurthy, S. V., Acampora, A.S., and Zorzi, M. 2001. Polling based medium access control protocols for use with smart adaptive array antennas. *IEEE/ACM Transactions on Networking*, 9(2):148–161.

Kurose, J.F. and Ross, K.W. 2003. *Computer Networking: A Top-Down Approach Featuring the Internet*, 2nd ed. Addison-Wesley, Boston, MA.

Lawson, S. 2003. IEEE approves 802.11g standard. *Computerworld*, June 12, 2003. (http://www.computerworld.com/hardwaretopics/hardware/story/0,10801,82068,00.html).

Li, S.-Y.R. 2001. *Algebraic Switching Theory and Broadband Applications.* Academic Press, San Diego, CA.

Liberti, J.C. and Rappaport, T.S. 1999. *Smart Antennas for Wireless Communications: IS-95 and Third Generation CDMA Applications.* Prentice Hall PTR, Upper Saddle River, NJ.

Ma, H. and Shin, K.G. 2002. Multicast video-on-demand services. *ACM SIGCOMM Computer Communication Review*, 32(1):31–43.

Metcalfe, B. 1996. From the Ether: The Internet is collapsing; the question is who's going to be caught in the fall. *InfoWorld*, November 18, 1996.

Meyer, E.E., Jr. 1970. ARPA network protocol notes (Network Working Group Request for Comments 46). Internet Engineering Task Force (http://www.ietf.org/).

Oie, Y., Suda, T., Murata, M., and Miyahara, H. 1990. Survey of switching techniques in high-speed networks and their performance. In *Proceedings IEEE INFOCOM '93, Volume 3.* IEEE Press, Picataway, NJ.

Ortiz, S. 2000. Broadband fixed wireless travels the last mile. *IEEE Computer*, 33(7):18–21.

Peden, M. and Young, G. 2001. From voice-band modems to DSL technologies. *International Journal of Network Management*, 11(5):265–276.

Perlman, R. 1999. *Interconnections: Bridges, Routers, Switches, and Internetworking Protocols*, 2nd ed., Addison-Wesley Longman Publishing Co., Inc., Boston, MA.

Peterson, L.L. and Davie, B.S. 2000. *Computer Networks: A Systems Approach*, 2nd ed., Morgan Kaufmann Publishers, San Francisco, CA.

Robertazzi, T.G. (Ed.). 1998. *Performance Evaluation and High Speed Switching Fabrics and Networks: ATM, Broadband ISDN, and MAN Technology.* Wiley-IEEE Press.

Rosenberg, J. and Schulzrinne, H. 2000. A framework for telephony routing over IP (Network Working Group Request for Comments 2871). Internet Engineering Task Force (http://www.ietf.org/).

Stallings, W. 2000. *Data and Computer Communications*, 6th ed., Prentice Hall, Upper Saddle River, NJ.

Stein, Jonathan (Y). 2000. *Digital Signal Processing: A Computer Science Perspective.* John Wiley & Sons, Inc., New York.

Tanenbaum, A.S. 1996. *Computer Networks*, 3rd ed., Prentice Hall, Upper Saddle River, NJ.

Turner, J. and Yamanaka, N. 1998. Architectural choices in large scale ATM switches. *IEICE Transactions on Communications*, E81-B(2):120–137.

Walrand, J. and Varaiya, P. 2000. *High-Performance Communication Networks*, 2nd ed., Morgan Kaufmann Publishers, San Francisco, CA.

Yeh, Y., Hluchyj, M.G., and Acampora, A.S. 1987. The knockout switch: a simple, modular architecture for high-performance packet switching. *IEEE J. Sel. Areas Commun.*, 5(8):1274–1283.

Further Information

A number of recent textbooks provide excellent surveys of the architecture of modern networks. In addition to those already mentioned, the books by Stallings [2000], and Peterson and Davie [2000] are worth examining.

It is nearly impossible to study computer networking without examining the telecommunications industry, because so much of networking, particular wide area networking, depends on the existing communication infrastructure. The book by Anttalainen [1999] gives a good overview of communications technologies, and concludes with chapters on data communication networks. Walrand and Varaiya [2000] also achieve a very smooth integration of networking and communication concepts.

A large number of professional organizations sponsor sites on the World Wide Web devoted to networking technologies. The IEEE's pilot "Get IEEE 802" program makes the 802 LAN standards documents freely available electronically (http://standards.ieee.org/getieee802/). In addition, the IEEE's Communications Society publishes a number of useful magazines and journals, including *IEEE Transactions on Communications, IEEE/ACM Transactions on Networking,* and *IEEE Transaction on Wireless Communications.*

The ATM Forum (http://www.atmforum.org) is a particularly well-organized and useful site, offering free standards documents, tutorials, and White Papers in all areas of Asynchronous Transfer Mode and broadband technology.

25

Fault Tolerance

25.1   Introduction 25-1
25.2   Failures, Errors, and Faults 25-2
    Failures • Errors • Faults
25.3   Metrics and Evaluation 25-4
    Metrics • Evaluation
25.4   System Failure Response 25-6
    Error on Output • Error Masking • Fault Secure
    Techniques • Fail-Safe Techniques
25.5   System Recovery 25-11
25.6   Repair Techniques 25-12
    Built-in Self-Test and Diagnosis • Fail-Soft Techniques
    • Self-Repair Techniques
25.7   Common-Mode Failures and Design Diversity 25-14
25.8   Fault Injection 25-14
25.9   Conclusion 25-15
25.10 Further Reading 25-16

Edward J. McCluskey
Stanford University

Subhasish Mitra
Stanford University

25.1   Introduction

Fault tolerance is the ability of a system to continue correct operation after the occurrence of hardware or software failures or operator errors. The intended system application is what determines the system reliability requirement. Since computers are used in a vast variety of applications, reliability requirements differ tremendously. For very low-cost systems, such as digital watches, calculators, games, or cell phones, there are minimal requirements: the products must work initially and should continue to operate for a reasonable time after purchase. Failures of these systems are easily discovered by the user. Any repair may be uneconomical. At the opposite extreme are systems in which errors can cause loss of human life. Examples are nuclear power plants and active control systems for civilian aircraft. The reliability requirement for the computer system on an aircraft is specified to be a probability of error less than 10^{-9} per hour [hissa.nist.gov/chissa/SEI_Framework/framework_7.html].

More typical reliability requirements are those associated with commercial computer installations. For such systems, the emphasis is on designing system features to permit rapid and easy recovery from failures. Major factors influencing this design philosophy are the reduced cost of commercial off-the-shelf (COTS) hardware and software components, the increasing cost and difficulty of obtaining skilled maintenance personnel, and applications of computers in banking, on-line reservations, networking, and also in harsh environments such as automobiles, industrial environments with noise sources, nuclear power plants, medical facilities, and space applications. For applications such as banking, on-line reservations, or e-commerce, the economic impact of computer system outages is significant. For applications such as space missions or satellites, computer failures can have a huge economic impact and cause missed opportunities to record valuable data that may be available for only a short period of time. Computer failures in industrial environments, automobiles, and nuclear power plants can cause serious health hazards or loss of human life.

Fault tolerance generally includes detection of a system malfunction followed by identification of the faulty unit or units and recovery of the system from the failure. In some cases, especially applications with very short mission times or real-time applications (e.g., control systems used during aircraft landing), a fault-tolerant system requires correct outputs during the short period of mission time. After the mission is completed, the failed part of the system is identified and the system is repaired. Failures that cause a system to stop or crash are much easier to detect than failures that corrupt the data being processed by the system without any apparent irregularities in the system behavior. Techniques to recover a system from failure include system reboot, reloading the correct system state, and repairing or replacing a faulty module in the system.

This discussion is restricted mainly to techniques to tolerate hardware failures. The applicability of the described techniques to software failures will also be discussed.

25.2 Failures, Errors, and Faults

25.2.1 Failures

Any deviation from the expected behavior is a *failure*. Failures can happen due to incorrect functions of one or more physical components, incorrect hardware or software design, or incorrect human interaction.

Physical failures can be either permanent or temporary. A *permanent failure* is a failure that is always present. Permanent failures are caused by a component that breaks due to a mechanical rupture or some wearout mechanism, such as metal electromigration, oxide defects, corrosion, time-dependent dielectric breakdown, or hot carriers [Ohring 98, Blish 00]. Usually, permanent failures are localized. The occurrence of permanent failures can be minimized by careful design, reliability verification, careful manufacture, and screening tests. They cannot be eliminated completely.

A *temporary failure* is a failure that is not present all the time for all operating conditions. Temporary failures can be either transient or intermittent. Examples of causes of transient failures include externally induced signal perturbation (usually due to electromagnetic interference), power-supply disturbances [Cortes 86], and radiation due to alpha particles from packaging material and neutrons from cosmic rays [Ziegler 96, Blish 00].

An *intermittent failure* causes a part to produce incorrect outputs under certain operating conditions and occurs when there is a weak component in the system. For example, some circuit parameter may degrade so that the resistance of a wire increases or drive capability decreases. Incorrect signals occur when particular patterns occur at internal leads. Intermittent failures are generally sensitive to the temperature, voltage, and frequency at which the part operates. Often, intermittent failures cause the part to produce incorrect outputs at the rated frequency for a particular operating condition but to produce correct outputs when operated at a lower frequency of operation. Not all intermittent failures may be due to inaccuracies in manufacture. Intermittent failures can be caused by design practices leading to incorrect or marginal designs. This category includes cross-talk failures caused by capacitive coupling between signal lines and failures caused by excessive power-supply voltage drop. The occurrence of intermittent failures is minimized by careful design, reliability verification, and stress testing of chips to eliminate weak parts.

Major causes of software failures include incorrect software design (referred to as *design bugs*) and incorrect resource utilization such as references to undefined memory locations in data structures, inappropriate allocation of memory resoures and incorrect management of data structures, especially those involving dynamic structures such as pointers. Bugs are also common in hardware designs, and can involve millions of dollars when big semiconductor giants are involved [PC World 99, Bentley 01].

Failures due to human error involve maintenance personnel and operators and are caused by incorrect inputs through operator–machine interfaces [Toy 86]. Incorrect documentation in specification documents and user's manuals, and complex and confusing interfaces are some potential causes of human errors.

25.2.2 Errors

The function of a computing system is to produce data or control signals. An *error* is said to have occurred when incorrect data or control signals are produced. When a failure occurs in a system, the effect may be to cause an error or to make operation of the system impossible. In some cases, the failure may be benign and have no effect on the system operation.

25.2.3 Faults

A *fault model* is the representation of the effect of a failure by means of the change produced in the system signals. The usefulness of a fault model is determined by the following factors:

Effectiveness in detecting failures
Accuracy with which the model represents the effects of failures
Tractability of design tools that use the fault model

Fault models often represent compromises between these frequently conflicting objectives of accuracy and tractability. A variety of models are used. The choice of a model or models depends on the failures expected for the particular technology and the system design; it also depends on design tools available for the various models.

The most common fault model is the *single stuck-at fault*. Exactly one of the signal lines in a circuit described as a network of elementary gates (AND, OR, NAND, NOR, and NOT gates) is assumed to have its value fixed at either a logical 1 or a logical 0, independent of the logical values on the other signal lines. The single stuck-at fault model has gained wide acceptance in connection with manufacturing test. It has been shown that although the single stuck-at fault model is not a very accurate representation of manufacturing defects, test patterns generated using this fault model are very effective in detecting defective chips [McCluskey 00]. Radiation from cosmic rays that cause transient errors on signal lines have effects similar to a transient stuck-at fault which persists for one clock cycle.

Another variation of the stuck-at fault model is the *unidirectional fault model*. In this model, it is assumed that one or more stuck-at faults may be present, but all the stuck signals have the same logical value, either all 0 or all 1. This model is used in connection with storage media whose failures are appropriately represented by such a fault. Special error-detecting codes [Rao 89] are used in such situations.

More complex fault models are the *multiple stuck-at* and *bridging fault* models. In a *bridging fault* model, two or more distinct signal lines in a logic network are unintentionally shorted together to create wired logic [Mei 74]. It is shown in McCluskey [00] that a manufacturing defect can convert a combinational circuit into a sequential circuit. This can happen due to either *stuck-open faults* created by any failure that leaves one or more of the transistors of CMOS logic gates in a nonconducting state, or *feedback bridging faults* in which two signal lines, one dependent on the other, are shorted so that wired logic occurs [Abramovici 90].

The previous fault models all involve signals having incorrect logic values. A different class of fault occurs when a signal does not assume an incorrect value, but instead fails to change value soon enough. These faults are called *delay faults*. Some of the delay fault models are the transition fault model [Eichelberger 91], gate delay fault model [Hsieh 77], and path delay fault model [Shedletsky 78a, Smith 85].

The previously discussed fault models also have the convenient property that their effects are independent of the signals present in the circuit. Not all failure modes are adequately modeled by such faults. For example, *pattern sensitive faults* are used in the context of random–access memory (RAM) testing, in which the effect of a fault depends on the contents of RAM cells in the vicinity of the RAM cell to be tested. Failures occurring due to cross-talk effects also belong to this category [Breuer 96].

The fault models described until now generally involve signal lines or gates in a logic network. Many situations occur in which a less detailed fault model at a higher level of abstraction is the most effective choice. In designing a fault-tolerant system, the most useful fault model may be one in which it is assumed that any single module can fail in an arbitrary fashion. This is called a *single module fault*. The only restriction placed on the fault is the assumption that at most one module will be bad at any given time.

There have been several attempts to develop fault models for software failures [Siewiorek 98]. However, there is no clear consensus about the effectiveness of any of these models — or even whether fault models can be developed for software failures at all.

25.3 Metrics and Evaluation

25.3.1 Metrics

Several metrics are used to quantify how reliable a system is. These are reliability, availability, safety, maintainability, performability, and testability.

The *reliability* of a system at time t is the probability that the system will produce correct outputs up to time t, provided it was producing correct outputs to start with. The concept of reliability directly applies situations in which the system must produce correct outputs for a given period of time. Examples include aircraft control, fly-by-wire systems, automobiles, nuclear reactors, and military and space applications.

The *availability* of a system at time t is defined as the probability that the system is operational at time t. Unlike a reliable system, a highly available system may not be operational for a very short while but must be quickly repaired so that it becomes operational again very quickly. Examples include servers and mainframes, network equipment, and telephone systems. For telephone networks, the availability requirement is 99.999% [http://www.iec.org/tutorials/five-nines/topic01.html], which roughly translates to 3 to 5 minutes of downtime per year.

The *safety* of a system at time t is the probability that the system either will be operating correctly or will fail in a "safe" manner. The way a system can fail in a "safe" manner depends on the actual application. For example, a safe failing state of a traffic light is when it is stays red and keeps blinking.

The *performability* of a system at time t is the probability that the system is operating correctly at full throughput or operating at a reduced throughput greater than or equal to a given value. The concept of performability, first introduced in Beudry [77] in terms of performance-related reliability measures, is useful in the context of systems with graceful degradation. As a simple example, consider a network of computers in which one or more network links or computers may not be working. In that case, some of the links and computers will have more traffic and higher workloads. As a result, the system throughput will be less than the full system throughput with all working network links and computers. Depending on the application, such a performance degradation may or may not be acceptable.

The *maintainability* of a system, denoted by $M(t)$, is the probability that it takes t units of time to restore a failed system to normal operation. Depending on the situation, a simple reload of the system state may be sufficient, or the faulty unit in the system (also referred to as the *field replaceable unit* or *FRU*) must be identified to be replaced or repaired. Maintainability is important; if it takes a long time to bring a failed system back to normal state, then the downtime increases and system availability suffers.

The *testability* of a system is the ease with which the system can be tested — this includes the ease with which test patterns can be generated and applied to the system. Test pattern generation cost depends on either the computer time required to run the test pattern generation program plus the (prorated) capital cost of developing the program, or the number of man-hours required for a person to write the test patterns plus the increase in system development time caused by the time taken to develop tests. Test application cost is determined by the cost of the test equipment plus the time required applying the test (sometimes called *socket time*). Attempts to understand circuit attributes that influence testability have produced the concepts of observability (visibility) and controllability (control). *Observability* refers to the ease with which the state of internal signals can be determined at the circuit outputs. *Controllability* refers to the ease of producing a specific internal signal value by applying signals to the circuit inputs [McCluskey 86].

25.3.2 Evaluation

Suppose that we want to estimate the reliability of a component (which can be a system itself) over time. An experimental approach to measure reliability is to take a large number N of these components and test them continuously. Suppose that at any time t, $G(t)$ is the number of components that are operating correctly up to time t and $F(t)$ is the number of components that failed from the time the experiment started up to time t. Note that $G(t) + F(t) = N$. Thus, the reliability $R(t)$ at time t is estimated to be $(G(t)/N)$, assuming that all components had equal opportunity to fail. The rate at which components fail is $(dF(t)/dt)$. The *failure rate* per component at time t, $\lambda(t)$, is $(1/G(t))(dF(t)/dt)$. Substituting $F(t) = N - G(t)$ and $R(t) = (G(t)/N)$, we obtain $\lambda(t) = -(1/R(t))(dR(t)/dt)$. When the failure rate is constant over time, represented by a constant failure rate λ, reliability $R(t)$ can be derived to be equal to $e^{-\lambda t}$ by solving the previous differential equation. This model of constant failure rate and exponential relationship between reliability and failure rate is widely used. Several other distributions that are used in the context of evaluation of reliable systems are hypoexponential, hyperexponential, gamma, Weibull, and lognormal distributions [Klinger 90, Trivedi 02].

Another metric very closely related to reliability is the *mean time to failure* (MTTF), which is the expected time that a component will operate correctly before failing. The MTTF is equal to $\int_{-\infty}^{\infty} t R(t) dt$. For a system with constant failure rate λ, the MTTF is equal to $\int_{-\infty}^{\infty} t e^{-\lambda t} dt = (1/\lambda)$.

The failure rate is generally estimated from field data. For hardware systems, accelerated life testing is also useful in failure rate estimation [Klinger 90]. Figure 25.1 shows how the failure rate of a system varies with time. To start with, the failure rate is high for integrated circuits (ICs) for the first 20 weeks or so; during this time, weak parts that were not identified as defective during manufacturing testing typically fail in the system. After that, the failure rate stays more or less constant over time (in fact, it decreases a little bit), until the system lifetime is reached. After that, the failure rate increases again due to wearout. This dependence of failure rate on time is often referred to as the *bathtub curve*.

When a system fails, it must be repaired before it can be put into operation again. Depending on the extent of the failure, the time required to repair a system may vary. In general, it is assumed that the repair rate of the system is constant (generally represented by μ) and the *mean time to repair* (MTTR) is equal to $(1/\mu)$. For real situations, the assumption of constant repair rate may not be justified; however, this assumption makes the associated mathematics simple and manageable.

Once we know the system failure rate and the system repair rate, the system availability can be calculated using a simple Markov chain [Trivedi 02]. The average availability over a long period of time is given by $(MTTF/MTTF + MTTR)$. The *mean time between failures* (MTBF) is equal to MTTF + MTTR. In general, MTTR should be very small compared to MTTF so that the average availability is very close to 1.

As a hypothetical example, consider system A, with MTTF = 10 hours and MTTR = 1 hour, and system B, with MTTF = 10 days and MTTR = 1 day. The average availability over a long period of time is the same for both systems. While MTTR should be brought down as much as possible, having a system with MTTF = 10 hours may not be acceptable because of frequent disruptions and system outage. Thus, average availability does not model all aspects of system availability. Other availability metrics include instantaneous availability and interval availability [Klinger 90].

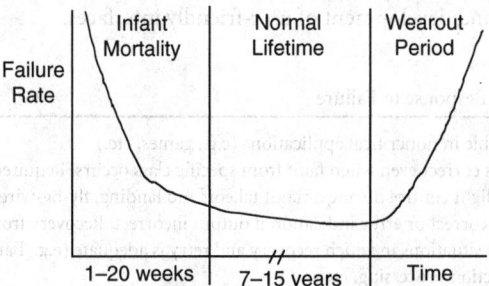

FIGURE 25.1 Failure rate vs. time: bathtub curve. The numbers on the time axis are relevant for integrated circuits.

Several techniques are used to evaluate the previous metrics for reliable system design [Trivedi 02, Iyer 96]. These include analytical techniques based on combinatorial methods and Markov modeling, simulation of faulty behaviors using fault injection, and experimental evaluation.

25.4 System Failure Response

The major reason for introducing fault tolerance into a computer system is to limit the effects of malfunctions on the system outputs. The various possibilities for system responses to failures are listed in Table 25.1.

25.4.1 Error on Output

With no fault tolerance present, failures can cause system outputs to be in error. For many systems, this is acceptable and the cost of introducing fault tolerance is not justified. Such systems are generally small and are used in noncritical applications such as computer games, word processing, etc. The important reliability attribute of such a system is the MTTF. This is an example of a product for which the best approach to reliability is to design it in such a way that it is highly unlikely that it does not fail — a technique called *fault avoidance*. For hardware systems, fault avoidance techniques include

Robust design (e.g., special circuit design techniques, radiation hardened designs, shielding, etc.) [Kang 86, Calin 96, Wang 99]

Design validation verification (e.g., simulation, emulation, fault injection, and formal verification)

Reliability verification (to evaluate the impact of cross-talk, electromigration, hot carriers, careful manufacturing)

Thorough production testing [Crouch 99, McClusley 86, Needham 91, Perry 95]

At present, it is not possible to manufacture quantities of ICs that are all free of defects. Defective chips are eliminated by testing all chips during production. Some parts may produce correct outputs only for certain operating conditions but may not be detected because production testing is not perfect. It is not economically feasible to test each part for all operating conditions. In addition, there are weak parts that may produce correct outputs for all specified operating conditions right after manufacture but will fail early in the field (within a few weeks) — much earlier than the other parts. Reliability screens are used to detect these parts. Techniques include burn-in [Jensen 82, Hnatek 87, Ohring 98], very low voltage (VLV) testing [Hao 93], and SHOrt Voltage Elevation (SHOVE) tests [Chang 97], and Iddq testing [Gulati 93]. (The applicability of Iddq testing is questionable for deep-submicron technologies.) The cost of IC testing is a significant portion of the manufacturing cost.

Depending on the application, fault avoidance techniques may be very expensive. For example, radiation hardening of hardware for space applications has a very limited market and is very expensive. Moreover, the radiation hardened designs are usually several generations behind the highest-performance designs.

Fault avoidance techniques for software systems include techniques to ensure correct specifications, validation, and testing. Fault avoidance techniques for human errors include adequate training, proper review of user documents, and development of user-friendly interfaces.

TABLE 25.1   System Output Response to Failure

Error on output	Acceptable in noncritical applications (e.g., games, etc.)
Errors masked	Outputs correct even when fault from specific class occurs. Required in critical control applications (e.g., flight control during aircraft takeoff and landing, fly-by-wire systems, etc.)
Fault secure or data integrity	Output correct or error indication if output incorrect. Recovery from failure required. Useful for critical situations in which recovery and retry is adequate (e.g., banking, telephony, networking, transaction processing, etc.)
Fail safe	Output correct or at "safe value." Useful for situations in which one class of output error is acceptable (e.g., flashing red for faulty traffic control light)

TABLE 25.2   Masking Techniques

Hardware	Voted logic — Each module is replicated. The outputs of all copies of a module are connected to a voter.
	Error correcting codes — Information has extra bits added. Some errors are corrected. Used in RAM and serial memory devices.
	Quadded (interwoven) logic — Each gate is replaced by four gates. Faults are automatically corrected by the interconnection pattern used.
	Fail safe — Output correct or at "safe value." Useful for situations in which one class of output error is acceptable (e.g., flashing red for faulty traffic control light).
Software	N-version programs — Execute a number of different programs written independently to implement the same specification. Vote on the final result.
	Recovery block — Execute acceptance test program upon completion of procedure. Execute alternate procedure if acceptance test is failed.

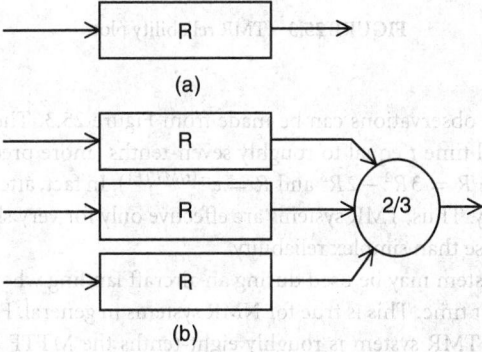

FIGURE 25.2   Voted logic. (a) Simplex, (b) TMR.

25.4.2   Error Masking

Several systems require that no errors are produced at their outputs. This is referred to as *error masking*. Such systems are usually real-time control systems in which outputs cause direct actions. Examples include air or spacecraft control surfaces, control for automobiles, medical systems, and nuclear and manufacturing controls. For such systems, the appropriate reliability parameter is the *fault tolerance*: the total number of failed elements that can be present without causing output errors. For some designs, it is important to consider both the total number of failed elements and the number of simultaneously failing elements that can be tolerated. Major error masking techniques are listed in Table 25.2.

Triple modular redundancy (TMR) is a widely used error working technique. Figure 25.2 illustrates this scheme, in which each module is replaced by three modules whose outputs are passed through a voting network before being transmitted as outputs. The three modules may be identical or different implementations of the same logic function. The concept of TMR was first developed in Von Neumann [56], in the context of models of neurons in the human brain.

One major advantage of TMR is its flexibility in choosing the module that forms the basic unit of replication. This can be as small as a single gate or as large as an entire computer. Any fault that affects only a single module will be masked. A direct extension of the TMR technique is *N-modular redundancy* (NMR), in which each module is replaced by N modules and voting is performed on the outputs of all modules. Several approaches to voting in TMR are discussed in Johnson [96], and Mitra [00a].

Suppose the reliability of each individual module of a TMR system is R; this is also referred to as the reliability of a simplex system or *simplex reliability*. The reliability of a TMR system is equal to $R^3 + 3R^2(1 - R) = 3R^2 - 2R^3$. Here, we assume that the voter reliability is 1, which is not generally the case. In that case, the TMR reliability must be multiplied by voter reliability. Figure 25.3 shows plots of simplex reliability and TMR reliability. It is assumed that the failure rate is constant, in which case R is equal to $e^{-(t/MTTF)}$, where MTTF is the mean time to failure of a simplex system.

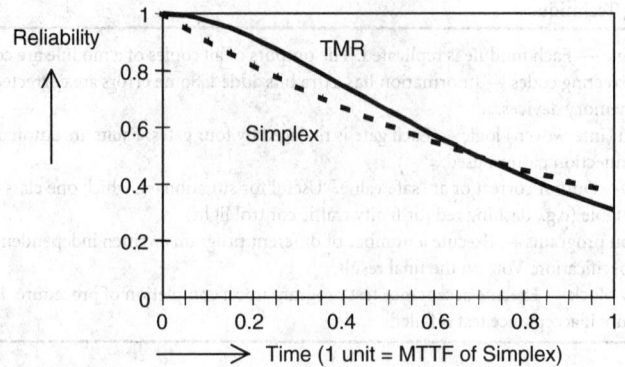

FIGURE 25.3 TMR reliability plot.

The following interesting observations can be made from Figure 25.3. The TMR reliability is greater than simplex reliability until time t equal to roughly seven-tenths (more precisely $log_e 2$) of the simplex MTTF (by solving for t when $R = 3R^2 - 2R^3$ and $R = e^{-(t/MTTF)}$). In fact, after that point TMR reliability is less than simplex reliability. Thus, TMR systems are effective only for very short mission times, beyond which their reliability is worse than simplex reliability.

As an example, a TMR system may be used during an aircraft landing when the system must produce correct outputs for that short time. This is true for NMR systems in general. For a system with a constant failure rate, the MTTF of a TMR system is roughly eight-tenths the MTTF of a simplex system. Thus, if MTTF is used as the only measure of reliability, then it will seem that the TMR is less reliable than a simplex system. However, this is not true for very short mission times (Figure 25.3). A TMR–simplex system [Siewiorek 98], which switches from a TMR to a simplex system, can be used to overcome the problem related to reliability of TMR for longer mission times. There are several examples of commercial systems using TMR [Riter 95, http://www.resilience.com]. Another example of hardware error masking is interwoven redundancy based on the concept of quadded logic [Tryon 62].

Error correcting codes (ECCs) are commonly used for error masking in RAMs and disks. Additional bits are appended to information stored or transmitted. Any faulty bit patterns within the capability of the used code are corrected, so that only correct information is provided at the outputs [Rao 89]. It relies on error correction circuitry to change faulty information bits and is thus effective when this circuitry is fault-free.

Two major software techniques for fault masking are *N*-version programming [Chen 78] and recovery blocks [Randall 75]. *N-version programming* requires that several versions of a program be written independently. Each program is run on the same data, and the outputs are obtained by voting on the outputs from the individual programs. This technique is claimed to be effective in detecting failures in writing a program.

The other software method, *recovery blocks*, also requires that several programs be written. However, the extra programs are run only when an error has been detected. Upon completion of a procedure, an acceptance test is run to check that no errors are present. If an error is detected, an alternate program for the procedure is run and checked by the acceptance test. One of the difficulties with this technique is the determination of suitable acceptance tests. A classification of acceptance tests into accounting checks, reasonableness tests, and run-time checks is discussed in Hecht [96].

25.4.3 Fault Secure Techniques

Fault secure techniques ensure that the output is correct unless an error indication is present. Errors must be detected but need not be corrected. Thus, the major objective is to preserve system data integrity so that any data corruption due to faults from a given class is detected. Error detection during normal system

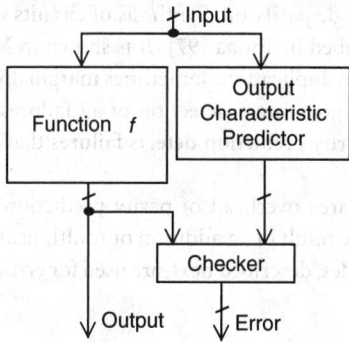

FIGURE 25.4 General architecture of a CED scheme.

operation is also referred to as *concurrent error detection* (CED) or *on-line checking*. CED techniques can be classified mainly into two major groups:

Hardware redundancy

Time redundancy

The basic principle behind CED based on hardware redundancy is shown in Figure 25.4 [Mitra 00b]. The system realizes function f and produces output sequence $f(i)$ for any input sequence i. A CED scheme generally contains another unit which independently predicts some special characteristic of the system output $f(i)$ for every input sequence i. Finally, a checker unit checks whether the special characteristic of the actual output produced by the system in response to sequence i matches the predicted characteristic and signals *error* when a mismatch occurs. Some examples of characteristics of $f(i)$ used for error detection purposes are $f(i)$ itself (duplication), its parity (parity prediction), 1s count, 0s count, transition count, etc. Coding schemes using parity and cyclic redundancy check (CRC) [Rao 89] are generally used for error detection in memories and data packets transmitted through computer networks.

A CED scheme is characterized by the class of failures in the presence of which data integrity is guaranteed. For the CED example in Figure 25.4, if the error signal is stuck-at-0, then an error will never be signaled. Thus, the CED system must be able to detect errors in the checking circuitry. An overview of self-checking checker designs can be obtained from Wakerly [78], and McCluskey [90]. Several CED techniques have been developed and used commercially for designing reliable systems [Sellers 68, Hsiao 81, Kraft 81, Chen 92, Webb 97, Spainhower 99]. Some of the important CED techniques are described next.

25.4.3.1 Duplication

Duplication is an example of a classic CED scheme in which there are two implementations, identical or different, of the same logic function. A self-checking comparator based on two-rail checkers is used to check whether the outputs from the two implementations agree. Duplication guarantees data integrity when one of the two implementations produces errors (single-module faults). Duplication has been used in several commercial systems [Kraft 81, Pradhan 96, Siewiorek 98, Webb 97, Spainhower 99].

25.4.3.2 Parity Prediction

The odd parity function indicates whether the number of 1s in a set of binary digits is odd. CED techniques for datapath circuits [Nicolaidis 97] and general combinational and sequential circuits based on parity prediction have been described in De [94], Touba [97], and Zeng [99]. For a design with parity prediction using a single parity bit, the circuit is designed in such a way that there is no sharing among the logic cones generating the circuit outputs. Thus, if a single fault affects a logic cone, at most one output will be erroneous, so that the error is detected by a parity checker. A self-checking parity checker design is given in McCluskey [90]. The restriction of no logic sharing among different logic cones may result in large area overhead for circuits with a single parity bit. Synthesis of circuits with multiple parity bits and some sharing among logic cones is described in Touba [97]. It is shown in Mitra [00b] that the area overhead of parity prediction is comparable to duplication, sometimes marginally better, for general combinational circuits. Unlike duplication, which guarantees detection of all failures that cause nonidentical errors at the outputs of the two modules, parity prediction detects failures that cause errors at an odd number of outputs.

For datapath logic circuits, the area overhead of parity prediction becomes very high, because it is difficult to calculate the parity of the result of an addition or multiplication operation from the individual parities of the operands. Residue codes, described next, are used for error detection in these cases [Langdon 70, Avizienis 71].

25.4.3.3 Residue Codes

Given an n-bit binary word, which is the binary representation of the decimal number x, the *residue* modulo b is defined as equal to $y = x \bmod b$. The recommended value of b is of the form $2^m - 1$ for high error detection coverage and low cost of implementation. For example, when $b = 3$, we need two bits to represent y. Suppose we add two numbers x_1 and x_2 and we have the residues of these two numbers for error detection along the datapath: that is, $y_1 = x_1 \bmod b$, and $y_2 = x_2 \bmod b$. The residue of the sum $x_3 = x_1 + x_2$ is given by $y_3 = (y_1 + y_2) \bmod b$; the residue of the product $x_4 = x_1 \times x_2$ is given by $y_4 = (y_1 \times y_2) \bmod b$. Hence, addition and multiplication operations are said to preserve the residues of its operands. The overall structure of error detection using residue codes is shown in Figure 25.5.

25.4.3.4 Application-Specific Techniques

The overhead of error detection can be reduced significantly if the specific characteristics of an application are utilized. For example, for some functions it is very easy to compute the inverse of that function; that is, given the output it is very easy to compute the input. For such functions, a cost-effective error detection scheme is to compute the inverse of a particular output and match the computed inverse with the actual input. The LZ-based compression algorithm is an example of such a function, where computation of the inverse is much easier than computation of the actual function. Compression is a fairly complex process involving string matching; however, decompression is much simpler, involving a memory lookup. Hence, for error detection in compression hardware, the compressed word can be decompressed to check whether it matches the original word. This makes error detection for compression hardware very simple [Huang 00]. Other examples of application-specific techniques for error detection with low hardware overhead are presented in Jou [88] and Austin [99].

Error detection based on time redundancy uses some form of repetition of actual operations on a piece of hardware. For detection of temporary errors, simple repetition may be enough. For detection of permanent faults, it must be ensured that the same faulty hardware is not used in exactly the same way during repetition. Techniques such as alternate data retry [Shedletsky 78b], RESO [Patel 82], and ED^{4}I

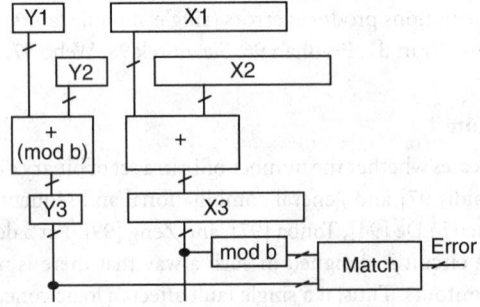

FIGURE 25.5 Error detection using residue codes.

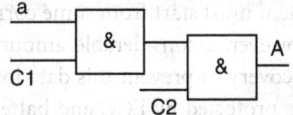

FIGURE 25.6   A simple fail-safe network.

[Oh 02a] are based on this idea. The performance overhead of error detection based on time redundancy may not be acceptable. Techniques such as multithreading [Saxena 00, Rotenberg 99, Reinhardt 00] or proper scheduling of instructions [Oh 02b] may be used reduce this overhead.

Some error detection techniques rely on a combination of hardware and software resources. Techniques include hardware facilities to check whether a memory access is to an area for which the executing program has authorization and also to verify that the type of access is allowed. Examples of other exception checks are given in Toy [86] and Siewiorek [98]. In a multiprocessor system, it is possible to run the same program on two different processors and compare the results. A more economical technique is to execute a reduced version of the main program on a separate device called a *watchdog processor* [Lu 82, Mahmood 88]. The watchdog processor is used to check the correct control flow of the program executed on the main processor, a technique called *control flow checking*. When the separate device is a simple counter, it is called a *watchdog timer*. Only faults that change the execution time of the main program are detected. Heartbeat signals that indicate the progress of a system are often used to ensure that a system is alive.

For signal processing applications involving matrix operations in a multiprocessor system, the number of processors required to check the errors in results can be significantly fewer than the number of processors required for the application itself, using a technique called *algorithm-based fault tolerance* (ABFT) [Huang 84].

Error detection using only software is very common and effective, but it must be developed for each individual program. Whenever some property of the outputs that ensures correctness can be identified, it is possible to execute a program that checks the property to detect errors. A simple example of checking output properties is the use of daily balance checks in the banking industry. The checked properties are also known as *assertions* [Leveson 83, Mahmood 83, Mahmood 85, Boley 95]. Automated software techniques for error detection without any hardware redundancy are described in Shirvani [00], Lovellette [02], Oh [02a], Oh [02b], and Oh [02c].

25.4.4   Fail-Safe Techniques

Fail-safe mechanisms cause the output state to remain in or revert to a safe state when a fault occurs. They typically involve replicated control signals and cause authorization to be denied when the control signals disagree. A simple example is a check that requires multiple signatures. Any missing or invalid signature causes the check not to be honored.

A simple fail-safe network is shown in Figure 25.6 Signals C1 and C2 have the same value in a fault-free situation. When they are both 0, the output is 0. If a fault causes them to disagree, the output is held at the safe value 0. This type of network is used in the fault-tolerant multiprocessor to control access to buses. No single fault can cause an incorrect bus access [Hopkins 78].

25.5   System Recovery

The rate of temporary failures is generally much higher than the rate of permanent faults [Lin 88]. Hence, when an error is detected, it is usually assumed that the error is due to some temporary failure. This temporary failure is expected to be gone when the operation is retried. If the failure persists even after a predetermined number of retries, it is assumed that the failure is a permanent one and repair routines are invoked.

Before a retry is performed, the system must start from some correct state. One solution is to initialize the entire system by rebooting it. However, a considerable amount of data may be lost as a result. A more systematic method of system recovery to prevent this data loss is checkpointingand rollback. The system state is stored in stable storage protected by ECC and battery backup at regular, predetermined intervals of time. This is referred to as *checkpointing*. Upon error detection, the system state is restored to the last checkpoint and the operation is retried; this is referred to as *rollback*. A particular application determines what system data must be checkpointed. For example, in a microprocessor checkpointing may be performed by copying all register and cache contents modified since the last checkpoint.

Techniques for roll-forward recovery in distributed systems that reduce the performance overhead of rollback recovery are described in Pradhan [94]. In a real-time system, rollback recovery may not be feasible because of the strict timing constraints. In that scenario, roll-forward recovery using TMR-based systems is applicable [Yu 01]. This technique also enables the use of TMR systems for temporary failures in long mission times.

Another useful recovery technique is called *scrubbing*. If the system memory is not used very frequently, then an error affecting a memory bit may not be corrected, because that memory word was not read out. As a result, errors may start accumulating, and used ECC codes may not be able to detect or correct the errors when the memory word is actually read out. Scrubbing techniques can be used to read the memory contents, check errors, and write back correct contents, even when the memory is not used. Scrubbing may be implemented in hardware or as a software process scheduled during idle cycles [Shirvani 00].

25.6 Repair Techniques

Unless a system is to be discarded when failures cause improper operation, failed components must be removed, replaced, or repaired. System repair may be manual or automatic. In the case of component removal, the system is generally reconfigured to run with fewer components, called *failed soft* or *graceful degradation*. If the faulty function can be automatically replaced or repaired, the system is called *self-repairing*. Self-repair techniques are useful for unmanned environments where manual repair is impossible or extremely costly. Examples include satellites, space missions, and remote or dangerous locations.

25.6.1 Built-in Self-Test and Diagnosis

The first error due to a fault typically occurs after a delay. This time is called the *error latency* [Shedletsky 76]. It is present because the output depends on the fault site logic value only for a subset of the possible input values. The error latency can be very long — so long that another fault (e.g., a temporary fault) occurs before the first fault is detected. Because of the limitations of the error-detection circuitry, the double error may be undetectable. In order to avoid this situation, provision is often made to test explicitly for faults that have not caused an output error. This technique is called *built-in self-test*. Built-in self-test is also useful in identifying the faulty unit after error detection.

Built-in self-test is generally executed periodically by interrupting the normal operation or when the system is idle. It is usually implemented by using test routines or diagnostic programs for programmed systems or by using built-in test hardware to apply a sequence of test vectors to the functional circuitry and checking its response [Bardell 87, Mitra 00c].

25.6.2 Fail-Soft Techniques

The most common form of automatic failure management involves disconnecting a failed component from the rest of the system and reconfiguring to continue operation until the bad device can be replaced. This can happen for as small a component as a portion of memory or for an entire processor, in the case of a multiprocessor system. The failed component is generally identified by means of built-in testing. Because no attempt is made to replace the disconnected component automatically, system operation may be degraded. This technique is sometimes called *graceful degradation*.

25.6.3 Self-Repair Techniques

Systems that must operate unattended for long periods require that the faulty components be replaced after being configured out of the system.

25.6.3.1 Standby Spares

One of the earliest self-repair techniques included a number of unused modules, *standby spares*, which could be used to replace failed components. The possibility of not applying power to a spare module until it is connected to the system is attractive for situations in which power is a critical resource or when the failure rate is much higher for powered than for unpowered circuits. Extensive hardware facilities are required to implement this technique, which is sometimes called *dynamic redundancy*. There are several major issues related to switching in a spare module without interrupting system operation.

25.6.3.2 Hybrid Redundancy

Fault masking and self-repair are obtained by combining TMR with standby spares, as shown in Figure 25.7 [Siewiorek 98]. Initially, the three top modules are connected through the switch to the voter inputs. The disagreement detector compares the voter output with each of the module outputs. If a module output disagrees with the voter output, the switch is reconfigured to disconnect the failed module from the voter and to connect a standby module to the voter instead. As long as at least two modules have correct outputs, the voter output will be correct. A failed module will automatically be replaced by a good module, as long as there are good modules available and the reconfiguration circuitry is fault-free. Thus, this system will continue to operate correctly until all spares are exhausted, the reconfiguration circuitry fails, or more than one of the on-line modules fail.

An advantage of this technique is the possibility of not applying power to a spare until it is connected to the voter. A disadvantage is the complexity of the reconfiguration circuitry. The complexity increases the cost of the system and limits its reliability [Ogus 75].

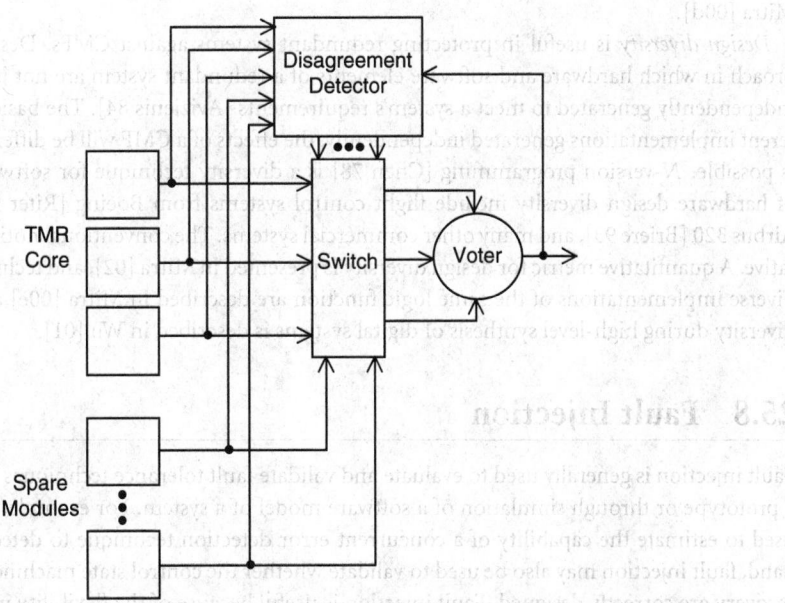

FIGURE 25.7 Hybrid redundancy with a TMR core and standby spares.

25.6.3.3 Self-Purging Redundancy

A self-purging system provides fault masking and self-repair with less complexity than hybrid redundancy [Losq 76]. It requires that all modules are powered until disconnected from the output due to a failure. Initially, all modules have their outputs connected to the voter inputs. The voter is a threshold network whose output agrees with the majority of voter inputs. Whenever one of the module outputs disagrees with the voter output, the module output is disconnected from the voter.

25.6.3.4 Reconfigurable Systems Self-Repair

In a reconfigurable system, such as a system with field programmable gate arrays (FPGAs), no spare modules are required for system self-repair. These systems can be configured multiple times by loading an appropriate configuration which implements logic functions in programmable logic blocks and inter-connections among the logic blocks, controlled by switches implanted using pass transistors. After error detection, the failing portion (which can be a programmable logic block or a failing switch) is identified, and a configuration that does not use the failed resource is loaded. For practical purposes, such an approach does not cause any significant reliability or performance degradation of the repaired system. Self-repair and fault isolation techniques for reconfigurable systems are described in Lach [98], Abramovici [99], Huang [01a], Huang [01b], and Mitra [04a, 04b].

25.7   Common-Mode Failures and Design Diversity

Most fault-tolerance techniques are based on a single fault assumption. For example, for a TMR system or for error detection based on duplication, it is assumed that one of the modules will be faulty. However, in an actual scenario, several failure sources may affect more than one module of a redundant system at the same time, generally due to a common cause. These failures are referred to as *common-mode failures* (CMFs) [Lala 94]. Some examples of CMFs include design faults, power-supply disturbances, a single source of radiation creating multiple upsets, etc. For a redundant system with identical modules, it is likely that a CMF will have identical error effects in the individual modules [Tamir 84]. For example, if a CMF affects both modules of a system using duplicated modules for error detection, then data integrity is not guaranteed — both modules can produce identical errors. CMFs are surveyed in Lala [94] and Mitra [00d].

Design diversity is useful in protecting redundant systems against CMFs. Design diversity is an approach in which hardware and software elements of a redundant system are not just replicated; they are independently generated to meet a system's requirements [Avizienis 84]. The basic idea is that, with different implementations generated independently, the effects of a CMF will be different, so error detection is possible. *N*-version programming [Chen 78] is a diversity technique for software systems. Examples of hardware design diversity include flight control systems from Boeing [Riter 95], the space shuttle, Airbus 320 [Briere 93], and many other commercial systems. The conventional notion of diversity is qualitative. A quantitative metric for design diversity is presented in Mitra [02], and techniques for synthesizing diverse implementations of the same logic function are described in Mitra [00e] and Mitra [01]. Use of diversity during high-level synthesis of digital systems is described in Wu [01].

25.8   Fault Injection

Fault injection is generally used to evaluate and validate fault tolerance techniques by injecting faults into a prototype or through simulation of a software model of a system. For example, fault injection may be used to estimate the capability of a concurrent error detection technique to detect errors; on the other hand, fault injection may also be used to validate whether the control state machines designed to perform recovery are correctly designed. Fault injection is useful because of the flexibility in injecting faults at the areas of interest in a controlled fashion, unlike in an actual environment. Also, faults can be injected at a much higher frequency than in an actual scenario, where the failure rate may be very low. Thus, statistical information about the capabilities of fault-tolerance techniques can be obtained in a much shorter time than in a real environment, where it may take several years to collect the same information.

However, it is questionable whether the injected faults represent actual failures. For example, results from test chip experiments demonstrate that very few manufacturing defects actually behave as single stuck-at faults [McCluskey 00]. Hence, experiments must be performed to validate the effectiveness of fault models used during fault injection. We briefly describe some of the commonly used fault injection techniques [Iyer 96, Shirvani 98].

Disturb signals on the pins of an IC — The signal values at the pins of an IC can be changed at random, or there may be some sequence and timing dependence between error events on the pins [Scheutte 87]. The problem with this approach is that there is no control over errors at the internal nodes of a chip not directly accessible through the pins.

Radiation experiments — Soft errors due to cosmic rays in the internal nodes of a chip can be created by putting the chip under heavy-ion radiation or a high-energy proton beam in radiation facilities. The angle of incidence and the amount of radiation can be controlled very precisely. This technique is used widely in industry.

Power-supply disturbance — Errors can be caused in the system by disturbing the power supply [Cortes 86, Miremadi 95]. This technique models errors due to disturbances in the power supply and in industrial environments.

Simulation techniques — Errors can be introduced at selected nodes of a design, and then the system can be simulated to understand its behavior in the presence of these errors. This technique is applicable at all levels of abstraction: transistor level, gate–level, or even a high-level description language level of the system. The choice of appropriate fault models is very important in this context. One of the drawbacks of simulation-based fault injection is the slow speed of simulation compared to the actual operation speed of a circuit. Several fault injection tools based on simulation have been developed, such as Kanawati [95] and Goswami [97].

25.9 Conclusion

Computing and communication systems constitute an inseparable part of our everyday lives — home appliances, communication, transportation, banking and finance management, embedded computing in space applications, and biomedical applications, including automated remote surgery, ICs implanted in human bodies (the pacemaker being a very simple example), and health monitoring systems. Malfunctions of computing and communication systems have also become an inseparable part of our daily lives. Such malfunctions can cause massive financial losses or loss of human lives for critical applications, such as servers, automobiles, avionics, space vehicles, and biomedical equipment. Security violations due to lapses in security protocol designs or unprecedented attacks by external agents are already a major cause of concern that will significantly grow with our increased reliance on computers and networks.

There is general consensus in the industry that these failure sources will become more pronounced. We need robust system designs ensuring high system availability, maintainability, quality of service, security, and efficient recovery from failures. These systems must be available even during bursts of heaviest demands, must detect errors in processed transactions, and must be able to repair themselves from failures very quickly. They also must be transparent to the user, through retry, reconfiguration, or reorganization through graceful degradation, for example. Hence, design techniques for self-organizing, self-repairing, and self-maintaining adaptive systems are required.

This chapter has surveyed techniques for evaluating and designing such robust systems. We have demonstrated that the issues of error control and repair are distinct and appropriate specifications for each must be matched to the intended applications. We also have presented techniques for satisfying these aspects of the system specifications and discussed fault injection techniques to evaluate the effectiveness of these techniques and validate their design. Some of these techniques are applicable in the context of designing robust, adaptive, self-repairing architectures for molecular and quantum computing systems.

25.10 Further Reading

The reader is encouraged to study the following topics related to this chapter:

  Security and survivability [http://www.iaands.org/iaands2002/index.html]
  Autonomic computing [http://www.research.ibm.com/autonomic]
  Software rejuvenation [http://www.software-rejuvination.com]
  Efforts related to dependability benchmarking [http://www.ece.cmu.edu/~koopman/ifip_wg_10_4_sigdeb]
  Defect and fault tolerance in molecular computing systems [http://www.hpl.hp.com/research/qsr]

References

[Abramovici 90] Abramovici, M., M.A. Breuer, and A.D. Friedman, *Digital Systems Testing and Testable Design*, IEEE Press, 1990.

[Abramovici 99] Abramovici, M., C. Stroud, C. Hamilton, C. Wijesuriya, and V. Verma, "Using Roving Stars for On-line Testing and Diagnosis of FPGAs," *Proc. Intl. Test Conf.*, pp. 973–982, 1999.

[Austin 99] Austin, T., "DIVA: A Reliable Substrate for Deep Submicron Microarchitecture Design," *Micro-32*, 1999.

[Avizienis 71] Avizienis, A., "Arithmetic Error Codes: Cost and Effectiveness Studies for Applications in Digital System Design," *IEEE Trans. Computers*, Vol. C-20, No. 11, pp. 1322–1331, Nov. 1971.

[Avizienis 84] Avizienis, A., and J.P.J. Kelly, "Fault Tolerance by Design Diversity: Concepts and Experiments," *IEEE Computer*, pp. 67–80, Aug. 1984.

[Bardell 87] Bardell, P.H., W.H. McAnney, and J. Savir, *Built-in Test for VLSI: Pseudorandom Techniques*, John Wiley & Sons, 1987.

[Bentley 01] Bentley, B., "Validating the Intel Pentium 4 Microprocessor," *Proc. Intl. Conf. Dependable Systems and Networks*, pp. 493–498, 2001.

[Beudry 77] Beudry, M.D., "Performance Related Reliability Measures for Computing Systems," *Proc. Intl. Symp. Fault-Tolerant Computing*, pp. 16–21, 1977.

[Blish 00] Blish, R., and N. Durrant, Eds., "Semiconductor Device Reliability Failure Models," *Tech. Transfer #00053955A-XFR, International Sematech*, May 31, 2000.

[Boley 95] Boley, D., G.H. Golub, S. Makar, N. Saxena, and E.J. McCluskey, "Floating Point Fault-Tolerance with Backward Error Assertions," *IEEE Trans. Computers, Special Issue on Fault-Tolerant Computing*, pp. 302–311, Feb. 1995.

[Briere 93] Briere, D., and P. Traverse, "Airbus A320/A330/A340 Electrical Flight Controls: A Family of Fault-Tolerant Systems," *Proc. Intl. Conf. Fault Tolerant Computing*, pp. 616–623, 1993.

[Breuer 96] Breuer, M.A., and S.K. Gupta, "Process Aggravated Noise (PAN): New Validation and Test Problems," *Proc. Intl. Test Conf.*, pp. 914–923, 1996.

[Calin 96] Calin, T., M. Nicolaidis, and R. Velazco, "Upset Hardererned Memory Design for Submicron CMOS Technology," *IEEE Trans. Nuclear Science*, Vol. 43, No. 6, pp. 2874–2878, Dec. 1996.

[Chang 97] Chang, T.Y.J., and E.J. McCluskey, "SHOrt Voltage Elevation (SHOVE) Test for Weak CMOS ICs," *Proc. IEEE VLSI Test Symp.*, pp. 446–451, 1997.

[Chen 78] Chen, L., and A. Avizienis, "N-Version Programming: A Fault-Tolerance Approach to Reliability of Software Operation," *Proc. Intl. Symp. Fault-Tolerant Computing*, pp. 3–9, 1978.

[Chen 92] Chen, C.L., et al., "Fault-Tolerance Design of the IBM Enterprise System/9000 Type 9021 Processors," *IBM Journal Res. And Dev.*, Vol. 36, No. 4, pp. 765–779, July 1992.

[Cortes 86] Cortes, M., and E.J. McCluskey, "Modeling Power-Supply Disturbances in Digital Circuits," *Intl. Solid State Circuits Conf.*, pp. 164–165, 1986.

[Crouch 99] Crouch, A. L., *Design-for-Test for Digital ICs and Embedded Core Systems*, Prentice Hall, 1999.

[De 94] De, K., C. Natarajan, D. Nair, and P. Banerjee, "RSYN: A System for Automated Synthesis of Reliable Multilevel Circuits," *IEEE Trans. VLSI*, Vol. 2, pp. 186–195, June 1994.

[Eichelberger 91] Eichelberger, E.B., E. Lindbloom, J.A. Waicukauski, and T.W. Williams, *Structured Logic Testing*, Prentice Hall, Englewood Cliffs, NJ, 1991.

[Goswami 97] Goswami, K.K., R.K. Iyer, and L.Y. Young, "DEPEND: A Simulation-Based Environment for System Level Dependability Analysis," *IEEE Trans. Computers*, Vol. 46, No. 1, pp. 60–74, Jan. 1997.

[Gulati 93] Gulati, R.K., and C.F. Hawkins, Eds., *Iddq Testing of VLSI Circuits*, Kluwer Academic Publishers, 1993.

[Hao 93] Hao, H., and E.J. McCluskey, "Very-Low-Voltage Testing for Weak CMOS Logic ICs," *Proc. Intl. Test Conf.*, pp. 275–284, 1993.

[Hecht 96] Hecht, H., and M. Hecht, "Fault Tolerance in Software," *Fault-Tolerant Computer System Design*, D.K. Pradhan, Ed., pp. 428–477, Prentice Hall, Upper Saddle River, NJ, 1996.

[Hnatek 87] Hnatek, E. R., *Integrated Circuit Quality and Reliability*, Marcel Dekker Inc., 1987.

[Hopkins 78] Hopkins, A.L., T.B. Smith, and J.H. Lala, "FTMP — A Highly Fault-Tolerant Multiprocessor for Aircraft," *Proc. IEEE*, Vol. 66, No. 10, pp. 1221–1239, Oct. 1978.

[Hsiao 81] Hsiao, M-Y., W.C. Carter, J.W. Thomas, and W.R. Stringfellow, "Reliability, Availability and Serviceability of IBM Computer Systems: A Quarter Century of Progress," *IBM Journal Res. And Dev.*, Vol. 25, No. 5, pp. 453–469, Sept. 1981.

[Hsieh 77] Hsieh, E.P., R.A. Rasmussen, L.J. Vidunas, and W.T. Davis, "Delay Test Generation," *Proc. Design Automation Conf.*, pp. 486–491, 1977.

[Huang 84] Huang, K.H., and J.A. Abraham, "Algorithm Based Fault Tolerance for Matrix Operations," *IEEE Trans. Computers*, Vol. C-33, No. 6, pp. 518–528, June 1984.

[Huang 00] Huang, W.J., N.R. Saxena, and E.J. McCluskey, "A Reliable LZ Data Compressor on Reconfigurable Coprocessors," *Proc. IEEE Field Programmable Custom Computing Machines (FCCM)*, 2000.

[Huang 01a] Huang, W.J., and E.J. McCluskey "Column-Based Precompiled Configuration Techniques for FPGA Fault Tolerance," *Proc. IEEE Field Programmable Custom Computing Machines (FCCM)*, 2001.

[Huang 01b] Huang, W.J., S. Mitra, and E.J. McCluskey, "Fast Run-Time Fault Location for Dependable FPGA-based Applications," *Proc. Intl. Symp. Defect and Fault-Tolerance of VLSI Systems*, pp. 206–214, 2001.

[IRPS 02] Special Workshop on "Soft Errors," *Intl. Reliability Physics Symp.*, 2002.

[Iyer 96] Iyer, R.K., and D. Tang, "Experimental Analysis of Computer System Dependability," *Fault-Tolerant Computer System Design*, D.K. Pradhan, Ed., pp. 282–392, Prentice Hall, Upper Saddle River, NJ, 1996.

[Jensen 82] Jensen, F. and N.E. Petersen, *Burn-In: On Engineering Approach to the Design and Analysis of Burn-In Procedurese*, Wiley, 1982.

[Johnson 96] Johnson, B.W., "An Introduction to the Design and Analysis of Fault-Tolerant Systems," *Fault-Tolerant Computer System Design*, D.K. Pradhan, Ed., pp. 1–88, Prentice Hall, Upper Saddle River, NJ, 1996.

[Jou 88] Jou, J.Y., and J.A. Abraham, "Fault-Tolerant FFT Networks," *IEEE Trans. Computers*, Vol. 37, No. 5, pp. 548–561, May 1988.

[Kanawati 95] Kanawati, G.A., N.A. Kanawati, and J.A. Abraham, "FERRARI: A Flexible Software-Based Fault and Error Injection System," *IEEE Trans. Computers*, Vol. 44, No. 2, pp. 248–260, Feb. 1995.

[Kang 86] Kang, S.M., and D. Chu, "CMOS Circuit Design for Prevention of Single Event Upset," *Proc. Intl. Conf. Computer Design*, pp. 385–388, 1986.

[Klinger 90] Klinger, D.J., Y. Nakada, and M.A. Menendez, Eds., *AT&T Reliability Manual*, Van Nostrand Reinhold, New York, 1990.

[Kraft 81] Kraft, G.D., and W.N. Toy, *Microprogrammed Control and Reliable Design of Small Computers*, Prentice Hall, Englewood Cliffs, NJ, 1981.

[Lach 98] Lach, J., W.H. Manglone-Smith, and M. Potkonjak, "Efficiently Supporting Fault-Tolerance in FPGAs," *Proc. Intl. Symp. FPGAs*, pp.105–115, 1998.

[Lala 94] Lala, J.H., and R.E. Harper, "Architectural Principles for Safety-Critical Real-Time Applications," *Proc. IEEE*, Vol. 82, No. 1, pp. 25–40, Jan. 1994.

[Langdon 70] Langdon, G.G., and C.K. Tang, "Concurrent Error Detection for Group Look-ahead Binary Adders," *IBM Journal Res. and Dev.*, pp. 563–573, Sept. 1970.

[Leveson 83] Leveson, N.G., and P.R. Harvey, "Analyzing Software Safety," *IEEE Trans. Software Eng.*, SE-9(5), pp. 569–579, Sept. 1983.

[Lin 88] Lin, T.Y. and D.P. Siewiorek, "Error Log Analysis: Statistical Modeling and Heuristic Trend Analysis," *IEEE Trans. Reliability* Vol. 39, No. 4, pp. 419–432, Oct. 1990.

[Losq 76] Losq, J., "A Highly Efficient Redundancy Scheme: Self-Purging Redundancy," *IEEE Trans. Computers*, Vol. 12, No. 6, pp. 569–578, June 1976.

[Lovellette 02] Lovellette, M.N., et al., "Strategies for Fault-Tolerant, Space-Based Computing: Lessons Learned from the ARGOS Testbed," *Proc. IEEE Aerospace Conf.*, Vol. 5, pp. 2109–2119, 2002.

[Lu 82] Lu, D.J., "Watchdog Processors and Structural Integrity Checking," *IEEE Trans. Computers*, Vol. C-31, No. 7, pp. 681–685, July 1982.

[Mahmood 83] Mahmood, A., E.J. McCluskey, and D.J. Lu, "Concurrent Fault Detection Using a Watchdog Processor and Assertions," *Proc. Intl. Test Conf.*, pp. 622–628, 1983.

[Mahmood 85] Mahmood, A., E. Ersoz, and E.J. McCluskey, "Concurrent System-Level Error Detection Using a Watchdog Processor," *Proc. Intl. Test Conf.*, pp. 145–152, 1985.

[Mahmood 88] Mahmood, A., and E.J. McCluskey, "Concurrent Error Detection Using Watchdog Processors — A Survey," *IEEE Trans. Computers*, Vol. 37, No. 2, pp. 160–174, Feb. 1988.

[McCluskey 86] McCluskey, E.J., *Logic Design Principles with Emphasis on Testable Semicustom Circuits*, Prentice Hall, Englewood Cliffs, NJ, 1986.

[McCluskey 90] McCluskey, E.J., "Design Techniques for Testable Embedded Error Checkers," *IEEE Computer*, Vol. 23, No. 7, pp. 84–88, July 1990.

[McCluskey 00] McCluskey, E.J., and C.W. Tseng, "Stuck-Fault Tests vs. Actual Defects," *Proc. Intl. Test Conf.*, pp. 336–343, 2000.

[Mei 74] Mei, K.C.Y., "Bridging and Stuck-at Faults," *IEEE Trans. Computers*, C-23, No. 7, pp. 720–727, July 1974.

[Miremadi 95] Miremadi, G., and J. Torin, "Evaluating Processor-Behavior and Three Error-Detection Mechanisms Using Physical Fault-Injection," *IEEE Trans. Reliability*, Vol. 44, No. 3, pp. 441–453, Sept. 1995.

[Mitra 00a] Mitra, S., and E.J. McCluskey, "Word-Voter: A New Voter Design for Triple Modular Redundant Systems," *Proc. IEEE VLSI Test Symp.*, pp. 465–470, 2000.

[Mitra 00b] Mitra, S., and E.J. McCluskey, "Which Concurrent Error Detection Scheme to Choose?" *Proc. Intl. Test Conf.*, pp. 985–994, 2000.

[Mitra 00c] Mitra, S., and E.J. McCluskey, "Fault Escapes in Duplex Systems," *Proc. IEEE VLSI Test Symp.*, pp. 453–458, 2000.

[Mitra 00d] Mitra, S., N.R. Saxena, and E.J. McCluskey, "Common-Mode Failures in Redundant VLSI Systems: A Survey," *IEEE Trans. Reliability*, Special Section on Fault-Tolerant VLSI Systems, Vol. 49, No. 3, pp. 285–295, Sept. 2000.

[Mitra 00e] Mitra, S., and E.J. McCluskey, "Combinational Logic Synthesis for Diversity in Duplex Systems," *Proc. Intl. Test Conf.*, pp. 179–188, 2000.

[Mitra 01] Mitra, S., and E.J. McCluskey, "Design of Redundant Systems Protected Against Common-Mode Failures," *Proc. IEEE VLSI Test Symp.*, pp. 190–195, 2001.

[Mitra 02] Mitra, S., N.R. Saxena, and E.J. McCluskey, "A Design Diversity Metric and Analysis of Redundant Systems," *IEEE Trans. Computers*, Vol. 51, No. 5, pp. 498–510, May 2002.

[Mitra 04a] Mitra, S., W.J. Huang, N.R. Sexena, S.Y. Yu, and E.J. McCluskey, "Dependable Reconfigurable Computing: Reliability Obtained by Adaptive Reconfiguration," *ACM Trans. Embedded Computing Systems*, To appear.

[Mitra 04b] Mitra, S., W.J. Huang, N.R. Saxena, S.Y. Yu, and E.J. McCluskey, "Reconfigurable Architecture for Autonomous Self-Repair," *IEEE Design & Test of Computers*, Special Issue on Design for Yield and Reliability, May-June, 2004.

[Needham 91] Needham, W.M., *Designer's Guide to Testable ASIC Devices*, Van Nostrand Reinhold, 1991.

[Nicolaidis 97] Nicolaidis, M., R.O. Duarte, S. Manich, and J. Figueras, "Fault-Secure Parity Prediction Arithmetic Operators," *IEEE Design and Test of Computers*, Vol. 14, No. 2, pp. 60–71, 1997.

[Ogus 75] Ogus, R.C., "Reliability Analysis of Hybrid Redundant Systems with Nonperfect Switches," *Technical Report, Computer Systems Lab., Stanford University*, CSL-TR-65, 1975.

[Oh 02a] Oh, N., S. Mitra, and E.J. McCluskey, "ED^{4}I: Error Detection by Diverse Data and Duplicated Instructions," *IEEE Trans. Computers, Special Issue on Fault-Tolerant Embedded Systems*, Vol. 51, No. 2, pp. 180–199, Feb. 2002.

[Oh 02b] Oh, N., P.P. Shirvani, and E.J. McCluskey, "Error Detection by Duplicated Instructions in Super-Scalar Processors," *IEEE Trans. Reliability*, Vol. 51, No. 1, pp. 63–75, March 2002.

[Oh 02c] Oh, N., P.P. Shirvani, and E.J. McCluskey, "Control-Flow Checking by Software Signatures," *IEEE Trans. Reliability*, Vol. 51, No. 1, pp. 111–122, March 2002.

[Ohring 98] Ohring, M., *Reliability and Failure of Electronic Materials and Devices*, Academic Press, 1998.

[Patel 82] Patel, J.H., and L.Y. Fung, "Concurrent Error Detection in ALUs by Recomputing with Shifted Operands," *IEEE Trans. Computers*, Vol. C-31, No. 7, pp. 589–595, July 1982.

[PC World 99] "Vendor Settles Suit over Alleged Problems in Floppy Disk Drives," *PC World*, Nov. 10, 1999.

[Perry 95] Perry, 6., *Digital Testing Course*, 1995, (http://www.soft-test.com).

[Pradhan 94] Pradhan, D.K., and N.H. Vaidya, "Roll-Forward Checkpointing Scheme: A Novel Fault-Tolerant Architecture," *IEEE Trans. Computers*, Vol. 43, No. 10, pp. 1163–1174, Oct. 1994.

[Pradhan 96] Pradhan, D.K., *Fault-Tolerant Computer System Design*, Prentice Hall, Upper Saddle River, NJ, 1996.

[Randall 75] Randall, B., "System Structure for Software Fault Tolerance," *IEEE Trans. Software Engineering*, Vol. SE-1, No. 2, pp. 220–232, June 1975.

[Rao 89] Rao, T.R.N., and E. Fujiwara, *Error-Control Coding for Computer Systems*, Prentice Hall, 1989.

[Reinhardt 00] Reinhardt, S., and S. Mukherjee, "Transient Fault Detection via Simultaneous Multi-threading," *Intl. Symp. Computer Architecture*, pp. 25–36, 2000.

[Riter 95] Riter, R., "Modeling and Testing a Critical Fault-Tolerant Multi-Process System," *Proc. Intl. Symp. Fault-Tolerant Computing*, pp. 516–521, 1995.

[Rotenberg 99] Rotenberg, E., "AR-SMT: A Microarchitectural Approach to Fault Tolerance in Microprocessor," *Proc. Fault-Tolerant Computing Symposium*, 1999.

[Saxena 00] Saxena, N.R., S. Fernandez Gomez, W.J. Huang, S. Mitra, S.Y. Yu, and E.J. McCluskey, "Dependable Computing and On-line Testing in Adaptive and Reconfigurable Systems," *IEEE Design and Test of Computers*, pp. 29–41, Jan–Mar 2000.

[Scheutte 87] Scheutte, M.A., and J.P. Shen, "Processor Control Flow Monitoring Using Signatured Instruction Streams," *IEEE Trans. Computers*, Vol. C-36, No. 3, pp. 264–276, March 1987.

[Sellers 68] Sellers, F., M.Y. Hsiao, and L.W. Bearnson, *Error Detection Logic for Digital Computers*, McGraw-Hill, 1968.

[Shedletsky 76] Shedletsky, J.J., and E.J. McCluskey, "The Error Latency of a Fault in a Sequential Circuit, *IEEE Trans. Computers*, C-25, pp. 655–659, June 1976.

[Shedletsky 78a] Shedletsky, J.J., "Delay Testing LSI Logic," *Proc. Intl. Symp. Fault Tolerant Computing*, pp. 159–164, 1978.

[Shedletsky 78b] Shedletsky, J.J., "Error Correction by Alternate-Data Retry," *IEEE Trans. Computers*, Vol. C-27, No. 2, pp. 106–112, Feb. 1978.

[Shirvani 98] Shirvani, P.P., and E.J. McCluskey, "Fault-Tolerant Systems in a Space Environment: The CRC Argos Project," *Technical Report, Center for Reliable Computing, Stanford University, CRC-TR-98-2, CSL-TR-98-774*, Dec. 1998.

[Shirvani 00] Shirvani, P.P., N. Saxena, and E.J. McCluskey, "Software-Implemented EDAC Protection Against SEUs," *IEEE Trans. Reliability, Special Section on Fault-Tolerant VLSI Systems*, Vol. 49, No. 3, pp. 273–284, Sept. 2000.

[Siewiorek 98] Siewiorek, D.P., and R.S. Swarz, *Reliable Computer Systems Design and Evaluation 3rd Ed.*, A.K. Peters, Massachusetts, 1998.

[Smith 85] Smith, G.L., "Model for Delay Faults," *Proc. Intl. Test Conf.*, pp. 342–349, 1985.

[Spainhower 99] Spainhower, L., and T.A. Gregg, "S/390 Parallel Enterprise Server G5 Fault Tolerance," *IBM Journal Res. and Dev.*, Vol. 43, pp. 863–873, Sept.–Nov., 1999.

[Tamir 84] Tamir, Y., and C.H. Sequin, "Reducing Common Mode Failures in Duplicate Modules," *Proc. Intl. Conf. Computer Design*, pp. 302–307, 1984.

[Touba 97] Touba, N.A., and E.J. McCluskey, "Logic Synthesis of Multilevel Circuits with Concurrent Error Detection," *IEEE Trans. CAD*, Vol. 16, No. 7, pp. 783–789, July 1997.

[Toy 86] Toy, W., and B. Zee, *Computer Hardware/Software Architecture*, Prentice Hall, Englewood Cliffs, NJ, 1986.

[Trivedi 02] Trivedi, K.S., *Probability and Statistics with Reliability, Queuing and Computer Science Applications, 2nd Ed.*, John Wiley & Sons, New York, 2002.

[Tryon 62] Tryon, J.G., "Quadded Logic," *Redundancy Techniques for Computing Systems*, Wicox and Mann, Eds., pp. 205–208, Spartan Books, Washington D.C., 1962.

[Von Neumann 56] Von Neumann, J., "Probabilistic Logics and the Synthesis of Reliable Organisms from Unreliable Components," *Annals of Mathematical Studies*, Vol. 34, Eds. C.E. Shannon and J. McCarthy, pp. 43–98, 1956.

[Wakerly 78] Wakerly, J.F., *Error Detecting Codes, Self-Checking Circuits and Applications*, Elsevier North-Holland, New York, 1978.

[Wang 99] Wang, J.J., et al., "SRAM-based Reprogrammable FPGA for Space Applications," *IEEE Trans. Nuclear Science*, Vol. 46, No. 6, pp. 1728–1735, Dec. 1999.

[Webb 97] Webb, C.F., and J.S. Liptay, "A High Frequency Custom S/390 Microprocessor," *IBM Journal Res. and Dev.*, Vol. 41, No. 4/5, pp. 463–474, 1997.

[Wu 01] Wu, K., and R. Karri, "Algorithm Level Recomputing with Allocation Diversity: A Register Transfer Level Time Redundancy Based Concurrent Error Detection Technique," *Proc. Intl. Test Conf.*, pp. 221–229, 2001.

[Yu 01] Yu, S.Y., and E.J. McCluskey, "On-line Testing and Recovery in TMR Systems for Real-Time Applications," *Proc. Intl. Test Conf.*, pp. 240–249, 2001.

[Zeng 99] Zeng, C., N.R. Saxena, and E.J. McCluskey, "Finite State Machine Synthesis with Concurrent Error Detection," *Proc. Intl. Test Conf.*, pp. 672–680, 1999.

[Ziegler 96] Ziegler, J.F., et al., "IBM Experiments in Soft Fails in Computer Electronics (1978–1994)," *IBM Journal Res. and Dev.*, No. 1, Jan. 1996.

III

Computational Science

Computational Science unites computational simulation, scientific experimentation, geometry, mathematical models, visualization, and high performance computing to address some of the "grand challenges" of computing in the sciences and engineering. Advanced graphics and parallel architectures enable scientists to analyze physical phenomena and processes, providing a virtual microscope for inquiry at an unprecedented level of detail. These chapters provide detailed illustrations of the computational paradigm as it is used in specific scientific and engineering fields, such as ocean modeling, chemistry, astrophysics, structural mechanics, and biology.

26 **Geometry-Grid Generation**   *Bharat K. Soni and Nigel P. Weatherill* **26**-1
Introduction • Underlying Principles • Best Practices • Grid Systems
• Research Issues and Summary

27 **Scientific Visualization**   *William R. Sherman, Alan B. Craig,*
M. Pauline Baker, and Colleen Bushell **27**-1
Introduction • Historic Overview • Underlying Principles • The Practice of Scientific
Visualization • Research Issues and Summary

28 **Computational Structural Mechanics**   *Ahmed K. Noor* **28**-1
Introduction • Classification of Structural Mechanics Problems • Formulation of
Structural Mechanics Problems • Steps Involved in the Application of Computational
Structural Mechanics to Practical Engineering Problems • Overview of Static, Stability,
and Dynamic Analysis • Brief History of the Development of Computational Structural
Mechanics Software • Characteristics of Future Engineering Systems and Their
Implications on Computational Structural Mechanics • Primary Pacing Items and
Research Issues

29 **Computational Electromagnetics**   *J. S. Shang* **29**-1
Introduction • Governing Equations • Characteristic-Based Formulation
• Maxwell Equations in a Curvilinear Frame • Eigenvalues and Eigenvectors
• Flux-Vector Splitting • Finite-Difference Approximation • Finite-Volume
Approximation • Summary and Research Issues

30 **Computational Fluid Dynamics**   *David A. Caughey* **30**-1
Introduction • Underlying Principles • Best Practices • Research Issues and Summary

31 **Computational Ocean Modeling**   *Lakshmi Kantha and Steve Piacsek* **31**-1
Introduction • Underlying Principles • Best Practices • Nowcast/Forecast in the Gulf
of Mexico (a Case Study) • Research Issues and Summary

32 Computational Chemistry   *Frederick J. Heldrich, Clyde R. Metz,*
Henry Donato, Kristin D. Krantzman, Sandra Harper,
Jason S. Overby, and Gamil A. Guirgis **32**-1
Introduction • Computational Chemistry in Education • Computational Aspects of
Chemical Kinetics • Molecular Dynamics Simulations • Modeling Organic
Compounds • Computational Organometallic and Inorganic Chemistry • Use of *Ab Initio*
Methods in Spectroscopic Analysis • Research Issues and Summary

33 Computational Astrophysics   *Jon Hakkila, Derek Buzasi, and*
Robert J. Thacker **33**-1
Introduction • Astronomical Databases • Data Analysis
• Theoretical Modeling • Research Issues and Summary

34 Computational Biology   *David T. Kingsbury* **34**-1
Introduction • Databases • Imaging, Microscopy, and Tomography • Determination of
Structures from X-Ray Crystallography and NMR • Protein Folding • Genomics

26

Geometry-Grid Generation

26.1 Introduction 26-1
Structured Grids • Unstructured Grids
• Generation Process
26.2 Underlying Principles 26-5
Terminology and Grid Characteristics • Geometry Preparation
• Structured Grid Generation • Unstructured Grid Generation
26.3 Best Practices 26-15
Structured Grid Generation • The Delaunay Algorithm
• Hybrid Grid Generation
26.4 Grid Systems 26-25
26.5 Research Issues and Summary 26-26

Bharat K. Soni
Mississippi State University

Nigel P. Weatherill
University of Wales Swansea

26.1 Introduction

With the advent and rapid development of supercomputers and high-performance workstations, computational field simulation (CFS) is rapidly emerging as an essential analysis tool for science and engineering problems. In particular, CFS has been extensively utilized in analyzing fluid mechanics, heat and mass transfer, biomedics, geophysics, electromagnetics, semiconductor devices, atmospheric and ocean science, hydrodynamics, solid mechanics, civil engineering related transport phenomena, and other physical field problems in many science and engineering firms and laboratories.

The basic equations governing the general physical field can be represented as a set of nonlinear partial differential equations pursuant to a particular set of boundary conditions. For computational simulation, the field is decomposed into a collection of elemental areas (2-D)/volumes (3-D). The governing equations associated with the field under consideration are then approximated by a set of algebraic equations on these elemental volumes and are numerically solved to get discrete values which approximate the solution of the pertinent governing equations over the field. This discretization of the field (domain, region) into finite-elemental areas/volumes is referred to as grid generation, and the collection of discretized elemental areas/volumes is called the grid.

The numerical solution process associated with general CFS applications first involves discretization of the integral or differential form of the governing set of partial differential equations (PDEs) formulated in continuum. The discretization of these equations is usually influenced by the grid strategy under consideration and the solution strategy to be employed. In general, the solution strategies are classified as: finite difference, finite volume, and finite element. In the case of finite differences, the derivatives in the PDEs are represented by algebraic difference expressions obtained by performing Taylor series expansions of the associated solution variables at several neighbors of the point of evaluation [Thompson and Mastin

1-58488-360-X/$0.00+$1.50
© 2004 by CRC Press, LLC

1983]. The differential forms of the governing equations are utilized in this case. However, the integral forms of the governing equations are used in the cases of finite-element and finite-volume strategies. Here the solution process involves representation of the solution variables over the cell in terms of selected functions, and then these functions are integrated over the volume (in case of finite-element) or the associated fluxes through cell sides (edges) are balanced (in case of finite volume).

The finite-element approach itself comes in two basic forms — the *variational*, where the PDEs are replaced by a more fundamental integral variational principle (from which they arise through the calculus of variations), or the *weighted residual* (Galerkin) approach in which the PDEs are multiplied by certain functions and then integrated over the cell.

In the finite-volume approach the fluxes through the cell sides (which separate discontinuous solution values) are best calculated with a procedure which represents the dissolution of such a discontinuity during the time step (Riemann solver).

The finite-difference approach, using the discrete points, is associated by many with rectangular Cartesian grids since such a regular lattice structure provides easy identification of neighboring points to be used in the representation of derivatives, whereas the finite-element approach has always been, by the nature of its construction on discrete cells, considered well-suited for irregular regions since a network of cells can be made to fill any arbitrarily shaped region and each cell is an entity unto itself, the representation being on a cell, not across cells. In view of the discretization strategy employed, the grids can be classified as structured, unstructured, or hybrid.

26.1.1 Structured Grids

Let $\mathbf{r} = (x, y, z)$ and $\Omega = (\xi, \eta, \zeta)$ denote the coordinates in the physical and computational space. The structured grid is presented by a network of lines of constant ξ, η, and ζ such that a one-to-one mapping can be established between physical and computational space. The computational space is made up of uniformly distributed points within a square in two dimensions or a cube in three dimensions as demonstrated in Figure 26.1. The structured grid involving identity transformation between physical and computational space (that is, $x = \xi, y = \eta, z = \zeta$) is called a Cartesian grid. However, the body-fitted grid generated by utilizing discrete/analytic arbitrary transformations between physical and computational space is classified as a curvilinear grid. A grid around a cylinder demonstrating the Cartesian grid and curvilinear two-dimensional grid demonstrating O, C, and H type strategies and their respective correspondence with the computational domain are displayed in Figure 26.2a through Figure 26.2d.

The curvilinear grid points conform to the solid surfaces/boundaries and hence provide the most economical and accurate way for specifying boundary conditions. For example, in the O-type grid the boundary of the cylinder is specified at $\eta = 1$ boundary, and $\xi = 1$ and $\xi = \xi_{\max}$ boundaries represent the same physical boundary (commonly referred to as a cut line). In the C-type grid the cylinder boundary is mapped into only a part of the ξ boundary, as shown in the Figure 26.2c. Here, the boundary segment in the front of the airfoil in the computational domain and at the tail of the airfoil represent the cut line

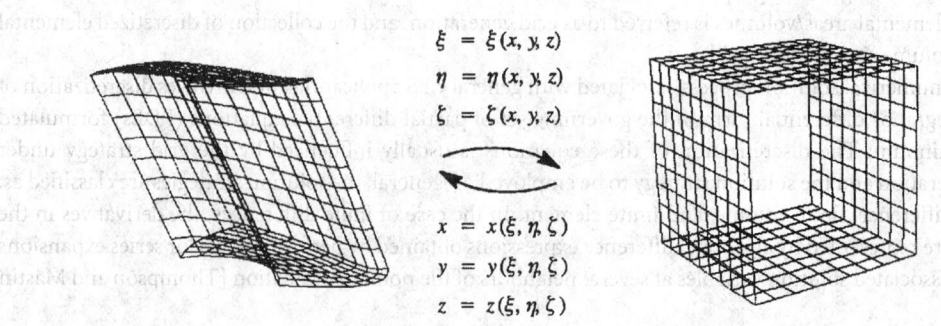

$$\xi = \xi(x, y, z)$$
$$\eta = \eta(x, y, z)$$
$$\zeta = \zeta(x, y, z)$$

$$x = x(\xi, \eta, \zeta)$$
$$y = y(\xi, \eta, \zeta)$$
$$z = z(\xi, \eta, \zeta)$$

FIGURE 26.1   Physical to computational space mapping.

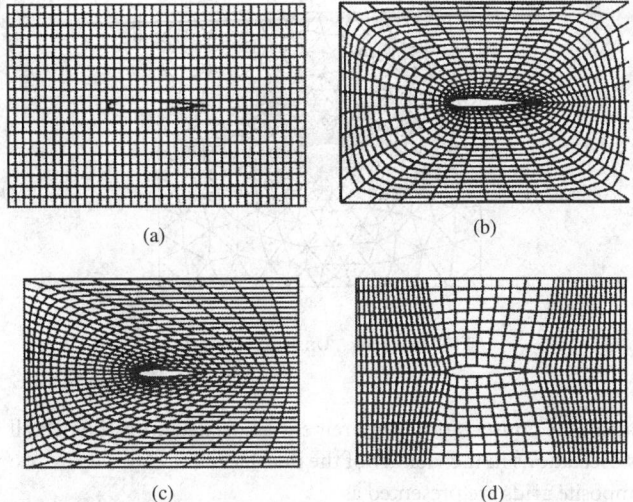

FIGURE 26.2 (a) Cartesian grid, (b) O-type grid, (c) C-type grid, and (d) H-type grid.

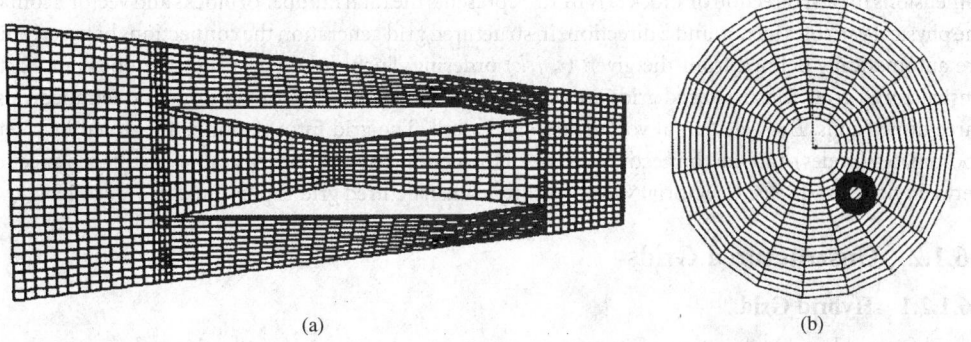

FIGURE 26.3 (a) Multiblock grid showing matching, nonmatching, and overlapping strategies and (b) overlaid chimera grid.

boundary. However, in the H-type grid an airfoil is mapped into the middle of the computational domain as shown in Figure 26.2d.

For complicated geometrical configurations, the physical region is divided into subregions, within each of which a structured grid is generated. These subgrids are commonly referred to as composite grids, and the generation process is referred to as domain decomposition [Thompson 1987b]. The resulting subgrids may be patched together at common interfaces, may be overlapped, or may be overlaid. Self-explanatory pictorial views of domain decomposition strategies allowing patched blocks, overlapping blocks, and overlaid blocks are presented in Figure 26.3a and Figure 26.3b. Overlaid grids are also called chimera grids after the composite monster of Greek mythology.

The use of composite grids allowing patched and overlapped blocks is very common in computational fluid dynamics where geometrical configurations representative of the physical space are extremely complex and/or distinct sets of underlying partial differential equations are to be simulated in different regions. An application of a chimera (overlaid) grid simplifies an overall grid generation and solution process when temporally deforming/moving geometrical components are involved in the simulation. The transfer of solution information at the block interface is very critical for successful simulation. In case of chimera and nonmatching blocks, the interface treatment involves interpolation, and these interpolated physical

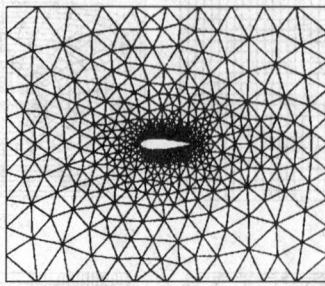

FIGURE 26.4   Unstructured grid.

variables may not satisfy the conservation requirement associated with the overall simulation process, resulting in spurious oscillations in the vicinity of the interface.

In general, the composite grids are presented as

$$\mathbf{r}_{ijkl}(i = 1, \ldots, I\text{MAX}(l), \quad j = 1, \ldots, J\text{MAX}(l), \quad k = 1, \ldots, K\text{MAX}(l), \quad l = 1, \ldots, N\text{BLKS})$$

where i, j, and k identify three curvilinear coordinates, $I\text{MAX}(l)$, $J\text{MAX}(l)$, $K\text{MAX}(l)$ denote the grid dimensions in each direction of block l. $N\text{BLKS}$ represents the total number of blocks and vector $\mathbf{r}$ contains the physical variables in x, y, and z direction. In structured grid generation the connections between points are automatically defined from the given (i, j, k) ordering. Such ordering (structure) does not exist in unstructured grids. Unstructured grids are composed of triangles in two dimensions and tetrahedrons in three dimensions. Each elemental volume is called a cell. The grid information is presented by a set of coordinates (nodes) and the connectivity between the nodes. The connectivity table specifies connections between nodes and cells. A pictorial view of a simple unstructured grid is presented in Figure 26.4.

26.1.2   Unstructured Grids

26.1.2.1   Hybrid Grid

A grid formed by a combination of structured–unstructured grids and/or allowing polygonal cells with different numbers of sides is called a hybrid or generalized grid. An usual practice is to generate structured grids near solid components up to desired distance and fill in remaining regions with unstructured (triangular/tetrahedron) grids. A typical hybrid grid is displayed in Figure 26.5.

26.1.3   Generation Process

Regardless of which grid strategy is being considered, creation of a computational grid requires the following:

1. *Computational mapping:* Establishing an appropriate mapping from physical to computational space allowing proper multiblock strategies in case of structured and hybrid grids or establishing an ordering of nodes in case of unstructured grids and hybrid grids.
2. *Geometry generation:* Defining an accurate numerical description of all solid components (surfaces) in conjunction with associated computational mapping criteria and a desired distribution of points.
3. *Computational modeling:* Generating an *appropriate* grid around these surfaces according to some criteria, usually with a specified multiblock strategy, point distribution, smoothness, and orthogonality in case of structured grids and desired background mesh representative of point distribution in case of unstructured grids.

The relationship of geometry to the grid generation process is analogous to the relationship between boundary conditions and the solution of the governing fluid flow equations. An accurate construction of

FIGURE 26.5 Hybrid grid.

the geometry with the proper distribution of points usually consumes 85 to 90 percent of the total time spent on the grid generation process. The geometry specification associated with grid generation involves the following:

1. Determine the desired distribution of grid points. This depends upon the expected flow characteristics.
2. Evaluate boundary segments and surface patches to be defined in order to resolve an accurate mathematical description of the geometry in question.
3. Select the geometry tools to be utilized to define these boundary segments/surface patches.
4. Follow an appropriate logical path to blend the aforementioned tasks obtaining the desired discretized mathematical description of the geometry with properly distributed points.

The accuracy of the numerical algorithm depends not only on the formal order of approximations but also on the distribution of grid points in the computational domain. The grid employed can have a profound influence on the quality and convergence rate of the solution. Grid adaption offers the use of excessively fine, computationally expensive, grids by directing the grid points to congregate so that a functional relationship on these points can represent the physical solution with sufficient accuracy.

Underlying principles and methodologies for grid generation and adaption follow.

26.2 Underlying Principles

26.2.1 Terminology and Grid Characteristics

The differencing and solution techniques involving Cartesian (regular) grids are well developed and well understood. The use of structured curvilinear grids (nonorthogonal in most cases) in the numerical solution of PDEs is not, in principle, any more difficult than using Cartesian grids. This is accomplished by transforming the pertinent PDEs from the physical space to computational space. The following notations will be utilized in the development of structured grids and these transformations; detailed mathematical analysis can be found in Thompson et al. [1985]:

$$
\begin{aligned}
\mathbf{r} &= (x, y, z) \equiv (x_1, x_2, x_3) & &= \text{physical space} \\
\mathbf{\Omega} &= (\xi, \eta, \zeta) \equiv (\xi^1, \xi^2, \xi^3) & &= \text{computational space} \\
\mathbf{a}_i &= \mathbf{r}_{\xi^i}, \ i = 1, 2, 3 & &= \text{covariant base vectors} \\
\mathbf{a}^i &= \nabla \xi^i, \ i = 1, 2, 3 & &= \text{contravariant base vectors} \\
g_{ij} &= \mathbf{a}_i \cdot \mathbf{a}_j \ (i = 1, 2, 3), \ (j = 1, 2, 3) & &= \text{covariant metric tensor components} \\
g^{ij} &= \mathbf{a}^i \cdot \mathbf{a}^j \ (i = 1, 2, 3), \ (j = 1, 2, 3) & &= \text{contravariant metric tensor components} \\
\sqrt{g} &= a_1 \cdot (a_2 \times a_3) & &= \text{Jacobian of transformation}
\end{aligned}
$$

Also it can be shown that for any tensor **u**

$$\nabla \cdot \mathbf{u} = \frac{1}{\sqrt{g}} \sum_{i=1}^{3} [(\mathbf{a}_j \cdot \mathbf{a}_k) \cdot \mathbf{u}]_{\xi^i}, \quad (i, j, k) \text{ cyclic} \tag{26.1}$$

and

$$\nabla \cdot \mathbf{u} = \frac{1}{\sqrt{g}} \sum_{i=1}^{3} (\mathbf{a}_j \cdot \mathbf{a}_k) \cdot \mathbf{u}_{\xi^i}, \quad (i, j, k) \text{ cyclic} \tag{26.2}$$

Although Equation 26.1 and Equation 26.2 are equivalent expressions, the numerical representations of these two forms may not be equivalent. The form given by Equation 26.1 is called conservative form and that of Equation 26.2 is called the nonconservative form. Equation 26.1 or Equation 26.2 is utilized for transforming derivatives from physical to computational space. With moving grids the time derivatives also must be transformed using the following equation:

$$\left(\frac{\partial \mathbf{u}}{\partial t} \right)_{\mathbf{r}} = \left(\frac{\partial \mathbf{u}}{\partial t} \right)_{\Omega} - \dot{\mathbf{r}} \cdot \nabla \mathbf{u} \tag{26.3}$$

where $\dot{\mathbf{r}}$ indicates the associated grid speed.

The discretization process associated with the finite-volume scheme employed in unstructured or hybrid grids is also very straightforward. Here, the integral equations are utilized. The edge-based data structure is used for connectivity information and can be easily utilized to compute areas of cells and associated fluxes. The area, for example, of a region bounded by a two-dimensional boundary $\partial\Omega$ is

$$A = \int_{\partial\Omega} x \, dy \tag{26.4}$$

which can be approximated to

$$A = \sum_{\text{edges}} x \Delta y \tag{26.5}$$

where x and Δy are interpreted as edge quantities. Also, the governing equations are discretized in similar fashion. For example, consider an integral form of the Navier-Stokes equation without body force as follows:

$$\frac{\partial}{\partial t} \int_{v} \mathbf{Q} \, dv + \oint_{\partial\Omega} \mathbf{F}(Q) \cdot \mathbf{n} \, ds = \oint_{\partial\Omega} \mathbf{F}^{v}(Q) \cdot \mathbf{n} \, ds \tag{26.6}$$

where **n** is the outward normal to the control volume with components n_x and n_y in the x and y directions. The domain of interest is discretized into a set of nonoverlapping polygons (unstructured or hybrid grid), and the cell-averaged variables are stored at the cell center. For each cell the semidiscretized form of the governing Equation 26.6 can be written as

$$V_i \frac{\partial Q}{\partial t} = -\sum_{j=1}^{k} F_{ij} \cdot n_j \, ds + \sum_{j=1}^{k} F_{ij}^{v} \cdot n_j \, ds \tag{26.7}$$

where j varies over the cell faces, and F_{ij} and F_{ij}^{v} are the convective and viscous part of the numerical flux at jth edge of cell i. Detailed analysis and description of this discretization process can be found in Weatherill [1990] and Koomullil et al. [1996b].

On structured grids the definition of an order of a difference representation is integrally tied to point distribution functions, commonly referred to as stretching functions. The order is determined by the error behavior as the spacing varies with the points fixed in a certain distribution, either by increasing the number of points or by changing a parameter in the distribution. Actually, a global order is not meaningful in case of nonuniform structured grids. The order is relevant only locally in regions where the spacing does in fact decrease as the point distribution changes. Looking at the truncation error analysis involving nonuniform structured curvilinear grids, the following grid requirements can be outlined: (1) The structured grid system must be either a right-handed system [$\sqrt{g} > 0$ for all (i, j, k)] or a left-handed system [$\sqrt{g} < 0$ for all (i, j, k)]. (2) Application of the distribution function with bounds on higher order derivatives does not change the formal order of approximation. The following functions involving exponential and hyperbolic tangent or hyperbolic sine stretching functions are widely utilized as distribution functions for distributing N points:

$$x(\xi) = \frac{e^{\alpha(s-1)} - 1}{e^{\alpha} - 1} \quad \text{or} \quad x(\xi) = 1 - \frac{\tanh(\alpha(1 - s))}{\tanh(\alpha)} \tag{26.8}$$

where

$$s = (\xi - 1)/(N - 1) \quad 1 \le \xi \le N.$$

(3) Numerical derivative evaluation of the distribution function is preferred, i.e., instead of analytical definition of x_ξ, where $x_\xi = [(x_{i+1} - x_{i-1})/2]$. (4) The truncation error is inversely proportional to the sine of the angle between grid lines. This in turn indicates that the mildly skewed grid does not increase truncation error significantly. In fact, as a rule of thumb, nonorthogonality resulting in an angle between grid lines not less than $45°$ does not increase truncation errors significantly.

26.2.2   Geometry Preparation

The geometry preparation is the most critical and labor intensive part of the overall grid generation process. Most of the geometrical configurations of interest to practical engineering problems are designed in the computer-aided design/computer-aided manufacturing (CAD/CAM) system. There are numerous geometry output formats which require the designer to spend a great deal of time manipulating geometrical entities in order to achieve a useful sculptured geometrical description with appropriate distribution for grid points. Also, there is a danger of loosing fidelity of the underlying geometry in this process. The desired point distribution on the boundary segment/surface patch is achieved by computing a concentration array of unit length. The concentration array is computed using specified spacing and by selecting an exponential or hyperbolic tangent stretching function. The geometry preparation involves the discrete-sculptured definitions of all outer boundaries/surfaces associated with the domain of interest. In case of unstructured or hybrid grids, the geometry preparation involves the definition of discrete points on the boundaries associated with the domain. The following definitions will be utilized in this development.

Definition 26.1   Given a set of points on a curve with physical Cartesian coordinates $(x_i, y_i, z_i), i = 1i, i1 + 1, \ldots, i2$. A number sequence $r = (r_1, r_2, \ldots, r_n)$, with $0 \le r_j \le 1, r_1 = 0, r_n = 1, r_i \le r_j$ for all $i < j$ represents the distribution of points, such that there exists a one-to-one correspondence between the element r_j of r and the triplet (x_j, y_j, z_j). This number sequence r is called a curve distribution. For example, the normalized chord length with $r_{i1} = 0$ and

$$r_i = \frac{\sum_{u=i1+1}^{i} \sqrt{(x_u - x_{u-1})^2 + (y_u - y_{u-1})^2 + (z_u - z_{u-1})^2}}{\sum_{u=i1+1}^{i2} \sqrt{(x_u - x_{u-1})^2 + (y_u - y_{u-1})^2 + (z_u - z_{u-1})^2}} \tag{26.9}$$

where $i = i1 + 1, \ldots, i2$ satisfies the definition of a curve distribution.

Definition 26.2   Given a set of points on a surface with physical Cartesian coordinates $(x_{ij}, y_{ij}, z_{ij}), i = 1i, i1 + 1, \ldots, i2; \ j = j1, j1 + 1, \ldots, j2$. A mesh $(s_{ij}, t_{ij}), \ i = 1i, i1 + 1, \ldots, i2; \ j = j1, j1 + 1, \ldots, j2$, is called a surface distribution mesh if $(s_{i1j}, \ldots, s_{i2j})$ for $j = j1, \ldots, j2$ represents the curve distribution for the curve $((x_{ij}, y_{ij}, z_{ij}), \ i = i1, i1 + 1, \ldots, i2)$ for $j = j1, j1, \ldots, j2)$ and $(t_{ij1}, \ldots, t_{ij2})$ for $i = i1, \ldots, i2$ represents the curve distribution for the curve $((x_{ij}, y_{ij}, z_{ij}), \ j = j1, j1 + 1, \ldots, j2)$, for $i = i1, i1 + 1, \ldots, i2)$.

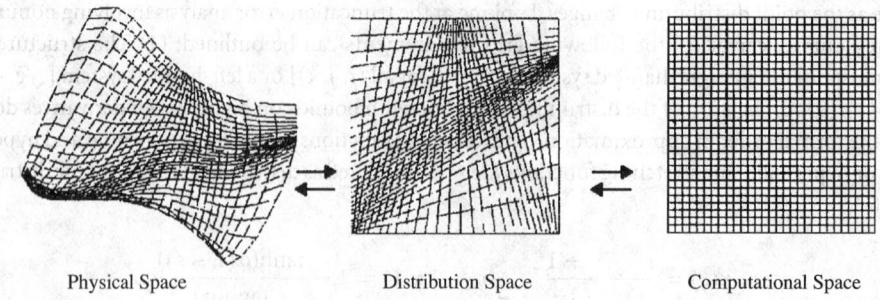

Physical Space   Distribution Space   Computational Space

FIGURE 26.6   Relationship between physical space, distribution mesh, and computational space.

Also, there exists a one-to-one correspondence between the physical domain and the surface distribution mesh and between the surface distribution mesh and the computational domain. These relations are demonstrated in Figure 26.6.

Definition 26.3   Let $(X_{ijk}, Y_{ijk}, Z_{ijk})$, $i = 1, 2, \ldots, N$; $j = 1, 2, \ldots, M$; $k = 1, 2, \ldots, L$ be a single block three-dimensional structured grid. A mesh $(s_{ijk}, t_{ijk}, q_{ijk})$, $i = 1, 2, \ldots, N$; $j = 1, 2, \ldots, M$; $k = 1, 2, \ldots, L$ is called a volume distribution mesh if $(\bar{s}_{ij\bar{k}}, \bar{t}_{ij\bar{k}})$ $i = 1, 2, \ldots, N$; $j = 1, 2, \ldots, M$ represents the surface distribution mesh for the surface $(\overline{X}_{ij\bar{k}}, Y_{ij\bar{k}}, Z_{ij\bar{k}})$, $\bar{k} = 1, 2, \ldots, L$, $(t_{\bar{i}jk}, q_{\bar{i}jk})$ represents the surface distribution mesh for the surface $(X_{\bar{i}jk}, \overline{Y}_{\bar{i}jk}, \overline{Z}_{\bar{i}jk})$, for all $\bar{i}$ and $(s_{i\bar{j}k}, q_{i\bar{j}k})$ represents the surface distribution mesh for the surface $(\overline{X}_{i\bar{j}k}, \overline{Y}_{i\bar{j}k}, \overline{Z}_{i\bar{j}k})$, for all $\bar{j}$.

26.2.3 Structured Grid Generation

26.2.3.1 Algebraic Generation Methods

An algebraic 3-D generation system based on transfinite interpolation (using either Lagrange or Hermite interpolation) is widely utilized for grid generation [Gordon and Thiel 1982, Soni 1992b]. The interpolation, in general complete transfinite interpolation from all boundaries, can be restricted to any combination of directions or lesser degrees of interpolation, and the form (Lagrange, Hermite, or incomplete Hermite) can be different in different directions or in different blocks. The blending functions can be linear or, more appropriately, based on the distribution surface/volume mesh. Hermite interpolation, based on cubic blending functions, allows orthogonality at the boundary. Incomplete Hermite uses quadratic functions and, hence, can give orthogonality at any one of two opposing boundaries, whereas Lagrange, with its linear functions, does not give orthogonality.

The transfinite interpolation is accomplished by the appropriate combination of 1-D projectors F for the type of interpolation specified. (Each projector is simply the 1-D interpolation in the direction indicated.) For interpolation from all sides of the section, if all three directions are indicated and the section is a volume, this interpolation is from all six sides, and the combination of projectors is the Boolean sum of the three projectors,

$$F_1 \oplus F_2 \oplus F_3 \equiv F_1 + F_2 + F_3 - F_1F_2 - F_2F_3 - F_3F_1 + F_1F_2F_3 \qquad (26.10)$$

With interpolation in only the two directions j and k, or if the section is a surface on which ξ^i is constant, the combination is the Boolean sum of F_j and F_k

$$F_j \oplus F_k \equiv F_j + F_k - F_jF_k, \quad (i, j, k) \text{ cyclic} \qquad (26.11)$$

With interpolation in only a single direction i, or if the section is a line on which ξ^i varies, the interpolation is between the two sides on which ξ^i is constant using only the single projector F_i.

Blocks can be divided into subblocks for the purpose of generation of the algebraic grid. Point distributions on the sides of the subblocks can either be specified or automatically generated by transfinite interpolation from the edge of the side. This allows additional control over the grid in general configurations and is particularly useful in cases where point distributions need to be specified in the interior of a block or to prevent grid overlap in highly curved regions. This also allows points in the interior of the field to be excluded if desired, e.g., to represent holes in the field.

26.2.3.2 Elliptic Generation Method

An elliptic grid generation system [Thompson 1987b] commonly used is

$$\sum_{i=1}^{3}\sum_{l=1}^{3} g^{il} r_{\xi^i\xi^l} + \sum_{l=1}^{3} g^{ll} P_l r_{\xi^l} = 0 \tag{26.12}$$

where the g^{il}, the elements of the contravariant metric tensor, can be evaluated as

$$g^{il} = \frac{1}{g}(g_{jm}g_{kn} - g_{jn}g_{km}), \quad (i = 1, 2, 3), \quad (l = 1, 2, 3) \tag{26.13}$$

$$(i, j, k) \text{ cyclic}, \quad (l, m, n) \text{ cyclic}$$

where g is the square of the Jacobian.

The P_n are the *control functions*, which serve to control the spacing and orientation of the grid lines in the field. The first and second coordinate derivatives are normally calculated using second-order central differences. One-sided differences dependent on the sign of the control function P_n (backward for $P_n < 0$ and forward for $P_n > 0$) are useful to enhance convergence with very strong control functions. The control functions are evaluated either directly from the initial algebraic grid and then smoothed or by interpolation from the boundary point distributions.

The three components of the elliptic grid generation system provide a set of three equations that can be solved simultaneously at each point for the three control functions $P_n(n = 1, 2, 3)$, with the derivatives here represented by central differences.

The elliptic generation system is solved by point successive over relaxation (SOR) iteration using a field of locally optimum acceleration parameters. These optimum parameters make the solution robust and capable of convergence with strong control functions.

Control functions can also be evaluated on the boundaries using the specified boundary point distribution in the generation system, with certain necessary assumptions (orthogonality at the boundary) to eliminate some terms, then they can be interpolated from the boundaries into the field. More general regions can, however, be treated by interpolating elements of the control functions separately. Thus, control functions on a line on which ξ^n varies can be expressed as

$$P_n = A_n + \frac{S_n}{\varrho_n} \tag{26.14}$$

where A_n is the logarithmic derivative of the arc length, S_n is the arc length spacing, and ϱ_n is the radius of curvature of the surface on which ξ^n is constant.

A second-order elliptic generation system allows either the point locations on the boundary or the coordinate line slope at the boundary to be specified, but not both. It is possible, however, to iteratively adjust the control functions in the generation system until not only a specified line slope, but also the spacing of the first coordinate surface off the boundary is achieved, with the point locations on the boundary specified. The extent of the orthogonality into the field can also be controlled. This orthogonality feature is also applicable on specified grid surfaces within the field, allowing grid surfaces in the field to be kept fixed while retaining continuity of slope of the grid line crossing the surface. This is quite useful in controlling the skewness of grid lines in some troublesome areas. Alternatively, boundary orthogonality can be achieved through Neumann boundary conditions, which allow the boundary points to move over a surface spline.

The boundary locations are located by Newton iteration on the spline to be at the foot of normals to the adjacent field points.

26.2.3.3 Hyperbolic Generation Method

It is also possible to base a grid generation system on hyperbolic PDEs rather than elliptic equations. In this case, the grid is generated by numerically solving a hyperbolic system [Steger and Chaussee 1980], marching in the direction of one curvilinear coordinate between two boundary curves in two dimensions or between two boundary surfaces in three dimensions. The hyperbolic system, however, allows only one boundary to be specified and is therefore of interest only for use in calculation on physically unbounded regions where the precise location of a computational outer boundary is not important. The hyperbolic grid generation system has the advantage of being generally faster than elliptic generation systems, but, as just noted, is applicable only to certain configurations. Hyperbolic generation systems can be used to generate orthogonal grids. In two dimensions, the condition of orthogonality is simply $g_{12} = 0$.

If either the cell area $\sqrt{g}$ or the cell diagonal length (squared) $g_{11} + g_{22}$ is a specified function of the curvilinear coordinates, i.e.,

$$\sqrt{g} = F(\xi, \eta) \qquad \text{or} \qquad g_{11} + g_{22} = F(\xi, \eta) \tag{26.15}$$

then the system consisting of $g_{12} = 0$ and either of the preceding two equations is hyperbolic. Since the system is hyperbolic, a noniterative marching solution can be constructed proceeding in one coordinate direction away from a specified boundary.

26.2.3.4 Multiblock Systems

Although in principle it is possible to establish a correspondence between any physical region and a single empty logically rectangular block for general 3-D configurations, the resulting grid is likely to be much too skewed and irregular to be usable when the boundary geometry is complicated. A better approach with complicated physical boundaries is to segment the physical region into contiguous subregions, each bounded by six curved sides (four in 2-D) and each of which transforms to a logically rectangular block in the computational region. Each subregion has its own curvilinear coordinate system, irrespective of that in the adjacent subregions (see Figure 26.7). This then allows both the grid generation and numerical solutions on the grid to be constructed to operate in a logically rectangular computational region, regardless of the shape or complexity of the full physical region. The full region is treated by performing the solution

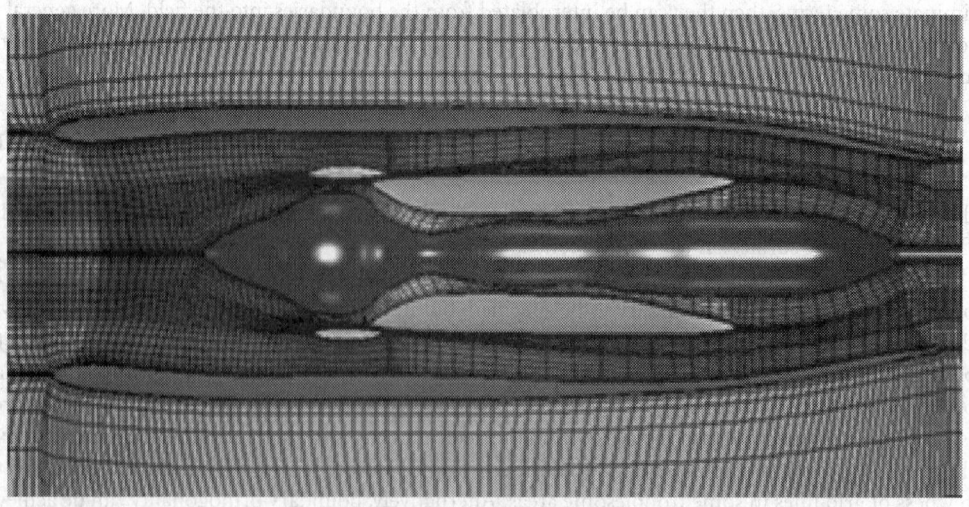

FIGURE 26.7   Multiblock grid.

operations in all of the logically rectangular computational blocks. With the composite framework, PDE solution procedures written to operate on logically rectangular regions can be incorporated into a code for general configurations in a straightforward manner, since the code only needs to treat a rectangular block. The entire physical field then can be treated in a loop over all the blocks. Transformation relations for PDEs are covered in detail in Thompson et al. [1985]. Discretization error related to the grid is covered in Thompson and Mastin [1983]. The evaluation and control of grid quality is an ongoing area of active research [Gatlin et al. 1991].

Grid lines at the interfaces may meet with complete continuity, with or without slope continuity, or may not meet at all. Complete continuity of grid lines across the interface requires that the interface [Thompson 1987a] be treated as a branch cut on which the generation system is solved just as it is in the interior of blocks. The interface locations are then not fixed, but are determined by the grid generation system. This is most easily handled in coding by providing an extra layer of points surrounding each block. Here, the grid points on an interface of one block are coincident in physical space with those of another interface of the same or another block, and also the grid points on the surrounding layer outside the first interface are coincident with those just inside the other interface, and vice versa. This coincidence can be maintained during the course of an iterative solution of an elliptic generation system by setting the values on the surrounding layers equal to those at the corresponding interior points after each iteration. All of the blocks are thus iterated to convergence, so that the entire composite grid is generated at once. The same procedure is followed by PDE solution codes on the block-structured grid.

26.2.3.5 Chimera Grids

The *chimera* (overlaid) grids [Belk 1995, Meakin 1991, Benek et al. 1985] are composed of completely independent component grids which may even overlap other component boundary elements, creating holes in the component grids. This requires flagging procedures to locate grid points that lie out of the field of computation, but such holes can be handled even in tridiagonal solvers by placing 1s at the corresponding positions on the matrix diagonal and all 0s off the diagonal. These overlaid grids also require interpolation to transfer data between grids, and that subject is the principal focus of effort with regard to the use of this type of composite grid.

26.2.3.6 Adaptive Grid Generation

With structured grids, the adaptive strategy based on redistribution is by far the most simple to implement, requiring only the regeneration of the grid and interpolation of flow properties at the new grid points at each adaptive stage without modification of the flow solver unless time accuracy is desired. Time accuracy can be achieved, as far as the grid is concerned, by simply transforming the time derivatives, thus adding convectivelike terms that do not alter the basic conservation form of the PDEs.

Adaptive redistribution of points traces its roots to the principle of equidistribution of error [Brackbill 1993, Soni et al. 1993] by which a point distribution is set so as to make the product of the spacing and a weight function constant over the points

$$w \, \Delta x = \text{const}$$

With the point distribution defined by a function ξ_i, where ξ varies by a unit increment between points, the equidistribution principle can be expressed as

$$w x_\xi = \text{const} \tag{26.16}$$

This one-dimensional equation can be applied in each direction in an alternating fashion. A direct extension to multiple dimensions using algebraic, variational, and elliptic systems has been developed.

The weight function is usually formulated by utilizing scaled gradients and curvatures of the solution variables considered for adaption.

The control of the characteristics and distribution of a grid system can be achieved by varying the values of the control functions P_l in Equation 26.12. The application of the one-dimensional form of

Equation 26.12 with Equation 26.16 results in the definition of control functions in three dimensions,

$$P_i = \frac{W_{\xi^i}}{W} \quad (i = 1, 2, 3) \tag{26.17}$$

These control functions were generalized by Eiseman [1985] as

$$P_i = \sum_{j=1}^{3} \frac{g^{ij}}{g^{ii}} \frac{(W_i)_{\xi^i}}{W_i} \quad (i = 1, 2, 3) \tag{26.18}$$

In order to conserve the geometrical characteristics of the existing grid the definition of control functions is extended as

$$P_i = (P_{\text{initial geometry}})_i + c_i(P_{wt}) \quad (i = 1, 2, 3) \tag{26.19}$$

where $(P_{\text{initial geometry}})$ is the control function based on initial grid geometry, P_{wt} is the control function based on gradient of flow parameter, and c_i is the constant weight factors.

These control functions are evaluated based on the current grid at the adaptation step. This can be formulated as

$$P_i^{(n)} = P_i^{(n-1)} + c_i(P_{wt})^{(n-1)} \quad (i = 1, 2, 3) \tag{26.20}$$

where

$$P_i^{(1)} = (P_{\text{initial geometry}})_i^{(0)} + c_i(P_{wt})^{(0)} \quad (i = 1, 2, 3) \tag{26.21}$$

26.2.4 Unstructured Grid Generation

26.2.4.1 The Delaunay Triangulation

Dirichlet, in 1850, first proposed a method whereby a domain could be systematically decomposed into a set of packed convex polyhedra. For a given set of points in space, $\{P_k\}, k = 1, \ldots, K$, the regions $\{V_k\}, k = 1, \ldots, K$, are the territories which can be assigned to each point P_k, such that V_k represents the space closer to P_k than to any other point in the set. Clearly, these regions satisfy

$$V_k = \{P_i : |p - P_i| < |p - P_j|\} \forall j \neq i \tag{26.22}$$

This geometrical construction of tiles is known as the Dirichlet tessellation or Voronoi [1908] diagram. This tessellation of a closed domain results in a set of nonoverlapping convex polyhedra, called Voronoi regions, covering the entire domain. If all point pairs which have some segment of a Voronoi boundary in common are joined, the result is a triangulation of the convex hull of the set of points $\{P_k\}$. This triangulation is known as the *Delaunay triangulation* [Baker 1990, George and Hermeline 1992]. The definition is valid for n-dimensional space.

From the preceding discussion, it is apparent that in two dimensions a line segment of the Voronoi diagram is equidistant from the two points it separates. Hence, the vertices of the Voronoi diagram must be equidistant from each of the three nodes which form the Delaunay triangles. Clearly, it is possible to construct a circle, centered at a Voronoi vertex, which passes through the three points, which form a triangle. Furthermore, it is evident that, given the definition of Voronoi line segments and regions, no circle can contain any point. This latter condition is referred to as the in-circle criterion.

26.2.4.2 Advancing Front Procedure

The advancing front procedure is based on the method originally proposed in Peraire et al. [1987] for two dimensions and then extended to three dimensions in [Peraire et al. 1988, 1990]. The advocated approach is regarded as a generalization of the advancing front technique [George 1971, Lo 1985] with the distinctive feature that elements, i.e., triangles or tetrahedra, and points are generated simultaneously. This enables the generation of elements of variable size and stretching and differs from the approach followed in tetrahedral generators which are based on Delaunay concepts [Baker 1990, Cavendish et al. 1985], which generally connect grid points which have already been distributed in space.

The generation problem consists of subdividing an arbitrarily complex domain into a consistent assembly of elements. The consistency of the generated mesh is guaranteed if the generated elements cover the entire domain and the intersection between elements occurs only on common points, sides, or triangular faces in the three-dimensional case. The final mesh is constructed in a bottom-up manner. The process starts by discretising each boundary curve. Nodes are placed on the boundary curve components and then contiguous nodes are joined with straight line segments. In later stages of the generation process, these segments will become sides of some triangles. The length of these segments must therefore, be consistent with the desired local distribution of mesh size. This operation is repeated for each boundary curve in turn.

The next stage consists of generating triangular planar faces. For each two-dimensional region or surface to be discretized, all of the edges produced when discretizing its boundary curves assembled into the so-called initial front. The relative orientation of the curve components with respect to the surface must be taken into account in order to give the correct orientation to the sides in the initial front. The front is a dynamic data structure which changes continuously during the generation process. At any given time, the front contains the set of all of the sides which are currently available to form a triangular face. A side is selected from the front and a triangular element is generated. This may involve creating a new node or simply connecting to an existing one. After the triangle has been generated, the front is updated and the generation proceeds until the front is empty. The size and shape of the generated triangles must be consistent with the local desired size and shape of the final mesh. In the three-dimensional case, these triangles will become faces of the tetrahedra to be generated later.

The geometrical characteristics of a general mesh are locally defined in terms of certain mesh parameters. If $N = (2 \text{ or } 3)$ is the number of dimensions, then the parameters used are a set of N mutually orthogonal directions $\boldsymbol{\alpha}_i, i = 1, \ldots, N$, and N associated element sizes $\delta_i, i = 1, \ldots, N$. Thus, at a certain point, if all N element sizes are equal, the mesh in the vicinity of that point will consist of approximately equilateral elements. To aid the mesh generation procedure, a transformation T which is a function of $\boldsymbol{\alpha}_i$ and δ_i is defined. This transformation is represented by a symmetric $N \times N$ matrix and maps the physical space onto a space in which elements, in the neighborhood of the point being considered, will be approximately equilateral with unit average size. This new space will be referred to as the normalized space. For a general mesh this transformation will be a function of position. The transformation T is the result of superimposing N scaling operations with factors $1/\delta_i$ in each $\boldsymbol{\alpha}_i$ direction. Thus,

$$T(\boldsymbol{\alpha}_i, \delta_i) = \sum_{i=1}^{N} \frac{1}{\delta_i} \boldsymbol{\alpha}_i \otimes \boldsymbol{\alpha}_i \tag{26.23}$$

where $\otimes$ denotes the tensor product of two vectors.

26.2.4.3 Grid Adaption Methods

For the solution adaptive grid generation procedure, an error indicator is required that detects and locates appropriate features in the flowfield. In order to provide flexibility in isolating varying features, multiple error indicators are used. Each can isolate a particular type of feature. The error indicators are usually set to the negative and positive components of the gradient in the direction of the velocity vector as given by

$$e_1 = \min[\mathbf{V} \cdot \nabla(\mathbf{u}), 0]$$
$$e_2 = \max[\mathbf{V} \cdot \nabla(\mathbf{u}), 0] \tag{26.24}$$

and the magnitude of the gradient in all directions normal to the velocity vector as given by

$$e_3 = |\nabla \mathbf{u} - \mathbf{V}(\mathbf{V} \cdot \nabla \mathbf{u})/\mathbf{V} \cdot \mathbf{V}| \tag{26.25}$$

Where $\mathbf{V}$ is the velocity vector and $\mathbf{u}$ is any suitable flow property. Typically, density is used as the basis for the error indicator. The first two error indicators represent expansions and compressions in the flow

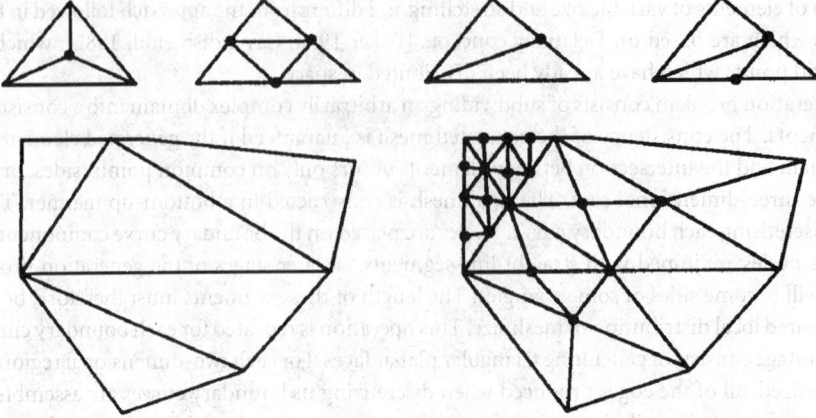

FIGURE 26.8   Types of h-refinement in two dimensions.

direction and the third represents gradients normal to the flow direction. The indicators can be scaled by the relative element size. Length scaling can improve detection of weak features on a coarse grid with the present procedure. Each error indicator is treated independently, allowing particular features in the flowfield to be isolated. For each error indicator, an error is determined from

$$e_{\lim} = e_m + c_{\lim} \cdot e_s \tag{26.26}$$

where $e_{\lim}$ is the error limit, e_m is the mean of the error indicator, e_s is the standard deviation of the error indicator, and $c_{\lim}$ is a constant. Typically a value near 1 is used for the constant. The error indicators are used to control the local reduction in relative element size during grid generation.

One of the advantages of an unstructured grid is that it provides a natural environment for grid adaptation using h-refinement or mesh enrichment. Points can be added to the mesh with the consequence that new elements are formed and only local modifications to the connectivity matrix need to be made. In addition, no modification or special treatments are required within the solution algorithm provided that on enriching the mesh the distribution of elements and points remains smooth. Once the regions for enrichment are determined and individual elements are identified there are a number of strategies for adding points. The most suitable methods attempt to ensure smoothness of the enriched mesh, and in this respect local refinement strategies can prove to be useful. Some examples of point enrichment are given in Figure 26.8.

In addition to h-refinement, node movement has been found to be necessary for an efficient implementation of grid adaptation. Node movement can be applied in the form

$$\mathbf{r}_0^{n+1} = \mathbf{r}_0^n + \omega_i \frac{\sum_{i=1}^{M} C_{i0} \left(\mathbf{r}_i^n - \mathbf{r}_0^{ni} \right)}{\sum_{i=1}^{M} c_{i0}} \tag{26.27}$$

where $\mathbf{r} = (x, y)$, $\mathbf{r}_0^{n+1}$ is the position of node 0 at relaxation level $n+1$, C_{i0} is the adaptive weight function between nodes i and 0, and ω is the relaxation parameter. An adaptive weight function C_{i0} is used which takes the form

$$C_{i0} = k_1 + k_2 \left| \frac{\phi_i - \phi_0}{\phi_i + \phi_0} \right| \tag{26.28}$$

where ϕ is the driving variable e.g., pressure, density, Mach number, etc.; k_1 and k_2 are constants, k_1 acts to damp out noise; and k_2 amplifies the gradients along edges. In practice, this is implemented in a form which guarantees positive area cells after movement, even in regions close to a wall, which for viscous grids can have very small volumes.

26.3 Best Practices

In the last few years, numerical grid generation has evolved as an essential tool in obtaining the numerical solution of the partial differential equations of fluid mechanics. A multitude of techniques and computer codes have been developed to support multiblock structured and unstructured grid generation associated with complex configurations. Structured grid generation methodologies can be grouped in two main categories: direct methods, where algebraic interpolation techniques are utilized, and indirect methods, where a set of partial differential equations is solved. Both of these techniques are utilized either separately or in combination, to efficiently generate grids in the aforementioned codes.

In algebraic methods the most widely used technique is transfinite interpolation. Historically, application of algebraic methods in grid generation has progressed as follows. In the 1970s (and early 1980s) the algebraic methods based on Lagrange and Hermite (mostly cubic) interpolation methods in tensor product form or Coon's patching, commonly referred to as transfinite interpolation technique form, and parametric spline (mostly cubic natural splines) were utilized to construct an initial grid for the iterative grid evaluation associated with a set of partial differential equations (mostly elliptic equations). In the 1980s (and 1990–1991), the development of high-powered graphics workstations along with the application of Bezier B-spline curve/surface definition in a control point form revolutionized the grid generation process with graphically interactive generation strategies and grid quality (smoothness-orthogonality with precise distribution improvements) with fast and efficient parametric curve/surface description based on basis splines. The parametric based nonuniform rational B-spline (NURBS) is a widely utilized representation for geometrical entities in computer-aided geometric design (CAGD) and CAD/CAM systems [Yu 1995]. The convex hull, local support, shape preserving forms, and variation diminishing properties of NURBS are extremely attractive in engineering design applications and in geometry-grid generation. In fact, the NURBS representation is becoming the de facto standard for the geometry description in most of the modern grid generation systems. Recently, the research concentration in algebraic grid generation is placed on utilizing CAGD techniques for efficient and accurate geometric modeling (boundary/surface grid generation). The development of NASA initial graphics exchange specification (IGES) standard [Blake and Chou 1992]; NURBS data structure in the National Grid Project [Thompson 1993]; DTNURBS library and its implementation in various grid systems; the grid systems NGP, ICEM, GRIDGEN, IGG, GENIE++; and computer aided grid interface (CAGI) system is just a partial list of the outcome of this research concentration.

The best practice in grid generation is to transform all geometrical entities associated with the complex configuration under consideration into parametric control points based on NURBS representation allowing standard data structure. The grid generation algorithms are then tailored to exhibit NURBS representation in the generation process. An overall generation process is usually based on utilizing the best features of direct and indirect methods in case of structured grids and Delaunay triangulation and advancing front methods in case of unstructured grids.

26.3.1 Structured Grid Generation

The following observations and evaluations on the structured grid generation methodologies are important to consider before the development of overall grid generation process: algebraic systems are fast and economical; precise spacing control (a well-distributed grid) is always achieved with algebraic systems; grid generation by elliptic systems is always smooth; algebraic systems may cause grids to overlap; however, elliptic systems resist grid line overlapping; the control functions can be formulated to achieve boundary orthogonality and spacing control (near solid boundary surfaces) by elliptic generation systems; the control functions can be formulated to accomplish field orthogonality in a given computational direction (ξ, ς, or η) and spacing control by elliptic generation systems by iteratively updating various terms in the generation system. This is very time consuming especially in three-dimensional problems. Algebraic systems require a high degree of understanding and visual user interaction. However, elliptic systems can be readily adaptable for generalization. This is extremely useful in grid adaptations. The hyperbolic systems preserve the orthogonality at the solid boundary and the point distribution in the field. However, their applicability is restricted to external flows where the accurate geometrical shape of the outer boundaries/surfaces is not important as long as their location is a certain distance away from the body. Also in three-dimensional applications of hyperbolic systems the grid quality is directly influenced by the characteristics of the surfaces associated with the computational domain.

26.3.1.1 Transfinite Interpolation Method

In general, the algebraic methods are based on utilizing tensor product form of interpolation (in case of surface generation) and transfinite interpolation (in case of 2-D or full 3-D volume grid generation). Define a one-dimensional interpolation projector as follows:

$$T[r(s)] = \sum_{k=0}^{q} \sum_{j=0}^{p} \phi_{j,k}(s) r_j^{(k)} \qquad (26.29)$$

where the parameter s is such that $0 \leq s \leq 1$, $\phi_{j,k}$ are the blending functions, and $r_j(k)$ is the kth derivative of the variable r at parametric location $s_j (0 = s_0 \leq s_1 \leq s_2 \cdots \leq s_p = 1)$. The following example clarifies Equation 26.29.

Example 26.1

Let

$$q = 0, \qquad p = 1$$
$$\phi_{0,0}(s) = (1-s), \phi_{1,0}(s) = s$$

then

$$T[r(s)] = \phi_{0,0}(s) r_0^0 + \phi_{1,0}(s) r_1^0 = (1-s) r_0^0 + s r_1^0$$

defines a straight line between two points r_0^0 and r_1^0.

Example 26.2

Let

$$q = 1, \qquad p = 1$$
$$\phi_{0,0}(s) = 2s^3 - 3s^2 + 1$$
$$\phi_{1,0}(s) = -2s^3 + 3s^2$$
$$\phi_{0,1}(s) = s^3 - 2s^2 + 1 \qquad (26.30)$$
$$\phi_{1,1}(s) = s^3 - s^2$$
$$T[r(s)] = \phi_{0,0}(s) r_0^0 + \phi_{1,0}(s) r_1^0 + \phi_{0,1}(s) r_0^1 + \phi_{1,1}(s) r_1^1$$

defines a Hermite cubic polynomial between two points r_0^0 and r_1^0 with specified slopes r_0^1 and r_1^1 at the respective end points. The linear interpolation and Hermite cubic interpolation are widely applied in grid generation.

For 2-D grid generation the transfinite interpolation (TFI) method is defined as follows:

$$T[r(s_{ij}, t_{ij})] = T_I[r(s_{ij}, t_{ij})] \oplus T_J[r(s_{ij}, t_{ij})]$$

where T_I is a one-dimensional interpolation projector applied in the $i(\xi)$ direction keeping t_{ij} fixed and T_J is the one-dimensional interpolation projector applied in the $j(\eta)$ direction keeping s_{ij} fixed. $T[r] = (T_I + T_J - T_I T_J)[r]$, and (s_{ij}, t_{ij}) is the distribution mesh for the 2-D grid configuration under consideration. The $T_I T_J$ represents the tensor product of interpolation in both the I and J directions.

The transfinite interpolation method for 3-D grid generation can be defined as

$$T[r(s_{ijk}, t_{ijk}, q_{ijk})] = T_I[r(s_{ijk}, t_{ijk}, q_{ijk})] \oplus T_J[r(s_{ijk}, t_{ijk}, q_{ijk})] \oplus T_K[r(s_{ijk}, t_{ijk}, q_{ijk})] \quad (26.31)$$

where T_I is a one-dimensional interpolation projector applied in the $t(\xi)$ direction keeping t_{ijk} and q_{ijk} fixed, T_J and T_K are similarly defined. Here $(s_{ijk}, t_{ijk}, q_{ijk})$ represents volume distribution mesh associated with 3-D grid generation.

For example, if (s_{ij}, t_{ij}) represents an $N \times M$ size distribution mesh, $((X_{i1}, Y_{i1}), (X_{iM}, Y_{iM}))$, $i = 1, 2, \ldots, N$; and $((X_{1j}, Y_{1j}), (X_{Nj}, Y_{Nj}))$, $j = 1, 2 \ldots, M$ boundaries are known, T_I is selected as a linear interpolation projector and T_J is selected as a Hermite interpolation projector and then

$$T_I[r(s_{ij}, t_{ij})] = (1 - s_{ij})r_{1j} + s_{ij}r_{Nj} \quad (26.32)$$

and

$$T_J[r(s_{ij}, t_{ij})] = \Phi_{00}(t_{ij})r_{i1} + \Phi_{1,0}(t_{ij})r_{iM} + \Phi_{0,1}(t_{ij})\frac{\partial r}{\partial \eta}\bigg|_{r_{i1}} + \Phi_{1,1}(t_{ij})\frac{\partial r}{\partial \eta}\bigg|_{r_{iM}} \quad (26.33)$$

and the respective 2-D grid can be evaluated as

$$r_{ij} = T_I + T_J - T_I T_J$$

where

$$T_I T_J[r(s_{ij}, t_{ij})] = (1 - s_{ij}) \left[\Phi_{0,0}(t_{ij})\gamma_{11} + \Phi_{1,0}(t_{ij})\gamma_{1M} + \Phi_{0,1}(t_{ij})\frac{\partial r}{\partial \eta}\bigg|_{r_{11}} + \Phi_{1,1}(t_{ij})\frac{\partial r}{\partial \eta}\bigg|_{1M} \right]$$

$$+ \left[s_{ij} \Phi_{0,0}(t_{ij})r_{N1} + \Phi_{1,0}(t_{ij})r_{NM} + \Phi_{0,0}(t_{ij})\frac{\partial r}{\partial \eta}\bigg|_{r_{NM}} + \Phi_{1,1}(t_{ij})\frac{\partial r}{\partial \eta}\bigg|_{NM} \right] \quad (26.34)$$

An important factor in applying the Hermite interpolation projectors in TFI formulation is the evaluation of slopes and twist vectors (cross derivatives). The slope vector $\bar{r}_\xi$ can be evaluated by solving

$$\mathbf{r}_\xi \cdot \mathbf{r}_\eta = (\sqrt{g_{11}g_{22}}) \cos(\theta_1)$$
$$\|\mathbf{r}_\xi \times \mathbf{r}_\zeta\| = A \quad (26.35)$$

where θ_1 is the desired angle between grid lines ξ and η, and A is the desired area of the cell. The metric terms g_{11} and g_{22} can be evaluated using the desired change in arc length or from the appropriate algebraic grid (precise spacing control property of the algebraic grid can be exploited here). The system (26.35) can be uniquely solved to evaluate $\mathbf{r}_\xi$ and $\mathbf{r}_\eta$, and $\mathbf{r}_\zeta$ can be evaluated similarly. The twist vectors $\mathbf{r}_{\xi\eta}$ and $\mathbf{r}_{\xi\xi}$ can be evaluated by solving

$$(\mathbf{r}_\xi \cdot \mathbf{r}_\eta)_\eta = [(\sqrt{g_{11}g_{22}}) \cos(\theta_1)]_\eta$$
$$\|\mathbf{r}_\xi \times \mathbf{r}_\eta\|_\eta = (A)_\eta \quad (26.36)$$

and

$$(\mathbf{r}_\xi \cdot \mathbf{r}_\eta)_\xi = [(\sqrt{g_{11}g_{22}}) \cos(\theta_1)]_\xi$$
$$\|\mathbf{r}_\xi \times \mathbf{r}_\xi\|_\eta = (A)_\xi \quad (26.37)$$

respectively. The other cross derivatives can be evaluated similarly. Observe that if orthogonality is desired, that is, $\theta_1 = 90°$, then the right-hand side will have zeros except for A, A_ξ, and A_η where A_ξ and A_η represent the change in desired volumes in all areas in the ξ and η directions. This concept can be easily extended to three-dimensional configurations.

26.3.1.2   Elliptic Grid Generation

A multitude of general purpose elliptic generation systems here appeared [Thompson 1987b]. Most of these algorithms are based on an iterative adjustment of control functions to achieve boundary orthogonality. The following analysis is provided to illustrate this development.

Consider

$$(g_{ij})_{\xi^k} \equiv (\text{derivative of } g_{ij} \text{ with respect } \xi^k)$$
$$\equiv \mathbf{r}_{\xi^i \xi^k} \cdot \mathbf{r}_{\xi^j} + \mathbf{r}_{\xi^i} \cdot \mathbf{r}_{\xi^j \xi^k} \tag{26.38}$$
$$i = 1, 2, 3; \qquad j = 1, 2, 3; \qquad \text{and} \qquad k = 1, 2, 3$$

Using Equation 26.12, the following statement can be obtained:

$$\mathbf{r}_{\xi^i \xi^j} \cdot \mathbf{r}_{\xi^k} = \frac{(g_{ik})_{\xi^j} - (g_{ij})_{\xi^k} + (g_{jk})_{\xi^i}}{2}$$
$$i = 1, 2, 3; \qquad j = 1, 2, 3; \qquad k = 1, 2, 3$$

The three-dimensional elliptic grid generation system presented in Eq. (26.12) can be rewritten by taking the dot product with r_{ξ^q}, $q = 1, 2, 3$ as

$$\sum_{i=1}^{3} \sum_{j=1}^{3} g^{ij} \mathbf{r}_{\xi^i \xi^j} \cdot \mathbf{r}_{\xi^q} + \sum_{k=1}^{3} \Phi_k g^{kk} \mathbf{r}_{\xi^k} \cdot \mathbf{r}_{\xi^q} = 0 \tag{26.39}$$

This can be written in terms of metric terms and their derivatives as

$$\sum_{i=1}^{3} \sum_{j=1}^{3} g^{ij} \frac{((g_{iq})_{\xi^j} - (g_{ij})_{\xi^q} + (g_{jq})_{\xi^i})}{2} + \sum_{k=1}^{3} \Phi_k g^{kk} g_{kq} = 0 \quad q = 1, 2, 3 \tag{26.40}$$

Now

$$g_{ii} = \bar{\mathbf{r}}_{\xi^i} \cdot \bar{\mathbf{r}}_{\xi^i} = \|\bar{\mathbf{r}}_{\xi^i}\|^2$$

represents an increment of arc length on a coordinate line along which ξ^i varies and

$$g_{ij} = \bar{\mathbf{r}}_{\xi^i} \cdot \bar{\mathbf{r}}_{\xi^j} = |\bar{\mathbf{r}}_{\xi^i}| \cdot |\bar{\mathbf{r}}_{\xi^j}| \cdot \cos\theta \quad i \neq j$$

represents a measure of orthogonality between grid lines along which ξ^i and ξ^j varies. These quantities can be evaluated if the desired increment in the arc length and desired angles between grid lines are known. Looking at the precise control of spacing property of the algebraic grid [Soni 1992a, b], the quantities g_{ij} can be evaluated from the well-defined algebraic grid, and using

$$g_{ij} = (\sqrt{g_{ii}})(\sqrt{g_{jj}}) \cos\theta \tag{26.41}$$

where θ is the desired angle between ξ^i, ξ^j grid lines, the quantities g_{ij}, $i \neq j$, can be evaluated. Once all g_{ij} are known, then Eq. (26.40) can be solved for the forcing functions Φ_k, $k = 1, 2, 3$.

If orthogonality is enforced, i.e., $\theta = 90°$ or $g_{ij} = 0$ for $i \neq j$, then Φ_k, $k = 1, 2, 3$, can be formulated as

$$\hat{\Phi}_k = \frac{1}{2} \frac{d}{d\xi^k} \left(\ell n \frac{g_{kk}}{g_{ii} g_{jj}} \right), \qquad (i, j, k) \text{ cyclic} \quad k = 1, 2, 3 \tag{26.42}$$

and

$$\Phi_k = -\frac{(g_{ii})(g_{jj}) \hat{\Phi}_k}{g}, \qquad (i, j, k) \text{ cyclic} \quad k = 1, 2, 3 \tag{26.43}$$

A usual practice is to utilize Equation 26.43 in the following form:

$$\Phi_k = -\frac{\mathbf{r}_{\xi^k} \cdot \mathbf{r}_{\xi^k\xi^k}}{|\mathbf{r}_{\xi^k}|^2} \frac{\mathbf{r}_{\xi^k} \cdot \mathbf{r}_{\xi^i\xi^i}}{|\mathbf{r}_{\xi^i}|^2} \frac{\mathbf{r}_{\xi^k} \cdot \mathbf{r}_{\xi^j\xi^j}}{|\mathbf{r}_{\xi^j}|^2} \tag{26.44}$$

In fact, the firm term of the definition of Φ_k provides the distribution control and the remaining two terms contribute toward the curvature control.

To understand the iterative evaluation of the control function, consider a two-dimensional elliptic system:

$$g_{22}(\mathbf{r}_{\xi\xi} - \phi\mathbf{r}_{\xi}) - 2g_{12}\mathbf{r}_{\xi\eta} + g_{11}(\mathbf{r}_{\eta\eta} - \psi\mathbf{r}_{\eta}) = 0 \tag{26.45}$$

The control functions ϕ and ψ can be formulated as

$$\Phi = -\frac{\mathbf{r}_{\xi} \cdot \mathbf{r}_{\xi\xi}}{|\mathbf{r}_{\xi}|^2} - \frac{\mathbf{r}_{\xi} \cdot \mathbf{r}_{\eta\eta}}{|\mathbf{r}_{\eta}|^2} \tag{26.46}$$

and

$$\Psi = -\frac{\mathbf{r}_{\eta} \cdot \mathbf{r}_{\eta\eta}}{|\mathbf{r}_{\eta}|^2} - \frac{\mathbf{r}_{\eta} \cdot \mathbf{r}_{\xi\xi}}{|\mathbf{r}_{\xi}|^2} \tag{26.47}$$

During the evaluation of ϕ: (1) the quantities $\mathbf{r}_{\xi}$ and $\mathbf{r}_{\xi\xi}$ can be evaluated by utilizing appropriate finite-difference approximations, (2) the $\mathbf{r}_{\eta}$ are evaluated by solving

$$\mathbf{r}_{\xi} \cdot \mathbf{r}_{\eta} = 0 \quad \text{and} \quad \|\mathbf{r}_{\xi} \times \mathbf{r}_{\eta}\| = (\Delta A) \tag{26.48}$$

where (ΔA) is the desired cell area, and (3) the $\mathbf{r}_{\eta\eta}$ quantities are calculated using the finite-difference approximation on the current grid. These quantities are updated at every iteration. Another approach is to utilize well-distributed algebraic grid characteristics to solve the following equations in order to evaluate $\mathbf{r}_{\eta\eta}$:

$$(\mathbf{r}_{\xi} \cdot \mathbf{r}_{\eta})_{\xi} = 0 \quad \text{and} \quad \|\mathbf{r}_{\xi} \times \mathbf{r}_{\eta}\|_{\xi} = (\Delta A)_{\xi} \tag{26.49}$$

and

$$(\mathbf{r}_{\xi} \cdot \mathbf{r}_{\eta})_{\eta} = 0 \quad \text{and} \quad \|\mathbf{r}_{\xi} \times \mathbf{r}_{\eta}\|_{\eta} = (\Delta A)_{\eta} \tag{26.50}$$

where $(\Delta A)_{\xi}$ and $(\Delta A)_{\eta}$ represent the change of cell area in the ξ and η directions, respectively (they can be computed using finite-difference approximation from a well-distributed algebraic grid or by utilizing desired cell areas on the boundaries). The control functions are usually evaluated on the boundaries and then interpolated in the interior. The distribution mesh can be utilized as a parametric space for doing this interpolation.

26.3.2 The Delaunay Algorithm

The algorithm widely utilized to generate the Delaunay triangulation is based upon the work of Bowyer. The algorithm is based on the in-circle criterion, and is a sequential process with each point introduced into an existing Delaunay satisfying structure, which is broken and then reconnected to form a new Delaunay triangulation [Baker 1990, George and Hermeline 1992]. The algorithm is applicable in two- and three-dimensions and in step-by-step format as follows:

Algorithm I.

1. *Define a set of points which forms a convex hull within which all points will lie.*
2. *Introduce a new point anywhere within the convex hull.*
3. *Determine all vertices of the Voronoi diagram to be deleted. A point which lies within a circle (sphere), centered at a vertex of the Voronoi diagram and which passes through its three (four) forming points, results in the deletion of that vertex. This follows from the in-circle criterion of the Voronoi construction.*

4. *Find the forming points of all the deleted Voronoi vertices. These are the contiguous points to the new point.*

5. *Determine the neighboring Voronoi vertices to the deleted vertices which have not themselves been deleted.*

6. *Determine the forming points of the new Voronoi vertices. The forming points of new vertices must include the new point together with two (three) points which are contiguous to the new point and form an edge (face) of a neighboring Voronoi diagram data structure, overwriting the entries of the deleted vertices.*

7. *Repeat steps (2–6) for the next point.*

In the preceding algorithm, the interpretation for three dimensions is included in parentheses. This algorithm has been used for the construction of the triangulation in two and three dimensions. It does not differ in content from that used in earlier work, but its implementation has made use of highly efficient search procedures and, hence, the computational time is considerably less than that used in earlier work.

For grid generation purposes the boundary of the domain is defined by points and associated connectivities. It will be assumed that the grid points on the boundary reflect appropriate variations in geometrical slope and curvature. Ideally any method which automatically creates points should ensure that the boundary point distribution is extended into the domain in a spatially smooth manner. An algorithm which achieves this in both two and three dimensions is the following:

Algorithm II.

1. *Compute the point distribution function for each boundary point $r_0 = (x, y, z)$ (i.e., for point 0)*

$$dp_0 = \frac{1}{M} \sum_{i=1}^{M} |r_1 - r_0|$$

   where || is the Euclidean distance and it is assumed that point 0 is surrounded by M points, $i = 1, M$.

2. *Generate the Delaunay triangulation of the boundary points.*

3. *Initialize the number of interior field points created, $N = 0$.*

4. *For all tetrahedra within the domain:*

   a. *Define a prospective point Q to be at the centroid of the tetrahedron.*

   b. *Derive the point distribution dp for the point Q, by interpolating the point distribution function from the nodes of the tetrahedron, dp_m, $m = 1, \ldots, 4$.*

   c. *Compute the distances d_m, $m = 1, \ldots, 4$, from the prospective point, Q, to each of the four points of the tetrahedron.*
      If $\{d_m < a\,dp_m\}$ for any $m = 1, \ldots, 4$ then
      *   reject the point :- Return to the beginning of step 4.*
      If $\{d_m > a\,dp_m\}$ for any $m = 1, \ldots, 4$ then
      *   compute the distance s_j, $(j = 1, \ldots, N)$, from the prospective point Q, to other points to*
      *   be inserted, P_j, $j = 1, N$.*
      *   If $\{s_j < \beta\,dp_m\}$ then*
      *   reject the point :- Return to the beginning of step 4.*
      *   If $\{s_j > \beta\,dp_m\}$ then*
      *   accept the point Q for insertion by the Delaunay triangulation algorithm and include Q in*
      *   the list P_j, $j = 1, N$.*

   d. *Assign the interpolated value of the point distribution function dp to the new node P_N.*

   e. *Next tetrahedra.*

5. *If $N = 0$ go to step 7.*

6. *Perform the Delaunay triangulation of the derived points, P_j, $j = 1, N$. Go to step 3.*

7. *Smooth the mesh.*

In the preceding algorithm, the term tetrahedron and triangle are interchangeable. The coefficient α controls the grid point density by changing the allowable shape of formed tetrahedra, whereas β has an influence on the regularity of the triangulation by not allowing points within a specified distance of each other to be inserted in the same sweep of the tetrahedra within the field. The effects of the parameters α and β are demonstrated in the following examples, which for convenience are presented for domains in two dimensions.

The interpolation of the boundary point distribution function is linear throughout the field. This can be modified to provide a weighting toward the boundaries so as to ensure greater point density in such regions. The implementation of such a procedure involves a scaling of the point distribution of the nodes, which form an element on the boundary. It should be noted that this point creation algorithm can be implemented very efficiently within the Delaunay triangulation procedure. In particular, if a point is accepted for insertion, then in the Delaunay algorithm a tetrahedron is known which contains this point, since by the very nature of the procedure the tetrahedron from which the point was created is known. However, after the insertion of one point the tetrahedron numbering can be changed, and if the tetrahedra formed from the inserted points overlap, then the tetrahedron numbers which have been flagged for each new point can be then incorrect. However, the exclusion zone, controlled by the parameter β, ensures that the points created from one sweep through the tetrahedra are sufficiently spatially separated that on the insertion of each point the resulting tetrahedra do not overlap and, hence, the original tetrahedron numbers associated with each new point are valid. Hence, in this way β improves the regularity of the tetrahedra and also ensures that no search is required to find a circle which includes the point.

The procedure outlined creates points consistent with the point distribution on the boundaries. Simple modifications provide greater flexibility.

26.3.2.1 Point Creation by the Use of Sources

In somewhat of an analogous way to point sources used as control functions with elliptic partial differential equations, it is possible to define line and point sources to provide grid control for unstructured meshes. Local point spacing, at position r, can be defined as

$$dp(r) = A_j e^{B_j |R_j - r|}$$

where A_j and B_j are the user specified amplification and decay parameters of the sources j, $j = 1, M$, and R_j is the position of each point source. Grid point creation is then performed as outlined in Algorithm II but in step 4b the appropriate point distribution function at the centroid is determined by Equation 26.2. Various forms of implementation of this can be devised. One simple modification is to define the point spacing as

$$dp(r) = \min \left(dp_{\text{boundary}}, A_j e^{B_j |R_j - r|} \right)$$

In this case dp_{boundary} is the point distribution from the boundary spacing. This then provides the desired point clustering in regions influenced by the sources but farther away the boundary point distribution has a dominant effect.

26.3.2.2 Point Creation Controlled by a Background Mesh

Another way to control the point spacing in the domain is to use a background mesh. A mesh is overlaid over the domain and at each node a point spacing is specified. To encompass this approach within the framework of Algorithm II, the point distribution function dp for a prospective point is obtained from the interpolated spacing from the background mesh. Within Algorithm II step 4b is replaced by the interpolation of dp from the background mesh. The effect is similar to that achieved by sources.

26.3.2.3 Boundary Integrity

This problem is widely recognized as a problem inherent to the generation of boundary conforming grids using the Delaunay triangulation. Several approaches to its solution have been proposed. The procedure followed here is an extension of earlier work and is closely related to the work of George. Both these approaches involve the addition of points to *block* the penetration of tetrahedra through the boundary surface. In the former approach, points were added on the boundary, which were then connected to the surrounding points using the Delaunay triangulation procedure. In two dimensions and in some cases in three dimensions, this approach proves to be adequate. Hence, the approach used in two dimensions uses this technique. However, in three dimensions, for some severe cases, it proves to be difficult to completely ensure the reconstruction of surface triangles. Hence, a different approach has been devised for three dimensions, which also involves the addition of points, but these are connected directly to the tetrahedral construction rather than using the Delaunay triangulation. A finite number of direct connections can be formulated for all types of tetrahedral penetration and hence in the proposed procedure the recovery of the surface can be guaranteed.

The necessary and sufficient conditions for a face with nodes (P, Q, R) to be present in the triangulation $\{\tau_l\}, l = 1, T$ are as follows:

1. The nodes P, Q, and R exist in the tetrahedral construction $\{\tau_l\}, l = 1, T$.
2. The cyclic combinations of P, Q, and R that is, PQ, QR, RP occur in one of the tetrahedra in the construction $\{\tau_l\}, l = 1, T$.
3. The combination (P, Q, R) exists in one of the tetrahedra in $\{\tau_l\}$, $l = 1, T$.

Hence, to recover an arbitrary set of triangular faces, these three conditions must be met for each face. The first condition appears to be self-evident.

If a boundary face is not in τ, this is because edges and faces of the tetrahedra $\{\tau\}$ intersect the required face. Since a face is formed from edges, and it is assumed that the points P, Q and R are present, it is necessary to firstly recover edges PQ, QR, RP and then the face (P, Q, R). This is achieved, for a given face PQR, by firstly finding the tetrahedra which are intersected by the edges PQ, QR, and RP. These tetrahedra are then modified and new tetrahedra created so that the required edges are present. Once edges are present, a similar procedure follows to recover the face. If the edges PQ, QR, and RP exist but the face (P, Q, R) does not exist, then all tetrahedra which possess at least one edge which intersects the face (P, Q, R) are determined. These tetrahedra are then modified accordingly to recover the missing faces.

26.3.2.4 Edge Swapping

In circumstances where it is required to have a complete surface, although there is not a fixed constraint that a given set of faces is recovered, it is possible to ensure boundary faces are coincident with faces within the tetrahedral construction by swapping edges within the boundary surface triangulation. If in the tetrahedral construction faces ABC and BCD exist, but in the surface triangulation faces ACD and ABD exist, then the two can be made to agree if in the surface triangulation edge AD is replaced by BC. There are conditions under which such a transformation is not allowed and these must be checked.

Edge swapping can be incorporated as an option. However, if it can be used, it can greatly reduce the amount of work to be carried out in the edge and face recovery routines.

26.3.2.5 Boundary Edge Recovery

The procedure to recover a missing edge of a boundary face involves two steps. First, it is necessary to identify the faces, edges, and points of tetrahedra which the edge intersects. Second, local transformations involving tetrahedra are performed to recover the edge. The intersections of edges with tetrahedra can be readily computed. Once this has been performed, a set of transformations are used to recover edges.

26.3.2.6 Boundary Face Recovery

After the recovery of all boundary face edges, the boundary faces can then be recovered. Clearly, although the edges of all boundary faces are present, this does not imply that boundary faces are present, since other

FIGURE 26.9   Volume grid of Ford Explorer.

FIGURE 26.10   Surface grid of Ford Explorer interior.

tetrahedra can penetrate the interior of a face but not the boundary face edges. If a boundary face is not present, it is necessary to determine all tetrahedra which intersect the face. One, two, three, or four edges and associated segments of a tetrahedra can intersect a face. Hence, for each missing face, all tetrahedra which have an edge or edges which intersect the face are determined and each of the tetrahedra are then classified accordingly.

26.3.2.7   Removal of Added Points

Most of the transformations used to recover the edges and faces in both 2-D and 3-D grids involve the creation of one or more points. These added surface points are used purely as part of the boundary recovery procedure and are removed after the boundary is complete. The mechanics of node removal involve taking each added point in turn and finding all elements connected to it. These elements are deleted leaving an empty polyhedron, which is then triangulated in a direct manner by finding point connections which lead to the optimum-shaped tetrahedral construction. This is a rapid process since this operation is performed locally for a relatively small number of points. A pictorial view of an unstructured grid for automotive application is presented in Figure 26.9 and Figure 26.10. The grid is generated by advancing front local reconnection algorithm [Marcum 1995].

26.3.3   Hybrid Grid Generation

A hybrid grid consists of structured grid in part of the physical domain and unstructured grid in rest of the field. In general, a hybrid grid can be defined as an agglomeration of cells having polygons with a different number of sides. This necessitates a hybrid grid generation process as a combination of structured and unstructured grid methodologies. The truncation error of a triangular cell is inversely proportional to

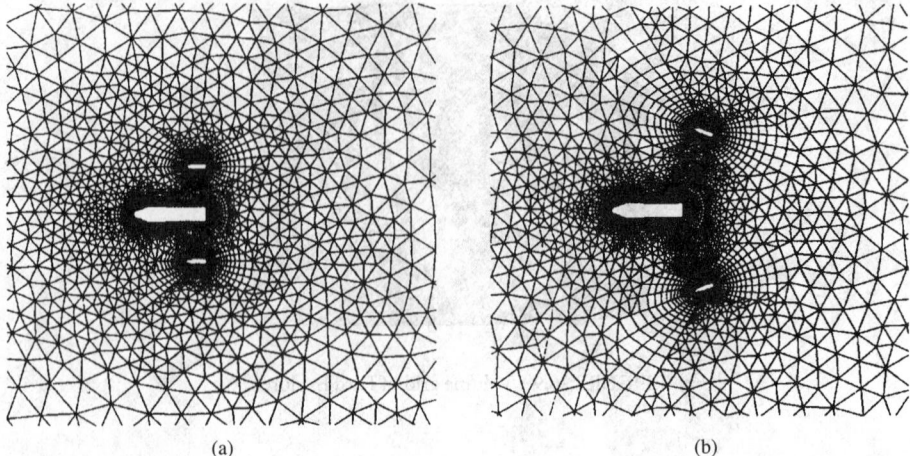

FIGURE 26.11 Hybrid grid around a launch vehicle with boosters: (a) grid at an initial state and (b) grid after booster separation.

the sine of the minimum angle of the triangle. Therefore, to reduce the truncation error the number of triangles in the boundary layer has to increase and, consequently, it will increase the total number of cells in the field. This will lead to more memory and central processing unit (CPU) time for the flow solver. The structured grid in the boundary layer will help to overcome these difficulties.

The basic steps involved in hybrid grid generation are the following. The first step is to decompose the complex geometries into simple geometric entities. A structured grid is generated around these geometric entities using an advancing layer-type method based on the surface normals together with the application of a local elliptic solver [Koomullil et al. 1996a]. The second step involves the trimming of the overlapping structured grid from different solid bodies in the domain, by comparing the aspect ratio. Cells having aspect ratio less than unity are removed. In the third step the void in the physical domain after the trimming of the structured grid is filled with unstructured grid.

The hybrid grid for dynamic motion-type geometries can be generated quickly and efficiently using the approach present in Koomullil et al. [1996a]. An example of a hybrid grid for the moving geometry problem is illustrated in Figure 26.11. The strap-on separation from the main launch vehicle booster is shown in this figure. The relative position of the strap-ons and booster rockets at different times and the hybrid grid around this configuration is shown in Figures 26.11a and b.

26.3.3.1 Grid Adaption: Construction of Weight Functions

Application of the equidistribution law results in grid spacing inversely proportional to the weight function, and hence, the weight function determines the grid point distribution. Ideally, the weight would be the local truncation error ensuring a uniform distribution of error. Determination of this function is one of the most challenging areas of adaptive grid generation. The overall solution is only as accurate as the least accurate region. Thus, excessive resolution in certain regions does not increase the accuracy of the overall solution.

Evaluation of higher order derivatives from discrete data is progressively less accurate and subject to noise. However, lower order derivatives must be nonzero in regions of wide variations of higher order derivatives, and are proportional to the rate of variation. Therefore, it is possible to employ lower order derivatives as a proxy for the truncation error.

Analysis of the weight functions explored to date indicates that density or velocity derivatives are not independently sufficient to represent the different types and strengths of flow features. Density, or pressure for that manner, varies insufficiently in the boundary layer to be used to construct weight functions for representation of these features. Whereas velocity derivatives for viscous flows by themselves are dominated by the boundary layer, additional variables must be included to represent other flow features. The weight function [Thornburg et al. 1996] consists of relative derivatives of density and the three conservative velocities,

$$W_{ijk}^{k} = 1.0 + \frac{|\hat{q}_{ijk}|}{|\hat{q}_{ijk} + e|_{\max}} + \frac{|(\hat{q}_{\xi^k})_{ijk}/\hat{q}_{ijk}|}{|(\hat{q}_{\xi^k})_{ijk}/\hat{q}_{ijk} + e|_{\max}} + \frac{|(\hat{q}_{\xi^k\xi^k})_{ijk}/\hat{q}_{ijk}|}{|(\hat{q}_{\xi^k\xi^k})_{ijk}/\hat{q}_{ijk} + e|_{\max}}$$

where $\hat{q} = (\varrho, \varrho u, \varrho v, \varrho w)$. The relative derivatives are necessary to detect features of varying intensity, so that weaker, but important structures such as vortices are accurately reflected in the weight function. One-sided differences are used at boundaries, and no-slip boundaries require special treatment since the velocity is zero. This case is handled in the same manner as zero velocity regions in the field. A small value, epsilon in the preceding equation, is added to all normalizing quantities. Also it appears that the Boolean sum construction method of Thornburg et al. [1998] would balance the weight functions more evenly, as several features are reflected in multiple variables, whereas some are reflected in only one.

26.4 Grid Systems

A multitude of general purpose grid generation codes to address complex three-dimensional structured–unstructured grid generation needs are newly available in the public domain or as proprietary commercial code. A brief description of the widely utilized candidate general-purpose codes follows.

The National Grid Project (NGP) system of Mississippi State University is an interactive geometry and grid generation system for block-structured and tetrahedral grids. The system reads CAD data via IGES and converts all surface patches to NURBS. A carpet, composed of interfacing NURBS patches, is then laid over the CAD patches to correct for gaps and overlaps. The system also has internal CAD capability for the construction or repair of surfaces. Surface grids are generated on the NURBS carpet, and can be projected onto the original CAD patches. Both the surface grids and the subsequent volume grid can be generated as block structured via elliptic, hyperbolic, or TFI methods, or as unstructured via Delaunay or advancing front procedure. A pictorial view demonstrating the graphical interface is displayed in Figure 26.12.

The ICEM–CFD [Akdag and Wulf 1992] system is a commercial code which offers block-structured grids, tetrahedral grids, and unstructured hexahedral grids. The system interfaces with numerous CAD systems and has been connected to a number of flow solvers.

The GRIDGEN [Steinbrenner and Chawner 1993] system is a graphically interactive block-structured commercial code. The user constructs curves, which are in turn used to build the topological surface and

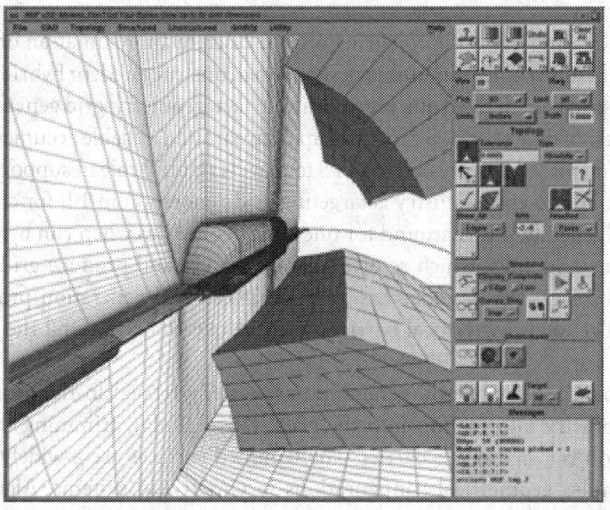

FIGURE 26.12 NGP code interface demonstrating multiblock grid.

volume components. The user then selects curves as the boundaries of surface grids, and finally surfaces as the boundaries of volume grids (blocks). With this system, grid generation is a user-in-the-loop task.

EAGLEView [Soni et al. 1992] is a graphical system that allows interactive construction of geometry and block-structured grids with journaling capability.

GridPro/az3000 [Eiseman 1995] is a commercial block-structured code topology input language (TIL) to define both the surface and the block-structured grid. The language includes components (objects) that can be invoked, and therefore admits the formation of element libraries.

CFD-GEOM [Hufford et al. 1995] is an interactive geometric modeling and grid generation system for block structured grids, tetrahedral (advancing front) grids, and hybrid grids. All elements are linked so that updates are propagated throughout the database. The geometry is NURBS based, reads IGES files, and has some internal CAD capabilities. The system also has macrolibrary capability.

The GEMS [Dener et al. 1994] block-structured grid generation system of SAMTEK-ITC in Turkey is based on object-oriented programming and C++ that uses case-based reasoning and reinforcement learning to capture CFD expertise. The system selects the case that is best suited for a particular geometry from among known ones.

The 3-DGRAPE/AL [Sorenson and Alta 1995] system of NASA Ames is a block-structured grid generator that now includes the specification of arbitrary intersection angles at boundary surfaces, as well as the orthogonality pioneered by Steger–Sorenson.

The GENIE++ [Soni et al. 1992] block-structured grid generation system of Mississippi State was also introduced in the late 1980s and has been continually enhanced over the years. This system uses TFI with elliptic smoothing and includes various splining methods.

VGRID [Parikh and Pirzadeh 1992] of NASA Langley is a tetrahedral grid generator which uses advancing front with a Cartesian background grid to control resolution.

TGrid of Fluent is a tetrahedral grid generator based on the Delaunay approach.

The first general-purpose-domain connectivity codes for chimera grids were the PEGSUS (from the Air Force Arnold Engineering Development Center) and CMPGRD (from IBM) codes in the late 1980s [Meakin 1995], which continue to be enhanced. Advances in CMPGRD are detailed in Henshaw et al. [1992]. Later codes are DCF3D of NASA Ames and Overset Methods (MEAKIN 1991) and BEGGAR of the Air Force Wright Laboratory at Eglin [Belk 1995, Maple and Belk 1994]. A detailed description of these codes and a comprehensive review of existing codes and technology can be found in Thompson [1996].

26.5 Research Issues and Summary

The first step of the CFS process is the construction of a discrete approximation of the general region of interest. This representation could be multiblock structured, unstructured, or hybrid. Only for the simplest of applications can a grid be generated quickly or easily. In fact, geometry-grid generation is by far the most time consuming aspect of the entire CFS process. Rapid turnaround, reliable accuracy, and affordability are the three key requirements to be addressed for CFS to play its rightful role in supporting multidisciplinary design environment. To this end, industry is targeting grid generation in 1 h for complex configuration, i.e., reliable 1-h grid generation turnaround for one-time geometries when run by designers. The system must include CAD-to-grid links, which resolve tolerance issues and produce grids with a quality good enough for the CFS solver. The designer has to feel that the grid generation process is under control and is predictable. The following critical barriers must be overcome to fulfill the aforestated industrial requirements.

The time-consuming aspects of grid generation are usually related to the geometry definition, i.e., the input of the geometry information into the grid system. Today, trimming the surfaces associated with their intersection is a significant barrier. The trimmed surfaces are a widely utilized entity in the construction of complex geometrical configurations in CAD/CAM systems. Algorithms which utilize surface triangulation techniques and solid modeling schemes need to be developed. Reliable methodologies for representing triangulated surfaces into a single surface represented by the NURBS are needed. The

CAD/CAM technology and methodology evolution based on solid modeling will also reduce this present barrier of addressing geometries undergoing design perturbations.

Automatic (noninteractive) algorithms for domain decomposition for the development of multiblock structured grids pose a barrier for addressing multidisciplinary design applications involving geometry optimization. The solution grid adaptive algorithms, at present, are limited to simple three-dimensional configurations. Techniques are needed to enhance the applicability of adaptive schemes pertaining to complex configurations. Parallel and distributed processing of grid generation algorithms is also essential for these multidisciplinary applications. Algorithms need to be developed to improve the quality of unstructured surface grids since they highly influence the quality of unstructured volume grids. Hybrid-generalized grid techniques are promising, especially for multidisciplinary CFS applications that include dynamic motion. This technology does not exist for full three-dimensional configurations.

References

Akdag, V. and Wulf, A. 1992. Integrated geometry and grid generation system for complex configurations, p. 161. In *Proc. Software Syst. Surface Modeling Grid Generation Workshop*. R. E. Smith, ed. NASA Conf. Pub. 3143, NASA Langley Research Center, Hampton, VA.

Baker, T. J. 1990. Unstructured mesh generation by a generalized Delaunay algorithm, pp. 20.1–20.10. In *Appl. Mesh Generation to Complex 3-D Configurations*. AGARD Conf. Proc. No. 464.

Belk, D. M. 1995. The role of overset grids in the development of the general purpose CFD code, p. 193. In *Proc. Surface Modeling, Grid Generation Related Issues Comput. Fluid Dyn. Workshop*, NASA Conf. Publ. 3291, NASA Lewis Research Center, Cleveland, OH, May.

Benek, J. A., Buning, P. G., and Steger, J. L. 1985. A 3-D chimera grid embedding technique. AIAA Paper 85–1523.

Blake, M. W. and Chou, J. J. 1992. The NASA-IGES geometry data exchange standard. *Proc. Workshop Sponsored by NASA*. Washington, DC, Langley Research Center, Hampton, VA, April.

Brackbill, J. U. 1993. An adaptive grid with directional control. *J. Comput. Physics*. 108:38.

Cavendish, J. C., Field, D. A., and Frey, W. H. 1985. An approach to automatic three dimensional finite element mesh generation. *Int. J. Num. Methods Eng.* 21:329–348.

Dener, C., Koc, E., and Sirin, I. 1994. Extentions to GEMS for automatic grid generation and intelligent topology definition. *Numerical Grid Generation in Computational Field Simulation and Related Fields*. N. P. Weatherill, P. R. Eiseman, J. Hauser, and J. F. Thompson, Eds., p. 453. Proc. 4th Int. Grid Con. Pineridge Press, Ltd. Swansea, Wales, UK.

Dirichlet, G. L. 1850. Uber die Reduction der positiven quadratischen formen mit drei understimmten, ganzen Zahlen. *J. Reine Angew. Math.* 40(3):209–227.

Eiseman, P. R. 1985. Grid generation for fluid mechanics computations. *Annu. Rev. Fluid Mech.* 17:487–522.

Eiseman, P. R. 1995. Multiblock grid generation with automatic zoning, p. 143. In *Proc. Surface Modeling, Grid Generation Related Issues Comput. Fluid Dyn. Workshop*, NASA Conf. Pub. 3291, NASA Lewis Research Center, Cleveland, OH, May.

Gatlin, B., Thompson, J. F., Yoon, Y.-H., Luong, P. V., Ganapathiraju, D., and Wolverton, M. K. 1991. Extensions to the EAGLE grid code for quality control and efficiency. *29th AIAA Aerospace Sci. Meeting*, AIAA Paper 91-0148, Reno, NV, Jan.

George, A. J. 1971. *Computer Implementation of the Finite Element Method*. Ph.D. Thesis, Stanford University, STAN-CS-71-208.

George, P. L. and Hermeline, F. 1992. Delaunay's mesh of a conven polyhedron in dimension D; application for arbitrary polyhedra. *Int. J. Num. Methods Eng.* 33:975–995.

Gordon, W. J. and Thiel, L. C. 1982. Transfinite mappings and their application to grid generation. In *Numerical Grid Generation*. J. F. Thompson, Ed. North Holland, Amsterdam.

Henshaw, W. D., Chessire, G., and Henderson, M. E. 1992. On constructing three-dimensional overlapping grids with CMPGRD, p. 415. In *Proc. Software Systems Surface Modeling Grid Generation Workshop*, R. E. Smith, Ed. NASA Conf. Pub. 3143, NASA Langley Research Center, Hampton, VA.

Hufford, G. S., Harrand, V. J., Patel, B. C., and Mitchell, C. R. 1995. Evaluation of grid generation technologies from an applied perspective, p. 401. In *Proc. Surface Modeling, Grid Generation Related Issues Comput. Fluid Dyn. Workshop,* NASA Conf. Pub. 3291, NASA Lewis Research Center, Cleveland, OH, May.

Koomullil, R. P., Soni, B. K., and Huang, C. 1996a. Flow simulations on generalized grids, pp. 527–536. In *5th Int. Conf. Num. Grid Generation Comput. Fluid Dyn. Related Fields.* B. K. Soni, J. F. Thompson, J. Hauser and P. Eiseman, Eds. Mississippi State University, April 1–5.

Koomullil, R. P., Soni, B. K., and Huang, C. 1996b. Navier–Stokes Simulation on Hybrid Grids. *34th Aerospace Sci. Meeting,* AIAA Paper 96-767, Reno, NV, Jan. 15–18.

Lo, S. H. 1985. A new mesh generation scheme for arbitrary planar domains. *Int. J. Num. Methods in Eng.* 21:1403–1426.

Lohner, R. and Parikh, P. 1988. Three-dimensional grid generation by the advancing-front method. *Int. J. Num. Methods Fluids* 8:1135–1149.

Maple, R. C. and Belk, D. M. 1994. Automated setup of blocked, patched and embedded grids in the beggar flow solver. In *Numerical Grid Generation in Computational Field Simulations and Related Fields,* N. P. Weatherill, P. R. Eiseman, J. Hauser, and J. F. Thompson, Eds., p. 151. Proc. 4th Int. Grid Conf. Pineridge Press Limited. Swansea, Wales, UK.

Marcum, D. L. 1995. Generation of unstructured grids for viscous flow applications. *33rd Aerospace Sci. Meeting and Exhibit,* AIAA Paper 95-0212, Reno, NV, Jan. 9–12.

Meakin, R. L. 1991. A new method for establishing intergrid communication among systems of overset grids. In *10th AIAA Comput. Fluid Dyn. Conf.,* AIAA Paper 91-1586, Honolulu, HI, June.

Meakin, R. L. 1995. Grid related issues for static and dynamic geometry problems using systems of overset structured grids, p. 181. In *Proc. Surface Modeling, Grid Generation Related Issues Comput. Fluid Dyn. Workshop.* NASA Conf. Pub. 3291, NASA Lewis Research Center, Cleveland, OH, May.

Parikh, P. and Pirzadeh, S. 1992. Recent advanced in unstructured grid generation, p. 435. In *Proc. Software Syst. Surface Modeling Grid Generation Workshop.* R. E. Smith, Ed. NASA Conf. Pub. 3143, NASA Langley Research Center, Hampton, VA.

Peraire, J., Morgan, K., and Peiro, J. 1990. Unstructured finite element mesh generation and adaptive procedures for CFD, pp. 18.1–18.12. In *Appl. Mesh Generation Complex 3-D Configurations,* AGARD Conf. Proc. No. 464.

Peraire, J., Peiro, J., Formaggia, L., Morgan, K., and Zeinkiewica, O. C. 1988. Finite element euler computations in three dimensions. *Int. J. Num. Methods Eng.* 26.

Peraire, J., Vahdati, M., Morgan, K., and Zienkiewicz, O. C. 1987. Adaptive remeshing for compressible flow computations. *J. Complex Physics* 72:449–466.

Soni, B. K. 1992a. Grid generation: algebraic and partial differential equations techniques revisited. In *Proc. Comput. Fluid Dyn. '92,* C. Hirsch, J. Periaux, and W. Kordulla, Eds. Vol. 2, pp. 929–936, Sept.

Soni, B. K. 1992b. Grid generation for internal flow configurations. *Comput. Math. Appl.* 24(5/6):191–201.

Soni, B. K., Thompson, J. F., Stokes, M. L., and Shih, M.-H. 1992. GENIE^{++}, EAGLEView and TIGER: general and special purpose graphically interactive grid systems. *30th AIAA Aerospace Sci. Meeting,* AIAA Paper 92-0071, Reno, NV, Jan.

Soni, B. K., Weatherill, N. P., and Thompson, J. F. 1993. Grid adaptive strategies in CFD. In *Advances in Hydro–Science & Engineering.* S. S. Y. Wang, Ed., pp. 1.A:201–208. University of Mississippi Press, Jackson, MS.

Sorenson, R. L. and Alta, S. J. 1995. 3-D GRAPE/AL: the Ames-Langley technology upgrade, p. 447. In *Proc. Surface Modeling, Grid Generation Related Issues Comput. Fluid Dyn. Workshop.* NASA Conf. Pub. 3291, NASA Lewis Research Center, Cleveland, Ohio, May.

Steger, J. L. and Chaussee, D. S. 1980. Generation of body-fitted coordinates using hyperbolic pratial differential equations. *SIAM J. Sci. and Stat. Comput.* 1:431–443.

Steinbrenner, J. P. and Chawner, J. R. 1993. Incorporation of a hierarchical grid component structure into GRIDGEN. *31st AIAA Aerospace Sci. Meeting.* AIAA Paper 93-0429. Reno, NV. Jan.

Thompson, J. F. 1985. A survey of dynamically adaptive grids in numerical solution of partial differential equations. *Appl. Numerical Math.* 1:3–27.

Thompson, J. F. 1987a. A composite grid generation code for general 3-D regions. *25th AIAA Aerospace Sci. Meeting.* AIAA Paper 87-0275. Reno, NV, Jan. 1987.

Thompson, J. F. 1987b. A general three-dimensional elliptic grid generation system on a composite block structure. *Comput. Methods Appl. Mech. Eng.* 64:377–411.

Thompson, J. F. 1993. The national grid project. *Comput. Syst. Eng.* 3(1–4):393–399.

Thompson, J. F. 1996. A reflection on grid generation in the 90's: trends, needs, and influences. In *Int. Num. Grid Generation Comput. Field Simulations.* B. K. Soni, J. F. Thompson, J. Hauser, and P. R. Eiseman, Eds., 1029. Proc. 5th Int. Grid Generation Conf. ERC Press.

Thompson, J. F. and Mastin, C. W. 1983. Order of difference expressions curvilinear coordinate systems. *J. Fluids Eng.* 50:215.

Thompson, J. F., Warsi, Z. U. A., and Mastin, C. W. 1985. *Numerical Grid Generation: Foundations and Applications.* North Holland, Amsterdam.

Thornburg, H., Soni, B., and Boyalakuntla, K. 1998. A structured based solution–adaptive technique for complex separated flows. *Appl. Math. Comput.* 89:199–211.

Voronoi, G. 1908. Nouvelles applications des parametres continus a la theorie des formes quadratiques, Rescherches sur les parallelloedres primitifs. *J. Reine Angew. Math.* 134.

Weatherill, N. P. 1990. Mixed structured-unstructured meshes for aerodynamic flow simulation. *Aeronautical J.* 94(934):111–123.

Yu, T.-U. 1995. *CAGD Techniques in Grid Generation.* Ph.D. dissertation, Computational Engineering Program, Mississippi State University.

Further Information

For complete in-depth literature on grid generation, the reader is referred to Thompson et al. [1985] and five proceedings associated with the 1985, 1988, 1992, 1994, and 1996 International Conferences on Numerical Grid Generation in Computational Fluid Dynamics and Related Areas. (The first four proceedings were published by Pineridge Press and the fifth conference proceedings was published by the Engineering Research Center at Mississippi State University.) The literature on surface grid generation and practical applications can also be found in the NASA conference *Proceedings on Surface Modeling, Grid Generation, and Related Issues in Computational Fluid Dynamics Solutions* of 1992 and 1995.

In view of the importance of and worldwide interest in grid generation, the organizing committee of the 5th International Conference has proposed an establishment of the International Society of Grid Generation (ISGG). The ISGG will be a focal point for assimilating the progress and advances realized in grid generation by publishing a quarterly electronic journal of grid generation and a newsletter and by maintaining a grid generation Internet index to the grid generation literature, researchers, test cases, and information on public domain and commercial geometry-grid systems. The ISGG can also be the focal point for organizing future grid generation related workshops and conferences. The organization committee feels that the time is right for the emergence of a formal society and journal for grid generation. Additional information on the ISGG can be obtained from Bharat Soni, NSF Engineering Research Center, Mississippi State University, by e-mailing: bsoni@erc.msstate.edu.

27

Scientific Visualization

William R. Sherman
National Center for Supercomputing Applications

Alan B. Craig
National Center for Supercomputing Applications

M. Pauline Baker
National Center for Supercomputing Applications

Colleen Bushell
National Center for Supercomputing Applications

27.1　Introduction 27-1
27.2　Historic Overview................................27-1
　　　The Motivation for Computer-Generated Visualization
　　　• The Process of Computational Science in Relation
　　　to Visualization • What Exactly Is Scientific Visualization?
　　　• Other Modes of Presenting Information
　　　• Application Areas • Evolution of Scientific Visualization
27.3　Underlying Principles 27-5
　　　The Goal of Scientific Visualization • The Basic Steps of the
　　　Scientific Visualization Process
27.4　The Practice of Scientific Visualization 27-8
　　　Representation Techniques • The Visualization Process
　　　• Visualization Tools • Examples of Scientific Visualizations
　　　• Visualizing Smog
27.5　Research Issues and Summary 27-29

27.1　Introduction

The field of scientific visualization is broad and requires technical knowledge and an understanding of many communication issues. This chapter provides information about its evolution, its uses in computational science, and the creative process involved. Also included are descriptions of various software tools currently available and examples of work which illustrate various visualization techniques. Relevant concerns, such as visual perception, representation, audience communication, and information design, are discussed throughout the chapter and are referenced for further investigation. An overview of current research efforts provides insight into the future directions of this field.

27.2　Historic Overview

Visualization did not begin after the advent of computers. There has always been a need for people to visualize information. At the dawn of human history, humans began spreading pigment on surfaces to convey events that took place and later to indicate quantities of goods. From that time on, the medium of choice for representing such information has continued to evolve (Figure 27.1).

In general, visualization efforts required that the creators of the image represent their data by hand. Often this was a painstaking process that involved an artistic ability to mentally envision a pictorial representation of a phenomenon and the manual skills required to transpose the mental image into a suitable medium.

1-58488-360-X/$0.00+$1.50
© 2004 by CRC Press, LLC

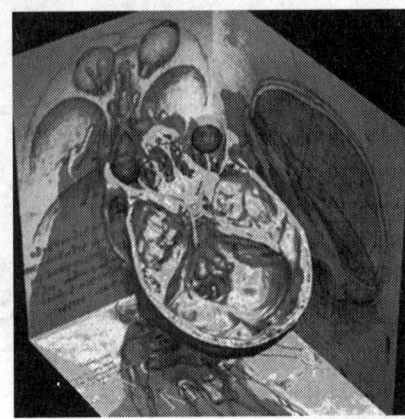

FIGURE 27.1 A combination of scientific representations from the fifteenth century and today demonstrates that the craft of visualization has been practiced for many years. (Courtesy of U. Tiede, T. Schiemann, and K. H. Hohne, IMDM, University of Hamburg, Germany, 1996.)

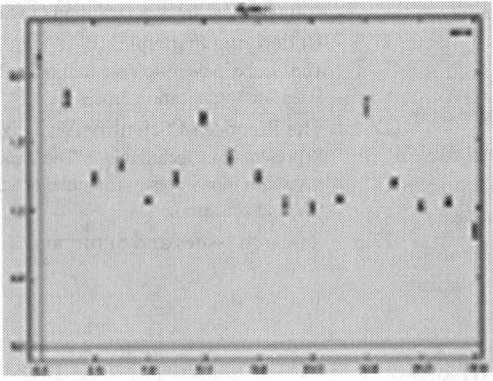

FIGURE 27.2 An XY plot with error bars.

The researcher had to be a capable artist and craftsman as well as a scientist. Usually, the visualizer would render the representation onto paper. However, other media for visualization were used as well.

As the scientific method developed, certain forms of visualization became accepted practices. As a scientist observed a phenomenon, it could be recorded onto an XY plot, representing the relationship between two quantities. A line was often drawn through the data to show the probable continuous pattern. Error bars were added to represent uncertainty in the data (Figure 27.2). We can now render detailed, data-based visual images by machine to show both quantitative relationships and qualitative overviews. How this process is accomplished, and its value to the scientific method, demand investigation.

27.2.1   The Motivation for Computer-Generated Visualization

The advent of the digital computer brought about the ability to collect, create, and store more information than previously. It also brought about a new method of science: *computational* [Kaufmann and Smarr 1993].

Computational science is the process of simulating a relevant subset of the laws of nature using a supercomputer. The laws of nature are described by a set of equations, which yield numeric solutions regarding the science being studied. Because these computational simulations of nature produce vasts amounts of numeric information, a scientist may not be able to see, much less interpret, all the results. Fortunately, as the computational power of computers has increased, allowing these complex simulations to be calculated, so has the graphical power of computers increased. Thus, we also have access to a medium capable of creating and presenting all this information in a way useful to the researcher.

There are tradeoffs in how numeric information can be visualized. Interactive visualization gives the researcher the ability to control specific portions of the dataset to examine and to control the type and parameters of the visual output. Interactivity, though, may limit the percentage of the data that can be examined at a given moment and limit the types of representation available. Alternatively, the data display may be created as a batch process, allowing complex representations which are not possible in real time. The researcher may take advantage of both methods by beginning with interactive exploration and, when an interesting region of the data is located, producing a detailed animation.

Another consideration when creating visualization is whether to render a view of the entire dataset (a qualitative overview) or to precisely represent a subset of the data that the scientist can analyze (a quantitative study). Both are important in computational science. The qualitative overview can give the scientist a sense of the entire simulation, which can help in comparisons with observed nature. Because it gives an overall understanding of the dataset, it provides a sense of context when looking at the details.

The details are in the quantitative representation provided by a precise mathematical description. Qualitative information is helpful to this process, but high-resolution quantitative displays are essential. Representations such as contour plots and two-dimensional (2-D) vector diagrams are precise and aid in data analysis. Quantitative representations, such as these, provide the ability to pore over a particular subset of the data, even to the point of measuring phenomenon from the display. Because there is a limit to the amount of useful information that can be rendered on a screen, focusing on a subset allows the data to be more completely displayed.

27.2.2   The Process of Computational Science in Relation to Visualization

To understand the role of scientific visualization in the computational science process, we must first review the scientific process itself. Figure 27.3 depicts the steps involved in the computational science process [Arrott and Latta 1992]. Computational science begins with observations of some natural phenomenon,

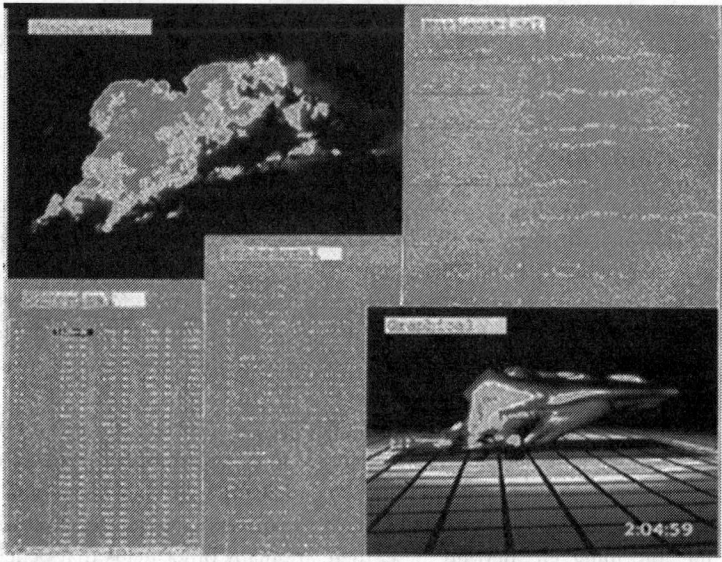

FIGURE 27.3 Computational science consists of observations, equations, algorithms, numerical solutions, and graphical representations. (Courtesy of Matthew Arrott; NCSA.)

in this particular case the formation of a severe thunderstorm. The scientist then expresses the observations in mathematics — the language of science. These equations can be manipulated by the researcher, though the problems are generally of a complexity sufficient to require solution on a supercomputer. In today's computing environment, the mathematical representation of the phenomenon is not a suitable means of input to typical computing systems. The computer requires the phenomenon be simulated in discrete steps of space and time, whereas mathematics allows a continuous representation. The mathematical representation is therefore translated into a programming language, implementing appropriate algorithms for a discrete numerical solution.

The resultant solution is typically a numeric value or set of values — a **dataset**. While the scientist may be able to gain insight from these numeric representations, a more intuitive visual form often aids in understanding. Also, others are often much more likely to understand the scientist's work when it is presented visually than as numbers only.

27.2.3 What Exactly Is Scientific Visualization?

Although we have previously discussed methods by which scientific data were represented before the advent of computers, for our purposes, **scientific visualization** is the use of data-driven computer graphics to aid in the understanding of scientific information. We will refer to an artist's rendering of a concept as illustration, or more specifically **scientific illustration**.

Is scientific visualization just computer graphics, then? Computer graphics is the medium in which modern visualization is practiced; however, visualization is more than simply computer graphics. Visualization uses graphics as the tool to represent data and concepts from computer simulations and collected data. Visualization is actually the process of selecting and combining representations and packaging this information in understandable presentations.

27.2.4 Other Modes of Presenting Information

There are ways to present information other than visually, of course. The primary of these is aurally, but one also might imagine the use of haptic (force, texture, temperature, etc.) display [Brooks et al. 1990], or even display to our other senses. Currently, sound (sonification) is increasingly recognized as an important method of information display for special types of data [Scaletti and Craig 1991].

Perhaps the field of representing information would be more appropriately termed **perceptualization**. The goal is, after all, to increase the information observer's perception of what is taking place in the data.

27.2.5 Application Areas

The use of scientific visualization to represent data is as broad as science itself. It spans the range of scales from the atomic and subatomic worlds to the vastness of the universe. It encompasses the study of complicated molecules and the building of complicated machinery. It looks at dynamic systems of living creatures and at the dynamics of whole ecosystems. Each of the areas touched by scientific visualization has representations that are particular to itself. Yet, there is much overlap in the techniques used by visualization developers due to commonality in the underlying mathematical expressions of natural systems.

Often a variety of what would seem unrelated sciences share similar or identical computational techniques. For example, computational fluid dynamics is used to study atmospheric effects, ocean currents, cosmology, mixing, injection molding, blood flow, and aeronautics. Finite-element analysis is used for solid and structural mechanics, fracture mechanics, crash-worthiness, heat transfer, electromagnetic fields, soil mechanics, metal forming, etc. Beyond these, there are many other fields that benefit from similar computational algorithms, including molecular modeling, population dynamics, diffusion, wave theory, and *n*-body problems.

27.2.6   Evolution of Scientific Visualization

The representation of the numeric output of simulations has developed from the simple printing of characters on paper, to vector display and plotter graphics, to three-dimensional (3-D) static images, to animated 2-D and then 3-D renderings of a simulation over time. The level of interaction has increased from the creation of visualization animations in batch mode, to real-time viewpoint control of fixed geometries, to interactive rendering allowing modification of the simulation in real time, and now to interacting with the simulation and representations in an immersive virtual environment.

As the underlying tools have improved, so have the idioms for representing information. New idioms are developed, and old ones are used in new ways. Many advances in representation are able to occur because of the advance of computing power and of computer input and output enhancements. Faster computing means more graphical computations can be done to create images, and higher-resolution displays allow for more detail to be presented.

These advances bring higher expectations. When 3-D pictures, animated 2-D pictures, and then finely rendered 3-D animations were first used by researchers to present their work, the visualization might have been considered the highlight of the presentation rather than the underlying science. The broader the audience, the more likely for this to be the case, giving rise to a situation where the scientists' credibility to the audience may be more correlated with the beauty of their images than with the underlying scientific theory. If scientists do not use the latest methods of computer graphics and animation to present their work, then it may not receive the attention it deserves. We are not arguing that this is how it *should* work, but it is important for scientists to be aware of the impact that visualization has on the communication of their research.

27.3   Underlying Principles

In this section, we will look at the various reasons for using scientific visualization and the effect they have on how a visualization is produced. We will also examine the basic concepts of visualization production and some of the considerations a producer should think about. Why are these important? Because scientific visualization is a means of communication. Sometimes the communication is between the raw numeric data and the researcher, and sometimes it is between the researcher and a group of people. Either way, for effective communication, it is important that both the producer of the visuals (or the tools used to create the visuals) and the audience have a grasp on what happens to the information as it passes from numbers to pictures.

27.3.1   The Goal of Scientific Visualization

Recall that the reason scientific visualization is employed as a tool is to more readily gain insight into a natural process. There are other similar goals that may be accomplished with scientific visualization. For instance, the goal might be to demonstrate a scientific concept to others, in which case the medium, representation, and the degree of detail chosen vary with the audience. A presentation shown to other scientists in the field will differ widely from one shown to the public or to government bodies.

The amount and the level of explanation required in a visualization is based on the experience of the intended audience. Also, design choices should be made which determine the amount of interactivity possible for wider audiences. For example, presentations designed for a mass audience are typically designed as noninteractive video animations. Alternatively, by utilizing computer delivered media, such as CD-ROMs, or multimedia presentations over the Internet, a limited dataset can be presented with a limited selection of visualization options, allowing for some experimentation by the audience.

When the audience is only the individual scientist, the primary goal is to uncover patterns in the data. Still, the goals of the study can vary. The goal might be to compare the patterns in the simulated data with patterns observed in nature. The closer these patterns match, the more confidence the scientist has in the theory expressed in the computational algorithm. Or the goal might be to discover new patterns that give clues to a better mathematical expression of the process. This is more frequently the case when the process is less well understood and the data are collected rather than simulated. For example, in analysis of the stock market, researchers might look for patterns that give rise to the ability to determine profitable opportunities.

27.3.2 The Basic Steps of the Scientific Visualization Process

Though it is possible to jump right into using visualization as a tool, there are several important steps that occur in the process of creating an effective visualization. At one level, scientific visualization can be thought of analytically as simply a transfer function between numbers and images, bearing in mind that this transfer may be irreversible and cause distortion of meaning.

At another level, visualization involves a barrage of procedures, each of which influences the final outcome and its ability to convey meaningful information. That is, the process of visualization includes consideration for data filtering, representation, potential inaccuracy, and human perception.

27.3.2.1 Data Filtering

Seldom is it possible to make pictures straight from the data source (a computational simulation or data acquisition system). Typically, work needs to be done on the data before they can be appropriately visualized. This can include cleaning up the data and performing operations on them that yield more useful data. Examples of cleaning up data are removing noise, replacing missing values, clamping values to be within the range of interest, etc. New numeric forms of the data often are derived in order to produce a dataset which will lead to greater insight. For example, the vertical vorticity of a fluid flow can be calculated from the horizontal wind velocity.

The medium of presentation may also be the cause of data filtering. From a practical point of view, it is often necessary to adjust the data so that they can be conveniently produced within the constraints of a particular medium. For example, a standard NTSC video device displays at a rate of 30 frames per second. Thus, the dataset must be adjusted to produce 30 images for each second of the animation. This is done by interpolating the data over time. To fill spatial gaps in the visual imagery, interpolation is also frequently done over space.

No matter what the medium, whenever any form of interpolation or other data-filtering operation is used, there are problems that can arise which may cause the imagery to be misleading to viewers who don't know or understand what has happened to the data.

27.3.2.2 Representation Issues

As noted at the beginning of this chapter, computational science involves choosing appropriate representations of the phenomenon being studied (e.g., numeric, mathematical, etc.). The representations of the data appropriate for computer manipulation are not necessarily the most conducive to human understanding. Representing numeric data so that they can be more readily perceived by the audience involves mapping those numbers to a geometric form, sonic waves, etc.

Visual representation of information requires a certain literacy on the part of the developer and the viewer [Keates 1996]. Minimal information will be communicated if the viewer is not able to understand the visual language the developer has chosen. The language elements (symbols) of visualization come primarily from the adoption and evolution of symbols used in other visual domains, particularly scientific illustration. When new symbols are created, care must be taken to ensure that adequate explanation is given.

Beyond representing the numerical output of the simulation, it is desirable to indicate information about the simulation itself. This includes items such as a representation of the grid of the computational domain, the coordinate system, scale information, and the resolution of the computation.

When choosing which representational idiom to use, it is important to consider several issues before coming to a decision. What type of information is being investigated? Is the primary goal to convey quantitative or qualitative information? How detailed should the representations be? Making these determinations depends to a great extent on the characteristics of the audience for which the visualization is intended. The resolution of the display affects the ability to present quantitative information and thus is a factor in which idioms can be chosen.

The goal of the visualization limits the medium of delivery. The medium, in turn, puts constraints on the possible choice of representation and interaction. So, for example, if motion is important to show some aspect of the data, then a medium that can support time-varying imagery needs to be used (e.g., film, videotape, interactive computer graphics). If the delivery medium is constrained to be a single image, then one must find a means to represent motion statically.

The ability to communicate with the audience relies on a well-designed presentation of information. A common problem is to give equal visual importance to all the elements in a scene, making it difficult to comprehend. We can learn techniques for making effective imagery from experts in information presentation such as graphic designers and instructional designers.

27.3.2.3 Accuracy

It is good practice for scientists to question the accuracy and validity of any information they are presented. All too often, compelling visualizations are used without the audience really questioning what they are seeing. Today, visualizations sometimes accompany peer-reviewed publications without being subjected to the same critical examination as the paper.

Where does inaccuracy come from, and what forms does it take? Whenever data change representation, the possibility for the introduction of error exists. Illusions are a danger in any medium. This is especially true when representing three-dimensional imagery on a two-dimensional display. For example, parallax can lead to misreading sizes of objects. The bias of the visualization producer often can affect the accuracy of the presentation. This does not necessarily imply that they might deliberately misrepresent the information, but many of the choices made during the production can add up to a presentation that gives an inaccurate view of the results. Some of these might include the choice of which representation to focus on, and which to leave in the background, or the selection of viewpoint or color and lighting that can make objects look ominous or insignificant.

High production values often lead to a sense of quality. The quality of the imagery does not necessarily reflect the quality of the underlying science or representations. The computer graphics techniques used should not get more emphasis than the science (i.e., should not have "glitz" merely for its own sake). Adding glitz can make a visualization appealing but can also occlude the important elements of the presentation. High production values and accuracy are two separate factors and should not be confused. The overuse of glitz in visualization is satirically treated in the animation *The Dangers of Glitziness and Other Visualization Faux Pas* (Figure 27.4) [Lytle 1993].

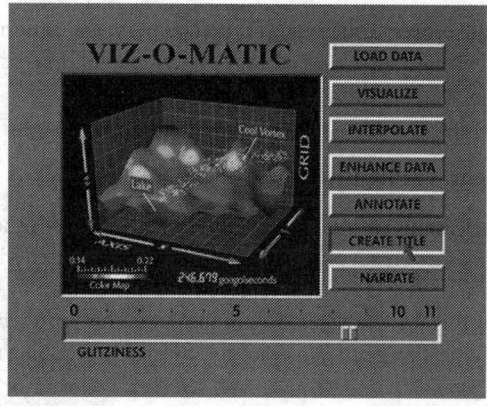

FIGURE 27.4 The Viz-o-matic animation pokes fun at the tendency to overuse graphic elements at the expense of accurate portrayal of the data. (Courtesy of Wayne Lytle; CTC.)

Labels can be used in any visualization, both as a tool for showing features of the visual representations and as a means to help clarify potentially confusing or unclear items. By adding labels, the viewer can be shown the size of the domain, the range of the values, the coarseness of the simulation, etc. Labels make the visualization more clear, understandable, and, therefore, a more useful means of communication.

27.3.2.4 Human Perception

One filter that will always have an effect on how data are viewed is the human perceptual system [Weintraub and Walker 1969]. Our perception of the world does not exactly match with physical reality. In fact, there are many elements of the real world that we cannot directly perceive at all. For example, human vision covers only a small range of the electromagnetic spectrum; human hearing perceives only air pressure changes within a specific range of frequencies; human olfactory nerves have limited precision and no real ability to determine directionality of smells. Each sense has its own perceptual anomalies.

It is likely that some animals are able to perceive phenomena that humans cannot consciously perceive. For example, many species of birds can sense the earth's magnetic field and use it as a navigational aid (magnetoperception). Fortunately, we are able to use instruments that can sense elements of the physical world that we are unable to sense. These instruments then translate the sensory input into a display that humans are able to interpret. Visualization often involves the mapping of information from imperceptible forms to something we can interpret and analyze.

The field of *human factors* studied the relationship between people and machines and has a deep body of research to draw upon. It will be beneficial for anyone creating visualizations to familiarize themselves with the work of this field [Wickens et al. 1994].

All human perception, however, suffers from the additional problem that each of our brains interprets the incoming signals differently. Our experiences through life have trained our perceptual systems uniquely. Many biases are fairly consistent throughout a culture because the individuals within that culture will have had many similar experiences. Symbols in a culture lead to an understanding of what lies ahead or within; a skull and crossbones has a specific meaning in western culture, as does a railroad crossing sign. Colors are also culturally biased. Red can mean danger or stop; yellow, caution; and green, go. But for some, green also implies money or envy. In science, colors often are used to represent a scale of information, but this might not be understood by the rest of the culture. Which colors should represent high and low values, and which should signify the interesting data?

It is important to take human perceptual biases into account when designing an information display, but one also must recognize that these biases are not necessarily universal. Thus, a scale or legend should always be used to illustrate how information is being mapped.

27.4 The Practice of Scientific Visualization

In this section, we examine the process and visual components of visualization, we discuss several types of tools available for producing visualization, and we look at examples of how visualization has been applied to particular sciences and evolved over the course of several years.

27.4.1 Representation Techniques

One of the most important elements in creating scientific visualizations is the choice of visual representation, or **visual idioms.** This section surveys a variety of commonly used visual idioms, each of which is appropriate in different situations. The visual representation is created by combining the elements of form, color, and motion that together show features of the data. This representation can range from very realistic renderings of real physical objects to abstract glyphs used to combine many pieces of information into a single idiom.

The traditional forms of data display should not be dismissed. Quantitative data can readily be retrieved from such representations as the *XY* plot, the contour map, and the bar chart (Figure 27.5).

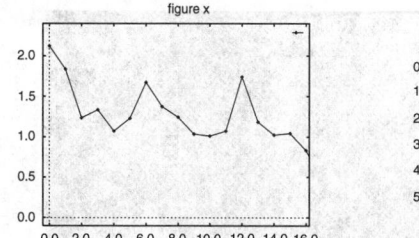 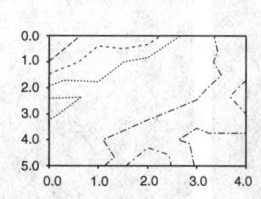 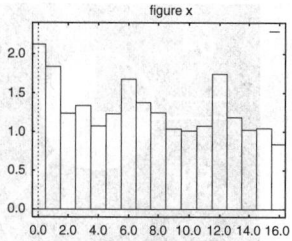

FIGURE 27.5 Simple 2-D graph representations convey both quantitative and qualitative information.

FIGURE 27.6 A realistic graphic representation of a front-end loader rendered from CAD data. (Courtesy of Caterpillar; Mark Bajuk; NCSA.)

27.4.1.1 Realism Continuum

Accurate representation of information does not necessitate that the display be realistic. Information can be represented in a very abstract form, and its contents can be read accurately by someone that knows what the form symbolizes. Sometimes it is useful, though not always possible, to show a realistic representation of what the data symbolize, as in many architectural visualizations and prototype designs (Figure 27.6).

In contrast, it is often more useful to create an abstract representation. This is especially true when there is not a direct physical counterpart to the concept being displayed. This is a convenient way of representing many variables simultaneously. The drawback with using very abstract representational schemes is that it takes time for the viewer to learn and become fluent with the symbols. In Figure 27.7, symbols (or glyphs) represent different aspects of current weather conditions.

27.4.1.2 Color

Color adds to the appeal of a visualization. More importantly, color is used to convey information about data. Some fields, including atmospheric sciences, chemistry, and seismic analysis, have developed common conventions about color use within their application areas. Colors must be chosen carefully, with a view to the goal of the visual analysis. For example, visual displays often are used to identify the quantitative value at a point. Color can be used for this task by assigning colors to specific data values. The number of colors should be limited to about seven, and the colors should be easily distinguished from each other. Alternatively, the visualization task might be to determine the overall structure of the data slice, so that a color map that is continuously varying can be more effective.

Figure 27.8 shows the use of a variety of color palettes to color the same 2-D slice of data. The top two color maps use a wide range of hues — these are variations of a "rainbow" palette. These color maps

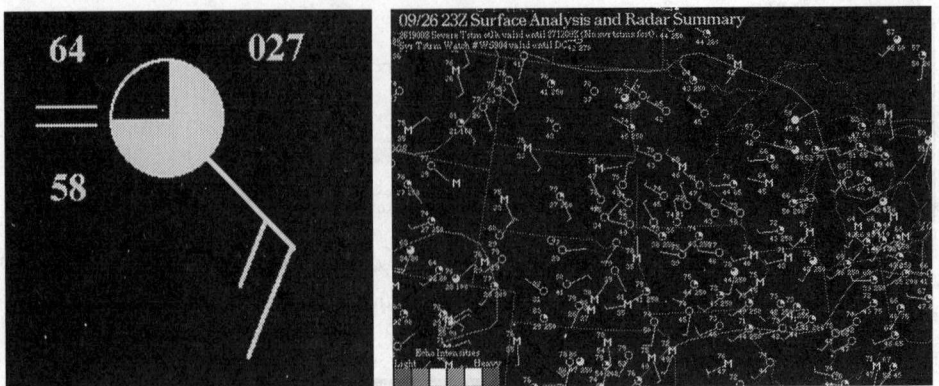

FIGURE 27.7   Weather maps often include an abstract symbol which depicts wind speed and direction, cloud coverage, etc.

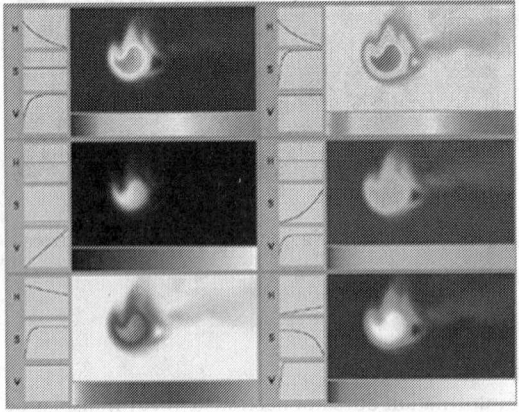

FIGURE 27.8   (See Plate 27.8 in the color insert following page 29-22.) A variety of color palettes used to color the the same two-dimensional slice of data. The top two color maps use a wide range of hues and introduce discontinuities into the image that may not be present in the data. If we are looking for overall structure, a smoothly changing palette with a restricted hue - such as the "fire" palette shown in the lower right - is more appropriate. (Courtesy of NCSA, Yale University.)

introduce discontinuities into the image that may not be present in the data. For example, the edge from yellow to green or yellow to red suggests a distinct change in the data, which may or may not be there. The distinct bands may be useful if we're trying to isolate all the data values within a particular range. However, if we're looking for overall structure, a smoothly changing palette with a restricted hue, such as the "fire" palette shown in the lower right, is more appropriate.

27.4.1.3   Form

27.4.1.3.1   *Two-Dimensional Plane*

The simplest form to present on a sheet of paper or a computer screen is a 2-D plane. Depending on the dimensionality of the data, there can be tradeoffs between representing data purely in two dimensions and giving a 3-D view of the data. In a 2-D representation, we do not have to deal with perspective projection or other potentially confusing 3-D cues, thus making it much easier to make quantitative determinations from the rendering. Of course, this works best for data which inherently lie on a 2-D grid. Higher-dimensional datasets can be examined a slice at a time by cutting 2-D planes through it.

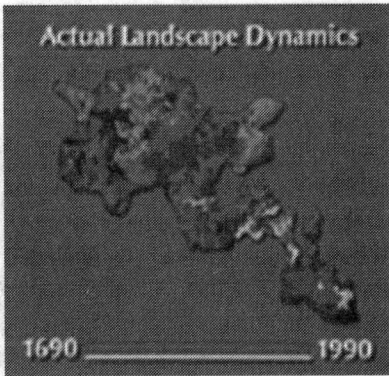

FIGURE 27.9 (See Plate 27.9 in the color insert following page **29**-22.) Discrete color mapping is used to depict the age of the forest in Yellowstone National Park [Kovacic et al. 1990]. (Courtesy of D. Kovacic, A. Craig, and R. Patterson; UIUC, NCSA.)

One can represent data on a 2-D raster image (also known as a false-color map). An example of this is the visualization of a 2-D fluid flow problem involving a jet of high-density material into a lower-density material.

The range of values is mapped to a range of colors. The choice of colors requires considerable thought in order to produce a meaningful image. Choosing colors arbitrarily may lead to false impressions of what the data indicate. It is not always necessary to employ a continuous range of color values. Sometimes a few colors chosen to indicate discrete states can convey information more effectively. In Figure 27.9, discrete shades of yellow and green represent the age of the forest since the last fire in the region, red indicates a current fire, and black represents areas of nonvegetation.

There are several other idioms for representing information on a 2-D plane, many of which extend into 3-D forms. Contour lines and vector plots are two examples. Contour lines can be drawn on a 2-D plane of data indicating where specific values of the data are located. Typically, multiple contours are drawn in a single visualization to show the range of values. Vector values on a 2-D plane of data often are represented as arrows that point in the direction of the vector, with a length proportional to the magnitude. To view the entire domain, arrows are placed at the location of each data element.

27.4.1.3.2 Height
If a goal is to see correlations between two variables in the dataset, one can add the element of height to the aforementioned color mapping idiom. One variable is mapped to the color, and the other is used to determine the height of the surface (e.g., Figure 27.10). One must be careful in viewing such representations so as to avoid inaccuracies due to poor viewpoint selection. For example, in the extreme case, viewing a height–color rendering from directly above causes the height information to be lost.

27.4.1.3.3 Volume Rendering
The technique of color-mapping a single variable of a 2-D dataset can be extended to a 3-D dataset using a computer graphics technique called volume rendering [Drebin et al. 1988]. In looking at 3-D data, a viewpoint must be selected, and then an image is rendered by traversing through the elements of the dataset and assigning color and transparency values. A common usage is to volume-render medical data, such as the dog heart in Figure 27.11. In this example, the less dense material is assigned a high transparency value, so we can see through to the higher density material of the muscle and bone [Moran and Potter 1992].

27.4.1.3.4 Vectors: Arrows, Tracers, Streamlines, etc.
There are several common idioms used for displaying vector data. A standard representation of a 2-D vector field is to simply draw an arrow from each cell in the data in the direction of the vector, with the length proportional to the magnitude (Figure 27.12).

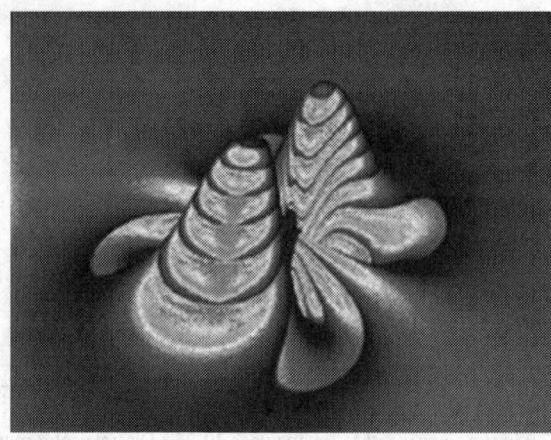

FIGURE 27.10   Visualization of gravitational effects of colliding black holes. (Courtesy of Mark Bajuk, Edward Seidel; NCSA [Anninos et al. 1993].)

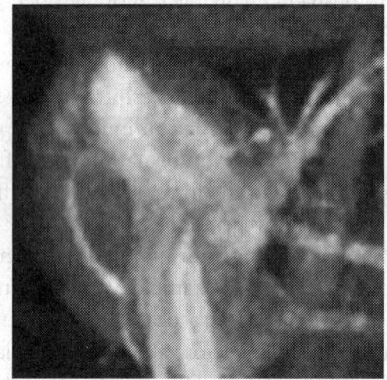

FIGURE 27.11   (See Plate 27.11 in the color insert following page **29**-22.) Volume rendering of a dog's heart. (Courtesy of E. Hoffman, P. Moran, C. Potter; NCSA.)

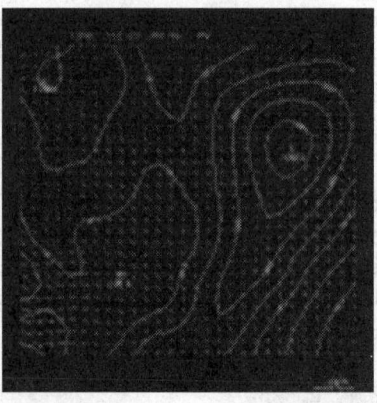

FIGURE 27.12   Wind vector field and pressure contours.

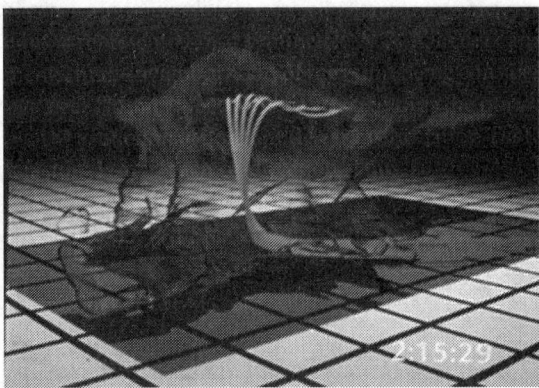

FIGURE 27.13 (See Plate 27.13 in the color insert following page **29**-22.) Tracers clarify the wind flow within a simulation of a severe thunderstorm. (Courtesy of Robert Wilhelmson et al. NCSA.)

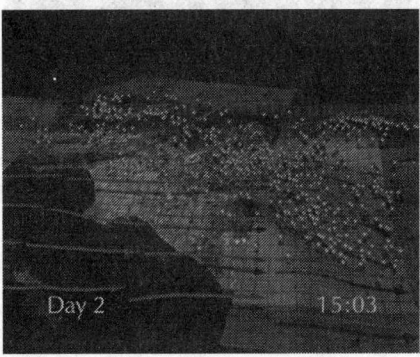

FIGURE 27.14 (See Plate 27.14 in the color insert following page **29**-22.) Streaklines and particles depict smog in Los Angeles. Each green particle represents 10 tons of reactive organic gases; each yellow particle, carbon monoxide; and each red particle, nitrogen oxide. (Courtesy of W. Sherman, M. McNeill et al. NCSA.)

Just as techniques such as colored smoke are used to visualize fluid flows in the laboratory, computer graphics equivalents of smoke may be used to display flowfields in computational simulations [Merzkirch 1987]. There are several idioms that give an effective qualitative view of the data. The simplest utilizes particles to indicate the flow. By adding lines showing the history of the paths taken by the particles, we derive a different technique often called **tracers** (Figure 27.13). These are also referred to as **streamlines**, especially when the particle itself is not represented. A similar idiom, the **streakline**, shows a path through the vector field for a static moment of time (Figure 27.14). In other words, it shows a portion of the path a particle would take if it were traveling infinitely fast.

As with many geometric representations, additional information (age, velocity, temperature, etc.) can be mapped onto a tracer, streamline, or streakline. This information can be represented with color, transparency, texture, or twist. Figure 27.15 shows a good example of representing the streamwise vorticity of a fluid flow simulation. In this representation, the twist along a streamline is proportional to the amount of vorticity at the particular location of the domain.

27.4.1.3.5 Isosurfaces

Extending the isoline (or contour line) to three dimensions gives us the isosurface [Lorensen and Cline 1987]. An **isosurface** is a surface of constant value in three dimensions. All points inside the surface have values below the threshold level, and all outside have values above the threshold level, or vice versa. For

FIGURE 27.15   (See Plate 27.15 in the color insert following page **29**-22.) A ribbon's rate of twist indicates the streamwise vorticity. (Courtesy of Robert Wilhelmson et al. NCSA.)

FIGURE 27.16   Isosurfaces depict regions of electron-density change. (Courtesy of Jeffrey Thingvold; NCSA.)

example, the isosurfaces shown in the cloud surface shown in Figure 27.16 represent changes in electron density.

27.4.1.3.6 *Forms from Traditional Science*

It often is desirable to choose representational idioms with which the audience is already familiar. For example, chemists are familiar with representing molecular structure with ball-and-stick models. These can be replicated in computer graphics. Using computer-generated imagery, however, removes certain physical constraints, and additional information may be given. For example, Figure 27.17 shows a picture of a leukotriene molecule in which colored, orthographic shadow views are projected onto a three-walled stage to give the viewer a better clue to the 3-D shape.

27.4.1.3.7 *Cutting Plane*

Sometimes it is useful when looking at a 3-D dataset to cut the 3-D cube with what is called a cutting plane, revealing data along a single 2-D slice of that cube. Figure 27.18 shows a cutting plane, together with the 3-D isosurface. The black line in the isosurface indicates where the plane intersects it. The scale alongside the cutting place shows the mapping between the level of water content and the colors.

27.4.1.3.8 *Alpha Shapes*

A technique called **alpha shapes** allows one to represent the concept of shape applied to a collection of points in space [Edelsbrunner and Mucke 1994]. Alpha is the name of the parameter that, when varied,

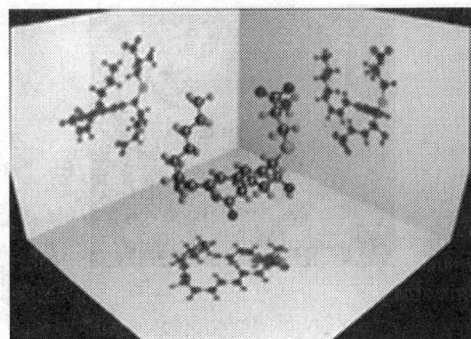

FIGURE 27.17 (See Plate 27.17 in the color insert following page **29**-22.) A computer-generated ball-and-stick model of a leukotriene molecule. Colored shadows aid in perceiving the 3-D shape. (Courtesy of D. Herron, Eli Lilly & Co.; J. Thingvold, W. Sherman, NCSA.)

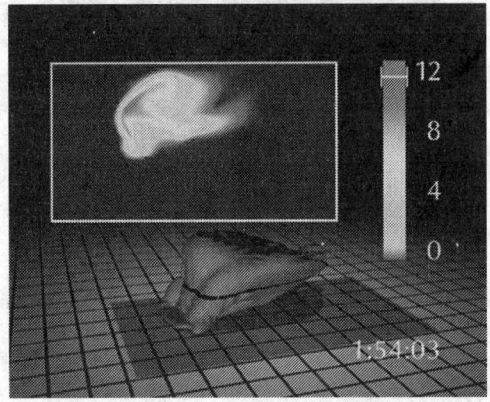

FIGURE 27.18 (See Plate 27.18 in the color insert following page **29**-22.) A cutting plane displays additional detail of a particular slice of a 3-D dataset. (Courtesy of Robert Wilhelmson et al. NCSA.)

affects the complexity of the resultant shape. Large values of alpha produce a simple representation of the shape — a convex hull. As the value of alpha is continuously decreased, more complex shapes are created, revealing cavities, tunnels, and voids inherent in the dataset. For example, Figure 27.19 shows a geometric representation of a molecule of gramicidin A, the first clinically used antibiotic (predating penicillin).

27.4.1.4 Motion

Motion can be a very important cue to the viewer for understanding the data being portrayed. It provides a qualitative overview of dynamic data by showing how the system evolves over time. Of course, the medium in which the imagery is presented affects the ability to indicate motion. In a book, for example, motion must be indicated by techniques such as motion blur, tracers, or series of still images from a time sequence (small multiples) [Tufte 1996a] (Figure 27.31).

Motion also can be used to aid the viewer in discerning the three-dimensionality of the object. Figure 27.20 shows an example of a 3-D set of molecules. The image in this book looks flat. However, when the molecules are rotating on a screen, a strong sense of the 3-D structure is conveyed. This representation also demonstrates how motion through the simulation can be represented in a single static geometry. In this visualization, the collection of small spheres represents the positions the atoms take over the course of the entire simulation.

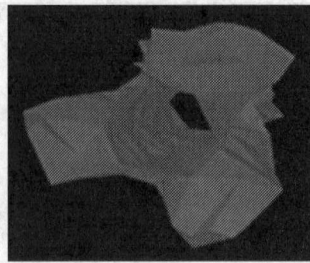

FIGURE 27.19   Alpha-shape representation of gramicidin A molecule. (Courtesy of H. Edelsbrunner, P. Fu, UIUC, NCSA.)

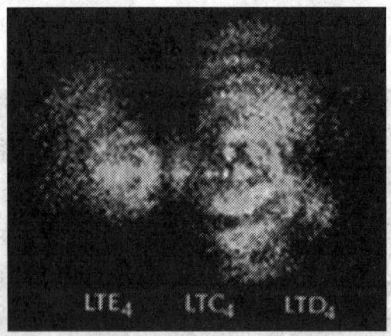

FIGURE 27.20   This image shows all of simulation time in a single geometric form and uses motion in the animation to aid the viewer in discerning the 3-D structure. (Courtesy of D. Herron, Eli Lilly & Co.; J. Thingvold, W. Sherman, NCSA.)

27.4.1.5   Transparency

Sometimes it is desirable to show something that's inside or behind something else in the scene. By increasing the transparency of the geometry, the visualizer can allow internal structure or hidden objects to be viewed. When using transparency, often many of the shadings and other cues that indicate shape are less compelling. In Figure 27.15, a transparency technique was used to allow the twisting ribbons inside the cloud structure to be seen. If the cloud had been made uniformly transparent, it would be difficult to see its shape. Here, the transparency is dependent on how nearly perpendicular the surface is to the viewpoint. This makes it easy to see the twisting ribbons in the center, while maintaining a well-defined edge to the surface.

27.4.1.6   Combining Techniques

Any of the above idioms may be combined to illustrate different aspects of the dataset (e.g., see Figure 27.21). The benefit of this is the ability to show correlations between various parameters. When combining idioms, however, care must be taken not to overload the display or occlude important information.

27.4.2   The Visualization Process

This subsection provides an overview of the process of creating various kinds of visualizations. We begin by looking at basic still images. We then add features such as choreography, real-time interaction, and, finally, simulation control from within the visualization process.

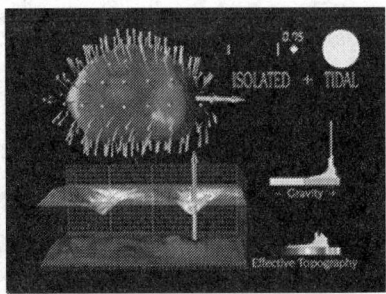

FIGURE 27.21 Multiple techniques combine to show gravity pull on the surface of Mars's moon Phobos. (Courtesy of Wayne Lytle, CTC.)

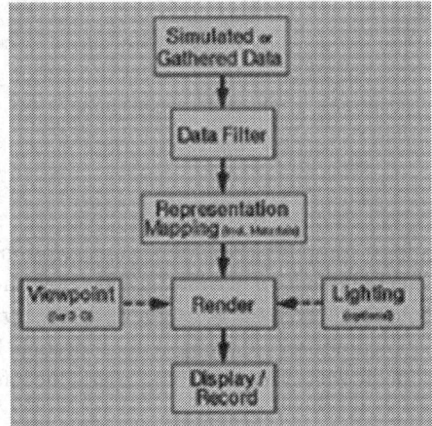

FIGURE 27.22 The basic visualization process. (Courtesy of W. Sherman.)

27.4.2.1 Still Imagery

The basic method for transforming numeric data into visual output is depicted in Figure 27.22. In this diagram, information flow follows the arrows, with dashed lines indicating optional parameters. At the top, data are provided from some numeric source such as a mathematically based computational model, a conceptually based computational model, or a collection of observed values. Data filtering involves a wide range of operations. Data must be translated into a form that is required by the visualization software. Additionally, the filtering process is used to sift out the most relevant aspects and discard unnecessary values.

After the data are filtered, they are mapped to some geometric form. At that point, decisions must be made about the materials and the characteristics of what these objects will look like — are they shiny, are they smooth, are they rough, are they dull, are they red, are they green, are they semitransparent? These parameters can also be driven by the computational model within the constraints imposed by the visualization software. Lighting information — i.e., how many lights, how diffuse they are, their relative location, etc. — is then combined with the viewpoint and information about the geometry by a computer program called the renderer. The renderer takes this information and computes the 2-D image which the eyes see. For a stereoscopic view, two images are rendered, from the point of view of each eye. The resultant images are then either recorded on film or video tape or displayed on a computer screen or other output device.

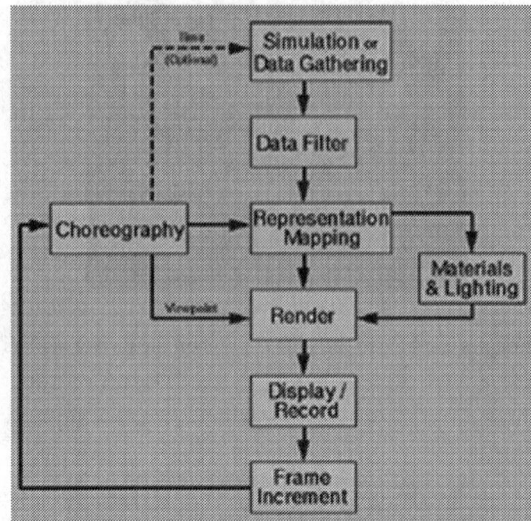

FIGURE 27.23   Process of creating a scientific visualization animation. (Courtesy of W. Sherman.)

27.4.2.2 Animation

The basic methodology for creating a computer animation is an extension of the process used to create the still image. The major extension is the concept of choreography. Now, the same steps are followed as in Figure 27.22, but the viewpoint information can be choreographed, along with the objects, their representations, and the passage of simulation time itself, resulting in the diagram in Figure 27.23. Animation is the result of rendering the scene repeatedly while varying the choreographic information and then displaying the images in rapid succession.

Choreography is that part of the process which controls the viewpoint of the scene and the movement of objects within the scene. Choreography can also be used to control how representations of the data change during an animation, such as the transparency of a material of a particular idiom.

Another important element of the dataset that can be controlled by the choreographer is *time*. Usually, time is considered to be constantly moving, at least as long as the animation is playing. However, there are two notions of time in a visualization animation. One is the passage of time of the viewer while watching the animation. The other, for visualizations of time-varying datasets, is the notion of time within the data. There is no requirement that data time move at a steady pace through the animation. It often can be insightful to hold time constant while other parameters of the representation are choreographed, revealing additional information. Moving time at different rates may reveal features that otherwise would be hidden.

27.4.2.3 Interaction

Animated sequences are made of several still images displayed in rapid succession. Often these images take longer to create than the fraction of a second that they will be displayed. This results in a situation where the choreography for all the action is planned ahead of time and is then unchangeable once the animation is complete. This is not generally the optimal method for a researcher who wants to probe a dataset interactively.

To have a more insightful experience with the data, scientists want to be able to spend more time looking at specific pieces. They want to be able to look at them from different angles, change representation parameters, and watch them at different rates of speed. To allow this kind of interaction, the images that are animated on the screen must be created as fast as they are displayed (i.e., in *real time*). Thus, the rendering stage in our visualization figure becomes a real-time (RT) process, and the display–record stage is now display only, with the option of recording (Figure 27.24).

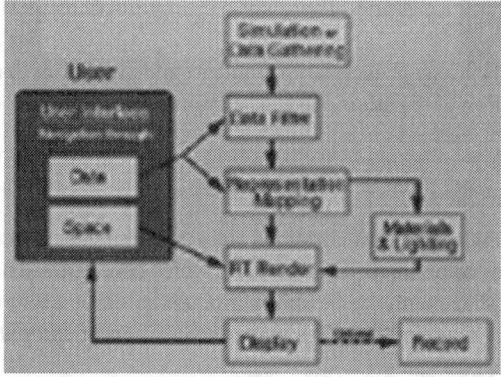

FIGURE 27.24 Interactive visualization allows the user to control the data filter, the representation, and the viewpoint. (Courtesy of W. Sherman.)

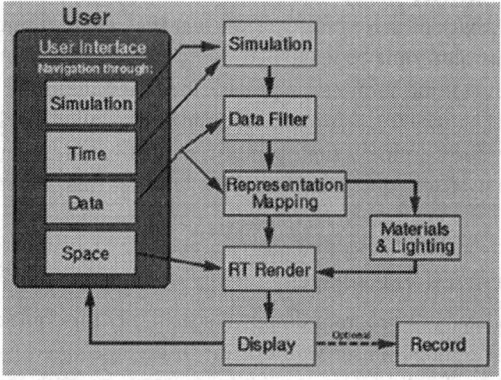

FIGURE 27.25 Interactive steering allows the user to manipulate the simulation in addition to the visualization. (Courtesy of W. Sherman.)

The most significant change in the interactive visualization process is that the choreography now happens in real time too; instead of an animator creating the choreography, the scientists themselves choreograph the scene through the package's user interface (UI). At this level of interaction, users are able to guide themselves through space and time and control the representation of the data (Figure 27.24).

27.4.2.4 Interactive Steering

Interactive steering is the ability for the user to alter the course of the simulation in real time [Haber 1989]. Figure 27.25 further extends the visualization flow chart to include user control of the simulation — including direct control over passage of time. It is vital that both the simulation and graphics system provide real-time performance and allow for user input. The user can then interactively steer the simulation by modifying variables and data in the simulation, as shown in the figure. The user can still steer the processing of the data into graphical representations and steer the viewpoint of the scene.

27.4.3 Visualization Tools

A number of tools are available for creating visualizations of information. The choices have improved with the evolution of computer technology. Visualization tools provide varying elements of the visualization processes shown in the figures. They can be categorized as follows: plotting libraries, turnkey packages, dataflow packages, and animation packages. There is also the option of creating the visualization by writing the software in-house.

Each package (and type of package) has certain advantages and limitations that may make one more suitable for a particular application or situation. We will examine some examples of each type of visualization tool.

27.4.3.1 Plotting Libraries

Historically, computer-graphics-based visualization originated with the use of alphanumeric printouts to mimic a bar graph, a shaded "color" map, or values from which contours can be hand-plotted. Computer visualization began to take shape with the flexibility brought about by new output devices, such as pen plotters and vector displays. These new devices made the creation of computer-generated contour and vector plots more feasible.

Software libraries were developed that enabled researchers to generate charts, graphs, and plots without the need for reinventing the graphics themselves. These packages have evolved as the computers and computer I/O have improved and are still available today. Since these packages are basically programming interfaces, they do not typically involve any form of interaction other than through programming and (perhaps) a command-line interface.

The limits on the interactivity of these tools suggest that they are of limited functionality. Although this is partially true, there are also some benefits. Because scientists often want to work with quantitative representations of their data, plotting libraries can provide the easiest means for producing the appropriate output. Print output has much higher resolution than screen output, and these libraries can produce high-resolution plots specifically for output to such printers. Screen-based interactive packages often limit their resolution to what is available on the screen, thus limiting their ability to produce quantitative visualization.

27.4.3.2 Turnkey Visualization Packages

A turnkey visualization package is a program designed specifically for doing visualization and contains controls (widgets) for most options a user would want to exercise when visualizing the data. This is accomplished through the use of pulldown menus or popup windows with control panels.

A number of commercial and no-cost visualization software packages are available. Most packages are available for Unix workstations, and many are also available for PC compatibles and Apple Macintoshes. Commercial packages include Fluent's CFD software package [Fluent], NASA/Sterling Software's FAST [FAST], and Fortner Research's visualization suite [Fortner]. Freely available packages include VIS-5D from the University of Wisconsin [Vis5D], Sci-An from Florida State University [Sci-An], and Gnu-Plot from Dartmouth College [Gnuplot].

There are definite advantages to using turnkey visualization software. The primary advantage is that one can often read one's data directly into the packages and immediately begin manipulating the controls to look at various aspects of them. This assumes that the data are initially in a format readable by the package and that the controls are intuitive enough for someone to experiment with and get good results.

These packages are designed as interactive tools; their ability to manipulate the view and the visualization parameters is key to putting the visualization process into the hands of individual researchers. Also, because the target user is a scientist, the user interface is usually designed for ease of learning and use.

Off-the-shelf packages also have some limitations, however. Foremost is that their flexibility is often limited. For instance, the user won't be able to add visualization representations that are not provided by the software. Also, if too many representations are included, then the user interface may suffer from an overwhelming number of control selections, making it hard for the user to find the control needed to perform a specific operation.

The desire for interactivity also imposes limits on these packages. To be interactive, data must be easily retrievable, and the representational forms should not be rapidly changing. This means that displaying features of time-dependent data may be too slow to be very interactive or may not be allowed by the package at all. To have interactive real-time rendering, many constraints are put on the amount of complexity the images can have. Generally, very beautiful imagery requires more complexity than can be achieved in real time. One solution for this is to allow the user to experiment with representations and choreography of the data; when a satisfactory situation is selected, information can be sent to a noninteractive rendering system to produce nicer images.

27.4.3.3 Dataflow Packages

Like the turnkey visualization packages above, dataflow visualization packages are designed as tools to be used directly by the scientist. However, dataflow packages are much more modular, with each stage of the visualization process represented as an independent unit. These units are then connected in the appropriate manner to allow data to be passed (or *flow*) from one unit to the next. This style of interface is inherently more flexible and also provides a map of the visualization process the system is using. These packages also are designed to be more extensible, allowing the user to add features and functionality that are not provided off the shelf.

Dataflow software currently requires the availability of a Unix workstation. Most packages now run across different brands of workstations. Despite the narrowing gap between PC and workstation, not many run on PCs. Three factors that contribute to this situation are the requirement for doing a considerable amount of processing of the data, the use of large amounts of memory, and the need for large-screen displays with reasonable graphics performance.

The dataflow concept consists in breaking down tasks into small programs, each of which does one thing. Each task is represented as a *module* (a software building block, or "black box") that performs the specified operation. Each module has a defined set of required inputs and outputs for passing information between modules. Figure 27.26 shows a simple connection network of some modules. In this example, data are retrieved with the Read HDF module and flow to the Isosurface module, which creates a geometric representation of the data, passing the new information to the Render module, which renders a pixel map, which in turn is passed to the Display module, which puts the image on the screen.

Though the dataflow concept had been described for doing other tasks (such as image manipulation), AVS (Application Visualization System) was probably the first package to apply the dataflow concept to the task of doing visualization [Upson et al. 1989, AVS]. AVS is available on most Unix workstations. IRIS Explorer [IRIS Explorer], originally developed for Silicon Graphics workstations, is also now available on other workstations. IBM's Data Explorer [IBM Data Explorer] is another example of commercially available software. Khoros [Khoros] is one freely available package of this nature. SCIrun, a powerful new implementation of the dataflow paradigm, is also freely available to some institutions [Johnson and Parker 1995]. The user interfaces for most of these programs are surprisingly similar, and familiarity with one package makes learning another easier.

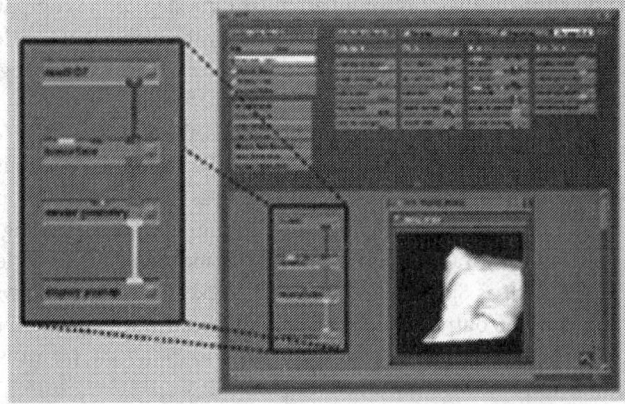

FIGURE 27.26 (See Plate 27.26 in the color insert following page **29-22**.) Dataflow packages such as AVS connect modules to customize a visualization. (Courtesy of W. Sherman, NCSA.)

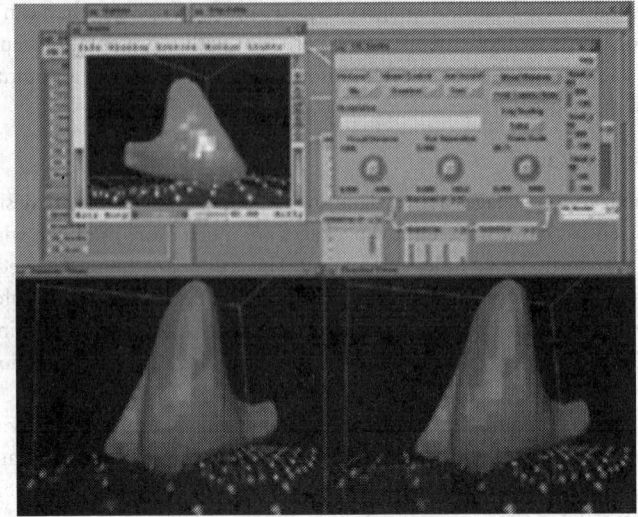

FIGURE 27.27   (See Plate 27.27 in the color insert following page **29**-22.) Customization capabilities allow the package to be extended to new display paradigms, such as in this stereo image linked to a virtual-reality viewing device. (Courtesy of W. Sherman, NCSA.)

Advantages of systems designed around the dataflow paradigm are numerous. As with turnkey packages, real-time rendering coupled with user control of the viewpoint and visualization parameters allows researchers to experience their data first hand. Flexibility is enhanced by the user's ability to connect modules together, creating a network of modules specific to the task at hand.

Flexibility is raised another degree by having the option to create new modules that perform operations not provided by the base collection of modules provided with the system. New modules may be written in common programming languages, such as C or Fortran. By utilizing routines provided with the system, the new modules will exhibit the same user interface and data handling as the standard modules. Since many people typically share the same desire for specific extensions, sharing of modules is common practice, thus making these packages even more valuable.

There are some disadvantages to this form of visualization package. For many, the primary disadvantage is the fact that most dataflow packages are available only on Unix workstations. Another is the difficulty with which this type of package handles data that change over time as part of a real-time visualization. Typically, data flowing down the network is for a particular time step of the computational simulation, and the modules that provide the representation of particle flow through a vector field will not be able to get the data fast enough to give smooth-flowing animation of those particles in a changing dataset (though they generally can handle particles flowing through a time-static dataset).

And last, the dataflow interface *is* a type of programming; thus, the end user at one level must learn how to program the system. While easier than writing code in C or Fortran, this process is more complicated than in the turnkey approach. The degree of difficulty increases as the networks become complex and the application has greater requirements, such as looping time-varying data that are flowing through the system.

The modularity of the dataflow paradigm also allows for the addition of a new style of interfacing with the user. New input devices such as Spaceballs™ and 3-D trackers can allow the system to control the viewpoint of the environment. New output devices can provide a head-tracked, stereo view, allowing *virtual reality* to become a potential method of interacting with a visualization [Sherman 1990] (Figure 27.27).

27.4.3.4 Animation

Today, presentation-quality visualization animations typically use standard computer animation packages [Fangmeier 1988]. Until recently, computer animation packages were the only off-the-shelf software available that provided 3-D rendering of any sort. Thus, these were often used to provide the rendering stage of the visualization process.

Computer animation packages are designed for all types of computer animation, not just for use in scientific visualization. However, it can be said that whatever they are used for, they are used to show (or visualize) the world the animator has conceptualized. For scientific visualization, the objects in the scene are typically derived from data. An animation package typically has separate components that allow the modeling of objects; the choreography of the objects, lights, and viewpoint; and the ability to design visual material qualities for each of the objects. These are all tied together into an image or series of images by the rendering component.

The primary advantage of using animation packages is that they are extremely flexible. The user is able to create objects of any type or size to represent the subject of the visualization and any supporting objects that might be required to give context to the representations. In a visualization environment, most of the objects are created by writing specialized programs that convert data from their native format to a format suitable for the animation package. This flexibility does have limits, however, such as when there is a need to represent the data using a method not available in the package. For example, a volume rendering technique may be required, and the renderer for the package may support only polygonal techniques.

Another advantage of animation packages is the ability to create very high-quality imagery. It is important to note, though, that the image quality is also dependent on the work done by the visualization creator.

The flexibility and complexity of animation packages also can be a disadvantage. These tools can be sufficiently complex that to use them as a visualization tool requires an outside expert to bring the visualization to fruition. Such an expert would be able to write the custom programs that convert data from their original state to an appropriate format. The fact that the consultant doing the visualization work is usually not an expert in the field of study being represented also can be a disadvantage.

Because each image in an animation can be very detailed, rendered using computationally expensive techniques that give rise to high-quality scenes, the time required to create the image is unlikely to be short enough to allow for what is referred to as real-time rendering. In fact, it can take several minutes to several hours to produce an image that will be viewed for one-thirtieth of a second (at video rates). The time required is a disadvantage even in situations where months are available to produce the animation, because of the delay between the time a small change is made and the time the scientist can view the resulting work.

Another disadvantage of using an animation package is the increased potential loss of data integrity that can occur in the process of translating the data from their original form to a form suitable for the animation package. The likelihood of data integrity problems is higher in this environment for two reasons: more custom work is done to translate the data, and the animation package may have options to make the output look good at the expense of giving an accurate view of the object being rendered. In an off-the-shelf visualization package, the techniques used to bring data to the screen are often well-known ones that a user can research and learn how they may affect the resulting image. Custom code used with animation packages may not be as heavily scrutinized if it is used only once for a particular application. Also, animation packages have features such as automatic smoothing that take the rough spots out of the data. This changed representation probably will not give an accurate view of the original information.

27.4.3.5 WYOS: Write Your Own Software

Before visualization and animation packages were available, tools to look at data were custom programmed for the task at hand. This is still done for certain visualizations — especially ones that handle large amounts of time-varying data or require new computer graphics techniques that have not been implemented in off-the-shelf software (e.g., certain volume visualization techniques). Now, however, users generally choose to use off-the-shelf software, perhaps with a few modifications.

Some examples of visualizations requiring custom renderers include *L.A. — The Movie* [Hussey et al. 1986], which used special techniques for overlaying satellite imagery onto a topological map of the Los Angeles region, and *The Deluge* [Sims 1989], which brings a still image to life by using a renderer written specifically for that task.

The only advantage to writing your own application software is in situations where that is the only way you can obtain the visualization you need.

27.4.3.6 Tools Summary

In summary, many tools have been developed which aid in visualizing scientific data. These tools are improving, making visualization an easier task to perform. However, even with a very easy user interface, the skill of the visualization developer is the overriding issue. No tool, no matter how easy to use, can replace the skill and insight of a visualization expert.

27.4.4 Examples of Scientific Visualizations

We have explained the components of scientific visualization and examined how these are used to create output that the researcher can use. In order to have a better understanding of how visualization is typically used in practice, and to look at how representational idioms have evolved, we will describe a selection of visualizations used as a means for scientific study.

27.4.4.1 The Study of a Severe Thunderstorm

The primary example of scientific visualization we will examine in this chapter is the computational simulation of severe thunderstorms. The simulation of meteorological processes has a long history in the computational sciences and consequently can be used to demonstrate the evolution of many visualization techniques. In particular, we will look at the research performed by Robert Wilhelmson's research group at NCSA and the Department of Atmospheric Sciences at UIUC and the visualizations done by his group and the NCSA Visualization Group.

27.4.4.1.1 Evolution of Atmospheric Simulations and Visualization

Wilhelmson's group has been studying the process of thunderstorm development computationally for 25 years. Over this period, several developments have been made in scientific visualization, and these are reflected in the many visualizations made of his work.

In the 1950s, the simulations run on digital computers were so simple that the researchers could literally examine every number produced. In the 1960s, digital computers had become fast enough to be used to execute 2-D simulations of atmospheric conditions over time. At first, the data were examined simply by printing the numbers in a 2-D array on paper. The researcher could then actually draw contours through the data by hand based on the printed numbers. An alternative early technique was to print a 2-D array of letters, with each letter representing a range of data values (this was often referred to as "gray-shading," with each character effectively representing a different level of gray) (Figure 27.28). About this time, a package for producing standard visualizations of 2-D data was released by the National Center for Atmospheric Research (NCAR). This package, known as *NCAR-graphics* [NCAR] could produce representations such as the image of contour levels within the developing storm front shown in Figure 27.29. In general, all the visual output of the simulations were of a static slice of time.

The 1970s saw the true advent of 3-D simulation models for all areas of atmospheric science. With the expansion into three dimensions, it became more difficult to interpret and explain the results. By this time, NCAR-graphics was the primary visualization package used by atmospheric sciences. With the addition of the ability to produce 3-D isocontours (i.e., isosurfaces) (Figure 27.30), Wilhelmson was able to view storms from a real-world perspective and from a radar perspective, among others.

Other techniques that began to be used in the late 1970s include the use of color to visually separate different representations, giving the ability to view them simultaneously. Color also was used to draw different layers of isosurfaces within a 3-D structure. In addition to color, Wilhelmson began to look at the change of the system over time; initially, this was limited to an alternation between two images on a raster-frame storage device.

In the early 1980s, the use of flow tracers (particles) to visualize flow through a field began in earnest — though some had been done in the late 1970s. Computing particle location was a new way to represent

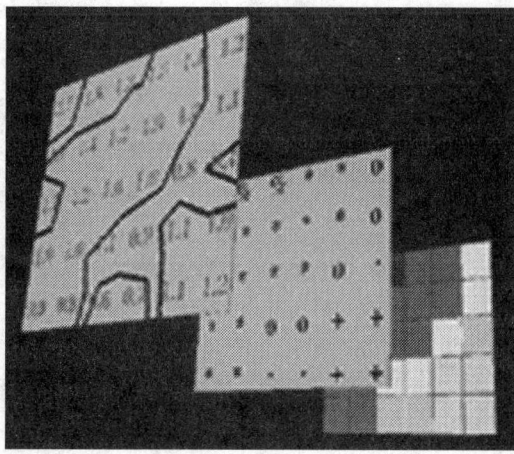

FIGURE 27.28 Early visualization attempts included hand-drawn contours and shaded images constructed from a judicious choice of ASCII characters.

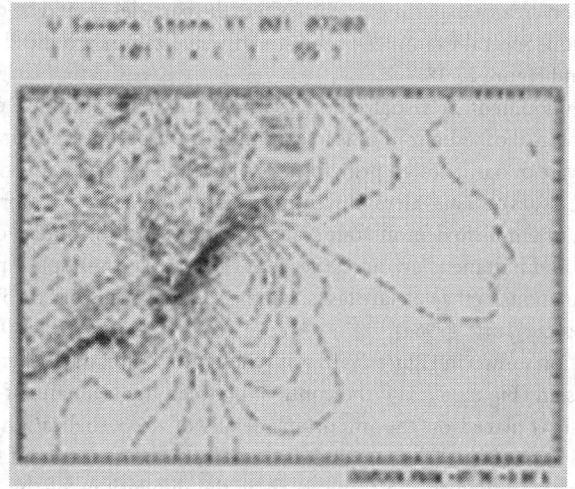

FIGURE 27.29 Developing storm front.

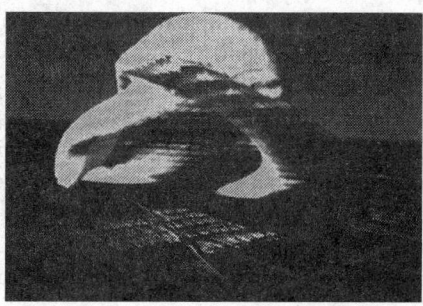

FIGURE 27.30 An early isosurface representing the late stage of the storm.

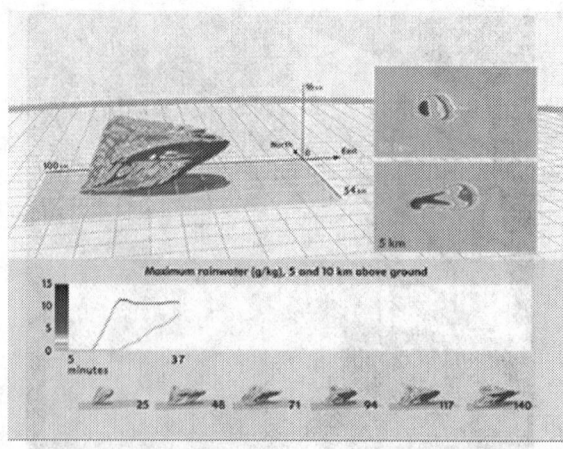

FIGURE 27.31   The thunderstorm visualization was redesigned with less contrast between the grid and the ground and small multiples added to represent changes in time. (Courtesy of Yale University, NCSA.)

flow through a fluid substance, providing the ability to view a large region of the simulation at one time. In practice, the visualization creator experiments with where the particles should be released to show specific aspects of the simulation. Several examples of the storm visualization produced in 1989 can be seen in Figures 27.13, 27.15, and Figure 27.18.

In line with the development of visualization technology is the continued exploration of effective information design. The goal of visualization is to communicate aspects of the data clearly. To achieve this clarity, one must be aware of basic design principles and have the ability to effectively critique the design quality of visualization. Issues of visual hierarchy, composition, choreography, color theory, and typography cannot be disregarded. Attempts to demonstrate scale, provide a sense of context, combine qualitative and quantitative information simultaneously, direct the viewer's eye, create multiple representations, and point out intentional manipulations in the data are tasks that the creator must work through in order to produce quality information design [Tufte 1996a].

The storm visualization shown in Figure 27.18 was redesigned to illustrate several issues of information design. In the new design (Figure 27.31), the contrast between the elements of the scene was adjusted so that visual emphasis is placed on the important details of the visualization. The grid floor in the original animation is a strong visual element and distracts from the dynamics of the cloud development. Reference information was added to the scene, such as the dimensions of the computational domain and the diagrammatic clock at the bottom. This clock includes a stripe of small images that illustrate the entire duration of the visualization. This gives contextual information by indicating which portion of the animation one is viewing and in addition shows the numerical time step for specific reference. The two-dimensional cutting planes in the scene in conjunction with the three-dimensional cloud, and the quantitative scale at the bottom, provide multiple representations of the data for comparison and clarification [Baker and Bushell 1995, Tufte 1996b].

Today, we are reaching the point where the flow of large numbers of particles can be computed and visualized. Too many particles, however, can make it hard to see what is going on, so the technique of selectively chosing to display specific particles is being explored. In the picture in Figure 27.32, particles are used to represent the flow of a tornado. To highlight just the tornado, only particles with high vertical vorticity, within a certain region, are shown. The amount of data and objects required to represent it also brings about a logistical problem of where all this information is stored. Large disk drives are required to create visualizations of massive numbers of particles, and files must be carefully managed to stay within the constraints of the storage.

Virtual reality (VR) also is being experimented with as a means of gaining insights from simulated scientific experiments [Baker and Wickens 1995]. VR experiences have been designed for the VROOM

FIGURE 27.32   Thousands of particles show the shape of a tornado. (Courtesy of Matthew Arrott.)

[Brown 1994] and SuperVROOM [Korab and Brown 1995] venues at SIGGRAPH '94 and SuperComputing '95, respectively. In these experiences, the domain of the data is represented by a bounding box, with various tools available to represent the data, including isosurfaces, and particle flow. In a CAVE™ [Cruz-Neira et al. 1992] VR system, the user can easily navigate the data space and has the ability to control the representations. In the particle flow representation, the user can control the release of particles with a hand-held device. As the user releases the particles, a flow simulation program is begun that updates the particle positions. This program (and any of the representation modules) can execute on a separate larger supercomputer for better interactivity.

It is hard to judge how much additional insight can be gained by immersive interactive visualizations (i.e., virtual reality). This stems from the deep knowledge atmospheric scientists already have of their simulations. Major concerns of the scientist in using virtual reality include: Would I use a device that requires me to be standing or walking for an extended period? Analyzing the data requires knowing where one is with respect to the cloud, which is easier when viewing the cloud from the outside — in which case, what is the advantage over the desktop? To be truly immersed, the user should feel the wind, hear the thunder, see the lightning, none of which is accurately representable with current VR technology.

Scientists have become proficient at deciphering the 3-D content from flat 2-D displays from years of experience. Because of this, they see less need for the improved 3-D cues that VR provides. However, they are continuing to explore the use of virtual reality as a visualization tool. Since it is a new medium, there are many technological and user interface improvements ahead; VR still may prove to provide unique insight not obtainable through other means.

The World Wide Web (Figure 27.33) is viewed as a very useful tool for disseminating research information. It can be used by the public to access weather forecast information or by other atmospheric researchers to directly examine the simulation data themselves. It can be the complete tool for a researcher — the notebook that contains all the diary entries over the course of a study, including explanations, data, images, animations, and interactive visualization tools. It can be made available not only to the researcher but also to colleagues, advisors, and (if desired) the scientific community [Jewett and Wilhelmson 1995].

27.4.4.1.2   *What Does the Scientist Use Today?*

With all that is available to the scientist today, what tools are actually used? To be specific, everything. Line graphs, 2-D contours, and 2-D vector plots still provide the basis for closely investigating the results of a simulation. Three-dimensional stills and animations are excellent tools for illustrating concepts. Interactive visualization tools help to find out what's in the dataset and to locate interesting regions that demand close inspection.

Animation, while a good tool for presenting the overall contents of a research study, poses problems when it comes to publishing the study. Scientific publishing still exists primarily as a print medium, and

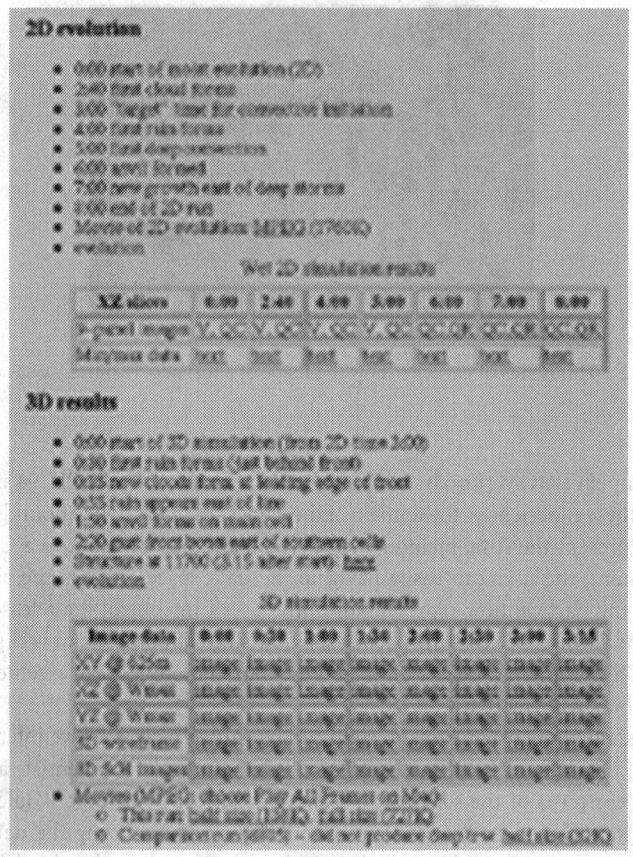

FIGURE 27.33 Portion of a WWW page used to document a research project. (Courtesy of Brian Jewett; NCSA.)

thus it is difficult to include animation. Indeed, even submitting an animation for inclusion in the review process is difficult. The World Wide Web is beginning to provide a solution to these problems. The entire research submission, including animations and interactive tools, can be included as one comprehensive document for the review. Once reviewed, such a submission can be published on CD-ROM, while still remaining available online where it can be periodically updated.

27.4.5 Visualizing Smog

Smog: Visualizing the Components is another animation visualization in the domain of atmospheric science. However, there are some interesting information presentation features that are highlighted in this animation.

One segment of the *Smog* animation has particles that represent different forms of pollutants flowing through the atmosphere, based on air movement data. Unlike most particle flow animations where particles are released from locations within the simulation to highlight specific aspects of flow, the particles in this animation emanate from actual locations of pollution sources, at a rate based on measured releases. This animation also demonstrates a slightly different form of the wind-tracing ribbons known as streaklines (Figure 27.14).

An additional representation of the data in this animation was made through the use of sound. A sirenlike tone reinforces the information displayed in the XY plot (Figure 27.34). It also experimented with mapping the ozone level to the repetition rate of digital recordings of coughs.

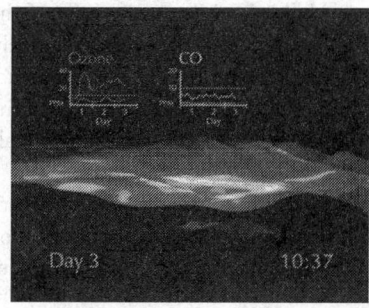

FIGURE 27.34   Los Angeles smog visualization. (Courtesy of NCSA.)

27.5   Research Issues and Summary

Though great strides have been taken in developing scientific visualization as a tool in computational science, a multitude of issues remain for further research and development. As computer power increases, the need for effective visualization tools increases due to the larger volume of information being produced. Fortunately, the increase in computer power also allows us to develop more compelling visualization tools with more flexibility, better performance, and higher fidelity.

When studying the field of visualization, it is important to recognize that visualization is being done effectively in a variety of fields outside of the realm of computational science. Medical imaging, factory automation, and financial researchers, to name a few, are taking advantage of the power of visual representation of information.

Additionally, there is a growing wealth of knowledge about the visual representation of information, and how we perceive and interpret visual information, being published by researchers in fields such as cognitive psychology, perceptual psychology, human factors, etc. As visualization researchers and tool developers incorporate new findings, we may take better advantage of visualization as a powerful analysis and communication tool. The art and design community also is demonstrating new ways of presenting and representing information. The use of sound in representing scientific data is becoming more important. Research is being done to find what types of information are best represented in audio.

In order to move beyond the current level of visualization tools, we must continue to develop the basic technologies and algorithms involved in the image-generation aspects of the process — for example, improved rendering algorithms, improved user interface, improved representations, etc.

In the ideal world, the computational scientist would not be required to think about human visual perception, optical illusions, how cultural biases affect the interpretation of imagery, or how color is understood. If research in the appropriate areas can be integrated into the tools which scientists use, we can raise the level of effectiveness of the visual analysis and presentation of science and, indeed, the science itself. Research is currently under way [Baker 1994] to develop tools which do automatic idiom selection. Based on input from the scientists regarding the type of data, the insight they are trying to achieve, etc., the tool will generate representations that have a high likelihood of being useful.

Visualization is rapidly moving beyond the researcher's workstation and videotape presentations. New media and communications systems are allowing the scientist to collaborate remotely over networks and share data, imagery, etc. The Internet and the World Wide Web are allowing the scientist to publish findings, including visual representations, to a worldwide audience. These new media bring a variety of new concerns and research issues. Some examples include the size and resolution of imagery that it is possible to share. Current network speeds render it impractical to share multiple large, high-resolution images. Animations also are just beginning to be practical over the network. As data compression techniques and sharing of executable code over the network improve, the scientist will truly be able to explore this new medium.

Current work in VR techniques is enabling scientists to see and interact with their research in many new ways. Stereoscopic imagery, wide fields of view, and tracking of the viewer's head enable the scientist to become immersed within the simulation and become a part of the system. Current systems suffer from a lack of resolution, tracker lag, etc. As these improve, virtual reality promises to be a powerful tool for computational scientists. Many scientists who have currently abandoned virtual reality systems with their clumsy, low-resolution head-mounted displays are finding that the less intrusive, higher-resolution projection-based displays such as the CAVE™ are transforming virtual reality from a novelty to a useful tool for analysis and display.

In summary, the field of scientific visualization is sufficiently mature to allow the scientist to harness the power of the eye–brain connection for data analysis and presentation. At the same time, it is breaking new ground with respect to online collaborative computing, worldwide publication, and fully immersive interaction through virtual reality.

Acknowledgments

We would like to thank Robert Wilhelmson for sharing his wealth of experience and historical perspective on visualization in computational science; also, from the Atmospheric Science Group, Crystal Shaw and Brian Jewett for additional support. Thanks also to the reviewers and editors for helpful comments and suggestions to improve this chapter.

Many of the ideas in this chapter were the result of our interactions with a variety of scientists. We have had the opportunity to work on a wide variety of challenging visualization problems with numerous world-class researchers.

We gained a much greater insight into the field of scientific visualization as a direct result of interaction with the participants of the Representation Project, including the guest speakers from a wide variety of disciplines and institutions. In particular, the members of the former Visualization Group at NCSA including Boss Dan Brady, Matthew Arrott, Mark Bajuk, Ingrid Kallick, Mike Krogh, Mike McNeill, Gautam Mehrotra, Jeff Thingvold, and Deanna Spivey. This group was intellectually stimulating, an endless source of talent and ideas, and most of all, our friends. We feel honored to have been part of such a magical team.

Defining Terms

Alpha shapes: A technique that allows one to represent the concept of shape applied to a collection of points in space.

Choreography: In computer animation, the timing and sequencing of activity and representation.

Computer graphics: The medium in which modern visualization is practiced.

Dataset: A value or set of values that is the input or output of a computational simulation.

Glyph: Generally, a symbol used to represent information. For example, in visualizing vector fields, an arrow often is used to show the direction and magnitude of the vector value at each location.

Interactive steering: The practice of dynamically modifying the parameters of a running simulation, guided by a real-time visualization of the simulation's progress.

Isosurface: The shape defined within a volume of scalar values on which all the values are equal to some constant.

Parallax: The difference in the apparent position of an object caused by a change in the point of observation.

Perceptualization: A term perhaps more suitable than "visualization," in recognition of the efficacy of using auditory and tactile techniques for representing and communicating about scientific data.

Scientific illustration: The traditional use of graphics created by an artist to show scientific concepts.

Scientific visualization: The use of computer-generated graphics, often animated, interactive, and three-dimensional, to represent scientific data and concepts.

Simulation: A computer model of natural phenomena.

Streakline: A line showing the path taken by all particles that pass through a given location in a vector field.

Streamline: A line drawn in a vector field such that, at any instant, the tangent to the line at any point on the line is the direction of the flow. Often restricted to fields with steady flow, in which case the streamline shows the path of a tracer particle.

Tracer: An animated symbol, usually a sphere, showing the path that would be taken by a particle in a vector field.

Vector field: An n-dimensional collection of vector values arranged in space, such as the wind velocity over a two-dimensional surface.

Virtual reality: A medium composed of highly interactive computer simulations that utilizes data about the participant's position and replaces or augments one or more of their senses — giving the feeling of being immersed, or being present, in the simulation [Sherman and Craig 1996].

Visual idiom: A technique for representing scientific data that has a commonly accepted interpretation.

References

Anninos, P., Bajuk, M., Bernstein, Seidel, E., Smarr, L., and Hobill, D. 1993. The evolution of distorted black holes. Physics Computing '93.

Arrott, M. and Latta, S. 1992. Perspectives on visualization, pp. 61–65. *IEEE Spectrum*, Sept.

AVS. AVS Home Page, http://www.avs.com/.

Bajuk, M. 1992. Camera evidence: visibility analysis through a multi-camera viewpoint,

Baker, M. P. 1994. KnowVis: an experiment in automating visualization, p. 456. In *Proc. Decision-Support 2001*, Toronto, Ontario, Sept.

Baker, M. P. and Bushell, C. 1995. After the storm: considerations for information visualization. *IEEE Trans. Comput. Graphics Appl.* 15(3):12–15 (May).

Baker, M. P. and Wickens, C. D. 1995. Human factors in virtual environments for the visual analysis of scientific data. *NCSA Tech. Rep.* 032, Aug.

Brooks, F. P., Jr., Ming O.-Y., Batter, J. J., and Kilpatrick, P. J. 1990. Project GROPE — haptic displays for scientific visualization. *Comput. Graphics (Proc. SIGGRAPH)* 24(4):177–185 (Aug.).

Brown, M., ed. 1994. *Comput. Graphics*, SIGGRAPH '94 Visual Proceedings, Aug., Orlando, FL.

Cruz-Neira, C., Sandin, D., DeFanti, T, Kenyon, R., and Hart, J. 1992. The CAVE audio visual experience automatic virtual environment. *Comm. ACM* 35(6):64–72. (June). URL: http://www.ncsa.uiuc.edu/EVL/docs/html/CAVE.html.

Drebin, R. A., Carpenter, L., and Hanrahan, P. 1988. Volume rendering. *Comput. Graphics (Proc. SIGGRAPH)* 22(4):64–75 (Aug.).

Edelsbrunner, H. and Mucke, E. P. 1994. Three-dimensional alpha shapes. *ACM Trans. Graphics* 13:43–72.

Fangmeier, S. M. 1988. The scientific visualization process, pp. 26–38. SIGGRAPH '88 Course Notes, Course 20: Computer Graphics in Science.

FAST. Flow Analysis Software Toolkit (FAST) home page. http://www.nas.nasa.gov/ NAS/FAST/fast.html. See also Sterling Software WWW Home Page. http://www. sterling.com/.

Fluent. Fluent Incorporated Home Menu Page. Fluent CFD software. http://www.fluent.com/.

Fortner. Fortner Research LLC. http://www.langsys.com/langsys/.

Gnuplot. Gnuplot. http://www.cs.dartmouth.edu/gnuplot_info.html.

Haber, R. B. 1989. Scientific visualization and the rivers project at the National Center for Supercomputing Applications. *Computer* 22(8):84–89 (Aug.).

Herron, D. K., Bollinger, N. G., Chaney, M. O., Varshavsky, A. D., Yost, J. B., Sherman, W. R., and Thingvold, J. A. 1995. Visualization and comparison of molecular dynamics simulations or leukotriene C(4), leukotriene D(4), and leukotriene E(4). *J. Mol. Graphics* 13:337–341, Elsevier Science, New York.

Hussey, K., Mortensen, B., and Hall, J. 1986. Jet Propulsion Lab Animation: *L.A. — The Movie*. Visualization in scientific computing. *ACM SIGGRAPH Video Review*, No. 28.

IBM Data Explorer. IBM Visualization Data Explorer (DX). http://www-i.almaden.ibm.com/dx/.

IRIS Explorer. IRIS Explorer Center. http://www.nag.co.uk:70/1h/Welcome_IEC.

Jewett, B. F. and Wilhelmson, R. B. 1995. Use of HTML and web tools in atmospheric sciences research. URL: http://redrock.ncsa.uiuc.edu/AOS/publications/IIPS96/web-atmossci.html.

Johnson, C. R. and Parker, S. G. 1995. Applications in computational medicine using SCIRun: a computational steering programming environment, H. W. Meuer, ed., pp. 2–19. In *Supercomputing '95*. URL: http://www.cs.utah.edu/~sci/.

Kaufmann, W. J., III, and Smarr, L. L. 1993. *Supercomputing and Science*. Scientific American Library, New York.

Keates, J. S. 1996. *Understanding Maps*, Hallstead Press. p. 39.

Khoros. The Khoral Research Home Page. http://www.khoros.unm.edu/home.html.

Korab, H. and Brown, M., eds. 1995. Virtual environments and distributed computing at SC '95. ACM/IEE allstead Press. Supercomputing '95, Dec., San Diego, CA.

Kovacic, D., Craig, A., Patterson, R., Romme, W., and Despain, D. 1990. *Fire Dynamics in the Yellowstone Landscape, 1690–1990: An Animation*. Model driven visual simulation, Proc. Resource Technology 90, Second International Symposium on Advanced Technology in Natural Resources Management.

Lorensen, W. E., and Cline, H. 1987. Marching cubes: a high resolution 3-D surface construction algorithm, *Comput. Graphics (Proc. SIGGRAPH)* 21(4):163–169 (July).

Lytle, W. 1993. The dangers of glitziness and other visualization faux pas. Animation, Cornell Theory Center, *SIGGRAPH 93 Electronic Theater*. No. 91.

Marshall, R., Kempf, J., and Dyer, S. 1990. Visualization methods and simulation steering for a 3-D turbulence model of Lake Erie. Symposium on Interactive 3-D Graphics. *ACM SIGGRAPH* 24(2):89–97 (Mar.).

Merzkirch, W. 1987. *Flow Visualization*, 2nd ed. Academic Press, Orlando, FL.

Moran, P. J. and Potter, C. S. 1992. Tiller: a tool for analyzing 4-d data. In *Visual Data Interpretation*, Proc. SPIE, 1668:124–128.

NCAR. NCAR Graphics Home Page. http://ngwww.ucar.edu/.

Scaletti, C. and Craig, A. B. 1991. Using sound to extract meaning from complex data. In *Extracting Meaning from Complex Data: Processing, Display, Interaction II*, Proc. SPIE, 1459:207–219.

SciAn. SciAn — Scientific Visualization and Animation Package. http://www.scri. fsu.edu/~mimi/scian.html.

Sherman, W., McNeill, M., Arrott, M., Bajuk, M., and Corson, A. 1990. Animation: Smog: Visualizing the Components. SIGGRAPH '90 Film & Video Theater. *ACM SIGGRAPH Video Rev.*, No. 62.

Sherman, W. R. 1990. Integrating virtual environments into the dataflow paradigm. Fourth Eurographics Workshop on Visualization in Scientific Computing, Abingdon, U.K., Apr.

Sherman, W. R. and Craig, A. B. 1997. *Working with Virtual Reality*. Morgan Kaufmann, San Francisco.

Sims, C. 1989. Animation: Leonardo's Deluge SIGGRAPH '89 Computer Graphics Theater. *ACM SIGGRAPH Video Rev.*, No. 52.

Tiede, U., Schiemann, T., and Hohne, K. H. 1996. Visualizing the visible human. In *IEEE Computer Graphics & Applications* 16(1).

Tufte, E. R. 1996a. *Envisioning Information*. Graphics Press, Cheshire, CT.

Tufte, E. R. 1996b. *Visual Explanations*. Graphics Press, Cheshire, CT.

Upson, C., Faulhaber, T., Kamins, D., Laidlaw, D., Vroom, J., Gurwitz, R., and van Dam, A. 1989. The application visualization system: a computational environment for scientific visualization. *IEEE Trans. Comput. Graphics Appl.* 9(4):30–42 (July).

Vis5D. Vis5D Home Page. http://www.ssec.wisc.edu/~billh/vis5d.html.

Weintraub, D. J. and Walker, E. H. 1969. *Perception*. Brooks/Cole, Belmont, CA.

Wickens, C., Merwin, D., and Lin, E. 1994. The human factors implications of graphics enhancements for the visualization of scientific data. In *1994 Human Factors*, Vol. 36. Also in *Proceedings of the 38th Annual Meeting of the Human Factors Society (1994)*, pp. 44–61.

Wickens, C. and Seidler, K. 1995. Information access and utilization. In *Emerging Needs and Opportunities for Human Factors Research*, R. Nickerson, Ed., National Academy Press, Washington, D.C.

Wilhelmson, R., 2nd. 1993. PATHFINDER: Probing ATmospHeric Flows in an INteractive and Distributed EnviRonment. In *Proc. 9th Conf. on Interactive Information and Processing Systems for Meteorology, Oceanography, and Hydrology*, Anaheim, CA, Jan.

Further Information

Brown, J. R., Earnshaw, R., Jern, M., and Vince, J. 1995. *Visualization: Using Computer Graphics to Explore Data and Present Information*. Wiley, New York.

Chambers, J., Cleveland, W., Kleiner, B., and Tukey, P. 1983. *Graphical Methods for Data Analysis*. Wadsworth International Group, Belmont, CA.

Dent, B. D. 1990. *Cartography: Thematic Map Design*. William C. Brown, New York.

Dondis, D. A. 1973. *A Primer of Visual Literacy*. MIT Press, Cambridge, MA.

Friedhoff, R. M. and Benzon, W. 1989. *Visualization*. W. H. Freeman, New York.

Gallager, R. S., ed. 1995. *Computer Visualization*. CRC Press.

Hearn, D. and Baker, M. P. 1994. *Computer Graphics*, 2nd ed. Addison–Wesley.

Huff, D. 1954. *How to Lie with Statistics*. Norton, New York.

Kaufmann, W. J., III, and Smarr, L. L. 1993. *Supercomputing and Science*. Scientific American Library, New York.

Keates, J. F. *Understanding Maps*.

Keller, P. R. and Keller, M. M. 1993. *Visual Cues*. IEEE Press.

Lauer, D. A. 1990. *Design Basics*. Harcourt Brace Jovanovich.

MacEachren, A. M. 1995. *How Maps Work*. Guilford Press, New York.

McCormick, B. H., DeFanti, T. A., and Brown, M. D. 1987. Visualization in scientific computing. *Comput. Graphics* 21(6): (Nov.).

Tufte, E. R. 1983. *The Visual Display of Quantitative Information*. Graphics Press, Cheshire, CT.

Wickens, C. 1992. *Engineering Psychology and Human Performance*, 2nd ed. HarperCollins, New York.

28

Computational
Structural Mechanics

28.1 Introduction **28**-1
28.2 Classification of Structural Mechanics Problems **28**-2
      Structural Characteristics and Source Variables
      • Different Classes of Structural Mechanics Problems
      • Deterministic and Nondeterministic Methods
28.3 Formulation of Structural Mechanics Problems **28**-3
      Different Formulations of Structural Mechanics Problems
      • Description of the Motion of a Structure
28.4 Steps Involved in the Application of Computational
      Structural Mechanics to Practical Engineering
      Problems **28**-5
      Major Steps in the Application of Computational Structural
      Mechanics • Selection of the Mathematical Models
      • Discretization Techniques • Model and Mesh Generation
      • Quality Assessment and Control of Numerical Solutions
28.5 Overview of Static, Stability, and Dynamic Analysis... **28**-9
      Static Analysis • Dynamic Analysis • Energy Balance in
      Transient Dynamic Analysis • Stability Analysis • Eigenvalue
      Problems • Sensitivity Analysis • Strategies and Numerical
      Algorithms for New Computing Systems
28.6 Brief History of the Development of Computational
      Structural Mechanics Software **28**-18
28.7 Characteristics of Future Engineering Systems
      and Their Implications on Computational
      Structural Mechanics **28**-19
28.8 Primary Pacing Items and Research Issues **28**-20
      High-Fidelity Modeling of the Structure • Failure and Life
      Prediction Methodologies • Hierarchical, Integrated Multiple
      Methods and Adaptive Modeling Techniques
      • Nondeterministic Analysis, Modeling, and Risk Assessment
      • Validation of Numerical Simulations • Multidisciplinary
      Analysis and Design Optimization • Related Tasks

Ahmed K. Noor
Old Dominion University

28.1 Introduction

Structural mechanics deals with (1) the idealization of actual structures and their environments and (2) prediction of response, failure, life, and performance of structures. In the last three decades the discipline of computational structural mechanics (CSM) has emerged as an insightful blend between structural mechanics, on the one hand, and other disciplines such as computer science, numerical analysis, and

1-58488-360-X/$0.00+$1.50
© 2004 by CRC Press, LLC

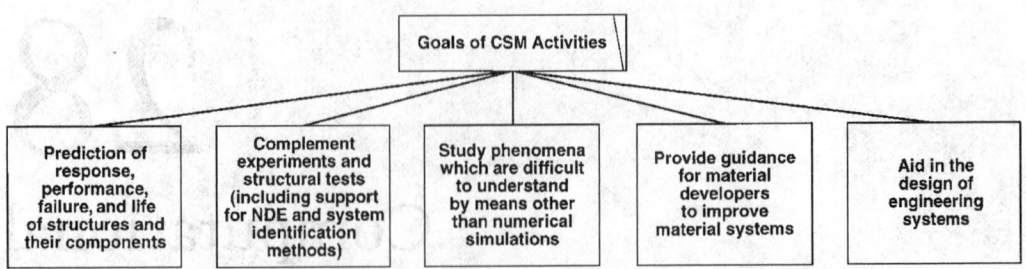

FIGURE 28.1   Five major goals of CSM activities (NDE refers to nondestructive evaluation techniques).

approximation theory, on the other hand. This rapidly evolving discipline is having a major impact on the development of structures technology, as well as on its application to various engineering systems. Development of the modern finite-element method during the 1950s marks the beginning of CSM. Finite-element technology is the backbone of many structural analysis software systems which are widely used by government, academia, and industry to solve complex structures problems.

The five major goals of CSM activities are shown in Figure 28.1. In support of these goals, current activities of CSM cover the study of phenomena, through numerical simulations, at a wide range of length scales ranging from the microscopic level to the structural level. Today, no important design can be completed without CSM, nor can any new theory be validated without it.

A number of survey papers and monographs have been written on various aspects of CSM (see, for example, Noor and Atluri [1987], Noor and Venneri [1990, 1995], and Ju [1995]). Also, a number of workshops and symposia have been devoted to CSM and proceedings have been published (for example, Grandhi et al. [1989], Noor et al. [1992], Oñaté et al. [1992], Ladevéze and Zienkiewicz [1992], Stein [1993], and Storaasli and Housner [1994]). The present chapter attempts to present, in a concise manner, the broad spectrum of problems covered by CSM along with the basic principles, formulations, and solution techniques for these problems. A brief history is given of the development of software systems used for the modeling and analysis of structures. The research areas in CSM, which have high potential for meeting future technological needs, are identified.

28.2   Classification of Structural Mechanics Problems

28.2.1   Structural Characteristics and Source Variables

The functions which govern the response, failure, life, and performance of structures can be grouped into four categories, namely:

Kinematic variables: e.g., displacements, velocities, strains, and strain rates
Kinetic variables: e.g., stresses and internal forces (or stress resultants)
Material characteristics: e.g., material stiffness and compliance coefficients
Source variables: which include environmental effects and external forces (e.g., mechanical, aerodynamic, thermal, optical, and electromagnetic forces)

The relations between the external forces and response quantities are shown in Table 28.1.

28.2.2   Different Classes of Structural Mechanics Problems

A number of classifications can be made for structural mechanics problems depending on: (1) presence or absence of uncertainties about the structural characteristics and source variables, (2) nature of the functions which govern the behavior of the structure (e.g., time dependence or independence), (3) the functional

TABLE 28.1 Relations Between External Forces and Response Quantities

Quantities	Relation	Type of Relation
External forces	Balance equations	Physical (conservation) laws
Stresses	Constitutive relations	Semi-empirical based on experiments
Strains Displacements	Strain-displacement relations	Geometric-based on logic

form of the relations between the source variables and response quantities (e.g., linear or nonlinear), and (4) the geometric characteristics of the structure and its components. A general classification, incorporating the aforementioned factors, is shown in Figure 28.2. Additional classifications can be made based on: (1) material response (e.g., homogeneous or nonhomogeneous, isotropic or anisotropic), (2) nature of source variables (e.g., conservative or nonconservative), and (3) coupling or noncoupling between source variables and response quantities (e.g., whether the changes in the source variables with structural deformations are pronounced or not).

28.2.3 Deterministic and Nondeterministic Methods

Deterministic methods of structural mechanics have become quite elaborate and include sophisticated mathematical models, highly refined computational methods, and optimization techniques. However, there is a growing realization among engineers during the past 25 years that unavoidable uncertainties in geometry, material properties, boundary conditions, loading, and operational environment must be taken into account to produce meaningful designs.

Three possible approaches for handling uncertainty can be identified, depending on the type of uncertainty and the amount of information available about the structural characteristics and the operational environment. The three approaches are: *probabilistic analysis, fuzzy-logic approach,* and *set theoretical, convex* (or antioptimization) *approach.* A discussion of the three approaches and their combinations is given in Elishakoff [1995]. In the probabilistic analysis, the structural characteristics and/or the source variables are assumed to be random variables (or functions), and the joint probability density functions of these variables are selected. The main objective of the analysis is the determination of the reliability of the system. Reliability is defined as the probability that the structure will adequately perform its intended mission for a specified interval of time, when operating under specified environmental conditions.

If the uncertainty is due to vague definition of structural and/or operational characteristics, imprecision of data and subjectivity of opinion or judgment, then fuzzy logic-based treatment is appropriate. The distinction between randomness and fuzziness is the fact that whereas randomness describes the uncertainty in the occurrence of an event (e.g., damage or failure of the structure), fuzziness describes the ambiguity of the event (e.g., imprecisely defined criteria for failure or damage — see Ross [1995]).

When the information about the structural and/or operational characteristics is fragmentary (e.g., only a bound on a maximum possible response function is known), then convex modeling can be used. Convex modeling produces the maximum or least favorable response, and the minimum or most favorable response, of the structure under the constraints within the set-theoretical description.

28.3 Formulation of Structural Mechanics Problems

28.3.1 Different Formulations of Structural Mechanics Problems

Several classifications can be made of the different formulations used for structural mechanics problems. Two of the major classifications are discussed next.

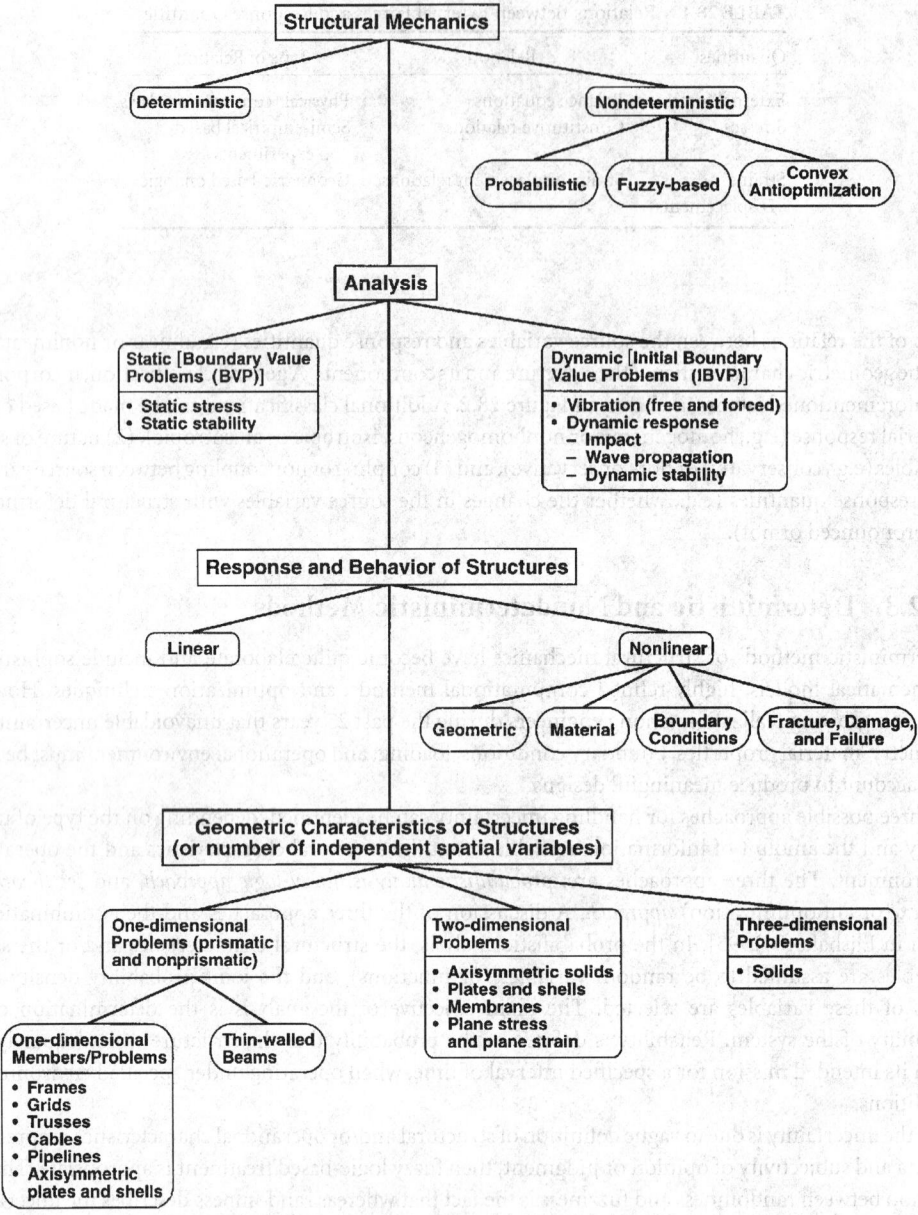

FIGURE 28.2 Classification of structural mechanics problems.

First, with respect to the approach used in deriving governing equations:

1. Differential equation formulation. This is the classical approach which is referred to as the *strong statement* of the problem.
2. Variational formulation. This is based on application of the principle of virtual displacements, and is referred to as the *weak statement* of the problem. A variety of variational principles for static and dynamic problems can be derived from the principle of virtual displacements (see, for example, Washizu [1982] and Geradin [1980]).
3. Integral, integro-differential or boundary integral equation formulation.

Some of the discretization techniques can be applied directly to the strong statement of the problem. The most notable example is the finite-difference method. Other techniques, such as finite elements, are typically used with the weak statement. Boundary element methods are generally used with the third formulation.

Second, with respect to fundamental unknowns in the governing equations or the governing functional:

1. Single-field formulation in which the fundamental unknowns belong to one field. The most commonly used formulation is the displacement formulation.
2. Multifield formulation in which the fundamental unknowns belong to more than one field, i.e., stresses and displacements. A detailed discussion of multifield formulations is given in Noor [1980].

28.3.2 Description of the Motion of a Structure

For nonlinear structural problems, three basic choices need to be made, namely, the approach used for describing the motion, the kinematic description (i.e., deformation measure), and the kinetic description (i.e., stress measure). Typically, the choice of the deformation measure determines the stress measure since the two have to be work conjugates. The two most commonly used approaches for describing the motion are as follows.

First is a *total Lagrangian description,* in which all the kinematic and kinetic variables are referred to the initial (undeformed) configuration. The Green–Lagrange strain tensor is used as the measure of deformation, and the associated second Piola–Kirchhoff stress tensor is used as the stress measure. For large-rotation problems, it is useful to separate the rigid-body movement from stretching by using *local corotational frames* and the polar decomposition method. This eliminates the problems associated with approximating finite rotations using trigonometric functions (or series expansions thereof).

Second is an *updated Lagrangian description,* in which the kinematic and kinetic variables are referred to the current configuration. For small-strain problems, the Almansi strain is used as the measure of deformation and the associated Cauchy stress is used as the stress measure. For large-strain problems, it is convenient to use the velocity strain (or rate of deformation) and the Jaumann stress rate as the deformation and stress measures.

A detailed discussion of the aforementioned formulations and their combinations, along with the appropriate deformation and stress measures is given in Bathe [1996] and Cescotto et al. [1979].

28.4 Steps Involved in the Application of Computational Structural Mechanics to Practical Engineering Problems

28.4.1 Major Steps in the Application of Computational Structural Mechanics

The application of CSM to contemporary structural problems involves a sequence of five steps, namely (Figure 28.3):

- Observation of response phenomena of interest.
- Development of computational models for the numerical simulation of these phenomena. This in turn includes: (1) selecting mathematical models which represent the environmental effects and external forces and describe the phenomena, analyzing the models to ensure that the problem is properly formulated, and testing the range of validity of the models; and (2) development of a discrete model, computational strategy, and numerical algorithms to approximate the mathematical model. Successful computational models for structures are based on a thorough familiarity with the response phenomena being simulated and a good understanding of the mathematical models available to describe them.
- Development and assembly of software and/or hardware to implement the computational model.
- Postprocessing and interpretation of the predictions of the computational model.
- Utilization of the computational model in the analysis and design of engineering structures.

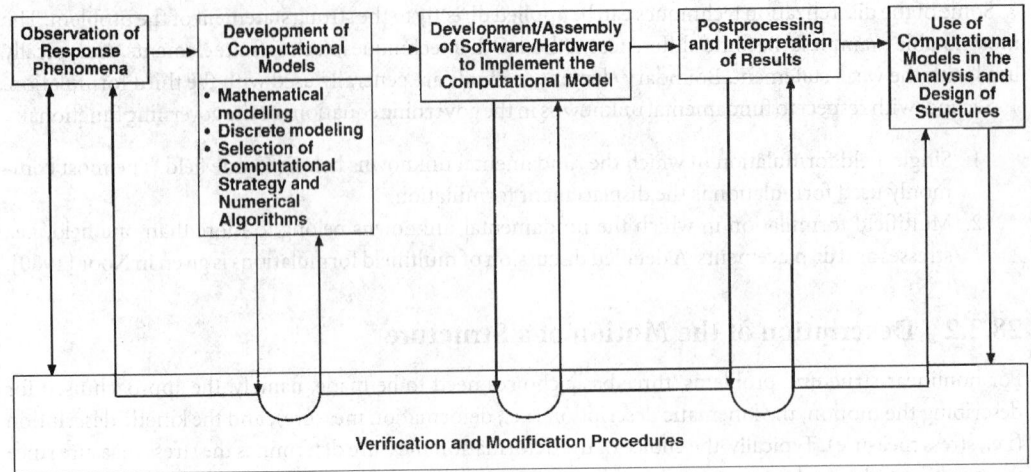

FIGURE 28.3 Application of CSM to practical structural problems.

Within this general framework, CSM is being used today in a broad range of practical applications. To date, large structural calculations are performed which account for complicated geometry, complex loading history, and material behavior. The applications span several industries, including aerospace, automotive, naval, nuclear, and microelectronics.

28.4.2   Selection of the Mathematical Models

The mathematical models are idealized representations of the real structure and its environment. Proper selection of the mathematical models is strongly influenced by the goals of the computation. The models will represent reality only if they take into consideration all factors which affect the conclusions drawn from them. The mathematical models are described by their governing equations (in one of the forms described in the previous section).

It is useful to view a particular mathematical model as one in a sequence (or hierarchy) of models of progressively increasing complexity. The effect of the model selection on the accuracy of the response predictions for simple structures, boundary conditions, and joints is discussed in Szabo and Babuska [1991]. A number of comparisons can be made between the predictions of the model being formulated and more elaborate models, i.e., models which account for a greater number of potentially relevant phenomena.

28.4.3   Discretization Techniques

Because of the complexity of the governing equations of the mathematical models used in representing real structures, exact solutions can only be obtained in very few cases, and one has to resort to approximate or numerical discretization techniques. A variety of approximate and numerical discretization techniques have been applied to structural mechanics problems. Two possible classifications of these techniques are shown in Figure 28.4 and are based on: (1) the formulation used, namely, differential equation, variational, or integral-equation formulation; and (2) modifications made in the form of the governing equations (replacement of the governing equations by an equivalent form).

The finite-element method is the most commonly used discretization technique to date. Extensive literature exists on various aspects of the finite-element method. A list of monographs, books, and conference proceedings on the method is given in Noor [1991]. The boundary element method is a computational tool for the boundary integral equation formulation. The method works with values of the dependent

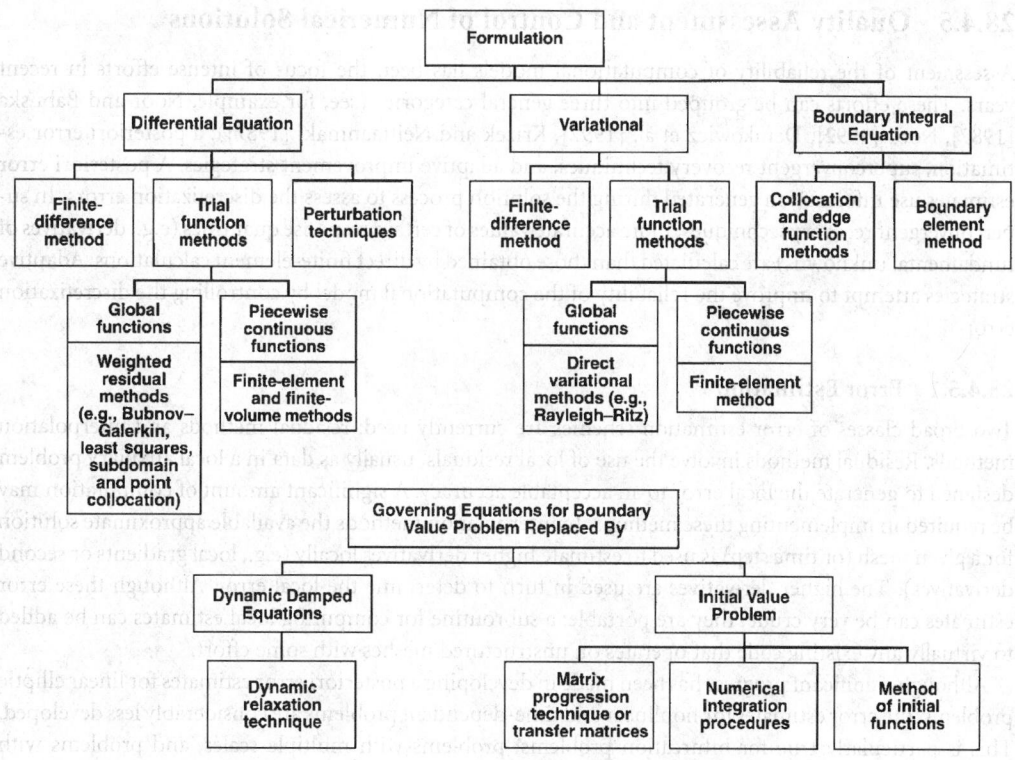

FIGURE 28.4 Approximate and numerical discretization techniques for structural mechanics problem.

variables on the boundary only, and therefore, is well suited to problems involving a large volume to surface ratio (Kane et al. [1993], Banerjee [1994]). In a number of applications, hybrid combinations of analytical and numerical discretization techniques were shown to be more effective than individual techniques, and resulted in dramatic savings in the computational effort (see, for example, Noor and Andersen [1992] and Noor [1994]).

28.4.4 Model and Mesh Generation

The reliability of the predictions of numerical discretization techniques (e.g., finite differences, finite elements, finite volumes, and boundary elements), and the computational effort involved in obtaining them, is very much influenced by the selection of the mesh and the procedure for generating it. Considerable effort has been devoted to the development of robust mesh generation procedures capable of producing controlled meshes over domains of arbitrary complex geometry. Finite-element mesh generation activities have been approached from both geometric modeling and adaptive finite-element viewpoints. Overviews and classifications of finite-element mesh generation methods are given in Shephard and Wentorf [1994], Shephard [1988], Ho-Le [1988], George [1991], Sabin [1991], and Mackerle [1993].

Among the recent activities on model generation are: (1) application of knowledge-based analysis-assistance tools, which allow a simple description of the analysis objectives and generate the corresponding discrete models appropriate for these objectives (Turkiyyah and Fenves [1996]); and (2) the development of paving and plastering techniques for automated generation of quadrilateral and hexahedral finite-element grids. These techniques generate well-formed elements (with reasonably small distortion metric), and are based on iteratively layering or paving rows of elements to the interior of a region's boundary (see Blacker and Stephenson [1991]).

28.4.5 Quality Assessment and Control of Numerical Solutions

Assessment of the reliability of computational models has been the focus of intense efforts in recent years. These efforts can be grouped into three general categories (see, for example, Noor and Babuska [1987], Noor [1992], Demkowicz et al. [1992], Krizek and Neittaanmaki [1987]): a posteriori error estimation, superconvergent recovery techniques, and adaptive improvement strategies. A posteriori error estimates use information generated during the solution process to assess the discretization errors. In superconvergent recovery techniques, more accurate values of certain response quantities (e.g., derivatives of fundamental unknowns) are calculated than those obtained by direct finite-element calculations. Adaptive strategies attempt to improve the reliability of the computational model by controlling the discretization error.

28.4.5.1 Error Estimation

Two broad classes of error estimation schemes are currently used: residual methods and interpolation methods. Residual methods involve the use of local residuals, usually as data in a local auxiliary problem designed to generate the local error to an acceptable accuracy. A significant amount of computation may be required in implementing these methods. In interpolation methods the available approximate solution for a given mesh (or time step) is used to estimate higher derivatives locally (e.g., local gradients or second derivatives). The higher derivatives are used in turn to determine the local error. Although these error estimates can be very crude, they are portable: a subroutine for computing local estimates can be added to virtually any existing code that operates on unstructured meshes with some effort.

Although significant progress has been made in developing a posteriori error estimates for linear elliptic problems, the error estimates for nonlinear and time-dependent problems are considerably less developed. This is particularly true for bifurcation problems, problems with multiple scales, and problems with resonance. Work on error estimation for highly nonlinear problems has mainly been a subject of ad hoc experimentation.

28.4.5.2 Superconvergent Recovery Techniques

Superconvergent recovery techniques refer to simple postprocessing techniques which provide increased accuracy of the sought quantities at some isolated points (e.g., Gauss-Legendre, Jacobi, or Lobatto); in a subdomain; or even in the whole domain (Krizek and Neittaanmaki [1987]). In the latter two cases, the techniques are referred to as local- and global-superconvergent recovery techniques, respectively. Recent work included development of local-superconvergent patch derivative techniques for both interior and boundary (or material interface) points (Babuska and Miller [1984a, 1984b], Zienkiewicz and Zhu [1992], Zienkiewicz et al. [1993], Tabbara et al. [1994]). It was shown in Zienkiewicz and Zhu [1992] and Zienkiewicz et al. [1993] that the superconvergent recovery technique can be used to obtain a posteriori error estimates for the finite-element solution.

28.4.5.3 Adaptive Strategies

Different strategies have been used for adaptive improvement of the numerical solutions, including: (1) mesh refinement (or derefinement) schemes, h methods; (2) moving mesh (node redistribution) schemes, r methods; (3) subspace enrichment schemes (selection of the local order of approximation), p methods; (4) mesh superposition schemes (overlapping local finite-element meshes on the global one), s methods (see, for example, Fish and Markolefas [1993, 1994]); and (5) hybrid (or combined) schemes. Examples of these schemes are: (1) simultaneous selection of the meshes and local order of approximation, h–p methods; recent theoretical results have shown that the fastest possible convergence rates can be attained by optimally decreasing the mesh size h and increasing the degree of the polynomial degree p in a special way; and (2) simultaneous selection of the meshes and node redistribution, h–r method. These methods can be effective in shock problems since an r-method might align the mesh along discontinuities prior to a mesh refinement.

28.5 Overview of Static, Stability, and Dynamic Analysis

Flow charts for the basic components of the solution methods for static stability and dynamic problems are given in Figure 28.5 and Figure 28.6. In this section a brief summary is given of the fundamental equations and solution techniques used in static, stability, and dynamic analysis. A single-field displacement formulation is used, and the spatial discretization of the structure is assumed to have been performed. The external load vector and associated displacement vector will henceforth be denoted by $\{F^{ext}\}$ and $\{X\}$, respectively. For linear problems $\{X\}$ is a linear function of the components of $\{F^{ext}\}$, and for nonlinear problems $\{X\}$ is a nonlinear function of these components.

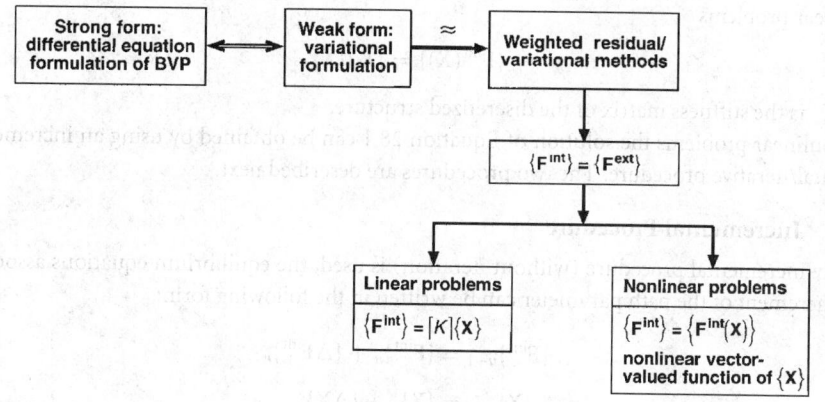

FIGURE 28.5 Basic components of solution methods for static problems.

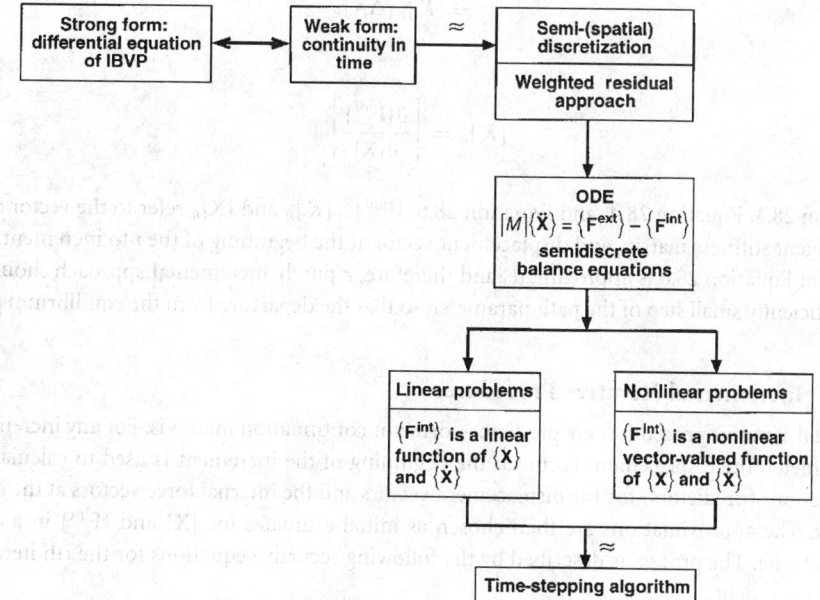

FIGURE 28.6 Basic components of solution methods for dynamic problems.

28.5.1 Static Analysis

The equilibrium equations for the discretized structure can be written in the following form:

$$\{\mathbf{F}^{int}(\mathbf{X})\} = \{\mathbf{F}^{ext}\} \tag{28.1}$$

where $\{\mathbf{F}^{int}(\mathbf{X})\}$ is the vector of internal forces, which is a vector-valued function of the displacements $\{\mathbf{X}\}$. For conservative loading $\{\mathbf{F}^{ext}\}$ is independent of $\{\mathbf{X}\}$, and for nonconservative loading it is a function of $\{\mathbf{X}\}$.

The equilibrium path is usually expressed in terms of one or more variable path parameters (typically taken as load, displacement, or arc-length parameters in the solution space). For simplicity, in subsequent discussion the loading is assumed to be conservative and proportional to a single path parameter p. The displacement vector is therefore a function of p, i.e., $\{\mathbf{X}\} = \{\mathbf{X}(p)\}$.

For linear problems

$$\{\mathbf{F}^{int}(\mathbf{X})\} = [K]\{\mathbf{X}\} \tag{28.2}$$

where $[K]$ is the stiffness matrix of the discretized structure.

For nonlinear problems the solution of Equation 28.1 can be obtained by using an incremental or an incremental/iterative procedure. The two procedures are described next.

28.5.1.1 Incremental Procedure

If a purely incremental procedure (without iteration) is used, the equilibrium equations associated with the nth increment of the path parameter can be written in the following form:

$$\{\mathbf{F}^{int}\}_{n+1} = \{\mathbf{F}^{int}\}_n + \{\Delta\mathbf{F}^{int}\}_n \tag{28.3}$$

$$\{\mathbf{X}\}_{n+1} = \{\mathbf{X}\}_n + \{\Delta\mathbf{X}\}_n \tag{28.4}$$

where

$$\{\Delta\mathbf{F}^{int}\}_n = \{\mathbf{F}^{ext}\}_{n+1} - \{\mathbf{F}^{int}\}_n \tag{28.5}$$

$$\cong [K]_n\{\Delta\mathbf{X}\}_n \tag{28.6}$$

and

$$[K]_n = \left[\frac{\partial\{\mathbf{F}^{int}\}}{\partial\{\mathbf{X}\}}\right]_n \tag{28.7}$$

In Equation 28.3, Equation 28.4, and Equation 28.6, $\{\mathbf{F}^{int}\}_n$, $[K]_n$ and $\{\mathbf{X}\}_n$ refer to the vector of internal forces, tangent stiffness matrix, and displacement vector at the beginning of the nth increment.

Note that Equation 28.6 is approximate, and therefore, a purely incremental approach should be used with a sufficiently small step of the path parameter, so that the departure from the equilibrium position is small.

28.5.1.2 Incremental-Iterative Procedures

Incremental-iterative procedures are predictor-corrector continuation methods. For any increment of the path parameter, the displacement vector at the beginning of the increment is used to calculate suitable approximations (predictors) for the displacement vectors and the internal force vectors at the end of that increment. The approximations are then chosen as initial estimates for $\{\mathbf{X}\}$ and $\{\mathbf{F}^{int}\}$ in a corrective-iterative scheme. The process is described by the following recursive equations for the ith iteration cycle of the nth increment.

$$\{\mathbf{R}\}^{(i)} = \{\mathbf{F}^{ext}\}_{n+1} - \{\mathbf{F}^{int}\}_{n+1}^{(i-1)} \tag{28.8}$$

$$= [K]\{\Delta\mathbf{X}\}_n^{(i)} \tag{28.9}$$

and

$$\{\mathbf{X}\}_{n+1}^{(i)} = \{\mathbf{X}\}_{n+1}^{(i-1)} + \{\Delta\mathbf{X}\}_n^{(i)} \tag{28.10}$$

where $\{\mathbf{R}\}$ is the residual force vector.

The iterational process is continued until convergence. As a test for convergence, a number of error norms can be used. Two of the error norms are described next. First is the modified Euclidean (spectral) norm,

$$|e| = \frac{1}{n}|\Delta\mathbf{X}|/|\mathbf{X}| \leq \text{tolerance} \tag{28.11}$$

where n is the total number of displacement parameters in the model. Second is the energy norm,

$$|e| = \frac{[\{\Delta\mathbf{X}\}^{(i)}]^T\left[\{\mathbf{F}^{\text{ext}}\}_{n+1} - \{\mathbf{F}^{\text{int}}\}_{n+1}^{(i-1)}\right]}{[\{\Delta\mathbf{X}\}^{(1)}]^T\left[\{\mathbf{F}^{\text{ext}}\}_{n+1} - \{\mathbf{F}^{\text{int}}\}_n\right]} \leq \text{tolerance} \tag{28.12}$$

In Equation 28.8 to Equation 28.10 superscripts $(i - 1)$ and (i) refer to the values of the vectors at the beginning and end of the ith iteration cycle and $[K]$ is an approximation to the tangent stiffness matrix. A number of different iterative processes are distinguished by the choice of $[K]$.

28.5.1.3 Newton–Raphson Technique

The matrix $[K]$ is selected to be the tangent stiffness matrix based on the solution at the end of iteration cycle $(i - 1)$, i.e.,

$$[K] = \left[\frac{\partial\{\mathbf{F}^{\text{int}}\}}{\partial\{\mathbf{X}\}}\right]_{n+1}^{(i-1)} = [K]_{n+1}^{(i-1)} \tag{28.13}$$

28.5.1.4 Modified Newton Method

The matrix $[K]$ is selected to be the tangent stiffness matrix associated with increment m of the path parameter, where $m \leq n$, i.e.,

$$[K] = [K]_m \tag{28.14}$$

28.5.1.5 Broyden–Fletcher–Goldfarb–Shanno (BFGS) Method

A secant approximation to the stiffness matrix is used in successive iterations, through updating the inverse of the stiffness matrix using vector products. This is equivalent to updating the stiffness matrix, based on iteration history, as follows:

$$[K] = [K]_{n+1}^{(i-1)} \tag{28.15}$$

28.5.2 Dynamic Analysis

The balance equations of the discretized structure can be written in the form of a system of ordinary differential equations in time as follows:

$$[M]\{\ddot{\mathbf{X}}\} = \{\mathbf{F}^{\text{ext}}\} - \{\mathbf{F}^{\text{int}}\} \tag{28.16}$$

where $[M]$ is the global mass matrix and $\{\ddot{\mathbf{X}}\}$ is the acceleration vector. The explicit characterization of $[M], \{\ddot{\mathbf{X}}\}$, and $\{\mathbf{F}^{\text{int}}\}$ depends on the particular structure under consideration. When damping or viscous effects exist, the vector $\{\mathbf{F}^{\text{int}}\}$ is a function of both the displacement vector and the velocity vector, i.e.,

$$\{\mathbf{F}^{\text{int}}\} = \{\mathbf{F}^{\text{int}}(\mathbf{X}, \dot{\mathbf{X}})\} \tag{28.17}$$

For linear problems, $\{\mathbf{F}^{\text{int}}\}$ is a linear function of both $\{\mathbf{X}\}$ and $\{\dot{\mathbf{X}}\}$, i.e.,

$$\{\mathbf{F}^{\text{int}}\} = [K]\{\mathbf{X}\} + [C]\{\dot{\mathbf{X}}\} \tag{28.18}$$

where $[C]$ is the global damping matrix.

For nonlinear problems $\{\mathbf{F}^{\text{int}}\}$ is a nonlinear function of $\{\mathbf{X}\}$ and $\{\dot{\mathbf{X}}\}$. At any time instant, the tangent stiffness and tangent damping matrices are given by

$$[K] = \left[\frac{\partial\{\mathbf{F}^{\text{int}}\}}{\partial\{\mathbf{X}\}}\right] \tag{28.19}$$

$$[C] = \left[\frac{\partial\{\mathbf{F}^{\text{int}}\}}{\partial\{\dot{\mathbf{X}}\}}\right] \tag{28.20}$$

The approaches used for obtaining the response-time history of the structure [solution of Equation 28.16] can be divided into two general categories, namely, direct integration techniques and modal superposition methods. The application of the two approaches to nonlinear dynamic analysis is described next.

28.5.2.1 Direct Integration Techniques

Direct temporal integration techniques are time-stepping (or step-by-step) strategies in which the response vectors at an initial time are used to generate the corresponding vectors at subsequent times. The techniques are based on: (1) satisfying the balance equations only at discrete time intervals and (2) assuming the functional dependence of the response vectors within each time interval. A variety of approximations for the response vectors within each time interval have been applied to structural dynamics problems. The approximations for the velocity and displacement vectors in the nth time step can be expressed as follows:

$$\{\dot{\mathbf{X}}\}_{n+1} = \frac{\alpha_1}{\Delta t}\{\ddot{\mathbf{X}}\}_{n+1} + L(\{\dot{\mathbf{X}}\}_n, \{\ddot{\mathbf{X}}\}_n, \ldots) \tag{28.21}$$

$$\{\mathbf{X}\}_{n+1} = \frac{\alpha_2}{(\Delta t)^2}\{\ddot{\mathbf{X}}\}_{n+1} + M(\{\mathbf{X}\}_n, \{\dot{\mathbf{X}}\}_n, \{\ddot{\mathbf{X}}\}_n, \ldots) \tag{28.22}$$

where Δt is the time-step size, α_1 and α_2 are coefficients used in the approximation, L and M are functions of the response vectors, and subscripts n and $n + 1$ refer to the values of the vectors at the beginning and end of the nth time step.

Temporal integration techniques can be classified into two general categories, namely, explicit and implicit techniques. In explicit methods the response at the end of a time step is evaluated using the balance equations at the beginning of the time step [α_1 and α_2 in Equation 28.21 and Equation 28.22 are both zero]. By contrast, implicit methods use the balance equations at the end of the time step, with either α_1 and/or α_2 in Equation 28.21 and Equation 28.22 as nonzero.

Explicit techniques generally require fewer computations per time step than implicit techniques and can easily handle complex nonlinearities. However, the time-step size must often be very small to ensure numerical stability. By contrast, for linear problems, the time-step size in implicit techniques is only restricted by accuracy requirements. However, based on available information, unconditional stability of implicit methods in linear problems does not necessarily extend to nonlinear problems. Explicit and implicit techniques can also be classified into *single-step* and *multistep methods* according to whether the response at any instant is related to the response at one or more previous times. Detailed discussions and assessments of different explicit and implicit techniques are given in Belytschko and Hughes [1983] and Belytschko et al. [1987]. Mixed explicit-implicit techniques are described in Belytschko and Hughes [1983].

Herein, the application of two of the widely used temporal integration techniques, central difference explicit scheme and the Newmark implicit scheme is described. Both are single-step methods.

28.5.2.1.1 Central Difference Explicit Scheme

The central difference scheme is based on the following approximations for the velocity and the acceleration vectors:

$$\{\dot{\mathbf{X}}\}_{n+\frac{1}{2}} = \frac{1}{\Delta t}(\{\mathbf{X}\}_{n+1} - \{\mathbf{X}\}_n) \tag{28.23}$$

$$\{\ddot{\mathbf{X}}\}_n = \frac{1}{(\Delta t)^2}[\{\mathbf{X}\}_{n-1} - 2\{\mathbf{X}\}_n + \{\mathbf{X}\}_{n+1}] \tag{28.24}$$

where Δt is the time-step size, and subscripts $n, n + \frac{1}{2}$, and $n + 1$ refer to the values of the vectors at the beginning, middle, and end of the nth time step.

Based on Equation 28.25 and Equation 28.26, the update formulas for the velocity and displacement vectors are

$$\{\dot{\mathbf{X}}\}_{n+\frac{1}{2}} = \{\dot{\mathbf{X}}\}_{n-\frac{1}{2}} + \Delta t[M]^{-1}\left(\{\mathbf{F}^{\text{ext}}\}_n - \{\mathbf{F}^{\text{int}}\}_n\right) \tag{28.25}$$

$$\{\mathbf{X}\}_{n+1} = \{\mathbf{X}\}_n + \Delta t\{\dot{\mathbf{X}}\}_{n+\frac{1}{2}} \tag{28.26}$$

If a lumped (diagonal) mass matrix is used, then Equation 28.25 uncouples.

Initially (at $t = 0$), Equation 28.25 and Equation 28.26 are replaced by

$$\{\dot{\mathbf{X}}\}_{\frac{\Delta t}{2}} = \{\dot{\mathbf{X}}\}_0 + \frac{\Delta t}{2}[M]^{-1}\left(\{\mathbf{F}^{\text{ext}}\}_0 - \{\mathbf{F}^{\text{int}}\}_0\right) \tag{28.27}$$

and

$$\{\mathbf{X}\}_{\Delta t} = \{\mathbf{X}\}_0 + \frac{\Delta t}{2}\{\dot{\mathbf{X}}\}_{\frac{\Delta t}{2}} \tag{28.28}$$

where subscript 0 in Equation 28.27 and Equation 28.28 refers to the value of the vector at $t = 0$.

Note that the central difference scheme is only conditionally stable. Therefore, the time step Δt must be smaller than a critical value Δt_{cr}, which for linear problems is calculated from the smallest period of vibration $T_{\min}$. Specifically,

$$\Delta t \leq \Delta t_{\text{cr}} = \frac{T_{\min}}{\pi} = \frac{2}{\omega_{\max}}$$

For nonlinear problems experience has shown that a 10–20% reduction in the time step is usually sufficient to maintain stability.

28.5.2.1.2 Newmark's Method

Newmark's method is based on the following approximations for the displacement and velocity vectors:

$$\{\mathbf{X}\}_{n+1} = \{\mathbf{X}\}_n + \Delta t\{\dot{\mathbf{X}}\}_n + (\Delta t)^2\left[\left(\frac{1}{2} - \beta\right)\{\ddot{\mathbf{X}}\}_n + \beta\{\ddot{\mathbf{X}}\}_{n+1}\right] \tag{28.29}$$

$$\{\dot{\mathbf{X}}\}_{n+1} = \{\dot{\mathbf{X}}\}_n + \Delta t[(1 - \gamma)\{\ddot{\mathbf{X}}\}_n + \gamma\{\ddot{\mathbf{X}}\}_{n+1}] \tag{28.30}$$

where β and γ are free parameters of Newmark's method. The particular choice $\beta = 1/6, \gamma = 1/2$ corresponds to a linear approximation of the acceleration over the nth time step. The choice $\beta = 1/4, \gamma = 1/2$ corresponds to a constant average acceleration (trapezoidal rule), which is unconditionally stable for linear systems. The central difference scheme corresponds to $\beta = 0, \gamma = 1/2$.

If Equation 28.29 and Equation 28.30 are combined with the balance equations, Equation 28.16, a set of simultaneous algebraic equations result at each time step. For nonlinear problems, the equations are nonlinear and are expressed by the following process for the *n*th time step:

1. Predict velocities and displacements using the explicit approximations

$$\{\tilde{\dot{X}}\}_{n+1} = \{\dot{X}\}_n + \Delta t(1 - \gamma)\{\ddot{X}\}_n \tag{28.31}$$

$$\{\tilde{X}\}_{n+1} = \{X\}_n + \Delta t\{\dot{X}\}_n + (\Delta t)^2\left(\frac{1}{2} - \beta\right)\{\ddot{X}\}_n \tag{28.32}$$

where a tilde ($\sim$) refers to the predicted value of the vector.

2. Obtain corrections to the displacements, velocities, and accelerations by the following iterative process:

$$[\overset{*}{K}]\{\Delta X\}_{n+1}^{(i)} = \{R\}_{n+1}^{(i)} \tag{28.33}$$

$$\{X\}_{n+1}^{(i+1)} = \{X\}_{n+1}^{(i)} + \{\Delta X\}_{n+1}^{(i)} \tag{28.34}$$

$$\{\dot{X}\}_{n+1}^{(i+1)} = \{\dot{X}\}_{n+1}^{(i)} + \gamma \Delta t\{\ddot{X}\}_{n+1}^{(i+1)} \tag{28.35}$$

$$\{\ddot{X}\}_{n+1}^{(i+1)} = \frac{1}{\beta(\Delta t)^2}\left[\{X\}_{n+1}^{(i+1)} - \{\tilde{X}\}_{n+1}\right] \tag{28.36}$$

For $i = 0$, the following values are used for the displacement, velocity, and acceleration vectors

$$\{X\}_{n+1}^{(0)} = \{\tilde{X}\}_{n+1}$$

$$\{\dot{X}\}_{n+1}^{(0)} = \{\tilde{\dot{X}}\}_{n+1}$$

$$\{\ddot{X}\}_{n+1}^{(0)} = 0$$

where superscript (0) refers to the value of the vector associated with $i = 0$, $[\overset{*}{K}]$ and $\{R\}$ in Equation 28.33 are the effective stiffness matrix and the residual vector given by

$$[\overset{*}{K}] = [K] + \frac{\gamma}{\beta \Delta t}[C] + \frac{1}{\beta(\Delta t)^2}[M] \tag{28.37}$$

$$\{R\}_{n+1}^{(i)} = \{F^{\text{ext}}\}_{n+1} - [M]\{\ddot{X}\}_{n+1}^{(i)} - \{F^{\text{int}}\}_{n+1}^{(i)} \tag{28.38}$$

Superscripts i and $i+1$ refer to the values of the vectors at the beginning and end of the *i*th iteration cycle. As for nonlinear static problems, the choice of $[\overset{*}{K}]$ depends on the iterative procedure used.

28.5.2.1.3 *Modal Superposition Method*

In this method the structural response at any time is expressed as a linear combination of a number of preselected modes (or basis vectors). This is expressed by the following transformation:

$$\{X\} = [\Gamma]\{\psi\} \tag{28.39}$$

where $[\Gamma]$ is the transformation matrix whose columns are the modes (or basis vectors) and $\{\psi\}$ is a vector of unknown coefficients (amplitudes of the preselected modes).

The number of modes (or basis vectors) is usually selected to be much smaller than the number of degrees of freedom of the discretized structure (components of the vector $\{X\}$). A Bubnov–Galerkin technique is used to approximate the balance equations of the discretized structure, Equation 28.16, by a much smaller system of equations in $\{\psi\}$. The resulting equations have the following form:

$$[\Gamma]^T[M][\Gamma]\{\ddot{\psi}\} = [\Gamma]^T\{F^{\text{ext}}\} - [\Gamma]^T\{F^{\text{int}}(\psi)\} \tag{28.40}$$

For linear problems, if the basis vectors are selected to be the free vibration modes of the structure, Equation 28.40 uncouples in the components of $\{\psi\}$. For nonlinear problems the free vibration modes and frequencies of the structure change with time.

The mode superposition technique is effective when the response can be adequately represented by few modes and the time integration has to be carried out over many time steps (e.g., earthquake loading), and the cost of calculating the required modes is reasonable.

28.5.3 Energy Balance in Transient Dynamic Analysis

Energy balance (conservation) in transient analysis is a reflection of the accuracy of the time integrator. Furthermore, conservation properties are intimately related to stability. The construction of stable integrators is often approached by enforcing conservation laws.

In linear problems, instability can be easily detected by the spurious growth in velocities. On the other hand, in large structural problems with material nonlinearities the energy generated by an instability may be rapidly dissipated as the material becomes inelastic and the erroneousness of the results does not become obvious to the user. This kind of instability has been termed *arrested instability* in Belytschko and Hughes [1983] and the energy balance check described subsequently was suggested to detect it.

Let $\{\Delta \mathbf{X}\}_n$ be the increment of the displacement vector from time t_n to time t_{n+1}, and let the internal energy, and the increment of external work be approximated by the trapezoidal rule as follows:

$$U_j = \sum_{l=1}^{j-1} \frac{1}{2}\{\Delta \mathbf{X}\}_l^T \left(\{\mathbf{F}^{\text{int}}\}_l + \{\mathbf{F}^{\text{int}}\}_{l+1} \right) \tag{28.41}$$

$$\Delta W_j = \frac{1}{2}\{\Delta \mathbf{X}\}_j^T \left(\{\mathbf{F}^{\text{ext}}\}_j + \{\mathbf{F}^{\text{ext}}\}_{j+1} \right) \tag{28.42}$$

The kinetic energy T_j at time t_j is given by

$$T_j = \frac{1}{2}\{\dot{\mathbf{X}}\}_j^T [M]\{\dot{\mathbf{X}}\}_j \tag{28.43}$$

If the mass matrix is positive definite, then the energy and the displacements of the structure at time t_{n+1} are bounded if the following inequality is satisfied:

$$T_{n+1} + U_{n+1} \leq (1+\varepsilon)(T_n + U_n) + \Delta W_n \tag{28.44}$$

where ε is a small number. Equation 28.44 provides an energy balance check provided $U_n \geq 0$.

Detailed discussion of the stability, convergence, and decay of energy is given in Hughes [1976]. Examples of the construction of integrators through energy conservation are given in Hughes et al. [1978] and Haug et al. [1977].

28.5.4 Stability Analysis

A number of instability phenomena can occur in structures depending on the structural characteristics and operational environment. These instabilities can be either time dependent or time invariant. Time-dependent instability analysis can be performed by using Lyapunov's method (see Kratzig and Eller [1992] and Kounadis and Kratzig [1995]). Time-invariant instabilities can be in the form of bifurcation points (simple or multiple) or limit points in the solution path. The *critical loads* associated with these points are referred to as bifurcation buckling and limit loads, respectively.

The algorithmic tools for *time-invariant instability analysis* encompass three distinct aspects: (1) nonlinear (or linear) analysis of the prebuckling state, using the nonlinear equations — Equation 28.1 [or the linearized version, Equation 28.2]; (2) determination of the critical points (e.g., bifurcation and/or limit points) on the equilibrium path; and (3) tracing the postcritical response. Critical points can be detected by the singularity of the tangent stiffness matrix, Equation 28.19. Bifurcation loads associated with linear prebuckling state can be determined by solving a linear eigenvalue problem as described in the succeeding subsection. Special numerical algorithms are available for overcoming the difficulties associated with commonly used iterative techniques near critical points (see, for example, Riks [1984] and Crisfield [1983]).

28.5.5 Eigenvalue Problems

Undamped free vibration and linear (bifurcation) buckling problems of structures can be represented by the general linear matrix eigenvalue equation

$$[K]\{\mathbf{X}\} = \lambda[B]\{\mathbf{X}\} \tag{28.45}$$

where $[K]$ is the stiffness matrix of the discretized structure, $\{\mathbf{X}\}$ is the displacement vector, $[B]$ is the mass matrix for free vibration problems or the negative of the geometric stiffness matrix for buckling problems, and λ is the eigenvalue square of vibration frequency or buckling load parameter. Typically, most of the elements of $[K]$ and $[B]$ are zero and only a few pairs $(\lambda, \{\mathbf{X}\})$ are wanted.

Although the governing equations for both the free vibration and linear buckling problems are similar, the properties of the two matrices $[K]$ and $[B]$ are different. For vibration problems, the stiffness matrix can be positive definite, positive semidefinite, or indefinite. For an unsupported structure the stiffness matrix is indefinite. If the structure is stable (except for rigid motions), the stiffness matrix will be positive semidefinite. Also, some lumping procedures can produce mass matrices, which are positive semidefinite because they have zero mass elements on the diagonal. The mass matrix can be positive definite or semidefinite (if some of the diagonal mass elements are zero, due to the lumping procedure used).

For buckling problems, the stiffness matrix is positive definite, provided the deformation state is stable (which may be assessed by a buckling analysis). The geometric stiffness matrix may be indefinite.

The preferred eigenvalue extraction techniques in use to date are sampling techniques. They create a linear operator (or matrix) and apply it to a sequence of carefully constructed vectors. From these transformed vectors the dominant eigenvectors and their eigenvalues can be approximated. Examples of these techniques are subspace iteration (or simultaneous iteration) and Lanczos techniques. The details of these methods are described in Bathe [1996], Parlett [1987], and Hughes [1987].

The eigensolver algorithms typically generate a set of eigenpairs (eigenvalues and associated eigenvectors) for the lowest eigenstates of the system, i.e., for the smallest eigenvalues (in absolute order). Often, it is necessary to compute eigenpairs for cases other than the set of smallest eigenvalues. This may be performed by introducing a shift α, which defines the shifted eigenvalue

$$\lambda = \overline{\lambda} + \alpha$$

The shifted eigenvalue problem is

$$[[K] - \alpha[B]]\{\mathbf{X}\} = \overline{\lambda}[B]\{\mathbf{X}\} \tag{28.46}$$

Equation 28.46, with a nonzero α, can be used to compute the eigenvalues for an unsupported structure (one for which $[K]$ is singular).

28.5.6 Sensitivity Analysis

Sensitivity analysis refers to methods of calculating the rates of change of: (1) response quantities (e.g., displacements, stresses, vibration frequencies, and buckling loads) with respect to changes in the structure characteristics (e.g., geometric and material parameters of the structure) and (2) the optimum design variable values with respect to changes in the structure parameters (e.g., applied loads and allowable stresses). The two types of calculations are usually designated by response and optimum design sensitivity analysis, and the rates of change referred to as sensitivity coefficients.

A number of techniques have been developed for evaluating the sensitivity coefficients of response quantities using either the governing discrete equations or continuum equations of the structure. The techniques used with the discrete equations can be grouped into three categories: *analytical, direct differentiation methods; finite difference methods;* and *semianalytical or quasianalytical methods* (see Kleiber and Hisada [1993] and Hinton and Sienz [1994]).

Methods for computing sensitivity coefficients for linear structural response have been developed for over 20 years (see, for example, Haftka and Adelman [1989], Haber et al. [1993], and Choi [1993]). However, only recently have attempts been made to extend the domain of sensitivity analysis to (1) nonlinear structural response and to path-dependent problems, for which the sensitivity coefficients depend also on the deformation history (e.g., viscoplastic response and frictional contact — see, for example, Kleiber et al. [1994], Kulkarni and Noor [1995], and Karaoglan and Noor [1995]); and (2) structural systems exhibiting probabilistic uncertainties.

Because of the importance of its role in structural optimization and in assessing the effect of uncertainties in the input parameters on the structural response, some commercial software systems have incorporated response sensitivity analysis into their systems. Also, an automatic differentiation facility has been developed for evaluating the derivatives of functions defined by computer programs exactly to within machine precision. The facility, automatic differentiation of Fortran (ADIFOR), is described in Chinchalkar [1994] and Carle et al. [1994]. The use of ADIFOR to evaluate the sensitivity coefficients from incremental/iterative forms of three-dimensional fluid flow problems is discussed in Sherman et al. [1994], and the additional facilities needed for ADIFOR to become competitive with hand-differentiated codes are listed in Carle et al. [1994].

28.5.7 Strategies and Numerical Algorithms for New Computing Systems

In recent years, intense efforts have been devoted to the development of efficient computational strategies and numerical algorithms which exploit the capabilities of new computing systems (in particular, the vector and parallel processing of the powerful high-performance computers — see, for example, Oñate et al. [1992]). Efficient direct and iterative numerical algorithms have been developed for solution of large sparse linear systems of equations (see Wang and Bruch [1993], Law and Mackay [1993], Qin and Mackay [1993], Vaughan [1991], and Papadrakakis [1993]).

Most parallel strategies are related to the *divide-and-conquer* paradigm based on breaking a large problem into a number of smaller subproblems which may be solved separately on individual processors. The degree of independence of the subproblems is a measure of the effectiveness of the algorithm since it determines the amount and frequency of communication and synchronization. The numerical algorithms developed for structural analysis can be classified into three major categories: namely, *elementwise algorithms, nodewise algorithms, and domainwise algorithms.*

The elementwise parallel algorithms include element-by-element equation solvers and parallel frontal equation solvers. The nodewise parallel equation solvers include node-by-node iterative solvers as well as column-oriented direct solvers. The domainwise algorithms include nested dissection-based (substructuring) techniques and domain-decomposition methods. The first two categories of numerical algorithms allow only small granularity of the parallel tasks and require frequent communication among the processors. By contrast, the third category allows a larger granularity, which can result in improved performance for the algorithm.

Nested dissection ordering schemes have been found to be effective in reducing both the storage requirements and the total computational effort required of direct factorization [Noor et al. 1978]. The performance of nested dissection-based linear solvers depends on balancing the computational load across processors in a way that minimizes interprocessor communication. Several nested dissection ordering schemes have been developed which differ in the strategies used in partitioning the structure and selecting the separators. Among the proposed partitioning strategies are: recursive bisection strategies (e.g., spectral graph bisection, recursive coordinate bisection, and recursive graph bisection [Pothen et al. 1990, Hendrickson and Leland 1992]); combinatorial and design-optimization based strategies (e.g., simulated annealing algorithm, genetic algorithm and neural-network-based techniques [Khan and Topping 1993]); and heuristic strategies (e.g., methods based on geometric projections and mappings; and algorithms based on embedding the problem in Euclidean space [Bui and Jones 1993]). For highly irregular and/or three-dimensional structures the effectiveness of nested dissection-based schemes may be reduced. However, this is also true for most other parallel numerical algorithms. Scalable parallel computational strategies for nonlinear, postbuckling, and dynamic contact/impact problems are presented in Watson and Noor [1996a, 1996b].

28.6 Brief History of the Development of Computational Structural Mechanics Software

Development of CSM software spans a period of less than 40 years, and may be divided into four stages, each stage lasting approximately 8–10 years. In the first stage (during the 1950s and 1960s), the aircraft industry pioneered development of in-house finite-element programs. These programs were generally based on the force method of analysis and were used to automate analysis of highly redundant structural components. Subsequent efforts in industry and academia led to development of simple two- and three-dimensional finite elements based on the displacement formulation. The variational process for formulating the elemental matrices was also introduced in this period.

In the second stage, general-purpose finite-element programs, such as NASTRAN, ASKA, ANSYS, STARDYNE, MARC, SAP, SESAM, and SAMCEF, were released for public use in the U.S. and Europe. These programs brought a significant technology base that led to development of numerous commercial finite-element software systems. This development included mixed and hybrid finite-element models with the fundamental unknowns consisting of stress and displacement parameters, efficient numerical algorithms for the solution of algebraic equations and extraction of eigenvalues, and substructuring and modal synthesis techniques for handling large problems. The finite-element method's success in linear static problems has encouraged bolder applications to nonlinear and transient response problems.

The third stage involved refining the commercial software codes and expanding their technology base. Design optimization techniques were also developed in this stage, as were pre- and postprocessing software and computer-aided design systems. The technology development included singular elements for fracture mechanics applications, boundary-element techniques, coupling of finite elements with other discretization techniques such as boundary elements, and quality assurance methods for both software and finite-element models.

The fourth stage included in the adaptation of CSM software to new computing systems (vector, multiprocessor, and massively parallel machines), development of efficient computational strategies and numerical algorithms for these machines, widespread availability of CSM software on personal computers and workstations, and the addition of substantial nonstructural modeling and analysis capabilities to CSM software systems such as MSC/NASTRAN and ANSYS. The latter capabilities were added because future flight vehicles and high-performance engineering systems (e.g., health monitoring aircraft and microsized spacecraft) will require significant interactions between CSM and other disciplines such as aerodynamics, controls, acoustics, electromagnetics, and optics.

Technology development in the fourth stage included introduction of advanced material models, development of stochastic models and probabilistic methods for structural analysis and design, and development of facilities for quality assessment and control of numerical simulations.

The four stages of CSM software development parallel the four stages of the computing environment's evolution: noncentralized mainframes, centralized mainframes, mainframe computing with timesharing, and distributed computing and networking. A summary of the major characteristics of currently used finite-element systems is given in Mackerle and Fredriksson [1988], and a guide to information sources on finite-element methods is given in Mackerle [1991]. Commercial finite-element programs for structural analysis have a rich variety of elements, and are widely used for performing structural calculations on large components and/or entire structures (see, for example, Figure 28.7 to Figure 28.9).

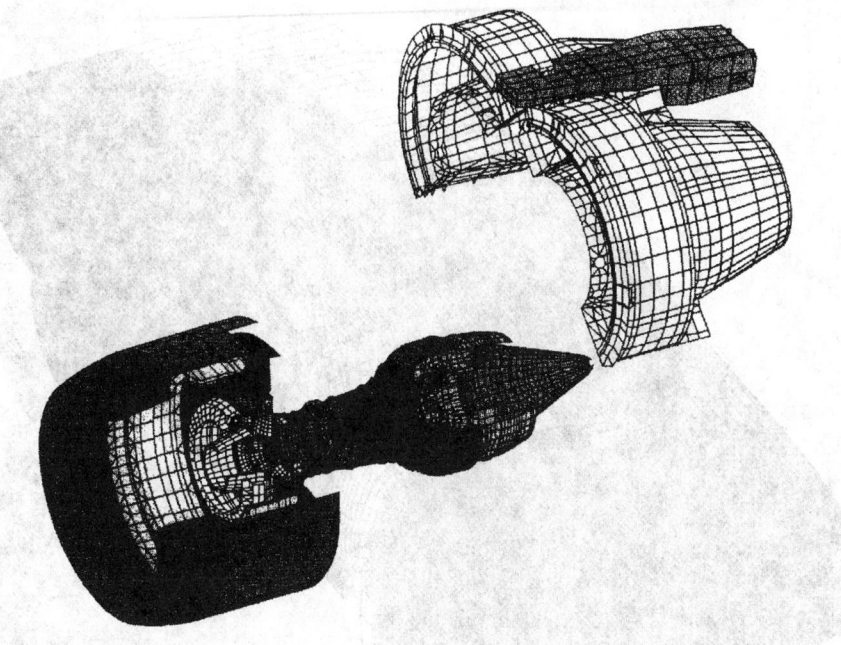

FIGURE 28.7   MSC/NASTRAN finite-element model of a G.E. Engine: 180,000 degrees of freedom. (Courtesy of GE Aircraft Engines.)

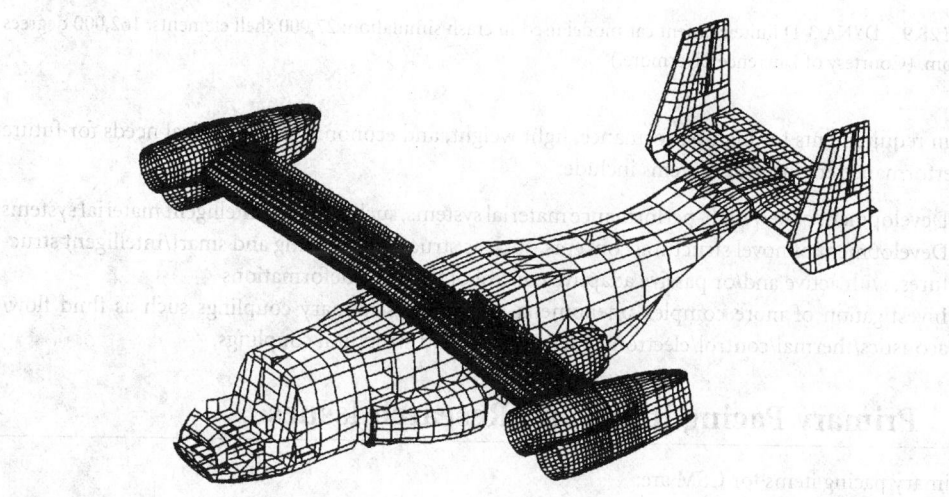

FIGURE 28.8   MSC/NASTRAN finite-element dynamics model of the V-22 Osprey Tiltrotor: 134,982 degrees of freedom (22,497 grid points), 44,006 elements. (Courtesy of Bell-Boeing.)

28.7   Characteristics of Future Engineering Systems and Their Implications on Computational Structural Mechanics

The demands that future high-performance engineering systems place on CSM differ somewhat from those of current systems. The radically different and more unpredictable operational environments for many of the systems (e.g., future flight vehicles) are one reason for this difference. Another is the stringency

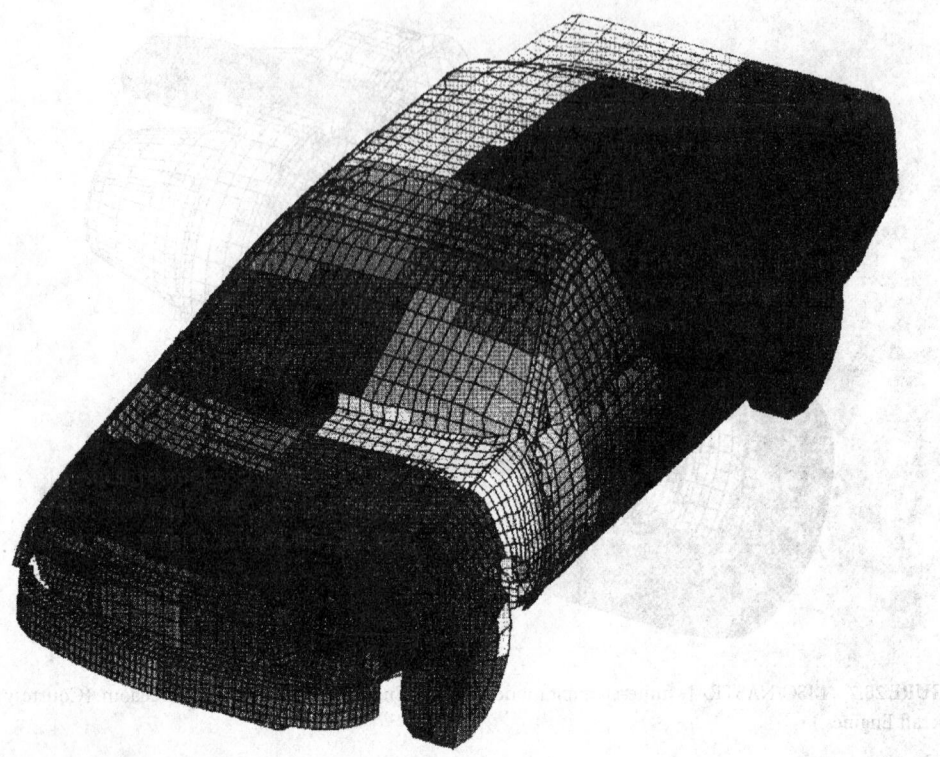

FIGURE 28.9   DYNA 3-D finite element car model used in crash simulation: 27,000 shell elements, 162,000 degrees of freedom. (Courtesy of Lawrence Livermore.)

of design requirements for high performance, light weight, and economy. The technical needs for future high-performance engineering systems include:

1. Development of new high-performance material systems, such as smart/intelligent material systems
2. Development of novel structural concepts, such as structural tailoring and smart/intelligent structures, with active and/or passive adaptive control of dynamic deformations
3. Investigation of more complex phenomena and interdisciplinary couplings such as fluid flow/ acoustics/thermal/control/electromagnetic/optics, and structural couplings.

28.8   Primary Pacing Items and Research Issues

The primary pacing items for CSM are:

1. High-fidelity modeling of the structure and its components
2. Failure and life prediction methodologies
3. Hierarchical, integrated methods and adaptive modeling techniques
4. Nondeterministic analysis, modeling methods, and risk assessment
5. Validation of numerical simulations
6. Multidisciplinary analysis and design optimization

For each of the aforementioned items, attempts should be made to exploit the major characteristics of high-performance computing technologies, as well as the future computing environment. The six primary pacing items are described subsequently. Note that some of the tasks within the pacing items are of a generic nature. Others are specific to certain components of future engineering systems (e.g., propulsion systems or airframes).

28.8.1   High-Fidelity Modeling of the Structure

The reliability of the predictions of response, failure, and life of structures is critically dependent on: (1) the accurate characterization and modeling of material behavior and (2) high-fidelity modeling of the critical details of the structure and its components (e.g., joints, damping, and for large deformations, frictional contact between the different parts of the structure). The simple material models used to date are inadequate for many of the future applications, especially those involving severe environment (e.g., high temperatures). Needed work on material modeling can be grouped in two general areas: (1) modeling the response and damage of advanced material systems in the actual operating environment of future engineering systems and (2) numerical simulation of manufacturing (fabrication) processes.

Advanced material systems include new polymer composites, metal-matrix composites, ceramic composites, carbon/carbon, and advanced metallics. The length scale selected in the model must be adequate for capturing the response phenomena of interest (e.g., micromechanics, mesomechanics, and macromechanics). For materials used in high-temperature applications, work is needed on the modeling of damage accumulation and propagation to fracture; modeling of thermoviscoplastic response, thermal-mechanical cycling, and ratcheting; and prediction of long-term material behavior from short-term data, which are particularly important.

28.8.2   Failure and Life Prediction Methodologies

Practical numerical techniques are needed for predicting the life, as well as the failure initiation and propagation in structural components made of new, high-performance materials in terms of measurable and controllable parameters. Examples of these materials are high-temperature materials; piezoelectric composites; and electronic, optical, and smart materials. For some of the materials, accurate constitutive descriptions, failure criteria, damage theories, and fatigue data are needed, along with more realistic characterization of interface phenomena (such as contact and friction). The constitutive descriptions may require investigations at the microstructure level or even the atomic level, as well as carefully designed and conducted experiments. Failure and life prediction of structures made of these materials is difficult and numerical models often constructed under restricting assumptions may not capture the dominant and underlying physical failure mechanisms. Moreover, material failure and structural response (such as instability) often couple in the failure mechanism.

28.8.3   Hierarchical, Integrated Multiple Methods and Adaptive Modeling Techniques

The effective use of numerical simulations for predicting the response, life, performance, and failure of future engineering systems requires strategies for treating phenomena occurring at disparate spatial and time scales, using reasonable computer resources. The strategies are based on using multiple mathematical models in different regions of the structure to take advantage of efficiencies gained by matching the model to the expected response in each region. To achieve the full potential of hierarchical modeling, there should be minimal reliance on a priori assumptions about the response. This is accomplished by adding adaptivity to the strategy. The key tasks of the research in this area are: (1) simple design-oriented models for use in the early stages of the design process; (2) rational selection of a set of nested mathematical models for different regions, and discretization techniques for use in conjunction with the mathematical models, which in turn requires the availability of a capability for holistic modeling from micro to structural response with varying degrees of accuracy; (3) simulation of local phenomena through global/local methodologies; (4) automated (or semiautomated) coupling of different mathematical/discrete models; (5) error estimation and adaptive modeling strategies; (6) efficient methods for coupling different components (e.g., engine airframe and rotor/engine frame); and (7) sensitivity analysis to assess the sensitivity of the response to each of the parameters neglected in the current mathematical model.

28.8.4 Nondeterministic Analysis, Modeling, and Risk Assessment

The new methodology developed for treating different types of uncertainties in geometry, material properties, boundary conditions, loading, and operational environment in the structural analysis formulation of structural components needs to be extended to the design and risk assessment of engineering systems. The ability to quantify inherent uncertainties in the response of the structure is obviously of great advantage. However, the principal benefit of using any nondeterministic method consists of the insights into engineering, safety, and economics that are gained in the process of arriving at those quantitative results and carrying out reliability analyses. As future engineering structures become more complicated, modeling of failure mechanisms will account for uncertainties from the beginning of the design process, and potential design improvements will be evaluated to assess their effects on reducing overall risk. The results combined with economic considerations will be used in systematic cost-benefit analyses (perhaps also done on a nondeterministic basis) to determine the structural design with the most acceptable balance of cost and risk.

28.8.5 Validation of Numerical Simulations

In addition to selecting a benchmark set of structures for assessing new computational strategies and numerical algorithms, a high degree of interaction and communication is needed between computational modelers and experimentalists. This is done on four different levels, namely:

1. Laboratory tests on small specimens to obtain material data
2. Component tests to validate computational models
3. Full-scale tests to validate the modeling of details, and for flight vehicles
4. Flight tests to validate the entire modeling process

28.8.6 Multidisciplinary Analysis and Design Optimization

The realization of new complex engineering systems requires integration between the structures discipline and other traditionally separate disciplines such as aerodynamics, propulsion, and control. This is mandated by significant interdisciplinary interactions and couplings which need to be accounted for in predicting response, as well as in the optimal design of these structures. Examples are the couplings between the aerodynamic flow field, structural heat transfer, and structural response of high-speed aircraft and propulsion systems and the couplings between the control system and structural response in control-configured aircraft and spacecraft. This activity also includes design optimization with multiobjective functions (e.g., performance, durability, integrity, reliability, and cost), and multiscale structural tailoring (micro, local, and global levels). For propulsion systems, it also includes design with damping for high-cycle fatigue, low-cycle fatigue, and acoustic fatigue.

Typically, in the design process questions arise regarding influence of design variable changes on system behavior. Answers to these questions, quantified by the derivatives of behavior with respect to the design variables or by parametric studies, guide design improvements toward a better overall system. In large applications, this improvement process is executed by numerical optimization, combined with symbolic/artificial intelligence (AI) techniques, and human judgment aided by data visualization. Efficiency of the computations that provide data for such a process is decisive for the depth, breadth, and rate of progress achievable, and hence, ultimately, is critical for the final product quality.

28.8.7 Related Tasks

For CSM to impact the design process, the following tasks need to be addressed by the research community: (1) development of automated or semiautomated model (and mesh) generation facilities; (2) pre- and postdata processing and use of advanced visualization technology, including multimedia and virtual reality facilities; and (3) adaptation of object-oriented technology and AI tools (knowledge-based expert systems and neural networks) to CSM.

CSM's wide acceptance can affect the design and operation of future engineering systems and structures in three ways. It can provide a better understanding of the phenomena associated with response, failure, and life, thereby identifying desirable structural design attributes. CSM can verify and certify designs and also allow low-cost modifications to be made during the design process. Finally, it can improve the design team's productivity and allow major improvements and innovations during the design phase, enabling fully integrated design in an integrated product and process development (IPPD) environment. Such an environment allows computer simulation of the entire life cycle of the engineering system, including material selection and processing, multidisciplinary design, automated manufacturing and fabrication, quality assurance, certification, operation, health monitoring and control, retirement, and disposal.

Acknowledgment

The present work is partially supported by NASA Cooperative Agreement NCCW-0011 and by Air Force Office of Scientific Research Grant AFOSR-F49620-93-1-0184.

References

Babuska, I. and Miller, A. 1984a. The postprocessing approach in the finite element method — part 1: calculation of displacements, stresses and other higher derivatives of the displacements. *Int. J. Num. Methods. Eng.* 20:1085–1109.

Babuska, I. and Miller, A. 1984b. The postprocessing approach in the finite element method — part 2: the calculation of stress intensity factors. *Int. J. Num. Methods. Eng.* 20:1110–1129.

Banerjee, P. K. 1994. *The Boundary Element Methods in Engineering*, 2nd ed. McGraw–Hill, New York.

Bathe, K. J. 1996. *Finite Element Procedures*. Prentice–Hall, Englewood Cliffs, NJ.

Belytschko, T., Engelmann, B. E., and Liu, W. K. 1987. A review of recent developments in time integration. In *State-of-the-Art Surveys on Computational Mechanics*, A. K. Noor and J. T. Oden, Eds., pp. 185–199. American Society of Mechanical Engineers, New York.

Belytschko, T. and Hughes, T. J. R. 1983. *Computational Methods for Transient Analysis*. Elsevier Science, Amsterdam.

Blacker, T. D. and Stephenson, M. B. 1991. Pacing: a new approach to automated quadrilateral mesh generation. *Int. J. Num. Methods. Eng.* 32:811–847.

Bui, T. N. and Jones, C. 1993. A heuristic for reducing fill-in in sparse matrix factorization. *Proc. 6th SIAM Conf. Parallel Process. Sci. Comput.*

Carle, A., Green, L. L., Bischof, C. H., and Newman, P. A. 1994. Application of automatic differentiation in CFD. *Proc. 25th AIAA Fluid Dynamics Conf.* Colorado Springs, CO, June 20–23, AIAA Paper 94-2197.

Cescotto, S., Frey, F., and Fonder, G. 1979. Total and updated Lagrangian descriptions in nonlinear structural analysis: a unified approach. In *Energy Methods in Finite Element Analysis*, R. Glowinski, E. Y. Rodin, and O. C. Zienkiewicz, Eds., pp. 283–296. Wiley, New York.

Chinchalkar, S. 1994. The application of automatic differentiation to problems in engineering analysis. *Comp. Methods. Appl. Mech. Eng.* 118:197–207.

Choi, K. K. 1993. Design sensitivity of nonlinear structures — II. In *Structural Optimization: Status and Promise*, M. P. Kamat, Ed., pp. 407–446. American Institute of Aeronautics and Astronautics, Washington, DC.

Crisfield, M. A. 1983. An arc-length method including line searches and accelerations. *Int. J. Num. Methods Eng.* 19:1269–1289.

Demkowicz, L., Oden, J. T., and Babuska, I. 1992. *Reliability in Computational Mechanics. Comput. Methods Appl. Mech. Eng.* Spec. issue 101(1–3).

Elishakoff, I. 1995. Essay on uncertainties in elastic and viscoelastic structures: from A. M. Freudenthal's criticisms to modern convex modeling. *Comput. Struct.* 56(6):871–895.

Fish, J. and Markolefas, S. 1993. Adaptive s-method for linear elastostatics. *Comput. Methods. Appl. Mech. Eng.* 104:363–396.

Fish, J. and Markolefas, S. 1994. Adaptive global-local refinement strategy based on the interior error estimates of the h-method. *Int. J. Num. Methods Eng.* 37:827–838.

George, P. L. 1991. *Automatic Mesh Generation. Application to Finite Element Methods.* Wiley, New York.

Geradin, M. 1980. Variational methods of structural dynamics and their finite element implementation. In *Advanced Structural Dynamics,* J. Donea, Ed., pp. 1–41. Applied Science, Ltd., London.

Grandhi, R. V., Stroud, W. J., and Venkayya, V. B., eds. 1989. *Computational Structural Mechanics and Multidisciplinary Optimization,* AD Vol. 16, American Society of Mechanical Engineers, New York.

Haber, R. B., Tortorelli, D. A., and Vidal, C. A. 1993. Design sensitivity analysis of nonlinear structures — I: large deformation hyperelasticity and history dependent material response. In *Structural Optimization: Status and Promise,* M. P. Kamat, Ed., pp. 369–406. American Institute of Aeronautics and Astronautics, Washington, DC.

Haftka, R. T. and Adelman, H. M. 1989. Recent developments in structural sensitivity analysis. *Struct. Optimization* 1(3):137–151.

Haug, E., Nguyen, Q. S., and de Rouvray, A. L. 1977. An improved energy conserving implicit time integration algorithm for nonlinear dynamic structural analysis. In *Trans. 4th Conf. Struct. Mech. Reactor Tech.,* T. A. Jaeger and B. A. Boley, Eds. IASMiRT, Paper No. M-5/3.

Hendrickson, B. and Leland, R. 1992. An Improved Spectral Graph Partitioning Algorithm for Mapping Parallel Computations. *Sandia National Lab.* TR SAND92-1460.

Hinton, E. and Sienz, J. 1994. Aspects of adaptive finite-element analysis and structural optimization. In *Advances in Structural Optimization,* B. H. V. Topping and M. Papadrakakis, Eds., pp. 1–25. Civil-Comp, Ltd., Edinburgh, Scotland.

Ho-Le, K. 1988. Finite element mesh generation methods: review and classification. *Comput. Aided Design* 1(20):27–38.

Hughes, T. J. R. 1976. Stability, convergence and growth and decay of energy of the average acceleration method in nonlinear structural dynamics. *Comput. Struct.* 6:313–324.

Hughes, T. J. R. 1987. *The Finite Element Method.* Prentice–Hall, Englewood Cliffs, NJ.

Hughes, T. J. R., Caughey, T. K., and Liu, W. K. 1978. Finite element methods for nonlinear elastodynamics which conserve energy. *J. Appl. Mech.* 45:366–370.

Ju, J. W., ed. 1995. *Numerical Methods in Structural Mechanics,* ASME Press, New York.

Kane, J. H., Maier, G., Tosaka, N., and Atluri, S. N., Eds. 1993. *Advances in Boundary Element Techniques.* Springer–Verlag, New York.

Karaoglan, L. and Noor, A. K. 1995. Dynamic sensitivity analysis of frictional contact/impact response of axisymmetric composite structures. *Comp. Methods Appl. Mech. Eng.* 128(1–2):169–190.

Khan, A. I. and Topping, B. H. V. 1993. Subdomain generation for parallel finite element analysis. *Comput. Syst. Eng.* 4(4–6):473–488.

Kleiber, M., Hien, T. D., and Postek, E. 1994. Incremental finite-element sensitivity analysis for nonlinear mechanics applications. *Int. J. Num. Methods. Eng.* 37(19):3291–3308.

Kleiber, M. and Hisada, T., eds. 1993. *Design-Sensitivity Analysis.* Atlanta Technology, Atlanta, GA.

Kounadis, A. N. and Kratzig, W. B. 1995. *Nonlinear Stability of Structures: Theory and Computational Techniques.* Springer–Verlag, Vienna.

Kratzig, W. B. and Eller, C. 1992. Numerical algorithms for unstable dynamic shell responses. *Comput. Struct.* 44(1–2):263–271.

Krizek, M. and Neittaanmaki, P. 1987. On superconvergence techniques. *Acta Applicandae Mathematicae* 9:175–198.

Kulkarni, M. and Noor, A. K. 1995. Sensitivity analysis of the nonlinear dynamic viscoplastic response of two-dimensional structures with respect to material parameters. *Int. J. Num. Methods. Eng.* 38(2):183–198.

Ladevéze, P. and Zienkiewicz, O. C., eds. 1992. *New Advances in Computational Structural Mechanics (Proc. Eur. Conf. New Advances Comput. Struc. Mechanics).* Giens, France, April 2–5, 1991. Elsevier, Amsterdam.

Law, K. H. and Mackay, D. R. 1993. A parallel row-oriented sparse solution method for finite element structural analysis. *Int. J. Num. Methods. Eng.* 36:2895–2919.

Mackerle, J. 1991. *Finite Element Methods, A Guide to Information Sources.* Elsevier, Amsterdam.

Mackerle, J. 1993. Mesh generation and refinement for FEM and BEM — A bibliography. *Finite Elements Anal. Design* 15:177–188.

Mackerle, J. and Fredriksson, B. 1988. *Handbook of Finite Element Software.* Studentlitteratur, Lund, Sweden.

Noor, A. K. 1980. Mixed methods of analysis. In *Structural Mechanics Software Series,* N. Perrone and W. Pilkey, Eds., Vol. III, pp. 263–305. University Press of Virginia, Charlottesville.

Noor, A. K. 1991. Bibliography of books and monographs on finite element technology. *Appl. Mech. Rev.,* 44(6):307–317.

Noor, A. K., ed. 1992. *Adaptive, Multilevel and Hierarchical Computational Strategies.* Proc. Symp. ASME Winter Annu. Meeting, Nov. 8–13, Anaheim, CA. AMD Vol. 157, American Society of Mechanical Engineers, New York.

Noor, A. K. 1994. Recent advances and applications of reduction methods. *Appl. Mech. Rev.* 47(5): 125–146.

Noor, A. K. and Andersen, C. M. 1992. Hybrid analytical techniques for the nonlinear analysis of curved beams. *Comput. Struct.* 43(5):823–830.

Noor, A. K. and Atluri, S. N. 1987. Advances and trends in computational structural mechanics. *AIAA J.* 25(7):977–995.

Noor, A. K. and Babuska, I. 1987. Quality assessment and control of finite element solutions. *Finite Elements Anal. Design* 3:1–26.

Noor, A. K., Housner, J. M., Starnes, J. H., and Hopkins, D. A. compilers. 1992. *Computational Structures Technology for Airframes and Propulsion Systems.* NASA CP-3142.

Noor, A. K., Kamel, H. A., and Fulton, R. E. 1978. Substructuring techniques — status and projections. *Comput. Struct.* 8:621–632.

Noor, A. K. and Venneri, S. L. 1990. Advances and trends in computational structural technology. *Comput. Syst. Eng.* 1(1):23–36.

Noor, A. K. and Venneri, S. L., Eds. 1995. Computational structures technology. *Flight-Vehicle Materials, Structures and Dynamics,* Vol. 6. American Society of Mechanical Engineers, New York.

Onaté, E., Periaux, J., and Samuelsson, A., eds. 1992. *The Finite Element Method in the Nineteen Ninety's: A Book Dedicated to O. C. Zienkiewicz.* Springer–Verlag, New York.

Papadrakakis, M., ed. 1993. *Solving Large-Scale Problems in Mechanics: The Development and Application of Computational Solution Methods.* Wiley, Chichester, UK.

Parlett, B. N. 1987. The state-of-the-art in extracting eigenvalues and eigenvectors in structural mechanics. In *State-of-the-Art Surveys on Computational Mechanics.* A. K. Noor and J. T. Oden, Eds., pp. 201–218. American Society of Mechanical Engineers, New York.

Pothen, A., Simon, H. D., and Liou, K. 1990. Partitioning sparse matrices with eigenvectors of graphs. *SIAM J. Matrix Anal. App.* 11(3):430–452.

Qin, J. and Mackay, D. R. 1993. A new parallel-vector finite element analysis software on distributed memory computers. *Proc. 34th AIAA/ASME/ASCE/AHS/ASC Struct. Structural Dynamics Mater. Conf.* La Jolla, CA, April 15–22.

Riks, E. 1984. Bifurcation and stability. A numerical approach. In *Proc. Int. Conf. Innovative Methods Nonlinear Probl.* W. K. Liu, T. Belytschko, and K. C. Park, Eds., pp. 313–344. Pineridge Press, Swansea, UK.

Ross, T. 1995. *Fuzzy Logic with Engineering Applications.* McGraw–Hill, New York.

Sabin, M. 1991. Criteria for comparison of automatic mesh generation methods. *Advances Eng. Software & Workstations* 13(5–6):220–225.

Shephard, M. S. 1988. Approaches to the automatic generation and control of finite element meshes. *Appl. Mech. Rev.* 41:169–185.

Shephard, M. S. and Wentorf, R. 1994. Toward the implementation of automated analysis idealization control. *Appl. Num. Math.* 14(1–3).

Sherman, L. L., Taylor, A. C., III, Green, L. L., Newman, P. A., Hou, G. J.-W., and Korivi, V. M. 1994. First-
and second-order aerodynamic sensitivity derivatives via automatic differentiation with incremental
iterative methods. *5th AIAA/USAF/NASA/ISSMO Symp. Multidisciplinary Anal. Optimization*, AIAA
Paper 94-4262. Panama City Beach, FL. Sept. 7–9.

Stein, E., ed. 1993. *Progress in Computational Analysis of Inelastic Structures*. Springer–Verlag, Vienna.

Storaasli, O. O. and Housner, J. M., Eds. 1994. *Large-Scale Structural Analysis for High-Performance Computers and Workstations, Comput. Syst. Eng.* Spec. issue 5(4–6).

Szabo, B. and Babuska, I. 1991. *Finite Element Analysis*. Wiley, New York.

Tabbara, M., Blacker, T., and Belytschko, T. 1994. Finite element derivative recovery by moving least square
interpolants. *Comput. Meth. Appl. Mech. Eng.* 117(1–2):211–223.

Turkiyyah, G. M. and Fenves, S. J. 1996. Knowledge-based assistance for finite element modeling. *IEEE
Expert Intell. Syst. Their Appl.* (June)11(3):23–32.

Vaughan, C. T. 1991. Structural analysis on massively parallel computers. *Comput. Systems Eng.* 2(2/3):261–
267.

Wang, K. P. and Bruch, J. C. 1993. A highly efficient iterative parallel computational method for finite
element systems. *Eng. Comput.* 10:195–204.

Washizu, K. 1982. *Variational Methods in Elasticity and Plasticity,* 3rd ed. Pergamon Press, Oxford, UK.

Watson, B. C. and Noor, A. K. 1996a. Large-scale contact/impact simulation and sensitivity analysis on
distributed-memory computers. *Comp. Methods Appl. Mech. Eng.* 65:6.

Watson, B. C. and Noor, A. K. 1996b. Sensitivity analysis for large-deflection and postbuckling responses
on distributed-memory computers. *Comp. Methods. Appl. Mech. Eng.* 129:393–409.

Zienkiewicz, O. C. and Zhu, J. Z. 1992. The superconvergent patch recovery and *a posteriori* error
estimates — part 1: the recovery technique. *Int. J. Num. Methods. Eng.* 33:1331–1364.

Zienkiewicz, O. C., Zhu, J. Z., and Wu, J. 1993. Superconvergent patch recovery techniques — some further
tests. *Commun. Num. Methods Eng.* 9(3):251–258.

Further Information

Information about CSM software is available on the Internet including structural analysis and design
programs, and commercial finite-element programs. A number of publications on finite-element practice
is available from the National Agency for Finite Element Methods and Standards (NAFEMS), Department of
Trade and Industry, National Engineering Laboratory, East Kilbride, Glasgow G75 0QU, United Kingdom,
including *A Finite Element Primer* (1986) and *Guidelines to Finite Element Practice* (1984). A finite element
bibliography, including books and conference proceedings published since 1967, is available on the World
Wide Web. The WWW address is: http://ohio.ikp.liu.se/fe/index.html.

29

Computational Electromagnetics

29.1	Introduction	29-1
29.2	Governing Equations	29-3
29.3	Characteristic-Based Formulation	29-5
29.4	Maxwell Equations in a Curvilinear Frame	29-7
29.5	Eigenvalues and Eigenvectors	29-10
29.6	Flux-Vector Splitting	29-11
29.7	Finite-Difference Approximation	29-14
29.8	Finite-Volume Approximation	29-15
29.9	Summary and Research Issues	29-16

J. S. Shang
Air Force Research

Nomenclature

B	=	Magnetic flux density
C	=	Coefficient matrix of flux-vector formulation
D	=	Electric displacement
E	=	Electric field strength
F	=	Flux vector component
H	=	Magnetic flux intensity
i, j, k	=	Index of discretization
J	=	Electric current density
n	=	Index of temporal level of solution
S	=	Similar matrix of diagonalization
t	=	Time
U	=	Dependent variables
V	=	Elementary-cell volume
x, y, z	=	Cartesian coordinates
Δ	=	Forward difference operator
λ	=	Eigenvalue
ξ, η, ζ	=	Transformed coordinates
∇	=	Gradient, backward difference operator

29.1 Introduction

Computational electromagnetics (CEM) is a natural extension of the analytical approach in solving the Maxwell equations. In spite of the fundamental difference between representing the solution in a continuum and in a discretized space, both approaches satisfy all pertaining theorems rigorously. The analytic approach to electromagnetics is elegant, and the results can describe the specific behavior as well as the general patterns of a physical phenomenon in a given regime. However, exact solutions to the Maxwell equations are usually unavailable. Some of the closed-form results that exist have restrictive underlying assumptions that limit their range of validity. Solutions of CEM generate only a point value for a specific simulation, but complexity of the physics or of the field configuration is no longer a limiting factor. The numerical accuracy of CEM is an issue to be addressed. Nevertheless, with the advent of high-performance computing systems, CEM is becoming a mainstay for engineering applications.

CEM in the present context is focused on simulation methods for solving the Maxwell equations in the time domain. First of all, time dependence is the most general form of the Maxwell equations, and the dynamic electromagnetic field is not confined to a time-harmonic phenomenon. Therefore, CEM in the time domain has the widest range of engineering applications. In addition, several new numerical algorithms for solving the first-order hyperbolic partial differential equations, as well as coordinate transformation techniques, were introduced recently to the CEM community. These finite-difference and finite-volume numerical algorithms were devised specifically to mimic the physics involving directional wave propagation. Meanwhile, very complex shapes associated with the field can be easily accommodated by incorporating a coordinate transformation technique. These methodologies have the potential to radically change future research in electromagnetics.

In order to use CEM effectively, it will be beneficial to understand the fundamentals of numerical simulation and its limitations. The inaccuracy incurred by a numerical simulation is attributable to the mathematical model for the physics, the numerical algorithm, and the computational accuracy. In general, differential equations in computational electromagnetics consist of two categories: the first-order divergence–curl equations and the second-order curl–curl equations [Elliott 1966, Harrington 1961, 1968]. In specific applications, further simplifications into the frequency-domain or the Helmholtz equations and the potential formulation have been accomplished. Poor numerical approximations to physical phenomena can result, however, from solving overly simplified governing equations. Under these circumstances, no meaningful quantification of errors for the numerical procedure can be achieved. Physically incorrect values and inappropriate implementation of initial and/or boundary conditions are another major source of error. The placement of the far-field boundary and the type of initial or boundary conditions have also played an important role. These concerns are easily appreciated in the light of the fact that the governing equations are identical, but the different initial/boundary conditions generate different solutions.

Numerical accuracy is also controlled by the algorithm and computing system adopted. The error induced by the discretization consists of the roundoff and the truncation error. The roundoff error is contributed by the computing system and is problem-size-dependent. Since this error is random, it is the most difficult to evaluate. One anticipates that this type of error will be a concern for solution procedures involving large-scale matrix manipulation such as the method of moments and the implicit numerical algorithm for finite-difference or finite-volume methods. The truncation error for time-dependent calculations appears as dissipation and dispersion, which can be assessed and alleviated by mesh-system refinements.

Finally, numerical error can be the consequence of a specific formulation. The error becomes pronounced when a special phenomenon is investigated or when a discontinuous and distinctive stratification of the field is encountered, such as a wave propagating through the interface between media of different characteristic impedances, for which the solution is piecewise continuous. Only in a strongly conservative formulation can the discontinuous phenomenon be adequately resolved. Another example is encountered in radar cross-section simulation, where the scattered-field formulation has been shown to be superior to the total-field formulation.

The Maxwell equations in the time domain consist of a first-order divergence–curl system and are difficult to solve by conventional numerical methods. Nevertheless, the pioneering efforts by Yee and others have attained impressive achievements [Yee 1966, Taflove 1992]. Recently, numerical techniques in CEM have been further enriched by development in the field of computational fluid dynamics (CFD). In CFD, the Euler equations, which are a subset of the Navier–Stokes equations, have the same partial-differential-system classification as that of the time-dependent Maxwell equations. Both are hyperbolic systems and constitute initial-value problems [Sommerfeld 1949]. For hyperbolic partial differential equations, the solutions need not be analytic functions. More importantly, the initial values together with any possible discontinuities are continued along a time–space trajectory, which is commonly referred to as the characteristic. A series of numerical schemes have been devised in the CFD community to duplicate the directional information-propagation feature. These numerical procedures are collectively designated as the characteristic-based method, which in its most elementary form is identical to the Riemann problem [Roe 1986]. The characteristic-based method when applied to solve the Maxwell equations in the time domain has exhibited many attractive attributes. A synergism of the new numerical procedures and scalable parallel-computing capability will open up a new frontier in electromagnetics research. For this reason, a major portion of the present chapter will be focused on introducing the characteristic-based finite-volume and finite-difference methods [Shang 1993, Shang and Gaitonde 1993, Shang and Fithen 1994].

29.2 Governing Equations

The time-dependent Maxwell equations for the electromagnetic field can be written as [Elliott 1966, Harrington 1961]:

$$\frac{\partial \mathbf{B}}{\partial t} + \nabla \times \mathbf{E} = 0 \tag{29.1}$$

$$\frac{\partial \mathbf{D}}{\partial t} - \nabla \times \mathbf{H} = -\mathbf{J} \tag{29.2}$$

$$\nabla \cdot \mathbf{B} = 0 \tag{29.3}$$

$$\nabla \cdot \mathbf{D} = \rho \tag{29.4}$$

The only conservation law for electric charge and current densities is

$$\frac{\partial \rho}{\partial t} + \nabla \cdot \mathbf{J} = 0 \tag{29.4}$$

where ρ and $\mathbf{J}$ are the charge and current density, respectively, and represent the source of the field. The constitutive relations between the magnetic flux density and intensity and between the electric displacement and field strength are $\mathbf{B} = \mu \mathbf{H}$ and $\mathbf{D} = \epsilon \mathbf{E}$. Equation 29.4 is regarded as a fundamental law of electromagnetics, derived from the generalized Ampere's circuit law and Gauss's law. Since Equation 29.1 and Equation 29.2 contain the information on the propagation of the electromagnetic field, they constitute the basic equations of CEM.

The above partial differential equations also can be expressed as a system of integral equations. The following expression is obtained by using Stoke's law and the divergence theorem to reduce the surface and volume integrals to a circuital line and surface integrals, respectively [Elliott 1966]. These integral relationships hold only if the first derivatives of the electric displacement $\mathbf{D}$ and the magnetic flux density $\mathbf{B}$ are continuous throughout the control volume:

$$\oint \mathbf{E} \cdot d\mathbf{L} = -\iint \frac{\partial \mathbf{B}}{\partial t} \cdot d\mathbf{S} \tag{29.5}$$

$$\oint \mathbf{H} \cdot d\mathbf{L} = \iint \left(\mathbf{J} + \frac{\partial \mathbf{D}}{\partial t} \right) \cdot d\mathbf{S} \tag{29.6}$$

$$\iint \mathbf{D} \cdot d\mathbf{S} = \iiint \rho \, d\mathbf{V} \tag{29.7}$$

$$\iint \mathbf{B} \cdot d\mathbf{S} = 0 \tag{29.8}$$

The integral form of the Maxwell equations is rarely used in CEM. They are, however, invaluable as a validation tool for checking the global behavior of field computations.

The second-order curl–curl form of the Maxwell equations is derived by applying the curl operator to get

$$\nabla \times \nabla \times \mathbf{E} + \frac{1}{c^2}\frac{\partial^2 \mathbf{E}}{\partial t^2} = -\frac{\partial(\mu \mathbf{J})}{\partial t} \qquad (29.9)$$

$$\nabla \times \nabla \times \mathbf{B} + \frac{1}{c^2}\frac{\partial^2 \mathbf{B}}{\partial t^2} = \nabla \times (\mu \mathbf{J}) \qquad (29.10)$$

The outstanding feature of the curl–curl formulation of the Maxwell equations is that the electric and magnetic fields are decoupled. The second-order equations can be further simplified for harmonic fields. If the time-dependent behavior can be represented by a harmonic function $e^{i\omega t}$, the separation-of-variables technique will transform the Maxwell equations into the frequency domain [Elliott 1966, Harrington 1961]. The resultant partial differential equations in spatial variables become elliptic:

$$\nabla \times \nabla \times \mathbf{E} - k^2 \mathbf{E} = -i\omega(\mu \mathbf{J})$$
$$\nabla \times \nabla \times \mathbf{B} - k^2 \mathbf{B} = \nabla \times (\mu \mathbf{J}) \qquad (29.11)$$

where $k = \omega/c$ is called the propagation constant or the wave number, and ω is the angular frequency of a component of a Fourier series or a Fourier integral [Elliott 1966, Harrington 1961]. The above equations are frequently the basis for finite-element approaches [Rahman et al. 1991].

In order to complete the description of the differential system, initial and/or boundary values are required. For Maxwell equations, only the source of the field and a few physical boundary conditions at the media interfaces are pertinent [Elliott 1966, Harrington 1961]:

$$\mathbf{n} \times (\mathbf{E}_1 - \mathbf{E}_2) = 0$$
$$\mathbf{n} \times (\mathbf{H}_1 - \mathbf{H}_2) = \mathbf{J}_s$$
$$\mathbf{n} \cdot (\mathbf{D}_1 - \mathbf{D}_2) = \rho_s \qquad (29.12)$$
$$\mathbf{n} \cdot (\mathbf{B}_1 - \mathbf{B}_2) = 0$$

where the subscripts 1 and 2 refer to media on two sides of the interface, and $\mathbf{J}_s$ and ρ_s are the surface current and charge densities, respectively.

Since all computing systems have finite memory, all CEM computations in the time domain must be conducted on a truncated computational domain. This intrinsic constraint requires a numerical far-field condition at the truncated boundary to mimic the behavior of an unbounded field. This numerical boundary unavoidably induces a reflected wave to contaminate the simulated field. In the past, absorbing boundary conditions at the far-field boundary have been developed from the radiation condition [Sommerfeld 1949, Enquist and Majda 1977, Higdon 1986, Mur 1981]. In general, a progressive order-of-accuracy procedure can be used to implement the numerical boundary conditions with increasing accuracy [Enquist and Majda 1977, Higdon 1986]. On the other hand, the characteristic-based methods which satisfy the physical domain of influence requirement can specify the numerical boundary condition readily. For this formulation, the reflected wave can be suppressed by eliminating the undesirable incoming numerical data. Although the accuracy of the numerical far-field boundary condition depends on the coordinate system, in principle this formulation under ideal circumstances can effectively suppress artificial wave reflections.

29.3 Characteristic-Based Formulation

The fundamental idea of the characteristic-based method for solving the hyperbolic system of equations is derived from the eigenvalue–eigenvector analyses of the governing equations. For Maxwell equations in the time domain, every eigenvalue is real but not all of them are distinct [Shang 1993, Shang and Gaitonde 1993, Shang and Fithen 1994]. In a time–space plane, the eigenvalue actually defines the slope of the characteristic or the phase velocity of the wave motion. All dependent variables within the time–space domain bounded by two intersecting characteristics are completely determined by the values along these characteristics and by their compatibility relationship. The direction of information propagation is also clearly described by these two characteristics [Sommerfeld 1949]. In numerical simulation, the well-posedness requirement on initial or boundary conditions and the stability of a numerical approximation are also ultimately linked to the eigenvalues of the governing equation [Anderson et al. 1984, Richtmyer and Morton 1967]. Therefore, characteristic-based methods have demonstrated superior numerical stability and accuracy to other schemes [Roe 1986, Shang 1993]. However, characteristic-based algorithms also have an inherent limitation in that the governing equation can be diagonalized only in one spatial dimension at a time. The multidimensional equations are required to split into multiple one-dimensional formulations. This limitation is not unusual for numerical algorithms, such as the approximate factored and the fractional-step schemes [Shang 1993, Anderson et al. 1984]. A consequence of this restriction is that solutions of the characteristic-based procedure may exhibit some degree of sensitivity to the orientation of the coordinate selected. This numerical behavior is consistent with the concept of optimal coordinates.

In the characteristic formulation, data on the wave motion are first split according to the direction of phase velocity and then transmitted in each orientation. In each one-dimensional time–space domain, the direction of the phase velocity degenerates into either a positive or a negative orientation. They are commonly referred to as the right-running and the left-running wave components [Sommerfeld 1949, Roe 1986]. The sign of the eigenvalue is thus an indicator of the direction of signal transmission. The corresponding eigenvectors are the essential elements for diagonalizing the coefficient matrices and for formulating the approximate Riemann problem [Roe 1986]. In essence, knowledge of eigenvalues and eigenvectors of the Maxwell equations in the time domain becomes the first prerequisite of the present formulation.

The system of governing equations cast in the flux-vector form in the Cartesian frame becomes [Shang 1993, Shang and Gaitonde 1993, Shang and Fithen 1994]

$$\frac{\partial U}{\partial t} + \frac{\partial F_x}{\partial x} + \frac{\partial F_y}{\partial y} + \frac{\partial H_z}{\partial z} = -J \tag{29.13}$$

where U is the vector of dependent variables. The flux vectors are formed by the inner product of the coefficient matrix and the dependent variable: $F_x = C_x U$, $F_y = C_y U$, and $F_z = C_z U$, with

$$U = \{B_x, B_y, B_z, D_x, D_y, D_z\}^T \tag{29.14}$$

and

$$F_x = \{0, -D_z/\epsilon, D_y/\epsilon, 0, B_z/\mu, -B_y/\mu\}^T$$

$$F_y = \{D_z/\epsilon, 0, -D_x/\epsilon, -B_z/\mu, 0, B_x/\mu\}^T \tag{29.15}$$

$$F_z = \{-D_y/\epsilon, D_x/\epsilon, 0, B_y/\mu, -B_x/\mu, 0\}^T$$

The coefficient matrices, or the Jacobians of the flux vectors C_x, C_y, and C_z are [Shang 1993]:

$$C_x = \begin{bmatrix} 0 & 0 & 0 & 0 & 0 & 0 \\ 0 & 0 & 0 & 0 & 0 & -\frac{1}{\epsilon} \\ 0 & 0 & 0 & 0 & \frac{1}{\epsilon} & 0 \\ 0 & 0 & 0 & 0 & 0 & 0 \\ 0 & 0 & \frac{1}{\mu} & 0 & 0 & 0 \\ 0 & -\frac{1}{\mu} & 0 & 0 & 0 & 0 \end{bmatrix} \tag{29.16a}$$

$$C_y = \begin{bmatrix} 0 & 0 & 0 & 0 & 0 & \frac{1}{\epsilon} \\ 0 & 0 & 0 & 0 & 0 & 0 \\ 0 & 0 & 0 & -\frac{1}{\epsilon} & 0 & 0 \\ 0 & 0 & -\frac{1}{\mu} & 0 & 0 & 0 \\ 0 & 0 & 0 & 0 & 0 & 0 \\ \frac{1}{\mu} & 0 & 0 & 0 & 0 & 0 \end{bmatrix} \tag{29.16b}$$

$$C_z = \begin{bmatrix} 0 & 0 & 0 & 0 & -\frac{1}{\epsilon} & 0 \\ 0 & 0 & 0 & \frac{1}{\epsilon} & 0 & 0 \\ 0 & 0 & 0 & 0 & 0 & 0 \\ 0 & \frac{1}{\mu} & 0 & 0 & 0 & 0 \\ -\frac{1}{\mu} & 0 & 0 & 0 & 0 & 0 \\ 0 & 0 & 0 & 0 & 0 & 0 \end{bmatrix} \tag{29.16c}$$

where ϵ and μ are the permittivity and permeability, which relate the electric displacement to the electric field intensity and the magnetic flux density to the magnetic field intensity, respectively.

The eigenvalues of the coefficient matrices C_x, C_y, and C_z in the Cartesian frame are identical and contain multiplicities [Shang 1993, Shang and Gaitonde 1993]. Care must be exercised to ensure that all associated eigenvectors

$$\lambda = \left\{ +\frac{1}{\sqrt{\epsilon\mu}}, -\frac{1}{\sqrt{\epsilon\mu}}, 0, +\frac{1}{\sqrt{\epsilon\mu}}, -\frac{1}{\sqrt{\epsilon\mu}}, 0 \right\} \tag{29.17}$$

are linearly independent. The linearly independent eigenvectors associated with each eigenvalue are found by reducing the matrix equation, $(C - \bar{\bar{I}}\lambda)U = 0$, to the Jordan normal form [Shang 1993, Shang and Fithen 1994].

Since the coefficient matrices C_x, C_y, and C_z can be diagonalized, there exist nonsingular similar matrices S_x, S_y, and S_z such that

$$\Lambda_x = S_x^{-1} C_x S_x$$
$$\Lambda_y = S_y^{-1} C_y S_y \tag{29.18}$$
$$\Lambda_z = S_z^{-1} C_z S_z$$

where the Λs are the diagonalized coefficient matrices. The columns of the similar matrices S_x, S_y, and S_z are simply the linearly independent eigenvectors of the coefficient matrices C_x, C_y, and C_z, respectively.

The fundamental relationship between the characteristic-based formulation and the Riemann problem can be best demonstrated in the Cartesian frame of reference. For the Maxwell equations in this frame of reference and for an isotropic medium, all the similar matrices are invariant with respect to temporal and spatial independent variables. In each time–space plane (x–t, y–t, and z–t), the one-dimensional governing equation can be given just as in the x–t plane:

$$\frac{\partial U}{\partial t} + C_x \frac{\partial U}{\partial x} = 0 \tag{29.19}$$

Substitute the diagonalized coefficient matrix to get

$$\frac{\partial U}{\partial t} + S_x \Lambda_x S_x^{-1} \frac{\partial U}{\partial x} = 0 \tag{29.20}$$

Since the similar matrix in the present consideration is invariant with respect to time and space, it can be brought into the differential operator. Multiplying the above equation by the left-hand inverse of S_x, S_x^{-1}, we have

$$\frac{\partial \left(S_x^{-1} U \right)}{\partial t} + \Lambda_x \frac{\partial \left(S_x^{-1} U \right)}{\partial x} = 0 \tag{29.21}$$

One immediately recognizes the group of variables $S_x^{-1} U$ as the characteristic, and the system of equations is decoupled [Shang 1993]. In scalar-variable form and with appropriate initial values, this is the Riemann problem [Sommerfeld 1949, Courant and Hilbert 1965]. This differential system is specialized to study the breakup of a single discontinuity. The piecewise continuous solutions separated by the singular point are also invariant along the characteristics. Equally important, stable numerical operators can now be easily devised to solve the split equations according to the sign of the eigenvalue. In practice it has been found if the multidimensional problem can be split into a sequence of one-dimensional equations, this numerical technique is applicable to those one-dimensional equations [Roe 1986, Shang 1993].

The gist of the characteristic-based formulation is also clearly revealed by the decomposition of the flux vector into positive and negative components corresponding to the sign of the eigenvalue:

$$\Lambda = \Lambda^+ + \Lambda^-, \qquad F = F^+ + F^- \tag{29.22}$$

$$F^+ = S\Lambda^+ S^{-1}, \qquad F^- = S\Lambda^- S^{-1} \tag{29.23}$$

where the superscripts $+$ and $-$ denote the split vectors associated with positive and negative eigenvalues, respectively.

The characteristic-based algorithms have a deep-rooted theoretical basis for describing the wave dynamics. They also have however an inherent limitation in that the diagonalized formulation is achievable only in one dimension at a time. All multidimensional equations are required to be split into multiple one-dimensional formulations. The approach yields accurate results so long as discontinuous waves remain aligned with the computational grid. This limitation is also the state-of-the-art constraint in solving partial differential equations [Roe 1986, Shang 1993, Anderson et al. 1984].

29.4 Maxwell Equations in a Curvilinear Frame

In order to develop a versatile numerical tool for computational electromagnetics in a wide range of applications, the Maxwell equations can be cast in a general curvilinear frame of reference [Shang and Gaitonde 1993, Shang and Fithen 1994]. For efficient simulation of complex electromagnetic field configurations, the adoption of a general curvilinear mesh system becomes necessary. The system of equations in general curvilinear coordinates is derived by a coordinate transformation [Anderson et al. 1984, Thompson 1982]. The mesh system in the transformed space can be obtained by numerous grid generation procedures [Thompson 1982]. Computational advantages in the transformed space are also realizable. For a body-oriented coordinate system, the interface between two different media is easily defined by one of the coordinate surfaces. Along this coordinate parametric plane, all discretized nodes on the interface are precisely prescribed without the need for an interpolating procedure. The outward normal to the interface, which is essential for boundary-value implementation, can be computed easily by $n = \nabla S / \|\nabla S\|$. In the transformed space, computations are performed on a uniform mesh space, but the corresponding physical spacing can be highly clustered to enhance the numerical resolution.

As an illustration of the numerical advantage of solving the Maxwell equations on nonorthogonal curvilinear, body-oriented coordinates, a simulation of the scattered electromagnetic field from a re-entry vehicle has been performed [Shang and Gaitonde 1994]. The aerospace vehicle, X24C-10D, has

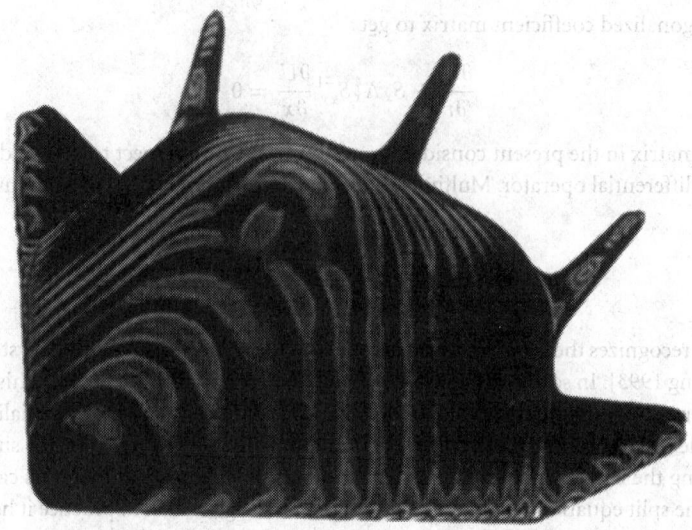

FIGURE 29.1 Radar-wave fringes on X24C-10D, grid $181 \times 59 \times 162$, TE excitation, $L/\lambda = 9.2$.

a complex geometrical shape (Figure 29.1). In addition to a blunt leading-edge spherical nose and a relatively flat delta-shaped underbody, the aft portion of the vehicle consists of five control surfaces — a central fin, two middle fins, and two strakes. A body-oriented, single-block mesh system enveloping the configuration is adopted. The numerical grid system is generated by using a hyperbolic grid generator for the near-field mesh adjacent to the solid surface, and a transfinite technique for the far field. The two mesh systems are merged by the Poisson averaging technique [Thompson 1982, Shang and Gaitonde 1994]. In this manner, the composite grid system is orthogonal in the near field but less restrictive in the far field. All solid surfaces of the X24C-10D are mapped onto a parametric surface in the transformed space, defined by $\eta = 0$. The entire computational domain is supported by a $181 \times 59 \times 162$ grid system, where the first coordinate index denotes the number of cross sections in the numerical domain. The second index describes the number of cells between the body surface and the far-field boundary, while the third index gives the cells used to circumscribe each cross-sectional plane. The electromagnetic excitation is introduced by a harmonic incident wave traveling along the x-coordinate. The fringe pattern of the scattered electromagnetic waves on the X24C-10D is presented in Figure 29.1 for a characteristic-length-to-wavelength ratio $L/\lambda = 9.2$. A salient feature of the scattered field is brought out by the surface curvature: the smaller the radius of surface curvature, the broader the diffraction pattern. The numerical result exhibits highly concentrated contours at the chine (the line of intersection between upper and lower vehicle surfaces) of the forebody and the leading edges of strakes and fins.

For the most general coordinate transformation of the Maxwell equations in the time domain, a one-to-one relationship between two sets of temporal and spatial independent variables is required. However, for most practical applications, the spatial coordinate transformation is sufficient:

$$\xi = \xi(x, y, z)$$
$$\eta = \eta(x, y, z) \tag{29.24}$$
$$\zeta = \zeta(x, y, z)$$

The governing equation in the strong conservation form is obtained by dividing the chain-rule-differentiated equations by the Jacobian of coordinate transformation and by invoking metric identities [Shang and Gaitonde 1993, Anderson et al. 1984]. The time-dependent Maxwell equations on a general curvilinear frame of reference and in the strong conservative form are

$$\frac{\partial U}{\partial t} + \frac{\partial F_\xi}{\partial \xi} + \frac{\partial F_\eta}{\partial \eta} + \frac{\partial F_\zeta}{\partial \zeta} = -J \tag{29.25}$$

where the dependent variables are now defined as

$$U = U(B_x V, B_y V, B_z V, D_x V, D_y V, D_z V) \tag{29.26}$$

Here V is the Jacobian of the coordinate transformation and is also the inverse local cell volume. If the Jacobian has nonzero values in the computational domain, the correspondence between the physical and the transformed space is uniquely defined [Anderson et al. 1984, Thompson 1982]. Since systematic procedures have been developed to ensure this property of coordinate transformations, detailed information on this point is not repeated here [Anderson et al. 1984, Thompson 1982]. We have

$$V = \begin{bmatrix} \xi_x & \eta_x & \zeta_x \\ \xi_y & \eta_y & \zeta_y \\ \xi_z & \eta_z & \zeta_z \end{bmatrix} \tag{29.27}$$

and ξ_x, η_x, ζ_x, etc. are the metrics of coordinate transformation and can be computed easily from the definition given by Equation 29.24. The flux-vector components in the transformed space have the following form:

$$F_\xi = \begin{bmatrix} 0 & 0 & 0 & 0 & -\frac{\xi_z}{\epsilon V} & \frac{\xi_y}{\epsilon V} \\ 0 & 0 & 0 & \frac{\xi_z}{\epsilon V} & 0 & -\frac{\xi_x}{\epsilon V} \\ 0 & 0 & 0 & -\frac{\xi_y}{\epsilon V} & \frac{\xi_x}{\epsilon V} & 0 \\ 0 & \frac{\xi_z}{V\mu} & -\frac{\xi_y}{V\mu} & 0 & 0 & 0 \\ -\frac{\xi_z}{V\mu} & 0 & \frac{\xi_x}{V\mu} & 0 & 0 & 0 \\ \frac{\xi_y}{V\mu} & -\frac{\xi_x}{V\mu} & 0 & 0 & 0 & 0 \end{bmatrix} \begin{Bmatrix} B_x \\ B_y \\ B_z \\ D_x \\ D_y \\ D_z \end{Bmatrix} \tag{29.28}$$

$$F_\eta = \begin{bmatrix} 0 & 0 & 0 & 0 & -\frac{\eta_z}{\epsilon V} & \frac{\eta_y}{\epsilon V} \\ 0 & 0 & 0 & \frac{\eta_z}{\epsilon V} & 0 & -\frac{\eta_x}{\epsilon V} \\ 0 & 0 & 0 & -\frac{\eta_y}{\epsilon V} & \frac{\eta_x}{\epsilon V} & 0 \\ 0 & \frac{\eta_z}{V\mu} & -\frac{\eta_y}{V\mu} & 0 & 0 & 0 \\ -\frac{\eta_z}{V\mu} & 0 & \frac{\eta_x}{V\mu} & 0 & 0 & 0 \\ \frac{\eta_y}{V\mu} & -\frac{\eta_x}{V\mu} & 0 & 0 & 0 & 0 \end{bmatrix} \begin{Bmatrix} B_x \\ B_y \\ B_z \\ D_x \\ D_y \\ D_z \end{Bmatrix} \tag{29.29}$$

$$F_\zeta = \begin{bmatrix} 0 & 0 & 0 & 0 & -\frac{\zeta_z}{\epsilon V} & \frac{\zeta_y}{\epsilon V} \\ 0 & 0 & 0 & \frac{\zeta_z}{\epsilon V} & 0 & -\frac{\zeta_x}{\epsilon V} \\ 0 & 0 & 0 & -\frac{\zeta_y}{\epsilon V} & \frac{\zeta_x}{\epsilon V} & 0 \\ 0 & \frac{\zeta_z}{V\mu} & -\frac{\zeta_y}{V\mu} & 0 & 0 & 0 \\ -\frac{\zeta_z}{V\mu} & 0 & \frac{\zeta_x}{V\mu} & 0 & 0 & 0 \\ \frac{\zeta_y}{V\mu} & -\frac{\zeta_x}{V\mu} & 0 & 0 & 0 & 0 \end{bmatrix} \begin{Bmatrix} B_x \\ B_y \\ B_z \\ D_x \\ D_y \\ D_z \end{Bmatrix} \tag{29.30}$$

After the coordinate transformation, all coefficient matrices now contain metrics which are position-dependent, and the system of equations in the most general frame of reference possesses variable coefficients. This added complexity of the characteristic formulation of the Maxwell equations no longer permits the system of one-dimensional equations to be decoupled into six scalar equations and reduced to the true Riemann problem [Shang 1993, Shang and Gaitonde 1993, Shang and Fithen 1994] like that on the Cartesian form.

29.5 Eigenvalues and Eigenvectors

As previously mentioned, eigenvalue and the eigenvector analyses are the prerequisites for characteristic-based algorithms. The analytic process to obtain the eigenvalues and the corresponding eigenvectors of the Maxwell equations in general curvilinear coordinates is identical to that in the Cartesian frame. In each of the temporal–spatial planes t–ξ, t–η, and t–ζ, the eigenvalues are easily found by solving the sixth-degree characteristic equation associated with the coefficient matrices [Sommerfeld 1949, Courant and Hilbert 1965]

$$\lambda_\xi = \left\{ -\frac{\alpha}{V\sqrt{\epsilon\mu}}, -\frac{\alpha}{V\sqrt{\epsilon\mu}}, \frac{\alpha}{V\sqrt{\epsilon\mu}}, \frac{\alpha}{V\sqrt{\epsilon\mu}}, 0, 0 \right\} \tag{29.31}$$

$$\lambda_\eta = \left\{ -\frac{\beta}{V\sqrt{\epsilon\mu}}, -\frac{\beta}{V\sqrt{\epsilon\mu}}, \frac{\beta}{V\sqrt{\epsilon\mu}}, \frac{\beta}{V\sqrt{\epsilon\mu}}, 0, 0 \right\} \tag{29.32}$$

$$\lambda_\zeta = \left\{ -\frac{\gamma}{V\sqrt{\epsilon\mu}}, -\frac{\gamma}{V\sqrt{\epsilon\mu}}, \frac{\gamma}{V\sqrt{\epsilon\mu}}, \frac{\gamma}{V\sqrt{\epsilon\mu}}, 0, 0 \right\} \tag{29.33}$$

where $\alpha = \sqrt{\xi_z^2 + \xi_y^2 + \xi_x^2}$, $\beta = \sqrt{\eta_z^2 + \eta_y^2 + \eta_x^2}$, and $\gamma = \sqrt{\zeta_z^2 + \zeta_y^2 + \zeta_x^2}$.

One recognizes that the eigenvalues in each one-dimensional time–space plane contain multiplicities, and hence the eigenvectors do not necessarily have unique elements [Shang 1993, Courant and Hilbert 1965]. Nevertheless, linearly independent eigenvectors associated with each eigenvalue still have been found by reducing the coefficient matrix to the Jordan normal form [Shang 1993, Shang and Fithen 1994]. For reasons of wide applicability and internal consistency, the eigenvectors are selected in such a fashion that the similar matrices of diagonalization reduce to the same form as in the Cartesian frame. Furthermore, in order to accommodate a wide range of electromagnetic field configurations such as antennas, wave guides, and scatterers, the eigenvalues are no longer identical in the three time–space planes. This complexity of formulation is essential to facilitate boundary-condition implementation on the interfaces of media with different characteristic impedances.

From the eigenvector analysis, the similarity transformation matrices for diagonalization in each time–space plane are formed by using eigenvectors as the column arrays as shown in the following equations. For an example, the first column of the similar matrix of diagonalization,

$$\left[-\frac{\sqrt{\mu}\xi_y}{\sqrt{\epsilon}\alpha}, \frac{\sqrt{\mu}(\xi_x^2 + \xi_z^2)}{\sqrt{\epsilon}\xi_x\alpha}, \frac{\sqrt{\mu}\xi_y\xi_z}{\sqrt{\epsilon}\xi_x\alpha}, \frac{\xi_y}{\xi_x}, 0, 1 \right]$$

in the t–ξ plane is the eigenvector corresponding to the eigenvalue $\lambda_\xi = -\alpha/V\sqrt{\epsilon\mu}$. We have

$$S_\xi = \begin{bmatrix} -\dfrac{\sqrt{\mu}\xi_y}{\sqrt{\epsilon}\alpha} & \dfrac{\sqrt{\mu}\xi_z}{\sqrt{\epsilon}\alpha} & \dfrac{\sqrt{\mu}\xi_y}{\sqrt{\epsilon}\alpha} & -\dfrac{\sqrt{\mu}\xi_z}{\sqrt{\epsilon}\alpha} & 1 & 0 \\[2mm] \dfrac{\sqrt{\mu}(\xi_x^2+\xi_z^2)}{\sqrt{\epsilon}\xi_x\alpha} & \dfrac{\sqrt{\mu}\xi_y\xi_z}{\sqrt{\epsilon}\xi_x\alpha} & -\dfrac{\sqrt{\mu}(\xi_x^2+\xi_z^2)}{\sqrt{\epsilon}\xi_x\alpha} & -\dfrac{\sqrt{\mu}\xi_y\xi_z}{\sqrt{\epsilon}\xi_x\alpha} & \dfrac{\xi_y}{\xi_x} & 0 \\[2mm] -\dfrac{\sqrt{\mu}\xi_y\xi_z}{\sqrt{\epsilon}\xi_x\alpha} & -\dfrac{\sqrt{\mu}(\xi_x^2+\xi_y^2)}{\sqrt{\epsilon}\xi_x\alpha} & \dfrac{\sqrt{\mu}\xi_y\xi_z}{\sqrt{\epsilon}\xi_x\alpha} & \dfrac{\sqrt{\mu}(\xi_x^2+\xi_y^2)}{\sqrt{\epsilon}\xi_x\alpha} & \dfrac{\xi_z}{\xi_x} & 0 \\[2mm] -\dfrac{\xi_z}{\xi_x} & -\dfrac{\xi_y}{\xi_x} & -\dfrac{\xi_z}{\xi_x} & -\dfrac{\xi_y}{\xi_x} & 0 & 1 \\[2mm] 0 & 1 & 0 & 1 & 0 & \dfrac{\xi_y}{\xi_x} \\[2mm] 1 & 0 & 1 & 0 & 0 & \dfrac{\xi_z}{\xi_x} \end{bmatrix} \tag{29.34}$$

$$S_\eta = \begin{bmatrix} -\dfrac{(\eta_y^2+\eta_z^2)\sqrt{\mu}}{\sqrt{\epsilon}\,\eta_y\beta} & -\dfrac{\eta_x\eta_z\sqrt{\mu}}{\sqrt{\epsilon}\,\eta_y\beta} & \dfrac{(\eta_y^2+\eta_z^2)\sqrt{\mu}}{\sqrt{\epsilon}\,\eta_y\beta} & \dfrac{\eta_x\eta_z\sqrt{\mu}}{\sqrt{\epsilon}\,\eta_y\beta} & \dfrac{\eta_x}{\eta_y} & 0 \\[2ex] \dfrac{\eta_x\sqrt{\mu}}{\sqrt{\epsilon}\beta} & -\dfrac{\eta_z\sqrt{\mu}}{\sqrt{\epsilon}\beta} & -\dfrac{\eta_x\sqrt{\mu}}{\sqrt{\epsilon}\beta} & \dfrac{\eta_z\sqrt{\mu}}{\sqrt{\epsilon}\beta} & 1 & 0 \\[2ex] \dfrac{\eta_x\eta_z\sqrt{\mu}}{\sqrt{\epsilon}\,\eta_y\beta} & \dfrac{(\eta_x^2+\eta_y^2)\sqrt{\mu}}{\sqrt{\epsilon}\,\eta_y\beta} & -\dfrac{\eta_x\eta_z\sqrt{\mu}}{\sqrt{\epsilon}\,\eta_y\beta} & -\dfrac{(\eta_x^2+\eta_y^2)\sqrt{\mu}}{\sqrt{\epsilon}\,\eta_y\beta} & \dfrac{\eta_z}{\eta_y} & 0 \\[2ex] 0 & 1 & 0 & 1 & 0 & \dfrac{\eta_x}{\eta_y} \\[2ex] -\dfrac{\eta_z}{\eta_y} & -\dfrac{\eta_x}{\eta_y} & -\dfrac{\eta_z}{\eta_y} & -\dfrac{\eta_x}{\eta_y} & 0 & 1 \\[2ex] 1 & 0 & 1 & 0 & 0 & \dfrac{\eta_z}{\eta_y} \end{bmatrix} \tag{29.35}$$

$$S_\zeta = \begin{bmatrix} \dfrac{\sqrt{\mu}(\zeta_y^2+\zeta_z^2)}{\sqrt{\epsilon}\,\zeta_z\gamma} & \dfrac{\sqrt{\mu}\zeta_x\zeta_y}{\sqrt{\epsilon}\,\zeta_z\gamma} & -\dfrac{\sqrt{\mu}(\zeta_y^2+\zeta_z^2)}{\sqrt{\epsilon}\,\zeta_z\gamma} & -\dfrac{\sqrt{\mu}\zeta_x\zeta_y}{\sqrt{\epsilon}\,\zeta_z\gamma} & \dfrac{\zeta_x}{\zeta_z} & 0 \\[2ex] -\dfrac{\sqrt{\mu}\zeta_x\zeta_y}{\sqrt{\epsilon}\,\zeta_z\gamma} & -\dfrac{\sqrt{\mu}(\zeta_x^2+\zeta_z^2)}{\sqrt{\epsilon}\,\zeta_z\gamma} & \dfrac{\sqrt{\mu}\zeta_x\zeta_y}{\sqrt{\epsilon}\,\zeta_z\gamma} & \dfrac{\sqrt{\mu}(\zeta_x^2+\zeta_z^2)}{\sqrt{\epsilon}\,\zeta_z\gamma} & \dfrac{\zeta_y}{\zeta_z} & 0 \\[2ex] -\dfrac{\sqrt{\mu}\zeta_x}{\sqrt{\epsilon}\gamma} & \dfrac{\sqrt{\mu}\zeta_y}{\sqrt{\epsilon}\gamma} & \dfrac{\sqrt{\mu}\zeta_x}{\sqrt{\epsilon}\gamma} & -\dfrac{\sqrt{\mu}\zeta_y}{\sqrt{\epsilon}\gamma} & 1 & 0 \\[2ex] 0 & 1 & 0 & 1 & 0 & \dfrac{\zeta_x}{\zeta_z} \\[2ex] 1 & 0 & 1 & 0 & 0 & \dfrac{\zeta_y}{\zeta_z} \\[2ex] -\dfrac{\zeta_y}{\zeta_z} & -\dfrac{\zeta_x}{\zeta_z} & -\dfrac{\zeta_y}{\zeta_z} & -\dfrac{\zeta_x}{\zeta_z} & 0 & 1 \end{bmatrix} \tag{29.36}$$

Since the similar matrices of diagonalization, S_ξ, S_η, and S_ζ, are nonsingular, the left-hand inverse matrices S_ξ^{-1}, S_η^{-1}, and S_ζ^{-1} are easily found. Although these left-hand inverse matrices are essential to the diagonalization process, they provide little insight for the following flux-vector splitting procedure. The rather involved results are omitted here, but they can be found in Shang and Fithen [1994].

29.6 Flux-Vector Splitting

An efficient flux-vector splitting algorithm for solving the Euler equations was developed by Steger and Warming [1987]. The basic concept is equally applicable to any hyperbolic differential system for which the solution need not be analytic [Sommerfeld 1949, Courant and Hilbert 1965]. In most CEM applications, discontinuous behavior of the solution is associated only with the wave across an interface between different media, a piecewise continuous solution. Even if a jump condition exists, the magnitude of the finite jump across the interface is much less drastic than the shock waves encountered in supersonic flows. Nevertheless, the salient feature of the piecewise continuous solution domains of the hyperbolic partial differential equation stands out: The coefficient matrices of the time-dependent, three-dimensional Maxwell equations cast in the general curvilinear frame of reference contain metrics of coordinate transformation. Therefore, the equation system no longer has constant coefficients even in an isotropic and homogeneous medium. Under this circumstance, eigenvalues can change sign at any given field location due to the metric variations of the coordinate transformation. Numerical oscillations have appeared in results calculated using the flux-vector splitting technique when eigenvalues change sign. A refined flux-difference splitting algorithm has been developed to resolve fields with jump conditions [Van Leer 1982, Anderson et al. 1985]. The newer flux-difference splitting algorithm is particularly effective at locations where the eigenvalues vanish. Perhaps more crucial for electromagnetics, the polarization of the medium, making the basic equations become nonlinear, occurs only in the extremely high-frequency range [Elliott 1966, Harrington 1961]. In general the governing equations are linear; at most, the coefficients of the differential system are dependent on physical location and phase velocity. For this reason, the difference between the flux-vector splitting [Steger and Warming 1987] and flux-difference splitting [Van Leer 1982, Anderson et al. 1985] schemes, when applied to the time-dependent Maxwell equations, is not of great importance.

The basic idea of the flux-vector splitting of Steger and Warming is to process data according to the direction of information propagation. Since diagonalization is achievable only in each time–space plane, the direction of wave propagation degenerates into either the positive or the negative orientation. This designation is consistent with the notion of the right-running and the left-running wave components. The flux vectors are computed from the point value, including the metrics at the node of interest. This formulation for solving hyperbolic partial differential equations not only ensures the well-posedness of the differential system but also enhances the stability of the numerical procedure [Roe 1986, Shang 1993, Anderson et al. 1984, Richtmyer and Morton 1967]. Specifically, the flux vectors F_ξ, F_η, and F_ζ will be split according to the sign of their corresponding eigenvalues. The split fluxes are differenced by an upwind algorithm to allow for the zone of dependence of an initial-value problem [Roe 1986, Shang 1993, Shang and Gaitonde 1993, Shang and Fithen 1994].

From the previous analysis, it is clear that the eigenvalues contain multiplicities, and hence the split flux of the three-dimensional Maxwell equations is not unique [Shang and Gaitonde 1993, Shang and Fithen 1994]. All flux vectors in each time–space plane are split according to the signs of the local eigenvalues:

$$F_\xi = F_\xi^+ + F_\xi^-$$
$$F_\eta = F_\eta^+ + F_\eta^-$$
$$F_\zeta = F_\zeta^+ + F_\zeta^-$$

$$(29.37)$$

The flux-vector components associated with the positive and negative eigenvalues are obtainable by a straightforward matrix multiplication:

$$F_\xi^+ = S_\xi \lambda_\xi^+ S_\xi^{-1} U$$
$$F_\xi^- = S_\xi \lambda_\xi^- S_\xi^{-1} U$$
$$F_\eta^+ = S_\eta \lambda_\eta^+ S_\eta^{-1} U$$
$$F_\eta^- = S_\eta \lambda_\eta^- S_\eta^{-1} U$$
$$F_\zeta^+ = S_\zeta \lambda_\zeta^+ S_\zeta^{-1} U$$
$$F_\zeta^- = S_\zeta \lambda_\zeta^- S_\zeta^{-1} U$$

$$(29.38)$$

It is also important to recognize that even if the split flux vectors in each time–space plane are non-unique, the sum of the split components must be unambiguously identical to the flux vector of the governing Equation 29.25. This fact is easily verifiable by performing the addition of the split matrices to reach the identities in Equation 29.28, Equation 29.29, and Equation 29.30. In addition, if one sets the diagonal elements of metrics, ξ_x, η_y, and ζ_z equal to unity and the off-diagonal elements equal to zero, the coefficient matrices will recover the Cartesian form:

$$
F_\xi^+ =
\begin{bmatrix}
\frac{\xi_y^2+\xi_z^2}{2\sqrt{\epsilon\mu}V\alpha} & \frac{-\xi_x\xi_y}{2\sqrt{\epsilon\mu}V\alpha} & \frac{-\xi_x\xi_z}{2\sqrt{\epsilon\mu}V\alpha} & 0 & \frac{-\xi_z}{2\epsilon V} & \frac{\xi_y}{2\epsilon V} \\
\frac{-\xi_x\xi_y}{2\sqrt{\epsilon\mu}V\alpha} & \frac{\xi_x^2+\xi_z^2}{2\sqrt{\epsilon\mu}V\alpha} & \frac{-\xi_y\xi_z}{2\sqrt{\epsilon\mu}V\alpha} & \frac{\xi_z}{2\epsilon V} & 0 & \frac{-\xi_x}{2\epsilon V} \\
\frac{-\xi_x\xi_z}{2\sqrt{\epsilon\mu}V\alpha} & \frac{-\xi_y\xi_z}{2\sqrt{\epsilon\mu}V\alpha} & \frac{\xi_x^2+\xi_y^2}{2\sqrt{\epsilon\mu}V\alpha} & \frac{-\xi_y}{2\epsilon V} & \frac{\xi_x}{2\epsilon V} & 0 \\
0 & \frac{\xi_z}{2V\mu} & \frac{-\xi_y}{2V\mu} & \frac{\xi_y^2+\xi_z^2}{2\sqrt{\epsilon\mu}V\alpha} & \frac{-\xi_x\xi_y}{2\sqrt{\epsilon\mu}V\alpha} & \frac{-\xi_x\xi_z}{2\sqrt{\epsilon\mu}V\alpha} \\
\frac{-\xi_z}{2V\mu} & 0 & \frac{\xi_x}{2V\mu} & \frac{-\xi_x\xi_y}{2\sqrt{\epsilon\mu}V\alpha} & \frac{\xi_x^2+\xi_z^2}{2\sqrt{\epsilon\mu}V\alpha} & \frac{-\xi_y\xi_z}{2\sqrt{\epsilon\mu}V\alpha} \\
\frac{\xi_y}{2V\mu} & \frac{-\xi_x}{2V\mu} & 0 & \frac{-\xi_x\xi_z}{2\sqrt{\epsilon\mu}V\alpha} & \frac{-\xi_y\xi_z}{2\sqrt{\epsilon\mu}V\alpha} & \frac{\xi_x^2+\xi_y^2}{2\sqrt{\epsilon\mu}V\alpha}
\end{bmatrix}
\begin{Bmatrix}
B_x \\ B_y \\ B_z \\ D_x \\ D_y \\ D_z
\end{Bmatrix}
\quad (29.39)
$$

$$F_\xi^- = \begin{bmatrix} \dfrac{-(\xi_y^2+\xi_z^2)}{2\sqrt{\epsilon\mu}V\alpha} & \dfrac{\xi_x\xi_y}{2\sqrt{\epsilon\mu}V\alpha} & \dfrac{\xi_x\xi_z}{2\sqrt{\epsilon\mu}V\alpha} & 0 & \dfrac{-\xi_z}{2\epsilon V} & \dfrac{\xi_y}{2\epsilon V} \\[3mm] \dfrac{\xi_x\xi_y}{2\sqrt{\epsilon\mu}V\alpha} & \dfrac{-(\xi_x^2+\xi_z^2)}{2\sqrt{\epsilon\mu}V\alpha} & \dfrac{\xi_y\xi_z}{2\sqrt{\epsilon\mu}V\alpha} & \dfrac{\xi_z}{2\epsilon V} & 0 & \dfrac{-\xi_x}{2\epsilon V} \\[3mm] \dfrac{\xi_x\xi_z}{2\sqrt{\epsilon\mu}V\alpha} & \dfrac{\xi_y\xi_z}{2\sqrt{\epsilon\mu}V\alpha} & \dfrac{-(\xi_x^2+\xi_y^2)}{2\sqrt{\epsilon\mu}V\alpha} & \dfrac{-\xi_y}{2\epsilon V} & \dfrac{\xi_x}{2\epsilon V} & 0 \\[3mm] 0 & \dfrac{\xi_z}{2V\mu} & \dfrac{-\xi_y}{2V\mu} & \dfrac{-(\xi_y^2+\xi_z^2)}{2\sqrt{\epsilon\mu}V\alpha} & \dfrac{\xi_x\xi_y}{2\sqrt{\epsilon\mu}V\alpha} & \dfrac{\xi_x\xi_z}{2\sqrt{\epsilon\mu}V\alpha} \\[3mm] \dfrac{-\xi_z}{2V\mu} & 0 & \dfrac{\xi_x}{2V\mu} & \dfrac{\xi_x\xi_y}{2\sqrt{\epsilon\mu}V\alpha} & \dfrac{-(\xi_x^2+\xi_z^2)}{2\sqrt{\epsilon\mu}V\alpha} & \dfrac{\xi_y\xi_z}{2\sqrt{\epsilon\mu}V\alpha} \\[3mm] \dfrac{\xi_y}{2V\mu} & \dfrac{-\xi_x}{2V\mu} & 0 & \dfrac{\xi_x\xi_z}{2\sqrt{\epsilon\mu}V\alpha} & \dfrac{\xi_y\xi_z}{2\sqrt{\epsilon\mu}V\alpha} & \dfrac{-(\xi_x^2+\xi_y^2)}{2\sqrt{\epsilon\mu}V\alpha} \end{bmatrix} \begin{Bmatrix} B_x \\ B_y \\ B_z \\ D_x \\ D_y \\ D_z \end{Bmatrix} \qquad (29.40)$$

$$F_\eta^+ = \begin{bmatrix} \dfrac{\eta_y^2+\eta_z^2}{2\sqrt{\epsilon\mu}\beta V} & \dfrac{-\eta_x\eta_y}{2\sqrt{\epsilon\mu}\beta V} & \dfrac{-\eta_x\eta_z}{2\sqrt{\epsilon\mu}\beta V} & 0 & \dfrac{-\eta_z}{2\epsilon V} & \dfrac{\eta_y}{2\epsilon V} \\[3mm] \dfrac{-\eta_x\eta_y}{2\sqrt{\epsilon\mu}\beta V} & \dfrac{\eta_x^2+\eta_z^2}{2\sqrt{\epsilon\mu}\beta V} & \dfrac{-\eta_y\eta_z}{2\sqrt{\epsilon\mu}\beta V} & \dfrac{\eta_z}{2\epsilon V} & 0 & \dfrac{-\eta_x}{2\epsilon V} \\[3mm] \dfrac{-\eta_x\eta_z}{2\sqrt{\epsilon\mu}\beta V} & \dfrac{-\eta_y\eta_z}{2\sqrt{\epsilon\mu}\beta V} & \dfrac{\eta_x^2+\eta_y^2}{2\sqrt{\epsilon\mu}\beta V} & \dfrac{-\eta_y}{2\epsilon V} & \dfrac{\eta_x}{2\epsilon V} & 0 \\[3mm] 0 & \dfrac{\eta_z}{2V\mu} & \dfrac{-\eta_y}{2V\mu} & \dfrac{\eta_y^2+\eta_z^2}{2\sqrt{\epsilon\mu}\beta V} & \dfrac{-\eta_x\eta_y}{2\sqrt{\epsilon\mu}\beta V} & \dfrac{-\eta_x\eta_z}{2\sqrt{\epsilon\mu}\beta V} \\[3mm] \dfrac{-\eta_z}{2V\mu} & 0 & \dfrac{\eta_x}{2V\mu} & \dfrac{-\eta_x\eta_y}{2\sqrt{\epsilon\mu}\beta V} & \dfrac{\eta_x^2+\eta_z^2}{2\sqrt{\epsilon\mu}\beta V} & \dfrac{-\eta_y\eta_z}{2\sqrt{\epsilon\mu}\beta V} \\[3mm] \dfrac{\eta_y}{2V\mu} & \dfrac{-\eta_x}{2V\mu} & 0 & \dfrac{-\eta_x\eta_z}{2\sqrt{\epsilon\mu}\beta V} & \dfrac{-\eta_y\eta_z}{2\sqrt{\epsilon\mu}\beta V} & \dfrac{\eta_x^2+\eta_y^2}{2\sqrt{\epsilon\mu}\beta V} \end{bmatrix} \begin{Bmatrix} B_x \\ B_y \\ B_z \\ D_x \\ D_y \\ D_z \end{Bmatrix} \qquad (29.41)$$

$$F_\eta^- = \begin{bmatrix} \dfrac{-(\eta_y^2+\eta_z^2)}{2\sqrt{\epsilon\mu}\beta V} & \dfrac{\eta_x\eta_y}{2\sqrt{\epsilon\mu}\beta V} & \dfrac{\eta_x\eta_z}{2\sqrt{\epsilon\mu}\beta V} & 0 & \dfrac{-\eta_z}{2\epsilon V} & \dfrac{\eta_y}{2\epsilon V} \\[3mm] \dfrac{\eta_x\eta_y}{2\sqrt{\epsilon\mu}\beta V} & \dfrac{-(\eta_x^2+\eta_z^2)}{2\sqrt{\epsilon\mu}\beta V} & \dfrac{\eta_y\eta_z}{2\sqrt{\epsilon\mu}\beta V} & \dfrac{\eta_z}{2\epsilon V} & 0 & \dfrac{-\eta_x}{2\epsilon V} \\[3mm] \dfrac{\eta_x\eta_z}{2\sqrt{\epsilon\mu}\beta V} & \dfrac{\eta_y\eta_z}{2\sqrt{\epsilon\mu}\beta V} & \dfrac{-(\eta_x^2+\eta_y^2)}{2\sqrt{\epsilon\mu}\beta V} & \dfrac{-\eta_y}{2\epsilon V} & \dfrac{\eta_x}{2\epsilon V} & 0 \\[3mm] 0 & \dfrac{\eta_z}{2V\mu} & \dfrac{-\eta_y}{2V\mu} & \dfrac{-(\eta_y^2+\eta_z^2)}{2\sqrt{\epsilon\mu}\beta V} & \dfrac{\eta_x\eta_y}{2\sqrt{\epsilon\mu}\beta V} & \dfrac{\eta_x\eta_z}{2\sqrt{\epsilon\mu}\beta V} \\[3mm] \dfrac{-\eta_z}{2V\mu} & 0 & \dfrac{\eta_x}{2V\mu} & \dfrac{\eta_x\eta_y}{2\sqrt{\epsilon\mu}\beta V} & \dfrac{-(\eta_x^2+\eta_z^2)}{2\sqrt{\epsilon\mu}\beta V} & \dfrac{\eta_y\eta_z}{2\sqrt{\epsilon\mu}\beta V} \\[3mm] \dfrac{\eta_y}{2V\mu} & \dfrac{-\eta_x}{2V\mu} & 0 & \dfrac{\eta_x\eta_z}{2\sqrt{\epsilon\mu}\beta V} & \dfrac{\eta_y\eta_z}{2\sqrt{\epsilon\mu}\beta V} & \dfrac{-(\eta_x^2+\eta_y^2)}{2\sqrt{\epsilon\mu}\beta V} \end{bmatrix} \begin{Bmatrix} B_x \\ B_y \\ B_z \\ D_x \\ D_y \\ D_z \end{Bmatrix} \qquad (29.42)$$

$$F_\zeta^+ = \begin{bmatrix} \dfrac{\zeta_y^2+\zeta_z^2}{2\sqrt{\epsilon\mu}V\gamma} & \dfrac{-\zeta_x\zeta_y}{2\sqrt{\epsilon\mu}V\gamma} & \dfrac{-\zeta_x\zeta_z}{2\sqrt{\epsilon\mu}V\gamma} & 0 & \dfrac{-\zeta_z}{2\epsilon V} & \dfrac{\zeta_y}{2\epsilon V} \\[3mm] \dfrac{-\zeta_x\zeta_y}{2\sqrt{\epsilon\mu}V\gamma} & \dfrac{\zeta_x^2+\zeta_z^2}{2\sqrt{\epsilon\mu}V\gamma} & \dfrac{-\zeta_y\zeta_z}{2\sqrt{\epsilon\mu}V\gamma} & \dfrac{\zeta_z}{2\epsilon V} & 0 & \dfrac{-\zeta_x}{2\epsilon V} \\[3mm] \dfrac{-\zeta_x\zeta_z}{2\sqrt{\epsilon\mu}V\gamma} & \dfrac{-\zeta_y\zeta_z}{2\sqrt{\epsilon\mu}V\gamma} & \dfrac{\zeta_x^2+\zeta_y^2}{2\sqrt{\epsilon\mu}V\gamma} & \dfrac{-\zeta_y}{2\epsilon V} & \dfrac{\zeta_x}{2\epsilon V} & 0 \\[3mm] 0 & \dfrac{\zeta_z}{2V\mu} & \dfrac{-\zeta_y}{2V\mu} & \dfrac{\zeta_y^2+\zeta_z^2}{2\sqrt{\epsilon\mu}V\gamma} & \dfrac{-\zeta_x\zeta_y}{2\sqrt{\epsilon\mu}V\gamma} & \dfrac{-\zeta_x\zeta_z}{2\sqrt{\epsilon\mu}V\gamma} \\[3mm] \dfrac{-\zeta_z}{2V\mu} & 0 & \dfrac{\zeta_x}{2V\mu} & \dfrac{-\zeta_x\zeta_y}{2\sqrt{\epsilon\mu}V\gamma} & \dfrac{\zeta_x^2+\zeta_z^2}{2\sqrt{\epsilon\mu}V\gamma} & \dfrac{-\zeta_y\zeta_z}{2\sqrt{\epsilon\mu}V\gamma} \\[3mm] \dfrac{\zeta_y}{2V\mu} & \dfrac{-\zeta_x}{2V\mu} & 0 & \dfrac{-\zeta_x\zeta_z}{2\sqrt{\epsilon\mu}V\gamma} & \dfrac{-\zeta_y\zeta_z}{2\sqrt{\epsilon\mu}V\gamma} & \dfrac{\zeta_x^2+\zeta_y^2}{2\sqrt{\epsilon\mu}V\gamma} \end{bmatrix} \begin{Bmatrix} B_x \\ B_y \\ B_z \\ D_x \\ D_y \\ D_z \end{Bmatrix} \qquad (29.43)$$

$$
F_\zeta^- = \begin{bmatrix}
\dfrac{-(\zeta_y^2+\zeta_z^2)}{2\sqrt{\epsilon\mu}V\gamma} & \dfrac{\zeta_x\zeta_y}{2\sqrt{\epsilon\mu}V\gamma} & \dfrac{\zeta_x\zeta_z}{2\sqrt{\epsilon\mu}V\gamma} & 0 & \dfrac{-\zeta_z}{2\epsilon V} & \dfrac{\zeta_y}{2\epsilon V} \\[2ex]
\dfrac{\zeta_x\zeta_y}{2\sqrt{\epsilon\mu}V\gamma} & \dfrac{-(\zeta_x^2+\zeta_z^2)}{2\sqrt{\epsilon\mu}V\gamma} & \dfrac{\zeta_y\zeta_z}{2\sqrt{\epsilon\mu}V\gamma} & \dfrac{\zeta_z}{2\epsilon V} & 0 & \dfrac{-\zeta_x}{2\epsilon V} \\[2ex]
\dfrac{\zeta_x\zeta_z}{2\sqrt{\epsilon\mu}V\gamma} & \dfrac{\zeta_y\zeta_z}{2\sqrt{\epsilon\mu}V\gamma} & \dfrac{-(\zeta_x^2+\zeta_y^2)}{2\sqrt{\epsilon\mu}V\gamma} & \dfrac{-\zeta_y}{2\epsilon V} & \dfrac{\zeta_x}{2\epsilon V} & 0 \\[2ex]
0 & \dfrac{\zeta_z}{2V\mu} & \dfrac{-\zeta_y}{2V\mu} & \dfrac{-(\zeta_y^2+\zeta_z^2)}{2\sqrt{\epsilon\mu}V\gamma} & \dfrac{\zeta_x\zeta_y}{2\sqrt{\epsilon\mu}V\gamma} & \dfrac{\zeta_x\zeta_z}{2\sqrt{\epsilon\mu}V\gamma} \\[2ex]
\dfrac{-\zeta_z}{2V\mu} & 0 & \dfrac{\zeta_x}{2V\mu} & \dfrac{\zeta_x\zeta_y}{2\sqrt{\epsilon\mu}V\gamma} & \dfrac{-(\zeta_x^2+\zeta_z^2)}{2\sqrt{\epsilon\mu}V\gamma} & \dfrac{\zeta_y\zeta_z}{2\sqrt{\epsilon\mu}V\gamma} \\[2ex]
\dfrac{\zeta_y}{2V\mu} & \dfrac{-\zeta_x}{2V\mu} & 0 & \dfrac{\zeta_x\zeta_z}{2\sqrt{\epsilon\mu}V\gamma} & \dfrac{\zeta_y\zeta_z}{2\sqrt{\epsilon\mu}V\gamma} & \dfrac{-(\zeta_x^2+\zeta_y^2)}{2\sqrt{\epsilon\mu}V\gamma}
\end{bmatrix}
\begin{Bmatrix} B_x \\ B_y \\ B_z \\ D_x \\ D_y \\ D_z \end{Bmatrix}
\tag{29.44}
$$

29.7 Finite-Difference Approximation

Once the detailed split fluxes are known, formulation of the finite-difference approximation is straightforward. From the sign of an eigenvalue, the stencil of a spatially second- or higher-order-accurate windward differencing can be easily constructed to form multiple one-dimensional difference operators [Shang 1993, Anderson et al. 1984, Richtmyer and Morton 1967]. In this regard, the forward difference and the backward difference approximations are used for the negative and the positive eigenvalues, respectively. The split flux vectors are evaluated at each discretized point of the field according to the signs of the eigenvalues. For the present purpose, a second-order accurate procedure is given:

$$
\begin{aligned}
\text{If} \quad \lambda < 0, \quad & \Delta U_i = [-3U_i + 4U_{i+1} - U_{i+2}]/2 \\
\text{If} \quad \lambda > 0, \quad & \nabla U_i = [3U_i - 4U_{i-1} + U_{i-2}]/2
\end{aligned}
\tag{29.45}
$$

The necessary metrics of the coordinate transformation are calculated by central-differencing, except at the edges of computational domain, where one-sided differences are used. Although the fractional-step or the time-splitting algorithm [Shang 1993, Anderson et al. 1984, Richtmyer and Morton 1967] has demonstrated greater efficiency in data storage and a higher data-processing rate than predictor–corrector time integration procedures [Shang 1993, Shang and Gaitonde 1993, Shang and Fithen 1994], it is limited to second-order accuracy in time. With respect to the fractional-step method, the temporal second-order result is obtained by a sequence of symmetrically cyclic operators [Shang 1993, Richtmyer and Morton 1967]:

$$
U^{n+2} = L_\xi L_\eta L_\zeta L_\zeta L_\eta L_\xi U^n
\tag{29.46}
$$

where L_ξ, L_η, and L_ζ are the difference operators for one-dimensional equations in the ξ, η, and ζ coordinates, respectively.

In general, second-order and higher temporal resolution is achievable through multiple-time-step schemes [Anderson et al. 1984, Richtmyer and Morton 1967]. However, one-step schemes are more attractive because they have less memory requirements and don't need special startup procedures [Shang and Gaitonde 1993, Shang and Fithen 1994]. For future higher-order accurate solution development potential, the Runge–Kutta family of single-step, multistage procedures is recommended. This choice is also consistent with the accompanying characteristic-based finite-volume method [Shang and Gaitonde 1993].

In the present effort, the two-stage, formally second-order accurate scheme is used:

$$U_0 = U_n$$
$$U_1 = U_0 - \Delta U\,(U_0)$$
$$U_2 = U_0 - 0.5\,(\Delta U\,(U_1) + \Delta U\,(U_0)) \tag{29.47}$$
$$U_{n+1} = U_2$$

where ΔU comprises the incremental values of dependent variables during each temporal sweep. The resultant characteristic-based finite-difference scheme for solving the three-dimensional Maxwell equations in the time domain is second-order accurate in both time and space.

The most significant feature of the flux-vector splitting scheme lies in its ability to easily suppress reflected waves from the truncated computational domain. In wave motion, the compatibility condition at any point in space is described by the split flux vector [Shang 1993, Shang and Gaitonde 1993, Shang and Fithen 1994]. In the present formulation, an approximated no-reflection condition can be achieved by setting the incoming flux component equal to zero:

$$\text{either} \quad \lim_{r \to \infty} F^+(\xi, \eta, \zeta) = 0 \quad \text{or} \quad \lim_{r \to \infty} F^-(\xi, \eta, \zeta) = 0 \tag{29.48}$$

The one-dimensional compatibility condition is exact when the wave motion is aligned with one of the coordinates [Shang 1993]. This unique attribute of the characteristic-based numerical procedure in removing a fundamental dilemma in computational electromagnetics will be demonstrated in detail later.

29.8 Finite-Volume Approximation

The finite-volume approximation solves the governing equation by discretizing the physical space into contiguous cells and balancing the flux-vectors on the cell surfaces. Thus in discretized form, the integration procedure reduces to evaluation of the sum of all fluxes aligned with surface-area vectors

$$\frac{\Delta U}{\Delta t} + \frac{\Delta F}{\Delta \xi} + \frac{\Delta G}{\Delta \eta} + \frac{\Delta H}{\Delta \zeta} - J = 0 \tag{29.49}$$

In the present approach, the continuous differential operators have been replaced by discrete operators. In essence, the numerical procedure needs only to evaluate the sum of all flux vectors aligned with surface-area vectors [Shang and Gaitonde 1993, Shang and Fithen 1994, Van Leer 1982, Anderson et al. 1985]. Only one of the vectors is required to coincide with the outward normal to the cell surface, and the rest of the orthogonal triad can be made to lie on the same surface. The metrics, or more appropriately the direction cosines, on the cell surface are uniquely determined by the nodes and edges of the elementary volume. This feature is distinct from the finite-difference approximation. The shape of the cell under consideration and the stretching ratio of neighbor cells can lead to a significant deterioration of the accuracy of finite-volume schemes [Leonard 1988].

The most outstanding aspect of finite-volume schemes is the elegance of its flux-splitting process. The flux-difference splitting for Equation 29.25 is greatly facilitated by a locally orthogonal system in the transformed space [Van Leer 1982, Anderson et al. 1985]. In this new frame of reference, eigenvalues and eigenvectors as well as metrics of the coordinate transformation between two orthogonal systems are well known [Shang 1993, Shang and Gaitonde 1993]. The inverse transformation is simply the transpose of the forward mapping. In particular, the flux vectors in the transformed space have the same functional form as that in the Cartesian frame. The difference between the flux vectors in the transformed and the Cartesian coordinates is a known quantity and is given by the product of the surface outward normal and the cell volume, $V(\nabla S / \|\nabla S\|)$ [Shang and Gaitonde 1993]. Therefore, the flux vectors can be split in the transformed space according to the signs of the eigenvalues but without detailed knowledge of the associated eigenvectors in the transformed space. This feature of the finite-volume approach provides a tremendous advantage over the finite-difference approximation in solving complex problems in physics.

The present formulation adopts Van Leer's kappa scheme in which solution vectors are reconstructed on the cell surface from the piecewise data of neighboring cells [Van Leer 1982, Anderson et al. 1985]. The spatial accuracy of this scheme spans a range from first-order to third-order upwind biased approximations,

$$U_{i+\frac{1}{2}}^{+} = U_i + \frac{\phi}{4} \left[(1-\kappa)\nabla + (1+\kappa)\Delta \right] U_i$$

$$U_{i+\frac{1}{2}}^{-} = U_i - \frac{\phi}{4} \left[(1+\kappa)\nabla + (1-\kappa)\Delta \right] U_{i+1}$$

(29.50)

where $\Delta U_i = U_i - U_{i-1}$ and $\nabla U_i = U_{i+1} - U_i$ are the forward and backward differencing discretizations. The parameters ϕ and κ control the accuracy of the numerical results. For $\phi = 1$, $\kappa = -1$ a two-point windward scheme is obtained. This method has an odd-order leading truncation-error term; the dispersive error is expected to dominate. If $\kappa = \frac{1}{3}$, a third-order upwind-biased scheme will emerge. In fact both upwind procedures have discernible leading phase error. This behavior is a consequence of using the two-stage time integration algorithm, and the dispersive error can be alleviated by increasing the temporal resolution. For $\phi = 1$, $\kappa = 0$ the formulation recovers the Fromm scheme [Van Leer 1982, Anderson et al. 1985]. If $\kappa = 1$, the formulation yields the spatially central scheme. Since the fourth-order dissipative term is suppressed, the central scheme is susceptible to parasitic odd–even point decoupling [Anderson et al. 1984, 1985].

The time integration is carried out by the same two-stage Runge–Kutta method as in the present finite-difference procedure [Shang 1993, Shang and Gaitonde 1993]. The finite-volume procedure is therefore second-order accurate in time and up to third-order accurate in space [Shang and Gaitonde 1993, Shang and Fithen 1994]. For the present purpose, only the second-order upwinding and the third-order upwind biased options are exercised. The second-order windward schemes in the form of the flux-vector splitting finite-difference and the flux-difference splitting finite-volume scheme are formally equivalent [Shang and Gaitonde 1993, Shang and Fithen 1994, Van Leer 1982, Anderson et al. 1985, Leonard 1988].

29.9 Summary and Research Issues

The technical merits of the characteristic-based methods for solving the time-dependent, three-dimensional Maxwell equations can best be illustrated by the following two illustrations. In Figure 29.2, the exact electrical field of a traveling wave is compared with numerical results. The numerical results were generated at the maximum allowable time-step size defined by the Courant–Friedrichs–Lewy (CFL) number of 2 ($\lambda \Delta x/\Delta t = 2$) [Anderson et al. 1984, Richtmyer and Morton 1967]. The numerical solutions presented are at instants when a right-running wave reaches the midpoint of the computational domain and exits the numerical boundary respectively. For this one-dimensional simulation, the characteristic-based scheme using the single-step upwind explicit algorithm exhibits the shift property, which indicates a perfect translation of the initial value in space [Anderson et al. 1984]. As the impulse wave moves through the initially quiescent environment, the numerical result duplicates the exact solution at each and every discretized point, including the discontinuous incoming wavefront. Although this highly desirable property of a numerical solution is achievable only under very restrictive conditions and is not preserved for multidimensional problems [Anderson et al. 1984, Richtmyer and Morton 1967], the ability to simulate the nonanalytic solution behavior in the limit is clearly illustrated.

In Figure 29.3, another outstanding feature of the characteristic-based method is highlighted by simulating the oscillating electric dipole. For the radiating electric dipole, the depicted temporal calculations are sampled at the instant when the initial pulse has traveled a distance of 2.24 wavelengths from the dipole. The numerical results are generated on a $48 \times 48 \times 96$ mesh system with the second-order scheme. Under that condition each wavelength is resolved by 15 mesh points and the difference between numerical results by the finite-volume and the finite-difference method is negligible. However, on an irregular

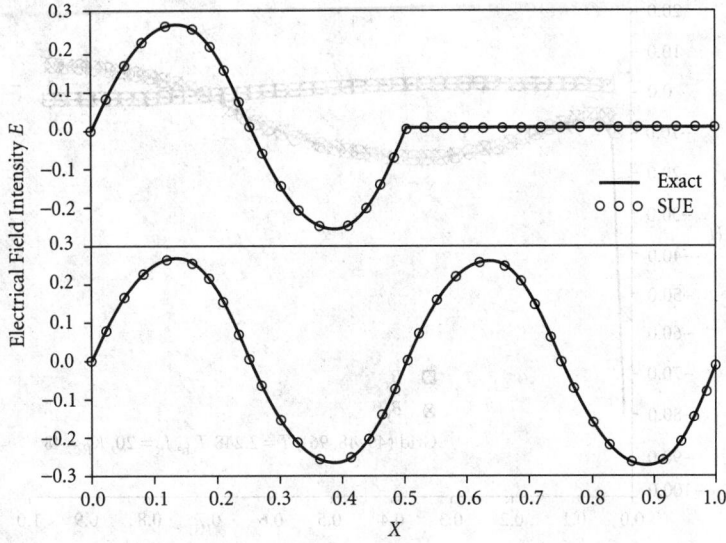

FIGURE 29.2 Perfect-shift property of a one-dimensional wave computation, CFL = 2.

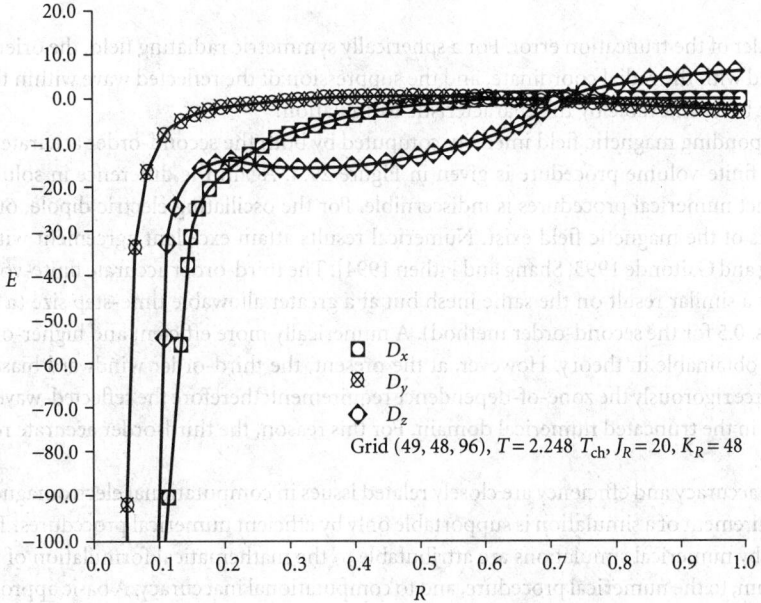

FIGURE 29.3 Instantaneous distributions of oscillating-dipole electric field.

mesh like the present spherical polar system, the finite-volume procedure has shown a greater numerical error than the finite-difference procedure on highly stretched grid systems [Shang and Fithen 1994, Anderson et al. 1985]. Under the present computational conditions, both numerical procedures yield uniformly excellent comparison with the theoretical result. Significant error appeared only in the immediate region of the coordinate origin. At that location, the solution has a singularity that behaves as the inverse cube of the radial distance. The most interesting numerical behavior, however, is revealed in the truncated far field. The no-reflection condition at the numerical boundary is observed to be satisfied

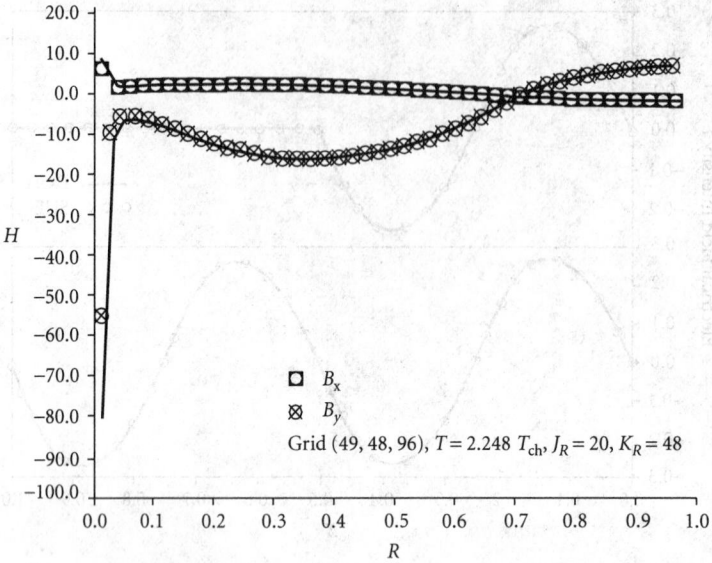

FIGURE 29.4 Instantaneous distributions of oscillating-dipole magnetic field.

within the order of the truncation error. For a spherically symmetric radiating field, the orientation of the wave is aligned with the radial coordinate, and the suppression of the reflected wave within the numerical domain is the best achievable by the characteristic formulation.

The corresponding magnetic field intensity computed by both the second-order accurate finite-difference and the finite-volume procedure is given in Figure 29.4. Again the difference in solution between the two distinct numerical procedures is indiscernible. For the oscillating electric dipole, only the x and y components of the magnetic field exist. Numerical results attain excellent agreement with theoretical values [Shang and Gaitonde 1993, Shang and Fithen 1994]. The third-order accurate finite-volume scheme also produces a similar result on the same mesh but at a greater allowable time-step size (a CFL value of 0.87 is used vs. 0.5 for the second-order method). A numerically more efficient and higher-order accurate simulation is obtainable in theory. However, at the present, the third-order windward biased algorithm cannot reinforce rigorously the zone-of-dependence requirement; therefore the reflected-wave suppression is incomplete in the truncated numerical domain. For this reason, the third-order accurate results are not included here.

Numerical accuracy and efficiency are closely related issues in computational electromagnetics. A high-accuracy requirement of a simulation is supportable only by efficient numerical procedures. The inaccuracies incurred by numerical simulations are attributable to the mathematical formulation of the problem, to the algorithm, to the numerical procedure, and to computational inaccuracy. A basic approach to relieve the accuracy limitation must be derived from using high-order schemes or spectral methods. The numerical efficiency of CEM can be enhanced substantially by using scalable multicomputers [Shang et al. 1993]. The effective use of a distributed-memory, message-passing homogeneous multicomputer still requires a judicious tradeoff between a balanced work load and interprocessor communication. A characteristic-based finite-volume computer program has been successfully mapped onto distributed-memory systems by a rudimentary domain decomposition strategy [Shang et al. 1993]. For example, a square waveguide, at five different frequencies up to the cutoff, was simulated.

Figure 29.5 displays the x-component of the magnetic field intensity within the waveguide. The simulated transverse electric mode, $TE_{1,1}$, $E_x = 0$, which has a half period of π along the x and y coordinates, is generated on a $24 \times 24 \times 128$ mesh system. Since the entire field is described by simple harmonic functions, the remaining field components are similar and only half the solution domain along the z-coordinate is

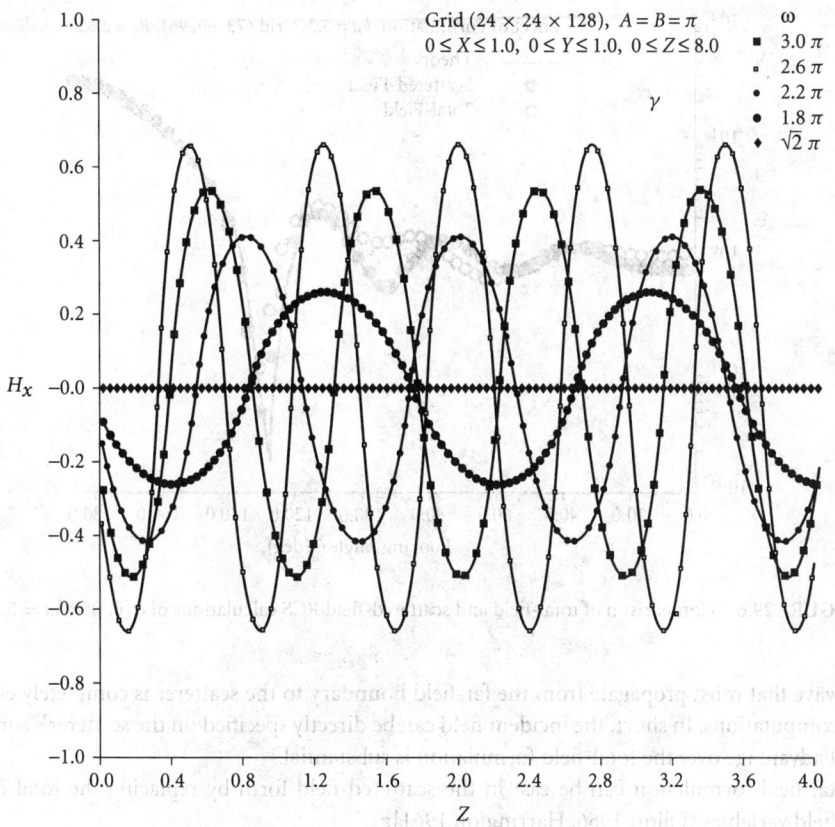

FIGURE 29.5 Cutoff frequency of a square waveguide, $TE_{1,1}$.

presented to minimize repetition. In short, the agreement between the closed-form and numerical solutions is excellent at each frequency. In addition, the numerical simulations duplicate the physical phenomenon at the cutoff frequency, below which there is no phase shift along the waveguide and the wave motion ceases [Elliott 1966, Harrington 1961]. For simple harmonic wave motion in an isotropic medium, the numerical accuracy can be quantified. At a grid-point density of 12 nodes per wavelength, the L2 norm [Richtmyer and Morton 1967] has a nearly uniform magnitude of order 10^{-4}. The numerical results are fully validated by comparison with theory. However, further efforts are still required to substantially improve the parallel and scalable numerical efficiency. In fact, this is the most promising area in CEM research.

The pioneering efforts in CEM usually employed the total-field formulation on staggered-mesh systems [Yee 1966, Taflove 1992]. That particular combination of numerical algorithm and procedure has been proven to be very effective. In earlier RCS calculations using the present numerical procedure, the total-field formulation was also utilized [Shang and Gaitonde 1994, Shang et al. 1993]. However, for three-dimensional scatterer simulation, the numerical accuracy requirement for RCS evaluations becomes extremely stringent. In the total-field formulation, the dynamic range of the field variables has a substantial difference from the exposed and the shadow region, and the incident wave must also traverse the entire computation domain. Both requirements impose severe demands on the numerical accuracy of simulation. In addition, the total field often contains only the residual of partial cancelations of the incident and the diffracted waves. The far-field electromagnetic energy distribution becomes a secular problem — a small difference between two variables of large magnitude. An alternative approach via the scattered-field formulation for RCS calculations appears to be very attractive. Particularly in this formulation, the numerical dissipation of the

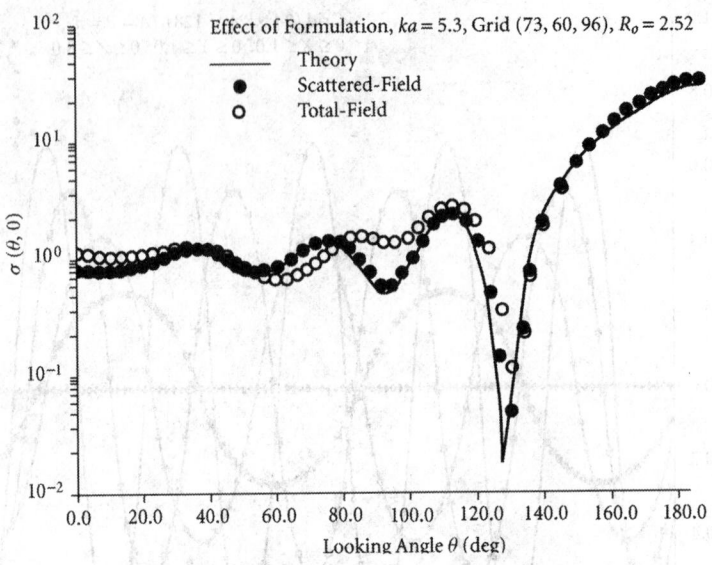

FIGURE 29.6   Comparison of total-field and scattered-field RCS calculations of $\sigma(\theta, 0°)$, $ka = 5.3$.

incident wave that must propagate from the far-field boundary to the scatterer is completely eliminated from the computations. In short, the incident field can be directly specified on the scatterer's surface. The numerical advantage over the total-field formulation is substantial.

The total-field formulation can be cast in the scattered-field form by replacing the total field with scattered field variables [Elliott 1966, Harrington 1961]:

$$U_s = U_t - U_i \tag{29.51}$$

Since the incident field U_i must satisfy the Maxwell equations identically, the equations of the scattered field remain unaltered from the total-field formulation. Thus, the scattered-field formulation can be considered as a dependent-variable transform of the total-field equations. In the present approach, both formulations are solved by a characteristic-based finite-volume scheme.

The comparison of horizontal polarized RCS of a perfect electrically conducting (PEC) sphere, $\sigma(\theta, 0.0)$, from the total-field and scattered-field formulations at $ka = 5.3$ (where $k =$ wave number and $a =$ diameter of the sphere) is presented in Figure 29.6. The validating datum is the exact solution for the scattering of a plane electromagnetic wave by a PEC sphere, which is commonly referred to as the Mie series [Elliott 1966]. Both numerical results are generated under identical computational conditions. The location of the truncated far-field boundary is prescribed at 2.5 wavelengths from the center of the PEC sphere. Numerical results of the total-field formulation reveal far greater error than for the scattered-field formulation. The additional source of error is incurred when the incident wave must propagate from the far-field boundary to the scatterer. In the scattered-field formulation, the incident field data are described precisely by the boundary condition on the scatterer surface. Since the far-field electromagnetic energy distribution is derived from the near-field parameters [Taflove 1992, Sommerfeld 1949, Shang and Gaitonde 1993], the advantage of describing the data incident on a scatterer without error is tremendous. Numerical errors of the total-field calculations are evident in the exaggerated peaks and troughs over the entire viewing-angle displacement.

In Figure 29.7, the vertically polarized RCS $\sigma(\theta, 90.0°)$ of the $ka = 5.3$ case substantiates the previous observation. In fact, the numerical error of the total-field calculation is excessive in comparison with the result of the scattered-field formulation. Since the results are not obtained for the optimal far-field placement for RCS computation, the results of the scattered-field formulation overpredict the theoretical values by 2.7 percent. The deviation of the total-field result from the theory, however, exceeds

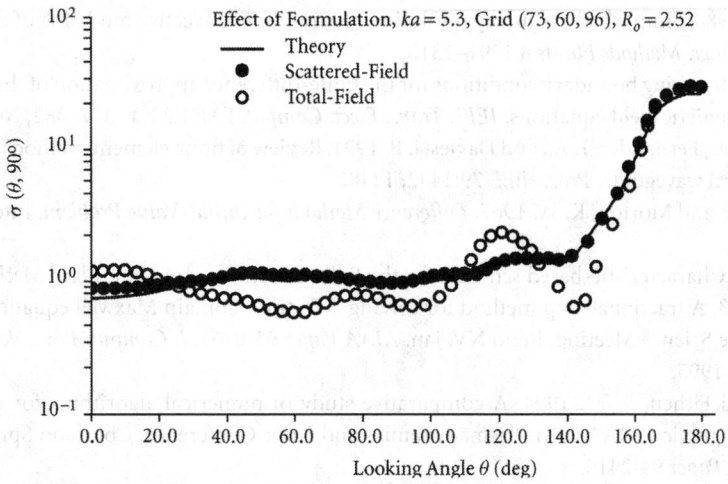

FIGURE 29.7 Comparison of total-field and scattered-field RCS calculations of $\sigma(\theta, 90°)$, $ka = 5.3$.

25.6 percent and becomes unacceptable. In addition, computations by the total-field formulation exhibit a strong sensitivity to placement of the far-field boundary. A small perturbation of the far-field boundary placement leads to a drastic change in the RCS prediction: a feature resembling the ill-posedness condition, which is highly undesirable for numerical simulation. Since there is very little difference in computer coding for the two formulations, the difference in computing time required for an identical simulation is insignificant. On the Cray C90, 1,505.3 s at a data-processing rate of 528.8 Mflops and an average vector length of 62.9 is needed to complete a sampling period. At the present, the most efficient calculation on a distributed memory system, Cray T3-D, has reduced the processing time to 1,204.2 s using 76 computing nodes.

In summary, recent progress in solving the three-dimensional Maxwell equations in the time domain has opened a new frontier in electromagnetics, plasmadynamics, and optics, as well as the interface between electrodynamics and quantum mechanics [Taflove 1992]. The progress in microchip and interconnect network technology has led to a host of high-performance distributed memory, message-passing parallel computer systems. The synergism of efficient and accurate numerical algorithms for solving the Maxwell equations in the time domain with high-performance multicomputers will propel the new interdisciplinary simulation technique to practical and productive applications.

References

Anderson, D. A., Tannehill, J. C., and Pletcher, R. H. 1984. *Computational Fluid Mechanics and Heat Transfer*. Hemisphere, New York.

Anderson, W. K., Thomas, J. L., and Van Leer, B. 1985. A comparison of finite-volume flux splittings for the Euler equations., AIAA 23rd Aerospace Science Meeting, Reno NV, Jan. *AIAA Paper 85-0122*.

Courant, R. and Hilbert, D. 1965. *Methods of Mathematical Physics*, Vol. II. Interscience, New York.

Elliott, R. A. 1966. *Electromagnetics*, Ch. 5. McGraw–Hill, New York.

Enquist, B. and Majda, A. 1977. Absorbing boundary conditions for the numerical simulation of waves. *Math Comp.* 31:629-651, July.

Harrington, R. F. 1961. *Time-Harmonic Electromagnetic Fields*. McGraw–Hill, New York.

Harrington, R. F. 1968. *Field Computation by Moment Methods*, 4th ed. Robert E. Krieger, Malabar, FL.

Higdon, R. 1986. Absorbing boundary conditions for difference approximation to multidimensional wave equation. *Math Comp.* 47(175):437–459.

Leonard, B. P. 1988. Simple high-accuracy resolution program for convective modeling of discontinuities. *Int. J. Numer. Methods Fluids* 8:1291–1318.

Mur, G. 1981. Absorbing boundary conditions for the finite-difference approximation of the time-domain electromagnetic-field equations. *IEEE Trans. Elect. Compat.* EMC-23(4):377–382, Nov.

Rahman, B. M. A., Fernandez, F. A., and Davies, J. B. 1991. Review of finite element methods for microwave and optical waveguide. *Proc. IEEE* 79:1442, 1448.

Richtmyer, R. D. and Morton, K. W. 1967. *Difference Methods for Initial-Value Problem.* Interscience, New York.

Roe, P. L. 1986. Characteristic-based schemes for the Euler equations. *Ann. Rev. Fluid Mech.* 18:337–365.

Shang, J. S. 1993. A fractional-step method for solving 3-D, time-domain Maxwell equations. AIAA 31st Aerospace Science Meeting, Reno NV, Jan. *AIAA Paper* 93-0461; *J. Comput. Phys.* Vol. 118(1):109–119, Apr. 1995.

Shang, J. S. and Fithen, R. M. 1994. A comparative study of numerical algorithms for computational electromagnetics. AIAA 25th Plasmadynamics and Laser Conference, Colorado Springs, June 20–23, *AIAA Paper* 94-2410.

Shang, J. S. and Gaitonde, D. 1993. Characteristic-based, time-dependent Maxwell equation solvers on a general curvilinear frame. AIAA 24th Fluid Dynamics, Plasmadynamics, and Laser Conference, Orlando FL, July. *AIAA Paper* 93-3178; *AIAA J.* 33(3):491–498, Mar. 1995.

Shang, J. S. and Gaitonde, D. 1994. Scattered electromagnetic field of a reentry vehicle. AIAA 32nd Aerospace Science Meeting, Reno NV, Jan. *AIAA Paper* 94-0231; *J. Spacecraft and Rockets* 32(2):294–301, Mar.–Apr. 1995.

Shang, J. S., Hill, K. C., and Calahan, D. 1993. Performance of a characteristic-based, 3-D, time-domain Maxwell equations solver on a massively parallel computer. AIAA 24th Plasmadynamics & Lasers Conference, Orlando FL, July 6–9 *AIAA Paper* 3179; *Appl. Comput. Elect. Soc.* 10(1):52–62, Mar. 1995.

Sommerfeld, A. 1949. *Partial Differential Equations in Physics*, Ch. 2. Academic Press, New York.

Steger, J. L. and Warming, R. F. 1987. Flux vector splitting of the inviscid gas dynamics equations with application to finite difference methods. *J. Comput. Phys.* 20(2):263–293, Feb.

Taflove, A. 1992. Re-inventing electromagnetics: supercomputing solution of Maxwell's equations via direct time integration on space grids. *Comput. Systems Eng.* 3(1–4):153–168.

Thompson, J. F. 1982. *Numerical Grid Generation.* Elsevier Science, New York.

Van Leer, B. 1982. Flux-vector splitting for the Euler equations. *TR* 82-30, ICASE, Sept., pp. 507–512, *Lecture Notes in Physics*, Vol. 170.

Yee, K. S. 1966. Numerical solution of initial boundary value problems involving Maxwell's equations. In *Isotropic Media, IEEE Trans. Ant. Prop.* 14(3):302–307.

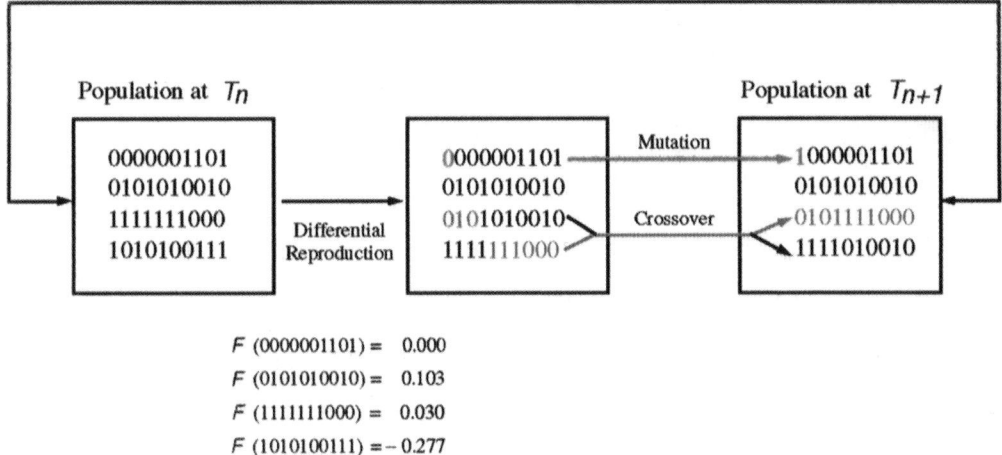

$$F(0000001101) = 0.000$$
$$F(0101010010) = 0.103$$
$$F(1111111000) = 0.030$$
$$F(1010100111) = -0.277$$

PLATE 14.1 Genetic algorithm overview: A population of four individuals is shown. Each is assigned a fitness value by the function $F(x,y) = yx^2 - x^4$. On the basis of these fitnesses, the selection phase assigns the first individual (0000001101) one copy, the second (0101010010) two, the third (1111111000) one copy, and the fourth (1010100111) zero copies. After selection, the genetic operators are applied probabilistically; the first individual has its first bit mutated from a 0 to a 1, and crossover combines the last two individuals into two new ones. The resulting population is shown in the box labeled $T_{(N+1)}$.

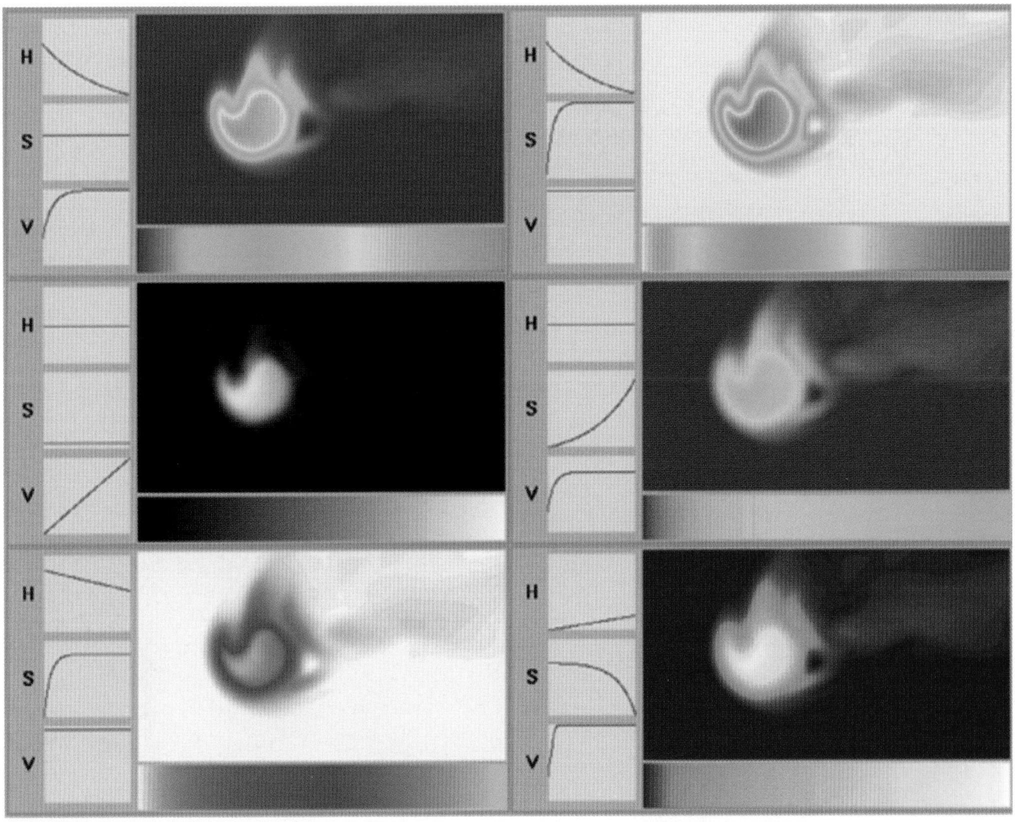

PLATE 27.8 A variety of color palettes used to color the same two-dimensional slice of data. The top two color maps use a wide range of hues and introduce discontinuities into the image that may not be present in the data. If we are looking for overall structure, a smoothly changing palette with a restricted hue — such as the "fire" palette shown in the lower right — is more appropriate. (Courtesy of NCSA, Yale University).

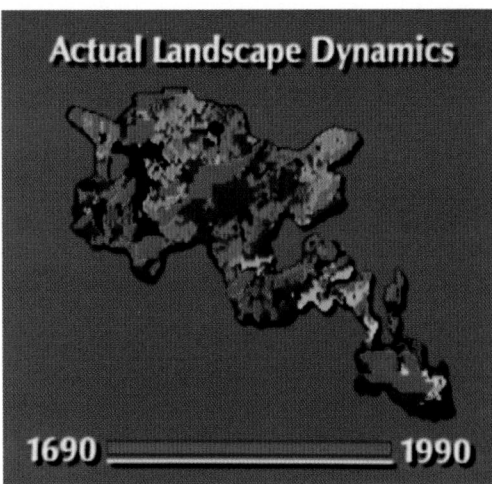

PLATE 27.9 Discrete color mapping is used to depict the age of the forest in Yellowstone National Park [Kovacic et al. 1990]. (Courtesy of D. Kovacic, A. Craig, and R. Patterson; UIUC, NCSA.)

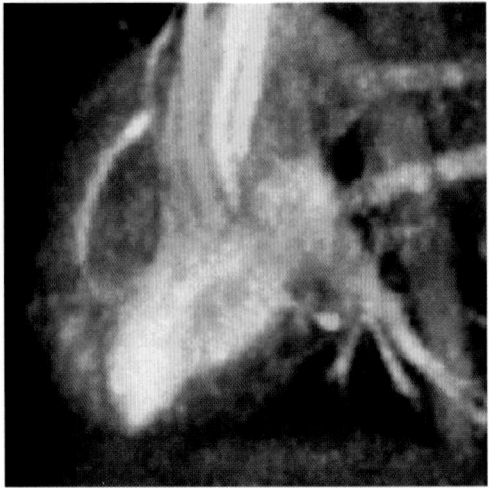

PLATE 27.11 Volume rendering of a dog's heart. (Courtesy of E. Hoffman, P. Moran, and C. Potter; NCSA.)

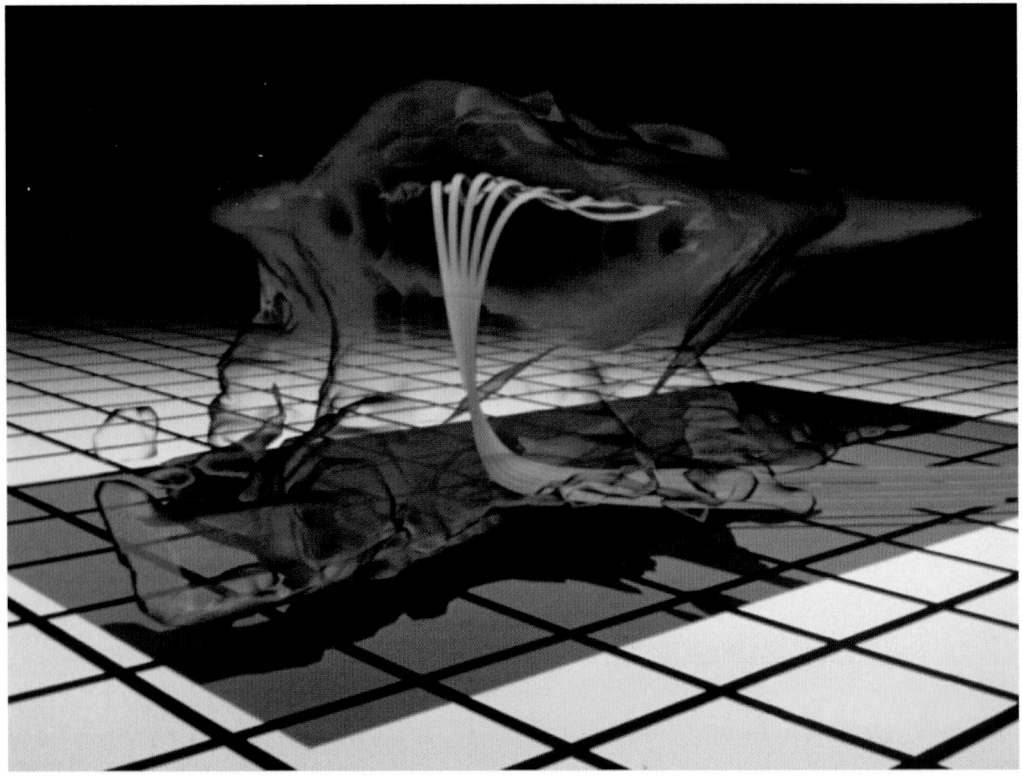

PLATE 27.13 Tracers clarify the wind flow within a simulation of a severe thunderstorm. (Courtesy of Robert Wilhelmson et al., NCSA.)

PLATE 27.14   Streaklines and particles depict smog in Los Angeles. Each green particle represents ten tons of reactive organic gases; each yellow particle, carbon monoxide; and each red particle, nitrogen oxide. (Courtesy of W. Sherman, M. McNeill et al., NCSA.)

PLATE 27.15   A ribbon's rate of twist indicates the streamwise vorticity. (Courtesy of Robert Wilhelmson et al., NCSA.)

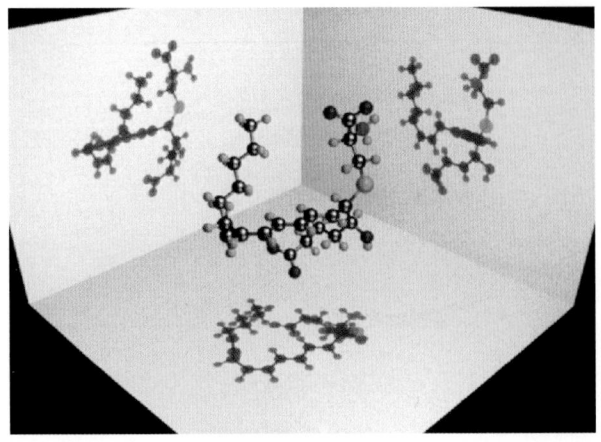

PLATE 27.17 A computer-generated ball-and-stick model of a leukotriene molecule. Colored shadows aid in perceiving the 3-D shape. (Courtesy of D. Herron, Eli Lilly & Co.; J. Thingvold and W. Sherman, NCSA.)

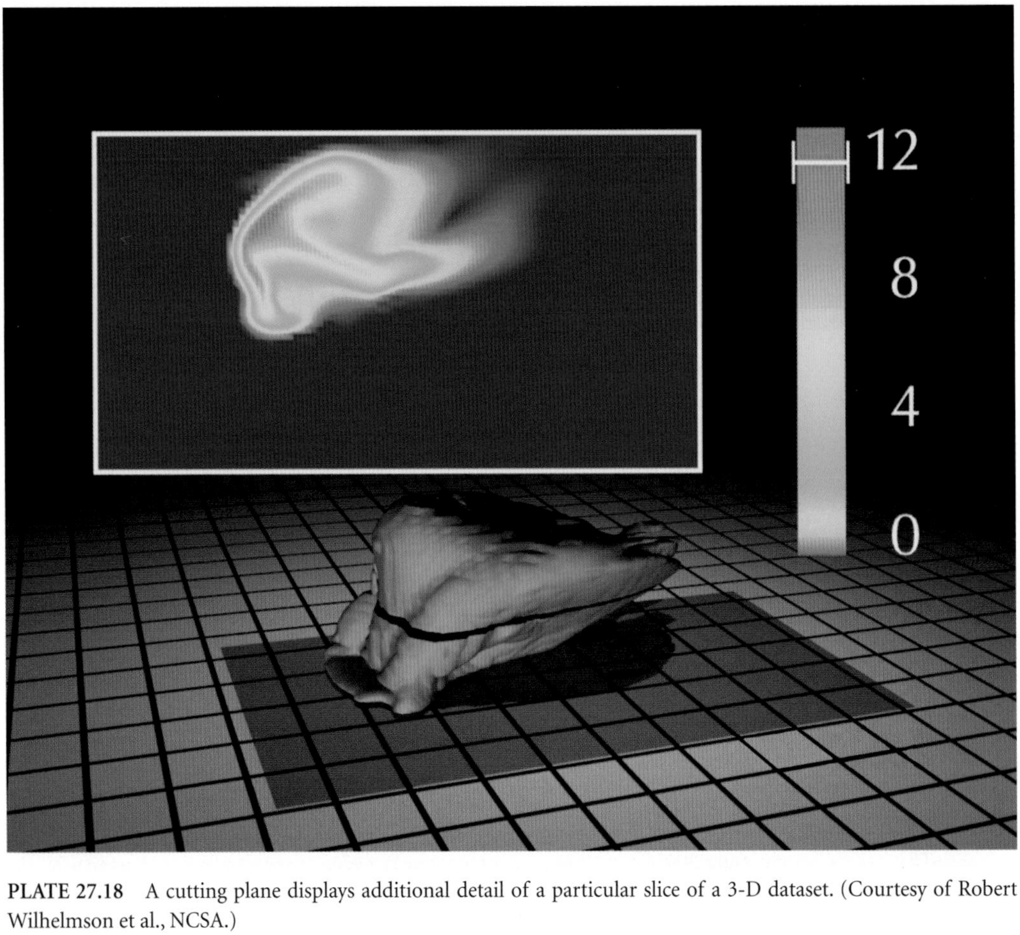

PLATE 27.18 A cutting plane displays additional detail of a particular slice of a 3-D dataset. (Courtesy of Robert Wilhelmson et al., NCSA.)

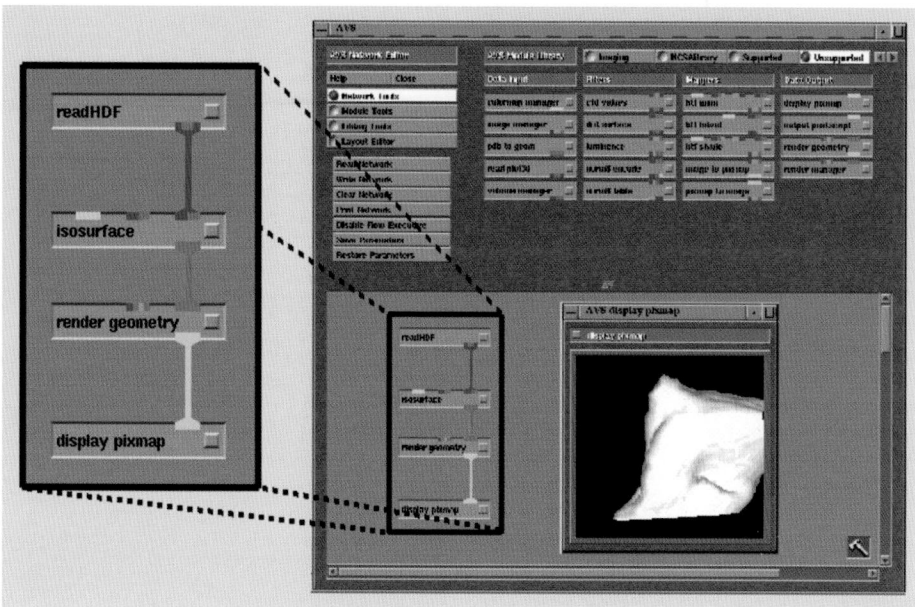

PLATE 27.26 Dataflow packages such as AVS connect modules to customize a visualization. (Courtesy of W. Sherman, NCSA.)

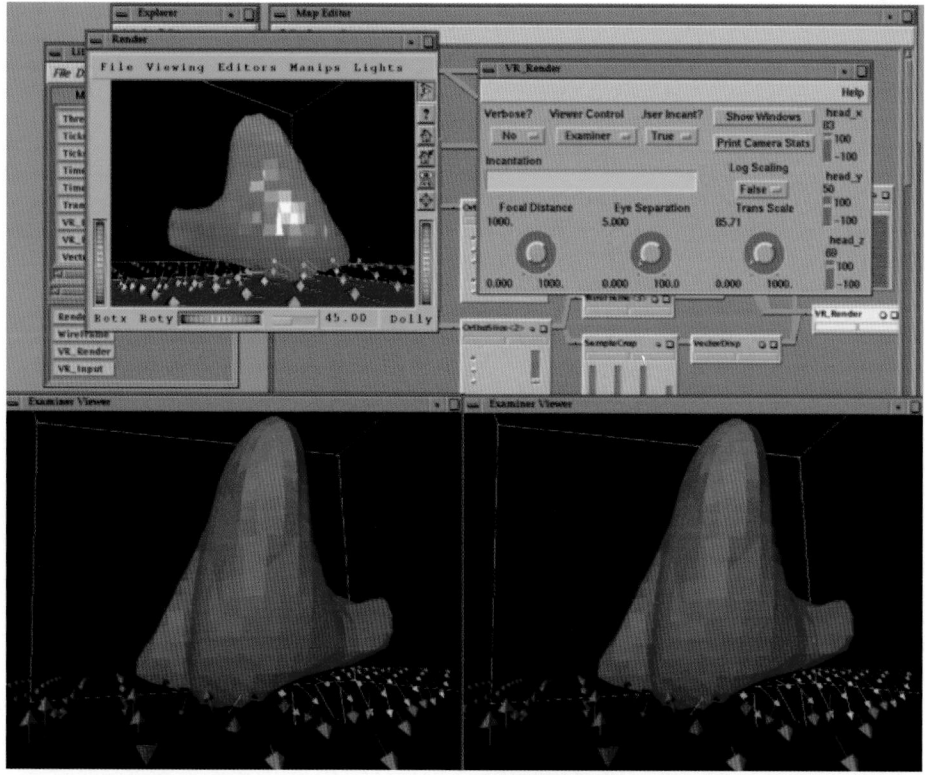

PLATE 27.27 Customization capabilities allow the package to be extended to new display paradigms, such as in this stereo image linked to a virtual-reality viewing device. (Courtesy of W. Sherman, NCSA.)

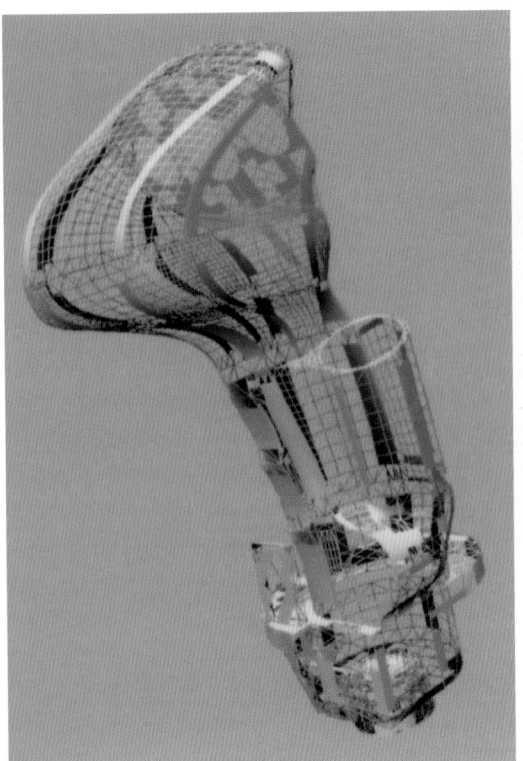
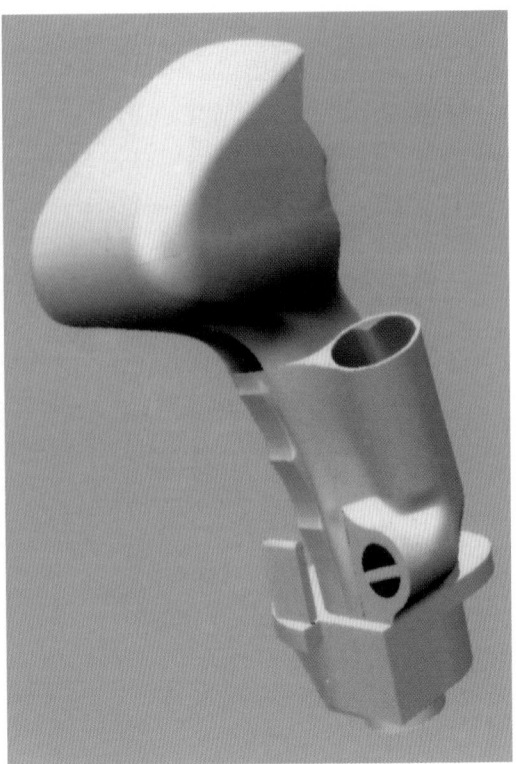

PLATE 36.5 A model composed of simple triangular facets (polygon model).

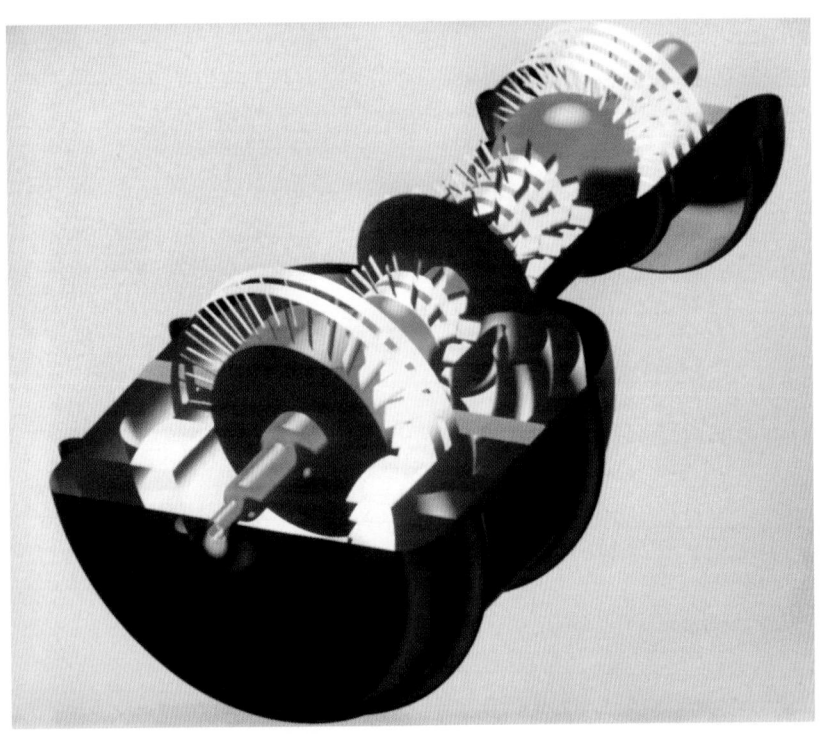

PLATE 36.18 Turbine engine modeled by B-spline surfaces.

PLATE 37.2 A fractal terrain model by Ken Musgrave. (© 1992 *Slickrock*. F. Kenton Musgrave. Used by permission).

PLATE 37.4 Horse chestnut tree created with a modified L-system that takes into account branch competition for light. (© 1995 by R. Mech and P. Prusinkiewicz, University of Calgary, Canada. Used by permission).

PLATE 37.5 Steam rising from a teacup. (© 1991 by David S. Ebert).

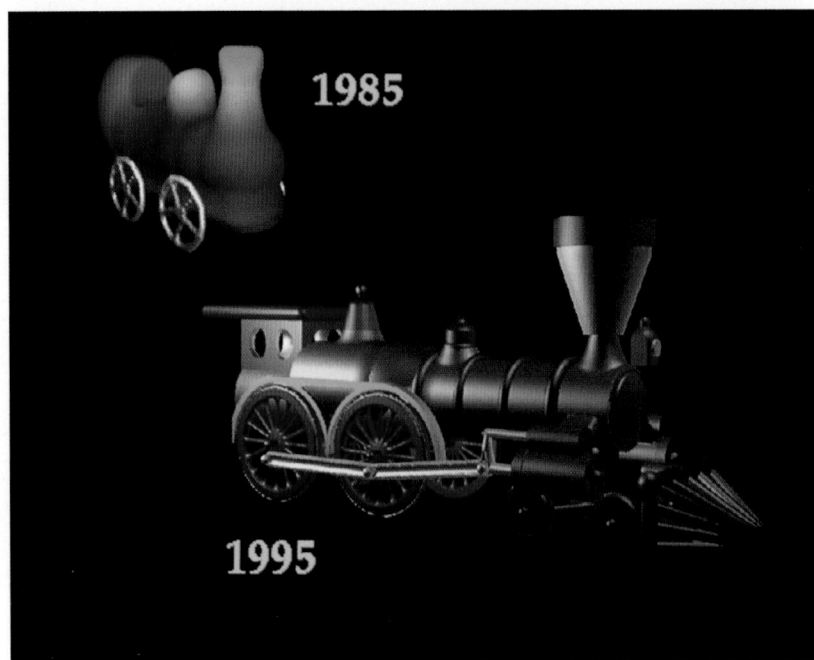

PLATE 37.8 Ten years in implicit surface modeling. The locomotive labeled 1985 shows a more traditional soft object created by implicit surface techniques. The locomotive labeled 1995 shows the results achievable by incorporating constructive solid geometry techniques with implicit surface models. (© 1995 by Brian Wyvill. Used by permission).

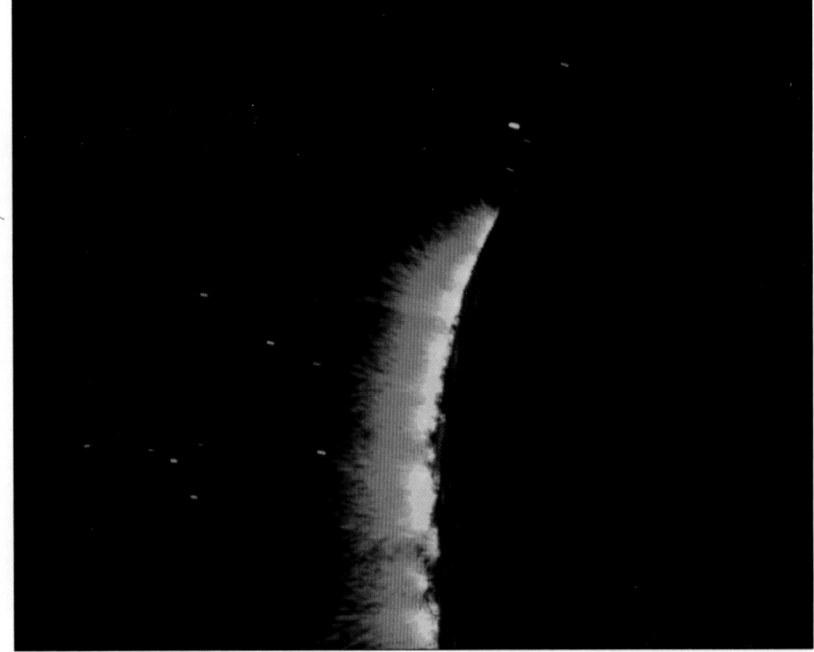

PLATE 37.9 An image from *Star Trek II: The Wrath of Khan* showing a wall of fire created with a particle system. (© 1987 Pixar. Used by permission).

30

Computational Fluid Dynamics

30.1 Introduction 30-1
30.2 Underlying Principles 30-2
    Fluid-Dynamical Background • Treatment of Geometry
30.3 Best Practices 30-8
    Panel Methods • Nonlinear Methods • Navier–Stokes
    Equation Methods
30.4 Research Issues and Summary 30-15

David A. Caughey
Cornell University

30.1 Introduction

The use of computer-based methods for the prediction of fluid flows has seen tremendous growth in the past several decades. Fluid dynamics has been one of the earliest, and most active fields for the application of numerical techniques. This is due to the essential nonlinearity of most fluid flow problems of practical interest — which makes analytical, or closed-form, solutions virtually impossible to obtain — combined with the geometrical complexity of these problems. In fact, the history of computational fluid dynamics can be traced back virtually to the birth of the digital computer itself, with the pioneering work of John von Neumann and others in this area. Von Neumann was interested in using the computer not only to solve engineering problems, but to understand the fundamental nature of fluid flows themselves. This is possible because the complexity of fluid flows arises, in many instances, not from complicated or poorly understood formulations, but from the nonlinearity of partial differential equations that have been known for more than a century. A famous paragraph written by von Neumann in 1946 serves to illustrate this point. He wrote [Goldstine and von Neumann 1963]:

> Indeed, to a great extent, experimentation in fluid mechanics is carried out under conditions where the underlying physical principles are not in doubt, where the quantities to be observed are completely determined by known equations. The purpose of the experiment is not to verify a proposed theory but to replace a computation from an unquestioned theory by direct measurements. Thus, wind tunnels are, for example, used at present, at least in part, as computing devices of the so-called analogy type . . . to integrate the nonlinear partial differential equations of fluid dynamics.

The present article provides some of the basic background in fluid dynamics required to understand the issues involved in numerical solution of fluid flow problems, then outlines the approaches that have been successful in attacking problems of practical interest.

30.2   Underlying Principles

In this section, we will provide the background in fluid dynamics required to understand the principles involved in the numerical solution of the governing equations. The formulation of the equations in generalized, curvilinear coordinates and the geometrical issues involved in the construction of suitable grids also will be discussed.

30.2.1   Fluid-Dynamical Background

As can be inferred from the quotation of von Neumann presented in the introduction, fluid dynamics is fortunate to have a generally accepted mathematical framework for describing most problems of practical interest. Such diverse problems as the high-speed flow past an aircraft wing, the motions of the atmosphere responsible for our weather, and the unsteady air currents produced by the flapping wings of a housefly all can be described as solutions to a set of partial differential equations known as the Navier–Stokes equations. These equations express the physical laws corresponding to conservation of mass, Newton's second law of motion relating the acceleration of fluid elements to the imposed forces, and conservation of energy, under the assumption that the stresses in the fluid are linearly related to the local rates of strain of the fluid elements. This latter assumption is generally regarded as an excellent approximation for everyday fluids such as water and air — the two most common fluids of engineering and scientific interest.

We will describe the equations for problems in two space dimensions, for the sake of economy of notation; here, and elsewhere, the extension to problems in three dimensions will be straightforward unless otherwise noted. The Navier–Stokes equations can be written in the form

$$\frac{\partial \mathbf{w}}{\partial t} + \frac{\partial \mathbf{f}}{\partial x} + \frac{\partial \mathbf{g}}{\partial y} = \frac{\partial \mathbf{R}}{\partial x} + \frac{\partial \mathbf{S}}{\partial y} \tag{30.1}$$

where $\mathbf{w}$ is the vector of conserved quantities

$$\mathbf{w} = \{\rho, \rho u, \rho v, e\}^T \tag{30.2}$$

where ρ is the fluid density, u and v are the fluid velocity components in the x and y directions, respectively, and e is the total energy per unit volume. The inviscid flux vectors $\mathbf{f}$ and $\mathbf{g}$ are given by

$$\mathbf{f} = \{\rho u, \rho u^2 + p, \rho u v, (e+p)u\}^T \tag{30.3}$$

$$\mathbf{g} = \{\rho v, \rho u v, \rho v^2 + p, (e+p)v\}^T \tag{30.4}$$

where p is the fluid pressure, and the flux vectors describing the effects of viscosity are

$$\mathbf{R} = \{0, \tau_{xx}, \tau_{xy}, u\tau_{xx} + v\tau_{xy} - q_x\}^T \tag{30.5}$$

$$\mathbf{S} = \{0, \tau_{xy}, \tau_{yy}, u\tau_{xy} + v\tau_{yy} - q_y\}^T \tag{30.6}$$

The viscous stresses appearing in these terms are related to the derivatives of the components of the velocity vector by

$$\tau_{xx} = 2\mu \frac{\partial u}{\partial x} - \frac{2}{3}\mu \left\{ \frac{\partial u}{\partial x} + \frac{\partial v}{\partial y} \right\} \tag{30.7}$$

$$\tau_{xy} = \mu \left\{ \frac{\partial u}{\partial y} + \frac{\partial v}{\partial x} \right\} \tag{30.8}$$

$$\tau_{yy} = 2\mu \frac{\partial v}{\partial y} - \frac{2}{3}\mu \left\{ \frac{\partial u}{\partial x} + \frac{\partial v}{\partial y} \right\} \tag{30.9}$$

and

$$q_x = -k \frac{\partial T}{\partial x} \tag{30.10}$$

$$q_y = -k \frac{\partial T}{\partial y} \tag{30.11}$$

represent the x and y components of the heat flux vector, according to Fourier's law. In these equations μ and k represent the coefficients of viscosity and thermal conductivity, respectively.

If the Navier–Stokes equations are nondimensionalized by normalizing lengths with respect to a representative length L and velocities by a representative velocity V_∞, and normalizing the fluid properties (such as density and coefficient of viscosity) by their values in the freestream, an important nondimensional parameter, the *Reynolds number*

$$\mathbf{Re} = \frac{\rho_\infty V_\infty L}{\mu_\infty} \tag{30.12}$$

appears as a parameter in the equations. In particular, the viscous stress terms on the right-hand side of Equation 30.1 are multiplied by the reciprocal of the Reynolds number. Physically, the Reynolds number can be interpreted as an inverse measure of the relative importance of the contributions of the viscous stresses to the dynamics of the flow; i.e., when the Reynolds number is large, the viscous stresses are small almost everywhere in the flowfield.

The computational resources required to solve the complete Navier–Stokes equations are enormous, particularly when the Reynolds number is large and regions of **turbulent flow** must be resolved. In 1970 Howard Emmons of Harvard University estimated the computer time required to solve a simple turbulent pipe-flow problem, including direct computation of all turbulent eddies containing significant energy [Emmons 1970]. For a computational domain consisting of approximately 12 diameters of a pipe of circular cross section, the computation of the solution at a Reynolds number based on the pipe diameter of $\mathbf{Re}_d = 10^7$ would require approximately 10^{17} seconds on a 1970s mainframe computer. Of course, much faster computers are now available, but even at a computational speed of 100 gigaflops — i.e., 10^{11} floating-point operations per second — such a calculation would require more than 3000 *years* to complete.

Because the resources required to solve the complete Navier–Stokes equations are so large, it is common to make approximations to bring the required computational resources to a more modest level for problems of practical interest. Expanding slightly on a classification introduced by Chapman [1979], the sequence of fluid-mechanical equations can be organized into the hierarchy shown in Table 30.1.

Table 30.1 summarizes the physical assumptions and time periods of most intense development for each of the stages in the fluid-mechanical hierarchy. Stage IV represents an approximation to the Navier–Stokes equations in which only the largest, presumably least isotropic scales of turbulence are resolved; the subgrid scales are modeled. Stage III represents an approximation in which the solution is decomposed into time-averaged or ensemble-averaged and fluctuating components for each variable. For example, the

TABLE 30.1 The Hierarchy of Fluid-Mechanical Approximations, Following Chapman [1979]

Stage	Model	Equations	Time Frame
I	Linear potential	Laplace's equation	1960s
IIa	Nonlinear potential	Full potential equation	1970s
IIb	Nonlinear inviscid	Euler equations	1980s
III	Modeled turbulence	Reynolds-averaged Navier–Stokes equations	1990s
IV	Large-eddy simulation (LES)	Navier–Stokes equations with subgrid turbulence model	1980s–1990s
V	Direct numerical simulation (DNS)	Fully resolved Navier–Stokes equations	1980s–1990s

velocity components and pressure can be decomposed into

$$u = U + u' \tag{30.13}$$
$$v = V + v' \tag{30.14}$$
$$p = P + p' \tag{30.15}$$

where U, V, and P are the average values of u, v, and p, respectively, taken over a time interval that is long compared to the turbulence time scales, but short compared to the time scales of any nonsteadiness of the averaged flowfield. If we let $\langle u \rangle$ denote such a time average of the u-component of the velocity, then, e.g.,

$$\langle u \rangle = U \tag{30.16}$$

When a decomposition of the form of Equation 30.13 to Equation 30.15 is substituted into the Navier–Stokes equations and the equations are averaged as described above, the resulting equations describe the evolution of the mean-flow quantities (such as U, V, P, etc.). These equations, called the **Reynolds-averaged Navier–Stokes equations**, are nearly identical to the original Navier–Stokes equations written for the mean-flow variables because terms that are linear in $\langle u' \rangle$, $\langle v' \rangle$, $\langle p' \rangle$, etc., are identically zero. The nonlinearity of the equations, however, introduces terms that depend upon the fluctuating components. In particular, terms proportional to $\langle \rho u' v' \rangle$, $\langle \rho u'^2 \rangle$, and $\langle \rho v'^2 \rangle$ appear in the averaged equations. Dimensionally, these terms are equivalent to stresses; in fact, these quantities are called the Reynolds stresses. Physically, the Reynolds stresses are the turbulent counterparts of the molecular viscous stresses, and appear as a result of the transport of momentum by the turbulent fluctuations. The appearance of these terms in the equations describing the mean flow means that the mean flow cannot be determined without some knowledge of these fluctuating components.

In order to solve the Reynolds-averaged Navier–Stokes equations, the Reynolds stresses must be related to the mean flow at some level of approximation using a phenomenological model. The simplest procedure is to try to relate the Reynolds stresses to the local mean-flow properties. Since the turbulence that is responsible for these stresses is a function not only of the local mean-flow state, but also of the flow history as well, such an approximation cannot have broad generality, but such local, or algebraic, turbulence models, based on the analogy of the mixing length from the kinetic theory of gases, are useful for many flows of engineering interest, especially for flows in which boundary-layer separation does not occur.

A more general procedure is to develop partial differential equations for the Reynolds stresses themselves, or for quantities that can be used to determine the scales of the turbulence, such as the turbulent kinetic energy k and the dissipation rate ε. The latter, so-called k–ε model is widely used in engineering analyses of turbulent flows. These differential equations can be derived by taking higher-order moments of the Navier–Stokes equations. For example, if the x-momentum equation is multiplied by v and then averaged, the result will be an equation for the Reynolds-stress component $\langle \rho u' v' \rangle$. Again because of the nonlinearity of the equations, however, yet higher moments of the fluctuating components (e.g., terms proportional to $\langle \rho u'^2 v' \rangle$, $\langle \rho u' v'^2 \rangle$, etc.) will appear in the equations for the Reynolds stresses. This is an example of the problem of closure of the equations for the Reynolds stresses; i.e., to solve the equations for the Reynolds stresses, these third-order quantities must be known. Equations for the third-order quantities can be derived by taking yet higher-order moments of the Navier–Stokes equations, but these equations will involve fourth-order moments. Thus, at some point the taking of moments must be terminated and models must be developed for the unknown higher-order quantities. It is hoped that more nearly universal models may be developed for these higher-order quantities, but the superiority of second-order models, in which the equations for the Reynolds-stress components are solved, subject to modeling assumptions for the required third-order quantities, has yet to be established for most practical problems.

For many design purposes, especially in aerodynamics, it is sufficient to represent the flow as that of an inviscid, or ideal, fluid. This is appropriate for flows at high Reynolds numbers that contain only negligibly small regions of separated flow. The equations describing inviscid flows can be obtained as a simplification of the Navier–Stokes equations in which the viscous terms are neglected altogether. This results in the

Euler equations of inviscid, compressible flow,

$$\frac{\partial \mathbf{w}}{\partial t} + \frac{\partial \mathbf{f}}{\partial x} + \frac{\partial \mathbf{g}}{\partial y} = 0 \tag{30.17}$$

This approximation corresponds to stage IIb in the hierarchy described in Table 30.1.

The Euler equations constitute a hyperbolic system of partial differential equations, and their solutions contain features that are absent from the Navier–Stokes equations. In particular, while the viscous diffusion terms appearing in the Navier–Stokes equations guarantee that solutions will remain smooth, the absence of these dissipative terms from the Euler equations allows them to have solutions that are discontinuous across surfaces in the flow. Solutions to the Euler equations must be interpreted within the context of generalized (or weak) solutions, and this theory provides the mathematical framework for developing the properties of any discontinuities that may appear. In particular, jumps in dependent variables (such as density, pressure, and velocity) across such surfaces must be consistent with the original conservation laws upon which the differential equations are based. For the Euler equations, these jump conditions are called the Rankine–Hugoniot conditions.

Solutions of the Euler equations for flows containing shock waves can be computed using either **shock fitting** or **shock capturing** methods. In the former, the shock surfaces must be located and the Rankine–Hugoniot jump conditions enforced explicitly. In shock capturing methods, **artificial viscosity** terms are added in the numerical approximation to provide enough dissipation to allow the shocks to be captured automatically by the scheme, with no special treatment in the vicinity of the shock waves. The numerical viscosity terms usually act to smear out the discontinuity over several grid cells. Numerical viscosity also is used when solving the Navier–Stokes equations for flows containing shock waves, because usually it is impractical to resolve the shock structure defined by the physical dissipative mechanisms.

In many cases, the flow can further be approximated as steady and irrotational. In these cases, it is possible to define a velocity potential Φ such that the velocity vector $\mathbf{V}$ is given by

$$\mathbf{V} = \nabla \Phi \tag{30.18}$$

and the fluid density is given by the isentropic relation

$$\rho = \left(1 + \frac{\gamma - 1}{2} \mathbf{M}_\infty^2 [1 - (u^2 + v^2)] \right)^{\frac{1}{\gamma - 1}} \tag{30.19}$$

where

$$u = \frac{\partial \Phi}{\partial x} \tag{30.20}$$

$$v = \frac{\partial \Phi}{\partial y} \tag{30.21}$$

and $\mathbf{M}_\infty$ is a reference Mach number corresponding to the freestream state in which $\rho = 1$ and $u^2 + v^2 = 1$. The steady form of the Euler equations then reduces to the single equation

$$\frac{\partial \rho u}{\partial x} + \frac{\partial \rho v}{\partial y} = 0 \tag{30.22}$$

Equation (30.19) can be used to eliminate the density from Eq. (30.22), which then can be expanded to the form

$$(a^2 - u^2)\frac{\partial^2 \Phi}{\partial x^2} - 2uv\frac{\partial^2 \Phi}{\partial x\, \partial y} + (a^2 - v^2)\frac{\partial^2 \Phi}{\partial y^2} = 0 \tag{30.23}$$

where a is the local speed of sound, which is a function only of the fluid speed $V = \sqrt{u^2 + v^2}$. Thus, Equation 30.23 is a single equation that can be solved for the velocity potential Φ.

Equation 30.23 is a second-order quasilinear partial differential equation whose type depends on the sign of the discriminant $1 - \mathbf{M}^2$, where $\mathbf{M}$ is the *local* Mach number. The equation is elliptic or hyperbolic according as the Mach number is less than or greater than unity. Thus, the nonlinear potential equation contains a mathematical description of the physics necessary to predict the important features of transonic flows. It is capable of changing type, and the conservation form of Equation 30.22 allows surfaces of discontinuity, or shock waves, to be computed. Solutions at this level of approximation, corresponding to stage IIa, are considerably less expensive to compute than solutions of the full Euler equations, since only one dependent variable need be computed and stored. The jump relation corresponding to weak solutions of the potential equation is different than the Rankine–Hugoniot relations, but is a good approximation to the latter when the shocks are not too strong.

Finally, if the flow can be approximated by small perturbations to some uniform reference state, Equation 30.23 can further be simplified to

$$\left(1 - \mathbf{M}_\infty^2\right)\frac{\partial^2 \phi}{\partial x^2} + \frac{\partial^2 \phi}{\partial y^2} = 0 \tag{30.24}$$

where ϕ is the *perturbation* velocity potential defined according to

$$\Phi = x + \phi \tag{30.25}$$

if the uniform velocity in the reference state is assumed to be parallel to the x-axis and be normalized to have unit magnitude. Further, if the flow can be approximated as incompressible, i.e., in the limit as $\mathbf{M}_\infty \to 0$, Equation 30.23 reduces to

$$\frac{\partial^2 \Phi}{\partial x^2} + \frac{\partial^2 \Phi}{\partial y^2} = 0 \tag{30.26}$$

Since Equation 30.24 and Equation 30.26 are linear, superposition of elementary solutions can be used to construct solutions of arbitrary complexity. Numerical methods to determine the singularity strengths for aerodynamic problems are called **panel methods**, and constitute stage I in the hierarchy of approximations.

It is important to realize that, even though the time periods for development of some of the different models overlap, the applications of the various models may be for problems of significantly differing complexity. For example, DNS calculations were performed as early as the 1970s, but only for the simplest flows — homogeneous turbulence — at very low Reynolds numbers. Flows at higher Reynolds numbers and nonhomogeneous flows are being performed only now, whereas calculations for three-dimensional flows with modeled turbulence were performed as early as the mid 1980s.

30.2.2 Treatment of Geometry

For all numerical solutions to partial differential equations, including those of fluid mechanics, it is necessary to discretize the boundaries of the flow domain. For the panel method of stage I, this is all that is required, since the problem is linear and superposition can be used to construct solutions satisfying the boundary conditions. For nonlinear problems, it is necessary to discretize the entire flow domain. This can be done using either structured or unstructured grid systems. *Structured grids* are those in which the grid points can be ordered in a regular Cartesian structure; i.e., the points can be given indices (i, j) such that the nearest neighbors of the (i, j) point are identified by the indices $(i \pm 1, j \pm 1)$. The grid cells for these meshes are thus quadrilateral in two dimensions and hexahedral in three dimensions. *Unstructured grids* have no regular ordering of points or cells, and a connectivity table must be maintained to identify which points and edges belong to which cells. Unstructured grids most often consist of triangles (in two dimensions), tetrahedra (in three dimensions), or combinations of these and quadrilateral and hexahedral cells, respectively. In addition, grids having purely quadrilateral cells may also be unstructured, even though they have a locally Cartesian structure — e.g., when multilevel grids are used for adaptive refinement.

Implementations on structured grids are generally more efficient than those on unstructured grids, since indirect addressing is required for the latter, and efficient implicit methods often can be constructed that take advantage of the regular ordering of points (or cells) in structured grids. A great deal of effort has been expended to generate both structured and unstructured grid systems, much of which is closely related to the field of computational geometry.

30.2.2.1 Structured Grids

A variety of techniques are used to generate structured grids for use in fluid-mechanical calculations. These include relatively fast algebraic methods, including those based on transfinite interpolation and conformal mapping, as well as more costly methods based on the solution of either elliptic or hyperbolic systems of partial differential equations for the grid coordinates. These techniques are discussed in a review article by Eiseman [1985].

For complex geometries, it often is not possible to generate a single grid that conforms to all the boundaries. Even if it is possible to generate such a grid, it may have undesirable properties, such as excessive skewness or a poor distribution of cells, which could lead to poor stability or accuracy in the numerical algorithm. Thus, structured grids for complex geometries are generally constructed as combinations of simpler grid blocks for various parts of the domain. These grids may be allowed to overlap, in which case they are referred to as *Chimera* grids, or be required to share common surfaces of intersection, in which case they are identified as *multiblock* grids. In the latter case, grid lines may have varying degrees of continuity across the interblock boundaries, and these variations have implications for the construction and behavior of the numerical algorithm in the vicinity of those boundaries. Grid generation techniques based on the solution of systems of partial differential equations are described by Thompson et al. [1985].

Numerical methods to solve the equations of fluid mechanics on structured grid systems are implemented most efficiently by taking advantage of the ease with which the equations can be transformed to a generalized coordinate system. The expression of the system of conservation laws in the new, body-fitted coordinate system reduces the problem to one effectively of Cartesian geometry. The transformation will be described here for the Euler equations.

Consider the transformation of Equation 30.17 to a new coordinate system (ξ, η). The local properties of the transformation at any point are contained in the Jacobian matrix of the transformation, which can be defined as

$$J = \begin{Bmatrix} x_\xi & x_\eta \\ y_\xi & y_\eta \end{Bmatrix} \tag{30.27}$$

for which the inverse is given by

$$J^{-1} = \begin{Bmatrix} \xi_x & \xi_y \\ \eta_x & \eta_y \end{Bmatrix} = \frac{1}{h}\begin{Bmatrix} y_\eta & -x_\eta \\ -y_\xi & x_\xi \end{Bmatrix} \tag{30.28}$$

where $h = x_\xi y_\eta - y_\xi x_\eta$ is the determinant of the Jacobian matrix. Subscripts in Equation 30.27 and Equation 30.28 denote partial differentiation.

It is natural to express the fluxes in conservation laws in terms of their contravariant components. Thus, if we define

$$\{\mathbf{F}, \mathbf{G}\}^T = J^{-1}\{\mathbf{f}, \mathbf{g}\}^T \tag{30.29}$$

then the transformed Euler equations can be written in the compact form

$$\frac{\partial h\mathbf{w}}{\partial t} + \frac{\partial h\mathbf{F}}{\partial \xi} + \frac{\partial h\mathbf{G}}{\partial \eta} = 0 \tag{30.30}$$

if the transformation is independent of time (i.e., if the grid is not moving). If the grid is moving or deforming with time, the equations can be written in a similar form, but additional terms must be included that allow for the fluxes induced by the motion of the mesh.

The Navier–Stokes equations can be transformed in a similar manner, although the transformation of the viscous contributions is somewhat more complicated and will not be included here. Since the nonlinear potential equation is simply the continuity equation (the first of the equations that comprise the Euler equations), the transformed potential equation can be written as

$$\frac{\partial}{\partial \xi}(\rho h U) + \frac{\partial}{\partial \eta}(\rho h V) = 0 \tag{30.31}$$

where

$$\begin{Bmatrix} U \\ V \end{Bmatrix} = J^{-1} \begin{Bmatrix} u \\ v \end{Bmatrix} \tag{30.32}$$

are the contravariant components of the velocity vector.

30.2.2.2 Unstructured Grids

Unstructured grids generally have greater geometric flexibility than structured grids, because of the relative ease of generating triangular or tetrahedral tesselations of two- and three-dimensional domains. Advancing-front methods and Delaunay triangulations are the most frequently used techniques for generating triangular/tetrahedral grids. Unstructured grids also are easier to adapt locally so as to better resolve localized features of the solution.

30.3 Best Practices

30.3.1 Panel Methods

The earliest numerical methods used widely for making fluid-dynamical computations were developed to solve linear potential problems, described as stage I calculations in the previous section. Mathematically, panel methods are based upon the fact that Equation 30.26 can be recast as an integral equation giving the solution at any point (x, y) in terms of the freestream speed U (here assumed unity) and angle of attack α and the line integrals of singularities distributed along the curve C representing the body surface:

$$\Phi(x, y) = x \cos \alpha + y \sin \alpha + \int_C \sigma \ln\left(\frac{r}{2\pi}\right) ds + \int_C \delta(\partial/\partial n) \ln\left(\frac{r}{2\pi}\right) ds \tag{30.33}$$

In this equation, σ and δ represent the source and doublet strengths distributed along the body contour, respectively, r is the distance from the point (x, y) to the boundary point, and n is the direction of the outward normal to the body surface. When the point (x, y) is chosen to lie on the body contour C, Equation 30.33 can be interpreted as giving the solution Φ at any point on the body in terms of the singularities distributed along the surface. This effectively reduces the dimension of the problem by one (i.e., the two-dimensional problem considered here becomes essentially one-dimensional, and the analogous procedure applied to a three-dimensional problem results in a two-dimensional equation requiring the evaluation only of integrals over the body surface).

Equation 30.33 is approximated numerically by discretizing the boundary C into a collection of panels (line segments in this two-dimensional example) on which the singularity distribution is assumed to be of some known functional form, but of an as yet unknown magnitude. For example, for a simple nonlifting body, the doublet strength δ might be assumed to be zero, while the source strength σ is assumed to be constant on each segment. The second integral in Equation 30.33 is then zero, while the first integral can be written as a sum over all the elements of integrals that can be evaluated analytically as

$$\Phi(x, y) = x \cos \alpha + y \sin \alpha + \sum_{i=1}^{N} \sigma_i \int_{C_i} \ln\left(\frac{r}{2\pi}\right) ds \tag{30.34}$$

where C_i is the portion of the body surface corresponding to the ith segment and N is the number of segments, or panels, used.

The source strengths σ_i must be determined by enforcing the boundary condition

$$\frac{\partial \Phi}{\partial n} = 0 \tag{30.35}$$

which specifies that the component of velocity normal to the surface be zero (i.e., that there be no flux of fluid across the surface). This is implemented by requiring that Equation 30.35 be satisfied at a selected number of control points. For the example of constant-strength source panels, if one control point is chosen on each panel, the requirement that Equation 30.35 be satisfied at each of the control points will result in N equations of the form

$$\sum_{i=1}^{N} A_{i,j}\sigma_i = \mathbf{U} \cdot \hat{\mathbf{n}}, \quad j = 1, 2, \ldots, N \tag{30.36}$$

where $A_{i,j}$ are the elements of the influence-coefficient matrix that give the normal velocity at control point j due to sources of unit strength on panel i, and $\mathbf{U}$ is a unit vector in the direction of the freestream. Equation 30.35 and Equation 30.36 constitute a system of N linear equations that can be solved for the unknown source strengths σ_i. Once the source strengths have been determined, the velocity potential, or the velocity itself, can be computed directly at any point in the flowfield using Equation 30.34. A review of the development and application of panel methods is provided by Hess [1990].

A major advantage of panel methods, relative to the more advanced methods required to solve the nonlinear problems of stages II–V, is that it is necessary to describe (and to discretize into panels) only the surface of the body. While linearity is a great advantage in this regard, it is not clear that the method is computationally more efficient than the more advanced nonlinear field methods. This results from the fact that the influence-coefficient matrix in the system of equations that must be solved to give the source strengths for each panel is not sparse; i.e., the velocities at each control point are affected by the sources on all the panels. In contrast, the solution at each mesh point in a finite-difference formulation (or each mesh cell in a finite-volume formulation) is related to the values of the solution at only a few neighbors, resulting in a very sparse matrix of influence coefficients that can be solved very efficiently using iterative methods. Thus, the primary advantage of the panel method is the geometrical one associated with the reduction in dimension of the problem. For nonlinear problems the use of finite-difference, finite-element, or finite-volume methods requires discretization of the entire flowfield, and the associated mesh generation task has been a major pacing item limiting the application of such methods.

30.3.2 Nonlinear Methods

For nonlinear equations, superposition of elementary solutions is no longer a valid technique for constructing solutions, and it becomes necessary to discretize the solution throughout the entire domain, not just the boundary surface. In addition, since the equations are nonlinear, some sort of iteration must be used to compute successively better approximations to the solution. This iterative process may approximate the physics of an unsteady flow process or may be chosen to provide more rapid convergence to the solution for steady-state problems. Solutions for both the nonlinear potential and Euler equations are generally determined using a finite-difference, finite-volume, or finite-element method. In any of these techniques a grid, or network of points and cells, is distributed throughout the flowfield. In a finite-difference method the derivatives appearing in the original differential equation are approximated by discrete differences in the values of the solution at neighboring points (and times, if the solution is unsteady), and substitution of these into the differential equation yields a system of algebraic equations relating the values of the solution at neighboring grid points. In a finite-volume method, the unknowns are taken to be representative of the values of the solution in the control volumes formed by the intersecting grid surfaces, and the equations are constructed by balancing fluxes across the bounding surfaces of each control volume with the rate of change of the solution within the control volume. In a finite-element method, the solution is represented using simple interpolating functions within each mesh cell, or element, and equations for the nodal values are obtained by integrating a variational or residual formulation of the equations over the elements. The algebraic equations relating the values of the solution in neighboring cells can be very similar (or even identical) in appearance for all three methods, and the choice of method often is primarily a matter of taste. Stable and efficient finite-difference and finite-volume methods were developed earlier than finite-element methods for compressible flow problems, but finite-element methods capable of treating very complex compressible flows now are available. A review of recent progress in the development of finite-element methods for compressible flows and remaining issues is given by Glowinski and Pironneau [1992].

Since the nonlinear potential, Euler, and Navier–Stokes equations all are nonlinear, the algebraic equations resulting from these discretization procedures also are nonlinear, and a scheme of successive approximation usually is required to solve them. As mentioned earlier, however, these equations tend to be highly local in nature, and efficient iterative methods have been developed to solve them in many cases.

30.3.2.1 Nonlinear Potential-Equation Methods

The primary advantage of solving the potential equation rather than the Euler (or Navier–Stokes) equations derives from the fact that the flowfield can be described completely in terms of a single scalar function, the velocity potential Φ. The formulation of numerical schemes to solve Equation 30.22 is complicated by the fact that, as noted earlier, the equation changes type according to whether the local Mach number is subsonic or supersonic. Differencing schemes for the potential equation must, therefore, be type-dependent — i.e., they must change their form depending on whether the local Mach number is less than or greater than unity. These methods usually are based upon central, or symmetric, differencing formulas that are appropriate for the elliptic case (corresponding to subsonic flows); they are then modified by adding an upwind bias to maintain stability in hyperbolic regions (where the flow is supersonic). This directional bias can be introduced into the difference equations either by adding an artificial viscosity proportional to the third derivative of the velocity potential Φ in the streamwise direction, or by replacing the density at each point in supersonic zones with its value at a point slightly upstream of the actual point. Mathematically, these two approaches can be seen to be equivalent, since the upwinding of the density evaluation also effectively introduces a correction proportional to the third derivative of the potential in the flow direction.

It is important to introduce such artificial viscosity (or compressibility) terms in such a way that their effect vanishes in the limit as the mesh is refined. In this way, the numerical approximation will approach the differential equation in the limit of zero mesh spacing, and the method is said to be consistent with the original differential equation. In addition, for flows with shock waves, it is important that the numerical approximation be **conservative**; this guarantees that the properties of discontinuous solutions will be consistent with the jump relations of the original conservation laws. The shock jump relations corresponding to Equation 30.22, however, are different from the Rankine–Hugoniot conditions describing shocks within the framework of the Euler equations. Since entropy is everywhere conserved in the potential theory, and since there is a finite entropy jump across a Rankine–Hugoniot shock, it is clear that the jump relations must be different. For weak shocks, however, the differences are small and the economies afforded by the potential formulation make computations based on this approximation attractive for many transonic problems.

Perturbation techniques can be used to demonstrate that the effect of entropy jump across the shocks is more important than the rotationality introduced by these weak shocks. Thus, it makes sense to develop techniques that allow for the entropy jump, but are still within the potential formulation. Such techniques have been developed (see, e.g., Hafez [1985]), but have been relatively little used, as developments in techniques to solve the Euler equations have overtaken these approaches.

The nonlinear difference equations resulting from discrete approximations to the potential equation generally are solved using iterative, or relaxation, techniques. The equations are linearized by computing approximations to all but the highest (second) derivatives from the preceding solution in an iterative sequence, and a correction is computed at each mesh point in such a way that the equations are identically satisfied. It is useful in developing these iterative techniques to think of the iterative process as a discrete approximation to a continuous time-dependent process [Garabedian 1956]. Thus, the iterative process approximates an equation of the form

$$\beta_0 \frac{\partial \Phi}{\partial t} + \beta_1 \frac{\partial^2 \Phi}{\partial \xi \, \partial t} + \beta_2 \frac{\partial^2 \Phi}{\partial \eta \, \partial t} = \frac{a^2}{\rho} \left\{ \frac{\partial}{\partial \xi}(\rho h U) + \frac{\partial}{\partial \eta}(\rho h V) \right\} \tag{30.37}$$

The parameters β_0, β_1, and β_2, which are related to the mix of old and updated values of the solution used in the difference equations, can then be chosen to ensure that the time-dependent process converges to a steady state in both subsonic and supersonic regions. The formulation of transonic potential flow problems and their solution is described by Caughey [1982].

Even when the values of the parameters are chosen to provide rapid convergence, many hundreds of iterations may be necessary to achieve convergence to acceptable levels, especially when the mesh is very fine. This slow convergence is a characteristic of virtually all iterative schemes, and is a result of the fact that the representation of the difference equations must be highly local if the scheme is to be computationally efficient. As a result of this locality, the reduction of the low-wave-number component of the error to acceptable levels often requires many iterations. A powerful technique for circumventing this difficulty with the iterative solution of numerical approximations to partial differential equations, called the multigrid technique, has been applied with great success to problems in fluid mechanics.

The multigrid method relies for its success on the fact that after a few cycles of any good iterative technique, the error remaining in the solution is relatively smooth, and can be represented accurately on a coarser grid. Application of the same iterative technique on the coarser grid soon makes the error on this grid smooth as well, and the grid-coarsening process can be repeated until the grid contains only a few cells in each coordinate direction. The corrections that have been computed on all coarser levels are then interpolated back to the finest grid, and the process is repeated. The accuracy of the converged solution on the fine grid is determined by the accuracy of the approximation on that grid, since the coarser levels are used only to effect a more rapid convergence of the iterative process.

A particularly efficient multigrid technique for steady transonic potential-flow problems has been developed by Jameson [1979]. It uses a generalized alternating-direction implicit smoothing algorithm in conjunction with a full-approximation-scheme multigrid algorithm. In theory, for a wide class of problems, the work (per mesh point) required to solve the equations using a multigrid approach is independent of the number of mesh cells. In many practical calculations, 10 or fewer multigrid cycles may be required even on very fine grids.

30.3.2.2 Euler-Equation Methods

As noted earlier, the Euler equations constitute a hyperbolic system of partial differential equations, and numerical methods for solving them rely heavily upon the rather well-developed mathematical theory of such systems. As for the nonlinear potential equation, discontinuous solutions corresponding to flows with shock waves play an important role. Shock-capturing methods are much more widely used than shock-fitting methods, and for these methods it is important to use schemes that are conservative.

As mentioned earler, it is necessary to add artificial, or numerical, dissipation to stabilize the Euler equations. This can be done by adding dissipative terms explicitly to an otherwise nondissipative central difference scheme or by introducing upwind approximations in the flux evaluations. Both approaches are highly developed. The most widely used central difference methods are those modeled after the approach of Jameson et al. [1981]. This approach introduces an adaptive blend of second and fourth differences of the solution in each coordinate direction; a local switching function, usually based on a second difference of the pressure, is used to reduce the order of accuracy of the approximation to first order locally in the vicinity of shock waves and to turn off the fourth differences there. The second-order terms are small in smooth regions where the fourth-difference terms are sufficient to stabilize the solution and ensure convergence to the steady state. More recently, Jameson [1992] has developed improved symmetric limited positive (SLIP) and upstream limited positive (USLIP) versions of these blended schemes (see also Tatsumi et al. [1995]).

Much effort has been directed toward developing numerical approximations for the Euler equations that capture discontinuous solutions as sharply as possible without overshoots in the vicinity of the discontinuity. For purposes of exposition of these methods, we consider the one-dimensional form of the

Euler equations, which can be written

$$\frac{\partial \mathbf{w}}{\partial t} + \frac{\partial \mathbf{f}}{\partial x} = 0 \tag{30.38}$$

where $\mathbf{w} = \{\rho, \rho u, e\}^T$ and $\mathbf{f} = \{\rho u, \rho u^2 + p, (e + p)u\}^T$. Not only is the exposition clearer for the one-dimensional form of the equations, but the implementation of these schemes for multidimensional problems also generally is done by dimensional splitting in which one-dimensional operators are used to treat variations in each of the mesh directions.

For smooth solutions, Equation 30.38 are equivalent to the quasilinear form

$$\frac{\partial \mathbf{w}}{\partial t} + \mathbf{A} \frac{\partial \mathbf{w}}{\partial x} = 0 \tag{30.39}$$

where $\mathbf{A} = \{\partial \mathbf{f}/\partial \mathbf{w}\}$ is the Jacobian of the flux vector with respect to the solution vector. The eigenvalues of $\mathbf{A}$ are given by $\lambda = u, u + a, u - a$, where $a = \sqrt{\gamma p/\rho}$ is the speed of sound. Thus, for subsonic flows, one of the eigenvalues will have a different sign than the other two. For example, if $0 < u < a$, then $u - a < 0 < u < u + a$. The fact that various eigenvalues have different signs in subsonic flows means that simple one-sided difference methods cannot be stable. One way around this difficulty is to split the flux vector into two parts, the Jacobian of each of which has eigenvalues of only one sign. Such an approach has been developed by Steger and Warming [1981]. They used a relatively simple splitting that has discontinuous derivatives whenever an eigenvalue changes sign; an improved splitting has been developed by van Leer [1982] that has smooth derivatives at the transition points.

Each of the characteristic speeds can be identified with the propagation of a wave. If a mesh surface is considered to represent a discontinuity between two constant states, these waves constitute the solution to a Riemann (or shock-tube) problem. A scheme developed by Godunov [1959] assumes the solution to be piecewise constant over each mesh cell, and uses the fact that the solution to the Riemann problem can be given in terms of the solution of algebraic (but nonlinear) equations to advance the solution in time. Because of the assumption of piecewise constancy of the solution, Godunov's scheme is only first-order accurate. Van Leer [1979] has shown how it is possible to extend these ideas to a second-order monotonicity-preserving scheme using the so-called monotone upwind scheme for systems of conservation laws (MUSCL) formulation. The efficiency of schemes requiring the solution of Riemann problems at each cell interface for each time step can be improved by the use of approximate solutions to the Riemann problem [Roe 1986].

More recent ideas to control oscillation of the solution in the vicinity of shock waves include the concept of total-variation-diminishing (TVD) schemes, first introduced by Harten (see, e.g., Harten [1983, 1984]), and essentially nonoscillatory (ENO) schemes, introduced by Osher and his coworkers (see, e.g., Harten et al. [1987] and Shu and Osher [1988]).

Hyperbolic systems describe the evolution in time of physical systems undergoing unsteady processes governed by the propagation of waves. This feature frequently is used in fluid mechanics, even when the flow to be studied is steady. In this case, the unsteady equations are solved for long enough times that the steady state is reached asymptotically — often to within roundoff error. To maintain the hyperbolic character of the equations, and to keep the numerical method consistent with the physics it is trying to predict, it is necessary to determine the solution at a number of intermediate time levels between the initial state and the final steady state. Such a sequential process is said to be a time marching of the solution.

The simplest practical methods for solving hyperbolic systems are **explicit** in time. The size of the time step that can be used to solve hyperbolic systems using an explicit method is limited by a constraint known as the Courant–Friedrichs–Lewy or **CFL condition**. Broadly interpreted, the CFL condition states that the time step must be smaller than the time required for a signal to propagate across a single mesh cell. Thus, if the mesh is very fine, the allowable time step also must be very small, with the result that many time steps must be taken to reach an asymptotic steady state.

Multistage, or Runge–Kutta, methods have become extremely popular for use as explicit time-stepping schemes. After discretization of the spatial operators, using finite-difference, finite-volume, or finite-element approximations, the Euler equations can be written in the form

$$\frac{d\mathbf{w}_i}{dt} + Q(\mathbf{w}_i) + D(\mathbf{w}_i) = 0 \tag{30.40}$$

where $\mathbf{w}_i$ represents the solution at the ith mesh point, or in the ith mesh cell, and Q and D are discrete operators representing the contributions of the Euler fluxes and numerical dissipation, respectively. An m-stage time integration scheme for these equations can be written in the form

$$\mathbf{w}_i^{(k)} = \mathbf{w}_i^{(0)} - \alpha_k \, \Delta t \left[Q\big(\mathbf{w}_i^{(k-1)}\big) + D\big(\mathbf{w}_i^{(k-1)}\big) \right], \quad k = 1, 2, \ldots, m \tag{30.41}$$

with $\mathbf{w}_i^{(0)} = \mathbf{w}_i^n$, $\mathbf{w}_i^{(m)} = \mathbf{w}_i^{n+1}$, and $\alpha_m = 1$. The dissipative and dispersive properties of the scheme can be tailored by the sequence of α_i chosen; note that, for nonlinear problems, this method may be only first-order accurate in time, but this is not necessarily a disadvantage if one is interested only in the steady-state solution. The principal advantage of this formulation, relative to versions that may have better time accuracy, is that only two levels of storage are required regardless of the number of stages used. This approach was first introduced for problems in fluid mechanics by Graves and Johnson [1978], and has been further developed by Jameson et al. [1981]. In particular, Jameson and his group have shown how to tailor the stage coefficients so that the method is an effective smoothing algorithm for use with multigrid (see, e.g., Jameson and Baker [1983]).

Implicit techniques also are highly developed, especially when structured grids are used. Approximate factorization of the implicit operator usually is required to reduce the computational burden of solving a system of linear equations for each time step. Methods based on alternating-direction implicit (ADI) techniques date back to the pioneering work of Briley and McDonald [1974] and Beam and Warming [1976]. An efficient diagonalized ADI method has been developed within the context of the multigrid method by Caughey [1987], and a lower-upper symmetric Gauss–Seidel method has been developed by Yoon and Kwak [1994]. The multigrid implementations of these methods are based on the work of Jameson [1983].

30.3.3 Navier–Stokes Equation Methods

As described earlier, the relative importance of viscous effects is characterized by the value of the Reynolds number. If the Reynolds number is not too large, the flow remains smooth, and adjacent layers (or laminae) of fluid slide smoothly past one another. When this is the case, the solution of the Navier–Stokes equations is not too much more difficult than solution of the Euler equations. Greater resolution is required to resolve the large gradients in the boundary layers near solid boundaries, especially as the Reynolds number becomes large, so more mesh cells are required to achieve acceptable accuracy. In most of the flowfield, however, the flow behaves as if it were nearly inviscid, so methods developed for the Euler equations are appropriate and effective. The equations must, of course, be modified to include the additional terms resulting from the viscous stresses, and care must be taken to ensure that any artificial dissipative effects are small relative to the physical viscous dissipation in regions where the latter is important. The solution of the Navier–Stokes equations for laminar flows, then, is somewhat more costly in terms of computer resources, but not significantly more difficult from an algorithmic point of view, than solution of the Euler equations. Unfortunately, most flows of engineering interest occur at large enough Reynolds numbers that the flow in the boundary layers near solid boundaries becomes turbulent.

30.3.3.1 Turbulence Models

Solution of the Reynolds-averaged Navier–Stokes equations requires modeling of the Reynolds stress terms. The simplest models, based on the original mixing-length hypothesis of Prandtl, relate the Reynolds stresses to the local properties of the mean flow. The Baldwin–Lomax model [Baldwin and Lomax 1978] is the most widely used model of this type and gives good correlation with experimental measurements for wall-bounded shear layers so long as there are no significant regions of separated flow.

The most complete commonly used turbulence models include two additional partial differential equations that determine characteristic length and time scales for the turbulence. The most widely used of these techniques is called the k–ε model, since it is based on equations for the turbulence kinetic energy (usually given the symbol k) and the turbulence dissipation rate (usually given the symbol ε). This method has grown out of work by Launder and Spaulding [1972]. Another variant, based on a formulation by Kolmogorov, calculates a turbulence frequency ω instead of the dissipation rate (and hence is called a k–ω model). More complete models that compute all elements of the Reynolds stress tensor are in an active state of development. The common base for most of these models is the work of Launder et al. [1975], with more recent contributions by Lumley [1978], Speziale [1987], and Reynolds [1987]. These models, and their implementation within the context of CFD, are described in the book by Wilcox [1993].

More limited models, based on a single equation for a turbulence scale, also have been developed. These include the models of Baldwin and Barth [1991] and of Spalart and Allmaras [1992]. These models are applicable principally to boundary-layer flows of aerodynamic interest.

30.3.3.2 Large-Eddy Simulations

The difficulty of developing generally applicable phenomenological turbulence models on the one hand, and the enormous computational resources required to resolve all scales in turbulent flows at large Reynolds number on the other, have led to the development of large-eddy simulation (LES) techniques. In this approach, the largest length and time scales of the turbulent motions are resolved, but the smaller (subgrid) scales are modeled. This is an attractive approach because it is the largest scales that contain the preponderance of turbulent kinetic energy and that are responsible for most of the mixing. At the same time, the smaller scales are believed to be more nearly isotropic and independent of the larger scales, and thus are less likely to behave in problem-specific ways — i.e., it should be easier to develop universal models for these smaller scales.

A filtering technique is applied to the Navier–Stokes equations which results in equations having a form similar to the original equations, but with additional terms representing a subgrid-scale tensor that describes the effect of the modeled terms on the larger scales. The solution of these equations is not well posed if there is no initial knowledge of the subgrid scales; the correct statistics are predicted for the flow, but it is impossible to reproduce a specific realization of the flow without this detailed initial condition. Fortunately, for most engineering problems, it is only the statistics that are of interest.

LES techniques date back to the pioneering work of Smagorinsky [1963], who developed an eddy-viscosity subgrid model for use in meteorological problems. The model turned out to be too dissipative for large-scale meteorological problems in which large-scale, predominantly two-dimensional motions are affected by three-dimensional turbulence. Smagorinsky's model finds wide application in engineering problems, however. Details of the LES approach are discussed by Rogallo and Moin [1984], and more recent developments and applications are described by Lesieur and Métais [1996].

30.3.3.3 Direct Numerical Simulations

Direct numerical simulations generally use spectral or pseudospectral approximations for the spatial discretization of the Navier–Stokes equations (see, e.g., Gottlieb and Orszag [1977] or Hussaini and Zang [1987]). The difference between spectral and pseudospectral methods is in the way that products are computed; the advantage of spectral methods, which are more expensive computationally, is that aliasing errors are removed exactly [Orszag 1972].

A description of the issues involved and some results are given by Rogallo and Moin [1984]. Direct numerical simulations are particularly valuable for the insight that they provide into the fundamental nature of turbulent flows. The computational resources required for DNS calculations grow so rapidly with increasing Reynolds number that they are unlikely to be directly useful for engineering predictions, but they will remain an invaluable tool providing insight needed to construct better turbulence models for use in LES and with the Reynolds-averaged equations.

30.4   Research Issues and Summary

Computational fluid dynamics continues to be a field of intense research activity. The development of accurate algorithms based on unstructured grids for problems involving complex geometries, and the increasing application of CFD techniques to unsteady problems, including aeroelasticity and acoustics, are examples of areas of great current interest. Algorithmically, there continues to be fruitful work on the incorporation of adaptive grids that automatically increase resolution in regions where it is required to maintain accuracy, and on the development of inherently multidimensional high-resolution schemes. Finally, the continued expansion of computational capability allows the application of DNS and LES methods to problems of higher Reynolds number and increasingly realistic flow geometries.

Defining Terms

Artificial viscosity: Terms added to a numerical approximation that provide artificial — i.e., nonphysical — dissipative mechanisms to stabilize a solution or to allow shock waves to be captured automatically by the numerical scheme.

CFL condition: The Courant–Friedrichs–Lewy (CFL) condition is a stability criterion that limits the time step of an explicit time-marching scheme for hyperbolic systems of differential equations. In the simplest one-dimensional case, if Δx is the physical mesh spacing and $\rho(\mathbf{A})$ is the spectral radius of the Jacobian matrix $\mathbf{A}$ corresponding to the fastest wave speed for the problem, then the time step for explicit schemes must satisfy the constraint $\Delta t \leq K \, \Delta x / \rho(\mathbf{A})$, where K is a constant. For the simplest explicit schemes, $K = 1$, which implies that the time step must be less than that required for the fastest wave to cross the cell.

Conservative: A numerical approximation is said to be conservative if it is based on the conservation (or divergence) form of the differential equations, and the net flux across a cell interface is the same when computed from either direction; in this way the properties of discontinuous solutions are guaranteed to be consistent with the jump relations for the original integral form of the conservation laws.

Direct numerical simulation (DNS): A solution of the complete Navier–Stokes equations in which all length and time scales, down to those describing the smallest eddies containing significant turbulent kinetic energy, are fully resolved.

Euler equations: The equations describing the inviscid flow of a compressible fluid. These equations constitute a hyperbolic system of differential equations; the Euler equations are nondissipative, and weak solutions containing discontinuities (which can be viewed as approximations to shock waves) must be allowed for many practical problems.

Explicit method: A method in which the solution at each point for the new time level is given explicitly in terms of values of the solution at the previous time level; to be contrasted with an *implicit method*, in which the solution at each point at the new time level also depends on the solution at one or more neighboring points at the new time level, so that an algebraic system of equations must be solved at each time step.

Finite-difference method: A numerical method in which the solution is computed at a finite number of points in the domain; the solution is determined from equations that relate the solution at each point to its values at selected neighboring points.

Finite-element method: A numerical method for solving partial differential equations in which the solution is approximated by simple functions within each of a number of small elements into which the domain has been divided.

Finite-volume method: A numerical method for solving partial differential equations, especially those arising from systems of conservation laws, in which the rate of change of quantities within each mesh volume is related to fluxes across the boundaries of the volume.

Implicit method: See **explicit method**.

Large-eddy simulation: A numerical solution of the Navier–Stokes equations in which the largest, energy-carrying eddies are completely resolved, but the effects of the smaller eddies, which are more nearly isotropic, are accounted for by a subgrid model.

Mach number: The ratio $\mathbf{M} = V/a$ of the fluid velocity V to the speed of sound a. This nondimensional parameter characterizes the importance of compressibility in the dynamics of the fluid motion.

Mesh generation: The generation of mesh systems suitable for the accurate representation of solutions to partial differential equations.

Panel method: A numerical method to solve Laplace's equation for the velocity potential of a fluid flow. The boundary of the flow domain is discretized into a set of nonoverlapping facets, or panels, on each of which the strength of some elementary solution is assumed constant. Equations for the normal velocity at control points on each panel can be solved for the unknown singularity strengths to give the solution. In some disciplines this approach is called the *boundary integral element method* (BIEM).

Reynolds-averaged Navier–Stokes equations: Equations for the mean quantities in a turbulent flow obtained by decomposing the fields into mean and fluctuating components and averaging the Navier–Stokes equations. Solution of these equations for the mean properties of the flow requires knowledge of various correlations (the Reynolds stresses), of the fluctuating components.

Shock capturing: A numerical method in which shock waves are treated by smearing them out with artificial dissipative terms in a manner such that no special treatment is required in the vicinity of the shocks; to be contrasted with *shock fitting* methods in which shock waves are treated as discontinuities with the jump conditions explicitly enforced across them.

Shock fitting: See **shock capturing**.

Shock wave: Region in a compressible flow across which the flow properties change almost discontinuously; unless the density of the fluid is extremely small, the shock region is so thin relative to other significant dimensions in most practical problems that it is a good approximation to represent the shock as a surface of discontinuity.

Turbulent flow: Flow in which unsteady fluctuations play a major role in determining the effective mean stresses in the field; regions in which turbulent fluctuations are important inevitably appear in fluid flow at large Reynolds numbers.

Upwind method: A numerical method for CFD in which upwinding of the difference stencil is used to introduce dissipation into the approximation, thus stabilizing the scheme. This is a popular mechanism for the Euler equations, which have no natural dissipation, but is also used for Navier–Stokes algorithms, especially those designed to be used at high Reynolds number.

Visualization: The use of computer graphics to display features of solutions to CFD problems.

References

Abid, R., Vatsa, V. N., Johnson, D. A., and Wedan, B. W. 1990. Prediction of separated transonic wing flows with non-equilibrium algebraic turbulence models. *AIAA J.* 28:1426–1431.

Baldwin, B. S. and Barth, T. J. 1991. A one-equation turbulence transport model for high Reynolds number wall-bounded flows. *AIAA Paper* 91-0610, 29th Aerospace Sciences Meeting, Reno, NV.

Baldwin, B. S. and Lomax, H. 1978. Thin layer approximation and algebraic model for separated turbulent flows. *AIAA Paper* 78-257, 16th Aerospace Sciences Meeting, Huntsville, AL.

Beam, R. M. and Warming, R. F. 1976. An implicit finite-difference algorithm for hyperbolic systems in conservation law form. *J. Comput. Phys.* 22:87–110.

Briley, W. R. and McDonald, H. 1974. Solution of the three-dimensional compressible Navier–Stokes equations by an implicit technique. In *Lecture Notes in Physics*, Vol. 35, pp. 105–110. Springer–Verlag, New York.

Caughey, D. A. 1982. The computation of transonic potential flows. *Ann. Rev. Fluid Mech.* 14:261–283.

Caughey, D. A. 1987. Diagonal implicit multigrid solution of the Euler equations. *AIAA J.* 26:841–851.

Chapman, D. R. 1979. Computational aerodynamics: review and outlook. *AIAA J.* 17:1293–1313.

Eiseman, P. R. 1985. Grid generation for fluid mechanics computations. *Ann. Rev. Fluid Mech.* 17:487–522.

Emmons, H. W. 1970. Critique of numerical modeling of fluid-mechanics phenomena. *Ann. Rev. Fluid Mech.* 2:15–36.

Garabedian, P. R. 1956. Estimation of the relaxation factor for small mesh size. *Math. Tables Aids Comput.* 10:183–185.

Glowinski, R. and Pironneau, O. 1992. Finite element methods for Navier–Stokes equations. *Ann. Rev. Fluid Mech.* 24:167–204.

Godunov, S. K. 1959. A finite-difference method for the numerical computation of discontinuous solutions of the equations of fluid dynamics. *Mat. Sb.* 47:357–393.

Goldstine, H. H. and von Neumann, J. 1963. On the principles of large scale computing machines. In *John von Neumann, Collected Works*, A. H. Taub, ed. Vol. 5, p. 4. Pergamon Press, New York.

Gottlieb, D. and Orszag, S. A. 1977. *Numerical Analysis of Spectral Methods: Theory and Application*, CBMS-NSF Reg. Conf. Ser. Appl. Math. 26. SIAM, Philadelphia.

Graves, R. A. and Johnson, N. E. 1978. Navier–Stokes solutions using Stetter's method. *AIAA J.* 16:1013–1015.

Hafez, M. M. 1985. Numerical algorithms for transonic, inviscid flow calculations. In *Advances in Computational Transonics*, W. G. Habashi, Ed., pp. 23–58. Pineridge Press, Swansea.

Harten, A. 1983. High resolution schemes for hyperbolic conservation laws. *J. Comput. Phys.* 49:357–393.

Harten, A. 1984. On a class of total-variation stable finite-difference schemes. *SIAM J. Numer. Anal.* 21:1–23.

Harten, A., Engquist, B., Osher, S., and Chakravarthy, S. 1987. Uniformly high order accurate, essentially non-oscillatory schemes III. *J. Comput. Phys.* 71:231–323.

Hess, J. L. 1990. Panel methods in computational fluid dynamics. *Ann. Rev. Fluid Mech.* 22:255–274.

Hussaini, M. Y. and Zang, T. A. 1987. Spectral methods in fluid dynamics. *Ann. Rev. Fluid Mech.* 19:339–367.

Jameson, A. 1979. A multi-grid scheme for transonic potential calculations on arbitrary grids, pp. 122–146. In *Proc. AIAA 4th Comput. Fluid Dynamics Conf.* Williamsburg, VA.

Jameson, A. 1983. Solution of the Euler Equations by a Multigrid Method. *MAE Rep.* 1613, Department of Mechanical and Aerospace Engineering, Princeton University.

Jameson, A. 1992. *Computational Algorithms for Aerodynamic Analysis and Design. Tech. Rep. INRIA 25th Anniversary Conference on Computer Science and Control*, Paris. Princeton University Rep. MAE 1966, Princeton, NJ, Dec.

Jameson, A. and Baker, T. J. 1983. Solution of the Euler equations for complex configurations, pp. 293–302. In *Proc. AIAA Comput. Fluid Dynamics Conf.*

Jameson, A., Schmidt, W., and Turkel, E. 1981. Numerical solution of the Euler equations by finite volume methods using Runge–Kutta time stepping schemes. *AIAA Paper* 81-1259, AIAA Fluid and Plasma Dynamics Conf., Palo Alto, CA.

Launder, B. E., Reese, G. J., and Rodi, W. 1975. Progress in the development of a Reynolds-stress turbulence closure. *J. Fluid Mech.* 68(3):537–566.

Launder, B. E. and Spaulding, D. B. 1972. *Mathematical Models of Turbulence.* Academic Press, London.

Lesieur, M. and Métais, O. 1996. New trends in large-eddy simulations of turbulence. *Ann. Rev. Fluid Mech.* 28:45–82.

Lumley, J. L. 1978. Computational modeling of turbulent flows. *Adv. Appl. Mech.* 18:123–176.

Orszag, S. A. 1972. Comparison of pseudo-spectral and spectral approximation. *Stud. Appl. Math.* 51:253–259.

Reynolds, W. C. 1987. Fundamentals of turbulence for turbulence modeling and simulation, pp. 1–66. In *Lecture Notes for von Karman Institute*, AGARD Lecture Series No. 86, NATO, New York.

Roe, P. L. 1986. Characteristic-based schemes for the Euler equations. *Ann. Rev. Fluid Mech.* 18:337–365.

Rogallo, R. S. and Moin, P. 1984. Numerical simulation of turbulent flows. *Ann. Rev. Fluid. Mech.* 16:99–137.

Shu, C. and Osher, S. 1988. Efficient implementation of essentially non-oscillatory shock-capturing schemes. *J. Comp. Phys.* 77:439–471.

Smagorinsky, J. 1963. General circulation experiments with the primitive equations. *Mon. Weather Rev.* 91:99–164.

Spalart, P. R. and Allmaras, S. R. 1992. A one-equation turbulence model for aerodynamic flows. *AIAA Paper* 92-0439, 30th Aerospace Sciences Meeting, Reno, NV.

Speziale, C. G. 1987. On nonlinear k–ℓ and k–ε models of turbulence. *J. Fluid Mech.* 178:459–475.

Steger, J. L. and Warming, R. F. 1981. Flux vector splitting of the inviscid gasdynamic equations with application to finite-difference methods. *J. Comput. Phys.* 40:263–293.

Tatsumi, S., Martinelli, L., and Jameson, A. 1995. Flux-limited schemes for the compressible Navier–Stokes equations. *AIAA J.* 33:252–261.

Thompson, J. F., Warsi, Z. U. A., and Mastin, C. W. 1985. *Numerical Grid Generation*. North Holland, New York.

van Leer, B. 1974. Towards the ultimate conservative difference scheme, II. Monotonicity and conservation combined in a second-order accurate scheme. *J. Comput. Phys.* 14:361–376.

van Leer, B. 1979. Towards the ultimate conservative difference scheme, V. A second-order sequel to Godunov's scheme. *J. Comput. Phys.* 32:101–136.

van Leer, B. 1982. Flux-vector splitting for the Euler equations. In *Lecture Notes in Phys.* 170:507–512.

Wilcox, D. C. 1993. *Turbulence Modeling for CFD*. DCW Industries, La Cañada, CA.

Yoon, S. and Kwak, D. 1994. Multigrid convergence of an LU scheme, pp. 319–338. In *Frontiers of Computational Fluid Dynamics — 1994*, D. A. Caughey and M. M. Hafez, eds. Wiley-Interscience, Chichester, U.K.

Further Information

Several organizations sponsor regular conferences devoted completely, or in large part, to computational fluid dynamics. The American Institute of Aeronautics and Astronautics (AIAA) sponsors the AIAA Computational Fluid Dynamics Conferences in odd-numbered years, usually in July; the proceedings of this conference are published by AIAA. In addition, there typically are many sessions on CFD and its applications at the AIAA Aerospace Sciences Meeting, held every January, and the AIAA Fluid and Plasma Dynamics Conference, which is held every summer, in conjunction with the AIAA CFD Conference in those years when the latter is held. The Fluids Engineering Conference of the American Society of Mechanical Engineers, held every summer, also contains sessions devoted to CFD. In even-numbered years, the International Conference on Numerical Methods in Fluid Dynamics is held, alternating between Europe and America; the proceedings of this conference are published in the *Lecture Notes in Physics* series by Springer-Verlag. The International Symposium on Computational Fluid Dynamics, sponsored by the CFD Society of Japan, is held in odd-numbered years, alternating between the U.S. and Asia; in September 1997 this meeting will be held in Beijing, China.

The *Journal of Computational Physics* contains many articles on CFD, especially covering algorithmic issues. The *AIAA Journal* also has many articles on CFD, including aerospace applications. The *International Journal for Numerical Methods in Fluids* contains articles emphasizing the finite-element method applied to problems in fluid mechanics. The journals *Computers and Fluids* and *Theoretical and Computational Fluid Dynamics* are devoted exclusively to CFD, the latter journal emphasizing the use of CFD to elucidate basic fluid-mechanical phenomena. The first issue of the *CFD Review*, which attempts to review important developments in CFD, was published in 1995. The *Annual Review of Fluid Mechanics* also contains a number of review articles on topics in CFD.

The following textbooks provide excellent coverage of many aspects of CFD:

Anderson, D. A., Tannehill, J. C., and Pletcher, R. H. 1984. *Computational Fluid Mechanics and Heat Transfer*. Hemisphere, Washington.

Hirsch, C. 1988, 1990. *Numerical Computation of Internal and External Flows, Vol. I: Fundamentals of Numerical Discretization, Vol. II: Computational Methods for Inviscid and Viscous Flows*. Wiley, Chichester.

Wilcox, D. C. 1993. *Turbulence Modeling for CFD*. DCW Industries, La Cañada, CA.

An up-to-date summary on algorithms and applications for high-Reynolds-number aerodynamics is found in Caughey, D. A. and Hafez, M. M., eds. *Frontiers of Computational Fluid Dynamics — 1994*. J. Wiley Chichester, U.K.

31
Computational Ocean Modeling

Lakshmi Kantha
University of Colorado

Steve Piacsek
Naval Research Laboratory

31.1 Introduction **31**-1
31.2 Underlying Principles **31**-2
Global or Regional • Deep Basin or Shallow Coastal
• Rigid Lid or Free Surface • Comprehensive or Purely
Dynamical • With Applications to Short-Term Simulations or
Long-Term Climate Studies • Quasigeostrophic (QG) or
Primitive-Equation-Based • Barotropic or Baroclinic
• Purely Physical or Physical–Chemical–Biological
• Process-Studies-Oriented or Application-Oriented
• With and Without Coupling to Sea Ice • Coupled to the
Atmosphere or Uncoupled
31.3 Best Practices **31**-8
Barotropic Models • z-Level Models • Sigma-Coordinate
Models • Layered Models • Isopycnal Models
• Data Assimilation • Computational Issues
31.4 Nowcast/Forecast in the Gulf of Mexico
(a Case Study) **31**-19
31.5 Research Issues and Summary **31**-21

31.1 Introduction

Oceanography is a relatively young field, barely a century old; major discoveries — such as the reason for the western intensification of currents such as the Gulf Stream and Kuroshio, and the existence of a deep sound channel in which acoustic energy can travel for thousands of kilometers with little attenuation — were not made till the 1940s. Even today, our knowledge of the circulation in the global oceans is rather sketchy. Computational ocean modeling is even younger; the very first comprehensive numerical global ocean model was formulated by Kirk Bryan [1969] in the late 1960s. However, the advent of supercomputers has led to a phenomenal growth in the field, especially in the last decade. In a brief chapter such as this, it is impossible to provide a detailed account of the many different versions of the ocean models that exist at present. Instead we will attempt to provide a bird's-eye view of the field and a detailed account of a selected few. The objective is to provide a road map that enables an interested reader to consult appropriate sources for details of a particular approach.

Oceans act as thermal flywheels and moderate our long-term weather. They are also huge reservoirs of CO_2 and have long memory and therefore play a crucial role in determining the climatic conditions on our planet on a variety of time scales. A better understanding of the oceans is also important for other reasons, including defense and commerce needs of nations. They are a source of protein, and might be able to supply part of our energy and mineral needs in the coming century. However, the oceans are data-poor in general. It is only in the last decade or so that satellite-borne sensors such as infrared radiometers, microwave imagers, and altimeters have begun to fill in the data gaps, especially in the poorly explored southern-hemisphere oceans. Since collection of in situ data in the oceans is quite expensive, and since satellite-borne sensors provide information mostly on the near-surface layers of the ocean, it is often thought that ocean models are central to understanding the way the oceans function. The hope is that comprehensive ocean models in combination with sparse in situ and relatively abundant remotely sensed data will provide the best means of studying and monitoring the oceans. Herein lies the importance of ocean modeling. For prediction purposes, of course, numerical ocean models are quite indispensable.

31.2 Underlying Principles

The choice of a particular ocean model or modeling approach depends very much on the intended application and on the computational and pre- and postprocessing capabilities available. A judicious compromise is essential for success. With this in mind, numerical ocean models can be classified in many different ways.

31.2.1 Global or Regional

The former necessarily requires supercomputing capabilities, whereas it may be possible to run the latter on powerful modern workstations. Even then, the resolution demanded (grid sizes in the horizontal and vertical) is critical. A doubling of the resolution in a three-dimensional model often requires an order of magnitude increase in computing (and analysis) resources. It is therefore quite easy to overwhelm even the most modern supercomputer (or workstation), whether it be a coarse-grained multiple CPU vector processor such as a Cray C-90 or a modern massively parallel machine such as a CM5 or Cray T3-D, irrespective of whether the model is global or regional. Regional models have to contend with the problem of how to inform the model about the state of the rest of the ocean, in other words, of prescribing suitable conditions along the open boundaries. Often the best solution is to nest the fine-resolution regional model in a coarse-resolution model of the basin.

31.2.2 Deep Basin or Shallow Coastal

The prevailing physical processes and the underlying driving mechanisms are essentially different for the two cases. Circulation in shallow coastal regions is highly variable, driven primarily by synoptic wind and other rapidly changing surface forcing (and, near river outflows, by buoyancy differences between the fresh river water and saline ambient shelf water). Wind mixing at the surface and benthic processes are impor-tant, and a numerical model that has reliable mixing physics and that resolves the benthic boundary layer is therefore better suited for coastal applications. A model such as the one developed at Princeton University, which employs a bottom-following, sigma-coordinate vertical grid and incorporates an advanced turbu-lence closure [Blumberg and Mellor 1987; Kantha and Piacsek 1993], may be essential for such applications.

Deep basins, on the other hand, are comparatively sluggish, and the horizontal density gradients, especially below the wind-mixed upper layers, are a dominant factor in the circulation. The upper mixed layer can often be modeled less rigorously, especially for applications that do not require consideration of air–sea interaction processes. The popular z-level Geophysical Fluid Dynamics Laboratory Modular Ocean Model (MOM), with or without an upper mixed layer, is a good candidate for modeling the ocean basins (and deep marginal and semienclosed seas) on a variety of time scales ranging from synoptic to climate. The very first global ocean model [Bryan 1969], on which many modern global models are based [Semtner and Chervin 1992], was a z-level model without an upper mixed layer.

In a z-level model, a number of horizontal levels are defined in the water column, and the equations are written for the oceanic variables at each level and each point on the model grid in the horizontal and solved. This is an Eulerian approach. Another equally viable approach is a semi-Lagrangian approach [Hurlburt and Thompson 1980] that divides the ocean vertically into a number of layers and models the variation in properties such as the thickness and density of each layer at each grid point on the horizontal grid. More recently developed isopycnal models [Oberhuber 1993, Bleck and Smith 1990] belong to this category. Since mixing in the deep oceans is primarily along **isopycnals** (density surfaces), isopycnal models perform a better job of depicting interior mixing and are ideal for long-term simulations of circulation in the global oceans.

31.2.3   Rigid Lid or Free Surface

Oceanic response to surface forcing can often be divided into two parts: fast barotropic response mediated by external Kelvin and gravity waves on the sea surface, and relatively slow baroclinic adjustment via internal gravity and other waves. On long time scales, it is the internal adjustment that is important to model and it is possible to suppress the external gravity waves by imposing a *rigid lid* on the free surface. This permits larger time-stepping of the model, and models used for climatic-type simulations are usually the rigid-lid kind. The very first global ocean model [Bryan 1969] was a rigid-lid model. At each time step an elliptic Poisson equation for the stream function has to be solved. This is difficult to carry out efficiently on vector and parallel processors and for complicated basin shapes (including islands). Also, under synoptic forcing, the convergence of the iterative solver slows down. For these reasons, free-surface models are becoming more popular for nonclimatic simulations. A mode-splitting technique must then be employed to circumvent the severe limitation on time-stepping that would otherwise be imposed. For shallow-water applications, such as storm-surge and tide modeling, free-surface dynamics must be retained. To diminish the drawbacks of a rigid-lid model, Dietrich et al. [1987] and Dukowicz and Smith [1994] have developed versions in which one works with the pressure on the rigid lid, rather than the barotropic stream function, leaving the domain multiply connected and with better matrix inversion characteristics.

31.2.4   Comprehensive or Purely Dynamical

Since the density gradients are overwhelmingly important and the density below the upper layers in the global oceans changes very slowly, it is often possible to ignore completely the changes in density with time. The model then becomes purely dynamical and can be used to explore the consequences of changing wind forcing at the surface. Purely dynamical layered models, originally developed at Florida State University [Metzger et al. 1992; Wallcraft 1991], belong to this category and are essentially isopycnal models without the thermodynamic component [Hurlburt and Thompson 1980]. Their principal advantage is that it is often possible to select a limited number of layers in the vertical (as few as two) and still include salient dynamical processes. This enables very high horizontal resolutions necessary for resolving mesoscale eddies in the oceans to be afforded. The highest-resolution global model at present is the Naval Research Laboratory $\frac{1}{8}^\circ$ eddy-resolving global model [Metzger et al. 1992] that needs a 16-processor Cray C-90 for multiyear simulations.

Even with modern computing power, it is necessary to sacrifice either vertical or horizontal resolution for many (especially global) simulations. The layered (and isopycnal) models sacrifice vertical resolution, whereas the z-level models, employing large numbers of levels in the vertical, are necessarily comparatively coarse-grained in the horizontal. The highest-resolution dynamical–thermodynamic z-level model at present is the $\frac{1}{6}^\circ$ Semtner model at Los Alamos (Dukowicz and Smith 1994; see Semtner and Chervin 1992 for a description of the basic model) and it stretches the capability of a 256-processor CM5 to the limit.

31.2.5   With Applications to Short-Term Simulations or Long-Term Climate Studies

On climatic time scales, it is extremely important to model correctly the thermohaline circulation driven by the formation of dense deep-water masses during strong wintertime cooling in subpolar seas, especially in the Atlantic. These water masses flow along the ocean bottom, and several centuries are needed for a water particle that sank (say) in the north Atlantic to surface again in the Indian or the Pacific Ocean.

Because of this long memory of the deep oceans, it is necessary to make multicentury simulations, and, irrespective of whether isopycnal or z-level models are employed, the horizontal and vertical resolutions that can be afforded are necessarily coarse. Accurate ocean simulations on climatic time scales belong to the category of grand challenge problems requiring a teraflop (10^{12} floating-point operations per second) computing capability that has been the holy grail of the computer industry.

31.2.6 Quasigeostrophic (QG) or Primitive-Equation-Based

In the 1970s and early 1980s, the limited computing power available led some to explore simplifications to the governing equations to be solved. QG models assume that there is a near-balance between the Coriolis acceleration and pressure gradient in the dynamical equations in the rotating coordinate frame of reference in which most ocean models are formulated. The resulting simplification enables higher vertical and horizontal resolutions to be achieved. QG models have strong limitations with respect to the accuracy with which physical processes are depicted and are becoming obsolete in the modern high-computing-power environment. We will not discuss QG models in this article, but instead refer the reader to Holland [1986]. Neither shall we discuss intermediate models, which are in between QG and primitive equation (PE) models in complexity.

31.2.7 Barotropic or Baroclinic

In the former, the density gradients are neglected and therefore the currents become independent of the depth in the water column. Many phenomena, such as tidal sea-surface elevation fluctuations and storm surges, can be simulated quite adequately by a **barotropic** model, which is a two-dimensional (in the horizontal) model based on the vertically integrated equations of motion. The advantage is that a barotropic model requires an order of magnitude less computing resources than a comparable baroclinic model. However, when it is important to model the vertical structure of currents, or the density field, a fully three-dimensional **baroclinic** model is necessary.

31.2.8 Purely Physical or Physical–Chemical–Biological

Often there is a need to model the fate of chemical and biological constituents in the ocean, and to do so it is essential to include not only the dynamical equations governing the circulation and other physical variables, but also the conservation equations for chemical and biological variables. Modeling the fate of inorganic CO_2 in the oceans (a problem germane to global warming) and modeling the primary production in the upper layers of the ocean are two such examples. The former requires solving for at least two more variables, the total CO_2 and alkalinity, whereas the simplest biological model must solve for at least three additional quantities, the nutrient, phytoplankton, and zooplankton concentrations (the so-called NPZ model). The governing equations are transport equations with appropriate source and sink terms, whose parametrization is not always quite straightforward. This not only implies additional complexity but also requires considerably more computing (and data) resources.

31.2.9 Process-Studies-Oriented or Application-Oriented

Models that are used to study some salient processes (for example western boundary currents and gyre circulation) can often be considerably simplified and hence made less computationally (and observational data-) intensive, since it is often possible to isolate and retain only the relevant physical process and ignore the rest in the model. Also, such models may not require extensive observational data for model initialization or forcing. They are most often run free in a predictive (or prognostic) mode. On the other hand, application-oriented models such as those used for ocean prediction purposes require extensive observational data for realistic initialization, forcing, and data assimilation. Assimilation into the model by one means or other of observational data is indispensable for **nowcast**, forecast, and **hindcast** applications, and data-assimilative models often employ approaches very much similar to those employed by numerical weather prediction (NWP) models in the atmosphere.

31.2.10 With and Without Coupling to Sea Ice

Global ocean models do not at present include sea ice. Ice-ocean-coupled basin models of the Arctic exist, however; for coupling to a z-level ocean model, see Hibler and Bryan [1987], and to an isopycnal model, see Oberhuber [1993]. Sea ice insulates the ocean from the cold atmosphere and mediates the exchange of heat and momentum between the two, and therefore such models involve solving dynamical and thermodynamic equations for the sea ice cover and its coupling to the underlying ocean.

31.2.11 Coupled to the Atmosphere or Uncoupled

Finally, for accurate simulation of long-time-scale processes, it is essential to couple ocean models with atmospheric models. Such coupled models are being increasingly used for such things as forecasting **El Niño** events. Most often, either the atmosphere or the ocean is highly simplified in such models, although modern supercomputers are enabling comprehensive atmospheric general circulation models to be coupled to global ocean models, at least for simulations of interannual variability. Truly comprehensive coupled models with applications to long-term climate studies require teraflop computing capability that might be routinely available in the coming century.

All numerical ocean models solve one form or other of the same governing equations for oceanic motions, written in a rotating coordinate frame of reference. These equations are essentially Navier–Stokes equations (or more appropriately **Reynolds-averaged** equations for mean term quantities, since the flow is invariably turbulent), but with the buoyancy and Coriolis force terms (fictitious accelerations due to the noninertial nature of the rotating coordinate frame) prominent in the dynamical balance. In addition, an equation of state relating the density of seawater to its temperature and salinity, and conservation equations for temperature and salinity, are also solved. In those models which model turbulence explicitly, equations for turbulence quantities such as the turbulence velocity scale (or equivalently turbulence kinetic energy) and turbulence macroscale are also solved. If chemical or biological components are included, conservation equations for relevant species with appropriate source and sink terms are solved as well.

Global and basin-scale models are formulated in spherical coordinates, but regional models are usually cast in rectangular coordinates instead. For simplicity we present the governing equations in rectangular Cartesian coordinates (spherical coordinate version can be found for example in Semtner [1986]). The x_1-axis is usually taken to be in the zonal direction (positive to the east), the x_2-axis is in the meridional direction (positive to the north), and the z-axis is in the vertical direction, positive upwards, with the origin located at the sea surface. Using tensorial notation and treating the horizontal coordinates separately from the vertical (indices take values 1 and 2 only), the governing equations consist of the continuity equation

$$\frac{\partial U_k}{\partial x_k} + \frac{\partial W}{\partial z} = 0 \tag{31.1}$$

where U_k denotes the horizontal components of velocity and W the vertical, and the momentum equations

$$\frac{\partial U_j}{\partial t} + \frac{\partial}{\partial x_k}(U_k U_j) + \frac{\partial}{\partial z}(W U_j) + f \varepsilon_{j3k} U_k = -\frac{\partial P}{\partial x_j} + \frac{\partial \Phi}{\partial x_j} + \frac{\partial}{\partial z}\left(K_M \frac{\partial U_j}{\partial z}\right) + F_j \tag{31.2}$$

$$\frac{\partial P}{\partial z} = -\frac{\rho}{\rho_0} g \tag{31.3}$$

where f is the Coriolis parameter ($2\,\Omega \cos \phi$), Ω the earth's angular velocity, ϕ the latitude, P the kinematic pressure, K_M the vertical mixing coefficient, ρ the in situ density, ρ_0 the reference density, g the gravitational acceleration, and Φ the gravitational potential.

The transport equations for **potential temperature** Θ and salinity S are

$$\frac{\partial \Theta}{\partial t} + \frac{\partial}{\partial x_k}(U_k \Theta) + \frac{\partial}{\partial z}(W\Theta) = \frac{\partial}{\partial z}\left(K_H \frac{\partial \Theta}{\partial z}\right) + S_\Theta + F_\Theta \tag{31.4}$$

$$\frac{\partial S}{\partial t} + \frac{\partial}{\partial x_k}(U_k S) + \frac{\partial}{\partial z}(WS) = \frac{\partial}{\partial z}\left(K_H \frac{\partial S}{\partial z}\right) + F_s \tag{31.5}$$

where K_H is the vertical mixing coefficient for scalar quantities, and S_Θ denotes a volumetric heat source such as due to penetrative solar heating. The equation of state is given by

$$\rho = \rho(\Theta, S) \tag{31.6}$$

Various simplifications have been made in deriving the above equations. The ocean is considered incompressible, a very good approximation for most applications. It is also considered to be in hydrostatic balance in the vertical, and hence the only terms remaining in the vertical component of the momentum equations are the buoyancy and pressure gradient terms. The hydrostatic approximation involves neglecting the vertical acceleration and regarding the fluid as essentially motionless in the vertical. In addition, the Boussinesq approximation has been used. This involves considering the ocean to be of constant density except when buoyancy forces are computed, thus assuming $\rho = \rho_0$ in all except the terms involving gravitation. Also, terms containing the horizontal component of rotation are neglected. The resulting equations are sufficiently accurate for most ocean circulation modeling. The equation of state employed is the so-called UNESCO equation [Pond and Pickard 1979] and is in the form of polynomial expansions in temperature and salinity. The pressure P is given by

$$P = g\eta + \frac{g}{\rho_0}\int_z^0 \rho(x_j, z', t)\,dz' \tag{31.7}$$

where η is the sea surface height. The terms F_j, F_Θ, F_S are horizontal mixing terms corresponding to unresolved subgrid-scale processes and are most often parametrized simply as Laplacian diffusion terms:

$$F_{j,\Theta,S} = A_{M,H}\frac{\partial^2}{\partial x_k \partial x_k}(U_j, \Theta, S) \tag{31.8}$$

where $A_{M,H}$ are horizontal mixing coefficients. A more rigorous form for these terms can be found in Blumberg and Mellor [1987]. The values for these coefficients are most often chosen as constants in a rather ad hoc manner based on purely numerical considerations. While the vertical mixing coefficients K_M and K_H can be rigorously modeled by turbulence closure theories [for example, Kantha and Clayson 1994, Mellor and Yamada 1982], there does not exist a similar approach for these terms. One approach, widely used in atmospheric modeling, is that due to Smagorinsky [1963], which is similar to the classical mixing-length theory of turbulence. Here the mixing coefficient is assumed to be proportional to the mean strain rate, so that

$$A_M = C(\Delta x_1)(\Delta x_2)\left[\left(\frac{\partial U_i}{\partial x_j} + \frac{\partial U_j}{\partial x_i}\right)\left(\frac{\partial U_i}{\partial x_j} + \frac{\partial U_j}{\partial x_i}\right)\right]^{1/2}, \quad A_H \sim A_M \tag{31.9}$$

where C is the Smagorinsky coefficient, with a value of around 0.04, and Δx_1 and Δx_2 are the grid sizes. This approach assumes that the subgrid scales fall within the Kolmogoroff inertial subrange, an assumption not always satisfied. A practical consequence of using this model is that strong horizontal shear is accompanied by strong horizontal mixing, which tends to smear out thermal fronts. A more general approach is to assume that the mixing coefficients are a sum of a constant background value and the Smagorinsky value given by Equation 31.9 and to choose the values assigned to each appropriately [for example, Kantha 1995]. Another approach is to assign a constant cell Reynolds number $R_N = |U_j||\Delta x_j|/A_M$ and determine the value of the mixing coefficient thus [Choi and Kantha 1995]. Some modelers have used biharmonic form

$$\frac{\partial^2}{\partial x_k \partial x_k} \frac{\partial^2}{\partial x_k \partial x_k}$$

to model these terms [O'Brien 1985]. In this form, the terms serve principally to control the so-called 2 Δx noise in the numerical solutions. Suffice it to say that modeling horizontal diffusion terms is still rather ad hoc.

The oceans are driven by momentum, heat, and salt fluxes at the air–sea interface. The boundary conditions at the sea surface ($z = \eta$) are therefore

$$K_M\left(\frac{\partial U_j}{\partial z}\right) = \tau_{0j}$$

$$K_H\left(\frac{\partial}{\partial z}(\Theta, S)\right) = (q_H, q_S) \tag{31.10}$$

$$W = \frac{\partial \eta}{\partial t} + U_j \frac{\partial \eta}{\partial x_j}$$

where τ_{0j} is the kinematic shear stress acting at the free surface due to the action of winds and waves (taken mostly as equal to the kinematic wind stress) and $q_{H,S}$ are the kinematic heat and salt fluxes. The value of q_H is determined by the net heat balance at the air–sea interface due to short-wave and long-wave solar heating, back radiation by the ocean surface, and the turbulent sensible and latent heat exchanges, and that of q_S is determined by the difference between evaporation and precipitation. Accurate parametrization of these air–sea fluxes has been the subject of intense research for several decades (for example, the 1992 multinational Tropical Ocean Global Atmosphere/Coupled Ocean Atmosphere Response Experiment).

The conditions at the ocean bottom ($z = -H$) are of no mass transfer through the bottom

$$W = -U_j \frac{\partial H}{\partial x_j}$$

and

$$K_M\left(\frac{\partial U_j}{\partial z}\right) = \tau_{bj}$$

$$K_H\left(\frac{\partial}{\partial z}(\Theta, S)\right) = (0, 0) \tag{31.11}$$

The last of the above conditions implies no heat or salt transfer through the ocean bottom. The bottom stresses are usually parametrized using a quadratic drag law with $c_d \sim 0.0025$:

$$\tau_{bj} = c_d |U_{bj}| U_{bj} \tag{31.12}$$

or by assuming that the lowest model grid point falls within the logarithmic-law region and using the well-known logarithmic relationship between the mean velocity and the friction velocity to derive the drag coefficient [Blumberg and Mellor 1987, Kantha and Piacsek 1993, for example].

When additional quantities such as turbulence velocity and length scales are modeled explicitly, corresponding conservation equations need to be solved along with appropriate boundary conditions at the ocean surface and bottom [Blumberg and Mellor 1987, Kantha and Clayson 1994, Kantha and Piacsek 1993]. The same holds for modeling of chemical and biological quantities.

If the lateral boundary is closed, then it is straightforward to apply the lateral boundary condition; the component of velocity perpendicular to the boundary is zero, and there is no lateral mass, heat, or salt flux through the boundary. If it is open, on the other hand, as is usual in regional models of coastal and marginal seas, then it is necessary to prescribe open boundary conditions. This is a difficult problem, since complete information on various flow properties must be specified, and this depends to a large extent on how well the flow at the boundary is known. The best strategy is to nest the model in a coarser-resolution model of the basin. In many cases, this is not feasible, and hence it is not possible to inform the model about what the rest of the ocean is doing. The best under these conditions is some form of Sommerfeld radiation boundary condition on dynamical quantities, which ensures that disturbances approaching the boundary from the inside are radiated out and not bottled up [Blumberg and Kantha 1985, Kantha et al. 1990, Roed and Cooper 1986]. This is usually of the form

$$\frac{\partial \zeta}{\partial t} + C \frac{\partial \zeta}{\partial x_n} = 0 \tag{31.13}$$

where ζ is a variable such as the sea surface height, n denotes direction normal to the boundary, and C is the phase speed of the approaching disturbance. Proper prescription of C is important to the success of the radiative boundary condition and has been the subject of much research [Blumberg and Kantha 1985, Orlonski 1976].

If there is inflow at the lateral boundary, temperature and salinity of the incoming flow must be prescribed. If there is outflow, on the other hand, these quantities are simply advected out:

$$\frac{\partial (\Theta, S)}{\partial t} + U_n \frac{\partial (\Theta, S)}{\partial x_n} = 0 \tag{31.14}$$

31.3 Best Practices

In many applications, ocean simulation is an initial-value problem and therefore the initial state of the modeled ocean must be prescribed as accurately as possible. This is difficult to do in practice, since observational data are insufficient to specify the state of the ocean at any given point in time. The best alternative is to prescribe some sort of a climatological average as the initial state and spin up the ocean from that state under prescribed surface forcing. Here appeal is made to databases such as [Levitus 1982] (updated in 1995) and the U.S. Navy GDEM that contain distributions of climatological average temperature and salinity in the global oceans derived from historical archives of in situ observations.

Prescription of surface forcing is itself a major problem. Both momentum and buoyancy fluxes need to be specified at the sea surface, and both are determined by processes in the adjacent atmospheric boundary layer. For determining the long-term average state of the oceans, it is once again possible to appeal to climatological databases such as that due to Hellerman and Rosenstein [1983] and the Comprehensive Ocean-Atmosphere Data Set (COADS) database (the database has been updated to 1990s recently), which provide gridded monthly average values of wind stress over the global oceans derived from historical marine surface observations. However, for many applications, including ocean prediction, it is necessary to provide surface forcing on a daily or even a multihourly basis. This is impossible to do from observations alone (although satellite-sensed air–sea fluxes may help in the future), and one has little choice but to appeal to six-hourly analyses and predictions of NWP centers, even though the accuracy of the surface forcing so derived depends very much on the skill of the particular NWP model and methodology.

Finally, the governing equations are discretized and the resulting finite-difference equations (or a set of algebraic equations) are solved to determine the evolution with time of the various oceanic variables such as temperature, salinity, and velocity at each model grid point. There are two principal means of discretization in the horizontal direction: finite differences [for example, Blumberg and Mellor 1987, Choi and Kantha 1995, Semtner and Chervin 1992] and finite elements [Le Provost et al. 1994], the latter being quite popular with civil engineers. The principal advantage of the latter is the flexibility the finite-element grid provides in assigning localized high resolution where needed. The reader is referred to Kantha [1995] for an example of a finite-difference approach to modeling the barotropic equations for global tides and to Le Provost et al. [1994] for a finite-element approach to the same problem. Advanced discretization techniques such as adaptive grids and nonorthogonal coordinate systems that are quite routinely used in conventional computational fluid dynamics are still in the developmental stage in ocean modeling; their efficacy is still largely unproven.

The finite-difference grid can be staggered or nonstaggered, with the former being more accurate. There are several possibilities [Mesinger and Arakawa 1976], including the so-called Arakawa C-grid, where the velocity component U_1 is displaced half a grid to the west and U_2 half a grid to the south of the grid center where all scalar quantities such as η, Θ, and S reside. The C-grid has better wave propagation characteristics if the grid size is smaller than the Rossby radius of deformation. The Arakawa B-grid, where both velocities are displaced half a grid point to the west, is better if it is larger [Semtner 1986]. With increased computing power and hence finer resolution, C-grid is becoming more popular.

Explicit or implicit methods can be used for time-stepping the equations; the latter are more efficient but require more complex simultaneous solution at all model grid points. The former are more easily adapted to massively parallel processors and are being increasingly used despite the limitation imposed by numerical stability considerations. The maximum time step that can be taken in an explicit scheme (for a staggered grid) is given by the Courant–Friedrichs–Lewy (CFL) condition, which is of the form

$$\Delta t \le 0.5(\Delta x_e / C_e) \tag{31.15}$$

where

$$\Delta x_e = \left(1/(\Delta x_1)^2 + 1/(\Delta x_2)^2\right)^{-1/2}$$

is the effective grid size, which is smaller than the grid size in the individual directions, and C_e is the effective gravity-wave speed, which is the sum of the gravity-wave speed and the advection velocity. In the barotropic problem, for example, $C_e = \max[|U_j| + \sqrt{gH}]$.

Explicit inclusion of the free-surface dynamics in a model requires that a mode-splitting technique [Blumberg and Mellor 1987, Kantha and Piacsek 1993, Madala and Piacsek 1977] be employed to overcome the severe limitations on the solution due to stability considerations imposed by fast-moving external gravity waves on the free surface. This technique consists essentially of splitting the solution into barotropic and baroclinic modes, with the barotropic part solved at the time step dictated by external gravity waves and the baroclinic part at a much larger time step, 20 to 50 times larger. This approach takes into account the fact that internal baroclinic adjustments are much slower.

It is the discretization of the vertical coordinate that is the most distinguishing feature of various ocean models. Several choices are possible, including that of no discretization (for a barotropic model). We will describe these next.

31.3.1 Barotropic Models

If the density gradients are neglected in the governing equations, or alternatively the ocean is considered to be of uniform density, the current distribution in the vertical becomes independent of depth (away from regions of frictional influence such as the surface and the bottom). Under these conditions, it is possible to ignore the transport equations for Θ and S and integrate the governing equations for continuity and momentum over the water column to arrive at a vertically integrated set of equations that govern the sea-surface elevation η and the vertically averaged velocity components $\overline{U}_j$:

$$\frac{\partial \eta}{\partial t} + \frac{\partial}{\partial x_k}(\overline{U}_k D) = 0$$

$$\frac{\partial}{\partial t}(\overline{U}_j D) + \frac{\partial}{\partial x_k}(\overline{U}_j \overline{U}_k D) + f\varepsilon_{j3k}(\overline{U}_k D) = -gD\frac{\partial \eta}{\partial x_j} - D\frac{\partial P_a}{\partial x_j} \tag{31.16}$$

$$+ gD\frac{\partial \xi}{\partial x_j} + (\tau_{0j} - \tau_{bj}) + D\overline{F}_j$$

where $D = H + \eta$ is the total depth of the water column.

The bottom friction is now determined using the column average velocity $\overline{U}_j$. Note the presence of tidal potential terms involving ξ on the right-hand side of the momentum equations that contain astronomical forcing terms due to the gravitational forces of the moon and the sun. Note also the terms due to atmospheric pressure and wind stress forcing. The astronomical forcing can be prescribed a priori from a knowledge of the ephemerides of the sun and the moon [see, for example, Kantha 1995, Schwiderski 1980]. The atmospheric forcing terms are also known and can be prescribed as a function of time during the model run. This set of equations can be used to solve for the sea surface height (SSH) and depth-averaged currents due to phenomena such as tides and storm surges.

Figure 31.1 shows an example of the application of barotropic equations to the problem of deducing the tidal SSH in the global oceans. The reader is referred to Kantha [1995] for details, but, briefly, the equations are cast in spherical coordinates, and the tidal potential terms are expressed as a sum of a series containing various tidal components such as the semidiurnal M_2, with a period of 12.42 h, and the diurnal K_1, with a period of 23.93 h (the atmospheric forcing terms are zeroed out for this application). The resulting equations are solved on a $\frac{1}{5}^\circ$ latitude–longitude C-grid covering the global oceans (excluding the Arctic) for each tidal component. The bottom depths over the model grid are derived from a digital database (ETOP05 from NOAA) containing world topography at $\frac{1}{12}^\circ$ resolution. However, for the results to be accurate enough for certain applications such as altimetry, inevitable errors that result from inaccurate knowledge of bottom depths and friction coefficients have to be overcome by data assimilation. Tidal SSHs can be derived in the deeper parts of the oceans quite accurately from measurements of SSH fluctuations by a satellite-borne microwave altimeter. The tidal SSH data derived from the currently operational NASA/CNES TOPEX/Poseidon precision **altimeter** [Desai and Wahr 1995] have been assimilated into the model as well as those from coastal tide gauges around the world's coastlines. A simple data assimilation scheme has been used where, at each time step, the model-predicted SSH is replaced by a weighted sum of the model SSH and the observed SSH, with weights determined a priori. The result is tidal SSH that is accurate to within a few centimeters over the global oceans, including shallow coastal and semienclosed seas. This information is useful for many applications, such as an accurate determination of the subtidal SSH variability from altimetric data, gravimetry, and determination of tidal dissipation. Figure 31.1 shows the M_2 **coamplitude and cophase** (with respect to Greenwich) distributions of the tidal SSH and the tidal-current ellipses over the global oceans. Figure 31.2 shows the accuracy attained by this data-assimilative tidal model in the form of scatterplots of comparison of modeled and observed tides from an independent set of accurate tide and bottom-pressure gauges over the global oceans, whose locations are also shown.

Barotropic models such as these can also be used to study the response of the SSH to atmospheric pressure forcing [Kantha et al. 1994, Ponte 1994]. It is often assumed that the ocean responds instantaneously to pressure forcing as an inverse barometer with roughly one centimeter of increase (decrease) for every millibar of drop (rise) in atmospheric pressure. This is not always true, and the departures from the inverse-barometer response are quite important to satellite ocean altimetry [Kantha et al. 1994].

Finally, a very important application of barotropic models is for prediction of storm surge effects along a coastline due to approaching hurricanes. The strong hurricane-force winds (augmented by the pressure drop in the eye of the hurricane) pile up water against the coast that often leads to an increase in sea level of several meters and consequent inundation of structures along the coastline. Hurricane Camille in 1969 caused a storm surge of nearly 8 m along the Mississippi coast, leading to widespread destruction and devastation. Provided the local bathymetry is known accurately and the characteristics of the hurricane (such as the wind stress distribution and forward velocity) can be deduced reasonably well from NWP forecasts, it is possible to predict the resulting storm surge quite accurately using a barotropic model driven by the wind stress and atmospheric pressure terms on the right-hand side.

31.3.2 z-Level Models

The Bryan–Cox–Semtner z-level model [Bryan 1969, Cox 1985, Semtner and Chervin 1992] is the oldest and the most popular global ocean model. Several versions exist, including the Modular Ocean Model (MOM) from the Geophysical Fluid Dynamics Laboratory in Princeton, New Jersey, the latest version

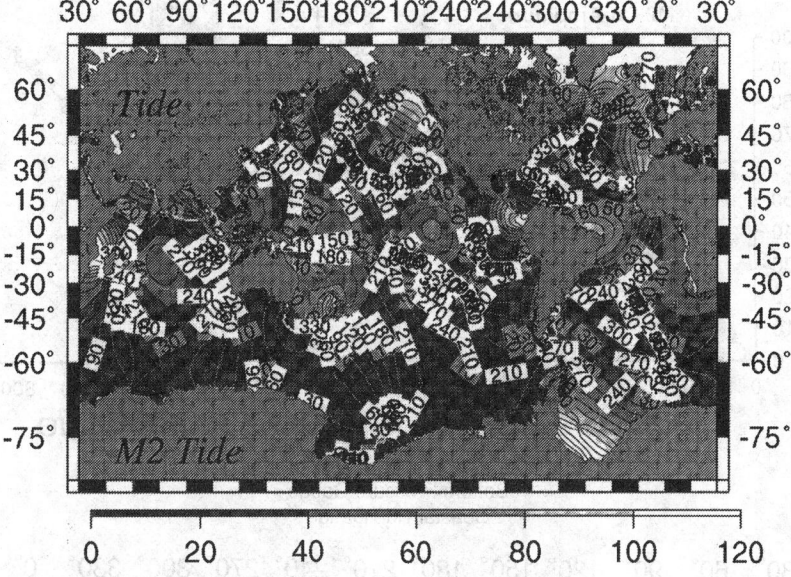

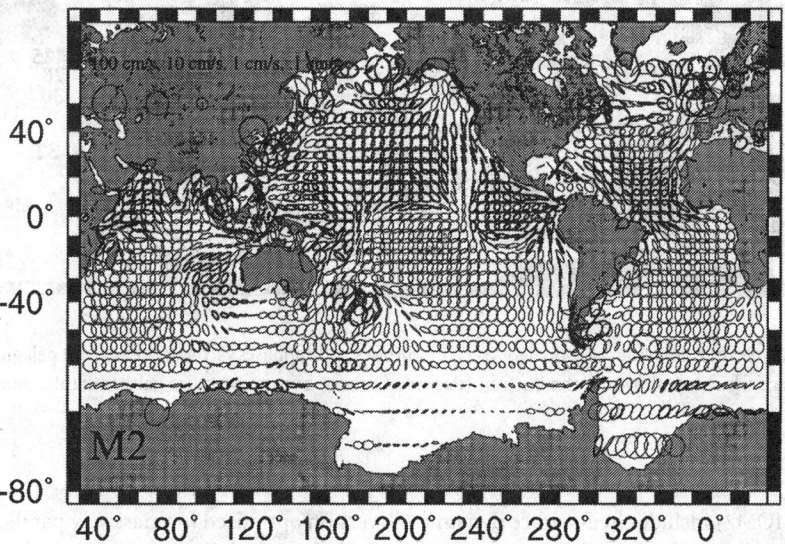

FIGURE 31.1 A map of the distribution of coamplitude and cophase (top), and tidal-current ellipses plotted every 25th point in each direction (bottom) for the M_2 tidal component in the global oceans. Note the logarithmic scale for ellipses.

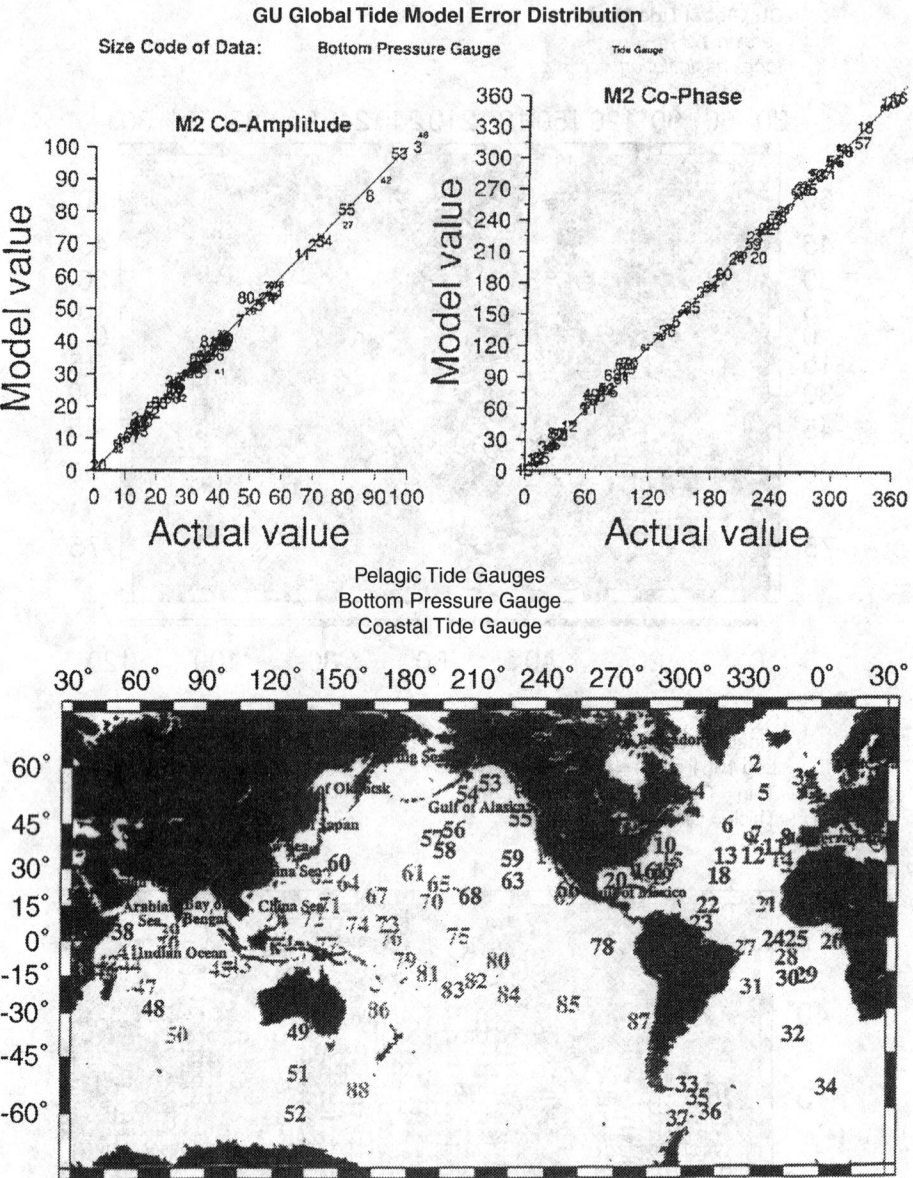

FIGURE 31.2   Scatterplots (top) of modeled M_2 coamplitudes and cophases vs. those observed at pelagic tidal stations, the locations of which are shown at the bottom (darker numbers: bottom pressure gauges; lighter ones: coastal tide gauges).

of which (MOM2) includes free-surface dynamics. A version optimized for massively parallel processors, called POPS (Parallel Ocean Prediction System), is available from Los Alamos Scientific Laboratory [Smith et al. 1992]. Recent improvements include inclusion of a free surface in a variety of versions around the world [Killworth et al. 1991, Dukowicz and Smith 1994], and adoption of a C-grid. A recent review of the current state of ocean modeling using z-level models can be found in Semtner [1995].

Imposition of a rigid lid (via the boundary condition $W = 0$ at $z = 0$) means that the pressure at $z = 0$ enters as an unknown. It is eliminated from Equation 31.2 by cross-differentiation, and an equation for the stream function ψ for vertically integrated transport in the water column is derived (for details, see

Semtner [1986]). This is an elliptic equation and is solved subject to conditions imposed on lateral ocean boundaries, which are in general multiply connected. Herein lies the principal problem with rigid-lid models. While they are efficient, the solution technique is more complicated and not easily adapted to vector and parallel processors. The problem is alleviated somewhat by not cross-differentiating to derive the stream function, but working with the pressure on the rigid lid (Section 31.2, "Rigid Lid or Free Surface").

Numerous applications of the z-level Bryan–Cox–Semtner model and its various versions can be found in the literature (for example in the *Journal of Physical Oceanography* and the *Journal of Geophysical Research, Oceans*). It has been used extensively to study the seasonal, interannual, and climatic variations in the global oceans. It is also a central part of the ocean analysis system for the tropical oceans [Leetma and Ji 1989], where a best estimate of the state of these oceans is determined by assimilation of observational data into a tropical-ocean version of the model. The most recent application can be found in [Semtner and Chervin 1992]. The highest-resolution global z-level model at present is the $\frac{1}{6}^\circ$ POPS model at Los Alamos that is run on a 256-node CM5 (A. Semtner, personal communication).

31.3.3 Sigma-Coordinate Models

Governing Equations 31.1 to 31.5 can be cast in a bottom-topography-following coordinate system by defining a new variable $\sigma = (z-\eta)/(H+\eta)$ and transforming the equations to the new coordinate system [Blumberg and Mellor 1987]; (see also Kantha and Piacsek [1993] for the general orthogonal curvilinear coordinate form):

$$\frac{\partial \eta}{\partial t} + \frac{\partial (U_k D)}{\partial x_k} + \frac{\partial \omega}{\partial \sigma} = 0 \tag{31.17}$$

$$\frac{\partial (U_j D)}{\partial t} + \frac{\partial}{\partial x_k}(U_k U_j D) + \frac{\partial}{\partial \sigma}(\omega U_j) + f\varepsilon_{j3k}U_k D = -D\frac{\partial P}{\partial x_j} + D\frac{\partial \Phi}{\partial x_j} + \frac{\partial}{\partial \sigma}\left(\frac{K_M}{D}\frac{\partial U_j}{\partial \sigma}\right) + DF_j \tag{31.18}$$

$$\frac{\partial P}{\partial \sigma} = -\frac{\rho}{\rho_0}gD \tag{31.19}$$

$$\frac{\partial (\Theta D)}{\partial t} + \frac{\partial}{\partial x_k}(U_k \Theta D) + \frac{\partial}{\partial \sigma}(\omega \Theta) = \frac{\partial}{\partial \sigma}\left(\frac{K_H}{D}\frac{\partial \Theta}{\partial \sigma}\right) + DS_\Theta + DF_\Theta \tag{31.20}$$

$$\frac{\partial (SD)}{\partial t} + \frac{\partial}{\partial x_k}(U_k SD) + \frac{\partial}{\partial \sigma}(\omega S) = \frac{\partial}{\partial \sigma}\left(\frac{K_H}{D}\frac{\partial S}{\partial \sigma}\right) + DF_S \tag{31.21}$$

where $D = H + \eta$, the total depth of the water column, and ω is the pseudo vertical velocity in the new coordinate system, zero at the ocean surface ($\sigma = 0$) and the bottom ($\sigma = -1$).

These equations, along with corresponding conservation relations for turbulence quantities, form the basis of the popular **sigma-coordinate** Princeton model developed by George Mellor's group at Princeton University [Blumberg and Mellor 1987; see also Mellor 1991, Kantha and Piacsek 1993]. In this coordinate system, the number of levels is the same everywhere in the ocean, irrespective of the depth of the water column. It is therefore possible to resolve the bottom boundary layer where needed. This set of equations is best suited to modeling the shallow coastal oceans, although there is no inherent barrier to its application to deep basins. The principal problem is in applying it over sharply changing topography such as the continental slope separating the shelf from the deep basin. Here, unless the topographic gradients are suitably reduced by a nonlinear smoother, the errors in the calculation of pressure gradients induced by horizontal gradients of density can lead to spurious along-slope currents [Haney 1991]. While the problem due to strong topographic changes manifests itself in one form or another in all ocean models, the problem is particularly serious in sigma-coordinate models.

Many applications of this model can be found in the literature (for example in the *Journal of Physical Oceanography* and the *Journal of Geophysical Research, Oceans*). An application of a modified version developed at the University of Colorado, incorporating an improved mixed-layer formulation [Kantha and

Clayson 1994] and involving assimilation of altimetric data, is given in Section 31.4. This version has also been converted to CM5 and applied to the Straits of Sicily, and its Cray T3-D version is being applied to the North Pacific Ocean.

31.3.4 Layered Models

In layered models, the ocean is divided into several (N) layers in the vertical, and Equation 31.1 to Equation 31.3 are integrated over each layer ($n = 1, \ldots, N$) to obtain expressions for the thickness of and velocity in each layer. For example, Wallcraft [1991] obtains

$$\frac{\partial h^n}{\partial t} + \frac{\partial}{\partial x_k}\left(h^n U_k^n\right) = w^n - w^{n-1}$$

$$\frac{\partial \left(h^n U_j^n\right)}{\partial t} + \left[\frac{\partial}{\partial x_k}\left(h^n U_k^n\right) + U_k^n \frac{\partial}{\partial x_k}\right]U_j^n + f\varepsilon_{j3k}h^n U_k^n$$

$$= -h^n \sum_{k=1}^{N} G_k^n \frac{\partial}{\partial x_k}\left(h^n - h_0^n\right) + \left(\tau_j^{n-1} - \tau_j^n\right) + A_M \frac{\partial^2}{\partial x_k \partial x_k}\left(h^n U_j^n\right) \qquad (31.22)$$

$$+ \max\left(0, -w^{n-1}\right)U_j^{n-1} + \max(0, w^k)U_j^{n+1} - [\max(0, -w^n)$$

$$+ \max(0, w^{n-1})]U_j^n + \max(0, -c_{de}w^{n-1})\left(U_j^{n-1} - U_j^n\right)$$

$$+ \max(0, -c_{de}w^n)\left(U_j^{n+1} - U_j^n\right)$$

where h^n is the thickness and U_j^n the velocity of the nth layer, w^k is the vertical velocity at the kth interface, and h_0 is the layer thickness at rest. The Nth layer contains the model basin topography, and its thickness is the total depth of the water column minus the sum of the thicknesses of the remaining layers. Finally,

$$G_k^n = \begin{cases} g, & k \geq n \\ g\left[1 - \dfrac{\rho^n - \rho^k}{\rho_0}\right], & k < n \end{cases}$$

$$\tau_j^n = \begin{cases} \tau_w, & n = 0 \\ c_{dn}\left|U_j^n - U_j^{n+1}\right|\left(U_j^n - U_j^{n+1}\right), & n = 1, \ldots, N-1 \\ c_{db}\left|U_j^N\right|U_j^N, & n = N \end{cases} \qquad (31.23)$$

The factor c_d is the drag coefficient, c_{de} is the drag due to entrainment of fluid from one layer to the adjacent one, τ_w is the wind stress, and ρ^n is the density of the nth layer. Note that the layer densities do not change with time, only their thickness does at each model grid point. The conditional statements have to do with entrainment and detrainment at each interface between two adjacent layers, the details of which can be found in Wallcraft [1991].

The thinning of a layer to vanishing thickness is a major problem in layered models that leads to numerical difficulties. The traditional solution has been to make each layer thick enough, but this distorts the representation of the oceanic vertical structure. An alternative solution is to entrain fluid into the thinning layer from below to thicken it. Such entrainment has to be balanced by global detrainment in the layer so as to keep the density of each layer constant in space and time. For details of this and the model numerics, see Wallcraft [1991].

It is essential to select the number and rest thicknesses of layers carefully in layered models. Since topographic variations are contained in the bottommost layer only, these models are generally incapable of simulating circulation in coastal and shallow seas. They are, however, excellent at capturing the important lowest-order dynamics of the basin circulation and are therefore widely used for process-oriented studies. They are also being increasingly used for a variety of applications. One example is the six-layer, $\frac{1}{8}°$ global model at the Naval Research Laboratory at Stennis Space Center, Mississippi, the SSH from which is shown in two parts, the Atlantic and Indian Oceans in Figure 31.3a and the Pacific Ocean in Figure 31.3b. Realistic depiction of mesoscale activity, especially in regions of strong ocean currents — such as the Gulf Stream in the Atlantic, Kuroshio in the Pacific, the Brazil/Malvinas Current off Brazil, the Agulhas Current off Africa, and the Circumpolar Current around the continent of Antarctica — are noteworthy. The SSH variability from a layered model like this, driven by synoptic winds from a NWP center such as Fleet Numerical Meteorology and Oceanography Center, compares well with the variability indicated by altimeters such as the U.S. Navy's GEOSAT.

A simple subset of the layered model is the so-called reduced-gravity model (also called $1\frac{1}{2}$-layer model), where the water column is assumed to consist of two layers: an active top layer of thickness H and a quiescent bottom layer of infinite thickness, with a density interface between the two of intensity $\Delta\rho$. It is remarkable that this very simple model often captures the essential dynamics of the circulation; for example, a reduced-gravity model of the Gulf of Mexico demonstrated conclusively that it is the instability of the Loop Current that is responsible for the shedding of the Loop Current eddies [Hurlburt and Thompson 1980]. The governing equations are identical to the barotropic Equation 31.15, except that the gravity parameter g is replaced by $g' = g\left(\Delta\rho/\rho_0\right)$, the reduced gravity (whose value is two orders of magnitude smaller than g; hence the name reduced-gravity model), with H now denoting the rest thickness of the upper layer and η denoting the deflection of the interface.

31.3.5 Isopycnal Models

Isopycnal models are similar to the layered models discussed above but are fully dynamical and thermodynamic. Despite the numerical problems associated with surfacing and vanishing of layers, they are well suited to simulate basin dynamics. Considerable progress has been made over the last decade in isopycnal modeling, and with the inclusion of adequate upper-mixed-layer physics they are also becoming quite practical. Examples of applications can be found in Oberhuber [1993] and Bleck and Smith [1990]. Since they principally deal with isopycnals (surfaces of equal density) and do not consider temperature and salinity separately, but instead treat density as the prognostic variable, they are not well suited to handling situations where temperature and salinity must be computed separately. A linear equation of state and identical diffusion characteristics for temperature and salinity are implicit in these models. This is valid over a majority of the global oceans, if one excludes regions such as those near river outflows and sea-ice formation.

31.3.6 Data Assimilation

Inevitable errors in initial conditions and imperfect parametrization of physical processes make a model ocean diverge rapidly from the real ocean. This is simply due to the extreme sensitivity of this system to even minute changes in initial conditions, typical of chaotic nonlinear systems. It is therefore essential to employ observational data in ocean models to retain the modeled ocean state close to the real state. The situation is no different from that in modeling the state of the atmosphere for NWP purposes, except that the time scales for loss of predictability is weeks for the oceans compared to days for the atmosphere. The process of employing observed data from the real ocean (atmosphere) to keep the modeled ocean (atmosphere) realistic is called data assimilation [for example, Anderson and Moore 1986] and consists of combining the modeled fields with observed data at various points in the domain to produce the best possible estimate of the real state of the ocean over the entire model domain. Exactly how this is best done has been the subject of considerable research in the atmospheric community [Bengtsson et al. 1981] over the past few decades, and more recently in the oceanic community as well [Haidvogel and Robinson 1989].

NWP centers use predominantly the so-called analysis–forecast cycle of assimilation. Here, the current state of the modeled atmosphere as predicted by the previous forecast is combined with observations of the atmosphere by radiosondes and surface stations all over the world, by an analysis–initialization process, to produce initial fields of various model variables suitable for describing the initial state for the next model

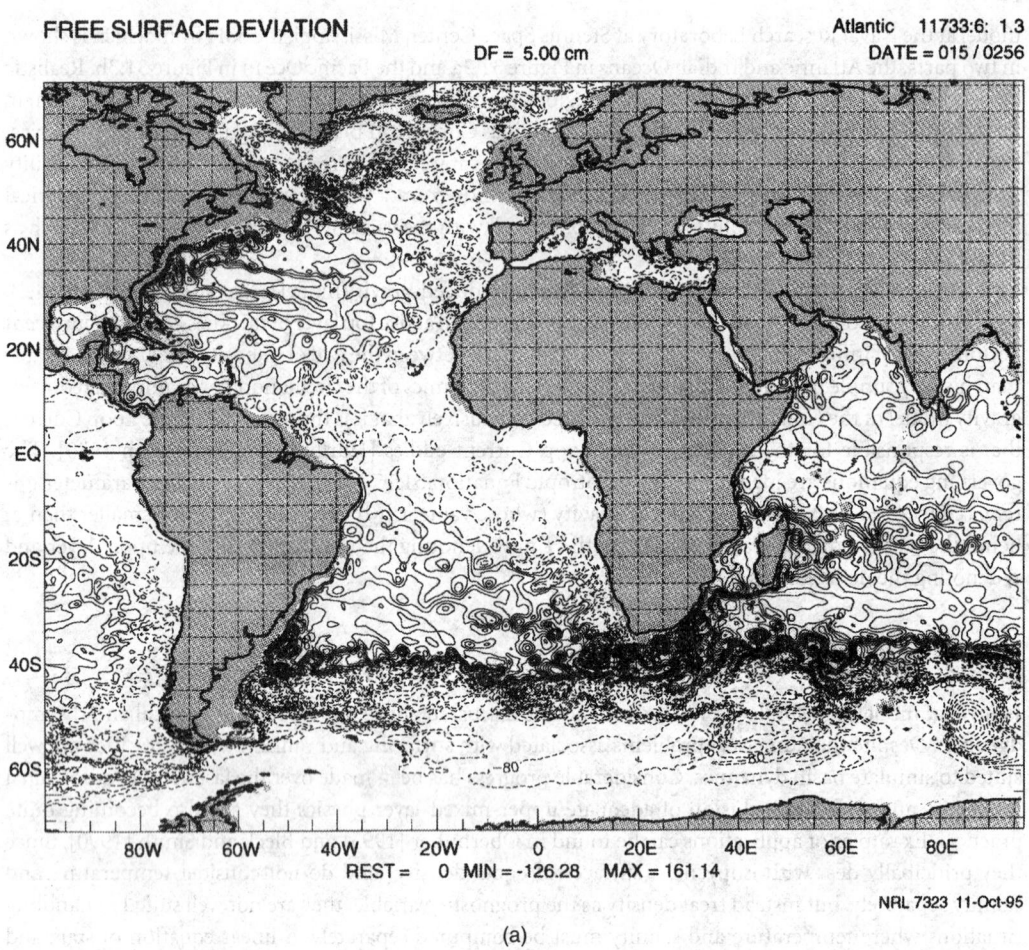

FREE SURFACE DEVIATION   Atlantic 11733:6: 1.3
                    DF = 5.00 cm   DATE = 015 / 0256

REST =   0 MIN = -126.28 MAX = 161.14

NRL 7323 11-Oct-95

(a)

FIGURE 31.3 Sea surface from the six-layer Naval Research Laboratory $\frac{1}{8}°$ global model for the Atlantic and the Indian Oceans (a) and for the Pacific Ocean (b). Note the eddy-resolving capability of this model displayed in the realistic mesoscale activity in regions of strong currents such as the Aghulas around Africa.

forecast. The forecast skill depends very much on the accuracy of the initial state so derived, since errors in the latter tend to get magnified with time during the forecast.

Another possible assimilation method is the so-called continuous assimilation [Bengtsson 1981], where a numerical model is kept running and current by assimilation of observed data as and when they become available. A forecast can then be initiated by a similar model running forward *free* (without any data assimilation). The principal advantage of this method over the analysis–forecast cycle is that the model derives benefit from all past data as opposed to a single set of observations at a particular time. Also, the shock of data insertion due to inevitable mismatch between the model and observed states is less severe. Since such a mismatch can lead to severe noise superimposed on the true state of the forecast atmosphere (ocean), often making the forecast worthless, considerable effort has been expended in devising means to minimize such a mismatch, resulting in a procedure called *initialization* in NWP terminology. Continuous assimilation tends to reduce this shock and is therefore often preferable. The reader is referred to Bengtsson et al. [1981] and Haidvogel and Robinson [1989] for a discussion of assimilation philosophies.

The method of combining data into a model can vary from the simplest one (called data insertion), in which the model-predicted values are just replaced by observed values, to Kalman filters [Gelb 1988] (which blend the model and observed values optimally, taking into account the model error and observational

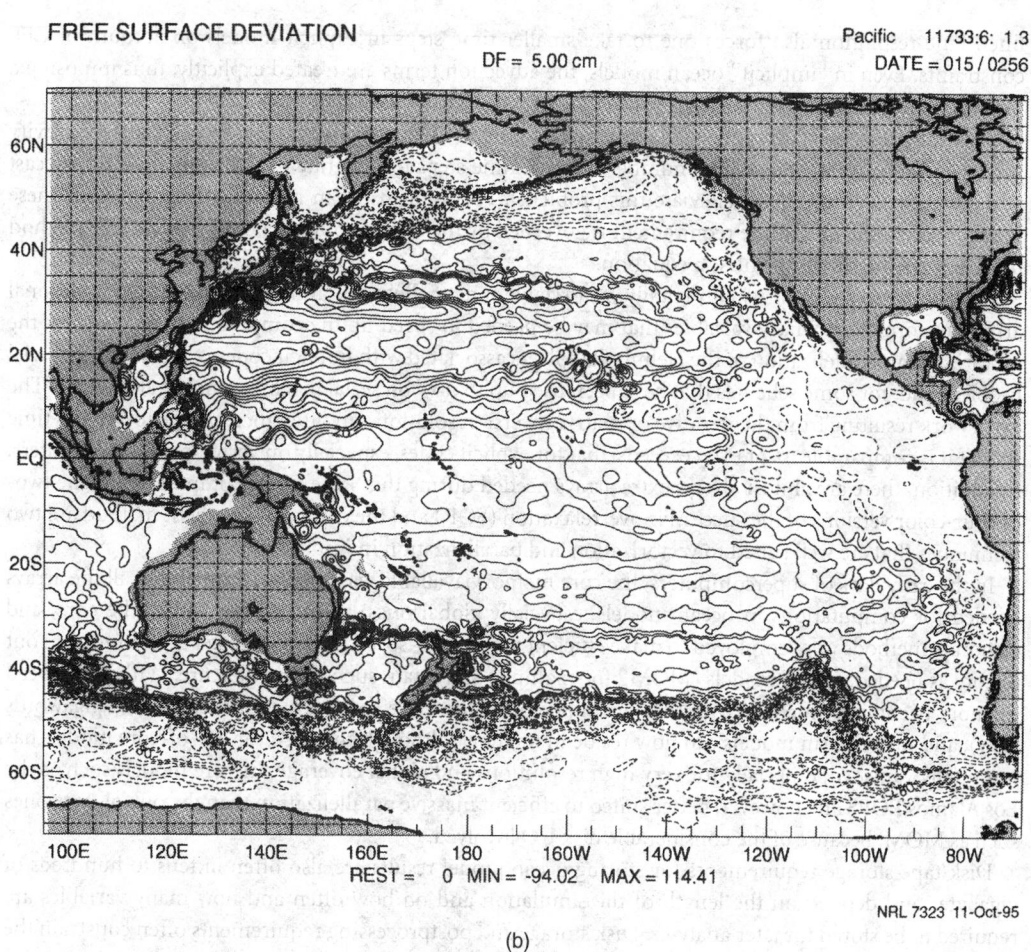

FREE SURFACE DEVIATION   Pacific   11733:6: 1.3
DF = 5.00 cm   DATE = 015 / 0256

REST =   0   MIN = -94.02   MAX = 114.41

NRL 7323 11-Oct-95

(b)

FIGURE 31.3 *continued*

error statistics), adjoint techniques [Thacker and Long 1987], and variational methods [Derber and Rosati 1989]. It is also possible to use nudging techniques in which appropriate Newtonian damping terms that damp the variable to the observed value with a predetermined time scale are introduced into the governing equations. The most commonly employed method is optimal interpolation [see Choi and Kantha 1995, for example], since methods such as Kalman filters and adjoint techniques are computationally expensive and at present still impractical for applications in NWP and ocean prediction.

It is beyond the scope of this article to go into details of data assimilation methods. Instead, the reader is referred to the above references (and more recent work in the literature, especially on NWP), with a reminder that most assimilation methods replace the model-predicted values by a weighted combination of model-predicted value and observed values during the assimilation step, with the weight either determined a priori by statistical methods such as optimal interpolation or updated at each assimilation step by a method such as Kalman filtering. For examples of oceanic data assimilation, the reader is referred to Derber and Rosati [1989], Glenn and Robinson [1995], and Choi and Kantha [1995].

31.3.7 Computational Issues

Ocean models make a large demand on computer resources, CPU time, core memory, and disk storage, because ocean eddies are much smaller than weather systems, and the resolution needed is therefore much finer. Fine resolution also forces one to take smaller time steps in explicit models on account of CFL constraints. Even in "implicit" ocean models, the advection terms are treated explicitly, thus imposing a time-step limitation.

For explicit free-surface models, the time step is limited by the step of the fast-moving surface gravity waves, and one has to take a large number of small time steps to integrate over a simulation or forecast period (mode splitting helps alleviate this problem). For implicit ocean models, which filter out these gravity modes, the CPU-time requirement is governed by the rate of convergence of the iterative method used to solve the resulting Poisson equation.

Explicit model codes are usually readily vectorizable and parallelizable and generally need few additional arrays to store the auxiliary variables that may be needed to speed up the computations. In contrast, the vectorization/parallelization of the Helmholtz solvers associated with implicit codes is usually a nontrivial problem and, for some schemes that have been used up to now on serial machines, not at all feasible. The extra work resulting from the iterative or matrix inversion solution can often increase the total CPU time so that it is comparable to, or even exceeds, that for explicit codes, especially on vector/parallel computers. In addition, there are almost always extra arrays needed during this stage of the computations. The two- or four-color versions of the successive overrelaxation (SOR) and the conjugate-gradient method are two techniques that are well suited to vectorization and parallelization in implicit codes.

In the early days of supercomputers, the core memory available was usually so small that all the arrays needed for computations in ocean models, especially global ones, would not fit within the core, and elaborate methods were employed to make efficient use of high-speed disks to transfer arrays into and out of core as needed. GFDL models (MOM2 for example) still retain such an architecture. With high-speed memory becoming much cheaper, modern supercomputers have core memories measured in gigawords (Gw), and many ocean models can now reside in memory, although the need for out-of-core models has not totally vanished, especially for very high resolutions and global coverage. In-core models such as the Los Alamos POPS are, however, better suited to efficient massive parallelization than the out-of-core ones such as MOM, because of the considerable disk I/O involved.

Disk/tape storage requirements for storing ocean model results are also often in tens to hundreds of gigabytes and depend on the length of the simulation and on how often and how many variables are required to be stored for later analyses. Disk storage and postprocessing requirements often constrain the temporal resolution and the details of the analyses carried out on the results of an ocean model.

Data-assimilative ocean models require even more resources than the free-running ones, with the additional memory and CPU-time requirements depending very much on the method of assimilation. It is not unusual for assimilation to more than double the CPU time requirements, even for simple OI-type schemes. Methods such as Kalman filters and adjoint methods are even more demanding. Generally, data assimilation on massively parallel computers requires considerable investment of time and effort for efficient implementation. We will give some typical CPU-time and memory requirements for large ocean models and for diverse computers, to cover a spectrum of configurations and to familiarize the reader with resource requirements of computational ocean modeling. The $\frac{1}{8}^\circ$ six-layer NRL global model ($2051 \times 1145 \times 6$ grid) requires 1.8 Gw of memory and 2 CPU hours per month of simulation on a 256-node CM5-E. The $\frac{1}{16}^\circ$ Pacific model ($1977 \times 1313 \times 6$ grid) requires 385 Megawords (Mw) and 17 single-processor CPU hours per month on a Cray C-90. The $\frac{1}{5}^\circ$ global explicit barotropic tidal model discussed in Section 31.3 under "Barotropic Models" (1801×729 grid) employs a time step of 13 s, assimilates 4000 data points every time step, and requires 65 Mw of memory and 22 single-processor CPU hours for a 10-day simulation on a 16-processor Cray C-90, assimilating 4000 data points every time step. A 15-level northern hemisphere Arctic ice–ocean model ($360 \times 360 \times 15$ grid) requires 40 Mw of memory and 25 CPU hours per month on a Cray C-90. A 30-level sigma-coordinate model of the eastern Pacific ($163 \times 229 \times 30$ grid) requires 42 Mw of memory and 4 CPU hours per month on Cray C-90. The small Gulf of Mexico sigma-coordinate nowcast–forecast model ($85 \times 86 \times 22$ grid) discussed in the next section requires 30 Mw of memory and 6 CPU hours for a month-long simulation in the nowcast mode, and 4 CPU hours in the forecast mode, the additional time requirements for the simple OI-based data assimilation being in this case about 50%. An idea of the storage requirements can be obtained from the fact that even this small model required 2 Gbytes to store the model output at 5-day intervals for a 10-year-long simulation without any data assimilation, and postprocessing of this output required numerous hours on a powerful Sun Sparc workstation.

31.4 Nowcast/Forecast in the Gulf of Mexico (a Case Study)

An important application of ocean models is in prediction of the current (nowcast) and future (forecast) state of the ocean. Given the fact that more than 50% of the burgeoning human population lives within 100 miles of a coastline and hence uses/abuses the coastal oceans, such predictions, especially in the coastal and marginal seas, might be particularly useful for societal needs such as sea-level predictions, mapping of currents, and pollution tracking. We will provide one such example from a marginal semienclosed sea in the north Atlantic, the Gulf of Mexico. The offshore oil fields of this "mini-ocean basin" account for roughly half the U.S. domestic oil production, and the Louisiana–Texas (LATEX) continental shelf is dotted with thousands of oil platforms. Exploration and production are expanding steadily into deeper waters, waters as deep as 1000 m.

The major oceanic phenomenon in the Gulf is the so-called Loop Current variability. Every second, about 28 million cubic meters of subtropical waters enter the Gulf through the Yucatan Straits between Mexico and Cuba and leave it through the Florida Straits between Florida and Cuba to eventually become the Gulf Stream. The extent of penetration of this so-called Loop Current into the Gulf is highly variable. Occasionally the Loop Current becomes unstable and sheds off a huge anticyclonic (clockwise) eddy, anywhere from 100 to 350 km in diameter, that pinches off the Loop Current and moves into the western Gulf. This Loop Current eddy (LCE) is the principal mechanism for renewal of waters in the western Gulf [Hurlburt and Thompson 1980]. The path of LCEs is also highly variable, and occasionally a LCE traverses the Gulf in close proximity to the LATEX continental shelf. Because of the strong currents (often as much as 4 knots, 2 m s^{-1} in magnitude) associated with LCEs, this is the second major source of concern (the first being hurricanes in late summer and fall) to production and exploration activities in the Gulf. A capability to accurately forecast the movement of a LCE is valuable to the oil industry.

A forecast of the path an LCE takes is possible with the use of a numerical model of the Gulf. However, accurate information on the initial location of the LCE once it is shed and the corresponding Gulf-wide oceanic state is crucial to the forecast skill. An accurate nowcast is therefore essential, and this requires a data-assimilative numerical model. Since in situ data, even in the Gulf, are sparse and often nonexistent, remotely sensed data need to be relied upon for this purpose. Since the sea surface temperature from IR sensors is not always useful in locating a LCE (especially in summer) and since altimetry can almost always detect such an eddy if it happens to straddle its track, altimetric SSH anomalies can be assimilated into the ocean model to provide a reliable nowcast of not only the eddy location but also the initial state of the Gulf. Forecasts can then be made from this nowcast and the path of the LCE predicted.

The methodology employed by Choi and Kantha [1995] for producing a nowcast (in a hindcast mode, that is, prediction of past events for which data are available for verification) is called the *continuous* or *four-dimensional* assimilation method and has its origins in NWP. The model is run from a time in the past to the present, assimilating altimetric data track by track [see Choi and Kantha 1995 for details]. Altimetric SSH anomalies derived from NASA/CNES TOPEX and ESA ERS1 altimeters are converted to anomalies in the temperature in the water column and assimilated into the model using simple optimal interpolation. In the particular example shown here, the model was run starting at the beginning of January 1993 (day 1 corresponds to Jan. 1) to produce a nowcast for day 240, at which a LCE pinches off and separates from the Loop Current. The model is then run forward free (without assimilation) to produce a forecast over the next 40 days, assuming nonchanging winds over the period. The forecast skill is assessed by comparison with the nowcast, which was also carried out over the rest of 1993.

Figure 31.4 shows a comparison of the forecast and nowcast SSH fields over the Gulf at days 260 (top) and 280 (bottom). The forecasts are shown on the left and the corresponding nowcasts on the right.

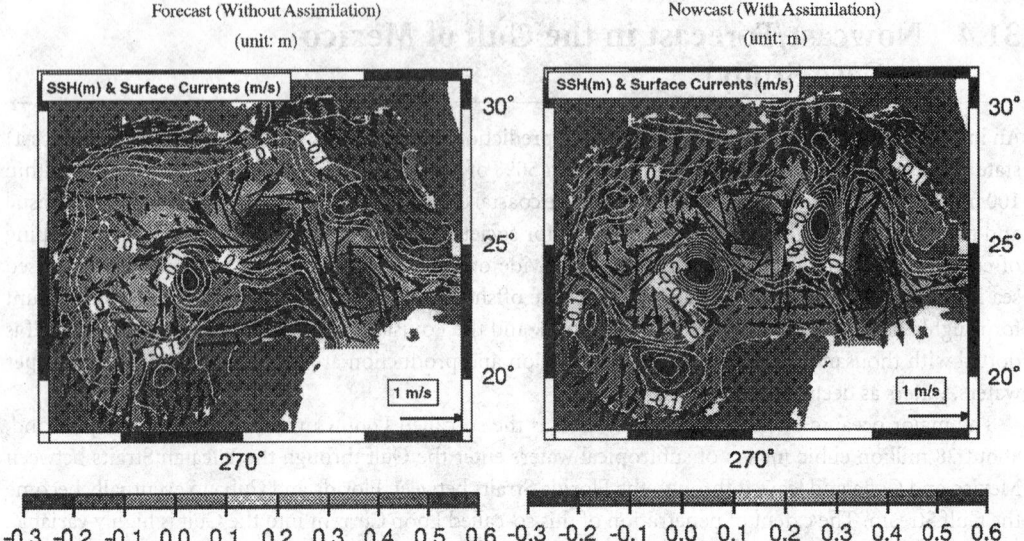

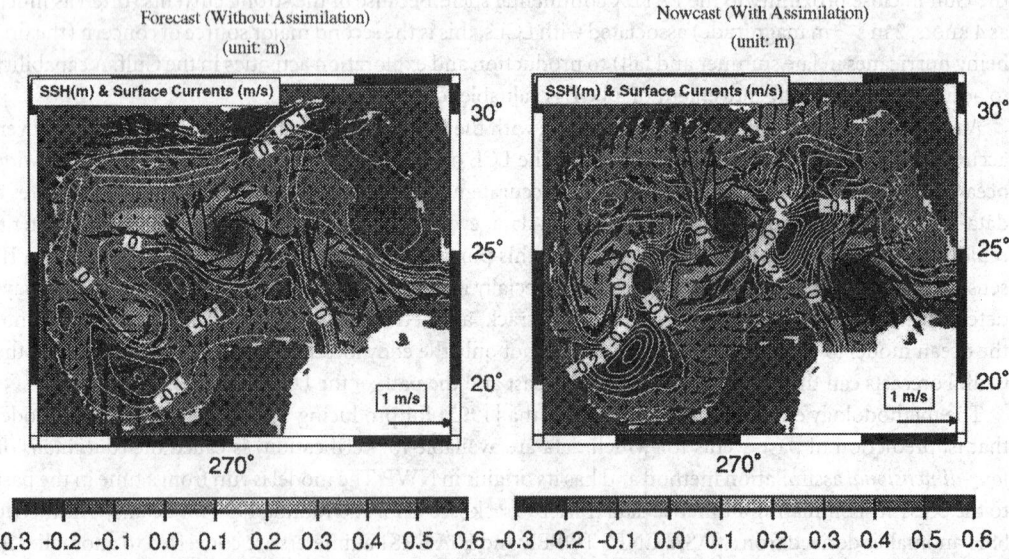

FIGURE 31.4 The sea surface elevation and currents from the forecast (left) and the nowcast (right) at days 260 (top) and 280 (bottom) from a three-dimensional circulation model of the Gulf of Mexico assimilating altimetric data from TOPEX and ERS1. The forecast was started at day 240. Compare the forecast with the corresponding nowcast to assess the model skill.

There is a close correspondence between forecast LCE position and the nowcast position, suggesting that the LCE path is being predicted reasonably well. The error, however, between forecast and nowcast LCE positions is larger at day 280 than at day 260. This particular experiment suggests that the forecast has some skill to about 30 days or so, beyond which the predicted path (forecast) deviates increasingly from the actual path (nowcast for the corresponding day). Since altimetric data are available within several days of their collection by the sensor, this experiment suggests that with some skill forecasts can be made two to three weeks in advance. If this is proven correct, this nowcast–forecast capability might be useful to drilling/exploration activities in the Gulf. It is in applications such as this that an ocean model, acting in concert with routine ocean monitoring via satellite-borne sensors, can prove useful.

31.5   Research Issues and Summary

We have provided a thumbnail sketch of ocean modeling as it is practiced today. As we said earlier, the field has undergone a phenomenal growth in recent years, and it is impossible to do justice to the subject in a short review like this. The reader is encouraged to pursue a particular model or approach of interest via the references cited.

The major issue in ocean modeling is the dearth of data for model initialization, forcing, assimilation, and of course verification or skill assessment. In situ data are rather sparse and, given the cost of ship time, likely to remain so. Therefore, increasing reliance will be placed on remote sensing to fill in gaps. However, this approach itself has limitations, and it is not clear what might fill the gap. Smart autonomous vehicles, a product of the Cold War, roaming the world oceans, and buoys sprinkled into the global oceans, telemetering data via communication satellites, may one day provide more in situ data than we currently acquire. Combined with multiteraflop computing capabilities of the coming century, an ocean observing and monitoring system consisting of satellites, moored arrays, buoys, and autonomous vehicles might one day finally enable us to set up realistic ocean prediction systems to satisy the needs of the coming generations. Ocean modeling will play a central role in all this.

Acknowledgments

Lakshmi Kantha acknowledges with pleasure the support provided by The Minerals Management Service of the Department of the Interior through an interagency agreement with the U.S. Navy through contract N00014-92C-6011, administered by Walter Johnson of MMS and Donald Johnson of the Naval Research Laboratory. Lakshmi Kantha was also supported by the NOMP program of the Office of Naval Research under contract N00014-95-1-0343, administered by Tom Curtin, and by the Coastal Sciences Section of the Office of Naval Research under contract N00014-92-J-1766, administered by Thomas Kinder.

Defining Terms

Altimeter: A microwave device measuring the time delay between an emitted microwave signal and its return by reflection from the sea surface. When the position of the instrument in space is independently determined, it enables sea surface topography to be measured to an accuracy of a few centimeters along the satellite track.

Baroclinic: Conditions in which the vertical shear is generated because of the horizontal gradients of density.

Barotropic: Conditions in which there are no variations in currents in the vertical direction.

Coamplitude and cophase: Lines of maximum tidal amplitude and lines of the time of occurrence of maximum tide, referred to either local or universal time.

Coriolis force: A fictitious force needed to allow for the noninertial nature of a rotating coordinate system.

Data assimilation: The process of blending observational data into numerical models.

El Niño: A frequent phenomenon in the tropical Pacific, occurring at 3–7-year intervals, when the eastern Pacific gets anomalously warm and sets off changes in the tropical atmosphere that affect weather all over the globe.

Gravimetry: The science of precise measurement of the earth's gravity.

Hindcast: A forecast excercise conducted for a period in the past to take account of the availability of accurate observational data for forcing, assimilation, and verification.

Inverse barometer effect: The effect where the changes in the atmospheric pressure are compensated exactly by the ocean by inverse changes in its height so that no oceanic motions are induced.

Isopycnal: A surface on which the density is constant.

Kelvin waves: Waves that run along the ocean margins (with the coast to the right in the northern hemisphere) at the speed of the shallow-water gravity wave. These waves are important for oceanic adjustment to changing surface forcing.

Nowcast: An estimate of the present state, often by an optimal blend of model and data.

Potential temperature: The temperature attained by a parcel of water brought adiabatically to a reference depth.

Reynolds averaging: The process of obtaining equations for mean quantities in a turbulent flow by considering each quantity to consist of a mean and a fluctuating component and taking averages over time or realizations.

Sigma coordinates: A coordinate system where the vertical coordinate is normalized by local depth; it is bottom-fitting or topographically conformal.

Synoptic forcing: Multihourly forcing from atmospheric models run at NWP centers, obtained in the past from a synopsis of weather charts.

Western intensification (boundary current): Strong currents found at the western boundaries of the ocean basins or eastern sides of continents because the effect of Earth's rotation variation with latitude (the so-called β-effect).

References

Andersen, O. B., Woodworth, P. L., and Flather, R. A. 1995. Intercomparison of recent ocean tidal models. *J. Geophys. Res.* 100:25261–25282.

Anderson, D. L. T. and Moore, A. M. 1986. Data assimilation. In *Advanced Physical Oceanographic Numerical Modelling*, J. J. O'Brien, Ed., pp. 437–464. Reidel, Dordrecht.

Bengtsson, L., Ghil, M., and Kallen, E., ed. 1981. *Dynamic Meteorology: Data Assimilation Methods*, p. 330. Springer–Verlag, New York.

Bleck, R. and Smith, L. T. 1990. A wind-driven isopycnic coordinate model of the north and equatorial Atlantic Ocean. Part I: Model development and supporting experiments. *J. Geophys. Res.* 95:3273–3285.

Blumberg, A. F. and Kantha, L. H. 1985. Open boundary conditions for circulation models. *J. Hydraulic Eng.* 111:237–255.

Blumberg, A. F. and Mellor, G. L. 1987. A description of a three-dimensional coastal ocean circulation model. In *Three-dimensional Coastal Ocean Models*, N. Heaps, Ed., pp. 1–16. American Geophysical Union, Washington, DC.

Bryan, K. 1969. A numerical model for the study of the circulation of the world oceans. *J. Comput. Phys.* 4:347–359.

Choi, J.-K. and Kantha, L. H. 1995. A nowcast/forecast experiment using TOPEX/ Poseidon and ERS-1 altimetric data assimilation into a three-dimensional circulation model of the Gulf of Mexico. Abstract, XXI IAPSO Meeting, Hawaii, Aug. 5–12.

Cox, M. D. 1985. An eddy-resolving numerical model of the ventilated thermocline. *J. Phys. Oceanogr.* 15:1312–1324.

Derber, J. and Rosati, A. 1989. A global oceanic data assimilation system. *J. Phys. Oceanogr.* 19:1333–1347.

Desai, S. D. and Wahr, J. M. 1995. Empirical ocean tide models estimated from TOPEX/POSEIDON altimetry. *J. Geophys. Res.* 100:25205–25228.

Dietrich, D. E., Marietta, M. G., and Roach, P. J. 1987. An ocean modeling system with turbulent boundary layers and topography. *Int. J. Numer. Methods Fluids* 7:833–855.

Dukowicz, J. K. and Smith, R. D. 1994. Implicit free-surface model for the Bryan–Cox–Semtner ocean model. *J. Geophys. Res.* 99:7991.

Dukowicz, J. K., Smith, R. D., and Malone, R. C. 1993. A reformulation and implementation of the Bryan–Cox–Semtner ocean model on the Connection Machine. *J. Atmos. Ocean. Technol.* 10:195.

Gelb, A., ed. 1988. *Applied Optimal Estimation*, p. 374. MIT Press, Cambridge, MA.

Gill, A. E. 1982. *Atmosphere–Ocean Dynamics*. p. 666. Academic Press, New York.

Glenn, S. M. and Robinson, A. R. 1995. Verification of an operational Gulf Stream forecasting model. In *Quantitative Skill Assessment for Coastal Ocean Models*, D. R. Lynch and A. M. Davies, Eds., pp. 469–499. American Geophysical Union, Washington, DC.

Haidvogel, D. B. and Robinson, A. R. 1989. In *Data Assimilation*, Special issue, *Dyn. Atmos. Oceans*. 13:171–515.

Haney, R. L. 1991. On the pressure gradient force over steep topography in sigma-coordinate ocean models. *J. Phys. Oceanogr.* 21:610–619.

Heaps, N., ed. 1987. *Three-Dimensional Coastal Ocean Models*. p. 208. American Geophysical Union, Washington, DC.

Hellerman, S. and Rosenstein, M. 1983. Normal monthly wind stress over the world ocean with error estimates. *J. Phys. Oceanogr.* 13:1093–1104.

Hibler, W. D., III and Bryan, K. 1987. Diagnostic ice–ocean model. *J. Phys. Oceanogr.* 17:987–1015.

Holland, W. R. 1986. Quasi-geostrophic modeling of eddy-resolved ocean circulation. In *Advanced Physical Oceanographic Numerical Modeling*, J. J. O'Brien, Ed., pp. 203–231. Reidel, Dordrecht.

Hurlburt, H. E. and Thompson, J. D. 1980. A numerical study of Loop Current intrusions and eddy-shedding. *J. Phys. Oceanogr.* 10:1611.

Kantha, L. H. 1995. Barotropic tides in the global oceans from a nonlinear tidal model assimilating altimetric tides. 1. Model description and results. *J. Geophys. Res.* 100:25283–25308.

Kantha, L. H., Blumberg, A. F., and Mellor, G. L. 1990. Computing phase speeds at an open boundary. *J. Hydraulic Eng.* 116:592–597.

Kantha, L. H. and Clayson, C. A. 1994. An improved mixed layer model for geophysical applications. *J. Geophys. Res.* 99:25235–25266.

Kantha, L. H. and Piacsek, S. A. 1993. Ocean models. In *Computational Science Education Project*, Oak Ridge Nat. Lab., Dept. Energy Rep. CSEP.

Kantha, L., Whitmer, K., and Born, G. 1994. The inverted barometer effect in altimetry: a study in the North Pacific. *TOPEX/Poseidon Res. News* 2:18–23.

Killworth, P. D., Stainforth, D., Webb, D. J., and Paterson, S. M. 1991. The development of a free surface Bryan–Cox–Semtner ocean model. *J. Phys. Oceanogr.* 21:1333–1348.

Kowalik, Z. and Murty, T. S. 1993. *Numerical Modeling of Ocean Dynamics*, p. 481. World Scientific, Singapore.

Le Provost, C., Genco, M. L., Lyard, F., Vincent, P., and Canceil, P. 1994. Spectroscopy of the world tides from a finite element hydrodynamical model. *J. Geophys. Res.* 99:24777–24797.

Levitus, S. 1982. Climatological atlas of the world ocean. *NOAA Professional Paper* 13, Geophys. Fluid Dyn. Lab., Princeton, NJ. 173 pp.

Madala, R. V. and Piacsek, S. A. 1977. A model for baroclinic oceans. *J. Comput. Phys.* 22:167.

Mellor, G. L. 1991. User's guide for a three-dimensional, primitive equation, numerical ocean model. *Princeton University Rep.* 91. 35 pp.

Mellor, G. L. and Yamada, T. 1982. Development of a turbulence closure model for geophysical fluid problems. *Rev. Geophys.* 20:851–875.

Mesinger, F. and Arakawa, A. 1976. *Numerical Methods Used in Atmospheric Models*, Vol. 1. Global Atmospheric Research Program Publication 17. 64 pp.

Metzger, E. J., Hurlburt, H. E., Kindle, J. C., Serkes, Z., and Pringle, J. M. 1992. Hindcasting of wind-driven anomalies using a reduced-gravity global ocean model. *Mar. Technol. Soc. J.* 26:23–32.

Oberhuber, J. M. 1993. Simulation of the Atlantic circulation with a coupled sea ice–mixed layer–isopycnal general circulation model. Part I: Model description. *J. Phys. Oceanogr.* 23:808–829.

O'Brien, J. J. 1985. *Advanced Physical Oceanographic Numerical Modeling.* Reidel, New York.

Orlonski, I. 1976. A simple boundary condition for unbounded hyperbolic flows. *J. Comput. Phys.* 21: 251–269.

Pond, S. and Pickard, G. L. 1979. *Introductory Dynamical Oceanography*, 2nd ed. p. 329. Pergamon Press, New York.

Ponte, R. M. 1994. Understanding the relation between wind driven sea level variability and atmospheric pressure. *J. Geophys. Res.* 99:8033–8040.

Roed, L. P. and Cooper, C. K. 1986. Open boundary conditions in numerical ocean models. In *Advanced Physical Oceanographic Numerical Modeling*, J. J. O'Brien, Ed., pp. 411–436. Reidel, Dordrecht.

Schwiderski, E. W. 1980. On charting global ocean tides. *Rev. Geophys.* 18:243–268.

Semtner, A. J. 1986. Finite-difference formulation of a world ocean model. In *Advanced Physical Oceanographic Numerical Modeling*, J. J. O'Brien, Ed., pp. 187–202. Reidel, Dordrecht.

Semtner, A. J. 1995. Modeling ocean circulation. *Science* 269:1379–1385.

Semtner, A. J., Jr. and Chervin, R. M. 1992. Ocean general circulation from a global eddy-resolving model. *J. Geophys. Res.* 97:5493–5550.

Smagorinskiy, J. 1963. General circulation experiments with primitive equations: I. The basic experiment. *Mon. Weather Rev.* 91:99–164.

Smith, R. D., Dukowicz, J. K., and Malone, R. C. 1992. Parallel ocean general circulation modeling. *Phys. D* 60:38.

Thacker, W. C. and Long, R. B. 1987. Fitting dynamics to data. *J. Geophys. Res.* 93:1227–1240.

Wallcraft, A. J. 1991. The Navy layered ocean model users guide. *NOARL Rep.* 35, 21 pp.

Warren, B. A. and Wunsch, C., Eds. 1981. *Evolution of Physical Oceanography.* p. 623. MIT Press, Cambridge, MA.

Further Information

This review chapter has been necessarily sketchy. The reader is therefore encouraged to consult the various references cited for more details. The monograph on numerical ocean modeling edited by James O'Brien [1985] is still the best starting point, especially since the models described there have remained essentially unchanged, undergoing only small evolutionary changes such as adaptation to massively parallel computers and inclusion of better mixing algorithms. A good starting point in coastal ocean modeling is the American Geophysical Union (AGU) volume edited by N. Heaps [1987]. Kowalik and Murty [1993] is an excellent "cookbook" for details of numerics such as finite-differencing and the split-mode technique. Reference can be made to Haidvogel and Robinson [1989] for a good description of data assimilation methods. Textbooks by Pond and Pickard [1979] and Gill [1982] are good starting points for exploring the dynamics of the oceans. The Henry Stommel 60th Birthday volume on physical oceanography [Warren and Wunsch 1981] is a good followup.

There is no specific journal for ocean modeling; instead, modeling advances are published in journals such as the *Journal of Physical Oceanography* of the American Meteorological Society (AMS) and *Journal of Geophysical Research (Oceans)* of the American Geophysical Union (AGU). The *Journal of Hydraulic Engineering* of the American Society of Civil Engineers publishes modeling papers mostly related to coastal and estuarine studies. The *Journal of Continental Shelf Research* specializes in coastal research, including coastal modeling. Purely computational advances often appear in journals such as the *Journal of Computational Physics*. Semiyearly meetings of the AGU, meetings of the AMS and biennial Ocean Sciences, and quadrennial meetings of the International Union of Geodesy and Geophysics (IUGG) are examples of venues where latest advances in ocean modeling are presented and critiqued.

The GFDL *z*-level deep-water-basin model (ftp.gfdl.gov; directory pub/ GFDL_MOM3), Princeton sigma-coordinate shallow-water coastal model (ftp.gfdl.gov; directory pub/slm), and University of Miami isopycnal model (http://oceanmodeling.rsmas.miami.edu/micom) are all available. Readers are encouraged to offload the model codes and experiment with them. A good starting point for hands-on ocean modeling is the Ocean Models chapter of the Computational Science Education Project [Kantha and Piacsek 1993], available on the World Wide Web at http://csepl.phy.oral.gov/csep.html. It contains model code, graphics, animation, and exercises on simple ocean models that serve as a good introduction to the field.

32

Computational
Chemistry

32.1 Introduction 32-1
Underlying Principles

32.2 Computational Chemistry in Education 32-2
Journal of Chemical Education • Project SERAPHIM
• JCE Software

32.3 Computational Aspects of Chemical Kinetics 32-4
Numerical Solution of Differential Equations
• Monte Carlo Methods

32.4 Molecular Dynamics Simulations 32-5
The Methodology of Molecular Dynamics Simulations
• Applications of MD Simulations • Concluding Comments
on MD Simulations

32.5 Modeling Organic Compounds 32-9
Empirical Solutions • Semiempirical Methods
• *Ab Initio* Methods

32.6 Computational Organometallic and Inorganic
Chemistry 32-13
Semiempirical Methods • *Ab Initio* Methods

32.7 Use of *Ab Initio* Methods in Spectroscopic Analysis ...32-16
Hartree–Fock Approximation • Electron Correlations
• Gaussian Basis Functions • Notation • Vibrational State
and Spectra

32.8 Research Issues and Summary 32-19

Frederick J. Heldrich
College of Charleston

Clyde R. Metz
College of Charleston

Henry Donato
College of Charleston

Kristin D. Krantzman
College of Charleston

Sandra Harper
College of Charleston

Jason S. Overby
College of Charleston

Gamil A. Guirgis
College of Charleston

32.1 Introduction

The use of computational methods in the study of chemistry touches upon every area of chemical inquiry. Indeed, the art and the science of computation are a natural fit with the study of chemistry. From the earliest times, beginning even with alchemy, chemists have used models to render comprehensible the abstract theories and concepts of their field. It is only logical, therefore, that chemists would use the power of modern computational methods to extend and explore their understanding of chemical compounds and processes.

Computational applications in chemistry are so vast and varied that it would be impossible to cover the entire field within the confines of this chapter. Instead, we will provide an overview of the types of computational applications in the field of chemistry, followed by a more detailed presentation in a few areas to show how chemists integrate computation into their discipline.

Anyone who has taken an introductory course in chemistry will remember using calculations to solve chemical problems — at least equilibrium, kinetics, and stoichiometry problems. Chemists have become adept at using computational tools to solve such mathematical problems and in using those tools to model, and thereby test, their understanding of chemical phenomena and processes. Tools such as spreadsheets, math programs, graphing calculators, and iconic modeling programs have replaced the slide rule of 40 years ago. These tools bring greater predictive power, better understanding, and the potential to solve ever more complex problems.

Chemists describe compounds at the most fundamental level as a reflection of the nature of the atoms, the bonds between those atoms, and the electrons that comprise them. In fact, since Schrödinger's development of functions in the 1920s to describe electrons as waves, chemists have attempted to describe chemical compounds and processes, with increasing sophistication, in purely mathematical terms. The limitations in this approach are both theoretical (the conceptual framework for our understanding of chemistry is not perfect) and practical (the mathematical and computational tools are not perfect, either). Since this approach was initiated, however, great strides have been and continue to be made in both the theoretical and practical arenas.

This overview of computational chemistry begins with a presentation of how students are introduced to computation, followed by a description of how mathematical modeling is used to comprehend chemical processes that do not require detailed understanding of the chemicals involved. The chapter concludes with a description of how chemists use computational methods to understand the structure of compounds and the nuances of chemical reactions.

32.1.1 Underlying Principles

For many areas of computational chemistry, mathematicians, physicists, and chemists (such as Schrödinger, Hartree, and Pauling) laid the theoretical foundation in the 1920s and 1930s. The power of today's desktop computers allow experimental chemists to bring these principles into practice. As these theories are tested and as the comparison of experimental and computational results reveals the theories' limitations, advances in theory continue to be made.

32.2 Computational Chemistry in Education

In the chemistry classroom, computation ranges from various forms of modeling (molecular and mathematical modeling, solving complex simulation problems, and molecular animation) to text and class supplements (homework and testing, computational tools, demonstrations and animations, and interactive figures). Computation appears in the chemistry laboratory as prelaboratory assignments (discussion of theoretical concepts and the proper use of equipment), simulated experiments and instruments, and the use of computational tools for data analysis.

Through the *Journal of Chemical Education* (*JCE*), the Division of Chemical Education of the American Chemical Society publishes articles on the theory and application of computational chemistry, summarizes symposia from national meetings, and reviews software programs. In addition, the division makes available inexpensive, high-quality software to instructors and students from pre–high school through graduate school through *JCE Software*, Project SERAPHIM, and the various Web-based services available to *JCE* subscribers. Software capabilities have paralleled advances in operating systems, from various flavors of Apple II and MS/PC-DOS to Macintosh and Windows.

32.2.1 Journal of Chemical Education

For more than 75 years, *JCE* has served as the primary source of information for chemical educators. Today, *JCE* offers monthly published issues, software through *JCE Software*, online access, and various printed materials. In addition to the general articles in *JCE*, several feature columns are useful to computational chemistry:

    Reviews of books, media, and software
    Computer Bulletin Board
    JCE Online

JCE Software

Molecular Modeling

Only@JCE Online, featuring *JCE* WebWare, Mathcad in the Chemistry Curriculum, and WWW site reviews

Teaching with Technology

The journal is fully searchable, with approximately 15 index keywords related to computational chemistry. To illustrate the current quality of coverage, a search using the keywords *computation chemistry* for the year 2002 resulted in nearly 200 hits.

Many general articles and Only@JCE Online features carry a symbol resembling the capital letter *W*. This symbol indicates that supplementary material (such as software, live spreadsheets and worksheets, additional data and exercises, laboratory instructions, animations, and video) is available to subscribers online. For example, one can link to the programs needed to analyze a kinetics simulation model for a drug poisoning victim, the software for finding the irreducible representations in a reducible representation for hybridization of orbitals and molecular vibrational motion, or Mathcad worksheets for Hückel theory calculations.

A relatively new addition to *JCE* is the regular feature JCE WebWare, which presents various Web-based applications suitable for computational chemistry in the laboratory, the classroom, or at home. Typically, the WebWare consists of small software programs, add-ins for spreadsheets and other standard programs, animations, movies, Java applets, or static and dynamic HTML pages. Recently, JCE WebWare included offerings for acid–base equilibria, nomenclature, games, chemical formatting add-ins for MS Word and Excel, spreadsheet analysis for first-order kinetics, a determinant solver for Hückel theory calculations, mechanism-based kinetics simulations, data analysis tools, and point group calculations. Each month, links are provided for fully interactive Chime-based models of some of the molecules discussed in the general articles in *JCE*.

Mathcad, a symbolic math software program, has become important in chemistry and, in particular, in physical chemistry. The *JCE* regular feature, Mathcad in the Chemistry Curriculum, presents abstracts of submitted documents that are useful for various chemistry courses. Recent worksheets include Hückel theory calculations, Bohr correspondence principle, NMR, modeling pH in natural waters, and variational treatment of a harmonic oscillator.

32.2.2 Project SERAPHIM

Project SERAPHIM began over 20 years ago as a clearinghouse for software and materials for computational chemistry. Membership fees support the clearinghouse, teacher workshops, and summer fellowships for software development. Software was distributed at a modest cost to members. A catalog listing several hundred different computer programs is available at the SERAPHIM Web site (http://ice.chem.wisc.edu/seraphim). These programs include

Databases — NMR library and experiments published in *JCE*

Data analysis tools — Significant figures, least squares analysis, dimensional analysis, spreadsheet calculations, and integration techniques

Chemical clip art

Teaching support — Quiz preparation and computer testing

Simulation problems — Wastewater treatment and water pollution

Tutorials and games

Laboratory — Experiment simulations, analytical techniques, and instrument interfacing

Core curriculum topics include states of matter (gases and crystals), nomenclature and formulas, stoichiometry (chemical equations, titrations, limiting reactants, and equilibria), thermochemistry, atomic structure (**electron configurations**, orbitals, and **SCF** calculations), chemical bonding (**molecular orbitals**, **HF** calculations, **Hückel calculations**, and **group theory**), spectroscopy (atomic, **NMR**, and **ESR**), dynamics (distribution of gas molecular speeds and chemical reaction kinetics), and descriptive chemistry (acid–base, redox and electrochemistry, complexes, qualitative analysis, reaction prediction, organic molecules, biochemistry, polymers, and industrial chemistry). Currently, all Project SERAPHIM materials are available as free downloads from the Web site.

32.2.3   JCE Software

JCE Software resulted from collaboration between *JCE* and Project SERAPHIM with initial support from the Dreyfus Foundation. The motto of the journal is "*JCE Software* is not *about* software, it *is* software." Originally, the journal contained three series: for Apple II, Macintosh, and MS/PC-DOS users; later, a fourth series was added for Windows users. The software and corresponding printed materials were sent to subscribers twice a year. Currently, in addition to various video materials, *JCE Software* offers over 15 special issue software collections, covering laboratory and supplementary classroom materials for Macintosh and Windows users:

General and advanced chemistry collections — Student-designed collections featuring animations, simulations, and computational tools for acid–base chemistry, equilibria, spectroscopy, crystal structure, and **quantum mechanics**

Chemistry Comes Alive! collections — Movies, pictures, and animations of reaction types, stoichiometry, states of matter, thermodynamics and electrochemistry, organic chemistry and biochemistry, and laboratory techniques

Collections on specific topics — Periodic table, laboratory techniques, NMR, solid-state surfaces, material science, crystallography, and problem-based learning

Many of the software programs are Web-ready, and appropriate licensing is available for local intranets. Much of the older software for MS/PC-DOS is available to subscribers as free downloads at the *JCE* Web site.

32.3   Computational Aspects of Chemical Kinetics

Chemical reaction sequences (CRSes) are ubiquitous in chemistry and biochemistry. CRSes are used to describe models of phenomena as diverse as the sequence of elementary steps occurring in a single chemical or biochemical reaction, a metabolic sequence of reactions, and the complex chemical processes occurring in environmental systems, such as the atmosphere. It is almost always of interest to understand the evolution of these systems in time. Writing the differential rate equations for each elementary reaction in the sequence conveniently expresses the theoretical temporal evolution. If the rate constant for each elementary step is known, then, in principle, one has a complete description of the temporal evolution of that CRS. However, since experimentally one can usually measure the concentration of one or more of the species involved in a CRS at different times, comparing the theoretical model and the experimental data involves one of the following:

Differentiating the experimental data — Analysis of enzyme kinetics using the **Michaelis–Menton** equation

Integrating the coupled set of differential rate equations associated with the CRS — Determining the order of a chemical reaction by plotting $\ln([\text{Reactant}])$ vs. time, $1/[\text{Reactant}]$ vs. time, etc.

A great deal of effort has gone into developing computational procedures that can convert the theoretical description of CRS into concentration vs. time information, which may then be compared directly to experimental results. There are two major approaches used to accomplish that goal: the numerical solution of differential equations and the **Monte Carlo** approach. Each has its own set of advantages and disadvantages, and software packages are available for each.

32.3.1   Numerical Solution of Differential Equations

CRS can be described by coupled sets of differential equations. Integrating these equations analytically is often not possible, so numerical techniques must be used. The various numerical techniques have been described and sample programs have been presented in the literature (e.g., [Press et al., 1992]). Many CRSes of interest are **stiff**, that is, they contain rate constants that span many orders of magnitude. In order to analyze these systems effectively, implicit numerical methods, such as the one developed by Gear [Gear, 1971], must be used. The following programs, some of which may be downloaded free or for a modest fee, accomplish integration of coupled sets of stiff differential rate equations:

Gespasi [Mendes, 1993, 1997]
KINSIM [Frieden, 1993]
BerkeleyMadonna
Kintecus

Many of these software packages offer the capability to optimize the CRS under consideration. That is, the set of kinetic constants that bring the model in closest agreement with the kinetic data can be found [Mendes and Kell, 1998]. It is also possible to simulate stiff problems using Mathcad in conjunction with VisSim. Other software packages, which do not handle stiff differential equations (e.g., STELLA), have been used for less demanding applications. One of the most dramatic uses of stiff differential equation solvers to study CRS is the study of the ozone chemistry in the atmosphere. There, laboratory studies of individual atmospheric elementary reactions, coupled with atmospheric models, led to the decision to stop using CFCs. The 1995 Nobel lectures of Rowland, Molina, and Crutzen summarize this research [Crutzen, 1996; Molina, 1996; Rowland, 1996].

32.3.2 Monte Carlo Methods

Using an entirely different approach, researchers have developed methods to model the stochastic events of CRS rather than find numerical solutions to the coupled differential equations describing the CRS. An early report by Kibby [Kibby, 1969], followed by a complete theoretical exposition by Gillespie [Gillespie, 1976, 1977], describes **Monte Carlo** methods for simulating the time evolution of molecular events occurring in CRS. The probability of a particular reaction event occurring is the product of the intrinsic probability of such a reaction event (given by the rate constant) and the number of possible reaction events (given by the numbers of reacting molecules). The time interval is chosen in which the next reaction event is likely to occur, and then the particular reaction event that occurs is chosen from all possible reaction events. After adjusting the number of molecules for that event, the process is repeated.

This is how the actual stochastic events in a CRS are simulated. While the simulation can involve a very large number of events, stiff CRSes are simulated in exactly the same way as nonstiff CRSes. Furthermore, the method applies to very small-volume systems, such as a living cell or cellular organelle. This makes the method attractive to biochemists, in whose work it may not be appropriate to treat concentrations of molecules as a continuously varying quantity that changes deterministically over time [McAdams and Arkin, 1997]. A software package developed at IBM is available for free download that implements stochastic simulations of CRS.

32.4 Molecular Dynamics Simulations

In general chemistry, students are introduced to Dalton's intuitive picture of chemical reactions, in which atoms collide and rearrange to form new combinations. The computational method of molecular dynamics (MD) simulations maintains this classical view by treating atoms as billiard balls, with forces between them akin to springs [Garrison et al., 2000]. MD has a breadth of applications in chemistry, ranging from the determination of the lowest energy configuration of a protein to the study of mechanisms of chemical reactions. A distinct advantage of MD over methods based on electronic structure theory is that MD can be applied to large systems composed of thousands of atoms. Quantities can be calculated from the simulations, which can be compared with experimental values. Animations of atomic motions as a function of time are created, which can contribute mechanistic insights on a microscopic level unobtainable by experiment.

This section focuses on the application of MD simulations to study the high-energy bombardment of organic targets with atomic and polyatomic projectiles [Garrison et al., 2000; Zaric et al., 1998; Townes et al., 1999; Nguyen et. al., 2000], which is important in secondary ion mass spectrometry (**SIMS**). In SIMS, a primary ion beam bombards the surface with a low enough dose that each impact samples a fresh, undamaged portion of the surface. Secondary ions ejected from the surface are detected by a mass spectrometer.

SIMS is a widely used analytical technique, which is being developed for applications to molecular-specific imaging on a submicron scale [Berry et al., 2001]. Experiments have demonstrated that the secondary ion yield depends nonlinearly on the number of atoms in the projectile, and the nonlinear enhancement increases with the number of atoms [Van Stipdonk, 2001]. Therefore, polyatomic projectiles have the potential to improve significantly the sensitivity of SIMS. Sensitivity is the limiting factor in imaging applications, in which the maximum amount of analytical information must be obtained from a limited number of target molecules on the surface.

The objective of the simulations is to understand the mechanisms by which polyatomic projectiles enhance the secondary ion yield and to determine the optimum conditions for the use of polyatomic projectiles. The model systems used in these studies are composed of a thin organic film that is physisorbed to an atomic substrate. The simulations have compared the effect of Cu_n clusters with the same kinetic energy per atom [Zaric et al., 1998]. Simulations with SF_5 and Xe, which have the same mass, have been compared at the same bombarding energy. An illustration of the bombardment process is shown in Figure 32.1. Here, an SF_5 projectile impacts a monolayer of biphenyl molecules on a silicon substrate [Townes et al., 1999]. The energetic particle strikes the surface and dissipates its kinetic energy through the solid. Collision cascades develop, and molecules are lifted from the surface into the vacuum by the underlying substrate atoms.

32.4.1 The Methodology of Molecular Dynamics Simulations

A microcrystallite composed of N atoms is constructed that models the experimental system of interest. The nuclear motions of the atoms are assumed to obey the laws of classical mechanics:

$$\mathbf{F}_i = m_i \mathbf{a}_i = m_i \frac{d^2 \mathbf{r}_i}{dt^2} \text{ which can be expressed as } \frac{d\mathbf{v}_i}{dt} = \frac{\mathbf{F}_i}{m} \text{ and } \frac{d\mathbf{r}_i}{dt} = \mathbf{v}_i \qquad (32.1)$$

The force is obtained as the gradient of the **potential energy function** that describes the interactions between the atoms in the system:

$$\mathbf{F}_i = -\nabla V (\mathbf{r}_1, \mathbf{r}_2, \ldots, \mathbf{r}_N) \qquad (32.2)$$

After the initial positions and velocities are chosen, Hamilton's equations of motion are numerically integrated to determine the position and velocity of each atom as a function of time. From the final positions and velocities of the atoms, the identity and kinetic energy of all ejected species can be calculated. Additionally, the atomic motions leading to the ejection of the molecule can be analyzed [Garrison et al., 2000; Garrison, 2001].

A limitation of MD is that atoms are assumed to obey the laws of classical mechanics. Therefore, quantum effects such as electronic excitation and ionization are not included. The questions that one is hoping to answer should be chosen so that quantum mechanical effects are not essential. Furthermore, it should be kept in mind how the assumption of classical behavior may affect the results [Garrison et al., 2000; Garrison, 2001].

The chemical interactions between atoms are modeled by potential energy functions that describe how the energy depends on their relative positions. Ideally, the potential energy function would be the solution to the electronic Schrödinger equation within the **Born–Oppenheimer approximation**. However, such a solution is not possible in simulations with thousands of atoms. Instead, a suitable mathematical function that models the physical behavior is used. Parameters in the equation are fit to available experimental data [Garrison et al., 2000; Garrison, 2001].

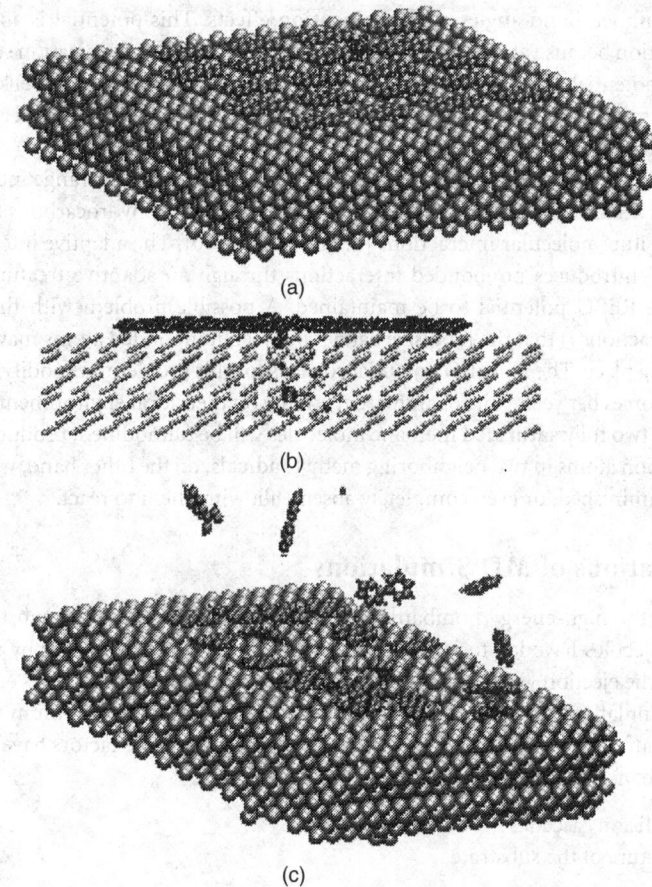

FIGURE 32.1 Collision cascade and ejection occurring for SF$_5$ bombardment of a monolayer of biphenyl molecules on Si{100}-(2x1). The Si atoms are represented by silver spheres, the carbon and hydrogen atoms are represented by dark gray spheres, and the S and F atoms are represented by black spheres. (a) An early snapshot in time that shows the SF$_5$ projectile as it moves toward the surface. (b) At 150 fs, the SF$_5$ projectile has penetrated into the open lattice and has broken up within the substrate. In this frame, the radii of the silicon spheres are drawn smaller so that the projectile can be seen within the substrate. (c) At 1500 fs, collision cascades initiated by the breakup of the projectile within the surface lead to the ejection of biphenyl molecules and molecular fragments.

Initially, the only available potential energy functions were **pair potentials** that depend on the distance between two atoms, such as the **Morse potential** and the **Lennard–Jones potential**. The total potential energy is calculated as a sum of the pairwise potential between each pair of atoms in the system. The **pairwise additive assumption** cannot take into account the effect of neighboring atoms. Therefore, this approach is not successful for modeling extended solids. Pairwise potentials are especially limited for chemistry applications, because they cannot accurately describe polyatomic molecules. For example, the pairwise additive approach would predict that H$_3$ is more stable than H$_2$ [Garrison, 2001]. The potential energy between each pair of atoms in a molecule depends on the coordination number of the central atom, the bond angle, and the torsion angle. In order to model chemical reactions, changes in hybridization must also be taken into account.

In the last decade, a great stride in these simulations was taken with the development of **many-body potentials** that include the effect of neighboring atoms on the interaction between a pair of atoms [Garrison et al., 2000; Garrison, 2001]. This development produced accurate potential energy functions for modeling face-centered cubic metals and silicon. Of particular interest to chemists is Brenner's reactive empirical bond-order (REBO) potential [Brenner, 1990], which varies the bond strength depending on the coordination number, bond angles, and **conjugation** effects. This potential is able to model bond breaking and formation because atoms may change neighbors and their hybridization state. These sophisticated many-body potentials are blended with empirical pairwise potentials to model keV bombardment of organic films on metal surfaces [Garrison et al., 2000; Zaric et al., 1998; Townes et al., 1999; Nguyen et al., 2000; Garrison, 2001].

A limitation of the Brenner REBO potential is that it cannot describe long-range interactions between molecules. Recently, Stuart et al. have developed a reactive potential for hydrocarbons that includes both **covalent** bonds and intermolecular interactions [Stuart et al., 2000]. The adaptive intermolecular REBO (AIREBO) potential introduces nonbonded interactions through an adaptive treatment, which allows the reactivity of the REBO potential to be maintained. A possible problem with the introduction of intermolecular interactions is that the repulsive barrier between nonbonded atoms may prevent chemical reactions from taking place. The AIREBO potential corrects for this problem by modifying the strength of the intermolecular forces between pairs of atoms, depending on their local environment. For example, the interaction between two fully **saturated** methane molecules will be unmodified, producing a large barrier to reaction. The carbon atoms in two neighboring methyl **radicals**, on the other hand, will have a repulsive interaction that is diminished, or even completely absent, allowing them to react.

32.4.2   Applications of MD Simulations

MD simulations of the high-energy bombardment of organic films on atomic substrates with atomic and polyatomic projectiles have led to interesting insights about the mechanisms by which polyatomic projectiles enhance the ejection yield [Garrison et al., 2000; Zaric et al., 1998; Townes et al., 1999; Nguyen et al., 2000]. The simulations also have contributed information about the optimum conditions for the effectiveness of polyatomic projectiles. As a result of the simulations, three factors have been identified as important to the enhancement in yield with polyatomic projectiles:

> Collaborating collision cascades
> Open lattice structure of the substrate
> Mass matching

First, molecules that have multiple contact points to the surface are ejected intact when several substrate atoms hit different parts of the molecule from below. Polyatomic projectiles enhance the emission yield by increasing the probability of adjacent collision cascades in the substrate, which can collaborate to lift the molecule gently from the surface, as shown in Figure 32.2.

Second, the nature of the substrate is a critical factor in the effectiveness of polyatomic projectiles [Townes et al., 1999; Nguyen et al., 2000]. The enhancement in yield is greater on a substrate with a more open lattice structure, such as silicon, than on a more closely packed substrate, such as copper. When SF_5 bombards an organic layer on copper, a densely packed solid, the polyatomic projectile breaks apart as it hits the surface and is reflected toward the vacuum. With the silicon substrate, on the other hand, the entire SF_5 projectile is able to penetrate the surface and break apart within the substrate, as shown in Figure 32.1. The breakup of the cluster within the lattice initiates upward-moving collision cascades that work together to lift intact molecules from the surface.

Third, polyatomic projectiles are most effective when there is mass matching between the atoms in the projectile and the atoms in the target solid [Townes et al., 1999; Nguyen et al., 2000]. When the mass of the atom (or atoms) of the projectile is much larger than the mass of the substrate atoms, the projectile passes through the solid without transferring much energy to the atoms in the top surface layers. When the projectile atom (or atoms) has much less mass than the substrate atoms, the projectile reflects from the surface, retaining much of its initial kinetic energy.

32.4.3   Concluding Comments on MD Simulations

The mechanistic insights obtained from MD simulations are significant for the further development of SIMS as an analytical tool. The simulations predict that the greatest enhancements with polyatomic

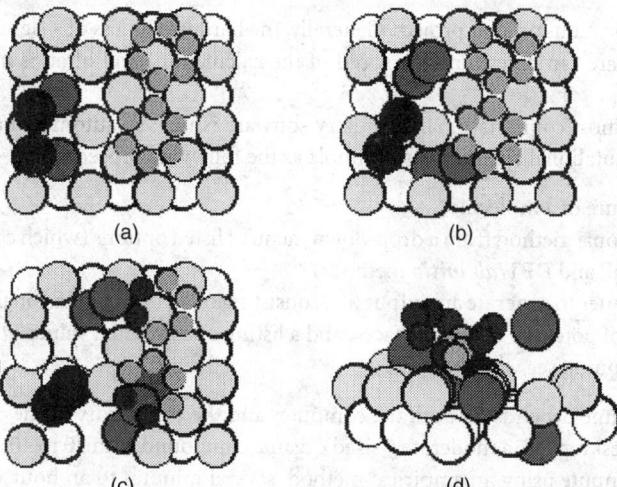

FIGURE 32.2 Schematic diagram illustrating the mechanism for the ejection of a biphenyl molecule with a diatomic projectile at 0.200 keV or 0.100 keV per atom. The incoming cluster atoms are black, and the biphenyl molecule is shaded gray. As atoms become part of the collision sequence leading to ejection of the molecule, they are shaded a darker gray. The two atoms in the dimer act collaboratively to initiate two adjacent collision cascades that lead to hitting carbon atoms in each ring of the molecule. (a) 45 fs, top view. (b) 63 fs, top view. (c) 84 fs, top view. (d) 104 fs, top view.

projectiles such as SF_5^+ and C_{60}^+ will be on organic solids, which have an open-lattice structure and are composed of light atoms. The development of focused polyatomic ion beams would have a significant impact on molecular specific imaging experiments, for which sensitivity is presently the limiting factor [Berry et al., 2001]. With the recent development of the AIREBO potential, the next challenge is to perform MD simulations of the high-energy bombardment of molecular solids, in which both short-range intramolecular forces and long-range intermolecular forces are present.

32.5 Modeling Organic Compounds

Organic chemistry encompasses all aspects of carbon-containing compounds, including the study of their chemical and physical properties, reactions, synthesis, and various uses. Organic chemistry is the foundation of biochemistry, polymer chemistry, materials science, medicinal chemistry, colorants, fragrances, the pulp and paper industry, and many other areas. The history of organic chemistry has been relatively brief: its formal beginnings were in 1826, with Friedrich Wöhler's preparation of urea that debunked the hypothesis of a **vital force**. Since that time, the use of models has been fundamental to the field. Organic chemistry is driven largely by the modeling tools that organic chemists use. Computational chemistry is currently one of the most powerful, and most rapidly developing, of those tools.

Two key models for chemical bonding guide the work of organic chemists: **valence bond theory** and molecular orbital theory. Organic computational chemistry is founded in both theories. Organic chemists use computational chemistry as a tool to solve three general types of problems:

Predicting or rationalizing chemical or physical properties
Predicting or rationalizing the stabilities of compounds
Predicting or rationalizing the reactions of compounds

The computational software used by particular organic chemists depends on personal preferences and abilities, and it varies in robustness and hardware availability. There are many sources — some free — for computational chemistry software. Because not all software packages use exactly the same programs or parameters, the publisher of the software and the specific program(s) used for each research project must be stated in any publications on that project. Generally, the hardware is less of a significant limiting factor (although the hardware can determine the speed of the calculation) and often is not mentioned in the publication.

The front end of most computational chemistry software is now so automated and polished that its effective use in computational chemistry is as simple as the following procedure:

Draw a 2-D structure of a molecule

Select a computational method from a drop-down menu of listed options (which can include **empirical**, **semiempirical**, and **DFT/*ab initio*** methods)

Wait for the computer to generate an output file, consisting of a three-dimensional structural representation, maps of potential energy surfaces, and a listing of computed values

Look over the output to see what it reveals

The computational time depends on both the computer and the complexity of the computation. For example, on a robust desktop PC, a moderately sized organic compound with 20 to 40 atoms might take less than a minute to compute using an empirical method, several minutes to an hour with a semiempirical method, and a week or more for a DFT/*ab initio* process. Unless the chemist is a specialist in computation, most of the personnel time is spent on the correct selection of the computational method to be performed (definition of the model), and then on the analysis of the output. End users are indebted to those computational chemists who have pioneered these techniques and made the tools of computational chemistry available to all chemists. For example, Professor Norman L. Allinger at the University of Georgia led a group that developed MM3, one of the more popular methods currently in use [Allinger et al., 1989].

32.5.1 Empirical Solutions

For many organic chemists, empirical methods, known collectively as **molecular mechanics** (MM) routines or *force-field methods*, are the easiest, fastest, and most generally used computational tools. This is not surprising, because these programs were originally designed for application to organic compounds. Conceptually, these methods fit easily into the historical development of valence bonding theory, and they are rooted mathematically in expressions of Hooke's law. By summing the energies of all bonded atoms, an estimate of the relative energy of the entire structure is derived. The force-field methods are, in a sense, mathematical extensions of valence bond theory's physical constructs of bonds and molecules, which contributed to early discoveries such as the alpha helix and **conformations** of cyclic and acyclic structures.

In practice, the proper selection of both the force-field routine and the parameter sets used for the class of compound being evaluated is crucial to getting a reliable result. Most bench chemists use an unmodified parameter set provided with the software. An empirical program will typically evolve over several years. The identities of these programs may be designated by year (e.g., MM3[92] or MM3[96]), by special characters (e.g., MM2*), or by a generic description of change (e.g., MM2 augmented). A problem with any computational result, but especially with an empirical calculation, is that the result occurs even if it has no practical validity. Thus, the user must know the limitations of the method and verify the reliability of any calculation.

Despite this limitation, the development of force-field programs over the years has created an ability to model reliably many classes of organic compounds, making these programs the first choice of many chemists. The basic set of equations used to model an organic compound computationally partitions and sums the contribution of the several factors to total energy. Those factors are related to the mass of the atoms in the molecule, the known preferences for bond angles and bond lengths between those atoms, **van der Waals forces** between atoms, interaction of bonds with dihedral relationships, and electrostatic interactions between atoms. In one commercially available software package, CAChe, the MM3 augmented routine includes bond stretch, bond angle, dihedral angle, improper **torsion**, torsion stretch, bend bend, van der Waals forces, electrostatics, and hydrogen bond terms. In comparison, the augmented MM2 routine in the same package does not include the bend bend interaction.

Parameters developed to describe a term in one force field are not necessarily useful in other force fields to describe the same term. It takes significant effort to develop and validate new parameters when a new routine is developed to solve a new problem or when an existing force field is modified to deal more effectively with a new functional group [Woods et al., 1992; Todebush et al., 2002]. In most MM programs, the molecule is modeled without molecular solvent or other interacting molecules, effectively modeling the preferences of a compound as a gas. Because organic chemistry is often performed in the solid state or solution phase, the influences of these states must be considered separately. As previously mentioned, the major limitation of MM methods is that they do not describe electronic properties of molecules.

Case Study

To illustrate the use of computational tools, we review the application of molecular mechanics in the development of the synthesis of a vasopressin receptor antagonist, SR 121463 A, as presented in [Venkatesan et al., 2001].

A key step in this synthesis involves the stereoselective reduction of a ketone, which results in the formation of two different alcohol products, designated **syn** and **anti** isomers. (The *syn/anti* nomenclature describes the relative position of the alcohol to the amide carbonyl in the product.) While reduction with a commonly employed reagent (LiAlH$_4$) gives acceptable initial results (4:1 syn:anti), Venkatesan et al. sought improved selectivity. They wanted to avoid loss of product by minimizing the amount of the anti product formed. They also wanted to reduce the effort needed to separate the syn from the anti alcohol.

The researchers used MM routines in MacroModel, a software package that allows for incorporation of solvent effects by means of a continuum model. They demonstrated computationally that, in the preferred conformation for the syn alcohol in solution, the alcohol was in an **equatorial** position. (This position is preferred over an alternative structure with the alcohol in the **axial** position by 2.0 kcal/mol using MM3*, by 1.1 kcal/mol using **MMFF** and, as determined for comparative purposes here, by 1.5 kcal/mol using MM3 in CAChe without solvent parameterization). This was in agreement with known general stabilities for equatorial and axial alcohols. However, upon examination of a solid state X-ray structure of the syn alcohol, it was clear that the alcohol in the syn compound was axial. The researchers rationalized that the rather small energy preference for the equatorial alcohol in the syn compound was easily outweighed in the solid state by the increased stabilization from intermolecular hydrogen bonding in the solid state when the alcohol was axial.

To explain the increased preference for production of the syn alcohol when using lithium cation-derived reagents (as opposed to sodium cation reagents, which gave syn:anti product ratios of only 3:1), the starting material was modeled when complexed with a cationic replacement (ammonium ion) for the lithium cation (because the parameters for Li atom were not in the programs used). If the cation is coordinated to both the ketone and the amide carbonyls, the modeled minimum energy structure adopts a **twist boat conformation**, which is only 2.7 kcal/mol higher in energy than the normally expected **chair** structure. If the compound were to react exclusively from the di-coordinated twist boat structure, then the expected product would be the syn alcohol. They also determined that the barrier for conversion from the chair to the twist boat (which requires adoption of a higher energy structure, known as a **half chair** conformer) was only 4.0 kcal/mol. This represents a significant increase in the stability of the twist boat structure compared to the normally more stable chair structure. In the absence of other influences, the chair form is generally 5.5 kcal/mol more stable and the barrier for interconversion (via the half chair) is about 10 kcal/mol. While this does not entirely account for the increased preference for forming the syn alcohol as one varies the reducing reagent from NaBH$_4$ to LiAlH$_4$ to L-selectride (which gave a 66:1 syn:anti preference), it aids in understanding the process, which is important if this process of stereoselective reduction is to be extended to use in other systems.

32.5.2 Semiempirical Methods

Organic chemists often use more complex methods than empirical force-field methods in order to accurately predict chemical process, especially if the nature of the electronic interactions is a controlling factor. In computational organic chemistry, molecular orbital theory provides an effective methodology. On a simple level, many problems can be evaluated by taking into consideration only the electrons most intimately involved in a chemical process. Semiempirical methods do this, by applying approximations based on empirically derived data to the **Hamiltonian** equations used to model the compound.

For example, an estimate of electronic transitions in ultraviolet-visible spectrophotometric measurements is modeled and understood by looking at the **frontier molecular orbitals** (FMO) of the **pi** system undergoing a transition, rather than including all electrons in the molecule (nonbonded and **sigma** bonded). A semiempirical method, such as **ZINDO**, is often used for this purpose. Other semiempirical methods make different approximations to simplify the computational task, and their approximations are included as look-up parameters from experimentally determined data. Three that are commonly found in software packages are the modified neglect of diatomic overlap (MNDO), the Austin models (AM1, AM3), and the third parametization of the MNDO model (PM3). The best strategy for selecting a semiempirical method is often simply to use the one that comes closest to fitting the experimental truth.

Although FMO had been used qualitatively for many years by organic chemists to rationalize chemical reactions [Fleming, 1976], the semiempirical methods, which may or may not provide accurate quantitative results, often fail even at a qualitative level.

32.5.3 *Ab Initio* Methods

The most robust computational methods employ full quantum mechanical methods. Even here, approximations are made, but they are minor compared to semiempirical methods. Such methods are often referred to as *ab initio* calculations, because they derive the energy of the molecule entirely from its native collected electrons.

There are different ways to perform *ab initio* calculations, allowing the electrons to express differing degrees of sophistication and freedom. It is possible to include all electrons, or to consider only the valence electrons, treating the nonvalence electrons as the so-called *frozen core*. Borden and Davidson elucidate the need to include all electrons with full electron correlation in complex computational chemistry problems [Borden and Davidson, 1996]. Although they take more time and employ more sophisticated calculations based on application of the Schrödinger equation, these *ab initio* methods (and DFT-type *ab initio* calculations) are required to evaluate **transition states** of **pericyclic processes**, chemistry of the **excited state**, and studies on radical structures or processes that have potential radical intermediates. Problems that must be addressed by DFT/*ab initio* methods are generally evaluated in stages: using empirical methods to get an initial structure, using semiempirical methods to get a refined structure, then crunching out the DFT/*ab initio* calculations.

Case Study

Here, we summarize the use of DFT methods to ascribe the stable conformers of a set of similar molecules, each of which undergoes an intramolecular **Diels–Alder** reaction, and then to ascertain the transition states of the Diels–Alder reactions [Bur et al., 2002].

Bur et al. set out to determine why four seemingly similar molecules required such vastly different experimental conditions to undergo an intramolecular Diels–Alder reaction. In the first part of the analysis, they identified a problem in the use of empirical methods (MM2* or MMFF using MacroModel7.0) for determining of the lowest energy conformation of the starting materials. They employed Monte Carlo searching in conjunction with applied force fields to identify the lowest energy structures. However, the lowest energy output placed two hydrogen atoms too close to each other in the molecule (in one case at a distance of only 2.40 Å, which implies contact between the two).

To resolve this issue, the researchers employed DFT calculations with **B3LYP**/3-21G* **basis sets** using Gaussian 98, followed by further refinement using 6-31G* to obtain stable conformers that had reasonable interatomic distances. They then used a 6-31G* basis set to model the Diels–Alder transition states and found that the results were qualitatively useful (the compounds requiring external heat in excess of 100°C to bring about a reaction also had computed activation energies of about 4.7 kcal/mol over the reactions known to occur at room temperature). However, the authors recognized that the results were not quantitative. By comparing the calculated activation energies for the two slower reactions (requiring heating to 145°C or 110°C, respectively), they saw that the energy difference of the transition states, only 0.3 kcal/mol, could not explain the temperature differences needed to bring about reaction.

Recognizing that the difference in transition state energy alone would not account for the observed discrepancy in the reactivities of the systems, they reexamined the conformational profiles of the starting materials for other factors. They identified two computationally modeled factors for the conformations of the reactants that corresponded to the reaction rates. The first was the influence of the carbonyl linkage (present only in the faster-reacting compound) between the 2-pi electron system and the 4-pi electron system, which resulted, as one might expect, in a more favorable alignment in the structure. The second accelerating feature was a preferential rotation about the C–N bond in the linkage that again favored a conformation of prealignment between the 2-pi and 4-pi systems in the faster-reacting materials.

Computational analysis is now central to the practice of organic chemistry (as spectroscopic analysis had become in the middle of the 20th century). The importance of computational chemistry for organic chemists is likely to increase as they find more useful tools for predicting and rationalizing chemical processes — and as computational chemists continue to advance and refine their science.

32.6   Computational Organometallic and Inorganic Chemistry

The application of modern computational techniques to inorganic and organometallic chemistry has truly undergone a renaissance during the past decade. Stoichiometric and catalytic metal reactions have attracted great interest for their many applications in industrial and synthetic processes. Metal reactions are critical in many thermodynamically feasible processes, because they accelerate the reaction by opening a lower **activation energy** pathway. These metal-centered reactions consist of one or more elementary reactions, such as substitution, oxidative addition, reductive elimination, migratory insertion, hydrogen exchange, β-hydrogen transfer, σ-bond **metathesis**, and **nucleophilic** addition.

However, unlike organic compounds, inorganic compounds are unsuited to the application of empirical molecular mechanics calculations. As mentioned in the case study in Section 32.4.1, parameters for many of the most common elements (e.g., Li) are not included in these programs, and the terms used were not developed to handle issues relevant to inorganic compounds (e.g., expanded octets, multiple stable valences, etc.).

Recent progress in computational chemistry has shown that many important chemical and physical properties of the species involved in these reactions can be predicted. Calculated values, such as predicted geometries, **vibrational frequencies**, bond dissociation energies, and other chemically important properties, have become reliable enough to complement and sometimes even challenge experimental data, especially in those cases in which experimental results are difficult to obtain. The most challenging aspect of this area has undoubtedly been to model the **metallic** elements reliably and efficiently. Metals, particularly the *d*- and *f*-**block elements**, typically present three main challenges for modeling:

The large number of orbitals, many of which are core orbitals

The **electron correlation** problem, which is accentuated by the presence of low-energy excited states

Relativistic effects for the heaviest metals

32.6.1 Semiempirical Methods

A range of quantum chemical methodologies can be used to study inorganic and organometallic compounds. The semiempirical quantum mechanics methods have great latitude in the number and type of approximations made to the full Schrödinger equation that involve the replacement of quantities that are difficult to determine with experimental or theoretical estimates or the removal of interactions, like electron interactions, which are thought to be of lesser importance. The trade-off for such approximate methods is accuracy vs. computational efficiency. For many inorganic or organometallic materials, the sacrifice of accuracy for speed is problematic because of the challenges listed previously.

The use or extension of approximate semiempirical methods necessitates a parameterization phase. Here, it is necessary to determine those parameters that maintain computational efficiency, while maximizing the model's descriptive and predictive power. Ideally, the parameterization process should incorporate the full range of motifs that characterize a chemical family. Therefore, one major issue for parameterizing metal-containing compounds is the development of a robust parameterization to handle a diverse set of influences. Such chemical diversity may be defined as the ability of metals to stabilize distinct bonding environments involving different bond types, bound-atom types, **spin** and **formal oxidation** states, **coordination numbers**, and geometries. For these reasons, it is difficult to use semiempirical methods when the calculations involve metal atoms.

32.6.2 *Ab Initio* Methods

At the other end of quantum chemical methodology spectrum lie *ab initio* methods. These techniques employ computations derived from theoretical principles without inclusion of experimental data. While *ab initio* methods such as **Hartree–Fock** (HF) and **Møller–Plesset** (MP2) have been utilized extensively for organic compounds, their application to metal-containing systems is limited. Results from such calculations involving metals have been more-or-less useless: bond lengths were wrong by tenths of angstroms, the relative energies of **isomers** were often wrong, and the relative energies of possible spin multiplicities were wrong, to name a few instances. For organometallic compounds, the primary problem with application of *ab initio* methods was founded in the change in the error with bond length that exceeds the change in the actual energy. This swamped other errors that result from use of simplified models for complicated **ligands**, neglect of the solvent, and neglect of relativistic corrections.

However, two techniques have been developed to deal with these challenges: quasi-relativistic **effective core potentials** (ECP) (also commonly referred to as *pseudopotentials*) and **density functional theory** (DFT). There is general agreement that DFT methods are superior to classical *ab initio* methods at the HF and MP2 levels for the calculation of metal compounds. This is because the accuracy of the DFT results is similar to, or in some cases even better than, MP2 data, and the computational time costs are significantly less. DFT combined with ECPs are now considered the standard operating procedure for handling inorganic and organometallic compounds. Indeed, computational inorganic and organometallic chemistry is almost synonymous with DFT for medium-sized molecules. This is due in part to computational improvements but also, perhaps more importantly, to the inclusion of these techniques in powerful, yet relatively user-friendly computational chemistry packages, such as Gaussian, Spartan, Jaguar, GAMESS, MOLPRO, CAChe, Hyperchem and ADF.

Powerful DFT methods incorporating ECPs allow many topics of interest to inorganic and organometallic chemists to be studied. A significant body of literature has been produced concerning theoretical studies of **transition metal**-catalyzed chemical reactions including industrially relevant processes, such as Ziegler–Natta polymerization, copolymerization, hydroformylation, and the water-gas shift reaction [Torrent et al., 2000]. Other catalytic systems studied include hydrogenation, epoxidation, dihydroxylation, and thioboration [Torrent et al., 2000]. It remains to be seen whether there is a catalytic system unamenable to computational study, such as **heterogeneous** catalysts, for which it is difficult to characterize intermediates experimentally.

Another topic of interest to inorganic and organometallic chemists is the nature of bonding. Included in bonding studies is the multiple bonding to ligands and other metals, both transition and main group

[Frenking and Fröhlich, 2000; Cundari, 2000]. The aim of most theoretical investigations of the chemical bond is to find a correlation between the chemical behavior of a molecule or its physical observables and calculated data, such as charge distribution or orbital structure. When including metals in theoretical models, a considerable problem is understanding the energetic contributions to the chemical bonds in terms of what the electrons are doing. Much progress has been made in understanding the nature of bonding to metals, and the future of this area is robust with possibilities.

The advent of faster computers and better algorithms has made possible new areas of theoretical work with metals. Until recently, there has been a paucity of work with **actinide** complexes, undoubtedly due to the experimental and computational difficulties in handling these elements. In addition to the problems mentioned earlier, the application of theoretical electronic structure methods to actinide complexes has long been deterred by several well known challenges. The ability to accommodate f orbitals with computational efficiency; the incorporation of a larger number of valence electrons; dynamic electron correlation effects; the large number of low-lying, near-degenerate states; the severe relativistic effects caused by the large atomic numbers of the actinides; and the overall large size of actinide complexes place extremely high demands on the choice of suitable basis sets, efficient numerical algorithms, and computational resources.

With DFT methods, actinide complexes are no longer ignored [Li and Bursten, 2001]. DFT methodology allows experimentally important properties (such as the geometry, vibrational frequencies, and infrared absorption intensities) to be calculated, even for large actinide systems. Many aspects of actinide chemistry are experimentally challenging, so reliable theoretical calculations provide a valuable adjunct to experimental studies and can provide theoretical interpretations of experimental results.

Accurate quantum chemical treatment of transition metal complexes in biochemical systems is a relatively new area [Siegbahn and Blomberg, 2000]. With each passing discovery of an important metalloenzyme or metal-mediated biological process, the ability to predict *in vivo* function of metals in biological systems is becoming increasingly important. There are two striking differences in the application of computational chemistry to biological systems, in contrast with other chemical systems such as **homogeneous** catalysis. The first difference is that the overall chemistry that affects biochemical complexes is considerably more complicated than in a typical catalytic cycle of a laboratory process. The second difference is that the amount of experimental information on biochemical systems is immense. Often, decades of experimental investigations are done by researchers focusing on only one system. To make significant contributions, the large amount of biochemical information must be put in the context of a quantum chemical modeling, which is not a trivial matter.

As with inorganic and organometallic systems, metal-containing biological systems must be studied with methods that incorporate correlation (e.g., DFT), because significant errors occur in **thermodynamic** properties, which are at least an order of magnitude larger. It is safe to predict that almost all future studies of biological transition metal systems will be accomplished using more accurate methods, such as gradient-corrected DFT. Systems that have been treated quantum-mechanically include accurate geometric determinations of the metal binding site in blue copper proteins, mechanisms of methane monooxygenase, and water oxidation in photosystem II. The number of applications is still not large, but this area will grow rapidly in the future. It is increasingly common in the computational study of large biological systems to use hyphenated methodologies, such as QM-MM. Quantum mechanical (QM) methods are employed to study the inner working of the active site or the metal center of the metalloprotein (where electron correlations are of paramount importance); empirical methods (MM) are used to define the carbon skeleton scaffolding that comprises the bulk of the system. The interface of the quantum mechanical region and the empirical region of the system is difficult, and discussion of how to address that problem is treated elsewhere [Carloni et al., 2002; Monard and Merz, 1999].

Finally, while recent work has focused on the use of DFT, it must be noted that such methods do not solve all problems. The magnetism and spin states of multimetal clusters and solids are not well determined by DFT; these are most easily described by an empirical Heisenberg Hamiltonian. Detailed potential curves and states of very small molecules to spectroscopic accuracy require configuration interaction (CI) methods. Solvent effects on spectra remain a difficult problem as well. While DFT certainly has revolutionized inorganic and organometallic computational chemistry, it is not a panacea for all problems facing the discipline. However, the growth and development of computational inorganic chemistry has been unprecedented over the past decade and will likely continue to expand in coming years.

32.7 Use of *Ab Initio* Methods in Spectroscopic Analysis

It is now possible to carry out molecular orbital *ab initio* calculations on a reasonably complex molecular structure. These calculations can yield details on a number of important molecular properties, such as the following:

Geometry in ground and excited states
Atomic charge and dipole moment
Molecular energy
Conformational stability and structure of macromolecules and biomolecules
Vibrational frequencies
Infrared intensities
Raman activities
Electrostatic potential energy
Force constants
Dynamics of molecular collision
Rate constants of elementary reaction
Simulation of molecular motions
Thermodynamic properties

The challenge here is determining how to obtain this information and what tools are needed to do so. As previously noted, much *ab initio* computational chemistry is based on the Schrödinger equation, $H\Psi = E\Psi$, developed in 1926 by the Austrian mathematician and physicist Erwin Schrödinger. This is a single equation, whose solution is the wave function for the system under consideration and describes the spatial motion of all particles of the molecular system. In order for the equation to work, the wave function must satisfy certain properties. Unfortunately, the exact solution of this equation can be used to calculate the energy of only a single atom: the one-electron hydrogen atom. Exact solutions are not possible for even the two-electron helium atom, or for any other elements or compounds. Still, the Schrödinger equation is utilized for larger systems consisting of many interacting electrons and nuclei. To do this, a number of mathematical methods that use approximation techniques such as the **variational** theorem, self-consistent field theory (Hartree–Fock approximation), and **linear combination of atomic orbitals** (LCAO) are applied to solve this equation and to describe the atomic and molecular structures.

32.7.1 Hartree–Fock Approximation

The central difficulty in solving the Schrödinger equation is dealing with electron–electron interactions. The energy of a particular electron depends on the electric field generated by the nuclei of the elements and all the other electrons in the system. According to the **variational principle**, we can approach an accurate solution of many electron wave functions once a set of adjustable coefficients is used with each orbital to minimize the total energy of the system. All the contributing one-electron functions are then varied until the energy obtained is at its lowest value. Such orbitals are referred to as *self-consistent field* or **Hartree–Fock** molecular orbitals.

The advantage of this method is that it allows us to interpret the molecular properties as derived from those of the constituent atoms. Further, this method assumes that the electrons are moving in an environment that is an average potential of the other electrons. Thus, the instantaneous position of an electron is not influenced by the presence of neighboring electrons. It should be noted that the HF approach fails to take into consideration the Pauli exclusion principle that two electrons cannot have the same quantum numbers, so HF does not account for electron–electron repulsion. Several methods are used to represent this electron correlation, as mentioned in the following section.

32.7.2 Electron Correlations

The correlation of electrons is crucial in studying the optimization of structural parameters, energies of conformational stabilities, and vibrational frequencies, each of which is essential to spectroscopic analysis. In HF calculations, as mentioned earlier, every electron is affected by the average of the other electrons in the atoms but is insensitive to any individual electron–electron interaction. Several electron correlation methods are used, such as Møller–Plesset to the nth order of correlations (MPn), configuration interaction (CI), multiconfigurational self-consistent field (MCSCF), generalized valence bond theory (GVB), and coupled cluster theory (CC). Including correlation functions in the calculations results in more accurate computational energies and structural parameters.

32.7.3 Gaussian Basis Functions

In 1930, Slater defined a particular set of functions associated with the molecular configuration. This set, which depends only on the nuclear charge, results in what are known as Slater-type atomic orbitals (STO), which have exponential radial components represented as $e^{-\xi r}$. Slater functions are of limited use because such integrals must be solved by numerical methods and are not well suited to numerical calculations in molecular systems. In 1950, a suggestion was made by S.F. Boys that the atomic orbitals should be expressed in terms of Gaussian-type orbitals (GTOs), in which the exponential radial parts are represented as $e^{-\alpha r^2}$, instead of $e^{-\xi r}$ as in Slater-type atomic orbitals. The exponential terms ξ and α are constants that determine the size of the orbital.

The advantage to using GTOs is that they do not require numerical integration, and the product of two Gaussian functions is another Gaussian function. The disadvantage to using Gaussians is that the atomic orbital is not well represented by a simple Gaussian function but by a sum of several Gaussian functions. That is, a Gaussian function decays too slowly and thus lacks a cusp at the origin, as is required by STO representations. (See Figure 32.3). Although fast, modern computers make the use of multiple Gaussian functions feasible, there is still some debate over the intrinsic value of using multiple Gaussians. One school of thought holds that increasing the number of basis functions will improve the model for the calculation of structural parameters. Others believe, however, that this will not improve the model but give rise to erroneous results.

In any event, the basis functions for multiple Gaussians are further modified by adding polarization and diffusion functions for hydrogen and heavy atoms. The choice of basis set affects the computation time to perform the calculation to the order N^4, where N is the number of basis functions. The smallest basis set used in the calculations is called the *minimal basis set*, for example, STO-3G. Despite this cost in computational time, polarization functions are still used because they often produce more accurately computed geometries and frequencies.

32.7.4 Notation

In the literature, there are two popular types of basis sets: designated STO-nG and the Pople or split valence set. The latter is designated a-bcG, where n, a, b, and c correspond to number of Gaussian functions used to form each Slater-type orbital. In STO-3G set, the 3 corresponds to the use of three Gaussian functions to mimic Slater orbitals; STO-4G indicates that four Gaussian functions are used, and so on. In split valence sets, such as 3-21G (which is the minimal basis set in this family), three primitive Gaussian functions are used to describe the core shell (nonvalence orbital). The valence shell is described by the linear combination of two primitive Gaussian functions and one primitive Gaussian function, so that a total of six GTOs are utilized to mimic the Slater-type orbital. In 6-31G, the core orbital is represented by six primitive Gaussian functions, and the valence shell is described by two functions that are linearly combined: one with three primitive Gaussian functions and the other with one.

As stated, these two basis sets (3-21G and 6-31G) do not allow for the polarization of the orbitals. This means that the electrons are not allowed to occupy orbitals other than those they would occupy based

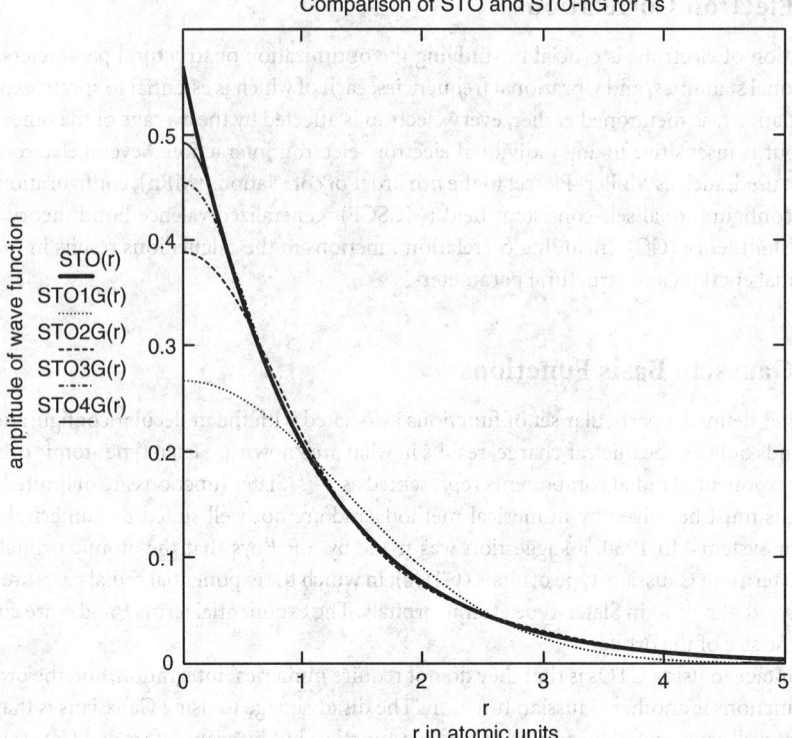

FIGURE 32.3 Representation of Slater-type orbital (heavy solid line) using various Gaussian-type orbitals.

on standard electron configuration principles. Polarization allows orbitals to change shape (e.g., from a spherically symmetrical s shape to a dumbbell p shape), and this is an important feature of orbitals that can be achieved by applying a polarization function to the basis set. This is designated by adding one or two asterisks to the basis set: for example, 6-31G* or 6-31G**. The polarization functions give the **wave function** flexibility to change its shape by adding another set of primitives. Polarization functions are crucial in the computation of structural parameters and vibrational frequencies. In the first case (6-31G* or 3-21G*), the single asterisk means that the basis set has a polarization function on the heavy atoms or nonhydrogen atoms in the molecule. Addition of a second asterisk (6-31G** or 3-21G**) indicates the use of a polarization function for the hydrogen atoms, so that polarization of all orbitals is manifest. A diffusion function can also be added to the basis set, indicated by a plus sign (as in 6-31$^+$G and 6-31^{++}G). This is important for calculations involving **anions**. A single plus sign indicates that the diffusion function has been added to the heavy atoms (nonhydrogen atoms); adding two plus signs indicates that the diffusion function is added to all atoms.

32.7.5 Vibrational State and Spectra

In order to carry out the quantum mechanical treatment of molecular vibration, it is necessary to introduce a new set of coordinates, $Q_k, k = 1, 2, 3 \ldots 3N$, called *normal coordinates*. Each normal mode of vibration can be characterized by a single normal coordinate Q, which varies periodically. A given normal coordinate is a measure of the amplitude of a specific normal mode of vibration. The energy levels of a harmonic oscillator are given by the expression $W = (v + \frac{1}{2})h\nu$, where v is the quantum number, ν is the classical frequency of the system, and h is Planck's constant. Consequently, the vibrational energy of the molecule with several classical frequencies ν_k is described as shown in Equation 32.3.

$$W = \left(\nu_1 + \tfrac{1}{2}\right)h\nu_1 + \left(\nu_2 + \tfrac{1}{2}\right)h\nu_2 + \cdots + \left(\nu_{3N-6} + \tfrac{1}{2}\right)h\nu_{3N-6} \qquad (32.3)$$

In other words, every frequency coordinate Q_k has associated with it a quantum number ν_k and a normal frequency ν_k, which is the classical normal frequency of vibration.

In order to obtain a more complete description of the molecular motions involved in the normal modes of a molecule, a normal coordinate analysis is performed. The force field in Cartesian coordinates can be obtained by the Gaussian program from the calculations of Hartree–Fock or Møller–Plesset perturbation level of theory utilizing any basis set. The force constants that result from the *ab initio* calculations are then used to obtain the vibrational frequencies for infrared intensities and Raman activities. Initially, a scaling factor of 1.0 is applied to produce the pure *ab initio* calculated frequencies, called *unscaled frequencies*. Since they are the result of *ab initio* calculations, the predicted frequency is always higher than the observed frequency (in accordance with the variational principle). To compensate, scaling factors of 0.88 for the CH stretches, 0.9 for the CH bends and heavy atom stretches, and 1.0 for all the other modes, such as torsional modes (rotation about single bonds), are used to calculate the scaled frequencies and the potential energy distributions.

The calculated Raman spectra are simulated from the *ab initio* calculations to generate scaled predicted frequencies and Raman scattering activities. The Raman scattering cross section, $\partial_j/\partial\Omega$, which is proportional to the Raman intensity, can be calculated from the scattering activities and the predicted frequencies for each normal mode [Amos, 1986; Polavarapu, 1990]. To obtain the polarized Raman scattering cross sections, the polarizabilities are incorporated into S_j by S_j $[(1 - \rho_j)/(1 + \rho_j)]$, where ρ_j is the depolarization ratio of the jth normal mode. The Raman scattering cross sections and the calculated scaled frequencies are used with a Lorenzian function to obtain the calculated spectrum. Infrared intensities are calculated based on the dipole moment derivatives with respect to the Cartesian coordinates. The derivatives are taken from the *ab initio* calculations and transformed to normal coordinates by

$$\left(\frac{\partial\mu_\mu}{\partial Q_i}\right) = \sum_j \left(\frac{\partial\mu_\mu}{\partial X_j}\right)L_{ji} \qquad (32.4)$$

where Q_i is the ith normal coordinate, X_j is the jth Cartesian displacement coordinate, and L_{ji} is the transformation matrix between the Cartesian displacement coordinates and the normal coordinates.

The infrared intensities are then calculated by

$$I_i = \frac{N\pi}{3c^2}\left[\left(\frac{\partial\mu_x}{\partial Q_i}\right)^2 + \left(\frac{\partial\mu_y}{\partial Q_i}\right)^2 + \left(\frac{\partial\mu_z}{\partial Q_i}\right)^2\right] \qquad (32.5)$$

The literature contains several representative examples of predicted Raman and infrared spectra derived from representative compounds [Guirgis et al., 2002; Guirgis et al., 2001; Mohamed et al., 1999].

32.8   Research Issues and Summary

In many ways, computational chemistry is just beginning to show its potential as a tool for explaining reaction processes. However, the greatest potential applications for computational chemistry seem likely to come from predictions for probable outcomes in complex chemical systems. Experimental chemistry will always be important, since every prediction must be confirmed experimentally. However, theoretical predictions already serve to guide experiment, and computational chemistry raises the power of that predictive ability to a new level.

As an example of the future predictive power of computational chemistry, consider the evidence presented in Section 32.3.3 and Section 32.5.2.1 of this chapter. As a further example of the promise for the predictive power of computational chemistry and the increasingly important role that computational chemistry is likely to play in directing experimental work, we close with the following case study.

Case Study

This case study involves stereocartography [Lipkowitz et al., 2002]. The issue of **chiral** control in chemical processes is of enormous importance. The FDA now requires biological testing of both **enantiomers** of chiral agents. The synthesis of a chiral material adds significant cost and difficulty to any preparative procedure. For these reasons and others, the ability to control the chirality of a chemical process is crucial. Moreover, the ability to achieve that control in a catalytic manner has obvious financial implications to the chemical industry. A major limitation in catalytic processes is obtaining a fundamental understanding of how the catalyst achieves its effect of accelerating a process, and, in the case of a chiral reaction, how the catalyst differentiates (and thus accelerates) the rate of one chiral pathway from another.

Many catalysts are metal-based systems. From a computational standpoint, this makes the modeling situation very complex. The working premise of this particular protocol was the readily believable but previously unproven hypothesis that an effective catalysis in a chiral system is one in which the chirality of the catalyst is as close as possible to the reactive site of the catalyst. With this assumption, a computational paradigm was needed to evaluate the effectiveness of known chiral catalysts. This paradigm maps the "stereodiscriminating regions around a chiral catalyst — hence the term stereocartography."

The first step was to make the center of mass of the catalyst at 0,0,0 on the Cartesian coordinate system, surrounded by a uniform three-dimensional grid. The second step was to put a transition state structure at grid points and compute the intermolecular energy (between transition state and catalyst) using molecular mechanics. The transition state–catalyst interaction was modeled 1,728 times at each grid point, using different alignments each time. For a typical analysis, this resulted in 950 million calculations! This is clearly a situation that calls for use of a highly parallel computational configuration, known as a *Beowulf cluster*.

To perform the analysis, parallel computing was employed "using a loosely networked cluster of SGI machines and on a small cluster of 26 AMD Athlon processors running a Linux operating system." The software used for semiempirical calculations of the transition state structures was Spartan 5.0, available from Wavefunction, Inc., which uses PM3 with a transition metal parameter set. **AMBER**, in MacroModel 7.0, was used for MM calculations.

The catalyst structures were obtained from the Cambridge Structural Database (Wavefunction), with PM3 optimization when necessary (i.e., when the database structure was not identical to the catalyst). To achieve chiral discrimination, the **R** and **S** transition states both were modeled within the grid, and the difference map (by subtraction) of the **electrostatic potentials** of the two **diastereomeric** complexes was examined. In the difference map, stereodiscrimination was revealed by the presence of electrostatic potential — if that potential were close to the site of the chemical transformation, the hypothesis would be confirmed.

Using this method, the authors examined 18 catalysts that were well described experimentally. Strikingly, 17 of the 18 known catalysts were shown to obey the Lipkowitz hypothesis. This level of success will assuredly lead many experimenters to model their catalysts with the Lipkowitz method before going to the difficulty of synthesis and experimental evaluation of their effectiveness. The authors appear to have made great progress toward their goal of quantifying the factors that influence the chiral induction of the catalytic system "so that ligands for use as **asymmetric** catalysts could be improved upon, or better yet, be designed *de novo*."

Defining Terms

Ab initio The computational construction of a molecule from its constituent atoms without any prior assumptions other than the identity, quantum mechanical properties, and predetermined connectivity of those atoms.

Actinide A series of f-block elements with increasing numbers of nuclear protons (atomic numbers 89–104) from actinium to rutherfordium.

Activation energy The energy needed to pass over the transition state when going from reactants to products.

AMBER A force-field method name that is an acronym for *assisted model building with energy refinement*.

Anions Negatively charged species, in which the total number of electrons exceeds the total number of protons.

Anti The disposition of two objects in reference to a plane so that the two objects are on opposite sides of the plane.

Asymmetric Without symmetry; *asymmetric* and *chiral* are synonyms.

Axial The location of a group that is directly attached to a six-membered ring structure and lies in a plane roughly perpendicular to that of the six-membered ring.

B3LYP An acronym for *Becke 3 term, Lee, Yang, and Parr*, an advanced variant of the density functional method that includes gradient correction factors for electron correlation potential energy.

Basis sets A set or collection of functions used to describe the atomic orbitals that are combined in the generation of a molecular orbital description of a compound.

Born–Oppenheimer approximation The assumption that because electron motion far exceeds nuclear motion, the electron motion is essentially independent of nuclear motion, so that the wave function is separated into two parts: one for the nucleus and one for the electrons.

Chair structure A conformation of the global energy minimum structure for cyclohexane (C_6H_{12}) that resembles a lounge chair (with a headrest and footrest), in which all opposing sides of the structure are parallel.

Chime A program developed by MDL for visualization of structural models on the Web.

Chiral A synonym of *asymmetric*; an object that is not identical to its mirror image.

Conformations Different structural representations of the same compound that differ only in the twist and turn of bonds (without breaking the bonds).

Conjugation An interaction of electron density in two or more adjacent systems. In a valence bond approach, adjacent pi bonds (or nonbonded electrons with a pi bond), with the pi electrons located in parallel, coplanar orbitals that can overlap each other, allow for interaction of the pi systems. Such an interaction imparts greater thermodynamic stability to the system than would be manifest if the adjacent electronic systems did not overlap (interact with) each other.

Coordination number The number of ligands that surround and are bonded to the central metal atom in a complexed structure.

Covalent bonds These occur when the electrons are shared evenly (but not necessarily equally) by each of two bonded atoms.

d-block elements The metallic elements in which highest-energy electrons are placed in the d orbitals of the atom.

De novo A *de novo* design is the construction of an entirely new structure based on new insights (such as computational models), rather than on the modification of an existing structure.

Density functional theory (DFT) Those *ab initio* methods that deal with the total electron density of the molecule.

Diastereomeric relationship A relationship between two objects that are not mirror images of each other but are still stereoisomers, meaning that they have different locations of atoms in space where the differences are not conformational. Diastereomers are expected to have different chemical and physical properties.

Diels–Alder reaction   A reaction named after the two chemists who discovered it. It is a pericyclic process resulting in the formation of a cyclohexene structure from two species, one having a 2-pi electron system and the other having a conjugated 4-pi electron system.

Diffusion function   A model for treatment of electrons that are at a considerable distance from the nucleus of the atom.

Effective core potential (ECP)   A potential function that represents the nonvalence (nonbonding) electrons of an atom.

Electron configuration   The manner in which electrons are added to atoms of increasing atomic number. The same protocol for electron configurations is applied to filling electrons into atomic and molecular orbitals.

Electron correlation   The avoidance behavior that characterizes electrons; the electrons do not want to be near each other.

Electron spin resonance (ESR)   A technique that uses the interaction between the nuclear magnetic spin of nuclei that exist as single electron (radical) species with an applied magnetic field to obtain spectroscopic data. ESR can be used to characterize the environment in which the atoms reside.

Electrostatic potential map   The result of an interaction between a charge and the surface of a molecule, which reveals positions of affinity for both negative and positive species within the molecule.

Empirical methods   Methods based on parameters assumed to be representative of atoms, bonds, and interactions in the target structure, employing classical mathematical relationships using those parameters to solve for structural energies.

Enantiomer   The named relationship between two objects that have the same composition and connectivity but are mirror images of each other and are not identical. Enantiomers have different locations of atoms in space where the differences are not conformational. A classic example of an enantiomeric relationship is the set of right and left hands.

Equatorial   Pertaining to the location of a group that is directly attached to a six-membered ring and lies in a plane roughly approximating that of the carbon atoms of that ring.

Excited state   A state in which the ground state (low-energy) electronic configuration of the molecule has been altered so that one electron in the structure is elevated to a higher level of energy.

f-block elements   The lanthanide and actinide metallic elements in which highest energy electrons are placed into the f orbitals of the atom.

Formal oxidation   When atoms react by either giving up or accepting electrons, the formal accounting of that process is denoted by the change in formal oxidation of an element relative to the number of electrons that the neutral atom would contain in a nonbonded state.

Frontier molecular orbitals   The highest-energy molecular orbitals of a compound, typically taken as those that are integral to a conjugated system.

Group theory   Mathematical methods commonly used to characterize the symmetry of compounds.

Half chair structure   The global energy maximum structure for a cyclohexane (C_6H_{12}) compound, which resembles a chair with one end of the structure, either the headrest or the footrest, flattened out in a planar arrangement.

Hamiltonian   The term H from the Schrödinger equation that serves as the operator for energy.

Hartree–Fock (HF) method   A self-consistent *ab initio* procedure that models the total number of electrons in a system N by N wave functions. In Hartree–Fock, electron correlation is not considered, and the variation principle is assumed.

Heterogeneous   Pertaining to a system of two or more different phases, such as a mixture of an insoluble solid and a liquid (e.g., sand in water) or a mixture of two insoluble liquids (e.g., oil and water).

Homogeneous   Pertaining to a system of only one phase, such as a mixture of a soluble solid and a liquid (e.g., NaCl in water) or a mixture of two soluble liquids (e.g., ethanol and water).

Hückel calculations   The mathematical treatment developed by Hückel to deal with compounds in which the electron interactions can be described as a matrix, with the sum of the squares of the coefficients of each column in the matrix representing electron density on one atom, and of each row the electron density on one orbital.

Hybridization A construct developed by Linus Pauling that became a central tool of valence bond theory to adopt blended atom orbitals (e.g., spherically symmetrical *s* orbitals and dumbbell-shaped *p* orbitals) to create hypothetical mixed atom orbitals (e.g., sp^3, which denotes a mixing of one part *s* and three parts *p*) in order to account for observed bond angles and lengths.

Isomers Compounds that have the same elemental composition but differ from each other in some other respect. The three major types of isomers are constitutional (atoms have different connectivity), conformational, and configurational (atoms occupy different locations in space, and those differences cannot be resolved without breaking and remaking bonds).

Kinetics The study of the rates of chemical reactions.

Lennard–Jones 12-6 potential This is used to describe weaker interactions, such as van der Waals forces, and has the functional form $V_{LJ}(r) = 4\varepsilon[(\frac{\sigma}{r})^{12} - (\frac{\sigma}{r})^6]$. The $\sim 1/r^{12}$ is a repulsive term that dominates at short distances, and the $\sim 1/r^6$ is an attractive term that dominates at large distances. The parameter ε is the depth of the well, and σ is related to the equilibrium bond distance $\sigma = 2^{1/6} r_e$.

Ligand An atom or a bonded grouping of atoms that forms a bond to a centrally located atom in a complex by donating its electrons to that atom.

Linear combination of atomic orbitals (LCAO) The process of combining the atomic basis functions to generate a wave function.

Many-body potential A potential energy function that includes many-body effects by changing the pair potential to incorporate interactions from nearby atoms.

Mechanism The description of bond-making and bond-breaking in a chemical process.

Merck molecular force field (MMFF) A version of a MM program developed by Merck Pharmaceutical Company to evaluate organic compounds of pharmacological potential.

Metallic Pertaining to elements that have a proclivity to give up valence electrons to other elements when combined with those elements to form compounds.

Metathesis Any reaction in which an atom (or group of atoms) in one reactant switches with another atom (or group of atoms) from another reactant in a chemical transformation.

Michaelis–Menton equation The expression of kinetics in biochemical systems is most often treated with the Michaelis–Menton equation in which a substrate (S) and an enzyme (E) react to form a product (P) by intermediate steps described as enzyme–substrate complexation (ES) and enzyme–product decomplexation, so that E + S are in equilibrium with ES, which is in equilibrium with EP, which is in equilibrium with E + P.

Molecular dynamics (MD) An application of a force-field method with classic equations of motion in a series of time steps. MD is useful to describe processes such as SIMS simulation or conformation searching.

Molecular mechanics (MM) A method that primarily uses a classical mechanics ball-and-spring approach to evaluate the structure of a compound.

Molecular orbitals Concept used to describe the distribution of electrons in a molecule when assuming that the orbitals no longer belong to the specific atoms of which the molecule is composed, but rather to the entire molecule.

Møller–Plesset method Electron correlation in *ab initio* methods that is based on perturbation theory.

Monte Carlo method A randomization of the three-dimensional locations of objects, generally used to seek a lower energy region that may not be accessible or evident by more regular manipulations of the same objects in a more orderly or logical manner.

Morse potential An empirical pair potential that describes the stretching of a chemical bond. The functional form is $V(r) = D_e \left(1 - e^{-\beta(r-r_e)}\right)^2$, where r_e is the equilibrium bond distance, D_e is the depth of the well, and β is related to the curvature near the minimum.

Nuclear magnetic resonance (NMR) A technique that uses the interaction of nuclear magnetic spin of certain nuclei (like ^{1}H and ^{13}C) with an applied magnetic field to obtain spectroscopic data that can be used to characterize the environment in which the atoms reside.

Nucleophilic behavior The preference for an electron-rich species to bond to the positive nucleus of another species, other than hydrogen.

Orbital A three-dimensional volume of space which is the probable location for finding electrons. If the orbital belongs to an atom, it is called an *atomic orbital*; if it belongs to the molecule, it is called a *molecular orbital*.

Pair potential A potential energy function that describes how the potential energy between a pair of atoms depends on the distance between them.

Pairwise additive assumption The assumption that the interaction between each pair of atoms is independent of the other atoms in the system. The total potential energy of the system is assumed to be equal to the sum of the interaction between each pair of atoms: $V_{tot}(r_1, r_2, \ldots, r_n) = \sum_i \sum_{j>i} V(r_{ij})$.

Pericyclic processes Processes (including the Diels–Alder reaction) with cyclic transition states that reflect the reorganization of sigma and pi bonds in the chemical transformation.

Pi electrons Those electrons positioned in orbitals that allow for side-to-side overlap. In valence bond theory, p orbitals are used most commonly to construct pi systems.

Potential energy function A mathematical function with a functional form that describes how the potential energy depends on the relative position of each atom in the system. The function represents the actual solution to the Schrödinger equation within the Born–Oppenheimer approximation for the physical system. The parameters of the functional form are fit to experimental data and also to data from *ab initio* calculations.

Quantum mechanics The physics of the very small. The rules of quantum mechanics govern atomic and molecular phenomena and determine the relative probabilities of electron locations.

R Denotes the spatial orientation of groups about a central atom in one of the pair of enantiomeric orientations, R and S.

Radical species Having a single, non-bonded electron.

Raman spectroscopy A technique for measuring a molecule's vibrational, rotational, or electronic energy, which depends largely on polarizability.

Relativistic effects As electrons get closer to the nucleus, their speeds approach the speed of light and the theory of relativity must be applied to them. This is especially important for atoms with larger atomic numbers.

S Denotes the spatial orientation of groups about a central atom in one of the pair of enantiomeric orientations, R and S.

Saturated A carbon compound is saturated if the carbons in the compound are bonded to as many hydrogen atoms as possible. Using the formula $C_n H_{2n+2}$, where n is the number of carbons, $2n + 2$ is the number of hydrogen atoms required for the compound to be classified as saturated.

Schrödinger equation Erwin Schrödinger (also spelled *Schroedinger*) developed the model of the hydrogen atom using wave functions upon which *ab initio* methods are founded. In this model, it is possible to solve for an electron's energy with great precision, but the electrons' locations can only be described probabilistically.

Secondary ion mass spectrometry (SIMS) Procedure that bombards a sample with primary ions, causing the sample to eject molecules and molecular fragments from the sample surface. The charged ejected particles (i.e., the secondary ions) are then analyzed by mass spectrometry.

Self-consistent field (SCF) A method of iterative refinement, starting from an arbitrary initial value and performing calculations of new values, replacing the initial value with the new values until the new (lower energy) and initial (higher energy) converge to an acceptable degree.

Semiempirical method A method that includes both quantum mechanics and empirical parameters to compute molecular structures and properties.

Sigma electrons Those electrons in a valence bond system that are positioned in end-to-end, overlapping atomic orbitals between two atoms. Typically, the atomic orbitals in such a bond have at least some percentage composition of s-type (symmetrically disposed around the atom) character.

Spin Electrons are described by many characteristic quantities, one of which is the spin quantum number, which can be either $+1/2$ or $-1/2$.

Stiff equations Sets of differential equations whose solutions vary over so great a span of values that their simultaneous integration is problematic by normal methods, such as Euler or Runge–Kutta.

Syn The disposition of two objects in reference to a plane so that the two objects are on the same side of the plane.

Thermodynamic Pertaining to the study or characteristics of energy flux in any chemical process.

Torsion The interaction of two bonds or groups about a central connecting bond (e.g., interactions of bonds A–B and C–D or groups A and D in a structure such as A-B-C-D).

Transition metals The collection of d- and f- block elements (including the actinides and lanthanides).

Transition states The high-energy structures (where the bonds are partially made or broken) that represent the lowest energy barrier that must exist between starting materials and products in reactions that create bonds.

Twist boat conformation A local energy minimum structure for cyclohexane (C_6H_{12}) that resembles a boat (with a bow, stern, and transom) whose sides are not parallel.

Valence bond theory A theory that describes molecules as a series of bonds made by the overlap and sharing of spin-paired electrons in the atomic orbitals between the atoms.

Van der Waals force The force between objects that arises from temporary electrostatic interactions of their surrounding electrons.

Variational principle Since an estimated wave function will never be equal to or lower than the actual energy, the result of a wave function calculation is used repeatedly as the approximation in successive calculations, until the approximate value converges with the calculated value.

Vibrational frequency The frequency of the vibration of atoms, which depends on the masses of the atoms that are bonded and the strength of the bond between them.

Vital force The theory that compounds from living systems, called *organic compounds*, can only be made by living systems, *in vivo* (in life), and not by humans, *in vitro* (in glassware).

Wave function The description of the probable distribution of electrons as a function of their x, y, z coordinates and spin.

ZINDO A program developed by Zerner that uses semiempirical quantum mechanical values to compute molecular spectra.

References

Allinger, N.L., Yuh, Y.H., Lii, J.-H. 1989. Molecular mechanics — the MM3 force-field for hydrocarbons. 1. *J. Am. Chem. Soc.*, 111: 8551–8566.

Amos, R.D. 1986. Calculation of polarizability derivatives using analytic gradient methods, *Chem. Phys. Lett.*, 124: 376–381.

Berry, J.I., Ewing, A.E., and Winograd, N. 2001. Biological Systems. In *ToF-SIMS: Surface Analysis by Mass Spectrometry*, Vickerman, J.C. and Briggs, D., Eds., SurfaceSpectraLtd and IMPublications, London, 595–626.

Borden, W.T., and Davidson, E.R. 1996. The importance of including dynamic electron correlation in *ab initio* calculations, *Acc. Chem. Res.*, 29: 67–75.

Brenner, D.W. 1990. Empirical potential for hydrocarbons for use in simulations of the chemical vapor deposition of diamond films, *Phys. Rev. B*, 42: 9458–9471.

Bur, S.K., Lynch, S.M., and Padwa, A. 2002. Influence of ground state conformations on the intramolecular Diels–Alder reaction, *Org. Lett.*, 4(4): 473–476.

Carloni, P., Rothlisberger, U., and Parrinello, M. 2002. The role and perspective of *ab initio* molecular dynamics in the study of biological systems, *Acc. Chem. Res.*, 35: 455–464.

Crutzen, P.J. 1996. My life with O3, NOx, and other YZOx compounds (Nobel Lecture), *Angew. Chem., Int. Ed. Engl.*, 35: 1758–1777.

Cundari, T.R. 2000. Computational studies of transition metal–main group multiple bonding, *Chem. Rev.*, 100: 807–818.

Fleming, I. 1976. *Frontier Molecular Orbitals and Organic Chemical Reaction*, John Wiley & Sons, New York.

Frenking, G., and Fröhlich, N. 2000. The nature of the bonding in transition-metal compounds, *Chem. Rev.* 100: 717–774.

Frieden, C. 1993. Numerical integration of rate equations by computer, *Trends Biochem. Sci.* 18: 58–60.

Garrison, B.J. 2001. Molecular Dynamics Simulations, the Theoretical Partner to State SIMS. In *ToF-SIMS: Surface Analysis by Mass Spectrometry*, Vickerman, J.C. and Briggs, D., Eds., SurfaceSpectraLtd and IMPublications, London, 233–257.

Garrison, B.J., Delcorte, A., and Krantzman, K.D. 2000. Molecule liftoff from surfaces, *Accts. Chem. Res.* 33: 69–77.

Gear, C.W. 1971. *Numerical Initial Value Problems in Ordinary Differential Equations*, Prentice Hall, Englewood Cliffs, NJ.

Gillespie, D.T. 1976. A general method for numerically simulating the stochastic time evolution of coupled chemical reactions, *J. Comp. Phys.*, 22: 403–434.

Gillespie, D.T. 1977. Exact stochastic simulation of coupled chemical reactions, *J. Phys. Chem.*, 81: 2340–2361.

Guirgis, G.A., Bell, S., Zheng, C., Durig, J.R. 2002. Infrared and Raman spectra, conformational stability, vibrational assignment, *ab initio* calculations and r0 structural parameters for N-methylpropargylamine, *Phys. Chem. Chem. Phys.*, 4: 1438–1450.

Guirgis, G.A., Zhu, X., Bell, S., Durig, J.R. 2001. Conformational analysis, barriers to internal rotation, *ab initio* calculations, and vibrational assignment of 4-fluoro-1-butyne, *J. Phys. Chem. A*, 105: 363–373.

Kibby, M.R. 1969. Stochastic method for the simulation of biochemical systems on a digital computer, *Nature*, 222: 298–299.

Li, J., and Bursten, B.E. 2001. The Electronic Structure of Organoactinide Complexes via Relativistic Density Functional Theory: Applications to the Actinocene Complexes $An(\eta^8\text{-}C_8H_8)_2$ (An = Th-Am). In *Computational Organometallic Chemistry*, Cundari, T. R., Ed., Marcel Dekker, New York.

Lipkowitz, K.B., D'Hue, C.A., Sakamoto, T., and Stack, J.N. 2002. Stereocartography: a computational mapping technique that can locate regions of maximum stereoinduction around chiral catalysts, *J. Am. Chem. Soc.*, 124: 14255–14267.

McAdams, H.H., and Arkin, A. 1997. Stochastic mechanisms in gene expression, *Proc. Natl. Acad. Sci.* 94: 814–819.

Mendes, P. 1993. GEPASI: a software package for modeling the dynamics, steady states and control of biochemical and other systems, *Comput. Applic. Biosci.*, 9: 563–571.

Mendes, P. 1997. Biochemistry by numbers: simulation of biochemical pathways with GEPASI 3, *Trends Biochem. Sci.*, 22: 361–363.

Mendes, P., and Kell, D.P. 1998. Non-linear optimization of biochemical pathways: applications to metabolic engineering and parameter estimation, *Bioinformatics*, 14: 869–883.

Mohamed, T.A., Guirgis, G.A., Nashed, Y.E., Durig, J.R. 1999. Spectra and structure of silicon-containing compounds. XXV. Raman and infrared spectra, r0 structural parameters, vibrational assignment, and *ab initio* calculations of ethyl chlorosilane-Si-d2, *Struct. Chem.*, 10: 333–348.

Molina, M.J. 1996. Polar ozone depletion (Nobel Lecture), *Angew. Chem., Int. Ed. Engl.*, 35: 1778–1785.

Monard, G., and Merz, K.M. 1999. Combined quantum mechanical/molecular mechanical methodologies applied to biomolecular systems, *Acc. Chem. Res.*, 32: 904–911.

Nguyen, T.C., Ward, D.W., Townes, J.A., White, A.K., Krantzman, K.D., and Garrison, B.J. 2000. A theoretical investigation of the yield-to-damage enhancement with polyatomic projectiles in organic SIMS, *J. Phys. Chem. B*, 104: 8221–8228.

Polavarapu, P.L. 1990. *Ab initio* vibrational Raman and Raman optical activity spectra, *J. Phys. Chem.*, 94: 8106–8112.

Press, W.H., Teukolsky, S.A., Vetterling, W.T., and Flannery, B.P. 1992. Integration of Ordinary Differential Equations. In *Numerical Recipes in Fortran, 2nd Ed.,* Cambridge University Press, Cambridge, Chapter 19.

Rowland, F.S. 1996. Stratospheric ozone depletion by chlorofluorocarbons (Nobel Lecture), *Angew. Chem., Int. Ed. Engl.,* 35: 1786–1798.

Siegbahn, P.E.M., and Blomberg, M.R.A. 2000. Transition-metal systems in biochemistry studied by high-accuracy quantum chemical methods, *Chem. Rev.* 100: 421–437.

Stuart, S., Tutein, A.B, and Harrison, J.A. 2000. A reactive potential for hydrocarbons with intermolecular interactions, *J. Chem. Phys,* 112: 6472–6468.

Todebush, P.M., Liang, G., and Bowen, J.P. 2002. Molecular mechanics (MM4) force field development for phosphine and its alkyl derivatives, *Chirality,* 14: 220–231.

Torrent, M., Solà, M., and Frenking, G. 2000. Theoretical studies of some transition-metal mediated reactions of industrial and synthetic importance, *Chem. Rev.* 100: 439–493.

Townes, J.A., White, A.K., Wiggins, E.N., Krantzman, K.D., Garrison, B.J., and Winograd, N. 1999. Mechanism for increased yield with the SF_5^+ projectile in organic SIMS: the substrate effect, *J. Phys. Chem. A,* 24: 4587–4589.

Van Stipdonk, M.J. 2001. Polyatomic Cluster Beams. In *ToF-SIMS: Surface Analysis by Mass Spectrometry,* Vickerman, J.C. and Briggs, D., Eds., SurfaceSpectraLtd and IMPublications, London, 309–346.

Venkatesan, H., Davis, M.C., Altas, Y., Snyder, J.P., and Liotta, D.C. 2001. Total synthesis of SR 121463 A, a highly potent and selective vasopressin V2 receptor antagonist, *J. Org. Chem.,* 66: 3653–3661.

Woods, R.J., Andrews, C.W., and Bowen, J.P. 1992. Molecular mechanical investigations of the properties of oxocarbenium ions. 1. Parameter development, *J. Am. Chem. Soc.,* 114: 850–858.

Zaric, R., Person, B., Krantzman, K.D., Garrison, B.J. 1998. Molecular dynamics simulations to explore the effect of projectile size on the ejection of organic targets from metal surfaces, *Int. J. Mass Spectrom. Ion Processes,* 174: 155–166.

Further Information

General Reviews

Cramer, C.J. 2002. *Essentials of Computational Chemistry: Theories and Models,* John Wiley & Sons, New York.

Leach, A.R. 2001. *Molecular Modelling: Principles and Applications, 2nd Ed.,* Pearson Education, Essex, England.

Lipkowitz, K.B., and Boyd, D.B., Eds. 1990–2002. *Reviews in Computational Chemistry,* Vols 1–18, John Wiley & Sons, New York.

Schleyer, P.R. (Editor-in-Chief) 1998. *Encyclopedia of Computational Chemistry,* 5 vols., John Wiley & Sons, New York.

Young, D. 2001. *Computational Chemistry: A Practical Guide for Applying Techniques to Real World Problems,* John Wiley & Sons, Inc, New York.

Web Sites

BerkeleyMadonna: http://www.berkeleymadonna.com

Chemical kinetics simulator (CKS): http://www.almaden.ibm.com/st/msim

Encyclopedia of Computational Chemistry: http://www.mrw.interscience.wiley.com/ecc/

Gepasi biochemical simulation: http://www.gepasi.org/

Journal of Chemical Education (JCE): http://JChemEd.chem.wisc.edu

KINSIM: http://www.wistl.edu/cflab/message.html

Kintecus: http://www.kintecus.com

NIH Center for Molecular Modeling: http://cmm.info.nih.gov/modeling/

Project SERAPHIM: http://ice.chem.wisc.edu/seraphim

Reviews in Computational Chemistry index: http://chem.iupui.edu/~boyd/rcc.html

STELLA: http://www.hps-inc.com/

33

Computational Astrophysics

33.1 Introduction 33-1
33.2 Astronomical Databases 33-1
Electronic Information Dissemination • Data Collection
• Accessing Astronomical Databases • Data File Formats
33.3 Data Analysis 33-4
Data Analysis Systems • Data Mining • Multi-Wavelength
Studies • Specific Examples
33.4 Theoretical Modeling 33-10
The Role of Simulation • The Gravitational n-body Problem
• Hydrodynamics • Magnetohydrodynamics and Radiative
Transfer • Planetary and Solar System Dynamics • Stellar
Astrophysics • Star Formation and the Interstellar Medium
• Cosmology • Galaxy Clusters, Galaxy Formation, and
Galactic Dynamics • Numerical Relativity • Compact
Objects • Parallel Computation in Astrophysics
33.5 Research Issues and Summary 33-16

Jon Hakkila
College of Charleston

Derek Buzasi
U.S. Air Force Academy

Robert J. Thacker
McMaster University

33.1 Introduction

Modern **astronomy/astrophysics** is a computationally driven discipline. During the 1980s, it was said that an astronomer would choose a computer over a telescope if given the choice of only one tool. However, just as it is impossible to separate "astronomy" from astrophysics, most astrophysicists would no longer be able to separate the computational components of astrophysics from the processes of data collection, data analysis, and theory. The links between astronomy and computational astrophysics are so close that a discussion of computational astrophysics is essentially a summary of the role of computation in all of astronomy. We have chosen to concentrate on a few specific areas of interest in computational astrophysics rather than attempt the monumental task of summarizing the entire discipline. We further limit the context of this chapter by discussing astronomy rather than related disciplines such as planetary science and the engineering-oriented aspects of space science.

33.2 Astronomical Databases

Physics and astronomy have been leaders among the sciences in the widespread use of and access to online access and data retrieval. This has most likely occurred because the relatively small number of astronomers is broadly distributed so that few astronomers typically reside in any one location; it also perhaps results because astronomy is less protective of data rights than other disciplines driven more by commercial spin-offs. Astronomers regularly enjoy online access to journal articles, astronomical catalogs/databases, and software.

33.2.1 Electronic Information Dissemination

All major astrophysical journals are available electronically [Boyce et al. 2001]. Several electronic preprint servers are also available [Ginsparg 1996, Hanisch 1999]. Furthermore, the Astrophysical Data Service (ADS) [Kurtz et al. 2000] is the electronic index to all major astrophysical publications; it is accessible from a variety of mirror sites worldwide. The ADS is a data retrieval tool that allows users to index journal articles by title, keyword, author, astronomical object and even by text search within an article. The ADS is a citation index as well as a tool capable of accessing the complete text of articles and in many cases the original data, although it has not served its original purpose (as designed by NASA) of allowing for the integrated data management of all astrophysics missions.

Electronic dissemination plays an important role in the collection of astronomical data. Many variable sources (such as high-energy transients and peculiar **galaxies**) often exhibit changes on extremely short timescales. Follow-up observations require the community to rapidly disseminate source locations, brightnesses, and other pertinent information electronically (notification sometimes must go out to other observatories on timescales as short as seconds). There are a variety of electronic circulars available to the observational community. The primary source of these is the International Astronomical Union (http://www.iau.org/).

33.2.2 Data Collection

Astronomy is generally an observational rather than experimental science (although there are laboratory experiments that emulate specific astronomical conditions). Modern astronomical observations are made using computer-controlled telescopes, balloon instruments, and/or satellites. Many of these are controlled remotely, and some are robotic.

All telescopes need electronic guidance because they are placed on moving platforms; even the Earth is a moving platform when studying the heavens. From the perspective of the terrestrial observer, the sky appears to pivot overhead about the Earth's equatorial axis; an electronic drive is needed to rotate the telescope with the sky to keep the telescope trained on the object. However, the great weight of large telescopes is easier to mount altazimuthally (perpendicular to the horizon) than equatorially. A processor is needed to make real-time transformations from equatorial to altazimuth coordinates so that the telescope can track objects; it is also needed to accurately and quickly point the telescope. Precession of the equinoxes must also be taken into account; the direction of the Earth's rotational axis slowly changes position relative to the sky. Computers have thus been integrated into modern telescope control systems.

Additional problems are present for telescopes flown on balloons, aircraft, and satellites. The telescopes must either be pointed remotely or at least have the directions they were pointed while observing accessible after the fact. Flight software is generally written in machine code to run in real-time. Stability in telescope pointing is required because astronomical sources are distant; long observing times are needed to integrate the small amount of light received from sources other than the sun, moon, and planets.

As an example, we mention computer guidance on the Hubble Space Telescope (HST). HST has different sensor types that provide feedback to maintain a high degree of telescope pointing accuracy; the Fine Guidance Sensors have been key since their installation in 1997. The three Fine Guidance Sensors provide an error signal so that the telescope has the pointing stability needed to produce high-resolution images. The first sensor monitors telescope pitch and yaw, the second monitors roll, and the third serves both as a scientific instrument and as a backup. Each sensor contains lenses, mirrors, prisms, servos, beam splitters, and photomultiplier tubes. Software coordinates the pointing of the sensors onto entries in an extremely large star catalog. The guidance sensors lock onto a star and then deviations in its motion to a 0.0028-arcsecond accuracy. This provides HST with the ability to point at the target to within 0.007

arcseconds of deviation over extended periods of time. This level of stability and precision is comparable to being able to hold a laser beam constant on a penny 350 miles away.

To complicate matters, many telescopes are designed to observe in **electromagnetic spectral** regimes other than the visible. An accurate computer guidance system is needed to ensure that the telescope is correctly pointing even when a source cannot be identified optically, and that the detectors are integrated with the other electronic telescope systems. Many bright sources at nonvisible wavelengths are extremely faint in the visible spectral regime. Furthermore, telescopes must be programmed to avoid pointing at bright objects that can burn out photosensitive equipment, and sometimes must avoid collecting data when conditions arise that are dangerous to operation. For example, orbital satellites must avoid operating when over the South Atlantic Ocean — in a region known as the South Atlantic Anomaly, where the Earth's magnetic field is distorted — because high-energy ions and electrons can interact with the satellite's electronic equipment. This can cause faulty instrument readings and/or introduce additional noise.

There are other examples where computation is necessary to the data collection process. In particular, we mention the developing field of **adaptive optics**. This process requires the fast, real-time inversion of large matrices [Beckers 1993]. *Speckle imaging techniques* (e.g., [Torgerson and Tyler 2002]) also remove atmospheric distortion by simultaneously taking short-exposure images from multiple cameras.

Detectors and electronics must often be integrated by one computer system. Many interesting specific data collection problems exist that require modern computer-controlled instrumentation. For example, radio telescopes with large dishes cannot move, and sources pass through the telescope's field-of-view as the Earth rotates. Image reconstruction is necessary from the data stream because all sources in the dish's field of view are visible at any given time. Another interesting data collection problem occurs in infrared astronomy. The sky is itself an infrared emitter, and strong source signal-to-noise can only be obtained by constantly subtracting sky flux from that of the observation. For this reason, infrared telescopes are equipped with an oscillating secondary mirror that wobbles back and forth, and flux measurements alternate between object and sky. A third example concerns the difficulty in observing x-ray and gamma-ray sources. X-ray and gamma-ray sources emit few photons, and these cannot be easily focused due to their short wavelengths. Computational techniques such as Monte Carlo analysis are needed to deconvolve photon energy, flux, and source direction from the instrumental response. Very often, this analysis must be done in real-time, requiring a fast processor. In addition, the telescope guidance system must be coordinated with onboard visual observations because the visual sky still provides the basis for telescope pointing and satellite navigation.

33.2.3 Accessing Astronomical Databases

Database management techniques have allowed astronomers to address the increasing problem of storing and accurately cross-referencing astronomical observations. Historically, bright stars were given catalog labels based on their intensities and on the constellation in which they were found. Subsequent catalogs were sensitive to an increased number of fainter objects and thus became larger while simultaneously assigning additional catalog numbers or identifiers to the same objects. The advent of photographic and photoelectric measurement techniques and the development of larger telescopes dramatically increased catalog sizes. Additional labels were given to stars in specialty catalogs (e.g., those containing bright stars, variable stars, and binary stars). Solar system objects such as asteroids and comets are not stationary and have been placed in catalogs of their own.

In 1781, Charles Messier developed a catalog of fuzzy objects that were often confused with comets. Subsequent observations led to identification of these extended objects as star clusters, galaxies, and gaseous nebulae. Many of these extended astronomical sources (particulary regions of the **interstellar medium**) do not have easily identified boundaries (and the observed boundaries are often functions of the **passband** used); this inherent fuzziness presents a problem in finding unique identifiers as well as a database management problem. Furthermore, sources are often extended in the radial direction (sources are three-dimensional), which presents additional problems because distances are among the most difficult astrophysical quantities to measure.

As astronomy entered the realm of multi-wavelength observations in the 1940s, observers realized the difficulty in directly associating objects observed in different spectral regimes. An x-ray emitter might be undetectable or appear associated with an otherwise unexciting stellar source when observed in the optical. Angular resolution is a function of wavelength, so it is not always easy to directly associate objects observed in different passbands.

Temporal variability further complicates source identification. Some objects detected during one epoch are absent in observations made during other epochs. Signal-to-noise ratios of detectors used in each epoch contribute to the source identification problem. Additionally, gamma-ray and x-ray sources tend to be more intrinsically variable due to the violent, nonthermal nature of their emission. Examples of sources requiring access via their temporal characteristics include supernovae, gamma-ray bursts, high-energy transient sources, and some variable stars and extra-galactic objects.

There are tremendous numbers of catalogs available to astronomers, and many of these are found online. Perhaps the largest single repository of online catalogs and metadata links is VisieR (http://vizier.u-strasbg.fr/viz-bin/VizieR) [Ochsenbein et al. 2000]. Online catalogs also exist at many other sites, including HEASARC (High Energy Astrophysics Science Archive Research Center at http://heasarc.gsfc.nasa.gov/), HST (Hubble Space Telescope at http://www.stsci.edu/resources/), and NED (NASA/IPAC Extragalactic Database at http://nedwww.ipac.caltech.edu/).

Large astronomical databases exist for specific ground-based telescopes and orbital satellites. Some of these databases are large enough to present information retrieval problems. Examples of these databases are 2MASS (Two Micron All Sky Survey) [Kleinmann et al. 1994]; DPOSS (Digitized Palomar Observatory Sky Survey) [Djorgovski et al. 2002]; SDSS (Sloan Digital Sky Survey) [York et al. 2000]; and NVSS (The NRAO VLA Sky Survey) [Condon et al. 1998]. Databases span the range of astronomical objects from stars to galaxies, from active galactic nuclei to the interstellar medium, and from gamma-ray bursts to the cosmic microwave background. Databases are often specific to observations made in predefined spectral regimes rather than specific to particular types of sources; this reflects the characteristics of the instrument making the observations.

33.2.4 Data File Formats

The astronomic community has evolved a standard data format for the transfer of data. The Flexible Image Transport System (FITS) has broad general acceptance within the astronomic community and can be used for transferring images, spectroscopic information, time series, etc. [Hanisch et al. 2001]. It consists of an ASCII text header with information describing the data structure that follows. Although originally defined for nine-track, half-inch magnetic tape, FITS format has evolved to be generic to different storage media. The general FITS structure has undergone incremental improvements, with acceptance determined by vote of the International Astronomical Union.

Despite the general acceptance of FITS file format, other methods are also used for astronomical data transfer. This is not surprising, given the large range of data types and uses. Some data types have been difficult to characterize in FITS formats (such as solar magnetospheric data). Satellite and balloon data formats are often specific to each instrument and/or satellite.

Due to the large need for storing astronomical images, a wide range of image compression techniques have been applied to astronomy. These include fractal, wavelets, pyramidal median, and JPEG (e.g. [Louys et al. 1999]).

33.3 Data Analysis

Mathematical and statistical analyses are the driving forces behind the use of computation in astrophysics.

Data analysis and theoretical software can be accessed at a variety of sites worldwide. A few of the most well-known sites include the Astrophysics Source Code Library (http://ascl.net/), the UK Starlink site (http://star-www.rl.ac.uk/), and the Astronomical Software and Documentation Service at STScI, (http://asds.stsci.edu). Data analysis tools written in RSI's proprietary IDL programming language are available at the IDL Astronomy User's Library (http://idlastro.gsfc.nasa.gov/homepage.html). IDL has become a computing language of choice by many astronomers because it has been designed as an image-processing language, it has been written with mathematical and statistical uses in mind, it can handle multidimensional arrays easily, and it has many built-in data visualization tools.

33.3.1 Data Analysis Systems

Some 20 years ago, a myriad of different data analysis systems existed within astronomy. Typically, each institution (or sometimes individual groups within an institution) had its own data analysis package, and compatibility between these packages was limited or nonexistent. In the fall of 1981, however, astronomers at Kitt Peak National Observatory began development of the Image Reduction and Analysis Facility (IRAF), intended to serve as a general-purpose, flexible, extendable data reduction package. IRAF has grown beyond its original use primarily by ground-based optical astronomers to encompass space-based experiments as well at wavelengths ranging from x-ray to infrared.

IRAF is currently a mature system, with new releases occurring roughly annually, and is operated by about 5000 users at 1500 distinct sites. It is supported under a number of different computer architectures running UNIX or UNIX-like (e.g., Linux) operating systems. Oversight of IRAF development is formalized, with a Technical Working Group and various User's Committees overseeing evolution of the software. Numerous large astronomical projects have adopted IRAF as their data analysis suite of choice, typically by providing extensions to the basic system. These projects include the x-ray "Great Observatory" Chandra, the Hubble Space Telescope, PROS (ROSAT XRAY Data Analysis System), and FTOOLS (a FITS utility package from HEASARC).

The IRAF core system provides data I/O tools, interactive graphics and image display tools, and a variety of image manipulation and statistical tools. Commonly available "packages" that are part of the standard installation include tools for basic CCD data reduction and photometry, and support one-dimensional, two-dimensional, echelle, and fiber spectroscopy. Most tasks can be operated in either interactive or batch mode.

IRAF supports the FITS file format, as well as its own internal data type. In addition, extensions exist to handle other data types. Currently, those include STF format (used for HST data), QPOE format (used for event list data such as from the x-ray and EUV satellites), and PLIO for pixel lists (used to flag individual pixels in a region of interest). IRAF uses text files as database or configuration files, and provides a number of conversion tools to produce images from text-based data. Binary tables can be manipulated directly using the TABLES package.

Other general-purpose data reduction packages coexist with IRAF. These include the Astronomical Image Processing System (AIPS), and its successor (AIPS++), developed at the U.S. National Radio Astronomy Observatory (NRAO), and is still the image analysis suite of choice for radio astronomy. Image processing has been a very important subdiscipline within computational astrophysics, and we mention a number of image reconstruction methods with special applications to astronomy: the Maximum Entropy Method (e.g., [Lasenby et al. 2001]; the Pixon Method [Piña and Puetter 1993]; and Massive Inference [Skilling 1998]). XIMAGE and its relatives (XSPEC and XRONOS) serve a similar function for the x-ray astronomy community. A significant number of optical astronomers use Figaro, developed at the Anglo-Australian telescope. Figaro is particularly popular throughout Australia and the United Kingdom. Recently, Figaro has been adapted to run within the IRAF system, allowing users to have the best of both worlds. While there is no formal software system operating under IDL, the profusion of astronomical programs available in that proprietary language makes it worthy of mention.

One of the most significant drivers for development of mission-independent data analysis software within astronomy has been NASA's Applied Information Systems Research Program (AISRP). This program has encouraged the development of software to serve community-wide needs and has also fought the recurrent tendency for each project to develop its own software system, but instead to write software "packages" within the IRAF or IDL architecture. Of course, a general data analysis system is not the best solution for all specialized needs, particularly for space-based astronomy. In these cases, many missions have developed their own packages, and not all of these exist within an IRAF/AIPS/XIMAGE/IDL framework.

33.3.2 Data Mining

For many years, both NASA and ESA (the European Space Agency) have collected and preserved data from observatories in space. Similar activities are underway (although with varying degrees of success) at ground-based observatories. Most of these archives are available online, although the heterogenous nature of user interfaces and metadata formats — even when constrained by HTML and CGI — can make combining data from multiple sources an unnecessarily involved process. In addition, astronomical archives are growing at rates unheard of in the past: terabytes (TB) per year are now typical for most significant archives. Two recently established databases that illustrate the trend are the MACHO database (8 TB) and the Sloan Digital Sky Survey (15 TB).

The simplest kinds of questions one might ask of these kinds of data sets involve resource discovery; for example, "Find all the sources within a given circle on the sky," or "List all the observations of a given object at x-ray wavelengths." Facilities such as SIMBAD [Wenger et al. 2000]; VisieR [Ochsenbein et al. 2000]; NED [Mazzarella et al. 2001]; and ASTROBROWSE [Heikkila et al. 1999] permit these kinds of queries, although they are still far from comprehensive in their selection of catalogs and/or data sets to be queried. One problem arises from the nature of the FITS format, which is poorly defined as far as metadata are concerned; XML may provide a solution here, but astronomers have yet to agree on a consistent, common set of metadata tags and attributes.

Difficulties due to heterogeneous formats and user interfaces are being addressed by a number of so-called **virtual observatory** projects, such as the U.S. National Virtual Observatory (NVO), the European Astrophysical Virtual Observatory (AVO), and the British Astrogrid Consortium. The most basic intent of all these projects, which are coordinated with one another at some level, is to deliver a form of integrated access to the vast and disparate collection of existing astronomical data. In addition, all intend to provide data visualization and mining tools.

In its most encompassing form, astronomical data mining involves combining data from disparate data sets involving multiple sensors, multiple spectral regimes, and multiple spatial and spectral resolutions. Sources within the data are frequently time-variable. In addition, the data are contaminated with ill-defined noise and systematic effects caused by varying data sampling rates and gaps and instrumental characteristics. Finally, astronomers typically wish to compare observations with the results of simulations, which may be performed with mesh scales dissimilar to that of the observations and which suffer from systematic effects of their own. It has been observed [Page 2001] that data mining in astronomy presently (and in the near future) focuses on the following functional areas:

1. Cross-correlation to find association rules
2. Finding outliers from distributions
3. Sequence analysis
4. Similarity searches
5. Clustering and classification

Despite the difficulties outlined above, the nature of astronomical data — which are generally freely available and in computer-accessible form — has led to numerous early applications of data mining techniques in the field. As far back as 1981, FOCAS (Faint Object Classification and Analysis System) [Jarvis and Tyson 1981] was developed for the detection and classification of images on astronomical plates for the automatic assembly of catalogs of faint objects. Neural nets and decision trees have also been applied for the purposes of discriminating between galaxies and stars in image data [Odewahn 1992, Fayyad 1996] and for morphological classification of galaxies [Storrie-Lombardi 1992].

More recently, projects such as JPL's Diamond Eye [Roden et al. 1999] have begun to experiment with more general data mining enterprises. In this particular case, users interact with data mining servers via a Java applet (and thus need not have any particular data mining expertise themselves). Algorithms being tested include adaptive recognition (and its application to dynamic events) and ad hoc queries. Another current program is SKICAT (http://www-aig.jpl.nasa.gov/public/mls/skicat/skicat_home.html), which is an integrated software system applying image processing, database management, and AI classification to large image database analysis. The basic image processing routines detect objects and measure a set of features (surface brightness, extent, morphology, etc.) for each object. Algorithms such as GID3*, O-BTree, and RULER are used to produce decision trees and classification rules based on training data, and these classifiers are then applied to the new data.

The greatest near-term successes of data mining are likely to arise from its application to large but coherent data sets. In this case, the data has a common format and noise characteristics, and data mining applications can be planned from the beginning. Perhaps the most ambitious of such ongoing projects are the Sloan Digital Sky Survey (SDSS) [York et al. 2000] and 2MASS (e.g. [Nikolaev et al. 2000]). SDSS uses a dedicated 2.5-meter telescope to gather images of approximately 25% of the sky (some 10^8 objects), together with spectra of approximately 10^5 objects of cosmological significance. Pipelines have been developed to convert the raw images into astrometric, spectroscopic, and photometric data that will be stored in a common science database, which are indexed in a hierarchical manner. The science database is accessible via a specialized query engine. The SDSS has required Microsoft to put new features into its SQL server; in particular, Microsoft has added a tree structure to allow rapid processing of queries with geographic parameters. The Two Micron All Sky Survey (2MASS) is another high-resolution infrared sky survey that contains massive amounts of data for discrete as well as nebulous sources.

The key problem in astronomical data mining will most likely revolve around interpretation of discovered classes. Astronomy is ruled by instrumental and sampling biases; these systematic effects can cause data to cluster in ways that mimic true source populations statistically. Because data mining tools are capable of finding classes that are only weakly defined statistically, these tools enhance the user's ability to find "false" classes. Subtle systematic biases are present (but minimized) even in controlled cases when data are collected from only one instrument. The astronomical community must be careful to test the hypothesis so that class structures are not produced by instrumental and/or sampling biases before accepting the validity of newly discovered classes.

33.3.3   Multi-Wavelength Studies

Multi-wavelength studies have become increasingly important in astronomy, as new spectral regimes (x-ray, extreme UV, and gamma ray) have opened to practitioners, and astronomers have become aware that most problems cannot be adequately addressed by studies within any one spectral band. Such studies are now recognized as essential to the understanding of objects ranging from the Sun and stars, through x-ray bursters and classical novae, to the interstellar medium and active galactic nuclei. The computational requirements peculiar to multi-wavelength astrophysics essentially fall into one of two categories, where the distinction is in the time rather than the spectral domain.

In the first case, we have situations in which the timescales associated with the phenomena under study are long compared with typical observational timescales. One such case is in studies of the interstellar medium (ISM), which is important in astrophysics because galactic gas and dust clouds are the source of new generations of stars. The ISM becomes more enriched as generations of stars die and return heavy elements to it. In the study of the ISM, one can identify three characteristic temperature scales: <100 K, $\approx 10^4$ K, and $\approx 10^7$ K; these temperatures necessitate observations at radio/infrared, optical/UV, and x-ray wavelengths, respectively (e.g., [Zhang et al. 2001]). In each case, the strength of its emission or absorption depends on the local density, composition, metallicity, temperature, and distribution of ambient photons. Because this causes the interstellar medium (and to a lesser extent the intergalactic medium) to interfere with observations of other galactic and extragalactic sources, stellar, galactic, and extragalactic astronomers are interested in knowing the radiative properties of the interstellar medium, which is by definition nebulous and three-dimensional, as well as knowing where this material is located. An example of code used to locate the interstellar medium in the visual spectral regime can be found at http://ascl.net/extinct.html [Hakkila et al. 1997]. Because the characteric evolutionary timescale of the ISM is long by human standards, simultaneous (and even contemporaneous) multi-wavelength observations are unnecessary. In this milieu, catalogs and databases come into their own, and archival multi-wavelength research is possible, focusing on spatial rather than temporal correlations.

An idea of the numerous databases and collections of online data available on the ISM can be found at http://adc.gsfc.nasa.gov/adc/quick_ref/ref_ism.html.

A different situation occurs with variable sources such as gamma-ray bursters, and extremely short-duration and high-energy events occurring at cosmological distances. The source of the bursts is as yet unknown, and they may be created by mergers of a pair of neutron stars or black holes, or by a hypernova, a type of exceptionally violent exploding star. Gamma-ray bursts and their afterglows have been detected across the electromagnetic spectrum, and further study of these objects clearly calls for multi-wavelength studies (e.g. [Galama 1999]). However, unlike the case obtaining for the ISM, gamma-ray bursts have timescales ranging from milliseconds up to approximately 10^3 seconds, and thus coordinated and simultaneous (or near-simultaneous) multi-wavelength observations are essential. In this case, the computational demand is more on rapid deployment of various computer-controlled telescopes (both on the ground and in space) than on correlation analyses of existing databases. Thus, systems such as GCN (GRB Coordinates Network) have been developed [Barthelmy et al. 2001] to distribute locations of GRBs to observers in real-time or near-real-time, and to distribute reports of the resulting follow-up observations.

33.3.4 Specific Examples

33.3.4.1 Cosmic Microwave Background (CMB) Data Analysis

One of the main goals of CMB data analysis is to derive the power spectrum of the temperature fluctuations, which correlates directly to fluctuations in the density of matter in the early Universe. The precise spectrum of matter perturbations depends acutely on cosmological parameters and, hence, CMB data is an excellent diagnostic for determining the parameters of cosmological models. However, the task of constructing the power spectrum is daunting because the raw CMB signal has instrument noise, as well as noise from the interstellar medium and other astrophysical objects, imposed upon it. To further complicate matters, all these sources of noise are often correlated.

Analysis proceeds by first creating a physical map from the time series of instrument pointings. A pixel map is first constructed by dividing up the area of sky surveyed. For the COBE experiment [Smoot et al. 1992], only a few thousand pixels were necessary; while for the PLANCK Surveyor satellite (http://astro.estec.esa.nl/SA-general/Projects/Planck/), tens of millions of pixels are required. Thus, the main step in creating the map is the separation of noise, which can be done under the assumption that the time-series of noise signals is drawn from a Gaussian distribution. Linear algebra methods are used to calculate the pixel-pixel noise correlation, which is then used to construct the map. Brute-force methods are inefficient and exploiting sparse matrix methods is the only way to do the calculation efficiently. Nonetheless, the calculation is sufficiently large that massively parallel computing is necessary, and a general analysis package (MADCAP) has been developed [Borrill 1999].

Having constructed the map, the next step is to calculate the power spectrum. This is a significantly more difficult process than map creation, because each pixel contains information about the signal on all scales within the map. Further, the power spectrum that produces the signal must be derived from a maximum likelihood analysis, which in turn requires an iterative process. The steps involved in the iterative process are extremely computationally costly because correlation matrices and numerical derivatives must be calculated multiple times. Estimates of the total number of flops required to analyze the PLANCK data set are close to 10^{24}, which would take almost 1000 years on the Earth Simulator alone. Even smaller data sets, such as MAXIMA-2, require tens of petaflops. Ultimately, analysis of the larger data sets requires the development of new analysis methods.

33.3.4.2 Gamma-Ray Burst Data Analysis

Gamma-ray bursts (GRBs) are short bursts of primarily gamma-radiation having fluxes and spectra that evolve on short timescales. Evidence suggests that GRBs are produced when relativistic shock waves collide with each other and with the ambient interstellar medium. The source of the shocks is currently presumed to be a *hypernova*; a supernova variant in which a significant portion of the collapsing stellar core's energy is focused into shocked bipolar accelerated particle beams. GRBs exhibit a wide variety of complex behaviors and yet there is evidence of multiple GRB classes. Class identification is as important in astrophysics as in other sciences; the properties of distinct GRB classes can lead to a better general understanding of GRB physics as well as to a better understanding of the different sources and environments producing the classes. However, class identification is not helpful if the mechanisms responsible for the producing the classes cannot be determined.

Data mining techniques have proven useful in the study of GRB classes. Data mining techniques are needed to identify clusters in the GRB attribute space because individual GRB behaviors overlap. Some overlap results from the large intrinsic range of GRB behaviors, some is due to distinctly different properties of GRB classes, and some is due to observational and instrumental bias. Bias can cause phantom classes to appear by creating clustering where no distinct source populations exist.

A reference to data mining techniques applied to GRBs can be found in [Hakkila et al. 2003]. GRB data mining has thus far been used primarily with the large GRB database collected by BATSE (the Burst And Transient Source Experiment on NASA's Compton Gamma-Ray Observatory) because this experiment has well-documented properties [Paciesas et al. 1999]. Data mining tools identify three distinct GRB classes rather than two known historically. Detailed analyses find that the newest (third) GRB class does not represent a separate source population; it is, instead, produced by observational biases resulting primarily from low signal-to-noise observations and from the instrumental trigger characteristics [Hakkila et al. 2003]. This successful result indicates that scientific data mining can be used not just to determine that classes exist, but also to determine *why* they exist.

33.3.4.3 Time Series Analysis

Astronomers make use of time series analysis techniques for a variety of purposes, including studies of variable stars and cataclysmic variables [Gilliland et al. 1998, Kiss et al. 2001]; pulsating or oscillating stars [Buzasi et al. 2000, Poretti et al. 2002]; asteroid rotation rates, active galactic nuclei [Pronik et al. 1999]; and detection of extrasolar planets [Brown et al. 2001]. Typically, astronomical time series suffer from relatively low signal-to-noise ratios, uneven sampling, and numerous gaps, at times leading to severe aliasing at the $1 \, \text{day}^{-1}$ frequency and difficulties in estimating a traditional autocorrelation function. In addition, some applications (e.g., AGN studies and planetary detection) are dominated by nonsinusoidal signals or pulses.

Historically, the primary tools used to support these efforts have been discrete Fourier transforms and periodograms [Scargle 1982]. Both of these estimators of the autocorrelation function can be defined in such a way as to satisfactorily represent unevenly sampled data, and small gaps in the time series can be handled using clever binning or interpolation techniques, but these necessarily distort the data and lead to the loss of information. Early efforts to deal with the difficulties raised by irregular sampling and gaps focused on applying the CLEAN algorithm [Roberts et al. 1987] to apply a nonlinear deconvolution in the frequency domain. Unfortunately, CLEAN, originally developed for use by radio astronomers, suffers from nonuniqueness as well as the tendency to fail when aliasing problems are severe. In some cases, aliasing can be minimized or eliminated by experimental design (e.g., GONG), but more often astronomers are confronted with this fundamental problem and this has driven numerous recent efforts in the area.

Recent developments in this area focus on the application of nonlinear, nonstationary techniques and Bayesian methods. The wavelet transformation shows great promise [Abry et al. 1995; Scargle 1997] because, unlike the DFT, it can be used to construct power spectrum estimators that are nearly independent of the signal shape and amplitude in the presence of noise. Such estimators are likely to find increasing use in the planetary-detection and AGN modeling communities. Perhaps an even more useful application of wavelets is in the denoising of power spectra, a technique pioneered by the solar physics community. Bayesian methods are also increasing in popularity, as they are well-suited to finding change points in long series of time-tagged data such as is typically obtained from high-energy astrophysics experiments [Scargle 1998].

A rapidly growing difficulty is the size of power spectra produced by astronomical experiments. Helioseismological observations can easily give rise to time series with in excess of 10^7 points, and asteroseismological observations are rapidly approaching this level [Schou and Buzasi 2001]. Achieving the maximum time resolution inherent in data such as time-tagged x-ray photon lists can require estimators with in excess of 10^9 points [Ransom et al. 2002]. Furthermore, upcoming space-based experiments such as Eddington and Kepler will give rise to data sets that are so large as to mandate the use of automated techniques.

33.4 Theoretical Modeling

Prior to the advent of computers, theoretical modeling consisted largely of solving idealized systems of equations for a given problem. Very often, to make problems tractable, simplifying assumptions such as spherical symmetry and linearization of the problem would be necessary. Rapid numerical solutions of equations avoid the need for simplifying assumptions, but at a cost: one can no longer achieve an elegant formula for the solution to the problem, and insight often derived from manually solving the equations is lost.

33.4.1 The Role of Simulation

Although computation is commonplace in theoretical modeling, perhaps the most heavily computationally biased aspect is *simulation*. Simulation can be viewed as an extension of finding numerical solutions to a given equation set; however, the set of equations is often enormous in size (such as that produced by the gravitational interaction problem between a large number of bodies). Simulation also often involves visualizing the "data set" to help understand the phenomenon being studied, and the systems under investigation are almost always cast as an Initial Value Problem.

The roots of simulation in astrophysics can be traced back to at least the 1940s. Driven by a desire to understand the clustering of galaxies, Holmberg built an analog computer consisting of light sources and photocells to simulate the mutual interaction of two galaxies via gravity. Today, almost all simulation is conducted on digital computers. Development of fast, efficient algorithms for solving complex equation sets can often lead to programs containing tens of thousands of lines of code. Although it has been the tradition for individual researchers to develop codes in isolation, the past few years have seen the appearance of collaborations of researchers who work together on large coding projects. This trend is likely to continue and it is probable that in the near future researchers will converge to using a handful of readily available simulation packages (e.g., NEMO, http://bima.astro.umd.edu/nemo/).

33.4.2 The Gravitational *n*-body Problem

Although Newton's Law of Universal Gravitation has been supplanted by **General Relativity**, Newton's Law remains highly accurate for a very large number of astrophysical problems. However, solving the interaction problem for any number of bodies (*n* bodies) is difficult because at first appearances, the number of operations scales as n^2. However, provided that small errors in the force calculation are acceptable (RMS errors typically less than 0.5%), then approximate solutions can be found using order $n \log n$ operations. Roughly speaking, the algorithms used by researchers fall into two categories: treecodes [Barnes and Hut 1986] and grid (FFT) methods [Hockney and Eastwood 1981]. Treecodes are usually about an order of magnitude slower than grid codes for homogeneous distributions of particles, but are potentially much faster for very inhomogeneous distributions. To date, the largest gravitational simulations conducted contain approximately 1 billion particles, and have been used to coarsely simulate volumes representing as much as 10% of the entire visible universe [Evrard et al. 2002].

33.4.3 Hydrodynamics

Hydrodynamic modeling — or equivalently, computational fluid dynamics — plays an extremely important role in astrophysics. Although most astrophysical **plasmas** are not fluids in the everyday sense, the physical description of them is the same. Modern hydrodynamic methods fall into two main groups: Eulerian (fixed) descriptions and Lagrangian (moving) descriptions. Eulerian descriptions can be broadly decomposed into finite difference and finite element methods. In astrophysics, the finite difference method is by far the most common approach. Lagrangian descriptions can be decomposed into (1) "moving mesh" methods where the grid deforms with the flow in the fluid, and (2) particle methods, for which Smoothed Particle Hydrodynamics (SPH) is a popular example [Gingold and Monaghan 1977].

Because shocks play an important role in the evolution of stars and the ISM, a significant amount of research has focused on "shock capturing methods." Most early approaches to shock capturing, and indeed a number of methods still in use today, provide stability by using an artificial viscosity to smooth out flow discontinuities (shocks). Although these methods work well, they often introduce additional, unwanted dissipation into the simulation. Perhaps the best alternative approach is the Godunov's Method [Godunov 1959], which is a simple example of a first-order method where the Riemann shock tube problem is solved at the interface of each grid cell. More modern algorithms have extended this idea to higher-order integration schemes, such as the Piecewise-Parabolic Method [Collela and Woodward 1984].

33.4.4 Magnetohydrodynamics and Radiative Transfer

Magnetic fluid dynamic modeling (MHD) is the focus of a large amount of research in computational astrophysics [Falgarone and Passot 2003]. The system of equations for MHD is that of hydrodynamics plus the addition of coupling terms corresponding to magnetic and electric forces and Maxwell's equations that constrain and evolve the magnetic and electric fields. Because of the severe complexities arising from the divergenceless nature of the magnetic field, most MHD methods are finite difference; and although particle methods have been used, quite often they produce significant integration errors.

Modern methods, as in hydrodynamics, often cast the problem in terms of a system of conservation laws. It is usual to look at variation along a given axis direction and to recast the problem in terms of "characteristic variables" that are constructed from eigenvalues and the primitive variables, such as density, pressure, and flow speed. Such recasting aids the development of the numerical method because timestepping can be viewed as propagating the system an infinitesimal amount along a characteristic. This formulation often allows development of stable integration schemes that produce accurate numerical solutions even when large time steps are used.

Radiative transfer (RT), the study of how radiation interacts with gaseous plasmas, is an extremely difficult problem. It bears parallels to gravity in that all points within a system can usually affect all others, but is further complicated by the possibility of objects along any given direction producing non-isotropic attenuation. The radiation intensity is a function of position, two angles for direction of propagation, and frequency — a total of six independent variables. There any many different approaches to solving RT, ranging from explicit ray tracing to Monte Carlo methods, as well as characteristic methods [Peraiah 2001]. Much of the modern research effort focuses on deriving useful approximation methods that ease the computational effort required.

33.4.5 Planetary and Solar System Dynamics

Recent advances in telescope instrumentation have led to a cascade of discoveries of extra-solar planets and, at the time of writing, more than 100 extra-solar planets are known. Consequently, there is now a large amount of interest in studying planet and solar-system formation. Solar-system formation occurs during star formation, and the inherent differences in the planets are due to a differentiation process that enables different elements to condense out of the solar nebula at different radii.

Planets form by hierarchical merging processes within the disk of the solar nebula. Dust grains form the first level of the hierarchy and planets the last, while objects of all mass scales and sizes exist in between. It should be noted that representing this variation of masses and sizes within a simulation is impossible because resolution is always limited by the available computing power and memory. Theoretical models of the agglomeration process must include hydrodynamics and gravity, although currently there is debate about the role of hydrodynamics in gas giant planet formation. At present, theory can be roughly divided into two approaches: (1) the study of stability properties of the solar nebula disk from an analytic perspective, and (2) the simulation of the process from a first principles perspective. Simulations with a million mass elements, designed to follow the agglomeration process in the inner part of the solar system, were conducted in 1998 [Richardson et al. 2000]. More recent hydrodynamic simulations [Mayer et al. 2002] using the SPH technique have shown that gas giant planets can form extremely rapidly because of instabilities in protoplanetary disks.

The realization that our Solar System contains many small asteroids and meteorites capable of causing severe damage to the Earth has renewed interest in solar system dynamics. Calculating accurate orbits for these systems is difficult because they often have chaotic orbits. Chaotic systems place great demands on numerical integrations because truncation errors can rapidly pollute the integration. Thus, the integration schemes used must be highly accurate (often quadruple precision is used), and much effort has been devoted to "long-term" integrators (such as "sympletic integrators," see [Clarke and West 1997]) that preserve numerical stability over long-periods of simulation time. The chaos observed in long-term simulations of the solar system inspired a new theory [Murray and Holman 1999] that demonstrates that the Solar System is chaotic (Uranus could possibly be ejected) but the timescale for this is extremely long (10^{17} years).

33.4.6 Stellar Astrophysics

Among the first astrophysical problems to be addressed using modern computational methods were models of stellar interiors [Henyey et al. 1959; Cox et al. 1960] and atmospheres [Kurucz 1969]. One simplifying assumption needed for early stellar codes was that of Local Thermodynamic Equilibrium, which meant that stellar structural variations were not expected to occur on short timescales. Such assumptions are no longer necessary: stellar interior and atmosphere codes have become increasingly complex as computers and computational techniques have evolved. Theoreticians have been able to study rapid evolutionary phases and complex atmospheric processes in stars. Some of the difficult problems currently being addressed include the evolution of rotating stars [Meynet and Maeder 2002]; radial and nonradial stellar pulsations [Buchler et al. 1997; Crsico and Benvenuto 2002]; stellar magnetospheres [Wade et al. 2001]; evolution of stars in binary systems [Beer and Podsiadlowski 2002]; and supernovae [Woosley et al. 2002].

33.4.7 Star Formation and the Interstellar Medium

One of the greatest challenges in astrophysics is understanding the star formation process. Star formation is an enormously difficult problem because it encompasses gravity, hydrodynamics, radiative transfer, and magnetic fields. Further, the difference in density between between the initial gas cloud from which the star forms and the final star itself is 21 orders of magnitude, or equivalently, a change in physical scale of 7 orders of magnitude.

One of the most significant questions in this field is: Why do most stars form in binary systems? To address this question, numerical simulations have been run that follow the fragmentation of a large cloud of gas. The methods used have been primarily Lagrangian ones (such as SPH), although Adaptive Mesh Refinement (AMR) techniques [Berger and Collela 1989] are becoming more popular. The reason for the growth of interest in AMR is that recent results have demonstrated a severe error in a large body of numerical simulations of the cloud fragmentation process: they lacked resolution to adequately follow the balance between gravitational forces and local pressure forces [Truelove et al. 1997]. Simulations currently suggest that turbulent fragmentation plays a critical role in determining the formation of multiple star systems and that a filamentary structure is the main mechanism for transferring mass to the protostellar disk [Klein et al. 2000]. A similar process seems to govern formation of the first stars in the Universe [Abel et al. 2002].

Studying the interstellar medium presents different challenges. Traditionally, the ISM is understood as having a series of distinct phases that determine local star formation [McKee and Ostriker 1977], with regulation of the phases provided by heating and cooling mechanisms. Stellar winds and supernovae are the primary heating mechanism, while radiative cooling is the dominant cooling mechanism for the hot gas phases. The supernovae and winds also constantly stir the ISM, which in combination with rapid radiative cooling, serve to make it a highly turbulent medium. Turbulent media are difficult to understand because motions on very large scales can quickly couple to motions on much smaller scales, and thus accurate modeling requires resolution of large and small scales [Mac Low 2000]. Because of this range of scales, achieving sufficient resolution to be able to accurately model turbulence is difficult, and a number of researchers rely on two-dimensional models to provide sufficient dynamic range. Recent simulations have shown that self-gravity alone, without the stirring provided from supernovae explosions, is sufficient to produce the spectrum of perturbations expected from analytical descriptions of turbulence [Wada et al. 2002]. In the near future, three-dimensional simulations with a similar resolution to two-dimensional models will be possible, although the incorporation of MHD turbulence makes large-scale three-dimensional simulations a formidable challenge.

33.4.8 Cosmology

The study of the Big Bang and quantum gravity epoch is still largely conducted analytically, although some aspects of this research lend themselves to computer algebra. Following these earliest moments, the Universe undergoes a series of phase transitions (or "symmetry breaking") as the forces of nature separate out of the "Unified Field" [Kolb and Turner 1993]. Computation has been used to examine the nature of the phase transitions that occur as each of the forces separates. For example, the Electroweak phase transition has been extensively examined using lattice calculations to explore whether the phase transition is first (most probable) or second order [Kajantie et al. 1993]. Numerical simulations have also investigated how non-uniform symmetry breaking can lead to the formation of defects [Ach'ucarro et al. 1999].

Computation is used extensively in the study of **Big Bang Nucleosynthesis** (BBN) and the relic Cosmic Microwave Background (CMB). However, at present, CMB data analysis probably represents the biggest challenge computationally. Theoretical modeling of BBN dates back to the 1940s [Alpher et al. 1948], and a very detailed numerical approach to solving the coupled set of equations describing the reaction network was developed comparatively early [Wagoner et al. 1967]. Currently, there are a number of BBN codes available, and considerable effort has been put into reconciling results from different codes. CMB modeling is comparatively straightforward because the equations describing the evolution of a thermal spectrum of radiation in an expanding Universe are not overly complex. However, because the CMB spectrum we measure has foreground effects superimposed upon it (such as clusters of galaxies), a large amount of effort is expended simulating the effect of foreground pollution [Bond et al. 2002].

The theoretical modeling of large-scale structure in the Universe has relied heavily on computation. Because on large-scales "dark matter" dominates dynamics, only gravity need be included, and a Newtonian approximation can be used without significant error. Particle-based algorithms are used to evolve an initially smooth distribution of particles into a clustered state representative of the Universe at its current epoch. The first simulations with moderate resolution (3×10^5 mass elements) of the distribution of galaxies were conducted in the early 1980s [Efstathiou and Eastwood 1981]. Simulations have played a leading role in establishing the accuracy of the **Cold Dark Matter** (CDM) model of structure formation [Blumenthal et al. 1984]. In this cosmological model, structures are formed via a hierarchical merging process. Simulation has also shown that dark matter tends to form cuspy halos that have a universal core profile [Navarro et al. 1997], while the large-scale distribution of matter is dominated by filamentary structures.

33.4.9 Galaxy Clusters, Galaxy Formation, and Galactic Dynamics

The modeling of clusters of galaxies and galaxy formation relies on the same codes as the study of large-scale structure, with the addition of hydrodynamics to model the gas that condenses to form stars and nebulae within galaxies. Typically, the hydrodynamic methods used are either Eulerian grid-based algorithms or Lagrangian particle-based methods [Frenk et al. 1999], although, as in star formation, AMR methods are being adopted. Modeling of galaxy clusters is comparatively straightforward because the intracluster gas tends toward hydrostatic equilibrium. However, simulations have shown that the gas in galaxy clusters shows evidence of an epoch of preheating [Eke et al. 1998].

Galaxy formation is an exceptional difficult problem to study numerically because the evolution of the gas is strongly affected by supernovae explosions that occur on scales smaller than the best simulations can currently simulate [White 1997]. The physics is also technically challenging because galaxy formation occurs in the very nonlinear regime of gravitational collapse while simultaneously being a radiation hydrodynamics problem (although an optically thin approximation for the gas works very well). Only within the past few years has a sufficient understanding evolved to enable simulations of galaxy formation to produce moderate facsimiles of the galaxies we observe [Thacker and Couchman 2001]. Nonetheless, the very best simulations continue to lack both important physics and sufficient resolution to describe the galaxy formation process in great detail.

The first large-scale numerical studies of galactic dynamics were conducted in the 1960s [Hockney 1967]. At least initially, and to a large extent today, most n-body simulations are used to confirm analytic solutions derived from idealized models of galactic disks [Binney and Tremaine 1987]. Typically, these simulations begin with a model of a given galaxy, which usually consists of a disk of stars and an extended dark halo, which is then perturbed in some fashion to mimic the phenomenon under study. Recently, it has been highlighted that accurate modeling of galactic dynamics in CDM universes is exceptionally difficult due to coupling between the substructure in the larger galactic halo and the galactic disk [Weinberg and Katz 2002]. Traditionally, researchers believed that approximately 1 million particles were sufficient to model galaxies reasonably; however, these new results have pushed that estimate at least an order of magnitude higher. Although researchers in this field use codes similar to those in large-scale structure, specialist codes, which are designed to reduce numerical noise, have been developed (e.g., the Self Consistent Field code of Hernquist and Ostriker [1992]).

33.4.10 Numerical Relativity

General relativity calculations are extremely computationally demanding because not only is the theory exceptionally nonlinear, but there are a number of elliptic constraint equations that must be satisfied at each iteration. Further, the rapid change of scales that can accompany collapse problems often requires adaptive methods to resolve. There are also other subtleties related to the boundary conditions around black holes that present severe intellectual challenges. Currently, the strongest science driver behind these calculations is the need to calculate the gravitational wave signal of cataclysmic events (such as binary black hole coalescence), which may be detectable with the LIGO gravitational wave detector (http://www.ligo.caltech.edu/). Because of the large amount of computation involved in computing space-time geometry, and the comparatively low amount of communication between processors, numerical relativity is an ideal candidate for Grid-based computation. The CACTUS framework has been developed to aid such calculations (http://www.cactuscode.org).

Relativity calculations are most often mesh based (although spectral methods are used occasionally). Before determining the evolution equations for the space-time, a gauge must first be decided upon, and the most common gauge is the so called "3 + 1 formalism," where the space-time is sliced into a one-parameter family of space-like slices. Other formulations exist, such as the characteristic formalism, and in general the gauge is chosen to suit the problem being studied. Initial conditions for the space-time are provided and then the simulation is integrated forward, with suitable boundary conditions being applied. Building upon this body of research, the first calculation of the gravitational waveform from binary black-hole coalescence was performed in 2001 [Alcubierre et al. 2001].

33.4.11 Compact Objects

The study of compact objects such as white dwarf and neutron stars presents a formidable theoretical challenge. These systems exhibit extreme density, in turn requiring detailed nuclear physics as well as relativistic descriptions. Compact objects are widely believed to be the source of energy behind GRBs, with energetic scenarios, such as sudden mergers, driving a highly relativistic "fireball" shock wave that produces an extreme amount of gamma-ray radiation during collisions with other shock waves or the interstellar medium [van Putten 2001].

Neutron star collisions have been simulated for similar reasons to black holes: the calculation of their gravitational wave spectrum [Calder and Yang 2002]. Neutron stars are also the beginning point of core collapse (Type II) supernovae, and the simulation of the ignition process has attracted much attention. Fully general relativistic models are now beginning to appear [Bruenn et al. 2001]. Of particular interest is how the neutrinos drive a wind shortly after collapse begins [Burrows et al. 1995].

Type I supernovae occur when white dwarfs accrete sufficient mass to exceed the Chandrasekhar limit and subsequently undergo collapse. The physics is challenging because the process occurs far from equilibrium and entails radiative transfer as well as hydrodynamic instabilities. As in studies of Type II supernovae, to date most calculations use a two-dimensional approximation [Niemeyer et al. 1996] and Eulerian approaches; however, some explorations have used SPH in three-dimensions [Bravo and Garcia-Senz 1995]. The push toward full, high-resolution three-dimensional calculations is gathering momentum [Reinecke et al. 2002]. Such simulations are necessary to fully understand instabilities and include more accurate physics. However, the computational challenge is significant and, ultimately, progress awaits the development of 100-Teraflop computers.

33.4.12 Parallel Computation in Astrophysics

Parallel computing in astrophysics is often used to examine problems that simply cannot be addressed on a desktop computer, regardless of how long one could wait. The primary driver in this case is the large amount of memory available in parallel computers as compared to serial ones: the largest parallel computations simply do not fit into a desktop. The secondary use of parallel computation is to speed up data analysis, which involves performing the same analysis on many subsets of data. In this case, parallel computers significantly help in lowering the data reduction time.

More than 20% of the total cycles at the U.S. National Center for Supercomputing Applications are devoted to astrophysics, which is second only to materials science in terms of resource usage. Astrophysicists have a history of developing unique and ingenious parallel algorithms to solve the problems they face. Indeed, a number of research problems in astrophysics, such as binary black-hole coalescence and the formation of galaxies, are considered to be computational "Grand Challenges" by the National Science Foundation. These problems have computing demands that are similar to the nuclear ignition codes in the Accelerated Strategic Computing Initiative, which is part of the U.S. Government Stockpile Stewardship Program.

Parallel codes for distributed memory platforms are most often developed using the Message Passing Interface (MPI). Prior to the standardization of MPI, the Parallel Virtual Machine (PVM) API was very popular, and PVM remains the most common mechanism for parallelizing simple codes. Higher-performance APIs, such as the remote direct memory access provided provided by MPI-2, are yet to receive significant attention, primarily because vendors have failed to provide full support for this emerging standard. All of these APIs can lead to a significant increase in the size of a parallel program compared to the serial one. It is not uncommon for parallel codes to be over twice the length of serial ones. Codes written using these APIs have the potential to scale to many hundreds of processors.

Shared memory parallel codes are most often developed using the standardized OpenMP API. This API is particularly simple to use because it enables simple parallelization of codes using "pragmas" that are inserted into the code before iterative loops. The iterations within the loop, provided that they meet certain data independence requirements, can then be distributed to different CPUs, thereby speeding up execution times. The OpenMP API often does not lead to significantly longer parallel programs, but is limited in terms of scalability by the requirement of running on shared memory computers, which typically have a maximum of around 32 processors.

Over the past few years it has become increasingly apparent that although the physics being simulated by two codes is often quite different, the underlying data structures being used, such as grids or trees, are quite similar. This has led to the development of skeleton packages in which researchers need only add the numerical implementation of their equations and the communication between processors is handled by the package. CACTUS and PARAMESH are examples of this type of framework. However, most researchers seem reluctant to rely on these packages and instead develop an optimized communication layers themselves.

It is unclear what role "The Grid" will play in the development of theoretical modeling. Grid technologies have an inherently large latency, which makes simulation of elliptic-like problems (where the solution at one point depends on all others) particularly difficult. Relativistic problems appear to be comparatively well-suited to a grid environment due to the exceptionally large amount of computation involved in tensor calculations. So far, demonstration calculations run on a Grid environment using CACTUS are the best example of an effective Grid application. The potential of The Grid in astrophysics will probably be realized through extensive data analysis (akin to the SETI@home model) and data mining.

33.5 Research Issues and Summary

In this section we first highlight a number of key areas in computational astrophysics that will receive increasing attention in the future. No significance should be attached to the ordering of the items.

1. *Development of a Virtual Observatory.* A Virtual Observatory is a geographically distributed, Web-based repository of space-based and ground-based digital sky surveys, observatory and mission archives, and astronomy data and literature services. Virtual Observatories will provide powerful data analysis, visualization, and data mining toolkits for the effective and rapid scientific exploration of the resulting massive data sets. Virtual Observatories are to be enabled by technology, but driven by science.

2. *Planet formation.* Models need to include both hydrodynamics and dust to better understand the formation process in both the inner and outer solar system. Increasing resolution and space-borne optical interferometry will require detailed theoretical support from simulations to analyze observations.

3. *Star formation.* Following the full collapse of a gas cloud through to nuclear ignition remains an enormously challenging theoretical proposition. There are also a large number of questions about the formation of stars in groups, especially globular clusters, that will require key insights from simulations.

4. *Galaxy formation.* Progress is limited by both a lack of numerical resolution and an incomplete understanding of the physics of the process. It is currently unclear how to model the extremely important effect of heating from supernovae on protogalactic collapse. New discoveries will depend heavily on the growth of understanding of the star formation process.

5. *CMB data analysis.* New approximation algorithms are required to reduce the enormity of the analysis calculations for high-resolution data sets. Brute force and larger computers are not the answer, as even Petaflop computers will take over a year to analyze upcoming data sets.

6. *Black-hole coalescence.* With the growth of gravitational wave astronomy, calculating wave forms for binary systems will become increasingly important. These calculations have the potential to exploit the growth of "The Grid" computing environment.

7. *Supernovae ignition.* Current models can only simulate with adequate resolution in two-dimensions. A 100-fold improvement in computational power is needed to enable full three-dimensional calculations.

8. *Visualization.* The increasing complexity of data sets makes analysis more and more difficult. Haptic interfaces and/or multidimensional representation will be necessary to understand the phenomenological aspects of the next generation of astronomical data sets and simulations.

9. *Data mining tools that incorporate measurement error.* Efforts are being undertaken to include measurement errors in the performance of data mining tools. Pattern recognition algorithms currently define classes without regard to data quality; poorly measured data can smear out true clusters that would be easily visible if the data were limited to that of high quality. In some cases, poor-quality measurements can cause data to artificially cluster (e.g., when an upper flux limit is used in lieu of a measurement). Because the scientific method is statistical in nature, computational tools for science are needed that better reflect statistical uncertainties.

10. *Autonomous spacecraft/instrumentation operation.* Autonomous spacecraft are being designed to carry out day-to-day operations independent of ground control. Some of these operations include navigation and the scheduling and execution of observations and experiments. Support technologies for autonomous spacecraft include robotics, artificial intelligence, and control theory. This approach will reduce mission operation costs while simultaneously allowing orbital satellites to work dynamically rather than passively.

Astrophysics is gradually and irreversibly becoming a computational science. Astronomical data are being stored in and retrieved from progressively larger databases; data and metadata are being used by larger audiences. The analyses of such data are aided by pattern recognition algorithms and improved data visualization tools. Theoretical modeling has developed beyond the point where elegant calculations using relatively simple assumptions suffice; detailed models with many physical parameters are often required to adequately explain the detailed observations made by the newest orbital and ground-based telescopes spanning the electromagnetic spectrum. Parallel computation is often used in carrying out time-consuming calculations. High precision is needed to accurately calculate model parameters, and computationally intensive statistical tools are required to evaluate theoretical model efficacies. Astrophysics is thus evolving, and the new generation of astrophysicist is increasingly well-versed in the use of computational tools. This can only help us better understand the structure, evolution, and nature of the universe.

Defining Terms

Adaptive optics: An active form of observing light (rather than a passive one) in which lasers are fired at the sky from the telescope site, and the reflected laser light is used to differentially correct observed starlight for atmospheric refraction and convection.

Astrophysics: The study of the physical properties, structure, kinematics, dynamics, and evolution of celestial objects.

Big Bang nucleosynthesis: The epoch of the early Universe when the nuclei of atoms were formed. Theories of this epoch reproduce observations of elemental abundances exceptionally well, strongly supporting the Hot Big Bang theory.

Cold dark matter: Hypothesized sub-atomic particles that contribute roughly 30% of the mass in the Universe. Although in the early Universe these particles interacted vigourously, their influence is now seen only via gravitation. Theories that include the effects of these particles match observations exceptionally accurately.

Cosmic microwave background: Relic electromagnetic radiation left over from 300,000 years after the Big Bang. The distribution of radiation is very close to being homogeneous and isotropic, but carries small perturbations on it that correspond to perturbations of the matter density in the early Universe.

Cosmology: The branch of astrophysics dealing with the structure and evolution of the Universe on the largest scales.

Electromagnetic spectrum: Classification of light by wavelength (which is inversely related to energy). Ranging from light with the shortest wavelength (and largest energy) to that with the longest wavelength (and smallest energy), the electromagnetic spectrum includes gamma-ray, x-ray, ultraviolet, visible, infrared, and radio.

Galaxy: A large, gravitationally bound ensemble of stars, gas, and dust. Galaxies are traditionally classified as spiral, elliptical, irregular, or peculiar, based on their morphological structure. Our *Milky Way* galaxy is a large spiral galaxy containing more than a hundred billion stars.

General relativity: Einstein's theory of gravitation, which casts gravity not as a force, but rather as a consequence of warping in the local space-time continuum. The theory relies strongly upon geometrical concepts of distance and curvature.

Interstellar medium (ISM): The material between the stars. It is composed primarily of hydrogen and helium gas, with 1 to 2% heavy elements chemically mixed to form larger molecules ("dust"). Dense, cool molecular clouds are the seeds of star formation.

Passband: An electromagnetic regime defined by the spectral response of a particular filter and/or instrument.

Plasma: An ionized gas. A plasma behaves differently than a normal gas (which can often be modeled as a fluid) because a treatment of electromagnetic theory is needed to address the electrical charges found within it.

Supernova: A massive stellar explosion during which a star can briefly brighten to almost a billion times its original luminosity. Such events can be seen at enormous distances, although maximimum luminosity usually lasts only for tens of days.

Virtual observatory: Proposed publicly accessible metadatabase of archived ground-based, balloon, and satellite astronomical observations. The database will also be associated with a variety of search engines and data mining tools.

References

Abel, T., Bryan G., and Norman, M. L. 2002. The Formation of the First Star in the Universe. *Science* 295:93–98.

Abry, P. Goncalves, P., and Flandrin, P. 1995. *Lect. Notes in Statistics 103, Wavelets and Statistics* 15, A. Antoniadis and G. Oppenheim, Editors (New York: Springer).

Alcubierre, M. et al. 2001. 3-D Grazing Collision of Two Black Holes. *Phys. Rev. Lett.* 87:271103.

Alpher R., Bethe, H., and Gamov, G., 1948. The Origin of Chemical Elements. *Phys. Rev.* 73:803–804.

Ach'ucarro, A., Borrill, J., and Liddle, A. R. 1999. The Formation Rate of Semilocal Strings. *Phys. Rev. Lett.* 82:3742–3749.

Barnes, J. and Hut, P., 1986. A Hierarchical O(NlogN) Force-Calculation Algorithm. *Nature* 324:446–449.

Barthelmy, S., Cline, T., and Butterworth, P. 2001. GRB Coordinates Network (GCN): A Status Report, *Gamma 2001: Gamma-Ray Astrophysics: AIP Conference Proceedings, Vol. 587*, 213; S. Ritz, N. Gehrels, and C. R. Shrader., Editors (Melville, NY: American Institute of Physics).

Beckers, J. M. 1993. Adaptive Optics for Astronomy — Principles, Performance, and Applications. *Annu. Rev. Astron. Astrophys.* 31:13–62.

Beer, M. E. and Podsiadlowski, P. A. 2002. General Three-Dimensional Fluid Dynamics Code for Stars in Binary Systems. *Mon. Not. Roy. Astron. Soc.* 335(2):358–368.

Binney, J. and Tremaine, S., 1987. Galactic Dynamics. (Princeton Unviersity Press: Princeton NJ).

Blumenthal, G. R., Faber, S. M., Primack, J. R., and Rees M. J. 1984. Formation of Galaxies and Large-Scale Structure with Cold Dark Matter. *Nature* 311:517–525.

Bond, J. R. et al. 2002. The Sunyaev-Zeldovich Effect in CMB-Calibrated Theories Applied to the Cosmic Background Imager Anisotropy Power at l > 2000. http://arxiv.org/abs/astro-ph/0205386.

Borrill, J. D., 1999. MADCAP — The Microwave Anisotropy Dataset Computational Analysis Package. *Proceedings of the 5th European SGI/Cray MPP Workshop.* http://xxx.lanl.gov/abs/astro-ph/9911389.

Boyce, P. B., Tenopir, C., and Milkey, R. W. 2001. Electronic Journal Usage Patterns in Astronomy. *Bull. Am. Astron. Soc.* 199:1004B.

Bravo, E. and Garcia-Senz, D., 1995. Smooth Particle Hydronamics Simulations of Deflagrations in Supernovae. *Astrophys. J.* 450:L17–L21.

Brown, T. M., Charbonneau, D., Gilliland, R. L., Noyes, R. W., and Burrows, A. 2001. Hubble Space Telescope Time-Series Photometry of the Transiting Planet of HD 209458. *Astrophys. J.* 552:699.

Bruenn, S. W., DeNisco, K. R., and Mezzacappa, A. 2001. General Relativistic Effects in the Core Collapse Supernova Mechanism. *Astrophys. J.* 560:326–338.

Buchler, J. R., Kollath, Z., and Marom, A. 1997. An Adaptive Code for Radial Stellar Model Pulsations. *Astrophys. Space Sci.* 253(1):139–160.

Burrows, A. Hayes, J., and Fryxell, B., 1995. On the Nature of Core-Collapse Supernova Explosions. *Astrophys. J.* 450:830–850.

Buzasi, D. Catanzarite, J., Laher, R., Conrow, T., Shupe, D., Gautier, T. N., III, Kreidl, T., and Everett, D. 2000. The Detection of Multimodal Oscillations on Alpha Ursae Majoris. *Astrophys. J.* 532:133.

Calder, A. C. and Yang, E. Y. M. Numerical Models of Binary Neutron Star System Mergers. II. Coalescing Models with Post-Newtonian Radiation Reaction Forces. *Astrophys. J.* 570:303–313.

Clarke, D. A. and West, M. J. 1997. *ASP Conf. Ser. 123: Computational Astrophysics; 12th Kingston Meeting on Theoretical Astrophysics.*

Collela, P. and Woodward, P. 1984. The Piecewise Parabolic Method for Gasdynamical Simulation. *J. Comp. Phys.* 54:174.

Condon, J. J. et al. 1998. The NRAO VLA Sky Survey. *Astron. J.* 115(5):1693–1716.

Córsico, A. H. and Benvenuto, O. G. 2002. A New Code for Nonradial Stellar Pulsations and its Application to Low-Mass, Helium White Dwarfs. *Astrophys. Space Sci.* 279(3):281–300.

Cox, A. N., Bowers, D. L., and Brownlee, R. R. 1960. A Method of Computing Stellar Interior Models. *Astron. J.* 65:487.

Crane, P. C. 2001. Applications of the DFT/CLEAN Technique to Solar Time Series. *Solar Phys.* 203:381.

Djorgovski, S.G. et al. 2002. The Digital Palomar Observatory Sky Survey (DPOSS): General Description and the Public Data Release. *Bull. Am. Astron. Soc.* 200:6006D.

Efstathiou, E. and Eastwood, J. W. 1981. On the Clustering of Particles in an Expanding Universe. *Mon. Not. R. Astr. Soc.* 194:503–525.

Eke, V. R., Navarro, J. F., and Frenk, C. S. 1998. The Evolution of X-Ray Clusters in a Low-Density Universe. *Astrophys. J.* 503:569–592.

Evrard A. E. et al. 2002. Galaxy Clusters in Hubble Volume Simulations: Cosmological Constraints from Sky Survey Populations. *Astrophys. J.* 573:7–36.

Falgarone, E. and Passot, T. 2003. Turbulence and Magnetic Fields in Astrophysics. *Lecture Notes in Physics, Vol. 614* (Spinger-Verlag: Heidelberg).

Galama, T. J. 1999. The Effect of Magnetic Fields on Gamma-Ray Bursts Inferred from Multi-Wavelength Observations of the Burst of 23 January 1999. *Nature* 398:394.

Gilliland, R. L., Bono, G., Edmonds, P. D., Caputo, F., Cassisi, S., Petro, L. D., Saha, A., and Shara, M. M. 1998. Oscillating Blue Stragglers in the Core of 47 Tucanae. *Astrophys. J.* 507:818.

Gingold, R. A. and Monaghan, J. J., 1977. Smoothed Particle Hydrodynamics — Theory and Applications to Non-Spherical Stars. *Mon. Not. R. Astr. Soc.* 204:715–733.

Ginsparg, P. 1996. Winners and Losers in the Global Research Village. Conference held at UNESCO HQ, Paris. http://arXiv.org/blurb/pg96unesco.html.

Hakkila, J., Myers, J. M., Stidham, B. J., and Hartmann, D. H. 1997. A Computerized Model of Large-Scale Visual Interstellar Extinction. *Astron. J.* 114:2043.

Hakkila, J., Giblin, T. W., Roiger, R. J., Haglin, D. J., Paciesas, W. S., and Meegan, C. A. 2003. How Sample Completeness Affects Gamma-Ray Burst Classification. *Astrophys. J.* 582:320–329.

Hanisch, R. 1999. Electronic Preprints. *Bull. Am. Astron. Soc.* 31:1519.

Hanisch, R. J., Farris, A., Greisen, E. W., Pence, W. D., Schlesinger, B. M., Teuben, P. J., Thompson, R. W., and Warnock, A., III 2001. Definition of the Flexible Image Transport System (FITS). *Astron. Astrophys.* 376:359–380.

Heikkila, C. W., McGlynn, T. A., and White, N. E. 1999. Astrobrowse: A Web Agent for Querying Astronomical Databases. *ASP Conf. Ser. 172: Astronomical Data Analysis Software and Systems VIII.* 8:221. D. M. Mehringer, R. L. Plante, and D. A. Roberts, Editors.

Henyey, L. G., Lelevier, R., and Levee, R. D. 1959. Evolution of Main-Sequence Stars. *Astrophys. J.* 129:2.

Hernquist, L. and Ostriker, J. P. 1992. A Self-Consistent Field Method for Galactic Dynamics. *Astrophys. J.* 386:375–397.

Hockney, R. W., 1967. Gravitational Experiments with a Cylindrical Galaxy. *Astrophys. J.* 150:797–806.

Hockney, R. W. and Eastwood, J. W., 1981. Computer Simulation Using Particles. (New York: McGraw-Hill).

Kajantie, K., Rummukainen, K., and Shaposhnikov, M. 1993. A Lattice Monte Carlo Study of the Hot Electroweak Phase Transition. *Nucl. Phys.* B407:356–372.

Kiss, L. L. Szabó, G. M., Sziládi, K., Furész, G., Sárneczky, K., and Csák, B. 2001. A Variable Star Survey of the Open Cluster M37. *Astron. Astrophys.* 376:561–567.

Klein, R. I., Fisher, R., and McKee C. F. 2001. The Formation of Binary Stars. *Proceedings of IAU Symp. 2001*, held 10–15 April 2000, in Potsdam, Germany ASP. H. Zinnecker and R.D. Mathieu. editors.

Kleinmann, S.G. et al. 1994. The Two Micron All Sky Survey. *Experimental Astronomy* 3:65–72.

Kolb, E. W. and Turner, M. S. 1993. The Early Universe. (Perseus Publishing: Cambridge).

Kurtz, M. J. et al. 2000. The NASA Astrophysics Data System: Overview. *Astron. Astrophys. Suppl.* 143:41–59.

Kurucz, R.L. 1969. A Matrix Method for Calculating the Source Function, Mean Intensity, and Flux in a Model Atmosphere. *Astrophys. J.* 156:235–240.

Lasenby, A., Barreiro, B., and Hobson, M. 2001. Regularization and Inverse Problems. *Mining the Sky, Proceedings of the MPA/ESO/MPE Workshop*, held at Garching, Germany, 31 July–4 August, 2000, 15. A. J. Banday, S. Zaroubi, and M. Bartelmann, Editors (Heidelberg: Springer-Verlag).

Louys, M., Starck, J. L., Mei, S., Bonnarel, F., and Murtagh, F. 1999. Astronomical Image Compression. *Astron. Astrophys. Suppl.* 136:579–590.

Mac Low, M.-M. 2000. The Dynamical Interstellar Medium: Insights from Numerical Models. *Stars, Gas, and Dust in Galaxies ASP Conference Series: San Francisco* D. Alloin, K. Olsen, and G. Galez, Editors.

Mayer, L., Quinn, T., Wadsley, J., and Stadel, J. 2002. Formation of Giant Planets by Fragmentation of Protoplanetary Disks. *Science* 298:1756.

Mazzarella, J. M., Madore, B. F., and Helou, G. 2001. Capabilities of the NASA/IPAC Extragalactic Database in the Era of a Global Virtual Observatory. *Proc. SPIE 4477: Astronomical Data Analysis.* 20–34, J.-L. Starck and F. D. Murtagh, Editors.

McKee, C. F. and Ostriker, J. P. 1977. A Theory of the Interstellar Medium — Three Components Regulated by Supernova Explosions in an Inhomogeneous Substrate. *Astrophys. J.* 218:148–169.

McGlynn, T., Scollick, K., and White, N. 1998. SKYVIEW: The Multi-Wavelength Sky on the Internet. *New Horizons from Multi-Wavelength Sky Surveys, Proceedings of the 179th Symposium of the International Astronomical Union*, held in Baltimore, USA August 26–30, 1996, 465. Kluwer Academic Publishers, B. J. McLean, D. A. Golombek, J. J. E. Hayes, and H. E. Payne, Editors.

Meynet, G. and Maeder, A. 2002. Stellar Evolution with Rotation. VIII. Models at $Z = 10^{-5}$ and CNO Yields for Early Galactic Evolution. *Astron. Astrophys.* 390:561–583.

Murray, J. and Holman, M., 1999. The Origin of Chaos in the Outer Solar System. *Science* 283:1877.

Ochsenbein, F., Bauer, P., and Marcout, J. 2000. The VizieR Database of Astronomical Catalogues. *Astron. Astrophys. Suppl.* 143:23–32.

Navarro, J. F., Frenk. C. S., and White, S. D. M. 1997. A Universal Density Profile from Hierarchical Clustering. *Astrophys. J.* 490:493–508.

Niemeyer, J. C., Hillebrandt, W., and Woosley, S. E., 1996. Off-Center Deflagrations in Chandrasekhar Mass Type IA Supernova Models. *Astrophys. J.* 471:903–914.

Nikolaev, S., Weinberg, M. D., Skrutskie, M. F., Cutri, R. M., Wheelock, S. L., Gizis, J. E., and Howard, E. M. 2000. A Global Photometric Analysis of 2MASS Calibration Data. *Astron. J.* 120(6):3340–3350.

Paciesas, W. S. et al. 1999. The Fourth BATSE Gamma-Ray Burst Catalog (Revised). *Astrophys. J. Suppl.* 122:465–495.

Peraiah, A. 2001. An Introduction to Radiative Transfer. (Cambridge University Press: Cambridge).

Piña, R. K. and Puetter, R. C. 1993. Bayesian Image Reconstruction — The Pixon and Optimal Image Modeling. *Publ. Astron. Soc. Pac.* 105(688):630–637.

Piro, L. et al. 2002. The Bright Gamma-Ray Burst of 2000 February 10: A Case Study of an Optically Dark Gamma-Ray Burst. *Astrophys. J.* 577:680–690.

Poretti, E., Buzasi, D., Laher, R., Catanzarite, J., and Conrow, T. 2002. Asteroseismology from Space: The Delta Scuti Star Theta 2 Tauri Monitored by the WIRE Satellite. *Astron. Astrophys.* 382:157–163.

Pronik, I. I., Merkulova, N. I., and Metik, L. P. 1999. Characteristics of the Variability of the Nucleus of NGC 1275 in the Optical in 1982–1994. *Astron. Astrophys.* 351:21–30.

Ransom, S. M., Eikenberry, S. S., and Middleditch, J. 2002. Fourier Techniques for Very Long Astrophysical Time-Series Analysis. *Astron. J.* 124:1788–1809.

Reinecke, M., Hillebrandt., W., and Neimeyer, J. 2002. Three-dimensional Simulations of Type Ia Supernovae. *Astron. Astrophys.* 391:1167–1172.

Richardson, D. C., Quinn, T., Stadel, J., Lake, G. 2000. Direct Large-Scale N-Body Simulations of Planetesimal Dynamics. *Icarus* 143:45–59.

Roberts, D. H., Lehar, J., and Dreher, J. W. 1987. Time Series Analysis with Clean — Part One — Derivation of a Spectrum. *Astron. J.* 93:968.

Roden, J., Burl, M. C., and Fowlkes, C. 1999, The Diamond Eye Image Mining System. In *Demo for the Scientific and Statistical Database, Management Conf.*, (Cleveland, OH), June 1999.

Scargle, J. D. 1982. Studies in Astronomical Time Series Analysis. II. Statistical Aspects of Spectral Analysis of Unevenly Spaced Data. *Astrophys. J.* 263:835.

Scargle, J. 1997. Astronomical Time Series Analysis: New Methods for Studying Periodic and Aperiodic Systems. *Applications of Time Series Analysis in Astronomy and Metrology*, 215. (London: Chapman and Hall).

Scargle, J. 1998. Studies in Astronomical Time Series Analysis. V. Bayesian Blocks, a New Method to Analyze Structure in Photon Counting Data. *Astrophys. J.* 504:405.

Schou, J. and Buzasi, D. L. 2001. Observations of P-modes in Alpha; Cen. *Proceedings of the SOHO 10/GONG 2000 Workshop: Helio- and Asteroseismology at the Dawn of the Millennium*, EAS SP-464:391, A. Wilson, Ed., (Noordwijk: ESA Publications Division).

Skilling, J. 1998. Massive Inference and Maximum Entropy. Maximum Entropy and Bayesian Methods. *Proceedings of the 17th International Workshop on Maxiumum Entropy and Bayesian Methods of Statistical Analysis*, held in Boise, Idaho, 1997, 1. G. J. Erickson, J. T. Rychert, and C. R. Smith, Editors (Dordrecht/Boston/London:Kluwer).

Smoot, G. F. et al. 1992. Structure in the COBE Differential Microwave Radiometer First-Year Maps. *Astrophys. J.* 396:L1–L5.

Thacker, R. J. and Couchman, H. M. P. 2001. Star Formation, Supernova Feedback, and the Angular Momentum Problem in Numerical Cold Dark Matter Cosmogony: Halfway There? *Astrophys. J.* 555:L17–L20.

Torgersen, T. C. and Tyler, D. W. 2002. Practical Considerations in Restoring Images from Phase-Diverse Speckle Data. *Pub. Astron. Soc. Pac.* 114(796):671–685.

van Putten, M. H. P. M. 2001. Gamma-Ray Bursts: LIGO/VIRGO Sources of Gravitational Radiation. *Phys. Rep.* 345:1–59.

Wada, K., Meurer, G., and Norman, C. A. 2002. Gravity-driven Turbulence in Galactic Disks. *Astrophys. J.* 577:197–205.

Wade, G. A., Bagnulo, S., Kochukhov, O., Landstreet, J. D., Piskunov, N., and Stift, M. J. 2001. LTE Spectrum Synthesis in Magnetic Stellar Atmospheres. The Interagreement of Three Independent Polarised Radiative Transfer Codes. *Astron. Astrophys.* 374:265–279.

Wagoner, R. V., Fowler, W. A., and Hoyle, F., 1967. On the Synthesis of the Elements at Very High Temperatures. *Astrophys. J.* 148:3–50.

Weinberg, M. D. and Katz, N. 2002. Bar-Driven Dark Halo Evolution: A Resolution of the Cusp-Core Controversy. *Astrophys. J.* 580:627–633.

Wenger, M. et al. 2000. The SIMBAD Astronomical Database. The CDS Reference Database for Astronomical Objects. *Astron. Astrophys. Suppl.* 142:9–22.

White, S. D. M. 1997. Formation and Evolution of Galaxies. Cosmology and Large Scale Structure: Les Houches Session LX. 349–430. R. Schaeffer et al, Editors.

Woosley, S. E., Heger, A., and Weaver, T. A. 2002. *Rev. Mod. Phys.* 74:1015–1071.

York, D.G. et al. 2000. The Sloan Digital Sky Survey: Technical Summary. *Astron. J.* 120(3):1579–1587.

Zhang, Q., Fall, S. M., and Whitmore, B. C. 2001. A Multiwavelength Study of the Young Star Clusters and Interstellar Medium in the Antennae Galaxies. *Astrophys. J.* 561:727–750.

34

Computational Biology

34.1 Introduction 34-1
34.2 Databases..34-3
    Access and Communication • Representation/Data Modeling
34.3 Imaging, Microscopy, and Tomography 34-6
34.4 Determination of Structures from X-Ray
    Crystallography and NMR..........................34-7
    Determination of Macromolecular Structures from NMR Data
    • X-Ray Structure Determination of Macromolecules
34.5 Protein Folding 34-10
34.6 Genomics..34-11
    Genetic Mapping • Sequence Assembly • Sequence Analysis

David T. Kingsbury

Gordon and Betty Moore Foundation

34.1 Introduction

The past decade has witnessed the emergence of computational biology, in many forms, as a discipline in its own right. The application of mathematical and computational tools to all areas of biology is producing exciting results and providing insights into biological problems too complex for traditional analysis. In all areas of the life sciences, computational tools — from databases to computational models — have become commonplace in all laboratory settings. There is not a pharmaceutical or biotechnology company that does not have a computational biology or informatics group. Likewise, computational biology has emerged as an established academic field. Talent shortages and the inherent interdisciplinary nature of the field, however, have limited the rate of development of the academic sector.

The emergence of large-scale biological research efforts, such as the Drosophila and Human Genome Projects among other genome efforts, has contributed to the continued landslide of complex data. What sets biology apart from other data-rich fields is the *complexity* of its data, and the emergence of the fields of proteomics and systems biology has brought even more focus to that problem. Whereas a few years ago, biology was generally viewed as a scientific "cottage industry," with data being generated in a highly distributed mode, the recent move to large-scale projects has accelerated the generation of widely shared data. Still, biology has failed to agree on standard formats or syntax, leading to significant losses of data utility and requiring extraordinary efforts to use shared information. The new paradigm of *Discovery Science*, as contrasted with the traditional *Hypothesis-Driven* investigation, is seriously limited by the lack of widely used standards.

Thus, all areas of the biological sciences have urgent needs for the development of organized and accessible storage of biological data which incorporates the use of standardized syntax and file formats. The emergence of XML as a self-defining and common data format has been embraced by many in the life sciences. This has been accompanied by the use of **SOAP** and other tools to deploy data exchange via

Web Services. While this may be considered a form of biological database development, this approach is an attempt to enhance traditional database technology such as transaction-oriented relational database systems, which are often inadequate to serve many areas of the biological sciences. It is clear that collaboration between computer scientists and biologists continues to be necessary to design information platforms which accommodate the needs for variation in the representation of biological data, the distributed nature of the data acquisition system, the variable demands placed on different data sets, and the absence of adequate algorithms for data comparison, all of which are characteristic of biological science.

The continued advances in commercially available hardware and a greater emphasis on hardware clusters and grids are beginning to have a profound effect on computational biology. While in the past, traditional general-purpose hardware was inadequate for many of the most computationally intense problems, this condition has essentially disappeared as high-performance general-purpose instruments, commodity clusters, and grids have become more widely available. However, not only hardware limitations have affected the productivity of the computational biologist. There is a continuing need for new algorithm development to cover many tasks, especially real-time data analysis and comparisons between objects and images. Imaging technology is central to almost all of biology, and data representation through image construction remains an elusive but astoundingly powerful tool.

During the past decade there were dramatic advances in instrumentation and related methodologies for both light and electron microscopy. The advances lie not simply in higher resolution, but rather in a broader size range of structures that can be analyzed, more powerful methods for putting together the pieces of three-dimensional puzzles of cell form, and the addition of dynamic details of biological form and function, ranging from the subcellular to the physiological level. These approaches are computationally demanding. Computational resources have been expanding to meet these needs but remain inadequate for dealing with the massive data flow. As new experiential and computational approaches emerge in a few laboratories around the world, and as the ability to combine data from multiple instruments expands, the demand for better hardware and software continues. However, it is important that new software be developed within the context of the experimental research driving the needs; that is, there must be close collaboration between those developing the software and the groups carrying out research on static and dynamic structures.

X-ray crystallography and *NMR* are the major experimental methods for deducing macromolecular structures at atomic resolution. Both methods produce extremely large amounts of data and are entirely dependent upon the availability of powerful computers and sophisticated processing algorithms for the interpretation of raw data. In addition, there are fundamental scientific problems in both areas that require major computational advances. Substantial opportunities exist for combining structural information from several experimental techniques. This may provide the basis for a structural solution where only partial data is available from any single technique. With improved computational tools, combining physical data from a variety of sources has become more common. These developments will allow solutions to be obtained for structural problems which would otherwise be intractable. Analysis of errors in structures based upon experimental data from several sources also represents a significant computational challenge.

Advances in x-ray and NMR data analysis will lead directly to rapid developments in the field of protein-folding and structure-function prediction, which will be synergistic with developments in other areas of biology itself, and especially computational biology. Common problems of data representation, search strategy, pattern recognition, and data visualization appear in many fields. There is a particularly exciting synergistic relationship between the protein-folding field and the fields of structure determination by x-ray crystallography and 2-D NMR. Each field will benefit from rapid advances in the others. Improved folding algorithms provide a new way to attack the phase problem in crystallography, and new, more carefully refined protein structures provide rich new insights into protein folding.

Computational neurobiology is one of the most rapidly developing areas of computational biology. It gives us the hope of interpreting the mass of anatomical and physiological information about the nervous system, much of it derived from diagnostic testing, that is now available in functional terms. Better integration and interpretation of these data will permit neurobiology to make contact with other fields such as psychology and artificial intelligence. This work is making specific, testable predictions in the areas of sensory perception (visual, olfactory, and auditory), memory, learning, and motor control. Above all, it will lead to the integration of all these aspects to provide an eventual understanding of the total functioning of the nervous system. Such integration can be expected to provide new insights that will lead to improvements in the treatment of diseases of the nervous system at all levels, from neuropharmacology to psychotherapy.

The area of *genome analysis* has been, and continues to be, a major focal point in computational biology, and much progress has been made over the past few years. The sequencing of the human, mouse, rat, and fruit fly genomes in the past few years has challenged computational biologists and statisticians in many areas. While robust approaches to both **linkage mapping** and **physical mapping** were developed in the past, a new set of genetic challenges emerged from the genome sequencing effort. The human genome contains a wide variety of gene variations, or *polymorphisms*, that are responsible for the unique character of each individual, including inherited disease. A single human genome contains millions of these polymorphisms, and the detection and statistically valid association of a particular polymorphism with a given trait is now a major problem in computational genomics.

In many cases this analysis requires the ability to analyze tens of thousands of markers in family pedigrees. To be fully useful in a meaningful quantitative sense, this analysis will require powerful computer simulation and modeling. Major algorithmic advances have been made in the area of sequence assembly and clone assembly in physical mapping. As biologists continue to pursue the rapid sequencing of many genomes, the most common strategy is to use "shotgun sequencing," which relies on the assembly of random fragments. While powerful assembly algorithms have been developed, many in the field still consider this an important problem.

Common to all of the problem areas mentioned is the need for good visualization of data. Visualization is necessary because the map and sequence analysis phase for a molecular biologist is equivalent to exploratory analysis for a statistician. It is at this point that the experimentalist gains the feeling for, and understanding of, a physical or linkage map or sequence, which may then guide many months of experimental work. The complexity inherent in biological systems is so great that very sophisticated methods of analysis are required. These are the tools that must be readily accessible to molecular and cellular biologists untrained in computer technology.

Ecology and evolutionary biology encompass a broad range of levels of biological organization, from the organism through the population to communities and whole ecosystems. This complexity demands computational solutions. The need for enhanced computational ability is most evident when one attempts to couple large numbers of individual units into highly interactive and largely parallel networks, whether at the tissue, community, or ecosystem level of organization. The proliferation of information from remote sensing introduces the need for geographical information systems that provide a framework for classifying information, spatial statistics for analyzing patterns, and dynamic simulation models that allow the integration of information across multiple spatial, temporal, and organizational scales.

What follows is an examination of several specific areas of computational biology, with particular emphasis on those areas related to molecular biology, and a short development of the experimental paradigm and highlights of the current computational challenges regarding the requirements for further development of that area. There are common themes that appear in several of the sections, and these themes deserve special attention because they appear to be limiting the development of the entire field, regardless of the specific area of research. (This review will not attempt to cover the important areas of computational neuroscience and ecology and evolutionary biology in any further detail. Both fields are rich in computational challenges and theory and, like some of the areas covered here, deserve a chapter of their own.)

34.2 Databases

Biology is inherently information-rich because of the complexity and variety of living systems. Understanding these systems requires information about their organization, structure, and function at a multitude of levels, from the macroscopic to the molecular. Moreover, each species (and in many species, each organism) represents the potential for a unique solution to the problems of life processes and the organism's interaction with the environment. The full understanding of biology requires extending organismic complexity to include the relationships of species and organisms in their ecological niches, as well as the evolution of the biosphere over time.

Historically, much of this information was accessible to scientific inquiry only at high levels of abstraction, so that the inherent richness of information was not reflected in the volume of data available. This situation has changed dramatically in a number of biological fields, among them molecular biology, neurobiology, ecology, and taxonomy. This change has been made especially dramatic by the ubiquitous use of the World Wide Web (WWW). Emerging scientific paradigms will require even more data, organized into large and dynamic databases to support ongoing biological research. Indeed, some aspects of modern biology (e.g., the various genome projects, systems biology or protein structure–function studies) are now utterly dependent upon database and computer technology. In many cases, current data collections are not well organized for ease of retrieval or error correction but remain central to work in a given field. For example, several important macromolecular databases are maintained as flat files, poorly delimited and not accessible to ad hoc queries because of the absence of well-structured fields. Because of the importance of such data, investigators around the world struggle to find solutions to ready access and query. The reliance of modern molecular biology on databases is exemplified by the heavy dependence on the collection of DNA-sequence data (GenBank, EMBL, and DDBJ). These databases contain the definitive public domain DNA sequence data. However, because of the difficulty in the use of such data, and the large-scale generation of proprietary data, a new group of commercial data sets has become available (e.g., the Celera Discovery System, Incyte's LifeSeq, Gene Logic's GeneExpress, etc.).

As a result of this need, a substantial number of people now devote their careers to data management and computational analysis in biological disciplines. However, despite significant and vigorous efforts, the present generation of biological databases will fail in the next decade without significant continued development. They were not (and could not have been) designed to deal with the volume, complexity, and diversity of the data that will need to be accessible for future biological research.

The explosion of data is derived, in large part, from the desire of increasing numbers of investigators to have shared data repositories. The pressure on databases is severe in a quantitative sense, but is equally daunting in terms of the diversity and the interrelationships that must be represented among the data. These problems are further complicated by the way data is generated in biological research, which is geographically dispersed and, more importantly, lacks any meaningful standardization. Taken together, these problems pose unique transdisciplinary challenges for database design. Indeed, it is important to note that a single technology is not yet in hand that can support the design of adequate databases for much of the biology of the next decade. In fact, the application of the term "database" itself tends to be misleading. Database technology and theory as it currently exists is an inadequate paradigm for what is needed now and in the future to represent and organize biological information. For example, biological data includes mixtures of measurements, images, and interpretation, including extensive collections of metadata and derived data. Ideally, each of these data types would be available to ad hoc queries. At present, only a few extended relational systems have addressed these needs, and without complete success. The most recent focus has been to define this data in an XML format that enables the development of a variety of solutions to access and analysis, including access through Web services models.

Biologists have become increasingly sophisticated in their use of computers and in their abilities to state their research requirements in terms of informational and computational strategies. Biological science is now posing questions that not only require computational solutions, but also provide problems of fundamental interest to computer science researchers, thus creating the possibility of effective interdisciplinary work. Fortunately, there is a growing community of transdisciplinary workers whose expertise is centered at the interface of biology and computing, and who can provide much of the insight into how the two fields can interact productively.

One of the principal challenges in scientific computing in the next decade will be the development of database systems that can handle the inherent complexity of biological information and the marriage of those systems with visualization tools that enable laboratory scientists to interpret and understand the data and analytical results. The existence and availability of such databases will transform the way the science is done, and make possible completely new paradigms of biological research. Meeting this challenge will require the construction of databases in fields where none are available, significant research and development in database and knowledge-base technology, and the provision of a robust and widely available computational infrastructure for biological science through new algorithmic approaches to data analysis and tools to embed database access into analysis and modeling.

34.2.1 Access and Communication

The emergence of the World Wide Web has revolutionized access to biological databases. Because the subdomains of biology are fragmented, it has been necessary to develop many distinct and customized databases. The fact that there is little semantic consistency between these databases has raised a significant barrier to linking them through standard query mechanisms. Several investigators have built hypertext-based linkages between a number of different databases, bringing together a richer data resource. One representative of the several available systems is the Biology Workbench developed at the San Diego Supercomputer Center and the University of California, San Diego (http://workbench.sdsc.edu/). Another approach to database linking through a common interface was developed by the European Molecular Biology Laboratory (EMBL) and has been commercialized and further refined by Lion Biosciences (http://www.lionbioscience.com/solutions/products/srs). The tool, termed SRS (sequence retrieval system), enables a scientist to use common terms to query a variety of databases and then to establsih links between them based on common features and cross references. However, it does not enable *ad hoc* SQL queries across all data sources.

As powerful as the WWW-based systems are, they still lack the potential for supporting complex *ad hoc* queries that would be achieved through a true federation of biological databases. The need for such a federation has been recognized most acutely in the genomics community, and the outlines for such a federation were developed several years ago [Robbins 1994]. It was suggested that for minimum technical linkage, all of the participating databases present similar **APIs** (application programming interfaces) to the Net. All of the databases in the federation should also be relational systems that support SQL queries. Ideally, these databases should (1) be self-documenting, (2) be stable, and (3) conform to agreed-upon federation-wide semantics. The problem with this strategy is that in many cases it places the goals of the federation in conflict with the rapidly changing nature of biological research and the needs of the specialty user community. To cope with these problems, several systems have been developed to integrate heterogeneous databases. The two most common are DiscoveryLink developed by IBM and discoveryHub developed by GeneticXchange. Both approaches utilize an intelligent broker architecture with a series of wrappers that describe the specific databases they are linking. Neither has fully solved the problem of the semantic inconsistency of biological databases; however, discoveryHub has attempted to approach this problem through a semantic translator.

34.2.2 Representation/Data Modeling

To enhance the continued development of shared data resources, we must increase database expressiveness, study representation of biological knowledge, and automate modeling of database schemas. Biological knowledge is extremely rich and diverse; it includes raw experimental data (images, numbers, symbols); interpretations of experimental data (descriptions of biological objects with complex properties and internal structures); descriptions of experimental methods (complex procedures); and theoretical knowledge (documents, equations, descriptions of processes such as gene expression). Existing database technology provides a small number of simple representational primitives (such as relations); allowing databases to capture only a small piece of this rich biological semantics. Research is needed to provide richer representational capabilities, and to study how to represent the wide range of biological knowledge. Further, because biological databases will model a large number of complex entities, biological database schemas will be correspondingly complex. Researchers must investigate automated methods of managing database schemas to increase the efficiency of the database design process.

34.3 Imaging, Microscopy, and Tomography

Image reconstruction with light and electron microscopy and other imaging techniques is a powerful tool for the characterization of biological structures in three dimensions over a wide range of scales. Biological structures amenable to one or more techniques of imaging include single macromolecules and macromolecular assemblies, subcellular organelles, whole cells, and tissues. Characterization of the spatial organization of biological structures is critical for determining the functionality of these structures. Many fundamental problems in biology are open to study by microscopy and other imaging techniques. The advances in computer-controlled scanning microscopies, high-resolution atomic-force microscopy, and cryoelectron microscopy have substantially expanded the level of detail that can be attained by these methods.

The continuing advances and widespread applications of **transmission electron microscopy** at conventional, intermediate, and high voltages (with and without energy filtration), as well as confocal light microscopy, are leading to the production of vast amounts of data that need to be processed to extract the structural information and biologically important details. Each of these techniques can produce three-dimensional images of biologically important structures. Successes in these studies do not imply that the technical problems have been solved; currently, there remain substantial computational challenges to meet before achieving the goal of making efficient use of these techniques. These challenges include developing improved image processing and reconstruction algorithms.

In transmission electron microscopy, there have evolved three distinct methodologies for producing three-dimensional information: (1) electron crystallography of two-dimensional lattices and the processing of images of symmetric structures; (2) the analysis of multiple images of isolated asymmetric macromolecular assemblies; and (3) electron tomography, or three-dimensional reconstruction from a tilt series of images obtained from a single structure. Confocal light microscopy (CLM), a tool used in cell biology, produces three-dimensional images by scanning light focused to a single point over a three-dimensional grid in the specimen, and then imaging the light onto a point detector. Compared to conventional microscopies, CLM has a somewhat higher resolution and a much smaller depth of field, minimizing mixing of image data from different depths in the specimen.

The success of these methods has been substantial (Fernandez et al., 2002). This has stimulated the development of a comprehensive user-oriented facility, the National Center for Microscopy and Imaging Research (NCMIR). Generally known as the *Telescience Portal*, it is used by molecular and cell biologists, electron microscopists, and computer scientists around the world. The portal links instruments and computers at laboratories in Europe, North America, and Asia over a dedicated IPv6 network. NCMIR's central instrument is a state-of-the-art 400,000-volt intermediate-voltage transmission electron microscope equipped with charge-coupled detectors (CCDs) and can be operated under complete computer control and via the Internet. Because of its design, the instrument is able to penetrate thicker samples than conventional electron microscopes and is therefore particularly good at 3-D reconstruction. In collaboration with Japanese scientists, the portal also provides network-base access to a 3-MeV ultrahigh-voltage electron microscope. These resources are also linked to high-performance computing facilities in Taiwan and San Diego. Thus, the Portal is an applications environment supplying centralized access to all of the tools necessary for high-level electron tomography. A full description of the Telescience Portal and a related facility (the Biomedical Informatics Research Network (BIRN)) is available at http://www.ncmir.ucsd.edu.

Use of these imaging techniques entails three fundamental problems. The first is the vast quantity of data being produced that requires complex processing. This is a problem common to all imaging techniques within and outside biology. For example, cryoelectron-microscopic analysis of icosahedral viruses involves sample preparation followed by microscopy followed further by a series of computational steps, many of which are very computation-intensive [Cheng et al. 1995]. Because of the limited computing power available, only a limited number of independent images can be analyzed. However, three-dimensional images are intrinsically complex, and the amount of information required to characterize an image in detail is very high. For example, construction of the three-dimensional image of a molecule from electron-microscope images of single particles routinely requires thousands of particle images. This means that the actual time to produce a three-dimensional image is weeks to months, due in large part to the user-mediated steps that remain in the analysis. For efforts like this to prove maximally fruitful, it is crucial that each step be automated as much as possible.

The second fundamental problem common to all imaging techniques is the existence of a point spread function due to the instrumental broadening that is intrinsic to each form of imaging. For example, transmission electron microscopy loses a cone of data in the Fourier transform of the image, and the restoration of this cone represents a difficult, open problem. In **scanning confocal microscopy**, the point spread function is greatly extended in the direction parallel to the optical axis, and narrowing it could improve the resulting three-dimensional images.

Third, the results of any imaging method must be quantitated and displayed. The problems of image enhancement and visualization are completely general, although each technique may benefit more from specific display modes than others. Quantitative comparison of images also provides substantial challenges, especially in the presence of noise. Comparison of two images of flexible objects (e.g., cells or chromosomes) represents a substantially greater challenge.

One of the recurring problems common to all imaging techniques is the existence of artifacts due to incomplete data collection. These artifacts may seriously interfere with the interpretation of the recon-struction, and may even lead to incorrect conclusions. This problem is most serious in electron microscopy, where a full range of viewing angles is not usually accessible, and data corresponding to a cone or wedge in Fourier space cannot be collected. In confocal microscopy, the resolution in the z-direction (parallel to the optical axis) is much lower than in the x and y directions, as reflected by a nonisotropic instrumental point spread function.

Although several restoration algorithms have been in existence for some time, only the recent increases in computational speed have made their practical implementation possible. Two algorithms with potential application to signal restoration have attracted special attention due to their success in other fields — pro-jection onto convex sets (POCS) [Bellon and Lanzavecchia 1995] and maximum entropy (ME) [Schmeider et al. 1995]. Both methods are extremely computation-intensive, making their application to realistic-sized image volumes ($64 \times 64 \times 64$ or $128 \times 128 \times 128$) extremely demanding. The full development of these algorithms into something useful for detailed images of biological systems will require many computation cycles, to allow many different parameter values to be tested (ME) or a variety of different constraints to be used (POCS). Thus, a serious attempt to make three-dimensional image restoration viable for biological images will always require the highest available computational speed, along with advances in the theory and design of algorithms.

34.4 Determination of Structures from X-Ray Crystallography and NMR

The three-dimensional structures of proteins and nucleic acids are essential elements in the pathway relating gene sequence to function. The problem of predicting three-dimensional structure from a sequence remains unsolved at present. X-ray crystallography and NMR are the major experimental methods for deducing these structures at atomic resolution. While the rate at which these structures are being determined has increased dramatically, it lags significantly behind the rate at which new sequences are being accumulated. Each newly determined structure increases our knowledge in two ways. First, it adds to the database of known structures that can be used in knowledge-based methods in subsequent structure determinations of other macromolecules. Equally important, each new structure provides insights into the fundamental biological processes that support all life.

NMR and x-ray crystallography both produce extremely large amounts of raw data and are entirely dependent upon the availability of powerful computers and sophisticated processing algorithms for the interpretation of those data. In addition, there are fundamental scientific problems in both areas that require major computational advances. Both NMR and crystallography make use of constraint refinement to optimize the fit of experimental data to working models, and both fields use large scale-simulations to correlate the molecular models to known biological properties. Like investigators in three-dimensional microscopy, crystallographers and NMR spectroscopists are using maximum-entropy reconstruction as a major tool in computational analysis.

34.4.1   Determination of Macromolecular Structures from NMR Data

NMR methods for determining three-dimensional structures of macromolecules in solution have become increasingly important over the past several years. Major limitations on the speed and ease of analyzing the NMR data include the difficulty of assigning individual resonances in the spectra to particular protons in the molecule and then of calculating the full three-dimensional structure using distance geometry, molecular dynamics, or algorithms. For example, when determining the structure of a 100–150-residue protein, it may take as much as two weeks of NMR spectrometer time to collect the raw data, two or three days to calculate the spectra, and months or years to fully interpret the results. The complexity of this process depends on both the size of the protein and the extent of peak overlap within the spectra. Because this is the critical bottleneck in obtaining the structural information, it limits the size of molecules that can be considered. One critical element is the solubility of biological macromolecules. The balance between solubility and the sensitivity of modern NMR equipment places the current lower limit on concentration at around 0.5 mM. Many proteins, especially those of high molecular mass, aggregate at such high concentrations, leading to broad spectral lines of no value in structural studies. The solution to this problem lies with the development of enhanced computational approaches, such as maximum-entropy reconstruction of specially collected data sets. This approach requires approximately 100 times the computational work of traditional discrete-Fourier-transform processing. The refinement of this approach will require the continued collaboration of structural biologists and computer scientists [Schmieder et al. 1995]. Future advances in algorithms, software development, and the availability of more powerful computers will make a major impact on the time required to interpret NMR data.

A critical step in interpreting the data is to use the relationships between protons signified by two-dimensional (or three-dimensional) crosspeaks to assign the resonances to particular protons in the molecule. Assignment is frequently the rate-limiting step in structure determination. There are a number of different strategies for assignment, and approaches to automate this process are under active investigation. It is clear that several approaches will be necessary to deal with the problems associated with ambiguous data. Both the **sequential** and **mainchain-directed** assignment procedures make use of patterns of J-correlated and distance-correlated relationships. In both cases, the procedure is still largely manual, although there have been recent attempts to automate parts of the analysis. The development of computer-assisted or fully automatic pattern recognition techniques would make a major impact on the time required to make the assignments, as well as the size of molecules that can be studied. This is particularly true as the complexity of the original NMR data increases.

Once protons have been assigned, a three-dimensional structure can be estimated using the peak areas from the **NOESY** spectrum to determine distance constraints. Extensive computing is required to calculate structures using distance geometry, molecular dynamics, or **Kalman filtering** techniques. Several refinements of the structure are needed, and a family of structures is usually generated. Recently, back-calculation of the NOESY spectrum has been used to try to refine the structure. The value of this procedure and the effect of using different techniques to calculate NMR structures still need to be established. However, both the development and application of this technique require major computing power.

34.4.2   X-Ray Structure Determination of Macromolecules

X-ray crystallography provides a fine example of how the availability of affordable supercomputers has dramatically sped up the rate at which x-ray structures can be determined. There are four major phases in crystal structure determination: crystal growth, solution of the phase problem, interpretation of the resulting electron density in terms of an atomic model; and refinement of the atomic model. Up until very recently, the refinement of protein structures using x-ray data required substantial manual intervention and took one or more years to get to a satisfactory stage. With the incorporation of **simulated annealing** methods into the refinement procedure, the time required for this phase continues to decrease as more computational power becomes available. It should be noted that while the established protocols for simulated annealing can now be run fairly comfortably on workstations, the original development and testing of the method required the availability of supercomputers such as the Cray.

The fundamental problem in protein crystallography is the **phase problem**, which, with the improvement in refinement techniques, has become the major bottleneck in the structure determination process. The current method of multiple isomorphous replacement (MIR) relies on measuring data using crystals of the protein soaked in different metal compounds. The soaking procedure does not always yield suitable crystals, and in some cases the lack of metal derivatives delays the determination of the structure. Attempts to solve the phase problem without resorting to MIR or other experimental techniques fall into two classes:

1. *Ab initio phasing.* This does not rely on any additional experimental information other than that provided by the x-ray data set on the native protein crystal. Methods that have been successful for small molecules break down in the range of 100 atoms or so, well below the range of 1000 atoms or more in protein molecules. The availability of more computing power is gradually leading to the extension of some of these methods to larger polypeptides, and intensive computational effort may yield success for the smaller proteins (approximately 50 amino acids). The *ab initio* solution of virus structures may be aided by taking advantage of their high degree of noncrystallographic symmetry. If an adequate starting model is available, then these techniques allow the extension of phases directly over a wide resolution range [Rossman et al. 1985]. Novel approaches, such as those based on maximum entropy, may also provide computational tools that will have a substantial impact on obtaining phases for diffraction data.

2. *Use of prior knowledge.* In cases where the three-dimensional structure of a closely related protein is known, the phase problem can be solved using the known structure as a starting point for refinement *(molecular replacement)*. Computation-intensive methods such as simulated annealing have proved particularly valuable here, because significant distortions may have to be introduced into the starting structure and the optimization procedure needs to escape from the local minimum of the starting structure. The inclusion of prior knowledge into phase determination is likely to become extremely important in the near future, because it is becoming clear that proteins are built up from smaller subunits or segments that are commonly shared between large numbers of proteins. As the database of known structures increases, it should be possible to use segments of structures in search procedures to solve the phase problem. There is a need here to develop and apply novel search procedures and pattern recognition algorithms, many of which will also be applicable in the next stage of the structure determination.

Noisy electron density maps are difficult to interpret manually because of false or missing connections in the electron density. It is not uncommon for a structure determination project to arrive at this stage rapidly, once crystals are obtained, and then to spend a long time getting a better electron density map (the phase problem). Computing is essential in two ways. First, phase bias from the model must be minimized by optimizing relative contributions from calculated and observed diffraction data. Second, rapid interactive refinement of difficult-to-interpret regions must be performed to help to solve the problem. This requires high-speed supercomputing and a fast network link between the supercomputer and a high-performance graphics workstation.

Even in the case of a well-defined electron density map, fitting an atomic model to these maps is a time-consuming process. The problem lies in automatic recognition of characteristic patterns of macromolecular structure, such as alpha helices, that cannot be achieved by existing algorithms. Because of the complexity of the three-dimensional pattern recognition, further advances in methodology and computing speed are required.

34.5 Protein Folding

Protein folding, recognized for many years as one of biology's core problems, remains a focus of much work and attention. Folding converts the linear, one-dimensional information of the amino acid sequence into the biologically active three-dimensional structure. Folding may be thought of as a final unsolved aspect of the genetic code, and it is clear that progress on the folding problem will have tremendous theoretical and practical implications for biology. As described above, there have been dramatic advances in crystallography and two-dimensional NMR, and the virtual explosion of protein sequence data reemphasizes the importance of working on the folding problem. Even limited progress in this area could have a tremendous payoff, especially because of the recognition that such diseases as cystic fibrosis, Alzheimer's, Creutzfeldt-Jacob (human form of Mad Cow), and many others are protein folding-related problems [Dobson 1999, Fink 1998, Hammarstrm et al. 2003]. The *protein folding problem* can be considered either as:

- The problem of understanding the actual kinetic pathway by which a protein folds, or
- The problem of predicting the final folded conformation.

Obviously, a detailed structural understanding of folding intermediates would lead to a prediction of the final folded structure. However, experimental studies of folding intermediates have been very difficult because the intermediates are present in vanishingly small amounts. The rich database of known structures provides an excellent guide regarding the final folded conformation and can serve to guide theoretical work. In addition, there is an evolving literature regarding fruitful experimental approaches [Hammarström et al. 2003].

At first glance, the protein folding problem may seem to have a tantalizing simplicity (a given string of 100 amino acids contains all the data needed to determine the final folded structure), but the problem is extraordinarily complex. If each residue in a polypeptide chain can adopt ten distinct conformations, then the protein could adopt 10^{100} distinct structures, which leads to an extraordinarily difficult search problem. Many different strategies have been used in approaching the folding problem. Some methods have relied on detailed physical models of the polypeptide chain and have tried to carefully simulate the interactions (hydrogen bonds, van der Waals contacts, electrostatic interactions, etc.) that stabilize the chain. Other methods rely on the structural database that has accumulated over the past decades. In some sense, it appears that "structure is more conserved than sequence," so that the structural database is a useful guide when modeling new proteins.

Current approaches to the folding problem can generally be placed into one of two methods: direct determination of the folded confirmations, or a template-based method. Direct methods seek to determine the lowest acceptable energy point in a suitably defined conformational space. Template-based methods compare the sequence of the unknown with a collection of solved structures and select a limited number of highly scored possibilities [Luthy et al. 1992, Sali et al. 1994].

One core problem with direct methods is the difficulty of searching through the astronomical number of possible structures. A naive calculation may suggest that there are 10^{100} conformations, yet it is clear that only "a few" of these are of low enough energy to be plausible structures or plausible folding intermediates. The multiple-minima problem arises repeatedly in studies of folding. Both physical approaches (based on a detailed molecular potential surface) and pattern recognition schemes (based on analogy) encounter the same problems with multiple minima.* Stochastic search algorithms have proved especially useful in handling problems with multiple minima, and the method of **simulated annealing** is frequently used. This method corresponds to a simulation of the molecular dynamics under the influence of random thermal forces. Other search algorithms involve buildup or stochastic buildup based on genetic algorithms. To estimate the difficulty of the multiple-minima problem in protein folding, it is possible to draw upon

---

*The same search problem occurs in other parts of computational biology, including sequence comparison problems in genomics, neural networks, and immune-network modeling.

some parts of statistical physics for help. Simple lattice models have been used to estimate excluded-volume effects after polymer collapse, and these models indicate that the number of allowed conformations may be far smaller than suggested by the initial naive estimates. This result is very encouraging because it suggests that the search problem can focus on a much smaller region of conformational space.

Detailed atomic models have been used to study protein folding and stability. The models are based upon well-understood principles of physical chemistry and have been used in conjunction with molecular dynamics and Monte Carlo methods to study the underlying forces that determine the stability of folded proteins. The application of free-energy perturbation theory has been a particularly exciting development. These approaches are beginning to provide a much better understanding of the key forces involved in protein folding and stability, such as the true role of electrostatic interactions and the origin of the hydrophobic effect. Continued close interactions between experimental biologists and computational/theoretical researchers have also been extremely important for this field. Theory can help design new experiments, and better data can allow the refinement of basic physical models.

Although not directly linked to the protein folding problem, an extremely important area of computational biology involves the molecular modeling of protein function. A molecular understanding of enzyme catalysis can now be approached through a combination of molecular dynamics and quantum chemistry. There have been many applications of this approach, and important insights about the mechanism of triose phosphate isomerase have recently come from such simulations. Another exciting area is the recent development of computational approaches to modeling electron transfer. This a fundamental and inherently quantum-mechanical process involved in energy transduction and photosynthesis. Signal transduction is another extremely important and active area of research. This involves problems of docking and protein–protein recognition. Allosteric transitions are a frequent consequence of such interactions, and new methods for studying protein motions on long timescales are being developed [Gilson et al. 1994].

Much information has been derived from the ability to clone and express a protein of choice, followed by deliberate mutation of individual residues or larger segments. Perhaps the simplest application of folding that can be envisioned would be to predict the structural perturbation caused by a single-residue mutational change in a protein of known structure [Hammarström et al. 2003]. However, frequently, such mutations do not result in major rearrangements of the chainfold, as shown by the work of Kossiakoff [Eigenbrot et al. 1993] and others. While the structural effect of single-site changes has been successfully predicted in some cases [Desjarlais and Berg 1992], there are many conspicuous failures, and clearly more work is needed. At the level of larger segments, attempts to predict antibody hypervariable loops have also met with partial success [Tramontano and Lesk 1995; Pan et al. 1995]. Recent experiments suggest that deletion of entire loops can be tolerated while partial deletions of the same segments are not. Such results suggest the existence of quasi-independent modules, which would simplify calculations.

34.6 Genomics

Genomics is the study of DNA and its products at the genome level. This includes both experimental and theoretical aspects of the problem. For the computer scientist, the interest lies in the discrete domains and output of the system; for the biologist, the goal is to reveal the function of a sequence, either of the DNA or, more commonly, of the protein gene product. The biological effort includes genetic mapping, physical mapping (restriction maps, ordered clones, x-ray diffraction maps, cytogenetic maps, and sequence assembly), and sequence analysis including polymorphism identification and location. There is a natural division of the work into two major computational areas: support for the construction and representation of various maps, and analysis of the data produced. This overlaps at the boundary with computation involved in the analysis of protein folding and interacts with work to assign function to gene products. The systematic effort to map genomes in the presence of variability and error also mandates the use of new computational techniques to assist in the design of efficient experiments. Beyond this, there are four major areas in which the use of computers is indispensable: database searching for DNA sequence analysis, maximum-likelihood calculations for genetic linkage mapping, DNA sequence assembly, and general bookkeeping (e.g., laboratory information management systems).

34.6.1 Genetic Mapping

Genetic mapping deals with the inheritance of certain genetic **markers** within the pedigree of families. These markers can be genes, sequences associated with genetic disease (polymorphic regions, often single nucleotide changes), or arbitrary probes determined to be of significance [e.g., single nucleotide polymorphisms (SNPs), sequence tagged sites (STSs), or expressed site tags (ESTs)]. The sequence of such markers and their probabilistic distance (traditionally measured in centimorgans and now in many cases in megabase pairs) along the genome can often be determined with fair accuracy by the use of maximum-likelihood methods. Inheritance of traits across the pedigrees of multiple families is determined by a number of techniques that essentially hybridize each family member's genome against the predetermined probes. Eventually, the genetic map most likely to produce the observed data is constructed. Only a few years ago, the knowledge of the mathematics involved and the computational complexity of algorithms based on that mathematics limited analysis to no more than five or six markers. As knowledge of approximations to the formulas and likelihood estimation has improved, software capable of producing maps for 60 markers or more [Magness et al. 1993, Matise et al. 1994] has been developed. Early advances in this area include the identification of a large number of SNP and EST probes as well as software capable of tractably producing maps based on hybridization pedigree data [Cinkosky and Fickett 1993]. Further progress in this area has used mathematical methods such as combinatorics, graph theory, and statistics, and computer science methods such as search theory. Although significant progress has been made over the past few years, considerable effort is still required to make genetic linkage maps effective tools for genetic research.

The human genome contains a wide variety of gene variations that are responsible for the unique character of each individual, including inherited disease and genetic susceptibility to disease. These variations also predict a variety of responses to external factors such as chemicals and drugs. These variations are of several types and two individuals may vary by as many as 5 million locations. The association of these variations with specific traits is a significant mathematical and computational challenge. Furthermore, tools to address more complex situations (such as manic-depressive disease, which is likely to involve multiple genes) are badly needed.

34.6.2 Sequence Assembly

The recent successes in sequencing the fruit fly [Adams et al. 2000], mouse [Waterston et al. 2002] and human [Lander et al. 2001, Venter, et al. 2001] genomes were derivative of a rapid sequencing and **sequence assembly** technique that was intensely computational. The assembly technique is referred to as a "shotgun assembly" process. Shotgun assembly is an example of an inverse problem: starting with a set of sequence reads randomly sampled from a target, reconstruct the order and position of those fragments in the original target. In this case, fragments of a genetic sequence are randomly sampled, and then experiments are performed on the fragments to determine which pairs of fragments come from overlapping regions [Iris 1994]. This overlap information is then used to piece together, or assemble, the fragments into an ordered layout of the fragments that covers the original genetic sequence (see [Venter et al., 2001] pp. 1308–1319 for a complete description and an example). This problem can be solved using multiple strategies but relies on the existence of multiple data types to establish a physical scaffold onto which the assembly is placed. For example, the Celera whole genome assembler (WGA) has a five-stage process: Screener, Overlapper, Unitigger, Scaffolder, and Repeat Resolver. This is a very computer-intensive process, taking days with current state-of-the-art hardware. Every overlap computed was statistically a 1-in-10^{17} event and, therefore, not a coincidental event.

The sequence assembly problem has been extensively examined and several solutions have been applied. Venter et al. [2001] assembled the human genome using the Celera WGA as described above, while a somewhat different assembly approach was taken by Kent and colleagues for the public version of the genome [Lander et al. 2001]. As additional large genomes are solved, the assembly problem will continue to be further refined and computational efficiencies achieved. This continues to be an exciting and productive interface between computer scientists, mathematicians, algoritmists, and bioinformaticists.

The assembly problem is further compounded by the fact that all the data is inaccurate (e.g., digest lengths are off by up to 10%, electrophoretically determined sequences contain 0.5% incorrect base assignments, etc.). One approach to dealing with these problems has been developed by Phil Green, the author of the Phred [Ewing et al. 1998, Ewing and Green 1998] and Phrap programs (http://www.phrap.org). Phred reads DNA sequencer trace data, calls bases, assigns quality values to the bases, and writes the base calls and quality values to output files. Phrap is a program for assembling shotgun DNA sequence data. Key features are that it allows the use of entire reads (not just trimmed high-quality parts); uses a combination of user-supplied and internally computed data quality information to improve accuracy of assembly in the presence of repeats; constructs contig sequences as a mosaic of the highest quality parts of reads (rather than a consensus); and, provides extensive information about assembly (including quality values for contig sequences) to assist in troubleshooting. It is able to handle very large datasets and was influential in the subsequent development of more complex assemblers.

Additionally, the data is partial, not all regions of the original are represented in the sample, and the orientation of fragments is frequently unknown. Other issues involve the amount and type of information gathered to infer overlaps. More information implies more confidence in the veracity of an overlap, but some false positives will occur by chance. Consequently, it is likely that when building a scaffold from ordered clone libraries, a variety of different experiments yielding heterogeneous types of information will be performed and must be used simultaneously to detect overlaps. A key analysis problem is to accurately assess the statistical significance of overlaps under some stochastic model. Another issue involves how much to sample. Statistical and biological considerations show that for large problems with moderate overlap information, one will never achieve coverage without an impractical amount of sampling. This is the gambler's dilemma: at some point one must stop rolling the dice and move to another strategy to complete an assembly project.

While preliminary solutions to genome analysis have been of great use, many challenges remain, including:

1. The scale of the problems is such that they are computationally demanding. Better algorithms and the exploitation of parallelism are required.
2. Each assembly problem has a somewhat different combinatorial structure due to variations in the experimental methods used to infer overlaps. There is clearly a central generalized assembly problem, but each variation requires its own optimization to best lever the combinatorial structure. However, many of these projects are one-time efforts. The challenge is to build software that is both general and efficient.
3. The resulting assemblages are large, and software is needed that permits one to visualize and navigate a solution. Further engineering is required to allow investigators to manipulate solutions according to their expert knowledge, and to maintain versions of the data and a record of the work.
4. Finally, the central assembly problem involves NP-hard combinatorial problems for which heuristics work well on typical data. But as the scale and number of the problems to be solved increase, we will need ever more trustworthy solutions [Goldberg et al. 1995].

34.6.3 Sequence Analysis

Database searching has become an essential part of modern molecular biology. Database searching is our most effective means of identifying a potential biochemical function of a newly sequenced protein or gene. It is hard to imagine the number of hours of trial and error that are eliminated by the advent of this technique.

34.6.3.1 Matching a Defined Pattern

The need for speed in database searches has led to using heuristic methods. Current research topics in this area include the improvement of the sensitivity of searching, which can be severely reduced by the overall nucleotide composition of the genomes involved. Database searches are generally conducted with a global alignment algorithm that finds the "best" alignment over the entire length of both sequences. Recent advances in our understanding of domain structure of proteins and the intron–exon organization of genes has made it very desirable to develop a fast, sensitive local alignment search algorithm to identify regions within a pair of sequences that show the highest similarity. Currently, the most promising approach to this is an implementation of one of the rigorous dynamic programming local alignment algorithms on very highly parallel hardware [Lim et al. 1993].

Several other searching techniques are widely used by experimental molecular biologists as aids for guiding and interpreting experiments. These include things as simple as finding the highly hydrophilic and hydrophobic regions in a protein sequence or regions capable of forming helices on the surface of a protein. Finding specific patterns of codon usage in a newly sequenced gene can yield insights into its expression, and finding the intron–exon junction is necessary before the correct protein sequence can be derived from a genomic DNA sequence. There are several important research topics in searching for signals. First is the need for procedures for easily and clearly specifying very arbitrary, complex patterns in a sequence. Faster algorithms for finding these patterns are needed as well. In many cases, the patterns, or biological signals, that are being sought are too complex to be readily identified by visual inspection of sequences. This is especially true if some variation is permissible in the signal pattern. Thus, another important research topic is better algorithms for selecting the most likely patterns to be associated with a signal in a sequence. For example, given several genes known to contain exons, from a single organism, how do we discover the pattern that signals the beginning and end of each exon? A variety of tools have been applied in the large genome projects [Adams et al. 2000, Lander et al. 2001, Venter et al., 2001] and further refinement is still needed.

34.6.3.2 Alignment

Sequence alignment is an important type of searching. It is, basically, the search for the most similar juxtaposition of sequences or regions within sequences. Good alignments are necessary if our inferences about the homology of genes are to be accurate. Even more crucially dependent on good alignments are methods for reconstructing phylogenies based on sequence data [Steel et al. 1994]. Finally, some problems in identifying signal patterns are appreciably simplified given well-aligned sequences. Fortunately, the state-of-the-art in pairwise sequence alignment is well-advanced. There are rigorous algorithms for both global and local alignments of pairs of sequences. These algorithms allow both flexible and sophisticated treatment of insertion/deletion gaps. One possible topic for research in this area is the context-sensitive treatment of these gaps. This would include cases where an insertion/deletion would change the reading frame of a coding region in a gene sequence or change the relative positions of amino acids known to be essential to protein function.

Multiple sequence alignment techniques are not as far advanced as the pairwise techniques. The rigorous pairwise algorithm can be, and has been, extended to multiple sequence problems. However, this approach requires computer time and memory proportional to the product of the lengths of the sequences. Recent advances have reduced this requirement by a large constant factor. However, even with this improvement, this approach soon exhausts even the fastest present-day computers. If a phylogeny of the sequences is available independent of the sequences themselves, it can be used to convert a sequence alignment problem into a series of pairwise problems. However, because a frequent reason for doing a multiple sequence alignment is to generate a phylogeny, this is not a general solution. Thus, a variety of heuristic algorithms are used for most multiple sequence alignment problems.

Several kinds of research are needed here. First, faster and more rigorous algorithms are required. Where algorithms are not completely rigorous, we need to characterize their performance so that we know how close to an optimal solution we can come. We also need to identify what sequence features might cause an algorithm to perform badly.

Defining Terms

API: Application programming interface. An API provides a wide and varied set of functions, messages, and structures that give applications access to the features and capabilities of the underlying operating system. The API consists of a set of standard interfaces to such features as window management, graphics device interface, system services, multimedia services, and remote procedure calls.

Clone assembly: The essence of experimental physical mapping is the ordering and placement of fragmented regions of chromosomes in an overlapping contiguous stretch of DNA that covers the entire chromosomal interval. The DNA fragments are derived from the original DNA and each fragment propagated independently (cloned) to obtain working quantities. Fragment overlaps are determined by hybridization experiments with unique DNA sequences that will permit the unique identification of a chromosomal region. This is a computationally difficult problem.

Kalman filtering: A digital-image averaging procedure for the enhancement of the signal-to-noise ratio in noisy images. The Kalman filter is a recursive version of the true averaging procedure, and takes the form $y_1 = (i - 1)y_i - 1/i$, in which the filter parameters are not constant but vary with the frame number i so that the latest averaged image y_n always equals $(x_1 + x_2 + \cdots + x_n)/n$. This gives a straightforward average over n frames, with a maximum signal-to-noise ratio, without having to prespecify n.

Linkage and linkage mapping: The proximity of two or more markers (e.g., genes, DNA sequences) on a chromosome; the closer the markers are, the lower the probability that they will be separated during repair or replication, and therefore, the greater the probability they will be inherited together. A map of the relative positions of genetic markers on a chromosome, determined on the basis of how often they are inherited together, is referred to as a linkage map.

Marker: An identifiable physical location on a chromosome (e.g., a specific identifiable sequence such as a restriction-enzyme cutting site, or a gene) whose inheritance can be monitored. Markers can be expressed regions of DNA (genes) or a segment of DNA with no known coding function but whose inheritance can be followed with molecular techniques.

Mainchain-directed alignment: A technique for predicting protein structure from NMR data, based on aligning only mainchain residues and ignoring sidechain structures at the primary alignment stage. This technique ignores the complexity of the sidechain packing but suffers from distortions based on the effects of sidechain groups.

NOESY: Nuclear Overhauser Effect (NOE) Enhancement Spectroscopy. The NOE is a consequence of relaxation of dipole-coupled spins induced in magnetic resonance. The enhancement factor in NMR is the deviation from thermal equilibrium induced in one of a pair of dipole-coupled spins following the selective radiation of one of the pairs.

Phase problem: In x-ray diffraction structure determination, a collimated beam of monochromatic x-rays is directed through an object. The x-rays are scattered in all directions by the electrons of every atom in the object. The magnitude of the scattering is proportional to the size of the electron complement of atoms in the target. Because of the variable composition of the target, the scattering of x-rays appears continuous. However, when regular structures are radiated, and known heavy-metal ions are included, it is possible through a series of least-squares calculations to reassemble the phased waves of the diffracted x-rays and compute a clean electron density map.

Physical map: A map of the locations of identifiable regions on DNA (e.g., specific identifiable sequences or genes), *regardless of inheritance*. Distance is measured in base pairs. The physical map with the highest possible resolution is the entire nucleotide sequence of the chromosome.

Scanning confocal microscopy: A technique that — unlike conventional light microscopy, which focuses an optical image of a specimen on the image plane of a collection system — involves the physical scanning of the specimen with a diffraction-limited point of light. In some cases, this may be a one- or two-dimensional array of points. In such microscopy, the result of the interaction of the scanned light beam(s) with successive regions of the specimen is measured and recorded. With such instruments, digital signal processing may significantly enhance the signal-to-noise ratio, yielding three-dimentional images of extraordinary quality.

Sequence assembly: One of the most common methods of sequencing DNA is "shotgun" sequencing, in which a DNA strand is read as a series of random substrings of length 350 to 500. The reconstruction of the sequence of the whole molecule from these random strings is referred to as sequence assembly. There is great complexity in this process, because the reactions to produce the sequencing substrate may be obtained from either strand. Therefore, when comparing two fragments, one must take into account that they could be derived from the same strand or from different strands; in the latter case, it is necessary to take the reverse complement of one of them before making the comparison.

Sequential alignment: A technique for evaluating the degree of secondary structure alignment by sequential evaluation of the root-mean-square deviation between backbone atoms. This involves identification of secondary structure, application of a clustering function to locate collections of such structures, and examining the more extended alignments outside of the initial regions.

Simulated annealing: This technique is a derivative of the Metropolis algorithm and applies statistical mechanics to optimization to many-body systems. In brief, the Metropolis algorithm is used to provide a simulation of a number of atoms in equilibrium at a given temperature. In each step of the algorithm, an atom is given a small random displacement, and the resulting change ΔE in the energy of the system is computed. If the $\Delta E \leq 0$, the displacement is accepted and the value is the basis for the next round. Through a series of probabilistic and cost functions, the algorithm optimizes at a given temperature. The simulated annealing procedure applies this function to a system that has been "melted," and the temperature is lowered until the system is frozen. It is essential that at each stage the system proceed long enough to reach a steady state.

SOAP: SOAP is a lightweight protocol for exchange of information in a decentralized, distributed environment. It is an XML-based protocol that consists of three parts: an envelope that defines a framework for describing what is in a message and how to process it, a set of encoding rules for expressing instances of application-defined datatypes, and a convention for representing remote procedure calls and responses. SOAP can potentially be used in combination with a variety of protocols.

Stochastic search algorithm: A form of optimization searching based on random sampling (Monte Carlo) methods where points v from a set V are chosen at random with probability $1/|V|$. The minimum values of $f(v)$ are recorded as the random sampling proceeds, and the sampling does not terminate arbitrarily, as might occur in a deterministic search. Simulated annealing is an example of a stochastic algorithm.

Transmission electron microscopy (TEM): A technique in which a suitably prepared sample is placed in a beam of electrons being controlled in an electric field. A moving electron has a wavelength that is inversely proportional to its momentum (mass times velocity). Therefore, the higher the accelerating voltage of the TEM, the smaller the wavelength and the higher the resolution. The modern TEM consists of an electron source, an imaging lens, and an image-recording system, all housed in a column under high vacuum. Electrons are emitted from a heated tungston filament held at a large negative potential and accelerated through voltages greater than 80 kV. The column is equipped with condenser electromagnetic lenses for focusing.

Web Services: A Web service is a software system identified by a URI (Uniform Resource Identifiers, a.k.a. URLs, are short strings that identify resources on the Web), whose public interfaces and bindings are defined and described using XML. Its definition can be discovered by other software systems. These systems can then interact with the Web service in a manner prescribed by its definition, using XML-based messages conveyed by Internet protocols.

References

Adams, M.H. et al. 2000. The genome sequence of *Drosophila melanogaster*. *Science*, 287:2185–2195.

Bellon, P.L. and Lanzavecchia, S. 1995. A direct Fourier method (DFM) for x-ray tomographic reconstructions and the accurate simulations of sinograms. *Int. J. Bio-Med. Comput.*, 38(1):55–69.

Cheng, R.H., Kuhn, R.J., Olson, N.H., Rossmann, M.G., Choi, H.K., Smith, T.J., and Baker, T.S. 1995. Nucleocapsid and glycoprotein organization in an enveloped virus. *Cell*, 80(4):621–630.

Cinkosky, M.J. and Fickett, J.W. 1993. *SIGMA User Manual*. Los Alamos National Laboratory, Los Alamos, NM.

Desjarlais, J.R. and Berg, J.M. 1992. Redesigning the DNA-binding specificity of a zinc finger protein: a database guided approach. *Proteins: Struct. Funct. Genet.*, 12:101–104.

Dobson, C.M. 1999. Protein misfolding, evolution and disease. *Trends Bichem. Sci.*, 24:329–332.

Dyer, M., Frieze, A., and Suen, S. 1995. Ordering clone libraries in computational biology. *J. Comput. Biol.*, 2:207–218.

Eigenbrot, C., Randal, M., Presta, L., Carter, P., and Kossiakoff, A.A. 1993. X-ray structures of the antigen-binding domains from three variants of humanized anti-p185HER2 antibody 4D5 and comparison with molecular modeling. *J. Mol. Biol.*, 229:969–995.

Ewing, B., Hillier, L., Wendl, M.C., Green, P. 1998. Base-calling of automated sequencer traces using phred. I. Accuracy assessment. *Genome Research*, 8:175–185.

Ewing, B. and Green, P. 1998. Base-calling of automated sequencer traces using phred. II. Error probabilities. *Genome Research*, 8:186–194.

Fernandez, J.-J., Lawrence, A.F., Roca, J., García, I., Ellisman, M.H., and Carazo, J-M. 2002. High-performance electron tomography of complex biological specimens. *J. Struct. Biol.*, 138:6–20.

Fink, A.L. 1998. Protein aggregation: folding aggregates, inclusion bodies and amyloid. *Fold. Dis.*, 3:R9–23.

Gilson, M.K., Straatsma, T.P., McCammon, J.A., Ripcoll, D.R., Faerman, C.H., Axelsen, P.H., Silman, I., and Sussman, J.L. 1994. Open "back door" in a molecular dynamics simulation of acetylcholinesterase. *Science* 263:1276–1278.

Goldberg, P.W., Golumbic, M.C., Kaplan, H., and Shamir, R. 1995. Four strikes against physical mapping of DNA. *J. Comput. Biol.*, 2:139–152.

Hammarström, P., Wiseman, R.L., Powers, E.T., and Kelly, J.W. 2003. Prevention of Transthyretin Amyloid Disease by changing protein misfolding energetics. *Science*, 299:713-716.

Iris, F.J.M. 1994. Optimized methods for large-scale shotgun sequencing in Alu-rich genomic regions, pp. 199–210. In M.D. Adams, C. Fields, and J.C. Venter, Eds. *Automated DNA Sequencing and Analysis*, Academic Press, London.

Lander, E.S. et al., 2001. Initial sequencing and analysis of the human genome. *Nature*, 409:860–921.

Lim, H.A., Fickett, J.W., Cantor, C.R., and Robbins, R.J., Eds. 1993. In *Proc. 2nd Int. Conf. on Bioinformatics, Supercomputing and Complex Genome Analysis*, St. Petersburg, FL, July 1992. World Scientific, Singapore.

Luthy, R., Bowie, J.U., and Eisenberg, D. 1992. Assessment of protein models with three-dimensional profiles. *Nature (London)*, 356:83–85.

Magness, C., Xu, Y., and Green, P. 1993. *SEGMAP — a program for computing and displaying YAC-based STS-content maps.* Washington University School of Medicine, St. Louis.

Matise, T.C., Perlin, M., and Chakravarti, A. 1994. Automated construction of genetic linkage maps using an expert system (MultiMap): a human genome linkage map. *Nature Genet.*, 6:384–390.

Pan, Y., Yuhasz, S.C., and Amzel, L.M. 1995. Anti-idiotypic antibodies: biological function and structural studies. *FASEB J.*, 9:43–49.

Robbins, R.J., Ed. 1994. Report on the Invitational DOE Workshop on Genome Informatics, 26–27 April 1993, Baltimore, MD. Genome Informatics I: Community Databases. *J. Comput. Biol.*, 1:173–190.

Rossman, M.G., Arnold, E., Erickson, J.W., Frankenberger, E.A., Griffith, J.P., Hecht, H.J., Johnson, J.E., Kamer, G., Luo, M., Mosser, A.G., Rueckert, R.R., Sherry, B., and Vriend, G. 1985. Structure of a human common cold virus and functional relationship to other picornaviruses. *Nature (London)*, 317:145–153.

Sali, A., Shakhnovich, E.I., and Karplus, M. 1994. How does a protein fold? *Nature (London)*, 369:248–251.

Schmieder, P., Hoch, J.C., Stern, A.S., and Wagner, G. 1995. Maximum entropy reconstruction of non-linearly sampled data. Keystone Symposium on Frontiers of NMR in Molecular Biology — IV, Keystone, Colorado, Apr. 3–9, 1995. *J. Cell. Biochem. Suppl.*, 21b, p. 76.

Steel, M.A., Szekely, L.A., and Hendy, M.D. 1994. Reconstructing trees when sequence sites evolve at variable rates. *J. Comput. Biol.*, 1:153–163.

Tramontano, A. and Lesk, A.M. 1995. Common features of the conformations of antigen-binding loops in immunoglobulins and application to modeling loop conformations. *Proteins*, 13:231–245.

Uberbacher, E. and Mural, R. 1991. Locating protein coding regions in human DNA sequences by a multiple sensor–neural network approach. *Proc. Natl. Acad. Sci. USA*, 88:11261–11265.

Venter, J.C. et al. 2001. The sequence of the human genome. *Science*, 291:1304–1351.

Waterston, R.H. et al., 2002. Initial sequencing and comparative analysis of the mouse genome. *Nature*, 420:520–562.

Further Information

The Journal of Computational Biology is a regular source of computational molecular biology. Additional journals related to computational biology are in the late planning stages. The *Journal of Structural Biology* and *Proteins: Structure, Function and Genetics* is also a source of current work. For a comprehensive treatment of the mathematics of molecular biology, the reader is directed to *Introduction to Computational Biology: Maps, Sequences and Genomes*, by Michael Waterman, published by Chapman & Hall, 1995.

IV

Graphics and Visual Computing

Graphics and Visual Computing is the study and realization of complex processes for representing physical and conceptual objects visually on a computer screen. Fundamental to all graphics applications are the processes of modeling objects abstractly and rendering them on a computer screen. Also important are object identification, light, color, shading, projection, and animation. The reconstruction of scanned images and the virtual simulation of reality are also of major research interest, as is the ultimate goal of simulating human vision itself.

35 **Overview of Three-Dimensional Computer Graphics** *Donald H. House* **35**-1
Introduction · Organization of a Three-Dimensional Computer Graphics
System · Research Issues and Summary

36 **Geometric Primitives** *Alyn P. Rockwood* **36**-1
Introduction · Screen Specification · Simple Primitives · Wireframes · Polygons
· The Triangular Facet · Implicit Modeling · Parametric Curves · Parametric
Surfaces · Standards · Research Issues and Summary

37 **Advanced Geometric Modeling** *David S. Ebert* **37**-1
Introduction · Fractals · Grammar-Based Models · Procedural Volumetric
Models · Implicit Surfaces · Particle Systems · Research Issues and Summary

38 **Mainstream Rendering Techniques** *Alan Watt and Steve Maddock* **38**-1
Introduction · Rendering Polygon Mesh Objects · Rendering Using Ray
Tracing · Rendering Using the Radiosity Method · The (Irresistible) Survival of
Mainstream Rendering · An OpenGL Example

39 **Sampling, Reconstruction, and Antialiasing** *George Wolberg* **39**-1
Introduction · Sampling Theory · Reconstruction · Reconstruction
Kernels · Aliasing · Antialiasing · Prefiltering · Example: Image Scaling · Research
Issues and Summary

40 **Computer Animation** *Nadia Magnenat Thalmann and Daniel Thalmann* **40**-1
Introduction · Underlying Principles and Best Practices · Physics-based
Methods · Behavioral Methods · Crowds and Groups · Facial
Animation · Algorithms · Research Issues and Summary

41 Volume Visualization   *Arie Kaufman and Klaus Mueller* **41**-1
Introduction • Volumetric Data • Rendering via Geometric Primitives • Direct Volume
Rendering: Prelude • Volumetric Function Interpolation • Volume Rendering
Techniques • Acceleration Techniques • Classification and Transfer
Functions • Volumetric Global Illumination • Special-Purpose Rendering
Hardware • General-Purpose Rendering Hardware • Irregular
Grids • High-Dimensional and Multivalued Data • Volume Graphics • Conclusions

42 Virtual Reality   *Steve Bryson* **42**-1
Introduction • Underlying Principles • Best Practices • Software
Architectures • Environment Design Concepts • Distributed Virtual
Reality • Application Evaluation and Design • Case Studies • Research
Issues • Summary

43 Computer Vision   *Daniel Huttenlocher* **43**-1
Introduction • Low-Level Vision • Middle-Level Vision • High-Level Vision

35

Overview of Three-Dimensional Computer Graphics

Donald H. House
Texas A&M University

35.1 Introduction 35-1
35.2 Organization of a Three-Dimensional
Computer Graphics System.........................35-1
Scene Specification • Rendering • Storage and Display
35.3 Research Issues and Summary 35-17

35.1 Introduction

The name *three-dimensional computer graphics* has been used freely in the computer graphics community for many years now [Foley et al. 1990, Glassner 1995, Hill 1990, Rogers 1985, Watt and Watt 1992]. It is something of a misnomer, because the graphics themselves are not in any sense three-dimensional (3-D). Rather, the way that the graphics are generated is dependent upon the construction of a virtual 3-D model in the computer, which is then imaged via a virtual camera, usually implying a simulation of a real physical illumination process. The term *three-dimensional* merely emphasizes the fact that a simulation of a 3-D world underlies the image-making process and also that the images produced often display the kinds of foreshortening distortions apparent in photographs or perspective drawings of real 3-D scenes. This chapter is devoted to outlining the various aspects of the process of generating 3-D computer graphic images. It is meant to give the reader an overview, or "big picture," that can be filled in by reading Chapter 36 through Chapter 43 of the handbook, which provide more detailed information on specific aspects of the process.

35.2 Organization of a Three-Dimensional Computer Graphics System

A three-dimensional graphics system can be thought of as having three major components, each of which performs a distinct and clearly defined key role in the process of image generation. These three components are responsible for *scene specification, rendering,* and *image storage and display*. Figure 35.1 gives a schematic view of the process used in 3-D graphics, showing the role that each of these components plays. Each of these major components can itself be broken down into groups of important subcomponents.

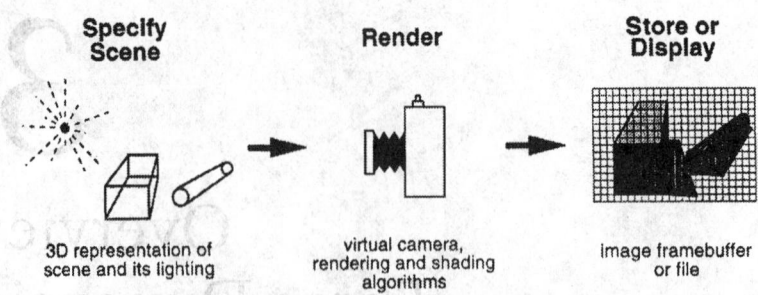

FIGURE 35.1   The 3-D graphics process.

35.2.1   Scene Specification

The *scene specification* section of a 3-D graphics system is responsible for providing an internal representation of the virtual scene that is eventually to be imaged. This requires both an interface to allow user specification and modification of the scene definition and a set of internal data structures that store and organize the scene so that it can be accessed by the user interface and the rendering system. This can be a highly interactive program, providing access to a variety of modeling tools via an interactive user interface; it may be script-driven, providing a scene description language that the user communicates in; or it can be as simple as a program that reads basic geometric information from a tightly formatted scene description file. In any case, the scene specification system will need to support some concept of a geometric coordinate system and provide some way of describing the geometry of the scene to be imaged. Scene description systems also will provide a way in which the user can specify what (virtual) materials objects are made of and how the scene is lit.

35.2.1.1   Coordinate Systems

Key to the geometric structure of a 3-D graphics system is a compact means for storing and utilizing descriptions of **local coordinate systems**. The local coordinate systems are used in the definition of the various components of a model describing the geometry and other characteristics of the scene, much as the local coordinates used on a plan are used in describing the design of a real object. For example, the coordinates on the plan for a complete airplane will necessarily be much different from the coordinates used on the plan for the airplane's wheel assembly.

Consistent with the usual representation of 3-D coordinates in mathematics and engineering, most current books, articles, and implementations of 3-D graphics systems use right-handed coordinate systems [Foley et al. 1990]. This gives a natural organization with respect to the display screen, with the x-coordinate measuring horizontal distance across the screen, the y-coordinate measuring vertical distance up the screen, and the z-coordinate providing the third spatial dimension as distance in front of the screen. However, in the early development of computer graphics, coordinate systems were often left-handed [Foley and van Dam 1982]. In screen space, the difference is that the positive z or depth coordinate is measured into the screen. Of course, for modeling, a local coordinate system can be positioned and oriented anywhere in space and is not usually aligned with the screen. Figure 35.2 shows the ordering of right-handed and left-handed coordinate systems.

A local coordinate system is usually defined in terms of a small set of intuitive geometric operations —
the **affine transformations**. These are:

1. Translation — a change in the position of the origin of the local system
2. Scaling — a change in the scale of measurement in the local system
3. Rotation — a change in the orientation of the local system
4. Shear — transformation from an orthogonal coordinate system to a nonorthogonal system or vice versa via shearing deformations

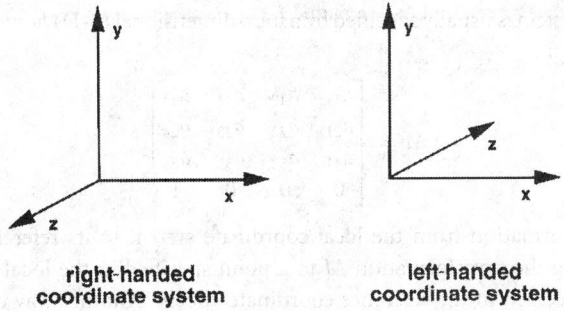

FIGURE 35.2 Right- and left-handed coordinate systems.

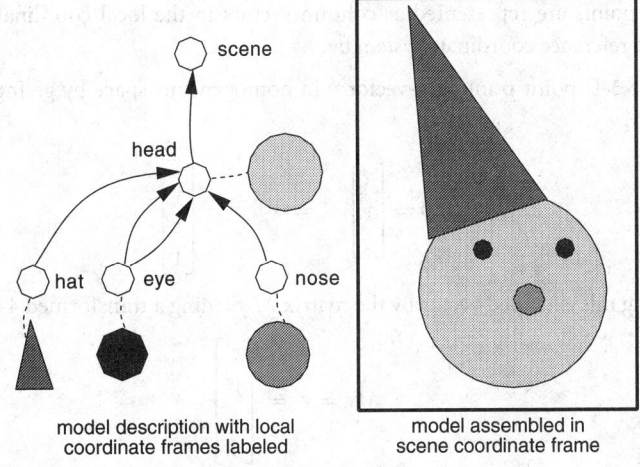

FIGURE 35.3 A clown described by a hierarchy of local coordinate frames.

All of these elements of the local-coordinate-system definition are specified with respect to the origin, scale, orientation, and shear of some external reference system, which might itself be specified relative to some other external system. In this way, local coordinates can be nested within each other, providing the possibility for models to be described in a **hierarchical** fashion. For example, the simple clown model of Figure 35.3 is described in terms of a hierarchy of coordinate frames, which allows for the design and modeling of the head, eye, nose, and hat in their own separate local coordinate frames but then places the two eyes, the nose, and the hat on the head with respect to the head's frame. Finally, the assembled head is placed and oriented in the scene with respect to a local reference frame for the scene. The reference frame at the top of such a hierarchy is usually referred to as the **global coordinate system**.

The basic geometric unit is the 3-D point, which is typically represented in a 3-D graphics system as a 3-vector and stored as a an array of three elements, representing the x, y, and z components of the point. Orientation vectors, like normals to surfaces and directions in space, are also represented by 3-vectors. Thus, the point (x, y, z) is given by the vector

$$\begin{bmatrix} x \\ y \\ z \end{bmatrix}$$

(or in some systems by its transpose $[x \ \ y \ \ z]$).

A local-coordinate-system is usually specified by a four-dimensional (4-D) *homogeneous transformation matrix* of the form

$$M = \begin{bmatrix} a_{11} & a_{12} & a_{13} & a_{14} \\ a_{21} & a_{22} & a_{23} & a_{24} \\ a_{31} & a_{32} & a_{33} & a_{34} \\ 0 & 0 & 0 & 1 \end{bmatrix}$$

which specifies a transformation from the local coordinate system to its reference coordinate system. In other words, applying the transformation M to a point specified in the local coordinate system will yield the same point specified in the reference coordinate system. Another way of thinking of the same transformation matrix M is that when applied to the reference coordinate system it aligns it with and scales it to the local coordinate system. The 4-D homogeneous form of the transformation matrix M allows the unification of translation with scaling, rotation, and shear in a single matrix representation.

The transformation implied by matrix M is actually implemented by a three-step process. Assuming that 3-D geometric points are represented as column vectors in the local coordinate system, they are transformed into the reference coordinate system by:

1. Extending the 3-D point $\mathbf{p}$ into a 4-vector $\mathbf{v}$ in homogeneous space by giving it a fourth, or w, coordinate of 1:

$$\mathbf{p} = \begin{bmatrix} x \\ y \\ z \end{bmatrix} \implies \mathbf{v} = \begin{bmatrix} x \\ y \\ z \\ 1 \end{bmatrix}$$

2. Premultiplying this extended vector by the matrix M yielding a transformed 4-vector $\mathbf{v}'$:

$$M\mathbf{v} = \mathbf{v}' = \begin{bmatrix} x' \\ y' \\ z' \\ 1 \end{bmatrix}$$

3. Converting the resulting 4-vector $\mathbf{v}'$ into the transformed 3-D point $\mathbf{p}'$ by discarding its w-coordinate:

$$\mathbf{v}' = \begin{bmatrix} x' \\ y' \\ z' \\ 1 \end{bmatrix} \implies \mathbf{p}' = \begin{bmatrix} x' \\ y' \\ z' \end{bmatrix}$$

Inspection of matrix M will show that it is defined to always send the original w-coordinate to itself, thus making the third step legitimate. (In earlier computer graphics systems, it was usual for points to be represented by row vectors instead of column vectors and for step 2 to be done by *postmultiplying* the homogeneous row vector $\mathbf{v}$ by the transpose of matrix M.)

The basic transformations of translation, rotation, scaling, and shear are given by the following matrices, which assume that points are represented as column vectors in a right-handed coordinate system and that transformations will be done by premultiplication of the vector (extended to homogeneous coordinates) by the matrix. Use of left-handed coordinates instead of right-handed coordinates will affect the rotations only as indicated below. If row vectors and postmultiplication are being used to represent points and their transforms, the matrices must be transposed.

- *Translation* by Δx in the x-direction, Δy in the y-direction, and Δz in the z-direction:

$$T(\Delta x, \Delta y, \Delta z) = \begin{bmatrix} 1 & 0 & 0 & \Delta x \\ 0 & 1 & 0 & \Delta y \\ 0 & 0 & 1 & \Delta z \\ 0 & 0 & 0 & 1 \end{bmatrix}, \qquad T(\Delta x, \Delta y, \Delta z) \begin{bmatrix} x \\ y \\ z \end{bmatrix} = \begin{bmatrix} x + \Delta x \\ y + \Delta y \\ z + \Delta z \end{bmatrix}$$

- *Scaling* of s_x in the x-direction, s_y in the y-direction, and s_z in the z-direction:

$$S(s_x, s_y, s_z) = \begin{bmatrix} s_x & 0 & 0 & 0 \\ 0 & s_y & 0 & 0 \\ 0 & 0 & s_z & 0 \\ 0 & 0 & 0 & 1 \end{bmatrix}, \qquad S(s_x, s_y, s_z) \begin{bmatrix} x \\ y \\ z \end{bmatrix} = \begin{bmatrix} s_x x \\ s_y y \\ s_z z \end{bmatrix}$$

- *Rotation* through angle θ around the x, y, or z axis, with right-handed rotation around the axis taken as a positive rotation (i.e., aligning the thumb of the right hand with the axis, the fingers grasp the axis in the direction of positive rotation; note that if left-handed coordinates are being used, the signs of the *sine* terms in R_x and R_y should be reversed, but R_z is unaffected):

$$R_x(\theta) = \begin{bmatrix} 1 & 0 & 0 & 0 \\ 0 & \cos\theta & -\sin\theta & 0 \\ 0 & \sin\theta & \cos\theta & 0 \\ 0 & 0 & 0 & 1 \end{bmatrix}, \qquad R_x(\theta) \begin{bmatrix} x \\ y \\ z \end{bmatrix} = \begin{bmatrix} x \\ y\cos\theta - z\sin\theta \\ y\sin\theta + z\cos\theta \end{bmatrix}$$

$$R_y(\theta) = \begin{bmatrix} \cos\theta & 0 & \sin\theta & 0 \\ 0 & 1 & 0 & 0 \\ -\sin\theta & 0 & \cos\theta & 0 \\ 0 & 0 & 0 & 1 \end{bmatrix}, \qquad R_y(\theta) \begin{bmatrix} x \\ y \\ z \end{bmatrix} = \begin{bmatrix} x\cos\theta + z\sin\theta \\ y \\ -x\sin\theta + z\cos\theta \end{bmatrix}$$

$$R_z(\theta) = \begin{bmatrix} \cos\theta & -\sin\theta & 0 & 0 \\ \sin\theta & \cos\theta & 0 & 0 \\ 0 & 0 & 1 & 0 \\ 0 & 0 & 0 & 1 \end{bmatrix}, \qquad R_z(\theta) \begin{bmatrix} x \\ y \\ z \end{bmatrix} = \begin{bmatrix} x\cos\theta - y\sin\theta \\ x\sin\theta + y\cos\theta \\ z \end{bmatrix}$$

- *Shear* parallel to the (x, y) plane as a function of z, or parallel to the (y, z) plane as a function of x, or parallel to the (z, x) plane as a function of y:

$$H_{xy}(h_{xz}, h_{yz}) = \begin{bmatrix} 1 & 0 & h_{xz} & 0 \\ 0 & 1 & h_{yz} & 0 \\ 0 & 0 & 1 & 0 \\ 0 & 0 & 0 & 1 \end{bmatrix}, \qquad H_{xy}(h_{xz}, h_{yz}) \begin{bmatrix} x \\ y \\ z \end{bmatrix} = \begin{bmatrix} x + h_{xz}z \\ y + h_{yz}z \\ z \end{bmatrix}$$

$$H_{yz}(h_{yx}, h_{zx}) = \begin{bmatrix} 1 & 0 & 0 & 0 \\ h_{yx} & 1 & 0 & 0 \\ h_{zx} & 0 & 1 & 0 \\ 0 & 0 & 0 & 1 \end{bmatrix}, \qquad H_{yz}(h_{yx}, h_{zx}) \begin{bmatrix} x \\ y \\ z \end{bmatrix} = \begin{bmatrix} x \\ y + h_{yx}x \\ z + h_{zx}x \end{bmatrix}$$

$$H_{zx}(h_{xy}, h_{zy}) = \begin{bmatrix} 1 & h_{xy} & 0 & 0 \\ 0 & 1 & 0 & 0 \\ 0 & h_{zy} & 1 & 0 \\ 0 & 0 & 0 & 1 \end{bmatrix}, \qquad H_{zx}(h_{xy}, h_{zy}) \begin{bmatrix} x \\ y \\ z \end{bmatrix} = \begin{bmatrix} x + h_{xy}y \\ y \\ z + h_{zy}y \end{bmatrix}$$

More complex coordinate transformations, involving combinations of the basic transformations, can be obtained by specifying them as a sequence of operations, each described by a 4-D transformation matrix. The product of these matrices will yield a compound transformation matrix that has the same effect as each transformation applied separately. For example, a rotation of 30° around the x axis followed by a translation to the point $(10, 20, -10)$ would be given by

$$M = T(10, 20, -10) R_x(30°)$$

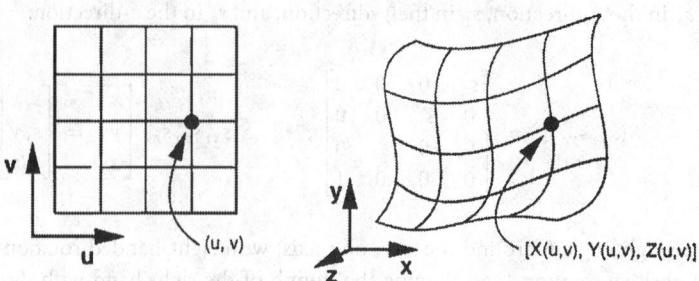

FIGURE 35.4   A biparametric surface.

35.2.1.2 Geometric Modeling

Virtually all 3-D graphics systems provide the ability to work with simple geometric primitives that can be specified as lists of 3-D points. These primitives include points, lines, and polygons. Points can be arranged together to indicate a *sampled* surface, lines to form a *wireframe* representation, and polygons to form *polyhedral* surfaces. More sophisticated modelers will provide **parametric surfaces**, which are defined via an underlying piecewise polynomial formulation [Rogers and Adams 1990, Bartels et al. 1987]. Polynomial coefficients are adjusted to give the surface a specific shape, and these coefficients are often given intuitive form by encoding them via simple geometric devices, such as control polyhedra.

A typical surface formulation is a *biparametric surface*, which describes a surface in three spatial dimensions (x, y, z) via a set of three functions of two parameters u and v:

$$x = X(u, v), \qquad y = Y(u, v), \qquad z = Z(u, v)$$

This concept is illustrated in Figure 35.4. The rectangular grid on the left of the figure, defined in the (u, v) parametric coordinate system, is mapped, via the functions X, Y, and Z, into a 3-D surface like that shown on the right. A set of points on a parametric surface can be obtained algorithmically by looping over a collection of sample points on the (u, v) plane. Simple geometric primitives and parametric surfaces are described more fully in Chapter 36.

Implicit surfaces are a common alternative to parametric surfaces. Here, surfaces are defined as the set of points satisfying a mathematical expression of the form

$$F(x, y, z) = 0$$

Thus, these surfaces are defined implicitly. Any point (x, y, z) in 3-D space can be tested to determine whether or not it is above $[F(x, y, z) > 0]$, below $[F(x, y, z) < 0]$, or on $[F(x, y, z) = 0]$ the surface. However, it is not generally easy to algorithmically generate a set of points guaranteed to be on the surface, without resorting to iterative search or relaxation techniques. Implicit formulations are especially useful for defining solids, where the implicit equation can be evaluated to determine whether a point is inside, outside, or on the surface of the solid. For example, the well-known equation for a sphere of radius r centered at the point (x_0, y_0, z_0) can be written implicitly as

$$(x - x_0)^2 + (y - y_0)^2 + (z - z_0)^2 - r^2 = 0$$

Implicit techniques can be generalized to describe surfaces or solids defined algorithmically in the form of a programmed function. This technique is useful in describing a variety of natural-looking forms [Ebert 1994]. Functional techniques are described in Chapter 37.

Some geometric data come naturally in the form of a set of scalar values distributed in a 3-D field. For example, data from medical scanners, such as MR or CT devices, consist of a set of material density values distributed on a regular lattice within a volume of space. This type of data has its own specialized set of modeling and visualization techniques that are described thoroughly in Chapter 41.

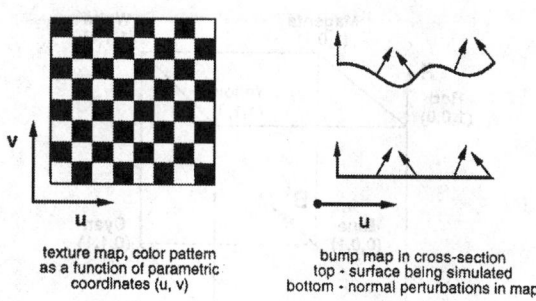

texture map, color pattern
as a function of parametric
coordinates (u, v)

bump map in cross-section
top - surface being simulated
bottom - normal perturbations in map

FIGURE 35.5 Texture and bump maps.

35.2.1.3 Materials

In the context of a 3-D graphics system, a **material** is an attribute of a geometric object that provides a description of how the surface of the object will appear when viewed from a particular direction under a particular illumination. In physical terms, what we need to define here is how a surface reflects (or transmits) light as a function of incident angle, reflection (or refraction angle), and wavelength. A function providing these relationships is known as the material's **bidirectional reflectance distribution function** or BRDF.

In practical terms for computer graphics applications, it is usually enough to approximate the BRDF for a material with a collection of parameters and maps. A usual material specification system will provide parameters for the specification of a material's color, diffuse reflectance factor, specular reflectance factor, specularity, transmissivity, and refraction index. These factors and their use in lighting calculations are described in detail in Chapter 38.

A material specification will also often include the capability to provide both **texture maps** and **bump maps.** A texture map provides a pattern of color that is to be applied to the surface of an object during the rendering process. These can be anything from a digital image that will be projected onto the surface, to a regular geometric pattern like a checker-board or polka-dot design. A bump map is a pattern of perturbations to the normal vector to a surface that simulates the effect that bumps would have on the appearance of the surface. Figure 35.5 illustrates texture and bump maps. Some systems also provide for **displacement maps,** which are used to locally perturb both the surface normal and the surface itself. It is usual to relate texture and other map coordinates directly to the parametric coordinates of a parametric surface. For nonparametric surfaces, specific texture coordinates must be provided by the user or by some algorithmic technique. For example, many modelers allow the user to provide texture coordinates along with 3-D geometric coordinates for each vertex in a polygonal surface.

Within the field of computer graphics, color is becoming as complex a topic as it is in physics, psychology, and art. However, from the point of view of usual practice, color is most often represented in 3-D graphics systems in one of two related color systems, the *RGB* and the *HSV* systems.

The RGB or "red–green–blue" system is the usual system used for storage or display of color. This is because this color system relates directly to the three-electron-gun RGB organization of color CRT displays. An RGB color is stored as a triple of three numbers giving the relative amount of each of the three color primaries — red, green, and blue. Because the RGB system organizes color into three *primaries,* and allows us to scale each primary independently, we can think of all of the colors that are represented by the system as being organized in the shape of a cube, as shown in Figure 35.6. We call this the RGB color cube or the RGB color space (when we add coordinate axes to measure R, G, and B levels). Note that the corners of the RGB color cube represent pure black and pure white; the three primary colors red, green, and blue; and the three secondary colors yellow, cyan, and magenta. The diagonal from the black corner to the white corner represents all of the gray levels. Other locations within the cube correspond with all of the other colors that can be displayed.

The HSV or "hue–saturation–value" color system represents colors using three measures that relate directly to how artists often think about color. It provides separate measures of *hue* (corresponding to

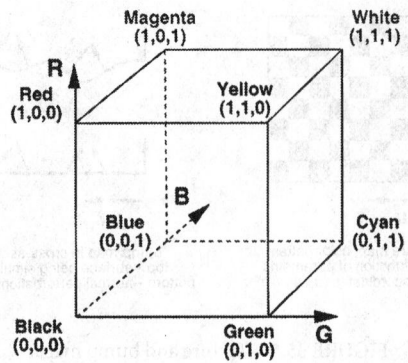

FIGURE 35.6   RGB color cube.

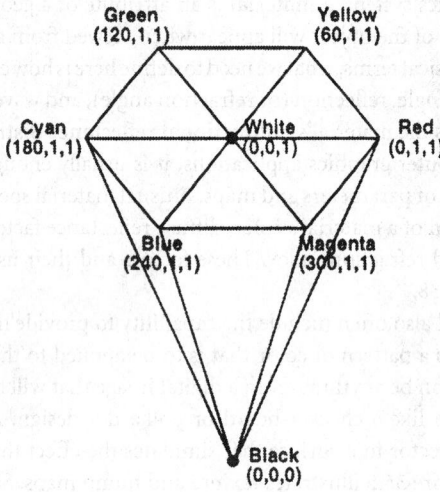

FIGURE 35.7   HSV color cone.

dominant color name), *saturation* (purity of color), and *value* (brightness on gray scale). Its structure is derived directly from the RGB system, and in fact there is a simple translation from RGB to HSV and back. If the RGB color cube of Figure 35.6 is viewed along its white–black diagonal, it presents a hexagonal silhouette. "Peeling" off layers of the face of the RGB cube visible along the white–black diagonal and projecting these cross sections onto a flat surface result in a series of smaller and smaller hexagonal cross sections. If these cross sections are then stacked up, they form a hexagonal cone. The HSV system is the cone-shaped space derived in this way, shown in Figure 35.7. Figure 35.8 shows how the coordinates of the HSV color space are organized. The hue or h coordinate is an angular measurement (usually in degrees) around the face of the cone, with red at $0°$, green at $120°$, and blue at $240°$. The saturation or s-coordinate is measured from the center of the face of the cone out to its perimeter, with 0 at the center corresponding with gray and 1 at the perimeter corresponding with a fully saturated color of the chosen hue. The value or v-coordinate is measured from the apex of the cone to its face along the central axis, with 0 at the apex corresponding with no illumination (black) and 1 at the face corresponding with full illumination.

35.2.1.4   Lights

The purpose of lights in a 3-D graphics system is to provide the illumination source for the simulated shading calculations done by the renderer in making an image. Thus, all light sources must define a

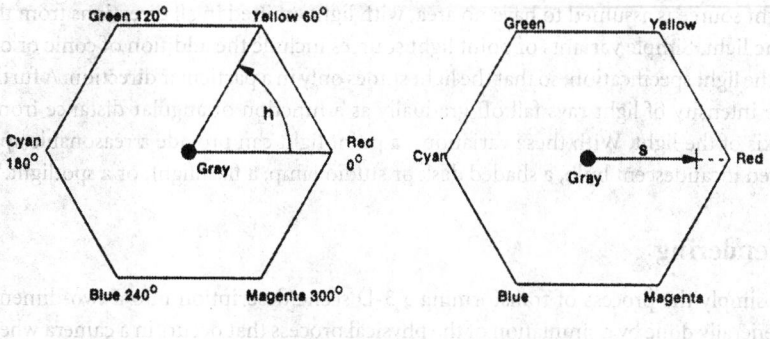

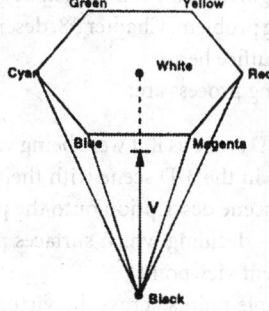

FIGURE 35.8   HSV parameterization of hue, saturation, and value.

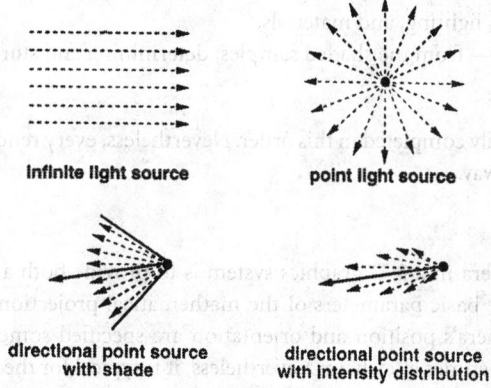

FIGURE 35.9   Infinite and point light sources and variants.

color of the illumination that they provide, usually specified in RGB or HSV coordinates. Note that this illumination color combines the notions of the intensity of the light and its chromatic attributes. Lights are arranged in a scene along with geometric objects but usually carry no geometric properties other than their position and, for directional lights, a direction of orientation. Some rendering algorithms work with light sources with other geometric properties such as shape and area, but most 3-D graphics systems work with only two types of lights — **infinite lights** and **point lights**. Figure 35.9 illustrates infinite and point light sources and some of their variants.

An infinite light is one that is so far away from the geometry being illuminated that light rays can be assumed to be parallel to each other, like the rays of light from the sun illuminating the earth. For this type of light, no position needs to be specified. All that matters geometrically is the direction of the light rays.

A point light source is assumed to have no area, with light emitted in all directions from the geometric position of the light. Simple variants of point light sources include the addition of conic or other shading devices with the light specification, so that the light shines only in a particular direction. A further variation is to have the intensity of light rays fall off gradually as a function of angular distance from the central directional axis of the light. With these variations, a point light can provide a reasonable approximation to an unshaded incandescent bulb, a shaded desk or studio lamp, a flashlight, or a spotlight.

35.2.2 Rendering

Rendering is simply the process of transforming a 3-D scene description into a two-dimensional (2-D) image. It is generally done by a simulation of the physical process that occurs in a camera when a picture is recorded on film. Making this process algorithmic, so that it can be simulated efficiently and accurately in a computer, is the essence of the rendering problem. Chapter 38, describes practical approaches to rendering in some detail, so a brief synopsis will suffice here.

Briefly, the main steps in the rendering process are:

1. Point of view — orienting the 3-D scene as if it were being viewed from a particular point in space,
2. Projection — associating points in the 3-D scene with their images on a 2-D virtual image plane or screen by projecting the 3-D scene description onto the plane,
3. Visible-surface determination — deciding which surfaces projected onto the image plane would actually be visible from the present viewpoint,
4. Sampling — fixing a set of sample points across the virtual image plane, usually corresponding in some way with the pixels that will be used to store the calculated image, and associating these sample points with visible points on the scene's 3-D geometry,
5. Shading calculation — determining, for these sample points, what color would be reflected or transmitted to the viewpoint from the geometry visible at the sample point, taking into account the scene's geometry, lighting, and materials,
6. Image construction — from the shaded samples, determining and storing colors for each pixel in the output image.

These steps are not necessarily completed in this order. Nevertheless, every renderer will have to solve each of these problems in some way.

35.2.2.1 Camera

The role of the virtual camera in a 3-D graphics system is to provide both a point of view from which to render an image and the basic parameters of the mathematical projection that will be used to form the virtual image. The camera's position and orientation are specified sometimes as part of the scene description system and sometimes elsewhere. Nevertheless, it is typical for the camera to be positioned in the global coordinate system, usually with some positioning controls that correspond to the operation of a real studio camera. Often, a camera is positioned by one set of controls and aimed by another set. Aiming is usually made easier by the option to select a **center of interest**, which is a reference point in the scene toward which the camera will orient itself.

Theoretically, cameras can have any projection characteristics, corresponding to the entire variety of lens types, either real or imagined. However, practical 3-D graphics implementations usually implement only the standard parallel or perspective projections that are common in architectural and design drafting.

A perspective projection is one in which all light rays coming from the scene converge at a common point, known as the **center of projection**. If a projection plane is interposed between the scene and the center of projection, the point at which a ray from the scene through the center of projection intersects the projection plane is the *image* of that point. Figure 35.10 shows the geometry of this projection. The parallel projections can be considered as special cases of the perspective projections, where the center of projection is infinitely far from the scene and virtual screen. Thus, rays between points on the scene and the center of projection are parallel to each other when intersecting the screen.

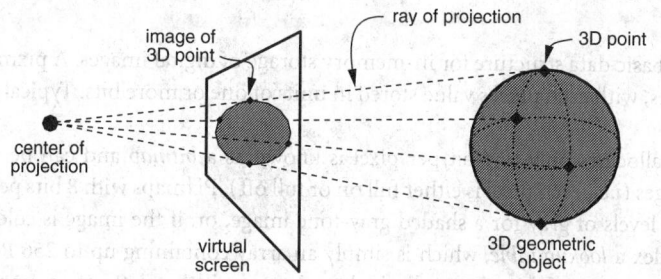

FIGURE 35.10 Geometry of a perspective projection.

In order to completely specify the perspective projection, the position of the camera (i.e., the center of projection), the direction in which the camera is aimed (i.e., its central ray of projection), the camera's up direction, the distance of the virtual screen from the center of projection along the central projection ray, and the width and height of the virtual screen must be known. This assumes that the virtual screen is centered on the central ray of projection, with its surface normal aligned with the central ray (i.e., it is perpendicular to the central ray). It is also possible to build fancier cameras, where the screen can be moved off center and oriented skewed to the central ray.

35.2.2.2 Renderer

The renderer in a 3-D graphics system is essentially the engine that drives the picture-making process. We can think of the renderer as viewing the scene through the lens of the virtual camera and constructing an image of what it sees, by first sampling points on the scene geometry and calling on the shader to calculate colors for each sample, and then combining these sampled colors into the pixels of the image. There are so many approaches to rendering, and the subject is so complex, that we will direct the interested reader to Chapter 38 for more information.

35.2.2.3 Shader

The shader is the algorithm that uses the information collected by the renderer about a point sample on the scene geometry, its material attributes, and the available lighting to calculate a color for the sample. Generally this is done by a more or less approximate physical simulation of how light is reflected toward the camera from the position on the surface at which the sample is being taken. Again, the reader is referred to Chapter 38 for more detailed information on shading and how it is done in a typical graphics system.

35.2.2.4 Image Construction

The final step in the rendering process is the construction of a digital image from the set of shaded samples across the virtual screen. This is done in any of a variety of ways, all of which are forms of low-pass filtering and resampling, providing a smooth blending and interpolation of the color samples into image pixel values [Wolberg 1990]. In practical terms, the digital image pixel grid is superimposed over the virtual screen, so that its pixels become associated with locations on the screen. Then the color of each pixel in the grid is calculated by taking a weighted average of the shaded samples in the vicinity of the pixel.

35.2.3 Storage and Display

For a 3-D graphics system to be useful, there must be a way to turn the results of calculations into tangible images that can be both viewed and stored for archiving and transmission. Thus, a 3-D graphics system is organized around a model of a digital image data structure, one or more image file formats, and a notion of the kind of display device that will be used to view images.

35.2.3.1 Image

The *pixmap* is the basic data structure for in-memory storage of digital images. A pixmap is simply a 2-D array of pixel values, with each pixel's value stored in units of one or more bits. Typical pixel sizes are 1, 8, 24, and 32 bits.

A pixmap that allocates only one bit per pixel is known as a *bitmap* and can be used to store only monochrome images (i.e., each pixel is either full on or full off). Pixmaps with 8 bits per pixel can be used to store up to 256 levels of gray for a shaded gray-tone image, or, if the image is colored, the 8 bits are usually used to index a *lookup table*, which is simply an array containing up to 256 RGB colors that are used in the image. Pixmaps of this type are limited to pictures with a palette of no more than 256 distinct colors, although these colors can be drawn from a much larger set of possible colors. The size of this set is determined by the number of bits per entry in the lookup table. This scheme is often supported by hardware as described below in the discussion of *framebuffers*.

Pixmaps with 24 bits per pixel normally allocate 8 bits, or one byte, to each of the three RGB color primaries, giving a color resolution of 256 levels per primary, or 16,777,216 distinct colors. This color resolution is well beyond the ability of the human eye to distinguish color differences, so that even color gradations in these images appear to be as smooth as they would be in a continuous-tone color image.

On a high-end graphics computer, it is not unusual to allocate more than 24 bits per pixel in a pixmap. The extra space can be used to store colors at higher than 8-bit resolution, which is often handy to avoid roundoff errors in image-processing operations. A common configuration is a 32-bit pixel, where only 8 bits are used for each color primary, and the additional 8 bits are used to store an *alpha* value. The alpha value is used in image-compositing operations as a measure of pixel opacity. For purposes of compositing images together, pixels with high alpha values are treated as if they were opaque and pixels with low alpha values are treated as if they were transparent. Other uses for extra bits in a pixmap are to store aspects of the geometric information of the original model, such as surface normal, object id, material type, 3-D position, etc. This information can be used in postprocessing of the image to do things like modify shading, or to add embellishments to the image that give a notion of the underlying form and structure of the geometry of the imaged objects.

35.2.3.2 Display Devices

The display device most frequently used in conjunction with a 3-D graphics system is the *CRT* or *cathode-ray tube*. A CRT works on exactly the same principle as a simple vacuum tube. A schematic diagram of the organization of a monochrome CRT is shown in Figure 35.11. Electrons traveling from the negatively charged cathode toward the positively charged plate are focused into a beam by focusing coils. The plate end of the CRT is a glass screen coated with phosphor. The grid control voltage adjusts the intensity of the beam and thus determines the brightness of the glowing phosphor dot where the beam hits the screen.

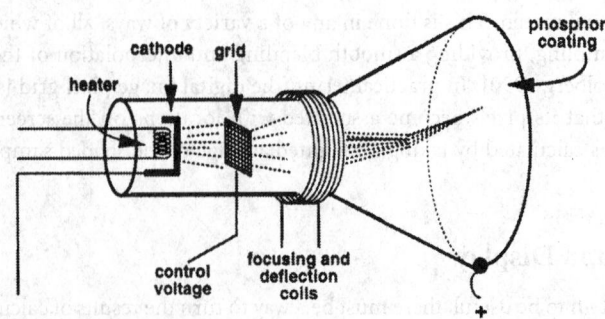

FIGURE 35.11 Schematic diagram of a CRT.

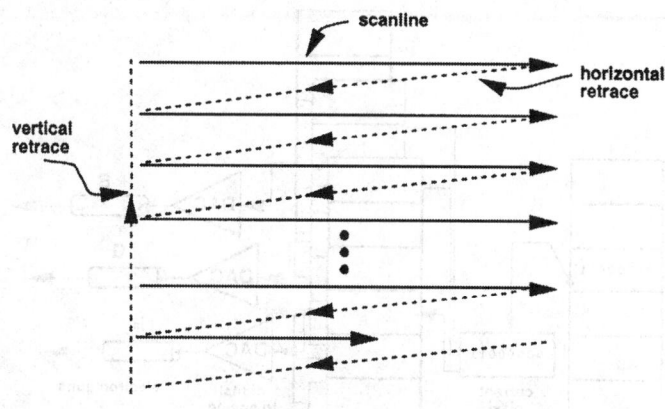

FIGURE 35.12   Raster scan pattern.

Steering or deflection coils push the beam left/right and up/down so that it can be precisely directed to any desired spot on the screen.

A color CRT works like a monochrome CRT, but the tube has three separately controllable electron beams or *guns*. The screen has dots of red-, green-, and blue-colored phosphors, and each of the three beams is calibrated to illuminate only one of the phosphor colors. Thus, even though beams of electrons have no color, we can think of the CRT as having red, green, and blue electron guns. Colors are made using the RGB system, as optical mixing of the colors of the tiny adjacent dots takes place in the eye. Typically, the colored phosphors are arranged in triangular patterns known as *triads*, and an opaque *shadow mask* is positioned between the electron guns and the phosphors to ensure that each gun excites only the phosphors of the appropriate color.

A CRT can be used to display a picture in two different ways. The electron beam can be directed to "draw" a line-drawing on the screen — much like a high-speed electronic Etch-a-Sketch. The picture is drawn over and over on the screen at very high speed, giving the illusion of a permanent image. This type of device is known as a *vector display* and was quite popular for use in computer graphics and computer-aided design up until the early 1980s. By far the most popular type of CRT-based display device today is the *raster display*. These work by scanning the electron beam across the screen in a regular pattern of *scanlines* to "paint" out a picture, as shown in Figure 35.12. The resulting pattern of scanlines is known as a **raster**. As a scanline is traced across the screen by the beam, the beam is modulated proportional to the intended brightness of the corresponding point on the picture. After a scanline is drawn, the beam is turned off and brought back to the starting point of the next scanline. As opposed to a vector display, which essentially makes a line drawing on the screen, a raster display can be used to paint out a shaded image.

The NTSC broadcast TV standard that is used throughout most of America uses 585 scanlines with 486 of these in the visible raster. The extra scanlines are used to transmit additional information, like control signals and closed-caption titling. The NTSC standard specifies a *framerate* of 30 frames per second, with each *frame* (single image) broadcast as two *interlaced fields*. The first of each pair of fields contains every even-numbered scanline, and the second contains every odd-numbered scanline. In this way, the screen is *refreshed* 60 times every second, giving the illusion of a solid flicker-free image. Actually, most of the screen is blank (or dark) most of the time. High-quality color CRTs for computer graphics greatly exceed the resolution and framerate of the NTSC standard, offering noninterlaced framerates of 60 or more frames per second with 1000 or more scanlines per frame.

35.2.3.3   Framebuffers

A *framebuffer* is the hardware interface between the pixmap data structure of a digital image and a CRT display. It is simply an array of computer memory, large enough to hold the color information for one

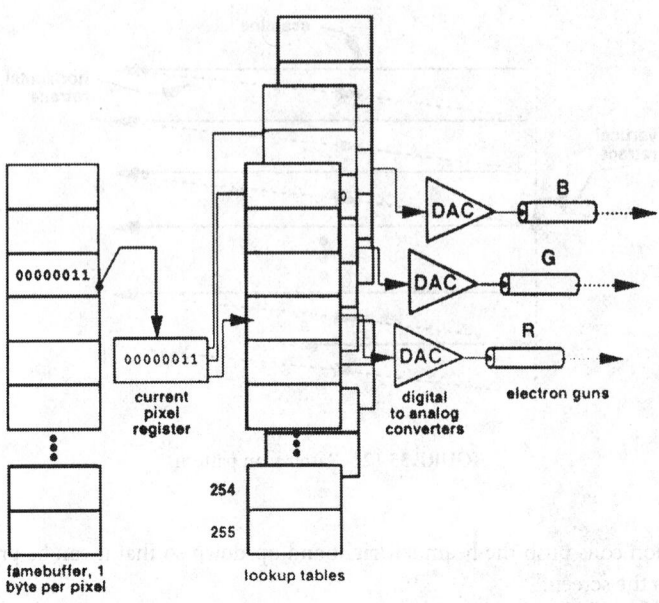

FIGURE 35.13   8-bit color framebuffer with lookup table.

or more frames (i.e., screenful) and display hardware to convert the current frame into control signals to drive a CRT. The color framebuffer schematized in Figure 35.13 holds an 8-bit per pixel color image in a pixmap. The circuitry that controls the electron gun on the CRT loops through each row of the image array, fetching each pixel value in turn and using it to index an array of 256×3 high-speed hardware registers arranged as a lookup table. The values fetched from each of the three indexed registers are converted to voltages by digital-to-analog converters (DACs) and used to control the grid voltages of the CRT's three RGB electron guns. The timing has to be such that the memory fetches and conversion to grid voltages are synchronized exactly with the trace of the beam across the corresponding screen scanline, so that the correct position on the screen is associated with the appropriate pixel from the framebuffer.

A full-color-resolution framebuffer is shown in schematic form in Figure 35.14. This type of device will have at least 24 bits per pixel (8 bits per color primary), driving three color guns, either directly or (as shown in the figure) through a separate lookup table per color primary. In this case, the lookup table is not used to increase the color resolution but instead can be used to correct nonlinearities or to obtain certain effects like overlay planes or pseudocoloring. Higher-end framebuffers may have more than 24 bits allocated per pixel, for hardware handling of such tasks as image compositing, depth buffering for hidden-surface resolution, double buffering for real-time animated display, and overlays.

35.2.3.4   Image Files

Due to the potential for using huge amounts of space, image file storage is a very important issue in computer graphics. A TV-resolution image has about $\frac{1}{3}$ million pixels — so a full-color RGB TV image will contain $3 \times \frac{1}{3} = 1$ million bytes of color information. Now, at 1800 frames (or images) per minute in a computer animation, we can expect to use up most of a 2-gigabyte disk for each minute of animation. Fortunately, we can do somewhat better than this by various file compression techniques, but disk storage space remains a crucial issue. Related to the space issue is the speed-of-access issue — that is, the bigger an image file, the longer it takes to read, write, and display.

For purposes of this overview, we will look closely at only a very simple, but very widely used image file format that has no intrinsic notion of compression. The *PPM*, or *portable pixmap*, format was devised to be an intermediate format for use in developing file-format conversion systems [Murray and vanRyper

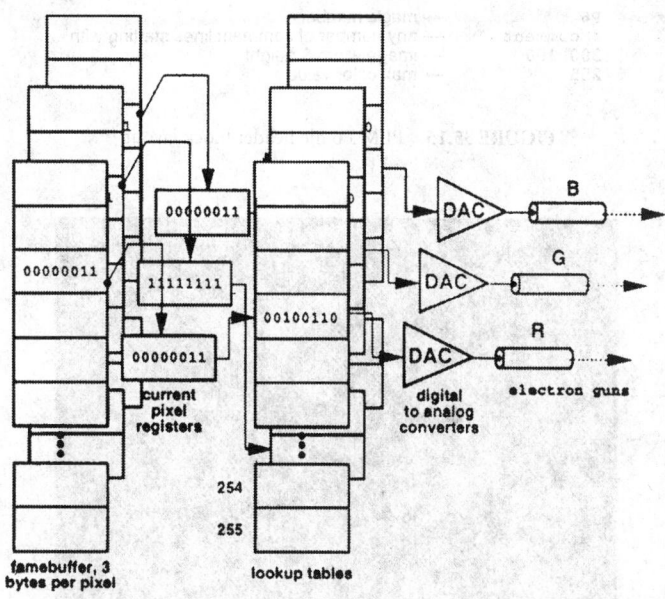

FIGURE 35.14 24-bit color framebuffer with three lookup tables.

1994]. Anyone familiar with computer graphics knows that the number of popular image file formats is immense. These include GIF, Targa, RLA, SGI, PICT, RLE, RLB, and many more. Converting images from one format to another is one of the common tasks in computer graphics, because different software packages and hardware units require different file formats. If there were N different file formats, and we wanted to be able to convert any one of these formats into any of the other formats, we would have to have $N \times (N - 1)$ conversion programs. The PPM idea is that we have one format that serves as a common source or target for all format conversions. We then write only N programs to convert all other formats into PPM and N more programs to convert PPM files into all other formats. In this way, we need write only $2 \times N$ programs to build a complete image conversion library.

The PPM format is not intended to be an archival format, so it does not need to be too storage-efficient. Although it is one of the simplest formats, PPM will nevertheless serve to illustrate some common features of image files. Most file formats are variants of the following organization. The file will typically contain some indication of the file type (which has come to be known as the format's *magic number*), a block of header or control information, and the image description data. Most (but not all) formats have a magic number, which identifies the file type. Often the magic number is not a number at all but rather a string of characters. The header block contains various descriptive information necessary to interpret the data in the image data block or can contain such archival information as creation date, image name, etc. The image data block is some form of encoding of the pixmap that describes the image. Some formats are much more complex than this, but this is the basic layout of a high percentage of formats.

In the PPM format, the magic number is either P1, P2, P3, P4, P5, or P6. P1 and P4 indicate that the image data are in a bitmap. These files are called PBM (portable bitmap) files. P2 and P5 are used to indicate gray-scale images or PGM (portable graymap) files. P3 and P6 are used to indicate full-color PPM (portable pixmap) files. The lower numbers — P1, P2, P3 — indicate that the image data are stored as ASCII characters; i.e., all numbers are stored as character strings. This has the advantage that you can read the file in a text editor. The higher numbers — P4, P5, P6 — indicate that image data are stored in a more compact binary encoding. We will look here only at P6-type files.

The header for a PPM file consists of the information shown in Figure 35.15, stored as ASCII characters in consecutive bytes in the file. In the header, all whitespace is ignored, so the program that writes the file can freely intersperse spaces and line breaks. The image width and height determine the length of a

```
P6                  — magic number
# comment           — any number of comment lines starting with
200  300            — image width & height
255                 — max color value
```

FIGURE 35.15   PPM P6 file header block layout.

(a) Red cube on a midgrey (0 5 0.5 0.5) background

```
000000  5036 0a33 3030 2032 3030 0a32 3535 0a7f   P6!300 200!255!!
000010  7f7f 7f7f 7f7f 7f7f 7f7f 7f7f 7f7f 7f7f   !!!!!!!!!!!!!!!!

009870  886d 6d92 5959 8a69 6984 7676 817d 7d80   'mm'YY'ii!vv']]'
009880  7f7f 7f7f 7f7f 7f7f 7f7f 7f7f 7f7f 7f7f   !'!!!!'!!!''!!!'!

009b10  7f80 7e7e 9d41 41b5 09D9 b211 11a9 2424   !!~.!AA!!!!!!$$
009c00  9f3b 3b94 5454 8b68 6886 7272 827b 7b80   !:::!TT!hh!rr!{}!
009c10  7f71 7f7f 7f7f 7f7f 7f7f 7f7f 7f7f 7f7f   !!!!!!!!!!!!!!!!

009170  7f71 7f71 7f82 7979 a72b 2bb9 0D0D ba0D   !!!!!yy!+.!!!!!
009180  00ba 0000 b9D1 01b7 0606 b201 01a1 1d1d   !!!!!!!!!!!!!!!!
009190  a532 3297 4d4d 8d62 6286 7272 827a 7a80   !22!MM!bb!rr!zz!
009fa0  7e7e 7f71 7f7f 7f7f 7f71 7f7f 7f7f 7f7f   --!!!!!!!!!!!!!!
009fb0  7f7f 7f71 7f7f 7f7f 7f71 7f71 7f7f 7f7f   !'!!'!!!!!'!!!'!

00a210  7f71 7f71 7f71 7f7f 7f8a 6969 b014 14ba   !!!!!!!!!! ii !!!!
00a300  0D00 ba00 00ba 0000 ba00 00ba 0000 ba00   !!!!!!!!!!!!!!!!
00a310  00ba 0000 ba00 00b9 0505 b60d 0daf 1d1d   !!!!!!!!!!!!!!!!
00a320  a62f 2f9d 4141 915b 5b88 6d6d 8279 7980   !//'AA![['mm!yy'
00a330  7e7e 7f71 7f7f 7f7f 7f71 7f71 7f7f 7f7f   --!!!!'!!'!!'!!'!
00a340  7f71 7f71 7f7f 7f7f 7f71 7f71 7f71 7f7f   !!!!!!!!!!!!!!!!
```

(b) Dump of start of PPM P6 red-cube image file

FIGURE 35.16   Red cube image and corresponding PPM P6 image file data: (a) red cube on a mid-gray (0.5 0.5 0.5) background, (b) dump of start of PPM P6 red-cube image file.

scanline and the number of scanlines. The maximum color value cannot exceed 255, but it may be less if less than 8 bits of color information per primary are available.

The PPM P6 data block begins with the first pixel of the top scanline of the image (upper left-hand corner), and pixel data are stored in scanline order from left to right in 3-byte chunks giving the R, G, B values for each pixel, encoded as binary numbers. There is no separator between scanlines, and none is needed, as the image width given in the header block exactly determines the number of pixels per scanline. Figure 35.16a shows a red cube on a mid-gray background, and Figure 35.16b gives the first several lines of a hexadecimal dump of the contents of the corresponding PPM file. Each line of this dump has the hexadecimal byte count on the left, followed by 16 bytes of hexadecimal information from the file, and ends with the same 16 bytes of information displayed in ASCII (nonprinting characters are displayed as !). Except for the first line of the dump, which contains the file header information, the ASCII information is meaningless, because the image data in the file are binary encoded. A line in the dump containing only a * indicates a sequence of lines all containing exactly the same information as the line above.

35.3 Research Issues and Summary

In this chapter, we have taken a quick, broad-brush look at 3-D graphics systems, from scene specification to image display and storage. The attempt has been to lay a basic foundation and to provide certain essential details necessary for further study. A practical example of the implementation and use of a 3-D graphics system appears in Chapter 43.

As research in graphics tends to be specialized, readers are directed to the research issues and summary sections of Chapter 37 through Chapter 42 of this handbook for information on important and interesting open research areas. However, we note here some broad areas of research that are both timely and important to the development of the field. In the area of rendering, there are two areas that seem to be generating much current interest: extending solutions to the global illumination problem to handle a wider variety of material types and developing nonphotorealistic techniques. The entire fields of virtual reality and volumetric modeling are just getting off the ground and promise to be strong research areas for many years. Within the subfield of computer animation, three areas are of very strong current interest. These are physically based modeling and simulation [Barzel 1992], artificial-intelligence and artificial-life techniques for character animation and choreography, and higher-order interactive techniques that exploit the capabilities of new 3-D position and motion tracking devices. Finally, in the area of modeling, there is much room for improvement in interactive modeling tools and techniques, again possibly exploiting 3-D position and motion trackers. And there is a continuing search for compact, powerful ways to represent natural forms.

Defining Terms

Affine transformation: A coordinate-system transformation where each transformed coordinate is a linear sum of the three original coordinates plus a constant.

Bidirectional reflectance distribution function: A function defined for a reflecting material that, given arrival and departure directions, gives the fraction of light energy of a particular wavelength arriving at a surface from the arrival direction that is reflected from the surface in the departure direction.

Bump map: A pattern of surface normal displacements, simulating the undulations of a bumpy surface, that is to be mapped onto a geometric surface during rendering.

Center of interest: A point in space toward which the virtual camera is always aimed.

Center of projection: The point in space at which all rays of projection for a perspective projection converge. This often is considered to be the position of the virtual camera.

Displacement map: A pattern of surface position displacements to create the undulations of a bumpy surface that is to be mapped onto a geometric surface during rendering.

Global coordinate system: A coordinate system with respect to which all other coordinate systems in the definition of a 3-D scene are defined.

Hierarchical model: A geometric model defined within a set of nested reference coordinate frames, whose references form a directed tree or acyclic graph from a set of leaf frames, where basic geometric objects are defined, to a root frame in which the entire scene is defined.

Homogeneous coordinate system: In a 3-D graphics system, this is a 4-D coordinate system into which points are transformed before being multiplied by a homogeneous transformation matrix. The system is called homogeneous because all 3-D points lie on the same hyperplane perpendicular to the fourth coordinate axis. Typically, this coordinate is 1 for all points.

Implicit surface: A surface defined implicitly as the set of points satisfying an equation of the form $F(x, y, z) = 0$.

Infinite light: A light source taken to be infinitely far from the model being illuminated, so that all of its rays reaching the model can be considered to be parallel.

Local coordinate system: A coordinate system defined with respect to some reference coordinate system, usually used to define a part or subassembly within a scene definition.

Material: The collective set of properties of the surface of an object that determines how it will reflect or transmit light.

Parametric surface: A surface defined explicitly by a set of functions of the form $X(\mathbf{u})$, $Y(\mathbf{u})$, $Z(\mathbf{u})$ that returns (x, y, z) coordinates of a point on the surface as a function of the set of parameters $\mathbf{u}$. Most commonly, $\mathbf{u} = (u, v)$, which yields a biparametric surface parametrized by the parametric coordinates u and v.

Pixel: A square or rectangular uniformly colored area on a raster display that forms the basic unit or picture element of a digital image.

Point light: A light source that radiates light uniformly in all directions from a single geometric point in space.

Projection: A transformation typically from a higher-dimensional space to a lower-dimensional space. In computer graphics the most commonly used projection is the camera projection, which projects 3-D scene geometry onto the plane of a 2-D virtual screen, as one of the key steps in the rendering process.

Raster: An array of scanlines, painted across a CRT screen, which taken together form a rectangular 2-D image. Often the term raster is used to refer to the 2-D array of pixel values stored digitally in a framebuffer.

Surface normal: A vector perpendicular to the tangent plane to a surface at a point. If the surface is planar, the three coefficients of the surface normal vector are the three scaling coefficients of the plane equation.

Texture map: A pattern of color to be mapped onto a geometric surface during rendering.

Transformation matrix: For a 3-D system, this is a 4×4 matrix that, when multiplied by a point in homogeneous coordinates, gives the coordinates of the point in a transformed homogeneous coordinate system. The 4-D homogeneous form of the matrix allows the unification of 3-D translation, rotation, scaling, and shear into one operator.

References

Bartels, R., Beatty, J., and Barsky, B. 1987. *An Introduction to Splines for Use in Computer Graphics and Geometric Modeling*. Morgan Kaufmann, Los Altos, CA.

Barzel, R. 1992. *Physically-Based Modeling for Computer Graphics*. Academic Press, San Diego.

Ebert, D. S., ed. 1994. *Texturing and Modeling: A Procedural Approach*. AP Professional, Boston.

Foley, J. and van Dam, A. 1982. *Fundamentals of Interactive Computer Graphics*. Addison–Wesley, Reading, MA.

Foley, J., van Dam, A., Feiner, S., and Hughes, J. 1990. *Computer Graphics Principles and Practice*. Addison–Wesley, Reading, MA.

Glassner, A., ed. 1990. *Graphics Gems*. Academic Press, San Diego.

Glassner, A. 1995. *Principles of Digital Image Synthesis*. Morgan Kaufmann, San Francisco.

Gonzalez, R. and Woods, R. 1992. *Digital Image Processing*. Addison–Wesley, Reading, MA.

Hill, F. 1990. *Computer Graphics*. Macmillan, New York.

Murray, J. and vanRyper, W. 1994. *Encyclopedia of Graphics File Formats*. O'Reilly, Sebastopol, CA.

Press, W., Teukolsky, S., Vetterling, W., and Flannery, B. 1992. *Numerical Recipes in C, The Art of Scientific Computing*, 2nd ed. Cambridge University Press, Cambridge.

Rogers, D. 1985. *Procedural Elements of Computer Graphics*. McGraw–Hill, New York.

Rogers, D. and Adams, J. 1990. *Mathematical Elements of Computer Graphics*. McGraw–Hill, New York.

Russ, J. 1992. *The Image Processing Handbook*. CRC Press, Boca Raton, FL.

Watt, A. and Watt, M. 1992. *Advanced Animation and Rendering Techniques*. Addison–Wesley, Reading, MA.

Wolberg, G. 1990. *Digital Image Warping*. IEEE Computer Society Press, Los Alamitos, CA.

Further Information

The reader seeking further information on three-dimensional graphics systems should refer first to Chapters 36 through 43 of this Handbook, which provide much of the detail that this overview intentionally skips. The Further Information sections of these chapters provide pointers to the best source books on each of the specialized topics covered.

Beyond this volume, the primary source book for a broad coverage of the field is *Computer Graphics Principles and Practice* by Foley, van Dam, Feiner, and Hughes, published by Addison–Wesley. For a host of practical information and implementation tips, the five-volume series *Graphics Gems*, published by Academic Press, is an invaluable source. Information on image file formats can be found in the *Encyclopedia of Graphics File Formats* by Murray and vanRyper, published by O'Reilly. Fine practical guides to image-processing techniques are *The Image Processing Handbook* by Russ, published by CRC Press, and *Digital Image Processing* by Gonzalez and Woods, published by Addison–Wesley. The mathematics of computer graphics is given a very lucid treatment in *Mathematical Elements of Computer Graphics* by Rogers and Adams, published by McGraw–Hill, and there is no better reference to practical approaches to the implementation of numerical algorithms than *Numerical Recipes* (in various computer-language editions) by Press, Teukolsky, Vetterling, and Flannery, published by Cambridge University Press. Finally, the recent two-volume set *Principles of Digital Image Synthesis* by Glassner, published by Morgan Kaufmann, provides an excellent comprehensive coverage of the theoretical groundings of the field.

Persons interested in keeping up with the latest research in the field should turn to the ACM SIGGRAPH Conference Proceedings, published each year as a special issue of the ACM journal *Computer Graphics*. Other important conferences with published proceedings are *Eurographics* sponsored by the European Association for Computer Graphics, *Graphics Interface* sponsored by the Canadian Human–Computer Communications Society, and *Computer Graphics International* sponsored by the Computer Graphics Society. The IEEE journal *Computer Graphics and Applications* provides an applications-oriented view of recent developments, as well as publishing news and articles of general interest. ACM's *Transactions on Graphics* carries significant research papers, often with a focus on geometric modeling. Other important journals include *The Visual Computer*, IEEE's *Transactions on Visualization and Computer Graphics*, and the *Journal of Visualization and Computer Animation*.

36

Geometric Primitives

36.1	Introduction 36-1
36.2	Screen Specification 36-2
36.3	Simple Primitives 36-2
	Text • Lines and Polylines • Elliptical Arcs
36.4	Wireframes 36-4
36.5	Polygons 36-4
36.6	The Triangular Facet 36-5
36.7	Implicit Modeling 36-6
	Implicit Primitives • CSG Objects
36.8	Parametric Curves 36-8
	Bezier Curves • B-Spline Curves
36.9	Parametric Surfaces 36-12
	Bezier Surfaces • B-Spline Surfaces
36.10	Standards 36-15
36.11	Research Issues and Summary 36-16

Alyn P. Rockwood

Colorado School of Mines

36.1 Introduction

Geometric primitives are rudimentary for creating the sophisticated objects seen in computer graphics. They provide uniformity and standardization in addition to enabling hardware support.

Initially, definition of geometric primitives was driven by the capabilities of the hardware. Only simple primitives were available, e.g., points, line segments, triangles. In addition to the hardware constraints, other driving forces in the development of a geometric primitive have been either its general applicability to a broad range of needs or its satisfying ad hoc, but useful applications. The triangular facet is an example of a primitive that is simple to generate, easy to support in hardware, and widely used to model many graphics objects. An example of a specific primitive can be drawn from flight simulation in the case of *light strings*, which are instances of variable-intensity, directional points of light used to model airport and city lights at night. It is not a common primitive, but it is supported by a critical and profitable application.

As hardware and CPU increased in capability, the sophistication of the primitives grew as well. The primitives became somewhat less dependent on hardware; software primitives became more common, although for raw speed hardware primitives still dominate.

One direction for the sophistication of graphics primitives has been in the geometric order of the primitive. Initially, primitives were discrete or first-order approximations of objects, that is, collections of points, lines, and planar polygons. This has been extended to higher-order primitives represented by polynomial or rational curves and surfaces in any dimension.

The other direction for the sophistication of primitives has been in attributes that are assigned to the geometry. Color, transparency, associated bitmaps and textures, surface normals, and labels are examples of attributes attached to a primitive and used in its display.

1-58488-360-X/$0.00+$1.50
© 2004 by CRC Press, LLC

This summary of graphics primitives is in rough chronological order of development which basically corresponds to increasing complexity. It concentrates on common primitives. It is beyond the scope of this review to include anything but occasional allusions to the plentiful special-purpose developments.

36.2   Screen Specification

To locate the graphics primitive in the viewing window, a local coordinate system is defined. By convention the origin is at the bottom left corner of the window. The positive x-axis extends horizontally from it, while the positive y-axis extends vertically. For 3-D objects, the z-axis is imagined to extend into the screen away from the viewer. In the 3-D case, it is necessary to transform the object to the screen via a set of viewing transformations (see Chapter 35).

Unlike pen and paper, we cannot draw a straight line between two points. The screen is a discrete grid; individual pixels must be illuminated in some pattern to indicate the desired line segment or other graphics primitive. A screen has from 80 to 120 pixels per inch, with high-resolution screens exhibiting 300 per inch. Most screens are about 1024 pixels wide by 780 pixels high. The problem of rendering a graphics primitive on a raster screen is called *scan conversion* and is discussed in Chapter 38. It is mentioned because it is closely related to the geometry of the object drawn and related drawing attributes. It is the scan conversion method that is embedded in hardware to accelerate the display of the graphics in a system. The expense and efficacy of graphics hardware depend on careful selection of the primitives for the facility desired.

36.3   Simple Primitives

36.3.1   Text

There are two standard ways to represent textual characters for graphics purposes. The first method is to save a representation of the letter as a bitmap, called a *font cache*. This method allows fast reproduction of the character on screen. Usually the font cache has more resolution than needed, and the character is downsampled to the display pixels. Even on high-resolution devices such as a quality laser printer, the discrete nature of the bitmap can be apparent, creating jagged edges. When transformations are applied to bitmaps such as rotations or shearing, aliasing problems can also be apparent. See the example in Figure 36.1.

To improve the quality of transformed characters and to compress the amount of data needed to transfer text, a second method of representing characters was developed using polygons or curves. When the text is displayed, the curves or polygons are scan-converted; thus the quality is constant regardless of the transformation (Figure 36.2). The transformation is applied to the curve or polygon basis before scan conversion. Postscript™ is a well-known product for text transferal. In a "Postscript" printer, for instance, the definition of the fonts resides in the printer where it is scan-converted. Transfer across the network requires only a few parameters to describe font size, type, style, etc. Those printers which do not have resident Postscript databases and interpreters must transfer bitmaps with resulting loss of quality and time.

Postscript is based on parametrically defined curves called Bezier curves (see Section 36.8).

Fonts are designed in the bitmap case by simply scanning script from existing print, while special font design programs exist to design fonts with curves.

FIGURE 36.1   Bitmap characters. Note jagged edges and sampling artifacts. Compare to Figure 36.2.

MQ MQ

FIGURE 36.2 Curved representation of characters scan-converted.

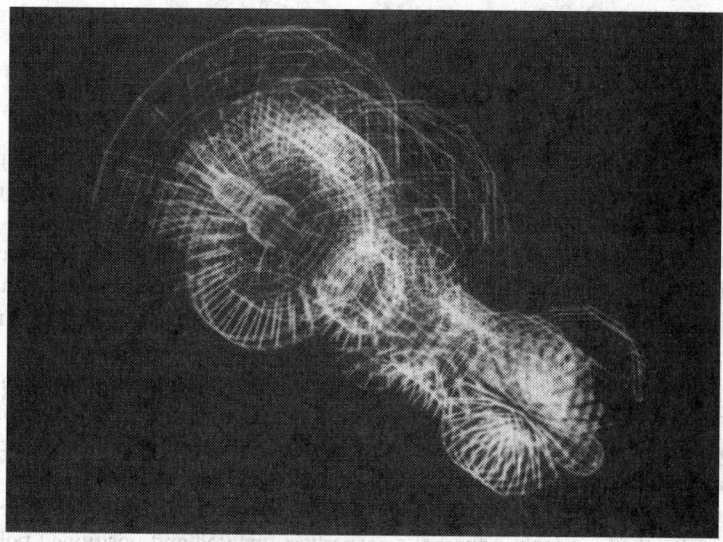

FIGURE 36.3 A polyline display of a turbine engine (Created on an Evans and Sutherland Multi-Picture System in 1986).

36.3.2 Lines and Polylines

The first and still one of the most prevalent of the graphics primitives is the *line* (actually a line segment), typically specified by "move to (*x1, y1*)" and then "draw to (*x2, y2*)" commands, where (*x1, y1*) and (*x2, y2*) are the end points of the line segments. The endpoints can also be given as 3-D points and then projected to the screen through viewing transforms (see Chapter 35).

The line can be easily extended to a *polyline* graphics primitive, that is to a piecewise polygonal path. It is usually stored as an array of 2-D or 3-D points. The first element of the array tells how many points are in the polyline. The following array elements are then the arguments of the line drawing commands. The first entry invokes the "move to" command; successive array entries use the "draw to" command as an operator.

Attributes for lines and polylines are used to modify their appearance without changing position. Popular attributes include *thickness, color,* and *style* (i.e., continuous, dotted, dashed, and variable dash lengths). In more advanced systems each vertex of a polyline is associated to a color and the line segment is drawn by linearly blending the one color to the other along the segment. This is called *color blending*.

Attributes are set either by extending the array to include attribute fields (polylines), or they are set *modally* (lines and polylines); that is, there is an independent command that sets the mode of the line drawing until changed explicitly by another command. Default values control the attributes that are not explicitly set. "Continuous" is the default for style, for example.

Impressive pictures can be produced from these simple line-based primitives. Figure 36.3 shows a turbine engine done with polylines.

Because of the discrete, pixel-based nature of the raster display screen, lines are prone to alias (see Chapter 39); that is, they may break apart or merge with one another to form distracting moire patterns.

They are also susceptible to jagged edges, a signature of older graphics displays. Serious line-drawing systems found it therefore important to add hardware that would "antialias" their lines while drawing them. See Chapter 39 for more details. Figure 36.3 used hardware antialiasing.

A *marker* is another primitive; it is either a point or a small square, triangle, or circle, often placed at the vertices of a polyline to indicate specific details such as distinguishing between lines of a chart. Markers are themselves usually made by predetermined lines or polylines. This unifies the display technology for the hardware, making them faster to draw. Even the point is often a very short line. A single command is then used to center the marker. Style attributes don't usually apply to markers, but color and thickness attributes can be used to advantage.

36.3.3 Elliptical Arcs

Elliptical arcs in 2-D may be specified in many equivalent ways. One way is to give the center position, major and minor axes, and start and end angles. Both angles are given counterclockwise from the major axis. Elliptical arcs may have all the attributes given to lines.

Such an arc may be a closed figure if the angles are properly chosen. In this case, it makes sense to have the ability to *fill* the ellipse with a given pattern or color using a scan conversion algorithm. Even in the case of partial arcs, the object is closed by a line between the end points of the arc so that it may be filled, if wished.

In 3-D the plane of the elliptical arc must also be specified by the unit normal of the plane. While 2-D arcs are common, 3-D arcs are limited to high-end systems that can justify the cost of the hardware. Viewing transforms must compute the arc that is the image on the screen and then scan-convert that arc (usually elliptical, since perspective takes conics to conics).

It should be mentioned that arcs can also be represented by a polyline with enough segments; thus a software primitive for the arc which induces the properly segmented and positioned polyline may be a cost-effective *macro* for elliptical arcs. This macro should consider the effects of zoom and perspective to avoid revealing the underlying polygonality of the arc. It is surprising how small a number of line segments, properly chosen, can give the impression of a smooth arc.

One of the most commonly used elliptical arcs is, of course, the circular arc and its closure, the circle.

36.4 Wireframes

Given hardware or software macro arc primitives in a system, complex curves can be generated by piecing the arcs and lines end to end. Several computer aided design (CAD) systems exist that can generate a rich set of line-based models in both 2-D and 3-D. They are called **wireframe** models. Figure 36.3 is a wireframe model of a turbine engine.

Wireframe models are popular in engineering applications, for instance, because of their visual precision. Drafting is also a natural application. They have other advantages in 3-D because of the ability to see through them to parts of the object that are behind.

Too many lines can be confusing, however; to improve wireframe models a hidden-line routine may be employed (see Chapter 38). This requires derivation of a surface implied by the line. Yet line models can be ambiguous. Figure 36.4 shows a classical example of an object for which the implied surface can be legitimately interpreted in many ways.

Finally, wireframe objects do not support realism. Most objects have a surface that is colored and reflects light according to physical laws of irradiance. This leads us to the next type of primitive.

36.5 Polygons

Closing a polyline by matching start and end points creates a **polygon.** It is probably the most commonly employed primitive in graphics, because it is easy to define an associated surface. Not all polygons have surfaces (they may be a primitive used in wireframe modeling), but most systems that admit polygons will

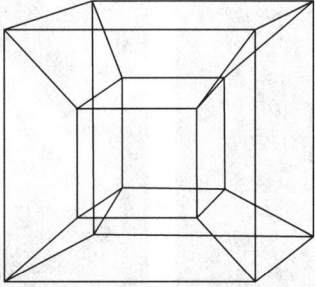

FIGURE 36.4 An ambiguous wireframe model. Where do the surfaces belong?

support both *filled* and *empty* polygons. Filling a polygon is an important example of scan conversion. There are many ways to do this (see Chapter 38). The different ways to fill become attributes of the polygon. The filled polygon is the basis for the numerous hidden-surface algorithms (see Chapter 38). Because of their usefulness for defining surfaces, polygons almost always appear as 3-D primitives which subsume the 2-D case.

The most sophisticated polygon primitive allows nonconvex polygons that contain other polygons, called *holes* or *islands* depending on whether they are filled or not. The scan conversion routine selectively fills the appropriate portions of the polygon depending on whether they are holes or islands. This complex polygon is probably made as a macro out of simple polygon primitives. Triangulation routines exist, for example, that reduce such polygons to simple triangular facets.

36.6 The Triangular Facet

If the polygon is the most popular primitive, then the simple triangle is certainly the most popular polygon. Figure 36.5 shows an object simple object composed of triangular facets.

As mentioned, the triangular facet often supports more complex polygons. It is fast to draw and supports many diverse and powerful methods (see Chapter 38). Another major advantage is that they are always flat; they do not leave the plane due to numerical problems, data errors, or nonplane-preserving transformations. At this writing, graphics workstations exist that can render 10 million triangular facets per second with hidden surfaces removed and smooth shaded display between neighboring triangles implemented. It is certain that this number will continue to increase. With such capacity several dozens of the objects in Figure 36.5 could be smoothly animated.

In order to increase the speed of polygon rendering and decrease the size of the database, triangular and quadrilateral facets can be stored and processed as meshes; i.e., collections of facets that share edges and vertices. The triangular mesh, for instance, is given by defining three vertices for the "lead" triangle, and then giving a new vertex for each successive triangle (Figure 36.6). The succeeding triangles use the last two vertices in the list with the newly given one to form the next triangle. Such a mesh contains about one-third the data and uses the shared edges to increase processing speed. Most graphics objects have large contiguous areas that can take advantage of meshing. The quadrilateral mesh requires two additional vertices be used, with the last two given, and therefore has less savings.

Because of its benefits and ubiquity, many other primitives are defined in software as configurations of the triangular facet. One example is the faceted sphere. With enough facets the sphere will look smooth. It is very commonly encountered in engineering applications and molecular modeling.

Most of the attributes associated with polygons, and triangular facets in particular, deal with the rendering of the facet (see Chapter 38), e.g., color blending between vertices, pointers into a bitmap for texturing, normal vectors (from an underlying surface), reflectance parameters, transparency, and color.

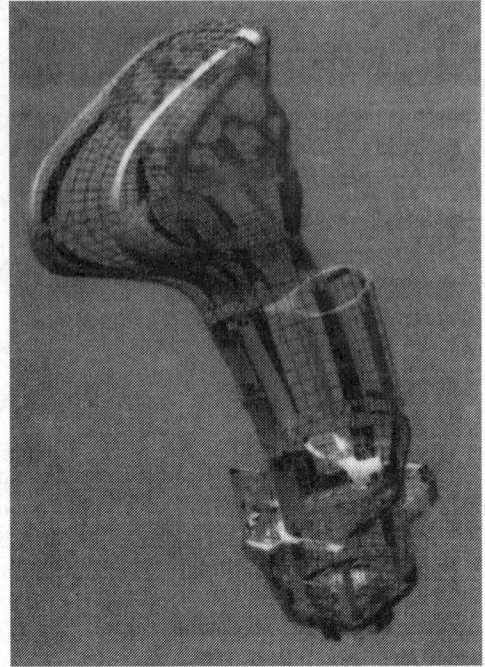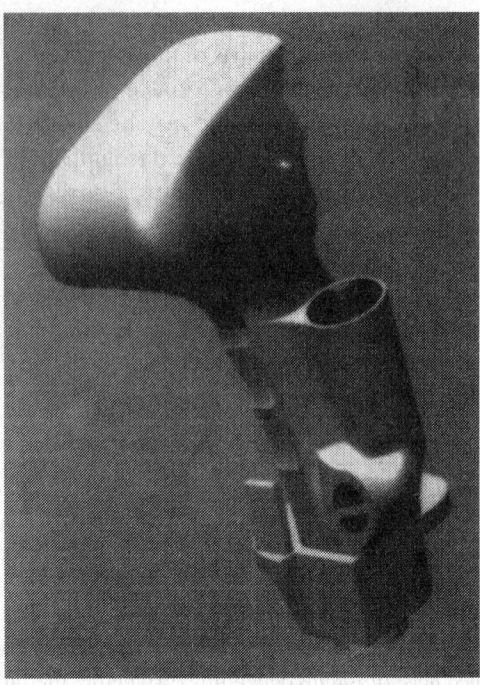

FIGURE 36.5   (See Plate 36.5 in color insert following page **29**-22.) A model composed of simple triangular facets (polygon model).

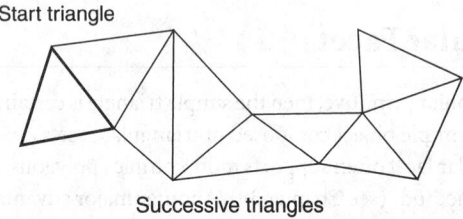

FIGURE 36.6   A mesh of triangular facets.

36.7   Implicit Modeling

36.7.1   Implicit Primitives

The need to model mechanical parts drove the definition of implicitly defined primitives. These modeling primitives naturally became graphics primitives to serve the design industry. They have since been used in other fields such as animation.

An implicit function f maps a point $\mathbf{x}$ in $\mathbf{R}^n$ to a real number, i.e., $f(\mathbf{x}) = r$. Usually $n = 3$. By absorbing r into the function we can view the implicit function as mapping to zero, i.e., $g(\mathbf{x}) = f(\mathbf{x}) - r = 0$. The importance of implicit functions in modeling is that the function divides space into three parts: $f(\mathbf{x}) < 0$, $f(\mathbf{x}) > 0$, and $f(\mathbf{x}) = 0$. The last case defines the surface of an **implicit object**. The other two cases define respectively the *inside* and *outside* of the implicit object. Hence implicit objects are useful in *solid modeling* where it is necessary to distinguish the inside and outside of an object. Modeling objects by polygons does not, for instance, distinguish between the inside and outside of the object. It is, in fact, quite possible to describe an object in which inside and outside are ambiguous. This facility to determine inside and outside is further enhanced by using Boolean operations on the implicit objects to define more complex objects (see the subsection on CSG objects below).

Another advantage of implicit objects is that the surface normal of an implicitly defined surface is given simply as the gradient of the function: $N(\mathbf{x}) = \nabla f(\mathbf{x})$. Furthermore, many common engineering surfaces are easily given as implicit surfaces; thus the *plane* (not the polygon) is defined by $\mathbf{n} \cdot \mathbf{x} + d = 0$, where $\mathbf{n}$ is the surface normal and d is the perpendicular displacement of the plane to the origin. The *ellipsoid* is defined by $(x/a)^2 + (y/b)^2 + (z/c)^2 - r^2 = 0$, where $\mathbf{x} = (x, y, z)$. General *quadrics*, which include ellipsoids, paraboloids, and hyperboloids, are defined by $\mathbf{x}^T M \mathbf{x} + \mathbf{b} \cdot \mathbf{x} + d = 0$, where M is a 3-by-3 matrix and $\mathbf{b}$ and $\mathbf{d}$ vectors in $\mathbf{R}^3$. The quadrics include such important forms as the cylinder, cone, sphere, paraboloids, and hyperboloids of one and two sheets. Other implicit forms used are the *torus*, *blends* (transition surfaces between other surfaces), and *superellipsoids* defined by $(x/a)^k + (y/b)^k + (z/c)^k - R = 0$ for any integer k.

36.7.2 CSG Objects

An important extension to implicit modeling arises from applying set operations such as union, intersection, and difference to the sets defined by implicit objects. The intersection of six half spaces defines a cube, for example. This method of modeling is called *constructive solid geometry* (CSG) [Foley 1990]. All set operations can be reduced to some combination of just union, intersection, and complementation. Because these create an algebra on the sets that is isomorphic to Boolean algebra, corresponding to multiply, add, and negate, respectively, the operations are often referred to as *Booleans*.

A convenient form for visualizing and storing a CSG object is to use a tree in which the nodes are implicit objects and the branches indicate the operation. Traversal of the tree indicates the order of the binary operations and to which sets they pertain. Figure 36.7 shows a CSG tree for a simple model.

Figure 36.8 demonstrates an object made exclusively from Boolean parts of plane quadrics, a part torus, and blended transition surfaces. For any point in space it is straightforward to determine whether it is

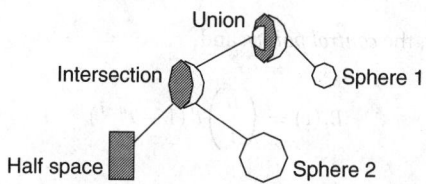

FIGURE 36.7 A CSG model and its tree.

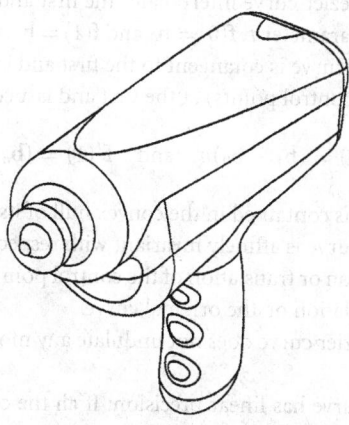

FIGURE 36.8 A solid model of a drill.

inside, outside, or on the surface of the object. This is important in determining volume, center of mass, and moments and for performing Boolean operations needed by CSG models.

Unfortunately, implicit forms tend to be difficult to render, except for ray tracing. Algorithms for polygonizing implicits and CSG models tend to be quite complex [Bloomenthal 1988]. The implicit object gives information about a point relative to the surface in space, but no information is given as to where on the surface a point is located; there is no local coordinate system. This makes it difficult to tessellate into rendering elements such as triangular facets. In the case of ray tracing, however, the parametric form of a ray $\mathbf{x}(t) = (x(t), y(t), z(t))$ composed with implicit function $f(\mathbf{x}(t)) = 0$ leads to a root-finding solution of the intersection points on the surface which are critical points in the ray-tracing algorithm. Determining whether points are part of a CSG model is simply exclusion testing on the CSG tree.

36.8 Parametric Curves

An important class of geometric primitives are formed by **parametrically defined curves and surfaces.** These constitute a flexible set of modeling primitives that are locally parametrized; thus in space the curve is given by $\mathbf{x}(t) = (x(t), y(t), z(t))$ and the surface by $\mathbf{s}(u, v) = (x(u, v), y(u, v), z(u, v))$. In this section and the next one, we will give examples of only the most popular types of parametric curves and surfaces. There are many variations on the parametric forms (see Farin [1992]).

36.8.1 Bezier Curves

The general form of a Bezier curve of degree n is

$$\mathbf{f}(t) = \sum_{i=0}^{n} \mathbf{b}_i B_i^n(t) \tag{36.1}$$

where $\mathbf{b}_i$ are vector coefficients, the *control points*, and

$$B_i(t) = \binom{n}{i} t^i (1 - t^{n-i})$$

where $\binom{n}{i}$ is the binomial coefficient. The $B_i^n(t)$ are called *Bernstein functions*. They form a basis for the set of polynomials of degree n. Bezier curves have a number of characteristics which are derived from the Bernstein functions and which define their behavior.

End-point interpolation: The Bezier curve interpolates the first and last control points, $\mathbf{b}_0$ and $\mathbf{b}_n$. In terms of the interpolation parameter t, $\mathbf{f}(0) = \mathbf{b}_0$ and $\mathbf{f}(1) = \mathbf{b}_n$.

Tangent conditions: The Bezier curve is cotangent to the first and last segments of the *control polygon* (defined by connecting the control points) at the first and last control points; specifically

$$\mathbf{f}'(0) = (\mathbf{b}_1 - \mathbf{b}_0)n \quad \text{and} \quad \mathbf{f}'(1) = (\mathbf{b}_n - \mathbf{b}_{n-1})n$$

Convex hull: The Bezier curve is contained in the convex hull of its control points for $0 \le t \le 1$.

Affine invariance: The Bezier curve is affinely invariant with respect to its control points. This means that any linear transformation or translation of the control points defines a new curve which is just the transformation or translation of the original curve.

Variation diminishing: The Bezier curve does not undulate any more than its control polygon; it may undulate less.

Linear precision: The Bezier curve has linear precision: If all the control points form a straight line, the curve also forms a line.

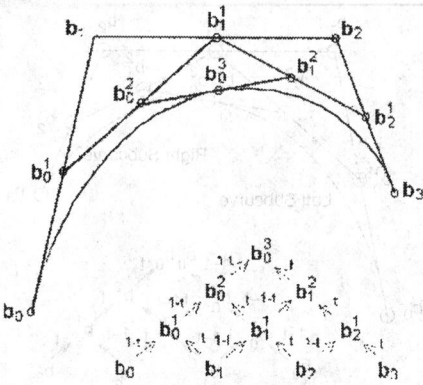

FIGURE 36.9 Bezier curve, control polygon, and de Casteljau algorithm.

Figure 36.9 shows a Bezier curve with its control polygon. Notice how it follows the general shape. This together with the other properties makes it desirable for shape design.

Evaluation of the Bezier curve function at a given value t produces a point $\mathbf{f}(t)$. As t varies from 0 to 1, the point $\mathbf{f}(t)$ traces out the curve segment. One way to evaluate Equation 36.1 is by direct substitution. This is probably the worst way. There are several better methods available for evaluating the Bezier curve. One method is the *de Casteljau algorithm*. This method not only provides a general, relatively fast and robust algorithm, but it gives insight into the behavior of Bezier curves and leads to several important operations on the curves.

To formalize de Casteljau's algorithm we need to use a recursive scheme. The control points are the input. Thereafter each point is superscripted by its level of recursion. Finally, for any point,

$$\mathbf{b}_i^j(t) = (1 - t)\mathbf{b}_i^{j-1} + t\mathbf{b}_{i+1}^{j-1} \quad \text{for } i = 0, \ldots, n, \quad j = 0, \ldots, n - i$$

Note that $\mathbf{b}_0^n(t) = \mathbf{f}(t)$; it is a point on the curve.

One of the most important devices for evaluating curves is the systolic array. It is a triangular arrangement of vectors in which each row reflects the levels of recursion of the de Casteljau algorithm. The first row consists of the Bezier control points. Each successive row corresponds to the points produced by iterating with de Casteljau's algorithm.

The point $\mathbf{b}_0^3$ is the point on the curve for some value of the parameter t. Any point in the systolic array may be computed by linearly interpolating the two points in the preceding row with the parameter t; thus for example:

$$\mathbf{b}_1^2 = \mathbf{b}_1^1(1 - t) + \mathbf{b}_2^1 t$$

One of the most important operations on a curve is that of subdividing it. The de Casteljau algorithm not only evaluates a point on the curve, it also subdivides a curve into two parts as a bonus. The control points of the two new curves are given as the legs of the systolic array. In Figure 36.10 is a cubic Bezier curve after three iterations of the de Casteljau algorithm, with the parameter $t = 0.5$. By using the left and right legs of the systolic array as control points, we obtain two separate Bezier curves which together replicate the original. We have subdivided the curve at $t = 0.5$.

Subdivision permits existing designs to be refined and modified; for example, by incorporating additional curves into an object. One method of intersecting a Bezier curve with a line is to recursively subdivide the curve, testing for intersections of the curve's control polygons with the line. This process is continued until a sufficiently fine intersection is attained.

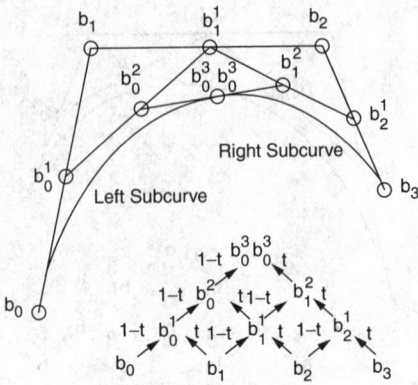

FIGURE 36.10   Subdividing the curve.

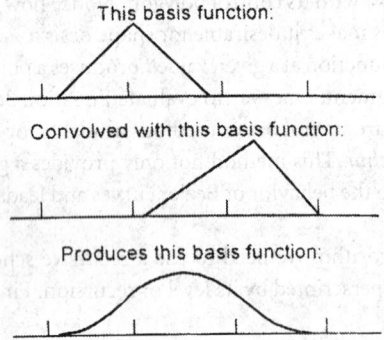

FIGURE 36.11   Defining basis functions by convolution.

36.8.2   B-Spline Curves

A single *B-spline* curve segment is defined much like a Bezier curve. It looks like

$$\mathbf{d}(t) = \sum_{i=0}^{n} \mathbf{d}_i N_i(t)$$

where $\mathbf{d}_i$ are control points, called *de Boor points*. The $N_i(t)$ are the basis functions of the B-spline curve. The degree of the curve is n.

Note that the basis functions used here are different from the Bernstein basis polynomials. Schoenberg first introduced the B-spline in 1949. He defined the basis functions using integral convolution (the "B" in B-spline stands for "basis"). Higher-degree basis functions are given by convolving multiple basis functions of one degree lower. Linear basis functions are just "tents" as shown in Figure 36.11. When convolved together they make piecewise parabolic "bell" curves.

The tent basis function (which has a degree of one) is non-zero over two intervals, the parabola is nonzero over three intervals, and so forth. This gives the region of influence for different degree B-spline control points. Notice also that each convolution results in higher-order continuity between segments of the basis function. When the control points (de Boor points) are weighted by these basis functions, the B-spline curve results.

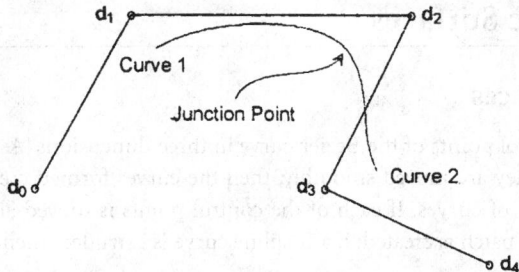

FIGURE 36.12 B-spline curve with two segments.

The major advantage of the B-spline form is in piecing curves together to form a *spline*. If two B-spline curve segments share $n - 1$ control points, they will fit together at the junction point with C^{n-1} continuity. The picture in Figure 36.12 shows the case for a cubic B-spline with two segments:

- The first curve has control points d_0, d_1, d_2, and d_3.
- The second curve has control points d_1, d_2, d_3, and d_4.

Instead of integrating to evaluate the basis functions, a recursive formula has been derived:

$$N_i^n(u) = \frac{u - u_{i-1}}{u_{i+n-1} - u_{i-1}} N_i^{n-1}(u) + \frac{u_{i+n} - u}{u_{i+n} - u_i} N_{i+1}^{n-1}(u)$$

where

$$N_i^0(u) = \begin{cases} 1 & \text{if } u_{i-1} \leq u \leq u_i \\ 0 & \text{otherwise} \end{cases}$$

The terms in u represent a *knot sequence*, the spans over which the de Boor points influence the B-spline.

More control points can be added to make a longer and more elaborate spline curve. As seen in Figure 36.12 neighboring curve segments share n control points.

It can be seen that for any parameter value u only four basis functions are nonzero; thus only four control points affect the curve at u. If a control point is moved it influences only a limited portion of the curve. This *local support* property is important for modeling.

If the first n and last n control points are made to correspond, then the curve's end points will match; it will form a closed curve. This is called a *periodic B-spline*.

Any point on the curve is a convex combination of the control points, i.e., it must be in the convex hull of the control points associated with the nonzero basis functions.

Like the Bezier curve, the B-spline curve also satisfies a variation-diminishing property, and is affinely invariant. Linear precision follows, as in the Bezier-curve case, from the convex-hull property.

The recursive form for evaluating B-splines via basis functions is seldom used in practice. The best way to evaluate a B-spline curve is to use the de Boor algorithm.

Formally, the de Boor algorithm is written as

$$d_i^k(u) = \frac{u_{i+n-k} - u}{u_{i+n-k} - u_{i-1}} d_{i-1}^{k-1}(u) + \frac{u - u_{i-1}}{u_{i+n-k} - u_{i-1}} d_i^{k-1}(u)$$

We see that the de Boor points form a systolic array; each point is defined in terms of preceding points. Thus we may write an iterative procedure to evaluate a point on a B-spline curve in much the same way as de Casteljau's algorithm above evaluates a point on a Bezier curve. Only the weighting factors differ. The last point produced in the method is the point on the curve.

36.9 Parametric Surfaces

36.9.1 Bezier Surfaces

Imagine moving the control points of the Bezier curve in three dimensions. As they move in space, new curves are generated. If they are moved smoothly, then the curves formed create a surface, which may be thought of as a bundle of curves. If each of the control points is moved along a Bezier curve of its own, then a Bezier surface patch is created; if a B-spline curve is extruded, then a B-spline surface results (Figure 36.13).

In the Bezier case this can be written by changing the control points in the Bezier formula into Bezier curves; thus a surface is defined by

$$s(u, v) = \sum_{i=0}^{n} \mathbf{b}_i(u) B_i(v) \tag{36.2}$$

Notice that we have one parameter for the control curves and one for the "swept" curve. It is convenient to write the control curves as Bezier curves of the same degree. If we let the ith control curve have control points $\mathbf{b}_{ij}$, then the surface given in Equation 36.2 above can be written as

$$s(u, v) = \sum_{i=0}^{n} \left(\sum_{j=0}^{m} \mathbf{b}_{ij} B_j(u) \right) B_i(v) \tag{36.3}$$

where m is the degree of the control curves.

A surface can always be thought of as nesting one set of curves inside another. From this simple characteristic we derive many properties and operations for surfaces. Simple algebra changes Equation 36.3 above into

$$s(u, v) = \sum_{i=0}^{n} \left(\sum_{j=0}^{m} \mathbf{b}_{ij} B_j(u) \right) B_i(v) = \sum_{i=0}^{n} \sum_{j=0}^{m} \mathbf{b}_{ij} B_i(v) B_j(u) \tag{36.4}$$

That is, even though we started with one curve and swept it along the other, there is no preferred direction. The surface patch could have been written as:

$$s(u, v) = \sum_{i=0}^{n} \mathbf{b}_j(v) B_j(u), \quad \text{where} \quad \mathbf{b}_j(v) = \sum_{i=0}^{n} \mathbf{b}_{ij} B_i(v)$$

The curve is simply swept in the other direction.

The set of control points forms a rectangular control mesh. A 3-by-3 (bicubic) control mesh is shown in Figure 36.14 with the surface. There are 16 control points in the bicubic control mesh. In general there

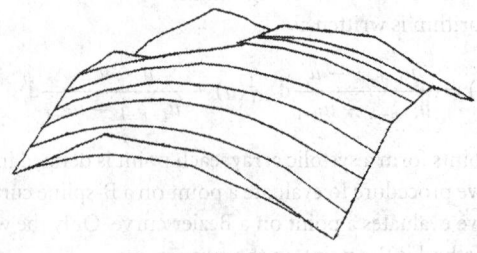

FIGURE 36.13 Sweeping a curve to make the surface.

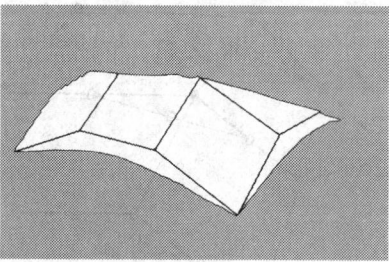

FIGURE 36.14   Bicubic surface with control mesh.

will be $(n + 1) \times (m + 1)$ control points. By convention we associate the i-index with the u-parameter, and the j-index with the v-parameter. Hence if we take:

$$\mathbf{b}_{i0}, \quad i = 1, \ldots, n$$

we get the Bezier curve

$$\mathbf{b}(u, 0) = \sum_{i=0}^{n} \mathbf{b}_{i0} B_i^n(u)$$

Each marginal set of control points defines a Bezier curve (the four border curves), and each of these curves is a boundary of the Bezier surface patch. Such a surface is shown in white in Figure 36.14.

Many of the properties of the Bezier surface are derived directly from those of the Bezier curve, especially those curves that form the boundaries of the patch:

End-point interpolation: The Bezier surface patch passes through all four corner control points.

Tangent conditions: The four border curves of the Bezier surface patch are cotangent to the first and last segments of each border control polygon, at the first and last control points. The normal to the surface patch at each vertex may be found from the cross product of the tangents.

Convex hull: The Bezier surface patch is contained in the convex hull of its control mesh for $0 < u < 1$ and $0 < v < 1$.

Affine invariance: The Bezier surface patch is affinely invariant with respect to its control mesh. This means that any linear transformation or translation of the control mesh defines a new patch which is just the transformation or translation of the original patch.

36.9.1.1   Evaluation of the Bezier Surface Patch

As with the properties described above, the evaluation of a Bezier surface patch can also be derived from the Bezier curve. If we want to evaluate a point on the patch at parameter value (u, v), we apply the de Casteljau algorithm in a nested fashion to Equation 36.3. That is, we first evaluate the control curves in the u direction, which reduce to control points in the v direction. These points are again evaluated with de Casteljau's algorithm.

36.9.1.2   Subdivision of the Bezier Surface Patch

Again, as with the Bezier curve, we can apply the de Casteljau algorithm in nested fashion to subdivide a Bezier surface patch. When a surface patch is subdivided, it yields four subpatches that share a corner at the (u, v) subdivision point. Recall that when a curve was subdivided, the new curve's control points appeared as the legs of a systolic array. In the surface case we subdivide each row of the control mesh, producing points of the systolic array for each. Each point on each leg of every row's systolic array now becomes a control point for a columnar set. A biquadratic case is shown in Figure 36.15. Here we see that three points in each row produce five after subdivision. Now we consider the points in columns, subdividing the

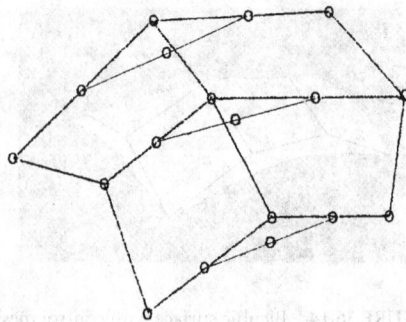

FIGURE 36.15 First-level subdivision.

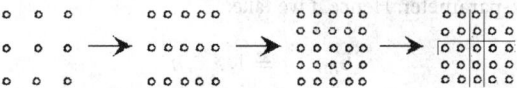

FIGURE 36.16 Control points: progression for subdivision.

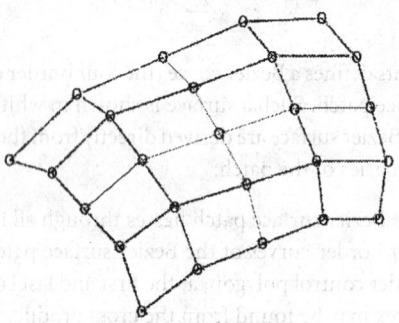

FIGURE 36.17 The subdivided patch.

columns with de Casteljau's algorithm. The points in the legs of their systolic arrays become the control points of the new subpatches. In our example rows with three control points produce five "leg" points, i.e., five columns of three points. Each column then produces five control points, so a 3-by-3 grid generates a 5-by-5 grid after subdivision. The control meshes of the four new patches are produced as shown in Figure 36.16. The central row and column of control points are shared by each 3-by-3 subpatch as shown in Figure 36.17. Note that the order of the scheme does not matter. Columns might have been taken first, and then rows.

Subdivision is a basic operation on surfaces. Many "divide and conquer" algorithms are based on it. To clip a surface to a viewing window we can use the convex-hull property and subdivision, for example. The convex-hull test can determine if a patch is entirely contained in the window. If not, it is subdivided, and the subpatches are then tested. Recursion is applied until the pixel level.

36.9.2 B-Spline Surfaces

As with the Bezier surface, the B-spline surface is defined as a nested bundle of curves, thus yielding

$$s(u,v) = \sum_{i=0}^{L+n-1} \sum_{j=0}^{M+m-1} d_{ij} N_i(u) N_j(v)$$

FIGURE 36.18   (See Plate 36.18 in color insert following page **29**-22.) Turbine engine modeled by B-spline surfaces.

where

  n, m are the degrees of the B-splines,
  L, M are the number of segments, so there are $L \times M$ patches.

All operations used for B-spline curves carry over to the surface via the nesting scheme implied by its definition. We recall that B-spline curves are especially convenient for obtaining continuity between polynomial segments. This convenience is even stronger in the case of B-spline surfaces; the patches meet with higher-order continuity depending on the degree of the respective basis functions. B-splines define quilts of patches with "local support" design flexibility. Finally, since B-spline curves are more compact to store than Bezier curves, the advantage is "squared" in the case of surfaces.

These advantages are tempered by the fact that operations are typically more efficient on Bezier curves. Conventional wisdom says that it is best to design and represent surfaces as B-splines, and then convert to Bezier form for operations. Figure 36.18 shows an object made of many surface patches.

36.10   Standards

Several movements have occurred to standardize primitives. Some of the standards grew out of the mechanical CAD/CAM industry, because of the obvious connection between it and graphics. IGES is a popular one. Although it cannot be said that IGES has become commonplace in the graphics world, it is quite often important to translate between graphics models and these standards when dealing with CAD/CAM applications. GKS [ANSI 1985], PHIGS, and PHIGS+ [ANSI 1988] are standards developed by a broad consortium of academics and industrial users. In spite of major efforts and backing, they have not been as directly successful as the standards evolved in the industry. They have, however, been an intellectual force that has influenced many of the industrial efforts.

In industry each company has developed and pushed for its particular set of graphics standards. Perhaps the most successful at this time is GL (for Graphics Library), which was developed by Silicon Graphics,

Inc., a company which based its computer workstation product on high-powered graphics. It has been licensed by IBM and many other companies. It has evolved into Open GL, which is supported by many manufacturers and threatens to become the standard.

36.11   Research Issues and Summary

There have been efforts to cast higher-order primitives like parametric surfaces into graphics hardware, but the best approach seems to be to convert these into polygons and then render [Rockwood et al. 1990]. There may be hardware support for this process, but the polygon processing remains at the heart of graphics primitives. This trend is likely to continue into the future if for no other reason than its own inertia. Special needs will continue to drive the development of specialized primitives.

One new trend that may affect development is that of volume rendering (see Chapter 41). Although it is currently quite expensive to render, hardware improvements and cheaper memory costs should make it increasingly more viable. As a technique it subsumes many of the current methods, usually with better quality, as well as enabling the visualization of volume-based objects. Volume-based primitives, i.e., tetrahedra, cuboids, and curvilinear volume cubes, will receive more attention and research.

Defining Terms

Implicit objects: Defined by implicit functions, they define solid objects of which outside and inside can be distinguished. Common engineering forms such as the plane, cylinder, sphere, torus, etc. are defined simply by implicit functions.

Parametrically defined curves and surfaces: Higher-order surface primitives used widely in industrial design and graphics. Parametric surfaces such as B-spline and Bezier surfaces have flexible shape attributes and convenient mathematical representations.

Polygon: A closed object consisting of vertices, lines, and usually an interior. When pieced together it gives a piecewise (planar) approximation of objects with a surface. Triangular facets are the most common form and form the basis of most graphics primitives.

Wireframe: Simplest and earliest form of graphics model, consisting of line segments and possibly elliptical arcs that suggest the shape of an object. It is fast to display and has advantages in precision and "see through" features.

References

Adobe Systems Inc. 1985. *Postscript Language Reference Manual*. Addison–Wesley, Reading, MA.
ANSI (American National Standards Institute). 1985. *American National Standards for Information Processing Systems — Computer Graphics — Graphical Kernel System (GKS) Functional Description*. ANSI X3.124-1985. ANSI, New York.
ANSI (American National Standards Institute). 1988. *American National Standards for Information Processing Systems — Programmer's Hierarchical Interactive Graphics Systems (PHIGS) Functional Description*. ANSI X3.144-1988. ANSI, New York.
Bezier, P. 1974. Mathematical and practical possibilities of UNISURF. In *Computer Aided Geometric Design*, R. E. Barnhill and R. Riesenfeld, Eds. Academic Press, New York.
Bloomenthal, J. 1988. Polygonisation of implicit surfaces. *Comput. Aided Geom. Des.* 5:341–345.
Boehm, W., Farin, G., and Kahman, J. 1984. A survey of curve and surface methods in CAGD. *Comput. Aided Geom. Des.* 1(1):1–60, July.
Farin, G. 1992. *Curves and Surfaces for Computer Aided Geometric Design*. Academic Press, New York.
Faux, I. D. and Pratt, M. J. 1979. *Computational Geometry for Design and Manufacture*. Wiley, New York.
Foley, J. D. 1990. *Computer Graphics: Principles and Practice*. Addison–Wesley, Reading, MA.
Rockwood, A., Davis, T., and Heaton, K. 1990. *Real Time Rendering of Trimmed Surfaces*, Computer Graphics 23,3.

37

Advanced Geometric Modeling

David S. Ebert

Purdue University

37.1   Introduction **37**-1
37.2   Fractals **37**-2
37.3   Grammar-Based Models...........................**37**-4
37.4   Procedural Volumetric Models **37**-6
37.5   Implicit Surfaces **37**-12
37.6   Particle Systems..................................**37**-14
37.7   Research Issues and Summary **37**-16

37.1   Introduction

Geometric modeling techniques in computer graphics have evolved significantly as the field matured and attempted to portray the complexities of nature. Earlier geometric models, such as polygonal models, patches, points, and lines, are insufficient to represent the complexities of natural objects and intricate man-made objects in a manageable and controllable fashion. Higher-level modeling techniques have been developed to provide an *abstraction* of the model, encode classes of objects, and allow high-level control and specification of the models. Most of these advanced modeling techniques can be considered *procedural* modeling techniques: code segments or algorithms are used to abstract and encode the details of the model, instead of explicitly storing vast numbers of low-level primitives. The use of algorithms unburdens the modeler/animator from low-level control, provides great flexibility, and allows amplification of their efforts through parametric control: a few parameters to the model yield large amounts of geometric detail (Smith [1984] referred to this as *database amplification*). This amplification allows a savings in storage of data and user specification time. The modeler has the flexibility to capture the *essence* of the object or phenomena being modeled without being constrained by the laws of physics and nature. He can include as much physical accuracy, and also as much artistic expression, as he wishes in the model.

This survey examines several types of procedural advanced geometric modeling techniques, including **fractals, grammar-based models, volumetric procedural models, implicit surfaces**, and **particle systems**. Most of these techniques are used to model natural objects and phenomena because the inherent complexity of nature renders traditional modeling techniques impractical. These techniques can also be classified into **surface-based modeling** techniques and **volumetric modeling** techniques. Fractals, grammar-based models, and implicit surfaces* are surface-based modeling techniques. Volumetric procedural models and particle systems are volumetric modeling techniques.

---

*Implicit surfaces are rendered as surfaces, although the actual model is volumetric.

37.2   Fractals

Fractals and chaos theory have rapidly grown in popularity since the early 1960s [Peitgen et al. 1992]. Mathematicians in the late 19th century and early 20th century, including Cantor, Sierpiński, and von Koch, "discovered" fractal mathematics, but considered these formulas to be "mathematical monsters" that defied normal mathematical principles. Benoit Mandelbrot, who coined the term *fractal*, was the first person to realize that these mathematical formulas were a geometry for describing nature.

Fractals [Peitgen et al. 1992] have a precise mathematical definition, but in computer graphics their definition has been extended to generally refer to models with a large degree of *self-similarity*: subpieces of the object appear to be scaled down, possibly translated and rotated versions of the original object.* Along these lines, Musgrave [Ebert et al. 2002] defines a fractal as "a geometrically complex object, the complexity of which arises through the repetition of form over some range of scale." Many natural objects exhibit this characteristic, including mountains, coastlines, trees, plants (e.g., cauliflower), water, and clouds. In describing fractals, the amount of "roughness," "detail," or amount of space filled by the fractal can be mathematically characterized by its *fractal dimension* (self-similarity dimension), D. The fractal dimension is related to the common integer dimensionality of geometric objects: a line is 1-D, a plane is 2-D, a sphere is 3-D. Fractal objects have noninteger dimensionality. An easy way to explain fractal dimension is to define it in terms of the recursive subdivision technique usually used to create simple fractals. If the original object is subdivided into a pieces using a reduction factor of s, the dimension D is related by the power law [Peitgen et al. 1992]

$$a = \frac{1}{s^D}$$

which yields

$$D = \frac{\log a}{\log (1/s)}$$

Normal geometric objects produce fractal (self-similarity) dimensions that are integers. Fractals produce noninteger, fractional, fractal dimensions. The following examples will illustrate this. If a cube is subdivided into 27 equal pieces, each one is scaled down by a factor of one third, yielding

$$D = \frac{\log 27}{\log \left(1/\frac{1}{3}\right)} = \frac{\log 27}{\log 3} = 3$$

Conversely, a fractal object such as the *von Koch snowflake*** has a noninteger fractal dimension. The von Koch snowflake can be constructed by taking each side of an equilateral triangle, recursively dividing it into three equal pieces, and replacing the middle piece with two equal length pieces rotated to form two sides of an equilateral triangle as illustrated in Figure 37.1. Analyzing the self-similarity dimension of this object yields a noninteger value:

$$D = \frac{\log 4}{\log \left(1/\frac{1}{3}\right)} = \frac{\log 4}{\log 3} \approx 1.2618595$$

The von Koch curve has a fractal dimension between one and two, indicating that it is more space-filling than a line, but not as much as a two-dimensional object. This property is characteristic of fractal curves. By definition, a fractal has a noninteger self-similarity dimension.

Fractals can generally be classified as deterministic and nondeterministic (also called random fractals), depending on whether they contain randomness. In computer graphics, deterministic fractals are closely

---

*Mathematically speaking, the self-similarity must be infinite for the set to be a true fractal.
**Named for the mathematician Helga von Koch, who "discovered" it in 1904.

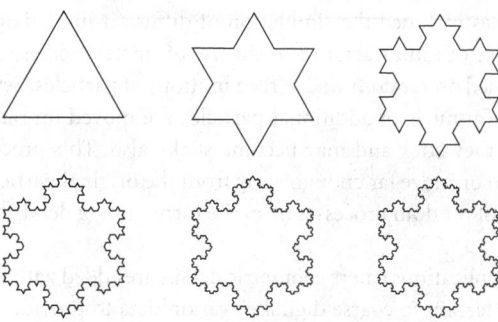

FIGURE 37.1   The von Koch snowflake after 0, 1, 2, 3, 4, and 5 iterations.

FIGURE 37.2   (See Plate 37.2 in the color insert following page **29**-22.) A fractal terrain model by Ken Musgrave. (*Slickrock* © 1992 F. Kenton Musgrave.)

related to the grammar-based L-systems described in the following section. Random fractals have been used extensively in computer graphics to model self-similar, complex, natural objects, including terrain, mountains, clouds, water, and even entire planets [Ebert et al. 2002]. Indeed, fractals are the most common technique used in graphics for modeling mountains. Most fractal terrain generation algorithms work through recursive subdivision and pseudorandom perturbation. An original surface is defined and divided equally into subparts. New vertices are added and pseudorandomly displaced from the original surface, with a displacement magnitude that decreases at each iteration. Therefore, the first iteration gives the large peaks on the surface, and later subdivisions add smaller-scale detail. A common algorithm for mountains by Fournier, Fussell, and Carpenter uses this technique by starting with a triangle, adding new vertices at the midpoint of each edge with a random height displacement, generating new triangles from these points, and repeating this subdivision with decreased height displacements at each iteration. Typically, only parameters for controlling the random number generator, the level of subdivision, and the "roughness" of the surface are needed to define extremely complex mountains and terrain. Others have used simulations of fractional Brownian motion (fBm) and statistical simulations of noise to produce the height displacements for the mountains. Musgrave [Ebert et al. 2002] uses a nonrecursive algorithm to displace the vertices of a regular grid with values iteratively calculated based on fBm and noise functions (at each iteration, the frequency is increased and the magnitude decreased) to create realistic terrain models. Recent work on fractal terrain generation has included eroding the fractal terrain to simulate natural erosion processes and the use of *multifractal* models. Multifractal terrain models allow the fractal dimension of the terrain to vary across the surface. This variation allows rougher areas and smoother areas, providing more realistic natural terrain. An example of the realistic landscapes that can be produced by these techniques can be seen in Figure 37.2.

Recent work in fractals has included the simulation of diffusion-limited aggregation (DLA) models, the previously mentioned use of multifractals, and the use of fractal models to add complex details into models. DLA is a process based on random walks (fBm motion) of particles. Several initial sticky particles are placed in space. A large number of additional particles are moved on random walks; if they touch one of the sticky particles, they stick and may become sticky also. This process continues until all the additional particles attach to or move far enough away from the original particles. DLA models are being used to model a wide range of random processes from the formation of dendrite clusters to the formation of galaxies.

There are two common applications where geometric details are added with fractals. One is the addition of realistic, detailed, fractal terrain to coarse digital elevation data to provide realistic, higher-resolution terrain models. Another is the use of fractals to add small levels of geometric detail to standard geometric models. This allows less geometry in the model, with the procedural fractal functions being applied at rendering time to add an appropriate level of detail [Hart 1995].

There are many open areas of research in fractal modeling. Better erosion models that take into account different rock hardnesses, better rain distribution models, deposition of material in addition to erosion, wind erosion, and the use of nonheight fields to allow rock overhangs will improve the realism of fractal terrains. Multifractals, diffusion-limited aggregation, and fractal detail addition are active areas of research that show great potential for geometric modeling.

37.3 Grammar-Based Models

Grammar-based models also allow natural complexity to be specified with a few parameters. Smith [1984] introduced grammar-based models to graphics, calling them *graftals*. The most commonly used grammar-based model, an L-system (named for Aristid Lindenmayer), was originally developed as a mathematical theory of plant development [Prusinkiewicz and Lindenmayer 1990]. An L-system is a formal language, a parallel graph grammar, where all the rules are applied in parallel to provide a final "word" describing the object. This parallel application of the production rules distinguishes L-systems from Chomsky grammars. Like Chomsky grammars, there are context-free L-systems (0L) and context-sensitive L-systems (1L and 2L).

Grammar-based models have been used by many authors, including Fowler, Lindenmayer, Prusinkiewicz, and Smith, to produce remarkably realistic models and images of trees, plants, and seashells. These models describe natural structures algorithmically and are closely related to deterministic fractals in their self-similarity, but fail to meet the precise mathematical definition of a fractal. Many deterministic fractals can be defined with L-systems, but not all L-systems meet the definition of a fractal.

As with most formal languages, an L-system can be described by an alphabet for the grammar, the grammar production rules, and an initial axiom. In plant modeling, alphabet symbols represent botanical structures (usually letters) and branching commands (usually "[ ]" denotes the beginning and end of a branch). We can add denotation for angular movement by introducing a "+" for clockwise rotations and "−" for counterclockwise rotations. The following simple L-system can produce a basic tree:

$$
\begin{aligned}
&\textit{Alphabet:} && a, [,], +, - \\
&\textit{Production Rule:} && a \to a[+a]a[-a-a]a \\
&\textit{Initial Axiom:} && a
\end{aligned}
$$

In the L-system, each terminal symbol represents a part of the object (e.g., a branch element, internode, apex) or a directional command (e.g., turn left 30 deg) to be interpreted by a three-dimensional drawing mechanism (turtle graphics). A "word" for a tree would contain subwords describing each branch, its length, size, and branching angle, when it develops, and its connection in the tree. For example, if we interpret each "[+a]" as defining a right branch at 30 deg, each "[−a]" as defining a left branch at 30 deg,

FIGURE 37.3 Trees produced after 1, 2, and 3 derivations with the PGF software by Prusinkiewicz.

and each "a" as a tree segment (internode or apex), the above grammar can be interpreted graphically as the trees in Figure 37.3 after 1, 2, and 3 derivations and symbolically as the following:

Derivation	Word
0	a
1	$a[+a]a[-a-a]a$
2	$a[+a]a[-a-a]a[+a[+a]a[-a-a]a]a[+a]a[-a-a]a[-a[+a]a[-a-a]a$
	$-a[+a]a[-a-a]a]a[+a]a[-a-a]a$

To allow more complex plant and plant growth models, L-systems have been extended to include context sensitivity, word age information, and stochastic rule evaluation. Context sensitivity allows the relationships between parts of plants to be incorporated into the model. 1L-systems have one-sided contexts: either a right or a left context, yielding production rules of the form

$$a_l < a \to F$$

and

$$a > a_r \to F$$

The production rule is applied if and only if either its right context, aa_r, is satisfied or its left context, $a_l a$, is satisfied (either if a is preceded by a_l or if a is followed by a_r). 2L-systems have both a left context and a right context, yielding production rules of the form

$$a_l < a > a_r \to F$$

Stochastic L-systems assign a probability to the application of a rule. This allows randomness into the creation of the plant, so that each plant is slightly different. Deterministic L-systems produce identical plants each time they are evaluated and would, therefore, create an unrealistic field of identical flowers. Stochastic rule evaluation is added into the grammar by associating a probability, p, with each production rule as follows:

$$a \xrightarrow{p} F$$

Parametric L-systems allow word age information and conditional expressions based on parameter values to be included into the model, permitting developmental growth models for plants. A parametric L-system has rules of the form

$$F(t) : t > 2 \to F(t-1)H(t+7)$$

where H is another parametric production rule. Parametric L-systems can be combined with stochastic L-systems and context-sensitive L-systems to achieve a powerful, flexible grammar.

Recent work in L-systems allows better developmental models, more advanced biologically based growth models, incorporation of more growth parameters, and environmental effects. Extensions of L-systems allow the death of the buds, dropping of leaves, and cutting of branches with the inclusion of erasing and

FIGURE 37.4   (See Plate 37.4 in the color insert following page **29**-22.) Horse chestnut tree created with a modified L-system that takes into account branch competition for light. (© 1995 R. Měch and P. Prusinkiewicz.)

cutting operators. Botanically based flowering structures (inflorescences), branching structures (sympodial and monopodial), and arrangement of lateral plant organs (phyllotaxis) [Fowler et al. 1992] have been incorporated into L-systems to more accurately model plants. **Tropism** effects (gravity, wind, growth toward light), pruning, amount of light, and availability of nutrients have also been incorporated into these grammars. These natural effects and growth processes not only affect the structure (topology) of the tree, but also affect the branching angles, petal and seed location and shape, and thickness of each branch segment. When generating the geometric plant model from the L-system grammar, these effects need to be included in determining the geometry and size of each structure in the plant. Figure 37.4 shows a realistic image of a horse chestnut tree generated by a modified L-system that takes into account the competition of branches for light.

Ongoing L-systems research includes environmentally-sensitive L-systems [Prusinkiewicz et al. 1995], the modeling of entire ecosystems [Deussen et al. 1998], modeling feathers [Chen et al. 2002], procedural modeling of cities [Parish and Miller 2001], and the use of L-systems for modeling other growth processes and artificial life. Additionally, better developmental models can be simulated and more accurate modeling of natural growth factors can be included.

37.4   Procedural Volumetric Models

Another procedural modeling technique, procedural volumetric modeling (also called hypertextures, volume density functions, and fuzzy blobbies), uses algorithms to define and animate three-dimensional volumetric objects and natural phenomena [Ebert et al. 2002]. These techniques have been used to model natural phenomena such as fire (Stam and Inakage), gases such as smoke, clouds, and fog (Ebert, Perlin, Sakas, Stam), and water (Ebert, Perlin). The volumetric procedures take as input a point location in space, time (if animating), and parameters that describe the object being modeled, and return the density and color of the object for that location in space. Complex volumetric phenomena can therefore be described with a few parameters.

An extremely simple procedural volumetric model for a spherical volume object is given below. This function, **spherical_pvm**, defines a sphere of radius **outer_radius**, with a solid center of radius **inner_radius**. The region between **inner_radius** and **outer_radius** is a semi-solid area that increases in density as the inner radius is approached. The following function, suggested by Perlin to create a *soft sphere* [Ebert et al. 2002], returns a density value when a point in space and the sphere definition are given as input:

```
typedef struct rgb_td { float r,g,b} rgb_td;
typedef struct xyz_td { float x,y,z} xyz_td;
typedef struct vol_td
    {
    float  density;
    rgb_td color;
    } vol_td;

vol_td spherical_pvm(xyz_td pnt,xyz_td center,float inner_radius,
              float outer_radius, float density_factor)
{
    float outer_radius_2, inner_radius_2, distance_2;
    vol_td vol;

    distance_2 = (pnt.x - center.x)*(pnt.x - center.x)
               + (pnt.y - center.y)* (pnt.y - center.y)
               + (pnt.z - center.z)*(pnt.z - center.z);
    /* compute outer and inner radius squared values */
    outer_radius_2 = outer_radius * outer_radius;
    inner_radius_2 = inner_radius * inner_radius;

  if (distance_2 < inner_radius_2)
    { /* inside inner radius */
    vol.density =1.0*density_factor;
    vol.color.r = 1.0; vol.color.g = 0.0; vol.color.b = 0.0;
    }
  else if (distance_2 > outer_radius_2)
    { /* outside of the sphere */
    vol.density = 0.0;
    vol.color.r = 0.0; vol.color.g = 0.0; vol.color.b = 0.0;
    }
  else
    { /* in the soft area of the sphere */
    vol.density = density_factor*(distance-inner_radius_2)/
        (outer_radius_2-inner_radius_2);
    vol.color.r=vol.color.g = 1.0*vol.density;vol.color.b=0;
    }
  return(vol);
}
```

Many authors have used these techniques to describe a wide range of natural objects. Perlin has successfully created realistic rock arches, woven fabric, smoke, and fur [Ebert et al. 2002], basing his procedures on a statistical simulation of turbulence and random noise to give natural-looking complexity to the objects. Perlin's turbulence function has been used as a building block for volumetric procedural objects, solid textures, and many other applications in computer graphics. This turbulence function defines a three-dimensional turbulence space by summing octaves of random noise, increasing the frequency and decreasing the amplitude at each step. The C function below is one implementation of Perlin's turbulence function:

```
float turbulence(xyz_td pnt, float pixel_size)
{
  float t, scale;
  t=0;
  for(scale=1.0; scale >pixel_size; scale/=2.0)
    {
      pnt.x = pnt.x/scale; pnt.y = pnt.y/scale;
      pnt.z = pnt.z/scale;
      t+=calc_noise(pnt) * scale;
    }
  return(t);
}
```

This function takes as input a three-dimensional point location in space, **pnt**, and an indication of the number of octaves of noise to sum, **pixel_size**,* and returns the turbulence value for that location in space. This function has a fractal characteristic in that it is self-similar and sums the octaves of random noise, doubling the frequency while halving the amplitude at each step. The heart of the **turbulence** function is the **calc_noise** function used to simulate uncorrelated random noise. Many authors have used various implementations of the noise function (see [Ebert et al. 2002] for several possible implementations). One implementation is the **calc_noise** function given below, which uses linear interpolation of a $64 \times 64 \times 64$ grid of random numbers**:

```
#define SIZE     64
#define SIZE_1   65
double drand48();
float calc_noise();
float noise[SIZE+1][SIZE+1][SIZE+1];

/*
 ****************************************************************
 *                        Calc_noise
 ****************************************************************
 * This is basically how the trilinear interpolation works:
 * interpolate down left front edge of the cube first, then the
 * right front edge of the cube(p_l, p_r). Next, interpolate down
 * the left and right back edges (p_l2, p_r2). Interpolate across
 * the front face between p_l and p_r (p_face1) and across the
 * back face between p_l2 and p_r2 (p_face2). Finally, interpolate
 * along line between p_face1 and p_face2.
 ****************************************************************
 */
```

---

*This variable name is used in reference to the projected area of the pixel in the three-dimensional turbulence space for antialiasing.

**The actual implementation uses a 65^3 table with the 64th entry equal to the 0th entry for quicker interpolation.

```
float
calc_noise(xyz_td pnt)
{
  float t1;
  float p_l,p_l2, /* value lerped down left side of face1 &
                   * face 2 */
        p_r,p_r2, /* value lerped down right side of face1 &
                   * face 2 */
        p_face1, /* value lerped across face 1 (x-y plane ceil
                  * of z) */
        p_face2, /* value lerped across face 2 (x-y plane floor
                  * of z) */
        p_final; /* value lerped through cube (in z) */

  extern float noise[SIZE_1][SIZE_1][SIZE_1];
  float        tnoise;
  register int x, y, z,px,py,pz;
  int          i,j,k, ii,jj,kk;
  static int   firstime =1;

  /* During first execution, create the random number table of
   * values between 0 and 1, using the Unix random number
   * generator drand48(). Other random number generators may be
   * substituted. These noise values can also be stored to a
   * file to save time.
   */
  if (firsttime)
    { for (i=0; i<SIZE; i++)
       for (j=0; j<SIZE; j++)
         for (k=0; k<SIZE; k++)
           {
             noise[i][j][k] = (float)drand48();
             /* A crude way to make element[64]=element[0] for
              * easier linear interpolation */
             if(i==0) noise[SIZE][j][k] = noise[i][j][k];
             if(j==0) noise[i][SIZE][k] = noise[i][j][k];
             if(k==0) noise[i][j][SIZE] = noise[i][j][k];
           }
       firsttime=0;
    }

  px = (int)pnt.x;
  py = (int)pnt.y;
  pz = (int)pnt.z;
  x = px &(SIZE-1); /* make sure the values are in the table */
  y = py &(SIZE-1); /* Effectively, replicates the table
                     * throughout space */
  z = pz &(SIZE-1);

  t1 = pnt.y - py;
  p_l = noise[x][y][z+1] +t1*(noise[x][y+1][z+1]
        - noise[x][y][z+1]);
```

```
p_r = noise[x+1][y][z+1] +t1*(noise[x+1][y+1][z+1]
      - noise[x+1][y][z+1]);
p_12 = noise[x][y][z] +t1*(noise[x][y+1][z] - noise[x][y][z]);
p_r2 = noise[x+1][y][z] +t1*(noise[x+1][y+1][z]
      - noise[x+1][y][z]);
t1 = pnt.x - px;
p_face1 = p_1 + t1 * (p_r - p_1);
p_face2 = p_12 + t1 * (p_r2 -p_12);
t1 = pnt.z - pz;
p_final = p_face2 + t1*(p_face1 -p_face2);
return(p_final);
}
```

Ebert et al. [2002] have used similar functions to model and animate steam, fog, smoke, clouds, and solid marble. The turbulence function and random noise functions allow a simple simulation of turbulent flow processes. To simulate steam rising from a teacup, a volume of gas is placed over the teacup. The basic gas is defined by the following function:

```
float basic_gas(xyz_td pnt, float density, float density_scalar,
                float exponent)
{
    float turb, density;
    turb =turbulence(pnt);
    density = pow(turb*density_scalar, exponent);
    return(density);
}
```

This function creates a three-dimensional gas space controlled by the values **density_scalar** and **exponent**. **density_scalar** controls the denseness of the gas, while **exponent** controls the sharpness and sparseness of the gas (from continuously varying to sharp individual plumes). This function is then shaped to create steam over the center of the teacup by spherically attenuating the density toward the edge of the cup and linearly attenuating the density as the distance from the top of the cup increases, simulating the gas dissipation as it rises. The following procedure will produce an image of steam rising from a teacup, as in Figure 37.5.

FIGURE 37.5   Steam rising from a teacup. (See Plate 37.5 in the color insert following page 29-22.) (© 1991 David S. Ebert.)

```
float steam(xyz_td pnt, xyz_td pnt_world, float exponent,
            float density_scalar,xyz_td tea_center,float radius)
    {
    float    turb, dist, dist_sq, density_max, fall_off, offset;
    xyz_td   diff;

    turb = turbulence(pnt);
    density = pow(turb*density_scalar, exponent);

    /* determine distance from center of the teacup squared. */
    XYZ_SUB(diff,tea_center, pnt_world);
    dist_sq = DOT_XYZ(diff,diff);
    /* calculate relative distance from center with some
     * randomness */
    density_max = dist_sq/(radius*radius);
    density_max += .2*noise(pnt);

    /* Use a cosine function to spherically attenuate the
     * density */
    if(density_max >= .25) /* ramp off if > 25% from center */
      { /* get table index 0:RAMP_SIZE-1 */
          density_max = MAX((density_max -.25)*4/3, 1.0);
          fall_off = (cos(density_max*M_PI)+1.0)/2.0;
          density *=fall_off;
      }
    /* Use exponential attenuation to decrease density as steam
     * rises */
    dist = pnt_world.y - tea_center.y;
    if(dist > 0.0)
      { dist = (dist +noise(pnt)*.1)/radius.y;
        if(dist > .05)
          { offset = (dist -.05)*1.111111;
            offset = 1 - (exp(offset)-1.0)/1.718282;
            density = density*offset;
          }
      }
    return(density);
    }
```

These procedural techniques can be easily animated by adding time as a parameter to the algorithm [Ebert et al. 2002]. They allow the use of simple simulations of natural complexity (noise, turbulence) to speed computation, but also allow the incorporation of physically based parameters where appropriate and feasible. This flexibility is one of the many advantages of procedural techniques.

Procedural volumetric models require volume rendering techniques to create images of these objects. Traditionally, most authors have used a modification of volume ray tracing. Several authors have incorporated physically based models for shadowing and illumination into rendering algorithms for these models [Ebert and Parent 1990, Stam and Fiume 1995]. However, with the advent of programmable graphics hardware that allow user-defined programs to be executed for each polygonal pixel-sized fragment, these function can potentially be evaluated at interactive rates by the graphics hardware. Ebert et al. [2002] describe methods to adapt procedural volumetric modeling techniques to modern graphics hardware.

Procedural volumetric modeling is still an active area of research and has many research problems to address. Efficient rendering of these models is still an important issue, as is the development of a

FIGURE 37.6   A close-up of a fly through of a procedural volumetric cloud incorporating volumetric implicit models into the procedural volumetric model. (©2002 David S. Ebert.)

larger toolbox of useful primitive functions. The incorporation of more physically based models will increase the accuracy and realism of the water, gas, and fire simulations. Finally, the development of an interactive procedural volumetric modeling system will speed the development of procedural volumetric modeling techniques. The procedural interfaces in the latest commercial modeling, rendering, and animation packages are now allowing the specification of procedural models, but the user control is still lacking.

Combining traditional volumetric procedural models with implicit functions, described below, creates a model that has the advantages of both techniques. Implicit functions have been used for many years as a modeling tool for creating solid objects and smoothly blended surfaces [Bloomenthal et al. 1997]. However, only a few researchers have explored their potential for modeling volumetric density distributions of semi-transparent volumes (e.g., [Nishita et al. 1996, Stam and Fiume 1991, Stam and Fiume 1993, Stam and Fiume 1995, Ebert 1997]). Ebert's early work on using volume rendered implicit spheres to produce a fly-through of a volumetric cloud was described in [Ebert et al. 1997]. This work has been developed further to use implicits to provide a natural way of specifying and animating the global structure of the cloud, while using more traditional procedural techniques to model the detailed structure. More details on the implementation of these techniques can be found in [Ebert et al. 2002]. An example of a procedural volumetric cloud modeled using the above turbulence-based techniques combined with volumetricly evaluated implicit spheres, can be seen in Figure 37.6.

37.5   Implicit Surfaces

While previously discussed techniques have been used primarily for modeling the complexities of nature, implicit surfaces [Bloomenthal et al. 1997] (also called blobby molecules [Blinn 1982], metaballs [Nishimura et al. 1985], and soft objects [Wyvill et al. 1986]) are used in modeling organic shapes, complex man-made shapes, and "soft" objects that are difficult to animate and describe using more traditional techniques. Implicit surfaces are surfaces of constant value, **isosurfaces**, created from blending primitives (functions or skeletal elements) represented by implicit equations of the form $F(x, y, z) = 0$, and were first introduced into computer graphics by Blinn [1982] to produce images of electron density clouds. A simple example of an implicit surface is the sphere defined by the equation

$$F(x, y, z) : x^2 + y^2 + z^2 - r^2 = 0$$

Implicit surfaces are a more concise representation than parametric surfaces and provide greater flexibility in modeling and animating soft objects.

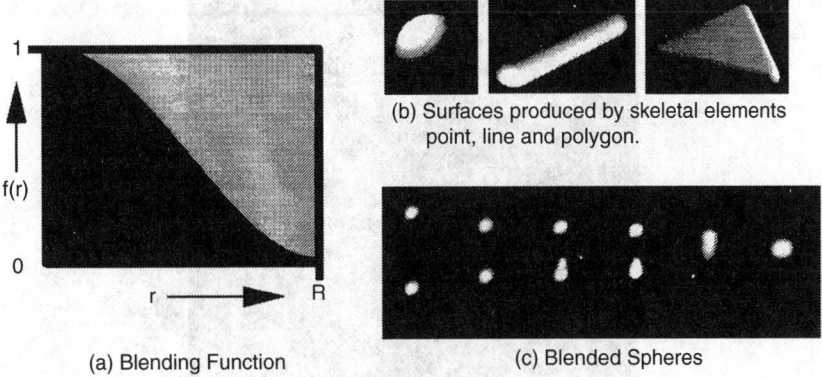

FIGURE 37.7 (a) Blending function, (b) surfaces produced by skeletal elements point, line and polygon, and (c) blended spheres. (© 1995 Brian Wyvill.)

For modeling complex shapes, several basic implicit surface primitives are smoothly blended to produce the final shape. For the blending function, Blinn used an exponential decay of the field values, whereas Wyvill [Wyvill et al. 1986, 1993] uses the cubic function

$$F_{\text{cub}}(r) = -\frac{4}{9}\frac{r^6}{R^6} + \frac{17}{9}\frac{r^4}{R^4} - \frac{22}{9}\frac{r^2}{R^2} + 1$$

This cubic blending function, whose values range from 1 when $r = 0$ to 0 at $r = R$, has several advantages for complex shape modeling. First, its value drops off quickly to zero (at the distance R), reducing the number of primitives that must be considered in creating the final surface. Second, it has zero derivatives at $r = 0$ and $r = R$ and is symmetrical about the contour value 0.5, providing for smooth blends between primitives. Finally, it can provide volume-preserving primitive blending. Figure 37.7(a) shows a graph of this blending function, and Figure 37.7(c) shows the blending of two spheres using this function. A good comparison of blending functions can be found in [Bloomenthal et al. 1997].

For implicit surface primitives, Wyvill uses procedures that return a functional (field) value for the field defined by the primitive. Field primitives, such as lines, points, polygons, circles, splines, spheres, and ellipsoids, are combined to form a basic skeleton for the object being modeled. The surfaces resulting from these skeletal elements can be seen in Figure 37.7(b). The object is then defined as an offset (isosurface) from this series of blended skeletal elements. Skeletons are an intuitive representation and are easily displayed and animated.

Modeling and animation of implicit surfaces is achieved by controlling the skeletal elements and blending functions, providing complex models and animations from a few parameters (another example of data amplification). Deformation, path following, warping, squash and stretch, gravity, and metamorphosis effects can all be easily achieved with implicit surfaces. Very high-level animation control is achieved by animating the basic skeleton, with the surface defining the character following naturally. The animator does not have to be concerned with specifying the volume-preserving deformations of the character as it moves.

There are two common approaches to rendering implicit surfaces. One approach is to directly ray-trace the implicit surfaces, requiring the modification of a standard ray tracer. The second approach is to polygonalize the implicit surfaces [Ning and Bloomenthal 1993, Wyvill et al. 1993] and then use traditional polygonal rendering algorithms on the result. Uniform-voxel space polygonization can create large numbers of unnecessary polygons to accurately represent surface details. More complicated tessellation and shrinkwrapping algorithms have been developed which create appropriately sized polygons [Wyvill et al. 1993].

Recent work in implicit surfaces [Wyvill and Gascuel 1995, Wyvill et al. 1999] has extended their use to character modeling and animation, human figure modeling, and representation of rigid objects through

FIGURE 37.8   (See Plate 37.8 in the color insert following page **29**-22.) Ten years in implicit surface modeling. The locomotive labeled 1985 shows a more traditional soft object created by implicit surface techniques. The locomotive labeled 1995 shows the results achievable by incorporating constructive solid geometry techniques with implicit surface models. (© 1995 Brian Wyvill.)

the addition of constructive solid geometry (CSG) operators. Implicit surface modeling techniques have advanced significantly in the past 10 years, as can be seen by comparing the locomotives in Figure 37.8. The development of better blending algorithms, which solve the problems of unwanted primitive blending and surface bulging, is an active area of research [Bloomenthal 1995]. Advanced animation techniques for implicit surfaces, including higher-level animation control, surface collision detection, and shape metamorphosis animation, are also active research areas. Finally, the development of interactive design systems for implicit surfaces will greatly expand the use of this modeling technique. The use of implicit functions have also expanded to compact representations of surface objects [Turk and O'Brien 2002].

37.6   Particle Systems

Particle systems are different from the previous four techniques in that their abstraction is in control of the animation and specification of the object. A particle-system object is represented by a large collection (cloud) of very simple geometric particles that change stochastically over time. Therefore, particle systems do use a large database of geometric primitives to represent natural objects ("fuzzy objects"), but the animation, location, birth, and death of the particles representing the object are controlled algorithmically. As with the other procedural modeling techniques, particle systems have the advantage of database amplification, allowing the modeler/animator to specify and control this extremely large cloud of geometric particles with only a few parameters.

Particle systems were first used in computer graphics by Reeves [1983] to model a wall of fire for the movie *Star Trek II: The Wrath of Khan* (see Figure 37.9). Because particle systems are a volumetric modeling technique, they are most commonly used to represent volumetric natural phenomena such as fire, water, clouds, snow, and rain [Reeves 1983]. An extension of particle systems, *structured particle systems,* has also been used to model grass and trees [Reeves and Balu 1985].

A particle system is defined by both a collection of geometric particles and the algorithms that govern their creation, movement, and death. Each geometric particle has several attributes, including its initial position, velocity, size, color, transparency, shape, and lifetime.

To create an animation of a particle system object, the following steps are performed at each time step [Reeves 1983]:

1. New particles are generated and assigned their attributes.
2. Particles that have existed in the system past their lifetime are removed.

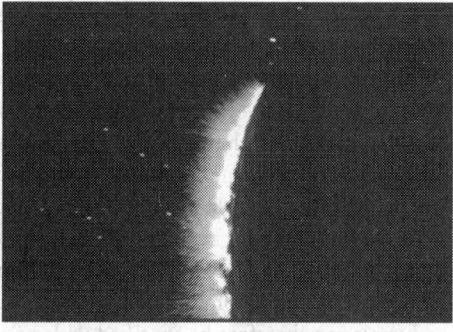

FIGURE 37.9 (See Plate 37.9 in the color insert following page **29**-22.) An image from *Star Trek II: The Wrath of Khan* showing a wall of fire created with a particle system. (© 1987 Pixar.)

3. Each remaining particle is moved and transformed by the particle-system algorithms as prescribed by their individual attributes.
4. These particles are rendered, using special-purpose rendering algorithms, to produce an image of the particle system.

The creation, death, and movement of particles are controlled by stochastic procedures, allowing complex, realistic motion to be created with a few parameters. The creation procedure for particles is controlled by parameters defining either the mean number of particles created at each time step and its variance, or the mean number of particles created per unit of screen area at each time step and its variance.* The actual number of particles created is stochastically determined to be within *mean + variance* and *mean − variance*. The initial color, velocity, size, and transparency are also stochastically determined by mean and variance values. The initial shape of the particle system is defined by an origin, a region about this origin in which new generated particles are placed, angles defining the orientation of the particle system, and the initial direction of movement for the particles.

The movement of particles is also controlled by stochastic procedures (stochastically determined velocity vectors). These procedures move the particles by adding their velocity vector to their position vector. Random variations can be added to the velocity vector at each frame, and acceleration procedures can be incorporated to simulate effects such as gravity, vorticity, and conservation of momentum and energy. The simulation of physically based forces allows realistic motion and complex dynamics to be displayed by the particle system, while being controlled by only a few parameters. In addition to the movement of particles, their color and transparency can also change dynamically to give more complex effects. The death of particles is controlled very simply by removing particles from the system whose lifetimes have expired or that have strayed more than a given distance from the origin of the particle system.

An example of the effects achievable by such a particle system can be seen in Figure 37.9, an image from the Genesis Demo sequence from *Star Trek II: The Wrath of Khan*. In this image, a two-level particle system was used to create the wall of fire. The first-level particle system generated concentric, expanding rings of particle systems on the planet's surface. The second-level particle system generated particles at each of these locations, simulating explosions. During the Genesis Demo sequence, the number of particles in the system ranged from several thousand initially to over 750,000 near the end.

Reeves extended the use of particle systems to model fields of grass and forests of trees, calling this new technique structured particle systems [Reeves and Blau 1985]. In structured particle systems, the particles are no longer an independent collection of particles, but rather form a connected, cohesive three-dimensional object and have many complex relationships among themselves. Each particle represents an

---

*These values can be varied over time as well.

element of a tree (e.g., branch, leaf) or part of a blade of grass. These particle systems are therefore similar to L-systems and graftals, specifically probabilistic, context-sensitive L-systems. Each particle is similar to a letter in an L-system alphabet, and the procedures governing the generation, movement, and death of particles are similar to the production rules. However, they differ from L-systems in several ways. First, the goal of structured particle systems is to model the visual appearance of whole collections of trees and grass, and not to correctly model the detailed geometry of each plant. Second, they are not concerned with biological correctness or modeling growth of plants. Structured particle systems construct trees by recursively generating subbranches, with stochastic variations of parameters such as branching angle, thickness, and placement within a value range for each type of tree. Additional stochastic procedures are used for placement of the trees on the terrain, random warping of branches, and bending of branches to simulate tropism. A forest of such trees can therefore be specified with a few parameters for distribution of tree species and several parameters defining the mean values and variances for tree height, width, first branch height, length, angle, and thickness of each species.

Both regular particle systems and structured particle systems pose special rendering problems because of the large number of primitives. Regular particle systems have been rendered simply as point light sources (or linear light sources for antialiased moving particles) for fire effects, accumulating the contribution of each particle into the frame buffer and compositing the particle system image with the surface rendered image (as in Figure 37.9). No occlusion or interparticle illumination is considered. Structured particle systems are much more difficult to render, and specialized probabilistic rendering algorithms have been developed to render them [Reeves and Blau 1985]. Illumination, shadowing, and hidden-surface calculations need to be performed for the particles. Because stochastically varying objects are being modeled, approximately correct rendering will provide sufficient realism. Probabilistic and approximate techniques are used to determine the shadowing and illumination of each tree element. The particle's distance into the tree from the light source determines its amount of **diffuse shading** and probability of having **specular highlights**. Self-shadowing is simulated by exponentially decreasing the **ambient illumination** as the particle's distance within the tree increases. External shadowing is also probabilistically calculated to simulate the shadowing of one tree by another tree. For hidden-surface calculations, an initial depth sort of all trees and a **painter's algorithm** is used. Within each tree, again, a painter's algorithm is used, along with a back-to-front bucket sort of all the particles. This will not correctly solve the hidden-surface problem in all cases, but will give realistic, approximately correct images.

Efficient rendering of particle systems is still an open research problem (e.g., [Etzmuss et al. 2002]). Although particle systems allow complex scenes to be specified with only a few parameters, they sometimes require rather slow, specialized rendering algorithms. Simulation of fluids [Miller and Pearce 1989], cloth [Breen et al. 1994, Baraff and Witkin 1998, Plath 2000], and surface modeling with oriented particle systems [Szeliski and Tonnesen 1992] are recent, promising extensions of particle systems. Sims [1990] demonstrated the suitability of highly parallel computing architectures to particle-system simulation. Particle systems, with their ease of specification and good dynamical control, have great potential when combined with other modeling techniques such as implicit surfaces [Witkin and Heckbert 1994] and volumetric procedural modeling.

Particle systems provide a very nice, powerful animation system for high-level control of complex dynamics and can be combined with many of the procedural techniques described in this chapter. For example, turbulence functions are often combined with particle systems, such as Ebert's use of particle systems for animating cloud dynamics [Ebert et al. 2002].

37.7 Research Issues and Summary

Advanced modeling techniques will continue to play an important role in computer graphics. As computers become more powerful, the complexity that can be rendered will increase; however, the capability of humans to specify more geometric complexity (millions of primitives) will not. Therefore, procedural techniques, with their capability to amplify the user's input, are the only viable alternative. These techniques will evolve in their capability to specify and control incredibly realistic and detailed models with a small number of user-specified parameters. More work will be done in allowing high-level control and specification of models in user-understandable terms, while more complex algorithms and improved physically based simulations will be incorporated into these procedures. Finally, the automatic generation of procedural models through artificial evolution techniques, similar to those of Sims [1994], will greatly enhance the capabilities and uses of these advanced modeling techniques.

Defining Terms

Ambient illumination: An approximation of the global illumination on the object, usually modeled as a constant amount of illumination per object.

Diffuse shading: The illumination of an object where light is reflected equally in all directions, with the intensity varying based on surface orientation with respect to the light source. This is also called Lambertian reflection because it is based on Lambert's law of diffuse reflection.

Fractal: Generally refers to a complex geometric object with a large degree of self-similarity and a non-integer fractal dimension that is not equal to the object's topological dimension.

Grammar-based modeling: A class of modeling techniques based on formal languages and formal grammars where an alphabet, a series of production rules, and initial axioms are used to generate the model.

Implicit surfaces: Isovalued surfaces created from blending primitives that are modeled with implicit equations.

Isosurface: A surface defined by all the points where the field value is the same.

L-system: A parallel graph grammar in which all the production rules are applied simultaneously.

Painter's algorithm: A hidden-surface algorithm that sorts primitives in back-to-front order, then "paints" them into the frame buffer in this order, overwriting previously "painted" primitives.

Particle system: A modeling technique that uses a large collection (thousands) of particles to model complex natural phenomena, such as snow, rain, water, and fire.

Phyllotaxis: The regular arrangement of plant organs, including petals, seeds, leaves, and scales.

Procedural volumetric models: Use algorithms to define the three-dimensional volumetric representation of an object.

Specular highlights: The bright spots or highlights on objects caused by angular-dependent illumination. Specular illumination depends on the surface orientation, the observer location, and the light source location.

Surface-based modeling: Refers to techniques for modeling the three-dimensional surfaces of objects.

Tropism: An external directional influence on the branching patterns of trees.

Volumetric modeling: Refers to techniques that model objects as three-dimensional volumes of material, instead of being defined by surfaces.

References

Baraff, D. and Witkin, A. 1998. Large Steps in Cloth Simulation. *Proceedings of SIGGRAPH*, 98, 43–54, ACM Press.

Blinn, J. F. 1982. A generalization of algebraic surface drawing. *ACM Trans. Graphics*, 1(3):235–256.

Bloomenthal, J. 1995. Skeletal Design of Natural Forms. Ph.D. thesis, Department of Computer Science, University of Calgary.

Bloomenthal, J., Bajaj, C., Blinn, J., Cani-Gascuel, M.P., Rockwood, A., Wyvill, B., and Wyvill, G. 1997. *Introduction to Implicit Surfaces*, Morgan Kaufman Publishers.

Breen, D.E., House, D.H., and Wozny, M.J. 1994. Predicting the drape of woven cloth using interacting particles, pp. 365–372. In *Proc. SIGGRAPH '94 (Orlando, Florida, July 24–29, 1994)*, A. Glassner, Ed., Computer Graphics Proceedings, Annual Conf. Ser. ACM SIGGRAPH, ACM Press.

Chen, Y., Xu, Y., Guo, B., and Shum, H. 2002. Modeling and Rendering of Realistic Feathers. *ACM Transactions on Graphics*, 21(3):630–636.

Deussen, O., Hanrahan, P., Lintermann, B., Mech, R., Pharr, M., and Prusinkiewicz, P. 1998. Realistic Modeling and Rendering of Plant Ecosystems. *Proceedings of SIGGRAPH '98*, 275–286.

Ebert, D. and Parent, R. 1990. Rendering and animation of gaseous phenomena by combining fast volume and scanline a-buffer techniques. *Comput. Graphics (Proc. SIGGRAPH)*, 24:357–366.

Ebert, D. 1997. Volumetric Modeling with Implicit Functions: A Cloud Is Born, *SIGGRAPH 97 Visual Proceedings (Technical Sketch)*, 147, ACM SIGGRAPH.

Ebert, D., Musgrave, F.K., Peachey, D., Perlin, K., and Worley, S. 2002. *Texturing and Modeling: A Procedural Approach, third edition.* Morgan Kaufman Publishers, San Francisco, CA. Professional, Boston.

Etzmuss, O., Eberhardt, B., and Hauth, M. 2000. Implicit-Explicit Schemes for Fast Animation with Particle Systems. *Computer Animation and Simulation 2000*, 138–151.

Foley, J.D., van Dam, A., Feiner, S.K., and Hughes, J.F. 1990. *Computer Graphics: Principles and Practices*, 2nd ed. Addison-Wesley, Reading, MA.

Fowler, D.R., Prusinkiewicz, P., and Battjes, J. 1992. A collision-based model of spiral phyllotaxis. *Comput. Graphics (Proc. SIGGRAPH)*, 26:361–368.

Hart, J. 1995. Procedural models of geometric detail. In *SIGGRAPH '95: Course 33 Notes.* ACM SIGGRAPH.

Mandelbrot, B.B. 1983. *The Fractal Geometry of Nature.* W. H. Freeman, New York.

Miller, G. and Pearce, A. 1989. Globular dynamics: a connected particle system for animating viscous fluids. *Comput. and Graphics*, 13(3):305–309.

Ning, P. and Bloomenthal, J. 1993. An evaluation of implicit surface tilers. *IEEE Comput. Graphics Appl.*, 13(6):33–41.

Nishimura, H., Hirai, A., Kawai, T., Kawata, T., Kawa, I.S., and Omura, K. 1985. Object modelling by distribution function and a method of image generation (in Japanese). In *Journals of Papers Given at the Electronics Communication Conference '85*, J68-D(4).

Nishita, T., Nakamae, E., and Dobashi, Y. 1996. Display of Clouds and Snow Taking into Account Multiple Anisotropic Scattering and Sky Light, *SIGGRAPH 96 Conference Proceedings*, 379–386. ACM SIGGRAPH.

Parish, Y. and Miller, P. 2001. Procedural Modeling of Cities. *Proceedings of ACM SIGGRAPH 2001*, 301–308.

Peitgen, H.-O., Jürgens, H., and Saupe, D. 1992. *Chaos and Fractals: New Frontiers of Science.* Springer-Verlag, New York.

Plath, J. 2000. Realistic modelling of textiles using interacting particle systems. *Computers & Graphics*, 24(6): 897–905.

Prusinkiewicz, P. and Lindenmayer, A. 1990. *The Algorithmic Beauty of Plants.* Springer-Verlag.

Prusinkiewicz, P., Hammel, M., and Mech, R. 1995. The artificial life of plants. In *SIGGRAPH '95: Course Notes.* ACM SIGGRAPH.

Reeves, W.T. 1983. Particle systems — a technique for modeling a class of fuzzy objects. *ACM Trans. Graphics*, 2:91–108.

Reeves, W.T. and Blau, R. 1985. Approximate and probabilistic algorithms for shading and rendering structured particle systems. *Comput. Graphics (Proc. SIGGRAPH)*, 19:313–322.

Sims, K. 1990. Particle animation and rendering using data parallel computation. *Comput. Graphics (Proc. SIGGRAPH)*, 24:405–413.

Sims, K. 1994. Evolving virtual creatures, pp. 15–22. *Proc. of SIGGRAPH '94*, Computer Graphics Proc., Annual Conf. Series. ACM SIGGRAPH, ACM Press.

Smith, A.R. 1984. Plants, fractals and formal languages. *Computer Graphics (Proc. SIGGRAPH)*, 18:1–10.

Stam, J. and Fiume, E. 1991. A multiple-scale stochastic modeling primitive, *Proceedings Graphics Interface '91.*

Stam, J. and Fiume, E. 1993. Turbulent wind fields for gaseous phenomena, *Computer Graphics (SIGGRAPH '93 Proceedings)*, 27:369–376.

Stam, J. and Fiume, E. 1995. Depicting fire and other gaseous phenomena using diffusion processes, pp. 129–136. In *Proc. SIGGRAPH '95*, Computer Graphics Proc., Annual Conf. Series. ACM SIGGRAPH, ACM Press.

Szeliski, R. and Tonnesen, D. 1992. Surface modeling with oriented particle systems. *Comput. Graphics (Proc. SIGGRAPH)*, 26:185–194.

Turk, G. and O'Brien, J. 2002. Modelling with implicit surfaces that interpolate. *ACM Transactions on Graphics*, 21(4):855–873.

Watt, A. and Watt, M. 1992. *Advanced Animation and Rendering Techniques: Theory and Practice*. Addison-Wesley, Reading, MA.

Witkin, A.P. and Heckbert, P.S. 1994. Using particles to sample and control implicit surfaces, pp. 269–278. In *Proc. SIGGRAPH '94*, Computer Graphics Proc. Annual Conf. Series. ACM SIGGRAPH, ACM Press.

Wyvill, B. and Gascuel, M.-P., 1995. *Implicit Surfaces '95, The First International Workshop on Implicit Surfaces*. INRIA, Eurographics.

Wyvill, G., McPheeters, C., and Wyvill, B. 1986. Data structure for soft objects. *The Visual Computer*, 2(4):227–234.

Wyvill, B., Bloomenthal, J., Wyvill, G., Blinn, J., Hart, J., Bajaj, C., and Bier, T. 1993. Modeling and animating implicit surfaces. In *SIGGRAPH '93: Course 25 Notes*.

Wyvill, B., Galin, E., and Guy, A. 1999. Extending The CSG Tree. Warping, Blending and Boolean Operations in an Implicit Surface Modeling System, *Computer Graphics Forum*, 18(2):149–158.

Further Information

There are many sources of further information on advanced modeling techniques. Two of the best resources are the proceedings and course notes of the annual ACM SIGGRAPH conference. The SIGGRAPH conference proceedings usually feature a section on the latest, and often best, results in modeling techniques. The course notes are a very good source for detailed, instructional information on a topic. Several courses at SIGGRAPH '92, '93, '94, and '95 contained notes on procedural modeling, fractals, particle systems, implicit surfaces, L-systems, artificial evolution, and artificial life.

Standard graphics texts, such as *Computer Graphics: Principles and Practice* by Foley, van Dam, Feiner, and Hughes [Foley et al. 1990] and *Advanced Animation and Rendering Techniques* by Watt and Watt [1992], contain introductory explanations to these topics. The reference list contains references to excellent books and, in most cases, the most comprehensive sources of information on the subject. Additionally, the book entitled *The Fractal Geometry of Nature*, by Mandelbrot [1983], is a classic reference for fractals. For implicit surfaces, the book by Bloomenthal, Wyvill, et al. [1997] is a great reference.

Another good source of reference material is specialized conference and workshop proceedings on modeling techniques. For example, the proceedings of the Eurographics '95 Workshop on Implicit Surfaces contains state-of-the-art implicit surfaces techniques.

38

Mainstream Rendering Techniques

Alan Watt
University of Sheffield

Steve Maddock
University of Sheffield

38.1 Introduction 38-1
38.2 Rendering Polygon Mesh Objects 38-2
Introduction • Viewing and Clipping • Clipping and Culling
• Projective Transformation and Three-Dimensional Screen
Space • Shading Algorithm • Hidden-Surface Removal
38.3 Rendering Using Ray Tracing 38-18
Intersection Testing
38.4 Rendering Using the Radiosity Method 38-21
Basic Theory • Form-Factor Determination • Problems with
the Basic Method
38.5 The (Irresistible) Survival of Mainstream
Rendering 38-26
38.6 An OpenGL Example 38-26

38.1 Introduction

Rendering is the name given to the process in three-dimensional graphics whereby a geometric description of an object is converted into a two-dimensional image–plane representation that looks real.

Three methods of rendering are now firmly established. The first and most common method is to use a simulation of light–object interaction in conjunction with **polygon mesh** objects; we have called this approach *rendering polygon mesh objects*. Although the light–object simulation is independent of the object representation, the combination of empirical light–object interaction and polygon mesh representation has emerged as the most popular rendering technique in computer graphics. Because of its ubiquity and importance, we shall devote most of this chapter to this approach.

This approach to rendering suffers from a significant disadvantage. The reality of light–object interaction is simulated as a crude approximation — albeit an effective and cheap simulation. In particular, objects are considered to exist in isolation with respect to a light source or sources, and no account is taken of light interaction between objects themselves. In practice, this means that although we simulate the reflection of light incident on an object from a light source, we resolutely ignore the effects that the reflected light has on the scene when it travels onward from its first reflection to encounter, perhaps, other objects, and so on. Thus, common phenomena that depend on light reflecting from one object onto another, like shadows and objects reflecting in each other, cannot be produced by such a model. Such defects in straightforward polygon mesh rendering have led to the development of many and varied enhancements that attempt to address its shortcomings. Principal among these are mapping techniques (texture mapping, environment mapping, etc.) and various shadow algorithms.

Such models are called **local reflection models** to distinguish them from **global reflection models**, which attempt to follow the adventures of light emanating from a source as it hits objects, is reflected, hits other objects, and so on. The reason local reflection models work — in the sense that they produce visually acceptable, or even impressive, results — is that in reality the reflected light in a scene that emanates from first-hit incident light predominates. However, the subtle object–object interactions that one normally encounters in an environment are important. This motivation led to the development of the two global reflection models: **ray tracing** and **radiosity**.

Ray tracing simulates global interaction by explicitly tracking infinitely thin beams, or rays, of light as they travel through the scene from object to object. *Radiosity*, on the other hand, considers light reflecting in all directions from the surface of an object and calculates how light radiates from one surface to another as a function of the geometric relationship between surfaces — their proximity, relative orientation, etc. Ray tracing operates on points in the scene, radiosity on finite areas called patches.

Ray tracing and radiosity formed popular research topics in the 1980s. Both methods are much more expensive than polygon mesh rendering, and a common research motivation was efficiency, particularly in the case of ray tracing.

For reasons that will become clear later, ray tracing and radiosity each can simulate only one aspect of global interaction. Ray tracing deals with specular interaction and is fine for scenes consisting of shiny, mutually reflective objects. On the other hand, radiosity deals with diffuse or dull surfaces and is used mostly to simulate interiors of rooms. In effect, the two methods are mutually exclusive: ray tracing cannot simulate diffuse interaction, and radiosity cannot cope with specular interaction. This fact led to another major research effort, which was to incorporate specular interaction in the radiosity method.

Whether radiosity and ray tracing should be categorized as mainstream is perhaps debatable. Certainly the biggest demand for three-dimensional computer graphics is real-time rendering for computer games. Ray tracing and radiosity cannot be performed in real time on consumer equipment and, unless used in precalculation mode, are excluded from this application. However, radiosity in particular is used in professional applications, such as computer-aided architectural design.

38.2 Rendering Polygon Mesh Objects

38.2.1 Introduction

The overall process of rendering polygon mesh objects can be broken down into a sequence of geometric transformations and pixel processes that have been established for at least two decades as a *de facto* standard. Although they are not the only way to produce a shaded image of a three-dimensional object, the particular processes we shall describe represent a combination of popularity and ease of implementation. There is no established name for this group of processes, which has emerged for rendering objects represented by a polygon mesh — by far the most popular form of representation. The generic term *rendering pipeline* applies to any set of processes used to render objects in three-dimensional graphics.

Ignoring any transformations that are involved in positioning many objects to make up a scene — modeling transformations — we can summarize these processes as follows:

Viewing transformation — A process that is invoked to generate a representation of the object or scene as seen from the viewpoint of an observer positioned somewhere in the scene and looking toward some aspect of it. This involves a simple transformation that changes the object from its database representation to one that is represented in a coordinate system related to the viewer's position and viewing direction. It establishes the size of the object, according to its distance from the viewer, and the parts of it seen from the viewing direction.

Clipping — The need for clipping is easily exemplified by considering a viewpoint that is embedded among objects in the scene. Objects and parts of objects, for example, behind the viewer must be eliminated from consideration. Clipping is nontrivial because, in general, it involves removing parts of polygons and creating new ones. It means "cutting chunks" off the objects.

Projective transformation — This transformation generates a two-dimensional image on the image or viewing plane from the three-dimensional view-space representation of the object.

Shading algorithm — The orientation of the polygonal facets that represent the object are compared with the position of a light source (or sources), and a reflected light intensity is calculated for each point on the surface of the object. In practice, "each point on the surface" means those pixels onto which the polygonal facet projects. Thus, it is convenient to calculate the set of pixels onto which a polygon projects and to drive this process from pixel space — a process that is usually called *rasterization*. Shading algorithms use a local reflection model and an *interpolative method* to distribute the appropriate light intensity among pixels inside a polygon. The computational efficiency and visual efficacy of the shading algorithm have supported the popularity of the polygon mesh representation. (The polygon mesh representation has many drawbacks — its major advantage is simplicity.)

Hidden-surface removal — Those surfaces that cannot be seen from the viewpoint need to be removed from consideration. In the 1970s, much research was carried out on the best way to remove hidden surfaces, but the Z-buffer algorithm, with its easy implementation, is the *de facto* algorithm, with others being used only in specialized contexts. However, it does suffer from inefficiency and produces aliasing artifacts in the final image.

The preceding processes are not carried out in a sequence but are merged together in a way that depends on the overall rendering strategy. The use of the Z-buffer algorithm, as we shall see, conveniently allows polygons to be fetched from the database in any order. This means that the units on which the whole rendering process operates are single polygons that are passed through the processes one at a time. The entire process can be seen as a black box, with a polygon input as a set of vertices in three-dimensional world space. The output is a shaded polygon in two-dimensional screen space as a set of pixels onto which the polygon has been projected.

Although, as we have implied, the processes themselves have become a kind of standard, rendering systems vary widely in detail, particularly in differences among subprocesses such as rasterization and the kind of viewing system used.

The marriage of interpolative shading with the polygon mesh representation of objects has served, and continues to serve, the graphics community well. It does suffer from a significant disadvantage, which is that **antialiasing** measures are not easily incorporated in it (except by the inefficient device of calculating a virtual image at a resolution much higher than the final screen resolution). Antialiasing measures are described elsewhere in this text.

The first two processes, viewing transformation and clipping, are geometric processes that operate on the vertex list of a polygon, producing a new vertex list. At this stage, polygons are still represented by a list of vertices where each vertex is a coordinate in a three-dimensional space with an implicit link between vertices in the list. The projective transformation is also a geometric process, but it is embedded in the pixel-level processes. The shading algorithm and hidden-surface removal algorithm are pixel-level processes operating in screen space (which, as we shall see, is considered for some purposes to possess a third dimension). For these processes, the polygon becomes a set of pixels in two-dimensional space. However, some aspects of the shading algorithm require us to return to three-dimensional space. In particular, calculating light intensity is a three-dimensional calculation. This functioning of the shading algorithm in both screen space and a three-dimensional object space is the source of certain visual artifacts. These arise because the projective transformation is nonlinear. Such subtleties will not be considered here, but see [Watt and Watt, 1992] for more information on this point.

38.2.2   Viewing and Clipping

When viewing transformations are considered in computer graphics, an analogy is often made with a camera, and the term **virtual camera** is employed. There are certainly direct analogies to be made between a camera, which records a two-dimensional projection of a real scene on a film, and a computer graphics system. However, keep in mind that these concern external attributes, such as the position of the camera and the direction in which it is pointing. There are implementations in a computer graphics system (notably

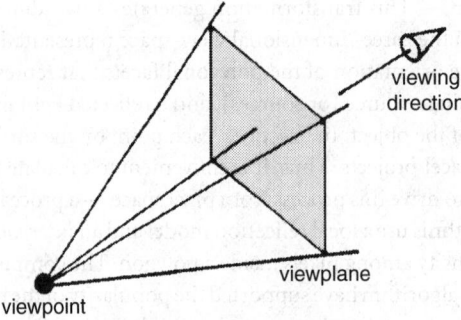

FIGURE 38.1 The three basic attributes required in a viewing system.

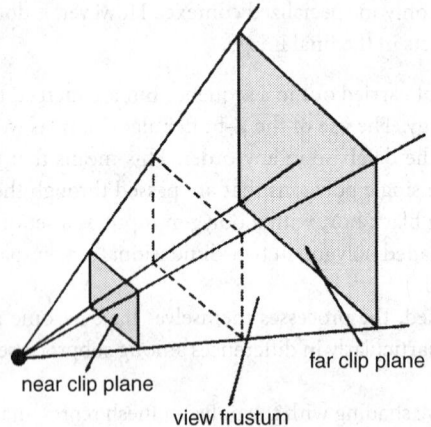

FIGURE 38.2 View frustum formed by near and far planes.

the near and far clip planes) that are not available in a camera and facilities in a camera that are not usually imitated in a computer graphics system (notably depth of field and lens distortion effects). The analogy is a general one, and its utility disappears when details are considered.

The facilities incorporated into a viewing system can vary widely. In this section, we will look at a system that will suffice for most general-purpose rendering systems. The ways in which the attributes of the viewing system are represented, which have ramifications for the design of a user interface for a viewing system, also vary widely.

We can discuss our requirements by considering just three attributes (see Figure 38.1). First, we establish a viewpoint and a viewing direction. A *viewpoint*, which we can also use as a center of projection, is a single point in world space, and a *viewing direction* is a vector in this space.

Second, we position a view plane somewhere in this space. It is convenient to constrain the view plane normal to be the viewing direction and its position to be such that a line from the viewpoint, in the viewing direction, passes through the center of the view plane. The distance of the viewpoint from the object being viewed and the distance of the view plane from the viewpoint determine the size of the projection of the object on the view plane. Also, these two distances determine the degree of perspective effect in the object projection, since we will be using a perspective projection. This arrangement defines a new three-dimensional space, known as *view space*. Normally, we would take the origin of this space to be the viewpoint, and the coordinate axes are oriented by the viewing direction.

Usually, we assume that the view plane is of finite extent and rectangular, because its contents will eventually be mapped onto the display device. Additionally, we can add *near* and *far clip planes* (Figure 38.2) to constrain further those elements of the scene that are projected onto the view plane — a caprice of computer graphics not available in a camera.

Such a setup, as can be seen in Figure 38.2, defines a so-called *view volume*, and consideration of this gives the motivation for clipping. Clipping means that the part of the scene that lies outside the view frustum should be discarded from the rendering process. We perform this operation in three-dimensional view space, clipping polygons to the view volume. This is a nontrivial operation, but it is vital in scenes of any complexity where only a small proportion of the scene will finally appear on the screen. In simple single-object applications, where the viewpoint will not be inside the bounds of the scene and we do not implement a near and a far clip plane, we can project all the scene onto the view plane and perform the clipping operation in two-dimensional space.

Now we are in a position to define **viewing** and **clipping** as those operations that transform the scene from world space into view space, at the same time discarding that part of the scene or object that lies outside the view frustum.

We will deal separately with the transformation into the view space and clipping. First, we consider the viewing transformation. A useful practical facility that we should consider is the addition of another vector to specify the rotation of the view plane about its axis (the view-direction vector). Returning to our camera analogy, this is equivalent to allowing the user to rotate the camera about the direction in which it is pointing. A user of such a system must specify the following:

1. A viewpoint or camera position **C**, which forms the origin of view space. This point is also the center of projection (see Section 38.2.4).
2. A viewing direction vector **N** (the positive z-axis in view space) — this is a vector normal to the view plane.
3. An "up" vector **V** that orients the camera about the view direction.
4. An optional vector **U**, to denote the direction of increasing x in the eye coordinate system. This establishes a right- or left-handed coordinate system (**UVN**). This system is represented in Figure 38.3.

The transformation required to take an object from world space into view space, T_{view}, can be split into a translation T and a change of basis B:

$$T_{\text{view}} = TB$$

where

$$T = \begin{bmatrix} 1 & 0 & 0 & -C_x \\ 0 & 1 & 0 & -C_y \\ 0 & 0 & 1 & -C_z \\ 0 & 0 & 0 & 1 \end{bmatrix}$$

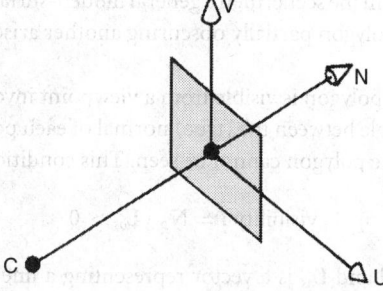

FIGURE 38.3   UVN coordinate system embedded in the view plane.

It can be shown [Fiume, 1989] that B is given by

$$B = \begin{bmatrix} U_x & U_y & U_z & 0 \\ V_x & V_y & V_z & 0 \\ N_x & N_y & N_z & 0 \\ 0 & 0 & 0 & 1 \end{bmatrix}$$

The only problem now is specifying a user interface for the system and mapping whatever parameters are used by the interface into $\mathbf{U}$, $\mathbf{V}$, and $\mathbf{N}$. A user needs to specify $\mathbf{C}$, $\mathbf{N}$, and $\mathbf{V}$. $\mathbf{C}$ is easy enough. $\mathbf{N}$, the viewing direction or view plane normal, can be entered, say, by using two angles in a spherical coordinate system. $\mathbf{V}$ is more problematic. For example, a user may require "up" to have the same sense as "up" in the world coordinate system. However, this cannot be achieved by setting

$$\mathbf{V} = (0, 0, 1)$$

because $\mathbf{V}$ must be perpendicular to $\mathbf{N}$. A useful strategy is to allow the user to specify, through a suitable interface, an approximate value for $\mathbf{V}$, having the program alter this to a correct value.

38.2.3 Clipping and Culling

Clipping and culling mean discarding, at an early stage in the rendering process, polygons or parts of polygons that will not appear on the screen. Polygons that must be discarded fall into three categories:

1. Complete objects that lie outside the view volume should be removed entirely without invoking any tests at the level of individual polygons. This can be done by comparing a bounding volume, such as a sphere, with the view-volume extents and removing (or not) the entire set of polygons that represent an object.
2. Complete polygons that face away from a viewer need not invoke the expense of a clipping procedure. These are called *back-facing polygons*, and in any (convex) object they account, on average, for 50% of the object polygons. These can be eliminated by a simple geometric test, which is termed **culling** or *back-face removal*.
3. Polygons that straddle a view-volume plane must be clipped against the view frustum and the resulting fragment, which comprises a new polygon, passed on to the remainder of the process.

Clipping is carried out in the three-dimensional domain of view space; culling is performed in this space, also. Culling is a pure geometric operation that discards polygons on the basis of the direction of their surface normal compared to the viewing direction. Clipping is an algorithmic operation, because some process must be invoked that produces a new polygon from the polygon that is clipped by one of the view-volume planes.

We deal with the simple operation of culling first. If we are considering a single convex object, then culling performs complete hidden-surface removal. If we are dealing with objects that are partially concave, or if there is more than one object in the scene, then a general hidden-surface removal algorithm is required. In these cases, the event of one polygon partially obscuring another arises — a situation impossible with a single convex object.

Determining whether a single polygon is visible from a viewpoint involves a simple geometric test (see Figure 38.4). We compare the angle between the (true) normal of each polygon and a line-of-sight vector. If this is greater than $90°$, then the polygon cannot be seen. This condition can be written as

$$\text{visibility} := \mathbf{N}_p \cdot \mathbf{L}_{os} > 0$$

where $\mathbf{N}_p$ is the polygon normal and $\mathbf{L}_{os}$ is a vector representing a line from one vertex of the polygon to the viewpoint. A polygon normal can be calculated from any three (noncollinear) vertices by taking a cross product of vectors parallel to the two edges defined by the vertices.

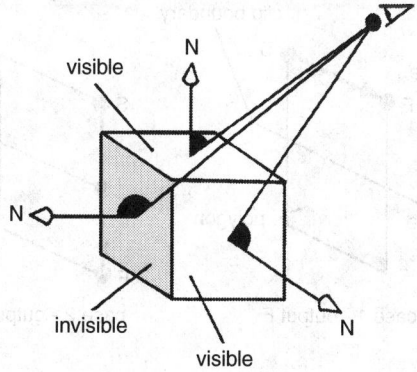

FIGURE 38.4 Back-face removal or culling.

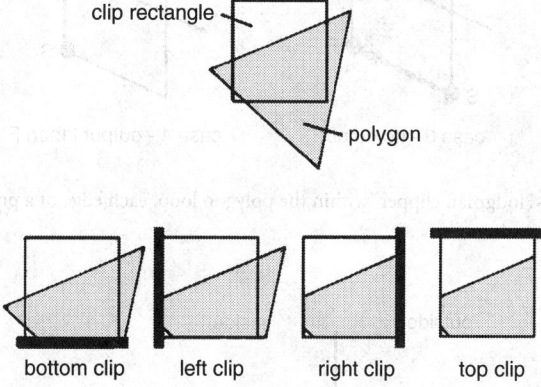

FIGURE 38.5 Sutherland–Hodgman clipper clips each polygon against each edge of each rectangle.

The most popular clipping algorithm, like most of the algorithms used in rendering, goes back over 25 years and is the Sutherland–Hodgman reentrant polygon clipper [Sutherland and Hodgman, 1974]. We will describe, for simplicity, its operation in two-dimensional space, but it is easily extended to three dimensions. A polygon is tested against a clip boundary by testing each polygon edge against a single infinite clip boundary. This structure is shown in Figure 38.5.

We consider the innermost loop of the algorithm, where a single edge is being tested against a single clip boundary. In this step, the process outputs zero, one, or two vertices to add to the list of vertices defining the clipped polygon. Figure 38.6 shows the four possible cases. An edge is defined by vertices **S** and **F**. In the first case, the edge is inside the clip boundary and the existing vertex **F** is added to the output list. In the second case, the edge crosses the clip boundary and a new vertex **I** is calculated and output. The third case shows an edge that is completely outside the clip boundary. This produces no output. (The intersection for the edge that caused the excursion outside is calculated in the previous iteration, and the intersection for the edge that causes the incursion inside is calculated in the next iteration.) The final case again produces a new vertex, which is added to the output list.

To calculate whether a point or vertex is inside, outside, or on the clip boundary, we use a dot-product test. Figure 38.7 shows clip boundary **C** with an outward normal $\mathbf{N}_c$ and a line with end points **S** and **F**. We represent the line parametrically as

$$\mathbf{P}(t) = \mathbf{S} + (\mathbf{F} - \mathbf{S})t \tag{38.1}$$

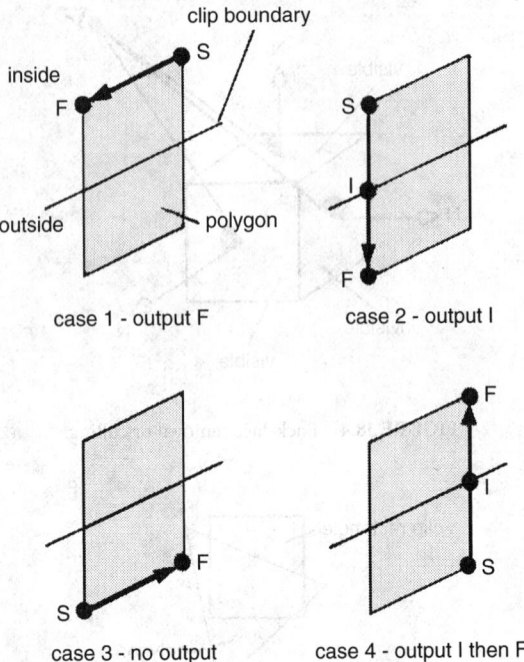

FIGURE 38.6   Sutherland–Hodgman clipper: within the polygon loop, each edge of a polygon is tested against each clip boundary.

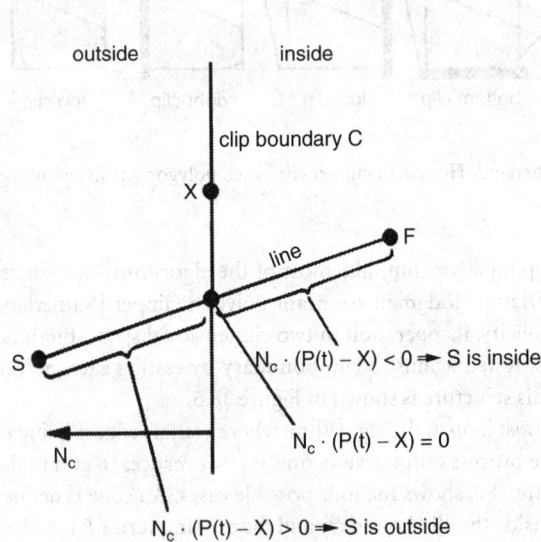

FIGURE 38.7   Dot-product test determines whether a line is inside or outside a clip boundary.

where

$$0 \leq t \leq 1$$

We define an arbitrary point on the clip boundary as $\mathbf{X}$ and consider a vector from $\mathbf{X}$ to any point on the line. The dot product of this vector and the normal allows us to distinguish whether a point on the line is outside, inside, or on the clip boundary. In the case shown in Figure 38.7,

$$\mathbf{N}_c \cdot (\mathbf{S} - \mathbf{X}) > 0 \Rightarrow \mathbf{S} \text{ is outside the clip region}$$

$$\mathbf{N}_c \cdot (\mathbf{F} - \mathbf{X}) < 0 \Rightarrow \mathbf{F} \text{ is inside the clip region}$$

and

$$\mathbf{N}_c \cdot (\mathbf{P}(t) - \mathbf{X}) = 0$$

defines the point of intersection of the line and the clip boundary. Solving Equation 38.1 for t enables the intersecting vertex to be calculated and added to the output list.

In practice, the algorithm is written recursively. As soon as a vertex is output, the procedure calls itself with that vertex, and no intermediate storage is required for the partially clipped polygon. This structure makes the algorithm eminently suitable for hardware implementation.

A projective transformation takes the object representation in view space and produces a projection on the view plane. This is a fairly simple procedure, somewhat complicated by the fact that we must retain a depth value for each point for eventual use in the hidden-surface removal algorithm. Sometimes, therefore, the space of this transformation is referred to as *three-dimensional screen space*.

38.2.4 Projective Transformation and Three-Dimensional Screen Space

A perspective projection is the more popular or common choice in computer graphics because it incorporates foreshortening. In a perspective projection, relative dimensions are not preserved, and a distant line is displayed smaller than a nearer line of the same length. This familiar effect enables human beings to perceive depth in a two-dimensional photograph or a stylization of three-dimensional reality. A perspective projection is characterized by a point known as the *center of projection*, the same point as the viewpoint in our discussion. The projection of three-dimensional points onto the view plane is the intersection of the lines from each point to the center of projection. This familiar idea is shown in Figure 38.8.

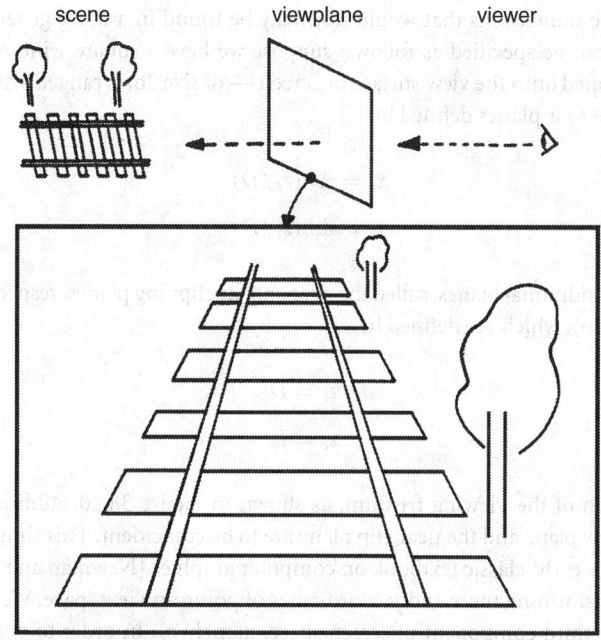

FIGURE 38.8 The perspective effect.

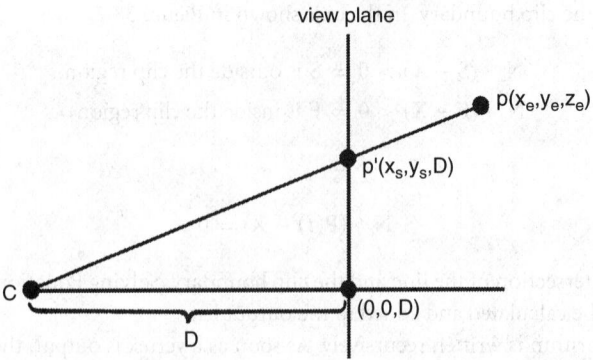

FIGURE 38.9 Perspective projection.

Figure 38.9 shows how a perspective projection is derived. Point $P(x_e, y_e, z_e)$ is a three-dimensional point in the view coordinate system. This point is to be projected onto a view plane normal to the z_e-axis and positioned at distance D from the origin of this system. Point P' is the projection of this point in the view plane. It has two-dimensional coordinates (x_s, y_s) in a view-plane coordinate system, with the origin at the intersection of the z_e-axis and the view plane. In this system, we consider the view plane to be the view surface or screen.

Similar triangles give

$$x_s = D(x_e/z_e)$$

$$y_s = D(y_e/z_e)$$

Screen space is defined to act within a closed volume — the viewing frustum that delineates the volume of space to be rendered. For the purposes of this chapter, we will consider a simplified view volume that constrains some of the dimensions that would normally be found in a more general viewing system. A simple view volume can be specified as follows: suppose we have a square window — that area of the view plane that is mapped onto the view surface or screen — of size 2h, arranged symmetrically about the viewing direction. The four planes defined by

$$x_e = \pm h(z_e/D)$$

$$y_e = \pm h(z_e/D)$$

together with the two additional planes, called the near and far clipping planes, respectively (perpendicular to the viewing direction), which are defined by

$$z_e = D$$

$$z_e = F$$

make up the definition of the viewing frustum, as shown in Figure 38.10. Additionally, we invoke the constraint that the view plane and the near clip plane are to be coincident. This simple system is based on a treatment given in an early, classic textbook on computer graphics [Newman and Sproull, 1973].

This deals with transforming the x and y coordinates of points in view space. We shall now discuss the transformation of the third component of screen space, namely z_e. In order to perform hidden-surface calculations (in the Z-buffer algorithm), depth information must be generated on arbitrary points, in practice pixels, within the polygon by interpolation. This is possible in screen space only if, in moving

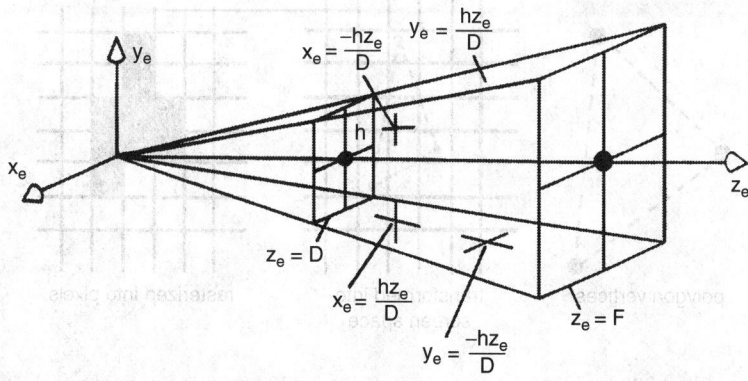

FIGURE 38.10 The six planes that define the view frustum.

from view space to screen space, lines transform into lines and planes transform into planes. It can be shown [Newman and Sproull, 1973] that these conditions are satisfied, provided the transformation of z takes the form

$$z_s = A + (B/z_e)$$

where A and B are constants. These constants are determined from the following constraints:

1. Choosing $B < 0$ so that as z_e increases, so does z_s. This preserves depth. If one point is behind another, then it will have a larger z_e-value; if $B < 0$, it will also have a larger z_s-value.
2. Normalizing the range of z_s-values so that the range z_e in [D, F] maps into the range z_s in [0, 1]. This is important to preserve accuracy, because a pixel depth will be represented by a fixed number of bits in the Z-buffer.

The full perspective transformation is then given by

$$x_s = D(x_e/(hz_e))$$
$$y_s = D(y_e/(hz_e))$$
$$z_s = F(1 - D/z_e)/(F - D)$$

where the additional constant h appearing in the transformation for x_s and y_s ensures that these values fall in the range $[-1, 1]$ over the square screen.

It is instructive to consider the relationship between z_e and z_s a little more closely; although, as we have seen, they both provide a measure of the depth of a point, interpolating along a line in eye space is not the same as interpolating along this line in screen space. As z_e approaches the far clipping plane, z_s approaches 1 more rapidly. Objects in screen space thus get pushed and distorted toward the back of the viewing frustum. This difference can lead to errors when interpolating quantities, other than position, in screen space.

38.2.5 Shading Algorithm

This part of the process calculates a light intensity for each pixel onto which a polygon projects. The input information to this process is the vertex list of the polygon, transformed into eye or view space, and the process splits into two subprocesses:

1. First we find the set of pixels that make up our polygon. We change the polygon from a vertex list into a set of pixels in screen space — through *rasterization*.

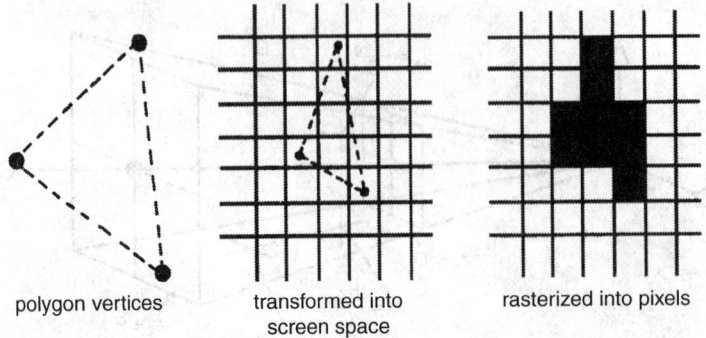

FIGURE 38.11   Different representations of a polygon in the rendering process.

2. Second we find a light intensity associated with each pixel. This is done by associating a local reflection model with each vertex of the polygon and calculating an intensity for the vertex. These vertex intensities are then used in a *bilinear interpolation* scheme to find the intensity of each pixel. By this process, we are finding the intensity of that part of the object surface that corresponds to the pixel area in screen space. The particular way in which this is done leads to the (efficiency–quality) hierarchy of **flat shading**, **Gouraud shading**, and **Phong shading**.

38.2.5.1   Rasterization

Rasterization, or finding the set of polygons onto which the polygon projects, must be done carefully because adjacent polygons must fit together accurately after they have been mapped into pixel sets. If it is not done accurately, holes can result in the image — probably the most common defect seen in rendering software.

As shown in Figure 38.11, the precise geometry of the polygon will map into a set of fully and partially covered pixels. We must decide which of the partially covered pixels are to be considered part of the polygon. Deciding on the basis of the area of coverage is extremely expensive and would wipe out the efficiency advantage of the bilinear interpolation scheme that is used to find a pixel intensity. It is better to map the vertices in some way to the nearest pixel coordinate and set up a consistent rule for deciding the fate of partially covered pixels. This is the crux of the matter. If the rules are not consistent and carefully formulated, then the rounding process will produce holes or unfilled pixels between polygons that share the same scan line. Note that the process will cause a shape change in the polygon, which we ignore because the polygon is already an approximation, and to some extent an arbitrary representation of the "real" surface of the object.

Sometimes called *scan-line conversion*, rasterization proceeds by moving a horizontal line through the polygon in steps of a pixel height. For a current scan line, interpolation (see Section 38.2.5.2) between the appropriate pairs of polygon vertices will yield x_{start} and x_{end}, the start and end points of the portions of the scan line crossing the polygons (using real arithmetic). The following scheme is a simple set of rules that converts these values into a run of pixels:

1. Round x_{start} up.
2. Round x_{end} down.
3. If the fractional part of x_{end} is 0, then subtract 1 from it.

Applying the rasterization process to a complete polygon implies embedding this operation in a structure that keeps track of the edges that are to be used in the current scan-line interpolation. This is normally implemented using a linked-list approach.

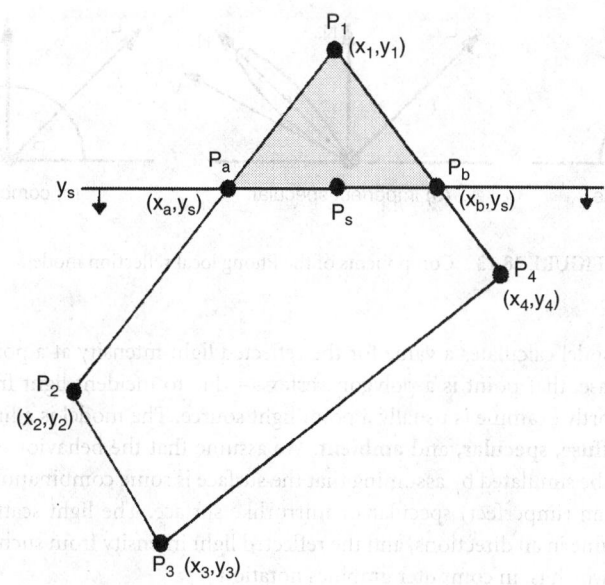

FIGURE 38.12 Notation used in property interpolation within a polygon.

38.2.5.2 Bilinear Interpolation

As we have already mentioned, light intensity values are assigned to the set of pixels that we have now calculated, not by individual calculation but by interpolating from values calculated only at the polygon vertices. At the same time, we interpolate depth values for each pixel to be used in the hidden-surface determination. So in this section, we consider the interpolation of a pixel property from vertex values independent of the nature of the property. Referring to Figure 38.12, the interpolation proceeds by moving a scan line down through the pixel set and obtaining start and end values for a scan line by interpolating between the appropriate pair of vertex properties. Interpolation along a scan line then yields a value for the property at each pixel. The interpolation equations are

$$p_a = \frac{1}{y_1 - y_2} \left[p_1(y_s - y_2) + p_2(y_1 - y_s) \right]$$

$$p_b = \frac{1}{y_1 - y_4} \left[p_1(y_s - y_4) + p_4(y_1 - y_s) \right] \tag{38.2}$$

$$p_s = \frac{1}{x_b - x_a} \left[p_a(x_b - x_s) + p_b(x_s - x_a) \right]$$

These would normally be implemented using an incremental form, the final equation, for example, becoming

$$p_s := p_s + \Delta p$$

with the constant value Δp calculated once per scan line.

38.2.5.3 Local Reflection Models

Given that we find pixel intensities by an interpolation process, the next thing to discuss is how to find the reflected light intensity at the vertices of a polygon, those values from which the pixel intensity values are derived. This is done by using a simple local reflection model — the one most commonly used is the Phong reflection model [Phong, 1975]. (This reflection model is not to be confused with Phong shading, which is a vector interpolation scheme.)

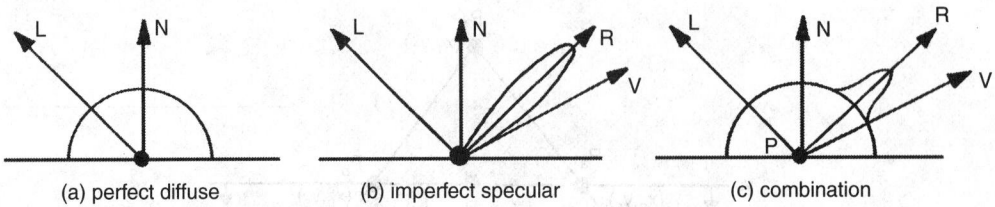

(a) perfect diffuse   (b) imperfect specular   (c) combination

FIGURE 38.13   Components of the Phong local reflection model.

A local reflection model calculates a value for the reflected light intensity at a point on the surface of an object — in this case, that point is a polygon vertex — due to incident light from a source, which for reasons we will shortly examine is usually a point light source. The model is a linear combination of three components: **diffuse, specular**, and **ambient**. We assume that the behavior of reflected light at a point on a surface can be simulated by assuming that the surface is some combination of a perfect diffuse surface together with an (imperfect) specular or mirrorlike surface. The light scattered from a perfect diffuse surface is the same in all directions, and the reflected light intensity from such a surface is given by Lambert's cosine law, which is, in computer graphics notation

$$I_d = I_i k_d \mathbf{L} \cdot \mathbf{N}$$

where $\mathbf{L}$ is the light direction vector and both $\mathbf{L}$ and $\mathbf{N}$ are unit vectors, as shown in Figure 38.13a; k_d is a diffuse reflection coefficient; and I_i is the intensity of a (point) light source. The specular contribution is a function of the angle between the viewing direction $\mathbf{V}$ and the mirror direction $\mathbf{R}$

$$I_s = I_i k_s (\mathbf{R} \cdot \mathbf{V})^n$$

where n is an index that simulates surface roughness and k_s is a specular reflection coefficient.

For a perfect mirror, n would be infinity and reflected light would be constrained to the mirror direction. For small integer values of n, a reflection lobe is generated, where the thickness of the lobe is a function of the surface roughness (see Figure 38.13b). The effect of the specular reflection term in the model is to produce a so-called highlight on the rendered object. This is basically a reflection of the light source spread over an area of the surface to an extent that depends on the value of n. The color of the specularly reflected light is different from that of the diffuse reflected light — hence the term *highlight*. In simple models of specular reflection, the specular component is assumed to be the color of the light source. If, say, a green surface were illuminated with white light, then the diffuse reflection component would be green, but the highlight would be white.

Adding the specular and diffuse components gives a very approximate imitation to the behavior of reflected light from a point on the surface of an object. Consider Figure 38.13c. This is a cross section of the overall reflectivity response as a function of the orientation of the view vector $\mathbf{V}$. The cross section is in a plane that contains the vector $\mathbf{L}$ and the point $\mathbf{P}$; thus, it slices through the specular bump. The magnitude of the reflected intensity, the sum of the diffuse and specular terms, is the distance from $\mathbf{P}$ along the direction $\mathbf{V}$ to where $\mathbf{V}$ intersects the profile.

An ambient component is usually added to the diffuse and specular terms. Such a component illuminates surfaces that, because we generally use a point light source, would otherwise be rendered black. These are surfaces that are visible from the viewpoint but not from the light source. Essentially, the ambient term is a constant that attempts to simulate the global interreflection of light between surfaces.

Adding the diffuse, specular, and ambient components (Equation 38.3), we have

$$I = k_a + I_i(k_d \mathbf{L} \cdot \mathbf{N} + k_s (\mathbf{R} \cdot \mathbf{V})^n) \tag{38.3}$$

where k_a is the constant ambient term. The expense of Equation 38.3 can be considerably reduced by making some geometric assumptions and approximations. First, if the light source and the viewpoint

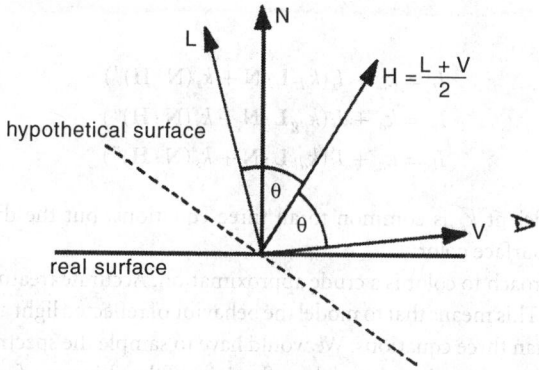

FIGURE 38.14 **H** is the unit normal to a hypothetical surface oriented in a direction that bisects the angle between **L** and **V**.

are considered to be at infinity, then **L** and **V** are constant over the domain of the scene. The vector **R** is expensive to calculate; although Phong gives an efficient method for calculating **R**, it is better to use a vector **H**. This appears to have been first introduced by Blinn [1977]. The specular term then becomes a function of **N·H** rather than **R·V**. **H** is the unit normal to a hypothetical surface that is oriented in a direction halfway between the light direction vector **L** and the viewing vector **V** (see Figure 38.14).

$$\mathbf{H} = (\mathbf{L} + \mathbf{V})/2$$

Together with a shading algorithm, this simple model is responsible for the look of most shaded computer graphics images, and it has been in use constantly since 1975. Its main disadvantages are that objects look as if they were made from some kind of plastic material, which is either shiny or dull. Also, in reality the magnitude of the specular component is not independent, as Equation 38.3 implies, of the direction of the incoming light. Consider, for example, the glare from a (nonshiny) road surface when you are driving in the direction of the setting sun: this does not occur when the sun is overhead.

We should now return to the term *local*. The reflection model is called a local model because it considers that the point on the surface under consideration is illuminated directly by the light source in the scene. No other (indirect) source of illumination is taken into account. Light reflected from nearby objects is ignored, so we see no reflections of neighboring objects in the object under consideration. It also means that shadows, which are areas that cannot "see" the light source and which receive their illumination indirectly from another object, cannot be modeled. In a scene using this model, objects are illuminated as if they were floating in a dark space illuminated only by the light source.

When shadows are added into rendering systems that use local reflection models, these are purely geometric. That is, the area that the shadow occupies on the surface of an object, due to the intervention of another object between it and the light source, is calculated. The reflected light intensity within this area is then arbitrarily reduced. When using local reflection models, there is no way to calculate how much indirect light illuminates the shadowed area. The visual consequences of this should be considered when including shadows in an add-on manner. These may detract from the real appearance of the final rendered image rather than add to it. First, because shadows are important to us in reality, we easily spot shadows in computer graphics that have the wrong intensity. This is compounded by the hard-edged shadow boundaries calculated by geometric algorithms. In reality, shadows normally have soft, subtle edges.

Finally, we briefly consider the role of color. For colored objects, the easiest approach is to model the specular highlights as white (for a white light source) and to control the color of the objects by appropriate setting of the diffuse reflection coefficients. We use three intensity equations to drive the monitor's red, green, and blue inputs:

$$I_r = k_a + I_i(k_{dr}\mathbf{L} \cdot \mathbf{N} + k_s(\mathbf{N} \cdot \mathbf{H})^n)$$
$$I_g = k_a + I_i(k_{dg}\mathbf{L} \cdot \mathbf{N} + k_s(\mathbf{N} \cdot \mathbf{H})^n)$$
$$I_b = k_a + I_i(k_{db}\mathbf{L} \cdot \mathbf{N} + k_s(\mathbf{N} \cdot \mathbf{H})^n)$$

where the specular coefficient k_s is common to all three equations, but the diffuse component varies according to the object's surface color.

This three-sample approach to color is a crude approximation. Accurate treatment of color requires far more than three samples. This means that to model the behavior of reflected light accurately, we would have to evaluate many more than three equations. We would have to sample the spectral energy distribution of the light source as a function of wavelength and the reflectivity of the object as a function of wavelength and apply Equation 38.3 at each wavelength sample. The solution then obtained would have to be converted back into three intensities to drive the monitor. The colors that we would get from such an approach would certainly be different from the three-sample implementation. Except in very specialized applications, this problem is completely ignored.

We now discuss shading options. These options differ in where the reflection model is applied and how calculated intensities are distributed among pixels. There are three options: flat shading, Gouraud shading, and Phong shading, in order of increasing expense and increasing image quality.

38.2.5.4 Flat Shading

Flat shading is the option in which we invoke no interpolation within a polygon and shade each pixel within the polygon with the same intensity. The reflection model is used once only per polygon. The (true) normal for the polygon (in eye or view space) is inserted into Equation 38.3, and the calculated intensity is applied to the polygon. The efficiency advantages are obvious — the entire interpolation procedure is avoided, and shading reduces to rasterization plus a once-only intensity calculation per polygon. The (visual) disadvantage is that the polygon edges remain glaringly visible, and we render not the surface that the polygon mesh represents but the polygon mesh itself. As far as image quality is concerned, this is more disadvantageous than the fact that there is no variation in light intensity among the polygon pixels. Flat shading is used as a fast preview facility.

38.2.5.5 Gouraud Shading

Both Gouraud and Phong shadings exhibit two strong advantages — in fact, these advantages are their *raison d'être*. Both use the interpolation scheme already described and so are efficient, and they diminish or eliminate the visibility of the polygon edges. In a Gouraud- or Phong-shaded object, these are now visible only along silhouette edges. This elegant device meant their enduring success; the idea originated by Gouraud [1971] and cleverly elaborated by Phong [1975] was one of the major breakthroughs in three-dimensional computer graphics.

In Gouraud shading, intensities are calculated at each vertex and inserted into the interpolation scheme. The trick is in the normals used at a polygon vertex. Using the true polygon normal would not work, because all the vertex normals would be parallel and the reflection model would evaluate the same intensity at each. What we must do is calculate a normal at each vertex that somehow relates back to the original surface. Gouraud vertex normals are calculated by considering the average of the true polygon normals of those polygons that contribute to the vertex (see Figure 38.15). This calculation is normally regarded as part of the setting up of the object, and these vectors are stored as part of the object database (although there is a problem when polygons are clipped: new vertex normals then must be calculated as part of the rendering process). Because polygons now share vertex normals, the interpolation process ensures that there is no change in intensity across the edge between two polygons; in this way, the polygonal structure of the object representation is rendered invisible. (However, an optical illusion, known as *Mach banding*, persists along the edges with Gouraud shading.)

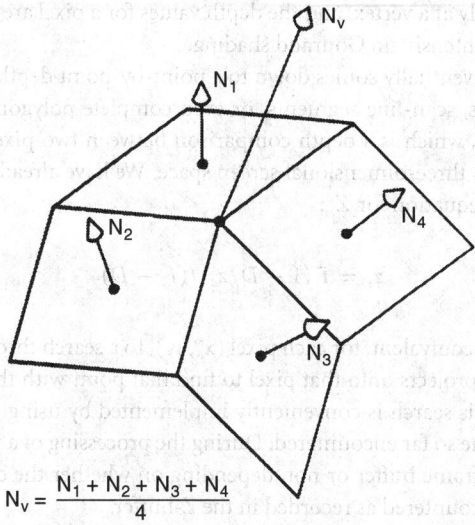

$$N_v = \frac{N_1 + N_2 + N_3 + N_4}{4}$$

FIGURE 38.15 The concept of a vertex normal.

Gouraud shading is used extensively and gives excellent results for the diffuse component. However, calculating reflected light intensity only at the vertices leads to problems with the specular component. The easiest case to consider is that of a highlight which, if it were visible, would be within the polygon boundaries — meaning it does not extend to the vertices. In this case, the Gouraud scheme would simply miss the highlight completely.

38.2.5.6 Phong Shading

Phong shading [Phong, 1975] was developed to overcome the problems of Gouraud shading and specular highlights. In this scheme, the property to be interpolated is the vertex normals themselves, with each vector component now inserted into three versions of Equation 38.3. It is a strange hybrid, with an interpolation procedure running in pixel or screen space controlling vector interpolation in three-dimensional view space (or world space). But it works very well.

We estimate the normal to the surface at a point that corresponds to the pixel under consideration in screen space, or at least estimate it to within the limitations and approximations that have been imposed by the polygonal representation and the interpolation scheme. We can then apply the reflection model at each pixel, and a unique reflected light intensity is now calculated for each pixel. We may end up with a result that is different from what would be obtained if we had access to the true surface normal at the point on the real surface that corresponded to the pixel, but it does not matter, because the quality of Phong shading is so good that we cannot perceive any erroneous effects on the monitor.

Phong shading is much slower than Gouraud shading because the interpolation scheme is three times as lengthy, and also the reflection model (Equation 38.3) is now applied at each pixel. A good rule of thumb is that Phong shading has five times the cost of Gouraud shading.

38.2.6 Hidden-Surface Removal

As already mentioned, we shall describe the Z-buffer as the *de facto* hidden-surface removal algorithm. That it has attained this status is due to its ease of implementation — it is virtually a single *if* statement — and its ease of incorporation into a polygon-based renderer. Screen space algorithms (the Z-buffer falls into this category) operate by associating a depth value with each pixel. In our polygon renderer, the depth values are available only at a vertex, and the depth values for a pixel are obtained by using the same interpolation scheme as for intensity in Gouraud shading.

Hidden-surface removal eventually comes down to a point-by-point depth comparison. Certain algorithms operate on area units, scan-line segments, or even complete polygons, but they must contain a provision for the worst case, which is a depth comparison between two pixels. The Z-buffer algorithm performs this comparison in three-dimensional screen space. We have already defined this space and we repeat, for convenience, the equation for Z_s:

$$z_s = F(1 - D/z_e)/(F - D)$$

The Z-buffer algorithm is equivalent, for each pixel (x_s, y_s), to a search through the associated z-values of every polygon point that projects onto that pixel to find that point with the minimum z-value — the point nearest the viewer. This search is conveniently implemented by using a Z-buffer, which holds for each pixel the smallest z-value so far encountered. During the processing of a polygon, we either write the intensity of a pixel into the frame buffer or not, depending on whether the depth of the current pixel is less than the depth so far encountered as recorded in the Z-buffer.

Apart from its simplicity, another advantage of the Z-buffer is that it is independent of object representation. Although we see it used most often in the context of polygon mesh rendering, it can be used with any representation: all that is required is the ability to calculate a z-value for each point on the surface of an object. If the z-values are stored with pixel values, separately rendered objects can be merged into a multiple-object scene using Z-buffer information on each object.

The main disadvantage of the Z-buffer is the amount of memory it requires. The size of the Z-buffer depends on the accuracy to which the depth value of each point (x, y) is to be stored, which is a function of scene complexity. Usually, 20 to 32 bits is deemed sufficient for most applications. Recall our previous discussion of the compression of z_s-values. This means that a pair of distinct points with different z_e-values can map into identical z_s-values. Note that for frame buffers with less than 24 bits per pixel, say, the Z-buffer will in fact be *larger* than the frame buffer. In the past, Z-buffers have tended to be part of the main memory of the host processor, but now graphics cards are available with dedicated Z-buffers. This represents the best solution.

38.3 Rendering Using Ray Tracing

Ray tracing is a simple and elegant algorithm whose appearance in computer graphics is usually attributed to Whitted [1980]. It combines in a single algorithm

Hidden-surface removal
Reflection due to direct illumination (the same factor we calculated in the previous method using a local model)
Reflection due to indirect illumination (i.e., reflection due to light striking) — the object which itself has been reflected from another object
Transmission of light through transparent or partially transparent objects
Shading due to object–object interaction (global illumination)
The computation of (hard-edged) shadows

It does this by tracing rays — infinitesimally thin beams of light — in the reverse direction of light propagation; that is, it traces light rays from the eye into the scene and from object to object. In this way, it "discovers" the way in which light interacts between objects and can produce visualizations such as objects reflecting in other objects and the distortion of an object viewed through another (transparent or glass) object due to refraction. Rays are traced from the eye or viewpoint, because we are interested only in those rays that pass through the view plane. If we traced rays from the light source, then theoretically we would have to trace an infinity of rays.

Ray-tracing algorithms exhibit a strong visual signature because a basic ray tracer can simulate only one aspect of the global interaction of light in an environment: specular reflection and specular transmission. Thus ray-traced scenes always look ray-traced, because they tend to consist of objects that exhibit mirrorlike reflection, in which you can see the perfect reflections of other objects. Simulating nonperfect specular reflection is computationally impossible with the normal ray-tracing approach, because this means that at a hit point a single incoming ray will produce a multiplicity of reflected rays instead of just one. The same argument applies to transparent objects. A single incident ray can produce only a single transmitted or refracted ray. Such behavior would happen only in a perfect material that did not scatter light passing through it. With transparent objects, the refractive effect can be simulated, but the material looks like perfect glass. Thus, perfect surfaces and perfect glass, behavior that does not occur in practice, betray the underlying rendering algorithm.

A famous development, called *distributed ray tracing* [Cook et al., 1984], addressed exactly this problem, using a Monte Carlo approach to simulate the specular reflection and specular transmission spread without invoking a combinatorial explosion. The algorithm produces shiny objects that look real (i.e., their surfaces look rough or imperfect), blurred transmission through glass, and blurred shadows. The modest cost of this method involved initiating 16 rays per pixel instead of one. This is still a considerable increase in an already expensive algorithm, and most ray tracers still utilize the perfect specular interaction model.

The algorithm is conceptually easy to understand and is also easy to implement using a recursive procedure. A pictorial representation is given in Figure 38.16. The algorithm operates in three-dimensional

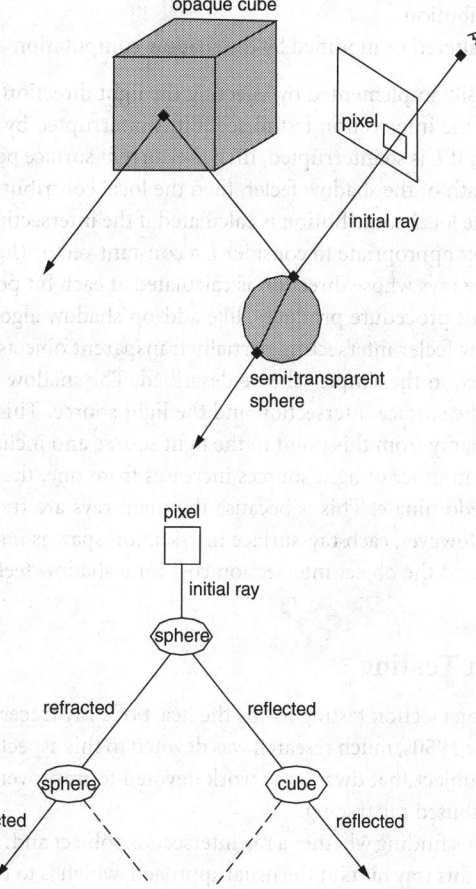

FIGURE 38.16 A representation of a ray-tracing algorithm.

world space, and for each pixel in screen space we calculate an initial ray from the viewpoint through the center of the pixel. The ray is injected into the scene and will either hit an object or not. (In the case of a closed environment, some object will always be encountered by an initial ray, even if it is just the background, such as a wall.) When the ray hits an object, it spawns two more rays: a reflected ray and a transmitted ray, which refracts into the object if the object is partially transparent. These rays travel onward and themselves spawn other rays at their next hits. The process is sometimes represented as a binary tree, with a light–surface hit at each node in the tree.

This process can be implemented as a recursive procedure, which for each ray invokes an intersection test that spawns a transmitted and a reflected ray by calling itself twice with parameters representing the reflected and the transmitted or refracted directions of the new rays. At the heart of the recursive control procedure is an intersection test. This procedure is supplied with a ray, compares the ray geometry with all objects in the scene, and returns the nearest surface that the ray intersects. If the ray is an initial ray, then this effectively implements hidden-surface removal. Intersection tests account for most of the computational overheads in ray tracing, and much research effort has gone into how to reduce this cost.

Grafted onto this basic recursive process, which follows specular interaction through the scene, are the computation of direct reflection and shadow computation. At each node or surface hit, we calculate these two contributions. Direct reflection is calculated by applying, for each light source, a Phong reflection model (or some other local model) at the node under consideration. The direct reflection contribution is diminished if the point is in shadow with respect to a light source. Thus, at any hit point or node, there are three contributions to the light intensity passed back up through the recursion:

A reflected-ray contribution
A transmitted-ray contribution
A local contribution unaltered or modified by the shadow computation

Shadow computation is easily implemented by injecting the light direction vector, used in the local contribution calculation, into the intersection test to see if it is interrupted by any intervening objects. This ray is called a *shadow feeler*. If L is so interrupted, then the current surface point lies in shadow. If a wholly opaque object lies in the path of the shadow feeler, then the local contribution is reduced to the ambient value. An attenuation in the local contribution is calculated if the intersecting object is partially transparent. Note that it is no longer appropriate to consider L a constant vector (light source at infinity) and the so-called shadow feelers are rays whose direction is calculated at each hit point. Because light sources are normally point sources, this procedure produces, like add-on shadow algorithms, hard-edged shadows. (Strictly speaking, a shadow feeler intersecting partially transparent objects should be refracted. It is not possible to do this, however, in the simple scheme described. The shadow feeler is initially calculated as the straight line between the surface intersection and the light source. This is an easy calculation, and it would be difficult to trace a ray from this point to the light source and include refractive effects.)

Finally note that, as the number of light sources increases from one, the computational overheads for shadow testing rapidly predominate. This is because the main rays are traced only to an average depth of between one and two. However, each ray-surface intersection spawns n shadow feelers (where n is the number of light sources), and the object intersection cost for a shadow feeler is exactly the same as for a main ray.

38.3.1 Intersection Testing

We have mentioned that intersection testing forms the heart of a ray tracer and accounts for most of the cost of the algorithm. In the 1980s, much research was devoted to this aspect of ray tracing. There is a large body of literature on the subject that dwarfs the work devoted to improvements in the ray-traced image such as, for example, distributed ray tracing.

Intersection testing means finding whether a ray intersects an object and, if so, the point of intersection. Expressing the problem in this way hints at the usual approach, which is to try to cut down the overall cost by using some scheme, like a bounding volume, that prevents the intersection test from searching through all the polygons in an object if the ray cannot hit the object.

The cost of ray tracing and the different possible approaches depend much on the way in which objects are represented. For example, if a voxel representation is used and the entire space is labeled with object occupancy, then discretizing the ray into voxels and stepping along it from the start point will reveal the first object that the ray hits. Contrast this with a brute-force intersection test, which must test a ray against every object in the scene to find the hit nearest to the ray start point.

38.4 Rendering Using the Radiosity Method

The *radiosity method* arrived in computer graphics in the mid-1980s, a few years after ray tracing. Most of the early development work was carried out at Cornell University under the guidance of D. Greenberg, a major figure in the development of the technique. The emergence of the **hemicube** algorithm and, later, the progressive refinement algorithm, established the method and enabled it to leave research laboratories and become a practical rendering tool. Nowadays, many commercial systems are available, and most are implementations of these early algorithms.

The radiosity method provides a solution to diffuse interaction, which, as we have discussed, cannot easily be incorporated in ray tracing, but at the expense of dividing the scene into large patches (over which the radiosity is constant). This approach cannot cope with sharp specular reflections. Essentially, we have two global methods: ray tracing, which simulates global specular reflection, and transmission and radiosity, which simulate global diffuse interaction.

In terms of the global phenomena that they simulate, the methods are mutually exclusive. Predictably, a major research bias has involved the unification of the two methods into a single global solution. Research is still actively pursued into many aspects of the method — particularly form-factor determination and scene decomposition into elements or patches.

38.4.1 Basic Theory

The radiosity method works by dividing the environment into largish elements called *patches*. For every pair of patches in the scene, a parameter F_{ij} is evaluated. This parameter, called a *form factor*, depends on the geometric relationship between patches i and j. This factor is used to determine the strength of diffuse light interaction between pairs of patches, and a large system of equations is set up which, on solution, yields the radiosity for each patch in the scene.

The radiosity method is an object–space algorithm, solving for a single intensity for each surface patch within an environment and not for pixels in an image-plane projection. The solution is thus independent of viewer position. This complete solution is then injected into a renderer that computes a particular view by removing hidden surfaces and forming a projection. This phase of the method does not require much computation (intensities are already calculated), and different views are easily obtained from the general solution. The method is based on the assumption that all surfaces are perfect diffusers or ideal Lambertian surfaces.

Radiosity, B, is defined as the energy per unit area leaving a surface patch per unit time and is the sum of the emitted and the reflected energy:

$$B_i \, dA_i = E_i \, dA_i + R_i \int_j B_j F_{ji} \, dA_j$$

Expressing this equation in words, we have for a single patch i

$$\text{radiosity} \; x \; \text{area} = \text{emitted energy} + \text{reflected energy}$$

E_i is the energy emitted from a patch, and emitting patches are, of course, light sources. The reflected energy is given by multiplying the incident energy by R_i, the reflectivity of the patch. The incident energy is the energy that arrives at patch i from all other patches in the environment; that is, we integrate over the environment, for all $j(j \neq i)$, the term $B_j F_{ji} \, dA_j$. This is the energy leaving each patch j that arrives at patch i.

For a discrete environment, the integral is replaced by a summation and constant radiosity is assumed over small discrete patches. It can be shown that

$$B_i = E_i + R_i \sum_{j=1}^{n} B_j F_{ij}$$

Such an equation exists for each surface patch in the enclosure, and the complete environment produces a set of n simultaneous equations of the form.

$$\begin{bmatrix} 1 - R_1 F_{11} & -R_1 F_{12} & \cdots & -R_1 F_{1n} \\ -R_2 F_{21} & 1 - R_2 F_{22} & \cdots & -R_2 F_{2n} \\ \cdots & \cdots & \cdots & \cdots \\ \cdots & \cdots & \cdots & \cdots \\ \cdots & \cdots & \cdots & \cdots \\ -R_n F_{n1} & -R_n F_{n2} & \cdots & 1 - R_n F_{nn} \end{bmatrix} \begin{bmatrix} B_1 \\ B_2 \\ \cdots \\ \cdots \\ \cdots \\ B_n \end{bmatrix} \begin{bmatrix} E_1 \\ E_2 \\ \cdots \\ \cdots \\ \cdots \\ E_n \end{bmatrix}$$

The E_is are nonzero only at those surfaces that provide illumination, and these terms represent the input illumination to the system. The R_is are known, and the F_{ij}s are a function of the geometry of the environment. The reflectivities are wavelength-dependent terms, and the previous equation should be regarded as a monochromatic solution, a complete solution being obtained by solving for however many color bands are being considered. We can note at this stage that $F_{ii} = 0$ for a plane or convex surface — none of the radiation leaving the surface will strike itself. Also, from the definition of the form factor, the sum of any row of form factors is unity.

Since the form factors are a function only of the geometry of the system, they are computed once only. Solving this set of equations produces a single value for each patch. This information is then input to a modified Gouraud renderer to give an interpolated solution across all patches.

38.4.2 Form-Factor Determination

A significant early development was a practical method to evaluate form factors. The algorithm is both an approximation and an efficient method of achieving a numerical estimation of the result. The form factor between patches i and j is defined as

$$F_{ij} = \frac{\text{radiative energy leaving surface } A_i \text{ that strikes } A_j \text{ directly}}{\text{radiative energy leaving } A_i \text{ in all directions in the hemispherical space surrounding } A_i}$$

This is given by

$$F_{ij} = \frac{1}{A_i} \int_{A_i} \int_{A_j} \frac{\cos \phi_i \cos \phi_j}{\pi r^2} dA_j dA_i$$

where the geometric conventions are illustrated in Figure 38.17. Now, it can be shown that this patch-to-patch form factor can be approximated by the differential-area-to-finite-area form factor

$$F_{dA_i A_j} = \int_{A_j} \frac{\cos \phi_i \cos \phi_j}{\pi r^2} dA_j$$

where we are now considering the form factor between the elemental area da_i and the finite area A_j. This approximation is calculated by the hemicube algorithm [Cohen and Greenberg, 1985]. The factors that enable the approximation to a single integral and its veracity are quite subtle and are outside the scope of this treatment.

A good intuition of the workings of the algorithm can be gained from a pictorial visualization (see Figure 38.18). Figure 38.18a is a representation of the property known as the *Nusselt analog*. In the example, patches A, B, and C all have the same form factor with respect to patch i. Patch B is the projection of A onto a hemicube, centered on patch i, and C is the projection of A onto a hemisphere. This property is the

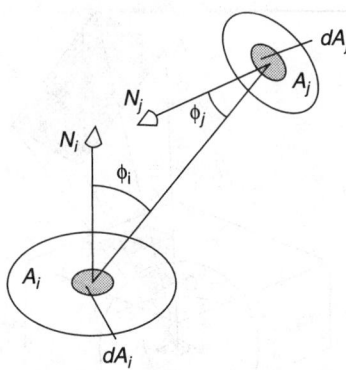

FIGURE 38.17 Parameters used in the definition of a form factor.

foundation of the hemicube algorithm, which places a hemicube on each patch i — (Figure 38.18b). The hemicube is subdivided into elements; associated with each element is a precalculated *delta form factor*. The hemicube is placed on patch i, and then patch j is projected onto it. In practice, this involves a clipping operation, because in general a patch can project onto three faces of the hemicube. Evaluating F_{ij} involves simply summing the values of the delta form factors onto which patch j projects (Figure 38.18c).

Another aspect of form-factor determination solved by the hemicube algorithm is the intervening-patch problem. Normally, we cannot evaluate the form-factor relationship between a pair of patches independently of one or more patches that happen to be situated between them. The hemicube algorithm solves this by making the hemicube a kind of Z-buffer in addition to its role as five projection planes. This is accomplished as follows. For the patch i under consideration, every other patch in the scene is projected onto the hemicube. For each projection, the distance from patch i to the patch being projected is compared with the smallest distance associated with previously projected patches, which is stored in hemicube elements. If a projection from a nearer patch occurs, then the identity of that patch and its distance from patch i are stored in the hemicube elements onto which it projects. When all patches are projected, the form factor F_{ij} is calculated by summing the delta form factors that have registered patch j as a nearest projection.

Finally, consider Figure 38.19, which gives an overall view of the algorithm. This emphasizes the fact that there are three entry points into the process for an interactive program. Changing the geometry of the scene means an entire recalculation, starting afresh with the new scene. However, if only the wavelength-dependent properties of the scene are altered (reflectivities of objects and colors of light sources), then the expensive part of the process — the form-factor calculations — is unchanged. Because the method is view-independent, changing the position of the viewpoint involves only the final process of interpolation and hidden-surface removal. This enables real-time, interactive walkthroughs using the precalculated solution, a popular application of the radiosity technique in computer-aided design (see Figure 38.20). The high-quality imagery gives designers a better feel for the final product than would be possible with simple rendering packages.

38.4.3 Problems with the Basic Method

Several problems occur if the method is implemented without elaboration. Here, we will restrict ourselves to identifying these and giving a pointer to the solutions that have emerged. The reader is referred to the appropriate literature for further details.

The first problem emerges from consideration of how to divide the environment into patches. Dividing the scene equally into large patches, as far as the geometry of the objects allows, will not suffice. The basic radiosity solution calculates a constant radiosity over the area of a patch. Larger patches mean fewer patches and a faster solution, but there will be areas in the scene that will exhibit a fast change in reflected

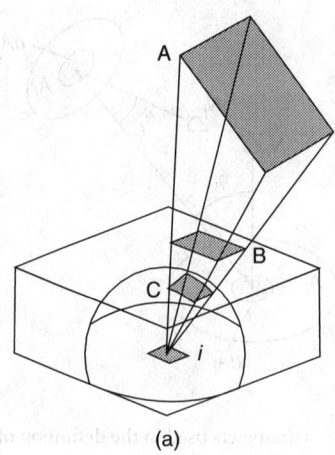

(a)

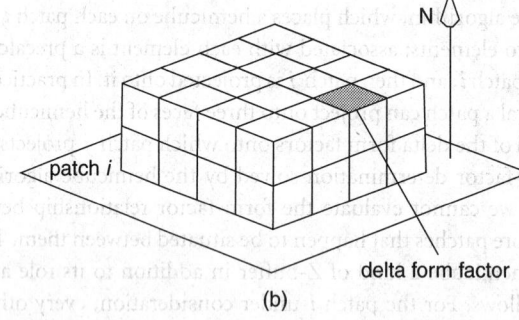

patch i

delta form factor

(b)

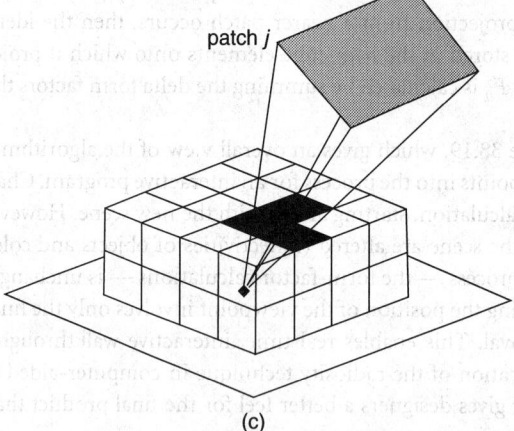

patch j

(c)

FIGURE 38.18 Visualization of the properties used in the hemicube algorithm for form-factor evaluation. (a) Nusselt analogue: patches A, B, and C have the same form factor with respect to patch i. (b) Delta form factors are precalculated for the hemicube. (c) The hemicube is positioned over patch i. Each patch j is projected onto the hemicube. F_{ij} is calculated by summing the delta form factors of the hemicube elements onto which j projects.

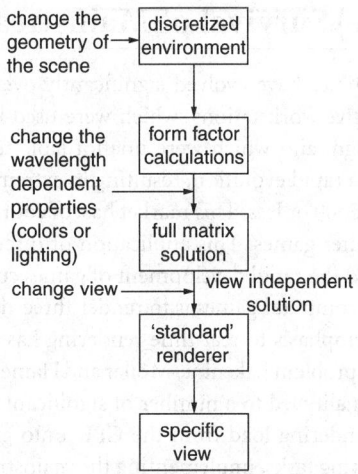

FIGURE 38.19 Processes and interactive entry points in a radiosity algorithm.

FIGURE 38.20 The interior of a car rendered using the radiosity method. (Image rendered using LightWorks, courtesy of LightWork Design.)

light per unit area. Shadow boundaries, for example, exhibit a shape that depends on the obscuring object, and small patches are required in the area of these, which will in general be curves. If the shadowed surface is not sufficiently subdivided, then the shadow boundary will exhibit steps in the shape of the (square or rectangular) patches. However, we do not know until the solution is computed where such areas are in the scene. A common approach to this problem is to incorporate a variable subdivision into the solution, subdividing as a solution emerges until the difference in radiosity between adjacent patches falls below a threshold [Cohen et al., 1986].

Another problem is the time taken by the algorithm — approximately one order of magnitude longer than ray tracing. An ingenious reordering of the method for solving the equation set, called the *progressive refinement algorithm* [Cohen et al., 1988], ameliorates this problem and provides a user with an early (approximate) image which is progressively refined, that is, it becomes more and more accurate. In a normal solution, each patch progresses toward a final value in turn. In the progressive refinement algorithm, each and every patch is updated for each iteration, and all patches proceed toward their final value simultaneously. The scene is at first dark and then gets lighter and lighter as the patches increase their radiosity. Early, approximate solutions are quickly computed by distributing an arbitrary ambient component among all patches. The early solutions are "fully" lit but incorrect. Progressive refinement means displaying more and more accurate radiosities on each patch and reducing the approximate ambient component at the same time. A user can thus see from the start some solution, which then proceeds to get more and more accurate. The process can then be interrupted at some stage if the image is not as required or has some visually obvious error.

38.5   The (Irresistible) Survival of Mainstream Rendering

Three-dimensional computer graphics have evolved significantly over the past three decades. We saw at first the development of expensive workstations, which were used in expensive applications, such as computer-aided architectural design, and which were unaffordable to home consumers. Recently, PC graphics hardware has undergone a rapid evolution, resulting in extremely powerful graphics cards being available to consumers at a cost of $500 or less. This market has, of course, been spurred by the apparently never-ending popularity of computer games. This application of three-dimensional computer graphics has dwarfed all others and hastened the rapid development of games consoles.

The demand from the world of computer games is to render three-dimensional environments, at high quality, in real time. The shift of emphasis to real-time rendering has resulted in many novel rendering methods that address the speed-up problem [Akenine-Möller and Haines, 2002; Watt and Policarpo, 2001]. The demand for ever-increasing quality led to a number of significant stages in rendering methodology. First, there was the shift of the rendering load from the CPU onto graphics cards. This add-on hardware performed most of the rendering tasks, implementing the mainstream polygon rendering described previously. Much use was made of texture mapping hardware to precalculate light maps, which enabled rendering to be performed in advance for all static detail in the environment.

At the time of writing (2003), consumer hardware is now available which is powerful enough to enable all rendering in real time, obviating the need for precalculation. This makes for better game content, with dynamic and static objects having access to the same rendering facilities.

Although graphics cards were at first simply viewed as black boxes that received a massive list of polygons, recent developments have seen cards with functionality exposed to the programmer. Such functionality has enabled the per-pixel (or per-fragment) programming necessary for real increases in image quality. Thus, we have the polygon processing returned to the control of the programmer and the need for an expansion of graphics APIs (such as OpenGL) to enable programmers to exploit this new functionality. Companies such as NVIDIA and ATI have thrived on offering such functionality.

One of the enduring facts concerning the history of this evolution is the inertia of the mainstream rendering methodology. Polygon meshes have survived as the most popular representation, and the rendering of polygons using interpolative shading and a set of light sources seems as firmly embedded as ever.

38.6   An OpenGL Example

We complete this chapter with a simple example in OpenGL that renders a single object: a teapot (see Figure 38.21). Two points are worth noting. First, polygons become triangles. Although we have consistently used the word *polygon* in this text, most graphics hardware is optimized to deal with triangles. Second, there is no exposure to the programmer of processes such as rasterization. This pixel process is invisible, although we briefly discussed in Section 38.5 that new graphics cards are facilitating access to pixel processing.

FIGURE 38.21   A teapot rendered using OpenGL.

```
// a1.cpp : Defines the entry point for the console application.
//

#include "stdafx.h"
#include <stdio.h>
#include <windows.h>
#include "math.h"

#include <gl/gl.h>
#include <gl/glu.h>
#include <gl/glut.h>

#define WINDOW_WIDTH 500
#define WINDOW_HEIGHT 500

#define NOOF_VERTICES 546
#define NOOF_TRIANGLES 1008

float vertices[NOOF_VERTICES][3] = {...};
int triangles[NOOF_TRIANGLES][3] = {...};
float vertexNormals[NOOF_VERTICES][3] = {...};

void loadData() {
  // load data from file
}

// *************************************************

void initLight0(void) {
  GLfloat light0Position[4] = {0.8,0.9,1.0,0.0};
                        // Since the final parameter is 0, this is a
                        // directional light at 'infinity' in the
                        // direction (0.8,0.9,1.0) from the world
                        // origin. A parameter of 1.0 can be used to
                        // create a positional (spot) light, with
                        // further commands to set the direction and
                        // cut-off angles of the spotlight.
  GLfloat whiteLight[4] = {1.0,1.0,1.0,1.0};
  GLfloat ambient[4] = {0.6,0.6,0.6,1.0};
  glLightfv(GL_LIGHT0, GL_POSITION, light0Position);
  glLightfv(GL_LIGHT0, GL_AMBIENT, ambient);
  glLightfv(GL_LIGHT0, GL_DIFFUSE, whiteLight);
  glLightfv(GL_LIGHT0, GL_SPECULAR, whiteLight);
}

void init(void) {
  glClearColor(0, 0, 0, 1);  // Screen will be cleared to black.
  glEnable(GL_DEPTH_TEST);   // Enables screen depth tests using z buffer.
  glEnable(GL_CULL_FACE);    // Enable culling.
  glCullFace(GL_BACK);       // Back-facing polygons will be culled.
                             // Note that by default the vertices of a
                             // polygon are considered to be listed in
                             // counterclockwise order.
  glShadeModel(GL_SMOOTH);   // Use smooth shading, not flat shading.
  glPolygonMode(GL_FRONT_AND_BACK, GL_FILL);
                             // Both 'sides' of a polygon are filled.
```

```
                              // Thus the 'inside' of enclosed objects could
                              // be viewed.

  glEnable(GL_LIGHTING);      // Enables lighting calculations.
  GLfloat ambient[4] = {0.2,0.2,0.2,1.0};
  glLightModelfv(GL_LIGHT_MODEL_AMBIENT, ambient);
                              // Set a global ambient value so that some
                              // 'background' light is included in all
                              // lighting calculations.
  glEnable(GL_LIGHT0);        // Enable GL_LIGHT0 for lighting calculations.
                              // OpenGL defines 8 lights that can be set and
                              // enabled.
  initLight0();               // Call function to initialise GL_LIGHT0 state.
}

void displayObject(void) {
                              // First the material properties of the object
                              // are set for use in lighting calculations
                              // involving the object.
  GLfloat matAmbientDiffuse[4] = {0.9,0.6,0.3,1.0};
                              // A mustard color.
  GLfloat matSpecular[4] = {1.0,1.0,1.0,1.0};
  GLfloat matShininess[1] = {64.0};
  GLfloat noMat[4] = {0.0,0.0,0.0,1.0};
  glMaterialfv(GL_FRONT, GL_AMBIENT_AND_DIFFUSE, matAmbientDiffuse);
                              // Ambient and diffuse can be set separately,
                              // but it is common to use
                              // GL_AMBIENT_AND_DIFFUSE to set them to the
                              // same value.
  glMaterialfv(GL_FRONT, GL_SPECULAR, matSpecular);
  glMaterialfv(GL_FRONT, GL_SHININESS, matShininess);
  glMaterialfv(GL_FRONT, GL_EMISSION, noMat);

  glPushMatrix();
  glRotatef(-90,1,0,0);
  glBegin(GL_TRIANGLES);      // The object is defined using triangles,
                              // rather than GL_QUADS
  for (int i=0; i<NOOF_TRIANGLES; i++)
                              // NOOF_TRIANGLES is a constant but could be a
                              // variable set when the data is read from
                              // file.
// Here we send three vertices for each triangle, although
// there are OpenGL commands to deal with triangle strips
// and with sending a display list of data.
  for (int j=0; j<3; j++) {
                              // The attributes for a vertex are sent before
                              // the vertex itself.
                              // Here we send the vertex normal, but texture
                              // coordinates could also be sent.
    glNormal3-D(vertexNormals[triangles[i][j]][0], // x of vertex normal
                vertexNormals[triangles[i][j]][1], // y of vertex normal
                vertexNormals[triangles[i][j]][2]);// z of vertex normal
    glVertex3-D(vertices[triangles[i][j]][0],      // x of vertex
                vertices[triangles[i][j]][1],      // y of vertex
                vertices[triangles[i][j]][2]);     // z of vertex
  }
```

```
    glEnd();                       // Finish sending triangles
    glPopMatrix();
}

void viewingSystem(int w, int h) {
    glViewport(0, 0, (GLsizei)w,(GLsizei)h);
                               // The viewpoort is set to occupy the whole
                               // screen window display area,
                               // OpenGL maintains two matrices for viewing
                               // and model transformation.
                               // First we deal with the projection matrix.
    glMatrixMode(GL_PROJECTION);
                               // The GL_PROJECTION matrix becomes current.
    glLoadIdentity();          // It is set to the identity matrix.
    gluPerspective(60.0, (GLfloat)w/(GLfloat)h, 1.0, 20.0);
                               // gluPerspective is a routine defined in the
                               // glu utility library that is included in
                               // OpenGL distributions. It creates a symmetric
                               // perspective-view frustum and multiplies the
                               // current (GL_PROJECTION) matrix by it.
                               // Parameter one is the angle of the field of
                               // view. Parameter two is the aspect ratio of
                               // the frustum. Parameter three is the distance
                               // of the near clipping plane from the
                               // viewpoint. Parameter four is the distance of
                               // the far clipping plane from the viewpoint.
                               // Screen coordinate (0,0) will be at the
                               // center of the viewport.

    glMatrixMode(GL_MODELVIEW);
                               // The GL_MODELVIEW matrix is now made current.
    glLoadIdentity();          // It is set to the identity matrix.
    gluLookAt(3.0,5.0,4.0, 0.0,0.0,0.0, 0.0,1.0,0.0);
                               // The utility routine gluLookAt defines a
                               // viewing transformation matrix and
                               // multiples it to the right of the current
                               // (GL_MODELVIEW) matrix.
                               // The eye position is set at (3.0,5.0,4.0).
                               // The next three parameters specify a point
                               // along the desired line of
                               // sight - here we are looking at the origin of
                               // the world coordinate system.
                               // The final three parameters indicate the
                               // direction that is 'up' in the
                               // viewing volume - these values are
                               // automatically adjusted so that the
                               // up direction is perpendicular to the line of
                               // sight.
}

void display(void) {
    glClear(GL_COLOR_BUFFER_BIT|GL_DEPTH_BUFFER_BIT);
                               // Each time the display function is called the
                               // relevant buffers must be cleared.

    displayObject();           // The object (or set of objects) is displayed.
```

```
    glFlush();                       // Forces previously issued OpenGL commands to
                                     // begin execution.
}


// The following main method makes use of the glut utility library to open
// a simple console window for display purposes. glut is a window-system-
// independent toolkit, written by Mark Kilgard, which is commonly used for
// simple OpenGL applications. For further information, see Kilgard, M.,
// "OpenGL Programming for the X Window System," Addison-Wesley, 1996.

int main(int argc, char** argv) {
    loadData();                      // Load data from file into vertices and
                                     // triangles arrays. Could also be used to
                                     // calculate vertex normals which are needed
                                     // for subsequent lighting calculations.
    glutInit(&argc, argv);           // Initializes glut.

    glutInitDisplayMode(GLUT_SINGLE | GLUT_RGBA | GLUT_DEPTH);
                                     // Decide on display modes needed. Here, we
                                     // specify that we want a single-buffered
                                     // window, rgba color mode, and a depth buffer.
    glutInitWindowSize(WINDOW_WIDTH,WINDOW_HEIGHT);
                                     // Size in pixels of the window.
    glutInitWindowPosition(0,0);
                                     // Screen location of upper left corner of
                                     // window.
    glutCreateWindow(argv[0]);// Creates a window with an OpenGL context.
    Init();                          // Initialize our use of OpenGL.
    glutDisplayFunc(display); // The function display will be called whenever
                                     // glut determines that the contents of
                                     // the screen window need to be redisplayed,
                                     // e.g., when a window is uncovered after being
                                     // covered by another window.
    glutReshapeFunc(viewingSystem);
                                     // If the window is resized, e.g., enlarged, the
                                     // viewingSystem function is called.
    glutMainLoop();                  // Last thing that is done. Now the window is
                                     // shown on screen and event processing begins,
                                     // i.e., continuous loop processing events such
                                     // as resize window until program is
                                     // terminated.
    return 0;
}
```

Defining Terms

Ambient: The constant term used in local reflection models to simulate global illumination. It illuminates parts of a surface which cannot be seen from the light source but which are visible from the viewpoint.

Antialiasing: The term given to measures designed to eliminate defects in computer graphics images, the most common being the effect produced when a curved edge in the image is displayed as jagged due to the finite extent of the pixels. The term is somewhat inappropriate, as *aliasing* is a specific signal-processing term and many computer graphics defects are not aliases in that sense.

Clipping: It is common to define a view frustum, a volume emanating from the viewpoint, against which objects are clipped. For example, objects behind the viewer must be eliminated from consideration.

Culling: A process to remove whole polygons that cannot be seen from the viewpoint and therefore do not need to be considered by the hidden-surface removal algorithm.

Diffuse: Local reflection models separately evaluate a diffuse, a specular, and an ambient component. The diffuse component is the light that is reflected from a point equally in all directions, simulating a matte or plastic-like surface.

Flat shading: A shading option in which all the pixels for a polygon are allocated the same shade — there is no variation within a polygon. This makes the underlying polygonal structure visible.

Global reflection models: Models that attempt to model indirect illumination. At a surface point, they consider both light coming directly from light sources and light that has been reflected from other objects.

Gouraud shading: An interpolative shading method for polygons in which the parameter that is interpolated is the reflected light intensity at the vertices.

Hemicube: An algorithmic device that enables an efficient calculation of the form-factor values in the radiosity method. A hemicube is an efficient approximation to a hemisphere.

Hidden-surface removal: The general algorithm that removes hidden surfaces and deals with such cases where one object partially obscures another. In general, a hidden-surface algorithm will eliminate those fragments of a polygon that are not visible because they are behind another object.

Local reflection models: Models that simulate the reflection of light incident directly on an object from a light source. Unlike global models, they take no account of indirect light reflected from another object.

Phong shading: An interpolative shading method in which the parameter that is interpolated is the vertex normal at the vertices. This produces an interpolated normal for every pixel onto which the polygon projects, and a reflection model is evaluated at each pixel in the image plane.

Polygon mesh: The most common form of object representation is to build a set of planar facets or polygons that (approximately, in general) represent the surface of an object.

Progressive refinement: An elaboration of the original radiosity method that enables a visualization of the solution to emerge as the equations are being solved. The solution, originally approximated with an ambient term, is gradually made more and more accurate.

Projective transformation: Usually the final geometric transformation, it produces the perspective foreshortening desired in most applications.

Radiosity: A global reflection model that divides the scene into large elements called patches, calculates a parameter that reflects the geometric relationship between all pairs of patches, and sets up a system of equations whose solution yields a constant radiosity for each patch. These radiosity values are input to a Gouraud-type interpolation to produce a rendered image. The geometric extent of the patches means that the basic method can deal only with diffuse interaction.

Ray tracing: A global reflection model that casts infinitesimally thin rays into the scene from each pixel and follows or traces these as they are reflected and transmitted by objects that they encounter. Such a ray tracer can find out only about specular interaction.

Shading algorithm: A general term that describes that part of the process that makes a geometric description look like a solid, three-dimensional object.

Specular: The specular component is the light reflected from a point in the mirror direction, as if the surface were a perfect mirror. In practice, this component is empirically spread to simulate a practical glossy surface.

Viewing transformation: Generates a representation of the object or scene as seen from the viewpoint of an observer positioned somewhere in the scene and looking toward some aspect of it.

Virtual camera: A common analogy used for the series of geometric transformations that form a two-dimensional image on the view surface. These transformations have the same effect as a (perfect) pinhole camera. The analogy is particularly useful in animation where the camera is to be choreographed.

References

Akenine-Möller, T., and Haines, E. 2002. *Real-time Rendering, 2nd Ed*. A.K. Peters, Ltd.

Blinn, J. 1977. Models of light reflection for computer synthesized pictures. pp. 192–198. *Comput. Graphics (Proc. SIGGRAPH)*.

Cohen, M.F., and Greenberg, D.P. 1985. A radiosity solution for complex environments. pp. 31–40. *Comput. Graphics (Proc. SIGGRAPH)*.

Cohen, M.F., Greenberg, D.P., and Immel, D.S. 1988. An efficient radiosity approach for realistic image synthesis. *IEEE Computer Graphics and Applications,* 6(2): 26–35.

Cohen, M.F., Chen, S.E., Wallace, J.R., and Greenberg, D.P. 1988. A progressive refinement approach to fast radiosity image generation. pp. 75–84. *Comput. Graphics (Proc. SIGGRAPH)*.

Cook, R.L., Porter, T., and Carpenter, L. 1984. Distributed ray tracing. pp. 137–145. *Comput. Graphics (Proc. SIGGRAPH)*.

Fiume, E.L. 1989. *The Mathematical Structure of Computer Graphics*. Academic Press, San Diego, CA.

Gouraud, H. 1971. Illumination for computer generated pictures. *Commun. ACM* 18(60): 628–678.

Newman, W., and Sproull, R. 1973. *Principles of Interactive Computer Graphics*. McGraw-Hill, New York.

Phong, B.T. 1975. Illumination for computer generated pictures. *Commun. ACM* 18(6): 311–317.

Sutherland, I.E., and Hodgman, G.W. 1974. Re-entrant polygon clipping. *Commun. ACM* 17(1): 32–42.

Watt, A., and Policarpo, F. 2001. *3-D Games, Real-time Rendering and Software Technology: Volume 1*. Addison-Wesley, Reading, MA.

Watt, A., and Watt, M. 1992. *Advanced Animation and Rendering Techniques*. ACM Press, New York.

Whitted, T. 1980. An improved illumination model for shaded display. *Commun. ACM* 26(6): 342–349.

Woo, M., Neider, J., Davis, T., and Shreiner, D. 1999. *OpenGL Programming Guide, 3rd Ed*. Addison-Wesley, Reading, MA.

Further Information

The References section comprises mostly the original sources of the algorithms that are commonly incorporated in rendering engines. A would-be implementer, however, would be best directed to a general textbook, such as [Watt and Watt, 1992], or the encyclopedic *Computer Graphics: Principles and Practice* by Foley et al.

Undoubtedly, the best source of rendering techniques and their development is the annual ACM SIGGRAPH conference (proceedings published by ACM Press). Browsing through past proceedings gives a feel for the fascinating development and history of image synthesis. Indeed, in 1998, ACM SIGGRAPH published the book *Seminal Graphics: Pioneering Efforts that Shaped the Field*, edited by Wolfe, which includes many of the pioneering rendering papers listed in the References section.

39

Sampling, Reconstruction, and Antialiasing

39.1 Introduction 39-1
39.2 Sampling Theory 39-2
    Sampling
39.3 Reconstruction 39-4
    Reconstruction Conditions • Ideal Low-Pass Filter • Sinc
    Function • Nonideal Reconstruction
39.4 Reconstruction Kernels 39-8
    Box Filter • Triangle Filter • Cubic Convolution
    • Windowed Sinc Function • Hann and Hamming Windows
    • Blackman Window • Kaiser Window • Lanczos Window
39.5 Aliasing 39-13
39.6 Antialiasing 39-14
    Point Sampling • Area Sampling • Supersampling
    • Adaptive Supersampling
39.7 Prefiltering 39-18
    Pyramids • Summed-Area Tables
39.8 Example: Image Scaling 39-21
39.9 Research Issues and Summary 39-25

George Wolberg
City College of New York

39.1 Introduction

This chapter reviews the principal ideas of sampling theory, reconstruction, and antialiasing. Sampling theory is central to the study of sampled-data systems, e.g., digital image transformations. It lays a firm mathematical foundation for the analysis of sampled signals, offering invaluable insight into the problems and solutions of sampling. It does so by providing an elegant mathematical formulation describing the relationship between a continuous signal and its samples. We use it to resolve the problems of image reconstruction and aliasing. Reconstruction is an interpolation procedure applied to the sampled data. It permits us to evaluate the discrete signal at any desired position, not just the integer lattice upon which the sampled signal is given. This is useful when implementing geometric transformations, or warps, on the image. Aliasing refers to the presence of unreproducibly high frequencies in the image and the resulting artifacts that arise upon undersampling.

Together with defining theoretical limits on the continuous reconstruction of discrete input, sampling theory yields the guidelines for numerically measuring the quality of various proposed filtering techniques.

1-58488-360-X/$0.00+$1.50
© 2004 by CRC Press, LLC

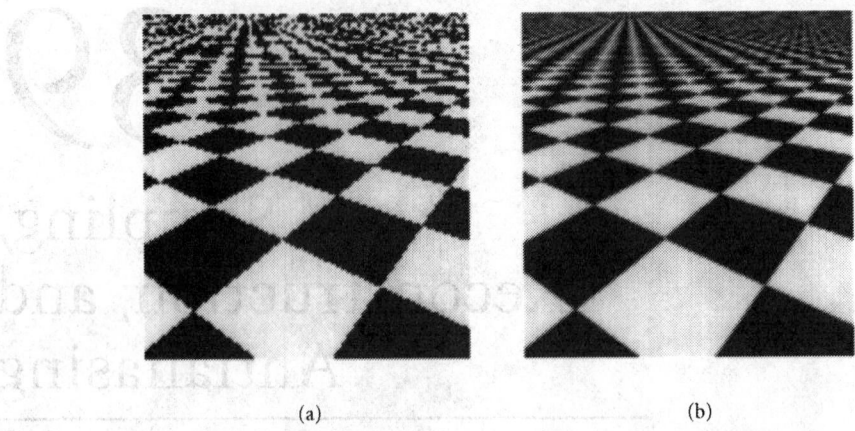

FIGURE 39.1   Oblique checkerboard: (a) unfiltered, (b) filtered.

This proves most useful in formally describing reconstruction, aliasing, and the filtering necessary to combat the artifacts that may appear at the output.

In order to better motivate the importance of sampling theory and filtering, we demonstrate its role with the following examples. A checkerboard texture is shown projected onto an oblique planar surface in Figure 39.1. The image exhibits two forms of artifacts: jagged edges and moiré patterns. Jagged edges are prominent toward the bottom of the image, where the input checkerboard undergoes magnification. It reflects poor reconstruction of the underlying signal. The moiré patterns, on the other hand, are noticeable at the top, where minification (compression) forces many input pixels to occupy fewer output pixels. This artifact is due to aliasing, a symptom of undersampling.

Figure 39.1(a) was generated by projecting the center of each output pixel into the checkerboard and sampling (reading) the value of the nearest input pixel. This point sampling method performs poorly, as is evident by the objectionable results of Figure 39.1(a). This conclusion is reached by sampling theory as well. Its role here is to precisely quantify this phenomenon and to prescribe a solution. Figure 39.1(b) shows the same mapping with improved results. This time the necessary steps were taken to preclude artifacts. In particular, a superior reconstruction algorithm was used for interpolation to suppress the jagged edges, and antialiasing filtering was carried out to combat the symptoms of undersampling that gave rise to the moiré patterns.

39.2   Sampling Theory

Both reconstruction and antialiasing share the twofold problem addressed by sampling theory:

1. Given a continuous input signal $g(x)$ and its sampled counterpart $g_s(x)$, are the samples of $g_s(x)$ sufficient to exactly describe $g(x)$?
2. If so, how can $g(x)$ be reconstructed from $g_s(x)$?

The solution lies in the frequency domain, whereby spectral analysis is used to examine the spectrum of the sampled data.

The conclusions derived from examining the reconstruction problem will prove to be directly useful for resampling and indicative of the filtering necessary for antialiasing. Sampling theory thereby provides an elegant mathematical framework in which to assess the quality of reconstruction, establish theoretical limits, and predict when it is not possible.

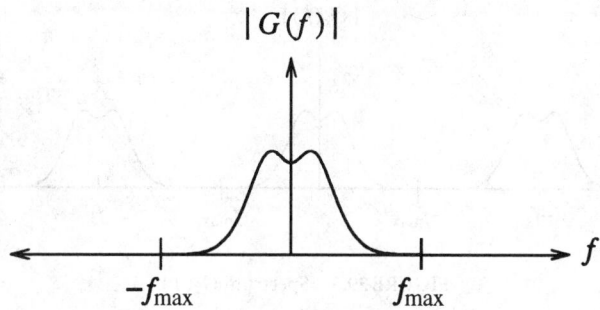

FIGURE 39.2 Spectrum $G(f)$.

39.2.1 Sampling

Consider a 1-D signal $g(x)$ and its spectrum $G(f)$, as determined by the Fourier transform:

$$G(f) = \int_{-\infty}^{\infty} g(x)e^{-i2\pi fx}\,dx \tag{39.1}$$

Note that x represents spatial position and f denotes spatial frequency.

The magnitude spectrum of a signal is shown in Figure 39.2. It shows the frequency content of the signal, with a high concentration of energy in the low-frequency range, tapering off toward the higher frequencies. Since there are no frequency components beyond f_{max}, the signal is said to be **bandlimited** to frequency f_{max}.

The continuous output $g(x)$ is then digitized by an ideal impulse sampler, the comb function, to get the sampled signal $g_s(x)$. The ideal 1-D sampler is given as

$$s(x) = \sum_{n=-\infty}^{\infty} \delta(x - nT_s) \tag{39.2}$$

where δ is the familiar impulse function and T_s is the sampling period. The running index n is used with δ to define the impulse train of the comb function. We now have

$$g_s(x) = g(x)s(x) \tag{39.3}$$

Taking the Fourier transform of $g_s(x)$ yields

$$G_s(f) = G(f) * S(f) \tag{39.4}$$

$$= G(f) * \left[\sum_{n=-\infty}^{n=\infty} f_s \delta(f - nf_s) \right] \tag{39.5}$$

$$= f_s \sum_{n=-\infty}^{n=\infty} G(f - nf_s) \tag{39.6}$$

where f_s is the sampling frequency and $*$ denotes convolution. The above equations make use of the following well-known properties of Fourier transforms:

1. Multiplication in the spatial domain corresponds to convolution in the frequency domain. Therefore, Equation 39.3 gives rise to a convolution in Equation 39.4.
2. The Fourier transform of an impulse train is itself an impulse train, giving us Equation 39.5.
3. The spectrum of a signal sampled with frequency f_s ($T_s = 1/f_s$) yields the original spectrum replicated in the frequency domain with period f_s Equation 39.6.

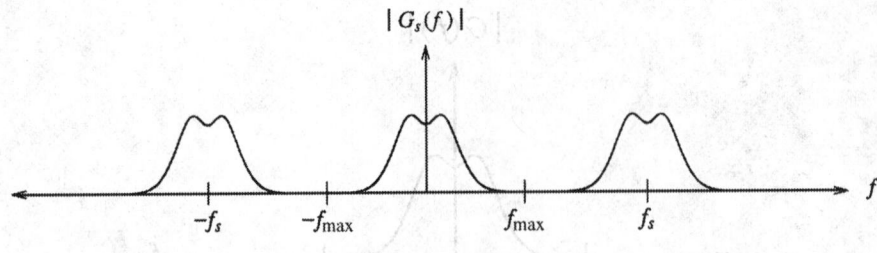

FIGURE 39.3   Spectrum $G_s(f)$.

This last property has important consequences. It yields a spectrum $G_s(f)$, which, in response to a sampling period $T_s = 1/f_s$, is *periodic in frequency* with period f_s. This is depicted in Figure 39.3. Notice, then, that a small sampling period is equivalent to a high sampling frequency, yielding spectra replicated far apart from each other. In the limiting case when the sampling period approaches zero ($T_s \to 0$, $f_s \to \infty$), only a single spectrum appears — a result consistent with the continuous case.

39.3   Reconstruction

The above result reveals that the sampling operation has left the original input spectrum *intact*, merely replicating it periodically in the frequency domain with a spacing of f_s. This allows us to rewrite $G_s(f)$ as a sum of two terms, the low-frequency (baseband) and high-frequency components. The *baseband* spectrum is exactly $G(f)$, and the high-frequency components, $G_{high}(f)$, consist of the remaining replicated versions of $G(f)$ that constitute harmonic versions of the sampled image:

$$G_s(f) = G(f) + G_{high}(f) \tag{39.7}$$

Exact signal reconstruction from sampled data requires us to discard the replicated spectra $G_{high}(f)$, leaving only $G(f)$, the spectrum of the signal we seek to recover. This is a crucial observation in the study of sampled-data systems.

39.3.1   Reconstruction Conditions

The only provision for exact **reconstruction** is that $G(f)$ be undistorted due to overlap with $G_{high}(f)$. Two conditions must hold for this to be true:

1. The signal must be bandlimited. This avoids spectra with infinite extent that are impossible to replicate without overlap.
2. The sampling frequency f_s must be greater than twice the maximum frequency f_{max} present in the signal. This minimum sampling frequency, known as the **Nyquist rate**, is the minimum distance between the spectra copies, each with bandwidth f_{max}.

The first condition merely ensures that a sufficiently large sampling frequency exists that can be used to separate replicated spectra from each other. Since all imaging systems impose a bandlimiting filter in the form of a point spread function, this condition is always satisfied for images captured through an optical system. Note that this does not apply to synthetic images, e.g., computer-generated imagery.

The second condition proves to be the most revealing statement about reconstruction. It answers the problem regarding the sufficiency of the data samples to exactly reconstruct the continuous input signal. It states that exact reconstruction is possible only when $f_s > f_{Nyquist}$, where $f_{Nyquist} = 2f_{max}$. Collectively, these two conclusions about reconstruction form the central message of sampling theory, as pioneered by Claude Shannon in his landmark papers on the subject [Shannon 1948, 1949].

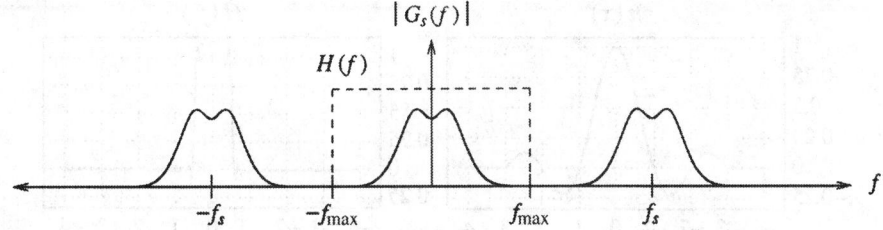

FIGURE 39.4   Ideal low-pass filter $H(f)$.

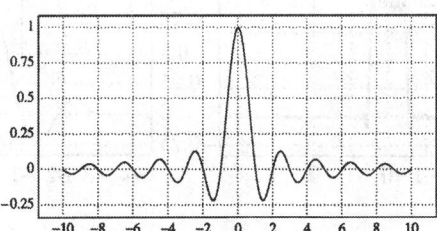

FIGURE 39.5   The sinc function.

39.3.2   Ideal Low-Pass Filter

We now turn to the second central problem: Given that it is theoretically possible to perform reconstruction, how may it be done? The answer lies with our earlier observation that sampling merely replicates the spectrum of the input signal, generating $G_{high}(f)$ in addition to $G(f)$. Therefore, the act of reconstruction requires us to completely suppress $G_{high}(f)$. This is done by multiplying $G_s(f)$ with $H(f)$, given as

$$H(f) = \begin{cases} 1, & |f| < f_{max} \\ 0, & |f| \geq f_{max} \end{cases} \tag{39.8}$$

$H(f)$ is known as an *ideal low-pass filter* and is depicted in Figure 39.4, where it is shown suppressing all frequency components above f_{max}. This serves to discard the replicated spectra $G_{high}(f)$. It is ideal in the sense that the f_{max} cutoff frequency is strictly enforced as the transition point between the transmission and complete suppression of frequency components.

39.3.3   Sinc Function

In the spatial domain, the ideal low-pass filter is derived by computing the inverse Fourier transform of $H(f)$. This yields the sinc function shown in Figure 39.5. It is defined as

$$\text{sinc}(x) = \frac{\sin(\pi x)}{\pi x} \tag{39.9}$$

The reader should note the reciprocal relationship between the height and width of the ideal low-pass filter in the spatial and frequency domains. Let A denote the amplitude of the sinc function, and let its zero crossings be positioned at integer multiples of $1/2W$. The spectrum of this sinc function is a rectangular pulse of height $A/2W$ and width $2W$, with frequencies ranging from $-W$ to W. In our example above, $A = 1$ and $W = f_{max} = 0.5$ cycles/pixel. This value for W is derived from the fact that digital images must not have more than one half cycle per pixel in order to conform to the Nyquist rate.

The sinc function is one instance of a large class of functions known as cardinal splines, which are interpolating functions defined to pass through zero at all but one data sample, where they have a value

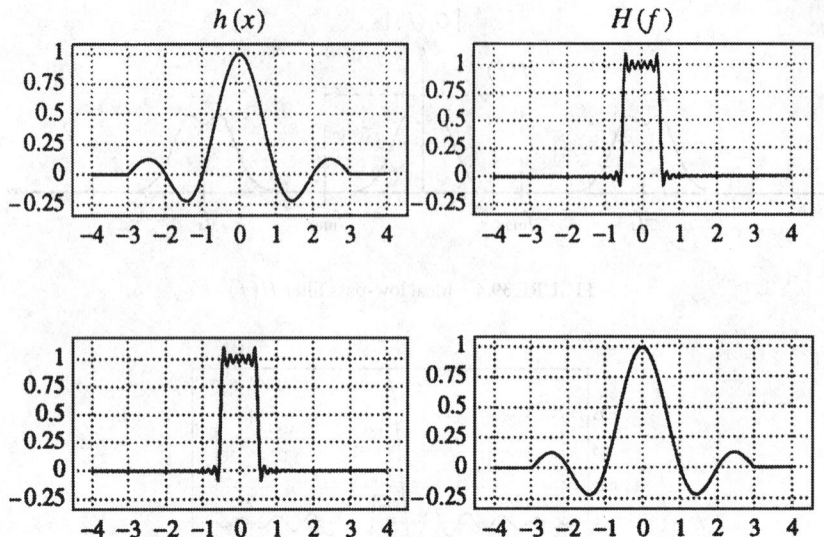

FIGURE 39.6   Truncation in one domain causes ringing in the other domain.

of one. This allows them to compute a continuous function that passes through the uniformly spaced data samples.

Since multiplication in the frequency domain is identical to convolution in the spatial domain, sinc (x) represents the convolution kernel used to evaluate any point x on the continuous input curve g given only the sampled data g_s:

$$g(x) = \text{sinc}(x) * g_s(x) = \int_{-\infty}^{\infty} \text{sinc}(\lambda)\, g_s(x - \lambda)\, d\lambda \qquad (39.10)$$

Equation 39.10 highlights an important impediment to the practical use of the ideal low-pass filter. The filter requires an infinite number of neighboring samples (i.e., an infinite filter support) in order to precisely compute the output points. This is, of course, impossible owing to the finite number of data samples available. However, truncating the sinc function allows for approximate solutions to be computed at the expense of undesirable "ringing," i.e., ripple effects. These artifacts, known as the **Gibbs phenomenon**, are the overshoots and undershoots caused by reconstructing a signal with truncated frequency terms. The two rows in Figure 39.6 show that truncation in one domain leads to ringing in the other domain. This indicates that a truncated sinc function is actually a poor reconstruction filter because its spectrum has infinite extent and thereby fails to bandlimit the input.

In response to these difficulties, a number of approximating algorithms have been derived, offering a tradeoff between precision and computational expense. These methods permit local solutions that require the convolution kernel to extend only over a small neighborhood. The drawback, however, is that the frequency response of the filter has some undesirable properties. In particular, frequencies below $f_{\max}$ are tampered, and high frequencies beyond $f_{\max}$ are not fully suppressed. Thus, nonideal reconstruction does not permit us to exactly recover the continuous underlying signal without artifacts.

39.3.4   Nonideal Reconstruction

The process of nonideal reconstruction is depicted in Figure 39.7, which indicates that the input signal satisfies the two conditions necessary for exact reconstruction. First, the signal is bandlimited, since the replicated copies in the spectrum are each finite in extent. Second, the sampling frequency exceeds the Nyquist rate, since the copies do not overlap. However, this is where our ideal scenario ends. Instead of

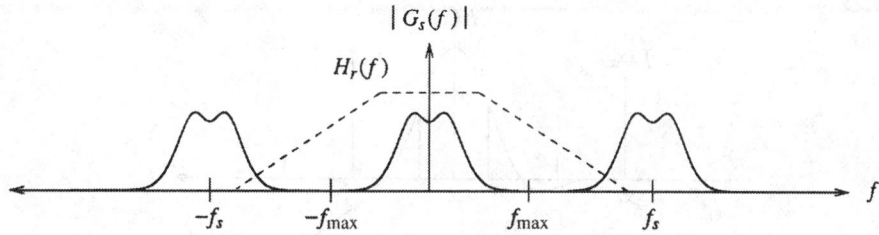

FIGURE 39.7 Nonideal reconstruction.

using an ideal low-pass filter to retain only the baseband spectrum components, a nonideal reconstruction filter is shown in the figure.

The filter response $H_r(f)$ deviates from the ideal response $H(f)$ shown in Figure 39.4. In particular, $H_r(f)$ does not discard all frequencies beyond f_{max}. Furthermore, that same filter is shown to attenuate some frequencies that should have remained intact. This brings us to the problem of assessing the quality of a filter.

The accuracy of a reconstruction filter can be evaluated by analyzing its frequency-domain characteristics. Of particular importance is the filter response in the passband and stopband. In this problem, the **passband** consists of all frequencies below f_{max}. The **stopband** contains all higher frequencies arising from the sampling process.

An ideal reconstruction filter, as described earlier, will completely suppress the stopband while leaving the passband intact. Recall that the stopband contains the offending high frequencies that, if allowed to remain, would prevent us from performing exact reconstruction. As a result, the sinc filter was devised to meet these goals and serve as the ideal reconstruction filter. Its kernel in the frequency domain applies unity gain to transmit the passband and zero gain to suppress the stopband.

The breakdown of the frequency domain into passband and stopband isolates two problems that can arise due to nonideal reconstruction filters. The first problem deals with the effects of imperfect filtering on the passband. Failure to impose unity gain on *all* frequencies in the passband will result in some combination of image smoothing and image sharpening. Smoothing, or blurring, will result when the frequency gains near the cutoff frequency start falling off. Image sharpening results when the high-frequency gains are allowed to exceed unity. This follows from the direct correspondence of visual detail to spatial frequency. Furthermore, amplifying the high passband frequencies yields a sharper transition between the passband and stopband, a property shared by the sinc function.

The second problem addresses nonideal filtering on the stopband. If the stopband is allowed to persist, high frequencies will exist that will contribute to aliasing (described later). Failure to fully suppress the stopband is a condition known as **frequency leakage**. This allows the offending frequencies to fold over into the passband range. These distortions tend to be more serious, since they are visually perceived more readily.

In the spatial domain, nonideal reconstruction is achieved by centering a finite-width kernel at the position in the data at which the underlying function is to be evaluated, i.e., reconstructed. This is an interpolation problem which, for equally spaced data, can be expressed as

$$f(x) = \sum_{k=0}^{K-1} f(x_k)h(x - x_k) \qquad (39.11)$$

where h is the reconstruction kernel that weighs K data samples at x_k. Equation 39.11 formulates interpolation as a convolution operation. In practice, h is nearly always a symmetric kernel, i.e., $h(-x) = h(x)$. We shall assume this to be true in the discussion that follows.

The computation of one interpolated point is illustrated in Figure 39.8. The kernel is centered at x, the location of the point to be interpolated. The value of that point is equal to the sum of the values of the

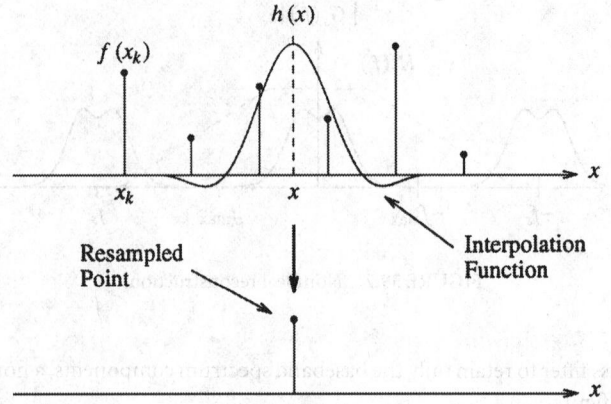

FIGURE 39.8   Interpolation of a single point.

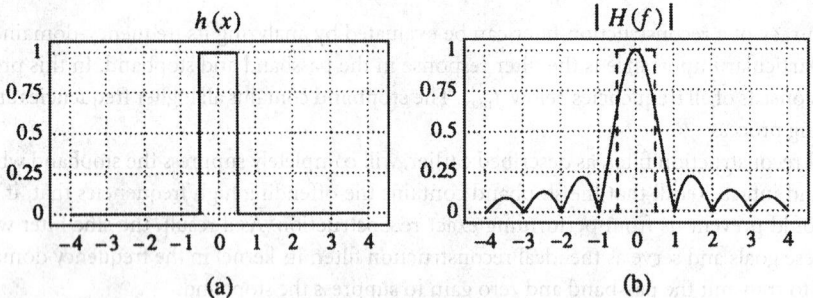

FIGURE 39.9   Box filter: (a) kernel, (b) Fourier transform.

discrete input scaled by the corresponding values of the reconstruction kernel. This follows directly from the definition of convolution.

39.4   Reconstruction Kernels

The numerical accuracy and computational cost of reconstruction are directly tied to the convolution kernel used for low-pass filtering. As a result, filter kernels are the target of design and analysis in the creation and evaluation of reconstruction algorithms. They are subject to conditions influencing the tradeoff between accuracy and efficiency. This section reviews several common nonideal reconstruction filter kernels in the order of their complexity: box filter, triangle filter, cubic convolution, and windowed sinc functions.

39.4.1   Box Filter

The box filter kernel is defined as

$$h(x) = \begin{cases} 1, & -0.5 < x \le 0.5 \\ 0, & \text{otherwise} \end{cases} \tag{39.12}$$

Various other names are used to denote this simple kernel, including the sample-and-hold function and Fourier window. The kernel and its Fourier transform are shown in Figure 39.9.

Convolution in the spatial domain with the rectangle function h is equivalent in the frequency domain to multiplication with a sinc function. Due to the prominent side lobes and infinite extent, a sinc function

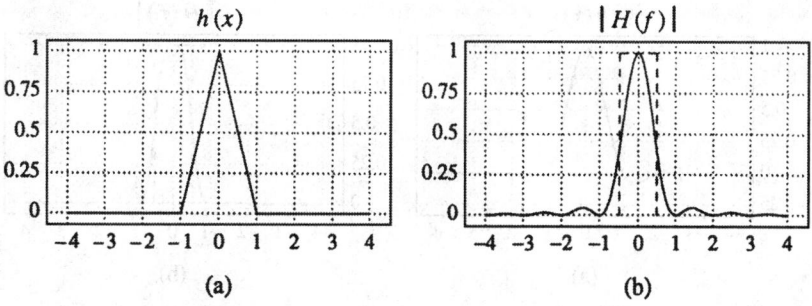

FIGURE 39.10 Triangle filter: (a) kernel, (b) Fourier transform.

makes a poor low-pass filter. Consequently, this filter kernel has a poor frequency-domain response relative to that of the ideal low-pass filter. The ideal filter, drawn as a dashed rectangle, is characterized by unity gain in the passband and zero gain in the stopband. This permits all low frequencies (below the cutoff frequency) to pass and all higher frequencies to be suppressed.

39.4.2 Triangle Filter

The triangle filter kernel is defined as

$$h(x) = \begin{cases} 1 - |x|, & 0 \le |x| < 1 \\ 0, & 1 \le |x| \end{cases} \tag{39.13}$$

The kernel h is also referred to as a tent filter, roof function, Chateau function, or Bartlett window.

This kernel corresponds to a reasonably good low-pass filter in the frequency domain. As shown in Figure 39.10, its response is superior to that of the box filter. In particular, the side lobes are far less prominent, indicating improved performance in the stopband. Nevertheless, a significant amount of spurious high-frequency components continues to leak into the passband, contributing to some aliasing. In addition, the passband is moderately attenuated, resulting in image smoothing.

39.4.3 Cubic Convolution

The cubic convolution kernel is a third-degree approximation to the sinc function. It is symmetric, space-invariant, and composed of piecewise cubic polynomials:

$$h(x) = \begin{cases} (a+2)|x|^3 - (a+3)|x|^2 + 1, & 0 \le |x| < 1 \\ a|x|^3 - 5a|x|^2 + 8a|x| - 4a, & 1 \le |x| < 2 \\ 0, & 2 \le |x| \end{cases} \tag{39.14}$$

where $-3 < a < 0$ is used to make h resemble the sinc function.

Of all the choices for a, the value -1 is preferable if visually enhanced results are desired. That is, the image is sharpened, making visual detail perceived more readily. However, the results are not mathematically precise, where precision is measured by the order of the Taylor series. To maximize this order, the value $a = -0.5$ is preferable. A cubic convolution kernel with $a = -0.5$ and its spectrum are shown in Figure 39.11.

39.4.4 Windowed Sinc Function

Sampling theory establishes that the sinc function is the ideal interpolation kernel. Although this interpolation filter is exact, it is not practical, since it is an infinite impulse response (IIR) filter defined by a slowly converging infinite sum. Nevertheless, it is perfectly reasonable to consider the effects of using a truncated, and therefore finite, sinc function as the interpolation kernel.

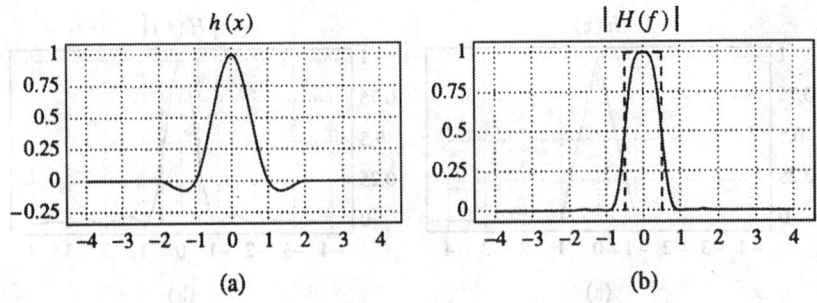

FIGURE 39.11   Cubic convolution: (a) kernel ($a = -0.5$), (b) Fourier transform.

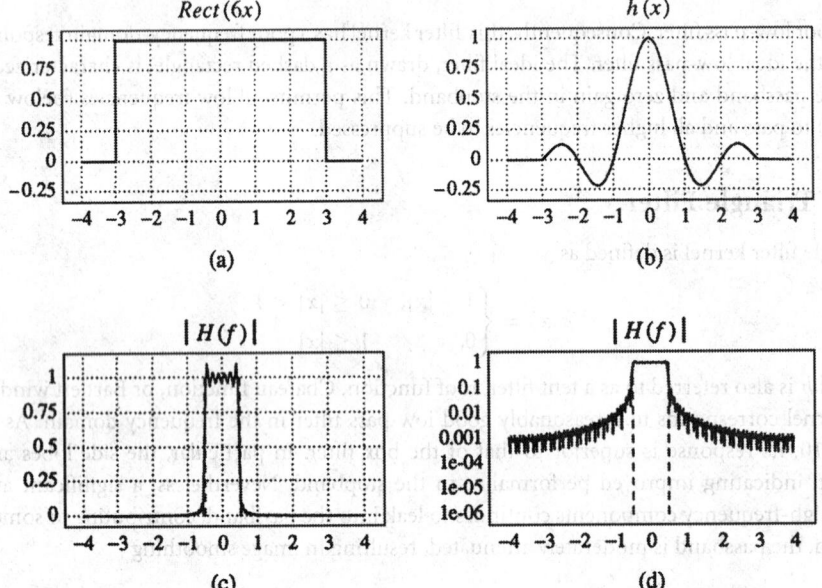

FIGURE 39.12   (a) Rectangular window; (b) windowed sinc; (c) spectrum; (d) log plot.

The results of this operation are predicted by sampling theory, which demonstrates that truncation in one domain leads to ringing in the other domain. This is due to the fact that truncating a signal is equivalent to multiplying it with a rectangle function Rect(x), defined as the box filter of Equation 39.12. Since multiplication in one domain is convolution in the other, truncation amounts to convolving the signal's spectrum with a sinc function, the transform pair of Rect(x). Since the stopband is no longer eliminated, but rather attenuated by a ringing filter (i.e., a sinc), the input is not bandlimited and aliasing artifacts are introduced. The most typical problems occur at step edges, where the Gibbs phenomena becomes noticeable in the form of undershoots, overshoots, and ringing in the vicinity of edges.

The Rect function above served as a window, or kernel, that weighs the input signal. In Figure 39.12(a), we see the *Rect* window extended over three pixels on each side of its center, i.e., Rect($6x$) is plotted. The corresponding windowed sinc function $h(x)$ is shown in Figure 39.12(b). This is simply the product of the sinc function with the window function, i.e., sinc (x) Rect($6x$). Its spectrum, shown in Figure 39.12(c), is nearly an ideal low-pass filter. Although it has a fairly sharp transition from the passband to the stopband, it is plagued by ringing. In order to more clearly see the values in the spectrum, we use a logarithmic scale for the vertical axis of the spectrum in Figure 39.12(d). The next few figures will use this same four-part format.

Ringing can be mitigated by using a different windowing function exhibiting smoother falloff than the rectangle. The resulting windowed sinc function can yield better results. However, since slow falloff requires larger windows, the computation remains costly.

Aside from the rectangular window mentioned above, the most frequently used window functions are Hann, Hamming, Blackman, and Kaiser. These filters identify a quantity known as the ripple ratio, defined as the ratio of the maximum side-lobe amplitude to the main-lobe amplitude. Good filters will have small ripple ratios to achieve effective attenuation in the stopband. A tradeoff exists, however, between ripple ratio and main-lobe width. Therefore, as the ripple ratio is decreased, the main-lobe width is increased. This is consistent with the reciprocal relationship between the spatial and frequency domains, i.e., narrow bandwidths correspond to wide spatial functions.

In general, though, each of these smooth window functions is defined over a small finite extent. This is tantamount to multiplying the smooth window with a rectangle function. While this is better than the Rect function alone, there will inevitably be some form of aliasing. Nevertheless, the window functions described below offer a good compromise between ringing and blurring.

39.4.5 Hann and Hamming Windows

The Hann and Hamming windows are defined as

$$
\text{Hann/Hamming}(x) = \begin{cases} \alpha + (1 - \alpha) \cos \frac{2\pi x}{N-1}, & |x| < \frac{N-1}{2} \\ 0, & \text{otherwise} \end{cases} \tag{39.15}
$$

where N is the number of samples in the windowing function. The two windowing functions differ in their choice of α. In the Hann window $\alpha = 0.5$, and in the Hamming window $\alpha = 0.54$. Since they both amount to a scaled and shifted cosine function, they are also known as the raised cosine window. The Hann window is illustrated in Figure 39.13. Notice that the passband is only slightly attenuated, but the stopband continues to retain high-frequency components in the stopband, albeit less than that of $\text{Rect}(x)$.

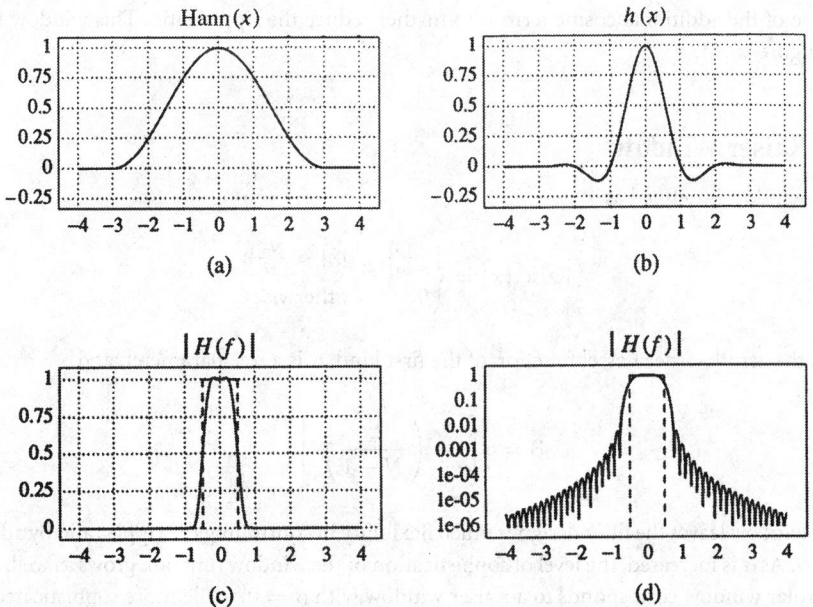

FIGURE 39.13 (a) Hann window; (b) windowed sinc; (c) spectrum; (d) log plot.

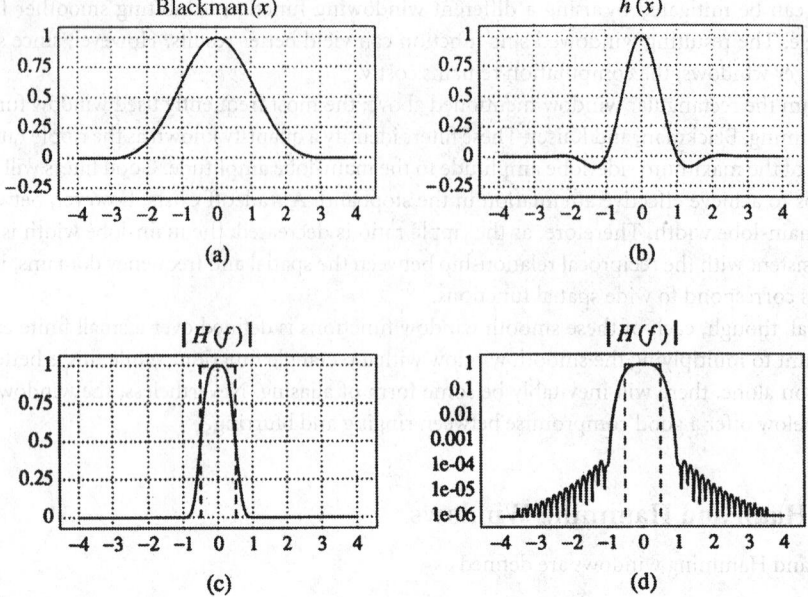

FIGURE 39.14   (a) Blackman window; (b) windowed sinc; (c) spectrum; (d) log plot.

39.4.6 Blackman Window

The Blackman window is similar to the Hann and Hamming windows. It is defined as

$$\text{Blackman}(x) = \begin{cases} 0.42 + 0.5 \cos \frac{2\pi x}{N-1} + 0.08 \cos \frac{4\pi x}{N-1}, & |x| < \frac{N-1}{2} \\ 0, & \text{otherwise} \end{cases} \tag{39.16}$$

The purpose of the additional cosine term is to further reduce the ripple ratio. This window function is shown in Figure 39.14.

39.4.7 Kaiser Window

The Kaiser window is defined as

$$\text{Kaiser}(x) = \begin{cases} \frac{I_0(\beta)}{I_0(\alpha)}, & |x| < \frac{N-1}{2} \\ 0, & \text{otherwise} \end{cases} \tag{39.17}$$

where I_0 is the zeroth-order Bessel function of the first kind, α is a free parameter, and

$$\beta = \alpha \left[1 - \left(\frac{2x}{N-1} \right)^2 \right]^{1/2} \tag{39.18}$$

The Kaiser window leaves the filter designer much flexibility in controlling the ripple ratio by adjusting the parameter α. As α is increased, the level of sophistication of the window function grows as well. Therefore, the rectangular window corresponds to a Kaiser window with $\alpha = 0$, while more sophisticated windows such as the Hamming window correspond to $\alpha = 5$.

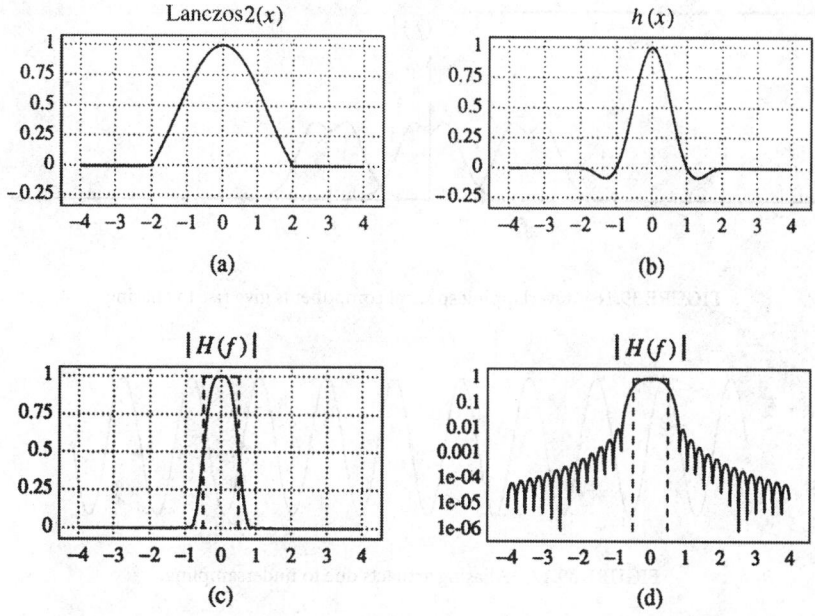

FIGURE 39.15 (a) Lanczos2 window; (b) windowed sinc; (c) spectrum; (d) log plot.

39.4.8 Lanczos Window

The Lanczos window is based on the sinc function rather than cosines as used in the previous methods. The two-lobed Lanczos window function is defined as

$$\text{Lanczos2}(x) = \begin{cases} \frac{\sin(\pi x/2)}{\pi x/2}, & 0 \le |x| < 2 \\ 0, & 2 \le |x| \end{cases} \tag{39.19}$$

The Lanczos2 window function, shown in Figure 39.15, is the central lobe of a sinc function. It is wide enough to extend over two lobes of the ideal low-pass filter, i.e., a second sinc function. This formulation can be generalized to an N-lobed window function by replacing the value 2 in Equation 39.19 with the value N. Larger N results in superior frequency response.

39.5 Aliasing

If the two reconstruction conditions outlined earlier are not met, sampling theory predicts that exact reconstruction is *not* possible. This phenomenon, known as **aliasing**, occurs when signals are not bandlimited or when they are undersampled, i.e., $f_s \le f_{\text{Nyquist}}$. In either case there will be unavoidable overlapping of spectral components, as in Figure 39.16. Notice that the irreproducible high frequencies fold over into the low frequency range. As a result, frequencies originally beyond f_{max} will, upon reconstruction, appear in the form of much *lower* frequencies. In comparison with the spurious high frequencies retained by nonideal reconstruction filters, the spectral components passed due to undersampling are more serious, since they actually corrupt the components in the original signal.

Aliasing refers to the higher frequencies becoming aliased and indistinguishable from the lower-frequency components in the signal if the sampling rate falls below the Nyquist frequency. In other words, undersampling causes high-frequency components to appear as spurious low frequencies. This is depicted in Figure 39.17, where a high-frequency signal appears as a low-frequency signal after sampling it too sparsely. In digital images, the Nyquist rate is determined by the highest frequency that can be

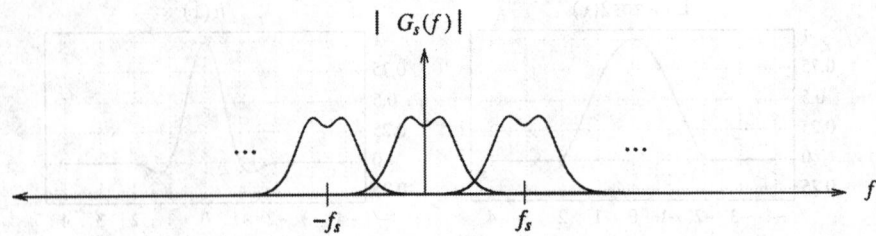

FIGURE 39.16 Overlapping spectral components give rise to aliasing.

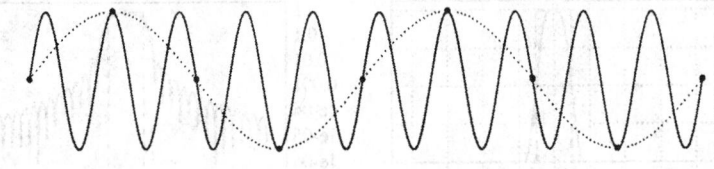

FIGURE 39.17 Aliasing artifacts due to undersampling.

displayed: one cycle every two pixels. Therefore, any attempt to display higher frequencies will produce similar artifacts.

There is sometimes a misconception in the computer graphics literature that jagged (staircased) edges are always a symptom of aliasing. This is only partially true. Technically, jagged edges arise from high frequencies introduced by inadequate reconstruction. Since these high frequencies are not corrupting the low-frequency components, no aliasing is actually taking place. The confusion lies in that the suggested remedy of increasing the sampling rate is also used to eliminate aliasing. Of course, the benefit of increasing the sampling rate is that the replicated spectra are now spaced farther apart from each other. This relaxes the accuracy constraints for reconstruction filters to perform ideally in the stopband, where they must suppress all components beyond some specified cutoff frequency. In this manner, the same nonideal filters will produce less objectionable output.

It is important to note that a signal may be densely sampled (far above the Nyquist rate) and continue to appear jagged if a zero-order reconstruction filter is used. Box filters used for pixel replication in real-time hardware zooms are a common example of poor reconstruction filters. In this case, the signal is clearly not aliased but rather poorly reconstructed. The distinction between reconstruction and aliasing artifacts becomes clear when we notice that the appearance of jagged edges is improved by blurring. For example, it is not uncommon to step back from an image exhibiting excessive blockiness in order to see it more clearly. This is a defocusing operation that attenuates the high frequencies admitted through nonideal reconstruction. On the other hand, once a signal is truly undersampled, there is no postprocessing possible to improve its condition. After all, applying an ideal low-pass (reconstruction) filter to a spectrum whose components are already overlapping will only blur the result, not rectify it.

39.6 Antialiasing

The filtering necessary to combat aliasing is known as **antialiasing**. In order to determine corrective action, we must directly address the two conditions necessary for exact signal reconstruction. The first solution calls for low-pass filtering *before* sampling. This method, known as **prefiltering**, bandlimits the signal to levels below f_{max}, thereby eliminating the offending high frequencies. Notice that the frequency at which the signal is to be sampled imposes limits on the allowable bandwidth. This is often necessary when the output sampling grid must be fixed to the resolution of an output device, e.g., screen resolution. Therefore, aliasing is often a problem that is confronted when a signal is forced to conform to an inadequate resolution due to physical constraints. As a result, it is necessary to bandlimit, or narrow, the input spectrum to conform to the allotted bandwidth as determined by the sampling frequency.

The second solution is to point-sample at a higher frequency. In doing so, the replicated spectra are spaced farther apart, thereby separating the overlapping spectra tails. This approach theoretically implies sampling at a resolution determined by the highest frequencies present in the signal. Since a surface viewed obliquely can give rise to arbitrarily high frequencies, this method may require extremely high resolution. Whereas the first solution adjusts the bandwidth to accommodate the fixed sampling rate f_s, the second solution adjusts f_s to accommodate the original bandwidth. Antialiasing by sampling at the highest frequency is clearly superior in terms of image quality. This is, of course, operating under different assumptions regarding the possibility of varying f_s. In practice, antialiasing is performed through a combination of these two approaches. That is, the sampling frequency is increased so as to reduce the amount of bandlimiting to a minimum.

39.6.1 Point Sampling

The naive approach for generating an output image is to perform **point sampling**, where each output pixel is a single sample of the input image taken independently of its neighbors (Figure 39.18). It is clear that information is lost between the samples and that aliasing artifacts may surface if the sampling density is not sufficiently high to characterize the input. This problem is rooted in the fact that intermediate intervals between samples, which should have some influence on the output, are skipped entirely.

The Star image is a convenient example that overwhelms most resampling filters due to the infinitely high frequencies found toward the center. Nevertheless, the extent of the artifacts is related to the quality of the filter and the actual spatial transformation. Figure 39.19 shows two examples of the moiré effects that can appear when a signal is undersampled using point sampling. In Figure 39.19(a), one out of every two pixels in the Star image was discarded to reduce its dimension. In Figure 39.19(b), the artifacts of undersampling are more pronounced, as only one out of every four pixels is retained. In order to see the small images more clearly, they are magnified using cubic spline reconstruction. Clearly, these examples show that point sampling behaves poorly in high-frequency regions.

Aliasing can be reduced by point sampling at a higher resolution. This raises the Nyquist limit, accounting for signals with higher bandwidths. Generally, though, the display resolution places a limit on the highest frequency that can be displayed, and thus limits the Nyquist rate to one cycle every two pixels. Any attempt to display higher frequencies will produce aliasing artifacts such as moiré patterns and jagged edges. Consequently, antialiasing algorithms have been derived to bandlimit the input *before* resampling onto the output grid.

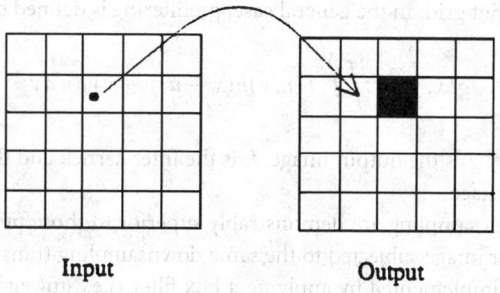

Input   **Output**

FIGURE 39.18 Point sampling.

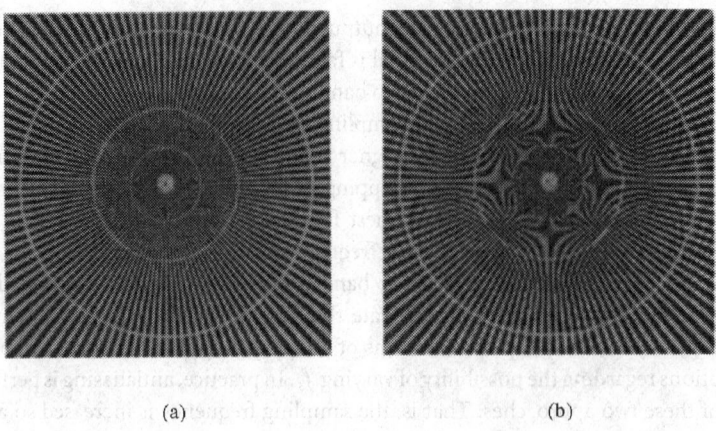

(a)   (b)

FIGURE 39.19 Aliasing due to point sampling: (a) 1/2 and (b) 1/4 scale.

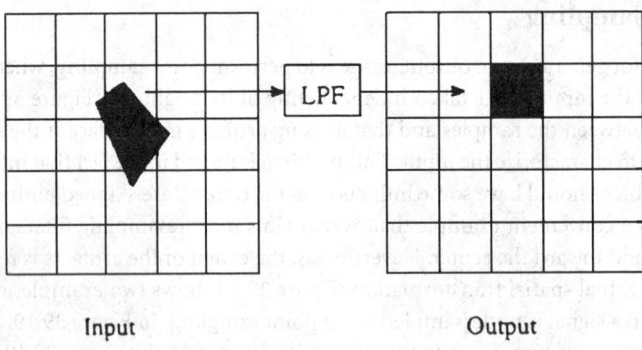

Input   Output

FIGURE 39.20 Area sampling.

39.6.2 Area Sampling

The basic flaw in point sampling is that a discrete pixel actually represents an area, not a point. In this manner, each output pixel should be considered a window looking onto the input image. Rather than sampling a point, we must instead apply a low-pass filter (LPF) upon the projected area in order to properly reflect the information content being mapped onto the output pixel. This approach, depicted in Figure 39.20, is called **area sampling**, and the projected area is known as the **preimage**. The low-pass filter comprises the prefiltering stage. It serves to defeat aliasing by bandlimiting the input image prior to resampling it onto the output grid. In the general case, prefiltering is defined by the convolution integral

$$g(x, y) = \iint f(u, v) h(x - u, y - v) \, du \, dv \qquad (39.20)$$

where f is the input image, g is the output image, h is the filter kernel, and the integration is applied to all $[u, v]$ points in the preimage.

Images produced by area sampling are demonstrably superior to those produced by point sampling. Figure 39.21 shows the Star image subjected to the same downsampling transformation as that in Figure 39.19. Area sampling was implemented by applying a box filter (i.e., unweighted averaging) to the Star image before point sampling. Notice that antialiasing through area sampling has traded moiré patterns for some blurring. Although there is no substitute for high-resolution imagery, filtering can make lower resolution less objectionable by attenuating aliasing artifacts.

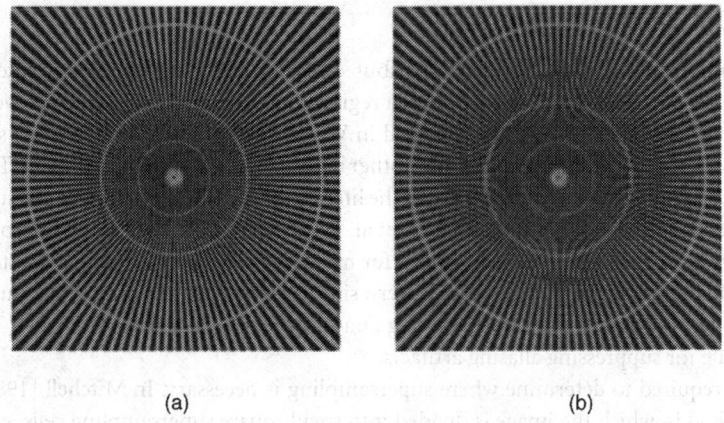

FIGURE 39.21 Aliasing due to area sampling: (a) 1/2 and (b) 1/4 scale.

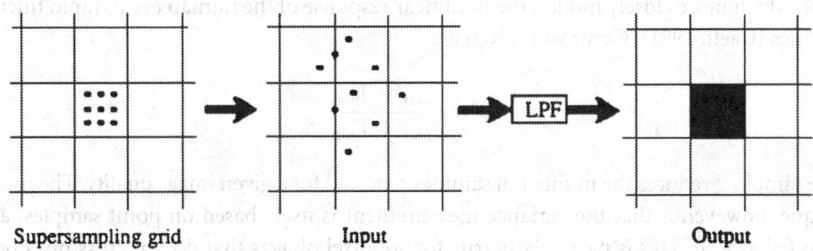

Supersampling grid   Input   Output

FIGURE 39.22 Supersampling.

39.6.3 Supersampling

The process of using more than one regularly spaced sample per pixel is known as **supersampling**. Each output pixel value is evaluated by computing a weighted average of the samples taken from their respective preimages. For example, if the supersampling grid is three times denser than the output grid (i.e., there are nine grid points per pixel area), each output pixel will be an average of the nine samples taken from its projection in the input image. If, say, three samples hit a green object and the remaining six samples hit a blue object, the composite color in the output pixel will be one-third green and two-thirds blue, assuming a box filter is used.

Supersampling reduces aliasing by bandlimiting the input signal. The purpose of the high-resolution supersampling grid is to refine the estimate of the preimages seen by the output pixels. The samples then enter the prefiltering stage, consisting of a low-pass filter. This permits the input to be resampled onto the (relatively) low-resolution output grid without any offending high frequencies introducing aliasing artifacts. In Figure 39.22 we see an output pixel subdivided into nine subpixel samples which each undergo inverse mapping, sampling the input at nine positions. Those nine values then pass through a low-pass filter to be averaged into a single output value.

Supersampling was used to achieve antialiasing in Figure 39.1 for pixels near the horizon. There are two problems, however, associated with straightforward supersampling. The first problem is that the newly designated high frequency of the prefiltered image continues to be fixed. Therefore, there will always be sufficiently higher frequencies that will alias. The second problem is cost. In our example, supersampling will take nine times longer than point sampling. Although there is a clear need for the additional computation, the dense placement of samples can be optimized. Adaptive supersampling is introduced to address these drawbacks.

39.6.4 Adaptive Supersampling

In **adaptive supersampling**, the samples are distributed more densely in areas of high intensity variance. In this manner, supersamples are collected only in regions that warrant their use. Early work in adaptive supersampling for computer graphics is described in Whitted [1980]. The strategy is to subdivide areas between previous samples when an edge, or some other high-frequency pattern, is present. Two approaches to adaptive supersampling have been described in the literature. The first approach allows sampling density to vary as a function of local image variance [Lee et al. 1985, Kajiya 1986]. A second approach introduces two levels of sampling densities: a regular pattern for most areas and a higher-density pattern for regions demonstrating high frequencies. The regular pattern simply consists of one sample per output pixel. The high density pattern involves local supersampling at a rate of 4 to 16 samples per pixel. Typically, these rates are adequate for suppressing aliasing artifacts.

A strategy is required to determine where supersampling is necessary. In Mitchell [1987], the author describes a method in which the image is divided into small square supersampling cells, each containing eight or nine of the low-density samples. The entire cell is supersampled if its samples exhibit excessive variation. In Lee et al. [1985], the variance of the samples is used to indicate high frequency. It is well known, however, that variance is a poor measure of visual perception of local variation. Another alternative is to use contrast, which more closely models the nonlinear response of the human eye to rapid fluctuations in light intensities [Caelli 1981]. Contrast is given as

$$C = \frac{I_{\max} - I_{\min}}{I_{\max} + I_{\min}} \tag{39.21}$$

Adaptive sampling reduces the number of samples required for a given image quality. The problem with this technique, however, is that the variance measurement is itself based on point samples, and so this method can fail as well. This is particularly true for subpixel objects that do not cross pixel boundaries. Nevertheless, adaptive sampling presents a far more reliable and cost-effective alternative to supersampling.

39.7 Prefiltering

Area sampling can be accelerated if constraints on the filter shape are imposed. Pyramids and preintegrated tables are introduced to approximate the convolution integral with a constant number of accesses. This compares favorably against direct convolution, which requires a large number of samples that grow proportionately to preimage area. As we shall see, though, the filter area will be limited to squares or rectangles, and the kernel will consist of a box filter. Subsequent advances have extended their use to more general cases with only marginal increases in cost.

39.7.1 Pyramids

Pyramids are multiresolution data structures commonly used in image processing and computer vision. They are generated by successively bandlimiting and subsampling the original image to form a hierarchy of images at ever decreasing resolutions. The original image serves as the base of the pyramid, and its coarsest version resides at the apex. Thus, in a lower-resolution version of the input, each pixel represents the average of some number of pixels in the higher-resolution version.

The resolution of successive levels typically differs by a power of two. This means that successively coarser versions each have one-quarter of the total number of pixels as their adjacent predecessors. The memory cost of this organization is modest: $1 + 1/4 + 1/16 + \cdots = 4/3$ times that needed for the original input. This requires only 33% more memory.

To filter a preimage, one of the pyramid levels is selected based on the size of its bounding square box. That level is then point sampled and assigned to the respective output pixel. The primary benefit of this approach is that the cost of the filter is constant, requiring the same number of pixel accesses independent of the filter size. This performance gain is the result of the filtering that took place while creating the

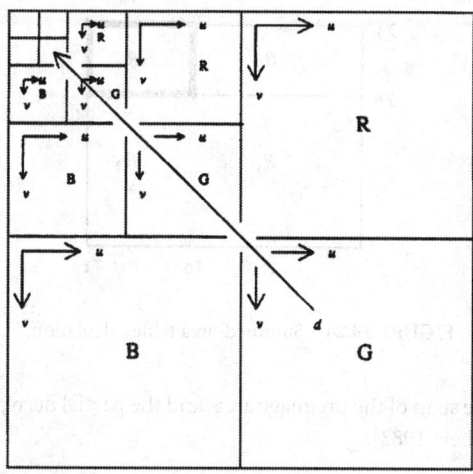

FIGURE 39.23 Mip-map memory organization.

pyramid. Furthermore, if preimage areas are adequately approximated by squares, the direct convolution methods amount to point-sampling a pyramid. This approach was first applied to texture mapping in Catmull [1974] and described in Dungan et al. [1978].

There are several problems with the use of pyramids. First, the appropriate pyramid level must be selected. A coarse level may yield excessive blur, while the adjacent finer level may be responsible for aliasing due to insufficient bandlimiting. Second, preimages are constrained to be squares. This proves to be a crude approximation for elongated preimages. For example, when a surface is viewed obliquely the texture may be compressed along one dimension. Using the largest bounding square will include the contributions of many extraneous samples and result in excessive blur. These two issues were addressed in Williams [1983] and Crow [1984], respectively, along with extensions proposed by other researchers.

Williams proposed a pyramid organization called *mip map* to store color images at multiple resolutions in a convenient memory organization [Williams 1983]. The acronym "mip" stands for "multum in parvo," a Latin phrase meaning "much in little." The scheme supports trilinear interpolation, where both intra- and interlevel interpolation can be computed using three normalized coordinates: u, v, and d. Both u and v are spatial coordinates used to access points within a pyramid level. The d coordinate is used to index, and interpolate between, different levels of the pyramid. This is depicted in Figure 39.23.

The quadrants touching the east and south borders contain the original red, green, and blue (RGB) components of the color image. The remaining upper left quadrant contains all the lower-resolution copies of the original. The memory organization depicted in Figure 39.23 clearly supports the earlier claim that the memory cost is 4/3 times that required for the original input. Each level is shown indexed by the $[u, v, d]$ coordinate system, where d is shown slicing through the pyramid levels. Since corresponding points in different pyramid levels have indices which are related by some power of two, simple binary shifts can be used to access these points across the multiresolution copies. This is a particularly attractive feature for hardware implementation.

The primary difference between mip maps and ordinary pyramids is the trilinear interpolation scheme possible with the $[u, v, d]$ coordinate system. Since they allow a continuum of points to be accessed, mip maps are referred to as pyramidal parametric data structures. In Williams's implementation, a box filter was used to create the mip maps, and a triangle filter was used to perform intra- and interlevel interpolation. The value of d must be chosen to balance the tradeoff between aliasing and blurring. Heckbert suggests

$$d^2 = \max\left(\left(\frac{\partial u}{\partial x}\right)^2 + \left(\frac{\partial v}{\partial x}\right)^2, \left(\frac{\partial u}{\partial y}\right)^2 + \left(\frac{\partial v}{\partial y}\right)^2\right) \tag{39.22}$$

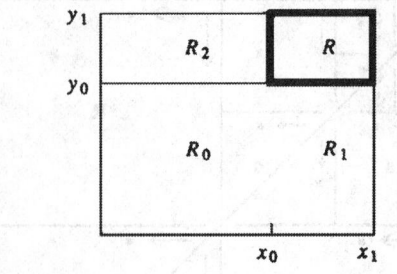

FIGURE 39.24   Summed-area table calculation.

where d is proportional to the span of the preimage area, and the partial derivatives can be computed from the surface projection [Heckbert 1983].

39.7.2 Summed-Area Tables

An alternative to pyramidal filtering was proposed by Crow [1984]. It extends the filtering possible in pyramids by allowing rectangular areas, oriented parallel to the coordinate axes, to be filtered in constant time. The central data structure is a preintegrated buffer of intensities, known as the **summed-area table**. This table is generated by computing a running total of the input intensities as the image is scanned along successive scanlines. For every position P in the table, we compute the sum of intensities of pixels contained in the rectangle between the origin and P. The sum of all intensities in any rectangular area of the input may easily be recovered by computing a sum and two differences of values taken from the table. For example, consider the rectangles R_0, R_1, R_2, and R shown in Figure 39.24. The sum of intensities in rectangle R can be computed by considering the sum at $[x1, y1]$, and discarding the sums of rectangles R_0, R_1, and R_2. This corresponds to removing all areas lying below and to the left of R. The resulting area is rectangle R, and its sum S is given as

$$S = T[x1, y1] - T[x1, y0] - T[x0, y1] + T[x0, y0] \qquad (39.23)$$

where $T[x, y]$ is the value in the summed-area table indexed by coordinate pair $[x, y]$.

Since $T[x1, y0]$ and $T[x0, y1]$ both contain R_0, the sum of R_0 was subtracted twice in Equation 39.23. As a result, $T[x0, y0]$ was added back to restore the sum. Once S is determined, it is divided by the area of the rectangle. This gives the average intensity over the rectangle, a process equivalent to filtering with a Fourier window (box filtering).

There are two problems with the use of summed-area tables. First, the filter area is restricted to rectangles. This is addressed in Glassner [1986], where an adaptive, iterative technique is proposed for obtaining arbitrary filter areas by removing extraneous regions from the rectangular bounding box. Second, the summed-area table is restricted to box filtering. This, of course, is attributed to the use of unweighted averages that keeps the algorithm simple. In Perlin [1985] and Heckbert [1986], the summed-area table is generalized to support more sophisticated filtering by repeated integration.

It is shown that by repeatedly integrating the summed-area table n times, it is possible to convolve an orthogonally oriented rectangular region with an nth-order box filter (B-spline). The output value is computed by using $(n + 1)^2$ weighted samples from the preintegrated table. Since this result is independent of the size of the rectangular region, this method offers a great reduction in computation over that of direct convolution. Perlin called this a selective image filter because it allows each sample to be blurred by different amounts.

Repeated integration has rather high memory costs relative to pyramids. This is due to the number of bits necessary to retain accuracy in the large summations. Nevertheless, it allows us to filter rectangular or elliptical regions, rather than just squares as in pyramid techniques. Since pyramid and summed-area tables both require a setup time, they are best suited for input that is intended to be used repeatedly, i.e., stationary background scenes or texture maps. In this manner, the initialization overhead can be amortized over each use.

39.8   Example: Image Scaling

In this section, we demonstrate the role of reconstruction and antialiasing in image scaling. The resampling process will be explained in one dimension rather than two, since resampling is carried out on each axis independently. For example, the horizontal scanlines are first processed, yielding an intermediate image, which then undergoes a second pass of interpolation in the vertical direction. The result is independent of the order: processing the vertical lines before the horizontal lines gives the same results.

A skeleton of a C program that resizes an image in two passes is given below. The input image is assumed to have **INheight** rows and **INwidth** columns. The first pass visits each row and resamples them to have width **OUTwidth**. The second pass visits each column of the newly formed intermediate image and resamples them to have height **OUTheight**:

```
INwidth   =  input image width (pixels/row);
INheight  =  input image height (rows/image);
OUTwidth  =  output image width (pixels/row);
OUTheight =  output image height (rows/image);
filter    =  convolution kernel to use to filter image;
offset    =  inter-pixel offset (stride);

allocate an intermediate image of size OUTwidth by INheight;

offset = 1;
for(y=0; y<INheight; y++) {        /* process rows */
    src = pointer to row y of input image;
    dst = pointer to row y of intermediate image;
    resample1-D(src, dst, INwidth, OUTwidth, filter, offset);
}

offset = OUTwidth;
for(x=0; x<w; x++) {              /* process columns */
    src = pointer to column x of intermediate image;
    dst = pointer to column x of output image;
    resample1-D(src, dst, INheight, OUTheight, filter, offset);
}
```

Function **resample1-D** is the workhorse of the resizing operation. The inner workings of this function will be described later. In addition to the input and output pointers and dimensions, **resample1-D** must be passed **filter**, an integer code specifying which convolution kernel to apply. In order to operate on both rows and columns, the parameter **offset** is given to denote the distance between successive pixels in the scanline. Horizontal scanlines (rows) have **offset = 1** and vertical scanlines (columns) have **offset = OUTwidth**.

There are two operations which **resample1-D** must be able to handle: magnification and minification. As mentioned earlier, these two operations are closely related. They both require us to project each output sample into the input, center a kernel, and convolve. The only difference between magnification and minification is the shape of the kernel. The magnification kernel is fixed at $h(x)$, whereas the minification kernel is $ah(ax)$, for $a < 1$. The width of the kernel for minification is due to the need for a low-pass filter to perform antialiasing. That filter now has a narrower response than that of the interpolation function. Consequently, we exploit the following well-known Fourier transform pair:

$$h(ax) \longleftrightarrow \frac{1}{a} H\left(\frac{f}{a}\right) \tag{39.24}$$

This equation expresses the reciprocal relationship between the spatial and frequency domains. Notice that multiplying the spatial axis by a factor of a results in dividing the frequency axis and the spectrum values by that same factor. Since we want the spectrum values to be left intact, we use $ah(ax)$ as the convolution kernel for blurring, where $a > 1$. This implies that the shape of the kernel changes as a function of scale factor when we are downsampling the input. This was not the case for magnification.

A straightforward method to perform 1-D resampling is given below. It details the inner workings of the **resample1-D** function outlined earlier. In addition, a few interpolation functions are provided. More such functions can easily be added by the user:

```
#define PI            3.1415926535897931160E0
#define SGN(A)        ((A) > 0 ? 1 : ((A) < 0 ? -1 : 0 ))
#define FLOOR(A)      ((int) (A))
#define CEILING(A)    ((A)==FLOOR(A) ? FLOOR(A) : SGN(A)+FLOOR(A))
#define CLAMP(A,L,H)  ((A)<=(L) ? (L) : (A)<=(H) ? (A) : (H))

resample1-D(IN, OUT, INlen, OUTlen, filtertype, offset)
unsigned char *IN, *OUT;
int INlen, OUTlen, filtertype, offset;
{
    int i;
    int left, right;   /* kernel extent in input */
    int pixel;         /* input pixel value */
    double u, x;       /* input (u) , output (x) */
    double scale;      /* resampling scale factor */
    double fwidth;     /* filter width (support) */
    double fscale;     /* filter amplitude scale */
    double weight;     /* kernel weight */
    double acc;        /* convolution accumulator */

    scale = (double) OUTlen / INlen;

    switch(filtertype) {
    case 0: filter = boxFilter; /* box filter (nearest
                                    neighbor) */
        fwidth = .5;
        break;
    case 1: filter = triFilter; /* triangle filter (linear
                                    intrp) */
        fwidth = 1;
        break;
    case 2: filter = cubicConv; /* cubic convolution
                                    filter */
        fwidth = 2;
        break;
    case 3: filter = lanczos3;  /* Lanczos3 windowed sinc
                                    function */
        fwidth = 3;
        break;
    case 4: filter = hann4;     /* Hann windowed sinc
                                    function */
        fwidth = 4;             /* 8-point kernel */
        break;
    }

    if(scale < 1.0) {           /* minification: h(x) -> h(x*scale)*
                                    scale */
```

```
                   fwidth = fwidth / scale;    /* broaden filter */
                   fscale = scale;             /* lower amplitude */
              } else      fscale = 1.0;

        /* project each output pixel to input, center kernel, and
           convolve */
            for(x=0; x<OUTlen; x++) {
                /* map output x to input u: inverse mapping */
                u = x / scale;

                /* left and right extent of kernel centered at u */
                if(u - fwidth < 0) {
                        left = FLOOR (u - fwidth);
                else left = CEILING(u - fwidth);
                right = FLOOR(u + fwidth);

                /* reset acc for collecting convolution products */
                acc = 0;

                /* weigh input pixels around u with kernel */
                for(i=left; i <= right; i++) {
                        pixel = IN[ CLAMP(i, 0, INlen-1)*offset];
                        weight = (*filter)((u - i) * fscale);
                        acc += (pixel * weight);
                }

                /* assign weighted accumulator to OUT */
                OUT[x*offset] = acc * fscale;
        }
}

/* ~~~~~~~~~~~~~~~~~~~~~~~~~~~~~~~~~~~~~~~~~~~~~~~~~~~~~~~~~~~~~~~~~~
 * boxFilter:
 *
 * Box (nearest neighbor) filter.
 */
double
boxFilter(t)
double t;
{
        if((t > -.5) && (t <= .5)) return(1.0);
        return(0.0);
}

/* ~~~~~~~~~~~~~~~~~~~~~~~~~~~~~~~~~~~~~~~~~~~~~~~~~~~~~~~~~~~~~~~~~~
 * triFilter:
 *
 * Triangle filter (used for linear interpolation).
 */
double
triFilter(t)
double t;
{
        if(t < 0) t = -t;
        if(t < 1.0) return(1.0 - t);
```

```
          return(0.0);
}
/* ~~~~~~~~~~~~~~~~~~~~~~~~~~~~~~~~~~~~~~~~~~~~~~~~~~~~~~~~~~~~~~~~~~
 * cubicConv:
 *
 * Cubic convolution filter.
 */
double
cubicConv(t)
double t;
{
      double A, t2, t3;

      if(t < 0) t = -t;
      t2 = t * t;
      t3 = t2 * t;

      A = -1.0;           /* user-specified free parameter */
      if(t < 1.0) return((A+2)*t3 - (A+3)*t2 + 1);
      if(t < 2.0) return(A*(t3 - 5*t2 + 8*t - 4));
      return(0.0);
}
/* ~~~~~~~~~~~~~~~~~~~~~~~~~~~~~~~~~~~~~~~~~~~~~~~~~~~~~~~~~~~~~~~~~
 * sinc:
 *
 * Sinc function.
 */
double
sinc(t)
double t;
{
      t *= PI;
      if(t != 0) return(sin(t) / t);
      return(1.0);
}
/* ~~~~~~~~~~~~~~~~~~~~~~~~~~~~~~~~~~~~~~~~~~~~~~~~~~~~~~~~~~~~~~~~~~
 * lanczos3:
 *
 * Lanczos3 filter.
 */
double
lanczos3(t)
double t;
{
      if(t < 0) t = -t;
      if(t < 3.0) return(sinc(t) * sinc(t/3.0));
      return(0.0);
}
/* ~~~~~~~~~~~~~~~~~~~~~~~~~~~~~~~~~~~~~~~~~~~~~~~~~~~~~~~~~~~~~~~~~~
 * hann:
```

```
    *
    * Hann windowed sinc function. Assume N (width) = 4.
    */
    double
    hann4(t)
    double t;
    {
        int N = 4; /* fixed filter width */

        if(t < 0) t = -t;
        if(t < N) return(sinc(t) * (.5 + .5*cos(PI*t/N)));
        return(0.0);
    }
```

There are several points worth mentioning about this code. First, the filter width **fwidth** of each of the supported kernels is initialized for use in interpolation (for magnification). We then check to see if the scale factor **scale** is less than one to rescale **fwidth** accordingly. Furthermore, we set **fscale,** the filter amplitude scale factor, to 1 for interpolation, or **scale** for minification. We then visit each of **OUTlen** output pixels, and project them back into the input, where we center the filter kernel. The kernel overlaps a range of input pixels from **left** to **right**. All pixels in this range are multiplied by a corresponding kernel value. The products are added in an accumulator **acc** and assigned to the output buffer.

Note that the **CLAMP** macro is necessary to prevent us from attempting to access a pixel beyond the extent of the input buffer. By clamping to either end, we are effectively replicating the border pixel for use with a filter kernel that extends beyond the image.

In order to accommodate the processing of rows and columns, the variable **offset** is introduced to specify the interpixel distance. When processing rows, **offset** = 1. When processing columns, **offset** is set to the width of a row.

This code can accommodate a polynomial transformation by making a simple change to the evaluation of **u.** Rather than computing **u** = **x**/scale , we may let u be expressed by a polynomial. The method of forward differences is recommended to simplify the computation of polynomials [Wolberg 1990].

The code given above suffers from three limitations, all dealing with efficiency:

1. A division operation is used to compute the inverse projection. Since we are dealing with a linear mapping function, the new position at which to center the kernel may be computed incrementally. That is, there is a constant offset between each projected output sample. Accordingly, **left** and **right** should be computed incrementally as well.
2. The set of kernel weights used in processing the first scanline applies equally to all the remaining scanlines as well. There should be no need to recompute them each time. This matter is addressed in the code supplied by Schumacher [1992].
3. The kernel weights are evaluated by calling the appropriate filter function with the normalized distance from the center. This involves a lot of run-time overhead, particularly for the more sophisticated kernels that require the evaluation of a sinc function, division, and several multiplies.

Additional sophisticated algorithms to deal with these issues are given in Wolberg [1990].

39.9   Research Issues and Summary

The computer graphics literature is replete with new and innovative work addressing the demands of sampling, reconstruction, and antialiasing. Nonuniform sampling has become important in computer graphics because it facilitates variable sampling density and it allows us to trade structured aliasing for noise. Recent work in adaptive sampling and nonuniform reconstruction is discussed in Glassner [1995]. Excellent surveys in nonuniform reconstruction, which is also known as scattered-data interpolation, can be found in Franke and Nielson [1991] and Hoschek and Lasser [1993]. These problems are also of direct consequence to image compression. The ability to determine a unique minimal set of samples to completely represent a signal within some specified error tolerance remains an active area of research. The solution must be closely coupled with a nonuniform reconstruction method. Although traditional reconstruction methods are well understood within the framework described in this chapter, the analysis of nonuniform sampling and reconstruction remains challenging.

We now summarize the basic principles of sampling theory, reconstruction, and antialiasing that have been presented in this chapter. We have shown that a continuous signal may be reconstructed from its samples if the signal is bandlimited and the sampling frequency exceeds the Nyquist rate. These are the two necessary conditions for image reconstruction to be possible. Since sampling can be shown to replicate a signal's spectrum across the frequency domain, ideal low-pass filtering was introduced as a means of retaining the original spectrum while discarding its copies. Unfortunately, the ideal low-pass filter in the spatial domain is an infinitely wide sinc function. Since this is difficult to work with, nonideal reconstruction filters are introduced to approximate the reconstructed output. These filters are nonideal in the sense that they do not completely attenuate the spectra copies. Furthermore, they contribute to some blurring of the original spectrum. In general, poor reconstruction leads to artifacts such as jagged edges.

Aliasing refers to the phenomenon that occurs when a signal is undersampled. This happens if the reconstruction conditions mentioned above are violated. In order to resolve this problem, one of two actions may be taken. Either the signal can be bandlimited to a range that complies with the sampling frequency, or the sampling frequency can be increased. In practice, some combination of both options is taken, leaving some relatively unobjectionable aliasing in the output.

Examples of the concepts discussed thus are concisely depicted in Figure 39.25 through 39.27. They attempt to illustrate the effects of sampling and low-pass filtering on the quality of the reconstructed signal and its spectrum. The first row of Figure 39.25 shows a signal and its spectrum, bandlimited to 0.5 cycle/pixel. For pedagogical purposes, we treat this signal as if it were continuous. In actuality, though, it is a 256-sample horizontal cross section taken from a digital image. Since each pixel has four samples contributing to it, there is a maximum of two cycles per pixel. The horizontal axes of the spectrum account for this fact.

The second row shows the effect of sampling the signal. Since $f_s = 1$ sample/pixel, there are four copies of the baseband spectrum in the range shown. Each copy is scaled by $f_s = 1$, leaving the magnitudes intact. In the third row, the 64 samples are shown convolved with a sinc function in the spatial domain. This corresponds to a rectangular pulse in the frequency domain. Since the sinc function is used here for image reconstruction, it must have an amplitude of unity value in order to interpolate the data. This forces the height of the rectangular pulse in the frequency domain to vary in response to f_s.

A few comments on the reciprocal relationship between the spatial and frequency domains are in order here, particularly as they apply to the ideal low-pass filter. We again refer to the variables A and W as the sinc amplitude and bandwidth. As a sinc function is made broader, the value $1/2W$ is made to change, since W is decreasing to accommodate zero crossings at larger intervals. Accordingly, broader sinc functions cause more blurring, and their spectra reflect this by reducing the cutoff frequency to some smaller W. Conversely, narrower sinc functions cause less blurring, and W takes on some larger value. In either case, the amplitude of the sinc function or its spectrum will change. That is, we can fix the amplitude of the sinc function so that only the rectangular pulse of the spectrum changes height $A/2W$ as W varies. Alternatively, we can fix $A/2W$ to remain constant as W changes, forcing us to vary A. The choice depends on the application.

When the sinc function is used to interpolate data, it is necessary to fix A to 1. Therefore, as the sampling density changes, the positions of the zero crossings shift, causing W to vary. This makes the amplitude of the spectrum's rectangular pulse change. On the other hand, if the sinc function is applied to bandlimit, not interpolate, the input signal, then it is important to fix $A/2W$ to 1 so that the passband frequencies remain intact. Since W is once again varying, A must change proportionately to keep $A/2W$ constant. Therefore, this application of the ideal low-pass filter requires the amplitude of the sinc function to be responsive to W.

In the examples presented below, our objective is to interpolate (reconstruct) the input, and so $A = 1$ regardless of the sampling density. Consequently, the height of the spectrum of the reconstruction filter changes. To make the Fourier transforms of the filters easier to see, we have not drawn the frequency

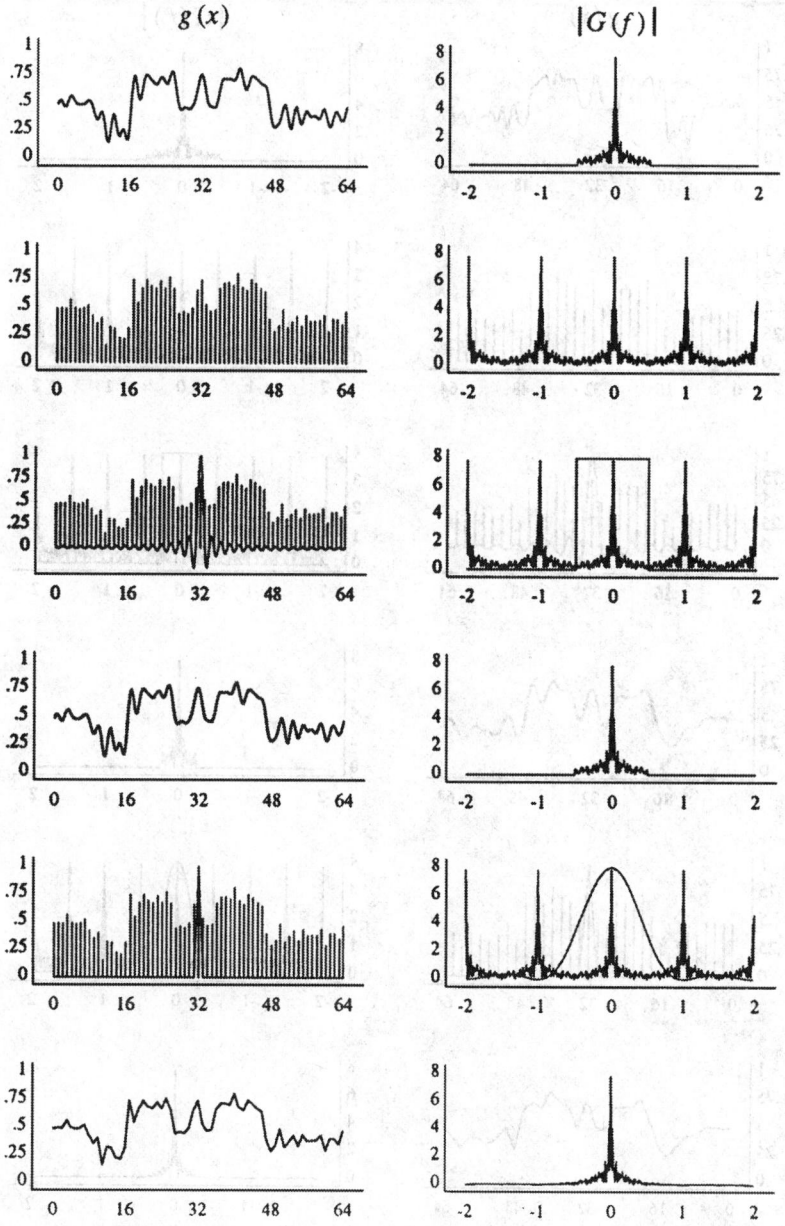

FIGURE 39.25 Sampling and reconstruction (with an adequate sampling rate).

response of the reconstruction filters to scale. Therefore, the rectangular pulse function in the third row of Figure 39.25 actually has height $A/2W = 1$. The fourth row of the figure shows the result after applying the ideal low-pass filter. As sampling theory predicts, the output is identical to the original signal. The last two rows of the figure illustrate the consequences of nonideal reconstruction filtering. Instead of using a sinc function, a triangle function corresponding to linear interpolation was applied. In the frequency domain this corresponds to the square of the sinc function. Not surprisingly, the spectrum of the reconstructed signal suffers in both the passband and the stopband.

The identical sequence of filtering operations is performed in Figure 39.26. In this figure, though, the sampling rate has been lowered to $f_s = 0.5$, meaning that only one sample is collected for every

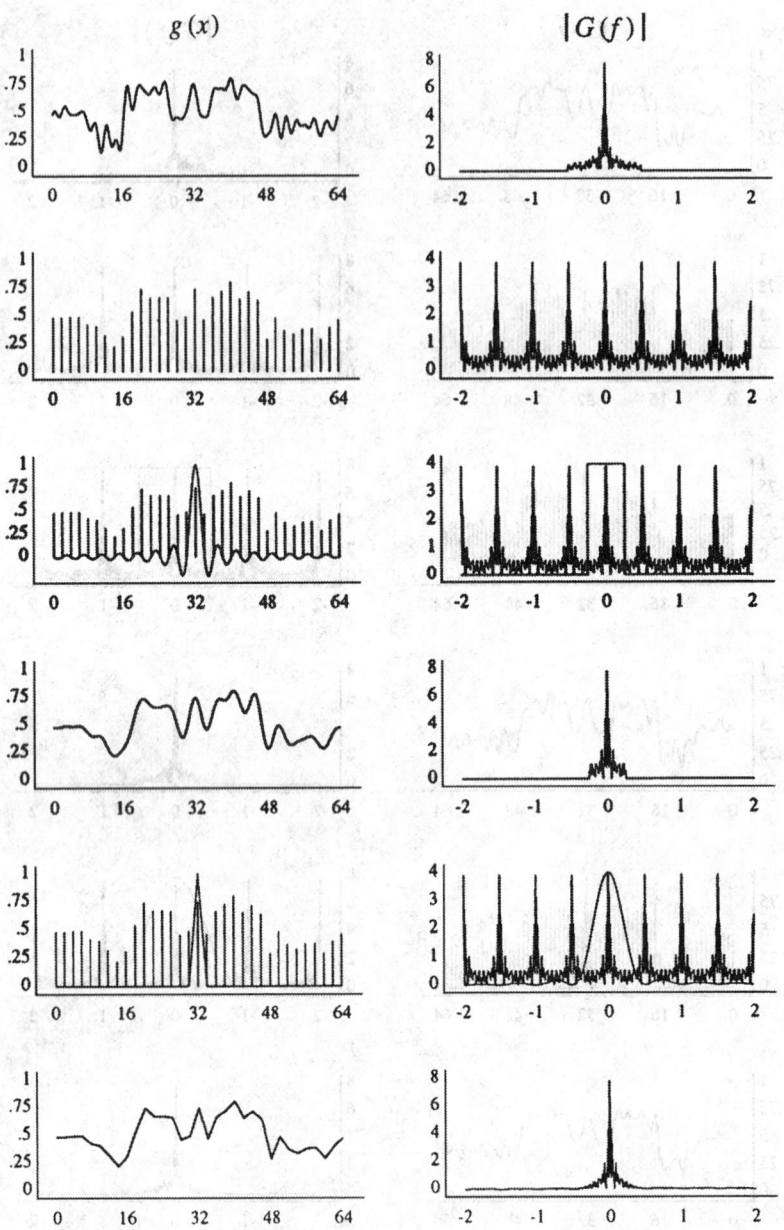

FIGURE 39.26   Sampling and reconstruction (with an inadequate sampling rate).

two output pixels. Consequently, the replicated spectra are multiplied by 0.5, leaving the magnitudes at 4. Unfortunately, this sampling rate causes the replicated spectra to overlap. This, in turn, gives rise to aliasing, as depicted in the fourth row of the figure. Applying the triangle function to perform linear interpolation also yields poor results.

In order to combat these artifacts, the input signal must be bandlimited to accommodate the low sampling rate. This is shown in the second row of Figure 39.27, where we see that all frequencies beyond $W = 0.25$ are truncated. This causes the input signal to be blurred. In this manner we have traded aliasing for blurring, a far less objectionable artifact. Sampling this function no longer causes the replicated copies to overlap. Convolving with an ideal low-pass filter now properly isolates the bandlimited spectrum.

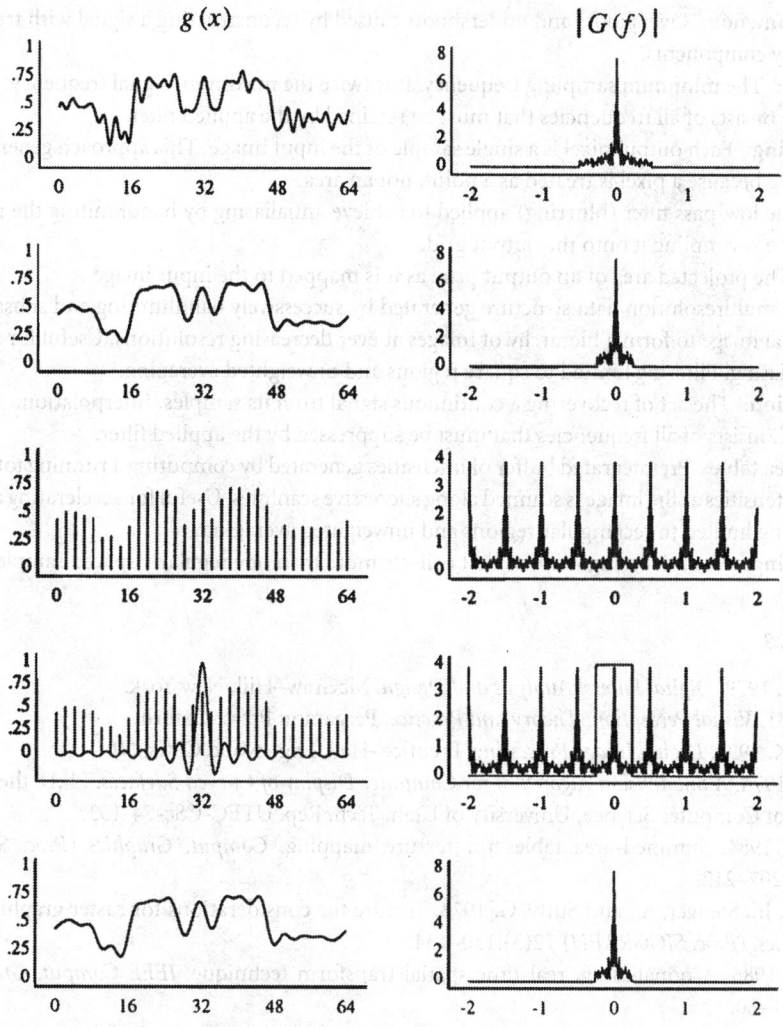

FIGURE 39.27 Antialiasing filtering, sampling, and reconstruction stages.

Defining Terms

Adaptive supersampling: Supersampling with samples distributed more densely in areas of high intensity variance.

Aliasing: Artifacts due to undersampling a signal. This condition prevents the signal from being reconstructed from its samples.

Antialiasing: The filtering necessary to combat aliasing. This generally requires bandlimiting the input before sampling to remove the offending high frequencies that will fold over in the frequency spectrum.

Area sampling: An antialiasing method that treats a pixel as an area, not a point. After projecting the pixel to the input, all samples in the preimage are averaged to compute a representative sample.

Bandlimiting: The act of truncating all frequency components beyond some specified frequency. Useful for antialiasing, where offending high frequencies must be removed to prevent aliasing.

Frequency leakage: A condition in which the stopband is allowed to persist, permitting it to fold over into the passband range.

Gibbs phenomenon: Overshoots and undershoots caused by reconstructing a signal with truncated frequency components.

Nyquist rate: The minimum sampling frequency. It is twice the maximum signal frequency.

Passband: Consists of all frequencies that must be retained by the applied filter.

Point sampling: Each output pixel is a single sample of the input image. This approach generally leads to aliasing because a pixel is treated as a point, not an area.

Prefilter: The low-pass filter (blurring) applied to achieve antialiasing by bandlimiting the input image prior to resampling it onto the output grid.

Preimage: The projected area of an output pixel as it is mapped to the input image.

Pyramid: A multiresolution data structure generated by successively bandlimiting and subsampling the original image to form a hierarchy of images at ever decreasing resolutions. Useful for accelerating antialiasing. Filtering limited to square regions and unweighted averaging.

Reconstruction: The act of recovering a continuous signal from its samples. Interpolation.

Stopband: Consists of all frequencies that must be suppressed by the applied filter.

Summed-area table: Preintegrated buffer of intensities generated by computing a running total of the input intensities as the image is scanned along successive scanlines. Useful for accelerating antialiasing. Filtering limited to rectangular regions and unweighted averaging.

Supersampling: An antialiasing method that collects more than one regularly spaced sample per pixel.

References

Antoniou, A. 1979. *Digital Filters: Analysis and Design*. McGraw–Hill, New York.

Caelli, T. 1981. *Visual Perception: Theory and Practice*. Pergamon Press, Oxford.

Castleman, K. 1996. *Digital Image Processing*. Prentice–Hall, Englewood Cliffs, NJ.

Catmull, E. 1974. *A Subdivision Algorithm for Computer Display of Curved Surfaces*. Ph.D. thesis, Department of Computer Science, University of Utah. Tech. Rep. UTEC-CSc-74-133.

Crow, F. C. 1984. Summed-area tables for texture mapping. *Comput. Graphics (Proc. SIGGRAPH)* 18(3):207–212.

Dungan, W., Jr., Stenger, A., and Sutty, G. 1978. Texture tile considerations for raster graphics. *Comput. Graphics. (Proc. SIGGRAPH)* 12(3):130–134.

Fant, K. M. 1986. A nonaliasing, real-time spatial transform technique. *IEEE Comput. Graphics Appl.* 6(1):71–80.

Franke, R. and Nielson, G. M. 1991. Scattered data interpolation and applications: a tutorial and survey. In *Geometric Modelling: Methods and Their Application*. H. Hagen and D. Roller, eds., pp. 131–160. Springer–Verlag, Berlin.

Glassner, A. 1986. Adaptive precision in texture mapping. *Comput. Graphics (Proc. SIGGRAPH)* 20(4):297–306.

Glassner, A. 1995. *Principles of Digital Image Synthesis*. Morgan Kaufmann, San Francisco.

Gonzalez, R. C. and Woods, R. 1992. *Digital Image Processing*. Addison–Wesley, Reading, MA.

Heckbert, P. 1983. Texture mapping polygons in perspective. Tech. Memo 13, NYIT Computer Graphics Lab.

Heckbert, P. 1986. Filtering by repeated integration. *Comput. Graphics (Proc. SIGGRAPH)* 20(4):315–321.

Hoschek, J. and Lasser, D. 1993. *Computer Aided Geometric Design*. A K Peters, Wellesley, MA.

Jain, A. K. 1989. *Fundamentals of Digital Image Processing*. Prentice–Hall, Englewood Cliffs, NJ.

Kajiya, J. 1986. The Rendering Equation. *Comput. Graphics Proc., SIGGRAPH* 20(4):143–150.

Lee, M., Redner, R. A., and Uselton, S. P. 1985. Statistically optimized sampling for distributed ray tracing. *Comput. Graphics (Proc. SIGGRAPH)* 19(3):61–67.

Mitchell, D. P. 1987. Generating antialiased images at low sampling densities. *Comput. Graphics (Proc. SIGGRAPH)* 21(4):65–72.

Mitchell, D. P. and Netravali, A. N. 1988. Reconstruction filters in computer graphics. *Comput. Graphics (Proc. SIGGRAPH)* 22(4):221–228.

Perlin, K. 1985. Course notes. SIGGRAPH'85 State of the Art in Image Synthesis Seminar Notes.

Pratt, W. K. 1991. *Digital Image Processing*, 2nd ed. J. Wiley, New York.

Russ, J. C. 1992. *The Image Processing Handbook*. CRC Press, Boca Raton, FL.

Schumacher, D. 1992. General filtered image rescaling. In *Graphics Gems III*. David Kirk, Ed., Academic Press, New York.

Shannon, C.E. 1948. A mathematical theory of communication. *Bell System Tech. J.* 27:379–423 (July 1948), 27:623–656 (Oct. 1948).

Shannon, C. E. 1949. Communication in the presence of noise. *Proc. Inst. Radio Eng.* 37(1):10–21.

Whitted, T. 1980. An improved illumination model for shaded display. *Commun. ACM* 23(6):343–349.

Williams, L. 1983. Pyramidal parametrics. *Comput. Graphics (Proc. SIGGRAPH)* 17(3):1–11.

Wolberg, G. 1990. *Digital Image Warping*. IEEE Comput. Soc. Press, Los Alamitos, CA.

Further Information

The material contained in this chapter was drawn from Wolberg [1990]. Additional image processing texts that offer a comprehensive treatment of sampling, reconstruction, and antialiasing include Castleman [1996], Glassner [1995], Gonzalez and Woods [1992], Jain [1989], Pratt [1991], and Russ [1992].

Advances in the field are reported in several journals, including *IEEE Transactions on Image Processing, IEEE Transactions on Signal Processing, IEEE Transactions on Acoustics, Speech, and Signal Processing*, and *Graphical Models and Image Processing*. Related work in computer graphics is also reported in *Computer Graphics* (ACM SIGGRAPH Proceedings), *IEEE Computer Graphics and Applications*, and *IEEE Transactions on Visualization and Computer Graphics*.

40

Computer Animation

40.1 Introduction 40-1
   Classification of Methods
40.2 Underlying Principles and Best Practices 40-2
   Geometric and Kinematic Methods • Character Deformations
40.3 Physics-based Methods 40-6
   Balance Control • Dynamic Simulation
40.4 Behavioral Methods 40-8
   Artificial and Virtual Life • Virtual Sensors
40.5 Crowds and Groups 40-9
40.6 Facial Animation 40-10
40.7 Algorithms 40-12
   Kochanek–Bartels Spline Interpolation • Principle of
   Behavioral Animation
40.8 Research Issues and Summary 40-14

Nadia Magnenat Thalmann
University of Geneva

Daniel Thalmann
*Swiss Federal Institute of Technology
(EPFL)*

40.1 Introduction

The main goal of computer animation is to synthesize the desired motion effect, a mix of natural phenomena, perception, and imagination. The animator designs the object's dynamic behavior with a mental representation of causality. The animator imagines how it moves, becomes misshapen, or reacts when pushed, pressed, pulled, or twisted. So, the animation system must provide the user with motion control tools able to translate his or her wishes from his or her own language.

Computer animation methods may also aid understanding of physical laws by adding motion control to data in order to show evolution over time. Visualization has become an important way to validate new models created by scientists. When the model evolves over time, computer simulation is generally used to obtain the evolution of time, and computer animation is a natural way to visualize the results obtained from the simulation. Computer animation may be defined as a technique in which the illusion of movement is created by displaying on a screen, or recording on a recording device, a series of individual states of a dynamic scene. Formally, any computer animation sequence may be defined as a set of objects characterized by state variables evolving over time. For example, a human character is normally characterized using its joint angles as state variables.

To produce a computer animation sequence, the animator has two principal techniques available. The first is to use a model that creates the desired effect. A good example is the growth of a green plant. The second is used when no model is available. In this case, the animator produces by hand the real-world motion to be simulated. Initially, most computer-generated films were produced using the second approach: motion capture, traditional computer animation techniques like keyframe animation, spline interpolation, etc. Then, animation languages, scripted systems, and director-oriented systems were developed. More recently,

1-58488-360-X/$0.00+$1.50
© 2004 by CRC Press, LLC

motion has been planned at a task level and computed using physical laws. Nowadays, researchers develop models of behavioral animation and simulation of autonomous creatures using artificial intelligence (AI) and agent technology.

40.1.1 Classification of Methods

We will start with the classification introduced by Magnenat Thalmann and Thalmann [1991], based on the method of controlling motion and according to characters' interactions. A *motion control method* specifies how a character is animated and may be characterized according to the type of information to which it is privileged in animating the character. For example, in a keyframe system for an articulated body, the privileged information to be manipulated is joint angles. In a forward dynamics–based system, the privileged information is a set of forces and torques; of course, in solving dynamic equations, joint angles are also obtained in this system, but we consider these to be derived information. In fact, any motion control method must eventually deal with geometric information (typically joint angles), but only geometric motion control methods are explicitly privileged to this information at the level of animation control.

The nature of privileged information for the motion control of characters falls into three categories: geometric, physical, and behavioral. These categories give rise to three corresponding categories of motion control method:

> The first approach corresponds to methods on which the animator relies heavily: motion capture, shape transformation, parametric keyframe animation. *Animated objects are locally controlled.* Methods are normally driven by geometric data. Typically, the animator provides a lot of geometric data corresponding to a local definition of the motion.

> The second approach guarantees realistic motion by using physical laws, especially dynamic simulation. The problem with this type of animation is controlling the motion produced by simulating the physical laws that govern motion in the real world. The animator should provide physical data corresponding to the complete definition of a motion. The motion is obtained by the dynamic equations of motion relating the forces, torques, constraints, and the mass distribution of objects. As trajectories and velocities are obtained by solving the equations, we may consider *the animated objects as globally controlled.* Functional methods based on biomechanics are also part of this class.

> The third type of animation is called **behavioral animation** and takes into account the relationships between each object and the other objects. Moreover, the control of animation may be performed at a task level, but we may also consider *the animated objects as autonomous creatures.*

40.2 Underlying Principles and Best Practices

40.2.1 Geometric and Kinematic Methods

40.2.1.1 Introduction

In this group of methods, the privileged information is of a geometric or kinematic nature. Typically, motion is defined in terms of coordinates, angles, and other shape characteristics, or it may be specified using velocities and accelerations, but no force is involved. Among the techniques based on geometry and kinematics, we will discuss motion capture, motion retargeting, keyframing, inverse kinematics (IK), and procedural animation. Although these methods have been mainly concerned with determining the displacement of objects, they may also be applied in calculating deformations of objects.

40.2.1.2 Motion Capture

Motion capture consists of measuring and recording direct actions of a real person or animal for immediate or delayed analysis and playback. The technique is especially used today in production environments for 3-D character animation, like movies and games. It involves mapping measurements onto the motion of the digital character.

FIGURE 40.1 Motion capture.

We may distinguish two main kinds of systems: optical and magnetic.

Optical motion capture systems — These are based on small reflective sensors, called *markers*, attached to an actor's body and on several infrared cameras focused on performance space. By tracking the positions of the markers, one can get locations for corresponding key points in the animated model. For example, we attach markers at a person's joints and record the position of markers from several different directions. We then reconstruct the 3-D position of each key point at each time. The main advantage of this method is freedom of movement; it does not require any cabling. There is, however, one main problem: occlusion, that is, a lack of data resulting from hidden markers. An example is when the performer lies on his or her back. Another problem comes from the lack of an automatic way to distinguish reflectors when they get very close to each other during motion. These problems may be minimized by adding more cameras — but at a higher cost, of course. Most optical systems operate with four or six cameras, but for high-end applications like movies, 16 cameras are usual. A good example of an optical system is the Vicon system. Researchers in computer vision try to develop similar systems based on simple video cameras.

Magnetic motion capture systems — These require the real actor to wear a set of sensors (see Figure 40.1), which are capable of measuring their spatial relationship to a centrally located magnetic transmitter. The position and orientation of each sensor is then used to drive an animated character. One problem is the need to synchronize receivers. The data stream from the receivers to a host computer consists of 3-D positions and orientations for each receiver. For human body motion, eleven sensors are generally needed: one on the head, one on each upper arm, one on each hand, one in the center of chest, one on the lower back, one on each ankle, and one on each foot. To calculate the rest of the necessary information, the most common method is inverse kinematics.

Motion capture methods offer advantages and disadvantages. Let us consider the case of human walking. A walking motion may be recorded and then applied to a computer-generated 3-D character. It will provide a very good motion, because it comes directly from reality. However, motion capture does not bring any new concept to animation methodology. For any new motion, it is necessary to record the reality again. Moreover, motion capture is not appropriate in real-time simulation activities, where the situations and actions of people cannot be predicted, and in dangerous situations, where one cannot involve a human actor.

40.2.1.3 Motion Retargeting

There is a great interest in recording motion using motion capture systems (magnetic or optical), and then attempting to alter such a motion to create a sense of individuality. This process is tedious, and there is no reliable method at this stage. Even if it is fairly easy to correct one posture by modifying its angular parameters (with an inverse kinematics engine, for instance), it becomes a difficult task to perform this over the whole motion sequence while ensuring that some spatial constraints are respected over a certain time range and that no discontinuities arise. When one tries to adapt a captured motion to a different character, the constraints are usually violated, leading to problems such as feet going into the ground or a hand being unable to reach an object that the character should grab. The problem of adaptation and adjustment is usually referred to as the motion retargeting problem.

Witkin and Popovic [1995] proposed a technique for editing motions by modifying the motion curves through warping functions, producing some of the first interesting results. In a more recent paper [Popovic and Witkin, 1999], they have extended their method to handle physical elements, such as mass and gravity, and described how to use characters with different numbers of degrees of freedom. Their algorithm is based on the reduction of the character to an abstract character, which is much simpler and only contains the degrees of freedom that are useful for a particular animation. The edition and modification are then computed on this simplified character and mapped again onto the end-user skeleton.

Bruderlin and Williams [1995] have described some basic facilities to change the animation, by modifying the motion parameter curves. The user can define a particular posture at time t, and the system is then responsible for smoothly blending the motion around t. These researchers also introduced the notion of the motion displacement map, which is an offset added to each motion curve.

The term *motion retargeting problem* was introduced by Gleicher [1998]. He designed a space–time constraints solver, into which every constraint is added, leading to a big optimization problem. He mainly focused on optimizing his solver, to avoid enormous computation time, and achieved very good results. Bindiganavale and Badler [1998] also addressed the motion retargeting problem, introducing new elements: using the zero-crossing of the second derivative to detect significant changes in the motion, using visual attention tracking (and the way to handle the gaze direction), and applying inverse kinematics to enforce constraints by defining six subchains (the two arms and legs, the spine, and the neck). Finally, Lee and Shin [1999] used in their system a coarse-to-fine hierarchy of B-splines to interpolate the solutions computed by their inverse kinematics solver. They also reduced the complexity of the IK problem by analytically handling the degrees of freedom for the four human limbs.

40.2.1.4 Keyframe

This is an old technique consisting of the automatic generation of intermediate frames, called *inbetweens*, based on a set of keyframes supplied by the animator. Originally, the inbetweens were obtained by interpolating the keyframe images themselves. As linear interpolation produces undesirable effects — such as lack of smoothness in motion, discontinuities in the speed of motion, and distortions in rotations — spline interpolation methods are used. *Splines* can be described mathematically as piecewise approximations of cubic polynomial functions. Two kinds of splines are very popular: interpolating splines, with C1 continuity at knots, and approximating splines, with C2 continuity at knots. For animation, the most interesting splines are the interpolating splines: cardinal splines, Catmull-Rom splines, and Kochanek-Bartels [1984] splines (see Section 40.7).

A way to produce better images is to interpolate parameters of the model of the object itself. This technique is called **parametric keyframe animation** and is used in most commercial animation systems. In a parametric model, the animator creates keyframes by specifying the appropriate set of parameter values. Parameters are then interpolated and images are finally constructed individually from the interpolated parameters. Spline interpolation is generally used for the interpolation.

40.2.1.5 Inverse Kinematics

This direct kinematics problem consists of finding the position of end-point positions (e.g., hand, foot) with respect to a fixed-reference coordinate system as a function of time without regard to the forces or the moments that cause the motion. Efficient and numerically well behaved methods exist for the transformation of position and velocity from joint-space (joint angles) to Cartesian coordinates (end of the limb). Parametric keyframe animation is a primitive application of direct kinematics.

Basically, the problem is to determine a joints configuration for which a desired task, usually expressed in Cartesian space, is achieved. For example, the shoulder, elbow, and wrist configurations must be determined so that the hand precisely reaches a position in space. The equations that arise from this problem are generally nonlinear and difficult to solve. In addition, a resolution technique must also deal with the difficulties described below. For the positioning and animation of articulated figures, the weighting strategy is the most frequent: some typical examples are given by Badler et al. [1987] and Zhao et al. [1994] for posture manipulation, and by Phillips et al. [1990, 1991] for achieving smooth solution blending.

40.2.1.6 Procedural Animation

Procedural animation corresponds to the creation of a motion by a procedure specifically describing the motion. Procedural animation should be used when the motion can be described by an algorithm or a formula. For example, consider the case of a clock based on the pendulum law:

$$\alpha = A \sin(\omega t + \phi)$$

A typical animation sequence may be produced using a program such as the following:

```
create CLOCK (...);
for FRAME:=1 to NB_FRAMES
  TIME:=TIME+1/25;
  ANGLE:=A*SIN (OMEGA*TIME+PHI);
  MODIFY (CLOCK, ANGLE);
  draw CLOCK;
  record CLOCK
  erase  CLOCK
```

40.2.2 Character Deformations

Modeling and deformation of 3-D characters, and especially human bodies, during the animation process is an important but difficult problem. Researchers have devoted significant efforts to the representation and deformation of the human body's shape. Broadly, we can classify their models into two categories: the surface model and the multilayered model.

The *surface model* [Magnenat Thalmann and Thalmann, 1987] is conceptually simple, containing a skeleton and an outer skin layer. The envelope is composed of planar or curved patches. One problem with this model is that it requires the tedious input of the significant points or vertices that define the surface. Another problem is that it is hard to control the realistic evolution of the surface across joints. Surface singularities or anomalies can easily be produced.

The *multilayered model* [Chadwick et al., 1989] contains a skeleton layer, intermediate layers which simulate the physical behavior of muscle, bone, fat tissue, etc., and a skin layer. Since the overall appearance of a human body is very much influenced by its internal muscle structures, the multilayered model is the most promising for realistic human animation. The key advantage of the layered methodology is that once the layered character is constructed, only the underlying skeleton need be scripted for an animation; consistent yet expressive shape deformations are generated automatically. Henne [1990] represents the skin by a mesh of bicubic surface patches, whose control points deform at joints in accordance with several constraints such as elasticity, area preservation, and implicit repulsive force fields that mimic bones and muscles. Deformations that occur away from joints are ignored by this approach. Yoshomito [1992] showed that implicit formulations like **metaballs** provide efficient ways of creating beautiful virtual humans at a reduced storage cost. Thalmann et al. [1996] extended this approach by combining implicit surfaces and B-spline patches. Ellipsoids and ellipsoidal metaballs represent the gross shape of bone, muscle, fat tissue. The motion/deformation of each primitive with respect to underlying joints is specified via a graphical interface. A skin of fixed topology is extracted by casting rays from the skeleton segments in a star-shaped manner and using the intersection points as control points of B-spline patches.

More recent work aims at mimicking more closely the actual anatomy of humans or animals. Wilhelms [1997] developed an interactive tool for designing and animating monkeys and cats. In her system, ellipsoids or triangular meshes represent bones and muscle. Each muscle is a generalized cylinder made up of a certain number of cross-sections that consist, in turn, of a certain number of points. The muscles show a relative incompressibility when deformed. In their work on anatomically modeling the human musculature, Scheepers et al. [1997] stressed the role of underlying components (muscles, tendons, etc.) on the form. They use three volume-preserving geometric primitives for three different types of muscles:

Ellipsoids are used for rendering fusiform muscles.
Multibelly muscles are represented by a set of ellipsoids positioned along two spline curves.
Tubular-shaped bicubic patches provide a general muscle model.

Isometric contraction is handled by introducing scaling factors and tension parameters. The skin is obtained by fitting bicubic patches to an implicit surface created from the geometric primitives.

Porcher-Nedel and Thalmann [1998] introduced the idea of abstracting muscles by an action line (a polyline in practice), representing the force produced by the muscle on the bones, and a surface mesh deformed by an equivalent mass-spring mesh. In order to smooth out mesh discontinuities, they employ special springs, termed *angular springs*, which tend to restore the initial curvature of the surface at each vertex. However, angular springs cannot deal with local inversions of the curvature. Aubel and Thalmann [2001] also use an action line and a muscle mesh. The action line, represented by a polyline with any number of vertices, is moved for each posture using a predefined behavior and a simple, physically based simulation. It is then used as a skeleton for the surface mesh, and the deformations are produced in a usual way. Seo et al. [2000] propose a very fast method of deformation for MPEG-4–based applications.

40.3 Physics-based Methods

40.3.1 Balance Control

Balance control is an essential problem for the realistic computer animation of articulated figures, and of humans in particular. While most people take for granted the action of keeping balance, it is a challenge for neurophysiologists to understand the mechanisms of balance in human and animal beings [Roberts, 1995]. Here, we focus on techniques for balance control in static equilibrium developed in the computer graphics community and well suited to the postural control problem. It requires only the additional information of the body mass distribution. Clearly, more advanced methods are required in dynamic situations.

The center of mass is an important characteristic point of a figure. In the computer graphics community, Phillips and Badler [1991] offered the first control of the center of mass by constraining the angular values of the ankle, knee, and hip joints of the leg that supports most of the weight. A more general approach to the control of the center of mass, called inverse kinetics, has been proposed by Boulic et al. [1994, 1996]. The constraint on the position of the center of mass is solved at the differential level with a special-purpose Jacobian matrix that relates differential changes of the joint coordinates to differential changes of the Cartesian coordinates of the center of mass. Recently, Baerlocher and Boulic [2001] extended the control of mass properties to the moments of inertia of the articulated structure.

40.3.2 Dynamic Simulation

A great deal of work exists on the dynamics of articulated bodies [Huston, 1990], and efficient direct dynamics algorithms have been developed in robotics for structures with many degrees of freedom [Featherstone, 1986; SDFAST, 1990]. In computer animation, these algorithms have been applied to the dynamic simulation of the human body [MacKenna and Zeltzer, 1996]. Given a set of external forces (like gravity or wind) and internal forces (due to muscles) or joint torques, these algorithms compute the motion of the articulated body according to the laws of rigid body dynamics. Impressive animations of passive structures like falling bodies on stairs can be generated in this way with little input from the animator. On the other hand, such an approach can turn adjustments into a tedious task, because the animator has only indirect control over the animation. More generally, it is impractical for an animator to specify a temporally coordinate sequence of force/torque activation to generate a desired active behavior [Multon et al., 1999]. This is the *inverse dynamics problem*, which has been addressed for walking in [Ko and Badler, 1996]. Impressive results have also been achieved by Hodgins et al. [1995] in the simulation of dynamic human activities such as running and jumping. Some controller architectures have been proposed by Lazlo et al. [1996] to generate a wide range of gaits and to combine elementary behaviors [Faloutsos et al., 2001]. The use of constraints to avoid the direct specification of torques has also been researched. For example, in the computer graphics community, the satisfaction of space–time constraints has been proposed by Witkin and Kass [1988] with the minimization of an objective function, such as the total energy expenditure.

40.3.2.1 Collision Detection and Response

In computer animation, collision detection and response are obviously more important. Some works have addressed collision detection and response. Hahn [1988] presented bodies in resting contact as a series of frequently occurring collisions. Baraff [1989] presented an analytical method for finding forces between contacting polyhedral bodies, based on linear programming techniques. He also proposed a formulation of the contact forces between curved surfaces that are completely unconstrained in their tangential movement. Bandi and Thalmann [1995] introduced an adaptive spatial subdivision of the object space based on octree structure and presented a technique for efficiently updating this structure periodically during the simulation. Volino and Magnenat Thalmann [1994] described a new algorithm for detecting self-collisions on highly discretized moving polygonal surfaces. This technique is based on geometrical shape regularity properties that permit avoiding many useless collision tests. Vassilev and Spanlang [2001] proposed to use the Z-buffer for collision detection to generate depth and normal maps. Computation time of their collision detection does not depend on the complexity of the body. However, the maps must be precomputed before simulation, restricting real-time application.

40.3.2.2 Cloth Animation

Weil [1986] pioneered cloth animation using an approximated model based on relaxation of the surface. Kunii and Godota [1990] used a hybrid model incorporating physical and geometrical techniques to model garment wrinkles. Aono [1990] simulated wrinkle propagation on a handkerchief using an elastic model. Terzopoulos et al. [1987] applied their general elastic to a wide range of objects, including cloth. Lafleur et al. [1991] and Carignan et al. [1992] have described complex interaction of clothes with synthetic actors in motion, which marked the beginning of a new era in cloth animation. However, there were still a number of restrictions on the simulation conditions of the geometrical structure and the mechanical situations, imposed by the simulation model or the collision detection. For this reason, Volino et al. [1995] proposed a mechanical model to deal with any irregular triangular meshes, handle high deformations despite rough discretization, and cope with complex interacting collisions. Figure 40.2 shows examples of cloth animation.

Recently, research has focused on the real-time aspect of the animation of deformable objects using physical simulation. Baraff and Witkin [1998] have used the implicit Euler integration method to compute cloth simulation in real time. They stated that the bottleneck of real-time cloth simulation is the fact that the time step must be small in order to avoid instability. They described a method that can stably take large time steps, suggesting the possibility of real-time animation of simple objects. Meyer et al. [2000] and Desbrun et al. [1999] have used a hybrid explicit/implicit integration algorithm to animate real-time clothes; integrated with this is a voxel-based collision detection algorithm. Their method seems to be limited by the maximum number of polygons they can animate in real time. James and Pai [1999] also have worked on real-time simulation; their paper describes the boundary integral equation formulation of static linear elasticity and the related boundary element method discretization technique. Their model is not dynamic but a collection of static postures, which limits its potential applications. Debunne et al. [2000]

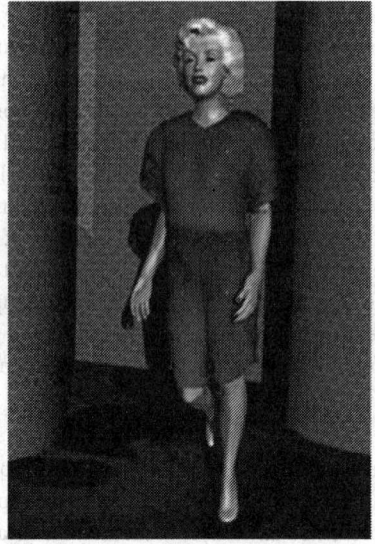 

FIGURE 40.2   Cloth animation.

have recently introduced a technique for animating soft bodies in real time. However, their method works on volumetric meshes. Therefore, it is not applicable to thin objects such as cloth.

Recently, for cloth animation on moving characters, Cordier et al. [2002] have proposed a real-time approach based on a classification into several categories, depending on how the garment is laid on and whether it sticks to, or flows on, the body surface. For instance, a tight pair of trousers will mainly follow the movement of the legs, whereas a skirt will float around the legs. Based on this approach, the MIRACloth system can be used for building and animating the garments on virtual actors. MIRACloth is a general animation framework in which different types of animation can be associated with the different objects: static, rigid, and deformable (keyframe, mechanical simulation). The methodology for building garments relies on the traditional garment design in real life. The 2-D patterns are created through a polygon editor, which are then taken to the 3-D simulator and placed around the body of a virtual actor. The process of seaming brings the patterns together, and the garment can then be animated with a moving virtual actor.

40.4   Behavioral Methods

40.4.1   Artificial and Virtual Life

This kind of research is strongly related to the research efforts in behavioral animation introduced by Reynolds's study [1987] in his distributed behavioral model simulating flocks of birds, herds of land animals, and schools of fish. For birds, the simulated flock is an elaboration of a particle system, with the simulated birds being the particles. A flock is assumed to be the result of the interactions between the behaviors of individual birds. Working independently, the birds try both to stick together and avoid collisions with one another and with other objects in their environment.

In a module of behavioral animation, the positions, velocities, and orientations of the actors are known from the system at any time. The animator may control several global parameters: for example, weight of the obstacle avoidance component, weight of the convergence to the goal, weight of the centering of the group, maximum velocity, maximum acceleration, minimum distance between actors. The animator provides data about the leader trajectory and the behavior of other birds relative to the leader. Wilhelms [1990] proposes a system based on a network of sensors and effectors. Ridsdale [1990] proposes a method that guides lower-level motor skills from a connectionist model of skill memory, implemented as collections of trained neural networks. Another approach to behavioral animation is based on timed and parameterized **L-systems** [Noser et al., 1992] with conditional and pseudostochastic productions. With this production-based approach, a user may create any realistic or abstract shape, play with fascinating tree structures, and generate any concept of growth and life development in the resulting animation.

40.4.2 Virtual Sensors

40.4.2.1 Perception through Virtual Sensors

In a typical behavioral animation scene, the actor perceives the objects and the other actors in the environment, providing information on their nature and position. This information is used by the behavioral model to decide the action to take, resulting in a motion procedure. In order to implement perception, virtual humans should be equipped with visual, tactile, and auditory sensors. These **virtual sensors** should be used as a basis for implementing everyday human behavior, such as visually directed locomotion, handling objects, and responding to sounds and utterances. For synthetic audition [Noser and Thalmann, 1995], one must model a sound environment where the synthetic actor can directly access positional and semantic sound source information of audible sound events. Simulating the haptic system corresponds roughly to a collision detection process.

But the most important perceptual subsystem is the vision system. The concept was first introduced by Renault et al. [1990] as a main information channel between the environment and the virtual actor. Reynolds [1993] more recently described an evolved, vision-based behavioral model of coordinated group motion. Tu and Terzopoulos [1994] also proposed artificial fish with perception and vision. In the Renault method, each pixel of vision input has the semantic information that gives the object projected on this pixel, and numerical information that gives the distance to this object. So it is easy to know, for example, that there is a table straight ahead at 3 meters. The synthetic actor perceives the environment from a small window in which the environment is rendered from the actor's point of view. Noser et al. [1995] proposed the use of an octree as the internal representation of the environment seen by an actor, because it can represent the visual memory of an actor in a 3-D environment with static and dynamic objects.

40.5 Crowds and Groups

Animating crowds [Musse and Thalmann, 2001] is challenging both in character animation and in virtual city modeling. Although different textures and colors may be used, the similarity of the virtual people would be soon detected even by nonexperts: for example, "Everybody walks the same in this virtual city!" Hence, it is useful to have a fast and intuitive way to generate motions with different personalities (depending on gender, age, emotions, etc.) from a sample motion, such as a genuine walking motion. The basic problem is to be able to generate variety among a finite set of motion requests and then to apply it either to an individual or to a member of a crowd. Very good tools are also required to tune the motion [Emering et al., 1998].

Bouvier et al. [1996, 1997] used a combination of particle systems and transition networks to model human crowds in visualizations of urban spaces. Lower-level behavior enabled people to avoid obstacles using attraction and repulsion forces analogous to physical electric forces. Higher-level behavior is modeled by transition networks, with transitions depending on timing, visiting of certain points, changes of local densities, and global events. Brogan and Hodgins [1994] simulated group behavior for systems with significant dynamics. They presented an algorithm for controlling the movements of creatures traveling as a group. The algorithm has a two steps: first, a perception model determines the creatures and obstacles visible to each individual, and then a placement algorithm determines the desired position for each individual, given the locations and velocities of perceived creatures and obstacles. Simulated systems included groups of legged robots, bicycle riders, and point-mass systems.

FIGURE 40.3 Crowd simulation.

Musse and Thalmann's [2001] proposed solution addresses two main issues: crowd structure and crowd behavior. Considering crowd structure, our approach deals with a hierarchy composed of crowd, groups, and agents, in which the groups are the most complex structure, containing the information to be distributed among the individuals. Concerning crowd behavior, the virtual agents are endowed with different levels of autonomy. They can either act according to an innate and scripted crowd behavior (programmed behavior), react as a function of triggered events (reactive or autonomous behavior), or be guided by an interactive process during simulation (guided behavior) [Musse et al., 1998]. Figure 40.3 shows a crowd guided by a leader.

Intelligence, memory, intention, and perception are focalized in the group structure. Also, each group can obtain one leader. This leader can be chosen randomly by the crowd system, be defined by the user, or emerge from the sociological rules.

For emergent crowds, Ulicny and Thalmann [2001] proposed a behavior model based on combinations of rules [Rosenblum et al., 1998] and finite state machines [Cremer et al., 1995] for controlling agents' behavior using a layered approach. The first layer deals with the selection of higher-level complex behavior appropriate to the agent's situation; the second layer implements these behaviors using low-level actions provided by the virtual human [Boulic et al., 1997]. At the higher level, rules select complex behaviors (such as *flee*) according to the agent's state (constituted by attributes) and the state of the virtual environment (conveyed by events). In rules, it is specified for whom (e.g., a particular agent or agents in particular group) and when (e.g., at a defined time, after receiving an event, or when some attribute reaches a specified value) the rule is applicable, and what is the consequence of rule firing (e.g., change of agent's high-level behavior or attribute). At the lower level, complex behaviors are implemented by hierarchical, finite state machines.

40.6 Facial Animation

The goal of facial animation systems has always been to obtain a high degree of realism using optimum-resolution facial mesh models and effective deformation techniques. Various muscle-based facial models with appropriate parameterized animation systems have been developed for facial animation [Parke, 1982;

FIGURE 40.4 Facial expressions.

Waters, 1987; Terzopoulos et al., 1990]. The facial action coding system [Friesen, 1978] defines high-level parameters for facial animation; several other systems are based on this one. Most facial animation systems typically use the following steps:

Define an animation structure on a facial model by parameterization.

Define building blocks or basic units of the animation in terms of these parameters, such as static expressions and visemes (visual counterparts of phonemes).

Use these building blocks as keyframes and define various interpolation and blending functions on the parameters to generate words and sentences from visemes and emotions from expressions (see Figure 40.4). The interpolation and blending functions contribute to the realism of a desired animation effect.

Generate the mesh animation from the interpolated or blended keyframes. Given the tools of parameterized face modeling and deformation, the most challenging task in facial animation is the design of realistic facial expressions and visemes.

The complexity of the keyframe-based facial animation system increases when we incorporate natural effects, such as coarticulation for speech animation and blending among a variety of facial expressions during speech. The use of speech synthesis systems and the subsequent application of coarticulation to the available temporized phoneme information is a widely accepted approach [Grandstrom, 1999; Hill et al., 1988].

Coarticulation is a phenomenon observed during fluent speech, in which facial movements corresponding to one phonetic or visemic segment are influenced by those corresponding to the neighboring segments. Two main approaches taken for coarticulation are those of Pelachaud [1991] and Cohen and Massaro [1993]. Both these approaches are based on the classification of phoneme groups and their observed interaction during speech pronunciation. Pelachaud arranged the phoneme groups according to the deformability and context dependence in order to decide the influence of the visemes on each other. Muscle contraction and relaxation times were also considered, and the facial action units were controlled accordingly.

For facial animation, the MPEG-4 standard is particularly important [MPEG-4]. The facial definition parameter (FDP) set and the facial animation parameter (FAP) set are designed to encode facial shape and animation of faces, thus reproducing expressions, emotions, and speech pronunciation. The FDPs are defined by the locations of the feature points and are used to customize a given face model to a particular face. They contain 3-D feature points, such as mouth corners and contours, eye corners, eyebrow centers, etc. FAPs are based on the study of minimal facial actions and are closely related to muscle actions. Each FAP value is simply the displacement of a particular feature point from its neutral position, expressed in terms of facial animation parameter units (FAPUs). The FAPUs correspond to fractions of distances between key facial features (e.g., the distance between the eyes). For example, the facial animation engine developed at MIRALab uses the MPEG-4 facial animation standard [Kshishagar et al., 1999] for animating 3-D facial models in real time. This parameterized model is capable of displaying a variety of facial expressions, including speech pronunciation, with the help of 66 low-level FAPs.

Recently, efforts in the field of phoneme extraction have resulted in software systems capable of extracting phonemes from both synthetic and natural speech and generating lip-synchronized speech animation from these phonemes. This creates a complete talking head system. It is possible to mix emotions with speech in a natural way, thus imparting to the virtual character an emotional behavior. Ongoing efforts concentrate on imparting emotional autonomy to the virtual face, enabling a dialogue between real and virtual humans with natural emotional responses. Kshirsagar and Magnenat Thalmann [2001] use a statistical analysis of the facial feature point movements. As the data is captured for fluent speech, the analysis reflects the dynamics of facial movements related to speech production. The results of the analysis were successfully applied for a more realistic speech animation. This has enabled us to blend various facial expressions and speech easily. Use of MPEG-4 feature points for data capture and facial animation enabled us to restrict the quantity of data being processed, at the same time offering more flexibility with respect to the facial model. We would like to improve the effectiveness of expressive speech further by the use of various time envelopes for expressions that may be linked to the meaning of the sentence. Kshirsagar and Magnenat Thalmann [2002] have also developed a system incorporating a personality model, for an emotionally autonomous virtual human.

40.7 Algorithms

40.7.1 Kochanek–Bartels Spline Interpolation

The method consists of interpolating splines with three parameters for local control: tension, continuity, and bias. Consider a list of points P_i and the parameter t along the spline to be determined. A point V is obtained from each value of t from only the two nearest given points along the curve (one behind P_i, one in front of P_{i+1}). But the tangent vectors D_i and D_{i+1} at these two points are also necessary. This means that we have

$$\mathbf{V} = THC^T \tag{40.1}$$

where T is the matrix $[t^3, t^2, t, 1]$, H is the Hermite matrix, and C is the matrix $[P_i, P_{i+1}, D_i, D_{i+1}]$. The Hermite matrix is given by

$$H = \begin{bmatrix} 2 & -2 & 1 & 1 \\ -3 & 3 & -2 & -1 \\ 0 & 0 & 1 & 0 \\ 1 & 0 & 0 & 0 \end{bmatrix} \tag{40.2}$$

This equation shows that the tangent vector is the average of the source chord $P_i - P_{i-1}$ and the destination chord $P_{i+1} - P_i$. Similarly, the source derivative (tangent vector) DS_i and the destination derivative (tangent vector) DD_i may be considered at any point P_i.

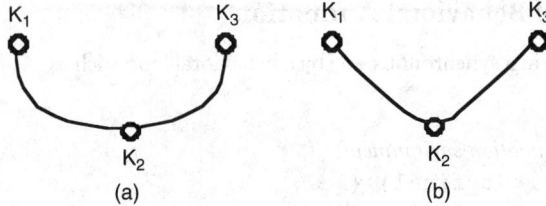

FIGURE 40.5 Variation of tension: the interpolation in b is more tense than the interpolation in a.

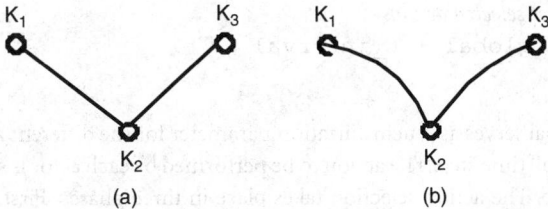

FIGURE 40.6 Variation of continuity: the interpolation in b is more discontinuous than the interpolation in a.

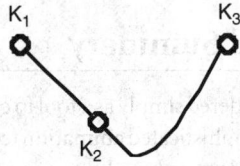

FIGURE 40.7 A biased interpolation at K_2.

Using these derivatives, Kochanek and Bartels [1984] propose the use of three parameters to control the splines: tension, continuity, and bias.

The tension parameter t controls how sharply the curve bends at a point P_i. As shown in Figure 40.5, in certain cases a wider, more exaggerated curve may be desired; in other cases, the desired path may be much tighter.

The continuity c of the spline at a point P_i is controlled by the parameter c. Continuity in the direction and speed of motion is not always desirable. Animating a bouncing ball, for example, requires the introduction of a discontinuity in the motion of the point of impact, as shown in Figure 40.6.

The direction of the path as it passes through a point P_i is controlled by the bias parameter b. This feature allows the animator to have a trajectory anticipate or overshoot a key position by a certain amount, as shown in Figure 40.7.

Equations combining the three parameters may be obtained:

$$\mathbf{DS}_i = 0.5[(1-t)(1+c)(1-b)(\mathbf{P}_{i+1} - \mathbf{P}_i) + (1-t)(1-c)(1+b)(\mathbf{P}_i - \mathbf{P}_{i-1})] \quad (40.3)$$

$$\mathbf{DD}_i = 0.5[(1-t)(1-c)(1-b)(\mathbf{P}_{i+1} - \mathbf{P}_i) + (1-t)(1+c)(1+b)(\mathbf{P}_i - \mathbf{P}_{i-1})] \quad (40.4)$$

A spline is then generated using Equation 1, with $\mathbf{DD}_i$ and $\mathbf{DS}_{i+1}$ instead of $\mathbf{D}_i$ and $\mathbf{D}_{i+1}$.

40.7.2   Principle of Behavioral Animation

A simulation is produced in a synchronous way by a behavioral loop such as

```
t_global  ←0.0
code to initialize the animation environment
while (t_global < t_final) {
code to update the scene
   for each actor
code to realize the perception of the environment
code to select actions based on sensorial input, actual state and specific behavior
   for each actor
code executing the above selected actions
   t_global ← t_global + t_interval
}
```

The global time t_global serves as synchronization parameter for the different actions and events. Each iteration represents a small time step. The action to be performed by each actor is selected by its behavioral model for each time step. The action selection takes place in three phases. First, the actor perceives the objects and the other actors in the environment, which provides information on their nature and position. This information is used by the behavioral model to decide the action to take, which results in a motion procedure with its parameters (e.g., grasp an object or walk with a new speed in a new direction). Finally, the actor performs the motion.

40.8   Research Issues and Summary

Computer animation should not be considered simply as a tool to enhance spatial perception by moving the virtual camera or rotating objects. More sophisticated animation techniques than keyframe animation must be widely used. Computer animation tends to be based more and more on physics and behavioral models. In the future, the application of computer animation to the scientific world will become common in many scientific areas: fluid dynamics, molecular dynamics, thermodynamics, plasma physics, astrophysics, etc. Real-time complex animation systems will be developed, taking advantage of virtual reality (VR) devices and simulation methods. An integration between simulation methods and VR-based animation will lead to systems allowing the user to interact with complex, time-dependent phenomena, providing interactive visualization and interactive animation. Moreover, real-time synthetic actors will be part of virtual worlds, and people will communicate with them through broadband multimedia networks. Such applications will only become possible through the development of new approaches to real-time motion, based on artificial intelligence and agent technology.

Defining Terms

Behavioral animation: Behavior of objects is driven by providing high-level directives indicating a specific behavior without any other stimulus.

L-systems: Production-based approach to generate any concept of growth and life development in the resulting animation.

Metaballs: Shapes based on implicit surfaces; they are employed to simulate the gross behavior of bone, muscle, and fat tissue.

Parametric keyframe animation: The animator creates keyframes by specifying the appropriate set of parameter values. Parameters are then interpolated and, finally, images are individually constructed from the interpolated parameters.

Performance animation: Consists of measuring and recording direct actions of a real person or animal for immediate or delayed analysis and playback.

Procedural animation: Corresponds to the creation of a motion by a procedure specifically describing the motion. Procedural animation may be specified using a programming language or an interactive system.

Space–time constraints: Method for creating automatic character motion by specifying *what* the character has to be, *how* the motion should be performed, what the character's *physical structure* is, and what physical *resources* are available to the character to accomplish the motion.

Virtual sensors: Used as a basis for implementing everyday human behavior, such as visually directed locomotion, handling objects, and responding to sounds and utterances. Virtual humans should be equipped with visual, tactile, and auditory sensors.

References

Aono, M. 1990. A wrinkle propagation model for cloth, *Proc. Computer Graphics International '90*, pp.96–115, Springer, Tokyo.

Aubel, A., and Thalmann, D. 2001. Interactive modeling of the human musculature, *Proc. Computer Animation*, Seoul, Korea.

Badler, N.I., Manoochehri, K.H., and Walters, G. 1987. Articulated figure positioning by multiple constraints, *IEEE Computer Graphics and Applications* 17(6): 28–38.

Baerlocher, P., and Boulic, R. 2001. Parametrization and range of motion of the ball-and-socket joint, *Deformable Avatars*, pp.180–190, Kluwer Academic Publishers.

Bandi, S., and Thalmann, D. 1995. An adaptive spatial subdivision of the object space for fast collision of animated rigid bodies, *Proc. Eurographics '95*, pp.259–270, Maastricht.

Baraff, D., and Witkin, A. 1998. Large steps in cloth simulation, in *SIGGRAPH '98 Conference Proceedings*, Michael Cohen, Ed., Annual Conference Series, pp. 43–54. ACM SIGGRAPH, Addison-Wesley, Reading, MA.

Baraff, D. 1989. Analytical methods for dynamic simulation of non-penetrating rigid bodies, *Proc. SIGGRAPH '89, Computer Graphics* 23(3): 223–232.

Bindiganavale, R., and Badler, N.I. 1998. Motion abstraction and mapping with spatial constraints. In *Modeling and Motion Capture Techniques for Virtual Environments, Lecture Notes in Artificial Intelligence*, N. Magnenat Thalmann and D. Thalmann, Eds., pp. 70–82, Springer. Held in Geneva, Switzerland, November, 1998.

Boulic, R., Mas, R., and Thalmann, D. 1996. A robust approach for the center of mass position control with inverse kinetics, *Journal of Computers and Graphics* 20(5): 693–701.

Boulic, R., Mas, R., and Thalmann, D. 1994. Inverse kinetics for center of mass position control and posture optimization, in *Race Workshop on Combined Real and Synthetic Image Processing for Broadcast and Video Production (Monalisa Project)*, Y. Parker and S. Wilbur, Eds., Workshop in Computing Series, Springer-Verlag, Hamburg.

Boulic, R., Becheiraz, P., Emering, L., and Thalmann, D. 1987. Integration of motion control techniques for virtual human and avatar real-time animation, *Proc. VRST '97*, pp. 111–118, ACM Press.

Bouvier, E., Cohen, E., and Najman, L. 1997. From crowd simulation to airbag deployment: particle systems, a new paradigm of simulation, *Journal of Electrical Imaging* 6(1):94–107.

Bouvier, E., and Guilloteau, P. 1996. Crowd simulation in immersive space management, *Proc. Eurographics Workshop on Virtual Environments and Scientific Visualization '96*, pp. 104–110, Springer-Verlag.

Brogan, D., and Hodgins, J. 1994. Robot herds: group behaviors for systems with significant dynamics, *Proc. Artificial Life IV*, pp.319–324.

Bruderlin, A., and Williams, 1995. L. Motion signal processing, in *SIGGRAPH '95 Conference Proceedings, Annual Conference Series*, Robert Cook, Ed., pp. 97–104. ACM SIGGRAPH, Addison-Wesley, Reading, MA. Held in Los Angeles, CA, August, 1995.

Carignan, M., Yang, Y., Magnenat Thalmann, N., and Thalmann, D. 1992. Dressing animated synthetic actors with complex deformable clothes, *Proc. SIGGRAPH '92, Computer Graphics* 26(4): 99–104.

Chadwick, J., Haumann, D., and Parent R. 1989. Layered construction for deformable animated characters, *Computer Graphics (SIGGRAPH '89 Proceedings)*, pp. 243–252.

Cohen, M.M., and Massaro, D.W. 1993. Modelling coarticulation in synthetic visual speech, in *Models and Techniques in Computer Animation*, N. M. Thalmann and D. Thalmann, Eds., pp. 139–156, Springer-Verlag.

Cordier, F., and Magnenat Thalmann, N. 2002. Real-time animation of dressed virtual humans, *Eurographics Conference Proceedings*, Blackwell Publishers.

Cremer, J., Kearney, J., and Papelis, Y. 1995. HCSM: framework for behavior and scenario control in virtual environments, *ACM Transactions on Modeling and Computer Simulation* 5(3): 242–267.

Debunne, G., Desbrun, M., Cani, M.P., and Barr, A. 2000. Adaptive simulation of soft bodies in real-time, *Computer Animation 2000, Annual Conference Series*, IEEE Press.

Desbrun, M., Schröder, P., and Barr, A.H. 1999. Interactive animation of structured deformable objects. In *Graphics Interface '99 Proceedings*, pp. 1–8.

Emering, L., Boulic, R., Molet, T., and Thalmann, D. 1998. Versatile tuning of humanoid agent activity, *Computer Graphics Forum*.

Faloutsos, P., van de Panne, M., and Terzopoulos D. 2001. Composable controllers for physics-based character animation. *Proceedings of ACM SIGGRAPH 2001*. Held in Los Angeles, August, 2001.

Featherstone, R. 1986. *Robot Dynamics Algorithms*, Kluwer Academic Publishers.

Friesen, E.W.V. 1978. *Facial Action Coding System: A Technique for the Measurement of Facial Movement*. Consulting Psychologists Press, Palo Alto, CA.

Gleicher, M. 1998. Retargeting motion to new characters, in *SIGGRAPH '98 Conference Proceedings, Annual Conference Series*, Michael Cohen, Ed., pp. 33–42. ACM SIGGRAPH, Addison-Wesley, Reading, MA.

Grandstrom B. Multi-modal speech synthesis with applications, in G. Chollet, M. Di

Hahn, J.K. 1988. Realistic animation of rigid bodies, *Proc. SIGGRAPH '88, Computer Graphics* 22(4): 299–308.

Henne, M. 1990. *A Constraint-Based Skin Model For Human Figure Animation*. Master's thesis, University of California, Santa Cruz.

Hill, D.R., Pearce, A., and Wyvill B. 1988. Animating speech: an automated approach using speech synthesized by rule, *The Visual Computer* 3: 277–289.

Hodgins, J., Wooten, W.L., Brogan, D.C., and O'Brien, J.F. 1995. Animating human athletics, *Proc. of SIGGRAPH '95*, pp. 71–78.

Huston, R.L. 1990. *Multibody Dynamics*, Butterworth-Heinemann, Stoneham.

James, D., and Pai, D. 1999. Accurate real-time deformable objects. *SIGGRAPH '99 Conference Proceedings, Annual Conference Series*, pp. 65–72. ACM SIGGRAPH, Addison-Wesley, Reading, MA.

Ko, H., and Badler, N.I. 1996. Animating human locomotion with inverse dynamics, *IEEE Comput. Graph. Appl.* 16: 50–59.

Kochanek, D.H., and Bartels, R.H. 1984. Interpolating splines with local tension, continuity, and bias control, *Proc. SIGGRAPH '84, Computer Graphics* 18: 33–41.

Kshirsagar, S., Garchery, S., and Thalmann, N. 2000. Feature point based mesh deformation applied to MPEG-4 facial animation, *Proceedings Deform 2000*.

Kshirsagar, S., and Magnenat Thalmann, N. 2001. Principal components of expressive speech animation, *Proc. Computer Graphics International 2001*, pp. 38–44, IEEE Computer Society.

Kshirsagar, S., and Magnenat Thalmann, N. 2002. Virtual humans personified, *Proceedings Autonomous Agents Conference (AAMAS) 2002*, pp. 356–359.

Kunii, T.L., and Gotoda, H. 1990. Modeling and animation of garment wrinkle formation processes, *Proc. Computer Animation '90*, pp.131–147, Springer, Tokyo.

Lafleur, B., Magnenat Thalmann, N., and Thalmann, D. 1991. Cloth animation with self-collision detection, *Proc. IFIP Conference on Modeling in Computer Graphics*, pp.179–187, Springer, Tokyo.

Lazlo, J., Van De Panne, M., and Fiume E. 1996. Limit cycle control and its application to the animation of balancing and walking, *Proc. SIGGRAPH '96*, pp. 155–162.

Lee, S., and Shin, S.Y. 1995. Warp generation and transition control in image morphing, in *Interactive Computer Animation*, N. Magnenat Thalmann and D. Thalmann D., Eds., Prentice Hall, .

MacKenna, M., and Zeltzer D. 1996. Dynamic simulation of a complex human figure model with low level behavior control, *Presence* 5(4): 431–456.

Magnenat Thalmann, N., and Thalmann, D. 1991. Complex models for animating synthetic actors, *IEEE Computer Graphics and Applications*,11 (September):.

Magnenat Thalmann, N., and Thalmann, D. 1987. The direction of synthetic actors in the film *Rendez-vous à Montréal*, *IEEE CG&A* 7(12): 9–19.

Meyer, M., Debunne, G., Desbrun, M., and Barr, A.H. 2000. Interactive animation of cloth-like objects in virtual reality, *Journal of Visualization and Computer Animation 2000*, John Wiley & Sons.

Monzani, J.S., Baerlocher, P., Boulic, R., and Thalmann D. 2000. Using an intermediate skeleton and inverse kinematics for motion retargeting, *Proc. Eurographics 2000*, pp.11–19.

MPEG-4 standard, Moving Picture Experts Group, http://www.cselt.it/mpeg.

Multon, F., France, L., Cani-Gascuel, M.-P., and Debunne, G. 1999. Computer animation of human walking: a survey, *J. Visual. Comput. Animat.* 10: 39–54.

Musse, S.R., and Thalmann, D. 2001. A behavioral model for real-time simulation of virtual human crowds, *IEEE Transactions on Visualization and Computer Graphics* 7(2): 152–164.

Musse, S.R., Babski, C., Capin, T., and Thalmann, D. 1998. Crowd modelling in collaborative virtual environments, *ACM VRST '98*, Taiwan.

Noser, H., and Thalmann, D. 1995. Synthetic vision and audition for digital actors, *Proc. Eurographics '95*, pp. 325–336, Maastricht, August.

Noser, H., Renault, O., Thalmann, D., and Magnenat Thalmann, N. 1995. Navigation for digital actors based on synthetic vision, memory and learning, *Computers and Graphics* 19(1): 7–19.

Noser, H., Turner, R., and Thalmann, D. 1992. Interaction between L-systems and vector force-field, *Proc. Computer Graphics International '92*, pp.747–761, Tokyo.

Parke, F.I. 1982. Parameterized models for facial animation, *IEEE Computer Graphics and Applications*, 2(19): 61–68.

Pelachaud, C. 1991. *Communication and Coarticulation in Facial Animation*, Ph.D. thesis, University of Pennsylvania.

Phillips, C.B., Zhao, J., and Badler, N.I. 1990. Interactive real-time articulated figure manipulation using multiple kinematic constraints *Computer Graphics* 24(2): 245–250.

Phillips, C.B., and Badler, N. 1991. Interactive behaviors for bipedal articulated figures, *Computer Graphics* 25(4): 359–362.

Popovic, Z., and Witkin, A. 1999. Physically based motion transformation. *Proceedings of SIGGRAPH '99*, pp. 11–20, Held in Los Angeles, CA, August, 1999.

Porcher-Nedel, L., and Thalmann, D. 1998. Real time muscle deformations using mass-spring systems, *Proc. CGI '98*, IEEE Computer Society Press.

Renault, O., Magnenat Thalmann, N., and Thalmann, D. 1990. A vision-based approach to behavioral animation, *Visualization and Computer Animation Journal* 1(1): 18–21.

Reynolds, C. 1987. Flocks, herds, and schools: a distributed behavioral model, *Proc. SIGGRAPH '87, Computer Graphics* 21(4): 25–34.

Reynolds, C.W. 1993. An evolved, vision-based behavioral model of coordinated group motion, in *From Animals to Animats, Proc. 2nd International Conf. on Simulation of Adaptive Behavior*, J.A. Meyer et al., Eds., pp.384–392, MIT Press, Cambridge, MA.

Ridsdale, G. 1990. Connectionist modelling of skill dynamics, *Journal of Visualization and Computer Animation* 1(2): 66–72.

Roberts T. 1995. *Understanding Balance: The Mechanics of Posture and Locomotion*, Chapman and Hall.

Rosenbloom, P.S., Laird, J.E., and Newell, A. 1993. *The Soar Papers: Research on Artificial Intelligence*, MIT Press, Cambridge, MA.

Scheepers, F., Parent, R., Carlson, W., and May, S. 1997. Anatomy-based modeling of the human musculature, *Computer Graphics (SIGGRAPH '97 Proceedings)*, pp. 163–172.

SDFAST User Manual. 1990.

Seo, H., Cordier, F., Philippon, and Magnenat Thalmann, N. 2000. *Interactive Modelling of MPEG-4 Deformable Human Body Models, Postproceedings Deform*, pp. 120–131, Kluwer Academic Publishers.

Terzopoulos, D., and Waters, K. 1990. Physically based facial modelling, analysis and animation, *Journal of Visualization and Computer Animation* 1(2): 73–90.

Terzopoulos, D., Platt, J.C., Barr, A.H., and Fleischer, K. 1987. Elastically deformable models, *Proc. SIGGRAPH '87, Computer Graphics* 21(4): 205–214.

Thalmann, D., Shen, J., and Chauvineau, E. 1996. Fast realistic human body deformations for animation and VR applications, *Computer Graphics International '96*. Held in Pohang, Korea, June, 1996.

Tu, X., and Terzopoulos, D. 1994. Artificial fishes: physics, locomotion, perception, behavior, *Proc. SIGGRAPH '94, Computer Graphics*, pp.42–48.

Ulicny, B., and Thalmann, D. 2001. Crowd simulation for interactive virtual environments and VR training systems, *Proc. Eurographics Workshop on Animation and Simulation '01*, Springer-Verlag.

Vassilev, T., and Spanlang, B. 2001. Fast cloth animation on walking avatars, *Eurographics*, September, 2001.

Volino, P., Courchesnes, M., and Magnenat Thalmann, N. 1995. Versatile and efficient techniques for simulating cloth and other deformable objects, *Proc. SIGGRAPH '95*.

Volino, P., and Magnenat Thalmann, N. 1994. Efficient self-collision detection on smoothly discretised surface animations using geometrical shape regularity, *Proc. Eurographics '94, Computer Graphics Forum* 13(3): 155–166.

Waters, K. 1987. A muscle model for animating three-dimensional facial expression, *Proc. SIGGRAPH '87, Computer Graphics* 21(4): 17–24.

Weil, J. 1986. The synthesis of cloth objects, *SIGGRAPH '86 Conference Proceedings, Computer Graphics*, Annual Conference Series 20: 49–54. ACM SIGGRAPH, Addison-Wesley, Reading, MA.

Wilhelms, J. 1990. A "notion" for interactive behavioral animation control, *IEEE Computer Graphics and Applications* 10(3): 14–22.

Wilhelms, J., and Van Gelder, A. 1997. Anatomically based modeling, *Computer Graphics (SIGGRAPH '97 Proceedings)*, pp. 173–180.

Williams, L. 1990. Performance driven facial animation, *Proc. SIGGRAPH '90*, pp. 235–242.

Witkin, A., and Popovic, Z. 1995. Motion warping. *Proceedings of SIGGRAPH '95*, pp. 105–108. Held in Los Angeles, CA, August, 1995.

Witkin, A., and Kass, M. 1988. Spacetime constraints, *Proc. SIGGRAPH '88, Computer Graphics* 22(4): 159–168.

Yoshimito, S. 1992. Ballerinas generated by a personal computer, *Journal of Visualization and Computer Animation* 3: 85–90.

Zhao, J., and Badler, N.I. 1994. Inverse kinematics positioning using nonlinear programming for highly articulated figures, *ACM Transactions on Graphics* 13(4): 313–336.

Further Information

Several textbooks on computer animation have been published:

Capin, T., Pandzic, I., and Magnenat Thalmann, N. 1999. *Avatars in Networked Virtual Environments*, John Wiley & Sons, New York.

Magnenat Thalmann, N., and Thalmann, D., Eds. 1996. *Interactive Computer Animation*, Prentice Hall, Englewood Cliffs, NJ.

Parent, R. 2001. *Computer Animation: Algorithms and Techniques*, Morgan Kaufmann, San Francisco.

Vince, J. 1992. *3-D Computer Animation*, Addison-Wesley, Reading, MA.

There is one journal dedicated to computer animation: *The Journal of Visualization and Computer Animation*. This journal has been published by John Wiley & Sons, in Chichester, UK, since 1990. In January 2004, this journal changed its name to *Computer Animation and Virtual Worlds*.

Although computer animation is always represented in major computer graphics conferences like SIG-GRAPH, Computer Graphics International (CGI), Pacific Graphics, and Eurographics, there are only two annual conferences dedicated to computer animation:

Computer Animation, organized each year in Geneva by the Computer Graphics Society. Proceedings are published by the IEEE Computer Society Press.

Symposium on Computer Animation, organized by SIGGRAPH and Eurographics.

41

Volume Visualization

41.1 Introduction **41**-1
41.2 Volumetric Data **41**-2
41.3 Rendering via Geometric Primitives **41**-2
41.4 Direct Volume Rendering: Prelude **41**-3
41.5 Volumetric Function Interpolation **41**-3
41.6 Volume Rendering Techniques...................... **41**-6
    Image-Order Techniques • Object-Order Techniques
    • Hybrid Techniques • Domain Volume Rendering
41.7 Acceleration Techniques **41**-13
41.8 Classification and Transfer Functions **41**-16
41.9 Volumetric Global Illumination **41**-18
41.10 Special-Purpose Rendering Hardware **41**-19
41.11 General-Purpose Rendering Hardware **41**-19
41.12 Irregular Grids **41**-20
41.13 High-Dimensional and Multivalued Data **41**-22
41.14 Volume Graphics................................. **41**-22
    Voxelization • Fundamentals of 3-D Discrete Topology
    • Binary Voxelization • Antialiased Voxelization
    • Block Operations and Constructive Solid Modeling
    • Texture Mapping and Solid Texturing • Amorphous
    Phenomena • Natural Phenomena • Volume Sculpting
41.15 Conclusions **41**-31

Arie Kaufman
State University of New York at Stony Brook

Klaus Mueller
State University of New York at Stony Brook

41.1 Introduction

Volume visualization is a method of extracting meaningful information from volumetric data using interactive graphics and imaging. It is concerned with volume data representation, modeling, manipulation, and rendering [20] [91] [92] [154]. Volume data are 3-D (possibly time-varying) entities that may have information inside them, might not consist of tangible surfaces and edges, or might be too voluminous to be represented geometrically. They are obtained by sampling, simulation, or modeling techniques. For example, a sequence of 2-D slices obtained from magnetic resonance imaging (MRI), computed tomography (CT), functional MRI (fMRI), or positron emission tomography (PET), is 3-D reconstructed into a volume model and visualized for diagnostic purposes or for planning of treatment or surgery. The same technology is often used with industrial CT for nondestructive inspection of composite materials or mechanical parts. Similarly, confocal microscopes produce data which are visualized to study the morphology of biological structures. In many computational fields, such as computational fluid dynamics, the results of simulations typically running on a supercomputer are often visualized as volume data for analysis and verification. Recently, the subarea of *volume graphics* [96] has been expanding, and many traditional geometric computer graphics applications, such as CAD and flight simulation, have been exploiting the advantages of volume techniques.

Over the years, many techniques have been developed to render volumetric data. Because methods for displaying geometric primitives were already well established, most of the early methods involved approximating a surface contained within the data using geometric primitives. When volumetric data are visualized using a surface rendering technique, a dimension of information is essentially lost. In response to this, volume rendering techniques were developed that attempt to capture all the 3-D data in a single 2-D image. Volume rendering conveys more information than surface rendering images, but at the cost of increased algorithm complexity and, consequently, increased rendering times. To improve interactivity in volume rendering, many optimization methods, both for software and for graphics accelerator implementations, as well as several special-purpose volume rendering machines, have been developed.

41.2 Volumetric Data

A volumetric data set is typically a set V of samples (x, y, z, v), also called *voxels*, representing the value v of some property of the data, at a 3-D location (x, y, z). If the value is simply 0 or an integer i within a set I, with a value of 0 indicating background and the value of i indicating the presence of an object O_i, then the data is referred to as binary data. The data may instead be multivalued, with the value representing some measurable property of the data, including, for example, color, density, heat or pressure. The value v may even be a vector, representing, for example, velocity at each location; results from multiple scanning modalities, such as anatomical (CT, MRI) and functional imaging (PET, fMRI); or color (RGB) triples, such as the Visible Human cryosection data set [82]. Finally, the volume data may be time-varying, in which case V becomes a 4-D set of samples (x, y, z, t, v).

In general, the samples may be taken at purely random locations in space, but in most cases the set V is isotropic, containing samples taken at regularly spaced intervals along three orthogonal axes. When the spacing between samples along each axis is a constant, but there may be three different spacing constants for the three axes, the set V is anisotropic. Since the set of samples is defined on a regular grid, a 3-D array (also called the *volume buffer*, 3-D *raster*, or simply the *volume*) is typically used to store the values, with the element location indicating position of the sample on the grid. For this reason, the set V will be referred to as the array of values $V(x, y, z)$, which is defined only at grid locations.

Alternatively, either rectilinear, curvilinear (structured), or unstructured grids are employed (e.g., [203]). In a *rectilinear* grid, the cells are axis-aligned, but grid spacings along the axes are arbitrary. When such a grid has been nonlinearly transformed while preserving the grid topology, the grid becomes *curvilinear*. Usually, the rectilinear grid defining the logical organization is called *computational space*, and the curvilinear grid is called *physical space*. Otherwise, the grid is called *unstructured* or *irregular*. An unstructured or irregular volume data set is a collection of cells whose connectivity must be specified explicitly. These cells can be of an arbitrary shape, such as tetrahedra, hexahedra, or prisms.

41.3 Rendering via Geometric Primitives

Several techniques have been developed that approximate a surface contained within the volumetric data by ways of geometric primitives, most commonly triangles, which can then be rendered using conventional graphics accelerator hardware. A surface can be defined by applying a binary segmentation function $B(v)$ to the volumetric data, where $B(v)$ evaluates to 1 if the value v is considered part of the object, and evaluates to 0 if the value v is part of the background. The surface is then contained in the region where $B(v)$ changes from 0 to 1.

Most commonly, $B(v)$ is either a step function

$$B(v) = 1, \forall v \geq v_{iso}$$

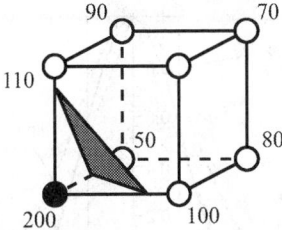

FIGURE 41.1 A grid cell with voxel values as indicated, intersected by an iso-surface (iso-value = 125). This is base case # 1 of the Marching Cube algorithm: a single triangle separating surface interior (black vertex) from exterior (white vertices). The positions of the triangle vertices are estimated by linear interpolation along the cell edges.

(where v_{iso} is called the *isovalue*) or an interval $[v_1, v_2]$ in which

$$B(v) = 1, \forall v \in [v_1, v_2]$$

(where $[v_1, v_2]$ is called the *isointerval*).

For the former, the resulting surface is called the *isosurface*; for the latter, the resulting structure is called the *isocontour*. Several methods for extracting and rendering isosurfaces have been developed; a few are briefly described here. The *Marching Cubes* algorithm [127] was developed to approximate an isovalued surface with a triangle mesh. The algorithm breaks down the ways in which a surface can pass through a grid cell into 256 cases, based on the $B(v)$ membership of the eight voxels that form the cell's vertices. By way of symmetry, the 256 cases reduce to 15 base topologies, although some of these have duals, and a technique called *asymptotic decider* [165] can be applied to select the correct dual case and thus prevent the incidence of holes in the triangle mesh. For each of the 15 cases (and their duals), a generic set of triangles representing the surface is stored in a lookup table. Each cell through which a surface passes maps to one of the base cases, with the actual triangle vertex locations being determined using linear interpolation of the cell vertices on the cell edges (see Figure 41.1). A normal value is estimated for each triangle vertex, and standard graphics hardware can be utilized to project the triangles, resulting in a relatively smooth shaded image of the isovalued surface.

41.4 Direct Volume Rendering: Prelude

Representing a surface contained within a volumetric data set using geometric primitives can be useful in many applications. However, there are several main drawbacks to this approach. First, geometric primitives can only approximate surfaces contained within the original data. Adequate approximations may require an excessive amount of geometric primitives. Therefore, a trade-off must be made between accuracy and space requirements. Second, because only a surface representation is used, much of the information contained within the data is lost during the rendering process. For example, in CT scanned data, useful information is contained not only on the surfaces, but also within the data. Also, amorphous phenomena, such as clouds, fog, and fire, cannot be represented adequately using surfaces, and therefore must have a volumetric representation and must be displayed using volume rendering techniques.

Before moving to techniques that visualize the data directly without going through an intermediate surface extraction step, we discuss some of the general principles that govern the theory of discretized functions and signals, such as the discrete volume data. We also present some specialized theoretical concepts relevant to the context of volume visualization.

41.5 Volumetric Function Interpolation

The volume grid V only defines the value of some measured property $f(x, y, z)$ at discrete locations in space. If one requires the value of $f(x, y, z)$ at an off-grid location (x, y, z), a process called interpolation must be employed to estimate the unknown value from the known grid samples $V(x, y, z)$. There are many

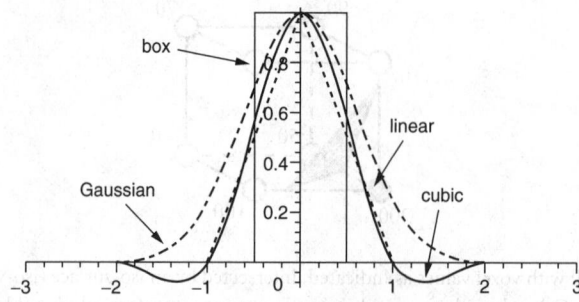

FIGURE 41.2 Popular filters in the spatial domain: box, linear, and Gaussian.

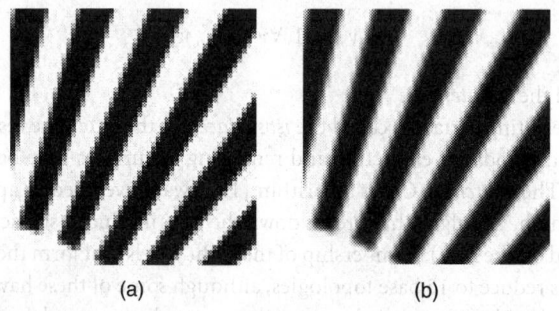

FIGURE 41.3 Magnification via interpolation with (a) a box filter; and (b) a bi-linear filter. The latter gives a much more pleasing result.

possible interpolation functions (also called *filters* or *filter kernels*). The simplest interpolation function is known as zero-order interpolation, which is actually just a nearest-neighbor function. That is, the value at any location (x, y, z) is simply that of the grid sample closest to that location:

$$f(x, y, z) = V(round(x), round(y), round(z)) \qquad (41.1)$$

This gives rise to a box filter (the black curve in Figure 41.2). With this interpolation method, there is a region of constant value around each sample in V. The human eye is very sensitive to the jagged edges and unpleasant staircasing that result from a zero-order interpolation. Therefore, this kind of interpolation generally gives the poorest visual results. See Figure 41.3a.

Linear or first-order interpolation (the magenta curve in Figure 41.2) is the next-best choice, and its 2-D and 3-D versions are called bilinear and trilinear interpolation, respectively. In 3-D, it can be written as three stages of seven linear interpolations, because the filter function is separable in higher dimensions. The first four linear interpolations are along x:

$$f(u, v_{0,1}, w_{0,1}) = (1 - u)(V(0, v_{0,1}, w_{0,1}) + uV(1, v_{0,1}, w_{0,1})) \qquad (41.2)$$

Using these results, two linear interpolations along y follow:

$$f(u, v, w_{0,1}) = (1 - v)f(u, 0, w_{0,1}) + vf(u, 1, w_{0,1}) \qquad (41.3)$$

One final interpolation along z yields the interpolation result:

$$f(x, y, z) = f(u, v, w) = (1 - w)f(u, v, 0) + wf(u, v, 1) \qquad (41.4)$$

Here the u, v, w are the distances (assuming a cell of size 1^3, without loss of generality) of the sample at (x, y, z) from the lower, left, rear voxel in the cell containing the sample point (e.g., the voxel with value 50 in Figure 41.1). A function interpolated with a linear filter no longer suffers from staircase artifacts; see Figure 41.3b. However, it has discontinuous derivatives at cell boundaries, which can lead to noticeable banding when the visual quantities change rapidly from one cell to the next.

A second-order interpolation filter that yields a $f(x, y, z)$ with a continuous first derivative is the cardinal spline function (see the blue curve in Figure 41.2), whose 1-D function is given by

$$
h(u) = \begin{pmatrix} (a+2)|u|^3 - (a+3)|u|^2 + 1 & 0 \leq |u| < 1 \\ a|u|^3 - 5a|u|^2 + 8a|u| - 4a & 1 \leq |u| \leq 2 \\ 0 & |u| > 2 \end{pmatrix} \tag{41.5}
$$

Here, u measures the distance of the sample location to the grid points that fall within the extent of the kernel, and $a = -0.5$ yields the Catmull–Rom spline, which interpolates a discrete function with the lowest third-order error [101]. The 3-D version of this filter $h(u, v, w)$ is separable, that is, $h(u, v, w) = h(u)h(v)h(w)$, and therefore interpolation in 3-D can be written as a three-stage nested loop.

A more general form of the cubic function has two parameters. The interpolation results obtained with different settings of these parameters have been investigated by Mitchell and Netravali [146]. In fact, the choice of filters and their parameters always presents trade-offs between the sensitivity to noise, sampling frequency ripple, aliasing (see discussion later in this section), ringing, and blurring, and there is no optimal setting that works for all applications. Marschner and Lobb [136] extended the filter discussion to volume rendering and created a challenging volumetric test function with a uniform frequency spectrum that can be employed to observe visually the characteristics of different filters. Finally, Möller et al. [149] applied a Taylor series expansion to devise a set of optimal nth order filters that minimize the $(n+1)$th order error.

Generally, higher filter quality comes at the price of wider spatial extent (compare Figure 41.2) and therefore larger computational effort. The best filter possible in the numerical sense is the *sinc* filter, but it has infinite spatial extent and also has noticeable ringing [146]. Sinc filters make excellent, albeit expensive, interpolation filters when used in truncated form and multiplied by a window function [136] [215], possibly adaptive to local detail [131]. In practice, first-order or linear filters give satisfactory results for most applications, providing good cost–quality trade-offs, but cubic filters are also used. Zero-order filters give acceptable results when the discrete function has already been sampled at a very high rate, for example, in high-definition function lookup tables [239].

All filters presented thus far are grid-interpolating filters, that is, their interpolation yields $f(x, y, z) = V(x, y, z)$ at grid points [217]. When presented with a uniform grid signal, they also interpolate a uniform $f(x, y, z)$ everywhere. This is not the case with a Gaussian filter function (the red curve in Figure 41.2), which can be written as

$$
h(u, v, w) = b \cdot e^{-a(u^2 + v^2 + w^2)} \tag{41.6}
$$

Here, a determines the width of the filter and b is a scale factor. The Gaussian has infinite continuity in the interpolated function's derivative, but it introduces a slight ripple (about 0.1%) into an interpolated uniform function. The Gaussian is most popular when a radially symmetric interpolation kernel is needed [237] [162] and for grids that assume that the frequency spectrum of $f(x, y, z)$ is radially band-limited [216] [161].

It should be noted that interpolation cannot restore sharp edges that may have existed in the original function $f_{org}(x, y, z)$ prior to sampling into the grid. Filtering will always smooth or somewhat *lowpass* the original function. Nonlinear filter kernels [83] or transformations of the interpolated results [158] are needed to re-create sharp edges, as we shall see later.

A frequent artifact that can occur is *aliasing*. Aliasing results from inadequate sampling and gives rise to strange patterns that did not exist in the sampled signal. Proper prefiltering (band-limiting) must be performed whenever a signal is sampled below its *Nyquist limit* (i.e., twice the maximum frequency that occurs in the signal). The interested reader may consult standard texts, such as [246] and [47], for more detail.

The gradient of $f(x, y, z)$ is also of great interest in volume visualization, mostly for the purpose of estimating the amount of light reflected from volumetric surfaces toward the eye (e.g., strong gradients indicate stronger surfaces and therefore stronger reflections). There are three popular methods to estimate a gradient from the volume data [148]. The first computes the gradient vector at each grid point via a process called *central differencing*:

$$
\begin{bmatrix} g_x \\ g_y \\ g_z \end{bmatrix} = \begin{bmatrix} V(x-1, y, z) \\ V(x, y-1, z) \\ V(x, y, z-1) \end{bmatrix} - \begin{bmatrix} V(x+1, y, z) \\ V(x, y+1, z) \\ V(x, y, z+1) \end{bmatrix} \tag{41.7}
$$

It then interpolates the gradient vectors at (x, y, z) using any of the filters described above. The second method also uses central differencing, but it does this at (x, y, z) by interpolating the required support samples on the fly. The third method is the most direct and employs a gradient filter [8] in each of the three axis directions to estimate the gradients. These three gradient filters could be simply the (u, v, w) partial derivatives of the filters described previously, or they could be a set of optimized filters [148].

The third method gives the best results because it only performs one interpolation step, whereas the other two methods have lower complexity and often have practical application-specific advantages. An important observation is that gradients are much more sensitive to the quality of the interpolation filter because they are used in illumination calculations, which consist of higher-order functions that involve the normal vectors, which in turn are calculated from the gradients via normalization [149].

41.6 Volume Rendering Techniques

The next subsections explore various fundamental volume rendering techniques. Volume rendering is the process of creating a 2-D image directly from 3-D volumetric data; hence it is often called direct volume rendering. Although several of the methods described in these subsections render surfaces contained within volumetric data, these methods operate on the actual data samples, without generating the intermediate geometric primitive representations used by the algorithms in the previous section.

Volume rendering can be achieved using an object-order, an image-order, or a domain-based technique. Hybrid techniques have also been proposed. *Object-order* volume rendering techniques use a forward mapping scheme in which the volume data is mapped onto the image plane. In *image-order* algorithms, a backward mapping scheme is used in which rays are cast from each pixel in the image plane through the volume data to determine the final pixel value. In a *domain-based* technique, the spatial volume data are first transformed into an alternative domain, such as compression, frequency, or wavelet, and then a projection is generated directly from that domain.

41.6.1 Image-Order Techniques

There are four basic volume rendering modes: x-ray rendering, maximum intensity projection (MIP), isosurface rendering, and full volume rendering. The third mode is just a special case of the fourth. These four modes share two common operations:

They all cast rays from the image pixels, sampling the grid at discrete locations along their paths.
They all obtain the samples via interpolation, using the methods described earlier.

The modes differ, however, in how the samples taken along a ray are combined. In x-ray, the interpolated samples are simply summed, giving rise to a typical image obtained in projective diagnostic imaging (Figure 41.4a). In MIP, only the interpolated sample with the largest value is written to the pixel (Figure 41.4b). In full volume rendering (Figure 41.4c and Figure 41.4d), on the other hand, the interpolated samples are further processed to simulate the light transport within a volumetric medium according to one of many possible models. In the remainder of this section, we shall concentrate on the full volume rendering mode because it provides the greatest degree of freedom, although rendering algorithms have been proposed that merge the different modes into a hybrid image generation model [74].

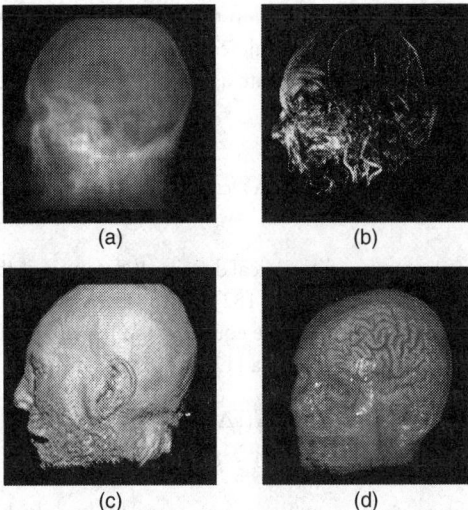

(a)   (b)

(c)   (d)

FIGURE 41.4 Ct head rendered in the four main volume rendering modes: (a) x-ray; (b) MIP; (c) Iso-surface; and (d) Translucent.

The fundamental element in full volume rendering is the volume rendering integral. In this section, we shall assume the *low-albedo* scenario, in which a certain light ray only scatters once before leaving the volume. The low-albedo *optical model* was first described by [10] and [89], and then formally derived by [137]. It computes, for each cast ray, the quantity $I_\lambda(x, r)$, which is the amount of light of wavelength λ coming from ray direction r that is received at point x on the image plane:

$$I_\lambda(x, r) = \int_0^L C_\lambda(s)\mu(s)\exp\left(-\int_0^s \mu(t)dt\right) ds \qquad (41.8)$$

Here, L is the length of ray r. We can think of the volume as being composed of particles with certain mass density values μ (sometimes called *light extinction values* [137]). These values, as well as the other quantities in this integral, are derived from the interpolated volume densities $f(x, y, z)$ via some mapping function. The particles can contribute light to the ray in three different ways: via emission [189], transmission, and reflection [219]. Thus, $C_\lambda(s) = E_\lambda(s) + T_\lambda(s) + R_\lambda(s)$. The latter two terms, T_λ and R_λ, transform light received from surrounding light sources, whereas the former, E_λ, is due to the light-generating capacity of the particle. The reflection term takes into account the specular and diffuse material properties of the particles. To account for the higher reflectivity of particles with larger mass densities, one must weight $C_{\lambda g}$ by μ. In low-albedo, we track only the light that is received on the image plane. Thus, in Equation 41.8, C_λ is the portion of the light of wavelength λ available at location s that is transported in the direction of r. This light then gets attenuated by the mass densities of the particles along r, according to the exponential attenuation function.

$R_\lambda(s)$ is computed via the standard illumination equation [47]:

$$R(s) = k_a C_a + k_d C_i C_o(s)(N(s) \cdot L(s)) + k_s C_l(N(s) \cdot H(s))^{ns} \qquad (41.9)$$

where we have dropped the subscript λ for reasons of brevity. Here, C_a is the ambient color, k_a is the ambient material coefficient, C_l is the color of the light source, C_o is the color of the object (determined by the density–color mapping function), k_d is the diffuse material coefficient, N is the normal vector (determined by the gradient), L is the light direction vector, k_s is the specular material coefficient, H is the half-vector, and ns is the Phong exponent.

The analytic volume rendering integral cannot, in the general case, be computed efficiently. However, an approximation of Equation 41.8 can be formulated using a Taylor expansion of the exponential and a discrete Riemann sum, where the rays interpolate a set of samples, most commonly spaced apart by a distance Δs:

$$I_\lambda(x,r) = \sum_{i=0}^{L/\Delta s-1} C_\lambda(i\Delta s)\alpha(i\Delta s) \cdot \prod_{j=0}^{i-1}(1-\alpha(j\Delta s)) \tag{41.10}$$

Here, α is the material *opacity*, a measure of its optical density. It determines the percentage of light allowed to pass and to be emitted at a given sample point s [182] (α- 1.0-transparency, assuming values in the range $\{0.0, 1.0\}$). Note that Equation 41.10 is a recursive equation in $(1-\alpha)$. It can be conveniently computed via the recursive *front-to-back compositing* formula [121] [182]:

$$c = C(i\Delta s)\alpha(i\Delta s)(1-\alpha)+c$$
$$\alpha = \alpha(i\Delta s)(1-\alpha)+\alpha \tag{41.11}$$

Thus, a practical implementation of volumetric ray casting would traverse the volume from front to back, calculating colors and opacities at each sampling site, weighting these colors and opacities by the current accumulated transparency $(1-\alpha)$, and adding these terms to the accumulated color and transparency to form the terms for the next sample along the ray. An attractive property of the front-to-back traversal is that a ray can be stopped once α approaches 1.0, which means that light originating from structures farther back is completely blocked by the cumulative opaque material in front. This provides for accelerated rendering and is called *early ray termination*. An alternative form of Equation 41.11 is the *back-to-front compositing* equation:

$$c = c(1-\alpha(i\Delta s))+C(i\Delta s)$$
$$\alpha = \alpha(1-\alpha(i\Delta s))+\alpha(i\Delta s) \tag{41.12}$$

Back-to-front compositing is a generalization of the Painter's algorithm. It does not enjoy the speed-up opportunities of early ray termination and is therefore less frequently used.

Equation 41.10 assumes that a ray interpolates a volume that stores at each grid point a color vector (usually a [red, green, blue] = RGB triple), as well as an α value [121] [122]. The mapping of voxel densities to colors C_o (in Equation 41.9) and α is implemented as a set of mapping functions, often implemented as 2-D tables, called *transfer functions*. By way of the transfer functions, users can interactively change the properties of the volume data set. Most applications give access to four mapping functions: $R(d)$, $G(d)$, $B(d)$, $A(d)$, where d is the value of a grid voxel, typically in the range of $[0,255]$ for 8-bit volume data. Thus, users can specify semitransparent materials by mapping their densities to opacities <1.0, which allows rays to acquire a mix of colors that is due to all traversed materials. More advanced applications also give users access to transfer functions that map $k_s(d)$, $k_d(d)$, $ns(d)$, and others. Wittenbrink pointed out that the colors and opacities at each voxel should be multiplied *prior* to interpolation to avoid artifacts on object boundaries [245].

The model in Equation 41.10 is called the *preclassified model*, because voxel densities are mapped to colors and opacities prior to interpolation. This model cannot resolve high-frequency detail in the transfer functions (see Figure 41.5 for an example), and also typically gives blurry images under magnification [158]. An alternative model that is more often used is the *post-classified model*. Here, the raw volume values are interpolated by the rays, and the interpolation result is mapped to color and opacity:

$$I_\lambda(x,r) = \sum_{i=0}^{L/\Delta s-1} C_\lambda(f(i\Delta s),g(i\Delta s))\alpha(f(i\Delta s)) \prod_{j=0}^{i-1}(1-\alpha(f(j\Delta s))) \tag{41.13}$$

The function value $f(i\Delta s)$ and the gradient vector $g(i\Delta s)$ are interpolated from $f_d(x,y,z)$ using a 3-D interpolation kernel, and C_λ and α are now the transfer and shading functions that translate the

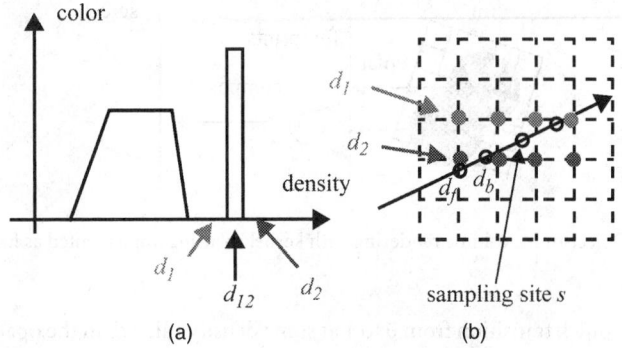

FIGURE 41.5 Transfer function aliasing. When the volume is rendered pre-classified, then both the top row (density d_1) and the bottom row (density d_2) voxels receive a color of zero, according to the transfer function shown on the left. At ray sampling this voxel neighborhood at s would then interpolate a color of zero as well. On the other hand, in post-classified rendering, the ray at s would interpolate a density close to density d_{12} (between d_1 and density d_2) and retrieve the strong color associated with density d_{12} in the transfer function.

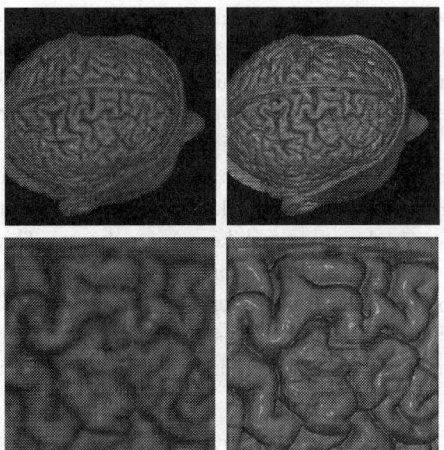

FIGURE 41.6 Pre-classified (left column) versus post-classified rendering (right column). The latter yields sharper images since the opacity and color classification is performed after interpolation. This eliminates the blurry edges introduced by the interpolation filter.

interpolated volume function values into color and opacity. This generates considerably sharper images (see Figure 41.6).

Post-classified rendering eliminates only some of the problems that come with busy transfer functions. Consider again Figure 41.5a, now assuming a very narrow peak in the transfer function at d_{12}. With this kind of transfer function, a ray that is point-sampling the volume at s may easily miss interpolating d_{12}. However, the ray may have interpolated d_{12}, had it just sampled the volume at $s + \delta s$. Preintegrated transfer functions [44] [186] solve this problem by precomputing a 2-D table that stores the analytical volume rendering integration for all possible density pairs (d_f, d_b). This table is then indexed during rendering by each ray sample pair (d_b, d_f), interpolated at sample locations Δs apart; see Figure 41.5b. The preintegration assumes a piecewise linear function within the density pairs, thus guaranteeing that no transfer function detail falling between two interpolated (d_f, d_b) fails to be considered in the discrete ray integration.

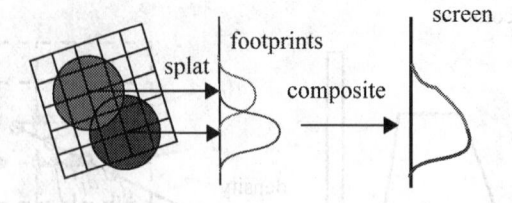

FIGURE 41.7   Object-order volume rendering with kernel splatting implemented as footprint mapping.

Finally, note that a quick transition from 0 to 1 at some density value d_i in the opacity transfer function selects the isosurface $d_{iso} = d_i$. Thus, isosurface rendering is merely a subset of full volume rendering, in which the ray hits a material with $d = d_{iso}$ and then immediately becomes opaque and terminates.

41.6.2   Object-Order Techniques

Object-order techniques decompose the volume into a set of basis elements or *basis functions*, which are individually projected onto the screen and assembled into an image. If the volume rendering mode is x-ray or MIP, then the basis functions can be projected in any order, because in x-ray and MIP the volume rendering integral degenerates to a commutative sum or MAX operation. In contrast, depth ordering is required when solving for the generalized volume rendering integral (Equation 41.8). Early work represented the voxels as disjointed cubes, which gave rise to the *cuberille* representation [61] [76] [77]. Because a cube is equivalent to a nearest neighbor kernel, the rendering results were inferior. Therefore, more recent approaches have turned to kernels of higher quality.

To understand better the issues associated with object-order projection, it helps to view the volume as a field of basis functions h, with one such basis kernel located at each grid point where it is modulated by the grid point's value (see Figure 41.7, where two such kernels are shown). This ensemble of modulated basis functions then makes up the continuous object representation. That is, one could interpolate a sample anywhere in the volume simply by adding up the contributions of the modulated kernels that overlap at the location of the sample value. Hence, one could still traverse this ensemble with rays and render it in image-order. However, a more efficient method emerges when realizing that the contribution of a voxel j with value d_j is given by

$$d_j \cdot \int h(s)ds$$

where s follows the line of kernel integration along the ray. Further, if the basis kernel is radially symmetric, then the integration

$$\int h(s)ds$$

is independent of the viewing direction. Therefore, one can perform a preintegration of

$$\int h(s)ds$$

and store the result into a lookup table. This table is called the kernel *footprint*, and the kernel projection process is referred to as *kernel splatting* or simply *splatting*. If the kernel is a Gaussian, then the footprint is also a Gaussian. Because the kernel is identical for all voxels, we can use it for all voxels. We can generate an image by going through the list of object voxels in depth order and performing the following steps for each (see again Figure 41.7):

1. Calculate the screen-space coordinate of the projected grid point.
2. Center the footprint around that point and stretch it according to the image magnification factor.

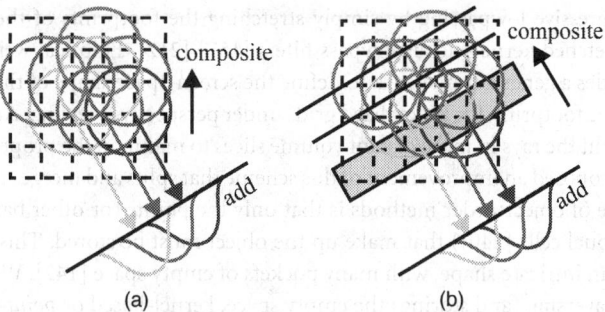

FIGURE 41.8 Sheet-buffered splatting: (a) axis-aligned — the entire kernel within the current sheet is added; (b) image-aligned — only slices of the kernels intersected by the current sheet-slab are added.

3. Rasterize the footprint to the screen, using the preintegrated footprint table and multiplying the indexed values by the voxel's value [237] [238] [239]. This rasterization can either be performed via fast DDA procedures [130] [153] or, in graphics hardware, by texture-mapping the footprint (basis image) onto a polygon [31].

There are three types of splatting: composite-only, axis-aligned sheet-buffered and image-aligned sheet-buffered. The *composite-only* method was proposed first [238] and is the most basic one (see Figure 41.7). Here, the object points are traversed in either front-to-back or back-to-front order. Each is first assigned a color and opacity using the shading equation (Equation 41.9) and the transfer functions. Then, each point is splatted into the screen's color and opacity buffers, and the result is composited with the present image (Equation 41.11). In this approach, color bleeding and slight sparkling artifacts in animated viewing may be noticeable, because the interpolation and compositing operations cannot be separated due to the preintegration of the basis (interpolation) kernel [239].

An attempt to solve this problem led to the *axis-aligned sheet-buffered* splatting approach [237] (see Figure 41.8a). Here, the grid points are organized into sheets (basically, the volume slices most parallel to the image plane), assigned a color and opacity, and splatted into the sheet's color and opacity buffers. The important difference is that now all splats within a sheet are added and not composited, whereas only subsequent sheets are composited. This prevents potential color bleeding of voxels located in consecutive sheets, due to the more accurate reconstruction of the opacity layer. The fact that the voxel sheets must be formed by the volume slices most parallel to the viewing axis leads to a sudden switch of the compositing order when the major viewing direction changes and an orthogonal stack of volume slices must be used to organize the voxels. This causes noticeable popping artifacts, where some surfaces suddenly reflect less light and others more [152].

The solution to this problem is to align the compositing sheet with the image plane at all times, which gives rise to the *image-aligned sheet-buffered* splatting approach [152] (see Figure 41.8b). Here, a slab is advanced across the volume, and all kernels that intersect the slab are sliced and projected. Kernel slices can also be preintegrated into footprints. Thus, this sheet-buffered approach differs from the original one in that each voxel must be considered more than once. The image-aligned splatting method provides the most accurate reconstruction of the voxel field prior to compositing and eliminates both color bleeding and popping artifacts. It is also best suited for post-classified rendering, because the density (and gradient) field is reconstructed accurately in each sheet. However, it is more expensive due to the multiple splatting of a voxel.

The divergence of rays under perspective viewing causes undersampling of the volume portions farther away from the viewpoint. This leads to aliasing in these areas, especially noticeable as disturbing flickering of small features in interactive roaming and time-animated viewing. Lowpassing can eliminate the artifacts caused by aliasing and replace them by visually more pleasing blur [157] [212]. For perspective rendering, the amount of required lowpassing increases with distance from the viewpoint. Kernel-based approaches can achieve this progressive lowpassing by simply stretching the footprints of the voxels as a function of depth, because stretched kernels act as lowpass filters [157] [212]. Elliptical weighted average (EWA) splatting [255] provides a general framework to define the screen-space shape of the footprints, and their mapping into a generic footprint, for generalized grids under perspective viewing. An equivalent approach for ray casting is to split the rays in more distant volume slices to maintain the proper sampling rate [170]. Kreeger et al. [113] proposed an improvement of this scheme that splits and merges rays in an optimal way.

A major advantage of object-order methods is that only the points (or other basis primitives, such as tetrahedral or hexagonal cells [240]) that make up the object must be stored. This can be advantageous when the object has an intricate shape, with many pockets of empty space [142]. While ray casting would spend much effort traversing (and storing) the empty space, kernel-based or *point-based* objects will not consider the empty space, neither during rendering nor for storage. However, there are trade-offs, because the rasterization of a footprint takes more time than the commonly used trilinear interpolation of ray samples. This is because the radially symmetric kernels employed for splatting must be larger than the trilinear kernels to ensure proper blending. Hence, objects with compact structure are more favorably rendered with image-order methods or hybrid methods (see Section 41.6.3). Another disadvantage of object-order methods is that early ray termination is not available to cull occluded material early from the rendering pipeline. The object-order equivalent is early point elimination, which is more difficult to achieve than early ray termination. Finally, image-order methods allow the extension of ray casting to ray tracing, in which secondary and higher-order rays are spawned at reflection sites. This facilitates mirroring on shiny surfaces, interreflections between objects, and soft shadows.

There are a number of ways to store and manage point-based objects. These schemes are distinguished mainly by their ability to exploit spatial coherence during rendering. The least spatial coherence results from storing the (nonair) points sorted by density [30]. It is best suited to sparse objects, and it allows fast isocontour selection via binary search. The method also requires that the points be depth-sorted during rendering or, at least, tossed into depth bins [155]. A compromise is struck by Ihm and Lee [86] who sort points by density within volume slices only, which gives implicit depth-ordering when used in conjunction with an axis-aligned sheet-buffer method. A number of approaches exist that organize the points into run length-encoded (RLE) lists, which allow the spatial coordinates to be computed incrementally when traversing the runs [102] [161].

A number of surface-based splatting methods have also been described. These do not provide the flexibility of volume exploration via transfer functions, because the original volume is discarded after the surface has been extracted. They only allow a fixed geometric representation of the object that can be viewed at different orientations and with different shadings. A popular method is *shell-rendering* [220], which extracts from the volume (possibly with a sophisticated segmentation algorithm) a certain thin or thick surface or contour and represents it as a closed shell of points. Shell-rendering is fast, because the number of points is minimized and its data structure has high cache coherence.

41.6.3 Hybrid Techniques

Hybrid techniques seek to combine the advantages of the image-order and object-order methods. That is, they use object-centered storage for fast selection of relevant material (a hallmark of object-order methods), and they use early ray termination for fast occlusion culling (a hallmark of image-order methods).

The shear-warp algorithm [116] is such a hybrid method. In shear-warp, the volume is rendered by a simultaneous traversal of RLE voxel and pixel runs, where opaque pixels and transparent voxels are efficiently skipped during these traversals [184]. Further speed comes from the fact that sampling only occurs in the volume slices via bilinear interpolation and that the ray grid resolution matches that of the volume slices. Therefore, the same bilinear weights can be used for all rays within a slice. The caveat is that the image must first be rendered from a sheared volume onto a so-called base plane, that is, aligned with the volume slice most parallel to the true image plane. After completing the base-plane rendering, the base plane image must be warped onto the true image plane, and the resulting image is displayed. All of this combined enables frame rates in excess of 10 frames/s on a current PC, for a 128^3 volume.

41.6.4 Domain Volume Rendering

In domain rendering, the spatial 3-D data is first transformed into another domain, such as compression, frequency, and wavelet domain, and then a projection is generated directly from that domain or with the help of information from that domain. *Frequency* domain rendering applies the Fourier slice projection theorem, which states that a projection of the 3-D data volume from a certain view direction can be obtained by extracting a 2-D slice perpendicular to that view direction out of the 3-D Fourier spectrum, and then by inverse Fourier transforming it. This approach obtains the 3-D volume projection directly from the 3-D spectrum of the data. Therefore, it reduces the computational complexity for volume rendering from $O(N^3)$ to $O(N^2 log(N))$ [41] [132] [218]. A major problem of frequency domain volume rendering is that the resulting projection is a line integral along the view direction, which does not exhibit any occlusion or attenuation effects. Totsuka and Levoy [218] proposed a linear approximation to the exponential attenuation [189] and an alternative shading model to fit the computation within the frequency domain rendering framework.

Compression domain rendering performs volume rendering from compressed scalar data without decompressing the entire data set, and therefore reduces the storage, computation, and transmission overhead of otherwise large volume data. For example, Ning and Hesselink [167] [168] first applied vector quantization in the spatial domain to compress the volume and then directly rendered the quantized blocks using regular spatial domain volume rendering algorithms. Fowler and Yagel [50] combined differential pulse-code modulation and Huffman coding, and developed a lossless volume compression algorithm, but their algorithm is not coupled with rendering. Yeo and Liu [253] applied a discrete cosine transform–based compression technique to overlapping blocks of the data. Chiueh et al. [21] applied the 3-D Hartley transform to extend the JPEG still-image compression algorithm [223] to the compression of subcubes of the volume. They performed frequency domain rendering on the subcubes before compositing the resulting subimages in the spatial domain. Each of the 3-D Fourier coefficients in each subcube was then quantized, linearly sequenced through a 3-D zigzag order, and then entropy encoded. In this way, they alleviated the problem of lack of attenuation and occlusion in frequency domain rendering while achieving high compression ratios, fast rendering speed (compared to spatial volume rendering), and improved image quality over conventional frequency domain rendering techniques. More recently, Guthe et al. [65] and Sohn et al. [202] have used principles from MPEG encoding to render time-varying data sets in the compression domain.

The *wavelet transform* [22] [37] has been used by a number of researchers to reduce the storage of volume data sets before rendering [63] [159] [235]. Guthe and Strasser [66] have recently used the wavelet transform to render very large volumes at interactive frame rates on texture mapping hardware. They employ a wavelet pyramid encoding of the volume to reconstruct, on the fly, a decomposition of the volume into blocks of different resolutions. Here, the resolution of each block is chosen based on the local error committed and the resolution of the screen area onto which the block is projected. Each block is rendered individually with 3-D texture-mapping hardware, and the block decomposition can be used for a number of frames, which amortizes the work spent on the inverse wavelet transform to construct the blocks.

41.7 Acceleration Techniques

The high computational complexity of volume rendering has led to a great variety of approaches to its acceleration. Acceleration techniques generally seek to take advantage of properties of the data, such as empty space, occluded space, and entropy, as well as properties of the human perceptual system, such as its insensitivity to noise over structural artifacts [56].

A number of techniques have been proposed to accelerate the grid traversal of rays in image-order rendering. Examples are the 3-D digital differential analyzer (DDA) method [1] [53], in which new grid positions are calculated by fast integer-based incremental arithmetic, and the template-based method [250], in which templates of the ray paths are precomputed and used during rendering to identify quickly the voxels to visit. Early ray termination can be sophisticated into a Russian roulette scheme [35] in which some rays terminate with lower and others with higher accumulated opacities. This capitalizes on the human eye's tolerance of error masked as noise [129]. In the object-order techniques, fast differential techniques to determine the screen-space projection of the points and to rasterize the footprints [130] [153] are also available.

Most of the object-order approaches deal well with empty space — they simply do not store and process it. In contrast, ray casting relies on the presence of the entire volume grid, because it requires it for sample interpolation and address computation during grid traversal. Although opaque space is quickly culled via early ray termination, the fast leaping across empty space is more difficult. A number of techniques are available to achieve this. The simplest form of space leaping is facilitated by enclosing the object in a set of boxes, possibly hierarchical, and quickly determining and testing the rays' intersection with each of the boxes before engaging in a more time-consuming volumetric traversal of the material within [99]. A better geometrical approximation is obtained by a polyhedral representation, chosen crudely enough to maintain ease of intersection. In fact, one case utilizes conventional graphics hardware to perform the intersection calculation, in which one projects the polygons twice to create two Z- (depth) buffers. The first Z-buffer is the standard closest-distance Z-buffer; the second is a farthest-distance Z-buffer. Because the object is completely contained within the representation, the two Z-buffer values for a given image plane pixel can be used as the starting and ending points of a ray segment on which samples are taken. This algorithm has been known as *PARC* (polygon-assisted ray casting) [201]. It is part of the *VolVis* volume visualization system [3] [4], which also provides a multialgorithm progressive refinement approach for interactivity. By using available graphics hardware, the user is given the ability to manipulate interactively a polyhedral representation of the data. When the user is satisfied with the placement of the data, light sources, and viewpoint, the Z-buffer information is passed to the PARC algorithm, which produces a ray-cast image.

A different technique for empty-space leaping was devised by Zuiderveld et al. [254] and by Cohen and Shefer [24], who introduced the concept of *proximity clouds*. Proximity clouds employ a distance transform of the object to accelerate the rays in regions far from the object boundaries. In fact, since the volume densities are irrelevant in empty volume regions, one can simply store the distance transform values in their place and, therefore, storage is not increased. Because the proximity clouds are the isodistance layers around the object's boundaries, they are insensitive to the viewing direction. Thus, rays that ultimately miss the object are often still slowed down. To address this shortcoming, Sramek and Kaufman [204] proposed a view-sensitive extension of the proximity clouds approach. Wan et al. [224] place a sphere at every empty voxel position, where the sphere radius indicates the closest nonempty voxel. They apply this technique to navigation inside hollow volumetric objects, as occurring in virtual colonoscopy [81], and reduce a ray's space traversal to just a few hops until a boundary wall is reached. Finally, Meissner et al. [143] suggested an algorithm that quickly recomputes the proximity cloud when the transfer function changes.

Proximity clouds only handle the quick leaping across empty space, but methods are also available that traverse occupied space faster when the entropy is low. These methods generally utilize a hierarchical decomposition of the volume in which each nonleaf node is obtained by lowpass filtering its children. Commonly, this hierarchical representation is formed by an octree [139], because these are easy to traverse and store. An *octree* is the 3-D extension of a quadtree [190], which is the 2-D extension of a binary tree. Most often, a nonleaf node stores the average of its children, which is synonymous with a box filtering of the volume, but more sophisticated filters are possible. Octree do not have to be balanced [241], or fully expanded into a single-root node or single-voxel leaf nodes. The last two give rise to a brick-of-bricks decomposition, in which the volume is stored as a flat hierarchy of bricks of size n^3 to improve cache coherence in the volume traversal. Parker et al. [172] [173] utilize this decomposition for the ray casting of very large volumes. They also give an efficient indexing scheme to find quickly the memory address of the voxels located in the 8-neighborhood required for trilinear interpolation.

When octrees are used for entropy-based rendering, nonleaf nodes store either an entropy metric of their children, such as standard deviation [35], minimum–maximum range [241], or Lipschitz range [208], or a measure of the error committed when the children are not rendered, such as the root mean square or the absolute error [66]. The idea is to have the user specify a tolerable error before the frame is rendered or to make the error dependent on the time maximally allowed to render the frame, which is known as *time-critical rendering*. In either case, the rays traversing the volume will advance across the volume but also will transcend up and down the octree, based on the metric used. This will either accelerate or decelerate the rays on their path. A method called *β-acceleration* will also make this traversal sensitive to the rays' accumulated opacity so far. The philosophy here is that the observable error from using a coarser node will be relatively small when it is weighted by a small transparency in Equation 41.11. Note, however, that the interpolated opacity must be normalized to unit stepsize before it is used in the compositing equation (see Chapter 6 in [126]).

Octrees are also easily used with object-order techniques, such as splatting. Laur and Hanrahan [117] have proposed an implementation that approximates nonleaf octree nodes by kernels of a radius that is twice the radius of the children's kernels, which gives rise to a magnified footprint. They store the children's average, and an error metric based on their standard deviation, in each parent node and use a preset error to select the nodes during rendering. This work was later generalized to hierarchies of elliptical kernels, found via wavelet analysis [234]. Although both of these works use nonleaf nodes during rendering, other splatting approaches only exploit them for fast occlusion culling. Lee and Ihm [118] and Mora et al. [151] store the volume as a set of bricks, which they render in conjunction with a dynamically computed hierarchical occlusion map to cull voxels quickly within occluded bricks from the rendering pipeline. An alternative scheme, which performs occlusion culling on a finer scale than the box-basis of an octree decomposition, is to calculate an occlusion map in which each pixel represents the average of all pixels within the box-neighborhood covered by a footprint [155].

Another form of acceleration (and volume compression) is to employ more efficient sampling grids, such as the body-centered Cartesian (BCC) lattice [28]. In 3-D, BCC grids reduce the number of grid samples by 30%, without loss of frequency content or accuracy. BCC grids are particularly attractive for splatting, because they reduce the number of points that must be rasterized and rendered [216]. Neophytou and Mueller [161] extended the use of these grids to 4-D volume rendering, where they reduce the data set to 50% of the original number of samples.

A comprehensive system for accelerated software-based volume rendering is the UltraVis system devised by Knittel [110]. It can render 256^3 volume at 10 frames/s. It achieves this by optimizing cache performance during both volume traversal and shading. This concept is rooted in the fact that good cache management is key to achieve fast volume rendering, since the data are so massive. As we have mentioned before, this was also realized by Parker et al. [172] [173], and it plays a key role in both custom and commodity hardware approaches, as we shall see later. The UltraVis system manages the cache by dividing it into four blocks: one block each for volume bricks, transfer function tables, image blocks, and temporary buffers. Because the volume can only map into a private cache block, it can never be swapped out by a competing data structure, such as a transfer function table or an image tile array. This requires that the main memory footprint of the volume is four times higher, because no volume data may be stored in an address space that would map outside the volume's private cache slots. By using a bricked volume decomposition in conjunction with a flock of rays traced simultaneously across the brick, the brick's data must only be brought in once before it can be discarded, when all rays have finished its traversal. A number of additional acceleration techniques give further performance.

Another type of acceleration is achieved by breaking the volume integral of Equation 41.10 or Equation 41.13 into segments and storing the composited color and opacity for each partial ray in a data structure [11] [156]. The idea is then to recombine these partial rays into complete rays for images rendered at viewpoints near the one for which the partial rays were originally obtained. This saves the cost of fully integrating all rays for each new viewpoint and reduces it to the much lower expense of compositing a few partial segments per ray. An alternative method, which uses a precomputed triangle mesh to achieve similar goals for isosurface volume rendering, was proposed by Chen et al. [18]. Yagel and Shi [252] warped complete images to nearby viewpoints, aided by a depth buffer.

41.8   Classification and Transfer Functions

In volume rendering, we seek to explore volumetric data using visuals. This exploration process aims to discover and emphasize interesting structures and phenomena embedded in the data, while deemphasizing or completely culling away occluding structures that are currently not of interest. Clipping planes and more general clipping primitives [233] provide geometric tools to remove or displace occluding structures in their entirety. On the other hand, transfer functions, which map the raw volume density data to color and transparencies, can alter the overall look and feel of the data set in a continuous fashion.

The exploration of a volume via transfer functions constitutes a navigation task, which is performed in a 4-D transfer function space, assuming three axes for RGB color and one for transparency (or opacity). It is often easier to specify colors in HSV (hue, saturation, value) color space, because this provides separate mappings for color and brightness. Simple algorithms exist to convert the HSV values into the RGB triples used in volume rendering [47].

Given the large space of possible settings, choosing an effective transfer function can be a daunting task. It is generally more convenient to gather more information about the data before the exploration via transfer functions begins. The easiest presentation of support data is in the form of 1-D histograms, which are data statistics collected as a function of raw density or some other quantity. A histogram of density values can be a useful indicator to point out dominant structures with narrow density ranges. A fuzzy classification function [39] can then be employed to assign different colors and opacities to these structures (see Figure 41.9). This works well if the data are relatively noise-free, if the density ranges of the features are well isolated, and if few distinct materials (e.g., bone, fat, and skin) are present. In most applications, however, this is not the case. In these settings, it helps to include the first and second derivatives into the histogram-based analysis [103].

The magnitude of the first derivative (the gradient strength) is useful because it peaks at densities where interfaces between different features exist (see Figure 41.10). Plotting a histogram of first derivatives over density yields an arc that peaks at the interface density. Knowing the densities at which feature boundaries exist narrows the transfer function exploration task considerably. One may now visualize these structures by assigning different colors and opacities within a narrow interval around these peaks. Levoy [121] showed that a constant width of (thick) surface can be obtained by making the width of the chosen density interval a linear function of the gradient strength. Kindlmann and Durkin [103] proposed a technique that uses the first and second derivatives to generate feature-sensitive transfer functions automatically. This method provides a segmentation of the data, in which the segmentation metric is a histogram of the first and second derivatives. Tenginakai et al. [214] extended the arsenal of metrics to higher-order moments, and computed from them additional measures, such as kurtosis and skew, in small neighborhoods. These measures can provide better delineations of features in histogram space.

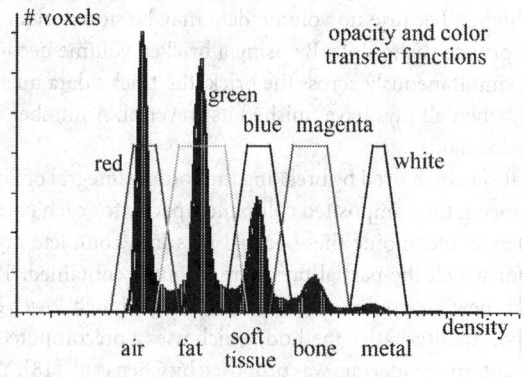

FIGURE 41.9   Histogram and a fuzzy classification into different materials, along with distinguishing colors and opacities.

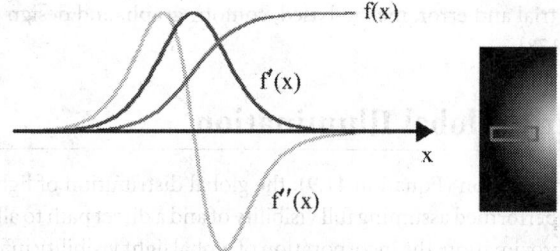

FIGURE 41.10 The relationship of densities and their first and second derivatives at an object boundary (shown as the box in the picture on the right).

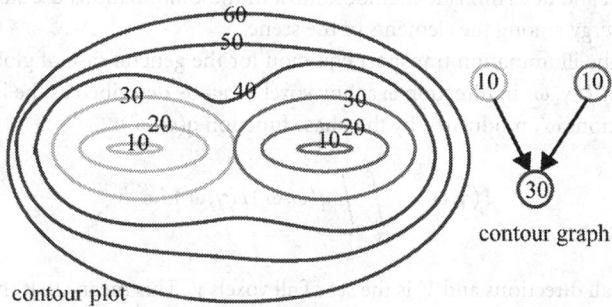

FIGURE 41.11 Simple contour graph. The first topological event occurs when the two inner contours are born at an iso-value of 10. The second topological events occurs at the iso-value at which the two inner contours just touch and give way to a single contour at iso-value = 30.

Another proposed analysis method is based on maxima in cumulative Laplacian-weighted density histograms [174].

Automated segmentation of images and volumes remains a difficult task and is, in many cases, observer- and task-dependent. In this regard, semisupervised segmentation algorithms, in which users guide the segmentation process in an interactive fashion, have a competitive edge. There are two examples of such systems: the PAVLOV architecture, which implements an interactive region-grow to delineate volumetric features of interest [112], and the dual-domain approach of Kniss et al. [106] [107]. These researchers embed Kindlemann's algorithm into an interactive segmentation application. Here, users work simultaneously within two domains (i.e., the histogram-coupled transfer function domain and the volume rendering domain) to bring out certain features of interest. For this to be effective, an interactive (hardware-based) volume renderer is required.

Another way to analyze the data is to look for topological changes in the isocontours of the volume, such as a merge of split of two contours (see Figure 41.11). These events are called *critical points*. By topologically sorting the critical points as a function of density, one can construct a contour graph, contour tree, or hyper Reeb graph, which yields a roadmap for an exploration of the volume [7] [17] [54] [115] [195] [213]. One can use the contour graph to come up with an automatic transfer function (simply position an isosurface between two nodes), or one can use it to guide users in the volume exploration process. A large number of critical points are potentially generated, especially when the data are noisy.

All of the methods presented so far base the transfer function selection on a prior analysis of the volume data. Another suggested strategy has been to render a large number of images with arbitrary transfer function settings and present these to the user, who then selects a subset of these for further refinement by application of genetic algorithms. This approach has been taken by the Design Galleries project [135], which is based in part on the method published by He et al. [75]. A good sample of all of the existing approaches (interactive trial and error, metric-based, contour graph, and design galleries) was presented in a symposium panel [178].

41.9   Volumetric Global Illumination

In the local illumination equation (Equation 41.9), the global distribution of light energy is ignored and shading calculations are performed assuming full visibility of and a direct path to all light sources. Although this is useful as a first approximation, the incorporation of global light visibility information (shadows, one instance of global illumination) adds a great deal of intuitive information to the image. This low-albedo [89] [200] lighting simulation has the ability to cast soft shadows by volume density objects. Generous improvements in realism are achieved by incorporating a high-albedo lighting simulation [89] [200], which is important in a number of applications (e.g., clouds [137], skin [68], and stone [38]). Although some use hierarchical and deterministic methods, most of these simulations use stochastic techniques to transport lighting energy among the elements of the scene.

We wish to solve the illumination transport equation for the general case of global illumination. The reflected illumination $I(\gamma, \omega)$ in direction ω at any voxel γ can be described as the integral of all incident radiation from directions ω', modulated by the phase function $q(\omega, \omega')$:

$$I(\gamma, \omega) = \int\limits_{V} \int\limits_{\Gamma} q(\omega, \omega') I(\gamma, \omega') d\omega' dv \qquad (41.14)$$

where Γ is the set of all directions and V is the set of all voxels v. This means that the illumination at any voxel is dependent on the illumination at every other voxel. In practice, this integral equation is solved by finite repeated projection of energy among voxels. This leads to a finite energy transport path, which is generally sufficient for visual fidelity.

In physics, equations of this sort are solved via *Monte Carlo simulations*. A large set of rays is cast from the energy sources into the volume, and at each voxel a "die is rolled" to determine how much energy is absorbed and how much energy is scattered and in what direction. After much iteration, the simulation is stopped and a final scattering of the radiosity volume is performed toward an arbitrarily positioned eye point.

A practical implementation of this process is volumetric backprojection. Backprojection is usually performed on a voxel-by-voxel basis, because this is the most obvious and direct method of computation. For example, in volumetric ray tracing [200], as illumination is computed for a volume sample, rays are cast toward the light sources, sampling the partial visibility of each. In computing high-albedo scattering illumination, Max [137] used the method of discrete ordinates to transport energy from voxel to voxel. For calculations of volumetric radiosity, voxels are usually regarded as discrete elements in the usual radiosity calculation on pairs of elements, thereby computing on a voxel-by-voxel basis [188] [200]. Particle tracing methods for global illumination track paths of scattered light energy through space, starting at the light sources [87].

In many cases, the backprojection can be reorganized into a single sweep through the volume, processing slice by slice. Because sunlight travels in parallel rays in one direction only, Kajiya and Herzen [89] calculated the light intensity of a cloudlike volume, one horizontal slice at a time. A similar technique was demonstrated as part of the Heidelberg ray-tracing model [140], in which shadow rays were propagated simultaneously, slice by slice and in the same general direction as rendering. Dachille et al. [32] described a backprojection approach that scatters the energy in the volume by a multipass, slice-by-slice sweep at random angles. He also devised a custom hardware architecture for a cache-efficient implementation of this algorithm.

Kniss et al. [108] [109] proposed a single-pass algorithm that approximates the scattering of light within a volume by a recursive slice-blurring operation, starting at the light source. The profile of the blurring filter is determined by the user-specified phase function. The method exploits 3-D texture-mapping hardware in conjunction with a dual image buffer, and runs at interactive frame rates. One buffer, the repeatedly blurred (light) buffer, contains the transported and scattered light energy on its path away from the source;

the other (frame) buffer holds the energy headed for the eye and is attenuated by the densities along the path to the eye. At each path increment, energy is transferred from the light buffer to the frame buffer.

41.10 Special-Purpose Rendering Hardware

The high computational cost of direct volume rendering makes it difficult for sequential implementations and general-purpose computers to deliver the targeted level of performance, although recent advances in commodity graphics hardware have started to blur these boundaries (as we shall see in Section 41.11). This situation is aggravated by the continuing trend toward higher and higher resolution data sets. For example, to render a data set of 1024^3 16-bit voxels at 30 frames per second requires 2 GB of storage, a memory transfer rate of 60 GB per second, and approximately 300 billion instructions per second, assuming 10 instructions per voxel per projection.

In the same way that the special requirements of traditional computer graphics led to high-performance graphics engines, volume visualization naturally lends itself to special-purpose volume renderers that separate real-time image generation from general-purpose processing. Most recent research focuses on accelerators for ray casting of regular data sets. Ray casting offers room for algorithmic improvements while still allowing for high image quality. More recent architectures [78] include VOGUE [111], VIRIM [64], Cube-4 [180] [181], VIZARD I and II [144] [145], and the commercial VolumePro500 board [179] (an ASIC implementation of the Cube-4 architecture) and VolumePro1000 [249] board.

Cube-4 [180] [181] has only simple and local interconnections, thereby allowing for easy scalability of performance. Instead of processing individual rays, Cube-4 manipulates a group of rays at a time. As a result, the rendering pipeline is directly connected to the memory. Accumulating compositors replace the binary compositing tree. A pixel-bus collects and aligns the pixel output from the compositors. Cube-4 is easily scalable to very high resolutions of 1024^3 16-bit voxels and true real-time performance implementations of 30 frames per second. The VolumePro500 board was the final design, in the form of an ASIC, and was released to market by Mitsubishi Electric in 1999 [179]. VolumePro has hardware for gradient estimation, classification, and per-sample Phong illumination. It is a hardware implementation of the shear-warp algorithm, but with true trilinear interpolation, which affords very high quality. The final warp is performed on the PC's graphics card. VolumePro streams the data through four rendering pipelines, maximizing memory throughput by using a two-level memory block- and bank-skewing mechanism to take advantage of the burst mode of its SDRAMs. No occlusion culling or voxel skipping is performed. Advanced features, such as gradient magnitude modulation of opacity and illumination, supersampling, cropping, and cut planes, are also available. The system renders 500 million interpolated, Phong-illuminated, composited samples per second, which is sufficient to render volumes with up to 16 million voxels (e.g., 256^3 volumes) at 30 frames per second.

41.11 General-Purpose Rendering Hardware

Another opportunity to accelerate volume rendering is to utilize the texture mapping capability of graphics hardware. The first such implementation was devised by Cabral et al. [14] and ran on SGI Reality Engine workstations. There are two ways to go about this. One represents the volume either as a stack of 2-D textures, one texture per volume slice, or as one single 3-D texture, which requires more sophisticated hardware. In the former case, three texture stacks are needed: one for each major viewing direction. An image is then rendered by choosing the stack that is most parallel to the image plane and rendering the textured polygons to the screen in front-to-back or back-to-front order. If the machine has 3-D texture capabilities, then one specifies a set of slicing planes parallel to the screen and composites the interpolated textures in depth order. The 3-D texturing approach generally provides better images, because an arbitrarily small slice distance can be chosen and no popping caused by texture stack switching can occur. Although the early approaches did not provide any shading, Gelder and Kim [58] introduced a fast technique to preshade the volume on the fly and then slice and composite an RGB volume to obtain an image with shading effects. Westermann and Ertl [236] introduced a fast multipass approach to display shaded isosurfaces.

The emergence of advanced PC graphics hardware has made texture-mapped volume rendering accessible to a much broader community, at less than 2% of the cost of the workstations that were previously required. However, the decisive factor triggering the revolution that currently dominates the field was the decision of manufacturers (e.g., NVidia, ATI, and 3-DLabs) to make two of the main graphics pipeline components programmable. These two components are the vertex shaders, which are the units responsible for vertex transformations, and the fragment shaders, which are the units that take over after the rasterizer. The first implementation that used these new commodity GPUs for volume rendering was published by Rezk-Salama et al. [185], who used the stack-or-textures approach, because 3-D texturing was not supported at that time. They overcame the undersampling problems associated with the large interslice distance at off-angles by interpolating, on the fly, intermediate slices, using the register combiners in the fragment shader compartment. Engel et al. [44] replaced this technique by the use of preintegrated transfer function tables (see Section 41.8).

To compute the gradients required for shading, one must also load a gradient volume into the texture memory. The interpolation of a gradient volume without subsequent normalization is generally incorrect, but the artifacts are not always visible. Meissner and Guthe [141] use a shading cube texture instead, which eliminates this problem. Even the most recent texture-mapping hardware cannot reach the performance of the specialized volume rendering hardware, such as VolumePro500 and the VolumePro 1000, at least not when volumes are rendered by brute force. Therefore, current research efforts have concentrated on reducing the load for the fragment shaders. Level-of-detail methods have been devised that rasterize lower-resolution texture blocks whenever the local volume detail or projected resolution allows them to do so [66] [120]. Li and Kaufman [124] [125] proposed an alternative approach that approximates the object by a set of texture boxes, which efficiently clips empty space from the rasterization.

41.12 Irregular Grids

All the algorithms discussed previously handle only regular gridded data. Irregular gridded data come in a large variety [203], including curvilinear data and unstructured (scattered) data, where no explicit connectivity is defined between cells (one can even be given a scattered collection of points that can be turned into an irregular grid by interpolation [166] [138]). Figure 41.12 illustrates the most prominent grid types, and Figure 41.13 shows an example of a translucent rendering of an irregular grid data set.

One approach to rendering irregular grids is the use of feed-forward (or projection) methods, where the cells are projected onto the screen one by one, accumulating their contributions incrementally to the final image [242] [138] [240] [196]. The projection algorithm that has gained popularity is the projected tetrahedra (PT) algorithm by Shirley and Tuchman [196]. It uses the projected profile of each

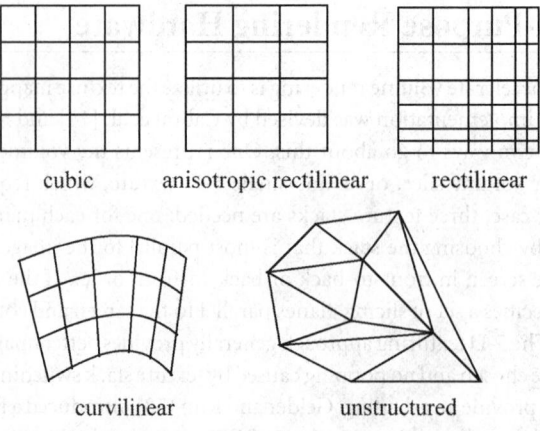

FIGURE 41.12   Various grid types in 2-D.

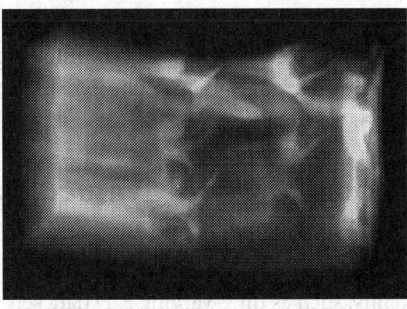

FIGURE 41.13   Visualization of an irregular grid dataset: combustion chamber.

tetrahedron with respect to the image plane to decompose it into a set of triangles. The color and opacity values for each triangle vertex are approximated using ray integration through the thickest point of the tetrahedron. The resulting semitransparent triangles are sorted in depth order and then rendered and composited using polygonal graphics hardware. Stein et al. [209] sort the cells before they are split into tetrahedra, and they utilize 2-D texture-mapping hardware to accelerate opacity interpolation and provide the correct per-pixel opacity values to avoid artifacts. Although their method can only handle linear transfer functions without artifacts, Röttger et al. [187] introduced the concept of preintegrated volume rendering to allow for arbitrary transfer functions. They created a 3-D texture map to provide hardware support in interpolating along the ray between the front and back faces of a tetrahedral cell. In this texture map, two of the three coordinates correspond to values at the cell entry and exit points, with the third coordinate mapping to the distance through the cell. This texture map is then approximated using 2-D texture mapping.

Cell projection methods require a sorted list of cells to be passed to the hardware. Starting with the work of Max et al. [138] and Williams [243], there has been substantial progress in the development of accurate visibility ordering algorithms [198] [27]. Graphics hardware architecture was also proposed, but not yet realized, by King et al. [104], which can both rasterize and sort tetrahedral meshes in hardware.

An alternative technique to visualize irregular grids is ray casting [57] [221]. Ray-casting methods tend to be more exact than projective techniques because they are able to "stab" or integrate the cells in depth order, even in the presence of cycles. This is generally not possible in cell-by-cell projection methods. Many ray-casting approaches employ the plane sweep paradigm, which is based on processing geometric entities in an order determined by passing a line or a plane over the data. Pioneered by Giertsen [59] for use in volume rendering, the method is based on a sweep plane that is orthogonal to the viewing plane (e.g., orthogonal to the y-axis). Events in the sweep are determined by vertices in the data set and by values of y that correspond to the pixel rows. When the sweep plane passes over a vertex, an active cell list (ACL) is updated accordingly, so that it stores the set of cells currently intersected by the sweep plane. When the sweep plane reaches a y-value that defines the next row of pixels, the current ACL is used to process that row, casting rays corresponding to the values of x that determine the pixels in the row, through a regular grid (hash table) that stores the elements of the ACL. This method has three major advantages:

It is unnecessary to store explicitly the connectivity between the cells.

It replaces the relatively expensive operation of 3-D ray casting with a simpler 2-D regular-grid ray casting.

It exploits coherence of the data between scanlines.

Since the work of Giertsen, a number of works have employed the sweep paradigm, most using a sweep plane parallel to the image plane. Some of these methods are assisted by hardware [251] [236]; others are pure software implementations [13] [46] [197]. The ZSweep [46] algorithm is very fast and has excellent memory efficiency. It sweeps the plane from front to back and rasterizes the cell faces as they are encountered by the sweep plane. This keeps the memory footprint low, because only the active cell set must held in memory. Finally, Hong and Kaufman [79] [80] have proposed a very fast ray-casting technique, which exploits the special topology of curvilinear grids, and Mao et al. [133] [134] first resample an irregular grid into an arrangement of ellipsoidal kernels, which can then be quickly projected from any viewpoint using splatting.

41.13   High-Dimensional and Multivalued Data

The following important extensions to the 3-D scalar data sets have been discussed so far:

Data sets of higher dimensionality, such as time-varying 3-D data sets or general n-D data.
Data volumes composed of field vectors (such as flow and strain), or attribute vectors.

The latter can either be *multichannel*, such as the RGB color volumes obtained by cryosectioning the Visible Human [82], or *multimodal*, that is, spatially colocated volumes acquired by scanning an object with multiple modalities, such as MRI, PET, and CT.

There have been significant developments in the rendering of time-varying volumetric data sets. These typically exploit time coherency for compression and acceleration [2] [66] [128] [194] [211] [235], but other methods have also been designed that allow general viewing [6] [9] [69] [70] [71] [100] [161] [232] of high-dimensional (n-D) data sets and require a more universal data decomposition.

The rendering of multimodal volumes requires the mixing of the data at some point in the rendering pipeline. There are at least three locations at which this can happen [15]:

At the image level (after rendering of the individual volumes)
At the accumulation level (before compositing)
At the illumination level (before shading and transfer function lookup)

Multichannel data, such as RGB data obtained by photographing slices of real volumetric objects, have the advantage that there is no longer a need to search for suitable color transfer functions to reproduce the original look of the data. On the other hand, the photographic data do not provide easy mapping to densities and opacities, which are required to compute normals and other parameters needed to bring out structural object detail in surface-sensitive rendering. One can overcome the perceptual nonlinearities of the RGB space by computing gradients and higher derivatives in the perceptually uniform color space L*u*v* [42]. In this method, the RGB data are first converted into the L*u*v* space, and the color distance between two voxels is calculated by their Euclidian distance in that color space.

41.14   Volume Graphics

Volume graphics, [20] [96] an emerging subfield of volume visualization, is concerned with the synthesis, modeling, manipulation, and rendering of volumetric geometric objects, stored in a volume buffer of voxels. Unlike the discussion so far, which has focused mainly on sampled and computed data sets, volume graphics is concerned primarily with modeled geometric scenes and commonly with those that are represented in a regular volume buffer. As an approach, volume graphics has the potential to greatly advance the field of 3-D graphics by offering a comprehensive alternative to traditional surface graphics.

Although the 3-D raster representation seems more natural for empirical imagery than for geometric objects, due to its ability to represent interiors and digital samples, the advantages of this representation are also attracting traditional surface-based applications that deal with the modeling and rendering of synthetic scenes made from geometric models. The geometric model is *voxelized* (3-D scan-converted) into a set of voxels that "best" approximate the model. Each of these voxels is then stored in the volume buffer, together with the voxel precomputed view-independent attributes. The voxelized model can be either binary (see [45] [84] [93] [94] [98]) or volume sampled (see [33] [51] [205] [226]), which generates alias-free density voxelization of the model. Some surface-based application examples are the rendering of fractals [169], hypertextures [176], fur [90], foliage, grass, and hair [163], gases [43], clouds [36] [137], and other complex models [199], including CAD models and terrain models for flight simulators [25]

[62] [96] [183] [225] [248]. Furthermore, in many applications involving sampled data, such as medical imaging, the data must be visualized along with synthetic objects that may not be available in digital form, such as scalpels, prosthetic devices, injection needles, radiation beams, and isodose surfaces. These geometric objects can be voxelized and intermixed with the sampled organ in the voxel buffer [97]. An alternative is to leave these geometric objects in a polygonal representation and render the assembly of volumetric and polygonal data in a hybrid rendering mode [114] [123] [200].

In the following subsections, we describe the volumetric approach to several common volume graphics modeling techniques. We describe the generation of object primitives from geometric models (voxelization) and images (reconstruction), 3-D antialiasing, solid texturing, modeling of amorphous and natural phenomena, modeling by block operations, constructive solid modeling, volume sculpting, volume deformation, and volume animation.

41.14.1 Voxelization

An indispensable stage in volume graphics is the synthesis of voxel-represented objects from their geometric representation. This stage, called *voxelization*, is concerned with converting geometric objects from their continuous geometric representation into a set of voxels that "best" approximates the continuous object. As this process mimics the scan-conversion process that pixelizes (rasterizes) 2-D geometric objects, it is also referred to as 3-D *scan-conversion*. In 2-D rasterization, the pixels are drawn directly onto the screen to be visualized, and filtering is applied to reduce the aliasing artifacts. However, the voxelization process does not render the voxels but merely generates a database of the discrete digitization of the continuous object.

A voxelization algorithm for any geometric object should meet several criteria:

Separability criterion — First, it must be efficient and accurate, and it must generate discrete surfaces that are thick enough so that they cannot be penetrated by a crossing line [23].

Minimality criterion — Second, the discrete surfaces should only contain those voxels indispensable to satisfying the separability requirement, such that a faithful delineation of the object's shape is warranted [23].

Smoothness criterion — Third, the generated discrete object should have smooth boundaries to ensure the antialiased gradient estimation necessary for high-quality volume rendering [226].

One practical meaning of separation is apparent when a voxelized scene is rendered by casting discrete rays from the image plane into the scene. The penetration of the background voxels (which simulate the discrete ray traversal) through the voxelized surface causes the appearance of a hole in the final image of the rendered surface. Another type of error might occur when a 3-D flooding algorithm is employed either to fill an object or measure its volume, surface area, or other properties. In this case, the nonseparability of the surface causes a leakage of the flood through the discrete surface. Unfortunately, the extension of the 2-D definition of separation to the third dimension and to voxel surfaces is not straightforward, because voxelized surfaces cannot be defined as an ordered sequence of voxels and a voxel on the surface does not have a specific number of adjacent surface voxels. Furthermore, there are important topological issues, such as the separation of both sides of a surface, which cannot be well defined by employing 2-D terminology. The theory that deals with these topological issues is called 3-D *discrete topology*. We next sketch some basic notions and informal definitions used in this field.

41.14.2 Fundamentals of 3-D Discrete Topology

The 3-D discrete space is a set of integral grid points in 3-D Euclidean space defined by their Cartesian coordinates (x, y, z). A voxel is the unit cubic volume centered at the integral grid point. The voxel value is mapped onto $\{0,1\}$: the voxels assigned 1 are called the *black* voxels, representing opaque objects, and those assigned 0 are the *white* voxels, representing the transparent background. Besides this binary representation, there are also nonbinary approaches in which the voxel value is mapped onto the interval

[0,1], representing either partial coverage, variable densities, or graded opacities. Due to its larger dynamic range of values, this approach supports 3-D antialiasing and thus supports higher-quality rendering.

Two voxels are 26-*adjacent* if they share a vertex, an edge, or a face. Every voxel has 26 such adjacent voxels: eight share a vertex (corner) with the center voxel, twelve share an edge, and six share a face. Accordingly, face-sharing voxels are defined as 6-*adjacent*, and edge-sharing and face-sharing voxels are defined as 18-*adjacent*. Here, we shall use the prefix N to define the adjacency relation, where $N = 6, 18$, or 26. A sequence of voxels having the same value (e.g., black) is called an N-*path* if all consecutive pairs are N-adjacent. A set of voxels W is N-*connected* if there is an N-path between every pair of voxels in W. An N-*connected component* is a maximal N-connected set.

Given a 2-D discrete 8-connected black curve, there are sequences of 8-connected white pixels (8-component) that pass from one side of the black component to the other without intersecting it. This phenomenon is a discrete disagreement with the continuous case, where there is no way of penetrating a closed curve without intersecting it. To avoid such a scenario, it has been the convention to define *opposite* types of connectivity for the white and black sets. Opposite types in 2-D space are 4 and 8; in 3-D space, 6 is opposite to 26 or to 18.

Assume that a voxel space, denoted by Σ, includes one subset of black voxels S. If $\Sigma - S$ is not N-connected, that is, if $\Sigma - S$ consists of at least two white N-connected components, then S is said to be N-*separating* in Σ. Loosely speaking, in 2-D, an 8-connected black path that divides the white pixels into two groups is 4-separating, and a 4-connected black path that divides the white pixels into two groups is 8-separating.

There are no analogous results in 3-D space. Let W be an N-separating surface. A voxel p E W is said to be an N-*simple voxel* if $W - p$ is still N-separating. An N-separating surface is called N-*minimal* if it does not contain any N-simple voxel. A *cover* of a continuous surface is a set of voxels such that every point of the continuous surface lies in a voxel of the cover. A cover is said to be a *minimal cover* if none of its subsets is also a cover. The cover property is essential in applications that employ space subdivision for fast ray tracing [60]. The subspaces (voxels) that contain objects must be identified along the traced ray. Note that a cover is not necessarily separating; on the other hand, as mentioned previously, it may include simple voxels. In fact, even a minimal cover is not necessarily N-minimal for any N [23].

41.14.3 Binary Voxelization

An early technique for the digitization of solids was spatial enumeration, which employs point or cell classification methods either in an exhaustive fashion or by recursive subdivision [119]. However, subdivision techniques for model decomposition into rectangular subspaces are computationally expensive and thus inappropriate for medium- or high-resolution grids. Instead, objects should be directly voxelized, preferably generating an N-separating, N-minimal, and covering set, where N is application-dependent. The voxelization algorithms should follow the same paradigm as the 2-D scan-conversion algorithms; they should be incremental and accurate, use simple arithmetic (preferably integer only), and have a complexity that is not more than linear with the number of voxels generated.

The literature on 3-D scan-conversion is relatively small. Danielsson [34] and Mokrzycki [147] independently developed similar 3-D curve algorithms in which the curve is defined by the intersection of two implicit surfaces. Voxelization algorithms have been developed for 3-D lines, 3-D circles, and a variety of surfaces and solids, including polygons, polyhedra, and quadric objects [98]. Efficient algorithms have been developed for voxelizing polygons using an integer-based decision mechanism embedded within a scan-line filling algorithm [93]; for parametric curves, surfaces, and volumes using an integer-based forward differencing technique [94]; and for quadric objects, such as cylinders, spheres, and cones using "weaving" algorithms by which a discrete circle or line sweeps along a discrete circle or line [26]. While these pioneering attempts focused more on efficiency and accuracy, later algorithms focused also on adherence to the topological requirements (i.e., to the separability and minimality criteria). Huang et al. [84] devised such an algorithm for the voxelization of polygon meshes, employing a geometric measure for each candidate voxel to determine its N-simplicity.

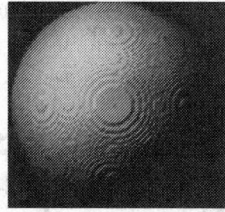

FIGURE 41.14 Binary sphere yields jagged surfaces when rendered.

41.14.4 Antialiased Voxelization

The previous subsection discussed binary voxelization, which generates topologically and geometrically consistent models but exhibits object space aliasing, caused by the binary classification of voxels into the {0,1} set. Therefore, the resolution of the 3-D raster ultimately determines the precision of the discrete model, and imprecise modeling results in jagged surfaces, known as *object space aliasing*. This leads to image space aliasing during the rendering (see Figure 41.14). To avoid the aliasing, one must employ object-space prefiltering, in which scalar-valued voxels represent the percentage of spatial occupancy of a voxel [227], an extension of the 2-D line antialiasing method of Gupta and Sproull [67]. The scalar-valued voxels determine a fuzzy set such that the boundary between inclusion and exclusion is smooth. Direct visualization from such a fuzzy set avoids image aliasing.

Some research on voxelization and debinarization of sampled volume data sets has focused on generating a distance volume for subsequent use in manipulation [12] or rendering [51]. The latter also employed an elastic surface wrap, called *surface nets*, to enable the generation of smoother distance fields. By means of the distance volume, one can then estimate smooth gradients and achieve pleasing renderings without jagged surfaces. Sramek and Kaufman [205] [206] showed that the optimal sampling filter for central difference gradient estimation in areas of low curvature is a one-dimensional box filter of width $2\sqrt{3}$ voxel units, oriented perpendicular to the surface. Since most volume rendering implementations utilize the central difference gradient estimation filter and trilinear sample interpolation, the oriented box filter is well suited for voxelization. Furthermore, this filter is an easily computed linear function of the distance from the triangle. Binary parametric surfaces and curves can be antialiased using a (slower) 3-D splatting technique.

Later methods have focused on providing more efficient algorithms for antialiased triangle voxelization, suitable for both software [33] [88] and hardware implementations [33] [45]. Because conventional graphics hardware only rasterizes points, lines, and triangles, higher-order primitives must be expressed as combinations of these basic primitives, most often as triangles. To voxelize solid objects, one can first voxelize the boundary as a set of triangles, then fill the interior using a volumetric filling procedure. A commodity hardware-based voxelization algorithm was proposed by Fang and Chen [45], which performs the voxelization on a per-volume sheet basis by slicing the polymesh (with antialiasing turned on) and storing the result in a 3-D (volumetric) texture map.

Dachille and Kaufman [33] devised a more accurate software method (in terms of the antialiasing), that employs fast, incremental arithmetic for rapid voxelization of polymeshes on a per-triangle basis. Figure 41.15 depicts the boundary region affected by the antialiased voxelization of a triangle and the profile of its voxelization. All voxels within the translucent surface, which is at a constant distance from the triangle, must be updated during the voxelization and assigned values corresponding to the distance to the triangle surface. The general idea of the algorithm is to voxelize a triangle by scanning a bounding box of the triangle in raster order. For each voxel in the bounding box, a filter equation (similar to that of [205]) is evaluated and the result stored in memory. The value of the equation is a linear function of the distance from the triangle. The result is stored using a fuzzy algebraic union operator — the *max* operator. A similar algorithm was also implemented on the VolumePro volume rendering board [33].

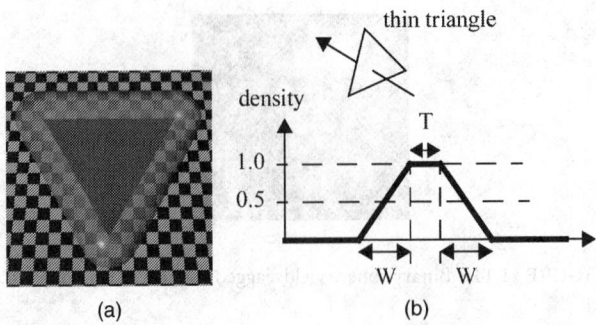

FIGURE 41.15   (a) The 3-D region of influence around a triangle, (b) the density profile of the oriented box filter along a line perpendicular to the triangle surface primitive. Here, T is the width of the triangle (usually very close to 0) and W is the width of the filter profile. The anti-aliased voxelization will maintain this profile everywhere within the red region of the triangle shown in (a). It is assumed that the iso-surface is positioned at a density value of 0.5, in the center of the profile. This ensures that the central difference operator meets a smooth boundary.

41.14.5   Block Operations and Constructive Solid Modeling

An intrinsic characteristic of the volume buffer is that adjacent objects in the scene are also represented by neighboring memory cells. Therefore, rasters lend themselves to various meaningful grouping-based operations, such as *bitblt* in 2-D, or *voxblt* in 3-D [95]. These include transfer of volume buffer rectangular blocks (cuboids) while supporting voxel-by-voxel operations between source and destination blocks. Block operations add a variety of modeling capabilities, which aid in the task of image synthesis and form the basis for the efficient implementation of a 3-D "room manager": the extension of window management to the third dimension.

Constructive solid geometry (CSG) is one of the most important modeling methods in computer graphics and computer-aided design (CAD). It allows complicated objects to be built as variously ordered unions, intersections, and differences of simpler objects, which may be bounded primitives or half-spaces. It is supported by Boolean algebra and a set of well understood, regularized set operations. Because the volume buffer lends itself to Boolean operations that can be performed on a voxel-by-voxel basis during the voxelization stage, it is advantageous to use CSG as the modeling paradigm with volumetric objects. Subtraction, union, and intersection operations between two discretized 3-D objects are accomplished at the voxel level, thereby reducing the original problem of evaluating a CSG tree during rendering time down to a 1-D Boolean operation between pairs of voxels during a preprocessing stage.

The volume buffer also allows a major extension of the traditional CSG paradigm, because it can also be employed to manipulate physical properties associated with objects, such as scalar or even vector and tensor fields. In *constructive volume geometry* (CVG) [19], the combinational operations, mostly defined in the real domain, can subsequently be used to model complex interior structures of objects and amorphous phenomena in a constructive manner, with a generalization of the well known CSG operators.

In CVG, combinational operators are defined on unbounded spatial objects and are constructed from simple arithmetic operations on scalars through a series of operational decompositions. The operations on scalars are normally defined in the real domain. The basic CVG operators include union, intersection, difference, and blending. With the flexibility and accuracy of the real domain, complex operators, such as those for data filtering and volume deformation, can easily be specified. CVG operates on both the interior and the exterior of objects and, therefore, preserves the main geometrical properties in volumetric data sets, such as volume density and multiple isosurfaces. Physical properties, such as colors, are defined and manipulated in the same way as geometry. CVG accommodates objects that are defined mathematically by scalar fields and those built from digitized volumetric data sets.

For two point-sampled binary objects, the Boolean operations of CSG or *voxblt* are trivially defined. However, the Boolean operations applied to volume-sampled models are analogous to those of fuzzy set

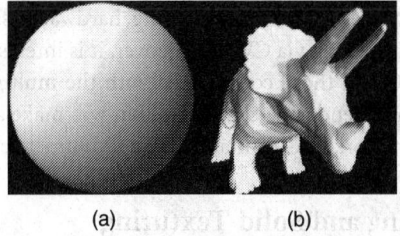

(a)   (b)

FIGURE 41.16   Voxelized objects with anti-aliased boundaries.

theory (see [40]). The volume-sampled model is a density function $d(x)$ over R^3, where d is 1 inside the object, 0 outside the object, and $0 < d < 1$ within the soft region of the filtered surface. Some of the common operations — intersection, complement, difference, and union — between two objects A and B are defined as follows:

$$
\begin{aligned}
d_{A \cap B} &= min(d_A(x), d_B(x)) \\
d_{\bar{A}} &= 1 - d_A(x) \\
d_{A-B} &= min(d_A(x), 1 - d_B(x)) \\
d_{A \cup B} &= max(d_A(x), d_B(x))
\end{aligned}
\tag{41.15}
$$

By performing the CVG operations in Equation 41.15 between sampled volumes, as obtained with 3-D scanners, complex geometric models can be generated. Volume-sampled models can also function as matte volumes [39] for various matting operations, such as performing cut-aways and merging multiple volumes into a single volume using the union operation.

The only law of set theory that is no longer true is the excluded-middle law:

$$ A \cap \bar{A} \neq \emptyset $$

and

$$ A \cup \bar{A} \neq Universe $$

The use of the *min* and *max* functions causes discontinuity at the region where the soft regions of the two objects meet, because the density value at each location in the region is determined solely by one of the two overlapping objects. In order to preserve continuity on the cut-away boundaries between the material and the empty space, one could use an alternative set of Boolean operators based on algebraic sum and algebraic product [40] [176]:

$$
\begin{aligned}
d_{A \cap B} &= d_A(x) d_B(x) \\
d_{\bar{A}} &= 1 - d_A(x) \\
d_{A-B} &= d_A(x) - d_A(x) d_B(x) \\
d_{A \cup B} &= d_A(x) + d_B(x) - d_A(x) d_B(x)
\end{aligned}
\tag{41.16}
$$

Unlike the *min* and *max* operators, algebraic sum and product operators result in

$$ A \cup A \neq A $$

which is undesirable. A consequence, for example, is that during modeling via sweeping, the resulting model is sensitive to the sampling rate of the swept path [227].

Once a CVG model has been constructed in voxel representation, it is rendered in the same way as any other volume buffer. This makes, for example, volumetric ray tracing [200] or splatting [150] [193] of constructive solid models straightforward. Texture-mapping, hardware-assisted rendering approaches will further promote the interactive modeling via CVG. Moreover, it is interesting to observe that the volume compositions generated via CVG and those constructed with the multimodal or multivalued data sets discussed earlier share a number of rendering challenges that will make attractive a common rendering, compositing, and modeling framework, most suitably using a volumetric scenegraph [160] [244].

41.14.6 Texture Mapping and Solid Texturing

One type of object complexity involves objects that are enhanced with texture mapping, photo mapping, environment mapping, or solid texturing. *Texture mapping* is commonly implemented during the last stage of the rendering pipeline, and its complexity is proportional to the object complexity. In volume graphics, however, texture mapping is performed during the voxelization stage, and the texture color is stored in each voxel in the volume buffer.

In *photo mapping*, six orthogonal photographs of the real object are projected back onto the voxelized object. Once this mapping is applied, it is stored with the voxels themselves during the voxelization stage and, therefore, does not degrade the rendering performance. Texture and photo mapping are also viewpoint-independent attributes, implying that once the texture is stored as part of the voxel value, texture mapping need not be repeated. This important feature is exploited, for example, by voxel-based flight simulators (see Figure 41.17) and in CAD systems.

A central feature of volumetric representation is that, unlike surface representation, it is capable of representing inner structures of objects, which can be revealed and explored with appropriate manipulation and rendering techniques. This capability is essential for the exploration of sampled or computed objects. Synthetic objects are also likely to be solid rather than hollow. One method for modeling various solid types is *solid texturing*, in which a procedural function or a 3-D map models the color of the objects in 3-D. During the voxelization phase, each voxel belonging to the objects is assigned a value by the texturing function or the 3-D map. This value is then stored as part of the voxel information. On the other hand, if solid texturing is to be used as a means to enrich a volume data set with more detail, without increasing the stored resolution of the data set, then the texturing function can also be evaluated during rendering time, at the ray-sampling locations. The statistical invariance under translation makes solid textured objects appear "carved out" of the simulated material (e.g., a voxelized chair or a CT head made of wood) [191].

The most important solid texturing basis functions are noise [175], turbulence, and nth closest [247]. Perlin's *noise* function [175] returns a pseudorandom value, in the range of $(-1, 1)$, by interpolating the gradient vector between predetermined lattice points. The *turbulence* basis function gives the impression

FIGURE 41.17   A satellite image, photo-mapped into voxelized terrain.

of Brownian motion (or turbulent flow) by summing noise values at decreasing frequencies, introducing a self-similar $1/f$ pattern, where f is the frequency of the noise. The *n*th-*closest* basis function [247] places feature points at random locations in R^3 and calculates the distance from a surface point to each of the *n*th-closest feature points. Combinations of these distances can then be used to index a color spline.

41.14.7 Amorphous Phenomena

Solid texturing produces objects that have simple surface definitions. However, many objects, such as fur, have surface definitions that are much more complex. Others, such as clouds, fire, and smoke, have no well defined surface at all. Although translucent objects can be represented by surface methods, these methods cannot efficiently support the modeling and rendering of amorphous phenomena, which are volumetric in nature and lack any tangible surfaces. A common modeling and rendering approach is based on a function that, for any input point in 3-D, calculates some object features, such as density, reflectivity, or color. These functions can then be rendered by ray casting, which casts a ray from each pixel into the function domain. Along the passage of the ray, at constant intervals, the function is evaluated to yield a sample. All samples along each ray are combined to form the pixel color.

Perlin and Hoffert [176] introduced a technique, called *hypertextures*, that allows for the production of such complex textures through the manipulation of surface densities. That is, rather than just coloring an object's surface with a texture map, its surface structure is changed (during rendering) using a 3-D texture function. Hypertextures introduce the idea of *soft objects*: objects with a large boundary region, modeled using an object density function $D(x)$. As with solid textures, combinations of noise and turbulence — together with two new density modulation functions, bias (controls the density variation across the soft region) and gain (controls the rate at which density changes across the midrange of the soft region) — are used to manipulate $D(x)$ to create hypertextured objects. Satherley and Jones [191] showed that nongeometric data sets, such as volumes, can be augmented with hypertextures by first performing a distance transform on them and then applying the hypertexture framework on the resulting distance volume, within the soft region of the object.

The modeling of amorphous detail via volumetric techniques has found a number of applications, including the texel approach introduced by Kajiya and Kay [90] for the rendering of fur, which was later extended by Neyret [163] for the rendering of foliage, grass, and hair. Other researchers have used volumetric representations to model and render fractals [73], gases [43] [48], and clouds [36] [108] [137].

Instead of using procedural functions based on noise functions, better renditions of physical, time-varying behavior can be obtained by modeling the actual underlying physical processes, such as smoke, fire, and water flow, via application of the Navier–Stokes equations [49] [164] [207] or lattice propagation methods [72] [229]. Although this requires much larger computational effort, recent advances in graphics hardware have yielded powerful SIMD processors that can run the required numerical solvers or lattice calculations at speedups of an order of magnitude or more, compared to traditional CPUs. For reasons of efficiency, the flow calculations are often performed on relatively coarse grids. Therefore, global illumination algorithms, such as photon maps [87], Monte Carlo volume renderers, or splats that are texture mapped with phenomena detail [105] [229], are often used to visually enhance the level of detail for visualization.

41.14.8 Natural Phenomena

Natural phenomena, such as the processes of thawing, natural weathering, and melting, are also inherently volumetric processes and are suitably modeled with volume graphics methods. Fuijishiro and Aoki [55] have used a mathematical morphology operator to simulate the effects of thawing. Here, the volume model is thought of as being made of ice and is left in the open (warm) air to thaw. The mathematical morphology operator is a phenomenological modeling operator and is shown in their work to provide a good approximation of the physical model. Further, to simulate the relegation of water on the base of volumetric ice statues, a cellular automata mechanism is employed.

Dorsey et al. [38] model the weathering of stone by employing a simulation of the flow of moisture through the surface into the stone. Here, the model governs the erosion of material from the surface, and the weathering process is confined to a thick crust on the surface of the volume. Ozawa and Fujishiro [171] also use the mathematical morphology technique for the weathering of stone. By applying a spatially variant structuring element for the morphology, they are able to simulate the stochastic nature of real weathering phenomena. Other researchers have used physically based methods, such as Navier–Stokes solvers [16] or advanced cellular automata methods (Lattice–Boltzmann) [230] [231], to simulate the process of melting and flowing of viscous materials, as well as sand, mud, and snow [210]. Varadhan and Mueller [222] proposed a physically based method for the simulation of ablation on volumetric models. They demonstrated the visual effect of ablative processes, such as a beam of heat emitted from a blow torch. Users can control ablative properties, such as energy propagation, absorption, and material evaporation, via a simple transfer function interface.

41.14.9 Volume Sculpting

Surface-based sculpting has been studied extensively (e.g., [29] [192]), while volume sculpting was introduced later for clay or waxlike sculptures [85] and for comprehensive detailed sculpting [228]. The latter approach is a free-form interactive modeling technique based on the metaphor of sculpting and painting a voxel-based solid material, such as a block of marble or wood. There are two motivations for this approach. First, modeling topologically complex and highly detailed objects is still difficult in most CAD systems. Second, sculpting has proved useful in volumetric applications. For example, scientists and physicians often need to explore the inner structures of their simulated or sampled data sets by gradually removing material.

Real-time human interaction could be achieved in this approach. Because the actions of sculpting (e.g., carving, sawing) and painting are localized in the volume buffer, a localized rendering can be employed to reproject only affected pixels. *Carving* is the process of taking a preexisting volume-sampled tool to chip or chisel the object bit by bit. Because both the object and the tool are represented as independent volume buffers, the process of sculpting involves positioning the tool with respect to the object and performing a Boolean subtraction between the two volumes. *Sawing* is the process of removing a whole chunk of material at once, much like a carpenter sawing off a portion of wood. Unlike carving, sawing requires generating the volume-sampled tool on the fly, using a user interface. To prevent object space aliasing and to achieve interactive speed, 3-D splatting is employed.

An important issue in digital sculpting is the data structure used to represent the sculpted object during and after the sculpting process. Although earlier systems employed flat 3-D arrays, multiresolution data structures, such as octrees, are better suited to capture a high level of detail where needed and at the same time to keep the memory requirements low in large homogeneous regions. Bærentzen [5] suggested the use of octrees with dynamic resolution, where nodes are inserted at different levels in the octree. Hence, there is no longer a predefined leaf level, because leaf nodes may be inserted at any level. This has two important implications:

Nonempty homogeneous regions may be grouped together and represented by a voxel at a lower level of subdivision, thereby storing the volume more efficiently.

Fine details may be added at a high level of subdivision, whenever they are needed to capture the local detail of the sculpted object.

A related data structure was introduced by Frisken et al. [52], who extended the work on distance volumes (see Section 41.14.1) to adaptively sampled distance fields (ADFs) for use as a fundamental graphical data structure. A *distance field* is a scalar field that specifies the minimum distance to a shape, where the distance may be signed to distinguish between the inside and outside of the shape. In ADFs, distance fields are adaptively sampled according to local detail and stored in a spatial hierarchy for efficient processing. They subsequently used an improved version of ADFs as a data structure within the Kizamu sculpting system [177], targeted to design digital characters for the entertainment industry.

41.15 Conclusions

Many of the important concepts and computational methods of volume visualization have been presented. Briefly described were surface rendering algorithms for volume data, in which an intermediate representation of the data is used to generate an image of a surface contained within the data. Object-order, image-order, domain-based, and hardware-based rendering techniques were presented for generating images of surfaces within the data, as well as volume-rendered images that attempt to capture in the 2-D image all 3-D embedded information, thus enabling a comprehensive exploration of the volumetric data sets. Several optimization techniques that aim at decreasing the rendering time for volume visualization and realistic global illumination rendering were also described.

Although volumetric representations and visualization techniques seem more natural for sampled or computed data sets, their advantages are also attracting traditional geometric-based applications. This trend implies an expanding role for volume visualization, which has the potential to revolutionize the field of computer graphics as a whole, by providing an alternative to surface graphics, called volume graphics. The emerging interactive volume rendering capabilities on GPUs and specialized hardware will only accelerate this trend.

Acknowledgments

This work has been partially supported by NSF grants CAREER ACI-0093157, CCR-0306438, IIS-0097646, ONR grant N000110034, DOE grant MO-068, NIH grant CA82402, and grants from NYSTAR and the Center for Advanced Technology in Biotechnology, Stony Brook.

References

[1] J. Amanatides and A. Woo, "A fast voxel traversal algorithm for ray tracing," *Eurographics '87*, 1987.

[2] K. Anagnostou, T. Atherton, and A. Waterfall, "4-D volume rendering with the shear warp factorization," *Symp. Volume Visualization and Graphics '00*, pp. 129–137, 2000.

[3] R. Avila, L. Sobierajski, and A. Kaufman, "Towards a comprehensive volume visualization system," *Proc. of IEEE Visualization '92*, pp. 13–20, 1992.

[4] R. Avila, T. He, L. Hong, A. Kaufman, H. Pfister, C. Silva, L. Sobierajski, and S. Wang, "VolVis: a diversified system for volume visualization," *Proc. of IEEE Visualization '94*, 1994.

[5] A. Bærentzen, "Octree-based volume sculpting," *Late Breaking Hot Topics, IEEE Visualization '98*, pp. 9–12, October 1998.

[6] C. Bajaj, V. Pascucci, and D. Schikore, "The contour spectrum," *Proc. IEEE Visualization '97*, pp. 167–175, 1997.

[7] C. Bajaj, V. Pascucci, G. Rabbiolo, and D. Schikore, "Hypervolume visualization: a challenge in simplicity," *Proc. 1998 Symposium on Volume Visualization '98*, pp. 95–102, 1998.

[8] M. Bentum, B.B.A. Lichtenbelt, and T. Malzbender, "Frequency analysis of gradient estimators in volume rendering," *IEEE Trans. on Visualization and Computer Graphics '96*, vol. 2, no. 3, pp. 242–254, 1996.

[9] P. Bhaniramka, R. Wenger, and R. Crawfis, "Isosurfacing in higher dimensions," *Proc. of IEEE Visualization '00*, pp. 267–273, 2000.

[10] F. Blinn, "Light reflection functions for simulation of clouds and dusty surfaces," *Proc. of SIGGRAPH '82*, pp. 21–29, 1982.

[11] M. Brady, K. Jung, H. Nguyen, and T. Nguyen, "Two-phase perspective ray casting for interactive volume navigation," *Visualization '97*, pp. 183–189, 1997.

[12] D. Breen, S. Mauch, and R. Whitaker, "3-D scan conversion of CSG models in distance volumes," *1998 Volume Visualization Symposium*, pp. 7–14, October 1998.

[13] P. Bunyk, A. Kaufman, and C. Silva, "Simple, fast, and robust ray casting of irregular grids," *Scientific Visualization*, pp. 30–36, 1997.

[14] B. Cabral, N. Cam, and J. Foran, "Accelerated volume rendering and tomographic reconstruction using texture mapping hardware," *Symp. on Volume Visualization '94*, pp. 91–98, 1994.

[15] W. Cai and G. Sakas, "Data intermixing and multi-volume rendering," *Computer Graphics Forum, (Eurographics '99)*, vol. 18, no. 3, 1999.

[16] M. Carlson, P. Mucha, R. Brooks Van Horn III, and G. Turk, "Melting and flowing," *ACM SIGGRAPH Symposium on Computer Animation*, 2002.

[17] H. Carr, J. Snoeyink, and U. Axen, "Computing contour trees in all dimensions," *Computational Geometry Theory and Applications '02*, vol. 24, no. 2, pp. 75–94, 2003.

[18] B. Chen, A. Kaufman, and Q. Tang, "Image-based rendering of surfaces from volume data," *Workshop on Volume Graphics '01*, pp. 279–295, 2001.

[19] M. Chen and J. Tucker, "Constructive volume geometry," *Computer Graphics Forum*, vol.19, no.4, pp. 281–293, 2000.

[20] M. Chen, A. Kaufman, and R. Yagel, Eds., *Volume Graphics*, Springer, London, February, 2000.

[21] T. Chiueh, T. He, A. Kaufman, and H. Pfister, "Compression domain volume rendering," *Tech. Rep.* 94.01.04, Computer science, SUNY Stony Brook, 1994.

[22] C. Chui, *An Introduction to Wavelets*, Academic Press, New York,1992.

[23] D. Cohen and A. Kaufman, "Fundamentals of surface voxelization," *CVGIP: Graphics, Models, Image Processing*, vol. 56, no. 6, pp. 453–461, 1995.

[24] D. Cohen and Z. Shefer, "Proximity clouds — an acceleration technique for 3-D grid traversal," *The Visual Computer*, vol. 10, no. 11, pp. 27–38, 1994.

[25] D. Cohen and A. Shaked, "Photo-realistic imaging of digital terrain," *Computer Graphics Forum*, vol. 12, no. 3, pp. 363–374, 1993.

[26] D. Cohen and A. Kaufman, "Scan-conversion algorithms for linear and quadratic objects," *Symposium on Volume Visualization '91*, pp. 280–301, 1991.

[27] J. Comba, J. Klosowski, N. Max, J. Mitchell, C. Silva, and P Williams, "Fast polyhedral cell sorting for interactive rendering of unstructured grids," *Computer Graphics Forum*, vol. 18, no. 3, pp. 369–376, 1999.

[28] J. Conway and N. Sloane, *Sphere Packings, Lattices and Groups, 2nd edition*, Springer-Verlag, 1993.

[29] S. Coquillart, "Extended free-form deformation: a sculpting tool for 3-D geometric modeling," *Computer Graphics*, vol. 24, no. 4, pp. 187–196, 1990.

[30] R. Crawfis, "Real-time slicing of data space," *Proc. of IEEE Visualization '96*, pp. 271–277, 1996.

[31] R. Crawfis, and N. Max, "Texture splats for 3-D scalar and vector field visualization," *Proc. of IEEE Visualization '93*, pp. 261–266, 1993.

[32] F. Dachille, K. Mueller, and A. Kaufman, "Volumetric backprojection," *Volume Visualization Symposium '00*, pp. 109–117, 2000.

[33] F. Dachille and A. Kaufman, "Incremental triangle voxelization," *Graphics Interface 2000*, pp. 205–212, May 2000.

[34] P. Danielsson, "Incremental curve generation," *IEEE Transactions on Computers*, C-19, pp. 783–793, 1970.

[35] J. Danskin and P. Hanrahan, "Fast algorithms for volume ray tracing," *Workshop on Volume Visualization*, pp. 91–98, 1992.

[36] Y. Dobashi, K. Kaneda, H. Yamashita, T. Okita, and T. Nishita, "A simple, efficient method for realistic animation of clouds," *Proc. of SIGGRAPH '00*, pp.19–28, 2000.

[37] I. Daubechies, "Ten lectures on wavelets," *CBMS-NSF Reg. Conf. Ser. Appl. Math. SIAM.* 1992.

[38] J. Dorsey, A. Edelman, H. Jensen, J. Legakis, and H. Pederson, "Modeling and rendering of weathered stone," *Proc. of SIGGRAPH '99*, pp. 225–234, 1999.

[39] R. Drebin, L. Carpenter, and P. Hanrahan, "Volume rendering," *Proc. of SIGGRAPH '88*, vol. 22, no. 4, pp. 65–74, 1988.

[40] D. Dubuis and H. Prade, *Fuzzy Sets and Systems: Theory and Applications*, Academic Press, New York, 1980.

[41] S. Dunne, S. Napel, and B. Rutt, "Fast reprojection of volume data," *Proc. of IEEE Visualization in Biomed. Comput,* pp. 11–18, 1990.

[42] D. Ebert, C. Morris, P. Rheingans, and T. Yoo, "Designing effective transfer functions for volume rendering from photographics volumes," *IEEE Trans. on Visualization and Computer 3-D Graphics,* vol. 8, no. 2, pp. 183–197, 2002.

[43] D. Ebert and R. Parent, "Rendering and animation of gaseous phenomena by combining fast volume and scanline A-buffer techniques," *Computer Graphics,* vol. 24, no. 4, pp. 357–366, 1990.

[44] K. Engel, M. Kraus, and T. Ertl, "High-quality pre-integrated volume rendering using hardware-accelerated pixel shading," *Proc. of SIGGRAPH Graphics Hardware Workshop '01,* pp. 9–16, 2001.

[45] S. Fang and H. Chen, "Hardware accelerated voxelization," *Computers and Graphics,* no. 24, vol. 3, pp. 433–442, June 2000.

[46] R. Farias, J. Mitchell, and C. Silva, "ZSWEEP: an efficient and exact projection algorithm for unstructured volume rendering," *ACM/IEEE Volume Visualization and Graphics Symposium,* pp. 91–99, 2000.

[47] J. Foley, A. Dam, S. Feiner, and J. Hughes, *Computer Graphics: Principles and Practice, 2nd edition.* Addison-Wesley, Reading, MA, 1996.

[48] N. Foster and D. Metaxas, "Modeling the motion of hot, turbulent gas," *Proc. of SIGGRAPH '97,* pp. 181–188, 1997.

[49] N. Foster and R. Fedkiw, "Practical animation of liquids," *Proc. of SIGGRAPH '01,* pp. 15–22, 2001.

[50] J. Fowler and R. Yagel, "Lossless compression of volume data," *Symp. of Volume Visualization '94,* pp. 43–50, 1994.

[51] S. Frisken Gibson, "Using distance maps for accurate surface representation in sampled volumes," *ACM Symposium on Volume Visualization '98,* pp. 23–30, 1998.

[52] S. Frisken, R. Perry, A. Rockwood, and T. Jones, "Adaptively sampled distance fields: a general representation of shape for computer graphics," *Proc. of ACM SIGGRAPH '00,* pp. 249–254, July 2000.

[53] A. Fujimoto, T. Tanaka, and K. Iwata, "ARTS: accelerated ray-tracing system," *IEEE Computer Graphics and Applications,* vol. 6, no. 4, pp. 16–26, 1986.

[54] I. Fujishiro, Y. Takeshima, T. Azuma, and S. Takahashi, "Volume data mining using 3-D field topology analysis," *IEEE Computer Graphics and Applications,* vol. 20, no. 5, pp. 46–51, 2000.

[55] I. Fujishiro and E. Aoki, "Volume graphics modeling of ice thawing," *Proc. Volume Graphics Workshop 2001,* pp. 69–81, 2001.

[56] A. Gaddipati, R. Machiraju, and R. Yagel, "Steering image generation using wavelet based perceptual metric," *Computer Graphics Forum (Proc. of Eurographics '97),* vol. 16, no. 3, pp. 241–251, 1997.

[57] M. Garrity, "Raytracing irregular volume data," *Computer Graphics,* pp. 35–40, November 1990.

[58] V. Gelder and K. Kim, "Direct volume rendering via 3-D texture mapping hardware," *Proc. of Vol. Rend. Symp.'96,* pp. 23–30, 1996.

[59] C. Giertsen, "Volume visualization of sparse irregular meshes," *IEEE Computer Graphics and Applications,* vol. 12, no.2, pp. 40–48, March 1992.

[60] A. Glassner, "Space subdivision for fast raytracing," *IEEE Computer Graphics & Applications,* vol. 4, no. 10, pp. 15–22, 1984.

[61] D. Gordon and R. Reynolds, "Image-space shading of 3-dimensional objects," *Computer Vision, Graphics, and Image Processing,* 29, pp. 361–376, 1985.

[62] A. Gosh, P. Prabhu, A. Kaufman, and K. Mueller, "Hardware assisted multichannel volume rendering," (to be presented) *Computer Graphics International '03,* 2003.

[63] M. Gross, R. Koch, L. Lippert, and A. Dreger, "A new method to approximate the volume rendering equation using wavelet bases and piecewise polynomials," *Computers & Graphics,* vol. 19 no. 1, pp. 47–62, 1995.

[64] T. Guenther, C. Poliwoda, C. Reinhard, J. Hesser, R. Maenner, H. Meinzer, and H. Baur, "VIRIM: a massively parallel processor for real-time volume visualization in medicine," *Proc. of the 9th Eurographics Hardware Workshop,* pp. 103–108, 1994.

[65] S. Guthe, M. Wand, J. Gonser, and W. Strasser, "Interactive rendering of large volume datasets," *Proc. of IEEE Visualization '02*, pp. 53–60, 2002.

[66] S. Guthe, and W. Strasser, "Real-time decompression and visualization of animated volume data," *Proc. of IEEE Visualization '01*, 2001.

[67] S. Gupta and R. Sproull, "Filtering edges for grayscale displays," *Computer Graphics*, vol. 15, no. 3, pp. 1–5, August 1981.

[68] P. Hanrahan and W. Krueger, "Reflection from layered surfaces due to subsurface scattering," *Computer Graphics (Proc. of SIGGRAPH '93)*, pp. 165–174, 1993.

[69] A. Hanson and P. Heng, "Four-dimensional views of 3-D scalar fields," *Proc. of IEEE Visualization '92*, pp. 84–91, 1992.

[70] A. Hanson and P. Heng, "Illuminating the fourth dimension," *IEEE Computer Graphics and Applications*, vol. 12, no. 4, pp. 54–62, 1992.

[71] A. Hanson and R. Cross, "Interactive visualization methods for four dimensions," *Proc. of IEEE Visualization '93*, pp. 196–203, 1993.

[72] M. Harris, G. Coombe, T. Scheuermann, and A. Lastra, "Physically-based visual simulation on graphics hardware," *Proc. of 2002 SIGGRAPH/Eurographics Workshop on Graphics Hardware*, 2002.

[73] J. Hart, D. Sandin, L. Kaufman, "Raytracing deterministic 3-D fractals," *Computer Graphics*, vol. 23, no. 3, pp. 289–306, 1989.

[74] H. Hauser, L. Mroz, G. Bischi, and M. Gröller, "Two-level volume rendering-flushing MIP and DVR," *Proc. of IEEE Visualization '00*, pp. 211–218, 2000.

[75] T. He, L. Hong, A. Kaufman, and H. Pfister, "Generation of transfer functions with stochastic search techniques," *Proc. of IEEE Visualization '96*, pp. 227–234, 1996.

[76] G. Herman and H. Liu, "Three-dimensional display of human organs from computed tomograms," *Comput. Graphics Image Process*, vol. 9, pp. 1–21, 1979.

[77] G. Herman and J. Udupa, "Display of three-dimensional discrete surfaces," *Proceedings SPIE*, 283, pp. 90–97, 1981.

[78] J. Hesser, R. Maenner, G. Knittel, W. Strasser, H. Pfister, and A. Kaufman, "Three architectures for volume rendering," *Computer Graphics Forum*, vol. 14, no. 3, pp. 111–122, 1995.

[79] L. Hong and A. Kaufman, "Accelerated ray-casting for curvilinear volumes," *Proc. of IEEE Visualization '98*, pp. 247–253, 1998.

[80] L. Hong and A. Kaufman, "Fast projection-based ray-casting algorithm for rendering curvilinear volumes," *IEEE Transactions on Visualization and Computer Graphics*, vol. 5, no. 4, pp. 322–332, 1999.

[81] L. Hong, S. Muraki, A. Kaufman, D. Bartz, and T. He, "Virtual voyage: interactive navigation in the human colon," *Proc. of ACM SIGGRAPH '97*, pp. 27–34, August 1997.

[82] http://www.nlm.nih.gov/research/visible/visible_human.html.

[83] J. Huang, R. Crawfis, and D. Stredney, "Edge preservation in volume rendering using splatting," *IEEE Volume Vis. '98*, pp. 63–69, 1998.

[84] J. Huang, R. Yagel, V. Fillipov, and Y. Kurzion, "An accurate method to voxelize polygon meshes," *ACM/IEEE Symposium on Volume Visualization '98*, Chapel Hill, NC, October 1998.

[85] J. Hughes and T. Galyean, "Sculpting: an interactive volumetric modeling technique," *Computer Graphics*, vol. 25, no. 4, pp. 267–274, 1991.

[86] I. Ihm and R. Lee, "On enhancing the speed of splatting with indexing," *Proc. of IEEE Visualization '95*, pp. 69–76, October 1995.

[87] H. Jensen and P. Christensen, "Efficient simulation of light transport in sciences with participating media using photon maps," *Proc. of SIGGRAPH '98*, pp. 311–320, 1998.

[88] M. Jones, "The production of volume data from triangular meshes using voxelisation," *Computer Graphics Forum*, vol. 15, no. 5, pp. 311–318, December 1996.

[89] J. Kajiya and B. Herzen, "Ray tracing volume densities," *Proc. of SIGGRAPH '84*, pp. 165–174, 1994.

[90] L. Kajiya and T. Kay, "Rendering fur with three-dimensional textures," *Computer Graphics*, vol. 23, no. 3, pp. 271–280, 1989.

[91] A. Kaufman, *Volume Visualization*, IEEE Computer Society Press Tutorial, Los Alamitos, CA.

[92] A. Kaufman, "Volume visualization," *ACM Computing Surveys*, vol. 28, no. 1 pp. 165–167, 1996.

[93] A. Kaufman, "An algorithm for 3-D scan conversion of polygons," *Proc. Eurographics '87*, pp. 197–208, Amsterdam, Netherlands, August 1996.

[94] A. Kaufman, "Efficient algorithms for 3-D scan conversion of parametric curves, surfaces, and volumes," *Computer Graphics*, vol. 21, no. 4, pp. 171–179, 1987.

[95] A. Kaufman, "The *voxblt* engine: a voxel frame buffer processor," *Advances in Graphics Hardware III*, A. Kujik, Ed., pp. 85102, Springer-Verlag, Berlin, 1992.

[96] A. Kaufman, D. Cohen, and R. Yagel, "Volume graphics," *IEEE Computer*, vol. 26, no. 7, pp. 51–64, 1993.

[97] A. Kaufman, R. Yagel, and D. Cohen, "Intermixing surface and volume rendering," *3-D Imaging in Medicine: Algorithms, Systems, Applications*, pp. 217–227, June 1990.

[98] A. Kaufman and E. Shimony, "3-D scan-conversion algorithms for voxel-based graphics," *Proc. ACM Workshop Interactive 3-D Graphics*, pp. 45–76, Chapel Hill, NC, October 1986.

[99] T. Kay and J. Kajiya, "Ray tracing complex scenes," *Proc. of SIGGRAPH '86*, pp. 269–278, 1986.

[100] Y. Ke and E. Panduranga, "A journey into the fourth dimension," *Proc. of IEEE Visualization'89*, pp. 219–229, 1989.

[101] R. Keys, "Cubic convolution interpolation for digital image processing," *IEEE Transactions on Acoustics, Speech, Signal Processing*, vol. 29, no. 6, pp. 1153–1160, 1981.

[102] S. Kilthau and T. Möller, "Splatting Optimizations," *Technical Report, School of Computing Science, Simon Fraser University*, (SFU- CMPT-04/01-TR2001-02), April 2001.

[103] G. Kindlmann and J. Durkin, "Semi-automatic generation of transfer functions for direct volume rendering," *Symp. Volume Visualization '98*, pp. 79–86, 1998.

[104] D. King, C. Wittenbrink, and H. Wolters, "An architecture for interactive tetrahedral volume rendering," *International Workshop on Volume Rendering '01*, 2001.

[105] S. King, R. Crawfis, and W. Reid, "Fast volume rendering and animation of amorphous phenomena," *Volume Graphics Workshop 1999*, Springer, London, 2000.

[106] J. Kniss, G. Kindlmann, and C. Hansen, "Interactive volume rendering using multidimensional transfer functions and direct manipulation widgets," *Proc. of IEEE Visualization '01*, pp. 255–262, 2001.

[107] J. Kniss, G. Kindlmann, and C. Hansen, "Multidimensional transfer functions for interactive volume rendering," *IEEE Transactions on Visualization and Computer Graphics*, vol. 8, no. 3, pp. 270–285, 2002.

[108] J. Kniss, S. Premoze, C. Hansen, P. Shirley, and A. McPherson, "Interactive volume light transport and procedural modeling," *IEEE Transactions on Visualization and Computer Graphics*, 2003.

[109] J. Kniss, S. Premoze, C. Hansen, and D. Ebert, "Interactive translucent volume rendering and procedural modeling," *Proc. of IEEE Visualization '02*, pp. 109–116, 2002.

[110] G. Knittel, "The ULTRAVIS system," *Proc. of Volume Visualization and Graphics Symposium '00*, pp. 71–80, 2000.

[111] G. Knittel and W. Strasser, "A compact volume rendering accelerator," *Volume Visualization Symposium Proceedings*, pp. 67–74, 1994.

[112] K. Kreeger and A. Kaufman, "Interactive volume segmentation with the PAVLOV architecture," *Proc. of Parallel Visualization and Graphics Symposium '99*, 1999.

[113] K. Kreeger, I. Bitter, F. Dachille, B. Chen, and A. Kaufman, "Adaptive perspective ray casting," *Volume Visualization Symposium '98*, pp. 55–62. 1998.

[114] K. Kreeger and A. Kaufman. "Mixing translucent polygons with volumes," *Proc. of IEEE Visualization '99*, October 1999.

[115] M. Kreveld, R. Oostrum, C. Bajaj, V. Pascucci, and D. Schikore, "Contour trees and small seed sets for isosurface traversal," *Proc. of the 13th ACM Symposium on Computational Geometry*, pp. 212–220, 1997.

[116] P. Lacroute and M. Levoy, "Fast volume rendering using a shear-warp factorization of the viewing transformation," *Proc. of SIGGRAPH '94*, pp. 451–458, 1994.

[117] D. Laur and P. Hanrahan, "Hierarchical splatting: a progressive refinement algorithm for volume rendering," *Proc. of SIGGRAPH '91*, pp. 285–288, 1991.

[118] R. Lee and I. Ihm, "On enhancing the speed of splatting using both object and image space coherence," *Graphical Models and Image Processing*, vol. 62, no. 4, pp 263–282, 2000.

[119] Y. Lee and A. Requicha, "Algorithms for computing the volume and other integral properties of solids: I — known methods and open issues; II — a family of algorithms based on representation conversion and cellular approximation," *Communications of the ACM*, vol. 25, no. 9, pp. 635–650, 1982.

[120] J. Leven, J. Corso, S. Kumar, and J. Cohen, "Interactive visualization of unstructured grids using hierarchical 3-D textures,"*Proc. of Symposium on Volume Visualization and Graphics '02*, pp. 33–40, 2002.

[121] M. Levoy, "Display of surfaces from volume data," *IEEE Comp. Graph. & Appl.*, vol. 8, no. 5, pp. 29–37, 1988.

[122] M. Levoy, "Efficient ray tracing of volume data," *ACM Trans. Comp. Graph.*, vol. 9, no. 3, pp. 245–261, 1990.

[123] M. Levoy. "A hybrid ray tracer for rendering polygon and volume data." *IEEE Computer Graphics & Applications*, vol. 10, no. 2, pp. 33–40, March 1990.

[124] W. Li and A. Kaufman, "Accelerating volume rendering with texture hulls," *IEEE/SIGGRAPH Symposium on Volume Visualization and Graphics 2002 (VolVis '02)*, pp. 115–122, 2002.

[125] W. Li and A. Kaufman, "Texture partitioning and packing for accelerating texture-based volume rendering," *Graphics Interface '03*, 2003.

[126] B. Lichtenbelt, R. Crane, and S. Naqvi, *Volume Rendering*, Prentice Hall, 1998.

[127] E. Lorensen and H. Cline, "Marching cubes: a high resolution 3-D surface construction algorithm," *Proc. of SIGGRAPH '87*, pp. 163–169, 1987.

[128] E. Lum, K. Ma, and J. Clyne, "Texture hardware assisted rendering of time-varying volume data," *Proc. of IEEE Visualization '01*, pp. 262–270, 2001.

[129] R. Machiraju, A. Gaddipati, and R. Yagel, "Detection and enhancement of scale coherent structures using wavelet transform products," *Proc. of the Technical Conference on Wavelets in Signal and Image Processing V*, SPIE Annual Meeting, pp. 458–469, 1997.

[130] R. Machiraju and R. Yagel, "Efficient feed-forward volume rendering techniques for vector and parallel processors," *SUPERCOMPUTING '93*, pp. 699–708, 1993.

[131] R. Machiraju and R. Yagel, "Reconstruction error and control: a sampling theory approach," *IEEE Transactions on Visualization and Graphics*, vol. 2, no. 3, December 1996.

[132] T. Malzbender and F. Kitson, "A Fourier technique for volume rendering," *Focus on Scientific Visualization*, pp. 305–316, 1991.

[133] X. Mao, "Splatting of non rectilinear volumes through stochastic resampling," *IEEE Transactions on Visualization and Computer Graphics*, vol. 2, no. 2, pp. 156–170, 1996.

[134] X. Mao, L. Hong, and A. Kaufman, "Splatting of curvilinear volumes," *Proc. IEEE Visualization 1995*, pp. 61–68, 1995.

[135] J. Marks, B. Andalman, P.A. Beardsley, W. Freeman, S. Gibson, J. Hodgins, T. Kang, B. Mirtich, H. Pfister, W. Rum, "Design galleries: a general approach to setting parameters for computer graphics and animation," *Proc. of SIGGRAPH '97*, pp. 389–400, 1997.

[136] S. Marschner and R. Lobb, "An evaluation of reconstruction filters for volume rendering," *Proc. of IEEE Visualization '94*, pp. 100–107, 1994.

[137] N. Max, "Optical models for direct volume rendering," *IEEE Trans. Vis. and Comp. Graph.*, vol. 1, no. 2, pp. 99–108, 1995.

[138] N. Max, P. Hanrahan, and R. Crawfis, "Area and volume coherence for efficient visualization of 3-D scalar functions," *Computer Graphics*, vol. 24, no. 5, pp. 27–33, 1990.

[139] D. Meagher, "Geometric modeling using octree encoding," *Computer Graphics and Image Processing*, vol. 19, no. 2, pp. 129–147, 1982.

[140] H. Meinzer, K. Meetz, D. Scheppelmann, U. Engelmann, and H. Baur, "The Heidelberg raytracing model," *IEEE Computer Graphics and Applications*, vol. 11, no. 6, pp. 34–43, 1991.

[141] M. Meissner and S. Guthe, "Interactive lighting models and pre-integration for volume rendering on PC graphics accelerators," *Graphics Interface '02*, 2002.

[142] M. Meissner, J. Huang, D. Bartz, K. Mueller, and R. Crawfis, "A practical comparison of popular volume rendering algorithms," *Symposium on Volume Visualization and Graphics 2000*, pp. 81–90, 2000.

[143] M. Meissner, M. Doggett, U. Kanus, and J. Hirche, "Efficient space leaping for ray casting architectures," *Proc. of the 2nd Workshop on Volume Graphics*, 2001.

[144] M. Meissner, U. Kanus, and W. Strasser, "VIZARD II: a PCI-card for real-time volume rendering," *Proc. of SIGGRAPH/Eurographics Workshop on Graphics Hardware '98*, pp. 61–67, 1998.

[145] M. Meissner, U. Kanus, G. Wetekam, J. Hirche, A. Ehlert, W. Strasser, M. Doggett, and R. Proksa, "A reconfigurable interactive volume rendering system," *Proc. of SIGGRAPH/Eurographics Workshop on Graphics Hardware '02*, 2002.

[146] D. Mitchell and A. Netravali, "Reconstruction filters in computer graphics," *Proc. of SIGGRAPH '88*, pp. 221–228, 1988.

[147] W. Mokrzycki, "Algorithms for discretization of algebraic spatial curves on homogeneous cubical grids," *Computer Graphics*, vol. 12, no. 3/4, pp. 477–487, 1988.

[148] T. Möller, R. Machiraju, K. Mueller, and R. Yagel, "A comparison of normal estimation schemes," *Proc. of IEEE Visualization '97*, pp. 19–26, 1997.

[149] T. Möller, R. Machiraju, K. Mueller, and R. Yagel, "Evaluation and design of filters using a Taylor series expansion," *IEEE Transactions on Visualization and Computer Graphics*, vol. 3, no. 2, pp. 184–199, 1997.

[150] C. Montani and R. Scopigno, "Ray tracing CSG trees using the sticks representation scheme." *Computers & Graphics*, vol. 14, no. 3, pp. 481–490, 1990.

[151] B. Mora, J. Jessel, and R. Caubet, "A new object-order raycasting algorithm," *Proc. of IEEE Visualization '02*, pp. 203–210, 2002.

[152] K. Mueller and R. Crawfis, "Eliminating popping artifacts in sheet buffer-based splatting," *Proc. of IEEE Visualization '98*, pp. 239–245, 1998.

[153] K. Mueller and R. Yagel, "Fast perspective volume rendering with splatting by using a ray-driven approach," *Proc. of IEEE Visualization '96*, pp. 65–72, 1996.

[154] K. Mueller, M. Chen, and A. Kaufman, Eds. *Volume Graphics '01*, Springer, London, 2001.

[155] K. Mueller, N. Shareef, J. Huang, and R. Crawfis, "High-quality splatting on rectilinear grids with efficient culling of occluded voxels," *IEEE Transactions on Visualization and Computer Graphics*, vol. 5, no. 2, pp. 116–134, 1999.

[156] K. Mueller, N. Shareef, J. Huang, and R. Crawfis, "IBR assisted volume rendering," *Proc. of IEEE Visualization '99*, pp. 5–8, 1999.

[157] K. Mueller, T. Moeller, J.E. Swan, R. Crawfis, N. Shareef, and R. Yagel, "Splatting errors and antialiasing," *IEEE Transactions on Visualization and Computer Graphics*, vol. 4, no. 2, pp. 178–191, 1998.

[158] K. Mueller, T. Möller, and R. Crawfis, "Splatting without the blur," *Proc. of IEEE Visualization '99*, pp. 363–371, 1999.

[159] S. Muraki, "Volume data and wavelet transform," *IEEE Comput. Graphics Appl.*, vol. 13, no. 4, pp. 50–56, 1993.

[160] D. Nadeau, "Volume scene graphs," *Symposium on Volume Visualization and Graphics '00*, October 2000.

[161] N. Neophytou and K. Mueller, "Space-time points: 4-D splatting on efficient grids," *Symposium on Volume Visualization and Graphics '02*, pp. 97–106, 2002.

[162] N. Neophytou and K. Mueller, "Post-convolved splatting," *Joint Eurographics–IEEE TCVG Symposium on Visualization '03*, 2003.

[163] F. Neyret, "Modeling, animating, and rendering complex scenes using volumetric textures," *IEEE Transactions on Visualization and Computer Graphics*, vol. 4, no. 1, 1998.

[164] D. Nguyen, R. Fedkiw, and H. Jensen, "Physically based modeling and animation of fire," *Proc. of SIGGRAPH '02*, pp. 721–728, 2002.

[165] G. Nielson and B. Hamann, "The asymptotic decider: resolving the ambiguity in marching cubes," *Proc. of IEEE Visualization '91*, pp. 29–38, 1991.

[166] M. Nielson, "Scattered data modeling," *IEEE Computer Graphics and Applications*, vol. 13, no. 1. pp. 60–70, January 1993.

[167] P. Ning and L. Hesselink, "Fast volume rendering of compressed data," *Proc. of IEEE Visualization '93*, pp. 11–18, 1993.

[168] P. Ning and L. Hesselink, "Vector quantization for volume rendering," *Proc. of IEEE Visualization '92*, pp. 69–74, 1992.

[169] V. Norton, "Generation and rendering of geometric fractals in 3-D," *Computer Graphics*, vol. 16, no. 3, pp. 61–67, 1982.

[170] L. Novins, F.X. Sillion, and D.P. Greenberg, "An efficient method for volume rendering using perspective projection," *Computer Graphics*, vol. 24, no. 5, pp. 95–100, 1990.

[171] N. Ozawa and I. Fujishiro, "A morphological approach to volume synthesis of weathered stone," *Proc. Volume Graphics Workshop 1999*, pp. 367–378, 1999.

[172] S. Parker, M. Parker, Y. Livnat, P. Sloan, C. Hansen, and P. Shirley, "Interactive ray tracing for volume visualization," *IEEE Transactions on Visualization and Computer Graphics*, vol. 5, no. 3, pp. 238–250, 1999.

[173] S. Parker, P. Shirley, Y. Livnat, C. Hansen, and P. Sloan, "Interactive ray tracing for isosurface rendering," *Proc. of IEEE Visualization '98*, pp. 233–238, 1998.

[174] V. Pekar, R. Wiemker, and D. Hempel, "Fast detection of meaningful isosurfaces for volume data visualization," *Proc. of IEEE Visualization '01*, pp. 223–230, 2001.

[175] K. Perlin, "An image synthesizer," *Computer Graphics (SIGGRAPH '85 Proceedings)*, vol. 19, no. 3, pp. 287–296, July 1985.

[176] K. Perlin and E. Hoffert, "Hypertexture," *Computer Graphics*, vol. 23, no. 3, pp. 253–261, 1989.

[177] R. Perry and S. Frisken, "Kizamu: a system for sculpting digital characters," *Proc. of ACM SIGGRAPH '01*, pp. 47–56, August 2001.

[178] H. Pfister, B. Lorensen, C. Bajaj, G. Kindlmann, W. Schroeder, L. Avila, K. Martin, R. Machiraju, and J. Lee, "The transfer function bake-off," *Proc. of IEEE Computer Graphics and Applications*, vol. 21, no. 3, pp. 16–22, 2001.

[179] H. Pfister, J. Hardenbergh, J. Knittel, H. Lauer, and L. Seiler, "The VolumePro real-time raycasting system," *Proc. of SIGGRAPH '99*, pp. 251–260, 1999.

[180] H. Pfister, A. Kaufman, and F. Wessels, "Toward a scalable architecture for real-time volume rendering," *Proc. of 10th Eurographics Workshop on Graphics Hardware '95*, 1995.

[181] H. Pfister and A. Kaufman, "Cube-4: a scalable architecture for real-time volume rendering," *Proc. of Volume Visualization Symposium '96*, pp. 47–54, 1996.

[182] T. Porter and T. Duff, "Compositing digital images," *Computer Graphics (Proc. SIGGRAPH '84)*, pp. 253–259, 1984.

[183] H. Qu, N. Zhang, F, Qin, A. Kaufman, and M. Wan, "Ray tracing height fields," *Proc. of Computer Graphics International '03*, Tokyo, July 2003.

[184] R. Reynolds, D. Gordon, and L. Chen, "A dynamic screen technique for shaded graphics display of slice-represented objects," *Computer Graphics and Image Processing*, vol. 38, pp. 275–298, 1987.

[185] C. Rezk-Salama, K. Engel, M. Bauer, G. Greiner, and T. Ertl, "Interactive volume rendering on standard PC graphics hardware using multi-textures and multi-stage-rasterization," *Proc. of SIGGRAPH/Eurographics Workshop on Graphics Hardware '00*, pp. 109–118, 2000.

[186] S. Roettger and T. Ertl, "A two-step approach for interactive pre-integrated volume rendering of unstructured grids," *Proc. of VolVis '02*, pp. 23–28, 2002.

[187] S. Roettger, M. Kraus, and T. Ertl, "Hardware-accelerated volume and isosurface rendering based on cell-projection," *Proc. of IEEE Visualization '00*, pp. 109–116, 2000.

[188] H. Rushmeier and E. Torrance, "The zonal method for calculating light intensities in the presence of a participating medium," *Computer Graphics*, vol. 21, no. 4, pp. 293–302, July 1987.

[189] P. Sabella, "A rendering algorithm for visualizing 3-D scalar fields," *ACM SIGGRAPH Computer Graphics*, vol. 22, no. 4, pp. 51–58, 1988.

[190] H. Samet, *Application of Spatial Data Structures*, Addison-Wesley, Reading, MA,1990.

[191] R. Satherley and M. Jones, "Extending hypertextures to non-geometrically definable volume data," *Proc. of Volume Graphics Workshop 1999*, Springer-Verlag, pp. 211–225, 2000.

[192] T. Sederberg and S. Parry, "Free-form deformation of solid geometric models, *Computer Graphics*, vol. 20, no. 4, pp. 151–159, 1986.

[193] N. Shareef and R. Yagel, "Rapid previewing via volume-based solid Modeling," *Proc. of Solid Modeling '95*, pp. 281–292, May 1995.

[194] H. Shen, L. Chiang, and K. Ma, "A fast volume rendering algorithm for time-varying fields using a time-space partitioning tree," *Proc. of IEEE Visualization '99*, pp. 371–377, 1999.

[195] Y. Shinagawa and T. Kunii, "Constructing a Reeb graph automatically from cross sections," *IEEE Computer Graphics and Applications*, vol. 11, no. 6, pp.45–51, 1991.

[196] P. Shirley and A. Tuchman, "A polygonal approximation to direct scalar volume rendering," *Computer Graphics*, vol. 24, no. 5, pp. 63–70, 1990.

[197] C. Silva and J. Mitchell, "The lazy sweep ray casting algorithm for rendering irregular grids," *IEEE Transactions on Visualization and Computer Graphics*, vol. 3, no. 2, April–June 1997.

[198] C. Silva, J. Mitchell, and P. Williams, "An exact interactive time visibility ordering algorithm for polyhedral cell complexes," *Volume Visualization Symposium '98*, pp. 87–94, October 1998.

[199] J. Snyder, and A. Barr, "Ray tracing complex models containing surface tessellations," *Proc. of SIGGRAPH '87*, pp. 119–128, 1987.

[200] L. Sobierajski and A. Kaufman, "Volumetric raytracing," *Symposium on Volume Visualization '94*, pp. 11–18, 1994.

[201] L. Sobierajski and R. Avila, "A hardware acceleration method for volumetric ray tracing," *Proc. of IEEE Visualization '95*, pp. 27–35, 1995.

[202] B. Sohn, C. Bajaj, and V. Siddavanahalli, "Feature based volumetric video compression for interactive playback," *VolVis '02*, pp. 89–96, 2002.

[203] D. Spearey and S. Kennon, "Volume probes: interactive data exploration on arbitrary grids," *Computer Graphics*, vol. 25, no. 5, pp. 5–12, 1990.

[204] M. Sramek and A. Kaufman, "Fast ray-tracing of rectilinear volume data using distance transforms," *IEEE Transactions on Visualization and Computer Graphics*, vol. 3, no. 6, pp. 236–252, 2000.

[205] M. Sramek and A. Kaufman, "Alias-free voxelization of geometric objects," *IEEE Transactions on Visualization and Computer Graphics*, vol. 3, no. 5, pp. 251–266, 1999.

[206] M. Sramek and A. Kaufman, "vxt: a C++ class library for object voxelization," *International Workshop on Volume Graphics*, pp. 295–306, March 1999.

[207] J. Stam, "Stable fluids," *Proc. of SIGGRAPH '99*, pp. 121–128, 1999.

[208] B. Stander and J. Hart. "A Lipschitz method for accelerated volume rendering," *Proc. of the Symposium on Volume Visualization '94*, pp. 107–114, 1994.

[209] C. Stein, B. Becker, and N. Max. "Sorting and hardware assisted rendering for volume visualization," *Symposium on Volume Visualization '94*, pp. 83–90, 1994.

[210] R. Sumner, J. O'Brien, and J. Hodgins, "Animating sand, mud, and snow," *Computer Graphics Forum*, vol. 18, no. 1, pp. 3–15, 1999.

[211] P. Sutton and C. Hansen, "Isosurface extraction in time-varying fields using a temporal branch-on-need tree (T-BON)," *Proc. of IEEE Visualization '99*, pp. 147–153, 1999.

[212] E. Swan, K. Mueller, T. Moller, N. Shareef, R. Crawfis, and R. Yagel, "An anti-aliasing technique for splatting," *Proc. of IEEE Visualization '97*, pp. 197–204, 1997.

[213] S. Takahashi, T. Ikeda, Y. Shinagawa, T.L. Kunii, and M. Ueda, "Algorithms for extracting correct critical points and constructing topological graphs from discrete geographical elevation data," *Computer Graphics Forum*, vol. 14, no. 3, pp. 181–192, 1995.

[214] S. Tenginakai, J. Lee, and R. Machiraju, "Salient iso-surface detection with model-independent statistical signatures," *Proc. of IEEE Visualization '01*, pp. 231–238, 2001.

[215] T. Theussl, H. Hauser, and M. Gröller, "Mastering windows: improving reconstruction," *Proc. of IEEE Symposium on Volume Visualization '00*, 2000.

[216] T. Theussl, T. Möller, and E. Gröller, "Optimal regular volume sampling,"*Proc. of IEEE Visualization '01*, 2001.

[217] P. Thévenaz, T. Blu, and M. Unser, "Interpolation revisited," *IEEE Transactions on Medical Imaging*, vol. 19, no. 7, pp. 739–758, 2000.

[218] T. Totsuka and M. Levoy, "Frequency domain volume rendering," *Proc. of SIGGRAPH '93*, pp. 271–278, 1993.

[219] H. Tuy and L. Tuy, "Direct 2-D display of 3-D objects," *IEEE Computer Graphics and Applications*, vol. 4, no. 10, pp. 29–33, 1984.

[220] J. Udupa and D. Odhner, "Shell rendering," *IEEE Computer Graphics and Applications*, vol. 13, no. 6, pp. 58–67, 1993.

[221] S. Uselton, "Volume rendering for computational fluid dynamics: initial results," *Tech Report RNR-91-026*, NASA Ames Research Center, 1991.

[222] H. Varadhan and K. Mueller, "Volume ablation rendering," *Volume Graphics Workshop 2003*, pp. 53–60, July 2003.

[223] G. Wallace, "The JPEG still picture compression standard," *Communications of the ACM*, vol. 34, no. 4, pp. 30–44, 1991.

[224] M. Wan, Q. Tang, A. Kaufman, Z. Liang, and M. Wax, "Volume rendering based interactive navigation within the human colon," *Proc. of IEEE Visualization '99*, pp.397–400, 1999.

[225] M. Wan, H. Qu, and A. Kaufman, "Virtual flythrough over a voxel-based terrain," *ACM Symposium on Volume Visualization*, pp. 53–60, 1999.

[226] S. Wang and A. Kaufman, "Volume-sampled voxelization of geometric primitives," *Proc. IEEE Visualization '93*, pp. 78–84, October 1993.

[227] S. Wang and A. Kaufman, "Volume-sampled 3-D modeling," *IEEE Computer Graphics and Applications*, vol. 14, no. 5, pp. 26–32, 1994.

[228] S. Wang and A. Kaufman, "Volume sculpting," *ACM Symposium on Interactive 3 − D Graphics*, pp. 151–156, 1995.

[229] X. Wei, W. Li, K. Mueller, and A. Kaufman, "Simulating fire with texture splats," *IEEE Visualization '02*, pp. 227–234, 2002.

[230] X. Wei, W. Li, and A. Kaufman, "Melting and flowing of highly viscous volumes in virtual environments," (poster) *IEEE Virtual Reality 2003*.

[231] X. Wei, W. Li and A. Kaufman, "Melting and flowing of viscous volumes," *Conference on Computer Animation and Social Agents*, 2003.

[232] C. Weigle and D. Banks, "Extracting iso-valued features in 4-dimensional datasets," *Proc. of IEEE Visualization '98*, pp. 103–110, 1998.

[233] D. Weiskopf, K. Engel, and T. Ertl, "Volume clipping via per-fragment operations in texture-based volume visualization," *Proc. of IEEE Visualization '02*, pp. 93–100, 2002.

[234] T. Welsh and K. Mueller, "A frequency-sensitive point hierarchy for images and volumes," *Proc. IEEE Visualization '03*, October 2003.

[235] R. Westermann, "Compression domain rendering of time-resolved volume data," *Proc. of IEEE Visualization '95*, pp. 168–174, 1995.

[236] R. Westermann and T. Ertl, "Efficiently using graphics hardware in volume rendering applications," *Proc. of SIGGRAPH '99*, pp.169–177, 1999.

[237] L. Westover, "Footprint evaluation for volume rendering," *Proc. of SIGGRAPH '90*, pp. 367–376, 1990.

[238] L. Westover, "Interactive volume rendering," *Chapel Hill Volume Visualization Workshop*, pp. 9–16, 1989.

[239] L. Westover, "SPLATTING: a parallel, feed-forward volume rendering algorithm," *Ph.D. Dissert.* UNC–Chapel Hill, 1991.

[240] J. Wilhelms and A. Gelder, "A coherent projection approach for direct volume rendering," *Proc. of SIGGRAPH '91*, vol. 25, no. 4, pp. 275–284, 1991.

[241] J. Wilhelms and A. Gelder, "Octrees for faster isosurface generation," *ACM Transactions on Graphics*, vol. 11, no. 3, pp. 201–227, 1992.

[242] P. Williams, "Interactive splatting of nonrectilinear volumes," *Proc. of IEEE Visualization '92*, pp. 37–44, 1992.

[243] P. Williams, "Visibility ordering meshed polyhedra," *ACM Transactions on Graphics*, vol. 11, no. 2, pp. 103–125, 1992.

[244] A. Winter and M. Chen, "vlib: a volume graphics API," *Volume Graphics Workshop 2001*, pp. 133–147, June 2001.

[245] C. Wittenbrink, T. Malzbender, and M. Goss, "Opacity-weighted color interpolation for volume sampling," *Symposium on Volume Visualization '98*, pp. 135–142, 1998.

[246] G. Wolberg, *Digital Image Warping*, IEEE Computer Society Press, Los Alamitos, CA, 1990.

[247] S. Worley and J. Hart, "Hyper-rendering of hyper-textured surfaces," *Proc. of Implicit Surfaces '96*, pp. 99–104, October 1996.

[248] J. Wright and J. Hsieh, "A voxel-based forward projection algorithm for rendering surface and volumetric data," *Proc. IEEE Visualization '92*, pp. 340–348, October 1992.

[249] Y. Wu, V. Bhatia, H. Lauer, and L. Seiler, "Shear-image ray casting volume rendering," *ACM SIGGRAPH Symposium on Interactive 3-D Graphics '03*, 2003.

[250] R. Yagel and A. Kaufman, "Template-based volume viewing," *Computer Graphics Forum, Proc. of EUROGRAPHICS '92*, vol. 11, no. 3, pp. 153–167, 1992.

[251] R. Yagel, D. Reed, A. Law, P.-W. Shih, and N. Shareef, "Hardware assisted volume rendering of unstructured grids by incremental slicing," *Volume Visualization Symposium '96*, pp. 55–62. 1996.

[252] R. Yagel and Z. Shi, "Accelerating volume animation by space-leaping," *Proc. of IEEE Visualization '93*, pp. 62–69, 1993.

[253] B. Yeo and B. Liu, "Volume rendering of DCT-based compressed 3-D scalar data," *IEEE Trans. Visualization Comput. Graphics*, vol. 1, no. 1, pp. 29–43, 1995.

[254] K. Zuiderveld, A. Koning, and M. Viergever, "Acceleration of ray-casting using 3-D distance transforms," *Visualization in Biomedical Computing '92*, pp. 324–335, 1992.

[255] M. Zwicker, H. Pfister, J. Baar, and M. Gross, "EWA volume splatting," *Proc. of IEEE Visualization '01*, 2001.

42

Virtual Reality

42.1 Introduction **42**-1
42.2 Underlying Principles **42**-2
42.3 Best Practices **42**-5
    Display of the Virtual Environment • Position Tracking
42.4 Software Architectures **42**-11
    Polling vs. Events • Navigation • Virtual Objects
42.5 Environment Design Concepts **42**-15
42.6 Distributed Virtual Reality **42**-16
42.7 Application Evaluation and Design **42**-16
42.8 Case Studies **42**-18
    Architectural Walkthrough • The Virtual Wind Tunnel
42.9 Research Issues.................................. **42**-18
42.10 Summary **42**-19

Steve Bryson
NASA Ames Research Center

42.1 Introduction

Virtual reality, also known as virtual environments or virtual worlds, is a new paradigm in computer–human interaction, in which three-dimensional computer-generated worlds, called virtual environments, are created which have the effect of containing objects that have their own location in three-dimensional space. The user's perception of this computer-generated world is as similar to the perception of the real world as the technology will allow, providing appropriate depth and three-dimensional structure cues. User perception in virtual reality can be via a variety of senses, including sight, sound, touch, and force. Virtual environments are often, but not necessarily, immersive, providing the effect of surrounding the user with virtual objects. Objects in the virtual environment are often autonomous and/or interactive. The user interacts with the virtual environment using several interaction techniques, with a stress on direct manipulation in three-dimensional space via interface metaphors from the real world, such as grab and point, where appropriate. In order to create the effect of interactive three-dimensional objects, the virtual environment must be processed and presented at a near-real-time rate of 10 frames/s or greater. The three-dimensional perception and interaction in the virtual environment, its real-world-like interface, and its inherently near-real-time response property make virtual reality a natural interface for three-dimensional applications, including training for real-world tasks.

More precisely, we define **virtual reality** as the use of computer systems and interfaces to create the effect of an interactive three-dimensional environment, called the virtual environment, which contains objects which have spatial presence. By **spatial presence**, we mean that objects in the environment effectively have the property of spatial location relative to and independent of the user in three-dimensional space. We call the effect of creating a three-dimensional environment which contains objects with a sense of spatial presence the virtual reality effect. The essence of virtual reality can be summed up in the idea of three-dimensional "things" in the virtual environment rather than (possibly animated) "pictures of things."

We are defining virtual reality as an interface: there is no statement of content in this definition. In particular there is nothing in the definition of virtual reality which implies an attempt to mimic or otherwise create the illusion of the real world in the computer-generated environment. While some applications such as real-world task training tasks may require mimicking the real world, other applications such as entertainment or scientific visualization use environments which do not attempt to duplicate the real world.

There has been some confusion about the meaning of the phrase "virtual reality," which some people take to be an oxymoron. "Virtual" means "having the effect of being something without actually being that thing," while the definition of "reality" appropriate for our purposes is "having the property of concrete existence." Thus the phrase "virtual reality" translates as "having the effect of concrete existence without actually having concrete existence." This definition is, to some extent, actually achieved in virtual reality systems and distinguishes virtual reality from conventional computer graphics.

Virtual reality is a young, interdisciplinary, growing research field. It is not possible to survey all interesting activities in virtual reality in this short chapter, nor is it possible to detail particular technologies without this chapter rapidly going out of date. I will therefore only survey the issues that arise in the design of a virtual reality system, with an emphasis on application development. Many of the results and principles described in this article are the result of experience rather than careful study.

42.2 Underlying Principles

The virtual reality effect is attained through the use of a combination of computer and human–computer interface technologies. While the virtual reality effect can be created using any one or combination of sensory modalities including sight, sound, touch, and force, we shall illustrate the virtual reality effect using the visual modality as an example. Later we shall comment on how the same effect is created using other sensory modalities.

The effect of the spatial presence of an object is provided for the visual sense by rendering the object using conventional three-dimensional computer graphics from the current point of view of the user. **Head tracking** technology is used to measure the position and orientation of the user's head. The computer uses this head information to compute the position and orientation of the center of each of the user's eyes. These eye data are used to set the point of view and perspective projection for the rendering of the three-dimensional virtual scene, once for each eye to create a stereoscopic effect. The graphics is displayed to the user, optionally in stereo, in a way which allows the user to move around to get varying views of the virtual environment.

Consider the case of a single motionless virtual object in the virtual environment: while the image presented to the user changes as the user moves about, that change approximately matches the expectations of the user's cognitive system. The result is that the user's cognitive system interprets the changing images as a motionless object being viewed from a moving point of view. We shall call this effect **spatial constancy**. It is the spatial constancy of virtual objects which results in the object's sense of spatial presence. The virtual scene must be rendered from the user's current point of view with a rate sufficient to provide spatial information about objects in the environment. For purposes of spatial constancy, a rate of 10 frames/s is sufficient, though a higher frame rate is required for interaction with environments which contain rapidly moving objects. Note that at 10 frames/s the environment still appears jerky and the frames are clearly discontinuous. Experience has shown, however, that 10 frames/s is sufficient for the effect of spatial constancy. For lower frame rates the effect of spatial constancy fails and the user perceives the scene as a succession of disconnected images. In general, the human cognitive system is very forgiving about endowing objects with the property of spatial constancy: slight delays in rendering, slightly inaccurate tracking, noticeably jerky motion, and low-quality graphics do not interfere with the effect of spatial constancy.

Interaction in the virtual environment is performed through the measurement of the user's body motions and the interpretation of that motion as intentions by the computer system. Interaction in virtual environments is very different from the transactional, discrete interaction in conventional human–computer interfaces: interaction in virtual environments is inherently continuous, typically involving spatial manual tasks such as picking, placing, and tracking. This continuous interaction requires accurate tracking and very fast response from the computer system in order to provide the user with appropriate feedback as to the state of the interaction.

The virtual reality effect is critically dependent on the virtual reality system providing a view of the virtual environment which corresponds as closely as possible with the user's head position and orientation as the user moves about. This requirement implies that there must be a minimal graphics frame rate and that the image presented to the user must correspond as closely as possible to the user's current head position and orientation, which implies a short delay between when that position and orientation are sampled and when the resulting rendered scene appears to the user. Thus there are two performance issues critical to the success of a virtual reality system: frame rate (analogous to bandwidth or throughput) and delay (analogous to latency).

Two considerations determine the required frame rate:

- Experience has shown that for the effect of spatial constancy to operate, the virtual scene must be rendered from the user's point of view with a frame rate of at least 10 frames/s. Failure to meet the 10 frame/s requirement will result in the failure of the virtual reality effect.
- If the virtual environment contains moving objects, the Shannon–Nyquist limit requires that the user "sample," in other words see, that virtual object with a frequency at least twice that of the highest frequency of motion of the object. In actual practice the display rate should be at least four times that of the highest frequency of motion. This puts an application-dependent lower limit on the acceptable frame rate. Further, low frame rates have a noticeable impact on the ability to perform spatial manipulation tasks (Bryson 1993, Burdea and Coiffet 1994). Failure to meet this frame-rate requirement will result in an impaired ability the user to correctly perceive motions and interact with objects in the environment.

Two considerations determine the acceptable delay:

- Delays in head tracking can result in motion or simulator sickness, as the images seen by the user do not correspond with head motion as sensed by the user's vestibular system. How strongly a given delay induces motion sickness is determined by many factors, including head motion frequency and field of view of the virtual reality display. Larger fields of view and high-frequency head motions induce greater motion sickness for a given delay. Experience has indicated that delays of 0.1 s or less are acceptable for head motions limited by reasonable frequencies and wide fields of view.
- Delays in response to hand tracking impair the ability to perform manual tasks such as pick and place and tracking. The highest allowable delay is determined by the accuracy with which tasks must be performed and the frequency of motion of objects with which the user must interact. For example, the accuracy of tracking tasks, where the user's hand must track a nonperiodic target, has been shown to depend linearly on both the delay and the target object's frequency of motion.

These considerations are summarized in the *virtual reality performance requirements*:

- The graphical frame rate (animation rate) must be greater than two to three times the highest frequency of motion of objects in the environment and in all cases must be greater than 10 frames/s.
- The end-to-end delay in response to user input must be small for interaction with objects which have high-frequency motion and in all cases must be less than 0.1 s.

The displays used in virtual reality come in several forms. The most famous form is that of the **head-mounted display**, which places a display in front of the eyes, usually a separate display for each eye. Head-mounted displays are usually worn on the head and move with the user, so the user always sees the virtual environment (though such displays are sometimes externally supported), providing a strong sense of presence. Issues that arise with head-mounted displays include comfort, image quality, and field of view and will be discussed in more detail below. *Stationary displays* are fixed displays for the virtual environment. Stationary displays may be conventional workstation screens or large projection screens, and may be vertical or horizontal or surround the user for a sense of immersion. Stereo display is typically provided in stationary displays via a time-multiplexed stereo signal, where a polarized image for each eye is displayed alternately, with polarized glasses worn by the user determining which image gets seen by which eye.

The virtual reality effect is attained through the use of head tracking and displays to provide various cues about the user's position and orientation in a three-dimensional environment. The human factors of visual perception, particularly depth perception and personal motion cues, are critical in the successful design of the virtual environment. Human depth cues include the following:

- *Head-motion parallax:* The relative motion of objects at various depths as the user's head moves about in the environment. Head-motion parallax is performed via head tracking.

- *Plane of focus (accommodation):* The location of the focus plane in the environment. As this chapter is written, plane of focus is not supported as a depth cue by any available virtual reality system.

- *Stereopsis:* The differences in relative positions of objects in the virtual environment as seen by the user's two eyes. Stereopsis is supported by providing a different image to each eye, either using a separate display for each eye or single display with images for each typically eye appearing sequentially in time.

- *Occlusion:* If one object blocks the view of another object, the first object is closer. Occlusion is supported by conventional three-dimensional computer graphics systems via hidden surface rendering algorithms.

- *Perspective:* The relative location of an object in the user's field of view as determined by classical perspective transformations. Perspective includes such cues as apparent size and the fact that objects that appear lower in the field of view are perceived to be closer. Perspective is supported by conventional three-dimensional computer graphics systems. Wide-angle displays significantly enhance this cue.

- *Textures:* The appearance of a known texture at different depths gives strong depth cues. Textures are supported by higher-end conventional three-dimensional computer graphics systems.

- *Atmospheric effects:* Blurring and fog effects due to distance. Atmospheric effects are supported by higher-end conventional three-dimensional computer graphics systems.

Self-location and *self-motion* cues are dominated by perception of the object's motion in the user's peripheral field. Thus wide-angle displays significantly enhance the sense of location and the accurate detection of self-motion in the virtual environment.

Hand tracking supports interaction in the virtual environment. Hand tracking takes two forms: position and orientation tracking and gesture (command) tracking. Position and orientation tracking is performed with much the same technology as that used in head tracking. Gesture tracking is performed via a variety of technologies: buttons are often used for a small number of gestures, while measurement of the user's finger joint angles can be used to infer the hand gesture that the user is performing. This hand gesture can then be interpreted as a command to the system. An example is the interpretation of a closed fist as a "grab and move" command. The user's finger joint angles are measured via an instrumented glove-type device.

An understanding of the human factors of manual interaction is critical in the successful design of a virtual environment. One basic result is *Fitts' law*, which states that the shortest time to reach an object is proportional to the log of the ratio of the object's distance to its size. Another result is in the study of manual tracking of a randomly moving target. If error is measured as the mean square distance from the target to the location of the user's cursor, that tracking error is:

- Linearly dependent on the frequency of motion of the target, with target motion frequencies greater that 5 Hz being essentially untrackable

- Linearly dependent on end-to-end delays between the user's motions and the resulting motions of the user's cursor

- Linearly dependent (according to preliminary results) on the inverse of the frame rate of the display, even when the display is completely up to date when first shown

Thus for manual tracking errors depend on both delay and frame rate. As any realistic virtual reality system will have both delays and a finite frame rate, applications which require manual tracking should have frame rates which are as high as possible and delays which are as small as possible.

The virtual environment may appear at many scales: Application requirements entirely determine the scale at which objects in the environment should appear. Some applications, such as real-world simulations or training applications, will have a naturally fixed scale. Other applications, such as a molecular structure application, will naturally have the scale set so that very small objects will appear very large. Other applications will have no natural environment scale at all, leaving the scale setting up to the user.

The above observation about scale generalizes to all aspects of the virtual environment: while virtual reality uses metaphors from the real world in the design of environments, there is no need beyond application requirements for behavior in the virtual environment to match behavior in the real world. Virtual reality applications can be tailored to perform tasks which would be either more difficult or impossible in the real world. Thus effective performance of application tasks should be the guiding principle in virtual environment application design. This focus on the application task as the guiding design principle has led to the abandonment of a conventional interface layer such as the menus and sliders in conventional graphical user interfaces. As this chapter is written it remains to be seen if another layer of conventionality will appear in virtual reality beyond the basic "pick up and move" metaphor of the direct manipulation interface. The desire for a conventional interface is at odds with the opportunity for the creation of application-specific objects which also act as interface objects.

A significant variation on virtual reality is *augmented reality*. In augmented reality the virtual environment is superimposed on the real world. Augmented reality uses either see-through displays, which place the virtual environment in a semitransparent window on the real world, or mixing of video images of the real world with computer-generated images of the virtual world. The dominant application of augmented reality is *information overlays*, which display information about the real world in the virtual environment. These information overlays may match the three-dimensional position and orientation of real-world objects. This matching of virtual and real objects requires very high accuracy in the tracking of the user's position and orientation as well as models of real-world behaviors. In this chapter we shall treat augmented reality as a subset of virtual reality.

42.3 Best Practices

42.3.1 Display of the Virtual Environment

One of the guiding principles of virtual reality is to provide the same types of information about the virtual environment as are available for the real world. Thus there are several sensual modalities used to present the virtual environment, including visual, auditory, and haptic (touch and force) displays.

42.3.1.1 Visual Display

Visual display is one of the most important components of a virtual reality system. There are a number of quality considerations which arise in the selection of a visual display for virtual reality. In addition to the usual display quality considerations from conventional graphics such as color, contrast, brightness, and refresh rate, the following issues arise in virtual reality displays:

Resolution: Defined as the angle subtended by a pixel as viewed by the user. Note that this sense of resolution is very different from the conventional sense used in computer graphics. Resolution will be determined by the pixel spacing on the screen and field of view as determined by the optics. A rough estimate of the angular resolution of a pixel can be obtained by dividing the horizontal field of view by the number of pixels in the horizontal direction.

Pixel spacing: Many screen technologies, such as shadow-masked color cathode-ray tubes and liquid-crystal displays, have significant space between the pixels. This space is magnified by wide-field optics and can significantly deteriorate the image.

Field of view: A strong determinant of the immersiveness of a display. The field of view is determined by the physical screen size and the optics.

Optical quality: Because head-coupled displays are very close to the eyes, they require optics to provide a focused image. These optics also determine the field of view and may induce strong optical distortion.

There are a variety of display technologies available which are appropriate to virtual reality. They fall roughly into two classes: **head-coupled**, which move with the user's head as the user moves about, and stationary, which do not. Head-coupled displays place a screen in front of each of the user's eyes, and are often head-mounted, rigidly attached to the user's head. Head-coupled displays provide a strong sense of immersion, because the user sees the virtual scene no matter which way the user's head is turned. Head-mounted displays require focusing optics, which can often provide a wide field of view. Some head-mounted display designs use folded optics to provide a more balanced design. Augmented reality displays often use a half-silvered mirror for display, allowing the virtual scene to be overlaid on the real world.

A significant issue for displays is ergonomics, particularly user comfort. Head-mounted displays can be uncomfortable, leading to rejection by the user community. Stationary displays provide both enhanced comfort and a sharable experience, but at the cost of immersiveness. Which display is chosen will generally depend on the application requirements.

42.3.1.2 Audio Displays

Sound output is an important modality of display in virtual reality. There are roughly two levels of sophistication of sound output: non-spatially-localized (possibly stereo) and three-dimensional spatially localized sound display.

Non-spatially-localized sound display involves using conventional sound rendering techniques, such as sampling and waveform synthesis, to provide nonspatial sound cues such as those found in conventional graphics applications. Nonspatial sounds are typically used to provide feedback as to the occurrence of events such as object collision in the virtual environment and (potentially) data display by varying the characteristics of a continuous sound.

Three-dimensional spatial sound uses various techniques to render sound whose source has a perceived location in three-dimensional space. This sound source location may or may not correspond to a visual or haptic object at the same location in three-dimensional space. Spatially localized sound provides the cues used by the human auditory perceptual system, which are interpreted as the sound source being located in three-dimensional space. These cues include variations in volume from ear to ear, phase differences due to the time difference between when the sound arrives at the two ears, and the distortion of the sound due to the structures on the exterior of the human ear (pinnae).

The simplest method of providing three-dimensional spatially located sound is to surround the user with an array of speakers. A particular sound source is given a perceived spatial location by appropriately balancing the volume of that sound from each speaker, taking into consideration the user's head position as measured by the head-tracking technology. The use of multiple external speakers provides appropriate volume to the user's ears, though the other cues are not well provided.

A more sophisticated method provides spatially localized sound via headphones and appropriate signal processing of the sound source (Begault 1993). Careful measurements are made of the differences in volume, phase, and sound distortion by placing a small microphone in a human ear while moving a sound source about in space. These measured parameters are stored, associated with a locaiton in space, and used to construct a convolution function as a function of three-dimensional position relative to the user. When a new sound is generated and associated with a three-dimensional location in the virtual environment, the convolution function modifies that sound for each ear in the same way that the sound would be modified as it reached the ear from a real-world sound source.

42.3.1.3 Haptic Displays

Haptics refers to the senses of force and touch. Haptic displays use various types of hardware to provide force or touch feedback when the user encounters an object in the virtual environment. At the time this chapter is written, haptic displays are highly experimental, with few commercial products available. Haptic displays are covered more thoroughly in Burdea and Coiffet (1994). The comments in this section are highly provisional and reflect an active research topic.

The primary purpose of haptic displays is to give the user the effect of "touching" objects in the virtual environment. The experience of touching an object in the real world is extemely complex, involving the texture of, temperature of, vibration (if any) of, and forces exerted by the object. Reflecting this complextiy, haptic displays fall into two classes:

- *Surface displays:* including texture, vibration, and temperature displays
- *Deep displays:* including force displays such as pneumatic robotics, compressed air actuators, and "memory metal" devices

Surface displays have been very difficult to build. The most common texture displays have typically involved small pin-type actuators, which are raised or lowered to give the effect of a smooth or rough surface, much like a graphical bitmap. Small vibrators and heat elements mounted in the fingertips of gloves have been used as a substitute for texture display with marginal success.

Deep displays involve exerting some kind of force on the user, the technology for which is well developed in the field of robotics. Thus deep displays are a good deal more mature than surface displays. As a general rule, the larger the force and the volume over which that force is to be exerted, the more difficult and cumbersome the haptic technology will be. As of this writing, a commercial product is available which delivers good force feedback to the fingertip over a volume about 0.5 m on a side.

There are two dominant technologies for force displays:

- Pneumatic actuators, used for whole-arm or whole-body forces. Pneumatic actuators tend to be very powerful but very cumbersome.
- Stepper-motor-based actuators, with the motors connected either directly to the actuator's lever arm or to a collection of strings or wires. Motor-based actuators tend to be smaller and more usable but have a smaller range of forces and a smaller working volume.

42.3.2 Position Tracking

There are two classes of tracking technologies: position and/or orientation trackers and angle trackers. Position and/or orientation trackers typically provide the position and/or orientation of an object at a single point in three-dimensional space. Angle trackers provide the angle between two different objects. Position and orientation sensors are typically used to track the user's head and hand, while angle sensors are usually used to track the angle of bend of a body joint such as the user's fingers.

42.3.2.1 Position Tracker Data

The data returned by position and orientation trackers fall into two classes: position and orientation. Position data are usually a three-dimensional vector relative to some known reference location. Orientation data are somewhat more complex, owing to the fact that three numbers are not sufficient to describe all orientations in three-dimensional space. Three numbers are sufficient to describe all but two orientatiNs (which two depend on the type and coordinate system of the orientation data), while four numbers are required to describe all orientations. Another option is to use the matrix description of an orientation, which requires nine numbers. There are three common methods of describing orientations:

Euler angles use an ordered triple (roll, pitch, yaw) of numbers: *roll* = rotation around the front-facing axis; *pitch* = rotation around the (new) horizontal axis; and *yaw* = rotation around the (new) vertical axis. Note that these rotations are concatenated. Because they describe rotations using three numbers, Euler angles do not describe all orientations: there are two orientations, characterized by pitch $= \pm 90°$, which are not correctly described by Euler angles. When the tracker is placed in one of these orientations, the tracker is said to be in *gimbal lock*.

Rotation matrices are 3×3 matrices, which describe the rotation of an object. Rotation matrices are standard in conventional computer graphics and are described in (Foley et al. 1990).

Quaternions are an ordered quadruple (w, x, y, z) of numbers which describe all orientations. Physically, quaternions represent a rotation of an angle given by arcsin $(w/2)$ around the axis specified by (x, y, z). In actual use, quaternions are usually translated to rotation matrices using the following formulas:

$$\begin{pmatrix} 1 - 2y^2 - 2z^2 & 2xy + 2wz & 2xz - 2wy \\ 2xy - 2wz & 1 - 2x^2 - 2z^2 & 2yz + 2wx \\ 2xz + 2wy & 2yz - 2wx & 1 - 2x^2 - 2y^2 \end{pmatrix}$$

Quaternions avoid the gimbal-lock problem and require smaller amounts of data storage than rotation matrices.

42.3.2.2 Position Tracker Technologies

There are a variety of tracker technologies which return position and/or orientation in three-dimensional space. Each of these technologies has its strengths and weaknesses:

- *Electromagnetic trackers:* Electromagnetic trackers use a source containing three orthogonal coils to sequentially produce three oriented radio-frequency electromagnetic fields, each component of which is read by three orthogonal coils in a sensor. These measurements result in nine numbers providing the strength of each of the three fields in three directions, which are used to reconstruct the position and orientation of the sensor relative to the source. Electromagnetic trackers do not require a clear line of sight from the source to the sensor, making them useful for both head and hand tracking, and are readily available commercially. Electromagnetic trackers have limited range (typically 1 to 3 m as of 1996) and are susceptible to electromagnetic distortion and noise in the physical environment, particularly from display monitors.

- *Acoustic trackers:* Acoustic trackers use an ultrasonic sound pulse, which is picked up by an array of receivers. The time of flight of the sound pulse from the source to each receiver is used to reconstruct the position and orientation of the receiver array relative to the source. Acoustic trackers are inexpensive and commercially available. They require a clear line of sight from the source to each receiver, limiting their operational envelope and making them potentially inappropriate for hand tracking. They are, however, appropriate for head tracking when using a stationary desktop display, where the user is physically always looking at a desktop display monitor so that line of sight is assured. Acoustic trackers have a limited range (1 to 2 m) and are very susceptible to acoustic noise and echoes in the physical environment.

- *Mechanical linkage trackers:* Mechanical linkage trackers use a jointed physical structure for position and orientation tracking. These structures have appropriately situated joints, each of which has an angle sensor. By measuring the angle of each joint and knowing the length of each segment, the position and orientation of the end relative to the base of the jointed structure can be determined. Mechanical linkage trackers have the advantage of very high accuracy, allowing a minimum of filtering, resulting in low delays. They have the disadvantage of a usually cumbersome physical structure which can get in the way of task performance.

- *Video tracking:* Video tracking uses multiple video cameras to track objects in the physical space, usually targets placed on the user's head and hands. The location of these targets is identified on the video images, and these image locations are used to reconstruct the three-dimensional position of the targets relative to the cameras. Orientation can be inferred by using multiple targets. Video tracking has the advantage of providing a very fast, accurate signal. It has the disadvantage of requiring a clear line of sight from camera to target, and so is more suited for head tracking than hand tracking in general circumstances. Other disadvantages include the complexity of the camera and video processing setup.

- *Inertial tracking:* Inertial tracking uses gyroscopes and accelerometers to measure the accelerations of the tracker, which are integrated to provide the position, orientation, and velocity. The primary advantage of inertial tracking is that it does not rely on the proximity of source and sensor and so can in principle track over very large volumes. The disadvantage is that errors in the acceleration measurements accumulate over time. As of 1996 practical accelerometers result in errors which become unacceptable within 30 to 60 s, so inertial tracking will not be further discussed here.

- *The global positioning system:* The global positioning system (GPS) is a system of satellites used for navigation. While the simplest civilian use of GPS results in position inaccuracies of a few meters, the use of *differential GPS*, which uses a GPS receiver in a known location to correct for errors in other, nearby moving GPS receivers, can provide accuracies of a few millimeters. This technology is experimental as of 1996 but shows great potential for many virtual reality applications.

42.3.2.3 Position Tracker Errors

There are two types of errors associated with trackers:

Static error: The difference between the actual position of the tracker and the position returned by the tracking system. Static error occurs in both position and orientation and is typically a function of the position of the tracker. A position error datum is typically a three-dimensional vector. Limits on the magnitude of this vector are usually provided by the tracker manufacturer. Orientation error data are usually expressed in terms of Euler angles, with limits on the error components provided by the tracker manufacturer. The source of static error will depend on the tracker technology.

Dynamic error: The difference between the history of the actual tracker position over time and the position data returned by the tracker system. The dominant sources of dynamic error are delays due to the time required to process the hardware tracker data and due to filtering to eliminate noise in the tracker signal. A second source of dynamic error is the suppression of high frequencies of motion due to the filtering. The hand and head, however, rarely have significant frequencies of motion above 5 Hz, and most filters do not suppress frequencies in this range.

42.3.2.4 Error Correction

Tracker errors can be corrected when a model of what the signal should be exists. The specific methods depend on whether the correction is to static or dynamic error.

- Static-error correction can be performed when the error is approximately static over time by directly measuring the error by taking tracker data at a known location and building a lookup table of error values indexed by measured position. The error at a given measured position can then be interpolated from the error table and added to the measured position.

- Dynamic-error correction requires a model of the motion of the tracker sensor. Such models are available for head tracking and are based on models of head motion supplemented by a noise factor. Such models are usually implemented via Kalman filters.

42.3.2.5 Using Head Tracker Data

Head tracker data are usually converted into a 4×4 homogeneous matrix containing both position and orientation information, inverted and multiplied onto the transformation stack of the geometry engine rendering the virtual environment. If V is a vertex to be rendered, and M_{head} is the 4×4 position and orientation matrix describing the user's head, then the transform of V is given by $M_1 M_2 M_3 \cdots M_n M_{\text{head}}^{-1} V$, where $M_1, M_2, M_3, \ldots, M_n$ are the various local transformations of V.

42.3.2.6 Using Hand Tracker Data for Direct Manipulation

Hand tracker data are used for either selecting or picking up and manipulating an object. There are several methods of selecting an object using hand tracker data:

- *Pointwise collision:* The user's hand position is represented by a simple point, the location of the hand tracker. Alternatively, when finger joint angle information is available, the point may be the tip of one of the user's fingers. An object is selected when that point is within a specified distance from the object.

- *Geometrical collision:* A geometrical model of the user's hand is constructed and used to detect polygon-by-polygon collisions with objects in the virtual environment. This type of selection is more compute-intensive but can more accurately mimic real-world object grasping.

- *Raycasting:* A ray is drawn from the user's hand position in a direction determined by the orientation of the user's hand and finger joint angles (if they are measured). If this ray intersects an object in the environment, then that object is considered selected.

Once an object is selected, it may be "picked up" and moved about by the user's hand. Following Robinett and Holloway (1992), this picking-up operation is performed by constraining the object's local transformation to be held constant relative to the user's hand transformation as that hand transformation changes. Let the hand transformation relative to the world coordinate system be M_{hand}, and let M_{object} be the object transformation relative to the world coordinate system. Then the orientatin of the object relative to the hand is $M_{hand}^{-1} M_{object}$. If M^{old} is the matrix from the last measured transformation of the hand or object, then, given a new hand transformation, the new object transformation is given by

$$M_{object} = M_{hand} M_{hand}^{old}{}^{-1} M_{object}^{old}$$

42.3.2.7 Gesture Recognition

Many virtual reality systems measure the finger joint angles of the user's hand. These finger joint angle data can be used to infer user commands based on hand gestures. There are two types of hand gestures: static and dynamic. In both cases gesture recognition is a problem of pattern recognition: recent or current measured hand data are matched against values which define a gesture.

A static gesture is defined as a particular configuration of the finger joint angles. The configuration is identified by demanding that each finger joint angle fall within a specified range. For example, the "fist" gesture may be defined as all finger joint angles greater than, say, $60°$. When detected, the application may interpret the fist gesture as a "grab and move" command. The use of many static gestures implies that the matching ranges are small, which makes recognition more sensitive to errors in the joint angle data.

Dynamic gestures are defined by the recent history of the finger angles and the hand position and orientation. The pattern recognition problem for dynamic gestures is complex and subtle and is best addressed by neural network techniques.

The measurement of finger joint angles takes place via a variety of technologies, including optical fibers which have been degraded so that the transmitted light is attenuated when the fiber bends, strain sensors which provide the strain on the sensor as the sensor is bent, and direct angle measurement via optical encoders or variable resistors. All of these technologies require calibration, which will depend on the details of the technology and the way in which the sensors are mounted on the hand. The sensor mountings vary from individual to individual and vary over time as the user's hand moves, introducing errors of as much as $10°$ into the calibration.

Significant issues arise in the definition and interpretation of gestures. While gestures are common in the real world, they are mostly nonconventional with different individuals making the "same" gestures differently. One guiding design principle is to reserve hand gestures for manual tasks such as pick and place, implying two conventional gestures: "fist" for picking up and moving objects, and "point with index finger" for selecting objects. Such use of small numbers of gestures allows large pattern-matching ranges and relative insensitivity to joint angle measurement errors.

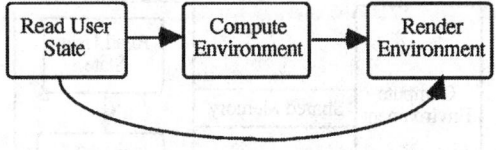

FIGURE 42.1 Classification of tasks.

42.4 Software Architectures

Because virtual environments are computer-generated, the software architecture used to maintain and operate the environment is of very high importance. The virtual reality performance requirements of frame rates greater than 10 frames/s and delays less that 0.1 s demand very high performance in terms of all tasks that must take place for the presentation of the virtual environment. We shall classify these tasks as shown in Figure 42.1.

First the user state is read, typically measuring head and hand position and orientation and command gesture. Then the environment state is computed. In some applications such as simple environment walk-throughs with static environment contents, very little computation may take place. Other applications may require a great deal of computation, which may include data access from mass storage or network communications. One example is a complex walkthrough with moving environment objects, which requires scene culling and other graphics optimization calculations. Another example requiring large amounts of computations is a visualization application, in which the large amounts of computation may result from the movement of an object in the environment, and may include access to large amounts of data from mass storage or a network.

After the environment is computed, it is rendered from the user's point of view as measured by the head tracking technology. In applications which contain a large amount of computation, the computation times may not satisfy the virtual reality frame-rate requirement of greater than 10 frames/s. In addition, the user's head may move, requiring a rerendering of the virtual environment from the new point of view even if the environment contents have not changed. User data obtained at the start of the computation phase will be out of date by the time the rendering phase starts if the rendering waits for the computation. This delay may violate the virtual reality delay requirement of less than 0.1 s. The use of out-of-date head tracking data for rendering will result in a swimming image, which typically induces motion sickness.

For these reasons, it is highly desirable to decouple the computation and rendering phases of the virtual environment, allowing the rendering process to run as fast as it can using the most recent head tracker data. This is accomplished by using one process for the rendering and another for the computation. The most efficient architecture is to have these processes on a single hardware platform implemented as lightweight processes communicating via shared memory, an architecture supported by many workstation vendors. Another option is to have the processes on separate platforms communicating over a network. The rendering and computation processes may themselves be multiple processes. The user data may be read in the rendering process, communicating the user data to the compute process, or the user data may be read in an additional process. An example using shared memory and reading the user state in one process is shown in Figure 42.2.

42.4.1 Polling vs. Events

There are two types of user input for the control of the virtual environment: discrete commands, and continuous motions such as those that arise when directly manipulating an object. This distinction is reflected in the choice of software architecture used to access the data from the interaction hardware. Discrete commands have the property that they must be executed in the order in which they were given by the user. Continuous commands for direct manipulation have the property that they should minimize delay, so that,

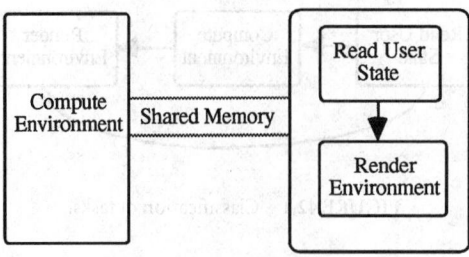

FIGURE 42.2   An example using shared memory and reading the user state in one process.

for example, the position of the user's hand is reflected as accurately as possible (up to specified coordinate transformations) by the position of the environment object showing the position of the user's hand.

We define the *event* interaction architecture as a queue of commands which is read by the application program. Commands may be put on this cue in response to hardware interrupts, polling by a deeper level of the software, or some other method. For our purposes the principal property of an event queue is that it preserves the history of the command sequence, ensuring that commands appear on the queue in the order in which they were given by the user. For a wide class of discrete commands, event queues are critical for the correct execution of user commands. Event queues, however, introduce delays and have no built-in sense of simultaneity.

Direct manipulation requires that the scene presented to the user reflect, as closely as possible, the state of the user at the present time, with no sense of history. Polling offers an effective method of taking a "snapshot" of the user's current state. Sampling the user data just before they are used minimizes delays. Simultaneity in the virtual environment can be defined as a single poll of all interaction devices.

42.4.1.1   Time-Critical Structures

Meeting the virtual reality frame-rate requirement can be a challenging aspect of virtual reality application design. This challenge is made more difficult by the often unpredictable nature of the virtual environment contents and activities, particularly in application environments which allow many user options. In many cases, applications may naturally take more computation or rendering time than the virtual reality frame-rate requirements would allow. In such cases algorithms must be built into the application which determines how much time is available and perform the application task as well as possible in that time. The result is typically a "graceful degradation" of the quality of the application for the purpose of maintaining virtual reality frame rates and responsiveness. While at first sight such degradation may seem to be highly undesirable, many users have indicated a clear preference for increased responsiveness at the cost of quality, so long as the loss of quality is appropriately understood. Designing algorithms which meet a specified time budget and a software architecture which supports these algorithms is known as the problem of *time-critical structures*. Time-critical structures falls naturally into two classes: *time-critical rendering* and *time-critical computing*.

42.4.1.2   Time-Critical Rendering

The time requirements of a rendering process depend on the method of rendering used, but for most graphics architectures this time has two components: the vertex projection rate, or the time required to compute the screen coordinates of a three-dimensional vertex, and the pixel fill rate, or the time required to fill in a solid object in screen space such as a polygon. Given a particular graphics architecture, the vertex projection time is determined primarily by the number of vertices that must be rendered, which is related to the scene complexity and the way in which the vertices are rendered (an object built out of disconnected triangles will have to render more vertices than the same object built out of triangle strips). Conventional methods of reducing the number of vertices to be projected include optimization techniques such as scene culling, so that only vertices that are potentially visible to the user are sent to the graphics system, and the use of triangle and rectangle strips. These techniques do not degrade the quality of the virtual environment and should be used regardless of the time budget, so they are not considered time-critical. The pixel fill rate is determined in part by the shading algorithm used, including the lighting conditions. Phong shading, for example, results in a slower rasterization than Gouraud shading, which is slower than flat shading.

Time-critical rendering techniques may use, for example, changes in scene complexity to limit the number of vertices and changes in lighting or in shading algorithm to affect the pixel fill rate. All of these choices can result in variations in scene quality, so they must be designed and implemented with great care. The appropriate choice of these parameters can be made through the use of cost and benefit functions for the rendering of an object (Funkhouser and Sequin 1993). The cost is roughly measured by the time required to render the object, while the benefit is related to such parameters as the object's size on the screen, whether the object appears in the center of the user's field of view is measured by head tracking, the state of motion of the object, and direct indications by the user. The benefit function measures the desirability of a certain object complexity, measured in terms of polygons, in a particular circumstance. The characterization of the benefit function is an area of ongoing research. Given cost and benefit functions, time-critical rendering becomes the problem of assigning a time budget to each object so that the ratio of total benefit to total cost for all time-critical objects in the environment is maximized. The most common way of varying the cost of an object is by reducing the number of vertices in the object, usually performed by selecting among several precomputed representations of the object at varying levels of detail.

42.4.1.3 Time-Critical Computing

Time-critical computing is the problem of designing computational algorithms which meet a specified time budget by degrading the quality of the output of the computation. As the computations that take place in an application are very application-dependent, time-critical algorithms will be very application-dependent. We will therefore confine ourselves to general comments, illustrated by a few examples.

When the results of the computation do not change very much from frame to frame, the results of a given frame's computation can be used to estimate the cost and benefit of an object in the next frame. Great care must be used, however, when determining what to do if the computation is too costly. For example, objects are of interest if they appear in the user's field of view and are not of interest if they are not. If the computation determines the motion of the object, that computation must be accurately performed in all frames to determine if the object appears in a later frame.

The cost of a computation can be controlled in a variety of ways depending on the computation algorithm: solutions of differential equations can be chosen for higher speed at the cost of accuracy; the computation of extended objects can be truncated after an allotted time; and linear approximations to more complex behavior can be implemented.

42.4.2 Navigation

At its most basic, navigation is the problem of controlling the user's point of view in the virtual environment. Head tracking is used for navigation over small distances and for controlling the orientation of the user's viewpoint. The effective envelope of head tracking is determined by the technology used to track the user's head. Many virtual environments are, however, considerably larger than the useful envelope of the head tracker. In these cases a method must be devised to allow the user to "move about" over large distances.

Movement over distances greater than those supported by head tracking is typically implemented by the addition of an additional transformation which determines the relationship of the head tracking coordinate system to the virtual environment's graphical world coordinate system. Changing this new transformation has the effect of moving the user's point of view as if the user were in a vehicle, so we call this new transformation a *vehicle transformation*. As navigation via head tracking is a well-understood problem, we define the problem of navigation as that of controlling the vehicle transformation. Several methods of controlling the vehicle transformation have appeared in various virtual reality systems:

- *Point and fly:* A three-dimensional orientation tracking device, usually a hand tracker but sometimes the head tracker, is used to indicate a direction in space. When the appropriate command is given, usually through a hand gesture or a button press, the vehicle transformation is translated in the indicated direction, resulting in the effect of flying through the scene. Variations on this method include using a continuous parameter such as the distance from the hand to the body or the angle of bend of a finger when using a dataglove-type device to control the speed of motion. Joystick-type devices can also be used to control both the speed and direction of motion. The primary advantage of the point-and-fly navigation method is its intuitive nature. Its primary disadvantages are poor control and the time required to travel large distances.

- *Teleportation:* When the desired location is known, the vehicle transformation can simply be set to that location. This location can be accessed from a list, indicated by a direct command, or indicated using a miniature representation of the virtual environment. This method has the advantage of speed and the disadvantage of a practical limit on the number of available target locations.

- *Direct manipulation of the environment space:* Rather than creating the effect of the user moving through the environment, limited navigation can be accomplished by providing the effect of the user moving the environment itself. One implementation is to use a hand tracking device to "grab open space," which results in the vehicle transformation being changed by the inverse of hand manipulations. This move may allow the user to control both translation and orientation of the environment. Repeated moves allow the user to manipulate the environment over moderate distances. The advantage of this method is accurate control over the destination of the navigation move and easy control over the orientation of the environment. The disadvantages include a limited operating range.

- *Variable scale:* This method solves the navigation problem by using a scale factor in the vehicle tranformation to shrink the virtual environment so that everything in the environment is within the effective envelope of the head and hand trackers. The hand tracker can then be used to indicate a new desired location, which determines the origin of subsequent scale operations. When the scale is increased again, it is around the new origin at the desired destination, so the user has the experience of ending up at the desired location. Variable scaling is best suited for applications which do not have a preferred scale. This method has the advantages of rapid operation over potentially very large distance scales, and fine control at a user-selectable scale. The disadvantages include the complexity of the navigation operation (scale, select new location, scale).

- *Manipulation of miniatures:* This method uses a miniature model of the environment which contains a representation of the user. Direct manipulation of the user's representation in the model controls the vehicle transformation, resulting in the effect of the user moving to the representation's location in the full-scale environment. It has been found (Pausch et al. 1995) that the most effective and least disorienting way of setting the vehicle transformation using the miniature is to interpolate the vehicle transformation between its original scale, orientation, and translation and that of the representation in the model. This results in the user's experience of the modeling expanding to become the full-scale environment seen from the desired new location and orientation (or equivalently the experience of zooming into the model). This method is appropriate for applications which have a preferred scale, such as an architectural walkthrough. Manipulation of miniatures has the advantage of an intuitive and simple control interface. The disadvantages include the requirement of the model of the environment and the limitations to effective navigation over varying scales using the limited scale of the model.

42.4.3 Virtual Objects

Careful design of the objects in the virtual environment is critical to implementation of a successful application. Such objects fall roughly into two classes:

- *Application objects:* Objects which are directly related to application tasks. These objects will have properties and behaviors directly relevant to the application task. Interaction with these objects will depend on the individual object and its task.

- *Interface objects (widgets):* Objects which exist solely for user control of the environment. Interface objects can provide a layer of a conventional interface between applications. Extensions of conventional graphical user interfaces such as menus and sliders inserted into the virtual environment are examples of interface objects. Using raycasting to select menu items (Jacoby and Ellis 1993) has been shown to be a fruitful implementation of menus in the virtual environment.

Some interface objects will be intimately connected to the application, so the distinction between application and interface objects can become very weak.

The extensive use of direct manipulation in the virtual environment raises the issue of arm fatigue: constantly holding the hand in front of the face when performing extended tasks can be tiring. For non-see-through displays, where the user's physical hand is not visible, one solution to the fatigue problem is to offset the hand cursor from the actual hand position so that when the hand is held low by the user's side the cursor appears in front of the user's face in the virtual environment. Experience has shown that people adapt very quickly to such offsets and these offsets do not impair task performance.

Many objects in the virtual environment will be autonomous, with behavior reflecting such things as simulated physics, information displays, or (in the case of simulated humans) complex volition. The maintenance of autonomous objects and their interactions with the user and other objects can place significant computational burdens on the virtual environment. Such autonomous behaviors will be highly application-dependent and will take advantage of the behavioral modeling literature from conventional computer graphics (Foley et al. 1990).

42.5 Environment Design Concepts

The ability to tailor a virtual environment for an application combined with the lack of a conventional interface layer makes the design of the virtual environment a challenging task. The use of metaphor in the design process has proven to be a useful guide. For our purposes a *metaphor* is a mapping from an application task to a task which is well understood by the user. The design of the virtual environment should be driven by the application, with metaphors from the application domain as the guiding design principle.

Metaphors in the virtual environment can appear at several levels:

- *Overall environment metaphor:* What is the driving metaphor of the application? How does that metaphor determine the overall appearance of the environment? The overall environment metaphor may include user navigation metaphors.
- *Object interaction metaphors:* What is the metaphor for interaction with objects in the virtual environment? There may be several classes of objects, with each class having its own metaphor. For example, an environment may have application objects which are picked up and moved and fall when they are released. The same environment may also have interaction objects which can also be picked up and moved but do not fall when released.
- *Individual object metaphors:* Individual objects in the environment may have their own metaphors which determine their appearance and behavior. A data display object may have a numerical appearance, while a training simulation object may faithfully mimic a real-world object.

When considering all of these levels of metaphor, the following questions should be asked:

- *Is there a metaphor intrinsic to the application?* An example of an intrinsic overall environment metaphor is "walking about the interior of a building" for an architectural walkthrough application. An example of an intrinsic object metaphor is a bat in a virtual baseball application.
- *Is there a metaphor from the language of the target user community?* The user community for the application may have conventions which can be the basis for design metaphors. These metaphors need not be understood by people outside the target user community. A common example of such conventions is the use of menus, sliders, and buttons for user interfaces.

42.6 Distributed Virtual Reality

The implementation of a virtual reality application distributed across several hardware platforms is an important capability, allowing physically separated users to share an environment in a collaborative setting and allowing remote access to large, high-capability systems. The virtual reality performance requirements of greater than 10 frames/s and delays of less than 0.1 s put extreme demands on the network systems involved in such distribution.

Several issues arise in the design of a distributed virtual environment:

- *Network capability:* Primarily the network's bandwidth and latency characteristics. Local area networks (LANs) have typically higher capability than wide area networks (WANs), as LANs usually have a higher bandwidth and lower latency. WANs will have latencies which will generally increase as more nodes are traversed.
- *Minimizing network traffic:* The amount of traffic involved in the operation of the virtual environment should be minimized. This is usually accomplished by transmitting only changes in the environment. Further minimization can be attained when environment changes can be modeled by each system. For example, when the position and velocity of, and forces acting on, an object are known, that object's motion can be predicted so long as no other force acts on that object. In this case only changes in the forces acting on the object need to be transmitted, avoiding constant messages indicating changes in position of the object.
- *Scaling with number of participants:* Some applications are designed for only a small number of participants, while others are designed for an unlimited number of participants. Architectures such as client–server and peer-to-peer unicast that require sequential maintenance of each participant will not have good scaling behavior. Multicast network protocols alleviate this problem somewhat. The NPSNET system (Macedonia et al. 1995) takes advantage of locality information, building multicast groups out of participants which are near each other in the virtual environment. In this case participants who are not near each other are assumed to be far apart.

There are several models of distribution that can be used for virtual reality:

- *Client–server:* In a client–server architecture the environment is maintained on a single computer system (which may itself be distributed over several components), with the state of the environment sent to client workstations which display the environment to the users. As the environment is maintained by a single system, issues of consistency are easily dealt with. User interactions are transmitted to the server, where they are interpreted and change the state of the environment.
- *Peer-to-peer:* In the peer-to-peer architecture, all systems maintain a model of the environment and changes in the environment are sent to each of the other systems. The messages may be sent to the other systems individually by address or via a multicast network message.

A natural issue that arises in any distributed application is that of standards. The lack of standard user interfaces and behaviors in virtual environment design impedes the implementation of standards for distributed virtual environments. Standards have been adapted for particular applications, most notably the Distributed Interactive Simulation standard developed for military simulators. Standards for three-dimensional graphics have also been applied to virtual reality. The Virtual Reality Modeling Language (VRML)(Pesce 1995) is one notable example of a graphics standard, which at the time of this writing is being extended to include simple behaviors.

42.7 Application Evaluation and Design

Virtual reality is based on a paradigm of highly responsive, inherently three-dimensional interaction and display. While this virtual reality paradigm can be highly advantageous when appropriately used, meeting the virtual reality performance requirements can be very difficult to achieve in a real application setting. Further, the new interaction concepts of virtual reality combined with the unusual and often flawed hardware interfaces can impede the acceptance of a virtual reality application by the target user community. Thus careful consideration must be given to the following two questions: "Is virtual reality valuable to the application?" and "Is virtual reality viable for the application?" We shall briefly consider approaches to these questions. As with all application design processes, we strongly recommend that input from the target user community be used when addressing these issues.

Is virtual reality valuable for the application? Virtual reality interfaces use inherently three-dimensional interaction and display technologies and concepts. Therefore a virtual reality interface is appropriate to an application if that application has a three- (or more-) dimensional spatial aspect. Classes of examples include: simulations of real-world activities, objects, experiences, or phenomena; abstract simulations such as scientific visualization; abstract experiences for artistic or entertainment purposes; or information displays which can be usefully mapped into a three-dimensional environment such as networking visualization. Examples of classes of applications which would probably not directly benefit from a virtual reality interface include: text-based applications; inherently two-dimensional applications such as image processing (though some image-processing techniques benefit from a three-dimensional representation); and applications which have no inherent spatial display content.

The value of a virtual reality interface for a particular application can be estimated in more detail by considering display and interaction separately:

- *Display: How would the three-dimensional display be used? What would be the role of head tracking?* These questions can be addressed by sketching out the expected appearance of the application environment as experienced by the user, including considerations of how the user would move about in the environment.

- *Interaction: How would the three-dimensional interaction capabilities be used? What is the role of direct manipulation?* These questions can be addressed by identifying representative application tasks that require three-dimensional interaction and sketching out how these tasks are performed.

If the answers to these questions are positive, then a virtual reality interface is probably useful to the application.

Is virtual reality viable for the application? Once the value of a virtual reality interface for an application has been established, it must be established that it is possible to implement that application in a way consistent with both the virtual reality performance requirements and with the current state of the art in both the computational and graphics capabilities and the virtual reality interface hardware. These issues fall into several categories:

- *Can the graphics for the application environment be rendered with a frame rate of 10 frames/s or higher?* Are there more polygons in the scene than can be rendered with available hardware in 0.1 s? Does the rendering require computationally intensive techniques such as volume rendering, raycasting, or sophisticated lighting techniques such as radiosity? While some high-end hardware platforms can perform some of these tasks with the performance required, will such platforms be available to the target user community? Failure to meet the graphics frame-rate requirement may result in a system which does not support head tracking and so does not take advantage of the virtual reality interface.

- *Can direct-manipulation-based application tasks be performed in less than 0.1 s?* Direct-manipulation-based tasks are those that occur in response to a user's continuous motion. Examples include the movement of an object in the virtual environment, which may trigger complex collision detection computations, or a data probe in a scientific visualization environment, where motion triggers the computation of new data displays. These computations may require access to data stored on disk or information over a network. Simple estimates of the computation time required can be made by counting the number of objects required for a collision detection computation or counting floating-point operations in a data display application. If it is found that the application task probably cannot be performed in 0.1 s or less, possibilities of limiting the task, changing how data are represented, or time-critical techniques which quickly provide approximations can be examined.

- *Do the currently available virtual reality interface hardware devices provide the required accuracy or qualtiy?* At any given time virtual reality interface devices provide only a particular level of quality. Do the available/affordable display devices provide sufficient image quality, resolution, field of view, and/or level of comfort required by this application? Do the trackers appropriate for the application provide the required accuracy over an appropriate range in the environment in which they will actually be used? Is the interface hardware convenient to use? Does the interface provide sufficient benefit to justify any inconvenience? Accuracy issues are particularly critical for augmented reality applications.

42.8   Case Studies

42.8.1   Architectural Walkthrough

An architectural walkthrough is an application which simulates walking through a building. This application is usually used to evaluate the building before it is actually built. The overall environment metaphor is that of being in a building, and there may or may not be individual interactive objects in the building. The navigation metaphor is largely determined by the available interface hardware, and is typically the point-and-fly paradigm. Collision with walls is not typically implemented. Time-critical graphics has been used in these walkthroughs to maintain a constant frame rate.

42.8.2   The Virtual Wind Tunnel

The virtual wind tunnel (Bryson et al. 1995) is an application of virtual reality to the visualization of simulated airflow. The overall environment metaphor is an aircraft body with airflow around it. Various tools are available to visualize the airflow, including streamlines, isosurfaces, and cutting planes. These tools are fully interactive, with movement by the user causing the recomputation of the visualization geometry. The recomputation and display of the visualization geometry allows real-time exploration of complex airflows, providing rapid understanding of the simulation. The virtual wind tunnel uses two asynchronous process groups, one for the computation of the visualization geometry and the other for rendering. Visualization geometry computation uses parallel processing when available. These two processes are typically in the same lightweight process group and communicate via shared memory, but they may be on separate systems communicating over a network in a client–server mode. Several users may be connected to the same server, providing the virtual wind tunnel with a shared-use mode.

   Time-critical computation in the virtual wind tunnel is implemented to maintain the responsiveness of the visualization tools when they are moved and to maintain animation rates for time-varying flows. The time-critical computation at a given frame is based on the assignment of a time budget to each visualization geometry computation. The ratio of the total computation time for all visualization geometry computations from the previous frame to the allotted time then multiplies these time budgets, providing a new time budget for each visualization geometry computation. It is then up to the individual computation to decide how to meet the new time budget, either by reducing the size of the visualization computed or by reducing the accuracy of the visualization by switching to a simpler, faster algorithm.

42.9   Research Issues

Essentially all aspects of virtual reality involve research issues. Critical research issues fall into roughly the following categorizations:

- *Human factors:* What is the impact on task perfomance of tracker inaccuracies, frame rates, end-to-end delays, the lack of full sensory feedback, and other technological distortions of experience in virtual reality? What is the appropriate approach to three-dimensional interfaces for task performance? What is the classification of tasks which are appropriate for virtual reality? What enhancements to the virtual environment can be inserted to aid task performance? How do we resolve the tension between application-specific and standard conventional interfaces?

- *Software:* Time-critical rendering and computation concepts, operating systems which support high-resolution scheduling and time-critical operations, and software structures for the rapid design of environments including complex behaviors need to be more fully developed.
- *Hardware:* Technologies need to be developed which deliver immersive visual displays with acceptable ergonomics, with the ideal being the form factor of sunglasses. While three-dimensional computer workstations, networks, and mass storage systems are a mature technology, they are usually developed to maximize throughput with little attention to latency. Developing hardware systems with both high throughput and low latency is a critical need for virtual reality.
- *Design:* Design methodologies for virtual environment applications are currently lacking. A classification of applications and useful approaches for virtual environments is required.

42.10 Summary

Virtual reality is an interface paradigm which relies on the effect of presenting the user with a computer-generated three-dimensional world which contains interactive objects that have three-dimensional locations independent of the user. Virtual reality is an inherently three-dimensional approach to human–computer interfaces which stresses responsive interaction with the virtual environment, which mimics real-world interaction. The three-dimensional aspects and high-performance requirements of virtual reality provide a platform for simulating real-world tasks, interactive entertainment, and exploration of complex data. These requirements also place significant stresses on the performance characteristics of the systems supporting the virtual environment.

Interface hardware for virtual reality involves several sensory modalities including visual, audio, touch, and force. The visual and audio display technologies are relatively mature, with the touch and force technologies requiring further development as of 1996. Virtual reality systems rely on tracking the user in three dimensions using technology which introduces errors and limitations. Working with these errors and limitations is one of the primary challenges of virtual reality application development.

Software systems for virtual reality are typically oriented toward specific applications or application domains. The high-performance requirements of virtual reality place unusual demands on the computation, rendering, and data management software that underlies a virtual reality application. Effective use of multiple processes and optimized, carefully written codes are critical for the success of a virtual reality application. Time-critical structures which gracefully degrade quality in order to maintain performance are important components in virtual reality software systems.

Virtual environment design raises further opportunities and challenges. The appropriate use of three-dimensional display and interface for a specific application must be approached on a task-by-task basis. The use of interface metaphors from the real world and the application domain naturally leads to the design of application-specific interfaces. The implementation and design of standard interfaces in this context is a challenging and open issue.

In spite of its difficulties, virutal reality allows the development of environments tailored to the best way for a user to perform an application task.

Defining Terms

Head-coupled display: A display which moves with the user's head, typically allowing the user to see the screen regardless of the position and orientation of the user's head.

Head-mounted display: A display which is rigidly mounted on the user's head. A subset of head-coupled displays.

Head tracking: The measurement of the position and orientation of the user's head, usually used for rendering the three-dimensional scene from the current point of view.

Spatial constancy: The property of having a spatial location when viewed from a moving point of view.

Spatial presence: The property of having a spatial location relative to and independent of a viewer in three-dimensional space.

Virtual reality: Use of computer systems and interfaces to create the effect of an interactive environment, called the virtual environment, which contains objects which have spatial presence. Also known as virtual environments or virtual worlds.

References

Begault, D. R. 1993. *3-D Sound for Virtual Reality and Multimedia*. Academic Press Professional, Cambridge, MA.

Bryson, S. 1993. Impact of lag and frame rate on various tracking tasks. In *Proc. SPIE Conf. on Stereoscopic Displays and Applications*, San Jose, CA.

Bryson, S., Johan, S., Globus, A., Meyer, T., and McEwen, C. 1995. Initial user reaction to the virtual windtunnel. AIAA 95-0114. In *33rd AIAA Aerospace Sciences Meeting and Exhibit*, Reno, NV.

Burdea, G. and Coiffet, P. 1994. *Virtual Reality Technology*. Wiley, New York.

Foley, J., van Dam, A., Feiner, S. K., and Hughes, J. 1990. *Computer Graphics: Principles and Practice*. Addison–Wesley, Reading, MA.

Funkhouser, T. A. and Sequin, C. H. 1993. Adaptive display algorithm for interactive frame rates during visualization of complex virtual environments. In *ACM SIGGRAPH '93 Conf. Proc.* Anaheim, CA, Aug.

Jacoby, R. H. and Ellis, S. R. 1993. Using virtual menus in a virtual environment. In *Proc. Symp. Electronic Imaging Science & Technology*, Vol. 1668. International Society for Optical Engineering/Society for Imaging Science & Technology.

Macedonia, M. R., Zyda, M. J., Pratt, D. R., and Barham, P. T. 1995. Exploiting reality with multicast groups: a network architecture for large scale virtual environments. In *Proc. 1995 IEEE Virtual Reality Annual Int. Symp.* IEEE Computer Society Press, Research Triangle Park, NC, Mar.

Pausch, R., Burnette, T., Broackway, D., and Weiblen, M. E. 1995. Navigation and locomotion in virtual worlds via flight into hand-held miniatures. In *ACM SIGGRAPH '95 Conf. Proc.* Los Angeles, CA, July.

Pesce, M. 1995. *VRML: Browsing and Building Cyberspace*. New Riders Publishing, Indianapolis. IN.

Robinett, W. and Holloway, R. 1992. Implementation of flying, scaling and grabbing in virtual worlds. In *Proc. 1992 Symp. on Interactive 3-D Graphics*, Boston, MA.

Further Information

The following books provide surveys of the virtual reality field:

Durlach, N. and Mavor, A. S., Eds. 1995. *Virtual Reality: Scientific and Technological Challenges*. National Academy Press, Washington, DC.

Burdea, G. and Coiffet, P. 1994. *Virtual Reality Technology*. Wiley, New York.

Badler, N. I., Phillips, C. B., and Webber, B. L. 1993. *Simulating Humans: Computer Graphics Animation and Control*. Oxford University Press, New York.

Kalawsky, R. S. 1993. *The Science of Virtual Reality and Virtual Environments*. Addison–Wesley, Wokingham, England.

The following books provide a summary of human factors issues:

Boff, K. R., Kaufman, L., and Thomas, J. P. 1986. *Handbook of Perception and Human Performance*, Vols. 1, 2. Wiley, New York.

Ellis, S. R, Kaiser, M, and Grunwald, A. J., Eds. 1993. *Pictorial Communications in Real and Virtual Environments*, 2nd ed. Taylor and Francis, Bristol, PA.

The following proceedings contain many important research papers of interest:

Proc. IEEE 1993 Symp. on Research Frontiers in Virtual Reality. IEEE Computer Society Press, San Jose CA, Oct. 1993.

Proc. 1993 IEEE Virtual Reality Annual Int. Symp. IEEE Press, Seattle, WA, Sept. 1993.

Proc. 1995 IEEE Virtual Reality Annual Int. Symp. IEEE Computer Society Press, Research Triangle Park, NC, Mar. 1995.

Proc. 1996 IEEE Virtual Reality Annual Int. Symp. IEEE Computer Society Press, Santa Clara, CA, Mar. 1996.

Singh, G., Feiner, S. K., and Thalmann, D., Eds. 1994. *Virtual Reality Software and Technology 1994 Proc.* World Scientific, Singapore.

43

Computer Vision

43.1 Introduction **43**-1
43.2 Low-Level Vision **43**-2
  Local Edge Detectors • Image Smoothing and Filtering
  • The Canny Edge Operator • Multiscale Processing
  • Visual Motion and Optical Flow
43.3 Middle-Level Vision............................... **43**-9
  Stereopsis • Structure from Motion • Snakes: Active
  Contour Models
43.4 High-Level Vision **43**-14
  Object Recognition • Correspondence Search: Interpretation
  Tree • Transformation Space Search • k-Tuple Search:
  Alignment and Linear Combinations • Invariants, Indexing,
  and Geometric Hashing • Dense Feature Matching: Hausdorff
  Distances • Appearance-Based Matching: Subspace Methods

Daniel Huttenlocher
Cornell University

43.1 Introduction

The goal of computer vision is to extract information from images. For example, **structure from motion** methods can recover a three-dimensional model of an object from a sequence of views, for use in robot grasping, medical imaging, and graphical modeling; **model-based recognition** methods can determine the best matches of stored models to image data, for use in visual inspection and image database searches; and *visual* **motion analysis** can recover image motion patterns for use in vehicle guidance and processing **digital video**. Computer vision is closely related to the field of image processing. In computer vision the focus is on extracting information from image data, whereas in image processing the focus is on transforming images. For instance, extracting a three-dimensional model from two-dimensional images is more of a computer vision problem than an image processing one, whereas image enhancement is more of an image processing problem than a computer vision one.

Computer vision is an interdisciplinary area, which falls primarily within the field of computer science, but also draws heavily on a number of other areas including image processing, differential and combinatorial geometry, numerical methods, and statistics. Some research in computer vision also has ties with biology and psychophysics; however, computer vision tends to be more concerned with building artificial vision systems than with accurately modeling human or animate systems. One of the main challenges for students of computer vision is developing adequate depth across such a wide range of areas. There are a number of books that provide relatively broad coverage of the field, with more advanced treatments in Faugeras [1993], Haralick and Shapiro [1992], and Horn [1986].

Human visual perception appears to be nearly effortless, in contrast with cognition, which can require substantial conscious effort. However, visual perception tasks are arguably at least as difficult as cognitive ones. For instance, computers can now beat all but the best human chess players and computational mathematics systems are routinely used to solve calculus problems that are too involved for people to do. Yet computational vision systems have only achieved human levels of performance in very restricted domains, such as automated parts inspection (under controlled lighting conditions). One particularly successful area of visual information processing is the development of systems for recognizing printed text. However, such optical character recognition (OCR) systems still make mistakes that a grade school student would not make, even if the child did not know the particular words being recognized. The main problem is that artificial vision systems are brittle, in the sense that small variations in the input may cause enormous changes in the output. Developing vision systems that degrade gracefully is a major challenge of computer vision.

Computer vision systems operate on **digital images**. A digital image is quantized into discrete values, both in space and in intensity. The discrete spatial locations are called **pixels**, and are generally arranged on a square grid, spaced equally apart (although the area covered by each pixel may not actually be square). Each pixel takes on a range of integer values. For a *gray-level* (or *intensity*) image these values are generally between 0 and 255 (8 b). For a *binary* image (or *bitmap*) the values are just 0 and 1. Color images can be represented in several ways; commonly three intensity images are used, one for each of three color channels (e.g., red, green, and blue). Digital images are large: a single gray-level frame from a video camera is about $\frac{1}{3}$ of a megabyte, a 24-b color image of a page scanned at 400 dots/in is about 44 megabytes, and an uncompressed 24-b color video stream is about 30 megabytes/s.

Computer vision methods are often classified into low, middle, and high levels. Although these classes are by no means universal, they still provide a useful way of categorizing computer vision problems. We will consider the following definitions:

- Low-level vision techniques are those that operate directly on images and produce outputs that are other images in the same coordinate system as the input. For example, an **edge detection** algorithm takes an intensity image as input and produces a binary image indicating where edges are present.
- Middle-level vision techniques are those that take images or the results of low-level vision algorithms as input and produce outputs that are something other than pixels in the image coordinate system. For example, a structure from motion algorithm takes as input sets of image features and produces as output the three-dimensional coordinates of those features.
- High-level vision techniques are those that take the results of low- or middle-level vision algorithms as input and produce outputs that are abstract data structures. For example, a model-based recognition system can take a set of image features as input and return the geometric transformations mapping models in its database to their locations in the image.

There are many applications of computer vision techniques. Traditionally, most computer vision systems have been designed for military and industrial applications. Common military applications include target recognition, visual guidance for autonomous vehicles, and interpretation of reconaissance imagery. Common industrial applications include parts inspection and visual control of automated systems. Over the past few years a number of new applications have emerged in medical imaging and multimedia systems. In medical applications computer vision methods are being used to register preoperative scans with a patient in the operating room. Computer vision techniques are also being used for realistic rendering and virtual reality applications, as well as image database retrieval.

In this chapter we will discuss a few computer vision problems in enough detail to give the reader an idea of some of the issues and to illustrate the kinds of techniques that are used to solve them. The presentation is divided according to low-, middle-, and high-level vision.

43.2 Low-Level Vision

Low-level vision computations operate directly on images and produce outputs that are pixel based and in the image coordinate system. Low-level vision computations include finding intensity edges in an image, representing images at multiple scales based on smoothing the image with different filters, computing visual motion fields, and analyzing the color information in images. In order to illustrate low-level vision methods, we will consider the problems of edge detection and image smoothing in more detail.

43.2.1 Local Edge Detectors

The primary goal of edge detection is to extract information about the geometry of an image for use in higher-level processing. There are many physical events in the world that cause intensity changes, or edges, in an image. Only some of these are geometric: object boundaries produce intensity changes due to a discontinuity in depth or difference in surface color and texture, surface boundaries produce intensity changes due to a difference in surface orientation. Other intensity changes do not directly reflect geometry (though it may be possible to derive some geometric information from them): specular reflections produce sharp intensity changes due to direct reflection of light; shadows and interreflections produce intensity changes due to other objects or parts of the same object.

We will refer to a gray-level image as $I(x, y)$, which denotes intensity as a function of the image coordinate system. Intensity edges correspond to rapid changes in the value of $I(x, y)$; thus it is common to use local differential properties such as the squared **gradient magnitude**,

$$\|\nabla I\|^2 = \left(\frac{\partial I}{\partial x}\right)^2 + \left(\frac{\partial I}{\partial y}\right)^2$$

Simplistically speaking, where the squared gradient magnitude is large, there is an edge. Another local differential operator that is used in edge detection is the Laplacian (see Horn [1986, Ch. 8]),

$$\nabla^2 I = \frac{\partial^2 I}{\partial x^2} + \frac{\partial^2 I}{\partial y^2}$$

This second derivative operator preserves information about which side of an edge is brighter. The zero crossings (sign changes) of $\nabla^2 I$ correspond to intensity edges in the image, and the sign on each side of a zero crossing indicates which side is brighter.

The images used in computer vision systems are digitized both in space and in intensity, producing an array $I[j, k]$ of discrete intensity values. Thus, in order to compute local differential operators, finite difference approximations are used to estimate the derivatives. For a discrete one-dimensional sampled function, represented as a vector of values $F[j]$, the derivative dF/dx can be approximated as $F[j + 1] - F[j]$, and the second derivative $d^2 F/dx^2$ can be approximated as $F[j - 1] - 2F[j] + F[j + 1]$. The squared gradient magnitude, $\|\nabla I\|^2 = (\partial I/\partial x)^2 + (\partial I/\partial y)^2$, can be approximated (at the center of a 2×2 grid of pixels) as

$$\left(\frac{\partial I}{\partial x}\right)^2 + \left(\frac{\partial I}{\partial y}\right)^2 \approx (I[j + 1, k + 1] - I[j, k])^2 + (I[j, k + 1] - I[j + 1, k])^2 \tag{43.1}$$

The Laplacian, $\nabla^2 I$, can be computed in a similar manner using the approximation to the second derivative (see Horn [1986, Ch. 8]).

In practice edges cannot be computed reliably using these kinds of local operators, which consider just a 2×2 or 3×3 window of pixel values. The high degree of variability in images causes such operators to both report edge points where there are none and to miss edge points. For example, Figure 43.1b shows the result of running a local gradient magnitude edge detector on the image shown in Figure 43.1a. This detector simply finds local maxima in the gradient magnitude which are larger than some threshold. Note the broken edges and large number of isolated edge points. In contrast, Figure 43.1c shows the edges detected using a nonlocal (or less local) gradient magnitude computation, which is described later in the section on the **Canny edge operator**.

A number of local edge operators have been developed, which can mainly be understood in terms of directional first and second derivatives. A more detailed discussion of some of these operators can be found

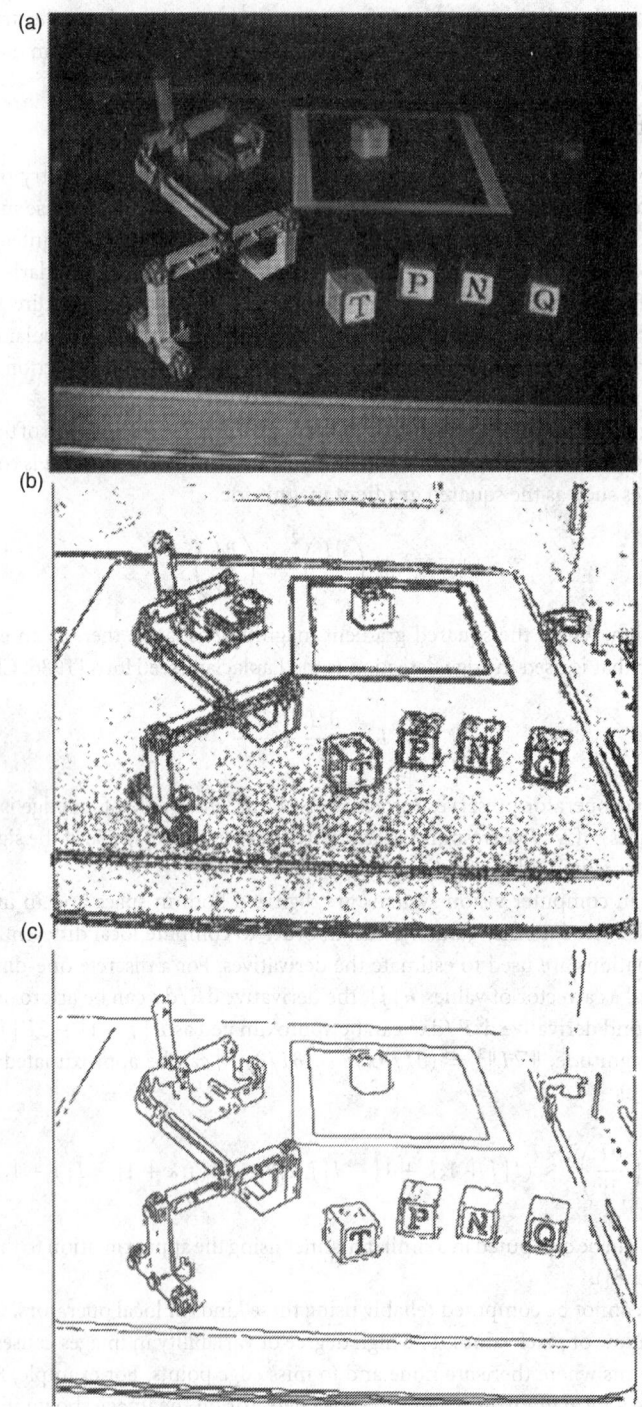

FIGURE 43.1 An example of edge detection: (a) an image, (b) a local gradient magnitude edge operator, (c) the Canny edge operator.

in Haralick and Shapiro [1992]. For example, the Sobel operator is a directional first derivative, based on the approximation $F[j+1] - F[j-1]$ (as opposed to $F[j+1] - F[j]$ as used earlier). The Sobel operator also uses a simple form of local smoothing. These local edge operators (which use 4×4 or 5×5 windows of pixel values) work slightly better in practice than the local gradient magnitude or Laplacian operators. The main reason is that they do local averaging (or weighted smoothing) of the image as part of the processing. We now turn to a discussion of local smoothing operations, and then put these operations together with the Laplacian and gradient magnitude to obtain edge detectors that work better than local methods such as Sobel.

43.2.2 Image Smoothing and Filtering

The basic operation used to smooth images in computer vision (and to filter images in general) is convolution. Consider the following function $g(x, y)$, defined in terms of $f(x, y)$ and $h(x, y)$,

$$g(x, y) = \int_{-\infty}^{\infty} \int_{-\infty}^{\infty} f(x - \xi, y - \eta) h(\xi, \eta) \, d\xi \, d\eta$$

We say that g is the convolution of f and h, which is written as $g = f \otimes h$. This function can be difficult to understand at first, because the value of g at a given point (x, y) depends on the values of f and h at all points. Convolution is commutative and associative, which allows computations to be rearranged in whatever fashion is most convenient (or efficient).

In the discrete approximation to the convolution, the sum of products can be expressed as four nested loops over two arrays that represent the (sampled) functions f and h. Let $h[i, j]$ be an $m \times m$ array and $f[x, y]$ be an $n \times n$ array, where $n > m$ and both arrays are indexed from 0. Then the code fragment in Table 43.1 computes the discrete convolution of the sampled functions f and h. The notation $\lfloor x \rfloor$ denotes the integer portion of x. Note that the iteration variables x and y cannot simply range between 0 and $n - 1$ as this would cause array references outside of f. These boundary cases can be handled in several ways and are important in any implementation.

Convolution can be used to smooth, or low-pass filter, an image in order to handle the problem of high-frequency variation (differences from one pixel to the next). In computer vision, the **Gaussian** is the most commonly used function for low-pass filtering. In one dimension, the Gaussian is given by

$$G_\sigma(x) = \frac{1}{\sqrt{2\pi}\sigma} e^{\frac{-x^2}{2\sigma^2}}$$

This is the canonical bell-shaped or normal distribution as used in statistics. The maximum value is attained at $G_\sigma(0)$, the function is symmetric about 0, and the area of the function is 1. The parameter σ controls the width of the curve: the larger the value of σ, the wider the bell. In two dimensions, the Gaussian can be defined as

$$G_\sigma(x, y) = \frac{1}{2\pi\sigma^2} e^{-\frac{(x^2+y^2)}{2\sigma^2}}$$

In the discrete case the values at integral steps over some range (generally $\pm 4\sigma$) are used as approximations. These values are normalized so that they sum to 1 (just as in the continuous case where the integral is 1).

TABLE 43.1 Four Nested Loops Which Compute the Discrete Convolution of f and h

```
for  x ← x min to x max
  do for  y ← y min to y max
    do sum = 0
      for  i ← 0 to m − 1
        do for  j ← 0 to m − 1
          do sum ← sum + h[i, j] f[x − ⌊m/2⌋ + i,  y − ⌊m/2⌋ + j]
      g[x, y] = sum
```

A direct implementation of discrete convolution as in Table 43.1 requires $O(m^2n^2)$ operations for an $m \times m$ mask representing the Gaussian and an $n \times n$ image. In the case of Gaussians, the operator is *separable* into the product of two one-dimensional Gaussians, and we can use this fact to speed up the convolution to $O(mn^2)$. The underlying idea is to do a one-dimensional convolution in the x-direction followed by a one-dimensional convolution in the y-dimension. The reason that this works is beyond the scope of this chapter. This is a significant savings, both theoretically and in practice, over the direct implementation. Any smoothing method (or edge detector) that uses separable filtering operators should be implemented in this manner. Gaussian smoothing can be made even faster using an approximation which is nearly independent of σ and thus the size of the mask, m. These methods are $O(n^2 + m)$, and use a form of the central limit theorem to approximate a Gaussian by several repeated convolutions with functions of constant height. Such convolutions can be performed in a constant number of operations for each image pixel (see Wells [1986]). In practice for Gaussian smoothing with σ of more than about 1 the repeated convolution approximation method is the fastest. For smaller values of σ the two one-dimensional convolutions are faster.

43.2.3 The Canny Edge Operator

The Canny [1986] edge detector is based on computing the squared gradient magnitude of the Gaussian smoothed image. Local maxima of the gradient are identified using a process known as *non-maximum suppression* (NMS). The NMS operation enables thin, connected chains of edge pixels to be identified. Conceptually, it is much like following the ridge lines in a mountain range, rather than just finding the (isolated) peaks. This is done by defining local maxima with respect to the gradient direction (the direction of steepest change) rather than in all directions. The NMS operation still leaves many local maxima that are not very large. These are then thresholded based on the gradient magnitude (or strength of the edge) to remove the small maxima. The maxima that pass this threshold are classified as edge pixels. Canny uses a thresholding operation with two thresholds, lo and hi. Any local maximum for which the gradient magnitude $m(x, y)$ is larger than hi is kept as an edge pixel. Moreover, any local maximum for which $m(x, y) >$ lo and some neighbor is an edge pixel is also kept as an edge pixel. Note that this is a recursive definition: any pixel that is above the low threshold and adjacent to an *edge* pixel is itself an *edge* pixel.

The steps of the Canny method are as follows:

1. Gaussian smooth the image, $I_s = G_\sigma \otimes I$.
2. Compute the gradient ∇I_s and the squared magnitude $\|\nabla I_s\|^2$ as in Equation 43.1.
3. Perform NMS. Let $(\delta\mathbf{x}, \delta\mathbf{y})$ be the unit vector in the gradient direction, ∇I_s. Compare $\|\nabla I_s(x, y)\|^2$ with $\|\nabla I_s(x + \delta\mathbf{x}, y + \delta\mathbf{y})\|^2$ and $\|\nabla I_s(x - \delta\mathbf{x}, y - \delta\mathbf{y})\|^2$ to see if it is a local maximum.
4. Threshold strong local maxima using $\|\nabla I_s\|^2$ as measure of edge strength, with two thresholds on edge strength, lo and hi, as previously described.

The edges in Figure 43.1(c) are from the Canny edge detector. In practice, this edge operator or variants of it (see Faugeras [1993, Ch. 4]) are the most useful and widely used.

43.2.4 Multiscale Processing

One problem with image filtering is choosing the *scale* of smoothing (e.g., the value of σ to use for the Gaussian). As the scale increases less of the detailed information in an image is preserved (and spatial localization gets worse). It is often desirable to be able to process an image at multiple scales; for instance, in order to determine the significance of edges by finding the range of scales over which they occur. Witkin [1983] developed the idea of multiscale signals resulting from smoothing with a Gaussian at different scales, which he called the **scale space** representation. For example, given an image $I(x, y)$, the corresponding scale-space function is

$$\mathcal{I}(\S, \dagger, \sigma) = \mathcal{I}(\S, \dagger) \otimes \mathcal{G}_\sigma(\S, \dagger)$$

where the scale parameter σ ranges from 0 to ∞ [at σ = 0 the scale-space function is the original function $I(x, y)$].

The use of multiscale representations and the characterization of signals from their edges at multiple scales can be viewed in terms of **wavelet** theory (cf. Mallat and Zhong [1992]). A wavelet function is a function whose integral is zero,

$$\int_{-\infty}^{\infty} h(x) = 0$$

and which has a scaling property

$$h_s(x) = \frac{1}{s} h\left(\frac{x}{s}\right)$$

The wavelet transform of a function f at scale s is then defined as $W_s^h f(x) = f(x) \otimes h_s(x)$, where h is a wavelet function (has integral zero and the scaling property). The derivative of a Gaussian has both the zero integral and scaling properties and thus is a wavelet function. Therefore, the multiscale edge representation of an image can be viewed as a wavelet transform. Wavelets have been used for a number of applications in image processing and analysis, such as image compression.

Multiscale representations using oriented filters are also common in computer vision [Perona 1992]. The gradient magnitude of the Gaussian smoothed image is not sensitive to orientation; it responds equally to edges in all directions. It is possible to design filters that are sensitive primarily to edges at a particular orientation, such as vertical or horizontal edges. The visual systems of many animals perform such orientation-sensitive filtering. For example, in many environments large horizontal edges are of considerable interest (especially moving ones, which could be predators).

43.2.5 Visual Motion and Optical Flow

Given a sequence of images taken over time (e.g., a digitized video stream at 30 frames/s), the goal of visual motion analysis is to recover information about the motion of objects in the image. The true motion of points in the world is not directly observable in an image sequence, only the changes in image intensity or the **optical flow**. For example, consider a rotating uniform sphere; in this case there is no optical flow (no change in brightness patterns in the image) but there is motion of the object. On the other hand, consider a shadow moving across a static scene. In this case there is optical flow, but there is no motion of the objects in the scene.

The optical flow is the vector field that tells us where each point in the image at time t "moved to" at time $t + \delta t$. Let $\mathbf{u}(x, y)$ and $\mathbf{v}(x, y)$ denote the components of this vector field; in other words the vector $(\mathbf{u}(x, y), \mathbf{v}(x, y))$ denotes the instantaneous motion of each point (x, y). One class of methods for computing the optical flow are based on measuring local intensity changes in the image with respect to time (see Horn [1986, Ch. 12]). The *optical flow constraint equation*

$$\frac{\partial I}{\partial x} \mathbf{u} + \frac{\partial I}{\partial y} \mathbf{v} + \frac{\partial I}{\partial t} = 0$$

shows how the components of the optical flow, $\mathbf{u}$ and $\mathbf{v}$, can be computed from local derivatives of the image with respect to space and time (x, y, and t). This equation also illustrates a fundamental problem with measuring motion from local intensity change. The equation can be rewritten in vector form as $(\partial I/\partial x, \partial I/\partial y) \cdot (\mathbf{u}, \mathbf{v}) = -\partial I/\partial t$. In other words, only the component of the optical flow in the direction of the brightness gradient $(\partial I/\partial x, \partial I/\partial y)$ is recoverable. This problem is known as the *aperture problem*; intuitively if a line segment is viewed through a small local window it is possible to tell what the motion of the segment is only in the direction normal to the segment.

There are a number of ways to provide additional constraints so that the flow problem can be solved. One approach is to assume that there is a single body moving rigidly [Horn 1986, Ch. 12]. This is powerful but generally overly restrictive. A second approach is to assume that the motion field varies smoothly in most parts of the image, which allows for nonrigid deformations and multiple objects, but has the drawback of performing poorly at motion boundaries because it blurs together multiple motions. A third approach is to use robust statistical techniques to combine local motion estimates [Black and Anandan 1993]. Another approach is to determine the motion of edge contours, rather than using local intensity differences with respect to time (see Horn [1986, Ch. 9]). As long as the contour is not a line segment, it is possible to extract the complete two-dimensional motion (because there are at least two distinct normals to the edge, which thus span the plane).

In practice, the best techniques for computing the optical flow are **area-based** methods, rather than computing the flow from discrete approximations to partial derivatives. Area-based methods operate by considering a small area, or _window_, around each pixel of an image. For each location (x, y) in the image at one time, I_t, a window of the image is matched against a set of windows in a local neighborhood of the next image I_{t+1}. The best matching window of I_{t+1}, using some match criterion, specifies the motion of the point (x, y). For example, if the best match for the window at (x, y) in I_t is the window at (w, z) in I_{t+1}, then the motion is $\mathbf{u}(x, y) = w - x$ and $\mathbf{v}(x, y) = z - y$. Area-based techniques generally use a simple matching measure for comparing windows, such as the **sum-squared difference** (SSD) of the corresponding pixels in the two windows. There is no ideal choice of window size for computing the optical flow. As the window gets smaller the flow estimate becomes noisy because the windows are not distinctive enough. As the window gets larger the motion boundaries become inaccurate because there are multiple motions in a window.

The pseudocode fragment in Table 43.2 illustrates the area-based computation of $\mathbf{u}$ and $\mathbf{v}$ for two images **img1** and **img2**, where the match window is of width m (size $(2m + 1) \times (2m + 1)$) and the search neighborhood is of width n (size $(2n + 1) \times (2n + 1)$). The window of **img1** centered at each (x, y) location is compared against the windows in the search neighborhood of **img2**, to find the best match for each window using the function **ssd**. The function **ssd** computes the sum-squared difference (or L_2 norm) of two images. The estimates of the optical flow resulting from a simple computation like this must generally be processed further in order to be useful.

As with local differential methods, the motions computed with area-based methods are generally processed by smoothing or aggregating the motion over local regions. Preprocessing of the images can also be used to improve the performance of area-based methods, particularly at motion boundaries where there are different motions in the same window. One such preprocessing method is based on transforming

TABLE 43.2   Area-Based Computation of the Optical Flow

$(\mathbf{u}(x, y), \mathbf{v}(x, y))$

```
for x ← x min to x max
  do for y ← y min to y max
    do min ← ∞
      for i ← −n to n
        do for j ← −n to n
          do diff ← SSD(img1, x, y, img2, x + i, y + j, m)
            if diff < min
              then min ← diff
                    u min ← i
                    v min ← j
      u(x, y) = u min
      v(x, y) = v min

SSD(img1, x1, y1, img2, x2, y2, w)
  sum ← 0
  for i ← −w to w
    do for j ← −w to w
      do sum ← sum + (img1(x1 + i, y1 + i) − img2(x2 + i, y2 + i))²
  return sum
```

the images using local nonparametric measures [Zabih and Woodfill 1994]. Two of these transforms are known as the *rank* and *census* transforms. The idea of both transforms is to replace the image intensities with measures based on order statistics in a local neighborhood. These measures are relatively insensitive to overall changes in intensity and thus are more reliable for use in area-based matching. In the rank transform, each pixel is replaced with the rank of its intensity over a local neighborhood. For example, when the neighborhood is size 15×15, if the point at the center of the neighborhood is brighter than any of the others its rank is 1, if it is darker than any of its neighbors its rank is 225, and if it is the median intensity its rank is 113. In the census transform a bit vector is used to encode information about which of the neighboring pixels are brighter or darker than the center pixel in the window.

43.3 Middle-Level Vision

Recall that we consider middle-level vision techniques to be those that take images or the results of low-level vision algorithms as inputs and produce some output other than pixels in the image coordinate system. One of the goals of middle-level vision is to extract three-dimensional geometric information from images. Extracting three-dimensional geometry from images is often referred to as *shape-from-x*, because there are a number of different sources of information that can be used to recover the three-dimensional structure, or shape, of a scene from two-dimensional images. Shading in an image reveals information about three-dimensional shapes (see Horn [1986, Ch. 11]). For instance, much of the way that the shape of a sphere in a photograph is perceived as being a solid rather than a disk is due to the uniform change in brightness away from the light source. Specular reflections can also provide information about the three-dimensional shapes of objects [Blake and Brelstaff 1988]. Another source of three-dimensional shape information is provided by the change in location of an object from one image to another in a set of two or more images. The main techniques for extracting image shape from multiple images are **stereopsis** and structure from motion.

Another goal of middle level vision is to extract structural descriptions of images. Active contour models or **snakes** are often used to fit models to data [Amini et al. 1990, Kass et al. 1988]. Relations between image structures can be identified using *perceptual grouping* methods. Grouping methods are generally concerned with recovering nonaccidental alignments of image primitives, such as colinear line segments or cocircular arcs [Lowe 1985]. We will not discuss grouping methods further in this chapter. First we will consider the shape-recovery methods of stereopsis and structure from motion, and then the fitting of contours using snakes.

43.3.1 Stereopsis

In the basic stereo vision paradigm, there are two cameras observing a scene. The central idea is that objects that are closer to the cameras will move more between the two views than will objects that are farther away. This phenomenon is readily observable by alternately blinking the two eyes. For a given point in the scene, the magnitude of its motion between the two views is referred to as the *disparity*. If the camera system is calibrated in world coordinates then the actual distance to a point can be determined from its disparity. The disparity between two images is often displayed as an intensity image with brighter points corresponding to larger intensities (things that are closer to the cameras). Figure 43.2 shows two intensity images and the resulting disparities from an area-based stereo matcher (discussed further subsequently).

A simple *pinhole* camera model consists of a focal point (or optical center), o, through which all rays of light pass, and an image plane I onto which these rays are projected. The *optical axis* of the camera is the line perpendicular to the image plane I and through the focal point o. The *focal length* f is the distance from the optical center o to the image plane I. For stereo vision there are two cameras at fixed relative positions. We will consider a simple stereo camera geometry in which the optical axes of the two cameras are parallel to one another and are perpendicular to the *baseline* that connects the two camera centers (which are denoted by o_l and o_r). We will also assume that the focal length f of the two cameras is the same. Denote the length of the baseline (distance between the camera centers) by b, and place the origin

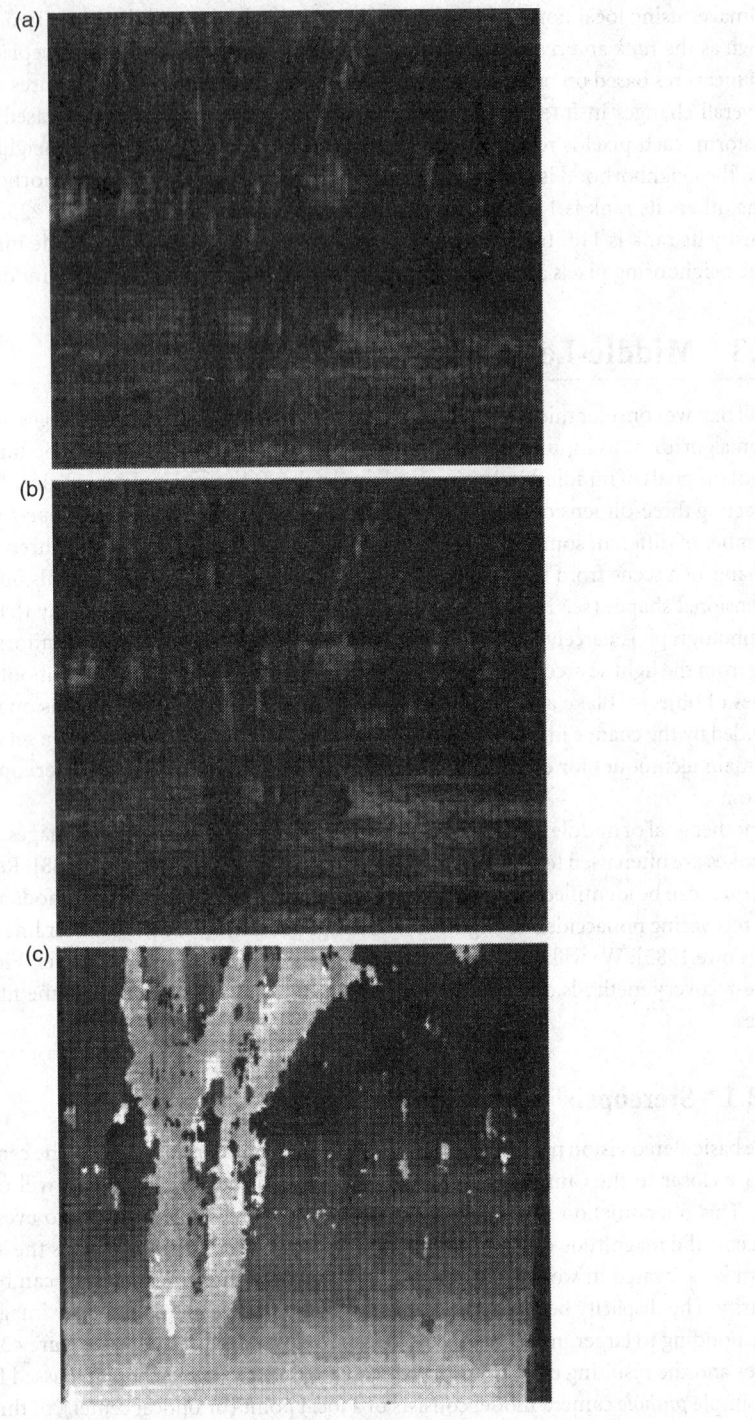

FIGURE 43.2 An example of stereo matching: (a) the left image, (b) the right image, (c) a disparity map with brighter points being larger disparity (closer to the cameras).

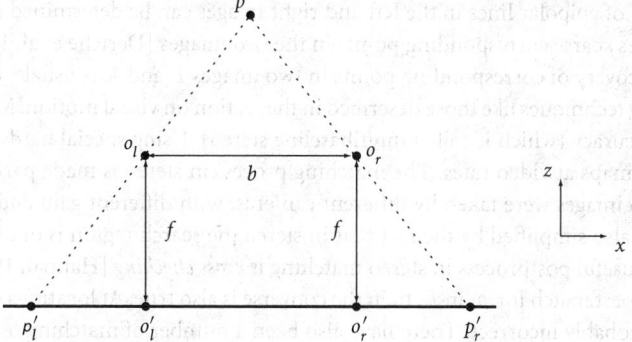

FIGURE 43.3 A simple stereo camera geometry.

of the world coordinate frame along the baseline, at the point equidistant between the two camera centers (at distance $b/2$ from each o_l and o_r). Let the origin of the coordinate system for the left image plane L be the projection of its optic axis o_l' (and similarly the origin of the right image plane R is at o_r'). This simple camera model is illustrated in Figure 43.3.

Consider a point $p = (x, y, z)$ in the world which is projected into L at location $p_l' = (x_l', y_l')$ and into R at location $p_r' = (x_r', y_r')$. From the geometry of the two cameras, it can be seen that

$$\frac{x_l'}{f} = \frac{x + b/2}{z}$$

$$\frac{x_r'}{f} = \frac{x - b/2}{z}$$

$$\frac{y_l'}{f} = \frac{y_r'}{f} = \frac{y}{z}$$

Note that in this simple camera geometry, only the x location of a projected point differs between the left and right images. The y location of a given point in space is the same for both images.

The disparity is defined as the distance between $p_l' = (x_l', y_l')$ and $p_r' = (x_r', y_r')$, which in this case is just $x_l' - x_r'$. From the preceding equations, we see that

$$\frac{x_l' - x_r'}{f} = \frac{b}{z}$$

Thus, if b and f are known, the depth z of the point p can be computed from the disparity. If b and f are unknown then the *relative* depths of points can be computed, but not their absolute distance from the camera.

The basic computation in stereopsis is that of finding corresponding pairs of points, p_l' in L and p_r' in R, which are images of the same point p in the world. Note that a given point p need not have an image in both L and R; there may be some other point in the scene that hides p from view in the left or right image, causing there to be no correspondence. This search for corresponding pairs is simplified by the fact that corresponding pairs of points must lie along certain lines in the two images, known as **epipolar lines**, rather than anywhere in the image. For the simple camera geometry illustrated in Figure 43.3, the corresponding epipolar lines are those with the same y-coordinates in the two images. Thus, given $p_l' = (x_l', y_l')$, it is necessary only to search for points p_r' in R that have $y_r' = y_l'$. In general, the corresponding epipolar lines are not parallel to one another but the search is still one dimensional. The epipolar lines in general form a pencil of lines in each image plane (a pencil is the set of lines through a given point). The point common to all of the epipolar lines in each image is the intersection of the image with the baseline connecting the two camera centers (for the simple camera geometry this point is at infinity so the epipolar lines are parallel).

The correspondence of epipolar lines in the left and right images can be determined through an iterative process that identifies sparse corresponding points in the two images [Deriche et al. 1994].

In practice the recovery of corresponding points in two images L and R is usually accomplished using area-based matching techniques like those described in the section on visual motion. Multiple cameras can be used for more accuracy (which is called multibaseline stereo). Using special hardware, these methods can compute depth maps at video rates. The matching process in stereo is made particularly difficult by the fact that the two images were taken by different cameras, with different gain and bias. However, the matching process is also simplified by the fact that in stereo the search region is one dimensional (along an epipolar line). A useful postprocess in stereo matching is *cross checking* [Hannah 1989], which ensures that if p_r in R is the best match for p_l in L, then the converse is also true. At locations where this is not the case, the match is probably incorrect. There have also been a number of matching techniques developed specifically for stereo, which directly implement constraints such as the fact that the disparity should in general not change quickly, except at object boundaries where there may be depth discontinuities (see Horn [1986, Ch. 6]).

In order to derive surface models from the depth (or disparity) information recovered from stereo, a surface interpolation process is often applied to the data (cf. Grimson [1983] and Terzopoulos [1988]). Such interpolation methods have broad applicability beyond computer vision.

43.3.2 Structure from Motion

Determining the three-dimensional structure, or shape, of an object in the world from a sequence of two-dimensional views is a problem that has been studied extensively (starting with Ullman [1979]). It is generally assumed that some feature points have been extracted and *tracked* through successive frames, so that the correspondence between points in each frame is known. The task is to recover the three-dimensional structure of the feature points in the world and the trajectory of the camera. Much of the early work on structure from motion was concerned with identifying the minimum number of points and frames that are needed to determine the three-dimensional structure of the points. We will not discuss these results here, because in practice it is necessary to have many more than the theoretical minimum number of observations in order to obtain a reliable solution.

To illustrate structure from motion techniques we will consider the case of orthographic projection (where all of the light rays are perpendicular to the image plane rather than going through a single focal point) and all of the points are visible in all of the frames. We will also limit the problem to a static scene and a moving camera, so that the goal is to recover the locations of P points in space and the positions of the camera at each of F time frames. The data provides $2PF$ observations (the x and y coordinates of the image points in each frame). The key issue is to use the redundancy of the observations to cancel out any (unbiased) noise in the locations of the image points and thereby recover the true structure and motion. There are a number of possible techniques for doing this; the approach of Tomasi and Kanade [1992] has been particularly effective.

Let (u_{fp}, v_{fp}) be the image coordinates of the pth point in the fth frame, $1 \leq p \leq P, 1 \leq f \leq F$, where the origin of each frame is the centroid of the P points in that frame. Given the assumption that all of the points are visible in each frame, using the centroid as the origin of each frame omits the need to consider translations of the points. Let $W = [UV]^T$ be a matrix containing the data points, where U and V are the $F \times P$ matrices of the u and v coordinates of each observed point. If there were no error in the observed image data, then the measurement matrix W could be expressed as the matrix product $W = MS$, where $M = [\mathbf{i}_1 \ldots \mathbf{i}_F \mathbf{j}_1 \ldots \mathbf{j}_F]^T$ represents the camera motion [each $(\mathbf{i}_f, \mathbf{j}_f)$ is a pair of orthogonal unit vectors specifying the orientation of the image plane at frame f] and $S = [s_1 \ldots s_P]$ is the shape matrix (the locations of the points in space, with their centroid as the origin). In other words the observed image data, as represented by W, is the product of two parts, the camera motion and the three-dimensional positions of the points.

As M is a $2F \times 3$ matrix and S is a $3 \times P$ matrix, if the measurements were exact $W = MS$ would be of rank at most 3. In reality, the problem is to recover M and S given W despite the fact that there is some error in the observed data. Of course, in the case of even small errors in the locations of the image points, the rank of W will not be 3. However, the singular value decomposition (SVD) can be used to determine the approximate rank of W, which we expect to be 3 (see a numerical methods book such as Press et al. [1988] for a description of the SVD).

The SVD of the measurement matrix is $W = L \Sigma R$ where L is a $2F \times P$ matrix, Σ is a $P \times P$ diagonal matrix of singular values and R is a $P \times P$ matrix. The information of interest corresponds to the three greatest singular values, so the best approximation is

$$\hat{W} = L' \Sigma' R' = \left(L' \Sigma'^{\frac{1}{2}} \right) \left(\Sigma'^{\frac{1}{2}} R' \right) = \hat{M} \hat{S}$$

where L' is the $2F \times 3$ matrix corresponding to the first three columns of L, Σ' is the 3×3 diagonal matrix corresponding to the upper left part of Σ, and R' is the $3 \times P$ matrix corresponding to the first three rows of R. One problem is that this is not a unique factorization as any linear transformation of $\hat{M} = L' \Sigma'^{\frac{1}{2}}$ and $\hat{S} = \Sigma'^{\frac{1}{2}} R'$ yields a valid result. It is possible to solve for the correct motion and shape transformations by noting that the true motion matrix M is composed of unit vectors, and the first F rows are orthogonal to the remaining rows (because each pair of rows $\mathbf{i}$ and $\mathbf{i} + F$ correspond to the two orthogonal unit vectors defining the image plane at frame $\mathbf{i}$). This specifies a unique solution up to a rotational ambiguity, which corresponds to the initial position of the camera with respect to the world. In other words, the overall orientation of the points in the world can be recovered only relative to the initial orientation of the camera.

A number of structures from motion methods have considered the problem of reconstructing the shape of objects up to certain transformations of space (such as affine or projective transformations). A nice book on the subject of solid shape is Koenderink [1990].

43.3.3 Snakes: Active Contour Models

Active contours, or *snakes*, are useful for applications that involve fitting models to image data. An active contour model is a curve that seeks to minimize both internal and external forces which control its shape. The internal forces are generally related to the smoothness of the curve, as reflected by measures such as first and second derivatives of the contour. The external forces are generally based on image measurements, but may also be due to user inputs in interactive curve-fitting applications. The image forces are often based on the gradient magnitude, so that the contour is attracted to edges. In most applications a contour is initially placed near some image structure, and then the constraint forces act on the snake to make it fit the image structure. An iterative update procedure is used to find the position of the snake which (locally) minimizes the forces.

The active contour model was defined in Kass et al. [1988] as a curve $\mathbf{v}(s) = (x(s), y(s))$ with an associated energy functional,

$$\int_0^1 E_{\text{int}}(s) + E_{\text{ext}}(s) \, ds \qquad (43.2)$$

The internal energy is composed of first- and second-order terms that measure the smoothness of the curve, and which make the contour act like a membrane or a thin plate,

$$E_{\text{int}}(s) = (\alpha(s)|\mathbf{v}_s(s)|^2 + \beta(s)|\mathbf{v}_{ss}(s)|^2)/2$$

where the subscripts denote the first and second derivatives of the curve with respect to arclength, and the functions α and β are weights (in practice these functions are often constant). The external energy is related to some image measure such as the gradient magnitude $E_{\text{ext}}(s) = -|\nabla I(x, y)|^2$. The curve that minimizes Equation 43.2 can be found using variational methods, by deriving a pair of Euler equations in x and y, discretizing them and then solving the discrete equations iteratively until they converge. This technique was developed in Kass et al. [1988]. Related methods have been developed for fitting three-dimensional energy-minimizing models to two- or three-dimensional data.

There are several difficulties with the variational approach to minimizing the energy of an active contour model. First, there is no constraint for the distance between points on the contour, so that many points of the contour can cluster near or on top of one another. More generally it is not possible to specify hard constraints, such as a minimum distance between points on the contour, because the energy functional must be differentiable. It is also difficult to choose the parameters of the minimization, and to determine to what degree the external energy term must be smoothed in order to produce a stable solution.

A different approach to the energy minimization is taken in Amini et al. [1990] where a discrete dynamic programming method is used. One of the main advantages of this method is that it allows the incorporation of hard constraints such as a minimum distance between points on the contour. One of the main disadvantages is that it is fairly computationally intensive. In this approach, the internal energy terms are discretized, $E_{\text{int}}(i) = (\alpha_i |\mathbf{v}_i - \mathbf{v}_{i-1}|^2 + \beta_i |\mathbf{v}_{i+1} - 2\mathbf{v}_i + \mathbf{v}_{i-1}|^2)/2$ and $E_{\text{ext}}(i) = -|\nabla I(x_i, y_i)|$. The energy over all of the contour points $(\mathbf{v}_0, \ldots, \mathbf{v}_{n-1})$, $\mathbf{v}_i = (x_i, y_i)$ is then

$$\sum_{i=0}^{n-1} E_{\text{int}}(i) + E_{\text{ext}}(i)$$

The minimization of this sum can be performed using $O(n)$ separate stages, where each stage considers only a local neighborhood around each of three successive contour points, because only the variables indexed by i, $i - 1$, and $i + 1$ must be considered simultaneously. This is used to develop a dynamic programming solution that runs in time $O(m^3 n)$ where m is the size of the local neighborhood around each point. In practice, the dynamic programming methods are easier to implement and more numerically stable than the variational ones.

43.4 High-Level Vision

High-level vision methods are those that make abstract decisions or categorizations based on visual data (generally using the outputs of low- or middle-level vision algorithms). For instance, *object recognition* systems can be used to determine whether or not particular objects are present in a scene, as well as to recover the locations of objects with respect to the camera. *Object tracking* systems can be used to follow a moving object in a video sequence and thus to guide a mobile robot or an autonomous vehicle (cf. Dickmanns and Graefe [1988] and Thorpe et al. [1988]). In this section we will discuss some approaches to object recognition.

43.4.1 Object Recognition

Object recognition systems generally operate by comparing an unknown image to stored object models in order to determine whether any of the models are present in the image. Many systems perform both recognition and localization, identifying an object and recovering its location in the image or in the world. The location of an object is often referred to as its **pose**, and it is generally specified by a transformation mapping the model coordinate system to the image or world coordinate system. One way of categorizing object recognition systems is in terms of the kind of problems they solve. The simplest recognition problems involve identifying two-dimensional objects which are completely unoccluded (i.e., none of the object is hidden from view), appear against a uniform background, and where the lighting conditions are controlled (e.g., there are no shadows or reflections). Many industrial inspection problems fall into this category and can be handled quite accurately with commercially available systems. Recognition problems become more difficult when there are many objects in a scene, when objects may be touching and occluding one another, when the background is highly textured, and when the lighting conditions are unknown. Recognition problems also become more difficult as the number of object models increases. Finally, recognizing three-dimensional objects in a two-dimensional image is more difficult than recognizing two-dimensional objects.

We will primarily consider approaches that can handle images with multiple objects, partly occluded objects, and some amount of background clutter. Most methods that address these kinds of recognition problems are based on extracting geometric information such as intensity edges from an image. This two-dimensional image geometry can then be compared with three-dimensional geometric models. For instance, Kriegman and Ponce [1990] use silhouettes for recognizing objects from intensity edges. A different approach is to compute invariant representations of the image geometry, which remain unchanged under changes in viewpoint [Mundy and Zisserman 1992]. These representations can be used to index into a library of object models. One of the central issues in geometric recognition systems is that of determining which portions of an image correspond to a given object. The recognition problem is often framed as that of recovering a correspondence between local features of an image and an object model. Three major classes of methods can be identified based on how they search for possible matches between model and image features: (1) *correspondence methods* consider the space of possible corresponding features, (2) *transformation space methods* consider the space of possible transformations mapping the model to the image, and (3) *hypothesize and test* methods consider k-tuples of model and data features. A more detailed treatment of geometric search methods can be found in Grimson [1990].

43.4.2 Correspondence Search: Interpretation Tree

Given a set of image features $S = \{s_1, \ldots, s_n\}$ and a set of model features $M = \{m_1, \ldots, m_r\}$, an *interpretation* of the image is a set of pairs $N = \{(m_i, s_j), \ldots\}$ specifying which model features correspond to which image features. If model features may be occluded and there may be extraneous image features, then in principle the set N can be any subset of the set of all pairs of model and image features. The **interpretation tree** approach (see Grimson [1990, Ch. 3]) is a pruned search of this exponential-sized space of possible interpretations (pairings of model and image features). The main idea is to use pairwise relations between features to prune a tree of possible model and image feature pairs, where paths in the tree correspond to interpretations. For concreteness, we consider the case where the features are points in the plane, and the transformation mapping a model to an image is a rigid motion (translation and rotation) plus some allowable error tolerance. As distances are preserved under rigid motion, the distance between pairs of points can be used as a constraint for pruning the search. In order for an interpretation to contain the pairs (m_i, s_j) and (m_p, s_q), the distances $\|m_p - m_i\|$ and $\|s_q - s_j\|$ must be equal (up to the allowable error tolerance).

The search for interpretations is structured as a pruned tree search, in the following manner. Each level of the tree (other than the root) corresponds to a given image feature. Each branch at a given node corresponds to a given model feature or to a special branch called the *null face*. Thus each node of the tree specifies a pair of an image feature (the one at that level of the tree) and a model feature or the null feature (the branch that was taken from the previous level). The search is depth first from the root, and at each node the null face branch is expanded last in the search. A given node is expanded only if it is pairwise consistent with all of the nodes along the path from the current node back to the root of the tree. That is, a given node is paired with each node along the path back to the root, and for each such pair the distance constraint is checked. Only when all of these pairs satisfy the constraint is the node expanded. (The null face branch is always consistent.)

A path from the root to a leaf node of the tree is an interpretation that accounts for zero or more model features (any of the branches may be null branches that do not account for a model feature). A path that accounts for k model features is called a k-*interpretation*. A threshold k_0 is used to filter out any hypotheses that do not account for enough model features (i.e., the matcher should report only those interpretations for which $k > k_0$). Note that a k-interpretation is guaranteed only to be *pairwise* consistent. Thus an additional step of model verification is performed, which estimates the best transformation for each k-interpretation and checks that this transformation brings each model feature within some error range ϵ of each corresponding image feature.

The HYPER and LFF systems (see Grimson [1990, Ch. 7]) similarly structure recognition as a search for consistent sets of model and image features, using pairwise constraints to prune the search. HYPER starts by matching a privileged model feature against compatible image features. A privileged feature is one that is believed a priori to be more reliable (e.g., for line segment features, longer segments might be considered more reliable). LFF searches for maximal cliques in a graph structure formed from pairwise consistent features.

43.4.3 Transformation Space Search

Transformation space (or pose space) methods are based on searching the space of possible transformations mapping a model to the image or world coordinate system. The idea underlying these methods is to accumulate independent pieces of evidence for a match. Pairs of model and image features, which are part of the same correct match, will specify approximately the same transformation, whereas random pairs of model and image features will tend to result in randomly distributed transformations. Therefore, pairs of features which result in a cluster of similar transformations are assumed to correspond to a match of the model to the data. The validity of this assumption, however, depends on there being a low likelihood that random clusters will be as large as those clusters resulting from correct matches (see Grimson [1990, Ch. 11]). Transformation space search methods compute the transformations that are consistent with each pair of model and image features (or in general each k-tuple of feature pairs). Then the space of possible transformations is searched to find clusters of similar transformations. The exact means of searching the space and identifying clusters depends on the particular transformation space search method.

The generalized **Hough transform** (see Grimson [1990, Ch. 11]) is a transformation search method that operates by voting for buckets in a discrete transformation space. For example, a rigid motion of the plane can be represented using three parameters x, y, and θ corresponding to two translations and a rotation. Each of these is broken into discrete ranges, forming a three-dimensional array of buckets, which tile the space of transformations. Every pair of model and image features then votes for those buckets containing transformations that map the given model feature to the given image feature, up to the allowable sensing error. A bucket that gets many votes corresponds to a possible transformation of the model to the image. There are a number of practical issues with the generalized Hough transform, such as what parameterization of the transformations to use, how to break the space up into a reasonable number of buckets (hierarchical schemes are often used rather than forming an explicit array of buckets), what kind of weighting scheme to use when a given feature pair votes for multiple buckets, and what to do about clusters that occur near bucket boundaries (because then votes may be spread over neighboring buckets instead of all occurring in the same bucket, possibly causing matches to be missed).

Another class of transformation search methods is based on precisely characterizing the regions of transformation space that are specified by each k-tuple of model and image feature pairs. The arrangement of these regions is then searched to find those cells where a large number of regions overlap. For example, consider the case of matching two point sets under translation, where a positional uncertainty of ϵ is allowed for each point. For each pair of model and image points there is a circle of translations of radius ϵ, which places the model point within ϵ of the image point. Translations at which many of these circles overlap are good potential matches of the model and image. Recognition methods that search the arrangement of these transformation space regions (e.g., Cass [1992]) make heavy use of techniques from computational geometry.

43.4.4 *k*-Tuple Search: Alignment and Linear Combinations

In the absence of sensor uncertainty, k pairs of model and image features exactly determine the transformation mapping a model to an image (where k depends on the kind of feature and the kind of transformation). For example, under translation a single pair of model and image points specifies the transformation. The idea underlying k-tuple search methods is to use the transformations specified by each k-tuple as an hypothesis about the pose of a model in an image (see Grimson [1990, Ch. 7]). If the transformation specified by k pairs of model and image features corresponds to a correct hypothesis, then it will map the other model features onto image features. In this case we say that the transformation *aligns* the model with the image. If the transformation is incorrect then other model features will in general not be mapped onto image features. When there is sensor uncertainty, it is necessary to account for the fact that the transformations computed from k-tuples will not in general bring other model features precisely into correspondence with image features.

Many applications of k-tuple search have considered the case of affine transformations of the plane. An affine transformation of the plane can be represented as $A(x) = Lx + b$, where L is a nonsingular 2×2 matrix, and b is a two-dimensional translation. Three pairs of corresponding points uniquely define such a transformation (which maps any triangle to any other triangle). Thus under an affine transformation each triple of model and image features defines a possible alignment of a model with an image. An affine transformation mapping three model points to three image points also constrains the three-dimensional location of an object. Under an orthographic projection camera model (where all of the light rays are parallel rather than going through an optical center), the three-dimensional position and orientation can be recovered up to a reflective ambiguity [Huttenlocher and Ullman 1990].

The most basic k-tuple search method is called the *alignment* technique, because it simply considers k-tuples of model and image feature pairs, checking the resulting transformations to find those that align a large number of model features with image features. In order to find all possible matches of a model to an image each ordered k-tuple of model and image features must be considered, although in some methods the search may terminate after one or more matches are found. For affine transformations of the plane, each ordered triple of model points and ordered triple of image points defines a basis set which specifies a possible transformation (or two transformations for a three-dimensional object). For each such basis set the transformation mapping the model into the image is computed, and the transformation is evaluated by using it to map the remaining model features into the image. The quality of a transformation is measured by the number of transformed model features for which there are nearby image features. The size of the region to search for each transformed model feature depends on the degree of error in sensing the image points, the spatial configuration of the basis triples, and the location of the given point with respect to the basis points.

An interesting extension of alignment techniques is the linear combinations method [Ullman and Basri 1991], which is based on the idea of forming two-dimensional images of a three-dimensional model as combinations of two-dimensional views of the object. That is, an object is modeled as a set of two-dimensional views, with known correspondence between the points in different views. A new view is recognized by determining whether it is a linear combination of a small number of stored two-dimensional views. This method assumes an orthographic projection camera model and is developed both for point sets and for objects with smooth bounding contours.

43.4.5 Invariants, Indexing, and Geometric Hashing

The central idea underlying the use of **geometric invariants** in model-based recognition is to develop representations of objects that remain unchanged as the viewpoint changes. Such invariants can be found in classical geometry and can also be derived using algebraic techniques (cf. Mundy and Zisserman [1992]). The invariant representations of objects can be used as keys to index into a table of stored objects. In principle it is then possible to look up an object in a large library of stored models in time that is essentially independent of the number of models. One of the key issues that separates such model indexing from hashing in general is the fact that different instances of the same object in various images will not generate exactly the same key or invariant signature, due to sensing uncertainty. Thus in practice exact numerical values cannot be used to hash into a data structure.

A number of methods have been developed for using invariants in recognition, based on geometric, differential, photometric, and even thermal properties of images (see Mundy and Zisserman [1992]). Structural indexing methods use combinations of simple features such as points and segments. Other geometric techniques use invariant properties of curves such as coplanar conics. The main advantage of using curve features is lower combinatorial complexity, and the main disadvantage is the need to extract these features from noisy, cluttered images. Photometric (intensity) information provides a richer description of an image than do geometric features such as points and line segments. The use of photometric invariants in recognition has been investigated in Nayar and Bolle [1993]. In order to illustrate the use of invariants in recognition we consider the **geometric hashing** approach [Lamdan and Wolfson 1988]. As in the previous section we examine the case of two-dimensional affine transformations. The fundamental observation underlying affine-invariant geometric hashing is the fact that three points define a coordinate system or basis with respect to which other points can be encoded in an invariant manner.

The geometric hashing method consists of two basic stages: (1) the construction of a model hash table and (2) the matching of the models to an image. The hash table is used to store a redundant, transformation-invariant representation of each object. Each model is entered into a hash table prior to recognition. For a given model, each ordered triple of model points m_1, m_2, m_3 forms an affine basis with origin $o = m_1$ and axes $u = m_2 - m_1$, $v = m_3 - m_1$. For each such basis every additional model point m_i is rewritten as (α_i, β_i) such that $m_i - o = \alpha_i u + \beta_i v$. The basis triple (o, u, v) and the point m_i are then stored in the hash table using the affine invariant indices (α_i, β_i). This results in a table with $O(r^4)$ entries for r model points. The table is generally formed using buckets rather than using a hashing scheme, as in practice the sensing uncertainty in real data makes it impossible to use exact values for retrieval. The issue of how to determine appropriate bucket sizes is somewhat complicated.

At recognition time, the hash table is used to determine which models are present in the image. The idea is that when the image points are rewritten in terms of an *image basis* that corresponds to an instance of the model, then the same *model basis* will be retrieved from the table many times. Each ordered triple of image points s_1, s_2, s_3 is used to form a basis, with origin $O = s_1$ and axes $U = s_2 - s_1$, $V = s_3 - s_1$. For a given image basis, each additional image point s_i is rewritten as (α'_i, β'_i) such that $s_i - O = \alpha'_i U + \beta'_i V$. The indices (α'_i, β'_i) are used to retrieve the corresponding entries from the hash table. For each model basis retrieved from the table, a corresponding counter is incremented in a histogram. Once all of the image points have been considered for a given image basis, the histogram contains votes for those model bases that could correspond to the current image basis, (O, U, V). If the peak in the histogram for a given model basis, (o, u, v), is sufficiently high, then this basis is selected as a possible match. When a new image basis is chosen the histogram counts are cleared.

This basic method often does not work well in practice due to the effects of sensing uncertainty on the locations of image features. Several weighted schemes have been developed which work well in practice. These methods enter each basis triple into multiple buckets in the table based on the sensing uncertainty [Rigoutsos and Hummel 1992].

43.4.6 Dense Feature Matching: Hausdorff Distances

The methods considered so far are primarily useful when there are relatively small numbers of model and image features, because they are based on considering subsets of features. A different approach is taken in Huttenlocher et al. [1993], which is based on computing distances between point sets rather than finding correspondences of points in two sets. These methods can be used for large sets of points, such as entire edge maps. The methods are similar to the template matching techniques used in some commercial recognition systems, but they use a new measure of image similarity based on the **Hausdorff** distance, and provide efficient algorithms for searching cluttered images.

Given two point sets $\mathcal{P}$ and $\mathcal{Q}$, with m and n points, respectively, and a fraction, $0 \leq f \leq 1$, the generalized Hausdorff measure is defined in Huttenlocher et al. [1993] and Rucklidge [1995] as

$$h_f(\mathcal{P}, \mathcal{Q}) = \{{}^{\text{th}}_{\sqrt{} \in \mathcal{P}} \min_{\coprod \in \mathcal{Q}} \|\sqrt{} - \coprod\|$$

where $f^{\text{th}}_{p \in \mathcal{P}} g(p)$ denotes the fth quantile of $g(p)$ over the set $\mathcal{P}$. For example, the 1st quantile is the maximum (the largest element), and the $\frac{1}{2}$th quantile is the median. This generalizes the classical Hausdorff distance, which *maximizes* over $p \in \mathcal{P}$. Hausdorff-based measures are asymmetric; for example, $h_f(\mathcal{P}, \mathcal{Q})$

and $h_f(\mathcal{Q}, \mathcal{P})$ can attain very different values as there may be points of $\mathcal{P}$ that are not near any points of $\mathcal{Q}$, or vice versa. This asymmetry is useful in recognition problems, where a hypothesize-and-test paradigm is often employed.

The generalized Hausdorff measure has been used for a number of matching and recognition problems. There are two complementary ways in which the measure has been employed. The first approach is to specify a fixed fraction f, and then determine the distance $d = h_f(\mathcal{P}, \mathcal{Q})$. In other words, find the smallest distance d, such that $k = \lceil fm \rceil$ of the points of $\mathcal{P}$ are within d of points of $\mathcal{Q}$. This has been termed finding the distance for a given fraction. Intuitively, it measures how well the best subset of size $k = \lceil fm \rceil$ of $\mathcal{P}$ matches $\mathcal{Q}$, with smaller distances being better matches. The second approach is to specify a fixed distance d, and then determine the resulting fraction of points that are within that distance. In other words, find the largest f such that $h_f(\mathcal{P}, \mathcal{Q}) \leq \lceil$. Intuitively, this measures what portion of $\mathcal{P}$ is near $\mathcal{Q}$ for some fixed neighborhood size d. This has been termed "finding the fraction for a given distance." It measures how well two sets match, with larger fractions being better matches.

Most applications of the measure are based on the second of the approaches, computing the *Hausdorff fraction*, because in most visual matching problems there is a reasonable prior estimate of the uncertainty in the positional location of image features. For example, a positional error of one pixel is generally introduced by the digitization process. If the feature points are edge features, then there is an uncertainty based on the degree of smoothing of the image. Efficient methods for finding the transformations of one point set such that the Hausdorff fraction is above some threshold (and the distance below some threshold) have been developed for affine transformations of the plane [Huttenlocher et al. 1993, Rucklidge 1995]. When the transformations are restricted to translations the fastest methods use dilation and correlation, whereas for full affine transformations the fastest methods use a hierarchical decomposition of the parameter space. The initial methods for computing Hausdorff distances were combinatorial algorithms using techniques from computational geometry, but current practice does not use these combinatorial techniques.

43.4.7 Appearance-Based Matching: Subspace Methods

Appearance-based recognition methods using **subspace techniques** have proven successful in a number of visual matching and recognition systems (e.g., Murase and Nayar [1995]). Appearance-based methods differ from those that we have already considered in that they operate by directly comparing images or image regions rather than by matching sets of local geometric features. The main advantage of appearance-based methods is that they are useful for tasks in which there is a large database of objects to be searched. The main disadvantage is that, in general, they do not work well with occlusion or with complex scenes and cluttered backgrounds. The most effective applications have been to problems such as the recognition of faces from mug shots, where the faces are generally about the same size and location in each image, and the background is a fixed color [Pentland et al. 1994].

The central idea underlying subspace methods is to represent images in terms of their projection into a relatively low-dimensional space, which captures the important characteristics of the set of objects to be recognized. This low-dimensional space is generally formed using the predominant eigenvectors (or principal components) of a set of known model images, where the k largest eigenvectors serve as the coordinate axes of the space (see Press et al. [1988] for more on eigendecomposition). Each model image is represented in terms of the k coefficients that result from projecting the image into the space. An unknown image is recognized by projecting it into the subspace and finding the closest model(s). The projection of an image can be thought of as a summary, which represents the image using just k numbers.

Subspace methods are attractive for problems in which there is a relatively large database of known objects, because the set of model images can be represented using just k coefficients in each, rather than the thousands of pixels in each image. This both saves storage and speeds the process of finding the closest matching images in the database. Moreover, to the extent that a subspace captures the important characteristics of a given set of images while omitting the unimportant characteristics, it can be used to generalize from a set of models. The main limitation of subspace methods is that when extraneous information from the background of an unknown image is projected into the subspace, it tends to cause incorrect recognition results. This is analogous to the problem that occurs when using the SSD to compare two image windows in motion or stereo, where background pixels included in a matching window can significantly alter the value of the SSD and cause incorrect matches.

Let I denote a two-dimensional image with N pixels, and let $\mathbf{x}$ be its representation as a (column) vector in scan line order. Given a set of training or model images, $\{I_m\}$, $1 \le m \le M$, define the matrix $X = [\mathbf{x}_1 - c, \ldots, \mathbf{x}_M - c]$, where $\mathbf{x}_m$ denotes the representation of I_m as a vector, and c is the average of the $\mathbf{x}_m$. The average image is subtracted from each $\mathbf{x}_m$ so that the predominant eigenvectors of XX^T will capture the maximal variation of the original set of images. Generally the $\mathbf{x}_m$ are also normalized prior to forming X, such as making $\|\mathbf{x}_m\| = 1$, to prevent the overall brightness of the images from affecting the results.

The eigenvectors of XX^T are an orthogonal basis in terms of which the $\mathbf{x}_m$ can be rewritten (and other, unknown, images as well). Let λ_i, $1 \le i \le N$, denote the ordered (from largest to smallest) eigenvalues of XX^T and let $\mathbf{e}_i$ denote each corresponding eigenvector. Define E to be the matrix of eigenvectors $[\mathbf{e}_1, \ldots, \mathbf{e}_N]$. Then $g_m = E^T(\mathbf{x}_m - c)$ is the rewriting of $\mathbf{x}_m - c$ in terms of the orthogonal basis defined by the eigenvectors of XX^T. It is straightforward to show that $\|\mathbf{x}_m - \mathbf{x}_n\|^2 = \|g_m - g_n\|^2$ [Murase and Nayar 1995], because distances are preserved under an orthonormal change of basis. That is, the SSD can be computed using the squared distance between the eigenspace representations of the two images.

The central idea underlying the use of subspace methods is to *approximate* the $\mathbf{x}_m$, and thus the SSD, using just those eigenvectors corresponding to the few largest eigenvalues. That is, $\mathbf{x}_m \approx \sum_{i=1}^k g_{m_i} \mathbf{e}_i + c$, where $k \ll N$. This low-dimensional representation is intended to capture the important characteristics of the set of training images. As this representation uses just the k predominant eigenvectors, it is not necessary to compute all N eigenvalues and eigenvectors of XX^T (which would be quite impractical as N is usually many thousands). Each model I_m is just represented by the k coefficients $(g_{m_1}, \ldots, g_{m_k})$, so comparing it with an unknown image requires only k rather than N comparisons. Generally k is considerably smaller than the number of models M, which is in turn much smaller than the number of pixels N.

Defining Terms

Appearance-based recognition: Recognizing objects based on views, generally using properties such as surface reflectance patterns; often used in contrast with model-based recognition.

Area-based matching: A means of identifying corresponding points in two images, for motion or stereo, by comparing small areas or windows of the two images.

Canny edge operator: An edge detector based on finding local maxima (peaks and ridges) of the gradient magnitude.

Correlation: The product of one function and shifted versions of another function; the maximum of the correlation can be used to find the best relative shift of two functions.

Digital image: Sampling of an image into discrete units in both space and intensity (or color) for processing with a digital computer; an array of pixels.

Digital video: A sequence of digital images representing a sampled video signal.

Edge detection: Locating significant changes in an intensity (gray-level) image, generally using local differential image properties.

Epipolar geometry: The constraint that the locations of points in one image must lie along a particular line in order to correspond to the same scene point as a given point in another image.

Gaussian smoothing: Convolution of a signal (an image) with a Gaussian function in order to remove high spatial frequency changes from the image (a type of low pass filtering).

Geometric hashing: A model-based recognition technique using a highly redundant representation that is invariant to certain geometric transformations.

Geometric invariants: Properties of a geometric model (e.g., a set of points) that remain unchanged under specified types of geometric transformations (e.g., distances between points under rigid motion).

Gradient magnitude: The magnitude of the gradient vector, or equivalently the square root of the sums of the squares of the local directional derivatives.

Hausdorff matching: A geometric technique for comparing binary images based on computing a variant of the classical Hausdorff distance from point set topology.

Hough transform: A technique used in model-based recognition to accumulate independent pieces of evidence for a match, using local features to vote for possible transformations mapping a model to an image.

Interpretation tree: A model-based recognition technique that uses a pruned exponential tree search to find corresponding sets of model and image features.

Model-based recognition: Approaches to recognizing objects in images that are based on comparing sets of stored model features against features extracted from an unknown image (generally geometric features).

Motion analysis: The recovery of information about how objects are moving in the image or in the world, or about the shape of objects, based on changes in brightness patterns over time.

Optical flow: The local change in image brightness as a function of time.

Pixels: The discrete spatial units of a digital image, generally obtained by conversion of an analog image signal from a camera or scanner; each pixel can generally take on a range of integer values (e.g., 0–255).

Pose recovery: Determining the position and orientation of an object in the world with respect to the camera coordinate system.

Scale space: Representation of a signal (an image) at multiple scales of smoothing.

Snakes: Energy-minimizing contours that combine internal constraints on their shape, such as smoothness of the contour, and external constraints from the image, such as brightness or gradient magnitude.

Stereopsis: The recovery of depth (or relative depth) by finding corresponding points in two or more images of the same scene.

Structure from motion: The recovery of the three-dimensional structure of an object based on a sequence of views as the camera moves with respect to the object.

Subspace methods: Reducing the dimensionality of image matching or recognition problems by representing the images in terms of their projection into a lower-dimensional space.

Sum-squared difference (SSD): A measure used to find the best relative shift of two images (or functions) based on the squared L_2 distance; similar to correlation, but based on minimizing a distance rather than maximizing a product.

Wavelets: Functions with a scaling and zero-sum property that are used to form a multiscale representation of an image (or function).

References

Amini, A. A., Weymouth, T. E., and Jain, R. C. 1990. Using dynamic programming for solving variational problems in vision. *IEEE Trans. Pattern Anal. Machine Intelligence* 12(9):855–867.

Black, M. J. and Anandan, P. 1993. A framework for the robust estimation of optical flow. In *Proc. Int. Conf. Comput. Vision*, pp. 231–236.

Blake, A. and Brelstaff, G. 1988. Geometry from specularities. In *Proc. Int. Conf. Comput. Vision*, pp. 394–403.

Canny, J. 1986. A computational approach to edge detection. *IEEE Trans. Pattern Anal. Machine Intelligence* 8(6):679–697.

Cass, T. A. 1992. Polynomial-time object recognition in the presence of clutter, occlusion, and uncertainty. In *Proc. Eur. Conf. Comput. Vision*, pp. 834–842.

Deriche, R., Zhang, Z., Luong, Q. T., and Faugeras, O. 1994. Robust recovery of the epipolar geometry for an uncalibrated stereo rig. *Proc. Eur. Conf. Comput. Vision* A:567–576.

Dickmanns, E. and Graefe, V. 1988. Dynamic monocular machine vision. *Machine Vision Appl.* 1:223–240.

Faugeras, O. 1993. *Three Dimensional Computer Vision, A Geometric Viewpoint.* MIT Press, Cambridge,MA.

Grimson, W. E. L. 1983. An implementation of a computational theory of visual surface interpolation. *Comput. Vision, Graphics, Image Proc.* 22(1):39–69.

Grimson, W. E. L. 1990. *Object Recognition by Computer: The Role of Geometric Constraints.* MIT Press, Cambridge, MA.

Hannah, M. J. 1989. A system for digital stereo image matching. *Photogrammetric Eng. Remote Sensing* 55:1765–1770.

Haralick, R. M. and Shapiro, L. G. 1992. *Computer and Robot Vision.* Addison–Wesley, Reading, MA.

Horn, B. K. P. 1986. *Robot Vision.* McGraw–Hill, New York.

Huttenlocher, D. P., Klanderman, G. A., and Rucklidge, W. J. 1993. Comparing images using the Hausdorff distance. *IEEE Trans. Pattern Anal. Machine Intelligence* 15(9):850–863.

Huttenlocher, D. P. and Ullman, S. 1990. Recognizing solid objects by alignment with an image. *Int. J. Comput. Vision* 5(2):195–212.

Kass, M., Witkin, A., and Terzopoulos, D. 1988. Snakes: active contour models. *Int. J. Comput. Vision* 1(3):321–331.

Koenderink, J. J. 1990. *Solid Shape.* MIT Press, Cambridge, MA.

Kriegman, D. J. and Ponce, J. 1990. On recognizing and positioning curved 3-d objects from image contours. *IEEE Trans. Pattern Anal. Machine Intelligence* 12:1127–1137.

Lamdan, Y. and Wolfson, H. J. 1988. Geometric hashing: a general and efficient model-based recognition scheme. In *Proc. Int. Conf. Comput. Vision,* pp. 238–249.

Lowe, D. G. 1985. *Perceptual Organization and Visual Recognition.* Kluwer Academic. New York.

Mallat, S. G. and Zhong, S. 1992. Characterization of signals for multiscale edges. *IEEE Trans. Pattern Anal. Machine Intelligence* 14(7):710–732.

Mundy, J. L. and Zisserman, A. 1992. *Geometric Invariants In Computer Vision.* MIT Press, Cambridge, MA.

Murase, H. and Nayar, S. K. 1995. Visual learning and recognition of 3-d objects from appearance. *Int. J. Comput. Vision.* 14:5–24.

Nayar, S. K. and Bolle, R. M. 1993. Reflectance ratio: a photometric invariant for object recognition. In *Proc. Int. Conf. Comput. Vision,* pp. 280–285.

Pentland, A., Moghaddam, B., and Starner, T. 1994. View-based and modular eigenspaces for face recognition. In *Proc. IEEE Conf. Comput. Vision Pattern Recognition,* pp. 84–91.

Perona, P. 1992. Steerable-scalable kernels for edge detection and junction analysis. In *Proc. Eur. Conf. Comput. Vision,* pp. 3–18.

Press, W. H., Flannery, B. P., Teukolsky, S. A., and Vetterling, W. T. 1988. *Numerical Recipes: The Art of Scientific Computing.* Cambridge University Press, Cambridge, England.

Rigoutsos, I. and Hummel, R. 1992. Massively parallel model matching: geometric hashing on the connection machine. *Computer* (Feb.):33–41.

Rucklidge, W. J. 1995. Locating objects using the Hausorff distance. In *Proc. Int. Conf. Comput. Vision,* pp. 457–464.

Terzopoulos, D. 1988. The computation of visible-surface representations. *IEEE Trans. Pattern Anal. Machine Intelligence* 10(4):417–438.

Thorpe, C. E., Hebert, M., Shafer, S. A., and Kanade, T. 1988. Vision and navigation for the Carnegie-Mellon navlab. *IEEE Trans. Pattern Anal. Machine Intelligence* 10(3):361–372.

Tomasi, C. and Kanade, T. 1992. Shape and motion from image streams under orthography: a factorization method. *Int. J. Comput. Vision* 9(2):137–154.

Ullman, S. 1979. *The Interpretation of Visual Motion.* MIT Press, Cambridge, MA.

Ullman, S. and Basri R. 1991. Recognition by linear combinations of models. *IEEE Trans. Pattern Anal. Machine Intelligence* 13(10):992–1006.

Wells, W. M. 1986. Efficient synthesis of Gaussian filters by cascaded uniform filters. *IEEE Trans. Pattern Anal. Machine Intelligence* 8:234–239.

Witkin, A. P. 1983. Scale-space filtering. In *Proc. Int. J. Conf. Artif. Intelligence*, pp. 1019–1022.
Zabih R. and Woodfill, J. 1994. Non-parametric local tranforms for computing visual correspondence. *Proc. Eur. Conf. Comput. Vision* B:151–158.

Further Information

Much of the material in computer vision is in the form of original research articles, a few of which are cited in the References. In-depth coverage of the field can be found in the books by Faugeras [1993] and Haralick and Shapiro [1992], and good coverage of low-level vision is provided by Horn [1986]. The area of model-based object recognition is covered in Grimson [1990], and the geometric invariants approach to recognition is in Mundy and Zisserman [1992].

V

Human–Computer Interaction

The subject area of Human–Computer Interaction is concerned with improving the quality and effectiveness of technology, its development, and its interactions with people. This includes the analysis, design, implementation, and evaluation of computing systems with a keen interest in user interfaces and user performance. With the recent global deployment of computers and software, both organizational and cultural issues have become critical factors in the design of interfaces that people can use in increasingly diverse professional and personal settings.

44 The Organizational Contexts of Development and Use *Jonathan Grudin and M. Lynne Markus* **44**-1
Introduction • The Need for Organizational Analysis in Human–Computer Interface Design • Organizations and Their Components • Organizational Modeling, Formal and Informal • Organizational Contexts of Development • Organizational Contexts of Use • What Are the Organizational Issues in Interactive System Use? • Conclusions

45 Usability Engineering *Jakob Nielsen* **45**-1
Introduction • Know the User • Competitive Analysis • Goal Setting • Coordinating the Total Interface • Heuristic Evaluation • Prototyping • User Testing • Iterative Design • Follow-Up Studies of Installed Systems

46 Task Analysis and the Design of Functionality *David Kieras* **46**-1
Introduction • Principles • Research and Application Background • Best Practices: How to Do a Task Analysis • Using GOMS Task Analysis in Functionality and Interface Design • Research Issues and Concluding Summary

47 Human-Centered System Development *Jennifer Tucker and Abby Mackness* **47**-1
Introduction • Underlying Principles • Best Practices • Research Issues and Summary

48 Graphical User Interface Programming *Brad A. Myers* **48**-1
Introduction • Importance of User Interface Tools • Models of User Interface Software • Technology Transfer • Research Issues • Conclusions

49 Multimedia *James L. Alty* **49**-1
Introduction: Media and Multimedia Interfaces • Types of Media • Multimedia Hardware Requirements • Distinct Application of Multimedia Techniques • The ISO Multimedia Design Standard • Theories about Cognition and Multiple Media • Case Study — An Investigation into the Effects of Media on User Performance • Authoring Software for Multimedia Systems • The Future of Multimedia Systems

50 Computer-Supported Collaborative Work   *Fadi P. Deek and James A. McHugh* **50**-1

Introduction • Media Factors in Collaboration • Computer-Supported Processes and Productivity • Information Sharing • Groupware • Research Issues and Summary

51 Applying International Usability Standards   *Wolfgang Dzida* **51**-1

Introduction • Underlying Principles • Best Practices • Research Issues and Summary

44

The Organizational Contexts of Development and Use

44.1 Introduction **44**-1
44.2 The Need for Organizational Analysis
      in Human–Computer Interface Design **44**-2
44.3 Organizations and Their Components **44**-2
      Organizations in Context • Organizational Components
      • Ways in Which Organizations Differ
44.4 Organizational Modeling, Formal and Informal **44**-5
44.5 Organizational Contexts of Development **44**-6
      The Emergence of Distinct Development Contexts in the U.S.
44.6 Organizational Contexts of Use **44**-9
44.7 What Are the Organizational Issues
      in Interactive System Use? **44**-11
      Initiation Phase • Acquisition • Implementation
      (Introduction) and Use • Impacts and Performance
44.8 Conclusions **44**-13

Jonathan Grudin
Microsoft Research

M. Lynne Markus
Bentley College

44.1 Introduction

Human–computer interaction has focused on individual users and their relationships to systems. Much of the progress in the field has come about by looking for commonalities across the increasing number and diversity of computer-supported tasks. The personal computer (PC) of the 1980s was the perfect laboratory for this effort, and the initial difficulty in networking PCs together only helped shield PC users from group and organizational influences.

The large, expensive systems that preceded the PC had few resources to devote to usability and less reason to worry about it. The users of these systems were people who used the output, typically paper reports. They did not interact directly with computers: that task was left to programmers and computer operators, who acquired the necessary technical competence. With spreadsheets and word processors on PCs, however, to a much greater extent user and operator were synonymous. These users did not see themselves as computer professionals. They had less desire to master technical aspects of systems, and the emerging shrinkwrap software market allowed them to seek out more usable software.

Today, PCs and workstations are networked, intranetworked, and internetworked; once again, all computer use is being carried out in organizational contexts. The implications of this move — from the three key elements of human–computer interaction: user, system, and use, to larger contexts — are described in other chapters. In this chapter, we examine what has always been a fundamental unit: organization. Organizations affect human–computer interaction in two important ways. First, systems and applications are developed in organizations, and the context of development influences the development process. Second, interactive systems and applications are used in organizations, and successful use is often affected by a range of organizational factors, which has implications for those developing, introducing, and using systems.

44.2   The Need for Organizational Analysis in Human–Computer Interface Design

A case study by Markus and Keil [1994] illustrates the benefit of a careful organizational analysis by demonstrating that a well-executed interface design project can produce a highly usable system that is not useful, and not used. The setting is "CompuSys," a major computer company and employer of leading interface designers. The project was initiated to redesign the interface to a system developed for internal use by the sales organization, an expert system of the sort that achieved prominence in the mid-1980s through the success of Digital's XCON [Barker and O'Connor 1989]. Sales representatives frequently made errors working out details of complex customer system configurations, such as omitting minor but necessary components. CompuSys swallowed the cost of repairing these errors. An expert system for product configuration was built; it was accurate, but was only used with a fraction of orders.

Costly errors continued. Why wasn't the system used? The sales force complained about its usability. A project employing many advanced interface design techniques, including iterative design and user feedback, led to a major redesign of the clearly awkward interface. Users from five pilot sites were trained on the new system. Millions of dollars later, a new system with a much improved interface was introduced. But the new system was not used much more than the one it replaced.

Why? The system design was based on the following model of a typical sales process: (1) a customer identifies system requirements, (2) the sales representative works out a system configuration to meet the requirements, (3) the price is calculated. This seemed logical to the designers, but it was wrong. More often, customers had a fuzzy sense of their problem and a concrete budget for the system. They indicated how much they had to spend and the sales representative would try to identify an adequate system that could be acquired for that amount. The expert system did not support reasoning back from price to configuration — it reasoned only in one direction, from configuration to price. The new interface made it much easier to work from configuration to the price, but it missed the point. The system was usable but not useful. At the end of the project, the developers did not understand why the usable system was neglected. An organizational analysis — actually an interorganizational analysis — uncovered the *counterintuitive* work process.

To avoid such costly mistakes requires greater analysis or awareness of the organizational context and actual work processes than these designers had, even after direct interviews with the sales force. The awareness might be obtained in different ways: through a more sophisticated survey, ethnographic observation, contextual inquiry, or more extensive participation by users. Intuitions could be enhanced through familiarity with the literature on organizational theory and practice. The next section of this chapter summarizes some of the insights from this literature, drawing from Grudin and Markus [in press].

44.3   Organizations and Their Components

Organizations are often defined as collections of people with a common purpose or task. How does this differ from groups? Is it only a matter of size and structure — are organizations (usually) groups of groups?

Organizations arguably have a longer lifespan than groups. Loss or change in one or two members often changes a group completely, and it would seem odd to consider a group the same following entire replacement of its personnel, but organizations (such as Ford Motor Company or the University of California) can continue despite extensive internal reorganization or change. A group may be more than its members, but an organization is much more so.

Organizations also differ from groups in that they often have distinct public and legal identities and engage in a range of activities as a result, whereas groups are more likely to have a single focus. Although rock groups may be anomalous in some ways, we could argue that the Rolling Stones as a group play music; the Rolling Stones as an organization plans concert tours, handles arrangements, invests money, and so forth. People invest in organizations, the annual reports of many organizations are scrutinized, organizations often engage in more competitive activities than we associate with groups. And for these reasons, organizations have possibly been studied more intensely than groups (except for rock groups!).

If we are trying to understand a human–computer dialogue, we would first move one level up and learn what the purpose and context of that interaction was. Only then would we consider the components: an individual with a display, keyboard, and mouse. Similarly, to understand an organization, we first examine the whole of which it is a part — the role that the organization plays in networks of organizations. Then we examine its component parts: groups and individuals with structured relationships and dynamic interactions.

44.3.1 Organizations in Context

Organizations operate in a larger societal context or environment than can usefully be considered a network of organizations. Consider the Boeing Airplane Company. We think of Boeing as making airplanes, but in fact Boeing manufactures very little of the aircraft apart from the wings. Boeing designs the plane, it assembles the plane, and the components are made by scores of organizations around the world. Much of Boeing's work consists of managing this network of organizations, which includes vendors in most countries to which Boeing sells planes. This of course helps ingratiate Boeing to their governments.

Governments are another kind of organization with which the company interacts, as are airlines, financial institutions, unions, passengers, competitors, and so forth. Each organization in this network of interacting organizations performs a different role. Each organization has its own goals or interests that are sometimes mutually compatible and sometimes in conflict. Some organizations are more powerful than the others, controlling more resources, and thus are more likely to win when there are conflicts of interests. If the Chinese government decides to make a foreign policy point by canceling plans for a large order, there is little Boeing can do.

Similarly, the software industry is a network of organizations with different roles involved in the production, sale, support, and use of various products and services. The 1980s saw a proliferation of organizations mediating between developers and users. These include consulting companies, standards organizations, value-added resellers, third-party developers, subcontractors, advertising agencies, professional organizations, magazine companies, and others.

There are opportunities for conflict as well as cooperation in these relationships. Some organizations both cooperate and compete, as in well-known joint ventures between Apple and IBM, for example. Organizations define success differently: for some it is the bottom line, the annual return to stockholders; for other organizations, such as universities, hospitals, governments, social-service agencies, and voluntary organizations, success may mean knowledge creation and transfer, health, social welfare, or member satisfaction. The interaction of organizations with different goals and interests leads to complex, rich dynamics.

44.3.2 Organizational Components

Apart perhaps from quite small organizations, an organization consists of people deployed in various work groups, which may be organized into departments, divisions, or other structural units. A computer manufacturer can be organized by product category (PCs, workstations, minicomputers) or by function (hardware, systems software, applications software, support); they can have sales organized by region or by product line, and so on.

There is no general agreement on optimal organizational structure. Although in some industries, virtually all organizations in the same size class have a similar organizational structure, others are marked by a great deal of local variation. Large organizations often shift from one structure to another, based perhaps on shifts in the external environment, or as a result of the personnel at hand, or perhaps even as a way of releasing adrenaline and enhancing vitality. Choices may not be fixed, but they can determine what the organization can do easily and well. For instance, a computer manufacturer organized by function may have a harder time delivering new PC models to market than a similar company with a product organization, but may have an easier time establishing consistency and interoperability across all product lines.

A specific example of particular interest to many in the human–computer interaction field is the deployment of usability specialists or human factors engineers in a very large research and development division. Should they be grouped together in a central usability laboratory, where they have the benefits of pooled resources and management that understands their roles? One drawback is possible isolation in a usability ghetto. Or should each specialist become a member of a product team, gaining the benefit of close interaction throughout the product cycle and a team that comes to understand and trust them? One drawback is that a usability specialist, managed as a member of a software team by someone with little understanding of usability, may become a software engineer. Both approaches and others are tried.

An ambitious analysis of organizational components has been developed by Mintzberg [1984, 1989]. Mintzberg identifies five basic organizational components or groups: the strategic apex (executives), the middle line (management), the operating core (production workers, whether surgeons in a hospital or assembly line workers), the technostructure (those who define the work processes in an organization), and the support staff (technical support, custodial, kitchen, and so on). Mintzberg argues that these functions will to some degree compete for control, and that five organizational forms can be found, each marking the dominance of one of these groups, along with some hybrids. Mintzberg extends his analysis to look at a wealth of accompanying and interacting factors, such as the approach to measuring output, the organizational environment, and so on. One of us once worked for a 500-person organization that was created with only a loose mission, and it was fascinating to see the correspondence of behavior to Mintzberg's model as each of the five groups jockeyed for control in this uniquely undefined setting.

Thus even when an organization as a whole has a clear set of goals and interests, individuals, groups, and subunits within the organization may not share them fully. Newspaper reporters and newspaper editors may differ, hardware and software engineers may engage in rivalry that is friendly or not so friendly — and both groups may have less than full respect for their colleagues in marketing or upper management. In many user organizations, the information systems staff differs in age and outlook from other groups, which can lead to breakdowns in communication or cooperation.

Employees may be subject to different performance measures and rewards. For example, a software unit may be evaluated on development schedules and budgets, whereas a hardware group may be evaluated largely on manufacturing cost. If so, it may be hard for the units to cooperate to achieve the total organization's goals. In the case of CompuSys, errors in computer system configuration originated with the sales department, which was rewarded for sales volume, but the manufacturing unit bore the costs associated with fixing incorrect orders. Given these incentives and rewards, the sales department devoted little time and energy to ensuring that orders were correct at the outset.

Good management, good communication of organizational purpose, an appropriate set of measures, rewards, and punishments might keep everyone singing the same tune. However, the complexity of large organizations makes it difficult to design efficient and effective management structures. Generally speaking, in organizations of any size and structural complexity, subunits exhibit a fair degree of goal displacement, pursuing goals and values that make sense to the subunit, but that do not necessarily advance the overall interests of the organization. And there is not agreement as to the appropriate level of conflict within an organization; one ideal would be to eliminate internal conflict and competition, but others endorse a management style that fosters a level of internal jockeying for resources.

Groups within organizations can develop distinct cultures (assumptions, beliefs, and language systems) that render clear communication with members of other groups and units more difficult. The same terms are often used in very different ways in different parts of an organization, creating problems for efforts to create a common database. Mark and Mambrey [1996] describe an example in which typists wanted documents categorized by author and completion date, whereas authors wanted categorization by topic. An example noted by Grudin and Markus [in press]: It is a sale to sales when a customer verbally commits to an order, but it is not a sale to legal until a contract is signed; it is not a sale to manufacturing until a purchase order is entered into the manufacturing control system, and the accounting department only acknowledges a sale when an invoice has been prepared. Language differences contribute to widely differing points of view on key organizational decisions. Conflict as well as cooperation occurs in organizational decision making, producing behavioral dynamics inside organizations that are as rich as those observable when organizations interact.

44.3.3 Ways in Which Organizations Differ

It is useful to think of at least three levels of reality operating simultaneously in organizations. The first can be called rational, technical, or economic reality. It takes a stated goal as given and looks for an efficient and effective means of achieving it. The second level of reality can be called socioemotional task reality. People have social needs and organizations are one place in which they attempt to meet them. In addition, people habituate to particular ways of thinking and behaving due in part to their membership in organizational groups and subunits. These ways of thinking and acting may differ dramatically from what an outside observer would see as the rational optimum. The third level of reality is structural/political, focusing on goals and interests created by resources and positions in units and task chains (often called business processes) or interorganizational networks. Keeping all three realities in mind helps in seeing and understanding the complex dynamics of organizational behavior.

This has been a highly simplified overview of organizational behavior. Next, we show how these organizational issues can have enormous implications for the adoption, deployment, use, and consequences of information technology in organizations. However, first we will review some major categories of differences between organizations and their component work groups and subunits that can significantly shape the use of information technology:

- Headcount
- Economic resources, particularly *slack* (uncommitted resources)
- Geography (scope of operations) and space (e.g., in buildings)
- Age of organization, demographic profile, experience including experience working together
- Stated or implicit goals (e.g., least cost producer vs. product innovator)
- Structure/basis of organization (product, geographic, function, technology, time)
- Culture (beliefs, assumptions, language systems, characteristic behavior patterns)
- Management style (measures, rewards, promotion patterns, etc.)
- Information technology infrastructure (prior investments, commitments, governance)

With so many variable factors, it is not surprising that organizations and subunits react quite differently to a given technology.

44.4 Organizational Modeling, Formal and Informal

When a system or application is developed for use within a single organization, the software is likely to better mesh with the behavior of the organization if pertinent aspects of the organization can be formally modeled; that is, represented in a form the computer can manipulate. Formal modeling of organizational processes and data have been a major concern in the development of large systems.

Before addressing this history, though, it is useful to contrast the development of a system for a single organization with other development situations. In particular, when developing a shrinkwrap software product, intended for use in a range of organizations, the initial developers have less motivation for organizational modeling. Rather, those tailoring a system for a given customer might consider organizational modeling. This is significant because commercial software has been the focus of most work in human–computer interaction (HCI). Most in HCI have no familiarity with organizational modeling; many have little experience with databases.

In the area of groupware and computer-supported cooperative work, formal modeling is appearing. Some is at the level of group behavior rather than organizational behavior, but workflow management systems (e.g., Abbott and Sarin [1994] and Marshak [1994]) involve a broad enough context to be considered organizational. The workflow tool developers do not model organizations; it is assumed the customer (or a consultant) will carry out the modeling.

Workflow management systems are often considered in the context of business process re-engineering (BPR). Both are predicated on the idea of creating detailed models of organizational processes. BPR looks to rationalize such processes; workflow management systems look to incorporate them in software and support them. Workflow management systems appear to be useful for high-volume, relatively routine business activities; whether current systems have the flexibility to support other activity is unclear. Bowers et al. [1995] is a nice study of a workflow system.

When we consider *user* organizations and the design and development of interactive software, formal modeling and its limitations arise. In considering the effects of *development* organizations on the design and development of interactive software, formal modeling does not arise, because our examination there deals with the need for people — designers and developers — to contend with the organizational environment, not a program. Fostering *awareness* of organizational influences is the objective, not modeling it formally. Awareness is a kind of informal modeling, of course, a point that arises in contexts of system use as well.

44.5 Organizational Contexts of Development

Developers are well aware that organizations constrain them through time pressures, approval processes, formal specifications, and other practices. What is often less evident is that these constraints differ markedly, and systematically, across organizations. To quote Mahoney [1988], "we speak of the computer industry as if it were a monolith rather than a network of interdependent industries with separate interests and concerns." An organization's structure and practices can have effects on the human–computer interfaces that it produces, and these may be major effects or subtle effects. Differences across segments of the industry affect what techniques can or should be applied, and what tools will or will not be useful. The history of segmentation within the field left traces in systems development practices, and the history differs in North America and Europe. (The following account draws on Grudin [1991a], Grudin and Poltrock [1995], and Grudin and Markus [in press], as well as the sources cited in the text.)

To illustrate the effects of organizational context, consider a key relationship in the design of interactive systems: the relationship between the developers and the users. Adapted from Grudin [1991a], Figure 44.1 shows, for three development contexts, the times at which development teams are identified and the actual users are identified for a new application. From the top, an organization putting out a project for competitive bidding must produce a preliminary design to give possible contractors a specification to bid on. The users are identified well before the development team is known, and to prevent any favoritism toward a particular contractor, interaction between user and development organization may be curtailed or prohibited. For in-house development of a system by the information systems group within a user organization, both parties are often identifiable from the outset. The developers of a novel commercial product work under still different conditions: After marketing or management has done some analysis and high-level specification, the development team is formed. But the users are not truly known until the product is marketed. As we will see, these and other differences in conditions can greatly affect the process of developing interactive systems.

44.5.1 The Emergence of Distinct Development Contexts in the U.S.

An examination of the historical emergence of development contexts illuminates their differences and explains aspects of systems development practices that can adversely affect human–computer interfaces.

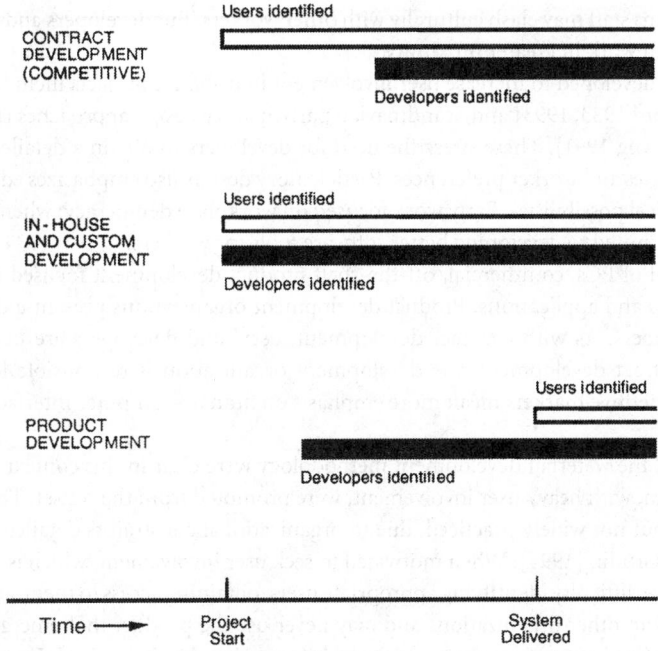

FIGURE 44.1 Identification of users and developers in different development contexts.

Most major early systems were contracted by the U.S. government. The government defined the requirements, and geographically distant organizations bid and were awarded contracts. Different contracts might be awarded for design, development, and maintenance. Legal restrictions governed their communication. This created tremendous organizational barriers to direct contact among developers and users. However, this was of less import because little software was interactive. Most computers were batch processors: a program and data were loaded and the computer computed, and perhaps produced a printed output.

Stage models of systems development evolved naturally in this environment. In the U.S., the waterfall model of Royce [1970], appearing at the dawn of the discipline of software engineering, was enormously influential. This design-it-first, specification-based approach, with little provision for feedback or iteration, became the dominant development methodology.

The design of usable human–computer dialogues requires abandoning the design-it-first approach. Boehm [1988] introduced spiral models explicitly to incorporate prototyping and iterative design for interactive software development. Nevertheless, the separation of developers and users remains a major organizational obstacle to interface design in contract development contexts. Winkler and Buie [1995] provide a good summary of the challenges and possible approaches, such as developing standards that can be included in requests for proposals (RFPs). The primary focus, however, is on raising awareness in user and developer organizations. Perhaps more important than trying to specify the interface in detail is to specify an interface development *process*. Winkler and Buie identify the primary interface development problems in government contracting to be organizational, not technical, in nature.

As business use of computers expanded, in-house development of software in large user organizations became the principal focus of system development. (In Europe, in-house development has always predominated.) Most processing remained batch rather than interactive, with end users primarily engaged in data entry and system operator tasks, but the waterfall development method approach was challenged from the outset in this organizational context, more effectively in Europe, where the stage models did not have the same prominence [Friedman 1989, Hirschheim et al. 1995].

Barriers to interaction among users and developers are less compelling for internal development. Organizational and cultural barriers still hinder user involvement in development — young, highly paid information systems staff may clash culturally with other workers. But developers and users have the same employer and often work in greater proximity.

Methodologies developed to increase user involvement in in-house projects include the sociotechnical approach [Mumford 1983, 1993] and Scandinavian participatory design approaches [Bjerknes et al. 1987, Greenbaum and Kyng 1991]. These stress the need for developers to obtain a detailed understanding of actual work processes and worker preferences. Participatory design also emphasizes educating prospective users about technical possibilities. Early work focused on workplace democracy; when interactive systems arrived, the focus shifted to developing better software tools for workers.

With the arrival of PCs, commercial, off-the-shelf product development focused attention on highly interactive systems and applications. Product development organizations present a different set of constraints on developers. As with contract development, users and developers are in different organizations. Unlike contract development, the development organization is responsible for defining the requirements. Competitive markets mean more emphasis on human–computer interfaces and greater time pressures.

Inadequacies of the waterfall development methodology were clear in this context. Rapid prototyping and iterative design, with heavy user involvement, were promoted from the outset. These principles were widely accepted, but not widely practiced, due to organizational constraints detailed in Grudin [1991b] and Poltrock and Grudin [1994]. When motivated to seek user involvement, which is not always the case, developers can have difficulty identifying appropriate users, obtaining access to them, and motivating users who after all work in other organizations and may never use the product that emerges. If these barriers are overcome, there are often obstacles to using the information that is received. Development schedules rarely provide time for iteration, and changes in the interface can be visible and unsettling to management, as well as to those working on product documentation and training, which are tied closely to the interface.

To overcome these obstacles, product developers have sought to adapt and extend participatory design and related techniques. A notable example is contextual inquiry and modeling [Holtzblatt and Beyer 1993, Holtzblatt and Jones 1993, Beyer and Holtzblatt 1995].

A fourth organizational context for development, custom development, blends characteristics of each of the three previously discussed. The development organization is distinct from the user organization and works under contract. The contract is not necessarily competitively bid and the development organization may be selected based on geographic proximity. Rather than prohibited, long-term, close involvement can serve to mutual advantage, infusing some aspects of internal development into the process. The development organization often focuses on a market niche, where similar custom jobs will amortize their investment. The more successful they are at finding customers with the same needs, the closer their situation comes to resemble product development; in fact, their software may evolve into a commercial product.

Custom development seems likely to thrive as demands for software become increasingly diverse and specialized. Grudin [1996] argues that custom development could be particularly promising for HCI tools that focus on documenting upstream design process for subsequent use in redesign or maintenance, because, for example, it is a context in which design, redesign, and maintenance are likely to be done by the same people.

Figure 44.2 summarizes responses to stage or waterfall models arising in contract development (software engineering), in-house development (information systems), and product or package development (human–computer interaction). For a richer view of the history, see Hirschheim et al. [1995].

This only suggests differences among development contexts and the effects of organizations. There exist organizations to mediate between a developer and a user organization. This activity takes place in economic, cultural, political, and societal conditions to which organizations contribute. Other organizational factors influence how technology is developed and used: structures and processes, size, intended market, geographical placement, age, application novelty, function, culture, environment. In a small startup company, all employees may see one another regularly; in a large organization, the software, documentation, and training developers may work in different states or countries. In the U.S., court decisions regarding *look and feel* copyrights or patents can affect the process and product of development; in Europe, codetermination laws can affect the development process by requiring user involvement.

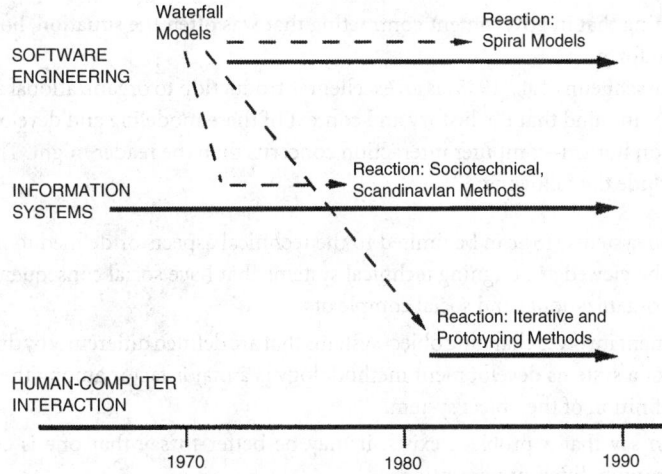

FIGURE 44.2 Approaches to systems development in different development contexts.

This list is not exhaustive, but it serves as an indicator that organizational factors within a development environment have a strong bearing on systems development, and thoughtful developers will consider those factors carefully. Knowledge of organizational influences can alert developers to obstacles and biases that can result in painful mistakes. It can also guide them to techniques that are suited to their context. Keil and Carmel [1995] have closely analyzed successful projects in two contexts, corresponding to what we have called in-house and product development. They report, for example, on 11 different approaches to establishing links among customers and developers and find that the methods used and their levels of effectiveness vary across contexts.

44.6 Organizational Contexts of Use

We now examine efforts to model, understand, or use intermediaries to represent user organizations. The emergence of different contexts of system use and development is relevant here as well. A very good treatment of this topic is found in Hirschheim et al. [1995]. This work to some degree builds on Friedman [1989] which is equally absorbing.

Both of these works focus virtually exclusively on internal or in-house systems development, with perhaps some attention to custom development. This is understandable: in-house development dominated systems development for decades, and this has been even more true in Europe. In the U.S., competitive contract development was more influential in the early days and commercial off-the-shelf software has attracted considerable attention recently.

It is important to keep in mind that close attention to human–computer interaction bloomed in commercial software development, not in in-house systems development. In the latter context, it makes intuitive sense to try to model aspects of the organization, but the focus has not been on improving human–computer interaction.

Hirschheim et al. [1995] note that those applications were growing more complex over time, but do not note the dramatic shifts in the user populations, the specific new demands that highly interactive systems introduce, or strong ties between the features of certain systems development methodologies and the development contexts in which they originate.

For example, they note that at one time user requirements elicitation was considered a largely noncontroversial task of asking users what they needed in their jobs. This may always have been a questionable assumption, but it was less so for batch processing than for interactive systems. Elsewhere, they note that software engineering focused on development steps "under the assumption that systems requirements are given" without noting that in government contracting that was often the situation, however problematic it might be. And so forth.

Nevertheless, Hirschheim et al. [1995] is an excellent introduction to organizational and data modeling, as long as one keeps in mind that the history and context of these modeling and development approaches placed less weight on human–computer interaction concerns than the reader might. The distinctions that are introduced include the following:

- Information systems (IS) can be limited to the technical aspects or defined to include humans. IS design can be viewed as designing technical systems that have social consequences or as primarily addressing organizational and social complexity.

- IS development involves changing object systems that are defined differently by different individuals; the choice of a systems development methodology is a major factor among the many that affect a person's definition of the object system.

- Rather than say that a problem exists, it may be better to say that one is constructed among stakeholders with different perceptions.

- Information systems development can be marked by uncertainty regarding the means employed, the effects that will result, or whether the right problem was attacked. Systems development methodologies can be categorized as process oriented or data oriented. Object systems can be perceived and modeled as static, dynamic, or hybrid.

- User participation in system design can be seen as expedient in collecting needed information or overcoming resistance to change, as a prerequisite for creating shared meanings necessary for design, or as a moral right.

- Most methodologies focus on one application; only a few embrace organization-wide planning.

The authors use these and other distinctions to provide alternative views of a range of methodologies. These include structured analysis and design methodologies, which focus on modeling organizational processes and data flows, and data-oriented analysis and design approaches, which represent organizations as a structured collection of facts and associations among them. Object-oriented analysis and design define software units that encapsulate methods and data. In parallel with the evolution of IS development, the authors consider the evolution of data modeling approaches emerging from the techniques that led to the development of database technologies.

The authors identify seven generations of information system design methodologies that replaced unstructured, ad hoc approaches to design. The sequencing is not strictly chronological or convincing, but it is a useful exercise. They are: (1) formal life cycle approaches (from mid-1960s); (2) structured approaches (mid-1970s), emphasizing user requirements; (3) prototyping and evolutionary approaches (late 1970s), tied here to greater competition but not especially to greater prominence of the human–computer interface; (4) sociotechnical and participative approaches (late 1970s, early 1980s) in which users take at least some control and redesign work situations; (5) sense making and problem formulation approaches (early 1980s), involving users and developers from very early in a project; (6) trade union led approaches (mid-1970s through mid-1980s), to support democratic planning and later to design tools to augment workers' skills; and (7) emancipatory approaches (future), a rather vaguely defined approach devised by the authors.

What emerges from this examination of modeling, interpreted through a sense of the contexts in which these approaches have been introduced, is a sense of the trajectory of change and the pressures driving it. When information systems consist largely of entering, organizing, storing, and transforming data, and when the data contain records such as employee names with their ages, salaries, managers, departments, and so forth, then data-centered IS design might flourish in ways that are less likely when the system focuses on real-time computer-mediated communication. A system designed to compute projectile trajectories might give rise to wide agreement on the definition of terms and relevant concepts, whereas a workflow system that must model a group of people and objects that have not previously been considered together will inevitably confront conflicting views of work, conflicting terminology, and other issues. Of particular relevance here, software applications in which over 50% of the program code is devoted to the human–computer interface inevitably introduce a range of considerations and challenges that did not exist for methodologies that often consciously ignored or relegated to a single process substep the establishment of the interface.

Although one does not leave the survey of current practice with a prescription for organizational modeling, one does leave it with a sense of the richness and importance of the problem, and an awareness of the steps taken in the direction of solutions. The growing focus on sense making, on recognizing the multitude of perspectives, and the constant shift in the sense of the target suggests that formal modeling of groups and organizations will encounter limits. And yet as people strive to build systems to support organizations through contact with most or all members, the effort to flexibly model the organizations will inevitably continue.

44.7 What Are the Organizational Issues in Interactive System Use?

This discussion is organized along broad system life cycle phases used by Grudin and Markus [in press]: initiation (idea origination, project funding), acquisition (system acquired or built), implementation and use (internal deployment, use or rejection), and impacts and consequences (system effects experienced; steps taken to augment, mitigate, or otherwise manage them).

44.7.1 Initiation Phase

The source of a technology investment idea can have important influences on downstream events. The source can be external or internal; if internal, it can be from high-ranking line management* (possibly responding to advertising or business fads) or lower, perhaps from the technical staff. Approval may require consent of one line manager or an internal technology approval board. Justification can serve a positive purpose of building shared understanding, but the understanding may be inaccurate if distortion is needed to meet preset criteria. See Dean [1987] for a discussion of the technology justification process and Cohen et al. [1972] for a classic paper on decision making in organizations. Suboptimal outcome possibilities include underutilized systems as well as systems rejected because they benefited few users (e.g., only the decision maker [Grudin 1994]). Kling [1978] describes a welfare case tracking system that was adopted despite its lack of operational benefits because the system made the agency look good to potential funders. This was considered a success.

During the initiation phase, the project schedule, funding level, and allocation of funds to different project activities are usually set. Often these decisions have lasting and not entirely positive downstream influences. For instance, an organizational decision to limit access to Lotus Notes to one particular work unit can create difficulties later for activities that cross the boundaries between work units. Or the decision makers may allocate enough resources to acquire or develop a technology but significantly underfund training and support for users, creating predictably negative consequences when the technology is up and running (see Walton [1989]).

In short, the initiation phase of the life cycle can involve political negotiations between people and groups who propose new technologies and people and groups who control essential resources (both technical and financial). These negotiations can result in the selection of inappropriate solutions to organizational problems, or they can result in perceptions and decisions that subsequently have negative effects on system features, use, and impacts.

---

*Line refers to managers in the core businesses or functions of the organization, as opposed to those in staff functions. In most organizations, IS is viewed as a staff function. This is frequently true even in software development firms, where the IS people who provide support for the internal operations of the business are organizationally separate from line software developers who work on the products the company sells.

44.7.2   Acquisition

A project leader is usually appointed to oversee acquisition of a software product. Several difficulties can arise in this phase.

As noted earlier, the project team may define the capabilities required by the interactive system in purely technical terms [Stinchcombe 1990]. The project team may focus on software features, hardware, and networking requirements, but neglect to provide for the training and support of users and necessary changes in organizational aspects such as their job descriptions, their performance evaluations, and their rewards. Likely result: limited user acceptance and poor quality use [Walton 1989].

The project team is often subjected to influence attempts by interested parties: some may desire to build the technology in-house, even with perfectly acceptable packages available. Some may wish to customize a package to unique organizational needs rather than to change the way the organization works. Project team members may have different personal preferences for a technology vendor, a software package, or a platform, or demand that the solution incorporate certain features. To meet schedule and budgets, certain aspects of the project included in the original proposal may be postponed or dropped altogether. In the acquisition phase the innovation becomes more real and more resistant to change. Errors made in initiation can be fixed, or a well-reasoned decision steered astray [Markus and Soh 1993, Soh and Markus 1995].

Finally, another outcome of the acquisition phase is a new set of social relations among various groups inside and outside the organization who will work with or support the technology during the *use* phase: linkages among technology vendors, third-party developers, in-house developers, in-house computer operations and support personnel, or an outsourcing firm, and users and their managers. Thorny support issues may remain unresolved due to the incentives built into outsourcer's service level agreements as negotiated during the acquisition phase. User skills may stagnate or decline, because a one-shot training program did not address the need for advanced training or for initial training for new hires. In short, what results from the acquisition phase may or may not be adequate for the organization's future needs.

44.7.3   Implementation (Introduction) and Use

Activities occurring during the implementation and use phase can quite substantially alter (for better or worse) the capabilities previously acquired. Technology that is made or bought and thrown "over the wall" to users and their managers usually results in a failure of the technology to yield appropriate organizational benefits. Sometimes line managers and users see some potential in the technologies tossed at them and adopt (and support) these technologies as their own.

There is no guarantee that technology will be used in ways consistent with either the initiator's or the project team's vision. The average user of even a modestly complex information technology like the digital telephone system uses only a small fraction of the technology's features [Manross and Rice 1986]. Once they acquire some level of proficiency, users often stop learning new features unless a new release, a conversion, or a major change in work requirements demands new learning [Tyre and Orlikowski 1994]. Often users will enact time consuming and inefficient workarounds [Gasser 1986], rather than invest the time, energy, and pain required to learn a more efficient procedure.

Sometimes users take simple information technologies and *overuse* them, getting more out of them than designers ever intended. Use unanticipated when the technology was first acquired, rather than the initially planned uses, often results in what are subsequently called organizational *transformations:* radical improvements in business processes, first-in-the-world new products and services, and so forth.

The technical term for the process we have been describing, in which users take a technology and redefine it, using it differently than developers, initiators, and implementors intended, is *reinvention* [Rogers 1995] or *emergence* [Markus and Robey 1988]. Reinvention is significant because it means that the use and hence the impacts of a technology can never fully be determined during the acquisition phase. Even when the project team involves users in design and fashions a careful technology implementation plan, both users and developers may fail to see the organizational implications of a technology until the technology itself is real, installed, and running in an organization.* What happens when users get their hands on a technology can never be fully predicted or controlled. Sometimes what emerges during use is much less than vendors, initiators, and implementors had hoped; sometimes it is much more.

44.7.4   Impacts and Performance

Some experts estimate that most benefits obtained from an innovation come from subsequent modifications and enhancements, rather than from the initial change itself [Stinchcombe 1990]. Thus, for example, Frito-Lay did not reap full advantages from its hand-held computer project until it developed analytic tools to improve product promotion decision making and changed the organizational level at which promotion decisions were made [Harvard 1993].

On the other hand, a major reason for the failure of information technology (IT) investments to pay off in terms of improved organizational performance is a tendency for organizations to make nonvalue added improvements in their IT environments [Baily 1986, Baily and Gordon 1988].** The people at VeriFone note, "If you're just using e-mail, there's no reason to have a Pentium. You don't need a Ferrari to drive to the supermarket" [Harvard 1994].

Positive organizational impacts from information technology are said to fall into four categories: new products and services enabled by IT, improved business processes, better organizational decision making attributable to databases and analytic tools, and increased organizational flexibility attributable to communication, collaboration, and coordination technologies [Sambamurthy and Zmud 1992]. However, two issues regarding the impacts of technology must be borne in mind.

First, positive organizational impacts due to technology investments do not always result in improved organizational performance, measured in terms important to various organizational stakeholders [Soh and Markus 1995]. Lack of performance improvement despite positive impacts can occur if the innovation is quickly duplicated by competitors or if the improvements only bring company performance up to existing customer expectations [Arthur 1990, Clemons 1991].

Second, positive organizational impacts are almost invariably accompanied by negative impacts on some dimensions of organizational life [Pool 1983, Rogers 1995]. For instance, the improved organizational efficiency and flexibility attributed to electronic communications technologies such as e-mail may be accompanied by depersonalization, stress, overload, and accountability politics [Sproull and Kiesler 1991, Markus 1994b]. And no matter how much people in an organization value the improvements in organizational functioning, they may still mourn the passing of traditional ways that gave meaning and quality to their working lives.

44.8   Conclusions

Human–computer interaction could postpone reckoning with organizational issues when PCs were computational islands. Today we are networked and on the Internet: The day of reckoning has arrived. Designers, developers, acquirers, users, and researchers must all be cognizant of group and organizational issues to a degree previously unnecessary. The HCI and IS fields are quickly merging. Organizational issues will affect many of us working on human–computer interaction, through the organizational contexts of system introduction and adoption, and the organizational contexts of system development. With this knowledge, frustration often gives away to challenge, and challenge evolves into adventure.

---

*Social scientists repeatedly warn that user participation in design does not ensure success, since participation can lead to incrementalism or recreation of the status quo [Walton 1989, Leonard-Barton 1988, 1990, Markus and Keil 1994].

**This issue and its relationship to usability is explored in detail by Landauer [1995].

References

Abbott, K. R. and Sarin, S. K. 1994. Experiences with workflow management: issues for the next generation, pp. 113–120. *Proc. CSCW '94*.

Arthur, W. B. 1990. Positive feedback in the economy. *Sci. Am.* (Feb.):92–99.

Baily, M. N. 1986. What has happened to productivity growth? *Science* 234:443–451.

Baily, M. N. and Gordon, R. J. 1988. The productivity slowdown, measurement issues and the explosion of computer power. In *Brookings Papers on Economic Activity*. W. C. Brainard and G. L. Perry, Eds. The Brookings Institute, Washington, DC.

Barker, V. and O'Connor, D. 1989. Expert systems for configuration at Digital: XCON and beyond. *Commun. ACM* 32(3):298–318.

Beyer, H. and Holtzblatt, K. 1995. Apprenticing with the customer. *Commun. ACM* 38(5):45–52.

Bjerknes, G., Ehn, P., and Kyng, M., Eds. 1987. *Computers and Democracy — a Scandinavian Challenge*. Gower, Aldershot, UK.

Boehm, B. 1988. A spiral model of software development and enhancement. *IEEE Comput.* 21(5):61–72.

Bowers, J., Button, G., and Sharrock, W. 1995. Workflow from within and without: technology and cooperative work on the print industry shopfloor, pp. 51–66. *Proc. ECSCW '95*.

Bridges, W. 1991. *Managing Transitions*. Addison–Wesley, Reading, MA.

Clemons, E. K. 1991. Evaluation of strategic investments in information technology. *Commun. ACM* 34(1):22–36.

Cohen, M. D., March J. G., and Olsen, J. P. 1972. A garbage can model of organizational choice. *Adm. Sci. Q.* 17:1–25.

Dean, J. W., Jr. 1987. Building for the future: the justification process for new technology. In *New Technology as Organizational Innovation*, J. M. Pennings and A. Buitendam, eds., pp. 35–58. Ballinger, Cambridge, MA.

Friedman, A. L. 1989. *Computer Systems Development: History, Organization and Implementation*. Wiley, Chichester, UK.

Gasser, L. 1986. The integration of computing and routine work. *ACM Trans. Office Inf. Syst.* 4(3):205–225.

Greenbaum, J. and Kyng, M., Eds. 1991. *Design at Work: Cooperative Design of Computer Systems*. Lawrence Erlbaum Associates, Hillsdale, NJ.

Grudin, J. 1991a. Interactive systems: bridging the gaps between developers and users. *IEEE Comput.* 24(4):59–69; republished in *Readings in Human–Computer Interaction: Toward the Year 2000*. R. M. Baecker, J. Grudin, W. A. S. Buxton, and S. Greenberg, Eds. Morgan Kaufmann, San Mateo, CA, 1995.

Grudin, J. 1991b. Systematic sources of suboptimal interface design in large product development organizations. *Hum.–Comput. Interaction* 6(2):147–196.

Grudin, J. 1994. Groupware and social dynamics: eight challenges for developers. *Commun. ACM* 37(1): 92–105.

Grudin, J. 1996. Evaluating opportunities for design capture. In *Design Rationale: Concepts, Techniques, and Use*. T. Moran and J. Carroll, Eds., pp. 453–470. Lawrence Erlbaum, Hillsdale, NJ.

Grudin, J. and Markus, M. L. 1997. Organizational issues in development and implementation of interactive systems. In *Handbook of Human–Computer Interaction*. M. Helander and T. Landauer, Eds. 2nd ed. Springer–Verlag.

Grudin, J. and Poltrock, S. 1995. Software engineering and the CHI & CSCW communities. In *Software Engineering and Human–Computer Interaction*. R. N. Taylor and J. Coutaz, eds. Lecture notes in computer science 896, pp. 93–112. Springer–Verlag, Berlin.

Harvard. 1993. Frito-Lay, Inc.: A Strategic Transition Case (D) 9-193-004. *Harvard Business School*. Cambridge, MA.

Harvard. 1994. VeriFone: The Transaction Automation Company, Case 9-195-088. *Harvard Business School*. Cambridge, MA.

Hirschheim, R., Klein, H. K., and Lyytinen, K. 1995. *Information Systems Development and Data Modeling: Conceptual and Philosophical Foundations*. Cambridge University Press, Cambridge, U.K.

Holtzblatt, K. and Beyer, H. 1993. Making customer-centered design work for teams. *Commun. ACM* 36(10):92–103.

Holtzblatt, K. and Jones, S. 1993. Contextual inquiry: a participatory technique for system design. In *Participatory Design: Principles and Practices*. D. Schuler and A. Namioka, eds. Lawrence Erlbaum Associates, Hillsdale, NJ.

Keil, M. and Carmel, E. 1995. Customer-developer links in software development. *Commun. ACM* 38(5):33–44.

Kling, R. 1978. Automated welfare client-tracking and service integration: the political economy of computing. *Commun. ACM* 21(6):484–493.

Kling, R. and Scacchi, W. 1982. The Web of Computing: Computer Technology as Social Organization. In *Advances in Computers*. M. C. Yovits, ed., pp. 1–89. Academic Press, Orlando, FL.

Landauer, T. K. 1995. *The Trouble with Computers: Usefulness, Usability, and Productivity*. MIT Press, Cambridge, MA.

Leonard-Barton, D. 1988. Implementation as mutual adaptation of technology and organization. *Res. Policy*. 17:251–267.

Leonard-Barton, D. 1990. Implementing new production technologies: exercises in corporate learning. In *Managing Complexity in High Technology Organizations*. Mary Ann Von Glinow and Susan Albers Mohrman, Eds., pp. 160–215. Oxford University Press, New York.

Mahoney, M. S. 1988. The history of computing in the history of technology. *Ann. Hist. Comput.* 10:113–125.

Manross, G. G. and Rice, R. E. 1986. Don't hang up: organizational diffusion of the intelligent telephone. *Inf. Manage.* 10(3):161–175.

Mark, G. and Mambrey, P. 1997. Models and metaphors in groupware: toward a group-centered design. *Interact 1997*. 477–484.

Markus, M. L. 1994a. Electronic mail as the medium of managerial choice. *Organ. Sci.* 5(4):502–527.

Markus, M. L. 1994b. Finding a happy medium: explaining the negative effects of electronic mail on social life at work. *ACM Trans. Inf. Sys.* 12(2):119–149.

Markus, M. L. and Keil, M. 1994. If we build it they will come: designing information systems that users want to use. *Sloan Manage. Rev.* (Summer):11–25.

Markus, M. L. and Robey, D. 1988. Information technology and organizational change: causal structure in theory and research. *Manage. Sci.* 34(5):583–598.

Markus, M. L. and Soh, C. 1993. Banking on information technology: converting IT spending into firm performance. In *Perspectives on the Strategic and Economic Value of Information Technology Investment*. R. D. Banker, R. J. Kauffman, and M. A. Mahmood, Eds., pp. 364–392. Idea Group, Middletown, PA.

Marshak, D. S. 1994. Workflow white paper: an overview of workflow software, pp. 15–42. *Proc. Workflow '94*.

Mintzberg, H. 1984. A typology of organizational structure. In *Organizations: A Quantum View*. D. Miller and P. H. Friesen, eds., pp. 68–86. Prentice–Hall, Englewood Cliffs, NJ.

Mintzberg, H. 1989. *Mintzberg on Management*. Free Press, New York.

Mumford, E. 1983. *Designing Human Systems*. Manchester Business School.

Mumford, E. 1993. The participation of users in systems design: an account of the origin, evolution, and use of the ETHICS method. In *Participatory Design: Principles and Practice*. D. Schuler and A. Namioka, Eds., pp. 257–270. Lawrence Erlbaum Associates, Hillsdale, NJ.

Poltrock, S. E. and Grudin, J. 1994. Organizational obstacles to interface design and development: two participant observer studies. *ACM Trans. Comput.–Hum. Interaction* 1(1):52–80.

Pool, I. d. S. 1983. *Forecasting the Telephone: A Retrospective Technology Assessment of the Telephone*. Ablex, Norwood, NJ.

Rogers, E. M. 1995. *Diffusion of Innovations*, 4th ed. Free Press, New York.

Royce, W. W. 1970. Managing the development of large software systems: concepts and techniques, pp. 1–9. *Proc. IEEE Wescon*.

Sambamurthy, V. and Zmud, R. W. 1992. Managing IT for Success: The Empowering Business Partnership, Financial Executives Research Foundation, Morristown, NY.

Soe, L. L. 1994. *Substitution and Complementarity in the Diffusion of Multiple Electronic Communication Media: An Evolutionary Approach.* Unpublished Ph.D. dissertation. University of California, Los Angeles.

Soh, C. and Markus, M. L. 1995. How IT creates business value: a process theory synthesis. *Proc. Int. Conf. Inf. Sys.*, pp. 29–41. Amsterdam, The Netherlands.

Sproull, L. and Kiesler, S. 1991. *Connections: New Ways of Working in the Networked Organization.* MIT Press, Cambridge, MA.

Stinchcombe, A. L. 1990. *Information and Organizations.* University of California Press, Berkeley, CA.

Tyre, M. J. and Orlikowski, W. J. 1994. Windows of opportunity: temporal patterns of technological adaptation in organizations. *Organ. Sci.* 5(1).

Walton, R. E. 1989. *Up and Running: Integrating Information Technology and the Organization.* Harvard Business School Press, Boston, MA.

Winkler, I. and Buie, E. 1995. HCI challenges in government contracting. *SIGCHI Bull.* 27(4):35–37.

45

Usability Engineering

45.1   Introduction **45**-1
45.2   Know the User **45**-2
         Individual User Characteristics • Task Analysis
         • Functional Analysis • International Use
45.3   Competitive Analysis **45**-5
45.4   Goal Setting **45**-6
         Parallel Design • Participatory Design
45.5   Coordinating the Total Interface **45**-8
45.6   Heuristic Evaluation **45**-9
45.7   Prototyping...................................... **45**-11
45.8   User Testing **45**-13
         The Test Users • Test Tasks • Role of Observers
         • Test Stages • Ethical Issues • Severity Ratings
         • Usability Laboratories
45.9   Iterative Design **45**-18
45.10   Follow-Up Studies of Installed Systems............. **45**-19

Jakob Nielsen
Nielsen Norman Group

45.1   Introduction

Usability engineering [Nielsen 1994b] is not a one-shot event where the user interface is fixed up before the release of a product. Rather, usability engineering is a set of activities that ideally take place throughout the lifecycle of the product, with significant activities happening at the early stages before the user interface has even been designed. The need to have multiple usability engineering stages supplement each other was recognized early in the field, though not always followed by development projects [Gould and Lewis 1985].

Usability cannot be seen in isolation from the broader corporate product development context where one-shot projects are fairly rare. Indeed, usability applies to the development of entire product families and extended projects where products are released in several versions over time. In fact, this broader context only strengthens the arguments for allocating substantial usability engineering resources as early as possible, since design decisions made for any given product have ripple effects due to the need for subsequent products and versions to be backward compatible. Consequently, some usability engineering specialists [Grudin et al. 1987] believe that "human factors involvement with a particular product may ultimately have its greatest impact on future product releases." Of course, having to plan for future versions is also a primary reason to follow up the release of a product with field studies of its actual use.

Table 45.1 shows a summary of the lifecycle stages discussed in this chapter. It is important to note that a usability engineering effort can still be successful even if it does not include every possible refinement at all of the stages.

The lifecycle model emphasizes that one should not rush straight into design. The least expensive way for usability activities to influence a product is to do as much as possible before design is started, since it will

TABLE 45.1   Stages of the Usability
Engineering Lifecycle

1. Know the user
   a. Individual user characteristics
   b. The user's current and desired tasks
   c. Functional analysis
   d. International use
2. Competitive analysis
3. Setting usability goals
4. Parallel design
5. Participatory design
6. Coordinated design of the total interface
7. Heuristic evaluation
8. Prototyping
9. User testing
10. Iterative design
11. Collect feedback from field use.

then not be necessary to change the design to comply with the usability recommendations. Also, usability work done before the system is designed may make it possible to avoid developing unnecessary features. Several of the predesign usability activities might be considered part of a market research or product planning process as well, and may sometimes be performed by marketing groups. However, traditional market research does not usually employ all of the methods needed to properly inform usability design, and the results are often poorly communicated to developers. But there should be no need for duplicate efforts if management successfully integrates usability and marketing activities [Wichansky et al. 1988]. One outcome of such integration could be the consideration of product usability attributes as features to be used by marketing to differentiate the product. Also, marketing efforts based on usability studies can sell the product on the basis of its benefits as perceived by users (*what* it can do that they want) rather than its features as perceived by developers (*how* does it do it).

45.2   Know the User

The first step in the usability process is to study the intended users and use of the product. At a minimum, developers should visit a customer site so that they have a feel for how the product will be used. Individual user characteristics and variability in tasks are the two factors with the largest impact on usability, so they need to be studied carefully. When considering users, one should keep in mind that they often include installers, maintainers, system administrators, and other support staff in addition to the people who sit at the keyboard. The concept of *user* should be defined to include everybody whose work is affected by the production in some way, including the users of the system's end product or output even if they never see a single screen.

Even though "know the user" is the most basic of all usability guidelines, it is often difficult for developers to get access to users. Grudin [1990, 1991a, 1991b] analyzes the obstacles to such access, including:

- The need for the development company to protect its developers from being known to customers, since customers may bypass established technical support organizations and call developers directly, sidetracking them from their main job
- The reluctance of sales representatives to let anybody else from the company talk to their customers, fearing that the developers or usability people may offend the customer or create dissatisfaction with the current generation of products
- User organizations only making users available for a short time, either because they are highly paid executives or because they are unionized and dislike being studied

All of these issues are real and need to be addressed when trying to get to know the user. No universal solutions are available, except to recommend an explicit effort to get direct access to representative users and not be satisfied with indirect access and hearsay. It is amazing how much time is wasted on certain development projects by arguing over what users *might* be like or what they *may* want to do. Instead of discussing such issues in a vacuum, it is much better (and actually less time consuming) to get hard facts from the users themselves.

45.2.1 Individual User Characteristics

It is necessary to know the class of people who will be using the system. In some situations this is easy since it is possible to identify these users as concrete individuals. This is the case when the product is going to be used in a specific department in a particular company. For other products, users may be more widely scattered such that it is possible to visit only a few, representative customers. Alternatively, the products might be aimed toward the entire population or a very large subset.

By knowing the users' work experience, educational levels, ages, previous computer experience, and so on, it is possible to anticipate their learning difficulties to some extent and to better set appropriate limits for the complexity of the user interface. Certainly one also needs to know the reading and language skills of the users. For example, very young children have no reading ability, so an entirely nontextual interface is required. Also, one needs to know the amount of time users will have available for learning and whether they will have the opportunity for attending training courses: The interface must be made much simpler if users are expected to use it within minimum training.

The users' work environment and social context also need to be known. As a simple example, the use of audible alarms, beeps, or more elaborate sound effects may not be appropriate for users in open office environments. In a field interview I once did, a secretary complained strongly that she wanted the ability to shut off the beep because she did not want others to think that she was stupid because her computer beeped at her all the time.

A great deal of the information needed to characterize individual users may come from market analysis or as a side benefit of the observational studies one may conduct as part of the task analysis. One may also collect such information directly through questionnaires or interviews. In may case, it is best not to rely totally on written information since new insights are almost always achieved by observing and talking to actual users in their own working environment.

45.2.2 Task Analysis

A task analysis [Diaper 1989, Fath and Bias 1992, Johnson 1992] is extremely important as early input to the system design. The users' overall goals should be studied as well as how they currently approach the task, what their information needs are, and how they deal with exceptional circumstances or emergencies. For example, systematic observation of users talking to their clients may reveal input and output needs for a transactions-processing system. Sometimes, interviewing or observing the users' clients or others who interact with them can also provide additional task analysis insights [Garber and Grunes 1992].

The users' model of the task should also be identified, since it can be used as a source for metaphors for the user interface. Also, seek out and observe especially effective users and user strategies and *workarounds* as hints of what a new system could support. Such lead users are often a major source of innovations [von Hippel 1988]. Finally, one should try to identify the weaknesses of the current situation: points where users fail to achieve goals, spend excessive time, or are made uncomfortable. These weaknesses present opportunities for improvements in the product being developed.

A typical outcome of a task analysis is a list of all of the things users want to accomplish with the system (the goals), all of the information they will need to achieve these goals (the preconditions), the steps that need to be performed and the interdependencies between these steps, all of the various outcomes and reports that need to be produced, the criteria used to determine the quality and acceptability of these results, and finally the communication needs of the users as they exchange information with others while performing the task or preparing to do so.

When interviewing users for the purpose of collecting task information, it is always a good idea to ask them to show concrete examples of their work products rather than keeping the discussion on an abstract level. Also, it is preferable to supplement such interviews with observations of some users working on real problems, since users will often rationalize their actions or forget about important details or exceptions when they are interviewed.

Often, a task analysis can be decomposed in a hierarchical fashion [Greif 1991], starting with the larger tasks and goals of the organization and breaking each of them down into smaller subtasks, that can again be further subdivided. Typically, each time a user says, "then I do this," an interviewer could ask two questions: "*Why* do you do it?" (to relate the activity to larger goals) and "*How* do you do it?" (to decompose the activity into subtasks that can be further studied). Other good questions to ask include, "Why do you not do this in such and such a manner?" (mentioning some alternative approach you can think of), "Do errors ever occur when doing this?," and "How do you discover and correct these errors?" [Nielsen et al. 1986].

Finally, users should be asked to describe exceptions from their normal work flow. Even though users cannot be expected to remember *all* of the exceptions that have ever occurred, and even though it will be impossible to predict all of the future exceptions, there is considerable value to having a list indicating the *range* of exceptions that must be accommodated. Users should also be asked for remarkable instances of notable successes and failures, problems, what they liked best and least, what changes they would like, what ideas they have for improvements, and what currently annoys them. Even though not all such suggestions may be followed in the final design, they are a rich source of inspiration.

45.2.3 Functional Analysis

A new computer system should not be designed simply to propagate suboptimal ways of doing things that may have been instituted because of limitations in previous technologies. Therefore, one should not analyze just the way users currently do the task, but also the underlying functional reason for the task: What is it that really needs to be done, and what are merely surface procedures which can, and perhaps should, be changed.

As a simple example, initial observations of people reading printed manuals could show them frequently turning pages to move through the document. A naive implementation of on-line documentation might take this observation to indicate the need for really good and fast paging or scrolling mechanisms. A functional analysis would show that manual users really turn pages this much because they want to find specific information, but they have a hard time locating the correct page. Based on this analysis, one could design an on-line documentation interface that first allowed users to specify their search needs, then used an outline of the document to show locations with high search scores, and finally allowed users to jump directly to these locations, highlighting their search terms to make it easier to judge the relevance of the information [Egan et al. 1989]. Of course, there is a limit to how drastically one can change the way users currently approach their task, so the functional analysis should be coordinated with a task analysis.

45.2.4 International Use

A final point related to knowing the users is to plan for any international use of the product from the very beginning of the usability engineering lifecycle [del Galdo and Nielsen 1996]. Some products are only intended for use in a single country, but many development projects need to consider foreign users. Traditionally, internationalization and localization was done after shipping the domestic version of the product, but true international usability requires that international users are considered throughout the lifecycle. Consider, for example, the design of the addressing feature in an e-mail program. If the program is going to be used in a country with an extremely strong sense of hierarchy in the workplace, then customers may require that the addressing feature sort the message recipients by rank, so fitting the program to that culture cannot be done simply by translating the menu items.

45.3   Competitive Analysis

Much of what you need to learn about user interface design for a new product can be gleaned from studies of your competitors' products [Nielsen 1995]. The best prototype of your next product is your own old product since you will presumably want to repeat everything that was good about it and avoid everything that was bad about it. The second-best prototypes are the competing products. Your competitors have invested significant resources in designing and implementing what they believe to be good user interfaces. You should take advantage of those investments.

Please note that I am not suggesting that you violate copyright by cloning your competitors' interface designs. What I *do* suggest is that you can learn a lot by analyzing other products that are designed to solve the same (or related) problems as your own future product. You can see what works and what does not work in these other designs, and you can learn how users approach the tasks by seeing how they work with the competing products. Competitive usability analysis should be performed very early in the usability engineering lifecycle. I would recommend performing competitive usability analysis after the first stages of customer visits, requirements gathering, and defining the product vision, but before you move on to actually designing and prototyping your own user interface.

For competitive usability analysis, I normally recommend acquiring the three or four leading competing products. Often, these products can be bought at a nominal price on the open market, especially in the case of PC software. Even if you are developing for high-end workstations or mainframes, much can still be learned from the design of lower end software even if there is some difference in the supported feature set. To design and develop a prototype yourself will normally take at least a week of engineering time, even for quite low-fidelity prototypes, so a home-grown prototype will cost a minimum of $4000 if the loaded cost of an engineer (whether usability engineer or development engineer) is $100 per hour. Normally one can buy at least 10 commercial software packages for the same money, and these packages will be very high-fidelity prototypes since they are fully functional (even if the features are not exactly the same as the ones you want in your product).

Some classes of products are substantially more expensive than PC software, but it is still often possible to buy evaluation copies, single-user licenses, or other cheap versions of high-end systems. Considering that a competitive usability analysis normally consists of having three or four users use the system for 1 or 2 h each, there is no need to buy the most elaborate version of the competing systems.

If even the cheapest versions of the competing systems are too expensive, you can rely on paper prototyping. Briefly, this method consists of showing users paper printouts of some of the screens from a user interface and asking them to describe what they would do with each screen. A selection of screendumps can usually be acquired from your competitors' sales brochures and so a few hours at a good trade show should suffice to collect more than enough material for an informative usability test.

Often you will have too many competitors to perform a usability study of them all. I usually find that one learns the most from studying four or five competing user interfaces. Three criteria should be used to select the systems that will be subjected to a competitive usability analysis:

- What products have an especially good reputation for good user interface design?
- What products show examples of interesting features or design ideas that you want for your own product?
- Who is the market leader?

Furthermore, you can consider pragmatic issues such as the price of a product and the difficulty of installing and running it on the equipment in your competitive analysis laboratory. If you do not have a competitive analysis laboratory, I highly recommend getting one: buy a computer from each of the major platform families, making sure that it is a high-end model with plenty of memory, a large hard disk, and a compact disc–read-only memory (CD-ROM) drive. You do not want to save on equipment purchases for the competitive analysis laboratory because you will not have the expertise to make each model run optimally: just buy nice big models in vendor-supported configurations. Also buy a good screendump utility for each machine because you will want to include shots of the competing user interfaces in your internal reports and presentations.

The first steps of a competitive usability analysis simply consist of familiarizing yourself with the products and checking how they have designed the features you are contemplating for your product. You can also make lists of user interface elements (commands, features, and attributes) to make sure that you do not overlook something important when designing your own interface.

The next step in competitive usability analysis is a brief usability test where a small number of users are exposed to the various products and asked to perform a few sample tasks. As always with user testing it is important to recruit test users who are representative of the intended user population (that is, the actual end users and not their managers or information systems (IS) support staff — unless, of course, the product is *intended* to have IS personnel as its main users). The tasks should also be chosen to represent the intended usage of the product. For competitive usability analysis, you should select users and tasks that are representative for your future product and not users and tasks that are representative for the other products. After all, the goal is not to evaluate whether the other companies have done a good job designing for *their* customer base but to see what you can learn from their efforts when applied to *your* customer base.

45.4   Goal Setting

Usability is not a one-dimensional attribute of a system. Usability comprises several components, including learnability, efficiency of use, user error rates, and subjective satisfaction, that can sometimes conflict. Normally, not all usability aspects can be given equal weight in a given design project, and so you will have to make your priorities clear on the basis of your analysis of the users and their tasks. For example, learnability would be especially important if new employees were constantly being brought in on a temporary basis, and the ability of infrequent to return to the system would be especially important for a reconfiguration utility that was used once every three or four months.

The different usability parameters can be operationalized and expressed in measurable ways. Before starting the design of a new interface, it is important to discuss the usability metrics of interest to the project and to specify the goals of the user interface in terms of measured usability [Chapanis and Budurka 1990]. One may not always have the resources available to collect statistically reliable measures of the usability metrics specified as goals, but it is still better to have some idea of the level of usability to strive for.

For each usability attribute of interest, several different levels of performance can be specified as part of a goal-setting process [Whiteside et al. 1988]. One would at least specify the minimum level which would be acceptable for release of the product, but a more detailed goal specification can also include the planned level one is aiming for as well as the current level of performance. Additionally, it can help to list the current value of the usability attribute as measured for existing or competing interfaces, and one can also list the theoretically best possible value, even though this value will typically not be attained. Figure 45.1 shows one possible notation, called a *usability goal line*, for representing the range of specification levels for one usability goal.

In the example in Figure 45.1, the number of user errors per hour is counted. When using the current system, users make an average of 4.5 errors/h and the planned number of user errors is 2.0/h. Furthermore,

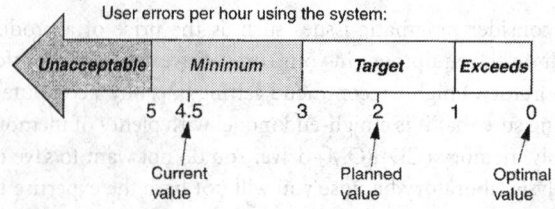

FIGURE 45.1   An example of a usability goal line in a notation similar to that used by Rideout [1991].

the theoretical optimum is obviously to have no errors at all. If the new interface is measured at anything between 1.0 and 3.0 user errors/h, it will be considered on target with respect to this usability goal. A performance in the interval of 3–5 would be a danger signal that the usability goal was not met, even though the new interface could still be released on a temporary basis since a minimal level of usability had been achieved. It would then be necessary to develop a plan to reduce user errors in future releases. Finally, more than 5.0 user errors/h would make this particular product sufficiently unusable to make a release unacceptable.

Usability goals are reasonably easy to set for new versions of existing systems or for systems that have a clearly defined competitor on the market. The minimum acceptable usability would normally be equal to the current usability level, and the target usability could be derived as an improvement that was sufficiently large to induce users to change systems. For completely new systems without any competition, usability goals are much harder to set. One approach is to define a set of sample tasks and ask several usability specialists how long it ought to take users to perform them. One can also get an idea of the minimum acceptable level by asking the users, but unfortunately users are notoriously fickle in this respect; countless projects have failed because developers believed users' claims about what they wanted, only to find that the resulting product was not satisfactory in real use.

45.4.1 Parallel Design

It is often a good idea to start the design with a parallel design process, in which several different designers work out preliminary designs [Nielsen et al. 1993, 1994, Nielsen and Faber 1996]. The goal of parallel design is to explore different design alternatives before one settles on a single approach that can then be developed in further detail and subjected to more detailed usability activities. Figure 45.2 is a conceptual illustration of the relation between parallel and iterative design.

Typically, one can have three or four designers involved in parallel design. For critical products, some large computer companies have been known to devote entire teams to developing multiple alternative designs almost to the final product stage, before upper management decided on which version to release. In general, though, it may not be necessary for the designers to spend more than a few hours or at the most one or two days on developing their initial designs. Also, it is normally better to have designers work individually rather than in teams, since parallel design only aims at generating rough drafts of the basic design ideas.

In parallel design, it is important to have the designers (or the design teams) work independently, since the goal is to generate as much diversity as possible. Therefore, the designers should not discuss their designs with each other until after they have produced their draft interface designs.

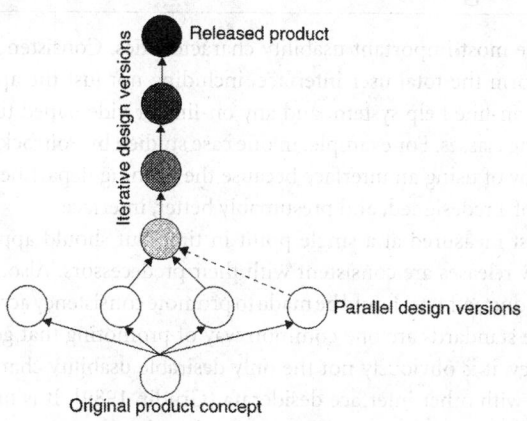

FIGURE 45.2 Conceptual illustration of the relation between parallel and iterative design. Normally, the first prototype would be based on ideas from several of the parallel design sketches.

When the designers have completed the draft designs, one will often find that they have approached the problem in at least two drastically different ways that would give rise to fundamentally different user interface models. Even those designers who are basing their designs on the same basic approach almost always have different details in their designs. Usually, it is possible to generate new combined designs after having compared the set of initial designs, taking advantage of the best ideas from each design. If several fundamentally different designs are available, it is preferable to pursue each of the main lines of design a little further in order to arrive at a small number of prototypes that can be subjected to usability evaluation before the final approach is chosen.

A variant of parallel design is called *diversified parallel design* and is based on asking the different designers to concentrate on different aspects of the design problem. For example, one designer could design an interface that was optimized for novice users, at the same time as another designer designed an interface optimized for expert users and a third designer explored the possibilities of producing an entirely nonverbal interface. By explicitly directing the design approach of each designer, diversified parallel design drives each of these approaches to their limit, leading to design ideas that might never have emerged in a unified design. Of course, some of these diversified design ideas may have to be modified to work in a single, integrated design.

It is especially important to employ parallel design for novel systems where little guidance is available for what interface approaches work the best. For more traditional systems, where competitive products are available, the competitive analysis previously discussed can serve as initial parallel designs, but it might still be advantageous to have a few designers create additional parallel designs to explore further possibilities.

The parallel design method might at first seem to run counter to the principle of cost-effective usability engineering, since most of the design ideas will have to be thrown away without even being implemented. In reality, though, parallel design is a very cheap way of exploring the design space, exactly *because* most of the ideas will not need to be implemented, the way they might be if some of them were not tried until later as part of the iterative design. The main financial benefit of parallel design is its parallel nature, which allows several design approaches to be explored at the same time, thus compressing the development schedule for the product and bringing it to market more rapidly. Studies have shown that about a third of the profits are lost when products ship as little as half a year late [House and Price 1991], and so anything that can speed up the development process should be worth the small additional cost of designing in parallel rather than in sequence.

45.4.2 Participatory Design

Participatory design is discussed further in Chapter 66 and elsewhere and will not be covered here.

45.5 Coordinating the Total Interface

Consistency is one of the most important usability characteristics. Consistency should apply across the different media which form the total user interface, including not just the application screens but also the documentation, the on-line help system, and any on-line or videotaped tutorials [Perlman 1989] as well as traditional training classes. For example, in one case studied by Poltrock [1996], training materials described an obsolete way of using an interface because the training department had not been informed about the introduction of a redesigned, and presumably better, interface.

Consistency is not just measured at a single point in time but should apply over successive releases of a product so that new releases are consistent with their predecessors. Also, since very few companies produce only a single product, efforts should be made to promote consistency across entire product families. Corporate user interface standards are one common way of promoting that goal. In spite of the general desirability of consistency, it is obviously not the only desirable usability characteristic, and consistency may sometimes conflict with other interface desiderata [Grudin 1989]. It is necessary to maintain some flexibility so that bad design is not forced upon users for the sake of consistency alone.

To achieve consistency of the total interface it is necessary to have some centralized authority for each development project to coordinate the various aspects of the interface. Typically, this coordination can be done by a single person, but on very large projects or to achieve corporatewide consistency, a committee structure may be more appropriate. Also, interface standards are an important approach to achieving consistency. In addition to such general standards, a project can develop its own ad hoc standard with elements such as a dictionary of the appropriate terminology to be used in all screen designs as well as in the other parts of the total interface.

In addition to formal coordination activities, it is helpful to have a shared culture in the development groups with common understanding of what the user interface should be like. Many aspects of user interface design (especially the dynamics) are hard to specify in written documents but can be fairly easily understood from looking at existing products following a given interface style. Actually, prototyping also helps achieve consistency, since the prototype is an early statement of the kind of interface toward which the project is aiming. Having an explicit instance of parts of the design makes the details of the design more salient for developers and encourages them to follow similar principles in subsequent design activities [Bellantone and Lanzetta 1991].

Furthermore, consistency can be increased through technological means such as code sharing or a constraining development environment. When several products use the same code for parts of their user interface, then those parts of the interface automatically will be consistent. Even if identical code cannot be used, it is possible to constrain developers by providing development tools and libraries that encourage user interface consistency by making it easier to implement interfaces that follow given guidelines [Tognazzini 1989, Wiecha et al. 1989].

45.6 Heuristic Evaluation

Guidelines list well-known principles for user interface design which should be followed in the development project. In any given project, several different levels of guidelines should be used: *general guidelines* applicable to all user interfaces, *category-specific guidelines* for the kind of system being developed (e.g., guidelines for window-based administrative data processing or for voice interfaces accessed through telephone keypads), and *product-specific guidelines* for the individual product. All of these guidelines can be used as background for heuristic evaluation.

Heuristic evaluation [Nielsen and Molich 1990, Nielsen 1994a] is the most popular of the usability inspection methods [Nielsen and Mack 1994]. Heuristic evaluation is done by looking at an interface and trying to come up with an opinion about what is good and bad about the interface. Heuristic evaluation should be done as a systematic inspection of a user interface design for usability. The goal of heuristic evaluation is to find the usability problems in a user interface design so that they can be attended to as part of an iterative design process. Heuristic evaluation involves having a small set of evaluators examine the interface and judge its compliance with recognized usability principles (the heuristics). A recommended set of heuristics is listed in Table 45.2.

In principle, individual evaluators can perform a heuristic evaluation of a user interface on their own, but the experience from several projects indicates that any single evaluator will miss most of the usability problems in an interface. Averaged over six projects, single evaluators found only 35% of the usability problems in the interfaces. However, since different evaluators tend to find different problems, it is possible to achieve substantially better performance by aggregating the evaluations from several evaluators. There is a nice payoff from using more than one evaluator, and it is recommended to use about five evaluators, and certainly at least three. The exact number of evaluators to use would depend on a cost-benefit analysis, and more evaluators should obviously be used in cases where usability is critical or when large payoffs can be expected due to extensive or mission-critical use of a system.

Heuristic evaluation is performed by having each individual evaluator inspect the interface alone. Only after all evaluations have been completed are the evaluators allowed to communicate and have their findings aggregated. This procedure is important in order to ensure independent and unbiased evaluations from each evaluator. The results of the evaluation can be recorded either as written reports from each evaluator or by having an observer present during the evaluation sessions and having the evaluators vocalize their comments as they go through the interface. Written reports have the advantage of presenting a formal

TABLE 45.2   List of 10 Heuristics for Good User Interface Design

---

Visibility of system status: The system should always keep users informed about what is going on, through appropriate feedback within reasonable time.

Match between system and the real world: The system should speak the users' language, with words, phrases, and concepts familiar to the user, rather than system-oriented terms. Follow real-world conventions, making information appear in a natural and logical order.

User control and freedom: Users often choose system functions by mistake and will need a clearly marked emergency exit to leave the unwanted state without having to go through an extended dialogue. Support undo and redo.

Consistency and standards: Users should not have to wonder whether different words, situations, or actions mean the same thing. Follow platform conventions.

Error prevention: Even better than good error messages is a careful design which prevents a problem from occurring in the first place.

Recognition rather than recall: Make objects, actions, and options visible. The user should not have to remember information from one part of the dialogue to another. Instructions for use of the system should be visible or easily retrievable whenever appropriate.

Flexibility and efficiency of use: Accelerators — unseen by the novice user — may often speed up the interaction for the expert user such that the system can cater to both inexperienced and experienced users. Allow users to tailor frequent actions.

Aesthetic and minimalist design: Dialogues should not contain information which is irrelevant or rarely needed. Every extra unit of information in a dialogue competes with the relevant units of information and diminishes their relative visibility.

Help users recognize, diagnose, and recover from errors: Error messages should be expressed in plain language (no codes), precisely indicate the problem, and constructively suggest a solution.

Help and documentation: Even though it is better if the system can be used without documentation, it may be necessary to provide help and documentation. Any such information should be easy to search, focused on the users' tasks, list concrete steps to be carried out, and not be too large.

---

record of the evaluation, but require an additional effort from the evaluators and also need to be read and aggregated by an evaluation manager. Using on observer adds to be overhead of each evaluation session but reduces the workload on the evaluators and provides the opportunity for having the result of the evaluation available fairly soon after the last evaluation session since the observer only needs to understand and organize his or her own notes and not a set of reports written by others. Furthermore, the observer can assist the evaluators in operating the interface in case of problems with, e.g., an unstable prototype, and help if the evaluators have limited domain expertise and need to have certain aspects of the interface explained.

Typically, a heuristic evaluation session for an individual evaluator lasts one or two hours. Longer evaluation sessions might be necessary for larger or very complicated interfaces with a substantial number of dialogue elements, but it is likely that it would be better to split up the evaluation in several smaller sessions, each concentrating on a part of the interface.

During the evaluation session, the evaluator goes through the interface several times and inspects the various dialogue elements and compares them with a list of recognized usability principles. These heuristics are general rules that seem to describe common properties of usable interfaces and it is possible to apply lists of heuristics that are more specialized for the particular application domain than the general list given in Table 45.2. In addition to the checklist of heuristics to be considered for all dialogue elements, the evaluator obviously is also allowed to consider any additional usability principles or results that come to mind that may be relevant for any specific dialogue element.

In principle, the evaluators decide on their own how they want to proceed with evaluating the interface. A general recommendation would be that they go through the interface at least twice, however. The first pass would be intended to get a feel for the flow of the interaction and the general scope of the system. The second pass then allows the evaluator to focus on specific interface elements while knowing how they fit the larger whole.

Since the evaluators are not *using* the system as such (to perform a real task), it is possible to perform heuristic evaluation of user interfaces that exist on paper only and have not yet been implemented [Nielsen 1990]. This makes heuristic evaluation suitable for use early in the usability engineering lifecycle.

45.7 Prototyping

One should not start full-scale implementation efforts based on early user interface designs. Instead, early usability evaluation can be based on prototypes of the final systems that can be developed much faster and much more cheaply, and which can thus be changed many times until a better understanding of the user interface design has been achieved.

In traditional models of software engineering most of the development time is devoted to the refinement of various intermediate work products, and executable programs are produced at the last possible moment. A problem with this *waterfall* approach is that there will then be no user interface to test with real users until this last possible moment, since the intermediate work products do not explicitly separate out the user interface in a prototype with which users can interact. Experience also shows that it is not possible to involve the users in the design process by showing them abstract specifications documents, since they will not understand them nearly as well as concrete prototypes.

The entire idea behind prototyping is to save on the time and cost to develop something that can be tested with real users. These savings can only be achieved by somehow reducing the prototype compared with the full system: either cutting down on the number of features in the prototype or reducing the level of functionality of the features such that they *seem* to work but do not actually *do* anything.

Reducing the number of features is called *vertical prototyping* since the result is a narrow system that does include in-depth functionality, but only for a few selected features. A vertical prototype can thus only test a limited part of the full system, but it will be tested in depth under realistic circumstances with real user tasks. For example, for a test of a website, in-depth functionality would mean that a user would actually access a set of documents with real content from the information providers.

Reducing the level of functionality is called *horizontal prototyping* since the result is a surface layer that includes the entire user interface to a full-featured system but with no underlying functionality. A horizontal prototype is a simulation [Life et al. 1990] of the interface where no real work can be performed. In the Web example, this would mean that users should be able to execute all navigation and search commands but without retrieving any real documents as a result of these commands. Horizontal prototyping makes it possible to test the entire user interface, even though the test is of course somewhat less realistic, since users cannot perform any real tasks on a system with no functionality. The main advantages of horizontal prototypes are that they can often be implemented fast with the use of various prototyping and screen design tools and that they can be used to assess how well the entire interface hangs together and feels as a whole.

Finally, one can reduce both the number of features and the level of functionality to arrive at a scenario that is only able to simulate the user interface as long as the test user follows a previously planned path. Scenarios are extremely easy and cheap to build, while at the same time not being particularly realistic. Scenarios are discussed further in other chapters.

In addition to reducing the proportion of the system that is implemented, prototypes can be produced faster by the following.

- Placing less emphasis on the efficiency of the implementation. For example, it will not matter how much disk space the prototype uses since it will only be used for a short time. Similarly, test users may be able to cope with slow response times that would never be acceptable in the final product. Note, however, that response times are an important aspect of usability and that test users may get very frustrated and make errors if the prototype is *too* slow. Of course, efficiency measures of the users' performance will be invalid if the prototype slows them down too much, so inefficient prototypes are better suited for early evaluation of interface concepts than for measurement studies.
- Accepting less reliable or poorer quality code. Even though bugs and crashes do distract users during testing, they can often be compensated for by the experimenter.
- Using simplified algorithms that cannot handle all of the special cases (such as leap years) that normally require a disproportionately large programming effort to get right.

- Using a human expert operating behind the scenes to take over certain computer operations that would be too difficult to program. This approach is often referred to as the *Wizard of Oz technique* after the "pay no attention to that man behind the curtain" scene in this story. Basically, the user interacts normally with the computer, but the users' input is not relayed directly to the program. Instead, the input is transmitted to the wizard who, using another computer, transforms the users' input into an appropriate format. A famous early Wizard of Oz study was the listening typewriter [Gould et al. 1983] simulation of a speech recognition interface where the users' spoken input was typed into a word processor by a human typist located in another room. When setting up a Wizard of Oz simulation, experience with previously implemented systems is helpful in order to place realistic bounds on the wizard's abilities [Maulsby et al. 1993].

- Using a different computer system than the eventual target platform. Often, one will have a computer available that is faster or otherwise more advanced than the final system and which can therefore support more flexible prototyping tools and require less programming tricks to achieve the necessary response times.

- Using low-fidelity media [Virzi 1989] that are not as elaborate as the final interface but still represent the essential nature of the interaction. For example, a prototype hypermedia system could use scanned still images instead of live video for illustrations.

- Using fake data and other content. For example, a prototype of a hypermedia system that was intended to include heavy use of video could use existing video material, even though it did not exactly match the topic of the text, in order to get a feel for the interaction techniques needed to deal with live images. A similar technique is used in the advertising industry, where so-called ripomatics are used as rudimentary television commercials with existing shots from earlier commercials to demonstrate concepts to clients before they commit to pay for the shooting of new footage.

- Using paper mockups instead of a running computer system. Such mockups are usually based on printouts of screen designs, dialogue boxes, pop-up menus, etc., that have been drawn up in some standard graphics or desktop publishing package. They are made into functioning prototypes by having a human play computer and find the next screen or dialogue element from a big pile of paper whenever the user indicates some action. This human needs to be an expert in the way the program is intended to work since it is otherwise difficult to keep track of the state of the simulated computer system and find the appropriate piece of paper to respond to the users' stated input.

- Paper mockups have the further advantage that they can be shown to larger groups on overhead projectors [Rowley and Rhoades 1992] and used in conditions where computers may not be available, such as customer conference rooms. Portable computers with screen projection attachments confer some of the same advantages to computerized prototypes, but also increase the risk of something going wrong.

- Relying on a completely imaginary prototype where the experimenter describes a possible interface to the user orally, posing a series of "what if (the interface did this or that) ... " questions as the user steps though an example task. This verbal prototyping technique has been called *forward scenario simulation* [Cordingley 1989] and is more akin to interviews or brainstorming than a true prototyping technique.

A prototype is a form of design specification, and the final implementation of a user interface is often performed with the prototype as a major way of communicating the design to developers. Unfortunately, the prototype can be *over* specified in some aspects that are not really intended to be part of the design. Whenever something is made concrete, there is a need to instantiate a multitude of representational details that might not have been explicitly designed by anybody. For example, a screen design will have to use certain colors and fonts, even though the designer's focus may have been on the wording and positioning of the dialogue elements. Basically, one needs to be aware that not every aspect of the prototype should be replicated in the final system, and the designers should inform developers about which aspects of the prototype are intentional and which are arbitrary.

45.8 User Testing

The most basic advice with respect to interface evaluation is simply to *do it*, and especially to conduct some user testing. The benefits of employing some reasonable usability engineering methods to evaluate a user interface rather than releasing it without evaluation are much larger than the incremental benefits of using exactly the right methods for a given project.

User testing with real users is the most fundamental usability method and is in some sense irreplaceable, since it provides direct information about how people use computers and what their exact problems are with the concrete interface being tested. Even so, other usability engineering methods [Nielsen 1994b] can serve as good supplements to gather additional information or to gain usability insights at a lower cost. In particular, user testing can often be combined with heuristic evaluation (discussed previously) or other usability inspection methods for greater efficiency: the heuristic evaluation will find many usability problems that should be cleaned up before presenting the design to real users.

The three main rules of user testing are:

- Get real users
- Have them do real tasks
- Shut up while they are trying

45.8.1 The Test Users

Your test users should accurately represent the system's intended users. You cannot simply test with the engineer in the neighboring office, despite his or her suspicious resemblance to a real person. Actually, much can be learned from testing with other engineers and it is also possible to gain substantial insight into potential usability problems by having other experts inspect a design, but a user test should always involve real users who have no special software skills.

Sometimes, during development, the specific individuals who will use the completed system can be identified. This is typically the case when a company is developing a system internally for use by a given department. This makes representative users easy to find, although it may be difficult to have them spend time on user testing instead of their primary job. Internal test users are often recruited through the users' management, who agree to provide a certain number of people. Unfortunately, managers often tend to select their most able staff members for such tests, either to make their department look good or because these staff members have the most interest in new technology. Thus, you should explicitly ask managers to choose a broad test sample based on characteristics such as experience and seniority.

In other cases, the system is designed for a certain user type, such as lawyers, secretaries in a dental clinic, or warehouse managers in small manufacturing companies. These groups can be more or less homogeneous, although it still may be desirable to involve test users from several different locations. Sometimes, existing customers will help with the test because doing so gives them an early look at new software and improves the quality of the resulting product, which they will be using.

But sometimes, no existing customers will be available, making it difficult to gain access to representative users. Test users can then be recruited from temporary employment agencies, or students in the application domain may be attending a local university or trade school. You may also recruit users who are currently unemployed by placing a classified advertisement under job openings. Of course, it will be necessary to pay all users thus recruited.

45.8.2 Test Tasks

Your users should perform specific tasks and not just try out the system. The experience of doing real work is different from that of simply dabbling with the software. For example, a user who plays around and discovers a menu with lots of obscure choices may try one or two commands and will then sound pleasantly surprised if something nice happens. The same user will be completely baffled when viewing that menu in the context of having to perform a specific task that does not map to the menu structure presented.

Test tasks should closely represent the uses to which the installed system will be put. Also, the tasks should provide reasonable coverage of the most important parts of the interface. You can design the test tasks based on a task analysis or on a product-identity statement that lists the product's intended uses. Information that helps you learn how users actually use systems — such as logging the frequency of use for specific commands in existing, similar systems, or direct field observation — can also help construct more representative test task sets for user testing.

The tasks must be small enough to be completed within the time limits of your user test, but they should not be so small they become trivial. For example, a good test task for a spreadsheet might be to enter sales figures for six regions through each of four quarters, using the sample numbers given in the task description. A second test task could be to obtain totals and percentages from the data entered, and a third might be to construct a bar chart showing trends across the six regions.

You should give all test users written task descriptions. This ensures they all receive identical information and also lets them refer to the description during the experiment. After the user receives the task descriptions and has a chance to read them, the tester should allow questions. Normally, task descriptions are distributed in printed form, but they can also be shown on line in a help window. The latter approach works best in computer-paced tests that require users to perform many tasks.

45.8.3 Role of Observers

During testing, the tester should not interface with users, but should let them discover the solutions to problems on their own. Not only does this lead to more valid and interesting test results, it also prevents users from feeling that they are so stupid the tester must solve the problems for them. On the other hand, the tester should not let users struggle endlessly with a task if they are clearly frustrated. In such cases, the tester can gently provide a hint or two to keep the test moving.

Mainly, though, observers should follow one simple rule during a user test: shut up and let the user do the talking. It is common for observers to offer help too quickly. It is human to want to assist a person who is struggling with a system (especially if you designed it), but doing so ruins the study.

45.8.4 Test Stages

A usability test typically has four stages:

1. Preparation
2. Introduction
3. The test itself
4. Debriefing

In preparation for the test, the tester should make sure the test room is prepared, the computer system is in the start specified in the test plan, and that all test materials, instructions, and questionnaires are available. For example, all files needed for the test tasks should be restored to their original content, and any files created during earlier tests should be moved to another computer or at least to another directory. To minimize users' discomfort and confusion, this preparation should be completed before their arrival. Also, any screen savers should be switched off, as should any other system components — such as e-mail notifiers — that might otherwise interrupt the test.

During the introduction, the tester welcomes the test users, gives a brief explanation of the test's purpose, explains the computer setup if it is unfamiliar to the users, and introduces the test procedure. After the introduction, the tester distributes any written instructions for the test, including the first test task, then asks before the start of the test if there are any questions regarding the test procedure, the instructions, or tasks.

Normally, to obtain the most feedback, the tester asks users to think aloud continuously during the test itself. By verbalizing their thoughts, test users help us understand how they view the computer system, which makes it easy to identify their major misconceptions. You get a very direct understanding of what parts of the dialogue cause the most problems, because the thinking-aloud method shows how users interpret each interface item. Thinking aloud should not be used if the test aims at gathering performance data, however, because users may be slowed by having to verbalize.

Thinking aloud feels unnatural to most people, and some test users find it difficult to make a steady stream of comments as they use a system. The tester may need to continuously prompt the user to think aloud by asking questions like, "What are you thinking now?" or, when a user spends more than a second or two on a particular window or dialogue box, "What do you think that message means?"

If the user asks a question like, "Can I do such-and-such?" the tester should not answer, but should instead keep the user talking with a counter-question like, "What do you think will happen if you do so?" If the user acts surprised after a system action but does not otherwise say anything, the tester may prompt the user with a question like, "Is that what you expected would happen?" Of course, following the general principle of not interfering in the users' use of the system, the tester should not use prompts like, "What do you think the message on the bottom of the screen means?" if the user has not appeared to notice that message yet.

After the test, the tester debriefs users and asks them to fill out any user-satisfaction questionnaires. To eliminate tester comments influencing the results, the questionnaires should be distributed before any further discussion of the system. During debriefing, ask users for comments about the system and suggested improvements. Such suggestions may not always lead to specific design changes; you will often find that different users make completely contradictory suggestions, but overall, this type of user suggestion can serve as a rich source of additional ideas to consider in the redesign.

45.8.5   Ethical Issues

Although usability test subjects normally escape actual bodily harm — even from irate developers resenting the users' mistreatment of their beloved software — test participation can still be quite distressing. Users feel a tremendous pressure to perform, even when told the study's purpose is to test the system and not the user. Also, users inevitably make errors and are slow to learn the system, especially when testing early designs that may be burdened with severe usability problems. Users can easily feel inadequate or stupid as they experience these difficulties. Knowing they are being observed, and possibly recorded, makes the feeling of performing inadequately even more unpleasant. On rare occasions, users have been known to cry during usability testing.

The tester is responsible for making the users feel as comfortable as possible during and after the test. Specifically, the tester must never laugh at the users or in any way indicate they are slow at discovering how to operate the system. During the test introduction, the tester should stress that the system is being tested, not the user. To reinforce this, test users should never be referred to as subjects, guinea pigs, or similar terms. More appropriate terms include participant and test user.

45.8.6   Severity Ratings

From whatever evaluation methods are used, a major result will be a list of the usability problems in the interface as well as hints for features to support successful user strategies. It is normally not feasible to solve all of the problems, and so one will need to prioritize them. Priorities are best based on experimental data about the impact of the problems on user performance (e.g., how many people will experience the problem and how much time each of them will waste because of it), but sometimes it is necessary to rely on intuitions only.

Severity ratings are usually gathered by sending a group of usability specialists a list of the usability problems discovered in the interface and asking them to rate the severity of each problem. Sometimes, the severity raters are given access to use the system while making their estimates, and sometimes they are asked to judge the problems based only on written description. Note that the latter approach is possible because the severity raters are supposed to be usability specialists. They should therefore be able to visualize the

TABLE 45.3 Table to Estimate the Severity of Usability Problems Based on the Frequency with Which the Problem Is Encountered by Users and the Impact of the Problems on Those Users Who Do Encounter It

Impact of problem on the users who experience it	Proportion of users experiencing the problem	
	Few	*Many*
Small	Low severity	Medium severity
Large	Medium severity	High severity

interface based on the written description (and possibly some screendumps) in a way that regular users would normally not be able to do. Typically, evaluators need only spend about 30 min to provide their severity ratings, though more time may of course be needed if the list of usability problems is extremely long. It is important to note that each usability specialist should provide the individual severity ratings independently of the other evaluators.

Two common approaches to severity ratings are either to have a single scale or to use a combination of several orthogonal scales. A single rating scale for the severity of usability problems might be:

- 0 = This is not a usability problem at all.
- 1 = Cosmetic problem only; need not be fixed unless extra time is available on project.
- 2 = Minor usability problem; fixing this should be given low priority.
- 3 = Major usability problem; important to fix, so should be given high priority.
- 4 = Usability catastrophe; imperative to fix this before product can be released.

Alternatively, severity can be judged as a combination of the two most important dimensions of a usability problem: how many users can be expected to have the problem and what is the extent to which those users who do have the problem are hurt by it. A simple example of such a rating scheme is given in Table 45.3. Of course, both dimensions in the table can be estimated at a finer resolution, using more categories than the two shown here for each dimension. Both the proportion of users experiencing a problem and the impact of the problem can be measured directly in user testing. A fairly large number of test users would be needed to measure reliably the frequency and impact of rare usability problems, but from a practical perspective, these problems are less important than more commonly occurring usability problems, so it is normally acceptable to have lower measurement quality for rare problems.

If no user test data is available, the frequency and impact of each problem can be estimated heuristically by usability specialists, but such estimates are probably best when made on the basis of at least a small number of user observations.

One can add a further severity dimension by judging whether a given usability problem will be a problem only the first time it is encountered or whether it will persistently bother users. For example, consider a set of pulldown menus where all of the menus are indicated by single words in the menubar except for a single menu that is indicated by a small icon (as, for example, the Apple menu on the Macintosh). Novice users of such systems can often be observed not even trying to pull down this last menu, simply because they do not realize that the icon is a menu heading. As soon as somebody shows the users that there is a menu under the icon (or if they read the manual), they immediately learn to overcome this small inconsistency and have no problems finding the last menu in future use of the system. This problem is thus not a persistent usability problem and would normally be considered less severe than a problem that also reduced the usability of the system for experienced users.

45.8.7 Usability Laboratories

Many user tests and other usability engineering activities take place in specially equipped usability laboratories [Nielsen 1994c]. Figure 45.3 shows one of Sun Microsystem's usability laboratories: the participant room and the control room. I should stress from the beginning that special laboratories are a convenience

FIGURE 45.3 View of the control room in a usability laboratory. The participant room is visible through the one-way mirror.

but not an absolute necessity for usability testing. It is possible to convert a regular office temporarily into a usability laboratory, and it is possible to perform usability testing with no more equipment than a notepad.

In September 1993, I surveyed 13 usability laboratories from a variety of companies [Nielsen 1994c]. The median floor space of the laboratories was 63 m² (678 ft²), and the median size of the test rooms was 13 m² (144 ft²). The smallest laboratory was 35 m² (377 ft²) with only 9 m² (97 ft²) for the test user. The largest laboratory was 237 m² and had 7 rooms, allowing a variety of tests to take place simultaneously [Lund 1994]. The largest single test room was 40 m² (430 ft²) and was found in a telephone company with a need to test groupware interfaces with many users. Even though the survey was conducted in the end of 1993, it seems that more recently built usability laboratories have about the same characteristics as the ones in the survey.

Having a permanent usability laboratory decreases the overhead of usability testing (once it is set up, that is!) and may thus encourage increased usability testing in an organization. Having a special room and special equipment dedicated to usability testing means that there will be fewer scheduling problems associated with each test and also makes it possible to run tests without disturbing other groups.

Usability laboratories typically have soundproof, one-way mirrors separating the observation room from the test room to allow the experimenters, other usability specialists, and the developers to discuss user actions without disturbing the user. Users are not so stupid that they do not know that there are observers behind a wall with a large mirror in a test room, so one might as well briefly show the users the observation room before the start of the test. Knowing who and what are behind the mirror is much less stressful for the users than having to imagine it. People usually come to ignore unseen observers during the test, even though they know they are there.

Having an executive observation area in the back of the main observation area allows a third group of observers (e.g., the development team) to discuss the test without disturbing the primary experimenters and the usability specialists.

Typically, a usability laboratory is equipped with several video cameras under remote control from the observation room: the average number of cameras in each test room was 2.2 in my survey, with 2 cameras being the typical number and a few labs using 1 or 3. These cameras can be used to show an overview of the test situation and to focus in on the users' face, the keyboard, the manual and the documentation, and the screen. A producer in the observation room then typically mixes the signal from these cameras to a single video stream that is recorded, and possibly time stamped for later synchronization with an observation log entered into a computer during the experiment. Such synchronization makes it possible to later find the video segment corresponding to a certain interesting user event without having to review the entire videotape.

In many ways, the most important equipment in a usability laboratory is the "do not enter" sign on the door since it makes it possible to conduct the usability test without interruptions. As long as one has a room with a do-not-disturb sign, one can conduct usability tests without any further equipment (you won't even need a computer if you are doing paper prototyping!). The second most important piece of equipment may be high-quality microphones: since there is normally a good deal of background noise from the computer, it will be impossible to hear what the user is saying unless professional microphones are used and unless the user is actually wearing the microphone.

45.9   Iterative Design

Based on the usability problems and opportunities disclosed by the empirical testing, one can produce a new version of the interface. Some testing methods such as thinking aloud provide sufficient insight into the nature of the problems to suggest specific changes to the interface in many cases. Log files of user interaction sequences often help by showing where the user paused or otherwise wasted time, and what errors were encountered most frequently. It often also helps if one is able to understand the underlying cause of the usability problem by relating it to established usability principles such as those listed in Table 45.2. In other cases alternative potential solutions need to be designed based solely on knowledge of usability guidelines, and it may be necessary to test several possible solutions before making a decision. Familiarity with the design options, insight gained from watching users, creativity, and luck are all needed at this point.

Some of the changes made to solve certain usability problems may fail to solve the problems. A revised design may even introduce new usability problems [Bailey 1993]. This is yet another reason for combining iterative design and evaluation. In fact, it is quite common for a redesign to focus on improving one of the usability parameters (for example, reducing the users' error rate), only to find that some of the changes have adversely impacted other usability parameters (for example, transaction speed).

In some cases, solving a problem may make the interface worse for those users who do not experience the problem. Then a tradeoff analysis is necessary as to whether to keep or change the interface, based on a frequency analysis of how many users will have the problem compared to how many will suffer because of the proposed solution. The time and expense needed to fix a particular problem is obviously also a factor in determining priorities. Often, usability problems can be fixed by changing the wording of a menu item or an error message. Other design fixes may involve fundamental changes to the software (which is why they should be discovered as early as possible) and will only be implemented if they are judged to impact usability significantly.

Furthermore, it is likely that additional usability problems appear in repeated tests after the most blatant problems have been corrected. There is no need to test initial designs comprehensively since they will be changed anyway. The user interface should be changed and retested as soon as a usability problem has been detected and understood, so that those remaining problems that have been masked by the initial glaring problems can be found.

I surveyed four projects that had used iterative design and had tested at least three user interface versions [Nielsen 1993]. The median improvement in usability per iteration was 38%, though with extremely high variability. In fact, in 5 of the 12 iterations studied, there was at least one usability metric that had gotten *worse* rather than better. This result certainly indicates the need to keep iterating past such negative results and to plan for at least three versions, since version two may not be any good. Also, the study showed that considerable additional improvements could be achieved after the first iteration, again indicating the benefits of planning for multiple iterations.

During the iterative design process it may not be feasible to test each successive version with actual users. The iterations can be considered a good way to evaluate design ideas simply by trying them out in a concrete design. The design can then be subjected to heuristic analysis and shown to usability experts and consultants or discussed with expert users (or teachers in the case of learning systems). One should not waste users by performing elaborate tests of every single design idea, since test subjects are normally hard to come by and should therefore be conserved for the testing of major iterations. Also, users get worn out as appropriate test subjects as they get more experience with the system and stop being representative of novice users seeing the design for the first time. Users who have been involved in participatory design are especially inappropriate as test subjects, since they will be biased.

45.10 Follow-Up Studies of Installed Systems

The main objective of usability work after the release of a product is to gather usability data for the next version and for new, future products. In the same way that existing and competing products were the best prototypes for the product in the initial competitive analysis phase, a newly released product can be viewed as a prototype of future products. Studies of the use of the product in the field assess how real users use the interface for naturally occurring tasks in their real-world working environment and can therefore provide much insight that would not be easily available from laboratory studies.

Sometimes, field feedback can be gathered as part of standard marketing studies on an ongoing basis. As an example, an Australian telephone company collected customer satisfaction data on a routine basis and found that overall satisfaction with the billing service had gone up from 67% to 84% after the introduction of a redesigned bill printout format developed according to usability engineering principles [Sless 1991]. If the trend in customer satisfaction had been the opposite, there would have been reason to doubt the true usability of the new bill outside the laboratory, but the customer satisfaction survey confirmed the laboratory results.

Alternatively, one may have to conduct specific studies to gather follow-up information about the use of released products. Basically, the same methods can be used for this kind of field study as for other field studies and task analysis, especially including interviews, questionnaires, and observational studies. Furthermore, since follow-up studies are addressing the usability of an existing system, logging data from instrumented versions of the software becomes especially valuable for its ability to indicate how the software is being used across a variety of tasks.

In addition to field studies where the development organization actively seeks out the users, information can also be gained from the more passive technique of analyzing user complaints, modification requests, and calls to help lines. Even when a user complaint at first sight might seem to indicate a programming error (for example, data lost), it can sometimes have its real roots in a usability problem, causing users to operate the system in dangerous or erroneous ways. Defect-tracking procedures are already in place in many software organizations and may only need small changes to be useful for usability engineering purposes [Rideout 1991]. Furthermore, information about common learnability problems can be gathered from instructors who teach courses in the use of the system.

Finally, economic data on the impact of the system on the quality and cost of the users' work product and work life are very important and can be gathered through surveys, supervisors' opinions, and statistics for absenteeism, etc. These data should be compared with similar data collected before the introduction of the system.

References

Bailey, G. 1993. Iterative methodology and designer training in human–computer interface design, pp. 198–205. In *Proc. ACM INTERCHI'93 Conf.* Amsterdam, The Netherlands, April 24–29.

Bellantone, C. E. and Lanzetta, T. M. 1991. Works as advertised: observations and benefits of prototyping, pp. 324–327. In *Proc. Hum. Factors Soc. 35th Annu. Meet.*

Benel, D. C. R., Ottens, D., Jr., and Horst, R. 1991. Use of an eyetracking system in the usability laboratory, pp. 461–465. In *Proc. Hum. Factors Soc. 35th Annu. Meet.*

Chapanis, A. and Budurka, W. J. 1990. Specifying human–computer interface requirements. *Behav. Inf. Tech.* 9(6):479–492.

Cordingley, E. 1989. Knowledge elicitation techniques for knowledge based systems. In *Knowledge Elicitation: Principles, Techniques, and Applications*. D. Diaper, Ed., pp. 89–172. Ellis Horwood, Chichester, U.K.

del Galdo, E. and Nielsen, J., Eds. 1996. *International User Interfaces*. Wiley, New York.

Diaper, D. ed. 1989. *Task Analysis for Human–Computer Interaction*. Ellis Horwood, Chichester, U.K.

Egan, D. E., Remde, J. R., Gomez, L. M., Landauer, T. K., Eberhardt, J., and Lochbaum, C. C. 1989. Formative design-evaluation of SuperBook. *ACM Trans. Inf. Syst.* 7(1):30–57.

Fath, J. L. and Bias, R. G. 1992. Taking the task out of task analysis, pp. 379–383. In *Proc. Hum. Factors Soc. 36th Annu. Meet.*

Garber, S. R. and Grunes, M. B. 1992. The art of search: a study of art directors, pp. 157–163. In *Proc. ACM CHI'92 Conf.* Monterey, CA, May 3–7.

Gould, J. D., Conti, J., and Hovanyecz, T. 1983. Composing letters with a simulated listening typewriter. *Commun. ACM* 26(4):295–308.

Gould, J. D. and Lewis, C. H. 1985. Designing for usability: key principles and what designers think. *Commun. ACM* 28(3):300–311.

Greif, S. 1991. Organisational issues and task analysis. In *Human Factors for Informatics Usability*. B. Shackel and S. Richardson, Eds., pp. 247–266. Cambridge University Press, Cambridge, U.K.

Grudin, J. 1989. The case against user interface consistency. *Commun. ACM* 32(10):1164–1173.

Grudin, J. 1990. Obstacles to user involvement in interface design in large product development organizations, pp. 219–224. In *Proc. IFIP INTERACT'90 3rd Int. Conf. Hum.–Comput. Interaction.* Cambridge, U.K, Aug. 27–31.

Grudin, J. 1991a. Interactive systems: bridging the gaps between developers and systems. *IEEE Comput.* 24(4):59–69.

Grudin, J. 1991b. Systematic sources of suboptimal interface design in large product development organizations. *Hum.–Comput. Interaction* 6(2):147–196.

Grudin, J., Ehrlich, S. F., and Shriner, R. 1987. Positioning human factors in the user interface development chain, pp. 125–131. In *Proc. ACM CHI+GI'87 Conf.* Toronto, Canada, April 5–9.

House, C. H. and Price, R. L. 1991. The return map: tracking product teams. *Harvard Bus. Rev.* (Jan.–Feb.):92–100.

Johnson, P. 1992. *Human Computer Interaction: Psychology, Task Analysis and Software Engineering*, McGraw–Hill, London, U.K.

Life, M. A., Narborough-Hall, C. S., and Hamilton, W. I., Eds. 1990. *Simulation and the User Interface*. Taylor & Francis, London, U.K.

Lund, A. M. 1994. Ameritech's usability laboratory: from prototpye of final design. *Behav. Inf. Tech.* 13(1&2):67–80.

Maulsby, D., Greenberg, S., and Mander, R. 1993. Prototyping an intelligent agent through Wizard of Oz, pp. 277–284. In *Proc. ACM INTERCHI'93 Conf.* Amsterdam, The Netherlands. April 24–29.

Nielsen, J. 1990. Paper versus computer implementations as mockup scenarios for heuristic evaluation, pp. 315–320, In *Proc. IFIP INTERACT'90 3rd Int. Conf. Hum.–Comput. Interaction.* Cambridge, U.K, Aug. 27–31.

Nielsen, J. 1993. Iterative user interface design. *IEEE Comput.* 26(11):32–41.

Nielsen, J. 1994a. Heuristic evaluation. In *Usability Inspection Methods*. J. Nielsen and R. L. Mack, Eds., pp. 25–62. Wiley, New York.

Nielsen, J. 1994b. *Usability Engineering*, paperback ed. AP Professional, Boston, MA.

Nielsen, J. 1994c. Usability laboratories. *Behav. Inf. Tech.* 13(1&2):3–8.

Nielsen, J. 1995. A home-page overhaul using other Web sites. *IEEE Software* 12(3):75–78.

Nielsen, J., Desurvire, H., Kerr, R., Rosenberg, D., Salomon, G., Molich, R., and Stewart, T. 1993. Comparative design review: an exercise in parallel design, pp. 414–417. In *Proc. ACM INTERCHI'93 Conf.* Amsterdam, The Netherlands, April 24–29.

Nielsen, J. and Faber, J. M. 1996. Improving system usability through parallel design. *IEEE Comput.* 29(3):29–35.

Nielsen, J., Fernandes, T., Wagner, A., Wolf, R., and Ehrlich, K. 1994. Diversified parallel design: contrasting design approaches. In *ACM CHI'94 Conf. Companion*. Boston, MA, April 24–28.

Nielsen, J. and Mack, R. L. 1994. *Usability Inspection Methods*. Wiley, New York.

Nielsen, J., Mack, R. L., Bergendorff, K. H., and Grischkowsky, N. L. 1986. Integrated software in the professional work environment: evidence from questionnaires and interviews, pp. 162–167. In *Proc. ACM CHI'86 Conf.* Boston, MA, April 13–17.

Nielsen, J. and Molich, R. 1990. Heuristic evaluation of user interfaces, pp. 249–256. In *Proc. ACM CHI'90 Conf.* Seattle, WA, April 1–5.

Perlman, G. 1989. Coordinating consistency of user interfaces, code, online help, and documentation with multilingual/multitarget software specification. In *Coordinating User Interfaces for Consistency*, J. Nielsen, Ed., pp. 35–55. Academic Press, Boston, MA.

Poltrock, S. E. 1996. Participant-observer studies of user interface design and development. In *Human–computer Interface Design: Success Cases, Emerging Methods, and Real-World Context*, M. Rudisill, T. McKay, C. Lewis, and P. Polson, eds. Morgan Kaufmann, San Francisco, CA.

Rideout, T. 1991. Changing your methods from the inside. *IEEE Software* 8(3):99–100, 111.

Rowley, D. E. and Rhoades, D. G. 1992. The cognitive jogthrough: a fast-paced user interface evaluation procedure, pp. 389–395. In *Proc. ACM CHI'92 Conf.* Monterey, CA. May 3–7.

Sless, D. 1991. Designing a new bill for Telecom Australia. *Inf. Design J.* 6(3):255–257.

Tognazzini, B. 1989. Achieving consistency for the Macintosh. In *Coordinating User Interfaces for Consistency*, J. Nielsen, Ed., pp. 57–73. Academic Press, Boston, MA.

Virzi, R. A. 1989. What can you learn from a low-fidelity prototype? pp. 224–228. In *Proc. Hum. Factors Soc. 33rd Annu. Meet.* Denver, CO, Oct. 16–20.

von Hippel, E. 1988. *The Sources of Innovation*. Oxford University Press, New York.

Whiteside, J., Bennett, J., and Holtzblatt, K. 1988. Usability engineering: our experience and evolution. In *Handbook of Human–Computer Interaction*. M. Helander, ed., pp. 791–817. North-Holland, Amsterdam.

Wichansky, A. M., Abernethy, C. N., Antonelli, D. C., Kotsonis, M. E., and Mitchell, P. P. 1988. Selling ease of use: human factors partnerships with marketing, pp. 598–602. In *Proc. Hum. Factors Soc. 32nd Annu. Meet.*

Wiecha, C., Bennett, W., Boies, S., and Gould, J. 1989. Tools for generating consistent user interfaces. In *Coordinating User Interfaces for Consistency*, J. Nielsen, ed., pp. 107–130. Academic Press, Boston, MA.

Further Information

Usability engineering is the main topic of the annual meetings of the Usability Professionals' Association. For further information contact its office: Usability Professionals' Association, 190 N. Bloomingdale Rd, Bloomingdale, IL 60108. http://www.upassoc.org.

46

Task Analysis and the Design of Functionality

46.1 Introduction 46-1
46.2 Principles 46-2
       The Critical Role of Task Analysis and Design
       of Functionality
46.3 Research and Application Background 46-4
       The Contribution of Human Factors to Task Analysis
       • Contributions of Human–Computer Interaction
       to Task Analysis
46.4 Best Practices: How to Do a Task Analysis 46-8
       Collecting Task Data • Representing Systems and Tasks
       • Task Analysis at the Whole-System Level
       • Representing the User's Task
46.5 Using GOMS Task Analysis in Functionality
       and Interface Design 46-15
       High-Level GOMS Analysis • An Example of High-Level
       GOMS Analysis
46.6 Research Issues and Concluding Summary 46-22

David Kieras
University of Michigan

46.1 Introduction

Task analysis is the process of understanding the user's task thoroughly enough to help design a computer system that will effectively support users in doing the task. By **task** is meant the user's job or work activities, what the user is attempting to accomplish. By *analysis* is meant a relatively systematic approach to understanding the user's task that goes beyond unaided intuitions or speculations, and attempts to document and describe exactly what the task involves. The design of **functionality** is a stage of the design of computer systems in which the user-accessible functions of the computer system are chosen and specified. The basic thesis of this chapter is that the successful design of functionality requires a task analysis early enough in the system design to enable the developers to create a system that effectively supports the user's task. Thus, the proper goal of the design of functionality is to choose functions that are *useful* in the user's task, and which, together with a good **user interface**, result in a system that is *usable*, that is, easy to learn and easy to use.

The user's task is not just to interact with the computer, but to get a job done. Thus, understanding the user's task involves understanding the user's **task domain** and the user's larger job goals. Many systems are designed for ordinary people, who presumably lack specialized knowledge, so the designers might believe that they understand the user's task adequately well without any further consideration. This belief is often incorrect; the tasks of even ordinary people are often complex and poorly understood by developers. In contrast, many economically significant systems are intended for expert users, and understanding their tasks is absolutely critical. For example, a system to assist a petroleum geologist must be based on an understanding of the knowledge and goals of the petroleum geologist. To be useful, such a system will require functions that produce information useful to the geologist; to be usable, the system must provide these functions in a way that the frequent and most important activities of the geologist are well supported. Thus, for success, the developer must design not just the user interface, but also the functionality behind the interface.

The purpose of this chapter is to provide some background and beginning how-to information about conducting a task analysis and approaching the design of functionality. The next section of this chapter discusses why task analysis and the design of functionality are critical stages in software development, and how typical development processes interfere with these stages. Then will be presented some background on methods for human–machine system design that have developed in the field of **human factors** over the last few decades, including newer methods that attempt to identify the more cognitive components of tasks. The final section provides a summary of existing methods and a newly developing method that is especially suitable for computer system design. A general overview of the user interface design process, usability, and other specific aspects is provided in other chapters.

46.2 Principles

46.2.1 The Critical Role of Task Analysis and Design of Functionality

In many software development organizations, some group, such as a marketing department or government procurement agents, prepares a list of requirements for the system to be developed. Such requirements specify the system functions, at least in part. The designers and developers then further specify the functions, possibly adding or deleting functions from the list, and then begin to design an implementation for them. Typically, only at this point or later is the user interface design begun. Ideally, the design of the user interface will use appropriate techniques to arrive at a usable design, but these techniques normally are based on whatever conception of the user and the user's tasks have already been determined, and the interface is designed in terms of the functions that have already been specified. Thus, even following a usability process might well arrive at only a local optimum defined by an inadequate characterization of the user's needs and the corresponding functions.

In other words, the focus of usability methods tends to be on relatively low-level questions (such as menu structure) and on how to conduct usability tests to identify usability problems; the problem is posed as developing a usable interface to functions which have already been chosen. Typically, the system requirements have been prepared by a group who "throws them over the wall" to a development group, who arrives at the overall design and functional implementation, and then throws that over the wall to a usability group to "put a good interface on it."

If the initial requirements and system functions are poorly chosen, the rest of the development will probably fail to produce a usable product. It is a truism in human–computer interaction (HCI) that if customers need the functionality, they will buy and use even a clumsy, hard-to-use product; if the functionality is poorly chosen, no amount of effort spent on user interface design will result in a usable system — and it might not even be useful at all. This is by no means a rare occurrence; it is easy to find cases of poorly chosen functionality that undermine whatever usability properties the system otherwise possesses. Some examples follow.

The interface is often not the problem. An important article with this title by Goransson et al. [1987] presents several brief case studies that involve failures of functionality design masquerading as usability problems. The most painful is a business organization's database system that was considered too difficult to use. The interface was improved to make the system reasonably easy to use, but then it became clear that nobody in the organization needed the data provided by the system! Apparently, the original system development did not include an analysis of the needs of the organization or the system users. The best way to improve the usability of the system would have been simply to remove it.

Half a loaf is worse than none. The second major version of an otherwise easy-to-use basic word processing application included a multiple-column feature; however, it was not possible to mix the number of columns on a page. Documents having a uniform layout of two or three columns throughout do not exist in the real world; rather, real multicolumn documents always mix the number of columns on at least one page. For example, a common pattern is a title page with a single column for the title that spans the page, followed by the body of the document in two-column format. The application could produce such a document only if two separate documents were prepared, printed, and then physically cut and pasted together! In other words, the multiple-column feature of this second version was essentially useless for preparing real documents. A proper task analysis would have determined the kinds and structures of multiple-column documents that users would be likely to prepare. Using this information during product development would have led either to more useful functionality (like the basic page-layout features in the third major release of the product) or to a decision not to waste resources on a premature implementation of incomplete functionality.

Why doesn't it do that? A first-generation handheld "digital diary" device provided calendar and date book functions equivalent to paper calendar books, but included no clock, no alarm, and no awareness of the current date, although such functions would have been minor additions to the hardware. In addition, there was no facility for scheduling repeating meetings, making such scheduling remarkably tedious. The only short cut was to use a rather clumsy copy–paste function, but it did not work for the meeting time field in the meeting information. A task analysis of typical user's needs would have identified all of these as highly desirable functions. Including them would have made the first generation of these devices much more viable.

Progress is not necessarily monotonic. The second version of a personal digital assistant (PDA) also had a problem with recurring meetings. In the first version, a single interface dialog was used for specifying recurring meetings, and it was possible to select multiple days per week for the repeating meeting. Thus, the user could easily specify a weekly repeating meeting schedule of the sort common in academics, for example, scheduling a class that meets at the same time every Monday, Wednesday, and Friday for a semester. However, in the second version, which attempted many interface improvements, this facility moved down a couple of menu levels and became both invisible in the interface (unless a certain option were selected) and undocumented in the user manual. If any task analysis was done in connection with either the original or the second interface design, it did not take into account the type of repeating meeting patterns needed in academic settings — a major segment of the user population which includes many valuable early adopter customers.

46.2.1.1 The Role of Task Analysis in Development

Problems with misdefined functionality arise because first, there is a tendency to assume that the requirements specifications for a piece of software can and should contain all that is necessary to design and implement the software, and second, the common processes for preparing these specifications often fail to include a real task analysis. Usually, the requirements are simply a list of desirable features or functions, chosen haphazardly, and without critical examination of how they will fit together to support the user. Simply mentioning a user need in the requirements does not mean that the final system will include the right functions to make the system either useful or usable for meeting that need. The result is often a serious waste: pointless implementation effort and unused functionality. Thus, understanding the user's task is the most important step in system and interface design. The results of task analyses can be used in several different phases in the development process:

Development of requirements. This stage of development is emphasized in this chapter. A task analysis should be conducted before developing the system requirements to guide the choice and design of the system functionality; the ultimate usability of the product is actually determined at this stage. The goal of task analysis at this point is to find out what the user needs to accomplish, so that the functionality of the system can be designed so that the user can accomplish the required tasks easily. Although some later revision is likely to be required, these critical choices can be made before the system implementation or user interface is designed.

User interface design and evaluation. Task analysis results are needed during interface design to design and evaluate the user interface effectively. The usability process itself and user testing both require information about the user's tasks. Task analysis results can be used to choose benchmark tasks for user testing that will represent important uses of the system. Usage scenarios valuable during interface design can be chosen that are properly representative of user activities. A task analysis will help to identify the portions of the interface that are most important for the user's tasks. Once an interface is designed and is undergoing evaluation, the original task analysis can be supplemented with an additional analysis of how the task would be done with the proposed interface. This can suggest usability improvements either by modifying the interface or by improving the fit of the functionality to the more specific form of the user's task entailed by a proposed interface. In fact, some task analysis methods are very similar to user testing. The difference is that in user testing, one seeks to identify problems that the users have with an interface while performing selected tasks; in task analysis, one tries to understand how users will perform their tasks given a specific interface. Thus, a task analysis might identify usability problems, but task analysis does not necessarily require user testing.

Follow-up after installation. Task analysis can be conducted on fielded or in-place systems to compare systems or to identify potential problems or improvements. When a fully implemented system is in place, it is possible to conduct a fully detailed task analysis. The results could be used to compare the demands of different systems, identify problems that should be corrected in a new system, or determine properties of the task that should be preserved in a new system.

46.3 Research and Application Background

46.3.1 The Contribution of Human Factors to Task Analysis

Task analysis developed in the discipline of human factors, which has a long history of concern with the design process and how human-centered issues should be incorporated into system design. Much of human factors has been concerned with human participation in extremely large and complex systems, usually military systems. In fact, task analysis methods have been applied to a broad range of systems, ranging from handheld radios to radar consoles to whole aircraft, chemical process plant control rooms, and very complex multiperson systems such as a warship combat information center. The breadth of this experience is no accident. Historically, human factors is the only discipline with an extended record of involvement and concern with human–system design, long predating the newer field of HCI. These methods have been presented in comprehensive collections of task analysis techniques (e.g., Kirwan and Ainsworth [1992], Beevis et al. [1992], and Diaper and Stanton, [2004a]), and it remains a active area of research within human factors and allied fields [Annett and Stanton, 2000a; Schraagen et al., 2000]. One purpose of this chapter is to summarize some of these techniques and concepts. But there are some key differences between the interfaces of the traditional kinds of systems treated in the human factors experience and the interfaces to computer-based systems.

Task properties of both types of systems. At a general level, the task analysis of both traditional and computer-based interfaces involve collecting and representing the answers to the following primary overall questions:

1. What does the system do as a whole?
2. Where does the human operator fit into the system? What role will the operator play?
3. What specific tasks must the operator perform in order to play that role?

A large-scale general difference between the two kinds of systems shows up in response to these first three questions. Namely, the function or mission of the system as a whole often goes well beyond the immediate goals of the human operators, especially for high-complexity systems. For example, the mission of a system such as AEGIS is to defend a naval task force against all forms of airborne attack. While each human operator of course has an interest in the success of this mission, the individual operator has a much more constrained set of goals, such as monitoring and identifying radar targets. HCI has tended to emphasize the design problem only at the single-operator level, as if only the third question were important. This myopic focus is another manifestation of the problem discussed previously, that the choice and the design of functionality are often not considered a part of the process of designing for usability.

Task properties of traditional systems. At the level of the detailed design of the interface, much of human factors expertise involves the detailed design of traditional interfaces constructed using conventional precomputer technology, often termed knobs-and-dials interfaces. A key design convention of traditional interfaces is that a single display device (e.g., a pressure gauge) displays a single system parameter or sensor value, and a single control device (e.g., a pump power switch) controls a single parameter or component of the system. Such interfaces require many individual physical devices and so tend to be distributed in space, resulting in control rooms in which every surface is covered with displays and controls. Thus, many of the interface design issues involve the visual and spatial properties of the devices and their layout. For example, some contributors to problems with nuclear power plant control rooms are gauges that cannot be seen by the operator, and misleadingly symmetrical switch layouts that have mirror-imaged on and off positions [Woods et al., 1987]. On the other hand, a useful design principle is to place related controls and displays together and organize them so that the spatial arrangement corresponds to the order in which they are used (e.g., left to right). The entire array of controls and displays should be always or readily visible and accessible simply by looking or moving, and thus constitute a large external memory for the operator. But the contents of this external memory are usually at the level of individual system parameters and components. This constraint is a product of the available interface and control technology used in traditional systems.

Thus, in addition to the first three task analysis questions, a task analysis of a traditional system seeks answers to the following fourth question:

4. What system parameters and components must be accessible to allow the operator to perform the tasks?

Once these four preliminary questions are answered, an interface can be designed or evaluated. If an interface design does not already exist, then human factors guidelines and principles can be used to guide the creation of a design. Once an interface design is available, a further, more detailed task analysis can then determine how the operator would access the controls and displays in order to carry out the tasks. Due to the spatiality of such interfaces, visual and spatial issues receive considerable emphasis. The sequential or temporal properties of the procedures followed by the operators are thus closely related to the spatial layout of the interface.

Task properties of computer-based systems. In contrast to the myriad controls and displays of traditional interfaces, computer-based systems often have few "in-place objects" [see Beevis et al., 1992], meaning that the interface for even an extremely complex system can consist of only a single screen, keyboard, and mouse. The operator must thus control the system by sequentially organized activities with this small number of mechanical or visual objects that tend to stay in the same place and compete for the limited keyboard and display space. Time has been traded for space; temporal layout of activity has been traded for spatial layout. The operator often must remember more, because the limited display space means that only a little information is visible at any one time, and complicated procedures are often needed to bring up other information. The user's procedures have relatively little spatial content, and instead become relatively arbitrary sequences of actions on the small set of objects, such as typing different command strings on a keyboard. These procedures can become quite elaborate sequential, repeating, and hierarchical patterns.

On the other hand, the power of the computer behind the interface permits the displayed information to be potentially more useful and more relevant to the user's task than the single-sensor/single-display traditional approach. For example, the software can combine several sensor readings and show a derived value that is what the user really wants to know. Likewise, the software can control several system components in response to a user command, radically simplifying the operator's procedures. Thus, the available computer technology means that the interface displays and controls can be chosen much more flexibly than the traditional knob-and-dial technology. In conjunction with the greater complexity of such systems, it is now both possible and critically important to choose the system display and control functionality on the basis of what will work well for the user, rather than simply sorting through the system components and parameters to determine the relevant ones.

Thus, a task analysis for a computer-based system must articulate what services or functions the computer should provide the operator with, rather than on what fixed components and parameters the operator must have access to. This leads to the following, additional question for computer-based systems:

5. What display and control functions should the computer provide to support the operator in performing the tasks?

In other words, the critical step in computer-based system design is the *choice of functionality*. Once the functions are chosen, the constraints of computer interfaces mean that the procedural requirements of the interface are especially prominent; if the functions are well chosen, the procedures that the operator must follow will be simple and consistent. Thus, the focus on spatial layout in traditional systems is replaced by a concern with the choice of functionality and interface procedures in computer-based systems. The importance of this combination of task analysis, choice of functionality, and the predominance of the procedural aspects of the interface is the basis for the recommendations in this chapter.

46.3.2 Contributions of Human–Computer Interaction to Task Analysis

The field of HCI is a relatively new and highly interdisciplinary field, which still lacks consensus on scientific, practical, and philosophical foundations. Consequently, a variety of ideas have been discussed concerning how developers should approach the problem of understanding what a new computer system must do for its users. While many original researchers and practitioners in HCI had their roots in human factors, several other disciplines have had a strong influence on HCI theory and practice.

These disciplines fall roughly into two groups. The first is **cognitive psychology**, a branch of scientific psychology concerned with human cognitive abilities such as comprehension, problem solving, and learning. The second is a mixture of ideas from the social sciences, such as social-organizational psychology, ethnography, and anthropology. While the contribution of these fields has been important in developing the scientific basis of HCI, they have either little experience with humans in a work context, as is the case with cognitive psychology, or no experience with practical system design problems, as with the social sciences. On the other hand, human factors is almost completely oriented toward solving practical design problems in an *ad hoc* manner and is almost completely atheoretic in content. Thus, the disciplines with a broad and theoretical science base lack experience in solving design problems, and the discipline with this practical knowledge lacks a comprehensive scientific foundation.

The current state of task analysis in HCI is thus rather confused [Diaper and Stanton, 2004b]; there has been an unfortunate tendency to reinvent task analysis under a variety of guises, as each theoretical approach presents its own insights about how to understand a work situation and design a system to support it. Moreover, many designers and developers have apparently simply started over with experience-based approaches. One example is contextual design [Holtzblatt, 2003], which is a collection of task-analytic techniques and activities that proponents claim will lead one through the stages of understanding what users do, what they need, and what system and interface will help them do it. Despite few explicit connections with previous work, many of the suggestions are familiar task-analytic techniques.

As each scientific community spawned its own ways of analyzing tasks, and as many development groups invented their own experienced-based approaches, the resulting hodgepodge of newly minted ideas has become bewildering enough to HCI specialists, but it is impenetrably obscure to the software developer who merely wants to develop a better system. For this reason, this chapter focuses on the tried-and-true pragmatic methodologies from human factors and a closely related methodology based on the most clearly articulated of the newer theoretical approaches: the GOMS model, defined later in this section.

New social-science approaches to task analysis. Since much computer usage takes place in organizations in which individual users must cooperate and interact as part of their work, viewing computer usage as a social activity can attempt to capture the larger context of a computer user's task. Some relevant social-science concepts can be summarized (see Baecker et al. [1995] for a sampling and overview). A general methodological approach is *ethnography*, which is the set of methods used by anthropologists to immerse oneself in a culture and document its structure (see Blomberg et al. [2003]). Another approach based on anthropology, called *situated cognition*, emphasizes understanding human activity in its larger social context (see Nardi [1995] for an overview). Another theoretical approach is *activity theory* [Nardi, 1995; Turner and McEwan, 2004] which originated in the former Soviet Union as a comprehensive psychological theory, with some interesting differences from conventional western or American psychology.

The proponents of all of these social-science approaches have had some successes, apparently due to their insistence on observing and documenting what people are actually doing in a specific situation and in their work context. In the context of common computer-industry practice in system design, this insistence might seem novel and noteworthy, but such attention to the user's context is characteristic of all competent task analyses. The contribution of these approaches is an emphasis on levels of the user's context that can be easily overlooked if one's focus is too narrowly on how the user interacts with the technological artifacts.

Contributions from cognitive psychology. The contribution of cognitive psychology to HCI is both more limited and more successful within that limited scope. Cognitive psychology treats an individual human as an information-processor who acquires information from the environment, transforms it, stores it, retrieves it, and acts on it. This *information-processing* approach, also called the *computer metaphor*, has an obvious application to how humans interact with computer systems. In a cognitive approach to HCI, the interaction between human and computer is viewed as two interacting information-processing systems with different capabilities, in which one, the human, has goals to accomplish, and the other, the computer, is an artificial system that should be designed to facilitate the human's efforts. The relevance of cognitive psychology research is that it directly addresses two important aspects of usability: how difficult it is for the human to learn how to interact successfully with the computer and how long it takes the human to conduct the interaction.

The underlying topics of human learning, problem solving, and skilled behavior have been intensively researched for decades in cognitive psychology.

The application of cognitive psychology research results to human–computer interaction was first systematically presented by Card et al. [1983] at two levels of analysis. The lower-level analysis is the Model Human Processor, a summary of about a century's worth of research on basic human perceptual, cognitive, and motor abilities in the form of an engineering model that could be applied to produce quantitative analysis and prediction of task execution times. The higher-level analysis was the **GOMS model**, a description of the procedural knowledge involved in doing a task. The acronym *GOMS* stands for the following. The user has **G**oals that can be accomplished with the system. **O**perators are the basic actions, such as keystrokes performed in the task. **M**ethods are the procedures, consisting of sequences of operators, that will accomplish the goals. **S**election rules determine which method is appropriate for accomplishing a goal in a specific situation.

In the Card et al. formulation, the new user of a computer system will use various problem-solving and learning strategies to figure out how to accomplish tasks using the computer system, and then, with additional practice, these results of problem solving will become procedures that the user can routinely invoke to accomplish tasks in a smooth, skilled manner. The properties of the procedures will thus govern both the ease of learning and ease of use of the computer system. In the research program stemming from the original proposal, approaches to representing GOMS models based on cognitive psychology theory have been developed and validated empirically, along with the corresponding techniques and computer-based tools for representing, analyzing, and predicting human performance in human–computer interaction situations (see John and Kieras [1996a, b] for reviews).

The significance of the GOMS model for task analysis is that it provides a method to describe the task procedures in a way that has a theoretically rigorous and empirically validated scientific relationship to human cognition and performance. Space limitations preclude any further presentation of how GOMS can be used to express and evaluate a detailed interface design (see John and Kieras, [1996a, b] and Kieras 1997 [2004]). In Section 46.5, a technique based on GOMS will be used to couple task analysis with the design of functionality.

46.4   Best Practices: How to Do a Task Analysis

The basic idea of conducting a task analysis is to understand the user's activity in the context of the whole system, either an existing or a future system. Although understanding human activity is the subject of scientific study in psychology and the social sciences, the conditions under which systems must be designed usually preclude the kind of extended and intensive research necessary to document and account for human behavior in a scientific mode. Thus, a task analysis for system design must be rather more informal and primarily heuristic in flavor, compared to scientific research. The task analyst must do his or her best to understand the user's task situation well enough to influence the system design given the limited time and resources available. This does not mean that a task analysis is an easy job; large amounts of detailed information must be collected and interpreted, and experience in task analysis is valuable even in the most structured methodologies (e.g., see Annett [2004]).

The role of formalized methods for task analysis. Despite the fundamentally informal character of task analysis, many formal and quasi-formal systems for task analysis have been proposed and widely recommended. Several will be summarized. It is critical to understand that these systems do not in themselves analyze the task or produce an understanding of the task. Rather, they are ways to structure the task analysis process and notations for representing the results of task analysis. They have the important benefit of helping the analyst observe and think carefully about the user's actual task activity, specifying what kinds of task information are likely to be useful to analyze, and providing a heuristic test for whether the task has actually been understood. That is, a good test for understanding something is whether one can represent or document it. Constructing such a representation can be a good approach to trying to understand it. A formal representation of a task shows the results of the task analysis in a form that can help document the analysis, so that it can be inspected, criticized, and revised. Finally, some of the more formal representations can be used as the basis for computer simulations or mathematical analyses to obtain quantitative predictions of task performance, but such results are no more correct than the original, and informally obtained, task analysis underlying the representation.

An informal task analysis is better than none. Most of the task analysis methods to be surveyed require significant time and effort; spending these resources would usually be justified, given the near-certain failure of a system that fails to meet the actual needs of users. However, the current reality of software development is that developers often will not have adequate time and support to conduct a full-fledged task analysis. Under these conditions, what can be recommended? As pointed out in sources such as Gould [1988] and Grudin [1991], perhaps the most serious problem is that the developers often have no contact with actual users. Thus, if nothing more systematic is possible, the developers should spend some time in informal observation of real users actually doing real work. The developers should observe unobtrusively but ask for explanation or clarification as needed, perhaps trying to learn the job themselves. They should not, however, make any recommendations or discuss the system design. The goal of this activity is simply to try to gain some experience-based intuitions about the nature of the user's job, what real users do and why. See Gould [1988] for additional discussion. Such informal, intuition-building contact with users will provide tremendous benefits at relatively little cost. Approaches such as contextual design [Holtzblatt, 2003] and the more elaborate methods presented here provide more detail and more systematic documentation, and will permit more careful and exact design and evaluation than casual observation. Some informal observation of users, however, is infinitely better than no attempt at task analysis at all.

46.4.1 Collecting Task Data

Task analysis requires information about the user's situation and activities, but simply collecting data about the user's task is not necessarily a task analysis. In a task analysis, the goal is to understand the properties of the user's task that can be used to specify the design of a system; this requires synthesis and interpretation beyond the data. The data collection methods summarized here are those that have been found to produce useful information about tasks (see Kirwan and Ainsworth [1992] and Gould [1988]). The task-analytic methods summarized in Section 46.4.2 are approaches that help analysts perform the synthesis and interpretation.

Observation of user behavior. In this fundamental family of methods, the analyst observes actual user behavior, usually with minimal intrusion or interference, and describes what has been observed in a thorough, systematic, and documented way. This type of task data collection is most similar to user testing, except that, as discussed previously, the goal of task analysis is to understand the user's task, not just to identify problems that the user might have with a specific system design.

The setting for the user's activity can be the actual situation (e.g., in the field) or a laboratory simulation of the actual situation. All of the user's behavior can be recorded, or it can be sampled periodically to cover more time while reducing the data collection effort. The user's activities can be categorized, counted, and analyzed in various ways. For example, the frequency of different activities could be tabulated, or the total time spent in different activities could be determined. Both such measures contribute valuable information about which task activities are most frequent or time-consuming, and thus are important to address in the system design. Finer-grain recording and analysis can provide information on the exact timing and sequence of task activities, which can be important in the detailed design of the interface. Videotaping users is a simple recording approach that supports both very general and very detailed analysis at low cost; consumer-grade equipment is often adequate.

A more intrusive method of observation is to have users *think aloud* about a task while performing it, or to have two users discuss and explain to each other how to do the task while performing it. The verbalization can disrupt normal task performance, but such *verbal protocols* are believed to be a rich source of information about the user's mental processes, such as inferences and decision making. The pitfall for the inexperienced is that the protocols can be extremely labor-intensive to analyze, especially if the goal is to reconstruct the user's cognitive processes. The most fruitful path is to transcribe the protocols, isolate segments of content, and attempt to classify them into an informative set of categories.

A final technique in this class is *walkthroughs* and *talkthroughs*, in which the users or designers carry out a task and describe it as they do so. The results are similar to a think-aloud protocol, but with more emphasis on the procedural steps involved. An important feature is that the interface or system need not exist; the users or designers can describe how the task would or should be carried out.

Critical incidents and major episodes. Instead of attempting to observe or understand the full variety of activity in the task, the analyst chooses incidents or episodes that are especially informative about the task and the system, and attempts to understand what happens in these. This is basically a case-study approach. Often the critical incidents are accidents, failures, or errors, and the analysis is based on retrospective reports from the people involved and any records produced during the incident. An important extension of this approach is the *critical decision method* [Wong, 2004], which focuses on understanding the knowledge involved in making expert-level decisions in difficult situations. However, the critical incident might be a major episode of otherwise routine activity that serves especially well to reveal the problems in a system. For example, observation of a highly skilled operator performing a very specialized task revealed that most of the time was spent doing ordinary file maintenance; understanding why led to major improvements in the system [Brooks, personal communication].

Questionnaires. Questionnaires are a fixed set of questions that can be used quite economically to collect some types of user and task information on a large scale. The main problem is that the accuracy of the data is unknown compared to observation, and can be susceptible to memory errors and social influences. Despite the apparent simplicity of a questionnaire, designing and implementing a successful one is not easy, and can require an effort comparable to interviews or workplace observation. The newcomer should consult sources on questionnaire design before proceeding.

Structured interviews. Interviews involve talking to users or domain experts about the task. Typically, some unstructured interviews might be done first, in which the analyst simply seeks any and all kinds of comments about the task. Structured interviews can then be planned; a series of predetermined questions for the interview is prepared to ensure more systematic, complete, and consistent collection of information.

Interface surveys. An interface survey collects information about an existing, in-place, or designed interface. Several examples are: Control and Display surveys determine what system parameters are shown to the user and what components can be controlled. Labeling and Coding surveys can determine whether there are confusing labels or inconsistent color codes present in the interface. Operator Modifications surveys assess changes made to the interface by the users, such as added notes or markings, that can indicate problems in the interface. Finally, Sightline surveys determine what parts of the interface can be seen from the operator's position; such surveys have found critical problems in nuclear power plant control rooms. Sightlines would not seem important for computer interfaces, but a common interface design problem is that the information required during a task is not on the screen at the time it is required; an analog to a sightline survey would identify such problems.

46.4.2 Representing Systems and Tasks

Once the task data is collected, the problem for the analyst is to determine how to represent the task data, which requires decisions about what aspects of the task are important and how much detail to represent. The key function of a representation is to make the task structure visible or apparent in some way that supports the analyst's understanding of the task. By examining a task representation, an analyst hopes to identify problems in the task flow, such as critical bottlenecks, inconsistencies in procedures, excessive workloads, and activities that could be better supported by the system. Traditionally, a graphical representation, such as a flowchart or diagram, has been preferred, but as the complexity of the system and the operating procedures increases, diagrammatic representations lose their advantage.

Task decomposition. One general form of task analysis is often termed task decomposition. This is not a well defined method at all; it merely reflects a philosophy that tasks usually have a complex structure, and a major problem for the analyst will be to decompose the whole task situation into subparts for further analysis. Some of these subparts will be critical to the system design; others are possibly less important. For example, one powerful approach is to consider how a task might be decomposed into a hierarchy of subtasks and the procedures for executing them, leading to a popular form of analysis called (somewhat too broadly) *hierarchical task analysis* (HTA). Another approach would be to decompose the task situation into considerations of how the controls are labeled, how they are arranged, and how the displays are coded. This is also a task decomposition and might also have a hierarchical structure, but the emphasis is on describing aspects of the displays in the task situation. Obviously, depending on the specific system and its interface, some aspects of the user's task situation may be far more important to analyze than others. Developing an initial task decomposition can help to identify what is involved overall in the user's task, thus allowing the analyst to choose what aspects of the task merit intensive analysis.

Level of detail. The question of how much detail to represent in a task analysis is difficult to answer. At the level of whole tasks, Kirwan and Ainsworth [1992] suggest a probability × cost rule: if the probability of inadequate performance multiplied by the cost of inadequate performance is low, then the task is probably not worthwhile to analyze. But even if a task has been chosen as important, the level of detail at which to describe the particular task still must be chosen. Some terminology must be clarified at this point: task decompositions can be viewed as a standard inverted tree structure, with a single item, the overall task, at the top, and the individual actions (such as keystrokes or manipulating valves) or interface objects (switches, gauges) at the bottom. A *high-level* analysis deals only with the low-detail top parts of the tree; a *low-level* analysis includes all of the tree from the top to the high-detail bottom.

The cost of task analysis rises quickly as more detail is represented and examined. On the other hand, many critical design issues appear only at a detailed level. For example, at a high enough level of abstraction, the Unix operating system interface is essentially just like the Macintosh operating system interface; both interfaces provide the functionality for invoking application programs and copying, moving, and deleting files and directories. The notorious usability problems of Unix relative to other systems only appear at a level of detail that the cryptic, inconsistent, and clumsy command structure and generally poor feedback come to the surface. The devil is in the details. Thus, a task analysis capable of identifying usability problems in an interface design typically involves working at a low, fully detailed level that involves individual commands and mouse selections. The opposite consideration holds true for the design of functionality, as will be discussed more below. When choosing functionality, ask how the user will carry out tasks using a set of system functions, and it is important to avoid being distracted by the details of the interface.

46.4.3 Task Analysis at the Whole-System Level

When large systems are being designed, an important component of task analysis is to consider how the system, consisting of all the machines and all the humans, is supposed to work as a whole in order to accomplish the overall system goal. This kind of very high-level analysis can be done even with very large systems, such as military systems involving multiple machines and humans. The purpose of the analysis is to determine what role in the whole system the individual human operators will play. Various methods for whole-system analysis have been in routine use for some time. Briefly, these are as follows (see Beevis et al. [1992] and Kirwan and Ainsworth [1992]).

Mission and scenario analysis. Mission and scenario analysis is an approach to starting the system design from a description of what the system must do (the mission), especially using specific, concrete examples or scenarios. See Brooks (this volume) for a related discussion.

Function-flow diagrams. Function-flow diagrams are constructed to show the sequential or information-flow relationships of the functions performed in the system. Beevis et al. [1992] provide a set of large-scale examples, such as naval vessels.

Petri nets. Petri nets also represent the causal and sequential relationships between the functions performed in a system, but in a rigorous formalism. Various methodologies for modeling and simulating the system performance can also include timing information.

Function allocation. Function allocation is a set of fairly informal techniques for deciding which system functions should be performed by machines and which by people. Usually mentioned in this context is the *Fitts list*, which describes what kinds of activities can best be performed by humans vs. machines. However, according to surveys described in Beevis et al. [1992], this classic technique is rarely used in real design problems, because it is not specific enough to drive design decisions. Rather, functions are typically allocated in an *ad hoc* manner, often simply maintaining whatever allocation was used in the predecessor system or following the rule that whatever can be automated should be — even though it is known that automation often produces safety or vigilance problems for human operators.

In the military systems analyzed heavily in human factors, the overall system goal is normally rather larger in scale and well above the level of concerns of the human operators. For example, in designing a new naval fighter aircraft, the system goals might be stated in terms such as "enable air superiority in naval operations under any conditions of weather and all possible combat theaters through the year 2010." At this level, the users of the system as a whole are military strategists and commanders, not the pilot, and the system as a whole will involve not just the pilot, but other people in the cockpit (such as a radar intercept operator) and maintenance and ground operations personnel. Thus, the humans involved in the system will have a variety of goals and tasks, depending on their role in the whole system.

At first glance, this level of analysis would appear to have little to do with computer systems. We often think of computer users as isolated individuals carrying out their tasks by interacting with their individual computers. However, when considering the needs of an organization, the mission level of analysis is clearly important; the system is supposed to accomplish something as a whole, and the individual humans all play roles defined by their relationship with each other and with the machines in the system. HCI has begun to consider higher levels of analysis, as in the field of computer-supported collaborative work, but perhaps the main reason why the mission level of analysis is not common parlance in HCI is that HCI has a cultural bias that organizations revolve around the humans, with the computers playing only a supporting role. Such a bias would explain the movement mentioned earlier toward incorporating more social-science methodology into system design. In contrast, in military systems, the human operators are often viewed as parts in the overall system, whose ultimate user is the commanding officer, leading to a concern with how the humans and machines fit together.

Regardless of the perspective taken on the whole system, at some point in the analysis, the activities of the individual humans who actually interact directly with the equipment begin to appear. It is then both possible and essential to identify the goals that they, as individual operators, must accomplish. At this point, task analysis methodology begins to overlap with the concerns of computer user interface design.

46.4.4 Representing the User's Task

Once the whole system and the roles of the individual users and operators have been characterized, the main focus of task-analytic work is to identify more specific properties of the situation and activities of the human operator or user. These can be summarized as what the user must know, what the user must do, what the user sees and interacts with, and what the user might do wrong.

46.4.4.1 Representing What the User Must Know

The goal of this type of task analysis is to represent what knowledge the human must have in order to operate the system effectively. Clearly, the human must know how to operate the equipment; such procedural knowledge is treated under its own heading below. But the operator might need additional procedural knowledge that is not directly related to the equipment, as well as additional nonprocedural conceptual background knowledge. For example, a successful fighter aircraft pilot must know more than just the procedures for operating the aircraft and the on-board equipment; he or she must have additional procedural skills such as combat tactics, navigation, and communication protocols, and an understanding of the aircraft mechanisms and overall military situation is valuable in dealing with unanticipated and novel situations.

Information about what the user needs to know is clearly useful for specifying the content of operator training and operator qualifications. It can also be useful in choosing system functionality, in that large benefits can be obtained by implementing system functions that make it unnecessary for users to know concepts or skills that are difficult to learn. Such simplifications typically are accompanied by simplifications in operating procedures. Aircraft computer systems that automate navigation and fuel conservation tasks are an obvious example.

In some cases, where the user knowledge is mostly procedural in content, it can be represented in a straightforward way, such as decision–action tables that describe what interpretation should be made of a specific situation, as in an equipment trouble-shooting guide. However, the required knowledge can be extremely hard to identify if it does not have a direct and overt relationship to "what to do" operating procedures. An example is the petroleum geologist, who, after staring at a display of complex data for some time, comes up with a decision about where to drill, and who probably cannot provide a rigorous explanation for how the decision was made. Understanding how and why users make such decisions is difficult, because there is little or no observable behavior prior to producing the result; it is all in the head, a *purely cognitive task.*

Analyzing a purely cognitive task in complete detail is essentially a cognitive psychology research project, and so is not usually practical in a system design context. Furthermore, to the extent that a cognitive task can be completely characterized, it becomes a candidate for automation, making moot the design of the user interface. Expert systems technology, when successfully applied using knowledge acquisition methods (see Boose [1992]), is an example of task analyses carried out in enough detail that the need for a human performer of the task is eliminated.

Most cognitive tasks seem neither possible nor practical to analyze this thoroughly, but there is still a need to support the human performer with system functions and interfaces and with training programs and materials that improve performance in the task, even if that task is not completely understood. For example, Gott [1988] surveys cases in which an intensive effort to identify the knowledge required for tasks can produce large improvements in training programs for highly demanding cognitive tasks, such as electronics troubleshooting.

During the 1990s, there were many efforts to develop methods for **cognitive task analysis** (CTA). CTA emphasizes describing the knowledge required for a task and how it is used in decision making, situation recognition, or problem–solving, rather than just the procedures involved. This is true even though most task analysis procedures developed prior to that time included cognitive activities, such as decision making and problem solving, and were often conducted in order to identify the required background knowledge for operator training. For example, Shepherd [2000] summarizes how the first systematic task analysis method, hierarchical task analysis (see Section 46.4.4.2) normally includes various cognitive information. Furthermore, as argued by Chipman et al. [2000], it makes little sense to attempt to divide cognitive processes such as decision making from the actions and procedures involved in actual task performance. There is currently a wide variety of CTA techniques proposed or under development; see Chipman et al. [2000] for an overview, and Dubois and Shalin [2000] and Seamster et al. [2000] for useful summaries of some important methods. Most current CTA methods involve some form of interview technique for eliciting a subset of the most critical knowledge involved in a task, often focusing on critical incidents (e.g., Militello and Hutton [2000] and Wong [2004]).

There are many cases in which CTA resulted in dramatically improved training materials and programs (e.g., O'Hare et al. [2000] and Schaafstal and Schraagen [2000]. However, an earlier review by Essens et al. [1994] reported that while cognitive task analysis techniques have demonstrated successes in developing training programs, there were few demonstrations of successful design of computer systems for decision aiding. Landauer [1995] notes that decision-support systems have yet to produce a convincing track record of improving human productivity; this result would be expected, given the general lack of task analysis in system design and the great difficulty of task analysis in the case of heavily cognitive tasks.

Given that support for cognitively demanding tasks is touted as one of the main contributions of computers, it is critical that more progress be made on how to conduct task analysis and system design for purely cognitive tasks.

46.4.4.2 Representing What the User Must Do

A major form of task analysis is describing the actions or activities carried out by the human operator while tasks are being executed. Such analyses have many uses. The description of how a task is currently conducted, or would be conducted with a proposed design, can be used for prescribing training, for assisting in the identification of design problems in the interface, or as a basis for quantitative or simulation modeling to obtain predictions of system performance. Depending on the level of detail chosen for the analysis, the description might be very high-level, or it might be fully detailed, describing the individual valve operations or keystrokes needed to carry out a task. The following paragraphs describe the major methods for representing procedures.

Operational sequence diagrams. Operational sequence diagrams and related techniques show the sequence of the operations (actions) carried out by the user (or the machine) to perform a task, represented graphically as a flowchart using standardized symbols for the types of operations. Such diagrams are often partitioned, showing the user's actions on one side and the machine's on the other, to illustrate the pattern of operation between the user and the machine.

Timeline analysis. Timeline analyses simply display activities, or some characteristic of them, as a function of time during task execution. For example, a workload profile for an airliner cockpit would show a large variety and intensity of activities during landing and takeoff, but not during cruising. After constructing a timeline display, the analyst looks for workload peaks, such as the operator's having to remember too many things, or conflicts, such as the operator's having to use two widely separated controls at the same time.

Hierarchical task analysis. HTA involves describing a task as a hierarchy of tasks and subtasks, emphasizing the procedures that operators will carry out, using several specific forms of description. The term *hierarchical* is somewhat misleading, because many forms of task analysis produce hierarchical descriptions; a better term might be *procedure hierarchy task analysis.* The results of an HTA are typically represented either graphically, as a sort of annotated tree diagram of the task structure similar to the conventional diagram of a function-call hierarchy in programming, or in a more compact tabular form. This is the original form of systematic task analysis [Annett et al., 1971] and still the single most heavily used form of task analysis. See Annett [2004] for additional background and a procedural guide.

HTA descriptions involve goals, tasks, operations, and plans. A *goal* is a desired state of affairs (e.g., a chemical process proceeding at a certain rate). A *task* is a combination of a goal and a context (e.g., get a chemical process going at a certain rate given the initial conditions in the reactor). *Operations* are activities for attaining a goal (e.g., procedures for introducing reagents into the reactor, increasing the temperature, and so forth). *Plans* specify which operations should be applied under what conditions (e.g., what procedure to follow if the reactor is already hot). Plans usually appear as annotations to the tree-structure diagram, explaining which portions of the tree will be executed under what conditions. Each operation in turn might be decomposed into subtasks, leading to a hierarchical structure. The analysis can be carried out to any desired level of detail, depending on the requirements of the analysis.

GOMS models. GOMS models, introduced in Section 46.3.2, are closely related to hierarchical task analysis. In fact, Kirwan and Ainsworth [1992] include GOMS as a form of HTA. GOMS models describe a task in terms of a hierarchy of goals and subgoals; methods, which are a sequence of operators (actions) that, when executed, will accomplish the goals; and selection rules that choose which method should be applied to accomplish a particular goal in a specific situation. However, both in theory and in practice, GOMS models are different from HTA. The concept of GOMS models grew out of research on human problem solving and cognitive skill, whereas HTA appears to have originated from the pragmatic, common-sense observation that tasks often involve subtasks, and eventually involve carrying out sequences of actions. Because of their more principled origins, GOMS models are more disciplined than HTA descriptions. The contrast is perhaps most clear in the difficulty HTA descriptions have in expressing the flow of control: the procedural structure of goals and subgoals must be deduced from the plans, which appear only as annotations to the sequence of operations. In contrast, GOMS models represent plans and operations in a uniform format, using only methods and selection rules. An HTA plan would be represented as simply a higher-order method that carries out lower-level methods or actions in the appropriate sequence, along with a selection rule for when the higher-order method should be applied.

46.4.4.3 Representing What the User Sees and Interacts With

The set of objects with which the user interacts during task execution is clearly closely related to the procedures that the user must follow, in that a full procedural description of the user's task will (or should) refer to all objects in the task situation that the user must observe or manipulate. However, it can be useful to attempt to identify and describe the relevant objects and events independently of the procedures in which they are used. Such a task analysis can identify some potential serious problems or design issues quite rapidly. For example, studies of nuclear power plant control rooms [Woods et al., 1987] found that important displays were located in positions where they could not be read by the operator. A task decomposition can be applied to break the overall task situation into smaller portions; the interface survey technique mentioned previously can then determine the various objects in the task situation. Collecting additional information (e.g., from interviews or walkthroughs) can lead to an assessment of whether and under what conditions the individual controls or displays are required for task execution. There are a variety of guidelines in human factors for determining whether the controls and displays are adequately accessible.

A related form of task analysis is concerned with the layout in space of the displays, controls, or other people with whom the operator must interact. *Link analysis* is a straightforward methodology for tabulating the pairwise accessibility relationships between people and objects in the task environment and weighting them by the frequency and difficulty of access (e.g., greater distance). Alternative arrangements can easily be explored to minimize the difficulty of the most frequent access paths. For example, a link analysis revealed that a combat information center on a warship was laid out in such a way that the movement and communication patterns involved frequent crossing of paths, sightlines, and so forth. A simple rearrangement of workstation positions greatly reduced the amount of interference. An analog for computer interfaces would be analyzing the transitions between different portions of the interface, such as dialogs or screen objects. A design could be improved by making the most frequent transitions short and direct.

46.4.4.4 Representing What the User Might Do Wrong

Human factors practitioners and researchers have developed a variety of techniques for analyzing situations in which errors have happened or might happen. The goal is to determine whether human errors will have serious consequences, and to identify where these might occur and how likely they are to occur. The design of the system or the interface can then be modified to try to reduce the likelihood of human errors or mitigate their consequences. Some key techniques are summarized in the following paragraphs.

Event trees. In an event tree, the possible paths, or sequences of behaviors, through the task are shown as a tree diagram. Each behavior's outcome is represented either as success/failure or as a multiway branch (e.g., for the type of diagnosis made by an operation in response to a system alarm display). An event tree can be used to determine the consequences of human errors, such as misunderstanding an alarm. Each path can be given a predicted probability of occurrence based on estimates of the reliability of human operators at performing each step in the sequence. (These estimates are controversial; see Reason [1990] for a discussion.)

Failure modes and effects analysis. The analysis of human failure modes and their effects is modeled after a common hardware reliability assessment process. The analyst considers each step in a procedure and attempts to list all the possible failures an operator might commit, such as omitting the action or performing it too early, too late, too forcefully, and so forth. The consequences of each such failure mode can then be worked out, and again a probability of failure predicted.

Fault trees. In a fault-tree analysis, the analyst starts with a possible system failure and then documents the logical combination of human and machine failures that could lead to it. The probability of the fault occurring can then be estimated, and possible ways to reduce the probability can be determined.

Until recently, these techniques had not been applied in computer user interface design to any visible extent. At most, user interface design guides contained a few general suggestions for how interfaces could be designed to reduce the chance of human error. Recent work, such as Stanton [2003] and Wood [1999], shows promise in using task analysis as a basis for systematically examining how errors might be made and how they can be detected and recovered from. Future work along these lines will be an extraordinarily important contribution to system and interface design.

46.5 Using GOMS Task Analysis in Functionality and Interface Design

The task analysis techniques described previously work well enough to have been developed and applied in actual system design contexts. However, they have mainly developed in the analysis of traditional interface technologies rather than computer user interfaces. As discussed earlier, the general concepts of task analysis hold for computer interfaces, but there are some key differences and a clear need to address computer user interfaces more directly. In summary, the following problems exist with traditional analysis methods.

Representing a mass of procedural detail. Computer interface procedures tend to be complicated, repetitious, and hierarchical. The primarily graphical and tabular representations traditionally used for procedures become unwieldy when the amount of detail is large.

Representing procedures that differ in level of analysis and type. For example, in HTA, a plan is represented differently from a procedure, even though a plan is simply a kind of higher-order procedure.

Moving from a task analysis to a functional design to an interface design. To a great extent, human factors practice uses different representations for different stages of the design process (see Kirwan and Ainsworth [1992] and Beevis et al. [1992]). It would be desirable to have a single representation that spans these stages, even if it covers only part of the task analysis and design issues.

This section describes how GOMS models could be used to represent a high-level task analysis that can be used to help choose the desirable functionality for a system. Because GOMS models have a programming language–like form, they can represent large quantities of procedural detail in a uniform notation that works from a very high level down to the lowest level of the interface design.

46.5.1 High-Level GOMS Analysis

Using high-level GOMS models is an alternative to the conventional requirements of development and interface design process discussed in the introduction to this chapter. The approach is to drive the choice of functionality from the high-level procedures for doing the tasks, choosing functions that will produce simple procedures for the user. By considering the task at a high level, these decisions can be made independently of, and prior to, the interface design, thereby improving the chances that the chosen functionality will enable a highly useful and usable product after a good interface is developed. Key interface design decisions, such as whether a color display is needed, can be made explicit and given a well founded basis, such as how color coding could be used to make the task easier.

The methodology involves choosing the system functionality based on high-level GOMS analysis of how the task would be done using a proposed set of functions. The analyst can then begin to elaborate the design by making some interface design decisions and writing the corresponding lower-level methods. If the functionality design is sound, it should be possible to expand the high-level model into a more detailed GOMS model that also has simple and efficient methods. If desired, the GOMS model can be fully elaborated down to the keystroke level of detail, which can produce usability predictions (see Kieras [1997] and John and Kieras [1996a, b]).

GOMS models involve goals and operators at all levels of analysis, with the lowest level being the so-called *keystroke level*, of individual keystrokes or mouse movements. The lowest-level goals will have methods consisting of keystroke-level operators and might be basic procedures, such as moving an object on the screen or selecting a piece of text. However, in a high-level GOMS model, the goals may refer only to parts of the user's task that are independent of the specific interface, and they may not specify operations in the interface. For example, a possible high-level goal would be *Add a Footnote*, but not *Select INSERT FOOTNOTE from EDIT menu*. Likewise, the operators must be well above the keystroke level of detail, not specific interface actions. The lowest level of detail an operator may have is to invoke a system function or to perform a mental decision or action, such as choosing which files to delete or thinking of a file name. For example, an allowable operator would be *Invoke the database update function*, but not *Click on the UPDATE button*.

The methods in a high-level GOMS model describe the order in which mental actions or decisions, submethods, and invocations of system functions are executed. The methods should document what information the user must acquire in order to make any required decisions and to invoke the system functions. They also should represent where the user might detect errors and how these might be corrected with additional system functions. All too often, support for error detection and correction by the user either is missing or is a clumsy add-on to a system design. By including it in the high-level model for the task, the designer may be able to identify ways in which errors can be prevented, detected, and corrected, early and easily.

46.5.2 An Example of High-Level GOMS Analysis

The domain for this example is electronic circuit design and computer-aided design systems for electronic design (ECAD). A task analysis of this domain would reveal many very complex activities on the part of electronic circuit designers. Of special interest are several tasks for which computer support is feasible: after a circuit is designed, its correct functioning must be verified, and then its manufacturing cost estimated, power and cooling requirements determined, the layout of the printed circuit boards designed, automated assembly operations specified, and so forth.

This example involves computer support for the task of verifying the circuit design by using computer simulation to replace the traditional slow and costly "breadboard" prototyping. For many years, computer-based tools for this process have been available and undergoing development, based on techniques for simulating the behavior of the circuit using an abstract mathematical representation of the circuit. For purposes of this example, attention will be limited to the somewhat simpler domain of digital circuit design. Here, the components are black-box modules (i.e., integrated circuits) with known behaviors, and the circuit consists simply of these components with their terminals interconnected by wires.

The high-level functionality design of such a system will be illustrated with an ideal task-driven design example, describing how such systems *should* have been designed. Then the design of a typical actual system will be presented in terms of the high-level analysis and compared to the task-driven design.

46.5.2.1 Task-Driven Design Example

Given the basic functionality concept of using a simulation of a circuit, the top-level method to accomplish this goal is the first method shown in Figure 46.1, which accomplishes the goal *Verify circuit with ECAD system*. This first method needs some explanation of the notation, which is the so-called natural GOMS language (NGOMSL) described in Kieras [1997] for representing GOMS models in a readable format.

The first line introduces a method for accomplishing the top-level user goal. It will be executed whenever the goal is asserted, and it terminates when the *return with goal accomplished* operator in Step 4 is executed. Step 1 represents the user's black-box mental activity of thinking up the original idea for the circuit design. This *think-of* operator is just a place holder; no attempt is made to represent the extraordinarily complex cognitive processes involved. Step 2 asserts a subgoal to specify the circuit for the ECAD system; the method for this subgoal appears next in Figure 46.1. Step 3 of the top-level method is a high-level operator for invoking the functionality of the circuit simulator and getting the results; no commitment is made at this point as to how this will be done in the interface. Step 4 documents that, at this point, the user will decide whether the job is complete, based on the output of the simulator. Step 5 is another placeholder for a complex cognitive process of deciding what modification to make to the circuit. Step 6 invokes the functionality to modify the circuit, which would probably be much like that involved in Step 2, but which will not be further elaborated in this example. Finally, the loop in Step 7 shows that the top-level task is iterative.

The next step in the example is to consider the method for entering a circuit into the ECAD system. In this domain, schematic circuit drawings are the conventional representations for a circuit, so the basic functionality that must be provided is a tool for drawing circuit diagrams. This is reflected in the method for the goal *Enter circuit into ECAD system*.

This method starts with invoking the tool, and then has a simple iteration consisting of thinking of something to draw, accomplishing the goal of drawing it, and repeating until done. The method gets more interesting when the goal of drawing an object is considered, because in this domain there are some fundamentally different kinds of objects, and the information requirements for drawing them are different. Only two kinds of objects will be considered here.

A selection rule is needed in a GOMS model to choose what method to apply, depending on the kind of object, and then a separate method is needed for each kind of object. The selection rule set in Figure 46.1 thus accomplishes a general goal by asserting a more specific goal, which then triggers the corresponding method. The method for drawing a component requires a decision about the type of component (e.g., what specific multiplexer chip should be used) and where in the diagram the component should be placed to produce a visually clear diagram. Drawing a connecting wire requires deciding which two points in the circuit the wire should connect, and also how the wire should be routed to produce a clear appearance.

At this point, the analysis has documented some possibly difficult and time-consuming activities that the user must do. Candidates for additional system functions to simplify these activities could be considered.

```
Method for goal: Verify circuit with ECAD system
      Step 1. Think-of circuit idea.
      Step 2. Accomplish Goal: Enter circuit into ECAD system.
      Step 3. Run simulation of circuit with ECAD system.
      Step 4. Decide: If circuit performs correct function,
                     then return with goal accomplished.
      Step 5. Think-of modification to circuit.
      Step 6. Make modification with ECAD system.
      Step 7. Go to 3.

Method for goal: Enter circuit into ECAD system
      Step 1. Invoke drawing tool.
      Step 2. Think-of object to draw next.
      Step 3. If no more objects, then Return with goal accomplished.
      Step 4. Accomplish Goal: draw the next object.
      Step 5. Go to 2.


Selection rule set for goal: Drawing an object
      If object is a component, then accomplish Goal: draw a component.
      If object is a wire, then accomplish Goal: draw a wire.
      ...
      Return with goal accomplished

Method for goal: Draw a component
      Step 1. Think-of component type.
      Step 2. Think-of component placement.
      Step 3. Invoke component-drawing function with type and placement.
      Step 4. Return with goal accomplished.

Method for goal: Draw a wire
      Step 1. Think-of starting and ending points for wire.
      Step 2. Think-of route for wire.
      Step 3. Invoke wire drawing function with starting point,
                     ending point, and route.
      Step 4. Return with goal accomplished.
```

FIGURE 46.1   Preliminary high-level methods for an ECAD system.

For example, the step of thinking of an appropriate component might be quite difficult, due to the huge number of integrated circuits available. Likewise, thinking of the starting and ending points for the wire involves knowing which input or output function goes with each of the many pins on the chips. Some kind of on-line documentation or database to provide this information in a form that meshes well with the task might be valuable. Similarly, a welcome function might be some automation to choose a good routing for wires.

Although the analysis of the desirable functions has just begun, it is worthwhile to consider what errors the user might make and how the system functionality will support identifying and correcting them. In this domain, the errors the user might make can be divided into *semantic* errors, in which the specified circuit does not do what it is supposed to do, and *syntactic* errors, in which the specified circuit is invalid, regardless of the function. The semantic errors can only be detected by the user evaluating the behavior of the circuit and discovering that the idea for the circuit was incorrect or incorrectly specified. Notice that the iteration in the top-level method provides for detecting and correcting semantic errors. Syntactic errors arise in the digital circuit domain because there are certain connection patterns that are incorrect in terms of what is permissible and meaningful with digital logic circuits. Two common cases are disallowed

```
Method for goal: Enter circuit into ECAD system
     Step 1. Invoke drawing tool.
     Step 2. Think-of object to draw next.
     Step 3. Decide: If no more objects, then go to 6.
     Step 4. Accomplish Goal: draw the next object.
     Step 5. Go to 2.
     Step 6. Accomplish Goal: Proofread drawing.
     Step 7. Return with goal accomplished.

Method for goal: Proofread drawing
     Step 1. Find missing connection in drawing.
     Step 2. Decide: If no missing connection, return with goal accomplished.
     Step 3. Accomplish Goal: Draw wire for connection.
     Step 4. Go to 1.

Method for goal: Draw a wire
     Step 1. Think-of starting and ending points for wire.
     Step 2. Think-of route for wire.
     Step 3. Invoke wire drawing function with starting point,
                  ending point, and route.
     Step 4. Decide: If wire is not disallowed, return with goal accomplished.
     Step 5. Correct the wire.
     Step 6. Return with goal accomplished.
```

FIGURE 46.2 Revised methods incorporating error detection and correction steps.

connections (e.g., shorting an output terminal to ground) and missing connections (e.g., leaving an input terminal unconnected).

It is important to correct syntactic errors before running the simulator, because their presence will cause the simulator to fail or to produce invalid results. Functions to detect syntactic errors can be implemented fairly easily; the design question is exactly how this functionality should be defined so that it most helps the user. Note that disallowed connections appear at the level of individual wires, whereas missing connections only show up at the level of the entire drawing. Thus, the functions to assist in detecting and correcting disallowed connections should be operating while the user is drawing wires, and those for missing connections while the user is working on the whole drawing. Figure 46.2 shows the corresponding revisions and additions to some of the methods in Figure 46.1.

The method for entering the circuit now incorporates a method invoked in Step 6 to proofread the drawing, which is done by finding the missing connections in the drawing and adding a wire for each one. The method for drawing a wire now has Step 4, which checks for the wire being disallowed immediately after it is drawn. If there is a problem, the wire will be corrected before proceeding.

At this point in the design of the functionality, the analysis has documented what information the user must get about syntactic errors in the circuit diagram, and where this information can be used to detect and correct the error immediately. Some thought can now be given to what functionality might be useful to support the user in finding syntactically missing and disallowed connections. One obvious candidate is the use of color coding. For example, perhaps unconnected input terminals could start out as red on the display, and then turn green when validly connected. Likewise, perhaps as soon as a wire is connected at both ends, it could turn red if it is a disallowed connection or green if it is legal. This use of color should work very well, because properly designed color coding is known to be an extremely effective way to aid visual search. In addition, the color coding calls the user's attention to the problems, but without forcibly interrupting the user. This design rationale for using color is an interesting contrast to actual ECAD displays, which generally make profligate use of color in ways that lack any obvious value in performing the task. Figure 46.3 presents a revision of the methods to incorporate this preliminary interface design decision.

```
Method for goal: Proofread drawing
      Step 1. Find a red terminal in drawing.
      Step 2. Decide: If no red terminals, return with goal accomplished.
      Step 3. Accomplish Goal: Draw wire at red terminal.
      Step 4. Go to 1.

Method for goal: Draw a wire
      Step 1. Think-of starting and ending points for wire.
      Step 2. Think-of route for wire.
      Step 3. Invoke wire drawing function with starting point,
                      ending point, and route.
      Step 4. Decide: If wire is now green, return with goal accomplished.
      Step 5. Decide: If wire is red, think-of problem with wire.
      Step 6. Go to 1.
```

FIGURE 46.3   Methods incorporating color codes for syntactic drawing errors.

At this point, the functionality design also has clear implications for how the system implementation must be designed, in that the system must be able to perform the required syntax-checking computations on the diagram quickly enough update the display while the drawing is in progress. Thus, performing the task analysis for the design of the functionality has not only helped guide the design to a fundamentally more usable approach, it also has produced some critical implementation specifications very early in the design.

46.5.2.2 An Actual Design Example

The preceding example of how the design of functionality can be aided by working out a high-level GOMS model of the task seems straightforward and unremarkable. A good design is usually intuitively "right," and once presented, seems obvious. However, at least the first few generations of ECAD tools did not implement such an intuitively obvious design at all, probably because nothing was done that resembled the kind of task and functionality analysis just presented. Rather, a first version of the system was probably designed and implemented whose methods were the obvious ones shown in Figure 46.1: the user will draw a schematic diagram in the obvious way and then run the simulator on it. However, once the system was in use, it became obvious that errors could be made in the schematic diagram that would cause the simulation to fail or to produce misleading results. The solution was simply to provide a set of functions to check the diagram for errors, but to do so in an opportunistic, *ad hoc* fashion, involving minimum implementation effort, which failed to take into account the impact on the user's task. Figure 46.4 shows the resulting method, which was actually implemented in some popular ECAD systems.

The top level is the same as in the previous example, except for Step 3, which checks and corrects the circuit after the entire drawing is completed. The method for checking and correcting the circuit first involves invoking a checking function, which was designed to produce a series of error messages that the user would process one at a time. For ease of implementation, the checking function does not work in terms of the drawing, but in terms of the abstract circuit representation, the *netlist*, and so reports the site of the syntactically illegal circuit feature in terms of the name of the node in the netlist. However, the only way the user can examine and modify the circuit is in terms of the schematic diagram. So the method for processing each error message first involves locating the corresponding point in the circuit diagram, and then making a modification to the diagram. To locate the site of the problem on the circuit diagram, the user invokes an identification function and provides the netlist node name; the function then highlights the corresponding part of the circuit diagram, which the user can locate on the screen. In other words, to check the diagram for errors, the user must wait until the entire diagram is completely drawn and then invoke a function whose output must be manually transferred into another function, which finally identifies the location of the error!

```
Method for goal: Verify circuit with ECAD system
     Step 1. Think-of circuit idea.
     Step 2. Accomplish Goal: Enter circuit into ECAD tool.
     Step 3. Accomplish Goal: Check and correct circuit.
     Step 4. Run simulation of circuit with ECAD tool.
     Step 5. Decide: If circuit performs correct function,
                     then return with goal accomplished.
     Step 6. Think-of modification to circuit.
     Step 7. Make modification in ECAD tool.
     Step 8. Go to 3.

Method for goal: Check and correct circuit
     Step 1. Invoke checking function.
     Step 2. Look at next error message.
     Step 3. If no more error messages, Return with goal accomplished.
     Step 4. Accomplish Goal: Process error message.
     Step 5. Go to 2.

Method for goal: Process error message
     Step 1. Accomplish Goal: Locate erroneous point in circuit.
     Step 2. Think-of modification to erroneous point.
     Step 3. Make modification to circuit.
     Step 4. Return with goal accomplished.

Method for goal: Locate erroneous point in circuit
     Step 1. Read type of error, netlist node name from error message.
     Step 2. Invoke identification function.
     Step 3. Enter netlist node name into identification function.
     Step 4. Locate highlighted portion of circuit.
     Step 5. Return with goal accomplished.
```

FIGURE 46.4   Methods for an actual ECAD system.

Obviously, the functionality design in this version of the system will inevitably result in a far less usable system than the task-driven design. Instead of getting immediate feedback at the time and place of an error, the user must finish drawing the circuit and then engage in a convoluted procedure to identify the errors in the drawing. Although the interface has not yet been specified, the inferior usability of the actual design relative to the task-driven design is clearly indicated by the additional number of methods and method steps, and the time-consuming nature of many of the additional steps. In contrast to the task-driven design, this actual design seems preposterous and could be dismissed as a silly example — except for the fact that at least one major vendor of ECAD software used exactly this design.

In summary, the task-driven design was based on an analysis of how the user would do the task and what functionality would help the user do it easily. The result was that users could detect and correct errors in the diagram while drawing the diagram, and so they could always work directly with the natural display of the circuit structure. In addition, good use was made of color display capabilities, which often go to waste. The actual design probably arose because user errors were not considered until very late in the development process, and the response was minimal add-ons of functionality, leaving the initial functionality decisions intact. The high-level GOMS model clarifies the difference between the two designs by showing the overall structure of the interaction. Even at a very high level of abstraction, poor functionality design can result in task methods that are inefficient and clumsy. Thus, high-level GOMS models can capture critical insights from a task analysis to help guide the initial design of a system and its functionality.

46.6 Research Issues and Concluding Summary

The major research problem in task analysis is attempting to bring some coherence and theoretical structure to the field. Although psychology as a whole rather seriously lacks a single theoretical structure, the subfields most relevant to human–system interaction are potentially unified by work in cognitive psychology on cognitive architectures, which are computational modeling systems that attempt to provide a framework for explaining human cognition and performance (see Byrne [2003] for an overview). These architectures are directly useful in system design in two ways: First, because the architecture must be "programmed" to perform the task with task-specific knowledge, the resulting model contains the content of a full-fledged task analysis, both the procedural and cognitive components. Thus, constructing a cognitive-architectural model is a way to represent the results of a task analysis and verify its completeness and accuracy. Second, because the architecture represents the constants and constraints on human activity (such as the speed of mouse pointing movements and short-term memory capacity), the model for a task is able to predict performance on the task, and so can be used to evaluate a design very early in the development process (see Kieras [2003] for an overview).

The promise is that these comprehensive cognitive architectures will encourage the development of coherent theory in the science base for HCI, and also provide a high-fidelity way to represent how humans would perform a task. While some would consider such predictive models the ultimate form of task analysis [Annett and Stanton, 2000b], there is currently a gap between the level of detail required to construct such models and the information available from a task analysis, both in principle and in practice [Kieras and Meyer, 2000]. There is no clear pathway for moving from one of the well established task analysis methods to a fully detailed cognitive-architectural model. It should be possible to bridge this gap, because GOMS models, which can be viewed as a highly simplified form of cognitive architectural model [John and Kieras, 1996a, b], are similar enough to HTA that it is easy to move from this most popular task analysis to a GOMS model. Future work in this area should result in better methods for developing high-fidelity predictive models in the context of more sophisticated task analysis methods.

Another area of research concerns the analysis of team activities and team tasks. A serious failing of conventional psychology and the social sciences in general is a gap between the theory of humans as individual intellects and actors and the theory of humans as members of a social group or organization. This leaves HCI as an applied science without an articulated scientific basis for moving between designing a system that works well for an individual user and designing a system that meets the needs of a group. Despite this theoretical weakness, task analysis can be done for whole teams with some success, as shown by Zachary, et al. [2000], Essens et al. [2000], and Klein, [2000]. What is less convincing at this point is how such analyses can be used to identify an optimal team structure or interfaces to support team performance optimally. One approach will be to use computational modeling approaches that take individual human cognition and performance into account as the fundamental determiner of the performance of a team, as in preliminary work by Santoro and Kieras [2001].

The claim that a task analysis is a critical step in system design is well illustrated by the introductory examples, in which entire systems were seriously weakened by failure to consider what users actually need to do and what functionality is needed to support them. This claim is also supported by the final example, which shows how, as opposed to the usual *ad hoc* design of functionality, a task analysis can directly support a choice of functions, resulting in a useful and usable system. While there are serious practical problems in performing task analysis, the experience of human factors shows that these problems can be overcome, even for large and complex systems. The numerous methods developed by human factors for collecting and representing task data are ready to be adapted to the problems of computer interface design. The additional contributions of cognitive psychology have resulted in procedural task analyses that can help evaluate designs rapidly and efficiently. System developers thus have a powerful set of concepts and tools already available, and they can anticipate even more comprehensive task analysis methods in the future.

Acknowledgment

The concept of high-level GOMS analysis was developed in conjunction with Ruven Brooks, of Rockwell Automation, who also provided helpful comments on the first version of this chapter.

Defining Terms

Cognitive psychology: A branch of psychology concerned with rigorous empirical and theoretical study of human cognition, the intellectual processes having to do with knowledge acquisition, representation, and application.

Cognitive task analysis: A task analysis that emphasizes the knowledge required for a task and its application, such as decision making, and its background knowledge.

Functionality: The set of user-accessible functions performed by a computer system; the kinds of services or computations performed that the user can invoke, control, or observe the results of.

GOMS model: A theoretical description of human procedural knowledge in terms of a set of **G**oals, **O**perators (basic actions), **M**ethods (sequences of operators that accomplish goals), and **S**election rules, which select methods appropriate for goals. The goals and methods typically have a hierarchical structure. GOMS models can be thought of as programs that the user learns and then executes in the course of accomplishing task goals.

Human factors: Originating when psychologists were asked to tackle serious equipment design problems during World War II, this discipline is concerned with designing systems and devices so that they can be used effectively by humans. Much of human factors is concerned with psychological factors, but important other areas are biomechanics, anthropometrics, work physiology, and safety.

Task: This term is not very well defined and is used differently in different contexts, even within human factors and HCI. Here, it refers to purposeful activities performed by users, either a general class of activities or a specific case or type of activity.

Task domain: The set of knowledge, skills, and goals possessed by users that is specific to a kind of job or task.

Usability: The extent to which a system can be used effectively to accomplish tasks. A multidimensional attribute of a system, covering ease of learning, speed of use, resistance to user errors, intelligibility of displays, and so forth.

User interface: The portion of a computer system with which the user interacts directly, consisting not just of physical input and output devices, but also the contents of the displays, the observable behavior of the system, and the rules and procedures for controlling the system.

References

Annett, J. 2004. Hierarchical task analysis. In D. Diaper and N.A. Stanton, Eds., *The handbook of task analysis for human–computer interaction.* Mahwah, NJ: Lawrence Erlbaum Associates. 67–82.

Annett, J. Duncan, K.D., Stammers, R.B., and Gray, M.J. 1971. *Task analysis.* London: Her Majesty's Stationery Office.

Annett, J. and Stanton, N.A., Eds. 2000a. *Task analysis.* London: Taylor & Francis.

Annett, J. and Stanton, N.A. 2000b. Research and development in task analysis. In J. Annett and N.A. Stanton, Eds., *Task analysis.* London: Taylor & Francis. 3–8.

Baber, C. and Stanton, N.A. In press. Task analysis for error identification. In D. Diaper and N.A. Stanton Eds., *The handbook of task analysis for human–computer interaction.* Mahwah, NJ: Lawrence Erlbaum Associates.

Baecker, R.M., Grudin, J., Buxton, W.A.S., and Greenberg, S., Eds. 1995. *Readings in human–computer interaction: toward the year 2000.* San Francisco: Morgan Kaufmann.

Beevis, D., Bost, R., Doering, B., Nordo, E., Oberman, F., Papin, J.-P., Schuffel, I.H., and Streets, D. 1992. Analysis techniques for man–machine system design. (Report AC/243(P8)TR/7). Brussels, Belgium: Defense Research Group, NATO HQ.

Blomberg, J., Burrell, M., and Guest, G. 2003. An ethnographic approach to design. In J.A. Jacko and A. Sears, Eds., *The human–computer interaction handbook*. Mahwah, NJ:. 964–986.

Boose, J.H. 1992. Knowledge acquisition. In *Encyclopedia of artificial intelligence*, 2nd edition. New York: Wiley. 719–742.

Byrne, M.D. 2003. Cognitive architecture. In J.A. Jacko and A. Sears, Eds., *The human–computer interaction handbook*. Mahwah, NJ. 97–117.

Chipman, S.F., Schraagen, J.M., and Shalin, V.L. 2000. Introduction to cognitive task analysis. In J.M. Schraagen, S.F. Chipman, and V.L. Shalin, Eds., *Cognitive task analysis*. Mahwah, NJ: Lawrence Erlbaum Associates. 3–24.

Diaper, D. and Stanton, N.A., Eds. 2004a. *The handbook of task analysis for human–computer interaction*. Mahwah, NJ: Lawrence Erlbaum Associates.

Diaper, D. and Stanton, N.A. 2004b. Wishing on a sTAr: the future of task analysis. In D. Diaper and N.A. Stanton, Eds., *The handbook of task analysis for human–computer interaction*. Mahwah, NJ: Lawrence Erlbaum Associates. 603–620.

Dubois, D. and Shalin, V.L. 2000. Describing job expertise using cognitively oriented task analysis. In J.M. Schraagen, S.F. Chipman, and V.L. Shalin, Eds., *Cognitive task analysis*. Mahwah, NJ: Lawrence Erlbaum Associates. 41–56.

Essens, P.J.M.D., Fallesen, J.J., McCann, C.A., Cannon-Bowers, J., and Dorfel, G. 1994. COADE: a framework for cognitive analysis, design, and evaluation. *Technical Report, TNO Human Factors Research Institute*, Soesterberg, Netherlands.

Essens, P.J.M.D., Post, W.M., and Rasker, P.C. 2000. Modeling a command center. In J.M. Schraagen, S.F. Chipman, and V.L. Shalin, Eds., *Cognitive task analysis*. Mahwah, NJ: Lawrence Erlbaum Associates. 385–400.

Goransson, B., Lind, M., Pettersson, E., Sandblad, B., and Schwalbe, P. 1987. The interface is often not the problem. In *Proceedings of CHI + GI 1987*. New York: ACM.

Gott, S.P. 1988. Apprenticeship instruction for real-world tasks: the coordination of procedures, mental models, and strategies. In E.Z. Rothkopf, Ed., *Review of research in education*. Washington, DC: AERA.

Gould, J.D. 1988. How to design usable systems. In M. Helander, Ed., *Handbook of human–computer interaction*. Amsterdam: North-Holland. 757–789.

Grudin, J. 1991. Systematic sources of suboptimal interface design in large product development organizations. *Human–computer interaction*, **6**: 147–196.

Holtzblatt, K. 2003. Contextual design. In J.A. Jacko. and A. Sears, Eds., *The human–computer interaction handbook*. Mahwah, NJ:. 941–963.

John, B.E. and Kieras, D.E. 1996a. Using GOMS for user interface design and evaluation: which technique? *ACM transactions on computer–human interaction*, **3**: 287–319.

John, B.E. and Kieras, D.E. 1996b. The GOMS family of user interface analysis techniques: comparison and contrast. *ACM transactions on computer–human interaction*, **3**: 320–351.

Kieras, D.E. 1997. A guide to GOMS model usability evaluation using NGOMSL. In M. Helander, T. Landauer, and P. Prabhu, Eds., *Handbook of human–computer interaction*, 2nd edition. Amsterdam: North-Holland. 733–766.

Kieras, D.E. 2003. Model-based evaluation. In J.A. Jacko. and A. Sears, Eds., *The human–computer interaction handbook*. Mahwah, NJ:. 1139–1151.

Kieras, D.E. 2004. GOMS models and task analysis. In D. Diaper and N.A. Stanton, Eds., *The handbook of task analysis for human–computer interaction*. Mahwah, NJ: Lawrence Erlbaum Associates. 83–116.

Kieras, D.E. and Meyer, D.E. 2000. The role of cognitive task analysis in the application of predictive models of human performance. In J.M. Schraagen, S.E. Chipman, and V.L. Shalin, Eds., *Cognitive task analysis*. Mahwah, NJ: Lawrence Erlbaum. 237–260.

Kirwan, B. and Ainsworth, L.K. 1992. *A guide to task analysis*. London: Taylor & Francis.

Klein, G. 2000. Cognitive task analysis of teams. In J.M. Schraagen, S.F. Chipman, and V.L. Shalin, Eds., *Cognitive task analysis*. Mahwah, NJ: Lawrence Erlbaum Associates. 417–430.

Landauer, T. 1995. *The trouble with computers: usefulness, usability, and productivity.* Cambridge, MA: MIT Press.

Militello, L.G. and Hutton, R.J.B. 2000. Applied cognitive task analysis (ACTA): a practitioner's toolkit for understanding cognitive task demands. In J. Annett and N.A. Stanton, Eds., *Task analysis.* London: Taylor & Francis. 90–113.

Nardi, B., Ed. 1995. *Context and consciousness: activity theory and human–computer interaction.* Cambridge, MA: MIT Press.

O'Hare, D., Wiggins, M., Williams, A., and Wong, W. 2000. Cognitive task analysis for decision centered design and training. In J. Annett and N.A. Stanton, Eds., *Task analysis.* London: Taylor & Francis. 170–190.

Reason, J. 1990. *Human error.* Cambridge: Cambridge University Press.

Santoro, T. and Kieras, D. 2001. GOMS models for team performance. In J. Pharmer and J. Freeman (organizers), Complementary methods of modeling team performance. Panel presented at the 45th annual meeting of the Human Factors and Ergonomics Society, Minneapolis/St. Paul, MN.

Schaafstal, A. and Schraagen, J.M. 2000. Training of troubleshooting: a structured task analytical approach. In J.M. Schraagen, S.F. Chipman, and V.L. Shalin, Eds., *Cognitive task analysis.* Mahwah, NJ: Lawrence Erlbaum Associates. 57–70.

Schraagen, J.M., Chipman, S.F., and Shalin, V.L., Eds. 2000. *Cognitive task analysis.* Mahwah, NJ: Lawrence Erlbaum Associates.

Seamster, T.L., Redding, R.E., and Kaempf, G.L. 2000. A skill-based cognitive task analysis framework. In J.M. Schraagen, S.F. Chipman, and V.L. Shalin, Eds., *Cognitive task analysis.* Mahwah, NJ: Lawrence Erlbaum Associates. 135–146.

Shepherd, A. 2000. HTA as a framework for task analysis. In J. Annett and N.A. Stanton, Eds., *Task analysis.* London: Taylor & Francis. 9–23.

Stanton, N.A. 2003. Human error identification in human–computer interaction. In J.A. Jacko and A. Sears, Eds., *The human–computer interaction handbook.* Mahwah, NJ: 371–383.

Turner, P. and McEwan, T. 2004. Activity theory: another perspective on task analysis. In D. Diaper and N.A. Stanton, Eds., *The handbook of task analysis for human–computer interaction.* Mahwah, NJ: Lawrence Erlbaum Associates. 423–440.

Wong, W. 2004. Data analysis for the critical decision method. In D. Diaper and N.A. Stanton, Eds., *The handbook of task analysis for human–computer interaction.* Mahwah, NJ: Lawrence Erlbaum Associates. 327–346.

Wood, S.D. 1999. The application of GOMS to error-tolerant design. Paper presented at the 17th International System Safety Conference, Orlando, FL.

Woods, D.D, O'Brien, J.F., and Hanes, L.F. 1987. Human factors challenges in process control: the case of nuclear power plants. In G. Salvendy, Ed., *Handbook of human factors.* New York: Wiley.

Zachary, W.W., Ryder, J.M., and Hicinbotham, J.H. 2000. Building cognitive task analyses and models of a decision-making team in a complex real-time environment. In J.M. Schraagen, S.F. Chipman, and V.L. Shalin, Eds., *Cognitive task analysis.* Mahwah, NJ: Lawrence Erlbaum Associates. 365–384.

For Further Information

The reference list contains useful sources for following up this chapter. Landauer's book provides excellent economic arguments on how many systems fail to be useful and usable. The most useful sources on task analysis are the books by Kirwan and Ainsworth and by Diaper and Stanton, and the report by Beevis et al. A readable introduction to GOMS modeling is B. John's article, "Why GOMS?" in *Interactions* magazine, 1995, 2(4). The references by John and Kieras and by Kieras provide detailed overviews and methods.

47

Human-Centered System Development

Jennifer Tucker
Booz Allen Hamilton

Abby Mackness
Booz Allen Hamilton

47.1 Introduction 47-1
The Changing IT Landscape • Human Dynamics Research
47.2 Underlying Principles 47-3
IT System Development Trends: History and Impact
• IT Management — Process Challenges • IT Personality
47.3 Best Practices 47-18
IT Developers • IT Managers • IT Educators
47.4 Research Issues and Summary 47-23

47.1 Introduction

The last decade has seen a significant change in the information technology (IT) landscape, resulting in a fundamental shift in the activities and skills required of IT professionals and teams. The mystical stories of isolated technical programmers huddled in the basement generating code are not the driving reality. Instead, while still demanding deep technical expertise, the work of IT individuals and teams is also becoming more interdisciplinary and interpersonal in nature, requiring a broader range of skills to meet end user needs.

Today's IT professionals are no longer solely technical specialists; they are also educators, facilitators, and consultants, working as teams in conjunction with end users to solve business needs. Amidst these new demands and roles, IT teams are under increasing pressure to create and deliver products and services that are on time, within budget, and of high quality.

These realities force a reexamination of the factors influencing the work of IT teams from a human dynamics perspective. Today's setting requires IT professionals and teams to possess a wide range of communication and interpersonal skills, skills that have not always been taught and supported in the technically focused environment of IT. This chapter considers these issues of human dynamics in system development, describes the changing role of the IT professional and team as technology has evolved, explores the skills required in today's setting, and proposes best practices for managing the "human side" of IT.

47.1.1 The Changing IT Landscape

The advent of ubiquitous computing and the birth of e-terms (e.g., e-mail, e-business) are small indicators of the degree of change that an information-hungry culture has experienced over the past 10 to 15 years. For many end users, IT is now an enabler, rather than an enigma — and has become the province of a broad user base, rather than a select group of technical gurus.

This rate of technological advance reflects the extensive study and development of both technical and process dynamics and principles. Despite this progress, system development efforts still continue to face failure at a high rate. The CHAOS 2000 Report [Standish Group, 2001] estimated that 23% of application development efforts failed between 1994 and 2000; an additional 49% were "challenged" (i.e., completed over budget, past deadline, and with fewer features).

What factors contribute to this situation? Even with our best development methodologies and technological sophistication, the human dimension of systems and software development remains the key element to success. For example, the CHAOS 2000 Report concludes that "user involvement" is the second most important criterion for project success, falling only behind "executive support" [Standish Group, 2001]. At the most basic level, the output from any development project emerges from the conversations and collaboration among many individuals working together over time. As such, success may ultimately be influenced by the development team's ability to manage the following types of human dynamics:

Team technical diversity — The range of talents required of an IT development team today is both broad and deep. Ten or fifteen years ago, a development team might have consisted primarily of programmers; today, the team is also likely to include functional experts, analysts, architects, writers, network and systems administrators, and perhaps even a facilitator. In fact, in a recent study across 36 IT-focused organizations, only 12% of IT professionals surveyed reported their role as being programmer or developer [DAU, 2003]. The result of this trend is that IT teams, which once shared at least a common technical base for building relationships, are now coming together from separate specialties and backgrounds within their own fields. This diversity creates profound opportunities for collaboration, but it can also lead to miscommunication and "stove-piped" efforts if not managed effectively.

Team collaboration — With decreased development times and increased focus on product integration over development, information technology activities demand that IT team members are able to transfer knowledge internally and communicate effectively. Gartner [2002] notes, "teamwork is key to software life cycle planning," and recommends that managers "establish cross-functional teams to enable consistent and constant communication across multiple groups." A study of the Microsoft NT operating system development effort concluded that the *team* was the most vital operating level of that organization. "So rapid are technological developments that the core of the corporation is now the team, the only unit small enough to maintain its intellectual edge" [Zachary, 1998].

Interaction with user — The emergence of the Internet and ubiquitous computing has resulted in a new type of end user — one who is more IT savvy than in the past and who faces uncertain and highly volatile requirements as business needs shift and demands grow. This has led to development efforts that are more connected to the user, and more dependent on social and technical interactions as systems are iterated through collaborative development efforts.

47.1.2 Human Dynamics Research

Research in human factors and human–computer interaction has predominantly focused on interface design, object-oriented techniques, and other issues related to a person's interaction with a computer or system. In the world of computer science and engineering, human factors often avoid the personal — and profoundly messy — areas of human-to-human interaction inherent in the system development process. This work has been primarily the realm of industrial psychology and organizational development, often difficult to translate to the unique setting of the system development experience. One of the books first connecting the fields of psychology and computer science even acknowledged that "the idea of the programmer as a human being is not going to appeal to some people" [Weinberg, 1971].

During that period, the field of software psychology brought new focus to the programmer's interaction with computing problems and solutions. Still, with some exceptions, early research in this area primarily involved the programmer's interaction with the computing environment — focusing on cognitive problem-solving skills, rather than the ability to work within a team or with others.

Recent research suggests that the focus has shifted to IT management and process, viewing the problems of system development from a project or management science perspective, rather than a personal one. When IT professionalism is discussed, it usually includes the context of technical or human resource–related issues, such as skills development, recruiting, and retention incentives. Still, while some of these works note the need for effective teamwork and communication among IT workers and users, they rarely describe how these dynamics can be assessed, taught, or developed [Curtis et al., 1988].

Recent notable works that describe the human dimension of the IT development experience include *Death March* [Yourdon, 1999], *Peopleware* [DeMarco and Lister, 1999], *Managing Technical People* [Humphrey, 1997], and *Adaptive Software Development* [Highsmith, 2000]. These works are insightful and useful; however, more often than not, the messages are not based on empirical evidence related to IT dynamics, as they rely heavily on case studies or anecdotal evidence. One notable exception to this is work by DeMarco and Lister [1999], which describes team productivity as it relates to several work environment characteristics, such as floor space allocation, noise levels, and degree of interruption.

In the first edition of this Handbook, Rosson [1995] notes the need for more empirical studies of computer professionals working in the field. This type of research poses unique challenges, because the range of variables and potential interactions impacting the human process carries a complexity that is difficult to manage. As a result, such studies are best approached from an epidemiological viewpoint, rather than as an experimental study based on controllable variables. Later in this chapter, we report the results of such a study, conducted with 632 IT professionals working in 77 different teams across the IT profession. This study is unique because it contains data for more than 80 variables, generated through both self-report and observational methods, collected from IT professionals [DAU, 2003]. This research can be used as a basis for developing new, human-focused training standards for IT managers, teams, and educators to assist them in responding to the changing demands of the IT landscape. First, however, we describe this changing landscape.

47.2 Underlying Principles

47.2.1 IT System Development Trends: History and Impact

New technology innovations over the past 10 to 15 years have profoundly impacted both the community of IT developers and the community of IT users. In this section, we review four key areas that have influenced the human dynamics of system development: distributed systems, ubiquitous computing, the growth of off-the-shelf software, and the Internet revolution. For each area, we describe its recent history and its impact on the human dynamics of system development.

47.2.1.1 Distributed Systems

In its simplest form, a *distributed system* is a collection of autonomous computers interconnected by a network. The emergence of distributed systems occurred in the early 1990s, in response to many of the problems associated with mainframe or mini-computing environments prevalent in the 1980s.

Distributed systems rely on software to produce a virtual integrated computing capability. These systems share several key characteristics: resource sharing, greater fault tolerance, transparency, scalability, concurrency, and openness. Resource sharing helps to reduce the overall costs of running concurrent systems, by allowing key resources such as hardware, software, and data to be shared across the system. Fault tolerance on distributed systems dampens the impact of hardware, software, or communications failures. Distributed systems allow for hardware redundancy, and they introduced the idea of recovery plans for system software failures. The openness of a distributed system governs its extendibility and the degree to which it adheres to industry standards for hardware, software, and communications protocols. *Concurrency* is the ability of the system to handle many different users accessing the same hardware and software resources. It is also concerned with the coordination of multiple responses to client requests in a coherent manner. *Scalability* relates to a system's ability to grow. Growth in a distributed environment usually involves an increased number of users, platforms, or additional software applications. In order for a distributed system to be robust, it must be scalable.

The emergence of distributed systems fundamentally impacted the human dynamics of system development. In this landscape, system users should not have to know (or care) where their hardware or software is physically located. The beauty of distributed system architecture is that it allows the administration and upkeep of the system to be located virtually anywhere without impacting the user. While this flexibility is beneficial, it can also affect the relationship between user and IT professional — with distributed systems, the easier face-to-face interactions encouraged by a centralized environment may take a back seat to the convenience and potential physical remoteness of a more virtual reality, where interaction with the end user may occur more through phone and e-mail.

The move to distributed systems has also had a major impact on the variety of IT positions in the workforce, as well as the IT professional pecking order. Distributed processing suddenly placed more emphasis on network and the system administration skills needed to keep the large, distributed infrastructures running. Additionally, new IT specialties, such as open system architecture and system security, have emerged. With the addition of these new specialties and the IT roles that accompany them, the traditional programmers and database experts of the prior decade have been relegated to secondary roles. System development is no longer focused solely on programming, but has broadened to include a new set of skills — and the diversity and complexity of the IT system development team has grown in response.

47.2.1.2 Ubiquitous Computing

The history and evolution of the personal computer (PC) is steeped in both folklore and competition. Many trace the origins of today's PCs to the Xerox Alto, but not all agree. For example, the history of IBM's PC, especially the XT and its spinoffs, have little or nothing to do with research done by Xerox on the Alto. Many believe that the first true personal computer was the Altair, introduced in 1975 [Pres, 1993]. The story of the Altair coincides with the emergence of two key people in system development: Bill Gates and Paul Allen, who started Micro-soft (*sic*) Corporation and developed programs for the Altair.

In 1977, Steven Jobs and Stephen Wozniak introduced the Apple II personal computer, the flagship for the newly introduced Apple Computer, Inc. For the first time, personal computers became affordable and accessible to the general population. Apple also introduced the intuitive graphical interface, opening the world of computing to millions.

Before the early 1980s, Digital Research, the developers of the CP/M operating system, had yet to build a CP/M version that would work with Intel's 16-bit central processing units. A small hardware vendor named Seattle Computer, with programmer Tim Paterson, offered an operating system much like CP/M, but which used Intel's 8086 processor. Under tremendous time pressure to get it to market, Paterson wrote the code hastily. Subsequently, the 8086-DOS operating system was dubbed QDOS, for *quick and dirty operating system*. Bill Gates, aware of IBM's problems obtaining an Intel-compatible operating system, acquired a license for and later bought QDOS. He then licensed it to IBM, where it became known as PC-DOS. The PC was a market success, and third-party applications developers began developing a wealth of products that would run on this platform [Pres, 1993].

Despite the rich, colorful, and very human history of the PC, most computing historians agree that current desktop configurations stabilized around 1990 [Pres, 1993]. During this time, the workstation became part of most professional business environments and appeared in many homes. Between 1993 and 1998, the percentage of U.S. households with a computer jumped from 23% to 42%; by 2000, the percentage had reached 51% [U.S. Census Bureau, 2000].

The 1990s' explosion of PC usage would forever change the human dynamics of computing, as IT professionals faced a redefinition of the term *user*. The number and variety of users in the PC world had become far greater than those of the earlier mainframe or mini-computer world. Users became much more sophisticated in the ways of computing, as they started thinking of computers more as interactive tools to meet business needs and less as sophisticated filing cabinets. This led to an increase in user demands from their systems and to heightened expectations of the programmers designing their applications. The ability to code was no longer the sole priority for an IT professional. Other key talents were now required: the ability to design effective user interfaces, an understanding of the client's business setting and requirements, and the ability to communicate effectively with IT-savvy clients.

47.2.1.3 The Emergence of Commercial Off-the-Shelf Applications

Spreadsheets, which originated in the 1970s with the package VisiCalc, led the proliferation of software applications in the 1980s in the growing PC market. Today, the market is inundated with commercially available software products or *commercial off-the-shelf* (COTS) packages. Sales of the top 500 software vendors were estimated at $72 billion in 1996, and the packaged software market grew by an estimated 269% between 1984 and 1994 [Carmel and Sawyer, 1998].

COTS components, in general, have the following characteristics [Vigder et al., 1996]:

Products are designed to sell multiple copies to multiple customers with no customization.
Design improvements and enhancements come from success, failure, and complaints in the marketplace.
The software vendor is responsible for ongoing customer support.
No single user, or group of users, has control over evolution of the product.
There is usually no access to source code except through applications programming interfaces (APIs).
System documentation may not be accessible by the user community, but user-oriented documentation
and training is usually well developed.

COTS are typically designed to meet specific, widespread needs. Today, it is common to see *COTS integration* projects, in which different COTS packages are fused together by custom-developed "glue code." Why is COTS integration so popular? Many, including the federal government, believe that COTS integration helps to reduce risk and cost in the long term. Successful examples of COTS integration have included graphical information systems, primarily because the components are usually bundled as a procedural library, various tools such as graphical user interface (GUI) constructors, office automation software, e-mail and messaging systems, Web browsers, database management systems, and system software components such as operating systems, windowing systems, and device drivers [Vigder et al., 1996].

Despite this trend, the success of COTS integration has been mixed [Sledge and Carney, 1998]. Pitfalls and limitations have included attempts to integrate components not designed for interworking, unstable or unsupported APIs, unanticipated labor required to wrap COTS in custom code to achieve required functionality, and inadequate stability and support from the COTS vendor. Furthermore, maintenance and upgrade of an integrated COTS-based system are tricky at best. COTS vendors have different release schedules, which must be coordinated into the integrated system's upgrade schedule. Every time upgrades of the various components are made, the system requires regression testing and often must be recalibrated, with changes to the custom glue code. COTS products also require custom code service providers to invest significant resources to stay current with existing and emerging products and tools in the marketplace.

Building an integrated COTS system is like building a house from a prefab kit for which the components come from different companies. Some are easier to fit together than others. Such an endeavor requires an extraordinary system architect and an integration team well versed in the integrated products. Even with such an extraordinary team, integrated systems remain at the mercy of changes made to future versions of individual components — changes that usually influence the functionality of the integrated system.

As with the skills shift accompanying distributed systems, the desirable IT skills associated with COTS integration have moved the emphasis away from programming. In the world of COTS integration, IT management and business requirements analysis skills become key, as IT professionals help users articulate their requirements and preferences in ways not thought of before. IT professionals must also remain current with the COTS products available on the market, from both a functional and a technical perspective.

Requirements and product research skills are not the only challenge; developers must also understand the intangible human dynamics at work in the COTS integration process. For example, a requirement will often start with the user preferring an existing software package or familiar GUI. If research suggests that better alternatives exist, then the IT professional must communicate with and educate the end users about these options. Often, users will reject a new package — even after training — because it pushes them beyond their comfort level. IT professionals often ignore this dynamic, perhaps because they cannot understand why end users would not jump at any chance to increase their functionality. The choice of software is often an emotional issue for users, and ignoring this reality can create gaps in the developer–user relationship, ultimately impacting the final product.

IT engineering courses often fail to teach the interpersonal communication and change management skills vital to these efforts. Failures in COTS integration efforts are often caused not by technical or feature failures (although this is sometimes the case in poorly integrated systems), but because the system has not been properly socialized with the end users. This understanding on the part of IT professionals will go a long way toward increasing customer satisfaction with the systems developed from COTS.

47.2.1.4 The Internet Revolution

In August 1962, J.C.R. Licklider described his "galactic network" concept in a series of memos, focusing on the major premise that social interaction could be enabled through widespread networking. He envisioned global, interconnected computers that would facilitate access to data and programs from anywhere in the world. Licklider went on to become the first head of the computer research program at the Advanced Research Projects Agency (ARPA), now the Defense Advanced Research Projects Agency (DARPA). While there, he refined the theory of the galactic network concept with Ivan Sutherland, Bob Taylor, and Lawrence Roberts [Leiner et al., 2002].

In 1961, Leonard Kleinrock published a paper on packet switching theory, following it with a book in 1964. Working with Lawrence Roberts, Kleinrock explored the feasibility of communications using packets rather than circuits. This theoretical milestone would prove crucial to the future of computer networking. In 1965, Roberts worked with Thomas Merrill to connect the TX-2 computer in Massachusetts with the Q-32 computer in California by way of a low-speed, dial-up telephone line. This experiment created the first wide-area computer network. The success of the experiment came with the realization that circuit-switched telephone systems would be insufficient for this massive level of communications. This realization helped to confirm Kleinrock's belief in packet switching as the critical component for the future Internet.

Working at ARPA in 1966, Roberts conceived a plan for the ARPANET and presented it at a conference attended by other researchers examining the same issues. Collaborative discussions among these researchers yielded a common understanding of the word *packet* and a projection for the proposed ARPANET of a speed of 50 kbps. By August 1968, the structure and specifications for the ARPANET were solidified, and in 1969, four host computers were networked together to form the initial ARPANET (Stanford University, University of California at Los Angeles, University of California at Santa Barbara, and University of Utah). Over time, more computers were added to the ARPANET, and in 1970, the host-to-host protocol called the *network control protocol* was completed and implemented. With the protocol in place, users began to develop applications to run on the network.

In 1972, ARPANET was publicly demonstrated at the International Computer Communications Conference (ICCC). It was also the year that the first e-mail application was written, primarily to facilitate discussion among ARPANET members. ARPANET would grow into the Internet we enjoy today. A key step in this evolution was the development of open architecture networking, formulated by R.E. Kahn, working at DARPA in the early 1970s. By 1985, the Internet was supporting a large number of system developers and researchers and was starting to branch out to include other communities, as well.

By the mid-1970s, other networks started springing up, primarily in the federal government. In addition to developing and introducing new thinking and technologies, the expanding federal government networking population made fundamental policy decisions that enabled the growth of the true World Wide Web. These policy decisions included the following [Leiner et al., 2002]:

Shared development costs and coordinated network access points were established for common infrastructure among federal agencies.

The Federal Networking Council in the U.S. and RARE in Europe were established. These organizations, along with others, developed the Coordinating Committee on Intercontinental Research Networking, to coordinate Internet research worldwide.

A culture developed that fostered sharing and cooperation between agencies. For example, DARPA allowed CSNET and the National Science Foundation (NSF) to share ARPANET's infrastructure.

In addition, the NSF encouraged its regional NSFNET networks to develop commercial customers, expanding their services and facilities to accommodate them.

The NSF instituted an acceptable use policy, which limited use of the NSFNET backbone to research and educational use. Consequently, a market in providing private, competitive, long-haul networks, such as PSI and UUNET, was born.

In 1988, an NSF-commissioned report, *Toward a National Research Network* proved to be very influential, particularly for then-Senator Al Gore. This report, and the subsequent political attention it prompted, established the high-speed networks that would become the foundation of the future information superhighway.

In 1994, a National Research Council report, *Realizing the Information Future: The Internet and Beyond* was released to the public. This document became the blueprint for the information superhighway. This report anticipated and discussed several key issues for Internet governance, including intellectual property rights, ethics, and the regulation of the Internet.

From a human dynamics perspective, the Internet has far exceeded even Licklider's vision of a "galactic network" to facilitate social interaction. Between 1997 and 2000, the percentage of U.S. households with Internet access jumped from 18% to 42% [U.S. Census Bureau, 2000]. Today, the Internet can facilitate almost every aspect of human interaction, including voice, video, and text-based communication.

The Internet fundamentally changed the way IT professionals do business. Even systems designed to meet individual group needs must now be considered in a broader context. Information contained in a stand-alone system often must be shared or even sold over the Internet. Consequently, many new stand-alone systems are designed to allow Web-enabling in the future. As a result, IT professionals are often asked to predict the requirements for a user population far beyond the local institution.

The Internet also helped to create a far more computer savvy user population, with high expectations for both attractive user interfaces and effective functionality. The length of development life cycles also continued to shorten. Because users can create their own Web pages and make these instantly available on the Web, expectations from IT developers for the same type of rapid development increased. System owners and business experts now see how the Internet and associated technology enable them to reach out to their customers and stakeholders in a way never before possible.

All these developments suggest that the IT professional must tap into this IT-savvy population and harness the power of the human factor to help build better systems for new and emerging applications. In addition, the IT profession needs to understand better its own biases, blind spots, and strengths, and learn how to use this knowledge to benefit future developments.

47.2.2 IT Management — Process Challenges

As users became more sophisticated and development times decreased, new engineering life cycles emerged to help guide development in this new and dynamic environment, with its ever evolving requirements. This section describes the changing face of the process side of IT and will focus on two themes. First, as engineers, IT professionals have attacked the dynamic problems of system development with process and procedure, creating a plethora of standards, protocols, and maturity models. Second, the IT profession has failed to address adequately — except through problem-focused exercises such as usability engineering (see Chapter 45 of this Handbook) — the human dynamics of our users' perspectives. Classical engineering training does not equip IT professionals to think in these terms, preferring the comfortable realm of the "problem first, people second" approach.

47.2.2.1 System Development Methodologies

Every complex endeavor needs an organized plan to keep it on track. Software engineering is no exception. In the world of software applications development, user requirements are usually not well defined, yet most users want their applications done quickly. Consequently, a number of software engineering life cycle models have evolved, describing the activities that must occur. Here, we summarize the most commonly used models, focusing in each case on the human dynamics for both for the development team and the user. (See also Chapter 110 in this Handbook.)

Pure waterfall. The waterfall life cycle model, first developed in the 1970s using traditional engineering processes and discipline, forms the basis for almost every other software development model. In the waterfall model, development occurs through a series of sequential phases, with a major milestone review or audit marking the end of each phase. The waterfall life cycle progresses through seven basic steps: software concept phase, requirements analysis, high-level or general design, detailed design, coding/unit test/debug phase, system testing, and implementation.

The waterfall life cycle works well for low-speed, low-change projects, in which there is a very clear and stable definition of the requirements, the product, and the supporting technology. It is also widely used in large projects where outputs are combined with other engineered systems. A project that would fit this approach is porting an existing application to a new platform.

The waterfall model is likely to appeal to developers and users who have lower tolerance for ambiguity, enjoy the stability of long-term projects, and prefer concrete milestones and metrics measuring completion. Such a process can be frustrating for users and developers, however, since the product often is not seen until late in its development cycle.

Code and fix. Probably the least disciplined, but still widely used, software engineering methodology is the *code-and-fix* approach. Although most software engineers would argue that code and fix is not really a life cycle at all, its frequency of use cannot be ignored. Code and fix is, of course, an *ad hoc* methodology. A development team will typically start with a general understanding of what they want to build, and then will proceed through steps that combine informal design, coding, and debugging (and maybe, but rarely, some documentation). The end result is rarely robust, but code and fix is attractive because it shows users results very quickly.

Should code and fix be avoided at all costs? Not necessarily. Code and fix is a useful methodology in a few distinct cases. For example, code and fix can be effective for very limited proof-of-concept demonstrations or throwaway prototypes to help evolve requirements. It can also be useful in testing the feasibility of a certain approach before more serious development is pursued.

Code and fix is often appealing to programmers because development moves directly to coding, away from the more abstract tasks of design and integration. It also appeals to ego, in the sense that a few very good hackers can actually build a fairly good system without the overhead associated with more complex methods. However, there is a danger in this approach: developers less schooled in formal methods often mistake code and fix for more disciplined approaches, such as those that follow.

Spiral model. Almost every software engineering project carries risk of failure. Risks can result from poorly understood requirements, for example, or from problems with the underlying technology. The *spiral model* (see Chapter 110) is designed to minimize risk [Boehm, 1998]. While successive rings of the spiral include the waterfall phases, the iterative steps inside each spiral enable designers to identify and address risks while it is still reasonably inexpensive to do so. Each iteration involves six basic steps [McConnell, 1996]: determine objectives, alternatives, and constraints; identify and resolve known risks; examine alternatives; produce and verify iteration deliverables; plan for the next iteration; and agree to the approach for the next iteration.

The spiral model is attractive because it allows developers to tailor the model continually to meet project objectives. For example, developers can use the model to identify and clarify risk at the beginning of the project. Once all major risks are identified and addressed, they may move into a more waterfall-oriented development process. Of course, like any of the development methodologies, the spiral model has its limitations. It is complex and requires diligent management, including the ability to establish verifiable milestones and validate them before proceeding.

Evolutionary prototyping. Often, users can articulate general needs for a system, but they are unable to express these needs in the form of specific requirements. Such systems can be designed with *evolutionary prototyping*, which focuses on the most visible parts of the system, usually the GUI and the reporting module. Through successive sessions with the end users, the prototype is modified and enhanced until it achieves the end users' desired result. At that point, the prototype becomes a *de facto* as-built specification and is solidified by the development team. The back-end development team then designs a more robust foundation that incorporates the agreed-upon user interface.

Techniques such as rapid application development (RAD) and joint application development (JAD) are often used to facilitate these evolutions. RAD refers to a process that delivers functionality to users quickly by dividing system functionality into logical segments for rapid development, often with some combination of prototyping, JAD sessions, SWAT teams, CASE tools, and time-boxed deliverables. JAD sessions bring users together with the development team to elicit consensus-based user requirements. Roles required for the development team include the JAD facilitator, who asks a series of prepared questions to elicit decisions from the users; observers, who listen to the interactions; and documenters, who capture the session discussions and decisions.

Many of the iterative development life cycles used today are criticized because of the amount of overhead and time required to implement them. Consequently, newer, "lightweight" software methodologies have begun to attract some attention. One of these is *extreme programming* (XP). XP came on the scene in about 1997 to force a reexamination of standard software development practices and to provide an alternative that would reduce cost. The rules and practices around XP can be grouped into four categories: planning, designing, coding, and testing. Designed for use by IT teams of up to 10 people, XP emphasizes iterative development centered on "user stories" to elicit requirements and drive design. It also encourages rapid iterations, developer-created schedule estimates, programming in pairs, continuous testing, and active user involvement [Beck, 1999].

The greatest strength of evolutionary prototyping is its ability to facilitate the discovery of requirements. There are also secondary benefits. Intimate end user participation in the system's evolution supports user buy-in at the end of the project. In addition, a representative user group participating in this way trains itself, subsequently yielding in-house personnel familiar with how the system works. The greatest weakness of evolutionary prototyping is defining when the project is complete. This becomes a subjective judgment, unless ground rules are well established in advance (e.g., number of iterations before requirements are locked, budget or schedule constraints, etc.). In addition, as with all life cycles, the development team must be constantly vigilant about requirements creep, managing this in a way that neither offends the client nor compromises the development schedule.

Evolutionary prototyping requires a diverse programming team and a broad set of communication and technical skills. This method introduces the role of facilitator to the IT team for JAD sessions and requires the GUI developer to spend extensive time working directly with the user. One of the developer's key roles is expectation management — for example, explaining to users the time required for robust back-end development once the GUI requirements are complete and locked.

Staged delivery. In staged delivery, the user receives incremental or *staged* deliveries of the system. The biggest difference between the staged model and other evolutionary models is that a complete set of requirements is well documented and understood before the first release. Once the detailed design is complete, a schedule of delivered functionality is agreed upon.

Staged delivery begins much like a pure waterfall process, but by iterating the activities of detailed design, coding, debugging, testing, and implementation, the user receives functionality much earlier in the life cycle. One of the major constraints of staged delivery is that it requires careful planning from both management and technical viewpoints. Deliveries must yield products that are both tangible and useful to clients. At the same time, functionality must be grouped in such a way that a feature slated for early release is not dependent on a feature slated for later development. Although this sounds like common sense, it can become very complicated on large-scale projects. This requires that the development team has clear communication processes and coordination skills; it also requires careful documentation to support future stages of the development process.

Design to schedule or cost. Closely associated with a staged release life cycle, this method is used when users do not have a firm understanding of requirements but are constrained by either a firm deadline or cost limits. In this life cycle, there is constant rescoping of requirements to ensure that the system can be delivered on time and below a ceiling cost.

This approach is often used by developers needing to demonstrate functioning software for a particular event, like a trade show or year-end. This approach often works for auxiliary products that are not on the critical path. For example, Microsoft Windows includes applets like WordPad, Paint, etc. Microsoft might use a design to schedule methodology for releasing new applets, to ensure that the released version does not delay or interfere with the primary product.

Design to schedule or cost requires the development team to predict accurately the time it will take to complete a specific task and to understand the essentials of project management, so that the most effective trade-offs can be made. This method can be frustrating for developers, who may feel held back by the user's limitations and who are not always aware of the management constraints at work.

47.2.2.2 Process Improvement Initiatives

The emphasis on system development models has been accompanied by a number of process improvement models. These process initiatives, primarily based on the principles of Deming and Crosby [Ahern et al., 2001], stress the concepts of critical thinking, root cause analysis, repeatable process, and assessment metrics. The "theory in use" manifestation of these initiatives has taken several forms, ranging from the level of organizational culture to the level of individual programmer processes (see also several chapters in Section XI, Software Engineering). This section provides examples and considers the potential impacts on the human dynamics of the development team.

ISO 9000 standards. At the organizational level, the ISO 9000 family of standards is a basis for documenting an organization's management system (what it does to manage its processes) and its approach to quality. Key areas of ISO include customer focus, leadership, involvement of people, process approach, system approach to management, continual improvement, factual approach to decision making, and mutually beneficial supplier relationships (ISO 2002).

From a human dynamics perspective, the strength and the weakness of the ISO system lie in its generic nature. ISO standards delineate specific areas of consideration for an organization (the *what*), but purposefully avoid implementation guidance (the *how*). For example, the leadership principle is encouraged to achieve results such as "establishing trust and eliminating fear." This is a noble goal, but how can a system development manager proceed to achieve it, amidst looming deadlines and irate clients? Unless significant organizational investment is made to articulate and introduce ISO within an organization, its principles are not likely to be embraced formally as a systemic and guiding set of values. Furthermore, ISO standards are grounded in the concepts of standardization and repeatable process — ideal for stable systems engineering processes, but less immediately attainable for the dynamic world of iterative design. On the other hand, for organizations with the investment resources and time, determining how to apply ISO standards can yield meaningful dialogue, resulting in fundamental cultural change if managed effectively.

The ISO publication related to human-centered design (ISO 13407) specifically outlines methods for considering usability in the system development process. Key activities include planning the design process, understanding context of use, specifying user requirements, producing designs and prototypes, and completing user-based assessments [Maguire, 2001].

Personal and team process metrics. Some software process improvement efforts, such as the *personal software process* and the *team software process*, focus on the documentation of time spent on specific activities performed by system engineers during the planning, development, and testing phases of the system development process. The underlying premise of these efforts is that quantifying the daily human activities of system development allows the developer or team to predict more accurately timelines associated with design projects, leading to more effective management of individual and team time [Humphrey, 1997].

From a human dynamics perspective, there are both benefits and risks to this approach. For developers who prefer the structure and predictability of time management techniques, these methods provide a useful tool for organizing and structuring the design continuum. On the other hand, for developers preferring a more flexible and adaptable approach to accomplishing work (and who may actually be most efficient when allowed to move between activity bursts and lulls), forced documentation may discourage the natural flow of their efforts.

Another possible risk of introducing team metrics relates to the strict documentation of specific activities and associated timelines. Such tasks represent a strong emphasis on goal orientation and product achievement — achieving the end by measuring the means. While understanding one's own process in pursuing a goal is an important aspect of self-management, such a strict emphasis on goal-focused time management and personal efficiency may decrease the spontaneous and creative outcomes that may emerge if the developer simply experiments with new ideas, with no particular goal in mind. (See Apter [2001] for an overview of telic and paratelic motivational states.)

It is interesting to note the general absence of human interaction time built into the standard worksheets associated with these process tools. The interaction time that supports the development of a relationship between IT professional and user is not recognized as a formal activity in these methods, yet this relationship can fundamentally impact project success. Customer management often includes the social aspects of IT development. Not all contributors to quality can be measured — the personal connection made during the basic activities of human interaction is one such intangible contributor.

Process tools. In the problem-focused world of information technology, software process improvement initiatives have been accompanied by a full suite of software packages, tools, and methodologies. From requirements management techniques and databases to object-oriented design techniques, CASE tools, and debugging tools, the emergence of tools to assist in the system development process are far reaching and have streamlined many traditionally chaotic development processes and management efforts (see also Chapter 116).

Fancy tools can be seductive, and development teams can easily fall into a trap of "form over function" when using them. As Yourdon notes in *Death March*, "What the users understand is their own native language . . . and what most users are willing to read is a short document that summarizes the requirements for the system. The key point is that it's English, it's terse, and it's to the point. . . . We already have a tool for creating such requirements documents; it's called a word processor" [Yourdon, 1997]. The use of structured tools does not eliminate ambiguous requirements, nor does it replace the independent critical thinking that is required for good problem solving. When the final system is complete and delivered, little attention will be paid to the tools used to construct it.

Capability maturity model (CMM). Developed by the Software Engineering Institute (SEI) at Carnegie Mellon University, the CMM process improvement model outlines five levels of process maturity, ranging from a state of no control or documentation (level 1) to a state of continuous improvement at the organizational level (level 5). Each of the five levels is associated with key process areas, focusing on the following concepts [Curtis, 2002]:

Stability as a path to improvement — "A fundamental premise of the process maturity framework is that a practice cannot be improved if it cannot be repeated."

A foundation for common practice — "Until basic management control of daily work is established, no organization-wide practices have a chance of being deployed successfully, because no one has the time to master them."

Quantitative management — "The premise underlying quantitative management is that if a well-understood process is repeated, you should get essentially the same result."

The staged progression of the CMM program is a key advantage, for it allows organizations to target the level most appropriate to their needs and most attainable with their resources. Not every software development organization needs to operate at level 5. Bringing an organization even to CMM level 2 or 3 requires commitment throughout the organization — from top management to the junior programmer. Like the personal and team process metrics described previously, CMM comes with a clear requirement for effective planning and documentation, which may not appeal to the brilliant hacker still in code-and-fix mode.

Adopting process improvement initiatives. Research has shown that adopting new IT process models and tools can be a difficult process, highly dependent on developer acceptance. For example, a 1999 SEI research study found that IT process tool adoption can be linked to three key factors: the developer's perceived control over the work, the developer's perception of the new tool, and the developer's perception of the impacts of using the new tool or process [Green and Hevner, 1999]. Understanding the personality of the IT professional may yield valuable insights into process improvement efforts and their potential impacts on different systems and individuals. These human dynamics are addressed in the following section.

47.2.3 IT Personality

Recent changes in the IT landscape, and the resulting change in the developer–team–user relationships, force a fresh look at the IT personality and the psychology of computer programming. In the past, much of the literature on software psychology used cognitive and skills-based assessments, focusing on the IT individual's abilities and orientation with respect to problem solving and the task at hand. Equally vital in today's setting are tools that help IT professionals understand their relationships with other people. Today's development environment requires tools that enable individuals and teams to identify their personal preferences and styles and to manage better themselves, their interactions with team members, and their interactions with users. This section describes two tools that are often useful when working with IT teams.

Personality type: background and theory. Several writers in the IT field (e.g., Weinberg [1998], DeMarco and Lister [1999], Humphrey [1997], and Yourdon [1997]) have recognized the importance of personality or psychological characteristics on team performance and success. One of the most popular and useful tools in assessing these characteristics is the Myers–Briggs type indicator (MBTI), an assessment tool grounded in the theory of personality types proposed by Carl Jung [Myers et al., 1998]. The MBTI measures an individual's personality preferences on four distinct scales, each with two opposite sides:

Energy scale: Extraversion–Introversion (E/I) — How people prefer to gather energy
Perceiving scale: Sensing–Intuitive (S/N) — How people prefer to gather new information or data
   (perceptions)
Judging scale: Thinking–Feeling (T/F) — How people prefer to make decisions (judgments)
Orientation scale: Judging–Perceiving (J/P) — How people express their perceptions or judgments

The theory of personality type proposes that each individual has a preference for one side of each of these dichotomous scales. For example, people energized by the outer world of people, action, and things, who prefer to process new ideas with others, have a preference for Extraversion. Conversely, people energized more by their inner world of concepts and ideas, who prefer to process new ideas alone before sharing them with others, may have a preference for Introversion. Table 47.1 lists each of the MBTI scales and the two preferences associated with each [OKA, 2000].

One key point is worth emphasizing: personality type is about preferences, not absolutes. Although someone might prefer subjective decision making based on what is best for the people involved (Feeling judgments), this does not prevent that person from making objective decisions based on analysis of cause and effect. It simply means that this person accesses the subjective decision making process first and with more ease, whereas it might take more effort and feel less comfortable to make the impersonal judgments that some decisions require. As a result, this person might self-select into activities that allow use of the preferred styles — in this case, working with people directly, rather than with impersonal problems.

Once an individual's letter preference on each scale is selected, a personality type is determined. This *MBTI type* is the four-letter grouping of preferences along the scales. For example, someone with the type ISTJ has preferences for Introversion, Sensing, Thinking, and Judging. With four scales and two possible preferences along each scale, there are 16 possible personality types, each reflecting a unique combination of personality preferences.

TABLE 47.1 Myers-Briggs Type Indicator (MBTI) Personality Scale Descriptions

Scale	Scale Descriptions	
E/I — Energy sources: introversion/extraversion	Extravert (E) — Gains energy from interacting with outer world of people, action, and things. Quiet time can be draining. Applicable words: interactive, expressive, disclosing, "speak to think."	Introvert (I) — Gains energy from inner world of concepts and ideas. Extensive interaction can be draining. Applicable words: concentrating, internal, contained, reflective, "think to speak."
S/N — Perceiving mental function: data gathering (What do you first notice?)	Sensor (S) — Prefers to perceive the immediate, practical, real facts of experience and life, collecting information through use of the five senses.	Intuitive (N) — Prefers to perceive possibilities, patterns, and meanings of experience, relying on a sixth sense of hunches to gather information.
T/F — Judging mental function: decision making (How do you prefer to make decisions?)	Thinker (T) — Makes decisions objectively and impersonally, seeking clarity by detaching from the problem. Cause–effect oriented.	Feeler (F) — Makes decisions subjectively and personally, seeking harmony with inner values by placing him- or herself within the problem. Relationship oriented.
J/P — Orientation attitude: (What is the world most likely to see from you: data or decisions?)	Judger (J) — More likely to show the external world his or her decision-making (judgments). Behaviorally, prefers to live in a decisive, planned, orderly way, aiming to regulate and control events. Often appears closure-oriented, with a focus on the goal to be reached.	Perceiver (P) — More likely to show the external world his or her perceiving mental function, sharing data and perceptions rather than decisions. Behaviorally, prefers to live in a spontaneous, flexible way, aiming to understand life and adapt to it.

Four specific pairings of MBTI letter preferences also result in four unique *temperaments*, which map well to specific behavioral, learning, and leadership styles. These four temperament groups follow [OKA, 2000]:

Sensing Judgers (SJ) — Stabilizers who prefer structure, order, accountability, reliance on existing systems that work, policies and procedures, and the proven way of doing things

Intuitive Thinkers (NT) — Visionaries who prefer nonconformity, systems theory, conceptualization, independence, objective complexity, and change for the sake of change (if it produces learning)

Intuitive Feelers (NF) — Catalysts who prefer interpersonal support, relationships, possibilities for people, interaction, cooperation, imagination, and supportiveness

Sensing Perceivers (SP) — Troubleshooters who prefer hands-on action and experimentation, practical solutions, variety and change, immediacy, flexibility, and adaptation

Personality type: applications. One of the key benefits to using the MBTI is that an individual or team can complete the assessment and learn about the results and applications in approximately four hours. Understanding personality type reveals important human dynamics in the development team or process. For example, consider requirements gathering in light of the Intuition–Sensing scale described previously. End users and developers who prefer gathering information through Sensing (called *Sensors*) may communicate detailed specifications and develop requirements by addressing specific technical needs. Sensors often describe systems from the ground up. End users and developers who prefer gathering information by Intuition (called *Intuitives*) may start by painting a broad picture of a future system, focusing on the possibilities that could be achieved — requirements that begin with the big picture, with specifics added later.

Communication difficulties often can be traced directly to the preferences associated with personality type. As an example, the authors once conducted requirements interviews with two different users describing the same need: a document management system allowing the user to search on key words or phrases within the stored documents. One user started the interview with, "I need golden words; help me find the golden words!" This user was likely an Intuitive, starting by painting a figurative picture of the need. The other user started the interview by presenting several sample screen shots from other similar systems, displaying the documents requiring storage. This user was likely a Sensor, communicating requirements by presenting existing, tangible samples of specific needs. Which response is better? The answer may depend on the interviewer's own preference for Sensing or Intuition. In fact, each response brings a unique and valuable perspective.

Helping IT professionals to identify their personality preferences enables them to manage better themselves and their interactions with team members and users. Consider the advantage to the requirements analyst armed with an understanding of personality type before entering the interviews just described. The educated analyst will begin where the user begins, and then migrate to the other preference to obtain a more complete picture. This often means starting with the big picture and drilling down with Intuitives, or starting with the details and broadening up with Sensors. Analysts unaware of this technique may force the client to begin where the analyst does (the client's nonpreference if they are opposites) or may never reach the other level of information (if client and analyst share the same preference).

Interpersonal needs: background and theory. The MBTI helps individuals understand their own personality preferences and apply that knowledge for self-management. A second tool, the fundamental interpersonal relations orientation–behavior (FIRO-B) survey, is a personality instrument that measures how one typically behaves toward a team or a group of people and what behaviors are expected in return [Waterman and Rogers, 1996]. This tool assesses three different scales:

Inclusion — Needs related to community, belonging, involvement, participation, recognition, and distinction. Inclusion assesses the extent of contact and prominence that a person needs.

Control — Needs related to power, authority, influence, responsibility, and consistency. Control relates to decision making, influence, and persuasion between people.

Affection — Needs related to acceptance, feedback, personal ties, consensus, sensitivity, support, and openness. Affection relates to emotional connections between people and determines the extent of closeness that a person seeks.

Each of these three scales has two dimensions: expressed (how much a person needs to extend this to others) and wanted (how much a person needs others to extend this back). Table 47.2 describes the six components of a FIRO-B assessment [Waterman and Rogers, 1996].

The FIRO-B assessment provides feedback on each of these six components, with scores ranging from low (limited or highly selective need) to high (strong preference or need for the behavior). For example, a person with low expressed control needs and high wanted control needs probably feels a limited need to exert individual power or influence in many situations, preferring someone else to provide the structure and direction. A person with high expressed control and low wanted control scores may prefer to exercise the control and direct a situation but not wish others to exert that same control in return.

TABLE 47.2   FIRO-B Interpersonal Needs Scale Descriptions

	Inclusion	Control	Affection
Expressed	How much do you try to include others in activities? How hard do you try to belong to groups and be with others?	How much do you try to exert control and influence, and to direct others?	How much do you try to be close to people? What is your level of comfort in expressing personal feelings and supportiveness?
Wanted	How much do you want others to include you in activities? How much do you want others to invite you to belong?	How strong is your need to be in well defined situations? To what degree do you want others to take control?	How much warmth do you want from others? What is your level of enjoyment when people share feelings, and when they encourage efforts?

Interpersonal needs: applications. The FIRO-B is useful in learning about team dynamics and the range of possible needs within a group. For example, it has been proposed that IT managers should hold weekly team sessions outside the work environment to encourage communication among members [Yourdon, 1997]. Some managers may wish to temper this recommendation, based on a consideration of the FIRO-B. If most team members have a very selective need for both expressed and wanted inclusion, the attractiveness of such social gatherings becomes less marked — most team members may not consider that type of interaction a prerequisite for effectiveness.

The challenge for many teams comes when there is incompatibility between the expressed needs and the wanted needs on a specific scale. If a manager with high expressed control needs (prefers to provide structure and direction and to be in a position of power) manages an IT team with low wanted control needs (generally do not want others to exert power or control over them), then the potential exists for conflict within that group over these issues. Understanding FIRO-B helps teams to discuss sensitive issues such as power and feedback, which may help articulate roles, process, and structure in the IT development process.

47.2.3.1 Psychology of the IT Professional

With this background, it is useful to look at the personality preferences and interpersonal needs of IT professionals and to consider the potential impact of these characteristics in the team environment and with the user. Recent research led by the authors with a representative sample of more than 600 IT professionals in 77 different IT teams revealed interesting insights into the psychology of the IT professional [DAU, 2003]:

Three quarters (77%) of the IT professionals surveyed reported an MBTI preference for Thinking decision making, with only 23% preferring Feeling decision making. Given that the split in the general population is generally even between these preferences, this represents a significant overrepresentation of IT individuals with the Thinking preference. Thinkers, as they are termed, generally prefer logical, objective, impersonal decision making, focused on cause–effect relationships and the clarity that comes from objectivity (problem first, people second). The underrepresented Feelers, conversely, prefer to make decisions by placing themselves within a problem, using empathy to connect with the individuals involved (people first, problem second).

Almost half of the IT professionals surveyed (41%) reported as Introverted Thinkers (a combination of the introversion and thinking preferences). This is nearly twice the percentage seen among the general population. Introverted Thinkers often prefer a "lone gun" approach to much of their work, avoiding teams, collaborative efforts, and the training that support such structures. This group is least likely to engage and connect interpersonally with others, and may be reluctant to create personal bridges of trust and openness with colleagues. This finding was supported with FIRO-B results, which revealed that 55% of IT professionals have very low wanted inclusion scores (i.e., a low or highly selective need to be included in the activities of others).

The two most prevalent temperaments among IT professionals are the Intuitive Thinking (NT) and the Sensing Judging (SJ) temperaments. These are represented at 27% and 48%, respectively, in IT teams, compared to 13% and 39% in the general population. Interestingly, these are also the most dynamically opposed temperaments, with SJs typically finding fulfillment in belonging to meaningful institutions and proven systems, and with NTs typically preferring to reinvent systems to experiment with new way of doing things.

Despite the trends toward individualism and autonomy noted previously, IT professionals do have trends toward moderate needs for affection and expressed inclusion (FIRO-B), meaning that they are willing to involve others and to provide a sense of connection with others at a moderate rate.

These findings are generally consistent with earlier studies investigating MBTI types among computer professionals [Hildebrand, 1995; Westbrook, 1988; Lyons 1985], although we found a slightly higher representation of the ESTJ whole type than those studies, which showed a higher percentage of INTJs.

The preferences associated with some of these findings echo the legendary portrait of the secluded IT programmer, content to spend hours at the keyboard as long as the work is challenging and someone occasionally shoves pizza under the door. Likewise, the personality preference characteristics described previously appear less consistent with the new vision of the IT professional as facilitative consultant, working side by side with end users to articulate requirements and build systems. The learning point, however, revolves around the concept that if IT professionals understand their basic preferences, they can use this information to identify individual preferences on a team, potential blind spots (often associated with a nonpreference), and potential interaction dynamics with other team members and clients.

For example, the ISTJ programmer's personality type suggests preferences for organization, privacy, practicality, responsibility, objectivity, and individual attention to detail. Under no circumstances do these preferences preclude an ISTJ from effectively facilitating a JAD session or from working collaboratively with users during an evolutionary prototyping effort, even though these activities are more obviously associated with Extraversion or Feeling personality preferences. It does mean that an ISTJ programmer introduced to these roles for the first time may not prefer to engage in these activities and may have more difficulty in accessing the personality characteristics that best support such work. Well structured training can help bridge these gaps, to help IT professionals develop the interpersonal skills required to meet today's IT needs.

47.2.3.2 IT Team Dynamics

System development is becoming more and more a team sport. This requires that team members have not only a fundamental grasp of their own personality preferences, but also of the needs and preferences of other team members, managers, and end users. The same research that identified personality preferences among IT individuals [DAU, 2003] also reveals interesting human dynamics at the team level.

Team dynamics. As noted previously, a recent study revealed that almost half of IT team members can be characterized as "lone gun" professionals, preferring to contribute through individual efforts with a strong focus on independence and self-sufficiency. For the study, these preferences were also assessed at the team level using a social climate instrument called the *work environment scales* (WES) [Moos, 1994]. The WES measures group-level dynamics across 10 scales and gathers data related to what team members believe the work place is like vs. their perception of the ideal work place. The preferences toward individuality were (not surprisingly) echoed in this team-level assessment, with teams reporting higher desired levels of autonomy, self-sufficiency, and task focus than currently experienced.

Despite this desire, however, these teams also reported desiring a significantly higher level of peer cohesion than currently experienced. Results also revealed that although team members generally desire little control exerted from management (consistent with the FIRO-B low wanted control scores), they desire a high level of supervisor support and want to know what is expected in their work.

Factors influencing IT success. Many studies have assessed the factors contributing to the success of IT projects [Standish Group, 2001]. Recent research [DAU, 2003] also assessed this factor, focusing on the level of the team (rather than the specific project). In this case, IT managers and team members were asked to describe the factors contributing to their success or their turmoil, depending on how they classified their teams. Managers and teams generally agreed on the three most important factors, regardless of whether the team was successful or in turmoil. These included delivers (or does not deliver) products on time, delivers (or does not deliver) high-quality products, and works together effectively (or does not). Although each team may have a different perception of what working together effectively requires, the finding echoes the theme that collaboration at the team level is a valued contributor to success.

Team dynamics: applications. Here is an example of how personality indicators such as the MBTI and FIRO-B can be used with a development team. The authors recently worked with a development team led by an ESTJ manager with high inclusion and affection needs on the FIRO-B. These results were consistent with her frank, direct, outgoing nature; evidence of frequent interaction with both the client and the team; and frequent feedback sessions about project status and metrics. The manager was frustrated because she sensed resistance to her requests for feedback from team members, and the team expressed little interest in engaging socially with her or with each other. The technical work generated by the team was excellent, but there appeared to be little interest in whether it met the client's aesthetic preferences.

Completing the MBTI and FIRO-B in a half-day workshop revealed that 80% of the manager's team were Introverted Thinkers with collectively low FIRO-B scores. They, too, were frustrated, because there were too many meetings, the client continually changed his mind about screen layout and colors, and the manager seemed too focused on group sharing.

Sharing and discussing MBTI and FIRO-B scores gave this manager and team a neutral vocabulary for discussing frustrations. Seeing the differences between the manager and the team helped the group to identify specific sources of conflict and to discuss them in a nonconfrontational way. Ultimately, the team established new ground rules related to communication pathways and frequency, designed to meet both managerial and team needs.

Once the team had moved past these basic communication issues, they were also better able to predict the personality preferences of the client. This led to clearer ground rules related to requirements management and review procedures. Although one four-hour workshop cannot heal the wounds of an IT team in turmoil, it is a useful starting point for more effective self-management, and it can open the door for more constructive discussion about the people issues that the IT personality tends to avoid.

47.2.3.3 IT Meets the User

No developer sets out to design or develop a substandard system that will not meet his or her users' needs. Moreover, no user purposefully sets out to confuse or muddy the requirements that he or she is trying to articulate. Effective communication and a positive, trusting working environment between IT professionals and their end users is vital to ensure a successful outcome, but both sides are often ill equipped to create such an environment. The creation of this environment is becoming increasingly important as the evolution of COTS integration and the sophistication of end users continue to develop.

The results of our IT study [DAU, 2003] demonstrate that there are significant differences between IT professionals and the general population. By understanding these differences and by formulating strategies that can help turn these differences into strengths, we can help to establish best practices to ensure team success and better systems. Here is an overview of some key human dynamics facing the development team when working with an end user:

User requirements — One of the key challenges in today's development environment is balancing the elicitation of often uncertain and changing user requirements against the minimization of development risk to result in a final product. Although many tools are available to assist IT developers with the documentation and management of requirements, little focus has been spent on the specific conversations and communication tools and approaches developers can take during interviews and sessions with the end users.

User as contributor — A second challenge in working with users is understanding their current perceptions with respect to information technology, and any assumptions they have about the development process. If the user has preferences for specific COTS products or unspoken concerns about specific possible outcomes, the developer must understand these at the outset. For example, sophisticated users may understand the need for a large-scale relational database with a COTS reporting module to fulfill requirements, but they may fear the loss of personal power that the hands-on control over the resident data currently provides. Understanding the personal history that a user brings to the process may sound trivial; however, given the significant impacts that these issues may have as work proceeds, it is important that the developer be sensitive toward them. Training in effective communication skills and coaching techniques is a first step toward developing this awareness among technical professionals.

Process problems vs. system needs — An IT development project is usually initiated to help resolve a business problem or to improve or streamline an existing business process. Even with the sophistication of today's technology, IT is not a panacea for business process woes or organizational turf battles. IT developers in today's business environment need the capability to recognize the difference between process, organizational, and technical problems, so that process mapping and reengineering efforts can be introduced if necessary. Some organizations address process issues by introducing a new system to manage roles and responsibilities, process, and interactions between players. Recognizing the connection between existing or desired business processes and the articulated requirements requires personal insight on the part of the IT analyst and should be included in the analyst's training and development as a professional.

Service focus — Today's relationship between end user and IT team is often long term and consultative in nature. Databases require long-term administrative support; Web sites require maintenance; technology changes lead to application upgrades; and changing business needs require future iterative development. This reality means that IT professionals must able to build lasting relationships with their clients — clients who are more likely to desire and need that interpersonal connection than the typical IT professional.

47.3   Best Practices

Myriad best practices are available to the IT professional and manager. Most of these practices are either technical or process in nature, focusing on the technical and/or management skills and tasks required to successfully navigate through a system development effort. This section presents a new perspective on the area of best practices, focusing purely on the human level of the IT individual and team. These best practices are personal in nature and result from our experience and belief that the success of an IT development effort depends heavily on the daily conversations between people — from the day the project need is identified to the day the first user accesses the completed system.

47.3.1   IT Developers

Self-understanding and management benefits technical work. Many IT professionals tend to avoid training that focuses on the development of interpersonal and communication skills, believing that such investment of time and resources is not as beneficial as learning a new technical skill. At the same time, it is hard to find a technical specialist who has not experienced conflict with team members, misunderstandings with the boss, or confusion with end users or clients. Many of these interpersonal issues can be traced to a fundamental style or communication mismatch, rather than a technical deficiency. IT professionals who recognize these root causes are better prepared to manage problems effectively, leading to better technical work. Understanding what one brings to the team from a human dynamics standpoint, and how that might mesh or clash with coworkers, is a key starting point.

Imagine the following scenario, using the MBTI principles described in Section 47.2.3. You are an Introverted Judging developer in a prototype review session with an Extraverted Perceiving client. Recall, this means that you prefer introversion and judging, and you may generally prefer to spend your time in your inner world of ideas and concepts. When you do engage with the outer world, you prefer it to be in a structured way, aiming for brevity, closure, and the regulation of events. Because you prefer to communicate final thoughts in a decision-based format, you expect that others generally operate this way, as well. Conversely, your Extraverted Perceiving client may view such interaction periods as open brainstorming sessions, designed to explore new ideas, cover the possibilities, and think out loud. Comments from this client may be alternatives for consideration or triggers for more thought (at least in the client's eyes), although you generally hear them as decisions. The result is that you leave with a long list of new requirements, and the client leaves thinking that some interesting new possibilities have been discussed, no strings attached.

This scenario, which is neither exaggerated nor uncommon, brings a fresh perspective to the enduring problem of requirements creep, noted as a key issue facing the IT community. Requirements creep is usually described as a problem at the system or project level; however, every need ever articulated for a system came from a human user expressing an individual thought. How that thought is heard and managed is ultimately what leads to the project-level problem of requirements creep. This is only one example: consider the other personality preferences and interpersonal needs, and the complexity and potential impact of human dynamics in the development process becomes even more dramatic.

What is the best practice for developers? Invest the time and effort to learn about your own personality preferences, and learn how to recognize them in others. In our experience, IT professionals with this knowledge reported better client management skills and less frustration, because having a language with which to talk about differences makes them easier to resolve. This conclusion is also supported by evidence from a study of 40 software development teams, which concluded "how people work together is a stronger predictor of performance than the individual skills and abilities of team members" [Sawyer, 2001].

Take accountability for requirements elicitation. A common complaint in failing IT development efforts takes the form of the developer's lament: "Customers just do not know what they want." This comment, made acceptable in IT literature, is fundamentally unhelpful, because it assigns blame to the customer, fails to capture the real problem, and broadens the divide between the end user and the developer.

Usually, in fact, customers do know what they want, or they would not have called the IT developers in the first place. Each customer has a problem, and what is needed is a solution to that problem. The burden is on the developer to help the user express that problem in the form of a functional requirement or specification. This fundamental shift in accountability is both frightening and empowering: frightening because it forces IT professionals to come down from a high perch and interact with clients in their functional domain, and empowering because the potential for success is greatly enhanced.

Shifting the problem from a lack of user knowledge to a lack of developer understanding or communication makes that problem more manageable and solvable. The result is more effective dialogue between developer and customer. Deming said it best: "If you do not know how to ask the right question, you discover nothing." The end user's ability to articulate what is wanted depends heavily on the developer's ability to elicit useful and meaningful information. If this is done properly, requirements that are more closely linked to the problem will result.

Two best practices evolve from this observation. First, developers should not let requirements management overtake requirements elicitation. They should learn the arts of active listening, reflective dialogue, and coaching, and use these when working with end users. Second, when a project is in turmoil, developers should go back to the beginning and talk openly with clients about the core need or root cause of the problem. They should ask probing questions that get at the *why* driving the effort, or they should have the client complete a structured visioning exercise describing what success looks like. Done well, this raises the level of dialogue, moving back from the technical negotiations to the overarching mission, until common ground and understanding are once again established.

Consider personality preferences in interface design. Just as personality preferences influence discussions between IT developer and users, they also influence the end user's experience with the final product. Consider the *Contact Us* page from the Web site of the Internal Revenue Service (IRS) in light of the personality preferences for data gathering and decision making described previously (see Figure 47.1). This page and its contents are designed to appeal to a variety of user preferences and styles. Here are some examples:

The site allows users to search for information in three different ways: by drilling down through subject menus, by selecting target audience category, or by searching on specific key words. This approach covers a variety of user and information-gathering preferences.

The language of the site includes objective information (a detailed sequential publication outlining methods for tracking one's return) delivered with a personal touch. For example, one IRS menu item describes a connecting link with: "Get the lowdown on your refund now. Secure access anytime from anywhere. What a deal!" (www.irs.gov, 2003). For those struggling with taxes, this balanced presentation builds a connection with a variety of users, from the no-nonsense filer locating a document to the more subjective user appreciating the humor of the moment.

The specific page in Figure 47.1 offers users two options for interaction: on-line resources for those who prefer to receive help without human intervention, and Customer Assistance Centers for those who prefer the face-to-face approach. While this relates more to business process than design, it reflects the IRS's accessibility through multiple channels.

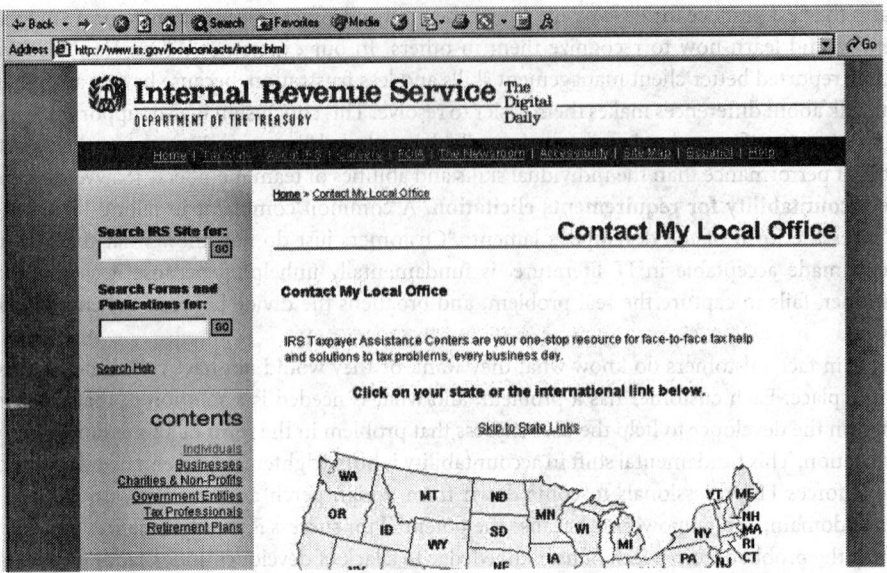

FIGURE 47.1   Example of effective user interface design.

This example suggests that developers test products with a variety of users early in the development process and consider different preferences when designing the product. Usability means different things to different people — a broad perspective with respect to user perceptions will add to the perceived value of the end product. (For other topics related to user interface design, see Chapter 45 and Chapter 48.)

Consider implementation issues early. In the heat of requirements, design, programming, and testing, it is easy to underestimate the logistics and politics of implementation once the system has been tested and accepted by the user. What business process will the system replace? Will roles and responsibilities within the sponsoring organization shift as a result? How will this be managed? What training will be required? What unforeseen changes may result from system introduction? Are any stakeholder groups likely to resist system use or introduction?

IT professionals should look ahead to the end game while still within the development process, so that potential production risks can be identified and mitigated early. Resulting activities may include business process reengineering, training, and an increased emphasis on socializing the ultimate product and its goals within affected stakeholder groups, particularly if these have not been included in the requirements process. (For other topics related to organizational contexts of development and use, see Chapter 45.)

Consider personal preferences when entering a project or role. Recognizing personality preferences and behavioral styles may help IT professionals select projects or roles that support personal values and areas of personal strength. Some people simply would rather work alone to generate a product that later can be joined with the efforts of others. Other people enjoy the shared nature of actively collaborative efforts. Some developers enjoy the iterative nature of evolutionary prototyping; others find the constant change and uncertainty more frustrating than freeing.

Developers should take the time to recognize their strengths, preferences, and potential blind spots when selecting jobs or projects. If a developer, for example, chafes against the requirements for structure and documentation that come with working on a CMM level 3 project, it is better to know this in advance. Self-knowledge is the key to making more effective decisions: decisions that will impact the individual, the development team, and ultimately the end user.

47.3.2 IT Managers

Early in this chapter, we highlighted an apparent conflict within IT teams: individual preferences for autonomy, coupled with the desire for improved effectiveness and cohesion at the team level. This leads to the key question: how can managers best support IT teams that clearly value effective team relationships, while also fulfilling strong needs related to objectivity, individual contribution, and independence?

Develop understanding of human dynamics within the team. Understanding the dynamics within the IT team and with the user empowers managerial leveraging of existing strengths and diversity, and the mitigation of weaknesses and blind spots. Whenever possible, analyze the client's dynamics and pinpoint risk or possible conflict areas early in the development life cycle. Insist on spending time on team dynamics as a necessary part of any development project. This includes training in people skills, such as communication and active listening, facilitation, coaching, and conflict management. This training allows IT professionals to develop their typically underused preferences, thereby expanding their flexibility with each other and with users. It is vital to note that the MBTI (and other assessments described here) are *not* predictors or indicators of success among computer programmers [Kerth et al., 1998] and should not be used to select people for participation on a project or team.

Welcome generalists and geeks. In today's development environment, a diverse team of functional specialists, analysts, programmers, architects, and administrators may be required to support multiple projects. In addition to a strong technical skills mix, a mix of communication, political, and interpersonal skills can also balance the strengths and weakness of a team.

IT managers should look for both technical and nontechnical gifts in each team member and utilize them in different project roles. It is tempting to hire those with the sharpest technical training and who share similar characteristics with other team members. Resist this urge by striving for balance, diversity — and even a bit of eccentricity. Klein et al. [2002] suggest that there are three unique perspectives that can support the development of an IT team: technical (project focused), end user (user focused), and sociopolitical (organizational system focused). Encouraging all three can mitigate risk and positively impact overall project success.

Understand dynamics of control and clarity. Faced with the complexity of a large development effort, many managers naturally respond by implementing control mechanisms to manage the process. In fact, most of the improvement models described in the previous section are designed for the goal of controlling process, so that it can be tracked, documented, and managed. This regulation comes at a cost. Most IT professionals report a low desire for wanted control on the FIRO-B (i.e., they dislike having others exert power and structure over them) and want more personal autonomy (i.e., individual decision making and self-sufficiency) than currently experienced. Furthermore, teams with control needs incompatible with the manager's (e.g., manager has high expressed control needs over a team with low wanted control needs) are more likely to report themselves as in turmoil. Conversely, IT professionals report wanting a significantly higher level of clarity in their work than currently experienced. This means that they want to know what to expect, and they want policies to be more explicitly communicated than they currently are.

Many managers have a difficult time separating the concepts of structure and policy from the concept of control, because there is a fine line between telling someone *what* needs to be done in an informative way and telling somehow *how* it shall be accomplished in a directive way. The first provides data to the team, whereas the second provides a decision. The distinction is not always obvious, particularly when the manager has a personality preference for Judging, which means that he or she naturally typically expresses both data and decisions in a closed-ended way. Judgers, as they are called, often sound like they are giving decisions and direction, when they may just be offering an opinion or impression. In this case, clarity may sound like control, even if the message in unintentional.

IT managers should consider the balance between *delineating* the road forward, describing all of its roadblocks and speed traps, and *directing* the team on how to drive in order to avoid them. For most managers, this requires a commitment to self-exploration and management, as well as requests from team members for frequent feedback on this issue.

Encourage healthy tension between the temperaments. Section 47.2.3.1 described the two most prevalent temperaments (i.e., behavioral preferences) among IT professionals: the Intuitive Thinking (NT) and Sensing Judging (SJ) temperaments, comprising 75% of the IT professionals surveyed. Tension between SJ and NT constituencies is common in organizations. Groups valuing the SJ approach may value established, "tried and true" policies and procedures, proven standards, chain of command accountability, and respect for organizational history and tradition. These groups may see the NTs as disrespectful of tradition, irreverent, and simply trying to stir up the pot by constantly reinventing the wheel. Groups valuing the NT approach may value systems that reward future-focused, innovative thinking and loose structure with minimal formal procedures and policies. These groups may see the SJs as the "ball and chain" traditionalists, who stifle creativity by their inability to think outside the box [Kroeger et al., 2002].

The temperament in power may drive which of these two approaches is more valued in the organization or team. At the IT team level, a manager who understands temperament and its impacts can apply both to the project's advantage. For example, SJs on the team may excel as administrators of systems requiring precision and organization, and prefer to focus on the specifics required for the work to be done today. On the other hand, NTs offer the challenge of questioning the established ways of doing things, with driving insights into the underlying principles of systems. Leveraging both should result in a better system overall, as different perspectives are considered during the process.

Engage users in the development process. A recurring theme in this chapter has been the need to work more effectively with users in the system development process. This discussion has focused on the level of the individual conversation, but it is also useful to review some of the structured methods by which this developer–user interaction can be encouraged [Maguire, 2001]:

User requirements interviews — Elicit individual views from a variety of users

Focus groups — Bring together stakeholder groups to discuss requirements

Task and process mapping — Defines the "as is" and "to be" states associated with envisioned system

Scenario visioning exercises — Help users to articulate what success looks like, leading to more definitive requirements and use cases

With effective planning and outreach, the IT manager can exert significant influence over the level of user involvement in system development efforts. Managers should consider introducing the methods described here as part of the work breakdown structure for development projects.

47.3.3 IT Educators

So far, this chapter has been primarily directed toward practicing IT professionals and managers, rather than the educators who prepare them for their profession. Here, we introduce suggestions for ways in which educators may introduce human dynamics into a range of educational programs, regardless of whether those programs are targeted toward computer science, systems engineering, or IT management.

Incorporate human uncertainty into assignments. Many educators use simulated exercises to allow programmers to practice programming skills. One common approach is to provide a team with an electronic kit containing both physical moving parts and a computer interface. The students are provided with specifications and must use their skills to program the robot to complete specific tasks.

We use a similar approach to investigate how IT teams approach ambiguous requirements [DAU, 2003]. We provide teams with a Lego Mindstorm Robot Kit and the following specifications: "Construct a robot that moves around a circle within 30 seconds, reverses, and moves around the circle in the opposite direction, also within 30 seconds. Creativity and elegance of design count." We purposely limit the time period allowed, so that a programmed automated solution is not possible. Instead, to be successful, the team must explore the terms *robot* and *moves around*. If the team agrees that the robot could be nonautomated (since we did not give them a better definition) and manually moves any object around the circle and back, they generally consider themselves successful. If they assume the term *robot* means *automated*, they generally make little headway. Usually, this exercise leads to interesting learning by the team with respect to the assumptions being made in the requirements process.

Introducing this type of requirements uncertainty at the beginning of an educational exercise offers the chance to explore this human dynamic of the development process. Educators should look for opportunities in the classroom to discuss IT professional–user interaction and to develop critical thinking skills in the face of ambiguity.

Encourage a broad curriculum. Students only have a short time to learn the skills required for success in today's IT environment. Furthermore, when a student leaves the educational setting, he or she may find a niche in programming, engineering, or management. As practitioners, we encourage educators to consider the following in planning a broad curriculum, focused on critical thinking and analysis:

Human and team dynamics — include as part of a well rounded software or computer engineering curriculum. Helping students to develop the habits of self-awareness and introspection will serve them well in any professional setting.

Functional areas outside the computer science and engineering disciplines — encourage students to explore. Students with an introduction to other subject areas may have more empathy for the challenges facing their end users. This also exposes them to different ways to approach the world.

Teach students to write — as practitioners, we have seen too many gifted programmers with underdeveloped writing skills. It is likely that students will eventually be required to write a strategic plan, document a technical architecture, write a requirements document, or write a critical analysis or product evaluation. Their education should include analytical writing skills. Again, this is a talent that will serve them well in any setting.

Balance theory and practice. Personality type theorists argue that the data-gathering function (assessed by the Sensing–Intuition scale of the MBTI) is tne most important in determining how people learn. Intuitives tend to learn best when given a theoretical framework within which to place new information. Sensors, conversely, often learn best when presented with practical applications or when they are able to interact with tools that allow practice and interaction with the new knowledge. This difference is particularly important in higher education, where it is estimated that 70% of all professors are Intuitive. On the other hand, the general population is only 30% Intuitive, with 70% Sensors. The result is an overrepresentation of Intuitive professors at the college level, teaching theory to classrooms filled predominantly with Sensors. Given this imbalance, educators should attempt to balance the emphasis between theory and practice in the classroom [Kroeger and Thuesen, 1988].

Continue research on the human dynamics of the IT profession. Research helps to support the work of practitioners by providing insight into the industry and its process. Given the subject of this chapter, we encourage more cross-disciplinary research to bridge the existing gaps between computer science and the social sciences. We also encourage educators to engage in collaborative research efforts with industry, to help unite today's curriculum with the industry challenges of tomorrow.

47.4   Research Issues and Summary

As practitioners, we approach the question of research direction from an applied viewpoint, searching for answers that will help us be better IT managers and professionals. Here, we suggest areas for consideration developed not only from our experience, but based on our applied research with IT teams:

We are intrigued by the finding that IT team members have a closer agreement about how the work setting or social climate *could be* than about what it actually is. Our research shows us that *not only* do IT professionals desire a greater degree of peer cohesion, task involvement, clarity, and physical comfort, but they also have more shared agreement as to what the ideal should look like than the reality currently experienced. If we can investigate these ideal views and the gaps that lie between here and there, we should be able to introduce improvements that will help to keep IT professionals working happily within their organizations, instead of moving on to perceived greener pastures. We selected proven psychological instruments (e.g., MBTI, FIRO-B, WES), popular and well documented in many other professional settings, for our research with IT teams. These instruments allow us to paint a broad portrait of IT teams. However, other areas were not addressed and deserve investigation. IT organizations can be quirky places, with casual dress as the rule, foosball tables in the break room, and 4:00 A.M. working sessions the norm. What impacts do these nontraditional workplace practices have on the team's success and productivity? There may be ways to capture these dynamics, and they may have direct influences on stress, creativity, and innovation. One promising tool in working with this population is the Apter motivational style profile (AMSP), based on *reversal theory*, a broad theory of personality, emotion, and motivation that emphasizes changeability and inconsistency of behavior [Apter, 2001]. Focused on four pairs of opposing motivational states, the AMSP helps explain, for example, why individuals who have been focused and "in the groove" of intense productivity suddenly need to shift into periods of playtime (this is the foosball effect). These issues of motivation, and the associated emotion, are areas ripe for research in the IT environment.

More empirical research is required to solidify the connection between effective human dynamics, communication, and conflict management and team/project success. DeMarco and Lister [1999], in particular, note success stories of the "jelled team." We believe that effective teams deliver results that exceed the traditional measures of time, quality, and cost; however, a broad range of evidence is currently lacking within the IT community. Future research should include benchmarking to assess before and after profiles of team and project success when human dynamics and communication training are actively introduced within the organization.

The fields of organizational development and information technology remain worlds far apart. There is a broad array of tools and training available from the organizational development community, but the translation of these into the IT setting can be difficult. If forced to attend "touchy-feely" training at all, IT professionals want it to be applicable to the work they do and the problems they face. As such, we recommend research into ways that IT-oriented applications and lessons learned can be integrated within the typically more generic programs currently offered by the organizational development and training communities, ones that provide objective evidence of the positive impacts such work will have on team effectiveness.

The IT landscape has shifted over the past decade. Today, IT professionals face faster development times, a broad range of skill requirements and development models, an increased emphasis on teams, and an increasingly IT-knowledgeable user base. These shifts signal the need for a new set of skills and tools for IT professionals themselves — skills that are less technical and more interpersonal in nature. In this chapter, we have described the fundamental paradigm shifts in human dynamics at work in the IT industry, suggested possible best practices for addressing them, and proposed areas for future work.

Every human endeavor involving more than one person is initiated with a single conversation between individuals. System development is no different. The tools and knowledge to build better conversations are available; they must be used by the next generation of IT professionals in order to improve the systems and applications they develop in the future.

Acknowledgment

The research for this chapter was assisted by contributions and support from Patrick Peck and Phyllis Griggs at Booz Allen Hamilton; Hile Rutledge at Otto Kroeger Associates (OKA); Dr. Philippe Gwet; and researchers at the Institute for Scientific Research (ISR): Rebecca Giorcelli, David Harris, Amy Jacquez, Anita Meinig, Robert Morgan, and Anthony Yancey. The authors are grateful for their contributions. We also extend special thanks to Congressman Alan B. Mollohan for his vision and support.

References

Ahern, D.M., Turner, R., and Clouse, A. 2001. *CMMI(SM) Distilled: A Practical Introduction to Integrated Process Improvement*. Addison-Wesley, Reading, MA.

Apter, M.J., Ed. 2001. *Motivational Styles in Everyday Life: A Guide to Reversal Theory*. American Psychological Association, Washington, D.C.

Beck, K. 1999. *Extreme Programming Explained: Embrace Change*. Addison-Wesley, Reading, MA.

Boehm, B.W. 1998. A spiral model of software development and enhancement. *IEEE Computer*. 21: 61–72.

Carmel, E., and Sawyer, S. 1998. Packed software development teams: what makes them different? *Information Technology and People*. 11(1): 7–19.

Ceruzzi, P. 2002. A view from 20 years. *IEEE Annals of the History of Computing*. Oct.–Dec.: 52–55.

Correia, J., and Eid, T. 2002. Teamwork is key to software life cycle planning. *Gartner Research Brief*, Gartner, Inc. May 14.

Curtis, W., Hefley, W.E., and Miller, S.A. 2001. *The People Capability Maturity Model: Guidelines for Improving the Workforce*. Addison-Wesley, Reading, MA.

Curtis, W., Krasner, H., and Iscoe, N. 1988. A field study of the software design process for large systems. *Communications of the ACM*. 31(11): 1268–1287.

Dafoulas, G.A. 2001. Facilitating group formation and role allocation in software engineering groups. *Proceedings of the ACS/IEEE International Conference on Computer Systems and Applications* (AICCSA'01).

Defense Acquisition University (DAU). 2003. *Team Dynamics in the Information Technology Industry*. Unpublished report prepared by Booz Allen Hamilton, the Institute for Scientific Research, and Otto Kroeger Associates.

DeMarco, T., and Lister, T. 1999. *Peopleware: Productive Projects and Teams, 2nd ed*. Dorset House.

Green, G., and Hevner, A.R. 1999. *Perceived Control of Software Developers and its Impact on the Successful Diffusion of Information Technology*. Carnegie Mellon University, Software Engineering Institute, Pittsburgh, PA. (CMU/SEI-98-SR-013).

Guinan, P.J., Cooprider, J.G., and Faraj, S. 1998. Enabling software development team performance during requirements definition: a behavioral versus technical approach. *Information Systems Research*. 9(2): 101–125.

Highsmith, J.A., III. 2000. *Adaptive Software Development: A Collaborative Approach to Managing Complex Systems*. Dorset House.

Hildebrand, C. 1995. I'm OK, you're really weird. *CIO Magazine*. Oct.: 86–96.

Hurley, F. 2002. Seven tips for keeping software development projects healthy. *IT Professional*. July/Aug.: 60–64.

Humphrey, W.S. 1997. *Managing Technical People: Innovation, Teamwork, and the Software Process*. Addison-Wesley Longman, Reading, MA.

Humphrey, W.S. 1996. *Introduction to the Personal Software Process*. Addison-Wesley, Reading, MA.

Kerth, N.L., Coplien, J., and Weinberg, J. 1998. Call for the rational use of personality indicators. *Computer*. 31: 146–147.

Klein, G., Jiang, J.J., and Tesch, D.B. 2002. Wanted: project teams with a blend of professional orientations. *Communications of the ACM*. 45(6): 81–87.

Kroeger, O., Thuesen, J., and Rutledge, H. 2002. *Type Talk at Work: How the 16 Personality Types Determine Your Success on the Job*. Dell Publishing, New York.

Kroeger, O., and Thuesen, J. 1988. *Type Talk: The 16 Personality Types That Determine How We Live, Love, and Work*. Dell Publishing, New York.

Leiner, B., Cerf, V., Clark, D., Kahn, R., Kleinrock, L., Lynch, D., Postel, J., Roberts, L., and Wolff, S. 2002. *All About the Internet: History of the Internet*. Internet Society.

Lyons, M.L. 1985. The DP psyche. *Datamation*. Aug.: 103–109.

Maguire, M. 2001. Methods to support human-centred design. *International Journal of Human–Computer Studies*. 55: 587–634.

McConnell, S. 1996. *Rapid Development*. Microsoft Press, Redmond, WA.

Moos, R.H. 1994. *Work Environment Scale Manual*. Consulting Psychologists Press, Palo Alto, CA.

Myers, I.B., McCaulley, M.H., Quenk, N.L., and Hammer, A.L. 1998. *MBTI Manual*. Consulting Psychologists Press, Palo Alto, CA.

Otto Kroeger Associates (OKA). 2000. *The Typewatching Toolkit, Version 2.0*. Otto Kroeger Associates, Fairfax, VA.

Pres, L. 1993. Before the Altair: the history of personal computing. *Communications of the ACM*. 36(9): 27–33.

Rosson, M.B. 1995. The human factor in programming and software development. In *Computer Science and Engineering Handbook, 1st Edition*, Ed. A.B. Tucker, pp. 1596–1618. CRC Press, Boca Raton, FL.

Sawyer, S. 2001. Effects of intra-group conflict on packaged software development team performance. *Information Systems Journal*. 11: 155–178.

Sledge, C., and Carney, D. 1998. *Case Study: Evaluating COTS Products for DoD Information Systems*. Carnegie Mellon University, Software Engineering Institute, Pittsburgh, PA.

Standish Group. 2001. *The CHAOS Report (2000)*. The Standish Group International, Inc.

United States Census Bureau. 2001. *Home Computers and Internet Use in the United States: August 2000*. U.S. Census Bureau, Current Population Survey, August 2000.

Vigder, M., Gentleman, W., and Dean, J. 1996. COTS software integration: state of the art. Presentation to the National Research Council, Canada, January 1996.

Waterman, J., and Rogers, J. 1996. *Introduction to the FIRO-B*. Consulting Psychologists Press, Palo Alto, CA.

Weinberg, G.M. 1998, 1971. *The Psychology of Computer Programming*. Dorset House.

Westbrook, P. 1988. Frequencies of MBTI types among computer technicians. *Journal of Psychological Type*. 15: 49–50.

Yourdon, E. 1997. *Death March: The Complete Software Developer's Guide to Mission Impossible Projects*. Prentice Hall, Upper Saddle River, NJ.

Zachary, G.P. 1998. Armed truce: software in an age of teams. *Information Technology and People*. 11(1): 62–65.

48

Graphical User Interface Programming

48.1 Introduction **48**-1
48.2 Importance of User Interface Tools **48**-2
Overview of User Interface Software Tools
• Tools for the World Wide Web
48.3 Models of User Interface Software **48**-20
48.4 Technology Transfer **48**-20
48.5 Research Issues **48**-20
New Programming Languages • Increased Depth
• Increased Breadth • End User Programming
and Customization • Application and User Interface
Separation • Tools for the Tools
48.6 Conclusions **48**-22

Brad A. Myers
Carnegie Mellon University

48.1 Introduction*

Almost as long as there have been user interfaces, there have been special software systems and tools to help design and implement the user interface software. Many of these tools have demonstrated significant productivity gains for programmers and have become important commercial products. Others have proved less successful at supporting the kinds of user interfaces people want to build. Virtually all applications today are built using some form of **user interface tool** [Myers 2000].

User interface (UI) software is often large, complex, and difficult to implement, debug, and modify. As interfaces become easier to use, they become harder to create [Myers 1994]. Today, direct-manipulation interfaces (also called GUIs for **graphical user interfaces**) are almost universal. These interfaces require that the programmer deal with elaborate graphics, multiple ways of giving the same command, multiple asynchronous input devices (usually a keyboard and a pointing device such as a mouse), a mode-free interface where the user can give any command at virtually any time, and rapid "semantic feedback" where determining the appropriate response to user actions requires specialized information about the objects in the program. Interfaces on handheld devices, such as a Palm organizer or a Microsoft PocketPC device, use similar metaphors and implementation strategies. Tomorrow's user interfaces will provide speech

---

*This chapter is revised from an earlier version: Brad A. Myers. 1995. "User Interface Software Tools," *ACM Transactions on Computer–Human Interaction.* 2(1): 64–103.

1-58488-360-X/$0.00+$1.50
© 2004 by CRC Press, LLC

recognition, vision from cameras, 3-D, intelligent agents, and integrated multimedia, and will probably be even more difficult to create. Furthermore, because user interface design is so difficult, the only reliable way to get good interfaces is to iteratively redesign (and therefore reimplement) the interfaces after user testing, which makes the implementation task even harder.

Fortunately, there has been significant progress in software tools to help with creating user interfaces. Today, virtually all user interface software is created using tools that make the implementation easier. For example, the MacApp system from Apple, one of the first GUI frameworks, was reported to reduce development time by a factor of four or five [Wilson 1990]. A study commissioned by NeXT claimed that the average application programmed using the NeXTStep environment wrote 83% fewer lines of code and took one-half the time, compared to applications written using less advanced tools, and some applications were completed in one-tenth the time. Over three million programmers use Microsoft's Visual Basic tool because it allows them to create GUIs for Windows significantly more quickly.

This chapter surveys UI software tools and explains the different types and classifications. However, it is now impossible to discuss all UI tools, because there are so many, and new research tools are reported every year at conferences such as the annual ACM User Interface Software and Technology Symposium (UIST) (see http://www.acm.org/uist/) and the ACM SIGCHI conference (see http://www.acm.org/sigchi/). There are also about three Ph.D. theses on UI tools every year. This article provides an overview of the most popular approaches, rather than an exhaustive survey. It has been updated from previous versions (e.g., [Myers 1995]).

48.2 Importance of User Interface Tools

There are many advantages to using user interface software tools. These can be classified into two main groups. First, the *quality* of the resulting user interfaces might be higher, for the following reasons:

Designs can be rapidly prototyped and implemented, possibly even before the application code is written. This, in turn, enables more rapid prototyping and therefore more iterations of iterative design, which is a crucial component of achieving high-quality user interfaces [Nielsen 1993b].

The reliability of the user interface will be higher, because the code for the user interface is created automatically from a higher-level specification.

Different applications are more likely to have consistent user interfaces if they are created using the same UI tool.

It will be easier for a variety of specialists to be involved in designing the user interface, rather than having the user interface created entirely by programmers. Graphic artists, cognitive psychologists, and usability specialists may all be involved. In particular, professional user interface designers, who may not be programmers, can be in charge of the overall design.

More effort can be expended on the tool than may be practical on any single user interface, because the tool will be used with many different applications.

Undo, Help, and other features are more likely to be available because the tools might support them.

Second, the UI code might be *easier and more economical* to create and maintain. This is because of the following:

Interface specifications can be represented, validated, and evaluated more easily.

There will be less code to write, because much is supplied by the tools.

There will be better modularization, due to the separation of the UI component from the application. This should allow the user interface to change without affecting the application, and a large class of changes to the application (such as changing the internal algorithms) should be possible without affecting the user interface.

The level of programming expertise of the interface designers and implementers can be lower, because the tools hide much of the complexity of the underlying system.

It will be easier to port an application to different hardware and software environments because the device dependencies are isolated in the UI tool.

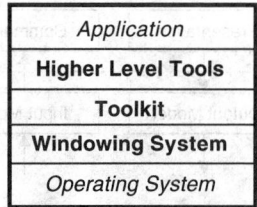

FIGURE 48.1 The components of user interface software.

48.2.1 Overview of User Interface Software Tools

Because user interface software is so difficult to create, it is not surprising that people have been working for a long time to create tools to help with it. Today, many of these tools and ideas have progressed from research into commercial systems, and their effectiveness has been amply demonstrated. Research systems also continue to evolve quickly, and the models that were popular five years ago have been made obsolete by more effective tools, changes in the computer market, and the emergence of new styles of user interfaces, such as handheld computing and multimedia.

48.2.1.1 Components of User Interface Software

As shown in Figure 48.1, UI software may be divided into various layers: the **windowing system**, the **toolkit**, and higher-level tools. Of course, many practical systems span multiple layers.

The windowing system supports the separation of the screen into different (usually rectangular) regions, called **windows**. The X system [Scheifler 1986] divides window functionality into two layers: the window system, which is the functional or programming interface, and the **window manager**, which is the user interface. Thus, the window system provides procedures that allow the application to draw pictures on the screen and get input from the user; the window manager allows the end user to move windows around and is responsible for displaying the title lines, borders, and **icons** around the windows. However, many people and systems use the name "window manager" to refer to both layers, because systems such as the Macintosh and Microsoft Windows do not separate them. This article will use the X terminology, and use the term *windowing system* to refer to both layers.

Note that Microsoft confusingly calls its entire system *Windows* (for example, *Windows 98* or *Windows XP*). This includes many different functions that here are differentiated into the operating system part (which supports memory management, file access, networking, etc.), the windowing system, and higher-level tools.

On top of the windowing system is the toolkit, which contains many commonly used **widgets** (also called *controls*) such as menus, buttons, scroll bars, and text input fields. On top of the toolkit might be higher-level tools, which help the designer to use the toolkit widgets. The following sections discuss each of these components in more detail.

48.2.1.2 Windowing Systems

A windowing system is a software package that helps the user monitor and control different contexts by separating them physically onto different parts of one or more display screens [Myers 1988b]. Although most of today's systems provide toolkits on top of the windowing systems, as will be explained later, toolkits generally only address the drawing of widgets such as buttons, menus, and scroll bars. Thus, when the programmer wants to draw application-specific parts of the interface and allow the user to manipulate these, the window system interface must be used directly. Therefore, the windowing system's programming interface has significant impact on most user interface programmers.

The first windowing systems were implemented as part of a single program or system. For example, the EMACs text editor [Stallman 1979], Smalltalk [Tesler 1981], and DLISP [Teitelman 1979] programming environments had their own windowing systems. Later systems implemented the windowing system as an

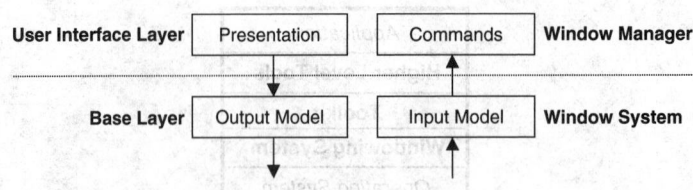

FIGURE 48.2 The windowing system can be divided into two layers, called the base (or window system) layer and the user interface (or window manager) layer. Each of these can be divided into parts that handle output and input.

integral part of the operating system, such as Sapphire for PERQs [Myers 1984], SunView for Suns, and the Macintosh and Microsoft Windows systems. In order to allow different windowing systems to operate on the same operating system, some windowing systems, such as X and Sun's NeWS [Gosling 1986], operate as a separate process and use the operating system's interprocess communication mechanism to connect to application programs.

48.2.1.2.1 *Structure of Windowing Systems*

A windowing system can be logically divided into two layers, each of which has two parts (see Figure 48.2). The window system, or base layer, implements the basic functionality of the windowing system. The two parts of this layer handle the display of graphics in windows (the output model) and the access to the various input devices (the input model), which usually includes a keyboard and a pointing device such as a mouse. The primary interface of the base layer is procedural and is called the windowing system's application programmer interface (API).

The other layer of windowing system is the window manager or user interface. This includes all aspects that are visible to the user. The two parts of the user interface layer are the presentation, which comprises the pictures that the window manager displays, and the commands, which are how the user manipulates the windows and their contents.

48.2.1.2.2 *Base Layer*

The base layer is the procedural interface to the windowing system. In the 1970s and early 1980s, there were a large number of different windowing systems, each with a different procedural interface (at least one for each hardware platform). People writing software found this to be unacceptable because they wanted to be able to run their software on different platforms, but they would have to rewrite significant amounts of code to convert from one window system to another. The X windowing system [Scheifler 1986] was created to solve this problem by providing a hardware-independent interface to windowing. X has been quite successful at this, and it drove all other windowing systems out of the workstation hardware market. X continues to be popular as the windowing system for Linux and all other UNIX implementations. In the rest of the computer market, most machines use some version of Microsoft Windows, with the Apple Macintosh computers having their own windowing system.

48.2.1.2.3 *Output Model*

The output model is the set of procedures that an application can use to draw pictures on the screen. It is important that all output be directed through the window system so that the graphics primitives can be clipped to the window's borders. For example, if a program draws a line that would extend beyond a window's borders, it must be clipped so that the contents of other, independent, windows are not overwritten. Most computers provide graphics hardware that is optimized to work efficiently with the window system.

In early windowing systems, such as Smalltalk [Tesler 1981] and Sapphire [Myers 1986], the primary output operation was BitBlt (also called RasterOp, and now sometimes CopyArea or CopyRectangle). These early systems primarily supported monochrome screens (each pixel is either black or white). BitBlt takes a rectangle of pixels from one part of the screen and copies it to another part. Various Boolean operations can be specified for combining the pixel values of the source and destination rectangles. For example, the source rectangle can simply replace the destination, or it might be XORed with the destination. BitBlt can be used to draw solid rectangles in either black or white, display text, scroll windows, and perform many other effects [Tesler 1981]. The only additional drawing operation typically supported by these early systems was drawing straight lines.

Later windowing systems, such as the Macintosh and X, added a full set of drawing operations, such as filled and unfilled polygons, text, lines, arcs, etc. These cannot be implemented using the BitBlt operator. With the growing popularity of color screens and nonrectangular primitives (such as rounded rectangles), the use of BitBlt has significantly decreased. Now, it is primarily used for scrolling and copying off-screen pictures onto the screen (e.g., to implement double-buffering).

A few windowing systems allowed the full PostScript imaging model [Adobe Systems Inc. 1985] to be used to create images on the screen. PostScript provides device-independent coordinate systems and arbitrary rotations and scaling for all objects, including text. Another advantage of using PostScript for the screen is that the same language can be used to print the windows on paper (because many printers accept PostScript). Sun created a version used in the NeWS windowing system, and then Adobe (the creator of PostScript) came out with an official version called Display PostScript, which was used in the NeXT windowing system. A similar imaging model is provided by Java 2-D [Sun Microsystems 2002], which works on top of (and hides) the underlying windowing system's output model.

All of the standard output models only contain drawing operations for two-dimensional objects. Extensions to support 3-D objects include PEX, OpenGL, and Direct3-D. PEX [Gaskins 1992] is an extension to the X windowing system that incorporates much of the PHIGS graphics standard. OpenGL [Silicon Graphics Inc. 1993] is based on the GL programming interface that has been used for many years on Silicon Graphics machines. OpenGL provides some machine independence for 3-D because it is available for various X and Windows platforms. Microsoft supplies its own 3-D graphics model, called Direct3-D, as part of Windows.

As shown in Figure 48.3, the earlier windowing systems assumed that a graphics package would be implemented using the windowing system. See Figure 48.3a. For example, the CORE graphics package was implemented on top of the SunView windowing system. Next, systems such as the Macintosh, X, NeWS, NeXT, and Microsoft Windows implemented a sophisticated graphics system as part of the windowing system. See Figure 48.3b and Figure 48.3c. Now, with Java2-D and Java3-D, as well as Web-based graphics systems such as VRML for 3-D programming on the Web [Web3-D Consortium 1997], we are seeing a return to a model similar to the one shown in Figure 48.3a, with the graphics on top of the windowing system. See Figure 48.3-D.

48.2.1.2.4 Input Model

The early graphics standards, such as CORE and PHIGS, provided an input model that does not support the modern, direct-manipulation style of interfaces. In those standards, the programmer calls a routine to request the value of a virtual device, such as a locator (pointing device position), string (edited text string), choice (selection from a menu), or pick (selection of a graphical object). The program would then pause, waiting for the user to take action. This is clearly at odds with the direct-manipulation mode-free style, in which the user can decide whether to make a menu choice, select an object, or type something.

With the advent of modern windowing systems, a new model was provided: a stream of event records is sent to the window that is currently accepting input. The user can select which window is getting events using various commands, described subsequently. Each event record typically contains the type and value of the event (e.g., which key was pressed), the window to which the event was directed, a timestamp, and the x and y coordinates of the mouse. The windowing system queues keyboard events, mouse button events, and mouse movement events together (along with other special events), and programs must dequeue the events and process them. It is somewhat surprising that, although there has been substantial progress in the output model for windowing systems (from BitBlt to complex 2-D primitives to 3-D), input is still

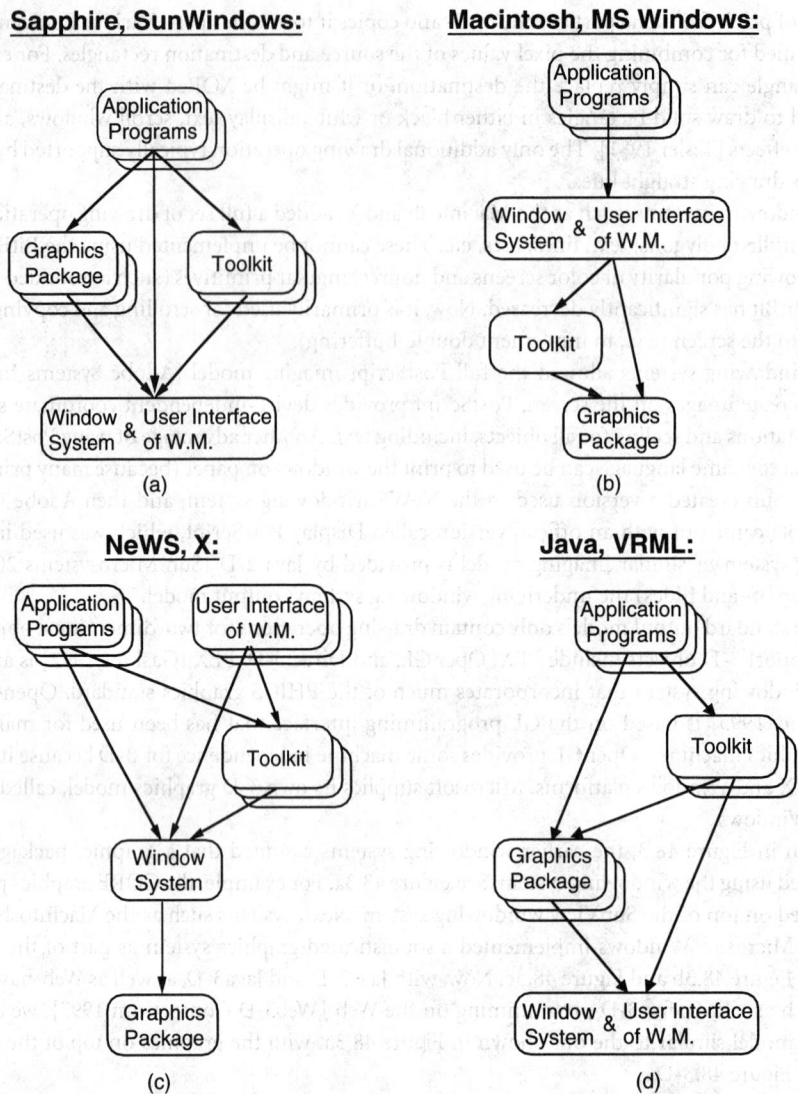

FIGURE 48.3 Various organizations that have been used by windowing systems. Boxes with extra borders represent systems that can be replaced by users. Early systems (a) tightly coupled the window manager and the window system, and assumed that sophisticated graphics and toolkits would be built on top. The next step in designs (b) was to incorporate into the windowing system the graphics and toolkits, so that the window manager itself could have a more sophisticated look and feel, and so applications would be more consistent. Other systems (c) allow different window managers and different toolkits, while still embedding sophisticated graphics packages. Newer systems (d) hark back to the original design (a) and implement the graphics and toolkit on top of the window system.

handled in essentially the same way today as it was in the original windowing systems, even though there are some well-known, unsolved problems with this model:

There is no provision for special stop-output (Ctrl+S) or abort (Ctrl+C, command-dot) events, so these will be queued with the other input events.

The same event mechanism is used to pass special messages from the windowing system to the application. When a window gets larger or becomes uncovered, the application must usually be notified so it can adjust or redraw the picture in the window. Most window systems communicate this by queuing special events into the event stream, which the program must then handle.

The application must always be willing to accept events in order to process aborts and redrawing requests. If not, then long operations cannot be aborted, and the screen may have blank areas while they are being processed.

The model is device-dependent, because the event record has fixed fields for the expected incoming events. If a 3-D pointing device or one with more than the standard number of buttons is used instead of a mouse, then the standard event mechanism cannot handle it.

Because the events are handled asynchronously, there are many race conditions that can cause programs to get out of synchronization with the window system. For example, in the X windowing system, if you press inside a window and release outside, under certain conditions the program will think that the mouse button is still depressed. Another example is that refresh requests from the windowing system specify a rectangle for the window that needs to be redrawn, but if the program is changing the contents of the window, the wrong area may be redrawn by the time the event is processed. This problem can occur when the window is scrolled.

Although these problems have been known for a long time, there has been little research on new input models (an exception is the Garnet Interactors model [Myers 1990b]).

48.2.1.2.5 Communication

In the X windowing system and NeWS, all communication between applications and the window system uses interprocess communication through a network protocol. This means that the application program can be on a different computer from its windows. In all other windowing systems, operations are implemented by directly calling the window manager procedures or through special traps into the operating system. The primary advantage of the X mechanism is that it makes it easier for a person to utilize multiple machines with all their windows appearing on a single machine. Another advantage is that it is easier to provide interfaces for different programming languages: for example, the C interface (called xlib) and the Lisp interface (called CLX) send the appropriate messages through the network protocol. The primary disadvantage is efficiency, because each window request will typically be encoded, passed to the transport layer, and then decoded, even when the computation and windows are on the same machine.

48.2.1.2.6 User Interface Layer

The user interface of the windowing system allows the user to control the windows. In X, the user can easily switch user interfaces, by killing one window manager and starting another. Some of the original window managers under X included uwm (with no title lines and borders), twm, mwm (the Motif window manager), and olwm (the OpenLook window manager). Newer choices include complete desktop environments that combine a window manager with a file browser and other GUI utilities (to better match the capabilities found in Windows and the Macintosh). Two popular desktop environments are KDE (K Desktop Environment — http://www.kde.org) with its window manager KWin, and Gnome (http://www.gnome.org), which provides a variety of window manager choices. X provides a standard protocol through which programs and the base layer communicate to the window manager, so that all programs continue to run without change when the window manager is switched. It is possible, for example, to run applications that use Motif widgets inside the windows controlled by the KWin window manager.

A discussion of the options for the user interfaces of window managers was previously published [Myers 1988b]. Also, the video *All the Widgets* [Myers 1990a] has a 30-minute segment showing many different forms of window manager user interfaces.

Some parts of the user interface of a windowing system, which is sometimes called its *look and feel*, can apparently be copyrighted and patented. Which parts is a highly complex issue, and the status changes with decisions in various court cases [Samuelson 1993].

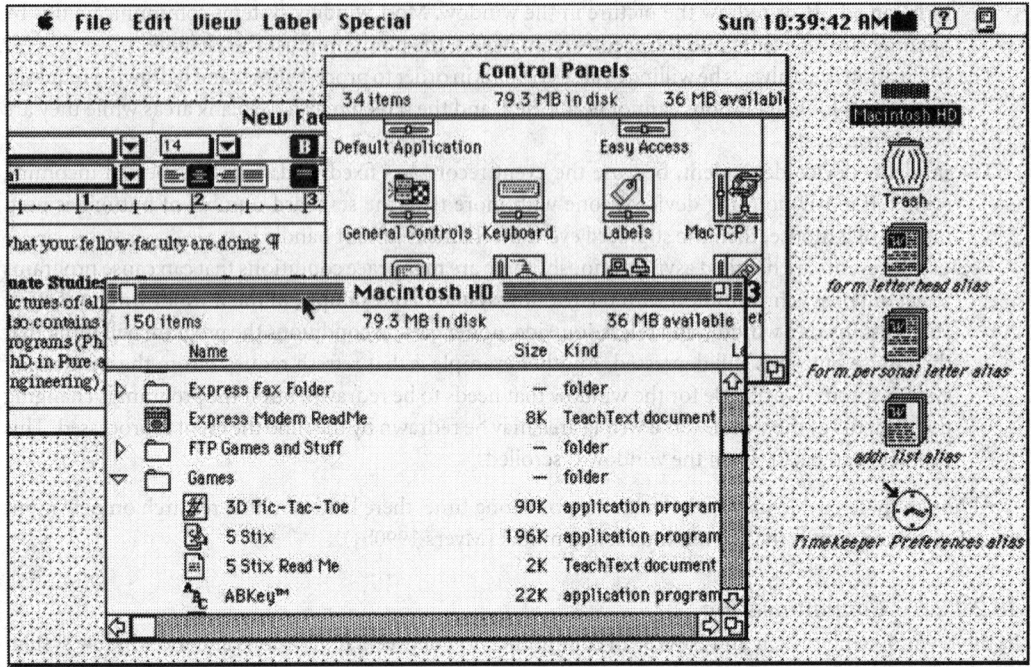

FIGURE 48.4 A screen from the original Macintosh showing three windows covering each other and some icons along the right margin.

48.2.1.2.7 *Presentation*

The presentation of the windows defines how the screen looks. One very important aspect of the presentation of windows is whether or not they can overlap. Overlapping windows, sometimes called covered windows, allow one window to be partially or totally on top of another window, as shown in Figure 48.4. This is also sometimes called the desktop metaphor, because windows can cover each other as pieces of paper can cover each other on a desk. There are usually other aspects to the desktop metaphor, however, such as presenting file operations in a way that mimics office operations, as originated in the Star office workstation [Smith 1982]. The alternative is tiled windows, which means that windows are not allowed to cover each other. Obviously, a window manager that supports covered windows can also allow them to be side by side, but not vice versa. Therefore, a window manager is classified as "covered" if it allows windows to overlap. The tiled style was popular for a while and was used by Cedar [Swinehart 1986] and by early versions of Star [Smith 1982], Andrew [Palay 1988], and even Microsoft Windows. A study even suggested that using tiled windows was more efficient for users [Bly 1986]. However, today tiled windows are rarely seen on conventional window systems, because users generally prefer overlapping.

Modern browsers for the World Wide Web, such as Netscape and Microsoft's Internet Explorer, provide a windowing environment inside the computer's main windowing system. Newer versions of browsers support frames containing multiple scrollable panes, which are a form of tiled window. In addition, if an application written in Java is downloaded (see Section 48.2.1.3.4), it can create multiple, overlapping windows like conventional GUI applications.

Another important aspect of the presentation of windows is the use of icons. These are small pictures that represent windows (or sometimes files). They are used because there would otherwise be too many windows to fit conveniently on the screen and to manage. Sapphire was the first window manager to group the icons into a window [Myers 1984], a format which was picked up by the Motif window manager. Now, the taskbar provides the icons and names of running and available processes in Windows and other modern window managers. Other aspects of the presentation include whether or not the window has a title line, what the background (where there are no windows) looks like, and whether the title and borders have control areas for performing window operations.

48.2.1.2.8 Commands

Because computers typically have multiple windows and only one mouse and keyboard, there must be a way for the user to control which window is getting keyboard input. This window is called the input (or keyboard) focus. Another term is the listener, because it is listening to the user's typing. Some systems called the focus the active window or current window, but these are poor terms because, in a multiprocessing system, many windows can be actively outputting information at the same time. Window managers provide various ways to specify and show which window is the listener. The most important options are the following:

Click-to-type — This means that the user must click the mouse button in a window before typing to it. This is used by the Macintosh and Microsoft Windows.

Move-to-type — This means that the mouse only has to move over a window to allow typing to it. This is usually faster for the user, but it may cause input to go to the wrong window if the user accidentally knocks the mouse.

Some X window managers (including the Motif window manager, mwm) allow the user to choose the desired method. However, the choice can have significant impact on the user interface of applications. For example, because the Macintosh requires click-to-type, it can provide a single menu bar at the top, and the commands can always operate on the focused window. With move-to-type, the user might have to pass through various windows (thus giving them the focus) on the way to the top of the screen. Therefore, Motif applications must have a menubar in each window so the commands will know which window to operate on.

All covered window systems allow the user to bring a window to the top (not covered by other windows), and some allow sending a window to the bottom (covered by all other windows). Other commands allow windows to be changed in size, moved, shrunk to an icon, made full-size, and destroyed.

48.2.1.3 Toolkits

A toolkit is a library of widgets that can be called by application programs. As mentioned previously, a widget (also called a control) is a way of using a physical input device to input a certain type of value. Typically, widgets in toolkits include menus, buttons, scroll bars, text type-in fields, etc. Figure 48.5 shows some examples of widgets. Creating an interface using a toolkit can only be done by programmers, because toolkits only have a procedural interface.

Using a toolkit has the advantage that the final UI will look and act similarly to other UIs created using the same toolkit, and each application does not have to rewrite the standard functions, such as menus. A problem with toolkits is that the styles of interaction are limited to those provided. For example, it is difficult to create a single slider that contains two indicators, which might be useful to input the upper and lower bounds of a range. In addition, the toolkits themselves are often expensive to create: "The primitives never seem complex in principle, but the programs that implement them are surprisingly intricate" [Cardelli 1985, p. 199]. Another problem with toolkits is that they are often difficult to use: they may contain hundreds of procedures, and it is often not clear how to use the procedures to create a desired interface.

As with the graphics package, the toolkit can be implemented either using or being used by the windowing system (see Figure 48.3). Early systems provided only minimal widgets (e.g., just a menu) and expected applications to provide others, as shown in Figure 48.3a. In the Macintosh and in Microsoft Windows, the toolkit is at a low level, and the window manager user interface is built using it. The advantage of this is that the window manager can then use the same sophisticated toolkit routines for its user interface. See Figure 48.3b. When the X system was being developed, the developers could not agree on a single toolkit, so they left the toolkit to be on top of the windowing system. In X, programmers can use a variety of toolkits (for example, the Motif, InterViews [Linton 1989], Amulet [Myers 1997], tcl/tk

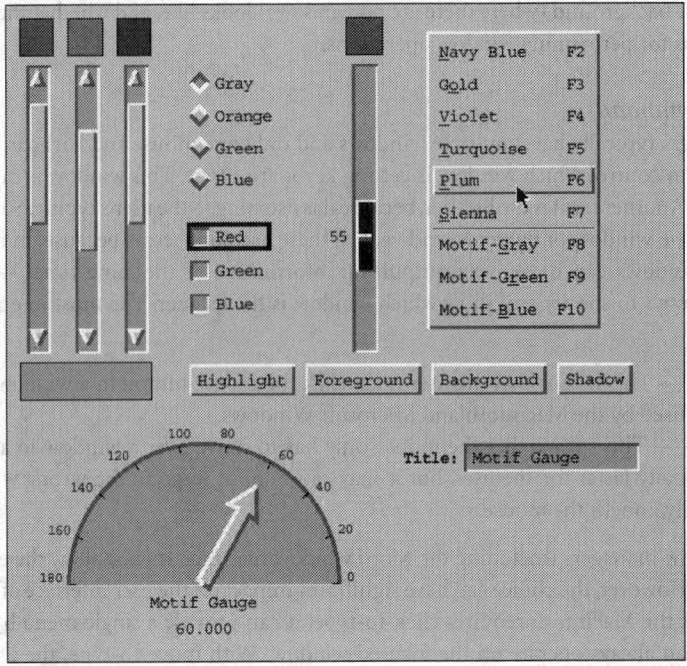

FIGURE 48.5 Some of the widgets with a Motif look and feel provided by the Garnet toolkit.

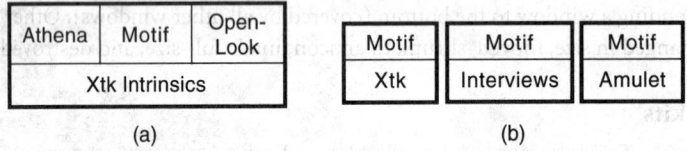

(a)   (b)

FIGURE 48.6 (a) At least three different widget sets that have different looks and feels were implemented on top of the Xt intrinsics. (b) The Motif look and feel has been implemented on many different intrinsics.

[Ousterhout 1991], and Gnome GTK+ [GNOME 2002] toolkits can be used on top of X), but the window manager must usually implement its user interface without using the toolkit, as in Figure 48.3c. The Java Swing toolkit is implemented on top of the Java 2-D graphics package, which in turn is on top of the windowing system. See Figure 48.3d.

Because the designers of X could not agree on a single look and feel, they created an **intrinsics** layer on which to build different widget sets, which they called Xt [McCormack 1988]. This layer provides the common services, such as techniques for object-oriented programming and layout control. The widget set layer is the collection of widgets implemented using the intrinsics. Multiple widget sets with different looks and feels can be implemented on top of the same intrinsics layer (Figure 48.6a), or else the same look and feel can be implemented on top of different intrinsics (Figure 48.6b).

48.2.1.3.1 Toolkit Intrinsics

Toolkits come in two basic varieties. The most conventional is simply a collection of procedures that can be called by application programs. Examples of this style include the SunTools toolkit for the SunView windowing system, and the original Macintosh Toolbox [Apple Computer Inc. 1985]. The second variety uses an object-oriented programming style that makes it easier for the designer to customize the interaction techniques. Examples include Smalltalk [Tesler 1981], Andrew [Palay 1988], InterViews [Linton 1989],

Xt [McCormack 1988], Amulet [Myers 1997], the Java Swing toolkit [Sun Microsystems 2003], and Gnome's GTK+ [GNOME 2002].

The advantages of using object-oriented intrinsics are that it is a natural way to think about widgets (the menus and buttons on the screen *seem* like objects), the widget objects can handle some of the chores that otherwise would be left to the programmer (such as refresh), and it is easier to create custom widgets (by subclassing an existing widget). The advantages of the older, procedural style are that it is easier to implement, no special object-oriented system is needed, and it is easier to interface to multiple programming languages.

To implement the objects, the toolkit might invent its own object system, as was done with Xt, Andrew, Amulet and GTK+, or it might use an existing object system, as was done in InterViews [Linton 1989], which uses C++; NeXTStep from NeXT, which uses Objective-C; and Swing, which uses Java.

The usual way that object-oriented toolkits interface with application programs is through the use of **call-back procedures**. These are procedures defined by the application programmer which are called when a widget is operated by the end user. For example, the programmer might supply a procedure to be called when the user selects a menu item. Experience has shown that real interfaces often contain hundreds of call-backs, which makes the code harder to modify and maintain [Myers 1992c]. In addition, different toolkits, even when implemented on the same intrinsics like Motif and OpenLook, have different call-back protocols. This means that code for one toolkit is difficult to port to a different toolkit. Therefore, research has been directed at reducing the number of call-backs in UI software [Myers 1991b].

Some research toolkits have added novel features to the toolkit intrinsics. For example, Garnet [Myers 1990d], Rendezvous [Hill 1994], Amulet [Myers 1997], and SubArctic [Hudson 1996] allow the objects to be connected using **constraints**, which are relationships that are declared once and then maintained automatically by the system. For example, the designer can specify that the color of a rectangle is constrained to be the value of a slider, and then the system will automatically update the color if the user moves the slider.

Many toolkits include a related capability for handling graphical layouts in a declarative manner. This is often called **geometry management**. Widgets can be specified to stay at the sides or the center of a container or to expand to fill up a specified space. This is particularly important when the size of objects might change, for example, in systems that can run on multiple architectures. An early example of this was in InterViews [Linton 1989], and layout managers are important parts of Motif and Java Swing.

48.2.1.3.2 Widget Set

Typically, the intrinsics layer is look and feel–independent, which means that the widgets built on top of it can have any desired appearance and behavior. However, a particular widget set must pick a look and feel. The video *All the Widgets* shows many examples of widgets that have been designed over the years [Myers 1990a]. For example, it shows 35 different kinds of menus. Like window manager UIs, the widgets' look and feel can be copyrighted and patented [Samuelson 1993].

As mentioned earlier, different widget sets (with different looks and feels) can be implemented on top of the same intrinsics. In addition, the same look and feel can be implemented on top of different intrinsics. For example, there are Motif look-and-feel widgets on top of the Xt, InterViews, and Amulet intrinsics. See Figure 48.6b. Although they all look and operate the same (and so would be indistinguishable to the end user), they are implemented quite differently and have completely different procedural interfaces for the programmer.

48.2.1.3.3 Specialized Toolkits

A number of toolkits have been developed to support specific kinds of applications or specific classes of programmers. For example, the SUIT system [Pausch 1992] (which contains a toolkit and an **interface builder**) is specifically designed to be easy to learn and is aimed at classroom instruction. Its successor, Alice [Pausch 1995], provides an easy way to program 3-D graphics and animation. Amulet [Myers 1997] provides high-level support for graphical, direct-manipulation interfaces and handles input as hierarchical command objects, making Undo easier to implement [Myers 1996a]. Rendezvous [Hill 1994], Visual Obliq

[Bharat 1994], and GroupKit [Roseman 1996] are designed to make it easier to create applications that support multiple users on multiple machines operating synchronously. Whereas most toolkits provide only 2-D interaction techniques, the Brown 3-D toolkit [Stevens 1994] and Silicon Graphics' Inventor toolkit [Wernecke 1994] provide preprogrammed 3-D widgets and a framework for creating others. Special support for animations has been added to Artkit [Hudson 1993], Amulet [Myers 1996b], and Alice [Pausch 1995]. Tk [Ousterhout 1991] is a popular toolkit for the X window system (and also for Windows) because it can be attached to a variety of interpretive languages, including the original, called tcl, Perl, and Python, which makes it possible to change the user interface dynamically. Tcl supports the UNIX style of programming, in which many small programs are glued together.

48.2.1.3.4 *Virtual Toolkits*

Although there are many small differences among the various toolkits, much remains the same. For example, all have some type of menu, button, scroll bar, text input field, etc. Although there are fewer windowing systems and toolkits than there were ten years ago, people are still finding it a lot of work to port software so that it works on the Macintosh, Microsoft Windows, and X on Linux.

Therefore, a number of systems, called **virtual toolkits**, have been developed that try to hide the differences among the various toolkits, by providing virtual widgets that can be mapped into the widgets of each toolkit. Another name for these tools is cross-platform development systems. The programmer writes the code once, using the virtual toolkit, and the code will run without change on different platforms and still look like it was designed with that platform's widgets. For example, the virtual toolkit might provide a single menu routine, which always has the same programmer interface but connects to a Motif menu, a Macintosh menu, or a Windows menu, depending on which machine the application is run on.

There are two styles of virtual toolkits. In the first, the virtual toolkit links to the different actual toolkits on the host machine. For example, XVT [XVT Software Inc. 1997] provides a C or C++ interface that links to the actual Motif, Macintosh, MS-Windows, and OS/2-PM toolkits (and also character terminals) and hides their differences. The Java AWT toolkit also works in this way and uses the underlying widget sets. The second style of virtual toolkit reimplements the widgets in each style. For example, Galaxy [Visix Software Inc. 1997], Amulet [Myers 1997], and Java Swing provide libraries of widgets that look like those on the various platforms. The advantage of the first style is that the user interface is more likely to be look and feel conformant (since it uses the real widgets). The disadvantage is that the virtual toolkit must still provide an interface to the graphical drawing primitives on the platforms. Furthermore, virtual toolkits tend only to provide functions that appear in all toolkits. Many of the virtual toolkits that take the second approach, for example, Java Swing, provide a sophisticated graphics package and complete sets of widgets on all platforms. However, with the second approach, there must always be a large run-time library, because in addition to the built-in widgets that are native to the machine, there is the reimplementation of these same widgets in the virtual toolkit's library.

All of the toolkits that work on multiple platforms can be considered virtual toolkits of the second type. For example, SUIT [Pausch 1992] and Garnet [Myers 1990d] work on X, Macintosh, and Windows. However, these use the same look and feel on all platforms (and therefore do not look the same as the other applications on that platform), so they are not classified as virtual toolkits.

48.2.1.4 Higher-Level Tools

Because programming at the toolkit level is quite difficult, there is a tremendous interest in higher-level tools that will make the UI software production process easier. These are discussed next.

48.2.1.4.1 *Phases*

Many higher-level tools have components that operate at different times. The design-time component helps the UI designer to design the user interface. For example, this might be a graphical editor that can lay out the interface, or a compiler to process a UI specification language. The next phase is when the end user is using the program. Here, the run-time component of the tool is used. This usually includes a toolkit but may also include additional software specifically for the tool. Because the run-time component is "managing" the user interface, the term **user interface management system** (UIMS) seems appropriate for tools with a significant run-time component.

There also may be an after-run-time component that helps with the evaluation and debugging of the user interface. Unfortunately, very few UI tools have an after-run-time component. This is partially because tools that have tried, such as MIKE [Olsen 1988], discovered that there are very few metrics that can be applied by computers. Some tools try to evaluate how people will interact with interfaces by automatically creating cognitive models from high-level descriptions of the user interface. For example, the GLEAN system generates quantitative predictions of a system's performance from a GOMS model [Kieras 1995].

48.2.1.4.2 *Specification Styles*

High-level UI tools come in a large variety of forms. One important way in which they can be classified is by the designer's specification of what the interface should be. Some tools require the programmer to program in a special-purpose language, some provide an **application framework** to guide the programming, some automatically generate the interface from a high-level model or specification, and others allow the interface to be designed interactively. Each of these types is discussed in the following sections. Of course, some tools use different techniques for specifying different parts of the user interface. These are classified by their predominant or most interesting feature.

48.2.1.4.3 *Language-Based*

With most of the older UI tools, the designer specifies the user interface in a special-purpose language. This language can take many forms, including context-free grammars, state transition diagrams, declarative languages, event languages, etc. The language is usually used to specify the syntax of the user interface, that is, the legal sequences of input and output actions. This is sometimes called the dialogue. Green [1986] provides an extensive comparison of grammars, state transition diagrams, and event languages, and Olsen [1992] surveys various older UIMS techniques.

48.2.1.4.3.1 *State Transition Networks*

Because many parts of user interfaces involve handling a sequence of input events, it is natural to think of using a state transition network to code the interface. A transition network consists of a set of states, with arcs out of each state labeled with the input tokens that will cause a transition to the state at the other end of the arc. In addition to input tokens, calls to application procedures and the output to display can also be put on the arcs in some systems. In 1968, Newman implemented a simple tool using finite state machines [Newman 1968] that handled textual input. This was apparently the first user interface tool. Many of the assumptions and techniques used in modern systems were present in Newman's tool: different languages for defining the user interface and the semantics (the semantic routines were coded in a normal programming language), a table-driven syntax analyzer, and device independence.

State diagram tools are most useful for creating user interfaces in which the user interface has a large number of modes (each state is really a mode). For example, state diagrams are useful for describing the operation of low-level widgets (e.g., how a menu or scroll bar works) or the overall global flow of an application (e.g., this command will pop up a dialogue box, from which you can get to these two dialogue boxes, and then to this other window, etc.). However, most highly interactive systems attempt to be mostly mode-free, which means that at each point, the user has a wide variety of choices of what to do. This requires a large number of arcs out of each state, so state diagram tools have not been successful for these interfaces. In addition, state diagrams cannot handle interfaces where the user can operate on multiple objects at the same time. Another problem is that they can be very confusing for large interfaces, because they get to be a "maze of wires," and off-page (or off-screen) arcs can be hard to follow.

Recognizing these problems, but still trying to retain the perspicuity of state transition diagrams, Jacob [1986] invented a new formalism, which is a combination of state diagrams and a form of event languages. There can be multiple diagrams active at the same time and flow of control transfers from one to another in a coroutine fashion. The system can create various forms of direct-manipulation interfaces. VAPS is a commercial system that uses the state transition model, and it eliminates the maze-of-wires problem by providing a spreadsheetlike table in which the states, events, and actions are specified [eNGENUITY 2002]. Transition networks have been thoroughly researched but have not proved particularly successful or useful, either as a research or a commercial approach.

48.2.1.4.3.2 Context-Free Grammars

Many grammar-based systems are based on parser generators used in compiler development. For example, the designer might specify the UI syntax using some form of BNF. Examples of grammar-based systems are Syngraph [Olsen 1983] and parsers built with YACC and LEX in UNIX.

Grammar-based tools, like state diagram tools, are not appropriate for specifying highly interactive interfaces, because they are oriented toward batch processing of strings with a complex syntactic structure. These systems are best for textual command languages; they have been mostly abandoned for user interfaces by researchers and commercial developers.

48.2.1.4.3.3 Event Languages

With event languages, the input tokens are considered to be events that are sent to individual event handlers. Each handler will have a condition clause that determines what types of events it will handle and when it is active. The body of the handler can cause output events, change the internal state of the system (which might enable other event handlers), or call application routines.

Sassafras [Hill 1986] is an event language in which the user interface is programmed as a set of small event handlers. The Elements-Events and Transitions (EET) language provides elaborate control over when the various event handlers are fired [Frank 1995]. In these earlier systems, the event handlers were global. In more modern systems, the event handlers are specific to particular objects. For example, the HyperTalk language, which is part of HyperCard for the Apple Macintosh, can be considered an event language. Microsoft's Visual Basic also contains event-language features; code is written to handle the response to events on objects.

The advantages of event languages are that they can handle multiple input devices active at the same time, and supporting nonmodal interfaces, where the user can operate on any widget or object, is straightforward. The main disadvantage is that it can be very difficult to create correct code, especially as the system gets larger, because the flow of control is not localized and small changes in one part can affect many different pieces of the program. It is also typically difficult for the designer to understand the code once it reaches a nontrivial size. However, the success of HyperTalk, Visual Basic, and similar tools shows that this approach is appropriate for small to medium-sized programs. The style of programming used by Java Swing and related systems, in which the programmer overrides methods that are called when events happen, is similar to the event style.

48.2.1.4.3.4 Declarative Languages

Another approach is to try to define a language that is declarative (stating what should happen) rather than procedural (stating how to make it happen). Cousin [Hayes 1985] and HP/Apollo's Open-Dialogue [Schulert 1985] both allow the designer to specify user interfaces in this manner. The user interfaces supported are basically forms, in which fields can be text that is typed by the user, or options selected using menus or buttons. There are also graphic output areas that the application can use in whatever manner is desired. The application program is connected to the user interface through variables, which can be set and accessed by both. As researchers have extended this idea to support more sophisticated interactions, the specification has grown into full application models, and newer systems are described in Section 48.2.1.4.5.

The layout description languages that come with many toolkits are also a type of declarative language. For example, Motif's user interface language (UIL) allows the layout of widgets to be defined. Because the UIL is interpreted when an application starts, users can (in theory) edit the UIL code to customize the interface. UIL is not a complete language, however, in the sense that the designer still must write C code for many parts of the interface, including any areas containing dynamic graphics and any widgets that change.

The advantage of using declarative languages is that the UI designer does not have to worry about the time sequence of events and can concentrate on the information that needs to be passed back and forth. The disadvantage is that only certain types of interfaces can be provided this way, and the rest must be programmed by hand in the "graphic areas" provided to application programs. The kinds of interactions available are preprogrammed and fixed. In particular, these systems provide no support for such things as dragging graphical objects, rubber-band lines, drawing new graphical objects, or even dynamically changing the items in a menu based on the application mode or context. However, these languages have been used as intermediate languages describing the layout of widgets (such as UIL) that are generated by interactive tools.

48.2.1.4.3.5 Constraint Languages

A number of UI tools allow the programmer to use constraints to define the user interface [Borning 1986b]. Early constraint systems include Sketchpad [Sutherland 1963], which pioneered the use of graphical constraints in a drawing editor, and Thinglab [Borning 1981], which used constraints for graphical simulation. Subsequently, Thinglab was extended to aid in the generation of user interfaces [Borning 1986b].

The previous discussion of toolkits mentions the use of constraints as part of the intrinsics of a toolkit. A number of research toolkits now supply constraints as an integral part of the object system (e.g., Garnet [Myers 1990d], Amulet [Myers 1997], and SubArctic [Hudson 1996]). In addition, some systems have provided higher-level interfaces to constraints. Graphical Thinglab [Borning 1986a] allows the designer to create constraints by wiring icons together, and NoPump [Wilde 1990] and C32 [Myers 1991a] allow constraints to be defined using a spreadsheet-like interface.

The advantage of constraints is that they are a natural way to express many kinds of relationships that arise frequently in user interfaces. For example, lines should stay attached to boxes; labels should stay centered within boxes, etc. A disadvantage with constraints is that they require a sophisticated run-time system to solve them efficiently, and it can be difficult for programmers to specify and debug constraint systems correctly. As yet, there are no commercial UI tools using general-purpose constraint solvers.

48.2.1.4.3.6 Database Interfaces

A very important class of commercial tools supports form-based or GUI-based access to databases. Major database vendors such as Oracle [Oracle Tools 1995] provide tools that allow designers to define the user interface for accessing and setting data. Often, these tools include interactive form editors (which are essentially interface builders) and special database languages. Fourth-generation languages (4GLs), which support defining the interactive forms for accessing and entering data, also fall into this category.

48.2.1.4.3.7 Visual Programming

Visual programs use graphics and 2-D (or more) layout as part of the program specification [Myers 1990c]. Many different approaches to using **visual programming** to specify user interfaces have been investigated. Most systems that support state transition networks use a visual representation. Another popular technique is to use dataflow languages. In these, icons represent processing steps, and the data flow along the connecting wires. The user interface is usually constructed directly by laying out prebuilt widgets, in the style of interface builders. Examples of visual programming systems for creating user interfaces include Labview [National Instruments 2003], which is specialized for controlling laboratory instruments, and Prograph [Pictorius 2002]. Using a visual language seems to make it easier for novice programmers, but large programs still suffer from the familiar maze-of-wires problem. Other papers (e.g., Myers [1990c]) have analyzed the strengths and weaknesses of visual programming in detail.

Another popular language is Visual Basic from Microsoft. However, this is more of a structure editor for Basic combined with an interface builder (see Section 48.2.1.4.6.3), and therefore does not really count as a visual language.

48.2.1.4.3.8 Summary of Language Approaches

In summary, many different types of languages have been designed for specifying user interfaces. One problem with all of these is that they can only be used by professional programmers. Some programmers have objected to the requirement of learning a new language for programming just the UI portion [Olsen 1987]. This has been confirmed by market research [X Business Group 1994]. Furthermore, it seems more natural to define the graphical part of a user interface using a graphical editor. However, it is clear that for the foreseeable future, much of the user interface must still be created by writing programs, so it is appropriate to continue investigations into the best language to use for this. Indeed, an entire book is devoted to investigating the languages for programming user interfaces [Myers 1992b].

48.2.1.4.4 *Application Frameworks*

After the Macintosh Toolbox had been available for a little while, Apple discovered that programmers had a difficult time figuring out how to call the various toolkit functions and how to ensure that the resulting interface met the Apple guidelines. They therefore created a software system that provides an overall application framework to guide programmers. This was called MacApp [Wilson 1990] and used the object-oriented language Object Pascal. Classes are provided for the important parts of an application, such as the main windows, the commands, etc. The programmer specializes these classes to provide the application-specific details, such as what is actually drawn in the windows and which commands are provided. MacApp was very successful at simplifying the writing of Macintosh applications. Today, there are multiple frameworks to help build applications for most major platforms, including the Microsoft Foundation Classes (MFC) for Windows and the portable Java Swing framework. A framework is a software architecture, often object-oriented, that guides the programmer so that implementing UI software is easier.

The Amulet framework [Myers 1997] is aimed at graphical applications, but due to its graphical data model, many of the built-in routines can be used without change (the programmer usually does not need to write methods for subclasses). Newer frameworks aim to help implement applications that take advantage of ubiquitous computing (also called pervasive computing) [Weiser 1993], multiple users (also called computer-supported cooperative work[CSCW]), various sensors that tell the computer where it is and who is around (called context-aware computing [Moran 2001]), and user interfaces that span multiple computers (called multicomputer user interfaces [Myers 2001]). For example, the BEACH framework [Tandler 2001] provides facilities to handle all of these kinds of user interfaces.

The component approach aims to replace today's large, monolithic applications with smaller pieces that attach together. For example, you might buy a separate text editor, ruler, paragraph formatter, spell checker, and drawing program, and have them all work together seamlessly. This approach was invented by the Andrew environment [Palay 1988], which provides an object-oriented document model that supports the embedding of different kinds of data inside other documents. These insets are unlike data that is cut and pasted in systems like the Macintosh, because they bring along the programs that edit them; therefore, they can always be edited in place. Furthermore, the container document does not need to know how to display or print the inset data because the original program that created it is always available. The designer creating a new inset writes subclasses that adhere to a standard protocol, so the system knows how to pass input events to the appropriate editor. Microsoft OLE [Petzold 1991], Apple's OpenDoc [Curbow 1995], and JavaBeans [JavaSoft 1996] use this approach. The Microsoft .Net initiative provides a component architecture for Web services.

All of these frameworks require the designer to write code, typically by creating application-specific subclasses of the standard classes provided as part of the framework.

48.2.1.4.5 *Model-Based Automatic Generation*

A problem with all of the language-based tools is that the designer must specify a great deal about the placement, format, and design of the user interfaces. To solve this problem, some tools use automatic generation so that the tool makes many of these choices from a much higher-level specification. Many of these tools, such as Mickey [Olsen 1989], Jade [Vander Zanden 1990], and DON [Kim 1993] have concentrated on creating menus and dialogue boxes. Jade allows the designer to use a graphical editor to edit the generated interface if it is not good enough. DON has the most sophisticated layout mechanisms and takes into account the desired window size, balance, symmetry, grouping, etc. Creating dialogue boxes automatically has been very thoroughly researched, but there are still no commercial tools that do this.

The **user interface design environment** (UIDE) [Sukaviriya 1993] requires that the semantics of the application be defined in a special-purpose language, and therefore might be included with the language-based tools. It is placed here instead because the language is used to describe the functions that the application supports and not the desired interface. UIDE is classified as a model-based approach because the specification serves as a high-level, sophisticated model of the **application semantics**. In UIDE, the description includes pre- and post-conditions of the operations, and the system uses these to reason about the operations, to automatically generate an interface, and to automatically generate help [Sukaviriya 1990].

The ITS system [Wiecha 1990] also uses rules to generate an interface. ITS was used to create the visitor information system for the EXPO 1992 World's Fair in Seville, Spain. Unlike the other rule-based systems, the designer using ITS is expected to write many of the rules, rather than just writing a specification on which the rules work. In particular, the design philosophy of ITS is that all design decisions should be codified as rules so that they can be used by subsequent designers, which hopefully will mean that interface designs will become easier and better as more rules are entered. As a result, the designer should never use graphical editing to improve the design, because then the system cannot capture the reason that the generated design was not sufficient.

Recently, there has been a resurgence of interest in model-based interfaces to try to provide interfaces that work on the many kinds of handheld devices. For example, the wireless access protocol (WAP) contains high-level descriptions of the information to be displayed, which the handhelds must convert to use specific layouts and interaction techniques. Research continues on ways to convert high-level specifications of appliances and other devices into appropriate remote control interfaces for handhelds, for use in "smart rooms" [Ponnekanti 2001], by the disabled [Zimmerman 2002], and for home appliances [Banavar 2000], [Nichols 2002].

48.2.1.4.6 *Direct Graphical Specification*

The tools described next allow the user interface to be defined, at least partially, by placing objects on the screen using a pointing device. This is motivated by the observation that the visual presentation of the user interface is of primary importance in graphical user interfaces, and a graphical tool seems to be the most appropriate way to specify the graphical appearance. Another advantage of this technique is that it is usually much easier for the designer to use. Many of these systems can be used by nonprogrammers. Therefore, psychologists, graphic designers, and UI specialists can more easily be involved in the UI design process when these tools are used.

These tools can be distinguished from those that use visual programming because with direct graphical specification, the actual user interface (or a part of it) is drawn, rather than being generated indirectly from a visual program. Thus, direct graphical specification tools have been called direct-manipulation programming, because the user is directly manipulating the UI widgets and other elements.

The tools that support graphical specification can be classified into four categories: prototyping tools, those that support a sequence of cards, interface builders, and editors for application-specific graphics.

48.2.1.4.6.1 *Prototyping Tools*

The goal of **prototyping tools** is to allow the designer to mock up quickly some examples of what the screens in the program will look like. Sometimes, these tools cannot be used to create the real user interface of the program; they just show how some aspects will look. This is the chief factor that distinguishes them from other high-level tools. Many parts of the interface may not be operative, and some of the things that look like widgets may just be static pictures. In many prototypers, no real toolkit widgets are used, which means that the designer must draw simulations that look like the widgets that will appear in the interface. The normal use is that the designer would spend a few days or weeks trying out different designs with the tool, and then completely reimplement the final design in a separate system. Most prototyping tools can be used without programming, so they can, for example, be used by graphic designers.

Note that this use of the term *prototyping* is different from the general phrase *rapid prototyping*, which has become a marketing buzzword. Advertisements for just about all UI tools claim that they support rapid prototyping, by which they mean that the tool helps create the UI software more quickly. In this chapter, the term *prototyping* is being used in a much more specific manner.

The first prototyping tool was probably Dan Bricklin's Demo program. This is a program for an IBM PC that allows the designer to create sample screens composed of characters and character graphics (where the fixed-size character cells can contain a graphic, such as a horizontal, vertical, or diagonal line). The designer can easily create the various screens for the application. It is also relatively easy to specify the actions (mouse or keyboard) that cause transitions from one screen to another. However, it is difficult to define other behaviors. In general, there may be some support for type-in fields and menus in prototyping tools, but there is little ability to process or test the results.

For GUIs, designers often use tools like Macromedia's Director [Macromedia 2003a], which is actually an animation tool. The designer can draw example screens and then specify that when the mouse is pressed in a particular place, an animation should start or a different screen should be displayed. Components of the picture can be reused in different screens, but again the ability to show behavior is limited. HyperCard and Visual Basic are also often used as prototyping tools. Research tools such as SILK [Landay 1995] and DENIM [Lin 2002] provide a quick sketching interface and then convert the sketches into actual interfaces.

The primary disadvantage of these prototyping tools is that sometimes the application must be recoded in a "real" language before the application is delivered. There is also the risk that the programmers who implement the real user interface will ignore the prototype.

48.2.1.4.6.2 Cards

Many graphical programs are limited to user interfaces that can be presented as a sequence of mostly static pages, sometimes called frames, cards, or forms. Each page contains a set of widgets, some of which cause transfer to other pages. There is usually a fixed set of widgets to choose from, which have been coded by hand.

An early example of this is Menulay [Buxton 1983a], which allows the designer to place text, graphical potentiometers, iconic pictures, and light buttons on the screen and see exactly what the end user will see when the application is run. The designer does not need to be a programmer to use Menulay.

Probably the most famous example of a card-based system is HyperCard from Apple. There are many similar programs, such as GUIDE [Owl International Inc. 1991], and ToolBook [Click2learn 1995]. In all of these, the designer can easily create cards containing text fields, buttons, etc., along with various graphic decorations. The buttons cause transfers to other cards. These programs provide a scripting language to offer more flexibility for buttons. HyperCard's scripting language is called HyperTalk and, as mentioned previously, is really an event language, because the programmer writes short pieces of code that are executed when input events occur. In its usual instantiation, the World Wide Web was represented as a sequence of pages, which were like cards, with embedded links that transfer to other pages that replace the previous page.

48.2.1.4.6.3 Interface Builders

An interface builder allows the designer to create dialogue boxes, menus, and windows that are to be part of a larger user interface. These are also called interface development tools (IDTs) or GUI builders. Interface builders allow the designer to select from a predefined library of widgets and place them on the screen using a mouse. Other properties of the widgets can be set using property sheets. Usually, there is also some support for sequencing, such as bringing up subdialogues when a particular button is hit. The Steamer project at BBN Technologies demonstrated many of the ideas later incorporated into interface builders and was probably the first object-oriented graphics system [Stevens 1983]. Other examples of research interface builders are DialogEditor [Cardelli 1988] and Gilt [Myers 1991b]. There are hundreds of commercial interface builders, including the resource editors that come with professional development environments such as Metrowerks CodeWarrior. Microsoft's Visual Basic is essentially an interface builder coupled with an editor for an interpreted language. Many of the tools discussed previously, such as the virtual toolkits, visual languages, and application frameworks, also contain interface builders.

Interface builders use the actual widgets from a toolkit, so they can be used to build parts of real applications. Most will generate C or C++ code templates that can be compiled along with the application code. Others generate a description of the interface in a language that can be read at run-time. It is sometimes important that the programmers not edit the output of the tools (such as the generated C code), or else the tool can no longer be used for later modifications.

Although interface builders make laying out the dialogue boxes and menus easier, this is only part of the UI design problem. These tools provide little guidance toward creating good user interfaces, because they give designers significant freedom. Another problem is that for any kind of program with a graphics area (such as drawing programs, CAD, visual language editors, etc.), interface builders do not help with the contents of the graphics pane. Also, they cannot handle widgets that change dynamically. For example, if the contents of a menu or the layout of a dialogue box changes based on program state, this must be programmed by writing code.

48.2.1.4.6.4 Data Visualization Tools

An important commercial category of tools is dynamic data visualization systems. These tools, which tend to be quite expensive, emphasize the display of dynamically changing data on a computer and are used as front ends for simulations, process control, system monitoring, network management, and data analysis. The interface to the designer is usually quite similar to an interface builder, with a palette of gauges, graphers, knobs, and switches that can be placed interactively. However, these controls usually are not from a toolkit and are supplied by the tool. Example tools in this category include DataViews [DataViews 2001] and SL-GMS [SL Corp. 2002].

48.2.1.4.6.5 Editors for Application-Specific Graphics

When an application has custom graphics, it would be useful if the designer could draw pictures of what the graphics should look like rather than having to write code for this. The problem is that the graphic objects usually need to change at run-time, based on the actual data and the end user's actions. Therefore, the designer can only draw an example of the desired display, which will be modified at run-time, and so these tools use demonstrational programming [Myers 1992a]. This distinguishes these programs from the graphical tools of the previous three sections, with which the full picture can be specified at design time. As a result of the generalization task of converting the example objects into parameterized prototypes that can change at run-time, these systems are still in the research phase.

Peridot [Myers 1988a] allows new, custom widgets to be created. The primitives, which the designer manipulates with the mouse, are rectangles, circles, text, and lines. The system generalizes from the designer's actions to create parameterized, object-oriented procedures such as those that might be found in toolkits. Experiments showed that Peridot could be used by nonprogrammers. Lapidary [Vander Zanden 1995] extends the ideas of Peridot to allow general application-specific objects to be drawn. For example, the designer can draw the nodes and arcs for a graph program. The DEMO system [Fisher 1992] allows some dynamic, run-time properties of the objects to be demonstrated, such as how objects are created. The Marquise tool [Myers 1993] allows the designer to demonstrate when various behaviors should happen and supports palettes that control the behaviors. With Pavlov [Wolber 1997], the user can demonstrate how widgets should control a car's movement in a driving game. Gamut [McDaniel 1999] has the user give hints to help the system infer sophisticated behaviors for games-style applications. Research continues on making these ideas practical.

48.2.1.4.6.6 Specialized Tools

For some application domains, there are customized tools that provide significant high-level support. These tend to be quite expensive, however (i.e., $20,000 to $50,000). For example, in the aeronautics and real-time control areas, there are a number of high-level tools, such as InterMAPhics [Gallium 1991].

48.2.2 Tools for the World Wide Web

Implementing user interfaces for the World Wide Web generally uses quite different tools than building GUIs and is covered in depth in other chapters of this volume. Furthermore, the technology and tools are changing quite rapidly. Therefore, this section just provides a brief overview.

Simple Web pages may be composed of static text and graphics with embedded links, and these can be authored by directly writing the underlying hypertext markup language (HTML). Alternatively, the author can use interactive tools, such as Microsoft FrontPage, which therefore serves as a kind of interface builder. FrontPage can also author pages that contain forms for filling in information (text fields, buttons, etc.). More dynamic pages can use a scripting language embedded in the HTML, such as JavaScript or VBScript (Visual Basic Script). Alternatively, a specialized animation language can be used, such as Macromedia's Flash language, which might be authored using an interactive tool such as Dreamweaver [Macromedia 2003b].

In all cases, the back-end that provides the pages, processes any input provided in form fields, and delivers new pages as a result, must be implemented using some kind of server-side scripting or database tool, which is typically quite different from the tools used to author the client-side pages that the user sees.

48.3   Models of User Interface Software

Because creating UI software is so difficult, there have been a number of efforts to describe the software organization at a very abstract level, by creating models of the software. The earliest attempts used the same levels that had been defined for compilers and talked about the semantic, syntactic, and lexical parts of the user interface, but this proved to be useful mostly for parser-based implementations [Buxton 1983b]. Another early model is the **Seeheim model** [Pfaff 1985], which separates the presentation aspects (output) from the dialogue management (what happens in what order, based on what input) from the application interface model (what the resulting changes in data are). This model does not work well with GUIs, because they deemphasize dialogue in favor of mode-free interaction. The **model-view-controller** concept [Krasner 1988] was first used by Smalltalk, and separates the output handling (view) from the input handling (controller). Both of these are separated from the underlying data (the model). Later systems, such as InterViews [Linton 1989], found it difficult to separate the view from the controller, and therefore used a simpler model–view organization, in which the view includes the controller. A new model for software organization — which tries to handle the various aspects required for multiple users, distributed processing and ubiquitous computing — is being developed as part of BEACH [Tandler 2002].

48.4   Technology Transfer

User interface tools are an area where research has had a tremendous impact on the current practice of software development [Myers 1998]. Of course, window managers and the resulting GUI style come from the seminal research at the Stanford Research Institute, Xerox Palo Alto Research Center (PARC), and MIT in the 1970s. Interface builders and card programs like HyperCard were invented in research laboratories at BBN, the University of Toronto, Xerox PARC, and others. Now, interface builders are widely used for commercial software development. Event languages, widely used in HyperTalk and Visual Basic, were first investigated in research laboratories. The current generation of environments, such as OLE and JavaBeans, are based on the component architecture that was developed in the Andrew environment from Carnegie Mellon University. Thus, whereas some early UIMS approaches, such as transition networks and grammars, may not have been successful, overall, UI tool research has changed the way that software is developed.

48.5   Research Issues

Although there are many UI tools, there are plenty of areas in which further research is needed. Previous reports discuss future research ideas for UI tools at length [Myers 2000, Olsen 1993]. Here, a few of the important ones are summarized.

48.5.1   New Programming Languages

The built-in input/output primitives in today's programming languages, such as printf/scanf or cout/cin, support a textual question-and-answer style of user interface that is modal and well known to be poor. Most of today's tools use libraries and interactive programs that are separate from programming languages.

However, many of the techniques, such as object-oriented programming, multiple-processing, and constraints, are best provided as part of the programming language. Even new languages, such as Java, make much of the user interface harder to program by leaving it in separate libraries. Furthermore, an integrated environment, where the graphical parts of an application can be specified graphically and the rest textually, would make the generation of applications much easier. How programming languages can be improved to better support UI software is the topic of a book by Myers [1992b].

48.5.2 Increased Depth

Many researchers are trying to create tools that will cover more of the user interface, such as application-specific graphics and behaviors. The challenge here is to allow flexibility to application developers while still providing a high level of support. Tools should also be able to support Help, Undo, and Aborting of operations.

Today's UI tools mostly help with the *generation* of the code of the interface, and assume that the fundamental UI design is complete. Also needed are tools to help with the generation, specification, and analysis of the *design* of the interface. For example, an important first step in UI design is task analysis, in which the designer identifies the particular tasks that the end user will need to perform. Research should be directed at creating tools to support these methods and techniques. These might eventually be integrated with the code generation tools, so that the information generated during early design can be fed into automatic generation tools, possibly to produce an interface directly from the early analyses. The information might also be used to generate documentation and run-time help automatically.

Another approach is to allow the designer to specify the design in an appropriate notation, and then provide tools to convert that notation into interfaces. For example, the UAN [Hartson 1990] is a notation for expressing the end user's actions and the system's responses.

Finally, much work is needed in ways for tools to help evaluate interface designs. Initial attempts, such as MIKE [Olsen 1988], have highlighted the need for better models and metrics against which to evaluate user interfaces. Research in this area by cognitive psychologists and other user interface researchers (e.g., Kieras [1995]) is continuing.

48.5.3 Increased Breadth

We can expect the user interfaces of tomorrow to be different from the conventional window-and-mouse interfaces of today, and tools will have to change to support the new styles. For example, already we are seeing tiny digital pagers and phones with embedded computers and displays, palm-sized computers such as the PalmOS devices, notebook-sized panel computers such as Microsoft's TabletPCs, and wall-sized displays. Furthermore, computing is appearing in more and more devices around the home and office. An important next wave will appear when the devices can all easily communicate with each other, probably using wireless radio technologies like 802.11 (Wi-Fi) or Bluetooth [Haartsen 1998]. Sound, video, and animations will increasingly be incorporated into user interfaces. New input devices and techniques will probably replace the conventional mouse and menu styles. For example, there will be substantially more use of techniques such as gestures, handwriting, and speech input and output. These are called recognition-based because they require software to interpret the input stream from the user to identify the content. In these "non-WIMP" applications [Nielsen 1993a] (WIMP stands for windows, icons, menus, and pointing devices), designers will also need better control over the timing of the interface, to support animations and various new media such as video. Although a few tools are directed at multiple-user applications, there are no direct graphical specification tools, and the current tools are limited in the styles of applications they support. A further problem is supporting *multiple* interfaces for the same application, so the application can run on small and large devices in a consistent manner. Also, sometimes a person might be using multiple devices *at the same time*, such as a PalmOS device *and* a big display screen [Myers 2001].

Another concern is supporting interfaces that can be moved from one natural language to another (like English to French). Internationalizing an interface is much more difficult than simply translating the text strings, and it includes different number, date, and time formats; new input methods; redesigned layouts; different color schemes; and new icons [Russo 1993]. How can future tools help with this process?

48.5.4   End User Programming and Customization

One of the most successful computer programs of all time is the spreadsheet. The primary reason for its success is that end users can program it (by writing formulas and macros). However, end-user programming is rare in other applications. Where it exists, it usually requires learning conventional programming. For example, AutoCAD provides Lisp for customization, and many Microsoft applications use Visual Basic. More effective mechanisms for users to customize existing applications and to create new ones are needed [Myers 1992b]. However, these should not be built into individual applications, as is done today, because this means that the user must learn a different programming technique for each application. Instead, the facilities should be provided at the system level, and therefore part of the underlying toolkit. Naturally, because this is aimed at end users, it will not be like programming in C, but rather at some higher level.

48.5.5   Application and User Interface Separation

One of the fundamental goals of UI tools is to allow better modularization and separation of UI code from application code. However, a survey reported that conventional toolkits actually make this separation more difficult, due to the large number of call-back procedures required [Myers 1992c]. Therefore, further research is needed into ways to better modularize the code and how tools can support this.

48.5.6   Tools for the Tools

It is very difficult to create the kinds of tools described in this chapter. Each one takes an enormous effort. Therefore, work is needed in ways to make the tools themselves easier to create. For example, the Garnet toolkit explored mechanisms specifically designed to make high-level graphical tools easier to create [Myers 1992d]. The Unidraw framework has also proved useful for creating interface builders [Vlissides 1991]. However, more work is needed.

48.6   Conclusions

Generally, research and innovation in *tools* trail innovation in user interface *design*, because it only makes sense to develop tools when you know what kinds of interfaces you are building tools for. Given the consolidation of the UI interaction style in the last 15 years, it is not surprising that tools have matured to the point where commercial tools have fairly successfully covered the important aspects of user interface construction. It is clear that the research on UI software tools has had enormous impact on the process of software development. Now, UI design is poised for a radical change, primarily brought on by the rise of the World Wide Web, ubiquitous computing, recognition-based user interfaces, handheld devices, wireless communication, and other technologies. Therefore, we expect to see a resurgence of interest in and research on UI software tools in order to support the new user interface styles.

Defining Terms

Application or application semantics: The part of the software that is *not* the user interface.
Application framework: A software architecture, often object-oriented, that guides the programmer so that implementing user interface software is easier.
Call-back procedures: Procedures defined by the application programmer that are called when a widget is operated by the end user.
Constraints: Relationships that are declared once and then maintained automatically by the system.

Geometry management: Part of the toolkit intrinsics that handles the placement and size of widgets.

Graphical user interface (GUI): A form of user interface that makes significant use of the direct-manipulation style using pointing with a mouse.

Icons: Small pictures that represent windows (or sometimes files) in window managers.

Interface builder: Interactive tool that lays out widgets to create dialogue boxes, menus, and windows that are to be part of a larger user interface. These are also called interface development tools and GUI builders.

Intrinsics: The layer of a toolkit on which different widgets are implemented.

Model-view-controller: Model of how user interface software might be organized, separating the application data (model), presentation (view), and input handling (controller) aspects.

Prototyping tools: These allow the designer to mock up quickly some examples of what the screens in the program will look like. Often, these tools cannot be used to create the real user interface of the program; they just show how some aspects will look.

Seeheim model: Model of how user interface software might be organized, separating the presentation, dialogue, and application aspects.

Toolkit: A library of widgets that can be called by application programs.

User interface (UI): The part of the software that handles the output to the display and the input from the person using the program.

User interface design environment (UIDE): General term for comprehensive user interface tools.

User interface management system (UIMS): An older term, not much used now. Sometimes used to cover all user interface tools, but usually limited to tools that handle the sequencing of operations (what happens after each event from the user).

User interface tool: Any software that helps to create user interfaces.

Virtual toolkits: Also called cross-platform development systems, these are programming interfaces to multiple toolkits that allow code to be easily ported to Macintosh, Microsoft Windows, and Unix environments.

Visual programming: Using graphics and two- (or more) dimensional layout as part of the program specification.

Widget: A way of using a physical input device to input a certain type of value. Typically, widgets in toolkits include menus, buttons, scroll bars, text type-in fields, etc.

Window: Region of the screen (usually rectangular) that can be independently manipulated by a program and/or user.

Window manager: The user interface of the windowing system. Can also refer to the entire windowing system.

Windowing system: Software that separates different processes into different rectangular regions (windows) on the screen.

References

[Adobe Systems Inc. 1985] Adobe Systems Inc. *PostScript Language Reference Manual*. Addison-Wesley, Reading, MA.

[Apple Computer Inc. 1985] Apple Computer Inc. *Inside Macintosh*. Addison-Wesley, Reading, MA.

[Banavar 2000] Guruduth Banavar, James Beck, Eugene Gluzberg, Jonathan Munson, Jeremy Sussman, and Deborra Zukowski. "Challenges: An Application Model for Pervasive Computing," *Sixth Annual ACM/IEEE International Conference on Mobile Computing and Networking (Mobicom 2000)*. http://www.research.ibm.com/PIMA.

[Bharat 1994] Krishna Bharat and Marc H. Brown. "Building Distributed, Multi-User Applications by Direct Manipulation," *ACM SIGGRAPH Symposium on User Interface Software and Technology*, Proceedings UIST'94. Marina del Rey, CA, Nov. 1994. pp. 71–81.

[Bly 1986] Sara A. Bly and Jarrett K. Rosenberg. "A Comparison of Tiled and Overlapping Windows," *Human Factors in Computing Systems*, Proceedings SIGCHI'86. Boston, MA, Apr. 1986. pp. 101–106.

[Borning 1981] Alan Borning. "The Programming Language Aspects of Thinglab: A Constraint-Oriented Simulation Laboratory," *ACM Transactions on Programming Languages and Systems.* **3**(4). pp. 353–387.

[Borning 1986a] Alan Borning. "Defining Constraints Graphically," *Human Factors in Computing Systems,* Proceedings SIGCHI'86. Boston, MA, Apr. 1986. pp. 137–143.

[Borning 1986b] Alan Borning and Robert Duisberg. "Constraint-Based Tools for Building User Interfaces," *ACM Transactions on Graphics.* **5**(4). pp. 345–374.

[Buxton 1983a] W. Buxton, M.R. Lamb, D. Sherman, and K.C. Smith. "Toward a Comprehensive User Interface Management System," *Computer Graphics,* Proceedings SIGGRAPH'83. Detroit, MI, July 1983. pp. 35–42.

[Buxton 1983b] William Buxton. "Lexical and Pragmatic Considerations of Input Structures," *Computer Graphics.* **17**(1). pp. 31–37.

[Cardelli 1988] Luca Cardelli. "Building User Interfaces by Direct Manipulation," *ACM SIGGRAPH Symposium on User Interface Software and Technology,* Proceedings UIST'88. Banff, Alberta, Canada, Oct. 1988. pp. 152–166.

[Cardelli 1985] Luca Cardelli and Rob Pike. "Squeak: A Language for Communicating with Mice," *Computer Graphics,* Proceedings SIGGRAPH'85. San Francisco, CA, July 22–26, 1985. pp. 199–204.

[Click2learn 1995] Click2learn. *ToolBook.* Click2learn, Inc. (formerly Asymetrix Corporation), Bellevue, WA. http://www.asymetrix.com/en/toolbook/index.asp.

[Curbow 1995] Dave Curbow, Elizabeth Dykstra-Erickson, Kerry Orteg, and Geoff Schuller. *Human Interface Specification for the Macintosh Implementation.* Apple Computer, Inc. OpenDoc Version 1.0, Specification Version 1.0.3. November 6, 1995.

[DataViews 2001] DataViews. GE Fanuc Automation NA. Albany, NY. www.dvcorp.com.

[eNGENUITY 2002] eNGENUITY. *VAPS.* eNGENUITY Technologies (formerly Virtual Prototypes, Inc.), Montreal. http://www.engenuitytech.com.

[Fisher 1992] Gene L. Fisher, Dale E. Busse, and David A. Wolber. "Adding Rule-Based Reasoning to a Demonstrational Interface Builder," *ACM SIGGRAPH Symposium on User Interface Software and Technology,* Proceedings UIST'92. Monterey, CA, Nov. 1992. pp. 89–97.

[Frank 1995] Martin R. Frank. *Model-Based User Interface by Demonstration and by Interview.* Computer Science Department, Georgia Institute of Technology. Ph.D. Thesis.

[Gallium 1991] Gallium. *InterMAPhics.* Gallium Software (formerly Prior Data Systems) Ottawa, Ontario Canada. http://www.gallium.com.

[Gaskins 1992] Tom Gaskins. *PEXlib Programming Manual.* O'Reilly and Associates, Inc., Sebastopol, CA.

[GNOME 2002] GNOME. *GNU Network Object Model Environment (GNOME).* http://www.gnome.org.

[Gosling 1986] James Gosling. *NeWS: A Definitive Approach to Window Systems.* Sun Microsystems Corp., Mountain View, CA.

[Green 1986] Mark Green. "A Survey of Three Dialog Models," *ACM Transactions on Graphics.* **5**(3). pp. 244–275.

[Haartsen 1998] Jaap Haartsen, Mahmoud Naghshineh, Jon Inouye, Olaf J. Joeressen, and Warren Allen. "Bluetooth: Vision, Goals, and Architecture," *ACM Mobile Computing and Communications Review.* **2**(4). pp. 38–45. www.bluetooth.com.

[Hartson 1990] H. Rex Hartson, Antonio C. Siochi, and Deborah Hix. "The UAN: A User-Oriented Representation for Direct Manipulation Interface Designs," *ACM Transactions on Information Systems.* **8**(3). pp. 181–203.

[Hayes 1985] Philip J. Hayes, Pedro A. Szekely, and Richard A. Lerner. "Design Alternatives for User Interface Management Systems Based on Experience with COUSIN," *Human Factors in Computing Systems,* Proceedings SIGCHI'85. San Francisco, CA, Apr. 1985. pp. 169–175.

[Hill 1986] Ralph D. Hill. "Supporting Concurrency, Communication and Synchronization in Human-Computer Interaction — The Sassafras UIMS," *ACM Transactions on Graphics.* **5**(3). pp. 179–210.

[Hill 1994] Ralph D. Hill, Tom Brinck, Steven L. Rohall, John F. Patterson, and Wayne Wilner. "The Rendezvous Architecture and Language for Constructing Multiuser Applications," *ACM Transactions on Computer-Human Interaction.* **1**(2). pp. 81–125.

[Hudson 1996] Scott E. Hudson and Ian Smith. "Ultra-Lightweight Constraints," *ACM SIGGRAPH Symposium on User Interface Software and Technology,* Proceedings UIST'96. Seattle, WA, Nov. 1996. pp. 147–155. http://www.cc.gatech.edu/gvu/ui/sub_arctic.

[Hudson 1993] Scott E. Hudson and John T. Stasko. "Animation Support in a User Interface Toolkit: Flexible, Robust, and Reusable Abstractions," *ACM SIGGRAPH Symposium on User Interface Software and Technology,* Proceedings UIST'93. Atlanta, GA, Nov. 1993. pp. 57–67.

[Jacob 1986] Robert J.K. Jacob. "A Specification Language for Direct Manipulation Interfaces," *ACM Transactions on Graphics.* **5**(4). pp. 283–317.

[JavaSoft 1996] JavaSoft. *JavaBeans.* Sun Microsystems. JavaBeans V1.0. December 4, 1996. http://java.sun. com/beans.

[Kieras 1995] David E. Kieras, Scott D. Wood, Kasen Abotel, and Anthony Hornof. "GLEAN: A Computer-Based Tool for Rapid GOMS Model Usability Evaluation of User Interface Designs," *Eighth Annual Symposium on User Interface Software and Technology,* Proceedings UIST'95. Pittsburgh, PA, Nov. 1995. pp. 91–100.

[Kim 1993] Won Chul Kim and James D. Foley. "Providing High-Level Control and Expert Assistance in the User Interface Presentation Design," *Human Factors in Computing Systems,* Proceedings INTERCHI'93. Amsterdam, The Netherlands, Apr. 1993. pp. 430–437.

[Krasner 1988] Glenn E. Krasner and Stephen T. Pope. "A Description of the Model-View-Controller User Interface Paradigm in the Smalltalk-80 system," *Journal of Object Oriented Programming.* **1**(3). pp. 26–49.

[Landay 1995] James Landay and Brad A. Myers. "Interactive Sketching for the Early Stages of User Interface Design," *Human Factors in Computing Systems,* Proceedings SIGCHI'95. Denver, CO, May 1995. pp. 43–50.

[Lin 2002] James Lin, Michael Thomsen, and James A. Landay. "A Visual Language for Sketching Large and Complex Interactive Designs," *ACM CHI'2002 Conference Proceedings: Human Factors in Computing Systems,* Minneapolis, MN, April 20–25, 2002. pp. 307–314.

[Linton 1989] Mark A. Linton, John M. Vlissides, and Paul R. Calder. "Composing User Interfaces with InterViews," *IEEE Computer.* **22**(2). pp. 8–22.

[Macromedia 2003a] Macromedia. *Director MX.* Macromedia, Inc., San Francisco, CA. http://www. macromedia.com/software/director.

[Macromedia 2003b] Macromedia. *Dreamweaver MX.* Macromedia, Inc., San Francisco, CA. http://www. macromedia.com/software/dreamweaver.

[McCormack 1988] Joel McCormack and Paul Asente. "An Overview of the X Toolkit," *ACM SIGGRAPH Symposium on User Interface Software and Technology,* Proceedings UIST'88. Banff, Alberta, Canada, Oct. 1988. pp. 46–55.

[McDaniel 1999] Richard G. McDaniel and Brad A. Myers. "Getting More out of Programming-by-Demonstration," *Human Factors in Computing Systems,* Proceedings CHI'99. Pittsburgh, PA, May 15–20, 1999. pp. 442–449.

[Moran 2001] Thomas P. Moran and Paul Dourish. "Special Issue on Context-Aware Computing," *HCI Journal.* **16**(2–4). pp. 87–419.

[Myers 1984] Brad A. Myers. "The User Interface for Sapphire," *IEEE Computer Graphics and Applications.* **4**(12). pp. 13–23.

[Myers 1986] Brad A. Myers. "A Complete and Efficient Implementation of Covered Windows," *IEEE Computer.* **19**(9). pp. 57–67.

[Myers 1988a] Brad A. Myers. *Creating User Interfaces by Demonstration.* Academic Press, Boston, MA.

[Myers 1988b] Brad A. Myers. "A Taxonomy of User Interfaces for Window Managers," *IEEE Computer Graphics and Applications.* **8**(5). pp. 65–84.

[Myers 1990a] Brad A. Myers. "All the Widgets," *SIGGRAPH Video Review*. **57.**

[Myers 1990b] Brad A. Myers. "A New Model for Handling Input," *ACM Transactions on Information Systems*. **8**(3). pp. 289–320.

[Myers 1990c] Brad A. Myers. "Taxonomies of Visual Programming and Program Visualization," *Journal of Visual Languages and Computing*. **1**(1). pp. 97–123.

[Myers 1991a] Brad A. Myers. "Graphical Techniques in a Spreadsheet for Specifying User Interfaces," *Human Factors in Computing Systems*, Proceedings SIGCHI'91. New Orleans, LA, Apr. 1991. pp. 243–249.

[Myers 1991b] Brad A. Myers. "Separating Application Code from Toolkits: Eliminating the Spaghetti of Call-Backs," *ACM SIGGRAPH Symposium on User Interface Software and Technology*, Proceedings UIST'91. Hilton Head, SC, Nov. 1991. pp. 211–220.

[Myers 1992a] Brad A. Myers. "Demonstrational Interfaces: A Step Beyond Direct Manipulation," *IEEE Computer*. **25**(8). pp. 61–73.

[Myers 1992b] Brad A. Myers, ed. *Languages for Developing User Interfaces*. Jones and Bartlett, Boston, MA.

[Myers 1994] Brad A. Myers. "Challenges of HCI Design and Implementation," *ACM Interactions*. **1**(1). pp. 73–83.

[Myers 1995] Brad A. Myers. "User Interface Software Tools," *ACM Transactions on Computer–Human Interaction*. **2**(1). pp. 64–103.

[Myers 1998] Brad A. Myers. "A Brief History of Human Computer Interaction Technology," *ACM Interactions*. **5**(2). pp. 44–54.

[Myers 2001] Brad A. Myers. "Using Hand-Held Devices and PCs Together," *Communications of the ACM*. **44**(11). pp. 34–41. http://www.cs.cmu.edu/~pebbles/papers/pebblescacm.pdf.

[Myers 1990d] Brad A. Myers, Dario A. Giuse, Roger B. Dannenberg, Brad Vander Zanden, David S. Kosbie, Edward Pervin, Andrew Mickish, and Philippe Marchal. "Garnet: Comprehensive Support for Graphical, Highly-Interactive User Interfaces," *IEEE Computer*. **23**(11). pp. 71–85.

[Myers 1996a] Brad A. Myers and David Kosbie. "Reusable Hierarchical Command Objects," *Proceedings CHI'96: Human Factors in Computing Systems*, Vancouver, BC, Canada, April 14–18, 1996. pp. 260–267.

[Myers 1993] Brad A. Myers, Richard G. McDaniel, and David S. Kosbie. "Marquise: Creating Complete User Interfaces by Demonstration," *Human Factors in Computing Systems*, Proceedings INTER-CHI'93. Amsterdam, The Netherlands, Apr. 1993. pp. 293–300.

[Myers 1997] Brad A. Myers, Richard G. McDaniel, Robert C. Miller, Alan Ferrency, Andrew Faulring, Bruce D. Kyle, Andrew Mickish, Alex Klimovitski, and Patrick Doane. "The Amulet Environment: New Models for Effective User Interface Software Development," *IEEE Transactions on Software Engineering*. **23**(6). pp. 347–365.

[Myers 1996b] Brad A. Myers, Robert C. Miller, Rich McDaniel, and Alan Ferrency. "Easily Adding Animations to Interfaces Using Constraints," *ACM SIGGRAPH Symposium on User Interface Software and Technology*, Proceedings UIST'96. Seattle, WA, Nov. 1996. pp. 119–128. http://www.cs.cmu.edu/~amulet.

[Myers 1992c] Brad A. Myers and Mary Beth Rosson. "Survey on User Interface Programming," *Human Factors in Computing Systems*, Proceedings SIGCHI'92. Monterey, CA, May 1992. pp. 195–202.

[Myers 1992d] Brad A. Myers and Brad Vander Zanden. "Environment for Rapid Creation of Interactive Design Tools," *The Visual Computer: International Journal of Computer Graphics*. **8**(2). pp. 94–116.

[Myers 2000] Brad Myers, Scott E. Hudson, and Randy Pausch. "Past, Present and Future of User Interface Software Tools," *ACM Transactions on Computer Human Interaction*. **7**(1). pp. 3–28.

[National Instruments 2003] National Instruments. *LabVIEW*. National Instruments Corporation, Austin, TX. http://www.ni.com.

[Newman 1968] William M. Newman. "A System for Interactive Graphical Programming," *AFIPS Spring Joint Computer Conference*, April 30–May 2, 1968. pp. 47–54.

[Nichols 2002] Jeffrey Nichols, Brad A. Myers, Michael Higgins, Joe Hughes, Thomas K. Harris, Roni Rosenfeld, and Mathilde Pignol. "Generating Remote Control Interfaces for Complex Appliances," *CHI Letters: ACM Symposium on User Interface Software and Technology, UIST'02,* Paris, France, Oct. 2002. pp. 161–170. http://www.cs.cmu.edu/~pebbles/papers/PebblesPUCuist.pdf.

[Nielsen 1993a] Jakob Nielsen. "Noncommand User Interfaces," *CACM*. **36**(4). pp. 83–99.

[Nielsen 1993b] Jakob Nielsen. *Usability Engineering.* Academic Press, Boston, MA.

[Olsen 1987] Dan R. Olsen, Jr. "Larger Issues in User Interface Management," *Computer Graphics.* **21**(2). pp. 134–137.

[Olsen 1989] Dan R. Olsen, Jr. "A Programming Language Basis for User Interface Management," *Human Factors in Computing Systems,* Proceedings SIGCHI'89. Austin, TX, Apr. 1989. pp. 171–176.

[Olsen 1992] Dan R. Olsen, Jr. *User Interface Management Systems: Models and Algorithms.* Morgan Kaufmann, San Mateo, CA.

[Olsen 1983] Dan R. Olsen, Jr., and Elizabeth P. Dempsey. "Syngraph: A Graphical User Interface Generator," *Computer Graphics,* Proceedings SIGGRAPH'83. Detroit, MI, July 25–29, 1983. pp. 43–50.

[Olsen 1993] Dan R. Olsen, Jr., James D. Foley, Scott E. Hudson, James Miller, and Brad Myers. "Research Directions for User Interface Software Tools," *Behaviour and Information Technology.* **12**(2). pp. 80–97.

[Olsen 1988] Dan R. Olsen, Jr., and Bradley W. Halversen. "Interface Usage Measurements in a User Interface Management System," *ACM SIGGRAPH Symposium on User Interface Software and Technology,* Proceedings UIST'88. Banff, Alberta, Canada, Oct. 1988. pp. 102–108.

[Oracle Tools 1995] *Oracle Tools.* Oracle Corporation, Redwood Shores, CA.

[Ousterhout 1991] John K. Ousterhout. "An X11 Toolkit Based on the Tcl Language," *Winter USENIX Technical Conference,* 1991. pp. 105–115.

[Owl International Inc. 1991] Owl International Inc. *Guide 2.* Owl International, Inc. Bellevue, WA.

[Palay 1988] Andrew J. Palay, Wilfred J. Hansen, Michael Kazar, Mark Sherman, Maria Wadlow, Tom Neuendorffer, Zalman Stern, Miles Bader, and Thom Peters. "The Andrew Toolkit: An Overview," *Proceedings Winter Usenix Technical Conference,* Dallas, TX, Feb. 1988. pp. 9–21.

[Pausch 1995] Randy Pausch, Tommy Burnette, A.C. Capehart, Matthew Conway, Dennis Cosgrove, Rob DeLine, Jim Durbin, Rich Gossweiler, Suichi Koga, and Jeff White. "Alice: A Rapid Prototyping System for 3-D Graphics," *IEEE Computer Graphics and Applications.* **15**(3). pp. 8–11.

[Pausch 1992] Randy Pausch, Matthew Conway, and Robert DeLine. "Lesson Learned from SUIT, the Simple User Interface Toolkit," *ACM Transactions on Information Systems.* **10**(4). pp. 320–344.

[Petzold 1991] C. Petzold. "Windows 3.1 — Hello to TrueType, OLE, and Easier DDE; Farewell to Real Mode," *Microsoft Systems Journal.* **6**(5). pp. 17–26.

[Pfaff 1985] Gunther R. Pfaff, ed. *User Interface Management Systems.* Springer-Verlag, Berlin.

[Pictorius 2002] Pictorius. *Prograph.* Pictorius Inc., Halifax, NS, Canada. http://www.pictorius.com.

[Ponnekanti 2001] S. R. Ponnekanti, B. Lee, A. Fox, P. Hanrahan, and T. Winograd. "ICrafter: A Service Framework for Ubiquitous Computing Environments," *UBICOMP 2001,* Atlanta, GA. pp. 56–75.

[Roseman 1996] M. Roseman and S. Greenberg. "Building Real Time Groupware with GroupKit, A Groupware Toolkit," *ACM Transactions on Computer Human Interaction.* **3**(1). pp. 66–106.

[Russo 1993] Patricia Russo and Stephen Boor. "How Fluent is Your Interface? Designing for International Users," *Human Factors in Computing Systems,* Proceedings INTERCHI'93. Amsterdam, The Netherlands, Apr. 1993. pp. 342–347.

[Samuelson 1993] Pamela Samuelson. "Legally Speaking: The Ups and Downs of Look and Feel," *CACM.* **36**(4). pp. 29–35.

[Scheifler 1986] Robert W. Scheifler and Jim Gettys. "The X Window System," *ACM Transactions on Graphics.* **5**(2). pp. 79–109.

[Schulert 1985] Andrew J. Schulert, George T. Rogers, and James A. Hamilton. "ADM-A Dialogue Manager," *Human Factors in Computing Systems,* Proceedings SIGCHI'85. San Francisco, CA, Apr. 1985. pp. 177–183.

[Silicon Graphics Inc. 1993] Silicon Graphics Inc. *Open-GL. 2011* Silicon Graphics Inc., Mountain View, CA.

[SL Corp. 2002] SL Corp. *SL-GMS*. SL Corp., Corte Madera, CA. http://www.sl.com.

[Smith 1982] David Canfield Smith, Charles Irby, Ralph Kimball, Bill Verplank, and Erik Harslem. "Designing the Star User Interface," *Byte*. **7**(4). pp. 242–282.

[Stallman 1979] Richard M. Stallman. *Emacs: The Extensible, Customizable, Self-Documenting Display Editor*. MIT Artificial Intelligence Lab. 519. Aug. 1979.

[Stevens 1983] Albert Stevens, Bruce Roberts, and Larry Stead. "The Use of a Sophisticated Graphics Interface in Computer-Assisted Instruction," *IEEE Computer Graphics and Applications*. **3**(2). pp. 25–31.

[Stevens 1994] Marc P. Stevens, Robert C. Zeleznik, and John F. Hughes. "An Architecture for an Extensible 3-D Interface Toolkit," *ACM SIGGRAPH Symposium on User Interface Software and Technology*, Proceedings UIST'94. Marina del Rey, CA, Nov. 1994. pp. 59–67.

[Sukaviriya 1990] Piyawadee Sukaviriya and James D. Foley. "Coupling a UI Framework with Automatic Generation of Context-Sensitive Animated Help," *ACM SIGGRAPH Symposium on User Interface Software and Technology*, Proceedings UIST'90. Snowbird, UT, Oct. 1990. pp. 152–166.

[Sukaviriya 1993] Piyawadee Sukaviriya, James D. Foley, and Todd Griffith. "A Second Generation User Interface Design Environment: The Model and the Runtime Architecture," *Human Factors in Computing Systems*, Proceedings INTERCHI'93. Amsterdam, The Netherlands, Apr. 1993. pp. 375–382.

[Sun Microsystems 2002] Sun Microsystems. *Java 2-D API*. http://java.sun.com/products/java-media/2-D.

[Sun Microsystems 2003] Sun Microsystems. *Java: Programming for the Internet*. http://java.sun.com.

[Sutherland 1963] Ivan E. Sutherland. "SketchPad: A Man-Machine Graphical Communication System," *AFIPS Spring Joint Computer Conference*. pp. 329–346.

[Swinehart 1986] Daniel Swinehart, Polle Zellweger, Richard Beach, and Robert Hagmann. "A Structural View of the Cedar Programming Environment," *ACM Transactions on Programming Languages and Systems*. **8**(4). pp. 419–490.

[Tandler 2001] Peter Tandler. "Software Infrastructure for Ubiquitous Computing Environments Supporting Synchronous Collaboration with Heterogeneous Devices," *UbiComp 2001*, Atlanta, GA, Sept. 30–Oct. 2, 2001. pp. 96–115. http://ipsi.fraunhofer.de/ambiente/paper/2001/UbiComp-2001-tandler.pdf.

[Tandler 2002] P. Tandler, N.A. Streitz, and Th. Prante. "Roomware — Moving toward Ubiquitous Computers," *IEEE Micro*. **22**(6). pp. 36–47. http://www2.darmstadt.gmd.de/ipsi/ambiente/abstract.asp?Pub_ID=293.

[Teitelman 1979] Warren Teitelman. "A Display Oriented Programmer's Assistant," *International Journal of Man-Machine Studies*. **11**(2). pp. 157–187. Also Xerox PARC Technical Report CSL-77-3, Palo Alto, CA, Mar. 8, 1977.

[Tesler 1981] Larry Tesler. "The Smalltalk Environment," *Byte*. **6**(8). pp. 90–147.

[Vander Zanden 1990] Brad Vander Zanden and Brad A. Myers. "Automatic, Look-and-Feel Independent Dialog Creation for Graphical User Interfaces," *Human Factors in Computing Systems*, Proceedings SIGCHI'90. Seattle, WA, Apr. 1990. pp. 27–34.

[Vander Zanden 1995] Brad Vander Zanden and Brad A. Myers. "Demonstrational and Constraint-Based Techniques for Pictorially Specifying Application Objects and Behaviors," *ACM Transactions on Computer-Human Interaction*. 1995. **2**(4). pp. 308–356.

[Visix Software Inc. 1997] Visix Software Inc. *Galaxy Application Environment*. (Company dissolved in 1998. Galaxy was bought by Ambiencia Information Systems, Inc., Campinas, Brazil. http://www.ambiencia.com).

[Vlissides 1991] John M. Vlissides and Steven Tang. "A Unidraw-Based User Interface Builder," *ACM SIGGRAPH Symposium on User Interface Software and Technology*, Proceedings UIST'91. Hilton Head, SC, Nov. 1991. pp. 201–210.

[Web3-D Consortium 1997] Web3-D Consortium. *The Virtual Reality Modeling Language*. ISO/IEC 14772-1:1997. http://www.web3-D.org/Specifications/VRML97.

[Weiser 1993] Mark Weiser. "Some Computer Science Issues in Ubiquitous Computing," *CACM*. **36**(7). pp. 74–83.

[Wernecke 1994] Josie Wernecke. *The Inventor Mentor*. Addison-Wesley Publishing Company, Reading, MA.

[Wiecha 1990] Charles Wiecha, William Bennett, Stephen Boies, John Gould, and Sharon Greene. "ITS: A Tool for Rapidly Developing Interactive Applications," *ACM Transactions on Information Systems*. **8**(3). pp. 204–236.

[Wilde 1990] Nicholas Wilde and Clayton Lewis. "Spreadsheet-Based Interactive Graphics: From Prototype to Tool," *Human Factors in Computing Systems*, Proceedings SIGCHI'90. Seattle, WA, Apr. 1990. pp. 153–159.

[Wilson 1990] David Wilson. *Programming with MacApp*. Addison-Wesley Publishing Company, Reading, MA.

[Wolber 1997] David Wolber. "An Interface Builder for Designing Animated Interfaces," *ACM Transactions on Computer-Human Interaction*. **4**(4). pp. 347–386.

[X Business Group 1994] X Business Group. *Interface Development Technology*. X Business Group, Fremont, CA.

[XVT Software Inc. 1997] XVT Software Inc. XVT, Boulder, CO. http://www.xvt.com.

[Zimmerman 2002] Gottfried Zimmerman, Gregg Vanderheiden, and Al Gilman. "Prototype Implementations for a Universal Remote Console Specification," *Human Factors in Computing Systems*, Extended Abstracts for CHI 2002. Minneapolis, MN, Apr. 1–6, 2002. pp. 510–511. See also http://www.ncits.org/tc_home/v2.htm.

49

Multimedia

49.1 Introduction: Media and Multimedia Interfaces **49**-1
49.2 Types of Media......................................**49**-3
    Output Media • Input Media • Wearable Computers
    and Ubiquitous Computing
49.3 Multimedia Hardware Requirements.................**49**-7
    Compact Disc Secondary Storage Technology
    • Video Storage and Manipulation • Animations
    • Audio Technology — Digital Audio and the Musical
    Instrument Digital Interface • Other Input, Output,
    or Combination Devices
49.4 Distinct Application of Multimedia
    Techniques..**49**-12
49.5 The ISO Multimedia Design Standard **49**-12
49.6 Theories about Cognition and Multiple Media **49**-13
    The Technology Debate • Theories of Cognition
49.7 Case Study — An Investigation into the
    Effects of Media on User Performance **49**-16
    The Laboratory Task • The Media Investigated
    • The Effect of Warnings on Performance • The Effects
    of Sound • Effects of Mental Coding • Do Multimedia
    Interfaces Improve Operator Performance?
49.8 Authoring Software for Multimedia Systems **49**-23
49.9 The Future of Multimedia Systems..................**49**-24

James L. Alty
Loughborough University

49.1 Introduction: Media and Multimedia Interfaces

In order to communicate information to other human beings, we must disturb the environment around us in such a way that the disturbances can be detected by the people with whom we wish to communicate. Furthermore, we must have previously agreed upon the meanings of such disturbances so that the messages can be understood. In other words, we must establish *a medium of communication* between ourselves and our target audience. It is in this sense that computer designers talk about *media*.

Some media of communication are very simple in nature — for example, the doorbell of a house. At this simple level, a ring means there is only one message: "Someone is at the door." However, one could imagine this medium being developed further in certain circumstances. For example, one ring could mean person A is at the door, two rings could indicate person B, etc. One could even transmit quite complex messages through the doorbell using Morse code. Why anyone would want to do this is not immediately obvious, but such a system might be useful for someone who was severely disabled and could not easily get to the door.

This simple example illustrates the essential components of a medium:

Basic tokens or symbols (such as the ringing of the bell)

Agreed-upon structures built from these elements (such as the number of consecutive soundings of the bell, the length of silences, the maximum number of rings in a structure)

Assigned meaning to the different structural elements (e.g., Morse code)

These three components are often termed the symbols (or lexicon), syntax (or structural rules), and semantics (meaning) of the medium. The parallel with language is obvious, and complex media are essentially a form of communication language. The fourth element of language, pragmatics, is also often present. Pragmatics are concerned with conventions and common usage (e.g., in the doorbell example, all communications might start with two rapid bells to gain attention).

The preceding doorbell example also illustrates multiple uses of media. The huge leap in communicative power that is released when the bell is used for Morse code results from developing a mapping between the simple bell and another very powerful medium of communication: our spoken or written language. The system is, in one sense, illustrating the use of multiple media because it can be used to alert the householder, to transmit a more complex message, or to do both at the same time.

Multimedia communication is the simultaneous (or sequential) use of more than one medium of communication to transmit information. For human beings, using multiple media is the normal way to communicate. A human being will usually, in parallel, employ spoken language, body language, gesture, and touch to transmit a message to another person. If they are constrained (e.g., by having their hands tied), human beings often find it more difficult to communicate. It is interesting to note that when people are observed talking on the telephone, they still use extensive gestures, even though no one is observing them.

Just as human beings need media to communicate, so computers (or the designers who write the programs) must employ media to communicate with their users. This is done through the human–computer interface (HCI). To communicate with users, *output media* are employed. These media must be comprehended by human beings and often involve words, pictures, or sounds. To interpret what the users are communicating to the computer system, *input media* are used. Such media often require special skills (such as typing skills) to be employed effectively. A medium is therefore a specific way of presenting information to a user or presenting information to a computer.

The simultaneous use of media to communicate with users on the human–computer interface is termed a *multimedia interface* and can refer to the input media, the output media, or both. Coupling media can often result in the creation of new media. For example, the coupling of moving video and audio into films or television has created new entertainment media.

Multimedia interfaces are more than a set of interesting new ways to use new technology. Many people believe that such communication is natural and corresponds closely to how the brain has developed. Marmollin [1992] has described multimedia as exercising "the whole mind." In this viewpoint, the human brain is seen as having evolved in a multisensory environment, where simultaneous input on different channels was essential for survival. Thus, the processing of the human brain has been fine-tuned to allow simultaneous sampling and comparison between different channels. When channels agree, a sense of safety and well-being is felt. When channels degrade, input from one channel can be used to compensate another. Thus, input channel redundancy (within limits) is thought to be desirable and is an agreeable experience.

It is important to realize that multimedia design is not just about choosing the most obvious medium for a particular communication requirement. Deliberately presenting information in an unusual medium can deliver new and interesting insights into a problem. For example, musical harmony is normally presented through the audio channel, yet new insights into harmonic progressions can be obtained by displaying harmony in a visual manner. A nice example of this is the HarmonySpace application of Holland [1994]. This tool offers both experts and beginners the opportunity to explore harmony by allowing them to use spatial attributes in their exploration (e.g., nearness, centrality, and shape similarity). Similarly, music can be used to assist in the understanding of computer algorithms or physical processes such as turbulence [Blattner et al., 1992].

One final and important point about multimedia interfaces is their significance for disabled users. The current high emphasis on visual output media can be severely disadvantaging to blind or partially sighted individuals. Designers should exploit the new presentation opportunities offered by the multimedia approach, but they must not forget that their interfaces may be used by someone unable to assimilate all the channels. Therefore, designers should allow sufficient redundancy on channels so that the partial loss of one medium does not fatally affect communication in other media. On the other hand, designers ought also to take advantage of the new aural media by offering specially adapted interfaces for the partially sighted. Some progress has already been made in this area. Edwards has created a word processor that uses musical tones and synthesized speech [Edwards, 1989]. The approach adapts visual interfaces so that blind users can use them. The system provides auditory windows, which signify their position by unique tones when the cursor enters them. Spoken menus are activated from these areas. The system, called Soundtrack, can be controlled solely through the audio channel. However, the interface also has a visual manifestation, and this redundancy can be utilized by a partially sighted person.

The importance of matching media appropriately with the capabilities or limitations of the user population is now high on the political agenda in many countries. Strong legislation is now either in force or due to come into force to ensure that designers take proper account of people's limitations so that sections of the community are not disadvantaged through the use of information technology. This has raised many questions with respect to the design and usability of multimedia interfaces. Here are some examples:

How can we design ubiquitous interfaces so that users with a range of disabilities can all use them effectively?

Are different combinations of media more effective for different user cognitive learning styles? For example, it is well known that users with dyslexia process information in different ways than nondyslexic users.

Can we exploit the properties of multimedia interfaces to overcome particular types of disabilities?

It is important that media designers acknowledge that they have a responsibility to think carefully about the usability of their interfaces for users with different types of disabilities.

49.2 Types of Media

As previously stated, media can be subdivided into input and output media. These can then be divided according to the sense used to detect them — visual, aural, and haptic (meaning touch) media — which can then be subdivided further, into language and graphics for visual output media, or into sound and music for aural media. Table 49.1, which is not intended to be exhaustive, gives some examples of common media.

TABLE 49.1 Some Common Media

	Aural	Visual	Haptic
Input Media	Natural sound	Video camera	Keyboard
	Spoken word	Text scan	Mouse
	Synthesized sound	Diagram scan	Trackball
		Gesture recognition	Data glove
		Eye tracking	Touch screen
			Foot pedal
			Breathing tube
Output Media	Natural sound	Written text	Data glove
	Music	Static graphics	Braille pad
	Synthesized sound	Animation	
	Spoken word	Still video	
		Moving video	

Currently, haptic media dominate the input media area, and visual media dominate the output media field. Aural media are still not fully exploited, particularly for input, where voice recognition could offer a flexible and natural interface.

In Section 49.2.3, wearable and ubiquitous computers are discussed. Such developments considerably extend our conventional ideas about media.

49.2.1 Output Media

Many current output media are reasonably well tuned to human capabilities. They are based on media that have been used among human beings for many years: text, graphics, pictures, video, and sound. Although normal-sized VDU screens do not have quite the same properties as standard-sized office paper, the correspondence is close. Most users therefore have little trouble in understanding well designed visual or aural output media. The problems of designing effective output using these media are essentially the same as those in traditional media design (e.g., books, movies, or diagrams). This does not mean that designing such media is necessarily straightforward, but well known design techniques exist.

One output medium, however, which has still not been fully utilized in traditional computer applications, is the auditory medium. Although most computers can support quite sophisticated aural output (e.g., music), this is rarely exploited. It is now a number of years since Gaver [1986] suggested the use of *auditory icons* — well known, usually natural sounds with common associations (such as a police car's siren). Blattner et al. [1989] have further suggested the use of structured *earcons*, based on simple musical motifs. Such motifs (or jingles) are often used in public address systems to precede messages and alert listeners. Alty [1995] has investigated mappings between computer algorithms and music. The internal workings of an algorithm are mapped to musical structures. Thus, in an audiolization of the Bubble Sort Algorithm, moving up the list, the swapping of elements, and the current state of the list are all mapped into different instruments and rhythmic structures. Using stereo output further assists disambiguation. It appears that people can understand the algorithms from the musical output alone, but more work is needed on which types of musical mappings are most appropriate. Musical mappings have also been suggested to aid computer program debugging [Vickers and Alty, 2000a, 2000b, 2000c].

Traditional entertainment media usually consist of output media only (films and TV programs are examples of these). In computing applications, however, the user is normally able to interact and control the progress of the interaction. This is termed *interactive multimedia*. In interactive media environments, linking between different media and within media can both be very important, and there has been considerable recent research work on how to link different elements within the various media.

Output text has been transformed through the creation of *hypertext* structures [Nielsen, 1990]. Hypertext techniques transform traditional sequential text into a cross-linked structure. Certain words in the text are made "active" in that, when selected, they will transfer the reader to another section of text. In extreme cases, the sequential nature of the text is lost, and the text becomes a structure with many paths through it, selectable by the reader. Hypertext linkages across communication networks have now become commonplace, an obvious example being the World Wide Web. Most textual information on the Web is now retrieved as hypertext (using a language called HTML). Such Web text contains links to other systems on the Web, and exceedingly long and complex chains of linkages can be followed.

The term *hypermedia* is often used to describe the creation and support of linkages between different media. Elements of text may link to photographs, movies, or even sound sequences, either on local systems or across the communication network. One of the current problems in hypertext and hypermedia structures is navigation. Users can easily become lost in hypermedia space. Nondynamic links can also inhibit exploratory learning.

One area where media are now being extensively employed in new ways is virtual reality, where input and output media are used to create virtual environments. For output media, this involves creating an impression of *immersion*. A key aspect of immersion is to avoid the current interface situation, where the interface is a small part of the visual or aural field. Thus, the whole field (visual or auditory) is completely filled. Visually, this can be done in two ways. First, very large curved screens (or even domes) can be used to fill the field of vision. Second, the user can wear glasses, which consist of two small computer output screens. Distinct presentations to the two screens can give an impression of immersive 3-D. The second technique is clearly less expensive and is economical in space usage. However, it requires more software effort to create the 3-D effects and to maintain stability in the environment when the head moves.

A related output medium, which has potential but is underexploited, is 3-D vision. The problem is, of course, the present requirement for special glasses. The third dimension has obvious applications in displaying 3-D molecules or architectural structures, but it can also be use to improve presentation of other data. Three-dimensional presentation has also been used in displaying information in databases. Because 3-D display is usually essential in virtual reality systems, it is expected that rapid developments will take place in this area.

49.2.2 Input Media

Current input media, in comparison with output media, are cumbersome and unnatural. They require skills (such as keyboard skills) to be used effectively. Furthermore, input media are unusual in the fact that they must be coupled with some form of output medium to be useful. For example, keyboard input is not effective unless the user receives simultaneous output of what is being typed. In a similar manner, input using a mouse requires visual feedback to be effective. This complicates the analysis of input media.

Recently, there has been more active research on new input media. Developments have been reported in voice recognition (now beginning to reach acceptable levels of performance), gesture and pointing (where the actual visual gestures are tracked by video cameras and interpreted), eye movement (the actual movement of the eye is tracked and can be used as a selection device), lip motion (to assist in speech recognition), facial expression, and handwriting. The research is driven by the current primitive state of input media in contrast to human–human communication.

An interesting feature of many input media is their impreciseness. Voice recognition is often a difficult process because of other extraneous noise; gesture is often vague and ambiguous; lip motion is not read accurately by most human beings. Human beings process such media effectively because they are usually processed in parallel with other media (e.g., gesture and lip movement usually accompany speech). The human processing system exploits the redundancy across these channels, comparing input from different channels for confirmation or seeking support for the interpretation of one channel from another input channel. Some experiments have even suggested that human beings combine acoustic and visual information before classifying them separately [Braida, 1991]. In other words, separate channel decisions are not made on each input. It is not surprising that much recent input media research work has been concerned with investigating the simultaneous use of a number of input media, and experiments have been carried out to see whether recognition can be improved through the simultaneous use of touch, speech input, and gesture.

Experiments on input media have been reported that involve the combination of speech recognition with lip reading, gesture with speech, and speech with handwriting [Waibel et al., 1995]. One experiment concerned the simultaneous input of lip reading and voice. The acoustic input performance was measured in clean and noisy environments. When the acoustic input was clean, word accuracies in excess of 90% were attained. The lip-reading performance, on its own, varied between 32% and 47% accuracy, and, when used in parallel with the acoustic input, had minimal effect on overall accuracy. When the noisy acoustic input was used, acoustic recognition on its own fell to around 50%, but with lip reading added in parallel, performance improved to over 70%. Thus, adding the lip-reading input (which had a relatively poor recognition rate on its own) boosted the recognition rates of acoustic input in noisy environments.

Waibel et al. [1995] also examined gestural input in some detail. Their input gestures were created by stylus moves on a 2-D graphics tablet. They report on the great variability in the way users make gestures: "No matter how many tokens we put in the training database to cover the different gestures used that mean 'delete text' for example, there may always be totally different gestures that are not yet part of the gesture vocabulary." Bordegoni and Hemmje [1993] have constructed a "dynamic gesture machine," providing graphical feedback in a 3-D interface. The system is based on a simple gesture language.

More novel input devices have been reported. Bordegoni and Hemmje [1993] report on the use of a "force-input" device. This device replaces the mouse in a 3-D virtual environment and is based on a space trackball, which not only can act as a normal trackball but, in addition, can detect the pressure exerted on the ball for providing 3-D movement. A data glove is used, in addition, to detect the position of the hand. Data gloves on their own offer interesting new possibilities, particularly in virtual reality applications [Zimmermann et al., 1987].

49.2.3   Wearable Computers and Ubiquitous Computing

These two concepts are in one sense quite different and in another rather similar. *Ubiquitous comput-ing* involves the embedding of computers into everyday objects such as desks, tables, chairs, and walls. When users move around in this environment, they are (perhaps unconsciously) interacting with these embedded computers, which might then respond. For example, a so-called "smart" room might keep track of the people in it by reading identification tags, and the configuration of the room might change depending on those currently present. The heating might be reduced when no one is in the room and air circulation increased as the room population increases. Laptop computers might automatically link up with each other and also with some central room computer when in the same room. There have al-ready been a number of experiments using badge readers [Want et al., 1992]. A major issue is that of privacy.

Wearable computers start at the other end of the spectrum from ubiquitous computing [Mann, 1996, 1997]. The body of the user carries the computing elements and all the sensors. In the extreme, this should be all that is necessary, because human beings have been able to work this way effectively for a long time! Wearable computers and ubiquitous computers both have been used to maintain personal diaries, where the privacy issue is likely to be less of a problem.

Wearable computers have their own constraints. They must be comfortable (ideally not noticeable). This was a problem for the first generation of wearables. They clearly must be portable. They must have a large number of sensors with extensive communication facilities and be switched on all the time. They should ideally allow the user to operate hands-free and eyes-free when possible; otherwise, the presence of the wearable will seriously intrude on the user's activities. Network connections will enable the user to retrieve data and perhaps to compare it with the world currently in view.

The wearables concept challenges our ideas of media. For example, most of our current input media are simply not suitable. One proposed input device is the *twiddler* — a one-handed, chorded keyboard and mouse with over 4000 combinations. It is claimed that up to 60 words a minute can be input using the twiddler, but this requires extensive training. Speeds of up to 10 words per minute can be learned during a weekend. Video input typically uses a tiny LED screen (720×280 monochrome pixels), viewable by one eye through a reflector, enabling the outside world to be seen at the same time. It looks like a normal 15-inch display viewed from about 2 feet away.

Ultimately, of course, the objective will be to make direct connections between the brain, the nervous system, and the computing systems. At present, this may sound a little futuristic, but already some progress has been reported. StartleCam [Healey and Picard, 1998] is a wearable system with a video camera, which is partially controlled by inputs from the body's physiological system. A skin conductivity sensor detects what is termed a *startled* response. The skin conductivity is measured by applying a small voltage, and this is continuously monitored. The palms of the hands and soles of the feet are preferred sites for the sensors. The startled response is a typical survival response, although it is also triggered by less threatening events, such as a light being turned on or an expression of anxiety. At the time of writing, such research is in its infancy, but if success is achieved, most of our ideas about media will need significant revision.

49.3 Multimedia Hardware Requirements

Multimedia systems inevitably require considerable hardware resources. Such systems need large amounts of memory (both primary and secondary), and any transmission of multimedia data over a network usually requires high bandwidths.

The most common multimedia platforms currently in use are based upon Sun workstations, PCs, or Apple Macintosh systems. Minimum multimedia enhancements will include a sound board, a super VGA graphics card, a CD read/write device, and a DVD reader.

49.3.1 Compact Disc Secondary Storage Technology

Optical storage media, as typified by the CD-ROM, have provided the much needed increased storage capability required for multimedia applications. The CD-ROM is rather like a traditional long-playing record but, in contrast to the traditional record player, the track is read at a constant linear velocity. This means that the rotation speed must change, depending upon the position of the track relative to the center. The data are divided into blocks to provide some direct access capability. The approach allows high data volumes to be stored, but it reduces direct access possibilities because the data are stored in a long spiral, starting at the inside. Originally, CD-ROMs were read-only. However, they are now available in two different technologies — CD-R (CD-recordable) and CD-RW (CD-rewritable). CD-R discs used to be called WORM drives (write once, read many times). Today, cheap CD recorders are available for writing either CD-R or CD-RW discs. They are not identical to the traditional CD (mass produced and bought in shops), and there may be problems in playing recordable media in older CD drives. Most recent CD players will, however, play all media types. For example, audio CDs can be created on a CD-R and then read on a car CD.

The time required to burn a CD depends on the speed of the drive and the amount of data to be stored. The base line is 75 minutes for 650 MB of data on a 1× drive, but this would reduce to about 19 minutes on a 4× speed drive.

The CD-ROM was developed from the audio CD (CD technologies are defined in colored books: audio technology is in the Red book, data CD technology is in the Yellow book, and recordable CD technology is defined in the Orange book). These standards define the basic hardware and storage mechanisms. However, there is also a need to standardize the ways in which operating systems access the information. The *ISO 9660* standard (originally known as *High Sierra*) achieves this and allows different operating systems to access the same CD-ROM. One improvement provided by the ISO standard is the concept of a *session*. A CD-ROM may be built up progressively as sessions are added. When the disc is read, by default, the last session is accessed. This session can access data written in previous sessions or prevent that information from being accessed (e.g., an update). This allows the designer to write information to the disk in different sessions (i.e., at different times) and to update earlier data.

CD-ROM technology stores data in a manner similar to normal filing systems, using a directory structure to locate the contents. A typical CD-ROM can store up to 650 MB of data, which corresponds to 250,000 pages of A4 text, 7,000 full-screen images, 72 minutes of full-bandwidth animation or full-screen video, or 75 minutes of uncompressed high-quality audio. It is important to realize that the high-volume storage capability of CD-ROM discs (> 600 MB) is achieved at the expense of access times (the worst case is about 150 KB/s, although it is proportionally higher on 2× or 4 × drives, etc.). The amount of data contained on a CD-ROM is large, and the data must be properly managed. Although careful placement of files can be used to improve efficiency, most management takes place outside the CD-ROM.

The introduction of low-cost write–read CD-ROM creation systems has revolutionized the production process for CD-ROMs. Previously, a designer had to create a binary image of the CD-ROM on magnetic media and deliver this to a CD-ROM manufacturer. From the binary file, a master disc was made, from which multiple discs were then pressed. This made low-volume production uneconomical. Now, a designer can buy cheap CD recorders, which can be attached to a PC and can master individual discs. The availability of this new hardware has expanded the application areas for CDs. They are now commonly used as removable media to back up files.

49.3.2 Video Storage and Manipulation

Handling video data places very high demands on the multimedia computer system. A typical full-screen still picture requires about 1 to 2 MB of memory. If this were to form part of a movie, the system would need to transfer about 30 MB/s to give the illusion of motion. Thus, one minute of video could occupy nearly a gigabyte of storage. In addition, huge transfer rates would be required to refresh the memory with new picture. In the previous decade, hard disk transfer rates were a limiting factor. Although transfer rates from the latest hard disks are now approaching these speeds, it is still vital to compress information in order to ensure cost-effective delivery over broadcast networks and the Internet.

One of the most common video compression systems used for still images is Joint Photographic Experts Group (JPEG). It can usually achieve compression ratios of about 30:1 without loss. This means that the 30 MB/s transfer rate reduces to 1 MB/sec — well within the capabilities of present hard disk drives — and an image can be stored in about 40 kilobytes. The compression speeds are high.

For moving pictures, the Moving Picture Experts Group (MPEG-2) compression technique is used. This achieves compression ratios of about 50:1 without degradation, and the compression and decompression can occur in real time. Using MPEG-2, a movie can be played from a CD-ROM. Higher compression is possible, but this results in a loss of image quality. MPEG-2 is also used for digital television and DVDs. Work continues on more efficient video compression; for example, MPEG-4 is currently mainly aimed at low-data streaming applications but could be used for future DVDs. A new compression standard, JVT/H.26L, can reduce bit rates by a further factor of 3 compared to MPEG-2, and similar results are achieved using the latest Microsoft Media Player.

The DVD (digital video disc, but now usually just called DVD) has provided dramatically improved levels of data storage. DVDs were specifically aimed at the home entertainment market and are now fully established in the marketplace. At present, most DVDs are read-only, but DVD writers are now available and are reducing in price. They store over two hours of high-quality video and audio data and with additional features, such as double-sides or dual layers, this can be increased to over eight hours. They can hold eight tracks of digital audio, providing multilanguage tracks, subtitles, and simple interactive features. The resultant video is much better than CD quality (though, of course, the quality depends upon the compression techniques used — MPEG-2 is essential). At the time of writing, the discs are still quite expensive, about $20, but this should slowly reduce. DVDs are now available for audio only (DVD-audio). The discs are actually available in two sizes, 12 cm and 8 cm, with a storage ratio of about 3:1. Capacity varies from an 8 cm single-sided/single-layer DVD (1.36 GB) to 12 cm double-sided/double-layer DVD (15.9 GB). With compression, these can store from half an hour of video to eight hours of video.

Many discs contain *regional codes*, and players are designed so that they will only play discs that correspond to their regional code. One of the main reasons for these codes is to control the release of DVDs in different countries. There are eight regional codes that span the world (Japan, USA/Canada, Australasia, Europe, etc.). Furthermore, DVDs are recorded in two incompatible formats corresponding to the two main TV formats (PAL and NTSC). Most PAL players can read and perform conversion on NTSC DVDs, but the reverse is not true.

In the short term, DVDs are likely to replace CD-ROMs; in the longer term, they will replace video recorders.

49.3.3 Animations

Animation covers any *related artificially created* moving images on the screen, not movement or change itself. So, screen changes are not animation, but moving banners and moving images of animals, cartoon characters, or human shapes clearly are. Movies are not animations because they are not artificially created, although some very specialized movies might be thought of as animations. There are also gray areas. For example, are mouse-over images animations?

Animations are important because the human physiological system is tuned to picking up movement, particularly peripheral movement. The evolutionary reasons are obvious — anything moving poses a threat (or could be food). Our visual system is therefore tuned to almost involuntary response to moving images, and movement will affect anything else we are doing at the time. Thus, animation provides two challenges to human–computer interaction:

Movement can be used to direct the attention of the user.

Irrelevant movement can seriously distract a user from the current task.

We should therefore exploit the former and minimize the latter. However the interface is designed, animation should add content or assist the user in using the interface by directing his or her attention.

There are a number of ways in which animations can be created using software:

A set of GIF (or JPEG) images can be created. Such images are essentially bit-mapped pictures with each succeeding picture slightly changed — like the pages of a flipbook. A good aspect of this approach is that it is not browser-dependent. However, the files created are very large, so download times are a problem. Because of the size, GIF animation can be rather jerky.

The alternative is to use vector graphics (the main product on the market being FLASH .swf files). These are relatively small (requiring one-tenth the memory of a GIF), but files are browser-dependent and require a plug-in. Another problem with using FLASH is that many browser functions are suspended (e.g., Stop). However, it has been estimated that 200 million people are equipped to view it. Shockwave vector graphics incorporates FLASH, 3-D, XML, and other data types.

JavaScript animation gives complete control over animation. A common use is JavaScript rollover on buttons.

Good animation, particularly when accompanied by voice-over explanations, can be very effective. It can also be used to direct and control what the user is looking at. When flashing or blinking items are on the screen, users are forced to direct attention toward the moving images. However, animation is often improperly used, particularly in Web pages, where irrelevant animations frequently distract and annoy users.

49.3.4 Audio Technology — Digital Audio and the Musical Instrument Digital Interface

Much of the audio activity in multimedia applications is concerned with sound, rather than with speech. Most books on multimedia applications hardly mention speech synthesis or voice recognition at all. Voice recognition is not discussed because it is still at a fairly early stage of development and is not usually reliable enough for many existing applications. Voice synthesis, on the other hand, works reasonably well and is considered straightforward. Most personal computers offer standard voice synthesis packages, which use text strings as a source.

Sound is stored in computer systems as digitized audio. Any sounds can be digitized: music, the singing voice, or natural sounds. The input analog signal is sampled at a regular rate and stored. Sampling frequencies vary from 11 to 48 KHz, and the sample size can be 8 or 16 bits. Audio quality requires a sampling rate of 48 KHz, whereas 22 KHz will store speech effectively. Sound editors are available which can work on the digitized waveform. In this way, the designer can add reverberation and chorus effects, fade in and out, and alter parameters such as volume.

Digital audio, at an appropriate sampling rate (about four times the maximum required audio frequency), stores a nearly faithful representation of the analog waveform. The musical instrument digital interface (MIDI) system, on the other hand, offers quite distinct facilities [Moog, 1986]. In contrast to digital audio, this system stores descriptions of a musical score and communicates this information along logical channels to other MIDI devices. Thus, a MIDI keyboard will record note-on and note-off actions, the note value, the pressure on the key, as well as other control signals which alter volume, define the timbre (instrument), and define the stereo profile. These signals can then be sent to a MIDI synthesizer, which will reproduce the analog signals necessary to play the score. Each MIDI instrument usually has MIDI-in, MIDI-thru, and MIDI-out ports. A typical MIDI setup is shown in Figure 49.1.

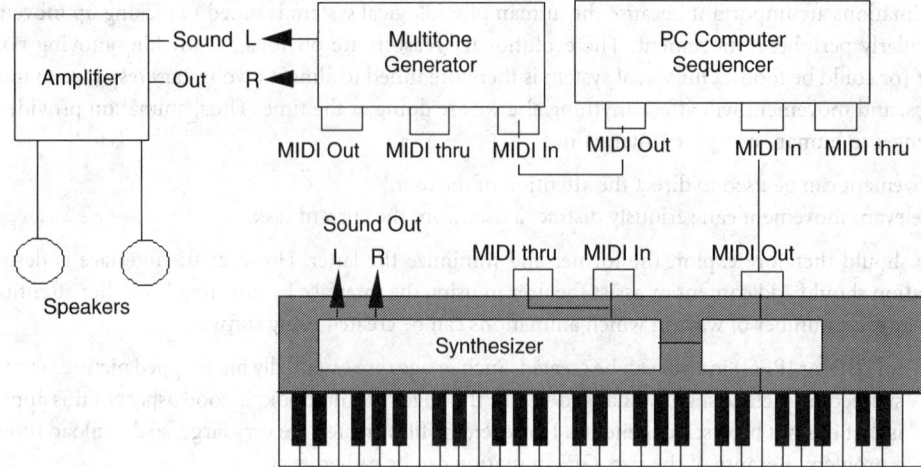

FIGURE 49.1   Typical MIDI connections.

The keyboard can be used to create basic MIDI instructions (note-on, note-off, etc.), in which case the output is taken from MIDI-out port and processed elsewhere. Alternatively, the internal synthesizer of the keyboard can be used to create sounds from the MIDI signals created by the keyboard. This results in audio output signals from sound out. The MIDI-in port can also be used by another keyboard, which would send in MIDI signals to be synthesized by the internal synthesizer.

If the MIDI-out port is used from the keyboard, this can either go directly into the MIDI-in port of the multitone generator to produce sounds at the sound out (in this case, simply using the multitone generator as a synthesizer instead of the internal synthesizer of the keyboard; connection not shown in diagram). Or it could be fed into the MIDI-in port of a PC that is running a MIDI sequencing software package. The resultant MIDI combined output will be stored on the disk of the PC and can then be played at any time, sending the output through the MIDI-out port of the PC to the MIDI-in port of the multitone generator, which will simultaneously create the necessary sounds at sound out. Finally, these sounds will be amplified and sent to stereo speakers in the normal way.

The MIDI-thru ports are used to connect to multiple devices. They allow the signals at a MIDI-in port to be presented elsewhere at the same time. For example, the MIDI-thru port on the multitone generator could be connected to the MIDI-in port of the keyboard to give an extra sound (but more likely to a different multitone generator).

It is important to note that MIDI signals are not digitized sounds; they are a sort of musical shorthand. As a result, MIDI files are far more compact — as much as 1000 times more compact — than digital audio files. The MIDI system can support up to 16 distinct channels simultaneously. Each channel can be assigned to a different instrument, and the signals can then be sent, in parallel, to a multitimbral tone generator that recreates the analog sounds. The sounds can be created in a sequencer, that is, a software system that allows a composer to build up the parallel descriptions of each part. Extensive editing facilities are available, and actual musical scores can be produced automatically from the MIDI description.

There are now a number of effective sequencing systems in the marketplace, such as Cubase [Steinberg, 2002] and Sibelius 2 [Sibelius, 2002]. Sibelius differs from Cubase in that the score is the "heart" of the system. To use it effectively, the user must be familiar with musical scoring notation. Cubase also provides a score, but creation and manipulation can also be carried out through a "piano roll"–type interface. Both packages offer professional-level music creation and manipulation facilities.

Using a system such as Cubase or Sibelius 2, the composer of the multimedia soundtrack can build up the separate channels and control instrument selection, volume, panning, and special effects. The timing of MIDI compositions can be accurately controlled so that they can be synchronized with other media, such as video. Cubase also offers extensive audio signal processing capabilities.

TABLE 49.2 Some General MIDI Sound Assignments

Identifier	Sound or Instrument
1	Acoustic grand piano
14	Xylophone
25	Acoustic electric guitar
41	Violin
61	French horn
72	Clarinet
98	Soundtrack
125	Telephone ring
128	Gunshot

Provided a standard mapping is used, the composer can also expect the desired instruments to be selected on playback in another MIDI device. There are a number of standard mappings, one of which is the general instrument sound map (general MIDI level 1), a small portion of which is given in Table 49.2.

Note that not only musical sounds are capable of being reproduced. Other general MIDI sounds include laughing, screaming, heartbeat, door slam, siren, dog, rain, thunder, wind, seashore, bubbles, and many others.

There are now a huge number of devices on the market for creating and manipulating sound. Auditory signals can be captured in digitized formats, such as .wav files. These can then be manipulated and edited in many ways. MIDI techniques can be used with digital files. For example, samplers can capture original sounds in a digitized format. These can then be stored on the hard disk for later use. Samplers exist that enable the user to buy CD-ROM collections of sampled sounds and load them into memory. These samples can then be assigned to MIDI channels and used to create the output from a sequencer. The result is very realistic sound. CD-ROMs are available that will provide, for example, samples from a complete orchestra playing string sounds, sampled over the complete chromatic scale and using different string techniques such as pizzicato, tremolo, and spicato. Mixers can then be used to adjust the final mix of sounds for performance and the final composition written to CD-ROM. The trend now is toward computer-based (i.e., software-driven) devices rather than stand-alone hardware.

49.3.5 Other Input, Output, or Combination Devices

In addition to the usual input devices (keyboard, mouse, tablet, etc.) multimedia systems may additionally have color scanners, voice input/output, and 3-D trackers (e.g., data gloves). Some of these devices combine input with feedback.

The flatbed scanner is like a photocopying machine. It is essential for adding artwork to multimedia presentations. The picture to be scanned is placed, face down, on a glass screen and is then scanned. Color scanners are normally used in multimedia work. They use three 8-bit values for each RGB component, providing a 24-bit image. As well as color resolution, a scanner has an optical resolution, measured in dots per inch (dpi). Typical values for optical resolution are 300, 400, or 600 dpi. Scanners require software to support them, which enables the user to process the image (e.g., to adjust brightness and contrast).

Three-dimensional trackers can be used to measure the absolute position and orientation of a sensor in space. The tracking is accomplished either by magnetic means or by acoustic means. The yaw, pitch, and roll may also be determined. Such measurements are important if human gesture is to be correctly interpreted. Trackers are often mounted on clothing, such as data gloves, helmets, or even body suits. The data glove worn on the hand can provide input data on hand position, finger movement, etc. Data gloves are particularly useful in virtual reality environments, where they can be used to select virtual objects by touch or grasping. An output feedback system can also be placed in the glove, which provides a sensation of pressure, making the grasping almost lifelike. An interesting example of such an application is the GROPE system [Brooks et al., 1990]. The GROPE system uses a device with force feedback to allow scientists to fit molecules into other molecules. The user can manipulate the computer-created objects and actually feel the molecules. Some improvement in performance was noted, and users reported that they obtained a better understanding of the physical process of the docking of molecules.

49.4 Distinct Application of Multimedia Techniques

Because media are at the heart of all human–computer interaction, an extreme viewpoint would regard all computer interfaces as multimedia interfaces, with single media interfaces (such as command languages) being special limiting cases. However, we will restrict the term *multimedia interface* to interfaces that employ two or more media: in series, in parallel, or both.

There is a major division in the way multimedia techniques are applied. First, the techniques can be used to front-end any computer application using the most appropriate media to transmit the required information to the user. Thus, different media might be used to improve a spreadsheet application, a database front-end retrieval system, an aircraft control panel, or the control desks of a large process control application. On the other hand, multimedia techniques have also created new types of computer application, particularly in educational and promotional areas. In the front-ending instances, multimedia techniques are enhancing existing interfaces, whereas in the latter fields, applications are being developed which were not previously viable. For example, educational programs analyzing aspects of Picasso's art or Beethoven's music were simply not possible with command-line interfaces. These new applications are quite distinct from more conventional interfaces. They have design problems that are more closely associated with those of movie or television production.

In both application areas, there is a real and serious lack of guidelines for how best to apply multimedia techniques. Just as very fast computers allowed programmers to make mistakes more quickly, so can the indiscriminate use of multiple media confuse users more effectively. The key issues related to the application of multimedia technology are therefore more concerned with design than with the technology per se. The past decade has been characterized more by bad design than by good design. There is no doubt, however, that good design practice will emerge eventually.

It is not difficult to understand why design will be a major issue. The early human–computer interfaces relied almost exclusively on text. Text has been with the human race for a few hundred years. We are all brought up on it, and we have all practiced communicating with it. When new graphics technology allowed programmers to expand their repertoire, the use of diagrams was not a large step, either. Human beings were already used to communicating in this way. When color became available, however, the first problems began to occur. Most human beings can appreciate a fine design which uses color in a clever manner. Few would, however, be able to create such a design. Thus, most human beings are not skilled in using color to communicate, so when programmers tried to add color to their repertoire, things went badly wrong. Many gaudy, overcolored interfaces were created before the advice of graphics designers was sought. More recently, the poor quality of most home video productions has shown that new skills are required in using video in interfaces. Such interfaces will require a whole new set of skills, which the average programmer does not currently possess.

49.5 The ISO Multimedia Design Standard

The International Organization for Standardization (ISO) has drafted a standard for multimedia user interface, ISO 14915, which consists at present of three parts:

Part 1 — Design principles and frameworks
Part 2 — Multimedia navigation and control
Part 3 — Media selection and combination

Specifications for these can be obtained from the ISO Web site [ISO, 2003] for a small cost. Media are classified first according to the sense used to detect them (auditory or visual), then by their realism (realistic or nonrealistic), then by their dynamism, (still or moving), and finally by their language base (verbal/nonverbal).

The standard develops media-neutral descriptions of information, which can then be mapped to media. Examples of information types include causal, descriptive, discrete action, event, physical, and relationship types (there are many more).

Mappings can then be made. For example, the audible click of an on–off switch can be mapped to audio (discrete action); graphs are nonrealistic still images (relationship); and a movie of a storm is a realistic moving image (physical).

The text of the ISO standard contains many interesting examples, and readers are encouraged to examine the standard.

49.6   Theories about Cognition and Multiple Media

49.6.1   The Technology Debate

Can the use of a technology such as multimedia enhance a student's learning experience? This is a question that has taxed the teaching profession for the last 50 years. There have been many discussions and false dawns in the application of technology to learning. For example, the initial introductions of radio and television were supposed to revolutionize teaching. More recently, the same claims are being made about the use of computer-based multiple media in the classroom. The stance of the antitechnology school was summed up in contributions by Clark [1983], who argued repeatedly in the 1980s and 1990s against the concentration of education research on technology and media. He claimed that pedagogy and teaching style were the main variables to be examined and that a concentration on media and their effects was a distraction.

For example, Clark wrote, "There is no cognitive learning theory that I have encountered where any media, media attribute, or any symbol system are included as variables that are related to learning" [Clark, 1983]. Clark's point was that many media are equally capable of delivering any instruction, so media choices are about cost and efficiency, but not about cognition and learning. However, Cobb has pointed out that "there may be no unique medium for any job, but this does not mean that one medium is not better than another, or that determining which is better is not an empirical question" [Cobb, 1997].

Cobb's argument shifts attention to cognitive efficiency and how choice of medium affects such efficiency. Clark's point is almost certainly valid in that the teacher and the pedagogy used should be the main focus. However, it is surely also true that, although any reasonable combination of media can be used to transfer knowledge, some media require more cognitive effort on the part of the learner than others. Musical appreciation can be taught using just musical scores. Actually being able to hear the music, as well, does make the task easier. The point is that appropriate choice of media can reduce the cognitive work required by the user, which might otherwise get in the way of the learning process.

49.6.2   Theories of Cognition

Although cognitive scientists would readily admit that they do not fully understand how human memory operates, they do have some useful models, which can be taken as reasonable approximations of particular aspects of memory behavior. Such models tend to be generalizations, but they can be used in a predictive manner. First, it is generally accepted that memory can be divided (at least logically) into a number of distinct units, two of which are often called working memory and long-term memory. Much of the initial evidence for the existence of some form of working memory was eloquently stated in Miller's well known paper [Miller, 1956]. Working memory provides a temporary store, which seems to be necessary for many cognitive skills. A great deal of experimental evidence has been collected to support the basic concept of working memory. There is also memory in the sensory devices themselves (eyes, etc.), but these are of very short duration.

Long-term memory is where all our durable memories are stored. It is durable and very large in size. In contrast, working memory is very limited in size and has been described as having a capacity of 7 ± 2 "chunks." Working memory is where human beings seem to make the connection between a stimulus from the outside world and the content of long-term memory. The word *forest* might trigger a picture of a forest we have read about in a book, the memory of a real forest with which we are familiar, or some stylized or prototypical generalization. This might then trigger a previous view or a journey that involved a forest and that we can actually "replay" in the mind.

There is also strong evidence that memories in both working memory and long-term memory exist in a number of forms that are similar to their original sensory stimulus — haptic, audio, and visual stimulations [Paivio, 1986]. In this view, music would be assumed to be stored as an auditory experience and a picture as a visual experience. Memories in long-term memory are much more than simple memories of faces, names, etc. They are thought to include structures called *schemas*, descriptive structures that can be triggered by particular external stimuli. For example, when a person walks into a shop, the "shopping" schema is immediately triggered. This creates expectations about what will happen and puts into context many other external inputs. However, if some external stimulus does not fit into the shopping schema pattern (say some goods fall off the shelf), the person suddenly becomes conscious of effort in working memory to resolve the situation.

Paivio [1986] has proposed what is called *dual coding theory*. In this theory (which has been extensively verified experimentally), items in memory are stored in the same modality as they were experienced (for a recent text, see Sadoski and Paivio [2001]). Thus, music is stored as some form of musical sequence (auditory) and pictures as visual representations. This view contrasts with other theories of memory, which claim that all sensory experiences are recoded into some common coding scheme. Furthermore, Paivio distinguishes between two distinct types of stored structure: *imagens* and *logogens*. Imagens and logogens can exist in any sensory form — visual, auditory, or haptic. For example, verbal utterances (logogens) can be stored as audio (words), written text (visual), or carvings in stone (haptic). Imagens can have auditory, visual, or haptic forms. There are strong referential connections between equivalent logogens and imagens. So, the logogen "table" can invoke the imagen "table," and vice versa.

Figure 49.2 illustrates the main aspects of the theory. The two subsystems exist to process the verbal and image representations independently. The imagens are quite different structures from the logogens, but there are referential links between them. Associative links also exist within each subsystem.

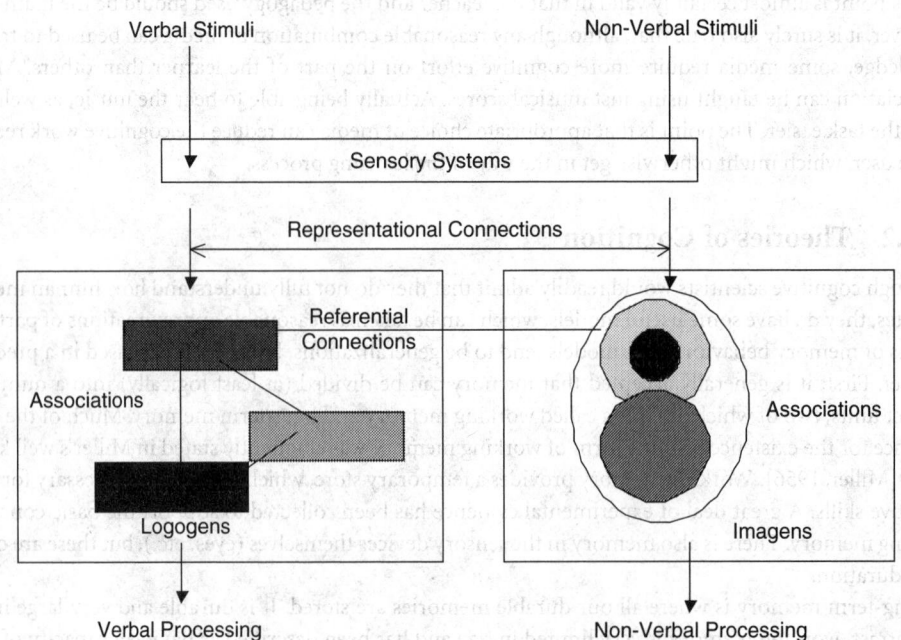

FIGURE 49.2 Processing in the dual coding theory.

Paivio suggests that the two types of storage are processed in fundamentally different ways. Imagens can be processed in parallel and from any viewpoint. Imagine your living room. You can view it in your mind, at will, from any angle. When you think of a face, you do not think of it as successively decomposed into lips, eyes, nose, nostrils, etc. This suggests that imagens are simultaneous structures. On the other hand, it is not easy to break into the middle of a logogen (e.g., a poem). It is very difficult to start in the middle and work backwards, because, Paivio argues, the structure in memory, a logogen, is essentially sequential. Music is more likely to be stored as a logogen than an imagen. For example, Minsky [1989] suggested that "[w]hen we enter a room, we seem to see it all at once: we are not permitted this illusion when listening to a symphony ... hearing has to thread a serial path through time, while sight embraces a space all at once."

The ideas of working memory and dual coding theory can now be connected. When a person sees an image, hears a word, or reads a set of words with which they are familiar, processing and recognition are almost instantaneous. This is called *automatic processing*. The incoming stimulus seems to trigger the right structure or schema in long-term memory. If an unusual image or new word is encountered, or if a set of words is read which is not understood, the automatic processing stops, and conscious processing is entered. There is a feeling of having to do work to understand the stimulus.

Rich schemas are what probably distinguish experts from novices, because an expert will have more higher-level schema to trigger. Novices, in contrast, must process unfamiliar schemas using their working memory, with a resultant high cognitive load. Sweller et al. [1998] have termed the procedure of connecting with an internal schema "automatic processing"; they call the process of converting a structure in working memory into a schema in long-term memory "schema acquisition." Schema acquisition puts a high load on working memory, which in itself is very limited. However, schema acquisition is rapid.

These theories can offer predictions about the use of multiple media. For example, Beacham et al. [2002] have presented similar material in three different forms to students and have measured the amount of material learned. Material was presented

As a voice-over with diagrams (Sound + D)
As written text with diagrams (Text + D)
As text only

Four different modules on statistical material were presented over four days to three groups of students. The student groups were given different presentation styles each day, to avoid any group bias effect. The Sound + D presentation style resulted in significantly higher recall than either of the other two methods for all four modules, even though the material was quite different in nature. For example, the first module (on the null hypothesis) was very discursive, whereas the modules on binomial probability distribution and normal distributions were highly mathematical in nature. The results of the recall of material for all four modules (presented on different days) are shown in Figure 49.3.

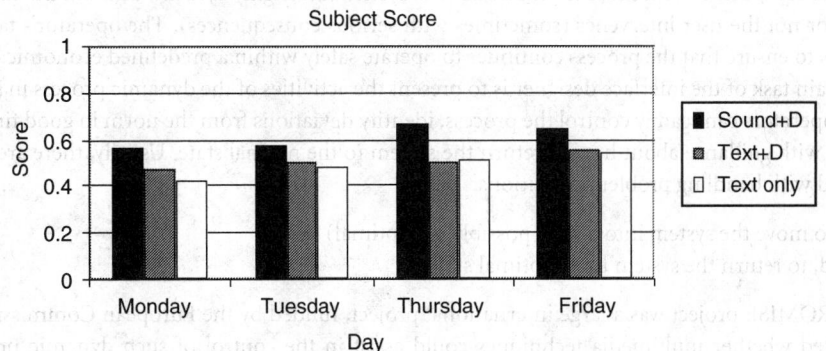

FIGURE 49.3 Recall of material for the three presentation styles.

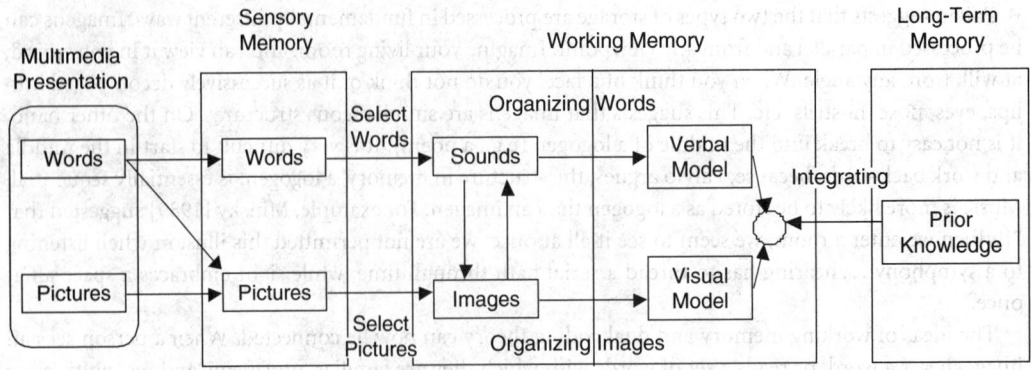

FIGURE 49.4   Mayer's cognitive theory of multimedia learning [Mayer, 2000].

Mayer [2000] has developed a cognitive theory of multimedia learning based on the work of Miller, Paivio, and Sweller. A diagrammatic version of Mayer's approach is given in Figure 49.4.

The multimedia presentation is represented on the left of the figure. For a brief period of time, the sounds or images are held in sensory memories and a selection is made, transferring material to working memory. In working memory, seen words can be converted to sounds and other correspondences made. The right-hand side corresponds to the dual coding approach, in which imagens and logogens are created and stored, being integrated with existing imagens and logogens in long-term memory.

The theory suggests a set of design principles:

The spatial contiguity principle — Place related words and pictures near each other.

The temporal contiguity principle — Present related words and pictures simultaneously, rather than separately.

The coherence principle — Exclude unnecessary or irrelevant words or pictures.

The modality principle — Present using different modalities. For example, animation is better supported by voice-overs than by text.

The reader is referred to Mayer [2000], which contains many references to experimental support.

49.7   Case Study — An Investigation into the Effects of Media on User Performance

The following case study is from the process control. This particular application area differs from the traditional computer system because process control systems are highly dynamic and will usually continue whether or not the user intervenes (sometimes with serious consequences). The operator's task in such systems is to ensure that the process continues to operate safely within a predefined economic envelope.

The main task of the interface designer is to present the activities of the dynamic process in such a way that the operators can readily control the process, identify deviations from the norm in good time, and be provided with guidance about how to return the system to the normal state. Usually, there are two goals associated with handling problem conditions:

First, to move the system into a safe (possibly suboptimal) state
Second, to return the system to its optimal state

The PROMISE project was a large international project, funded by the European Commission, which investigated whether multimedia techniques could assist in the control of such dynamic processes. A multimedia toolset was built and a series of experiments carried out to examine the effects of different media on operator performance. Some experiments were performed in the laboratory and others in a live

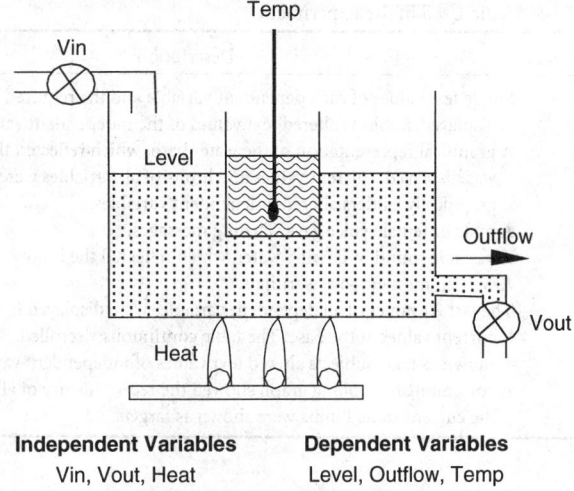

Independent Variables
Vin, Vout, Heat

Dependent Variables
Level, Outflow, Temp

FIGURE 49.5 Crossman's water bath.

chemical plant. Only the laboratory data will be reported here. Space precludes a complete analysis, so any reader interested in additional information is referred to Alty [1999] for the laboratory work and Alty et al. [1995] for the plant results.

49.7.1 The Laboratory Task

Process control tasks are characterized by having thousands of variables, so one of the main problems encountered when setting up process control laboratory studies is choosing a task that is reasonably representative of the domain under investigation, yet is reasonably straightforward to set up. It also must be learned relatively easily by subjects.

A task already existed that has these characteristics, and for which operator behavior had already been extensively studied. This task is Crossman's water bath [Crossman and Cooke, 1974], which has been studied on numerous occasions [Sanderson et al., 1989; Moray and Rotenberg, 1989]. The task is deceptively simple (see Figure 49.5).

Crossman's water bath is a simulated hydraulic system. A simple bath contains water, the *Level* of which can be altered by adding water through the in-valve *Vin* increasing the *Outflow* by opening the out-valve *Vout*, or both. At any time, there will be an outflow (*Outflow*) from the bath, which could be zero if *Vout* is closed. The bath is heated (*Heat*). Inside the bath is a further container with a fixed amount of water in it, whose temperature is continuously measured by *Temp*. Thus, changes to *Vin, Vout,* and *Heat* (the independent variables) cause changes to *Temp, Level,* and *Outflow* (the dependent variables). The subject is given the system in a particular state and asked to stabilize it at a new state (a set of limits within which the variables must lie after stabilization).

More than 50 engineering students used as subjects were given a brief introduction to the system. The task and the controls were explained, but without giving any principles of the underlying system. Each session consisted of two halves. In the first half, the subject had to complete 21 tasks of increasing difficulty (but all the tasks were relatively easy). In the second half, subjects had to complete a set of more difficult tasks, involving larger moves in the state space and narrower stabilization limits. Tasks were characterized by their *compatibility*, a task descriptor defined by Sanderson et al. [1989]. Tasks had a compatibility of 1 (easy) to 3 (most difficult). In the break between the two halves of the experiment, subjects were tested for their understanding of the state variables. This test was repeated at the end. All sessions were recorded, timing statistics were collected by the system, and subjects were asked to verbalize their current beliefs about the system and the reasons for their actions.

TABLE 49.3   The Basic Media Used in the Experiment

Medium	Description
Text only	Single text values of each dependent variable and the required limits were displayed. Subjects altered text values of the independent variables.
Graphics only	A graphical representation of the water bath, which reflected the current state, was displayed. Current values and limits of all variables were displayed graphically. Sliders altered independent variables.
Voice messages	A male or female voice gave warning messages.
Sound output	A variable sound of flowing water, which reflected the inflow rate, was presented.
Written messages	A written message gave warnings.
Scrolling text table	The last 20 values of all dependent variables were displayed in text with the current values at the base. The table continuously scrolled. The limits were shown as text. Subjects altered text values of independent variables.
Dynamic graph	A continuously scrolling graph showed the recent history of all the variables and the current state. Limits were shown as targets.

	Readings	Targets	
		Min	Max
Level (mm)	316	273	393
Temperature (°C)	25.44	10.0	34.61
Flow Rate (ml/sec)	1998	2204	3404

In Valve = 60   Out Valve = 75   Heater = 39

FIGURE 49.6   The text only interface.

49.7.2 The Media Investigated

A wide range of media were used to create different instances of the controlling interface. The basic media are outlined in Table 49.3.

An example of the text-only interface is given in Figure 49.6. The readings reflect the current state of the dependent variables. These can be compared with the targets. The user's task is to stabilize the system so that all the readings are within the target ranges. The values of the independent variables are at the base of the screen and can be changed by selection using the cursor. The scrolling text interface is shown in Figure 49.7, which also shows the sound outputs (these, and the written warnings, can be presented in any interface).

By comparing results from individual or combinations of elements, particular variables can be isolated. For example, a comparison of (text only and graphics only) with (text and sound and graphics and sound) can examine the effect of sound. The main measures taken during the experiment were as follows:

Time to complete the task
Total number of user actions
Number of warning situations entered
Subjective ratings of the interfaces by users

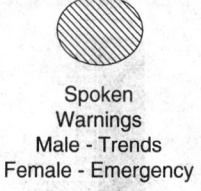

Spoken
Warnings
Male - Trends
Female - Emergency

Continuous
Sound
Indicating
Flow Rate

	Level (mm)	Temperature (C)	Flowrate (ml/sec)
	363	28.29	2350
	371	28.03	2409
	378	27.76	2463
	385	27.49	2512
	391	27.22	2557
	396	26.96	2598
	401	26.69	2635
Current	406	26.44	2669
Min:	345	10.00	1774
Max:	450	27.68	2974

FIGURE 49.7 The scrolling text interface.

49.7.3 The Effect of Warnings on Performance

Subjects showed no significant differences in performance as a result of receiving spoken or textual warnings, but they did rate spoken warnings as more important. A more detailed analysis of the data revealed that, although all subjects rated spoken warnings as important, this was not true for textual warnings. One group of users had rated them as more important than the other group. The group that rated textual warnings highly also found the tasks difficult. The other group found the tasks much easier. Written warnings require additional processing in comparison with verbal ones (because a switch must be made from the visual task at hand), and they can easily be missed. Therefore, it was likely that subjects who found the tasks difficult tended to check the written messages carefully and, therefore, rated them as more important. Spoken messages, on the other hand, could be processed in parallel with the visual perception of the screen and were rarely missed.

49.7.4 The Effects of Sound

The effects of the presence or absence of sound are shown in Figure 49.8, which again shows completion times, number of actions, and the number of warning situations entered. What is immediately striking about these results is an apparent detrimental effect of sound. Every effect deteriorates when sound is present — the task takes longer, there are more actions, and the number of warning situations entered is increased. A more careful analysis, separating the performance on the three different levels of task compatibility, however, reveals a much more interesting result. These results are shown in Figure 49.9.

A clear pattern emerges from these graphs, although the result does not reach significance. Sound appears to have an increasing effect with increasing task complexity, and one might speculate why this may be so. The moves in state space for category-3 tasks were much greater than for the other two. Thus, in general, larger changes occurred in the flowrate of category-3 tasks than for the simpler tasks. These large changes resulted in larger sound volume shifts, so that the sound volume may then have been more obvious and usable for category-3 tasks. In the easy tasks, the volume changes were not distinct enough to be useful.

It is obviously important to use media that can signal the required differences the human operator needs for successful control. In the case of sound, this was not happening in the easy tasks.

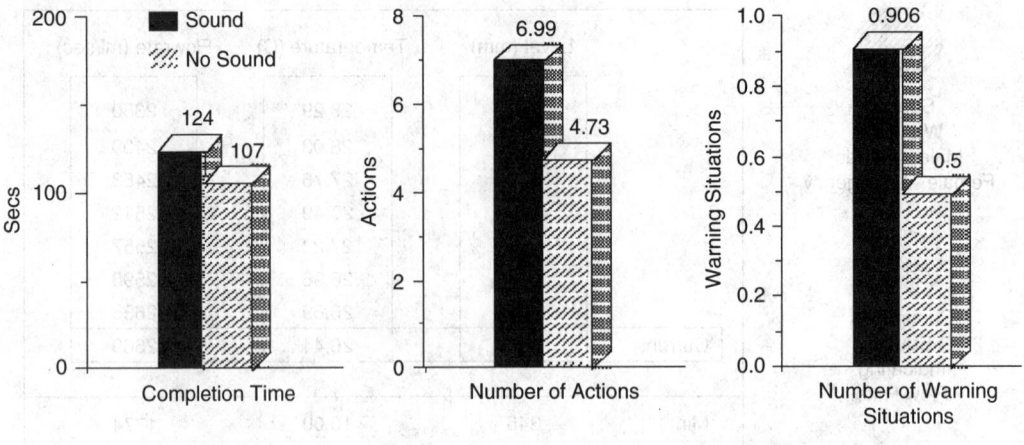

FIGURE 49.8   Effect of sound on performance.

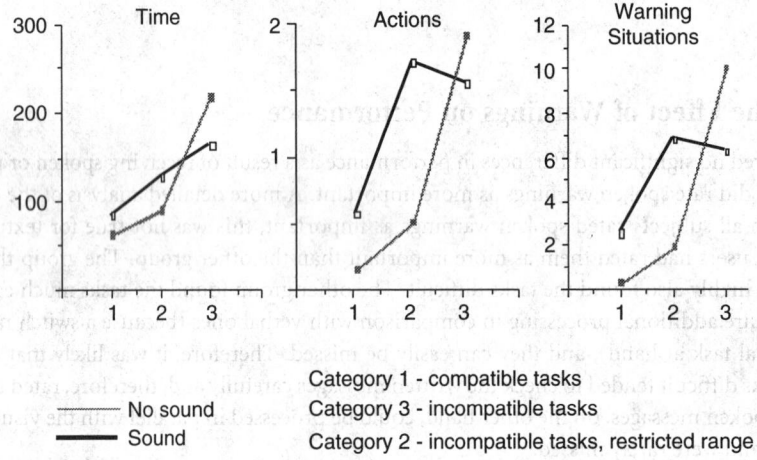

FIGURE 49.9   Effect of task compatibility and sound.

49.7.5   Effects of Mental Coding

Finally, the performance of subjects in the graphical conditions, compared with the textual conditions, was examined. Received wisdom would imply that subjects ought to perform better on spatial tasks when presented with information in graphical form. The independent variable being examined here is mental processing code (See Table 49.4).

The table shows that, from an efficiency point of view, the graphical interfaces yielded better performance than the textual ones. Only the difference in the number of actions approached significance, however ($p < 0.03$). The effect of graphical interfaces become more interesting when one analyzes the results again in terms of task difficulty. Figure 49.10 (similar to Figure 49.9) illustrates this point.

The trend is significant for both the number of actions ($p < 0.003$) and the warning situations ($p < 0.03$). There is therefore an indication that the graphical interfaces provided better information in a more appropriate form, particularly for more complex tasks. Thus, although the graphical representation might provide too much information for simple tasks, resulting in a low signal-to-noise ratio, as the task becomes more complex, the information provided becomes more relevant and essential to the task.

TABLE 49.4   Effect of Mental Processing Code on Performance

Condition	Completion Time	Number of Actions	Number of Warning Situations
Graphical	97	4.13	0.96
Textual	107	5.46	1.476

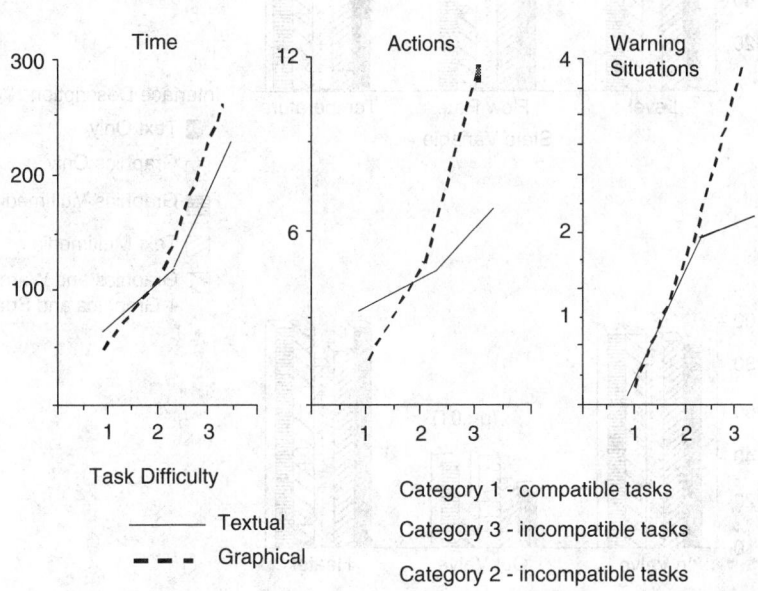

FIGURE 49.10   Effect of task compatibility and mental processing code.

49.7.6   Do Multimedia Interfaces Improve Operator Performance?

A basic assumption of multimedia designers is that multimedia interfaces do indeed make a difference to operator performance. But does presentation affect the way information is picked up and consequently reasoned about? Figure 49.11 illustrates the results from the knowledge questionnaires for users who had experienced the different multimedia interfaces.

The figure shows the results from the knowledge questionnaire for the six variables, broken down by medium type: text (T), graphics (G), graphics with warnings (GSP + GW), text multimedia (TMM), and graphics multimedia (GMM). A statistical analysis shows that, in the cases of *Outflow* and *Vout*, the result is highly significant ($p < 0.01$ in both cases). The other four variables, however, failed to yield a significant result. Thus, the type of medium chosen did affect comprehension in the cases of these two variables (although the analysis does not tell us how).

The figure exhibits some interesting features. Three of the variables have very high levels of comprehension — temperature, in-valve, and heater — and a fourth is reasonably well understood. The effect of the different media seems to be greatest when understanding is lowest (e.g., in the cases of outflow and out-valve).

The concepts of heater, temperature, in-valve, and level are easy to understand. The heater is increased, and the bath heats up. The in-valve is opened, and the flow into the bath increases. The inflow rate is directly proportional to the setting of the valve. The level variable is a little harder to understand, because it depends on the settings of both the in-valve and out-valve, but users seemed to be able to grasp the basic principles. The outflow and out-valve settings, however, are much harder to understand, and the reasons are in the physics. Although the out-valve may be set at a particular value, this does not result in a fixed outflow value. This is because the actual flowrate will depend on the setting of the out-valve *and* the

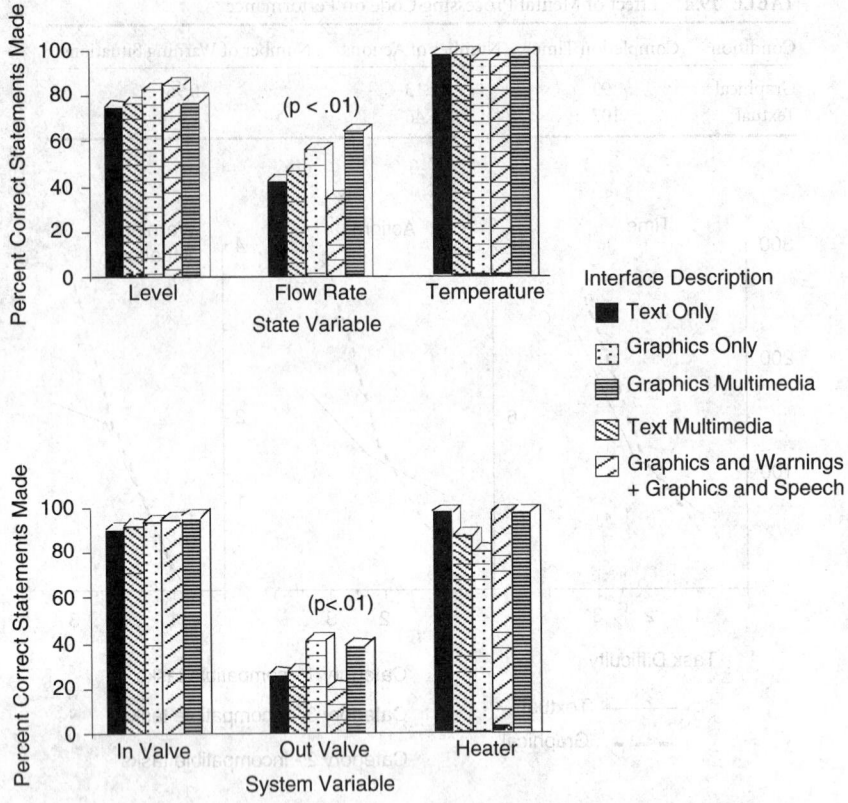

FIGURE 49.11 Overall effect of media on comprehension.

level in the bath (i.e., the pressure). Thus, setting the out-valve to a particular value does not guarantee a certain outflow rate. Moreover, this outflow rate will change as the bath empties or fills. The level of comprehension was quite low for both outflow (about 50%) and out-valve (about 30%).

Here is a possible explanation. The concepts of heater, temperature, level, and in-valve are well understood (probably even before the experiment). Thus, it does not matter what medium is used to portray their actions. Provided the basic information is exchanged, the users will understand. If the opposite extreme of this situation is considered, where the concepts being portrayed are not understood at all by the users (this condition was not present in this experiment), then one might reasonably postulate that it would not matter what medium was used to show their operation; the users would have great difficulty in understanding what was going on. In other words, the medium would not seriously affect comprehension, because there would be no point of contact. The intermediate situation is the interesting one. What if the users understand the basic elements of the domain but are operating on the edge of their understanding? In the cases of the outflow and the out-valve, the users did understand some aspects, but not all (they did not get zero marks in the comprehension test). In such cases, it appears that medium choice does make the difference in understanding.

If this explanation is true, it is an important result and is a strong argument for the appropriate use of multimedia elements in interface design. The intermediate area of understanding just described is an important one. In process control, it represents that area where the operators are trying to understand abnormal conditions in the behavior of the plant, or where users of computer applications are exploring new features. In more traditional applications, it represents the situation during initial exposure to the system or when users are exploring more advanced features of the applications.

49.8 Authoring Software for Multimedia Systems

In the case of promotional software, educational software, or games, the creation of a multimedia application strongly resembles the production of a film or television program, and one talks of multimedia productions. All the different aspects of the production, such as scanned photographs, movie sequences, musical sequences, and text, must be put together into a meaningful whole, all correctly synchronized. For such applications, software to organize and assist in bringing together the contributions from the different media is essential. Such software is often called *authoring software*. In addition, other software is required to create the components that make up the complete applications, such as sound sequences, animations, etc. Like most emerging software environments, the situation can change quite rapidly, and there are many options. However, there are now a number of reasonably well established tools on the market.

In the case of sound composition, tools like Cubase and Sibelius 2 have already been mentioned. These are sophisticated tools for creating sequences, orchestrations, and musical scores. In addition, Cubase can provide extensive digital wave editing and manipulation. There are still many hardware-based tools for sound recording and manipulation, but the trend is now rapidly moving toward computer- or software-based solutions.

For still pictures and animations, there are a number of highly developed packages. In the case of still pictures, packages like Photoshop [Adobe, 2002] offer full picture manipulation and editing. For animation and graphics, there are two basic techniques: raster and vector graphics. *Raster graphics* (or bit-maps) are composed of sets of individual pixels. The resolution is measured in dots per inch (dpi). *Vector graphics* are based on a totally different principle — line representations of images. This means that the components of a vector graphics–based image are calculated using mathematics. This has major implications for the size of the stored image, the processing power required to manipulate it, and its scalability. For example, a bit-map does not scale easily.

Authoring software enables the designer to create and edit different media, and then put them together in a desired sequence. As with most interface tools, different metaphors are used to ease the design task. One example of a common metaphor used is the timeline, in which all the events are laid out across the screen in a time sequence. Using a configurable time frame (usually about 1/30th of a second), elements are positioned on parallel time lines, and other events are cued along the timeline. These representations have tended to replace the card metaphor, rather like a card address file, in which a stack of cards formed the basis of the design. A second type of authoring tool is an existing visual programming environment, such as Visual Basic. Finally, other tools are icon based (similar to visual programming).

There are a number of authoring systems on the market. The situation is changing rapidly, so any detailed comments about tool facilities must be read with caution. The general principles, however, will apply. Tool-based systems include Visual Basic; timeline products include Macromedia Director, FLASH, Adobe Premiere, Action, and Animation Works; and an example of an icon-based authoring tool is Authorware.

All these are proprietary packages. In contrast, the synchronized multimedia integration language (SMIL) enables simple authoring of interactive audio-visual presentations. Using SMIL, a multimedia presentation is captured as a standardized XML application. Various graphical tools are appearing in the marketplace to create and manipulate SMIL presentations. A number of proprietary players can now also play SMIL presentations.

Which package to choose depends on how ambitious the planned multimedia project is. The icon-based tools are the easiest to use, and the timeline tools are probably the most complex. The tool-based systems may not provide all the facilities needed and may require further effort in mastering the tool itself. There are many tools on the market. A recent listing of multimedia products covered more than 90 different offerings, although not all of these are authoring tools. Prices for authoring tools vary from a few hundred dollars to over $20,000. As one might expect, the buyer often gets what has been paid for.

49.9   The Future of Multimedia Systems

It is clear that multimedia technology is already available that can greatly augment the choices open to the user interface designer. These choices will also be available for systems of all sizes, as computational speeds increase and hardware costs fall. A corollary of this is the ubiquity of high-bandwidth, worldwide networks capable of transmitting information in a variety of representational forms via multicast or point-to-point connections. For example, the use of the multimedia HTTP transfer protocol (as used in the World Wide Web) gives an interesting, albeit haphazard, view of multimedia interfaces of the future.

However, the lack of prevalent and wide-ranging design criteria still makes multimedia user interface design an ill defined and empirically lacking discipline. To counter this, user-centered, rather than technology-centered, research is focusing on the following areas:

Examining the effect of different media on human cognitive representations, particularly the construction of mental models of represented domains [Faraday, 1995; Williams, 1996]
Classifying media in linguistic terms by virtue of their expressiveness [Stenning and Oberlander, 1995]
Matching media to task descriptions [Maybury, 1993]
Developing cognitively based approaches to design [Mayer, 2000]

There is no doubt that if user interface designers are to fully utilize the technology on offer, suitable design methodologies must be developed. These must encompass both the cognitive and the goal-oriented aspects of the human–computer system. Without them, multimedia will remain a pragmatic area of HCI application, or worse, will only be fit for use in entertainment systems.

References

Adobe (2002) Adobe Systems Incorporated.

Alty, J.L. (1995) Can we use music in computer–human communication? In *People and Computers X*, M.A.R. Kirby, A.J. Dix, and J.E. Finlay, Eds., Proc. HCI '95: 409–423, Cambridge Univ. Press, Cambridge, UK.

Alty, J.L. (1999) Multimedia and process control: signals or noise? *Trans. of the Institute of Measurement and Control*, 21(4/5): 181–190.

Alty, J.L., Bergan, J., and Schepens, A. (1995) The design of the PROMISE multimedia system and its use in a chemical plant. In *Multimedia Systems and Applications*, R.A. Earnshaw and J.A. Vince, Eds., pp 53–78, Academic Press, London.

Beacham, N., Elliott, A., Alty, J.L., and Al-Sharrah, A. (2002) Media combinations and learning styles: a dual coding Approach, *Proc. Ed-Media 2002* (Denver), P. Barker and S. Rebelsky, Eds., pp 111–116.

Blattner, M., Greenberg, R., and Kamegai, M. (1992) Listening to turbulence: an example of scientific audiolisation. In *Multimedia Interface Design*, Chap. 6, M. Blattner and R. Dannenberg, Eds., pp 87–102, ACM Press, New York.

Blattner, M., Sumikawa, D., and Greenberg, R. (1989) Earcons and icons: their structure and common design principles, *Human–Computer Interaction*, 4(1): 11–44.

Bordegoni, M. and Hemmje, M. (1993) A dynamic gesture language and graphical feedback for interaction in a 3-D user interface, *Proc. EUROGR. '93*, pp 1–11.

Braida, L.D. (1991) Crossmodal integration in the identification of consonant segments, *J. Exp. Psych.* 43A(3): 647–677.

Brooks, Jr., F.P., Ouh-Young, M., Batter, J.J., and Kilpatrick, P. (1990) Project GROPE: haptic displays for scientific visualisation. *ACM Computer Graphics*, 24(4): 177–186.

Clark, R.E. (1983) Reconsidering research on learning from media, *Review of Educational Research*, 53(4): 445–459.

Cobb, T. (1997) Cognitive efficiency: toward a revised theory of media, *Educational Technology Research & Development*, 45(4): 21–35.

Crossman, E.R., and Cooke, F.W. (1974) Manual control of slow response systems. In *The Human Operator in Process Control*, E. Edwards and F. Lees, Eds., pp 51–64, Taylor & Francis, London,.

Edwards, A.D.N. (1989) Soundtrack: an auditory interface for blind users. *Human–Computer Interaction*, 4(1): 45–66.

Faraday, P. (1995) Evaluating multimedia presentations for comprehension. In *Doctoral Consortium, SIGCHI '95* (Denver, CO, April/May), pp 49–50, ACM Press, New York.

Gaver, W.W. (1986) Auditory icons: using sound in computer interfaces, *Human–Computer Interaction*, 2(1):167–177.

Healey, J., and Picard, R. (1998) StartleCam: a cybernetic wearable camera, *Proc. 2nd International Symposium on Wearable Computers*, pp 42–49.

Holland, S. (1994) Learning about harmony with harmony space: an overview. In *Music Education: An Artificial Intelligence Approach*, Springer-Verlag, London.

ISO (2003) ISO Copyright Office, Case Postale 56, CH-1211, Geneva 20.

Mann, S. (1996) Smart clothing: a shift to wearable computing, *Comm. of the ACM*, August 1996, pp 23–34.

Mann, S. (1997) Wearable computing, a first step toward personal imaging, *IEEE Computer*, 30(2): 25–31.

Marmollin H. (1992) Multimedia from the perspective of psychology. In *Multimedia: Systems Interactions and Applications*, L. Kjelldahl, Ed., pp 39–52, Springer-Verlag, Berlin.

Maybury, M.T. (1993) Planning multimedia explanations using communicative acts. In *Intelligent Multimedia Interface Design*, M. Maybury, Ed., pp 59–74, MIT Press, Cambridge, MA.

Mayer, R. (2000) *Multimedia Learning*, Cambridge University Press, Cambridge, UK.

Miller, G.A. (1956) The magic number seven plus or minus two: some limits on our capacity for processing information, *Psychological Review*, 63: 81–97.

Minsky, M. (1989) *The Music Machine: Selected Readings from the Music Journal*, C. Roads, Ed., pp 639–655.

Moog, R. (1986) MIDI: musical instrument digital interface, *J. Audio Eng. Soc.*, 34, pp 394–404.

Moray, N., and Rotenberg, I. (1989) Fault management in process control: eye movement and action. *Ergonomics*, 32(11): 1319–1342.

Nielsen, J. (1990) *Hypertext and Hypermedia*, Academic Press, London.

Paivio, A. (1986) *Mental Representations*, Oxford University Press, New York.

Sadoski, M., and Paivio, A. (2001) *Imagery and Text*, Lawrence Erlbaum Associates, NJ.

Sanderson, P.M., Verhage, A.G., and Fuld, R.B. (1989) Statespace and verbal protocol methods for studying the human operator in process control, *Ergonomics*, 32(11): 1343–1372.

Sibelius (2002) Sibelius Software Ltd., Cambridge, UK.

Steinberg (2002) Cubase SX, Steinberg Media Technologies, AG, Neuer Hoeltigbaum 22-32, 22143 Hamburg, Germany.

Stenning, K., and Oberlander, J. (1995) A cognitive theory of graphical and linguistic reasoning: logic and implementation, *Cognitive Sci.*, 19(1): 97–140.

Sweller, J., van Merrienboer J.J.G., and Paas, F.G.W.C. (1998) Cognitive architecture and instructional design, *Psychological Reviews*, 10(3): 251–296.

Vickers, P., and Alty, J.L. (2002a) When bugs sing, *Interacting with Computers*, 14(6): 793–819.

Vickers, P., and Alty, J.L. (2002b) Musical program auralisation: a structured approach to motif design, *Interacting with Computers*, 14(5): 456–484.

Vickers, P., and Alty, J.L. (2002c) Using music to communicate computing information, *Interacting with Computers*, 14(5): 434–455.

Waibel A., Tue Vo, M., Duchnowski, P., and Manke, S. (1995) Multimodal interfaces, *Artif. Intel. Rev.* pp 1–23.

Want, R., Hopper, A., Falcao, V., Gibbons, J. (1992) The active badge location system, *ACM Trans. on Information Systems*, 10(1): 91–102.

Williams, D.M. (1996) Multimedia, mental models, and complex domains, *Doctoral Consortium, SIGCHI '96* (Vancouver, Canada, April), pp 49–50, ACM Press: New York.

Zimmermann, T.G., Lanier, J., Blanchard, C., Bryson, S., and Harvill, Y. (1987) A hand gesture interface device, *Proc. ACM Conf. on Human Factors in Computing Systems and Graphics Interface*, pp 189–192.

General Texts That Provide Extended Information

Chapman, N. (2000) *Digital Multimedia*, Digital Multimedia, 582 pages.

Guerin, R. (2001) *CUBASE Power*, Muska and Lipman, 432 pages.

Jones S., Ed., (2002), *Encyclopedia of New Media: An Essential Reference to Communication and Technology*, Sage, Newbury Park, CA, 544 pages.

Labarge, R. (2001) *DVD Authoring and Production*, CMP Books, 496 pages.

Lee, W.L., and Owens, D.L. (2000) *Multimedia-Based Instructional Design: Computer-Based Training, Web-Based Training, and Distance Learning*, Jossey-Bass, San Francisco, 304 pages.

Neuschotz, N. (1999) *Introduction to Director and Lingo: Multimedia and Internet Applications*, Prentice Hall, Upper Saddle River, NJ, 617 pages.

Packer, R., and Jordan, K. (2001) *Multimedia: From Wagner to Virtual Reality*, W.W. Norton and Co., New York, 394 pages.

Taylor, J. (2000) *DVD Demystified, 2nd Edition*, McGraw-Hill, New York, 700 pages.

Vaughan, T. (2001) *Multimedia: Making It Work*, McGraw-Hill, New York, 600 pages.

Further general information on multimedia interfaces may be obtained from the World Wide Web. There are many FAQ (frequently answered questions) sites, which can be found by simply typing *DVD* or *CDROM* into a good search engine.

50

Computer-Supported Collaborative Work

50.1	Introduction	**50**-1
50.2	Media Factors in Collaboration	**50**-2
	Environmental Factors Affecting Collaboration	
	• Visual and Auditory Cues in Face-to-Face Collaboration	
	• Video vs. Audio-Only • Proxemic Effects • Dialog Structure	
	• Social Context Cues • Managerial Behavior and Information	
	Richness • Effects of I/O Rates and Asynchrony	
50.3	Computer-Supported Processes and Productivity......	**50**-9
	Process Gains and Losses • Production Blocking	
	• Anonymity and Free-Riding • Process and Task Structures	
	• Process Support Tools	
50.4	Information Sharing	**50**-12
	Information Availability • Opinion Formation in	
	Computer-Supported Groups	
50.5	Groupware	**50**-14
50.6	Research Issues and Summary	**50**-15

Fadi P. Deek
New Jersey Institute of Technology

James A. McHugh
New Jersey Institute of Technology

50.1 Introduction

The development of networked computing and the globalization of industry have dramatically increased the importance of what is called computer-supported collaborative work (CSCW). The Web has made geographically distributed collaborative systems feasible in a manner that was previously impossible [Deek and McHugh, 2003]. This chapter surveys various factors that affect collaboration, with an emphasis on the media characteristics of distributed computer systems, process management techniques, the information processing characteristics of groups, and the impact of organizational context on the use of collaborative systems.

We begin by considering, in Section 50.2, the effect of media characteristics of an interaction environment on collaboration under a variety of interaction modalities. Because remote groups lack physical copresence, computer-supported communication serves as a surrogate for a broad variety of physical factors, like visual and behavioral cues. These factors are fundamental to understanding the kind of infrastructure that computer-supported environments approximate, supplement, or substitute for.

Section 50.3 focuses on how computer mediation affects group productivity, and how productivity may be enhanced by appropriate computer-supported processes and by effectively structuring interactions. The view is based on the premise that group productivity is determined by the task to be solved, the resources available to the group, and the processes used to solve the task. The characteristics of these processes can increase or decrease group productivity. We consider several process-related effects that affect

1-58488-360-X/$0.00+$1.50
© 2004 by CRC Press, LLC

productivity, including production blocking, anonymous communication, and evaluation apprehension. We also consider techniques for structuring group interactions to make them more effective, such as interaction and task structuring techniques like templates and voting. Information exchange is a defining characteristic distinguishing individual and group problem solving.

Section 50.4 looks at factors that affect how groups handle information, including the availability characteristics of information, such as whether it is initially common, unique, or partially shared among its members, and issues related to opinion formation, like the role of information influence and normative influence. Section 50.5 focuses on impediments to organizational acceptance and issues that groupware developers should be alert to when determining the kinds of functionality appropriate to groupware and the effect of design presuppositions on groupware functionality.

50.2   Media Factors in Collaboration

The media we use to communicate profoundly influence the nature of those communications. This section focuses on the diverse and subtle impacts of the media characteristics of an **interaction environment** on collaboration. We consider a variety of interaction modalities: face-to-face, video-supported, audio-only, and synchronous and asynchronous computer-supported communication. We consider the characteristics of **collocated work**, environmental factors that affect the ability to establish a shared understanding of a problem, and the role of visual and auditory cues in communication. Although the cues in copresent collaboration are different from those in computer-supported communication, they can help us understand the characteristics of communicative exchanges in general. This knowledge can then be brought to bear on issues that arise in computer-supported collaboration. We compare video-mediated and audio-only communications, especially in the context of consensus generation, and consider the impact of so-called proxemic effects. We consider how conversational exchanges can be modeled and how interaction environments affect social context cues, leading to changes in behavior, as has been argued for the case of deindividuation and **group polarization**. We briefly consider which media environments may be preferred for different purposes, such as managerial or organizational objectives, and discuss the effects of I/O and asynchrony.

50.2.1   Environmental Factors Affecting Collaboration

The *interaction environment* for collaboration is defined as the modality or mechanism through which the group interaction occurs. In face-to-face collaboration, the interaction environment is the physical world. In computer-supported collaboration, the interaction environment is the computerized interface that supports the interaction [Whitworth et al., 2000]. A fundamental difference between face-to-face and computer-supported interactions is that the latter provide fewer of the cueing factors that facilitate shared cognitive attention, factors like gesturing, **deictic reference**, and nonverbal expression. A rich cueing environment makes it easier for a group to attain a common understanding or shared context for its discussions, letting everyone in a group be on the same page. The relative scarcity of such cues in computer-supported interactions represents a **process loss**, that is, an inefficiency associated with a process or environment which decreases the productivity of the process.

However, computer-supported interactions also exhibit **process gains**, that is, efficiencies that increase the productivity of the process. For example, computer-supported interactions are more documentable, more reviewable, and may be more precise than their face-to-face counterparts [Nunamaker et al., 1991a]. Some of the current technological limitations on the cueing characteristics of computer-supported environments will be mitigated over time by advances in technology, but other limitations are intrinsic to remote communication.

We begin by examining the simplest model of collaboration, collocated work, which refers to collaboration where participants are located at a common site, with workspaces separated by at most a short walk, say less than 30 meters, and where the work is done synchronously. We assume that there is a common or shared space where group members can meet and that all group members have convenient access to shared static media like blackboards, bulletin boards, boards for posting materials, etc. This kind of interactive environment is also referred to as *proximal interaction* [Olson and Olson, 2000]. Simple collocation and physical proximity are well known to have a strong positive effect on the initiation of collaboration [Kraut et al., 1990]. The interactions that can occur in a collocated synchronous work environment are characterized by the following factors:

Rapid feedback (allowing for quick correction)
Multiple information channels (vocal, facial, postural, gestural)
Coreference (such as deictic reference by gaze)
Spatiality of reference (participants are in the same physical space)
Opportunistic information exchange
Shared local context
Implicit cues
Easily nuanced information exchange

The discretized communication represented by text cannot match the continuum of nuances that occur in a proximal context. All the participants share the same local context, including the same time of day and circadian state. Having the same spatial frame of reference, so-called *coreference*, simplifies deictic reference, which refers to pointing to objects by gesturing and the ability to use the words *this* or *that* as references. There is also substantial individual control, with members able to focus discussion readily.

Most of these characteristics of collocated work derive from the shared spatiality of reference, which embeds in a single space both the individuals and the work products and artifacts they are working on. Each of these features of a collocated workspace has certain positive implications for collaboration. For example, the rapid, real-time feedback makes it easy to nip misunderstandings and errors in the bud. The instant personal identification of sources makes it easier to evaluate the information in the context of its source. The availability of spontaneous, unplanned interactions, which are less likely in a computer-supported context, permits opportunistic exchange of information and facilitates individual and group cohesion [Olson and Olson, 2000]. This is not to imply, however, that face-to-face interactions automatically always represent a gold standard for interactions [Olson and Olson, 2000], because remote collaborations may sometimes be more suitable.

Several factors facilitate collaboration and effective communication, the first being what is called **common ground** [Olson and Olson, 2000]. Common ground refers to the environmental factors or characteristics that facilitate or make possible a shared collaborative experience. Common ground characteristics include factors that enhance cueing, such as the copresence of participants, mutual visibility of participants, audibility, contemporality (which allows immediate reception of communications or messages), simultaneity (which allows participants to send and/or receive messages simultaneously), and sequentiality (which ensures speaker or communicator turns cannot get out of order). Other media characteristics relevant to common ground are factors that enhance the quality of communications or messages, including revisability of messages (messages can be revised by the sender before their final transmission) and reviewability of messages (messages from others can be reviewed after receipt) [Olson and Olson, 2000]. The richer the interaction environment or medium, the more cues it supports for establishing a commonly recognized, shared understanding. Copresence, for example, entails all the advantages that follow from convenient deictic reference.

Besides common ground, other factors that facilitate collaboration are the degree to which work is **tightly coupled**, the desire of participants to collaborate, and the technological preparedness of the group to work in a groupware environment. Another factor is the willingness of participants to collaborate remotely in the first place, which is tied to the incentive structure of the organization, defined as the system of rewards that encourage performance of an activity. Collaboratively sharing and seeking information from or with others requires an appropriate incentive structure. An example of an incentive structure hostile to collaboration (remotely or locally) is one that awards participants based on individual claims to ownership of ideas. In such a context, collaboration may cloud ownership and be perceived as contrary to an individual's benefit [Olson and Olson, 2000].

50.2.2 Visual and Auditory Cues in Face-to-Face Collaboration

Whittaker and O'Conaill [1997] give an overview of the role of visual cues in conversational communication and coordination. They observe that any communication, computer-supported or proximate, requires extensive coordination between speakers (senders) and listeners (receivers). First of all, there must be coordinating processes for initiating and terminating entire conversations (availability), as well as processes for coordinating how speakers and listeners alternate turns during an initiated conversation (turn-taking). In addition to process coordination, conversations also require content coordination, which refers to how participants in the conversation establish a shared understanding. The turn-taking process determines how transitions are negotiated and is remarkably fine-tuned. The process of recognizing availability concerns the initiation of entire conversations. Potential participants must identify when partners are available to converse and whether the moment is opportune to initiate a conversation. These processes require awareness and alertness by participants to cues signaling turn-taking and availability, including understanding the social protocols that signal readiness to begin a conversational process or to switch turns from speaker to listener.

Establishing shared understanding is more complex than merely navigating a conversation. For one thing, the literally expressed meaning of a communication underspecifies what the speaker fully intends to express. What is unspoken must be inferred by the listener from the discussion context, prior understandings, and contextual environmental cues in the physical environment. Some of this inferred understanding can be gathered from the preposited shared source called common ground. For example, common ground facilitates deictic reference, which lets participants easily identify artifacts in the environment. As part of establishing shared understanding, conversations require a feedback loop so speakers can confirm that listeners have correctly understood the intent of the communication [McGrath and Hollingshead, 1994]. The feedback loop helps maintain and extend the common ground or shared knowledge. In a proximate conversation, feedback information is generated on a real-time basis and occurs through a variety of cues. Whittaker and O'Conaill observe that in addition to informational exchange, participants must also be able to track dynamic changes in the affective states and interpersonal attitudes of conversational partners [Whittaker and O'Conaill, 1997].

Much of the information used to support conversations is based on visual cues available in face-to-face environments, including communicative cues about other participants in the interaction and cues about the shared environment itself. The communicative cues include gaze, facial expression, gestures, and posture. Communicative processes like turn-taking depend on information from multiple channels, such as gaze, gesture, and posture. It is worth emphasizing that "an important property of non-verbal signaling [is] that it can go on simultaneously with verbal communication without interrupting it" [Short et al., 1976]. The effects are subtle in terms of their impact on the conversational process. For example, a "negotiated mutual gaze" [Whittaker and O'Conaill, 1997] between the speaker and listener signals that the speaker is yielding the turn to the listener. Or the speaker can gaze at the listener to prompt attention on the listener's part. Gaze has many nuances and can be modulated to indicate the speaker's affective attitude to the listeners or the affective content of what the speaker is saying: trustworthy, sincere, skeptical, amicable, etc. The characteristics of a person's gaze behavior are also revealing. For example, an individual who looks at a conversational partner only a small part of the time will be evaluated as evasive, while a person with the opposite behavior may be interpreted as friendly or sincere. Speakers tend to gaze more when they are attempting to be persuasive or deceptive. Facial expressions are an even richer source of communicative information and feedback than gaze and range from head nod frequency to cross-cultural expressions of affect. A glance is another useful visual behavior that can allow a participant to determine whether another person is available for conversation, not present, or engaged in another activity. Glances also serve as useful "prompts to the identity of a potential participant" [Daly-Jones et al., 1998]. It is worth noting from the point of view of computer-supported communication that visual recognition of identity is possible with low bandwidth and that "more can be remembered about a person when prompted by their face than by their name" [Daly-Jones et al., 1998].

In addition to visible behavior, the shared visible environment itself supports common ground. For example, the presence of others can be automatically inferred, though less effectively so the larger the group. Direct knowledge of the proximity or activities of others can be used to initiate conversations and also affects how interruptions are handled. Visible cues about the availability of a person can affect conversation initiation. Conversely, dyadic conversations may be terminated or altered in content or mode by the arrival of a third person.

50.2.3 Video vs. Audio-Only

Early studies on audio-only conversations [Reid, 1977] concluded that audio channels significantly improved performance on tasks involving simple, objective information exchange, but there was no particular advantage from additional visual access. These studies also showed that audio-only collaboration led to the exchange of considerably more messages than purely written exchanges when solving problems for which there was a single correct solution. Face-to-face communications similarly led to considerably more messages than audio-only [Short et al., 1976].

Interestingly, very slight delays in auditory transmissions, such as can occur over transmission links or Internet connections, can significantly affect the interpretation of verbal communications. The classic study by Krauss and Brickner observed that increases in delay led to an increase in words used [Krauss and Bricker, 1966]. The task used consisted in identifying scrambled, random graphic symbols; each of two participants had identical sets of symbols. One participant had to characterize a symbol verbally to the other participant, who then had to recognize which symbol was being described. The number of so-called *free-standing utterances* (speech by one partner that was immediately preceded and followed by speech by the other partner) was measured, as was the length of the utterances. Delays beyond 1.8 seconds led participants to characterize partners as less attentive, and there were noticeable gender-related effects. The task used here is called a *referential communication task* [Krauss and Fussell, 1990]. These usually consist of a visual stimulus which one of the participants must describe to the other, who then must select the described object from a list. The object (referent) may be a nonsense figure, and the list of choices is called the *nonreferent array*. Although the tasks are limited and unlike ordinary conversational exchanges, reference experiments are useful for testing the effects of an environment on deixis and its communicative effectiveness.

The cuelessness model of Rutter et al. [1981], who analyzed dyadic exchanges over an audio-only link, proposed that, in comparison with face-to-face communications, audio-only interactions were both more issue/task oriented and more depersonalized, because the absence of visual cues "forfeited the regulatory information of non-verbal signals" [Daly-Jones et al., 1998]. O'Malley et al. [1996] used a map interpretation task to compare video, audio-only, and face-to-face communications. Each participant received a slightly different map of an area, and one participant had to tell the other how to follow a route, despite slight differences between the maps. The results indicated that video and audio-only communications were more alike than either was to face-to-face communication. In particular, high-quality video "has only small effects on the process of conversation and low-quality video can actually impede communication" [Daly-Jones et al., 1998].

50.2.4 Proxemic Effects

Visual proxemic effects illustrate the extent to which a sense of personal immediacy can be conveyed though computer-supported video. *Proxemic effects* refer to the apparent distance of one individual from another [Grayson and Coventry, 1998] and are one of the most primitive components of nonverbal communication. Social protocols govern the proximity behavior of people when they interact, with the rules depending not only on the context and relation between the individuals but also on culture and personality. Hall [1963] classified proximity as intimate, personal, social, or public space, with corresponding interactions associated with each. Thus, "talking to a close friend may occur within personal space (18 inches to 4 feet), whereas talking to a stranger will usually occur with the social space (4 to 12 feet)"

[Grayson and Coventry, 1998]. It is known that close proximity may increase perceived persuasiveness, although too-close proximity decreases persuasive influence.

The fact that video communication appears to enhance communication in situations that require negotiation may be related to proxemic effects. Grayson and Coventry [1998] examined proxemic effects in video connections, using dialog analysis to provide a standard decomposition of the conversation into backchannels, turns, overlaps, and utterances or words. *Backchannel responses* refer to communications that signal whether a listener is satisfied or not, agrees or not, is paying attention or not, understands or does not understand the current state of the discussion. Backchannels do not directly disrupt the speaker's flow of speech. They include body language signals, like nods or furled brows, and are used for state correlation but otherwise lack content.

In the Grayson and Conventry experiment, participants whose perceived distance was closer, that is, where the video image appeared closer, interacted more, with more turns taken and more words spoken. Because increased conversational interaction aids understanding more than mere listening, perceived proximity may enhance understanding. The putative explanation is that "conversation involves participants trying to establish the mutual belief that the listener has understood what the speaker meant . . . a collaborative process called grounding" [Grayson and Coventry, 1998]. This mutual view of conversational understanding contrasts with the autonomous view, whereby "merely hearing and seeing all that happens and having the same background knowledge is sufficient to understanding fully" [Grayson and Coventry, 1998]. Apparently, video conferencing conveys proxemic information, though at an attenuated level compared to face-to-face interaction.

50.2.5 Dialog Structure

Conversational dialog is the prototypical interaction, so it is critical to understand its structure. Models of discourse, like the conversation analysis of Sacks et al. [1974] and the interactionist model of Clark and Schaeffer [1989], can help analyze the structure of conversations. The model of Sacks et al. takes a regulatory view of dialog, emphasizing the role of turn-taking during discussions and how often speakers overlap; the lengths of gaps between speakers; and the lengths of individual turns, interruptions, and breakdowns in the dialog. The model tends to interpret breakdowns in conversational flow as failures of communication and interprets a smooth flow in turn-taking as indicating conversational success. According to the interactionist model, on the other hand, the issue is not regulating turn-taking but attaining shared understanding. The interactionist model interprets interruptions and overlaps in speech not as disruptions of a normative smooth sequence of turns, but as necessary to produce a common ground of shared understanding, which is the real objective of the dialog.

As the discussion of cues per Whittaker and O'Conaill [1997] indicates, human interaction has several components, including making contact in the first place, turn-taking disciplines, attention monitoring, comprehension feedback, and various kinds of deixis to objects or persons. Each of these components can be mediated by auditory or visual cues. Initiating contact, for example, uses a cooperative process called a *summons–answer sequence* in which "the caller seeks the attention of a desired recipient who in turn signals their availability and so precipitates an interaction" [Daly-Jones et al., 1998]. Both parties can use auditory or visual signals, with visual signals apparently being more important the larger the number of participants. Similarly, turn-taking can use either visual or auditory cues, the latter including effects like changes in vocal pitch, which serve as turn regulators. Semantic structures in conversation, like requests for attention or explicit questions that signal "entry points into the conversation" [Daly-Jones et al., 1998], also facilitate turn-taking via audition.

50.2.6 Social Context Cues

Social context cues are generally attenuated in computer-supported systems. Sproul and Kiesler [1986] claimed that this tends to result in behavior which is "more extreme, more impulsive, and less socially differentiated" than in a face-to-face context. The negative behaviors exhibited purportedly ranged from flaming to decreased inhibition about delivering bad news. They explain this behavior as resulting from reduced static and dynamic social cues. Static social cues refer to artifacts like a large desk (symbolizing elevated status) or appearance. Dynamic social cues refer to body expressions like nods of approval or frowns. Paper letters or memos reduce static cues to a minimum of "standardized format conventions" [Sproul and Kiesler, 1986] and provide no dynamic cues.

Spears and Lea present a very different analysis of the social ambience of computer-supported communications than is emphasized in the social cues and cuelessness approaches [Spears and Lea, 1992]. They claim that such communications are indeed more social, in the most fundamental sense of the word, than even face-to-face contact. Their use of the word *social* differs from that of theorists who emphasize social presence as the key manifestation of a social element. The latter tend to equate the social with interpersonal contact and immediacy, as mediated by the kind of social cues just described, and which are attenuated in a computer-supported environment. Spears and Lea use *social* in the sense of social identity theory, where it refers to the social categories or identities according to which individuals define themselves, with different categories being salient at different times and in different contexts.

The kind of information used to convey social cues in the interpersonal sense in a visual context are gestures, postures, backchannel signals, the tone of verbal exchanges, and so on. The kinds of cues used to communicate social information in the more fundamental sense defined in social identity theory include "information about the participants, [about] the context, and . . . relevant social category information" [Spears and Lea, 1992]. Interpersonal social cues are strongly affected by the communications media used. The categorical information required in the social identity sense is sparser and less dependent on the information-richness of the environment. Indeed, social categorical information "is often likely to be given or already inferred in the [computer-supported] context" [Spears and Lea, 1992]. Basic categorical information is of the type "generally supplied by message headers (e.g., name, giving cues to gender and ethnicity, organizational affiliation, distribution lists)" etc., [Spears and Lea, 1992]. In contrast, the understanding of social influence characteristic of the social cues line of research is strongly affected by the socio-emotional vs. task distinction proposed by Bales [1950]. While the gold standard for exchange of interpersonal information is the face-to-face environment, with the interpersonal taken as equivalent to the social, this interpersonal characterization of the social neglects the importance of "social categories, norms, and identifications which position communicators and define their relations to each other" [Spears and Lea, 1992].

Spears and Lea summarize the logic of the cuelessness, social presence, and reduced social cues models of interaction as basically asserting that fewer clues lead "to psychological distance, psychological distance leads to task-oriented and depersonalized content, and task-oriented depersonalized content leads in turn to a deliberate, unspontaneous style" [Spears and Lea, 1992]. Conversely, the richer face-to-face environment makes the interpersonal factor more salient. The impoverished social cue environment is then used to explain asocial behaviors in computer-supported groups, like decreased inhibition and group polarization. In fact, *group polarization* (or *risky shift*), which refers to the "tendency for the mean [in the sense of the average] attitudes or decisions of individuals to become more extreme in the direction of the already preferred pole of a given scale, as a result of discussion within a group" [Spears and Lea, 1992], is one of the most well-established characteristics of group behavior. The cuelessness model claims that computer-supported communications exacerbate group polarization because they "undermine the social and **normative influences** on individuals or groups, leading to a more deregulated and extreme (anti-normative) behavior" [Spears and Lea, 1992]. The classic explanation for this is that the egalitarian, uninhibited behavior of computer-supported groups increases the number of persuasive arguments supporting the general direction of the group opinion. Thus, the rationale offered by such models is that reduced social cues lead to more extensive exchanges of arguments and more extreme arguments.

Deindividuation and depersonalization also come into play. *Deindividuation,* defined as "the loss of identity and weakening of social norms and constraints associated with submergence in a group or crowd" [Spears and Lea, 1992], is encouraged by the **anonymity** and reduced feedback typical of computer-supported environments and which provoke less socially desirable behavior. *Depersonalization* occurs because the decrease in social cues redirects participant attention to the task or message context and away from the social context. These combined effects allegedly make the group less interpersonally and more information oriented, and so more affected by persuasive arguments, leading to group polarization. Spears and Lea contend, on the contrary, that people are more likely to be affected by group influence "under de-individuating conditions because the visual anonymity [provided by computer mediated environments] will further reduce perceived intra-group differences, thereby increasing the salience of the group" [Spears and Lea, 1992]. They explain group polarization not as a manifestation of a socially barren environment but, quite the opposite, as representing the convergence of the group "on an extremitized group norm" [Spears and Lea, 1992]. The applicable social psychology concept is referent informational influence according to which "social influence reflects conformity to the norm of the relevant group with which one identifies" [Spears and Lea, 1992]. The computer-supported environment's low information richness fosters the significance of social categorical information, decreasing the salience of interpersonal cues that undermine those social categories.

50.2.7 Managerial Behavior and Information Richness

Daft and Lengel [1984] introduced the concept of *richness* or *information richness*, defined as the "potential information-carrying capacity of data," a seemingly redundant definition clarified in their work. Lee [1994] surveys and critiques the idea of information richness. From a managerial viewpoint, a medium is richer the more rapidly it allows participants using the medium to disambiguate lack of clarity. Information richness emphasizes the effectiveness of a medium for producing shared understanding in a timely manner and how this is related to the medium's characteristics. Richness can also be thought of as a measure of the capacity of a medium to support learning. Consider how managers deal with the key issue of **equivocality**, which refers to uncertainty about the meaning of information when the information has multiple interpretations. Face-to-face communication, usually considered the standard for information richness, provides a broad range of cues to reduce equivocality. Whereas uncertainty about a situation (lack of information) can be reduced by obtaining additional information, equivocality is unaffected by further information and is only resolved or clarified through negotiation, which can be used to converge on a consensus interpretation that minimizes equivocality [Hohmann, 1997].

Negotiation seems to be facilitated by media richness [Dennis and Valacich, 1999]. The process of negotiation is highly social and interpersonal. As Kraut et al. [1990] observe, high equivocality leads decision makers into a fundamentally social interaction, because they are involved in a process of "generating shared interpretations of the problem, and enacting solutions based on those interpretations" [Kraut et al., 1990]. Indeed, Daft and Lengel [1984] (referring to Weick [1979]) claim organizations "are designed to reduce equivocality" and even that "organizations reduce equivocality through the use of sequentially less rich media down through the hierarchy." This is consistent with the notion that nonrich computer-supported communications are better suited to reducing uncertainty (lack of information) than to reducing equivocality (existence of multiple interpretations of information).

50.2.8 Effects of I/O Rates and Asynchrony

The relative speeds of keyboard typing, text reading, speaking, and listening affect the amount of information available to a group, which in turn affects process gains and losses [Nunamaker et al., 1991a]. McGrath and Hollingshead [1994] observe an interesting media asymmetry between output and input rates: "Most people can talk much faster than even very skilled typists can type," but "[m]ost people can read faster than they can listen, although the amount of that difference depends on type of material." In other words, the output in face-to-face communication (by speaking) is faster than output in computer-supported environments (by typing), but input in face-to-face communication (by listening) is slower than input in computer-supported environments (by reading). Adrianson and Hjelmquist [1991] also remark on the differing delay characteristics of computer-supported environments and face-to-face environments and how these affect communication. Thus, computer-supported environments rely primarily on the production and perception of textual messages. These messages can be composed without the need for the kind of relatively instantaneous generation required in a face-to-face environment. Furthermore, the messages can be reviewed and edited by the sender before transmission and reflected on by the receiver after reception. Dennis et al. [1990] also observe how the media difference characteristics of computer-supported environments can positively impact the efficiency of communications. For example, they may "dampen dysfunctional socializing and encourage people to be more succinct" [Dennis et al., 1990].

Another distinguishing characteristic of interaction environments is whether they are synchronous or asynchronous. Face-to-face communications are by nature synchronous, while computer-supported communications can be synchronous or asynchronous but are more often asynchronous. The different modes lead to distinct patterns of communication. McGrath and Hollingshead [1994] allude to various temporal effects of the different modes (also Hesse et al. [1990]). Synchronous communications are strongly constrained by temporal constraints, which tend to limit the length of communications, while asynchronous communication tends to encourage lengthier communications. This can lead to a greater number of simultaneous discussion topics in asynchronous exchanges and possibly more creative contributions, since asynchronous communications can be done over a longer period of time, rely on collective computer-retained memories of past discussions, and suffer no **production blocking** constraints (see Turoff [1984, 1991], Ocker et al. [1995], and Ocker [2001]).

The asynchronous mode may lead to coordination problems (for example, communications are subject to disruptions in sequence), but asynchrony also facilitates coordination. On the other hand, Hiltz et al. [2001] observe that in all their experiments, "even the most extreme asynchronous structures do not reduce the quality of the solutions when compared to the more classical coordination and group approaches." It is important to appreciate that asynchronous computer-supported communication is not just a poor substitute for immediate contact. As Turoff [1991] observes, it is widely misunderstood that asynchronous communication "is a problem, because it is not the sequential process that people use in the face-to-face mode." In fact, the real opportunities for improving group communications via asynchronous systems lie in capitalizing on the fact that such systems allow individuals to "deal with that part of the problem they can contribute to at a given time, regardless of where the other individuals are in the process" rather than trying "to maintain the sequential nature of the processes that groups go through in face-to-face settings" [Turoff, 1991].

50.3 Computer-Supported Processes and Productivity

Steiner's classic work [1972] on group problem solving proposed that group productivity is determined by the task to be solved, the resources available to the group, and the processes used to solve the task. The characteristics of the processes can either increase or decrease group productivity. This section considers process-related effects that affect productivity, including production blocking, anonymous communication, and evaluation apprehension. It also looks at techniques that have been proposed for structuring group interactions to make them more effective.

50.3.1 Process Gains and Losses

Computer-supported collaborative environments have many characteristics that affect productivity. For example, they allow parallel, remote, and relatively instantaneous communications and support a digital group memory. Dennis and Valacich [1993] examined whether the communication among the members of a real group improves performance or detracts from it by contrasting *real group* performance with that of *nominal groups*. The latter are statistical constructs: sets of noninteracting individuals mathematically viewed as a group for purposes of statistical comparison with a real group. Nominal groups are groups in name only, whereas in a real group, the members actually interact.

Process gains may be defined as factors that increase performance in a collaborative environment or efficiencies associated with the intrinsic characteristics of a process. For example, process gains could include the synergies and learning that can occur in a group environment [Nunamaker et al., 1991a] or the advantages associated with parallelism in a computer-supported environment. *Process losses* are inefficiencies associated with the intrinsic characteristics of a process or factors that decrease its performance [Dennis, 1996]. Nominal groups exhibit neither (communication) process gains nor losses, because there is no intermember communication. In contrast, there can be extensive communications among the members of a real group that can have positive or negative effects, leading to process gains or losses. Dennis and Valacich [1993] and Nunamaker et al. [1991b] focused on the gains and losses associated with communication and identified literally dozens of potential process losses. The most prominent of these process effects are production blocking, evaluation apprehension, and **free-riding**, each of which we examine in this section.

50.3.2 Production Blocking

Production blocking is the process loss that occurs in a face-to-face environment when more than one participant wishes to speak concurrently. Speaking requires mutually exclusive access to the floor, so only one person can speak at a time. Allocation of this air-time resource is managed by various social protocols. The delay caused by the associated access contention and its cognitive side-effects is a key source of productivity loss in face-to-face group problem solving. Computer-supported communications ameliorate this kind of production blocking by allowing simultaneous or parallel communication. The blocking is alleviated by a combination of parallel communications, which enable more than one participant to communicate at a time, and extensive logging of communications, which enables later access to these communications. Logged communications also provide a group memory, which reduces the need for members to keep abreast of and remember exchanges, reduces the cognitive effort required by listening, and facilitates reflection on what has been said.

Production blocking has many implications for group interaction. For example, participants who are not allowed to speak at a given point may subsequently forget what they were going to say when they are finally able to speak. Furthermore, after listening to subsequent discussion, participants may conclude that what they had intended to say is now less relevant, less original, or less compelling. The participant may not be accurate in this assessment, but in any case, the group does not know what they were thinking. This could in turn affect how the overall group discussion proceeds. Another effect of blocking is that individuals who are waiting to express an idea must concentrate on what they are about to say instead of listening to what others are saying or thinking productively about the problem. They may waste their cognitive resources trying to remember the idea rather than generating new ideas. Additionally, the act of listening to others speak may block individuals from generating alternatives [Nunamaker et al., 1991a]. Despite its benefits, there are also complications and costs associated with the parallel, synchronous, or asynchronous communications that alleviate production blocking. McGrath and Hollingshead [1994] observe that there is a potentially high cognitive load and possible sequencing complications to discourse caused by these communications, with the result that elimination of blocking can lead to information overload.

50.3.3 Anonymity and Free-Riding

Another phenomenon that affects the productivity of group communications is *evaluation apprehension*, a psychological and socially based effect, unlike production blocking, which is a physical effect based on the mutual exclusion required for speaking. Evaluation apprehension is the fear of being criticized for an opinion one expresses, causing an obvious process loss [Dennis and Valacich, 1993]. It especially affects timid or low-status individuals.

A simple **process structure** proposed to decrease evaluation apprehension is to impose a moratorium on criticism at points where individuals propose opinions, a policy that can easily be implemented in face-to-face and computer-supported environments. A standard mechanism for reducing evaluation apprehension is anonymity or anonymous communication. Anonymity allows individuals to present their ideas without fear of being openly embarrassed by critical comments and without having to challenge openly the opinions of others. The risk of embarrassment or even retaliation in an organizational environment is substantially decreased if no one can be sure who made a comment. Anonymity is straightforward to implement in a computer-supported environment, although there may be technical issues (related to system security and the trustworthiness of the software) in guaranteeing anonymity. On the other hand, if members of a small group know each other well, they may be able to guess the author of an anonymous exchange [Nunamaker et al., 1991b]. According to Nunamaker et al. [1996], however, such guesses are more often than not incorrect.

The effects of anonymity have been especially well studied for idea generation or brainstorming tasks, with interesting results obtained in a series of papers by Nunamaker, Dennis, Valacich, and Vogel (e.g., Nunamaker et al. [1991b]). Prior research found that anonymous groups generated more ideas during brainstorming than nonanonymous groups when using computer-supported communication, at least for tasks with low levels of conflict and under appropriate circumstances. Such groups also judged the interaction process to be both more effective and more satisfying, with similar results obtained in a variety of experimental settings. This occurred for groups with preexisting group histories and those without, groups of varying sizes, and groups from public and private organizations [Nunamaker et al., 1991b]. However, although so-called *negotiation support systems* increase idea productivity or generation of alternative solutions in low-conflict situations, these systems do not appear to have a significant impact on idea generation in high-conflict situations. Some have proposed using interaction processes that require a noncritical tone for group interactions to enhance idea generation, but in fact conflict and criticality of tone seem intrinsic to such processes (see, e.g., Connolly et al. [1993], which evaluated the effect of anonymity vs. evaluative tone).

The impact of anonymity on idea productivity is most pronounced in groups with high status differentials, with behavior changing as one migrates from a peer group to a status group. There is little impact from anonymity in laboratory-scale groups where there are no preexisting power structures, no preexisting vested interests in outcomes, and no fear of negative consequences for nonconformity, but there are significant anonymity effects in groups with existing organizational contexts [Nunamaker et al., 1991b].

Another important process loss is free-riding, the type of loss that occurs when some members of a group rely on others to do the group's work without making their own contribution or making minimum contributions [Dennis et al., 1990]. This loss is also called *social loafing* [Latane et al., 1979]. Free-riding can occur in any group environment. In a face-to-face environment, free-riding is exacerbated by physical group size, because size provides for a degree of anonymity in a physical environment. In a computer-supported environment, group size may promote free-riding, but anonymity is a more likely factor (see, e.g., Dennis and Valacich [1993] and Dennis et al. [1990]).

Computer-supported communications have both positive and negative impacts on free-riding. On the one hand, anonymous mediated communications can increase free-riding, because it may not be evident which members are contributing. That is, anonymity decreases accountability, thereby increasing free-riding. On the other hand, computer-supported communications lessen the barriers to participation in group exchanges.

50.3.4 Process and Task Structures

A process structure is an overall organization for group interaction. A variety of process structuring techniques has been developed, including dialectical inquiry, devil's advocacy, and the Delphi method (see Nunamaker et al. [1996] and Dennis et al. [1997]). *Dialectical inquiry* structures group interactions by splitting participants into subgroups, each of which argues for one alternative solution to a problem and against another. This tends to yield a more complete problem analysis. A related approach is *devil's advocacy*, in which one subgroup acts as a foil to dispute a solution proposed by another subgroup, thereby structuring the evaluation process. Similarly, the *Delphi method* is used "to develop the strongest pro and con arguments for alternative resolutions to a policy issue" [Turoff, 1991] using (but not restricted to) domain experts to represent alternatives and combining alternative generation and evaluation (see Hiltz and Turoff [1993] for an extensive review of this technique).

Process structures are *global* if they apply to the entire interaction process and *local* if they apply to only a single step or phase of the interaction. An example of a global process structure is an agenda or set of rules according to which a meeting will be conducted. DeGross et al. [1990], in a broad-ranging review, concluded that participants in computer-supported groups rated the process support provided by anonymity and the process structure provided by agendas as the most crucial elements of such systems. Agendas help ensure focus and proper allocation of time, so that pertinent issues are not overlooked and premature decisions are not made [DeGross et al., 1990]. The human facilitator frequently provided in electronic meeting systems (EMS) to direct the process represents a global process structure.

Information technology has usually supplied task support for accessing and integrating information, such as allowing group members to access a database. In contrast, task structure uses analytical techniques to improve group decisions and can be either qualitative or quantitative. Qualitative techniques include stakeholder analysis, value chain analysis, and assumption surfacing. Task structure information could be constructed by a human facilitator for an environment. An example is cognitive maps that depict the relations between the comments in a discussion for the purpose of clarifying the influence of one discussion factor on another or for depicting causal relations between factors [Nunamaker et al., 1996; Dennis et al., 1997].

Process structures are not organizationally neutral. Because they determine the patterns and timing of interactions — namely, what, when, how, and by whom things are done — they must address the question of roles: who does what. Consequently, process structures provided by groupware presuppose some kind of group organization. Kernaghan and Cooke [1990] identify three organizational styles typically used by groups for planning. In interacting groups, discussion is unbridled except as directed by the basic charge of the group and the time allowed for discussion. In leader-directed groups, the group leader is the individual recognized as most capable or who is able to identify the most talented members of the group. In nominal groups (a different use of the term than Steiner's), there is no formal group leader; each member states the problem and records responses and suggestions, with group decisions made on the basis of rankings and votes.

50.3.5 Process Support Tools

Templates are built-in sequences of problem-solving activities and so represent a type of global process structure. The templates a groupware environment provides to manage interactions may be hardwired into the interface or merely offered as options. As Nunamaker et al. [1996] observe, "Having a set of standard templates for processes can make it easier for a group to decide what tools to use and what processes to follow." Voting or group polling is useful for dynamically clarifying deliberations, because it helps a group to understand the nature of its disagreements, focusing the direction for the next round of group interactions. This method is also good for the standard purpose of recognizing when consensus has been reached.

Computer-supported environments transform the role of traditional consensus and approval methods like voting. In conventional interactions, voting typically occurs when discussions are nearing completion and is used to close, consummate, or approve conclusions. Computer-supported polling is much more dynamic and provides the group with the opportunity for real-time feedback on the collective state of mind of the group. "Teams find that [electronic] polling clarifies communication, focuses discussion, reveals patterns of consensus, and stimulates thinking"[Nunamaker et al., 1996]. Thus, voting becomes a problem-clarification tool, an alternative-selection tool, and a tool for facilitating consensus-based meeting management. Voting also supplements the information deficit caused by the reduced cues in a distributed meeting.

50.4 Information Sharing

The exchange of information is a defining difference between individual and group problem solving. Despite this, it is often done incompletely and exchanged information is often used ineffectively. This section looks at factors that affect how groups handle information, including the availability characteristics of information (such as whether it is initially common, unique, or partially shared among its members), and issues related to opinion formation, like the role of **information influence** and normative influence.

50.4.1 Information Availability

Information available to a group can be characterized as common, unique, or partially shared. **Common information** is known to every member before work begins. **Unique information** is initially known to only one member. Partially shared information is known to a subgroup prior to collaboration [Dennis, 1996].

The way groups react to shared information can be subtle. Results by Stasser et al. [1989] indicate that information unique to the presenting individual tends to be ignored by the group after its initial presentation more than common information. Group decisions tend to be based on preexisting, commonly held information rather than unique information exchanged in meetings [Gigone and Hastie, 1993]. There are also cognitive effects related to whether or not information supports an individual's preexisting preferences. For example, if information is exchanged, only some elements of which support aspects of an issue, participants tend to focus on those elements that support their preexisting preferences and discount those contrary to existing preferences, although this depends on the prevalence of a preference. Individuals also tend to assume the majority opinion is at least the correct frame of reference and interpret their own preferences in reference to it.

50.4.2 Opinion Formation in Computer-Supported Groups

Mere dissemination of information is not enough to establish or alter opinions. Doing so depends strongly on normative and information influence. *Normative influence* is based on the status of individuals proposing information or extraneous factors like the attractiveness of an advocate, the number of people supporting a position [Dennis, 1996], or sanctions imposed for nonconformity [Short et al., 1976; early work by Deutsch and Gerard, 1955]. This contrasts with *information influence*, which is based on the exchange of facts, ignoring social status clues. Information influence theory (also called persuasive arguments theory) proposes that individuals change their opinions primarily based on factual information [Dennis, 1996]. Normative influence theory (social comparison theory), on the other hand, claims that individuals adapt their opinions to conform to social norms, so that the preferences of the group are more critical than the information exchanged for decision making. To some extent, normative influence theory asserts that mere exposure to preference information is sufficient to alter preferences, although the evidence is inconclusive [Myers and Lamm, 1976; Dennis, 1996]. Normative influence theory also tends to imply that mere public assertion of a preference reinforces the acceptance of that preference.

Group support systems appear to increase the role of factual information or information influence relative to participant preferences or normative influence because of the process gains they provide. One reason is that the members of a computer-supported group can enter information more easily, because parallel input eliminates the input-driven production blocking characteristic of face-to-face environments. Furthermore, since parallelism removes the conceptual dependencies associated with sequential presentations, more themes or threads tend to emerge in such parallel environments than in face-to-face environments, whose sequential character makes them more subject to cognitive inertia [Dennis and Valacich, 1994]. The parallel communication also reduces the blocking in face-to-face environments related to the mutual exclusion between listening to information and processing that information. The group memory that group support systems provide also facilitates recalling common or shared information and unique information.

According to Dennis [1996], computer-supported environments increase information exchange relative to face-to-face groups, but the increased exchange does not appear to improve the group decisions or increase consensus. Computer-supported groups exhibit the typical predilection for common information over unique information exhibited by face-to-face groups and the same tendency for information that supports their preexisting preferences, rather than for neutral information or information that opposes *a priori* preferences. Such preferences, when expressed by a group leader, strongly affect both group discussions and outcomes [Janis, 1982]. Significantly, computer-supported groups are even less likely to use the information they exchange and take longer to make their decisions, even though those decisions are not, on average, better. The impact of group support systems on consensus change, defined as the difference between individuals' opinions before and after the group decision, reveals no significant differences between computer-supported and face-to-face groups. In most other respects, computer-supported groups exhibit the same cognitive failings that face-to-face groups do. In fact, members of computer-supported groups were even more likely than face-to-face groups to contribute information that supported their *a priori* preferences; this was true regardless of the kind of information exchanged. Members of computer-supported groups also recall less originally unknown information than members of face-to-face groups. Overall, while the availability of parallel communication, anonymous communication, and group memory substantially increases information exchange, the members of computer-supported groups appeared to not use the exchanged information to improve their decisions.

The information activities of a group consist in information exchange, information use, and information recall. Each of these places its own cognitive demands on the members of the group, each of whom has limited cognitive resources available for these activities. Petty and Cacioppo [1986a] have observed that processing preference information is less cognitively demanding than processing factual information. One consequence of this is that if too much factual information is shared or shared too rapidly, the individuals in the group tend to deal with the more easily processed preference information first. Since the very availability of information in computer-supported environments tends to flood participants with information for which the processing time is limited or unavailable, factual information may be processed inadequately, poorly integrated with existing information, or replaced altogether by the more readily processed *a priori* preference information. The same effect can occur on purely perceptual grounds if individuals merely perceive that extraction of available information from the computer-supported source is awkward or tedious. Failure to use information available in a computer-supported environment may also be due to unfamiliarity with how to use the system because of lack of prior experience. Other factors affecting the impact of information on a decision-making deliberation are the perceived importance, novelty, and credibility of the information [Dennis, 1996].

We previously discussed process gains associated with anonymous communication in computer-supported groups. Because these gains also affect how information is evaluated, they are not without associated process losses. For example, there is a trade-off between anonymity and credibility because credibility is negatively affected by anonymity [Dennis, 1996]. The credibility of a source of information critically influences its acceptability, particularly if the information is ambiguous or difficult to process [Petty and Cacioppo, 1986b]. When the source is anonymous, this credibility is harder to evaluate. Source anonymity also makes it harder to challenge the contributor. Thus, a key question becomes whether anonymity reduces source credibility to such an extent that it also reduces information processing. However, anonymity also reduces the impact of the dominant group preference [Nunamaker et al., 1991a, 1991b] and thereby of *a priori* preferences. In nonanonymous environments, members maintain their positions and contribute information to the group on the basis of whether it undermines alternatives presented by other members or supports their own preferences. This is less of an issue in an anonymous environment, because there is little need to save face in defense of previously announced public preferences, because those preferences are anonymous [Dennis, 1996].

50.5 Groupware

The context of a collaborative technology has a decisive influence on the success of that technology, yet the organizational, motivational, political, and economic factors that are central to group activity are infrequently addressed explicitly in the design of collaborative systems [Deek and McHugh, 2003]. Grudin [1994] identified various impediments that hindered developing and using groupware systems. To begin with, there is often a perceived disparity between the effort required to work in collaborative environments and the benefits perceived to accrue from their use. Collaborative tools require a critical mass of users to succeed, and thus must pass an initial adoption barrier before which the tool may not appear to be to the advantage of any single individual in a prospective collaborative group. Collaborative environments also tend to level the playing field, violating in-place social hierarchies. Groupware may not be flexible enough to accommodate the exception handling and improvisation required by group activity. Infrequently used features often obscure functionality by being integrated with less frequently used features. Groupware is also more difficult to evaluate than single-user systems, which are not affected by the backgrounds or personalities of group members. Groupware systems are often developed based on the needs of a subset of users or on experience from single-user applications. Developers often fail to recognize that groupware applications require participation from a range of users. With any computing system, organizational integration is critical, but especially careful efforts must be made to ensure that groupware is accepted on an organizationwide basis.

The success of e-mail provides an interesting reference point for understanding collaborative environments [Grudin, 1994]. In terms of effort vs. benefit (perceived disparity), e-mail provides a reasonable balance between sender and receiver, with the sender incurring somewhat more effort because the message must be composed and entered, while the receiver merely reads or scans a message. Critical mass is trivially attained, because even with only a single other user, e-mail can be useful; furthermore, its marginal utility increases monotonically with the number of other users. In terms of social practice, e-mail is natural and conversational. On the other hand, e-mail may upset or alter the typical communication patterns of organizational bureaucracies (disruption of social processes). Information flow between organizational units is usually vertical, in contrast to the lateral communication provided by e-mail. The asynchronous character of e-mail makes its use robust, and its basic features are easy to learn (unobtrusive availability). Evaluation of its costs and utility can be complex to determine, but its widespread use demonstrates its success. The vector of how e-mail use spread (adoption) is also notable, beginning in academic environments and spreading to business and popular uses, rather than through marketing.

Groupware applications tend to require additional work in order to enter the information the application requires for its tools or features to work. Because the benefits of the features may vary among group members, it may not be easy to get every individual to cooperate with this increased entry requirement, because of perceived and possibly real disparity in benefit. For example, scheduling tools primarily benefit one person, so the asymmetry or nonuniformity of the benefit vs. the cost in effort for the members may cause such tools to go unused, representing a so-called *misaligned benefit* [Grudin, 1994]. Testing or evaluating groupware is vastly more complicated than evaluating single-user systems (see Fjermestad and Hiltz [2000]). While laboratory-scale experiments can ferret out the perceptual, cognitive, and motor aspects of single-user applications in a relatively straightforward way, this is a labyrinthine task for group systems, which have inextricably embedded social, political, and motivational factors that influence their usability [Svanaes, 2001]. The acceptance of groupware is particularly sensitive to how it is introduced, because acceptance is required by all the individuals in the group expected to use the system (see Majchrzak et al. [2000]. Because groupware requires possibly substantial learning in order to extract its benefits, an incremental approach that simplifies learning and acceptance is to introduce groupware by adding or overlaying groupware functionality on existing systems.

These impediments challenge the design and acceptance of groupware. The perceived disparity and critical mass effects can be mitigated by educating potential users to the advantages of the systems, such as via training programs that enhance the sense of self-efficacy, defined in cognitive theory as the belief that one is able to use such technologies effectively [Compeau et al., 1999; Deek et al., 2003]. Nidamarthi et al. [2001] emphasize that collaborative environments should supplement, rather than replace, existing methods of communication. The systems should not destabilize traditional, effective means of communication, such as pencil-and-paper calculations. This criterion has implications for the specification of the technological implementation of collaborative systems and suggests the importance of a bottom-up view of groupware design, where systems are built on top of features that support the activities of individuals.

50.6 Research Issues and Summary

This chapter has selectively reviewed research concepts and principles in the area of computer-supported collaborative work, focusing especially on the following:

Media issues that arise in synchronous and asynchronous distributed collaboration
Basic characteristics of face-to-face and distributed interactions
The structure of interaction dialogs

Evaluation of the performance characteristics of computer-supported interactions

Challenges that arise in establishing the shared ground necessary for collaborative work

Various social–psychological effects, like group polarization, which manifest themselves in group environments and which may be either exacerbated or attenuated in computer-supported environments

Management information impacts, like the equivocality of information

The diverse effects and consequences of asynchronous vs. synchronous communications

We also address the impact of CSCW on the productivity of group work and specific types of process losses and process gains, like production blocking and anonymity, which arise in or may be ameliorated by computer-supported environments. We consider the impact of the distributed and partial awareness of information on group decision making and of social psychology factors, like normative influence, on opinion formation. Finally, we examine some of the organizational phenomena that can impede the adoption of groupware systems that support CSCW.

We have only touched the surface of the conceptual, perceptual, psychological, social, cognitive, technical, organizational, and process issues in collaboration. These issues have become increasingly important because of the rapid advances in computer interfaces and networked computing and the impact of these advances on CSCW. All of these areas are the subject of extensive research and development interest. For example, in the area of human cognition in distributed environments, an extensive body of theoretical and experimental knowledge has been developed, including quite sophisticated mathematical models and simulations of distributed cognition and also elaborations of classical models of individual cognition. There have been extensive empirical and statistical studies on the impact of computer support on groups' productivity, as well as much research on the correlative issue of statistical design (see Deek and McHugh [2003] for further discussion and references).

Defining Terms

Anonymity: Anonymous computer-mediated communications intended to reduce evaluation apprehension, especially in the presence of status differences or pressure to conform.

Collocated work: Collaboration where participants are at a common site with workspaces separated by a short walk (less than 30 meters).

Common ground factors: Environmental characteristics that facilitate establishing a shared collaborative experience. These characteristics include factors that enhance cueing (copresence, visibility of participants to each other, audibility, contemporality [immediate receipt of messages], simultaneity [all participants can send/receive messages simultaneously]), and factors that enhance message quality (reviewability [messages can be reviewed later] and revisability [messages can be revised before sending]).

Common information: Information that is known to all the members of a collaborative group prior to group discussion.

Deictic reference: Pointing to objects or ideas, gesturing, and ability to use *this* or *that* as references, for example, supported in a distributed environment by a telepointer.

Equivocality: This refers to uncertainty about the meaning of information. Consensus interpretation can be obtained by using negotiation to converge on an accepted meaning.

Free-riding: A type of process loss in which members of a group rely on others to achieve the task without their own contribution. In a noncomputerized environment, this is exacerbated by physical group size; in a computerized environment, this may be exacerbated by anonymity.

Group polarization: This phenomenon, also called *risky shift*, refers to the alleged tendency of groups to adopt more extreme positions or decisions than individuals, possibly because of normative influence.

Information influence: Support for an opinion derived from primary factors, such as the correctness, quality, or persuasiveness of information, rather than from social factors, such as the status or number of advocates of a position.

Interaction environment: The mechanism through which interaction occurs. In face-to-face interactions, it is the physical world. In computer-mediated environments, it is the technological or interface support system.

Normative influence: Support for an opinion derived from secondary factors, such as the number or the status of participants who hold a position. Normative influence also refers to the tendency of individuals to defer to what they perceive as the group opinion without the need for group pressure, coercion, or persuasion.

Process gains: Factors that increase performance in a collaborative environment or efficiencies associated with the intrinsic characteristics of a process, such as possible synergies and learning that can occur in a group environment.

Process losses: Inefficiencies associated with the intrinsic characteristics of a process, or factors that decrease performance. For example, in verbal communication, speakers must take turns speaking because only one person can have access to the floor at a time.

Process structure: Rules for directing the pattern, sequencing, or content of communications among group members. Group process structures include techniques such as dialectical inquiry (subgroups argue for different alternatives) and devil's advocacy (one subgroup acts as the foil to dispute a solution proposed by another subgroup).

Production blocking: Blocking associated with mutually exclusive access to a resource. For example, in a verbal exchange, only one person can speak at a time, so other participants are blocked in the meantime. Mitigated by simultaneous communication, such as that provided by groupware.

Tightly coupled work: Work that is not partitionable into subtasks requiring limited and less frequent communication among individuals. Tight coupling requires rapid and frequent communications, particularly for ambiguity resolution or repair, and may require geographical colocation. Design is typically tightly coupled, whereas a task like coauthoring is only moderately coupled.

Unique information: Information that is known only to a single member of a collaborative group prior to group discussion.

References

Adrianson, D., and Hjelmquist, E. 1991. Group process in face-to-face computer mediated communication. *Behavior and Information Technology*. 10(4): 281–296.

Bales, R. 1950. A set of categories for the analysis of small group interaction. *American Sociological Review*. 15: 257–263.

Clark, H.H., and Schaeffer, E. 1989. Contributing to discourse. *Cognitive Science*. 13: 259–294.

Compeau, D., Higgins, C.A., and Huff, S. 1999. Social cognitive theory and individual reactions to computing technology: a longitudinal study. *MIS Quarterly*. 23(2): 145–158.

Connolly, T., Routhieaux, R.L., and Schneider, S.K. 1993. On the effectiveness of group brainstorming: test of one underlying cognitive mechanism. *Small Group Research*. 24(4): 490–503.

Daft, R., and Lengel, R. 1984. Information richness: a new approach to managerial behavior and organizational design. In *Research on Organizational Behavior*, Ed. L. Cummings and B. Staw. JAI Press, Homewood, IL.

Daly-Jones, O., Monk, A., and Watts, L. 1998. Some advantages of video conferencing over high-quality audio conferencing: fluency and awareness of attentional focus. *International Journal of Human–Computer Studies*. 49: 21–58.

Deek, F.P., DeFranco-Tommarello, J., and McHugh, J. 2003. A model for a collaborative technologies in manufacturing. To appear in the *International Journal of Computer Integrated Manufacturing*.

Deek, F.P., and McHugh, A.M. 2003. *Computer-Supported Collaboration with Applications to Software Development*. Kluwer Academic Publishers, Boston, MA.

DeGross, J.I., Alavi, M., and Oppelland, H. 1990. *Proceedings of the 11th International Conference on Information Systems*. Copenhagen. 37–52.

Dennis, A.R. 1996. Information exchange and use in group decision making: you can lead a group to information, but you can't make it think. *Management Information Systems Quarterly*. 20(4): 433–457.

Dennis, A.R., Tyran, C.K., Vogel, D.R., and Nunamaker, J.F. 1997. Group support systems for strategic planning. *Journal of Management Information Systems*. 14(1): 155–184.

Dennis, A.R., and Valacich, J.S. 1993. Computer brainstorms: more heads are better than one. *Journal of Applied Psychology*. 78 (4): 531–537.

Dennis, A.R., and Valacich, J.S. 1994. Group, sub-group, and nominal group brainstorming: new rules for new media. *Journal of Management*. 20(4): 723–736.

Dennis, A.R., and Valacich, J.S. 1999. Rethinking media richness: towards a theory of media synchronicity. *Proceedings of the 32nd Hawaii International Conference on System Sciences*. IEEE, Hawaii.

Dennis, A.R., Valacich, J.S., and Nunamaker, J.F. 1990. An experimental investigation of the effects of group size in an electronic meeting environment. *IEEE Transactions on Systems, Man and Cybernetics*. 25(5): 1049–1057.

Deutsch, M., and Gerard, H.B. 1955. A study of normative and informational social influence upon individual judgment. *Journal of Abnormal Social Psychology*. 51: 629–636.

Fjermestad, J., and Hiltz, S.R. 2000. Group support systems: a descriptive evaluation of case and field studies. *Journal of Management Information Systems*. 17(3): 115–159.

Gigone, D., and Hastie, R. 1993. The common knowledge effect: information sharing and group judgment. *Journal of Personality and Social Psychology*. 65: 959–974.

Grayson, D., and Coventry, L. 1998. The effects of visual proxemic information in video mediated communication. *SIGCHI Bulletin*. 30(3): 30–39.

Grudin, J. 1994. Groupware and social dynamics: eight challenges for developers. *Communications of the ACM*. 37(1): 93–105.

Hall, E.T. 1963. A system for the notation of proxemic behavior. *American Anthropologist*. 65: 1003–1026.

Hiltz, S., Dufner, D., Fjermestad, J., Kim, Y., Ocker, R., Rana, A., and Turoff, M. 2001. Distributed group support systems: theory, development, and experimentation. In *Coordination Theory and Collaboration*, Ed. G. Olson, T. Malone, and J. Smith, 473–506. Lawrence Erlbaum Associates, Mahwah, NJ.

Hiltz, S., and Turoff, M. 1993. *The Network Nation*, revised edition. MIT Press, Cambridge, MA.

Hohmann, L. 1997. *Journey of the Software Professional*. Prentice Hall, Upper Saddle River, NJ.

Janis, I. 1982. *Groupthink: Psychological Studies of Policy Decisions and Fiascoes*. Houghton Mifflin, Boston, MA.

Jones, S., and Marsh, S. 1997. Human-computer-human interaction: trust in CSCW. *ACM SIGCHI Bulletin*. 29(3): 36–40.

Kernaghan, J.A., and Cooke, R.A. 1990. Teamwork in planning innovative projects: improving group performance by rational and interpersonal interventions in group process. *IEEE Transactions on Engineering Management*. 37(2): 109–116.

Krauss, R.M., and Bricker, P.D. 1966. Effects of transmission delay and access delay on the efficiency of verbal communication. *Journal of the Acoustical Society of America*. 41(2): 286–292.

Krauss, R.M., and Fussell, S.R. 1990. Mutual knowledge and communicative effectiveness. In *Intellectual Teamwork: Social and Technological Foundations of Cooperative Work*, Ed. J.R. Galegher, R.E. Kraut, and C. Egido, 111–146. Lawrence Erlbaum, Hillsdale, NJ.

Kraut, R.E., Egido, C., and Galegher, J. 1990. Patterns of contact and communication in scientific research collaborations. In *Intellectual Teamwork: Social and Technological Foundations of Cooperative Work*, Ed. J.R. Galegher, R.E. Kraut, and C. Egido, 149–171. Lawrence Erlbaum, Hillsdale, NJ.

Latane, B., Williams, K., and Harkins, S. 1979. Many hands light the work: the causes and consequences of social loafing. *Journal of Personality and Social Psychology*. 37: 822–832.

Lee, A.S. 1994. Electronic mail as a medium for rich communication: an empirical investigation using hermeneutic interpretation. *MIS Quarterly*. 18(2): 143–157.

Majchrzak, A., Rice, R., Malhotra, A., King, N., and Ba, S. 2000. Technology adaptation: the case of a computer-supported inter-organizational virtual team. *MIS Quarterly.* 24(4): 569–600.

McGrath, J.E., and Hollingshead, A.B. 1994. *Groups Interacting with Technology,* Sage Publications, Thousand Oaks, CA.

Myers, J.L., and Lamm, H. 1976. The group polarization phenomenon. *Psychological Bulletin.* 83: 602–627.

Nidamarthi, S., Allen, R.H., and Sriram, R.D. 2001. Observations from supplementing the traditional design process via Internet-based collaboration tools. *International Journal of Computer Integrated Manufacturing.* 14(1): 95–107.

Nunamaker, J.F., Briggs, R.O., Romano, N.C., and Mittleman, D. 1996. The virtual office workspace: group systems Web and case studies. In *Groupware: Collaborative Strategies for Corporate LANs and Intranets,* Ed. D. Coleman, Prentice Hall, Upper Saddle River, NJ.

Nunamaker, J.F., Dennis, A.R., Valacich, J.S., Vogel, D.R., and George, J.F. 1991a. Electronic meeting systems to support group work. *Communications of the ACM.* 34(7): 40–61.

Nunamaker, J.F., Dennis, A.R., Valacich, J.S., and Vogel, D.R. 1991b. Information technology for negotiating groups. *Management Science.* 37(10): 1325–1346.

Ocker, R. 2001. The relationship between interaction, group development, and outcome: a study of virtual communication. *Proceedings of the 33rd Hawaii International Conference on System Sciences.*

Ocker, R., Hiltz, S.R., Turoff, M., and Fjermestad, J. 1995. Computer support for distributed asynchronous software design teams: experimental results and creativity and quality. *Proceedings of the 28th Annual Hawaii International Conference on System Sciences.* IV: 4–13. IEEE Computer Society Press, Los Alamitos, CA.

Olson, G., and Olson, J. 2000. Distance matters. *Human–Computer Interactions.* 15: 139–178.

O'Malley, C., Langton, S., Anderson, A., Doherty-Sneddon, G., and Bruce, V. 1996. Comparison of face to face and video mediated interaction. *Interacting with Computers.* 8(2): 177–192.

Petty, R.E., and Cacioppo, J.T. 1986a. *Communication and Persuasion: Control and Peripheral Routes to Attitude Change.* Springer-Verlag, New York.

Petty, R.E., and Cacioppo, J.T. 1986b. The elaboration likelihood model of persuasion. In *Advances in Experimental Social Psychology,* Ed. L. Berkowitz, 123–205. Academic Press, New York.

Reid, A.A.L. 1977. Comparing telephone with face to face contact. In *The Social Impact of the Telephone,* Ed. I. Pool. MIT Press, Cambridge, MA.

Rutter, D.R., Stephenson, G.M., and Dewey, M.E. 1981. Visual communication and the content and style of conversation. *British Journal of Social Psychology.* 20(1): 41–52.

Sacks, H., Schegloff E.A., and Jefferson, G.A. 1974. A simplest systematics for the organization of turn-taking in conversation. *Language.* 50: 696–735.

Short, J., Williams, E., and Christie, B. 1976. *The Social Psychology of Telecommunications.* John Wiley & Sons, London.

Spears, R., and Lea, M. 1992. Social influences and influence of the "social" in computer-mediated communication. In *Contexts of Computer Mediated Communication,* Ed. M. Lea, 30–65. Harvester Wheatsheaf, Hemel Hempstead, NY.

Sproul, L., and Kiesler, S. 1986. Reducing social context cues: electronic mail in organizational communication. *Management Science.* 32(11): 1492–1512.

Stasser, G., Kerr, N.L., and Davis, J.H. 1989. Influence processes and consensus models in decision-making groups. In *Psychology of Group Influence,* Ed. P. Paul. Lawrence Erlbaum, Hillsdale, NJ.

Steiner, I.D. 1972. *Group Process and Productivity.* Academic Press, New York.

Svanaes, D. 2001. Context-aware technology: a phenomenological perspective. *Human–Computer Interaction.* 16: 379–400.

Turoff, M. 1984. Chapter 12. In *Human Factors and Interactive Computer Systems,* Ed. Y. Vassiliou. Ablex Publishing Corporation, Norwood, NJ.

Turoff, M. 1991. Computer-mediated communication requirements for group support. *Journal of Organizational Computing.* 1(1): 85–113.

Weick, K.E. 1979. *The Social Psychology of Organizing, 2nd edition,* Addison-Wesley, Reading, MA.

Whittaker, S., and O'Conaill, B. 1997. The role of vision in face-to-face and mediated communication. In *Video-Mediated Communication*, Ed. K. Finn, A. Sellen, and S. Wilbur. Lawrence Erlbaum Associates, Mahwah, NJ.

Whitworth, B., Gallupe, B., and McQueen, R. 2000. A cognitive three-process model of computer-mediated group interaction. *Group Decision and Negotiation.* 9(5): 431–456.

Further Information

The following journals are a good source for current research on computer-supported collaboration:

ACM Transactions on Computer–Human Interaction (TOCHI)
Journal of Management Information Systems
MIS Quarterly
Communications of the ACM
Small Group Research

The following conferences devote considerable attention to computer-supported collaborative issues:

Computer-Supported Collaborative Work (CSCW)
ACM Conference on Computer-Supported Cooperative Work
CHI Conference on Human Factors in Computing Systems
Hawaii International Conference on System Sciences
European Conference of Computer-Supported Cooperative Work

The authors' book, *Computer-Supported Collaboration with Applications to Software Development* (Deek and McHugh, Kluwer Academic Publishers, 2003) surveys research in this area and contains an extensive bibliography. Important handbooks are

Handbook of Applied Cognition (Durso, Ed., John Wiley & Sons, 1999)
Coordination Theory and Collaboration (Olson, Malone, Smith, Eds., Lawrence Erlbaum Associates, 2001)

51

Applying International Usability Standards

51.1 Introduction **51**-1
51.2 Underlying Principles **51**-3
User Interface Reference Models • The IFIP Reference Model
• Usability Test Criterion • Structure and Content of the
Usability Standards • Standards in Relation to the European
Council Directive
51.3 Best Practices **51**-11
Standards as Guidelines • User Participation • Analyzing the
Context of Use • Coping with the Uncertainty Principle in a
Design–Use Cycle • Conformity in Terms of Usability Test
Criteria • Conformance Testing vs. Heuristic Evaluations
51.4 Research Issues and Summary **51**-16

Wolfgang Dzida
Pro Context GmbH

51.1 Introduction

Two types of standards are distinguished in **user interface** technology: standard user interfaces and standards *for* user interfaces [Stewart 1990]. The first type establishes a *de facto* standard for user interface implementation, either provided as a corporate standard by a leading software producer (e.g., OPEN LOOK [Sun 1990] and Windows [Microsoft 1994]) or defined by consensus within the software industry (see OSF/Motif [OSF 1994; Berlage 1995]). The second type comprises a series of standards [ISO 9241 1999, Part 10 through Part 17] that are devoted to dialogue techniques of interactive systems, such as menu dialogues or **direct manipulation**. These standards provide design recommendations but do not include any guidance for implementation, nor do they involve toolboxes or programming interfaces. In addition to the product design standards, a process standard has been published [ISO 13407 1999] providing recommendations for user-centered design processes, such as context and requirements analyses. Meanwhile, usability engineering has been established as a discipline on its own right, like software engineering [Mayhew, 1999; Rosson and Carroll, 2002]. Both disciplines will set their own standards, as is typical with traditional engineering disciplines (e.g., civil engineering). Hence, the previously mentioned standards for designing products and processes can be called *usability standards* or *usability engineering standards*.

As is usual in the standardization of interface components for system-to-system interaction, the existing technologies of different companies are integrated by consensus — for example, so-called computer-aided design (CAD) frameworks to link CAD tools or protocols to ensure that different applications can interoperate at run-time. This kind of standardization is aimed at reducing development costs and time to market, an approach driven mainly by technology. One may therefore call this type of standard a *technical* standard. The second type of standard, the standards for user interfaces, may appropriately be called *ergonomic* standards. Hence, the design object is far more than purely technical, because the interface component is not located between technical components but between user and system. Standards for user interfaces hold for a variety of user interface products in all kinds of interactive systems and applications.

The major difference between the two types of standards is that ergonomic standards do not merely rest on the technical state of the art. Rather, their development is predominantly determined by the state of evidence obtained from ergonomic research [Dzida 1989]. The impetus for developing ergonomic standards is to achieve an acceptable level of ergonomic quality according to the current state of knowledge in the community of ergonomists.

Standards may also be distinguished by their scope of validity, which range from locally defined proprietary guidelines to international standards or directives. Within a software company, the user interface designer may be required to stick firmly to the corporate style guide and the associated toolkit, thereby assuring that similar interfaces are implemented for different applications [Pangalos 1992]. Consistency in design can be the main benefit for the user; however, designing consistency with the aid of user interface toolkits addresses only the tip of the iceberg [Berry 2000], that is, about 25% of usability [Travis 1997].

The international standards pursue more ambitious objectives than proprietary guidelines. They define a minimum level of usability that should be respected within an international market. Accordingly, the European Council Directive [ECD 1990] requires by law that interface design adhere to such a minimum level of ergonomic quality, so as to establish a harmonized level of work conditions across all member states of the European Union.

Many people are concerned about whether too much standardization in the user interface area might stifle the designer's creativity and competition between companies, in the same way that copyright might stifle them [Samuelson 1995]. Some of this criticism may apply to early approaches to ergonomic standardization and may still be valid for standard interfaces; for a discussion of further objections to standard interfaces, see Potosnak [1988]. However, the international ergonomic standards developed by the International Standards Organization [ISO TC 159/SC 4/WG 5, 2003] cannot be viewed in that way. ISO standards rely on consensus about the minimum level of quality to be required; above that level, there is freedom of design.

Sometimes standards have also been suspected of freezing a certain state of technical development. The innovative progress in user interface technology, however, should not be impeded by the ergonomic type of standards, because they are neutral toward any specific implementation. Nevertheless, standard user interfaces may have a conservative effect in the long run, unless new standard interfaces can outrival the established ones. The freezing of a state of technology may in fact be favored, if a standard interface is widely accepted as a reference for high quality.

This chapter focuses on international software ergonomic standards, increasingly referred to as usability standards. A methodology of testing products for compliance with international standards is of central concern, because a standard can contribute to improved design solutions only if the methods of requirement specification and testing are well accepted. Although usability standards suggest a procedure for requirements specification as well as for conformance testing, the "level of specification of the procedure is a matter of negotiation between the involved parties" [ISO 9241-14 1997, par. 4.3]. Relying only on such negotiations, however, will make it impossible to compare the quality of different products. Moreover, the negotiating partners must not specify usability requirements below the minimum level that is the defined in the usability standards.

The approach presented in this chapter has not yet achieved international consensus. However, a standard usability test has been agreed on at a national level. DATech [2003], the German body for the accreditation of test laboratories, is authorized to publish such a test in order to avoid letting a statement of conformity with a standard be an arbitrary interpretation of the standard. The DATech test procedures for usability of products and the maturity of processes are offered as downloads on the German DATech home page [DATech 2003].

51.2 Underlying Principles

Testing of **usability** has achieved the level of a standard procedure during the last decade. This chapter describes some methodological foundations that generally apply to usability testing of interactive systems and which may be of particular concern when testing products for conformity with standards. Key methodological questions include the following:

What portion of the user interface is under study?

What are the leading design objectives (principles of quality)?

How is usability embedded into a general quality model of software?

How are usability requirements structured in the international standards, so as to find them easily and apply them to design decisions or **conformity tests**?

Finally, the liability of ergonomic standards in an international market is addressed, especially from the perspective of the European Union.

51.2.1 User Interface Reference Models

To structure the rather complex user interface, through which the user interacts with the application program, one may specify areas of an interactive system particularly apt for usability design and evaluation. A number of human–computer interaction reference models have been published; for a survey, see Spring et al. [1993]. Some of the reference models are devoted to a functional specification, some describe system architectures, and others provide conceptual models [Norman 1988] for users. Conceptual models include a specific interface model, layer models [Fähnrich and Ziegler 1985], and linguistic models of interaction [Moran 1981; Myers 1989; Marcus and van Dam 1991]. An interface may be described by a set of rules (or attributes) that determine the interchange of data (information) between user and computer. The interface model was developed by the International Federation for Information Processing (IFIP) Working Group 6.5; for the original model see IFIP [1981], and for its formalized description see Dzida [1987, 1988].

Information designers of user interfaces confine their concept of an interface to only one component (i.e., the input/output interface), which is characterized by facilities of data input as well as attributes of data presentation (such as grouping and coding of data or echoing keystrokes and mouse clicks). The software designers' conceptual models (e.g., the MVC model [Goldberg 1990] or the PAC model [Coutaz 1987]) do not necessarily fit well with the users' conceptual model of an interface.

Incompatibility was uncovered, particularly for inheritance-based decompositions of interactive systems [Wegener 1995], which aimed at taking better account of the reusability of interface components. Abstract data types representing the input/output interface (e.g., push buttons, menus, icons) can be well separated as far as they are neutral toward the application. But data types for the **dialogue** have not found their way into the software designers' conceptual model as explicit abstractions; they are represented as part of input/output or application (sometimes both), but not separately. Separate data types for the dialogue thus appear to be irrelevant for software designers, although this concept remains significant from the user's perspective. As a consequence, the designer may develop an interaction between user and system with a concept of interface in mind that may be more restricted than the user's model. This may bring about a limited understanding of how the user wants to conduct a task.

51.2.2 The IFIP Reference Model

The IFIP user interface reference model (Figure 51.1) had some impact on the structure of international usability standards, insofar as ergonomic principles of design could be grouped according to the conceptual components of the user interface.

The user's interaction with the system is enabled by an interface, which can be structured into three interface components, each representing a separate aspect of interaction: input/output; the conduct of dialogue; and access to tools, services, or data. This structure can be amended by a fourth component for involving a specific type of interaction: information exchange in an organization or within a computer

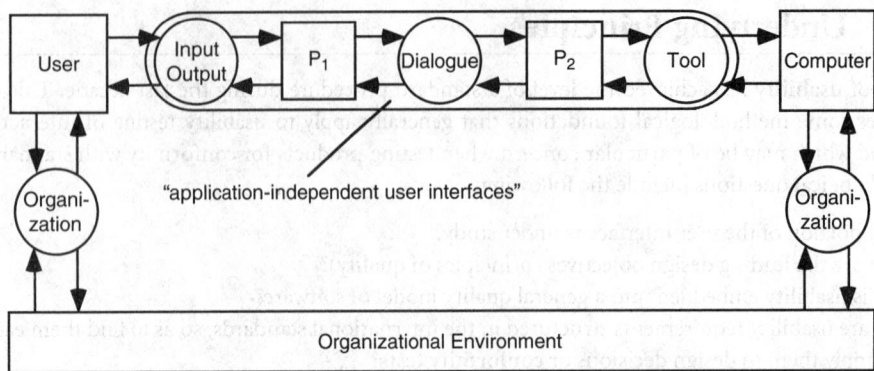

FIGURE 51.1 User interface reference model of IFIP WG 6.5. (From Dzida, W., *Psychological Issues of Human–Computer Interaction in the Work Place*, North-Holland, Amsterdam, 1987. With permission.)

network (see the organizational interface). Whereas the first three interface components are always available when using the computer as a tool, the fourth component is typically available when using the computer as a medium. An advantage of the model is its focus on four relatively independent concepts appropriate to structuring the interface and distinguishing between the computer as a tool and as a medium. The focus facilitates communication between users and designers on usability issues or on those issues that pertain to the design of a medium. Objectives for designing a medium may be incompatible with usability objectives. The Internet provides many examples demonstrating, for instance, the incompatibility between usability and marketing objectives.

In Figure 51.1, circles indicate interfaces, rectangles represent activities (processes), arrows point to the direction of dataflow.* From the user's point of view, the four interface components are of central concern, but the user may also be interested in the processes (e.g., P_1 and P_2) representing software components of the user interface provided by a user interface management system (UIMS) or application program. For instance, P_1 may realize a user input and then react by echoing the input; P_2 may prompt the user for further data input or signal an input error. P_1 and P_2 do not necessarily induce data processing by the application program, but they do change the display state. A separation of display-related interaction and task-related interaction can be introduced this way. This is a necessary conceptual separation, because the user is interested in the effects of an input (i.e., whether it solely changes the attributes of the screen or causes a transition from the current data state of a task at hand to a new data state).

A feature of the model that relies on the Petri net notation is the assimilation of three components into one interface, indicating that the user is virtually faced with one interface that presents three of the interface aspects simultaneously before or after running the application program. (See Figure 51.1) Note that the Petri net notation does not account for time relations among interface components; only causal relations are addressed.

Before design principles are discussed for the user interface components, their functions and attributes are described.

51.2.2.1 Input/Output Interface

This part is often referred to as the *surface* of the software. Rules for user input and system output govern the interface — for instance, movement and positioning of the cursor by a mouse or arrow keys, placement of a pop-up menu or an icon, size of a window, highlighting of a menu option, color coding, and information design. The design of this part of the user interface is predominantly captured by toolboxes or UIMSes. The notion of user interface is sometimes confined to this interface component, that is, to the tangible surface characteristics. This is, of course, too narrow a concept.

---

*The syntax of the model uses a type of Petri net called a *channel agency net* [Reisig 1985].

51.2.2.2 Dialogue Interface

Interaction with the system is dialogue-like: the user receives information and can control the system by means of an interface language. This language addresses the meaning of communication, including command names, letters of shortcut keys, symbols, direct manipulation, voice input, and gestures. Furthermore, the data exchange necessary for conducting a task characterizes the dialogue, which includes activities such as prompting, interrupting, switching to another window, and resuming an editing process. Also, data exchange is necessary after task accomplishment, such as recovering from errors in response to error messages, adhering to warnings, system messages, and help information. Characteristics of the dialogue (e.g., being sequential, asynchronous, or concurrent [Hartson 1989; Hartson and Hix 1989]) may also help to determine this interface.

51.2.2.3 Tool Interface or Application Interface

Rules or conventions govern the access to tools, data, and services. The user may want to undo the change of data, put into sequence a number of tools by a pipe (as in the UNIX system), concatenate tools in terms of a macro or a command procedure, or configure the set of tools actually necessary for use. Characteristics of tools determine this interface (e.g., generic or specific, elementary or compounded). There is no doubt about the impact of the tool interface characteristics on the usability of software. However, many of these characteristics are highly application-dependent compared with the attributes of dialogue or input/output. Consequently, the international standardization group restricted the scope of ergonomic requirements to application-independent characteristics. Nevertheless, the tool interface should not be ignored when evaluating the usability of software beyond the ergonomic standard requirements.

Notably, the IFIP user interface reference model has influenced the Seeheim model of interactive systems [Pfaff 1985], which is an approach to developing a system architecture that is effective in usability engineering for the development of UIMSes [Olsen 1992].

51.2.2.4 Principles of Design

The interface components of the reference model can help apply the principles of ergonomic design. They represent objectives of usability, the achievement of which is verified when a number of conceptually coherent user requirements have been satisfied.

The following principles of information design pertain to the input/output component of the interface [ISO 9241-12 1998]:

- **Clarity** — The information can be quickly conveyed.
- **Discriminability** — The information items can be accurately distinguished.
- **Conciseness** — Only necessary information is given.
- **Consistency** — The expected information is given in the same way.
- **Detectability** — Attention is directed to the information required.
- **Legibility** — The information is easy to read.
- **Comprehensibility** — The meaning is clearly understandable.

The following principles of ergonomic dialogue design are published in ISO 9241-10 [1996]; for the empirical basis of these principles, see Dzida et al. [1978]:

- **Suitability for the task** — Only relevant and task-related steps are required.
- **Self-descriptiveness** — The information is immediately clear or clarified on demand.
- **Controllability** — The user is in control of the dialogue steps required by the task.
- **Conformity with user expectations** — The dialogue fits well with conventions and user attributes.
- **Error tolerance** — Mismatches are prevented or can be managed with minimal effort.
- **Suitability for individualization** — The dialogue can be adapted to individual, special needs.
- **Suitability for learning** — Explorative system use is beneficial for becoming an advanced user.

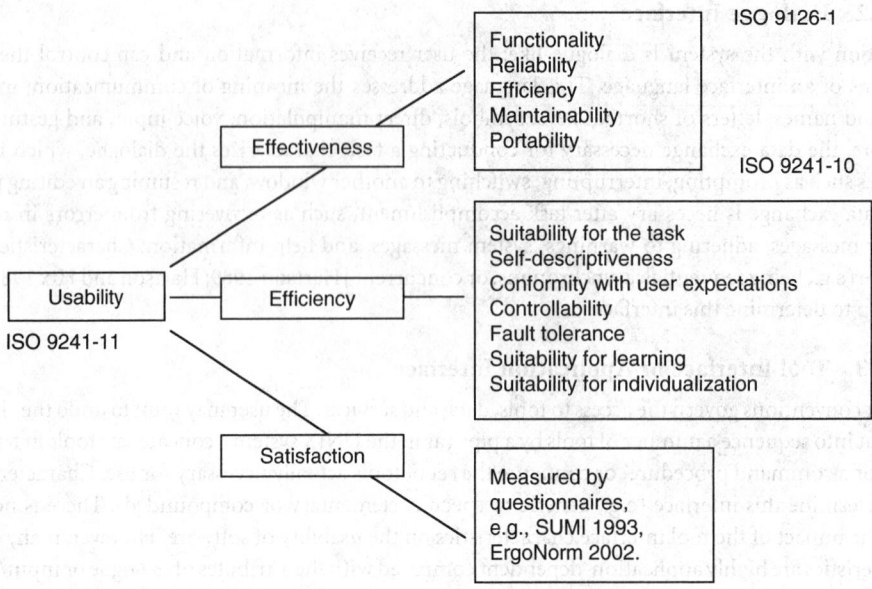

FIGURE 51.2   Usability quality model.

Although the international usability standards rarely involve ergonomic requirements for the design of the tool interface, a list of such principles is presented below as a suggestion; see also ISO/IEC 9126-1 [2001] and McCall et al. [1977]. In software development, it is indispensable to adhere to these principles, so that the user can use the software effectively (see effectiveness as a factor in the quality model, Figure 51.2; see also the definition of **usability**):

- **Functionality** — Functions suit user needs and provide accurate results.
- **Reliability** — System performance avoids faults or is fault tolerant.
- **Efficiency** — System performance provides appropriate processing time.
- **Maintainability** — Modifications include corrections, adaptations, and improvements.
- **Portability** — The software can be transferred to other environments.

Characteristics of the organizational interface are highly application-dependent and are therefore not an issue of ergonomic standardization; this interface will not be further discussed here. However, to be complete, as far as the design of tasks determines this interface (see the left part of the organizational interface in Figure 51.1), principles of task design [ISO 9241-2 1992] can help specify and improve the ergonomic quality of user performance. Regarding the organizational interface as a medium among computers of an organization (see the right part of the organizational interface in Figure 51.1), conventional rules of conduct may be described in terms of ergonomic principles of groupware design, which include suitability for cooperation, responsiveness, negotiability, and security [Herrmann et al. 1996]. As a medium, the organizational interface may also adhere to principles of marketing, especially in the Internet.

These principles help the designer to structure the short tradition of thought in usability engineering in abstract concepts, thereby developing a general understanding of user requirements. A specific requirement can be interpreted in terms of a principle, thereby guiding the designer and the user in achieving a common understanding. Principles may also help to clarify trade-offs and priorities among design proposals. Principles can also guide the reader of a standard to specify a requirement according to the needs implied in the product's context of use.

51.2.2.5 Usability Quality Model

Last but not least, principles of design contribute to the development of a quality model. A number of software engineering quality models have been discussed; see, for instance, Boehm et al. [1976] and McCall et al. [1977]. Results of this discussion have been crystallized in the international standards ISO/IEC 12119 [1994] and ISO/IEC 9126-1 [2001]. A usability quality model may analogously establish a framework of terminology for a growing international usability engineering community. After having structured the scope of usability design by means of principles, an attempt has been made to develop a usability quality model. But before presenting the suggested model (see Figure 51.2), the quality concept of usability must be defined.

Usability has been introduced as a general term for software quality; it replaces colloquial terms such as user friendliness or ease of use. In ISO 9241-11 [1998] usability is defined as the extent of **effectiveness**, **efficiency**, and **satisfaction** to which a product can be used to achieve specified goals in a particular **context of use**. Effectiveness, efficiency, and satisfaction can be viewed as the three quality factors of usability. Notably, high efficiency can only be achieved if effectiveness is given. Effectiveness is usually defined in terms of **task** results, and a degree of 100% effectiveness (complete and accurate result) is usually required. Given 100% effectiveness, the genuine ergonomic evaluation can take place, and the focus is then on the effort users must invest to achieve effectiveness. Effectiveness and efficiency are evaluated mostly by experts. But an expert may err, so the user's judgment about satisfaction is indispensable for usability evaluation. An observed dissatisfaction may help uncover a hidden shortcoming of the product. See Kirakowski and Corbett [1993] and also the ErgoNorm questionnaire [DATech 2002] for measuring subjective usability.

The usability quality model (Figure 51.2) introduces the concept of usability as one concept among many, but one that determines the overall quality of an interactive system. The rationale for setting usability as the ultimate quality objective is seen in the relation between validation and verification of quality. Adhering to software-technical principles just contributes to correctness, but a system being verified as correct is worthless to the user if it is invalid. The software-technical principles (e.g., reliability and functionality) determine the effectiveness with which a user can achieve a required task result. The usability-engineering principles (e.g., suitability for the task and controllability) determine the efficiency of user performance, with effectiveness included. (Note that efficiency of user performance and efficiency of system performance are distinguished.) The role of usability engineering in quality assurance is primarily appreciated due to its contribution to the validation of a product early in the manufacturing process [Dzida and Freitag 1998].

51.2.3 Usability Test Criterion

Testing an interactive system for usability requires that test criteria be specified. A usability **test criterion** is defined as a required level of measure, the achievement of which can be verified. To verify this level, the concept of usability is broken down into its constituent factors: effectiveness, efficiency, and satisfaction. As mentioned previously, effectiveness is assumed to be satisfied by the presence of all of the other technical quality concepts: reliability, portability, etc. Hence, a criterion of 100% effectiveness is postulated as a basis for an ergonomic specification of efficiency test criteria. Usually, these criteria are derived from subfactors (also referred to as *principles*) of efficiency. Two types of criteria should be specified during the construction of usability requirements:

- A **task performance** and its resulting effect, observable at the user interface or the outcome of a human cognitive process (user performance) accompanying a task performance
- A product attribute, which represents an appropriate design solution to enable the task performance or cognitive process

ISO 8402 [1994] defined a *requirement* as an expression of needs and/or their translation into stated requirements for the characteristics of an entity (par. 2.3). Although this definition is watered down in the newer standard ISO 9000 [2000], the good thinking in the original definition is that any requirement is twofold: a need and a corresponding product attribute. Hence, a usability requirement or test criterion is almost always expressed in terms of a task or user performance and a corresponding product attribute.

As an example of a required effect of task performance, consider the echoing of the selection of a menu option to the user. Note that, although echoing is an effect of task performance, it is not the intended (complete and accurate) final task result.

As an example of a required human cognitive outcome, consider the ability of a user to discriminate between active and nonactive menu options. Note here that discrimination is the required outcome.

As an example of a required product attribute, consider the use of brightness coding (highlighting) in menus to facilitate discrimination between active and nonactive menu options. Note that the highlighted menu option as an attribute is an appropriate design solution to enable a required user performance.

One may argue that echoing a user input is just as much a product attribute as highlighting a menu option. Indeed, an effect of task performance can be a user interface attribute. The difference, however, can be seen in the relation of an attribute to a user's activity (performance). When an attribute appears on the display as a consequence of such an activity, it is taken as an effect in the interaction. Effects are, for instance, echo, prompt, error or help message, alert, and cursor positioning. If an attribute is used to design for such an effect, it is taken as a product attribute. Highlighting, for instance, can be used as an attribute to design the system's echo (an effect) to the selection of a menu option (a user performance). Although the distinction between effects and attributes may appear artificial, it will become more useful when we deal with the evaluation of attributes that are task-related and those that are neutral (see Section 51.3.5). Effects are always task-related.

These examples illustrate how different kinds of usability criteria may complement each other in a design solution. Hence, when defining the user's list of requirements, it may be unsatisfactory to exclusively specify criteria in terms of product attributes because the designer may regard a required attribute as out of date or inconsistent with other design decisions. If the user simply requests a specific attribute, the designer may not know why. The level of measure to be specified for a test criterion, therefore, should consider the required effect of task performance at the user interface or the level of outcome of human performance. It should be up to the designer to select an adequate product attribute that fits well with the required levels of performance.

A user interface may provide numerous high-tech features, but the work performed at this interface may nevertheless be of low ergonomic quality. This may be caused by the fact that the tasks are designed without regard for basic ergonomic task requirements. The poor design of a task, of course, does not bring the usability of a product into discredit. However, a really human-centered approach considers not only the design of user interface attributes, but also the human conditions of work and organization. It is worth mentioning that the introduction of information technology can have effects on the content of jobs and individual interdependencies in an organization. These changes should be taken as an opportunity to redesign tasks and develop organizations according to ergonomic task requirements. ISO 9241-2 [1992] not only requires task design to facilitate tasks but also recommends that task design provide an appropriate degree of autonomy to the user in deciding on priority, pace, and procedure. The features of such a task can be mirrored by user interface attributes of controllability (see Section 51.2.2.4).

51.2.4 Structure and Content of the Usability Standards

The usability quality model (Figure 51.2) may clarify the structure of usability standards shown in Figure 51.3. Similar to the model, the standards are hierarchical, beginning with usability as the most general quality concept (Part 11), continuing with principles (Part 10 and Part 12), and ending with a series of standards (Part 13 through Part 17) devoted to specific forms of dialogue. With the quality model in mind, the reader can rapidly find an appropriate standard requirement at the intended level of abstraction. If a specific standard does not provide the guiding information to make a design proposal, the reader should ask for information in one of the two principal standards (dialogue or information presentation). The principles encourage freedom of design but remind the designer to meet design principles as closely as possible. In Figure 51.3, the principal standards, along with the specific standards on dialogue techniques, form the core of standard requirements. Their interpretation is primarily based on task requirements (Part 2) and the requirements derived from a context-of-use analysis (Part 11).

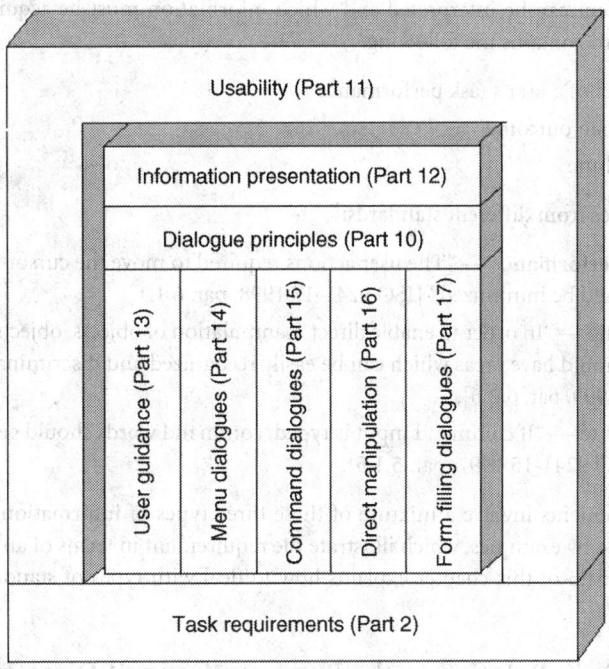

FIGURE 51.3 A structure of the usability parts of ISO 9241. (From Dzida, W. 1995. Standards for user-interfaces. *Comput. Stand. & Interfaces* 17: 89–97. With permission.)

TABLE 51.1 Parts of ISO 9241 (January 2003); Parts 10–17 are Software Usability Standards

Part	Title
1	General introduction
2	Guidance on task requirements
3	Visual display requirements
4	Keyboard requirements
5	Workstation layout and postural requirements
6	Environmental requirements
7	Display requirements with reflections
8	Requirements for displayed colors
9	Requirements for non-keyboard input devices
10	Dialogue principles
11	Guidance on usability
12	Presentation of information
13	User guidance
14	Menu dialogues
15	Command dialogues
16	Direct manipulation dialogues
17	Form filling dialogues

Source: From Dzida, W. 1995. Standards for user-interfaces. *Comput. Stand. & Interfaces* 17: 89–97. With permission.

Table 51.1 provides a list of all standards under the title ISO 9241 [1999]: Ergonomic requirements for office work with visual display terminals (VDTs). The specific usability parts of this standard are Part 10 through Part 17.

The differentiation between types of usability criteria helps to identify what type of information is provided by a specific standard requirement. Having the types of criteria in mind, the reader will immediately know how a requirement can be interpreted and which information must be acquired for checking its applicability. A standard contains the following:

- A required effect of a user's task performance
- A required human outcome
- A product attribute

Here are some examples from different standards:

- **Effect of task performance** — "The user actions required to move the cursor from one entry field to the next should be minimized" [ISO 9241-17 1998, par. 6.1.1].
- **Human outcome** — "In order to enable direct manipulation of objects, objects that can be directly manipulated should have areas which can be easily recognized and discriminated by the user ..." [ISO 9241-16 1999, par. 6.2.6].
- **Product attribute** — "If command input is typed, command words should generally not exceed 7 characters" [ISO 9241-15 1997, par. 5.1.5].

Some standard requirements involve a mixture of these three types of information. Nearly all requirements are accompanied by examples, which illustrate the requirement in terms of an implemented product attribute. Section 51.3 of this chapter explains how to deal with types of standard requirements in conformance testing.

51.2.5 Standards in Relation to the European Council Directive

To finish this section on principles of usability, we consider an issue of internationalization. The European Council Directive [ECD 1990] requires that "the principles of software ergonomics must be applied, in particular to human data processing" [ECD 1990, p. 18]. For an interpretation of this regulation, the notion of principles must be clarified. Further requirements in the list of minimum requirements (Section 3 of the Directive) point to the background information the authors probably had in mind when referring to principles. The minimum requirements are as follows:

"(a) software must be suitable for the task; (b) software must be easy to use and, where appropriate, adaptable to the operator's level of knowledge or experience; no quantitative or qualitative checking facility may be used without the knowledge of the workers; (c) systems must provide feedback to workers on their performance; (d) systems must display information in a format and at a pace which are adapted to operators ..." [ECD 1990, p.18].

The following is an interpretation of the requirements in terms of principles:

Point (a) corresponds to the dialogue principle suitability for the task, although the system's functionality is also addressed.

Point (b) is a twofold requirement: the required adaptability corresponds to the dialogue principle conformity with user expectations, and it also addresses maintainability.

Point (c) corresponds to the dialogue principle self-descriptiveness.

Point (d) corresponds to the dialogue principle controllability, although principles of information design are also addressed.

To sum up, when adhering to the principles of software ergonomics, the practitioner is well advised to adopt the ISO principles of dialogue [ISO 9241-10 1996] and the principles of information design [ISO 9241-12 1998].

An additional issue is the legal obligation of the European Council Directive previously listed. All member states of the European Union must transpose the Council Directive into national law. Certainly, national legislation will refer, at least indirectly, to these international standards, because the European minimum requirements correspond to the ISO minimum requirements. Although an ISO standard originally has the status of recommendation, all standards will be conceived as obligatory requirements within the European software market [Stewart 1992].

Concerning the software-producing companies outside of Europe, there is a question of how they will cope with the specific European standards if they want to deliver products to this market. Software producers will establish European service companies to intensify contact with customers as regards requirements specification, system adaptations, and conformance testing at the users' workplaces [Keil and Carmel 1995]. The next section, on best practices, is an attempt to address these issues in detail.

51.3 Best Practices

The series of software-ergonomic standards will be of no value if software designers and product assessors do not know how to apply them. Members of the standardization committee repeatedly asked software designers and usability assessors to judge the applicability of standards. Complaints about the difficulty of interpreting the standards were made. The main difficulties were in testing products for compliance with the standards. Another concern was raised about when which standard must be applied. This section provides help in reading and interpreting the usability standards in order to convert the recommendations given in the standards into valid test criteria.

51.3.1 Standards as Guidelines

Usability standards are mostly formulated as guidelines, rather than precise specifications. The problem with guidelines is that they may become too detailed and voluminous or too brief (and thus overly generic). The authors of the most comprehensive collection of guidelines predicted that designers may be disappointed if they look to guidelines for specific rules but find only general advice instead [Mosier and Smith 1986]. The standard guidelines represent the present state of an ergonomically accepted technology, but the product attributes involved in the standard lack exact quantitative values. This lack of exactness may make it difficult for the designer to obtain a compliant proposal for a specific design decision; it may also make it difficult for the assessor to check a product feature for compliance with a guideline.

Nevertheless, this characteristic of the guidelines need not be regarded as a drawback. The opposite may be true, in fact, because the guidelines are not at all unclearly stated for a reader who acquires usability requirements from the context of use before interpreting the guidelines. This is similar to interpreting another type of standards, national or international laws, which also require the reader to apply them to a specific state of affairs and its context. To judge conformity, the reader of a usability standard should not expect that a paragraph of a standard can easily be compared to a product attribute. Before introducing best practices for applying the usability standards, we outline why usability standards are like guidelines, what the advantages of guidelines are, and how test criteria can be determined so as to enable the designer or the assessor to apply a standard.

Usability standards are formulated as guidelines for several reasons:

- Freedom of design is warranted, that is, the standard does not specify any particular product attribute.
- The requirements do not imply any specific implementation.
- The requirements do not imply any specific user or user target group.
- The requirements do not presuppose any specific task or organizational setting.

Certainly, the standardization committee should not specify requirements for specific target groups, tasks, or contexts of use. This would produce a tremendous proliferation of standards that would be practically unmanageable. Software-producing companies would protest against standardized attributes, because it is the style of such attributes that should establish the appearance of a product as being unique to a specific company.

Using the standards as a source for specifying usability test criteria requires the usability specialist to acquire valid user data with the help of the user target group(s).

51.3.2 User Participation

The development of usability requirements evolves from a mature relationship between customer and manufacturer. Valid user data can only gathered in the customer organization at the users' workplaces. It may be a risk to the validity of a product when users are ignored in the cooperation of customer and manufacturer. What a customer wants is not always what the users really need. Users are experts at describing what is going on at their workplaces and which problems occur when using a product within its context. Nevertheless, what a user wants is not always what a user really needs. It happens that the latest advances in features and functions impress users. It is tempting to require them, and we should not prohibit users from doing so. The requirements engineer, however, should try to redirect the users' attention to the required task performance rather than technical attributes. For example, we know that almost every user seems to be an expert at designing color combinations. As far as the color attribute is a matter of taste, the user is always right. However, if the color attribute is a matter of usability (e.g., discriminability of items), then the user is respected as an expert on discriminability, but not on color design. Users are not good designers of product attributes, just as designers are not experts on user performance.

User participation is indispensable in usability engineering [ISO 13407 1999], in particular during context analysis and explorative prototyping. In the past, the usability specialists' common phrase was "Know the user." However, this turned out to be too narrow a focus, because only a few findings of mainstream psychology really helped to understand the computer user. While establishing usability research in the 1980s, the focus expanded toward "Know the user's task." However, even this focus was too narrow, because the initial understanding of task was that of a computerized task performance. Existing system features biased this understanding. User participation was confined to testing the user's acceptance of these features. In response to this immunization trap, the notion of *context of use* has been introduced [ISO 9241-11 1998], with the user, the actual key tasks, and organizational and social constraints being the source for creating a valid understanding of usability requirements.

51.3.3 Analyzing the Context of Use

The advantages of the guideline-like character of standards are achieved at the expense of readers who immediately want to apply them to design or testing. These readers first must define usability requirements to provide the basis for an interpretation of a guideline. ISO 9241-11 [1998] serves as a standard which guides the reader in analyzing the context of use of the product in order to specify the usability requirements. Additional guidance for context analysis can be retrieved from Bowden and Thomas [1995] and Bevan and Macleod [1994]. Essentially, this analysis provides a specification of user characteristics, their goals and tasks, the equipment (hardware, software, and materials), and the physical and social environments in which the product is or will be used. The specification of users' goals can be converted into a list of usability requirements. The DATech usability group [DATech 2003] has published a scenario-based analysis of the context of use and its application to specifying usability requirements. See Beyer and Holtzblatt [1998] and Rosson and Carroll [2002] for further approaches to context analysis. Information about the context of use provides the rationale for those usability requirements that are context-related. Usability requirements, which are independent of the context, must be further specified. Use of scenarios or prototypes can provide the rationale for these requirements [Carroll 1995; Rosson and Carroll 2002; DATech 2003]. The scenario-based techniques have been established as best practices of user participation.

The previously described practices of user participation in the analysis of a product's context of use contribute to the validation of usability requirements. Validation of requirements is thus introduced as a milestone in a mature usability engineering process [DATech 2003]. A narrow view of validation simply requires that the systems requirements be matched with the user's intention [Loukopoulos and Karakostas 1995]. In usability engineering, validation has been conceptualized as a consensus-formation process [Dzida and Freitag 1998], a process in the course of which proposed system attributes are evaluated against usability requirements to achieve a consensus among partners about usability requirements and corresponding design solutions. Validation is thus defined as a consensual process, rather than a simple analytic comparison. Validation becomes a constructive activity, which invites the user to intervene in the design process and invites the designer to refine an understanding of requirements. This process enables a shared understanding, that is, a valid understanding.

51.3.4 Coping with the Uncertainty Principle in a Design–Use Cycle

Partners in a project (i.e., manufacturer and customer) have shared responsibilities for specifying usability requirements before signing the contract. The ISO 9000 [2000] requires the manufacturer to focus on the customer's needs. The first quality management principle postulates that manufacturers depend on their customers and therefore should understand customer needs (par. 0.2.a). The customer's responsibility is laid down in the ISO 9001 [2000] standard: the customer shall select manufacturers based on their ability to supply products in accordance with the customer's requirements (par. 7.4.1). Hence, the partners should share responsibility for developing requirements throughout the project, even after installation of the product in its context of use.

From experiences in usability engineering, it became evident that the development of an interactive system can only be finished in the context of use. This experience fits well with Humphrey's uncertainty principle [1995], which states that the requirements will not be completely known until after the users have used the system. Therefore, in usability engineering, the so-called *software life cycle* is called **design–use cycle**. To meet the needs of users and customers, the model of the design–use cycle takes account of quality improvements to be evolutionarily achieved. For the software-engineering origin of the model, see Floyd et al. [1989].

The customer's responsibility (in the role of employer) is addressed in the European Council Directive [ECD 1990]. Article 3 requires an analysis of users' workstations to evaluate conditions that may result in complaints about physical problems, health problems, or mental stress. The test of a product for compliance with international standards could be used as a preventive measure, as the directive requires. The customer will require such a test to be performed in advance by the developer. However, subsequent use of a product can also uncover **defects** and **faults,** which would soon initiate an adaptation or a redesign of the product.

To manage these quality problems, "effective communications should be maintained to encourage users to discuss their concerns and to ensure timely and effective organization responses" [ISO 9241-2 1992, par. 5]. An organization may respond either by adapting the product or by adapting the product's context of use, so as to embed the system more properly. After the product has been adapted to user needs, the adaptation should be tested for conformity with the standards, taking into account that the customer will be responsible for making that test. From the evolutionary character of the design–use cycle, it is clear that a product's conformity with standards is not achieved once and forever but requires periodic retests during redesign and application.

51.3.5 Conformity in Terms of Usability Test Criteria

When testing a product for conformity with an international usability standard, one should not expect to conduct the test in the same way as style-guide compliance for specific product attributes, according to ISO/IEC 12119 [1994]. This kind of testing is the exception in usability conformance testing. The rule is that usability requirements derived from the context of use must be translated in terms of test criteria to prepare for the conformance test. The translation essentially involves an interpretation of a standard requirement in the light of the usability requirements. As previously outlined, this interpretation is necessary due to the fact that standard guidelines are formulated neutrally toward users, tasks, and environmental conditions. In principle, there are three possibilities to determine the test criteria:

> The standard requirement genuinely represents the test criterion (which is the exception). This holds for a required product attribute. For example, help explanations should vary in type and length [ISO 9241-10 1996]. Type and length are attributes, which can be inspected regardless of task and users, that is, a context of use analysis can be skipped.

The standard requirement must be interpreted with regard to the task at hand. Let us assume that the user wants to avoid the recurrent input of the same data, which instead should be easily available on the display after the first input. The standard [ISO 9241-10 1996] contains a corresponding requirement concerning default values, but this must be interpreted in the light of the real task. This task may be, for instance, the task of a CAD engineer who works on modeling the geometry of a steel girder and is repeatedly concerned with the values of its flange. The CAD system conforms to the standard if it presents the flange values as defaults.

The standard requirement must be interpreted with regard to task *and* user needs. This holds true for a required human outcome resulting from task performance. For example, "Explanations should assist the user in gaining a general understanding of the dialogue system. . . " [ISO 9241-10 1996, par. 3.3]; understanding can be assessed only in view of the user and the task. For a user who interacts with an integrated office system, the general level of understanding should be much higher than for a user who simply applies a form-filling dialogue. After having determined the required level of understanding, the test criterion is specified and the test for conformity with the standard is well prepared.

The usability standards, especially Part 13 through Part 17, introduce the concept of the conditional requirement, which is a special form of the criterion-oriented approach. A conditional requirement is a sentence formulated in terms of an *if-then* rule, thereby structuring the sentence into two components: a conditional part and a subsequent guideline part. For example, ISO 9241-14 [1997] contains the following conditional requirement for menu options: "If options can be arranged into conventional or natural groups known to users, options should be organized into levels and menus consistent with that order" (par. 5.1.1). The *if*-clause refers to the condition of applying the guideline. Actually, the context of use must be analyzed and the users must be interviewed in order to determine the test criterion, which is a specific interpretation of the guideline.

The criterion-oriented approach to conformance testing is organized as a process mainly determined by three activities (see Figure 51.4).

1. Derive the usability requirement from the context of use.
2. Specify the requirement in terms of a test criterion in view of the standards.
3. Test for conformity.

The second step determines the minimum level of usability. Finally, the test criterion is compared with relevant product attributes of the design solution. This step provides the test of the product for conformity with usability standard. In most cases, it suffices to consult ISO 9241-10 when a requirement is converted into a test criterion. The conformity test (step 3) acknowledges whether the design solution meets the

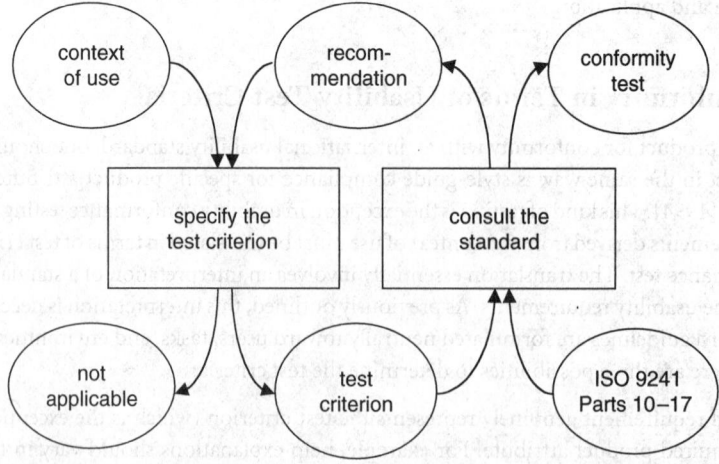

FIGURE 51.4   Conversion of a usability requirement in terms of a test criterion.

usability requirement, in principle. If a deviation from a principle is detected, then an appropriate standard within the series of ISO 9241 Part 12 through Part 17 should be consulted to specify a rationale for the minimum quality a design solution should satisfy.

It may occur that the design solution provides a higher level of quality than the standard requires. Conformity with the standard is achieved if the user or task performance enabled by the design solution equals or exceeds the minimum level of quality specified by the test criterion.

To avoid redundancy, some parts of ISO 9241 [1999] do not explicitly phrase requirements or recommendations in terms of an *if-then* clause. For example, the principle of suitability for the task recommends that "Help information should be task dependent" [ISO 9241-10 1996, par. 3.2]. Although no conditional clause is explicitly included, it is evident that this guideline can be applied only under the condition that the task has been investigated, so as to specify the test criterion prior to the conformance test. When adopting the criterion-oriented approach, conformance testing is possible for all guideline-like standards [Dzida 1995], regardless of what turn of phrase is used in the standard (conditional requirement or application of a principle).

Figure 51.4 summarizes the steps in interpreting a recommendation provided by a usability standard. It may be necessary to consult the usability standards repeatedly during the process of interpreting the appropriateness of recommendations before a test criterion is specified. As a final result, a checklist of usability test criteria can be achieved, which represents the scope of the conformance tests to be done. The major advantage of this checklist is that it can be made part of the test report, thereby explicitly confining the scope of conformity tested. The reproducibility of the test results is thus warranted. Documenting the conformity of a product in undifferentiated terms can be avoided. If a usability requirement changes during the application phase of a product, the assessor can easily pick up the corresponding test criteria that determine whether a subsequent conformance test is necessary.

The checklist containing usability test criteria differs from the typical type of checklist frequently applied in usability evaluation. Typically, checklists contain a mixture of required product attributes and arbitrarily stated user performance requirements. It seems to be quite convenient to adopt such a checklist in a variety of investigations, regardless of the product under study or the product's context of use. Opposite to this type of checklist, the criterion-oriented checklist is validated for a specific context of use and is legitimized as each item in the list has a defined linkage to a valid usability requirement. The major drawback of a conventional checklist, containing solely product attributes, can be seen in the unspecified linkage of these attributes to usability requirements.

51.3.6 Conformance Testing vs. Heuristic Evaluations

The criterion-oriented approach to usability testing should also be compared with other frequently applied methods; for a survey of methods, see Nielsen and Mack [1994]. In particular, the heuristic evaluation approach [Nielsen 1992] should be selected for a comparison, to contrast this method with the striking features of conformance testing. Nielsen classifies usability problems and allocates them to nine "principles" of good interface design, such as consistency, feedback, and shortcuts. All principles correspond to principles of the international standards; for instance, consistency issues are subsumed within the dialogue principle of conformity with user expectations [ISO 9241-10 1996].

Nevertheless, heuristic evaluation serves another purpose that is beyond the test for conformity with international standards. Heuristic evaluation is a cost-effective and rapid inspection method, like usability walkthroughs, with the main difference being that it is less formal than conformance testing. Inspection of the product can be taken as a *debugging* of usability defects, that is, an identification and diagnosis of the most striking and serious usability problems. Therefore, heuristic evaluation is particularly suited to the inspection of prototypes during the design cycle.

Just as conformance testing mostly includes an inspection of the product, so does heuristic evaluation. The major difference lies in the question of what is inspected. A conformance test is an inspection of a required product attribute, that is, a relevant attribute that fits well with a valid usability requirement. A heuristic evaluation inspects a usability problem. This approach requires the assessor to have the nine principles in mind as a vague quality definition to be matched with a number of arbitrarily selected product attributes. The product is not necessarily inspected for a selected set of relevant tasks (i.e., a specific context of use is not addressed and suitability for the task is not included in the list of heuristics). Hence, the output of heuristic evaluation is not a list of defects that violate specific test criteria but of those that indicate more or less serious usability problems. A shortcoming of heuristic evaluation is that the identified usability problems cannot be fixed for a constructive solution, because it will then be necessary to know a valid usability requirement. The advantage of heuristic evaluation, however, is its effectiveness in uncovering the major problems in a user interface. Thus, heuristic evaluation contributes to achieving the minimum level of quality, just as conformance tests do.

51.4   Research Issues and Summary

Many standard requirements need further development through empirical investigations. Research issues can easily be found in some of the usability standards (Part 13 through Part17) by looking at the annex that provides references to source documents for almost every requirement. Three types of sources are distinguished there: research studies, guidelines, and experts' consensus. For example, the following requirement was selected from guidelines (see Galitz [1985]) to be put into the menu standard [ISO 9241-14 1997, par. 5.3.6]: "If the frequency of option use is known … and option groups are small (eight or less), the most frequently used options should be placed first." This requirement may be reasonable for a pull-down menu, but it should not be generalized for pop-up menus without further empirical evidence. It may turn out that in a pop-up menu, the most frequently used option is more adequately placed at the end of the menu panel, in order to meet Fitts' law (i.e., to minimize the time necessary for moving the mouse to a frequently selected target option). Readers of the usability standards should be made aware of the rich potential for research issues that is raised by the need for empirical evidence to support standard requirements.

Usually, experts in usability design and evaluation will apply standards. A significant aspect is that expert judgment is based largely on accumulated practical experience rather than solid empirical data. Although for each requirement the empirical evidence is checked during the development of the standards, the practitioner will usually neglect the underlying source document. For many empirical studies, the related experimental results could not always be generalized as far as suggested by the standard requirement. Even if a requirement is based on solid data, a specific application may lead to a divergent interpretation because of the specific circumstances given in the context of use [Landauer and Galotti 1984]. The practitioner will not be concerned with such issues. For a professional researcher, however, the notion that design decisions may be based on judgment rather than on experimental data is repugnant [Smith 1986].

At the designer's workbench, design decisions cannot be postponed until supportive data have been obtained from empirical research. Therefore, many design decisions could be investigated *ex post facto* in experiments, the results of which then may influence the revision of design and standards. Unfortunately, professional researchers are not yet sufficiently integrated into this process. Also, they may be reluctant to take part in design decisions, thereby accompanying the designer when obtaining from this experience the impetus for further experimental studies. Even if a laboratory researcher is willing to take part in this process, there may be good reasons not to be too committed, because of the limitations of laboratory research to study usability [Eason 1984].

Thus, we may question whether solid empirical foundations for standard requirements can ever be achieved. This is why a design decision, as well as the underlying guideline, will rarely be absolutely true or false, but will be more or less appropriate relative to the usability requirements derived from a product's context of use.

With the publication of a standard concerning the usability design process [ISO 13407 1999], a new research issue has been identified. Every defect of a product can be traced back to a shortcoming in the manufacturing process. The process, as well as the product, can be assessed for its shortcomings and defects, thereby investigating the causal relationship of process and product quality. A challenging research issue is seen in a combination of assessment and subsequent improvement of both process and product, thus establishing a test and improvement cycle.

The series of usability standards [ISO 9241 1999, Part 10 through Part 17] provides a set of requirements for the minimum level of usability of interactive systems. Although the evidence for some of the requirements may be questioned and may need to be revised in a later review, the worth of the standards should not be underestimated. The set of requirements is based on a balance of different interests among partners in an international market and on a state-of-the-art knowledge in usability research, as well. Although the standards are as yet merely conceived as recommendations (except within the European Union), they should be respected throughout the world as a baseline of usability, beyond which software companies will have sufficient space for developing competitive, high-quality products. There may be regions of the world where ergonomic aspects of products and quality of work do not yet play an important role. With the advent of international standards, a worldwide harmonization of work conditions at computerized workplaces may become the most significant effect of this work in the long run.

Defining Terms

Conformity test: An operation that compares relevant attributes of a product with applicable standard requirements for determining the achievement of the level of quality required.

Context of use: The users, goals, tasks, equipment (hardware, software, and materials), and the physical and social environments in which a product is used [ISO 9241-11 1998].

Defect: An unintended attribute that impairs the efficient use of a product.

Design–use cycle: The evolutionary course of developmental improvements through which a product passes from its conception, through its use, to its redesign or the termination of its use.

Dialogue: A process in the course of which the user, to perform a given task, inputs data in one or more dialogue steps and receives for each step feedback with regard to the processing of the data concerned.

Direct manipulation: A dialogue technique by which the user acts directly on objects on the screen, for example, by pointing at them, moving them, or changing their physical characteristics (or values) via the use of an input device [ISO 9241-16 1999].

Effectiveness: The accuracy and completeness with which users achieve specified goals [ISO 9241-11 1998].

Efficiency: The accuracy and completeness, in relation to the resources expended, with which users achieve goals [ISO 9241-11 1998].

Fault: A missing attribute that impairs the effective use of a product.

Satisfaction: The comfort and acceptability of use [ISO 9241-11 1998].

Task: A specification including the intended result of an activity to be performed on an object (material) by a specified means (method).

Task performance: An activity carried out at a user interface and aimed at completing a task.

Test criterion: A measure of required level of quality against which attributes of an object (e.g., a product) or of a process (e.g., the design cycle) are judged, to assess the level of quality achieved.

Usability: The extent to which a product can be used by specified users to achieve specified goals of effectiveness, efficiency, and satisfaction in a specified context of use [ISO 9241-11 1998].

User interface: An interface that enables information to be passed between a human user and hardware or software components of a computer system [ISO/IEC 12119 1994].

User participation: The involvement of target users for developing and validating usability requirements and evaluating design proposals (or solutions) throughout the design–use cycle.

References

Berlage, T. 1995. OSF/Motif as a user interface standard. *Comput. Stand. & Interfaces* 17: 99–106.

Berry, D. 2000. The user experience: the iceberg analogy of usability. IBM Web site. http://www-106.ibm.com/developerworks/library/w-berry.

Bevan, N., and Macleod, M. 1994. Usability measurement in context. *Behaviour Inf. Tech.* 13: 132–145.

Beyer, H., and Holtzblatt, K. 1998. *Contextual Design: Defining Customer-Centered Systems*. Morgan Kaufmann, San Francisco.

Boehm, B.W., Brown, J.R., and Lipow, M. 1976. Quantitative evaluation of software quality. In *Proc. IEEE 2nd Int. Conf. Software Eng*, pp. 529–605. ACM Press, New York.

Bowden, R., and Thomas, C. 1995. *Usability Context Analysis Guide*, ver. 4. National Physical Laboratory, U.K.

Carroll, J.M., Ed. 1995. *Scenario-Based Design*. John Wiley & Sons, New York.

CEC 1990. *Social Europe, 2/90*, Ed. F. Gutman. Commission of the European Communities. Directorate-General V. Rue de la Loi 200, 1049 Brussels, Belgium.

Coutaz, J. 1987. PAC: an object oriented model for dialog design. In INTERACT '87 Conf. Proc. *Human–Computer Interaction*, H.J. Bullinger and B. Shackel, Eds., pp. 431–436. Elsevier, Amsterdam.

DATech 2003. *DATech-Prüfhandbuch Gebrauchstauglichkeit. Leitfaden für die software-ergonomische Evaluierung von Software auf der Grundlage von DIN EN ISO 9241, Teile 10 und 11*. Deutsche Akkreditierungsstelle Technik e.V., http://www.datech.de.

DATech 2003. Deutsche Akkreditierungsstelle Technik e.V., http://www.datech.de.

Dzida, W. 1987. On tools and interfaces. In *Psychological Issues of Human Computer Interaction in the Work Place*, M. Frese, E. Ulich, and W. Dzida, Eds., pp. 339–355. North-Holland, Amsterdam.

Dzida, W. 1988. Modellierung und Bewertung von Benutzerschnittstellen. *Software Kurier* 1: 13–28.

Dzida, W. 1989. The development of ergonomic standards. *SIGCHI Bull.* 20(3): 35–43.

Dzida, W. 1995. Standards for user-interfaces. *Comput. Stand. & Interfaces* 17: 89–97.

Dzida, W., Herda, S., and Itzfeldt, W.-D. 1978. User perceived quality of interactive systems. *IEEE Trans. Software Eng.* SE4(4): 270–276.

Dzida, W., and Freitag, R. 1998. Making use of scenarios for validating analysis and design. *IEEE Trans. Software Eng.* 24(12): 1182–1196.

Eason, K.D. 1984. Toward the experimental study of usability. *Behaviour Inf. Tech.* 3: 133–143.

ECD 1990. Council Directive on the minimum safety and health requirements for work with display screen equipment; fifth individual Directive within the meaning of Article 16(1) of Directive 87/391/EEC. No. 90/270/EEC of May 29. *Off. J. Eur. Communit.* L156: 14–18.

ErgoNorm 2002. User questionnaire. See DATech 2003.

Fähnrich, K.-P., and Ziegler, J. 1985. Workstations using direct manipulation as interaction mode — aspects of design, application and evaluation. In INTERACT '84 *Human–Computer Interaction*, B. Shackel, Ed., pp. 693–698. Elsevier, Amsterdam.

Floyd, C., Reisin, F.-M., and Schmidt, G. 1989. STEPS to software development with users. In *Proc. ESEC '89, 2nd Eur. Software Eng. Conf.*, C. Ghezzi and J.A. McDermid, Eds. Lecture Notes in Computer Science, 387, pp. 48–64. Springer, Heidelberg.

Galitz, W.O. 1985. *Handbook of Screen Format Design*. QED Information Sciences, Wellesley, MA.

Goldberg, A. 1990. Information models, views, and controllers. *Dr. Dobb's J.* 7: 54–61, 106–107.

Hartson, H.R. 1989. User-interface management control and communication. *IEEE Software* 1: 62–70.

Hartson, H.R., and Hix, D. 1989. Human–computer interface development: concepts and systems for its management. *ACM Comput. Surv.* 21(3): 5–92.

Herrmann, T., Wulf, V., and Hartmann, A. 1996. Requirements for a human-centered design of groupware. In *Design of Computer Supported Cooperative Work and Groupware Systems*, D. Shapiro, M. Tauber, and R. Traunmüller, Eds., pp. 77–100. Elsevier, Amsterdam.

Humphrey, W.S. 1995. *A Discipline for Software Engineering*. Addison-Wesley, Reading, MA.

IFIP 1981. Report of the 1st meeting of the European User Environment Subgroup of International Federation for Information Processing [IFIP] WG 6.5. GMD, 53754 Sankt Augustin, Germany, July.

ISO 8402 1994. Quality — Vocabulary.

ISO 9000 2000. Quality management systems — Fundamentals and vocabulary.

ISO 9001 2000. Quality management systems — Requirements.

ISO/IEC 9126-1 2001. Software engineering — Product quality — Quality model.

ISO/IEC 12119 1994. Information technology — Software product evaluation — Quality requirements and testing. 1st ed.

ISO 9241-2 1992. Ergonomic requirements for office work with display terminals (VDTs): guidance on task requirements.

ISO 9241-10 1996. Ergonomic requirements for office work with display terminals (VDTs): dialogue principles.

ISO 9241-11 1998. Ergonomic requirements for office work with display terminals (VDTs): guidance on usability.

ISO 9241-12 1998. Ergonomic requirements for office work with display terminals (VDTs): presentation of information.

ISO 9241-13 1998. Ergonomic requirements for office work with display terminals (VDTs): user guidance.

ISO 9241-14 1997. Ergonomic requirements for office work with display terminals (VDTs): menu dialogues.

ISO 9241-15 1997. Ergonomic requirements for office work with display terminals (VDTs): command dialogues.

ISO 9241-16 1999. Ergonomic requirements for office work with display terminals (VDTs): direct manipulation dialogues.

ISO 9241-17 1998. Ergonomic requirements for office work with display terminals (VDTs): form filling dialogues.

ISO 13407 1999. Human-centred design processes for interactive systems.

ISO TC 159/SC 4/WG 5 and WG 6 2003. Ergonomics of human system interaction. S. Krebs, Sec. DIN, 10772 Berlin, Germany.

Keil, M., and Carmel, E. 1995. Customer-developer links in software development. *Commun. ACM* 38: 33–44.

Kirakowski, J., and Corbett, M. 1993. SUMI: the software usability measurement inventory. *Brit. J. Educ. Tech.* 24(3): 210–212.

Landauer, T.K., and Galotti, K.M. 1984. What makes a difference when? Comments on Grudin and Barnard. *Human Factors* 26: 423–429.

Loukopoulos, K., and Karakostas, V. 1995. *Systems Requirements Engineering.* McGraw-Hill, London.

Mayhew, D.J. 1999. *The Usability Engineering Lifecycle: A Practitioner's Handbook for User Interface Design.* Morgan Kaufmann, San Francisco.

Marcus, A., and van Dam, A. 1991. User-interface developments for the nineties.*Computer* 24(9): 49–57.

McCall, J.A., Richards, P.K., and Walters, G.F. 1977. Factors in software quality, Vols. I, II, III. *U.S. Rome Air Development Center Rep.* NTIS AD/A - 049014, 015, 055.

Microsoft 1994. *Windows Software Development Kit. The Windows Interface: An Application Design Guide,* 2nd edition, Microsoft Press, Redmond, WA.

Moran, T.P. 1981. The command language grammar: a representation for the user interface of interactive computer systems. *Int. J. Man-Machine Stud.* 15: 3–50.

Mosier, J.N., and Smith, S.L. 1986. Application of guidelines for designing user interface software. *Behaviour Inf. Tech.* 5: 39–46.

Myers, B.A. 1989. Encapsulating interactive behaviours. In *Proc. CHI '89 Conf.,* pp. 319–324. ACM Press, New York.

Nielsen, J. 1992. Finding usability problems through heuristic evaluation. In *Proc. CHI '92 Conf.,* J. Bennett and G. Lynch, Eds., pp. 373–380. ACM Press, New York.

Nielsen, J., and Mack, R.L., Eds. 1994. *Usability Inspection Methods.* John Wiley & Sons, New York.

Norman, D.A. 1988. *The Psychology of Everyday Things.* Basic Books, New York.

Olsen, D.R. 1992. *User Interface Management Systems.* Morgan Kaufmann, New York.

OSF 1994. *OSF/Motif Style Guide.* Open System Foundation. Prentice Hall, Englewood Cliffs, NJ.

Pangalos, G.J. 1992. Standardization of the user interface. *Comput. Stand. & Interfaces* 14: 223–229.

Pfaff, G.E., Ed. 1985. *User Interface Management Systems.* Springer, Berlin.

Potosnak, K. 1988. What's wrong with standard user interfaces? *IEEE Software* 5(5): 91–92.

Reisig, W. 1985. *Petri Nets: An Introduction.* Springer, Heidelberg.

Rosson, M.B., and Carroll, J.M. 2002. *Usability Engineering: Scenario-Based Development of Human-Computer Interaction.* Morgan Kaufmann, San Francisco.

Samuelson, P. 1995. Software compatibility and the law. *Commun. ACM* 38: 15–22.

Smith, S.L. 1986. Standards versus guidelines for designing user interface software. *Behaviour Inf. Tech.* 5(1): 47–61.

Spring, M.B., Jamison, W., Fithen, K.T., Thomas, P.M, and Pavol, R.A. 1993. Models for a human-computer interaction. In *Encyclopedia of Microcomputers*, A. Kent and J.G. Williams, Eds. vol. 11, pp.189–218. Marcel Dekker, New York.

Stewart, T.F.M. 1990. SIOIS — standard interfaces or interface standards. In INTERACT '90, *Human-Computer Interaction,* D. Diaper et al., eds, pp. xxix–xxxiv. Elsevier, Amsterdam.

Stewart, T.F.M. 1992. The role of HCI standards in relation to the directive. *Displays* 13: 125–133.

SUMI 1993. User questionnaire. See Kirakowski and Corbett, 1993.

Sun 1990. *OPEN LOOK Graphical User Interface Application Style Guidelines.* Sun Microsystems. Addison-Wesley, Reading, MA.

Travis, D. 1997. Why GUIs fail. Web site of System-Concepts Ltd., http://www.system-concepts.com/articles/gui.html.

Wegener, H. 1995. The myth of the separable dialogue: software engineering vs. user models. In *Human–Computer Interaction*, K. Nordby et al. Eds., pp. 169–172. Chapman & Hall, London.

Further Information

Requests for information concerning international standards should be addressed to one of the ISO members listed here. The complete list can be obtained from the following address:

ISO Central Secretariat, 1, rue de Varembé, Case postale 56, 1211 Genève 20, Switzerland

Information about the current state of development of ISO standards can be retrieved from the Internet: http://www.iso.ch.

ISO members currently active in ISO TC 159/SC 4/WG 5 ("Ergonomics of human–system interaction") are as follows:

Canada (SCC) — Standards Council of Canada, 270 Albert Street, Suite 200, Ottawa, Ontario K1P 6N7

Denmark (DS) — Dansk Standard, Kollegievej 6, 2920 Charlottenlund

France (AFNOR) — Association Française de Normalisation, 11, avenue Francis de Pressensé, 93571 Saint-Denis La Plaine Cedex

Germany (DIN) — DIN Deutsches Institut für Normung, Burggrafenstraße 6, 10787 Berlin

Italy (UNI) — Ente Nationale Italiano di Unificazione, Via Battistotti Sassi 11/b, 20133 Milano

Japan (JISC) — Japanese Industrial Standards Committee, Ministry of International Trade and Industry, 1-3-1, Kasumigaseki, Chiyoda-ku, Tokyo 100

Netherlands (NEN) — Nederlands Normalisatie-instituut, Vlinderweg 6, 2623 AX Delft

Sweden (SIS) — Swedish Standards Institute, Sankt Paulsgatan 6, 11880 Stockholm

United Kingdom (BSI) — BSI British Standards Headquarters, 389 Chiswick High Road, London W4 4AL

United States (ANSI) — American National Standards Institute, 11 West 42nd Street, 13th floor, New York, NY 10036

More information on standards in computer science and engineering also can be found in the appendices of this Handbook.

Active national member bodies of ISO organize meetings of working groups, which mirror the previously mentioned ISO working group. The meetings are open to anyone who wants to contribute to the current ISO projects, either by proposals or by comments to committee drafts, or draft international standards.

Most standards need a period of 5 to 10 years from the first working document to the final version of the international standard. During this time, standards are commented on in national and international journals. A recognized source is the journal *Computer Standards & Interfaces*, published by Elsevier, Amsterdam.

VI

Information Management

Information Management is concerned with the collection, design, storage, organization, retrieval, and security of information in large databases. From a technology viewpoint, the emphasis is on the algorithms and structures underlying the databases. From an organizational viewpoint, the emphasis is on the relationship of information to business performance. Considerable research is also devoted to information management techniques that support the emerging global technological infrastructure. Particularly interesting here is the study of transaction processing in distributed computing environments, multimedia databases, and issues surrounding database security and privacy.

52 **Data Models**   *Avi Silberschatz, Henry F. Korth, and S. Sudarshan* **52**-1
   Introduction • The Relational Model • Object-Based Models • XML • Further Reading
53 **Tuning Database Design for High Performance**   *Dennis Shasha*
   and Philippe Bonnet **53**-1
   Introduction • Underlying Principles • Best Practices • Tuning the Application
   Interface • Monitoring Tools • Tuning Rules of Thumb • Summary and Research Results
54 **Access Methods**   *Betty Salzberg and Donghui Zhang* **54**-1
   Introduction • Underlying Principles • Best Practices • Research Issues and Summary
55 **Query Optimization**   *Yannis E. Ioannidis* **55**-1
   Introduction • Query Optimizer Architecture • Algebraic Space
   • Planner • Size-Distribution Estimator • Noncentralized Environments
   • Advanced Types of Optimization • Summary
56 **Concurrency Control and Recovery**   *Michael J. Franklin* **56**-1
   Introduction • Underlying Principles • Best Practices • Research Issues and Summary
57 **Transaction Processing**   *Alexander Thomasian* **57**-1
   Introduction • Secure Distributed Transaction Processing: Cryptography • Transaction
   Processing on the Web: Web Services • Concurrency Control for High-Contention
   Environments • Performance Analysis of Transaction Processing Systems • Conclusion
58 **Distributed and Parallel Database Systems**   *M. Tamer Özsu*
   and Patrick Valduriez **58**-1
   Introduction • Underlying Principles • Distributed and Parallel
   Database Technology • Research Issues • Summary

59 Multimedia Databases: Analysis, Modeling, Querying, and Indexing
   Vincent Oria, Ying Li, and Chitra Dorai **59**-1
   Introduction • Image Content Analysis • Video Content Analysis • Bridging the
   Semantic Gap in Content Management • Modeling and Querying Images • Modeling and
   Querying Videos • Multidimensional Indexes for Image and Video Features • Multimedia
   Query Processing • Emerging MPEG-7 as Content Description Standard • Conclusion

60 Database Security and Privacy   *Sushil Jajodia* **60**-1
   Introduction • General Security Principles • Access Controls • Assurance
   • General Privacy Principles • Relationship Between Security and Privacy Principles
   • Research Issues

52

Data Models

Avi Silberschatz
Yale University

Henry F. Korth
Lehigh University

S. Sudarshan
IIT Bombay

52.1 Introduction **52**-1
52.2 The Relational Model **52**-1
    Formal Basis • SQL • Relational Database Design
    • History
52.3 Object-Based Models................................ **52**-5
    The Entity-Relationship Model • Object-Oriented Model
    • Object-Relational Data Models
52.4 XML **52**-14
52.5 Further Reading.................................... **52**-17

52.1 Introduction

Underlying the structure of a database is the concept of a *data model*. A data model is a collection of conceptual tools for describing the real-world entities to be modeled in the database and the relationships among these entities. Data models differ in the primitives available for describing data and in the amount of semantic detail that can be expressed.

The various data models that have been proposed fall into three different groups: physical data models, record-based logical models, and object-based logical models. Physical data models are used to describe data at the lowest level. Physical data models capture aspects of database system implementation that are not covered in this article. Database system interfaces used by application programs are based on the logical data model; databases hide the underlying implementation details from applications.

This chapter focuses on logical data models, covering the relational data model, the E-R model, the object-oriented and object-relational data models, and XML.

52.2 The Relational Model

The **relational model** is currently the primary data model for commercial data-processing applications. It has attained its primary position because of its simplicity, which eases the job of the programmer, as compared to earlier data models.

A relational database consists of a collection of **tables**, each of which is assigned a unique name. An *instance* of a table storing customer information is shown in Table 52.1. The table has several rows, one for each customer, and several columns, each storing some information about the customer. The values in the *customer-id* column of the *customer* table serve to uniquely identify customers, while other columns store information such as the name, street address, and city of the customer.

The information stored in a database is broken up into multiple tables, each storing a particular kind of information. For example, information about accounts and loans at a bank would be stored in separate tables. Table 52.2 shows an instance of the *loan* table, which stores information about loans taken from the bank.

TABLE 52.1   The *Customer* Table

Customer-id	Customer-Name	Customer-Street	Customer-City
019-28-3746	Smith	North	Rye
182-73-6091	Turner	Putnam	Stamford
192-83-7465	Johnson	Alma	Palo Alto
244-66-8800	Curry	North	Rye
321-12-3123	Jones	Main	Harrison
335-57-7991	Adams	Spring	Pittsfield
336-66-9999	Lindsay	Park	Pittsfield
677-89-9011	Hayes	Main	Harrison
963-96-3963	Williams	Nassau	Princeton

TABLE 52.2   The *Loan* Table

Loan-Number	Amount
L-11	900
L-14	1500
L-15	1500
L-16	1300
L-17	1000
L-23	2000
L-93	500

TABLE 52.3   The *Borrower* Table

Customer-id	Loan-Number
019-28-3746	L-11
019-28-3746	L-23
244-66-8800	L-93
321-12-3123	L-17
335-57-7991	L-16
555-55-5555	L-14
677-89-9011	L-15
963-96-3963	L-17

In addition to information about "entities" such as customers or loans, there is also a need to store information about "relationships" between such entities. For example, the bank needs to track the relationship between customers and loans. Table 52.3 shows the *borrower* table, which stores information indicating which customers have taken which loans. If several people have jointly taken a loan, the same loan number would appear several times in the table with different customer-ids (e.g., loan number L-17). Similarly, if a particular customer has taken multiple loans, there would be several rows in the table with the customer-id of that customer (e.g., 019-28-3746), with different loan numbers.

52.2.1 Formal Basis

The power of the relational data model lies in its rigorous mathematical foundations and a simple user-level paradigm for representing data. Mathematically speaking, a **relation** is a subset of the Cartesian product of an ordered list of domains. For example, let E be the set of all employee identification numbers, D the set of all department names, and S the set of all salaries. An employment relation is a set of 3-tuples (e, d, s) where $e \in E$, $d \in D$, and $s \in S$. A tuple (e, d, s) represents the fact that employee e works in department d and earns salary s.

At the user level, a relation is represented as a table. The table has one column for each domain and one row for each tuple. Each column has a name, which serves as a column header, and is called an **attribute** of the relation. The list of attributes for a relation is called the **relation schema**. The terms "table" and "relation" are used synonymously, as are row and tuple, as also column and attribute.

Data models also permit the definition of *constraints* on the data stored in the database. For instance, *key constraints* are defined as follows. If a set of attributes L is specified to be a *super-key* for relation r, in any consistent ("legal") database, the set of attributes L would uniquely identify a tuple in r; that is, no two tuples in r can have the same values for all attributes in L. For instance, *customer-id* would form a super-key for relation *customer*. A relation can have more than one super-key, and usually one of the super-keys is chosen as a *primary key*; this key must be a minimal set, that is, dropping any attribute from the set would make it cease to be a super-key.

Another form of constraint is the *foreign key* constraint, which specifies that for each tuple in one relation, there must exist a matching tuple in another relation. For example, a foreign key constraint *from borrower referencing customer* specifies that for each tuple in *borrower*, there must be a tuple in *customer* with a matching *customer-id* value.

Users of a database system can query the data, insert new data, delete old data, or update the data in the database. Of these tasks, the task of querying the data is usually the most complicated. In the case of the relational data model, because data is stored as tables, a user can query these tables, insert new tuples, delete tuples, and update (modify) tuples. There are several languages for expressing these operations.

The tuple relational calculus and the domain relational calculus are nonprocedural languages that represent the basic power required in a relational query language. Both of these languages are based on statements written in mathematical logic. We omit details of these languages.

The relational algebra is a procedural query language that defines several operations, each of which takes one or more relations as input and returns a relation as output. For example:

- The **selection** operation is used to get a subset of tuples from a relation, by specifying a predicate. The selection operation $\sigma_P(r)$ returns the set of tuples of r that satisfy the predicate P.
- The **projection** operation $\Pi_L(r)$ is used to return a relation containing a specified set of attributes L of a relation r, removing the other attributes of r.
- The **union** operation $r \cup s$ returns the union of the tuples in r and s. The **intersection** and **difference** operations are similarly defined.
- The **natural join** operation $\bowtie$ is used to combine information from two relations. For example, the natural join of the relations *loan* and *borrower*, denoted *loan* $\bowtie$ *borrower* would be the relation defined as follows. First match each tuple in *loan* with each tuple in *borrower* that has the same values for the shared attribute *loan-number*; for each pair of matching tuples, the join operation creates a tuple containing all attributes from both tuples; the join result relation is the set of all such tuples.

  For instance, the natural join of the *loan* and *borrower* tables in Tables 52.2 and 52.3 contains tuples (L-17, 1000, 321-12-3123) and (L-17, 1000, 963-96-3963), since the tuple with loan number L-17 in the *loan* table matches two different tuples with loan number L-17 in the *borrower* table.

The relational algebra has other operations as well; for example, operations that can aggregate values from multiple tuples, for example by summing them up, or finding their average.

Because the result of a relational algebra operation is itself a relation, it can be used in further operations. As a result, complex expressions with multiple operations can be defined in the relational algebra.

Among the reasons for the success of the relational model are its basic simplicity, representing all data using just a single notion of tables, as well as its formal foundations in mathematical logic and algebra.

The relational algebra and the relational calculi are terse, formal languages that are inappropriate for casual users of a database system. Commercial database systems have, therefore, used languages with more "syntactic sugar." Queries in these languages can be translated into queries in relational algebra.

52.2.2 SQL

The SQL language has clearly established itself as *the* standard relational database language. The SQL language has a data definition component for specifying schemas, and a data manipulation component for querying data as well as for inserting, deleting, and updating data.

We illustrate some examples of queries and updates in SQL. The following query finds the name of the customer whose *customer-id* is 192-83-7465:

> **select**   *customer.customer-name*
> **from**   *customer*
> **where**   *customer.customer-id* = '192-83-7465'

Queries may involve information from more than one table. For example, the following query finds the amount of all loans owned by the customer with customer-id 019-28-3746:

> **select**   *loan.loan-number, loan.amount*
> **from**   *borrower, loan*
> **where**   *borrower.customer-id* = '019-28-3746' **and**
>   *borrower.loan-number* = *loan.loan-number*

If the above query were run on the tables shown earlier, the system would find that the loans L-11 and L-23 are owned by customer 019-28-3746, and would print out the amounts of the two loans, namely 900 and 2000.

The following SQL statement adds an interest of 5% to the loan amount of all loans with amounts greater than 1000.

> **update**   *loan*
> **set**   *amount* = *amount* * 1.05
> **where**   *amount* > 10000

Over the years, there have been several revisions of the SQL standard. The most recent is SQL:1999. QBE and Quel are two other significant query languages. Of these, Quel is no longer in widespread use, while QBE is used only in a few database systems such as Microsoft Access.

52.2.3 Relational Database Design

The process of designing a conceptual level schema for a relational database involves the selection of a set of relational schemas. There are several approaches to relational database design. One approach, which we describe in Section 52.3.1, is to create a model of the enterprise using a higher-level data model, such as the entity-relationship model, and then translate the higher-level model into a relational database design.

Another approach is to directly create a design, consisting of a set of tables and a set of attributes for each table. There are often many possible choices that the database designer might make. A proper balance must be struck among three criteria for a good design:

1. Minimization of redundant data
2. Ability to represent all relevant relationships among data items
3. Ability to test efficiently the data dependencies that require certain attributes to be unique identifiers

To illustrate these criteria for a good design, consider a database of employees, departments, and managers. Let us assume that a department has only one manager, but a manager may manage one or more departments. If we use a single relation *emp-info1(employee, department, manager)*, then we must repeat the manager of a department once for each employee. Thus we have **redundant** data.

We can avoid redundancy by *decomposing* (breaking up) the above relation into two relations *emp-mgr(employee, manager)* and *emp-dept(manager, department)*. However, consider a manager, Martin, who manages both the sales and the service departments. If Clark works for Martin, we cannot represent the fact that Clark works in the service department but not the sales department. Thus we cannot represent all relevant relationships among data items using the decomposed relations; such a decomposition is called a *lossy-join decomposition*. If instead, we chose the two relations *emp-dept*(*employee, department*) and *dept-mgr*(*department, manager*), we would avoid this difficulty, and at the same time avoid redundancy. With this decomposition, joining the information in the two relations would give back the information in *emp-info1*; such a decomposition is called a *lossless-join decomposition*.

There are several types of data dependencies. The most important of these are **functional dependencies**. A functional dependency is a constraint that the value of a tuple on one attribute or set of attributes determines its value on another. For example, the constraint that a department has only one manager could be stated as "department functionally determines manager." Because functional dependencies represent facts about the enterprise being modeled, it is important that the system check newly inserted data to ensure no functional dependency is violated (as in the case of a second manager being inserted for some department). Such checks ensure that the update does not make the information in the database inconsistent. The cost of this check depends on the design of the database.

There is a formal theory of relational database design that allows us to construct designs that have minimal redundancy, consistent with meeting the requirements of representing all relevant relationships, and allowing efficient testing of functional dependencies. This theory specifies certain properties that a schema must satisfy, based on functional dependencies. For example, a database design is said to be in a *Boyce-Codd normal form* if it satisfies a certain specified set of properties; there are alternative specifications, for instance the *third normal form*. The process of ensuring that a schema design is in a desired normal form is called **normalization**.

More details can be found in standard textbooks on databases; Ullman [Ull88], provides a detailed coverage of database design theory.

52.2.4 History

The relational model was developed in the late 1960s and early 1970s by E.F. Codd. The 1970s saw the development of several experimental database systems based on the relational model and the emergence of a formal theory to support the design of relational databases. The commercial application of relational databases began in the late 1970s but was limited by the poor performance of early relational systems. During the 1980s numerous commercial relational systems with good performance became available. Simultaneously, simple database systems based loosely on the relational approach were introduced for single-user personal computers. In the latter part of the 1980s, efforts were made to integrate collections of personal computer databases with large mainframe databases.

The relational model has since established itself as the primary data model for commercial data processing applications. Earlier generation database systems were based on the *network data model* or the *hierarchical data model*. Those two older models are tied closely to the data structures underlying the implementation of the database. We omit details of these models because they are now of historical interest only.

52.3 Object-Based Models

The relational model is the most widely used data model at the implementation level; most databases in use around the world are relational databases. However, the relational view of data is often too detailed for conceptual modeling. Data modelers need to work at a higher level of abstraction.

Object-based logical models are used in describing data at the conceptual level. The object-based models use the concepts of **entities** or **objects** and relationships among them rather than the implementation-based concepts of the record-based models. They provide flexible structuring capabilities and allow data constraints to be specified explicitly. Several object-based models are in use; some of the more widely known ones are:

- The entity-relationship model
- The object-oriented model
- The object-relational model

The entity-relationship model has gained acceptance in database design and is widely used in practice. The object-oriented model includes many of the concepts of the entity-relationship model, but represents executable code as well as data. The object-relational data model combines features of the object-oriented data model with the relational data model.

The semantic data model and the the functional data model are two other object-based data models; currently, they are not widely used.

52.3.1 The Entity-Relationship Model

The E-R data model derives from the perception of the world or, more specifically, of a particular enterprise in the world, as consisting of a set of basic objects called *entities*, and *relationships* among these objects. It facilitates database design by allowing the specification of an *enterprise schema*, which represents the overall logical structure of a database. The E-R data model is one of several semantic data models; that is, it attempts to represent the meaning of the data.

52.3.1.1 Basics

There are three basic notions that the E-R data model employs: entity sets, relationship sets, and attributes. An **entity** is a "thing" or "object" in the real world that is distinguishable from all other objects. For example, each person in the universe is an entity.

Each entity is described by a collection of features, called **attributes**. For example, the attributes *account-number* and *balance* may describe one particular account in a bank, and they form attributes of the *account* entity set. Similarly, attributes *customer-name*, *customer-street* address and *customer-city* may describe a *customer* entity.

The values for some attributes may uniquely identify an entity. For example, the attribute *customer-id* may be used to uniquely identify customers (because it may be possible to have two customers with the same name, street address, and city). A unique customer identifier must be assigned to each customer. In the United States, many enterprises use the social-security number of a person (a unique number the U.S. Government assigns to every person in the United States) as a customer identifier.

An entity may be concrete, such as a person or a book, or it may be abstract, such as a bank account, a holiday, or a concept.

An **entity set** is a set of entities of the same type that share the same properties (attributes). The set of all persons working at a bank, for example, can be defined as the entity set *employee*, and the entity John Smith may be a member of the *employee* entity set. Similarly, the entity set *account* might represent the set of all accounts in a particular bank. A database thus includes a collection of entity sets, each of which contains any number of entities of the same type.

Attributes are descriptive properties possessed by all members of an entity set. The designation of attributes expresses that the database stores similar information concerning each entity in an entity set; however, each entity has its own value for each attribute. Possible attributes of the *employee* entity set are *employee-name*, *employee-id*, and *employee-address*. Possible attributes of the *account* entity set are *account-number* and *account-balance*. For each attribute there is a set of permitted values, called the *domain* (or *value set*) of that attribute. The domain of the attribute *employee-name* might be the set of all text strings of a certain length. Similarly, the domain of attribute *account-number* might be the set of all positive integers.

Entities in an entity set are distinguished based on their attribute values. A set of attributes that suffices to distinguish all entities in an entity set is chosen, and called a **primary key** of the entity set. For the *employee* entity set, *employee-id* could serve as a primary key; the enterprise must ensure that no two people in the enterprise can have the same employee identifier.

A **relationship** is an association among several entities. Thus, an *employee* entity might be related by an *emp-dept* relationship to a *department* entity where that employee entity works. For example, there would be an *emp-dept* relationship between John Smith and the bank's credit department if John Smith worked in that department. Just as all *employee* entities are grouped into an *employee* entity set, all *emp-dept* relationship instances are grouped into a *emp-dept* **relationship set**. A relationship set may also have descriptive attributes. For example, consider a relationship set *depositor* between the *customer* and

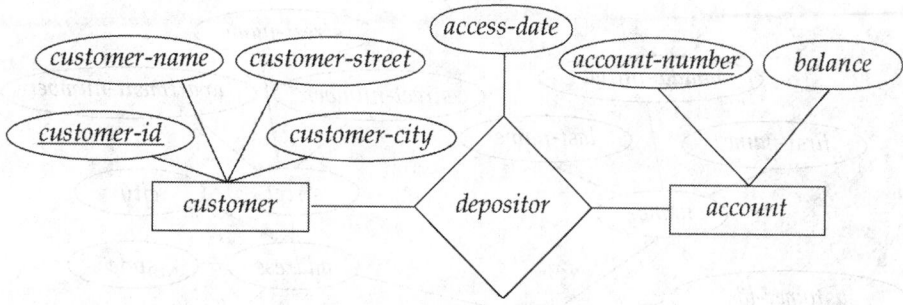

FIGURE 52.1   E-R diagram.

account entity sets. We could associate an attribute *last-access* to specify the date of the most recent access to the account. The relationship sets *emp-dept* and *depositor* are examples of a binary relationship set, that is, one that involves two entity sets. Most of the relationship sets in a database system are binary.

The overall logical structure of a database can be expressed graphically by an E-R **diagram**. Such a diagram consists of the following major components:

- **Rectangles**, which represent entity sets
- **Ellipses**, which represent attributes
- **Diamonds**, which represent relationship sets
- **Lines**, which link entity sets to relationship sets, and link attributes to both entity sets and relationship sets

An entity-relationship diagram for a portion of our simple banking example is shown in Figure 52.1. The primary key attributes (if any) of an entity set are shown underlined.

Composite attributes are attributes that can be divided into subparts (that is, other attributes). For example, an attribute *name* could be structured as a composite attribute consisting of *first-name*, *middle-initial*, and *last-name*. Using composite attributes in a design schema is a good choice if a user wishes to refer to an entire attribute on some occasions, and to only a component of the attribute on other occasions.

The attributes in our examples so far all have a single value for a particular entity. For instance, the *loan-number* attribute for a specific loan entity refers to only one loan number. Such attributes are said to be **single valued**. There may be instances where an attribute has a set of values for a specific entity. Consider an *employee* entity set with the attribute *phone-number*. An employee may have zero, one, or several phone numbers, and hence this type of attribute is said to be **multivalued**.

Suppose that the *customer* entity set has an attribute *age* that indicates the customer's age. If the *customer* entity set also has an attribute *date-of-birth*, we can calculate *age* from *date-of-birth* and the current date. Thus, *age* is a **derived attribute**. The value of a derived attribute is not stored, but is computed when required.

Figure 52.2 shows how composite, multivalued, and derived attributes can be represented in the E-R notation. Ellipses are used to represent composite attributes as well as their subparts, with lines connecting the ellipse representing the attribute to the ellipse representing its subparts. Multivalued attributes are represented using a double ellipse, while derived attributes are represented using a dashed ellipse.

Most of the relationship sets in a database system are *binary*, that is, they involve only two entity sets. Occasionally, however, relationship sets involve more than two entity sets. As an example, consider the entity sets *employee*, *branch*, and *job*. Examples of *job* entities could include manager, teller, auditor, and so on. Job entities may have the attributes *title* and *level*. The relationship set *works-on* among *employee*, *branch*, and *job* is an example of a **ternary relationship**. A ternary relationship among Jones, Perryridge, and manager indicates that Jones acts as a manager at the Perryridge branch. Jones could also act as auditor at the Downtown branch, which would be represented by another relationship. Yet another relationship could be among Smith, Downtown, and teller, indicating Smith acts as a teller at the Downtown branch.

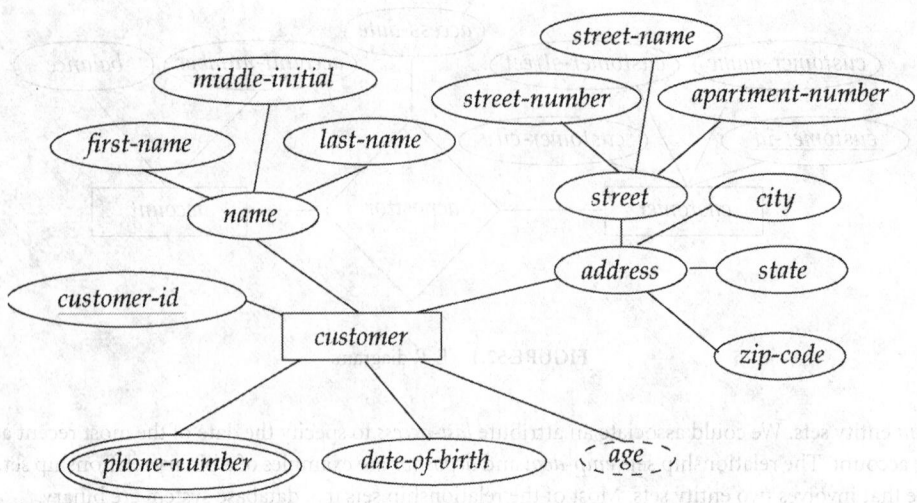

FIGURE 52.2   E-R diagram with composite, multivalued, and derived attributes.

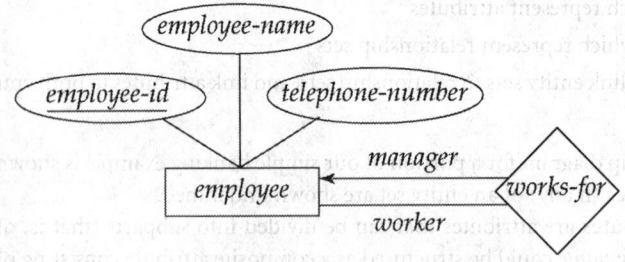

FIGURE 52.3   E-R diagram with role indicators.

Consider, for example, a relationship set *works-for* relating the entity set *employee* with itself. Each employee entity is related to the entity representing the manager of the employee. One employee takes on the role of *worker*, whereas the second takes on the role of *manager*. Roles can be depicted in E-R diagrams as shown in Figure 52.3.

Although the basic E-R concepts can model most database features, some aspects of a database may be more aptly expressed by certain extensions to the basic E-R model. Commonly used extended E-R features include specialization, generalization, higher- and lower-level entity sets, attribute inheritance, and aggregation. The notion of specialization and generalization are covered in the context of object-oriented data models in Section 52.3.2. A full explanation of the other features is beyond the scope of this chapter; we refer readers to the references listed at the end of this chapter for additional information.

52.3.1.2 Representing Data Constraints

In addition to entities and relationships, the E-R model represents certain constraints to which the contents of a database must conform. One important constraint is **mapping cardinalities**, which express the number of entities to which another entity can be associated via a relationship set. Therefore, relationships can be classified as many-to-many, many-to-one, or one-to-one. A many-to-many *works-for* relationship between *employee* and *department* exists if a department may have one or more employees and an employee may work for one or more departments. A many-to-one *works-for* relationship between *employee* and *department* exists if a department may have one or more employees but an employee must work for only department. A one-to-one *works-for* relationship exists if a department were required to have exactly one employee,

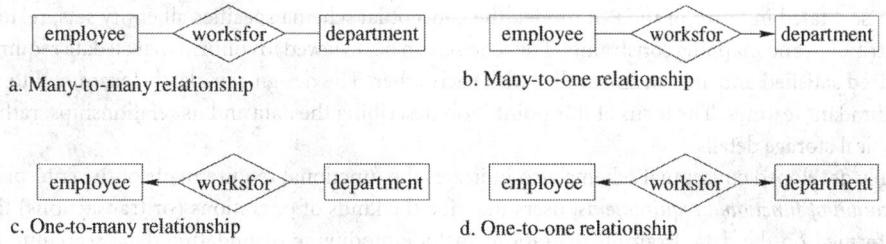

FIGURE 52.4 Relationship cardinalities.

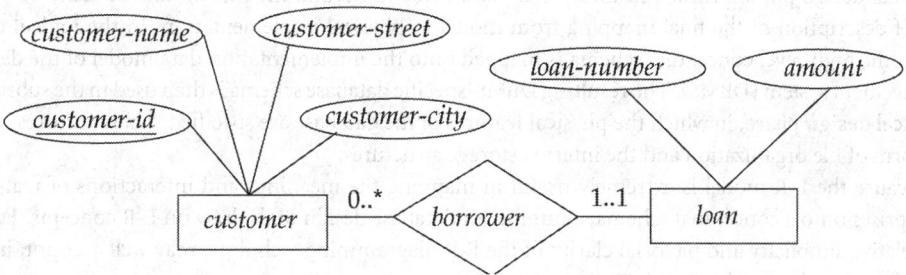

FIGURE 52.5 Cardinality limits on relationship sets.

and an employee was required to work for exactly one department. In an E-R diagram, an arrow is used to indicate the type of relationship, as shown in Figure 52.4.

E-R diagrams also provide a way to indicate more complex constraints on the number of times each entity participates in relationships in a relationship set. An edge between an entity set and a binary relationship set can have an associated minimum and maximum cardinality, shown in the form $l..h$, where l is the minimum and h the maximum cardinality. A maximum value of 1 indicates that the entity participates in at most one relationship, while a maximum value $*$ indicates no limit.

For example, consider Figure 52.5. The edge between *loan* and *borrower* has a cardinality constraint of 1..1, meaning the minimum and the maximum cardinality are both 1. That is, each loan must have exactly one associated customer. The limit 0..$*$ on the edge from *customer* to *borrower* indicates that a customer can have zero or more loans. Thus, the relationship *borrower* is one to many from *customer* to *loan*.

It is easy to misinterpret the 0..$*$ on the edge between *customer* and *borrower*, and think that the relationship *borrower* is many-to-one from *customer* to *loan* — this is exactly the reverse of the correct interpretation.

If both edges from a binary relationship have a maximum value of 1, the relationship is one-to-one. If we had specified a cardinality limit of 1..$*$ on the edge between *customer* and *borrower*, we would be saying that each customer must have at least one loan.

52.3.1.3 Use of E-R Model in Database Design

A high-level data model, such as the E-R model, serves the database designer by providing a conceptual framework in which to specify, in a systematic fashion, the data requirements of the database users and how the database will be structured to fulfill these requirements. The initial phase of database design, then, is to fully characterize the data needs of the prospective database users. The outcome of this phase will be a *specification of user requirements*. The initial specification of user requirements may be based on interviews with the database users, and the designer's own analysis of the enterprise. The description that arises from this design phase serves as the basis for specifying the logical structure of the database.

By applying the concepts of the E-R model, the user requirements are translated into a conceptual schema of the database. The schema developed at this **conceptual design** phase provides a detailed overview of the enterprise. Stated in terms of the E-R model, the conceptual schema specifies all entity sets, relationship sets, attributes, and mapping constraints. The schema can be reviewed to confirm that all data requirements are indeed satisfied and are not in conflict with each other. The design can also be examined to remove any redundant features. The focus at this point is on describing the data and its relationships, rather than on physical storage details.

A fully developed conceptual schema also indicates the functional requirements of the enterprise. In a *specification of functional requirements*, users describe the kinds of operations (or transactions) that will be performed on the data. Example operations include modifying or updating data, searching for and retrieving specific data, and deleting data. A review of the schema for meeting functional requirements can be made at the conceptual design stage.

The process of moving from a conceptual schema to the actual implementation of the database involves two final design phases. Although these final phases extend beyond the role of data models, we present a brief description of the final mapping from model to physical implementation. In the **logical design** phase, the high-level conceptual schema is mapped onto the implementation data model of the database management system (DBMS). The resulting DBMS-specific database schema is then used in the subsequent **physical design** phase, in which the physical features of the database are specified. These features include the form of file organization and the internal storage structures.

Because the E-R model is extremely useful in mapping the meanings and interactions of real-world enterprises onto a conceptual schema, a number of database design tools draw on E-R concepts. Further, the relative simplicity and pictorial clarity of the E-R diagramming technique may well account, in large part, for the widespread use of the E-R model.

52.3.1.4 Deriving a Relational Database Design from the E-R Model

A database that conforms to an E-R diagram can be represented by a collection of tables. For each entity set and each relationship set in the database, there is a unique table that is assigned the name of the corresponding entity set or relationship set. Each table has a number of columns that, again, have unique names. The conversion of database representation from an E-R diagram to a table format is the basis for deriving a relational database design.

The column headers of a table representing an entity set correspond to the attributes of the entity, and the primary key of the entity becomes the primary key of the relation. The column headers of a table representing a relationship set correspond to the primary key attributes of the participating entity sets, and the attributes of the relationship set. Rows in the table can be uniquely identified by the combined primary keys of the participating entity sets. For such a table, the primary keys of the participating entity sets are *foreign keys* of the table. The rows of the tables correspond to individual members of the entity or relationship set.

Table 52.1 through Table 52.3 show instances of tables that correspond, respectively, to the *customer* and *loan* entity sets, the *borrower* relationship set, of Figure 52.5.

52.3.2   Object-Oriented Model

The object-oriented data model is an adaptation of the object-oriented programming paradigm to database systems. The object-oriented approach to programming was first introduced by the language Simula 67, which was designed for programming simulations. More recently, the languages Smalltalk, C++, and Java have become the most widely known object-oriented programming languages. Database applications in such areas as computer-aided design and bio-informatics do not fit the set of assumptions made for older, data-processing-style applications. The object-oriented data model has been proposed to deal with some of these applications. The model is based on the concept of encapsulating data, and code that operates on that data, in an object.

52.3.2.1   Basics

Like the E-R model, the object-oriented model is based on a collection of objects. Entities, in the sense of the E-R model, are represented as **objects** with attribute values represented by *instance variables* within the object. The value stored in an instance variable may itself be an object. Objects can contain objects to an arbitrarily deep level of nesting. At the bottom of this hierarchy are objects such as integers, character strings, and other data types that are built into the object-oriented system and serve as the foundation of the object-oriented model. The set of built-in object types varies from system to system.

In addition to representing data, objects have the ability to initiate operations. An object may send a *message* to another object, causing that object to execute a *method* in response. Methods are procedures, written in a general-purpose programming language, that manipulate the object's local instance variables and send messages to other objects. Messages provide the only means by which an object can be accessed. Therefore, the internal representation of an object's data need not influence the implementation of any other object. Different objects may respond differently to the same message. This encapsulation of code and data has proven useful in developing higher modular systems. It corresponds to the programming language concept of abstract data types.

The only way in which one object can access the data of another object is by invoking a method of that other object. This is called *sending a message* to the object. Thus, the call interface of the methods of an object defines its externally visible part. The internal part of the object — the instance variables and method code — are not visible externally. The result is two levels of data abstraction.

To illustrate the concept, consider an object representing a bank account. Such an object contains instance variables *account-number* and *account-balance*, representing the account number and account balance. It contains a method *pay-interest*, which adds interest to the balance. Assume that the bank had been paying 4% interest on all accounts but now is changing its policy to pay 3% if the balance is less than $1000 or 4% if the balance is $1000 or greater. Under most data models, this would involve changing code in one or more application programs. Under the object-oriented model, the only change is made within the *pay-interest* method. The external interface to the object remains unchanged.

52.3.2.2 Classes

Objects that contain the same types of values and the same methods are grouped together into *classes*. A class may be viewed as a type definition for objects. This combination of data and code into a type definition is similar to the programming language concept of abstract data types. Thus, all *employee* objects may be grouped into an *employee* class. Classes themselves can be grouped into a hierarchy of classes; for example, the *employee* class and the *customer* classes may be grouped into a *person* class. The class *person* is a **superclass** of the *employee* and *customer* classes because all objects of the *employee* and *customer* classes also belong to the *person* class. Superclasses are also called **generalization**s. Correspondingly, the *employee* and *customer* classes are **subclasses** of *person*; subclasses are also called **specializations**.

The hierarchy of classes allows sharing of common methods. It also allows several distinct views of objects: an employee, for an example, may be viewed either in the role of person or employee, whichever is more appropriate.

52.3.2.3 The Unified Modeling Language UML

The **Unified Modeling Language** (UML) is a standard for creating specifications of various components of a software system. Some of the parts of UML are:

- **Class diagram**. Class diagrams play the same role as E-R diagrams, and are used to model data. Later in this section we illustrate a few features of class diagrams and how they relate to E-R diagrams.
- **Use case diagram**. Use case diagrams show the interaction between users and the system, in particular the steps of tasks that users perform (such as withdrawing money or registering for a course).
- **Activity diagram**. Activity diagrams depict the flow of tasks between various components of a system.
- **Implementation diagram**. Implementation diagrams show the system components and their interconnections, both at the software component level and the hardware component level.

We do not attempt to provide detailed coverage of the different parts of UML here; we only provide some examples illustrating key features of UML class diagrams. See the bibliographic notes for references on UML for more information.

UML class diagrams model objects, whereas E-R models entities. Objects are similar to entities, and have attributes, but additionally provide a set of functions (called methods) that can be invoked to compute values on the basis of attributes of the objects, or to update the object itself. Class diagrams can depict methods in addition to attributes.

We represent binary relationship sets in UML by drawing a line connecting the entity sets. We write the relationship set name adjacent to the line. We may also specify the role played by an entity set in a relationship set by writing the role name on the line adjacent to the entity set. Alternatively, we may write the relationship set name in a box, along with attributes of the relationship set, and connect the box by a dotted line to the line depicting the relationship set. This box can then be treated as an entity set, in the same way as an aggregation in E-R diagrams and can participate in relationships with other entity sets. UML 1.3 supports non-binary relationships, using the same diamond notation used in E-R diagrams.

Cardinality constraints are specified in UML in the same way as in E-R diagrams, in the form $l..h$, where l denotes the minimum and h the maximum number of relationships an entity can participate in. However, the interpretation here is that the constraint indicates the minimum/maximum number of relationships an object can participate in, *given that the other object in the relationship is fixed*. You should be aware that, as a result, the positioning of the constraints is exactly the reverse of the positioning of constraints in E-R diagrams, as shown in Figure 52.6. The constraint $0..*$ on the $E2$ side and $0..1$ on the $E1$ side means that each $E2$ entity can participate in, at most, one relationship, whereas each $E1$ entity can participate in many relationships; in other words, the relationship is many-to-one from $E2$ to $E1$.

Single values such as 1 or $*$ may be written on edges; the single value 1 on an edge is treated as equivalent to $1..1$, while $*$ is equivalent to $0..*$.

We represent generalization and specialization in UML by connecting entity sets by a line with a triangle at the end corresponding to the more general entity set. For instance, the entity set *person* is a generalization of *customer* and *employee*. UML diagrams can also represent explicitly the constraints of disjoint/overlapping on generalizations. For instance, if the *customer/employee*-to-*person* generalization is disjoint, it means that no one can be both a *customer* and an *employee*. An overlapping generalization allows a person to be both a *customer* and an *employee*. Figure 52.6 shows how to represent disjoint and overlapping generalizations of *customer* and *employee* to *person*.

52.3.2.4 Object-Oriented Database Programming Languages

There are two approaches to creating an object-oriented database language: the concepts of object orientation can be added to existing database languages, or existing object-oriented languages can be extended to deal with databases by adding concepts such as persistence and collections. Object-relational database systems take the former approach. Persistent programming languages follow the latter approach.

Persistent extensions to C++ and Java have made significant technical progress in the past decade. Several object-oriented database systems succeeded in integrating persistence fairly seamlessly and orthogonally with existing language constructs. The Object Data Management Group (ODMG) developed standards for integrating persistence support into several programming languages such as Smalltalk, C++, and Java. However, object-oriented databases based on persistent programming languages have faced significant hurdles in commercial adoption, in part because of their lack of support for legacy applications, and in part because the features provided by object-oriented databases did not make a significant different to typical data processing applications.

Object-relational database systems, which integrate object-oriented features with traditional relational support, have fared better commercially because they offer an easy upgrade path for existing applications.

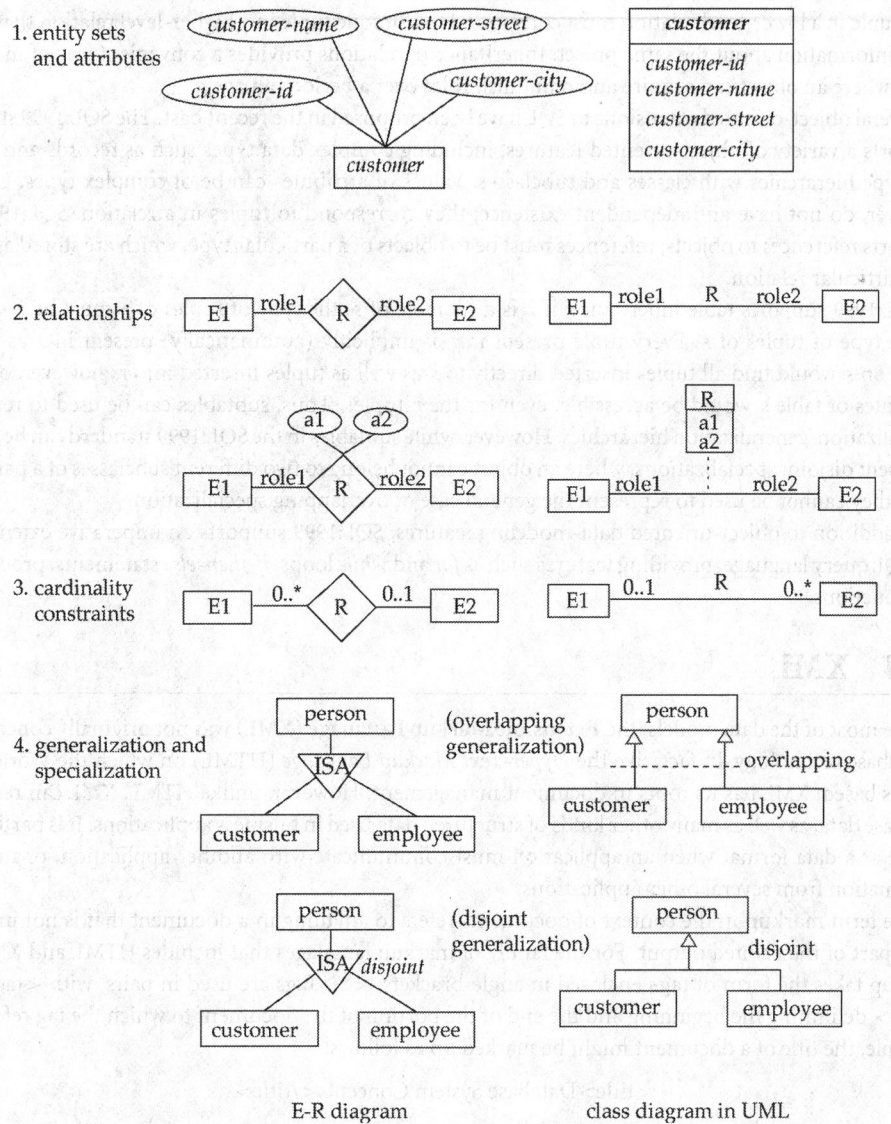

1. entity sets and attributes

2. relationships

3. cardinality constraints

4. generalization and specialization

E-R diagram   class diagram in UML

FIGURE 52.6 Correspondence of symbols used in the E-R diagram and UML class diagram notation.

52.3.3 Object-Relational Data Models

Object-relational data models extend the relational data model by providing a richer type system including object orientation. Constructs are added to relational query languages such as SQL to deal with the added data types. The extended type systems allow attributes of tuples to have complex types, including non-atomic values such as nested relations. Such extensions attempt to preserve the relational foundations, in particular the declarative access to data, while extending the modeling power.

Object-relational database systems (that is, database systems based on the object-relational model) provide a convenient migration path for users of relational databases who wish to use object-oriented features. Complex types such as nested relations are useful to model complex data in many applications. Object-relational systems combine complex data based on an extended relational model with object-oriented concepts such as object identity and inheritance. Relations are allowed to form an inheritance hierarchy;

each tuple in a lower-level relation must correspond to a unique tuple in a higher-level relation that represents information about the same object. Inheritance of relations provides a convenient way of modeling roles, where an object can acquire and relinquish roles over a period of time.

Several object-oriented extensions to SQL have been proposed in the recent past. The SQL:1999 standard supports a variety of object-oriented features, including complex data types such as records and arrays, and type hierarchies with classes and subclasses. Values of attributes can be of complex types. Objects, however, do not have an independent existence; they correspond to tuples in a relation. SQL:1999 also supports references to objects; references must be to objects of a particular type, which are stored as tuples of a particular relation.

SQL:1999 supports table inheritance; if r is a subtable of s, the type of tuples of r must be a subtype of the type of tuples of s. Every tuple present in r is implicitly (automatically) present in s as well. A query on s would find all tuples inserted directly to s as well as tuples inserted into r; however, only the attributes of table s would be accessible, even for the r tuples. Thus, subtables can be used to represent specialization/generalization hierarchies. However, while subtables in the SQL:1999 standard can be used to represent disjoint specializations, where an object cannot belong to two different subclasses of a particular class, they cannot be used to represent the general case of overlapping specialization.

In addition to object-oriented data-modeling features, SQL:1999 supports an imperative extension of the SQL query language, providing features such as *for* and *while* loops, *if-then-else* statements, procedures, and functions.

52.4 XML

Unlike most of the data models, the **Extensible Markup Language** (XML) was not originally conceived as a database technology. In fact, like the *Hyper-Text Markup Language* (HTML) on which the World Wide Web is based, XML has its roots in document management. However, unlike HTML, XML can represent database data, as well as many other kinds of structured data used in business applications. It is particularly useful as a data format when an application must communicate with another application, or integrate information from several other applications.

The term **markup** in the context of documents refers to anything in a document that is not intended to be part of the printed output. For the family of markup languages that includes HTML and XML, the markup takes the form of **tags** enclosed in angle-brackets, <>. Tags are used in pairs, with <tag> and </tag> delimiting the beginning and the end of the portion of the document to which the tag refers. For example, the title of a document might be marked up as follows:

<title>Database System Concepts</title>

Unlike HTML, XML does not prescribe the set of tags allowed, and the set may be specialized as needed. This feature is the key to XML's major role in data representation and exchange, whereas HTML is used primarily for document formatting.

For example, in our running banking application, account and customer information can be represented as part of an XML document as in Table 52.4. Observe the use of tags such as account and account-number. These tags provide context for each value and allow the semantics of the value to be identified. The contents between a start tag and its corresponding end tag is called an **element**.

Compared to storage of data in a database, the XML representation may be inefficient because tag names are repeated throughout the document. However, despite this disadvantage, an XML representation has significant advantages when it is used to exchange data, for example, as part of a message:

- The presence of the tags makes the message **self-documenting**; that is, a schema need not be consulted to understand the meaning of the text. We can readily read the fragment above, for example.
- The format of the document is not rigid. For example, if some sender adds additional information, such as a tag last-accessed noting the last date on which an account was accessed, the recipient of the XML data may simply ignore the tag. The ability to recognize and ignore unexpected tags allows the format of the data to evolve over time, without invalidating existing applications.

TABLE 52.4 XML Representation of Bank Information

```
<bank>
    <account>
        <account-number> A-101 </account-number>
        <branch-name> Downtown </branch-name>
        <balance> 500 </balance>
    </account>
    <account>
        <account-number> A-102 </account-number>
        <branch-name> Perryridge </branch-name>
        <balance> 400 </balance>
    </account>
    <account>
        <account-number> A-201 </account-number>
        <branch-name> Brighton </branch-name>
        <balance> 900 </balance>
    </account>
    <customer>
        <customer-name> Johnson </customer-name>
        <customer-street> Alma </customer-street>
        <customer-city> Palo Alto </customer-city>
    </customer>
    <customer>
        <customer-name> Hayes </customer-name>
        <customer-street> Main </customer-street>
        <customer-city> Harrison </customer-city>
    </customer>
    <depositor>
        <account-number> A-101 </account-number>
        <customer-name> Johnson </customer-name>
    </depositor>
    <depositor>
        <account-number> A-201 </account-number>
        <customer-name> Johnson </customer-name>
    </depositor>
    <depositor>
        <account-number> A-102 </account-number>
        <customer-name> Hayes </customer-name>
    </depositor>
</bank>
```

- Elements can be nested inside other elements, to any level of nesting. Table 52.5 shows a representation of the bank information from Table 52.4, but with account elements nested within customer elements.

  The nested representation permits representation of complex information within a single document. For instance, a purchase order element may have within it elements representing the supplier, customer, and each of the parts ordered. Each of these elements, in turn, may have subelements; for instance, a part element may have subelements for its part number, name, and price.

  Although the nested representation makes it easier to represent some information, it can result in redundancy; for example, the nested representation of the bank database would store an account element redundantly (multiple times) if it is owned by multiple customers.

  Nested representations are widely used in XML data interchange applications to avoid joins. For instance, a shipping application would store the full address of sender and receiver redundantly on a shipping document associated with each shipment, whereas a normalized representation may require a join of shipping records with a *company-address* relation to get address information.

- Because the XML format is widely accepted, a wide variety of tools are available to assist in its processing, including browser software and database tools.

TABLE 52.5   Nested XML Representation of Bank Information

```
<bank-1>
    <customer>
        <customer-name> Johnson </customer-name>
        <customer-street> Alma </customer-street>
        <customer-city> Palo Alto </customer-city>
        <account>
            <account-number> A-101 </account-number>
            <branch-name> Downtown </branch-name>
            <balance> 500 </balance>
        </account>
        <account>
            <account-number> A-201 </account-number>
            <branch-name> Brighton </branch-name>
            <balance> 900 </balance>
        </account>
    </customer>
    <customer>
        <customer-name> Hayes </customer-name>
        <customer-street> Main </customer-street>
        <customer-city> Harrison </customer-city>
        <account>
            <account-number> A-102 </account-number>
            <branch-name> Perryridge </branch-name>
            <balance> 400 </balance>
        </account>
    </customer>
</bank-1>
```

Just as SQL is the dominant *language* for querying relational data, XML is becoming the dominant *format* for data exchange.

In addition to elements, XML specifies the notion of an *attribute*. For example, the type of an account is represented below as an attribute named acct-type.

```
...
    <account acct-type= "checking" >
        <account-number> A-102 </account-number>
        <branch-name> Perryridge </branch-name>
        <balance> 400 </balance>
    </account>
...
```

The attributes of an element appear as *name=value* pairs before the closing ">" of a tag. Attributes are strings and do not contain markup. Furthermore, an attribute name can appear only once in a given tag, unlike subelements, which may be repeated.

Note that in a document construction context, the distinction between subelement and attribute is important — an attribute is implicitly text that does not appear in the printed or displayed document. However, in database and data exchange applications of XML, this distinction is less relevant, and the choice of representing data as an attribute or a subelement is often arbitrary.

The **document type definition** (**DTD**) is an optional part of an XML document. The main purpose of a DTD is much like that of a schema: to constrain and type the information present in the document. However, the DTD does not, in fact, constrain types in the sense of basic types like integer or string. Instead, it only constrains the appearance of subelements and attributes within an element. The DTD is primarily a list of rules for what pattern of subelements appear within an element.

For instance, the DTD for the XML data in Table 52.5 is shown below:

```
<!DOCTYPE bank [
    <!ELEMENT bank (customer*)>
    <!ELEMENT customer ( customer-name customer-street customer-city account+)>
    <!ELEMENT customer-name( #PCDATA )>
    <!ELEMENT customer-street( #PCDATA )>
    <!ELEMENT customer-city( #PCDATA )>
    <!ELEMENT account ( account-number branch-name balance )>
    <!ELEMENT account-number ( #PCDATA )>
    <!ELEMENT branch-name ( #PCDATA )>
    <!ELEMENT balance( #PCDATA )>
] >
```

The above DTD indicates that a bank may have zero or more customer subelements. Each customer element has a single occurrence of each of the subelements customer-name, customer-street, and customer-city, and one or more subelements of type account. These subelements customer-name, customer-street, and customer-city are declared to be of type #PCDATA, indicating that they are character strings with no further structure (PCDATA stands for "parsed character data"). Each account element, in turn, has a single occurrence of each of the subelements account-number, branch-name, and balance.

The following DTD illustrates a case where the nesting can be arbitrarily deep; such a situation can arise with complex parts that subparts that themselves have complex subparts, and so on.

```
<!DOCTYPE parts [
    <!ELEMENT part (name, subpartinfo*)>
    <!ELEMENT subpartinfo (part, quantity)>
    <!ELEMENT name ( #PCDATA )>
    <!ELEMENT quantity ( #PCDATA )>
] >
```

The above DTD specifies that a part element may contain within it zero or more subpart elements, each of which in turn contains a part element. DTDs such as the above, where an element type is recursively contained within an element of the same type, are called **recursive DTDs**. The **XMLSchema** language plays the same role as DTDs, but is more powerful in terms of the types and constraints it can specify.

The **XPath** and **XQuery** languages are used to query XML data. The XQuery language can be thought of as an extension of SQL to handle data with nested structure, although its syntax is different from that of SQL.

Many database systems store XML data by mapping them to relations. Unlike in the case of E-R diagram to relation mappings, the XML to relation mappings are more complex and done transparently. Users can write queries directly in terms of the XML structure, using XML query languages.

In summary, the XML language provides a flexible and self-documenting mechanism for modeling data, supporting a variety of features such as nested structures and multivalued attributes, and allowing multiple types of data to be represented in a single document. Although the basic XML model allows data to be arbitrarily structured, the schema of a document can be specified using DTDs or the XMLSchema language. Both these mechanisms allow the schema to be flexibly and partially specified, unlike the rigid schema of relational data, thus supporting **semi-structured data**.

52.5 Further Reading

- **The Relational Model.** The relational model was proposed by E.F. Codd of the IBM San Jose Research Laboratory in the late 1960s [Cod70]. Following Codd's original paper, several research projects were formed with the goal of constructing practical relational database systems, including System R at the IBM San Jose Research Laboratory (Chamberlin et al. [CAB+81]), Ingres at the

University of California at Berkeley (Stonebraker [Sto86b]), and Query-by-Example at the IBM T.J. Watson Research Center (Zloof [Zlo77]).

General discussion of the relational data model appears in most database texts, including Date [Dat00], Ullman [Ull88], Elmasri and Navathe [EN00], Ramakrishnan and Gehrke [RG02], and Silberschatz et al. [SKS02]. Textbook descriptions of the SQL-92 language include Date and Darwen [DD97] and Melton and Simon [MS93].

Textbook descriptions of the network and hierarchical models, which predated the relational model, can be found on the Web site http://www.db-book.com (this is the Web site of the text by Silberschatz et al. [SKS02])

- **The Object-Based Models.**

  - **The Entity-Relationship Model.** The entity-relationship data model was introduced by Chen [Che76]. Basic textbook discussions are offered by Elmasri and Navathe [EN00], Ramakrishnan and Gehrke [RG02], and Silberschatz et al. [SKS02]. Various data manipulation languages for the E-R model have been proposed, although none is in widespread commercial use. The concepts of generalization, specialization, and aggregation were introduced by Smith and Smith [SS77].

  - **Object-Oriented Models.** Numerous object-oriented database systems were implemented as either products or research prototypes. Some of the commercial products include ObjectStore, Ontos, Orion, and Versant. More information on these may be found in overviews of object-oriented database research, such as Kim and Lochovsky [KL89], Zdonik and Maier [ZM90], and Dogac et al. [DOBS94]. The ODMG standard is described by Cattell [Cat00].

    Descriptions of UML may be found in Booch et al. [BJR98] and Fowler and Scott [FS99].

  - **Object-Relational Models.** The nested relational model was introduced in [Mak77] and [JS82]. Design and normalization issues are discussed in [OY87, [RK87], and [MNE96]. POSTGRES ([SR86] and [Sto86a]) was an early implementation of an object-relational system. Commercial databases such as IBM DB2, Informix, and Oracle support various object-relational features of SQL:1999. Refer to the user manuals of these systems for more details.

    Melton et al. [MSG01] and Melton [Mel02] provide descriptions of SQL:1999; [Mel02] emphasizes advanced features, such as the object-relational features, of SQL:1999. Date and Darwen [DD00] describes future directions for data models and database systems.

- **XML.** The XML Cover Pages site (www.oasis-open.org/cover/) contains a wealth of XML information, including tutorial introductions to XML, standards, publications, and software. The World Wide Web Consortium (W3C) acts as the standards body for Web-related standards, including basic XML and all the XML-related languages such as XPath, XSLT, and XQuery. A large number of technical reports defining the XML related standards are available at www.w3c.org.

  A large number of books on XML are available in the market. These include [CSK01], [CRZ03], and [W+00].

Defining Terms

Attribute: 1. A descriptive feature of an entity or relationship in the entity-relationship model. 2. The name of a column header in a table, or, in relational-model terminology, the name of a domain used to define a relation.

Class: A set of objects in the object-oriented model that contains the same types of values and the same methods; also, a type definition for objects.

Data model: A collection of conceptual tools for describing the real-world entities to be modeled in the database and the relationships among these entities.

Element: The contents between a start tag and its corresponding end tag in an XML document.

Entity: A distinguishable item in the real-world enterprise being modeled by a database schema.

Foreign key: A set of attributes in a relation schema whose value identifies a unique tuple in another relational schema.

Functional dependency: A rule stating that given values for some set of attributes, the value for some other set of attributes is uniquely determined. X functionally determines Y if whenever two tuples in a relation have the same value on X, they must also have the same value on Y.

Generalization: A superclass; an entity set that contains all the members of one or more specialized entity sets.

Instance variable: attribute values within objects.

Key: 1. A set of attributes in the entity relationship model that serves as a unique identifier for entities. Also known as *superkey*. 2. A set of attributes in a relation schema that functionally determines the entire schema. 3. *Candidate key*: a minimal key. 4. *Primary key*: a candidate key chosen as the primary means of identifying/accessing an entity set, relationship set, or relation.

Message: The means by which an object invokes a method in another object.

Method: Procedures within an object that operate on the instance variables of the object and/or send messages to other objects.

Normal form: A set of desirable properties of a schema. Examples include the Boyce-Codd normal form and the third normal form.

Object: Data and behavior (methods) representing an entity.

Persistence: The ability of information to survive (persist) despite failures of all kinds, including crashes of programs, operating systems, networks, and hardware.

Relation: 1. A subset of a Cartesian product of domains. 2. Informally, a table.

Relation schema: A type definition for relations, consisting of attribute names and a specification of the corresponding domains.

Relational algebra: An algebra on relations; consists of a set of operations, each of which takes as input one or more relations and returns a relation, and a set of rules for combining operations to create expressions.

Relationship: An association among several entities.

Subclass: A class that lies below some other class (a superclass) in a class inheritance hierarchy; a class that contains a subset of the objects in a superclass.

Subtable: A table such that (a) its tuples are of a type that is a subtype of the type of tuples of another table (the supertable), and (b) each tuple in the subtable has a corresponding tuple in the supertable.

Specialization: A subclass; an entity set that contains a subset of entities of another entity set.

References

[BJR98] G. Booch, I. Jacobson, and J. Rumbaugh. *The Unified Modeling Language User Guide*. Addison-Wesley, 1998.

[CAB+81] D.D. Chamberlin, M.M. Astrahan, M.W. Blasgen, J.N. Gray, W.F. King, B.G. Lindsay, R.A. Lorie, J.W. Mehl, T.G. Price, P.G. Selinger, M. Schkolnick, D.R. Slutz, I.L. Traiger, B.W. Wade, and R.A. Yost. A history and evaluation of System R. *Communications of the ACM*, 24(10):632–646, October 1981.

[Cat00] R. Cattell, Editor. *The Object Database Standard: ODMG 3.0*. Morgan Kaufmann, 2000.

[Che76] P.P. Chen. The Entity-Relationship model: toward a unified view of data. *ACM Transactions on Database Systems*, 1(1):9–36, January 1976.

[Cod70] E.F. Codd. A relational model for large shared data banks. *Communications of the ACM*, 13(6): 377–387, June 1970.

[CRZ03] A.B. Chaudhri, A. Rashid, and R. Zicari. *XML Data Management: Native XML and XML-Enabled Database Systems*. Addison-Wesley, 2003.

[CSK01] B. Chang, M. Scardina, and S. Kiritzov. *Oracle9i XML Handbook*. McGraw-Hill, 2001.

[Dat00] C.J. Date. *An Introduction to Database Systems*. Addison-Wesley, 7th edition, 2000.

[DD97] C.J. Date and H. Darwen. *A Guide to the SQL Standard*. Addison-Wesley, 4th edition, 1997.

[DD00] C.J. Date and H. Darwen. *Foundation for Future Database Systems: The Third Manifesto*. Addison Wesley, 2nd edition, 2000.

[DOBS94] A. Dogac, M.T. Ozsu, A. Biliris, and T. Selis. *Advances in Object-Oriented Database Systems*, volume 130. Springer Verlag, 1994. Computer and Systems Sciences, NATO ASI Series F.

[EN00] R. Elmasri and S.B. Navathe. *Fundamentals of Database Systems*. Benjamin Cummings, 3rd edition, 2000.

[FS99] M. Fowler and K. Scott. *UML Distilled: A Brief Guide to the Standard Object Modeling Language*. Addison-Wesley, 2nd edition, 1999.

[JS82] G. Jaeschke and H.J. Schek. Remarks on the algebra of non first normal form relations. In *Proc. of the ACM SIGMOD Conf. on Management of Data*, pages 124–138, 1982.

[KL89] W. Kim and F. Lochovsky, Editors. *Object-Oriented Concepts, Databases, and Applications*. Addison-Wesley, 1989.

[Mak77] A. Makinouchi. A consideration of normal form on not-necessarily normalized relations in the relational data model. In *Proc. of the International Conf. on Very Large Databases*, pages 447–453, 1977.

[Mel02] J. Melton. *Advanced SQL: 1999 — Understanding Object-Relational and Other Advanced Features*. Morgan Kaufmann, 2002.

[MNE96] W.Y. Mok, Y.-K. Ng, and D.W. Embley. A normal form for precisely characterizing redundancy in nested relations. *ACM Transactions on Database Systems*, 21(1):77–106, March 1996.

[MS93] J. Melton and A.R. Simon. *Understanding the New SQL: A Complete Guide*. Morgan Kaufmann, 1993.

[MSG01] J. Melton, A.R. Simon, and J. Gray. *SQL: 1999 — Understanding Relational Language Components*. Morgan Kaufmann, 2001.

[OY87] G. Ozsoyoglu and L. Yuan. Reduced MVDs and minimal covers. *ACM Transactions on Database Systems*, 12(3):377–394, September 1987.

[RG02] R. Ramakrishnan and J. Gehrke. *Database Management Systems*. McGraw-Hill, 3rd edition, 2002.

[RK87] M.A. Roth and H.F. Korth. The design of ¬1nf relational databases into nested normal form. In *Proc. of the ACM SIGMOD Conf. on Management of Data*, pages 143–159, 1987.

[SKS02] A. Silberschatz, H.F. Korth, and S. Sudarshan. *Database System Concepts*. McGraw-Hill, 4th edition, 2002.

[SR86] M. Stonebraker and L. Rowe. The design of postgres. In *Proc. of the ACM SIGMOD Conf. on Management of Data*, 1986.

[SS77] J.M. Smith and D.C.P. Smith. Database abstractions: aggregation and generalization. *ACM Transactions on Database Systems*, 2(2):105–133, March 1977.

[Sto86a] M. Stonebraker. Inclusion of new types in relational database systems. In *Proc. of the International Conf on Data Engineering*, pages 262–269, 1986.

[Sto86b] M. Stonebraker, Editor. *The Ingres Papers*. Addison-Wesley, 1986.

[Ull88] J.D. Ullman. *Principles of Database and Knowledge-base Systems*, Volume 1. Computer Science Press, Rockville, MD, 1988.

[W+00] K. Williams (Editor) et al. *Professional XML Databases*. Wrox Press, 2000.

[Zlo77] M.M. Zloof. Query-by-example: a data base language. *IBM Systems Journal*, 16(4):324–343, 1977.

[ZM90] S. Zdonik and D. Maier. *Readings in Object-Oriented Database Systems*. Morgan Kaufmann, 1990.

53

Tuning Database Design for High Performance

53.1 Introduction 53-1
53.2 Underlying Principles 53-2
    What Databases Do • Performance Spoilers
53.3 Best Practices 53-4
    Tuning Hardware • Tuning the Operating System • Tuning
    Concurrency Control • Indexes • Tuning Table Design
53.4 Tuning the Application Interface 53-15
    Assemble Object Collections in Bulk • Cursors Cause Friction
    • The Art of Insertion
53.5 Monitoring Tools 53-18
53.6 Tuning Rules of Thumb 53-18
53.7 Summary and Research Results 53-19

Dennis Shasha
Courant Institute New York University

Philippe Bonnet
University of Copenhagen

53.1 Introduction

In fields ranging from arbitrage to tactical missile defense, speed of access to data can determine success or failure. Database tuning is the activity of making a database system run faster. Like optimization activities in other areas of computer science and engineering, database tuning must work within the constraints of its underlying technology. Just as compiler optimizers, for example, cannot directly change the underlying hardware, database tuners cannot change the underlying database management system.

The tuner can, however, modify the design of tables, select new indexes, rearrange transactions, tamper with the operating system, or buy hardware. The goals are to eliminate bottlenecks, decrease the number of accesses to disks, and guarantee response time, at least in a statistical sense.

Understanding how to do this well requires deep knowledge of the interaction among the different components of a database management system (Figure 53.1).

Further, interactions between database components and the nature of the bottlenecks change with technology. For example, inserting data in a table with a clustered index was a potential source of bottleneck using page locking; currently, all sytems support row locking, thus removing the risk of such bottlenecks. Tuning, then, is for well-informed generalists. This chapter introduces a principled foundation for tuning, focusing on principles that are likely to hold true for years to come.

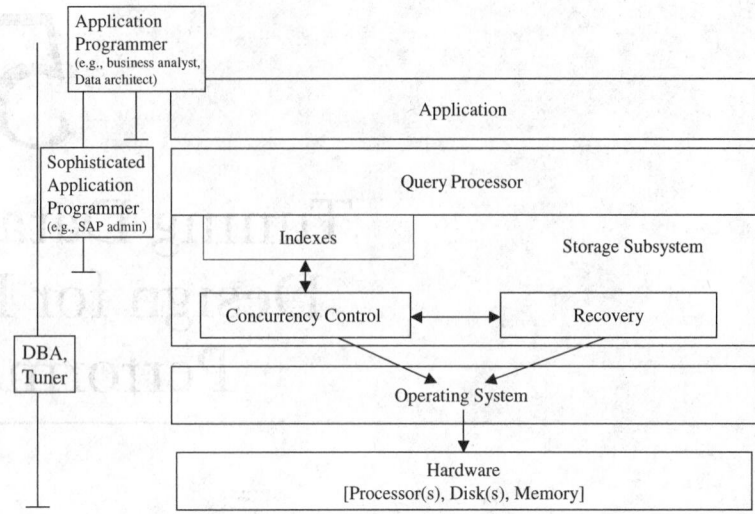

FIGURE 53.1 Database system architecture. Database tuning requires deep knowledge of the interaction among the different components and levels of a database system.

53.2   Underlying Principles

To understand the principles of tuning, you must understand the two main kinds of database applications and what affects performance.

53.2.1   What Databases Do

At a high level of abstraction, databases are used for two purposes: online transaction processing and decision support. **Online transaction processing** typically involves access to a small number of records, generally to modify them. A typical such transaction records a sale or updates a bank account. These transactions use indexes to access their few records without scanning through an entire table. **E-commerce** applications share many of these characteristics, especially the need for speed — it seems that potential e-customers will abandon a site if they have to wait more than 7 seconds for a response.

Decision support queries, by contrast, read many records often from a **data warehouse**, compute an aggregate result, and sometimes apply that aggregate back to an individual level. Typical decision support queries are "find the total sales of widgets in the last quarter in the northeast" or "calculate the available inventory per unit item." Sometimes, the results are actionable, as in "find frequent flyer passengers who have encountered substantial delays in their last few flights and send them free tickets and an apology."

53.2.2   Performance Spoilers

Having divided the database applications into two broad areas, we can now discuss what slows them down:

1. *Imprecise data searches.* These occur typically when a selection retrieves a small number of records from a large table, yet must search the entire table to find those data. Establishing an index may help in this case, although other actions, including reorganizing the table, may also have an effect (see Figure 53.2).
2. *Random vs. sequential disk accesses.* Sequential disk bandwidth is between one and two orders of magnitude larger than random-access disk bandwidth. In 2002, for mid-range disks, sequential

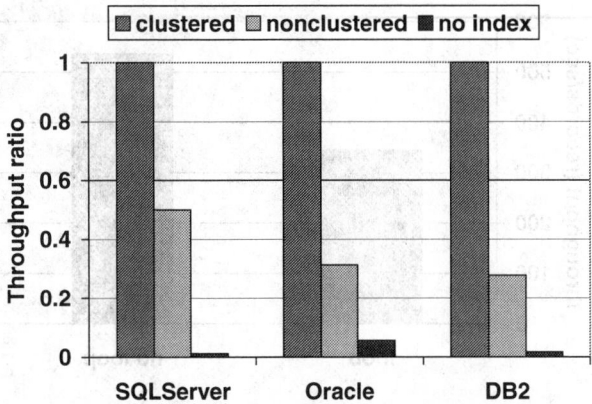

FIGURE 53.2   Benefits of clustering index. In this graph, each query returns 100 records out of the 1,000,000 that the table contains. For such a query, a clustering index is twice as fast as a non-clustering index and orders of magnitude faster than a full table scan when no index is used; clustering and non-clustering indexes are defined below. These experiments were performed on DB2 UDB V7.1, Oracle8i, and SQL Server 7 on Windows 2000.

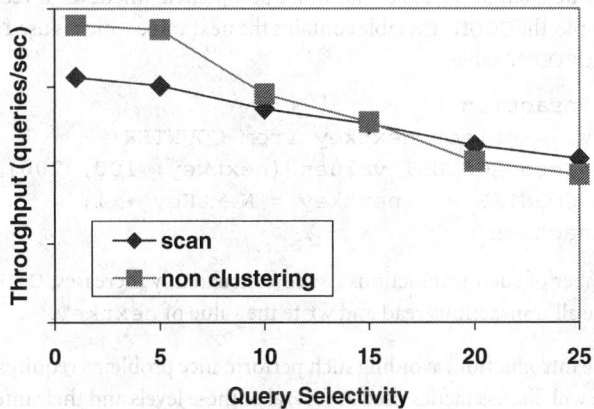

FIGURE 53.3   Index may hurt performances. We submit range queries selecting a variable portion of the underlying table and measure the performance using an index or a scan. We observe that the non-clustering index is better when the percentage of selected records is below a certain threshold. Above this threshold, a scan performs better because it is faster to sequentially access all records than to randomly access a relatively large portion of them (15% in this experiment). This experiment was performed using DB2 UDB V7.1 on Windows 2000.

> bandwidth was about 20 Mb/sec while random bandwidth was about 200 KB/sec. (The variation depends on technology and on tunable parameters, such as the degree of prefetching and size of pages.) Non-clustered index accesses tend to be random, whereas scans are sequential. Thus, removing an index may sometimes improve performance, because either the index is never used for reading (and therefore constitutes only a burden for updates) or the index is used for reading and behaves poorly (see Figure 53.3).
>
> 3. *Many short data interactions, either over a network or to the database.* This may occur, for example, if an object-oriented application views records as objects and assembles a collection of objects by accessing a database repeatedly from within a "for" loop rather than as a bulk retrieval (see Figure 53.4).

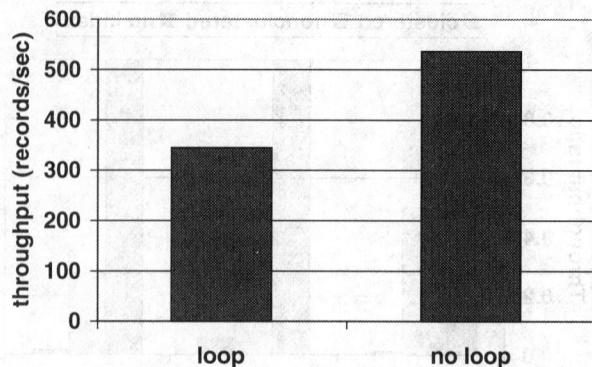

FIGURE 53.4 Loop constructs. This graph compares two programs that obtain 2000 records from a large table (line item from TPC-H). The *loop* program submits 200 queries to obtain this data, while the *no loop* program submits only one query and thus enjoys much better performance.

4. *Delays due to lock conflicts.* These occur either when update transactions execute too long or when several transactions want to access the same datum, but are delayed because of locks. A typical example might be a single variable that must be updated whenever a record is inserted. In the following example, the COUNTER table contains the next value which is used as a key when inserting values in the ACCOUNT table.

```
begin transaction
   NextKey := select nextkey from COUNTER;
   insert into ACCOUNT values (nextkey, 100, 200);
   update COUNTER set nextkey = NextKey + 1;
end transaction
```

When the number of such transactions issued concurrently increases, COUNTER becomes a bottleneck because all transactions read and write the value of nextkey.

As mentioned in the introduction, avoiding such performance problems requires changes at all levels of a database system. We will discuss tactics used at several of these levels and their interactions — hardware, concurrency control subsystem, indexes, and conceptual level. There are other levels, such as recovery and query rewriting, that we mostly defer to reference [4].

53.3 Best Practices

Understanding how to tune each level of a database system (see Figure 53.1) requires understanding the factors leading to good performance at that level. Each of the following subsections discusses these factors before discussing tuning tactics.

53.3.1 Tuning Hardware

Each processing unit consists of one or more processors, one or more disks, and some memory. Assuming a 1-GIPS (billion instructions per second) processor, disks will be the bottleneck for online transaction processing applications until the processor is attached to around 10 disks (counting 500,000 instructions per random I/O issued by the database system and 200 random I/O per second). Each transaction spends far more time waiting for head movement on disk than in the processor.

Decision support queries, by contrast, often entail massive scans of a table. In theory, a 1-GIPS processor is saturated when connected to three disks (counting 500,000 instructions per sequential I/O and 5000 I/O per second, considering 20 MB/sec and 64 KB per I/O). In practice, around three disks overflow the PCI

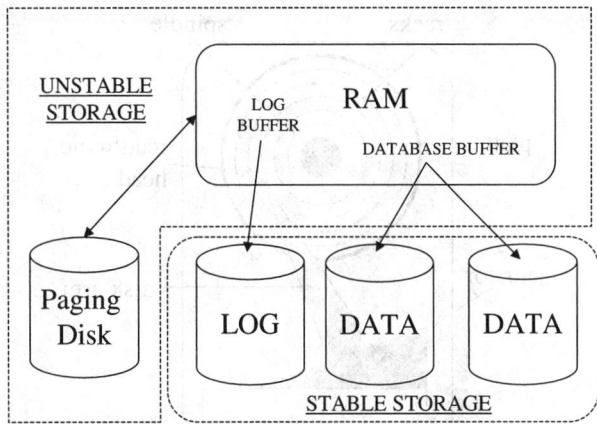

FIGURE 53.5 Buffer organization. The database buffer is located in virtual memory (i.e., RAM and paging file). Its greater part should be in RAM. It is best to have the paging file on its own disk.

bus of the server they are connected to. Thus, decision support sites may need fewer disks per processor than transaction processing sites for the purposes of matching aggregate disk bandwidth to processor speed.*

Solid-state random access memory (RAM) obviates the need to go to disk. Database systems reserve a portion of RAM as a *buffer*, whose logical role is illustrated in Figure 53.5. In all applications, the buffer usually holds frequently accessed pages (*hot* pages, in database parlance), including the first few levels of indexes. Increasing the amount of RAM buffer tends to be particularly helpful in online transaction applications where disks are the bottleneck.

The read **hit ratio** in a database is the portion of database reads that are satisfied by the buffer. Hit ratios of 90% or higher are common in online transaction applications but less common in decision support applications. Even in transaction processing applications, hit ratios tend to level off as you increase the buffer size if there are one or more tables that are accessed unpredictably and are much larger than available RAM (e.g., sales records for a large department store).

53.3.2 Tuning the Operating System

The operating system, in combination with the lower levels of the database system, determines such features as the layout of files on disk as well as the assignment and safe use of transaction priorities.

53.3.2.1 File Layout

File layout is important because of the moving parts on mechanical (as opposed to solid state) disks (Figure 53.6). Such a disk consists of a set of platters, each of which resembles a CD-ROM. A platter holds a set of tracks, each of which is a concentric circle. The platters are held together on a spindle, so that track *i* of one platter is in the same *cylinder* as track *i* of all other platters.

Accessing (reading or writing) a page on disk requires (1) moving the disk head over the proper track, say track *t*, an operation called **seeking** (the heads for all tracks move together, so all heads will be at cylinder *t* when the seek is done); (2) waiting for the appropriate page to appear under the head, a time

---

*This point requires a bit more explanation. There are two reasons you might need more disks: (1) for disk bandwidth (the number of bytes coming from the disk per second), or (2) for space. Disk bandwidth is usually the issue in online transaction processing. Airline reservations systems, for example, often run their disks at less than 50% utilization. Decision support applications tend to run into the space issue more frequently, because scanning allows disks to deliver their optimal bandwidth.

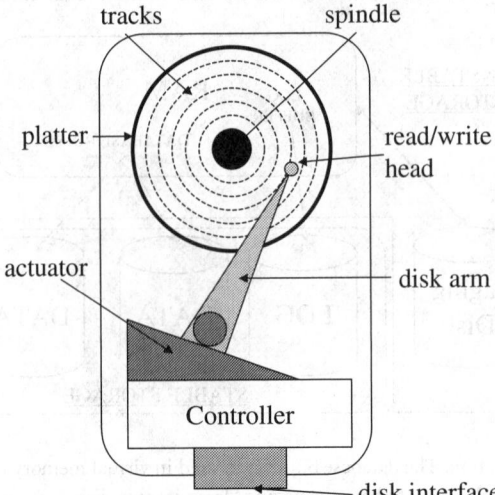

FIGURE 53.6   Disk organization. A disk is a collection of circular platters placed one on top of the other and rotating around a common axis (called a spindle). The concentric dashed circles are called tracks.

period called rotational delay; and (3) accessing the data. Mechanical disk technology implies that seek time > rotational delay > access time.

As noted above, if you could eliminate the overhead caused by seeks and rotational delay, the aggregate bandwidth could increase by a factor of 10 to 100. Making this possible requires laying out the data to be read sequentially along tracks.*

Recognizing the advantage of sequential reads on properly laid-out data, most database systems encourage administrators to lay out tables in relatively large *extents* (consecutive portions of disk). Having a few large extents is a good idea for tables that are scanned frequently or (like database recovery logs or history files) are written sequentially. Large extents, then, are a necessary condition for good performance, but not sufficient, particularly for history files. Consider, for example, the scenario in which a database log is laid out on a disk in a few large extents, but another hot table is also on that disk. The accesses to the hot table may entail a seek from the last page of the log; the next access to the log will entail another seek. So, much of the gain of large extents will be lost. For this reason, each log or history file should be the only hot file on its disk, unless the disk makes use of a large RAM cache to buffer the updates to each history file (Figure 53.7).

When accesses to a file are entirely random (as is the case in online transaction processing), seeks cannot be avoided. But placement can still minimize their cost, because seek time is roughly proportional to a constant plus the square root of the seek distance.

53.3.3   Tuning Concurrency Control

As the chapter on Concurrency Control and Recovery in this handbook explains, database systems attempt to give users the illusion that each transaction executes in isolation from all others. The ANSI SQL standard, for example, makes this explicit with its concept of degrees of isolation [3, 5]. Full isolation or **serializability** is the guarantee that each transaction that completes will appear to execute one at a time *except that its*

---

*Informed readers will realize that the physical layout of sequential data on tracks is not always contiguous — whether it is or not depends on the relative speed ratios of the controller and the disk. The net effect is that there is a layout that eliminates rotational and seek time delay for table scans.

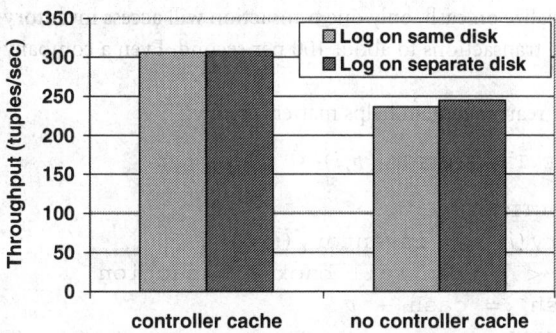

FIGURE 53.7 Impact of the controller cache. For this experiment, we use the line item table from the TPC-H benchmark and we issue 300,000 insert or update statements. This experiment was performed with Oracle8i on a Windows server with a RAID controller. This graph shows that the controller cache hides the performance penalty due to disk seeks.

performance may be affected by other transactions. This ensures, for example, that in an accounting database in which every update (e.g., sale, purchase, etc.) is recorded as a double-entry transaction, any transaction that sums assets, liabilities, and owners' equity will find that assets equal the sum of the other two. There are less stringent notions of isolation that are appropriate when users do not require such a high degree of consistency.

The concurrency-control algorithm in predominant use is two-phase locking, sometimes with optimizations for data structures. Two-phase locking has **read** (or *shared*) and **write** (or *exclusive*) locks. Two transactions may both hold a shared lock on a datum. If one transaction holds an exclusive lock on a datum, however, then no other transaction can hold any lock on that datum; in this case, the two transactions are said to **conflict**. The notion of datum (the basic unit of locking) is deliberately left unspecified in the theory of concurrency control because the same algorithmic principles apply regardless of the size of the datum, whether a page, a record, or a table. The performance may differ, however. For example, record-level locking works much better than page-level locking for online transaction processing applications.

53.3.3.1 Rearranging Transactions

Tuning concurrency control entails trying to reduce the number and duration of conflicts. This often entails understanding application semantics. Consider, for example, the following code for a purchase application of item i for price p for a company in bankruptcy (for which the cash cannot go below 0):

```
PURCHASE TRANSACTION (p,i)

    1    BEGIN TRANSACTION
    2        if cash < p then roll back transaction
    3        inventory(i) := inventory(i) + p
    4        cash := cash − p
    5    END TRANSACTION
```

From a concurrency-control-theoretical point of view, this code does the right thing. For example, if the cash remaining is 100, and purchase P1 is for item i with price 50, and purchase P2 is for item j with price 75, then one of these will roll back.

From the point of view of performance, however, this transaction design is very poor, because every transaction must acquire an exclusive lock on cash from the beginning to avoid deadlock. (Otherwise, many transactions will obtain shared locks on cash and none will be able to obtain an exclusive lock on cash.) That will make cash a bottleneck and have the effect of serializing the purchases. Because inventory is apt to be large, accessing inventory(i) will take at least one disk access, requiring about 10 ms. Because the transactions will serialize on cash, only one transaction will access inventory at a time. This will limit the number of purchase transactions to about 100 per second. Even a company in bankruptcy may find this rate unacceptable.

A surprisingly simple rearrangement helps matters greatly:

REDESIGNED PURCHASE TRANSACTION (p, i)

```
1    BEGIN TRANSACTION
2        inventory(i) := inventory(i) + p
3        if cash < p then roll back transaction
4        else cash := cash − p
5    END TRANSACTION
```

Cash is still a hot spot, but now each transaction will avoid holding cash while accessing inventory. Because cash is so hot, it will be in the RAM buffer. The lock on cash can be released as soon as the commit occurs.

Other techniques are available that "chop" transactions into independent pieces to shorten lock times further, but they are quite technical. We refer interested readers to [4].

53.3.3.2 Living Dangerously

Many applications live with less than full isolation due to the high cost of holding locks during user interactions. Consider the following full-isolation transaction from an airline reservation application:

AIRLINE RESERVATION TRANSACTION (p, i)

```
1    BEGIN TRANSACTION
2        Retrieve list of seats available.
3        Reservation agent talks with customer regarding availability.
4        Secure seat.
5    END TRANSACTION
```

The performance of a system built from such transactions would be intolerably slow because each customer would hold a lock on all available seats for a flight while chatting with the reservations agent. This solution does, however, guarantee two conditions: (1) no two customers will be given the same seat, and (2) any seat that the reservation agent identifies as available in view of the retrieval of seats will still be available when the customer asks to secure it.

Because of the poor performance, however, the following is done instead:

LOOSELY CONSISTENT AIRLINE RESERVATION TRANSACTION (p, i)

```
1    Retrieve list of seats available.
2    Reservation agent talks with customer regarding availability.
3    BEGIN TRANSACTION
4    Secure seat.
5    END TRANSACTION
```

This design relegates lock conflicts to the secure step, thus guaranteeing that no two customers will be given the same seat. It does allow the possibility, however, that a customer will be told that a seat is available, will ask to secure it, and will then find out that it is gone. This has actually happened to a particularly garrulous colleague of ours.

53.3.4 Indexes

Access methods, also known as indexes, are discussed in another chapter. Here we review the basics, then discuss tuning considerations. An **index** is a data structure plus a method of arranging the data tuples in the table (or other kind of collection object) being indexed. Let's discuss the data structure first.

53.3.4.1 Data Structures

Two data structures are most often used in practice: B-trees and Hash structures. Of the two, B-trees are used most often (one vendor's tuning book puts it this way: "When in doubt, use a B-tree"). Here, we review those concepts about B-trees most relevant to tuning.

A **B-tree** (strictly speaking, a B+ tree) is a balanced tree whose nodes contain a sequence of key–pointer pairs [2]. The keys are sorted by value. The pointers at the leaves point to the tuples in the indexed table.

B-trees are self-reorganizing through operations known as splits and merges (although occasional reorganizations for the purpose of reducing the number of seeks do take place). Further, they support many different query types well: equality queries (find the employee record of the person having a specific social security number), min–max queries (find the highest-paid employee in the company), and range queries (find all salaries between $70,000 and $80,000).

Because an access to disk secondary memory costs about 5 ms if it requires a seek (as index accesses will), the performance of a B-tree depends critically on the number of nodes in the average path from root to leaf. (The root will tend to be in RAM, but the other levels may or not be, and the farther down the tree the search goes, the less likely they are to be in RAM.) The number of nodes in the path is known as the number of levels. One technique that database management systems use to minimize the number of levels is to make each interior node have as many children as possible (1000 or more for many B-tree implementations). The maximum number of children a node can have is called its *fan-out*. Because a B-tree node consists of key–pointer pairs, the bigger the key is, the smaller the fan-out.

For example, a B-tree with a million records and a fan-out of 1000 requires three levels (including the level where the records are kept). A B-tree with a million records and a fan-out of 10 requires seven levels. If we increase the number of records to a billion, the numbers of levels increase to four and ten, respectively. This is why accessing data through indexes on large keys is slower than accessing data through small keys on most systems (the exceptions are those few systems that have good compression).

Hash structures, by contrast, are a method of storing key–value pairs based on a pseudorandomizing function called a *hash function*. The hash function can be thought of as the root of the structure. Given a key, the hash function returns a location that contains either a page address (usually on disk) or a directory location that holds a set of page addresses. That page either contains the key and associated record or is the first page of a linked list of pages, known as an *overflow chain* leading to the record(s) containing the key. (You can keep overflow chaining to a minimum by using only half the available space in a hash setting.)

In the absence of overflow chains, hash structures can answer equality queries (e.g., find the employee with Social Security number 156-87-9864) in one disk access, making them the best data structures for that purpose. The hash function will return arbitrarily different locations on key values that are close but unequal (e.g., Smith and Smythe). As a result, records containing such close keys will likely be on different pages. This explains why hash structures are completely unhelpful for range and min–max queries.

53.3.4.2 Clustering and Sparse Indexes

The data structure portion of an index has pointers at its leaves to either data pages or data records, as shown in Figure 53.8.

- If there is at most one pointer from the data structure to each data page, then the index is said to be **sparse**.
- If there is one pointer to each record in the table, then the index is said to be **dense**.

If records are small compared to pages, then there will be many records per data page and the data structure supporting a sparse index will usually have one less level than the data structure supporting a dense index. This means one less disk access if the table is large. By contrast, if records are almost as large as pages, then a sparse index will rarely have better disk access properties than a dense index.

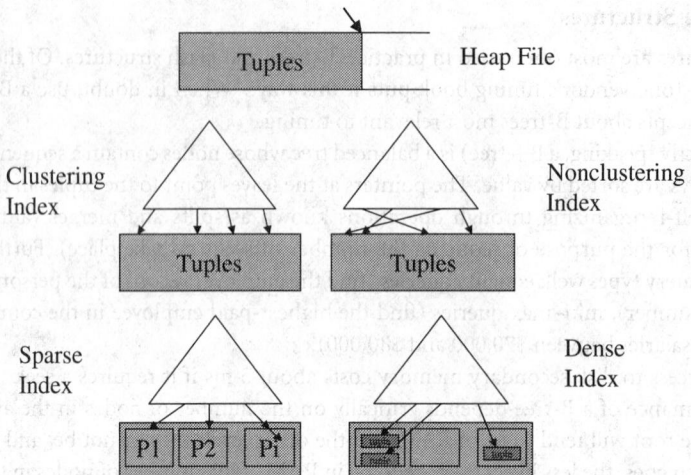

FIGURE 53.8 Data organization. This diagram represent various data organization: a heap file (records are always inserted at the end of the data structure), a clustering index (records are placed on disk according to the leaf node that points to them), a nonclustering index (records are placed on disk independently of the index structure), a sparse index (leaf-nodes point to pages), and a dense index (leaf nodes point to records). Note that a nonclustering index must be dense, while a clustering index might be sparse or dense.

The main virtue of dense indexes is that they can support certain read queries within the data structure itself, in which case they are said to **cover** the query. For example, if there is a dense index on the keywords of a document retrieval system, a query can count the records containing some term (e.g., "derivatives scandals,") without accessing the records themselves. (Count information is useful for that application because queriers frequently reformulate a query when they discover that it would retrieve too many documents.) A secondary virtue is that a query that makes use of several dense indexes can identify all relevant tuples before accessing the data records; instead, one can just form intersections and unions of pointers to data records.

A **clustering index** on an attribute (or set of attributes) X is an index that puts records close to one another if their X-values are *near* one another. What "near" means depends on the data structure. On B-trees, two X-values are near if they are close in their sort order. For example, 50 and 51 are near, as are Smith and Sneed. In hash structures, two X-values are near only if they are identical.

Sparse indexes must be clustering, but clustering indexes need not be sparse. In fact, clustering indexes are sparse in some systems (e.g., SQL Server, ORACLE hash structures) and dense in others (e.g., ORACLE B-trees, DB2). Because a clustering index implies a certain table organization and the table can be organized in only one way at a time, there can be at most one clustering index per table.

A **nonclustering index** (sometimes called a *secondary* index) is an index on an attribute (or set of attributes) Y that puts no constraint on the table organization. The table can be clustered according to some other attribute X or can be organized as a heap, as we discuss below. A nonclustering index must be dense, so there is one leaf pointer per record. There can be many nonclustering indexes per table. Throughput trade-offs among these various indexing strategies are illustrated in Figure 53.9.

A **heap** is the simplest table organization of all. Records are ordered according to their time of entry (Figure 53.8). That is, new insertions are added to the last page of the data structure. For this reason, inserting a record requires a single page access.

Nonclustering indexes are useful if each query retrieves significantly fewer records than there are pages in the file. We use the word "significant" for the following reason: a table scan can often save time by reading many pages at a time, provided the table is stored on contiguous tracks. Therefore, even if the scan and the index both read all the pages of the table, the scan may complete more than 10 times faster than if it reads one page at a time (see Figure 53.3).

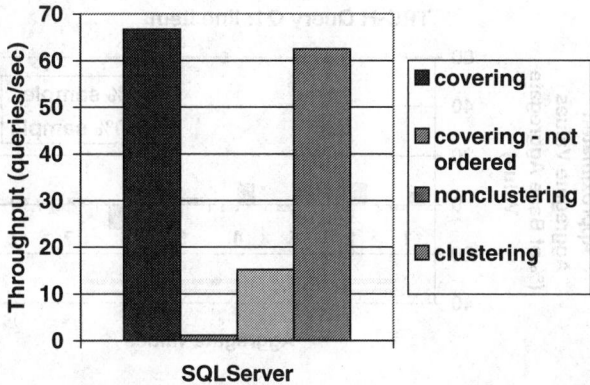

FIGURE 53.9   Covering index. This experiment illustrates that a covering index can be as good as or even better than a clustering index as long as (1) the query submitted is a prefix match query (a *prefix match query* on an attribute or sequence of attributes X is one that specifies only a prefix of X); and (2) the order of the attributes in the prefix match query matches the order in which the attributes have been declared in the index. If this is not the case, then the composite index does not avoid a full table scan on the underlying relation. A covering index is also significantly faster than a nonclustering index that is not covering because it avoids access to the table records. This experiment was performed with SQL Server 7 on Windows 2000 (i.e., the clustering index is sparse).

In summary, nonclustering indexes work best if they cover the query. Otherwise, they work well if the average query using the index will access far fewer records than there are data pages. Large records and high selectivity both contribute to the usefulness of nonclustering indexes.

53.3.4.3   Data Structures for Decision Support

Decision support applications often entail querying on several, perhaps individually unselective, attributes. For example, "Find people in a certain income range who are female, live in California, buy boating equipment, and work in the computer industry." Each of these constraints is unselective in itself, but together form a small result. The best all-around data structure for such a situation is the bitmap. A bitmap is a collection of vectors of bits. The length of each vector equals the length of the table being indexed and has a 1 in position i if the ith record of the table has some property. For example, a bitmap on state would consist of 50 vectors, one for each state. The vector for California would have a 1 for record i if record i pertains to a person from California. In our experiments, bitmaps outperform multidimensional indexes by a substantial margin.

Some decision support queries compute an aggregate, but never apply the result of the aggregate back to individuals. For example, you might want to find the approximate number of California women having the above properties. In that case, you can use approximate summary tables as a kind of indexing technique. The Aqua system [1], for example, proposes an approximation based on constructing a database from a random sample of the most detailed table T (sometimes known as the *fact table* in data warehouse parlance) and then joining that result with the reference tables R1, R2, . . . , Rn based on foreign key joins. In the TPC-H setting, for example, the fact table is lineitem. With even a 10% sample, a seven-fold improvement in performance is obtained at a decent accuracy. (See Figure 53.10 through Figure 53.12).

53.3.4.4   Final Remarks Concerning Indexes

The main point to remember is that the use of indexes is a two-edged sword: we have seen an index reduce the time to execute a query from hours to a few seconds in one application, yet increase batch load time by a factor of 80 in another application. Add them with care.

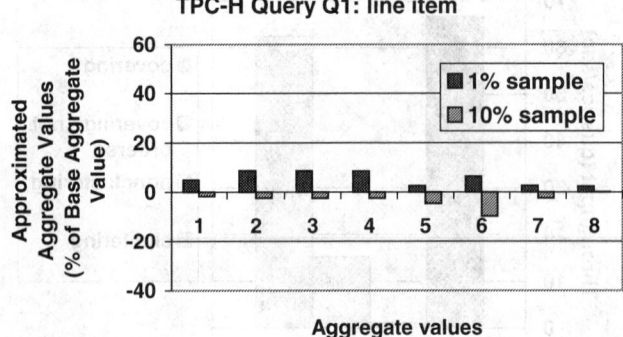

FIGURE 53.10   Approximation on one relation. We sample 1% and 10% of the line item table by selecting the top N records on an attribute which is not related to the attributes involved in the subsequent queries (here l_linenumber). That is, we are taking an approximation of a random sample. We compare the results of a query (Q1 in TPC-H) that accesses only records in the line item relation. The graph shows the difference between the aggregated values obtained using the base relations and our two samples. There are 8 aggregate values projected out in the select clause of this query. Using the 1% sample, the difference between the aggregated value obtained using base relations and sample relations is never greater than 9%; using a 10% sample, this difference falls to around 2% in all cases but one.

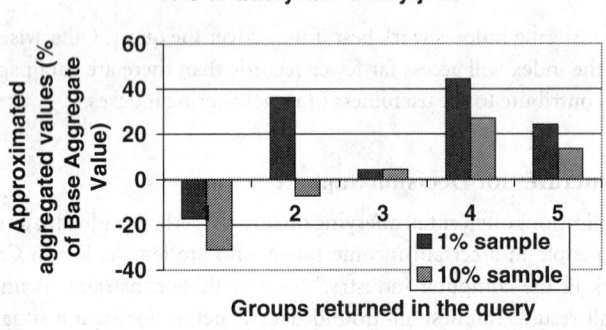

FIGURE 53.11   Approximation on a 6-way join. As indicated in this section, we take a sample of the line item table and join from there on foreign keys to obtain samples for all tables in the TPC-H schema. We run query Q5, which is a 6-way join. The graph shows the error for the five groups obtained with this query (only one aggregated value is projected out). For one group, using a 1% sample (of line item and using the foreign key dependencies to obtain samples on the other tables), we obtain an aggregated value which is 40% off the aggregated value we obtained using the base relations; and using a 10% sample, we obtain a 25% difference. As a consequence of this error, the groups are not ordered the same way using base relations and approximate relations.

53.3.5   Tuning Table Design

Table design is the activity of deciding which attributes should appear in which tables in a relational system. The *Conceptual Database Design* chapter discusses this issue, emphasizing the desirability of arriving at a **normalized** schema. Performance considerations sometimes suggest choosing a nonnormalized schema, however (see Figure 53.13). More commonly, performance considerations may suggest choosing one normalized schema over another or they may even suggest the use of redundant tables.

53.3.5.1   To Normalize or Not to Normalize

Consider the normalized schema consisting of two tables: *Sale*(sale_id, customer_id, product, quantity) and *Customer*(customer_id, customer_location).

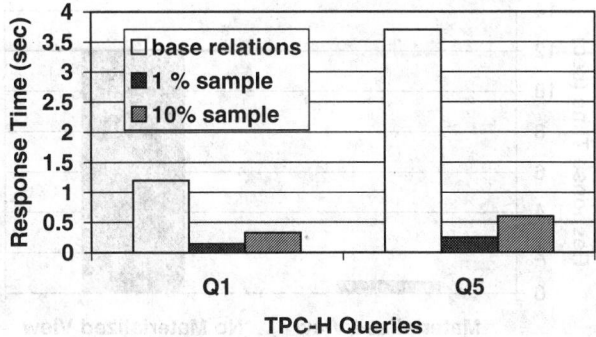

FIGURE 53.12 Response time benefits of approximate results. The benefit of using approximated relations that are much smaller than the base relations is, naturally, significant.

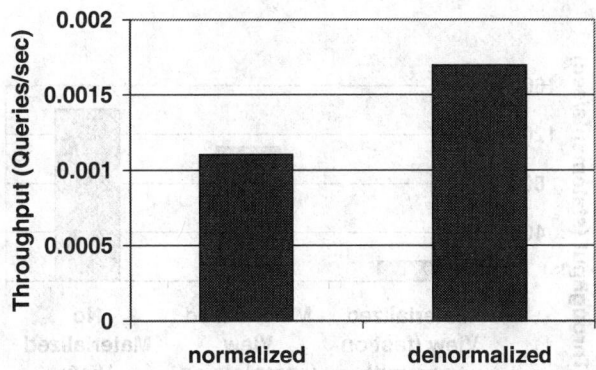

FIGURE 53.13 Denormalization. We use the TPC-H schema to illustrate the potential benefits of denormalization. This graph shows the performance of a query that finds all line items whose supplier is in Europe. With the normalized schema, this query requires a 4-way join between line item, supplier, nation, and region. If we denormalize line item and introduce the name of the region each item comes from, then the query is a simple selection on line item. In this case, denormalization provides a 30% improvement in throughput. This graph was obtained with Oracle 8i Enterprise Edition running on Windows 2000.

If we frequently want sales per customer location or sales per product per customer location, then this table design requires a join on customer_id for each of these queries. A denormalized alternative is to add customer_location to *Sale*, yielding *Sale*(sale_id, customer_id, product, quantity, customer_location) and *Customer*(customer_id, customer_location). In this alternative, we still would need the *Customer* table to avoid anomalies such as the inability to store the location of a customer who has not yet bought anything.

Comparing these two schemas, we see that the denormalized schema requires more space and more work on insertion of a sale. Typically, the data-entry operator would type in the customer_id, product, and quantity; the system would generate a sale_id and do a join on customer_id to get customer_location. On the other hand, the denormalized schema is much better for finding the products sold at a particular customer location.

The trade-off of space plus insertion cost vs. improved speeds for certain queries is the characteristic one in deciding when to use a denormalized schema. Good practice suggests starting with a normalized schema and then denormalizing sparingly.

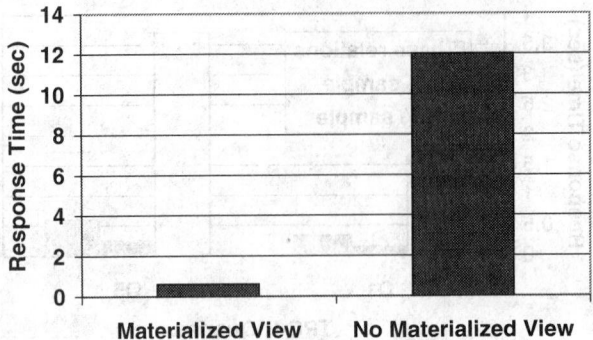

FIGURE 53.14   Aggregate maintenance with materialized views (queries). We implement redundant tables using materialized views in Oracle9i on Linux. Materialized views are transparently maintained by the system to reflect modifications on the base relations. The use of these materialized views is transparent when processing queries; the optimizer rewrites the aggregate queries to use materialized views if appropriate. The speed-up for queries is two orders of magnitude.

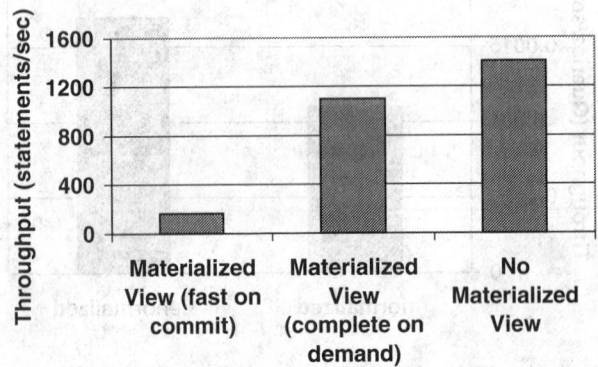

FIGURE 53.15   Aggregate maintenance with materialized views (insertions). There are two main parameters for the maintenance of materialized views as insertions/updates/deletions are performed on the base relations: (1) the materialized view can be updated in the transaction that performs the insertions (ON COMMIT) or it can be updated offline after all transactions are performed (ON DEMAND); (2) the materialized view is recomputed completely (COMPLETE) or only incrementally, depending on the modifications of the base tables (FAST). The graph shows the throughput when inserting 100,000 records in the orders relation for FAST ON COMMIT and COMPLETE ON DEMAND. On commit refreshing has a very significant impact on performance. On demand refreshing should be preferred if the application can tolerate that materialized views are not completely up-to-date or if insertions and queries are partitionned in time (as is the case in a data warehouse).

53.3.5.2 Redundant Tables

The previous example illustrates a special situation that we can sometimes exploit by implementing wholly redundant tables. Such tables store the aggregates we want. For example:

> *Sale*(sale_id, customer_id, product, quantity)
> *Customer*(customer_id, customer_location)
> *Customer_Agg* (customer_id, totalquantity)
> *Loc_Agg* (customer_location, totalquantity)

This reduces the query time but imposes an update time as well as a small space overhead (see Figure 53.14 and Figure 53.15). The trade-off is worthwhile in situations where many aggregate queries are issued (perhaps in a data warehouse situation) and an exact answer is required.

53.3.5.3 Tuning Normalized Schemas

Even restricting our attention to normalized schemas without redundant tables, we find tuning opportunities because many normalized schemas are possible. Consider a bank whose *Account* relation has the normalized schema (account_id is the key):

Account(account_id, balance, name, street, postal_code)

Consider the possibility of replacing this by the following pair of normalized tables:

AccountBal(account_id, balance)
AccountLoc(account_id, name, street, postal_code)

The second schema results from **vertical partitioning** of the first (all nonkey attributes are partitioned). The second schema has the following benefits for simple account update transactions that access only the id and the balance:

- A sparse clustering index on account_id of *AccountBal* may be a level shorter than it would be for the *Account* relation, because the name, street, and postal_code fields are long relative to account_id and balance. The reason is that the leaves of the data structure in a sparse index point to data pages. If *AccountBal* has far fewer pages than the original table, then there will be far fewer leaves in the data structure.
- More account_id–balance pairs will fit in memory, thus increasing the hit ratio. Again, the gain is large if *AccountBal* tuples are much smaller than *Account* tuples.

On the other hand, consider the further decomposition:

- *AccountBal*(account_id, balance)
- *AccountStreet*(account_id, name, street)
- *AccountPost*(account_id, postal_code)

Although still normalized, this schema probably would not work well for this application, because queries (e.g., monthly statements, account update) require both street and postal_code or neither. Vertical partitioning, then, is a technique to be used by users who have intimate knowledge of the application.

53.4 Tuning the Application Interface

A central tuning principle asserts *start-up costs are high; running costs are low*. When applied to the application interface, this suggests that you want to transfer as much necessary data as possible between an application language and the database per connection. Here are a few illustrations of this point.

53.4.1 Assemble Object Collections in Bulk

Object-oriented encapsulation allows the implementation of one class to be modified without affecting the rest of the application, thus contributing greatly to code maintenance. Encapsulation sometimes is interpreted as "the specification is all that counts." That interpretation can lead to horrible performance.

The problem begins with the fact that the most natural object-oriented design on top of a relational database is to make records (or sometimes fields) into objects. Fetching one of these objects then translates to a fetch of a record or a field. So far, so good.

But then the temptation is to build bulk fetches from fetches on little objects (the "encapsulation imperative"). The net result is a proliferation of small queries instead of one large query.

Consider, for example, a system that delivers and stores documents. Each document type (e.g., a report on a customer account) is produced according to a certain schedule that may differ from one document type to another. Authorization information relates document types to users. This gives a pair of tables of the form:

authorized(user, documenttype)
documentinstance(id, documenttype, documentdate)

When a user logs in, the system should say which document instances he or she can see. This can easily be done with the join:

```
select documentinstance.id, documentinstance.documentdate
from documentinstance, authorized
where documentinstance.documenttype = authorized.documenttype
and authorized.user = <input user name>
```

However, if each document type is an object and each document instance is another object, then one might be tempted to write the following code:

```
Authorized authdocs = new Authorized();
authdocs.init(<input user name>);
for (Enumeration e = authdocs.elements(); e.hasMoreElements();)
{
    DocInstance doc = new DocInstance();
    doc.init(e.nextElement());
    doc.print();
}
```

This application program will first issue one query to find all the document types for the user (within the init method of Authorized class):

```
select documentinstance.documenttype
from authorized
where authorized.user = <input user name>
```

and then for each such type t to issue the query (within the init method of DocInstance class):

```
select documentinstance.id, documentinstance.documentdate
from documentinstance
where documentinstance.documenttype = t
```

This is much slower than the previous SQL formulation. The join is performed in the application and not in the database server.

The point is not that object-orientation is bad. Encapsulation contributes to maintainability. The point is that programmers should keep their minds open to the possibility that accessing a bulk object (e.g., a collection of documents) should be done directly rather than by forming the member objects individually and then grouping them into a bulk object on the application side.

Figure 53.4 illustrates the performance penalty of looping over small queries rather than getting all necessary data at once.

53.4.2 Cursors Cause Friction

Programmers who have grown up with programming language loops find a familiar idiom in cursors. Unfortunately, the performance of cursors is horrible in almost all systems. Shasha once had the experience of rewriting an 8-hour query having nested cursors into a cursor-free query that took 15 seconds. We illustrate a less dramatic cursor penalty with a simple experiment, as shown in Figure 53.16.

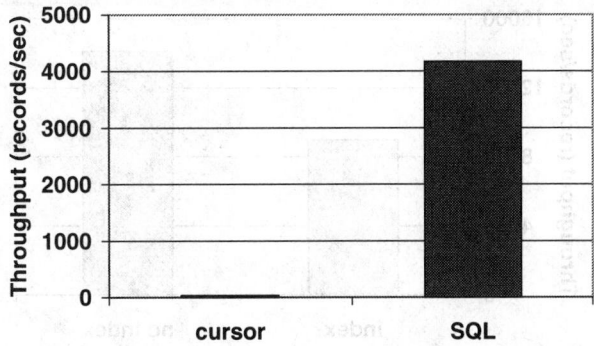

FIGURE 53.16 Cursors drag. This experiment consists of retrieving 200,000 rows from the table Employee (each record is 56 bytes), using both a set-oriented formulation (SQL) and using a cursor to iterate over the table contents (cursor). Using the cursor, records are transmitted from the database server to the application one at a time. The query takes a few seconds with the SQL formulation and more than an hour using a cursor. This experiment was run on SQL Server 2000 on Windows 2000.

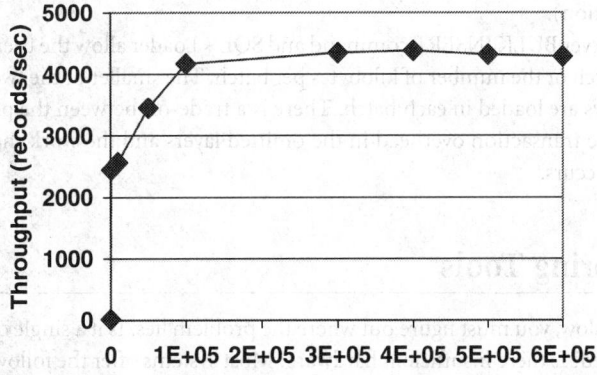

FIGURE 53.17 Batch size. We used the BULK INSERT command to load 600,500 tuples into the line item relation on SQL Server 2000 on Windows 2000. We varied the number of tuples loaded in each batch. The graph shows that throughput increases steadily until batch size reaches 100,000 tuples, after which there seems to be no further gain. This suggests that a satisfactory trade-off can be found between performance (the larger the batch, the better up to a certain point) and the amount of data that has to be reloaded in case of a problem when loading a batch (the smaller the batch, the better).

53.4.3 The Art of Insertion

We have spoken so far about retrieving data. Inserting data rapidly requires understanding the sources of overhead of putting a record into the database:

1. As in the retrieval case, the first source of overhead is an excessive number of round-trips across the database interface. This occurs if the batch size of your inserts is too small. In fact, up to 100,000 rows, increases in the batch size improve performance on some systems, as Figure 53.17 illustrates.
2. The second reason has to do with the ancillary overhead that an insert causes: updating all the indexes on the table. Even a single index can hurt performance as Figure 53.18 illustrates.
3. Finally, the layers of software within a database system can get in the way. Database systems provide bulk loading tools that achieve high performance by bypassing some of the database layers (mostly having to do with transactional recovery) that would be traversed if single row INSERT statements

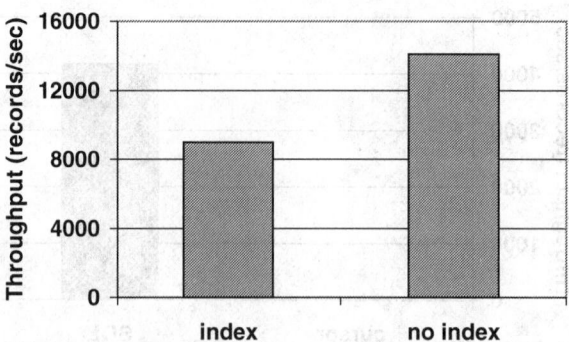

FIGURE 53.18   High index overhead for insertion. We insert 100,000 records in the table Order(ordernum, itemnum, quantity, purchaser, vendor). We measure throughput with or without a nonclustered index defined on the ordernum attribute. The presence of the index significantly impacts performance.

were used, as we show in Figure 53.18. For example, *SQL * Loader* is a tool that bulk loads data into Oracle databases. It can be configured to bypass the query engine of the database server (using the direct path option).

The SQL Server BULK INSERT command and SQL * Loader allow the user to define the number of rows per batch or the number of kilobytes per batch. The smaller of the two is used to determine how many rows are loaded in each batch. There is a trade-off between the performance gained by minimizing the transaction overhead in the omitted layers and the work that has to be redone in case a failure occurs.

53.5   Monitoring Tools

When your system is slow, you must figure out where the problem lies. Is it a single query? Is some specific resource misconfigured? Is there insufficient hardware? Most systems offer the following basic monitoring tools:

1. Event monitors (sometimes known as Trace Data Viewer or Server Profiler) capture usage measurements (processor usage ratio, disk usage, locks obtained, etc.) at the end of each query. You might then look for expensive queries.
2. If you have found an expensive query, you might look to see how it is being executed by looking at the query plan. These Plan Explainer tools tell you which indexes are used, when sorts are done, and which join ordering is chosen.
3. If you suspect that some specific resource is overloaded, you can check the consumption of these resources directly using operating system commands. This includes the time evolution of processor usage, disk queueing, and memory consumption.

53.6   Tuning Rules of Thumb

Often, tuning consists of applying the techniques cited above, such as the selection and placement of indexes or the splitting up of transactions to reduce locking conflicts. At other times, tuning consists of recognizing fundamental inefficiencies and attacking them.

1. Simple problems are often the worst. We have seen a situation where the database was very slow because the computer supporting the database was also the mail router. Offloading nondatabase applications is the simplest method of speeding up database applications.

2. Another simple problem having a simple solution concerns locating and rethinking specific queries. The authors have had the experience of reducing query times by a factor of ten by the judicious use of outer joins to avoid superlinear query performance [4].

3. The use of triggers can often result in surprisingly poor performance. Because procedural languages for triggers resemble standard programming languages, bad habits sometimes emerge. Consider, for example, a trigger that loops over all records inserted by an update statement. If the loop has an expensive multitable join operation, it is important to pull that join out of the loop if possible. We have seen a ten-fold speedup for a critical update operation following such a change.

4. There are many ways to partition load to avoid performance bottlenecks in a large enterprise. One approach is to distribute the data across many sites connected by wide area networks. This can result, however, in performance and administrative overheads unless networks are extremely reliable. Another approach is to distribute queries over time. For example, banks typically send out $\frac{1}{20}$ of their monthly statements every working day rather than send out all of them at the end of the month.

53.7 Summary and Research Results

Database tuning is based on a few principles and a body of knowledge. Some of that knowledge depends on the specifics of systems (e.g., which index types each system offers), but most of it is independent of version number, vendor, and even data model (e.g., hierarchical, relational, or object oriented). This chapter has attempted to provide a taste of the principles that govern effective database tuning.

Various research and commercial efforts have attempted to automate the database tuning process. Among the most successful is the tuning wizard offered by Microsoft's SQL server. Given information about table sizes and access patterns, the tuning wizard can give advice about index selection, among other features. Tuners would do well to exploit such tools as much as possible. Human expertise then comes into play only when deep application knowledge is necessary (e.g., in rewriting queries and in overall design) or when these tools do not work as advertised (the problems are all NP-complete).

Diagnosing performance problems and finding solutions may not require a good bedside manner, but good tuning can transform a slothful database into one full of pep.

Acknowledgment

The anonymous reviewer of this chapter improved the presentation greatly.

Defining Terms

B-tree: The most used data structure in database systems. A B-tree is a balanced tree structure that permits fast access for a wide variety of queries. In virtually all database systems, the actual structure is a B+-tree in which all key–pointer pairs are at the leaves.

Clustering index: A data structure plus an implied table organization. For example, if there is a clustering index based on a B-tree on last name, then all records with the last names that are alphabetically close will be packed onto as few pages as possible.

Conflict (between locks): An incompatibility relationship between two lock types. Read locks are compatible (nonconflicting) with read locks, meaning different transactions may have read locks on the same data item s. A write lock, however, conflicts with all kinds of locks.

Covering index: An index whose fields are sufficient to answer a query.

Decision support: Queries that help planners decide what to do next (e.g., which products to push, which factories require overtime, etc.).

Denormalization: The activity of changing a schema to make certain relations denormalized for the purpose of improving performance (usually by reducing the number of joins). Should not be used for relations that change often or in cases where disk space is scarce.

Dense index: An index in which the underlying data structure has a pointer to each record among the data pages. Clustering indexes can be dense in some systems (e.g., ORACLE). Nonclustering indexes are always dense.

E-commerce applications: Applications entailing access to a Web site and a back-end database system.

Hash structure: A tree structure whose root is a function, called the hash function. Given a key, the hash function returns a page that contains pointers to records holding that key or is the root of an overflow chain. Should be used when selective equality queries and updates are the dominant access patterns.

Heap: In the absence of a clustering index, the tuples of a table will be laid out in their order of insertion. Such a layout is called a heap. (Some systems, such as RDB, reuse the space in the interior of heaps, but most do not.)

Hit ratio: The number of logical accesses satisfied by the database buffer divided by the total number of logical accesses.

Index: A data organization to speed the execution of queries on tables or object-oriented collections. It consists of a data structure (e.g., a B-tree or hash structure) and a table organization.

Locking: The activity of obtaining and releasing read locks and write locks (see corresponding entries) for the purposes of concurrent synchronization (concurrency control) among transactions.

Nonclustering index: A dense index that puts no constraints on the table organization; also known as a secondary index. For contrast, *see* **clustering index**.

Normalized: A relation R is normalized if every functional dependency "X functionally determines A," where A and the attributes in X are contained in R (but A does not belong to X), has the property that X is the key or a superset of the key of R. X functionally determines A if any two tuples with the same X values have the same A value. X is a key if no two records have the same values on all attributes of X.

Online transaction processing: The class of applications where the transactions are short, typically ten disk I/Os or fewer per transaction; the queries are simple, typically point and multipoint queries; and the frequency of updates is high.

Read lock: If a transaction T holds a read lock on a data item x, then no other transaction can obtain a write lock on x.

Seek: Moving the read/write head of a disk to the proper track.

Serializability: The assurance that each transaction in a database system will appear to execute in isolation of all others. Equivalently, the assurance that a concurrent execution of committed transactions will appear to execute in serial order as far as their input/output behaviors are concerned.

Sparse index: An index in which the underlying data structure contains exactly one pointer to each data page. Only clustering indexes can be sparse.

Track: A narrow ring on a single platter of a disk. If the disk head over a platter does not move, then a track will pass under that head in one rotation. The implication is that reading or writing a track does not take much more time than reading or writing a portion of a track.

Transaction: A program fragment delimited by Commit statements having database accesses that are supposed to appear as if they execute alone on the database. A typical transaction can process a purchase by increasing inventory and decreasing cash.

Two-phase locking: An algorithm for concurrency control whereby a transaction acquires a write lock on x before writing x and holds that lock until after its last write of x; acquires a read or write lock on x before reading x and holds that lock until after its last read of x; and never releases a lock on any item x before obtaining a lock on any (perhaps different) item y. Two-phase locking can encounter deadlock. The database system resolves this by rolling back one of the transactions involved in the deadlock.

Vertical partitioning: A method of dividing each record (or object) of a table (or collection of objects) so that some attributes, including a key, of the record (or object) are in one location and others are in another location, possibly another disk. For example, the account id and the current balance may be in one location, and the account id and the address information of each tuple may be in another location.

Write lock: If a transaction T holds a write lock on a datum x, then no other transaction can obtain any lock on x.

Transaction: Unit of work within a database application that should appear to execute atomically (i.e., either all its updates should be reflected in the database or none should; it should appear to execute in isolation).

References

[1] S. Acharya, P.B. Gibbons, V. Poosala, and S. Ramaswamy. 1999. Join synopses for approximate query answering. In A. Delis, C. Faloutsos, and S. Ghandeharizadeh, Editors, *SIGMOD 1999, Proceedings ACM SIGMOD International Conference on Management of Data*, June 1–3, 1999, Philadephia, Pennsylvania, USA, pages 275–286. ACM Press.

[2] D. Comer. 1979. The ubiquitous B-tree. *ACM Comput. Surveys*, 11(2):121–137.

[3] J. Gray, and A. Reuter. 1993. *Transaction Processing: Concepts and Techniques.* Morgan Kaufmann, San Mateo, CA.

[4] D. Shasha, and P. Bonnet. 2002. *Database Tuning: Principles, Experiments, and Troubleshooting Techniques.* Morgan-Kaufmann Publishing Company, San Mateo, CA. Experiments may be found in the accompanying Web site: http://www.distlab.dk/dbtune/

[5] G. Weikum and G. Vossen. 2001. *Transactional Information Systems: Theory, Algorithms, and Practice of Concurrency Control and Recovery.* Morgan-Kaufmann, San Mateo, CA.

[6] Oracle Web site. http://otn.oracle.com/

[7] DB2 Web site. http://www.ibm.com/software/data/db2/

[8] SQL Server Web site, ongoing. http://www.microsoft.com/sql/

[9] A. Thomasian, and K. Ryu. 1991. Performance analysis of two-phase locking. *IEEE Trans. Software Eng.*, 17(5):68–76 (May).

[10] G. Weikum, C. Hasse, A. Moenkeberg, and P. Zabback. 1994. The COMFORT automatic tuning project. *Inf. Systems*, 19(5):381–432.

Further Information

Whereas the remarks of this chapter apply to most database systems, each vendor will give you valuable specific information in the form of tuning guides or administrator's manuals. The guides vary in quality, but they are particularly useful for telling you how to monitor such aspects of your system as the relationship between buffer space and hit ratio, the number of deadlocks, the disk load, etc.

A performance-oriented general textbook on databases is Pat O'Neil's book, *Database*, published by Morgan-Kaufmann.

Jim Gray has produced some beautiful viewgraphs of the technology trends and applications leading to parallel database architectures (http://research.microsoft.com/ gray/).

Our book, *Database Tuning: Principles, Experiments, and Troubleshooting Techniques*, published by Morgan Kaufmann goes into greater depth regarding all the topics in this chapter.

54

Access Methods

Betty Salzberg
Northeastern University

Donghui Zhang
Northeastern University

54.1 Introduction **54**-1
54.2 Underlying Principles **54**-2
54.3 Best Practices **54**-4
    The B^+-Tree • Hashing • Spatial Methods
    • Temporal Methods • Spatio-Temporal Methods
54.4 Research Issues and Summary **54**-24

54.1 Introduction

Although main memories are becoming larger, there are still many large databases that cannot fit entirely in main memory. In addition, because main memory is larger and processing is faster, new applications are storing and displaying image data as well as text, sound, and video. This means that the data stored can be measured in terabytes. Few main memories hold a terabyte of data. So data still have to be transferred from a magnetic disk to main memory. Such a transferral is called a **disk access**.

Disk access speeds have improved. However, they have not and cannot improve as rapidly as central processing unit (CPU) speed. Disk access requires mechanical movement. To move a **disk page** from a magnetic disk to main memory, first one must move the **arm** of the disk drive to the correct **cylinder**. A cylinder is the collection of **tracks** at a fixed distance from the center of the disk drive. The disk arm moves toward or away from the center to place the read/write **head** over the correct track on one of the disks. As the disks rotate, the correct part of the track moves under the head. Only then can the page be transferred to the main memory of the computer. A disk drive is illustrated in Figure 54.1.

The fastest disks today have an average access time of 5 ms. This is at a time when CPU operations are measured in nanoseconds. Therefore, the access of one disk page is at least one million times slower than adding two integers in the CPU. In addition, to request a disk page, the CPU has to perform several thousand instructions, and often the operating system must make a process switch. Thus, although the development of efficient access methods is not a new topic, it is becoming increasingly important.

In addition, new application areas are requiring more complex disk access methods. Some of the data being stored in large databases are multidimensional, requiring that the records stored in one disk page refer to points in two- or three-dimensional space, which are close to each other in that space. Data mining, or discovery of patterns over time, requires access methods that are sensitive to the time dimension. The use of video requires indexing that will allow retrieval by pictorial subject matter. The increasingly large amount of textual data being gathered electronically requires new thinking about information retrieval. Other chapters in this book look at video and text databases, but here we will treat spatial and temporal data as well as the usual linear business data.

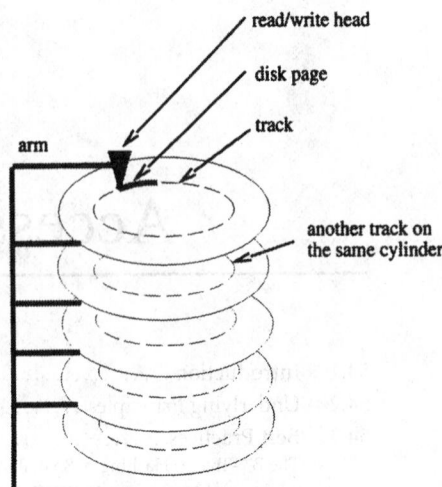

FIGURE 54.1   A disk drive.

54.2   Underlying Principles

Because disk access is so expensive in elapsed time and in CPU operations, and because the minimum number of bytes that can be transferred at one time is now 8 kilobytes (8192 bytes) or so, the main goal of access methods is to group or cluster on one disk page data which will be requested by an application in a short timeframe. Then after a page is placed in main memory, in order to read a requested item, there will be a good chance that when a subsequent item is requested, a separate disk access will not be necessary.

Because application logic cannot be predicted in general, the clustering will follow some kind of logical relationship or some kind of specific knowledge about the applications that are run. For example, if it is known that address labels are printed out in order of the zip code, one can attempt to store records with the same zip code in the same page.

In addition to disk page clustering, there are several other important principles of access methods. One is that disk space should be used wisely. Suppose an application accesses customer records individually and randomly by social security number. A very simple solution that would require only one disk access per customer would be to reserve one disk page for each possible social security number. With 1000 customers, 1×10^9 disk pages would be reserved. Because each disk page are 8 kilobytes, this are about 8 terabytes of disk space. Although this seems like a ridiculous example, papers have been written that propose access methods and do not take into account the amount of disk space used. Total disk space used should always be considered in proposing a disk access method.

Another principle regards the number of pages needed to be accessed to find one particular record. In general, one aims to minimize the number of pages accessed that contain *none* of the data wanted. This is one of the reasons why binary search trees are not used for large databases. Binary search trees can be very unbalanced, and the mapping from parts of the tree to disk pages is not defined. One could, for example, read in the page that had the root of binary search tree in it and then follow the tree in memory until it led to another disk page, and so forth. If one of the records that is accessed relatively often thus required reading 20 disk pages every time it is requested, this is a poor access method for this application. In general, if pagination of the access method cannot be specified, it will perform poorly.

Finally, insertion and deletion of records should modify as few disk pages as possible. Most good access methods modify only one page when a new record is inserted most of the time. That is, the record is inserted into a page that has empty space. This page is the only one modified. Occasionally, one or two other pages must be modified. For a good access method, it is *never* necessary to modify a large number of pages when one record is inserted or deleted.

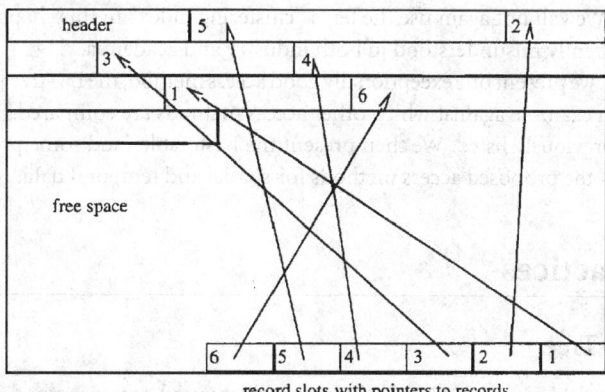

record slots with pointers to records

FIGURE 54.2   A typical disk page layout. New records are placed in the free space after old records (growing down) and new slot numbers and pointers are placed before the old ones (growing up). Variable-length records are accommodated.

We summarize the principles we have discussed in a list. We will often refer to these in the rest of the chapter. The **properties of good access methods** are as follows:

1. *Clustering:* Data should be clustered in disk pages according to anticipated queries.
2. *Space usage:* Total disk space usage should be minimized.
3. *Few answer-free pages in search:* Search should not touch many pages having no relevant data.
4. *Local insertion and deletion:* Insertion and deletion should modify only one page most of the time and occasionally two or three. Insertion and deletion should never modify a large number of pages.

Most access methods have **data pages** and also have pages that contain no data, which are called **index pages**. The index pages contain **references** to other index pages and/or references to data pages or references to records within data pages. A reference to a data page or an index page is a disk page address. A reference to a record may be only the disk page address of the page where the record is kept (and some other means must be used to locate the record in the page, usually the value of one of the attributes of the record), it may be the combination of a disk page address and a record slot number within that disk page (see Figure 54.2), or it may be some attribute value that can be used in another index to find the given record. (We freely exchange the words *index* and *access method*.)

If the index pages reference data pages but never individual records within those data pages, the access method is called a **sparse index** or sometimes a **primary index**. In this case, the location for insertion of a new record is determined by the access method and usually (but not always) depends on the value of a primary key for the record. (It could instead depend on some other attribute that is not primary, for example, the zip code in an address.) Thus, *primary indices* are not always associated with *primary keys*.

If there are references in the index to individual records, so that a data page with 20 data records in it has at least 20 separate references in the index, the index is called a **dense index** or a **secondary index**. Insertion of new records can then occur anywhere. Often, secondary indices are associated with data pages where the data is placed in order of insertion.

We will use the terms "primary index" and "secondary index." A data collection often has one primary index and several secondary indices. The primary index determines the placement of the records and the secondary indices allow lookup by attributes other than the ones used for placement. Most primary indices can be converted to secondary indices by replacing the data records with references to the same data records and actually storing the data records elsewhere. Thus, the question, "Is this access method a primary index or a secondary index?" does not always make sense.

Indices are sometimes called **clustering indices**. In commercial products, this usually means that the data is loaded into data pages initially in the order specified by a secondary index. When new data is inserted in the database, such an index loses the clustering property. Such an index clusters data *statically* and not *dynamically*. We will not again use the term "clustering index" in this chapter as we believe it has been abused and frequently misunderstood in both industry and academia.

Next in this chapter, we present one exceptionally good access method, the B^+-tree [Bayer and McCreight 1972], and use it as an example against which other access methods are compared. The B^+-tree has all of the good properties previously listed. We then present the hash table (and some proposed variants) and briefly review some of the proposed access methods for spatial and temporal data.

54.3   Best Practices

54.3.1   The B^+-Tree

The B^+-tree [Bayer and McCreight 1972] is the most widely used access method in databases today. A picture of a B^+-tree is shown in Figure 54.3. Each node of the tree is a disk page and, hence, contains 4096 bytes. The leaves of the tree when it is used as a primary index contain the data records or, in the case of a secondary B^+-tree, references to the data records that lie elsewhere. The leaves of the tree are all at the same level of the tree.

The index entries contain values and pointers. Search begins at the root. The search key is compared with the values in the index entries. In Figure 54.3, the pointer associated with the largest index entry value smaller than or equal to the search key is followed. To search for coconut, for example, first the pointer at the root associated with caramel is followed. Then at the next level, the pointer associated with chocolate is followed. Search for a single record visits only one node at each level of the tree.

In the remainder of this section, we assume the B^+-tree is being used as a primary index. This is certainly not always the case. Many commercial database management systems have no primary B^+-tree indices. Even those that do offer the B^+-tree as a primary index must offer the B^+-tree as a secondary index as well, as it will be necessary in most cases to have more than one index on each relation, and only one of them can be primary.

The main reasons that the B^+-tree is the most widely used index is (1) that it clusters data in pages in the order of one of the attributes (or a concatenation of attributes) of the records, and (2) it maintains that

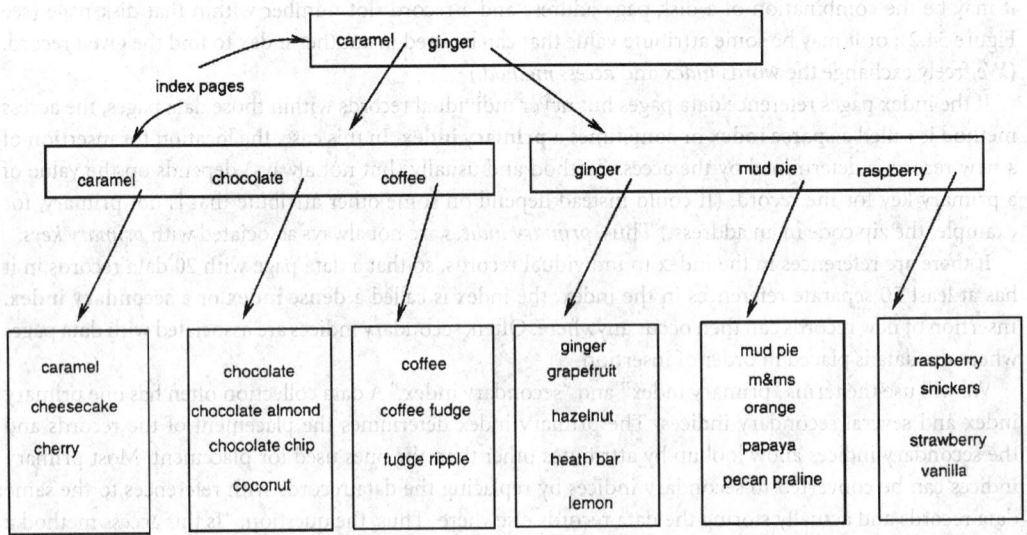

FIGURE 54.3   A B^+-tree.

clustering dynamically, never having to be reorganized to retain disk page clustering. Thus, the B^+-tree satisfies property 1 for good access methods: it clusters data for the anticipated queries.

Storage utilization for B^+-trees in both index and data pages is 69% on average [Yao 1978]. Total space usage is approximately 1/0.69 times the space needed to store the data records packed in pages. All of the data is stored in the leaves of the tree. The space needed above the level of the leaves is less than 1/100 of the space needed for the leaves. Thus, the B^+-tree satisfies property 2 for good access methods: it does not use too much total disk space.

In addition, B^+-trees have no pathological behavior. That is, no records require many accesses to be found. Searches for a record given its key follow one path from the root of the tree to a leaf visiting only one page at each level of the tree. If the record is in the database, it is in the leaf page visited. All of the leaves are at the same level of the tree.

All records can thus be found with, at most, as many disk accesses as the height of the tree and the height of the tree is always small. The reason some records can be found with less disk accesses is that the upper levels of the tree are often still in main memory when subsequent requests for data are made. Thus, B^+-trees satisfy property 3 for good access methods: few answer-free pages (in this case, index pages) are accessed in any search.

Insertion and deletion algorithms enable the B^+-tree to retain its clustering and to maintain the property that all leaves are at the same level of the tree. Thus, search for one record remains efficient and search for ranges remains efficient. Insertion in a leaf that has enough empty space requires writing only one disk page. (Although this seems obvious, we will see that the R-tree discussed subsequently does not have this property.) Deletion from a page that does not require it to become sparse also writes only one disk page. The probability that more than one existing index or data page must be updated is low. In the worst case, one page is updated and a new page created at each level of the tree. This is an extremely rare event. Thus, the B^+-tree satisfies property 4 for good access methods: the B^+-tree has local insertion and deletion algorithms.

54.3.1.1 Fan-Out Calculations

The height of the tree is small because the **fan-out** is large. The fan-out is the ratio of the size of the data page collection to the size of the index page collection. The secret of the B^+-tree is that each index page has hundreds of children.

Some data structure textbooks suggest a B-tree where data records are stored in all pages of the index. But when data records are stored high up in the tree, the fan-out is too small. A data record may have 100 or 200 bytes. This would limit the number of children each high-level page in the index could have. This variation has never been used for database management systems for this reason. It is also the reason we use the notation B^+-tree, which has historically stood for the variation of B-trees, which has all of the data in the leaves, the only variation of the B-tree in use.

A disk page address is usually 4 bytes. A key value may be about 8 bytes if it is an alphanumeric key. This means each index term is 12 bytes. But there are 4096 bytes in an index page. Theoretically, each page could have 4096/12 = 341 children. (Actually, there is some header information.) But when pages are full and they must be split as insertions are made, each new page is only half-full. An average case analysis [Yao 1978] showed that B^+-trees are, on average, $ln2$ full, or about 69% full. Thus, this would give our average node, allowing for some header space, about 230 children.

Let us do some calculations. If the root has 230 children and each of them has 230 children and each of them is a data page, then there are 52,900 data pages. Each data page is also 4 kilobytes. This is 211,600 kilobytes or, in round numbers, about 200 megabytes. Thus, with at most three disk accesses we can find any record in a 200-megabyte relation. However, we can also fix the root in the main memory so that every record can be found in two disk accesses. If we use 231 pages of main memory space (231×4 kilobytes, or 924 kilobytes of main memory), we can store the top two levels of the tree in main memory and we have a one-disk access method.

Even if we do not specify that index pages be stored in main memory, if we use a standard least recently used (LRU) page replacement algorithm to manage the memory, it is likely that upper levels of the tree are in memory. This is because all searches must start at the root and travel down the tree.

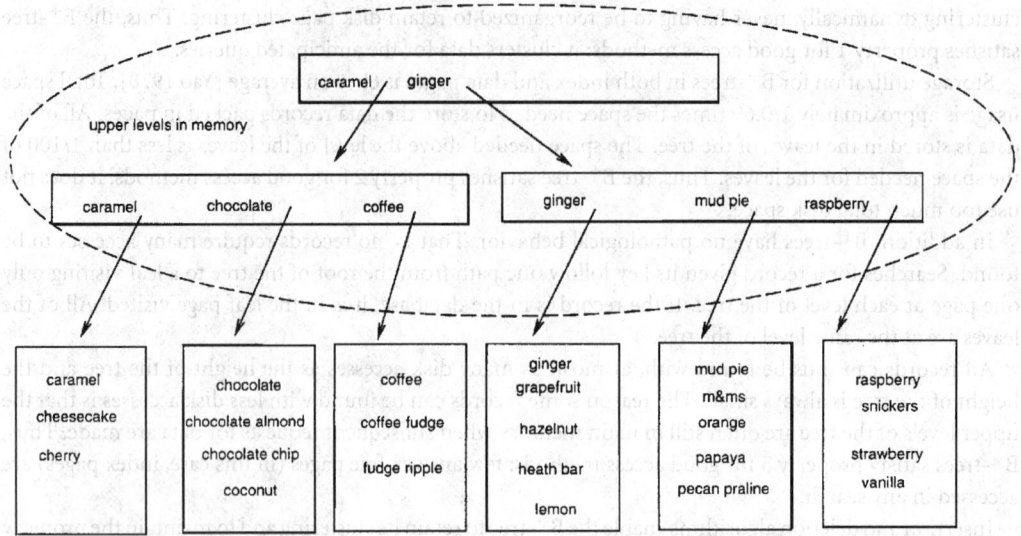

FIGURE 54.4 The number of disk accesses to find a record in a B$^+$-tree depends on how much of the tree can be kept in main memory. Adding one more level, as shown here, does not necessarily add one more disk access.

The number of levels above the leaves of a tree containing n leaves is ceiling($\log_{\text{fan-out}} n$). The height of the tree is one more than this number. Thus, if the fan-out is 230 and the number of leaves is 50,000, there are two ($= $ ceiling($\log_{230}$ 50,000)) levels above the leaves and the height of the tree is three. For a 4-gigabyte relation, with about 1 million leaves of 4 kilobytes each, a height of 4 is required. But in this case, the level right below the root may be small. Suppose the level below the root has only 20 pages. If each of these is a root of a subtree of height three with 50,000 leaves, we have 1 million leaves in total. But storing in memory these 20 pages at the level below the root plus the root takes up only 84 kilobytes (21×4 kilobytes). It is likely that these two levels will stay in main memory. Then, even with a height of 4, the number of disk accesses for search for one record is only two.

Thus, tree height is not the same as number of disk accesses required to find a record. The number of disk accesses required to find a record depends on what part of the top levels of the tree are already stored in main memory. This is illustrated in Figure 54.4, where we have shown the top two levels of the sample tree in main memory, making this B$^+$-tree a one-disk-access method.

54.3.1.2 Key Compression and Binary Search within Pages

Because the fan-out determines the height of the tree, which in turn limits the number of disk accesses for record search, it is worthwhile to enhance fan-out. This would be particularly desirable if the size of the attributes or concatenated attributes used as the key in the B$^+$-tree is large. A smaller key means a larger fan-out.

One technique often used for enhancing B$^+$-tree performance is key compression. The prefix B$^+$-tree [Bayer and Unterauer 1977] is used in many implementations. Here, what is stored in the index pages is not the full key but only a **separator**. A separator holds only enough characters to differentiate one page from the next. Binary search still can be used within the index pages. Separators are illustrated in Figure 54.5. (It is not true that separators must be changed when records are deleted from the database. It is probably not worthwhile to make a disk access to modify the separator when a record is deleted. The old separator still allows correct search, although it is not as short as it might be.)

Although schemes to omit some of the first characters of the key when they are shared by some of the entries have been considered, unless they are shared by *all* of the index entries in an index page, such schemes have not been used. This is because compressing prefixes of keys makes binary search impossible,

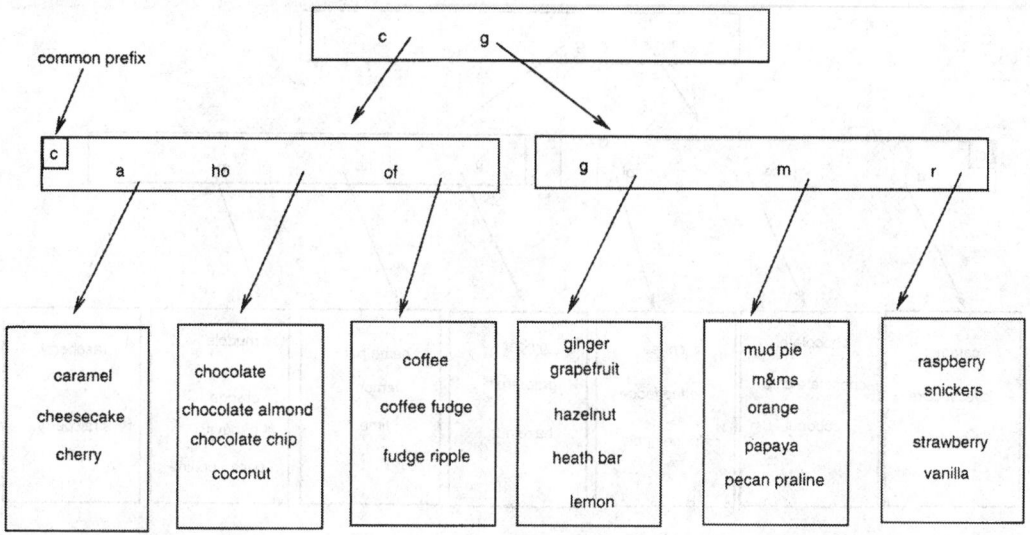

FIGURE 54.5 Separators in a prefix B$^+$-tree.

and search within an index page becomes linear. Because index page search is a frequent operation, prefix compression is not worthwhile. Also, it is predicted that disk page sizes will get larger, making linear search even worse. Figure 54.5 illustrates storing a common prefix (the first few characters used in *every key in a page*) only once.

Either the index terms can be organized within index pages using a binary search tree, or they can be kept in order using an array. The usual practice for index pages is to use an array, which saves space and allows more fan-out. Index pages are updated relatively infrequently. When a new index term is inserted, the array elements after the insertion must of course be moved to a new position in the page. Binary search is performed on the array in each index page visit during a record search operation.

In data pages, the keys can be stored separately from the rest of the record, with a pointer to the rest of the record. Then the keys also can be stored in order in an array, and binary search can be used.

54.3.1.3 Insertion and Deletion

Usually, insertion of a record simply searches for the correct leaf page in the tree and updates that leaf page. Occasionally [with a probability of 2/max(recs) where max(recs) is the maximum number of records in a leaf page], the leaf is full, and it must be split. A leaf split in a B$^+$-tree is illustrated in Figure 54.6. In Figure 54.6, the record with key lime is entered in the tree of Figure 54.5. In this example, we assume that only five records can reside in a leaf. Thus, we split the leaf where lime would be inserted.

Half of the records stay in the old leaf node and half of the records are copied to a newly allocated leaf node. An entry is placed in the parent to separate the old and new nodes. With a much lower probability, the parent may need to be split as well, and this can percolate up the tree on a path to the root. The probability that an insertion causes a split of both the leaf and its parent is $(2/ \max(\text{recs})) \times (2/ \max(\text{index entries}))$.

For example, if records are 200 bytes and there are 4 kilobyte pages, max(rec) might be 20 allowing for header space in the data page. Then, after a split, 10 more records must be inserted in the page before a split is needed again (assuming no deletes). Thus, the probability of a split is at worst 1/10. If max(index entries) is equal to 300, the probability of an insertion causing an index page split is $(1/10) \times (1/150)$, or 1/1500.

For record deletion, similarly, one need only remove the record from the data page in most cases. Sometimes, pages are considered sparse and are merged with their siblings. In fact, many commercial products do not consider any pages sparse unless they are completely empty [Mohan and Levine 1992]. An analysis of this issue, which concludes that node consolidation in B$^+$-trees is not useful unless the nodes are completely empty, can be found in Johnson and Shasha [1989].

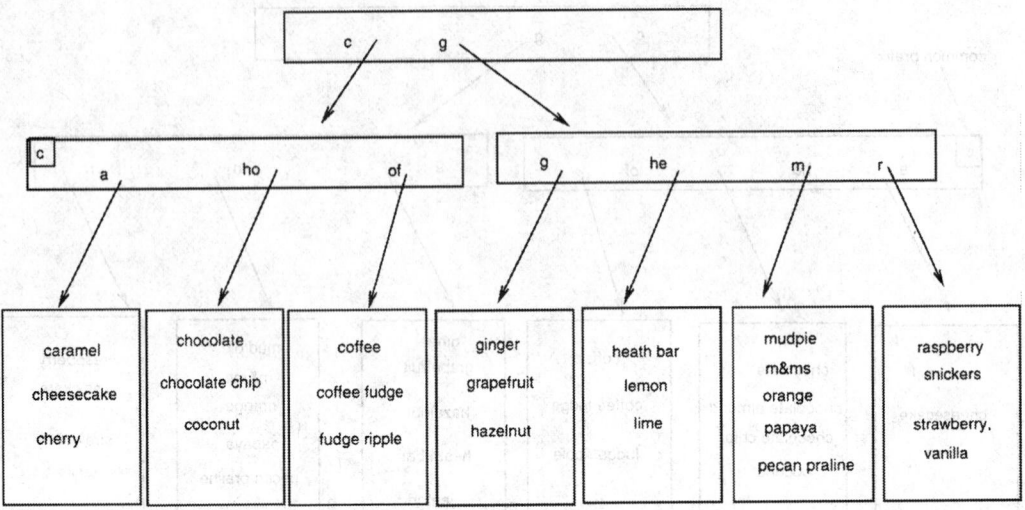

FIGURE 54.6   A leaf split in a B^+-tree.

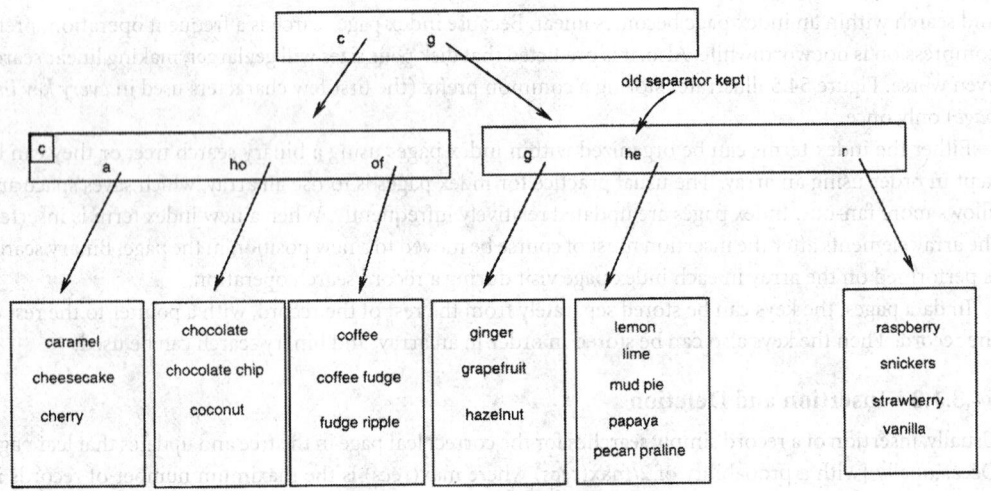

FIGURE 54.7   A B^+-tree node consolidation.

Most textbooks, however, consider anything less than 50% to be sparse in order to maintain the property that all nodes are at least 50% full. This is not a good idea in practice, even if one does not go so far as to wait until nodes are empty before consolidating them, because it may cause **thrashing**.

If the threshold for node sparseness is 50%, so that nodes are not allowed to ever fall below 50% full, and if a record is inserted into an overflowing page, then the page is *split* to get a 50% utilization, then a record is deleted from the page, the B^+-tree may needlessly thrash between splitting and consolidating the same node. This issue is discussed in detail in Maier and Salveter [1981].

A B^+-tree node consolidation is pictured in Figure 54.7. (The records with keys m&ms, orange, and heath bar have been removed from Figure 54.6.) We do not change the separator he, although a shorter separator could have been chosen. In some cases, search will be incorrect if a new separator is chosen. Also, using the old separator makes the node consolidation algorithm *simpler*, which is an important consideration for access methods.

The case pictured deallocates a page. It also is possible to move records from a full sibling to a sparse one. In both cases, the parent must be updated to reflect the change. In rare cases, sparse node consolidation could percolate to higher levels, because deletion of an index term could cause a parent to become sparse.

54.3.1.4 Range Searches and Reorganizing

B^+-trees are very good for small range searches. Each leaf of the tree contains all of the records whose keys are within a given range. If the range is small, it is worthwhile to use the B^+-tree to access only a few disk pages to get all of the records wanted. However, if the range is large, it might be better to read in all of the leaf pages at once and not use the B^+-tree to find out which ones satisfy the query.

The reason for this is that after there have been many splits of the leaves in a B^+-tree, the leaf pages are not stored on the disk in order. Within each page, the records are in key order. But two adjacent pages in terms of key order could be in very different places on the disk. The disk arm might need to move a long distance to access the pages that were required. On the other hand, sequential reading on the disk, where the arm sits over one cylinder and reads all of the tracks on that cylinder and then moves to the next cylinder and reads all of the tracks on it and so forth, is much faster than reading the same amount of data one page at a time, moving the disk arm for each page.

Calculations in Salzberg [1988] show that in circumstances where a fairly large-size range is required (say 10% of the data), it is better to read in all of the data in no particular order than to make separate disk accesses for pages that satisfy the query. This is even worse when the B^+-tree is used as a secondary index. The range must be quite small to make use of the index worthwhile. Here, each record in a range could be stored in a separate disk page.

Because primary B^+-trees get out of order in this sense, it has sometimes been considered worthwhile to reorganize them. This can be done while keeping the B^+ trees online, as shown in Smith [1990], a description of an algorithm written and programmed by F. Putzolu. One leaf is written at a time. Sometimes the data in two leaves is interchanged. This is all done as a background process, and it is done transactionally (i.e., with lock protection so that searches will be correct even when reorganization is in progress).

54.3.1.5 Bounded Disorder

Efforts have been made to improve the B^+-tree. Most of these have not been implemented in commercial systems because it is not considered worth the trouble. Mostly, the B^+-tree is good enough. However, we will outline a particularly nice attempt in this area [Lomet 1988].

The basic idea of bounded disorder is to keep a small B^+-tree in memory and then vary the size of the leaves, using consecutive disk pages for one leaf. To find a particular record within a leaf, a hash algorithm is used to find the correct **bucket** within the leaf holding that record. There is one hash overflow bucket stored with each leaf. In the variation we illustrate, buckets are two or three consecutive disk pages.

When a small leaf becomes full, it can be replaced with a larger leaf. (Only two different sizes for leaves are suggested, one one and one-half times bigger than the other, thus expanding the space by a factor of 1.5.) When a larger size leaf is full, it can be replaced with two smaller size leaves, thus expanding by a factor of 1.33. This gradual expansion makes the average space utilization better than that of the B^+-tree. This follows some ideas about partial expansion found in Larson [1980].

For single record search, bounded disorder is fast because it usually requires only one disk access; the B^+-tree in memory directs the search to the correct leaf and the hash algorithm to the bucket within the leaf. For large-range searches, it is fast because the leaves are large. For small-range searches, it is not bad; it may read in one or two large leaves and then, when they are in memory, search within each bucket for the records that are in the queried range.

Bounded disorder is illustrated in Figure 54.8. Here we have assumed a hash function of $h(key) = key \bmod 5$. First, the in-memory tree is consulted. It tells the location and the size of the leaf (two or three units). Then the hash function is used to determine the bucket. If the record is not there, the overflow buckets are searched.

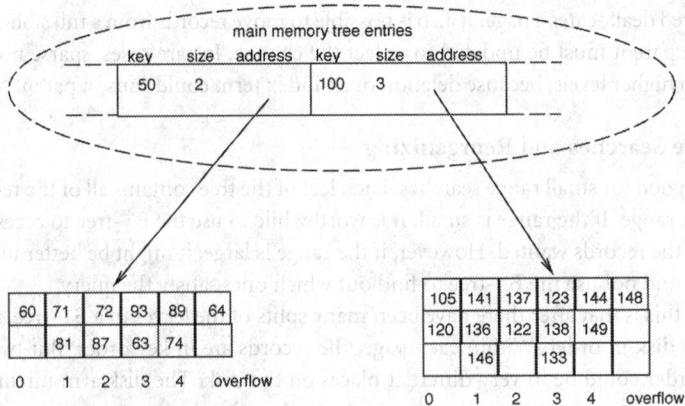

FIGURE 54.8 Bounded disorder. A small main memory B$^+$-tree directs the search to a large leaf. A hash function (here $h(key)=key$ mod 5) yields the bucket within the leaf.

54.3.1.6 Summary of the B$^+$-Tree

The B$^+$-tree is about as good an access method as there is. It dynamically maintains clustering, uses a reasonable total amount of disk space, never has pathological search cases, and has local insertion and deletion algorithms, usually modifying only one page. All other access methods we shall describe do not do as well.

54.3.2 Hashing

The second most used access method is **hashing**. In hashing, a function applied to a database key determines the location of the record. For example, if the set of keys were known to be the integers between 1 and 1000 and the data records all had the same size, contiguous disk space for 1000 records could be allocated, and the identity function would yield the offset from the beginning of the allocated space. However, usually database keys are not consecutive integers.

Hashing algorithms start with a function that maps a database key to a number. To make the result uniformly distributed, certain functions have been found to be especially effective. The best coverage of hashing functions can still be found in Knuth [1968].

Basically, the database key is transformed to a number N by some method such as adding together the ASCII (American Standard Code for Information Interchange) codes of the letters of the key. Then N is multiplied by a large prime number P_1 and then added to another number P_2, and the result is taken modulo P_3, where P_3 corresponds to the number of consecutive pages on the disk allocated for the **primary area** of the hash table. Thus, $f(N) = (N \times P_1 + P_2)$ mod P_3. Knuth [1968] discusses how the parameters are chosen.

For a primary hashing index, the data records themselves are stored in the primary area. For a secondary hashing index, the primary area is filled with addresses of data records which are stored elsewhere. In the remainder of our discussion, we assume all indices are primary indices.

A page (or sometimes a set of consecutive pages) corresponding to one value of the hash function is called a bucket. When all records fit into their correct bucket, hashing is a one-disk-access method. The hash value gives the bucket address as an offset from the beginning of the primary area. That bucket is accessed and the record is there. The same algorithm is used to insert a new record into a bucket as long as there is room for it.

When there is no longer room in a bucket, an overflow bucket is allocated from the **overflow area**. Let us assume each overflow bucket is a disk page. When an overflow bucket is allocated, its address is placed in the primary bucket to which it corresponds (or in the last overflow bucket allocated to this hash number), forming a chain of overflow buckets for each hash value). (Most overflow chains should be empty if the average search time is to be reasonable.)

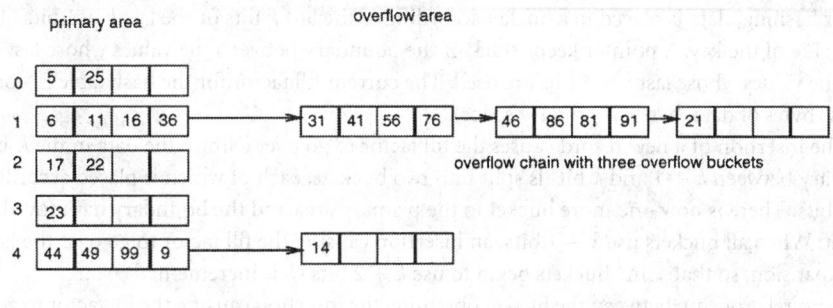

FIGURE 54.9   A hashing table with overflow buckets.

Then the search algorithm is as follows: Hash the database key to get the address of the correct bucket in the primary area. Search in the bucket. If the record is not there, search in the overflow bucket(s). A basic hashing index with overflow buckets is illustrated in Figure 54.9.

We define the **hash table fill factor** as the total space needed for the data divided by the space used by the primary area. As long as the hash table fill factor is a certain amount below one, there is not much overflow and search time is fast. For example, if buckets large enough for 50 records are used and the hash table fill factor is near 70% (which would be comparable to B^+-tree leaf space utilization), search is very close to one disk access on average [Knuth 1968].

However, one basic drawback of hashing is that the fill factor grows as the database grows. Hashing does not adjust well to growth of the database. The database can even become several times as large as the allocated primary area. In this case, the number of disk accesses needed on average can become quite high. Even without the database becoming large, hashing functions that produced very long chains of overflow buckets are possible and have occurred. Thus, hashing does not always satisfy property 3 for good access methods (efficient search).

If hashing must be reorganized and a new table constructed because the old table is too small (which is done in most systems that use hashing), property 4 is violated: insertion is not local when massive reorganization is required.

Further, by its nature, hashing does not cluster data that is related, unless all of the records that are related have the same key. Hashing functions that effectively create a uniform distribution are never order preserving. If such an order-preserving hashing function existed, it could do sorting of n elements in $O(n)$ time, by hashing each one into its proper place in an array [Lomet 1991]. Thus, hashing does not satisfy property 1: clustering. In fact, it is virtually useless to use a hashing structure for a range query. Each possible value in the range would have to be hashed.

Because hashing does not support range queries and because it does not adjust well to growth, it is not used as often as B^+-trees. When memories were small, a B^+-tree access from a B^+-tree of height 4 meant at least 3 and sometimes 4 disk accesses, because not even the root could always be kept in main memory. Now that memories are larger, single-record search in B^+-trees is more likely to be 1 or 2 disk accesses even when the height of the tree is 4. More importantly, for most relations, a B^+-tree of height 3 is likely to be a one-disk access method. This makes B^+-trees competitive with hashing for single-record search.

Before leaving this topic, we look at two proposed variations on hashing to support database growth. Although many papers have been written on this topic, most are refinements of these two. Most commercial systems that provide hashing have only hashing with overflow buckets.

54.3.2.1 Linear Hashing

Linear hashing [Litwin 1980] and extendible hashing [Fagin et al. 1979] both add one new bucket to the primary area at a time. Linear hashing (in the variation we explain here) adds a new bucket when the hash table fill factor becomes too large. Extendible hashing adds one whenever an overflow bucket would otherwise occur. We explain linear hashing first.

In linear hashing, data is placed in a bucket according to the last k bits or the last $k + 1$ bits of the hash function value of the key. A pointer keeps track of the boundary between the values whose last k bits are used and the values whose last $k + 1$ bits are used. The current fill factor for the hash table is stored as the two values (bytes of data, bytes of primary space).

When the insertion of a new record causes the fill factor to go over a limit, the data in the k-bucket on the boundary between $k + 1$ and k bits is split into two buckets, each of which is placed according to the last $k + 1$ bits. There is now one more bucket in the primary area and the boundary has moved down by one bucket. When all buckets use $k + 1$ bits, an insertion causing the fill factor to go over the limit starts another expansion, so that some buckets begin to use $k + 2$ bits (k is incremented).

There is no relationship between the bucket obtaining the insertion causing the fill factor to go over the limit and the bucket that is split. Linear hashing also has overflow bucket chains.

We give an example of linear hashing in Figure 54.10. We assume that k is 2. We assume that each bucket has room for three records and the limit for the fill factor is 0.667. We show how the insertion of one record causes the fill factor limit to be exceeded and how a new bucket is added to the primary area.

The main advantage of linear hashing is that insertion never causes massive reorganization. Search can still be long if long overflow chains exist. Range searches and clustering still are not enabled. There are two main criticisms of linear hashing: (1) some of the buckets are responsible for twice as many records on average as others, causing overflow chains to be likely even when the fill factor is reasonable, and (2) in order for the addressing system to work, massive amounts of consecutive disk pages must be allocated.

Actually, the second objection is not more of a problem here than for other access methods. File systems have to allocate space for growing data collections. Usually this is done by allocating an **extent** to a relation when it is created and specifying how large new extents should be when the relation grows. Extents are large amounts of consecutive disk space. The information about the extents should be a very small table that is kept in main memory while the relation is in use. (Some file systems are not able to do this and are unsuitable for large relations.)

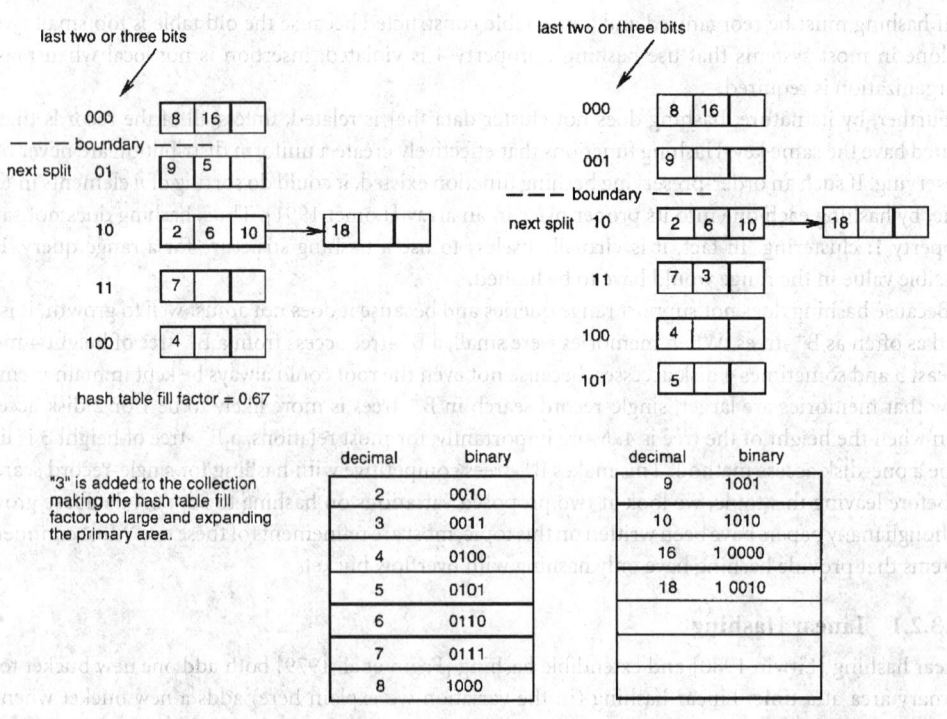

FIGURE 54.10   Linear hashing.

Many papers have been written about expanding linear hashing tables by a factor of less than 2. The basic idea here is that, for example (as was used in the bounded disorder method previously mentioned), at the first expansion, what took two units of space expands to three units of space. At the second expansion, what was in three buckets expands to four buckets. At this point, the file is twice as big as it was before the first expansion. In this way, no buckets are responsible for twice as much data as any other buckets; the factors are 1.5 or 1.33. This idea originated in Larson [1980].

54.3.2.2 Extendible Hashing

Another variant on hashing that also allows the primary area to grow one bucket at a time is called extendible hashing [Fagin et al. 1979]. Extendible hashing does not allow overflow buckets. Instead, when an insertion would cause a bucket to overflow, its contents are split between a new bucket and an old bucket and a table keeping track of where the data is updated.

The table is based on the first k bits of a hash number. A bucket B can belong to 2^j table entries, where $j < k$. In this case, all of the numbers whose first $k - j$ bits match those in the table will be in B. For example, if k is 3, there are eight entries in the table. They could refer to eight different buckets. Or two of the entries with the same two first bits could refer to the same bucket. Or four of the entries with the same first bit could refer to the same bucket. We illustrate extendible hashing in Figure 54.11.

The insertion of a new record that would cause an overflow either causes the table to double or else it causes some of the entries to be changed. For example, a bucket that was referred to by four entries might have its contents split into two buckets, each of which was referred to by two entries. Both cases are illustrated in Figure 54.11.

The advantage of extendible hashing is that it never has more than two disk accesses for any record. Often, the table will fit in memory, so that it becomes a one-disk-access method. There are no overflow chains to follow.

The main problems with this variation on hashing are total space utilization and the need for massive reorganization (of the table). Suppose the buckets can hold 50 records and there are 51 records with the identical first 13 bits in the hash number. Then there are at least $2^{14} = 16,384$ entries in the table. It does not matter how many other records there are in the database.

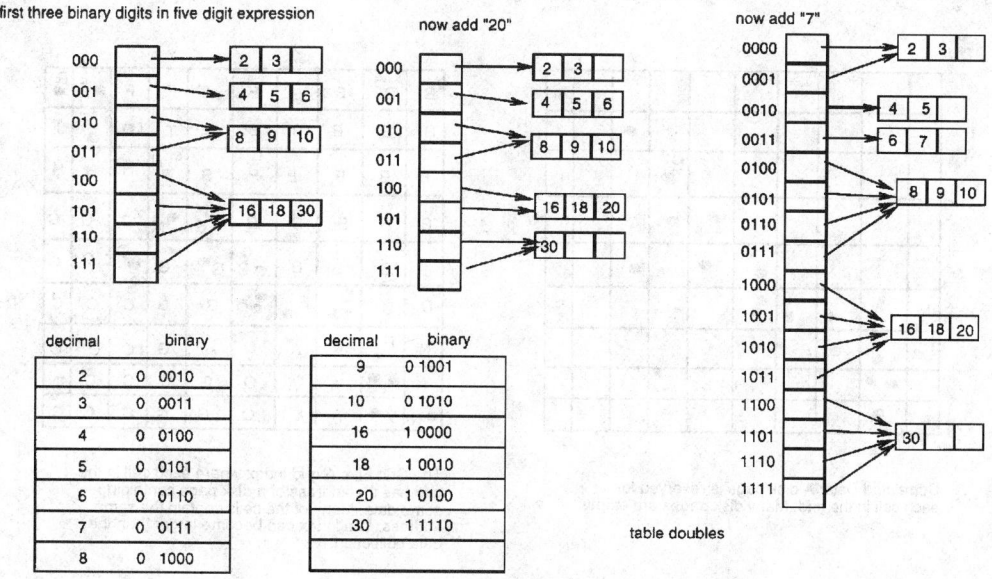

FIGURE 54.11 Extendible hashing.

Like the other variations on hashing, extendible hashing does not support range queries. All hashing starts with a *hashing function*, which will randomize the keys before applying the rest of the algorithm.

54.3.3 Spatial Methods

New application areas in geography, meteorology, astronomy, and geometry require spatial access and make nearest-neighbor queries. For spatial access methods, data should be clustered in disk pages by nearness in the application area space, for example, in latitude and longitude. Then the question: "find all of the desert areas in photographs whose center is within 100 miles of the equator" would have its answer in a smaller number of disk pages than if the data were organized alphabetically, for example, by name.

However, it is a difficult problem to organize data spatially and still maintain the four properties of good access methods. For example, one way to organize space is to make a grid and assign each cell of the grid to one disk page. However, if the data is correlated as in Figure 54.12, most of the disk pages will be empty. In this case, $O(n^k)$ disk space is needed for n records in k-dimensional space. Using a grid as an index has similar problems; only the constant in the asymptotic expression is changed. A grid index is also pictured in Figure 54.12.

A proposal was made for a grid index or *grid file* in Nievergelt et al. [1984]. Because it uses $O(n^k)$ space, in the worst case it can use too much total disk space for the index; thus, it violates property 2. Range searches can touch very many pages of the index just to find one data page; thus, it violates property 3. Insertion or deletion can cause massive reorganization; thus, it violates property 4. One major problem with the index is that it is not paginated (no specification of which disk pages correspond to which parts of the grid is made). Thus, a search over a part of it, which may even be small, can touch many disk pages of the index.

Over the past 25 years or so, researchers have proposed many spatial access methods, as surveyed in Gaede and Günther [1998]. In particular, the R-tree and Z-ordering have been used commercially, especially in geographic information systems.

54.3.3.1 R-Tree and R*-Tree

The R-tree [Guttman 1984] organizes the data in disk pages (nodes of the R-tree) corresponding to a brick or a rectangle in space. It was originally suggested for use with spatial objects. Each object is represented by the coordinates of its smallest enclosing rectangle (or brick) with sides parallel to the coordinate axes.

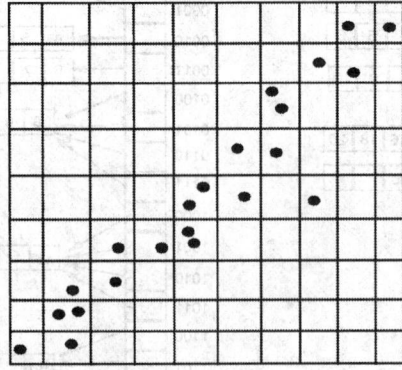

Correlated Data: A disk page is reserved for each cell in the grid. Many disk pages are empty.

The Grid File: A grid index where each cell in the grid has the address of a disk page containing some data. Many of the cells contain the same address. The index can become larger than the data collection.

FIGURE 54.12   A grid and a grid index.

Thus, any two-dimensional object is represented by four numbers: its lowest x-coordinate, its highest x-coordinate, its lowest y-coordinate, and its highest y-coordinate. Then when a number of such objects are collected, their smallest enclosing brick becomes the boundary of the disk page (a leaf node) containing the records (or in the case of a secondary index, pointers to the records).

At each level, the boundaries of the spaces corresponding to nodes of the R-tree can overlap. Thus, search for objects involves backtracking.

When a new item is inserted, at each level of the tree, the node where the insertion would cause the least increase in the corresponding brick is chosen. Often, the new item can be inserted in a data page without increasing the area at all.

But sometimes the area, and hence the boundaries, of nodes must change when a new data item is inserted. This is an example of a case where the insertion of an element into a leaf node, although there is room and no splits occur, causes updates of ancestors. For when the boundaries of a leaf node change, its entry in its parent must be updated. This also can affect the grandparent and so forth if further enclosing boundaries are changed. A deletion could also cause a boundary to be changed, although this could be ignored at the parent level without causing false search. The adjustment of boundaries can thus violate property 4, local insertion and deletion.

Node splits are similar to those of the B$^+$-tree because, when a node splits, some of its contents are moved to a newly allocated disk page. A parent index node obtains a new entry describing the boundaries of the new child and the entry referring to the old child is updated to reflect its new boundaries. An R-tree split is illustrated in Figure 54.13.

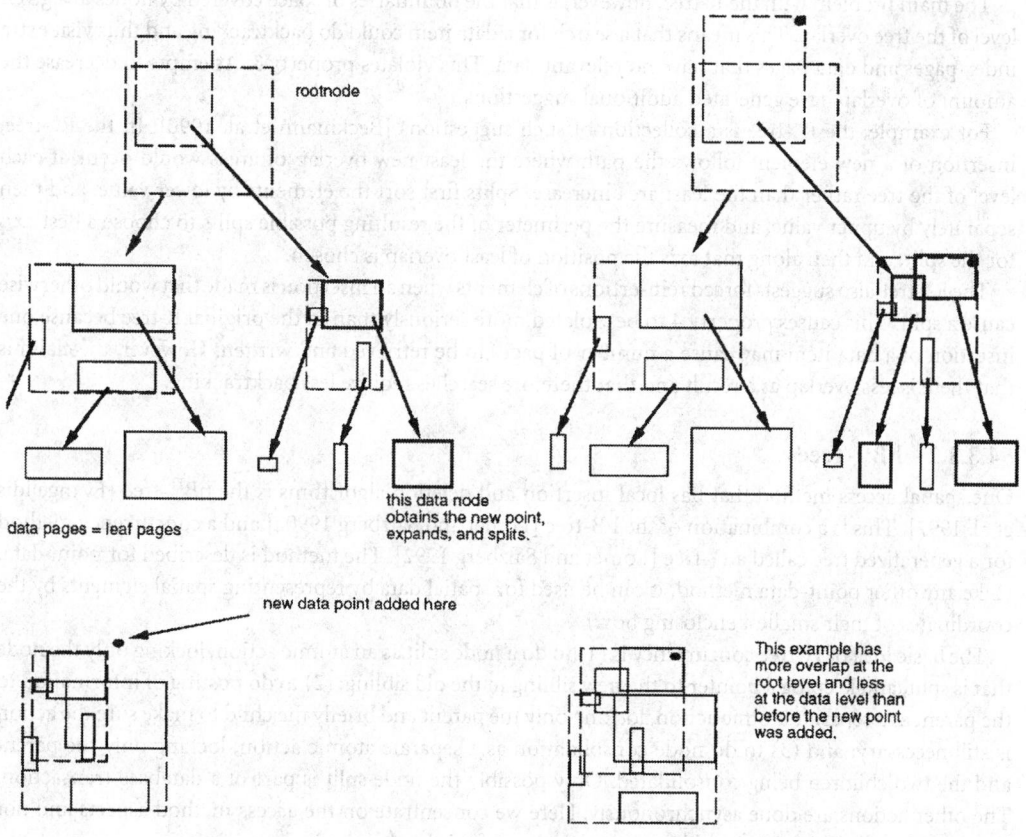

rootnode

this data node
obtains the new point,
expands, and splits.

data pages = leaf pages

new data point added here

This example has
more overlap at the
root level and less
at the data level than
before the new point
was added.

FIGURE 54.13 An R-tree split.

Because an R-tree node can be split in many different ways, an algorithm is presented that first chooses two *seed elements* in the node to be split. One seed element is to remain in the old node, and the other seed is to be placed in the new node. After that, each remaining element is tested to see in which of the two nodes its insertion would cause the greatest increase in area (or volume). These elements are placed in the node that causes the least increase in area, until one of the nodes has reached a threshold in the number of elements. Then all remaining elements are placed in the other node.

Sparse nodes are consolidated by deallocating the sparse node, removing its reference from its parent, adjusting upward boundaries, and reinserting the orphaned elements from the sparse node (each of which requires a search, an insertion, and a possible boundary adjustment). Thus, property 4 is violated by sparse node consolidation.

Another drawback of the R-tree is its sensitivity to dimension. This is not as severe as the problems of grid files; the space usage is $O(n^k)$ because each of the $2k$ coordinates of each child must be stored in its parent. This affects the fan-out.

Keeping boundaries does have the good property of stopping searches in areas that have no points. For example, with the information in the root of the R-tree, one knows the outer boundaries of the entire data collection. A search for a point not in this space can be stopped without accessing any lower-level pages. Because the root is likely to be in main memory, this means negative searches (searches that retrieve no data) can in some cases be very efficient. The trade-off between keeping boundaries and searching in space where there are no points is a general one, not confined to the R-tree. Every search method must make a decision whether to keep boundaries of existing data in index terms, making the index larger but making negative searches more efficient, or having a smaller index but risking accessing several pages when making searches in areas where there is no data.

The main problem with the R-tree, however, is that the boundaries of space covered by nodes at a given level of the tree overlap. This means that a search for a data item could do backtracking and thus visit extra index pages and data pages that have no relevant data. This violates property 3. Attempts to decrease the amount of overlap have generated additional suggestions.

For example, the R*-tree is a collection of such suggestions [Beckmann et al. 1990]. In the R*-tree, insertion of a new element follows the path where the least new overlap of areas would occur at each level of the tree rather than the least area increase. Splits first sort the elements by lower value, and then separately by upper value, and measure the perimeter of the resulting possible splits to choose a best *axis* for the split, and then along that axis the position of least overlap is chosen.

The R*-tree also suggests forced reinsertions of elements when an insertion is made that would otherwise cause a split. This causes property 4 to be violated more seriously than in the original R-tree because one insertion of a data item may cause a number of pages to be retrieved and written. However, the claim is that there is less overlap as a result and that therefore searches require less backtracking.

54.3.3.2 hB$^\Pi$-Trees

One spatial access method that has local insertion and deletion algorithms is the hB$^\Pi$-tree [Evangelidis et al. 1997]. This is a combination of the hB-tree [Lomet and Salzberg 1990a] and a concurrency method for a generalized tree called a Π-tree [Lomet and Salzberg 1992]. The method is described for point data. (Like any other point-data method, it can be used for spatial data by representing spatial elements by the coordinates of their smallest enclosing box.)

The basic idea of Π-tree concurrency is (1) to do a node split as an atomic action, locking only the node that is split, and keeping a pointer to the new sibling in the old sibling; (2) to do posting of information to the parent as a separate atomic action, locking only the parent and briefly the child to make sure the action is still necessary; and (3) to do node consolidation as a separate atomic action, locking only the parent and the two children being consolidated. Only possibly the node split is part of a database transaction. The other actions are done asynchronously. Here we concentrate on the access method aspects and not the concurrency, which is treated in another chapter of this book.

A k-d-tree [Bentley 1979] is used in index nodes to describe the spaces of its children and simultaneously to describe the spaces of the siblings it is pointing to. In data nodes, a k-d-tree also describes the space

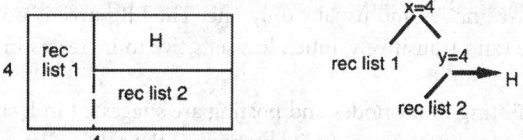

(a) data node organization in hB–Pi tree. A local kd–tree indicates
the spaces where there are record lists and the addresses of
siblings that have been extracted. Here, an upper corner of the
space has been split off and is in another node with disk address "H."

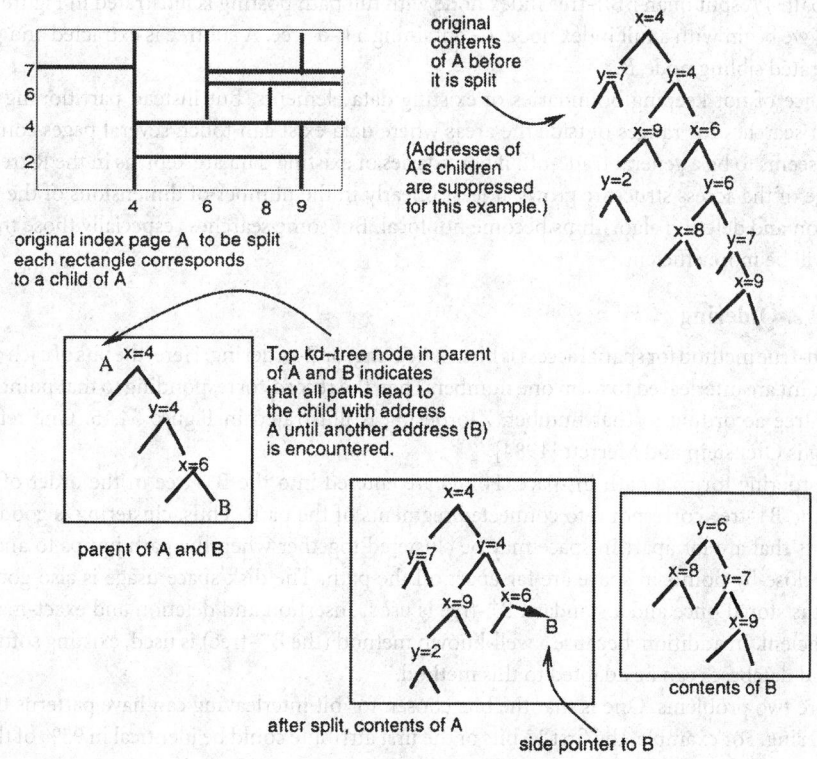

(b) hB–Pi tree index split with full path posted

FIGURE 54.14 An hB$^\Pi$-tree split.

now covered by a sibling the data node points to. An illustration of a data node in an hB$^\Pi$-tree is given in Figure 54.14a.

Search is as in the B$^+$-tree. There is exactly one path from the root to the leaves for any point in the space. There is no backtracking. Insertion of an element never writes on more than one page if there is room for the element in the page, unlike the R-tree.

The splitting discipline of the hB$^\Pi$-tree is similar to the B$^+$-tree or the R-tree. When a page becomes full, some of the contents are moved to a new sibling and an index term will be posted (in this case, asynchronously) in the parent. Unlike the R-tree, the spaces of the two siblings never overlap. At each level of the tree, the spaces corresponding to the pages partition the whole space.

Splitting data nodes is the same in the hB$^\Pi$-tree as in the hB-tree. Usually, one axis can be chosen and between one third and two thirds of the data is on one side of the hyperplane described by some one value on that axis. In Lomet and Salzberg [1990a], it is proved that, in any case, a corner can be chosen that contains between one third and two thirds of the data. The information posted to the parent consists of at most k k-d-tree nodes and usually only one. The hB$^\Pi$-tree is not as sensitive to increases in dimension as the R-tree (and transitively, much less sensitive to increases in dimension than the grid file).

Several variations on splitting index nodes and posting are suggested in Evangelidis et al. [1997]. We briefly describe *full paths* and *split anywhere*. Split anywhere is the split policy of the hB-tree [Lomet and Salzberg 1990a]. Index nodes contain k-d-trees. Index nodes contain no data. To split an index node, one follows the path from the root to a subtree having more than two thirds of the total contents of the node. The split is made there. The subtree is moved to a newly allocated disk page. All of the k-d-tree nodes on the path from the root of the original k-d-tree to the extracted subtree are copied to the parent(s). (This is the full path.) A split of an hB$^\Pi$-tree index node with full path posting is illustrated in Figure 54.14b. In this figure, we begin with a full index node A containing a k-d-tree. A subtree is extracted and placed in a newly allocated sibling node B.

The choice of not keeping boundaries of existing data elements, but instead partitioning the space, means that searches for ranges outside the areas where data exist can touch several pages containing no data. This seems to be a general trade-off: if boundaries of existing data are kept, as in the R-tree, the total space usage of the access structure grows at least linearly in the number of dimensions of the space, and the insertion and deletion algorithms become nonlocal. But some searches (especially those that retrieve no data) will be more efficient.

54.3.3.3 Z-Ordering

A tried-and-true method for spatial access is bit interleaving, or Z-ordering. Here, the bits of each coordinate of a data point are interleaved to form one number. Then the record corresponding to that point is inserted into a B$^+$-tree according to that number. Z-ordering is illustrated in Figure 54.15. One reference for Z-ordering is Orenstein and Merrett [1984].

The Z-ordering forms a path in space. Points are entered into the B$^+$-tree in the order of that path. Leaves of the B$^+$-tree correspond to connected segments of the path. Thus, clustering is good, although some points that are far apart in space may be clustered together when the path jumps to another area, and some close-by points in space are far apart on the path. The disk space usage is also good because each point is stored once and a standard B$^+$-tree is used. Insertion and deletion and exact-match search are also efficient. In addition, because a well-known method (the B$^+$-tree) is used, existing software in file systems and databases can be adapted to this method.

There are two problems. One is that the bits chosen for bit interleaving can have patterns that inhibit good clustering. For example, the first 13 bits of the first attribute could be identical in 95% of the records. Then, if two attributes are used, clustering is good only for the second attribute.

The other problem is the range query. Ranges correspond to many disjoint segments of the path and many different B$^+$-tree leaves. How can these segments be determined? In Orenstein and Merrett [1984], a recursive algorithm finds all segments completely contained in the search area and then obtains all B-tree leaves intersecting those segments. (This may require visiting a number of index pages and data pages that may, in fact, have no points in the search area, but this is true of all spatial access methods as pages whose space intersects the border of the query space may or may not contain answers to the query.)

54.3.4 Temporal Methods

To do *data mining*, for example, to discover trends in buying patterns, or for many legal and financial applications, all old versions of records are maintained. This is in contrast to the usual policy in databases of replacing records with their new versions, or *updating in place*. To maintain a database of current and old versions of records, special access structures are necessary.

In this chapter, we look only at record versions that are marked with a timestamp associated to the transaction that created the version. This is called *transaction time*. Transaction time has the interesting and useful property that it is monotonically increasing. Newly created record versions have more recent timestamps than older versions.

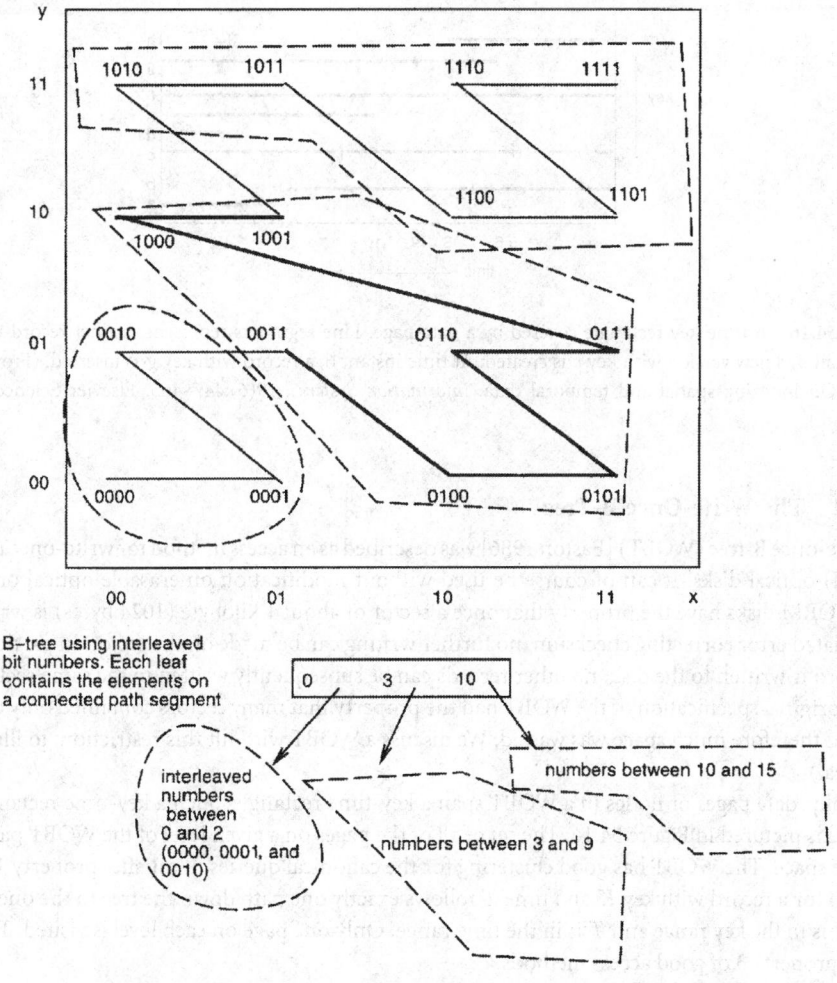

FIGURE 54.15   Z-ordering.

We also assume that each version of a record is assumed valid until a new version is created or until the record with that key is deleted from the database. Thus, to find the version of a record valid at time T, one finds the most recent version created at or before T.

As before, we discuss the indices as if they were primary indices, although, as usual, they can all be regarded as secondary indices if records are replaced with references to records. Thus, our indices will determine the placement of the records in disk pages.

We assume four canonical queries:

1. *Time slice:* Find all records as of time T.
2. *Exact match:* Find the record with key K at time T.
3. *Key range/time slice:* Find records with keys in range (K_1, K_2) valid at time T.
4. *Past versions:* Find all past versions of this record.

A survey of recent temporal access methods appears in Salzberg and Tsotras [1999]. Here, because we assume key ranges are of interest, we look only at access methods that cluster by time and by key range in disk pages. We will in addition restrict ourselves to methods that partition the time-key space rather than allowing overlapping of time-key rectangles. The access methods we outline here are all variations on the write-once B-tree.

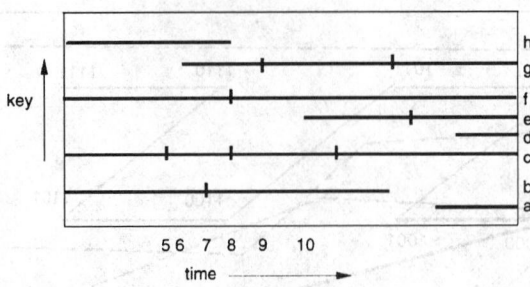

FIGURE 54.16   A time-key rectangle covered by a data page. Line segments represent distinct record versions. At time instant 5, a new version with key c is created. At time instant 6, a record with key g is inserted. (From Salzberg, B. 1994. On indexing spatial and temporal data. *Information Systems*, 19(6):447–465. Elsevier Science Ltd. With permission.)

54.3.4.1   The Write-Once B-Tree

The write-once B-tree (WOBT) [Easton 1986] was described as an access method for write-once read-many (WORM) optical disks. It can of course be used without modification on erasable optical or magnetic disks. WORM disks have the property that once a sector of about 1 kilobyte (1024 bytes) is written, with its associated error correcting checksum, no further writing can be made on the sector. In particular, if say one record is written to the disk, no other records can be subsequently written in the same sector.

(The original specification of the WOBT had the property that many sectors contained only one record each, and therefore much space was wasted. We discuss a WOBT without this restriction, to illustrate the basic idea.)

Basically, data pages or nodes in a WOBT span a key-time rectangle. Such a key-time rectangle on the data level is pictured in Figure 54.16. The set of all of the pages on a given level of the WOBT partition the key-time space. The WOBT has good clustering for the canonical queries. It satisfies property 1.

Search for a record with key K and time T follows exactly one path down the tree to the one data page where K is in the key range and T is in the time range. Only one page on each level is visited. The WOBT satisfies property 3 of good access methods.

The WOBT splits as a B^+-tree. Assume a new record version is to be inserted in a data page, including its key and the current time. All of the current record versions in that page are then copied to one or two newly allocated pages. Only one page is used if the number of current records in the page falls below a threshold. Two pages are used if the number of record versions in the page that are still valid is large. If two pages are used, they are distinguished by key range: all of the current records with key value over or equal to some value K_0 are placed in one page and those with key value less than K_0 are placed in the other page. WOBT data node splits are illustrated in Figure 54.17a (by time) and Figure 54.18a (by time, then by key).

Index nodes are split the same way. They refer to children with a time-key range rectangle. The current children references are copied to the one or two new index nodes. Note that this makes the WOBT a directed acyclic graph (DAG), and not a tree. In particular, current nodes have more than one parent.

When a current node splits, its split time is only in the time-key rectangle of its most recent parent. So the most recent parent is the only one that needs to be updated. The new index term indicates the split time and the address of the new node(s) containing the copies of the current records. In case of a (time-and-) key split, two index terms are posted. The WOBT satisfies property 4, local insertion and deletion algorithms.

However, old data cannot be moved because that would involve updating several parents referring to the old time-key rectangle. (Imagine a long-lived data page whose parents have split many times.) This means that older data cannot be moved to an archive.

In addition, sometimes it does not make sense to make a time split before every key split. This makes unnecessary copies of data. These two problems of the WOBT were addressed by the TSB-tree.

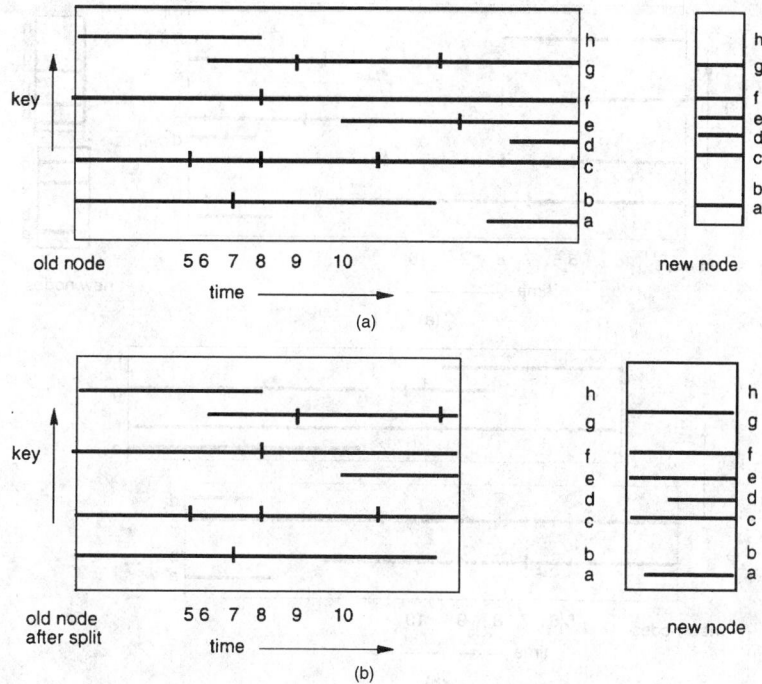

FIGURE 54.17 WOBT and TSB-tree time splits. (a) The WOBT splits at current time, copying current records into a new node. (b) The TSB tree can choose other times to split. (From Salzberg, B. 1994. On indexing spatial and temporal data. *Information Systems*, 19(6):447–465. Elsevier Science Ltd. With permission.)

54.3.4.2 The TSB-Tree

The time-split B-tree (TSB-tree) [Lomet and Salzberg 1989] modifies the splitting algorithms of the WOBT. It allows pure key splits (so that a node may split by key without splitting first by time) and it allows data nodes to split at any time, not just the current time. Data node splits of the TSB-tree are indicated in Figure 54.17b and Figure 54.18b.

In the TSB-tree, index nodes may split only by a time before or at the earliest begin time of any current child. This way, current children have only one parent and old data can be migrated to an archive. Key splits for index nodes must be by a key boundary of a current child. This is also needed to ensure that current children have only one parent. TSB-tree index node splits are illustrated in Figure 54.19.

The total space utilization of the TSB-tree is significantly smaller than that of the WOBT. This is important because all past versions of records are kept and because copies of records are made. An analysis of the space usage is in Lomet and Salzberg [1990b]. Both the TSB-tree and the WOBT use $O(N)$ space, where N is the number of distinct versions. The constant is smaller for the TSB-tree. However, this sometimes comes at a price. Because pure key splits are allowed, it is possible to split a data node so that in some of the earlier time instants not many record versions are valid in one or both of the new nodes. This is illustrated in Figure 54.18b.

Both TSB-tree and the WOBT have no node-consolidation algorithm. This problem is solved by two other extensions of the WOBT.

54.3.4.3 Other Extensions of the Write-Once B-Tree

The persistent B-tree [Lanka and Mays 1991] does time splits by current time only, as does the WOBT. It specifies a node consolidation algorithm by splitting a sparse current node and one of its siblings at current time and copying the combined current record versions to a new node. If there are too many of

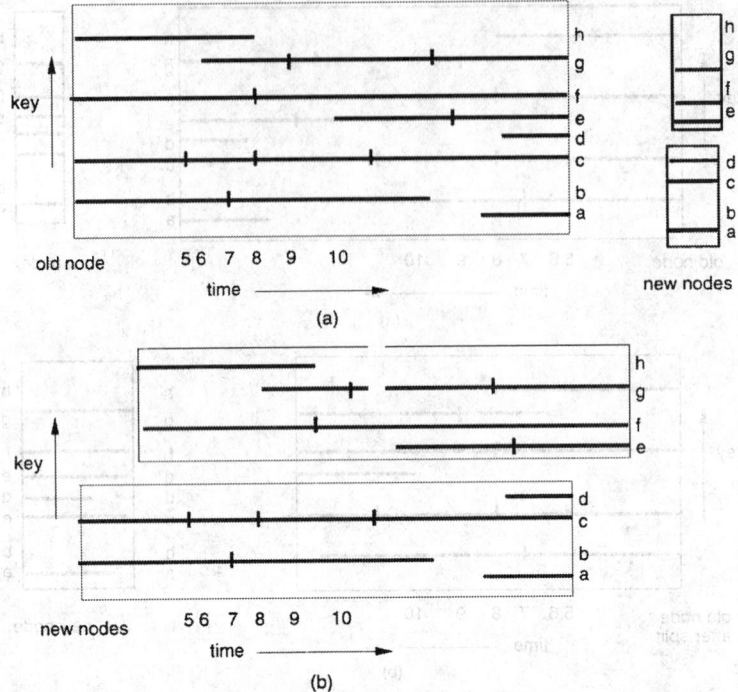

FIGURE 54.18 WOBT and TSB-tree key splits. (a) The WOBT splits data nodes first by time and then sometimes also by key. (b) The TSB-tree can split by key alone. (From Salzberg, B. 1994. On indexing spatial and temporal data. *Information Systems*, 19(6):447–465. Elsevier Science Ltd. With permission.)

them, two new nodes can be used instead. It makes the mistake we discussed earlier with the B^+-tree of having a 50% threshold for node consolidation, thereby allowing thrashing.

It also has two features that are not useful in general: (1) extra nodes on the path from the root to the leaves and (2) an unbalanced overall structure caused by having a *superroot* called *root**. Having a superroot usually increases the average height of the tree. (The WOBT, but not the TSB-tree, also has a superroot.)

The multiversion B-tree [Becker et al. 1993] eliminates the extra nodes in the search path and uses a smaller than 50% threshold for node consolidation but does not eliminate the superroot.

Both the persistent B-tree and the multiversion B-tree always split by current time. This implies that a lower limit on the number of record versions valid at a given time in a node's time interval can be guaranteed, but old record versions cannot be migrated to an archive.

A good compromise of all of the previous methods would be to allow pure-key splitting, as in the TSB-tree, but only when the resulting minimum number of valid records in the time intervals of each new node is above a certain level. Index nodes should be split by earliest begin time of current children, as in the TSB-tree. (This includes the case when they are split for node consolidation.) Node consolidation should be supported as in the persistent B-tree, but with a lower threshold to avoid thrashing as in the multiversion B-tree.

54.3.5 Spatio-Temporal Methods

Recently, with the advances in areas such as mobile computing, GPS technology, and cellular communications, the field of efficiently indexing and querying spatio-temporal objects has received much attention.

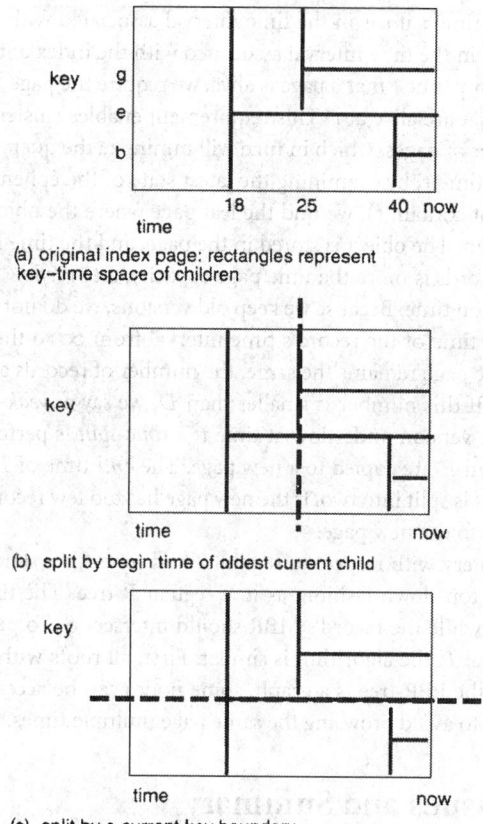

key g
e
b

18   25   40 now
time

(a) original index page: rectangles represent
key–time space of children

key

time   now

(b) split by begin time of oldest current child

key

time   now

(c) split by a current key boundary

FIGURE 54.19 TSB-tree index-node splits. (From Salzberg, B. 1994. On indexing spatial and temporal data. *Information Systems*, 19(6):447–465. Elsevier Science Ltd. With permission.)

An object has both spatial attributes and temporal attributes. Spatio-temporal access methods need to efficiently support the following selection queries — the *region-timeslice query*: "find all objects that are in region R at time t" and the *region-interval query*: "find all objects that are in region R at some time during time interval I."

To efficiently support the region-timeslice query, theoretically we could store a separate R-tree for every time instant. However, the space utilization is prohibitively expensive. Also, region-interval queries will not be supported efficiently because many R-trees need to be browsed.

The **Partially Persistent R-tree (PPR-tree)** [Kollios et al. 2001, Tao and Papadias 2001] is an access method that has asymptotically the same efficiency for timeslice queries as the theoretical approach described in the previous paragraph, while it has linear storage cost. The idea is as follows. Two consecutive versions of ephemeral R-trees are quite similar. Thus we combine the common parts of the ephemeral R-trees and only store separately the difference. This, of course, requires us to store a time interval along with each record specifying all the versions that share it. Here, an index is *partially persistent* if a query can be performed on any version while an update can be performed only on the current version.

The PPR-tree is a directed acyclic graph of pages. This graph embeds many R-trees and has a number of root pages. Each root is responsible for providing access to a subsequent part of the ephemeral R-tree data objects. Each record stored in the PPR-tree is thus extended to also include a time interval. This interval represents both the *start* time when the object was inserted into the database and the *end* time when the object was deleted.

A record is called *alive* at time t if t is in the time interval associated with the record. Similarly, a tree node is called *alive* at t if t is in the time interval associated with the index entry referencing the node. To ensure query efficiency, for any time t that a page is alive, we require the page (with the exception of root) to have at least D records that are alive at t. This requirement enables clustering of the alive objects at a given time in a small number of pages, which in turn will minimize the query I/O.

To insert a new object at time t, by examining the latest state of the ephemeral R-tree (omit all nodes whose time interval does not contain t), we find the leaf page where the object should be inserted using the R-tree insertion algorithm. The object is stored in the page, and the time interval associated with it is $[t, \infty)$. If the number of records is more than the page capacity, an *overflow* happens. Another case that needs special care is at deletion time. Because we keep old versions, we do not physically delete the object. Instead, we change the *end* time of the record's time interval from ∞ to the deletion time t. Although the number of records in the page remains the same, the number of records alive at t (or some later time instant) is reduced by one. If this number is smaller than D, we say a *weak-version underflow* happens. To handle overflow or weak-version underflow at time t, a *time-split* is performed on the target page P as follows. All alive records in P are copied to a new page. The *end* time of P is changed to t. If the new page has too many records, it is split into two. If the new page has too few records, alive records from some sibling page of P are copied to the new page.

To perform a timeslice query with respect to region R and time t, we start with the root page alive at t. The tree is searched in a top-down fashion, as in a regular R-tree. The time interval of every record traversed should contain t, while the record's MBR should intersect S. To perform a query with respect to region S and time interval I, the algorithm is similar. First, all roots with intervals intersecting I are found, and so on. Because the PPR-tree is a graph, some nodes can be accessed multiple times. We can keep a list of accessed pages to avoid browsing the same page multiple times.

54.4 Research Issues and Summary

Good access methods should cluster data according to anticipated queries, use only a reasonable amount of total disk space, be efficient in search, and have local insertion and deletion algorithms. We have seen that for one-dimensional data, usually used in business applications, the B^+-tree has all of these properties. As a result, it is the most used access method today.

Hashing can be very fast, especially for a large and nearly static collection of data. However, as hashed databases grow, they require massive reorganization to regain their good performance. In addition, they do not support range queries. Hashing is the second most used access method today.

The requirements of spatial and temporal indexing, on the other hand, lead to subtle problems. Grid-style solutions tend to take up too much space, especially in large dimensions. R-tree-like solutions with overlapping can have poor search performance due to backtracking. R-trees are also somewhat sensitive to larger dimensions as all boundary coordinates of each child are stored in their parent. Access methods based on interleaved bits depend on the bit patterns of the data. Methods such as the hB^Π-tree, which is not sensitive to increases in dimension but where index terms do not keep boundaries of existing data in children, may cause searches to visit too many data nodes without data in the query area.

Temporal methods (using transaction time) trade off total space usage (numbers of copies of records) with efficiency of search. Some variation of the WOBT, which allows pure key splits some of the time, does node consolidation, and splits index nodes by earliest begin time of current children, is a good compromise solution to the problem.

To index objects with both spatial attributes and temporal attributes, we can use the partially persistent R-tree. It can be thought as many R-trees, one for each time instant. However, by combining the common parts of adjacent versions into one record along with a time interval describing the versions, the partially persistent R-tree has good space utilization as well.

Defining Terms

Arm: The part of a disk drive that moves back and forth toward and away from the center of the disks.

Bucket: One or several consecutive disk pages corresponding to one value of a hashing function.

Clustering index: A commercial term often used to denote a secondary index that is used for data placement only when the data is loaded in the database. After initial loading, the index can be used for record placement only when there is still space left in the correct page. Records never move from the page where they are originally placed. Clustering indices tend not to be clustering after a number of insertions in the database. *This term is avoided in this chapter for this reason.*

Cylinder: The set of tracks on a collection of disks on a disk drive that are the same distance from the center of the disks. (One track on each side of each disk.) Reading information that is stored on the same cylinder of a disk drive is fast because the disk arm does not have to move.

Data page: A disk page in an access method that contains data records.

Dense index: A secondary index.

Disk access: The act of transferring information from a magnetic disk to the main memory of a computer, or the reverse. This involves the mechanical movement of the disk arm so that it is placed at the correct cylinder and the rotation of the disk so that the correct disk page falls under a read/write head on the disk arm.

Disk page: The smallest unit of transfer from (or to) a disk to (or from) main memory. In most systems today, this is 4 kilobytes, or 4096 bytes. It is expected that the size of a disk page will grow in the future so that many systems may begin to have 8-kilobyte disk pages, or even 32-kilobyte disk pages. The reason for the size increase is that main memory space and CPU speed are increasing faster than disk access speed.

Extent: The large amount of consecutive disk space assigned to a relation when it is created and subsequent such chunks of consecutive disk space assigned to the relation as it grows. Some file systems cannot assign extents and thus are unsuitable for many access methods.

Fan-out: The ratio of the size of the data page collection to the size of the index page collection. In B^+-tree N-like access methods, this is approximately the average number of children of an index node, and sometimes fan-out is used in this sense.

Hash table fill factor: The total space needed for the data divided by the space used by the primary area.

Hashing: In hashing, a function maps a database key to the location (address) of the record having that key. (A secondary hashing method maps the database key to the location containing the address of the record.)

Head: The head on a disk arm is where the bits are moved off and onto the disk (read and write). Much effort has been made to allow disk head placement to be more precise so that the density of bits on the disk (number of bits per track and number of tracks per disk of a fixed diameter) can become larger.

Index page: A disk page in an access method or indexing method that does not contain any data records.

Overflow area: That part of a hashing access method where records are placed when there is no room for them in the correct bucket in the primary area.

Primary area: That part of the disk that holds the buckets of a hashing method accessible with one disk access, using the hashing function.

Primary index: A primary index determines the physical placement of records. It does not contain references to individual records. A primary index can be converted to a secondary index by replacing all of the data records with references to data records. Many database systems have no primary indices.

Reference: A reference to an index page or to a data page is a disk address. A reference to a data record can be (1) just the address of the disk page where the record is, with the understanding that some other criteria will be used to locate the record; (2) a disk page address and a slot number within that disk page; or (3) some collection of attribute values from the record that can be used in another index to find the record.

Secondary index: An index that contains a reference to every data record. Secondary indices do not determine data record placement. Sometimes, secondary indices refer to data that is placed in the database in insertion order. Sometimes, secondary indices refer to data that is placed in the database according to a separate primary index based on other attributes of the record. Sometimes, secondary indices refer to data that is loaded into the database originally in the same order as specified by the secondary index but not thereafter. Many database management systems have only secondary indices.

Separator: A prefix of a (possibly former) database key that is long enough to differentiate one page on a lower level of a B^+-tree from the next. These are used instead of database keys in the index pages of a B^+-tree.

Sparse index: A primary index.

Thrashing: When repeated deletions and insertions of records cause the same data page to be repeatedly split and then consolidated, the access method is said to be thrashing. Commercial database systems prevent thrashing by setting the threshold for B^+-tree node consolidation at 0% full; only empty nodes are considered sparse and are consolidated with their siblings.

Track: A circle on one side of one disk with its center in the middle of the disk. Tracks tend to hold on the order of 100,000 bytes. The set of tracks at the same distance from the center, but on different disks or on different sides of the disk, form a cylinder on a given disk drive.

References

Bayer, R. and McCreight, E. 1972. Organization and maintenance of large ordered indices. *Acta Informatica*, 1(3):173–189.

Bayer, R. and Unterauer, K. 1977. Prefix B-trees. *ACM Trans. Database Syst.*, 2(1):11–26.

Becker, B., Gschwind, S., Ohler, T., Seeger, B., and Widmayer, P. 1993. On optimal multiversion access structures. In *Proc. Symp. Large Spatial Databases*. Lecture Notes in Computer Science 692, pp. 123–141. Springer-Verlag, Berlin.

Beckmann, N., Kriegel, H.-P., Schneider, R., and Seeger, B. 1990. The R*-tree: an efficient and robust access method for points and rectangles, pp. 322–331. In *Proc. ACM SIGMOD*.

Bentley, J.L. 1979. Multidimensional binary search trees in database applications. *IEEE Trans. Software Eng.*, 5(4):333–340.

Comer, D. 1979. The ubiquitous B-tree. *Comput. Surv.*, 11(4):121–137.

Easton, M.C. 1986. Key-sequence data sets on indelible storage. *IBM J. Res. Dev.*, 30(3):230–241.

Evangelidis, G., Lomet, D., and Salzberg, B. 1997. The hB^Π-tree: a multiattribute index supporting concurrency, recovery and node consolidation. *J. Very Large Databases*, 6(1).

Fagin, R., Nievergelt, J., Pippenger, N., and Strong, H.R. 1979. Extendible hashing — a fast access method for dynamic files. *Trans. Database Syst.*, 4(3):315–344.

Gaede, V. and Günther, O. 1998. Multidimensional Access Methods, *ACM Computing Surveys*, 30(2).

Guttman, A. 1984. R-trees: a dynamic index structure for spatial searching, pp. 47–57. In *Proc. ACM SIGMOD*.

Johnson, T. and Shasha, D. 1989. B-trees with inserts and deletes: why free-at-empty is better than merge-at-half. *J. Comput. Syst. Sci.*, 47(1):45–76.

Knuth, D.E. 1968. *The Art of Computer Programming*. Addison-Wesley, Reading, MA.

Kollios, G., Gunopulos, D., Tsotras, V.J., Delis, A., and Hadjieleftheriou, M. 2001. Indexing Animated Objects using Spatiotemporal Access Methods. *IEEE Trans. on Knowledge and Data Engineering (TKDE)*, 13(5).

Lanka, S. and Mays, E. 1991. Fully persistent B^+-trees, pp. 426–435. In *Proc. ACM SIGMOD*.

Larson, P. 1980. Linear hashing with partial expansions, pp. 224–232. In *Proc. Very Large Database*.

Litwin, W. 1980. Linear hashing: a new tool for file and table addressing, pp. 212–223. In *Proc. Very Large Database*.

Lomet, D. 1988. A simple bounded disorder file organization with good performance. *Trans. Database Syst.*, 13(4):525–551.

Lomet, D. 1991. Grow and post index trees: role, techniques and future potential. In *2nd Symp. Large Spatial Databases (SSD91), Advances in Spatial Databases.* Lecture Notes in Computer Science 525, pp. 183–206. Springer-Verlag, Berlin.

Lomet, D. and Salzberg, B. 1989. Access methods for multiversion data, pp. 315–324. In *Proc. ACM SIGMOD.*

Lomet, D. and Salzberg, B. 1990a. The hB-tree: A multiattribute indexing method with good guaranteed performance. *Trans. Database Syst.*, 15(4):625–658.

Lomet, D. and Salzberg, B. 1990b. The performance of a multiversion access method, pp. 353–363. In *Proc. ACM SIGMOD.*

Lomet, D. and Salzberg, B. 1992. Access method concurrency with recovery, pp. 351–360. In *Proc. ACM SIGMOD.*

Maier, D. and Salveter, S.C. 1981. Hysterical B-trees. *Inf. Process. Lett.*, 12:199–202.

Mohan, C. and Levine, F. 1992. ARIES/IM: an efficient and high concurrency index management method using write-ahead logging, pp. 371–380. In *Proc. ACM SIGMOD.*

Nievergelt, J., Hinterberger, H., and Sevcik, K.C. 1984. The grid file: an adaptable, symmetric, multikey file structure. *Trans. Database Syst.*, 9(1):38–71.

Orenstein, J.A. and Merrett, T. 1984. A class of data structures for associative searching, pp. 181–190. In *Proc. ACM SIGMOD/SIGACT Principles Database Syst. (PODS).*

Salzberg, B. 1988. *File Structures: An Analytic Approach.* Prentice Hall, Englewood Cliffs, NJ.

Salzberg, B. and Tsotras, V.J. 1999. Comparison of access methods for time-evolving data. *ACM Computing Surveys*, 31(2):158–221.

Smith, G. 1990. Online reorganization of key-sequenced tables and files. (Description of software designed and implemented by F. Putzolu.) *Tandem Syst. Rev.*, 6(2):52–59.

Tao, Y. and Papadias, D. 2001. MV3R-Tree: A Spatio-Temporal Access Method for Timestamp and Interval Queries. *Proc., VLDB.*

Yao, A.C. 1978. On random 2-3 trees. *Acta Informatica*, 9:159–170.

Further Information

To build a solid background on access methods on or before 1998, please refer to the survey paper on spatial access methods [Gaede and Günther 1998] and the survey paper on temporal access methods [Salzberg and Tsotras 1999].

The best information about new access methods for new applications can be obtained by attending top database conferences and reading their proceedings. The best database conferences include: (1) ACM International Conference for Special Interest Group on Management of Data (SIGMOD); (2) International Conference on Very Large Data Bases (VLDB); (3) IEEE International Conference on Data Engineering (ICDE); and (4) International Conference on Extending Database Technology (EDBT). Another top conference is the ACM Symposium on Principles of Database Systems (PODS), which is held annually together with SIGMOD. The papers in PODS generally are more theoretical. For more information on the conferences (e.g., to download research papers), one can visit the following URL: http://www.informatik.uni-trier.de/~ley/db/index.html.

Important database events (e.g., upcoming database conferences or submission deadlines) will be announced in the newsgroup dbworld. To subscribe to dbworld, send a message by e-mail to **listproc@cs.wisc.edu** with the words "subscribe dbworld" and your full name. The reason to attend conferences in person is that you will hear information that is not published and you will get an impression of which articles are the best ones to read in detail.

Although in the database field, conference papers are more important than journal papers, there are some journals worth reading: (1) *ACM Transactions on Database Systems (TODS)*, (2) *The VLDB Journal*; and (3) *IEEE Transactions on Knowledge and Data Engineering (TKDE)*.

There are several excellent textbooks that contain information on access methods: (1) *Transaction Processing: Techniques and Concepts* by Jim Gray and Andreas Reuter has chapters on file structures and access methods in a modern setting. This book was published in 1993 by Morgan Kaufmann. (2) The first author of this chapter, Betty Salzberg, has written a textbook entitled *File Structures: An Analytic Approach*. Many topics touched upon in this chapter are elaborated with exercises and examples. This book was published in 1988 by Prentice Hall. (3) *Database Principles, Programming, and Performance (second edition)* by Patrick E. O'Neil and Elizabeth J. O'Neil, published by Morgan Kaufmann in 2000. (4) *Database Management Systems (third edition)* by Raghu Ramakrishnan and Johannes Gehrke, published by McGraw-Hill in 2001.

55

Query Optimization

55.1 Introduction 55-1
55.2 Query Optimizer Architecture 55-4
      Overall Architecture • Module Functionality
      • Description Focus
55.3 Algebraic Space 55-6
55.4 Planner 55-8
      Dynamic Programming Algorithms • Randomized
      Algorithms • Other Search Strategies
55.5 Size-Distribution Estimator 55-13
      Histograms • Other Techniques
55.6 Noncentralized Environments 55-15
      Parallel Databases • Distributed Databases
55.7 Advanced Types of Optimization 55-16
      Semantic Query Optimization • Global Query Optimization
      • Parametric/Dynamic Query Optimization
55.8 Summary 55-18

Yannis E. Ioannidis
University of Wisconsin

55.1   Introduction

Imagine yourself standing in front of an exquisite buffet filled with numerous delicacies. Your goal is to try them all out, but you need to decide in what order. What order of tastes will maximize the overall pleasure of your palate?

Although much less pleasurable and subjective, that is the type of problem that query optimizers are called to solve. Given a query, there are many plans that a database management system (DBMS) can follow to process it and produce its answer. All plans are equivalent in terms of their final output but vary in their cost, i.e., the amount of time that they need to run. What is the plan that needs the least amount of time?

Such *query optimization* is absolutely necessary in a DBMS. The cost difference between two alternatives can be enormous. For example, consider the following database schema, which will be used throughout this chapter:

emp(name,age,sal,dno)
dept(dno,dname,floor,budget,mgr,ano)
acnt(ano,type,balance,bno)
bank(bno,bname,address)

Further, consider the following very simple SQL query:

select name, floor
from emp, dept
where emp.dno=dept.dno **and** sal> 100 K

TABLE 55.1   Characteristics of a Sample Database

Parameter Description	Parameter Value
Number of emp pages	20,000
Number of emp tuples	100,000
Number of emp tuples with sal>100 K	10
Number of dept pages	10
Number of dept tuples	100
Indices of emp	Clustered B+-tree on emp.sal (3 levels deep)
Indices of dept	Clustered hashing on dept.dno (average bucket length of 1.2 pages)
Number of buffer pages	3
Cost of one disk page access	20 ms

Assume the characteristics in Table 55.1 for the database contents, structure, and run-time environment: Consider the following three different plans:

P1 : Through the B+-tree find all tuples of emp that satisfy the selection on emp.sal. For each one, use the hashing index to find the corresponding dept tuples. (Nested loops, using the index on both relations.)

P2 : For each dept page, scan the entire emp relation. If an emp tuple agrees on the dno attribute with a tuple on the dept page and satisfies the selection on emp.sal, then the emp–dept tuple pair appears in the result. (Page-level nested loops, using no index.)

P3 : For each dept tuple, scan the entire emp relation and store all emp–dept tuple pairs. Then, scan this set of pairs and, for each one, check if it has the same values in the two dno attributes and satisfies the selection on emp.sal. (Tuple-level formation of the cross product, with subsequent scan to test the join and the selection.)

Calculating the expected I/O costs of these three plans shows the tremendous difference in efficiency that equivalent plans may have. P1 needs 0.32 s, P2 needs a bit more than an hour, and P3 needs more than a whole day. Without query optimization, a system may choose plan P2 or P3 to execute this query, with devastating results. Query optimizers, however, examine "all" alternatives, so they should have no trouble choosing P1 to process the query.

The path that a query traverses through a DBMS until its answer is generated is shown in Figure 55.1. The system modules through which it moves have the following functionality:

- The *Query Parser* checks the validity of the query and then translates it into an internal form, usually a relational calculus expression or something equivalent
- The *Query Optimizer* examines all algebraic expressions that are equivalent to the given query and chooses the one that is estimated to be the cheapest
- The *Code Generator* or the *Interpreter* transforms the access plan generated by the optimizer into calls to the query processor
- The *Query Processor* actually executes the query

Queries are posed to a DBMS by interactive users or by programs written in general-purpose programming languages (e.g., C/C++, Fortran, PL/I) that have queries embedded in them. An interactive (ad hoc) query goes through the entire path shown in Figure 55.1. On the other hand, an embedded query goes through the first three steps only once, when the program in which it is embedded is compiled (*compile time*). The code produced by the Code Generator is stored in the database and is simply

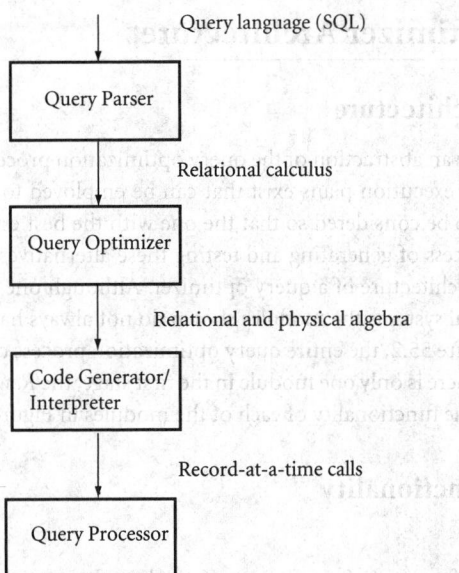

FIGURE 55.1 Query flow through a DBMS.

invoked and executed by the Query Processor whenever control reaches that query during the program execution (*run time*). Thus, independent of the number of times an embedded query needs to be executed, optimization is not repeated until database updates make the access plan invalid (e.g., index deletion) or highly suboptimal (e.g., extensive changes in database contents). There is no real difference between optimizing interactive or embedded queries, so we make no distinction between the two in this chapter.

The area of query optimization is very large within the database field. It has been studied in a great variety of contexts and from many different angles, giving rise to several diverse solutions in each case. The purpose of this chapter is to primarily discuss the core problems in query optimization and their solutions and only touch upon the wealth of results that exist beyond that. More specifically, we concentrate on optimizing a single *flat SQL query* with "and" as the only Boolean connective in its qualification (also known as *conjunctive query, select–project–join query*, or *nonrecursive Horn clause*) in a centralized relational DBMS, assuming that full knowledge of the run-time environment exists at compile time. Likewise, we make no attempt to provide a complete survey of the literature, in most cases providing only a few example references. More extensive surveys can be found elsewhere [Jarke and Koch 1984, Mannino et al. 1988].

The rest of the chapter is organized as follows. Section 55.2 presents a modular architecture for a query optimizer and describes the role of each module in it. Section 55.3 analyzes the choices that exist in the shapes of relational query access plans, and the restrictions usually imposed by current optimizers to make the whole process more manageable. Section 55.4 focuses on the dynamic programming search strategy used by commercial query optimizers and briefly describes alternative strategies that have been proposed. Section 55.5 defines the problem of estimating the sizes of query results and/or the frequency distributions of values in them and describes in detail histograms, which represent the statistical information typically used by systems to derive such estimates. Section 55.6 discusses query optimization in noncentralized environments, i.e., parallel and distributed DBMSs. Section 55.7 briefly touches upon several advanced types of query optimization that have been proposed to solve some hard problems in the area. Finally, Section 55.8 summarizes the chapter and raises some questions related to query optimization that still have no good answer.

55.2   Query Optimizer Architecture

55.2.1   Overall Architecture

In this section, we provide an abstraction of the query optimization process in a DBMS. Given a database and a query on it, several execution plans exist that can be employed to answer the query. In principle, all the alternatives need to be considered so that the one with the best estimated performance is chosen. An abstraction of the process of generating and testing these alternatives is shown in Figure 55.2, which is essentially a modular architecture of a query optimizer. Although one could build an optimizer based on this architecture, in real systems the modules shown do not always have boundaries so clear-cut as in Figure 55.2. Based on Figure 55.2, the entire query optimization process can be seen as having two stages: *rewriting* and *planning*. There is only one module in the first stage, the *Rewriter*, whereas all other modules are in the second stage. The functionality of each of the modules in Figure 55.2 is analyzed below.

55.2.2   Module Functionality

55.2.2.1   Rewriter

This module applies transformations to a given query and produces equivalent queries that are hopefully more efficient, e.g., replacement of views with their definition, flattening out of nested queries, etc. The transformations performed by the Rewriter depend only on the declarative, i.e., static, characteristics of queries and do not take into account the actual query costs for the specific DBMS and database concerned. If the rewriting is known or assumed to always be beneficial, the original query is discarded; otherwise, it is sent to the next stage as well. By the nature of the rewriting transformations, this stage operates at the *declarative* level.

55.2.2.2   Planner

This is the main module of the ordering stage. It examines all possible execution plans for each query produced in the previous stage and selects the overall cheapest one to be used to generate the answer of the original query. It employs a *search strategy*, which examines the space of execution plans in a particular fashion. This space is determined by two other modules of the optimizer, the *Algebraic Space* and the *Method–Structure Space*. For the most part, these two modules and the search strategy determine the cost, i.e., running time, of the optimizer itself, which should be as low as possible. The execution plans examined by the Planner are compared based on estimates of their cost so that the cheapest may be chosen. These costs are derived by the last two modules of the optimizer, the *Cost Model* and the *Size-Distribution Estimator*.

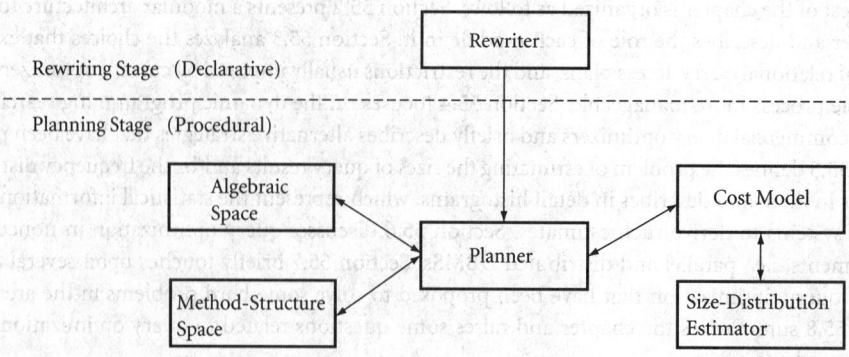

FIGURE 55.2   Query optimizer architecture.

55.2.2.3 Algebraic Space

This module determines the action execution orders that are to be considered by the Planner for each query sent to it. All such series of actions produce the same query answer but usually differ in performance. They are usually represented in relational algebra as formulas or in tree form. Because of the algorithmic nature of the objects generated by this module and sent to the Planner, the overall planning stage is characterized as operating at the *procedural* level.

55.2.2.4 Method–Structure Space

This module determines the implementation choices that exist for the execution of each ordered series of actions specified by the Algebraic Space. This choice is related to the available join methods for each join (e.g., nested loops, merge scan, and hash join), if supporting data structures are built on the fly, if/when duplicates are eliminated, and other implementation characteristics of this sort, which are predetermined by the DBMS implementation. This choice is also related to the available indices for accessing each relation, which is determined by the physical schema of each database stored in its catalogs. Given an algebraic formula or tree from the Algebraic Space, this module produces all corresponding complete execution plans, which specify the implementation of each algebraic operator and the use of any indices.

55.2.2.5 Cost Model

This module specifies the arithmetic formulas that are used to estimate the cost of execution plans. For every different join method, for every different index type access, and in general for every distinct kind of step that can be found in an execution plan, there is a formula that gives its cost. Given the complexity of many of these steps, most of these formulas are simple approximations of what the system actually does and are based on certain assumptions regarding issues like buffer management, disk–CPU overlap, sequential vs random I/O, etc. The most important input parameters to a formula are the size of the buffer pool used by the corresponding step, the sizes of relations or indices accessed, and possibly various distributions of values in these relations. While the first one is determined by the DBMS for each query, the other two are estimated by the Size-Distribution Estimator.

55.2.2.6 Size-Distribution Estimator

This module specifies how the sizes (and possibly frequency distributions of attribute values) of database relations and indices as well as (sub)query results are estimated. As mentioned above, these estimates are needed by the Cost Model. The specific estimation approach adopted in this module also determines the form of statistics that need to be maintained in the catalogs of each database, if any.

55.2.3 Description Focus

Of the six modules of Figure 55.2, three are not discussed in any detail in this chapter: the Rewriter, the Method–Structure Space, and the Cost Model. The Rewriter is a module that exists in some commercial DBMSs (e.g., DB2-Client/Server and Illustra), although not in all of them. Most of the transformations normally performed by this module are considered an advanced form of query optimization and not part of the core (planning) process. The Method–Structure Space specifies alternatives regarding join methods, indices, etc., which are based on decisions made outside the development of the query optimizer and do not really affect much of the rest of it. For the Cost Model, for each alternative join method, index access, etc., offered by the Method–Structure Space, either there is a standard straightforward formula that people have devised by simple accounting of the corresponding actions (e.g., the formula for the tuple-level nested-loop join) or there are numerous variations of formulas that people have proposed and used to approximate these actions (e.g., formulas for finding the tuples in a relation having a random value in an attribute). In either case, the derivation of these formulas is not considered an intrinsic part of the query optimization field. For these reasons, we do not discuss these three modules any further until Section 55.7, where some Rewriter transformations are described. The following three sections provide a detailed description of the Algebraic Space, the Planner, and the Size-Distribution Estimator modules, respectively.

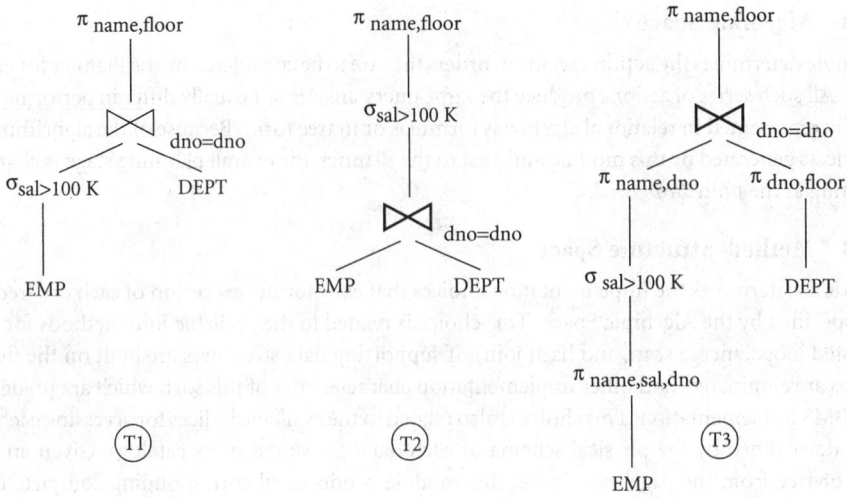

FIGURE 55.3   Examples of general query trees.

55.3   Algebraic Space

As mentioned above, a flat SQL query corresponds to a select–project–join query in relational algebra. Typically, such an algebraic query is represented by a *query tree* whose leaves are database relations and nonleaf nodes are algebraic operators like selections (denoted by σ), projections (denoted by π), and joins* (denoted by ⋈). An intermediate node indicates the application of the corresponding operator on the relations generated by its children, the result of which is then sent further up. Thus, the edges of a tree represent data flow from bottom to top, i.e., from the leaves, which correspond to data in the database, to the root, which is the final operator producing the query answer. Figure 55.3 gives three examples of query trees for the query

> **select** name, floor
> **from** emp, dept
> **where** emp.dno=dept.dno **and** sal>100 K

For a complicated query, the number of all query trees may be enormous. To reduce the size of the space that the search strategy has to explore, DBMSs usually restrict the space in several ways. The first typical restriction deals with selections and projections:

> *R1*: Selections and projections are processed on the fly and almost never generate intermediate relations. Selections are processed as relations are accessed for the first time. Projections are processed as the results of other operators are generated.

For example, plan P1 of Section 55.1 satisfies restriction R1: the index scan of emp finds emp tuples that satisfy the selection on emp.sal on the fly and attempts to join only those; furthermore, the projection on the result attributes occurs as the join tuples are generated. For queries with no join, R1 is moot. For queries with joins, however, it implies that all operations are dealt with as part of join execution. Restriction R1 eliminates only suboptimal query trees, since separate processing of selections and projections incurs additional costs. Hence, the Algebraic Space module specifies alternative query trees with join operators only, selections and projections being implicit.

---

*For simplicity, we think of the cross-product operator as a special case of a join with no join qualification.

Given a set of relations to be combined in a query, the set of all alternative join trees is determined by two algebraic properties of join: commutativity ($R_1 \bowtie R_2 \equiv R_2 \bowtie R_1$) and associativity $[(R_1 \bowtie R_2) \bowtie R_3 \equiv R_1 \bowtie (R_2 \bowtie R_3)]$. The first determines which relation will be inner and which outer in the join execution. The second determines the order in which joins will be executed. Even with the R1 restriction, the alternative join trees that are generated by commutativity and associativity are very large, $\Omega(N!)$ for N relations. Thus, DBMSs usually further restrict the space that must be explored. In particular, the second typical restriction deals with cross products.

> *R2:* Cross products are never formed, unless the query itself asks for them. Relations are combined always through joins in the query.

For example, consider the following query:

> **select** name, floor, balance
> **from** emp, dept, acnt
> **where** emp.dno=dept.dno **and** dept.ano=acnt.ano

Figure 55.4 shows the three possible join trees (modulo join commutativity) that can be used to combine the emp, dept, and acnt relations to answer the query. Of the three trees in the figure, tree T3 has a cross product, since its lower join involves relations emp and acnt, which are not explicitly joined in the query. Restriction R2 almost always eliminates suboptimal join trees due to the large size of the results typically generated by cross products. The exceptions are very few and are cases where the relations forming cross products are extremely small. Hence, the algebraic-space module specifies alternative join trees that involve no cross product. The exclusion of unnecessary cross products reduces the size of the space to be explored, but that still remains very large. Although some systems restrict the space no further (e.g., Ingres and DB2-Client/Server), others require an even smaller space (e.g., DB2/MVS). In particular, the third typical restriction deals with the shape of join trees:

> *R3:* The inner operand of each join is a database relation, never an intermediate result.

For example, consider the following query:

> **select** name, floor, balance, address
> **from** emp, dept, acnt, bank
> **where** emp.dno=dept.dno **and** dept.ano=acnt.ano **and** acnt.bno=bank.bno

Figure 55.5 shows three possible cross-product-free join trees that can be used to combine the emp, dept, acnt, and bank relations to answer the query. Tree T1 satisfies restriction R3, whereas trees T2 and T3 do not, since they have at least one join with an intermediate result as the inner relation. Because of their shape (Figure 55.5), join trees that satisfy restriction R3, e.g., tree T1, are called *left-deep*. Trees that have their outer relation always being a database relation, e.g., tree T2, are called *right-deep*. Trees with at least one join between two intermediate results, e.g., tree T3, are called *bushy*. Restriction R3 is of a more heuristic nature than R1 and R2 and may well eliminate the optimal plan in some cases. It has been claimed that

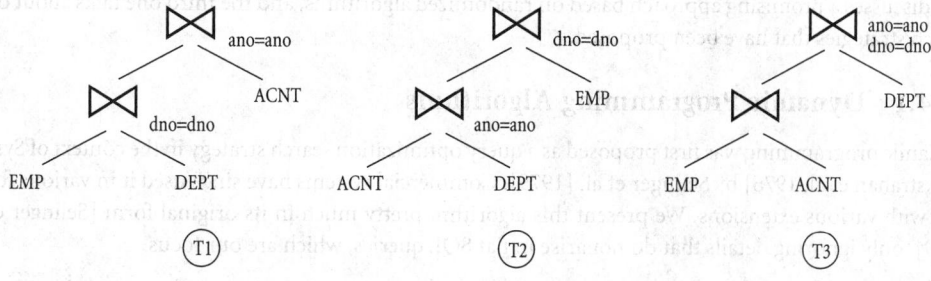

FIGURE 55.4 Examples of join trees; T3 has a cross product.

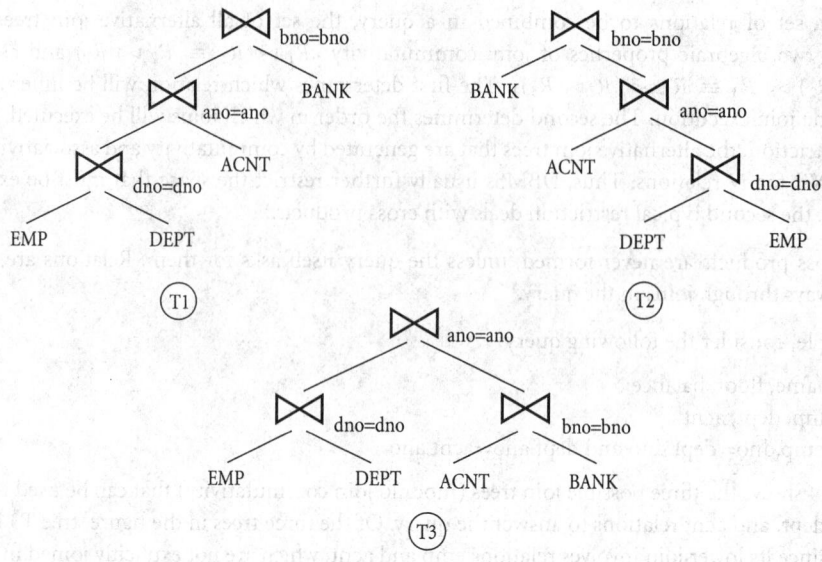

FIGURE 55.5 Examples of left-deep (T1), right-deep (T2), and bushy (T3) join trees.

most often the optimal left-deep tree is not much more expensive than the optimal tree overall. The typical arguments used are two:

- Having original database relations as inners increases the use of any preexisting indices.
- Having intermediate relations as outers allows sequences of nested-loops joins to be executed in a pipelined fashion.*

Both index usage and pipelining reduce the cost of join trees. Moreover, restriction R3 significantly reduces the number of alternative join trees to $O(2^N)$ for many queries with N relations. Hence, the Algebraic Space module of the typical query optimizer specifies only join trees that are left-deep.

In summary, typical query optimizers make restrictions R1, R2, and R3 to reduce the size of the space they explore. Hence, unless otherwise noted, our descriptions follow these restrictions as well.

55.4 Planner

The role of the Planner is to explore the set of alternative execution plans, as specified by the Algebraic Space and the Method–Structure Space, and find the cheapest one, as determined by the Cost Model and the Size-Distribution Estimator. The following three subsections deal with different types of search strategies that the Planner may employ for its exploration. The first one focuses on the most important strategy, dynamic programming, which is the one used by essentially all commercial systems. The second one discusses a promising approach based on randomized algorithms, and the third one talks about other search strategies that have been proposed.

55.4.1 Dynamic Programming Algorithms

Dynamic programming was first proposed as a query optimization search strategy in the context of System R [Astrahan et al. 1976] by Selinger et al. [1979]. Commercial systems have since used it in various forms and with various extensions. We present this algorithm pretty much in its original form [Selinger et al. 1979], only ignoring details that do not arise in flat SQL queries, which are our focus.

---

*A similar argument can be made in favor of right-deep trees regarding sequences of hash joins.

The algorithm is essentially a dynamically pruning, exhaustive search algorithm. It constructs all alternative join trees (that satisfy restrictions R1–R3) by iterating on the number of relations joined so far, always pruning trees that are known to be suboptimal. Before we present the algorithm in detail, we need to discuss the issue of *interesting order*. One of the join methods that is usually specified by the Method–Structure-Space module is *merge scan*. Merge scan first sorts the two input relations on the corresponding join attributes and then merges them with a synchronized scan. If any of the input relations, however, is already sorted on its join attribute (e.g., because of earlier use of a B+-tree index or sorting as part of an earlier merge-scan join), the sorting step can be skipped for the relation. Hence, given two partial plans during query optimization, one cannot compare them based on their cost only and prune the more expensive one; one has to also take into account the sorted order (if any) in which their result comes out. One of the plans may be more expensive but may generate its result sorted on an attribute that will save a sort in a subsequent merge-scan execution of a join. To take into account these possibilities, given a query, one defines its *interesting orders* to be orders of intermediate results on any relation attributes that participate in joins. (For more general SQL queries, attributes in order-by and group-by clauses give rise to interesting orders as well.) For example, in the query of Section 55.3, orders on the attributes emp.dno, dept.dno, dept.ano, acnt.ano, acnt.bno, and bank.bno are interesting. During optimization of this query, if any intermediate result comes out sorted on any of these attributes, then the partial plan that gave this result must be treated specially.

Using the above, we give below a detailed English description of the dynamic programming algorithm optimizing a query of N relations:

Step 1: For each relation in the query, all possible ways to access it, i.e., via all existing indices and including the simple sequential scan, are obtained. (Accessing an index takes into account any query selection on the index key attribute.) These partial (single-relation) plans are partitioned into equivalence classes based on any interesting order in which they produce their result. An additional equivalence class is formed by the partial plans whose results are in no interesting order. Estimates of the costs of all plans are obtained from the Cost Model module, and the cheapest plan in each equivalence class is retained for further consideration. However, the cheapest plan of the no-order equivalence class is not retained if it is not cheaper than all other plans.

Step 2: For each pair of relations joined in the query, all possible ways to evaluate their join using all relation access plans retained after step 1 are obtained. Partitioning and pruning of these partial (two-relation) plans proceeds as above.

⋮

Step i: For each set of $i - 1$ relations joined in the query, the cheapest plans to join them for each interesting order are known from the previous step. In this step, for each such set, all possible ways to join one more relation with it without creating a cross product are evaluated. For each set of i relations, all generated (partial) plans are partitioned and pruned as before.

⋮

Step N: All possible plans to answer the query (the unique set of N relations joined in the query) are generated from the plans retained in the previous step. The cheapest plan is the final output of the optimizer to be used to process the query.

For a given query, the above algorithm is guaranteed to find the optimal plan among those satisfying restrictions R1–R3. It often avoids enumerating all plans in the space by being able to dynamically prune suboptimal parts of the space as partial plans are generated. In fact, although in general still exponential, there are query forms for which it generates only $O(N^3)$ plans [Ono and Lohman 1990].

An example that shows dynamic programming in its full detail takes too much space. We illustrate its basic mechanism by showing how it would proceed on the simple query below:

select name, mgr
from emp, dept
where emp.dno=dept.dno **and** sal>30 K **and** floor=2

TABLE 55.2 Entire Set of Alternatives Appropriately Partioned

Relation	Interesting Order	Plan Description	Cost
emp	emp.dno	Access through B+-tree on emp.dno.	700
	—	Access through B+-tree on emp.sal.	200
		Sequential scan.	600
dept	—	Access through hashing on dept.floor.	50
		Sequential scan.	200

TABLE 55.3 Entire Set of Alternatives for the Last Step of the Algorithm

Join Method	Outer/Inner		Plan Description	Cost
Nested loops	emp/dept	•	For each emp tuple obtained through the B+-tree on emp.sal, scan dept through the hashing index on dept.floor to find tuples matching on dno.	1800
		•	For each emp tuple obtained through the B+-tree on emp.dno and satisfying the selection on emp.sal, scan dept through the hashing index on dept.floor to find tuples matching on dno.	3000
	dept/emp	•	For each dept tuple obtained through the hashing index on dept.floor, scan emp through the B+-tree on emp.sal to find tuples matching on dno.	2500
		•	For each dept tuple obtained through the hashing index on dept.floor, probe emp through the B+-tree on emp.dno using the value in dept.dno to find tuples satisfying the selection on emp.sal.	1500
Merge scan	—	•	Sort the emp tuples resulting from accessing the B+-tree on emp.sal into L_1.	2300
		•	Sort the dept tuples resulting from accessing the hashing index on dept.floor into L_2.	
		•	Merge L_1 and L_2.	
		•	Sort the dept tuples resulting from accessing the hashing index on dept.floor into L_2.	2000
		•	Merge L_2 and the emp tuples resulting from accessing the B+-tree on emp.dno and satisfying the selection on emp.sal.	

Assume that there is a B+-tree index on emp.sal, a B+-tree index on emp.dno, and a hashing index on dept.floor. Also assume that the DBMS supports two join methods: nested loops and merge scan. (Both types of information should be specified in the Method–Structure-Space module.) Note that, based on the definition, potential interesting orders are those on emp.dno and dept.dno, since these are the only join attributes in the query. The algorithm proceeds as follows:

Step 1: All possible ways to access emp and dept are found. The only interesting order arises from accessing emp via the B+-tree on emp.dno, which generates the emp tuples sorted and ready for the join with dept. The entire set of alternatives, appropriately partitioned, is shown in Table 55.2. Each partial plan is associated with some hypothetical cost; in reality, these costs are obtained from the Cost-Model module. Within each equivalence class, only the cheapest plan is retained for the next step, as indicated by the boxes surrounding the corresponding costs in the table.

Step 2: Since the query has two relations, this is the last step of the algorithm. All possible ways to join emp and dept are found, using both supported join methods and all partial plans for individual relation access retained from step 1. For the nested-loops method, which relation is inner and which is outer is also specified. Since this is the last step of the algorithm, there is no issue of interesting orders. The entire set of alternatives is shown in Table 55.3 in a way similar to step 1. Based on hypothetical costs for each of the plans, the optimizer produces as output the plan indicated by the box surrounding the corresponding cost in the table.

As the above example illustrates, the choices offered by the Method–Structure Space in addition to those of the Algebraic Space result in an extraordinary number of alternatives that the optimizer must search through. The memory requirements and running time of dynamic programming grow exponentially with query size (i.e., number of joins) in the worst case, since all viable partial plans generated in each step must be stored to be used in the next one. In fact, many modern systems place a limit on the size of queries that can be submitted (usually around fifteen joins), because for larger queries the optimizer crashes due to its very high memory requirements. Nevertheless, most queries seen in practice involve less than ten joins, and the algorithm has proved to be very effective in such contexts. It is considered the standard in query optimization search strategies.

55.4.2 Randomized Algorithms

To address the inability of dynamic programming to cope with really large queries, which appear in several novel application fields, several other algorithms have been proposed recently. Of these, randomized algorithms, i.e., algorithms that "flip coins" to make decisions, appear very promising.

The most important class of these optimization algorithms is based on plan transformations instead of the plan construction of dynamic programming, and includes algorithms like *Simulated Annealing*, *Iterative Improvement*, and *Two-Phase Optimization*. These are generic algorithms that can be applied to a variety of optimization problems and are briefly described below as adapted to query optimization. They operate by searching a graph whose nodes are all the alternative execution plans that can be used to answer a query. Each node has a cost associated with it, and the goal of the algorithm is to find a node with the globally minimum cost. Randomized algorithms perform *random walks* in the graph via a series of *moves*. The nodes that can be reached in one move from a node S are called the *neighbors* of S. A move is called *uphill* move (*downhill*) if the cost of the source node is lower (higher) than the cost of the destination node. A node is a *global minimum* if it has the lowest cost among all nodes. It is a *local minimum* if, in all paths starting at that node, any downhill move comes after at least one uphill move.

55.4.2.1 Algorithm Description

Iterative Improvement (II) [Nahar et al. 1986, Swami 1989, Swami and Gupta 1988] performs a large number of *local optimizations*. Each one starts at a random node and repeatedly accepts random downhill moves until it reaches a local minimum. II returns the local minimum with the lowest cost found.

Simulated Annealing (SA) performs a continuous random walk, accepting downhill moves always and uphill moves with some probability, trying to avoid being caught in a high-cost local minimum [Ioannidis and Kang 1990, Ioannidis and Wong 1987, Kirkpatrick et al. 1983]. This probability decreases as time progresses and eventually becomes zero, at which point execution stops. Like II, SA returns the node with the lowest cost visited.

The *Two-Phase Optimization* (2PO) algorithm is a combination of II and SA [Ioannidis and Kang 1990]. In phase 1, II is run for a small period of time, i.e., a few local optimizations are performed. The output of that phase, which is the best local minimum found, is the initial node of the next phase. In phase 2, SA is run starting from a low probability for uphill moves. Intuitively, the algorithm chooses a local minimum and then searches the area around it, still being able to move in and out of local minima, but practically unable to climb up very high hills.

55.4.2.2 Results

Given a finite amount of time, these randomized algorithms have performance that depends on the characteristics of the cost function over the graph and the connectivity of the latter as determined by the neighbors of each node. They have been studied extensively for query optimization, being mutually compared and also compared against dynamic programming [Ioannidis and Kang 1990, Ioannidis and Wong 1987, Kang 1991, Swami 1989, Swami and Gupta 1988]. The specific results of these comparisons vary depending on the choices made regarding issues of the algorithms' implementation and setup, but also choices made in other modules of the query optimizer, i.e., the Algebraic Space, the Method–Structure Space, and the Cost Model. In general, however, the conclusions are as follows. First, up to about ten joins, dynamic programming is preferred over the randomized algorithms because it is faster and it guarantees finding the optimal plan. For larger queries, the situation is reversed, and despite the probabilistic nature of the randomized algorithms, their efficiency makes them the algorithms of choice. Second, among randomized algorithms, II usually finds a reasonable plan very quickly, while given enough time, SA is able to find a better plan than II. 2PO gets the best of both worlds and is able to find plans that are as good as those of SA, if not better, in much shorter time.

55.4.3 Other Search Strategies

To complete the picture on search strategies we briefly describe several other algorithms that people have proposed in the past, deterministic, heuristic, or randomized. Ibaraki and Kameda were the ones that proved that query optimization is an NP-complete problem even if considering only the nested-loops join method [Ibaraki and Kameda 1984]. Given that result, there have been several efforts to obtain algorithms that solve important subcases of the query optimization problem and run in polynomial time. Ibaraki and Kameda themselves presented an algorithm (referred to as IK here) that takes advantage of the special form of the cost formula for nested loops and optimizes a tree query of N joins in $O(N^2 \log N)$ time. They also presented an algorithm that is applicable to even cyclic queries and finds a good (but not always optimal) plan in $O(N^3)$ time.

The KBZ algorithm uses essentially the same techniques, but it is more general and more sophisticated and runs in $O(N^2)$ time for tree queries [Krishnamurthy et al. 1986]. As with IK, the applicability of KBZ depends on the cost formulas for joins to be of a specific form. Nested loops and hash join satisfy this requirement but, in general, merge scan does not.

The AB algorithm mixes deterministic and randomized techniques and runs in $O(N^4)$ time [Swami and Iyer 1993]. It uses KBZ as a subroutine, which needs $O(N^2)$ time, and essentially executes it $O(N^2)$ times on randomly selected spanning trees of the query graph. Through an interesting separation of the cost of merge scan into a part that affects optimization and a part that does not, AB is applicable to all join methods despite the dependence on KBZ.

In addition to SA, II, and 2PO, *Genetic Algorithms* [Goldberg 1989] form another class of generic randomized optimization algorithms that have been applied to query optimization. These algorithms simulate a biological phenomenon: a random set of solutions to the problem, each with its own cost, represents an initial population; pairs of solutions from that population are matched (*cross over*) to generate offspring that obtain characteristics from both parents, and the new children may also be randomly changed in small ways (*mutation*); between the parents and the children, those with the least cost (*most fit*) survive in the next generation. The algorithm ends when the entire population consists of copies of the same solution, which is considered to be optimal. Genetic algorithms have been implemented for query optimization with promising results [Bennett et al. 1991].

Another interesting randomized approach to query optimization is pure, uniformly random generation of access plans [Galindo-Legaria et al. 1994]. Truly uniform generation is a hard problem but has been solved for tree queries. With an efficient implementation of this step, experiments with the algorithm have shown good potential, since there is no dependence on plan transformations or random walks.

In the artificial intelligence community, the A* heuristic algorithm is extensively used for complex search problems. A* has been proposed for query optimization as well and can be seen as a direct extension to the traditional dynamic programming algorithm [Yoo and Lafortune 1989]. Instead of proceeding in steps and using all plans with n relations to generate all plans with $n + 1$ relations together, A* proceeds by expanding one of the generated plans at hand at a time, based on its expected proximity to the optimal plan. Thus, A* generates a full plan much earlier than dynamic programming and is able to prune more aggressively in a branch-and-bound mode. A* has been proposed for query optimization and has been shown quite successful for not very large queries.

Finally, in the context of extensible DBMSs, several unique search strategies have been proposed, which are all rule-based. Rules are defined on how plans can be constructed or modified, and the Planner follows the rules to explore the specified plan space. The most representative of these efforts are those of Starburst [Haas et al. 1990, Lohman 1988] and Volcano/Exodus [Graefe and DeWitt 1987, Graefe and McKenna 1993]. The Starburst optimizer employs constructive rules, whereas the Volcano/Exodus optimizers employ transformation rules.

55.5   Size-Distribution Estimator

The final module of the query optimizer that we examine in detail is the Size-Distribution Estimator. Given a query, it estimates the sizes of the results of (sub)queries and the frequency distributions of values in attributes of these results.

Before we present specific techniques that have been proposed for estimation, we use an example to clarify the notion of frequency distribution. Consider the simple relation *OLYMPIAN* on the left in Table 55.4, with the frequency distribution of the values in its Department attribute on the right.

One can generalize the above and discuss distributions of frequencies of combinations of arbitrary numbers of attributes. In fact, to calculate/estimate the size of any query that involves multiple attributes from a single relation, multiattribute joint frequency distributions or their approximations are required. Practical DBMSs, however, deal with frequency distributions of individual attributes only, because considering all possible combinations of attributes is very expensive. This essentially corresponds to what is known as the *attribute value independence assumption*, and, although rarely true, it is adopted by all current DBMSs.

Several techniques have been proposed in the literature to estimate query result sizes and frequency distributions, most of them contained in the extensive survey by Mannino et al. [1988] and elsewhere [Christodoulakis 1989]. Most commercial DBMSs (e.g., DB2, Informix, Ingres, Sybase, Microsoft SQL server) base their estimation on *histograms*, so our description mostly focuses on those. We then briefly summarize other techniques that have been proposed.

55.5.1   Histograms

In a *histogram* on attribute a of relation R, the domain of a is partitioned into *buckets*, and a uniform distribution is assumed within each bucket. That is, for any bucket b in the histogram, if a value $v_i \in b$, then the frequency f_i of v_i is approximated by $\sum_{v_j \in b} f_j / |b|$. A histogram with a single bucket generates the same approximate frequency for all attribute values. Such a histogram is called *trivial* and corresponds to making the *uniform distribution assumption* over the entire attribute domain. Note that, in principle,

TABLE 55.4   The Relation Olympian with the Frequency Distribution of the Values in its Department Attribute

Name	Salary	Department	Department	Frequency
Zeus	100 K	General Management	General Management	2
Poseidon	80 K	Defense	Defense	2
Pluto	80 K	Justice	Education	1
Aris	50 K	Defense	Domestic Affairs	2
Ermis	60 K	Commerce	Agriculture	1
Apollo	60 K	Energy	Commerce	1
Hefestus	50 K	Energy	Justice	1
Hera	90 K	General Management	Energy	3
Athena	70 K	Education		
Aphrodite	60 K	Domestic Affairs		
Demeter	60 K	Agriculture		
Hestia	50 K	Domestic Affairs		
Artemis	60 K	Energy		

TABLE 55.5   An Example of Two Histograms on the Department Attribute

	Histogram H1		Histogram H2	
Department	Frequency in Bucket	Approximate Frequency	Frequency in Bucket	Approximate Frequency
Agriculture	[1]	1.5	[1]	1.33
Commerce	[1]	1.5	[1]	1.33
Defense	[2]	1.5	[2]	1.33
Domestic Affairs	[2]	1.5	(2)	2.5
Education	(1)	1.75	[1]	1.33
Energy	(3)	1.75	(3)	2.5
General Management	(2)	1.75	[2]	1.33
Justice	(1)	1.75	[1]	1.33

any arbitrary subset of an attribute's domain may form a bucket and not necessarily consecutive ranges of its natural order.

Continuing with the example of the *OLYMPIAN* relation, we present in Table 55.5 two different histograms on the Department attribute, both with two buckets. For each histogram, we first show which frequencies are grouped in the same bucket by enclosing them in the same shape (box or circle), and then we show the resulting approximate frequency, i.e., the average of all frequencies enclosed by identical shapes.

There are various classes of histograms that systems use or researchers have proposed for estimation. Most of the earlier prototypes, and still some of the commercial DBMSs, use trivial histograms, i.e., make the uniform distribution assumption [Selinger et al. 1979]. That assumption, however, rarely holds in real data, and estimates based on it usually have large errors [Christodoulakis 1984, Ioannidis and Christodoulakis 1991]. Excluding trivial ones, the histograms that are typically used belong to the class of *equiwidth* histograms [Kooi 1980]. In those, the number of consecutive attribute values or the size of the range of attribute values associated with each bucket is the same, independent of the frequency of each attribute value in the data. Since these histograms store a lot more information than trivial histograms (they typically have 10–20 buckets), their estimations are much better. Histogram H1 above is equiwidth, since the first bucket contains four values starting from A–D and the second bucket contains also four values starting from E–Z.

Although we are not aware of any system that currently uses histograms in any other class than those mentioned above, several more advanced classes have been proposed and are worth discussing. *Equidepth* (or *equiheight*) histograms are essentially duals of equiwidth histograms [Kooi 1980, Piatetsky-Shapiro and Connell 1984]. In those, the sum of the frequencies of the attribute values associated with each bucket is the same, independent of the number of those attribute values. Equiwidth histograms have a much higher worst-case and average error for a variety of selection queries than equidepth histograms. Muralikrishna and DeWitt [1988] extended the above work for multidimensional histograms that are appropriate for multiattribute selection queries.

In *serial* histograms [Ioannidis and Christodoulakis 1993], the frequencies of the attribute values associated with each bucket are either all greater or all less than the frequencies of the attribute values associated with any other bucket. That is, the buckets of a serial histogram group frequencies that are close to each other with no interleaving. Histogram H1 in Table 55.5 is not serial, as frequencies 1 and 3 appear in one bucket and frequency 2 appears in the other, while histogram H2 is. Under various optimality criteria, serial histograms have been shown to be optimal for reducing the worst-case and the average error in equality selection and join queries [Ioannidis 1993, Ioannidis and Christodoulakis 1993, Ioannidis and Poosala 1995].

Identifying the optimal histogram among all serial ones takes exponential time in the number of buckets. Moreover, since there is usually no order correlation between attribute values and their frequencies, storage of serial histograms essentially requires a regular index that will lead to the approximate frequency of every individual attribute value. Because of all these complexities, the class of *end-biased* histograms has been introduced. In those, some number of the highest frequencies and some number of the lowest frequencies in an attribute are explicitly and accurately maintained in separate individual buckets, and the remaining (middle) frequencies are all approximated together in a single bucket. End-biased histograms are serial, since their buckets group frequencies with no interleaving. Identifying the optimal end-biased histogram, however, takes only slightly over linear time in the number of buckets. Moreover, end-biased histograms require little storage, since usually most of the attribute values belong in a single bucket and do not have to be stored explicitly. Finally, in several experiments it has been shown that most often the errors in the estimates based on end-biased histograms are not too far off from the corresponding (optimal) errors based on serial histograms. Thus, as a compromise between optimality and practicality, it has been suggested that the optimal end-biased histograms should be used in real systems.

55.5.2 Other Techniques

In addition to histograms, several other techniques have been proposed for query result size estimation [Christodoulakis 1989, Mannino et al. 1988]. Those that, like histograms, store information in the database typically approximate a frequency distribution by a parametrized mathematical distribution or a polynomial. Although requiring very little overhead, these approaches are typically inaccurate because most often real data do not follow any mathematical function. On the other hand, those based on *sampling* primarily operate at run time [Haas and Swami 1992, 1995, Lipton et al. 1990, Olken and Rotem 1986] and compute their estimates by collecting and possibly processing random samples of the data. Although producing highly accurate estimates, sampling is quite expensive, and therefore its practicality in query optimization is questionable, especially since optimizers need query result size estimations frequently.

55.6 Noncentralized Environments

The preceding discussion focuses on query optimization for sequential processing. This section touches upon issues and techniques related to optimizing queries in noncentralized environments. The focus is on the Method–Structure-Space and Planner modules of the optimizer, as the remaining ones are not significantly different from the centralized case.

55.6.1 Parallel Databases

Among all parallel architectures, the shared-nothing and the shared-memory paradigms have emerged as the most viable ones for database query processing. Thus, query optimization research has concentrated on these two. The processing choices that either of these paradigms offers represent a huge increase over the alternatives offered by the Method–Structure-Space module in a sequential environment. In addition to the sources of alternatives that we discussed earlier, the Method–Structure-Space module offers two more: the number of processors that should be given to each database operation (*intraoperator parallelism*) and placing operators into groups that should be executed simultaneously by the available processors (*interoperator parallelism*, which can be further subdivided into *pipelining* and *independent parallelism*). The *scheduling* alternatives that arise from these two questions add at least another superexponential factor to the total number of alternatives and make searching an even more formidable task. Thus, most systems and research prototypes adopt various heuristics to avoid dealing with a very large search space. In the two-stage approach [Hong and Stonebraker 1991], given a query, one first identifies the optimal sequential plan for it using conventional techniques like those discussed in Section 55.4, and then one identifies the optimal parallelization/scheduling of that plan. Various techniques have been proposed in the literature for the second stage, but none of them claims to provide a complete and optimal answer to the scheduling question, which remains an open research problem. In the segmented execution model, one considers only schedules that process memory-resident right-deep segments of (possibly bushy) query plans one-at-a-time (i.e., no independent interoperator parallelism). Shekita et al. [1993] combined this model with a novel heuristic search strategy with good results for shared-memory. Finally, one may be restricted to deal with right-deep trees only [Schneider and DeWitt 1990].

In contrast to all the search-space reduction heuristics, Lanzelotte et al. [1993] dealt with both deep and bushy trees, considering schedules with independent parallelism, where all the pipelines in an execution are divided into phases, pipelines in the same phase are executed in parallel, and each phase starts only after the previous phase ended. The search strategy that they used was a randomized algorithm, similar to 2PO, and proved very effective in identifying efficient parallel plans for a shared-nothing architecture.

55.6.2 Distributed Databases

The difference between distributed and parallel DBMSs is that the former are formed by a collection of independent, semiautonomous processing sites that are connected via a network that could be spread over a large geographic area, whereas the latter are individual systems controlling multiple processors that are in the same location, usually in the same machine room. Many prototypes of distributed DBMSs have been implemented [Bernstein et al. 1981, Mackert and Lohman 1986], and several commercial systems are offering distributed versions of their products as well (e.g., DB2, Informix, Sybase, Oracle).

Other than the necessary extensions of the Cost-Model module, the main differences between centralized and distributed query optimization are in the Method–Structure-Space module, which offers additional processing strategies and opportunities for transmitting data for processing at multiple sites. In early distributed systems, where the network cost was dominating every other cost, a key idea was using semijoins for processing in order to only transmit tuples that would certainly contribute to join results [Bernstein et al. 1981, Mackett and Lohman 1986]. An extension of that idea is using Bloom filters, which are bit vectors that approximate join columns and are transferred across sites to determine which tuples *might* participate in a join so that only these may be transmitted [Mackett and Lohman 1986].

55.7 Advanced Types of Optimization

In this section, we attempt to provide a brief glimpse of advanced types of optimization that researchers have proposed over the past few years. The descriptions are based on examples only; further details may be found in the references provided. Furthermore, there are several issues that are not discussed at all due to lack of space, although much interesting work has been done on them, e.g., nested query optimization, rule-based query optimization, query optimizer generators, object-oriented query optimization, optimization with materialized views, heterogeneous query optimization, recursive query optimization, aggregate query optimization, optimization with expensive selection predicates, and query-optimizer validation.

55.7.1 Semantic Query Optimization

Semantic query optimization is a form of optimization mostly related to the Rewriter module. The basic idea lies in using integrity constraints defined in the database to rewrite a given query into *semantically equivalent* ones [King 1981]. These can then be optimized by the Planner as regular queries, and the most efficient plan among all can be used to answer the original query. As a simple example, using a hypothetical SQL-like syntax, consider the following integrity constraint:

    assert sal-constraint **on** emp:
        sal>100 K **where** job="Sr. Programmer".

Also consider the following query:

> **select** name, floor
> **from** emp, dept
> **where** emp.dno=dept.dno **and** job="Sr. Programmer".

Using the above integrity constraint, the query can be rewritten into a semantically equivalent one to include a selection on sal:

> **select** name, floor
> **from** emp, dept
> **where** emp.dno=dept.dno **and** job="Sr. Programmer" **and** sal>100 K.

Having the extra selection could help tremendously in finding a fast plan to answer the query if the only index in the database is a B+-tree on emp.sal. On the other hand, it would certainly be a waste if no such index exists. For such reasons, all proposals for semantic query optimization present various heuristics or rules on which rewritings have the potential of being beneficial and should be applied and which should not.

55.7.2 Global Query Optimization

So far, we have focused our attention to optimizing individual queries. Quite often, however, multiple queries become available for optimization at the same time, e.g., queries with unions, queries from multiple concurrent users, queries embedded in a single program, or queries in a deductive system. Instead of optimizing each query separately, one may be able to obtain a global plan that, although possibly suboptimal for each individual query, is optimal for the execution of all of them as a group. Several techniques have been proposed for global query optimization [Sellis 1988]. As a simple example of the problem of global optimization consider the following two queries:

> **select** name, floor
> **from** emp, dept
> **where** emp.dno=dept.dno **and** job="Sr. Programmer,"
> **select** name
> **from** emp, dept
> **where** emp.dno=dept.dno **and** budget>1 M.

Depending on the sizes of the emp and dept relations and the selectivities of the selections, it may well be that computing the entire join once and then applying separately the two selections to obtain the results of the two queries is more efficient than doing the join twice, each time taking into account the corresponding selection. Developing Planner modules that would examine all the available global plans and identify the optimal one is the goal of global/multiple query optimizers.

55.7.3 Parametric/Dynamic Query Optimization

As mentioned earlier, embedded queries are typically optimized once at compile time and are executed multiple times at run time. Because of this temporal separation between optimization and execution, the values of various parameters that are used during optimization may be very different during execution. This may make the chosen plan invalid (e.g., if indices used in the plan are no longer available) or simply not optimal (e.g., if the number of available buffer pages or operator selectivities has changed, or if new indices have become available). To address this issue, several techniques [Cole and Graefe 1994, Graefe and Ward 1989, Ioannidis et al. 1992] have been proposed that use various search strategies (e.g., randomized algorithms [Ioannidis et al. 1992] or the strategy of Volcano [Cole and Graefe 1994]) to optimize queries as much as possible at compile time, taking into account all possible values that interesting parameters may have at run time. These techniques use the actual parameter values at run time and simply pick the plan that was found optimal for them with little or no overhead. Of a drastically different flavor is the technique of Rdb/VMS [Antoshenkov 1993], where by dynamically monitoring how the probability distribution of plan costs changes, plan switching may actually occur during query execution.

55.8   Summary

To a large extent, the success of a DBMS lies in the quality, functionality, and sophistication of its query optimizer, since that determines much of the system's performance. In this chapter, we have given a bird's-eye view of query optimization. We have presented an abstraction of the architecture of a query optimizer and focused on the techniques currently used by most commercial systems for its various modules. In addition, we have provided a glimpse of advanced issues in query optimization, whose solutions have not yet found their way into practical systems, but could certainly do so in the future.

Although query optimization has existed as a field for more than twenty years, it is very surprising how fresh it remains in terms of being a source of research problems. In every single module of the architecture of Figure 55.2, there are many questions for which we do not have complete answers, even for the most simple, single-query, sequential, relational optimizations. When is it worthwhile to consider bushy trees instead of just left-deep trees? How can one model buffering effectively in the system's cost formulas? What is the most effective means of estimating the cost of operators that involve random access to relations (e.g., nonclustered index selection)? Which search strategy can be used for complex queries with confidence, providing consistent plans for similar queries? Should optimization and execution be interleaved in complex queries so that estimate errors do not grow very large? Of course, we do not even attempt to mention the questions that arise in various advanced types of optimization.

We believe that the next twenty years will be as active as the previous twenty and will bring many advances to query optimization technology, changing many of the approaches currently used in practice. Despite its age, query optimization remains an exciting field.

Acknowledgments

I would like to thank Minos Garofalakis, Joe Hellerstein, Navin Kabra, and Vishy Poosala for their many helpful comments. Partially supported by the National Science Foundation under Grants IRI-9113736 and IRI-9157368 (PYI Award) and by grants from DEC, IBM, HP, AT&T, Informix, and Oracle.

References

Antoshenkov, G. 1993. Dynamic query optimization in Rdb/VMS, pp. 538–547. In *Proc. IEEE Int. Conf. on Data Engineering*, Vienna, Austria, Mar.

Astrahan, M. M. et al. 1976. System R: a relational approach to data management. *ACM Trans. Database Sys.* 1(2):97–137, June.

Bennett, K., Ferris, M. C., and Ioannidis, Y. 1991. A genetic algorithm for database query optimization, pp. 400–407. In *Proc. 4th Int. Conf. on Genetic Algorithms*, San Diego, CA, July.

Bernstein, P. A., Goodman, N., Wong, E., Reeve, C. L., and Rothnie, J. B. 1981. Query processing in a system for distributed databases (SDD-1). *ACM Trans. Database Syst.* 6(4):602–625, Dec.

Christodoulakis, S. 1984. Implications of certain assumptions in database performance evaluation. *ACM Trans. Database Syst.* 9(2):163–186, June.

Christodoulakis, S. 1989. On the estimation and use of selectivities in database performance evaluation. *Research Report CS-89-24.* Dept. of Computer Science, University of Waterloo, June.

Cole, R. and Graefe G. 1994. Optimization of dynamic query evaluation plans, pp. 150–160. In *Proc. ACM-SIGMOD Conf. on the Management of Data*, Minneapolis, MN, June.

Galindo-Legaria, C., Pellenkoft, A., and Kersten, M. 1994. Fast, randomized join-order selection — why use transformations?, pp. 85–95. In *Proc. 20th Int. VLDB Conf.*, Santiago, Chile, Sept. Also available as *CWI Tech. Report CS-R9416*.

Goldberg, D. E. 1989. *Genetic Algorithms in Search, Optimization, and Machine Learning.* Addison–Wesley, Reading, MA.

Graefe, G. and DeWitt, D. 1987. The exodus optimizer generator, pp. 160–172. In *Proc. ACM-SIGMOD Conf. on the Management of Data*, San Francisco, CA, May.

Graefe, G. and McKenna, B. 1993. The Volcano optimizer generator: extensibility and efficient search. In *Proc. IEEE Data Engineering Conf.*, Vienna, Austria, Mar.

Graefe, G. and Ward, K. 1989. Dynamic query evaluation plans, pp. 358–366. In *Proc. ACM-SIGMOD Conference on the Management of Data*, Portland, OR, May.

Haas, L. et al. 1990. Starburst mid-flight: as the dust clears. *IEEE Trans. Knowledge and Data Eng.* 2(1):143–160, Mar.

Haas, P. and Swami, A. 1992. Sequential sampling procedures for query size estimation, pp. 341–350. In *Proc. 1992 ACM-SIGMOD Conf. on the Management of Data*, San Diego, CA, June.

Haas, P. and Swami, A. 1995. Sampling-based selectivity estimation for joins using augmented frequent value statistics. In *Proc. 1995 IEEE Conf. on Data Engineering*, Taipei, Taiwan, Mar.

Hong, W. and Stonebraker, M. 1991. Optimization of parallel query execution plans in xprs, pp. 218–225. In *Proc. 1st Int. PDIS Conf.*, Miami, FL, Dec.

Ibaraki, T. and Kameda, T. 1984. On the optimal nesting order for computing n-relational joins. *ACM Trans. Database Syst.* 9(3):482–502, Sept.

Ioannidis, Y. 1993. Universality of serial histograms, pp. 256–267. In *Proc. 19th Int. VLDB Conf.*, Dublin, Ireland, Aug.

Ioannidis, Y. and Christodoulakis, S. 1991. On the propagation of errors in the size of join results, pp. 268–277. In *Proc. 1991 ACM-SIGMOD Conf. on the Management of Data*, Denver, CO, May.

Ioannidis, Y. and Christodoulakis, S. 1993. Optimal histograms for limiting worst-case error propagation in the size of join results. *ACM Trans. Database Syst.* 18(4):709–748, Dec.

Ioannidis, Y. and Kang, Y. 1990. Randomized algorithms for optimizing large join queries, pp. 312–321. In *Proc. ACM-SIGMOD Conf. on the Management of Data*, Atlantic City, NJ, May.

Ioannidis, Y. and Poosala, V. 1995. Balancing histogram optimality and practicality for query result size estimation, pp. 233–244. In *Proc. 1995 ACM-SIGMOD Conf. on the Management of Data*, San Jose, CA, May.

Ioannidis, Y. and Wong, E. 1987. Query optimization by simulated annealing, pp. 9–22. In *Proc. ACM-SIGMOD Conf. on the Management of Data*, San Francisco, CA, May.

Ioannidis, Y., Ng, R., Shim, K., and Sellis, T. K. 1992. Parameteric query optimization, pp. 103–114. In *Proc. 18th Int. VLDB Conf.*, Vancouver, BC, Aug.

Jarke, M. and Koch, J. 1984. Query optimization in database systems. *ACM Comput. Surveys* 16(2):111–152, June.

Kang, Y. 1991. *Randomized Algorithms for Query Optimization.* Ph.D. thesis, University of Wisconsin, Madison, May.

King, J. J. 1981. Quist: a system for semantic query optimization in relational databases, pp. 510–517. In *Proc. 7th Int. VLDB Conf.*, Cannes, France, Aug.

Kirkpatrick, S., Gelatt, C. D., Jr., and Vecchi, M. P. 1983. Optimization by simulated annealing. *Science* 220(4598):671–680, May.

Kooi, R. P. 1980. *The Optimization of Queries in Relational Database.* Ph.D. thesis, Case Western Reserve University, Sept.

Krishnamurthy, R., Boral, H., and Zaniolo, C. 1986. Optimization of nonrecursive queries, pp. 128–137. In *Proc. 12th Int. VLDB Conf.*, Kyoto, Japan, Aug.

Lanzelotte, R., Valduriez, P., and Zait, M. 1993. On the effectiveness of optimization search strategies for parallel execution spaces, pp. 493–504. In *Proc. 19th Int. VLDB Conf.*, Dublin, Ireland, Aug.

Lipton, R. J., Naughton, J. F., and Schneider, D. A. 1990. Practical selectivity estimation through adaptive sampling, pp. 1–11. In *Proc. 1990 ACM-SIGMOD Conf. on the Management of Data*, Atlantic City, NJ, May.

Lohman, G. 1988. Grammar-like functional rules for representing query optimization alternatives, pp. 18–27. In *Proc. ACM-SIGMOD Conf. on the Management of Data.*, Chicago, IL, June.

Mackert, L. F. and Lohman, G. M. 1986. *R** validation and performance evaluation for distributed queries, pp. 149–159. In *Proc. 12th Int. VLDB Conf.*, Kyoto, Japan, Aug.

Mannino, M. V., Chu, P., and Sager, T. 1988. Statistical profile estimation in database systems. *ACM Comput. Surveys* 20(3):192–221, Sept.

Muralikrishna, M. and DeWitt, D. J. 1988. Equi-depth histograms for estimating selectivity factors for multi-dimensional queries, pp. 28–36. In *Proc. 1988 ACM-SIGMOD Conf. on the Management of Data*, Chicago, IL, June.

Nahar, S., Sahni, S., and Shragowitz, E. 1986. Simulated annealing and combinatorial optimization, pp. 293–299. In *Proc. 23rd Design Automation Conf.*

Olken, F. and Rotem, D. 1986. Simple random sampling from relational databases, pp. 160–169. In *Proc. 12th Int. VLDB Conf.*, Kyoto, Japan, Aug.

Ono, K. and Lohman, G. 1990. Measuring the complexity of join enumeration in query optimization, pp. 314–325. In *Proceedings of the 16th Int. VLDB Conf.*, Brisbane, Australia, Aug.

Piatetsky-Shapiro, G. and Connell, C. 1984. Accurate estimation of the number of tuples satisfying a condition, pp. 256–276. In *Proc. 1984 ACM-SIGMOD Conf. on the Management of Data*, Boston, MA, June.

Schneider, D. and DeWitt, D. 1990. Tradeoffs in processing complex join queries via hashing in multiprocessor database machines, pp. 469–480. In *Proc. of the 16th Int. VLDB Conf.*, Brisbane, Australia, Aug.

Selinger, P. G., Astrahan, M. M., Chamberlin, D. D., Lorie, R. A., and Price, T. G. 1979. Access path selection in a relational database management system, pp. 23–34. In *Proc. ACM-SIGMOD Conf. on the Management of Data*, Boston, MA, June.

Sellis, T. 1988. Multiple query optimization. *ACM Trans. Database Syst.* 13(1):23–52, Mar.

Shekita E., Young, H., and Tan, K.-L. 1993. Multi-join optimization for symmetric multiprocessors, pp. 479–492. In *Proc. 19th Int. VLDB Conf.*, Dublin, Ireland, Aug.

Swami, A. 1989. Optimization of large join queries: combining heuristics and combinatorial techniques, pp. 367–376. In *Proc. ACM-SIGMOD Conf. on the Management of Data*, Portland, OR, June.

Swami, A. and Gupta, A. 1988. Optimization of large join queries, pp. 8–17. In *Proc. ACM-SIGMOD Conf. on the Management of Data*, Chicago, IL, June.

Swami, A. and Iyer, B. 1993. A polynomial time algorithm for optimizing join queries. In *Proc. IEEE Int. Conf. on Data Engineering*, Vienna, Austria, Mar.

Yoo, H. and Lafortune, S. 1989. An intelligent search method for query optimization by semijoins. *IEEE Trans. Knowledge and Data Eng.* 1(2):226–237, June.

56

Concurrency Control and Recovery

56.1	Introduction	56-1
56.2	Underlying Principles	56-4
	Concurrency Control • Recovery	
56.3	Best Practices	56-8
	Concurrency Control • Recovery	
56.4	Research Issues and Summary	56-17

Michael J. Franklin
University of California at Berkeley

56.1 Introduction

Many service-oriented businesses and organizations, such as banks, airlines, catalog retailers, hospitals, etc., have grown to depend on fast, reliable, and correct access to their "mission-critical" data on a constant basis. In many cases, particularly for global enterprises, 7×24 access is required; that is, the data must be available seven days a week, twenty-four hours a day. *Database management systems* (DBMSs) are often employed to meet these stringent performance, availability, and reliability demands. As a result, two of the core functions of a DBMS are (1) to protect the data stored in the database and (2) to provide correct and highly available access to those data in the presence of concurrent access by large and diverse user populations, despite various software and hardware failures. The responsibility for these functions resides in the **concurrency control** and **recovery** components of the DBMS software. *Concurrency control* ensures that individual users see consistent states of the database even though operations on behalf of many users may be interleaved by the database system. *Recovery* ensures that the database is fault-tolerant; that is, that the database state is not corrupted as the result of a software, system, or media failure. The existence of this functionality in the DBMS allows applications to be written without explicit concern for concurrency and fault tolerance. This freedom provides a tremendous increase in programmer productivity and allows new applications to be added more easily and safely to an existing system.

For database systems, correctness in the presence of concurrent access and/or failures is tied to the notion of a **transaction**. A transaction is a unit of work, possibly consisting of multiple data accesses and updates, that must **commit** or **abort** as a single atomic unit. When a transaction *commits*, all updates it performed on the database are made permanent and visible to other transactions. In contrast, when a transaction *aborts*, all of its updates are removed from the database and the database is restored (if necessary) to the state it would have been in if the aborting transaction had never been executed. Informally, transaction executions are said to respect the **ACID properties** [Gray and Reuter 1993]:

Atomicity: This is the "all-or-nothing" aspect of transactions discussed above — either all operations of a transaction complete successfully, or none of them do. Therefore, after a transaction has completed (i.e., committed or aborted), the database will not reflect a partial result of that transaction.

Consistency: Transactions preserve the consistency of the data — a transaction performed on a database that is internally consistent will leave the database in an internally consistent state. Consistency is typically expressed as a set of declarative *integrity constraints*. For example, a constraint may be that the salary of an employee cannot be higher than that of his or her manager.

Isolation: A transaction's behavior is not impacted by the presence of other transactions that may be accessing the same database concurrently. That is, a transaction sees only a state of the database that could occur if that transaction were the only one running against the database and produces only results that it could produce if it was running alone.

Durability: The effects of *committed* transactions survive failures. Once a transaction commits, its updates are guaranteed to be reflected in the database even if the contents of volatile (e.g., main memory) or nonvolatile (e.g., disk) storage are lost or corrupted.

Of these four transaction properties, the concurrency control and recovery components of a DBMS are primarily concerned with preserving *atomicity, isolation*, and *durability*. The preservation of the *consistency* property typically requires additional mechanisms such as compile-time analysis or run-time triggers in order to check adherence to integrity constraints.* For this reason, this chapter focuses primarily on the A, I, and D of the ACID transaction properties.

Transactions are used to structure complex processing tasks which consist of multiple data accesses and updates. A traditional example of a transaction is a money transfer from one bank account (say account A) to another (say B). This transaction consists of a withdrawal from A and a deposit into B and requires four accesses to account information stored in the database: a read and write of A and a read and write of B. The data accesses of this transaction are as follows:

TRANSFER( )
    01 A_bal := **Read**(A)
    02 A_bal := A_bal − $50
    03 **Write**(A, A_bal)
    04 B_bal := **Read**(B)
    05 B_bal := B_bal + $50
    06 **Write**(B, B_bal)

The value of A in the database is read and decremented by $50, then the value of B in the database is read and incremented by $50. Thus, TRANSFER preserves the invariant that the sum of the balances of A and B prior to its execution must equal the sum of the balances after its execution, regardless of whether the transaction commits or aborts.

Consider the importance of the atomicity property. At several points during the TRANSFER transaction, the database is in a temporarily inconsistent state. For example, between the time that account A is updated (statement 3) and the time that account B is updated (statement 6) the database reflects the decrement of A but not the increment of B, so it appears as if $50 has disappeared from the database. If the transaction reaches such a point and then is unable to complete (e.g., due to a failure or an unresolvable conflict, etc.), then the system must ensure that the effects of the partial results of the transaction (i.e., the update to A) are removed from the database — otherwise the database state will be incorrect. The durability property, in contrast, only comes into play in the event that the transaction successfully commits. Once the user is notified that the transfer has taken place, he or she will assume that account B contains the transferred funds and may attempt to use those funds from that point on. Therefore, the DBMS must ensure that the results of the transaction (i.e., the transfer of the $50) remain reflected in the database state even if the system crashes.

Atomicity, consistency, and durability address correctness for **serial execution** of transactions, where only a single transaction at a time is allowed to be in progress. In practice, however, database management systems typically support **concurrent execution**, in which the operations of multiple transactions can

---

*In the case of triggers, the recovery mechanism is typically invoked to abort an offending transaction.

TRANSFER	REPORTSUM
01 A_bal := **Read**(A)	
02 A_bal := A_bal - $50	
03 **Write**(A,A_bal)	
	01 A_bal := **Read**(A) /* value is $250 */
	02 B_bal := **Read**(B) /* value is $200 */
	03 Print(A_bal + B_bal) /* result = $450 */
04 B_bal := **Read**(B)	
05 B_bal := B_bal + $50	
06 **Write**(B,B_bal)	

FIGURE 56.1 An incorrect interleaving of TRANSFER and REPORTSUM.

be executed in an interleaved fashion. The motivation for concurrent execution in a DBMS is similar to that for multiprogramming in operating systems, namely, to improve the utilization of system hardware resources and to provide multiple users a degree of fairness in access to those resources. The *isolation* property of transactions comes into play when concurrent execution is allowed.

Consider a second transaction that computes the sum of the balances of accounts A and B:

REPORTSUM( )
    01 A_bal := **Read**(A)
    02 B_bal := **Read**(B)
    03 Print(A_bal + B_bal)

Assume that initially, the balance of account A is $300 and the balance of account B is $200. If a REPORTSUM transaction is executed on this state of the database, it will print a result of $500. In a database system restricted to *serial execution* of transactions, REPORTSUM will also produce the same result if it is executed after a TRANSFER transaction. The atomicity property of transactions ensures that if the TRANSFER aborts, all of its effects are removed from the database (so REPORTSUM would see A = $300 and B = $200), and the durability property ensures that if it commits, then all of its effects remain in the database state (so REPORTSUM would see A = $250 and B = $250).

Under concurrent execution, however, a problem could arise if the isolation property is not enforced. As shown in Figure 56.1, if REPORTSUM were to execute after TRANSFER has updated account A but before it has updated account B, then REPORTSUM could see an inconsistent state of the database. In this case, the execution of REPORTSUM sees a state of the database in which $50 has been withdrawn from account A but has not yet been deposited in account B, resulting in a total of $450 — it seems that $50 has disappeared from the database. This result is not one that could be obtained in any serial execution of TRANSFER and REPORTSUM transactions. It occurs because in this example, REPORTSUM accessed the database when it was in a temporarily inconsistent state. This problem is sometimes referred to as the *inconsistent retrieval* problem. To preserve the isolation property of transactions the DBMS must prevent the occurrence of this and other potential anomalies that could arise due to concurrent execution. The formal notion of correctness for concurrent execution in database systems is known as **serializability** and is described in Section 56.2.

Although the transaction processing literature often traces the history of transactions back to antiquity (such as Sumerian tax records) or to early contract law [Gray 1981, Gray and Reuter 1993, Korth 1995], the roots of the transaction concept in information systems are typically traced back to the early 1970s and the work of Bjork [1973] and Davies [1973]. Early systems such as IBM's IMS addressed related issues, and a systematic treatment and understanding of ACID transactions was developed several years later by members of the IBM System R group [Gray et al. 1975, Eswaran et al. 1976] and others [e.g., Rosenkrantz et al. 1977, Lomet 1977]. Since that time, many techniques for implementing ACID transactions have been proposed and a fairly well accepted set of techniques has emerged. The remainder of this chapter contains an overview of the basic theory that has been developed as well as a survey of the more widely known implementation techniques for concurrency control and recovery. A brief discussion of work on extending the simple transaction model is presented at the end of the chapter.

It should be noted that issues related to those addressed by concurrency control and recovery in database systems arise in other areas of computing systems as well, such as file systems and memory systems. There are, however, two salient aspects of the ACID model that distinguish transactions from other approaches. First is the incorporation of both isolation (concurrency control) and fault-tolerance (recovery) issues. Second is the concern with treating arbitrary groups of *write* and/or *read* operations on multiple data items as atomic, isolated units of work. While these aspects of the ACID model provide powerful guarantees for the protection of data, they also can induce significant systems implementation complexity and performance overhead. For this reason, the notion of ACID transactions and their associated implementation techniques have remained largely within the DBMS domain, where the provision of highly available and reliable access to "mission critical" data is a primary concern.

56.2   Underlying Principles

56.2.1   Concurrency Control

56.2.1.1   Serializability

As stated in the previous section, the responsibility for maintaining the isolation property of ACID transactions resides in the concurrency-control portion of the DBMS software. The most widely accepted notion of correctness for concurrent execution of transactions is *serializability*. Serializability is the property that an (possibly interleaved) execution of a group of transactions has the same effect on the database, and produces the same output, as some serial (i.e., noninterleaved) execution of those transactions. It is important to note that serializability does not specify any *particular* serial order, but rather, only that the execution is equivalent to *some* serial order. This distinction makes serializability a slightly less intuitive notion of correctness than transaction initiation time or commit order, but it provides the DBMS with significant additional flexibility in the scheduling of operations. This flexibility can translate into increased responsiveness for end users.

A rich theory of database concurrency control has been developed over the years [see Papadimitriou 1986, Bernstein et al. 1987, Gray and Reuter 1993], and serializability lies at the heart of much of this theory. In this chapter we focus on the simplest models of concurrency control, where the operations that can be performed by transactions are restricted to *read*(x), *write*(x), *commit*, and *abort*. The operation *read*(x) retrieves the value of a data item from the database, *write*(x) modifies the value of a data item in the database, and *commit* and *abort* indicate successful or unsuccessful transaction completion respectively (with the concomitant guarantees provided by the ACID properties). We also focus on a specific variant of serializability called *conflict serializability*. Conflict serializability is the most widely accepted notion of correctness for concurrent transactions because there are efficient, easily implementable techniques for detecting and/or enforcing it. Another well-known variant is called *view serializability*. View serializability is less restrictive (i.e., it allows more legal schedules) than conflict serializability, but it and other variants are primarily of theoretical interest because they are impractical to implement. The reader is referred to Papadimitriou [1986] for a detailed treatment of alternative serializability models.

56.2.1.2   Transaction Schedules

Conflict serializability is based on the notion of a **schedule** of transaction operations. A schedule for a set of transaction executions is a partial ordering of the operations performed by those transactions, which shows how the operations are interleaved. The ordering defined by a schedule can be partial in the sense that it is only required to specify two types of dependencies:

- All operations of a given transaction for which an order is specified by that transaction must appear in that order in the schedule. For example, the definition of REPORTSUM above specifies that account A is read before account B.
- The ordering of all **conflicting operations** from different transactions must be specified. Two operations are said to conflict if they both operate on the same data item and at least one of them is a *write*( ).

The concept of a schedule provides a mechanism to express and reason about the (possibly) concurrent execution of transactions. A *serial* schedule is one in which all the operations of each transaction appear consecutively. For example, the serial execution of TRANSFER followed by REPORTSUM is represented by the following schedule:

$$r_0[A] \rightarrow w_0[A] \rightarrow r_0[B] \rightarrow w_0[B] \rightarrow c_0 \rightarrow r_1[A] \rightarrow r_1[B] \rightarrow c_1 \qquad (56.1)$$

In this notation, each operation is represented by its initial letter, the subscript of the operation indicates the *transaction number* of the transaction on whose behalf the operation was performed, and a capital letter in brackets indicates a specific data item from the database (for read and write operations). A transaction number (tn) is a unique identifier that is assigned by the DBMS to an execution of a transaction. In the example above, the execution of TRANSFER was assigned tn 0 and the execution of REPORTSUM was assigned tn 1. A right arrow ($\rightarrow$) between two operations indicates that the left-hand operation is ordered before the right-hand one. The ordering relationship is transitive; the orderings implied by transitivity are not explicitly drawn.

For example, the interleaved execution of TRANSFER and REPORTSUM shown in Figure 56.1 would produce the following schedule:

$$r_0[A] \rightarrow w_0[A] \rightarrow r_1[A] \rightarrow r_1[B] \rightarrow c_1 \rightarrow r_0[B] \rightarrow w_0[B] \rightarrow c_0 \qquad (56.2)$$

The formal definition of serializability is based on the concept of equivalent schedules. Two schedules are said to be *equivalent* ($\equiv$) if:

- They contain the same transactions and operations, and
- They order all conflicting operations of nonaborting transactions in the same way.

Given this notion of equivalent schedules, *a schedule is said to be serializable if and only if it is equivalent to some serial schedule.* For example, the following concurrent schedule is serializable because it is equivalent to Schedule 56.1:

$$r_0[A] \rightarrow w_0[A] \rightarrow r_1[A] \rightarrow r_0[B] \rightarrow w_0[B] \rightarrow c_0 \rightarrow r_1[B] \rightarrow c_1 \qquad (56.3)$$

In contrast, the interleaved execution of Schedule 56.2 is *not* serializable. To see why, notice that in any serial execution of TRANSFER and REPORTSUM either *both* writes of TRANSFER will precede *both* reads of REPORTSUM or vice versa. However, in schedule (56.2) $w_0[A] \rightarrow r_1[A]$ but $r_1[B] \rightarrow w_0[b]$. Schedule 56.2, therefore, is not equivalent to any possible serial schedule of the two transactions so it is not serializable. This result agrees with our intuitive notion of correctness, because recall that Schedule 56.2 resulted in the apparent loss of $50.

56.2.1.3 Testing for Serializability

A schedule can easily be tested for serializability through the use of a *precedence graph*. A precedence graph is a directed graph that contains a vertex for each *committed* transaction execution in a schedule (noncommitted executions can be ignored). The graph contains an edge from transaction execution T_i to transaction execution T_j ($i \neq j$) if there is an operation in T_i that is constrained to precede an operation of T_j in the schedule. A schedule is serializable if and only if its precedence graph is *acyclic*. Figure 56.2(a) shows the precedence graph for Schedule 56.2. That graph has an edge $T_0 \rightarrow T_1$ because the schedule

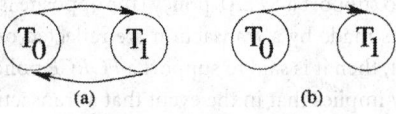

FIGURE 56.2 Precedence graphs for (a) nonserializable and (b) serializable schedules.

contains $w_0[A] \rightarrow r_1[A]$ and an edge $T_1 \rightarrow T_0$ because the schedule contains $r_1[B] \rightarrow w_0[b]$. The cycle in the graph shows that the schedule is nonserializable. In contrast, Figure 56.2(b) shows the precedence graph for Schedule 56.1. In this case, all ordering constraints are from T_0 to T_1, so the precedence graph is acyclic, indicating that the schedule is serializable.

There are a number of practical ways to implement conflict serializability. These and other implementation issues are addressed in Section 56.3. Before discussing implementation issues, however, we first survey the basic principles underlying database recovery.

56.2.2 Recovery

56.2.2.1 Coping with Failures

Recall that the responsibility for the atomicity and durability properties of ACID transactions lies in the recovery component of the DBMS. For recovery purposes it is necessary to distinguish between two types of storage: (1) **volatile storage**, such as main memory, whose state is lost in the event of a system crash or power outage, and (2) **nonvolatile storage**, such as magnetic disks or tapes, whose contents persist across such events. The recovery subsystem is relied upon to ensure correct operation in the presence of three different types of failures (listed in order of likelihood):

- *Transaction failure:* When a transaction that is in progress reaches a state from which it cannot successfully commit, all updates that it made must be removed from the database in order to preserve the atomicity property. This is known as *transaction rollback*.

- *System failure:* If the system fails in a way that causes the loss of volatile memory contents, recovery must ensure that: (1) the updates of all transactions that had committed prior to the crash are reflected in the database and (2) all updates of other transactions (aborted or in progress at the time of the crash) are removed from the database.

- *Media failure:* In the event that data are lost or corrupted on the nonvolatile storage (e.g., due to a disk-head crash), then the on-line version of the data is lost. In this case, the database must be restored from an archival version of the database and brought up to date using operation logs.

In this chapter we focus on the issues of rollback and crash recovery, the most frequent uses of the DBMS recovery subsystem. Recovery from media crashes requires substantial additional mechanisms and complexity beyond what is covered here. Media recovery is addressed in the recovery-related references listed at the end of this chapter.

56.2.2.2 Buffer Management Issues

The process of removing the effects of an incomplete or aborted transaction for preserving atomicity is known as *UNDO*. The process of reinstating the effects of a committed transaction for durability is known as *REDO*. The amount of work that a recovery subsystem must perform for either of these functions depends on how the DBMS buffer manager handles data that are updated by in-progress and/or committing transactions [Haerder and Reuter 1983, Bernstein et al. 1987]. Recall that the buffer manager is the DBMS component that is responsible for coordinating the transfer of data between main memory (i.e., volatile storage) and disk (i.e., nonvolatile storage). The unit of storage that can be written atomically to nonvolatile storage is called a *page*. Updates are made to copies of pages in the (volatile) buffer pool, and those copies are written out to nonvolatile storage at a later time. If the buffer manager allows an update made by an *uncommitted* transaction to overwrite the most recent committed value of a data item on nonvolatile storage, it is said to support a *STEAL* policy (the opposite is called *NO-STEAL*). If the buffer manager ensures that all updates made by a transaction are reflected on nonvolatile storage before the transaction is allowed to commit, then it is said to support a *FORCE* policy (the opposite is *NO-FORCE*).

Support for the STEAL policy implies that in the event that a transaction needs to be rolled back (due to transaction failure or system crash), UNDOing the transaction will involve restoring the values of any nonvolatile copies of data that were overwritten by that transaction back to their previous committed state.

In contrast, a NO-STEAL policy guarantees that the data values on nonvolatile storage are valid, so they do not need to be restored. A NO-FORCE policy raises the possibility that some committed data values may be lost during a system crash because there is no guarantee that they have been placed on nonvolatile storage. This means that substantial REDO work may be required to preserve the durability of committed updates. In contrast, a FORCE policy ensures that the committed updates *are* placed on nonvolatile storage, so that in the event of a system crash, the updates will still be reflected in the copy of the database on nonvolatile storage.

From the above discussion, it should be apparent that a buffer manager that supports the combination of NO-STEAL and FORCE would place the fewest demands on UNDO and REDO recovery. However, these policies may negatively impact the performance of the DBMS during normal operation (i.e., when there are no crashes or rollbacks) because they restrict the flexibility of the buffer manager. NO-STEAL obligates the buffer manager to retain updated data in memory until a transaction commits or to write those data to a temporary location on nonvolatile storage (e.g., a swap area). The problem with a FORCE policy is that it can impose significant disk write overhead during the critical path of a committing transaction. For these reasons, many buffer managers support the STEAL and NO-FORCE (**STEAL/NO-FORCE**) policies.

56.2.2.3 Logging

In order to deal with the UNDO and REDO requirements imposed by the STEAL and NO-FORCE policies respectively, database systems typically rely on the use of a **log**. A log is a sequential file that stores information about transactions and the state of the system at certain instances. Each entry in the log is called a **log record**. One or more log records are written for each update performed by a transaction. When a log record is created, it is assigned a **log sequence number** (LSN) which serves to uniquely identify that record in the log. LSNs are typically assigned in a monotonically increasing fashion so that they provide an indication of relative position in the log. When an update is made to a data item in the buffer, a log record is created for that update. Many systems write the LSN of this new log record into the page containing the updated data item. Recording LSNs in this fashion allows the recovery system to relate the state of a data page to logged updates in order to tell if a given log record is reflected in a given state of a page.

Log records are also written for transaction management activities such as the commit or abort of a transaction. In addition, log records are sometimes written to describe the state of the system at certain periods of time. For example, such log records are written as part of the **checkpointing** process. Checkpoints are taken periodically during normal operation to help bound the amount of recovery work that would be required in the event of a crash. Part of the checkpointing process involves the writing of one or more *checkpoint records*. These records can include information about the contents of the buffer pool and the transactions that are currently active, etc. The particular contents of these records depend on the method of checkpointing that is used. Many different checkpointing methods have been developed, some of which involve quiescing the system to a consistent state, while others are less intrusive. A particularly nonintrusive type of checkpointing is used by the ARIES recovery method [Mohan et al. 1992] that is described in Section 56.3.

For transaction update operations there are two basic types of logging: *physical* and *logical* [Gray and Reuter 1993]. Physical log records typically indicate the location (e.g., position on a particular page) of modified data in the database. If support for UNDO is provided (i.e., a STEAL policy is used), then the value of the item prior to the update is recorded in the log record. This is known as the *before image* of the item. Similarly the *after image* (i.e., the new value of the item after the update), is logged if REDO support is provided. Thus, physical log records in a DBMS with STEAL/NO-FORCE buffer management contain both the old and new data values of items. Recovery using physical log records has the property that recovery actions (i.e., UNDOs or REDOs) are *idempotent*, meaning that they have the same effect no matter how many times they are applied. This property is important if recovery is invoked multiple times, as will occur if a system fails repeatedly (e.g., due to a power problem or a faulty device).

Logical logging (sometimes referred to as *operational logging*) records only high-level information about operations that are performed, rather than recording the actual changes to items (or storage locations) in the database. For example, the insertion of a new tuple into a relation might require many physical changes to the database such as space allocation, index updates, and reorganization, etc. Physical logging would require log records to be written for all of these changes. In contrast, logical logging would simply log the fact that the insertion had taken place, along with the value of the inserted tuple. The REDO process for a logical logging system must determine the set of actions that are required to fully reinstate the insert. Likewise, the UNDO logic must determine the set of actions that make up the inverse of the logged operation.

Logical logging has the advantage that it minimizes the amount of data that must be written to the log. Furthermore, it is inherently appealing because it allows many of the implementation details of complex operations to be hidden in the UNDO/REDO logic. In practice however, recovery based on logical logging is difficult to implement because the actions that make up the logged operation are not performed atomically. That is, when a system is restarted after a crash, the database may not be in an *action consistent* state with respect to a complex operation — it is possible that only a subset of the updates made by the action had been placed on nonvolatile storage prior to the crash. As a result, it is difficult for the recovery system to determine which portions of a logical update are reflected in the database state upon recovery from a system crash. In contrast, physical logging does not suffer from this problem, but it can require substantially higher logging activity.

In practice, systems often implement a compromise between physical and logical approaches that has been referred to as *physiological logging* [Gray and Reuter 1993]. In this approach log records are constrained to refer to a single page, but may reflect logical operations on that page. For example, a physiological log record for an insert on a page would specify the value of the new tuple that is added to the page, but would not specify any free-space manipulation or reorganization of data on the page resulting from the insertion; the REDO and UNDO logic for insertion would be required to infer the necessary operations. If a tuple insert required updates to multiple pages (e.g., data pages plus multiple index pages), then a separate physiological log record would be written for each page updated. Physiological logging avoids the action consistency problem of logical logging, while reducing, to some extent, the amount of logging that would be incurred by physical logging. The ARIES recovery method is one example of a recovery method that uses physiological logging.

56.2.2.4 Write-Ahead Logging (WAL)

A final recovery principle to be addressed in this section is the **write-ahead logging** (WAL) protocol. Recall that the contents of volatile storage are lost in the event of a system crash. As a result, any log records that are not reflected on nonvolatile storage will also be lost during a crash. WAL is a protocol that ensures that in the event of a system crash, the recovery log contains sufficient information to perform the necessary UNDO and REDO work when a STEAL/NO-FORCE buffer management policy is used. The WAL protocol ensures that:

1. All log records pertaining to an updated page are written to nonvolatile storage before the page itself is allowed to be overwritten in nonvolatile storage
2. A transaction is not considered to be committed until all of its log records (including its commit record) have been written to stable storage

The first point ensures that UNDO information required due to the STEAL policy will be present in the log in the event of a crash. Similarly, the second point ensures that any REDO information required due to the NO-FORCE policy will be present in the nonvolatile log. The WAL protocol is typically enforced with special support provided by the DBMS buffer manager.

56.3 Best Practices

56.3.1 Concurrency Control

56.3.1.1 Two-Phase Locking

The most prevalent implementation technique for concurrency control is locking. Typically, two types of locks are supported, *shared* (S) locks and *exclusive* (X) locks. The compatibility of these locks is defined by the *compatibility matrix* shown in Table 56.1. The compatibility matrix shows that two different transactions

TABLE 56.1 Compatibility Matrix for S and X Locks

	S	X
S	y	n
X	n	n

are allowed to hold S locks simultaneously on the same data item, but that X locks cannot be held on an item simultaneously with any other locks (by other transactions) on that item. S locks are used for protecting *read* access to data (i.e., multiple concurrent readers are allowed), and X locks are used for protecting *write* access to data. As long as a transaction is holding a lock, no other transaction is allowed to obtain a conflicting lock. If a transaction requests a lock that cannot be granted (due to a lock conflict), that transaction is *blocked* (i.e., prohibited from proceeding) until all the conflicting locks held by other transactions are released.

S and X locks as defined in Table 56.1 directly model the semantics of conflicts used in the definition of conflict serializability. Therefore, locking can be used to enforce serializability. Rather than testing for serializability after a schedule has been produced (as was done in the previous section), the blocking of transactions due to lock conflicts can be used to prevent nonserializable schedules from *ever* being produced.

A transaction is said to be **well formed** with respect to *reads* if it always holds an S or an X lock on an item while reading it, and well formed with respect to *writes* if it always holds an X lock on an item while writing it. Unfortunately, restricting all transactions to be well formed is not sufficient to guarantee serializability. For example, a nonserializable execution such as that of Schedule 56.2 is still possible using well formed transactions. Serializability can be enforced, however, through the use of **two-phase locking** **(2PL)**. Two-phase locking requires that all transactions be well formed and that they respect the following rule: *Once a transaction has released a lock, it is not allowed to obtain any additional locks.* This rule results in transactions that have two phases:

1. A *growing phase* in which the transaction is acquiring locks
2. A *shrinking phase* in which locks are released

The two-phase rule dictates that the transaction shifts from the growing phase to the shrinking phase at the instant it first releases a lock.

To see how 2PL enforces serializability, consider again Schedule 56.2. Recall that the problem arises in this schedule because $w_0[A] \rightarrow r_1[A]$ but $r_1[B] \rightarrow w_0[B]$. This schedule could not be produced under 2PL, because transaction 1 (REPORTSUM) would be blocked when it attempted to read the value of A because transaction 0 would be holding an X lock on it. Transaction 0 would not be allowed to release this X lock before obtaining its X lock on B, and thus it would either abort or perform its update of B before transaction 1 is allowed to progress. In contrast, note that Schedule 56.1 (the serial schedule) would be allowed in 2PL. 2PL would also allow the following (serializable) interleaved schedule:

$$r_1[A] \rightarrow r_0[A] \rightarrow r_1[B] \rightarrow c_1 \rightarrow w_0[A] \rightarrow r_0[B] \rightarrow w_0[B] \rightarrow c_0 \qquad (56.4)$$

It is important to note, however, that two-phase locking is sufficient but not necessary for implementing serializability. In other words, there are schedules that are serializable but would not be allowed by two-phase locking. Schedule 56.3 is an example of such a schedule.

In order to implement 2PL, the DBMS contains a component called a *lock manager*. The lock manager is responsible for granting or blocking lock requests, for managing queues of blocked transactions, and for unblocking transactions when locks are released. In addition, the lock manager is also responsible for dealing with **deadlock** situations. A deadlock arises when a set of transactions is blocked, each waiting for another member of the set to release a lock. In a deadlock situation, none of the transactions involved can make progress. Database systems deal with deadlocks using one of two general techniques: avoidance or detection. Deadlock avoidance can be achieved by imposing an order in which locks can be obtained on data, by requiring transactions to predeclare their locking needs, or by aborting transactions rather than blocking them in certain situations.

Deadlock detection, on the other hand, can be implemented using *timeouts* or explicit checking. Timeouts are the simplest technique; if a transaction is blocked beyond a certain amount of time, it is assumed that a deadlock has occurred. The choice of a timeout interval can be problematic, however. If it is too short, then the system may infer the presence of a deadlock that does not truly exist. If it is too long, then deadlocks may go undetected for too long a time. Alternatively the system can explicitly check for deadlocks using a structure called a *waits-for graph*. A waits-for graph is a directed graph with a vertex for each active transaction. The lock manager constructs the graph by placing an edge from a transaction T_i to a transaction T_j $(i \neq j)$ if T_i is blocked waiting for a lock held by T_j. If the waits-for graph contains a cycle, all of the transactions involved in the cycle are waiting for each other, and thus they are deadlocked. When a deadlock is detected, one or more of the transactions involved is rolled back. When a transaction is rolled back its locks are automatically released, so the deadlock will be broken.

56.3.1.2 Isolation Levels

As should be apparent from the previous discussion, transaction isolation comes at a cost in potential concurrency. Transaction blocking can add significantly to transaction response time.* As stated previously, serializability is typically implemented using two-phase locking, which requires locks to be held at least until all necessary locks have been obtained. Prolonging the holding time of locks increases the likelihood of blocking due to data contention.

In some applications, however, serializability is not strictly necessary. For example, a data analysis program that computes aggregates over large numbers of tuples may be able to tolerate some inconsistent access to the database in exchange for improved performance. The concept of *degrees of isolation* or *isolation levels* has been developed to allow transactions to trade concurrency for consistency in a controlled manner [Gray et al. 1975, Gray and Reuter 1993, Berenson et al. 1995]. In their 1975 paper, Gray et al. defined four degrees of consistency using characterizations based on locking, dependencies, and anomalies (i.e., results that could not arise in a serial schedule). The degrees were named degree 0–3, with degree 0 being the least consistent, and degree 3 intended to be equivalent to serializable execution.

The original presentation has served as the basis for understanding relaxed consistency in many current systems, but it has become apparent over time that the different characterizations in that paper were not specified to an equal degree of detail. As pointed out in a recent paper by Berenson et al. [1995], the SQL-92 standard suffers from a similar lack of specificity. Berenson et al. have attempted to clarify the issue, but it is too early to determine if they have been successful. In this section we focus on the locking-based definitions of the isolation levels, as they are generally acknowledged to have "stood the test of time" [Berenson et al. 1995]. However, the definition of the degrees of consistency requires an extension to the previous description of locking in order to address the *phantom problem*.

An example of the phantom problem is the following: assume a transaction T_i reads a set of tuples that satisfy a query predicate. A second transaction T_j inserts a new tuple that satisfies the predicate. If T_i then executes the query again, it will see the new item, so that its second answer differs from the first. This behavior could never occur in a serial schedule, as a "phantom" tuple appears in the midst of a transaction; thus, this execution is anomalous. The phantom problem is an artifact of the transaction model, consisting of reads and writes to *individual* data that we have used so far. In practice, transactions include *queries* that dynamically define sets based on predicates. When a query is executed, all of the tuples that satisfy the predicate at that time can be locked as they are accessed. Such individual locks, however, do not protect against the later addition of further tuples that satisfy the predicate.

---

*Note that other, non-blocking approaches discussed later in this section also suffer from similar problems.

One obvious solution to the phantom problem is to lock predicates instead of (or in addition to) individual items [Eswaran et al. 1976]. This solution is impractical to implement, however, due to the complexity of detecting the overlap of a set of arbitrary predicates. Predicate locking can be approximated using techniques based on locking clusters of data or ranges of index values. Such techniques, however, are beyond the scope of this chapter. In this discussion we will assume that predicates can be locked without specifying the technical details of how this can be accomplished (see Gray and Reuter [1993] and Mohan et al. [1992] for detailed treatments of this topic).

The locking-oriented definitions of the isolation levels are based on whether or not read and/or write operations are well formed (i.e., protected by the appropriate lock), and if so, whether those locks are *long duration* or *short duration*. Long-duration locks are held until the end of a transaction (EOT) (i.e., when it commits or aborts); short-duration locks can be released earlier. Long-duration write locks on data items have important benefits for recovery, namely, they allow recovery to be performed using *before images*. If long-duration write locks are not used, then the following scenario could arise:

$$w_0[A] \rightarrow w_1[A] \rightarrow a_0 \qquad (56.5)$$

In this case restoring A with T_0's before image of it will be incorrect because it would overwrite T_1's update. Simply ignoring the abort of T_0 is also incorrect. In that case, if T_1 were to subsequently abort, installing its before image would reinstate the value written by T_0. For this reason and for simplicity, locking systems typically hold long-duration locks on data items. This is sometimes referred to as *strict* locking [Bernstein et al. 1987].

Given these notions of locks, the degrees of isolation presented in the SQL-92 standard can be obtained using different lock protocols. In the following, all levels are assumed to be well formed with respect to writes and to hold long duration *write* (i.e., exclusive) locks on updated data items. Four levels are defined (from weakest to strongest:)*

READ UNCOMMITTED: This level, which provides the weakest consistency guarantees, allows transactions to read data that have been written by other transactions that have not committed. In a locking implementation this level is achieved by being ill formed with respect to reads (i.e., not obtaining read locks). The risks of operating at this level include (in addition to the risks incurred at the more restrictive levels) the possibility of seeing updates that will eventually be rolled back and the possibility of seeing some of the updates made by another transaction but missing others made by that transaction.

READ COMMITTED: This level ensures that transactions only see updates that have been made by transactions that have committed. This level is achieved by being well formed with respect to reads on individual data items, but holding the read locks only as short-duration locks. Transactions operating at this level run the risk of seeing *nonrepeatable* reads (in addition to the risks of the more restrictive levels). That is, a transaction T_0 could read a data item twice and see two different values. This anomaly could occur if a second transaction were to update the item and commit in between the two reads by T_0.

REPEATABLE READ: This level ensures that reads to individual data items are repeatable, but does not protect against the phantom problem described previously. This level is achieved by being well formed with respect to reads on individual data items, and holding those locks for long duration.

SERIALIZABLE: This level protects against all of the problems of the less restrictive levels, including the phantom problem. It is achieved by being well formed with respect to reads on *predicates* as well as on individual data items and holding all locks for long duration.

---

*It should be noted that two-phase locks can be substituted for the long-duration locks in these definitions without impacting the consistency provided. Long-duration locks are typically used, however, to avoid the recovery-related problems described previously.

A key aspect of this definition of degrees of isolation is that as long as all transactions execute at the READ UNCOMMITTED level or higher, they are able to obtain at least the degree of isolation they desire without interference from any transactions running at lower degrees. Thus, these degrees of isolation provide a powerful tool that allows application writers or users to trade off consistency for improved concurrency. As stated earlier, the definition of these isolation levels for concurrency-control methods that are not based on locking has been problematic. This issue is addressed in depth in Berenson et al. [1995].

It should be noted that the discussion of locking so far has ignored an important class of data that is typically present in databases, namely, *indexes*. Because indexes are auxiliary information, they can be accessed in a non-two-phase manner without sacrificing serializability. Furthermore, the hierarchical structure of many indexes (e.g., B-trees) makes them potential concurrency bottlenecks due to high contention at the upper levels of the structure. For this reason, significant effort has gone into developing methods for providing highly concurrent access to indexes. Pointers to some of this work can be found in the Further Information section at the end of this chapter.

56.3.1.3 Hierarchical Locking

The examples in the preceeding discussions of concurrency control primarily dealt with operations on a single granularity of data items (e.g., tuples). In practice, however, the notions of conflicts and locks can be applied at many different granularities. For example, it is possible to perform locking at the granularity of a page, a relation, or even an entire database. In choosing the proper granularity at which to perform locking there is a fundamental tradeoff between potential concurrency and locking overhead. Locking at a fine granularity, such as an individual tuple, allows for maximum concurrency, as only transactions that are truly accessing the same tuple have the potential to conflict. The downside of such fine-grained locking, however, is that a transaction that accesses a large number of tuples will have to acquire a large number of locks. Each lock request requires a call to the lock manager. This overhead can be reduced by locking at a coarser granularity, but coarse granularity raises the potential for *false conflicts*. For example, two transactions that update different tuples residing on the same page would conflict under page-level locking but not under tuple-level locking.

The notion of hierarchical or multigranular locking was introduced to allow concurrent transactions to obtain locks at different granularities in order to optimize the above tradeoff [Gray et al. 1975]. In hierarchical locking, a lock on a granule at a particular level of the granularity hierarchy implicitly locks all items included in that granule. For example, an S-lock on a relation implicitly locks all pages and tuples in that relation. Thus, a transaction with such a lock can read any tuple in the relation without requesting additional locks. Hierarchical locking introduces additional lock modes beyond S and X. These additional modes allow transactions to declare their *intention* to perform an operation on objects at lower levels of the granularity hierarchy. The new modes are IS, IX, and SIX for *intention shared*, *intention exclusive*, and *shared with intention exclusive*. An IS (or IX) lock on a granule provides no privileges on that granule, but indicates that the holder intends to obtain S (or X) locks on one or more finer granules. An SIX lock combines an S lock on the entire granule with an IX lock. SIX locks support the common access pattern of scanning the items in a granule (e.g., tuples in a relation) and choosing to update a fraction of them based on their values.

Similarly to S and X locks, these lock modes can be described using a compatibility matrix. The compatibility matrix for these modes is shown in Table 56.2. In order for transactions locking at different granularities to coexist, all transactions must follow the same hierarchical locking protocol starting from the root of the granularity hierarchy. This protocol is shown in Table 56.3. For example, to read a single record, a transaction would obtain IS locks on the database, relation, and page, followed by an S lock on the specific tuple. If a transaction wanted to read all or most tuples on a page, then it could obtain IS locks on the database and relation, followed by an S lock on the entire page. By following this uniform protocol, potential conflicts between transactions that ultimately obtain S and/or X locks at different granularities can be detected.

A useful extension to hierarchical locking is known as *lock escalation*. Lock escalation allows the DBMS to automatically adjust the granularity at which transactions obtain locks, based on their behavior. If the

TABLE 56.2 Compatibility Matrix for
Regular and Intention Locks

	IS	IX	S	SIX	X
IS	y	y	y	y	n
IX	y	y	n	n	n
S	y	n	y	n	n
SIX	y	n	n	n	n
X	n	n	n	n	n

TABLE 56.3 Hierarchical Locking Rules

To Get	Must Have on All Ancestors
IS or S	IS or IX
IX, SIX, or X	IX or SIX

system detects that a transaction is obtaining locks on a large percentage of the granules that make up a larger granule, it can attempt to grant the transaction a lock on the larger granule so that no additional locks will be required for subsequent accesses to other objects in that granule. Automatic escalation is useful because the access pattern that a transaction will produce is often not known until run time.

56.3.1.4 Other Concurrency Control Methods

As stated previously, two-phase locking is the most generally accepted technique for ensuring serializability. Locking is considered to be a *pessimistic* technique because it is based on the assumption that transactions are likely to interfere with each other and takes measures (e.g., blocking) to ensure that such interference does not occur. An important alternative to locking is **optimistic concurrency control**. Optimistic methods [e.g., Kung and Robinson 1981] allow transactions to perform their operations without obtaining any locks. To ensure that concurrent executions do not violate serializability, transactions must perform a *validation phase* before they are allowed to commit. Many optimistic protocols have been proposed. In the algorithm of Kung and Robinson [1981], the validation process ensures that the reads and writes performed by a validating transaction did not conflict with any other transactions with which it ran concurrently. If during validation it is determined a conflict had occurred, the validating transaction is aborted and restarted.

Unlike locking, which depends on *blocking* transactions to ensure isolation, optimistic policies depend on transaction *restart*. As a result, although they don't perform any blocking, the performance of optimistic policies can be hurt by data contention (as are pessimistic schemes) — a high degree of data contention will result in a large number of unsuccessful transaction executions. The performance tradeoffs between optimistic and pessimistic have been addressed in numerous studies [see Agrawal et al. 1987]. In general, locking is likely to be superior in resource-limited environments because blocking does not consume cpu or disk resources. In contrast, optimistic techniques may have performance advantages in situations where resources are abundant, because they allow more executions to proceed concurrently. If resources are abundant, then the resource consumption of restarted transactions will not significantly hurt performance. In practice, however resources are typically limited, and thus concurrency control in most commercial database systems is based on locking.

Another class of concurrency control techniques is known as **multiversion concurrency control** [e.g., Reed 1983]. As updating transactions modify data items, these techniques retain the previous versions of the items on line. Read-only transactions (i.e., transactions that perform no updates) can then be provided with access to these older versions, allowing them to see a consistent (although possibly somewhat out-of-date) snapshot of the database. Optimistic, multiversion, and other concurrency control techniques (e.g., timestamping) are addressed in further detail in Bernstein et al. [1987].

56.3.2 Recovery

The recovery subsystem is generally considered to be one of the more difficult parts of a DBMS to design for two reasons: First, recovery is required to function in failure situations and must correctly cope with a huge number of possible system and database states. Second, the recovery system depends on the behavior of many other components of the DBMS, such as concurrency control, buffer management, disk management, and query processing. As a result, few recovery methods have been described in the literature in detail. One exception is the ARIES recovery system developed at IBM [Mohan et al. 1992]. Many details about the ARIES method have been published, and the method has been included in a number of DBMSs. Furthermore, the ARIES method involves only a small number of basic concepts. For these reasons, we focus on the ARIES method in the remainder of this section. The ARIES method is related to many other recovery methods such as those described in Bernstein et al. [1987] and Gray and Reuter [1993]. A comparison with other techniques appears in Mohan et al. [1992].

56.3.2.1 Overview of ARIES

ARIES is a fairly recent refinement of the write-ahead-logging (WAL) protocol. Recall that the WAL protocol enables the use of a STEAL/NO FORCE buffer management policy, which means that pages on stable storage can be overwritten at any time and that data pages do not need to be forced to disk in order to commit a transaction. As with other WAL implementations, each page in the database contains a log sequence number (LSN) which uniquely identifies the log record for the latest update that was applied to the page. This LSN (referred to as the *pageLSN*) is used during recovery to determine whether or not an update for a page must be redone. LSN information is also used to determine the point in the log from which the REDO pass must commence during restart from a system crash. LSNs are often implemented using the physical address of the log record in the log to enable the efficient location of a log record given its LSN.

Much of the power and relative simplicity of the ARIES algorithm is due to its REDO paradigm of *repeating history*, in which it redoes updates for *all* transactions — including those that will eventually be undone. Repeating history enables ARIES to employ a variant of the *physiological logging* technique described earlier: it uses *page-oriented REDO* and a form of *logical UNDO*. Page-oriented REDO means that REDO operations involve only a single page and that the affected page is specified in the log record. This is part of physiological logging. In the context of ARIES, logical UNDO means that the operations performed to undo an update do not need to be the exact inverses of the operations of the original update.

In ARIES, logical UNDO is used to support fine-grained (i.e., tuple-level) locking and high-concurrency index management. For an example of the latter issue, consider a case in which a transaction T1 updates an index entry on a given page P1. Before T1 completes, a second transaction T2 could split P1, causing the index entry to be moved to a new page (P2). If T1 must be undone, a physical, page-oriented approach would fail because it would erroneously attempt to perform the UNDO operation on P1. Logical UNDO solves this problem by using the index structure to find the index entry, and then applying the UNDO operation to it in its new location. In contrast to UNDO, page-oriented REDO can be used because the repeating history paradigm ensures that REDO operations will always find the index entry on the page referenced in the log record — any operations that had affected the location of the index operation at the time the log record was created will be replayed before that log record is redone.

ARIES uses a three-pass algorithm for restart recovery. The first pass is the *analysis* pass, which processes the log forward from the most recent checkpoint. This pass determines information about dirty pages and active transactions that is used in the subsequent passes. The second pass is the *REDO* pass, in which history is repeated by processing the log forward from the earliest log record that could require REDO, thus ensuring that all logged operations have been applied. The third pass is the *UNDO* pass. This pass proceeds backwards from the end of the log, removing from the database the effects of all transactions that had not committed at the time of the crash. These passes are shown in Figure 56.3. (Note that the relative ordering of the starting point for the REDO pass, the endpoint for the UNDO pass, and the checkpoint can be different than that shown in the figure.) The three passes are described in more detail below.

ARIES maintains two important data structures during normal operation. The first is the *transaction table*, which contains status information for each transaction that is currently running. This information

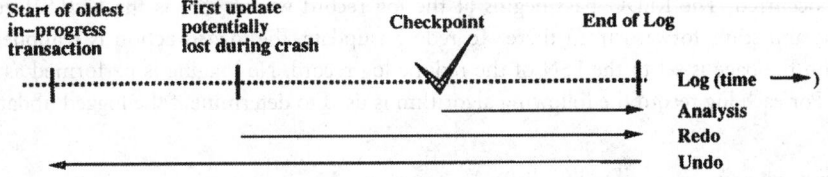

FIGURE 56.3   The three passes of ARIES restart.

includes a field called the *lastLSN*, which is the LSN of the most recent log record written by the transaction. The second data structure, called the *dirty-page table*, contains an entry for each "dirty" page. A page is considered to be dirty if it contains updates that are not reflected on stable storage. Each entry in the dirty-page table includes a field called the *recoveryLSN*, which is the LSN of the log record that caused the associated page to become dirty. Therefore, the *recoveryLSN* is the LSN of the earliest log record that might need to be redone for the page during restart. Log records belonging to the same transaction are linked backwards in time using a field in each log record called the *prevLSN* field. When a new log record is written for a transaction, the value of the *lastLSN* field in the transaction-table entry is placed in the *prevLSN* field of the new record and the new record's LSN is entered as the *lastLSN* in the transaction-table entry.

During normal operation, checkpoints are taken periodically. ARIES uses a form of fuzzy checkpoints which are extremely inexpensive. When a checkpoint is taken, a checkpoint record is constructed which includes the contents of the transaction table and the dirty-page table. Checkpoints are efficient, since no operations need be quiesced and no database pages are flushed to perform a checkpoint. However, the effectiveness of checkpoints in reducing the amount of the log that must be maintained is limited in part by the earliest *recoveryLSN* of the dirty pages at checkpoint time. Therefore, it is helpful to have a background process that periodically writes dirty pages to non-volatile storage.

56.3.2.2   Analysis

The job of the analysis pass of restart recovery is threefold: (1) it determines the point in the log at which to start the REDO pass, (2) it determines which pages could have been dirty at the time of the crash in order to avoid unnecessary I/O during the REDO pass, and (3) it determines which transactions had not committed at the time of the crash and will therefore need to be undone.

The analysis pass begins at the most recent checkpoint and scans forward to the end of the log. It reconstructs the transaction table and dirty-page table to determine the state of the system as of the time of the crash. It begins with the copies of those structures that were logged in the checkpoint record. Then, the contents of the tables are modified according to the log records that are encountered during the forward scan. When a log record for a transaction that does not appear in the transaction table is encountered, that transaction is added to the table. When a log record for the commit or the abort of a transaction is encountered, the corresponding transaction is removed from the transaction table. When a log record for an update to a page that is not in the dirty-page table is encountered, that page is added to the dirty-page table, and the LSN of the record which caused the page to be entered into the table is recorded as the *recoveryLSN* for that page. At the end of the analysis pass, the dirty-page table is a conservative (since some pages may have been flushed to nonvolatile storage) list of all database pages that could have been dirty at the time of the crash, and the transaction table contains entries for those transactions that will actually require undo processing during the UNDO phase. The earliest *recoveryLSN* of all the entries in the dirty-page table, called the *firstLSN*, is used as the spot in the log from which to begin the REDO phase.

56.3.2.2.1   REDO

As stated earlier, ARIES employs a redo paradigm called *repeating history*. That is, it redoes updates for *all* transactions, committed or otherwise. The effect of repeating history is that at the end of the REDO pass, the database is in the same state with respect to the logged updates that it was in at the time that the crash occurred. The REDO pass begins at the log record whose LSN is the *firstLSN* determined by analysis and scans forward from there. To redo an update, the logged action is reapplied and the *pageLSN* on the page is set to the LSN of the redone log record. No logging is performed as the result of a redo. For each log record the following algorithm is used to determine if the logged update must be redone:

- If the affected page is not in the dirty-page table, then the update does *not* require redo.
- If the affected page is in the dirty-page table, but the *recoveryLSN* in the page's table entry is *greater than* the LSN of the record being checked, then the update does *not* require redo.
- Otherwise, the LSN stored on the page (the *pageLSN*) must be checked. This may require that the page be read in from disk. If the *pageLSN* is *greater than or equal to* the LSN of the record being checked, then the update does *not* require redo. Otherwise, the update *must* be redone.

56.3.2.2.2 UNDO

The UNDO pass scans backwards from the end of the log. During the UNDO pass, all transactions that had not committed by the time of the crash must be undone. In ARIES, undo is an *unconditional* operation. That is, the *pageLSN* of an affected page is not checked, because it is always the case that the undo must be performed. This is due to the fact that the *repeating of history* in the REDO pass ensures that all logged updates have been applied to the page.

When an update is undone, the undo operation is applied to the page and is logged using a special type of log record called a *compensation log record* (CLR). In addition to the undo information, a CLR contains a field called the *UndoNxtLSN*. The *UndoNxtLSN* is the LSN of the next log record that must be undone for the transaction. It is set to the value of the *prevLSN* field of the log record being undone. The logging of CLRs in this fashion enables ARIES to avoid ever having to undo the effects of an undo (e.g., as the result of a system crash during an abort), thereby limiting the amount of work that must be undone and bounding the amount of logging done in the event of multiple crashes. When a CLR is encountered during the backward scan, no operation is performed on the page, and the backward scan continues at the log record referenced by the *UndoNxtLSN* field of the CLR, thereby jumping over the undone update and all other updates for the transaction that have already been undone (the case of multiple transactions will be discussed shortly). An example execution is shown in Figure 56.4.

In Figure 56.4, a transaction logged three updates (LSNs 10, 20, and 30) before the system crashed for the first time. During REDO, the database was brought up to date with respect to the log (i.e., 10, 20, and/or 30 were redone if they weren't on nonvolatile storage), but since the transaction was in progress at the time of the crash, they must be undone. During the UNDO pass, update 30 was undone, resulting in the writing of a CLR with LSN 40, which contains an *UndoNxtLSN* value that points to 20. Then, 20 was undone, resulting in the writing of a CLR (LSN 50) with an *UndoNxtLSN* value that points to 10. However, the system then crashed for a second time before 10 was undone. Once again, history is repeated during REDO, which brings the database back to the state it was in after the application of LSN 50 (the CLR for 20). When UNDO begins during this second restart, it will first examine the log record 50. Since the record is a CLR, no modification will be performed on the page, and UNDO will skip to the record whose LSN is stored in the *UndoNxtLSN* field of the CLR (i.e., LSN 10). Therefore, it will continue by undoing the

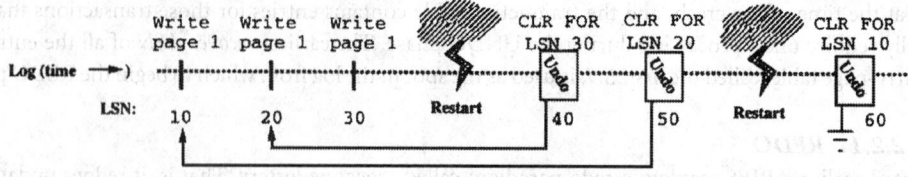

FIGURE 56.4   The use of CLRs for UNDO.

update whose log record has LSN 10. This is where the UNDO pass was interrupted at the time of the second crash. Note that no extra logging was performed as a result of the second crash.

In order to undo multiple transactions, restart UNDO keeps a list containing the next LSN to be undone for each transaction being undone. When a log record is processed during UNDO, the *prevLSN* (or *UndoNxtLSN*, in the case of a CLR) is entered as the next LSN to be undone for that transaction. Then the UNDO pass moves on to the log record whose LSN is the most recent of the next LSNs to be redone. UNDO continues backward in the log until all of the transactions in the list have been undone up to and including their first log record. UNDO for *transaction rollback* works similarly to the UNDO pass of the restart algorithm as described above. The only difference is that during transaction rollback, only a single transaction (or part of a transaction) must be undone. Therefore, rather than keeping a list of LSNs to be undone for multiple transactions, rollback can simply follow the backward chain of log records for the transaction to be rolled back.

56.4 Research Issues and Summary

The model of ACID transactions that has been described in this chapter has proven to be quite durable in its own right, and serves as the underpinning for the current generation of database and transaction processing systems. This chapter has focused on the issues of concurrency control and recovery in a centralized environment. It is important to note, however, that the basic model is used in many types of distributed and parallel DBMS environments and the mechanisms described here have been successfully adapted for use in these more complex systems. Additional techniques, however, are needed in such environments. One important technique is *two-phase commit*, which is a protocol for ensuring that all participants in a distributed transaction agree on the decision to commit or abort that transaction.

While the basic transaction model has been a clear success, its limitations have also been apparent for quite some time [e.g., Gray 1981]. Much of the ongoing research related to concurrency control and recovery is aimed at addressing some of these limitations. This research includes the development of new implementation techniques, as well as the investigation of new and extended transaction models.

The ACID transaction model suffers from a lack of flexibility and the inability to model many types of interactions that arise in complex systems and organizations. For example, in collaborative work environments, strict isolation is not possible or even desirable [Korth 1995]. Workflow management systems are another example where the ACID model, which works best for relatively simple and short transactions, is not directly appropriate. For these types of applications, a richer, multilevel notion of transactions is required.

In addition to the problems raised by complex application environments, there are also many computing environments for which the ACID model is not fully appropriate. These include environments such as mobile wireless networks, where large periods of disconnection are expected, and loosely coupled wide-area networks (the Net is an extreme example) in which the availability of systems is relatively low. The techniques that have been developed for supporting ACID transactions must be adjusted to cope with such highly variable situations. New techniques must also be developed to provide concurrency control and recovery in nontraditional environments such as heterogeneous systems, dissemination-oriented environments, and others.

A final limitation of ACID transactions in their simplest form is that they are a general mechanism, and hence, do not exploit the semantics of data and/or applications. Such knowledge could be used to significantly improve system performance. Therefore, the development of concurrency control and recovery techniques that can exploit application-specific properties is another area of active research.

As should be obvious from the preceding discussion, there is still a significant amount of work that remains to be done in the areas of concurrency control and recovery for database systems. The basic concepts, however, such as serializability theory, two-phase locking, write-ahead logging, etc., will continue to be a fundamental technology, both in their own right and as building blocks for the development of more sophisticated and flexible information systems.

Acknowledgment

Portions of this chapter are reprinted with permission from Franklin, M., Zwilling, M., Tan, C., Carey, M., and DeWitt, D., Crash recovery in client-server EXODUS. In *Proc. ACM Int. Conf. on Management of Data (SIGMOD'92)*, San Diego, June 1992. © 1992 by the Association for Computing Machinery, Inc. (ACM).

Defining Terms

Abort: The process of rolling back an uncommitted transaction. All changes to the database state made by that transaction are removed.

ACID properties: The transaction properties of atomicity, consistency, isolation, and durability that are upheld by the DBMS.

Checkpointing: An action taken during normal system operation that can help limit the amount of recovery work required in the event of a system crash.

Commit: The process of successfully completing a transaction. Upon commit, all changes to the database state made by a transaction are made permanent and visible to other transactions.

Concurrency control: The mechanism that ensures that individual users see consistent states of the database even though operations on behalf of many users may be interleaved by the database system.

Concurrent execution: The (possibly) interleaved execution of multiple transactions simultaneously.

Conflicting operations: Two operations are said to conflict if they both operate on the same data item and at least one of them is a *write( )*.

Deadlock: A situation in which a set of transactions is blocked, each waiting for another member of the set to release a lock. In such a case none of the transactions involved can make progress.

Log: A sequential file that stores information about transactions and the state of the system at certain instances.

Log record: An entry in the log. One or more log records are written for each update performed by a transaction.

Log sequence number (LSN): A number assigned to a log record, which serves to uniquely identify that record in the log. LSNs are typically assigned in a monotonically increasing fashion so that they provide an indication of relative position.

Multiversion concurrency control: A concurrency control technique that provides read-only transactions with conflict-free access to previous versions of data items.

Nonvolatile storage: Storage, such as magnetic disks or tapes, whose contents persist across power failures and system crashes.

Optimistic concurrency control: A concurrency control technique that allows transactions to proceed without obtaining locks and ensures correctness by validating transactions upon their completion.

Recovery: The mechanism that ensures that the database is fault-tolerant; that is, that the database state is not corrupted as the result of a software, system, or media failure.

Schedule: A schedule for a set of transaction executions is a partial ordering of the operations performed by those transactions, which shows how the operations are interleaved.

Serial execution: The execution of a single transaction at a time.

Serializability: The property that a (possibly interleaved) execution of a group transactions has the same effect on the database, and produces the same output, as some serial (i.e., non-interleaved) execution of those transactions.

STEAL/NO-FORCE: A buffer management policy that allows committed data values to be overwritten on nonvolatile storage and does not require committed values to be written to nonvolatile storage. This policy provides flexibility for the buffer manager at the cost of increased demands on the recovery subsystem.

Transaction: A unit of work, possibly consisting of multiple data accesses and updates, that must commit or abort as a single atomic unit. Transactions have the ACID properties of *atomicity, consistency, isolation,* and *durability*.

Two-phase locking (2PL): A locking protocol that is a sufficient but not a necessary condition for serializability. Two-phase locking requires that all transactions be well formed and that once a transaction has released a lock, it is not allowed to obtain any additional locks.

Volatile storage: Storage, such as main memory, whose state is lost in the event of a system crash or power outage.

Well formed: A transaction is said to be well formed with respect to reads if it always holds a shared or an exclusive lock on an item while reading it, and well formed with respect to writes if it always holds an exclusive lock on an item while writing it.

Write-ahead logging: A protocol that ensures all log records required to correctly perform recovery in the event of a crash are placed on nonvolatile storage.

References

Agrawal, R., Carey, M., and Livny, M. 1987. Concurrency control performance modeling: alternatives and implications. *ACM Trans. Database Systems* 12(4), Dec.

Berenson, H., Bernstein, P., Gray, J., Melton, J., Oneil, B., and Oneil, P. 1995. A critique of ANSI SQL Isolation Levels. In *Proc. ACM SIGMOD Int. Conf. on the Management of Data*, San Jose, CA, June.

Bernstein, P., Hadzilacos, V., and Goodman, N. 1987. *Concurrency Control and Recovery in Database Systems*. Addison–Wesley, Reading, MA.

Bjork, L. 1973. Recovery scenario for a DB/DC system. In *Proc. ACM Annual Conf.* Atlanta.

Davies, C. 1973. Recovery semantics for a DB/DC system. In *Proc. ACM Annual Conf.* Atlanta.

Eswaran, L., Gray, J., Lorie, R., and Traiger, I. 1976. The notion of consistency and predicate locks in a database system. *Commun. ACM* 19(11), Nov.

Gray, J. 1981. The transaction concept: virtues and limitations. In *Proc. Seventh International Conf. on Very Large Databases*, Cannes.

Gray, J., Lorie, R., Putzolu, G., and Traiger, I. 1975. Granularity of locks and degrees of consistency in a shared database. In *IFIP Working Conf. on Modelling of Database Management Systems*.

Gray, J. and Reuter, A. 1993. *Transaction Processing: Concepts and Techniques*. Morgan Kaufmann, San Mateo, CA.

Haerder, T. and Reuter, A. 1983. Principles of transaction-oriented database recovery. *ACM Comput. Surveys* 15(4).

Korth, H. 1995. The double life of the transaction abstraction: fundamental principle and evolving system concept. In *Proc. Twenty-First International Conf. on Very Large Databases*, Zurich.

Kung, H. and Robinson, J. 1981. On optimistic methods for concurrency control. *ACM Trans. Database Systems* 6(2).

Lomet, D. 1977. Process structuring, synchronization and recovery using atomic actions. *SIGPLAN Notices* 12(3), Mar.

Mohan, C. 1990. ARIES/KVL: a key-value locking method for concurrency control of multiaction transactions operating on B-tree indexes. In *Proc. 16th Int. Conf. on Very Large Data Bases*, Brisbane, Aug.

Mohan, C., Haderle, D., Lindsay, B., Pirahesh, H., and Schwarz, P. 1992. ARIES: a transaction method supporting fine-granularity locking and partial rollbacks using write-ahead logging. *ACM Trans. Database Systems* 17(1), Mar.

Papadimitriou, C. 1986. *The Theory of Database Concurrency Control*. Computer Science Press, Rockville, MD.

Reed, D. 1983. Implementing atomic actions on decentralized data. *ACM Trans. Comput. Systems* 1(1), Feb.

Rosenkrantz, D., Sterns, R., and Lewis, P. 1977. System level concurrency control for distributed database systems. *ACM Trans. Database Systems* 3(2).

Further Information

For many years, what knowledge that existed in the public domain about concurrency control and recovery was passed on primarily though the use of multiple-generation copies of a set of lecture notes written by Jim Gray in the late seventies ("Notes on Database Operating Systems" in *Operating Systems: An Advanced Course* published by Springer–Verlag, Berlin, 1978). Fortunately, this state of affairs has been supplanted by the publication of *Transaction Processing: Concepts and Techniques* by Jim Gray and Andreas Reuter (Morgan Kaufmann, San Mateo, CA, 1993). This latter book contains a detailed treatment of all of the topics covered in this chapter, plus many others that are crucial for implementing transaction processing systems.

An excellent treatment of concurrency control and recovery theory and algorithms can be found in *Concurrency Control and Recovery in Database Systems* by Phil Bernstein, Vassos Hadzilacos, and Nathan Goodman (Addison–Wesley, Reading, MA, 1987). Another source of valuable information on concurrency control and recovery implementation is the series of papers on the ARIES method by C. Mohan and others at IBM, some of which are referenced in this chapter. The book *The Theory of Database Concurrency Control* by Christos Papadimitriou (Computer Science Press, Rockville, MD, 1986) covers a number of serializability models.

The performance aspects of concurrency control and recovery techniques have been only briefly addressed in this chapter. More information can be found in the recent books *Performance of Concurrency Control Mechanisms in Centralized Database Systems* edited by Vijay Kumar (Prentice–Hall, Englewood Cliffs, NJ, 1996) and *Recovery in Database Management Systems*, edited by Vijay Kumar and Meichun Hsu (Prentice–Hall, Englewood Cliffs, NJ, in press). Also, the performance aspects of transactions are addressed in *The Benchmark Handbook: For Database and Transaction Processing Systems* (2nd ed.), edited by Jim Gray (Morgan Kaufmann, San Mateo, CA, 1993).

Finally, extensions to the ACID transaction model are discussed in *Database Transaction Models*, edited by Ahmed Elmagarmid (Morgan Kaufmann, San Mateo, CA, 1993). Papers containing the most recent work on related topics appear regularly in the ACM SIGMOD Conference and the International Conference on Very Large Databases (VLDB), among others.

57

Transaction Processing

57.1   Introduction 57-1
Transaction Processing Preliminaries • Transaction Processing
Infrastructure • Distributed Transaction Processing

57.2   Secure Distributed Transaction Processing:
Cryptography 57-7
The iKP Protocol • NetBill

57.3   Transaction Processing on the Web: Web Services 57-9
Introduction to Web Services • Components of Web Services
• Web Services Transactions (WS-Transactions)

57.4   Concurrency Control for High-Contention
Environments 57-12
Wait-Depth Limited Methods • Two-Phase Processing
Methods • Reducing Data Contention

57.5   Performance Analysis of Transaction
Processing Systems 57-16
Hardware Resource Contention • Performance Degradation
Due to Locking

57.6   Conclusion 57-22

Alexander Thomasian
New Jersey Institute of Technology

57.1   Introduction

"Six thousand years ago the Sumerians invented writing for transaction processing" [Gray and Reuter 1993], but the same goal can be accomplished today with a few clicks of the mouse.

The discussion of transaction processing concepts in this chapter is somewhat abbreviated, because aspects of transaction processing are covered in chapters on *Concurrency Control and Recovery* (Chapter 56) and *Distributed and Parallel Database Systems* (Chapter 58). Enough material is included here to make this chapter self-complete. In this section we review the fundamentals of transaction processing, its infrastructure, and distributed transaction processing.

Section 57.2 is an introduction to the cryptography required for e-commerce. The emphasis is on the protocols rather than the mathematical aspects of *cryptography*, which is the title of Chapter 9. This is followed by transaction processing on the Web, better known as Web Services, in Section 57.3.

Section 57.4 is a review of concurrency control methods to reduce the level of data contention in high-contention environments. This discussion is motivated by the increase in the volume of transactions made possible by electronic shopping. With the rapid increase in computing power, it is data rather than hardware resource contention that may become the bottleneck, unless the software is carefully designed not to thrash under high levels of lock contention. Such behavior was observed as part of a benchmarking study of an e-commerce application with a mySQL DBMS (DBMS = database management system) [Elnikety et al. 2003].

Section 57.5 discusses performance analysis for transaction processing, taking both hardware resource contention and data contention into account. Transactions tend to have stringent response time requirements, so it is important to understand the factors affecting transaction performance. We are not concerned here with software performance engineering, whose role is to predict performance as the software is being developed. Conclusions are given in Section 57.6.

57.1.1   Transaction Processing Preliminaries

Transactions can be described by four **ACID properties** as follows:

1. **Atomicity:** all or nothing property.
2. **Consistency:** a transaction's modification of a consistent database will lead to another consistent database state. Consistency can be enforced by **integrity constraints**, which provide a notification when a constraint is violated; for example, the salary of an employee, after a raise, exceeds his manager's salary (a company policy) [Ramakrishnan and Gehrke 2003].
3. **Isolation:** it is as if a transaction is executing by itself, although there are other transactions being processed concurrently.
4. **Durability:** the updates made by a **committed** transaction will endure system failures.

Transaction commit is the declaration that a transaction has completed successfully, so that it cannot be rolled back and that its updates have been made durable. It is reassuring to know this when making a bank deposit. Transaction **abort** can be self-induced due to an internal error condition, or the user who initiated a transaction can abort it in the middle, or the system can initiate the abort to resolve a deadlock (such transactions are automatically restarted by the system).

The ACID properties are required for pragmatic reasons. Transaction updates are first applied to database pages residing in the **database buffer** in volatile main memory, whose content may be lost due to power outages, but also system crashes, software failures, and operator errors. Most DBMSs use the **NO-FORCE** policy, so that the pages modified by a transaction do not have to be propagated to disk as part of transaction commit. This policy is a necessity in high-performance transaction processing systems, because writing to disk of certain frequently updated database pages would result in unnecessary disk traffic.

The **STEAL** policy allows dirty pages of *uncommitted* transactions to be written to disk. Atomicity and durability in a centralized database are implemented by the **recovery** component of the DBMS, which ensures both by **logging** appropriate information onto **non-volatile storage** (**NVS**). As far as the NO-FORCE policy is concerned, logging writes the **after-images** or modifications made by the transaction. As far as the STEAL policy is concerned, the logging process writes the **before-image** of modified data to disk. This allows the database to be returned to its original state in case the transaction which made the change is aborted.

Checkpointing, which is invoked periodically, writes (committed) dirty pages to disk with the intent of reducing **recovery time** when the system is restarted after a failure occurs. This is because database pages on disk may not be up-to-date because of the NO-FORCE policy. A recovery method, such as ARIES, uses the log to bring disk pages to a consistent state, so that they reflect updates by committed transactions and no modifications made by aborted transactions. This ensures transaction atomicity.

Concurrent execution of transactions or **multiprogramming** is a system requirement for transaction processing systems (see Section 57.5 for more details). **Concurrency control** methods are required to ensure that transaction updates do not result in an incorrect execution; for example, the updates of one transaction overwrite another's update. In fact, concurrency control methods ensure both consistency *and* isolation.

Strict two-phase locking (2PL) is the dominant concurrency control method in commercial DBMSs. Strict 2PL ensures **serializability**, that is, *the concurrent execution of a set of transactions is considered correct as long as it corresponds to a serial execution schedule*. While this definition is the *de facto* standard, certain specialized applications, such as stock trading, may have more stringent requirements.

Commercial DBMSs do not strictly adhere to strict 2PL, but rather provide options to suit application requirements. Transactions may specify variations in the **isolation level** required for their execution. A strict 2PL paradigm is unacceptable in some cases for pragmatic reasons. An example is the running of a read-only query to determine the total amount held in the checking accounts of a bank, when the records are stored in a relational table. Holding a shared lock on the table is unacceptable for the duration of the transaction, since it will block access to short online transactions, e.g., those generated by ATM (automatic teller machine) access. One solution is for the read-only transaction to lock one page at a time and release the lock immediately after it is done with the page. which is referred to as the **cursor-stability** isolation level. The shortcoming is that the query will only provide an approximation to the total amount.

The **timestamp-ordering** concurrency control method was proposed to deal with data conflicts in distributed databases using a local algorithm, so as to minimize the overhead associated with concurrency control. There are many variations of this method, but it is generally known to provide poor performance.

The **optimistic concurrency control (OCC)** method was originally proposed to deal with locking overhead in low data contention environments. We will discuss optimistic concurrency control methods in some detail in Section 57.4.

Two phase locking stipulates that transactions acquire **shared locks** on objects they read and **exclusive locks** on objects to be modified and that *no lock is released until all locks are acquired*. Transaction T_2 can access an object modified by T_1, as soon as T_1 releases its lock, but T_2 cannot commit until T_1 commits. If T_1 aborts we might have **cascading aborts**. **Strict 2PL** eliminates cascading aborts, at the cost of a reduced concurrency level in processing transactions, by requiring locks to be held until transaction commit time. For example, T_2 cannot access an object modified by T_1, until T_1 commits or aborts.

Most lock requests are successful, because most database objects are not locked most of the time. Otherwise, only shared locks are compatible with each other. A transaction encountering a lock conflict is blocked awaiting the release of the lock, until the transaction holding an exclusive lock or all transactions holding a shared lock commit or abort.

A **deadlock** occurs when an active transaction T_1 requests a lock held by a blocked transaction T_2, which is in turn waiting for T_1's completion (or abort). A deadlock may involve only one object, which occurs when two transactions holding a shared lock on an object need to upgrade their lock to exclusive mode. **Update locks**, which are not compatible with each other but are compatible with shared locks, were introduced to prevent the occurrence of such deadlocks.

So far, we have discussed **flat transactions**. The **nested transaction** paradigm offers "more decomposable execution units and finer grained control over concurrency and recovery than flat transactions" [Moss 1985]. This paradigm also supports the decomposition of a "unit of work" into subtasks and their appropriate distribution in a computer system.

Multilevel transactions are related to nested transactions, but are more specialized. Transactions hold two types of locks: (1) long-term object locks (e.g., locks on records) and (2) short-term locks are held by the subtransactions on database pages for the duration of operations on records, e.g., to increase record size by adding another field [Weikum and Vossen 2002]. Compensating operations for subtransactions are provided for rollback.

57.1.1.1   Further Reading

The reader is referred to Chapter 56 in this *Handbook* on "Concurrency Control and Recovery" especially Section 56.4.7 on nested transactions and Section 56.49 on multilevel transactions, as well as Ramakrishnan and Gehrke [2003], Lewis et al. [2002], and Gray and Reuter [1993].

57.1.2   Transaction Processing Infrastructure

Online transaction processing (OLTP) replaced **batch processing** to handle situations where delayed updates are unacceptable (e.g., tracking cash withdrawals from ATMs). Some nonrelational DBMSs, which were mainly used for batch processing, were extended with OLTP capabilities in the 1970s. The transition to

OLTP became possible with the advent of direct access storage devices (DASD), since it became possible to access the required data records in a few milliseconds (see Chapter 86 on "Secondary Storage Filesystems").

Transaction processing monitors (or **TP monitors**) can be considered specialized operating systems that can execute transactions concurrently using threads, although the operating systems can accomplish the same goal via multiprogramming. The advantages and disadvantages of the two approaches are beyond the scope of this discussion (see Gray and Reuter [1993]). IMS has two components — IMS/DB and IMS/DC (data communications) — where the latter is a TP monitor providing **message processing regions (MPRs)**. Transactions are classified into classes that are assigned to different MPRs to run. Smaller degrees of concurrency are required to maintain a lower level of lock contention, as discussed below.

Batch processing requires a very simple form of concurrency control, that is, the locking of complete files, allowing two applications accessing the same file need to be run serially. Concurrent execution of the two programs is possible by partitioning the files into subfiles, such that with perfect synchronization (possible in a perfect world), program one can update partition $i + 1$ while program two can read partition i. At best, the two programs can be merged to attain pipelining at the record level. This can be accomplished by global query optimization (see Chapter 55 on "Query Optimization").

OLTP led to a flurry of research and development activities in the area of concurrency control and recovery in the late 1970s, but many refinements followed later [Gray and Reuter 1993]. This heightened activity level coincided with the advent of **relational databases** and the introduction of sophisticated locking methods, such as **intent locks** to facilitate **hierarchical locking**; for example, to detect a conflict between a shared lock on a relational table and exclusive locks on records in that table.

Application programs communicate with the TP monitors via a library of functions or a language such as the **Structured Transaction Definition Language (STDL)** [STDL 1996]. There is a STDL compiler that translates STDL statements to API (application programming interface) calls to supported TP monitors. STDL supports transaction demarcation, exception handling, interfaces to resource managers, transaction workspaces, transactional presentation messaging, calls for high-level language programs, spawning independent transactions, concurrent execution of procedures within a transaction, enqueueing and dequeueing data, data typing, and multilingual messages [STDL 1996].

Up to some point, transaction processing involved "dumb terminals" and mainframes or servers, which ran all the software required to get the job done. Things became slightly more complicated with the advent of **client/server computing** [Orfali et al. 1996]. A **three-tiered architecture** has the **clients** or users with a **graphical user interface (GUI)** on their PCs at the lowest level. The next level is an **application server** with (application) programs that are invoked by a user. The application server uses **object request brokers (ORBs)**, such as **CORBA (common ORB architecture)** or the *de facto* standard **DCOM (distributed component object model)**. Web or Internet servers are yet another category that use HTTP and XML. Finally, there is a **data server**, or simply a server.

A **two-tiered architecture** is the result of combining the client and application tiers, leading to a client/server system with **fat clients**. Clients can then communicate directly with the data server via the **SQL Structured Query Language** embedded in a client/server communication protocol such as **ODBC (Open Database Connectivity)** or **JDBC**. ODBC and JDBC use indirection to achieve SQL code portability across various levels. A two-tiered architecture may consist of a **thin client**, with the application and data-server layers at the server.

57.1.3 Distributed Transaction Processing

Centralized (with independent software components) and distributed transaction processing both require a **two-phase commit (2PC)** protocol, which is discussed in Section 57.3.

The strict 2PL protocol is also utilized in distributed transaction processing. An SQL query is subdivided into subqueries, based on the location of the relational tables being accessed. The subqueries are processed in a distributed manner at the nodes where the data resides, acquiring appropriate locks locally.

We use **waits-for-graphs (WFGs)** to represent transaction blocking, such that the nodes of the graph represent transactions and directed edges represent the waits-for relationship. **Local deadlocks** are detected by checking for cycles in local WFGs, while **distributed deadlocks** can be detected by transmitting the WFGs at various nodes to a designated node, which builds the global WFG for deadlock detection. This can be costly in the number of messages involved, and phantom deadlocks (due to out-of-date WFGs) can result in unnecessary aborts.

The **wound-wait** and **wait-die** methods proposed for distributed transaction processing are deadlock-free [Bernstein et al. 1987]. Both methods associate a **timestamp** with each transaction based on its initiation time. The timestamp is attached to lock requests and used in lock conflict resolution, as follows. The wound-wait method blocks a transaction T_A requesting a lock held by T_B, if T_A is not older than T_B; otherwise, T_B is aborted. The wait-die method allows a younger transaction blocked by an older transaction to wait; otherwise, the transaction encountering the lock conflict is aborted.

Data access in distributed transaction processing can be accomplished according to one of the following methods [Thomasian 1996a]:

- **I/O request shipping.** An I/O request is sent to the node that holds the data on its disk or in its database buffer.

- **Data request (or call) shipping.** The query optimizer determines the location of the data by consulting the **distributed data directory** and then initiates SQL calls to appropriate sites.

- **Distributed transaction processing.** This is accomplished by **remote procedure calls (RPCs)**, which are similar to calls to (local) **stored procedures**. This approach has the advantage of minimizing the volume of the data to be transferred, because the procedure returns the answer, which may be very short.

Peer-to-peer programming allows more flexibility than RPC [Bernstein and Newcomer 1997]:

- **Flexible message sequences.** An RPC requires master–slave communication. This is a synchronous call-return model and all call-return pairs should be properly nested. Consider program A or P_A that calls P_B, that calls P_C. P_B cannot do anything until it hears from P_C, at which point it can initiate another call to P_C or return control to P_A.

- **Transaction termination.** All called programs must first announce the termination of their processing to the caller. In the above example, P_A cannot terminate until P_B terminates, and P_B cannot terminate until P_C terminates. Only then can P_A initiate commit processing. In the peer-to-peer model, any program may invoke termination via a **synchpoint** or **commit** operation. Transaction commit is delayed when some transactions are still running.

- **State of the transaction.** An RPC is **connectionless**. For the client and the server to share state, the server should return the state to the client. A **context handle** is used in some client/server systems for this purpose. In the peer-to-peer paradigm, communication programs share the transaction id and whether it is active, committed, or aborted. As far as **Web servers** are concerned the **http protocol** is stateless. Aside from maintaining the state in a middle tier, the state can be maintained via a **cookie**, which is a *(name,value)* pair. Cookies are perceived to be invasive and browsers may disallow cookies from being saved.

- **Communication mechanism.** Connection-oriented peer-to-peer communication protocols are favored in transaction processing. IBM's **Logical Unit (LU6.2)** is a transactional peer-to-peer or RPC protocol specification, and is a *de facto* standard supported by many TP monitors.

Queued transaction processing can be used to deal with failures, so when the server or the client is down, queueing the request (from the client to the server) and the reply (from the server to the client) can be used to ensure eventual delivery. *Receiver-initiated* load balancing, where an idle server can pick up requests from a shared queue, has been shown to outperform *sender-initiated* transaction routing, because the latter can result in an *unbalanced load* (e.g., a server is idle while other servers have a queue). IBM's MQSeries is one of the leading products in this area.

57.1.3.1 Data Replication

Data replication in a distributed system can be used to improve data availability and also performance, because data can be read from the "closest" node, which may be local. Updates are expensive if we use the **read-one write-all (ROWA)** paradigm or **synchronous replication**. When all nodes are not available, the **ROWA-A(available)** protocol updates all available nodes. The **Quorum Concensus Protocol** uses a write quorum Q_W and a read quorum Q_R, where $Q_W > N/2$ and $Q_R + Q_W > N$, and where the latter condition ensures that a read request will encounter at least one up-to-date copy of the data.

Alternatively, one of the nodes can be designated as the **primary copy**, so that all updates are first carried out at that node. Updates to the primary can be propagated to other nodes via **asynchronous update propagation** by sending the log records from the primary node to others.

To deal with network partitions, the **majority concensus** paradigm allows a subset of nodes with more than one half of the nodes to have a primary copy. The **quorum concensus** algorithm assigns different weights to nodes to deal with ties (i.e., when half of the nodes are in each partition).

57.1.3.2 Data Sharing or Shared Disk Systems

Computer systems, from the viewpoint of database applications, have been classified as:

- **Shared everything.** This is typically a **shared memory multiprocessor (SMP)**, where the processors share memory and disks. SMPs have become more powerful and can meet high-end transaction processing requirements.

- **Shared nothing.** Independent computers using message passing for communication. This configuration is most suitable for parallel database applications, such as in the case of the Teradata DBC/1012 computer or the NonStop SQL system from Tandem, which was also used for transaction processing (these are the original names of the computers and companies in the mid-1980s).

- **Shared disk or data sharing.** Data sharing can be used as a method to exceed the performance limits of a single computer, by allowing multiple computers access to data residing on a set of shared disks. **Cluster computers** with shared disk also belong to this category. Data sharing can be implemented on a shared nothing system with I/O request shipping.

A data sharing system in addition to concurrency control requires **coherency control** because more than one computer's buffer can hold the same database page. When such a page is updated by one computer, the other copies in the database buffers need to be invalidated or updated. The log is written as part of transaction commit, but the updating of disk data can wait because a NO-FORCE policy is in effect. The **broadcast invalidation** method, which is the preferred method for processor caches, will result in the purging of an updated page from the caches of other computers.

A transaction router determines the routing of transactions to the nodes of a data sharing system. The routing can be done with **load balancing** as a goal. The objective of **transaction affinity** is to reduce transaction processing requirements. This is accomplished by processing transactions in the same class or transactions that access the same datasets at the same node, so that the node buffers the pages associated with those datasets.

Primary copy locking (PCL), where each node is responsible for a database partition, is preferable to a centralized lock manager due to the following reasons [Rahm 1993]: (1) proper use of PCL results in a balanced load for processing lock requests; and (2) **affinity based routing (ABR)**, which matches transactions classes with database partitions, reduces inter-node communication. The node responsible for a part of the data is referred to as the **primary copy authority (PCA)**.

Integrated concurrency and coherency control uses **on-request-invalidation** or **check-on-access** to combine a lock request with checking the validity of a page cached locally. PCA caches at each node the current copy of recently accessed pages that belong to that partition. A node making a lock request attaches the version number in the database buffer of the page for which the lock is being requested. The PCA returns the current copy of the page if the version number is different, along with the granting of the lock. Conversely, lock releases attach the current version of the page for caching at the primary site.

Data sharing concurrency and coherency control methods have a lot in common with client/server methods, yet there are many subtle differences [Franklin et al. 1997].

57.1.3.3 Further Information

For a more thorough discussion of the topics in this section, the reader is referred to Chapter 58 on "Distributed and Parallel Database Systems," as well as Cellary et al. [1988], Ceri and Pelagatti [1984], Gray and Reuter [1993], Bernstein and Newcomer [1997], Orfali et al. [1996], Ozsu and Valduriez [1999], and Ramakrishnan and Gehrke [2003].

57.2 Secure Distributed Transaction Processing: Cryptography

Electronic commerce is growing at a rapid pace with respect to traditional commerce, and provides major application of distributed transaction processing. This is a rapidly evolving area and our goal here is to introduce the reader to one of the more important issues, **cryptography**, and to discuss two particular protocols: iKP and NetBill.

Encryption is used to stop an intruder from reading a message that is only to be read by a certain receiver. This is accomplished by encrypting the **plaintext** ($Text_{plain}$), using a *sender key* (K_s) to produce **cyphertext** ($Text_{cipher}$), which is then decrypted by the receiver using the *receiver key* (K_r). More formally, $Text_{cipher} = K_s Text_{plain}$ and, conversely, $Text_{plain} = K_r Text_{cipher}$.

Symmetric or **secret key (SK)** cryptography uses the same key for both encryption and decryption: $K_r = K_s$, so this key should be known to each sender–receiver pair. This is the classical approach to cryptography and has been realized in countless ways.

The popular **Data Encryption Standard (DES)** uses a sequence of *transpositions* and *substitutions* on 64-bit blocks of plaintext using a 56-bit key. A successor to DES is the **Advanced Encryption Standard (AES)**, which allows 128, 192, and 256-bit key sizes. The Kerberos protocol developed at MIT also uses SK cryptography to authenticate a client to a server and create session keys on demand.

Asymmetric or **public key (PK)** cryptography assigns different public and private keys to each person (c) K_{pb}^c and K_{pr}^c, respectively. Public keys are known to everybody (e.g., by being posted on an individual's Web page). Plaintext intended for a person with K_{pb}^c is encrypted with that key, but only that person can retrieve the original plaintext, since $T_{plain} = K_{pr}^c (K_{pb}^c T_{plain})$.

This algorithm relies on **one-way functions**, whose inverses are very expensive to compute and so the code is hard to break (this is related to computing the factors of a very large number). The most popular PK algorithm (RSA) uses modular exponentiation, so encryption, which uses a small exponent, is easy, while decryption with a large exponent is expensive. Public (resp. private) key operations were recently measured to be in tens (resp. thousands) of milliseconds as the key size is increased from 512 to 2048 bits [Menasce and Almeida 2000].

An electronic document can be signed using a **digital signature** based on PK encryption, except that K_{pr} is used for signing and P_{kp} for verification. A digital signature, similar to an ordinary signature, can be used to verify that a message received by B was indeed sent by A, rather than an imposter pretending to be A. This is accomplished by A first encrypting the message being sent using his *private key* and then encrypting it using B's public key. After receiving a message, B first applies his private key and then A's public key. The message received by B could only have come from A, because nobody else knows A's private key. This allows **non-repudiation** because A cannot claim that he did not send the message.

If it is just a matter of authenticating the sender, rather than hiding the contents of the message, then the aforementioned encryptions can be applied to a **message signature** or a **message digest**, rather than the message itself. The message signature, which is obtained by a one-way function (e.g., hashing), is much smaller than the message; therefore, it is much less expensive to encrypt. There are many techniques for generating digests, such as **MD5** and **SHA-1**, that generate 128- and 160-bit digests, respectively.

The recipient A applies the hash function to the received message and compares it with the signature (after decrypting it). If they match, then the message was indeed from B and it was not corrupted in passage (perhaps by an interceptor) because the signature serves as a checksum.

A customer interested in purchasing goods from a vendor encrypts his message with the vendor's public key, so that only that particular vendor can decrypt it. There is a problem with reliably finding out a vendor's public key because, otherwise, a customer's order may go to another vendor who is impersonating the intended one.

Netscape's **Secure Sockets Layer (SSL)** protocol uses certificates to support secure communication and authentication between clients (customers) and servers (of the vendors). The Kerberos protocol developed at MIT uses SK cryptography to authenticate a customer to a server and create *session keys* on demand (see Chapter 74 on "Network and Internet Security" in this *Handbook*). Kerberos is *not* suitable for the high volumes of transactions in an e-commerce environment because its server could become a bottleneck.

Vendors who want to be authenticated obtain the certificate from a **certification authority (CA)**, (e.g., Verisign); the CA issues certificates to vendors determined to be "reliable." A customer with a browser running the SSL protocol who wants to place an order with a particular vendor is first provided with the vendor's *certificate*. The browser has the public keys of all CAs, but it does not need to communicate with the CA for authentication. The **X.509 certificate**, which the vendor encrypts with its private key before sending to the customer, has the following fields: *name of the CA, vendor's name, url, public key, timestamp, and expiration time*. The browser uses the public key of the CA to decrypt the message and verifies that the URLs are the same.

The customer's browser next generates and sends to the vendor a **pre-master secret**, which is then used by the customer and the vendor, who use the same algorithm, to generate two **session keys** for communication each way during the same session. From this point on, the customer and the vendor communicate using a *symmetric* encryption protocol based on the session key. The reason for using symmetric rather than asymmetric encryption is that that the latter is much more expensive.

To authenticate the customer's identity, the customer may be required to first obtain an account with the vendor, at which point the vendor verifies the customer's identity. The customer is asked to log in after the session key is established using his userid and password.

Netscape's SSL protocol, which is invoked when the browser points to a URL starting with *https*, offers authentication, confidentiality, and non-repudiation, but has been superseded by **Transport Level Security (TLS)**, which is now an IETF RFC (Internet Engineering Task Force Request for Comments). TLS running between the http and TCP layers consists of a handshake protocol and record protocol. The handshake protocol selects the DES algorithm used for bulk encryption, the **Message Authentication Code (MAC)** used for message authentication, and the compression algorithm used by the record protocol.

The **Secure Electronic Transaction (SET)** protocol can ensure that the vendor does not have access to the customer's credit card number and cannot misuse it. This can be accomplished simply by encoding credit card–related information using the public key of the credit card company. In a more elaborate scheme, each customer has a certificate that contains his credit card number and expiration date, and this information properly encrypted can only be processed by the **payment gateway**, rather than the vendor.

We next discuss two protocols for electronic commerce: the iKP family of protocols [Bellare et al. 2000] and the NetBill security and transaction protocol [Cox et al. 1995].

57.2.1 The iKP Protocol

The iKP ($i = 1, 2, 3$) family of protocols are similar in that we are dealing with customers, merchants, and a gateway between the electronic world and the payment infrastructure. The following steps are followed:

1. A customer initiates the action by sending a message to the merchant expressing interest in a product.
2. The merchant sends an invoice.
3. The customer uses his credit card for the payment.

4. The merchant asks for authorization from the gateway.
5. The gateway (also called the *acquirer*) grants authorization (most of the time).
6. The merchant sends the confirmation and the goods to the customer.

iKP protocols vary in the number of public-key pairs utilized. In 1KP, only the gateway possesses public and private keys. Payment is authenticated by sending the credit card number and the associated PIN, encrypted using the gateway public key. A weakness of this protocol is that it does not offer non-repudiation (i.e., disputes about the non-authenticity of orders, etc.). In 2KP, in addition to the gateway, the merchants hold keys, such that the customers can be sure that they are dealing with the right merchant. 3KP requires customers to have a public key, which ensures non-repudiation.

57.2.2 NetBill

NetBill extends (distributed) transaction atomicity with additional concepts to suit e-commerce as follows [Tygar 1998]:

1. **Money atomic protocols:** money is not created or destroyed in funds transfer.
2. **Goods atomic protocols:** in addition to money atomicity, ensure the exact exchange of goods for money, which is similar to the cash-on-delivery (COD) protocol.
3. **Certified delivery:** in addition to money and goods atomicity, allows both the consumer and merchant to prove that electronic goods were delivered. With COD, it is as if the contents of the delivered parcel are recorded by a trusted third party.
4. **Anonymity:** consumers do not want anybody to know their identity, for example, to preserve their privacy.

A representative anonymous electronic commerce protocol works as follows: a customer withdraws money from a bank in the form of a cryptographic token. He makes the money untraceable by cryptographically transforming the token, but the merchant can still check its validity. When spending the money, the customer applies a transformation, inserting the merchant's identity, who ensures that he has not received the token previously before sending the goods to the customer and also deposits his token at the bank. The bank checks the token for uniqueness and the customer remains anonymous, unless the token has been reused.

If the customer is not sure that the token reached the merchant, he can return the token to the bank, but if the token was, in fact, received by the merchant, then there is a problem and the identity of the customer is revealed. If the customer does not return the token and the token was not received by the merchant, then the customer will lose his money without receiving the goods.

A trusted server in NetBill acts as an ultimate authority, but security failures are possible when this server is corrupted. Appropriate log records are required to ensure recovery.

The issue of transaction size is important because it is important to ensure that the cost of processing the transaction remains a small fraction of the amount involved.

57.2.2.1 Further Information

A more in-depth treatment of this material can be found in books on cryptography and network security, such as Kaufman et al. [2002], as well as some database textbooks, such as Ramakrishnan and Gehrke [2003] and Lewis et al. [2002].

57.3 Transaction Processing on the Web: Web Services

We start with a brief introduction to **Web services (WS)**, then discuss the WS building blocks, and conclude with a discussion of **WS-transactions**.

57.3.1   Introduction to Web Services

Transaction processing is tied to WS, which is a level of abstraction like the Internet that sits above application servers like (e.g., CORBA) [WS Architecture 2003]:

> A Web service is defined by a URI, whose public interfaces and bindings are defined and described using XML. Its definition can be discovered by other software systems. These systems may then interact with the Web service in a manner prescribed by its definition, using XML-based messages conveyed by Internet protocols.

The **URI (uniform resource identifier)** is better known as the **URL (uniform resource locator)**. **XML** stands for Extensible Markup Language.

Let's consider an example. A user specifies the following to a travel agent who handles vacation packages: preferences for vacation time, preferred locations ("one of the Hawaiian islands"), preferences for airlines, cars, and hotels, budget limits, etc. Some of this information may be unnecessary because the agent may be aware of the user's preferences (e.g., his participation in an airline's promotional program).

The travel agent interacts with service providers, which have posted information on the Web and can handle online reservations. Consumer payments are guaranteed by credit card companies. The user, who is *the only human in this scenario*, is, of course, interested in the best package at the lowest price. The travel agent is interested in a desirable package that will meet the user's approval and also maximize the commission. The service provider is interested in selling as many products as possible, while also minimizing cost (e.g., by routing the user via a route with excess capacity). The credit card company guarantees and makes payments for purchases.

In effect, we have a negotiation among **Web service agents** with vested interests. Before a requesting agent and a provider agent interact, there must be an agreement among the entities that own them. The travel agent uses **ontologies** (i.e., formal descriptions of a set of concepts and their relationships) to deal with the different services. Additional required technologies include: (1) trust maintenance, (2) reliability, (3) trust mechanisms, and (4) **orchestration** of services. A **choreography** is the "pattern of possible interactions between a set of services" and orchestration is a technique for realizing choreographies. [WS Architecture 2003].

This transaction can be carried out in multiple steps, with the agent getting user approval step by step. An impasse may be reached in some cases, requiring the undoing of some previous steps, as in the case where the flight is reserved but no hotel rooms are available at the destination. In case the airline reservation was made, this should be undone by canceling the reservation. Otherwise, it is possible that by the time the hotel reservation is made, the airline seat is no longer available and this step must then be repeated.

57.3.2   Components of Web Services

WS is based on XML, which is basically a markup language like **HTML (Hypertext Markup Language)**. HTML is applicable to text, while XML is also applicable to data. WS use URLs, but instead of an HTML file, an XML file is downloaded.

WSDL (Web Services Description Language) via appropriate abstractions provides an interface to WS. The two parties of a WS interaction share a common WSDL file, to generate messages at the sender end and to interpret them at the receiver.

SOAP (Simple Object Access Protocol) is a *de facto* standard for XML messaging. It provides lightweight communication capability for Web Services, but also **RPCs (remote procedure calls)**, publish and subscribe, and other communication styles. It can be utilized in conjunction with DBMSs, middleware systems, application servers, and deals with heterogeneity (e.g., J2EE and .NET) [SOAP V.1.2 P.0: Primer 2003].

UDDI (Universal Description, Discovery and Integration) is like a Yellow Pages directory and provides a repository for Web services descriptions. It is similar to Internet's DNS (domain name service) that translates host names into TCP addresses. Web services available at an Internet address are made available publicly in WSDL files.

These standards are used together in the following manner. The document submitted by a user to a Web Service is according to WSDL format. The sender's SOAP ensures that the data to be sent is appropriately converted to XML data types before being sent. The receiver's SOAP converts it to the format of the receiving computer. The receiver parses the XML message and validates it for consistency.

Distributed computing architectures, such as CORBA and DCOM, provide the same functionality as Web services. The difference is that there is a tight relationship between clients and servers, while the Web allows previously unknown connections to be made.

It is difficult to implement two-phase commit (2PC) for distributed transactions on top of HTTP, but the following connection-oriented messaging protocols for transaction coordination have been proposed: **Reliable HTTP (HTTPR)** by IBM and **Blocks Extensible Exchange Protocol (BEEP)** by IETF (Internet Engineering Task Force). The **Transaction Internet Protocol (TIP)** by IETF is then used for 2PC.

57.3.3   Web Services Transactions (WS-Transactions)

There are two categories of WS-transactions: atomic and business. WS-transactions rely on *Web services coordination* [WS Coordination 2002], whose functions are to:

1. Create a *coordination context* (*CC*) for a new atomic transaction at its coordinator.
2. Add interposed coordinators to existing transactions (if necessary).
3. Propagate CC in messages between WS.
4. Register for participation in coordination protocols.

An application sends a *Create Coordination Context* (*CCC*) message to its coordinator's *Application Service* (*AS*) and to register for coordination protocols to the *Registration Service* (*RS*). WS coordinators allow different coordination protocols, as discussed below.

We illustrate our discussion with the coordination of two applications App1 and App2 with their own coordinators CRa and CRb, with application services ASa and ASb, and registration services RSa and RSb The two CRs have a common protocol Y and protocol services Ya and Yb. The coordination proceeds in five steps.

1. App1 sends a CCC message for coordination type Q to ASa and gets back a *context* Ca, which consists of an *activity identifier* A1, the coordination type Q, and PortReference to RSa.
2. App1 then sends an application message to App2, including the context Ca.
3. App2 sends a CCC message to CRb's RSb with Ca as context. It gets back its context Cb with the same activity identifier and coordination type as Ca, but with its own registration service RSb.
4. App2 determines the protocol supported by coordination type Q and registers protocol Y at CRb.
5. CRb passes the registration to RSa (of CRa) registration service. It is agreed that protocol Y will be used.

The following commit protocols have been defined:

- **Completion protocol.** Completion is registered as a prelude to commit or abort.
- **PhaseZero.** This also precedes 2PC and is a notification to a transaction to FORCE outstanding cached data updates to disk.
- **Two-phase commit (2PC).** 2PC is briefly described below.
- **Outcome notification.** A transaction participant who wants to be notified about a commit-abort decision.

The WS coordinator messages first establish: (1) who are the participants, (2) the coordination protocol to be followed, and (3) the ports to be used. Web services WS-transactions use this infrastructure to carry out the commit protocol with relatively predictable message flows.

The 2PC protocol can be specified by a state transition diagram where transactions have six states: S0-active, S1-aborting, S2-preparing, S3-prepared, S4-committing, and S5-ended. We only specify a few important transitions.

The **transaction coordinator** initiates the protocol by issuing the **Prepare** message, which is a request for participants to vote. Three responses are possible:

1. **ReadOnly.** The node has not been engaged in writing any data. This is a vote to commit. The participant does not have to participate further.
2. **Aborted.** This is a vote not to commit. No further participation by the node is required.
3. **Prepared.** This is a vote to commit, and a Prepared status also indicates that the participant has logged information so that it can deal with subsequent commits or aborts.

A transaction coordinator, which receives Prepared messages from all participants, can decide to commit or abort the transaction. An appropriate message is sent after the coordinator has logged his decision. Logging is tantamount to permanently storing data onto non-volatile storage.

The **presumed commit protocol** has the following implications:

1. The coordinator need not log anything until the commit decision is made.
2. A participant can forget about a transaction after sending an Aborted or ReadOnly message.
3. When the outcome of a transaction is commit, then the coordinator has to remember it until all Committed Acks are received.

A one-phase-commit (1PC) protocol is also possible, in which the coordinator can issue a commit or abort messages, all by himself.

Detailed message flows for WS coordinations and transactions for App1 running at the Web server, App2 at the middleware server, and accessing the DBMS at the database server are given in WS Transactions [2002].

Web services security (WS-security) protocols can be used in conjunction with SOAP to ensure message integrity, confidentiality, and single message authentication. A brief review of techniques used for this purpose appear in the previous section of this chapter.

ebXML (electronic business XML) is a parallel effort to Web services and is geared toward enterprise users [Newcomer 2002].

57.3.3.1 Further Information

This is a rapidly evolving field, so that most of the more interesting and up-to-date information can be found on the Web. There are several detailed books discussing the technologies mentioned here. A book that ties everything together is Newcomer [2002]. A forthcoming book is *Web Services: Concepts, Techniques, and Examples* by S. Khoshafian, Morgan-Kaufmann Publishers, 2004. The reader is referred to http://www.w3.org for online material for most of the topics covered here, and to Brown and Haas [2002] for a glossary of terms.

57.4 Concurrency Control for High-Contention Environments

In this section we review concurrency control methods that have appeared in the research literature to deal with high data contention. More specifically, we are envisaging a system where data, rather than hardware resource contention, is the limiting factor for system throughput.

The standard locking method is strict 2PL (with the transaction blocking option upon a lock conflict), which limits the degree of transaction concurrency and is susceptible to thrashing, as discussed in the next section. The methods described here allow transaction aborts and restarts, which result in a reduction in the lock contention level. The "wasted" processing is expected to be tolerable with the advent of faster processors.

We first consider two classes of concurrency control methods: (1) **wait-depth limited (WDL)** methods, which limit the wait-depth of blocked transactions, while strict 2PL is the overall policy; and (2) **two-phase processing** methods, which are based on **access invariance**, that is, *a restarted transaction tends to access the same set of objects it accessed before* [Franaszek et al. 1992]. We also provide a quick overview of several methods for reducing data contention.

57.4.1 Wait-Depth Limited Methods

Before discussing the WDL method [Franaszek et al. 1992], we briefly review some other methods that also limit the wait depth.

An extreme WDL method is the **no waiting (NW)** or **immediate restart** policy, which disallows any waiting; that is, a transaction encountering a lock conflict is aborted and restarted immediately [Tay 1987]. Because an immediate restart will result in another lock conflict and abort, resulting in repeated wasted processing, **restart waiting** can be introduced to defer the restart of the aborted transaction until the transaction causing the conflict departs.

The **running priority (RP)** policy increases the degree of transaction concurrency and provides an approximation to **essential blocking**: *a transaction can be blocked only by an active transaction, which is also doing useful work*; that is, it will not be aborted in the future [Franaszek and Robinson 1985]. The approximation is due to the fact that it is not known in advance whether a transaction will commit successfully. Consider a transaction T_C that requests a lock held by T_B, which is blocked by an active transaction T_A. RP aborts T_B so that T_C can acquire its requested lock. There is a *symmetric* version of RP as well. T_C is blocked by T_B, which is initially active, but T_B is aborted when it becomes blocked by an active transaction T_A at a later time.

The **cautious waiting** aborts T_C when it becomes blocked by T_B [Hsu and Zhang 1992]. Although this policy limits the wait depth, it has the same deficiencies as the no-waiting policy.

A family of WDL methods are described in Franaszek et al. [1992]. We only consider WDL(1), which is similar to (symmetric) RP but takes into account the progress made by transactions involved in the conflict, including the active transaction, in selecting which transaction to abort. A transaction that has acquired a large number of locks and consumed a significant amount of system resources is not aborted in favor of a transaction that has made little progress, even if that transaction is active.

A simulation study of WDL methods shows that WDL(1) outperforms RP, which outperforms other methods [Thomasian 1997]. Simulation results of the WDL(1), which limits the wait depth to one, against two-phase and optimistic methods, shows that it outperforms others, unless "infinite" hardware resources are available [Franaszek et al. 1992].

Variations of the WDL are described for distributed databases in Franaszek et al. [1993]. Simulation studies show that this method outperforms strict 2PL and the wound-wait method. 2PC is the commit protocol in all cases.

57.4.2 Two-Phase Processing Methods

The first phase of transaction execution, without applying any concurrency control method, serves the role of prefetching the data required for transaction execution from disk. This phase is called **simulation** or **virtual execution** [Franaszek et al. 1992]. In the second phase, the transaction is executed according to a concurrency control method. The restarted transaction's execution time is very short in the second phase, because the buffer is primed and it can access the blocks required for its execution from the database buffer without incurring disk I/O.

What we described is a generalization of *pipelined processing* [Gray and Reuter 1993], where transactions are executed serially in the second phase so that there is no need for a concurrency control method, but this method has a limited throughput. A generalization of this method is to execute transactions in non-interfering classes on different processors of a multiprocessor system.

Rather than wasting the execution on the first phase, some concurrency control method such as RP can be adopted. If the transaction execution is successful, it commits, and the second execution phase is not required. A transaction can be aborted according to the RP paradigm, but instead of being restarted, it continues executing in virtual execution mode. A locking method is used in the second phase, but lock requests by second phase transactions preempt locks held by first phase transactions, forcing them into virtual execution. The standard RP method (with restart waiting) is adopted when a second phase transaction is aborted (by another second phase transaction) according to the RP paradigm.

Multiphase processing methods work better with **optimistic** concurrency control methods, which execute without requesting any locks on objects they access. A transaction instead posts *access entries* to identify objects that it has accessed into an appropriate hash class. These objects are also copied into the transaction's *private workspace* and modified locally if an update is required. Upon completing its first phase, which is called the **read phase**, a transaction enters its second or **validation phase**, during which it checks whether any of the objects accessed by the transaction have been modified since they were "read." If this is true, then the transaction is aborted, otherwise it can commit. Transaction commit includes the third or **write phase**, which involves externalizing the modified objects from the private workspace into the database buffer (after appropriate logging for recovery). Note that the three optimistic steps constitute a single phase in transaction processing.

A committing transaction can invalidate others, those that had accessed objects which it has modified. Validation then just involves checking whether a transaction was conflicted (in the past) or not. In fact, a conflicted transaction can be aborted right away, which is what is done according to the **optimistic kill** policy, but two-phase processing favors the **optimistic die policy**, so that a first phase transaction executes to the end, prefetching all data, and dying a natural death.

A transaction running according to the optimistic die policy is susceptible to fail its validation according to a **quadratic effect**, that is, the probability that a transaction is conflicted is proportional to the number of objects accessed (k) and the execution time of the transaction, which is also proportional to k [Franaszek et al. 1992]. In case a transaction with the optimistic die policy is restarted after failing its validation, its second execution phase can be very short, so that the quadratic effect is not a problem. The quadratic effect is a problem when the system is processing variable-size transactions. Larger transactions, which are more prone to conflict than shorter transactions, contribute heavily to wasted processing when they do so [Ryu and Thomasian 1987].

In fact, given that all of the objects required for the execution of a transaction have been prefetched, there is no advantage in running it to completion. The **optimistic kill** rather than the optimistic die policy should be used in the second and further phases, because doing so will reduce the wasted CPU processor and also the transaction response time.

An optimistic kill policy may result in more than two phases of execution. To minimize the number of executions, a locking method can be used in the second phase. On demand or **dynamic locking** is still susceptible to deadlocks, although we know that deadlocks tend to be rare. Because the identity of all objects required for the second phase is known, **lock preclaiming** or **static locking** can be used to ensure that the second execution phase is successful.

The optimistic die/lock preclaiming method can be utilized in a distributed system as long as the validation required for the first phase is carried out in the same order at all nodes [Thomasian 1998b]. Two-phase commit can be carried out by including lock requests as part of the *pre-commit message*. If any of the objects has been modified, its modified value is sent to the coordinator node at which the transaction executes. The transaction is reexecuted at most once because it holds lock on all required objects.

It was shown in Franaszek et al. [1992] that the performance of two-phase methods is quite similar, but with "infinite" hardware resources, they outperform WDL(1).

57.4.3 Reducing Data Contention

A short list of some interesting methods to reduce the level of lock contention is given at this point.

Ordered sharing allows a flexible lock compatibility matrix, as long as operations are executed in the same order as locks are acquired [Agrawal et al. 1994]. Thus, it introduces restrictions on the manner transactions are written. For example, a transaction T_2 can obtain a shared lock on an object locked in exclusive mode by T_1 (i.e., read the value written by T_1) but this will result in deferring T_1's commit to after T_2 is committed. Simulation results show that there is an improvement in performance with respect to standard locking.

Altruistic locking allows transactions to *donate* previously locked objects once they are done with them, but before the objects are actually unlocked at transaction completion time [Salem et al. 1994]. Another transaction may lock a donated object, but to ensure serializability, it should remain in the "wake" of the original transaction (i.e., accesses to objects should be ordered). Cascading aborts, which are a possibility when the donated object is locked in exclusive mode, can be prevented by restricting "donations" to objects held in shared mode only. This makes the approach more suitable for read-only queries or long-running transactions with few updates.

A method for allowing interleaved execution of **random batch transactions** and short update transactions is proposed in Bayer [1986]. The random batch transaction updates database records only once and the updates can be carried out in any order (e.g., giving a 5% raise to all employees). In effect, the batch transaction converts "old" records into "new" records. Because the blocking delay is not tolerable for short transactions, the batch transaction may update the *required* old records and make them available to short transactions, after taking *intermediate commit points*.

The **escrow method** [O'Neil 1986] is a generalization of the **field calls** approach in IMS FastPath [Gray and Reuter 1993]. The minimum, current, and maximum values of an aggregate variable, such as a bank balance, are made available to other transactions.

The **proclamation-based** model for cooperating transactions is described in Jagadish and Shmueli [1992]. In addition to its original motivation of transaction cooperation, it can be used to reduce the level of lock contention. This method is different from altruistic locking in that a transaction, before releasing its lock on an object (it is not going to modify again), proclaims one or a set of possible values for it. Trivially, the two values may be the original and modified value. Transactions interested in the object can proceed with their execution according to the *proclaimed* values.

The lock holding time by long-lived transactions can be reduced using intermediate commit points according to the **sagas** paradigm [Garcia-Molina and Salem 1987]. A long-lived transaction T_1 is viewed as a set of subtransactions $T_1, \ldots, T_n$, that are executed sequentially and can be committed individually at their completion. However, the abort of subtransaction T_j results in the undoing of the updates of all preceding subtransactions from a semantic point of view through compensating subtransactions $C_1, \ldots, C_{j-1}$. Compensating transactions consult the log to determine the parameters to be used in compensation.

A method for **chopping** larger transactions into smaller ones to reduce the level of lock contention and increase concurrency, while preserving correctness, is presented in Shasha et al. [1995].

Semantics-based concurrency control methods rely on the semantics of transactions or the semantics of operations on database objects. The former is utilized in Garcia-Molina [1983], where transactions are classified into types and a compatibility set is associated with different types. Semantics-based concurrency control methods for objects are based on the commutativity of operations. **Recoverability of operations** is an extension of this concept that allows an operation to proceed when it is recoverable with respect to an uncommitted operation [Badrinath and Ramamritham 1992]. Various operations on stacks and tables belong to this category. Two methods based on commutativity of operations are presented in Weihl [1988]; they differ in that one method uses intention lists and the other uses undo logs. This work is extended in Lynch et al. [1994].

Checkpointing at the transaction level can be used to reduce the wasted processing due to transaction aborts. The effect of checkpointing on performance has been investigated in the context of optimistic concurrency control with kill option [Thomasian 1995]. As previously discussed, data conflicts result in transaction abort and resumption of its execution from the beginning. A reduction in checkpointing cost is to be expected due to the private workspace paradigm. There is a trade-off between checkpointing overhead and the saved processing due to partial rollbacks, which allows a transaction to resume execution from the checkpoint preceding the data item causing the data conflict.

57.4.3.1   Further Information

A more detailed discussion of these topics appears in Ramamritham and Chrisanthis [1996], Thomasian [1996a], and Thomasian [1998a] and, of course, the original papers.

57.5 Performance Analysis of Transaction Processing Systems

The performance of a **transaction processing** system is affected by **hardware resource contention** as well as **data contention**. We provide a brief introduction to **queueing theory** and especially **queueing network models (QNMs)**, which provide solutions to the first problem. The description that follows is brief, yet self-complete.

The analysis of lock contention in databases is more academic in nature, but provides insight into the effect that lock contention has on system performance.

57.5.1 Hardware Resource Contention

QNMs have been used successfully in analyzing the performance of transaction processing systems. Crude QNMs can be used to roughly estimate performance, while more specialized QNM-based tools for **capacity planning** (such as BEST/1 and MAP [Lazowska et al. 1984]) can predict computer performance more accurately. Specialized tools are available to automate the extraction of QNM's input parameters.

The QNM of a computer system consists of nodes that correspond to **active resources** of the system, such as the processor and disks. The processors of a multiprocessor constitute a single node with multiple servers, which is referred to as the CPU (central processing unit). Each node in the QNM consists of a queue, which holds pending requests, and server(s). We will assume that the queueing discipline at all nodes is first-come, first serve (FCFS).

Transactions correspond to **jobs** being processed by a QNM. We consider transactions that belong to a **job class**, but multiple job classes can be defined based on their processing requirements at the resources of the computer system. Multiple transaction (or job) classes would be required if the system under consideration processed short online and long batch transactions, which definitely belong to different categories.

Transactions execute concurrently in a multiprogrammed environment to attain higher transaction throughput and resource utilization, that is, to reduce the cost per transaction per time unit. The main memory is a **passive resource** because it limits the degree of transaction concurrency or the **multiprogramming level (MPL)**, which is denoted by M. Transactions that cannot be activated due to the maximum MPL constraint (M_{max}) are held in the **memory queue**. More generally, transactions are delayed because they can only be processed by a limited number of threads.

The sharing of (active) hardware resources by concurrent transactions has an adverse effect on transaction response time. The analysis of the underlying QNM can then be used to determine the queueing delays at the active resources, as well as the maximum MPL constraint.

Transactions have multiple processing steps, where each step leads to an access to a database object. This is preceded by a lock request and acquisition. The transaction proceeds to its next step if its lock request is successful, but otherwise the transaction is blocked until the requested lock is released (this effect is discussed in the next section). The last step leads to transaction commit and the release of all locks. At this point, we assume that all lock requests are successful because we are only concerned with hardware resource contention. Lock contention is discussed in Section 57.5.2.

Objects accessed by online transactions tend to be small (e.g., a database record) so they are contained in database pages or blocks (say, 8 KBytes). Most DBMSs maintain a database buffer in main memory that caches recently accessed database pages. If the required page is found in the database buffer, then the transaction can proceed to the next step in its execution; otherwise, there is a page fault. The transaction encountering a page fault remains blocked until the I/O request is completed, at which point it resumes its execution or is enqueued in the processor queue if all processors are busy. The **miss ratio** of the buffer, which is the fraction of references to missing pages, has a major effect on transaction performance. This is because of the high disk access time overhead.

A QNM with external arrivals is called an **open QNM**, while a QNM with a fixed degree of transaction concurrency, where a completed transaction is immediately replaced by a new transaction, is called a **closed QNM**. We next discuss open and closed QNMs, followed by a solution method for "closed" QNMs subject to external arrivals.

57.5.1.1 Analysis of an Open QNM

A transaction arriving at the computer system (the CPU) behaves as a token, which transitions from one node to another, acquiring service, until it is completed and leaves the system. We consider transaction execution at a special QNM known as the **central server model (CSM)**, which was introduced for modeling multiprogrammed computer systems. The central server is the CPU (node 1) and the peripheral servers are the disks (nodes $2 - N$). Transactions alternate between CPU and disk processing. CPU processing ends when a transaction makes an I/O request to disk as a result of a page fault. A transaction resumes CPU processing after the disk is accessed and the page fault is resolved. Transactions leave the system after completing their last step at the CPU, which is just a matter of convention.

After completing a CPU burst, a transaction leaves the system with probability p_1 or with probability $p_n, 2 \leq n \leq N$, and accesses the nth disk, so that $\sum_{n=1}^{N} p_n = 1$. The probability of selecting the nth disk is $q_n = p_n/(1 - p_1), 2 \leq n \leq N$. The probability that a transaction makes k visits to the CPU is $P_{CPU}(k) = p_1(1 - p_1)^{k-1}, k \geq 1$, and the mean number of visits to the CPU is then $v_{CPU} = v_1 = 1/p_1$. The total number of visits to disks is $v_{disk} = 1/p_1 - 1$, so that the mean number of visits to the nth disk is $v_n = q_n v_{disk} = p_n/p_1, 2 \leq n \leq N$.

The **service demand** of a transaction at device n is the total service time it requires at this device: $D_n = v_n \bar{x}_n$, where $\bar{x}_n$ is the mean service time per visit. The hardware resource requirements of a transaction can be summarized as $(D_1, D_2, \ldots, D_N)$. In fact, detailed transition probabilities and service times are not required at all, unless one is interested in simulating the system. Data reduction applied to software measurement data can be used in estimating service demands required by QNM solvers.

An open QNM with Poisson arrivals is **product-form**; that is, an efficient solution for it is available if the service times at nodes with an FCFS queueing discipline are exponentially distributed. Poisson arrivals are random in time and transactions are generated from an "infinite" number of sources, so that the arrival rate is not affected by the number of transactions at the computer system.

The **utilization factors** of the nth node with m_n servers is given as $\rho_n = \Lambda D_n/m_n$, where $1 \leq n \leq N$. The utilization factor is the fraction of time that the servers at a node are busy and also the mean number of requests at one of the servers (provided that they have equal utilizations). Unequal server utilizations would be possible if a scheduler assigns servers in an ordered fashion, always starting with the lowest indexed server. It is intuitively clear that the condition for the stability of the system is $\rho_n < 1$ for all n. In this case, the system is *flow-balanced*, so the **system throughput** or the transaction completion rate equals the arrival rate Λ.

Little's result states that *the mean number of requests in a queueing system equals the product of the arrival rate of requests and the mean time they spend in the system* [Trivedi 2002]. The queueing system may be the combination of the queue and the servers, the queue alone, the service facility, or one of the servers, as discussed above.

The **bottleneck** node is defined as the node with the highest service demand; but in the presense of multiple servers, it is the node with the highest D_n/m_n. The maximum throughput is given as $\Lambda_{max} = m_n/D_n$ and the system will be *saturated* if $\Lambda > \Lambda_{max}$. Faster disks and processors can be used to reduce the service demand so that a higher Λ_{max} can be attained. The load of a bottleneck disk can be reduced by reallocating its files. At best, the service demands at all disks can be made equal.

Disk arrays utilize *striping* partition files into stripe units and allocate them a in round-robin manner on disks. In this way, "hot" files with high access rates are intermixed with "cool" files. See Chapter 21 on "Secondary Storage Systems" in this *Handbook* for more detail.

More efficient software will also result in a reduction of the service demands at the processor. Database tuning, such as implementing appropriate B-tree indices, where the higher levels of the index are held in main memory, can be used in reducing the number of disk I/Os. In effect, we have reduced the miss ratio of the database cache.

The mean residence time of a transaction at a disk is given as $R_n = D_n/(1 - \rho_n), 2 \leq n \leq N$; for example, D_n is expanded by a factor of two if $\rho_n = 0.5$. This formula applies to the CPU if it is a uniprocessor; but in the case of a multiprocessor with $m_1 = 2$, $R_1 = D_1/(1 - \rho^2)$, that is, $R_1 \approx 1.33$ for $\rho_1 = 0.5$.

The mean transaction response time (R) is the sum of its residence times at the nodes of the computer system: $R = \sum_{n=1}^{N} R_n$.

The mean number of transactions at the computer system $(\bar{N})$ according to Little's result is the product of the arrival rate of transactions (Λ) and the mean time transactions spend at the computer system (R); that is, $\bar{N} = \Lambda R$. This result holds for general interarrival and service time distributions and the individual nodes of a QNM: $\bar{N}_{server} = \lambda r$; for the queues: $\bar{N}_{queue} = \lambda w$; and for the servers: $\bar{N}_{servers} = \lambda \bar{x} = m\rho$, where λ is the arrival rate of requests to a node, $\bar{x}$, the mean service time (per visit), m the number of servers, w the mean waiting or queueing time, and $r = w + \bar{x}$, the mean residence time per visit.

Each node in an open QNM can be analyzed as an *M/M/1* queue, where the first *M* implies Poisson arrivals, the second *M* exponential service times, and there is one server (in fact, the arrivals to nodes with feedback are not Poisson). The mean waiting time (w) can be expressed as $w = \bar{N}_q \bar{x} + \rho \overline{x'}$, which is the sum of the mean delay due to requests in the queue and the request being served (if any). The probability that the server is busy is $\rho = \lambda \bar{x}$ and $\overline{x'}$ is the mean *residual service time* of the request being served at arrival time. This equality holds for Poisson arrivals because *Poisson arrivals see time averages (PASTA)*, due to the *memoryless property* of the exponential distribution $\overline{x'} = \bar{x}$. Noting that $\bar{N}_q = \lambda w$, we have $w = \rho \bar{x}/(1 - \rho)$. The mean response at a node per visit is $r = w + \bar{x} = \bar{x}/(1 - \rho)$ and the mean residence time at the node is then $R = vr$.

This analysis also applies to an M/G/1 queue with a general service, in which case $\overline{x'} = \overline{x^2}/(2\bar{x})$, where the numerator is the second moment of service time. This leads to $W = \lambda \overline{x^2}/(2(1 - \rho))$, which is the well-known Pollaczek-Khinchine formula for M/G/1 queues. Departures from an M/G/1 queue with FCFS are not Poisson, so that this queue cannot be included in a product-form QNM.

57.5.1.2 Analysis of a Closed QNM

A system with a maximum MPL M_{max} can be treated as an open QNM if the probability of exceeding this limit is very small. For example, the distribution of the number of jobs in M/M/1 queues is given by the geometric distribution [Trivedi 2002] $P(m) = (1 - \rho)\rho^m, m \geq 1$, so that the probability that the buffer capacity is exceeded is $P_{overflow} = \sum_{m > M_{max}} p(m)$.

The joint-distribution of the number of jobs in a product-form open QNM is

$$P(m_1, m_2, \ldots, m_K) = P(m_1)P(m_2) \cdots P(m_K)$$

where each term is $P(m_k) = (1 - \rho_k)\rho_k^{m_k}$ if the node is a single server (more complicated expressions for multiserver nodes are given in Trivedi [2002]). The distribution of probability when the total number of jobs is $N = \sum_{k=1}^{K} n_k$ can be easily computed. The closed QNM can be considered open if the probability of exceeding M_{max} is quite small. We also need to ascertain the throughput of the system running at M_{max}: $T(M_{max}) > \Lambda$.

In a closed QNM, a completed transaction is immediately replaced by a new transaction, so that the number of transactions remains fixed at M. A closed QNM can be succinctly defined by its MPL M and transaction service demands: $(D_n, 1 \leq n \leq N)$. The **throughput characteristic** $T(M), M \geq 1$ is a nondecreasing (and convex) function of M. As $M \to \infty$, $T_{max} = m_n/D_n$, where n is the index of the bottleneck resource.

The **convolution algorithm** or **mean value analysis (MVA)** can be used to determine the system throughput $T(M)$ or the mean transaction residence time $R(M)$, which are related by Little's result: $R(M) = M/T(M)$. Analysis of QNMs using MVA is specified in Lazowska et al. [1984], but at this point we provide the analysis of a **balanced QNM**, which consists of single server nodes with all service demands equal to D.

Due to symmetry there are M/N transactions at each node, on average. According to the *arrival theorem*, which is the closed system counterpart of PASTA, an arriving transaction encounters $(M - 1)/N$ transactions at each node, as if there is one transaction less in the system (the arriving transaction itself). This observation forms the basis of an iterative solution, but no iteration is required in this special case because the number of requests at each node is known. The mean residence time of transactions at node n is the sum of its service time and the queueing delay: $R_n(M) = D[1 + (M - 1)/N]$.

It follows that the mean residence time of a transaction in the system is $R(M) = NR_n(M) = (M + N - 1)D$ and that $T(M) = M/R(M) = M/[(M + N - 1)D]$, which means that as $M \to \infty$, $T_{max} = 1/D$.

Balanced job bounds, which utilize the solution to a balanced QNM, can be used to obtain upper and lower bounds to the throughput characteristic $(T(M), M \geq 1)$ of a computer system. Both bounds are obtained by assuming that the QNM is balanced, with the upper bound using $D_u = D_{avg} = \sum_{n=1}^{N} D_n/N$ and the lower bound $D_L = D = D_{max}$, where *max* is the index of the node with the largest service demand.

Asymptotic bounds are more robust, in that they are applicable to multiple server nodes. The maximum throughput is equal to m_{min}/D_{min}, where *min* is the node with the smallest such ratio. Another asymptote passes through the origin and the point $(M = 1, T(1) = 1/\sum_{n=1}^{N} D_n)$ of the throughput characteristic.

A rule of thumb (ROT) to determine M_{max} in a system with a single CPU and many disks (so that there is no queueing delay at the disks) is $M_{max} = D_{disk}/D_{CPU} + 1$. This ROT is based on the observation that with perfect synchronization of CPU and disk processing times, which could only be possible if they have a fixed value, M_{max} would utilize the processor 100% (with no queueing delays). If the CPU has $m_{CPU} = m_1$ processors, then $M_{max} = m_{CPU} D_{disk}/D_{CPU} + 1$.

57.5.1.3 Hierarchical Solution Method

We next consider the analysis of a transaction processing system with external arrivals, but with a constraint on the maximum MPL (M_{max}). As far as external arrivals are concerned, rather than assuming an infinite number of sources, we consider the more realistic case of a finite number of sources $I(>M_{max})$ so that queueing is possible at the memory queue. Each source has an exponentially distributed think time with mean $Z = 1/\lambda$, which is the time it takes the source to generate its next request.

A two-step **hierarchical solution method** is required, which substitutes the computer system, regardless of its complexity, with a **flow-equivalent service center (FESC)** specified by its throughput characteristic: $T(M), 1 \leq M \leq M_{max}$ and $T(M) = T(M_{max}), M \geq M_{max}$. This approximation has been shown to be very accurate by validation against simulation results.

The hierarchical solution method models the system by a one-dimensional **Markov chain**, where the state S_M designates that there are M transactions at the computer system. There are $I + 1$ states, since $0 \leq M \leq I$. The arrival rate of transactions at S_M is $(I - M)\lambda$, that is, the arrival rate decreases linearly with M. When $M \leq M_{max}$, all transactions will be activated and processed at a rate $T(M)$. Otherwise, when $M > M_{max}$, the number of transactions enqueued at the memory queue is $M - M_{max}$. The analysis, however, postulates that the FESC processes all transactions, but the maximum rate does not exceed $T(M_{max})$ even when $M > M_{max}$.

The MC is, in fact, a **birth-death process** because the transitions are only among neighboring states. The forward transitions $S_{M-1} \to S_M$ have a rate $(I - M + 1)\lambda$ and the backward transitions $S_M \to S_{M-1}$ are $T(M), 1 \leq M \leq I$. In equilibrium, *the rate of the forward transitions multiplied by the fraction of time the system spends in that state equals the same product for backward transitions*. Note that the fraction of time spent in a state can be expressed simply as the state probability $\pi(M)$ for S_M.

The **state equilibrium** or **steady-state** equations are given as follows:

$$(I - M + 1)\lambda\pi(M - 1) = T(M)\pi(M), \qquad 1 \leq M \leq M_{Max}$$

where all probabilities can be expressed as a function of $\pi(0)$, which in turn can be determined using the condition that probabilities sum to one.

Given $\pi(M)$, we can obtain the mean number of transactions at the computer system as $\bar{M} = \sum_{M=1}^{I} M\pi(M)$ and the system throughput as $\bar{T} = \sum_{M=1}^{I} T(M)\pi(M)$. The mean transaction response time, which includes the delay in the memory queue, is $R = \bar{M}/\bar{T}$; and the mean memory queue length is $\bar{M}_q = \sum_{M_{max}}^{I} (M - M_{max})\pi(M)$.

A shortcut solution is possible via **equilibrium point analysis** [Tasaka 1986], which obtains the intersection point of the throughput characteristic: $T(M), M \geq 1$ vs. M and the arrival rate: $\Lambda(M) = (I - M)\lambda$ vs. M. The intersection point at coordinates $(\bar{M}, T)$ is the equilbrium point because the arrival rate of requests equals the completion rate. The mean transaction response time is then $R = \bar{M}/T$. Alternatively, it follows from Little's result that $I = (R + Z)T$, so that $R = I/T - Z$.

The Transaction Processing Council's benchmarks (www.tpc.org) compare systems based on the maximum throughput (in processing one of its carefully specified benchmarks) as long as a certain percentile of transaction response time does not exceed a threshold of a few seconds.

57.5.2 Performance Degradation Due to Locking

Performance degradation due to locking is expected to be low because if it is high, steps are taken to eliminate the sources of lock contention. For example, the lock contention in a DBMS with page level locking can be reduced by introducing record level locking, that is, using a finer granularity of locking. This is at the cost of making recovery more complicated, as in the case of the ARIES recovery method [Ramakrishnan and Gehrke 2003].

There is little information about the level of lock contention in high-performance systems, although most DBMSs record the occurrence of lock conflicts. The very few studies of lock contention by analyzing lock traces (e.g., Singhal and Smith [1997]) have been carried out without a global view of the system: (1) transaction classes, their data access pattern and their frequencies; (2) the organization of the database, information about its tables and indexes, etc.; and (3) transaction scheduling, which serializes the execution of transactions in the same or conflicting classes to reduce the level of lock contention.

The solution methods provided in this section potentially can be used to predict the effect of increased arrival rates on system performance, but it should be mentioned that unlike validation studies of analytical models for hardware resource contention [Lazowska et al. 1984], we are not aware of validation studies for lock contention.

In addition to obtaining expressions for the probability of lock conflict and deadlock, we obtain the effect of locking on transaction response time. We also provide an analysis to explain the **thrashing phenomenon**, and conclude with the issue of load control and a more realistic model for lock contention.

57.5.2.1 Probability of Lock Conflict and Deadlock

A (validated) straw-man analysis of lock contention and deadlocks served as a starting point for analytic studies in this area [Gray and Reuter 1993].

We consider a closed system with M transactions that are processed according to the strict 2PL paradigm. Requested locks are uniformly distributed over the D database objects (e.g., database pages). This is, in fact, the effective database size, which can be estimated indirectly if we know (1) the probability of lock conflict, (2) that requested locks are uniformly distributed over active objects, and (3) the number of locks requested per transaction.

The effect of shared and exclusive locks and nonuniform lock distribution can be taken into account by utilizing an **effective database size**, which is larger (resp. smaller) than D from the first (resp. second) viewpoint [Tay 1987]. For example, if a fraction f_X of lock requests are exclusive, then $D_{eff} = D/[1 - (1 - f_X)^2]$, so that $D_{eff} \approx 2D$ when $f_X = 0.7$, and $D_{eff} = \infty$ when all requests are reads (i.e., no lock conflicts).

We consider fixed-size transactions of size k and variable-size transactions, where the fraction of transactions of size k is f_k, $1 \leq k \leq K_{max}$ with $\sum_{k=1}^{K_{max}} f_k = 1$. K_i denotes the ith moment of transaction size; that is, $K_i = \sum_{k=1}^{K_{max}} k^i f_k$.

A size k transaction consists of $k + 1$ steps with identically distributed execution times denoted by s, which has been characterized by a uniform and exponential distributions, but this has a secondary effect on the lock contention level. The mean step durations $s(\bar{M}_a)$ can be determined from the QNM (queueing network model) of the transaction processing, taking into account the mean number of active transactions. Different steps may have different service demands and durations, in which case we will need the mean number of active transactions in different steps, but these complications remain beyond the scope of this discussion.

The value of $\bar{M}_a$ is initially unknown, but the analysis yields the fraction of blocked transactions (β), which can be used to obtain the mean number of active and blocked transactions as $\bar{M}_a = (1 - \beta)M$ and

$\bar{M}_b = M - \bar{M}_a = \beta M$. The mean residence time of transactions in the system, while they are not blocked due to lock conflicts, is $r(\bar{M}_a) = (K_1 + 1)s(\bar{M}_a)$.

The first k steps of a transaction lead to a lock request and the final step leads to transaction commit and the release of all locks according to the strict 2PL paradigm. Transactions are blocked upon a lock conflict awaiting the release of the requested lock. The effect of deadlocks is ignored in this analysis because deadlocks are relatively rare, even in systems with a high level of lock contention.

The probability of lock conflict for this model is the ratio of the number of locks held by transactions to the total number of locks: $P_c \approx (M - 1)\bar{L}/D$; we use $M - 1$ because the lock requested by the target transaction, may conflict with locks held by the *other* $M - 1$ transactions, and $\bar{L}$ is the mean number of locks held by a transaction, which is the ratio of the time-space of locks (a step function, so that each acquired lock adds one unit and there is a drop to zero when all locks are released) and the execution time of the transaction. In the case of fixed-size transactions, $\bar{L} \approx k/2$ and in the case of variable size transactions, $\bar{L} \approx K_2/(2K_1)$ [Thomasian and Ryu 1991].

The probability that a transaction encounters a two-way deadlock can be approximated by $P_{D2} \approx (M - 1)k^4/(12D^2)$, which is very small because D tends to be very large [Thomasian and Ryu 1991].

57.5.2.2 Effect of Lock Contention on Response Time

The mean response time of fixed-size (of size k) and variable-size transactions is given as:

$$R_k(M) = (k + 1)s(\bar{M}_a) + kP_c W$$

$$R(M) = \sum_{k=l}^{K_{max}} R_k(M)f_k = (K_1 + 1)s(\bar{M}_a) + K_1 P_c W = r(\bar{M}_a) + K_1 P_c W$$

Note that $R(M)$ is a weighted sum of $R_k(M)$'s, based on transaction frequencies.

The fraction of blocked transactions can be expressed as the fraction of time transactions spend in the blocked state: $\beta = \bar{M}_b/M = K_1 P_c W/R(M)$ (the second ratio follows from the first by dividing both sides by $T(M)$). We have $R(M) = r(\bar{M}_a)/(1 - \beta)$, which indicates that the mean transaction residence time is expanded by the one's complement of the fraction of blocked transactions.

In a system with lower lock contention levels, most lock conflicts are with active transactions, in which case $W \approx W_1$. W_1 normalized by the mean transaction response time is $A = W_1/R \approx 1/3$; for fixed-size transactions and for variable-size transactions, we have $A \approx (K_3 - K_1)/(3K_1(K_2 + k_1))$ [Thomasian and Ryu 1991].

As far as transaction blocking is concerned, we have a forest of transactions with an active transaction at level zero (the root), transactions blocked by active transactions at level one, etc. The probability that a transaction is blocked by a level $i > 1$ transaction is $P_b(i) = \beta^i$ and, hence, $P_b(1) = 1 - \beta - \beta^2 - \cdots$. Approximating the mean waiting time of transactions blocked at level $i > 1$ by $W_i = (i - 0.5)W_1$, the mean overall transaction blocking time is

$$W = \sum_{I \geq 1} P_b(i)W_i = W_1 \left[1 - \sum_{I \geq 1} \beta^i + \sum_{i > 1} (i - 1/2)\beta^{i-1} \right]$$

Multiplying both sides by $K_1 P_c/R(M)$ and defining $\alpha = K_1 P_c A$ (with $A = W_1/R(M)$), we obtain:

$$\beta = \alpha(1 + 0.5\beta + 1.5\beta^2 + 2.5\beta^3 + \cdots)$$

Because $\alpha < 1$, we can obtain the following cubic equation:

$$\beta^3 - (1.5\alpha + 2)\beta^2 + (1, 5\alpha + 1)\beta - \alpha = 0$$

The cubic equation has three roots for $\alpha < 0.226$. Two of the roots are less than one and one root greater than one [Thomasian 1993]. Only the smallest root is meaningful. For $\alpha = 0.226$, we have $\beta = 0378$; but for $\alpha > 0.226$, there is only one root, which is greater than one. In other words, the model predicts that the system is thrashing.

The simulation program used to ascertain the accuracy of the approximate analysis of strict 2PL shows the analysis to be quite accurate. The same simulation program was used to show that very long runs are required in some cases to induce thrashing, but the duration of these runs decreases as the variability of transaction sizes increases.

As we increase M, the mean number of active transactions $\bar{M}_a$ is maximized at $\beta \approx 0.3$, or when 30% of transactions are blocked. Given that the throughput characteristic $T(M)$ increases with M, this also means that $T(0.7\bar{M})$ is the maximum system throughput.

57.5.2.3 More General Models

An analytic solution of variable-size transactions with variable step durations showed that the maximum throughput is attained at $\bar{M}_a \approx 0.7M$, as before. The analysis in this case involves a new parameter $\rho = \bar{L}_b/\bar{L}$, which is the ratio of the mean number of locks held by transactions in the blocked state to $\bar{L}$. In fact, ρ is related to the **conflict ratio** (c_r), which is the ratio of the total number of locks held by transactions and the total number of locks held by active transactions [Weikum et al. 1994]. It is easy to see that $\rho = 1 - 1/c_r$. Analysis of lock traces shows that the critical value for c_r is 1.3, which is in agreement with our analysis that $0.2 \leq \rho \leq 0.3$ or $1.25 \leq c_r \leq 1.33$.

A possible use of the aforementioned parameters (α and β, ρ and the conflict ratio c_r) is that they can be used as load control parameters to avoid thrashing. In fact, the fraction of blocked transactions β is the easiest to measure.

Since the above work was published, more realistic models of lock contention have been introduced. In these, the database is specified as multiple tables with different sizes and transactions are specified by the frequency of their accesses to different tables [Thomasian 1996b]. While it is not difficult to analyze these more complicated models, it is difficult to estimate the parameters.

Index locking can be a source of lock contention unless appropriate locking mechanisms are utilized. The reader is referred to Gray and Reuter [1993] and Weikum and Vossen [2002] for a description of algorithms for dealing with B+-trees. The analysis of several methods is reported in Johnson and Shasha [1993].

57.5.2.4 Further Information

Lazowska et al. [1984] is a good source for the material that we presented here on the issue of hardware resource contention. Other pertinent examples are given in Menasce and Almeida [2000]. Trivedi [2002] is an excellent textbook covering basic probability theory and random processes used in this section.

There are two monographs dealing with data contention. Tay [1987] gives an elegant analysis of the analysis of the no-waiting and blocking methods for strict 2PL. Thomasian [1996a] provides a monograph reviewing his work, while Thomasian [1998b] is a shortened version, which is much more readily available. An analysis of optimistic methods appears in Ryu and Thomasian [1987]. The analysis of locking methods with limited wait depth (WDL methods) appears in Thomasian [1998c], while simulation results are reported in Thomasian [1997].

57.6 Conclusion

Readers who have reached this point should have realized the vastness of this field. The study of transaction processing is a must in the area of Web services, where the field is evolving rapidly.

Due to space limitations, we have not discussed **workflow models**, which have some transactional properties. Business transactions discussed in Section 57.3.3 belong to this category. Several advanced transaction models are described in ElMagarmid [1992].

Data mining extracts information from the data items transacted by a transaction, such as the items purchased in a single visit to a supermarket. **Asociation rule mining** applied to such "market basket data" has discovered association rules such as the following (at a statistically significant level): "purchases of diapers are accompanied by purchases of beer." Transaction data is also compiled in summary tables and used for **OLAP (online analytical processing)** [Dunham 2003].

Acknowledgment

We acknowledge the support of the National Science Foundation through Grant 0105485 in Computer System Architecture.

Defining Terms

We refer the reader to the terms defined in Chapter 56 ("Concurrency Control and Recovery) and Chapter 58 ("Distributed and Parallel Database Systems") in this *Handbook*. The reader is also referred to a glossary of transaction processing terms in Gray and Reuter [1993], network security and cryptography terms in Kaufman et al. [2002], and Web services terms in Brown and Haas [2002].

References

[Agrawal et al. 1994] D. Agrawal, A. El Abbadi, and A.E. Lang. "The performance of protocols based on locks with ordered sharing," *IEEE Trans. Knowledge and Data Eng.*, 6(5):805–818 (1994).

[Badrinath and Ramamritham 1992] B.R. Badrinath and K. Ramamritham. "Semantics-based concurrency control: beyond commutativity," *ACM Trans. Database Systems*, 17(1):163–199 (1992).

[Bayer 1986] R. Bayer. "Consistency of transactions and the random batch," *ACM Trans. Database Systems*, 11(4):397–404 (1986).

[Bellare et al. 2000] M. Bellare et al. "Design, implementation and deployment of the iKP secure electronic payment protocol," *IEEE J. Selected Areas of Commun.*, 18(4): 611–627 (2000).

[Bernstein et al. 1987] P.A. Bernstein, V. Hadzilacos, and N. Goodman. *Concurrency Control and Recovery in Database Systems*, Addison-Wesley, 1987 (also available from http://www.research. microsoft.com/pubs/ccontrol/).

[Bernstein and Newcomer 1997] P.A. Bernstein and E. Newcomer. *Principles of Transaction Processing for the Systems Professional*, Morgan Kaufmann Publishers, 1997.

[Brown and Haas 2002] A. Brown and H. Haas (W3C Working Draft 14 Nov. 2002). *Web Services Glossary*, http://www.w3.org/TR/2002/WD-ws-gloss-20021114.

[Cellary et al. 1988] W. Cellary, E. Gelenbe, and T. Morzy. *Concurrency Control in Distributed Database Systems*, North-Holland, 1988.

[Ceri and Pelagatti 1984] S. Ceri and G. Pelagatti. *Distributed Databases: Principles and Systems*, McGraw-Hill, 1984.

[Cox et al. 1995] B. Cox, J.D. Tygar, and M. Sirbu. "NetBill security and transaction protocol," *Proc. 1st USENIX Workshop on Electronic Commerce*, 1995.

[Dunham 2003] M.H. Dunham. *Data Mining: Introductory and Advanced Topics*, Prentice Hall, 2003.

[ElMagarmid 1992] A.K. ElMagarmid (Ed.) *Database Transaction Models for Advanced Applications*, Morgan Kaufmann, 1992.

[Elnikety et al. 2003] S. Elnikety, E. Nahum, J. Tracey, and W. Zwaenepoel. "Admission control for e-commerce web sites," *Proc. 1st Workshop on Algorithms and Architectures for Self-Managing Systems*, 2003 (http://tesla.hpl.hp.com/self-manage03/).

[Franaszek and Robinson 1985] P.A. Franaszek and J.T. Robinson. "Limitations of concurrency in transaction processing," *ACM Trans. Database Systems*, 10(1): 1–28 (1985).

[Franaszek et al. 1992] P.A. Franaszek, J.T. Robinson, and A. Thomasian. "Concurrency control for high contention environments," *ACM Trans. Database Systems*, 17(2): 304–345 (1992).

[Franaszek et al. 1993] P.A. Franaszek, J.R. Haritsa, J.T. Robinson, and A. Thomasian. "Distributed concurrency control based on limited wait-depth," *IEEE Trans. Parallel and Distributed Systems*, 4(11):1246–1264 (1993).

[Franklin et al. 1997] M.J. Franklin. M.J. Carey, and M. Livny. "Transactional client-server cache consistency: alternatives and performance," *ACM Trans. Database Systems*, 22(3):315–363 (1997).

[Garcia-Molina 1983] H. Garcia-Molina. "Using semantic knowledge for transaction processing in a distributed database," *ACM Trans. Database Systems*, 8(2):186–213 (1983).

[Garcia-Molina and Salem 1987] H. Garcia-Molina and K. Salem. "Sagas," *Proc. ACM SIGMOD Int. Conf.*, 1987, pp. 249–259.

[Gray and Reuter 1993] J. Gray and A. Reuter. *Transaction Processing: Concepts and Techniques*, Morgan Kaufmann Publishers, 1993.

[Hsu and Zhang 1992] M. Hsu and B. Zhang. "Performance evaluation of cautious waiting," *ACM Trans. Database Systems*, 17(3):477–512 (1992).

[Jagadish and Shmueli 1992] H.V. Jagadish and O. Shmueli. "A proclamation based model for cooperating transactions," *Proc. 18th VLDB Conf.*, 1992, pp. 265–276.

[Johnson and Shasha 1993] T. Johnson and D. Shasha. "The performance of current B-tree algorithms," *ACM Trans. Database Systems*, 18(1):51–101 (1993).

[Kaufman et al. 2002] C. Kaufman, R. Perlman, and M. Speciner. *Network Security: Private Communication in a Public World*, Prentice Hall 2002.

[Kumar 1996] V. Kumar (Editor). *Performance of Concurrency Control Mechanisms in Centralized Database Systems*, Prentice Hall, 1996.

[Lazowska et al. 1984] E.D. Lazowska, J. Zoharjan, G.S. Graham, and K. Sevcik. *Quantitative System Performance*, Prentice Hall, 1984. Available online at: http://www.cs.washington.edu/homes/lazowska/qsp.

[Lewis et al. 2002] P.M. Lewis, A. Bernstein, and M. Kifer. *Databases and Transaction Processing*, Addison-Wesley, 2002.

[Lynch et al. 1994] N. Lynch, M. Merritt, W.E. Weihl, and A. Fekete. *Atomic Transactions*, Morgan Kaufmann, 1994.

[Menasce and Almeida 2000] D. Menasce and A.F. Almeida. *Scaling for E-Business: Technologies, Models, Performance, and Capacity Planning*, Prentice Hall, 2000.

[Moss 1985] J.E.B. Moss. *Nested Transactions: An Approach to Reliable Distributed Computing*, MIT Press, 1985.

[Newcomer 2002] E. Newcomer. *Understanding Web Services: XML, WSDL, SOAP, and UDDI*, Addison-Wesley, 2002.

[O'Neil 1986] P. E. O'Neil. "The escrow transaction method," *ACM Trans. Database Systems*, 11(4):405–430 (1986).

[Orfali et al. 1996] R. Orfali, D. Harkey, and J. Edwards. *The Essential Client/Server Survival Guide, 2nd ed.*, Wiley & Sons, 1996.

[Ozsu and Valduriez 1999] M.T. Ozsu and P. Valduriez. *Principles of Distributed Database Systems*, Prentice Hall, 1999.

[Rahm 1993] E. Rahm. "Empirical performance evaluation of concurrency and coherency control protocols for data sharing systems," *ACM Trans. Database Systems*, 18(2):333–377 (1993).

[Ramakrishnan and Gehrke 2003] R. Ramakrishnan and J. Gehrke. *Database Management Systems, 3rd ed.*, McGraw-Hill, 2003.

[Ramamritham and Chrisanthis 1996] K. Ramamritham and P.K. Chrisanthis. *Advances in Concurrency Control and Transaction Processing*, IEEE Computer Society Press, 1996.

[Ryu and Thomasian 1987] I.K. Ryu and A. Thomasian. "Performance analysis of centralized databases with optimistic concurrency control," *Performance Evaluation*, 7(3):195–211 (1987).

[Salem et al. 1994] K. Salem, H. Garcia-Molina, and J. Shands. "Altruistic locking," *ACM Trans. Database Systems*, 19(1):117–165 (1994).

[Shasha et al. 1995] D. Shasha, F. Llirbat, E. Simon, and P. Valduriez. "Transaction chopping: algorithms and performance studies," *ACM Trans. Database Systems*, 20(3):325–363 (1995).

[Singhal and Smith 1997] V. Singhal and A.J. Smith. "Analysis of locking behavior in three real database systems," *VLDB J.*, 6(1):40–52 (1997).

[SOAP V.1.2 P.0: Primer 2003] W3C Proposed Recommendation, *SOAP V.1.2 P.0 Primer 2003*, May 2003. http://www.w3.org/TR/soap12-part0/.

[STDL 1996] X/Open CAE Specification, *Structured Transaction Definition Language (STDL)*, 1996. http://www.opengroup.org/onlinepubs/009638899/toc.pdf.

[Tasaka 1986] S. Tasaka. *Performance Analysis of Multiple Access Protocols*, MIT Press, 1986.

[Tay 1987] Y.C. Tay. *Locking Performance in Centralized Databases*, Academic Press, 1987.

[Thomasian and Ryu 1991] A. Thomasian and I.K. Ryu. "Performance analysis of two-phase locking," *IEEE Trans. on Software Eng.*, 17(5):386–402 (1991).

[Thomasian 1993] A. Thomasian. "Two-phase licking performance and its thrashing behavior," *ACM Trans. Database Systems*, 18(4):579–625 (1993).

[Thomasian 1995] A. Thomasian. "Checkpointing for optimistic concurrency control methods," *IEEE Trans. Knowledge and Data Eng.*, 7(2):332–339 (1995).

[Thomasian 1996a] A. Thomasian. *Database Concurrency Control: Methods, Performance, and Analysis*, Kluwer Academic Publishers, 1996.

[Thomasian 1996b] A. Thomasian. "A more realistic locking model and its analysis," *Information Systems*, 21(5):409–430 (1996).

[Thomasian 1997] A. Thomasian. "A performance comparison of locking methods with limited wait-depth," *Trans. Knowledge and Data Eng.*, 9(3):421–434 (1997).

[Thomasian 1998a] A. Thomasian. "Concurrency control: methods, performance, and analysis," *ACM Computing Surveys*, 30(1):70–119 (1998).

[Thomasian 1998b] A. Thomasian. "Distributed optimistic concurrency control methods for high-performance transaction processing," *IEEE Trans. Knowledge and Data Eng.*, 10(1):173–189 (1998).

[Thomasian 1998c] A. Thomasian. "Performance analysis of locking methods with limited wait depth," *Performance Evaluation* 34(2): 69–89 (1998).

[Trivedi 2002] K.S. Trivedi. *Probability and Statistics and Reliability Queueing and Computer Science Applications, 2nd ed.*, Wiley 2002.

[Tygar 1998] J.D. Tygar. "Atomicity versus anonymity: Distributed transactions for electronic commerce," *Proc. 24th VLDB Conf.*, 1998, pp. 1–12.

[Weihl 1988] W.E. Weihl. "Commutativity based concurrency control for abstract data types," *IEEE Trans. Computers 37*(12): 1488–1505 (1988).

[Weikum et al. 1994] G. Weikum, C. Hasse, A. Menkeberg, and P. Zabback. "The COMFORT automatic tuning project (Invited Project Review)," *Information Systems*, 19(5):381–432 (1994).

[Weikum and Vossen 2002] G. Weikum and G. Vossen. *Transactional Information Systems*, Morgan Kaufmann, 2002.

[WS Architecture 2003] W3C Working Draft May 2003. *WS Architecture 2003*, http://www.w3.org/TR/ws-arch/.

[WS Architecture Usage Scenarios] W3C Working Draft, 2003. *Web Services Architecture Usage Scenarios*, http://www.w3.org/TR/ws-arch-scenarios/.

[WS Coordination 2002] F. Cabrera et al. *Web Services Coordination (WS-Coordination)*, August 2002. http://www.ibm.com/developerworks/library/ws-coor/.

[WS Transactions 2002] F. Cabrera et al. *Web Services Transaction (WS-Transaction)*, August 2002. http://www.ibm.com/developerworks/library/ws-transpec/.

Further Information

We have provided information on further reading at the end of each section. We refer the reader to Chapter 56 on "Concurrency Control and Recovery" by Michael J. Franklin and Chapter 58 on "Distributed and Parallel Database Systems" by M.T. Ozsu and P. Valduriez.

Conferences that publish papers in this area are ACM SIGMOD, the Very Large Data Base (VLDB), the International Conference on Data Engineering (ICDE), among others. Relevant journals are *ACM Transactions on Database Systems* (TODS), *The VLDB Journal, IEEE Transactions on Knowledge and Data Engineering* (TKDE), and *Informations Systems*.

58

Distributed and Parallel Database Systems

M. Tamer Özsu
University of Waterloo

Patrick Valduriez
INRIA and IRIN

58.1 Introduction **58-1**
58.2 Underlying Principles **58-1**
58.3 Distributed and Parallel Database Technology **58-4**
  Architectural Issues • Data Integration • Concurrency
  Control • Reliability • Replication • Data Placement
  • Query Processing and Optimization • Load Balancing
58.4 Research Issues.................................... **58-16**
  Mobile Databases • Large-Scale Query Processing
58.5 Summary **58-18**

58.1 Introduction

The maturation of database management system (DBMS) technology has coincided with significant developments in distributed computing and parallel processing technologies. The end result is the emergence of **distributed database management systems** and **parallel database management systems**. These systems have become the dominant data management tools for highly data-intensive applications. With the emergence of the Internet as a major networking medium that enabled the subsequent development of the World Wide Web (WWW or Web) and grid computing, distributed and parallel database systems have started to converge.

A parallel computer, or multiprocessor, is itself a distributed system composed of a number of nodes (processors and memories) connected by a fast network within a cabinet. Distributed database technology can be naturally revised and extended to implement parallel database systems, that is, database systems on parallel computers [DeWitt and Gray 1992, Valduriez 1993]. Parallel database systems exploit the parallelism in data management in order to deliver high-performance and high-availability database servers.

This chapter presents an overview of the distributed DBMS and parallel DBMS technologies, highlights the unique characteristics of each, and indicates the similarities between them. This discussion should help establish their unique and complementary roles in data management.

58.2 Underlying Principles

A distributed database (DDB) is a collection of multiple, logically interrelated databases distributed over a computer network. A distributed database management system (distributed DBMS) is then defined as the software system that permits the management of the distributed database and makes the distribution transparent to the users [Özsu and Valduriez 1999]. These definitions assume that each site logically consists of a single, independent computer. Therefore, each site has its own primary and secondary storage, runs its own operating system (which may be the same or different at different sites), and has the capability to execute applications on its own. The sites are interconnected by a computer network rather than a multiprocessor configuration. The important point here is the emphasis on loose interconnection between processors that have their own operating systems and operate independently.

The database is physically distributed across the data sites by **fragmenting** and **replicating** the data [Ceri et al. 1987]. Given a relational database schema, fragmentation subdivides each relation into horizontal or vertical partitions. *Horizontal fragmentation* of a relation is accomplished by a selection operation that places each tuple of the relation in a different partition based on a fragmentation predicate (e.g., an Employee relation can be fragmented according to the location of the employees). *Vertical fragmentation*, divides a relation into a number of fragments by projecting over its attributes (e.g., the Employee relation can be fragmented such that the Emp_number, Emp_name, and Address information is in one fragment, and Emp_number, Salary, and Manager information is in another fragment). Fragmentation is desirable because it enables the placement of data in close proximity to its place of use, thus potentially reducing transmission cost, and it reduces the size of relations that are involved in user queries.

Based on user access patterns, each of the fragments can also be replicated. This is preferable when the same data is accessed from applications that run at a number of sites. In this case, it may be more cost-effective to duplicate the data at a number of sites rather than continuously moving it between them.

In the context of the Web, the above process of defining a database schema that is then fragmented and distributed is not always possible. The data (and the databases) usually exist and one is faced with the problem of providing integrated access to this data. This process is known as *data integration* and is discussed further below.

When the architectural assumption of each site being a (logically) single, independent computer is relaxed, one gets a parallel database system. The differences between a parallel DBMS and a distributed DBMS are somewhat blurred. In particular, shared-nothing parallel DBMS architectures, which we discuss below, are quite similar to the loosely interconnected distributed systems. Parallel DBMSs exploit multiprocessor computer architectures to build high-performance and high-availability database servers at a much lower price than equivalent mainframe computers.

A parallel DBMS can be defined as a DBMS implemented on a multiprocessor computer. This includes many alternatives, ranging from the straightforward porting of an existing DBMS, which may require only rewriting the operating system interface routines, to a sophisticated combination of parallel processing and database system functions into a new hardware/software architecture. As always, we have the traditional trade-off between *portability* (to several platforms) and *efficiency*. The sophisticated approach is better able to fully exploit the opportunities offered by a multiprocessor at the expense of portability.

The solution, therefore, is to use large-scale parallelism to magnify the raw power of individual components by integrating these in a complete system along with the appropriate parallel database software. Using standard hardware components is essential in order to exploit the continuing technological improvements with minimal delay. Then, the database software can exploit the three forms of parallelism inherent in data-intensive application workloads. **Inter-query parallelism** enables the parallel execution of multiple queries generated by concurrent transactions. **Intra-query parallelism** makes the parallel execution of multiple, independent operations (e.g., select operations) possible within the same query. Both inter-query and intra-query parallelism can be obtained using *data partitioning*, which is similar to horizontal fragmentation. Finally, with **intra-operation parallelism**, the same operation can be executed as many sub-operations using *function partitioning* in addition to data partitioning. The set-oriented mode of database languages (e.g., SQL) provides many opportunities for intra-operation parallelism.

There are a number of identifying characteristics of the distributed and parallel DBMS technology:

1. *The distributed/parallel database is a database*, not some "collection" of files that can be individually stored at each node of a computer network. This is the distinction between a DDB and a collection of files managed by a distributed file system. To form a DDB, distributed data should be logically

related, where the relationship is defined according to some structural formalism (e.g., the relational model), and access to data should be at a high level via a common interface.

2. *The system has the full functionality of a DBMS.* It is neither, as indicated above, a distributed file system, nor is it a transaction processing system. Transaction processing is only one of the functions provided by such a system, which also provides functions such as query processing, structured organization of data, and others that transaction processing systems do not necessarily deal with.

3. *The distribution (including fragmentation and replication) of data across multiple sites/processors is not visible to the users.* This is called **transparency**. The distributed/parallel database technology extends the concept of *data independence*, which is a central notion of database management, to environments where data is distributed and replicated over a number of machines connected by a network. This is provided by several forms of transparency: *network* (and, therefore, *distribution*) *transparency*, *replication transparency*, and *fragmentation transparency*. Transparent access means that users are provided with a single logical image of the database although it may be physically distributed, enabling them to access the distributed database as if it were a centralized one. In its ideal form, full transparency would imply a query language interface to the distributed/parallel DBMS that is no different from that of a centralized DBMS.

Transparency concerns are more pronounced in the case of distributed DBMSs. There are two fundamental reasons for this. First, the multiprocessor system on which a parallel DBMS is implemented is controlled by a single operating system. Therefore, the operating system can be structured to implement some aspects of DBMS functionality, thereby providing some degree of transparency. Second, software development on parallel systems is supported by parallel programming languages, which can provide further transparency.

In a distributed DBMS, data and the applications that access that data can be localized at the same site, eliminating (or reducing) the need for remote data access that is typical of teleprocessing-based timesharing systems. Furthermore, because each site handles fewer applications and a smaller portion of the database, contention for resources and for data access can be reduced. Finally, the inherent parallelism of distributed systems provides the possibility of inter-query parallelism and intra-query parallelism.

If access to the distributed database consists only of querying (i.e., read-only access), then provision of inter-query and intra-query parallelism would imply that as much of the database as possible should be replicated. However, because most database accesses are not read-only, the mixing of read and update operations requires support for distributed transactions (as discussed in a later section).

High performance is probably the most important objective of parallel DBMSs. In these systems, higher performance can be obtained through several complementary solutions: database-oriented operating system support, parallelism, optimization, and load balancing. Having the operating system constrained and "aware" of the specific database requirements (e.g., buffer management) simplifies the implementation of low-level database functions and therefore decreases their cost. For example, the cost of a message can be significantly reduced to a few hundred instructions by specializing the communication protocol. Parallelism can increase transaction throughput (using inter-query parallelism) and decrease transaction response times (using intra-query and intra-operation parallelism).

Distributed and parallel DBMSs are intended to improve reliability because they have replicated components and thus eliminate single points of failure. The failure of a single site or processor, or the failure of a communication link that makes one or more sites unreachable, is not sufficient to bring down the entire system. Consequently, although some of the data may be unreachable, with proper system design users may be permitted to access other parts of the distributed database. The "proper system design" comes in the form of support for distributed transactions. Providing transaction support requires the implementation of distributed concurrency control and distributed reliability (commit and recovery) protocols, which are reviewed in a later section.

In a distributed or parallel environment, it should be easier to accommodate increasing database sizes or increasing performance demands. Major system overhauls are seldom necessary; expansion can usually be handled by adding more processing and storage power to the system.

Ideally, a parallel DBMS (and to a lesser degree a distributed DBMS) should demonstrate two advantages: **linear scaleup** and **linear speedup**. Linear scaleup refers to a sustained performance for a linear increase in both database size and processing and storage power. Linear speedup refers to a linear increase in performance for a constant database size, and a linear increase in processing and storage power. Furthermore, extending the system should require minimal reorganization of the existing database.

The price/performance characteristics of microprocessors and workstations make it more economical to put together a system of smaller computers with the equivalent power of a single, big machine. Many commercial distributed DBMSs operate on minicomputers and workstations to take advantage of their favorable price/performance characteristics. The current reliance on workstation technology has come about because most commercially distributed DBMSs operate within local area networks for which the workstation technology is most suitable. The emergence of distributed DBMSs that run on wide area networks may increase the importance of mainframes. On the other hand, future distributed DBMSs may support hierarchical organizations where sites consist of clusters of computers communicating over a local area network with a high-speed backbone wide area network connecting the clusters.

58.3   Distributed and Parallel Database Technology

Distributed and parallel DBMSs provide the same functionality as centralized DBMSs except in an environment where data is distributed across the sites on a computer network or across the nodes of a multiprocessor system. As discussed above, users are unaware of data distribution. Thus, these systems provide users with a *logically integrated* view of the *physically distributed* database. Maintaining this view places significant challenges on system functions. We provide an overview of these new challenges in this section. We assume familiarity with basic database management techniques.

58.3.1   Architectural Issues

There are many possible alternatives for distributing data and disseminating it to the users. We characterize the data delivery alternatives along three orthogonal dimensions: delivery modes, regularity of data delivery, and communication methods.

The alternative delivery modes are pull-only, push-only, and hybrid. In the *pull-only* mode of data delivery, the transfer of data from servers to clients is initiated by a client pull. When a client request is received at a server, the server responds by locating the requested information. The main characteristic of pull-based delivery is that the arrival of new data items or updates to existing data items are carried out at a server without notification to clients unless clients explicitly poll the server. Also, in pull-based mode, servers must be interrupted continuously to deal with requests from clients. Furthermore, the information that clients can obtain from a server is limited to when and what clients know to ask for. Conventional DBMSs (including relational and object-oriented ones) offer primarily pull-based data delivery.

In the *push-only* mode of data delivery, the transfer of data from servers to clients is initiated by a server push in the absence of any specific request from clients. The main difficulty in the push-based approach is in deciding which data would be of common interest, and when to send it to clients (periodically, irregularly, or conditionally). Thus, the usefulness of server push depends heavily on the accuracy of a server to predict the needs of clients. In push-based mode, servers disseminate information to either an unbounded set of clients (random broadcast) who can listen to a medium or a selective set of clients (multicast) who belong to some categories of recipients that may receive the data.

The *hybrid* mode of data delivery combines the client-pull and server-push mechanisms. For example, the transfer of information from servers to clients is first initiated by a client pull, and the subsequent transfer of updated information to clients is initiated by a server push [Liu et al. 1996].

There are three typical frequency measurements that can be used to classify the regularity of data delivery: *periodic*, *conditional*, and *ad-hoc* or *irregular*.

In periodic delivery, data is sent from the server to clients at regular intervals. The intervals can be defined by system default or by clients using their profiles. Both pull and push can be performed in periodic fashion.

Periodic delivery is carried out on a regular and prespecified repeating schedule. A client request for IBM's stock price every week is an example of a periodic pull. An example of periodic push is when an application can send out stock price listings on a regular basis, for example, every morning. Periodic push is particularly useful for situations in which clients might not be available at all times, or might be unable to react to what has been sent, such as in the mobile setting where clients can become disconnected.

In conditional delivery, data is sent from servers whenever certain conditions installed by clients in their profiles are satisfied. Such conditions can be as simple as a given time span or as complicated as event-condition-action rules. Conditional delivery is mostly used in the hybrid or push-only delivery systems. Using conditional push, data is sent out according to a prespecified condition, rather than any particular repeating schedule. An application that sends out stock prices only when they change is an example of conditional push. An application that sends out a balance statement only when the total balance is 5% below the predefined balance threshold is an example of hybrid conditional push. Conditional push assumes that changes are critical to the clients, and that clients are always listening and need to respond to what is being sent. Hybrid conditional push further assumes that missing some update information is not crucial to the clients.

Ad-hoc delivery is irregular and is performed mostly in a pure pull-based system. Data is pulled from servers to clients in an ad-hoc fashion whenever clients request it. In contrast, periodic pull arises when a client uses polling to obtain data from servers based on a regular period (schedule).

The third component of the design space of information delivery alternatives is the communication method. These methods determine the various ways in which servers and clients communicate for delivering information to clients. The alternatives are *unicast* and *one-to-many*. In unicast, the communication from a server to a client is one-to-one: the server sends data to one client using a particular delivery mode with some frequency. In one-to-many, as the name implies, the server sends data to a number of clients. Note that we are not referring here to a specific protocol; one-to-many communication can use a multicast or broadcast protocol.

Within this framework, we can identify a number of popular architectural approaches. The popular *client/server architecture* [Orfali et al. 1994] is a pull-only, ad hoc, and unicast data delivery system. The client/server DBMS, in which a number of client machines access a single database server, is the most straightforward one. In which can be called *multiple-client/single-server*, the database management problems are considerably simplified because the database is stored on a single server. The pertinent issues relate to the management of client buffers and the caching of data and (possibly) locks. The data management is done *centrally* at the single server.

A more distributed and more flexible architecture is the *multiple-client/multiple server* architecture, where the database is distributed across multiple servers that must communicate with each other in responding to user queries and in executing transactions. Each client machine has a "home" server to which it directs user requests. The communication of the servers among themselves is transparent to the users. Most current DBMSs implement one or the other type of client server architecture.

A truly distributed DBMS does not distinguish between client and server machines. Ideally, each site can perform the functionality of a client and a server. Such architectures, called *peer-to-peer* (P2P), require sophisticated protocols to manage the data distributed across multiple sites. P2P systems have started to become very popular with the emergence of the Internet as the primary long-haul communication medium and the development of applications such as Napster (for sharing music files) and Gnutella (for sharing general files). These first applications had a simple design with poor scaling and performance. However, newer P2P systems, using distributed hash tables to identify objects over the network, have good scaling characteristics [Balakrishnan et al. 2003]. P2P has many advantages over client server. First, there is no centralized access control. This avoids any kind of bottleneck and makes it possible to scale up to a very large number of sites. P2P systems on the Internet already claim millions of user sites. Second, by replicating data at different sites, P2P systems increase both data availability and performance through parallelism.

The primary objective of a parallel database system is to provide DBMS functions with a better cost/performance. Load balancing (i.e., the ability of the system to divide the workload evenly across all nodes) is crucial for high performance and is more or less difficult, depending on the architecture. Parallel system architectures range between two extremes — the **shared-memory** and the **shared-nothing** architectures — and a useful intermediate point is the **shared-disk** architecture. Hybrid architectures, such as Non-Uniform Memory Architecture (NUMA) and cluster, can combine the benefits of these architectures.

In the shared-memory approach, any processor has access to any memory module or disk unit through a fast interconnect. Examples of shared-memory parallel database systems include XPRS [Hong 1992], DBS3 [Bergsten et al. 1991], and Volcano [Graefe 1990], as well as portings of major commercial DBMSs on symmetric multiprocessors. Most shared-memory commercial products today can exploit inter-query parallelism to provide high transaction throughput and intra-query parallelism to reduce response time of decision-support queries. Shared-memory makes load balancing simple. But because all data access goes through the shared memory, extensibility and availability are limited.

In the shared-disk approach, any processor has access to any disk unit through the interconnect, but exclusive (non-shared) access to its main memory. Each processor can then access database pages on the shared disk and copy them into its own cache. To avoid conflicting accesses to the same pages, global locking and protocols for the maintenance of cache coherency are needed. Shared-disk provides the advantages of shared-memory with better extensibility and availability, but maintaining cache coherency is complex.

In the shared-nothing approach, each processor has exclusive access to its main memory and disk unit(s). Thus, each node can be viewed as a local site (with its own database and software) in a distributed database system. In particular, a shared-nothing system can be designed as a P2P system [Carey et al. 1994]. The difference between shared-nothing parallel DBMSs and distributed DBMSs is basically one of implementation platform; therefore, most solutions designed for distributed databases can be reused in parallel DBMSs. Shared-nothing has three important virtues: cost, extensibility, and availability. On the other hand, data placement and load balancing are more difficult than with shared-memory or shared-disk. Examples of shared-nothing parallel database systems include the Teradata's DBC and Tandem's NonStopSQL products, as well as a number of prototypes such as BUBBA [Boral et al. 1990], GAMMA [DeWitt et al. 1990], GRACE [Fushimi et al. 1986], and PRISMA [Apers et al. 1992].

To improve extensibility, shared-memory multiprocessors have evolved toward NUMA, which provides a shared-memory programming model in a scalable shared-nothing architecture [Lenoski et al. 1992, Hagersten et al. 1992, Frank et al. 1993]. Because shared-memory and cache coherency are supported by hardware, remote memory access is very efficient, only several times (typically 4 times) the cost of local access. Database techniques designed for shared-memory DBMSs also apply to NUMA [Bouganim et al. 1999].

Cluster architectures (sometimes called hierarchical architectures) combine the flexibility and performance of shared-disk with the high extensibility of shared-nothing [Graefe 1993]. A cluster is defined as a group of servers that act like a single system and enable high availability, load balancing, and parallel processing. Because they provide a cheap alternative to tightly coupled multiprocessors, large clusters of PC servers have been used successfully by Web search engines (e.g., Google). They are also gaining much interest for managing autonomous databases [Röhm et al. 2001, Gançarski et al. 2002]. This recent trend attests to further convergence between distributed and parallel databases.

58.3.2 Data Integration

Data integration involves the process by which information from multiple data sources can be integrated to form a single cohesive system. This problem has been studied for quite some time, and various names have been given to it over the years (e.g., heterogeneous databases, federated systems, multidatabases), currently settling on the more generic term "data integration," which recognizes that the data sources do not all have to be databases. These systems are identified by two main characteristics of the data sources: autonomy and heterogeneity. Autonomy indicates the degree to which the individual data sources can operate independently. Each source is free to join and leave the integrated environment and is free to share some or all (or none) of its data or execute some queries but not others. This is an increasingly important issue as Web data sources become more prevalent and as Web data integration gains importance. Heterogeneity can occur in various forms in distributed systems, ranging from hardware heterogeneity and differences in networking protocols to variations in data sources. The important ones from the perspective of this discussion relate to data sources, in particular their functionality. Not all data sources will be database managers, so they may not even provide the typical database functionality. Even when a number of database managers are considered, heterogeneity can occur in their data models, query languages, and implementation protocols. Representing data with different modeling tools creates heterogeneity because of the inherent expressive powers and limitations of individual data models. Heterogeneity in query languages not only involves the use of completely different data access paradigms in different data models (set-at-a-time access in relational systems vs. record-at-a-time access in network and hierarchical systems), but also covers differences in languages even when the individual systems use the same data model. Different query languages that use the same data model often select very different methods for expressing identical requests. Heterogeneity in implementation techniques and protocols raises issues as to what each systems can and cannot do.

In such an environment, building a system that would provide integrated access to diverse data sources raises challenging architectural, model, and system issues. The dominant architectural model is the *mediator architecture* [Wiederhold 1992], where a middleware system consisting of *mediators* is placed in between data sources and users/applications that access these data sources. Each middleware performs a particular function (provides domain knowledge, reformulates queries, etc.) and the more complex system functions may be composed using multiple mediators. An important mediator is one that reformulates a user query into a set of queries, each of which runs on one data source, with possible additional processing at the mediator to produce the final answer.

Each data source is "wrapped" by *wrappers* that are responsible for providing a common interface to the mediators. The sophistication of each wrapper varies, depending on the functionality provided by the underlying data source. For example, if the data source is not a DBMS, the wrapper may still provide a declarative query interface and perform the translation of these queries into code that is specific to the underlying data source. Wrappers, in a sense, deal with the heterogeneity issues.

To run queries over diverse data sources, a *global schema* must be defined. This can be done either in a bottom-up or top-down fashion. In the bottom-up approach, the global schema is specified in terms of the data sources. Consequently, for each data element in each data source, a data element is defined in the global schema. In the top-down approach, the global schema is defined independent of the data sources, and each data source is treated as a view defined over the global schema. These two approaches are called *global-as-view* and *local-as-view* [Lenzerini 2002]. The details of the methodologies for defining the global schema are outside the bounds of this chapter.

58.3.3 Concurrency Control

Whenever multiple users access (read and write) a shared database, these accesses must be synchronized to ensure database consistency. The synchronization is achieved by means of **concurrency control algorithms** that enforce a correctness criterion such as **serializability**. User accesses are encapsulated as **transactions** [Gray 1981], whose operations at the lowest level are a set of read and write operations to the database. Concurrency control algorithms enforce the **isolation** property of transaction execution, which states that the effects of one transaction on the database are isolated from other transactions until the first completes its execution.

The most popular concurrency control algorithms are **locking**-based. In such schemes, a lock, in either shared or exclusive mode, is placed on some unit of storage (usually a page) whenever a transaction attempts to access it. These locks are placed according to lock compatibility rules such that *read-write*, *write-read*, and *write-write* conflicts are avoided. It is a well-known theorem that if lock actions on behalf of concurrent transactions obey a simple rule, then it is possible to ensure the serializability of these transactions: "No lock on behalf of a transaction should be set once a lock previously held by the transaction is released." This is known as **two-phase locking** [Gray 1979], because transactions go through a growing phase when they obtain locks and a shrinking phase when they release locks. In general, releasing of locks prior to the end of a transaction is problematic. Thus, most of the locking-based concurrency control algorithms are *strict*, in that they hold on to their locks until the end of the transaction.

In distributed DBMSs, the challenge is to extend both the serializability argument and the concurrency control algorithms to the distributed execution environment. In these systems, the operations of a given transaction can execute at multiple sites where they access data. In such a case, the serializability argument is more difficult to specify and enforce. The complication is due to the fact that the serialization order of the same set of transactions may be different at different sites. Therefore, the execution of a set of distributed transactions is serializable if and only if:

1. The execution of the set of transactions at each site is serializable, *and*
2. The serialization orders of these transactions at all these sites are identical.

Distributed concurrency control algorithms enforce this notion of *global serializability*. In locking-based algorithms, there are three alternative ways of enforcing global serializability: centralized locking, primary copy locking, and distributed locking.

In *centralized locking*, there is a single lock table for the entire distributed database. This lock table is placed, at one of the sites, under the control of a single lock manager. The lock manager is responsible for setting and releasing locks on behalf of transactions. Because all locks are managed at one site, this is similar to centralized concurrency control and it is straightforward to enforce the global serializability rule. These algorithms are simple to implement but suffer from two problems: (1) the central site may become a bottleneck, both because of the amount of work it is expected to perform and because of the traffic that is generated around it; and (2) the system may be less reliable because the failure or inaccessibility of the central site would cause system unavailability.

Primary copy locking is a concurrency control algorithm that is useful in replicated databases where there may be multiple copies of a data item stored at different sites. One of the copies is designated as a primary copy, and it is this copy that has to be locked in order to access that item. The set of primary copies for each data item is known to all the sites in the distributed system, and the lock requests on behalf of transactions are directed to the appropriate primary copy. If the distributed database is not replicated, copy locking degenerates into a distributed locking algorithm. Primary copy locking was proposed for the prototype distributed version of INGRES.

In *distributed* (or *decentralized*) *locking*, the lock management duty is shared by all sites in the system. The execution of a transaction involves the participation and coordination of lock managers at more than one site. Locks are obtained at each site where the transaction accesses a data item. Distributed locking algorithms do not have the overhead of centralized locking ones. However, both the communication overhead to obtain all the locks and the complexity of the algorithm are greater. Distributed locking algorithms are used in System R* and in NonStop SQL.

One side effect of all locking-based concurrency control algorithms is that they cause **deadlocks**. The detection and management of deadlocks in a distributed system is difficult. Nevertheless, the relative simplicity and better performance of locking algorithms make them more popular than alternatives such as *timestamp-based algorithms* or *optimistic concurrency control*. Timestamp-based algorithms execute the conflicting operations of transactions according to their timestamps, which are assigned when the transactions are accepted. Optimistic concurrency control algorithms work from the premise that conflicts among transactions are rare and proceed with executing the transactions up to their termination, at which point a validation is performed. If the validation indicates that serializability would be compromised by the successful completion of that particular transaction, then it is aborted and restarted.

58.3.4 Reliability

We indicated earlier that distributed DBMSs are potentially more reliable because there are multiples of each system component, which eliminates single points of failure. This requires careful system design and the implementation of a number of protocols to deal with system failures.

In a distributed DBMS, four types of failures are possible: *transaction failures*, *site (system) failures*, *media (disk) failures*, and *communication line failures*. Transactions can fail for a number of reasons. Failure can be due to an error in the transaction caused by input data, as well as the detection of a present or potential deadlock. The usual approach to take in cases of transaction failure is to abort the transaction, resetting the database to its state prior to the start of the database.

Site (or system) failures are due to a hardware failure (e.g., processor, main memory, power supply) or a software failure (e.g., bugs in system or application code). The effect of system failures is the loss of main memory contents. Therefore, any updates to the parts of the database that are in the main memory buffers (also called **volatile database**) are lost as a result of system failures. However, the database that is stored in secondary storage (also called **stable database**) is safe and correct. To achieve this, DBMSs typically employ **logging protocols**, such as *Write-Ahead Logging*, that record changes to the database in system logs and move these log records and the volatile database pages to stable storage at appropriate times. From the perspective of distributed transaction execution, site failures are important because the failed sites cannot participate in the execution of any transaction.

Media failures refer to the failure of secondary storage devices that store the stable database. Typically, these failures are addressed by duplexing storage devices and maintaining archival copies of the database. Media failures are frequently treated as problems local to one site and therefore are not specifically addressed in the reliability mechanisms of distributed DBMSs.

The three types of failures described above are common to both centralized and distributed DBMSs. Communication failures, on the other hand, are unique to distributed systems. There are a number of types of communication failures. The most common ones are errors in the messages, improperly ordered messages, lost (or undelivered) messages, and line failures. Generally, the first two are considered the responsibility of the computer network protocols and are not addressed by the distributed DBMS. The last two, on the other hand, have an impact on the distributed DBMS protocols and therefore need to be considered in the design of these protocols. If one site is expecting a message from another site and this message never arrives, this may be because (1) the message is lost, (2) the line(s) connecting the two sites may be broken, or (3) the site that is supposed to send the message may have failed. Thus, it is not always possible to distinguish between site failures and communication failures. The waiting site simply timeouts and has to assume that the other site cannot communicate. Distributed DBMS protocols must deal with this uncertainty. One drastic result of line failures may be *network partitioning*, in which the sites form groups where communication within each group is possible but communication across groups is not. This is difficult to deal with in the sense that it may not be possible to make the database available for access while at the same time guaranteeing its consistency.

Two properties of transactions are maintained by reliability protocols: **atomicity** and **durability**. Atomicity requires that either all the operations of a transaction are executed or none of them are (all-or-nothing). Thus, the set of operations contained in a transaction is treated as one atomic unit. Atomicity is maintained even in the face of failures. Durability requires that the effects of successfully completed (i.e., committed) transactions endure subsequent failures.

The enforcement of atomicity and durability requires the implementation of *atomic commitment protocols* and *distributed recovery protocols*. The most popular atomic commitment protocol is **two-phase commit** (2PC). The recoverability protocols are built on top of the local recovery protocols, which are dependent upon the supported mode of interaction (of the DBMS) with the operating system.

Two-phase commit (2PC) is a very simple and elegant protocol that ensures the atomic commitment of distributed transactions. It extends the effects of local atomic commit actions to distributed transactions by insisting that all sites involved in the execution of a distributed transaction agree to commit the transaction before its effects are made permanent (i.e., all sites terminate the transaction in the same manner). If all the sites agree to commit a transaction, then all the actions of the distributed transaction take effect; if one of the sites declines to commit the operations at that site, then all the other sites are required to abort the transaction. Thus, the fundamental 2PC rule states that:

1. If even one site rejects to commit (which means it votes to abort) the transaction, the distributed transaction must be aborted at each site where it executes; *and*
2. If all the sites vote to commit the transaction, the distributed transaction is committed at each site where it executes.

The simple execution of the 2PC protocol is as follows. There is a *coordinator* process at the site where the distributed transaction originates, and *participant* processes at all the other sites where the transaction executes. Initially, the coordinator sends a "prepare" message to all the participants, each of which independently determines whether or not it can commit the transaction at that site. Those that can commit send back a "vote-commit" message, while those that are not able to commit send back a "vote-abort" message. Once a participant registers its vote, it cannot change it. The coordinator collects these messages and determines the fate of the transaction according to the 2PC rule. If the decision is to commit, the coordinator sends a "global-commit" message to all participants. If the decision is to abort, it sends a "global-abort" message to those participants that had earlier voted to commit the transaction. No message needs to be sent to those participants that had originally voted to abort because they can assume, according to the 2PC rule, that the transaction is going to be eventually globally aborted. This is known as the *unilateral abort* option of the participants.

There are two rounds of message exchanges between the coordinator and the participants; hence the name 2PC protocol. There are a number of variations of 2PC, such as linear 2PC and distributed 2PC, that have not found much favor among distributed DBMS implementations. Two important variants of 2PC are the *presumed abort 2PC* and *presumed commit 2PC* [Mohan and Lindsay 1983]. These are important because they reduce the message and I/O overhead of the protocols. Presumed abort protocol is included in the X/Open XA standard and has been adopted as part of the ISO standard for Open Distributed Processing.

An important characteristic of 2PC protocol is its *blocking* nature. Failures can occur during the commit process. As discussed above, the only way to detect these failures is by means of a timeout of the process waiting for a message. When this happens, the process (coordinator or participant) that times out follows a **termination protocol** to determine what to do with the transaction that was in the middle of the commit process. A non-blocking commit protocol is one whose termination protocol can determine what to do with a transaction in case of failures under any circumstance. In the case of 2PC, if a site failure occurs at the coordinator site and one participant site while the coordinator is collecting votes from the participants, the remaining participants cannot determine the fate of the transaction among themselves, and they have to remain blocked until the coordinator or the failed participant recovers. During this period, the locks that are held by the transaction cannot be released, which reduces the availability of the database.

Assume that a participant times out after it sends its commit vote to the coordinator, but before it receives the final decision. In this case, the participant is said to be in READY state. The termination protocol for the participant is as follows. First, note that the participant cannot unilaterally reach a termination decision. Because it is in the READY state, it must have voted to commit the transaction. Therefore, it cannot now change its vote and unilaterally abort it. On the other hand, it cannot unilaterally decide to commit the transaction because it is possible that another participant may have voted to abort it. In this case, the participant will remain blocked until it can learn from someone (either the coordinator or some other participant) the ultimate fate of the transaction. If we consider a centralized communication structure where the participants cannot communicate with one another, the participant that has timed out has to wait for the coordinator to report the final decision regarding the transaction. Because the coordinator has failed, the participant will remain blocked. In this case, no reasonable termination protocol can be designed.

If the participants can communicate with each other, a more distributed termination protocol can be developed. The participant that times out may simply ask all other participants to help it reach a decision. If during termination, all the participants realize that only the coordinator site has failed, they can elect a new coordinator that can restart the commit process. However, in the case where both a participant site and the coordinator site have failed, it is possible for the failed participant to have received the coordinator's decision and terminated the transaction accordingly. This decision is unknown to the other participants; thus, if they elect a new coordinator and proceed, there is the danger that they may decide to terminate the transaction differently from the participant at the failed site. The above case demonstrates the blocking nature of 2PC. There have been attempts to devise non-blocking commit protocols (e.g., three-phase commit) but the high overhead of these protocols has precluded their adoption.

The inverse of termination is recovery. When the failed site recovers from the failure, what actions does it have to take to recover the database at that site to a consistent state? This is the domain of distributed recovery protocols. Consider the recovery side of the case discussed above, in which the coordinator site recovers and the recovery protocol must now determine what to do with the distributed transaction(s) whose execution it was coordinating. The following cases are possible:

1. The coordinator failed before it initiated the commit procedure. Therefore, it will start the commit process upon recovery.
2. The coordinator failed while in the READY state. In this case, the coordinator has sent the "prepare" command. Upon recovery, the coordinator will restart the commit process for the transaction from the beginning by sending the "prepare" message one more time. If the participants had already terminated the transaction, they can inform the coordinator. If they were blocked, they can now resend their earlier votes and resume the commit process.
3. The coordinator failed after it informed the participants of its global decision and terminated the transaction. Thus, upon recovery, it does not need to do anything.

58.3.5 Replication

In replicated distributed databases,* each logical data item has a number of physical instances. For example, the salary of an employee (*logical data item*) may be stored at three sites (*physical copies*). The issue in this type of a database system is to maintain some notion of consistency among the copies. The most discussed consistency criterion is **one copy equivalence**, which asserts that the values of all copies of a logical data item should be identical when the transaction that updates it terminates.

If replication transparency is maintained, transactions will issue read and write operations on a logical data item x. The replica control protocol is responsible for mapping operations on x to operations on physical copies of x ($x_1, \ldots, x_n$). A typical replica control protocol that enforces one copy serializability is known as the **Read-Once/Write-All** (ROWA) protocol. ROWA maps each read on x [Read(x)] to a read on one of the physical copies x_i [Read(x_i)]. The copy that is read is insignificant from the perspective of the replica control protocol and may be determined by performance considerations. On the other hand, each write on logical data item x is mapped to a set of writes on *all* copies of x.

The ROWA protocol is simple and straightforward but requires that all copies of all logical data items that are updated by a transaction be accessible for the transaction to terminate. Failure of one site may block a transaction, reducing database availability.

A number of alternative algorithms have been proposed that reduce the requirement that all copies of a logical data item be updated before the transaction can terminate. They relax ROWA by mapping each write to only a subset of the physical copies.

This idea of possibly updating only a subset of the copies, but nevertheless successfully terminating the transaction, has formed the basis of quorum-based voting for replica control protocols. Votes are assigned to each copy of a logical data item and a transaction that updates that logical data item can successfully complete as long as it has a majority of the votes. Based on this general idea, an early **quorum-based voting algorithm** [Gifford 1979] assigns a (possibly unequal) vote to each copy of a replicated data item. Each operation then has to obtain a *read quorum* (V_r) or a *write quorum* (V_w) to read or write a data item, respectively. If a given data item has a total of V votes, the quorums must obey the following rules:

1. $V_r + V_w > V$ (a data item is not read and written by two transactions concurrently, avoiding the read-write conflict).
2. $V_w > V/2$ (two write operations from two transactions cannot occur concurrently on the same data item thus avoiding write-write conflict).

---

*Replication is not a significant concern in parallel DBMSs because the data is normally not replicated across multiple processors. Replication may occur as a result of data shipping during query optimization, but this is not managed by the replica control protocols.

The difficulty with this approach is that transactions are required to obtain a quorum even to read data. This significantly and unnecessarily slows down read access to the database. An alternative quorum-based voting protocol that overcomes this serious performance drawback has also been proposed [Abbadi et al. 1985]. However, this protocol makes unrealistic assumptions about the underlying communication system.

Single copy equivalence replication, often called eager replication, is typically implemented using 2PC. Whenever a transaction updates one replica, all other replicas are updated inside the same transaction as a distributed transaction. Therefore, mutual consistency of replicas and strong consistency are enforced. However, it reduces availability as all the nodes must be operational. In addition, synchronous protocols may block due to network or node failures. Finally, to commit a transaction with 2PC, the number of messages exchanged to control transaction commitment is quite significant and, as a consequence, transaction response times may be extended as the number of nodes increases. A solution proposed in Kemme and Alonso [2000] reduces the number of messages exchanged to commit transactions compared to 2PC, but the protocol is still blocking and it is not clear if it scales up. For these reasons, eager replication is less and less used in practice.

Lazy replication is the most widely used form of replication in distributed databases. With lazy replication, a transaction can commit after updating one replica copy at some node. After the transaction commits, the updates are propagated to the other replicas, and these replicas are updated in separate transactions. Different from eager replication, the mutual consistency of replicas is relaxed and strong consistency is not assured. A major virtue of lazy replication is its easy deployment because is avoids all the constraints of eager replication [Gray et al. 1996]. In particular, it can scale up to large configurations such as cluster systems.

In lazy replication, a primary copy is stored at a master node and secondary copies are stored in slave nodes. A primary copy that may be stored at and updated by different master nodes is called a multi-owner copy. These are stored in multi-owner nodes and a multi-master configuration consists of a set of multi-owner nodes on a common set of multi-owner copies. Several configurations such as lazy master, multi-master, and hybrid configurations (combining lazy master and multi-master) are possible. Multi-master replication provides the highest level of data availability: a node failure does not block updates on the replicas it carries. Replication solutions that assure strong consistency for lazy master are proposed in Pacitti et al. [1999], Pacitti et al. [2001], and Pacitti and Simon [2000]. A solution to provide strong consistency for fully replicated multi-master configurations, in the context of cluster systems, is also proposed in Pacitti et al. [2003].

58.3.6 Data Placement

In a parallel database system, proper data placement is essential for load balancing. Ideally, interference between concurrent parallel operations can be avoided by having each operation work on an independent dataset. These independent datasets can be obtained by the *partitioning* of the relations based on a function (hash function or range index) applied to some placement attribute(s), and allocating each partition to a different disk. As with horizontal fragmentation in distributed databases, partitioning is useful to obtain inter-query parallelism, by having independent queries working on different partitions; and intra-query parallelism, by having a query's operations working on different partitions. Partitioning can be single-attribute or multi-attribute. In the latter case [Ghandeharizadeh et al. 1992], an exact match query requiring the equality of multi-attributes can be processed by a single node without communication. The choice between hashing or range index for partitioning is a design issue: hashing incurs less storage overhead but provides direct support for exact-match queries only, while range index can also support range queries. Initially proposed for shared-nothing systems, partitioning has been shown to be useful for shared-memory designs as well, by reducing memory access conflicts [Bergsten et al. 1991].

Full partitioning, whereby each relation is partitioned across all the nodes, causes problems for small relations or systems with large numbers of nodes. A better solution is *variable partitioning*, where each relation is stored on a certain number of nodes as a function of the relation size and access frequency

[Copeland et al. 1988]. This can be combined with multirelation clustering to avoid the communication overhead of binary operations.

When the criteria used for data placement change to the extent that load balancing degrades significantly, dynamic reorganization is required. It is important to perform such dynamic reorganization online (without stopping the incoming of transactions) and efficiently (through parallelism). By contrast, existing database systems perform static reorganization for database tuning [Shasha and Bonnet 2002]. Static reorganization takes place periodically when the system is idle to alter data placement according to changes in either database size or access patterns. In contrast, dynamic reorganization does not need to stop activities and adapts gracefully to changes. Reorganization should also remain transparent to compiled programs that run on the parallel system. In particular, programs should not be recompiled because of reorganization. Therefore, the compiled programs should remain independent of data location. This implies that the actual disk nodes where a relation is stored or where an operation will actually take place can be known only at runtime.

Data placement must also deal with data replication for high availability. A naive approach is to maintain two copies of the same data, a primary and a backup copy, on two separate nodes. However, in case of a node failure, the load of the node having the copy may double, thereby hurting load balancing. To avoid this problem, several high-availability data replication strategies have been proposed [Hsiao and DeWitt 1991]. An interesting solution is Teradata's interleaved partitioning, which partitions the backup copy on a number of nodes. In failure mode, the load of the primary copy is balanced among the backup copy nodes. However, reconstructing the primary copy from its separate backup copies may be costly. In normal mode, maintaining copy consistency may also be costly. A better solution is Gamma's chained partitioning, which stores the primary and backup copy on two adjacent nodes. In failure mode, the load of the failed node and the backup nodes are balanced among all remaining nodes using both primary and backup copy nodes. In addition, maintaining copy consistency is cheaper. Fractured mirrors [Ramamurthy et al. 2002] go one step further in storing the two copies in two different formats, each as is and the other decomposed, in order to improve access performance.

58.3.7 Query Processing and Optimization

Query processing is the process by which a declarative query is translated into low-level data manipulation operations. SQL is the standard query language that is supported in current DBMSs. **Query optimization** refers to the process by which the "best" execution strategy for a given query is found from among a set of alternatives.

In centralized DBMSs, the process typically involves two steps: *query decomposition* and *query optimization*. Query decomposition takes an SQL query and translates it into one expressed in relational algebra. For a given SQL query, there are more than one possible algebraic queries. Some of these algebraic queries are "better" than others. The quality of an algebraic query is defined in terms of expected performance. The traditional procedure is to obtain an initial algebraic query by translating the predicates and the target statement into relational operations as they appear in the query. This initial algebraic query is then transformed, using algebraic transformation rules, into other algebraic queries until the "best" one is found. The "best" algebraic query is determined according to a cost function that calculates the cost of executing the query according to that algebraic specification. This is the process of query optimization.

In distributed DBMSs, two more steps are involved between query decomposition and query optimization: *data localization* and *global query optimization*.

The input to data localization is the initial algebraic query generated by the query decomposition step. The initial algebraic query is specified on global relations irrespective of their fragmentation or distribution. The main role of data localization is to localize the query's data using data distribution information. In this step, the fragments involved in the query are determined and the query is transformed into one that operates on fragments rather than global relations. As indicated earlier, fragmentation is defined through fragmentation rules that can be expressed as relational operations (horizontal fragmentation by selection, vertical fragmentation by projection). A distributed relation can be reconstructed by applying the inverse of the fragmentation rules. This is called a *localization program*. The localization program for a horizontally (vertically) fragmented query is the union (join) of the fragments. Thus, during the data localization step, each global relation is first replaced by its localization program, and then the resulting fragment query is simplified and restructured to produce another "good" query. Simplification and restructuring may be done according to the same rules used in the decomposition step. As in the decomposition step, the final fragment query is generally far from optimal; the process has only eliminated "bad" algebraic queries.

The input to the third step is a fragment query, that is, an algebraic query on fragments. The goal of query optimization is to find an execution plan for the query which is close to optimal. Remember that finding the optimal solution is computationally intractable. An execution plan for a distributed query can be described with *relational algebra operations* and *communication primitives* (send/receive operations) for transferring data between sites. The previous layers have already optimized the query — for example, by eliminating redundant expressions. However, this optimization is independent of fragment characteristics such as cardinalities. In addition, communication operations are not yet specified. By permuting the ordering of operations within one fragment query, many equivalent query execution plans may be found. Query optimization consists of finding the "best" one among candidate plans examined by the optimizer.* The query optimizer is usually seen as three components: a search space, a cost model, and a search strategy. The *search space* is the set of alternative execution plans to represent the input query. These plans are equivalent, in the sense that they yield the same result but they differ on the execution order of operations and the way these operations are implemented. The *cost model* predicts the cost of a given execution plan. To be accurate, the cost model must have accurate knowledge about the parallel execution environment. The *search strategy* explores the search space and selects the best plan. It defines which plans are examined and in which order.

In a distributed environment, the cost function, often defined in terms of time units, refers to computing resources such as disk space, disk I/Os, buffer space, CPU cost, communication cost, etc. Generally, it is a weighted combination of I/O, CPU, and communication costs. To select the ordering of operations, it is necessary to predict execution costs of alternative candidate orderings. Determining execution costs before query execution (i.e., static optimization) is based on fragment statistics and the formulas for estimating the cardinalities of results of relational operations. Thus, the optimization decisions depend on the available statistics on fragments. An important aspect of query optimization is *join ordering*, because permutations of the joins within the query may lead to improvements of several orders of magnitude. One basic technique for optimizing a sequence of distributed join operations is through use of the semijoin operator. The main value of the semijoin in a distributed system is to reduce the size of the join operands and thus the communication cost.

Parallel query optimization exhibits similarities with distributed query processing. It takes advantage of both intra-operation parallelism and inter-operation parallelism. Intra-operation parallelism is achieved by executing an operation on several nodes of a multiprocessor machine. This requires that the operands have been previously partitioned across the nodes. The set of nodes where a relation is stored is called its *home*. The *home of an operation* is the set of nodes where it is executed and it must be the home of its operands in order for the operation to access its operands. For binary operations such as join, this might imply repartitioning one of the operands. The optimizer might even sometimes find that repartitioning both the operands is useful. Parallel optimization to exploit intra-operation parallelism can make use of some of the techniques devised for distributed databases.

Inter-operation parallelism occurs when two or more operations are executed in parallel, either as a dataflow or independently. We designate as *dataflow* the form of parallelism induced by *pipelining*. *Independent* parallelism occurs when operations are executed at the same time or in arbitrary order. Independent parallelism is possible only when the operations do not involve the same data.

---

*The difference between an optimal plan and the best plan is that the optimizer does not, because of computational intractability, examine all of the possible plans.

There is a necessary trade-off between optimization cost and quality of the generated execution plans. Higher optimization costs are probably acceptable to produce "better" plans for repetitive queries, because this would reduce query execution cost and amortize the optimization cost over many executions. However, high optimization cost is unacceptable for ad hoc queries, which are executed only once. The optimization cost is mainly incurred by searching the solution space for alternative execution plans. In a parallel system, the solution space can be quite large because of the wide range of distributed execution plans. The crucial issue in terms of search strategy is the join ordering problem, which is NP-complete in the number of relations [Ibaraki and Kameda 1984]. A typical approach to solving the problem is to use dynamic programming [Selinger et al. 1979], which is a *deterministic* strategy. This strategy is almost exhaustive and assures that the best of all plans is found. It incurs an acceptable optimization cost (in terms of time and space) when the number of relations in the query is small. However, this approach becomes too expensive when the number of relations is greater than 5 or 6. For this reason, there is interest in *randomized* strategies, which reduce the optimization complexity but do not guarantee the best of all plans. Randomized strategies investigate the search space in a way that can be fully controlled such that optimization ends after a given optimization time budget has been reached. Another way to cut off optimization complexity is to adopt a heuristic approach. Unlike deterministic strategies, *randomized* strategies allow the optimizer to trade optimization time for execution time [Ioannidis and Wong 1987, Swami and Gupta 1988, Ioannidis and Kang 1990, Lanzelotte et al. 1993].

58.3.8 Load Balancing

Load balancing is distributing the amount of work the parallel system has to do between all nodes so that more work gets done in the same amount of time and, thus, all users get served faster. In a parallel database system, load balancing can be done at three levels: intra-operator, inter-operator, and inter-query.

Load balancing problems can appear with intra-operator parallelism because of the variation in partition size, namely *data skew*. The effects of skewed data distribution on parallel execution are classified in Walton et al. [1991]. *Attribute value skew (AVS)* is skew inherent in the dataset while *tuple placement skew (TPS)* is the skew introduced when the data is initially partitioned (e.g., with range partitioning). *Selectivity skew (SS)* is introduced when there is variation in the selectivity of select predicates on each node. *Redistribution skew (RS)* occurs in the redistribution step between two operators. It is similar to TPS. Finally, *join product skew (JPS)* occurs because the join selectivity may vary between nodes. Capturing these various skews in a static cost model is hard and error-prone. A more reasonable strategy is to use a dynamic approach, that is, redistribute the load dynamically in order to balance execution.

The negative impact of skew on intra-operator parallelism can be reduced with specific algorithms. A robust hash-join algorithm is proposed in Kitsuregawa and Ogawa [1990] for a shared-nothing architecture: the idea is to further partition the large hash buckets among the processors. Another solution [DeWitt et al. 1992] is to have several algorithms, each specialized for a different degree of skew, and to use a small sample of the relations to determine which algorithm is best. Distributed shared memory (implemented by software) can also help load balancing [Shatdal and Naughton 1993]. When a processor is idle, it steals work from a randomly chosen processor using distributed shared memory.

Load balancing at the inter-operator level is more involved [Wilshut et al. 1995]. First, the degree of parallelism and the allocation of processors to operators, decided in the query optimization phase, are based on a static cost model that may be inaccurate. Second, the choice of the degree of parallelism is subject to errors because both processors and operators are discrete entities. Finally, the processors associated with the latest operators in a pipeline chain may remain idle a significant time. This is called the pipeline delay problem. These problems stem from the fixed association between data, operators, and processors.

Thus, the solution is to differ the choice of these associations until runtime with more dynamic strategies. In Mehta and DeWitt [1995], the processors to run each operator are dynamically determined (just prior to execution) based on a cost model that matches the rate at which tuples are produced and consumed. Other load balancing algorithms are proposed in Rahm and Marek [1995] and Garofalakis and Ioanidis [1996] using statistics on processor usage.

In the context of hierarchical systems (i.e., shared-nothing systems with shared-memory nodes), load balancing is exacerbated because it must be addressed at two levels, locally among the processors of each shared-memory node and globally among all nodes. The *Dynamic Processing (DP)* execution model [Bouganim et al. 1996] proposes a solution for intra- and inter-operator load balancing. The idea is to break the query into self-contained units of sequential processing, each of which can be carried out by any processor. The main advantage is to minimize the communication overhead of inter-node load balancing by maximizing intra- and inter-operator load balancing within shared-memory nodes.

Parallel database systems typically perform load balancing at the operator level (inter- and intra-operator), which is the finest way to optimize the execution of complex queries (with many operators). This is possible only because the database system has full control over the data. However, cluster systems are now being used for managing autonomous databases, for instance, in the context of an application service provider (ASP). In the ASP model, customers' applications and databases are hosted at the provider site and should work as if they were local to the customers' sites. Thus, they should remain autonomous and unchanged after migration to the provider site's cluster. Using a parallel DBMS such as Oracle Rapid Application Cluster or DB2 Parallel Edition is not acceptable because it requires heavy migration and hurts application and database autonomy [Gançarski et al. 2002]. Given such autonomy requirements, the challenge is to fully exploit the cluster resources, in particular parallelism, in order to optimize performance. In a cluster of autonomous databases, load balancing can only be done at the coarser level of inter-query. In a shared-disk cluster architecture, load balancing is easy and can use a simple round-robin algorithm that selects, in turn, each processor to run an incoming query. The problem is more difficult in the case of a shared-nothing cluster because the queries need be routed to the nodes that hold the requested data. The typical solution is to replicate data at different nodes so that users can be served by any of the nodes, depending on the current load. This also provides high-availability because, in the event of a node failure, other nodes can still do the work. With a replicated database organization, executing update queries in parallel at different nodes can make replicas inconsistent. The solution proposed in Gançarski et al. [2002] allows the administrator to control the trade-off between consistency and performance based on users' requirements. Load balancing is then achieved by routing queries to the nodes with the required consistency and the least load. To further improve performance of query execution, query routing can also be cache-aware [Röhm et al. 2001]. The idea is to base the routing decision on the states of the node caches in order to minimize disk accesses. In a cluster of autonomous databases (with black-box DBMS components), the main issue is to estimate the cache benefit from executing a query at a node, without any possibility of access to the directory caches. The solution proposed in Röhm et al. [2001] makes use of predicate signatures that approximate the data ranges accessed by a query.

58.4 Research Issues

Distributed and parallel DBMS technologies have matured to the point where fairly sophisticated and reliable commercial systems are now available. As expected, there are a number of issues that have yet to be satisfactorily resolved. In this section we provide an overview of some of the more important research issues.

58.4.1 Mobile Databases

Mobile computing promises to access information anywhere on the network, anytime, and from any kind of mobile appliance (digital assistant, cellular phone, smartcard, etc.). Such information ubiquity makes it possible to envision new personal and professional applications that will have a strong impact on distributed information systems. For instance, traveling employees could access, wherever they are, their company's or co-workers' data. Distributed database technologies have been designed for fixed clients and servers connected by a wired network. The properties of mobile environments (low bandwidths of wireless networks, frequent disconnections, limited power of mobile appliances, etc.) change radically the assumptions underlying these technologies. Mobile database management (i.e., providing database functions in a mobile environment) is thus becoming a major research challenge [Helal et al. 2002]. In this section, we only mention the specific issues of distributed data management, and we ignore other important issues such as scaling down database techniques to build picoDBMS that fit in a very small device [Pucheral et al. 2001].

New distributed architectures must be designed that encompass the various levels of mobility (mobile user, mobile data, etc.) and wireless networks with handheld (lightweight) terminals, connected to much faster wire networks with database servers. Various distributed database components must be defined for the mobile environment, for example clients, data sources and wrappers, mediators, client proxies (representing mobile clients on the wire network), mediator proxies (representing the mediator on the mobile clients), etc.

Traditional client server and three-tier models are not well suited to mobile environments as servers are central points of failure and bottlenecks. Other models are better. For example, the publish-subscribe model enables clients to be notified only when an event corresponding to a subscription is published. Or a server could repeatedly broadcast information to clients who will eventually listen. These models can better optimize the bandwidth and deal with disconnections. One major issue is to scale up to very large numbers of clients.

Localizing mobile clients (e.g., cellular phones) and data accessed by the clients requires databases that deal with moving objects. Capturing moving objects efficiently is a difficult problem because it must take into account spatial and temporal dimensions, and impacts query processing.

Mobile computing suggests connected and disconnected working phases, with asynchronously replicated data. The replication model is typically symmetric (multi-master) between the mobile client and the database server, with copy divergence after disconnection. Copy synchronization is thus necessary after reconnection. Open problems are the definition of a generic synchronization model, for all kinds of objects, and the scaling up of the reconciliation algorithms.

Synchronization protocols guarantee a limited degree of copy consistency. However, there may be applications that need stronger consistency with ACID properties. In this case, transactions could be started on mobile clients and their execution distributed between clients and servers. The possible disconnection of mobile clients and the unbounded duration of disconnection suggests reconsideration of the traditional distributed algorithms to support the ACID properties.

58.4.2 Large-Scale Query Processing

Database technology is now mature enough to support all kinds of data, including complex structured documents and multimedia objects. Distributed and parallel technologies now make it possible to build very large-scale systems, for example, connecting millions of sites over the Web or thousands of nodes in shared-nothing cluster systems. These combined advances are creating the need for large-scale distributed database systems that federate very large numbers of autonomous databases. The number of potential users, some with handheld devices, is also growing exponentially. This makes query processing a very challenging problem.

Distributed query processing typically performs static query optimization, using a cost model, followed by query execution. We already discussed the problems of static query optimization in parallel database systems (skew data distributions, inaccurate estimates, etc.) that can make execution plans inefficient and some solutions. In a large-scale federated system, these problems are exacerbated by the fact that autonomous databases do not easily export cost-based information. Furthermore, workloads submitted by large communities of users are highly unpredictable. Another problem is the need to deal with data and programs in a more integrated way. In scientific applications, for example, a common requirement [Tanaka and Valduriez 2001] is the ability to process over the network data objects that can be very large (e.g., satellite images) by scientific user programs that can be very long running (e.g., image analysis). User programs can be simply modeled as expensive user-defined predicates and included in SQL-like queries. In this context, static query optimization does not work. In particular, predicates are evaluated over program execution results that do not exist beforehand. Furthermore, program inputs (e.g., satellite images) range within an unconstrained domain, which makes statistical estimation difficult.

To adapt to runtime conditions, fully dynamic query processing techniques are needed, which requires us to completely revisit more than two decades of query optimizer technology. The technique proposed in Bouganim et al. [2001] for mediator systems adapts to the variances in estimated selectivity and cost of expensive predicates and supports nonuniform data distributions. It also exploits parallelism. Eddies [Avnur and Hellerstein 2000] have also been proposed as an adaptive query execution framework where query optimization and execution are completely mixed. Data is directed toward query operators, depending on their actual consume/produce efficiency. Eddies can also be used to dynamically process expensive predicates. More work is needed to introduce learning capabilities within dynamic query optimization.

Another area that requires much more work is load balancing. Trading consistency for performance based on user requirements is a promising approach [Gançarski et al. 2002]. It should be useful in many P2P applications where consistency is not a prime requirement.

Finally, introducing query and query processing capabilities within P2P Systems presents new challenges [Harren et al. 2002]. Current P2P systems on the Web employ a distributed hash table (DHT) to locate objects in a scalable way. However, DHT is good only for exact-match. More work is needed to extend the current techniques to support complex query capabilities.

58.5 Summary

Distributed and parallel DBMSs have become a reality. They provide the functionality of centralized DBMSs, but in an environment where data is distributed over the sites of a computer network or the nodes of a multiprocessor system. Distributed databases have enabled the natural growth and expansion of databases by the simple addition of new machines. The price-performance characteristics of these systems are favorable, in part due to the advances in computer network technology. Parallel DBMSs are perhaps the only realistic approach to meet the performance requirements of a variety of important applications that place significant throughput demands on the DBMS. To meet these requirements, distributed and parallel DBMSs need to be designed with special consideration for the protocols and strategies. In this chapter, we have provided an overview of these protocols and strategies.

One issue that we omitted is distributed object-oriented databases. The penetration of database management technology into areas (e.g., engineering databases, multimedia systems, geographic information systems, image databases) that relational database systems were not designed to serve has given rise to a search for new system models and architectures. A primary candidate for meeting the requirements of these systems is the object-oriented DBMS [Dogac et al. 1994]. The distribution of object-oriented DBMSs gives rise to a number of issues generally categorized as distributed object management [Özsu et al. 1994]. We have ignored both multidatabase system and distributed object management issues in this chapter.

Defining Terms

Atomicity: The property of transaction processing whereby either all the operations of a transaction are executed or none of them are (all-or-nothing).

Client/server architecture: A distributed/parallel DBMS architecture where a set of client machines with limited functionality access a set of servers that manage data.

Concurrency control algorithm: An algorithms that synchronizes the operations of concurrent transactions that execute on a shared database.

Data independence: The immunity of application programs and queries to changes in the physical organization (physical data independence) or logical organization (logical data independence) of the database, and vice versa.

Deadlock: An occurrence where each transaction in a set of transactions circularly waits on locks that are held by other transactions in the set.

Distributed database management system: A database management system that manages a database that is distributed across the nodes of a computer network and makes this distribution transparent to the users.

Durability: The property of transaction processing whereby the effects of successfully completed (i.e., committed) transactions endure subsequent failures.

Inter-query parallelism: The parallel execution of multiple queries generated by concurrent transactions.

Intra-operation parallelism: The execution of one relational operation as many sub-operations.

Intra-query parallelism: The parallel execution of multiple, independent operations possible within the same query.

Isolation: The property of transaction execution that states that the effects of one transaction on the database are isolated from other transactions until the first completes its execution.

Linear scaleup: Sustained performance for a linear increase in both database size and processing and storage power.

Linear speedup: Linear increase in performance for a constant database size and linear increase in processing and storage power.

Locking: A method of concurrency control where locks are placed on database units (e.g., pages) on behalf of transactions that attempt to access them.

Logging protocol: The protocol that records, in a separate location, the changes that a transaction makes to the database before the change is actually made.

One copy equivalence: Replica control policy that asserts that the values of all copies of a logical data item should be identical when the transaction that updates that item terminates.

Parallel database management system: A database management system that is implemented on a tightly coupled multiprocessor.

Query optimization: The process by which the "best" execution plan for a given query is found from among a set of alternatives.

Query processing: The process by which a declarative query is translated into low-level data manipulation operations.

Quorum-based voting algorithm: A replica control protocol where transactions collect votes to read and write copies of data items. They are permitted to read or write data items if they can collect a quorum of votes.

Read-Once/Write-All protocol: The replica control protocol that maps each logical read operation to a read on one of the physical copies and maps a logical write operation to a write on all of the physical copies.

Serializability: The concurrency control correctness criterion that requires that the concurrent execution of a set of transactions should be equivalent to the effect of some serial execution of those transactions.

Shared-disk architecture: A parallel DBMS architecture where any processor has access to any disk unit through the interconnect but exclusive (non-shared) access to its main memory.

Shared-memory architecture: A parallel DBMS architecture where any processor has access to any memory module or disk unit through a fast interconnect (e.g., a high-speed bus or a cross-bar switch).

Shared-nothing architecture: A parallel DBMS architecture where each processor has exclusive access to its main memory and disk unit(s).

Stable database: The portion of the database that is stored in secondary storage.

Termination protocol: A protocol by which individual sites can decide how to terminate a particular transaction when they cannot communicate with other sites where the transaction executes.

Transaction: A unit of consistent and atomic execution against the database.

Transparency: Extension of data independence to distributed systems by hiding the distribution, fragmentation, and replication of data from the users.

Two-phase commit: An atomic commitment protocol that ensures that a transaction is terminated the
same way at every site where it executes. The name comes from the fact that two rounds of messages
are exchanged during this process.

Two-phase locking: A locking algorithm where transactions are not allowed to request new locks once
they release a previously held lock.

Volatile database: The portion of the database that is stored in main memory buffers.

References

[Abbadi et al. 1985] A.E. Abbadi, D. Skeen, and F. Cristian. "An Efficient, Fault-Tolerant Protocol for
Replicated Data Management," in *Proc. 4th ACM SIGACT–SIGMOD Symp. on Principles of Database
Systems*, Portland, OR, March 1985, pp. 215–229.

[Apers et al. 1992] P. Apers, C. van den Berg, J. Flokstra, P. Grefen, M. Kersten, and A. Wilschut.
"Prisma/DB: a Parallel Main-Memory Relational DBMS," *IEEE Trans. on Data and Knowledge Eng.*,
(1992), 4(6):541–554.

[Avnur and Hellerstein 2000] R. Avnur and J. Hellerstein. "Eddies: Continuously Adaptive Query Processing," *Proc. ACM SIGMOD Int. Conf. on Management of Data*, Dallas, May 2000, pp. 261–272

[Balakrishnan et al. 2003] H. Balakrishnan, M.F. Kaashoek, D. Karger, R. Morris, and I. Stoica. "Looking
Up Data in P2P Systems," *Commun. of the ACM*, (2003), 46(2):43–48.

[Bell and Grimson 1992] D. Bell and J. Grimson. *Distributed Database Systems*, Reading, MA: Addison-
Wesley, 1992.

[Bergsten et al. 1991] B. Bergsten, M. Couprie, and P. Valduriez. "Prototyping DBS3, a Shared-Memory
Parallel Database System," in *Proc. Int. Conf. on Parallel and Distributed Information Systems*, Miami,
December 1991, pp. 226–234.

[Bernstein and Newcomer 1997] P. A. Bernstein and E. Newcomer. *Principles of Transaction Processing*.
Morgan Kaufmann, 1997.

[Boral et al. 1990] H. Boral, W. Alexander, L. Clay, G. Copeland, S. Danforth, M. Franklin, B. Hart, M.
Smith, and P. Valduriez. "Prototyping Bubba, a Highly Parallel Database System," *IEEE Trans. on
Knowledge and Data Engineering*, (March 1990), 2(1):4–24.

[Bouganim et al. 1996] L. Bouganim, D. Florescu, and P. Valduriez. "Dynamic Load Balancing in Hierarchical Parallel Database Systems," in *Proc. 22th Int. Conf. on Very Large Data Bases*, Bombay,
September 1996, pp. 436–447.

[Bouganim et al. 1999] L. Bouganim, D. Florescu, and P. Valduriez. Multi-Join Query Execution
with Skew in NUMA Multiprocessors, *Distributed and Parallel Database Systems*, (1999),
7(1):99–121.

[Bouganim et al. 2001] L. Bouganim, F. Fabret, F. Porto, and P. Valduriez, "Processing Queries with Expensive Functions and Large Objects in Distributed Mediator Systems," In *Proc. 17th IEEE Int. Conf.
on Data Engineering*, Heidelberg, April 2001, pp. 91–98.

[Carey et al. 1994] M. Carey et al. "Shoring Up Persistent Applications," *Proc. ACM SIGMOD Int. Conf. on
Management of Data*, Minneapolis, June 1994, pp. 383–394.

[Ceri and Pelagatti 1984] S. Ceri and G. Pelagatti. *Distributed Databases: Principles and Systems*. New York:
McGraw-Hill, 1984.

[Ceri et al. 1987] S. Ceri, B. Pernici, and G. Wiederhold. "Distributed Database Design Methodologies,"
Proc. IEEE, (May 1987), 75(5):533–546.

[Copeland et al. 1988] G. Copeland, W. Alexander, E. Boughter, and T. Keller. "Data Placement in Bubba,"
In *Proc. ACM SIGMOD Int. Conf. on Management of Data*, Chicago, May 1988, pp. 99–108.

[DeWitt et al. 1990] D.J. DeWitt, S. Ghandeharizadeh, D.A. Schneider, A. Bricker, H.-I. Hsiao, and
R. Rasmussen. "The GAMMA Database Machine Project," *IEEE Trans. on Knowledge and Data
Eng.*, (March 1990), 2(1):44–62.

[DeWitt and Gray 1992] D. DeWitt and J. Gray. "Parallel Database Systems: The Future of High-Performance Database Systems," *Commun. of ACM*, (June 1992), 35(6):85–98.

[DeWitt et al. 1992] D.J. DeWitt, J.F. Naughton, D.A. Schneider, and S. Seshadri. "Practical Skew Handling in Parallel Joins," in *Proc. 22th Int. Conf. on Very Large Data Bases*, Vancouver, August 1992, pp. 27–40.

[Dogac et al. 1994] A. Dogac, M.T. Özsu, A. Biliris, and T. Sellis (Eds.). *Advances in Object-Oriented Database Systems*, Berlin: Springer-Verlag, 1994.

[Elmagarmid 1992] A.K. Elmagarmid (Ed.). *Transaction Models for Advanced Database Applications*. San Mateo, CA: Morgan Kaufmann, 1992.

[Frank et al. 1993] S. Frank, H. Burkhardt, and J. Rothnie. The KSR1: Bridging the Gap Between Shared-Memory and MPPs. In *Proc. CompCon'93*, San Francisco, February 1993.

[Freytag 1987] J.-C. Freytag. "A Rule-based View of Query Optimization," in *Proc. ACM SIGMOD Int. Conf. on Management of Data*, San Francisco, 1987, pp. 173–180.

[Freytag et al. 1993] J.-C. Freytag, D. Maier, and G. Vossen. *Query Processing for Advanced Database Systems*. San Mateo, CA: Morgan Kaufmann, 1993.

[Fushimi et al. 1986] S. Fushimi, M. Kitsuregawa, and H. Tanaka. "An Overview of the System Software of a Parallel Relational Database Machine GRACE," in *Proc. 12th Int. Conf. on Very Large Data Bases*, Kyoto, August 1986, pp. 209–219.

[Gançarski et al. 2002] S. Gançarski, H. Naacke, E. Pacitti, and P. Valduriez. "Parallel Processing with Autonomous Databases in a Cluster System," in *Proc. Confederated International Conferences, DOA, CoopIS and ODBASE*, Irvine, CA, 2002, pp. 410–428.

[Garcia-Molina and Lindsay 1990] H. Garcia-Molina and B. Lindsay. "Research Directions for Distributed Databases," *IEEE Q. Bull. Database Eng.* (December 1990), 13(4):12–17.

[Garofalakis and Ioanidis 1996] M.N. Garofalakis and Y.E. Ioannidis. "Multi-dimensional Resource Scheduling for Parallel Queries," in *Proc. ACM SIGMOD Int. Conf. on Management of Data*, Montreal, June 1996, pp. 365–376.

[Ghandeharizadeh et al. 1992] S. Ghandeharizadeh, D. DeWitt, and W. Quresh. "A Performance Analysis of Alternative Multi-Attributed Declustering Strategies," *ACM SIGMOD Int. Conf. on Management of Data*, San Diego, CA, June 1992, pp. 29–38.

[Gifford 1979] D.K. Gifford. "Weighted Voting for Replicated Data," in *Proc. 7th ACM Symp. on Operating System Principles*, Pacific Grove, CA, December 1979, pp. 150–159.

[Graefe 1990] G. Graefe. "Encapsulation of Parallelism in the Volcano Query Processing Systems," in *Proc. ACM SIGMOD Int. Conf.*, Atlantic City, NJ, May 1990, pp. 102–111.

[Graefe 1993] G. Graefe. "Query Evaluation Techniques for Large Databases," *ACM Comp. Surv.* (June 1993), 25(2):73–170.

[Gray 1979] J.N. Gray. "Notes on Data Base Operating Systems," In *Operating Systems: An Advanced Course*, R. Bayer, R.M. Graham, and G. Seegmüller (Eds.), New York: Springer-Verlag, 1979, pp. 393–481.

[Gray 1981] J. Gray. "The Transaction Concept: Virtues and Limitations," in *Proc. 7th Int. Conf. on Very Large Data Bases*, Cannes, France, September 1981, pp. 144–154.

[Gray and Reuter 1993] J. Gray and A. Reuter. *Transaction Processing: Concepts and Techniques*. San Mateo, CA: Morgan Kaufmann, 1993.

[Gray et al. 1996] J. Gray, P. Helland, P. O'Neil, and D. Shasha. "The Danger of Replication and a Solution," in *Proc. ACM SIGMOD Int. Conf. on Management of Data*, Montreal, 1996, pp. 173–182.

[Hagersten et al. 1992] E. Hagersten, E. Landin, and S. Haridi. "Ddm – a Cache-Only Memory Architecture," *IEEE Computer*, (September 1992), 25(9): pp. 44–54.

[Harren et al. 2002] M. Harren, J. Hellerstein, R. Huebsch, B. Thau Loo, S. Shenker, and Ion Stoica. "Complex Queries in DHT-Based Peer-to-Peer Networks," in *Peer-to-Peer Systems, 1st Int. Workshop*, Cambridge, 2002, pp. 242–259.

[Helal et al. 2002] A. Helal, B. Haskell, J. Carter, R. Brice, D. Woelk, and M. Rusinkiewicz. *Any Time, Anywhere Computing: Mobile Computing Concepts and Technology*. Kluwer Academic, 2002.

[Hong 1992] W. Hong. "Exploiting Inter-Operation Parallelism in XPRS,". in *Proc. ACM SIGMOD Int. Conf. on Management of Data*, San Diego, June 1992, pp. 19–28.

[Hsiao and DeWitt 1991] H.-I. Hsiao and D. DeWitt. "A Performance Study of Three High-Availability Data Replication Strategies," in *Proc. Int. Conf. on Parallel and Distributed Information Systems*, Miami, 1991, pp. 18–28.

[Ibaraki and Kameda 1984] T. Ibaraki and T. Kameda. "On the Optimal Nesting Order for Computing N-Relation Joins," *ACM Trans. Database Syst.*, (September 1984), 9(3):482–502.

[Ioannidis and Wong 1987] Y. Ioannidis and E. Wong. "Query Optimization by Simulated Annealing," in *Proc. of the ACM SIGMOD Int. Conf. on Management of Data*, 1987, pp. 9–22.

[Ioannidis and Kang 1990] Y. Ioannidis and Y.C. Kang. "Randomized Algorithms for Optimizing Large Join Queries," in *Proc. of the ACM SIGMOD Int. Conf. on Management of Data*, 1990, pp. 312–321.

[Kemme and Alonso 2000] B. Kemme and G. Alonso. "Don't Be Lazy Be Consistent: Postgres-R, a New Way to Implement Database Replication," in *Proc. 26th Int. Conf. on Very Large Databases*, Cairo, 2000, pp. 134–143.

[Kitsuregawa and Ogawa 1990] M. Kitsuregawa and Y. Ogawa. "Bucket Spreading Parallel Hash: A New, Robust, Parallel Hash Join Method for Data Skew in the Super Database Computer," in *Proc. 16th Int. Conf. on Very Large Data Bases*, Brisbane, Australia, 1990, pp. 210–221.

[Lanzelotte et al. 1993] R. Lanzelotte, P. Valduriez, and M. Zait. "On the Effectiveness of Optimization Search Strategies for Parallel Execution Spaces," in *Proc. 19th Int. Conf. on Very Large Data Bases*, Dublin, August 1993, pp. 493–504.

[Lenoski et al. 1992] D. Lenoski et al. "The Stanford Dash Multiprocessor," *IEEE Computer*, (March 1992), 25(3):63–79.

[Lenzerini 2002] M. Lenzerini. "Data Integration: A Theoretical Perspective," in *Proc. 21st ACM SIGMOD–SIGACT–SIGSART Symp. on Principles of Database Systems*, Madison, WI, June 2002, pp. 233–246.

[Liu et al. 1996] L. Liu, C. Pu, R. Barga, and T. Zhou. "Differential Evaluation of Continual Queries," in *Proc. IEEE Int. Conf. Dist. Comp. Syst.*, May 1996, pp. 458–465.

[Mehta and DeWitt 1995] M. Mehta and D. DeWitt. "Managing Intra-operator Parallelism in Parallel Database Systems," in *Proc. 21st Int. Conf. on Very Large Data Bases*, Zurich, September 1995, pp. 382–394.

[Mohan and Lindsay 1983] C. Mohan and B. Lindsay. "Efficient Commit Protocols for the Tree of Processes Model of Distributed Transactions," in *Proc. 2nd ACM SIGACT–SIGMOD Symp. on Principles of Distributed Computing*, 1983, pp. 76–88.

[Orfali et al. 1994] R. Orfali, D. Harkey, and J. Edwards. *Essential Client/Server Survival Guide*, New York: John Wiley, 1994.

[Özsu, 1994] M.T. Özsu. "Transaction Models and Transaction Management in Object-Oriented Database Management Systems," in *Advances in Object-Oriented Database Systems*, A. Dogac, M.T. Özsu, A. Biliris, and T. Sellis (Eds.), Berlin: Springer-Verlag, 1994, pp. 147–183.

[Özsu and Valduriez 1991] M.T. Özsu and P. Valduriez. "Distributed Database Systems: Where Are We Now?," *IEEE Computer*, (August 1991), 24(8):68–78.

[Özsu and Valduriez 1999] M.T. Özsu and P. Valduriez. *Principles of Distributed Database Systems*, 2nd edition, Englewood Cliffs, NJ: Prentice Hall, 1999.

[Özsu et al., 1994] M.T. Özsu, U. Dayal, and P. Valduriez (Eds.). *Distributed Object Management*, San Mateo CA: Morgan Kaufmann, 1994.

[Pacitti et al. 1999] E. Pacitti, P. Minet, and E. Simon. "Fast Algorithms for Maintaining Replica Consistency in Lazy Master Replicated Databases," in *Proc. 25th Int. Conf. on Very Large Data Bases*, Edimburg, 1999, pp. 126–137.

[Pacitti et al. 2001] E. Pacitti, P. Minet, and E. Simon. "Replica Consistency in Lazy Master Replicated Databases," *Distributed and Parallel Databases*, (2001), 9(3):237–267.

[Pacitti et al. 2003] E. Pacitti, T. Özsu, and C. Coulon. "Preventive Multi-Master Replication in a Cluster of Autonomous Databases," in *Proc. Int. Conf. on Parallel and Distributed Computing (Euro-Par)*, Klagenfurt, Austria, 2003, pp. 318–327.

[Pacitti and Simon 2000] E. Pacitti and E. Simon. "Update Propagation Strategies to Improve Freshness in Lazy Master Replicated Databases," *The VLDB Journal*, (2000), 8(3-4):305–318.

[Pucheral et al. 2001] P. Pucheral, L. Bouganim, P. Valduriez, and C. Bobineau. "PicoDBMS: Scaling Down Database Techniques for the Smartcard," *The VLDB Journal*, (2001) Special Issue on Best Papers from VLDB2000, 10(2-3).

[Rahm and Marek 1995] E. Rahm and R. Marek. "Dynamic Multi-Resource Load Balancing in Parallel Database Systems," in *Proc. 21st Int. Conf. on Very Large Data Bases*, Zurich, Switzerland, September 1995.

[Ramamurthy et al. 2002] R. Ramamurthy, D. DeWitt, and Q. Su. "A Case for Fractured Mirrors," in *Proc. 28th Int. Conf. on Very Large Data Bases*, Hong Kong, August 2002, pp. 430-441.

[Röhm et al. 2001] U. Röhm, K. Böhm, and H.-J. Schek. "Cache-Aware Query Routing in a Cluster of Databases", In *Proc. 17th IEEE Int. Conf. on Data Engineering*, Heidelberg, April 2001, pp. 641–650.

[Selinger et al. 1979] P.G. Selinger, M.M. Astrahan, D.D. Chamberlin, R.A. Lorie, and T.G. Price. "Access Path Selection in a Relational Database Management System," in *Proc. ACM SIGMOD Int. Conf. on Management of Data*, Boston, MA, May 1979, pp. 23–34.

[Shasha and Bonnet 2002] D. Shasha and P. Bonnet. *Database Tuning: Principles, Experiments, and Troubleshooting Techniques*, Morgan Kaufmann Publishing, 2002.

[Shatdal and Naughton 1993] A. Shatdal and J.F. Naughton, "Using Shared Virtual Memory for Parallel Join Processing," in *Proc. ACM SIGMOD Int. Conf. on Management of Data*, Washington, May 1993.

[Sheth and Larson 1990] A. Sheth and J. Larson. "Federated Databases: Architectures and Integration," *ACM Comput. Surv.*, (September 1990), 22(3):183–236.

[Stonebraker, 1989] M. Stonebraker. "Future Trends in Database Systems," *IEEE Trans. Knowledge and Data Eng.*, (March 1989), 1(1):33–44.

[Stonebraker et al. 1988] M. Stonebraker R. Katz, D. Patterson, and J. Ousterhout. "The Design of XPRS," in *Proc. 14th Int. Conf. on Very Large Data Bases*, Los Angeles, September 1988, pp. 318–330.

[Swami and Gupta 1988] A. Swami and A. Gupta. "Optimization of Large Join Queries," in *Proc. of the ACM SIGMOD Int. Conf. on Management of Data*, 1988, pp. 8–17.

[Tanaka and Valduriez 2001] A. Tanaka and P. Valduriez. "The Ecobase Environmental Information System: Applications, Architecture and Open Issues," in *ACM SIGMOD Record*, 30(3), 2001, pp. 70–75.

[Valduriez 1993] P. Valduriez. "Parallel Database Systems: Open Problems and New Issues," *Distributed and Parallel Databases*, (April 1993), 1(2):137–165.

[Walton et al. 1991] C.B. Walton, A.G. Dale and R.M. Jenevin. "A Taxonomy and Performance Model of Data Skew Effects in Parallel Joins," in *Proc. 17th Int. Conf. on Very Large Data Bases*, Barcelona, September 1991.

[Weihl 1989] W. Weihl. "Local Atomicity Properties: Modular Concurrency Control for Abstract Data Types," *ACM Trans. Prog. Lang. Syst.*, (April 1989), 11(2):249–281.

[Wiederhold 1992] G. Widerhold. "Mediators in the Architecture of Future Information Systems," *IEEE Computer*, (March 1992), 25(3):38–49.

[Wilshut et al. 1995] A. N. Wilshut, J. Flokstra, and P.G. Apers. "Parallel Evaluation of Multi-join Queries," in *Proc. ACM SIGMOD Int. Conf. on Management of Data*, San Jose, 1995.

Further Information

There are two current textbooks on distributed and parallel databases. One is our book [Özsu and Valduriez 1999] and the other book is Bell and Grimson [1992]. The first serious book on this topic was Ceri and Pelagatti [1984], which is now quite dated. Our paper [Özsu and Valduriez 1991], which is a companion to our book, discusses many open problems in distributed databases. Two basic papers on parallel database systems are DeWitt and Gray [1992] and Valduriez [1993].

There are a number of more specific texts. On query processing, Freytag et al. [1993] provide an overview of many of the more recent research results. Elmagarmid [1992] has descriptions of a number of advanced transaction models. Gray and Reuter [1993] provide an excellent overview of building transaction managers. Another classical textbook on transaction processing is Bernstein and Newcomer [1997]. These books cover both concurrency control and reliability.

59

Multimedia Databases: Analysis, Modeling, Querying, and Indexing

59.1	Introduction	59-1
59.2	Image Content Analysis	59-2
	Low-Level Image Content Analysis • Mid- to High-Level Image Content Analysis	
59.3	Video Content Analysis	59-5
	Visual Content Analysis • Audio Content Analysis • Video Abstraction	
59.4	Bridging the Semantic Gap in Content Management	59-12
	Computational Media Aesthetics	
59.5	Modeling and Querying Images	59-14
	An Example Object-Relational Image Data Model • An Example Object-Oriented Image Data Model	
59.6	Modeling and Querying Videos	59-15
	Segmentation-Based Models • Annotation-Based Models • Salient Object-Based Models	
59.7	Multidimensional Indexes for Image and Video Features	59-17
	Tree-Based Index Structures • Dimensionality Curse and Dimensionality Reduction	
59.8	Multimedia Query Processing	59-21
59.9	Emerging MPEG-7 as Content Description Standard	59-22
59.10	Conclusion	59-22

Vincent Oria
New Jersey Institute of Technology

Ying Li
IBM T.J. Watson Research Center

Chitra Dorai
IBM T.J. Watson Research Center

59.1 Introduction

With rapidly growing collections of images, news programs, music videos, movies, digital television programs, and training and education videos on the Internet and corporate intranets, new tools are needed to harness digital media for different applications ranging from image and video cataloging, media archival and search, multimedia authoring and synthesis, and smart browsing. In recent years, we have witnessed the growing momentum in building systems that can query and search video collections efficiently and accurately for desired video segments just in the manner text search engines on the Web have enabled easy retrieval of documents containing a required piece of text located on a server anywhere in the world. The digital video archival and management systems are also important to broadcast studios, post-production houses, stock footage houses, and advertising agencies working with large videotape and multimedia collections, to enable integration of content in their end-to-end business processes. Further, because the digital form of videos enables rapid content editing, manipulation, and synthesis, there is burgeoning interest in building cheap, personal desktop video production tools.

An image and video content management system must allow archival, processing, editing, manipulation, browsing, and search and retrieval of image and video data for content repurposing, new program production, and other multimedia interactive services. Annotating or describing images and videos manually through a preview of the material is extremely time consuming, expensive, and unscalable with formidable data accumulation. A content management system, for example, in a digital television studio serves many sets of people, ranging from the program producer who often needs to locate material from the studio archive, to the writer who needs to write a story about the airing segment, the editor who needs to edit in the desired clip, the librarian who adds and manages new material in the archive, and the logger who actually annotates the material in terms of its metadata such as medium ID, production details, and other pertinent information about the content that enables locating it easily. Therefore, a content management tool must be scalable and highly available, ensure integrity of content, and enable easy and quick retrieval of archived material for content reuse and distribution. Automatic extraction of image and video content descriptions is highly desirable to ease the pain of manual annotation and to result in a consistent language of content description when annotating large video collections.

To answer user queries during media search, it is crucial to define a suitable representation for the media, their metadata, and the operations to be applied to them. The aim of a data model is to introduce an abstraction between the physical level (data files and indexes) and the conceptual representation, together with some operations to manipulate the data. The conceptual representation corresponds to the conceptual level in the ANSI relational database architecture [1] where algebraic optimizations and algorithm selections are performed. Optimizations at the physical level (data files and indexes) consist of defining indexes and selecting the right access methods to be used in query processing.

This chapter surveys techniques used to extract descriptions of multimedia data (mainly image, audio and video) through automated analysis and current database solutions in managing, indexing, and querying of multimedia data. The chapter is divided in two parts: multimedia data analysis and database techniques for multimedia. The multimedia data analysis part is composed of Section 59.2, which presents common features used in image databases and the techniques to extract them automatically, and Section 59.3, which discusses audio and video analysis and extraction of audiovisual descriptions. Section 59.4 then describes the problem of semantic gap in multimedia content management systems and emerging approaches to address this critical issue. The second part comprises Section 59.5, which presents some image database models; Section 59.6, which discusses video database models; Section 59.7, which describes multidimensional indexes; and Section 59.8, which discusses issues in processing multimedia queries. Section 59.9 gives an overview of the emerging multimedia content description standard. Finally, Section 59.10 concludes the chapter.

59.2 Image Content Analysis

Existing work in image content analysis can be coarsely categorized into two groups based on the features employed. The first group indexes an image based on low-level features such as color, texture, and shape, while the second group attempts to understand the image's semantic content by using mid- to high-level features and by applying more complex analysis models. Representative work in both groups is surveyed below.

59.2.1 Low-Level Image Content Analysis

Research in this area proposes to index images based on low-level features that are easy to extract and fast to implement. Some well-known content-based image retrieval (CBIR) systems such as QBIC [2], MARS [3], WebSEEK [4], and Photobook [5] have employed these features for image indexing, browsing, and retrieval with reasonable performance achieved. However, due to the low-level nature of these features, there still exists a gap between the information revealed by these features and the real image semantics. Obviously, more high-level features are needed to truly understand the image content. Below we review some commonly used image features, including color, texture, and shape.

59.2.1.1 Color

Color is one of the most recognizable elements of image content, and is widely used as a feature for image retrieval because of its invariance to image scaling, translation, and rotation. Key issues in color feature extraction include the selection of color space and the choice of color quantization scheme.

A color space is a multidimensional space in which different dimensions represent different color components. Theoretically, any color can be represented by a linear combination of the three primary colors, the red (R), the green (G), and the blue (B). However, the RGB color space is not perceptually uniform; that is, equal distances in different areas and along different dimensions of this space do not correspond to equal perception of color dissimilarity. Therefore, some other color spaces such as the CIELAB and CIEL*u*v* have been proposed. Other widely used color spaces include YCbCr, YIQ, YUV, HSV, and Munsell spaces. The MPEG-7 standard [6], which is formally known as "Multimedia Content Description Interface," has adopted the RGB, YCbCr, HMMD, Monochrome, and HSV color spaces, as well as some linear transformation matrices with reference to RGB. Readers are referred to [7] for more detailed descriptions on color spaces.

Color quantization is used to reduce the color resolution of an image. Because a color space, where each color is represented by 24 bits, contains 2^{24} distinct colors, using a quantized color map can considerably decrease the computational complexity in color feature extraction. The commonly used color quantization schemes include uniform quantization, vector quantization, tree-structured vector quantization, and product quantization. MPEG-7 supports linear, nonlinear, and lookup table quantization types.

Three widely used color features (also called color descriptors in MPEG-7) for images are global color histogram, local color histogram, and dominant color. The global color histogram captures the color content of the entire image while ignoring information on the colors' spatial layout. Specifically, a global color histogram represents an image I by an N-dimensional vector $H(I) = [H(I, j), j = 1, 2, \ldots, N]$, where N is the total number of quantized colors and $H(I, j)$ is the number of pixels having color j.

In contrast, the local color histogram representation considers the position and size of each individual image region so as to describe the spatial structure of image colors. For instance, Stricker and Dimai [8] segmented each image into five non-overlapping spatial regions, from which color features were extracted and subsequently used for image matching. In [9], a scalable blob histogram was proposed, where the term *blob* denotes a group of pixels with a homogeneous color. This descriptor is able to distinguish images, which contain objects of different sizes and shapes, from each other without performing color segmentation.

The dominant color representation is considered one of the major color descriptors in MPEG-7 because of its simplicity and association with human perception. Various algorithms have been proposed to extract this feature. For instance, Ohm et al. took color clusters' means (i.e., the cluster centroids) as the image's dominant colors [10]. Considering that human eyes are more sensitive to the changes in smooth regions than those in detailed ones, Deng et al. proposed to extract dominant colors by hierarchically merging similar and unimportant color clusters while leaving distinct and more important clusters untouched [11].

Some other commonly used color feature representations include color moments and color sets, which have been adopted to overcome undesirable color quantization effects.

59.2.1.2 Texture

Texture refers to visual patterns with properties of homogeneity that do not result from the presence of only a single color or intensity. Tree barks, clouds, water, bricks, and fabrics are some examples. Typical texture features include contrast, uniformity, coarseness, roughness, frequency, density, and directionality, which contain important information about the structural arrangement of surfaces as well as their relationship to the surrounding environment. So far, many research efforts have been reported on texture analysis due to its usefulness and effectiveness in applications such as pattern recognition, computer vision, and image retrieval.

There are two basic types of texture descriptors: statistical model-based and transform-based. The first approach explores the gray-level spatial dependence of textures and extracts meaningful statistics as texture representation. For instance, Haralick et al. proposed to represent textures using a co-occurrence matrix [12], where the gray-level spatial dependence of texture was explored. Moreover, they also did Line-Angle-Ratio statistics by analyzing the spatial relationships of lines as well as the properties of their surroundings. Interestingly, Tamura et al. addressed this topic from a totally different viewpoint [13]. In particular, based on psychological measurements, they claimed that the six basic textural features should be coarseness, contrast, directionality, line-likeness, regularity, and roughness. Two well-known CBIR systems namely, the QBIC and the MARS systems, have adopted this representation. Some other work has chosen to use a subset of the above six features, such as the contrast, coarseness, and directionality, for texture classification and recognition purposes.

Some commonly used transforms for transform-based texture extraction include DCT (Discrete Cosine Transform), Fourier-Mellin transform, Polar Fourier transform, Gabor, and wavelet transform. Alata et al. [14] proposed to classify rotated and scaled textures using a combination of Fourier-Mellin transform and a parametric two-dimensional spectrum estimation method (Harmonic Mean Horizontal Vertical). In [15], Wan and Kuo reported their work on texture feature extraction for JPEG images based on the analysis of DCT-AC coefficients. Chang and Kuo [16] presented a tree-structured wavelet transform that provided a natural and effective way to describe textures that have dominant middle- to high-frequency subbands. Readers are referred to [7] for detailed descriptions of texture feature extraction.

59.2.1.3 Shape

Compared to color and texture, the shape feature is less developed due to the inherent complexity of representing it. Two major steps are required to extract a shape feature: object segmentation and shape representation.

Object segmentation has been studied for decades, yet it remains a very difficult research area in computer vision. Some existing image segmentation techniques include the global threshold-based approach, the region growing approach, the split and merge approach, the edge detection-based approach, the color- and texture-based approach, and the model-based approach. Generally speaking, it is difficult to achieve perfect segmentation results due to the complexity of individual object shapes, as well as the existence of shadows and noise.

Existing shape representation approaches could be categorized into the following three classes: the boundary-based representation, the region-based representation, and their combination.

The boundary-based representation emphasizes the closed curve that surrounds the shape. Numerous models have been proposed to describe this curve, which include the chain code, polygons, circular arcs, splines, explicit and implicit polynomials, boundary Fourier descriptor, and UNL descriptor. Because digitization noise can significantly affect this approach, some robust approaches have been proposed.

The region-based representation, on the other hand, emphasizes the area within the closed boundary. Various descriptors have been proposed to model the interior regions, such as the moment invariants, Zernike moments, morphological descriptor, and pseudo-Zernike moments. Generally speaking, region-based moments are invariant to an image's affine transformations. Readers are referred to [17] for more details.

Recent work in shape representation includes the finite element method (FEM), the turning function, and the wavelet descriptor. Moreover, in addition to the above work in two-dimensional shape representation, there are also some research efforts on three-dimensional shape representation. Readers are referred to [7] for more detailed discussions on shape features.

Each descriptor, whether boundary based or region based, is intuitively appealing and corresponding to a perceptually meaningful dimension. Clearly, they could be used either independently or jointly. Although the two representations are interchangeable in the sense of information content, the issue of which aspects of shape have been made explicit matters to the subsequent phases of the computation. Shape features represented explicitly will generally achieve more efficient retrieval when these particular features are queried [7].

59.2.2 Mid- to High-Level Image Content Analysis

Research in this area attempts to index images based on their content semantics such as salient image objects. To achieve this goal, various mid- to high-level image features, as well as more complex analysis models, have been proposed.

One good attempt was reported in [11], where a low-dimensional color indexing scheme was proposed based on homogeneous image regions. Specifically, it first applied a color segmentation approach, called JSEG, to obtain homogeneous regions; then colors within each region were quantized and grouped into a small number of clusters. Finally, color centroids as well as their percentages were used as features descriptors.

More recent work starts to understand image content by learning its semantic concepts. For instance, Minka and Picard [18] developed a system that first generated segmentations or groups of image regions using various feature combinations; then they learned from a user's input to decide which combinations best represented predetermined semantic categories. This system, however, requires supervised training for various parts of the image. In contrast, Li et al. proposed to detect salient image regions based on segmented color and orientation maps without any human intervention [19].

The Stanford SIMPLIcity system, presented in [20], applied statistical classification methods to group images into coarse semantic classes such as textured vs. non-textured and graph vs. photograph. This approach is, however, problem specific and does not extend directly to other domains. Targeting automatic linguistic indexing of pictures, Li and Wang introduced statistical modeling in their work [21]. Specifically, they first employed two-dimensional multi-resolution hidden Markov models (2-D MHMMs) to represent meaningful image concepts such as "snow," "autumn," and "people." Then, to measure the association between the image and concept, they calculated the image occurrence likelihood from its characterizing stochastic process. A high likelihood would then indicate a strong association.

Targeting a moderately large lexicon of semantic concepts, Naphade et al. proposed an SVM-based learning system for detecting 34 visual concepts, which include 15 scene concepts (e.g., outdoors, indoors, landscape, cityscape, sky, beach, mountain, and land) and 19 object concepts (e.g., face, people, road, building, tree, animal, text overlay, and train) [22]. Using TREC 2002 benchmark corpus for training and validation, this system has achieved reasonable performance with moderately large training samples.

59.3 Video Content Analysis

Video content analysis, which consists of both visual content analysis and audio content analysis, has attracted enormous interest in both academic and corporate research communities. This research appeal, in turn, further brings areas that are primarily built upon content analysis modules such as video abstraction, video browsing, and video retrieval, to be actively developed. In this section, a comprehensive survey of all these research topics is presented.

59.3.1 Visual Content Analysis

The first step in video content analysis is to extract its content structure, which could be represented by a hierarchical tree exemplified in Figure 59.1 [23]. As shown, given a continuous video bitstream, we first segment it into a series of cascaded video *shots*, where a shot contains a set of contiguously recorded image frames. Because the content within a shot is always continuous, in most cases, one or more frames, which are known as *keyframes*, can be extracted to represent its underlying content. However, while the shot forms the building block of a video sequence, this low-level structure does not directly correspond to the

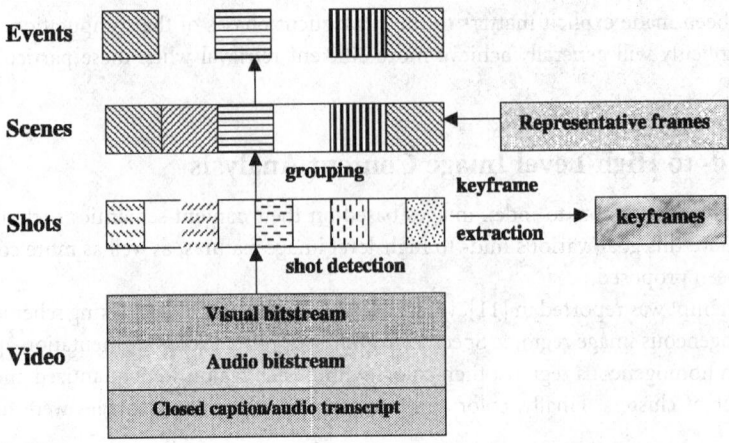

FIGURE 59.1   A hierarchical representation of video content.

video semantics. Moreover, this processing often leads to a far too fine segmentation of the video data in terms of its semantics.

Therefore, most recent work tends to understand the video semantics by extracting the underlying video scenes, where a *scene* is defined as a collection of semantically related and temporally adjacent shots that depicts and conveys a high-level concept or story. A common solution to video scene extraction is to group semantically related shots into a scene.

Nevertheless, not every scene contains a meaningful theme. For example, in feature films, there are certain scenes that are only used to establish story environment; thus, they do not contain any thematic topics. Therefore, it is necessary to find important scenes that contain specific thematic topics such as dialogs or sports highlights. Such a video unit is called an *event* in this chapter.

Previous work on the detection of video shots, scenes, and events is reviewed in this section.

59.3.1.1   Video Shot Detection

A shot can be detected by capturing camera transitions, which could be either abrupt or gradual. An abrupt transition is also called a camera break or cut, where a significant content change occurs between two consecutive frames. In contrast, a gradual transition is usually caused by some special effects such as dissolve, wipe, fade-in, and fade-out, where a smooth content change is observed over a set of consecutive frames.

Existing work in shot detection can be generally categorized into the following five classes: pixel based, histogram based, feature based, statistics based, and transform based. In particular, the pixel-based approach detects the shot change by counting the number of pixels that have changed from one frame to the next. While this approach gives the simplest way to detect the content change between two frames, it is too sensitive to object and camera motions. As a result, the histogram-based approach, which detects the content change by comparing the histogram of neighboring frames, has gained more popularity as histograms are invariant to image rotation, scaling, and transition. In fact, it has been reported that this approach can achieve good trade-off between the accuracy and speed. Many research efforts have been reported along this direction [24].

A feature-based shot detection approach was proposed in [25], where the intensity edges between two consecutive frames were analyzed. It was claimed by the authors that, during the cut and dissolve operations, new intensity edges would appear far away from the old ones; thus, by counting the new and old edge pixels, the shot transitions could be detected and classified. In [26], a visual rhythm-based approach was proposed where a visual rhythm is a special two-dimensional image reduced from a three-dimensional video such that its pixels along a vertical line are the pixels uniformly sampled along the diagonal line of a video frame. Some other technologies such as image segmentation and object tracking have also been employed to detect the shot boundary.

Kasturi and Jain developed a statistics-based approach in which the mean and standard deviations of pixel intensities were used as features for shot boundary detection [27]. To avoid manually determining the threshold, Boreczky and Wilcox built a Hidden Markov Model (HMM) to model shot transitions where audio cepstral coefficients and color histogram differences were used as features [28].

To accommodate the trend that an increasing amount of video data is currently stored and transmitted in compressed form, transform-based approaches have been proposed where video shots are directly detected in the compressed domain. In this case, the processing could be greatly sped up because no full-frame decompression is needed. Among reported work in this domain, the DCT (Discrete Cosine Transform) and wavelet transform are the two most frequently used approaches.

Compared to the large amount of work on cut detection, little work has been directed toward the gradual transition detection due to its complex nature. A "twin-comparison" algorithm was proposed in [29] where two thresholds were utilized to capture the minor content change during the shot transition. To detect the dissolve effect, past research efforts have mainly focused on finding the relations between the dissolve formula and the statistics of interpolated MBs (Macroblocks) in P- and B-frames. Similar work was also reported for wipe detection, yet with special considerations on various wipe shapes, directions, and patterns. Clearly, in the case of gradual transition detection, algorithms developed for one type of effect may not work for another.

A detailed evaluation and comparison of several popular shot detection algorithms can be found in [30], where both abrupt and gradual transitions have been studied.

59.3.1.2 Video Scene and Event Detection

Existing scene detection approaches can be classified into the following two categories: the model-based approach and the model-free approach. In the former case, specific structure models are usually built up to model specific video applications by exploiting their scene characteristics, discernible logos, or marks. For instance, in [31], temporal and spatial structures were defined to parse TV news, where the temporal structure was modeled by a series of shots, including anchorperson shots, news shots, commercial break shots, and weather forecast shots. Meanwhile, the spatial structure was modeled by four frame templates with each containing either two anchorpersons, one anchorperson, one anchorperson with an upper-right news icon, or one anchorperson with an upper-left news icon. Some other work along this direction has tried to integrate multiple media cues such as visual, audio, and text (closed captions or audio transcripts) to extract scenes from real TV programs.

The model-based approach has also been applied to analyze sports video because a sports video can be characterized by a predictable temporal syntax, recurrent events with consistent features, and a fixed number of views. For instance, Zhong and Chang proposed to analyze tennis and baseball videos by integrating domain-specific knowledge, supervised machine learning techniques, and automatic feature analysis at multiple levels [32].

Compared to the model-based approach, which has very limited application areas, the model-free approach can be applied to very generic applications. Work in this area can be categorized into three classes according to the use of visual, audio, or both audiovisual cues. Specifically, in visual-based approaches, the color or motion information is utilized to locate the scene boundary. For instance, Yeung et al. proposed to detect scenes by grouping visually similar and temporally close shots [33]. Moreover, they also constructed a Scene Transition Graph (STG) to represent the detected scene structure. Compressed video sequences were used in their experiments. Some other work in this area has applied the cophenetic dissimilarity criterion or a set of heuristic rules to determine the scene boundary.

Pure audio-based work was reported in [34], where the original video was segmented into a sequence of audio scenes such as speech, silence, music, speech with music, song, and environmental sound based on low-level audio features. In [35], sound tracks in films and their indexical semiotic usage were studied based on an audio classification system that could detect complex sound scenes as well as the constituent sound events in cinema. Specifically, it has studied the car chase and the violence scenes for action movies based on the detection of their characteristic sound events such as horns, sirens, car crashes, tires skidding, glass breaking, explosions, and gunshots.

However, due to the difficulty of precisely locating the scene boundaries based on pure audio cues, more recent work starts to integrate multiple media modalities for more robust results. For instance, three types of media cues, including audio, visual, and motion, were employed by [36] to extract semantic video scenes from broadcast news. Sundaram and Chang reported their work on extracting computable scenes in films by utilizing audiovisual memory models. Two types of scenes, namely, N-type and M-type, were considered, where the N-type scene was further classified into pure dialog, progressive, and hybrid [37]. A good integration of audio and visual cues was reported in [38], where audio cues, including ambient noise, background music, and speech, were cooperatively evaluated with visual features extraction in order to precisely locate the scene boundary. Special movie editing patterns were also considered in this work.

Compared to the large amount of work on scene detection, little attention has been paid to event detection. Moreover, because event is a subjectively defined concept, different work may assign it different meanings. For instance, it could be the highlight of a sports video or an interesting topic in a video document. In [39], a query-driven approach was presented to detect topics of discussion events by using image and text contents of query foils (slide) found in a lecture. While multiple media sources were integrated in their framework, identification results were mainly evaluated in the domain of classroom lectures/talks due to the special features adopted. In contrast, work on sports highlight extraction mainly focuses on detecting the announcer's speech, the audience ambient speech noise, the game-specific sounds (e.g., the baseball hits), and various background noise (e.g., the audience cheering). Targeting movie content analysis, Li et al. proposed to detect three types of events, namely, two-speaker dialogs, multi-speaker dialogs, and hybrid events, by exploiting multiple media cues and special movie production rules [40].

Contrary to all the work above on detecting interesting events, Nam and colleagues tried to detect undesired events (such as violence scenes) from movies [41]. In particular, violence-related visual cues, including spatio-temporal dynamic activity, flames in gunfire/explosion scenes, and splashed blood, were detected and integrated with the detection of violence-related audio cues such as abrupt loud sounds to help locate offensive scenes.

59.3.2 Audio Content Analysis

Existing research on content-based audio data analysis is still quite limited, and can be categorized into the following two classes.

59.3.2.1 Audio Segmentation and Classification

One basic problem in this area is the discrimination between speech and music, which are the two most important audio types. A general solution is to first extract various audio features such as the average zero-crossing rate (ZCR) and the short-time energy from the signals, and then distinguish the two sound types based on the feature values. For instance, 13 audio features calculated in time, frequency, and cepstrum domains were employed in [42] for classification purposes. It also examined and compared several popular classification schemes, including the multidimensional Gaussian maximum *a posteriori* estimator, the Gaussian mixture model, a spatial partitioning scheme based on k-d trees, and a nearest neighbor classifier. Generally speaking, a relatively high accuracy could be achieved in distinguishing speech and music because these two signals are quite different in both spectral distributions and temporal change patterns.

A more advanced classification algorithm usually takes more sound types into consideration. For instance, Wyse and Smoliar classified audio signals into three types, including "music," "speech," and "others" [43]. Specifically, music was first detected based on the average length of the interval during which peaks were within a narrow frequency band; then speech was separated out by tracking the pitches. Research in [44] was devoted to analyze the signal's amplitude, frequency, and pitch. It has also conducted simulations on human audio perception; the results were utilized to segment the audio data and recognize the music component. More recently, Zhang and Kuo presented an extensive feature extraction and classification system for audio content segmentation and classification purposes [45]. Five audio features, including energy, average zero-crossing rate, fundamental frequency, and spectral peak tracks, were extracted to fulfill the task. A two-step audio classification scheme was proposed in [46], where in the first step, speech and non-speech were discriminated based on KNN and LSP VQ schemes. In the second step, the non-speech signals were further classified into music, environment sounds, and silence based on a feature thresholding scheme.

59.3.2.2 Audio Analysis for Video Indexing

In this section, some purely audio-based work developed for the video indexing purpose is reviewed.

Five different video classes, including news report, weather report, basketball, football, and advertisement, were distinguished in [47] using both multilayer neural networks (MNN) and the Hidden Markov Model (HMM). Features such as the silence ratio, the speech ratio, and the subband energy ratio were extracted to fulfill this task. It was shown that while MNN worked well in distinguishing among reports, games, and advertisements, it had difficulty in classifying different types of reports or games. On the contrary, the use of HMM increased the overall accuracy but it could not well classify all five video types. In [48], features such as the pitch, the short-time average energy, the band energy ratio, and the pause rate were first extracted from the coded sub-band of an MPEG audio clip; then they were integrated to characterize the clip into either silence, music, or dialog. Another approach to index videos based on music and speech detection was proposed in [49], where image processing techniques were applied to the spectrogram of the audio signals. In particular, the spectral peaks of music were recognized by applying an edge-detection operator and the speech harmonics were detected with a comb filter.

59.3.3 Video Abstraction

Video abstraction, as the name implies, generates a short summary for a long video document. Specifically, a video abstract is a sequence of still or moving images that represents the video essence in a very concise way. Video abstraction is primarily used for video browsing, and is an inseparable part of a video indexing and retrieval system.

There are two fundamentally different kinds of video abstracts: still- and moving-image abstracts. The still-image abstract, also known as static storyboard or video summary, is a small collection of keyframes extracted or generated from the underlying video source. The moving-image abstract, also known as moving storyboard or video skim, consists of a collection of image sequences, as well as the corresponding audio abstract extracted from the original sequence. Thus, it is itself a video clip but is of considerably shorter length.

59.3.3.1 Video Skimming

There are basically two types of video skim: the summary sequence and the highlight. A summary sequence is used to provide users with an impression about the entire video content, while a highlight generally only contains some interesting video segments. A good example of a video highlight is the movie trailer, which only shows some very attractive scenes without revealing the story's end.

In the VAbstract system developed in [50], the most characteristic movie segments were extracted to generate a movie trailer. Specifically, scenes that contained important objects/people, had high actions, or contained speech dialogs were selected and organized in their original temporal order to form the movie trailer. In the improved version of VAbstract, which is called MoCA, special events such as closed-up shots of leading actors, explosion, and gunfire were detected to help determine important scenes.

Defining which video segments are highlights is actually a very subjective and difficult process, and it is also difficult to map human cognition into the automated abstraction process. Therefore, most current video-skimming work focuses on the generation of the summary sequence. One of the most straightforward approaches in this case would be to compress the original video by speeding up the playback. As studied in [51], using a time compression technology, a video could be watched in a fast playback mode without distinct pitch distortion. However, according to recent research results, the maximum time compression could only reach 1.5 to 2.5, beyond which the speech will become incomprehensible.

The Informedia Project [52] at the Carnegie Mellon University aimed at creating a short synopsis of the original video by extracting significant audio and video information. In particular, an audio skim was first created by extracting audio segments with respect to preextracted text keywords; then an image skim was generated by selecting video frames that satisfied predefined visual rules. Finally, the video skim was formed based on consideration of both word relevance and structures of prioritized audio and image skims. Toklu et al. [53] also proposed to generate video skim by integrating visual, audio, and text cues. Specifically, they first grouped video shots into story units based on detected "change of speaker" and "change of subject" markers. Then, audio segments corresponding to all generated story units were extracted and aligned with closed captions. Finally, a video skim was formed by integrating the audio and text information. Similar to the Informedia Project, this work also heavily depends on the text information.

Nam and Tewfik [54] proposed to generate video skims based on a dynamic sampling scheme. Specifically, they first decomposed a video sequence into sub-shots, and then computed a motion intensity index for each of them. Next, all indices were quantized into predefined bins, where each bin was assigned a different sampling rate. Finally, keyframes were sampled from each sub-shot based on its assigned rate. Recently, a segment-based video skimming system was presented in [55]. This system was specifically developed for documentary, presentation, and educational videos, where word frequency analysis was carried out to score segments, and segments that contained frequently occurring words were selected to form the final skim. A sophisticated user study was carried out, but it did not produce satisfactory results as expected.

Some other work in this area attempts to find solutions to domain-specific video data where special features can be employed. For example, the VidSum project developed at Xerox PARC employed the presentation structure, which was particularly designed for their regular weekly forum, to assist in mapping low-level signal events onto semantically meaningful events that could be used to assemble the summary [56]. He et al. [57] reported Microsoft Research's work on summarizing corporate informational talks. Some special knowledge of the presentation, such as the pitch and pause information, the slide transition points, as well as the information about the user access patterns, was utilized to generate the summary. Another work reported in [58] mainly focused on the summarization of home videos, which was more usage model-based than content-based. In this approach, all video shots were first clustered into five different levels based on the time and date they were taken. Then, a shot shortening process was applied to uniformly segment longer shots into 2-minute clips. Finally, clips that met certain sound pressure criteria were selected to form the final abstract.

59.3.3.2 Video Summarization

Compared to video skimming, video summarization has attracted much more research interest in recent years. Based on the way keyframes are extracted, existing work in this area can be categorized into the following three classes: sampling based, shot based, and segment based.

59.3.3.2.1 Sampling and Shot-Based Keyframe Extraction

Most of the earlier summarization work was sampling based, where keyframes were either randomly chosen or uniformly sampled from the original video. This approach gives the simplest way to extract keyframes, yet it fails to truly represent the video content.

More sophisticated work thus tends to extract keyframes by adapting to the dynamic video content. Because a shot is taken within a continuous capture period, a natural and straightforward way is to extract one or more keyframes from each shot. Based on the features used to select keyframes, we categorize the existing work into the following three classes: the color-based approach, the motion-based approach, and the mosaic-based approach.

One typical color-based approach was reported in [59], where keyframes were extracted in a sequential fashion. In particular, the first frame of each shot was always chosen as the first keyframe. Then, the next frame, which had a sufficiently large difference from the latest keyframe, was chosen as the next keyframe. Zhuang et al. [60] proposed to extract keyframes based on an unsupervised clustering scheme. Specifically, all video frames within a shot were first grouped into clusters; then the frames that were closest to cluster centroids were chosen as keyframes.

FIGURE 59.2 A mosaic image generated from a panning sequence.

Because the color histogram is invariant to image orientations and robust to background noises, color-based keyframe extraction algorithms have been widely used. However, because most of this work is heavily threshold dependent, the underlying video dynamics cannot be well captured when there are frequent camera or object motions.

The motion-based approaches are relatively better suited for controlling the number of keyframes when the video presents significant temporal dynamics. A general solution along this direction is to first measure the amount of motion contained in each frame based on calculated optical flows; then frames that have minimum motion activities are chosen as keyframes.

A domain-specific keyframe extraction method was proposed in [61], where sophisticated global motion and gesture analyses were carried out to generate a summary for videotaped presentations. Three different operation levels were suggested in [62], where at the lowest level, pixel-based frame differences were computed to generate a "temporal activity curve"; at level two, color histogram-based frame differences were computed to extract "color activity segments"; and at level three, sophisticated camera motion analysis was carried out to detect "motion activity segments." Keyframes were then selected from each segment and the necessary elimination was applied to obtain the final result.

A limitation of the above approaches is that it is not always possible to find keyframes that can well represent the entire video content. For example, given a camera panning/tilting sequence, even if multiple keyframes are selected, the underlying dynamics still cannot be well captured. In this case, a mosaic-based approach, which generates a synthesized panoramic image to cover the video content, can provide a better solution.

Mosaic, also known as salient still, video sprite, or video layer, is usually generated in the following two steps: (1) fitting a global motion model to the motion between each pair of successive frames; (2) compositing frames into a single panoramic image by warping them using estimated camera parameters. Some commonly used motion models, such as the translational model, rotation/scaling model, affine model, planar perspective model, and quadratic model, can be found in the MPEG-7 standard [6]. Figure 59.2 shows a mosaic generated from 183 video frames using an affine model. As we can see, this single still image can provide much more information than regular keyframes can do.

To capture both foreground and background, Irani and Anandan [63] developed two types of mosaics, namely, a static background mosaic and a synopsis mosaic. While the static mosaic could capture the background scene, the synopsis mosaic was constructed to provide a visual summary of the entire foreground dynamics by detecting object trajectories. The final mosaic image was then obtained by simply combining these two mosaics. In addition, to accommodate video sources with complex camera operations and frequent object motions, Taniguchi et al. proposed to interchangeably use either a regular keyframe or mosaic image, whichever is more suitable [64].

59.3.3.2.2 Segment-Based Keyframe Extraction

One major drawback of using one or more keyframes for each shot is that it does not scale well for long video. Therefore, recently people have begun to work on a higher-level video unit, which we call "segment" here. A video segment could be a scene, an event, or even the entire sequence. Some interesting research along this direction is summarized below.

In [65], a video sequence was first partitioned into segments; then an importance measure was computed for each segment based on its length and rarity. Finally, the frame closest to the center of each qualified

FIGURE 59.3   A video summary containing variable-sized keyframes.

segment was extracted as the representative keyframe with its size proportional to the importance index. Figure 59.3 shows one of their exemplary summaries.

Yeung and Yeo reported their work on summarizing video at a scene level [66]. In particular, it first grouped shots into clusters using a proposed "time-constrained clustering" approach; then, meaningful story units or scenes were subsequently extracted. Next, an R-image was extracted from each scene to represent its component shot clusters, whose dominance value was computed based on either the frequency count of visually similar shots or the shots' durations. Finally, all extracted R-images were organized into a predefined visual layout with their sizes being proportional to their dominance values. To allow users to freely browse the video content, a scalable video summarization scheme was proposed in [23], where the number of keyframes could be adjusted based on user preference. In particular, it first generated a set of default keyframes by distributing them among hierarchical video units, including scene, sink, and shot, based on their respective importance ranks. Then, more or less keyframes would be returned to users based on their requirements and keyframe importance indices.

Some other work in this category has attempted to treat video summarization task in a more mathematical way. For instance, some of them introduce fuzzy theory into the keyframe extraction scheme and others represent the video sequence as a curve in a high-dimensional feature space. The SVD (Singular Value Decomposition), PCA (Principle Component Analysis), mathematical morphology, and SOM (Self-Organizing Map) techniques are generally used during these processes.

59.4   Bridging the Semantic Gap in Content Management

We complete the survey by describing the problem of semantic gap in multimedia content management systems and emerging approaches to address this critical issue. Most of the approaches surveyed above for video content analysis can be broadly grouped together into two classes:

1. The first class of approaches seeks to extract as much frame-level information as possible from a video source. All further processing is carried out to merge frames and shots on the basis of various visual and aural similarity measures using low-level features. Shots that are similar in terms of the extracted features are considered semantically similar and labeled based on their common and dominant attributes.
2. The other class consist of approaches that target predefined genres of videos by describing a video in terms of its structure (e.g., news video) or events (sports). They exploit domain-specific constraints to carefully select low-level features and to analyze the patterns of their occurrences to result in higher-level descriptions of what is happening in videos. The goal is to generate descriptions from a domain-specific, finite, commonly accepted vocabulary for a specific task such as shot or scene labeling, genre discrimination, or sports events extraction.

However, it has become evident via real-world installations of content management systems that these fall far short of the expectations of users. A major problem is the gap between the descriptions that are computed by the automatic methods and those employed by users to describe an aspect of video, such as motion during their search. While users want to query in a way natural to them in terms of persons, events, topics, and emotions, actual descriptions generated by current techniques remain at a much lower level, closer to machine-speak than to the natural language. For example, in most systems, instead of being able to specify that one is looking for a clip where the U.S. President is limping to the left in a scene, one often needs to specify laboriously, "Object=human, identity=the US President, movement=left, rate of movement=x pixels per frame, etc.," using descriptive fields amenable to the computations of algorithms provided by the annotation systems. Further, even if we allow that some systems have recently begun to address the problem of object motion-based annotation and events, what is still missing is the capability to handle high-level descriptions of, not just what the objects are and what they do in a scene, but also of emotional and visual appeal of the content seen and remembered.

The other concern is that most video annotation and search systems ignore the *fallout* rate, which measures the number of nonmatching items that were not retrieved upon a given query. This measure is extremely important for video databases, because even as a simple measure, there are more than 100,000 frames in just a single hour of video with a frame rate of 30 fps. An important design criterion would therefore emphasize deriving annotation indices and search measures that are more discriminatory, less frame-oriented, and result in high fallout rates.

59.4.1 Computational Media Aesthetics

To bridge the semantic gap between the high-level meaning sought by user queries in search for media and the low-level features that we actually compute today for media indexing and search, one promising approach [67] is founded upon an understanding of media elements and their roles in synthesizing meaning, manipulating perceptions, and crafting messages, with a systematic study of media productions. Content creators worldwide use widely accepted conventions and cinematic devices to solve problems presented when transforming a written script to an audiovisual narration, be it a movie, documentary, or a training video.

This new approach, called computational media aesthetics, is defined as the algorithmic study of a variety of image and aural elements in media, founded on their patterns of use in film grammar, and the computational analysis of the principles that have emerged underlying their manipulation, individually or jointly, in the creative art of clarifying, intensifying, and interpreting some event for the audience [68]. The core trait of this approach is that in order to create effective tools for automatically understanding video, we must be able to interpret the data with its maker's eye.

This new research area has attracted computer scientists, content creators, and producers who seek to address the fundamental issues in spanning the data-meaning gulf by a systematic understanding and application of media production methods. Some of the issues that remain open for examination include:

- Challenges presented by semantic gap in media management
- Assessment of problems in defining and extracting high-level semantics from media
- Examination of high-level expressive elements relevant in different media domains
- New algorithms, tools, and techniques for extracting characteristics related to space, motion, lighting, color, sound, and time, and associated high-level semantic constructs
- Production principles for manipulation of affect and meaning
- Semiotics for new media
- Metrics to assess extraction techniques and representational power of expressive elements
- Case studies and working systems

Media semantics can lead to the development of shared vocabularies for structuring images and video, and serve as the foundation for media description interfaces. There is structure regardless of the particular media context but there may not be homogeneity, and therefore it helps to be guided by production knowledge in media analysis. New software models like this will enable technologies that can emulate human perceptual capabilities on a host of difficult tasks such as parsing video into sections of interest, making inferences about semantics, and about the perceptual effectiveness of the messages contained.

Once content descriptions are extracted from the multimedia data, the main questions that follow include: (1) What is the best representation for the data? and (2) What are the basic operations needed to manipulate the data and express "all" the user queries?

59.5 Modeling and Querying Images

An image database model to organize and query images is relatively new in image databases. Usually, visual feature vectors extracted from the images are directly maintained in a multidimensional index structure to enable similarity searches. The main problems with this approach include:

- *Flexibility.* The index is the database. In traditional database systems, indexes are hidden at the physical level and are used as access methods to speed up query processing. The database system can still deliver results to queries without any index. The only problem is that the query processing will take more time as the data files will be scanned. Depending on the type of queries posed against the database, different indexes can exist at the same time, on the same set of data.

- *Expressiveness.* The only type of queries that can be handled is the query supported by the index (Query by Examples in general).

- *Portability.* Similarity queries are based on some metrics defined on the feature vectors. Once the metric has been chosen, only a limited set of applications can benefit from the index because metrics are application dependent.

To address these issues, some image models are being proposed. The image models are built on top of existing database models, mainly object relational and object oriented.

59.5.1 An Example Object-Relational Image Data Model

In [69], an image is stored in a table $T(h : Integer, x_1 : X_1, \ldots, x_n : X_n)$ where h is the image identifier and x_i is an image feature attribute of domain (or type) X_i (note that classical attributes can be added to this minimal schema). The tuple corresponding to the image k is indicated by $T[k]$. Each tuple is assigned a score (ζ) which are real numbers such that $T[k].\zeta$ is a distance between the image k and the current query image. The value of ζ is assigned by a scoring operator $\Sigma_T(s)$ given a scoring function $s: (\Sigma_T(s))[k].\zeta = s(T[k].x_1, \ldots, T[k].x_n)$.

Because many image queries are based on distance measures, a set of distance functions ($d : X \times X \rightarrow [0, 1]$) are defined for each feature type X. Given an element $x : X$ and a distance function d defined on X, the scoring function s assigns $d(x)$, a distance from x to every element of X. In addition, a set of score combination operators $\diamondsuit : [0, 1] \times [0, 1] \rightarrow [0, 1]$ are defined.

New selection and join operators defined on the image table augmented with the scores allow the selection of n images with lowest scores, the images whose scores are less than a given score value ρ and the images from a table $T = T(h : Integer, x_1 : X_1, \ldots, x_n : X_n)$ that match images from a Table $Q = Q(h : Integer, y_1 : Y_1, \ldots, y_n : Y_n)$ based on score combination functions as follows:

- K-nearest neighbors: $\sigma_k^{\#}(\Sigma_T(s))$ returns the k rows of the table T with the lowest distance.
- Range query operator: $\sigma_\rho^<(\Sigma_T(s))$ returns all the rows of the table T with a distance less than ρ.
- $\diamondsuit join$: $T \bowtie Q$ joins the tables T and Q on their identifiers h and returns the table $W = W(h : Integer, x_1 : X_1, \ldots, x_n : X_n, y_1 : Y_1, \ldots, y_n : Y_n)$. The distance in the table W is defined as $W.d = T.d \diamondsuit Q.d$.

In [70] the same authors proposed a design model with four kind of feature dependencies that can be exploited for the design of efficient search algorithms.

59.5.2   An Example Object-Oriented Image Data Model

In the DISIMA model, an image is composed of physical salient objects (regions of the image) whose semantics are given by logical salient objects that represent real-world objects. Both images and physical salient objects can have visual properties. The DISIMA model uses an object-oriented concept and introduces three new types — *Image, Physical Salient Objects, Logical Salient Objects* — and operators to manipulate them.

Images and related data are manipulated through predicates and operators defined on images, physical and logical salient objects are used to query the images. They can be directly used in calculus-based queries to define formulas or in the definition of algebraic operators. Because the classical predicates $\{=, <, \leq, >, \geq\}$ are not sufficient for images, a new set of predicates were defined to be used on images and salient objects.

- *Contain predicate.* Let i be an image, o an object with a behavior *pso* that returns the associated set of physical salient objects $contains(i, o) \Longleftrightarrow \exists p \in o.pso \land p \in i.pso$.

- *Shape similarity predicates.* Given a shape similarity metric d_{shape} and a similarity threshold ϵ_{shape}, two shapes s and t are similar with respect to d_{shape} if $d_{shape}(s, t) \leq \epsilon_{shape}$. In other words: $shape\_similar(s, t, \epsilon_{shape}) \Longleftrightarrow d_{shape}(s, t) \leq \epsilon_{shape}$.

- *Color similarity predicates.* Given two color representations (c_1, c_2) and a color distance metric d_{color}, the color representations c_1 and c_2 are similar with respect to d_{color} if $d_{color}(c_1, c_2) \leq \epsilon_{color}$.

Based on the above-defined predicates, some operators are defined: *contains* or *semantic join* (to check whether a salient object is found in an image), and the *similarity join* that is used to match two images or two salient objects with respect to a predefined similarity metric on some low-level features (color, texture, shape, etc.), and *spatial join* on physical salient objects:

- *Semantic join.* Let S be a set of semantic objects of the same type with a behavior *pso* that returns, for a semantic object, the physical salient objects it describes. The *semantic join* between an image class extent I and the semantic object class extent S, denoted by $I \bowtie_{contains} S$, defines the elements of $I \times S$ where for $i \in I$, and $s \in S$, $contains(i, s)$.

- *Similarity join.* Given a similarity predicate *similar* and a threshold ϵ, the *similarity join* between two sets R and S of images or physical salient objects, denoted by $R \bowtie_{similar(r.i,s.j,\epsilon)} S$ for $r \in R$ and $s \in S$, is the set of elements from $R \times S$ where the behaviors i defined on the elements of R and j on the elements of S return some compatible metric data type T and $similar(r.i, s.j)$ (the behaviors i and j can be the behaviors that return color, texture or shape).

- *Spatial join.* The spatial join of the extent of two sets R and S, denoted by $R \bowtie_{r.i\theta s.j} S$, is the set of elements from $R \times S$ where the behaviors i defined on the elements of R and j on the elements of S return some spatial data type, θ is a binary spatial predicate, and $R.i$ stands in relation θ to $S.j$ (θ is a spatial operator like north, west, northeast, intersect, etc.)

The predicates and the operators are the basis of the declarative query languages MOQL [71] and Visual-MOQL [72].

59.6   Modeling and Querying Videos

After obtaining video structure from the content analysis module, low- and/or high-level features can be subsequently extracted from the underlying video units to facilitate the content indexing. For instance, Zhang et al. [24] extracted four image features (color, texture, shape, and edge) and two temporal features (camera operations and temporal brightness variation) from shots and keyframes, and used them to index and retrieve clips from a video database. Similar work was also reported in [31,73], where TV news was used to demonstrate the proposed indexing scheme. More sophisticated investigation of indexing TV broadcast news can be found in [74], where speech, speech transcript, and visual information were combined together in the proposed DANCERS system. Tsekeridou and Pitas [75] also reported their work on indexing TV news where, extracted faces, which could be talking or non-talking, and speaker identities were employed as indexing features.

A system called "PICTURESQUE" was proposed in [76], where object motions represented by its trajectory coordinates were utilized to index the video. A "VideoBook" system was presented in [77], where multiple features including motion, texture, and colorimetry cues were combined to characterize and index a video clip. A sports video indexing scheme was presented in [78], where speech under-standing and image analysis were integrated to generate meaningful indexing features. A comprehensive video indexing and browsing environment (ViBE) was discussed in [79] for a compressed video database. Specifically, given a video sequence, it first represented each shot with a hierarchical structure (shot tree). Then, all shots were classified into pseudo-semantic classes according to their contents, which were finally presented to end users in an active browsing environment. A generic framework of integrating existing low- and high-level indexing features was presented in [80], where the low-level features in-cluded color, texture, motion, and shape, while the high-level features could be video scenes, events, and hyperlinks.

In general, video database models can be classified into segmentation-based models, annotation-based models, and salient object-based models.

59.6.1 Segmentation-Based Models

In segmentation-based approaches [81–83], the video data model follows the video segmentation (events, scenes, shots, and frames) and keyframes extracted from shots and scenes are used to summarize the video content. The visual features extracted from the key frames are then used to index the video.

Mahdi et al. [83] proposed a temporal cluster graph (TCG) as a data model that combines visual similarity of shots and and semantic concepts such as sequence and scene. The scene construction method uses two main video features: spatial and temporal clues extracted and shot rhythms in the video. The shot rhythm is a temporal effect obtained from the duration of successive shots that is supposed to lead to a particular scene sensation. Shots are first clustered based on their color similarity. Then the clusters are grouped into sequences. A sequence is a narrative unity formed by one or several scenes. Sequences are linked to each other through an effect of gradual transition (dissolve, fade-in, or fade-out). The temporal cluster graph (TCG) is constructed to describe the clusters and their temporal relationships. A node is associated with each cluster, and the edges represent temporal relationships between clusters. Queries are directly posed against the graph.

59.6.2 Annotation-Based Models

In annotation-based approaches [84,85], the video stream is augmented with a description (annotation) layer. Each descriptor is associated with a logical video sequence and/or a physical object in the video.

Annotation-based video data models can be further classified into frame-based annotation and attribute-value pair structure. In frame-based annotation, the annotation is directly associated to the frame se-quences. The annotation itself can be made of keywords [84], a description in a natural language [85], or semantic icons [86]. Smith and Davenport [84] proposed a layered annotation representation model called the stratification model that segments contextual information of the video. The stratification model divides the video sequence into a set of overlapping strata. A stratum consists of a descriptive infor-mation with the corresponding keywords and boundaries. A stratum can be contained in another stra-tum and may encompass multiple descriptions. The content information of a stratum is obtained by the union of all the contextual descriptions associated with it. Based on the stratification model, Weiss et al. [87] introduced a video data model called the algebraic video model. The algebraic video model is composed of hierarchical composition of video expression with semantic description. The atomic video expression is a single window presentation from a raw video segmentation. These segments are defined by the name of the raw video data, and the starting and ending frames. Compound video expressions can be constructed from primitive video expression or other compound video expression using the algebraic operations. The video algebra operations falls into four categories: creation, composition, output and description.

An example of a natural language annotation is the VideoText model [85]. This model allows free text annotation of logical video segments. VideoText supports incremental, dynamic, and multiple creation of annotations. Basic information retrieval (IR) techniques are used to evaluate the queries and rank the results. To support interval queries based on temporal characteristics of videos, the set of classical IR operations is extended with some interval operators.

59.6.3   Salient Object-Based Models

In salient object-based approaches [88–90], salient objects (objects of interest in the video) are somehow identified and extracted, and some spatio-temporal operators are used to express events and concepts in the queries. Video data modeling based on segmentation employs image processing techniques and only deals with low-level video image features (color, shape, texture, etc). The entire modeling process can be automated. However, this solution is very limited as only query involving low-level features and shots can be posed. Chen et al. [90] proposed a model that combines segmentation and salient objects. The model extends the DISIMA model with a video block that models video following video segmentation. Each shot is represented by a set of keyframes that are treated as images following the DISIMA image data model and some new operators are defined for the videos.

59.7   Multidimensional Indexes for Image and Video Features

In the presence of a large database, a sequential scan of the database each time a query is posed is unacceptable. The aim of indexes is to filter the database and select a significantly small subset of the database that contains the result of the query. The multidimensional index structures, also known as *spatial access methods* (SAMs), are employed to index image visual features, such as average colors, color histograms, and textures, because these features are usually modeled as points in multidimensional spaces. SAMs usually organize the multidimensional points in a number of buckets, each of which corresponds to a disk block and to some subspace of the universal space. There are two categories of SAMs: tree-based methods and hashing-based methods.

Indexes are normally based on the total order of the space to be indexed and the main difficulty encountered in designing multidimensional indexing structures is that, unlike the case in one-dimensional space, there exists no total ordering among multidimensional points that preserves their proximity. In other words, there is no mapping from two- or higher-dimensional space into one-dimensional space, such that any two objects that are close in the higher-dimensional space are also close to each other in the one-dimensional sorted sequence. Multidimensional indexing structures usually employ "bucket methods" [91]. The points in the database are organized into a number of buckets, each of which corresponds to some subspace of the universal multidimensional space. The subspaces are often referred to as bucket regions, although their dimensionality may be greater than two. Through partitioning the universe into bucket regions, the multidimensional indexing structures potentially achieve clustering: separating objects that are far apart and grouping objects that are close to each other.

A typical representative for hashing-based methods is the grid file [92]. The grid file superimposes a multidimensional orthogonal grid on the universal space. One data bucket, which is stored as a disk block, is associated with one or more of these grid cells. The addresses of the data buckets are recorded in a grid directory. A common criticism of the grid file is that the directory expansion approaches an exponential rate as the data distribution becomes less uniform, and the problem is magnified by the number of dimensions of the space [93]. In [94] a multidimensional hashing method that controls the directory

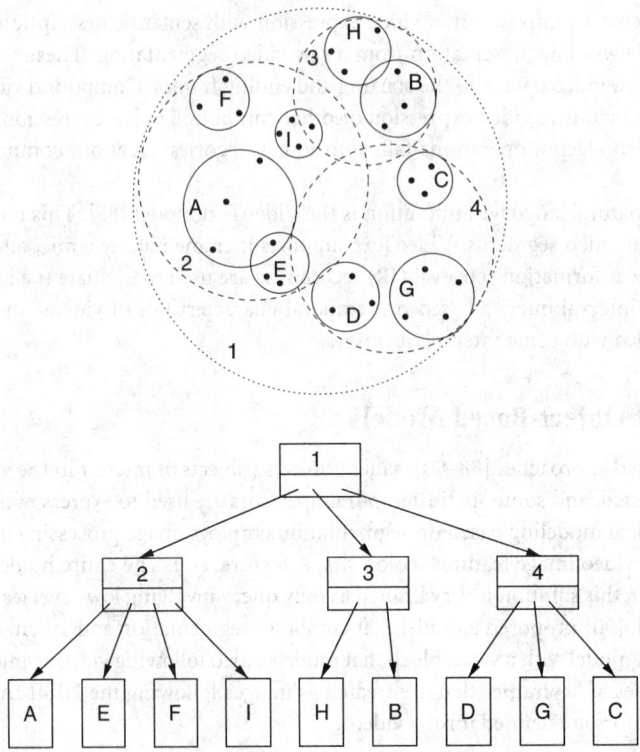

FIGURE 59.4   An SS-tree data space and the corresponding SS-tree (adapted from [97]).

expansion through a structure called mask was proposed. There has been more work on tree structures that we will summarize in the following subsection. Then we discuss the problem of dimensionality curse and the existing solutions.

59.7.1 Tree-Based Index Structures

The R-tree [95] was originally proposed to index *n*-dimensional rectangles but can be used to index *n*-dimensional points as well if points are considered as zero size rectangles. The R-tree [95] was among the first multidimensional index structures and has influenced most of work done subsequently on multidimensional indexes: R*-tree [96], SS-tree [97], and SR-tree [98] (there are others). The R-tree and the R*-tree are used for range queries and divide the universal space into multidimensional rectangles. The SS-tree is an improvement of the R*-tree by dividing the universal space into multidimensional spheres instead (better for similarity searches). The similarity search tree (SS-tree) [97] is specially designed to support similarity query in high-dimensional visual feature space. The SS-tree indexes high-dimensional vectors (points). The center of a sphere is the centroid of the underlying points. Figure 59.4 shows points in two-dimensional space followed by the corresponding SS-tree.

The SR-tree combines the use of rectangles and spheres, and it is reported that the SR-tree outperforms the R*-tree and SS-tree [98]. The SR-tree is a combination of the SS-tree and the R*-tree in that it uses both bounding spheres and bounding rectangles to group data points. Using spheres to bound data points is more suitable for similarity searches than using rectangles, because similarity search regions are usually spheres too. However, bounding spheres occupy much a larger volume than bounding rectangles when the dimensionality is high. Regions with larger volume tend to produce more overlap among themselves, which reduces similarity search efficiency. Katayama and Satoh propose the SR-tree [98] to solve this problem by integrating bounding spheres and bounding rectangles. The SR-tree specifies a region by the

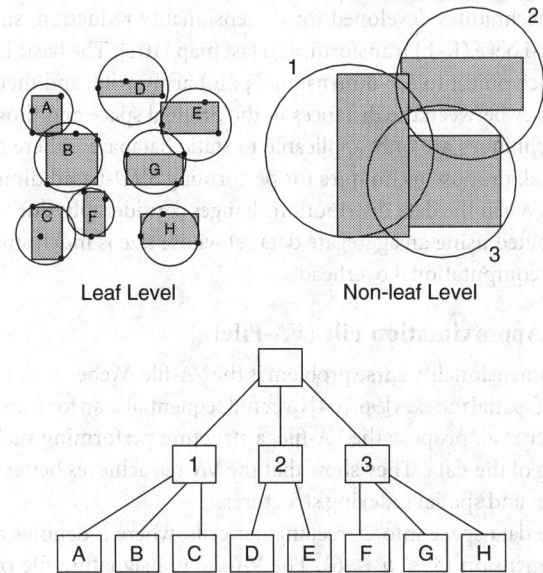

FIGURE 59.5   An SR-tree data space and the corresponding SR-tree (adapted from [98]).

intersection of a bounding sphere and a bounding rectangle (Figure 59.5). The introduction of bounding rectangles permits neighborhoods to be partitioned into smaller regions than the SS-tree and improves the disjointness among regions.

59.7.2   Dimensionality Curse and Dimensionality Reduction

The multimedia feature vectors usually have a high number of dimensions. For example, color histograms typically have at least 64 dimensions. However, it is well known that current multidimensional indexing structures suffer from "dimensionality curse," which refers to the phenomenon that the query performance of the indexing structures degrades as the data dimensionality increases. Moreover, Beyer et al. reported [99,100] a "clustering" phenomenon: as dimensionality increases, the distance to the nearest data point approaches the distance to the farthest data point. The "clustering" phenomenon can occur for as few as 10 to 15 dimensions. Under this circumstance, high-dimensional indexing is not meaningful: linear scan can outperform the R*-tree, SS-tree, and SR-tree [100]. Hence, developing more sophisticated multidimensional indexing structures is not a complete answer to the question of how to provide effective support for querying high-dimensional data. Different solutions have been proposed to address this problem: reducing the dimensionality of the data, applying a sophisticated filtering to sequential scan, and indexing the metric space.

59.7.2.1   Dimensionality Reduction

The dimensionality reduction problem is defined as: given a set of vectors in n-dimensional space, find the corresponding vectors in k-dimensional space ($k < n$) such that the distances between the points in the original space are maintained as well as possible. The following *stress* function gives the average relative error that a distance in k-dimensional space suffers from:

$$stress = \sqrt{\frac{\sum_{i,j}(d'_{ij} - d_{ij})^2}{\sum_{i,j} d_{ij}^2}},$$

where d_{ij} is the distance between objects i and j in their original n-dimensional space and d'_{ij} is their distance in the resulting k-dimensional space. Preserving distances means minimizing the stress.

There have been several techniques developed for dimensionality reduction, such as multidimensional scaling (MDS), Karhunen-Loève (K-L) transform, and fast map [101]. The basic idea of multidimensional scaling is to first assign each object to a k-dimensional point arbitrarily; and then try to move it in order to minimize the discrepancy between the distances in the original space and those in the resulting space. The above-mentioned techniques are only applicable to static databases where the set of data objects is known *a priori*. Kanth et al. propose techniques for performing SVD-based dimensionality reduction in dynamic databases [102]. When the data distribution changes considerably, due to inserts and deletes, the SVD transform is recomputed using an aggregate data set whose size is much smaller than the size of the database, in order to save computational overhead.

59.7.2.2 The Vector Approximation File (VA-File)

Another solution to the dimensionality curse problem is the VA-file. Weber et al. [103] report experiments showing little advantage of spatial indexes [96,104] over full sequential scan for feature vectors of ten or more dimensions. Hence, Weber et al. propose the VA-file, a structure performing such a scan combined with an intelligent pre-filtering of the data. They show that the VA-file achieves better performance compared to a simple sequential scan and spatial indexing structures.

The VA-file divides the data space into 2^b rectangular cells, where b denotes a user-specified number of bits to encode each dimension ($4 \leq b \leq 6$). The VA-file is a signature file containing a compressed approximation of the original data vectors. Each data vector is approximated by a bit-string encoding of the hypercube in which it lies. The hypercubes are generated by partitioning each data dimension into the number of bins representable by the number of bits used for that dimension. Typically, the compressed file is 10 to 15% of the size of the original data file. The maximum and minimum distances of a point to the hypercube provide upper and lower bounds on the distance between the query location and the original data point. In a K-nearest neighbor search, a filtering phase selects the possible K-NN points through a sequential scan of the VA-file. An approximated vector is selected if its lower bound is less than the current 5th closest upper bound. The second phase visits the candidates in ascending order of the lower bounds until the lower bound of the next candidate is greater than the actual distance to the current Kth nearest neighbor.

The pre-filtering of the VA-file requires each data point in the space to be analyzed, leading to linear complexity with a low constant.

59.7.2.3 Indexing Metric Spaces

In addition to indexing data objects in vector spaces, the indexing problem can be approached from a rather different perspective, that is, indexing in metric spaces. In metric spaces, how data objects are defined is not important (data objects may or may not be defined as vectors); what is important is the definition of the distance between data objects.

Berman proposes using triangulation tries [105] to index in metric spaces. The idea is to choose a set of *key objects* (key objects may or may not be in the datasets to be indexed), and for each object in the dataset, create a vector consisting of the ordered set of distances to the key objects. These vectors are then combined into a trie.

Space decomposition is another approach to indexing in metric spaces. An example is the *generalized hyper-plane decompositions* [106]. A generalized hyper-plane is defined by two objects o_1 and o_2 and consists of the set of objects p satisfying $d(p, o_1) = d(p, o_2)$. An object x is said to lie on the o_1-side of the plane if $d(p, o_1) < d(p, o_2)$. The generalized hyper-plane decomposition builds a binary tree. At the root node, two arbitrary objects are picked to form a hyper-plane. The objects that are on the one side of the hyper-plane are placed in one branch of the tree, and those on the other side of the hyper-plane are placed in the other branch. The lower-level branches of the tree are constructed recursively in the same manner.

The above decomposition methods build trees by a top-down recursive process, so the trees are not guaranteed to remain balanced in case of insertions and deletions. Furthermore, these methods do not consider secondary memory management, so they are not suitable for large databases that must be stored on disks. To address these problems, the M-tree [107,108] is proposed. The M-tree is a dynamic and balanced tree. Each node of the M-tree corresponds to a disk block. The M-tree uses sphere cuts to break up the metric space and is a multi-branch tree with a bottom-up construction. All data objects are stored in leaf nodes.

Metric space decompositions are made based on distance measures from some reference objects in data sets. The use of data set elements in defining partitions tends to permit exploitation of the distribution features of the data set itself, and thus may provide good query performance.

Indexing in metric spaces requires nothing to be known about the objects other than their pairwise distances. It only makes use of the properties of distance measures (symmetry, non-negativity, triangle inequality) to organize the objects and prune the search space. Thus, it can deal with objects whose topological relationships are unknown.

59.8 Multimedia Query Processing

The common query in multimedia is similarity search, where the object-retrieved are ordered according to some scores based on a distance function defined on a feature vector. In the presence of specialized indexes (e.g., an index for color features, an index for texture features), a similarity query involving the two or more features has to be decomposed into sub-queries and the sub-results integrated to obtain the final result. In relational database systems where sub-query results are not ordered, the integration is done using set operators (e.g., INTERSECTION, UNION, and DIFFERENCE). Because of the inherent order, a blind use of these set operators is not applicable in ordered sets (sequences).

The integration problem of N ranked lists has been studied for long time, both in IR and WWW research [109–112]. In both contexts, "integration" means find a scoring function able to aggregate the partial scores (i.e., the numbers representing the goodness of each returned object). However, all the proposed solutions make the assumption that a "sorted access" (i.e., a sequential scan) on the data has to exist based on some distance. In this way, it is possible to obtain the score for each data object accessing the sorted list and proceeding through such a list sequentially from the top. In other words, given a set of k lists, the problem, also named the "rank aggregation" problem, consists of finding a unique list that is a "good" consolidation of the given lists. In short, the problem consists of finding an aggregation function, such as *min*, *max*, or *avg*, that renders a consolidated distance.

Fagin [109,113] assumes that each multimedia object has a score for each of its attributes. Following the running example, an image object can have a color score and a texture score. For each attribute, a sorted list, which lists each object and its score under that attribute, sorted by score (highest score first) is available. For each object, an overall score is computed by combining the score attributes using a predefined monotonic aggregation function (e.g., average, min, and max). In particular, the algorithm uses upper and lower bounds on the number of objects that it is necessary to extract from a repository to meet the number of objects required in the consolidated list.

Fagin's approach, however, works only when the sources support sorted access of the objects. The same problem has been addressed in the Web context by Gravano et al. [110,111], in which the so-called "meta-ranking" concept is defined for data sources available on the Internet that queried separately and the results merged to compose a final result to a user query. In this work also, the authors assume the existence of scores returned together with the relevant objects.

It is known that linear combinations of scores favor correlated features. When the scores do not exist, or are not available, the integration of multimedia sub-query results cannot be performed following the above-mentioned approaches. This is the case for the Boolean models and search engines, for example. In a Boolean model, the sub-queries are logical constructs [114]. Given a query, the database is divided into a set of relevant and not relevant objects. The function is analogogous to a membership function on sets. Search engines usually do not disclose the scores given to the retrieved objects and the metrics used for evident commercial reasons. Instead, the objects are ranked.

59.9   Emerging MPEG-7 as Content Description Standard

MPEG-7, the Multimedia Content Description Interface, is an ISO metadata standard defined for the description of multimedia data. The MPEG-7 standard aims at helping with searching, filtering, processing, and customizing multimedia data through specifying its features in a "universal" format. MPEG-7 does not specify any applications but the format of the way the information contained within the multimedia data is represented, thereby supporting descriptions of multimedia made using many different formats. The objectives of MPEG-7 [115] include creating methods to describe multimedia content, manage data flexibly, and globalize data resources.

In creating methods to describe multimedia content, MPEG-7 aims to provide a set of tools for the various types of multimedia data. Usually there are four fundamental areas that can be addressed, depending on the data, so that the content is specified thoroughly. The first one in the basic fundamental areas is specifying the medium from which the document was created. This also includes the physical aspects of the medium such as what type of film it was originally shot on or information about the camera lenses. Another area concerns the physical aspects of the document. This type of information covers computational features that are not perceived by a person viewing the document. An example of such data includes the frequency of a sound in the document. Grouped with the perceptual area sometimes are the perceived descriptions. These descriptions specify the easily noticed features of the multimedia data such as the color or textures. Finally, the transcription descriptions control specifying the transcripts, or the textual representation of the multimedia information, within the MPEG-7.

MPEG-7 essentially provides two tools: the description definition language (MPEG-7 DDL) [116] for the definition of media schemes and an exhaustive set of media description schemes mainly for media low-level features. The predefined media description schemes are composed of visual feature descriptor schemes [117], audio feature descriptor schemes [118], and general multimedia description schemes [119]. Media description through MPEG-7 is achieved through three main elements: descriptors (Ds), description schemes (DSs), and a description definition language (DDL). The descriptors essentially describe a feature of the multimedia data, with a feature being a distinctive aspect of the multimedia data. An example of a descriptor would be a camera angle used in a video. The description scheme organizes the descriptions specifying the relationship between descriptors. Description schemes, for example, would represent how a picture or a movie would be logically ordered. The DDL is used to specify the schemes and allow modifications and extensions to the schemes.

The MPEG-7 DDL is a superset of XML Schema [120], the W3C schema definition language for XML documents. The extensions to XML Schema comprise support for array and matrix data types as well as additional temporal data types. MPEG-7 is commonly admitted as a multimedia content description tool and the number of MPEG-7 document available is increasing. With the increase of MPEG-7 documents, there will certainly be the need for suitable database support. Because MPEG-7 media descriptions are XML documents that conform to the XML Schema definition, it is natural to suggest XML database solutions for the management of MPEG-7 document as in [121]. Current XML database solutions are oriented toward text and MPEG-7 encodes nontextual data. Directly applying current XML database solutions to MPEG-7 will lower the expressive power because only textual queries will be allowed.

59.10   Conclusion

This chapter has surveyed multimedia databases from data analysis to querying and indexing. In multimedia databases, the raw multimedia is unfortunately not too useful because it is of large size and not meaningful by itself. Usually, data is coupled with descriptive data (low-level and possibly semantics) obtained from an analysis of the raw data.

The low-level features of an image are color, texture, and shape. Color is the most common feature used in images. There exist different color spaces that represent a color in three dimensions. The MPEG-7 standard has adopted the RGB, HSV, HMMD, Monochrome, and YCbCr color spaces. Texture captures the visual patterns that result from the presence of different color intensities and expresses the structural arrangement of the surfaces in the image. Examples of texture include tree barks, clouds, water, bricks, and fabrics. The common representation classifies textures into coarseness, contrast, directionality, line-likeness, regularity, and roughness. Object shapes are usually extracted by segmenting the image into homogeneous regions. A shape can be represented by the boundary, the region (area), and a combination of the first two representations. A video can be seen as a sequences of images and is often summarized by a sequence of keyframes (images). In addition, a video has a structure (e.g., event, scene, shot), an audio component that can be analyzed and embeds some movements. Although it is relatively easier to analyze images and videos for low-level features, it is not evident to deduce semantics from the low-level features because features do not intrinsically carry any semantics. The dichotomy between low-level features and semantics is known as the "semantic gap."

The multimedia data and the related metadata are normally stored in a database following a data model that defines the representation of the data and the operations to manipulate it. Due to the volume and the complexity of multimedia data, the analysis is performed at the acquisition of the data. Multimedia databases are commonly built on top of object or object-relational database systems. Multidimensional indexes are used to speed up query processing. The problem is that the low-level properties are represented as vectors of large size and it is well-known that beyond a certain number of dimensions, sequential scan outperforms multidimensional indexes. Current and future multimedia research is moving toward the integration of more semantics.

References

[1] R. Ramakrishnan and J. Gehrke. *Database Management Systems, 2nd ed.* McGraw-Hill, 2000.

[2] J. J. Ashley et al. Automatic and semiautomatic methods for image annotation and retrieval in query by image content (QBIC). *Proc. SPIE*, 2420:24–35, March 1995.

[3] T. S. Huang, S. Mehrotra, and K. Ramachandran. Multimedia analysis and retrieval system (MARS) project. *Proc. 33rd Annual Clinic on Library Application of Data Processing-Digital Image Access and Retrieval*, pp. 104–117, March 1996.

[4] J. R. Smith and S. F. Chang. Image and video search engine for the World Wide Web. *Proc. SPIE*, 3022:84–95, 1997.

[5] A. Pentland, R. W. Picard, W. Rosalind, and S. Sclaroff. Photobook: tools for content-based manipulation of image databases. *Proc. SPIE*, 2368:37–50, 1995.

[6] B.S. Manjunath, P. Salembier, T. Sikora, and P. Salembier. *Introduction to MPEG 7: Multimedia Content Description Language*. John Wiley & Sons, New York, June 2002.

[7] V. Castelli and L. D. Bergman. *Image Databases-Search and Retrieval of Digital Imagery*. John Wiley & Sons, New York, 2002.

[8] M. Stricker and A. Dimai. Color indexing with weak spatial constraints. *Proc. SPIE*, 2670: 29–40, 1996.

[9] Sharp Laboratories of America. Scalable blob histogram descriptor. *MPEG-7 Proposal 430, MPEG-7 Seoul Meeting*, Seoul, Korea, March 1999.

[10] J. R. Ohm, F. Bunjamin, W. Liebsch, B. Makai, and K. Muller. A set of descriptors for visual features suitable for MPEG-7 applications. *Signal Processing: Image Communication*, 16(1-2):157–180, 2000.

[11] Y. Deng, C. Kenney, M. Moore, and B. Manjunath. Peer group filtering and perceptual color quantization. *IEEE International Symposium on Circuits and Systems*, 4:21–24, 1999.

[12] R. M. Haralick, K. Shanmugam, and I. Dinstein. Texture features for image classification. *IEEE Trans. on Sys. Man. Cyb.*, SMC-3(6):1345–1350, 1973.

[13] H. Tamura, S. Mori, and T. Yamawaki. Texture features corresponding to visual perception. *IEEE Trans. Sys. Man. Cyb.*, SMC-8(6):780–786, 1978.

[14] O. Alata, C. Cariou, C. Ramannanjarasoa, and M. Najim. Classification of rotated and scaled textures using HMHV spectrum estimation and the Fourier-Mellin Transform. *ICIP'98*, pp. 53–56, 1998.

[15] X. Wan and C. C. Kuo. Image retrieval based on JPEG compressed data. *Proc. SPIE*, 2916:104–115, 1996.

[16] T. Chang and C.-C. Kuo. Texture analysis and classification with tree-structured wavelet transform. *IEEE Trans. on Image Processing*, 2(4):429–441, 1993.

[17] D. Kapur, Y. N. Lakshman, and T. Saxena. Computing invariants using elemination methods. *ICIP'95*, 3:335–341, 1995.

[18] T. P. Minka and R. W. Picard. Interactive learning using a society of models. *Pattern Recognition*, 30(3):565–581, 1999.

[19] Y. Li, Y. Ma, and H. Zhang. Salient region detection and tracking in video. *ICME'03*, 2003.

[20] J. Z. Wang. *Integrated Region-based Image Retrieval*. Kluwer Academic, 2001.

[21] J. Li and J. Z. Wang. Automatic linguistic indexing of pictures by a statistical modeling approach. *IEEE Trans. on Pattern Analysis and Machine Intelligence*, 25(9):1075–1088, 2003.

[22] M. R. Naphade, C. Lin, A. Natsev, B. Tseng, and J. Smith. A framework for moderate vocabulary semantic visual concept detection. *ICME'03*, 2003.

[23] Y. Li and C.-C. Kuo. *Video Content Analysis Using Multimodal Information: For Movie Content Extraction, Indexing and Representation*. Kluwer Academic Publishers, 2003.

[24] H. J. Zhang, C. Y. Low, S. W. Smoliar, and J. H. Wu. Video parsing, retrieval and browsing: an integrated and content-based solution. *ACM Multimedia'95*, pp. 15–24, November 1995.

[25] R. Zabih, J. Miller, and K. Mai. A feature-based algorithm for detecting and classifying scene breaks. *ACM Multimedia'95*, 1995.

[26] M. Chung, H. Kim, and S. Song. A scene boundary detection method. *ICIP'00*, 2000.

[27] R. Kasturi and R. Jain. Dynamic vision. *Computer Vision: Principles*, IEEE Computer Society Press, pp. 469–480, 1991.

[28] J. S. Boreczky and L. D. Wilcox. A hidden markov model framework for video segmentation using audio and image features. *ICASSP'98*, pages 3741–3744, Seattle, May 1998.

[29] H. J. Zhang, A. Kankanhalli, and S. W. Smoliar. Automatic partitioning of full-motion video. *Multimedia Systems*, 1(1):10–28, 1993.

[30] R. Lienhart. Comparison of automatic shot boundary detection algorithms. *Proc. SPIE*, 3656:290–301, January 1999.

[31] H. J. Zhang, S. Y. Tan, S. W. Smoliar, and G. Y. Hong. Automatic parsing and indexing of news video. *Multimedia Systems*, 2(6):256–266, 1995.

[32] D. Zhong and S. F. Chang. Structure analysis of sports video using domain models. *ICME'01*, 2001.

[33] M. Yeung, B. Yeo, and B. Liu. Extracting story units from long programs for video browsing and navigation. *IEEE Proc. Multimedia Computing & Systems*, pages 296–305, 1996.

[34] T. Zhang and C.-C. Jay Kuo. Audio-guided audiovisual data segmentation, indexing and retrieval. *Proc. SPIE*, 3656:316–327, 1999.

[35] S. Moncrieff, C. Dorai, and S. Venkatesh. Detecting indexical signs in film audio for scene interpretation. *ICME'01*, 2001.

[36] Q. Huang, Z. Liu, and A. Rosenberg. Automated semantic structure reconstruction and representation generation for broadcast news. *Proc. SPIE*, 3656:50–62, January 1999.

[37] H. Sundaram and S. F. Chang. Determining computable scenes in films and their structures using audio-visual memory models. *ACM Multimedia'00*, Marina Del Rey, CA, pp. 85–94, November 2000.

[38] Y. Li and C.-C. Kuo. A robust video scene extraction approach to movie content abstraction. *To appear in International Journal of Imaging Systems and Technology: Special Issue on Multimedia Content Description and Video Compression*, 2004.

[39] T. S. Mahmood and S. Srinivasan. Detecting topical events in digital video. *ACM Multimedia'00*, pages 85–94, Marina Del Rey, CA, November 2000.

[40] Y. Li, S. Narayanan, and C.-C. Kuo. Content-based movie analysis and indexing based on audiovisual cues. *To appear in IEEE Trans. on Circuits and Systems for Video Technology*, 2004.

[41] J. Nam, M. Alghoniemy, and A. H. Tewfik. Audio-visual content-based violent scene characterization. *ICIP'00*, 2000.

[42] E. Scheirer and M. Slaney. Construction and evaluation of a robust multifeature speech/music discrimination. *ICASSP'97*, 2:1331–1334, Munich, Germany, 1997.

[43] L. Wyse and S. Smoliar. Towards content-based audio indexing and retrieval and a new speaker discrimination technique. Downloaded from http://www.iss.nus.sg/People/lwyse/lwyse.html, Institute of Systems Science, National Univ. of Singapore, December 1995.

[44] S. Pfeiffer, S. Fischer, and W. Effelsberg. Automatic audio content analysis. *ACM Multimedia'96, Boston, MA*, PP. 21–30, November 1996.

[45] T. Zhang and C.-C. Kuo. Audio content analysis for on-line audiovisual data segmentation. *IEEE Transactions on Speech and Audio Processing*, 9(4):441–457, 2001.

[46] L. Lu, H. Jiang, and H. Zhang. A robust audio classification and segmentation method. *ACM Multimedia'01*, pp. 203–211, 2001.

[47] Z. Liu, J. Huang, and Y. Wang. Classification of TV programs based on audio information using hidden markov model. *Proc. IEEE 2nd Workshop on Multimedia Signal Processing*, pp. 27–32, December 1998.

[48] N. V. Patel and I. K. Sethi. Audio characterization for video indexing. *Proc. SPIE: Storage and Retrieval for Image and Video Databases IV*, 2670:373–384, San Jose, February 1996.

[49] K. Minami, A. Akutsu, and H. Hamada. Video handling with music and speech detection. *IEEE Multimedia*, pp. 17–25, Fall 1998.

[50] S. Pfeiffer, R. Lienhart, S. Fischer, and W. Effelsberg. Abstracting digital movies automatically. *Journal of Visual Communication and Image Representation*, 7(4):345–353, December 1996.

[51] N. Omoigui, L. He, A. Gupta, J. Grudin, and E. Sanocki. Time-compression: system concerns, usage and benefits. *Proc. ACM Conference on Computer Human Interaction*, pp. 136–143, 1999.

[52] M. Smith and T. Kanade. Video skimming for quick browsing based on audio and image characterization. Technical Report CMU-CS-95-186, Carnegie Mellon University, July 1995.

[53] C. Toklu, S. P. Liou, and M. Das. Videoabstract: a hybrid approach to generate semantically meaningful video summaries. *ICME'00*, New York, 2000.

[54] J. Nam and A. H. Tewfik. Video abstract of video. *IEEE 3rd Workshop on Multimedia Signal Processing*, pp. 117–122, September 1999.

[55] C. M. Taskiran, A. Amir, D. Ponceleon, and E. J. Delp. Automated video summarization using speech transcripts. *Proc. SPIE*, 4676:371–382, January 2002.

[56] D. M. Russell. A design pattern-based video summarization technique: moving from low-level signals to high-level structure. *Proc. 33rd Hawaii International Conference on System Sciences*, Vol. 3, January 2000.

[57] L. He, E. Sanocki, A. Gupta, and J. Grudin. Audio-summarization of audio-video presentation. *ACM Multimedia'99*, pp. 489–498, 1999.

[58] R. Lienhart. Dynamic video summarization of home video. *Proc. SPIE*, 3972:378–389, January 2000.

[59] H. J. Zhang, C. Y. Low, and S. W. Smoliar. Video parsing and browsing using compressed data. *SPIE conference on Multimedia Tools and Applications*, 1(1):89–100, 1995.

[60] Y. T. Zhuang, Y. Rui, T. S. Huang, and Sharad Mehrotra. Adaptive key frame extraction using unsupervised clustering. *ICIP'98*, 1998.

[61] S. X. Ju, M. J. Black, S. Minneman, and D. Kimber. Summarization of video-taped presentations: automatic analysis of motion and gestures. *IEEE Transactions on Circuits and Systems for Video Technology*, 8(5):686–696, 1998.

[62] C. Toklu and S. P. Liou. Automatic keyframe selection for content-based video indexing and access. *Proc. SPIE*, 3972:554–563, 2000.

[63] M. Irani and P. Anandan. Video indexing based on mosaic representation. *IEEE Trans. PAMI*, 86(5):905–921, May 1998.

[64] Y. Taniguchi, A. Akutsu, and Y. Tonomura. PanoramaExcerpts: extracting and packing panoramas for video browsing. *ACM Multimedia'97*, pp. 427–436, November 1997.

[65] S. Uchihashi, J. Foote, A. Girgensohn, and J. Boreczky. Video manga: generating semantically meaningful video summaries. *ACM Multimedia'99*, 1999.

[66] M. M. Yeung and B. L. Yeo. Video visualization for compact presentation and fast browsing of pictorial content. *IEEE Transactions on Circuits and Systems for Video Technology*, 7(5), October 1997.

[67] C. Dorai and S. Venkatesh. Computational Media Aesthetics: finding meaning beautiful. *IEEE Multimedia*, 8(4):10–12, October-December 2001.

[68] C. Dorai and S. Venkatesh, Editors. *Media Computing: Computational Media Aesthetics.* International Series in Video Computing. Kluwer Academic Publishers, June 2002.

[69] S. Santini and A. Gupta. An extensible feature management engine for image retrieval. In *Proc. SPIE Vol. 4676, Storage and Retrieval for Media Databases*, San Jose, CA, 2002.

[70] S. Santini and A. Gupta. Principles of schema design in multimedia databases. *IEEE Transactions on Multimedia Systems*, 4(2):248–259, 2002.

[71] J. Z. Li, M. T. Özsu, D. Szafron, and V. Oria. MOQL: a multimedia object query language. In *Proc. 3rd International Workshop on Multimedia Information Systems*, pp. 19–28, Como, Italy, September 1997.

[72] V. Oria, M. T. Özsu, B. Xu, L. I. Cheng, and P.J. Iglinski. VisualMOQL: The DISIMA visual query language. In *Proc. 6th IEEE International Conference on Multimedia Computing and Systems*, Vol. 1, pp. 536–542, Florence, Italy, June 1999.

[73] H. J. Zhang and S. W. Smoliar. Developing power tool for video indexing and retrieval. *Proc. SPIE*, 2185:140–149, 1994.

[74] A. Hanjalic, G. Kakes, R. Lagendijk, and J. Biemond. Indexing and retrieval of TV broadcast news using DANCERS. *Journal of Electronic Imaging*, 10(4):871–882, 2001.

[75] S. Tsekeridou and I. Pitas. Content-based video parsing and indexing based on audio-visual interaction. *IEEE Transactions on Circuits and Systems for Video Technology*, 11(4):522–535, 2001.

[76] S. Dagtas, A. Ghafoor, and R. L. Kashyap. Motion-based indexing and retrieval of video using object trajectories. *ICIP'00*, 2000.

[77] G. Iyengar and A. B. Lippman. VideoBook: an experiment in characterization of video. *ICIP'96*, 3:855–858, 1996.

[78] Y. L. Chang, W. Zeng, I. Kamel, and R. Alonso. Integrated image and speech analysis for content-based video indexing. *Proc. ICMCS*, pp. 306–313, September 1996.

[79] J. Y. Chen, C. Taskiran, A. Albiol, E. J. Delp, and C. A. Bouman. ViBE: a compressed video database structured for active browsing and search. *Proc. SPIE*, 3846:148–164, 1999.

[80] R. Tusch, H. Kosch, and L. Boszormenyi. VIDEX: an integrated generic video indexing approach. *ACM Multimedia'00*, pp. 448–451, 2000.

[81] H.J. Zhang, A. Kankanhalli, and S.W. Smoliar. Automatic partitioning of full-motion video. *ACM Multimedia Systems*, 1(1):10–28, 1993.

[82] B. Gunsel and A. M. Tekapl. Content-based video abstraction. In *Proceedings of the IEEE International Conference on Image Processing*, pp. 128–131, Chicago, IL, October 1998.

[83] W. Mahdi, M. Ardebilian, and L.M. Chen. Automatic video scene segmentation based on spatial-temporal clues and rhythm. *Networking and Information Systems Journal*, 2(5):1–25, 2000.

[84] T.G.A. Smith and G. Davenport. The stratification system: A design environment for random access video. In *Proc. Workshop on Networking and Operating System Support for Digital Audio and Video*, pp. 250–261, La Jolla, CA, November 1992.

[85] T. Jiang, D. Montesi, and A. K. Elmagarmid. Videotext database systems. In *Proc. IEEE International Conference on Multimedia Computing and Systems*, pp. 344–351, Ottawa, ON, Canada, June 1997.

[86] M. Davis. Videotext database systemmedia streams: an iconic visual language for video annotations. In *Proc. IEEE Symposium on Visual Languages*, pp. 196–202, Bergen, Norway, August 1993.

[87] R. Weiss, A. Duda, and D.K Gifford. Composition and search with a video algebra. *IEEE Multimedia Magazine*, 2(1):12–25, 1995.

[88] Y.F. Day, S. Dagtas, M. Iino, A. Khokha, and A. Ghafoor. Object-oriented conceptual modeling of video data. In *Proc. 11th IEEE International Conference on on Data Engineering*, pp. 401–408, Taipei, Taiwan, March 1995.

[89] M. Nabil, A. H.H. Ngu, and J. Shepherd. Modeling and retrieval of moving objects. *Multimedia Tools and Applications*, 13(1):35–71, 2001.

[90] L. Chen, M. T. Özsu, and V. Oria. Modeling video data for content-based queries: extending the DISIMA image data model. In *Proc. 9th International Conference on Multimedia Modeling (MMM'03)*, pp. 169–189, Taipei, Taiwan, January 2003.

[91] H. Samet. *The Design and Analysis of Spatial Data Structures*. Addison-Wesley Publishing, 1990.

[92] J. Nievergelt, H. Hinterberger, and K. C. Sevcik. The grid file: an adaptable, symmetric multikey file structure. *ACM Transactions on Database Systems*, 9(1):38–71, March 1984.

[93] M. Freeston. The BANG file: a new kind of grid file. In *Proc. ACM SIGMOD 1987 Annual Conference*, pp. 260–269, San Francisco, CA, May 1987.

[94] S. Lin, M. T. Özsu, V. Oria, and R. Ng. An extendible hash for multi-precision similarity querying of image databases. In *Proc. 27th VLDB Conference*, Rome, Italy, pp. 221–230, September 2001.

[95] A. Guttman. R-trees: a dynamic index structure for spatial searching. In *Proc. ACM SIGMOD 1984 Annual Meeting*, pp. 47–57, Boston, MA, June 1984.

[96] N. Beckmann, H. Kriegel, R. Schneider, and B. Seeger. The R*-tree: an efficient and robust access method for points and rectangles. In *Proc. 1990 ACM SIGMOD International Conference on Management of Data*, pp. 322–331, Atlantic City, NJ, May 1990.

[97] D. A. White and R. Jain. Similarity indexing with the SS-tree. In *Proc. 12th International Conference on Data Engineering*, pp. 516–523, New Orleans, LA, 1996.

[98] N. Katayama and S. Satoh. The SR-tree: an index structure for high-dimensional nearest neighbor queries. In *Proc. ACM SIGMOD International Conference on Management of Data*, pp. 369–380, Tucson, AZ, May 1997.

[99] K. S. Beyer, J. Goldstein, R. Ramakrishnan, and U. Shaft. When is "nearest neighbor" meaningful? Technical Report TR1377, Department of Computer Science, University of Wisconsin-Madison, June 1998.

[100] K. S. Beyer, J. Goldstein, R. Ramakrishnan, and U. Shaft. When is "nearest neighbor" meaningful? In *Proc. 7th International Conference on Database Theory*, pp. 217–235, Jerusalem, Israel, January 1999.

[101] C. Faloutsos and K. Lin. Fastmap: a fast algorithm for indexing, data-mining and visualization of traditional and multimedia datasets. In *Proc. 1995 ACM SIGMOD International Conference on Management of Data*, pp. 163–174, San Jose, CA, May 1995.

[102] K. V. R. Kanth, D. Agrawal, A. E. Abbadi, and A. K. Singh. Dimensionality reduction for similarity searching in dynamic databases. In *Proc. 1998 ACM SIGMOD International Conference on Management of Data*, pp. 166–176, Seattle, WA, June 1998.

[103] Roger Weber, Hans-Jörg Schek, and Stephen Blott. A quantitative analysis and performance study for similarity-search methods in high-dimensional spaces. In Ashish Gupta, Oded Shmueli, and Jennifer Widom, Editors, *VLDB'98, Proc. 24rd International Conference on Very Large Data Bases*, August 24–27, 1998, New York City, pp. 194–205. Morgan Kaufmann, 1998.

[104] Stefan Berchtold, Daniel A. Keim, and Hans-Peter Kriegel. The x-tree: an index structure for high-dimensional data. In T. M. Vijayaraman, Alejandro P. Buchmann, C. Mohan, and Nandlal L. Sarda, Editors, *VLDB'96, Proc. 22th International Conference on Very Large Data Bases*, September 3–6, 1996, Mumbai (Bombay), India, pp. 28–39. Morgan Kaufmann, 1996.

[105] A. P. Berman. A new data structure for fast approximate matching. Technical Report 1994-03-02, Department of Computer Science, University of Washington, 1994.

[106] J. K. Uhlmann. Satisfying general proximity/similarity queries with metric trees. *Information Processing Letters*, 40(4):175–179, November 1991.

[107] P. Zezula, P. Ciaccia, and F. Rabitti. M-tree: a dynamic index for similarity queries in multimedia databases. Technical Report 7, HERMES ESPRIT LTR Project, 1996. URL http://www.ced.tuc.gr/hermes/.

[108] P. Ciaccia, M. Patella, and P. Zezula. M-tree: an efficient access method for similarity search in metric spaces. In *Proc. 23rd International Conference on Very Large Data Bases*, pp. 426–435, Athens, Greece, 1997.

[109] R. Fagin. Combining fuzzy information from multiple systems. In *Proc. Fifteenth ACM SIGACT-SIGMOD-SIGART Symposium on Principles of Database Systems*, pp. 216–226, Montreal, Canada, June 1996.

[110] L. Gravano and H. García-Molina. Merging ranks from heterogeneous internet sources. In *Proc. 23rd International Conference on Very Large Data Bases (VLDB'97)*, pp. 196–205, Athens, Greece, August 1997.

[111] N. Bruno, L. Gravano, and A. Marian. Evaluating top-*k* queries over web-accessible databases. In *Proc. 18th International Conference on Data Engineering (ICDE'02)*, pp. 369–382, San Jose, CA, February 2002.

[112] R. Fagin, R. Kumar, and D. Sivakumar. Comparing top *k* lists. In *Proc. 2003 ACM SIAM Symposium on Discrete Algorithms (SODA'03)*, pp. 28–36, Baltimore, MD, January 2003.

[113] R. Fagin, A. Lotem, and M. Naor. Optimal aggregation algorithms for middleware. In *Proc. Twenteenth ACM SIGACT-SIGMOD-SIGART Symposium on Principles of Database Systems*, pp. 216–226, Santa Barbara, CA, May 2001.

[114] J. Fauqueur and N. Boujemaa. New image retrieval paradigm: logical composition of region categories to appear. In *Proc. IEEE International Conference on Image Processing (ICIP'2003)*, Barcelona, Spain, September 2003.

[115] MPEG Requirements Group. MPEG-7 context, objectives and technical roadmap. *Doc. ISO/MPEG N2861, MPEG Vancouver Meeting*, July 1999.

[116] ISO/IEC JTC 1/SC 29/WG 11. Information Technology Multimedia Content Description Interface. Part 2: Description Definition Language. International Organization for Standardization/International Electrotechnical Commission (ISO/IEC)ISO/IEC Final Draft International Standard 15938-2:2001, International Organization for Standardization/International Electrotechnical Commission, September 2001.

[117] ISO/IEC JTC 1/SC 29/WG 11. Information Technology Multimedia Content Description Interface. Part 3: Visual. International Organization for Standardization/International Electrotechnical Commission (ISO/IEC)ISO/IEC Final Draft International Standard 15938-2:2001, International Organization for Standardization/International Electrotechnical Commission, July 2001.

[118] ISO/IEC JTC 1/SC 29/WG 11. Information Technology Multimedia Content Description Interface. Part 4: Audio. International Organization for Standardization/International Electrotechnical Commission (ISO/IEC)ISO/IEC Final Draft International Standard 15938-4:2001, International Organization for Standardization/International Electrotechnical Commission, June 2001.

[119] ISO/IEC JTC 1/SC 29/WG 11. Information Technology Multimedia Content Description Interface. Part 5: Multimedia Description Schemes. International Organization for Standardization/International Electrotechnical Commission (ISO/IEC)ISO/IEC Final Draft International Standard 15938-5:2001, International Organization for Standardization/International Electrotechnical Commission, October 2001.

[120] H. Thompson, D. Beech, and M. Maloney. Xml Schema. Part 1: structures. W3C Recommendation, World Wide Web Consortium (W3C), May 2001.

[121] H. Kosch. Mpeg-7 and multimedia database systems. *ACM SIGMOD Record*, 31(2):34–39, 2002.

60

Database Security and Privacy

60.1 Introduction 60-1
60.2 General Security Principles 60-2
60.3 Access Controls 60-3
    Discretionary Access Controls • Limitation of Discretionary
    Access Controls • Mandatory Access Controls
60.4 Assurance 60-7
60.5 General Privacy Principles 60-8
60.6 Relationship Between Security
    and Privacy Principles 60-9
60.7 Research Issues 60-9
    Discretionary Access Controls • Mandatory Access Controls
    • Authorization for Advanced Database Management Systems

Sushil Jajodia
George Mason University

60.1 Introduction

With rapid advancements in computer and network technology, it is possible for an organization to collect, store, and retrieve vast amounts of data of all kinds quickly and efficiently. This, however, represents a threat to the organizations as well as individuals. Consider the following incidents of security and privacy problems:

- On November 2, 1988, Internet came under attack from a program containing a *worm*. The program affected an estimated 2000–3000 machines, bringing them to a virtual standstill.

- In 1986, a group of West German hackers broke into several military computers, searching for classified information, which was then passed to the KGB.

- According to a U.S. General Accounting Office study, authorized users (or *insiders*) were found to represent the greatest threat to the security of the Federal Bureau of Investigation's National Crime Information Center. Examples of misuse included insiders disclosing sensitive information to outsiders in exchange for money or using it for personal purposes (such as determining if a friend or a relative has a criminal record).

- Another U.S. General Accounting Office study uncovered improper accesses of taxpayer information by authorized users of the Internal Revenue Service (IRS). The report identified instances where IRS employees manipulated taxpayer records to generate unauthorized refunds and browsed tax returns that were unrelated to their work, including those of friends, relatives, neighbors, or celebrities.

The essential point of these examples is that databases of today no longer contain only data used for day-to-day data processing; they have become information systems that store everything, whether it is vital or not to an organization. Information is of strategic and operational importance to any organization; if the concerns related to security are not properly resolved, security violations may lead to losses of information that may translate into financial losses or losses whose values are obviously high by other measures (e.g., national security).

These large information systems also represent a threat to personal privacy since they contain a great amount of detail about individuals. Admittedly, the information collection function is essential for an organization to conduct its business; however, indiscriminate collection and retention of data can represent an extraordinary intrusion on the privacy of individuals.

To resolve these concerns, security or privacy issues must be carefully thought out and integrated into a system very early in its developmental life cycle. Timely attention to system security generally leads to effective measures at lower cost. A complete solution to security and privacy problems requires the following three steps:

- *Policy:* The first step consists of developing a security and privacy policy. The policy precisely defines the requirements that are to be implemented within the hardware and software of the computing system, as well as those that are external to the system such as physical, personnel, and procedural controls. The policy lays down broad goals without specifying how to achieve them. In other words, it expresses what needs to be done rather than how it is going to be accomplished.

- *Mechanism:* The security and privacy policy is made more concrete in the next step, which proposes the mechanism necessary to implement the requirements of the policy. It is important that the mechanism perform the intended functions.

- *Assurance:* The last step deals with the assurance issue. It provides guidelines for ensuring that the mechanism meets the policy requirements with a high degree of assurance. Assurance is directly related to the effort that would be required to subvert the mechanism. Low-assurance mechanisms may be easy to implement, but they are also relatively easy to subvert. On the other hand, high-assurance mechanisms can be notoriously difficult to implement.

Since most commercial database management systems (DBMSs) and database research have security rather than privacy as their main focus, we devote most of this chapter to the issues related to security. We conclude with a brief discussion of the issues related to privacy in database systems.

60.2 General Security Principles

There are three high-level objectives of security in any system:

- *Secrecy* aims to prevent unauthorized disclosure of information. The terms *confidentiality* or *nondisclosure* are synonyms for secrecy.
- *Integrity* aims to prevent unauthorized modification of information or processes.
- *Availability* aims to prevent improper denial of access to information. The term *denial of service* is often used as a synonym for denial of access.

These three objectives apply to practically every information system. For example, payroll system secrecy is concerned with preventing an employee from finding out the boss's salary; integrity is concerned with preventing an employee from changing his or her salary in the database; availability is concerned with ensuring that the paychecks are printed and distributed on time as required by law. Similarly, military command and control system secrecy is concerned with preventing the enemy from determining the target coordinates of a missile; integrity is concerned with preventing the enemy from altering the target coordinates; availability is concerned with ensuring that the missile does get launched when the order is given.

60.3 Access Controls

The purpose of access controls is to ensure that a user is permitted to perform certain operations on the database only if that user is authorized to perform them. Commercial DBMSs generally provide access controls that are often referred to as **discretionary access controls** (as opposed to the **mandatory access controls** which will be described later in the chapter).

Access controls are based on the premise that the user has been correctly identified to the system by some **authentication** procedure. Authentication typically requires the user to supply his or her claimed identity (e.g., user name, operator number, etc.) along with a password or some other authentication token. Authentication may be performed by the operating system, the DBMS, a special authentication server, or some combination thereof. Authentication is not discussed further in this chapter; we assume that a suitable mechanism is in place to ensure proper access controls.

60.3.1 Discretionary Access Controls

Most commercial DBMSs provide security by controlling modes of access by users to data. These controls are called discretionary since any user who has discretionary access to certain data can pass the data along to other users. Discretionary policies are used in commercial systems because of their flexibility; this makes them suitable for a variety of environments with different protection requirements.

There are many different administrative policies that can be applied to issue authorizations in systems that enforce discretionary protection. Some examples are *centralized* administration, where only a few privileged users may grant and revoke authorizations; *ownership-based* administration, where the creator of an object is allowed to grant and revoke accesses to the object; and *decentralized* administration, where other users, at the discretion of the owner of an object, may also be allowed to grant and revoke authorizations on the object.

60.3.1.1 Granularity and Modes of Access Control

Access controls can be imposed in a system at various degrees of granularity. In relational databases, some possibilities are the entire database, a single relation, or some rows or columns within a relation. Access controls are also differentiated by the operation to which they apply. For instance, among the basic SQL (Structured Query Language) operations, access control modes are distinguished as SELECT access, UPDATE access, INSERT access, and DELETE access. Beyond these access control modes, which apply to individual relations or parts thereof, there are also privileges which confer special authority on selected users. A common example is the DBA privilege for database administrators.

60.3.1.2 Data-Dependent Access Control

Database access controls can also be established based on the contents of the data. For example, some users may be limited to seeing salaries which are less than $30,000. Similarly, managers may be restricted to seeing the salaries only for employees in their own departments. *Views* and *query modification* are two basic techniques for implementing data-dependent access controls in relational databases.

60.3.1.3 Granting and Revoking Access

The granting and revocation operations allow users with authorized access to certain information to selectively and dynamically grant or restrict any of those access privileges to other users. In SQL, granting of access privileges is accomplished by means of the GRANT statement, which has the following general form:

```
GRANT     privileges
[ON       relation]
TO        users
[WITH     GRANT OPTION]
```

Possible privileges users can exercise on relations are *select* (select tuples from a relation), *insert* (add tuples to a relation), *delete* (delete tuples from a relation), and *update* (modify existing tuples in a relation). These access modes apply to a relation as a whole, with the exception of the update privilege, which can be further refined to refer to specific columns inside a relation. When a privilege is given with the grant option, the recipient can in turn grant the same privilege, with or without grant option, to other users.

The GRANT command applies to base relations within the database as well as views. Note that it is not possible to grant a user the grant option on a privilege without allowing the grant option itself to be further granted.

Revocation in SQL is accomplished by means of the REVOKE statement, which has the following general format:

> REVOKE   privileges
> [ON   relation]
> FROM   users

The meaning of REVOKE depends upon who executes it, as explained next.

A grant operation can be modeled as a tuple of the form $\langle s, p, t, ts, g, go \rangle$ stating that user s has been granted privilege p on relation t by user g at time ts. If $go = yes$, s has the grant option and, therefore, s is authorized to grant other users privilege p on relation t, with or without grant option. For example, tuple $\langle Bob, select, T, 10, Ann, yes \rangle$ indicates that Bob can select tuples from relation T, and grant other users authorizations to select tuples from relation T, and that this privilege was granted to Bob by Ann at time 10. Tuple $\langle C, select, T, 20, B, no \rangle$ indicates that user C can select tuples from relation T and that this privilege was granted to C by user B at time 20; this authorization, however, does not entitle user C to grant other users the select privilege on T.

The semantics of the revocation of a privilege from a user (revokee) by another user (revoker) is to consider as valid the authorizations that would have resulted had the revoker never granted the revokee the privilege. As a consequence, every time a privilege is revoked from a user, a recursive revocation may take place to delete all of the authorizations which would have not existed had the revokee never received the authorization being revoked.

To illustrate this concept, consider the sequence of grant operations for privilege p on relation t illustrated in Figure 60.1a, where every node represents a user, and an arc between node u_1 and node u_2 indicates that u_1 granted the privilege on the relation to u_2. The label of the arc indicates the time the privilege was

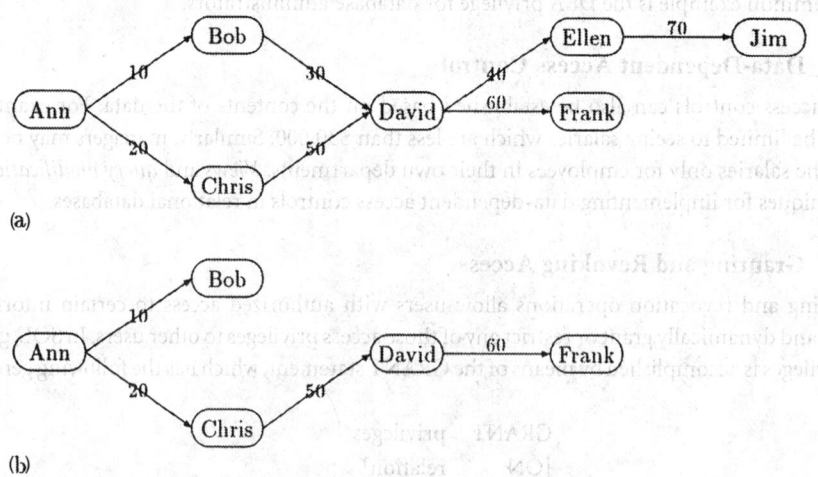

FIGURE 60.1 Bob revokes the privilege from David.

granted. For the sake of simplicity, we make the assumption that all authorizations are granted with the grant option. Suppose now that Bob revokes the privilege on the relation from David at some time later than 70. According to the semantics of recursive revocation, the resulting authorization state has to be as if David had never received the authorization from Bob, and the time of the original granting is the arbiter of this recursion. That is, if David had never received the authorization from Bob, he could not have granted the privilege to Ellen (his request would have been rejected by the system at time 40). Analogously, Ellen could not have granted the authorization to Jim. Therefore, the authorizations granted by David to Ellen and by Ellen to Jim must also be deleted. Note that the authorization granted by David to Frank does not have to be deleted since David could have granted it even if he had never received the authorization from Bob (because of the authorization from Chris at time 50). The set of authorizations holding in the system after the revocation is shown in Figure 60.1b.

60.3.2   Limitation of Discretionary Access Controls

Whereas discretionary access control mechanisms are adequate for preventing unauthorized disclosure of information to honest users, malicious users who are determined to seek unauthorized access to the data must be restricted by other devices. The main drawback of discretionary access controls is that although it allows an access only if it is authorized, it does not impose restrictions on further dissemination of information by a user once the user obtains it. This weakness makes discretionary controls vulnerable to **Trojan horse** attacks. A Trojan horse is a computer program with an apparent or actual useful function, but which contains additional *hidden* functions that surreptitiously exploit the access gained by legitimate authorizations of the invoking process. To understand how a Trojan horse can leak information to unauthorized users despite discretionary access control, consider the following example.

Suppose a user Burt (the bad guy) wants to access a file called **my_data** owned by Vic (the victim). To achieve this, Burt creates another file **stolen_data** and gives Vic the write authorization to **stolen_data** (Vic is not informed about this). Moreover, Burt modifies the code of an application generally used by Vic to include a Trojan horse containing two hidden operations, the first operation reads **my_data** and the second operation copies **my_data** into **stolen_data.** When Vic executes the application the next time, the application executes on behalf of Vic and, as a result, the personal information in **my_data** is copied to **stolen_data,** which can then be read by Burt.

This simple example illustrates how easily the restrictions stated by the discretionary authorizations can be bypassed and, therefore, the lack of assurance that results from the authorizations imposed by discretionary policies. For this reason discretionary policies are considered unsafe and not satisfactory for environments with stringent protection requirements.

To overcome this weakness further restrictions, beside the simple presence of the authorizations for the required operations, should be imposed on the accesses. To this end, the idea of *mandatory* (or *nondiscretionary*) access controls, together with a protection mechanism called the **reference monitor** for enforcing them, have been developed [Denning 1982].

60.3.3   Mandatory Access Controls

Mandatory access control policies provide a way to protect data against illegal accesses such as those gained through the use of the Trojan horse. These policies are mandatory in the sense that the accesses allowed are determined by the administrators rather than the owners of the data.

Mandatory access controls are usually based on the **Bell–LaPadula model** [Denning 1982], which is stated in terms of *subjects* and *objects*. An object is understood to be a data file, record, or a field within a record. A subject is an active process that can request access to an object. Every object is assigned a classification and every subject a clearance. Classifications and clearances are collectively referred to as *security* or *access* classes. A security class consists of two components: a hierarchical component (usually, *top secret, secret, confidential,* and *unclassified,* listed in decreasing order of sensitivity) together with a set (possibly empty) of nonhierarchical categories (e.g., NATO or Nuclear).*

Security classes are partially ordered as follows: Given two security classes L_1 and L_2, $L_1 \geq L_2$ if and only if the hierarchical component of L_1 is greater than or equal to that of L_2 and the categories in L_1 contain those in L_2. Since the set inclusion is not a total order, neither is $\geq$.

The Bell–LaPadula model imposes the following restrictions on all data accesses:

The simple security property: A subject is allowed a read access to an object only if the former's clearance is identical to or higher (in the partial order) than the latter's classification.

The ⋆-property: A subject is allowed a write access to an object only if the former's clearance is identical to or lower than the latter's classification.

These two restrictions are intended to ensure that there is no direct flow of information from high objects to low subjects.** The Bell–LaPadula restrictions are mandatory in the sense that the reference monitor checks security classes of all reads and writes and enforces both restrictions automatically. The ⋆-property is specifically designed to prevent a Trojan horse operating on behalf of a user from copying information contained in a high object to another object having a lower or incomparable classification.

60.3.3.1 Covert Channels

It turns out that a system may not be secure even if it always enforces the two Bell–LaPadula restrictions correctly. A secure system must guard against not only the direct revelation of data but also violations that do not result in the direct revelation of data yet produce illegal information flows. **Covert channels** fall into the violations of the latter type. They provide indirect means by which information by subjects within high-security classes can be passed to subjects within lower security classes.

To illustrate, suppose a distributed database uses two-phase commit protocol to commit a transaction. Further, suppose that a certain transaction requires a *ready-to-commit* response from both a secret and an unclassified process to commit the transaction; otherwise, the transaction is aborted. From a purely database perspective, there does not appear to be a problem, but from a security viewpoint, this is sufficient to compromise security. Since the secret process can send one bit of information by agreeing either to commit or not to commit a transaction, both secret and unclassified processes may cooperate to compromise security as follows: The unclassified process generates a number of transactions; it always agrees to commit a transaction, but the secret process by selectively causing transaction aborts can establish a covert channel to the unclassified process.

60.3.3.2 Polyinstantiation

The application of mandatory policies in relational databases requires that all data stored in relations be classified. This can be done by associating security classes with a relation as a whole, with individual tuples (rows) in a relation, with individual attributes (columns) in a relation, or with individual elements (attribute values) in a relation. In this chapter we assume that each tuple of a relation is assigned a classification.

The assignment of security classes to tuples introduces the notion of a *multilevel* relation. An example of a multilevel relation is shown in Table 60.1. Since the security class of the first tuple is secret, any user logged in at a lower security class will not be shown this tuple.

Multilevel relations suffer from a peculiar integrity problem known as **polyinstantiation** [Abrams et al. 1995]. Suppose an unclassified user (i.e., a user who is logged in at an unclassified security class) wants to enter a tuple in a multilevel relation in which each tuple is labeled either secret or unclassified. If the same key is already occurring in a secret tuple, we cannot prevent the unclassified user from inserting the

---

*Although this discussion is couched within a military context, it can easily be adapted to meet nonmilitary security requirements.

**The terms *high* and *low* are used to refer to two security classes such that the former is strictly higher than the latter in the partial order.

TABLE 60.1 A Multilevel Relation

STARSHIP	DESTINATION	SECURITY_CLASS
Voyager	Rigel	Secret
Enterprise	Mars	Unclassified

TABLE 60.2 A Polyinstantiated Multilevel Relation

STARSHIP	DESTINATION	SECURITY_CLASS
Voyager	Rigel	Secret
Voyager	Mars	Unclassified

unclassified tuple without leakage of one bit of information by inference. In other words the classification of the tuple has to be treated as part of the relation key. Thus unclassified tuples and secret tuples will always have different keys, since the keys will have different security classes.

To illustrate this further, consider the multilevel relation of Table 60.2, which has the key STARSHIP, SECURITY_CLASS. Suppose a secret user inserts the first tuple in this relation. Later, an unclassified user inserts the second tuple of Table 60.2 This later insertion cannot be rejected without leaking the fact to the unclassified user that a secret tuple for the Voyager already exists. The insertion is therefore allowed, resulting in the relation of Table 60.2. Unclassified users see only one tuple for the Voyager, viz., the unclassified tuple. Secret users see two tuples. There are two different ways these two tuples might be interpreted as follows:

- There are two distinct starships named Voyager going to two distinct destinations. Unclassified users know of the existence of only one of them, viz., the one going to Mars. Secret users know about both of them.
- There is a single starship named Voyager. Its real destination is Rigel, which is known to secret users. There is an unclassified cover story alleging that the destination is Mars.

Presumably, secret users know which interpretation is intended.

The main drawback of mandatory policies is their rigidity, which makes them unsuitable for many application environments. In particular, in most environments there is a need for a decentralized form of access control to designate specific users who are allowed (or who are forbidden) access to an object. Thus, there is a need for access control mechanisms that are able to provide the flexibility of discretionary access control and, at the same time, the high assurance of mandatory access control. The development of a high-assurance discretionary access control mechanism poses several difficult challenges. Because of this difficulty, the limited research effort that has been devoted to this problem has yielded no satisfactory solutions.

60.4 Assurance

In order that a DBMS meets the U.S. Department of Defense (DoD) requirements, it must also be possible to *demonstrate* that the system is secure. To this end, designers of secure DBMSs follow the concept of a **trusted computing base***(**TCB**) (also known as a *security kernel*), which is responsible for all security-relevant actions of the system. TCB mediates all database accesses and cannot be bypassed; it is small enough and simple enough so that it can be formally verified to work correctly; it is isolated from the rest of the system so that it is tamperproof.

DoD established a metric against which various computer systems can be evaluated for security. It developed a number of *levels*, A1, B3, B2, B1, C2, C1, and D, and for each level, it listed a set of requirements

---

*The reference monitor resides inside the trusted computing base.

that a system must have to achieve that level of security. Briefly, systems at levels C1 and C2 provide discretionary protection of data, systems at level B1 provide mandatory access controls, and systems at levels B2 or above provide increasing assurance, in particular against covert channels. The level A1, which is most rigid, requires verified protection of data. The D level consists of all systems which are not secure enough to qualify for any of levels A, B, or C.

Although these criteria were designed primarily to meet DoD requirements, they also provide a metric for the non-DoD world. Most commercial systems which implement security would fall into the C1 or D levels. The C2 level requires that decisions to grant or deny access can be made at the granularity of individual users. In principle, it is reasonably straightforward to modify existing systems to meet C2 or even B1 requirements. This has been successfully demonstrated by several operating system and DBMS vendors. It is not clear how existing C2 or B1 systems can be upgraded to B2 because B2 imposes modularity requirements on the system architectures. At B3 or A1 it is generally agreed that the system would need to be designed and built from scratch.

For obvious reasons the DoD requirements tend to focus on secrecy of information. Information integrity, on the other hand, is concerned with unauthorized or improper modification of information, such as caused by the propagation of viruses which attach themselves to executables. The commercial world also must deal with the problem of authorized users who misuse their privileges to defraud the organization. Many researchers believe that we need some notion of mandatory access controls, possibly different from the one based on the Bell–LaPadula model, in order to build high-integrity systems. Consensus on the nature of this mandatory access controls has been illusive.

60.5   General Privacy Principles

In this section, we describe the basic principles for achieving information privacy. These principles are made more concrete when specific mechanisms are proposed to support them:

- *Proper acquisition and retention* are concerned with what information is collected and after it is collected how long it is retained by an organization.
- *Integrity* is concerned with maintaining information on individuals that is correct, complete, and timely. The source of the information should be clearly stated, especially when the information is based on indirect sources.
- *Aggregation and derivation of data* are concerned with ensuring that any aggregations or derivations performed by an organization on its information are necessary to carry out its responsibilities. Aggregation is the combining of information from various sources. Derivation goes one step further; it uses different pieces of data to deduce or create new or previously unavailable information from the aggregates. Aggregation and derivation are important and desirable effects of collecting data and storing them in databases; they become a problem, however, when legitimate data are aggregated or used to derive information that is either not authorized by law or not necessary to the organizations. Aggregates and derived data pose serious problems since new information can be derived from available information in several different ways. Nonetheless, it is critical that data be analyzed for possible aggregation or derivation problems. With a good understanding of the ways problems may arise, it should be possible to take steps to eliminate them.
- *Information sharing* is concerned with authorized or proper disclosure of information to outside organizations or individuals. Information should be disclosed only when specifically authorized and used solely for the limited purpose specified. This information should be generally prohibited from being redisclosed by requiring that it be either returned or properly destroyed when no longer needed.
- *Proper access* is concerned with limiting access to information and resources to authorized individuals who have a demonstrable need for it in order to perform official duties. Thus, information should not be disclosed to those that either are not authorized or do not have a need to know (even if they are authorized).

Privacy protection is a personal and fundamental right of all individuals. Individuals have a right to expect that organizations will keep personal information confidential. One way to ensure this is to require that organizations collect, maintain, use, and disseminate identifiable personal information and data only as necessary to carry out their functions. In the U.S., Federal privacy policy is guided by two key legislations:

Freedom of Information Act of 1966: It establishes an openness in the Federal government by improving the public access to the information. Under this act, individuals may make written requests for copies of records of a department or an agency that pertain to them.

The Privacy Act of 1974: It provides safeguards against the invasion of personal privacy by the Federal government. It permits individuals to know what records pertaining to them are collected, maintained, used, and disseminated.

60.6 Relationship Between Security and Privacy Principles

Although there appears to be a large overlap in principle between security and privacy, there are significant differences between their objectives.

Consider the area of secrecy. Although both security and privacy seek to prevent unauthorized observation of data, security principles do not concern themselves with whether it is proper to gather a particular piece of information in the first place and, after it is collected, how long it should be retained. Privacy principles seek to protect individuals by limiting what is collected and, after it is collected, by controlling how it is used and disseminated. As an example, the IRS is required to collect only the information that is both necessary and relevant for tax administration and other legally mandated or authorized purposes. The IRS must dispose of personally identifiable information at the end of the retention periods required by law or regulation.

Security and privacy have different goals when new, more general information is deduced or created using available information. The objective of security controls is to determine the sensitivity of the derived data; any authorized user can access this new information. Privacy concerns, on the other hand, dictate that the system should not allow aggregation or derivation if the new information is either not authorized by law or not necessary to carry out the organization's responsibilities.

There is one misuse — denial of service — that is of concern to security but not privacy. In denial of service misuse, an adversary seeks to prevent someone from using features of the computer system by tying up the computer resources.

60.7 Research Issues

Current research efforts in the database security area are moving in three main directions. We refer the reader to Bertino et al. [1995] for a more detailed discussion and relevant citations.

60.7.1 Discretionary Access Controls

The first research direction concerns discretionary access control in relational DBMSs. Recent efforts are attempting to extend the capabilities of current authorization models so that a wide variety of application authorization policies can be directly supported. Related to these extensions is the problem of developing appropriate tools and mechanisms to support those models. Examples of these extension are models that permit negative authorizations, role-based and task-based authorization models, and temporal authorization models.

One extension introduces a new type of revoke operation. In the current authorization models, whenever an authorization is revoked from a user, a recursive revocation takes place. A problem with this approach is that it can be very disruptive. Indeed, in many organizations the authorizations users possess are related

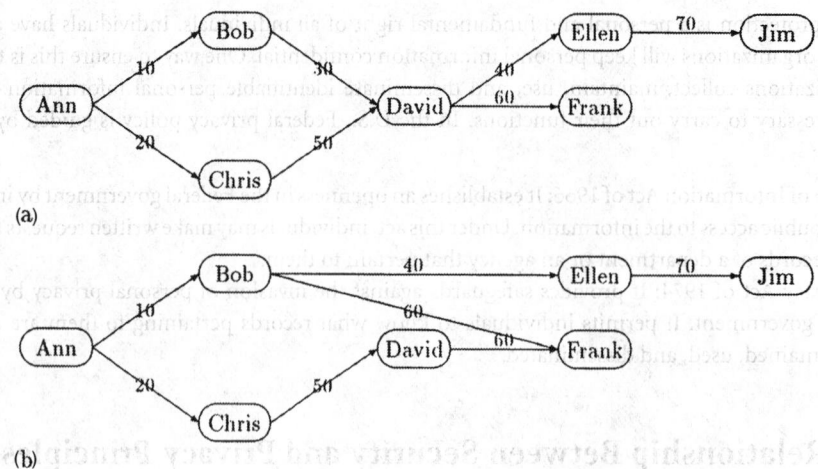

FIGURE 60.2   Bob revokes the privilege from David without cascade.

to their particular tasks or functions within the organization. If a user changes his or her task or function (for example, if the user is promoted), it is desirable to remove only the authorizations of this user, without triggering a recursive revocation of all of the authorizations granted by this user.

To support this concept, a new type of revoke operation, called *noncascading* revoke, has been introduced. Whenever a user, say Bob, revokes a privilege from another user, say David, a noncascading revoke operation would not revoke authorizations granted by David; instead, they are respecified as if they had been granted by Bob, the user issuing revocation. The semantics of the revocation without cascade is to produce the authorization state that would have resulted if the revoker (Bob) had granted the authorizations that had been granted by revokee (David).

To illustrate how noncascading revocation works, consider the sequence of authorizations shown in Figure 60.2a. Suppose now that Bob invokes the noncascading revoke operation to the privilege granted to David. Figure 60.2b illustrates the authorization state after revocation. The authorizations given by David to Ellen and Frank are respecified with Bob as the grantor and Jim retains the authorization given him by Ellen.

Another extension of current authorization models concerns negative authorizations. Most DBMSs use a *closed world* policy. Under this policy, the lack of an authorization is interpreted as a negative authorization. Therefore, whenever a user tries to access a table, if a positive authorization (i.e., an authorization permitting access) is not found in the system catalogs, the user is denied the access.

This approach has a major problem in that the lack of a given authorization for a user does not guarantee that he or she will not acquire the authorization any time in the future. That is, anyone possessing the right to administer an object can grant any user the authorization to access that object. The use of explicit negative authorizations can overcome this drawback. An explicit negative authorization expresses a *denial* for a user to access a table under a specified mode. Conflicts between positive and negative authorizations are resolved by applying the *denials-take-precedence* policy under which negative authorizations override positive authorizations. That is, whenever a user has both a positive and a negative authorization for a given privilege on the same table, the user is prevented from using that privilege on the table. The user is denied access even if a positive authorization is granted *after* a negative authorization has been granted. There are more flexible models in which negative authorizations do not always take precedence over positive authorizations [Bertino et al. 1995].

Negative authorizations can also be used for temporarily blocking possible positive authorizations of a user and for specifying exceptions. For example, it is possible to grant an authorization to all members of a group except one specific member by granting the group the positive authorization for the privilege on the table and the given member the corresponding negative authorization.

60.7.2 Mandatory Access Controls

The second research direction deals with extending the relational model to incorporate mandatory access controls. Several results have been reported for relational DBMSs, some of which have been applied to commercial products.

When dealing with multilevel secure DBMSs, there is a need to revise not only the data models but also the transaction processing algorithms. In this section, we show that the two most popular concurrency control algorithms, *two-phase locking* and *timestamp ordering*, do not satisfy the secrecy requirements.

Consider a database that stores information at two levels: low and high. Any low-level information is made accessible to all users of the database by the DBMS; on the other hand, high-level information is available only to a selected group of users with special privileges. In accordance with the mandatory security policy, a transaction executing on behalf of a user with no special privileges would be able to access (read and write) only low-level data elements, whereas a high-level transaction (initiated by a high user) would be given full access to the high-level data elements and read-only access to the low-level elements.

It is easy to see that the previous transaction rules would prevent direct access by unauthorized users to high-level data. However, there could still be ways for an ingenious saboteur to circumvent the intent of these rules, if not the rules themselves. Imagine a conspiracy of two transactions: T_L and T_H. T_L is a transaction confined to the low-level domain; T_H is a transaction initiated by a high user and, therefore, able to read all data elements. Suppose that a two-phase locking scheduler is used and that only these two transactions are currently active. If T_H requests to read a low-level data element d, a lock will be placed on d for that purpose. Suppose that next T_L wants to write d. Since d has been locked by another transaction, T_L will be forced by the scheduler to wait. T_L can measure such a delay, for example, by going into a busy loop with a counter. Thus, by selectively issuing requests to read low-level data elements, transaction T_H could modulate delays experienced by transaction T_L, effectively sending signals to T_L. Since T_H has full access to high-level data, by transmitting such signals, it could pass on to T_L the information that the latter is not authorized to see. The information channel thus created is known as a **signaling channel**.

Note that we can avoid a signaling channel by aborting the high transactions whenever a low-transaction wants to acquire a conflicting lock on a low data item. However, the drawback with this approach is that a malicious low transaction can starve a high transaction by causing it to abort repeatedly.

The standard timestamp-ordering technique also possesses the same secrecy-related flaw. Let T_L, T_H, and d be as before. Suppose that timestamps are used instead of locks to synchronize concurrent transactions. Let $ts(T_L)$ and $ts(T_H)$ be the (unique) timestamps of transactions T_L and T_H. Let $rts(d)$ be the read timestamp of data element d. (By definition, $rts(d) = \max(rts(d), ts(T))$, where T is the last transaction that read d.) Suppose that $ts(T_L) < ts(T_H)$ and T_H reads d. If, after that, T_L attempts to write d, then T_L will be aborted. Since a high-transaction can selectively cause a (cooperating) low transaction to abort, a signaling channel can be established.

Since there does not appear to be a completely satisfactory solution for single-version multilevel databases, researchers have been looking in alternative directions for solutions. One alternative is to maintain multiple versions of data instead of a single version. Using this alternative, transaction T_H will be given older versions of low-level data, thus eliminating both the signaling channels and starvations. The other alternative is to use correctness criteria that are weaker than serializability, yet they preserve database consistency in some meaningful way.

60.7.3 Authorization for Advanced Database Management Systems

A third direction concerns the development of adequate authorization models for advanced DBMSs, like object-oriented DBMSs or active DBMSs. These DBMSs are characterized by data models that are richer than the relational model. Advanced data models often include notions such as inheritance hierarchies, composite objects, versions, and methods. Therefore, authorization models developed for relational DBMSs must be properly extended to deal with the additional modeling concepts.

Authorization models developed in the framework of relational DBMSs need substantial extensions to be suitable for object-oriented DBMSs (OODBMSs). The main requirements driving such extensions can be summarized as follows. First, the authorization model must account for all semantic relationships which may exist among data (i.e., inheritance, versioning, or composite relationship). For example, in order to execute some operation on a given object (e.g., an instance), the user may need to have the authorization to access other objects (e.g., the class to which the instance belongs). Second, administration of authorizations becomes more complex. In particular, the ownership concept does not have a clear interpretation in the context of object-oriented databases. For example, a user can create an instance from a class owned by some other user. As a result, it is not obvious who should be considered the owner of the instance and administer authorizations to access the instance. Finally, different levels of authorization granularity must be supported. Indeed, in object-oriented database systems, objects are the units of access. Therefore, the authorization mechanism must allow users to associate authorizations with single objects. On the other hand, such fine granularity may decrease performance when accessing sets of objects, as in the case of queries. Therefore, the authorization mechanisms must allow users to associate authorizations with classes, or even class hierarchies, if needed. Different granularities of authorization objects are not required in relational DBMSs, where the tuples are always accessed in a set-oriented basis, and thus authorizations can be associated with entire relations or views.

Some of those problems have been been addressed by recent research. However, work in the area of authorization models for object-oriented databases is still at a preliminary stage. Of the OODBMSs, only Orion and Iris provide authorization models comparable to the models provided by current relational DBMSs.

With respect to mandatory controls, the Bell–LaPadula model is based on the subject-object paradigm. Application of this paradigm to object-oriented systems is not straightforward. Although this paradigm has proven to be quite effective for modeling security in operating systems as well as relational databases, it appears somewhat forced when applied to object-oriented systems. The problem is that the notion of an object in the object-oriented data model does not correspond to the Bell–LaPadula notion of an object. The former combines the properties of a passive information repository, represented by attributes and their values, with the properties of an active entity, represented by methods and their invocations. Thus, the object of the object-oriented data model can be thought of as the object and the subject of the Bell–LaPadula paradigm fused into one. Moreover, as with relational databases, the problem arises of assigning security classifications to information stored inside objects. This problem is made more complex by the semantic relationships among objects which must be taken into consideration in the classification. For example, the access level of an instance cannot be lower than the access level of the class containing the instance; otherwise, it would not be possible for a user to access the instance.

Some work has been performed on applying the Bell–LaPadula principles to object-oriented systems. A common characteristic to the various models is the requirement that objects must be single level (i.e., all attributes of an object must have the same security level). A model based on single-level objects has the important advantage of making the security monitor small enough that it can be easily verified. However, entities in the real world are often multilevel: some entities may have attributes with different levels of security. Since much modeling flexibility would be lost if multilevel entities could not be represented in the database, most of the research work on applying mandatory policies to object-oriented databases has dealt with the problem of representing these entities with single-level objects.

Defining Terms

Authentication: The process of verifying the identity of users.

Bell–LaPadula model: A widely used formal model of mandatory access control. It requires that the simple security property and the ★-property be applied to all subjects and objects.

Covert channel: Any component or feature of a system that is misused to encode or represent information for unauthorized transmission, without violating access control policy of the system.

Discretionary access controls: Means of restricting access. Discretionary refers to the fact that the users at their discretion can specify to the system who can access their files.

Mandatory access controls: Means of restricting access. Mandatory refers to the fact that the security restrictions are applied to all users. Mandatory access control is usually based on the Bell–LaPadula security model.

Polyinstantiation: A multilevel relation containing two or more tuples with the same primary key values but differing in security classes.

Reference monitor: Mechanism responsible for deciding if an access request of a subject for an object should be granted or not. In the context of multilevel security, it contains security classes of all subjects and objects and enforces two Bell–LaPadula restrictions faithfully.

Signaling channel: A means of information flow inherent in the basic model, algorithm, or protocol and, therefore, implementation invariant.

Trojan horse: A malicious computer program that performs some apparently useful function but contains additional hidden functions that surreptitiously leak information by exploiting the legitimate authorizations of the invoking process.

Trusted computing base: Totality of all protection mechanisms in a computer system, including all hardware, firmware, and software that is responsible for enforcing the security policy.

References

Abrams, M. D., Jajodia, S., and Podell, H. J., Eds. 1995. *Information Security: An Integrated Collection of Essays.* IEEE Computer Society Press.

Adam, N. R. and Wortmann, J. C. 1989. Security-control methods for statistical databases: a comparative study. *ACM Comput. Sur.* 21(4):515–556.

Amoroso, E. 1994. *Fundamentals of Computer Security Technology.* Prentice–Hall, Englewood Cliffs, NJ.

Bertino, E., Jajodia, S., and Samarati, P. 1995. Database security: research and practice. *Inf. Syst.* 20(7): 537–556.

Castano, S., Fugini, M., Martella, G., and Samarati, P. 1994. *Database Security.* Addison–Wesley, Reading, MA.

Cheswick, W. R. and Bellovin, S. M. 1994. *Firewalls and Internet Security.* Addison–Wesley, Reading, MA.

Denning, D. E. 1982. *Cryptography and Data Security.* Addison–Wesley, Reading, MA.

Kaufman, C., Perlman, R., and Speciner, M. 1995. *Network Security: Private Communication in a Public World.* Prentice–Hall, Englewood Cliffs, NJ.

Further Information

In this chapter, we have mainly focused on the security issues related to DBMSs. It is important to note, however, that the security measures discussed here constitute only a small aspect of overall security. As an increasing number of organizations become dependent on access to their data over the Internet, network security is also critical.

The most popular security measure these days is a *firewall* [Cheswick and Bellovin 1994]. A firewall sits between an organization's internal network and the Internet. It monitors all traffic from outside to inside and blocks any traffic that is unauthorized. Although firewalls can go a long way to protect organizations against the threat of intrusion from the Internet, they should be viewed only as the first line of defense. Firewalls are not immune to penetrations; once an outsider is successful in penetrating a system, firewalls typically do not provide any protection for internal resources. Moreover, firewalls do not protect against security violations from *insiders*, who are an organization's authorized users. Most security experts believe that insiders are responsible for a vast majority of computer crimes.

For general reference on computer security, refer to Abrams et al. [1995], Amoroso [1994], and Denning [1982]. Text by Castano et al. [1994] is specific to database security. Kaufman et al. [1995] deals with security for computer networks. Security in statistical databases is covered in Denning [1982] and in the survey by Adam and Wortmann [1989].

VII

Intelligent Systems

The study of Intelligent Systems, often called "artificial intelligence" (AI), uses computation as a medium for simulating human perception, cognition, reasoning, learning, and action. Current theories and applications in this area are aimed at designing computational mechanisms that process visual data, understand speech and written language, control robot motion, and model physical and cognitive processes. Fundamental to all AI applications is the ability to efficiently search large and complex information structures and to utilize the tools of logic, inference, and probability to design effective approximations for various intelligent behaviors.

61 Logic-Based Reasoning for Intelligent Systems *James J. Lu and Erik Rosenthal* ... **61**-1
Introduction • Underlying Principles • Best Practices • Research Issues and Summary

62 Qualitative Reasoning *Kenneth D. Forbus* **62**-1
Introduction • Qualitative Representations • Qualitative Reasoning
Techniques • Applications of Qualitative Reasoning • Research Issues and Summary

63 Search *D. Kopec, T.A. Marsland, and J.L. Cox* **63**-1
Introduction • Uninformed Search Methods • Heuristic Search Methods • Game-Tree
Search • Parallel Search • Recent Developments

64 Understanding Spoken Language *Stephanie Seneff and Victor Zue* **64**-1
Introduction • Underlying Principles • Best Practices • Research Issues and Summary

65 Decision Trees and Instance-Based Classifiers *J. Ross Quinlan* **65**-1
Introduction • Decision Trees • Instance-Based Approaches • Composite Classifiers

66 Neural Networks *Michael I. Jordan and Christopher M. Bishop* **66**-1
Introduction • Representation • Learning from Data • Graphical Models

67 Planning and Scheduling *Thomas Dean and Subbarao Kambhampati* **67**-1
Introduction • Classifying Planning Problems • Algorithms, Complexity, and
Search • Research Issues and Summary

68 Explanation-Based Learning *Gerald DeJong* **68**-1
Introduction • Background • Explanation-Based and Empirical
Learning • Explanation-Based Learning • Constructing
Explanations • Generalizing • The Utility Problem: Selecting a Concept
Descriptor • Annotated Summary of EBL Issues

69 Cognitive Modeling *Eric Chown* **69**-1
Introduction • Underlying Principles • Research Issues • Best Practices • Summary

70 Graphical Models for Probabilistic and Causal Reasoning *Judea Pearl* **70**-1
Introduction • Historical Background • Bayesian Networks as Carriers of Probabilistic
Information • Bayesian Networks as Carriers of Causal Information • Counterfactuals

71 Robotics *Frank L. Lewis, John M. Fitzgerald, and Kai Liu* **71**-1
Introduction • Robot Workcells • Workcell Command and Information
Organization • Commercial Robot Configurations and Types • Robot Kinematics,
Dynamics, and Servo-Level Control • End Effectors and End-of-Arm
Tooling • Sensors • Workcell Planning • Job and Activity Coordination
• Error Detection and Recovery • Human Operator Interfaces • Robot Workcell
Programming • Mobile Robots and Automated Guided Vehicles

61
Logic-Based Reasoning for Intelligent Systems

61.1 Introduction **61**-1

61.2 Underlying Principles **61**-2
Propositional Logic • Inference and Deduction • First-Order Logic

61.3 Best Practices **61**-4
Classical Logic • Resolution • The Method of Analytic Tableaux and Path Dissolution • Model Finding in Propositional Logic • Nonclassical Logics

61.4 Research Issues and Summary **61**-20

James J. Lu
Emory University

Erik Rosenthal
University of New Haven

61.1 Introduction

Modern interest in artificial intelligence (AI) is coincident with the development of high-speed digital computers. Shortly after World War II, many hoped that truly intelligent machines would soon be a reality. In 1950, Turing, in his now-famous article "*Computing Machinery and Intelligence*," which appeared in the journal *Mind*, predicted that machines would duplicate human intelligence by the end of the century. In 1956, at a workshop held at Dartmouth College, McCarthy introduced the term "*artificial intelligence*," and the race was on. The first attempts at mechanizing reasoning included Newell and Simon's 1956 computer program, the *Logic Theory Machine*, and a computer program developed by Wang that proved theorems in propositional logic.

Early on, it was recognized that **automated reasoning** is central to the development of machine intelligence, and central to automated reasoning is automated theorem proving, which can be thought of as mechanical techniques for determining whether a logical formula is satisfiable. The key to automated theorem proving is inference rules that can be implemented as algorithms. The first major breakthrough was Robinson's landmark paper in 1965, in which the resolution principle and the unification algorithm were introduced [Robinson 1965]. That paper marked the beginning of a veritable explosion of research in machine-oriented logics.

The focus of this chapter is on reasoning through mechanical inference techniques. The fundamental principles underlying a number of logics are introduced, and several of the numerous theorem proving techniques that have been developed are explored.

The underlying logics can generally be classified as **classical** — roughly, the logic described by Aristotle — or as **nonstandard** — logics that were developed somewhat more recently. Most of the alternative logics that have been proposed are extensions of classical logic, and inference methods for them are typically based on classical deduction techniques. Reasoning with uncertainty through **fuzzy logic** and nonmonotonic reasoning through **default logics**, for example, have for the most part been based on variants and extensions of classical proof techniques; see, for example, Reiter's paper in Bobrow [1980] and Lee [1972] and Lifschitz [1995].

This chapter touches on three disciplines: artificial intelligence, automated theorem proving, and symbolic logic. Many researchers have made important contributions to each of them, and it is impossible to describe all logics and inference rules that have been considered. We believe that the methodologies described are typical and should give the reader a basis for further exploration of the vast and varied literature on the subject.

61.2 Underlying Principles

We begin with a brief review of **propositional classical logic**. Propositional logic may not be adequate for many reasoning tasks, so we also examine first-order logic and consider some nonstandard logics. An excellent (and more detailed, albeit somewhat dated) exposition of the fundamentals of computational logic is the book by Chang and Lee [1973].

61.2.1 Propositional Logic

A *proposition* is a statement that is either true or false (but not both); *carbon is an element* and *Hercules is president* are both examples. A single proposition is often called an *atomic formula* or simply an *atom*. Logical formulas are built from a set $\mathcal{A}$ of atoms, a set of connectives, and a set of logical constants in the following way: atoms and constants are formulas; if Θ is an *n*-ary connective and if $\mathcal{F}_1, \mathcal{F}_2, \ldots, \mathcal{F}_n$ are formulas, then so is $\Theta(\mathcal{F}_1, \mathcal{F}_2, \ldots, \mathcal{F}_n)$. The expression

$$(A \vee \neg B) \leftrightarrow (B \rightarrow true)$$

is an example with one constant *true*, two atoms A and B, one unary connective $\neg$ (*negation*), and three binary connectives $\vee$ (*logical or*), $\leftrightarrow$ (*logical if and only if*), and $\rightarrow$ (*logical implication*). Negated atoms play a special role in most deduction techniques, and the word **literal** is used for an atom or for the negation of an atom.

The semantics (meaning) of a logical formula are characterized by truth values. In classical logic, the possible truth values are *true* and *false*; in general, the set of truth values may be any set.

Connectives can be viewed as function symbols appearing in the strings representing formulas, or, less formally, if Δ is the set of truth values, then an *n*-ary connective is a function from Δ^n to Δ. In classical logic, $\Delta = \{true, false\}$; there are four functions from Δ to Δ, and only one is an interesting connective: standard negation. There are 16 binary connectives (i.e., functions from $\Delta \times \Delta$ to Δ). Of particular interest are conjunction $\wedge$ and disjunction $\vee$.

Most **inference rules** assume that formulas have been normalized in some manner. A formula is in **negation normal form** (**NNF**) if conjunction and disjunction are the only binary connectives and if all negations are at the atomic level. A formula is in **conjunctive normal form** (**CNF**) if it is a conjunction of clauses, where a clause is a disjunction of literals; the term "**clause form**" refers to CNF. Observe that CNF is a special case of NNF.

An **interpretation** is a function from the atom set $\mathcal{A}$ to the set Δ of truth values. In practice, we use the word "interpretation" for a *partial interpretation* for a formula $\mathcal{F}$: an assignment of a truth value only to the atoms that occur in $\mathcal{F}$. Interpretations can be extended to complex formulas according to the functions represented by the connectives in the logic. A formula in classical logic is **satisfiable** if it evaluates to *true* under some interpretation. For other logics, where the set Δ of truth values is arbitrary, *satisfiability* is determined by a designated subset Δ^*. That is, an interpretation I satisfies a formula $\mathcal{F}$ if I maps $\mathcal{F}$ to a truth value in Δ^*. A formula C is said to be a **logical consequence** of $\mathcal{F}$ if every interpretation I that satisfies $\mathcal{F}$ also satisfies C; in that case, we write $\mathcal{F} \models C$.

61.2.2 Inference and Deduction

To paraphrase Hayes [1977], the meaning and the implementation of a logic meet in the notion of inference. Automated inference techniques can roughly be put into one of two categories: **inference rules** and **rewrite rules**. Inference rules are applied to a formula, producing a conclusion that is conjoined to the original formula. When a rewrite rule is applied to a formula, the result is a new formula in which the original formula may not be present. The distinction is really not that clear: the rewritten formula can be interpreted as a conclusion and conjoined to the original formula. We will consider examples of both. Resolution is an inference rule, and the **tableau method** and **path dissolution** can be thought of as rewrite rules.

Inference rules can be written in the following general form:

$$\frac{premise}{conclusion} \tag{61.1}$$

where *premise* is a set of formulas* and *conclusion* is a formula. A *deduction* of a formula C from a given set of formulas S is a sequence

$$C_0, C_1, \ldots, C_n$$

such that $C_0 \in S, C_n = C$, and for each $i, 1 \leq i \leq n, C_i$ satisfies one of the following conditions:

1. $C_i \in S$
2. There is an inference rule (*premise/conclusion*) such that *premise* $\subseteq \{C_0, C_1, \ldots, C_{i-1}\}$, and $C_i =$ *conclusion*

We use the notation $S \vdash C$ to indicate that there is a deduction of C from S.

A simple example of an inference rule is

$$\frac{premise}{chocolate\ is\ good\ stuff}$$

that is, *chocolate is good stuff* is inferred from any premise. We have several colleagues who can really get behind this particular rule, but it appears to lack something from the automated reasoning point of view. To avoid the possible problems inherent in this rule, there are two standards against which inference rules are judged:

1. **Soundness**: Suppose $\mathcal{F} \vdash C$. Then $\mathcal{F} \models C$.
2. **Completeness**: Suppose $\mathcal{F} \models C$. Then $\mathcal{F} \vdash C$.

Of the two properties, the first is the more important; the ability to draw valid (and only valid!) conclusions is more critical than the ability to draw all valid conclusions. In practice, many researchers are interested in *refutation* completeness, that is, the ability to verify that an unsatisfiable formula is, in fact, unsatisfiable. As we shall see when considering nonmonotonic reasoning, even soundness may not always be a desirable property.

61.2.3 First-Order Logic

Theorem proving often requires **first-order logic**. Typically, one starts with propositional inference rules and then employs some variant of Robinson's **unification algorithm** and the **lifting lemma** [Robinson 1965]. In this section we present the basics of first-order logic.

Atoms are usually called *predicates* in first-order logic, and predicates are allowed to have arguments. For example, if M is the predicate "is a man," then $M(x)$ may be interpreted as "x is a man." Thus, M(Socrates), $M(7)$, and (because function symbols are allowed) $M(f(x))$ are all well formed. In general, predicates can have any (finite) number of arguments, and any *term* can be substituted for any argument. Terms are

---

*In most settings — certainly in this chapter — a *set of formulas* is essentially the conjunction of the formulas in the set.

defined recursively as follows: variables and constant symbols are terms, and if $t_1, t_2, \ldots, t_n$ are terms and if f is an n-ary function symbol, then $f(t_1, t_2, \ldots, t_n)$ is a term.

First-order formulas are essentially the same as propositional formulas, with the obvious exception that the atoms that appear are predicates. However, first-order formulas can be **quantified**. In the following example, c is a constant, x is a *universally* quantified variable, and y is *existentially* quantified; the unquantified variable z is said to be *quantifier-free* or simply *free*,

$$\forall x \exists y (P(x, y) \vee \neg Q(y, z, c))$$

Interpretations at the first-order level are different because a **domain of discourse** over which the variables may vary must be selected. If $\mathcal{F}$ is a formula with n free variables, if I is an interpretation, and if $\mathcal{D}$ is the corresponding domain of discourse, then I maps $\mathcal{F}$ to a function from $\mathcal{D}^n$ to Δ. A *valuation* is an assignment of variables to elements of $\mathcal{D}$. Under interpretation I and valuation V, a formula $\mathcal{F}$ yields a truth value, and two formulas are said to be *equivalent* if they evaluate to the same truth value under all interpretations and valuations.

Of particular importance to the theoretical development of inference techniques in AI is the class of *Herbrand interpretations*. These are interpretations whose domain of discourse is the **Herbrand universe**, which is built from the variable-free terms in the given formula. It can be defined recursively as follows. Let $\mathcal{F}$ be any formula. Then, H_0 is the set of constants that appear in $\mathcal{F}$. If there are no constants, let a be any constant symbol, and let $H_0 = \{a\}$. For each nonnegative integer n, H_{n+1} is the union of H_n and the set of all terms of the form $f(t_1, t_2, \ldots, t_m)$, where $t_i \in H_n$ for $i = 0, 1, 2, \ldots, m$. Then the Herbrand universe is $H = \bigcup_{i=0}^{\infty} H_i$. The importance of Herbrand interpretations is made clear by the following theorem: *A formula $\mathcal{F}$ is unsatisfiable if and only if $\mathcal{F}$ is unsatisfiable under Herbrand interpretations.*

In general, it is possible to transform any first-order formula to an equivalent (i.e., truth preserving) **prenex normal form:** all quantifiers appear in the front of the formula. A formula $\mathcal{F}$ in prenex normal form can be further normalized to a satisfiability preserving **Skolem standard form** $\mathcal{G}$: All existentially quantified variables are replaced by constants or by functions of constants and the universally quantified variables. *Skolemizing* a formula in this manner preserves satisfiability: $\mathcal{F}$ is satisfiable if and only if $\mathcal{G}$ is.*

Because the quantifiers appearing in $\mathcal{G}$ are all universal, we can (and typically do) write $\mathcal{G}$ without quantifiers, it being understood that all variables are universally quantified.

A **substitution** is a function that maps variables to terms. Any substitution can be extended in a straightforward way to apply to arbitrary expressions. Given a set of expressions $E_1, \ldots, E_n$, each of which can be a term, an atom, or a clause, a substitution θ is a **unifier** for the set $\{E_1, \ldots, E_n\}$ if $\theta(E_1) = \theta(E_2) = \cdots = \theta(E_n)$. A unifier θ of a set of expressions E is called *the most general unifier* (mgu) if given any unifier γ of E, $\gamma \circ \theta = \gamma$. For example, the two expressions $P(a, y), P(x, f(z))$ are unifiable via the substitution θ_1, which maps y to $f(z)$ and x to a. They are also unifiable via the substitution θ_2, which maps y to $f(a)$, z to a, and x to a. The substitution θ_1 is more general than θ_2. When a substitution is applied to a formula, the resulting formula is called an *instance* of the given formula.

Robinson's unification algorithm [Robinson 1965] provides a means of finding the mgu of any set of unifiable expressions. Robinson proved the lifting lemma in the same paper, and the two together represent what may be the most important single advance in automated theorem proving.

61.3 Best Practices

61.3.1 Classical Logic

Not surprisingly, the most widely adopted logic in AI systems is classical (two-valued) logic. The truth value set Δ is {*true, false*}, and the designated truth value set Δ^* is {*true*}. Some examples of AI programs based on classical logic include problem solvers such as Green's program [Green 1969], theorem provers such

---

*Perhaps surprisingly, Skolemization does not, in general, preserve equivalence.

as OTTER [McCune 1992], Astrachan's METEOR (see Wrightson [1994]), the Boyer and Moore [1979] theorem prover, the *Rewrite Rule Laboratory* [Kapur and Zhang 1989], and a number of model finding systems for propositional logic [Moskewcz et al. 2001, Zhang and Stickel 2000, Selman et al. 1992]. There are several deduction-based programming languages such as Prolog; a good source is the book by Sterling and Shapiro [1986]. In this section we describe one inference rule (**resolution**) and two rewrite rules (the **tableau method** and its generalization, **path dissolution**). These methods are *refutation* complete; that is, they verify that an unsatisfiable formula is in fact unsatisfiable. In contrast, Section 61.3.4 on "Model Finding in Propositional Logic" examines several complete and incomplete techniques for finding *models*, that is, for finding satisfying interpretations of a formula, if any exist.

61.3.2 Resolution

Perhaps the most widely applied inference rule in all of AI is the resolution principle of Robinson [1965]. It assumes that each formula is in CNF (a conjunction of clauses). To define resolution for propositional logic, suppose we have a formula in CNF containing the two clauses in the premise; then the conclusion may be inferred,

$$\frac{(A_1 \vee A_2 \vee \cdots \vee A_m \vee L) \wedge (B_1 \vee B_2 \vee \cdots \vee B_n \vee \neg L)}{(A_1 \vee A_2 \vee \cdots \vee A_m \vee B_1 \vee B_2 \vee \cdots \vee B_n)} \tag{61.2}$$

The conclusion is called the *resolvent*, and the two clauses in the premise are called the *parent clauses*.

It is easy to see why resolution is sound. If an interpretation satisfies the formula, then it must satisfy every clause. Since L and $\neg L$ cannot simultaneously evaluate to *true*, one of the other literals must be *true*. Resolution is also complete; the proof is beyond the scope of this chapter.

The lifting lemma [Robinson 1965] enables the application of resolution to formulas in first-order logic. Roughly speaking, it says that if instances of two clauses can be resolved, then the clauses can be unified and resolved. The effect is that two first-order clauses can be resolved if they contain, respectively, positive and negative unifiable occurrences of the same predicate. To state the first-order resolution inference rule, let L_1 and L_2 be two occurrences of the same predicate (one positive, one negative) and let θ be the mgu of L_1 and L_2. Then,

$$\frac{(A_1 \vee A_2 \vee \cdots \vee A_m \vee L) \wedge (B_1 \vee B_2 \vee \cdots \vee B_n \vee \neg L)}{(\theta(A_1) \vee \theta(A_2) \vee \cdots \vee \theta(A_m) \vee \theta(B_1) \vee \theta(B_2) \vee \cdots \vee \theta(B_n))} \tag{61.3}$$

In practice, an implementation based on resolution alone has limited value because unrestricted resolution tends to produce an enormous number of inferences. There are several approaches to controlling the search space. One is the *set of support* strategy, which identifies a subset of the original set of clauses as its set of support, and then insists that at least one parent clause in every resolvent come from the set of support. Another strategy, one that is especially useful for logic programming, is the *linear restriction*, wherein one parent clause must be the most recent resolvent. These strategies may be thought of as control of the deduction process through meta-level restrictions on the search space.

As a simple example, consider the knowledge that "Tweety is a canary," "a canary is a bird," and "a bird flies," encoded as the following set of clauses:

$$Canary(I)$$

$$\neg Canary(x) \vee Bird(x)$$

$$\neg Bird(x) \vee Flies(x)$$

A linear resolution deduction of the fact that Tweety flies can be obtained as shown in Figure 61.1. The substitutions θ_1 and θ_2 both map x to the constant *Tweety*.

Bundy et al. [1988] argue that "[logic] provides only a low-level, step-by-step understanding, whereas a high-level, strategic understanding is also required," and there are restrictions on the search space that attempt to incorporate some kind of understanding. For example, equality and inequality have a special status in many theories. An axiom $a = b$ in a theory typically indicates that a and b can be used

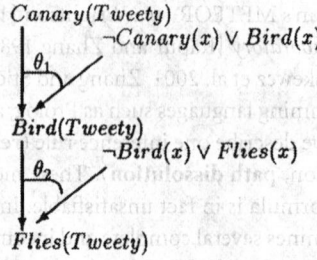

FIGURE 61.1   A deduction of *Flies(Tweety)*.

interchangeably in any context. Formally, the following equality axioms are implicitly assumed for theories requiring this property:

1. (Reflexivity) $x = x$.
2. (Symmetry) $(x = y) \rightarrow (y = x)$.
3. (Transitivity) $(x = y) \wedge (y = z) \rightarrow (x = z)$.
4. (Substitution 1) $(x_i = y) \wedge P(x_1, \ldots, x_i, \ldots, x_n) \rightarrow P(y_1, \ldots, y, \ldots, y_n)$ for $1 \leq i \leq n$, for each n-ary predicate symbol P.
5. (Substitution 2) $(x_i = y) \rightarrow f(x_1, \ldots, x_i, \ldots, x_n) = f(x_1, \ldots, y, \ldots, x_n)$ for $1 \leq i \leq n$, for each n-ary function symbol f.

The explicit incorporation of these axioms tends to drastically increase the search space, so Robinson and Wos [1969] proposed a specialized inference rule, **paramodulation**, for handling equality. Let $L[t]$ be a literal, let θ be the mgu of r and t, and let $\theta(L[s])$ be the literal obtained from $L[t]$ by replacing one occurrence of $\theta(t)$ in $L[t]$ with $\theta(s)$:

$$\frac{L[t] \vee D_1, (r = s) \vee D_2}{\theta(L[s]) \vee \theta(D_1) \vee \theta(D_2)} \tag{61.4}$$

The conclusion is called the *paramodulant* of the two clauses.

Using the Tweety example, suppose we have the additional knowledge that Tweety is known by the alias "Fred" (i.e., *Tweety = Fred*). Then the question, "Can Fred fly?" may be answered by extending the resolution proof shown in Figure 61.1 with the paramodulation inference, which substitutes the constant *Tweety* in the conclusion *Flies(Tweety)* with *Fred* to obtain *Flies(Fred)*.

An inference rule such as paramodulation is semantically based because its definition comes from unique properties of the predicate and function symbols. Paramodulation treats the equality symbol $=$ with a special status that enables it to perform larger inference steps. Other semantically based inference rules can be found in Slagle [1972], Manna and Waldinger [1986], Stickel [1985], and Bledsoe et al. [1985].

Controlling paramodulation in an implementation is difficult. One system designed to handle equality is the RUE* system of Digricoli and Harrison [1986]. Its goal-directed nature tends to produce better computational behavior than paramodulation. The essential idea, illustrated in the following example, is to build the two substitution axioms into resolution. Let S be the set of clauses

$$\{P(f(a)), \neg P(f(b)), (a = b)\}$$

and let $\mathcal{E}$ be the equality axioms; that is, $\mathcal{E}$ consists of the rules for reflexivity, transitivity, symmetry, and the following two substitution axioms:

$$(x = y) \wedge \neg P(x) \rightarrow P(y)$$
$$(x = y) \rightarrow (f(x) = f(y))$$

---

*Resolution with Unification and Equality.

As an equality theory, the set is unsatisfiable. That is, $\mathcal{S} \cup \mathcal{E}$ is not satisfiable; a straightforward resolution proof can be obtained as follows. Apply resolution to the first substitution axiom and the clause $\neg P(f(b))$ from $\mathcal{S}$; the resolvent is

$$x \neq f(b) \vee P(x)$$

Resolving now with the clause $P(f(a))$ yields the resolvent

$$f(a) \neq f(b)$$

Finally, this clause can be resolved with the second substitution axiom to produce the clause $a \neq b$. This resolves with $a = b$ from the set $\mathcal{S}$ to complete the proof.

RUE builds into the resolution inference the substitution axioms by observing that the two substitution axioms can be expressed equivalently as

$$P(y) \wedge \neg P(x) \rightarrow x \neq y$$
$$f(x) \neq f(y) \rightarrow x \neq y$$

The inference rule

$$\frac{P(y), \neg P(x)}{x \neq y}$$

is introduced from the first axiom, and from the second, we obtain the inference rule

$$\frac{f(x) \neq f(y)}{x \neq y}$$

RUE further optimizes its computation by allowing the application of both inference rules in a single step. Thus, from the clauses $P(f(a))$ and $\neg P(f(b))$, the RUE resolvent $a \neq b$ can be obtained in a single step. Note that if only the first inference rule is applied, then the RUE resolvent of $P(f(a))$ and $\neg P(f(b))$ would be $f(a) \neq f(b)$.

61.3.3 The Method of Analytic Tableaux and Path Dissolution

Most theorem proving techniques employ clause form; the tableau method and the more general path dissolution do not. Tableau methods were originally developed and studied by a number of logicians, among them Beth [1955], Hintikka [1955], and Smullyan [1995], who built on the work of Gentzen [1969]. It is probably Smullyan who is most responsible for popularizing these methods; his particularly elegant variation on these techniques is known as the *method of analytic tableaux*. More recently, tableau methods have been receiving considerable attention from researchers investigating both automated deduction and logics for artificial intelligence; this includes serious implementors and those whose focus is primarily theoretical. See Beckert and Posegga [1995] for a compact (eight lines of Prolog code!) implementation of the tableau method. *Path dissolution* operates on a complementary pair of literals within a formula by restructuring the formula in such a way that all paths through the link vanish. The tableau method restructures the formula so that the paths through the link are immediately accessible and then marks them closed, in effect deleting them. It does this by selectively expanding the formula toward disjunctive normal form. The sense in which path dissolution generalizes the tableau method is that dissolution need not distinguish between the restructuring and the closure operations.

61.3.3.1 Negation Normal Form

One way to classify deduction systems is by what normalization they require of the formulas upon which they operate. As we have seen, resolution uses conjunctive normal form. Path dissolution and the tableau method cannot be restricted to clause form; both restructure a formula in such a way that clause form may not be preserved. In essence, both methods work with formulas in negation normal form (NNF). It turns

$$((\overline{C} \wedge A) \vee D \vee E) \wedge (\overline{A} \vee (B \wedge C)) \quad \equiv \quad \begin{array}{c} \overline{C} \\ \wedge \quad \vee \quad D \quad \vee \quad E \\ A \\ \wedge \\ \qquad B \\ \overline{A} \quad \vee \quad \wedge \\ \qquad C \end{array}$$

FIGURE 61.2   Graphical representation of a formula.

out that NNF formulas can be far more complex than formulas in clause form, and a careful analysis of NNF will facilitate the introduction of these proof techniques.

Recall that a formula is in NNF if conjunction and disjunction are the only binary connectives and if all negations are at the atomic level. Propositional formulas in NNF can be described recursively as follows:

1. The constants t and f are NNF formulas.
2. The literals A and $\neg A$ are NNF formulas.
3. If $\mathcal{F}$ and $\mathcal{G}$ are formulas, then so are $\mathcal{F} \wedge \mathcal{G}$ and $\mathcal{F} \vee \mathcal{G}$.

Each formula used in the construction of an NNF formula is called an *explicit subformula*. When a formula contains occurrences of *true* and *false*, the obvious truth-functional reductions apply. For example, if $\mathcal{F}$ is any formula, then $t \wedge \mathcal{F} = \mathcal{F}$. Unless otherwise stated, we will assume that formulas are automatically so reduced.

It is often convenient to write formulas as two-dimensional graphs in a manner that can easily be understood by considering a simple example. In Figure 61.2, the formula on the left is displayed graphically on the right. For a more detailed exposition, see Murray and Rosenthal [1993].

Naturally, a literal A can occur more than once in a formula. As a result, we use the term "node" for a literal occurrence in a formula. If A and B are nodes in a formula $\mathcal{F}$, and if $\mathcal{F}$ contains the subformula $X \wedge Y$ with A in X and B in Y, then we say that A and B are *c-connected*; *d-connected* nodes are similarly defined. In Figure 61.2, C is c-connected to each of B, A, $\overline{C}$, D, and E is d-connected to $\overline{A}$.

Let $\mathcal{F}$ be a formula. A *partial c-path through* $\mathcal{F}$ is a set of nodes such that any two are c-connected, and a *c-path* through $\mathcal{F}$ is a partial c-path that is not properly contained in any partial c-path. The c-paths of the formula of Figure 61.2 are $\{\overline{C}, A, \overline{A}\}$, $\{\overline{C}, A, B, C\}$, $\{D, \overline{A}\}$, $\{D, B, C\}$, $\{E, \overline{A}\}$, and $\{E, B, C\}$. A d-path is similarly defined using d-connected nodes in place of c-connected nodes.

The c-paths of a formula are the clauses of one of its disjunctive normal form equivalents. Similarly, the d-paths correspond to the clauses of a CNF equivalent. It is easy to see that a formula is unsatisfiable if and only if every c-path in it is unsatisfiable, and a c-path is unsatisfiable if and only if it contains a *link* (i.e., a complementary pair of literals [an atom and its negative]). Most inference mechanisms operate on links; several are path based. The idea of the method of analytic tableaux is to isolate paths containing a link and then to eliminate them. Path dissolution accomplishes this without first isolating the paths in question. Other path-based methods include Andrews's work on matings [Andrews 1976] and Bibel's connection and connection graph methods (see Bibel [1987]).

61.3.3.2 The Tableau Method

The reader should be forewarned that there is a potentially misleading difference in emphasis between the typical descriptions of the tableau method and the one presented here. Tableau proofs are usually cast as tree structures in which paths can grow through the addition of new *lines* in the tree; the number of paths can increase due to a *splitting* or *branching* operation. The lines and branch points of a tableau proof tree are meta-linguistic representations of conjunction and disjunction, respectively. Here we strip them of their special status; a tableau proof tree then becomes merely a single formula. This simplifies the presentation, which is an advantage because of space limitations, and makes the relationship to path dissolution easier to see. Smullyan's [1995] book is an excellent source for the traditional description of the tableau method.

Defining the tableau method in terms of formulas requires three rules: *separation*, *dispersion*, and *closure*. It is also convenient to designate certain subformulas as *primary*; they form a tree that corresponds precisely to the proof tree maintained by the more traditional approach to the tableau method. Initially, the entire formula is the only primary subformula. A separation is performed on any primary subformula whose highest-level connective is a conjunction by removing the primary designation from it and bestowing that designation on its conjuncts. There is essentially no cost to this operation, and it can be regarded as automatic whenever a conjunction becomes primary.

A separation can also be performed on a disjunction that is a leaf in the primary tree. (Separating an interior disjunction is not allowed because such an operation would destroy the tree structure.) Separations of such disjunctions should not be regarded as automatic; this operation increases the number of paths in the tree (we call such paths *tree* paths to distinguish them from c-paths); thus, although there is no cost to the operation itself, there is a potential penalty from the extra paths.

The process of dispersing a primary subformula whose highest-level connective is a disjunction can now be defined precisely: a copy of the subformula is placed at the end of one path descending from it and separated. For example, suppose that $X = X_1 \vee X_2$ is a primary subformula, and that the leaf Y is a descendant of X. On the left, we show the original tree path from X to Y; on the right is the extension of that path produced by dispersing X:

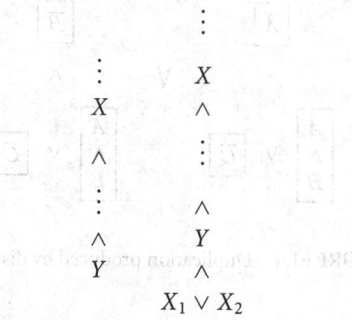

The subformulas X and Y remain primary, and X_1 and X_2 are now designated as primary. Note that if a subformula is eventually dispersed to the ends of every path descending from it, the original copy is no longer required.

The key operations in a tableau deduction are the closures, which close tree paths. Marking a tree path closed is equivalent to deleting it. In terms of the tree of primaries, a tree path can be closed when a separation or a dispersion makes primary a literal that forms a link with one of its ancestors. The literal and all of its descendants are deleted, and any leaves that result are, in turn, deleted. Because the tree path through the link is removed, the effect is to delete all c-paths through the link.

Consider, for example, the unsatisfiable formula $(((A \wedge B) \vee \overline{C}) \wedge C \wedge (\overline{A} \vee \overline{B}))$, pictured in Figure 61.3a. Boxes are used to designate primary subformulas; initially, the entire formula is the only one. Because it is a conjunction, this primary subformula is automatically separated; the result is Figure 61.3b.

Note that there is yet only one path in the tree. As a result, a dispersion can be performed by moving (without a duplicate copy) any primary to the leaf position and then separating. For simplicity, we separate the primary that is already a leaf (Figure 61.3c). The formula, as a formula, is unchanged. But by designating $\overline{A}$ and $\overline{B}$ as primary, the proof tree has split and now contains two tree paths. Only one operation can follow: dispersion of the upper primary. It can be dispersed twice, once for each tree path, allowing the deletion of the original copy at the root; the result is pictured in Figure 61.4.

The remainder of the proof is straightforward. Each of the primary subformulas $(A \wedge B)$ is separated into two primaries, and each of the four paths in the tree can be closed; two paths contain a $\{C, \overline{C}\}$ link, another contains $\{\overline{A}, A\}$, and the fourth contains $\{\overline{B}, B\}$.

A tree path is the conjunction of its nodes, that is, of its primary subformulas. We can also view a tree path as a collection of c-paths. For example, the single tree path in Figure 61.3c that contains $\overline{A}$ is a conjunction of three primaries and contains the c-paths $\{\{A, B, C, \overline{A}\}, \{\overline{C}, C, \overline{A}\}\}$. Note also that dispersion is the source

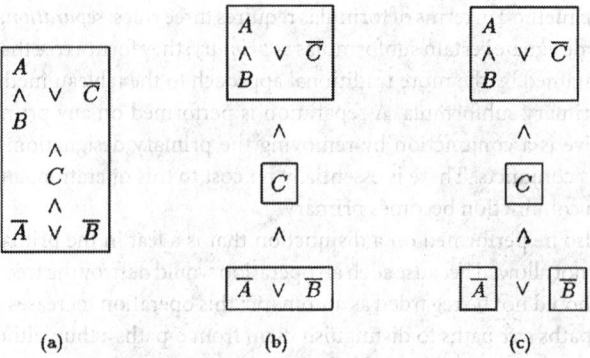

FIGURE 61.3   Separation of primary subformulas.

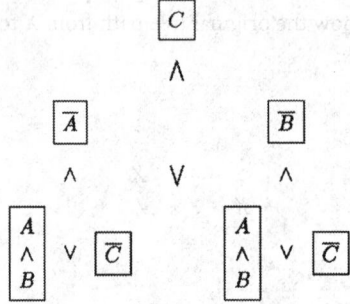

FIGURE 61.4   Duplication produced by dispersion.

of all literal duplication (the expensive part) with the tableau method. In Figure 61.4, for example, an extra copy (in this case only one) of $((A \wedge B) \vee \overline{C})$ has been created for all but the last descendant leaf to which it has been dispersed.

61.3.3.3   Path Dissolution

Path dissolution operates on a *link* — a complementary pair of literals — within a formula by restructuring the formula in such a way that all paths through the link vanish. The tableau method restructures a formula so that the paths through the link are immediately accessible and then marks them closed, in effect deleting them. It does this by selectively expanding the formula toward disjunctive normal form. The sense in which dissolution generalizes the tableau method is that dissolution need not distinguish between the restructuring and closure operations.

Path dissolution is, in general, applicable to collections of links; here we restrict attention to single links. Suppose then that we have complementary literals A and $\overline{A}$ residing in conjoined subformulas X and Y, respectively. Consider, for example, the link $\{A, \overline{A}\}$ on the left in Figure 61.5. Then the formula can be written $\mathcal{D} = (X \wedge Y)$, where

$$X = \begin{matrix} \overline{C} \\ \wedge \\ A \end{matrix} \vee D \vee E \quad \text{and} \quad Y = \overline{A} \vee \begin{matrix} B \\ \wedge \\ C \end{matrix}$$

The *c-path complement* of a node A with respect to X, written $CC(A, X)$, is defined to be the subformula of X consisting of all literals in X that lie on c-paths that do not contain A; the *c-path extension* of A with respect to X, written $CPE(A, X)$, is the subformula containing all literals in X that lie on paths that *do* contain A.

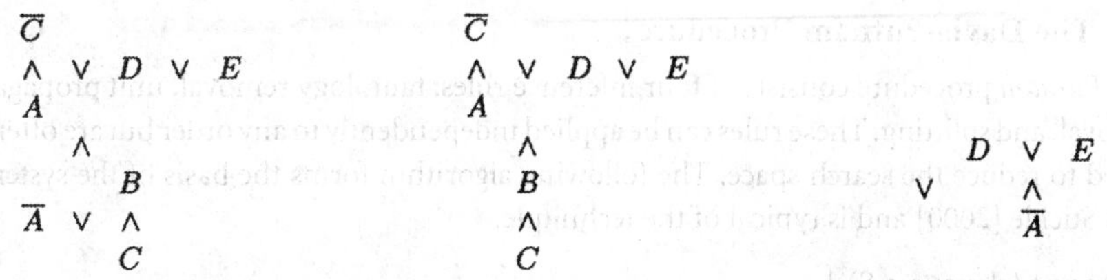

$$\overline{C} \qquad\qquad\qquad\qquad \overline{C}$$
$$\wedge \ \vee \ D \ \vee \ E \qquad\qquad \wedge \ \vee \ D \ \vee \ E$$
$$A \qquad\qquad\qquad\qquad A$$
$$\wedge \qquad\qquad\qquad\qquad \wedge \qquad\qquad\qquad D \ \vee \ E$$
$$B \qquad\qquad\qquad\qquad B \qquad\qquad\qquad \vee \ \wedge$$
$$\overline{A} \ \vee \ \wedge \qquad\qquad \wedge \qquad\qquad\qquad \overline{A}$$
$$C \qquad\qquad\qquad\qquad C$$

FIGURE 61.5   The right side is the path-dissolvent of $\{A, \overline{A}\}$ on the left.

In Figure 61.5, $CC(A, X) = (D \vee E)$; $CPE(A, X) = (\overline{C} \wedge A)$.

It is intuitively clear that the paths through $(X \wedge Y)$ that do not contain the link are those through $CPE(A, X) \wedge CC(\overline{A}, Y))$ plus those through $(CC(A, X) \wedge CPE(\overline{A}, Y))$ plus those through $(CC(A, X) \wedge CC(\overline{A}, Y))$. The reader is referred to Murray and Rosenthal [1993] for the formal definitions of CC and of CPE and for the appropriate theorems.

The *dissolvent* of the link $H = \{A, \overline{A}\}$ in $M = X \wedge Y$ is defined by

$$DV(H, M) = \begin{array}{ccccc} CPE(A, X) & & CC(A, X) & & CC(A, X) \\ \wedge & \vee & \wedge & \vee & \wedge \\ CC(\overline{A}, Y) & & CPE(\overline{A}, Y) & & CC(\overline{A}, Y) \end{array}$$

The c-paths of $DV(H, M)$ are exactly the c-paths of M that *do not* contain the link. Thus, M and $DV(H, M)$ are equivalent. In general, M need not be the entire formula; without being precise, M is the smallest part of the formula that contains the link. If $\mathcal{F}$ is the entire formula, then the *dissolvent of $\mathcal{F}$ with respect to H*, denoted Diss $(\mathcal{F}, H)$, is the formula produced by replacing M in $\mathcal{F}$ by $DV(H, M)$. If $\mathcal{F}$ is a propositional formula, then Diss $(\mathcal{F}, H)$ is equivalent. Because the paths of the new formula are all that appeared in $\mathcal{F}$ except those that contained the link, this formula has strictly fewer c-paths than $\mathcal{F}$. As a result, finitely many dissolutions (bounded above by the number of c-paths in the original formula) will yield a linkless equivalent formula. We can therefore say that path dissolution is a *strongly complete* rule of inference for propositional logic; that is, if a formula is unsatisfiable, any sequence of dissolution steps will eventually produce the empty clause.

A useful special case of dissolution arises when X consists of A alone; then $CC(A, X)$ is empty, and the dissolvent of the link $\{A, \overline{A}\}$ in the subformula $X \wedge Y$ is $X \wedge CC(\overline{A}, Y)$; that is, dissolving has the effect of replacing Y by $CC(\overline{A}, Y)$, which is formed by deleting $\overline{A}$ and anything directly conjoined to it. Hence, no duplications whatsoever are required. A tableau closure is essentially a dissolution step of this type. Observe that a separation in a tableau proof does not really affect the structure of the formula; it is a bookkeeping device employed to keep track of the primaries in the tree. A dispersion is essentially an application of the distributive laws, which of course can be used by any logical system. As a result, every tableau proof is a dissolution proof, but certainly not vice versa.

61.3.4   Model Finding in Propositional Logic

Model finding is the counterpart of determining unsatisfiability; it is the process of searching for an interpretation that satisfies a given formula. A number of high-performance model finding systems have been developed in recent years. They are able to handle formulas with thousands of variables and millions of clauses. These systems generally fall into one of two categories: *systematic* or *stochastic*. Roughly speaking, systematic methods explore the space of interpretations through variations of backtracking techniques. Given sufficient time and space, these methods are guaranteed to find a satisfying interpretation, if one exists. Systematic methods are thus complete. Stochastic methods are, in contrast, incomplete. They apply heuristic search techniques to prune large portions of the search space and often find satisfying interpretations much more quickly than systematic methods. The basis for most systematic systems is the Davis-Putnam procedure. The **GSAT** algorithm introduced by Selman, Levesque, and Mitchell [1992] is an example of a simple yet rather ingenious stochastic system.

61.3.4.1   The Davis-Putnam Procedure

The *Davis-Putnam* procedure consists of four inference rules: tautology removal, unit propagation, pure-literal removal, and splitting. These rules can be applied independently in any order but are often combined and ordered to reduce the search space. The following algorithm forms the basis of the system **SATO** by Zhang and Stickle [2000] and is typical of the technique.

```
bool DP-sat(clause set S) {
    repeat
        for each unit clause L in S do
            /* unit propagation */
            delete from S every clause containing L
            delete ¬L from every clause of S
        if S is empty return true
        else if S contains the empty clause return false
    until no changes occur in S
    /* splitting */
    choose a literal L in S
    if DP-sat(S ∪ {L})
        return true
    else if DP-sat(S ∪ {¬L})
        return true
    else
        return false
}
```

For example, consider the unsatisfiable input clause set,

$$\{A \lor B \lor \neg C, A \lor \neg B, \neg A, C, D\}$$

The algorithm repeatedly applies unit propagation until the empty clause is derived. Suppose A is selected first; then unit propagation produces

$$\{B \lor \neg C, \neg B, C, D\}$$

Selecting C next yields

$$\{B, \neg B, D\}$$

Finally, unit propagation on B produces a set that contains the empty clause (represented by $\square$):

$$\{\square, D\}$$

The presence of $\square$ signals that the original clause set is unsatisfiable.

Note that if the clause set is satisfiable, then the literals selected during the computation of the algorithm produce a model of the set, as the next example illustrates. Consider the input set

$$\{A \lor B, \neg B, \neg A \lor B \lor \neg C\}$$

Selecting $\neg B$ and applying unit propagation yields the clause set

$$\{A, \neg A \lor \neg C\}$$

Applying unit propagation on A then produces the set

$$\{\neg C\}$$

Finally, unit propagation on $\neg C$ results in the empty set. The model thus obtained for the initial clause set is $\{\neg B, A, \neg C\}$.

Literals selected for splitting can also contribute to the model. For example, no unit propagation is possible with the input set

$$\{A \vee \neg B, \neg A \vee B, B \vee \neg C, \neg B \vee \neg C\}$$

If A is selected for splitting, then the clause set passed to the recursive call of the algorithm is

$$\{A \vee \neg B, \neg A \vee B, B \vee \neg C, \neg B \vee \neg C, A\}$$

The only literal on which unit propagation is possible is A. Hence, if the clause set is satisfiable, the model produced will contain A. It is easy to verify that the model for this example is $\{A, B, \neg C\}$.

Different implementations of the basic Davis-Putnam procedure have been developed. These include the use of sophisticated heuristics for literal selection, complex data structures, and efficient conflict detection methods.

61.3.4.2 GSAT

The GSAT algorithm employs a hill-climbing heuristic to find models of propositional clauses. Given a set of clauses, the algorithm first randomly assigns a truth value to each propositional variable and records the number of satisfied clauses. The main loop of the algorithm then repeatedly toggles the truth value of variables to increase the number of satisfied clauses. The algorithm continues until either all clauses are satisfied, or a preset amount of time has elapsed. The latter can occur either with an unsatisfiable clause set or if a model for a satisfiable clause set has not been found. That is to say, GSAT is incomplete. Figure 61.6 outlines the GSAT algorithm in more detail.

61.3.5 Nonclassical Logics

Many departures from classical logic that have been formalized in the AI research program have been aimed at common-sense reasoning. Perhaps the two most widely addressed limits of classical logic are its inability to model either reasoning with uncertain knowledge or reasoning with incomplete information. A number of nonclassical logics have been proposed; here we consider **multiple-valued logics**, fuzzy logic, and default logic. Alternatives to uncertain reasoning include probabilistic reasoning — see Chapter 70

Input: A set of clauses C, MAX-FLIPS, MAX-TRIES

Output: An interpretation that satisfies C or "don't know"

1. for $i = 1$ to MAX-TRIES
   - T = randomly generate a truth assignment for the variables of C
   - for $j = 1$ to MAX-FLIPS
     (a) if T satisfies C return T
     (b) find a variable p such that a change in its truth value gives the largest increase in the total number of clauses of C satisfied by T
     (c) $T = T$ modified by the changing the truth assignment of p
2. return "don't know"

FIGURE 61.6 The GSAT algorithm.

TABLE 61.1   Truth Table

G	H	$G \wedge H$	$G \vee H$	$G \rightarrow H$	$\neg G$
0	0	0	0	1	1
0	1/2	0	1/2	1	1
0	1	0	1	1	1
1/2	0	0	1/2	1/2	1/2
1/2	1/2	1/2	1/2	1	1/2
1/2	1	1/2	1	1	1/2
1	0	0	1	0	0
1	1/2	1/2	1	1/2	0
1	1	1	1	1	0

and nonmonotonic formalisms for knowledge representation. Inference techniques in nonclassical logics remain important; a common approach is to designate a set of satisfying truth values and then to adapt classical inference techniques.

61.3.5.1   Multiple-Valued Logics

One of the drawbacks of classical logic is its restricted set Δ of truth values. A logic that generalizes Δ to be an arbitrary set is commonly referred to as a *multiple-valued* (or *many-valued*) logic (MVL). Although the precise boundary of what can be classified as a multiple-valued logic is not all that clear [Urquhart 1986], logics that are generally agreed to fit that description have been applied to reasoning with uncertainty, to reasoning with inconsistency, to natural language processing, and, to some extent, to nonmonotonic reasoning.

We examine extensions of classical inference techniques to MVLs based on the framework of **signs**: sets of truth values. The discussion of MVLs leads naturally to the examination of fuzzy logic, which, from a deduction point of view, can be viewed as an MVL.

An MVL Λ is more general than classical logic in one respect: the set Δ of truth values can be arbitrary. As with classical logic, an interpretation for Λ is a function from its atom set $\mathcal{A}$ to Δ; that is, an assignment of truth values to every atom in Λ. A connective Θ of arity n denotes a function $\Theta : \Delta^n \rightarrow \Delta$. Interpretations are extended in the usual way to mappings from the set of formulas to Δ.

Consider, for example, Łukasiewicz's [1970] three-valued logic. The set of truth values is $\Delta = \{0, 1/2, 1\}$; the binary connectives are $\wedge$, $\vee$, and $\rightarrow$; and the only unary connective is $\neg$. The connectives are defined by the truth table, Table 61.1.

The designated set of truth values of Łukasiewicz's logic is $\{1\}$. Thus, the consequence relation means that $\mathcal{F} \models C$ if whenever an interpretation I assigns the value 1 to $\mathcal{F}$, then I also assigns 1 to C.

Intuitively, the truth value 1/2 in Łukasiewicz's logic denotes possible. A variant of this three-valued logic is Kleene's, in which the truth value 1/2 corresponds to undefined [Kleene 1952]. In a more recent three-valued logic proposed by Priest, the value 1/2 may be thought of as denoting inconsistency.

A good reference for deduction techniques for MVLs is Hähnle's monograph, *Automated Deduction in Multiple-Valued Logics* [Hähnle 1994]. The key to the approach described there is the use of signs — subsets of the set Δ of truth values. If $\mathcal{F}$ is any formula and if S is any sign, the expression $S : \mathcal{F}$ can be interpreted as the assertion "$\mathcal{F}$ evaluates to a truth value in S." For each interpretation over the MVL Λ, this assertion is either true or false, so $S : \mathcal{F}$ can be treated as a proposition in classical logic. For example, let $\mathcal{F}$ be a formula in Łukasiewicz's logic, and suppose we are interested in determining whether a formula C is a logical consequence of $\mathcal{F}$. One strategy is to determine whether the following disjunction of signed formulas must evaluate to *true* under all interpretations over Λ:

$$\{0, 1/2\} : \mathcal{F} \vee \{1\} : C$$

The connective $\vee$ is classical. Thus, given any interpretation, for the disjunction to evaluate to 1, either $\mathcal{F}$ evaluates to 0 or to 1/2, or C evaluates to 1. In particular, if $\mathcal{F}$ evaluates to 1, then so must C, which is to say that C is a logical consequence of $\mathcal{F}$.

For example, let $\mathcal{F}$ be the formula $(p \wedge r) \wedge (p \rightarrow q)$. To show that q is a logical consequence of $\mathcal{F}$, we must determine whether the formula

$$\{0, 1/2\}:\mathcal{F} \vee \{1\} : q \qquad (61.5)$$

is a tautology. Using the tableau method, we attempt to find a closed tableau for the negation of Equation 61.5. It is useful to first drive the signs inward; thus, from the truth table:

$$\begin{aligned}
\{1\}:((p \wedge r) \wedge (p \rightarrow q)) \wedge \{0, 1/2\}:q &= \{1\}:p \wedge \{1\}:r \wedge \{1\}:(\neg p \vee q) \wedge \{0, 1/2\}:q \\
&= \{1\}:p \wedge \{1\}:r \wedge (\{1\}:\neg p \vee \{1\}:q) \wedge \{0, 1/2\}:q \\
&= \{1\}:p \wedge \{1\}:r \wedge (\{0, 1/2\}:p \vee \{1\}:q) \wedge \{0, 1/2\}:q
\end{aligned}$$

The last formula is the initial tableau tree; if the disjunction is dispersed, the tableau becomes

$$\boxed{\{1\} : p}$$
$$\wedge$$
$$\boxed{\{1\} : r}$$
$$\wedge$$
$$\boxed{\{0, 1/2\} : q}$$
$$\wedge$$
$$\boxed{\{0, 1/2\} : p} \quad \vee \quad \boxed{\{1\} : q}$$

To close a tree path, the path must contain a *generalized link*: signed atoms $S_1 : A$ and $S_2 : A$, where S_1 and S_2 are disjoint. There are two tree paths in this tableau; the left path contains the link $\{\{1\} : p, \{0, 1/2\} : p\}$, and the right contains the link $\{\{0, 1/2\} : q, \{1\} : q\}$. Thus, the tableau can be closed and the proof is complete.

Most other classical inference techniques — for example, resolution and path dissolution — can similarly be generalized to MVLs using signed formulas, which can be thought of as formalizing metalevel reasoning about an arbitrary MVL Λ. In Lu et al. [1998], signed formulas provide a framework for adapting most classical inference techniques. The idea is to treat each signed formula $S : \mathcal{F}$ as a proposition in classical logic. Then, classical inferences made by restricting attention to the Λ-*consistent interpretations* — those interpretations that assign true to a proposition $S : \mathcal{F}$ if and only if there is a corresponding interpretation over Λ that assigns some truth value in S to $\mathcal{F}$ — yield an inference about the MVL Λ.

The language of signed formulas is closely related to the system of annotated logic studied in Kifer and Lozinskii [1992] and in Blair and Subrahmanian [1989]. Those authors had the goal of developing reasoning systems capable of dealing with inconsistent information. An examination of the relationship between logic programming based on signs and annotations can be found in Lu [1996]. Automated reasoning systems that implement reasoning in MVLs based on signed formulas include the 3TAP program of Beckert et al. [1992].

61.3.5.2 Fuzzy Logic

In recent years, fuzzy logic, which was introduced by Zadeh in 1965, has received considerable attention, largely for engineering applications such as heuristic control theory. There are at least two other views of fuzzy logic — see Dubois and Prade [1995] and Gaines [1977] — both closer to mainstream AI. One is that fuzzy logic can be regarded as an extension of classical logic for handling *uncertain* propositions — propositions whose truth values are derived from the unit interval $[0, 1]$. But relatively little attention has been paid to inference techniques for fuzzy logic; examples of deduction-based systems for fuzzy logic include Baldwin [1986], Lee [1972], Mukaidono [1982], and Weigert et al. [1993].

In fuzzy logic, the set of truth values is $\Delta = [0, 1]$, and so interpretations assign each proposition a value in the interval $[0, 1]$. As usual, n-ary connectives are functions from $[0, 1]^n$ to $[0, 1]$. Conjunction $\wedge$ is usually the function min, disjunction $\vee$ is usually the function max, and $\neg$ is usually defined by $\neg\phi = 1-\phi$. There are several possibilities for the function $\rightarrow$; perhaps the most obvious is $A \rightarrow B \equiv \neg A \vee B$.

The designated set of truth values Δ^* in fuzzy logic is a subinterval of $[0, 1]$ of the form $[\alpha, 1]$ for some $\alpha \geq 0.5$. We call such an interval positive, and correspondingly call an interval of the form $[0, \alpha]$, where $\alpha \leq 0.5$, negative. For example, Weigert et al. [1993] defined a threshold of acceptability, $\tau \geq 0.5$, which in effect specifies Δ^* to be $[\tau, 1]$. On the other hand, Lee and Mukaidono do not explicitly define Δ^*. However, their systems implicitly adopt $\Delta^* = [0.5, 1]$. We begin this section by considering the fuzzy logic developed by Lee [1972] and extended by Mukaidono [1982], and then examine the more recent work of Weigert et al. [1993].

If we restrict attention to fuzzy formulas that use $\wedge$ and $\vee$ interpreted as min and max, respectively, as the only binary connectives and $\neg$ as defined as the only unary connective, then, as in the classical case, a formula can be put into an equivalent CNF. The keys are the observations

$$\neg(\mathcal{F} \wedge \mathcal{G}) = \neg\mathcal{F} \vee \neg\mathcal{G} \quad \text{and} \quad \neg(\mathcal{F} \vee \mathcal{G}) = \neg\mathcal{F} \wedge \neg\mathcal{G} \tag{61.6}$$

The resolution inference rule introduced by Lee is the obvious generalization of classical resolution. Let C_1 and C_2 be clauses (i.e., disjunctions of literals), and let L be an atom. Then the resolvent is defined by

$$\frac{L \vee C_1, \neg L \vee C_2}{C_1 \vee C_2} \tag{61.7}$$

Lee proved the following: Let C_1 and C_2 be two clauses, and let $R(C_1, C_2)$ be a resolvent of C_1 and C_2. If I is any interpretation, let $\max\{I(C_1), I(C_2)\} = b$ and $\min\{I(C_1), I(C_2)\} = a > 0.5$. Then $a \leq I(R(C_1, C_2)) \leq b$.

Mukaidono defines an inference to be *significant* if for any interpretation, the truth value of the conclusion is greater than or equal to the truth value of the minimum of the clauses in the premise. Lee's theorem can thus be interpreted to say that an inference using resolution is significant.

Weigert et al. [1993] built on the work of Lee and Mukaidono. They augmented the language by allowing infinitely many negation symbols that they call *fuzzy operators*. A formula is defined as follows: Let A be an atom, let $\mathcal{F}$ and $\mathcal{G}$ be fuzzy formulas, and let $\phi \in [0, 1]$.* Then:

1. ϕA is a fuzzy formula (also called a *fuzzy literal*)
2. $\phi(\mathcal{F} \wedge \mathcal{G})$ is a fuzzy formula
3. $\phi(\mathcal{F} \vee \mathcal{G})$ is a fuzzy formula

A simple example of a fuzzy formula is

$$\mathcal{F} = A \wedge 0.3(0.9B \vee 0.2C)$$

Several observations are in order. First, fuzzy operators are represented by real numbers in the unit interval. (That there are uncountably many fuzzy operators should not cause alarm. In practice, considering only rational fuzzy operators is not likely to be a problem. Indeed, with a computer implementation, we are restricted to a finite set of terminating decimals of at most n digits for some not very large n.) In particular, real numbers in the unit interval denote both truth values and fuzzy operators. Second, every formula and subformula is prefixed by a fuzzy operator; any subformula that does not have an explicit fuzzy operator prefix is understood to have 1 as its fuzzy operator.

The semantics of fuzzy operators are given via a kind of fuzzy product.

Definition 61.1   If $\phi, \delta \in [0, 1]$, then $\phi \otimes \delta = (2\phi - 1) \cdot \delta - \phi + 1$.

Observe that $\otimes$ is commutative and associative. Also observe that

$$\phi \otimes \delta = \phi \cdot \delta + (1 - \phi) \cdot (1 - \delta)$$

---

*Weigert et al. [1993] use λ for the fuzzy operator and Λ for the threshold of acceptability. We use ϕ and τ to avoid confusion with notation used in other parts of the chapter.

This last observation provides the intuition behind the fuzzy product $\otimes$: Were ϕ the probability that A_1 is true and were δ the probability that A_2 is true, then $\phi \otimes \delta$ would be the probability that A_1 and A_2 are both true or both false. (This probabilistic analogy is for intuition only; fuzzy logic is not based on probability.)

It turns out that the following generalization of Equation 61.6 holds: Let $\mathcal{F}$ and $\mathcal{G}$ be fuzzy formulas, and let ϕ be a fuzzy operator. If $\phi > 0.5$, then

$$\phi(\mathcal{F} \wedge \mathcal{G}) = \phi\mathcal{F} \wedge \phi\mathcal{G} \quad \text{and} \quad \phi(\mathcal{F} \vee \mathcal{G}) = \phi\mathcal{F} \vee \phi\mathcal{G}$$

If $\phi < 0.5$, then

$$\phi(\mathcal{F} \wedge \mathcal{G}) = \phi\mathcal{F} \vee \phi\mathcal{G} \quad \text{and} \quad \phi(\mathcal{F} \vee \mathcal{G}) = \phi\mathcal{F} \wedge \phi\mathcal{G}$$

In particular, every fuzzy formula is equivalent to one in which 1 is the only fuzzy operator applied to nonatomic arguments.

In addition to introducing fuzzy operators, Weigert et al. extended Lee and Mukaidono's work with the *threshold of acceptability*: a real number $\tau \in [0.5, 1]$. Then an interpretation I is said to τ-*satisfy* the formula $\mathcal{F}$ if $I(\mathcal{F}) \geq \tau$. Observe that the threshold of acceptability is essentially a redefinition of Δ^* to $[\tau, 1]$. That is, the threshold of acceptability provides a variable for the definition of the designated set of truth values.

The significance of the threshold can be made clear by looking at some simple examples. Let $\tau = 0.7$ and consider each of the following three formulas: $0.8A$, $0.2A$, and $0.6A$. Suppose A is 1; that is, $I(A) = 1$ for some interpretation I. Then $I(0.8A) = 0.8 \geq \tau$, so that the first formula is satisfied. The latter two evaluate to 0.2 and to 0.6, and so neither is satisfied. Now suppose A is 0. The first formula evaluates to $0.8 \otimes 0 = 0.2$, and the second evaluates to $0.2 \otimes 0 = 0.8$, so that the second formula is τ-satisfied. In effect, because the fuzzy operator 0.2 is less than $1 - \tau$, $0.2A$ is a negative literal and is τ-satisfied by assigning false to the atom A. The value of the third formula is now $0.6 \otimes 0 = 0.4$. Thus, in either case, the third formula is τ-unsatisfiable. Weigert et al., in fact, define a clause to be τ-*empty* if every fuzzy operator of every literal in the clause lies between $1 - \tau$ and τ; it is straightforward to prove that every τ-empty clause is τ-unsatisfiable.

The fuzzy resolution rule relies upon complementary pairs of literals. However, complementarity is a relative notion depending on the threshold τ. Two literals $\phi_1 A$ and $\phi_2 A$ are said to be τ-*complementary* if $\phi_1 \leq 1 - \tau$ and $\phi_2 \geq \tau$. Resolution for fuzzy logic, which Weigert et al. proved is sound and complete, can now be defined with respect to the threshold τ: Let $\phi_1 A$ and $\phi_2 A$ be τ-complementary, and let C_1 and C_2 be fuzzy clauses. Then,

$$\frac{\phi_1 A \vee C_1, \phi_2 A \vee C_2}{C_1 \vee C_2} \tag{61.8}$$

Suppose we now reconsider the example from the Resolution section, which encoded the knowledge that "Tweety is a canary," "A canary is a bird," and "A bird flies." In reality, not all birds fly. A perhaps more realistic representation of this knowledge is:

$$1 \ Canary(Tweety)$$
$$0 \ Canary(x) \vee 1 \ Bird(x)$$
$$0 \ Bird(x) \vee 0.8 \ Flies(x)$$

The first two clauses represent the facts that Tweety is a canary and that all canaries are birds. In other words, the fuzzy operator 1 represents true, and the fuzzy operator 0 represents false. The fuzzy operator 0.8 can be interpreted as *highly likely*; thus, the third clause expresses the notion that most birds fly. Using the threshold 0.7 again and applying the same resolution steps as in Figure 61.1, we can infer $0.8 \ flies(Tweety)$. Observe that this means that the truth value of *flies*(*Tweety*) must be at least 5/6 because $0.8 \otimes 5/6 = 0.7$.

61.3.5.3 Nonmonotonic Logics

Common-sense reasoning requires the ability to draw conclusions in the presence of incomplete information. Indeed, very few conclusions in our everyday thinking are based on knowledge of every piece
of relevant information. Typically, numerous assumptions are required. Even a simple inference such as,
if x is a bird, then x flies, is based on a host of assumptions regarding *x*; for instance, that *x* is not an
unusual bird such as an ostrich or a penguin. It follows that logics for common-sense reasoning must be
capable of modeling reasoning processes that permit incorrect (and reversible) conclusions based on false
assumptions. This observation has motivated the development of nonmonotonic logics, whose origins
can be traced to foundational works by Clark, McCarthy, McDermott and Doyle, and Reiter; see Gallaire
and Minker [1978] and Bobrow [1980].

The term **nonmonotonic logic** highlights the fundamental technical difference from classical logic,
which is monotonic in the sense that

$$\mathcal{F}_1 \vdash \gamma \quad \text{and} \quad \mathcal{F}_1 \subseteq \mathcal{F}_2 \quad \text{implies } \mathcal{F}_2 \vdash \gamma \tag{61.9}$$

That is, classical entailment dictates that the set of conclusions from a knowledge base is inviolable; the
addition of new knowledge never invalidates previously inferred conclusions. A classically based reasoning
agent will therefore never be able to retract a conclusion in light of new, possibly contradictory information.
Nonmonotonic logics, on the other hand, need not obey Equation 61.9.

The investigation of inference techniques for nonmonotonic logics has been limited, but there have
been several interesting attempts. Comprehensive studies of nonmonotonic reasoning and default logic
include Etherington [1988], Marek and Truszczyński [1993], Besnard [1989], and Moore's autoepistemic
logic [Moore 1985]. We will focus on the nonmonotonic formalism of Reiter known as *default logic*; see
his paper in Bobrow [1980]. Some of the other proposed systems contain technical differences, but the
essence of *nonmonotonicity* — the failure to obey Equation 61.9 — is adhered to by all.

A *default* is an inference (scheme) of the form

$$\frac{\alpha : M\beta_1, \ldots, M\beta_m}{\gamma} \tag{61.10}$$

where $\alpha, \beta_1, \ldots, \beta_m$, and γ are formulas. The formula α is the *prerequisite* of the default, $\{M\beta_1, \ldots, M\beta_m\}$ is
the *jusitification*, and γ is the *consequent*; the *M* in the justification serve merely to demark the justification.
A *default theory* is a pair (D, W), where W is a set of formulas and D is a set of defaults. Intuitively, W
can be thought of as the set of knowledge that is known to be true. A default theory (D, W) then enables
a reasoner to draw additional conclusions through the defaults in D.

As a motivating example, consider the default rule

$$\frac{Bird(x) : MFly(x)}{Fly(x)} \tag{61.11}$$

The prerequisite specifies *if an individual x is a bird*, and the justification specifies, *if it is consistent to assume
that x flies*, then we may infer by default that *x* flies. Suppose we have a theory consisting of the clause
set $W = \{Canary(Tweety), \neg Canary(x) \vee Bird(x)\}$. Using classical logic, one logical consequence of W is
the fact $Bird(Tweety)$. If $S = \{\alpha \mid W \vdash \alpha\}$ represents the reasoner's knowledge, then $S \cup \{Fly(Tweety)\}$
is consistent. That is, $S \cup \{Fly(Tweety)\}$ has a satisfying interpretation in classical logic. The default rule
Equation 61.11 then warrants the inference $Fly(Tweety)$.

It is important to note that formulas and consequences in default logics are classical in the sense that a
formula is either true or false. The difference from classical logic lies in the manner in which consequences
are derived from a given set of formulas. From W we can conclude $Fly(Tweety)$. However, suppose we add
to W the additional knowledge set A consisting of the clauses:

$$Broken\_Wings(Tweety)$$

$$Broken\_Wings(x) \rightarrow \neg Fly(x)$$

Then $S = \{\alpha \mid W \vdash \alpha\}$ contains, among other things, the fact $\neg Fly\ (Tweety)$. In this case, $Fly\ (Tweety)$ is inconsistent with S; that is, $S \cup \{Fly\ (Tweety)\}$ is not satisfiable. Thus, the condition for the justification part of the default rule (Equation 61.11) is not met, and hence the conclusion $Fly\ (Tweety)$ cannot be drawn. This provides a clear illustration of the nonmonotonic nature of default inference rules such as the default rule in Equation 61.11:

$$W \vdash_D Fly\ (Tweety), \quad \text{but} \quad W \cup A \nvdash_D Fly(Tweety)$$

where $\vdash_D$ means deduction based on classical inference or the default rules in D.

In the initial version of the example, the conclusion $Fly\ (Tweety)$ was obtained starting with the set S (which is the set of all classical consequences of W), and then applying the default rule according to the condition that $Bird\ (Tweety)$ is a consequence of S, and that $\{Fly\ (Tweety)\} \cup S$ is satisfiable. Now suppose a reasoner holds the following initial set of beliefs:

$$S_0 = \{Bird(Tweety), Fly(Tweety), Canary(Tweety), Canary(x) \rightarrow Bird(x)\}$$

This set contains S, and applications of the default rule (Equation 61.11) with respect to S_0 yield no additional conclusions. That is, because $Bird\ (Tweety)$ is a classical consequence of S_0, and $Fly\ (Tweety)$ $\in S_0$, so that $\{Fly\ (Tweety)\ \} \cup S_0$ is consistent, adding the conclusion $Fly\ (Tweety)$ from the consequent of the default rule produces no changes in the beliefs S_0. A set that has such a property holds special status in default logic and is called an *extension*. Intuitively, an extension E can be thought of as a set of formulas that *agree* with all the default rules in the logic; that is, every default whose prerequisites are in E and whose justification is consistent with E must have its consequence in E. Still another way to look at an extension E of W is as a superset of W that is closed under both classical and default inference.

To formally define extension, given a set of formulas E, let $th(E)$ denote the set of all classical consequence of E; that is, $th(E) = \{\alpha \mid E \vdash \alpha\}$. If (D, W) is a default theory, let $\Gamma(E)$ be the smallest set of formulas that satisfies the following conditions:

1. $W \subseteq \Gamma(E)$.
2. $\Gamma(E) = th(\Gamma(E))$.
3. Suppose $(\alpha : M\beta_1, \ldots, M\beta_m/\gamma) \in D, \alpha \in \Gamma(E)$, and $\neg\beta_1, \ldots, \neg\beta_m \notin E$. Then $\gamma \in \Gamma(E)$.

Then, E is an extension of (D, W) if $E = \Gamma(E)$.

Observe that the third part of the definition requires $\neg\beta_i \notin E$ for each i. This is, in general, a weaker notion than the requirement that β_i be consistent with E. That is, were E not deductively closed under inference in classical logic, it is possible that one $\neg\beta_i$ is not a member of but is a logical consequence of E; $E \cup \{\beta_i\}$ would then be inconsistent. However, in the case of an extension, which is closed under classical deduction, the two notions coincide.

The simple example just illustrated suggests a natural way of computing extensions; namely, begin with the formulas in W and repeatedly apply each default inference rule until no new inferences are possible. Of course, because default rules can be interdependent, the choice of which default rule to apply first can affect the extension that is obtained. For example, with the following two simple rules, the application of one prevents the application of the other by making the justification of the other inconsistent with the inferred fact:

$$\frac{true : MB}{\neg A} \quad \frac{true : MA}{\neg B}$$

Marek and Truszczyński introduced the operator R^D to compute extensions of a default logic. Let U be a set of formulas; then

$$R^D(U) = th\left(U \cup \left\{\gamma \mid \frac{\alpha}{\gamma} \in D \text{ and } \alpha \in U\right\}\right)$$

Observe that α/γ is justification-free. Thus, R^D amounts to the closure of U under classical and justification-free default inference.

Repeated applications of the operator R^D is merely function composition and can be written as follows:

$$R^D \uparrow 0(U) = U$$
$$R^D \uparrow (\alpha + 1)(U) = R^D(R^D \uparrow \alpha(U))$$
$$R^D \uparrow \lambda(U) = \bigcup \{R^D \uparrow \alpha(U) \mid \alpha < \lambda\} \qquad \text{for a limit ordinal } \lambda$$

Given a default theory (D, W) and a set of formulas S, the *reduct* of D with respect to S (D_S) is defined to be the set of justification-free inference rules of the form α/γ, where $(\alpha : M\beta_1, \ldots, M\beta_m/\gamma)$ is a default in D, and $\neg\beta_i \notin S$ for each i. The point is, once we know that a justification is satisfiable, then the corresponding justification-free rule is essentially equivalent.

Marek and Truszczyński showed that from the knowledge base W of a default theory, it is possible to use the operator R^D to determine whether a set of formulas is an extension.* More precisely, a set of formulas E is an extension of a default theory (D, W) if and only if $E = R^{D_E} \uparrow \omega(W)$.

Default logic is intimately connected with nonmonotonic logic programming. Analogs of the many results regarding extensions can be found in nonmonotonic logic programming. The problem of determining whether a formula is contained in some extension of a default theory is called the *extension membership problem*; in general, it is quite difficult because it is not semidecidable (as compared, for example, with first-order classical logic). This makes implementation of nonmonotonic reasoning much harder than the already-difficult task of implementing monotonic reasoning. Reiter speculated that a reasonable computational approach to default logic will necessarily allow for incorrect (unsound) inferences [Bobrow 1980]. This issue was also considered in Etherington [1988]. Some recent work on proof procedures for default logic can be found in Barback and Lobo [1995] (resolution based), in the work of Thielscher and Schaub (see Lifschitz [1995]), and in the tableau-based work of Risch and Schwind (see Wrightson [1994]). Work on general nonmonotonic deduction systems include Kraus et al. [1990].

61.4 Research Issues and Summary

Logic-based deductive reasoning has played a central role in the development of artificial intelligence. The scope of AI research has expanded to include, for example, vision and speech [Russell and Norvig 1995], but the importance of logical reasoning remains. At the heart of logic-based deductive reasoning is the ability to perform inference. In this chapter we have discussed but a few of the numerous inference rules that have been widely applied.

There are several directions that researchers commonly pursue in automated reasoning. One is the exploration of new logics. This line of research is more theoretical and intimately tied to the philosophical foundation of the reasoning processes of intelligent agents. Typically, the motivation behind newly proposed logics lies with some aspect of reasoning for which classical two-valued logic may not be adequate. Among the many examples are temporal logic, which attempts to deal with time-oriented reasoning [Allen 1991]; modal logics, which address questions of knowledge and beliefs [Fagin et al. 1992], and alternative MVLs [Ginsberg 1988].

Another area of ongoing research is the development of new inference techniques for existing logics, both classical and nonclassical; for example, Henschen [1979] and McRobbie [1991]. Such techniques might produce better general purpose inference engines or might be especially well-suited for some narrowly defined reasoning process.

Inference also plays an important role in *complexity theory*, which is, more or less, the analysis of the running time of algorithms; it is carefully described in other chapters. A fundamental question in complexity theory — indeed, a famous open question in all of computer science — is, "Does the class $\mathcal{NP}$ equal the class $\mathcal{P}$?" It has been shown [Cook 1971] that this question is equivalent to the question,

---

*Even for propositional default theory, this checking process is computationally very expensive because it is necessary to choose the set E non-deterministically.

"Is there a fast algorithm for determining whether a formula in classical propositional logic is satisfiable?" (Roughly speaking, *fast* means a running time that is polynomial in the size of the input.)

Implementation of deduction techniques continues to receive a great deal of attention from researchers. Considerable effort has gone into controlling the search space. In recent years, many authors have chosen to replace domain-independent, general-purpose control strategies with domain-specific strategies, for example, the work of Bundy, van Harmelen, Hesketh and Smaill, Smith, and Wos. Other implementation techniques for theorem provers include the use of discrimination trees by McCune, flatterms by Christian, and parallel representation by Fishman and Minker.

Defining Terms

Artificial intelligence: The field of study that attempts to capture apsects of human intelligence in machines.

Automated deduction, automated reasoning: Deduction techniques that can be mechanized.

Classical logic: The standard logic that employs the usual connectives and the truth values {*true, false*}.

Clause: A disjunction of literals.

Completeness: An inference or rewrite rule is complete if the rule can verify that an unsatisfiable formula is unsatisfable.

Conjunctive normal form (CNF): A conjunction of clauses.

Davis-Putnam procedure: A systematic, backtracking procedure for propositional model finding.

Default logic: A nonmonotonic logic.

Domain of discourse: The set of values to which a first-order variable can be assigned.

First-order logic: Logic in which predicates can have arguments and formulas can be quantified.

Fuzzy logic: An extension of classical logic for handling uncertain propositions.

GSAT: A stochastic procedure for propositional model finding.

Herbrand universe: A domain of discourse constructed from the constants and function symbols that appear in a logical formula.

Inference rule: A rule that, when applied to a formula, produces another formula.

Interpretation: A function from the atom set to the set of truth values.

Lifting lemma: A lemma for proving completeness at the first-order level from completeness at the propositional level.

Literal: An atom or the negation of an atom.

Logical consequence: A formula C is a logical consequence of $\mathcal{F}$ if every interpretation that satisfies $\mathcal{F}$ also satisfies C.

Multiple-valued logic: Any logic whose set of truth values is *not* restricted to {*true, false*}.

Negation normal form (NNF): A form for logical formulas in which conjunction and disjunction are the only binary connectives and in which all negations are at the atomic level.

Nonmonotonic logic: A logic in which the addition of new knowledge may invalidate previously inferrable conclusions.

Paramodulation: A specialized inference rule for handling equality.

Path dissolution: An inference mechanism that operates on formulas in NNF.

Prenex normal form: A form for first-order logical formulas in which all quantifiers appear in the front of the formula.

Propositional logic: Logic in which predicates cannot have variables as arguments.

Quantifier: A restriction on variables in a first-order formula.

Resolution: An inference rule that operates on sets of clauses.

Rewrite rule: A rule that modifies formulas.

Satisfiable: A formula in classical logic is satisfiable if it evaluates to *true* under some interpretation.

Sign: Any subset of the set of truth values.

Skolem standard form: A form for first-order logical formulas in which all existentially quantified variables are replaced by constants or by functions of constants and the universally quantified variables.

Soundness: An inference (or rewrite rule) is sound if every inferred formula is a logical consequence of the original formula.

Substitution: A function that maps variables to terms.

Tableau method: An inference mechanism that operates on formulas in NNF.

Unification algorithm: An algorithm that finds the most general unifier of a set of terms.

Unifier: A substitution that unifies — makes identical — terms in different predicate occurrences.

References

Allen, J.F. 1991. Time and time again: the many ways to represent time. *J. Intelligent Syst.*, 6:341–355.

Andrews, P.B. 1976. Refutations by matings. *IEEE Trans. Comput.*, C-25:801–807.

Baldwin, J. 1986. Support logic programming. In *Fuzzy Sets Theory and Applications*. A. Jones, A. Kaufmann, and H. Zimmermann., Eds., pp. 133–170. D. Reidel.

Barback, M. and Lobo, J. 1995. A resolution-based procedure for default theories with extensions. In *Nonmonotonic Extensions of Logic Programming*, J. Dix, L. Pereira, and T. Przymusinski., Eds., pp. 101–126. Springer-Verlag, Heidelberg.

Beckert, B., Gerberding, S., Hähnle, R., and Kernig, W. 1992. The tableau-based theorem prover 3TAP for multiple-valued logics. In *Proc. 11th Int. Conf. Automated Deduction*, pp. 758–760. Springer-Verlag, Heidelberg.

Beckert, B. and Posegga, J. 1995. leanTAP: Lean tableau-based deduction. *J. Automated Reasoning*, 15(3):339–358.

Besnard, P. 1989. *An Introduction to Default Logic*. Springer-Verlag, Heidelberg.

Beth, E.W. 1955. Semantic entailment and formal derivability. *Mededelingen van de Koninklijke Nederlandse Akad. van Wetenschappen, Afdeling Letterkunde, N.R.*, 18(3):309–342.

Bibel, W. 1987. *Automated Theorem Proving*. Vieweg Verlag, Braunschweig.

Blair, H.A. and Subrahmanian, V.S. 1989. Paraconsistent logic programming. *Theor. Comput. Sci.*, 68(2):135–154.

Bledsoe, W.W., Kunen, K., and Shostak, R. 1985. Completeness results for inequality provers. *Artificial Intelligence*, 27(3):255–288.

Bobrow, D.G., Ed. 1980. *Artificial Intelligence: Spec. Issue Nonmonotonic Logics*. 13.

Boyer, R.S. and Moore, J.S. 1979. *A Computational Logic*. Academic Press, New York.

Bundy, A., van Harmelen, F., Hesketh, J., and Smaill, A. 1988. Experiments with proof plans for induction. *J. Automated Reasoning*, 7(3):303–324.

Chang, C.L. and Lee, R.C.T. 1973. *Symbolic Logic and Mechanical Theorem Proving*. Academic Press, New York.

Cook, S.A. 1971. The complexity of theorem proving procedures, pp. 151–158. In *Proc. 3rd Annu. ACM Symp. Theory Comput.* ACM Press, New York.

Digricoli, V.J. and Harrison, M.C. 1986. Equality based binary resolution. *J. ACM*, 33(2):253–289.

Dubois, D. and Prade, H. 1995. What does fuzzy logic bring to AI? *ACM Comput. Surveys*, 27(3):328–330.

Etherington, D.W. 1988. *Reasoning with Incomplete Information*. Pitman, London, UK.

Fagin, R., Halpern, J.Y., and Vardi, M.Y. 1992. What can machines know? On the properties of knowledge in distributed systems. *J. ACM*, 39(2):328–376.

Fitting, M. 1990. *Automatic Theorem Proving*. Springer-Verlag, Heidelberg.

Gabbay, D.M., Hogger, C.J., and Robinson, J.A., Eds. 1993–95. *Handbook of Logic in Artificial Intelligence and Logic Programming*. Vols. 1–4, Oxford University Press, Oxford, U.K.

Gaines, B.R. 1977. Foundations of fuzzy reasoning. In *Fuzzy Automata and Decision Processes*. M.M. Gupta, G.N. Saridis, and B.R. Gaines, Eds., pp. 19–75. North-Holland.

Gallaire, H. and Minker, J., Eds. 1978. *Logic and Data Bases*. Plenum Press.

Genesereth, M.R. and Nilsson, N.J. 1988. *Logical Foundations of Artificial Intelligence*. Morgan Kaufmann, Menlo Park, CA.

Gent, I. and Walsh, T., Eds. 2000. *J. of Automated Reasoning* special issue: satisfiability in the Year 2000, 24(1–2).

Gentzen, G. 1969. Investigations in logical deduction. In *Studies in Logic*, M.E. Szabo, Ed., pp. 132–213. Amsterdam.

Ginsberg, M. 1988. Multivalued logics: a uniform approach to inference in artificial intelligence. *Comput. Intelligence*, 4(3):265–316.

Green, C. 1969. Application of theorem proving to problem solving, pp. 219–239. In *Proc. 1st Int. Conf. Artificial Intelligence*. Morgan Kaufmann, Menlo Park, CA.

Hähnle, R. 1994. *Automated Deduction in Multiple-Valued Logics*. Vol. 10. International Series of Monographs on Computer Science. Oxford University Press, Oxford, U.K.

Hayes, P. 1977. In defense of logic, pp. 559–565. In *Proc. 5th IJCAI*. Morgan Kaufman, Palo Alto, CA.

Henschen, L. 1979. Theorem proving by covering expressions. *J. ACM*, 26(3):385–400.

Hintikka, K.J.J. 1955. Form and content in quantification theory. *Acta Philosohica Fennica*, 8:7–55.

Kapur, D. and Zhang, H. 1989. An overview of RRL: rewrite rule laboratory. In *Proc. 3rd Int. Conf. Rewriting Tech. Its Appl.* LNCS 355:513–529.

Kautz, H. and Selman, B., Eds. 2001. *Proceedings of the LICS 2001 Workshop on Theory and Applications of Satisfiability Testing*, Electronic Notes in Discrete Mathematics, 9.

Kifer, M. and Lozinskii, E. 1992. A logic for reasoning with inconsistency. *J. Automated Reasoning*, 9(2):179–215.

Kleene, S.C. 1952. *Introduction to Metamathematics*. Van Nostrand, Amsterdam.

Kraus, S., Lehmann, D., and Magidor, M. 1990. Nonmonotonic reasoning, preferential models and cumulative logics. *Artificial Intelligence*, 44(1–2):167–207.

Lee, R.C.T. 1972. Fuzzy logic and the resolution principle. *J. ACM*, 19(1):109–119.

Lifschitz, V., Ed. 1995. *J. Automated Reasoning: Spec. Issue Common Sense Nonmonotonic Reasoning*, 15(1).

Loveland, D.W. 1978. *Automated Theorem Proving: A Logical Basis*. North-Holland, New York.

Lu, J.J. 1996. Logic programming based on signs and annotations. *J. Logic Comput.* 6(6):755–778.

Lu, J.J., Murray, N.V., and Rosenthal, E. 1998. A framework for automated reasoning in multiple-valued logics. *J. Automated Reasoning* 21(1):39–67.

Łukasiewicz, J. 1970. *Selected Works*, L. Borkowski, Ed., North-Holland, Amsterdam.

Manna, Z. and Waldinger, R. 1986. Special relations in automated deduction. *J. ACM*, 33(1):1–59.

Marek, V.W. and Truszczyński, M. 1993. *Nonmonotonic Logic: Context-Dependent Reasoning*. Springer-Verlag, Heidelberg.

McCune, W. 1992. Experiments with discrimination-tree indexing and path indexing for term retrieval. *J. Automated Reasoning*, 9(2):147–168.

McRobbie, M.A. 1991. Automated reasoning and nonclassical logics: introduction. *J. Automated Reasoning: Spec. Issue Automated Reasoning Nonclassical Logics*, 7(4):447–452.

Mendelson, E. 1979. *Introduction to Mathematical Logic*. Van Nostrand Reinhold, Princeton, NJ.

Moore, R.C. 1985. Semantical considerations on nonmonotonic logic. *Artificial Intelligence*, 25(1):27–94.

Moskewcz, J.W., Madigan, C.F., Zhao, Y., Zhang, L., and Malik, S. 2001. Chaff: Engineering an Efficient SAT Solver. In *Proc. of the 39th Design Automation Conference*. Las Vegas, pp. 530–535. ACM Press.

Mukaidono, M. 1982. Fuzzy inference of resolution style. In *Fuzzy Set and Possibility Theory*, R. Yager, Ed., pp. 224–231. Pergamon, New York.

Murray, N.V. and Rosenthal, E. 1993. Dissolution: making paths vanish. *J. ACM*, 40(3):502–535.

Robinson, J.A. 1965. A machine-oriented logic based on the resolution principle. *J. ACM*, 12(1):23–41.

Robinson, J.A. 1979. *Logic: Form and Function*. Elsevier North-Holland, New York.

Robinson, G. and Wos, L. 1969. Paramodulation and theorem proving in first-order theories with equality. In *Machine Intelligence*, Vol. IV, B. Melzer and D. Michie, Eds., pp. 135–150. Edinburgh University Press, Edinburgh, U.K.

Russell, S. and Norvig, P. 1995. *Artificial Intelligence: A Modern Approach*. Prentice Hall, Englewood Cliffs, NJ.

Selman, B., Levesque, H.J., and Mitchell, D.V. 1992. A new method for solving hard satisfiability problems. In *Proceedings of the Tenth National Conference on Aritificial Intelligence*, P. Rosenbloom and P. Szolovits, Eds. pp. 440–446. AAAI Press, Menlo Park, CA.

Slagle, J. 1972. Automatic theorem proving with built-in theories including equality, partial ordering, and sets. *J. ACM*, 19(1):120–135.

Smullyan, R.M. 1995. *First-Order Logic*, 2nd ed. Dover, New York.

Sterling, L. and Shapiro, E. 1986. *The Art of Prolog*. ACM Press, Cambridge, MA.

Stickel, M.E. 1985. Automated deduction by theory resolution. *J. Automated Reasoning*, 1(4):333–355.

Urquhart, A. 1986. Many-valued logic. In *Handbook of Philosophical Logic*, Vol. III, D. Gabbay and F. Guenthner, Eds., pp. 71–116. D. Reidel.

Weigert, T.J., Tsai, J.P., and Liu, X.H. 1993. Fuzzy operator logic and fuzzy resolution. *J. Automated Reasoning*, 10(1):59–78.

Wos, L., Overbeek, R., Lusk, E., and Boyle, J. 1992. *Automated Reasoning: Introduction and Applications*. 2nd ed. Prentice Hall, Englewood Cliffs, NJ.

Wrightson, G., Ed. 1994. *J. Automated Reasoning: Spec. Issues Automated Reasoning Analytic Tableaux* 13(2,3).

Zadeh, L.A. 1965. Fuzzy sets. *Inf. Control*, 8(3):338–353.

Zhang, H. and Stickel, M. 2000. Implementing the Davis-Putnam procedure. *J. Automated Reasoning*, 24(1–2):277–296.

Further Information

The *Journal of Automated Reasoning* is an excellent reference for current research and advances in logic-based automated deduction techniques for both classical and nonclassical logics. The International Conference on Automated Deduction (CADE) is the major forum for researchers focusing on logic-based deduction techniques; its proceedings are published by Springer-Verlag. Other conferences with an emphasis on computational logic and logic-based reasoning include the International Logic Programming Conference (ICLP), Logics in Computer Science (LICS), Logic Programming and Nonmonotonic Reasoning Conference (LPNMR), International Symposium on Multiple-Valued Logics (ISMVL), the International Symposium on Methodologies for Intelligent Systems (ISMIS), and the IEEE International Conference on Fuzzy Systems.

More general conferences on AI include two major annual meetings: the conference of the AAAI and the International Joint Conference on AI. Each of these conferences regularly publishes logic-based deduction papers. The *Artificial Intelligence* journal is an important source for readings on logics for common-sense reasoning and related deduction techniques.

Other journals of relevance include the *Journal of Logic and Computation*, the *Journal of Computational Intelligence*, the *Journal of Logic Programming*, the *Journal of Symbolic Computation*, *IEEE Transactions on Fuzzy Systems*, *Theoretical Computer Science*, and the *Journal of the Association of Computing Machinery*.

Most of the texts referenced in this chapter provide a more detailed introduction to the field of computational logic. They include Bibel [1987], Chang and Lee [1973], Fitting [1990], Loveland [1978], Robinson [1979], and Wos et al. [1992]. Good introductory texts for mathematical logic are Mendelson [1979] and Smullyan [1995].

Model finding and propositional deduction is an active area of research with special issues of journals and workshops dedicated to the topic. See Gent and Walsh [2000] and Kautz and Selman [2001] for recent advances in the field. Improvements on implementation techniques continue to be reported at a rapid rate. A useful source of information on the latest developments in the field can be found at http://www.satlive.org

62

Qualitative Reasoning

62.1 Introduction 62-1
62.2 Qualitative Representations 62-2
Representing Quantity • Representing Mathematical
Relationships • Ontology • State, Time, and Behaviors
• Space and Shape • Compositional Modeling, Domain
Theories, and Modeling Assumptions
62.3 Qualitative Reasoning Techniques 62-9
Model Formulation • Causal Reasoning • Simulation
• Comparative Analysis • Teleological Reasoning • Data
Interpretation • Planning • Spatial Reasoning
62.4 Applications of Qualitative Reasoning 62-13
Monitoring, Control, and Diagnosis • Design • Intelligent
Tutoring Systems and Learning Environments • Cognitive
Modeling
62.5 Research Issues and Summary 62-16

Kenneth D. Forbus
Northwestern University

62.1 Introduction

Qualitative reasoning is the area of artificial intelligence (AI) that creates representations for continuous aspects of the world, such as space, time, and quantity, which support reasoning with very little information. Typically, it has focused on scientific and engineering domains, hence its other name, qualitative physics. It is motivated by two observations. First, people draw useful and subtle conclusions about the physical world without equations. In our daily lives we figure out what is happening around us and how we can affect it, working with far less data, and less precise data, than would be required to use traditional, purely quantitative methods. Creating software for robots that operate in unconstrained environments and modeling human cognition require understanding how this can be done. Second, scientists and engineers appear to use qualitative reasoning when initially understanding a problem, when setting up more formal methods to solve particular problems, and when interpreting the results of quantitative simulations, calculations, or measurements. Thus, advances in qualitative physics should lead to the creation of more flexible software that can help engineers and scientists.

Qualitative physics began with de Kleer's investigation on how qualitative and quantitative knowledge interacted in solving a subset of simple textbook mechanics problems [de Kleer, 1977]. After roughly a decade of initial explorations, the potential for important industrial applications led to a surge of interest in the mid-1980s, and the area grew steadily, with rapid progress. Qualitative representations have made their way into commercial supervisory control software for curing composite materials, design, and **FMEA** (Failure Modes and Effects Analysis). The first product known to have been designed using qualitative physics techniques appeared on the market in 1994 [Shimomura et al., 1995]. Given its demonstrated utility in industrial applications and its importance in understanding human cognition, work in qualitative modeling is likely to remain an important area in artificial intelligence.

This chapter first surveys the state of the art in qualitative representations and in qualitative reasoning techniques. The application of these techniques to various problems is discussed subsequently.

62.2 Qualitative Representations

As with many other representation issues, there is no single, universal right or best qualitative representation. Instead, there exists a spectrum of choices, each with its own advantages and disadvantages for particular tasks. What all of them have in common is that they provide notations for describing and reasoning about continuous properties of the physical world. Two key issues in qualitative representation are resolution and **compositionality**. We discuss each in turn.

Resolution concerns the level of information detail in a representation. Resolution is an issue because one goal of qualitative reasoning is to understand how little information suffices to draw useful conclusions. Low-resolution information is available more often than precise information ("the car heading toward us is slowing down" vs. "the derivative of the car's velocity along the line connecting us is −28 km/hr/sec"), but conclusions drawn with low-resolution information are often ambiguous. The role of ambiguity is important: the prediction of alternative futures (i.e., "the car will hit us" vs. "the car won't hit us") suggests that we may need to gather more information, analyze the matter more deeply, or take action, depending on what alternatives our qualitative reasoning uncovers. High-resolution information is often needed to draw particular conclusions (i.e., a finite element analysis of heat flow within a notebook computer design to ensure that the CPU will not cook the battery), but qualitative reasoning with low-resolution representations reveals what the interesting questions are. Qualitative representations comprise one form of tacit knowledge that people, ranging from the person on the street to scientists and engineers, use to make sense of the world.

Compositionality concerns the ability to combine representations for different aspects of a phenomenon or system to create a representation of the phenomenon or system as a whole. Compositionality is an issue because one goal of qualitative reasoning is to formalize the modeling process itself. Many of today's AI systems are based on handcrafted knowledge bases that express information about a specific artifact or system needed to carry out a particular narrow range of tasks involving it. By contrast, a substantial component of the knowledge of scientists and engineers consists of principles and laws that are broadly applicable, both with respect to the number of systems they explain and the kinds of tasks they are relevant for. Qualitative physics is developing the ideas and organizing techniques for knowledge bases with similar expressive and inferential power, called **domain theories**.

The remainder of this section surveys the fundamental representations used in qualitative reasoning for quantity, mathematical relationships, **modeling assumptions**, causality, space, and time.

62.2.1 Representing Quantity

Qualitative reasoning has explored trade-offs in representations for continuous parameters ranging in resolution from sign algebras to the hyperreals. Most of the research effort has gone into understanding the properties of low-resolution representations because the properties of high-resolution representations tend to already be well understood due to work in mathematics.

The lowest resolution representation for continuous parameters is the status abstraction, which represents a quantity by whether or not it is *normal* [Abbott et al., 1987]. It is a useful representation for certain diagnosis and monitoring tasks because it is the weakest representation that can express the difference between something working and not working. The next step in resolution is the sign algebra, which represents continuous parameters as either −, +, or 0, according to whether the sign of the underlying continuous parameter is negative, positive, or zero. The sign algebra is surprisingly powerful: because a parameter's derivatives are themselves parameters whose values can be represented as signs, some of the main results of the differential calculus (e.g., the mean value theorem) can be applied to reasoning about sign values [de Kleer and Brown, 1984]. This allows sign algebras to be used for qualitative reasoning about dynamics, including expressing properties such as oscillation and stability. The sign algebra is the weakest representation that supports such reasoning.

Representing continuous values via sets of ordinal relations (also known as the **quantity space** representation) is the next step up in resolution [Forbus, 1984]. For example, the temperature of a fluid might be represented in terms of its relationship between the freezing point and boiling point of the material that comprises it. Like the sign algebra, quantity spaces are expressive enough to support qualitative reasoning about dynamics. (The sign algebra can be modeled by a quantity space with only a single comparison point, zero.) Unlike the sign algebra, which draws values from a fixed finite algebraic structure, quantity spaces provide variable resolution because new points of comparison can be added to refine values. The temperature of water in a kettle on a stove, for instance, will likely be defined in terms of its relationship with the temperature of the stove as well as its freezing and boiling points. There are two kinds of comparison points used in defining quantity spaces. **Limit points** are derived from general properties of a domain as applicable to a specific situation. Continuing with the kettle example, the particular ordinal relationships used were chosen because they determine whether or not the **physical processes** of freezing, boiling, and heat flow occur in that situation. The precise numerical value of limit points can change over time (e.g., the boiling point of a fluid is a function of its pressure). **Landmark** values are constant points of comparison introduced during reasoning to provide additional resolution [Kuipers, 1986]. To ascertain whether an oscillating system is overdamped, underdamped, or critically damped, for instance, requires comparing successive peak values. Noting the peak value of a particular cycle as a landmark value, and comparing it to the landmarks generated for successive cycles in the behavior, provides a way of making this inference.

Intervals are a well-known variable-resolution representation for numerical values and have been heavily used in qualitative reasoning. A quantity space can be thought of as partial information about a set of intervals. If we have complete information about the ordinal relationships between limit points and landmark values, these comparison points define a set of intervals that partition a parameter's value. This natural mapping between quantity spaces and intervals has been exploited by a variety of systems that use intervals whose endpoints are known numerical values to refine predictions produced by purely qualitative reasoning. Fuzzy intervals have also been used in similar ways, for example, in reasoning about control systems.

Order of magnitude representations stratify values according to some notion of scale. They can be important in resolving ambiguities and in simplifying models because they enable reasoning about what phenomena and effects can safely be ignored in a given situation. For instance, heat losses from turbines are generally ignored in the early stages of power plant design, because the energy lost is very small relative to the energy being produced. Several stratification techniques have been used in the literature, including hyperreal numbers, numerical thresholds, and logarithmic scales. Three issues faced by all these formalisms are (1) the conditions under which many small effects can combine to produce a significant effect, (2) the soundness of the reasoning supported by the formalism, and (3) the efficiency of using them.

Although many qualitative representations of number use the reals as their basis, another important basis for qualitative representations of number is finite algebras. One motivation for using finite algebras is that observations are often naturally categorized into a finite set of labels (i.e., very small, small, normal, large, very large). Research on such algebras is aimed at solving problems such as how to increase the compositionality of such representations (e.g., how to propagate information across different resolution scales).

62.2.2 Representing Mathematical Relationships

Like number, a variety of qualitative representations of mathematical relationships have been developed, often by adopting and adapting systems developed in mathematics. Abstractions of the analytic functions are commonly used to provide the lower resolution and compositionality desired. For example, **confluences** are differential equations over the sign algebra [de Kleer and Brown, 1984]. An equation such as $V = IR$ can be expressed as the confluence

$$[V] = [I] + [R]$$

where $[Q]$ denotes taking the sign of Q. Differential equations can also be expressed in this manner, for instance,

$$[F] = \partial_V$$

which is a qualitative version of $F = MA$ (assuming M is always positive). Thus, any system of algebraic and differential equations with respect to time can be described as a set of confluences.

Many of the algebraic operations taken for granted in manipulating analytic functions over the reals are not valid in weak algebras [Struss, 1988]. Because qualitative relationships most often are used to propagate information, this is not a serious limitation. In situations where algebraic solutions themselves are desirable, mixed representations that combine algebraic operations over the reals and move to qualitative abstractions when appropriate provide a useful approach [Williams, 1991].

Another low-resolution representation of equations uses monotonic functions over particular ranges; that is,

$$M + (\text{acceleration, force})$$

states that acceleration depends only on the force, and the function relating them is increasing monotonic [Kuipers, 1986]. Compositionality is achieved using qualitative proportionalities [Forbus, 1984] to express partial information about functional dependency; for example,

$$\text{acceleration} \propto_{Q+} \text{force}$$

states that acceleration depends on force and is increasing monotonic in its dependence on force, but may depend on other factors as well. Additional constraints on the function that determines force can be added by additional **qualitative proportionalities**; for example,

$$\text{acceleration} \propto_{Q-} \text{mass}$$

states that acceleration also depends on mass, and is decreasing monotonic in this dependence. Qualitative proportionalities must be combined via closed-world assumptions to ascertain all the effects on a quantity. Similar primitives can be defined for expressing relationships involving derivatives, to define a complete language of compositional qualitative mathematics for ordinary differential equations. As with confluences, few algebraic operations are valid for combining monotonic functions, mainly composition of functions of identical sign; that is,

$$M + (f, g) \wedge M + (g, h) \Rightarrow M + (f, h)$$

In addition to resolution and compositionality, another issue arising in qualitative representations of mathematical relationships is causality. There are three common views on how mathematical relationships interact with the causal relationships people use in common-sense reasoning. One view is that there is no relationship between them. The second view is that mathematical relationships should be expressed with primitives that also make causal implications. For example, qualitative proportionalities include a causal interpretation, that is, a change in force causes a change in acceleration, but not the other way around. The third view is that acausal mathematical relationships give rise to causal relationships via the particular process of using them. For example, confluences have no built-in causal direction, but are used in causal reasoning by identifying the flow of information through them while reasoning with a presumed flow of causality in the physical system they model. One method for imposing causality on a set of acausal constraint equations is by computing a causal ordering [Iwasaki and Simon, 1986] that imposes directionality on a set of equations, starting from variables considered to be exogenous within the system.

Each view of causality has its merits. For tasks where causality is truly irrelevant, ignoring causality can be a reasonable approach. To create software that can span the range of human common-sense reasoning, something like a combination of the second and third views appears necessary because the appropriate notion of causality varies [Forbus and Gentner, 1986]. In reasoning about chemical phenomena, for instance, changes in concentration are always caused by changes in the amounts of the constituent parts and never the other way around. In electronics, on the other hand, it is often convenient to consider voltage changes as being caused by changes in current in one part of a circuit and to consider current changes as being caused by changes in voltage in another part of the same circuit.

62.2.3 Ontology

Ontology concerns how to carve up the world, that is, what kinds of things there are and what sorts of relationships can hold between them. Ontology is central to qualitative modeling because one of its main goals is formalizing the art of building models of physical systems. A key choice in any act of modeling is figuring out how to construe the situation or system to be modeled in terms of the available models for classes of entities and phenomena. No single ontology will suffice for the span of reasoning about physical systems that people do. What is being developed instead is a catalog of ontologies, describing their properties and interrelationships and specifying conditions under which each is appropriate. While several ontologies are currently well understood, the catalog still contains gaps.

An example of ontologies will make this point clearer. Consider the representation of liquids. Broadly speaking, the major distinction in reasoning about fluids is whether one individuates fluid according to a particular collection of particles or by location [Hayes, 1985]. The former are called Eulerian, or piece of stuff, ontologies. The latter are called Lagrangian, or contained stuff, ontologies. It is the contained stuff view of liquids we are using when we treat a river as a stable entity, although the particular set of molecules that comprises it is changing constantly. It is the piece of stuff view of liquids we are using when we think about the changes in a fluid as it flows through a steady-state system, such as a working refrigerator. Ontologies multiply as we try to capture more of human reasoning. For instance, the piece of stuff ontology can be further divided into three cases, each with its own rules of inference: (1) molecular collections, which describe the progress of an arbitrary piece of fluid that is small enough to never split apart but large enough to have extensive properties; (2) slices which, like molecular collections, never subdivide but unlike them are large enough to interact directly with their surroundings; and (3) pieces of stuff large enough to be split into several pieces (e.g., an oil slick). Similarly, the contained stuff ontology can be further specialized according to whether or not individuation occurs simply by container (abstract contained stuffs) or by a particular set of containing surfaces (bounded stuffs). Abstract contained stuffs provide a low-resolution ontology appropriate for reasoning about system-level properties in complex systems (e.g., the changes over time in a lubricating oil subsystem in a propulsion plant), whereas bounded stuffs contain the geometric information needed to reason about the interactions of fluids and shape in systems such as pumps and internal combustion engines.

Cutting across the ontologies for particular physical domains are systems of organization for classes of ontologies. The most commonly used ontologies are the device ontology [de Kleer and Brown, 1984] and the process ontology [Forbus, 1984]. The device ontology is inspired by network theory and system dynamics. Like those formalisms, it construes physical systems as networks of devices whose interactions occur solely through a fixed set of ports. Unlike those formalisms, it provides the ability to write and reason automatically with device models whose governing equations can change over time.

The process ontology is inspired by studies of human mental models and observations of practice in thermodynamics and chemical engineering. It construes physical systems as consisting of entities whose changes are caused by physical processes. Process ontologies thus postulate a separate ontological category for causal mechanisms, unlike device ontologies, where causality arises solely from the interaction of the parts. Another difference between the two classes of ontologies is that in the device ontology the system of devices and connections is fixed over time, whereas in the process ontology entities and processes can come into existence and vanish over time. Each is appropriate in different contexts: for most purposes, an electronic circuit is best modeled as a network of devices, whereas a chemical plant is best modeled as a collection of interacting processes.

62.2.4 State, Time, and Behaviors

A qualitative state is a set of propositions that characterize a qualitatively distinct behavior of a system. A qualitative state describing a falling ball, for instance, would include information about what physical processes are occurring (e.g., motion downwards, acceleration due to gravity) and how the parameters of the ball are changing (e.g., its position is getting lower and its downward velocity is getting larger). A qualitative state can abstractly represent an infinite number of quantitative states: although the position and velocity of the ball are different at each distinct moment during its fall, until the ball collides with the ground, the qualitative state of its motion is unchanged.

Qualitative representations can be used to partition behavior into natural units. For instance, the time over which the state of the ball falling holds is naturally thought of as an interval, ending when the ball collides with the ground. The collision itself can be described as yet another qualitative state, and the fact that falling leads to a collision with the ground can be represented via a transition between the two states. If the ball has a nonzero horizontal velocity and there is some obstacle in its direction of travel, another possible behavior is that the ball will collide with that object instead of the ground. In general, a qualitative state can have transitions to several next states, reflecting ambiguity in the qualitative representations. Returning to our ball example, and assuming that no collisions with obstacles occur, notice that the qualitative state of the ball falling occurs again once the ball has reached its maximum height after the collision. If continuous values are represented by quantity spaces and the sources of comparisons are limit points, then a finite set of qualitative states is sufficient to describe every possible behavior of a system. A collection of such qualitative states and transitions is called an **envisionment** [de Kleer, 1977]. Many interesting dynamical conclusions can be drawn from an envisionment. For instance, oscillations correspond to cycles of states. Unfortunately, the fixed resolution provided by limit points is not sufficient for other dynamical conclusions, such as ascertaining whether or not the ball's oscillation is damped. If comparisons can include landmark values, such conclusions can sometimes be drawn, for example, by comparing the maximum height on one bounce to the maximum height obtained on the next bounce. The cost of introducing landmark values is that the envisionment no longer need be finite; every cycle in a corresponding fixed-resolution envisionment could give rise to an infinite number of qualitative states in an envisionment with landmarks.

A sequence of qualitative states occurring over a particular span of time is called a behavior. Behaviors can be described using purely qualitative knowledge, purely quantitative knowledge, or a mixture of both. If every continuous parameter is quantitative, the numerical aspects of behaviors coincide with the notion of trajectory in a state-space model. If qualitative representations of parameters are used, a single behavior can represent a family of trajectories through state space.

An idea closely related to behaviors is histories [Hayes, 1985]. Histories can be viewed as local behaviors, that is, how a single individual or property varies through time. A behavior is equivalent to a global history, that is, the union of all the histories for the participating individuals. The distinction is important for two reasons. First, histories are the dual of situations in the situation calculus; histories are bounded spatially and extended temporally, whereas situations are bounded temporally and global spatially. Using histories avoids the frame problem, instead trading it for the more tractable problems of generating histories locally and determining how they interact when they intersect in space and time. The second reason is that history-based simulation algorithms can be more efficient than state-based simulation algorithms because no commitments need to be made concerning irrelevant information.

In a correct envisionment, every possible behavior of the physical system corresponds to some path through the envisionment. Because envisionments reflect only local constraints, the converse is not true; that is, an arbitrary path through an envisionment may not represent a physically possible behavior. All such paths must be tested against global constraints, such as energy conservation, to ensure their physical validity. Because the typical uses of an envisionment are to test whether an observed behavior is plausible or to propose possible behaviors, this limitation is not serious. A more serious limitation is that envisionments are often exponential in the size of the system being modeled. This means that, in practice, envisionments often are not generated explicitly and, instead, possible behaviors are searched in ways similar to those used in other areas of AI.

Many tasks require integrating qualitative states with other models of time, such as numerical models. Including precise information (e.g., algebraic expressions or floating-point numbers) about the endpoints of intervals in a history does not change their essential character.

62.2.5 Space and Shape

Qualitative representations of shape and space play an important role in spatial cognition because they provide a bridge between the perceptual and the conceptual. By discretizing continuous space, they make it amenable to symbolic reasoning. As with qualitative representations of one-dimensional parameters, task constraints govern the choice of qualitative representation. However, problem-independent purely qualitative spatial representations suffice for fewer tasks than in the one-dimensional case, because of the increased ambiguity in higher dimensions [Forbus et al., 1991]. Consider, for example, deciding whether a protrusion can fit snugly inside a hole. If we have detailed information about their shapes, we can derive an answer. If we consider a particular set of protrusions and a particular set of holes, we can construct a qualitative representation of these particular protrusions and holes that would allow us to derive whether or not a specific pair would fit, based on their relative sizes. But if we first compute a qualitative representation for each protrusion and hole in isolation, in general the rules of inference that can be derived for this problem will be very weak. Work in qualitative spatial representations thus tends to take two approaches. The first approach is to explore what aspects do lend themselves to qualitative representations. The second approach is to use a quantitative representation as a starting point and compute problem-specific qualitative representations to reason with. We summarize each in turn.

There are several purely qualitative representations of space and shape that have proven useful. Topological relationships between regions in two-dimensional space have been formalized, with transitivity inferences similar to those used in temporal reasoning identified for various vocabularies of relations [Cohn and Hazarika, 2001]. The beginnings of rich qualitative mechanics have been developed. This includes qualitative representations for vectors using the sign of the vector's quadrant to reason about possible directions of motion [Nielsen, 1988] and using relative inclination of angles to reason about linkages [Kim, 1992].

The use of quantitative representations to ground qualitative spatial reasoning can be viewed as a model of the ways humans use diagrams and models in spatial reasoning. For this reason such work is also known as **diagrammatic reasoning** [Glasgow et al., 1995]. One form of diagram representation is the occupancy array that encodes the location of an object by cells in a (two- or three-dimensional) grid. These representations simplify the calculation of spatial relationships between objects (e.g., whether or not one object is above another), albeit at the cost of making the object's shape implicit. Another form of diagram representation uses symbolic structures with quantitative, for example, numerical, algebraic, or interval (cf. [Forbus et al., 1991]). These representations simplify calculations involving shape and spatial relationships, without the scaling and resolution problems that sometimes arise in array representations. However, they require a set of primitive shape elements that spans all the possible shapes of interest, and identifying such sets for particular tasks can be difficult. For instance, many intuitively natural sets of shape primitives are not closed with respect to their complement, which can make characterizing free space difficult.

Diagram representations are used for qualitative spatial reasoning in two ways. The first is as a decision procedure for spatial questions. This mimics one of the roles diagrams play in human perception. Often, these operations are combined with domain-specific reasoning procedures to produce an analog style of inference, where, for instance, the effects of perturbations on a structure are mapped into the diagram, the effect on the shapes in the diagram noted, and the results mapped back into a physical interpretation. The second way uses the diagram to construct a problem-specific qualitative vocabulary, imposing new spatial entities representing physical properties, such as the maximum height a ball can reach or regions of free space that can contain a motion. This is the **metric diagram/place vocabulary** model of qualitative spatial reasoning.

Representing and reasoning about kinematic mechanisms was one of the early successes in qualitative spatial reasoning. The possible motions of objects are represented by qualitative regions in configuration space representing the legitimate positions of parts of mechanisms [Faltings 1990]. Whereas, in principle, a single high-dimensional configuration space could be used to represent a mechanism's possible motions (each dimension corresponding to a degree of freedom of a part of the mechanism), in practice a collection of configuration spaces, one two-dimensional space for each pair of parts that can interact is used. These techniques suffice to analyze a wide variety of kinematic mechanisms [Joscowicz and Sacks, 1993].

Another important class of spatial representations concerns qualitative representations of spatially distributed phenomena, such as flow structures and regions in phase space. These models use techniques from computer vision to recognize or impose qualitative structure on a continuous field of information, gleaned from numerical simulation or scientific data. This qualitative structure, combined with domain-specific models of how such structures tie to the underlying physics, enables them to interpret physical phenomena in much the same way that a scientist examining the data would (cf. [Yip, 1991], [Nishida, 1994], [Huang and Zhao, 2000]).

An important recent trend is using rich, real-world data as input for qualitative spatial reasoning. For example, several systems provide some of the naturalness of sketching by performing qualitative reasoning on spatial data input as digital ink, for tasks like mechanical design (cf. [Stahovich et al., 1998]) and reasoning about sketch maps (cf. [Forbus et al., 2003]). Qualitative representations are starting to be used in computer vision as well, for example, as a means of combining dynamic scenes across time to interpret events (cf. [Fernyhough et al., 2000]).

62.2.6 Compositional Modeling, Domain Theories, and Modeling Assumptions

There is almost never a single correct model for a complex physical system. Most systems can be modeled in a variety of ways, and different tasks can require different types of models. The creation of a system model for a specific purpose is still something of an art. Qualitative physics has developed formalisms that combine logic and mathematics with qualitative representations to help automate the process of creating and refining models. The compositional modeling methodology [Falkenhainer and Forbus, 1991], which has become standard in qualitative physics, works like this: models are created from domain theories, which describe the kinds of entities and phenomena that can occur in a physical domain. A domain theory consists of a set of **model fragments**, each describing a particular aspect of the domain. Creating a model is accomplished by instantiating an appropriate subset of model fragments, given some initial specification of the system (e.g., the propositional equivalent of a blueprint) and information about the task to be performed. Reasoning about appropriateness involves the use of modeling assumptions. Modeling assumptions are the control knowledge used to reason about the validity or appropriateness of using model fragments. Modeling assumptions are used to express the relevance of model fragments. Logical constraints between modeling assumptions comprise an important component of a domain theory.

An example of a modeling assumption is assuming that a turbine is isentropic. Here is a model fragment that illustrates how this assumption is used:

```
(defEquation Isentropic-Turbine
  ((turbine ?g ?in ?out)(isentropic ?g))
  (= (spec-s ?in) (spec-s ?out)))
```

In other words, when a turbine is isentropic, the specific entropy of its inlet and outlet are equal. Other knowledge in the domain theory puts constraints on the predicate isentropic,

```
(for-all (?self (turbine ?self))
  (iff (= (nu-isentropic ?self) 1.0)
  (isentropic ?self)))
```

That is, a turbine is isentropic exactly when its isentropic thermal efficiency is 1. Although no real turbine is isentropic, assuming that turbines are isentropic simplifies early analyses when creating a new design. In later design phases, when tighter performance bounds are required, this assumption is retracted and the impact of particular values for the turbine's isentropic thermal efficiency is explored. The consequences of choosing particular modeling assumptions can be quite complex; the fragments shown here are less than one fourth the knowledge expressing the consequences of assuming that a turbine is isentropic in a typical knowledge base.

Modeling assumptions can be classfied in a variety of ways. An ontological assumption describes which onotology should be used in an analysis. For instance, reasoning about the pressure at the bottom of a swimming pool is most simply performed using a contained stuff representation, whereas describing the location of an oil spill is most easily performed using a piece of stuff representation. A perspective assumption describes which subset of phenomena operating in a system will be the subject. For example, in analyzing a steam plant, one might focus on a fluid perspective, a thermal perspective, or both at once. A grain assumption describes how much detail is included in an analysis. Ignoring the implementation details of subsystems, for instance, is useful in the conceptual design of an artifact, but the same implementation details may be critical for troubleshooting that artifact. The relationships between these classes of assumptions can be complicated and domain dependent; for instance, it makes no sense to include a model of a heating coil (a choice of granularity) if the analysis does not include thermal properties (a choice of perspective).

Relationships between modeling assumptions provide global structure to domain theories. Assumptions about the nature of this global structure can significantly impact the efficiency of model formulation, as discussed subsequently. In principle, any logical constraint could be imposed between modeling assumptions. In practice, two kinds of constraints are the most common. The first are implications, such as one modeling assumption requiring or forbidding another. For example,

```
(for-all (?s (system ?s))
  (implies (consider (black-box ?s))
           (for-all (?p (part-of ?p ?s)) (not (consider ?p)))))
```

says that if one is considering a subsystem as a black box, then all of its parts should be ignored. Similarly,

```
(for-all (?l (physical-object ?l))
  (implies (consider (pressure ?l))
           (consider (fluid-properties ?l))))
```

states that if an analysis requires considering something's pressure, then its fluid properties are relevant.

The second kind of constraint between modeling assumptions is assumption classes. An assumption class expresses a choice required to create a coherent model under particular conditions. For example,

```
(defAssumptionClass (turbine ?self)
  (isentropic ?self)
  (not (isentropic ?self)))
```

states that when something is modeled as a turbine, any coherent model including it must make a choice about whether or not it is modeled as isentropic. The choice may be constrained by the data so far (e.g., different entrance and exit specific entropies), or it may be an assumption that must be made in order to complete the model. The set of choices need not be binary. For each valid assumption class there must be exactly one of the choices it presents included in the model.

62.3 Qualitative Reasoning Techniques

A wide variety of qualitative reasoning techniques have been developed that use the qualitative representations just outlined.

62.3.1 Model Formulation

Methods for automatically creating models for a specific task are one of the hallmark contributions of qualitative physics. These methods formalize knowledge and skills typically left implicit by most of traditional mathematics and engineering.

The simplest model formulation algorithm is to instantiate every possible model fragment from a domain theory, given a propositional representation of the particular scenario to be reasoned about. This algorithm is adequate when the domain theory is very focused and thus does not contain much irrelevant information. It is inadequate for broad domain theories and fails completely for domain theories that include alternative and mutually incompatible perspectives (e.g., viewing a contained liquid as a finite object vs. an infinite source of liquid). It also fails to take task constraints into account. For example, it is possible, in principle, to analyze the cooling of a cup of coffee using quantum mechanics. Even if it were possible in practice to do so, for most tasks simpler models suffice. Just how simple a model can be and remain adequate depends on the task. If I want to know if the cup of coffee will still be drinkable after an hour, a qualitative model suffices to infer that its final temperature will be that of its surroundings. If I want to know its temperature within 5% after 12 min have passed, a macroscopic quantitative model is a better choice. In other words, the goal of model formulation is to create the simplest adequate model of a system for a given task.

More sophisticated model formulation algorithms search the space of modeling assumptions, because they control which aspects of the domain theory will be instantiated. The model formulation algorithm of Falkenhainer and Forbus [1991] instantiated all potentially relevant model fragments and used an assumption-based truth maintenance system to find all legal combinations of modeling assumptions that sufficed to form a model that could answer a given query. The simplicity criterion used was to minimize the number of modeling assumptions. This algorithm is very simple and general but has two major drawbacks: (1) full instantiation can be very expensive, especially if only a small subset of the model fragments is eventually used; and (2) the number of consistent combinations of model fragments tends to be exponential for most problems. The rest of this section describes algorithms that overcome these problems.

Efficiency in model formulation can be gained by imposing additional structure on domain theories. Under at least one set of constraints, model formulation can be carried out in polynomial time [Nayak, 1994]. The constraints are that (1) the domain theory can be divided into independent assumption classes; and (2) within each assumption class, the models can be organized by a (perhaps partial) simplicity ordering of a specific nature, forming a lattice of *causal approximations*. Nayak's algorithm computes a simplest model, in the sense of simplest within each local assumption class, but does not necessarily produce the globally simplest model.

Conditions that ensure the creation of *coherent* models, that is, models that include sufficient information to produce an answer of the desired form, provide powerful constraints on model formulation. For example, in generating "what-if" explanations of how a change in one parameter might affect particular other properties of the system, a model must include a complete causal chain connecting the changed parameter to the other parameters of interest. This insight can be used to treat model formulation as a best-first search for a set of model fragments providing the simplest complete causal chain [Rickel and Porter, 1994]. A novel feature of this algorithm is that it also selects models at an appropriate time scale. It does this by choosing the slowest time-scale phenomenon that provides a complete causal model, because this provides accurate answers that minimize extraneous detail.

As with other AI problems, knowledge can reduce search. One kind of knowledge that experienced modelers accumulate concerns the range of applicability of various modeling assumptions and strategies for how to reformulate when a given model proves inappropriate. Model formulation often is an iterative process. For instance, an initial qualitative model often is generated to identify the relevant phenomena, followed by the creation of a narrowly focused quantitative model to answer the questions at hand. Similarly, domain-specific error criterion can determine that a particular model's results are internally inconsistent, causing the reasoner to restart the search for a good model. Formalizing the decision making needed in iterative model formulation is an area of active research. Formalizing model formulation as a dynamic preference constraint satisfaction problem, where more fine-grained criteria for model preference than "simplest" can be formalized and exploited (cf. [Keppens and Shen, 2002]), is one promising approach.

62.3.2 Causal Reasoning

Causal reasoning explains an aspect of a situation in terms of others in such a way that the aspect being explained can be changed if so desired. For instance, a flat tire is caused by the air inside flowing out, either through the stem or through a leak. To refill the tire, we must both ensure that the stem provides a seal and that there are no leaks. Causal reasoning is thus at the heart of diagnostic reasoning as well as explanation generation.

The techniques used for causal reasoning depend on the particular notion of causality used, but they all share a common structure. First, causality involving factors within a state are identified. Second, how the properties of a state contribute to a transition (or transitions) to another state are identified, to extend the causal account over time. Because causal reasoning often involves qualitative simulation, we turn to simulation next.

62.3.3 Simulation

The new representations of quantity and mathematical relationships of qualitative physics expand the space of simulation techniques considerably. We start by considering varieties of purely qualitative simulation, and then describe several simulation techniques that integrate qualitative and quantitative information.

Understanding *limit analysis*, the process of finding state transitions, is key to understanding qualitative simulation. Recall that a qualitative state consists of a set of propositions, some of them describing the values of continuous properties in the system. (For simplicity in this discussion, we will assume that these values are described as ordinal relations, although the same method works for sign representations and representations richer than ordinals.) Two observations are critical: (1) the phenomena that cause changes in a situation often depend on ordinal relationships between parameters of the situation, and (2) knowing just the sign of the derivatives of the parameters involved in these ordinal relationships suffices to predict how they might change over time. The effects of these changes, when calculated consistently, describe the possible transitions to other states.

An example will make this more clear. Consider again a pot of water sitting on a stove. Once the stove is turned on, heat begins to flow to the water in the pot because the stove's temperature is higher than that of the water. The causal relationship between the temperature inequality and the flow of heat means that to predict changes in the situation, we should figure out their derivatives and any other relevant ordinal relationships that might change as a result. In this qualitative state, the derivative of the water's temperature is positive, and the derivative of the stove's temperature is constant. Thus, one possible state change is that the water will reach thermal equilibrium with the stove and the flow of heat will stop. That is not the only possibility, of course. We know that boiling can occur if the temperature of the water begins to rise above its boiling temperature. That, too, is a possible transition that would end the state. Which of these transitions occurs depends on the relationship between the temperature of the stove and the boiling temperature of water.

This example illustrates several important features of limit analysis. First, surprisingly weak information (i.e., ordinal relations) suffices to draw important conclusions about broad patterns of physical behavior. Second, limit analysis with purely qualitative information is fundamentally ambiguous: It can identify what transitions might occur but cannot by itself determine in all cases which transition will occur. Third, like other qualitative ambiguities, higher-resolution information can be brought in to resolve the ambiguities as needed. Returning to our example, any information sufficient to determine the ordinal relationship between the stove temperature and boiling suffices to resolve this ambiguity. If we are designing an electric kettle, for instance, we would use this ambiguity as a signal that we must ensure that the heating element's temperature is well above the boiling point; and if we are designing a drink warmer, its heating element should operate well below the boiling point.

Qualitative simulation algorithms vary along four dimensions: (1) their initial states, (2) what conditions they use to filter states or transitions, (3) whether or not they generate new landmarks, and (4) how much of the space of possible behaviors they explore. *Envisioning* is the process of generating an envisionment, that is, generating all possible behaviors. Two kinds of envisioning algorithms have been used in practice: *attainable* envisioners produce all states reachable from a set of initial states, and *total* envisioners produce a complete envisionment. *Behavior generation* algorithms start with a single initial state, generate landmark values, and use a variety of task-dependent constraints as filters and termination criteria (e.g., resource bounds, energy constraints).

Higher-resolution information can be integrated with qualitative simulation in several ways. One method for resolving ambiguities in behavior generation is to provide numerical envelopes to bound mathematical relationships. These envelopes can be dynamically refined to provide tighter situation-specific bounds. Such systems are called **semiquantitative simulators.**

A different approach to integration is to use qualitative reasoning to automatically construct a numerical simulator that has integrated explanation facilities. These *self-explanatory simulators* [Forbus and Falkenhainer, 1990] use traditional numerical simulation techniques to generate behaviors, which are also tracked qualitatively. The concurrently evolving qualitative description of the behavior is used both in generating explanations and in ensuring that appropriate mathematical models are used when applicability thresholds are crossed. Self-explanatory simulators can be compiled in polynomial time for efficient execution, even on small computers, or created in an interpreted environment.

62.3.4 Comparative Analysis

Comparative analysis answers a specific kind of "what-if" questions, namely, the changes that result from changing the value of a parameter in a situation. Given higher-resolution information, traditional analytic or numerical sensitivity analysis methods can be used to answer these questions; however (1) such reasoning is commonly carried out by people who have neither the data nor the expertise to carry out such analyses, and (2) purely quantitative techniques tend not to provide good explanations. Sometimes, purely qualitative information suffices to carry out such reasoning, using techniques such as *exaggeration* [Weld, 1990]. Consider, for instance, the effect of increasing the mass of a block in a spring-block oscillator. If the mass were infinite the block would not move at all, corresponding to an infinite period. Thus, we can conclude that increasing the mass of the block will increase the period of the oscillator.

62.3.5 Teleological Reasoning

Teleological reasoning connects the structure and behavior of a system to its goals. (By its goals, we are projecting the intent of its designer or the observer, because purposes often are ascribed to components of evolved systems.) To describe how something works entails ascribing a function to each of its parts and to explain how these functions together achieve the goals. Teleological reasoning is accomplished by a combination of abduction and recognition. Abduction is necessary because most components and behaviors can play several functional roles [de Kleer, 1984]. A turbine, for instance, can be used to generate work in a power generation system and to expand a gas in a liquefication system. Recognition is important because it explains patterns of function in a system in terms of known, commonly used abstractions. A complex power-generation system with multiple stages of turbines and reheating and regeneration, for instance, still can be viewed as a Rankine cycle after the appropriate aggregation of physical processes involved in its operation [Everett, 1999].

62.3.6 Data Interpretation

There are two ways that the representations of qualitative physics have been used in data interpretation problems. The first is to explain a temporal sequence of measurements in terms of a sequence of qualitative states; the second is to create a qualitative model of phase space by interpreting the results of successive numerical simulation experiments. The underlying commonality in these problems is the use of qualitative descriptions of physical constraints to formulate compatibility constraints that prune the set of possible interpretations. We describe each in turn.

In measurement interpretation tasks, numerical and symbolic data is partitioned into intervals, each of which can be explained by a qualitative state or sequence of qualitative states. Using precomputed envisionments or performing limit analysis online, possible transitions between states used as interpretations can be found for filtering purposes. Specifically, if a state S1 is a possible interpretation for interval I1, then at least one transition from S1 must lead to a state that is an interpretation for the next interval.

This compatibility constraint, applied in both directions, can provide substantial pruning. Additional constraints that can be applied include the likelihood of particular states occurring, the likelihood of particular transitions occurring, and estimates of durations for particular states. Algorithms have been developed that can use all these constraints to maintain a single best interpretation of a set of incoming measurements that operate in polynomial time [de Coste, 1991].

In phase space interpretation tasks, a physical experiment (cf. [Huang and Zhao, 2000]) or numerical simulation (cf. [Yip, 1991], [Nishida, 1994]) is used to gather information about the possible behaviors of a system given a set of initial parameters. The geometric patterns these behaviors form in phase space are described using vision techniques to create a qualitative characterization of the behavior. For example, initially, simulations are performed on a coarse grid to create an initial description of phase space. This initial description is then used to guide additional numerical simulation experiments, using rules that express physical properties visually.

62.3.7 Planning

The ability of qualitative physics to provide predictions with low-resolution information and to determine what manipulations might achieve a desired effect makes it a useful component in planning systems involving the physical world. A tempting approach is to carry out qualitative reasoning entirely in a planner, by *compiling* the domain theory and physics into operators and inference rules. Unfortunately, such straightforward translations tend to have poor combinatorics. A different approach is to treat actions as another kind of state transition in qualitative simulation. This can be effective if qualitative reasoning is interleaved with execution monitoring [Drabble, 1993] or used with a mixture of backward and forward reasoning with partial states.

62.3.8 Spatial Reasoning

Reasoning with purely qualitative representations uses constraint satisfaction techniques to determine possible solutions to networks of relationships. The constraints are generally expressed as transitivity tables. When metric diagrams are used, processing techniques adapted from vision and robotics research are used to extract qualitative descriptions. Some reasoning proceeds purely within these new qualitative representations, while other tasks require the coordination of qualitative and diagrammatic representations. Recently, the flow of techniques has begun to reverse, with vision and robotics researchers adopting qualitative representations because they are more robust to compute from the data and are more appropriate for many tasks (cf. [Kuipers and Byun, 1991, Fernyhough et al., 2000]).

62.4   Applications of Qualitative Reasoning

Qualitative physics began as a research enterprise in the 1980s, with successful fielded applications starting to appear by the early 1990s. For example, applications in supervisory process control (cf. [LeClair et al., 1989]) have been successful enough to be embedded in several commercial systems. Qualitative reasoning techniques were also used in the design of the Mita Corporation's DC-6090 photocopier [Shimomura et al., 1995], which came to market in 1994. By 2000, a commercial tool for FMEA in automobile electrical circuts had been adopted by a major automobile manufacturer [Price, 2000]. Thus, some qualitative reasoning systems are in routine use already, and more such applications are expected as research matures. Here we briefly summarize some of these research efforts.

62.4.1 Monitoring, Control, and Diagnosis

Monitoring, control, and diagnosis, although often treated as distinct problems, are in many applications deeply intertwined. Because these tasks also have deep theoretical commonalities, they are described together here. Monitoring a system requires summarizing its behavior at a level of description that is useful for taking action. Qualitative representations correspond to descriptions naturally applied by system operators and designers and thus can help provide new opportunities for automation. An important benefit of using qualitative representations is that the concepts the software uses can be made very similar to those of people who interact with software, thus potentially improving human–computer interfaces. Qualitative representations are important for control because qualitative distinctions provide criteria that make different control actions appropriate. Diagnosis tasks impose similar requirements. It is rarely beneficial to spend the resources required to construct a very detailed quantitative model of the way a particular part has failed when the goal is to isolate a problem. Qualitative models often provide sufficient resolution for fault isolation. Qualitative models also provide the framework for organizing fault detection (i.e., noticing that a problem has occurred) and for working around a problem, even when these tasks require quantitative information.

Operative diagnosis tasks are those where the system being monitored must continue being operated despite any faults. One example of operative diagnosis is diagnosing engine trouble in civilian commercial aircraft. FaultFinder [Abbott et al., 1987], under development at NASA Langley Research Center, is intended to detect engine trouble and provide easily understood advice to pilots, whose information processing load is already substantial. FaultFinder prototypes compare engine data with a numerical simulation to detect the onset of a problem. A causal model, using low-resolution qualitative information (essentially, working vs. not working), is used to construct failure hypotheses, to be communicated to the pilot in a combination of natural language and graphics.

Many alarm conditions are specified as thresholds, indicating when a system is approaching a dangerous mode of operation or when a component is no longer behaving normally. Alarms are insufficient for fault detection because they do not reflect the lack of normal behaviors. Experienced operators gain a feel for a system and can sometimes spot potential problems long before they become serious enough to trigger an alarm. Some of this expertise can be replicated using a combination of causal models and statistical reasoning over historical data concerning the system in question [Doyle, 1995].

One commercial success of qualitative reasoning has been in supervisory process control. It has been demonstrated that qualitative representations can be used to provide more robust control than statistical process control in curing composite parts [LeClair et al., 1989]. This technique is called *qualitative process automation* (QPA). In the early stage of curing a composite part, the temperature of the furnace needs to be kept relatively low because the part is outgassing. Keeping the furnace low during the entire curing process is inefficient, however, because lower temperatures means longer cure times. Therefore, it is more productive to keep temperature low until outgassing stops and then increase it to finish the cure process more quickly. Statistical process control methods use a combination of analytic models and empirical tests to figure out an optimal pattern of high/low cooking times. QPA incorporates a qualitative description of behavior into the controller, allowing it to detect the change in qualitative regime and control the furnace accordingly. The use of qualitative distinctions in supervisory control provided both faster curing times and higher yield rates than traditional techniques. QPA-inspired supervisory control techniques are now in regular use in curing certain kinds of composite components and have been incorporated into commercial control software.

Another use of qualitative representations is in describing control strategies used by machine operators, such as unloading cranes on docks. By recording the actions of crane operators, machine learning techniques can be used to reverse-engineer their strategies (cf. [Suc and Bratko, 2002]).

In some applications a small set of fault models can be pre-enumerated. A set of models, which includes the nominal model of the system plus models representing common faults, can then be used to track the behavior of a system with a qualitative or semiquantitative simulator. Any fault model whose simulation is inconsistent with the observed behavior can thus be ruled out. Relying on a pre-existing library of fault models can limit the applicability of automatic monitoring and diagnosis algorithms. One approach to overcoming this limitation is to create algorithms that require only models of normal behavior. Most consistency-based diagnosis algorithms take this approach. For example, in developing on-board diagnostics for Diesel engines, qualitative representations have been found to be useful as a robust level of description for reasoning and for abstracting away from sensor noise (cf. [Sachenbacher et al., 2000]).

One limitation with consistency-based diagnosis is that the ways a system can fail are still governed by natural laws, which impose more constraint than logical consistency. This extra constraint can be exploited by using a domain theory to generate explanations that could account for the problem, via abduction. These explanations are useful because they make additional predictions that can be tested and that also can be important for reasoning about safety in operative diagnosis (e.g., if a solvent tank's level is dropping because it is leaking, then where is the solvent going?). However, in many diagnosis tasks, this limitation is not a concern.

62.4.2 Design

Engineering design activities are divided into conceptual design, the initial phase when the overall goals, constraints, and functioning of the artifact are established; and detailed design, when the results of conceptual design are used to synthesize a constructable artifact or system. Most computer-based design tools, such as computer-aided design (CAD) systems and analysis programs, facilitate detailed design. Yet many of the most costly mistakes occur during the conceptual design phase. The ability to reason with partial information makes qualitative reasoning one of the few technologies that provides substantial leverage during the conceptual design phase. Qualitative reasoning can also help automate aspects of detailed design.

One example is Mita Corporation's DC-6090 photocopier [Shimomura et al., 1995]. It is an example of a *self-maintenance machine*, in which redundant functionality is identified at design time so that the system can dynamically reconfigure itself to temporarily overcome certain faults. An envisionment including fault models, created at design time, was used as the basis for constructing the copier's control software. In operation, the copier keeps track of which qualitative state it is in, so that it produces the best quality copy it can.

In some fields experts formulate general design rules and methods expressed in natural language. Qualitative representations can enable these rules and methods can be further formalized, so that they can automated. In chemical engineering, for instance, several design methods for distillation plants have been formalized using qualitative representations, and designs for binary distillation plants comparable to those in the chemical engineering research literature have been generated automatically.

Automatic analysis and synthesis of kinematic mechanisms have received considerable attention. Complex fixed-axis mechanisms, such as mechanical clocks, can be simulated qualitatively, and a simplified dynamics can be added to produce convincing animations. Initial forays into conceptual design of mechanisms have been made, and qualitative kinematics simulation has been demonstrated to be competitive with conventional approaches in some linkage optimization problems. Qualitative representations are also useful in case-based design, because they provide a level of abstraction that simplifies adaptation (cf. [Faltings, 2001]).

Qualitative reasoning also is being used to reason about the effects of failures and operating procedures. Such information can be used in failure modes and effects analysis (FMEA). For example, potential hazards in a chemical plant design can be identified by perturbing a qualitative model of the design with various faults and using qualitative simulation to ascertain the possible indirect consequences of each fault. Commercial FMEA software using qualitative simulation for electrical system design is now being used in automotive design [Price, 2000].

62.4.3 Intelligent Tutoring Systems and Learning Environments

One of the original motivations for the development of qualitative physics was its potential applications in intelligent tutoring systems (ITSs) and intelligent learning environments (ILEs). Qualitative representations provide a formal language for a student's mental models [Gentner and Stevens, 1983], and thus they facilitate communication between software and student. For example, a sequence of qualitative models can be designed that helps students learn complex domains such as electronics more easily. Student protocols can be analyzed in qualitative terms to diagnose misconceptions.

Qualitative representations are being used in software for teaching plant operators and engineers. They provide a level of explanation for how things work that facilitates teaching control. For example, systems for teaching the operation of power generation plants, including nuclear plants, are under construction in various countries. Teaching software often uses hierarchies of models to help students understand a typical industrial process and design controllers for it. Qualitative representations also can help provide teaching software with the physical intuitions required to help find students' problems. For instance, qualitative representations are used to detect physically impossible designs in an ILE for engineering thermodynamics.

Qualitative representations can be particularly helpful in teaching domains where quantitative knowledge is either nonexistent, inaccurate, or incomplete. For example, efforts underway to create ITSs for ecology in Brazil, to support conservation efforts, are using qualitative representations to explain how environmental conditions affect plant growth [Salles and Bredeweg, 2001]. For younger students, who have not had algebra or differential equations, the science curriculum consists of learning causal mental models that are well captured by the formalisms of qualitative modeling. By using a student-friendly method of expressing models, such as concept maps, software systems have been built which help students learn conceptual models [Forbus et al., 2001; Leelawong et al., 2001].

62.4.4 Cognitive Modeling

Since qualitative physics was inspired by observations of how people reason about the physical world, one natural application of qualitative physics is cognitive simulation, i.e., the construction of programs whose primary concern is accurately modeling some aspect of human reasoning, as measured by comparison with psychological results. Some research has been concerned with modeling scientific discovery, e.g., how analogy can be used to create new physical theories and modeling scientific discovery [Falkenhainer, 1990]. Several investigations suggest that qualitative representations have major role to play in understanding cognitive processes such as high-level vision [Fernyhough et al., 2000][Forbus et al., 2003]. Common sense reasoning appears to rely heavily on qualitative representations, although human reasoning may rely more on reasoning from experience than first-principles reasoning [Forbus and Gentner, 1997]. Understanding the robustness and flexibility of human common sense reasoning is an important scientific goal in its own right, and will provide clues as to how to build better AI systems. Thus potential use of qualitative representations by cognitive scientists may ultimately prove to be the most important application of all.

62.5 Research Issues and Summary

Qualitative reasoning is now a mature subfield with a mixture of basic and applied activities, including fielded applications. The substantial increases in available computing power, combined with the now urgent need to make software that is more articulate, suggests that the importance of qualitative reasoning will continue to grow.

Although there is a substantial research base to draw upon, there are many open problems and areas that require additional research. There are still many unanswered questions about purely qualitative representations (e.g., what is the minimum information that is required to guarantee that all predicted behaviors generated from an envisionment are physically possible?), but the richest vein of research concerns the integration of qualitative knowledge with other kinds of knowledge: numerical, analytic, teleological, etc. One outgrowth of such research is a new area, *hybrid systems*, dedicated to exploring mixtures of qualitative, numerical, and discrete models. The work on modeling to date, although a solid foundation, is still very primitive; better model formulation algorithms, well-tested conventions for structuring domain theories, and robust methods for integrating the results of multiple models are needed. Substantial domain theories for a broad range of scientific and engineering knowledge need to be created. And finally, there are many domains where traditional mathematics has intruded, but where the amount and/or precision of the data available has not enabled it to be very successful. These areas are ripe for qualitative modeling. Examples where such efforts are underway include medicine, organizational theory, economics, and ecology.

Defining Terms

Comparative analysis: A particular form of a what if question, i.e., how a physical system changes in response to the perturbation of one of its parameters.

Compositional modeling: A methodology for organizing domain theories so that models for specific systems and tasks can be automatically formulated and reasoned about.

Confluence: An equation involving sign values.

Diagrammatic reasoning: Spatial reasoning, with particular emphasis on how people use diagrams.

Domain theory: A collection of general knowledge about some area of human knowledge, including the kinds of entities involved and the types of relationships that can hold between them, and the mechanisms that cause changes (e.g., physical processes, component laws, etc.). Domain theories range from purely qualitative to purely quantitative to mixtures of both.

Envisionment: A description of all possible qualitative states and transitions between them for a system. *Attainable envisionments* describe all states reachable from a particular initial state; *total envisionments* describe all possible states.

FMEA: Failure Modes and Effects Analysis. Analyzing the possible effects of a failure of a component of a system on the operation of the entire system.

Landmark: A comparison point indicating a specific value achieved during a behavior, e.g., the successive heights reached by a partially elastic bouncing ball.

Limit point: A comparison point indicating a fundamental physical boundary, such as the boiling point of a fluid. Limit points need not be constant over time, e.g., boiling points depend on pressure.

Metric diagram: A quantitative representation of shape and space used for spatial reasoning, the computer analog to or model of the combination of diagram/visual apparatus used in human spatial reasoning.

Model fragment: A piece of general domain knowledge that is combined with others to create models of specific systems for particular tasks.

Modeling assumption: A proposition expressing control knowledge about modeling, such as when a model fragment is relevant.

Physical process: A mechanism that can cause changes in the physical world, such as heat flow, motion, and boiling.

Place vocabulary: A qualitative description of space or shape that is grounded in a quantitative representation.

Qualitative proportionality: A qualitative relationship expressing partial information about a functional dependency between two parameters.

Qualitative simulation: The generation of predicted behaviors for a system based on qualitative information. Qualitative simulations typically include branching behaviors due to the low resolution of the information involved.

Quantity space: A set of ordinal relationships that describes the value of a continuous parameter.

Semiquantitative simulation: A qualitative simulation that uses quantitative information, such as numerical values or analytic bounds, to constrain its results.

References

Abbott, K., Schutte, P., Palmer, M., and Ricks, W. 1987. Faultfinder: a diagnostic expert system with graceful degradation for onboard aircraft application. In *14th Int. Symp. Aircraft Integrated Monitoring Syst.*

Cohn, A.G. and Hazarika, S.M. 2001. Qualitative spatial representation and reasoning: an overview. *Fundamenta Informaticae*, 46(1-2): 1–29.

de Coste, D. 1991. Dynamic across-time measurement interpretation. *Artif. Intell.*, 51:273–341.

de Kleer, J. 1977. Multiple representations of knowledge in a mechanics problem solver, pp. 299–304. *Proc. IJCAI-77.*

de Kleer, J. 1984. How circuits work. *Artif. Intell.*, 24:205–280.

de Kleer, J. and Brown, J. 1984. A qualitative physics based on confluences. *Artif. Intell.*, 24:7–83.

Doyle, R. 1995. Determining the loci of anomalies using minimal causal models, pp. 1821–1827. *Proc. IJCAI-95.*

Drabble, B. 1993. Excalibur: a program for planning and reasoning with processes. *Artif. Intell.*, 62(1):1–40.

Everett, J. 1999. Topological inference of teleology: deriving function from structure via evidential reasoning. *Artif. Intell.*, 113(1-2): 149–202.

Falkenhainer, B. 1990. A unified approach to explanation and theory formation. In *Computational Models of Scientific Discovery and Theory Formation*. Shrager and Langley, Eds. Morgan Kaufmann, San Mateo, CA. Also in Sharlik and Dietterich (Eds.), *Readings in Machine Learning*. Morgan Kaufmann, San Mateo, CA.

Falkenhainer, B. and Forbus, K. 1991. Compositional modeling: finding the right model for the job. *Artif. Intell.*, 51:95–143.

Faltings, B. 1990. Qualitative kinematics in mechanisms. *Artif. Intell.*, 44(1):89–119.

Faltings, B. 2001. FAMING: Supporting innovative design using adaptation — a description of the approach, implementation, illustrative example and evaluation. In Chakrabarti (Ed.), *Engineering Design Synthesis*, Springer-Verlag.

Faltings, B. and Struss, P., Eds. 1992. *Recent Advances in Qualitative Physics*. MIT Press, Cambridge, MA.

Fernyhough, J., Cohn, A.G., and Hogg, D. 2000. *Image and Vision Computing*, 18, pp. 81–103.

Forbus, K. 1984. Qualitative process theory. *Artif. Intell.*, 24:85–168.

Forbus, K. and Falkenhainer, B. 1990. Self explanatory simulations: an integration of qualitative and quantitative knowledge, pp. 380–387. *Proc. AAAI-90.*

Forbus, K. and Gentner, D. 1986. Causal reasoning about quantities. *Proceedings of the Eighth Annual Conference of the Cognitive Science Society*, Amherst, MA, August.

Forbus, K. and Gentner, D. 1997. Qualitative mental models: Simulations or memories? *Proceedings of the Eleventh International Workshop on Qualitative Reasoning*, Cortona, Italy.

Forbus, K., Nielsen, P., and Faltings, B. 1991. Qualitative spatial reasoning: the CLOCK project. *Artif. Intell.*, 51:417–471.

Forbus, K., Carney, K., Harris, R., and Sherin, B. 2001. A qualitative modeling environment for middle-school students: A progress report. *Proceedings of the Fifteenth International Workshop on Qualitative Reasoning*, San Antonio, Texas, USA.

Forbus, K., Usher, J., and Chapman, V. 2003. Sketching for military courses of action diagrams. *Proceedings of IUI'03*, January, Miami, Florida.

Gentner, D. and Stevens, A. Eds. 1983. *Mental Models*. Erldaum, Hillsdale, NJ.

Glasgow, J., Karan, B., and Narayanan, N., Eds. 1995. *Diagrammatic Reasoning*. AAAI Press/MIT Press, Cambridge, MA.

Hayes, P. 1985. Naive physics 1: ontology for liquids. In *Formal Theories of the Commonsense World*, R. Hobbs and R. Moore, Eds. Ablex, Norwood, NJ.

Hollan, J., Hutchins, E., and Weitzman, L. 1984. STEAMER: an interactive inspectable simulation-based training system. *AI Mag.*, 5(2):15–27.

Huang, X. and Zhao, F. 2000. Relation based aggregation: Finding objects in large spatial datasets. *Intelligent Data Analysis*, 4:129–147.

Iwasaki, Y. and Simon, H. 1986. Theories of causal observing: reply to de Kleer and Brown. *Artif. Intell.*, 29(1):63–68.

Iwasaki, Y., Tessler, S., and Law, K. 1995. Qualitative structural analysis through mixed diagrammatic and symbolic reasoning. In *Diagrammatic Reasoning*. J. Glasgow, B. Karan, and N. Narayanan, Eds., pp. 711–729. AAAI Press/MIT Press, Cambridge, MA.

Joscowicz, L. and Sacks, E. 1993. Automated modeling and kinematic simulation of mechanisms. *Computer Aided Design* 25(2).

Keppens, J. and Shen, Q. 2002. On supporting dynamic constraint satisfiaction with order of magnitude preferences. *Proceedings of the Sixteenth International Workshop on Qualitative Reasoning (QR2002)*, pp. 75–82, Sitges, Spain.

Kim, H. 1992. Qualitative kinematics of linkages. In *Recent Advances in Qualitative Physics*. B. Faltings and P. Struss, Eds. MIT Press, Cambridge, MA.

Kuipers, B. 1986. Qualitative simulation. *Artif. Intell.*, 29:289–338.

Kuipers, B. 1994. *Qualitative Reasoning: Modeling and Simulation with Incomplete Knowledge*. MIT Press, Cambridge, MA.

Kuipers, B. and Byun, Y. 1991. A robot exploration and mapping strategy based on semantic hierarchy of spatial reasoning. *J. Robotics Autonomous Syst.*, 8:47–63.

Le Clair, S., Abrams, F., and Matejka, R. 1989. Qualitative process automation: self directed manufacture of composite materials. *Artif. Intell. Eng. Design Manuf.*, 3(2):125–136.

Leelawong, K., Wang, Y., Biswas, G., Vye, N., Bransford, J., and Schwartz, D. 2001. Qualitative reasoning techniques to support Learning by Teaching: The Teachable Agents project. *Proceedings of the Fifteenth International Workshop on Qualitative Reasoning*, San Antonio, Texas, USA.

Nayak, P. 1994. Causal approximations. *Artif. Intell.*, 70:277–334.

Nielsen, P. 1988. A qualitative approach to mechanical constraint. *Proc. AAAI-88*.

Nishida, T. 1994. Qualitative reasoning for automated explanation for chaos, pp. 1211–1216. *Proc. AAAI-94*.

Price, C. J. 2000. AutoSteve: Automated Electrical Design Analysis, in *Proceedings ECAI-2000*, pp. 721–725, August 2000.

Rickel, J. and Porter, B. 1994. Automated modeling for answering prediction questions: selecting the time scale and system boundary, pp. 1191–1198. *Proc. AAAI-94*.

Sachenbacher, M., Struss, P., and Weber, R. 2000. Advances in design and implementation of OBD functions for Diesel injection systems based on a qualitative approach to diagnosis. *SAE World Congress*, Detroit, USA.

Sallas, P. and Bredeweg, B. 2001. Constructing progressive learning routes through qualitative simulation models in ecology. Proceedings of the Fifteenth International Workshop on Qualitative Reasoning, San Antonio, Texas, USA.

Shimomura, Y., Tanigawa, S., Umeda, Y., and Tomiyama, T. 1995. Development of self-maintenance photocopiers, pp. 171–180. *Proc. IAAI-95*.

Stahovich, T.F., David, R., and Shrobe, H. 1998. Generating multiple new designs from a sketch. *Artificial Intelligence*, Vol 104., pp. 211–264.

Struss, P. 1988. Mathematical aspects of qualitative reasoning. *Int. J. Artif. Intell. Eng.*, 3(3):156–169.

Suc, D. and Bratko, I. 2002. Qualitative reverse engineering. *Proc. ICML'02 (Int. Conf. on Machine Learning)*, Sydney, Australia.

Weld, D. 1990. *Theories of Comparative Analysis*. MIT Press, Cambridge, MA.

Williams, B. 1991. A theory of interactions: unifying qualitative and quantitative algebraic reasoning. *Artif. Intell.*, 51(1–3):39–94.

Yip, K. 1991. *KAM: A System for Intelligently Guiding Numerical Experimentation by Computer*. Artificial intelligence series. MIT Press, Cambridge, MA.

Further Information

There are a variety of qualitative reasoning resources on the World Wide Web, including extensive bibliographies, papers, and software. A large number of edited collections have been published (cf. [Faltings and Struss, 1992]) An excellent textbook on the QSIM approach to qualitative physics is Kuipers [1994]. For an introduction to diagrammatic reasoning, see Glasgow *et al.* [1995].

Papers on qualitative reasoning routinely appear in *Artificial Intelligence, Journal of Artificial Intelligence Research (JAIR)*, and *IEEE Intelligent Systems*. Many papers first appear in the proceedings of the American Association for Artificial Intelligence (AAAI), the International Joint Conferences on Artificial Intelligence (IJCAI), and the European Conference on Artificial Intelligence (ECAI). Every year there is an International Qualitative Reasoning Workshop, whose proceedings document the latest developments in the area. Proceedings for a particular workshop are available from its organizers.

63

Search

Danny Kopec
Brooklyn College, CUNY

Tony A. Marsland
University of Alberta

J.L. Cox
Brooklyn College, CUNY

63.1 Introduction **63**-1
63.2 Uninformed Search Methods **63**-2
 Search Strategies • State-Space Search • Breadth-First Search
 • Depth-First Search • Bidirectional Search
63.3 Heuristic Search Methods **63**-6
 Hill Climbing • Best-First Search • The A* Algorithm
63.4 Game-Tree Search **63**-11
 The Alpha-Beta Algorithms • SSS* Algorithm • The MTD(f)
 Algorithm • Recent Developments
63.5 Parallel Search **63**-15
 Parallel Single-Agent Search • Adversary Games
63.6 Recent Developments **63**-21

63.1 Introduction

Efforts using artificial intelligence (AI) to solve problems with computers — which humans routinely handle by employing innate cognitive abilities, pattern recognition, perception, and experience — invariably must turn to considerations of search. This chapter explores search methods in AI, including both **blind** exhaustive methods and informed **heuristic** and optimal methods, along with some more recent findings. The search methods covered include (for non-optimal, uninformed approaches) **state-space search,** generate and test, means–ends analysis, problem reduction, AND/OR trees, depth-first search, and breadth-first search. Under the umbrella of heuristic (informed) methods, we discuss **hill climbing,** best-first search, bidirectional search, and the A* algorithm. Tree searching algorithms for games have proved to be a rich source of study and provide empirical data about heuristic methods. Included here are the SSS* algorithm, the use of iterative deepening, and variations on the alpha-beta minimax algorithm, including the recent MTD(f) algorithm.

Coincident with the continuing price–performance improvement of small computers is growing interest in reimplementing some of the heuristic techniques developed for problem solving and planning programs, to see whether they can be enhanced or replaced by more algorithmic methods. Because many of the heuristic methods are computationally intensive, the second half of the chapter focuses on parallel methods, which can exploit the benefits of parallel processing. The importance of parallel search is presented through an assortment of relatively recent algorithms, including the parallel iterative deepening algorithm (PIDA*), principal variation splitting (PVSplit), and the young brothers wait concept. In addition, dynamic tree-splitting methods have evolved for both shared memory parallel machines and networks of distributed computers. Here, the issues include load balancing, processor utilization, and communication overhead. For single-agent search problems, we consider not only **work-driven** dynamic parallelism, but also the more recent **data-driven parallelism** employed in transposition table–driven scheduling (TDS). In adversarial

1-58488-360-X/$0.00+$1.50
© 2004 by CRC Press, LLC

```
Repeat
      Generate a candidate solution
      Test the candidate solution
Until a satisfactory solution is found, or
      no more candidate solutions can be generated:
If an acceptable solution is found, announce it;
      Otherwise, announce failure.
```

FIGURE 63.1 Generate and test method.

games, tree pruning makes work load balancing particularly difficult, so we also consider some recent advances in dynamic parallel methods for game-tree search.

The application of raw computing power–while anathema to some — often provides better answers than is possible by reasoning or analogy. Thus, brute force techniques form a good basis against which to compare more sophisticated methods designed to mirror the human deductive process.

63.2   Uninformed Search Methods

63.2.1   Search Strategies

All search methods in computer science share in common three necessities:

A world model or database of facts, based on a choice of representation providing the current state, other possible states, and a goal state

A set of operators that defines possible transformations of states

A control strategy that determines how transformations among states are to take place by applying operators

Forward reasoning is one technique for identifying states that are closer to a goal state. Working backward from a goal to the current state is called *backward reasoning*. As such, it is possible to make distinctions between bottom-up and top-down approaches to problem solving. Bottom-up is often goal-oriented — that is, reasoning backward from a goal state to solve intermediary subgoals. Top-down or data-driven reasoning is based on reaching a state that is defined as closer to a goal. Often, application of operators to a problem state may not lead directly to a goal state, so some **backtracking** may be necessary before a goal state can be found [Barr and Feigenbaum, 1981].

63.2.2   State-Space Search

Exhaustive search of a problem space (or search space) is often not feasible or practical because of the size of the problem space. In some instances, however, it is necessary. More often, we are able to define a set of legal transformations of a state space (moves in a board game) from which those that are more likely to bring us closer to a goal state are selected and others are never explored further. This technique in problem solving is known as *split and prune*. In AI, the technique that emulates this approach is called **generate and test.** The basic method is shown in Figure 63.1.

Good generators are complete and will eventually produce all possible solutions, while not proposing redundant ones. They are also informed; that is, they will employ additional information to constrain the solutions they propose.

Means–ends analysis is another state-space technique whose purpose is to reduce the difference (distance) between a current state and a goal state. Determining "distance" between any state and a goal state can be facilitated by *difference-procedure tables*, which can effectively prescribe what the next state might be. To perform means–ends analysis, see Figure 63.2.

```
Repeat
        Describe the current state, the goal state,
            and the difference between the two.
        Use the difference between the current state and goal state,
            to select a promising transformation procedure.
        Apply the promising procedure and update the current state.
Until the GOAL is reached or
        no more procedures are available
If the GOAL is reached, announce success;
Otherwise, announce failure.
```

FIGURE 63.2 Means–ends analysis.

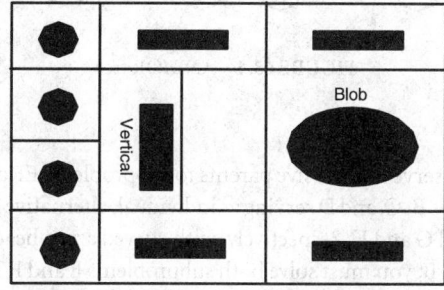

FIGURE 63.3 Problem reduction and the sliding block puzzle.

The technique of problem reduction is another important approach in AI. *Problem reduction* means to solve a complex or larger problem by identifying smaller, manageable problems (or subgoals), which you know can be solved in fewer steps.

For example, Figure 63.3 shows the "donkey" sliding block puzzle. It has been known for over 100 years. Subject to constraints on the movement of pieces in the sliding block puzzle, the task is to slide the blob around the vertical bar with the goal of moving it to the other side. The blob occupies four spaces and needs two adjacent vertical or horizontal spaces to be able to move; the vertical bar needs two adjacent empty vertical spaces to move left or right, or one empty space above or below it to move up or down. The horizontal bars' movements are complementary to the vertical bar. Likewise, the circles can move to any empty space around them in a horizontal or vertical line. A relatively uninformed state-space search can result in over 800 moves for this problem to be solved, with considerable backtracking required. Using problem reduction, resulting in the subgoal of trying the get the blob on the two rows above or below the vertical bar, it is possible to solve this puzzle in just 82 moves!

Another example of a technique for problem reduction is called **AND/OR trees**. Here, the goal is to find a solution path to a given tree by applying the following rules.

A node is solvable if

1. It is a terminal node (a primitive problem).
2. It is a non-terminal node whose successors are AND nodes that are all solvable.
3. OR it is a non-terminal node whose successors are OR nodes and least one of them is solvable.

Similarly, a node is unsolvable if

1. It is a non-terminal node that has no successors (a non-primitive problem to which no operator applies).
2. It is a non-terminal node whose successors are AND nodes and at least one of them is unsolvable.
3. OR it is a non-terminal node whose successors are OR nodes and all of them are unsolvable.

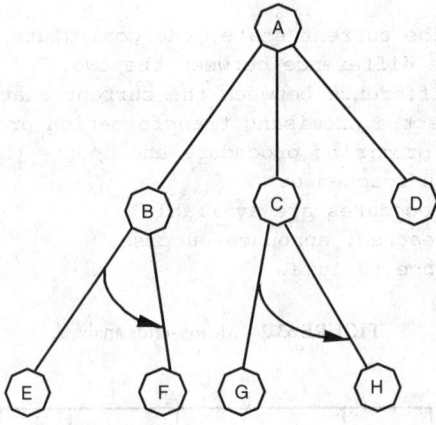

FIGURE 63.4   AND/OR tree.

In Figure 63.4, nodes B and C serve as exclusive parents to subproblems EF and GH, respectively. One way of viewing the tree is with nodes B, C, and D serving as individual, alternative subproblems representing OR nodes. Node pairs E and F and G and H, respectively, with curved arrowheads connecting them, represent AND nodes. To solve problem B, you must solve both subproblems E and F. Likewise, to solve subproblem C, you must solve subproblems G and H. Solution paths would therefore be {A-B-E-F}, {A-C-G-H}, and {A-D}. In the special case where no AND nodes occur, we have the ordinary graph occurring in a state-space search. However, the presence of AND nodes distinguishes AND/OR trees (or graphs) from ordinary state structures, which call for their own specialized search techniques. Typical problems tackled by AND/OR trees include games, puzzles, and other well defined state-space goal-oriented problems, such as robot planning, movement through an obstacle course, or setting a robot the task of reorganizing blocks on a flat surface.

63.2.3   Breadth-First Search

One way to view search problems is to consider all possible combinations of subgoals, by treating the problem as a tree search. **Breadth-first search** always explores nodes closest to the root node first, thereby visiting all nodes at a given layer first before moving to any longer paths. It pushes uniformly into the search tree. Because of memory requirements, Breadth-first search is only practical on shallow trees or those with an extremely low branching factor. It is therefore not much used in practice, except as a basis for such **best-first search** algorithms such as A* and SSS*.

63.2.4   Depth-First Search

Depth-first search (DFS) is one of the most basic and fundamental blind search algorithms. It is used for bushy trees (with a high branching factor), where a potential solution does not lie too deeply down the tree. That is, "DFS is a good idea when you are confident that all partial paths either reach dead ends or become complete paths after a reasonable number of steps." In contrast, "DFS is a bad idea if there are long paths, particularly indefinitely long paths, that neither reach dead ends nor become complete paths" [Winston, 1992]. To conduct a DFS, follow these steps:

1. Put the Start Node on the list called OPEN.
2. If OPEN is empty, exit with failure; otherwise, continue.
3. Remove the first node from OPEN and put it on a list called CLOSED. Call this node n.
4. If the depth of n equals the depth bound, go to **2**; otherwise, continue.

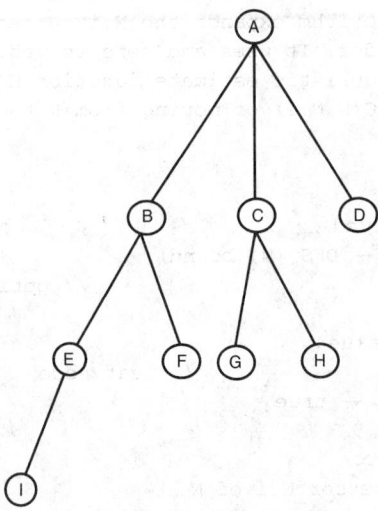

FIGURE 63.5   Tree example for depth-first and breadth-first search.

5. Expand node n, generating all immediate successors. Put these at the beginning of OPEN (in predetermined order) and provide pointers back to n.
6. If any of the successors are goal nodes, exit with the solution obtained by tracing back through the pointers; otherwise, go to **2**.

DFS always explores the deepest node to the left first, that is, the one farthest down from the root of the tree. When a dead end (terminal node) is reached, the algorithm backtracks one level and then tries to go forward again. To prevent consideration of unacceptably long paths, a *depth bound* is often employed to limit the depth of search. At each node, immediate successors are generated and a transition made to the leftmost node, where the process continues recursively until a dead end or depth limit is reached. In Figure 63.5, DFS explores the tree in the order I-E-b-F-B-a-G-c-H-C-a-D-A. Here, the notation using lowercase letters represents the possible storing of provisional information about the subtree. For example, this could be a lower bound on the value of the tree.

Figure 63.6 enhances depth-first search with a form of **iterative deepening** that can be used in a single-agent search like A*. DFS expands an immediate successor of some node N in a tree. The next successor to be expanded is (N.i), the one with lowest cost function. Thus, the expected value of node N.i is the estimated cost C(N,N.i) plus H(N), the known value of node N. The basic idea in iterative deepening is that a DFS is started with a depth bound of 1. This bound increases by one at each new iteration. With each increase in depth, the algorithm must reinitiate its depth-first search for the prescribed bound. The idea of iterative deepening, in conjunction with a memory function to retain the best available potential solution paths from iteration to iteration, is credited to Slate and Atkin [1977], who used it in their chess program. Korf [1985] showed how efficient this method is in single-agent searches, with his iterative deepening A* (IDA*) algorithm.

63.2.5 Bidirectional Search

To this point, all search algorithms discussed (with the exception of means–ends analysis and backtracking) have been based on forward reasoning. Searching backward from goal nodes to predecessors is relatively easy. Pohl [1971] combined forward and backward reasoning in a technique called **bidirectional search**. The idea is to replace a single search graph, which is likely to grow exponentially, with two smaller graphs — one starting from the initial state and one starting from the goal. The search terminates when the two graphs intersect. This algorithm is guaranteed to find the shortest solution path through a general state-space graph. Empirical data for randomly generated graphs shows that Pohl's algorithm expands

```
// The A* (DFS) algorithm expands the N.i successors of node N
// in best first order. It uses and sets solved, a global indicator.
// It also uses a heuristic estimate function H(N), and a
// transition cost C(N,N.i) of moving from N to N.i
//
IDA* (N) → cost
     bound ← H(N)
     while not solved
          bound ← DFS (N, bound)
     return bound                    // optimal cost


DFS (N, bound) → value
     if H(N) ≡ 0                     // leaf node
          solved ← true
          return 0
     new_bound ← ∞
     for each successor N.i of N
          merit ← C(N, N.i) + H(N.i)
          if merit ≤ bound
               merit ← C(N,N.i) + DFS (N.i, bound - C(N.i,N.i))
               if solved
                    return merit
          if merit < new_bound
               new_bound ← merit
     return new_bound
```

FIGURE 63.6 The A* DFS algorithm for use with IDA*.

only about one-quarter as many nodes as unidirectional search [Barr and Feigenbaum, 1981]. Pohl also implemented heuristic versions of this algorithm. However, determining when and how the two searches will intersect is a complex process.

Russell and Norvig [2003] analyze the bidirectional search and come to the conclusion that it is o $(b^{d/2})$ in terms of average case time and space complexity. They point out that this is significantly better than o (b^d), which would be the cost of searching exhaustively in one direction. Identification of subgoal states could do much to reduce the costs. The large space requirements of the algorithm are considered its weakness.

However, Kaindl and Kainz [1997] have demonstrated that the long-held belief that the algorithm is afflicted by the frontiers passing each other is wrong. They developed a new generic approach, which dynamically improves heuristic values but is only applicable to bidirectional heuristic. Their empirical results have found that the bidirectional heuristic search can be performed very efficiently, with limited memory demands. Their research has resulted in a better understanding of an algorithm whose practical usefulness has been long neglected, with the conclusion that it is better suited to certain problems than corresponding unidirectional searches. For more details, the reader should review their paper [Kaindl and Kainz, 1997]. The next section focuses on heuristic search methods.

63.3   Heuristic Search Methods

George Polya, via his wonderful book *How to Solve It* [1945], may be regarded as the father of heuristics. Polya's efforts focused on problem solving, thinking, and learning. He developed a short "heuristic dictionary" of heuristic primitives. Polya's approach was both practical and experimental. He sought to develop commonalities in the problem-solving process through the formalization of observation and experience.

Present-day notions of heuristics are somewhat different from Polya's [Bolc and Cytowski, 1992]. Current tendencies seek formal and rigid algorithmic solutions to specific problem domains, rather than the development of general approaches that could be appropriately selected and applied to specific problems.

The goal of a heuristic search is to reduce greatly the number of nodes searched in seeking a goal. In other words, problems whose complexity grows combinatorially large may be tackled. Through knowledge, information, rules, insights, analogies, and simplification — in addition to a host of other techniques — heuristic search aims to reduce the number of objects that must be examined. Heuristics do not guarantee the achievement of a solution, although good heuristics should facilitate this. Over the years, heuristic search has been defined in many different ways:

- It is a practical strategy increasing the effectiveness of complex problem solving [Feigenbaum and Feldman, 1963].
- It leads to a solution along the most probable path, omitting the least promising ones.
- It should enable one to avoid the examination of dead ends and to use already gathered data.

The points at which heuristic information can be applied in a search include the following:

1. Deciding which node to expand next, instead of doing the expansions in either a strict breadth-first or depth-first order
2. Deciding which successor or successors to generate when generating a node — instead of blindly generating all possible successors at one time
3. Deciding that certain nodes should be discarded, or pruned, from the search tree.

Bolc and Cytowski [1992] add:

"[U]se of heuristics in the solution construction process increases the uncertainty of arriving at a result ... due to the use of informal knowledge (rules, laws, intuition, etc.) whose usefulness have never been fully proven. Because of this, heuristic methods are employed in cases where algorithms give unsatisfactory results or do not guarantee to give any results. They are particularly important in solving very complex problems (where an accurate algorithm fails), especially in speech and image recognition, robotics and game strategy construction ....

"Heuristic methods allow us to exploit uncertain and imprecise data in a natural way .... The main objective of heuristics is to aid and improve the effectiveness of an algorithm solving a problem. Most important is the elimination from further consideration of some subsets of objects still not examined ...."

Most modern heuristic search methods are expected to bridge the gap between the completeness of algorithms and their optimal complexity [Romanycia and Pelletier, 1985]. Strategies are being modified in order to arrive at a quasi-optimal, rather than optimal, solution with a significant cost reduction [Pearl, 1984]. Games, especially two-person, zero-sum games of perfect information, like chess and checkers, have proved to be a very promising domain for studying and testing heuristics.

63.3.1 Hill Climbing

Hill climbing is a DFS with a heuristic measure that orders choices as nodes are expanded. The heuristic measure is the estimated remaining distance to the goal. The effectiveness of hill climbing is completely dependent upon the accuracy of the heuristic measure. To conduct a hill climbing search of a tree:

```
Form a one-element queue consisting of a zero-length path that contains
only the root node.

Repeat

  Remove the first path from the queue;

  Create new paths by extending the first path to all the neighbors
of the  terminal node.
```

```
If New Path(s) result in a loop Then

     Reject New Path(s).

Sort any New Paths by the estimated distances between their terminal
nodes and the GOAL.

  If any shorter paths exist Then

     Add them to the front of the queue.

Until the first path in the queue terminates at the GOAL node or

   the queue is empty

  If the GOAL node is found, announce SUCCESS, otherwise announce
FAILURE.
```

In this algorithm, *neighbors* refer to "children" of nodes that have been explored, and *terminal nodes* are equivalent to leaf nodes. Winston [1992] explains the potential problems affecting hill climbing. They are all related to issue of local vision vs. global vision of the search space. The *foothills problem* is particularly subject to local maxima where global ones are sought. The *plateau problem* occurs when the heuristic measure does not hint toward any significant gradient of proximity to a goal. The *ridge problem* illustrates its name: you may get the impression that the search is taking you closer to a goal state, when in fact you traveling along a ridge that prevents you from actually attaining your goal. **Simulated annealing** attempts to combine hill climbing with a random walk in a way that yields both efficiency and completeness [Russell and Norvig, 2003]. The idea is to temper the downhill process of hill climbing in order to avoid some of these pitfalls by increasing the probability of hitting important locations to explore. It is like intelligent guessing.

63.3.2   Best-First Search

Best-first search (Figure 63.7) is a general algorithm for heuristically searching any state-space graph — a graph representation for a problem that includes initial states, intermediate states, and goal states. In this sense, a directed acyclic graph (DAG), for example, is a special case of a state-space graph. Best-first search is equally applicable to data- and goal-driven searches and supports the use of heuristic evaluation functions. It can be used with a variety of heuristics, ranging from a state's "goodness" to sophisticated measures based on the probability of a state's leading to a goal that can be illustrated by examples of Bayesian statistical measures.

Similar to the depth-first and breadth-first search algorithms, best-first search uses lists to maintain states: OPEN to keep track of the current fringe of the search and CLOSED to record states already visited. In addition, the algorithm orders states on OPEN according to some heuristic estimate of their proximity to a goal. Thus, each iteration of the loop considers the most promising state on the OPEN list. According to Luger and Stubblefield [1993], best-first search improves at just the point where hill climbing fails with its short-sighted and local vision. The following description of the algorithm closely follows that of Luger and Stubblefield [1993, p. 121]:

"At each iteration, Best First Search removes the first element from the OPEN list. If it meets the goal conditions, the algorithm returns the solution path that led to the goal. Each state retains ancestor information to allow the algorithm to return the final solution path.

"If the first element on OPEN is not a goal, the algorithm generates its descendants. If a child state is already on OPEN or CLOSED, the algorithm checks to make sure that the state records the shorter of the

```
Procedure Best_First_Search (Start) → pointer
     OPEN ← {Start}                                            // Initialize
     CLOSED ← { }
     While OPEN ≠ { } Do                                       // States Remain
          remove the leftmost state from OPEN, call it X;
          if X ≡ goal then
               return the path from Start to X
          else
               generate children of X
               for each child of X do
               CASE
               the child is not on OPEN or CLOSED:
                    assign the child a heuristic value
                    add the child to OPEN
               the child is already on OPEN:
                    if the child was reached by a shorter path
                    then give the state on OPEN the shorter path
               the child is already on CLOSED:
                    if the child was reached by a shorter path then
                         remove the state from CLOSED
                         add the child to OPEN
               end_CASE
               put X on CLOSED;
               re-order states on OPEN by heuristic merit (best leftmost)
     return NULL                                               // OPEN is empty
```

FIGURE 63.7   The best-first search algorithm (based on Luger and Stubblefield [1993, p. 121]).

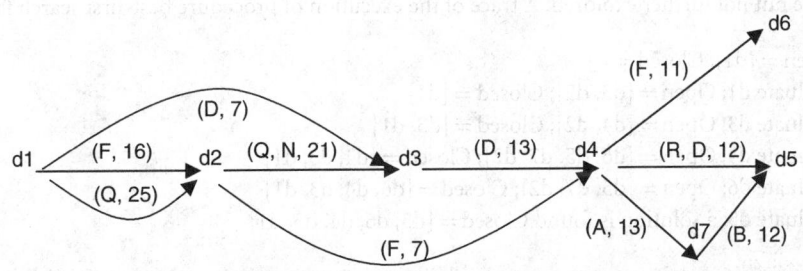

FIGURE 63.8   A state-space graph for a hypothetical subway system.

two partial solution paths. Duplicate states are not retained. By updating the ancestor history of nodes on OPEN and CLOSED, when they are rediscovered, the algorithm is more likely to find a quicker path to a goal.

"Best First Search then heuristically evaluates the states on OPEN, and the list is sorted according to the heuristic values. This brings the 'best' state to the front of OPEN. It is noteworthy that these estimates are heuristic in nature and therefore the next state to be examined may be from any level of the state space. OPEN, when maintained as a sorted list, is often referred to as a *priority queue*."

Here is a graph of a hypothetical search space. The problem is to find a shortest path from d1 to d5 in this directed and weighted graph (Figure 63.8), which could represent a sequence of local and express subway train stops. The F train starts at d1 and visits stops d2 (cost 16), d4 (cost 7), and d6 (cost 11). The D train starts at d1 and d3 (cost 7), d4 (cost 13), and d5 (cost 12). Other choices involve combinations of Q, N, R, and A trains with the F and/or D train. By applying the best-first search algorithm, we can find the shortest path from d1 to d5. Figure 63.9 shows a state tree representation of this graph.

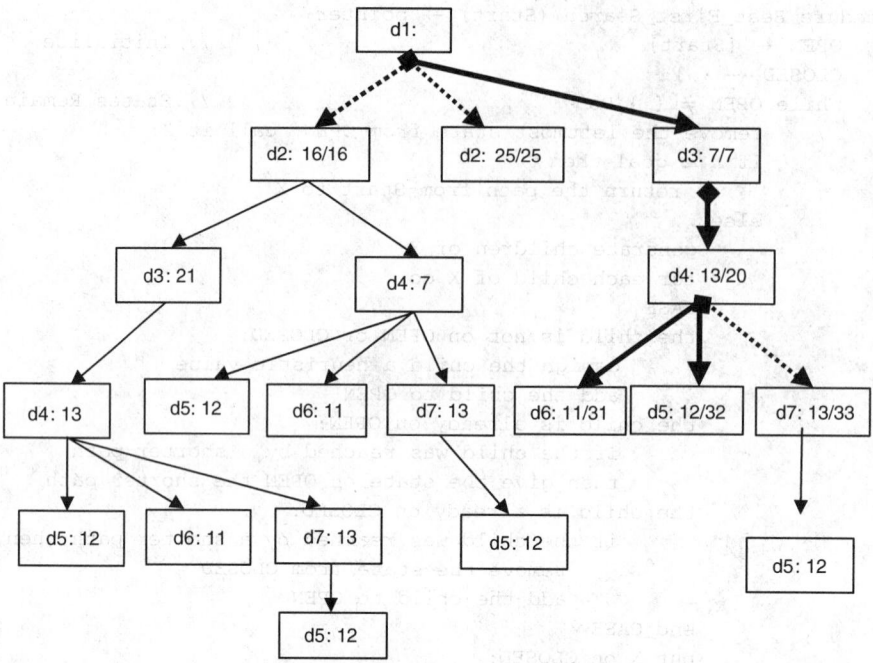

FIGURE 63.9   A search tree for the graph in Figure 63.8.

The thick arrowed path is the shortest path [d1, d3, d4, d5]. The dashed edges are nodes put on the open node queue but not further explored. A trace of the execution of procedure best-first search follows:

1. Open = [d1]; Closed = [ ]
2. Evaluate d1; Open = [d3, d2]; Closed = [d1]
3. Evaluate d3; Open = [d4, d2]; Closed = [d3, d1]
4. Evaluate d4; Open = [d6, d5, d7, d2]; Closed = [d4, d3, d1]
5. Evaluate d6; Open = [d5, d7, d2]; Closed = [d6, d4, d3, d1]
6. Evaluate d5; a solution is found Closed = [d5, d6, d4, d3, d1]

Note that nodes d6 and d5 are at the same level, so we do not take d6 in our search for the shortest path. Hence, the shortest path for this graph is [d1, d3, d4, d5]. After we reach our goal state d5, we can also find the shortest path from d5 to d1 by retracing the tree from d5 to d1.

When the best-first search algorithm is used, the states are sent to the open list in such a way that the most promising one is expanded next. Because the search heuristic being used for measurement of distance from the goal state may prove erroneous, the alternatives to the preferred state are kept on the open list. If the algorithm follows an incorrect path, it will retrieve the next best state and shift its focus to another part of the space. In the preceding example, children of node d2 were found to have poorer heuristic evaluations than sibling d3, so the search shifted there. However, the children of d3 were kept on open and could be returned to later, if other solutions were sought.

63.3.3   The A* Algorithm

The **A* algorithm**, first described by Hart et al. [1968], attempts to find the minimal cost path joining the start node and the goal in a state-space graph. The algorithm employs an ordered state-space search and an estimated heuristic cost to a goal state, f* (known as an *evaluation function*), as does the best-first search (see Section 63.3.2). It uniquely defines f*, so that it can guarantee an optimal solution path. The A* algorithm falls into the **branch and bound** class of algorithms, typically employed in operations research to find the shortest path to a solution node in a graph.

The evaluation function, $f^*(n)$, estimates the quality of a solution path through node n, based on values returned from two components: $g^*(n)$ and $h^*(n)$. Here, $g^*(n)$ is the minimal cost of a path from a start node to n, and $h^*(n)$ is a lower bound on the minimal cost of a solution path from node n to a goal node. As in branch and bound algorithms for trees, g^* will determine the single unique shortest path to node n.

For graphs, on the other hand, g^* can err only in the direction of overestimating the minimal cost; if a shorter path is found, its value readjusted downward. The function h^* is the carrier of *heuristic information*, and the ability to ensure that the value of $h^*(n)$ is less than $h(n)$ — that is, $h^*(n)$ is an underestimate of the actual cost, $h(n)$, of an optimal path from n to a goal node — is essential to the optimality of the A* algorithm. This property, whereby $h^*(n)$ is always less than $h(n)$, is known as the **admissibility condition**. If h^* is zero, then A* reduces to the blind uniform-cost algorithm. If two otherwise similar algorithms, A1 and A2, can be compared to each other with respect to their h^* function (i.e., $h1^*$ and $h2^*$) then algorithm A1 is said to be *more informed* than A2 if $h1^*(n) > h2^*(n)$, whenever a node n (other than a goal node) is evaluated. The cost of computing h^*, in terms of the overall computational effort involved and algorithmic utility, determines the *heuristic power* of an algorithm. That is, an algorithm that employs an h^* which is usually accurate, but sometimes inadmissible, may be preferred over an algorithm where h^* is always minimal but hard to effect [Barr and Feigenbaum, 1981].

Thus, we can summarize that the A* algorithm is a branch and bound algorithm augmented by the *dynamic programming principle*: the best way through a particular, intermediate node is the best way to that intermediate node from the starting place, followed by the best way from that intermediate node to the goal node. There is no need to consider any other paths to or from the intermediate node [Winston, 1992].

Stewart and White [1991] presented the *multiple-objective A* algorithm* (MOA*). Their research is motivated by the observation that most real-world problems have multiple, independent, and possibly conflicting objectives. MOA* explicitly addresses this problem by identifying the set of all non-dominated paths from a specified start node to given set of goal nodes in an OR graph. This work shows that MOA* is complete and is admissible, when used with a suitable set of heuristic functions.

63.4 Game-Tree Search

63.4.1 The Alpha-Beta Algorithms

To the human player of two-person games, the notion behind the **alpha-beta algorithm** is understood intuitively as follows:

> If I have determined that a move or a sequence of moves is bad for me (because of a refutation move or variation by my opponent), then I do not need to determine just how bad that move is. Instead, I can spend my time exploring other alternatives earlier in the tree.
> Conversely, if I have determined that a variation or sequence of moves is bad for my opponent, then I do not determine exactly how good it is for me.

Figure 63.10 illustrates some of these ideas. Here, the thick solid line represents the current solution path. This, in turn, has replaced a candidate solution, here shown with dotted lines. Everything to the right of the optimal solution path represents alternatives that are simply proved inferior. The path of the current solution is called the *principal variation* (PV), and nodes on that path are marked as PV nodes. Similarly, the alternatives to PV nodes are CUT nodes, where only a few successors are examined before a proof of inferiority is found. In time, the successor to a CUT node will be an ALL node, where everything must be examined to prove the cut-off at the CUT node. The number or bound value by each node represents the return to the root of the cost of the solution path.

In the 40 years since its inception, the alpha-beta minimax algorithm has undergone many revisions and refinements to improve the efficiency of its pruning and, until the recent invention of the MTD(f)

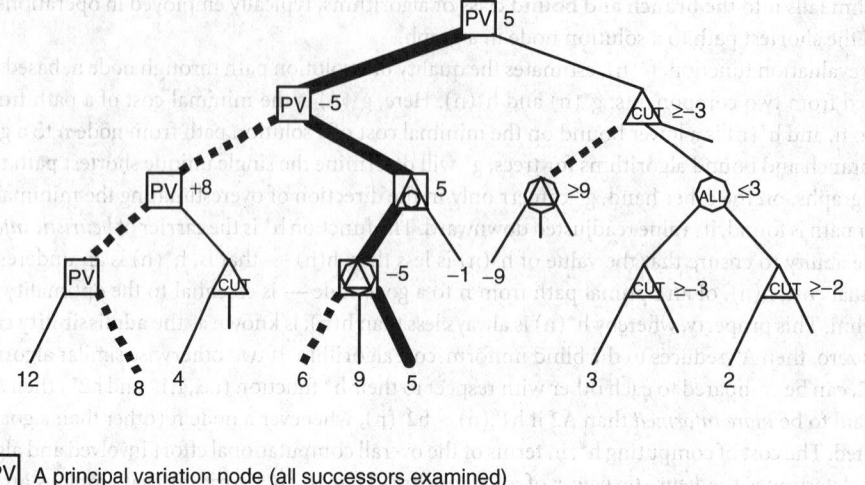

FIGURE 63.10 The PV, CUT, and ALL nodes of a tree, showing its optimal path (bold) and value (5).

variant, it has served as the primary search engine for two-person games. There have been many landmarks on the way, including Knuth and Moore's [1975] formulation in a negamax framework, Pearl's [1980] introduction of Scout, and the special formulation for chess with the principal variation search [Marsland and Campbell, 1982] and NegaScout [Reinefeld, 1983].

The essence of the method is that the search seeks a path whose value falls between two bounds called *alpha* and *beta*, which form a window. With this approach, one can also incorporate an artificial narrowing of the alpha-beta window, thus encompassing the notion of *aspiration search*, with a mandatory research on failure to find a value within the corrected bounds. This leads naturally to the incorporation of null window search (NWS) to improve on Pearl's test procedure. Here, the NWS procedure covers the search at a CUT node (Figure 63.10), where the cutting bound (beta) is negated and increased by 1 in the recursive call. This refinement has some advantage in the parallel search case, but otherwise NWS (Figure 63.11) is entirely equivalent to the minimal window call in NegaScout.

Additional improvements include the use of iterative deepening with *transposition tables* and other move-ordering mechanisms to retain a memory of the search from iteration to iteration. A transposition table is a cache of previously generated states that are typically hashed into a table to avoid redundant work. These improvements help ensure that the better subtrees are searched sooner, leading to greater pruning efficiency (more cut-offs) in the later subtrees. Figure 63.11 encapsulates the essence of the algorithm and shows how the first variation from a set of PV nodes, as well as any superior path that emerges later, is given special treatment. Alternates to PV nodes will always be CUT nodes, where a few successors will be examined. In a minimal game tree, only one successor to a CUT node will be examined, and it will be an ALL node where everything is examined. In the general case, the situation is more complex, as Figure 63.10 shows.

63.4.2 SSS* Algorithm

The **SSS* algorithm** was introduced by Stockman [1979] as a game-searching algorithm that traverses subtrees of the game tree in a best-first fashion similar to the A* algorithm. SSS* was shown to be superior to the original alpha-beta algorithm in the sense that it never looks at more nodes, while occasionally

```
ABS (node, alpha, beta, height) → tree_value
    if height ≡ 0
        return Evaluate(node)                            // a terminal node
    next ← FirstSuccessor (node)                         // a PV node
    best ← - ABS (next, -beta, -alpha, height -1)
    next ← SelectSibling (next)
    while next ≠ NULL do
        if best ≥ beta then
            return best                                  // a CUT node
        alpha ← max (alpha, best)
        merit ← - NWS (next, -alpha, height-1)
        if merit > best then
            if (merit ≤ alpha) or (merit ≥ beta) then
                best ← merit
            else best ← -ABS (next, -beta, -merit, height-1)
        next ← SelectSibling (next)
    end
    return best                                          // a PV node
end

NWS (node, beta, height) → bound_value
    if height ≡ 0 then
        return Evaluate(node)                            // a terminal node
    next ← FirstSuccessor (node)
    estimate ← - ∞
    while next ≠ NULL do
        merit ← - NWS (next, -beta+1, height-1)
        if merit > estimate then
            estimate ← merit
        if merit ≥ beta then
            return estimate                              // a CUT node
        next ← SelectSibling (next)
    end
    return estimate                                      // an ALL node
end
```

FIGURE 63.11   Scout/PVS version of alpha-beta search (ABS) in the negamax framework.

examining fewer [Pearl, 1984]. Roizen and Pearl [1983], the source of the following description of SSS*, state:

"...the aim of SSS* is the discovery of an optimal solution tree ... In accordance with the best-first split-and-prune paradigm, SSS* considers 'clusters' of solution trees and splits (or refines) that cluster having the highest upper bound on the merit of its constituents. Every node in the game tree represents a cluster of solution trees defined by the set of all solution trees that share that node ... the merit of a partially developed solution tree in a game is determined solely by the properties of the frontier nodes it contains, not by the cost of the paths leading to these nodes. The value of a frontier node is an upper bound on each solution tree in the cluster it represents ... SSS* establishes upper bounds on the values of partially developed solution trees by seeking the value of terminal nodes, left to right, taking the minimum value of those examined so far. These monotonically non-increasing bounds are used to order the solution trees so that the tree of highest merit is chosen for development. The development process continues until one solution tree is fully developed, at which point that tree represents the optimal strategy and its value coincides with the minimax value of the root...."

```
int MTDF ( node_type root, int f, int d)
{
        g = f;
        upperbound = +INFINITY;
        lowerbound = -INFINITY;
        repeat
                if (g == lowerbound)
                        beta = g + 1
                else
                        beta = g;
                g = AlphaBetaWithMemory(root, beta - 1, beta, d);
                if (g < beta)
                        then upperbound = g
                        else lowerbound = g;
        until (lowerbound >= upperbound);
    return g;
}
```

FIGURE 63.12   The MTD(f) algorithm pseudocode.

"The disadvantage of SSS* lies in the need to keep in storage a record of all contending candidate clusters, which may require large storage space, growing exponentially with search depth" [p. 245].

Heavy space and time overheads have kept SSS* from being much more than an example of a best-first search, but current research seems destined to relegate SSS* to a historical footnote. Plaat et al. [1995] have formulated the node-efficient SSS* algorithm into the alpha-beta framework using successive NWS search invocations (supported by perfect transposition tables) to achieve a memory-enhanced test procedure that provides a best-first search. With their introduction of the MTD(f) algorithm, Plaat et al. [1995] claim that SSS* can be viewed as a special case of the time-efficient alpha-beta algorithm, as opposed to the earlier view that alpha-beta is a k-partition variant of SSS*. MTD(f) is an important contribution that has now been widely adopted as the standard two-person game-tree search algorithm. It is described next.

63.4.3   The MTD(f) Algorithm

MTD(f) is usually run in an iterative deepening fashion, and each iteration proceeds by a sequence of minimal or NULL window alpha-beta calls. The search works by zooming in on the minimax value, as Figure 63.12 shows.

The bounds stored in *upperbound* and *lowerbound* form an interval around the true minimax value for a particular search depth d. The interval is initially set to $[-\infty, +\infty]$. Starting with the value f, returned from a previous call to MTD(f), each call to alpha-beta returns a new minimax value g, which is used to adjust the bounding interval and to serve as the pruning value for the next alpha-beta call. For example, if the initial minimax value is 50, alpha-beta will be called with the pruning values 49 and 50. If the new minimax value returned, g, is less than 50, upperbound is set to g. If the minimax value returned, g, is greater than or equal to 50, lowerbound is set to g. The next call to alpha-beta will use $g - 1$ and g for the pruning values (or g and $g + 1$, if g is equal to lowerbound). This process continues until upperbound and lowerbound converge to a single value, which is returned. MTD(f) will be called again with this newly returned minimax estimate and an increased depth bound, until the tree has been searched to a sufficient depth.

As a result of the iterative nature of MTD(f), the use of transposition tables is essential to its efficient implementation. In tests with a number of tournament game-playing programs, MTD(f) outperformed ABS (Scout/PVS, Figure 63.11). It generally produces trees that are 5% to 15% smaller than ABS [Plaat et al., 1996]. MTD(f) is now recognized as the most efficient variant of ABS and has been rapidly adopted as the new standard in minimax search.

63.4.4 Recent Developments

Sven Koenig has developed minimax learning real-time A* (Min-Max LRTA*), a real-time heuristic search method that generalizes Korf's [1990] earlier LRTA* to nondeterministic domains. Hence it can be applied to "robot navigation tasks in mazes, where robots know the maze but do not know their initial position and orientation (pose). These planning tasks can be modeled as planning tasks in non-deterministic domains whose states are sets of poses." Such problems can be solved quickly and efficiently with Min-Max LRTA*, requiring only a small amount of memory [Koenig, 2001].

Martin Mueller [2001] introduces the use of partial order bounding (POB) rather than scalar values for construction of an evaluation function for computer game playing. Propagation of partially ordered values through a search tree has been known to lead to many problems in practice. Instead, POB compares values in the leaves of a game tree and backs up Boolean values through the tree. The effectiveness of this method was demonstrated in examples of capture races in the game of GO [Mueller, 2001].

Schaeffer et al. [2001] demonstrate that the distinctions for evaluating heuristic search should not be based on whether the application is for single-agent or two-agent search. Instead, they argue that the search enhancements applied to both single-agent and two-agent problems for creating high-performance applications are the essentials. Focus should be on generality, for creating opportunities for reuse. Examples of some of the generic enhancements (as opposed to problem-specific ones) include the alpha-beta algorithm, transposition tables, and IDA*. Efforts should be made to enable more generic application of algorithms.

Hong et al. [2001] present a **genetic algorithm** approach that can find a good next move by reserving the board evaluation of new offspring in partial game-tree search. Experiments have proved promising in terms of speed and accuracy when applied to the game of GO.

The fast forward (FF) planning system of Hoffman and Nebel [2001] uses a heuristic that estimates goal distances by ignoring delete lists. Facts are not assumed to be independent. The system uses a new search strategy that combines hill climbing with systematic search. Powerful heuristic information is extended and used to prune the search space.

63.5 Parallel Search

The easy availability of low-cost computers has stimulated interest in the use of multiple processors for parallel traversals of decision trees. The few theoretical models of parallelism do not accommodate communication and synchronization delays that inevitably impact the performance of working systems. There are several other factors to consider, including the following:

How best to employ the additional memory and I/O resources that become available with the extra processors.

How best to distribute the work across the available processors.

How to avoid excessive duplication of computation.

Some important combinatorial problems have no difficulty with the last point, because every eventuality must be considered, but these tend to be less interesting in an AI context.

One problem of particular interest is game-tree search, where it is necessary to compute the value of the tree while communicating an improved estimate to the other parallel searchers as it becomes available. This can lead to an *acceleration anomaly* when the tree value is found earlier than is possible with a sequential algorithm. Even so, uniprocessor algorithms can have special advantages in that they can be optimized for best pruning efficiency, while a competing parallel system may not have the right information in time to achieve the same degree of pruning, and so do more work (suffer from search overhead). Further, the very fact that pruning occurs makes it impossible to determine in advance how big any piece of work (subtree to be searched) will be, leading to a potentially serious work imbalance and heavy synchronization (waiting for more work) delays.

Although the standard basis for comparing the efficiency of parallel methods is simply

$$\text{speedup} = \frac{\text{time taken by a sequential single-processor algorithm}}{\text{time taken by a P-processor system}}$$

this basis is often misused, because it depends on the efficiency of the uniprocessor implementation.

The exponential growth of the tree size (solution space) with depth of search makes parallel search algorithms especially susceptible to anomalous speedup behavior. Clearly, acceleration anomalies are among the welcome properties, but more commonly, anomalously bad performance is seen, unless the algorithm has been designed with care.

In game-playing programs of interest to AI, parallelism is not primarily intended to find the answer more quickly, but to get a more reliable result (e.g., based on a deeper search). Here, the emphasis lies on scalability instead of speedup. Although speedup holds the problem size constant and increases the system size to get a result sooner, scalability measures the ability to expand the sizes of both the problem and the system at the same time:

$$\text{scale-up} = \frac{\text{time taken to solve a problem of size s by a single-processor}}{\text{time taken to solve a (P} \times \text{s) problem by an P-processor system}}$$

Thus, scale-up close to unity reflects successful parallelism.

63.5.1 Parallel Single-Agent Search

Single-agent game-tree search is important because it is useful for several robot-planning activities, such as finding the shortest path through a maze of obstacles. It seems to be more amenable to parallelization than the techniques used in adversary games, because a large proportion of the search space must be fully seen — especially when optimal solutions are sought. This traversal can safely be done in parallel, since there are no cut-offs to be missed. Although move ordering can reduce node expansions, it does not play the same crucial role as in dual-agent game-tree search, where significant parts of the search space are often pruned away. For this reason, parallel single-agent search techniques usually achieve better speedups than their counterparts in adversary games.

Most parallel single-agent searches are based on A* or IDA*. As in the sequential case, parallel A* outperforms IDA* on a node count basis, although parallel IDA* needs only linear storage space and runs faster. In addition, cost-effective methods exist (e.g., parallel window search described in Section 63.5.1.3) that determine non-optimal solutions with even less computing time.

63.5.1.1 Parallel A*

Given P processors, the simplest way to parallelize A* is to let each machine work on one of the current best states on a global open list (a place holder for nodes that have not yet been examined). This approach minimizes the search overhead, as confirmed in practice by Kumar et al. [1988]. Their relevant experiments were run on a shared memory BBN Butterfly machine with 100 processors, where a search overhead of less than 5% was observed for the traveling salesperson (TSP) problem.

Elapsed time is more important than the node expansion count, however, because the global open list is accessed both before and after each node expansion, so memory contention becomes a serious bottleneck. It turns out that a centralized strategy for managing the open list is useful only in domains where the node expansion time is large compared to the open list access time. In the TSP problem, near linear time speedups were achieved with up to about 50 processors, when a sophisticated heap data structure was used to significantly reduce the open list access time [Kumar et al., 1988].

Distributed strategies using local open lists reduce the memory contention problem. But again, some communication must be provided to allow processors to share the most promising state descriptors, so that no computing resources are wasted in expanding inferior states. For this purpose, a global blackboard table can be used to hold state descriptors of the current best nodes. After selecting a state from its local open list, each processor compares its f-value (lower bound on the solution cost) to that of the states contained in the blackboard. If the local state is much better (or much worse) than those stored in the blackboard, then node descriptors are sent (or received), so that all active processors are exploring states of almost equal heuristic value. With this scheme, a 69-fold speedup was achieved on an 85-processor BBN Butterfly [Kumar et al., 1988].

Although a blackboard is not accessed as frequently as a global open list, it still causes memory contention with increasing parallelism. To alleviate this problem, Huang and Davis [1989] proposed a distributed heuristic search algorithm called *parallel iterative A** (PIA*), which works solely on local data structures. On a uniprocessor, PIA* expands the same nodes as A*; in the multiprocessor case, it performs a parallel best-first node expansion. The search proceeds by repetitive synchronized iterations, in which processors working on inferior nodes are stopped and reassigned to better ones. To avoid unproductive waiting at the synchronization barriers, the processors are allowed to perform speculative processing. Although Huang and Davis [1989] claim that "this algorithm can achieve almost linear speedup on a large number of processors," it has the same disadvantage as the other parallel A* variants, namely, excessive memory requirements.

63.5.1.2 Parallel IDA*

IDA* (Figure 63.6) has proved to be effective when excessive memory requirements undermine best-first schemes. Not surprisingly, it has also been a popular algorithm to parallelize. Rao et al. [1987] proposed PIDA*, an algorithm with almost linear speedup even when solving the 15-puzzle with its trivial node expansion cost. The 15-puzzle is a popular game made up of 15 tiles that slide within a 4 × 4 matrix. The object is to slide the tiles through the one empty spot until all tiles are aligned in some goal state. An optimal solution to a hard problem might take 66 moves. PIDA* splits the search space into disjoint parts, so that each processor performs a local cost-bounded depth-first search on its private portion of the state space. When a process has finished its job, it tries to get an unsearched part of the tree from other processors. When no further work can be obtained, all processors detect global termination and compute the minimum of the cost bounds, which is used as a new bound in the next iteration.

Note that more than a P-fold speedup is possible when a processor finds a goal node early in the final iteration. In fact, Rao et al. [1987] report an average speedup of 9.24 with nine processors on the 15-puzzle! Perhaps more relevant is the all-solution case, where no superlinear speedup is possible. Here, an average speedup of 0.93P with up to 30 (P) processors on a bus-based multiprocessor architecture (Sequent Balance 21000) was achieved. This suggests that only low multiprocessing overheads (locking, work transfer, termination detection, and synchronization) were experienced.

PIDA* employs a task attraction scheme like that shown in Figure 63.13 for distributing the work among the processors. When a processor becomes idle, it asks a neighbor for a piece of the search space. The donor then splits its depth-first search stack and transfers to the requester some nodes (subtrees) for parallel expansion. An optimal splitting strategy would depend on the regularity (uniformity of width and height) of the search tree, although short subtrees should never be given away. When the tree is regular (as in the 15-puzzle), a coarse-grained work transfer strategy can be used (e.g., transferring only nodes near the root); otherwise, a slice of nodes (e.g., nodes A, B, and C in Figure 63.13) should be transferred.

63.5.1.3 A Comparison with Parallel Window Search

Another parallel IDA* approach borrows from Baudet's [1978] parallel window method for searching adversary games (described subsequently). Powley and Korf [1991] adapted this method to single-agent search, under the title *parallel window search* (PWS). Their basic idea was to start simultaneously as many iterations as there are processors. This works for a small number of processors, which either expand the tree up to their given thresholds until a solution is found (and the search is stopped) or completely expand their search space. A global administration scheme then determines the next larger search bound, and node expansion starts over again.

Note that the first solution found by PWS need not necessarily be optimal. Suboptimal solutions are often found in searches of poorly ordered trees. There, a processor working with a higher cut-off bound finds a goal node in a deeper tree level, while other processors are still expanding shallower tree parts (which

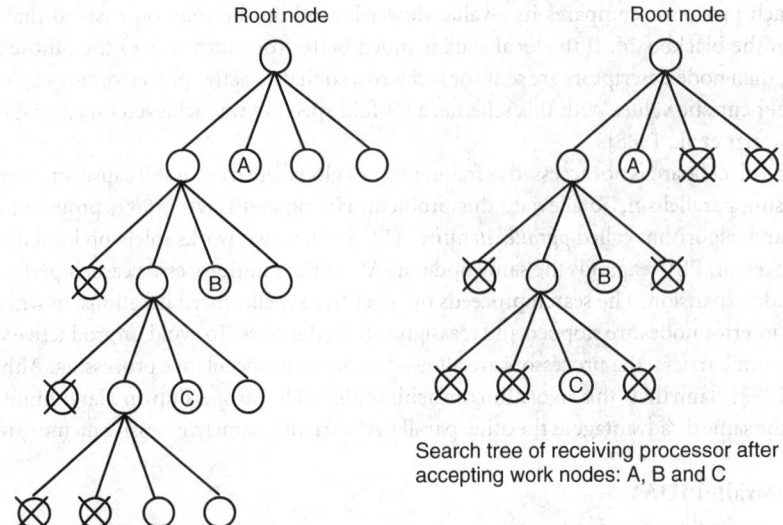

Search tree of receiving processor after
accepting work nodes: A, B and C

Search tree of sending processor before
transferring work nodes: A, B and C

FIGURE 63.13   A work distribution scheme.

may contain cheaper solutions). According to Powley and Korf [1991], PWS is not primarily meant to compete with IDA*, but it "can be used to find a nearly optimal solution quickly, improve the solution until it is optimal, and then finally guarantee optimality, depending on the amount of time available." Compared to PIDA*, the degree of parallelism is limited, and it remains unclear how to apply PWS in domains where the cost-bound increases are variable.

In summary, PWS and PIDA* complement each other, so it seems natural to combine them to form a single search scheme that runs PIDA* on groups of processors administered by a global PWS algorithm. The amount of communication needed depends on the work distribution scheme. A fine-grained distribution requires more communication, whereas a coarse-grained work distribution generates fewer messages (but may induce unbalanced work load). Note that the choice of the work distribution scheme also affects the frequency of good acceleration anomalies. Along these lines, perhaps the best results have been reported by Reinefeld [1995]. Using AIDA* (asynchronous parallel IDA*), near linear speedup was obtained on a 1024 transputer-based system solving 13 instances of the 19-puzzle. Reinefeld's paper includes a discussion of the communication overheads in both ring and toroid systems, as well as a description of the work distribution scheme.

63.5.2   Adversary Games

In the area of two-person games, early simulation studies with a **mandatory work first** (MWF) scheme [Akl et al., 1982] and the **PVSplit algorithm** [Marsland and Campbell, 1982] showed that a high degree of parallelism was possible, despite the work imbalance introduced by pruning. Those papers saw that in key applications (e.g., chess), the game-trees are well ordered, because of the wealth of move-ordering heuristics that have been developed [Slate and Atkin, 1977], so the bulk of the computation occurs during the search of the first subtree. The MWF approach uses the shape of the critical tree that must be searched. Because that tree is well defined and has regular properties, it is easy to generate.

In their simulation, Akl et al. [1982] considered the merits of searching the critical game tree in parallel, with the balance of the tree being generated algorithmically and searched quickly by simple tree splitting. Marsland and Campbell [1982], on the other hand, recognized that the first subtree of the critical game tree has the same properties as the whole tree, but its maximum height is one less. This so-called *principal variation* can be recursively split into parts of about equal size for parallel exploration. PVSplit, an algorithm based on this observation, was tested and analyzed by Marsland and Popowich [1985]. Even so, the static processor allocation schemes, like MWF and PVSplit, cannot achieve high levels of parallelism, although PVSplit does very well with up to half a dozen processors. MWF in particular ignores the true shape of the average game tree, and so is at its best with shallow searches, where the pruning imbalance from the so called *deep cut-offs* has less effect. Other working experience includes the first parallel chess program by Newborn, who later presented performance results [Newborn, 1988]. For practical reasons, Newborn only split the tree down to some pre-specified common depth from the root (typically 2), where the greatest benefits from parallelism can be achieved. This use of a common depth has been taken up by Hsu [1990] in his proposal for large-scale parallelism. Depth limits are also an important part of changing search modes and managing transposition tables.

63.5.2.1 Parallel Aspiration Window Search

In an early paper on parallel game-tree search, Baudet [1978] suggested partitioning the range of the alpha-beta window rather than the tree. In his algorithm, all processors search the whole tree, but each with a different, non-overlapping alpha-beta window. The total range of values is subdivided into P smaller intervals (where P is the number of processors), so that approximately one-third of the range is covered. The advantage of this method is that the processor having the true minimax value inside its narrow window will complete more quickly than a sequential algorithm running with a full window. Even the unsuccessful processors return a result: they determine whether the true minimax value lies below or above their assigned search window, providing important information for rescheduling idle processors until a solution is found.

Its low communication overhead and lack of synchronization needs are among the positive aspects of Baudet's approach. On the negative side, however, Baudet estimates a maximum speedup of between 5 and 6, even when using infinitely many processors. In practice, parallel window search can only be effectively employed on systems with two or three processors. This is because even in the best case (when the successful processor uses a minimal window), at least the critical game tree must be expanded. The critical tree has about the square root of the leaf nodes of a uniform tree of the same depth, and it represents the smallest tree that must be searched under any circumstances.

63.5.2.2 Advanced Tree-Splitting Methods

Results from fully recursive versions of PVSplit using the Parabelle chess program [Marsland and Popowich, 1985] confirmed earlier simulations and offered some insight into a major problem: in a P-processor system, P − 1 processors are often idle for an inordinate amount of time, thus inducing a high synchronization overhead for large systems. Moreover, the synchronization overhead increases as more processors are added, accounting for most of the total losses, because the search overhead (number of unnecessary node expansions) becomes almost constant for the larger systems.

This led to the development of variations that dynamically assign processors to the search of the principal variation. Notable is the work of Schaeffer [1989], which uses a loosely coupled network of workstations, and the independent implementation of Hyatt et al. [1989] for a shared-memory computer. These dynamic splitting works have attracted growing attention through a variety of approaches. For example, the results of Feldmann et al. [1990] show a speedup of 11.8 with 16 processors (far exceeding the performance of earlier systems), and Felten and Otto [1988] measured a 101 speedup on a 256-processor hypercube. This latter achievement is noteworthy because it shows an effective way to exploit the 256 times bigger memory that was not available to the uniprocessor. Use of the extra transposition table memory to hold results of search by other processors provides a significant benefit to the hypercube system, thus identifying clearly one advantage of systems with an extensible address space.

These results show a wide variation not only of methods but also of apparent performance. Part of the improvement is accounted for by the change from a static assignment of processors to the tree search (e.g., from PVSplit) to the dynamic processor reallocation schemes of Hyatt et al. [1989] and Schaeffer

[1989]. These later systems try to identify dynamically the ALL nodes of Figure 63.10 and search them in parallel, leaving the CUT nodes (where only a few successors might be examined) for serial expansion. In a similar vein, Ferguson and Korf [1988] proposed a bound-and-branch method that only assigned processors to the leftmost child of the tree-splitting nodes where no bound (subtree value) exists. Their method is equivalent to the static PVSplit algorithm and realizes a speedup of 12 with 32 processors for alpha-beta trees generated by Othello programs. This speedup result might be attributed to the smaller average branching factor of about 10 for Othello trees, compared to an average branching factor of about 35 for chess. If that uniprocessor solution is inefficient — for example, by omitting an important node-ordering mechanism like transposition tables [Reinefeld and Marsland, 1994] — the speedup figure may look good. For that reason, comparisons with a standard test suite from a widely accepted game are often done and should be encouraged.

Most of the working experience with parallel methods for two-person games has centered on the alpha-beta algorithm. Parallel methods for more node count-efficient sequential methods, like SSS*, have not been successful until recently, when the potential advantages of using heuristic methods like hash tables to replace the open list were exploited [Plaat et al., 1995].

63.5.2.3 Dynamic Distribution of Work

The key to successful large-scale parallelism lies in the dynamic distribution of work. There are four primary issues in dynamic search:

Search overhead — This measures the size of the tree searched by the parallel method with respect to the best sequential algorithm. As mentioned previously, in some cases superlinear speedup can occur when the parallel algorithm actually visits fewer nodes.

Synchronization overhead — Problems occur when processors are idle, waiting for results from other processors, thus reducing the effective use of the parallel computing power (processor utilization).

Load balancing — This reflects how evenly the work has been divided among available processors and similarly affects processor utilization.

Communication overhead — In a distributed memory system, this occurs when results must be communicated between processors via message passing.

Each of these issues must be considered in designing a dynamic parallel algorithm. The distribution of work to processors can either be accomplished in a work-driven fashion, whereby idle processors must acquire new work either from a blackboard or by requesting work from another processor.

The **young brothers wait concept** [Feldmann, 1993] is a work-driven scheme in which the parallelism is best described with the help of a definition: the search for a successor N.j of a node N in a game tree must not be started until after the leftmost sibling N.1 of N.j is completely evaluated. Thus, N.j can be given to another processor if and only if it has not yet been started and the search of N.1 is complete. This is also the requirement for the PVSplit algorithm. So how do the two methods differ and what are the trade-offs?

There are two significant differences. The first is at startup and the second is in the potential for parallelism. PVSplit starts much more quickly, because all the processors traverse the first variation (first path from the root to the search horizon of the tree) and then split the work at the nodes on the path as the processors back up the tree to the root. Thus, all the processors are busy from the beginning. On the other hand, this method suffers from increasingly large synchronization delays as the processors work their way back to the root of the game tree [Marsland and Popowich, 1985]. Thus, good performance is possible only with relatively few processors, because the splitting is purely static. In the work of Feldmann et al. [1990], the startup time for this system is lengthy, because initially only one processor (or a small group of processors) is used to traverse the first path. When that is complete, the right siblings of the nodes on the path can be distributed for parallel search to the waiting processors. For example, in the case of 1000 such processors, possibly less than 1% would initially be busy. Gradually, the idle processors are brought in to help the busy ones, but this takes time. However — and here comes the big advantage — the system is now much more dynamic in the way it distributes work, so it is less prone to serious synchronization loss. Further, although many of the nodes in the tree will be CUT nodes (which are a poor choice for parallelism because they generate high search overhead), others will be ALL nodes, where every successor must be examined, and they can simply be done in parallel. Usually CUT nodes generate a cut-off quite quickly, so by being cautious about how much work is initially given away once N.1 has been evaluated, one can keep excellent control of the search overhead, while getting full benefit from the dynamic work distribution that Feldmann's method provides.

On the other hand, **transposition table-driven scheduling** (TDS) for parallel single-agent and game-tree searches, proposed by Romein et al. [1999], is a data-driven technique that in many cases offers considerable improvements over work-driven scheduling on distributed memory architectures. TDS reduces the communication and memory overhead associated with the remote lookups of transposition tables partitioned among distributed memory resources. This permits lookup communication and search computation to be integrated. The use of the transposition tables in TDS, as in IDA*, prevents the repeated searching of previously expanded states.

TDS employs a distributed transposition table that works by assigning to each state a *home processor*, where the transposition entry for that state is stored. A signature associated with the state indicates the number of its home processor. When a given processor expands a new state, it evaluates its signature and sends it to its home processor without having to wait for a response, thus permitting the communication to be carried out asynchronously. In other words, the work is assigned to where the data on a particular state is stored, rather than having to look up a remote processor's table and wait for the results to be transmitted back. Alternatively, when a processor receives a node, it performs a lookup of its local transposition table to determine whether the node has been searched before. If not, the node is stored in the transposition table and added to the local work queue. Furthermore, since each transposition table entry includes a search bound, this prevents redundant processing of the same subtree by more than one processor. The resulting reduction in both communication and search overhead yields significant performance benefits. Speedups that surpass IDA* by a factor of more than 100 on 128 processors [Romein et al., 1999] have been reported in selected games.

Cook and Varnell [1998] report that TDS may be led into doing unnecessary work at the goal depth, however, and therefore they favor a hybrid combination of techniques that they term *adaptive parallel iterative deepening search*. They have implemented their ideas in the system called EUREKA, which employs machine learning to select the best technique for a given problem domain.

63.6 Recent Developments

Despite advances in parallel single-agent search, significant improvement in methods for game-tree search has remained elusive. Theoretical studies have often focused on showing that linear speedup is possible on worst-order game trees. While not wrong, they make only the trivial point that where exhaustive search is necessary and where pruning is impossible, even simple work distribution methods may yield excellent results. The true challenge, however, is to consider the case of average game trees, or even better, the strongly ordered model (where extensive pruning can occur), resulting in asymmetric trees with a significant work distribution problem and significant search overhead. The search overhead occurs when a processor examines nodes that would be pruned by the sequential algorithm but has not yet received the relevant results from another processor.

The intrinsic difficulty of searching game trees under pruning conditions has been widely recognized. Hence, considerable research has been focused on the goal of dynamically identifying when unnecessary search is being performed, thereby freeing processing resources for redeployment. For example, Feldmann et al. [1990] used the concept of making young brothers wait to reduce search overhead, and developed the *helpful master* scheme to eliminate the idle time of masters waiting for their slaves' results. On the other hand, young brothers wait can still lead to significant synchronization overhead.

Generalized depth-first searches are fundamental to many AI problems. In this vein, Kumar and Rao [1990] have fully examined a method that is well suited to doing the early iterations of single-agent IDA* search. The unexplored parts of the trees are marked and are dynamically assigned to any idle processor. In principle, this work distribution method (illustrated in Figure 63.13) could also be used for deterministic adversary game trees. Shoham and Toledo [2001] developed a parallel, randomized best-first minimax search (RBFM). RBFM expands randomly chosen terminal nodes, with higher probabilities assigned to nodes with better values. This method seems amenable to parallelism but also seems to suffer from speculative search overhead.

The advent of MTD(f) as the primary sequential search algorithm has greatly improved circumstances and seems likely to lead to significant advances in parallel game-tree search. The use of a minimal window, and hence a single bound for pruning the tree, can greatly reduce the search overhead. Local pruning is no longer sensitive to results obtained by different processors searching other parts of the game tree. Romein [2000] developed a parallel version of MTD(f) that achieved speedup over other parallel game tree algorithms. Since MTD(f) is iterative in nature, its use of transposition tables is especially critical to performance. Kishimoto and Schaeffer [2002] have developed an algorithm called TDSAB, which combines MTD(f) with the transposition table–driven scheduling of TDS to achieve significant performance gains in a number of selected games.

Many researchers feel that hardware advances will ultimately increase parallel processing power to the point where we will simply overwhelm the problem of game-tree search with brute force. They feel that this will make the need for continued improvement of parallel game tree algorithms less critical. Perhaps in time we will know whether massive parallelism solves all our game-tree search problems. However, the results of complexity theory should assure us that unless $P = NP$, improvements in search techniques and heuristics will remain both useful and desirable.

The advent of the World Wide Web in 1993 naturally led to interest in search techniques for the Internet, especially when intelligence can be applied. *Search engines* are computer programs that can automatically contact other network resources on the Internet, searching for specific information or key words, and report the results of their search. *Intelligent agents* are computer programs that help users to conduct routine tasks, search and retrieve information, support decision making, and act as domain experts. They do more than just search and match. There are intelligent agents for consumers that search and filter information. Intelligent agents have been developed for product and vendor finding, for negotiation, and for learning. They can help determine what to buy to satisfy a specific need by looking for specific product information and then critically evaluating the products. One example is Firefly, which uses a *collaborative filtering* process that can be described as word-of-mouth to build the profile. It asks a consumer to rate a number of products, then matches those ratings with the ratings of other consumers with similar tastes and recommends products that have not yet been rated by the consumer.

Intelligent agents for product and vendor finding can also find bargains. An example is Jango from NetBot/Excite. It originates requests from the user's site (rather than Jango's). Vendors have no way to determine whether the request is from a real customer or from the agent. Jango provides product reviews. Intelligent agents for consumers can act as negotiation agents, helping to determine the price and other terms of transactions. Kasbah, from MIT Lab, is for users who want to sell or buy a product. It assigns the task to an agent, which is then sent out to seek buyers or sellers proactively. Kasbah has multiple agents. Users create agents, which can exemplify three strategies: anxious, cool-headed, or frugal. Tete-@-Tete is an intelligent agent for considering a number of different parameters: price, warranty, delivery time, service contracts, return policy, loan options, and other value-added services. Upon request, it can even be argumentative.

Finally, there are learning agents capable of learning individuals' preferences and making suitable suggestions based on these preferences. Memory Agent from IBM and Learn Sesame from Open Sesame use learning theory for monitoring customers' interactions. The agent learns customers' interests, preferences, and behavior and delivers customized service accordingly [Deitel et al., 2001]. Expect many interesting developments in this arena, combining some of the theoretical findings we have presented with practical results.

Defining Terms

A* algorithm: A best-first procedure that uses an admissible heuristic estimating function to guide the search process to an optimal solution.

Admissibility condition: The necessity that the heuristic measure never overestimates the cost of the remaining search path, thus ensuring that an optimal solution will be found.

Alpha-beta algorithm: The conventional name for the bounds on a depth-first minimax procedure used to prune away redundant subtrees in two-person games.

AND/OR tree: A tree that enables the expression of the decomposition of a problem into subproblems; hence, alternate solutions to subproblems through the use of AND/OR node-labeling schemes can be found.

Backtracking: A component process of many search techniques whereby recovery from unfruitful paths is sought by backing up to a juncture where new paths can be explored.

Best-first search: A heuristic search technique that finds the most promising node to explore next by maintaining and exploring an ordered open node list.

Bidirectional search: A search algorithm that replaces a single search graph, which is likely to grow exponentially, with two smaller graphs — one starting from the initial state and one starting from the goal state.

Blind search: A characterization of all search techniques that are heuristically uninformed. Included among these would normally be state-space search, means–ends analysis, generate and test, depth-first search, and breadth-first search.

Branch and bound algorithm: A potentially optimal search technique that keeps track of all partial paths contending for further consideration, always extending the shortest path one level.

Breadth-first search: An uninformed search technique that proceeds level by level, visiting all the nodes at each level (closest to the root node) before proceeding to the next level.

Data-driven parallelism: A load-balancing scheme in which work is assigned to processors based on the characteristics of the data.

Depth-first search: A search technique that first visits each node as deeply and as far to the left as possible.

Generate and test: A search technique that proposes possible solutions and then tests them for their feasibility.

Genetic algorithm: A stochastic hill-climbing search in which a large population of states is maintained. New states are generated by mutation and crossover, which combines pairs of earlier states from the population.

Heuristic search: An informed method of searching a state space with the purpose of reducing its size and finding one or more suitable goal states.

Iterative deepening: A successive refinement technique that progressively searches a longer and longer tree until an acceptable solution path is found.

Mandatory work first: A static two-pass process that first traverses the minimal game tree and uses the provisional value found to improve the pruning during the second pass over the remaining tree.

Means–ends analysis: An AI technique that tries to reduce the "difference" between a current state and a goal state.

MTD(f) algorithm: A minimal window minimax search recognized as the most efficient alpha-beta variant.

Parallel window aspiration search: A method in which a multitude of processors search the same tree, each with different (non-overlapping) alpha-beta bounds.

PVSplit (principal variation splitting): A static parallel search method that takes all the processors down the first variation to some limiting depth, then splits the subtrees among the processors as they back up to the root of the tree.

Simulated annealing: A stochastic algorithm that returns optimal solutions when given an appropriate "cooling schedule."

SSS* algorithm: A best-first search procedure for two-person games.

Transposition table-driven scheduling (TDS): A data-driven, load-balancing scheme for parallel search that assigns a state to a processor based on the characteristics or signature of the given state.

Work-driven parallelism: A load-balancing scheme in which idle processors explicitly request work from other processors.

Young brothers wait concept: A dynamic variation of PVSplit in which idle processors wait until the first path of leftmost subtree has been searched before giving work to an idle processor.

References

Akl, S.G., Barnard, D.T., and Doran, R.J. 1982. Design, analysis and implementation of a parallel tree search machine. *IEEE Trans. on Pattern Anal. and Mach. Intell.*, 4(2): 192–203.

Barr, A., and Feigenbaum, E.A. 1981. *The Handbook of Artificial Intelligence V1.* William Kaufmann, Stanford, CA.

Baudet, G.M. 1978. *The Design and Analysis of Algorithms for Asynchronous Multiprocessors.* Ph.D. thesis, Dept. of Computing Science, Carnegie Mellon Univ., Pittsburgh, PA.

Bolc, L., and Cytowski, J. 1992. *Search Methods for Artificial Intelligence.* Academic Press, San Diego, CA.

Cook, D.J., and Varnell, R.C. 1998. Adaptive parallel iterative deepening search. *J. of AI Research*, 8: 139–166.

Deitel, H.M., Deitel, P.J., and Steinbuhler, K. 2000. *E-Business and e-Commerce for Managers.* Prentice Hall.

Feldmann, R., Monien, B., Mysliwietz, P., and Vornberger, O. 1990. Distributed game tree search. In V. Kumar, P.S. Gopalakrishnan, and L. Kanal, Eds., *Parallel Algorithms for Machine Intelligence and Vision,* pp. 66–101. Springer-Verlag, New York.

Feldmann, R. 1993. *Game Tree Search on Massively Parallel Systems.* Ph. D. thesis, University of Paderborn, Germany.

Felten, E.W., and Otto, S.W. 1989. A highly parallel chess program. *Procs. Int. Conf. on 5th Generation Computer Systems,* pp. 1001–1009.

Feigenbaum, E., and Feldman, J. 1963. *Computers and Thought.* McGraw-Hill, New York.

Ferguson, C., and Korf, R.E. 1988. Distributed tree search and its application to alpha-beta pruning. *Proc. 7th Nat. Conf. on Art. Intell.,* 1: 128–132, Saint Paul, MN. Morgan Kaufmann, Los Altos, CA.

Hart, P.E., Nilsson, N.J., and Raphael, B. 1968. A formal basis for the heuristic determination of minimum cost paths. *IEEE Transactions on SSC,* SSC-4: 100–107.

Hoffman, J., and Nebel, B. 2001. The FF planning system: fast plan generation through heuristic search. *J. of AI Research,* 14: 253–302.

Hong, T.P, Huang, K.Y., and Lin, W.Y. 2001. Adversarial search by evolutionary computation. *Evolutionary Computation,* 9(3): 371–385.

Hsu, F.-H. 1990. Large scale parallelization of alpha-beta search: an algorithmic and architectural study with computer chess. Technical Report CMU-CS-90-108, Carnegie-Mellon Univ., Pittsburgh, PA.

Huang, S., and Davis, L.R. 1989. Parallel iterative A* search: an admissible distributed search algorithm. *Procs. 11th Int. Joint Conf. on AI,* 1: 23–29, Detroit. Morgan Kaufmann, Los Altos, CA.

Hyatt, R.M., Suter, B.W., and Nelson, H.L. 1989. A parallel alpha-beta tree searching algorithm. *Parallel Computing* 10(3): 299–308.

Kaindl, H., and Kainz, G. 1997. Bidirectional heuristic search reconsidered. *J. of AI Research,* 7: 283–317.

Kishimoto, A., and Schaeffer, J. 2002. Distributed game-tree search using transposition table driven work scheduling. *International Conference on Parallel Processing (ICPP),* pp. 323–330.

Knuth, D., and Moore, R. 1975. An analysis of alpha-beta pruning. *Artificial Intelligence,* 6(4): 293–326.

Koenig, S. 2001. Minimax real-time heuristic search. *Artificial Intelligence,* 129: 165–195.

Korf, R.E. 1990. Real-time heuristic search. *Artificial Intelligence,* 42(2-3): 189–211.

Korf, R.E. 1985. Depth-first iterative-deepening: an optimal admissible tree search. *Artificial Intelligence*, 27(1): 97–109.

Kumar, V., Ramesh, K., and Nageshwara-Rao, V. 1988. Parallel best-first search of state-space graphs: a summary of results. *Procs. 7th Nat. Conf. on Art. Int., AAAI-88*, pp. 122–127, Saint Paul, MN. Morgan Kaufmann, Los Altos, CA.

Luger, J., and Stubblefield, W. 1993. *Artificial Intelligence: Structures and Strategies for Complex Problem Solving, 2nd ed.* Benjamin/Cummings, Redwood City, CA.

Luger, J. 2002. *Artificial Intelligence: Structures and Strategies for Complex Problem Solving, 4th ed.* Benjamin/Cummings, Redwood City, CA.

Marsland, T.A., and Campbell, M. 1982. Parallel search of strongly ordered game trees. *ACM Computing Surveys*, 14(4): 533–551.

Marsland, T.A., and Popowich, F. 1985. Parallel game-tree search. *IEEE Trans. on Pattern Anal. and Mach. Intell.*, 7(4): 442–452.

Mueller, M. 2001. Partial order bounding: a new approach to evaluation in game tree search. *Artificial Intelligence*, 129: 279–311.

Newborn, M.M. 1988. Unsynchronized iteratively deepening parallel alpha-beta search. *IEEE Trans. on Pattern Anal. and Mach. Intell.*, 10(5): 687–694.

Nilsson, N. 1971. *Problem-Solving Methods in Artificial Intelligence*. McGraw-Hill, New York.

Pearl, J. 1980. Asymptotic properties of minimax trees and game-searching procedures. *Artificial Intelligence*, 14(2): 113–38.

Pearl, J. 1984. *Heuristics: Intelligent Search Strategies for Computer Problem Solving*. Addison-Wesley, Reading, MA.

Plaat, A., Schaeffer, J., Pijls, W., and de Bruin, A. 1995. Best-first fixed-depth game-tree search in practice. *Proceedings of IJCAI-95*, pp. 273–279, Montreal, Canada, August. Morgan Kaufmann, Los Altos, CA.

Plaat, A., Schaeffer, J., Pijls, W., and de Bruin, A. 1996. Best-first fixed-depth minimax algorithms. *Artificial Intelligence*, 87(1-2): 1–38.

Pohl, I. 1971. Bi-directional search. In B. Meltzer and D. Michie, Eds., *Machine Intelligence 6*, pp. 127–140. American Elsevier, New York.

Polya, G. 1945. *How to Solve It*. Princeton University Press, Princeton, NJ.

Powley, C., and Korf, R.E. 1991. Single-agent parallel window search. *IEEE Trans. on Pattern Anal. and Mach. Intell.*, 13(5): 466–477.

Rao, V.N., Kumar, V. and Ramesh, K. 1987. A parallel implementation of Iterative-Deepening A*. *Procs. 6th Nat. Conf. on Art. Intell.*, pp. 178–182, Seattle.

Reinefeld, A. 1983. An improvement to the Scout tree-search algorithm. *Intl. Computer Chess Assoc. Journal*, 6(4): 4–14.

Reinefeld, A., and Marsland, T.A. 1994. Enhanced iterative-deepening search. *IEEE Trans. on Pattern Anal. and Mach. Intell.*, 16(7): 701–710.

Reinefeld, A. 1995. scalability of massively parallel depth-first search. In P.M. Pardalos, M.G.C. Resende, and K.G. Ramakrishnan, Eds., *Parallel Processing of Discrete Optimization Problems*, pp. 305–322. DIMACS Series in Discrete Mathematics and Theoretical Computer Science, Vol. 22, American Mathematical Society, Providence, RI.

Roizen, I., and Pearl, J. 1983. A minimax algorithm better than alpha-beta? Yes and no. *Artificial Intelligence*, 21(1–2): 199–220.

Romanycia, M., and Pelletier, F. 1985. What is heuristic? *Computational Intelligence* 1: 24–36.

Romein, J., Plaat, A., Bal, H., and Schaeffer, J. 1999. Transposition table driven work scheduling in distributed search. *Procs. AAAI National Conf.*, pp. 725–731, Orlando, FL.

Romein, J.W. 2000. Multigame — an environment for distributed game-tree search. Ph.D. thesis. ASCI dissertation series No. 53. Vrije Universiteit, Amsterdam.

Russell, S., and Norvig, P. 2003. *Artificial Intelligence: A Modern Approach, 2nd ed.* Prentice Hall, Englewood Cliffs, NJ.

Schaeffer, J. 1989. Distributed game-tree search. *J. of Parallel and Distributed Computing*, 6(2): 90–114.

Schaeffer, J., Plaat, A., and Junghanns, J. 2001. Unifying single-agent and two-player search. *Information Sciences*, 134(3–4): 151–175.

Shoham, Y., and Toledo, S. 2002. Parallel randomized best-first minimax search. *Artificial Intelligence*, 137: 165–196.

Slate, D.J., and Atkin, L.R. 1977. Chess 4.5 — the Northwestern University chess program. In P. Frey, Ed., *Chess Skill in Man and Machine*, pp. 82–118. Springer-Verlag, New York.

Stewart, B.S., and White, C.C. 1991. Multiobjective A*. *J. of the ACM*, 38(4): 775–814.

Stockman, G. 1979. A minimax algorithm better than alpha-beta? *Artificial Intelligence*, 12(2): 179–96.

Winston, P.H. 1992. *Artificial Intelligence, 3rd ed.* Addison-Wesley, Reading, MA.

Acknowledgment

The authors thank Islam M. Guemey for help with research; Erdal Kose, for the best-first example; and David Kopec for technical assistance and assistance with artwork.

For Further Information

The most regularly and consistently cited source of information for this chapter is the *Journal of Artificial Intelligence*. There are numerous other journals including, for example, *AAAI Magazine, CACM, IEEE Expert, ICGA Journal*, and the *International Journal of Computer Human Studies*, which frequently publish articles related to this subject. Also prominent have been the volumes of the *Machine Intelligence Series*, edited by Donald Michie with various others. An excellent reference source is the three-volume *Handbook of Artificial Intelligence* by Barr and Feigenbaum [1981].

In addition, there are numerous national and international conferences on AI with published proceedings, headed by the International Joint Conference on AI (IJCAI). Classic books on AI methodology include Feigenbaum and Feldman's *Computers and Thought* [1963] and Nils Nilsson's *Problem-Solving Methods in Artificial Intelligence* [1971]. There are a number of popular and thorough textbooks on AI. Two relevant books on the subject of search are *Heuristics* [Pearl, 1984] and the more recent *Search Methods for Artificial Intelligence* [Bolc and Cytowski, 1992]. An AI texts that has considerable focus on search techniques is George Luger's *Artificial Intelligence* [2002]. Particularly current is Russell and Norvig's *Artificial Intelligence: A Modern Approach* [2003].

64

Understanding Spoken Language

64.1 Introduction 64-1
    Defining the Problem • System Architecture and Research
    Issues

64.2 Underlying Principles 64-4
    Procedure for System Development • Data Collection
    • Speech Recognition • Language Understanding • Speech
    Recognition/Natural Language Integration • Discourse and
    Dialogue • Evaluation

64.3 Best Practices 64-11
    The Advanced Research Projects Agency Spoken Language
    System (SLS) Project • The SUNDIAL Program
    • Other Systems

64.4 Research Issues and Summary 64-14
    Working in Real Domains • The New Word Problem
    • Spoken Language Generation • Portability

Stephanie Seneff
Massachusetts Institute of Technology

Victor Zue
Massachusetts Institute of Technology

64.1 Introduction

Computers are fast becoming a ubiquitous part of our lives, and our appetite for information is ever increasing. As a result, many researchers have sought to develop convenient human–computer interfaces, so that ordinary people can effortlessly access, process, and manipulate vast amounts of information — any time and anywhere — for education, decision making, purchasing, or entertainment. A speech interface, in a user's own language, is ideal because it is the most natural, flexible, efficient, and economical form of human communication.

After many years of research, spoken input to computers is just beginning to pass the threshold of practicality. The last decade has witnessed dramatic improvement in **speech recognition** (SR) technology, to the extent that high-performance algorithms and systems are becoming available. In some cases, the transition from laboratory demonstration to commercial deployment has already begun. Speech input capabilities are emerging that can provide functions such as voice dialing (e.g., "call home"), call routing (e.g., "I would like to make a collect call"), simple data entry (e.g., entering a credit card number), and preparation of structured documents (e.g., a radiology report).

64.1.1 Defining the Problem

Speech recognition is a very challenging problem in its own right, with a well-defined set of applications. However, many tasks that lend themselves to spoken input, making travel arrangements or selecting a movie, as illustrated in Figure 64.1, are in fact exercises in interactive problem solving. The solution is

Book me a flight on American to Dallas tomorrow morning.
Send me all his x-rays taken over the past six months.
Wo ist das nächste Italienische Restaurant das die
    American Express Kreditkarte nimmt?
List all mergers and acquisitions in the entertainment and
    telecommunication industries in the last three months.
Quel temps fera-t-il à Washington?
Transfer $500 from my savings account to my checking
    account.
Dove sta la biblioteca vicino a Central Square?
What's that movie with Marilyn Monroe in which two men
    dressed as women?

FIGURE 64.1   An illustration of the types of queries a user is likely to produce for interactive problem solving.

often built up incrementally, with both the user and the computer playing active roles in the conversation. Therefore, several language-based input and output technologies must be developed and integrated to reach this goal. Regarding the former, speech recognition must be combined with **natural language** (NL) processing so that the computer can understand spoken commands (often in the context of previous parts of the dialogue). On the output side, some of the information provided by the computer, and any of the computer's requests for clarification, must be converted to natural sentences, perhaps delivered verbally.

This chapter describes the technologies that are utilized by computers to achieve spoken language understanding. Spoken language understanding by machine has been the focus of much research over the past 10 years around the world. In contrast, spoken **language generation** has not received nearly as much attention, even though it is a critical component of a fully interactive, conversational system. The remainder of this chapter will focus mainly on the input side. Although such an imbalance in treatment may be viewed as being inappropriate, it is, unfortunately, an accurate reflection of the research landscape.

Spoken language communication is an active process that utilizes many different sources of knowledge, some of them deeply embedded in the linguistic competence of the talker and the listener. For example, the two phrase, "lettuce spray" and "let us pray" differ in subtle ways at the acoustic level. In the first phrase, the /s/ phonemes ending the first word and starting the second merge into a longer acoustic segment than the /s/ in "us." Furthermore, the /p/ in "spray" is *unaspirated*, because it is embedded in a consonant cluster with the preceding /s/ in the syllable onset, thus sounding more like a /b/. Other very similar word sequences are more problematic, because in some cases the acoustic realizations can be essentially identical. For the pair "meter on Main Street," and "meet her on Main Street," *syntactic* constraints would suggest that the first one is not a well-formed sentence. For the contrastive pair, "is the baby crying," and "is the bay bee crying," the acoustic differences would be very subtle, and both phrases are *syntactically* legitimate. However, the second one could be ruled out as implausible on the basis of *semantic* constraints. A popular example among speech researchers is the pair, "recognize speech," and "wreck a nice beach." These two are surprisingly similar acoustically, but it is hard to imagine a **discourse** context that would support both. In practice, acoustics, syntax, semantics, and discourse context should all be utilized to contribute to a goodness score for all competing hypotheses.

Higher level linguistic knowledge can play an important role in helping to constrain the permissible word sequences. Thus, for example, the phoneme sequence /w ɛ r ɪ z ɪ t/ is linguistically much more likely to be "where is it" than "wear is it," simply because the first one makes more sense. On the other hand, if used too aggressively, syntactic and semantic constraints can cause a system to fail even though enough of the content words have been recognized correctly to infer a plausible action to take. Problems arise not only because people often violate syntactic rules in conversational speech, but also because recognition errors can lead to pathological syntactic forms. For instance, if the recognizer's top choice hypothesis is

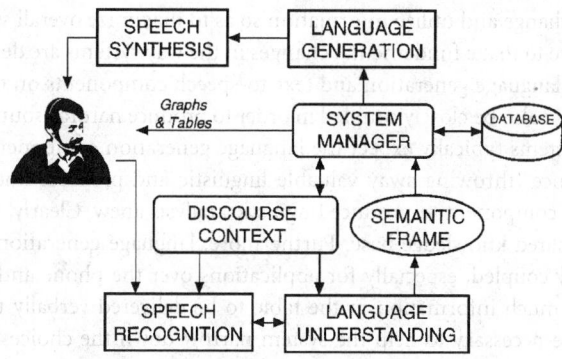

FIGURE 64.2   A generic block diagram for a typical spoken language system.

the nonsense phrase, "between four in five o'clock," a parser may fail to recognize the intended meaning. Three alternative and contrastive strategies to cope with such problems have been developed. The first would be to analyze "between four" and "five o'clock" as separate units, and then infer the relationship between them after the fact through plausibility constraints. A second approach would be to permit "in" to substitute for "and" at selected places in the grammar rules, based on the assumption that these are confusable pairs acoustically. The final, and intuitively most appealing method, is to tightly integrate the natural language component into the recognizer search, so that "and" is so clearly preferred over "in" in the preceding situation that the latter is never chosen. Although this final approach seems most logical, it turns out that researchers have not yet solved the problem of computational overload that occurs when a **parser** is used to predict the next word hypotheses. A compromise that is currently popular is to allow the recognizer to propose a list of *N* ordered theories, and have the linguistic analysis examine each theory in turn, choosing the one that appears the most plausible.

64.1.2   System Architecture and Research Issues

Figure 64.2 shows the major components of a typical **conversational system**. The spoken input is first processed through the speech recognition component. The natural language component, working in concert with the recognizer, produces a meaning representation. For information retrieval applications illustrated in this figure, the meaning representation can be used to retrieve the appropriate information in the form of text, tables, and graphics. If the information in the utterance is insufficient or ambiguous, the system may choose to query the user for clarification. Natural language generation and **text-to-speech synthesis** can be used to produce spoken responses that may serve to clarify the tabular information. Throughout the process, discourse information is maintained and fed back to the speech recognition and language understanding components, so that sentences can be properly understood in context.

The development of conversational systems offers a set of significant challenges to speech and natural language researchers, and raises several important research issues. First, the system must begin to deal with conversational, extemporaneously produced speech. Spontaneous speech is often extremely difficult to recognize and understand, since it may contain false starts, hesitations, and words and linguistic constructs unknown to the system.

Second, the system must have an effective strategy for coupling speech recognition with language understanding. Speech recognition systems typically implement linguistic constraints as a statistical grammar that specifies the probability of a word given its predecessors. Although these simple language models have been effective in reducing the search space and improving performance, they do not begin to address the issue of speech understanding. On the other hand, most natural language systems are developed with text input in mind; it is usually assumed that the entire word string is known with certainty. This assumption is clearly false for speech input, where many words are competing for the same time span, and some words may be more reliable than others because of varying signal robustness. Researchers in each discipline need to investigate how to exchange and utilize information so as to maximize overall system performance. In some cases, one may have to make fundamental changes in the way systems are designed.

Similarly, the natural language generation and text-to-speech components on the output side of conversational systems should also be closely coupled in order to produce natural-sounding spoken language. For example, current systems typically expect the language generation component to produce a textual surface form of a sentence (throwing away valuable linguistic and prosodic knowledge) and then require the text-to-speech component to produce linguistic analysis anew. Clearly, these two components would benefit from a shared knowledge base. Furthermore, language generation and dialogue modeling should be intimately coupled, especially for applications over the phone and without displays. For example, if there is too much information in the table to be delivered verbally to the user, a clarification subdialogue may be necessary to help the system narrow down the choices before enumerating a subset.

64.2 Underlying Principles

64.2.1 Procedure for System Development

Figure 64.3 illustrates the typical procedure for system development. For a newly emerging domain or language, an initial system is developed with some limited natural language capabilities, based on the inherent knowledge and intuitions of system developers. Once the system has some primitive capabilities, a **wizard** mode data collection episode can be initiated, in which a human wizard helps the system answer questions posed by naive subjects. The resulting data (both speech and text) are then used for further development and training of both the speech recognizer and the natural language component. As these components begin to mature, it becomes feasible to give the system increasing responsibility in later data collection episodes. Eventually, the system can stand alone without the aid of a wizard, leading to less costly and more efficient data collection possibilities. As the system evolves, its changing behaviors have a profound influence on the subjects' speech, so that at times there is a *moving target* phenomenon. Typically, some of the collected data are set aside for performance evaluation, in order to test how well the system can handle previously unseen material. The remainder of this section provides some background

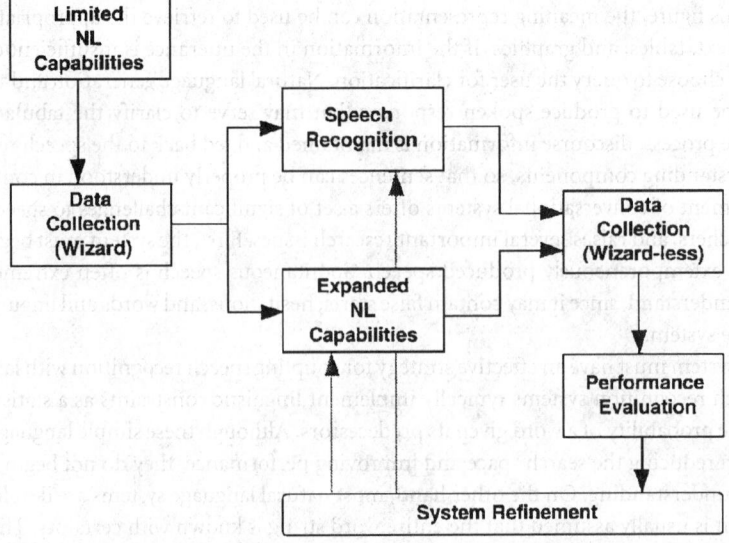

FIGURE 64.3 An illustration of the spoken language system development cycle.

information on speech recognition and language understanding components, as well as the data collection and performance evaluation procedures.

64.2.2 Data Collection

Development of spoken language systems is driven by the availability of representative training data, capturing how potential users of a system would want to talk to it. For this reason, data collection and evaluation have been important areas of research focus. Data collection enables application development and training of the recognizer and language understanding systems; evaluation techniques make it possible to compare different approaches and to measure progress. It is difficult to devise a way to collect realistic data reflecting how a user would use a spoken language system when there is no such system; indeed, the data are needed in order to *build* the system.

Most researchers in the field have now adopted an approach to data collection which uses a system in the loop to facilitate data collection and provide realistic data. At first, some limited natural language understanding capabilities are developed for the particular application. In early stages, the data are collected in a simulation mode, where the speech recognition component is replaced by an expert typist. An experimenter in a separate room types in the utterances spoken by the subject, typically after removing false starts and hesitations. The natural language component then translates the typed input into a query to the database, returning a display to the user, perhaps along with a verbal response clarifying what is being shown. In this way, data collection and system development are combined into a single tightly coupled cycle. Since only a transcriber is needed, not an expert wizard, this approach is quite cost effective, allowing data collection to begin quite early in the application development process, and permitting realistic data to be collected (see Figure 64.4.). As system development progresses, the simulated portions of the system can be replaced with their real counterparts, ultimately resulting in stand-alone data collection, yielding data that accurately reflect the way system would be used in practice.

Since the subjects brought in for data collection are not true users with clear goals, it is critical to provide a mechanism to help them focus their dialogue with the computer. A popular approach is to devise a set of short scenarios for them to solve. These are necessarily artificial, and the exact wording of the sentences in the scenarios often has a profound influence on the subjects' choices of linguistic constructs. An alternative is to allow the subjects complete freedom to design their own scenarios. This is perhaps somewhat more realistic, but subjects may wander from topic to topic because of a lack of a clearly defined problem.

As the system's dialogue model evolves, data previously collected can become somewhat obsolete, since the users' utterances are markedly influenced by the computer feedback. Hence it is problematic to achieve advances in the dialogue model without suffering from temporary inadequacies in recognition, until the system can bootstrap from new releases of training material.

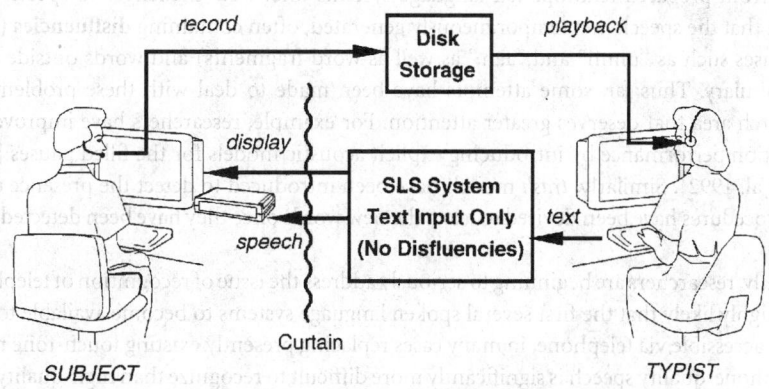

FIGURE 64.4   A person-behind-the-curtain, or wizard, paradigm for data collection.

64.2.3 Speech Recognition

The past decade has witnessed unprecedented progress in speech recognition technology. Word error rates continue to drop by a factor of 2 every two years while barriers to speaker independence, continuous speech, and large vocabularies have all but fallen. There are several factors that have contributed to this rapid progress. First, there is the coming of age of the stochastic modeling techniques known as **hidden Markov modeling (HMM)**. HMM is a doubly stochastic model, in which the generation of the underlying phoneme string and its surface acoustic realizations are *both* represented probabilistically as Markov processes [Rabiner 1986]. HMM is powerful in that, with the availability of training data, the parameters of the model can be trained automatically to give optimal performance. The systems typically operate with the support of an *n*-gram **(statistical) language model** and adopt either a **Viterbi** (time-synchronous) or an **A* (best fit) search** strategy. Although the application of HMM to speech recognition began nearly 20 years ago [Jelinek et al. 1974], it was not until the past few years that it has gained wide acceptance in the research community.

Second, much effort has gone into the development of large speech corpora for system development, training, and testing [Zue et al. 1990, Hirschman et al. 1992]. Some of these corpora are designed for acoustic phonetic research, whereas others are highly task specific. Nowadays, it is not uncommon to have tens of thousands of sentences available for system training and testing. These corpora permit researchers to quantify the acoustic cues important for phonetic contrasts and to determine parameters of the recognizers in a statistically meaningful way.

Third, progress has been brought about by the establishment of standards for performance evaluation. Only a decade ago, researchers trained and tested their systems using locally collected data and had not been very careful in delineating training and testing sets. As a result, it was very difficult to compare performance across systems, and the system's performance typically degraded when it was presented with previously unseen data. The recent availability of a large body of data in the public domain, coupled with the specification of evaluation standards [Pallett et al. 1994], has resulted in uniform documentation of test results, thus contributing to greater reliability in monitoring progress.

Finally, advances in computer technology have also indirectly influenced our progress. The availability of fast computers with inexpensive mass storage capabilities has enabled researchers to run many large-scale experiments in a short amount of time. This means that the elapsed time between an idea and its implementation and evaluation is greatly reduced. In fact, speech recognition systems with reasonable performance can now run in real time using high-end workstations without additional hardware — a feat unimaginable only a few years ago. However, recognition results reported in the literature are usually based on more sophisticated systems that are too computationally intensive to be practical in live interaction. An important research area is to develop more efficient computational methods that can maintain high-recognition accuracy without sacrificing speed.

Historically, speech recognition systems have been developed with the assumption that the speech material is read from prepared text. Spoken language systems offer new challenges to speech recognition technology in that the speech is extemporaneously generated, often containing **disfluencies** (i.e., unfilled and filled pauses such as "umm" and "aah," as well as word fragments) and words outside the system's working vocabulary. Thus far, some attempts have been made to deal with these problems, although this is a research area that deserves greater attention. For example, researchers have improved their system's recognition performance by introducing explicit acoustic models for the filled pauses [Ward 1990, Butzberger et al. 1992]. Similarly, *trash* models have been introduced to detect the presence of unknown words, and procedures have been devised to learn the new words once they have been detected [Asadi et al. 1991].

Most recently, researchers are beginning to seriously address the issue of recognition of telephone quality speech. It is highly likely that the first several spoken language systems to become available to the general public will be accessible via telephone, in many cases replacing presently existing touch-tone menu driven systems. Telephone-quality speech is significantly more difficult to recognize than high-quality recordings, both because the band-width has been limited to under 3.3 kHz and because noise and distortions are introduced in the line. Furthermore, the background environment could include disruptive sounds such as other people talking or babies crying.

64.2.4 Language Understanding

Natural language analysis has traditionally been predominantly syntax driven — a complete syntactic analysis is performed which attempts to account for *all* words in an utterance. However, when working with spoken material, researchers quickly came to realize that such an approach [Bobrow et al. 1990, Seneff 1992b], although providing some linguistic constraints to the speech recognition component and a useful structure for further linguistic analysis, can break down dramatically in the presence of unknown words, novel linguistic constructs, recognition errors, and spontaneous speech events such as false starts. Spoken language tends to be quite informal; people are perfectly capable of speaking, and willing to accept, sentences that are agrammatical.

Due to these problems, many researchers have tended to favor more semantic-driven approaches, at least for spoken language tasks in limited domains. In such approaches, a meaning representation or **semantic frame** is derived by *spotting* key words and phrases in the utterance [Ward 1990]. Although this approach loses the constraint provided by syntax, and may not be able to adequately interpret complex linguistic constructs, the need to accommodate spontaneous speech input has outweighed these potential shortcomings. At the present time, almost all viable systems have abandoned their original goal of achieving a complete syntactic analysis of every input sentence, favoring a more robust strategy that can still answer when a full parse is not achieved [Jackson et al. 1991, Seneff 1992a, Stallard and Bobrow 1992]. This can be achieved by identifying parsable phrases and clauses, and providing a separate mechanism for gluing them together to form a complete meaning analysis [Seneff 1992a]. Ideally, the parser includes a probabilistic framework with a smooth transition to parsing fragments when full linguistic analysis is not achievable. Examples of systems that incorporate such *stochastic* modeling techniques can be found in Pieraccini et al. [1992] and Miller et al. [1994].

64.2.5 Speech Recognition/Natural Language Integration

One of the critical research issues in the development of spoken language systems is the mechanism by which the speech recognition component interacts with the natural language component in order to obtain the correct meaning representation. At present, the most popular strategy is the so-called N-**best interface** [Soong and Huang 1990], in which the recognizer can propose its best N complete sentence hypotheses* one by one, stopping with the first sentence that is successfully analyzed by the natural language component. In this case, the natural language component acts as a filter on *whole sentence* hypotheses. However, it is still necessary to provide the recognizer with an inexpensive language model that can partially constrain the theories. Usually, a statistical language model such as a bigram is used, in which every word in the lexicon is assigned a probability reflecting its likelihood in following a given word.

In the N-best interface, a natural language component filters hypotheses that span the entire utterance. Frequently, many of the candidate sentences differ minimally in regions where the acoustic information is not very robust. Although confusions such as "an" and "and" are acoustically reasonable, one of them can often be eliminated on linguistic grounds. In fact, many of the top N sentence hypotheses could have been eliminated before reaching the end if syntactic and semantic analyses had taken place early on in the search. One possible control strategy, therefore, is for the speech recognition and natural language components to be tightly coupled, so that only the acoustically promising hypotheses that are linguistically meaningful are advanced. For example, partial theories are arranged on a stack, prioritized by score. The most promising partial theories are extended using the natural language component as a predictor of all possible next-word candidates; any other word hypotheses are not allowed to proceed. Therefore, any theory that completes

---

*N is a parameter of the system that can be set arbitrarily as a compromise between accuracy and computation.

is guaranteed to parse. Researchers are beginning to find that such a tightly coupled integration strategy can achieve higher performance than an *N*-best interface, often with a considerably smaller stack size [Goodine et al. 1991, Goddeau 1992, Moore et al. 1995]. The future is likely to see increasing instances of systems making use of linguistic analysis at early stages in the recognition process.

64.2.6 Discourse and Dialogue

Human verbal communication is a two-way process involving multiple, active participants. Mutual understanding is through direct and indirect speech acts, turn taking, clarification, and pragmatic considerations. An effective spoken language interface for information retrieval and interactive transactions must incorporate extensive and complex **dialogue modeling**: initiating appropriate clarification subdialogues based on partial understanding, and taking an active role in directing the conversation toward a valid conclusion. Although there has been some theoretical work on the structure of human–human dialogue [Grosz and Sidner 1990], this has not yet led to effective insights for building human–machine interactive systems.

Systems can maintain an active or a passive role in the dialogue, and each of these extremes has advantages and disadvantages. An extreme case is a system which asks a series of prescribed questions, and requires the user to answer each question in turn before moving on. This is analogous to the interactive voice response systems that are now available via the touch-tone telephone, and users are usually annoyed by their inflexibility. At the opposite extreme is a system that never asks any questions or gives any unsolicited advice. In such cases the user may feel uncertain as to what capabilities exist, and may, as a consequence, wander quite far from the domain of competence of the system, leading to great frustration because nothing is understood. Researchers are still experimenting with setting an important balance between these two extremes in managing the dialogue.

It is absolutely essential that a system be able to interpret a user's queries in context. For instance, if the user says, "I want to go from Boston to Denver," followed with, "show me only United flights," they clearly do not want to see *all* United flights, but rather just the ones that fly from Boston to Denver. The ability to inherit information from preceding sentences is particularly helpful in the face of recognition errors. The user may have asked a complex question involving several restrictions, and the recognizer may have misunderstood a single word, such as a flight number or an arrival time. If a good context model exists, the user can now utter a very short correction phrase, and the system will insert the correction for the misunderstood word correctly, preventing the user from having to reutter the entire sentence, running the risk of further recognition errors.

At this point, it is probably educational to give an example of a real dialogue between a spoken language system and a human. For this purpose, we have selected the Pegasus system, a system developed at Massachusetts Institute of Technology (MIT), which is capable of helping a user make flight plans [Zue et al. 1994]. Pegasus connects, via a modem over the telephone line, to the Eaasy Sabre flight database, offered by American Airlines. As a consequence, users can make explicit flight reservations on real flights using Pegasus Figure 64.5 and figure 64.6 contain an example of the log of an actual round-trip booking to illustrate the system's capability. This dialogue shows examples where the system asks directed questions, cases where a great deal of context information is carried over from one query to the next, "please wait" requests where the system is warning the user of possible delays, and instances where the system provides additional information that was not explicitly requested, such as the ticket summary.

64.2.7 Evaluation

Over the past few years, the speech recognition and natural language communities have placed strong emphasis on performance evaluation. With the use of common corpora and metrics, the relative merits of different approaches and systems can be assessed, using both subjective and objective measures.

Spoken language systems can be evaluated along several dimensions. First, the *accuracy* of the system and its various modules can be documented. Thus, for example, a given system's phonetic, word, and sentence accuracy can be measured, as well as linguistic and task completion accuracy. Second, the *coverage*

```
USER:      I WANT TO GO FROM BOSTON TO SAN FRANCISCO
PEGASUS:   What date will you be travelling on?
USER:      OCTOBER TWENTIETH
PEGASUS:   Here are the flights from Boston to San Francisco on Wednesday October
           twentieth.
              FLIGHT  FROM  TO   DEPARTURE    ARRIVAL    NSTOPS
           1  DL  565  BOS  SFO   7:00 A.M.   11:55 A.M.   1
           2  US   31  BOS  SFO   7:00 A.M.   12:03 P.M.   1
           3  NW  181  BOS  SFO   7:30 A.M.   10:58 A.M.   0
           4  UA   31  BOS  SFO   8:10 A.M.   11:23 A.M.   0
           5  NW  353  BOS  SFO   8:15 A.M.    1:18 P.M.   1
           6  UA   59  BOS  SFO   9:05 A.M.   12:34 P.M.   0
              . . .
           You can ask for more flights.
USER:      I'D LIKE A NONSTOP UNITED FLIGHT
PEGASUS:   Here are the non-stop United flights from Boston to San Francisco on
           Wednesday October twentieth.
              FLIGHT  FROM  TO   DEPARTURE    ARRIVAL    NSTOPS
           1  UA   31  BOS  SFO   8:10 A.M.   11:23 A.M.   0
           2  UA   59  BOS  SFO   9:05 A.M.   12:34 P.M.   0
           3  UA   93  BOS  SFO   5:10 P.M.    8:23 P.M.   0
              . . .
USER:      BOOK THE THIRD FLIGHT PLEASE
PEGASUS:   Please wait while I determine what fares are available for this flight.
           The following fares are available.  Which fare class would you like?
              ONE WAY   FARE CLASS       MIN-STAY  MAX-STAY  RETURN REQ'D
           1   $ 214.   THRIFT DISCOUNT    YES       30         YES
           2   $ 234.   COACH DISCOUNT     YES       30         YES
           3   $ 294.   THRIFT DISCOUNT    YES       NO         YES
           4   $ 304.   COACH DISCOUNT     YES       30         YES
              . . .
USER:      COACH DISCOUNT
PEGASUS:   The following fares are available.  Which fare class would you like?
              ONE WAY   FARE CLASS       MIN-STAY  MAX-STAY  RETURN REQ'D
           1   $ 234.   COACH DISCOUNT     YES       30         YES
           2   $ 304.   COACH DISCOUNT     YES       30         YES
           3   $ 336.   COACH DISCOUNT     31        NO         YES
              . . .
```

FIGURE 64.5 An example of an actual verbal booking dialogue using Pegasus. Due to space limitations, irrelevant parts of the system's responses have been omitted.

and *habitability* of the system can be evaluated. Coverage can be measured with respect to the lexicon, the language model, and the application back-end. Third, the system's *flexibility* can be established. For example, how easy is it to add new knowledge to the system? How easy is it for a naive user to learn how to communicate with the system? How difficult is it for researchers, and, more especially, for nonexperts, to port the system to a different application or language? Finally, the *efficiency* of the system can be assessed. One such measure may be the task completion time.

Whether one wants to evaluate the accuracy of a spoken language system in part or as a whole, one must first establish what the *reference* should be. For example, determining word accuracy for speech recognizers requires that the reference string of words first be transcribed. Similarly, assessing the appropriateness of a syntactic parse presupposes that we know what the correct parse is. In some cases, establishing the reference is relatively straightforward and can be done almost objectively. In other cases, such as specifying the correct system response, the process can be highly subjective. For example, should the correct answer to the query, "Do you know of any Chinese restaurants?" be simply, "Yes," or a list of the restaurants that the system knows?

```
USER:    THE CHEAPEST ONE
PEGASUS: I'll book United flight 93 from Boston to San Francisco on Wednesday
         October twentieth for you.  Your reservation has been entered.
         Now I'll help you with the return flight.
         What date will you be returning on?
USER:    ON SUNDAY
PEGASUS: Here are the United flights from San Francisco to Boston on Sunday
         October twenty-fourth.
             FLIGHT  FROM TO  DEPARTURE   ARRIVAL  NSTOPS
           1 UA 92   SFO  BOS  7:45 A.M.   4:06 P.M.  0
           2 UA 20   SFO  BOS  1:30 P.M.   9:54 P.M.  0
           3 UA 90   SFO  BOS 10:10 P.M.   6:32 A.M.  0
             ...
USER:    SHOW ME NONSTOP FLIGHTS LEAVING IN THE MORNING
PEGASUS: Here is United flight 92 from San Francisco to Boston on Sunday October
         twenty-fourth.
             FLIGHT  FROM TO  DEPARTURE   ARRIVAL  NSTOPS
           1 UA 92   SFO  BOS  7:45 A.M.   4:06 P.M.  0
USER:    I'LL TAKE THIS FLIGHT
PEGASUS: Please wait while I determine what fares are available for this flight.
         I'll book United flight 92 from San Francisco to Boston on Sunday
         October twenty-fourth for you.  Your reservation has been entered.
         Give me a moment to confirm your reservation.
         Here's your ticket summary.
         You saved a total of 894 dollars over regular coach fare.
         Based on fares in effect today, your airline tickets will cost:
         Each adult:   488.00 USD
         Fare codes:   QE7ONQ, QE7PNQ
         Adults:       1
         Total fare:   488.00 USD
         ... Please wait while I retrieve your itinerary...
                                              Price:   488.00
                                              Total:   488.00
         ------------------------------------------------------------
             Date    Flight   Class From  To    Leave   Arrive
           1 OCT20   UA 93    Q     BOS   SFO   510P    823P
           2 OCT24   UA 92    Q     SFO   BOS   745A    406P
         ------------------------------------------------------------
         Can I help you with something else?
```

FIGURE 64.6   Continuation of the example shown in Figure 64.5.

The outputs of the system modules naturally become more general at the higher levels of the system, since these outputs represent more abstract information. Unfortunately, this makes an automatic comparison with a reference output more difficult, both because the *correct* response may become more ambiguous and because the output representation must become more flexible. The added flexibility that is necessary to express more general concepts also allows a given concept to be expressed in many ways, making the comparison with a reference more difficult.

Objective evaluation of spoken language systems comes with large overhead costs, particularly if it is applied on a common evaluation dataset across a wide community. Researchers must first agree on formal definitions of the *correct* answers, which becomes particularly problematic when discourse context may lead to ambiguities. If a pool of systems are to be evaluated on a common set of dialogues, then the systems used to collect the dialogues must be extremely passive, since users' responses to system queries may not be interpretable in the absence of knowledge about the collection system's half of the conversation. Yet, if systems are not evaluated on common data, it becomes impossible to make objective comparisons of their performance.

A possible alternative is to utilize more subjective evaluations, where an evaluator examines a prior dialogue between a subject and a computer, and decides whether each exchange in the dialogue was effective. A small set of categories, such as correct, incorrect, partially correct, and out of domain, can be used to tabulate statistics on the performance. If the scenario comes with a single unique correct answer, then it is also straightforward to measure how many times users solved their problem successfully, as well as how long it took them to do so. The time is rapidly approaching when real systems will be accessible to the general public via the telephone line, and so the ultimate evaluation will be successful active use of such systems in the real world.

64.3   Best Practices

Spoken language systems are a relatively new technology, having first come into existence in the late 1980s. Prior to that time, computer processing and memory limitations precluded the possibility of real-time speech recognition making it difficult for researchers to conceive of interactive human computer dialogues. All of the systems focus within a narrowly defined area of expertise, and vocabulary sizes are generally limited to under 3000 words. Nowadays, these systems can typically run in real time on standard workstations with no additional hardware.

During the late 1980s, two major government-funded efforts involving multiple sites on two continents provided the momentum to thrust spoken language systems into a highly visible and exciting success story, at least within the computer speech research community. The two programs were the Esprit speech understanding and dialog (SUNDIAL) program in Europe [Peckham 1992] and the Advanced Research Projects Agency (ARPA) spoken language understanding program in the U.S. These two programs were remarkably parallel in that both involved database access for travel planning, with the European one including both flight and train schedules, and the American one being restricted to air travel. The European program was a multilingual effort involving four languages (English, French, German, and Italian), whereas the American effort was, understandably, restricted to English.

64.3.1   The Advanced Research Projects Agency Spoken Language System (SLS) Project

64.3.1.1   The Air Travel Information Service (ATIS) Common Task

The spoken language systems (SLS) program sponsored by ARPA of the Department of Defense in the U.S. has provided major impetus for spoken language system development. In particular, the program adopted the approach of developing the underlying technologies within a common domain called Air Travel Information Service (ATIS) [Price 1990]. ATIS permits users to verbally query for air travel information, such as flight schedules from one city to another, obtained from a small **relational database** excised from the Official Airline Guide. By requiring that all system developers use the same database, it has been possible to compare the performance of various spoken language systems based on their ability to extract the correct information from the database, using a set of prescribed training and test data, and a set of interpretation guidelines. Indeed, periodic common evaluations have occurred at regular intervals, and steady performance improvements have been observed for all systems. Figure 64.7 shows the error rates for the best ATIS systems, measured in several dimensions over the past four years. Many of the systems currently run in real time on high-end workstations with no additional hardware, although with some performance degradation.

As shown in Figure 64.7, the speech recognition performance has improved steadily over the past four years. Word error rate (WE) decreased by more than eightfold, while sentence error rate (SE) decreased more than fourfold in this period. In both cases, the reduction in error rate for spontaneous speech has followed the trend set forth for read speech, namely, halving the error every two years. In the most recent formal evaluation of the ARPA-SLS program in the ATIS domain, the best system achieved a word error rate of 2.3% and a sentence error rate of 15.2% [Pallett et al. 1994]. The vocabulary size was more than 2500 words, and the bigram and trigram language models had a **perplexity** of about 20 and 14, respectively.

TABLE 64.1   Examples Illustrating Particularly Difficult Sentences Within the ATIS Domain That Systems Are Capable of Handling

---

GIVE ME A FLIGHT FROM MEMPHIS TO LAS VEGAS AND NEW YORK CITY TO LAS VEGAS ON SUNDAY THAT ARRIVE AT THE SAME TIME

I WOULD LIKE A LIST OF THE ROUND TRIP FLIGHTS BETWEEN INDIANAPOLIS AND ORLANDO ON THE TWENTY SEVENTH OR THE TWENTY EIGHTH OF DECEMBER

I WANT A ROUND TRIP TICKET FROM PHOENIX TO SALT LAKE CITY AND BACK. I WOULD LIKE THE FLIGHT FROM PHOENIX TO SALT LAKE CITY TO BE THE EARLIEST FLIGHT IN THE MORNING AND THE FLIGHT FROM SALT LAKE CITY TO PHOENIX TO BE THE LATEST FLIGHT IN THE AFTERNOON.

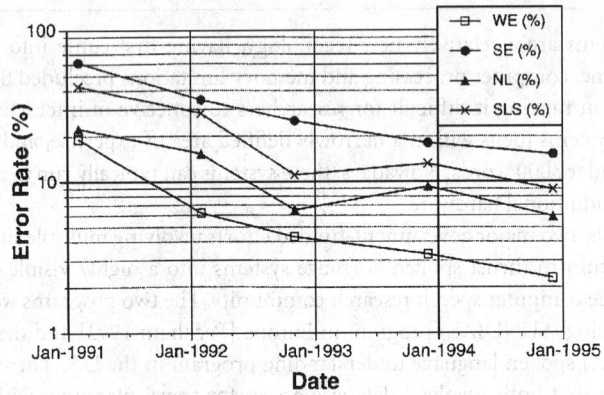

FIGURE 64.7   Best performance achieved by systems in the ATIS domain over the past four years. See text for a detailed description.

Note that all of the performance results quoted in this section are for the so-called *evaluable* queries, i.e., those queries that are within the ATIS domain and for which an appropriate answer is available from the database.

The ARPA-SLS community has carefully defined a common answer specification (CAS) evaluation protocol, whereby a system's performance is determined by comparing its output, expressed as a set of database tuples, with one or more predetermined reference answers [Bates et al. 1991]. The CAS protocol has the advantage that system evaluation can be carried out automatically, once the principles for generating the reference answers have been established and a corpus has been annotated accordingly. Since direct comparison across systems can be performed relatively easily with this procedure, the community has been able to achieve cross fertilization of research ideas, leading to rapid research progress. Figure 64.7 shows that language understanding error rate (NL) has declined by more than threefold in the past four years.* This error rate is measured by passing the transcription of the spoken input, after removing partial words, through the natural language component. In the most recent formal evaluation in the ATIS domain, the best natural language system achieved an understanding error rate of only 5.9% on all the evaluable sentences in the test set [Pallett et al. 1994]. Table 64.1 contains several examples of relatively complex sentences that some of the NL systems being evaluated are able to handle.

The performance of the entire spoken language system can be assessed using the same CAS protocol for the natural language component, except with speech rather than text as input. Figure 64.7 shows that this speech understanding error rate (SLS) has fallen from 42.6% to 8.9% over the four-year interval.

---

*The error rate for both text (NL) and speech (SLS) input increased somewhat in the 1993 evaluation. This was largely due to the fact that the database was increased from 11 cities to 46 that year, and some of the travel-planning scenarios used to collect the newer data were considerably more difficult.

It is interesting to note that this error rate is considerably less than the sentence recognition error rate, suggesting that a large number of sentences can be understood even though the transcription may contain errors.

64.3.1.2 Other Advanced Research Projects Agency Projects

The major ARPA sites are Carnegie Mellon University (CMU), MIT, SRI International, and Bolt, Beranek, and Newman (BBN) Systems and Technology. Most of these sites have developed other interesting spoken language systems besides the ATIS task. These systems are at varying stages of completion, and it is fair to say that none of them are as yet sufficiently robust to be deployable. However, each one has revealed interesting new research areas where gaps remain in our understanding of how to build truly practical spoken language systems. Active ongoing research in these domains should provide interesting new developments in the future.

Probably the most impressive early system was the MIT Voyager system, first assembled for English dialogues in 1989 [Zue et al. 1989]. Voyager can engage in verbal dialogues with users about a restricted geographical region within Cambridge, Massachusetts, in the U.S. The system can provide information about distances, travel times, or directions between landmarks located within this area (e.g., restaurants, hotels, banks, libraries, etc.) as well as handling specific requests for information such as address, phone number, or location on the map. Voyager has remained under active development since 1989, particularly in the dimension of multilingual capabilities. Voyager can now help a user solve ground travel planning problems in three languages: English, Japanese, and Italian [Glass et al. 1995]. The process of porting to new languages has motivated researchers to redesign the system so that language-dependent aspects can be contained in external tables and rules.

The CMU office manager system is designed to provide users with voice access to a set of application programs such as a calendar, an on-line Rolodex, a voice mail system, and a calculator for the office of the future [Rudnicky et al. 1991]. For speech recognition the office manager relies on a version of the Sphinx system [Lee 1989]. Language understanding support is provided either by a finite-state language model or a frame-based representation that extracts the meaning of an utterance by spotting key words and phrases [Ward 1990].

In early 1991, researchers at BBN demonstrated a spoken language interface [Bates et al. 1991] to the dynamic analysis and replanning tool (DART), which is a system for military logistical transportation planning. A spoken language interface to DART has the potential advantage of reducing task completion time by allowing the user to access information more efficiently and naturally than would be possible with the keyboard and mouse buttons. Their spoken language system demonstration centers around the task of database query and information retrieval by combining the BYBLOS speech recognition system [Chow et al. 1987] and the Delphi natural language system [Bates et al. 1990] for speech understanding.

An interesting realistic system, which grew out of the ARPA ATIS effort, is the previously mentioned Pegasus system, which is an extension of an existing ATIS system. The distinguishing feature of Pegasus is that it connects via a modem over the phone line to a real flight reservation system. The system has knowledge of flights to and from some 220 cities worldwide. Pegasus has a fairly extensive dialogue model to help it cope with difficult problems such as date restrictions imposed by discount fares or aborted flight plans due to selections being sold out. A *displayless* version of Pegasus is under development, which will ultimately enable users to make flight reservations by speaking with a computer over the telephone [Seneff et al. 1992].

In 1994, researchers at MIT started the development of Galaxy [Goddeau et al. 1994], a system that enables universal information access using spoken dialogue. Galaxy differs from current spoken language systems in a number of ways. First, it is distributed and decentralized: Galaxy uses a client-server architecture to allow sharing of computationally expensive processes (such as large vocabulary speech recognition), as well as knowledge intensive processes. Second, it is multidomain, intended to provide access to a wide variety of information sources and services while insulating the user from the details of database location and format. It is presently connected to many real, on-line databases, including the National Weather Services, the NYNEX Electronic Yellow Pages, and the World Wide Web. Users can query Galaxy in natural

English (e.g., "what is the weather forecast for Miami tomorrow," "how many hotels are there in Boston," and "do you have any information on Switzerland," etc.), and receive verbal and visual responses. Finally, it is extensible; new knowledge domain servers can be added to the system incrementally.

64.3.2 The SUNDIAL Program

Whereas the ARPA ATIS program in the U.S. emphasized competition through periodic common evaluations, the European SUNDIAL program [Peckham 1992] promoted cooperation and plug compatibility by requiring different sites to contribute distinct components to a single multisite system. Another significant difference was that the European program made dialogue modeling an integral and important part of the research program, whereas the American program was focused more strictly on speech understanding, minimizing the effort devoted to usability considerations. The common evaluations carried out in America led to an important breakthrough in forcing researchers to devise robust **parsing** techniques that could makes some sense out of even the most garbled spoken input. At the same time, the emphasis on dialogue in Europe led to some interesting advances in dialogue control mechanisms.

Although the SUNDIAL program formally terminated in 1993, some of the systems it spawned have continued to flourish under other funding resources. Most notable is the Philips Automatic Train Timetable Information System, which is probably the foremost real system in existence today [Eckert et al. 1993]. This system operates in a displayless mode and thus is capable of communicating with the user solely by voice. As a consequence, it is accessible from any household in Germany via the telephone line. The system is presently under field trial, and has been actively promoted through German press releases in order to encourage people to try it. Data are continuously collected from the callers, and can then be used directly to improve system performance. The system runs on a UNIX workstation, and has a vocabulary of 1800 words, 1200 of which are distinct railway station names. The dialogue relies heavily on confirmation requests to permit correction of recognition errors, but the overall success rate for usage is remarkably high.

64.3.3 Other Systems

There are a few other spoken language systems that fall outside of the ARPA ATIS and Esprit SUNDIAL efforts. A notable system is the Berkeley restaurant project (BeRP) [Jurafsky et al. 1994], which acts as a restaurant guide in the Berkeley area. This system is currently distinguished by its neural networks-based recognizer and its probabilistic natural language system. Another novel emergent system is the Waxholm system, being developed by researchers at KTH in Sweden [Blomberg et al. 1993]. Waxholm provides timetables for ferries in the Stockholm archipelago as well as port locations, hotels, camping sites, and restaurants that can be found on the islands. The Waxholm developers are designing a flexible, easily controlled dialogue module based on a scripting language that describes dialogue flow.

64.4 Research Issues and Summary

As we can see, significant progress has been made over the past few years in research and development of systems that can understand spoken language. To meet the challenges of developing a language-based interface to help users solve real problems, however, we must continue to improve the core technologies while expanding the scope of the underlying Human Language Technology (HLT) base. In this section, we outline some of the new research challenges that have heretofore received little attention.

64.4.1 Working in Real Domains

The rapid technological progress that we are witnessing raises several timely questions. When will this technology be available for productive use? What technological barriers still exist that will prevent large-scale HLT deployment? An effective strategy for answering these questions is to develop the underlying technologies within *real* applications, rather than relying on mockups, however realistic they might be, since this will force us to confront some of the critical technical issues that may otherwise elude our attention. Consider, for example, the task of accessing information in the Yellow Pages of a medium-sized metropolitan area such as Boston, a task that can be viewed as a logical extension of the Voyager system developed at MIT. The vocabulary size of such a task could easily exceed 100,000, considering the names of the establishments, street and city names, and listing headings. A task involving such a huge vocabulary presents a set of new technical challenges. Among them are:

- How can adequate acoustic and language models be determined when there is little hope of obtaining a sufficient amount of domain-specific data for training?
- What search strategy would be appropriate for very large vocabulary tasks? How can natural language constraints be utilized to reduce the search space while providing adequate coverage?
- How can the application be adapted and/or customized to the specific needs of a given user?
- How can the system be efficiently ported to a different task in the same domain (e.g., changing the geographical area from Boston to Washington D.C.), or to an entirely different domain (e.g., library information access)?

There are many other research issues that will surface when one is confronted with the need to make human language technology truly useful for solving real problems, some of which will be described in the remainder of this section. Aside from providing the technological impetus, however, working within real domains also has some practical benefits. While years may pass before we can develop unconstrained spoken language systems, we are fast approaching a time when systems with limited capabilities can help users interact with computers with greater ease and efficiency. Working on real applications thus has the potential benefit of shortening the interval between technology demonstration and its ultimate use. Besides, applications that can help people solve problems *will* be used by real users, thus providing us with a rich and continuing source of useful data.

64.4.2 The New Word Problem

Yet another important issue concerns unknown words. The traditional approach to spoken language recognition and understanding research and development is to define the working vocabulary based on domain-specific corpora [Hetherington and Zue 1991]. However, experience has shown that, no matter how large the size of the training corpora, the system will invariably encounter previously unseen words. This is illustrated in Figure 64.8. For the ATIS task, for example, a 100,000-word training corpus will yield a vocabulary of about 1,000 words. However, the probability of the system encountering an unknown word, is about 0.002. Assuming that an average sentence contains 10 words, this would mean that approximately 1 in 50 sentences will contain an unknown word.

In a *real* domain such as Electronic Yellow Pages, a much larger fraction of the words uttered by users will not be in the system's working vocabulary. This is unavoidable partly because it is not possible to anticipate all of the words that all users are likely to use, and partly because the database is usually changing with time (e.g., new restaurants opening up). In the past, we have not paid much attention to the unknown word problem because the tasks we have chosen assume a closed vocabulary. In the limited cases where the vocabulary has been open, unknown words have accounted for a small fraction of the word tokens in the test corpus. Thus researchers could either construct generic *trash word* models and hope for the best, or ignore the unknown word problem altogether and accept a small penalty on word error rate. In real applications, however, the system must be able to cope with unknown words simply because they will always be present, and ignoring them will not satisfy the user's needs; if a person wants to know how to go from MIT to Lucia's restaurant, they will not settle for a response such as, "I am sorry I don't understand you. Please rephrase the question." The system must be able not only to *detect* new words, taking into account acoustic, phonological, and linguistic evidence, but also to adaptively *acquire* them, both in terms of their orthography and linguistic properties. In some cases, fundamental changes in the problem formulation and search strategy may be necessary.

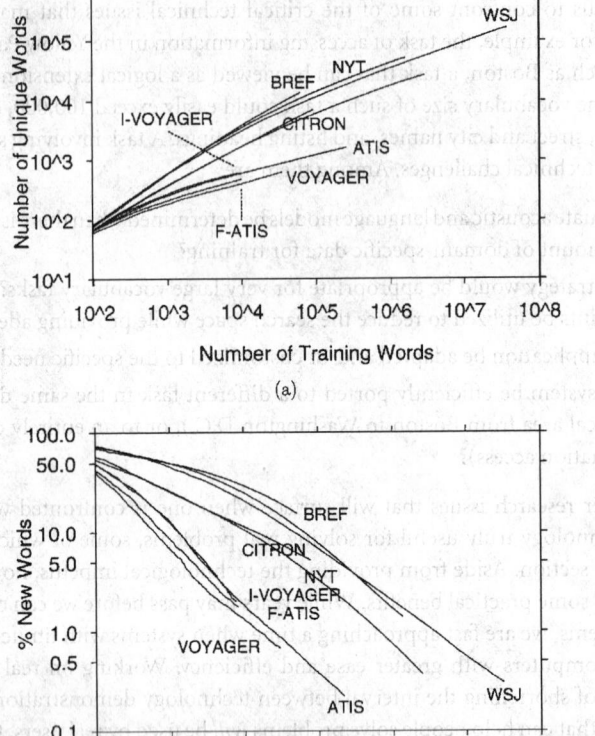

FIGURE 64.8   (a) The number of unique words (i.e., task vocabulary) as a function of the size of the training corpora, for several spoken language tasks and (b) the percentage of unknown words in previously unseen data as a function of the size of the training corpora used to determine the vocabulary empirically. The sources of the data are: F-ATIS = French ATIS, I-VOYAGER = Italian Voyager, BREF = French *La Monde*, NYT = *New York Times*, WSJ = *Wall Street Journal*, and CITRON = Directory Assistance.

64.4.3   Spoken Language Generation

With few exceptions [Zue et al. 1989, 1994], current research in spoken language systems has focused on the input side, i.e., the understanding of the input queries, rather than the *conveyance* of the information.

Spoken language generation is an extremely important aspect of the human–computer interface problem, especially if the transactions are to be conducted over a telephone. Models and methods must be developed that will generate natural sentences appropriate for spoken output, across many domains and languages [Glass et al. 1994]. In many cases, particular attention must be paid to the interaction between language generation and dialogue management; the system may have to initiate clarification dialogue to reduce the amount of information returned from the backend, in order not to generate unwieldy verbal responses. On the speech side, we must continue to improve speech synthesis capabilities, particularly with regard to the encoding of prosodic and paralinguistic information such as emotion and mood. As is the case on the input side, we must also develop integration strategies for language generation and speech synthesis. Finally, evaluation methodologies for spoken language generation technology must be developed, and comparative evaluation performed.

64.4.4 Portability

Currently, the development of speech recognition and language understanding technologies has been domain specific, requiring a large amount of annotated training data. However, it may be costly, or even impossible, to collect a large amount of training data for certain applications, such as Yellow Pages.

Therefore, we must address the problems of producing a spoken language system in a new domain given at most a small amount of domain-specific training data. To achieve this goal, we must strive to cleanly separate the algorithmic aspects of the system from the application-specific aspects. We must also develop automatic or semiautomatic methods for acquiring the acoustic models, language models, grammars, semantic structures for language understanding, and dialogue models required by a new application. The issue of portability spans across different acoustic environments, databases, knowledge domains, and languages. Real deployment of spoken language technology cannot take place without adequately addressing this issue.

Defining Terms

A* (best first) search: A search strategy for speech recognition in which the theories are prioritized by score, and the best scoring theory is incrementally advanced and returned to the stack. An estimated future score is included to normalize theories. The search is admissible if the estimated future score is an *upper-bound* estimate, in which case it can be guaranteed that the overall best-scoring theory will arrive at the end first.

Conversational system: A computer system that is able to carry on a spoken dialogue with a user in order to solve some problem. Usually there is a database of information that the user is attempting to access, and it may involve explicit goals such as making a reservation.

Dialogue modeling: The part of a conversational system that is concerned with interacting with the user in an effective way. This includes planning what to say next and keeping track of the state of completion of a task such as form filling. Important considerations are the ability to offer help at certain critical points in the dialogue or to recover gracefully from recognition errors. A good dialogue model can help tremendously to improve the usability of the system.

Discourse modeling: The part of a conversational system that is concerned with interpreting user queries in context. Often information that was mentioned earlier must be retained in interpreting a new query. The obvious cases are pronominal reference such as *it* or *this one*, but there are many difficult cases where inheritance is only implicit.

Disfluencies (false starts): Portions of a spoken sentence that are not fluent language. These can include false starts (a word or phrase that is abruptly ended prior to being fully uttered, and then verbally replaced with an alternative form), filled pauses (such as "umm" and "er"), or agrammatical constructs due to a changed plan midstream. Dysfluencies are particularly problematic for recognition systems.

Hidden Markov modeling (HMM): A very prevalent recognition framework that begins with an observation sequence derived from an acoustic waveform, and searches through a sequence of states, each of which has a set of hidden observation probabilities and a set of state transition probabilities, to seek an optimal solution. A distinguished *begin* state starts it off, and a distinguished *end* state concludes the search. In recognition, each phoneme is typically associated with an explicit state transition matrix, and each word is encoded as a sequence of specific phonemes. In some cases, phonological pronunciation rules may expand a word's phonetic realization into a set of alternate choices.

Language generation: The process of generating a well-formed expression in English (or some other language) that conveys appropriate information to a user based on diverse sources such as a database, a user query, a partially completed electronic form, and a discourse context (narrow definition for conversational systems).

Natural language understanding: The process of converting an utterance (text string) into a meaning representation (e.g., semantic frame).

N-best interface: An interface between a speech recognition system and a natural language system in which the recognizer proposes N whole-sentence hypotheses, and the NL system selects the most plausible alternative from among the N theories. In an alternative tightly coupled mode, the NL system is allowed to influence partial theories during the initial recognizer search.

n-Gram (statistical) language models: A powerful mechanism for providing linguistic constraint to a speech recognizer. The models specify the set of follow words with associated probabilities, based on the preceding $n - 1$ words. Statistical language models depend on large corpora of training data within the domain to be effective.

Parser: A program that can analyze an input sentence into a hierarchical structure (a parse tree) according to a set of prescribed rules (a grammar) as an intermediate step toward obtaining a meaning representation (semantic frame).

Perplexity: A measure associated with a statistical language model, characterizing the geometric mean of the number of alternative choices at each branching point. Roughly, it indicates the average number of words the recognizer must consider at each decision point.

Relational database: An electronic database in which a collection of tables contain database entries along with sets of attributes, such that the data can be accessed along complex dimensions using the standard query language (SQL). Such databases make it convenient to look up information based on specifications derived from a semantic frame.

Semantic frame: A meaning representation associated with a user query. For very restricted domains it could be a flat structure of (key: value) pairs. Parsers that retain the syntactic structure can produce semantic frames that preserve the clause structure of the sentence.

Speech recognition: The process of converting an acoustic waveform (digitally recorded spoken utterance) into a sequence of hypothesized words (an orthographic transcription).

Text-to-speech synthesis: The process of converting a text string representing a sentence in English (or some other language) into an acoustic waveform that appropriately expresses the phonetics of the text string.

Viterbi search: A search strategy for speech recognition in which all partial theories are advanced lock-stepped in time. Inferior theories are pruned prior to each advance.

Wizard-of-Oz paradigm: A procedure for collecting speech data to be used for training a conversational system in which a human wizard aids the system in answering the subjects' queries. The wizard may simply enter user queries verbatim to the system, eliminating recognition errors, or may play a more active role by extracting appropriate information from the database and formulating canned responses. As the system becomes more fully developed it can play an ever-increasing role in the data collection process, eventually standing alone in a wizardless mode.

References

Asadi, A., Schwartz, R., and Makhoul, J. 1991. Automatic modelling for adding new words to a large vocabulary continuous speech recognition system, pp. 305–308. In *Proc. ICASSP '91*.

Bates, M., Boisen, S., and Makhoul, J. 1990. Developing an evaluation methodology for spoken language systems, pp. 102–108. In *Proc. ARPA Workshop Speech Nat. Lang.*

Bates, M., Ellard, P., and Shaked, V. 1991. Using spoken language to facilitate military transportation planning, pp. 217–220. In *Proc. ARPA Workshop Speech Nat. Lang.* Morgan Kaufmann, San Mateo, CA.

Blomberg, M., Carlson, R., Elenius, K., Granstrom, B., Gustafson, J., Hunnicutt, S., Lindell, R., and Neovius, L. 1993. An experimental dialogue system: Waxholm, pp. 1867–1870. In *Proc. Eurospeech '93*. Berlin, Germany.

Bobrow, R., Ingria, R., and Stallard, R. 1990. Syntactic and semantic knowledge in the DELPHI unification grammar, pp. 230–236. In *Proc. DARPA Speech Nat. Lang. Workshop*.

Butzberger, J., Murveit, H., and Weintraub, M. 1992. Spontaneous speech effects in large vocabulary speech recognition applications, pp. 339–344. In *Proc. ARPA Workshop Speech Nat. Lang.*

Chow, Y. et al. 1987. BYBLOS: the BBN continuous speech recognition system, pp. 89–92. In *Proc. ICASSP.*

Eckert, W., Kuhn, T., Niemann, H., Rieck, S., Scheuer, A., and Schukat-Talamazzini, E. G. 1993. A spoken dialogue system for German intercity train timetable enquiries, pp. 1871–1874. In *Proc. Eurospeech '93.* Berlin, Germany.

Glass, J., Flammia, G., Goodine D., Phillips, M., Polifroni, J., Sakai, S., Seneff, S., and Zue, V. 1995. Multilingual spoken-language understanding in the MIT Voyager system. *Speech Commun.* 17:1–18.

Glass, J., Polifroni, J., and Seneff, S. 1994. Multilingual language generation across multiple domains, pp. 983–976. In *ICSLP '94.*

Goddeau, D. 1992. Using probabilistic shift-reduce parsing in speech recognition systems, pp. 321–324. In *Proc. Int. Conf. Spoken Lang. Process.*

Goddeau, D., Brill, E., Glass, J., Pao, C., Phillips, M., Polifroni, J., Seneff, S., and Zue, V. 1994. Galaxy: a human-language interface to on-line travel information, pp. 707–710. In *Proc. ICSLP '94.*

Goodine, D., Seneff, S., Hirschman, L., and Philips, M. 1991. Full integration of speech and language understanding in the MIT spoken language system, pp. 845–848. In *Proc. Eurospeech.*

Grosz, B. and Sidner, C. 1990. Plans for discourse. In *Intentions in Communication.* MIT Press, Cambridge, MA.

Hetherington, I. L. and Zue, V. 1991. New words: implications for continuous speech recognition. In *Proc. Eur. Conf. Speech Commun. Tech.* Berlin, Germany.

Hirschman, L. et al. 1992. Multi-site data collection for a spoken language corpus, pp. 903–906. In *Proc. Int. Conf. Spoken Lang. Process.*

Jackson, E., Appelt, D., Bear, J., Moore, R., and Podlozny, A. 1991. A template matcher for robust NL interpretation, pp. 190–194. In *Proc. DARPA Speech Nat. Lang. Workshop.*

Jelinek, F., Bahl, L., and Mercer, R. 1974. Design of a linguistic decoder for the recognition of continuous speech, pp. 255–266. In *Proc. IEEE Symp. Speech Recognition.*

Jurafsky, D., Wooters, C., Tajchman, G., Segal, J., Stolcke, A., Fosler, E., and Morgan, N. 1994. The Berkeley restaurant project, pp. 2139–2142. In *Proc. ICSLP '94.* Yokohama, Japan.

Lee, K. F. 1989. *Automatic Speech Recognition: The Development of the SPHINX System.* Kluwer Academic, Boston, MA.

Miller, S., Schwartz, R., Bobrow, R., and Ingria, R. 1994. Statistical language processing using hidden understanding models. *Proc. ARPA Speech Nat. Lang. Workshop.*

Moore, R., Appelt, D., Dowding, J., Gawron, J., and Moran, D. 1995. Combining linguistic and statistical knowledge sources in natural-language processing for ATIS, pp. 261–264. In *Proc. ARPA Spoken Language Systems Workshop.* Austin, TX.

Pallett, D., Fiscus, J., Fisher, W., Garafolo, J., Lund, B., Martin, A., and Pryzbocki, M. 1994. Benchmark tests for the ARPA spoken language program, pp. 5–36. In *Proc. ARPA Spoken Lang. Sys. Tech. Workshop.* Austin, TX.

Peckham, J. 1992. A new generation of spoken dialogue systems: results and lessons from the SUNDIAL project, pp. 33–40. In *Proc. Eurospeech.*

Pieraccini, R., Levin, E., and Lee, C. H. 1992. Stochastic representation of conceptual structure in the ATIS task, pp. 121–124. In *Proc. DARPA Speech Nat. Lang. Workshop.*

Price, P. 1990. Evaluation of spoken language systems: the ATIS domain, pp. 91–95. In *Proc. DARPA Speech Nat. Lang. Workshop.*

Rabiner, L. R. 1986. A tutorial on hidden Markov models and selected applications in speech recognition, pp. 257–285. In *Proc. IEEE* 77(2), February.

Rudnicky, A., Lunati, J.-M., and Franz, A. 1991. Spoken language recognition in an office management domain, pp. 829–832. In *Proc. ICASSP.*

Seneff, S. 1992a. Robust parsing for spoken language systems, pp. 189–192. In *Proc. ICASSP.*

Seneff, S. 1992b. TINA: A natural language system for spoken language applications. *Comput. Linguistics* 18(1):61–86.

Seneff, S., Meng, H., and Zue, V. 1992. Language modelling for recognition and understanding using layered bigrams, pp. 317–320. In *Proc. Int. Conf. Spoken Lang. Process.*

Seneff, S., Zue, V., Polifroni, J., Pao, C., Hetherington, L., Goddeau, D., and Glass, J. 1995. The preliminary development of a displayless Pegasus system, pp. 212–217. In *Proc. ARPA Spoken Lang. Tech. Workshop*. Austin, TX.

Soong, F. and Huang, E. 1990. A tree-trellis based fast search for finding the *N*-best sentence hypotheses in continuous speech recognition, pp. 199–202. In *Proc. ARPA Workshop Speech Nat. Lang.*

Stallard, D. and Bobrow, R. 1992. Fragment processing in the DELPHI system, pp. 305–310. In *Proc. DARPA Speech Nat. Lang. Workshop.*

Ward, W. 1989. Modelling non-verbal sounds for speech recognition, pp. 47–50. In *Proc. DARPA Workshop Speech Nat. Lang.*

Ward, W. 1990. The CMU air travel information service: understanding spontaneous speech, pp. 127–129. In *Proc. ARPA Workshop Speech Nat. Lang.* Morgan Kaufmann, San Mateo, CA.

Zue, V., Glass, J., Goodine, D., Leung, H., Phillips, M., Polifroni, J., and Seneff, S. 1989. The Voyager speech understanding system: a progress report, pp. 160–167. In *Proc. DARPA Speech Nat. Lang. Workshop.*

Zue, V., Seneff, S., and Glass, J. 1990. Speech database development at MIT: TIMIT and beyond. *Speech Commun.* 9(4):351–356.

Zue, V., Seneff, S., Polifroni, J., Phillips, M., Pao, C., Goddeau, D., Glass, J., and Brill, E. 1994. Pegasus: a spoken language interface for on-line air travel planning. *Speech Commun.* 15:331–340.

Further Information

Fundamentals of Speech Recognition, by Larry Rabiner and Bing-Huang Juang (Prentice–Hall, Englewood Cliffs, NJ, 1993) provides a good description of the basic speech recognition technology.

Natural Language Understanding, by James Allen (2nd ed., Benjamin Cummings, 1995) provides a good description of basic natural language technology.

Proceedings of ICASSP, Proceedings of Eurospeech, Proceedings of ICSLP, and *Proceedings of DARPA Speech and Natural Language Workshop* all provide excellent coverage of state-of-the-art spoken language systems.

65

Decision Trees and Instance-Based Classifiers

65.1 Introduction 65-1
      Attribute-Value Representation
65.2 Decision Trees 65-2
      Method for Constructing Decision Trees • Choosing Tests
      • Overfitting • Missing Attribute Values • Extensions
65.3 Instance-Based Approaches.........................65-10
      Outline of the Method • Similarity Metric, or Measuring
      Closeness • Choosing Instances to Remember • How Many
      Neighbors? • Irrelevant Attributes
65.4 Composite Classifiers 65-13

J. Ross Quinlan
University of New South Wales

65.1 Introduction

This chapter looks at two of the common learning paradigms used in artificial intelligence (AI), both of which are also well known in statistics. These methods share an approach to learning that is based on exploiting regularities among observations, so that predictions are made on the basis of similar previously encountered situations. The methods differ, however, in the way that similarity is expressed; trees make important shared properties explicit, whereas instance-based approaches equate (dis)similarity with some measure of distance.

65.1.1 Attribute-Value Representation

Decision tree and instance-based methods both represent each **instance** using a collection $\{A_1, A_2, \ldots, A_x\}$ of properties or **attributes**. Attributes are grouped into two broad types: *continuous* attributes have real or integer values, whereas *discrete* attributes have unordered nominal values drawn from a (usually small) set of possibilities defined for that attribute. Each instance also belongs to one of a fixed set of mutually exclusive **classes** $c_1, c_2, \ldots, c_k$. Both families of methods use a **training set** of classified instances to develop a mapping from attribute values to classes; this mapping can then be used to predict the class of a new instance from its attribute values.

Figure 65.1 shows a small collection of instances described in terms of four attributes. Attributes Outlook and Windy are discrete, with possible values {sunny, overcast, rain} and {true, false}, respectively, whereas the other two attributes have numeric values. Each instance belongs to one of the classes yes or no.

Outlook	Temp, –F	Humidity, %	Windy	Class
rain	70	96	false	yes
sunny	80	90	true	no
overcast	64	65	true	yes
sunny	75	70	true	yes
sunny	85	85	false	no
sunny	72	95	false	no
rain	75	80	false	yes
sunny	69	70	false	yes
overcast	83	78	false	yes
rain	65	70	true	no
overcast	72	90	true	yes
overcast	81	75	false	yes
rain	68	80	false	yes
rain	71	80	true	no

FIGURE 65.1 An illustrative training set of instances.

The x attributes define an x-dimensional **description space** in which each instance becomes a point. From this geometrical perspective, both instance-based and decision tree approaches divide the description space into regions, each associated with one of the classes.

65.2 Decision Trees

Methods for generating decision trees were pioneered by Hunt and his co-workers in the 1960s, although their popularity in statistics stems from the independent work of Breiman et al. [1984]. The techniques are embodied in software packages such as CART [Breiman et al. 1984] and C 4.5 [Quinlan 1993].

Decision tree learning systems have been used in numerous industrial applications, particularly diagnosis and control. In one early success, Leech [1986] learned comprehensible trees from data logged from a complex and imperfectly understood uranium sintering process. The trees pointed the way to improved control of the process with substantial gains in throughput and quality. Evans and Fisher [1994] describe the use of decision trees to prevent banding, a problem in high-speed rotogravure printing. The trees are used to predict situations in which banding is likely to occur so that preventive action can be taken, leading to a dramatic reduction in print delays. Several other tree-based applications are discussed in Langley and Simon [1995].

65.2.1 Method for Constructing Decision Trees

Decision trees are constructed by a recursive *divide-and-conquer* algorithm that generates a partition of the data. The tree for set D of instances is formed as follows:

- If D satisfies a specified **stopping criterion**, the tree for D is a **leaf** that identifies the most frequent class among the instances. The most common stopping criterion is that all instances of D belong to the same class, but some systems also stop when D contains very few instances.
- Otherwise, select some **test** T with mutually exclusive outcomes $T_1, T_2, \ldots, T_n$ and let D_i be the subset of D containing those instances with outcome T_i, $1 \leq i \leq n$. The decision tree for D then has T as its root with a subtree for each outcome T_i of T. If D_i is empty, the subtree corresponding to outcome T_i is a leaf that nominates the majority class in D; otherwise, the subtree for T_i is obtained by applying the same procedure to subset D_i of D.

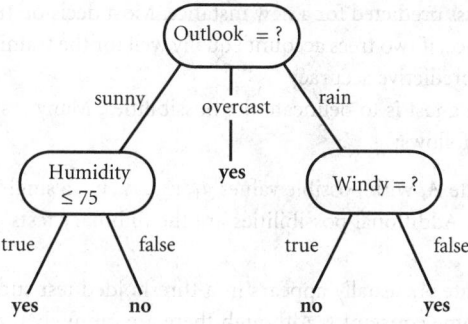

FIGURE 65.2   Decision tree for training instances of Figure 65.1.

Outlook	Temp, –F	Humidity, %	Windy	Class
sunny	75	70	true	yes
sunny	69	70	false	yes
sunny	80	90	true	no
sunny	85	85	false	no
sunny	72	95	false	no
overcast	72	90	true	yes
overcast	83	78	false	yes
overcast	64	65	true	yes
overcast	81	75	false	yes
rain	71	80	true	no
rain	65	70	true	no
rain	75	80	false	yes
rain	68	80	false	yes
rain	70	96	false	yes

FIGURE 65.3   Partition of the instances of Figure 65.1.

In the example of Figure 65.1, the test chosen for the root of the tree might be Outlook = ? with possible outcomes sunny, overcast, and rain. The subset of instances with outcome sunny might then be further subdivided by a test Humidity ≤ 75 with outcomes true and false. All instances with outlook overcast belong to the same class, so no further subdivision would be necessary. The instances with outlook rain might be further divided by a test Windy = ? with outcomes true and false. The resulting decision tree appears in Figure 65.2 and the corresponding partition of the training instances is in Figure 65.3.

The tree provides a mechanism for classifying any instance. Starting at the root, the outcome of the test for that instance is determined and the process continues with the corresponding subtree. When a leaf is encountered, the instance is predicted to belong to the class identified by the leaf. In the preceding example, a new instance

$$\text{Outlook} = \text{sunny}, \text{Temp} = 82, \text{Humidity} = 85, \text{Windy} = \text{true}$$

would follow the outcome sunny, then the outcome false before reaching a leaf labeled no.

65.2.2   Choosing Tests

The example does not explain how tests like Outlook = ? or Humidity ≤ 75 came to be used. Provided that the chosen test T always produces a nontrivial partition of the instances so that no subset D_i contains all of them, the process will terminate. Nevertheless, the choice of T determines the structure of the final tree and so can affect the class predicted for a new instance. Most decision tree systems are biased toward producing compact trees since, if two trees account equally well for the training instances, the simpler tree seems likely to have higher predictive accuracy.

The first step in selecting a test is to delineate the possibilities. Many systems consider only tests that involve a single attribute as follows:

- For a discrete attribute A_i with possible values $v_1, v_2, \ldots, v_m$, a single test $A_i = ?$ with m outcomes could be considered. Additional possibilities are the m binary tests $A_i = v_j$, each with outcomes true and false.

- A continuous attribute A_i usually appears in a thresholded test such as $A_i \leq t$ (with outcomes true and false) for some constant t. Although there are infinitely many possible thresholds t, the number of distinct values of A_i that appear in a set D of instances is at most $|D|$. If these values are sorted into an ascending sequence, say, $n_1 < n_2 < \cdots < n_l$, any value of t in the interval $[n_i, n_{i+1})$ will give the same partition of D, so only one threshold in each interval need be considered.

Most systems carry out an exhaustive comparison of simple tests such as those just described, although more complex tests (see "Extensions" subsections) may be examined heuristically.

Tests are evaluated with respect to some **splitting criterion** that allows the desirability of different tests to be assessed and compared. Such criteria are often based on the class distributions in the set D and subsets $\{D_i\}$ induced by a test. Two examples should illustrate the idea.

65.2.2.1 Gini Index and Impurity Reduction

Breiman et al. [1984] determine the *impurity* of a set of instances from its class distribution. If the relative frequency of instances belonging to class c_j in D is denoted by r_j, $1 \leq j \leq k$, then

$$Gini(D) = 1 - \sum_{j=1}^{k} (r_j)^2$$

The Gini index of a set of instances assumes its minimum value of zero when all instances belong to a single class.

Suppose now that test T partitions D into subsets $D_1, D_2, \ldots, D_n$ as before. The expected reduction in impurity associated with this test is given by

$$Gini(D) - \sum_{i=1}^{n} \frac{|D_i|}{|D|} \times Gini(D_i)$$

whose value is always greater than or equal to zero.

65.2.2.2 Gain Ratio

Criteria such as impurity reduction tend to improve with the number of outcomes n of a test. If possible tests have very different numbers of outcomes, such metrics do not provide a fair basis for comparison.

The gain ratio criterion [Quinlan 1993] is an information-based measure that attempts to allow for different numbers (and different probabilities) of outcomes. The residual uncertainty about the class to which an instance in D belongs can be expressed in a form similar to the preceding Gini index as

$$Info(D) = - \sum_{j=1}^{k} r_j \times \log_2(r_j)$$

and the corresponding information gained by a test T as

$$Info(D) - \sum_{i=1}^{n} \frac{|D_i|}{|D|} \times Info(D_i)$$

Like reduction in impurity, information gain focuses on class distributions. On the other hand, the potential information obtained by partitioning a set of instances is based on knowing the subset D_i into which an instance falls; this *split information* is given by

$$-\sum_{i=1}^{n} \frac{|D_i|}{|D|} \times \log_2 \left(\frac{|D_i|}{|D|} \right)$$

and tends to increase with the number of outcomes of a test. The gain ratio criterion uses the ratio of the information gain of a test T to its split information as the measure of its usefulness.

There have been numerous studies of the behavior of different splitting criteria, e.g., Liu and White [1994]. Some authors, including Breiman et al. [1984], see little operational difference among a broadly defined class of metrics.

65.2.3 Overfitting

Most data collected in practical applications involve some degree of *noise*. Values of continuous attributes are subject to measurement errors, discrete attributes such as color depend on subjective interpretation, instances are misclassified, and mistakes are made in recording.

When the divide-and-conquer algorithm is applied to such data, it often results in very large trees that *fit* the noise in addition to the meaningful structure in the task. The resulting over-elaborate trees are more difficult to understand and generally exhibit degraded predictive accuracy when classifying unseen instances.

Overfitting can be prevented either by restricting the growth of the tree, usually by means of significance tests of one form or another, or by **pruning** back the full tree to an appropriate size. The latter is generally preferred since it allows interactions of tests to be explored before deciding how much structure is justifiable; on the downside, though, growing and then pruning a tree requires more computation. Three common pruning strategies illustrate the idea.

65.2.3.1 Cost-Complexity Pruning

Breiman et al. [1984] describe a two-stage process in which a sequence of trees $Z_0, Z_1, \ldots, Z_z$ is generated, one of which is then selected as the final pruned tree. Consider a decision tree Z used to classify each of the $|D|$ instances in the training set from which it was constructed, and let e of them be misclassified. If $L(Z)$ is the number of leaves in Z, the *cost complexity* of Z is defined as the sum

$$\frac{e}{|D|} + \alpha \times L(Z)$$

for some value of the parameter α. Now, suppose we were to replace a subtree S of Z by a leaf identifying the most frequent class among the instances from which S was constructed. In general, the new tree would misclassify Δe more of the instances in the training set but would contain $L(S) - 1$ fewer leaves. This new tree would have the same cost complexity as Z if

$$\alpha = \frac{\Delta e}{|D| \times (L(S) - 1)}$$

The sequence of trees starts with Z_0 as the original tree. To produce Z_{i+1} from Z_i, each nonleaf subtree of Z_i is examined to find the minimum value of α. All subtrees with that value of α are then replaced by their respective best leaves, and the process continues until the final tree Z_z consists of a single leaf.

If a *pruning set* of instances separate from the training set D is available, a final tree can be selected simply by evaluating each Z_i on this set and picking the most accurate tree. If no pruning set is available, Breiman et al. [1984] employ a strategy based on **cross validation** that allows the **true error rate** of each Z_i to be estimated.

65.2.3.2 Reduced Error Pruning

The previous method considers only some subtrees of the original tree as candidates for the final pruned tree. Reduced error pruning [Quinlan 1987] presumes the existence of a separate pruning set and identifies among all subtrees of the original tree the one with the lowest error on the pruning set. This can be accomplished efficiently as follows.

Every instance in the pruning set is classified by the tree. The method records the number of errors at each leaf and also notes, for each internal node, the number of errors that would be made if that node were to be changed to a leaf. (As with a leaf, the class associated with an internal node is the most frequent class among the instances from which that subtree was constructed.) When all of these error counts have been determined, each internal node is investigated starting from the bottom levels of the tree. The number of errors made by the subtree rooted at that node is compared with the number of errors that would result from changing the node to a leaf and, if the latter is not greater than the former, the change is effected. Since the total number of errors made by a tree is the sum of the errors at its leaves, it is clear that the final subtree minimizes the number of errors on the pruning set.

65.2.3.3 Minimum Description Length Pruning

Rissanen's minimum description length (MDL) principle and Wallace and Boulton's similar minimum message length principle provide a rationale for offsetting fit on the training data against the complexity of the tree. The idea is to encode, as a single message, a theory (such as a tree) derived from training data together with the data given the theory. A complex theory that explains the data well might be expensive to encode, but the second part of the message should then be short. Conversely, a simple theory can be encoded cheaply but will not account for the data as well as a more complex theory, so that the second part of the message will require more bits. These principles advocate choosing a theory to minimize the length of the complete message; under certain measures of error or *loss functions*, this policy can be shown to maximize the probability of the theory given the data.

In this context, the alternative theories are pruned variants of the original tree. The scheme does not require a separate pruning set and is computationally simple, but its performance is sensitive to the encoding schemes used: the method for encoding a tree, for instance, implies different prior probabilities for trees of various shapes and sizes. The details would take us too far afield here, but Quinlan and Rivest [1989] and Wallace and Patrick [1993] discuss coding schemes and present comparative results.

65.2.4 Missing Attribute Values

Another problem often encountered with real-world datasets is that they are rarely complete; some instances do not have a recorded value for every attribute. This can impact decision tree methods at three stages:

- When comparing tests on attributes with different numbers of missing values
- When partitioning a set D on the outcomes of the chosen test, since the outcomes for some instances may not be known
- When classifying an unseen instance whose outcome for a test is again undetermined

These problems are usually handled in one of three ways:

- Filling in missing values. For example, if the value of a discrete attribute is not known, it can be assumed to be that attribute's most frequent value, and a missing numeric value can be replaced by the mean of the known values.
- Estimating test outcomes by some other means. Breiman et al. [1984] define the notion of a *surrogate split* for test T, viz., a test on a different attribute that produces a similar partition of D. When the value of a tested attribute is not known, the best surrogate split whose outcome is known is used to predict the outcome of the original test.
- Treating test outcomes probabilistically. Rather than determining a single outcome, this approach uses the probabilities of the outcomes as determined by their relative frequencies in the training

data. In the task of Figure 65.1, for instance, the probabilities of the outcomes sunny, overcast, and rain for the test Outlook $=$? are 5/14, 4/14, and 5/14, respectively. If the tree of Figure 65.2 is used to classify an instance whose value of Outlook is missing, all three outcomes are explored. The predicted classes associated with each outcome are then combined with the corresponding relative frequencies to give a probability distribution over the classes; this is straightforward, since the outcomes are mutually exclusive. Finally, the class with highest probability is chosen as the predicted class.

The approaches are discussed in more detail in Quinlan [1989] together with comparative trials of different combinations of methods.

65.2.5 Extensions

The previous sections sketch what might be called the fundamentals of constructing and using decision trees. We now look at extensions in various directions aimed at producing trees with higher predictive accuracies on new instances and/or reducing the computation required for learning.

65.2.5.1 More Complex Tests

Many authors have considered ways of enlarging the repertoire of possible tests beyond those set out in the section on choosing tests. More flexible tests allow greater freedom in dividing the description space into regions and so increase the number of classification functions that can be represented as decision trees.

65.2.5.1.1 Subset Tests

If an attribute A_i has numerous discrete values $v_1, v_2, \ldots, v_m$, a test $A_i =$? with one branch for every outcome will divide D into many small subsets. The ability to find meaningful structure in data depends on having sufficient instances to distinguish random and systematic association between attribute values and classes, so this *data fragmentation* generally makes learning more difficult.

One alternative to tests of this form is to group the values of A_i into a small number of subsets $S_1, S_2, \ldots, S_q$ ($q \ll m$), giving a test with outcomes $A_i \in S_j$, $1 \leq j \leq q$. Since there are $\sum_{q=2}^{m-1} q^{m-1}$ possible groupings of values, it is generally impossible to evaluate all of them.

In two-class learning tasks where the values are to be grouped into two subsets, Breiman et al. [1984] give the following algorithm for finding the subsets that optimize convex splitting criteria such as impurity reduction:

- For each value v_j, determine the proportion of instances with this value that belong to one of the classes (the majority class, say)
- Order the values on this proportion, giving $v'_1, v'_2, \ldots, v'_m$
- The optimal subsets are then $\{v'_1, v'_2, \ldots, v'_l\}$ and $\{v'_{l+1}, v'_{l+2}, \ldots, v'_m\}$ for some value of l in the range 1 to $m - 1$

This reduces the number of candidate subsets from 2^{m-1} to $m - 1$ and makes it feasible to find the true optimal grouping. This procedure depends on there being only two classes, but CART extends the idea to multiclass tasks by first assembling the classes themselves into two superclasses and then finding the optimal subsets with respect to the superclasses.

Another approach is to grow the subsets heuristically. C4.5 [Quinlan 1993] starts with each subset containing a single value and iteratively merges subsets. At each stage, the subsets to be merged are chosen so that the gain ratio of the new test is maximized; the process stops when merging any pair of subsets would lead to a lower value. Since this algorithm is based on greedy search, optimality of the final subsets cannot be guaranteed.

65.2.5.1.2 Linear Multiattribute Tests

If tests all involve a single attribute, the resulting regions in the description space are bounded by hyperplanes that are orthogonal to one of the axes. When the real boundaries are not so simple, the divide-and-conquer algorithm will tend to produce complex trees that approximate general boundaries by successions of small axis-orthogonal segments.

One generalization allows tests that involve a linear combination of attribute values, such as

$$w_0 + \sum_{i=1}^{x} w_i \times A_i \leq 0$$

with outcomes true and false. This clearly makes sense only when each attribute A_i has a numeric value. However, a discrete attribute with m values can be replaced by m binary-valued attributes, each having the value 1 when A_i has the particular value and 0 otherwise; when this is done, the linear test can also include multivalued discrete attributes. Systems such as LMDT [Utgoff and Brodley 1991] and OC1 [Murthy et al. 1994] that implement linear tests of this kind have been found to produce smaller trees, often with higher predictive accuracy on unseen instances. Brodley and Utgoff [1995] provide a summary of methods used to find the coefficients $w_0, w_1, \ldots, w_x$ and compare their performance empirically on several real-world datasets.

65.2.5.1.3 Symbolic Multiattribute Tests

One disadvantage of tests that compute a linear combination of attributes is that the tree can become more difficult to understand (although the complexity of the tests must be offset against the smaller overall tree size). Other multiattribute tests that do not use weights suffer less in this respect.

FRINGE [Pagallo and Haussler 1990] uses conjunctions of single-attribute tests. Consider the situation in which an instance belongs to a class if p Boolean conditions are satisfied. A conventional tree using p single-attribute tests would have to partition the training data into $p + 1$ subsets to represent this rule, risking the same problems with data fragmentation mentioned earlier. If a single test consisting of the conjunction of the p tests were used instead, the data would be split into only two subsets. FRINGE finds such conjunctions iteratively, starting with pairs of tests near the leaves of the tree and adding the conjunctions as new attributes for subsequent tree-building stages. In contrast, LFC [Ragavan and Rendell 1993] constructs conjunctive tests directly, in a manner reminiscent of the lookahead employed by the pioneering CLS system [Hunt et al. 1966].

Another form of combination is seen in the *m-of-n* test whose outcome is true if at least m of n single-attribute tests are satisfied and false otherwise. Tests of this kind are commonly used in biomedical domains but are extremely cumbersome to represent as trees with single-attribute tests. ID2-of-3 [Murphy and Pazzani 1991] constructs *m-of-n* tests at each node using a greedy search and often produces smaller trees as a result. Zheng [1995] generalized the idea to an *x-of-n* test that has one outcome for each possible number of conditions that can be satisfied, rather than just two outcomes based on a specified threshold number m of conditions being satisfied. Zheng shows that *x-of-n* tests are easier to construct than their *m-of-n* counterparts and have greater representational power.

65.2.5.2 Multiclass Problems

The effectiveness of decision tree methods is most easily seen in two-class learning tasks when each test contributes to discriminating one class from the other. When there are more than two classes, and especially when classes are very numerous, the goal of a test becomes less clear: should it try to separate one class from all of the others or one group of classes from other groups?

A task with k classes can also be viewed as k two-class tasks, each focusing on distinguishing a single class from all other classes. A separate decision tree can be grown for each class, and a new instance can be classified by looking at the predictions from all k trees. This poses a problem if two or more of the class trees claim the instance, or if none do; the procedure has to be augmented with conflict resolution and default strategies.

A similar idea motivates the *error-correcting output codes* of Dietterich and Bakiri [1995]. With each class is associated a pattern of d binary digits chosen so that the minimum Hamming distance h between any two patterns (i.e., the number of bits in which the patterns differ) is as large as possible. A separate tree is then learned to predict each bit of the class patterns. When a new instance is classified by the d trees, the d output bits may not correspond to the pattern for any class. However, if there are at most $(h - 1)/2$ errors in the output bits, the nearest class pattern will indicate the correct class. Case studies presented by Dietterich and Bakiri demonstrate that this technique can result in a large improvement in classification accuracy for domains with numerous classes.

65.2.5.3 Growing Multiple Trees

An interesting feature of the divide-and-conquer procedure is its sensitivity to the training data. Often two or more attributes will have nearly equal values of the splitting criterion and removing even a single instance from the training set will cause the selected attribute and the associated subtrees to change. (This can be seen as follows: Suppose that omitting one instance typically makes no difference to the learned tree. The correctness or otherwise with which the tree classifies this instance would then be the same whether or not the instance is included in the training set. That is, the **resubstitution error rate** of the tree on the training instances would be the same as the error rate determined by leave-one-out cross validation. However, the resubstitution error rate is known to be an extremely biased measure, substantially underestimating the error rate on new instances, whereas the error rate obtained from leave-one-out cross validation is unbiased.)

The greedy search employed by the divide-and-conquer algorithm will generate only one tree, but a more careful exploration of the space of possible trees will generally uncover many equally appealing trees, compounding the problem of selecting one of them. Buntine [1990] suggests a way around this difficulty that avoids choice altogether! Conceptually, Buntine's idea is to retain all trees that are strong candidates and, when an unseen instance is to be classified, to determine a consensus result by averaging over all predictions. Since it is computationally intractable to explore the complete space of candidate trees, Buntine approximates the process by the use of limited lookahead and by incorporating a small random component in the splitting criterion so that several trees can be constructed from the same set of instances D. Even this constrained search involves substantial increases in computation but, in domains for which classification accuracy is paramount, the additional effort seems justifiable. In a comparison of four methods over 10 learning task domains, Buntine found that the averaging approach using two-ply lookahead gave consistently more accurate predictions on unseen instances. Similarly, Breiman [1996] uses bootstrap samples from the original training set to generate different trees, leading to dramatic improvements on several real-world datasets.

65.2.5.4 Efficiency Issues

Divide-and-conquer is an efficient algorithm, and current decision tree systems require only seconds on a workstation to generate trees from thousands of instances. In some tasks, however, very large numbers of training instances or the need to constantly regrow trees makes even this computational requirement too demanding.

65.2.5.4.1 Determining Thresholds for Continuous Attributes

The most time-consuming operation in growing a tree is finding the possible thresholds for tests on continuous attributes since the values of the attribute that are present in the current set of instances must be sorted, a process of complexity $\Omega(|D| \times \log(|D|))$. This could be avoided, of course, if continuous attributes were thresholded once and for all before growing the tree, thus converting all continuous attributes to discrete attributes. Although this is an active research area, it is clear that continuous attributes will have to be divided into more than just two intervals; papers by Fayyad and Irani [1993] and Van de Merckt [1993] suggest algorithms for finding multiple thresholds.

65.2.5.4.2 Peepholing

Catlett [1991] investigates efficiency of induction from very large datasets. He first demonstrates that speedup cannot be achieved trivially by learning from samples of the data; in the several domains studied, the accuracy of the final classifier is always reduced when a significant fraction of the data is ignored. In Catlett's approach, a small subset of the training data is studied to determine which continuous-valued attributes can be eliminated from contention for the next test and, for the remainder, the interval in which a good threshold might lie. For the small overhead cost of processing the sample, this method allows the learning algorithm to avoid sorting on some attributes altogether and to sort only those values of the candidate attributes that lie within the indicated limits. As a result, the growth of learning time with the number of training instances is very much closer to linear.

65.2.5.4.3 *Incremental Tree Construction*

In some applications the data available for learning grow continually as new information comes to hand. The divide-and-conquer method is a batch-type process that uses all of the training instances to decide questions such as the choice of the next test. When the training set is enlarged, the previous tree must be discarded and the whole process repeated from scratch to generate a new tree. In contrast, Utgoff [1994] has developed incremental tree-growing algorithms that allow the existing tree to be modified as new training data arrive. Two key ideas are the retention of sufficient counting information at each node to determine whether the test at that node must be changed and a method of pulling up a test from somewhere in a subtree to its root. Utgoff's approach carries an interesting guarantee: the revised tree is identical to the tree that would be produced by divide-and-conquer using the enlarged training set.

65.3 Instance-Based Approaches

Although these approaches (usually under the name of *nearest neighbor* methods) have long interested researchers in pattern recognition, their use in the machine learning community has largely dated from Aha's influential work [Aha et al. 1991]. A useful summary of key developments from the perspective of someone outside AI is provided by the introductory chapter of Dasarthy [1991].

65.3.1 Outline of the Method

Recall that, in the geometrical view, attributes define a description space in which each instance is represented by a point. The fundamental assumption that underlies instance-based classification is that nearby instances in the description space will tend to belong to the same class, i.e., that closeness implies similarity. This does not suggest the converse (similarity implies closeness); there is no implicit assumption that instances belonging to a single class will form one cluster in the description space.

Unlike decision tree methods, instance-based approaches do not rely on a symbolic theory formed from the training instances to predict the class of an unseen instance. Instead, some or all of the training instances are remembered and a new instance is classified by finding instances that lie close to it in the description space and taking the most frequent class among them as the predicted class of the new instance. The central questions in this process are as follows:

- How should closeness in the description space be measured?
- Which training instances should be retained?
- How many neighbors should be used when making a prediction?

These are addressed in the following subsections.

65.3.2 Similarity Metric, or Measuring Closeness

65.3.2.1 Continuous Attributes

If all attributes are continuous, as was generally the case in early pattern recognition work [Nilsson 1965], the description space is effectively Euclidean. The square of the distance between two instances P and Q, described by their values for the x attributes ($P = \langle p_1, p_2, \ldots, p_x \rangle$ and $Q = \langle q_1, q_2, \ldots, q_x \rangle$) is

$$d^2(P, Q) = \sum_{i=1}^{x} (p_i - q_i)^2$$

and closeness can be equated with small distance. Alternatively, the attributes can be ascribed weights that reflect their relative magnitudes or importances, giving

$$d_w^2(P, Q) = \sum_{i=1}^{x} w_i^2 \times (p_i - q_i)^2$$

Common choices for weights to normalize magnitudes are as follows:

- $w_i = 1/\text{range}_i$. Here range_i is the difference between the largest and smallest values of attribute A_i observed in the training set.
- $w_i = 1/\text{sd}_i$. Here sd_i is the standard deviation of the values of A_i.

The former has the advantage that differences in values of an individual attribute range from 0 to 1, whereas the latter is particularly useful when attribute A_i is known to have a normal distribution.

65.3.2.2 Discrete Attributes

The difference between unordered values of a discrete attribute is more problematic. The obvious approach is to map the difference $p_i - q_i$ between two values of a discrete attribute A_i to 0 if p_i equals q_i and to 1 otherwise.

Stanfill and Waltz [1986] describe a significant improvement to this two-valued difference that takes account of the similarity of values with respect to the classes. Their *value difference metric* (VDM) first computes a weight for each discrete value of an instance and for each pair of discrete values. Let $n_i(v, c_j)$ denote the number of training instances that have value v for attribute A_i and also belong to class c_j, and let $n_i(v, \cdot)$ denote the sum of these over all classes. An attribute value is important to the extent that it differentiates among the classes. The weight associated with attribute A_i and instance P is taken as

$$w_i(P) = \sqrt{\sum_{j=1}^{k} \left(\frac{n_i(p_i, c_j)}{n_i(p_i, \cdot)} \right)^2}$$

The value difference between p_i and q_i is given by an analogous expression

$$\Delta v_i^2(P, Q) = \sum_{j=1}^{k} \left(\frac{n_i(p_i, c_j)}{n_i(p_i, \cdot)} - \frac{n_i(q_i, c_j)}{n_i(q_i, \cdot)} \right)^2$$

Combining these, the distance between instances P and Q becomes

$$d_{\text{VDM}}(P, Q) = \sum_{i=1}^{x} w_i(P) \times \Delta v_i^2(P, Q)$$

In the task of learning how to pronounce English words, Stanfill and Waltz [1986] found that VDM gave substantially improved performance over simple use of a 0–1 value difference.

Cost and Salzberg [1993] point out that VDM is not symmetric; $d_{\text{VDM}}(P, Q)$ is not generally equal to $d_{\text{VDM}}(Q, P)$ since only the first instance is used to determine the attribute weights. Their *modified value difference metric* (MVDM) drops the attribute weights in favor of an instance weight. They also prefer computing the value difference as the sum of the absolute values of the differences for each class rather that using the square of these differences. In summary,

$$d_{\text{MVDM}}(P, Q) = w(P) \times w(Q) \times \sum_{i=1}^{x} |\Delta v|_i(P, Q)$$

where

$$|\Delta v|_i(P, Q) = \sum_{j=1}^{k} \left| \frac{n_i(p_i, c_j)}{n_i(p_i, \cdot)} - \frac{n_i(q_i, c_j)}{n_i(q_i, \cdot)} \right|$$

The instance weights $w(P)$ and $w(Q)$ depend on their relative success in previous classification trials. If an instance P has been found to be closest to a test instance in t trials, in e of which the test instance belongs to a class different from P, the weight of P is

$$w(P) = \frac{t+1}{t-e+1}$$

This means that instances with a poor track record of classification will have a high weight and so appear to be more distant from (and thus less similar to) an unseen instance.

65.3.2.3 Mixed Continuous and Discrete Attributes

In learning tasks that involve attributes of both types, one strategy to measure distance would be simply to sum the different components as shown earlier, using the weighted square of distance (say) for continuous attributes and the MVDM difference for discrete attributes. Ting [1995] has found that instance-based learners employing nonuniform metrics of this kind have relatively poor performance. His experimental results suggest that it is preferable to convert continuous attributes to discrete attributes using thresholding (as discussed by Fayyad and Irani [1993] or Van de Merckt [1993]) and then to employ a uniform MVDM scheme throughout.

65.3.3 Choosing Instances to Remember

The performance of instance-based methods degrades in the presence of noisy training data. Dasarthy [1991, p. 4] states:

> [Nearest neighbor] classifiers perform best when the training data set is essentially noise free, unlike the other parametric and non-parametric classifiers that perform best when trained in an environment paralleling the operational environment in its noise characteristics.

Performance should improve, then, if noisy training instances are discarded or **edited**. Two approaches to selecting the instances to retain give a flavor of the methods.

IB3 [Aha et al. 1991] starts with training instances arranged in an arbitrary sequence. Each in turn is classified with reference to the (initially empty) pool of retained instances. Those that are classified correctly by the current pool are discarded, whereas misclassified instances are held as potential additions to the pool. Performance statistics for these potential instances are kept and an instance is pooled when a significance test indicates that it would lead to improved classification.

Cameron-Jones [1992] uses an MDL-based approach (see the section on minimum description length pruning). A subset of training instances is chosen heuristically, the goal being to minimize the number of bits in a message specifying the retained instances and the exceptions to the classes that they predict for the training data. This approach usually retains remarkably few instances and yet leads to excellent predictive accuracy.

65.3.4 How Many Neighbors?

Most instance-based approaches use a fixed number of neighbors when classifying a new instance. The size of the neighborhood is important for good classification performance: if it is too small, predictions will be unduly sensitive to the presence of misclassified training instances, whereas too large a value will cause regions of the description space containing fewer exemplars to be merged with surrounding regions. The number of neighbors is usually odd so as to minimize problems with tied class frequencies. Popular choices are one (e.g., Cost and Salzberg [1993]), three, five, and even more (Stanfill and Waltz [1986]).

It is also possible to determine an appropriate number of neighbors from the training instances themselves. A leave-one-out cross validation is performed: each instance in turn is classified using the remaining instances with various neighborhood sizes. The number of neighbors that gives the least number of errors over all instances is then chosen.

65.3.5 Irrelevant Attributes

Instance-based approaches are *parallel* classifiers that use the values of all attributes for each prediction, in contrast with *sequential* classifiers like decision trees that use only a subset of the attributes in each prediction [Quinlan 1994]. When some of the attributes are irrelevant, a random element is introduced to the measurement of distance between instances. Consequently, the performance of instance-based methods can degrade sharply in tasks that have many irrelevant attributes, whereas decision trees are more robust in this respect.

Techniques like MVDM go a long way toward relieving this problem. If a discrete attribute A_i is not related to the instances's classes, the ratio $n_i(v, c_j)/n_i(v, \cdot)$ should not change much for different attribute values v, so that $|\Delta v|_i$ should be close to zero. As a result, the contribution of A_i to the distance calculation should be slight, so that irrelevant attributes are effectively ignored.

Irrelevant attributes can also be excluded more directly by finding the subset of attributes that gives the highest accuracy on a leave-one-out cross validation. There are, of course, $2^x - 1$ nonempty subsets of x attributes, a number that can be too large to investigate if x is greater than 20 or so. Moore and Lee [1994] describe techniques called *racing* and *schemata search* that increase the efficiency of exploring large combinatorial spaces like this. The essence of racing is that competitive subsets are investigated in parallel and a subset is eliminated as soon as it becomes unlikely to win. Schemata search allows subsets of attributes to be described stochastically, using values 0, 1, and * to indicate whether each attribute is definitely excluded, definitely included, or included with probability 0.5. As it becomes clear that subsets including (or excluding) an attribute are performing better, the asterisks for this attribute are resolved in remaining schemata to 1 or 0, respectively.

65.4 Composite Classifiers

This short discussion of decision trees and instance-based methods should not leave the impression that they are solved problems; both are the subject of considerable research. One of the more interesting areas concerns the use of multiple approaches in classifier design. This is motivated by the observation that, even within a single task, there are likely to be regions of the description space in which one or another type of classifier has the edge. For example, instance-based approaches support more general region boundaries than the axis-orthogonal hyperplanes constructed by decision trees. In regions whose true boundaries are complex, the former should provide better models and so lead to more accurate predictions. Conversely, in regions where some attributes are irrelevant, decision trees are likely to prove more robust.

The most general scheme for combining classifiers is *stacking* [Wolpert 1992]. Suppose that y different learning methods are available. For each training instance in turn, y classifiers can be constructed from the remaining instances and used to predict the class of the training instance. This instance thus gives rise to a *first-level* instance with $x + y$ attributes, namely, all of the original attributes plus the y predictions. One of the learning methods can be used with this new dataset; its predictions may employ (selectively) the predictions made by other methods. The process can be repeated to form second-level data with $x + 2y$ attributes, and so on.

In contrast, Brodley [1993] uses hand-crafted rules to decide when a particular classification method is appropriate. One such rule relates to the use of single attribute versus multiattribute tests and can be paraphrased as: If the number of instances is less than the number of attributes, use a single-attribute test, otherwise prefer multiattribute tests.

Finally, Jordan [1994] generalizes the idea of a decision tree to one in which the outcomes of all tests are inherently fuzzy or probabilistic. Constructing a tree then involves not only determining its structure, but also learning a model for estimating the outcome probabilities at each node. Since the latter can involve techniques such as hidden Markov models, the resulting structure is a flexible hybrid.

Acknowledgments

I am most grateful for comments and suggestions from Nitin Indurkhya, Kai Ming Ting, Will Uther, and Zijian Zheng.

Defining Terms

Attribute: A property or feature of all instances. May have *discrete* (nominal) or *continuous* (numeric) values. In statistical terms, an independent variable.

Class: The nominal category to which an instance belongs. The goal of learning is to be able to predict an instance's class from its attribute values. In statistical terms, a dependent variable.

Cross validation: A method for estimating the true error rate of a theory learned from a set of instances. The data are divided into N (e.g., 10) equal-sized groups and, for each group in turn, a theory is learned from the remaining groups and tested on the hold-out group. The estimated true error rate is the total number of test misclassifications divided by the number of instances.

Description space: A conceptual space with one dimension for each attribute. An instance is represented by a point in this space.

Editing: A process of discarding instances from the training set.

Instance: A single observation or datum described by its values of the attributes.

Leaf: A terminal node of a decision tree; has a class label.

Pruning: A process of simplifying a decision tree; each subtree that is judged to add little to the tree's predictive accuracy is replaced by a leaf.

Resubstitution error rate: The misclassification rate of a learned theory on the data from which it was constructed.

Similarity metric: The method used to measure the closeness of two instances in instance-based learning.

Splitting criterion: The basis for selecting one of a set of possible tests.

Stopping criterion: The conditions under which a set of instances is not further subdivided.

Test: An internal node of a decision tree that computes an *outcome* as some function of the attribute values of an instance. A test node is linked to subtrees, one for every possible outcome.

Training set: The collection of instances with known classes that is given to a learning system.

True error rate: The misclassification rate of a theory on unseen instances.

References

Aha, D. W., Kibler, D., and Albert, M. K. 1991. Instance-based learning algorithms. *Machine Learning* 6(1):37–66.

Breiman, L. 1996. Bagging predictors. *Machine Learning* (to appear).

Breiman, L., Friedman, J. H., Olshen, R. A., and Stone, C. J. 1984. *Classification and Regression Trees.* Wadsworth, Belmont, CA.

Brodley, C. E. 1993. Addressing the selective superiority problem: automatic algorithm/model class selection. In *Proc. 10th Int. Conf. Machine Learning*, pp. 17–24. Morgan Kaufmann, San Francisco.

Brodley, C. E. and Utgoff, P. E. 1995. Multivariate decision trees. *Machine Learning* 19(1):45–77.

Buntine, W. L. 1990. *A Theory of Learning Classification Rules.* Ph.D. Thesis. School of Computing Sciences, University of Technology, Sydney, Australia.

Cameron-Jones, R. M. 1992. Minimum description length instance-based learning. In *Proc. 5th Australian J. Conf. Artif. Intelligence*, pp. 368–373. World Scientific, Singapore.

Catlett, J. 1991. *Megainduction.* Ph.D. Thesis. Basser Department of Computer Science, University of Sydney, Australia.

Cost, S. and Salzberg, S. 1993. A weighted nearest-neighbor algorithm for learning with symbolic features. *Machine Learning* 10(1):57–78.

Dasarthy, B. V., ed. 1991. *Nearest Neighbor Norms: NN Pattern Classification Techniques.* IEEE Computer Society Press, Los Alamitos, CA.

Dietterich, T. G. and Bakiri, G. 1995. Solving multiclass learning problems via error correcting output codes. *J. Artif. Intelligence Res.* 2:263–286.

Evans, R. and Fisher, D. 1994. Overcoming process delays with decision tree induction. *IEEE Expert* 9(1):60–66.

Fayyad, U. M. and Irani, K. B. 1993. Multi-interval discretization of continuous-valued attributes for classification learning. In *Proc. 13th Int. J. Conf. Artif. Intelligence,* pp. 1022–1027. Morgan Kaufmann, San Francisco.

Hunt, E. B., Marin, J., and Stone, P. J. 1966. *Experiments in Induction.* Academic Press, New York.

Jordan, M. I. 1994. A statistical approach to decision tree modeling. In *Proc. 11th Int. Conf. on Machine Learning,* pp. 363–370. Morgan Kaufmann, San Francisco.

Langley, P. and Simon, H. A. 1995. Applications of machine learning and rule induction. *Commun. ACM* 38(11):55–64.

Leech, W. J. 1986. A rule based process control method with feedback. In *Proc. Instrument Soc. of Am. Conf.,* pp. 169–175, Houston, TX.

Liu, W. Z. and White, A. P. 1994. The importance of attribute selection measures in decision tree induction. *Machine Learning* 15(1):25–41.

Moore, A. W. and Lee, M. S. 1994. Efficient algorithms for minimizing cross validation error. In *Proc. 11th Int. Conf. Machine Learning,* pp. 190–198. Morgan Kaufmann, San Francisco.

Murphy, P. M. and Pazzani, M. J. 1991. ID2-of-3: constructive induction of M-of-N concepts for discriminators in decision trees. In *Proc. 8th Int. Workshop Machine Learning,* pp. 183–187. Morgan Kaufmann, San Francisco.

Murthy, S. K., Kasif, S., and Salzberg, S. 1994. A system for induction of oblique decision trees. *J. Artif. Intelligence Res.* 2:1–32.

Nilsson, N. J. 1965. *Learning Machines.* McGraw–Hill, New York; 1990. Republished as *The Mathematical Foundations of Learning Machines.* Morgan Kaufmann, San Francisco.

Pagallo, G. and Haussler, D. 1990. Boolean feature discovery in empirical learning. *Machine Learning* 5(1):71–100.

Quinlan, J. R. 1987. Simplifying decision trees. *Int. J. Man-Machine Studies* 27:221–234.

Quinlan, J. R. 1989. Unknown attribute values in induction. In *Proc. 6th Int. Machine Learning Workshop,* pp. 164–168. Morgan Kaufmann, San Francisco.

Quinlan, J. R. 1993. *C4.5: Programs for Machine Learning.* Morgan Kaufmann, San Francisco.

Quinlan, J. R. 1994. Comparing connectionist and symbolic learning methods. In *Computational Learning Theory and Natural Learning Systems.* Vol. 1. S. J. Hanson, G. A. Drastal, and R. L. Rivest, Eds., pp. 445–456. MIT Press, Cambridge, MA.

Quinlan, J. R. and Rivest, R. L. 1989. Inferring decision trees using the minimum description length principle. *Inf. Comput.* 80(3):227–248.

Ragavan, H. and Rendell, L. 1993. Lookahead feature construction for learning hard concepts. In *Proc. 11th Int. Conf. Machine Learning,* pp. 252–259. Morgan Kaufmann, San Francisco.

Stanfill, C. and Waltz, D. 1986. Toward memory-based reasoning. *Commun. ACM* 29(12):1213–1228.

Ting, K. M. 1995. *Common Issues in Instance-Based and Naive Bayesian Classifiers.* Ph.D. Thesis. Basser Department of Computer Science, University of Sydney, Australia.

Utgoff, P. E. 1994. An improved algorithm for incremental induction of decision trees. In *Proc. 11th Int. Conf. Machine Learning,* pp. 318–325. Morgan Kaufmann, San Francisco.

Utgoff, P. E. and Brodley, C. E. 1991. Linear machine decision trees. University of Massachusetts, Amherst. COINS Tech. Rep. 91-10.

Van de Merckt, T. 1993. Decision trees in numerical attribute spaces. In *Proc. 13th Int. J. Conf. Artif. Intelligence,* pp. 1016–1021. Morgan Kaufmann, San Francisco.

Wallace, C. S. and Patrick, J. D. 1993. Coding decision trees. *Machine Learning* 11(1):7–22.

Wolpert, D. H. 1992. Stacked generalization. *Neural Networks* 5:241–259.

Zheng, Z. 1995. Constructing nominal X-of-N attributes. In *Proc. 14th Int. J. Conf. Artif. Intelligence*, pp. 1064–1070. Morgan Kaufmann, San Francisco.

Further Information

The principal computer science journals that report advances in learning techniques are *Machine Learning* (Kluwer), *Artificial Intelligence* (Elsevier), and *Journal of Artificial Intelligence Research*. The latter is an electronic journal; details are available at http://www.cs.washington.edu/research/jair/home.html or from jair-ed@ptolemy.arc.nasa.gov.

Papers on learning techniques are presented at the International Conferences in Machine Learning, the International Joint Conferences on Artificial Intelligence, the AAAI National Conferences on Artificial Intelligence, and the European Conferences on Machine Learning. Applications are not as easy to follow, although the Workshops and Conferences on Knowledge Discovery in Databases have relevant papers.

There are two moderated electronic newsletters that often contain relevant material: the *Machine Learning List* (http://www.ics.uci.edu/~mlearn) and *KDD Nuggets* (http://kddnuggets.com).

66

Neural Networks

66.1 Introduction 66-1
66.2 Representation 66-2
  Density Estimation • Linear Regression and Linear
  Discriminants • Nonlinear Regression and Nonlinear
  Classification • Decision Trees • General Mixture Models
66.3 Learning from Data 66-8
  Likelihood-Based Cost Functions • Gradients of the Cost
  Function • Optimization Algorithms • Hessian Matrices,
  Error Bars, and Pruning • Complexity Control • Bayesian
  Viewpoint • Preprocessing, Invariances, and Prior Knowledge
66.4 Graphical Models 66-16

Michael I. Jordan
University of California at Berkeley

Christopher M. Bishop
Microsoft Research

66.1 Introduction

Within the broad scope of the study of artificial intelligence (AI), research in neural networks is character-ized by a particular focus on pattern recognition and pattern generation. Many neural network methods can be viewed as generalizations of classical pattern-oriented techniques in statistics and the engineering areas of signal processing, system identification, and control theory. As in these parent disciplines, the notion of "pattern" in neural network research is essentially probabilistic and numerical. Neural network methods have had their greatest impact in problems where statistical issues dominate and where data are easily obtained.

A neural network is first and foremost a graph, with patterns represented in terms of numerical values at-tached to the nodes of the graph and transformations between patterns achieved via simple message-passing algorithms. Many neural network architectures, however, are also statistical processors, characterized by making particular probabilistic assumptions about data. As we will see, this conjunction of graphical algorithms and probability theory is not unique to neural networks but characterizes a wider family of probabilistic systems in the form of chains, trees, and networks that are currently being studied throughout AI [Spiegelhalter et al. 1993].

Neural networks have found a wide range of applications, the majority of which are associated with problems in pattern recognition and control theory. In this context, neural networks can best be viewed as a class of algorithms for statistical modeling and prediction. Based on a source of *training data*, the aim is to produce a statistical model of the process from which the data are generated, so as to allow the best predictions to be made for new data. We shall find it convenient to distinguish three broad types of statistical modeling problem, which we shall call **density estimation**, **classification**, and **regression**.

For density estimation problems (also referred to as *unsupervised learning* problems), the goal is to model the unconditional distribution of data described by some vector x. A practical example of the application of density estimation involves the interpretation of X-ray images (mammograms) used for breast cancer screening [Tarassenko 1995]. In this case, the training vectors x form a sample taken from normal (noncancerous) images, and a network model is used to build a representation of the density $p(x)$. When a new input vector x' is presented to the system, a high value for $p(x')$ indicates a normal image, whereas a low value indicates a novel input which might be characteristic of an abnormality. This is used to label regions of images that are unusual, for further examination by an experienced clinician.

For classification and regression problems (often referred to as *supervised learning* problems), we need to distinguish between *input* variables, which we again denote by x, and *target* variables, which we denote by the vector t. Classification problems require that each input vector x be assigned to one of C classes $C_1, \ldots, C_C$, in which case the target variables represent class labels. As an example, consider the problem of recognizing handwritten digits [LeCun et al. 1989]. In this case, the input vector would be some (preprocessed) image of the digit, and the network would have 10 outputs, one for each digit, which can be used to assign input vectors to the appropriate class (as discussed in Section 66.2).

Regression problems involve estimating the values of continuous variables. For example, neural networks have been used as part of the control system for adaptive optics telescopes [Sandler et al. 1991]. The network input x consists of one in-focus and one defocused image of a star and the output t consists of a set of coefficients that describe the phase distortion due to atmospheric turbulence. These output values are then used to make real-time adjustments of the multiple mirror segments to cancel the atmospheric distortion.

Classification and regression problems also can be viewed as special cases of density estimation. The most general and complete description of the data is given by the probability distribution function $p(x, t)$ in the joint input-target space. However, the usual goal is to be able to make good predictions for the target variables when presented with new values of the inputs. In this case, it is convenient to decompose the joint distribution in the form

$$p(x, t) = p(t \mid x) p(x) \tag{66.1}$$

and to consider only the conditional distribution $p(t \mid x)$, in other words the distribution of t *given* the value of x. Thus, classification and regression involve the estimation of *conditional* densities, a problem which has its own idiosyncracies.

The organization of the chapter is as follows. In Section 66.2 we present examples of network representations of unconditional and conditional densities. In Section 66.3 we discuss the problem of adjusting the parameters of these networks to fit them to data. This problem has a number of practical aspects, including the choice of optimization procedure and the method used to control network complexity. We then discuss a broader perspective on probabilistic network models in Section 66.4. The final section presents further information and pointers to the literature.

66.2   Representation

In this section we describe a selection of neural network architectures that have been proposed as representations for unconditional and conditional densities. After a brief discussion of density estimation, we discuss classification and regression, beginning with simple models that illustrate the fundamental ideas and then progressing to more complex architectures. We focus here on representational issues, postponing the problem of learning from data until the following section.

66.2.1   Density Estimation

We begin with a brief discussion of density estimation, utilizing the Gaussian **mixture model** as an illustrative model. We return to more complex density estimation techniques later in the chapter.

Although density estimation can be the main goal of a learning system, as in the diagnosis example mentioned in the Introduction, density estimation models arise more often as components of the solution to a more general classification or regression problem. To return to Equation 66.1, note that the joint density is composed of $p(t \mid x)$, to be handled by classification or regression models, and $p(x)$, the (unconditional) input density. There are several reasons for wanting to form an explicit model of the input density. First,

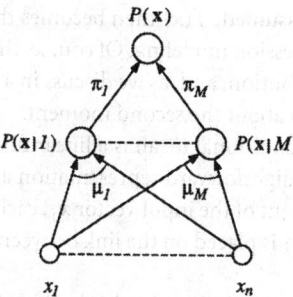

FIGURE 66.1 A network representation of a Gaussian mixture distribution. The input pattern x is represented by numerical values associated with the input nodes in the lower level. Each link has a weight μ_{ij}, which is the jth component of the mean vector for the ith Gaussian. The ith intermediate node contains the covariance matrix Σ_i and calculates the Gaussian conditional probability $p(x \mid i, \mu_i, \Sigma_i)$. These probabilities are weighted by the mixing proportions π_i and the output node calculates the weighted sum $p(x) = \sum_i \pi_i p(x \mid i, \mu_i, \Sigma_i)$.

real-life data sets often have missing components in the input vector. Having a model of the density allows the missing components to be filled in in an intelligent way. This can be useful both for training and for prediction (cf. Bishop [1995]). Second, as we see in Equation 66.1, a model of $p(x)$ makes possible an estimate of the joint probability $p(x, t)$. This in turn provides us with the necessary information to estimate the inverse conditional density $p(x \mid t)$. The calculation of such inverses is important for applications in control and optimization.

A general and flexible approach to density estimation is to treat the density as being composed of a set of M simpler densities. This approach involves modeling the observed data as a sample from a *mixture density*,

$$p(x \mid w) = \sum_{i=1}^{M} \pi_i \, p(x \mid i, w_i) \tag{66.2}$$

where the π_i are constants known as *mixing proportions*, and the $p(x \mid i, w_i)$ are the *component densities*, generally taken to be from a simple parametric family. A common choice of component density is the multivariate Gaussian, in which case the parameters w_i are the means and covariance matrices of each of the components. By varying the means and covariances to place and orient the Gaussians appropriately, a wide variety of high-dimensional, multimodal data can be modeled. This approach to density estimation is essentially a probabilistic form of clustering.

Gaussian mixtures have a representation as a network diagram, as shown in Figure 66.1. The utility of such network representations will become clearer as we proceed; for now, it suffices to note that not only mixture models, but also a wide variety of other classical statistical models for density estimation, are representable as simple networks with one or more layers of adaptive weights. These methods include *principal component analysis*, *canonical correlation analysis*, *kernel density estimation*, and *factor analysis* [Anderson 1984].

66.2.2 Linear Regression and Linear Discriminants

Regression models and classification models both focus on the conditional density $p(t \mid x)$. They differ in that in regression the target vector t is a real-valued vector, whereas in classification t takes its values from a discrete set representing the class labels.

The simplest probabilistic model for regression is one in which t is viewed as the sum of an underlying deterministic function $f(x)$ and a Gaussian random variable ϵ,

$$t = f(x) + \epsilon \tag{66.3}$$

If ϵ has zero mean, as is commonly assumed, $f(x)$ then becomes the *conditional mean* $E(t \mid x)$. It is this function that is the focus of most regression modeling. Of course, the conditional mean describes only the first moment of the conditional distribution, and, as we discuss in a later section, a good regression model will also generally report information about the second moment.

In a linear regression model, the conditional mean is a linear function of x: $E(t \mid x) = Wx$, for a fixed matrix W. Linear regression has a straightforward representation as a network diagram in which the jth input unit represents the jth component of the input vector x_j, each output unit i takes the weighted sum of the input values, and the weight w_{ij} is placed on the link between the jth input unit and the ith output unit.

The conditional mean is also an important function in classification problems, but most of the focus in classification is on a different function known as a **discriminant function**. To see how this function arises and to relate it to the conditional mean, we consider a simple two-class problem in which the target is a simple binary scalar that we now denote by t. The conditional mean $E(t \mid x)$ is equal to the probability that t equals one, and this latter probability can be expanded via Bayes rule

$$p(t = 1 \mid x) = \frac{p(x \mid t = 1)p(t = 1)}{p(x)} \tag{66.4}$$

The density $p(t \mid x)$ in this equation is referred to as the *posterior probability* of the class given the input, and the density $p(x \mid t)$ is referred to as the *class-conditional density*. Continuing the derivation, we expand the denominator and (with some foresight) introduce an exponential,

$$p(t = 1 \mid x) = \frac{p(x \mid t = 1)p(t = 1)}{p(x \mid t = 1)p(t = 1) + p(x \mid t = 0)p(t = 0)} \tag{66.5}$$

$$= \frac{1}{1 + \exp\left\{ -\ln\left[\frac{p(x \mid t=1)}{p(x \mid t=0)}\right] - \ln\left[\frac{p(t=1)}{p(t=0)}\right] \right\}}$$

We see that the posterior probability can be written in the form of the *logistic function*:

$$y = \frac{1}{1 + e^{-z}} \tag{66.6}$$

where z is a function of the likelihood ratio $p(x \mid t = 1)/p(x \mid t = 0)$, and the prior ratio $p(t = 1)/p(t = 0)$. This is a useful representation of the posterior probability if z turns out to be simple.

It is easily verified that if the class conditional densities are multivariate Gaussians with identical covariance matrices, then z is a linear function of x: $z = w^T x + w_0$. Moreover, this representation is appropriate for any distribution in a broad class of densities known as the exponential family (which includes the Gaussian, the Poisson, the gamma, the binomial, and many other densities). All of the densities in this family can be put in the following form:

$$g(x; \theta, \phi) = \exp\{(\theta^T x - b(\theta))/a(\phi) + c(x, \phi)\} \tag{66.7}$$

where θ is the *location parameter* and ϕ is the *scale parameter*. Substituting this general form in Equation 66.5, where θ is allowed to vary between the classes and ϕ is assumed to be constant between classes, we see that z is in all cases a linear function. Thus, the choice of a linear-logistic model is rather robust.

The geometry of the two-class problem is shown in Figure 66.2, which shows Gaussian class-conditional densities, and suggests the logistic form of the posterior probability.

The function z in our analysis is an example of a discriminant function. In general, a discriminant function is any function that can be used to decide on class membership [Duda and Hart 1973]; our analysis has produced a particular form of discriminant function that is an intermediate step in the calculation of a posterior probability. Note that if we set $z = 0$, from the form of the logistic function we obtain a probability of 0.5, which shows that $z = 0$ is a *decision boundary* between the two classes.

The discriminant function that we found for exponential family densities is linear under the given conditions on ϕ. In more general situations, in which the class-conditional densities are more complex

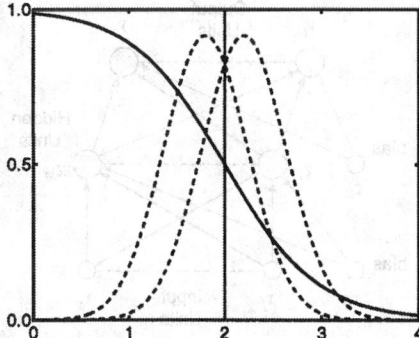

FIGURE 66.2   This shows the Gaussian class-conditional densities $p(x \mid C_1)$ (dashed curves) for a two-class problem in one dimension, together with the corresponding posterior probability $p(C_1 \mid x)$ (solid curve) which takes the form of a logistic sigmoid. The vertical line shows the decision boundary for $y = 0.5$, which coincides with the point at which the two density curves cross.

than a single exponential family density, the posterior probability will not be well characterized by the linear-logistic form. Nonetheless, it still is useful to retain the logistic function and focus on *nonlinear* representations for the function z. This is the approach taken within the neural network field.

To summarize, we have identified two functions that are important for regression and classification, respectively: the conditional mean and the discriminant function. These are the two functions that are of concern for simple linear models and, as we now discuss, for more complex nonlinear models as well.

66.2.3   Nonlinear Regression and Nonlinear Classification

The linear regression and linear discriminant functions introduced in the previous section have the merit of simplicity, but are severely restricted in their representational capabilities. A convenient way to see this is to consider the geometrical interpretation of these models. When viewed in the d-dimensional x-space, the linear regression function $w^T x + w_0$ is constant on hyperplanes which are orthogonal to the vector w. For many practical applications, we need to consider much more general classes of function. We therefore seek representations for nonlinear mappings which can approximate any given mapping to arbitrary accuracy. One way to achieve this is to transform the original x using a set of M nonlinear functions $\phi_j(x)$ where $j = 1, \ldots, M$, and then to form a linear combination of these functions, so that

$$y_k(x) = \sum_j w_{kj} \phi_j(x) \tag{66.8}$$

For a sufficiently large value of M, and for a suitable choice of the $\phi_j(x)$, such a model has the desired universal approximation properties. A familiar example, for the case of one-dimensional input spaces, is the simple polynomial, for which the $\phi_j(x)$ are simply successive powers of x and the w are the polynomial coefficients. Models of the form in Equation 66.8 have the property that they can be expressed as network diagrams in which there is a *single* layer of adaptive weights.

There are a variety of families of functions in one dimension that can approximate any continuous function to arbitrary accuracy. There is, however, an important issue which must be addressed, called the *curse of dimensionality*. If, for example, we consider an Mth-order polynomial then the number of independent coefficients grows as d^M [Bishop 1995]. For a typical medium-scale application with, say, 30 inputs, a fourth-order polynomial (which is still quite restricted in its representational capability) would have over 46,000 adjustable parameters. As we shall see in the section on complexity control, in order to achieve good generalization it is important to have more data points than adaptive parameters in the model, and this is a serious problem for methods that have a power law or exponential growth in the number of parameters.

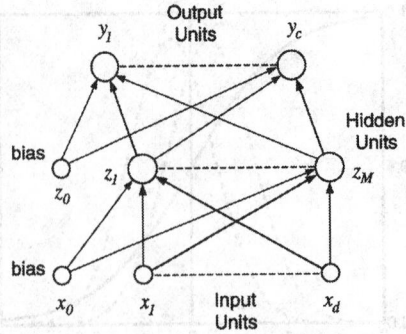

FIGURE 66.3 An example of a feedforward network having two layers of adaptive weights. The bias parameters in the first layer are shown as weights from an extra input having a fixed value of $x_0 = 1$. Similarly, the bias parameters in the second layer are shown as weights from an extra hidden unit, with activation again fixed at $z_0 = 1$.

A solution to the problem lies in the fact that, for most real-world data sets, there are strong (often nonlinear) correlations between the input variables such that the data do not uniformly fill the input space but are effectively confined to a subspace whose dimensionality is called the *intrinsic dimensionality* of the data. We can take advantage of this phenomenon by considering again a model of the form in Equation 66.8 but in which the basis functions $\phi_j(x)$ are *adaptive* so that they themselves contain weight parameters whose values can be adjusted in the light of the observed dataset. Different models result from different choices for the basis functions, and here we consider the two most common examples. The first of these is called the **multilayer perceptron** (MLP) and is obtained by choosing the basis functions to be given by linear-logistic functions Equation 66.6. This leads to a multivariate nonlinear function that can be expressed in the form

$$y_k(x) = \sum_{j=1}^{M} w_{kj} g \left(\sum_{i=1}^{d} w_{ji} x_i + w_{j0} \right) + w_{k0} \tag{66.9}$$

Here w_{j0} and w_{k0} are *bias* parameters, and the basis functions are called *hidden units*. The function $g(\cdot)$ is the logistic sigmoid function of Equation 66.6. This also can be represented as a network diagram as in Figure 66.3. Such a model is able to take account of the intrinsic dimensionality of the data because the first-layer weights w_{ji} can adapt and hence orient the surfaces along which the basis function response is constant. It has been demonstrated that models of this form can approximate to arbitrary accuracy any continuous function, defined on a compact domain, provided the number M of hidden units is sufficiently large. The MLP model can be extended by considering several successive layers of weights. Note that the use of nonlinear activation functions is crucial, because if $g(\cdot)$ in Equation 66.9 was replaced by the identity, the network would reduce to several successive linear transformations, which would itself be linear.

The second common network model is obtained by choosing the basis functions $\phi_j(x)$ in Equation 66.8 to be functions of the radial variable $x - \mu_j$ where μ_j is the *center* of the jth basis function, which gives rise to the **radial basis function (RBF) network** model. The most common example uses Gaussians of the form

$$\phi_j(x) = \exp \left\{ -\frac{1}{2} (x - \mu_j)^T \Sigma_j^{-1} (x - \mu_j) \right\} \tag{66.10}$$

Here both the mean vector μ_j and the covariance matrix Σ_j are considered to be adaptive parameters. The curse of dimensionality is alleviated because the basis functions can be positioned and oriented in input space such as to overlay the regions of high data density and hence to capture the nonlinear correlations between input variables. Indeed, a common approach to training an RBF network is to use a two-stage procedure [Bishop 1995]. In the first stage, the basis function parameters are determined using the input data alone, which corresponds to a density estimation problem using a mixture model in which the component densities are given by the basis functions $\phi_j(x)$. In the second stage, the basis function parameters are frozen and the second-layer weights w_{kj} are found by standard least-squares optimization procedures.

66.2.4 Decision Trees

MLP and RBF networks are often contrasted in terms of the support of the basis functions that compose them. MLP networks are often referred to as "global," given that linear-logistic basis functions are bounded away from zero over a significant fraction of the input space. Accordingly, in an MLP, each input vector generally gives rise to a distributed pattern over the hidden units. RBF networks, on the other hand, are referred to as "local," due to the fact that their Gaussian basis functions typically have support over a local region of the input space. It is important to note, however, that local support does not necessarily mean nonoverlapping support; indeed, there is nothing in the RBF model that prefers basis functions that have nonoverlapping support. A third class of model that does focus on basis functions with nonoverlapping support is the **decision tree** model [Breiman et al. 1984]. A decision tree is a regression or classification model that can be viewed as asking a sequence of questions about the input vector. Each question is implemented as a linear discriminant, and a sequence of questions can be viewed as a recursive partitioning of the input space. All inputs that arrive at a particular leaf of the tree define a polyhedral region in the input space. The collection of such regions can be viewed as a set of basis functions. Associated with each basis function is an output value which (ideally) is close to the average value of the conditional mean (for regression) or discriminant function (for classification; a majority vote is also used). Thus, the decision tree output can be written as a weighted sum of basis functions in the same manner as a layered network.

As this discussion suggests, decision trees and MLP/RBF neural networks are best viewed as being different points along the continuum of models having overlapping or nonoverlapping basis functions. Indeed, as we show in the following section, decision trees can be treated probabilistically as mixture models, and in the mixture approach the sharp discriminant function boundaries of classical decision trees become smoothed, yielding partially overlapping basis functions.

There are tradeoffs associated with the continuum of degree-of-overlap; in particular, nonoverlapping basis functions are generally viewed as being easier to interpret and better able to reject noisy input variables that carry little information about the output. Overlapping basis functions often are viewed as yielding lower variance predictions and as being more robust.

66.2.5 General Mixture Models

The use of mixture models is not restricted to density estimation; rather, the mixture approach can be used quite generally to build complex models out of simple parts. To illustrate, let us consider using mixture models to model a conditional density in the context of a regression or classification problem. A mixture model in this setting is referred to as a "mixtures of experts" model [Jacobs et al. 1991].

Suppose that we have at our disposal an elemental conditional model $p(t \mid x, w)$. Consider a situation in which the conditional mean or discriminant exhibits variation on a local scale that is a good match to our elemental model, but the variation differs in different regions of the input space. We could use a more complex network to try to capture this global variation; alternatively, we might wish to combine local variants of our elemental models in some manner. This can be achieved by defining the following probabilistic mixture:

$$p(t \mid x, w) = \sum_{i=1}^{M} p(i \mid x, v) p(t \mid x, i, w_i) \tag{66.11}$$

Comparing this mixture to the unconditional mixture defined earlier Equation 66.2, we see that both the mixing proportions and the component densities are now conditional densities dependent on the input vector x. The former dependence is particularly important: we now view the mixing proportion $p(i \mid x, v)$ as providing a probabilistic device for choosing different elemental models ("experts") in different regions of the input space. A learning algorithm that chooses values for the parameters v as well as the values for the parameters w_i can be viewed as attempting to find both a good partition of the input space and a good fit to the local models within that partition.

This approach can be extended recursively by considering mixtures of models where each model may itself be a mixture model [Jordan and Jacobs 1994]. Such a recursion can be viewed as providing a probabilistic interpretation for the decision trees discussed in the previous section. We view the decisions in the decision tree as forming a recursive set of probabilistic selections among a set of models. The total probability of target t given input x is the sum across all paths down the tree,

$$p(t \mid x, w) = \sum_{i=1}^{M} p(i \mid x, u) \sum_{j=1}^{M} p(j \mid x, i, v_i) \cdots p(t \mid x, i, j, \ldots, w_{ij} \cdots) \qquad (66.12)$$

where i and j are the decisions made at the first level and second level of the tree, respectively, and $p(t \mid x, i, j, \ldots, w_{ij} \cdots)$ is the elemental model at the leaf of the tree defined by the sequence of decisions. This probabilistic model is a conditional hierarchical mixture. Finding parameter values u, v_i, etc., to fit this model to data can be viewed as finding a nested set of partitions of the input space and fitting a set of local models within the partition.

The mixture model approach can be viewed as a special case of a general methodology known as *learning by committee*. Bishop [1995] provides a discussion of committees; we will also meet them in the section on Bayesian methods later in the chapter.

66.3 Learning from Data

The previous section has provided a selection of models to choose from; we now face the problem of matching these models to data. In principle, the problem is straightforward: given a family of models of interest we attempt to find out how probable each of these models is in the light of the data. We can then select the most probable model [a selection rule known as *maximum a posteriori* (MAP) estimation], or we can select some highly probable subset of models, weighted by their probability (an approach that we discuss in the section on Bayesian methods). In practice, there are a number of problems to solve, beginning with the specification of the family of models of interest. In the simplest case, in which the family can be described as a fixed structure with varying parameters (e.g., the class of feedforward MLPs with a fixed number of hidden units), the learning problem is essentially one of *parameter estimation*. If, on the other hand, the family is not easily viewed as a fixed parametric family (e.g., feedforward MLPs with a variable number of hidden units), then we must solve the *model selection* problem.

In this section we discuss the parameter estimation problem. The goal will be to find MAP estimates of the parameters by maximizing the probability of the parameters given the data $\mathcal{D}$. We compute this probability using Bayes rule,

$$p(w \mid \mathcal{D}) = \frac{p(\mathcal{D} \mid w) p(w)}{p(\mathcal{D})} \qquad (66.13)$$

where we see that to calculate MAP estimates we must maximize the expression in the numerator (the denominator does not depend on w). Equivalently we can minimize the negative logarithm of the numerator. We thus define the following **cost function** $J(w)$:

$$J(w) = -\ln p(\mathcal{D} \mid w) - \ln p(w) \qquad (66.14)$$

which we wish to minimize with respect to the parameters w. The first term in this cost function is a (negative) log **likelihood**. If we assume that the elements in the training set $\mathcal{D}$ are conditionally independent of each other given the parameters, then the likelihood factorizes into a product form. For density estimation we have

$$p(\mathcal{D} \mid w) = \prod_{n=1}^{N} p(x_n \mid w) \tag{66.15}$$

and for classification and regression we have

$$p(\mathcal{D} \mid w) = \prod_{n=1}^{N} p(t_n \mid x_n, w) \tag{66.16}$$

In both cases this yields a log likelihood which is the sum of the log probabilities for each individual data point. For the remainder of this section we will assume this additive form; moreover, we will assume that the log prior probability of the parameters is uniform across the parameters and drop the second term. Thus, we focus on *maximum likelihood* (ML) estimation, where we choose parameter values w_{ML} that maximize $\ln p(\mathcal{D} \mid w)$.

66.3.1 Likelihood-Based Cost Functions

Regression, classification, and density estimation make different probabilistic assumptions about the form of the data and therefore require different cost functions.

Equation 66.3 defines a probabilistic model for regression. The model is a conditional density for the targets t in which the targets are distributed as Gaussian random variables (assuming Gaussian errors ϵ) with mean values $f(x)$. We now write the conditional mean as $f(x, w)$ to make explicit the dependence on the parameters w. Given the training set $\mathcal{D} = \{x_n, t_n\}_{n=1}^{N}$, and given our assumption that the targets t_n are sampled independently (given the inputs x_n and the parameters w), we obtain

$$J(w) = \frac{1}{2} \sum_{n} \|t_n - f(x_n, w)\|^2 \tag{66.17}$$

where we have assumed an identity covariance matrix and dropped those terms that do not depend on the parameters. This cost function is the standard least-squares cost function, which is traditionally used in neural network training for real-valued targets. Minimization of this cost function is typically achieved via some form of gradient optimization, as we discuss in the following section.

Classification problems differ from regression problems in the use of discrete-valued targets, and the likelihood accordingly takes a different form. For binary classification the Bernoulli probability model $p(t \mid x, w) = y^t (1 - y)^{1-t}$ is natural, where we use y to denote the probability $p(t = 1 \mid x, w)$. This model yields the following log likelihood:

$$J(w) = -\sum_{n} [t_n \ln y_n + (1 - t_n) \ln(1 - y_n)] \tag{66.18}$$

which is known as the *cross-entropy* function. It can be minimized using the same generic optimization procedures as are used for least squares.

For multiway classification problems in which there are C categories, where $C > 2$, the multinomial distribution is natural. Define t_n such that its elements $t_{n,i}$ are one or zero according to whether the nth data point belongs to the ith category, and define $y_{n,i}$ to be the network's estimate of the posterior probability of category i for data point n; that is, $y_{n,i} \equiv p(t_{n,i} = 1 \mid x_n, w)$. Given these definitions, we obtain the following cost function:

$$J(w) = -\sum_{n} \sum_{i} t_{n,i} \ln y_{n,i} \tag{66.19}$$

which again has the form of a cross entropy.

We now turn to density estimation as exemplified by Gaussian mixture modeling. The probabilistic model in this case is that given in Equation 66.2. Assuming Gaussian component densities with arbitrary covariance matrices, we obtain the following cost function:

$$J(\mathbf{w}) = -\sum_n \ln \sum_i \pi_i \frac{1}{|\Sigma_i|^{1/2}} \exp\left\{ -\frac{1}{2} (\mathbf{x}_n - \mathbf{\mu}_i)^T \Sigma_i^{-1} (\mathbf{x}_n - \mathbf{\mu}_i) \right\} \tag{66.20}$$

where the parameters $\mathbf{w}$ are the collection of mean vectors $\mathbf{\mu}_i$, the covariance matrices Σ_i, and the mixing proportions π_i. A similar cost function arises for the generalized mixture models [cf. Equation 66.12].

66.3.2   Gradients of the Cost Function

Once we have defined a probabilistic model, obtained a cost function, and found an efficient procedure for calculating the gradient of the cost function, the problem can be handed off to an optimization routine. Before discussing optimization procedures, however, it is useful to examine the form that the gradient takes for the examples that we have discussed in the previous two sections.

The ith output unit in a layered network is endowed with a rule for combining the activations of units in earlier layers, yielding a quantity that we denote by z_i and a function that converts z_i into the output y_i. For regression problems, we assume linear output units such that $y_i = z_i$. For binary classification problems, our earlier discussion showed that a natural output function is the logistic: $y_i = 1/(1 + e^{-z_i})$. For multiway classification, it is possible to generalize the derivation of the logistic function to obtain an analogous representation for the multiway posterior probabilities known as the *softmax function* [cf. Bishop 1995]:

$$y_i = \frac{e^{z_i}}{\sum_k e^{z_k}} \tag{66.21}$$

where y_i represents the posterior probability of category i.

If we now consider the gradient of $J(\mathbf{w})$ with respect to z_i, it turns out that we obtain a single canonical expression of the following form:

$$\frac{\partial J}{\partial \mathbf{w}} = \sum_i (t_i - y_i) \frac{\partial z_i}{\partial \mathbf{w}} \tag{66.22}$$

As discussed by Rumelhart et al. [1995], this form for the gradient is predicted from the theory of generalized linear models [McCullagh and Nelder 1983], where it is shown that the linear, logistic, and softmax functions are (inverse) *canonical links* for the Gaussian, Bernoulli, and multinomial distributions, respectively. Canonical links can be found for all of the distributions in the exponential family, thus providing a solid statistical foundation for handling a wide variety of data formats at the output layer of a network, including counts, time intervals, and rates.

The gradient of the cost function for mixture models has an interesting interpretation. Taking the partial derivative of $J(\mathbf{w})$ in Equation 66.20 with respect to $\mathbf{\mu}_i$, we find

$$\frac{\partial J}{\partial \mathbf{\mu}_i} = \sum_n h_{n,i} \Sigma_i (\mathbf{x}_n - \mathbf{\mu}_i) \tag{66.23}$$

where $h_{n,i}$ is defined as follows:

$$h_{n,i} = \frac{\pi_i |\Sigma_i|^{-1/2} \exp\left\{ -\frac{1}{2} (\mathbf{x}_n - \mathbf{\mu}_i)^T \Sigma_i^{-1} (\mathbf{x}_n - \mathbf{\mu}_i) \right\}}{\sum_k \pi_k |\Sigma_k|^{-1/2} \exp\left\{ -\frac{1}{2} (\mathbf{x}_n - \mathbf{\mu}_k)^T \Sigma_k^{-1} (\mathbf{x}_n - \mathbf{\mu}_k) \right\}} \tag{66.24}$$

When summed over i, the quantity $h_{n,i}$ sums to one and is often viewed as the "responsibility" or "credit" assigned to the ith component for the nth data point. Indeed, interpreting Equation 66.24 using Bayes rule shows that $h_{n,i}$ is the posterior probability that the nth data point is generated by the ith component

Gaussian. A learning algorithm based on this gradient will move the ith mean μ_i toward the data point x_n, with the effective step size proportional to $h_{n,i}$.

The gradient for a mixture model will always take the form of a weighted sum of the gradients associated with the component models, where the weights are the posterior probabilities associated with each of the components. The key computational issue is whether these posterior weights can be computed efficiently. For Gaussian mixture models, the calculation (Equation 66.24) is clearly efficient. For decision trees there is a set of posterior weights associated with each of the nodes in the tree, and a recursion is available that computes the posterior probabilities in an upward sweep [Jordan and Jacobs 1994]. Mixture models in the form of a chain are known as **hidden Markov models**, and the calculation of the relevant posterior probabilities is performed via an efficient algorithm known as the Baum–Welch algorithm.

For general layered network structures, a generic algorithm known as backpropagation is available to calculate gradient vectors [Rumelhart et al. 1986]. Backpropagation is essentially the chain rule of calculus realized as a graphical algorithm. As applied to layered networks it provides a simple and efficient method that calculates a gradient in $O(W)$ time per training pattern, where W is the number of weights.

66.3.3 Optimization Algorithms

By introducing the principle of maximum likelihood in Section 66.1, we have expressed the problem of learning in neural networks in terms of the minimization of a cost function, $J(w)$, which depends on a vector, w, of adaptive parameters. An important aspect of this problem is that the gradient vector $\nabla_w J$ can be evaluated efficiently (for example, by backpropagation). Gradient-based minimization is a standard problem in unconstrained nonlinear optimization for which many powerful techniques have been developed over the years. Such algorithms generally start by making an initial guess for the parameter vector w and then iteratively updating the vector in a sequence of steps,

$$w^{(\tau+1)} = w^{(\tau)} + \Delta w^{(\tau)} \tag{66.25}$$

where τ denotes the step number. The initial parameter vector $w^{(0)}$ is often chosen at random, and the final vector represents a minimum of the cost function at which the gradient vanishes. Because of the nonlinear nature of neural network models, the cost function is generally a highly complicated function of the parameters and may possess many such minima. Different algorithms differ in how the update $\Delta w^{(\tau)}$ is computed.

The simplest such algorithm is called *gradient descent* and involves a parameter update which is proportional to the negative of the cost function gradient $\Delta = -\eta \nabla E$ where η is a fixed constant called the learning rate. It should be stressed that gradient descent is a particularly inefficient optimization algorithm. Various modifications have been proposed, such as the inclusion of a *momentum* term, to try to improve its performance. In fact, much more powerful algorithms are readily available, as described in standard textbooks such as Fletcher [1987]. Two of the best known are called *conjugate gradients* and *quasi-Newton* (or *variable metric*) methods. For the particular case of a sum-of-squares cost function, the *Levenberg–Marquardt* algorithm can also be very effective. Software implementations of these algorithms are widely available.

The algorithms discussed so far are called *batch* since they involve using the whole dataset for each evaluation of the cost function or its gradient. There is also a *stochastic* or *on-line* version of gradient descent in which, for each parameter update, the cost function gradient is evaluated using just one of the training vectors at a time (which are then cycled either in order or in a random sequence). Although this approach fails to make use of the power of sophisticated methods such as conjugate gradients, it can prove effective for very large datasets, particularly if there is significant redundancy in the data.

66.3.4 Hessian Matrices, Error Bars, and Pruning

After a set of weights have been found for a neural network using an optimization procedure, it is often useful to examine second-order properties of the fitted network as captured in the Hessian matrix $H = \partial^2 J / \partial w \, \partial w^T$. Efficient algorithms have been developed to compute the Hessian matrix in time $O(W^2)$

[Bishop 1995]. As in the case of the calculation of the gradient by backpropagation, these algorithms are based on recursive message passing in the network.

One important use of the Hessian matrix lies in the calculation of error bars on the outputs of a network. If we approximate the cost function locally as a quadratic function of the weights (an approximation which is equivalent to making a Gaussian approximation for the log likelihood), then the estimated variance of the ith output y_i can be shown to be

$$\sigma_{y_i}^2 = \left(\frac{\partial y_i}{\partial \mathbf{w}}\right)^T H^{-1} \left(\frac{\partial y_i}{\partial \mathbf{w}}\right) \tag{66.26}$$

where the gradient vector $\partial y_i / \partial \mathbf{w}$ can be calculated via backpropagation.

The Hessian matrix also is useful in pruning algorithms. A pruning algorithm deletes weights from a fitted network to yield a simpler network that may outperform a more complex, **overfitted** network (discussed subsequently) and may be easier to interpret. In this setting, the Hessian is used to approximate the increase in the cost function due to the deletion of a weight. A variety of such pruning algorithms is available [cf. Bishop 1995].

66.3.5   Complexity Control

In previous sections we have introduced a variety of models for representing probability distributions, we have shown how the parameters of the models can be optimized by maximizing the likelihood function, and we have outlined a number of powerful algorithms for performing this minimization. Before we can apply this framework in practice there is one more issue we need to address, which is that of model complexity. Consider the case of a mixture model given by Equation 66.2. The number of input variables will be determined by the particular problem at hand. However, the number M of component densities has yet to be specified. Clearly if M is too small the model will be insufficiently flexible and we will obtain a poor representation of the true density. What is not so obvious is that if M is too large we can also obtain poor results. This effect is known as *overfitting* and arises because we have a dataset of finite size. It is illustrated using the simple example of mixture density estimation in Figure 66.4. Here a set of 100 data points in one dimension has been generated from a distribution consisting of a mixture of two Gaussians (shown by the dashed curves). This dataset has then been fitted by a mixture of M Gaussians by use of the expectation-maximization (EM) algorithm. We see that a model with 1 component ($M = 1$) gives a poor representation of the true distribution from which the data were generated, and in particular is unable to capture the

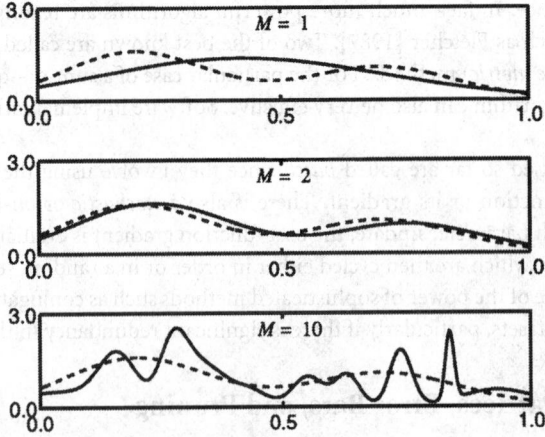

FIGURE 66.4   Effects of model complexity illustrated by modeling a mixture of two Gaussians (shown by the dashed curves) using a mixture of M Gaussians (shown by the solid curves). The results are obtained for 20 cycles of EM.

bimodal aspect. For $M = 2$ the model gives a good fit, as we expect since the data were themselves generated from a two-component Gaussian mixture. However, increasing the number of components to $M = 10$ gives a poorer fit, even though this model contains the simpler models as special cases.

The problem is a very fundamental one and is associated with the fact that we are trying to infer an entire distribution function from a finite number of data points, which is necessarily an ill-posed problem. In regression, for example, there are infinitely many functions which will give a perfect fit to the finite number of data points. If the data are noisy, however, the best generalization will be obtained for a function which does not fit the data perfectly but which captures the underlying function from which the data were generated. By increasing the flexibility of the model, we are able to obtain ever better fits to the training data, and this is reflected in a steadily increasing value for the likelihood function at its maximum. Our goal is to model the true underlying density function from which the data were generated since this allows us to make the best predictions for new data. We see that the best approximation to this density occurs for an intermediate value of M.

The same issue arises in connection with nonlinear regression and classification problems. For example, the number M of hidden units in an MLP network controls the model complexity and must be optimized to give the best generalization. In a practical application, we can train a variety of different models having different complexities, compare their generalization performance using an independent validation set, and then select the model with the best generalization. In fact, the process of optimizing the complexity using a validation set can lead to some partial overfitting to the validation data itself, and so the final performance of the selected model should be confirmed using a third independent data set called a *test* set.

Some theoretical insight into the problem of overfitting can be obtained by decomposing the error into the sum of bias and variance terms [Geman et al. 1992]. A model which is too inflexible is unable to represent the true structure in the underlying density function, and this gives rise to a high bias. Conversely, a model which is too flexible becomes tuned to the specific details of the particular data set and gives a high variance. The best generalization is obtained from the optimum tradeoff of bias against variance.

As we have already remarked, the problem of inferring an entire distribution function from a finite data set is fundamentally ill posed since there are infinitely many solutions. The problem becomes well posed only when some additional constraint is imposed. This constraint might be that we model the data using a network having a limited number of hidden units. Within the range of functions which this model can represent there is then a unique function which best fits the data. Implicitly, we are assuming that the underlying density function from which the data were drawn is relatively smooth. Instead of limiting the number of parameters in the model, we can encourage smoothness more directly using the technique of **regularization**. This involves adding penalty term Ω to the original cost function J to give the total cost function $\tilde{J}$ of the form

$$\tilde{J} = J + \nu\Omega \tag{66.27}$$

where ν is called a regularization coefficient. The network parameters are determined by minimizing $\tilde{J}$, and the value of ν controls the degree of influence of the penalty term Ω. In practice, Ω is typically chosen to encourage smooth functions. The simplest example is called *weight decay* and consists of the sum of the squares of all of the adaptive parameters in the model,

$$\Omega = \sum_i w_i^2 \tag{66.28}$$

Consider the effect of such a term on the MLP function (Equation 66.9). If the weights take very small values then the network outputs become approximately linear functions of the inputs (since the sigmoidal function is approximately linear for small values of its argument). The value of ν in Equation 66.27 controls the effective complexity of the model, so that for large ν the model is oversmoothed (corresponding to high bias), whereas for small ν the model can overfit (corresponding to high variance). We can therefore consider a network with a relatively large number of hidden units and control the effective complexity by changing ν. In practice, a suitable value for ν can be found by seeking the value which gives the best performance on a validation set.

The weight decay regularizer (Equation 66.28) is simple to implement but suffers from a number of limitations. Regularizers used in practice may be more sophisticated and may contain multiple regularization coefficients [Neal 1994].

Regularization methods can be justified within a general theoretical framework known as *structural risk minimization* [Vapnik 1995]. Structural risk minimization provides a quantitative measure of complexity known as the **VC dimension**. The theory shows that the VC dimension predicts the difference between performance on a training set and performance on a test set; thus, the sum of log likelihood and (some function of) VC dimension provides a measure of generalization performance. This motivates regularization methods (Equation 66.27) and provides some insight into possible forms for the regularizer Ω.

66.3.6   Bayesian Viewpoint

In earlier sections we discussed network training in terms of the minimization of a cost function derived from the principle of maximum a posteriori or maximum likelihood estimation. This approach can be seen as a particular approximation to a more fundamental, and more powerful, framework based on Bayesian statistics. In the maximum likelihood approach, the weights w are set to a specific value, w_{ML}, determined by minimization of a cost function. However, we know that there will typically be other minima of the cost function which might give equally good results. Also, weight values close to w_{ML} should give results which are not too different from those of the maximum likelihood weights themselves.

These effects are handled in a natural way in the Bayesian viewpoint, which describes the weights not in terms of a specific set of values but in terms of a probability distribution over all possible values. As discussed earlier (cf. Equation 66.13), once we observe the training dataset $\mathcal{D}$ we can compute the corresponding *posterior* distribution using Bayes' theorem, based on a *prior* distribution function $p(w)$ (which will typically be very broad), and a *likelihood* function $p(\mathcal{D} \mid w)$,

$$p(w \mid \mathcal{D}) = \frac{p(\mathcal{D} \mid w)\,p(w)}{p(\mathcal{D})} \tag{66.29}$$

The likelihood function will typically be very small except for values of w for which the network function is reasonably consistent with the data. Thus, the posterior distribution $p(w \mid \mathcal{D})$ will be much more sharply peaked than the prior distribution $p(w)$ (and will typically have multiple maxima). The quantity we are interested in is the predicted distribution of target values t for a new input vector x once we have observed the data set $\mathcal{D}$. This can be expressed as an integration over the posterior distribution of weights of the form

$$p(t \mid x, \mathcal{D}) = \int p(t \mid x, w)\,p(w \mid \mathcal{D})\,dw \tag{66.30}$$

where $p(t \mid x, w)$ is the conditional probability model discussed in the Introduction.

If we suppose that the posterior distribution $p(w \mid \mathcal{D})$ is sharply peaked around a single most-probable value w_{MP}, then we can write Equation 66.30 in the form:

$$p(t \mid x, \mathcal{D}) \simeq p(t \mid x, w_{MP}) \int p(w \mid \mathcal{D})\,dw \tag{66.31}$$

$$= p(t \mid x, w_{MP}) \tag{66.32}$$

and so predictions can be made by fixing the weights to their most probable values. We can find the most probable weights by maximizing the posterior distribution or equivalently by minimizing its negative logarithm. Using Equation 66.29, we see that w_{MP} is determined by minimizing a regularized cost function of the form in Equation 66.27 in which the negative log of the prior $-\ln p(w)$ represents the regularizer $\nu\Omega$. For example, if the prior consists of a zero-mean Gaussian with variance ν^{-1}, then we obtain the weight-decay regularizer of Equation 66.28.

The posterior distribution will become sharply peaked when the size of the dataset is large compared to the number of parameters in the network. For datasets of limited size, however, the posterior distribution has a finite width and this adds to the uncertainty in the predictions for t, which can be expressed in terms of error bars. Bayesian error bars can be evaluated using a local Gaussian approximation to the posterior distribution [MacKay 1992]. The presence of multiple maxima in the posterior distribution also contributes to the uncertainties in predictions. The capability to assess these uncertainties can play a crucial role in practical applications.

The Bayesian approach can also deal with more general problems in complexity control. This can be done by considering the probabilities of a set of alternative models, given the dataset

$$p(\mathcal{H}_i \mid \mathcal{D}) = \frac{p(\mathcal{D} \mid \mathcal{H}_i) p(\mathcal{H}_i)}{p(\mathcal{D})} \tag{66.33}$$

Here different models can also be interpreted as different values of regularization parameters as these too control model complexity. If the models are given the same prior probabilities $p(\mathcal{H}_i)$ then they can be ranked by considering the *evidence* $p(\mathcal{D} \mid \mathcal{H}_i)$, which itself can be evaluated by integration over the model parameters w. We can simply select the model with the greatest probability. However, a full Bayesian treatment requires that we form a linear combination of the predictions of the models in which the weighting coefficients are given by the model probabilities.

In general, the required integrations, such as that in Equation 66.30, are analytically intractable. One approach is to approximate the posterior distribution by a Gaussian centered on w_{MP} and then to linearize $p(t \mid x, w)$ about w_{MP} so that the integration can be performed analytically [MacKay 1992]. Alternatively, sophisticated Monte Carlo methods can be employed to evaluate the integrals numerically [Neal 1994]. An important aspect of the Bayesian approach is that there is no need to keep data aside in a validation set as is required when using maximum likelihood. In practical applications for which the quantity of available data is limited, it is found that a Bayesian treatment generally outperforms other approaches.

66.3.7 Preprocessing, Invariances, and Prior Knowledge

We have already seen that neural networks can approximate essentially arbitrary nonlinear functional mappings between sets of variables. In principle, we could therefore use a single network to transform the raw input variables into the required final outputs. However, in practice for all but the simplest problems the results of such an approach can be improved upon considerably by incorporating various forms of preprocessing, for reasons we shall outline in the following.

One of the simplest and most common forms of preprocessing consists of a simple normalization of the input, and possibly also target, variables. This may take the form of a linear rescaling of each input variable independently to give it zero mean and unit variance over the training set. For some applications, the original input variables may span widely different ranges. Although a linear rescaling of the inputs is equivalent to a different choice of first-layer weights, in practice the optimization algorithm may have considerable difficulty in finding a satisfactory solution when typical input values are substantially different. Similar rescaling can be applied to the output values, in which case the inverse of the transformation needs to be applied to the network outputs when the network is presented with new inputs. Preprocessing is also used to encode data in a suitable form. For example, if we have categorical variables such as red, green, and blue, these may be encoded using a 1-of-3 binary representation.

Another widely used form of preprocessing involves reducing the dimensionality of the input space. Such transformations may result in loss of information in the data, but the overall effect can be a significant improvement in performance as a consequence of the curse of dimensionality discussed in the complexity control section. The finite dataset is better able to specify the required mapping in the lower dimensional space. Dimensionality reduction may be accomplished by simply selecting a subset of the original variables but more typically involves the construction of new variables consisting of linear or nonlinear combinations of the original variables called *features*. A standard technique for dimensionality reduction is principal component analysis [Anderson 1984]. Such methods, however, make use only of the input data and ignore the target values and can sometimes be significantly suboptimal.

Yet another form of preprocessing involves correcting deficiencies in the original data. A common occurrence is that some of the input variables are missing for some of the data points. Correction of this problem in a principled way requires that the probability distribution $p(x)$ of input data be modeled.

One of the most important factors determining the performance of real-world applications of neural networks is the use of *prior knowledge*, which is information additional to that present in the data. As an example, consider the problem of classifying handwritten digits discussed in Section 66.1. The most direct approach would be to collect a large training set of digits and to train a feedforward network to map from the input image to a set of 10 output values representing posterior probabilities for the 10 classes. However, we know that the classification of a digit should be independent of its position within the input image. One way of achieving such *translation invariance* is to make use of the technique of *shared weights*. This involves a network architecure having many hidden layers in which each unit takes inputs only from a small patch, called a *receptive field*, of units in the previous layer. By a process of constraining neighboring units to have common weights, it can be arranged that the output of the network is insensitive to translations of the input image. A further benefit of weight sharing is that the number of independent parameters is much smaller than the number of weights, which assists with the problem of model complexity. This approach is the basis for the highly successful U.S. postal code recognition system of LeCun et al. [1989]. An alternative to shared weights is to enlarge the training set artificially by generating virtual examples based on applying translations and other transformations to the original training set [Poggio and Vetter 1992].

66.4 Graphical Models

Neural networks express relationships between variables by utilizing the representational language of graph theory. Variables are associated with nodes in a graph and transformations of variables are based on algorithms that propagate numerical messages along the links of the graph. Moreover, the graphs are often accompanied by probabilistic interpretations of the variables and their interrelationships. As we have seen, such probabilistic interpretations allow a neural network to be understood as a form of a probabilistic model and reduce the problem of learning the weights of a network to a problem in statistics.

Related graphical models have been studied throughout statistics, engineering, and AI in recent years. Hidden Markov models, Kalman filters, and path analysis models are all examples of graphical probabilistic models that can be fitted to data and used to make inferences. The relationship between these models and neural networks is rather strong; indeed, it is often possible to reduce one kind of model to the other. In this section, we examine these relationships in some detail and provide a broader characterization of neural networks as members of a general family of graphical probabilistic models.

Many interesting relationships have been discovered between graphs and probability distributions [Spiegelhalter et al. 1993, Pearl 1988]. These relationships derive from the use of graphs to represent conditional independencies among random variables. In an undirected graph, there is a direct correspondence between conditional independence and graph separation: random variables X_i and X_k are conditionally independent given X_j if nodes X_i and X_k are separated by node X_j (we use the symbol X_i to represent both a random variable and a node in a graph). This statement remains true for sets of nodes. [See Figure 66.5(a).] Directed graphs have a somewhat different semantics due to the ability of directed graphs to represent induced dependencies. An induced dependency is a situation in which two nodes which are marginally independent become conditionally dependent given the value of a third node. [See Figure 66.5(b).] Suppose, for example, that X_i and X_k represent independent coin tosses, and X_j represents the sum of X_i and X_k. Then X_i and X_k are marginally independent but are conditionally dependent given X_j. The semantics of independence in directed graphs is captured by a graphical criterion known as *d-separation* [Pearl 1988], which differs from undirected separation only in those cases in which paths have two arrows arriving at the same node [as in Figure 66.5(b)].

Although the neural network architectures that we have discussed until now all have been based on directed graphs, undirected graphs also play an important role in neural network research. Constraint satisfaction architectures, including the Hopfield network [Hopfield 1982] and the **Boltzmann machine** [Hinton and Sejnowski 1986], are the most prominent examples. A Boltzmann machine is an undirected probabilistic graph that respects the conditional independency semantics previously described

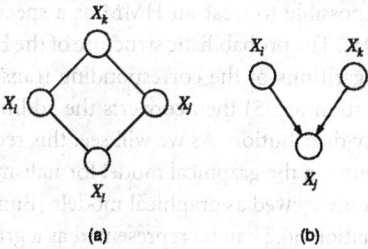

FIGURE 66.5   (a) An undirected graph in which X_i is independent of X_j given X_k and X_l, and X_k is independent of X_l given X_i and X_j. (b) A directed graph in which X_i and X_k are marginally independent but are conditionally dependent given X_j.

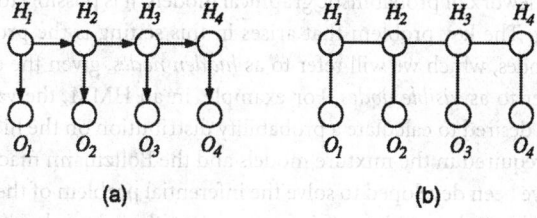

FIGURE 66.6   (a) A directed graph representation of an HMM. Each horizontal link is associated with the transition matrix A, and each vertical link is associated with the emission matrix B. (b) An HMM as a Boltzmann machine. The parameters on the horizontal links are logarithms of the entries of the A matrix, and the parameters on the vertical links are logarithms of the entries of the B matrix. The two representations yield the same joint probability distribution.

[cf. Figure 66.5(a)]. Each node in a Boltzmann machine is a binary-valued random variable X_i (or, more generally, a discrete-valued random variable). A probability distribution on the 2^N possible configurations of such variables is defined via an *energy function* E. Let J_{ij} be the weight on the link between X_i and X_j, let $J_{ij} = J_{ji}$, let α index the configurations, and define the energy of configuration α as follows:

$$E_\alpha = -\sum_{i<j} J_{ij} X_i^\alpha X_j^\alpha \tag{66.34}$$

The probability of configuration α is then defined via the Boltzmann distribution:

$$P_\alpha = \frac{e^{-E_\alpha/T}}{\sum_\gamma e^{-E_\gamma/T}} \tag{66.35}$$

where the *temperature* T provides a scale for the energy.

An example of a directed probabilistic graph is the hidden Markov model (HMM). An HMM is defined by a set of *state variables* H_i, where i is generally a time or a space index, a set of output variables O_i, a *probability transition matrix* $A = p(H_i \mid H_{i-1})$, and an *emission matrix* $B = p(O_i \mid H_i)$. The directed graph for an HMM is shown in Figure 66.6(a). As can be seen from considering the separatory properties of the graph, the conditional independencies of the HMM are defined by the following Markov conditions:

$$H_i \perp \{H_1, O_1, \ldots, H_{i-2}, O_{i-2}, O_{i-1}\} \mid H_{i-1}, \qquad 2 \leq i \leq N \tag{66.36}$$

and

$$O_i \perp \{H_1, O_1, \ldots, H_{i-1}, O_{i-1}\} \mid H_i, \qquad 2 \leq i \leq N \tag{66.37}$$

where the symbol $\perp$ is used to denote independence.

Figure 66.6(b) shows that it is possible to treat an HMM as a special case of a Boltzmann machine [Luttrell 1989, Saul and Jordan 1995]. The probabilistic structure of the HMM can be captured by defining the weights on the links as the logarithms of the corresponding transition and emission probabilities. The Boltzmann distribution (Equation 66.35) then converts the additive energy into the product form of the standard HMM probabilility distribution. As we will see, this reduction of a directed graph to an undirected graph is a recurring theme in the graphical model formalism.

General mixture models are readily viewed as graphical models [Buntine 1994]. For example, the unconditional mixture model of Equation 66.2 can be represented as a graphical model with two nodes — a multinomial hidden node, which represents the selected component, a visible node representing x, with a directed link from the hidden node to the visible node (hidden/visible distinction discussed subsequently). Conditional mixture models [Jacobs et al. 1991] simply require another visible node with directed links to the hidden node and the visible nodes. Hierarchical conditional mixture models [Jordan and Jacobs 1994] require a chain of hidden nodes, one hidden node for each level of the tree.

Within the general framework of probabilistic graphical models, it is possible to tackle general problems of inference and learning. The key problem that arises in this setting is the problem of computing the probabilities of certain nodes, which we will refer to as *hidden nodes*, given the observed values of other nodes, which we will refer to as *visible nodes*. For example, in an HMM, the variables O_i are generally treated as visible, and it is desired to calculate a probability distribution on the hidden states H_i. A similar inferential calculation is required in the mixture models and the Boltzmann machine.

Generic algorithms have been developed to solve the inferential problem of the calculation of posterior probabilities in graphs. Although a variety of inference algorithms have been developed, they can all be viewed as essentially the same underlying algorithm [Shachter et al. 1994]. Let us consider undirected graphs. A special case of an undirected graph is a *triangulated graph* [Spiegelhalter et al. 1993], in which any cycle having four or more nodes has a chord. For example, the graph in Figure 66.5(a) is not triangulated but becomes triangulated when a link is added between nodes X_i and X_j. In a triangulated graph, the cliques of the graph can be arranged in the form of a *junction tree*, which is a tree having the property that any node that appears in two different cliques in the tree also appears in every clique on the path that links the two cliques (the "running intersection property"). This cannot be achieved in nontriangulated graphs. For example, the cliques in Figure 66.5(a) are $\{X_i, X_k\}$, $\{X_k, X_j\}$, $\{X_j, X_l\}$, and it is not possible to arrange these cliques into a tree that obeys the running intersection property. If a chord is added, the resulting cliques are $\{X_i, X_j, X_k\}$ and $\{X_i, X_j, X_l\}$, and these cliques can be arranged as a simple chain that trivially obeys the running intersection property. In general, it turns out that the probability distributions corresponding to triangulated graphs can be characterized as *decomposable*, which implies that they can be factorized into a product of local functions (potentials) associated with the cliques in the triangulated graph.* The calculation of posterior probabilities in decomposable distributions is straightforward and can be achieved via a local message-passing algorithm on the junction tree [Spiegelhalter et al. 1993].

Graphs that are not triangulated can be turned into triangulated graphs by the addition of links. If the potentials on the new graph are defined suitably as products of potentials on the original graph, then the independencies in the original graph are preserved. This implies that the algorithms for triangulated graphs can be used for *all* undirected graphs; an untriangulated graph is first triangulated. (See Figure 66.7.) Moreover, it is possible to convert *directed* graphs to undirected graphs in a manner that preserves the probabilistic structure of the original graph [Spiegelhalter et al. 1993]. This implies that the junction tree algorithm is indeed generic; it can be applied to any graphical model.

The problem of calculating posterior probabilities on graphs is NP-hard; thus, a major issue in the use of the inference algorithms is the identification of cases in which they are efficient. Chain structures such as HMMs yield efficient algorithms, and indeed the classical forward-backward algorithm for HMMs is

---

*An interesting example is a Boltzmann machine on a triangulated graph. The potentials are products of $\exp(J_{ij})$ factors, where the product is taken over all (i, j) pairs in a particular clique. Given that the product across potentials must be the joint probability, this implies that the partition function (the denominator of Equation 66.35) must be unity in this case.

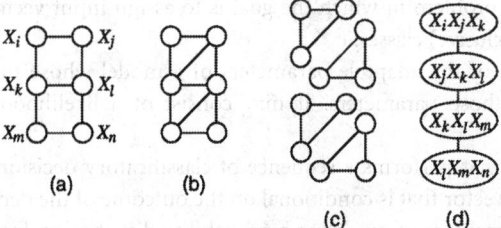

FIGURE 66.7 The basic structure of the junction tree algorithm for undirected graphs. The graph in (a) is first triangulated (b), then the cliques are identified (c), and arranged into a tree (d). Products of potential functions on the nodes in (d) yield probability distributions on the nodes in (a).

a special, efficient case of the junction tree algorithm [Smyth et al. 1996]. Decision tree structures such as the hierarchical mixture of experts yield efficient algorithms, and the recursive posterior probability calculation of Jordan and Jacobs [1994] described earlier is also a special case of the junction tree algorithm. All of the simpler mixture model calculations described earlier are therefore also special cases. Another interesting special case is the state estimation algorithm of the Kalman filter [Shachter and Kenley 1989]. Finally, there are a variety of special cases of the Boltzmann machine which are amenable to the exact calculations of the junction tree algorithm [Saul and Jordan 1995].

For graphs that are outside of the tractable categories of trees and chains, the junction tree algorithm often performs surprisingly well, but for highly connected graphs the algorithm can be too slow. In such cases, approximate algorithms such as Gibbs sampling are utilized. A virtue of the graphical framework is that Gibbs sampling has a generic form, which is based on the notion of a *Markov boundary* [Pearl 1988]. A special case of this generic form is the stochastic update rule for general Boltzmann machines.

Our discussion has emphasized the unifying framework of graphical models both for expressing probabilistic dependencies in graphs and for describing algorithms that perform the inferential step of calculating posterior probabilities on these graphs. The unification goes further, however, when we consider learning. A generic methodology known as the expectation-maximization algorithm is available for MAP and Bayesian estimation in graphical models [Dempster et al. 1977]. EM is an iterative method, based on two alternating steps: an *E step*, in which the values of hidden variables are estimated, based on the current values of the parameters and the values of visible variables, and an *M step*, in which the parameters are updated, based on the estimated values obtained from the E step. Within the framework of the EM algorithm, the junction tree algorithm can readily be viewed as providing a generic E step. Moreover, once the estimated values of the hidden nodes are obtained from the E step, the graph can be viewed as fully observed, and the M step is a standard MAP or ML problem. The standard algorithms for all of the tractable architectures described (mixtures, trees, and chains) are, in fact, instances of this general graphical EM algorithm, and the learning algorithm for general Boltzmann machines is a special case of a generalization of EM known as GEM [Dempster et al. 1977].

What about the case of feedforward neural networks such as the multilayer perceptron? It is, in fact, possible to associate binary hidden values with the hidden units of such a network (cf. our earlier discussion of the logistic function; see also Amari [1995]) and apply the EM algorithm directly. For N hidden units, however, there are 2^N patterns whose probabilities must be calculated in the E step. For large N, this is an intractable computation, and recent research has therefore begun to focus on fast methods for approximating these distributions [Hinton et al. 1995, Saul et al. 1996].

Defining Terms

Boltzmann machine: An undirected network of discrete valued random variables, where an energy function is associated with each of the links, and for which a probability distribution is defined by the Boltzmann distribution.

Classification: A learning problem in which the goal is to assign input vectors to one of a number of (usually mutually exclusive) classes.

Cost function: A function of the adaptive parameters of a model whose minimum is used to define suitable values for those parameters. It may consist of a likelihood function and additional terms.

Decision tree: A network that performs a sequence of classificatory decisions on an input vector and produces an output vector that is conditional on the outcome of the decision sequence.

Density estimation: The problem of modeling a probability distribution from a finite set of examples drawn from that distribution.

Discriminant function: A function of the input vector that can be used to assign inputs to classes in a classification problem.

Hidden Markov model: A graphical probabilistic model characterized by a state vector, an output vector, a state transition matrix, an emission matrix, and an initial state distribution.

Likelihood function: The probability of observing a particular data set under the assumption of a given parametrized model, expressed as a function of the adaptive parameters of the model.

Mixture model: A probability model that consists of a linear combination of simpler component probability models.

Multilayer perceptron: The most common form of neural network model, consisting of successive linear transformations followed by processing with nonlinear activation functions.

Overfitting: The problem in which a model which is too complex captures too much of the noise in the data, leading to poor generalization.

Radial basis function network: A common network model consisting of a linear combination of basis functions, each of which is a function of the difference between the input vector and a center vector.

Regression: A learning problem in which the goal is to map each input vector to a real-valued output vector.

Regularization: A technique for controlling model complexity and improving generalization by the addition of a penalty term to the cost function.

VC dimension: A measure of the complexity of a model. Knowledge of the VC dimension permits an estimate to be made of the difference between performance on the training set and performance on a test set.

References

Amari, S. 1995. The EM algorithm and information geometry in neural network learning. *Neural Comput.* 7(1):13–18.

Anderson, T. W. 1984. *An Introduction to Multivariate Statistical Analysis.* Wiley, New York.

Bengio, Y. 1996. *Neural Networks for Speech and Sequence Recognition.* Thomson Computer Press, London.

Bishop, C. M. 1995. *Neural Networks for Pattern Recognition.* Oxford University Press.

Breiman, L., Friedman, J. H., Olshen, R. A., and Stone, C. J. 1984. *Classification and Regression Trees.* Wadsworth International Group, Belmont, CA.

Buntine, W. 1994. Operations for learning with graphical models. *J. Artif. Intelligence Res.* 2:159–225.

Dempster, A. P., Laird, N. M., and Rubin, D. B. 1977. Maximum-likelihood from incomplete data via the EM algorithm. *J. R. Stat. Soc.* B39:1–38.

Duda, R. O. and Hart, P. E. 1973. *Pattern Classification and Scene Analysis.* Wiley, New York.

Fletcher, R. 1987. *Practical Methods of Optimization,* 2nd ed. Wiley, New York.

Geman, S., Bienenstock, E., and Doursat, R. 1992. Neural networks and the bias/variance dilemma. *Neural Comput.* 4:1–58.

Hertz, J., Krogh, A., and Palmer, R. G. 1991. *Introduction to the Theory of Neural Computation.* Addison–Wesley, Redwood City, CA.

Hinton, G. E., Dayan, P., Frey, B., and Neal, R. 1995. The wake-sleep algorithm for unsupervised neural networks. *Science* 268:1158–1161.

Hinton, G. E. and Sejnowski, T. 1986. Learning and relearning in Boltzmann machines. In *Parallel Distributed Processing: Vol. 1.* D. E. Rumelhart and J. L. McClelland, Eds., pp. 282–317. MIT Press, Cambridge, MA.

Hopfield, J. J. 1982. Neural networks and physical systems with emergent collective computational abilities. *Proc. Nat. Acad. Sci.* 79:2554–2558.

Jacobs, R. A., Jordan, M. I., Nowlan, S. J., and Hinton, G. E. 1991. Adaptive mixtures of local experts. *Neural Comput.* 3:79–87.

Jordan, M. I. and Jacobs, R. A. 1994. Hierarchical mixtures of experts and the EM algorithm. *Neural Comput.* 6:181–214.

LeCun, Y., Boser, B., Denker, J. S., Henderson, D., Howard, R. E., Hubbard, W., and Jackel, L. D. 1989. Backpropagation applied to handwritten zip code recognition. *Neural Comput.* 1(4): 541–551.

Luttrell, S. 1989. The Gibbs machine applied to hidden Markov model problems. *Royal Signals and Radar Establishment: SP Res. Note 99*, Malvern, UK.

MacKay, D. J. C. 1992. A practical Bayesian framework for back-propagation networks. *Neural Comput.* 4:448–472.

McCullagh, P. and Nelder, J. A. 1983. *Generalized Linear Models.* Chapman and Hall, London.

Neal, R. M. 1994. *Bayesian Learning for Neural Networks.* Unpublished Ph.D. thesis, Department of Computer Science, University of Toronto, Canada.

Pearl, J. 1988. *Probabilistic Reasoning in Intelligent Systems.* Morgan Kaufmann, San Mateo, CA.

Poggio, T. and Vetter, T. 1992. Recognition and structure from one 2-D model view: observations on prototypes, object classes and symmetries. *Artificial Intelligence Lab.*, AI Memo 1347, Massachusetts Institute of Technology, Cambridge, MA.

Rabiner, L. R. 1989. A tutorial on hidden Markov models and selected applications in speech recognition. *Proc. IEEE* 77:257–286.

Rumelhart, D. E., Durbin, R., Golden, R., and Chauvin, Y. 1995. Backpropagation: the basic theory. In *Backpropagation: Theory, Architectures, and Applications*, Y. Chauvin, and D. E. Rumelhart, Eds., pp. 1–35. Lawrence Erlbaum, Hillsdale, NJ.

Rumelhart, D. E., Hinton, G. E., and Williams, R. J. 1986. Learning internal representations by error propagation. In *Parallel Distributed Processing: Vol. 1.* D. E. Rumelhart and J. L. McClelland, Eds., pp. 318–363. MIT Press, Cambridge, MA.

Sandler, D. G., Barrett, T. K., Palmer, D. A., Fugate, R. Q., and Wild, W. J. 1991. Use of a neural network to control an adaptive optics system for an astronomical telescope. *Nature* 351:300–302.

Saul, L. K., Jaakkola, T., and Jordan, M. I. 1996. Mean field learning theory for sigmoid belief networks. *J. Artif. Intelligence Res.* 4:61–76.

Saul, L. K. and Jordan, M. I. 1995. Boltzmann chains and hidden Markov models. In *Advances in Neural Information Processing Systems 7*, G. Tesauro, D. Touretzky, and T. Leen, Eds. MIT Press, Cambridge, MA.

Shachter, R., Andersen, S., and Szolovits, P. 1994. Global conditioning for probabilistic inference in belief networks. In *Uncertainty in Artificial Intelligence: Proc. 10th Conf.*, pp. 514–522. Seattle, WA.

Shachter, R. and Kenley, C. 1989. Gaussian influence diagrams. *Management Sci.* 35(5):527–550.

Smyth, P., Heckerman, D., and Jordan, M. I. 1996 in press. Probabilistic independence networks for hidden Markov probability models. *Neural Computation.*

Spiegelhalter, D., Dawid, A., Lauritzen, S., and Cowell, R. 1993. Bayesian analysis in expert systems. *Stat. Sci.* 8(3):219–283.

Tarassenko, L. 1995. Novelty detection for the identification of masses in mammograms. *Proc. 4th IEE Int. Conf. Artif. Neural Networks* Vol. 4, pp. 442–447.

Vapnik, V. N. 1995. *The Nature of Statistical Learning Theory.* Springer–Verlag, New York.

Further Information

In this chapter we have emphasized the links between neural networks and statistical pattern recognition. A more extensive treatment from the same perspective can be found in Bishop [1995]. For a view of recent research in the field, the proceedings of the annual Neural Information Processing Systems (NIPS), MIT Press, conferences are highly recommended.

Neural computing is now a very broad field, and there are many topics which have not been discussed for lack of space. Here we aim to provide a brief overview of some of the more significant omissions, and to give pointers to the literature.

The resurgence of interest in neural networks during the 1980s was due in large part to work on the statistical mechanics of fully connected networks having symmetric connections (i.e., if unit i sends a connection to unit j then there is also a connection from unit j back to unit i with the same weight value). We have briefly discussed such systems; a more extensive introduction to this area can be found in Hertz et al. [1991].

The implementation of neural networks in specialist very large-scale integrated (VLSI) hardware has been the focus of much research, although by far the majority of work in neural computing is undertaken using software implementations running on standard platforms.

An implicit assumption throughout most of this chapter is that the processes which give rise to the data are stationary in time. The techniques discussed here can readily be applied to problems such as time series forecasting, provided this stationarity assumption is valid. If, however, the generator of the data is itself evolving with time, then more sophisticated techniques must be used, and these are the focus of much current research (see Bengio [1996]).

One of the original motivations for neural networks was as models of information processing in biological systems such as the human brain. This remains the subject of considerable research activity, and there is a continuing flow of ideas between the fields of neurobiology and of artificial neural networks. Another historical springboard for neural network concepts was that of adaptive control, and again this remains a subject of great interest.

67

Planning and Scheduling

67.1   Introduction 67-1
        Planning and Scheduling Problems • Distinctions
        and Disciplines

67.2   Classifying Planning Problems 67-3
        Representing Dynamical Systems • Representing Plans
        of Action • Measuring Performance • Categories of
        Planning Problems

67.3   Algorithms, Complexity, and Search 67-6
        Complexity Results • Planning with Deterministic Dynamics
        • Scheduling with Deterministic Dynamics • Improving
        Efficiency • Approximation in Stochastic Domains
        • Practical Planning

67.4   Research Issues and Summary 67-20

Thomas Dean
Brown University

Subbarao Kambhampati
Arizona State University

67.1 Introduction

In this chapter, we use the generic term **planning** to encompass both planning and scheduling problems, and the terms *planner* or *planning system* to refer to software for planning or scheduling. Planning is concerned with reasoning about the consequences of acting in order to choose from among a set of possible courses of action. In the simplest case, a planner might enumerate a set of possible courses of action, consider their consequences in turn, and choose one particular course of action that satisfies a given set of requirements.

Algorithmically, a planning problem has as input a set of possible courses of actions, a predictive model for the underlying dynamics, and a performance measure for evaluating courses of action. The output or solution to a planning problem is one or more courses of action that satisfy the specified requirements for performance. Most planning problems are combinatorial in the sense that the number of possible courses of actions or the time required to evaluate a given course of action is exponential in the description of the problem.

Just because there is an exponential number of possible courses of action does not imply that a planner has to enumerate them all in order to find a solution. However, many planning problems can be shown to be NP-hard, and, for these problems, all known exact algorithms take exponential time in the worst case. The computational complexity of planning problems often leads practitioners to consider approximations, computation time vs. solution quality tradeoffs, and heuristic methods.

67.1.1 Planning and Scheduling Problems

We use the travel planning problem as our canonical example of planning (distinct from scheduling). A *travel planning problem* consists of a set of travel options (airline flights, cabs, subways, rental cars, and shuttle services), travel dynamics (information concerning travel times and costs and how time and cost are affected by weather or other factors), and a set of requirements. The requirements for a travel planning problem include an itinerary (be in Providence on Monday and Tuesday, and in Phoenix from Wednesday morning until noon on Friday) and constraints on solutions (leave home no earlier than the Sunday before, arrive back no later than the Saturday after, and spend no more than $1000 in travel-related costs). Planning can be cast either in terms of **satisficing** (find some solution satisfying the constraints) or **optimizing** (find the least cost solution satisfying the constraints).

We use the job-shop scheduling problem as our canonical example of scheduling (distinct from planning). The specification of a *job-shop scheduling problem* includes a set of jobs, where each job is a partially ordered set of tasks of specified duration, and a set of machines, where each machine is capable of carrying out a subset of the set of all tasks. A *feasible solution* to a job-shop scheduling problem is a mapping from tasks to machines over specific intervals of time, so that no machine has assigned to it more than one task at a time and each task is completed before starting any other task that follows it in the specified partial order. Scheduling can also be cast in terms of either satisficing (find a feasible solution) or optimizing (find a solution that minimizes the total time required to complete all jobs).

67.1.2 Distinctions and Disciplines

To distinguish between planning and scheduling, we note that scheduling is primarily concerned with figuring out *when* to carry out actions whereas planning is concerned with *what* actions need to be carried out. In practice, this distinction often blurs and many real-world problems involve figuring out both what and when.

In real job shops, each task need not specify a rigid sequence of steps (the what). For example, drilling a hole in a casting may be accomplished more quickly if a positioning device, called a fixture, is installed on the machine used for drilling. However, the fixture takes time to install and may interfere with subsequent machining operations for other tasks. Installing a fixture for one task may either expedite (the next job needs the same fixture) or retard (the fixture is in the way in the next job) subsequent jobs. In this version of our canonical scheduling problem, planning can take on considerable importance.

We can, however, design a problem so as to emphasize either planning or scheduling. For example, it may be reasonable to let a human decide the what (e.g., the type of machine and specific sequence of machining steps for each task) and a computer program decide the when (e.g., the time and machine for each task). This division of labor has allowed the field of operations research to focus effort on solving problems that stress scheduling and finesse planning, as we previously distinguished the two. Restricting attention to scheduling has the effect of limiting the options available to the planner, thereby limiting the possible interactions among actions and simplifying the combinatorics. In addition, some scheduling problems do not allow for the possibility of events outside the direct control of the planner, so-called *exogenous* events. Planning researchers in artificial intelligence generally allow a wide range of options (specifying both what and when) resulting in a very rich set of interactions among the individual actions in a given course of action and between actions and exogenous events.

The travel planning problem includes as a special case the classic traveling salesperson problem, a problem of considerable interest in operations research. In the traveling salesperson problem, there is a completely connected graph with L vertices corresponding to L distinct cities, an $L \times L$ matrix whose entries encode the distance between each pair of cities, and the objective is to find a minimal-length tour of a specified subset of the cities. The classic traveling salesperson problem involves a very limited set of possible interactions (e.g., you must finish one leg of a tour before beginning the next, and the next leg of a tour must begin at the city in which the previous leg ended). In contrast, variants of the travel planning problem studied in artificial intelligence generally consider a much richer set of possible interactions (e.g., if you start on a multileg air trip it is generally more cost effective to continue with the same airline; travel that extends over a Saturday is less expensive than travel that does not).

Planning of the sort studied in artificial intelligence is similar in some respects to problems studied in a variety of other disciplines. We have already mentioned operations research; planning is also similar to the problem of synthesizing controllers in control theory or the problem of constructing decision procedures in various decision sciences. Planning problems of the sort considered in this chapter differ from those studied in other disciplines mainly in the details of their formulation. Planning problems studied in artificial intelligence typically involve very complex dynamics, requiring expressive languages for their representation, and encoding a wide range of knowledge, often symbolic, but invariably rich and multifaceted.

67.2 Classifying Planning Problems

In this section, we categorize different planning problems according to their inputs: the set of basic courses of action, the underlying dynamics, and the performance measure. We begin by considering models used to predict the consequences of action.

67.2.1 Representing Dynamical Systems

We refer to the environment in which actions are carried out as a **dynamical system**. A description of the environment at an instant of time is called the *state* of the system. We assume that there is a finite, but large set of states S, and a finite set of actions A, that can be executed.

States are described by a vector of *state variables*, where each state variable represents some aspect of the environment that can change over time (e.g., the location or color of an object). The resulting dynamical system can be described as a deterministic, nondeterministic, or stochastic finite-state machine, and time is isomorphic to the integers. In the case of a deterministic finite-state machine, the dynamical system is defined by a **state-transition function** f that takes a state $s_t \in S$ and an action $a_t \in A$ and returns the next state $f(s_t, a_t) = s_{t+1} \in S$.

If there are N state variables each of which can take on two or more possible values, then there are as many as 2^N states and the state-transition function is N dimensional. We generally assume each state variable at t depends on only a small number (at most M) of state variables at $t - 1$. This assumption enables us to *factor* the state-transition function f into N functions, each of dimension at most M, so that $f(s, a) = \langle g_1(s, a), \ldots, g_N(s, a) \rangle$ where $g_i(s, a)$ represents the ith state variable.

In most planning problems, a plan is constructed at one time and executed at a later time. The state-transition function models the evolution of the state of the dynamical system as a consequence of actions carried out by a *plan executor*. We also want to model the information available to the plan executor. The plan executor may be able to observe the state of the dynamical system, partial state information corrupted by noise, or only the current time. We assume that there is a set of possible observations $\mathcal{O}$ and the information available to the plan executor at time t is determined by the current state and the *output function* $h: S \to \mathcal{O}$, so that $h(s_t) = o_t$. We also assume that the plan executor has a clock and can determine the current time t.

Figure 67.1 depicts the general planning problem. The planner is notated as Γ; it takes as input the current observation o_t and has as output the current **plan** π_t. The planner need not issue a new plan on

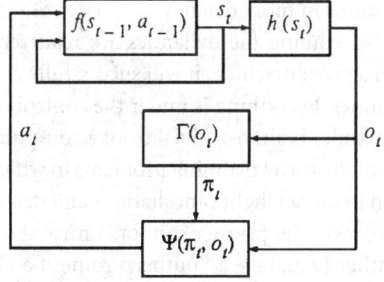

FIGURE 67.1 A block diagram for the general planning problem with state-transition function f, output function h, planner Γ, and plan executor Ψ.

every state transition and can keep a history of past observations if required. The plan executor is notated as Ψ; it takes as input the current observation o_t and the current plan π_t and has as output the current action a_t.

In the classic formulation of the problem, all planning is done prior to any execution. This formulation is inappropriate in cases where new information becomes available in the midst of execution and replanning is called for. The idea of the planner and plan executor being part of the specification of a planning problem is relatively new in artificial intelligence. The theory that relates to accounting for computations performed during execution is still in its infancy and is only touched upon briefly in this chapter.

Some physical processes modeled as dynamical systems evolve deterministically; the next state of the system is completely determined by the current state and action. Other processes, said to be *stochastic*, are subject to random changes or are so complex that it is often convenient to model their behavior in statistical terms; the next state of such a system is summarized by a distribution over the set of states.

If the state transitions are governed by a stochastic process, then the state-transition and output functions are random functions and we define the state-transition and output conditional probability distributions as follows:

$$Pr(f(s_t, a_t) \mid s_t, a_t)$$
$$Pr(h(s_t) \mid s_t)$$

In the general case, it requires $O(2^N)$ storage to encode these distributions for Boolean state variables. However, in many practical cases, these probability distributions can be factored by taking advantage of independence among state variables.

As mentioned earlier, we assume that the ith state variable at time t depends on a small subset (of size at most M) of the state variables at time $t-1$. Let Parents(i,s) denote the subset of state variables that the ith state variable depends on in s. We can represent the conditional probability distribution governing state transitions as the following product:

$$Pr(\langle g_1(s_t, a_t), \ldots, g_N(s_t, a_t)\rangle \mid s_t, a_t) = \prod_{i=1}^{N} Pr(g_i(s_t, a_t) \mid Parents(i, s_t), a_t)$$

This factored representation requires only $O(N2^M)$ storage for Boolean state variables, which is reasonable assuming that M is relatively small.

The preceding descriptions of dynamical systems provide the semantics for a planning system embedded in a dynamic environment. There remains the question of syntax, specifically: how do you represent the dynamical system? In artificial intelligence, the answer varies widely. Researchers have used first-order logic [Allen et al. 1991], dynamic logic [Rosenschein 1981], **state-space operators** [Fikes and Nilsson 1971], and factored probabilistic state-transition functions [Dean and Kanazawa 1989]. In the later sections, we examine some of these representations in more detail.

In some variants of job-shop scheduling the dynamics are relatively simple. We might assume, for example, that if a job is started on a given machine, it will successfully complete in a fixed, predetermined amount of time known to the planner. Everything is under the control of the planner, and evaluating the consequences of a given plan (schedule) is almost trivial from a computational standpoint.

We can easily imagine variants of the travel planning problems in which the dynamics are quite complicated. For example, we might wish to model flight cancellations and delays due to weather and mechanical failure in terms of a stochastic process. The planner cannot control the weather but it can plan to avoid the deleterious effects of the weather (e.g., take a Southern route if a chance of snow threatens to close Northern airports). In this case, there are factors not under control of the planner and evaluating a given travel plan may require significant computational overhead.

67.2.2 Representing Plans of Action

We have already introduced a set of actions $\mathcal{A}$. We assume that these actions are *primitive* in that they can be carried out by the hardware responsible for executing plans. Semantically, a plan π is a mapping from what is known at the time of execution to the set of actions. The set of all plans for a given planning problem is notated Π.

For example, a plan might map the current observation o_t to the action a_t to take in the current state s_t. Such a plan would be independent of time. Alternatively, a plan might ignore observations altogether and map the current time t to the action to take in state s_t. Such a plan is independent of the current state, or at least the observable aspects of the current state.

If the action specified by a plan is dependent on observations of the current state, then we say that the plan is *conditional*. If the action specified by a plan is dependent on the current time, then we say that the plan is *time variant*, otherwise we say it is *stationary*. If the mapping is one-to-one, then we say that the plan is deterministic, otherwise it is nondeterministic and possibly stochastic if the mapping specifies a distribution over possible actions.

Conditional plans are said to run in a **closed loop**, since they enable the executor to react to the consequences of prior actions. Unconditional plans are said to run in an **open loop**, since they take no account of exogenous events or the consequences of prior actions that were not predicted using the dynamical model.

Now that we have the semantics for plans, we can think about how to represent them. If the mapping is a function, we can use any convenient representation for functions, including decision trees, tabular formats, hash tables, or artificial neural networks. In some problems, an unconditional, time-variant, deterministic plan is represented as a simple sequence of actions. Alternatively, we might use a set of possible sequences of actions perhaps specified by a partially ordered set of actions to represent a nondeterministic plan [i.e., the plan allows any total order (sequence) consistent with the given partial order].

67.2.3 Measuring Performance

For a deterministic dynamical system in initial state s_0, a plan π determines a (possibly infinite) sequence of states $h_\pi = \langle s_0, s_1, \ldots \rangle$, called a **history or state-space trajectory**. More generally, a dynamical system together with a plan induces a probability distribution over histories, and h_π is a random variable governed by this distribution. A value function V assigns to each history a real value. In the deterministic case, the performance J of a plan π is the value of the resulting history, $J(\pi) = V(h_\pi)$. In the general case, the performance J of a plan is the expected value according to V over all possible histories, $J(\pi) = E[V(h_\pi)]$, where E denotes taking an expectation.

In artificial intelligence planning (distinct from scheduling), much of the research has focused on goal-based performance measures. A **goal** $\mathcal{G}$ is a subset of the set of states $\mathcal{S}$.

$$V(\langle s_0, s_1, \ldots \rangle) = \begin{cases} 1 & \text{if } \exists i, s_i \in \mathcal{G} \\ 0 & \text{otherwise} \end{cases}$$

Alternatively, we can consider the number of transitions until we reach a goal state as a measure of performance.

$$V(\langle s_0, s_1, \ldots \rangle) = \begin{cases} -\min_i s_i \in \mathcal{G} & \text{if } \exists i, s_i \in \mathcal{G} \\ -\infty & \text{otherwise} \end{cases}$$

In the stochastic case, the corresponding measure of performance is called *expected time to target*, and the objective in planning is to minimize this measure.

Generalizing on the expected-time-to-target performance measure, we can assign to each state a cost using the cost function C. This cost function yields the following value function on histories:

$$V(\langle s_0, s_1, \ldots \rangle) = -\sum_{i=0}^{\infty} C(s_i)$$

In some problems, we may wish to discount future costs using a discounting factor $0 \le \gamma < 1$,

$$V(\langle s_0, s_1, \ldots \rangle) = -\sum_{i=0}^{\infty} \gamma^i C(s_i)$$

This performance measure is called *discounted cumulative cost*. These value functions are said to be *separable* since the total value of a history is a simple sum or weighted sum (in the discounted case) of the costs of each state in the history.

It should be noted that we can use any of the preceding methods for measuring the performance of a plan to define either a satisficing criterion (e.g., find a plan whose performance is above some fixed threshold) or an optimizing criterion (e.g., find a plan maximizing a given measure of performance).

67.2.4 Categories of Planning Problems

Now we are in a position to describe some basic classes of planning problems. A planning problem can be described in terms of its dynamics, either deterministic or stochastic. We might also consider whether the actions of the planner completely or only partially determine the state of the environment.

A planning problem can be described in terms of the knowledge available to the planner or executor. In the problems considered in this chapter, we assume that the planner has an accurate model of the underlying dynamics, but this need not be the case in general. Even if the planner has an accurate predictive model, the executor may not have the necessary knowledge to make use of that model. In particular, the executor may have only partial knowledge of the system state and that knowledge may be subject to errors in observation (e.g., noisy, error-prone sensors).

We can assume that all computations performed by the planner are carried out prior to any execution, in which case the planning problem is said to be **off-line**. Alternatively, the planner may periodically compute a new plan and hand it off to the executor; this sort of planning problem is said to be **on-line**. Given space limitations, we are concerned primarily with off-line planning problems in this chapter.

Now that we have some familiarity with the various classes of planning problems, we consider some specific techniques for solving them. Our emphasis is on the design, analysis, and application of planning algorithms.

67.3 Algorithms, Complexity, and Search

Once we are given a set of possible plans and a *performance function* implementing a given performance measure, we can cast any planning or scheduling problem as a search problem. If we assume that evaluating a plan (applying the performance function) is computationally simple, then the most important issue concerns how we search the space of possible plans. Specifically, given one or more plans currently under consideration, how do we extend the search to consider other, hopefully better, performing plans? We focus on two methods for extending search: *refinement* methods and *repair* methods.

A refinement method takes an existing partially specified plan (schedule) and refines it by adding detail. In job-shop scheduling, for example, we might take a (partial) plan that assigns machines and times to k of the jobs, and extend it so that it accounts for $k + 1$ jobs. Alternatively, we might build a plan in chronological order by assigning the earliest interval with a free machine on each iteration.

A repair method takes a completely specified plan and attempts to transform it into another completely specified plan with better performance. In travel planning, we might take a plan that makes use of one airline's flights and modify it to use the flights of another, possibly less expensive or more reliable airline. Repair methods often work by first analyzing a plan to identify unwanted interactions or bottlenecks and then attempting to eliminate the identified problems.

The rest of this section is organized as follows. In "Complexity Results," we briefly survey what is known about the complexity of planning and scheduling problems, irrespective of what methods are used to solve them. In "Planning with Deterministic Dynamics," we focus on traditional search methods for generating plans of actions given deterministic dynamics. We begin with open-loop planning problems with complete knowledge of the initial state, progressing to closed-loop planning problems with incomplete knowledge of the initial state. In "Scheduling with Deterministic Dynamics," we focus on methods for generating schedules given deterministic dynamics. In both of the last two sections just mentioned we discuss refinement- and repair-based methods. In "Improving Efficiency," we mention related work in machine learning concerned with learning search rules and adapting previously generated solutions to planning and scheduling problems. In "Approximation in Stochastic Domains," we consider a class of planning problems involving stochastic dynamics and address some issues that arise in trying to approximate the value of conditional plans in stochastic domains. Our discussion begins with a quick survey of what is known about the complexity of planning and scheduling problems.

67.3.1 Complexity Results

Garey and Johnson [1979] provide an extensive listing of NP-hard problems, including a great many scheduling problems. They also provide numerous examples of how a hard problem can be rendered easy by relaxing certain assumptions. For example, most variants of job-shop scheduling are NP-hard. Suppose, however, that you can suspend work on one job in order to carry out a rush job, resuming the suspended job on completion of the rush job so that there is no time lost in suspending and resuming. With this assumption, some hard problems become easy. Unfortunately, most real scheduling problems are NP-hard. Graham et al. [1977] provide a somewhat more comprehensive survey of scheduling problems with a similarly dismal conclusion. Lawler et al. [1985] survey results for the traveling salesperson problem, a special case of our travel planning problem. Here again the prospects for optimal, exact algorithms are not good, but there is some hope for approximate algorithms.

With regard to open-loop, deterministic planning, Chapman [1987], Bylander [1991], and Gupta and Nau [1991] have shown that most problems in this general class are hard. Dean and Boddy [1988] show that the problem of evaluating plans represented as sets of partially ordered actions is NP-hard in all but the simplest cases. Bäckström and Klein [1991] provide some examples of easy (polynomial time) planning problems, but these problems are of marginal practical interest.

Regarding closed-loop, deterministic planning, Papadimitriou and Tsitsiklis [1987] discuss polynomial-time algorithms for finding an optimal conditional plan for a variety of performance functions. Unfortunately, the polynomial is in the size of the state space. As mentioned earlier, we assume that the size of the state space is exponential in the number of state variables. Papadimitriou and Tsitsiklis also list algorithms for the case of stochastic dynamics that are polynomial in the size of the state space.

From the perspective of worst-case, asymptotic time and space complexity, most practical planning and scheduling problems are computationally very difficult. The literature on planning and scheduling in artificial intelligence generally takes it on faith that any interesting problem is at least NP-hard. The research emphasis is on finding powerful heuristics and clever search algorithms. In the remainder of this section, we explore some of the highlights of this literature.

67.3.2 Planning with Deterministic Dynamics

In the following section, we consider a special case of planning in which each action deterministically transforms one state into another. Nothing changes without the executor performing some action. We assume that the planner has an accurate model of the dynamics. If we also assume that we are given complete information about the initial state, it will be sufficient to produce unconditional plans that are produced off-line and run in an open loop.

Recall that a state is described in terms of a set of state variables. Each state assigns to each state variable a value. To simplify the notation, we restrict our attention to Boolean variables. In the case of Boolean variables, each state variable is assigned either true or false. Suppose that we have three Boolean state variables: P, Q, and R. We represent the particular state s in which P and Q are true and R is false by the *state-variable assignment*, $s = \{P = \text{true}, Q = \text{true}, R = \text{false}\}$, or, somewhat more compactly, by $s = \{P, Q, \neg R\}$, where $X \in s$ indicates that X is assigned true in s and $\neg X \in s$ indicates that X is assigned false in s.

An action is represented as a *state-space operator* α defined in terms of *preconditions* ($\text{Pre}(\alpha)$) and *postconditions* (also called effects) ($\text{Post}(\alpha)$). Preconditions and postconditions are represented as state-variable assignments that assign values to subsets of the set of all state variables. Here is an example operator α_{eg}:

Operator α_{eg}

$$\text{Preconditions:} \quad P, \neg R$$

$$\text{Postconditions:} \quad \neg P, \neg Q$$

If an operator (action) is applied (executed) in a state in which the preconditions are satisfied, then the variables mentioned in the postconditions are assigned their respective values in the resulting state. If the preconditions are not satisfied, then there is no change in state.

In order to describe the state-transition function, we introduce a notion of consistency and define two operators $\oplus$ and $\ominus$ on state-variable assignments. Let φ and ϑ denote state-variable assignments. We say that φ and ϑ are *inconsistent* if there is a variable X such that φ and ϑ assign X different values; otherwise, we say that φ and ϑ are *consistent*. The operator $\ominus$ behaves like set difference with respect to the variables in assignments. The expression $\varphi \ominus \vartheta$ denotes a new assignment consisting of the assignments to variables in φ that have no assignment in ϑ (e.g., $\{P, Q\} \ominus \{P\} = \{P, Q\} \ominus \{\neg P\} = \{Q\} \ominus \{\} = \{Q\}$). The operator $\oplus$ takes two consistent assignments and returns their union (e.g., $\{Q\} \oplus \{P\} = \{P, Q\}$, but $\{P\} \oplus \{\neg P\}$ is undefined).

The state-transition function is defined as follows:

$$f(s, \alpha) = \begin{cases} s & \text{if } s \text{ and } Pre(\alpha) \text{ are inconsistent} \\ Post(\alpha) \oplus (s \ominus Post(\alpha)) & \text{otherwise} \end{cases}$$

If we apply the operator α_{eg} to a state where the variables P and Q are true, and R is false, we have

$$f(\{P, Q, \neg R\}, \alpha_{eg}) = \{\neg P, \neg Q, \neg R\}$$

We extend the state-transition function to handle sequences of operators in the obvious way,

$$f(s, \langle \alpha_1, \alpha_2, \ldots, \alpha_n \rangle) = f(f(s, \alpha_1), \langle \alpha_2, \ldots, \alpha_n \rangle)$$

$$f(s, \langle \rangle) = s$$

Our performance measure for this problem is goal based. Goals are represented as state-variable assignments that assign values to subsets of the set of all state variables. By assigning values to one or more state variables, we designate a set of states as the goal. We say that a state s *satisfies* a goal ϕ, notated $s \models \phi$, just in case the assignment ϕ is a subset of the assignment s. Given an initial state s_0, a goal ϕ, and a library of operators, the objective of the planning problem is to find a sequence of state-space operators $\langle \alpha_1, \ldots, \alpha_n \rangle$ such that $f(s_0, \langle \alpha_1, \ldots, \alpha_n \rangle) \models \phi$.

Using a state-space operator to transform one state into the next state is called **progression**. We can also use an operator to transform one goal into another, namely, the goal that the planner would have prior to carrying out the action corresponding to the operator. This use of an operator to transform goals is called **regression**. In defining regression, we introduce the notion of an impossible assignment, denoted $\bot$.

We assume that if you regress a goal using an operator with postconditions that are inconsistent with the goal, then the resulting regressed goal is impossible to achieve. Here is the definition of regression:

$$b(\phi, \alpha) = \begin{cases} \bot & \text{if } \phi \text{ and } Post(\alpha) \text{ are inconsistent} \\ Pre(\alpha) \oplus (\phi \ominus Post(\alpha)) & \text{otherwise} \end{cases}$$

67.3.2.1 Conditional Postconditions and Quantification

Within the general operator-based state-transition framework previously described, a variety of syntactic abbreviations can be used to facilitate compact action representation. For example, the postconditions of an action may be *conditional*. A conditional postcondition of the from $P \Rightarrow Q$ means that the action changes the value of the variable Q to true only if the value of P is true in the state where the operator is applied. It is easy to see that an action with such a conditional effect corresponds to two simpler actions, one which has a precondition P and the postcondition Q, and the other which has a precondition $\neg P$ and does not mention Q in its postconditions.

Similarly, when state variables can be typed in terms of objects in the domain to which they are related, it is possible to express preconditions and postconditions of an operator as quantified formulas. As an example, suppose in the travel domain, we have one state variable $loc\,(c)$ which is true if the agent is in city c and false otherwise. The action of flying from city c to city c' has the effect that the agent is now at city c', and the agent is not in any other city. If there are n cities, $c_1, \ldots, c_n$, the latter effect can be expressed either as a set of propositional postconditions $\neg loc\,(c_1), \ldots, \neg loc\,(c_{j-1}), \neg loc\,(c_{j+1}), \ldots, \neg loc\,(c_n)$ where $c' = c_j$, or, more compactly, as the quantified effect $\forall_{z:city\,(z)} \; z \neq c' \Rightarrow \neg loc\,(z)$. Since operators with conditional postconditions and quantified preconditions and postconditions are just shorthand notations for finitely many propositional operators, the transition function, as well as the progression and regression operations, can be modified in straightforward ways to accommodate them. For example, if a goal formula $\{W, S\}$ is regressed through an operator having preconditions $\{P, Q\}$ and postconditions $\{R \Rightarrow \neg W\}$, we get $\{\neg R, S, P, Q\}$. Note that by making $\neg R$ a part of the regressed formula, we ensure that $\neg W$ will not be a postcondition of the operator, thereby averting the inconsistency with the goals.

67.3.2.2 Representing Partial Plans

Although solutions to the planning problems can be represented by operator sequences, to facilitate efficient methods of plan synthesis, it is useful to have a more flexible representation for partial plans. A *partial plan* consists of a set of *steps*, a set of *ordering constraints* that restrict the order in which steps are to be executed, and a set of *auxiliary constraints* that restrict the value of state variables over particular intervals of time. Each step is associated with a state-space operator. To distinguish between multiple instances of the same operator appearing in a plan, we assign each step a unique integer i and represent the ith step as the pair (i, α_i) where α_i is the operator associated with the ith step.

Figure 67.2 shows a partial plan π_{eg} consisting of seven steps. The plan π_{eg} is represented as follows:

$$\langle \; \{(0, \alpha_0), (1, \alpha_1), (2, \alpha_2), (3, \alpha_3), (4, \alpha_4), (5, \alpha_5), (\infty, \alpha_\infty)\}, \\ \{(0 \preceq 1), (1 \prec 2), (1 \prec 4), (2 \prec 3), (3 \prec 5), (4 \prec 5), (5 \preceq \infty)\}, \\ \{(1 \overset{Q}{-} 2), (3 \overset{R}{-} \infty)\} \; \rangle$$

An ordering constraint of the form $(i \prec j)$ indicates that step i precedes step j. An ordering constraint of the form $(i \preceq j)$ indicates that step i is contiguous with step j, that is, step i precedes step j and no other steps intervene. The steps are *partially ordered* in that step 2 can occur either before or after step 4. An auxiliary constraint of the form $(i \overset{P}{-} j)$ is called an *interval preservation constraint* and indicates that P is to be preserved in the range between steps i and j (and therefore no operator with postcondition $\neg P$ should occur between steps i and j). In particular, according to the constraint $(3 \overset{R}{-} \infty)$, step 4 should not occur between steps 3 and ∞.

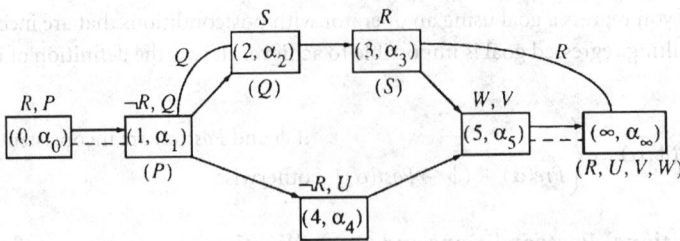

FIGURE 67.2 This figure depicts the partial plan π_{eg}. The postconditions (effects) of the steps are shown above the steps, whereas the preconditions are shown below the steps in parentheses. The ordering constraints between steps are shown by arrows. The interval preservation constraints are shown by arcs, whereas the contiguity constraints are shown by dashed lines.

The set of steps $\{\sigma_1, \sigma_2, \ldots, \sigma_n\}$ with contiguity constraints

$$\{(\sigma_0 \preceq \sigma_1), (\sigma_1 \preceq \sigma_2), \ldots, (\sigma_{n-1} \preceq \sigma_n)\}$$

is called the *header* of the plan π. The last element of the header σ_n is called the *head step*. The state defined by $f(s_0, \langle \alpha_{\sigma_1}, \ldots, \alpha_{\sigma_n} \rangle)$, where α_{σ_i} is the operator associated with σ_i, is called the *head state*. In a similar manner, we can define the *tail*, *tail step*, and *tail state*.

As an example, the partial plan π_{eg} shown in Figure 67.2 has the steps 0 and 1 in its header, with step 1 being the head step. The head state (which is the state resulting from applying α_1 to the initial state) is $\{P, Q\}$. Similarly, the tail consists of steps 5 and ∞, with step 5 being the tail step. The tail state (which is the result of regressing the goal conditions through the operator α_5) is $\{R, U\}$.

67.3.2.3 Refinement Search

A large part of the work on plan synthesis in artificial intelligence falls under the rubric of refinement search. Refinement search can be seen as search in the space of partial plans. The search starts with the empty partial plan, and adds details to that plan until a complete plan results. Semantically, a partial plan can be seen as a shorthand notation for the set of complete plans (action sequences) that are consistent with the constraints. A refinement strategy converts a partial plan π into a set of new plans $\{\pi_1, \ldots, \pi_n\}$ such that all of the potential solutions represented by π are represented by at least one of $\pi_1, \ldots, \pi_n$. Syntactically, this is accomplished by generating each of the children plans (refinements) by adding additional constraints to π.

The following is a general template for refinement search. The search starts with the null plan $\langle \{(0, \alpha_0), (\infty, \alpha_\infty)\}, \{(0 \prec \infty)\}, \{\} \rangle$, where α_0 is a dummy operator with no preconditions and postconditions corresponding to the initial state, and α_∞ is a dummy operator with no postconditions and preconditions corresponding the goal. For example, if we were trying to find a sequence of actions to transform the initial state $\{P, Q, \neg R\}$ into a state satisfying the goal $\{R\}$, then we would have $\text{Pre}(\alpha_0) = \{\}$, $\text{Post}(\alpha_0) = \{P, Q, \neg R\}$, $\text{Pre}(\alpha_\infty) = \{R\}$, and $\text{Post}(\alpha_\infty) = \{\}$.

We define a generic refinement procedure, $\text{Refine}(\pi)$, as follows [Kambhampati et al. 1995]:

1. If an action sequence $\langle \alpha_1, \alpha_2, \ldots, \alpha_n \rangle$ corresponds to a total order consistent with both ordering constraints and the auxiliary constraints of π, and is a solution to the planning problem, then terminate and return $\langle \alpha_1, \alpha_2, \ldots, \alpha_n \rangle$

2. If the constraints in π are inconsistent, then eliminate π from future consideration

3. Select a refinement strategy, and apply the strategy to π and add the resulting refinements to the set of plans under consideration

4. Select a plan π' from those under consideration and call $\text{Refine}(\pi')$

In step 3, the search selects a refinement strategy to be applied to the partial plan. There are several possible choices here, corresponding intuitively to different ways of splitting the set of potential solutions represented by the plan. In the following sections, we outline four popular refinement strategies employed in the planning literature.

67.3.2.4 State-Space Refinements

The most straightforward way of refining partial plans involves using progression to convert the initial state into a state satisfying the goal conditions, or using regression to convert a set of goal conditions into a set of conditions that are satisfied in the initial state. From the point of view of partial plans, this corresponds to growing the plan from either the beginning or the end.

Progression (or forward state-space) refinement involves advancing the head state by adding a step σ, such that the preconditions of α_σ are satisfied in the current head state, to the header of the plan. The step σ may be newly added to the plan or currently present in the plan. In either case, it is made contiguous to the current head step and becomes the new head step.

As an example, one way of refining the plan π_{eg} in Figure 67.2 using progression refinement would be to apply an instance of the operator α_2 (either the instance that is currently in the plan $(2, \alpha_2)$ or a new instance) to the head state (recall that it is $\{P, Q\}$). This is accomplished by putting a contiguity constraint between $(2, \alpha_2)$ and the current head step $(1, \alpha_1)$ (thereby making the former the new head step).

We can also define a refinement strategy based on regression, which involves regressing the tail state of a plan through an operator. For example, the operator α_3 is applicable (in the backward direction) through this tail state (which is $\{R, U\}$), whereas the operator α_4 is not (since its postconditions are inconsistent with the tail state). Thus, one way of refining π_{eg} using regression refinement would be to apply an instance of the operator α_3 (either the existing instance in step 3 or a new one) to the tail state in the backward direction. This is accomplished by putting a contiguity constraint between $(3, \alpha_3)$ and the current tail step.

From a search control point of view, one of the important questions is deciding which of the many refinements generated by progression and regression refinements are most likely to lead to a solution. It is possible to gain some focus by using *state difference heuristics*, which prefer the refinements where the set difference between the tail state and the head state is the smallest.

Although the state difference heuristic works well enough for regression refinements, it does not provide sufficient focus to progression refinements. The problem is that in a realistic planning problem, there potentially may be many operators that are applicable in the current head state, and only a few of them may be relevant to the goals of the problem. Thus, the strategy of generating all of the refinements and ranking them with respect to the state difference heuristic can be prohibitively expensive. We need a method of automatically zeroing on those operators which are possibly relevant to the goals.

One popular way of generating the list of relevant operators is to use *means-ends analysis*. The general idea is the following. Suppose we have an operator α whose postconditions match a goal of the problem. Clearly, α is a relevant operator. If the preconditions of α are satisfied in the head state of the current partial plan, we can apply it directly. Suppose they are not all satisfied. In such a case, we can consider the preconditions of α as subgoals, look for an operator α' whose postconditions match one of these subgoals, and check if it is applicable to the head state. This type of recursive analysis can be continued to find the set of relevant operators, and focus progression refinement.

67.3.2.5 Plan-Space Refinements

As we have seen previously in state-space refinements, partial plans are extended by adding new steps and new contiguity constraints. The contiguity constraints are required since without them the head state and tail state are not well defined. State-space refinements have the disadvantage that they completely determine the order and position of every step introduced into the plan. Although it is easy to see whether or not a given step is relevant to a plan, often the precise position at which a step must occur in the final plan is not apparent until all of the steps have been added. In such situations, state-space refinement can lead to premature commitment to the order of steps, causing extensive backtracking.

Plan-space refinement attempts to avoid this premature commitment. The main idea in plan-space refinement is to shift the attention from advancing the world state to establishing goals. A precondition P of a step (i, α_i) in a plan is said to be *established* if there is some step (j, α_j) in the plan that precedes i and

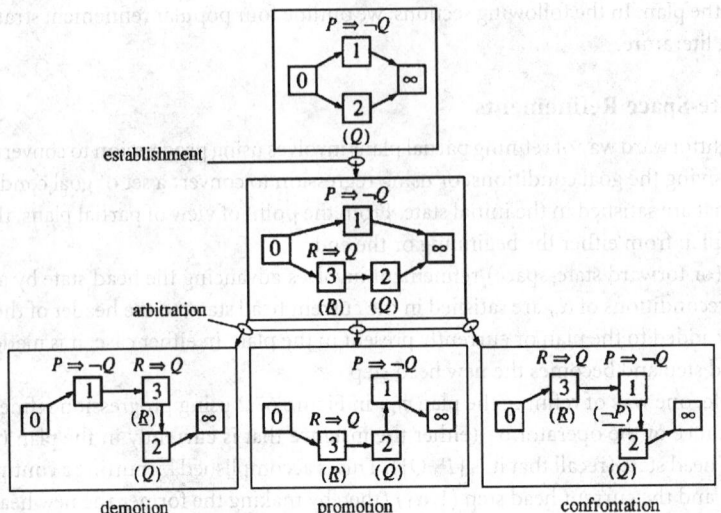

FIGURE 67.3 An example of precondition establishment. This diagram illustrates an attempt to establish Q for step 2. Establishing a postcondition can result in a potential conflict, which requires arbitration to avert the conflict. Underlined preconditions correspond to secondary preconditions.

causes P to be true, and no step that can possibly intervene between j and i has postconditions that are inconsistent with P. It is easy to see that if every precondition of every step in the plan is established, then that plan will be a solution plan. Plan-space refinement involves picking a precondition P of a step (i, α_i) in the partial plan, and adding enough additional step, ordering, and auxiliary constraints to ensure the establishment of P.

We illustrate the main ideas in precondition establishment through an example. Consider the partial plan at the top in Figure 67.3. Step 2 in this plan requires a precondition Q. To establish this precondition, we need a step which has Q as its postcondition. None of the existing steps have such a postcondition. Suppose an operator α_3 in the library has a postcondition $R \Rightarrow Q$. We introduce an instance of α_3 as step 3 into the plan. Step 3 is ordered to come before step 2 (and after step 0). Since α_3 makes Q true only when R is true before it, to make sure that Q will be true following step 3, we need to ensure that R is true before it. This can be done by posting R as a precondition of step 3. Since R is not a normal precondition of α_3, and is being posted only to guarantee one of its conditional effects, it is called a *secondary precondition* [Pednault 1988].

Now that we have introduced step 3 and ensured that it produces Q as a postcondition, we need to make sure that Q is not violated by any steps possibly intervening between steps 3 and 2. This phase of plan-space refinement is called *arbitration*. In our example, step 1, which can possibly intervene between steps 3 and 2, has a postcondition $P \Rightarrow \neg Q$ that is potentially inconsistent with Q. To avert this inconsistency, we can either order step 1 to come before step 3 (demotion), or order step 1 to come after step 2 (promotion), or ensure that the offending conditional effect will not occur. This last option, called confrontation, can be carried out by posting $\neg P$ as a (secondary) precondition of step 1. All these partial plans, corresponding to different ways of establishing the precondition Q at step 2 are returned as the refinements of the original plan.

One problem with this precondition-by-precondition establishment approach is that the steps added in establishing a precondition might unwittingly violate a previously established precondition. Although this does not affect the completeness of the refinement search, it can lead to wasted planning effort, and necessitate repeated establishments of the same precondition within the same search branch. Many variants of plan-space refinements avoid this inefficiency by *protecting* their establishments. Whenever a condition P of a step σ is established with the help of the effects of a step σ', an interval preservation constraint ($\sigma' \xrightarrow{P} \sigma$) is added to remember this establishment. If the steps introduced by later refinements violate this preservation constraint, those conflicts are handled much the same way as in the arbitration phase previously discussed. In the example shown in Figure 67.3, we can protect the establishment of precondition Q by adding the constraint $3 \xrightarrow{Q} 2$.

Although the order in which preconditions are selected for establishment does not have any effect on the completeness of a planner using plan-space refinement, it can have a significant impact on the size of the search space explored by the planner (and thereby its efficiency). Thus, any available domain specific information regarding the relative importance of the various types of preconditions can be gainfully exploited. As an example, in the travel domain, the action of taking a flight to go from one place to another may have as its preconditions having a reservation and being at the airport. To the extent that having a reservation is considered more critical than being at the airport, we would want to work on establishing the former first.

67.3.2.6 Task-Reduction Refinements

In both the state-space and plan-space refinements, the only knowledge that is available about the planning task is in terms of primitive actions (that can be executed by the underlying hardware), and their preconditions and postconditions. Often, one has more structured planning knowledge available in a domain. For example, in a travel planning domain, we might have the knowledge that one can reach a destination by either taking a flight or by taking a train. We may also know that taking a flight in turn involves making a reservation, buying a ticket, taking a cab to the airport, getting on the plane, etc. In such a situation, we can consider taking a flight as an abstract task (which cannot be directly executed by the hardware). This abstract task can then be reduced to a plan fragment consisting of other abstract or primitive tasks (in this case making a reservation, buying a ticket, going to the airport, getting on the plane). This way, if there are some high-level problems with the taking flight action and other goals (e.g., there is not going to be enough money to take a flight as well paying the rent), we can resolve them *before* we work on low-level details such as getting to the airport.

This idea forms the basis for task reduction refinement. Specifically, we assume that in addition to the knowledge about primitive actions, we also have some abstract actions, and a set of schemas (plan fragments) that can replace any given abstract action. Task reduction refinement takes a partial plan π containing abstract and primitive tasks, picks an abstract task σ, and for each reduction schema (plan fragment) that can be used to reduce σ, a refinement of π is generated with σ replaced by the reduction schema (plan fragment). As an example, consider the partial plan on the left in Figure 67.4. Suppose the operator α_2 is an abstract operator. The central box in Figure 67.4 shows a reduction schema for step 2, and the partial plan shown on the right of the figure shows the result of refining the original plan with this reduction schema. At this point any interactions between the newly introduced plan fragment and the previously existing plan steps can be resolved using techniques such as promotion, demotion, and confrontation discussed in the context of plan-space refinement. This type of reduction is carried out until all of the tasks are primitive.

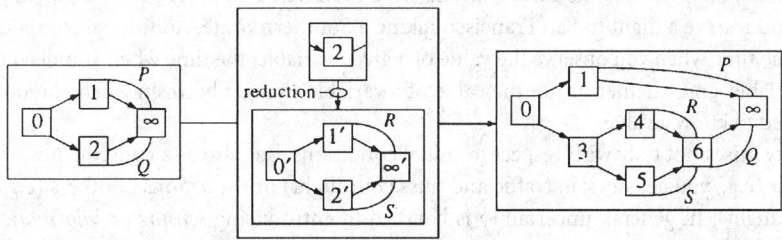

FIGURE 67.4 Step 2 in the partial plan shown on the left is reduced to obtain a new partial plan shown on the right. In the new plan, step 2 is replaced with the (renamed) steps and constraints specified in the reduction shown in the center box.

In some ways, task reduction refinements can be seen as macrorefinements that package together a series of state-space and plan-space refinements, thereby reducing a considerable amount of search. This, and the fact that in most planning domains, canned reduction schemas are readily available, have made task reduction refinement a very popular refinement choice for many applications.

67.3.2.7 Hybrid Refinements

Although early refinement planning systems tended to subscribe exclusively to a single refinement strategy, it is possible and often effective to use multiple refinement strategies. As an example, the partial plan π_{eg} shown in Figure 67.2 can be refined with progression refinement (e.g., by putting a contiguity constraint between step 1 and step 2), with regression refinement (e.g., by putting a contiguity constraint between step 3 and step 5), or plan-space refinement (e.g., by establishing the precondition S of step 3 with the help of the effect step 2). Finally, if the operator α_4 is a nonprimitive operator, we can also use task reduction refinement to replace α_4 with its reduction schema. There is some evidence that planners using multiple refinement strategies intelligently can outperform those using single refinement strategies [Kambhampati 1995]. However, the question as to which refinement strategy should be preferred when is still largely open.

67.3.2.8 Handling Incomplete Information

Although the refinement methods just described were developed in the context of planning problems where the initial state is completely specified, they can be extended to handle incompletely specified initial states. Incomplete specification of the initial state means that the values of some of the state variables in the initial state are not specified. Such incomplete specification can be handled as long as the state variables are observable (i.e., the correct value of the variable can be obtained at execution time).

Suppose the initial state is incomplete with respect to the value of the state variable ϕ. If ϕ has only a small number of values, K, then we can consider this planning problem to be a collection of K problems, each with the same goal and a complete initial state in which ϕ takes on a specific value. Once the K problems are solved, we can make a K-way conditional plan that gives the correct plan conditional given the observed value of the state variable ϕ. There exist methods for extending refinement strategies so that instead of working on K unconditional plans with significant overlap, a single, multithreaded conditional plan is generated [Peot and Shachter 1991].

Conditional planning can be very expensive in situations in which the unspecified variable ϕ has a large set of possible values or there are several unspecified variables. If there are U unspecified variables each with K possible values, then a conditional plan that covers all possible contingencies has to account for U^K possible initial states. In some cases, we can avoid a combinatorial explosion by performing some amount of on-line planning; first plan to obtain the necessary information, then, after obtaining this information, plan what to do next. Unfortunately, this on-line approach has potential problems.

In travel planning, for example, you could wait until you arrive in Boston's Logan Airport to check on the weather in Chicago in order to plan whether to take a Southern or Northern route to San Francisco. But, if you do wait and it is snowing in Chicago, you may find out that all of the flights taking Southern routes are already sold out. In this case, it would have been better to anticipate the possibility of snow in Chicago and reserve a flight to San Francisco taking a Southern route. Additional complications arise concerning the time when you observe the value of a given variable, the time when you need to know the value of a variable, and whether or not the value of a variable changes between when you observe it and when you need to know a value.

Uncertainty arises not only with respect to initial conditions, but also as a consequence of the actions of the planner (e.g., you get stuck in traffic and miss your flight) or the actions of others (e.g., the airline cancels your flight). In general, uncertainty is handled by introducing *sensing* or *information gathering* actions (operators). These operators have preconditions and postconditions similar to other operators, but some of the postconditions, those corresponding to the consequences of information gathering, are nondeterministic; we will not know the actual value of these postconditions until after we have executed the action [Etzioni et al. 1992].

The approach to conditional planning just sketched theoretically extends to arbitrary sources of uncertainty, but in practice search has to be limited to consider only outcomes that are likely to have a significant impact on performance. Subsequently we briefly consider planning using stochastic models that quantify uncertainty involving outcomes.

67.3.2.9 Repair Methods in Planning

The refinement methods for plan synthesis described in this section assume access to the complete dynamics of the system. Sometimes, the system dynamics are complex enough that using the full model during plan synthesis can be inefficient. In many such domains, it is often possible to come up with a simplified model of the dynamics that is approximately correct. As an example, in the travel domain, the action of taking a flight from one city to another has potentially many preconditions, including ones such as: having enough money to buy tickets and enough clean clothes to take on the travel. Often, most of these preconditions are trivially satisfied, and we are justified in approximating the set of preconditions to simply ensure that we have a reservation and are at the airport on time. In such problems, a simplified model can be used to drive plan generation using refinement methods, and the resulting plan can then be *tested* with respect to the complete dynamical model of the system. If the testing shows the plan to be correct, we are done. If not, the plan needs to be *repaired* or *debugged*. This repair process involves both adding and deleting constraints from the plan.

If the complete dynamical model is declarative (instead of being a black box), it is possible to extract from the testing phase an *explanation* of why the plan is incorrect (for example, in terms of some of the preconditions that are not satisfied, or are violated by some of the indirect effects of actions). This explanation can then be used to focus the repair activity [Simmons and Davis 1987, Hammond 1989]. Similar repair methods can also be useful in situations where we have probably approximately correct *canned* plans for generic types of goals, and we would like to solve planning problems involving collections of these goals by putting the relevant canned plans together and modifying them.

67.3.3 Scheduling with Deterministic Dynamics

As we mentioned earlier, scheduling is typically concerned with deciding when to carry out a given set of actions so as to satisfy various types of constraints on the order in which the actions need to be performed, and the ways in which different resources are consumed. Artificial intelligence approaches to scheduling typically declaratively represent and reason with the constraints.

Constraint-based schedulers used in a real applications generally employ sophisticated programming languages to represent a range of constraints. For example, many schedulers require temporal constraints that specify precedence, contiguity, duration, and earliest and latest start and completion times for tasks.

In some schedulers, temporal constraints are enforced rigidly, so that they never need to be checked during search. Many scheduling problems also manage a variety of resources. In the job-shop scheduling problem, machines are resources; only one task can be performed on a machine at a time. Other resources encountered in scheduling problems include fuel, storage space, human operators and crew, vehicles, and assorted other equipment. Tasks have constraints that specify their resource requirements and resources have capacity constraints that ensure that a schedule does not over allocate resources.

In addition to constraints on the time of occurrence and resources used by tasks, there are also constraints on the state of the world that are imposed by physics: the dynamics governing the environment. For example, a switch can be in the on or off position but not both at the same time. In some scheduling problems, the dynamical system is represented as a large set of state constraints.

67.3.3.1 Scheduling and Constraint Satisfaction

Scheduling problems are typically represented in terms of a set of variables and constraints on their values. A schedule is then represented as an assignment of values to all of the variables that satisfies all of the constraints. The resulting formulation of scheduling problems is called a *constraint satisfaction problem* [Tsang 1993].

Formally, a constraint satisfaction problem is specified by a set of n variables $\{x_1, \ldots, x_n\}$, their respective value domains $\Omega_1, \ldots, \Omega_n$, and a set of m constraints $\{C_1, \ldots, C_m\}$. A constraint C_i involves a subset $\{x_{i_1}, \ldots, x_{i_k}\}$ of the set of all variables $\{x_1, \ldots, x_n\}$ and is defined by a subset of the Cartesian product $\Omega_{i_1} \times \cdots \times \Omega_{i_k}$. A constraint C_i is *satisfied* by a particular assignment in which $x_{i_1} \leftarrow v_{i_1}, \ldots, x_{i_k} \leftarrow v_{i_k}$ just in case $\langle v_{i_1}, \ldots, v_{i_k} \rangle$ is in the subset of $\Omega_{i_1} \times \cdots \times \Omega_{i_k}$ that defines C_i. A solution is an assignment of values to all of the variables such that all of the constraints are satisfied. There is a performance function that maps every complete assignment to a numerical value representing the cost of the assignment. An optimal solution is a solution that has the lowest cost.

As an example, consider the following formulation of a simplified version of the job-shop scheduling problem as a constraint satisfaction problem. Suppose we have N jobs, $1, 2, \ldots, N$, each consisting of a single task, and M machines, $1, 2, \ldots, M$. Since there is exactly one task for each job, we just refer to jobs. Assume that each job takes one unit of time and there are T time units, $1, 2, \ldots, T$. Let $z_{ij} = 1$ if the jth machine can handle the ith job and $z_{ij} = 0$ otherwise. The z_{ij} are specified in the description of the problem. Let x_i for $1 \leq i \leq N$ take on values from $\{j \mid z_{ij} = 1\} \times \{1, 2, \ldots, T\}$, where $x_i = (j, k)$ indicates that the ith job is assigned to the jth machine during the kth time unit. The x_i are assigned values in the process of planning. There are $N(N-1)$ constraints of the form $x_i \neq x_j$, where $1 \leq i, j \leq N$, and $i \neq j$. We are searching for an assignment to the x_i that satisfies these constraints.

67.3.3.2 Refinement-Based Methods

A refinement-based method for solving a constraint satisfaction problem progresses by incrementally assigning values to each of the variables. A partial plan (schedule) π is represented as partial assignment of values to variables $\{x_{\pi_1} \leftarrow v_{\pi_1}, \ldots, x_{\pi_k} \leftarrow v_{\pi_k}\}$, where $\{x_{\pi_1}, \ldots, x_{\pi_k}\}$ is a subset of the set of all variables $\{x_1, \ldots, x_n\}$. The partial assignment π can be seen as a shorthand notation for all of the complete assignments that agree on the assignment of values to the variables in $\{x_{\pi_1}, \ldots, x_{\pi_k}\}$. A partial assignment π is said to be inconsistent if the assignment of values to variables in π already violates one or more constraints. If the partial assignment π is consistent, it can be refined by selecting a variable x_j that is not yet assigned a value in π and extending π to produce a set of refinements each of which assigns x_j one of the possible values from its domain Ω_j. Thus, the set of refinements of π is $\{\pi \cup \{x_j \leftarrow v\} \mid v \in \Omega_j\}$. In the case of satisficing scheduling, search terminates when a complete and consistent assignment is produced. In the case of optimizing scheduling, search is continued with *branch-and-bound* techniques until an optimal solution is found.

From the point of view of efficiency, it is known that the order in which the variables are considered and the order in which the values of the variables are considered during refinement have a significant impact on the efficiency of search. Considering variables with the least number of possible values first is known to provide good performance in many domains. Other ways of improving search efficiency include using *lookahead* techniques to prune inconsistent partial assignments ahead of time, to process the domains of the remaining variables so that any infeasible values are removed, or using dependency directed backtracking techniques to recover from inconsistent partial assignments intelligently. See Tsang [1993] for a description of these techniques and their tradeoffs.

67.3.3.3 Repair-Based Methods

A repair-based method for solving constraint satisfaction problems is to start with an assignment to all of the variables in which not all of the constraints are satisfied and reassign a subset of the variables so that more of the constraints are satisfied. Reassigning a subset of the variables is referred to as repairing an assignment. Consider the following repair method for solving the simplified job-shop scheduling problem.

We say that two variables x_i and x_j $(i \neq j)$ *conflict* if their values violate a constraint; in the simplified job-shop scheduling problem considered here, a constraint is violated if $x_i = x_j$. *Min-conflicts* is a heuristic for repairing an existing assignment that violates some of the constraints to obtain a new assignment that violates fewer constraints. The hope is that by performing a short sequence of repairs as determined by the min-conflicts heuristic we obtain an assignment that satisfies all of the constraints. The min-conflicts heuristic counsels us to select a variable that is in conflict and assign it a new value that minimizes the number of conflicts. See the Johnston and Minton article in [Zweben and Fox 1994] for more on the min-conflicts heuristic.

Min-conflicts is a special case of a more general strategy that proceeds by making *local* repairs. In the job-shop scheduling problem, a local repair corresponds to a change in the assignment of a single variable.

For the traveling salesperson problem, there is a very effective local repair method that works quite well in practice. Suppose that there are five cities A, B, C, D, E, and an existing tour (a path consisting of a sequence of edges beginning and ending in the same city) $(A, B), (B, C), (C, D), (D, E), (E, A)$. Take two edges in the tour, say (A, B) and (C, D), and consider the length of the tour $(A, C), (C, B), (B, D), (D, E), (E, A)$ that results from replacing (A, B) and (C, D) with (A, C) and (B, D). Try all possible pairs of edges [there are $O(L^2)$ such edges where L is the number of cities], and make the replacement (repair) that results in the shortest tour. Continue to make repairs in this manner until no improvement (reduction in the length of the resulting tour) is possible. Lin and Kernighan's algorithm, which is based on this local repair method, generates solutions that are within 10% of the length of the optimal tour on a large class of practical problems [Lin and Kernighan 1973].

67.3.3.4 Rescheduling and Iterative Repair Methods

Repair methods are typically implemented with iterative search methods; at any point during the scheduling process, there is a complete schedule available for use. This ready-when-you-are property of repair methods is important in applications that require frequent rescheduling, such as job shops in which change orders and new rush jobs are a common occurrence.

Most repair methods employ greedy strategies that attempt to improve the current schedule on every iteration by making local repairs. Such greedy strategies often have a problem familiar to researchers in combinatorial optimization. The problem is that many repair methods, especially those that perform only local repairs, are liable to converge to local extrema of the performance function and thereby miss an optimal solution. In many cases, these local extrema correspond to very poor solutions.

To improve performance and reduce the risk of becoming stuck in local extrema corresponding to badly suboptimal solutions, some schedulers employ stochastic techniques that occasionally choose to make repairs other than those suggested by their heuristics. Simulated annealing [Kirkpatrick et al. 1983] is one example of a stochastic search method used to escape local extrema in scheduling. In simulated annealing, there is a certain probability that the scheduler will choose a repair other than the one suggested by the scheduler's heuristics. These random repairs force the scheduler to consider repairs that at first may not look promising but in the long term lead to better solutions. Over the course of scheduling this probability is gradually reduced to zero. See the article by Zweben et al. in Zweben and Fox [1994] for more on iterative repair methods using simulated annealing.

Another way of reducing the risk of getting stuck in local extrema involves making the underlying search systematic (so that it eventually visits all potential solutions). However, traditional systematic search methods tend to be too rigid to exploit local repair methods such as the min-conflicts heuristic. In general, local repair methods attempt to direct the search by exploiting the local gradients in the search space. This guidance can sometimes be at odds with the commitments that have already been made in the current search branch. Iterative methods do not have this problem since they do not do any bookkeeping about the current state of the search. Recent work on partial-order dynamic backtracking algorithms [Ginsberg and McAllester 1994] provides an elegant way of keeping both systematicity and freedom of movement.

67.3.4 Improving Efficiency

Whereas the previous sections surveyed the methods used to organize the search for plans and discussed their relative advantages, as observed in the section on complexity results most planning problems are computationally hard. The only way we can expect efficient performance is to exploit the structure and idiosyncrasies of the specific applications. One attractive possibility involves dynamically customizing the performance of a general-purpose search algorithm to the structure and distribution of the application problems. A variety of machine learning methods have been developed and used for this purpose. We briefly survey some of these methods.

One of the simplest ways of improving performance over time involves caching plans for frequently occurring problems and subproblems, and reusing them in subsequent planning scenarios. This approach is called *case-based planning (scheduling)* [Hammond 1989, Kambhampati and Hendler 1992] and is motivated by similar considerations to those motivating task-reduction refinements. In storing a previous planning experience, we have two choices: store the final plan or store the plan along with the search decisions that lead to the plan. In the latter case, we exploit the previous experience by *replaying* the previous decisions in the new situation.

Caching typically involves only storing the information about the successful plan and the decisions leading to it. Often, there is valuable information in the search failures encountered in coming up with the successful plan. By analyzing the search failures and using *explanation-based learning* techniques, it is possible to learn *search control rules* that, for example, can be used to advise a planner as to which refinement or repair to pursue under what circumstances. For more about the connections between planning and learning see Minton [1992].

67.3.5 Approximation in Stochastic Domains

In this section, we consider a planning problem involving stochastic dynamics. We are interested in generating conditional plans for the case in which the state is completely observable [the output function is the identity $h(x_t) = x_t$] and the performance measure is expected discounted cumulative cost with discount γ. This constitutes an extreme case of closed-loop planning in which the executor is able to observe the current state at any time without error and without cost.

In this case, a plan is just a mapping from (observable) states to actions $\pi : \mathcal{S} \to \mathcal{A}$. To simplify the presentation, we notate states with the integers $0, 1, \ldots, |\mathcal{S}|$, where $s_0 = 0$ is the initial state. We refer to the performance of a plan π starting in state i as $J(\pi \mid i)$. We can compute the performance of a plan by solving the following set of $|\mathcal{S}| + 1$ equations in $|\mathcal{S}| + 1$ unknowns,

$$J(\pi \mid i) = C(i) + \gamma \sum_{j=0}^{|\mathcal{S}|} Pr(f(i, \pi(i)) = j \mid i, \pi(i)) J(\pi \mid j)$$

The objective in planning is to find a plan π from the set of all possible plans Πs such that for all $\pi' \in \Pi$, $J(\pi \mid i) \geq J(\pi' \mid i)$ for $0 \leq i \leq |\mathcal{S}|$.

As an aside, we note that the conditional probability distribution governing state transitions, $Pr(f(i, \pi(i)) = j \mid i, \pi(i))$, can be specified in terms of *probabilistic state space operators*, allowing us to apply the techniques of the section on planning with deterministic dynamics. A probabilistic state-space operator α is a set of triples of the form $\langle \phi, \rho, \omega \rangle$ where ϕ is a set of preconditions, ρ is a probability, and ω is a set of postconditions. Semantically, if ϕ is satisfied just prior to α, then with probability ρ the postconditions in ω are satisfied immediately following α. If a proposition is not included in ϕ, then it is assumed not to affect the outcome of α; if a proposition is not included in ω, then it is assumed to be unchanged by α. For example, given the following representation for α:

$$\alpha = \{\langle \{P\}, 1, \emptyset \rangle, \langle \{\neg P\}, 0.2, \{P\} \rangle, \langle \{\neg P\}, 0.8, \{\neg P\} \rangle\}$$

if P is true prior to α, nothing is changed following α; but if P is false, then 20% of the time P becomes true and 80% of the time P remains false. For more on planning in stochastic domains using probabilistic state-space operators, see Kushmerick et al. [1994].

There are well-known methods for computing an optimal plan for the problem previously described [Puterman 1994]. Most of these methods proceed using iterative repair-based methods that work by improving an existing plan π using the computed function $J(\pi \mid i)$. On each iteration, we end up with a new plan π' and must calculate $J(\pi' \mid i)$ for all i. If, as we assumed earlier, $|\mathcal{S}|$ is exponential in the number of state variables, then we are going to have some trouble solving a system of $|\mathcal{S}| + 1$ equations. In the rest of this section, we consider one possible way to avoid incurring an exponential amount of work in evaluating the performance of a given plan.

Suppose that we know the initial state s_0 and a bound $C_{\max}$ ($C_{\max} \geq \max_i C(i)$) on the maximum cost incurred in any state. Let π be any plan, $J_\infty(\pi) = J(\pi \mid 0)$ be the performance of π accounting for an infinite sequence of state transitions, and $J_K(\pi)$ the performance of π accounting for only K state transitions. We can bound the difference between these two measures of performance as follows (see Fiechter [1994] for a proof):

$$|J_\infty(\pi) - J_K(\pi)| \leq \gamma^K C_{\max}/(1 - \gamma)$$

These result implies that if we are willing to sacrifice a (maximum) error of $\gamma^K C_{\max}/(1-\gamma)$ in measuring the performance of plans, we need only concern ourselves with histories of length K. So how do we calculate $J_K(\pi)$? The answer is a familiar one in statistics; namely, we estimate $J_K(\pi)$ by sampling the space of K-length histories.

Using a factored representation of the conditional probability distribution governing state transitions, we can compute a random K-length history in time polynomial in K and N (the number of state variables), assuming that M (the maximum dimensionality of a state-variable function) is constant. The algorithm is simply, given s_0, for $t = 0$ to $K - 1$, determine s_{t+1} according to the distribution $Pr(s_{t+1} \mid s_t, \pi(s_t))$. For each history $\langle s_0, \ldots, s_K \rangle$ so determined, we compute the quantity $V(\langle s_0, \ldots, s_K \rangle) = \sum_{j=0}^{K} \gamma^j C(s_j)$ and refer to this as one *sample*.

If we compute enough samples and take their average, we will have an accurate estimate of $J_K(\pi)$. The following algorithm takes two parameters, ϵ and δ, and computes an estimate $\hat{J}_K(\pi)$ of $J_K(\pi)$ such that

$$Pr[J_K(\pi)(1 - \epsilon) \leq \hat{J}_K(\pi) \leq J_K(\pi)(1 + \epsilon)] > 1 - \delta$$

1. $T \leftarrow 0; Y \leftarrow 0$
2. $S \leftarrow 4 \log(2/\delta)(1 + \epsilon)/\epsilon^2$
3. *While* $Y < S$ do

    a. $T \leftarrow T + 1$
    b. *Generate a random history* $\langle s_0, \ldots, s_K \rangle$
    c. $Y \leftarrow Y + V(\langle s_0, \ldots, s_K \rangle)$

4. *Return* $J_K(\pi) = S/T$

This algorithm terminates after generating $E[T]$ samples, where

$$E[T] \leq 4 \log(2/\delta)(1 + \epsilon) \left(J_K(\pi)\epsilon^2\right)^{-1}$$

so that the entire algorithm for approximating $J_\infty(\pi)$ runs in expected time polynomial in $1/\delta$, $1/\epsilon$, $1/(1 - \gamma)$ (see Dagum et al. [1995] for a detailed analysis).

Approximating $J_\infty(\pi)$ is only one possible step in an algorithm for computing an optimal or near-optimal plan. In most iterative repair-based algorithms, the algorithm evaluates the current policy and then tries to improve it on each iteration. In order to have a polynomial time algorithm, we not only have to establish a polynomial bound on the time required for evaluation but also a polynomial bound on the total number of iterations. The point of this exercise is that when faced with combinatorial complexity, we need not give up but we may have to compromise. In practice, making reasonable tradeoffs is critical in solving planning and scheduling problems. The simple analysis demonstrates that we can trade time (the expected number of samples required) against the accuracy (determined by the ϵ factor) and reliability (determined by the δ factor) of our answers.

67.3.6 Practical Planning

There currently are no off-the-shelf software packages available for solving real-world planning problems. Of course, there do exist general-purpose planning systems. The SIPE [Wilkins 1988] and O-Plan [Currie and Tate 1991] systems are examples that have been around for some time and have been applied to a range of problems from spacecraft scheduling to fermentation planning for commercial breweries. In scheduling, several companies have sprung up to apply artificial intelligence scheduling techniques to commercial applications, but their software is proprietary and by no means turn-key. Moreover, these systems are rather large; they are really programming environments meant to support design and not necessarily to provide the basis for stand-alone products.

Why, you might ask, are there not convenient libraries in C, Pascal, and Lisp for solving planning problems much as there are libraries for solving linear programs? The answer to this question is complicated, but we can provide some explanation for why this state of affairs is to be expected. Before you can solve a planning problem you have to understand it and translate it into an appropriate language for expressing operators, goals, and initial conditions. Although it is true, at least in some academic sense, that most planning problems can be expressed in the language of propositional operators that we introduced earlier in this chapter, there are significant practical difficulties to realizing such a problem encoding. This is especially true in problems that require reasoning about geometry, physics, and continuous change.

In most problems, operators have to be encoded in terms of schemas and somehow generated on demand; such schema-based encodings require additional machinery for dealing with variables that draw upon work in automated theorem proving and logic programming. Dealing with quantification and disjunction, although possible in finite domains using propositional schemas, can be quite complex. Finally, in addition to just encoding the problem, it is also necessary to cope with the inevitable combinatorics that arise by encoding expert heuristic knowledge to guide search. Designing heuristic evaluation functions is more an art than a science and, to make matters worse, an art that requires deep knowledge of the particular domain.

The point is that the problem-dependent aspects of building planning systems are monumental in comparison with the problem-independent aspects that we have concentrated upon in this chapter. Building planning systems for real-world problems is further complicated by the fact that most people are uncomfortable turning over control to a completely automated system. As a consequence, the interface between humans and machines is a critical component in planning systems that we have not even touched upon in this brief overview.

To be fair, the existence of systems for solving linear programs does not imply off-the-shelf solutions to any real-world problems either. And, once you enter the realm of mixed integer and linear programs, the existence of systems for solving such programs is only of marginal comfort to those trying to solve real problems given that the combinatorics severely limit the effective use of such systems. The bottom line is that if you have a planning problem in which discrete-time, finite-state changes can be modeled as operators, then you can look for advice in books such as Wilkins's account of applying SIPE to real problems [Wilkins 1988] and look to the literature on heuristic search to implement the basic engine for guiding search given a heuristic evaluation function. But you should be suspicious of anyone offering a completely general-purpose system for solving planning problems. The general planning problem is just too hard to admit to quick off-the-shelf technological solutions.

67.4 Research Issues and Summary

In this chapter, we provide a framework for characterizing planning and scheduling problems that focuses on properties of the underlying dynamical system and the capabilities of the planning system to observe its surroundings. The presentation of specific techniques distinguishes between refinement-based methods that construct plans and schedules piece by piece, and repair-based methods that modify complete plans and schedules. Both refinement- and repair-based methods are generally applied in the context of heuristic search.

Most planning and scheduling problems are computationally complex. As a consequence of this complexity, most practical approaches rely on heuristics that exploit knowledge of the planning domain. Current research focuses on improving the efficiency of algorithms based on existing representations and on developing new representations for the underlying dynamics that account for important features of the domain (e.g., uncertainty) and allow for the encoding of appropriate heuristic knowledge. Given the complexity of most planning and scheduling problems, an important area for future research concerns identifying and quantifying tradeoffs, such as those involving solution quality and algorithmic complexity.

Planning and scheduling in artificial intelligence cover a wide range of techniques and issues. We have not attempted to be comprehensive in this relatively short chapter. Citations in the main text provide attribution for specifically mentioned techniques. These citations are not meant to be exhaustive by any means. General references are provided in the Further Information section at the end of this chapter.

Defining Terms

Closed-loop planner: A planning system that periodically makes observations of the current state of its environment and adjusts its plan in accord with these observations.

Dynamical system: A description of the environment in which plans are to be executed that account for the consequences of actions and the evolution of the state over time.

Goal: A subset of the set of all states such that a plan is judged successful if it results in the system ending up in one of these states.

History or state-space trajectory: A (possibly infinite) sequence of states generated by a dynamical system.

Off-line planning algorithm: A planning algorithm that performs all of its computations prior to executing any actions.

On-line planning algorithm: A planning algorithm in which planning computations and the execution of actions are carried out concurrently.

Open-loop planner: A planning system that executes its plans with no feedback from the environment, relying exclusively on its ability to accurately predict the evolution of the underlying dynamical system.

Optimizing: A performance criterion that requires maximizing or minimizing a specified measure of performance.

Plan: A specification for acting that maps from what is known at the time of execution to the set of actions.

Planning: A process that involves reasoning about the consequences of acting in order to choose from among a set of possible courses of action.

Progression: The operation of determining the resulting state of a dynamical system given some initial state and specified action.

Regression: The operation of transforming a given (target) goal into a prior (regressed) goal so that if a specified action is carried out in a state in which the regressed goal is satisfied, then the target goal will be satisfied in the resulting state.

Satisficing: A performance criterion in which some level of satisfactory performance is specified in terms of a goal or fixed performance threshold.

State-space operator: A representation for an individual action that maps each state into the state resulting from executing the action in the (initial) state.

State-transition function: A function that maps each state and action deterministically to a resulting state. In the stochastic case, this function is replaced by a conditional probability distribution.

References

Allen, J. F., Hendler, J., and Tate, A., Eds. 1990. *Readings in Planning*. Morgan Kaufmann, San Francisco, CA.

Allen, J. F., Kautz, H. A., Pelavin, R. N., and Tenenberg, J. D. 1991. *Reasoning about Plans*. Morgan Kaufmann, San Francisco, CA.

Bäckström, C. and Klein, I. 1991. Parallel non-binary planning in polynomial time, pp. 268–273. In *Proc. IJCAI 12*, IJCAII.

Bylander, T. 1991. Complexity results for planning, pp. 274–279. In *Proc. IJCAI 12*, IJCAII.

Chapman, D. 1987. Planning for conjunctive goals. *Artif. Intelligence* 32:333–377.

Currie, K. and Tate, A. 1991. O-Plan: the open planning architecture. *Artif. Intelligence* 51(1):49–86.

Dagum, P., Karp, R., Luby, M., and Ross, S. M. 1995. An optimal stopping rule for Monte Carlo estimation. In *Proc. 1995 Symp. Found. Comput. Sci.*

Dean, T., Allen, J., and Aloimonos, Y. 1995. *Artificial Intelligence: Theory and Practice*. Benjamin Cummings, Redwood City, CA.

Dean, T. and Boddy, M. 1988. Reasoning about partially ordered events. *Artif. Intelligence* 36(3):375–399.

Dean, T. and Kanazawa, K. 1989. A model for reasoning about persistence and causation. *Comput. Intelligence* 5(3):142–150.

Dean, T. and Wellman, M. 1991. *Planning and Control*. Morgan Kaufmann, San Francisco, CA.

Etzioni, O., Hanks, S., Weld D., Draper, D., Lesh, N., and Williamson, M. 1992. An approach to planning with incomplete information. In *Proc. 1992 Int. Conf. Principles Knowledge Representation Reasoning*.

Fiechter, C.-N. 1994. Efficient reinforcement learning, pp. 88–97. In *Proc. 7th Annu. ACM Conf. Comput. Learning Theory*.

Fikes, R. and Nilsson, N. J. 1971. Strips: a new approach to the application of theorem proving to problem solving. *Artif. Intelligence* 2:189–208.

Garey, M. R. and Johnson, D. S. 1979. *Computers and Intractibility: A Guide to the Theory of NP-Completeness*. W. H. Freeman, New York.

Georgeff, M. P. 1987. Planning. In *Annual Review of Computer Science*, Vol. 2, J. F. Traub, ed., Annual Review Incorp.

Ginsberg, M. L. and McAllester, D. 1994. GSAT and dynamic backtracking. In *Proc. 1994 Int. Conf. Principles Knowledge Representation and Reasoning*.

Graham, R. L., Lawler, E. L., Lenstra, J. K., Rinnooy, and Kan, A. H. G. 1977. Optimization and approximation in deterministic sequencing and scheduling: a survey. In *Proc. Discrete Optimization*.

Gupta, N. and Nau, D. S. 1991. Complexity results for blocks-world planning, pp. 629–633. In *Proc. AAAI-91*, AAAI.

Hammond, K. J. 1989. *Case-Based Planning*. Academic Press, New York.

Hendler, J., Tate, A., and Drummond, M. 1990. AI planning: systems and techniques. *AI Mag.* 11(2):61–77.

Kambhampati, S. 1995. A comparative analysis of partial-order planning and task-reduction planning. *ACM SIGART Bull* 6(1).

Kambhampati, S. and Hendler, J. 1992. A validation structure based theory of plan modification and reuse. *Artif. Intelligence* 55(2–3):193–258.

Kambhampati, S., Knoblock, C., and Yang, Q. 1995. Refinement search as a unifying framework for evaluating design tradeoffs in partial order planning. *Art. Intelligence* 76(1–2):167–238.

Kirkpatrick, S., Gelatt, C. D., and Vecchi, M. P. 1983. Optimization by simulated annealing. *Science* 220:671–680.

Kushmerick, N., Hanks, S., and Weld, D. 1994. An algorithm for probabilistic planning. In *Proc. AAAI-94*. AAAI.

Lawler, E. L., Lenstra, J. K., Rinnooy Kan, A. H. G., and Shmoys, D. B. 1985. *The Travelling Salesman Problem*. Wiley, New York.

Lin, S. and Kernighan, B. W. 1973. An effective heuristic for the travelling salesman problem. *Operations Res.* 21:498–516.

Minton, S., ed. 1992. *Machine Learning Methods for Planning and Scheduling*. Morgan Kaufmann, San Francisco, CA.

Papadimitriou, C. H. and Tsitsiklis, J. N. 1987. The complexity of Markov chain decision processes. *Math. Operations Res.* 12(3):441–450.

Pednault, E. P. D. 1988. Synthesizing plans that contain actions with context-dependent effects. *Comput. Intelligence* 4(4):356–372.

Penberthy, J. S. and Weld, D. S. 1992. UCPOP: a sound, complete, partial order planner for ADL, pp. 103–114. In *Proc. 1992 Int. Conf. Principles Knowledge Representation and Reasoning*.

Peot, M. and Shachter, R. 1991. Fusion and propagation with multiple observations in belief networks. *Artif. Intelligence* 48(3):299–318.

Puterman, M. L. 1994. *Markov Decision Processes*. Wiley, New York.

Rosenschein, S. 1981. Plan synthesis: a logical perspective, pp. 331–337. In *Proceedings IJCAI 7*, IJCAII.

Simmons, R. and Davis, R. 1987. Generate, test and debug: combining associational rules and causal models, pp. 1071–1078. In *Proceedings IJACI 10*, IJCAII.

Tsang, E. 1993. *Foundations of Constraint Satisfaction*. Academic, San Diego, CA.

Wilkins, D. E. 1988. *Practical Planning: Extending the Classical AI Planning Paradigm*. Morgan Kaufmann, San Francisco, CA.

Zweben, M. and Fox, M. S. 1994. *Intelligent Scheduling*. Morgan Kaufmann, San Francisco, CA.

Further Information

Research on planning and scheduling in artificial intelligence is published in the journals *Artificial Intelligence, Computational Intelligence,* and the *Journal of Artificial Intelligence Research*. Planning and scheduling work is also published in the proceedings of the International Joint Conference on Artificial Intelligence and the National Conference on Artificial Intelligence. Specialty conferences such as the International Conference on Artificial Intelligence Planning Systems and the European Workshop on Planning cover planning and scheduling exclusively.

Georgeff [1987] and Hendler et al. [1990] provide useful summaries of the state of the art. Allen et al. [1990] is a collection of readings that covers many important innovations in automated planning. Dean et al. [1995] and Penberthy and Weld [1992] provide somewhat more detailed accounts of the basic algorithms covered in this chapter. Zweben and Fox [1994] is a collection of readings that summarizes many of the basic techniques in knowledge-based scheduling. Allen et al. [1991] describe an approach to planning based on first-order logic. Dean and Wellman [1991] tie together techniques from planning in artificial intelligence, operations research, control theory, and the decision sciences.

68

Explanation-Based Learning

68.1	Introduction	68-1
68.2	Background	68-3
68.3	Explanation-Based and Empirical Learning	68-5
68.4	Explanation-Based Learning	68-7
68.5	Constructing Explanations	68-8
68.6	Generalizing	68-9

Irrelevant Feature Elimination • Identity Elimination
• Operationality Pruning • Disjunctive Augmentation

68.7	The Utility Problem: Selecting a Concept Descriptor	68-14
68.8	Annotated Summary of EBL Issues	68-16

Gerald DeJong
University of Illinois at Urbana-Champaign

68.1 Introduction

A machine learning system is one that automatically improves with experience, adapts to an external environment, or detects and extrapolates patterns. An appropriate machine learning technology could relieve the current economically dictated one-size-fits-all approach to application design. Help systems might specialize themselves to their users to choose an appropriate level of response-sophistication; portables might automatically prefer the most context-appropriate method to minimize power consumption; compilers may learn to optimize code to best exploit the processor, memory, and network resources of the machine on which they find themselves installed; and multimedia delivery systems might learn reliable time-varying patterns of available network bandwidth to optimize delivery decisions. The potential benefits to computer science of such abilities are immense and the opportunities ubiquitous. Machine learning promises to become the fractional horsepower motor of the information age.

Unfortunately, to date many formal results in machine learning and computational learning theory have been negative; they indicate that intractable numbers of training examples can be required for desirable real-world learning tasks. Such results are based upon statistical and information theoretic arguments and therefore apply to any algorithm. The reasoning can be paraphrased roughly along the following lines. When the desired concept is subtle and complex, a suitably flexible and expressive concept vocabulary must be employed to avoid cheating (i.e., directly encoding illegitimate preferences for the desired learning outcome). It follows that a great deal of evidence is required to tease apart the subtly different hypotheses. Each training example carries relatively little information and, thus, an inordinately large **training set** is required before we can be reasonably sure that an adequate concept has emerged. In fact, the numbers can be staggering: confidently acquiring good approximations to apparently simple everyday human-level concepts can require more training examples than there are molecules in the universe. We take this as an indication that current inductive approaches, which focus on empirical learning with an *a priori* bias, are missing some important aspects.

Explanation-based learning (EBL) offers an alternative. In it, the information embodied in the training examples is augmented through interaction with declarative prior knowledge. This prior knowledge (or **domain theory**) describes an expert's understanding of the relevant conceptual structure of the world. Interaction is managed by an inference engine that uses the prior domain knowledge to conjecture possible causal patterns underlying the training examples. That is, it suggests *explanations* for why a particular training example is assigned its training classification and actively looks for evidence to confirm or refute these conjectures. EBL represents one way to combine deductive inference with inductive inference. Traditionally, the deductive and inductive approaches are quite disparate but they have complementary strengths. In deductive logic, nonlinear interactions are easily specified and can be efficient to infer. Unfortunately, in many AI research systems, logic has proved too brittle to adequately model real-world phenomena. On the other hand, inductive or statistical inference has been found in many disciplines to be sufficiently robust to model real-world phenomena. However, nonlinear interactions can be exponentially expensive to infer from data. EBL uses the strengths of each to overcome the weaknesses of the other.

The deductive component of EBL embodies significant departures from conventional logic. Unlike the axiom set of conventional logic, the EBL domain theory need not perfectly capture world interactions. There is no micro-world or intended model on which to base a conventional possible-worlds semantics. Instead, we assume that the formal representations only approximately capture world interactions. As a result, an expression derivable from the domain theory, even using a sound inferencer, counts only as *evidence* for believing the expression. In conventional logic, any sound derivation is a theorem and, *ipso facto*, can be fully believed. Thus, the syntactic procedure of constructing explanations is deductive, but semantically it is more in line with Peirce's abduction.

Compared to conventional empirical learning, an EBL system needs to rely much less on the examples for selecting a concept to describe the underlying phenomenon. This greatly reduces the risk of **overfitting**.

To illustrate, consider learning a **classifier** or **concept descriptor** intended to predict cities where John would like to live. For training examples, John says that acceptable cities include San Francisco, Paris, Boston, and Toronto, and that he would not like to live in Peoria, Boise, Topeka, or El Paso. We might guess that he wants to live in a large city. Or perhaps John enjoys an active night life. As we discover John's opinion on additional training cities, the set of possible classification rules narrows.

Clearly, a plethora of reasonable classifiers are consistent with this set of city assignments. Interestingly, many other rules that are just as accurate on the training set seem far less reasonable. The positive examples all have poorer air quality than the negative ones. Perhaps John wants to live in a place with unhealthy air pollution levels. Perhaps John wishes to avoid cities with the letter "e" in their names. We as humans have a natural predisposition toward certain descriptors and away from others. This common-sense ability to avoid silly hypotheses is central to the computational tractability of human concept acquisition.

Common sense is intuitively compelling to humans but elusive to computers. The only mechanism in the framework of conventional inductive learning is to supply an *a priori* preference against such concepts. The problem is that such preferences should often not be *a priori* and absolute but contingent on our understanding, which can change with training. If we come to believe that John is an opera buff or a professor or that he may suffer from 'E'-phobia, then we should amend our prior preferences accordingly. The set of "reasonable" concepts changes with our beliefs about John.

An alternative analysis suggests that existing beliefs influence our interpretation of the training examples. If John is a human like us, we can infer a great deal about his desires and priorities. We cannot know exactly what kind of person John is, but knowing he *is* a person allows us to interpret training examples as direct evidence about his own individualistic traits. Once estimated, these human attributes can help us to prefer more reasonable concept descriptors. We can think of the domain theory as a mechanism by which input training examples can be embellished to magnify their information content. EBL offers a systematic way to incorporate approximate declarative and general prior world knowledge into an automated inductive concept learning system.

68.2 Background

The conventional formalization of induction requires two spaces: an **example space** and a **hypothesis space**. The example space is the set of all items of interest in the world. In the example, this is the set of all possible cities. Note that the example space is a mathematical construct and we need not explicitly represent all of its elements. It is sufficient to specify set membership rules and be able to assert properties about some small number of explicitly represented elements. What city properties need to be represented for individual examples? For this example, we might include the city's population, whether or not it has a prestigious university, number of parks, and so on. The example space must represent enough properties to support the distinctions required by the concept descriptor. The set of all expressible concept descriptors forms the hypothesis space.

Most often, each example is represented as a vector of conjoined feature values. The values themselves are ground predicate logic expressions. Thus, we might represent one city, CITY381, as

$$\text{name=Toronto} \wedge \text{population=697,000} \wedge \text{area=221.77} \wedge \ldots$$

An equal sign separates each feature name from its value. The symbol $\wedge$ indicates the logical AND connective. We interpret the expression as saying that CITY381 is equivalent to something whose name is Toronto, whose population is 697,000, and .... Thus, CITY381 is now defined by its features. As in other logical formalisms, the actual symbol used to denote this city is arbitrary. If CITY381 were replaced everywhere by CITY382, nothing would change. One can imagine including many additional features such as the city's crime rate or the average housing price.

Toronto is a city in which John would be happy. Thus, its classification is positive. For John, city classification is binary (each city is preferred by John or it is not). In other applications (e.g., identifying aircraft types from visual cues or diagnosing tree diseases from symptoms), many different classes can exist.

Figure 68.1 is a schematic depiction of the example space for John's city preference. San Francisco, Paris, Boston, and Toronto (the four positive examples in the training set) are represented by the symbol +; Peoria, Boise, Topeka, and El Paso are marked as −. For pedagogical purposes we will simply use two dimensions. In general, there may be many dimensions and they need not be ordered or metric. Importantly, distinct cities correspond to distinct points in the space.

A classifier or concept descriptor is any function that partitions the example space. Figure 68.2 depicts three sample concept descriptors for John's city preference. Each concept classifies some examples as positive (those within its boundary) and the others as negative. Concept C1 misclassifies four cities; one undesirable city is included and three desirable cities are excluded. Concepts C2 and C3 both correctly classify all eight cities of the training set but embody quite different partitionings of the example space. Thus, they venture different labelings for unseen cities. C2 might classify all large cities as desirable, whereas C3 might classify any city without the letter "e" in its name as desirable.

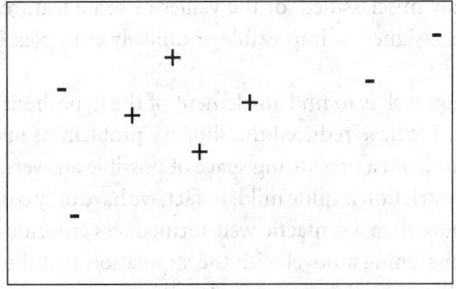

FIGURE 68.1 The training set.

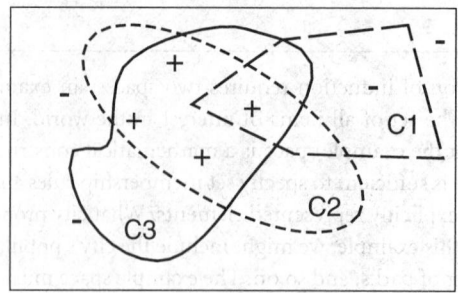

FIGURE 68.2   Three concept descriptors.

A concept descriptor can be represented using the same features employed to represent examples. Constrained variables augment ground expressions as feature values. Thus, a concept descriptor for large safe cities might be

$$\text{population} = ?x \land \text{crime-index} = ?y$$

with constraints $?x > 500,000$ and $?y < .03$

We might choose to allow disjunctions or other connectives. We might also allow more complex relations to serve as constraints among descriptor variables. By increasing the expressive power of the concept vocabulary, we allow more varied partitioning of the example space. The set of all expressible concept descriptors is the hypothesis space. Just as looking for the same needle is more challenging in a larger haystack, increasing the size of the hypothesis space by allowing a more expressive concept vocabulary can make the learning problem more difficult.

The unknown but objectively correct partitioning of the example space for the task at hand is called the **target concept**. In this example the target concept is embodied by John himself, who is the unimpeachable final authority on where he would like to live. Often, as in this example, the target concept is fundamentally unknowable. Most likely, we cannot fully capture John's city preferences. He might, for example, like large cities in general but have an unnatural fear of suspension bridges, a fondness for large groves of maple trees, or other subtle or peculiar personality quirks requiring distinctions not representable within our concept vocabulary. In machine learning, selecting a good approximate concept when the target concept is known not to be contained in the hypothesis space is known as agnostic learning. In the real world, it is the rule rather than the exception.

Even in those applications where we can be certain that within the hypothesis space lurks the true target concept, we may prefer an approximation. The target concept itself may be very complex, difficult to learn, or expensive to use, whereas a simpler approximation might be almost as accurate and much more efficient. Additionally, in most applications there is the possibility of noise. Some elements of the training set may be accidentally misclassified, or the values of some features may be incorrect, resulting in questionable training labels assigned or impossible or unlikely examples. These situations can often be treated in the agnostic regime.

Thus, the concept acquisition task is to find an element of the hypothesis space that is a good approximation of the target concept. We have reduced the slippery problem of inventing a new concept to the problem of finding a descriptor from a preexisting space of possible answers. Have we trivialized the concept formation task? No, this restriction is quite mild. In fact, we have only committed to a representational vocabulary. This is nothing more than a syntactic well-formedness criterion. Defining a hypothesis space is a bit like a publisher commissioning a novel with the stipulation that the manuscript be in English or requiring that it be typed with roman characters. Such restrictions are severe in the sense that they rule out far more items in our universe than they permit, but the author's creative latitude is not significantly diminished. So it is with properly defined hypothesis spaces. The commitment to a particular representation vocabulary for concept descriptors defines a space. It rules out far more things than it permits. But properly done, the hypothesis space still supports a sufficiently rich variety of concept descriptors so as not to trivialize the learning problem.

An empirical or conventional inductive learning algorithm searches the hypothesis space guided only by the training data. It entertains hypotheses from the hypothesis set until sufficient training evidence confirms one or indicates that no acceptable one is likely to be found. The search is seldom uniquely determined by the training set elements. The concept found is often a function of the learning algorithm's characteristics. Such concept preferences that go beyond the training data, including representational limitations, are collectively termed the **inductive bias** of the learning algorithm. It has been well established that inductive bias is inescapable in concept learning. Incidentally, among the important implications of this result is the impossibility of any form of Lockean *tabula rasa* learning.

Discipline is necessary in formulating the inductive bias. A concept vocabulary that is overly expressive can dilute the hypothesis space and greatly increase the complexity of the concept acquisition task. On the other hand, a vocabulary that is not expressive enough may preclude finding an adequate concept or (more often) may trivialize the search so that the algorithm is condemned to find a desired concept without relying on the training data as it should. Essentially, hypothesis design then functions as arcane programming to predispose the learner to do what the implementor knows to be the right thing for the learning problem.

68.3   Explanation-Based and Empirical Learning

Explanation-based learning is best viewed as a principled method for extracting the maximum information from the training examples. It works by constructing explanations for the training examples and using the explanations to guide the selection of a concept descriptor. The explanations interpret the examples, bringing into focus alternative coherent sets of features that might be important for correct classification. The explanations also augment the examples by inference, adding features deemed relevant to the classification task.

We now explore a brief and intuitive example illustrating the difference between the explanation-based and the conventional inductive approaches.

Suppose we are lost in the jungle with our pet gerbil. We have only enough food to keep ourselves alive and decide that bugs, which are plentiful and easy to catch, will have to suffice for the gerbil. Unfortunately, a significant number of insect-like jungle creatures are poisonous. To save our pet we must quickly acquire a descriptor that identifies nonpoisonous bugs.

Again, we represent examples as feature/value pairs. Features might include a bug's number of legs, the number of body parts, the average size, the bug's coloring, how many wings it has, what it seems to like to eat, what seems to like to eat it, where it lives, a measure of how social it is, etc. One insect we see might be represented as

$$\text{legs}=6 \wedge \text{body-parts}=3 \wedge \text{size}=2\text{cm} \wedge \text{color}=\text{bright-purple}$$
$$\wedge \text{ wings}=4 \wedge \text{wing-type}=\text{folding} \wedge \dots$$

Let us call this bug X7 for the 7th example of a bug that we catch. The bug representation vocabulary can also serve as the concept descriptor vocabulary as long as we allow constrained variables to be feature values. A concept descriptor for nonpoisonous bugs might be something like

$$\text{legs}=?\text{x1} \wedge \text{body-parts}=?\text{x1*2} \wedge \text{size}>1.5\text{cm} \wedge \text{color}=\text{purple}$$

which says that anything that has twice as many legs as body parts, is over 1.5 cm, and is purple will be considered nonpoisonous. This descriptor includes insects like X7 and long dark purple centipedes but excludes spiders (because they do not have enough body parts for their eight legs) and yellow butterflies (because they are the wrong color).

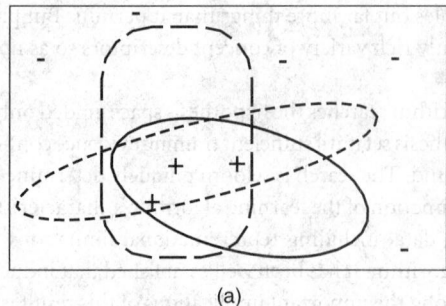 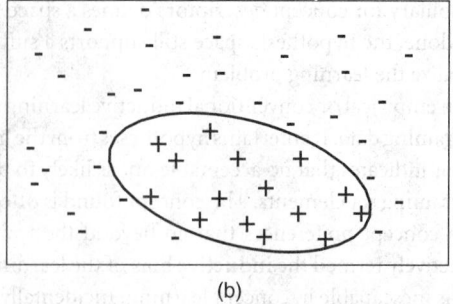

(a)   (b)

FIGURE 68.3   Training examples for an empirical learning approach: (a) hypothesized and (b) confirmed.

Empirical learning uncovers emergent patterns in the example space by viewing the training data as representative of all examples of interest. Patterns found within the training set are likely to be found in the world, provided the training set is a statistically adequate sample. In this example, we feed sample bugs to our gerbil to view the effect. An empirical system searches for an adequate concept descriptor by sampling broadly over the available bugs, revising estimates of the likelihood of various descriptors as evidence mounts. We may observe that the gerbil becomes nauseous and lethargic after eating X7, so we label it as poisonous. Next, X8, a reddish-brown many-legged worm that we find hidden on the jungle floor, is consumed with no apparent ill effects. It is labeled nonpoisonous. This process continues for several dozen bugs. A pattern begins to emerge. Figure 68.3a illustrates three of the potential concepts consistent with the first eight training examples. Assuming a reasonably expressive hypothesis space, there would be many such consistent concepts, and perhaps even an unboundedly large number.

We can be confident of the descriptor only after testing it on a statistically significant sample of bugs. This number can be quite large and depends on the expressivity of the hypothesis space. This can be quantified in a number of ways, the most popular being the Vapnik-Chervonenkis (or V-C) dimension. Perhaps after sampling several hundred or several thousand, we can statistically defend adopting the pattern that bugs that are either not bright purple or have more than ten legs and whose size is less than 3 cm are sufficiently plentiful and nonpoisonous to sustain our gerbil. This is illustrated in Figure 68.3b.

(legs>10 ∧ size<3cm) ∨ ¬ color=PURPLE

By contrast, the EBL approach searches for explanations of how gerbil poisoning/nonpoisoning can be connected to the observed bug's features using our background knowledge of the world. After observing that X7 makes our gerbil ill, we might recall the fact that poisonousness is a characteristic evolved to protect individuals of a species from being eaten. Furthermore, it is an effective deterrent only if the poisonous species itself is easily identifiable by its would-be predator. Thus, the bug should have a unique appearance, sharp patterns, bright colors, great regularity among individuals, etc. This explains the poisonousness of X7: the bright purple coloring is relevant, and the number of legs, the fact that it feeds on hemlock blossoms, etc. are irrelevant. The explanation also indicates that bugs that are plain-looking, camouflaged, or change their color to blend into their surroundings, are probably not poisonous. The palatability of X8 is justified by its plainness. Such an explanation provides evidence for and against the nonpoisonousness of many untested bugs. Figure 68.4a illustrates several possible positive and one negative explanation boundary. For a reasonable domain theory, there will be relatively few alternative ways of explaining positive or negative examples. Thus, if a few more test examples all turn out to be consistent with plainness, we can be quite confident that the rule of eating only plain-looking bugs will suffice for our gerbil. A confirmed EBL boundary is shown in Figure 68.4b. With EBL, the pattern emerges from the training data and explanation combined. Far fewer examples are required, which is fortunate for our gerbil.

Figures 68.3b and 68.4b show EBL and empirical approaches converging to the same concept boundaries in example space. This need not be the case. The particular concept selected may depend on many things,

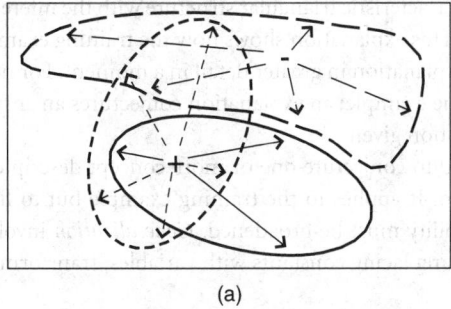 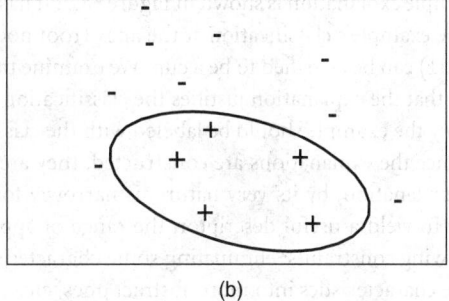

(a)   (b)

FIGURE 68.4   Training examples for an EBL concept: (a) hypothesized and (b) confirmed.

including the order of presentation of the training examples, and in EBL, aspects of the domain theory. For example, suppose the background knowledge supported this explanation of X7: bugs are very simple organisms; they are more likely to collect and concentrate poison from elsewhere than to manufacture it within their bodies. If we also know a good deal about which plants are poisonous, we might choose only those bugs that, like X8, are seen to prefer feeding on nonpoisonous plants. Alternatively or additionally, we might know that gerbils are similar to jungle mice. We notice that jungle mice feed almost exclusively on grubs of the mound-building beetles that construct colonies in jungle clearings. We might reason that mound-building beetle grubs are, therefore, likely to be edible for our gerbil. After a few sample grubs are successfully devoured, we might acquire a concept descriptor that fits only these grubs. Although quite different boundaries result, each concept descriptor is adequate for nourishing the gerbil. Any one or the disjunction of all can be used to keep our pet alive.

68.4   Explanation-Based Learning

As we have seen, by getting the most out of each training example, EBL can acquire a concept descriptor using relatively few training examples. The price of example efficiency is a domain theory. Acquiring a concept descriptor with EBL consists of three steps:

1. Constructing one or more explanations from training examples
2. Using the the explanation(s) and examples together to construct a set of hypothesized EBL concept descriptors
3. Selecting one or more of the descriptors to apply to the task at hand

The first step, constructing explanations, employs an inference mechanism over the domain theory. For our purposes, a domain theory is any set of prior beliefs about the world, and an inference mechanism is any procedure that suggests new beliefs by combining existing beliefs. Any such procedure is quite acceptable; inference by analogy and other unsound methods are perfectly consistent with explanation-based learning. However, the inference mechanism will usually be some sound logical procedure such as first-order resolution. If the domain theory is imperfect, the inferencer must not be of the refutational or indirect variety. This is because a flawed domain theory may already contain contradictions, and so discovering a contradiction after adding the negated goal may have nothing whatever to do with explaining it.

An **explanation** for a training example is any tree-structured graph with the following properties:

- Each leaf node is either a prior belief or a property of the example being explained.
- Each nonleaf node is the result of applying the inference procedure to prior nodes.
- The root node is the training classification assigned to the example.

A sample explanation is shown in Figure 68.7. It has a characteristic triangular structure with the inference of the example's classification at the apex (root node). This explanation shows how the training example (OBJ1) can be classified to be a cup. We examine this explanation in greater detail in a moment. For now, note that the explanation justifies the classification of the example; an explanation conjectures an answer to *why* the example should be labeled with the classification given.

Once the explanations are constructed, they are used to conjecture one or more concept descriptors. An explanation, by its very nature, is narrowly focused. It applies to the training example but to little else. To yield a useful descriptor, the range of applicability must be broadened. *Generalization* involves removing constraints: eliminating some characteristics, replacing constants with variables, transforming some characteristics into more abstract ones, etc.

68.5   Constructing Explanations

We now examine these steps in greater detail by way of an example. We consider learning a concept descriptor for a simplified drinking cup. A suitable domain theory for this task is given in Figure 68.5. The domain knowledge is represented as a collection of first-order Horn clauses. Horn theories are a popular formalism for knowledge representation in artificial intelligence (AI). They embody an effective compromise between expressiveness and computational tractability, but there is nothing particularly special about first-order Horn theories as far as EBL is concerned. The knowledge representation language must simply support the construction of an explanation that carries evidence for its conclusion.

This illustrates the difference in semantics of an EBL domain theory. Treated as a logical expression, R3 provides sufficient conditions to entail an object is "liftable." As a plausible domain theory expression, R3 suggests that "liftable" may be an important derived feature and that it might be adequately estimated from an object both being light and having a handle.

The domain theory: In our example, we provide the single training object, OBJ1, which is a known positive example of a cup. OBJ1 names the particular collection of properties shown in Figure 68.6a and Figure 68.6b. Two different representation schemes are presented. The first is a predicate calculus representation. It specifies a separate logical sentence for each of the relevant relations. The second is a semantic net representation. Here, each arrow or directed arc points from an object to a feature value that

R1: $\forall x$ drinkable-from$(x) \Rightarrow$ cup(x)

R2: $\forall x$ liftable$(x) \wedge$ open$(x) \Rightarrow$ drinkable-from(x)

R3: $\forall x \forall y$ weight$(x,$ LIGHT$) \wedge$ has-part$(x, y) \wedge$ isa$(y,$ HANDLE$) \Rightarrow$ liftable(x)

R4: $\forall x \forall y$ has-part$(x,y) \wedge$ isa$(y,$CONCAVITY$)] \Rightarrow$ open(x)

R5: $\forall x \forall y$ has-part$(x,y) \wedge$ isa$(y,$CONCAVITY$)] \wedge$ orientation$(x,$ UPWARD$)] \Rightarrow$ liquid-container(x)

R6: $\forall x \forall y$ has-part$(x,y) \wedge$ isa$(y,$FLAT-BOTTOM$)] \Rightarrow$ stable(x)

FIGURE 68.5   The CUP domain theory.

owner(OBJ1, HERMAN)   weight(OBJ1, LIGHT)

has-part(OBJ!, HAN31)   has-part(OBJ1, BOT7)

color(OBJ1, RED)   has-part(OBJ1, CONC12)

isa(CONC12, CONCAVITY)   orientation(CONC12, UPWARD)

shape(CONC12, CYLINDER)   isa(BOT7, FLAT-BOTTOM)

isa(HAN31, HANDLE)   shape(HAN31, CIRCLE)

FIGURE 68.6a   OBJ1, a cup, in predicate notation.

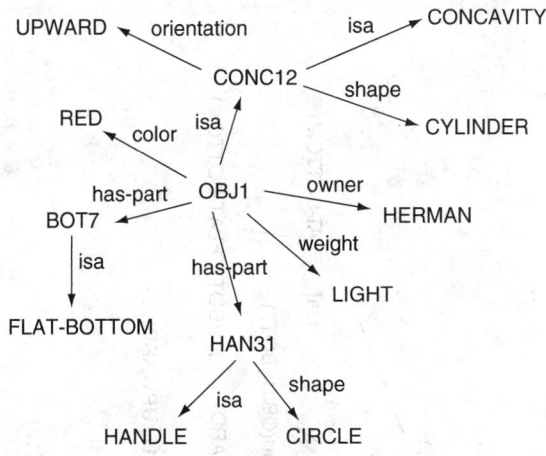

FIGURE 68.6b OBJ1 in semantic net notation.

the object possesses. The arrow is labeled with the name of the feature. Both representations specify that OJB1 has three parts, called CONC12, HAN31, and BOT7. OBJ1 is colored RED, is owned by HERMAN, and so on.

Given OBJ1 as a positive example, an EBL system attempts to construct an explanation for why OBJ1 is indeed a cup. Figure 68.7 shows such an explanation. Arrows denote the contribution of each background knowledge rule to the explanation. For clarity, the quantification variables (each rule's formal parameters) have been renamed to be unique. Arrows point from a rule's antecedents to its consequent. For example, rule R3 allows the consequent property liftable to be concluded from the antecedents weight, has-part, and isa. The three arrows on the lower left, which are labeled as R3, constitute an instantiation of this domain theory rule. Double lines show expressions across rules that must match for the explanation to hold. These are enforced by unifying the connected expressions. Thus, the first antecedent of R3,

$$\text{weight}(x_3, \text{LIGHT})$$

is unified with the known property from the training example

$$\text{weight}(\text{OBJ1}, \text{LIGHT})$$

This match is contingent upon the variable x_3, being bound to OBJ1 and variable y_3 being bound to HAN31. From the consequent we can infer that OBJ1 is liftable. This conclusion, along with others, supports the inference that OBJ1 has the property drinkable-from. By R1, we plausibly infer that OBJ1 is a cup, completing the explanation.

68.6 Generalizing

The next step produces a generalized EBL concept descriptor. There are several distinct kinds of generalization performed. But before describing them, it is important to be clear on the meaning of "generalization." In EBL, an explanation counts as derivational evidence for the classification label assigned to the training example. Generalizing means finding a larger set of examples where the same classification evidence applies. In the inductive learning literature, generalization can mean finding *any* larger set of examples that contains the example of interest. Simply removing a conjunct or adding a disjunct without regard for its effect on the explanation is not a well-formed EBL generalization.

The semantics of conventional first-order logic dictate that the sentences of the theory be consistent and correct (with respect to the intended interpretation) with a sound logical inference engine. Thus, if the

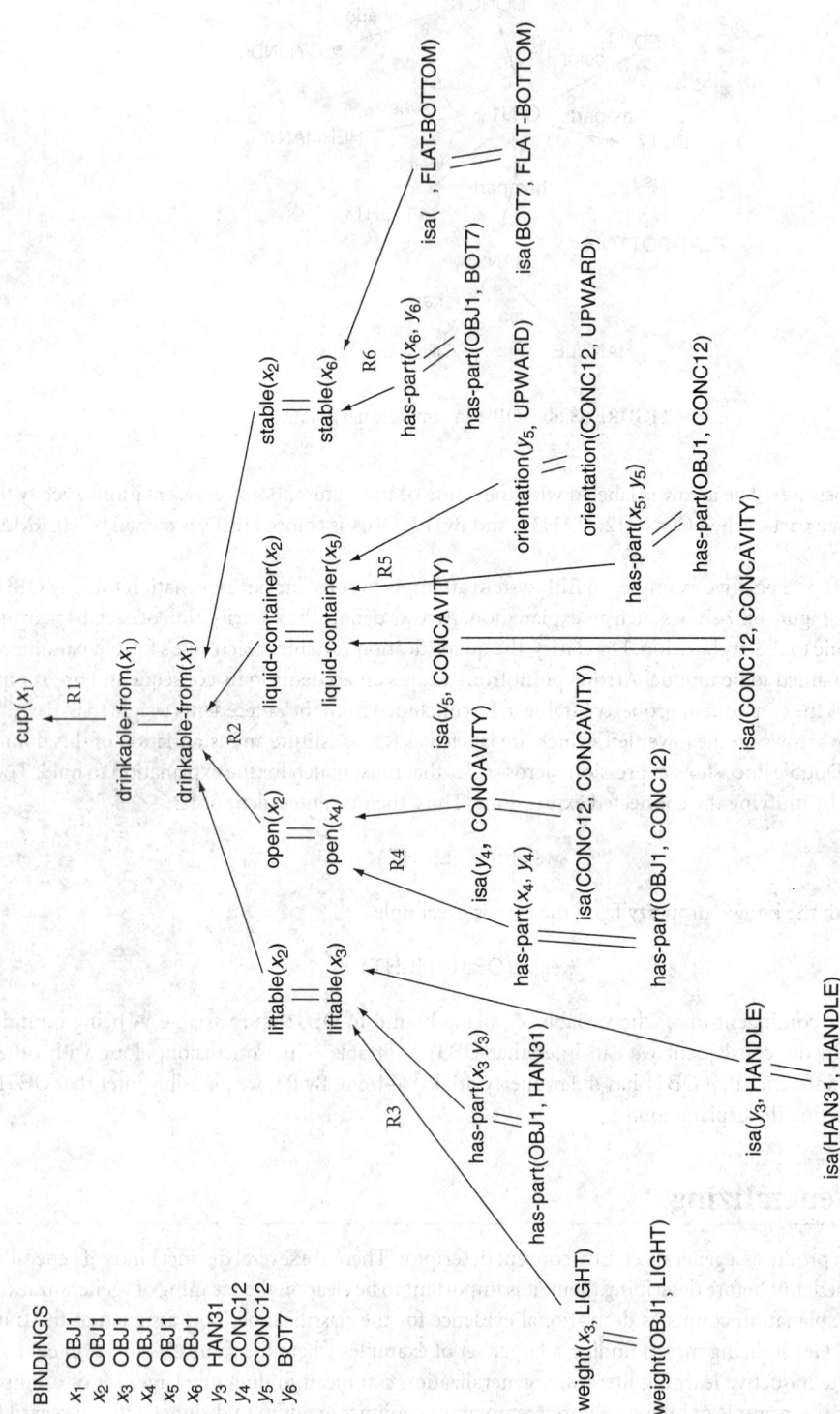

FIGURE 68.7 Explanation for the training example OBJ1.

foundation of the EBL system is conventional first-order logic, EBL-generalized descriptors are necessarily correct. This form of EBL is often termed *speed-up* learning. It is a common misconception in the literature that an EBL domain theory must be a conventional variety of logic. This has given rise to mistakenly *equating* EBL with speed-up learning, eliminating any **knowledge level change** in the performance system. This view precludes some of the most important strengths of EBL. Next we examine some individual types of EBL generalization.

68.6.1   Irrelevant Feature Elimination

The explanation provides sufficient grounds for believing that the training example (in our case, OBJ1) satisfies the goal specification (cupness). Clearly, any feature of OBJ1 not mentioned in the explanation could have a different value without affecting the explanation's veracity. The owner could have been George instead of Herman; the cup could be blue instead of red, etc. In general, most of an object's properties will not participate in the explanation. This is particularly true for more realistic representations. These might include many additional properties: its current location in the room, distances to other objects, what it is resting upon, how full it is, what it contains, the temperature of the contained liquid, whether it is clean or used and who drank from it last, where it was purchased, how valuable it is, etc.

Empirical learning systems can become overwhelmed in such situations. Coincidences abound when objects have many features, training data is limited, or the space of well-formed concepts is large. This is often the case in the real world. Coincidences, by their very nature, are not predictive of future examples. Empirical learners are vulnerable to such coincidences. The phenomenon of overfitting is related to this issue. Recall our first informal example of John's city preference. It is most likely a mere coincidence that testing for the letter "e" correctly classifies the training data. An empirical approach would have no legitimate mechanism to choose among descriptors that are equally complex and behave similarly on the training data. An EBL system, on the other hand, is unlikely to be sidetracked by such coincidences.

By irrelevant feature elimination, the example explanation of Figure 68.7 gives rise to the concept descriptor C1:

$$[\text{weight(OBJ1, LIGHT)} \wedge \text{has-part(OBJ1, HAN31)} \wedge \text{isa(HAN31, HANDLE)}$$
$$\wedge \text{has-part(OBJ1, CONC12)} \wedge \text{isa(CONC12, CONCAVITY)}$$
$$\wedge \text{orientation(CONC12,UPWARD)} \wedge \text{has-part(OBJ1, BOT7)}$$
$$\wedge \text{isa(BOT7, FLAT-BOTTOM)}] \Rightarrow \text{cup(OBJ1)}$$

This descriptor is not very general. Importantly, however, it is in the same form as the original background knowledge. That is, it provides an alternative method to infer the cupness of an object, and it states that weight, orientation, and the various isa and has-part relations are sufficient for this inference. No other features of the object are needed if these are present.

68.6.2   Identity Elimination

The second generalization type removes unnecessary mention of example characteristics. According to the explanation, OBJ1 is liftable by virtue of its having handle HAN31. Without a handle, the explanation's veracity is compromised. But the training example need not have had this particular handle. If it had some other handle (e.g., HAN32 or HAN4597), the conclusion of cupness would still be justified. Thus, HAN31 can be generalized into a predicate calculus variable. Likewise, OBJ1 can be generalized to a different predicate calculus variable. But we must ensure that required relations hold among these variables. The variable that replaces HAN31 must still be a handle; it must be a part of the variable that replaces OBJ1, and so on.

We can find the minimal general conditions required to conclude cupness by pruning the lowest portions of the explanation. Pruning involves breaking unification links within the explanation and keeping only the central structure (the subgraph that includes the goal justification). After this generalization step, we can obtain the descriptor C2:

$$\forall\, x_1, y_3, y_4, y_6\; [\text{weight}(x_1,\, \text{LIGHT}) \wedge \text{has-part}(x_1,\, y_3) \wedge \text{isa}(y_3,\, \text{HANDLE})$$
$$\wedge\, \text{has-part}(x_1,\, y_4) \wedge \text{isa}(y_4,\, \text{CONCAVITY}) \wedge \text{orientation}(y_4, \text{UPWARD})$$
$$\wedge\, \text{has-part}(x_1,\, y_6) \wedge \text{isa}(y_6,\, \text{FLAT-BOTTOM})] \Rightarrow \text{cup}(x_1)$$

This says that anything that has a handle, an upward pointing concavity, a flat bottom, and is light is a cup. This rule applies to many objects in addition to OBJ1 and can be used as a concept descriptor for cup classification. We can be confident of this descriptor's correctness although only one example has been seen. Identity elimination works because of generalities already built into the domain theory. These preexisting domain theory generalities are essential to EBL. If the domain theory were changed to include an alternate rule R3:

$$\text{R3a:}\; \forall x\; [\text{weight}(x, \text{LIGHT}) \wedge \text{has-part}(x, \text{HAN31}) \wedge \text{isa}(\text{HAN31}, \text{HANDLE})] \Rightarrow \text{liftable}(x)$$

then EBL generalization of HAN31 would not be possible although the conclusion of OBJ1's cupness would still be supported. For the sake of EBL, we would prefer to avoid domain rules such as R3a. Ideally, the role that an object plays in the domain theory is entirely determined by its properties, never by its identity. Philosophically, this has some interesting ramifications but it is uncontroversial so far as EBL is concerned. This has been termed the principle of no *function in form*. It is often adhered to in AI and usually results in a theory with fewer rules. This property is also important in the next generalization type.

68.6.3 Operationality Pruning

Easily reconstructable subexplanations of the original explanation can also be eliminated. For example, we could prune the substructure added by rule R6 by breaking the unification at the stable predicate. This results in a slightly different concept, C3:

$$\forall\, x_1, y_3, y_4, y_6\; [\text{weight}(x_1,\, \text{LIGHT}) \wedge \text{has-part}(x_1,\, y_3) \wedge \text{isa}(y_3,\, \text{HANDLE}) \wedge \text{has-part}(x_1,\, y_4)$$
$$\wedge\, \text{isa}(y_4,\, \text{CONCAVITY}) \wedge \text{orientation}(y_4, \text{UPWARD}) \wedge \text{stable}(x_1)] \Rightarrow \text{cup}(x_1)$$

This descriptor is syntactically simpler than the previous one. It has one fewer antecedent conjuncts. However, to determine that a new object is a cup, that object's stability must now be justified at the time the concept is applied. The previous descriptor only consults properties expressed directly in the definition of the test object. In that rule, the test object's stability is never an issue. Sufficient ancillary properties are tested to justify that the object is stable. In particular, all objects are required to have flat bottoms. In point of fact, many objects are stable although they do not have flat bottoms. Indeed, many cups do not possess flat bottoms. The zarf (Figure 68.8) is often used in Middle Eastern countries to support hot coffee

FIGURE 68.8   The zarf.

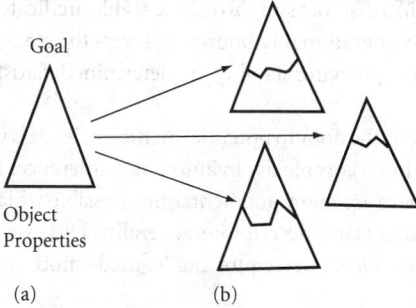

Goal

Object
Properties

(a)   (b)

FIGURE 68.9 Generality–operationality trade-off: (a) explanation and (b) operationality boundaries.

cups that have rounded bottoms. The zarf is a cylindrical chalice-like holder into which the rounded cup bottom is nestled. There are other common and not so common ways to achieve stability. While C3 is a more general concept descriptor, the price of this generality is that an inferencer must conclude stability when the concept descriptor is evaluated on an object. Thus, it is more expensive to use. We say it is less *operational*. Constructing a subexplanation on demand is typically a harder task than the straightforward lookup of several object properties.

Likewise, we could entertain concept descriptors that sever other of the explanation's unifications. Cups could be allowed to be liftable by other means than having a handle, which include Styrofoam cups and the like. Liquid containment could be other than by an open concavity, giving rise to covered travel mugs. The higher in the explanation that unifications are broken, the more general is the resulting concept descriptor but the more expensive it is to evaluate.

Thus, there is a trade-off between concept descriptor generality and operationality. The minimal generalization is the result of applying identity elimination. The maximal generalization is an uninformative repetition of the general domain rule:

$$\forall x_1 \ \text{drinkable-from}(x_1) \Rightarrow \text{cup}(x_1)$$

In between there are many alternative choices requiring progressively harder subexplanations to be constructed at the time of concept use.

Figure 68.9a shows a schematic explanation tree. An explanation's **operationality boundary** is any coherent choice of unification pruning traversing the explanation structure. Figure 68.9b shows several schematic choices for the operationality boundary.

In the following section on selecting a concept descriptor, we discuss different approaches that have been advanced for deciding how to choose the operationality boundary. For now it is sufficient to realize that even once a particular explanation has been constructed, there are many potential concept descriptors from which to choose.

68.6.4 Disjunctive Augmentation

Pruning requires a subexplanation to be constructed from scratch at the time the concept is used. This can be expensive. Disjunctive augmentation supports similar flexibility at a greatly reduced cost. Instead of simply pruning a subexplanation, we add one or more alternative subexplanations, any one of which can be used to support the requisite property. At the time the concept is used, the system need only choose among several pregeneralized subexplanations.

Suppose we give the learning system a second positive example of a cup. This cup's stability is achieved by a zarf as in Figure 68.8. We also suppose that the system has sufficient background theory to conclude that this cup is indeed stable. The explanation of this example's cupness will be similar to the explanation of Figure 68.7 except that stability is achieved by the zarf instead of the conventional flat bottom. Generalizing both explained examples results in a disjunct supporting the stable predicate. Note that this is quite different from operationality pruning. No operationality boundary severs the stable predicate from its support. In this case, two distinct methods of achieving stability are determined. Satisfying either qualifies the object as a cup.

Disjunctive augmentation opens the door to potential inefficiencies. It is well established that disjunction is a primary source of computational complexity in automated inference. It is possible, indeed likely, that in any interesting explanation there are many augmentations possible which make subtle distinctions and result in only marginal improvements in a descriptor's generality. Discovering alternative subexplanations is expensive. Similarly, use of the concept descriptor can be made more complex.

68.7   The Utility Problem: Selecting a Concept Descriptor

In the final step, EBL ensures that the acquired concept is useful. Concepts are typically acquired for some purpose. There is little point in acquiring a new concept unless an AI application program (a scheduling system, diagnosis system, database retrieval agent, etc.) works better with it. It is important to appreciate that performance improvement in an application system is the ultimate goal of any machine learning system. Ensuring that acquired concepts help rather than hurt performance is the **utility problem**.

Concept accuracy is no guarantee of concept utility. In the performance system there is some cost associated with possessing concepts. A newly acquired concept may be accurate so that it indeed saves work by simplifying the task. The benefit of applying that concept may not outweigh the overhead of including it in the system. To illustrate, consider a commuter who drives home every workday. The standard well-traveled route, Avenue A, takes 15 min. Suppose a learning system discovers two additional routes home. One follows Avenue B and takes only 10 min, whereas the other follows Avenue C and takes 11 min. Because they are poorer roads, they are often closed for repair; but because there is but one road crew, they are never simultaneously closed. The newly learned concept chooses between these routes, both of which are quicker than Avenue A. The concept must include an antecedent to ensure that the selected road be open because each route only works when there is no construction. The acquired concept might say something like, "if there is no maintenance on Avenue B, then take it and save 5 min, else take Avenue C and save 4 min." The overhead cost of a piece of knowledge in the performance system includes three components: (1) the cost of locating the knowledge in memory, (2) the cost of evaluating precondition antecedents in the situation, and (3) the slowdown of locating other knowledge for other purposes caused by the presence of this knowledge. Consider cost 2 in the context of our example. Suppose the only reliable way to determine whether or not there is maintenance is by placing a call to the Public Works department. If it happens that such calls are completed within, say, 2 min, then the newly acquired concept is a useful one: with the concept, the performance system saves 2 to 3 min over solving the problem without the concept. However, if the Public Works office is typically busy or inefficient so that the average call takes 6 min, then the performance system incurs a penalty of 1 to 2 min with each application. In either case, the concept is accurate; it adequately captures the relevant world knowledge, shortens driving time, and never causes a solvable problem to become unsolvable. However, its inclusion may result in poorer system performance.

The decision on what concepts to adopt can be quite subtle. A concept's utility can be strongly influenced by attributes far removed from it. Here the crucial feature in determining positive or negative concept utility (the efficiency of the Public Works telephone operator) would at first seem quite distant from the process of driving home. A concept's utility can also depend on the particular distribution of problems to which the performance system will be applied. It can be influenced by other concepts that have been previously adopted. Thus, a utility judgment may be difficult to make reliably.

The fact that correct knowledge can harm the performance system is particularly troublesome for EBL. Because EBL is more example efficient, a particular training set may yield many more concepts using an EBL approach. The performance system can quickly become swamped even if the average concept utility is only slightly negative. Each concept's utility must be evaluated before it can be given to the performance system. Only concepts with positive utilities should be added.

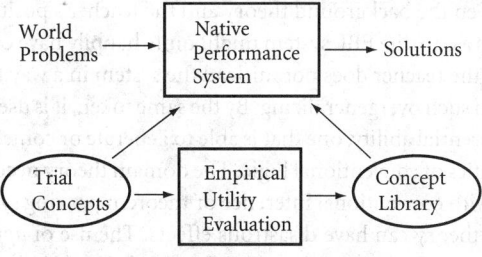

FIGURE 68.10   Empirical utility analysis.

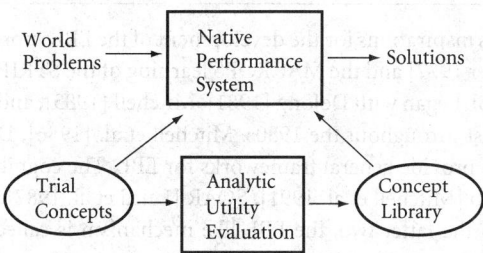

FIGURE 68.11   Analytic utility analysis.

There are two broad approaches to judging a concept's utility for a performance system. In the first or *empirical* approach, a hypothesized concept is temporarily added to the performance system to measure its effect, which is then taken as an estimate of its future utility. In the second or *analytic* approach, the decision to retain a new concept is inferred directly from the concept's definition.

Figure 68.10 illustrates the empirical approach. Newly hypothesized concepts from the generalizer are represented by the oval labeled Trial Concepts. The performance system, composed of the Native Performance System together with the Concept Library, processes problems from the world. The Empirical Utility Evaluation monitors the efficiency of the performance system augmented with individual trial concepts. As statistics are gathered and compared to baseline performance, evidence mounts for or against the utility of the currently conjectured concept. When there is sufficient evidence to warrant a confident judgment about the concept's utility, it is permanently added to the system's concept library or discarded.

The analytic approach is illustrated in Figure 68.11. The native performance system is not directly involved in the utility judgment. Rather, information on what constitutes efficient processing is represented in a theory of utility. Trial concepts are evaluated with respect to this *a priori* utility theory. Concepts that are judged likely to improve the performance system are added to the system's concept library; others are discarded.

In an empirical procedure, the definitions of the concept are not directly examined. The performance system simply continues its normal processing although it can now draw upon the conjectured concept. The empirical method quite naturally incorporates information implicit in the distribution of problems. The expected utility of a concept can vary even though the concept itself is static. The knowledge of how to control a high-speed car with a flat tire is of little use to a Bedouin who travels only by camel, but is crucial to a race car driver. Such phenomena do not preclude an analytic approach. The theory of utility may itself include information on estimated problem distributions, or the effect of problem distribution on utility may be insignificant in the face of other considerations. Hybrid approaches have also been examined.

Finally, we return to the issue of incorrect knowledge. One advantage of EBL is that the approach is quite forgiving of approximate knowledge representations. The functional specification of our cup (R1) is such a case. It allows a drinking glass to be a functional cup. However, concepts C1 through C3 do not apply to drinking glasses. The acquired concepts insist that the item in question have a handle. This robustness is due to the interplay between the background theory and the teacher's positive training examples. If the teacher had labeled a glass as a cup, the EBL system might quite happily have conjectured a rather different set of descriptors. Provided the teacher does not mislead the system in a way that admits explanation, the EBL system is often safe from such overgeneralizing. By the same token, it is useful to endow the EBL system with an overly powerful inferential ability, one that is able to generate or conjecture explanations that need not be true under the semantics of conventional logic. The domain theory need not be complete or correct. This is in striking contrast with conventional inference or theorem proving systems in which the presence of any contradiction in the theory can have disastrous effects. The use of *approximate*, *overly general*, or *plausible* background theories has been studied by several researchers and is the subject of ongoing research.

68.8   Annotated Summary of EBL Issues

Two early AI works served as inspirations for the development of the EBL approach. These were the notion of goal regression [Waldinger 1977] and the MACROPS learning of the STRIPS system [Fikes et al. 1972]. The modern approach to EBL began with DeJong [1981], Mitchell [1983], and Silver [1983]. It enjoyed an explosion of research interest throughout the 1980s. Mitchell et al. [1986], DeJong and Mooney [1986], and Segre and Elkan [1994] provide general frameworks for EBL. The cognitive architectures of Prodigy [Carbonell et al. 1990], Theo [Mitchell et al. 1991], SOAR [Laird et al. 1987], and ACT [Anderson 1983] employ versions of EBL. In the latter two, the EBL-like mechanism is called "chunking" and generally applies to preserving processing traces. Cohen [1994] has advanced an formalism built upon abduction.

Minton [1985] pointed out aspects of the utility problem. Tambe [1988] reported a similar phenomenon in SOAR. Minton [1988] provided a detailed empirical solution. Greiner and Jurisica [1992] and Gratch and DeJong [1992] refined the approach along decision theoretic lines. An alternative is to limit concept acquisition to those that can be shown *a priori* to be inexpensive to evaluate. This analytic approach is taken by Etzioni [1990], Subramanian and Feldman [1990], and Letovsky [1990], and SOAR [Tambe 1988].

EBL has been employed as a learning mechanism in many other AI areas. Illustrative examples include planning [Kambhampati 2000], theory revision [Hirsh 1987, Ourston and Mooney 1994], analogy understanding [Russell 1989], natural language processing [Bangalore and Joshi 1995, Neumann 1997], and reinforcement learning [Dietterich and Flann 1995, Laud and DeJong 2002].

The role of prior knowledge in human concept formation is well established [Murphy and Medin 1985]. There is ample psychological evidence that in knowledge-rich contexts adult concept learning is more in line with EBL than conventional empirical mechanisms [Ahn et al. 1987, Anderson 1987, Pazzani 1989]. Surprisingly, infants as young as 5.5 months exhibit EBL-like learning mechanisms [Kotovksy and Baillargeon 1998, Baillargeon 2003].

From its inception, integrating EBL with empirical induction was seen as an important direction [Lebowitz 1986, Danyluk 1987, Bergadano and Giordana 1988, Flann and Dietterich 1989, Kodratoff and Tecuci 1989, Pazzani 1989, Russell 1989]. Neural networks have served as a particularly effective integration substrate [Thrun and Mitchell 1993, Shavlik and Towell 1994].

Defining Terms

Classifier or concept descriptor: The declarative representation of a partitioning of the example space to reflect example classification.

Domain theory: Any information supplied to the concept acquisition system about the application task, the application system, the nature of expected inputs, etc. This is also known as background knowledge.

Example space: The set of all classifiable objects.

Explanation: A tree-structured graph such that each leaf node is either a prior belief or a property of the example being explained, each nonleaf node is the result of applying an inference procedure to prior nodes, the root node is the training classification assigned to the example.

Hypothesis space: The set of all well-formed concept descriptors.

Inductive bias: Any preference not due to the training set which is exhibited by a concept acquisition algorithm for one concept descriptor over another.

Knowledge level change: Any change to an AI system's representation that goes beyond the inferential closure of its previously represented knowledge.

Operationality boundary: In a generalized explanation, any division between the root subtree and the peripheral subtrees such that the root subtree yields a useful concept.

Overfitting: The selection of a concept descriptor that captures a pattern exhibited by the training set but not exhibited by the target concept. In Section 68.1, the concept that avoids cities with the letter "e" overfits the training data.

Target concept: The correct or desired partitioning of the example space.

Training set: A collection of examples whose classification is known. This is an input to a concept acquisition system from which a concept descriptor is induced.

Utility problem: The difficulty in ensuring that an acquired concept enhances the performance of the application system.

References

Ahn, W., Mooney, R. J., Brewer, W. F., and DeJong, G. F. 1987. Schema acquisition from one example: psychological evidence for explanation-based learning. In *9th Ann. Conf. Cognitive Sci. Soc.*, pp. 50–57. Lawrence Erlbaum Associates.

Anderson, J. 1983. *The Architecture of Cognition.* Harvard Universtiy Press.

Anderson, J. R. 1987. Causal analysis and inductive learning. In *4th Int. Workshop Machine Learning.*, pp. 288–299. Morgan Kaufmann.

Baillargeon, R. 2003. Infants' physical world. *Current Directions in Psychological Science.*

Bangalore, S. and Joshi, A. 1995. Some novel applications of explanation-based learning for parsing lexicalized tree-adjoining grammars. In *33rd Annual Meeting of the Association for Computational Linguistics*, pp. 268–275. Morgan Kaufmann.

Bergadano, F. and Giordana, A. 1988. A knowledge intensive approach to concept induction. In *5th Int. Conf. Machine Learning*, pp. 305–317. Morgan Kaufmann.

Carbonell, J., Knoblock, C. and Minton, S. 1990. Prodigy: An integrated architecture for planning and learning. In *Architectures for Intelligence*, K. VanLehn, Ed., pp. 241–278. Erlbaum.

Cohen, W. 1994. Incremental abductive EBL. *Machine Learning*, 15(1):5–24.

Danyluk, A. 1987. The use of explanations for similarity-based learning. In *10th Int. J. Conf. Artif. Intelligence*, pp. 274–276. Morgan Kaufmann.

DeJong, G. 1981. Generalizations based on explanations. In *7th Int. J. Conf. Artif. Intelligence*, pp. 67–70. IJCAI.

DeJong, G. and Mooney, R. 1986. Explanation-based learning: an alternative view. *Machine Learning*, 1(2):145–176.

Dietterich, T. and Flann, N. 1995. Explanation-based learning and reinforcement learning: a unified view. In *12th Int. Conf. Machine Learning*, pp. 176–184.

Etzioni, O. 1990. Why Prodigy/EBL works. In *8th Nat. Conf. Artif. Intelligence*, pp. 916–922. MIT Press.

Fikes, R., Hart, P., and Nilsson, N. 1972. Learning and executing generalized robot plans. *Artif. Intelligence*, 3(4):251–288.

Flann, N. S. and Dietterich, T. G. 1989. A study of explanation-based learning methods for inductive learning. *Machine Learning*, 4(2):187–226.

Gratch, J. and DeJong, G. 1992. COMPOSER: a probabilistic solution to the utility problem in speed-up learning. In *10th Nat. Conf. Artif. Intelligence*, pp. 235–240. MIT Press.

Greiner, R. and Jurisica, I. 1992. A statistical approach to solving the EBL utility problem. In *10th Nat. Conf. Artif. Intelligence*, pp. 241–248. MIT Press.

Hirsh, H. 1987. Explanation-based generalization in a logic-programming environment. In *10th Int. J. Conf. Artif. Intelligence*, pp. 221–227. Morgan Kaufmann.

Kambhampati, S. 2000. Planning graph as a (dynamic) CSP: exploiting EBL, DDB, and other search techniques in graphplan. *J. Artif. Intelligence Research*, 12:1–34.

Kodratoff, Y. and Tecuci, G. 1989. The central role of explanations in disciple. In *Knowledge Representation and Organization in Machine Learning*. K. Morik, Ed., pp. 135–147. Springer-Verlag.

Kotovsky, L. and Baillargeon, R. 1998. The development of calibration-based reasoning about collision events in young infants. *Cognition*, 67:311–351.

Laird, J., Newell, A., and Rosenbloom, P. 1987. SOAR: an architecture for general intelligence. *Artif. Intelligence*, 33(1):1–64.

Laud, A. and DeJong, G. 2002. Reinforcement learning and shaping: encouraging intended behaviors. In *19th Int. Conf. Machine Learning*, pp. 355–362. Morgan Kaufmann.

Lebowitz, M. 1986. Integrated learning: controlling explanation. *Cognitive Sci.*, 10(2):219–240.

Letovsky, S. 1990. Operationality criteria for recursive predicates. In *8th Nat. Conf. Artif. Intelligence*, pp. 936–941. AAAI/MIT Press.

Minton, S. 1985. Selectively generalizing plans for problem-solving. In *9th Int. J. Conf. Artif. Intelligence*, pp. 596–599. Morgan Kaufmann.

Minton, S. 1988. *Learning Search Control Knowledge: An Explanation-Based Approach*. Kluwer Academic.

Mitchell, T. 1983. Learning and problem solving. In *8th Int. J. Conf. Artif. Intelligence*, pp. 1139–1151. Morgan Kaufmann.

Mitchell, T., Keller, R., and Kedar-Cabelli, S. 1986. Explanation-based generalization: a unifying view. *Machine Learning*, 1(1):47–80.

Mitchell, T., Allen, J., Chalasani, P., Cheng, J., Etzioni, O., Ringuette, M., and Schlimmer, J., 1991. THEO: a framework for self-improving systems. In *Architectures for Intelligence*, K. VanLehn, Ed., pp. 323–355. Erlbaum.

Murphy, G. and Medin, D. 1985. The role of theories in conceptual coherence. *Psychological Rev.*, 92:289–316.

Neumann, G. 1997. Applying explanation-based learning to control and speeding-up natural language generation. In *35th Annual Meeting of the Association for Computational Linguistics and 8th Conference of the European Chapter of the Association for Computational Linguistics*, pp. 214–221. Morgan Kaufmann.

Ourston, D. and Mooney, R. 1994. Theory refinement combining analytical and empirical methods. *Artif. Intelligence*, pp. 273–309.

Pazzani, M. 1989. Creating high level knowledge structures from simple events. In *Knowledge Representation and Organization in Machine Learning*, K. Morik, Ed., pp. 258–287. Springer-Verlag.

Russell, S. 1989. *The Use of Knowledge in Analogy and Induction*. Morgan Kaufmann.

Segre, A. and Elkan, C. 1994. A high-performance explanation-based learning algorithm. *Artif. Intelligence*, 69(1–2):1–50.

Shavlik, J. and Towell, G. 1994. Knowledge-based artificial neural networks. *Artif. Intelligence*, 70(1–2):119–165.

Silver, B. 1983. Learning equation solving methods from worked examples. In *Proc. 1983 Int. Workshop Machine Learning*, pp. 99–104. CS Dept., University of Illinois.

Subramanian, D. and Feldman, R. 1990. The utility of EBL in recursive domain theories. In *8th Nat. Conf. Artif. Intelligence*, pp 942–949. AAAI/MIT Press.

Tambe, M. 1988. Some chunks are expensive. In *5th International Machine Learning Conference*, pp. 451–458. Morgan Kaufmann.

Thrun, S. and Mitchell, T. 1993. Integrating inductive neural network learning and explanation-based learning. In *Int. J. Conf. Artif. Intelligence*, pp. 930–936. Morgan Kaufmann.

Waldinger, R. 1977. Achieving several goals simultaneously. In *Machine Intelligence*, E. Elcock and D. Michie, Eds., pp. 94–136. Ellis Horwood.

69

Cognitive Modeling

69.1   Introduction 69-1
69.2   Underlying Principles 69-2
         Psychology • Neuroscience • Computer Science
         • Evolutionary and Environmental Psychology
69.3   Research Issues 69-5
         Are We Symbol Processors? • Grand Theories?
69.4   Best Practices 69-8
69.5   Summary 69-10

Eric Chown
Bowdoin College

69.1   Introduction

An important goal of cognitive science is to understand human cognition. Good models of cognition can be *predictive* — describing how people are likely to react in different scenarios — as well as *prescriptive* — describing limitations in cognition and potentially ways in which the limitations might be overcome. In a sense, the benefits of having cognitive models are similar to the benefits individuals accrue in building their own internal model. To quote Craik [1943]:

> If the organism carries a 'small-scale model' of external reality and of its own possible actions within its head, it is able to try out various alternatives, conclude which is the best of them, react to future situations before they arise, utilize the knowledge of past events in dealing with the present and future, and in every way to react in a much fuller, safer, and more competent manner to the emergencies which face it. (p. 61)

Among the important questions facing cognitive scientists are how such models are created and how they are represented internally. Craik emphasizes the importance of the predictive power of models, and it is the model's ability to make accurate predictions that is the ultimate measure of the model's value. One important value of computers in cognitive science is that computer simulations provide a means to instantiate theories and to concretely test their predictive power. Further, implementation of a theory in a computer model forces theoreticians to face practical issues that they may never have otherwise considered.

The role of computer science in cognitive modeling is not strictly limited to implementation and testing, however. A core belief of most cognitive scientists is that cognition is a form of computation (otherwise, computer modeling is a doomed enterprise) and the study of computation has long been a source of ideas (and material for debates) for building cognitive models. Computers themselves once served as the dominant metaphor for cognition. In more recent years, the influence of computers has been more in the area of computational paradigms, such as rule-based systems, neural network models, etc.

69.2   Underlying Principles

One of the dangers of cognitive modeling is falling under the spell of trying to apply computational models directly to cognition. A good example of this is the "mind as computer" metaphor that was once popular but has fallen into disfavor. Computer science offers a range of computational tools designed to solve problems, and it is tempting to apply these tools to psychological data and call the result a cognitive model. As McCloskey [1991] has pointed out, this falls far short of the criteria that could reasonably used to define a theory of cognition. One replacement for the "mind as computer" metaphor makes this point well. Neurally inspired models of cognition fell out of favor following Minsky and Papert's 1969 book *Perceptrons* that showed that the dominant neural models at the time were unable to model nonlinear functions (notably exclusive-or). The extension of these models by the PDP (Parallel Distributed Processing) group in 1986 [Rumelhart and McClelland, 1986] is largely responsible for the **connectionist** revolution of the past 25 years. The excitement generated by these models was twofold: (1) they were computationally powerful and simple to use, and (2) as neural-level models they appeared to be physiologically plausible.

A major difficulty for connectionist theory of the past 20 years has been that, despite the fact that the early PDP-style models (particularly models built upon **feed-forward back-propagation** networks) were proven to be implausible for both physiological and theoretical reasons (e.g., see [Lachter and Bever, 1988; Newell, 1990]), many cognitive models are still built using such discredited computational engines. The reason for this appears to be simple convenience. Back-propagation networks, for example, can approximate virtually any function and are simple to train. Because any set of psychological data can be viewed as a function that maps an input to a behavior, and because feed-forward back-propagation networks can approximate virtually any function, it is hardly surprising that such networks can "model" an extraordinary range of psychological phenomena. To put this another way, many cognitive models are written in computer languages like C. Although such models may accurately characterize a huge range of data on human cognition, no one would argue that the C programming language is a realistic model of human cognition. Feed-forward neural networks seem to be a better candidate for a cognitive model because of some of their features: they intrinsically learn, they process information in a manner reminiscent of neurons, etc. In any regard, this suggests that the criteria for judging the merits of a cognitive model must include many more constraints than whether or not the model is capable of accounting for a given data set. While issues such as how information is processed are useful for judging models, they are also crucial for constructing models.

There are a number of sources and types of constraints used in cognitive modeling. These break down relatively well by the disciplines that comprise the field. In practice most cognitive models draw constraints from some, but not all, of these disciplines. In broad terms, the data for cognitive models comes from psychology. "Hardware" constraints come from neuroscience. "Software" constraints come from computer science, which also provides methodologies for validation and testing. Two related fields that are relatively new, and therefore tend to provide softer constraints are evolutionary psychology and environmental psychology. The root idea of each of these fields is that the evolution process, and especially the environmental conditions that took place during evolution, are crucially important to the kind of brain that we now have. We will examine the impact of each field on cognitive modeling in turn.

69.2.1   Psychology

The ultimate test of any theory is whether or not it can account for, or correctly predict, human behavior. Psychology as a field is responsible for the vast majority of data on human behavior. Over the last century the source of this data has evolved from mere introspection on the part of theorists to rigorous laboratory experimentation. Normally the goal of psychological experiments is to isolate a particular cognitive factor; for example, the number of items a person can hold in short term memory. In general this isolation is used as a means of reducing complexity. In principle this means that cognitive theories can be constructed piecewise instead of out of whole cloth. It would be fair to say that the majority of work in cognitive science proceeds on this principle. A fairly typical paper in a cognitive science conference proceeding, for example, will present a set of psychological experiments on some specific area of cognition, a model to account for the data, and computer simulations of the model.

69.2.2 Neuroscience

The impact of neuroscience on cognitive science has grown dramatically in conjunction with the influence of neural models in the last 20 years. Unfortunately, terms such as "neurally plausible" have been applied fairly haphazardly in order to lend an air of credibility to models. In response, some critics have argued that neurons are not well understood enough to be productively used as part of cognitive theory. Nevertheless, though the low level details are still being studied, neuroscience can provide a rich source of constraints and information for cognitive modelers. Within the field there are several different types of architectural constraint available. These include:

1) *Information flow.*   We have learned from neuroscientists, for example, that the visual system is divided into two distinct parts, a "what" system for object identification, and a "where" system for determining spatial locations. This suggests computational models of vision should have similar properties. Further, these constraints can be used to drive cognitive theory as with the PLAN model of human cognitive mapping [Chown, et al. 1995]. In PLAN it was posited that humans navigate in two distinct ways, each corresponding to one of the visual pathways. Virtually all theories of cognitive mapping had previously included a "what" component based upon topological collections of landmarks, but none had a good theory of how more metric representations are constructed. The split in the visual system led the developers of PLAN to theorize that metric representations would have simple "where" objects as their basic units. This led directly to a representation built out of "scenes," which are roughly akin to snapshots.

2) *Modularity.*   A great deal of work in neuroscience goes towards understanding what kinds of processing is done by particular areas of the brain, such as the hippocampus. These studies can range from working with patients with brain damage to intentionally lesioning animal brains. More recently, imaging techniques such as fMRI (functional magnetic resonance imaging) have been used to gain information non-invasively. This work has provided a picture of the brain far more complex than the simple "right brain-left brain" distinction of popular psychology. The hippocampus, for example, has been implicated in the retrieval of long-term memories [Squire, 1992] as well as in the processing of spatial information [O'Keefe and Nadel, 1978]. In principle, discovering what each of the brain's different subsystems does is akin to determining what each function that makes up a computer program does.

Modularity in the brain, however, is not as clean as modularity in computer programs. This is largely due to the way information is processed in the brain, namely by neurons passing activity to each other in a massively parallel fashion. Items processed close together in the brain, for example, tend to interfere with each other because neural cells often have a kind of inhibitory surround. This fact is useful in understanding how certain perceptual processes work. Further it means that when one is thinking about a certain kind of math problem, it may be possible to also think about something unrelated like what will be for dinner, but it will be more difficult to simultaneously think about another math problem. The increased interference between similar items (processed close together) over items processed far apart has been called "the functional distance principle" by Kinsbourne [1982]. This suggests, among other things, that there may not be a clean separation of "modules" in the brain, and further that even within a module architectural issues impact processing.

3) *Mechanisms.*   Numerous data are simpler to make sense of in the context of neural processing mechanisms. A good example of this would be the **Necker cube**. From a pure information processing point of view, there is no reason that people would only be able to hold one view of the cube in their mind at a time. From a neural point of view, on the other hand, the perception of the cube can be seen as a competitive process with two mutually inhibitory outcomes. Perceptual theory is an area that has particularly benefited from a neural viewpoint.

4) *Timing.*   Perhaps the most famous constraint on cognitive processing offered by neuroscience is the "100 step rule." This rule is based upon looking at timing data of perception and the firing rate of

neurons. From these it has been determined that no perceptual algorithm could be more than 100 steps long (though the algorithm could be massively parallel as the brain itself is).

69.2.3 Computer Science

Aside from providing the means to implement and simulate models of cognition, computer science has also provided constraints on models through limits drawn from the theory of computation, and has been a source of algorithms for modelers.

One of the biggest debates in the cognitive modeling community is whether or not computers are even capable of modeling human intelligence. Critics, normally philosophers, point to the limitations on what is computable and have gone as far as suggesting that the mind may not be computational. While some find these debates interesting, they do not actually have a significant impact on the enterprise of modeling. On the other hand, there have been theoretical results from computational theory that have had a huge impact on the development of cognitive models. Probably the best example of this is the previously mentioned work done by Minsky and Papert on Perceptrons [1969]. They showed that perceptrons, which are a simple kind of neural network, are not capable of modeling nonlinear functions (including exclusive-or). This result effectively ended the majority of neural network research for more than a decade until the PDP group developed far more powerful neural network models [Rumelhart and McClelland, 1986].

69.2.4 Evolutionary and Environmental Psychology

In recent years two branches of psychology have come to prominence as providing alternate sources of constraints based upon evolution, and in particular the kinds of environments in which humans evolved. Evolutionary psychology is most often associated with the work of Tooby and Cosmides (e.g. [Tooby and Cosmides, 1992]) while environmental psychology is often associated with the work of Steve and Rachel Kaplan (e.g. [Kaplan & Kaplan, 1989]). What both of these fields have in common is a belief that the brain should not be studied in a vacuum, that some types of context are extremely meaningful.

In the case of evolutionary psychology the context is provided by evolution. As has been noted in many places, systems that are evolved rather than designed, tend to end up looking like the work of a "tinkerer." The eye is a well known example of a system that is poorly "designed" but is nonetheless functional and can be understood as a series of successive improvements, each adding functionality to the previous iteration [Dawkins, 1986]. This example captures the core tenets of evolutionary psychology, that evolution tends to work in piecemeal fashion with each change adding functionality to what existed previously. This does not tend to be a hard constraint on cognitive models, as it can be argued that the evolutionary story behind any particular theory simply has yet to be found. Nevertheless, the evolutionary view provides a powerful way to think about how pieces of the cognitive system came about and for what purpose.

As the name would suggest, work in environmental psychology focuses on the environment as a source of constraints. Most environmental psychologists focus their research on how people interact with different kinds of environments and how to use this knowledge to design better spaces. Another branch of the field, however, has noted that the environment adds additional meaningful context to evolutionary history. The evolutionary history of the brain is a story of information processing mechanisms that evolved to address the specific needs of our ancestors. The human ability to represent and reason about large-scale space, for example, allowed our ancestors to forage and hunt over large areas of savanna. In turn once these spatial abilities were in place they were available for the greater cognitive system and impact cognition of virtually every type [Chown, 1999]. The importance in understanding the evolutionary environments that the brain developed in is highlighted by the work of the Kaplans and their colleagues. The Kaplans have shown, for example, that people will recover more quickly in hospitals with views of nature, perform better in workplaces with views of trees, etc (for reviews see [Kaplan, 1993; Kaplan and Peterson, 1993]). This is a clear indicator that people do not treat information neutrally. As Kaplan and others [Chown, et al. 2002] have argued, the human emotional system can be understood in these terms. The argument is based upon the idea that human emotions address the need our ancestors had to make very fast decisions in encounters with other dangerous predators. In such cases it is usually better to act quickly than to pause and consider an optimal strategy. This view of cognition undercuts rationality approaches and helps explain many of the supposed shortcomings in human reasoning.

Some of the advantages of an evolutionary/environmental perspective have been clarified by work in robotics. Early artificial intelligence and cognitive mapping focused on reasoning, for example. This led to models that were too abstract to be implemented on actual robots. The move to using robots forced researchers to come at the problem from a far more practical point of view and to consider perceptual issues more directly.

69.3   Research Issues

Since so much about cognition is still not well understood, this section will focus on two of the key debates driving research in the field. These include: 1) is the brain a symbol processor, or does it need to be modeled in neural terms? 2) Should the field be working on grand theories of cognition, or is it better to proceed on a reductionist path?

69.3.1   Are We Symbol Processors?

The connectionist revolution brought a new way of thinking to cognitive science. The critical idea is rather simple — since the "hardware" of the brain is neural, then models of the brain should be described in neural terms. Lending credence to this position was a series of neural models that had a number of attractive properties that seemed notably lacking in symbolic models of the time (e.g. **content addressable memory**, **graceful degradation**, etc. [Rumelhart and McClelland, 1986]). On the face of it, the argument for neural models seems unassailable given that the brain is a neural system. Symbolists, notably Newell [1990] and Fodor and Plyshen [1988] have attacked these models on a number of grounds, however. Their arguments are based upon the idea that the brain, like any complex system, is hierarchical. Neural models, so the argument goes, provide appropriate computational descriptions for only the lowest levels of the cognitive hierarchy. From the point of view of the symbolists, these levels of cognition are also less well understood and not as interesting from a behavioral point of view as the so-called **cognitive band** [Newell, 1990]. From this point of view the operation of the cognitive band is nothing like a neural network, but is much more like a traditional symbol system. The argument is akin to finding the appropriate level at which to study computers. It is possible, and often necessary, to look at the performance of a computer from the point of view of gates. When trying to understand the performance of a complicated piece of software, however, it is much more appropriate to study the performance at the level of a high level programming language. Further, symbolists have effectively argued that current connectionist models are not capable of the full range of behaviors needed for cognitive modeling [Newell, 1990].

The argument for symbolic models comes from computational theory. It is based upon the idea of computational equivalence. Since symbolic models are Turing-complete they are equivalent computationally to any other Turing-complete model. Along these lines a number of efforts have been made to implement symbolic models in neural hardware. The case can then be made that this is exactly what the brain does. Symbolists see this equivalence as freeing them from the need to worry about mechanisms. Even so, the impact of connectionism and the capability of neural models for pattern recognition has led even strongly symbolic models like Soar [Laird, et al., 1987] to use neural networks as pattern recognizers to obtain the symbols.

The freeing up from the constraints of mechanisms has been something of a double-edged sword for symbolic models. On the one hand, symbolic models are easily implemented on computers and relatively easy to expand, debug, etc. Further, in terms of high-level behavior, symbolic models have been shown to be capable of a much wider range of behaviors than their current connectionist counterparts. For example, Soar agents are capable of modeling the flying behaviors of combat pilots [Jones et al., 1999]. On the other hand, critics have complained that systems like Soar are little more than symbolic programming languages. As noted earlier, while a computer running C may be Turing-complete, it would be ridiculous to call it a model of human cognition. Clark [2001] refers to this as "surface mimicry" and points out that the Soar model is far more homogeneous than the "Swiss Army Knife" model suggested by Tooby and Cosmides [1992]. The critics argue that symbolic models like Soar and ACT-R are under constrained, and therefore they cannot truly be called cognitive models. The lack of constraints goes even further than just mechanisms, symbolic models have also been attacked on evolutionary grounds. It is difficult to see how symbolic constructs like distal memory access could have evolved in any sort of piecemeal fashion.

There are several arguments that symbolists use against neural models. First is the "levels of modeling" argument. This argument posits that we simply do not understand the behavior of neurons well enough to construct credible models from them. To be fair, the majority of neural models do not model the behavior of individual neurons. Other attacks on connectionist models are based upon what current models cannot do. Popular connectionist models, such as feed-forward back-propagation networks, for example, exhibit a number of problems as memory models including "catastrophic forgetting" of old material when presented with new material [McCloskey and Cohen, 1989] . While it is certainly appropriate to attack the individual models on these grounds, it is less so to attack the entire connectionist position based on the failure of even its most notable examples. Similarly, other criticisms of connectionism have attacked the models as being nothing more than new versions of the discredited behaviorist position [Lachter and Bever, 1988]. Again, this is certainly the case with many connectionist models, and should rightfully lead to a search for better models, but it cannot undermine the general position.

A more interesting criticism focuses upon the way that connectionists have pursued cognitive modeling. The argument is the same as the one against some of the symbolic programs. As McCloskey [1991] put it, connectionists often pursue a path of "simulation in search of theory" (p. 388). McCloskey argues that many of these simulations are little better than "black boxes." Modelers do not provide insight into what aspects of the network are crucial to its performance with regard to the task. For example, if back-propagation was used to train the network, is that a crucial step, or could another training regime have been used? If back-propagation were crucial then it would undermine the model if back-propagation were found implausible on other grounds. In other words, connectionist systems are engaged in the same sort of "surface mimicry" as symbolic systems under the guise of "neural plausibility." McCloskey goes on to complain that connectionist models generally fall short in describing how their networks elucidate the cognitive processes they purport to model. To put it simply, connectionist networks are not well understood enough to tell us exactly how they manage to accomplish what they are trained for.

So far in discussing connectionist systems, we have avoided discussing connectionist "symbols." While it may be the case that connectionist units represent collections of neurons, they do not normally represent what most computer scientists would think of as symbols. There are, however, connectionist models that recognize the power of symbols as a basic unit of thought. Many of these models trace their lineage to the work of D.O. Hebb's book *The Organization of Behavior* [1949]. Hebb proposed that the "symbols" of thought were cell assemblies, tightly connected groups of neurons capable of functioning as a unit because of their strong interconnections. The problem with cell assemblies as originally formulated by Hebb was that all of the connections between neurons were positive and only became stronger through learning. Hebb omitted inhibition because there was still no hard evidence of it at the time of publication. A later simulation of the cell assembly construct [Rochester, et al., 1956] showed that without inhibition activity in the simulated brain quickly grew out of control. Unfortunately these results were sufficient to essentially stop research on cell assemblies for more than a decade even though the same paper showed that with the addition of inhibition the cell assembly construct was viable. In recent years, however, the cell assembly idea has undergone a revival as researchers from a number of domains have proposed models based upon Hebb's original idea, but modified with modern understandings of neuroscience [Kaplan et al., 1991; Amit, 1995].

Neural models based upon cell assemblies purport to contain the best of both symbolic and connectionist models. It is difficult to study sequence learning, for example, without something approximating a symbol to serve as a unit in the sequence. Further, many connectionist systems do not address temporal issues at all. Conversely, symbolic models do not ground the symbols in any physical mechanism, nor are symbolic systems able to take advantage of the properties of neural hardware. Chown [2002], for example, has shown that some learning results that have defied conventional modeling for nearly 40 years can be fairly easily explained when basic neural properties are accounted for in a cell assembly-based model.

69.3.2   Grand Theories?

In his book *Unified Theories of Cognition* Allen Newell [1990] called for a return to all-encompassing theories of mind called **UTCs** after the title of the book. Newell's reasoning echoes McCloskey's complaints about connectionism [1991], that by operating at too small a scale cognitive modelers have worked on under-constrained models. While it may be true, for example, that connectionist system X can model psychological data set Y, such models rarely address questions of how they would or could fit into a larger system. Modeling efforts such as these are sometimes attacked on the grounds that they are "doomed to succeed" in much the same way that models with too many parameters can fit all types of curves. Put another way, if a model is Turing complete, the question of whether or not it can be used to fit some data is not particularly interesting. The interesting question is whether or not there is actually evidence for it. McCloskey and others argue that for these reasons theory should drive simulation rather than the other way around.

Ironically, the two most notable examples of UTCs, Soar and ACT-R, have been attacked on virtually the same grounds — that they are under-constrained. Both systems are built on the same assumption, that at higher levels of cognition the brain is a rule-based system. At the heart of each system is a production system implementing the rule-base that serves as long-term memory. Since production systems are Turing-complete they are capable of modeling anything. To be fair, however, Soar does make some key theoretical commitments that can be used to judge its merits. First and foremost, Soar is a symbol manipulation system with all that that implies. Second, in Soar deliberative thought is equated to a search through a problem space. For example, Soar is equipped with all the basic weak search methods including breadth-first, depth-first, etc. At one time another key constraint associated with Soar was that all learning came as the result of a single mechanism called "chunking" [Laird, et al., 1984]. It is not clear, however, from recent work in Soar that this is still held as a central tenet of the system.

Soar and ACT-R have both been attacked for their commitment to production systems. This criticism actually pre-dates either system and is most famously associated with the philosopher Hubert Dreyfus [1972]. The Dreyfus position is that systems based upon rules are too brittle to account for the richness of human behavior. For example, Dreyfus discusses how knowledge about the health of a jockey's mother might influence how a bettor would make a wager. It seems unlikely that the bettor would have explicit rules dealing with such a situation, and yet humans are capable of dealing with such situations with ease. The response to this criticism has been to test it explicitly, as with Doug Lenat's CYC [Lenat and Feigenbaum, 1992] which aims to capture enough knowledge to perform common sense reasoning, or with the Soar program which builds more and more complex agents capable of difficult tasks such as flying jet airplanes in combat situations. In part, Dreyfus's criticisms can also be addressed by noting that rules need not all be specified at the same level of generality. For example, while a system for betting on horses might contain many specific rules concerning the records of horses and jockeys, a general cognitive system might reasonably be expected to include rules such as "when something traumatic happens to a person they will not perform at normal levels." Of course this raises further questions of how such rules are learned, how patterns such as "something traumatic" are recognized, etc. Dreyfus would argue that this leads to an endless cycle for any reasonably complex task.

There are also connectionist programs that work at the level of large theories of cognition. Steven Grossberg, for example, has produced a huge body of work that have never been explicitly put forth as a UTC, but which when viewed as a whole have many of the same principles. Probably the best example of this work is the ART model developed in conjunction with Gail Carpenter [1987]. The SESAME group, operating mainly out of the University of Michigan is also working on a cognitive architecture [Kaplan et al., 1991]. The SESAME architecture is based on the cell assembly and is also the only cognitive architecture to include a complete theory of spatial processing [Chown et al., 1995].

69.4   Best Practices

As the previous sections suggest, there are a number of pitfalls involved in putting together a cognitive model. History has shown that there are two problems that crop up again and again. The first is the danger of constructing a simulation without theoretically motivating the details. This is akin to the old saw that "if you have a big enough hammer everything looks like a nail." There is a related danger that once a simulation works (or at least models the data) it is often difficult to say why. Together these dangers suggest that there should be a close relationship between theory and the simulation process. The goal of a simulation should not be simply to model a dataset, but should also be to elucidate the theory. For example, some connectionist models propose a number of mechanisms as being central to understanding a particular process. These models can be systematically "damaged" by disabling the individual mechanisms. In many cases the damage to the model can be equated to damage to individuals. This provides a second dataset to model, and provides solid evidence of what the mechanism does in the simulation. Alternatively models can be built piecewise mechanism by mechanism. Each new piece of the simulation would correspond to a new theoretical mechanism aiming to address some shortcoming of the previous iteration. This motivates each mechanism and helps to clearly delineate its role in the overall simulation. In the SESAME group this style of simulation has been termed "the systematic exploitation of failure" by one of its members, Abraham Kaplan.

One of the earliest examples of this approach was done by Booker [1982] in an influential work that has helped shaped the adaptive systems paradigm. In an adaptive systems paradigm a simple creature is placed in a microworld where the goal is survival. Creatures are successively altered (and sometimes the environments are as well) by adding and subtracting mechanisms. In each case the success of the new mechanism can be judged by improvements in the survival rate of the organism. In addition to providing a way to motivate theoretical mechanisms, this paradigm is also essentially the same one used for the development of genetic algorithms.

The Soar architecture is probably the pre-eminent symbolic cognitive architecture. Soar is based upon a number of crucial premises that constrain all models written in Soar (which can be considered a kind of programming environment). First, Soar is a rule-based system implemented as a production system. In the Soar paradigm the production rules represent long-term memory and knowledge. One effect of a production firing in Soar can be to put new elements into working memory, Soar's version of short-term memory. For example, a Soar system might contain a number of perceptual productions that aim to identify different types of aircraft. When a production fires it might create a structure in working memory to represent the aircraft it identifies. This structure in turn might cause further productions to fire. Soar enforces a kind of hierarchy through the use of a subgoaling system. Productions can be written to apply generally, or might only match when a certain goal is active. The combination of goals and productions forms a problem space that provides the basic framework for any task. Finally, the Soar architecture contains a single mechanism for learning called "chunking." Essentially Soar systems learn when they reach an impasse generally created by not being able to match any productions. When impasses occur Soar can apply weak search methods to the problem space in order to discover what to do. Once a solution is found, a new production or "chunk" is created to apply to the situation.

Here is an example of a Soar production taken from the Soar tutorial [Laird, 2001]. In this example the agent is driving a tank in a battle exercise.

```
1    sp {wander*propose*move
2        (state <s> ∧ name wander
3            ∧io.input-linked-blocked.forward no)
4    - -
5        (<s> ∧operator <o> + = )
6        (<o> ∧name move
7            ∧ actions.move.direction forward) }
```

In the example "sp" stands for "Soar production" and starts every production. The production is named "wander*propose*move". The elements that come before the arrow represent the "if" part of the production. In this case the production fires only if the current subgoal is to wander and the forward direction is not blocked (input comes from a specialized structure tagged ∧io). The elements that come after the arrow represent the "then" parts of the production. In this case a new working memory element is created to represent the operator for moving forward. In a typical production cycle, productions are matched in parallel and can propose operators such as the move operator in this case. Then other productions can be used to select among the proposed operators. This selection can be based upon virtually any criteria; for example cognitive productions may be selected over more reactive productions. Production matching can be done in parallel to simulate the parallelism of the brain.

Both the Soar and ACT communities are engaged in programs of simulating more and more human behavior. These simulations can be done at the level of models of simple psychological experiments, or, as is increasingly the case, they can simulate human performance on complex tasks such as flying airplanes. The implicit argument is that if they can simulate anything that humans can do then they must be modeling human cognition. On one level this argument has merit, if either architecture can accurately simulate human performance then it can be used in a predictive fashion. Tac-air Soar [Jones et al., 1999], for example, simulates the performance of combat pilots and is used to train new pilots in a more cost-effective way than if experienced pilots had to be used. On the other hand equivalent functionality is not the same thing as equivalence. As noted previously, critics point out that both Soar and Act are essentially Turing-complete programming environments and therefore are capable of simulating any computable function given clever enough programmers. The fact that both systems still rely heavily on clever programming is still a major limitation with regard to being considered a fully realized model of human cognition. Although Soar's initial success was due in large part to its learning mechanism, for example, little progress has been made within the Soar community in building agents that exhibit any sort of developmental patterns. It is much simpler to build a Soar system that can fly planes than one that can learn to fly planes.

Gail Carpenter, Stephen Grosseberg and their associates have also attacked a wide range of problems, but have done so with much more of an eye towards cognitive theory than applications. While Carpenter and Grossberg have not explicitly developed a unified theory of cognition they have modeled a remarkable range of cognitive processes and has done so using an approach more sympathetic to a systems view of cognition than is typical in connectionist modelers. A good example of this approach can be found in their Adaptive Resonance Theory (ART) [Carpenter and Grossberg, 1987; Grossberg, 1987]. Superficially ART looks similar to many connectionist learning systems in that it is essentially a classification system, but it was developed to specifically address many of the shortcomings of such models. ART takes a feature vector as an input and uses it to provide a classification of the input. For example, a typical task would be to recognize hand-written numbers. The features would consist of the presence or absence of a pen stroke at different spatial locations.

One of the problems that ART was designed to address was what Grossberg [1987] referred to as the "stability-plasticity" dilemma. This is essentially a problem of how much new knowledge should impact what has been learned before. For example, a system that has been trained to recognize horses might have a problem when confronted with a zebra. The system could either change its representation of horses to include zebras, or it could create a separate representation for zebras. This is a significant issue for neural network models because they achieve a great deal of their power by having multiple representations share structure. Such sharing is useful for building compact representations and for automatic abstraction, but it also means that new knowledge tends to constantly overwrite what has come before. Among the problems this raises, is "catastrophic forgetting" as mentioned previously.

The stability-plasticity dilemma was addressed in part in ART through the introduction of a vigilance parameter that adaptively changed according to how well the system was performing. In some cases, for example, the system would be extremely vigilant and would require an unusually high degree of match before it would recognize an input as being familiar. In cases where inputs were not recognized as familiar, novel structure was created to form a new category or prototype. Such a new category would not share internal structure directly with previously learned categories. Essentially when vigilance is high the system creates "exemplars" or very specialized categories, whereas when vigilance is low ART will create "prototypes" that generalize across many instances. This makes ART systems attractive since they do not commit fully either to exemplar or prototype models, but can exhibit properties of both, as seems to be the case with human categorization.

In ART systems an input vector activates a set of feature cells within an attentional system, essentially storing the vector in short-term memory. These in turn activate corresponding pathways in a bottom-up process. The weights in these pathways represent long-term memory traces and act to pass activity to individual categories. The degree of activation of a category represents an estimate that the input is an example of the category. In the meantime the categories send top down information back to the features as a kind of hypothesis test. The vigilance parameter defines the criteria for whether the match is good enough. When a match is established the bottom up and top down signals are locked into a "resonant" state, and this in turn triggers learning, is incorporated into conciousness, etc.

It is important to note that ART, unlike many connectionist learning systems, is unsupervised — it learns the categories without any teaching signals.

ART has since been extended to a number of times, to models including ART1, ART2, ART3, and ARTMAP. Grossberg has also tied it to his FACADE model in a system called ARTEX [Grossberg and Williamson, 1999]. These models vary in features and complexity, but share intrinsic theoretical properties. ART models are self-organizing (i.e., unsupervised, though ARTMAP systems can include supervised learning) and consist of an attentional and an orienting subsystem. A fundamental property of any ART system (and many other connectionist systems) is that perception is a competitive process. Different learned patterns generate expectations that essentially compete against each other. Meanwhile, the orienting system controls whether or not such expectations sufficiently match the input — in other words it acts as a novelty detector.

The ART family of models demonstrate many of the reasons why working with connectionist models can be so attractive. Among them:

- The neural computational medium is natural for many processes including perception. Fundamental ideas such as representations competing against each other (including inhibiting each other) are often difficult to capture in a symbolic model. In a system like ART, on the other hand, a systemic property like the level of activation of a unit can naturally fill many roles from the straightforward transmission of information to providing different measures of the goodness of fit of various representations to input data.
- The architecture of the brain is a source of both constraints and ideas. Parameters, such as ART's vigilance parameter, can be linked directly to real brain mechanisms such as the arousal system. In this way what is known about the arousal system provides clues as to the necessary effects of the mechanism in the model and provides insight into how the brain handles fundamental issues such as the plasticity-stability dilemma.

69.5 Summary

Unlike many disciplines in computer science there are no provably correct algorithms for building cognitive models. Progress in the field is made through a process of successive approximation. Models are continually proposed and rejected; and with each iteration of this process the hope is that the models come closer to a true approximation of the underlying cognitive structure of the brain. It should be clear from the preceding sections that there is no "right" way to do this.

Improvements in cognitive models come from several sources. In many cases improvements result from an increased understanding of some aspect of cognition. For example, neuroscientists are constantly getting new data on how neurons work, how they are connected, what parts of the brain process what types of information, etc. In the meantime models are implemented on computers and on robots. These implementations provide direct feedback about model quality and shortcomings. This feedback often will lead to revisions in the models and sometimes may even drive further experimental work. Because of the complexity of cognition and the number of interactions amongst parts of the brain it is really the case that definitive answers can be found; which is not to say that cognitive scientists do not reach consensus on any issues. Over time, for example, evidence has accumulated that there are multiple memory systems operating at different time scales. While many models have been proposed to account for this there is general agreement on the kinds of behavior that those models need to be able to display. This represents real progress in the field because it eliminates whole classes of models that could not account for the different time scales. The constraints provided by data and by testing models work to continually narrow the field of prospective models.

Defining Terms

Back-propagation A method for training neural networks based upon gradient descent. An error signal is propagated backward from output layers toward the input layer through the network.

Cognitive band In Newell's hierarchy of cognition, the cognitive band is the level at which deliberate thought takes place.

Cognitive map A mental model. Often, but not exclusively, used for models of large-scale space.

Connectionist A term used to describe neural network models. The choice of the term is meant to indicate that the power of the models comes from the massive number of connections between units within the model.

Content addressable memory Memory that can be retrieved by descriptors. For example, people can remember a person when given a general description of the person.

Feed forward Neural networks are often constructed in a series of layers. In many models, information flows from an input layer toward an output layer in one direction. Models in which the information flows in both directions are called **recurrent**.

Graceful degradation The principle that small changes in the input to a model, or that result from damage to a model, should result in only small changes to the model's performance. For example, adding noise to a model's input should not break the model.

Necker cube A three-dimensional drawing of a cube drawn in such a way that either of the two main squares that comprise the drawing can be viewed as the face closest to the viewer.

UTC Unified Theory of Cognition.

References

Amit, D.J. (1995). The Hebbian paradigm reintegrated: local reverberations as internal representations. *Behavioral and Brain Sciences*, 18(4), 617–657.

Ballard, D.H. (1999) *An Introduction to Natural Computation*, Cambridge, MA: The MIT Press.

Booker, L.B. (1982). Intelligent Behavior as an Adaptation to the Task Environment. Ph.D. dissertation, The University of Michigan.

Carpenter, G.A. and Grossberg, S. (1987). A massively parallel architecture for a self-organizing neural pattern recognition machine. *Computer Vision, Graphics and Image Processing*, 37, 54–115.

Chown, E. (1999). Making predictions in an uncertain world: environmental structure and cognitive maps. *Adaptive Behavior*, 7(1), 1–17.

Chown, E. (2002). Reminiscence and arousal: a connectionist model. *Proceedings of the Twenty Fourth Annual Meeting of the Cognitive Science Society.* 234–239

Chown, E., Jones, R.M., and Henninger, A.E. (2002). An architecture for emotional decision-making agents. In *The proceedings of Autonomous Agents and Multi-Agent Systems '02*.

Chown, E., Kaplan, S., and Kortenkamp, D. (1995). Prototypes, location, and associative networks (PLAN): towards a unified theory of cognitive mapping. *Cognitive Science*, 19, 1–51.

Clark, A. (2001). *Mindware: An Introduction to the Philosophy of Cognitive Science*. New York: Oxford University Press.

Craik, K.J.W. (1943). *The Nature of Exploration*. London: Cambridge University Press.

Dawkins, R. (1986). *The Blind Watchmaker*. New York: W.W. Norton & Company.

Dreyfus, H. (1972). *What Computers Can't Do*. New York: Harper & Row.

Hebb, D.O. (1949). *The Organization of Behavior*. New York: John Wiley.

Fodor, J.A. and Pylyshyn, Z.W. (1988). Connectionism and cognitive architecture: a critical analysis. *Cognition*, 28, 3–71

Grossberg, S. (1987). Competitive learning: from interactive activation to adaptive resonance. *Cognitive Science*, 11, 23–63.

Grossberg, S. and Williamson, J.R. (1999). A self-organizing neural system for learning to recognize textured scenes. *Vision Research*, 39, 1385–1406.

Jones, R.M., Laird, J.E., Nielsen, P.E., Coulter, K.J., Kenny, P.G., and Koss, F., (1999). Automated intelligent pilots for combat flight simulation. *AI Magazine*, 20(1), 27–41.

Kaplan, R. (1993). The role of nature in the context of the workplace. *Landscape and Urban Planning*, 26, 193–201.

Kaplan, R. and Kaplan, S. (1989). *The Experience of Nature: A Psychological Perspective*. New York: Cambridge University Press.

Kaplan, S. and Peterson, C. (1993). Health and environment: a psychological analysis. *Landscape and Urban Planning*, 26, 17–23.

Kaplan, S., Sonntag, M., and Chown, E. (1991). Tracing recurrent activity in cognitive elements (TRACE): a model of temporal dynamics in a cell assembly. *Connection Science*, 3, 179–206.

O'Keefe, M.J. and Nadel, L. (1978). *The Hippocampus as a Cognitive Map*. Oxford: Clarendon Press.

Kinsbourne, M. (1982). Hemispheric specialization and the growth of human understanding. *American Psychologist*, 34, 411–420.

Lachter, J. and Bever, T. (1988). The relationship between linguistic structure and associative theories of language learning — A constructive critique of some connectionist teaching models. *Cognition*, 28, 195–247.

Laird, J.E. (2003). The Soar 8 Tutorial. http://ai.eecs.umich.edu/soar/tutorial.html.

Laird, J.E., Newell, A., and Rosenbloom, P.S. (1987). Soar: an architecture for general intelligence. *Artificial Intelligence*, 33, 1–64.

Laird, J.E., Rosenbloom, P.S., and Newell, A. (1984). Towards chunking as a general learning mechanism. *Proceedings of the AAAI'84 National Conference on Artificial Intelligence*. American Association for Artificial Intelligence, 188–192.

Lenat, D. and Feigenbaum, E. (1992). On the thresholds of knowledge. In D. Kirsh (Ed.), *Foundations of Artificial Intelligence*. MIT Press and Elsevier Science. 195–250.

McCloskey, M. (1991). Networks and theories: the place of connectionism in cognitive science. *Psychological Science*, 2(6), 387–395.

McCloskey, M. and Cohen, N.J. (1989). Catastrophic interference in connectionist networks: the sequential learning problem. In G.H. Bower, Ed. *The Psychology of Learning and Motivation*, Vol. 24, New York: Academic Press.

Newell, A. (1990). *Unified Theories of Cognition*. Harvard University Press: Cambridge, MA.

Rochester, N., Holland, J.H, Haibt, L.H., and Duda, W.L. (1956). Tests on a cell assembly theory of the action of the brain, using a large digital computer. *IRE Transactions on Information Processing Theory*, IT-2, 80–93.

Rumelhart, D.E. and McClelland, J.L., Eds. (1986). *Parallel Distributed Processing: Explorations in the Microstructure of Cognition*, The MIT Press: Cambridge, MA.

Squire, L.R. (1992). Memory and the hippocampus: a synthesis from findings with rats, monkeys, and humans. *Psychological Review*, 99, 195–231.

Tooby, J. and Cosmides, L. (1992). The psychological foundations of culture. In J. Barkow, L. Cosmides, and J. Tooby, Eds., *The Adapted Mind*, New York: Oxford University Press, 19–136.

Further Information

There are numerous journals and conferences on cognitive modeling. Probably the best place to start is with the annual conference of the Cognitive Science Society. This conference takes place in a different city each summer. The Society also has an associated journal, *Cognitive Science*. Information on the journal and the conference can be found at the society's homepage at http://www.cognitivesciencesociety.org.

Because of the lag-time in publishing journals, conferences are often the best place to get the latest research. Among other conferences, Neural Information Processing Systems (NIPS) is one of the best for work specializing in neural modeling. The Simulation of Adaptive Behavior conference is excellent for adaptive systems. It has an associated journal as well, *Adaptive Behavior*.

A good place for anyone interested in cognitive modeling to start is Allen Newell's book, *Unified Theories of Cognition*. While a great deal of the book is devoted to Soar, the first several chapters lay out the challenges and issues facing any cognitive modeler. Another excellent starting point is Dana Ballard's 1999 book, *An Introduction to Natural Computation*. Ballard emphasizes neural models, and his book provides good coverage on most of the major models in use. Andy Clark's 2001 book, *Mindware: An Introduction to the Philosophy of Cognitive Science*, covers much of the same ground as this article, but in greater detail, especially with regard to the debate between connectionists and symbolists.

70

Graphical Models for Probabilistic and Causal Reasoning

70.1 Introduction **70**-1
70.2 Historical Background **70**-2
70.3 Bayesian Networks as Carriers of Probabilistic Information **70**-3
Formal Semantics • Inference Algorithms • System's Properties • Recent Developments
70.4 Bayesian Networks as Carriers of Causal Information **70**-6
Causal Theories, Actions, Causal Effect, and Identifiability • Acting vs Observing • Action Calculus • Historical Remarks
70.5 Counterfactuals **70**-13
Formal Underpinning • Applications to Policy Analysis

Judea Pearl
University of California at Los Angeles

70.1 Introduction

This chapter surveys the development of graphical models known as Bayesian networks, summarizes their semantical basis, and assesses their properties and applications to reasoning and planning.

Bayesian networks are directed acyclic graphs (DAGs) in which the nodes represent variables of interest (e.g., the temperature of a device, the gender of a patient, a feature of an object, the occurrence of an event) and the links represent causal influences among the variables. The strength of an influence is represented by conditional probabilities that are attached to each cluster of parents–child nodes in the network.

Figure 70.1 illustrates a simple yet typical Bayesian network. It describes the causal relationships among the season of the year (X_1), whether rain falls (X_2) during the season, whether the sprinkler is on (X_3) during that season, whether the pavement would get wet (X_4), and whether the pavement would be slippery (X_5). All variables in this figure are binary, taking a value of either true or false, except the root variable X_1, which can take one of four values: spring, summer, fall, or winter. Here, the absence of a direct link between X_1 and X_5, for example, captures our understanding that the influence of seasonal variations on the slipperiness of the pavement is mediated by other conditions (e.g., the wetness of the pavement).

As this example illustrates, a Bayesian network constitutes a model of the environment rather than, as in many other knowledge representation schemes (e.g., logic, rule-based systems, and neural networks), a model of the reasoning process. It simulates, in fact, the causal mechanisms that operate in the environment and thus allows the investigator to answer a variety of queries, including associational queries, such as "Having observed A, what can we expect of B?"; abductive queries, such as "What is the most plausible

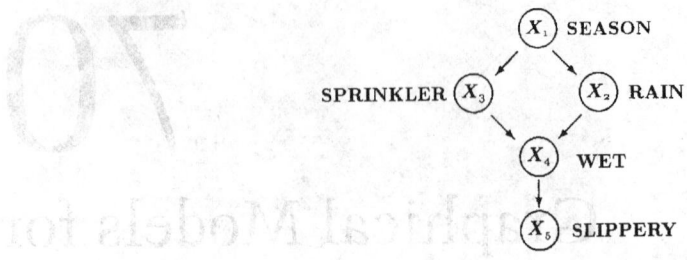

FIGURE 70.1 A Bayesian network representing causal influences among five variables.

explanation for a given set of observations?"; and control queries, such as "What will happen if we intervene and act on the environment?" Answers to the first type of query depend only on probabilistic knowledge of the domain, whereas answers to the second and third types rely on the causal knowledge embedded in the network. Both types of knowledge, associative and causal, can effectively be represented and processed in Bayesian networks.

The associative facility of Bayesian networks may be used to model cognitive tasks such as object recognition, reading comprehension, and temporal projections. For such tasks, the probabilistic basis of Bayesian networks offers a coherent semantics for coordinating top-down and bottom-up inferences, thus bridging information from high-level concepts and low-level percepts. This capability is important for achieving selective attention, that is, selecting the most informative next observation before actually making the observation. In certain structures, the coordination of these two modes of inference can be accomplished by parallel and distributed processes that communicate through the links in the network.

However, the most distinctive feature of Bayesian networks, stemming largely from their causal organization, is their ability to represent and respond to changing configurations. Any local reconfiguration of the mechanisms in the environment can be translated, with only minor modification, into an isomorphic reconfiguration of the network topology. For example, to represent a disabled sprinkler, we simply delete from the network all links incident to the node sprinkler. To represent a pavement covered by a tent, we simply delete the link between rain and wet. This flexibility is often cited as the ingredient that marks the division between deliberative and reactive agents and that enables the former to manage novel situations instantaneously, without requiring retraining or adaptation.

70.2 Historical Background

Networks employing DAGs have a long and rich tradition, starting with the geneticist Sewall Wright [1921]. He developed a method called *path analysis* [Wright 1934], which later became an established representation of causal models in economics [Wold 1964], sociology [Blalock 1971, Kenny 1979], and psychology [Duncan 1975]. Good [1961] used DAGs to represent causal hierarchies of binary variables with disjunctive causes. *Influence diagrams* represent another application of DAG representation [Howard and Matheson 1981]. Developed for decision analysis, they contain both event nodes and decision nodes. *Recursive models* is the name given to such networks by statisticians seeking meaningful and effective decompositions of contingency tables [Lauritzen 1982, Wermuth and Lauritzen 1983, Kiiveri et al. 1984].

The role of the network in the applications cited was primarily to provide an efficient description for probability functions; once the network was configured, all subsequent computations were pursued by symbolic manipulation of probability expressions. The potential for the network to work as a computational architecture, and hence as a model of cognitive activities, was noted in Pearl [1982], where a distributed scheme was demonstrated for probabilistic updating on tree-structured networks. The motivation behind this particular development was the modeling of distributed processing in reading comprehension [Rumelhart 1976], where both top-down and bottom-up inferences are combined to form a coherent interpretation. This dual mode of reasoning is at the heart of Bayesian updating and in fact motivated

Reverend Bayes's original 1763 calculations of posterior probabilities (representing explanations), given prior probabilities (representing causes), and likelihood functions (representing evidence).

Bayesian networks have not attracted much attention in the logic and cognitive modeling circles, but they did in expert systems. The ability to coordinate bidirectional inferences filled a void in expert systems technology of the late 1970s, and it is in this area that Bayesian networks truly flourished. Over the past 10 years, Bayesian networks have become a tool of great versatility and power, and they are now the most common representation scheme for probabilistic knowledge [Shafer and Pearl 1990, Shachter 1990, Oliver and Smith 1990, Neapolitan 1990]. They have been used to aid in the diagnosis of medical patients [Heckerman 1991, Andersen et al. 1989, Heckerman et al. 1990, Peng and Reggia 1990] and malfunctioning systems [Agogino et al. 1988]; to understand stories [Charniak and Goldman 1991]; to filter documents [Turtle and Croft 1991]; to interpret pictures [Levitt et al. 1990]; to perform filtering, smoothing, and prediction [Abramson 1991]; to facilitate planning in uncertain environments [Dean and Wellman 1991]; and to study causation, nonmonotonicity, action, change, and attention. Some of these applications are described in a tutorial article by Charniak [1991]; others can be found in Pearl [1988], Shafer and Pearl [1990], and Goldszmidt and Pearl [1996].

70.3 Bayesian Networks as Carriers of Probabilistic Information

70.3.1 Formal Semantics

Given a DAG G and a joint distribution P over a set $X = \{X_1, \ldots, X_n\}$ of discrete variables, we say that G *represents* P if there is a one-to-one correspondence between the variables in X and the nodes of G, such that P admits the recursive product decomposition

$$P(x_1, \ldots, x_n) = \prod_i P(x_i \mid pa_i) \tag{70.1}$$

where pa_i are the direct predecessors (called *parents*) of X_i in G. For example, the DAG in Figure 70.1 induces the decomposition

$$P(x_1, x_2, x_3, x_4, x_5) = P(x_1)P(x_2 \mid x_1)P(x_3 \mid x_1)P(x_4 \mid x_2, x_3)P(x_5 \mid x_4) \tag{70.2}$$

The recursive decomposition in Equation 70.1 implies that, given its parent set pa_i, each variable X_i is conditionally independent of all its other predecessors $\{X_1, X_2, \ldots, X_{i-1}\} \setminus pa_i$. Using Dawid's notation [Dawid 1979], we can state this set of independencies as

$$X_i \perp\!\!\!\perp \{X_1, X_2, \ldots, X_{i-1}\} \setminus pa_i \mid pa_i \quad i = 2, \ldots, n \tag{70.3}$$

Such a set of independencies is called *Markovian*, since it reflects the Markovian condition for state transitions: each state is rendered independent of the past, given its immediately preceding state. For example, the DAG of Figure 70.1 implies the following Markovian independencies:

$$X_2 \perp\!\!\!\perp \{0\} \mid X_1, \quad X_3 \perp\!\!\!\perp X_2 \mid X_1, \quad X_4 \perp\!\!\!\perp X_1 \mid \{X_2, X_3\}, \quad X_5 \perp\!\!\!\perp \{X_1, X_2, X_3\} \mid X_4 \tag{70.4}$$

In addition to these, the decomposition of Equation 70.1 implies many more independencies, the sum total of which can be identified from the DAG using the graphical criterion of d-separation [Pearl 1988]:

Definition 70.1 (d-separation.) Let a path in a DAG be a sequence of consecutive edges, of any directionality. A path p is said to be d-separated (or blocked) by a set of nodes Z iff:

1. p contains a chain $i \longrightarrow j \longrightarrow k$ or a fork $i \longleftarrow j \longrightarrow k$ such that the middle node j is in Z.
2. Or p contains an inverted fork $i \longrightarrow j \longleftarrow k$ such that neither the middle node j nor any of its descendants (in G) are in Z.

If X, Y, and Z are three disjoint subsets of nodes in a DAG G, then Z is said to d-separate X from Y, denoted $(X \parallel Y \mid Z)_G$, iff Z d-separates every path from a node in X to a node in Y.

In Figure 70.1, for example, $X = \{X_2\}$ and $Y = \{X_3\}$ are d-separated by $Z = \{X_1\}$; the path $X_2 \leftarrow X_1 \rightarrow X_3$ is blocked by $X_1 \in Z$, while the path $X_2 \rightarrow X_4 \leftarrow X_3$ is blocked because X_4 and all its descendants are outside Z. Thus, $(X_2 \parallel X_3 \mid X_1)_G$ holds in Figure 70.1. However, X and Y are not d-separated by $Z' = \{X_1, X_5\}$, because the path $X_2 \rightarrow X_4 \leftarrow X_3$ is rendered active by virtue of X_5, a descendant of X_4, being in Z'. Consequently $(X_2 \parallel X_3 \mid \{X_1, X_5\})_G$ does not hold; in words, learning the value of the consequence X_5 renders its causes X_2 and X_3 dependent, as if a pathway were opened along the arrows converging at X_4.

The d-separation criterion has been shown to be both necessary and sufficient relative to the set of distributions that are represented by a DAG G [Verma and Pearl 1990, Geiger et al. 1990]. In other words, there is a one-to-one correspondence between the set of independencies implied by the recursive decomposition of Equation 70.1 and the set of triples (X, Z, Y) that satisfies the d-separation criterion in G. Furthermore, the d-separation criterion can be tested in time linear in the number of edges in G. Thus, a DAG can be viewed as an efficient scheme for representing Markovian independence assumptions and for deducing and displaying all of the logical consequences of such assumptions.

An important property that follows from the d-separation characterization is a criterion for determining when two DAGs are observationally equivalent, that is, every probability distribution that is represented by one of the DAGs is also represented by the other:

Theorem 70.1 (Verma and Pearl 1990.)   *Two DAGs are observationally equivalent if and only if they have the same sets of edges and the same sets of v-structures, that is, head-to-head arrows with nonadjacent tails.*

The soundness of the d-separation criterion holds not only for probabilistic independencies but for any abstract notion of conditional independence that obeys the semigraphoid axioms [Verma and Pearl 1990, Geiger et al. 1990]. Additional properties of DAGs and their applications to evidential reasoning in expert systems are discussed in Pearl [1988, 1993a], Pearl et al. [1990], Geiger [1990] Lauritzen and Spiegelhalter [1988], and Spiegelhalter et al. [1993].

70.3.2   Inference Algorithms

The first algorithms proposed for probability updating in Bayesian networks used message-passing architecture and were limited to trees [Pearl 1982] and singly connected networks [Kim and Pearl 1983]. The idea was to assign each variable a simple processor, forced to communicate only with its neighbors, and to permit asynchronous back-and-forth message passing until equilibrium was achieved. Coherent equilibrium can indeed be achieved this way, but only in singly connected networks, where an equilibrium state occurs in time proportional to the diameter of the network.

Many techniques have been developed and refined to extend the tree-propagation method to general, multiply connected networks. Among the most popular are Shachter's [1988] method of node elimination, Lauritzen and Spiegelhalter's [1988] method of clique-tree propagation, and the method of loop-cut conditioning [Pearl 1988, Ch. 4.3].

Clique-tree propagation, the most popular of the three methods, works as follows. Starting with a directed network representation, the network is transformed into an undirected graph that retains all of its original dependencies. This graph, sometimes called a Markov network [Pearl 1988, Ch. 3.1], is then triangulated to form local clusters of nodes (cliques) that are tree structured. Evidence propagates from clique to clique by ensuring that the probability of their intersection set is the same, regardless of which of the two cliques is considered in the computation. Finally, when the propagation process subsides, the posterior probability of an individual variable is computed by projecting (marginalizing) the distribution of the hosting clique onto this variable.

Whereas the task of updating probabilities in general networks is NP-hard [Rosenthal 1977, Cooper 1990], the complexity for each of the three methods cited is exponential in the size of the largest clique found in some triangulation of the network. It is fortunate that these complexities can be estimated prior to actual processing; when the estimates exceed reasonable bounds, an approximation method such as stochastic simulation [Pearl 1987, Henrion 1988] can be used instead. Learning techniques have also been developed for systematic updating of the conditional probabilities $P(x_i \mid pa_i)$ so as to match empirical data [Spiegelhalter and Lauritzen 1990].

70.3.3 System's Properties

By providing graphical means for representing and manipulating probabilistic knowledge, Bayesian networks overcome many of the conceptual and computational difficulties of earlier knowledge-based systems [Pearl 1988]. Their basic properties and capabilities can be summarized as follows:

1. Graphical methods make it easy to maintain consistency and completeness in probabilistic knowledge bases. They also define modular procedures of knowledge acquisition that reduce significantly the number of assessments required [Pearl 1988, Heckerman 1991].
2. Independencies can be dealt with explicitly. They can be articulated by an expert, encoded graphically, read off the network, and reasoned about; yet they forever remain robust to numerical imprecision [Geiger 1990, Geiger et al. 1990, Pearl et al. 1990].
3. Graphical representations uncover opportunities for efficient computation. Distributed updating is feasible in knowledge structures which are rich enough to exhibit intercausal interactions (e.g., explaining away) [Pearl 1982, Kim and Pearl 1983]. And, when extended by clustering or conditioning, tree-propagation algorithms are capable of updating networks of arbitrary topology [Lauritzen and Spiegelhalter 1988, Shachter 1986, Pearl 1988].
4. The combination of predictive and abductive inferences resolves many problems encountered by first-generation expert systems and renders belief networks a viable model for cognitive functions requiring both top-down and bottom-up inferences [Pearl 1988, Shafer and Pearl 1990].
5. The causal information encoded in Bayesian networks facilitates the analysis of action sequences, their consequences, their interaction with observations, their expected utilities, and, hence, the synthesis of plans and strategies under uncertainty [Dean and Wellman 1991, Pearl 1993b, 1994b].
6. The isomorphism between the topology of Bayesian networks and the stable mechanisms, which operate in the environment, facilitates modular reconfiguration of the network in response to changing conditions and permits deliberative reasoning about novel situations.

70.3.4 Recent Developments

70.3.4.1 Causal Discovery

One of the most exciting prospects in recent years has been the possibility of using the theory of Bayesian networks to discover causal structures in raw statistical data. Several systems have been developed for this purpose [Pearl and Verma 1991, Spirtes et al. 1993], which systematically search and identify causal structures with hidden variables from empirical data. Technically, because these algorithms rely merely on conditional independence relationships, the structures found are valid only if one is willing to accept weaker forms of guarantees than those obtained through controlled randomized experiments: minimality and stability [Pearl and Verma 1991]. Minimality guarantees that any other structure compatible with the data is necessarily less specific, and hence less falsifiable and less trustworthy, than the one(s) inferred. Stability ensures that any alternative structure compatible with the data must be less stable than the one(s) inferred; namely, slight fluctuations in experimental conditions will render that structure no longer compatible with the data. With these forms of guarantees, the theory provides criteria for identifying genuine and spurious causes, with or without temporal information.

Alternative methods of identifying structure in data assign prior probabilities to the parameters of the network and use Bayesian updating to score the degree to which a given network fits the data [Cooper and Herskovits 1990, Heckerman et al. 1994]. These methods have the advantage of operating well under small sample conditions but encounter difficulties coping with hidden variables.

70.3.4.2 Plain Beliefs

In mundane decision making, beliefs are revised not by adjusting numerical probabilities but by tentatively accepting some sentences as true for all practical purposes. Such sentences, often named *plain beliefs*, exhibit both logical and probabilistic character. As in classical logic, they are propositional and deductively closed; as in probability, they are subject to retraction and to varying degrees of entrenchment [Spohn 1988, Goldszmidt and Pearl 1992].

Bayesian networks can be adopted to model the dynamic of plain beliefs by replacing ordinary probabilities with nonstandard probabilities, that is, probabilities that are infinitesimally close to either zero or one. This amounts to taking an order of magnitude approximation of empirical frequencies and adopting new combination rules tailored to reflect this approximation. The result is an integer-addition calculus, very similar to probability calculus, with summation replacing multiplication and minimization replacing addition. A plain belief is then identified as a proposition whose negation obtains an infinitesimal probability (i.e., an integer greater than zero). The connection between infinitesimal probabilities and nonmonotonic logic is described in Pearl [1994a] and Goldszmidt and Pearl [1996].

This combination of infinitesimal probabilities with the causal information encoded by the structure of Bayesian networks facilitates linguistic communication of belief commitments, explanations, actions, goals, and preferences and serves as the basis for current research on qualitative planning under uncertainty [Darwiche and Pearl 1994, Goldszmidt and Pearl 1992, Pearl 1993b, Darwiche and Goldszmidt 1994]. Some of these aspects will be presented in the next section.

70.4 Bayesian Networks as Carriers of Causal Information

The interpretation of DAGs as carriers of independence assumptions does not necessarily imply causation and will in fact be valid for any set of Markovian independencies along any ordering (not necessarily causal or chronological) of the variables. However, the patterns of independencies portrayed in a DAG are typical of causal organizations and some of these patterns can be given meaningful interpretation only in terms of causation. Consider, for example, two independent events, E_1 and E_2, that have a common effect E_3. This triple represents an intransitive pattern of dependencies: E_1 and E_3 are dependent, E_3 and E_2 are dependent, yet E_1 and E_2 are independent. Such a pattern cannot be represented in undirected graphs because connectivity in undirected graphs is transitive. Likewise, it is not easily represented in neural networks, because E_1 and E_2 should turn dependent once E_3 is known. The DAG representation provides a convenient language for intransitive dependencies via the converging pattern $E_1 \rightarrow E_3 \leftarrow E_2$, which implies the independence of E_1 and E_2 as well as the dependence of E_1 and E_3 and of E_2 and E_3. The distinction between transitive and intransitive dependencies is the basis for the causal discovery systems of Pearl and Verma [1991] and Spirtes et al. [1993]. (See subsection on causal discovery.)

However, the Markovian account still leaves open the question of how such intricate patterns of independencies relate to the more basic notions associated with causation, such as influence, manipulation, and control, which reside outside the province of probability theory. The connection is made in the mechanism-based account of causation.

The basic idea behind this account goes back to structural equations models [Wright 1921, Haavelmo 1943, Simon 1953] and it was adapted in Pearl and Verma [1991] for defining probabilistic causal theories, as follows. Each child–parents family in a DAG G represents a deterministic function

$$X_i = f_i(pa_i, \epsilon_i) \tag{70.5}$$

where pa_i are the parents of variable X_i in G, and $\epsilon_i, 0 < i < n$, are mutually independent, arbitrarily distributed random disturbances. Characterizing each child–parent relationship as a deterministic function, instead of the usual conditional probability $P(x_i \mid pa_i)$, imposes equivalent independence constraints on the resulting distributions and leads to the same recursive decomposition that characterizes DAG models (see Equation 70.1). However, the functional characterization $X_i = f_i(pa_i, \epsilon_i)$ also specifies how the resulting distributions would change in response to external interventions, since each function is presumed to represent a stable mechanism in the domain and therefore remains constant unless specifically altered. Thus, once we know the identity of the mechanisms altered by the intervention and the nature of the alteration, the overall effect of an intervention can be predicted by modifying the appropriate equations in the model of Equation 70.5 and using the modified model to compute a new probability function of the observables.

The simplest type of external intervention is one in which a single variable, say X_i, is forced to take on some fixed value x_i'. Such *atomic* intervention amounts to replacing the old functional mechanism $X_i = f_i(pa_i, \epsilon_i)$ with a new mechanism $X_i = x_i'$ governed by some external force that sets the value x_i'. If we imagine that each variable X_i could potentially be subject to the influence of such an external force, then we can view each Bayesian network as an efficient code for predicting the effects of atomic interventions and of various combinations of such interventions, without representing these interventions explicitly.

70.4.1 Causal Theories, Actions, Causal Effect, and Identifiability

Definition 70.2   A causal theory is a 4-tuple

$$T = \langle V, U, P(u), \{f_i\} \rangle$$

where:

1. $V = \{X_1, \ldots, X_n\}$ is a set of observed variables.
2. $U = \{U_1, \ldots, U_n\}$ is a set of unobserved variables which represent disturbances, abnormalities, or assumptions.
3. $P(u)$ is a distribution function over $U_1, \ldots, U_n$.
4. $\{f_i\}$ is a set of n deterministic functions, each of the form

$$X_i = f_i(PA_i, u) \quad i = 1, \ldots, n \tag{70.6}$$

where PA_i is a subset of V not containing X_i.

The variables PA_i (connoting parents) are considered the direct of X_i and they define a directed graph G, which may, in general, be cyclic. Unlike the probabilistic definition of parents in Bayesian networks (Equation 70.1) PA_i is selected from V by considering functional mechanisms in the domain, not by conditional independence considerations. We will assume that the set of equations in 70.6 has a unique solution for $X_i, \ldots, X_n$, given any value of the disturbances $U_i, \ldots, U_n$. Therefore, the distribution $P(u)$ induces a unique distribution on the observables, which we denote by $P_T(v)$.

We will consider concurrent actions of the form $do(X = x)$, where $X \subseteq V$ is a set of variables and x is a set of values from the domain of X. In other words, $do(X = x)$ represents a combination of actions that forces the variables in X to attain the values x.

Definition 70.3 (effect of actions.)   The effect of the action $do(X = x)$ on a causal theory T is given by a subtheory T_x of T, where T_x obtains by deleting from T all equations corresponding to variables in X and substituting the equations $X = x$ instead.

The framework provided by Definitions 70.2 and 70.3 permits the coherent formalization of many subtle concepts in causal discourse, such as causal influence, causal effect, causal relevance, average causal effect, identifiability, counterfactuals, exogeneity, and so on. Examples are as follows:

- X *influences* Y in context u if there are two values of X, x, and x', such that the solution for Y under $U = u$ and $do(X = x)$ is different from the solution under $U = u$ and $do(X = x')$.
- X *can potentially influence* Y if there exist both a subtheory T_z of T and a context $U = u$ in which X influences Y.
- Event $X = x$ *is the (singular) cause* of event $Y = y$ if (1) $X = x$ and $Y = y$ are true, and (2) in every context u compatible with $X = x$ and $Y = y$, and for all $x' \neq x$, the solution of Y under $do(X = x')$ is not equal to y.

The definitions are deterministic. Probabilistic causality emerges when we define a probability distribution $P(u)$ for the U variables, which, under the assumption that the equations have a unique solution, induces a unique distribution on the endogenous variables for each combination of atomic interventions.

Definition 70.4 (causal effect.)   Given two disjoint subsets of variables, $X \subseteq V$ and $Y \subseteq V$, the causal effect of X on Y, denoted $P_T(y \mid \hat{x})$, is a function from the domain of X to the space of probability distributions on Y, such that

$$P_T(y \mid \hat{x}) = P_{T_x}(y) \tag{70.7}$$

for each realization x of X. In other words, for each $x \in \text{dom}(X)$, the causal effect $P_T(y \mid \hat{x})$ gives the distribution of Y induced by the action $do(X = x)$.

Note that causal effects are defined relative to a given causal theory T, though the subscript T is often suppressed for brevity.

Definition 70.5 (identifiability.)   Let $Q(T)$ be any computable quantity of a theory T; Q is identifiable in a class M of theories if for any pair of theories T_1 and T_2 from M, $Q(T_1) = Q(T_2)$ whenever $P_{T_1}(v) = P_{T_2}(v)$.

Identifiability is essential for estimating quantities Q from P alone, without specifying the details of T, so that the general characteristics of the class M suffice. The question of interest in planning applications is the identifiability of the causal effect $Q = P_T(y \mid \hat{x})$ in the class M_G of theories that share the same causal graph G. Relative to such classes we now define the following:

Definition 70.6 (causal-effect identifiability.)   The causal effect of X on Y is said to be identifiable in M_G if the quantity $P(y \mid \hat{x})$ can be computed uniquely from the probabilities of the observed variables, that is, if for every pair of theories T_1 and T_2 in M_G such that $P_{T_1}(v) = P_{T_2}(v)$, we have $P_{T_1}(y \mid \hat{x}) = P_{T_2}(y \mid \hat{x})$.

The identifiability of $P(y \mid \hat{x})$ ensures that it is possible to infer the effect of action $do(X = x)$ on Y from two sources of information:

1. Passive observations, as summarized by the probability function $P(v)$.
2. The causal graph, G, which specifies, qualitatively, which variables make up the stable mechanisms in the domain or, alternatively, which variables participate in the determination of each variable in the domain.

Simple examples of identifiable causal effects will be discussed in the next subsection.

70.4.2 Acting vs Observing

Consider the example depicted in Figure 70.1. The corresponding theory consists of five functions, each representing an autonomous mechanism:

$$X_1 = U_1$$
$$X_2 = f_2(X_1, U_2)$$
$$X_3 = f_3(X_1, U_3) \tag{70.8}$$
$$X_4 = f_4(X_3, X_2, U_4)$$
$$X_5 = f_5(X_4, U_5)$$

To represent the action "turning the sprinkler ON," $do(X_3 = \text{ON})$, we delete the equation $X_3 = f_3(x_1, u_3)$ from the theory of Equation 70.8 and replace it with $X_3 = \text{ON}$. The resulting subtheory, $T_{X_3} = \text{ON}$, contains all of the information needed for computing the effect of the actions on other variables. It is easy to see from this subtheory that the only variables affected by the action are X_4 and X_5, that is, the descendant, of the manipulated variable X_3.

The probabilistic analysis of causal theories becomes particularly simple when two conditions are satisfied:

1. The theory is recursive, i.e., there exists an ordering of the variables $V = \{X_1, \ldots, X_n\}$ such that each X_i is a function of a subset PA_i of its predecessors

$$X_i = f_i(PA_i, U_i), \quad PA_i \subseteq \{X_1, \ldots, X_{i-1}\} \tag{70.9}$$

2. The disturbances $U_1, \ldots, U_n$ are mutually independent, that is,

$$P(u) = \prod_i P(u_i) \tag{70.10}$$

These two conditions, also called Markovian, are the basis of the independencies embodied in Bayesian networks (Equation 70.1) and they enable us to compute causal effects directly from the conditional probabilities $P(x_i \mid pa_i)$, without specifying the functional form of the functions f_i, or the distributions $P(u_i)$ of the disturbances. This is seen immediately from the following observations: The distribution induced by any Markovian theory T is given by the product in Equation 70.1

$$P_T(x_1, \ldots, x_n) = \prod_i P(x_i \mid pa_i) \tag{70.11}$$

where pa_i are (values of) the parents of X_i in the diagram representing T. At the same time, the subtheory $T_{x'_j}$, representing the action $do(X_j = x'_j)$ is also Markovian; hence, it also induces a product-like distribution

$$P_{T_{x'_j}}(x_1, \ldots, x_n) = \begin{cases} \prod_{i \neq j} P(x_i \mid pa_i) = \frac{P(x_1, \ldots, x_n)}{P(x_j \mid pa_j)} & \text{if } x_j = x'_j \\ 0 & \text{if } x_j \neq x'_j \end{cases} \tag{70.12}$$

where the partial product reflects the surgical removal of the

$$X_j = f_j(pa_j, U_j)$$

from the theory of Equation 70.9 (see Pearl [1993a]).

In the example of Figure 70.1, the pre-action distribution is given by the product

$$P_T(x_1, x_2, x_3, x_4, x_5) = P(x_1)P(x_2 \mid x_1)P(x_3 \mid x_1)P(x_4 \mid x_2, x_3)P(x_5 \mid x_4) \qquad (70.13)$$

whereas the surgery corresponding to the action $do(X_3 = \text{ON})$ amounts to deleting the link $X_1 \to X_3$ from the graph and fixing the value of X_3 to ON, yielding the postaction distribution

$$P_T(x_1, x_2, x_3, x_4, x_5 \mid do(X_3 = \text{ON})) = P(x_1)P(x_2 \mid x_1)P(x_4 \mid x_2, X_3 = \text{ON})P(x_5 \mid x_4) \qquad (70.14)$$

Note the difference between the action $do(X_3 = \text{ON})$ and the observation $X_3 = \text{ON}$. The latter is encoded by ordinary Bayesian conditioning, whereas the former by conditioning a mutilated graph, with the link $X_1 \to X_3$ removed. Indeed, this mirrors the difference between seeing and doing: after observing that the sprinkler is ON, we wish to infer that the season is dry, that it probably did not rain, and so on; no such inferences should be drawn in evaluating the effects of the deliberate action "turning the sprinkler ON." The amputation of $X_3 = f_3(X_1, U_3)$ from Equation 70.8 ensures the suppression of any abductive inferences from X_3, the action's recipient.

Note also that Equations 70.1 through Equation 70.14 are independent of T; in other words, the pre-action and postaction distributions depend only on observed conditional probabilities but are independent of the particular functional form of $\{f_i\}$ or the distribution $P(u)$, which generate those probabilities. This is the essence of identifiability as given in Definition 70.6, which stems from the Markovian assumptions 70.9 and 70.10. The next subsection will demonstrate that certain causal effects, though not all, are identifiable even when the Markovian property is destroyed by introducing dependencies among the disturbance terms.

Generalization to multiple actions and conditional actions are reported in Pearl and Robins [1995]. Multiple actions $do(X = x)$, where X is a compound variable result in a distribution similar to Equation 70.12, except that all factors corresponding to the variables in X are removed from the product in Equation 70.11. Stochastic conditional strategies [Pearl 1994b] of the form

$$do(X_j = x_j) \quad \text{with probability} \quad P^*(x_j \mid pa_j^*) \qquad (70.15)$$

where pa_j^* is the support of the decision strategy, also result in a product decomposition similar to Equation 70.11, except that each factor $P(x_j \mid pa_j)$ is *replaced* with $P^*(x_j \mid pa_j^*)$.

The surgical procedure just described is not limited to probabilistic analysis. The causal knowledge represented in Figure 70.1 can be captured by logical theories as well, for example,

$$
\begin{aligned}
x_2 &\Longleftrightarrow [(X_1 = \text{Winter}) \vee (X_1 = \text{Fall}) \vee ab_2] \wedge \neg ab_2' \\
x_3 &\Longleftrightarrow [(X_1 = \text{Summer}) \vee (X_1 = \text{Spring}) \vee ab_3] \wedge \neg ab_3' \\
x_4 &\Longleftrightarrow (x_2 \vee x_3 \vee ab_4) \wedge \neg ab_4' \\
x_5 &\Longleftrightarrow (x_4 \vee ab_5) \wedge \neg ab_5'
\end{aligned}
\qquad (70.16)
$$

where x_i stands for $X_i = \text{true}$, and ab_i and ab_i' stand, respectively, for triggering and inhibiting abnormalities. The double arrows represent the assumption that the events on the right-hand side of each equation are the *only* direct causes for the left-hand side, thus identifying the surgery implied by any action.

It should be emphasized though that the models of a causal theory are not made up merely of truth value assignments which satisfy the equations in the theory. Since each equation represents an autonomous process, the content of each individual equation must be specified in any model of the theory, and this can be encoded using either the graph (as in Figure 70.1) or the generic description of the theory, as in Equation 70.8. Alternatively, we can view a model of a causal theory to consist of a mutually consistent set of submodels, with each submodel being a standard model of a single equation in the theory.

70.4.3 Action Calculus

The identifiability of causal effects demonstrated in the last subsection relies critically on the Markovian assumptions 70.9 and 70.10. If a variable that has two descendants in the graph is unobserved, the disturbances in the two equations are no longer independent, the Markovian property 70.9 is violated, and identifiability may be destroyed. This can be seen easily from Equation 70.12; if any parent of the manipulated variable X_j is unobserved, one cannot estimate the conditional probability $P(x_j \mid pa_j)$, and the effect of the action $do(X_j = x_j)$ may not be predictable from the observed distribution $P(x_1, \ldots, x_n)$. Fortunately, certain causal effects are identifiable even in situations where members of pa_j are unobservable [Pearl 1993a] and, moreover, polynomial tests are now available for deciding when $P(x_i \mid \hat{x}_j)$ is identifiable and for deriving closed-form expressions for $P(x_i \mid \hat{x}_j)$ in terms of observed quantities [Galles and Pearl 1995].

These tests and derivations are based on a symbolic calculus [Pearl 1994b, 1995] to be described in the sequel, in which interventions, side by side with observations, are given explicit notation and are permitted to transform probability expressions. The transformation rules of this calculus reflect the understanding that interventions perform local surgeries as described in Definition 70.3, i.e., they overrule equations that tie the manipulated variables to their preintervention causes.

Let $X, Y,$ and Z be arbitrary disjoint sets of nodes in a DAG G. We denote by $G_{\overline{X}}$ the graph obtained by deleting from G all arrows pointing to nodes in X. Likewise, we denote by $G_{\underline{X}}$ the graph obtained by deleting from G all arrows emerging from nodes in X. To represent the deletion of both incoming and outgoing arrows, we use the notation $G_{\overline{X}\underline{Z}}$. Finally, the expression $P(y \mid \hat{x}, z) \triangleq P(y,z) \mid \hat{x}/P(z \mid \hat{x})$ stands for the probability of $Y = y$ given that $Z = z$ is observed and X is held constant at x.

Theorem 70.2   *Let G be the directed acyclic graph associated with a Markovian causal theory, and let $P(\cdot)$ stand for the probability distribution induced by that theory. For any disjoint subsets of variables $X, Y, Z,$ and W we have the following rules:*

- *Rule 1 (Insertion/deletion of observations):*

$$P(y \mid \hat{x}, z, w) = P(y \mid \hat{x}, w) \quad \text{if } (Y \perp\!\!\!\perp Z \mid X, W)_{G_{\overline{X}}} \tag{70.17}$$

- *Rule 2 (Action/observation exchange):*

$$P(y \mid \hat{x}, \hat{z}, w) = P(y \mid \hat{x}, z, w) \quad \text{if } (Y \perp\!\!\!\perp Z \mid X, W)_{G_{\overline{X}\underline{Z}}} \tag{70.18}$$

- *Rule 3 (Insertion/deletion of actions):*

$$P(y \mid \hat{x}, \hat{z}, w) = P(y \mid \hat{x}, w) \quad \text{if } (Y \perp\!\!\!\perp Z \mid X, W)_{G_{\overline{X},\overline{Z(W)}}} \tag{70.19}$$

where $Z(W)$ is the set of Z-nodes that are not ancestors of any W-node in $G_{\overline{X}}$.

Each of the inference rules follows from the basic interpretation of the $\hat{x}$ operator as a replacement of the causal mechanism that connects X to its pre-action parents by a new mechanism $X = x$ introduced by the intervening force. The result is a submodel characterized by the subgraph $G_{\overline{X}}$ (named *manipulated graph* in Spirtes et al. [1993]), which supports all three rules.

Corollary 70.1   A causal effect $q : P(y_1, \ldots, y_k \mid \hat{x}_1, \ldots, \hat{x}_m)$ is identifiable in a model characterized by a graph G if there exists a finite sequence of transformations, each conforming to one of the inference rules in Theorem 70.2, which reduces q into a standard (i.e., hat-free) probability expression involving observed quantities.

Although Theorem 70.2 and Corollary 70.1 require the Markovian property, they also can be applied to non-Markovian, recursive theories because such theories become Markovian if we consider the unobserved variables as part of the analysis, and represent them as nodes in the graph. To illustrate, assume that variable X_1 in Figure 70.1 is unobserved, rendering the disturbances U_3 and U_2 dependent since these terms now include the common influence of X_1. Theorem 70.2 tells us that the causal effect $P(x_4 \mid \hat{x}_3)$ is identifiable, because

$$P(x_4 \mid \hat{x}_3) = \sum_{x_2} P(x_4 \mid \hat{x}_3, x_2) P(x_2 \mid \hat{x}_3)$$

Rule 3 permits the deletion

$$P(x_2 \mid \hat{x}_3) = P(x_2), \quad \text{because } (X_2 \perp\!\!\!\perp X_3)_{G_{\overline{X}_3}}$$

whereas rule 2 permits the exchange

$$P(x_4 \mid \hat{x}_3, x_2) = P(x_4 \mid x_3, x_2), \quad \text{because } (X_4 \perp\!\!\!\perp X_3 \mid X_2)_{G_{\underline{X}_3}}$$

This gives

$$P(x_4 \mid \hat{x}_3) = \sum_{x_2} P(x_4 \mid x_3, x_2) P(x_2)$$

which is a hat-free expression, involving only observed quantities.

In general, it can be shown [Pearl 1995]:

1. The effect of interventions can often be identified (from nonexperimental data) without resorting to parametric models.
2. The conditions under which such nonparametric identification is possible can be determined by simple graphical tests.*
3. When the effect of interventions is not identifiable, the causal graph may suggest nontrivial experiments which, if performed, would render the effect identifiable.

The ability to assess the effect of interventions from nonexperimental data has immediate applications in the medical and social sciences, since subjects who undergo certain treatments often are not representative of the population as a whole. Such assessments are also important in artificial intelligence (AI) applications where an agent needs to predict the effect of the next action on the basis of past performance records, and where that action has never been enacted out of free will, but in response to environmental needs or to other agents' requests.

70.4.4 Historical Remarks

An explicit translation of interventions to *wiping out* equations from linear econometric models was first proposed by Strotz and Wold [1960] and later used in Fisher [1970] and Sobel [1990]. Extensions to action representation in nonmonotonic systems were reported in Goldszmidt and Pearl [1992] and Pearl [1993a]. Graphical ramifications of this translation were explicated first in Spirtes et al. [1993] and later in Pearl [1993b]. A related formulation of causal effects, based on event trees and counterfactual analysis, was developed by Robins [1986, pp. 1422–1425]. Calculi for actions and counterfactuals based on this interpretation are developed in Pearl [1994b] and Balke and Pearl [1994], respectively.

---

*These graphical tests offer, in fact, a complete formal solution to the *covariate-selection* problem in statistics: finding an appropriate set of variables that need be adjusted for in any study which aims to determine the effect of one factor upon another. This problem has been lingering in the statistical literature since Karl Pearson, the founder of modern statistics, discovered (1899) what in modern terms is called the Simpson's paradox; any statistical association between two variables may be reversed or negated by including additional factors in the analysis [Aldrich 1995].

70.5 Counterfactuals

A counterfactual sentence has the form:

If A were true, then C would have been true?

where A, the counterfactual antecedent, specifies an event that is contrary to one's real-world observations, and C, the counterfactual consequent, specifies a result that is expected to hold in the alternative world where the antecedent is true. A typical example is "If Oswald were not to have shot Kennedy, then Kennedy would still be alive," which presumes the factual knowledge of Oswald's shooting Kennedy, contrary to the antecedent of the sentence.

The majority of the philosophers who have examined the semantics of counterfactual sentences have resorted to some version of Lewis' *closest world* approach: "C if it were A" is true, if C is true in worlds that are closest to the real world yet consistent with the counterfactual's antecedent A [Lewis 1973]. Ginsberg [1986] followed a similar strategy. Whereas the closest world approach leaves the precise specification of the closeness measure almost unconstrained, causal knowledge imposes very specific preferences as to which worlds should be considered closest to any given world. For example, considering an array of domino tiles standing close to each other, the manifestly closest world consistent with the antecedent "tile i is tipped to the right" would be a world in which just tile i is tipped, and all of the others remain erect. Yet, we all accept the counterfactual sentence "Had tile i been tipped over to the right, tile $i + 1$ would be tipped as well" as plausible and valid. Thus, distances among worlds are not determined merely by surface similarities but require a distinction between disturbed mechanisms and naturally occurring transitions. The local surgery paradigm expounded in the beginning of Section 70.4 offers a concrete explication of the closest world approach which respects causal considerations. A world w_1 is closer to w than a world w_2 is, if the set of atomic surgeries needed for transforming w into w_1 is a subset of those needed for transforming w into w_2. In the domino example, finding tile i tipped and $i + 1$ erect requires the breakdown of two mechanism (e.g., by two external actions) compared with one mechanism for the world in which all j-tiles, $j > i$, are tipped. This paradigm conforms to our perception of causal influences and lends itself to economical machine representation.

70.5.1 Formal Underpinning

The structural equation framework offers an ideal setting for counterfactual analysis.

Definition 70.7 (context-based potential response.)   Given a causal theory T and two disjoint sets of variables, X and Y, the potential response of Y to X in a context u, denoted $Y(x, u)$ or $Y_x(u)$, is the solution for Y under $U = u$ in the subtheory T_x. $Y(x, u)$ can be taken as the formal definition of the counterfactual English phrase: "the value that Y would take in context u, had X been x."*

Note that this definition allows for the context $U = u$ and the proposition $X = x$ to be incompatible in T. For example, if T describes a logic circuit with input U it may well be reasonable to assert the counterfactual: "Given $U = u$, Y would be high if X were low," even though the input $U = u$ may preclude X from being low. It is for this reason that one must invoke some motion of intervention (alternatively, a theory change or a *miracle* [Lewis 1973]) in the definition of counterfactuals.

---

*The term *unit* instead of *context* is often used in the statistical literature [Rubin 1974], where it normally stands for the identity of a specific individual in a population, namely, the set of attributes u that characterizes that individual. In general, u may include the time of day, the experimental conditions under study, and so on. Practitioners of the counterfactual notation do not explicitly mention the notions of *solution* or *intervention* in the definition of $Y(x, u)$. Instead, the phrase "the value that Y would take in unit u, had X been x," viewed as basic, is posited as the definition of $Y(x, u)$.

If U is treated as a random variable, then the value of the counterfactual $Y(x, u)$ becomes a random variable as well, denoted as $Y(x)$ of Y_x. Moreover, the distribution of this random variable is easily seen to coincide with the causal effect $P(y \mid \hat{x})$, as defined in Equation 70.7, i.e.,

$$P((Y(x) = y) = P(y \mid \hat{x})$$

The probability of a counterfactual conditional $x \to y \mid o$ may then be evaluated by the following procedure:

- Use the observations o to update $P(u)$ thus forming a causal theory $T^o = \langle V, U, \{f_i\}, P(u \mid o) \rangle$
- Form the mutilated theory T_x^o (by deleting the equation corresponding to variables in X) and compute the probability $P_{T^o}(y \mid \hat{x})$ which T_x^o induces on Y

Unlike causal effect queries, counterfactual queries are not identifiable even in Markovian theories, but require that the functional-form of $\{f_i\}$ be specified. In Balke and Pearl [1994] a method is devised for computing sharp bounds on counterfactual probabilities which, under certain circumstances may collapse to point estimates. This method has been applied to the evaluation of causal effects in studies involving noncompliance and to the determination of legal liability.

70.5.2 Applications to Policy Analysis

Counterfactual reasoning is at the heart of every planning activity, especially real-time planning. When a planner discovers that the current state of affairs deviates from the one expected, a plan repair activity needs to be invoked to determine what went wrong and how it could be rectified. This activity amounts to an exercise in counterfactual thinking, as it calls for rolling back the natural course of events and determining, based on the factual observations at hand, whether the culprit lies in previous decisions or in some unexpected, external eventualities. Moreover, in reasoning forward to determine if things would have been different, a new model of the world must be consulted, one that embodies hypothetical changes in decisions or eventualities — hence, a breakdown of the old model or theory.

The logic-based planning tools used in AI, such as STRIPS and its variants or those based on situation calculus, do not readily lend themselves to counterfactual analysis, as they are not geared for coherent integration of abduction with prediction, and they do not readily handle theory changes. Remarkably, the formal system developed in economics and social sciences under the rubric structural equations models does offer such capabilities but, as will be discussed, these capabilities are not well recognized by current practitioners of structural models. The analysis presented in this chapter could serve both to illustrate to AI researchers the basic formal features needed for counterfactual and policy analysis and to call the attention of economists and social scientists to capabilities that are dormant within structural equation models.

Counterfactual thinking dominates reasoning in political science and economics. We say, for example, "If Germany were not punished so severely at the end of World War I, Hitler would not have come to power," or "If Reagan did not lower taxes, our deficit would be lower today." Such thought experiments emphasize an understanding of generic laws in the domain and are aimed toward shaping future policy making, for example, "defeated countries should not be humiliated," or "lowering taxes (contrary to Reaganomics) tends to increase national debt."

Strangely, there is very little formal work on counterfactual reasoning or policy analysis in the behavioral science literature. An examination of a number of econometric journals and textbooks, for example, reveals a glaring imbalance: although an enormous mathematical machinery is brought to bear on problems of estimation and prediction, policy analysis (which is the ultimate goal of economic theories) receives almost no formal treatment. Currently, the most popular methods driving economic policy making are based on so-called *reduced-form* analysis: to find the impact of a policy involving decision variables X on outcome variables Y, one examines past data and estimates the conditional expectation $E(Y \mid X = x)$, where x is the particular instantiation of X under the policy studied.

The assumption underlying this method is that the data were generated under circumstances in which the decision variables X act as exogenous variables, that is, variables whose values are determined outside the system under analysis. However, although new decisions should indeed be considered exogenous for the purpose of evaluation, past decisions are rarely enacted in an exogenous manner. Almost every realistic policy (e.g., taxation) imposes control over some endogenous variables, that is, variables whose values are determined by other variables in the analysis. Let us take taxation policies as an example. Economic data are generated in a world in which the government is reacting to various indicators and various pressures; hence, taxation is endogenous in the data-analysis phase of the study. Taxation becomes exogenous when we wish to predict the impact of a specific decision to raise or lower taxes. The reduced-form method is valid only when past decisions are nonresponsive to other variables in the system, and this, unfortunately, eliminates most of the interesting control variables (e.g., tax rates, interest rates, quotas) from the analysis.

This difficulty is not unique to economic or social policy making; it appears whenever one wishes to evaluate the merit of a plan on the basis of the past performance of other agents. Even when the signals triggering the past actions of those agents are known with certainty, a systematic method must be devised for selectively ignoring the influence of those signals from the evaluation process. In fact, the very essence of *evaluation* is having the freedom to imagine and compare trajectories in various counterfactual worlds, where each world or trajectory is created by a hypothetical implementation of a policy that is free of the very pressures that compelled the implementation of such policies in the past.

Balke and Pearl [1995] demonstrate how linear, nonrecursive structural models with Gaussian noise can be used to compute counterfactual queries of the type: "Given an observation set O, find the probability that Y would have attained a value greater than y, had X been set to x." The task of inferring causes of effects, that is, of finding the probability that $X = x$ is the cause for effect E, amounts to answering the counterfactual query: "Given effect E and observations O, find the probability that E would not have been realized, had X not been x." The technique developed in Balke and Pearl [1995] is based on probability propagation in dual networks, one representing the actual world and the other representing the counterfactual world. The method is not limited to linear functions but applies whenever we are willing to assume the functional form of the structural equations. The noisy OR-gate model [Pearl 1988] is a canonical example where such functional form is normally specified. Likewise, causal theories based on Boolean functions (with exceptions), such as the one described in Equation 70.16 lend themselves to counterfactual analysis in the framework of Definition 70.7.

Acknowledgments

The research was partially supported by Air Force Grant F49620-94-1-0173, National Science Foundation (NSF) Grant IRI-9420306, and Northrop/Rockwell Micro Grant 94-100.

References

Abramson, B. 1991. ARCO1: an application of belief networks to the oil market. In *Proc. 7th Conf. Uncertainty Artificial Intelligence*. Morgan Kaufmann, San Mateo, CA.

Agogino, A. M., Srinivas, S., and Schneider, K. 1988. Multiple sensor expert system for diagnostic reasoning, monitoring and control of mechanical systems. *Mech. Sys. Sig. Process.* 2:165–185.

Aldrich, J. 1995. Correlations genuine and spurious in Pearson and Yule. 10:364–376. *Stat. Sci.*

Andersen, S. K., Olesen, K. G., Jensen, F. V., and Jensen, F. 1989. Hugin — a shell for building Bayesian belief universes for expert systems, pp. 1080–1085. In *11th Int. Jt. Conf. Artificial Intelligence*.

Balke, A. and Pearl, J. 1994. Counterfactual probabilities: computational methods, bounds, and applications. In *Uncertainty in Artificial Intelligence 10*. R. Lopez de Mantaras and D. Poole, Eds., pp. 46–54. Morgan Kaufmann, San Mateo, CA.

Balke, A. and Pearl, J. 1995. Counterfactuals and policy analysis in structural models. In *Uncertainty in Artificial Intelligence 11*. P. Besnard and S. Hanks, Eds., pp. 11–18. Morgan Kaufmann, San Francisco, CA.

Blalock, H. M. 1971. _Causal Models in the Social Sciences._ Macmillan, London.

Charniak, E. 1991. Bayesian networks without tears. _AI Mag._ 12(4):50–63.

Charniak, E. and Goldman, R. 1991. A probabilistic model of plan recognition. In _Proceedings, AAAI-91._ AAAI Press/MIT Press, Anaheim, CA.

Cooper, G. F. 1990. Computational complexity of probabilistic inference using Bayesian belief networks. _Artificial Intelligence_ 42(2):393–405.

Cooper, G. F. and Herskovits, E. 1990. A Bayesian method for constructing Bayesian belief networks from databases, pp. 86–94. In _Proc. Conf. Uncertainty in AI._ Morgan Kaufmann.

Darwiche, A. and Goldszmidt, M. 1994. On the relation between kappa calculus and probabilistic reasoning. In _Uncertainty in Artificial Intelligence. 10,_ R. Lopez de Mantaras and D. Poole, Eds., pp. 145–153. Morgan Kaufmann, San Francisco, CA.

Darwiche, A. and Pearl, J. 1994. Symbolic causal networks for planning under uncertainty, pp. 41–47. In _Symp. Notes AAAI Spring Symp. Decision-Theoretic Planning._ Stanford, CA.

Dawid, A. P. 1979. Conditional independence in statistical theory. _J. R. Stat. Soc. Ser. A_ 41:1–31.

Dean, T. L. and Wellman, M. P. 1991. _Planning and Control._ Morgan Kaufmann, San Mateo, CA.

Duncan, O. D. 1975. _Introduction to Structural Equation Models._ Academic Press, New York.

Fisher, F. M. 1970. A correspondence principle for simultaneous equations models. _Econometrica_ 38:73–92.

Galles, D. and Pearl, J. 1995. Testing identifiability of causal effects. In _Uncertainty in Artificial Intelligence 11,_ P. Besnard and S. Hanks, Eds., pp. 185–195. Morgan Kaufmann, San Francisco, CA.

Geiger, D. 1990. _Graphoids: A Qualitative Framework for Probabilistic Inference._ Ph.D. Thesis. Department of Computer Science, University of California, Los Angeles.

Geiger, D., Verma, T. S., and Pearl, J. 1990. Identifying independence in Bayesian networks. In _Networks._ Vol. 20, pp. 507–534. Wiley, Sussex, England.

Ginsberg, M. L. 1986. Counterfactuals. _Artificial Intelligence_ 30:35–79.

Goldszmidt, M. and Pearl, J. 1992. Rank-based systems: a simple approach to belief revision, belief update, and reasoning about evidence and actions. In _Proc. 3rd Int. Conf. Knowledge Representation Reasoning._ B. Nobel, C. Rich, and M. Swartout, Eds., pp. 661–672. Morgan Kaufmann, San Mateo, CA.

Goldszmidt, M. and Pearl, J. 1996. Qualitative probabilities for default reasoning, belief revision, and causal modeling. _Artificial Intelligence_ 84(1–2):57–112.

Good, I. J. 1961. A causal calculus, I-II. _Br. J. Philos. Sci._ 11:305–318, 12:43–51.

Haavelmo, T. 1943. The statistical implications of a system of simultaneous equations. _Econometrica_ 11:1–12.

Heckerman, D. 1991. Probabilistic similarity networks. _Networks_ 20(5):607–636.

Heckerman, D., Geiger, D., and Chickering, D. 1994. Learning Bayesian networks: The combination of knowledge and statistical data, pp. 293–301. In _Proc. 10th Conf. Uncertainty Artificial Intelligence._ Seattle, WA, July. Morgan Kaufmann, San Mateo, CA.

Heckerman, D. E., Horvitz, E. J., and Nathwany, B. N. 1990. Toward normative expert systems: The pathfinder project. _Medical Comput. Sci. Group, Tech. Rep._ KSL-90-08, Section on Medical Informatics, Stanford University, Stanford, CA.

Henrion, M. 1988. Propagation of uncertainty by probabilistic logic sampling in Bayes' networks. In _Uncertainty in Artificial Intelligence 2._ J. F. Lemmer and L. N. Kanal, Eds., pp. 149–164. Elsevier Science, North-Holland, Amsterdam.

Howard, R. A. and Matheson, J. E. 1981. Influence diagrams. _Principles and Applications of Decision Analysis._ Strategic Decisions Group. Menlo Park, CA.

Kenny, D. A. 1979. _Correlation and Causality._ Wiley, New York.

Kiiveri H., Speed, T. P., and Carlin, J. B. 1984. Recursive causal models. _J. Australian Math. Soc._ 36:30–52.

Kim, J. H. and Pearl, J. 1983. A computational model for combined causal and diagnostic reasoning in inference systems, pp. 190–193. In _Proceedings IJCAI-83._ Karlsruhe, Germany.

Lauritzen, S. L. 1982. _Lectures on Contingency Tables,_ 2nd ed. University of Aalborg Press, Aalborg, Denmark.

Lauritzen, S. L. and Spiegelhalter, D. J. 1988. Local computations with probabilities on graphical structures and their application to expert systems (with discussion). _J. R. Stat. Soc. Ser. B_ 50(2):157–224.

Levitt, J. M., Agosta, T. S., and Binford, T. O. 1990. Model-based influence diagrams for machine vision. In M. Hension, R. D. Shachter, L. N. Kanal, and J. F. Lemmer, Eds., pp. 371–388. *Uncertainty in Artificial Intelligence* 5. North-Holland, Amsterdam.

Lewis, D. 1973. *Counterfactuals*. Basil Blackwell, Oxford, England.

Neapolitan, R. E. 1990. *Probabilistic Reasoning in Expert Systems: Theory and Algorithms*. Wiley, New York.

Oliver, R. M. and Smith, J. Q., Eds. 1990. *Influence Diagrams, Belief Nets, and Decision Analysis*. Wiley, New York.

Pearl, J. 1982. Reverend Bayes on inference engines: a distributed hierarchical approach, pp. 133–136. In *Proc. AAAI Nat. Conf. AI*. Pittsburgh, PA.

Pearl, J. 1987. Bayes decision methods. In *Encyclopedia of AI*, pp. 48–56. Wiley Interscience, New York.

Pearl, J. 1988. *Probabilistic Reasoning in Intelligence Systems*, rev. 2nd printing, Morgan Kaufmann, San Mateo, CA.

Pearl, J. 1993a. From Bayesian networks to causal networks, pp. 25–27. In *Proc. Adaptive Comput. Inf. Process. Semin*. Brunel Conf. Centre, London, Jan. See also *Stat. Sci*. 8(3):266–269.

Pearl, J. 1993b. From conditional oughts to qualitative decision theory. In *Proc. 9th Conf. Uncertainty Artificial Intelligence*. D. Heckerman and A. Mamdani, Eds., pp. 12–20. Morgan Kaufmann.

Pearl, J. 1994a. From Adams' conditionals to default expressions, causal conditionals, and counterfactuals. In *Probability and Conditionals*. E. Eells and B. Skyrms, Eds., pp. 47–74. Cambridge University Press, Cambridge, MA.

Pearl, J. 1994b. A probabilistic calculus of actions. In *Uncertainty in Artificial Intelligence 10*. R. Lopez de Mantaras and D. Poole, Eds., pp. 454–462. Morgan Kaufmann, San Mateo, CA.

Pearl, J. 1995. Causal diagrams for experimental research. *Biometrika* 82(4):669–710.

Pearl, J., Geiger, D., and Verma, T. 1990. The logic and influence diagrams. In *Influence Diagrams, Belief Nets and Decision Analysis*. R. M. Oliver and J. Q. Smith, Eds., pp. 67–87. Wiley, New York.

Pearl, J. and Robins, J. M. 1995. Probabilistic evaluation of sequential plans from causal models with hidden variables. In *Uncertainty in Artificial Intelligence 11*. P. Besnard and S. Hanks, Eds., pp. 444–453. Morgan Kaufmann, San Francisco, CA.

Pearl, J. and Verma, T. 1991. A theory of inferred causation. In *Principles of Knowledge Representation and Reasoning: Proc. 2nd Int. Conf*. J. A. Allen, R. Fikes, and E. Sandewall, Eds., pp. 441–452. Morgan Kaufmann, San Mateo, CA.

Peng, Y. and Reggia, J. A. 1990. *Abductive Inference Models for Diagnostic Problem-Solving*. Springer–Verlag, New York.

Robins, S. M. 1986. A new approach to causal inference in mortality studies with a sustained exposure period — applications to control of the healthy workers survivor effect. *Math. Model* 7:1393–1512.

Rosenthal, A. 1977. A computer scientist looks at reliability computations. In *Reliability and Fault Tree Analysis*. Barlow et al., Eds., pp. 133–152. SIAM, Philadelphia, PA.

Rubin, D. B. 1974. Estimating causal effects of treatments in randomized and nonrandomized studies. *J. Educ. Psych*. 66:688–701.

Rumelhart, D. E. 1976. Toward an interactive model of reading. *University of California Tech. Rep. CHIP-56*, University of California, La Jolla.

Shachter, R. D. 1986. Evaluating influence diagrams. *Op. Res*. 34(6):871–882.

Shachter, R. D. 1988. Probabilistic inference and influence diagrams. *Op. Res*. 36:589–604.

Shachter, R. D. 1990. Special issue on influence diagrams. *Networks: Int. J*. 20(5).

Shafer, G. and Pearl, J., Eds. 1990. *Readings in Uncertain Reasoning*. Morgan Kaufmann, San Mateo, CA.

Simon, H. A. 1953. Causal ordering and identifiability. In *Studies in Econometric Method*. W. C. Hood and T. C. Koopmans, Eds., pp. 49–74. Wiley, New York.

Sobel, M. E. 1990. Effect analysis and causation in linear structural equation models. *Psychometrika* 55(3):495–515.

Spiegelhalter, D. J. and Lauritzen, S. L. 1990. Sequential updating of conditional probabilities on directed graphical structures. *Networks* 20(5):579–605.

Spiegelhalter, D. J., Lauritzen, S. L., Dawid, P. A., and Cowell, R. G. 1993. Bayesian analysis in expert systems. *Stat. Sci.* 8:219–247.

Spirtes, P., Glymour, C., and Schienes, R. 1993. *Causation, Prediction, and Search.* Springer–Verlag, New York.

Spohn, W. 1988. A general non-probabilistic theory of inductive reasoning, pp. 315–322. In *Proc. 4th Workshop Uncertainty Artificial Intelligence.* Minneapolis, MN.

Strotz, R. H. and Wold, H. O. A. 1960. Causal models in the social sciences. *Econometrica* 28:417–427.

Turtle, H. R. and Croft, W. B. 1991. Evaluation of an inference network-based retrieval model. *ACM Trans. Inf. Sys.* 9(3).

Verma, T. and Pearl, J. 1990. Equivalence and synthesis of causal models. In *Uncertainty in Artificial Intelligence 6*, pp. 220–227. Elsevier Science, Cambridge, MA.

Wermuth, N. and Lauritzen, S. L. 1983. Graphical and recursive models for contingency tables. *Biometrika* 70:537–552.

Wold, H. 1964. *Econometric Model Building.* North-Holland, Amsterdam.

Wright, S. 1921. Correlation and causation. *J. Agric. Res.* 20:557–585.

Wright, S. 1934. The method of path coefficients. *Ann. Math. Stat.* 5:161–215.

71

Robotics

71.1 Introduction 71-2
71.2 Robot Workcells 71-2
71.3 Workcell Command and Information
Organization..................................... 71-4
Intelligent Control Architectures • Behaviors and Hybrid
Systems Design • Workcell Planning, Coordination, and
Control Structure
71.4 Commercial Robot Configurations and Types 71-7
Manipulator Performance • Common Kinematic
Configurations • Drive Types of Commercial Robots
• Commercial Robot Controllers
71.5 Robot Kinematics, Dynamics, and
Servo-Level Control 71-13
Kinematics and Jacobians • Robot Dynamics and Properties
• Robot Servo-level Motion Control • Robot Force/Torque
Servocontrol • Motion Trajectory Generation
71.6 End Effectors and End-of-Arm Tooling 71-21
Part Fixtures and Robot Tooling • Grippers and Fingers
• Robot Wrist Mechanisms • Robot/Tooling Process
Integration and Coordination
71.7 Sensors 71-25
The Philosophy of Robotic Workcell Sensors
• Types of Sensors • Sensor Data Processing
• Vision for Robotics
71.8 Workcell Planning 71-31
Workcell Behaviors and Agents • Task Decomposition and
Planning • Task Matrix Approach to Workcell Planning
• Path Planning
71.9 Job and Activity Coordination 71-41
Matrix Rule-Based Job Coordination Controller
• Process Integration, Digital I/O, and Job Coordination
Controller Implementation • Coordination of Multiple
Robots
71.10 Error Detection and Recovery 71-44
Error Detection • Error Recovery
71.11 Human Operator Interfaces 71-45
Levels of User Interface • Mechanisms for User Interface
71.12 Robot Workcell Programming 71-47
Robot Programming Languages • V+, A Representative
Robot Language
71.13 Mobile Robots and Automated Guided Vehicles 71-49
Mobile Robots • Automated Guided Vehicle Systems

Frank L. Lewis
University of Texas at Arlington

John M. Fitzgerald
Adept Technology

Kai Liu
Alcatel Telecom

1-58488-360-X/$0.00+$1.50
© 2004 by CRC Press, LLC

71.1   Introduction

The word *robot* was introduced by the Czech playright Karel Čapek in his 1920 play *Rossum's Universal Robots*. The word *robota* in Czech means simply work. In spite of such practical beginnings, science fiction writers and early Hollywood movies have given us a romantic notion of robots and expectations that they will revolutionize several walks of life including industry. However, many of the more far-fetched expectations from robots have failed to materialize. For instance, in underwater assembly and oil mining, teleoperated robots are very difficult to manipulate due to sea currents and low visibility, and have largely been replaced or augmented by automated smart quick-fit couplings that simplify the assembly task. However, through good design practices and painstaking attention to detail, engineers have succeeded in applying robotic systems to a wide variety of industrial and manufacturing situations where the environment is *structured* or predictable. Thus, the first successful commercial implementation of process robotics was in the U.S. automobile industry; the word *automation* was coined in the 1940s at Ford Motor Company, a contraction of automatic motivation.

As machines, robots have precise motion capabilities, repeatability, and endurance. On a practical level, robots are distinguished from other electromechanical motion equipment by their dexterous manipulation capability in that robots can work, position, and move tools and other objects with far greater dexterity than other machines found in the factory. The capabilities of robots are extended by using them as a basis for *robotic workcells*. Process robotic workcells are integrated functional systems with grippers, **end effectors**, sensors, and process equipment organized to perform a controlled sequence of jobs to execute a process. Robots must coordinate with other devices in the workcell such as machine tools, conveyors, part feeders, cameras, and so on. Sequencing jobs to correctly perform automated tasks in such circumstances is not a trivial matter, and robotic workcells require sophisticated planning, sequencing, and control systems.

Today, through developments in computers and artificial intelligence (AI) techniques (and often motivated by the space program), we are on the verge of another breakthrough in robotics that will afford some levels of autonomy in *unstructured environments*. For applications requiring increased autonomy it is particularly important to focus on the design of the data structures and command-and-control information flow in the robotic system. Therefore, this chapter focuses on the design of robotic workcell *systems*. A distinguishing feature of robotics is its multidisciplinary nature: to successfully design robotic systems one must have a grasp of electrical, mechanical, industrial, and computer engineering, as well as economics and business practices. The purpose of this chapter is to provide a background in these areas so that design of robotic systems may be approached from a position of rigor, insight, and confidence.

The chapter begins by discussing layouts and architectures for robotic workcell design. Then, components of the workcell are discussed from the bottom up, beginning with robots, sensors, and conveyors/part feeders, and progressing upwards in abstraction through task coordination, job sequencing, and resource dispatching, to task planning, assignment, and decomposition. Concepts of user interface and exception handling/fault recovery are included.

71.2   Robot Workcells

In factory automation and elsewhere it was once common to use layouts such as the one in Figure 71.1, which shows an assembly line with distinct workstations, each performing a dedicated function. Robots have been used at the workstation level to perform operations such as assembly, drilling, surface finishing, welding, palletizing, and so on. In the assembly line, parts are routed sequentially to the workstations by conveyors. Such systems are very expensive to install, require a cadre of engineering experts to design and program, and are extremely difficult to modify or reprogram as needs change. In today's high-mix low-volume (HMLV) manufacturing scenario, these characteristics tolled the death knell for such rigid antiquated designs.

In the assembly line, the robot is *restricted* by placing it into a rigid sequential system. Robots are versatile machines with many capabilities, and their potential can be significantly increased by using them as a basis

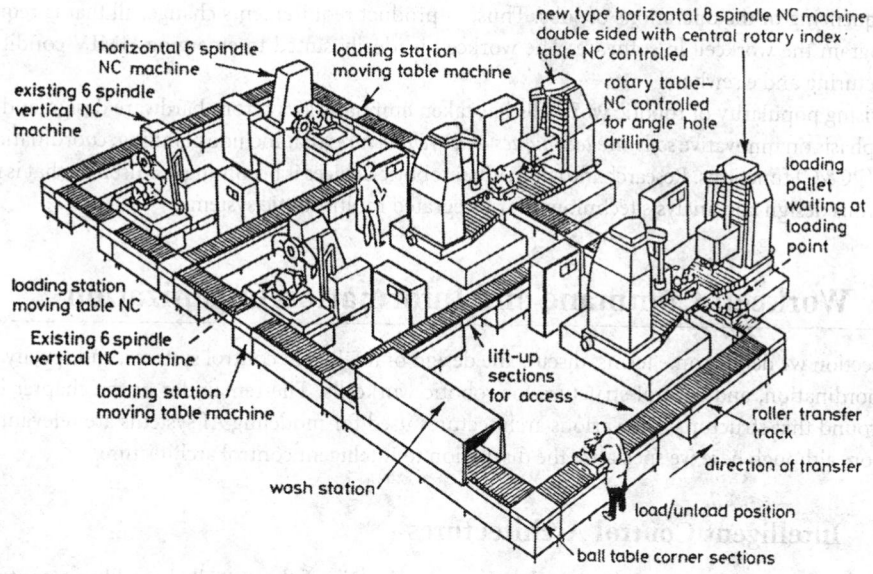

FIGURE 71.1 Antiquated sequential assembly line with dedicated workstations. (Courtesy of Edkins, M. 1983. Linking industrial robots and machine tools. In *Robotic Technology*. A. Pugh, Ed. Peregrinus, London.)

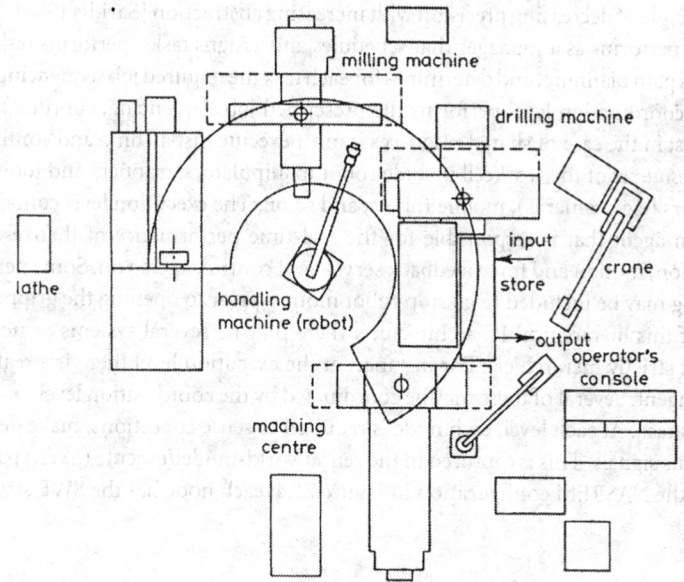

FIGURE 71.2 Robot workcell. (Courtesy of Edkins, M. 1983. Linking industrial robots and machine tools. In *Robotic Technology*. A. Pugh, Ed. Peregrinus, London.)

for robotic workcells such as the one in Figure 71.2 [Decelle 1988, Jamshidi et al. 1992, Pugh 1983]. In the robotic workcell, robots are used for part handling, assembly, and other process operations. The workcell is designed to make full use of the workspace of the robots, and components such as milling machines, drilling machines, vibratory part feeders, and so on are placed within the robots' workspaces to allow servicing by the robots. Contrary to the assembly line, the physical layout does not impose a priori a fixed sequencing of the operations or jobs. Thus, as product requirements change, all that is required is to reprogram the workcell in software. The workcell is ideally suited to emerging HMLV conditions in manufacturing and elsewhere.

The rising popularity of robotic workcells has taken emphasis away from hardware design and placed new emphasis on innovative *software techniques and architectures* that include planning, coordination, and control (PC&C) functions. Research into individual robotic devices is becoming less useful; what is needed are rigorous design and analysis techniques for integrated multirobotic systems.

71.3 Workcell Command and Information Organization

In this section we define some terms, discuss the design of intelligent control systems, and specify a planning, coordination, and control structure for robotic workcells. The remainder of the chapter is organized around that structure. The various architectures used for modeling AI systems are relevant to this discussion, although here we specialize the discussion to intelligent control architecture.

71.3.1 Intelligent Control Architectures

Many structures have been proposed under the general aegis of the so-called intelligent control (IC) architectures [Antsaklis and Passino 1992]. Despite frequent heated philosophical discussions, it is now becoming clear that most of the architectures have much in common, with apparent major differences due to the fact that different architectures focus on different aspects of intelligent control or different levels of abstraction. A general IC architecture based on work by Saridis is given in Figure 71.3, which illustrates the principle of decreasing precision with increasing abstraction [Saridis 1996]. In this figure, the organization level performs as a manager that schedules and assigns tasks, performs task decomposition and planning, does path planning, and determines for each task the required job sequencing and assignment of resources. The coordination level performs the prescribed job sequencing, coordinating the workcell agents or resources; in the case of shared resources it must execute dispatching and conflict resolution.

The *agents* or *resources* of the workcell include robot manipulators, grippers and tools, conveyors and part feeders, sensors (e.g., cameras), mobile robots, and so on. The execution level contains a closed-loop controller for each agent that is responsible for the real-time performance of that resource, including trajectory generation, motion and force feedback servo-level control, and so on. Some permanent built-in motion sequencing may be included (e.g., stop robot motion prior to opening the gripper).

At each level of this hierarchical IC architecture, there may be several systems or nodes. That is, the architecture is not strictly hierarchical. For instance, at the execution level there is a real-time controller for each workcell agent. Several of these may be coordinated by the coordination level to sequence the jobs needed for a given task. At each level, each node is required to sense conditions, make decisions, and give commands or status signals. This is captured in the sense/world-model/execute (SWE) paradigm of Albus [1992], shown in the NASREM configuration in Figure 71.4; each node has the SWE structure.

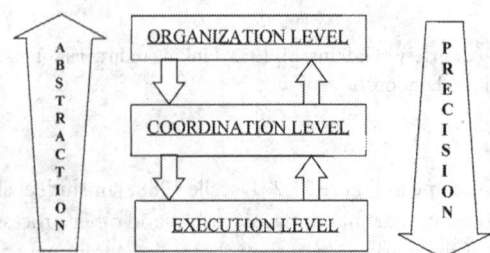

FIGURE 71.3   Three-level intelligent control architecture from work by Saridis.

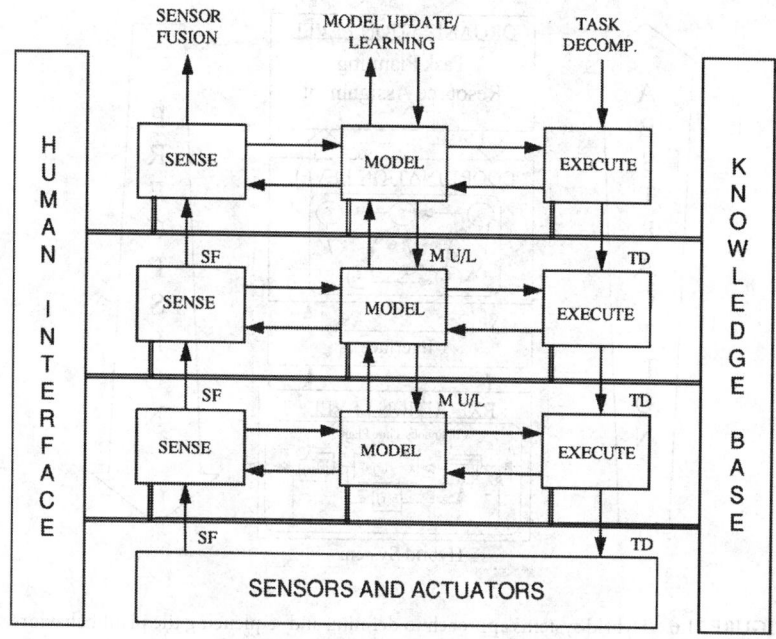

FIGURE 71.4   Three-element structure at all levels of the IC architecture: the NASREM paradigm.

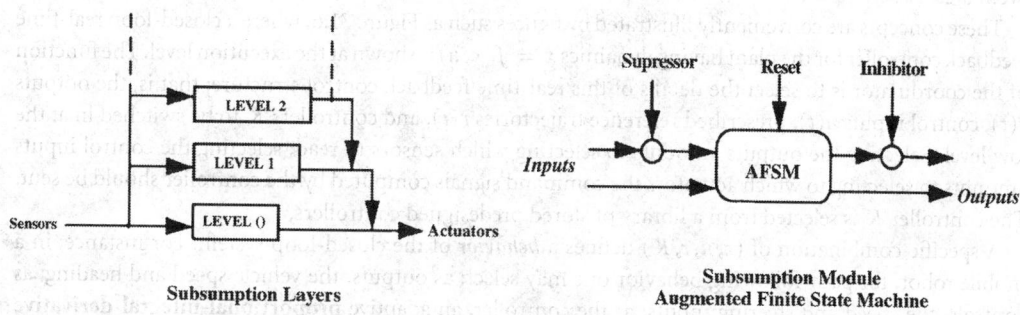

FIGURE 71.5   Behavior-based design after the subsumption technique of Brooks.

71.3.2   Behaviors and Hybrid Systems Design

In any properly designed IC system, the supervisory levels should not destroy the capabilities of the systems supervised. Thus, design should proceed in the manner specified by Brooks [1986], where *behaviors* are built in at lower levels, then selected, activated, or modified by upper-level supervisors. From the point of view of still higher level nodes, the composite performance appears in terms of new more complex or emergent behaviors. Such *subsumption* design proceeds in the manner of adding layers to an onion, as depicted loosely in Figure 71.5.

Near or slightly below the interfaces between the coordination level and the execution level one must face the transition between two fundamentally distinct worlds. Real-time servo-level controller design and control may be accomplished in terms of *state-space systems*, which are time-varying dynamical systems (either continuous time or discrete time) having continuous-valued states such as temperatures, pressures, motions, velocities, forces, and so on. On the other hand, the coordinator is not concerned about such details, but speaks in terms of *discrete events* such as "perform this job" or "check this condition." The

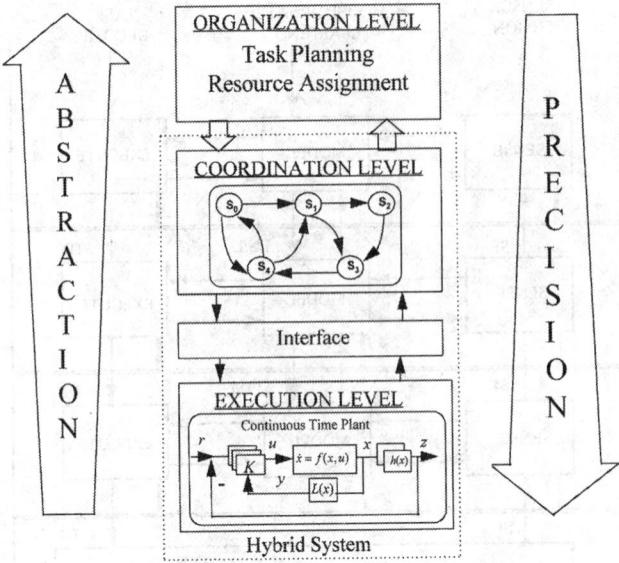

FIGURE 71.6   Hybrid systems approach to defining and sequencing the plant behaviors.

theory of *hybrid systems* is concerned with the interface between continuous-state systems and discrete event systems.

These concepts are conveniently illustrated by figures such as Figure 71.6, where a closed-loop real-time feedback controller for the plant having dynamics $\dot{x} = f(x, u)$ is shown at the execution level. The function of the coordinator is to select the details of this real-time feedback control structure; that is, the outputs $z(t)$, control inputs $u(t)$, prescribed reference trajectories $r(t)$, and controllers K to be switched in at the low level. Selecting the outputs amounts to selecting which sensors to read; selecting the control inputs amounts to selecting to which actuators the command signals computed by the controller should be sent. The controller K is selected from a library of stored predesigned controllers.

A specific combination of (z, u, r, K) defines a *behavior* of the closed-loop system. For instance, in a mobile robot, for path-following behavior one may select: as outputs, the vehicle speed and heading; as controls, the speed and steering inputs; as the controller, an adaptive **proportional-integral-derivative (PID) controller**; and as reference input, the prescribed path. For wall-following behavior, for instance, one simply selects as output the sonar distance from the wall, as input the steering command, and as reference input the prescribed distance to be maintained. These distinct closed-loop behaviors are sequenced by the coordinator to perform the prescribed job sequence.

71.3.3   Workcell Planning, Coordination, and Control Structure

A convenient planning, coordination, and control structure for robotic workcell design and operation is given in Figure 71.7, which is modified from the next generation controller (NGC) paradigm. This is an operational PC&C architecture fully consistent with the previous IC structures. In this figure, the term *virtual agent* denotes the agent plus its low-level servocontroller and any required built-in sequencing coordinators. For instance, a *virtual robot* includes the manipulator, its commercial controller with servo-level joint controllers and trajectory generator, and in some applications the gripper controller plus an agent internal coordinator to sequence manipulator and gripper activities. A *virtual camera* might include the camera(s) and framegrabber board, plus software algorithms to perform basic vision processing such as edge detection, segmentation, and so on; thus, the virtual camera could include a *data abstraction*, which is a set of data plus manipulations on that data.

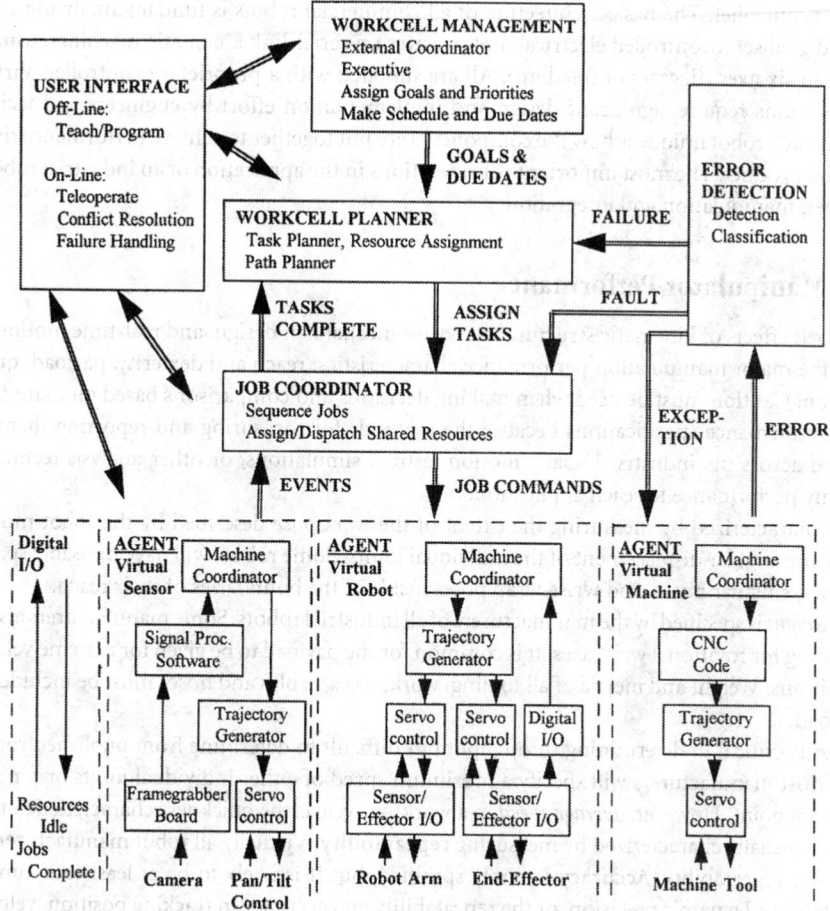

FIGURE 71.7 Robotic workcell planning, coordination, and control operational architecture.

The remainder of the chapter is structured after this PC&C architecture, beginning at the execution level to discuss robot manipulator kinematics, dynamics and control; end effectors and tooling; sensors; and other workcell components such as conveyors and part feeders. Next considered is the coordination level including sequencing control and dispatching of resources. Finally, the organization level is treated including task planning, path planning, workcell management, task assignment, and scheduling.

Three areas are particularly problematic. At each level there may be *human operator interfaces*; this complex topic is discussed in a separate section. An equally complex topic is *error detection and recovery*, also allotted a separate section, which occurs at several levels in the hierarchy. Finally, the strict NGC architecture has a component known as the *information* or *knowledge base*; however, in view of the fact that all nodes in the architecture have the SWE structure shown in Figure 71.4, it is clear that the knowledge base is distributed throughout the system in the world models of the nodes. Thus, a separate discussion on this component is not included.

71.4 Commercial Robot Configurations and Types

Robots are highly reliable, dependable, and technologically advanced factory equipment. The majority of the world's robots are supplied by established companies using reliable off-the-shelf component technologies. All commercial industrial robots have two physically separate basic elements, the manipulator arm and the controller. The basic architecture of all commercial robots is fundamentally the same, and consists of digital servocontrolled electrical motor drives on serial-link kinematic machines, usually with no more than six axes (degrees of freedom). All are supplied with a proprietary controller. Virtually all robot applications require significant design and implementation effort by engineers and technicians. What makes each robot unique is how the components are put together to achieve performance that yields a competitive product. The most important considerations in the application of an industrial robot center on two issues: manipulation and integration.

71.4.1 Manipulator Performance

The combined effects of kinematic structure, axis drive mechanism design, and real-time motion control determine the major manipulation performance characteristics: reach and dexterity, payload, quickness, and precision. Caution must be used when making decisions and comparisons based on manufacturers' published performance specifications because the methods for measuring and reporting them are not standardized across the industry. Usually motion testing, simulations, or other analysis techniques are used to verify performance for each application.

Reach is characterized by measuring the extent of the *workspace* described by the robot motion and *dexterity* by the angular displacement of the individual joints. Some robots will have unusable spaces such as dead zones, singular poses, and wrist-wrap poses inside of the boundaries of their reach.

Payload weight is specified by the manufacturers of all industrial robots. Some manufacturers also specify inertial loading for rotational wrist axes. It is common for the payload to be given for extreme velocity and reach conditions. Weight and inertia of all tooling, workpieces, cables and hoses must be included as part of the payload.

Quickness is critical in determining throughput but difficult to determine from published robot specifications. Most manufacturers will specify a maximum speed of either individual joints or for a specific kinematic tool point. However, *average speed* in a working cycle is the quickness characteristic of interest.

Precision is usually characterized by measuring **repeatability**. Virtually all robot manufacturers specify static position repeatability. **Accuracy** is rarely specified, but it is likely to be at least four times larger than repeatability. Dynamic precision, or the repeatability and accuracy in tracking position, velocity, and acceleration over a continuous path, is not usually specified.

71.4.2 Common Kinematic Configurations

All common commercial industrial robots are serial-link manipulators, usually with no more than six kinematically coupled axes of motion. By convention, the axes of motion are numbered in sequence as they are encountered from the base on out to the wrist. The first three axes account for the spatial positioning motion of the robot; their configuration determines the shape of the space through which the robot can be positioned. Any subsequent axes in the kinematic chain generally provide rotational motions to orient the end of the robot arm and are referred to as *wrist axes*. There are two primary types of motion that a **robot axis** can produce in its driven link — either **revolute** or **prismatic**. It is often useful to classify robots according to the orientation and type of their first three axes. There are four very common commercial robot configurations: articulated, type I selectively compliant assembly robot arm (**SCARA**), type II SCARA, and Cartesian. Two other configurations, cylindrical and spherical, are now much less common.

71.4.2.1 Articulated Arms

The variety of commercial articulated arms, most of which have six axes, is very large (Figure 71.8). All of these robot's axes are revolute. The second and third axes are parallel and work together to produce motion in a vertical plane. The first axis in the base is vertical and revolves the arm to sweep out a large work volume. Many different types of drive mechanisms have been devised to allow wrist and forearm drive motors and gearboxes to be mounted close to the first and second axis of rotation, thus minimizing the extended mass of the arm. The workspace efficiency of well-designed articulated arms, which is the degree

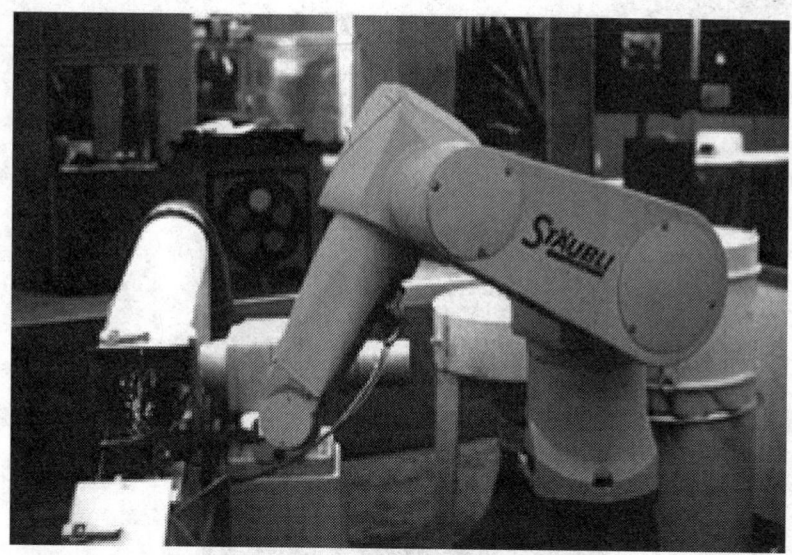

FIGURE 71.8   Articulated arm; six-axis arm grinding from a casting. (Courtesy of Staubli Unimation, Inc.)

of quick dexterous reach with respect to arm size, is unsurpassed by other arm configurations when five or more degrees of freedom are needed. A major limiting factor in articulated arm performance is that the second axis has to work to lift both the subsequent arm structure and the payload. Historically, articulated arms have not been capable of achieving accuracy as well as other arm configurations, as all axes have joint angle position errors which are multiplied by link radius and accumulated for the entire arm.

71.4.2.2   Type I SCARA

The type I SCARA (selectively compliant assembly robot arm) uses two parallel revolute joints to produce motion in the horizontal plane (Figure 71.9). The arm structure is weight-bearing but the first and second axes do no lifting. The third axis of the type I SCARA provides work volume by adding a vertical or z axis. A fourth revolute axis will add rotation about the z axis to control orientation in the horizontal plane. This type of robot is rarely found with more than four axes. The type I SCARA is used extensively in the assembly of electronic components and devices, and it is used broadly for the assembly of small- and medium-sized mechanical assemblies.

71.4.2.3   Type II SCARA

The type II SCARA, also a four-axis configuration, differs from type I in that the first axis is a long vertical prismatic z stroke, which lifts the two parallel revolute axes and their links (Figure 71.10). For quickly moving heavier loads (over approximately 75 lb) over longer distance (more than about 3 ft), the type II SCARA configuration is more efficient than the type I.

71.4.2.4   Cartesian Coordinate Robots

Cartesian coordinate robots use orthogonal prismatic axes, usually referred to as x, y, and z, to translate their end effector or payload through their rectangular workspace (Figure 71.11). One, two, or three revolute wrist axes may be added for orientation. Commercial robot companies supply several types of Cartesian coordinate robots with workspace sizes ranging from a few cubic inches to tens of thousands of cubic feet, and payloads ranging to several hundred pounds. Gantry robots, which have an elevated bridge structure, are the most common Cartesian style and are well suited to material handling applications where large areas and/or large loads must be serviced. They are particularly useful in applications such as arc welding, waterjet cutting, and inspection of large complex precision parts. Modular Cartesian robots are

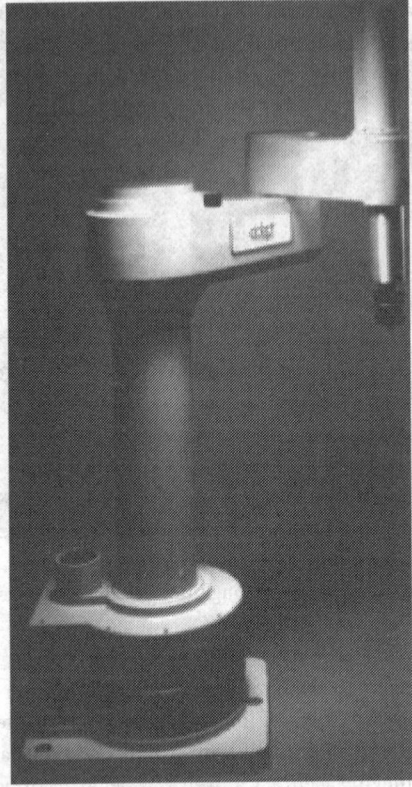

FIGURE 71.9 Type I SCARA arm. High-precision, high-speed midsized SCARA I. (Courtesy of Adept Technologies, Inc.)

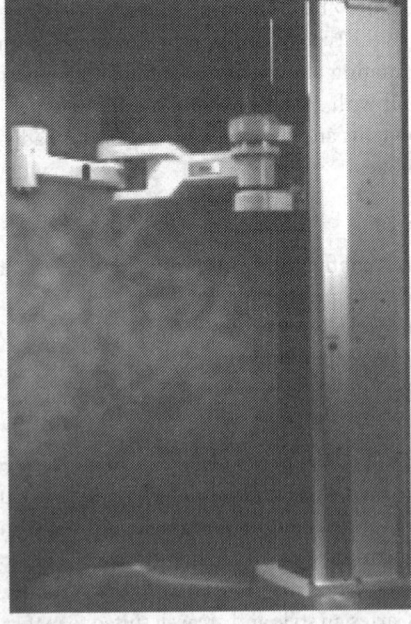

FIGURE 71.10 Type II SCARA. (Courtesy of Adept Technologies, Inc.)

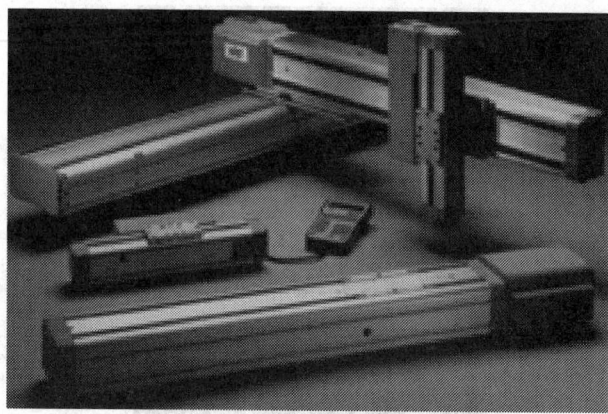

FIGURE 71.11 Cartesian robot. Three-axis robot constructed from modular single-axis motion modules. (Courtesy of Adept Technologies, Inc.)

also commonly available from several commercial sources. Each module is a self-contained completely functional single-axis actuator; the modules may be custom assembled for special-purpose applications.

71.4.2.5 Spherical and Cylindrical Coordinate Robots

The first two axes of the spherical coordinate robot are revolute and orthogonal to one another, and the third axis provides prismatic radial extension. The result is a natural spherical coordinate system with a spherical work volume. The first axis of cylindrical coordinate robots is a revolute base rotation. The second and third are prismatic, resulting in a natural cylindrical motion. Commercial models of spherical and cylindrical robots (Figure 71.12) were originally very common and popular in machine tending and material handling applications. Hundreds are still in use but now there are only a few commercially available models. The decline in use of these two configurations is attributed to problems arising from use of the prismatic link for radial extension/retraction motion; a solid boom requires clearance to fully retract.

71.4.3 Drive Types of Commercial Robots

The vast majority of commercial industrial robots use electric servomotor drives with speed reducing transmissions. Both ac and dc motors are popular. Some servohydraulic articulated arm robots are available now for painting applications. It is rare to find robots with servopneumatic drive axes. All types of mechanical transmissions are used, but the tendency is toward low- and zero-backlash type drives. Some robots use direct drive methods to eliminate the amplification of inertia and mechanical backlash associated with other drives. Joint angle position sensors, required for real-time servo-level control, are generally considered an important part of the drive train. Less often, velocity feedback sensors are provided.

71.4.4 Commercial Robot Controllers

Commercial robot controllers are specialized multiprocessor computing systems that provide four basic processes allowing integration of the robot into an automation system: motion trajectory generation and following, motion/process integration and sequencing, human user integration, and information integration.

71.4.4.1 Motion Trajectory Generation and Following

There are two important controller related aspects of industrial robot motion generation. One is the extent of manipulation that can be programmed, the other is the ability to execute controlled programmed motion. A unique aspect of each robot system is its real-time servo-level motion control. The details of real-time control are typically not revealed to the user due to safety and proprietary information secrecy reasons.

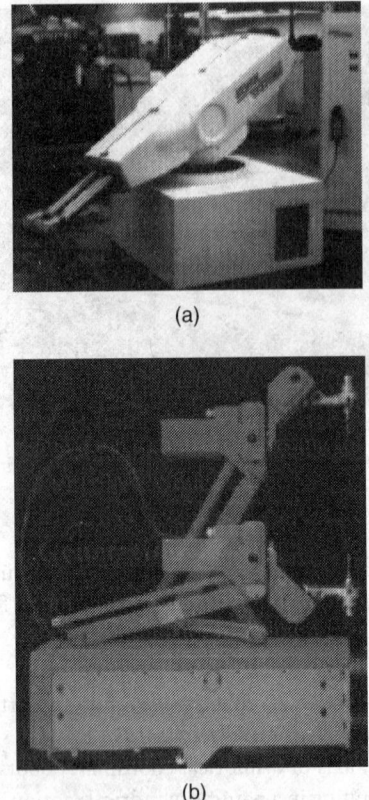

(a)

(b)

FIGURE 71.12   Spherical and cylindrical robots. (a) Hydraulic powered spherical robot. (*Source:* Courtesy of Kohol Systems, Inc. With permission.) (b) Cylindrical arm using scissor mechanism for radial prismatic motion. (Courtesy of Yamaha Robotics.)

Each robot controller, through its operating system programs, converts digital data from higher level coordinators into coordinated arm motion through precise computation and high-speed distribution and communication of the individual axis motion commands, which are executed by individual joint servocontrollers. Most commercial robot controllers operate at a sample period of 16 ms. The real-time motion controller invariably uses classical independent-joint proportional-integral-derivative control or simple modifications of PID. This makes commercially available controllers suitable for point-to-point motion, but most are not suitable for following continuous position/velocity profiles or exerting prescribed forces without considerable programming effort, if at all.

71.4.4.2 Motion/Process Integration and Sequencing

Motion/process integration involves coordinating manipulator motion with process sensors or other process controller devices. The most primitive process integration is through discrete digital input/output (I/O). For example, a machine controller external to the robot controller might send a 1-b signal indicating that it is ready to be loaded by the robot. The robot controller must have the ability to read the signal and to perform logical operations (if then, wait until, do until, etc.) using the signal. Coordination with sensors (e.g., vision) is also often provided.

71.4.4.3 Human Integration

The controller's human interfaces are critical to the expeditious setup and programming of robot systems. Most robot controllers have two types of human interface available: computer style CRT/keyboard terminals for writing and editing program code off line, and *teach pendants*, which are portable manual input terminals used to command motion in a telerobotic fashion via touch keys or joy sticks. Teach pendants are usually the most efficient means available for positioning the robot, and a memory in the controller makes it possible to play back the taught positions to execute motion trajectories. With practice, human operators can quickly teach a series of points which are chained together in playback mode. Most robot applications currently depend on the integration of human expertise during the programming phase for the successful planning and coordination of robot motion. These interface mechanisms are effective in unobstructed workspaces where no changes occur between programming and exceution. They do not allow human interface during execution or adaptation to changing environments.

71.4.4.4 Information Integration

Information integration is becoming more important as the trend toward increasing flexibility and agility impacts robotics. Many commercial robot controllers now support information integration functions by employing integrated personal computer (PC) interfaces through the communications ports (e.g., RS-232), or in some through direct connections to the robot controller data bus.

71.5 Robot Kinematics, Dynamics, and Servo-Level Control

In this section we shall study the kinematics, dynamics, and servocontrol of robot manipulators; for more details see Lewis et al. [1993]. The objective is to turn the manipulator, by proper design of the control system and trajectory generator, into an *agent with desirable behaviors*, which behaviors can then be selected by the job coordinator to perform specific jobs to achieve some assigned task. This agent, composed of the robot plus servo-level control system and trajectory genarator, is the *virtual robot* in Figure 71.7; this philosophy goes along with the subsumption approach of Brooks (Figure 71.5).

71.5.1 Kinematics and Jacobians

71.5.1.1 Kinematics of Rigid Serial-Link Manipulators

The kinematics of the robot manipulator are concerned only with relative positioning and not with motion effects.

71.5.1.1.1 Link A Matrices

Fixed-base serial-link rigid robot manipulators can be considered as a sequence of joints held together by links. Each joint i has a **joint variable** $\mathbf{q}_i$, which is an angle for revolute joints (units of degrees) and a length for prismatic or extensible joints (units of length). The *joint vector* of an n-link robot is defined as $\mathbf{q} = [\mathbf{q}_1 \ \mathbf{q}_2 \ \cdots \ \mathbf{q}_n]T \in \Re^n$; the joints are traditionally numbered from the base to the end effector, with link 0 being the fixed base. A robot with n joints has n degrees of freedom, so that for complete freedom of positioning and orientation in our 3-D space $\Re^3$ one needs a six-link arm.

For analysis purposes, it is considered that to each link is affixed a coordinate frame. The *base frame* is attached to the manipulator base, link 0. The location of the coordinate frame on the link is often selected according to the *Denavit–Hartenberg* (DH) convention [Lewis et al. 1993]. The relation between the links is given by the *A matrix for link i*, which has the form

$$A_i(\mathbf{q}_i) = \begin{bmatrix} R_i & \mathbf{p}_i \\ 0 & 1 \end{bmatrix} \tag{71.1}$$

where $R_i(\mathbf{q}_i)$ is a 3×3 rotation matrix ($R_i^{-1} = R_i^T$) and $\mathbf{p}_i(\mathbf{q}_i) = [x_i \ y_i \ z_i]^T \in \Re^3$ is a translation vector. R_i specifies the rotation of the coordinate frame on link i with respect to the coordinate frame on link $i - 1$; $\mathbf{p}_i$ specifies the translation of the coordinate frame on link i with respect to the coordinate frame on link $i - 1$. The 4×4 *homogeneous transformation* A_i thus specifies completely the orientation and translation of link i with respect to link $i - 1$.

The A matrix $A_i(\mathbf{q}_i)$ is a function of the joint variable, so that as $\mathbf{q}_i$ changes with robot motion, A_i changes correspondingly. A_i is also dependent on the parameters link twist and link length, which are fixed for each link. The A matrices are often given for a specific robot in the manufacturers handbook.

71.5.1.1.2 Robot T Matrix

The position of the end effector is given in terms of the base coordinate frame by the *arm T matrix* defined as the concatenation of A matrices

$$T(\mathbf{q}) = A_1(\mathbf{q}_1)A_2(\mathbf{q}_2)\cdots A_n(\mathbf{q}_n) \equiv \begin{bmatrix} R & \mathbf{p} \\ 0 & 1 \end{bmatrix} \qquad (71.2)$$

This 4×4 homogeneous transformation matrix is a function of the joint variable vector $\mathbf{q}$. The 3×3 cumulative rotation matrix is given by $R(\mathbf{q}) = R_1(\mathbf{q}_1)R_2(\mathbf{q}_2)\cdots R_n(\mathbf{q}_n)$.

71.5.1.1.3 Joint Space vs. Cartesian Space

An n-link manipulator has n degrees of freedom, and the position of the end effector is completely fixed once the joints variables $\mathbf{q}_i$ are prescribed. This position may be described either in joint coordinates or in Cartesian coordinates. The joint coordinates position of the end effector is simply given by the value of the n-vector $\mathbf{q}$. The Cartesian position of the end effector is given in terms of the base frame by specifying the orientation and translation of a coordinate frame affixed to the end effector in terms of the base frame; this is exactly the meaning of $T(\mathbf{q})$. That is, $T(\mathbf{q})$ gives the Cartesian position of the end effector.

The Cartesian position of the end effector may be completely specified in our 3-D space by a six vector; three coordinates are needed for translation and three for orientation. The representation of Cartesian translation by the arm $T(\mathbf{q})$ matrix is suitable, as it is simply given by $\mathbf{p}(\mathbf{q}) = [x \ y \ z]^T$. Unfortunately, the representation of Cartesian orientation by the arm T matrix is inefficient in that $R(\mathbf{q})$ has nine elements. More efficient representations are given in terms of quaternions or the *tool configuration vector*.

71.5.1.1.4 Kinematics and Inverse Kinematics Problems

The robot *kinematics problem* is to determine the Cartesian position of the end effector once the joint variables are given. This is accomplished simply by computing $T(\mathbf{q})$ for a given value of $\mathbf{q}$.

The **inverse kinematics** problem is to determine the required joint angles $\mathbf{q}_i$ to position the end effector at a prescribed Cartesian position. This corresponds to solving Equation 71.2 for $\mathbf{q} \in \Re^n$ given a desired orientation R and translation $\mathbf{p}$ of the end effector. This is not an easy problem, and may have more than one solution (e.g., think of picking up a coffee cup, one may reach with elbow up, elbow down, etc.). There are various efficient techniques for accomplishing this. One should avoid the functions arcsin, arccos, and use where possible the numerically well-conditioned arctan function.

71.5.1.2 Robot Jacobians
71.5.1.2.1 Transformation of Velocity and Acceleration

When the manipulator moves, the joint variable becomes a function of time t. Suppose there is prescribed a generally nonlinear transformation from the joint variable $\mathbf{q}(t) \in \Re^n$ to another variable $y(t) \in \Re^p$ given by

$$y(t) = h(\mathbf{q}(t)) \qquad (71.3)$$

An example is provided by the equation $y = T(\mathbf{q})$, where $y(t)$ is the Cartesian position. Taking partial derivatives one obtains

$$\dot{y} = \frac{\partial h}{\partial \mathbf{q}}\dot{\mathbf{q}} \equiv J(\mathbf{q})\dot{\mathbf{q}} \qquad (71.4)$$

where $J(\mathbf{q})$ is the *Jacobian* associated with $h(\mathbf{q})$. This equation tells how the joint velocities are transformed to the velocity $\dot{y}$.

If $y = T(\mathbf{q})$ the Cartesian end effector position, then the associated Jacobian $J(\mathbf{q})$ is known as the **manipulator Jacobian**. There are several techniques for efficiently computing this particular Jacobian; there are some complications arising from the fact that the representation of orientation in the homogeneous transformation $T(\mathbf{q})$ is a 3×3 rotation matrix and not a three vector. If the arm has n links, then the Jacobian is a $6 \times n$ matrix; if n is less than 6 (e.g., SCARA arm), then $J(\mathbf{q})$ is not square and there is not full positioning freedom of the end effector in 3-D space. The **singularities** of $J(\mathbf{q})$ (where it loses rank), define the limits of the robot workspace; singularities may occur within the workspace for some arms.

Another example of interest is when $y(t)$ is the position in a *camera coordinate frame*. Then $J(\mathbf{q})$ reveals the relationships between manipulator joint velocities (e.g., joint incremental motions) and incremental motions in the camera image. This affords a technique, for instance, for moving the arm to cause desired relative motion of a camera and a workpiece. Note that, according to the velocity transformation 71.4, one has that incremental motions are transformed according to $\Delta y = J(\mathbf{q})\Delta q$.

Differentiating Equation 71.4 one obtains the *acceleration transformation*

$$\ddot{y} = J\ddot{\mathbf{q}} + \dot{J}\dot{\mathbf{q}} \tag{71.5}$$

71.5.1.2.2 Force Transformation

Using the notion of virtual work, it can be shown that forces in terms of $\mathbf{q}$ may be transformed to forces in terms of y using

$$\tau = J^T(\mathbf{q})\mathbf{F} \tag{71.6}$$

where $\tau(t)$ is the force in joint space (given as an n-vector of torques for a revolute robot), and $\mathbf{F}$ is the force vector in y space. If y is the Cartesian position, then $\mathbf{F}$ is a vector of three forces $[\mathbf{f}_x\ \mathbf{f}_y\ \mathbf{f}_z]^T$ and three torques $[\tau_x\ \tau_y\ \tau_z]^T$. When $J(\mathbf{q})$ loses rank, the arm cannot exert forces in all directions that may be specified.

71.5.2 Robot Dynamics and Properties

The robot dynamics considers motion effects due to the control inputs and inertias, Coriolis forces, gravity, disturbances, and other effects. It reveals the relation between the control inputs and the joint variable motion $\mathbf{q}(t)$, which is required for the purpose of servocontrol system design.

71.5.2.1 Robot Dynamics

The dynamics of a rigid robot arm with joint variable $\mathbf{q}(t) \in \Re^n$ are given by

$$M(\mathbf{q})\ddot{\mathbf{q}} + V_m(\mathbf{q}, \dot{\mathbf{q}})\dot{\mathbf{q}} + \mathbf{F}(\mathbf{q}, \dot{\mathbf{q}}) + \mathbf{G}(\mathbf{q}) + \tau_d = \tau \tag{71.7}$$

where M is an inertia matrix, V_m is a matrix of Coriolis and centripetal terms, $\mathbf{F}$ is a friction vector, $\mathbf{G}$ is a gravity vector, and τ_d is a vector of disturbances. The n-vector $\tau(t)$ is the control input. The dynamics for a specific robot arm are not usually given in the manufacturer specifications, but may be computed from the kinematics A matrices using principles of Lagrangian mechanics.

The dynamics of any actuators can be included in the robot dynamics. For instance, the electric or hydraulic motors that move the joints can be included, along with any gearing. Then, as long as the gearing and drive shafts are noncompliant, the form of the equation with arm-plus-actuator dynamics has the same form as Equation 71.7. If the actuators are not included, the control τ is a torque input vector for the joints. If joint dynamics are included, then τ might be, for example, a vector of voltage inputs to the joint actuator motors.

The dynamics may be expressed in Cartesian coordinates. The *Cartesian dynamics* have the same form as Equation 71.7, but appearances there of $\mathbf{q}(t)$ are replaced by the Cartesian position $y(t)$. The matrices are modified, with the manipulator Jacobian $J(\mathbf{q})$ becoming involved. In the Cartesian dynamics, the control input is a six vector of forces, three linear forces and three torques.

71.5.2.2 Robot Dynamics Properties

Being a Lagrangian system, the robot dynamics satisfy many physical properties that can be used to simplify the design of servo-level controllers. For instance, the inertia matrix $M(\mathbf{q})$ is symmetric positive definite, and bounded above and below by some known bounds. The gravity terms are bounded above by known bounds. The Coriolis/centripetal matrix V_m is linear in $\dot{\mathbf{q}}$, and is bounded above by known bounds. An important property is the **skew-symmetric** property of rigid-link robot arms, which says that the matrix $(\dot{M} - 2V_m)$ is always skew symmetric.

This is a statement of the fact that the fictitious forces do no work, and is related in an intimate fashion to the *passivity* properties of Lagrangian systems, which can be used to simplify control system design. Ignoring passivity can lead to unacceptable servocontrol system design and serious degradations in performance, especially in teleoperation systems with transmission delays.

71.5.2.3 State-Space Formulations and Computer Simulation

Many commercially available controls design software packages, including MATLAB, allow the simulation of state-space systems of the form $\dot{\mathbf{x}} = f(\mathbf{x}, u)$ using, for instance, Runge–Kutta integration. The robot dynamics can be written in state-space form in several different ways. One state-space formulation is the position/velocity form

$$
\begin{aligned}
\dot{\mathbf{x}}_1 &= \mathbf{x}_2 \\
\dot{\mathbf{x}}_2 &= -M^{-1}(\mathbf{x}_1)[V_m(\mathbf{x}_1, \mathbf{x}_2)\mathbf{x}_2 + \mathbf{F}(\mathbf{x}_1, \mathbf{x}_2) + \mathbf{G}(\mathbf{x}_1) + \tau_d] + M^{-1}(\mathbf{x}_1)\tau
\end{aligned}
\tag{71.8}
$$

where the control input is $u = M^{-1}(\mathbf{x}_1)\tau$, and the state is $\mathbf{x} = [\mathbf{x}_1^T \ \mathbf{x}_2^T]^T$, with $\mathbf{x}_1 = \mathbf{q}$, and $\mathbf{x}_2 = \dot{\mathbf{q}}$ both n-vectors. In computation, one should not invert $M(\mathbf{q})$; one should either obtain an analytic expression for M^{-1} or use least-squares techniques to determine $\dot{\mathbf{x}}_2$.

71.5.3 Robot Servo-level Motion Control

The objective in robot servo-level motion control is to cause the manipulator end effector to follow a prescribed trajectory. This can be accomplished as follows for any system having the dynamics Equation 71.7, including robots, robots with actuators included, and robots with motion described in Cartesian coordinates. Generally, design is accomplished for robots including actuators, but with motion described in joint space. In this case, first, solve the inverse kinematics problem to convert the desired end effector motion $y_d(t)$ (usually specified in Cartesian coordinates) into a desired joint-space trajectory $\mathbf{q}_d(\mathbf{t}) \in \Re^n$ (discussed subsequently). Then, to achieve tracking motion so that the actual joint variables $\mathbf{q}(t)$ follow the prescribed trajectory $\mathbf{q}_d(t)$, define the *tracking error* $e(t)$ and *filtered tracking error* $r(t)$ as

$$
e(t) = \mathbf{q}_d(t) - \mathbf{q}(t)
\tag{71.9}
$$

$$
r(t) = \dot{e} + \Lambda e(t)
\tag{71.10}
$$

with Λ a positive definite design parameter matrix; it is common to select Λ diagonal with positive elements.

71.5.3.1 Computed Torque Control

One may differentiate Equation 71.10 to write the robot dynamics Equation 71.7 in terms of the filtered tracking error as

$$
M\dot{r} = -V_m r + f(\mathbf{x}) + \tau_d - \tau
\tag{71.11}
$$

where the *nonlinear robot function* is given by

$$
f(\mathbf{x}) = M(\mathbf{q})(\ddot{\mathbf{q}}_d + \Lambda\dot{e}) + V_m(\mathbf{q}, \dot{\mathbf{q}})(\dot{\mathbf{q}}_d + \Lambda e) + F(\mathbf{q}, \dot{\mathbf{q}}) + G(\mathbf{q})
\tag{71.12}
$$

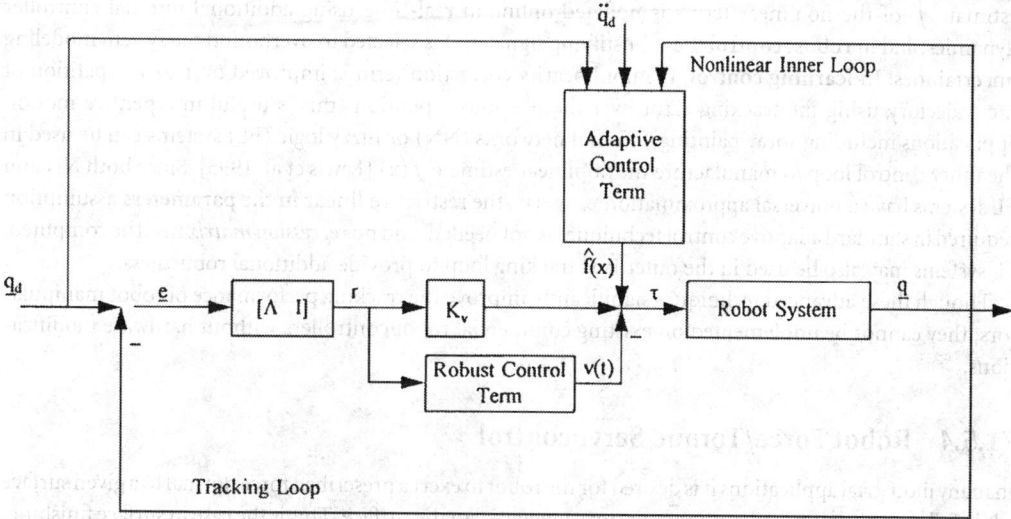

FIGURE 71.13 Robot servo-level tracking controller.

Vector $\mathbf{x}$ contains all of the time signals needed to compute $f(\cdot)$, and may be defined for instance as $\mathbf{x} \equiv [e^T \; \dot{e}^T \; \mathbf{q}_d^T \; \dot{\mathbf{q}}_d^T \; \ddot{\mathbf{q}}_d^T]^T$. It is important to note that $f(\mathbf{x})$ contains all the potentially unknown robot arm parameters including payload masses, friction coefficients, and Coriolis/centripetal terms that may simply be too complicated to compute.

A general sort of servo-level tracking controller is now obtained by selecting the control input as

$$\tau = \hat{f} + K_v r - v(t) \tag{71.13}$$

with $\hat{f}$ an *estimate* of the nonlinear terms $f(\mathbf{x})$, $K_v r = K_v \dot{e} + K_v \Lambda e$ an *outer proportional-plus-derivative (PD) tracking loop*, and $v(t)$ an auxiliary signal to provide robustness in the face of disturbances and modeling errors. The *multiloop control structure* implied by this scheme is shown in Figure 71.13. The nonlinear inner loop that computes $\hat{f}(x)$ provides *feedforward compensation* terms that improve the tracking capabilities of the PD outer loop, including an *acceleration feedforward* term $M(\mathbf{q})\ddot{\mathbf{q}}_d$, *friction compensation* $\mathbf{F}(\mathbf{q}, \dot{\mathbf{q}})$, and a *gravity compensation* term $\mathbf{G}(\mathbf{q})$.

This controller is a variant of **computed-torque control**, since the torque required for trajectory following is computed in terms of the tracking error and the additional nonlinear robot terms in $f(\mathbf{x})$. An integrator may be added in the outer tracking loop to ensure zero steady-state error, obtaining a PID outer loop.

71.5.3.2 Commercial Robot Controllers

Commercial robot controllers do not implement the entire computed torque law. Most available controllers simply use a PD or PID control loop around each joint, dispensing entirely with the inner nonlinear compensation loop $\hat{f}(\mathbf{x})$. It is not clear exactly what is going on in most commercially available controllers, as they are proprietary and the user has no way to modify the joint tracking loops. However, in some controllers (e.g., Adept Hyperdrive), there appears to be some inner-loop compensation, where some of the terms in $f(\mathbf{x})$ are included in $\tau(t)$. For instance, acceleration feedforward may be included. To implement nonlinear feedback terms that are not already built-in on commercial controllers, it is usually necessary to perform hardware modifications of the controller.

71.5.3.3 Adaptive and Robust Control

There are by now many advanced control techniques for robot manipulators that either estimate the nonlinear robot function or compensate otherwise for uncertainties in $f(\mathbf{x})$. In **adaptive control** the estimate $\hat{f}$ of the nonlinear terms is updated online in real-time using additional internal controller dynamics, and in **robust control** the robustifying signal $v(t)$ is selected to overbound the system modeling uncertainties. In **learning control**, the nonlinearity correction term is improved over each repetition of the trajectory using the tracking error over the previous repetition (this is useful in repetitive motion applications including spray painting). Neural networks (NN) or fuzzy logic (FL) systems can be used in the inner control loop to manufacture the nonlinear estimate $\hat{f}(\mathbf{x})$ [Lewis et al. 1995]. Since both NN and FL systems have a universal approximation property, the restrictive **linear in the parameters** assumption required in standard adaptive control techniques is not needed, and no *regression matrix* need be computed. FL systems may also be used in the outer PID tracking loop to provide additional robustness.

Though these advanced techniques significantly improve the tracking performance of robot manipulators, they cannot be implemented on existing commercial robot controllers without hardware modifications.

71.5.4 Robot Force/Torque Servocontrol

In many industrial applications it is desired for the robot to exert a prescribed force normal to a given surface while following a prescribed motion trajectory tangential to the surface. This is the case in surface finishing, etc. A hybrid position/force controller can be designed by extension of the principles just presented.

The robot dynamics with environmental contact can be described by

$$M(\mathbf{q})\ddot{\mathbf{q}} + V_m(\mathbf{q}, \dot{\mathbf{q}})\dot{\mathbf{q}} + \mathbf{F}(\mathbf{q}, \dot{\mathbf{q}}) + \mathbf{G}(\mathbf{q}) + \tau_d = \tau + J^T(\mathbf{q})\lambda \tag{71.14}$$

where $J(\mathbf{q})$ is a constraint Jacobian matrix associated with the contact surface geometry and λ (the so-called Lagrange multiplier) is a vector of contact forces exerted normal to the surface, described in coordinates relative to the surface.

The hybrid position/**force control** problem is to follow a prescribed motion trajectory $\mathbf{q}_{1_d}(t)$ tangential to the surface while exerting a prescribed contact force $\lambda_d(t)$ normal to the surface. Define the filtered motion error $r_m = \dot{e}_m + \Lambda e_m$, where $e_m = \mathbf{q}_{1_d} - \mathbf{q}_1$ represents the motion error in the plane of the surface and Λ is a positive diagonal design matrix. Define the force error as $\tilde{\lambda} = \lambda_d - \lambda$, where $\lambda(t)$ is the normal force measured in a coordinate frame attached to the surface. Then a hybrid position/force controller has the structure

$$\tau = \hat{f} + K_v L(\mathbf{q}_1)r_m + J^T[\lambda_d + K_f \tilde{\lambda}] - v \tag{71.15}$$

where $\hat{f}$ is an estimate of the nonlinear robot function 71.12 and $L(\cdot)$ is an extended Jacobian determined from the surface geometry using the implicit function theorem.

This controller has the basic structure of Figure 71.13, but with an additional *inner force control loop*. In the hybrid position/force controller, the nonlinear function estimate inner loop $\hat{f}$ and the robustifying term $v(t)$ can be selected using adaptive, robust, learning, neural, or fuzzy techniques. A simplified controller that may work in some applications is obtained by setting $\hat{f} = 0$, $v(t) = 0$, and increasing the PD motion gains $K_v \Lambda$ and K_v and the force gain K_f.

It is generally not possible to implement force control on existing commercial robot controllers without hardware modification and extensive low-level programming.

71.5.5 Motion Trajectory Generation

In the section on servo-level motion control it was shown how to design real-time servo-level control loops for the **robot joint** actuators to cause the manipulator to follow a prescribed joint-space trajectory $\mathbf{q}_d(t)$ and, if required by the job, to exert forces normal to a surface specified by a prescribed force trajectory $\lambda_d(t)$. Unfortunately, the higher level **path planner** and job coordinator in Figure 71.7 do not specify the position and force trajectories in the detail required by the servo-level controllers. Most commercial robot controllers operate at a sampling period of 16 ms, so that they require specific desired motion trajectories $\mathbf{q}_d(t)$ sampled every 16 ms. On the other hand, the path planner wishes to be concerned at the level of abstraction only with general path descriptions sufficient to avoid obstacles or accomplish desired high-level jobs (e.g., move to prescribed final position, then insert pin in hole).

71.5.5.1 Path Transformation and Trajectory Interpolation

71.5.5.1.1 Joint Space vs. Cartesian Space Prescribed Trajectories

The job coordinator in Figure 71.7 passes required path-following commands to the virtual robot in the form of discrete events to be accomplished, which could be in the form of commands to "move to a specified final position passing through prescribed *via points*." These prescribed path via points are given in Cartesian coordinates y, are usually not regularly spaced in time, and may or may not have required times of transit associated with them. Via points are given in the form $(y_i, \dot{y}_i, t_i)$, with y_i the required Cartesian position at point i and $\dot{y}_i$ the required velocity. The time of transit t_i may or may not be specified. The irregularly spaced Cartesian-space via points must be *interpolated* to produce joint-space trajectory points regularly spaced at every sampling instant, often every 16 ms. It should be clearly understood that the path and the joint trajectory are both prescribed for each coordinate: the path for three Cartesian position coordinates and three Cartesian orientation coordinates, and the trajectory for each of the n manipulator joints. If n is not equal to 6, there could be problems in that the manipulator might not be able to exactly reach the prescribed via points. Thus, in its planning process the path planner must take into account the limitations of the individual robots.

Two procedures may be used to convert prescribed Cartesian path via points into desired joint-space trajectory points specified every 16 ms. One may either: (1) use the arm inverse kinematics to compute the via points in joint-space coordinates and then perform trajectory interpolation in joint space, or (2) perform interpolation on the via points to obtain a Cartesian trajectory specified every 16 ms, and then perform the inverse kinematics transformation to yield the joint-space trajectory $q_d(t)$ for the servo-level controller. The main disadvantage of the latter procedure is that the full inverse kinematics transformation must be performed every 16 ms. The main disadvantage of the former procedure is that interpolation in joint space often has strange effects, such as unexpected motions or curvilinear swings when viewed from the point of Cartesian space; one should recall that the path planner selects via points in Cartesian space, e.g., to avoid obstacles, often assuming linear Cartesian motion between the via points. The latter problem may be mitigated by spacing the Cartesian path via points more closely together. Thus, procedure 1 is usually selected in robotic workcell applications.

71.5.5.1.2 Trajectory Interpolation

A trajectory specified in terms of via points, either in joint space or Cartesian space, may be interpolated to obtain connecting points every 16 ms by many techniques, including interpolation by cubic polynomials, second- or third-order splines, minimum-time techniques, etc. The interpolation must be performed separately for each coordinate of the trajectory (e.g., n interpolations if done in joint space). Cubic interpolation is not recommended as it can result in unexpected swings or overshoots in the computed trajectory.

The most popular technique for trajectory interpolation may be linear functions with parabolic blends (LFPB). Let us assume that the path via points are specified in joint space, so that the inverse kinematics transformation from the Cartesian path via points obtained from the path planner has already been performed. Then, the path is specified in terms of the via points $(q(t_i), \dot{q}(t_i), t_i)$; note that the time of transit t_i of point i is specified; the transit times need not be uniformly spaced. Within each path segment connecting two via points, one uses constant acceleration or deceleration to obtain the required transit velocity, then zero acceleration during the transit, then constant acceleration or deceleration to obtain the prescribed final position and velocity at the next via point. Sample LFPB trajectories are given in Figure 71.14. Note that LFPB results in quadratic motion, followed by linear motion, followed by quadratic motion. The maximum acceleration/deceleration is selected taking into account the *joint actuator torque limits*.

There are standard formulas available to compute the LFPB trajectory passing through two prescribed via points, for instance the following. In Figure 71.14 two design parameters are selected: the *blend time* t_b and the *maximum velocity* v_M. Then the joint-space trajectory passing through via points i and $(i+1)$,

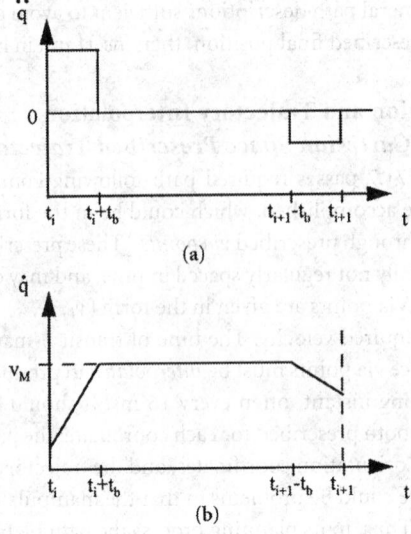

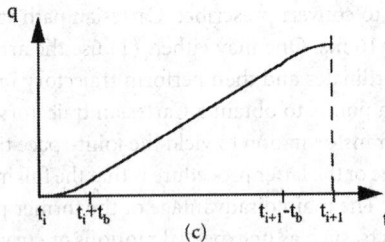

FIGURE 71.14   LFPB trajectory: (a) acceleration profile, (b) velocity profile, and (c) position profile.

shown in Figure 71.14(c), is given by

$$
\mathbf{q}_d(t) = \begin{cases} a + (t - t_i)b + (t - t_i)^2 c, & t_i \le t < t_i + t_b \\ d + v_M t, & t_i + t_b \le t < t_{i+1} - t_b \\ e + (t - t_{i+1})f + (t - t_{i+1})^2 g, & t_{i+1} - t_b \le t < t_{i+1} \end{cases} \tag{71.16}
$$

It is not difficult to determine that the coefficients required to pass through the ith and $(i+1)$st via points
are given by

$$
a = \mathbf{q}(t_i), \qquad b = \dot{\mathbf{q}}(t_i), \qquad c = \frac{v_M - \dot{\mathbf{q}}(t_i)}{2t_b}
$$

$$
d = \frac{\mathbf{q}(t_i) + \mathbf{q}(t_{i+1}) - v_M t_{i+1}}{2} \tag{71.17}
$$

$$
e = \mathbf{q}(t_{i+1}), \qquad f = \dot{\mathbf{q}}(t_{i+1})
$$

$$
g = \frac{v_M t_{i+1} + \mathbf{q}(t_i) - \mathbf{q}(t_{i+1}) + 2t_b[\dot{\mathbf{q}}(t_{i+1}) - v_M]}{2t_b^2}
$$

One must realize that this interpolation must be performed for each of the n joints of the robot. Then,
the resulting trajectory n-vector is passed as a prescribed trajectory to the servo-level controller, which
functions as in Robot Servo-level Motion Control subsection to cause trajectory-following arm motion.

71.5.5.2 Types of Trajectories and Limitations of Commercial Robot Controllers

The two basic types of trajectories of interest are motion trajectories and force trajectories. Motion specifications can be either in terms of motion from one prescribed point to another, or in terms of following a prescribed position/velocity/acceleration motion profile (e.g., spray painting).

In robotic assembly tasks point-to-point motion is usually used, without prescribing any required transit time. Such motion can be programmed with commercially available controllers using standard robot programming languages (Section 71.12). Alternatively, via points can usually be taught using a telerobotic teach pendant operated by the user (Section 71.11); the robot memorizes the via points, and effectively plays them back in operational mode. A speed parameter may be set prior to the motion that tells the robot whether to move more slowly or more quickly. Trajectory interpolation is automatically performed by the robot controller, which then executes PD or PID control at the joint servocontrol level to cause the desired motion. This is by far the most common form of robot motion control.

In point-to-point motion control the commercial robot controller performs trajectory interpolation and joint-level PD servocontrol. All of this is transparent to the user. Generally, it is very difficult to modify any stage of this process since the internal controller workings are proprietary, and the controller hardware does not support more exotic trajectory interpolation or servo-level control schemes. Though some robots by now do support following of prescribed position/velocity/acceleration profiles, it is generally extremely difficult to program them to do so, and especially to modify the paths once programmed. Various tricks must be used, such as specifying the Cartesian via points $(y_i, \dot{y}_i, t_i)$ in very fine time increments, and computing t_i such that the desired acceleration is produced.

The situation is even worse for force control, where additional sensors must be added to sense forces (e.g., wrist force-torque sensor, see Section 71.7), kinematic computations based on the given surface must be performed to decompose the tangential motion control directions from the normal force control directions, and then very tedious low-level programming must be performed. Changes in the surface or the desired motion or force profiles require time-consuming reprogramming. In most available robot controllers, hardware modifications are required.

71.6 End Effectors and End-of-Arm Tooling

End effectors and end-of-arm tooling are the devices through which the robot manipulator interacts with the world around it, grasping and manipulating parts, inspecting surfaces, and so on [Wright and Cutkosky 1985]. End effectors should not be considered as accessories, but as a major component in any workcell; proper selection and/or design of end effectors can make the difference between success and failure in many process applications, particularly when one includes reliability, efficiency, and economic factors. End effectors consist of the *fingers*, the *gripper*, and the *wrist*. They can be either standard commercially available mechanisms or specially designed tools, or can be complex systems in themselves (e.g., welding tools or dextrous hands). Sensors can be incorporated in the fingers, the gripper mechanism, or the wrist mechanism. All end effectors, end-of-arm tooling, and supply hoses and cables (electrical, pneumatic, etc.) must be taken into account when considering the manipulator payload weight limits of the manufacturer.

71.6.1 Part Fixtures and Robot Tooling

In most applications the end effector design problem should not be decoupled from the part fixturing design problem. One should consider the wrist, gripper, fingers, and part fixturing as a *single system*. Integrated design can often yield innovative solutions to otherwise intractable problems; nonintegrated design can often lead to unforseen problems and unexpected failure modes. Coordinated design of fixtures and end effectors can often avoid the use of high-level expensive sensors (e.g., vision) and/or complex feedback control systems that required overall coordinated control of the robot arm motion, the gripper action, and the part pose. An ideal example of a device that allows simplified control strategies is the **remote-center-of-compliance** (RCC) wrist in Figure 71.17(b), if correctly used.

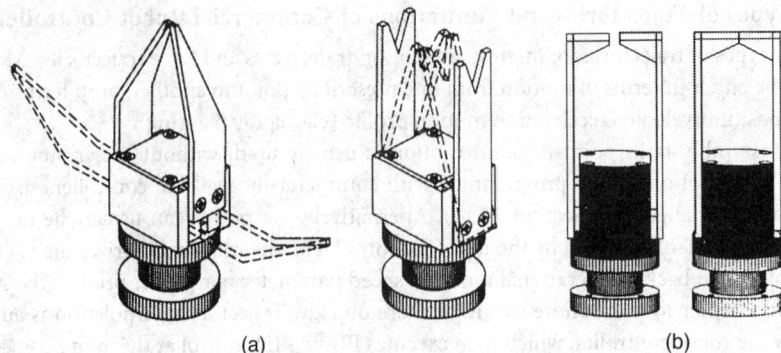

FIGURE 71.15 Angular and parallel motion robot grippers: (a) angular motion gripper and (b) parallel motion gripper, open and closed. (Courtesy of Robo-Tech Systems, Gastonia, NC.)

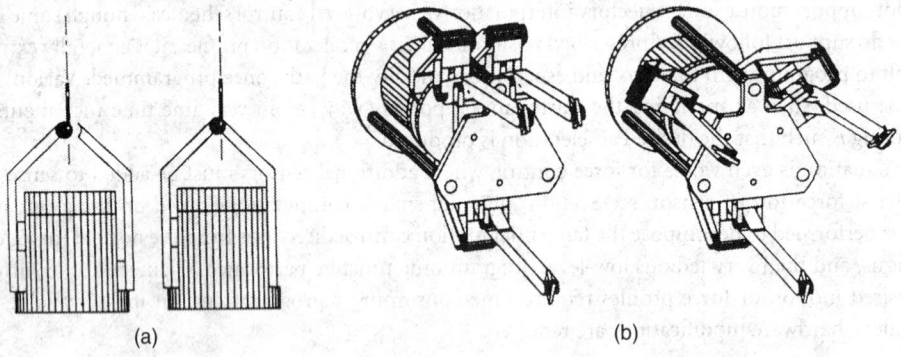

FIGURE 71.16 Robot grippers: (a) center seeking gripper showing part contact by first finger and final closure by second finger and (b) Versagrip III adjustable three-finger gripper. (Courtesy of Robo-Tech Systems, Gastonia, NC.)

71.6.2 Grippers and Fingers

Commercial catalogs usually allow one to purchase end effector components separately, including fingers, grippers, and wrists. Grippers can be actuated either pneumatically or using servomotors. Pneumatic actuation is usually either open or closed, corresponding to a binary command to turn the air pressure either off or on. Grippers often lock into place when the fingers are closed to offer failsafe action if air pressure fails. Servomotors often require analog commands and are used when finer gripper control is required. Available gripping forces span a wide range up to several hundred pounds force.

71.6.2.1 Gripper Mechanisms

Angular motion grippers, see Figure 71.15a, are inexpensive devices allowing grasping of parts either externally or internally (e.g., fingers insert into a tube and gripper presses them outward). The fingers can often open or close by 90°. These devices are useful for simple pick-and-place operations. In electronic assembly or tasks where precise part location is needed, it is often necessary to use *parallel grippers*, see Figure 71.15b, where the finger actuation affords exactly parallel closing motion. Parallel grippers generally have a far smaller range of fingertip motion that angular grippers (e.g., less than 1 in). In some cases, such as electronic assembly of parts positioned by wires, one requires *center seeking grippers*, see Figure 71.16a, where the fingers are closed until one finger contacts the part, then that finger stops and the other finger closes until the part is grasped.

There are available many grippers with advanced special-purpose mechanisms, including Robo-Tech's Versagrip III shown in Figure 71.16b, a 3-fingered gripper whose fingers can be rotated about a longitudinal axis to offer a wide variety of 3-fingered grasps depending on the application and part geometry. Finger rotation is affected using a fine motion servomotor that can be adjusted as the robot arm moves.

The gripper and/or finger tips can have a wide variety of sensors including binary part presence detectors, binary closure detectors, analog finger position sensors, contact force sensors, temperature sensors, and so on (Section 71.7).

71.6.2.2 The Grasping Problem and Fingers

The study of the *multifinger grasping problem* is a highly technical area using mathematical and mechanical engineering analysis techniques such as rolling/slipping concepts, friction studies, force balance and center of gravity studies, etc. [Pertin-Trocaz 1989]. These ideas may be used to determine the required gripper mechanisms, number of fingers, and finger shapes for a specific application. Fingers are usually specially designed for particular applications, and may be custom ordered from end-effector supply houses. Improper design and selection of fingers can doom to failure an application of an expensive robotic system. By contrast, innovative finger and contact tip designs can solve difficult manipulation and grasping problems and greatly increase automation reliability, efficiency, and economic return.

Fingers should not be thought of as being restricted to anthropomorphic forms. They can have vacuum contact tips for grasping smooth fragile surfaces (e.g., auto windshields), electromagnetic tips for handling small ferrous parts, compliant bladders or wraparound air bladders for odd-shaped or slippery parts, Bernoulli effect suction for thin fragile silicon wafers, or membranes covering a powder to distribute contact forces for irregular soft fragile parts [Wright and Cutkosky 1985].

Multipurpose grippers are advantageous in that a single end effector can perform multiple tasks. Some multipurpose devices are commercially available; they are generally expensive. The ideal multipurpose end effector is the *anthropomorphic dextrous hand*. Several dextrous robot hands are now available and afford potential applications in processes requiring active manipulation of parts or handling of many sorts of tooling. Currently, they are generally restricted to research laboratories since the problems associated with their expense, control, and coordination are not yet completely and reliably solved.

71.6.3 Robot Wrist Mechanisms

Wrist mechanisms couple the gripper to the robot arm, and can perform many functions. Commercial *adapter plates* allow wrists to be mounted to any commercially available robot arm. As an alternative to expensive multipurpose grippers, *quick change wrists* allow end effectors to be changed quickly during an application, and include quick disconnect couplings for mechanical, electrical, pneumatic and other connections. Using a quick change wrist, required tools can be selected from a magazine of available tools/end effectors located at the workcell. If fewer tools are needed, an alternative is provided by inexpensive *pivot gripper wrists*, such as the 2-gripper-pivot device shown in Figure 71.17a, which allows one of two grippers to be rotated into play. With this device, one gripper can unload a machine while the second gripper subsequently loads a new blank into the machine. Other *rotary gripper wrists* allow one of several (up to six or more) grippers to be rotated into play. With these wrists, the grippers are mounted in parallel and rotate much like the chamber of an old-fashioned western Colt 45 revolver; they are suitable if the grippers will not physically interfere with each other in such a parallel configuration.

Safety wrists automatically deflect, sending a fault signal to the machine or job coordinator, if the end-of-arm tooling collides with a rigid obstacle. They may be reset automatically when the obstacle is removed.

Part positioning errors frequently occur due to robot end effector positioning errors, part variations, machine location errors, or manipulator repeatibility errors. It is unreasonable and expensive to require the robot joint controller to compensate exactly for such errors. *Compliant wrists* offset positioning errors to a large extent by allowing small passive part motions in response to forces or torques exerted on the

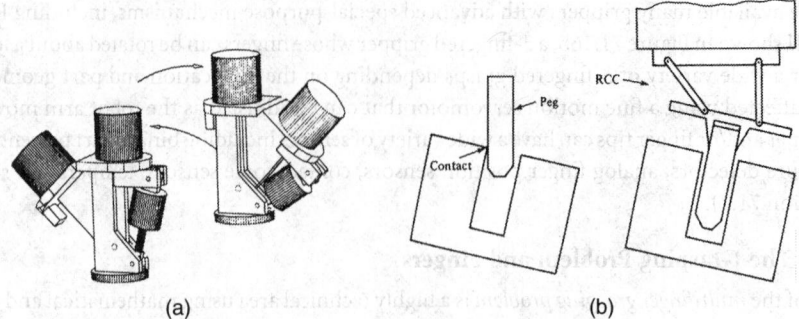

FIGURE 71.17   Robot wrists. (a) Pivot gripper wrist. (Courtesy of Robo-Tech Systems, Gastonia, NC.) (b) Remote-center-of-compliance (RCC) wrist. (Courtesy of Lord Corporation, Erie, PA.)

part. An example is pin insertion, where small positioning errors can result in pin breakage or other failures, and compensation by gross robot arm motions requires sophisticed (e.g., expensive) force-torque sensors and advanced (e.g., expensive) closed-loop feedback force control techniques. The compliant wrist allows the pin to effectively adjust its own position in response to sidewall forces so that it slides into the hole. A particularly effective device is the remote-center-of-compliance (RCC) wrist, Figure 71.17b, where the rotation point of the wrist can be adjusted to correspond, e.g., to the part contact point [Groover et al. 1986]. Compliant wrists allow successful assembly where vision or other expensive sensors would otherwise be needed.

The wrist can contain a wide variety of sensors, with possibly the most important class being the *wrist force-torque sensors* (Section 71.7), which are quite expensive. A general rule-of-thumb is that, for economic and control complexity reasons, robotic force/torque sensing and control should be performed at the lowest possible level; e.g., fingertip sensors can often provide sufficient force information for most applications, with an RCC wrist compensating for position inaccuracies between the fingers and the parts.

71.6.4   Robot/Tooling Process Integration and Coordination

Many processes require the design of sophisticated end-of-arm tooling. Examples include spray painting guns, welding tools, multipurpose end effectors, and so on. Indeed, in some processes the complexity of the tooling can rival or exceed the complexity of the robot arm that positions it. Successful coordination and sequencing of the robot manipulator, the end effector, and the end-of-arm tooling calls for a variety of considerations at several levels of abstraction in Figure 71.7.

There are two philosophically distinct points of view that may be used in considering the robot manipulator plus its end-of-arm tooling. In the first, the robot plus tooling is viewed as a *single virtual agent* to be assigned by an upper-level organizer/manager and commanded by a midlevel job coordinator. In this situation, all machine-level robot/tool coordination may be performed by the internal virtual robot machine coordinator shown in Figure 71.7. This point of view is natural when the robot must perform sophisticated trajectory motion during the task and the tool is unintelligent, such as in pick-and-place operations, surface finishing, and grinding. In such situations, the end effector is often controlled by simple binary on/off or open/close commands through digital input/output signals from the machine coordinator. Many commercially available robot controllers allow such communications and support coordination through their programming languages (Section 71.12).

In the second viewpoint, one considers the manipulator as a dumb platform that positions the tooling or maintains its relative motion to the workpiece while the tooling performs a job. This point of view may be taken in the case of processes requiring sophisticated tooling such as welding. In this situation, the robot manipulator and the tooling may be considered as *two separate agents* which are coordinated by the higher level job coordinator shown in Figure 71.7.

A variety of processes fall between these two extremes, such as assembly tasks which require some coordinated intelligence by both the manipulator and the tool (insert pin in hole). In such applications both machine-level and task-level coordination may be required. The decomposition of coordination commands into a portion suitable for machine-level coordination and a portion for task-level coordination is not easy. A rule-of-thumb is that any coordination that is invariant from process to process should be apportioned to the lower level (e.g., do not open gripper while robot is in motion). This is closely connected to the appropriate definition of robot/tooling behaviors in the fashion of Brooks [1986].

71.7 Sensors

Sensors and actuators [Tzou and Fukuda 1992] function as *transducers*, devices through which the workcell planning, coordination, and control system interfaces with the hardware components that make up the workcell. Sensors are a vital element as they convert states of physical devices into signals appropriate for input to the workcell PC&C control system; inappropriate sensors can introduce errors that make proper operation impossible no matter how sophisticated or expensive the PC&C system, whereas innovative selection of sensors can make the control and coordination problem much easier.

71.7.1 The Philosophy of Robotic Workcell Sensors

Sensors are of many different types and have many distinct uses. Having in mind an analogy with biological systems, *proprioceptors* are sensors internal to a device that yield information about the internal state of that device (e.g., robot arm joint-angle sensors). *Exteroceptors* yield information about other hardware external to a device. Sensors yield outputs that are either analog or digital; digital sensors often provide information about the status of a machine or resource (gripper open or closed, machine loaded, job complete). Sensors produce inputs that are required at all levels of the PC&C hierarchy, including uses for:

- Servo-level feedback control (usually analog proprioceptors)
- Process monitoring and coordination (often digital exteroceptors or part inspection sensors such as vision)
- Failure and safety monitoring (often digital, e.g., contact sensor, pneumatic pressure-loss sensor)
- Quality control inspection (often vision or scanning laser)

Sensor output data must often be processed to convert it into a form meaningful for PC&C purposes. The sensor plus required signal processing is shown as a *virtual sensor* in Figure 71.7; it functions as a *data abstraction*, that is, a set of data plus operations on that data (e.g., camera, plus framegrabber, plus signal processing algorithms such as image enhancement, edge detection, segmentation, etc.). Some sensors, including the proprioceptors needed for servo-level feedback control, are integral parts of their host devices, and so processing of sensor data and use of the data occurs within that device; then, the sensor data is incorporated at the servocontrol level or machine coordination level. Other sensors, often vision systems, rival the robot manipulator in sophistication and are coordinated by the job coordinator, which treats them as valuable shared resources whose use is assigned to jobs that need them by some priority assignment (e.g., dispatching) scheme. An interesting coordination problem is posed by so-called *active sensing*, where, e.g., a robot may hold a scanning camera, and the camera effectively takes charge of the coordination problem, directing the robot where to move it to effect the maximum reduction in entropy (increase in information) with subsequent images.

71.7.2 Types of Sensors

This section summarizes sensors from an operational point of view. More information on functional and physical principles can be found in Fraden [1993], Fu et al. [1987], Groover et al. [1986], and Snyder [1985].

71.7.2.1   Tactile Sensors

Tactile sensors [Nichols and Lee 1989] rely on physical contact with external objects. Digital sensors such as limit switches, microswitches, and vacuum devices give binary information on whether contact occurs or not. Sensors are available to detect the onset of slippage. Analog sensors such as spring-loaded rods give more information. Tactile sensors based on rubberlike carbon- or silicon-based *elastomers* with embedded electrical or mechanical components can provide very detailed information about part geometry, location, and more. Elastomers can contain resistive or capacitive elements whose electrical properties change as the elastomer compresses. Designs based on LSI technology can produce *tactile grid pads* with, e.g., 64×64 *force/* points on a single pad. Such sensors produce *tactile images* that have properties akin to digital images from a camera and require similar data processing. Additional tactile sensors fall under the classification of force sensors discussed subsequently.

71.7.2.2   Proximity and Distance Sensors

The noncontact proximity sensors include devices based on the Hall effect or inductive devices based on the electromagnetic effect that can detect ferrous materials within about 5 mm. Such sensors are often digital, yielding binary information about whether or not an object is near. Capacitance-based sensors detect any nearby solid or liquid with ranges of about 5 mm. Optical and ultrasound sensors have longer ranges.

Distance sensors include time-of-flight range finder devices such as sonar and lasers. The commercially available Polaroid sonar offers accuracy of about 1 in up to 5 ft, with angular sector accuracy of about 15°. For 360° coverage in navigation applications for mobile robots, both scanning sonars and ring-mounted multiple sonars are available. Sonar is typically noisy with spurious readings, and requires low-pass filtering and other data processing aimed at reducing the false alarm rate. The more expensive laser range finders are extremely accurate in distance and have very high angular resolution.

71.7.2.3   Position, Velocity, and Acceleration Sensors

Linear position-measuring devices include linear potentiometers and the sonar and laser range finders just discussed. Linear velocity sensors may be laser- or sonar-based Doppler-effect devices.

Joint-angle position and velocity proprioceptors are an important part of the robot arm servocontrol drive axis. Angular position sensors include potentiometers, which use dc voltage, and *resolvers*, which use ac voltage and have accuracies of ±15 min. Optical encoders can provide extreme accuracy using digital techniques. *Incremental optical encoders* use three optical sensors and a single ring of alternating opaque/clear areas, Figure 71.18a, to provide angular position relative to a reference point and angular velocity information; commercial devices may have 1200 slots per turn. More expensive *absolute optical encoders*, Figure 71.18b, have n concentric rings of alternating opaque/clear areas and require n optical sensors. They offer increased accuracy and minimize errors associated with data reading and transmission, particularly if they employ the *Grey code*, where only one bit changes between two consecutive sectors. Accuracy is $360°/2^n$, with commercial devices having $n = 12$ or so.

Gyros have good accuracy if repeatability problems associated with drift are compensated for. Directional gyros have accuracies of about ±1.5°; vertical gyros have accuracies of 0.15° and are available to measure multiaxis motion (e.g., pitch and roll). Rate gyros measure velocities directly with thresholds of 0.05°/s or so.

Various sorts of accelerometers are available based on strain gauges (next paragraph), gyros, or crystal properties. Commercial devices are available to measure accelerations along three axes.

71.7.2.4   Force and Torque Sensors

Various torque sensors are available, though they are often not required; for instance, the internal torques at the joints of a robot arm can be computed from the motor armature currents. Torque sensors on a drilling tool, for instance, can indicate when tools are dull. Linear force can be measured using load cells or strain gauges. A strain gauge is an elastic sensor whose resistance is a function of applied strain or deformation. The piezoelectric effect, the generation of a voltage when a force is applied, may also be

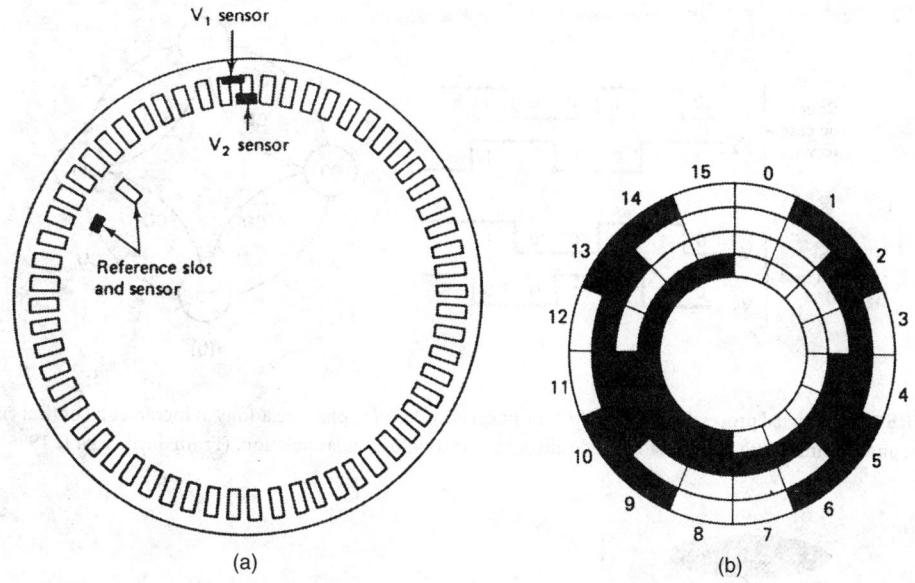

FIGURE 71.18 Optical encoders: (a) incremental optical encoder and (b) absolute optical encoder with $n = 4$ using Grey code (From Snyder, W. E. 1985. *Industrial Robots: Computer Interfacing and Control.* Prentice–Hall, Englewood Cliffs, NJ. With permission.)

used for force sensing. Other force sensing techniques are based on vacuum diodes, quartz crystals (whose resonant frequency changes with applied force), etc.

Robot arm force-torque wrist sensors are extremely useful in dextrous manipulation tasks. Commercially available devices can measure both force and torque along three perpendicular axes, providing full information about the Cartesian force vector **F**. Transformations such as Equation 71.6 allow computation of forces and torques in other coordinates. Six-axis force-torque sensors are quite expensive.

71.7.2.5 Photoelectric Sensors

A wide variety of photoelectric sensors are available, some based on fiber optic principles. These have speeds of response in the neighborhood of 50 μs with ranges up to about 45 mm, and are useful for detecting parts and labeling, scanning optical bar codes, confirming part passage in sorting tasks, etc.

71.7.2.6 Other Sensors

Various sensors are available for measuring pressure, temperature, fluid flow, etc. These are useful in closed-loop servocontrol applications for some processes such as welding, and in job coordination and/or safety interrupt routines in others.

71.7.3 Sensor Data Processing

Before any sensor can be used in a robotic workcell, it must be *calibrated*. Depending on the sensor, this could involve significant effort in experimentation, computation, and tuning after installation. Manufacturers often provide calibration procedures though in some cases, including vision, such procedures may not be obvious, requiring reference to the published scientific literature. Time-consuming recalibration may be needed after any modifications to the system.

Particularly for more complex sensors such as optical encoders, significant sensor signal conditioning and processing is required. This might include amplification of signals, noise rejection, conversion of data from analog to digital or from digital to analog, and so on. Hardware is usually provided for such purposes

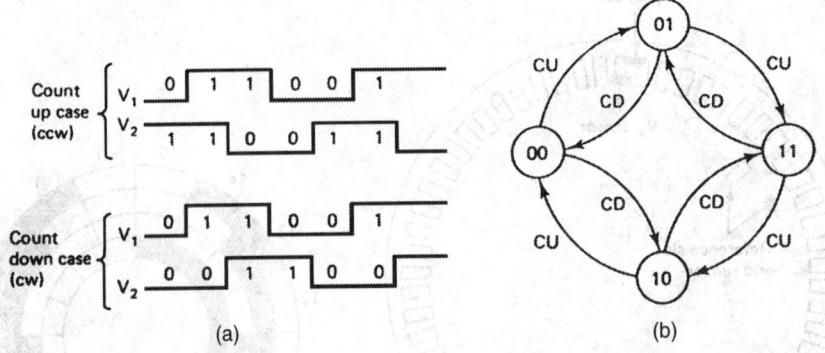

FIGURE 71.19   Signal processing using FSM for optical encoders: (a) phase relations in incremental optical encoder output and (b) finite state machine to decode encoder output into angular position. (From Snyder, W. E. 1985).

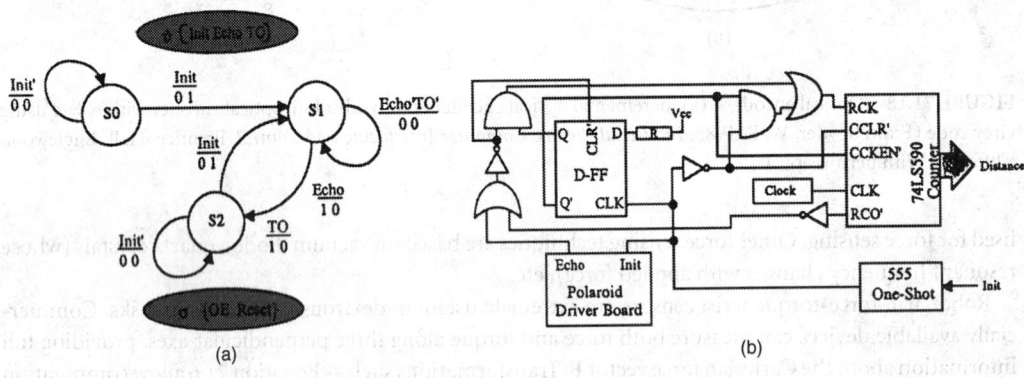

FIGURE 71.20   Hardware design from FSM: (a) FSM for sonar transducer control on a mobile robot and (b) sonar driver control system from FSM.

by the manufacturer and should be considered as part of the sensor package for robot workcell design. The sensor, along with its signal processing hardware and software algorithms may be considered as a data abstraction and is called the *virtual sensor* in Figure 71.7.

If signal processing does need to be addressed, it is often very useful to use *finite state machine* (FSM) *design*. A typical signal from an incremental optical encoder is shown in Figure 71.19a; a FSM for decoding this into the angular position is given in Figure 71.19b. FSMs are very easy to convert directly to hardware in terms of logical gates. A FSM for sequencing a sonar is given in Figure 71.20a; the sonar driver hardware derived from this FSM is shown in Figure 71.20b.

A particular problem is obtaining angular velocity from angular position measurements. All too often the position measurements are simply differenced using a small sample period to compute velocity. This is guaranteed to lead to problems if there is any noise in the signal. It is almost always necessary to employ a low-pass-filtered derivative where velocity samples v_k are computed from position measurement samples $\mathbf{p}_k$ using, e.g.,

$$v_k = \alpha v_{k-1} + (1 - \alpha)(\mathbf{p}_k - \mathbf{p}_{k-1})/T \tag{71.18}$$

where T is the sample period and α is a small filtering coefficient. A similar approach is needed to compute acceleration.

71.7.4 Vision for Robotics

Computer vision is covered in Chapter 43; the purpose of this section is to discuss some aspects of vision that are unique to robotics [Fu et al. 1987, Lee et al. 1991, 1994]. Industrial robotic workcells often require vision systems that are reliable, accurate, low cost, and rugged yet perform sophisticated image processing and decision making functions. Balancing these conflicting demands is not always easy. There are several commercially available vision systems, the most sophisticated of which may be the Adept vision system, which supports multiple cameras. However, it is sometimes necessary to design one's own system. Vision may be used for three purposes in robotic workcells: inspection and quality control, robotic manipulation, and servo-level feedback control. In quality control inspection systems the cameras are often affixed to stationary mounts while parts pass on a conveyor belt. In *active vision* inspection systems, cameras may be mounted as end effectors of a robot manipulator, which positions the camera for the required shots.

The operational phase of robot vision has six principal areas. *Low-level vision* includes *sensing* and *preprocessing* such as noise reduction, image digitization if required, and edge detection. Medium-level vision includes *segmentation, description*, and *recognition*. High-level vision includes *interpretation* and *decision making*. Such topics are disussed in Chapter 32. Prior to placing the vision system in operation, one is faced with several design issues including camera selection and illumination techniques, and the problem of system *calibration*.

71.7.4.1 Cameras and Illumination

Typical commercially available vision systems conform to the RS-170 standard of the 1950s, so that frames are acquired through a framegrabber board at a rate of 30 frames/s. Images are scanned; in a popular U.S. standard, each complete scan or *frame* consists of 525 lines of which 480 contain image information. This sample rate and image resolutions of this order are adequate for most applications with the exception of vision-based robot arm servoing (discussed subsequently). Robot vision system cameras are usually TV cameras: either the solid-state charge-coupled device (CCD), which is responsive to wavelengths of light from below 350 nm (ultraviolet) to 1100 nm (near infrared) and has peak response at approximately 800 nm, or the charge injection device (CID), which offers a similar spectral response and has a peak response at approximately 650 nm. Both *line-scan* CCD cameras, having resolutions ranging between 256 and 2048 elements, and *area-scan* CCD cameras are available. Medium-resolution area-scan cameras yield images of 256×256, though high-resolution devices of 1024×1024 are by now available. Line-scan cameras are suitable for applications where parts move past the camera, e.g., on conveyor belts. Framegrabbers often support multiple cameras, with a common number being four, and may support black-and-white or color images.

If left to chance, illumination of the robotic workcell will probably result in severe problems in operations. Common problems include low-contrast images, specular reflections, shadows, and extraneous details. Such problems can be corrected by overly sophisticated image processing, but all of this can be avoided by some proper attention to details at the workcell design stage. Illumination techniques include spectral filtering, selection of suitable spectral characteristics of the illumination source, diffuse-lighting techniques, backlighting (which produces easily processed silhouettes), structured-lighting (which provides additional depth information and simplifies object detection and interpretation), and directional lighting.

71.7.4.2 Coordinate Frames and Camera Perspective Transformation

A typical robot vision system is depicted in Figure 71.21, which shows a gimball-mounted camera. There are illustrated the base frame (or world frame) (X, Y, Z), the gimball platform, the camera frame (x, y, z), and the *image plane* having coordinates of (ξ, υ).

71.7.4.2.1 Image Coordinates of a Point in Base Coordinates

The primary tools for analysis of robot vision systems are the notion of *coordinate transforms* and the camera *perspective transformation*. Four-by-four *homogeneous transformations* (discussed earlier) are used, as they provide information on translations, rotations, scaling, and perspective.

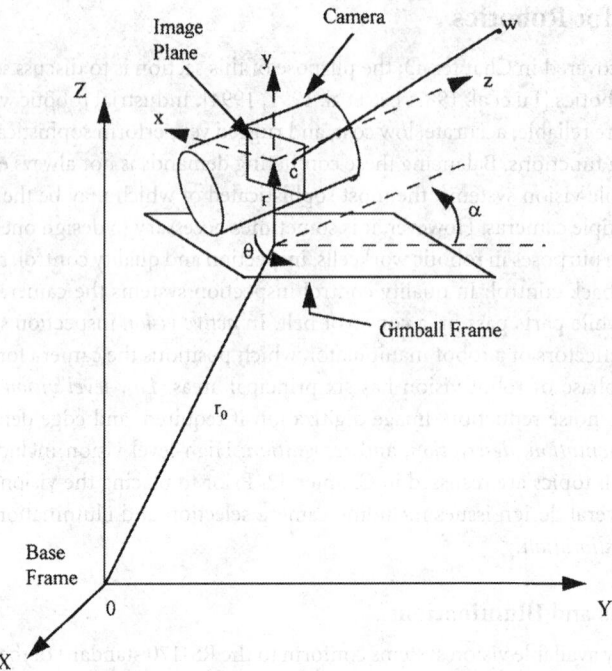

FIGURE 71.21   Typical robot workcell vision system.

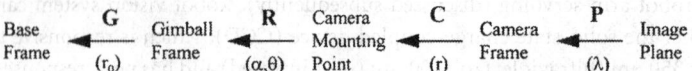

FIGURE 71.22   Homogeneous transformations associated with the robot vision system.

Four homogeneous transformations may be identified in the vision system, as illustrated in Figure 71.22. The gimball transformation G represents the base frame in coordinates affixed to the gimball platform. If the camera is mounted on a robot end effector, G is equal to T^{-1}, with T the robot arm T matrix detailed in earlier in Section 71.5; for a stationary-mounted camera G is a constant matrix capturing the camera platform mounting offset $r_0 = [X_0 \ Y_0 \ Z_0]^T$. The pan/tilt transformation R represents the gimball platform with respect to the mounting point of the camera. This rotation transformation is given by

$$R = \begin{bmatrix} \cos\theta & \sin\theta & 0 & 0 \\ -\sin\theta\cos\alpha & \cos\theta\cos\alpha & \sin\alpha & 0 \\ \sin\theta\sin\alpha & -\cos\theta\sin\alpha & \cos\alpha & 0 \\ 0 & 0 & 0 & 1 \end{bmatrix} \tag{71.19}$$

with θ the pan angle and α the tilt angle. C captures the offset $r = [r_x \ r_y \ r_z]^T$ of the camera frame with respect to the gimball frame. Finally, the perspective transformation

$$\begin{bmatrix} 1 & 0 & 0 & 0 \\ 0 & 1 & 0 & 0 \\ 0 & 0 & 1 & 0 \\ 0 & 0 & -\frac{1}{\lambda} & 1 \end{bmatrix} \tag{71.20}$$

projects a point represented in camera coordinates (x, y, z) onto a position (ξ, υ) in the image, where λ is the camera focal length.

In terms of these constructions, the image position of a point w represented in base coordinates as (X, Y, Z) is given by the *camera transform equation*

$$c = PCRGw \tag{71.21}$$

which evaluates in the case of a stationary-mounted camera to the image coordinates

$$\xi = \lambda \frac{(X - X_0) \cos \theta + (Y - Y_0) \sin \theta - r_x}{-(X - X_0) \sin \theta \sin \alpha + (Y - Y_0) \cos \theta \sin \alpha - (Z - Z_0) \cos \alpha + r_z + \lambda}$$
$$\upsilon = \lambda \frac{-(X - X_0) \sin \theta \cos \alpha + (Y - Y_0) \cos \theta \cos \alpha + (Z - Z_0) \sin \alpha - r_y}{-(X - X_0) \sin \theta \sin \alpha + (Y - Y_0) \cos \theta \sin \alpha - (Z - Z_0) \cos \alpha + r_z + \lambda} \tag{71.22}$$

71.7.4.2.2 Base Coordinates of a Point in Image Coordinates

In applications, one often requires the inverse of this transformation; that is, from the image coordinates (ξ, υ) of a point one wishes to determine its base coordinates (X, Y, Z). Unfortunately, the perspective transformation P is a *projection* which loses depth information z, so that the inverse perspective transformation P^{-1} is not unique. To compute unique coordinates in the base frame one therefore requires either two cameras, the ultimate usage of which leads to *stereo imaging*, or multiple shots from a single moving camera. Many techniques have been developed for accomplishing this.

71.7.4.2.3 Camera Calibration

Equation 71.22 has several parameters, including the camera offsets r_0 and r and the focal length λ. These values must be known prior to operation of the camera. They may be measured, or they may be computed by taking images of points w_i with known base coordinates (X_i, Y_i, Z_i). To accomplish this, one must take at least six points w_i and solve a resulting set of nonlinear simultaneous equations. Many procedures have been developed for accomplishing this by efficient algorithms.

71.7.4.3 High-Level Robot Vision Processing

Besides scene interpretation, other high-level vision processing issues must often be confronted, including decision making based on vision data, relation of recognized objects to stored CAD data of parts, recognition of faults or failures from vision data, and so on. Many technical papers have been written on all of these topics.

71.7.4.4 Vision-Based Robot Manipulator Servoing

In standard robotic workcells, vision is not often used for servo-level robot arm feedback control. This is primarily due to the facts that less expensive lower level sensors usually suffice, and reliable techniques for vision-based servoing are only now beginning to emerge. In vision-based servoing the standard frame rate of 30 ft/s is often unsuitable; higher frame rates are often needed. This means that commercially available vision systems cannot be used. Special purpose cameras and hardware have been developed by several researchers to address this problem, including the vision system in Lee and Blenis [1994].

Once the hardware problems have been solved, one has yet to face the design problem for real-time servocontrollers with vision components in the feedback loop. This problem may be attacked by considering the nonlinear dynamical system 71.7 with measured outputs given by combining the camera transformation 71.21 and the arm kinematics transformation 71.2 [Ghosh et al. 1994].

71.8 Workcell Planning

Specifications for workcell performance vary at distinct levels in the workcell planning, coordination, and control architecture in Figure 71.7. At the machine servocontrol level, motion specifications are given in terms of continuous trajectories in joint space sampled every 16 ms. At the job coordination level, motion specifications are in terms of Cartesian path via points, generally nonuniformly spaced, computed to achieve a prescribed task. Programming at these lower levels involves tedious specifications of points, motions, forces, and times of transit. The difficulties involved with such low-level programming have led to requirements for *task-level programming*, particularly in modern robot workcells which must be flexible and reconfigurable as products vary in response to the changing desires of customers. The function of the workcell planner is to allow task-level programming from the workcell manager by performing *task planning and decomposition* and *path planning*, thereby automatically providing the more detailed specifications required by the job coordinator and servo-level controllers.

71.8.1 Workcell Behaviors and Agents

The virtual machines and virtual sensors in the (PC&C) architecture of Figure 71.7 are constructed using the considerations discussed in previous sections. These involve commercial robot selection, robot kinematics and servo-level control, end effectors and tooling, and sensor selection and calibration. The result of design at these levels is a set of workcell *agents* — robots, machines, or sensors — each with a set of behaviors or *primitive actions* that each workcell agent is capable of. For instance, proper design could allow a robot agent to be capable of behaviors including accurate motion trajectory following, tool changing, force-controlled grinding on a given surface, etc. A camera system might be capable of identifying all Phillips screw heads in a scene, then determining their coordinates and orientation in the base frame of a robot manipulator.

Given the workcell agents with their behaviors, the higher level components in Figure 71.7 must be able to assign tasks and then decompose them into a suitable sequencing of behaviors. In this and the next section are discussed the higher level PC&C components of workcell planning and job coordination.

71.8.2 Task Decomposition and Planning

A *task* is a specific goal that must be accomplished, often by an assigned *due date*. These goals may include completed robotic processes (e.g., weld seam), finished products, and so on. Tasks are accomplished by a sequence of *jobs* that, when completed, result in the achievement of the final goal. *Jobs* are specific primitive activities that are accomplished by a well-defined set of resources or agents (e.g., drill hole, load machine). Once a resource has been assigned to a job it can be interpreted as a behavior. To attain the task goal, jobs must usually be performed in some *partial ordering*, e.g., tasks a and b are immediate prerequisties for task c. The jobs are not usually completely ordered, but have some possibility for concurrency.

At the workcell management level, tasks are assigned, along with their due dates, without specifying details of resource assignment or selection of specific agents. At the job coordination level, the required specifications are in terms of sequences of jobs, with resources assigned, selected to achieve the assigned goal tasks. The function of the workcell planner is to convert between these two performance specification paradigms. In the task planning component, two important transformations are made. First, assigned *tasks are decomposed into the required job sequences*. Second, workcell agents and *resources are assigned* to accomplish the individual jobs. The result is a task plan that is passed for execution to the job coordinator.

71.8.2.1 Task Plan

A *task plan* is a sequence of jobs, along with detailed resource assignments, that will lead to the desired goal task. Jobs with assigned resources can be interpreted as behaviors. Plans should not be overspecified — the required job sequencing is usually only a partial ordering, often with significant concurrency remaining among the jobs. Thus, some decisions based on real-time workcell status should be left to the job coordinator; among these are the final detailed sequencing of the jobs and any *dispatching* and *routing* decisions where shared resources are involved.

71.8.2.2 Computer Science Planning Tools

There are well-understood techniques in computer science that can be brought to bear on the robot task planning problem [Fu et al. 1987]. Planning and scheduling is covered in Chapter 67, decision trees in

Chapter 65, search techniques in Chapter 30, and decision making under uncertainty in Chapter 70; all of these are relevant to this discussion. However, the structure of the robotic workcell planning problem makes it possible to use some refined and quite rigorous techniques in this chapter, which are introduced in the next subsections.

Task planning can be accomplished using techniques from problem solving and learning, especially learning by analogy. By using *plan schema* and other *replanning* techniques, it is possible to modify existing plans when goals or resources change by small amounts. *Predicate logic* is useful for representing knowledge in the task planning scenario and many problem solving software packages are based on production systems.

Several task planning techniques use *graph theoretic* notions that can be attacked using search algorithms such as A^*. *State-space search techniques* allow one to try out various approaches to solving a problem until a suitable solution is found: the set of states reachable from a given initial state forms a graph. A plan is often represented as a finite state machine, with the states possibly representing jobs or primitive actions. *Problem reduction* techniques can be used to decompose a task into smaller subtasks; in this context it is often convenient to use AND/OR graphs. *Means–ends analysis* allows both forward and backward search techniques to be used, solving the main parts of a problem first and then going back to solve smaller subproblems.

For workcell assembly and production tasks, product data in CAD form is usually available. Assembly task planning involves specifying a sequence of assembly, and possibly process, steps that will yield the final product in finished form. *Disassembly planning techniques* work backwards from the final product, performing part disassembly transformations until one arrives at the initial raw materials. Care must be taken to account for part obstructions, etc. The relationships between parts should be specified in terms of symbolic spatial relationships between *object features* (e.g., place $block_1-face_2$ against $wedge_2-face_3$ and $block_1-face_1$ against $wedge_2-face_1$ or place pin in slot). *Constructive solid geometric* techniques lead to graphs that describe objects in terms of features related by set operations such as intersection, union, etc.

71.8.2.3 Industrial Engineering Planning Tools

In industrial engineering there are well-understood design tools used for product assembly planning, process planning, and resource assignment; they should be used in workcell task planning. The *bill of materials* (BOM) for a product is a computer printout that breaks down the various subassemblies and component parts needed for the product. It can be viewed as a matrix B whose elements $B(i, j)$ are set to 1 if subassembly j is needed to produce subassembly i. This matrix is known as *Steward's Sequencing Matrix;* by studying it one can decompose the assembly process into hierarchically interconnected subsystems of subassemblies [Warfield 1973], thereby allowing parallel processing and simplification of the assembly process. The *assembly tree* [Wolter et al. 1992] is a graphical representation of the BOM.

The *resource requirements matrix* is a matrix R whose elements $R(i, j)$ are set equal to 1 if resource j is required for job i. The resources may include machines, robots, fixtures, tools, transport devices, and so on. This matrix has been used by several workers for analysis and design of manufacturing systems; it is very straightforward to write down given a set of jobs and available resources. The *subassembly tree* is an assembly tree with resource information added.

71.8.3 Task Matrix Approach to Workcell Planning

Plans can often be specified as finite state machines. However, in the robot workcell case, FSM are neither general enough to allow versatile incorporation of workcell status and sensor information, nor specific enough to provide all of the information needed by the job coordinator.

A very general robot workcell task plan can be completely specified by four *task plan matrices* [Lewis and Huang 1995]. The *job sequencing matrix* and *job start matrix* are independent of resources and carry the job sequencing information required for task achievement. Resources are subsequently added by constructing the *resource requirements matrix* and the *resource release matrix*. The function of the task planner is to construct these four matrices and pass them to the job coordinator, who uses them for job coordination, sequencing, and resource dispatching. The task plan matrices are straightforward to construct and are

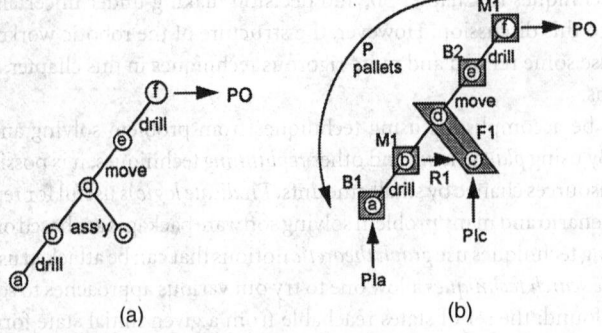

FIGURE 71.23 Product information for task planning: (a) assembly tree with job sequencing information and (b) subassembly tree with resource information added to the jobs.

easy to modify in the event of goal changes, resource changes, or failures; that is, they accommodate task planning as well as *task replanning*.

The task planning techniques advocated here are illustrated through an assembly design example, which shows how to select the four task plan matrices. Though the example is simple, the technique extends directly to more complicated systems using the notions of *block matrix (e.g., subsystem) design*. First, job sequencing is considered, then the resources are added.

71.8.3.1 Workcell Task Decomposition and Job Sequencing

In Figure 71.23(a) is given an assembly tree which shows the required sequence of actions (jobs) to produce a product. This sequence may be obtained from stored product CAD data through disassembly techniques, etc. The assembly tree contains information analogous to the BOM; it does not include any resource information. Part a enters the workcell and is drilled to produce part b, then assembled with part c to produce part d, which is again drilled (part e) to result in part f, which is the cell output (*PO* denotes "product out"). The assembly tree imposes only a *partial ordering* on the sequence of jobs. It is important not to overspecify the task decomposition by imposing additional temporal orderings that are not required for job sequencing.

71.8.3.1.1 Job Sequencing Matrix

Referring to Figure 71.23a, define the *job vector* as $\mathbf{v} = [\mathbf{a\,b\,c\,d\,e\,f}]^T$. The Steward's sequencing matrix F_v for the assembly tree in Figure 71.23a is then given by

$$
F_v = \begin{array}{c} \\ a \\ b \\ c \\ d \\ e \\ f \\ PO \end{array}
\begin{array}{c} \begin{array}{cccccc} a & b & c & d & e & f \end{array} \\
\left[\begin{array}{cccccc}
0 & 0 & 0 & 0 & 0 & 0 \\
1 & 0 & 0 & 0 & 0 & 0 \\
0 & 0 & 0 & 0 & 0 & 0 \\
0 & 1 & 1 & 0 & 0 & 0 \\
0 & 0 & 0 & 1 & 0 & 0 \\
0 & 0 & 0 & 0 & 1 & 0 \\
0 & 0 & 0 & 0 & 0 & 1
\end{array} \right] \end{array} . \tag{71.23}
$$

In this matrix, an entry of 1 in position (i, j) indicates that job j must be completed prior to starting job i. F_v is independent of available resources; in fact, regardless of the resources available, F_v will not change.

71.8.3.1.2 Sequencing State Vector and Job Start Equation

Define a *sequencing state vector* x, whose components are associated with the vector $[\mathbf{a\,b\,c\,d\,e\,f\,PO}]^T$, that checks the conditions of the rules needed for job sequencing. The components of $\mathbf{x}$ may be viewed as situated between the nodes in the assembly tree. Then, the *job start equation* is

$$
\mathbf{v}_s =
\begin{bmatrix}
1 & 0 & 0 & 0 & 0 & 0 & 0 \\
0 & 1 & 0 & 0 & 0 & 0 & 0 \\
0 & 0 & 1 & 0 & 0 & 0 & 0 \\
0 & 0 & 0 & 1 & 0 & 0 & 0 \\
0 & 0 & 0 & 0 & 1 & 0 & 0 \\
0 & 0 & 0 & 0 & 0 & 1 & 0
\end{bmatrix}
\begin{bmatrix}
\mathbf{x}_1 \\
\mathbf{x}_2 \\
\mathbf{x}_3 \\
\mathbf{x}_4 \\
\mathbf{x}_5 \\
\mathbf{x}_6 \\
\mathbf{x}_7
\end{bmatrix}
\equiv S_v \mathbf{x}
\tag{71.24}
$$

where $\mathbf{v}_s$ is the job start command vector. In the job start matrix S_v, an entry of 1 in position (i, j) indicates that job i can be started when component j of the sequencing state vector is active.

In this example, the matrix S_v has 1s in locations (i, i) so that S_v appears to be redundant. This structure follows from the fact that the assembly tree is an *upper semilattice*, wherein each node has a unique node above it; such a structure occurs in the manufacturing *re-entrant flowline with assembly*. In the more general *job shop* with variable part routings the semilattice structure of the assembly tree does not hold. Then, S_v can have multiple entries in a single column, corresponding to different routing options; nodes corresponding to such columns have more than one node above them.

71.8.3.2 Adding the Resources

To build a job dispatching coordination controller for shop-floor installation to perform this particular assembly task, the resources available must now be added. The issue of required and available resources is easily confronted as a *separate engineering design issue* from job sequence planning. In Figure 71.23b is given a *subassembly tree* for the assembly task, which includes resource requirements information. This information would in practice be obtained based on the resources and behaviors available in the workcell and could be assigned by a user during the planning stage using interactive software. The figure shows that part input *PIc* and part output (*PO*) do not require resources, pallets (*P*) are needed for part *a* and its derivative subassemblies, buffers (*B*1, *B*2) hold parts *a* and *e*, respectively, prior to drilling, and both drilling operations need the same machine (*M*1). The assembly operation is achieved by fixturing part *c* in fixture *F*1 while robot *R*1 inserts part *b*.

Note that drilling machine *M*1 represents a *shared resource*, which performs two jobs, so that dispatching decision making is needed when the two drilling jobs are simultaneously requested, in order to avoid possible problems with *deadlock*. This issue is properly faced by the job coordinator in real-time, as shown in Section 71.9, not by the task planner. Shared resources impose *additional temporal restrictions* on the jobs that are not present in the job sequencing matrix; these are concurrency restrictions of the form: both drilling operations may not be performed simultaneously.

71.8.3.2.1 Resource Requirements (RR) Matrix

Referring to Figure 71.23b, define the *resource vector* as $\mathbf{r} = [R1A\ F1A\ B1A\ B2A\ PA\ M1A]^T$, where A denotes available. In the RR matrix F_r, a 1 in entry (i, j) indicates that resource j is needed to activate sequencing vector component $\mathbf{x}_i$ (e.g., in this example, to accomplish job i). By inspection, therefore, one may write down the RR matrix

$$
F_r =
\left[
\begin{array}{ccccc|c}
0 & 0 & 1 & 0 & 1 & 0 \\
0 & 0 & 0 & 0 & 0 & 1 \\
0 & 1 & 0 & 0 & 0 & 0 \\
1 & 0 & 0 & 0 & 0 & 0 \\
0 & 0 & 0 & 1 & 0 & 0 \\
0 & 0 & 0 & 0 & 0 & 1 \\
0 & 0 & 0 & 0 & 0 & 0
\end{array}
\right]
\tag{71.25}
$$

Row 3, for instance, means that resource *F1A*, the fixture, is needed as a precondition for firing x_3; which matrix S_v associates with job c. Note that column 6 has two entries of 1, indicating that $M1$ is a shared resource that is needed for two jobs b and f. As resources change or machines fail, the RR matrix is easily modified.

71.8.3.2.2 Resource Release Matrix

The last issue to be resolved in this design is that of *resource release*. Thus, using manufacturing engineering experience and Figure 71.23b, select the *resource release matrix* S_r in the resource release equation

$$\mathbf{r}_s = \begin{bmatrix} R1A_s \\ F1A_s \\ B1A_s \\ B2A_s \\ PA_s \\ M1A_s \end{bmatrix} = \begin{bmatrix} 0 & 0 & 0 & 0 & 1 & 0 & 0 \\ 0 & 0 & 0 & 1 & 0 & 0 & 0 \\ 0 & 1 & 0 & 0 & 0 & 0 & 0 \\ 0 & 0 & 0 & 0 & 0 & 1 & 0 \\ 0 & 0 & 0 & 0 & 0 & 0 & 1 \\ 0 & 0 & 0 & 1 & 0 & 0 & 1 \end{bmatrix} \begin{bmatrix} x_1 \\ x_2 \\ x_3 \\ x_4 \\ x_5 \\ x_6 \\ x_7 \end{bmatrix} \equiv S_r \mathbf{x} \tag{71.26}$$

where subscript s denotes a command to the workcell to start resource release. In the resource release matrix S_r, a 1 entry in position (i, j) indicates that resource i is to be released when entry j of $\mathbf{x}$ has become high (e.g., in this example, on completion of job j). It is important to note that rows containing multiple ones in S_r correspond to columns having multiple ones in F_r. For instance, the last row of S_r shows that $M1A$ is a shared resource, since it is released after *either* x_4 is high or x_7 is high; that is, after either job b or job f is complete.

71.8.3.3 Petri Net from Task Plan Matrices

It will be shown in Section 71.9 that the four task plan matrices contain all of the information needed to implement a matrix-based job coordination controller on, for instance, a programmable logic workcell controller. However, there has been much discussion of uses of *Petri nets* in task planning. It is now shown that the four task plan matrices correspond to a Petri net (PN). The job coordinator would not normally be implemented as a Petri net; however, it is straightforward to derive the PN description of a manufacturing system from the matrix controller equations, as shown by the next result.

Theorem 71.1   *(Petri net from task plan matrices) Given the four task plan matrices F_v, S_v, F_r, S_r, define the activity completion matrix F and the activity start matrix S as*

$$F = [\, F_v \quad F_r \,], \qquad S = \begin{bmatrix} S_v \\ S_r \end{bmatrix} \tag{71.27}$$

Define X as the set of elements of sequencing state vector $\mathbf{x}$, and A (activities) as the set of elements of the job and resource vectors $\mathbf{v}$ and $\mathbf{r}$. Then (A, X, F, S^T) is a Petri net.

The theorem identifies F as the input incidence matrix and S^T as the output incidence matrix of a PN, so that the PN incidence matrix is given by

$$W = S^T - F = \begin{bmatrix} S_v^T - F_v & S_r^T - F_r \end{bmatrix} \tag{71.28}$$

Based on the theorem, the PN in Figure 71.24 is easily drawn for this example. In the figure, initial markings have been added; this is accomplished by determining the *number of resources available* in the workcell. The inputs u_{D1}, u_{D2} are required for dispatching the shared resource $M1$, as discussed in Section 71.9. This theorem provides a formal technique for constructing a PN for a workcell task plan and allows all of

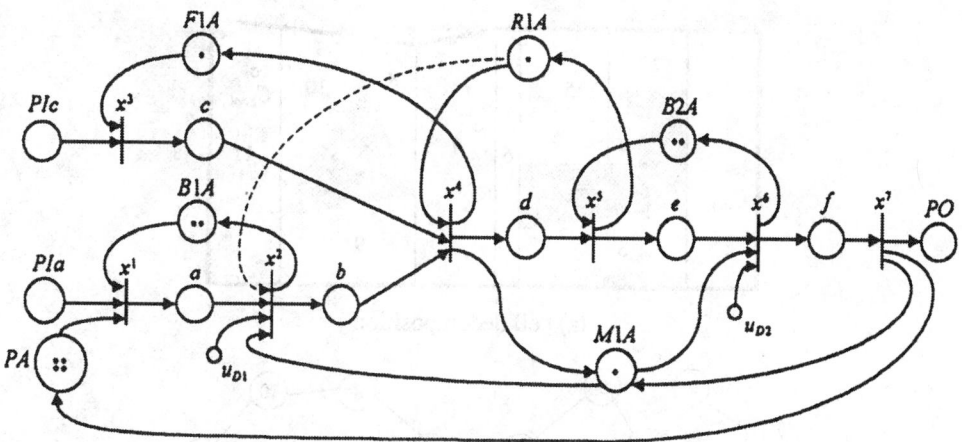

FIGURE 71.24   Petri net representation of workcell with shared resource.

the PN analysis tools to be used for analysis of the workcell plan. It formalizes some work in the literature (e.g., top-down and bottom-up design [Zhou et al. 1992]).

Behaviors. All of the PN transitions occur along the *job paths*. The places in the PN along the job paths correspond to jobs with assigned resources and can be interpreted as behaviors. The places off the task paths correspond to resource availability.

71.8.4   Path Planning

The path planning problem [Latombe 1991] may be decomposed into motion path planning, grasp planning, and error detection and recovery; only the first is considered here. *Motion path planning* is the process of finding a continuous path from an initial position to a prescribed final position or goal without collision. The output of the path planner for robotic workcells is a set of path *via points* which are fed to the machine trajectory generator (discussed previously). *Off-line path planning* can be accomplished if all obstacles are stationary at known positions or moving with known trajectories. Otherwise, *on-line or dynamic* path planning is required in real time; this often requires techniques of *collision or obstacle avoidance*. In such situations, paths preplanned off line can often be modified to incorporate collision avoidance. This subsection deals with off-line path planning except for the portion on the potential field approach, which is dynamic planning. See Zhou [1996] for more information.

Initial and final positions may be given in any coordinates, including the robot's joint space. Generally, higher level workcell components think in terms of Cartesian coordinates referred to some world frame. The Cartesian position of a robot end effector is given in terms of three position coordinates and three angular orientation coodinates; therefore, the general 3-D path planning problem occurs in $\Re^6$. If robot joint-space initial and final positions are given one may work in *configuration space*, in which points are specified by the joint variable vector $\mathbf{q}$ having coordinates $\mathbf{q}_i$, the individual joint values. For a six-degrees-of-freedom arm, configuration space is also isomorphic to $\Re^6$. Path planning may also be carried out for initial and final values of *force/torque*. In 3-D, linear force has three components and torque has three components, again placing the problem in $\Re^6$. Hybrid position/force planning is also possible. In this subsection path planning techniques are illustrated in $\Re^2$, where it is convenient to think in terms of planning paths for mobile robots in a plane.

If the number of degrees of freedom of a robot is less than six, there could be problems in that the manipulator may not be able to reach the prescribed final position and the via points generated in the planning process. Thus, the path planner must be aware of the limitations of the individual robots in its planning process; in fact, it is usually necessary to select a specific robot agent for a task prior to planning the path in order to take such limitations into account.

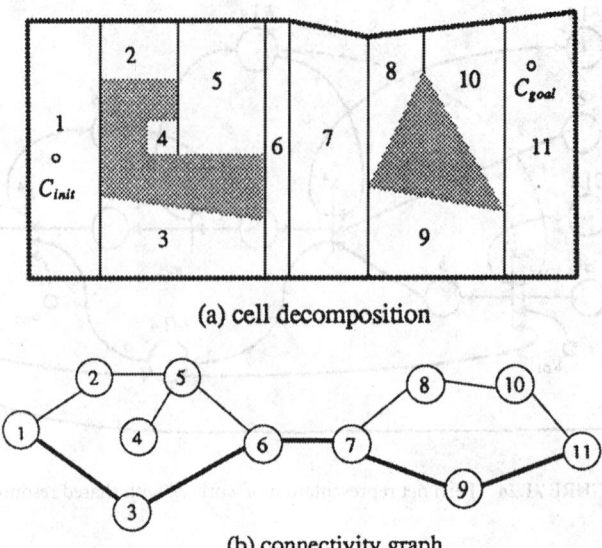

(a) cell decomposition

(b) connectivity graph

FIGURE 71.25 Cell decomposition approach to path planning: (a) free space decomposed into cells using the vertical-line-sweep method and (b) connectivity graph for the decomposed space.

71.8.4.1 Cell Decomposition Approach

In the *cell decomposition approach to path planning*, objects are enclosed in polygons. The object polygons are expanded by an amount equal to the radius of the robot to ensure collision avoidance; then, the robot is treated simply as a moving point. The free space is decomposed into simply connected free-space regions within which any two points may be connected by a straight line. When the Euclidean metric is used to measure distance, convex regions satisfy the latter requirement. A sample cell decomposition is shown in Figure 71.25. The decomposition is not unique; the one shown is generated by sweeping a vertical line across the space. Based on the decomposed space, a *connectivity graph* may be constructed, as shown in the figure. To the graph may be added weights or costs at the arcs or the nodes, corresponding to distances traveled, etc. Then, graph search techniques may be used to generate the shortest, or otherwise least costly, path.

71.8.4.2 Road Map Based on Visibility Graph

In the road map approach the obstacles are modeled as polygons expanded by the radius of the robot, which is treated simply as a moving point. A **visibility graph** is a nondirected graph whose nodes are the vertices of the polygons and whose links are straight line segments connecting the nodes without intersecting any obstacles. A *reduced visibility graph* does not contain links that are dominated by other links in terms of distance. Figure 71.26 shows a reduced visibility graph for the free space. Weights may be assigned to the arcs or nodes and graph search techniques may be used to generate a suitable path. The weights can reflect shortest distance, path smoothness, etc.

71.8.4.2.1 Road Map Based on Voronoi Diagram

A **Voronoi diagram** is a diagram where the path segment lines have equal distance from adjacent obstacles. In a polygonal space, the Voronoi diagram consists of straight lines and parabolas: when both adjacent object segments are vertices or straight lines, the equidistant line is straight, when one object is characterized by a vertex and the other by a straight line, the equidistant line is parabolic. In the Voronoi approach, generated paths are generally longer than in the visibility graph approach, but the closest point of approach (CPA) to obstacles is maximized.

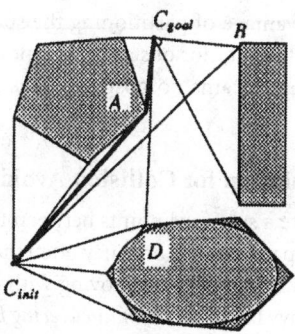

FIGURE 71.26 Road map based on visibility graph. (Courtesy of Zhou, C. 1996. Planning and intelligent control. In *CRC Handbook of Mechanical Engineering*. F. Kreith, Ed. CRC Press, Boca Raton, FL.)

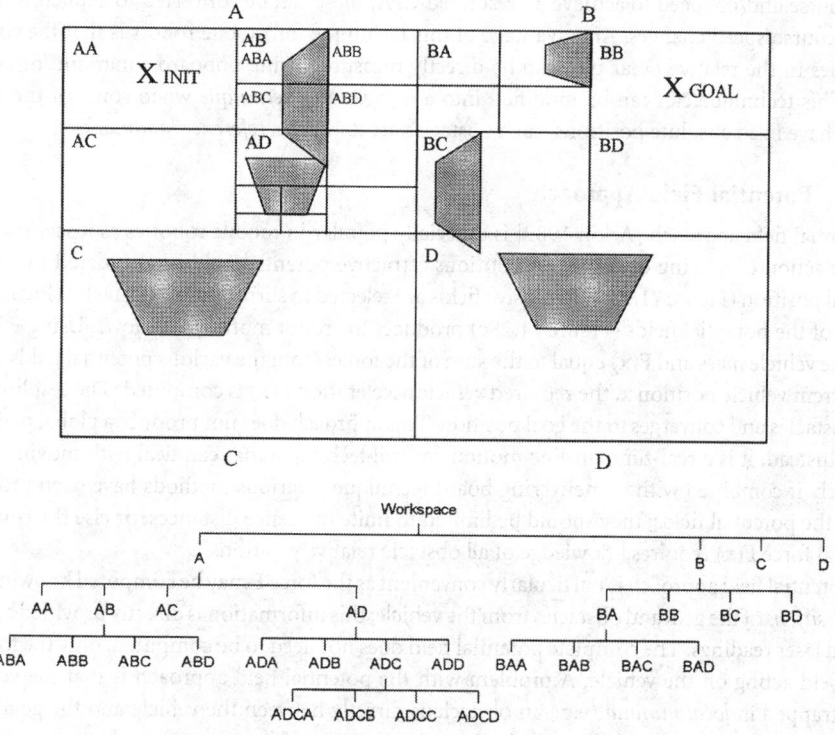

FIGURE 71.27 Quadtree approach to path planning: (a) quadtree decomposition of the work area and (b) quadtree constructed from space decomposition.

71.8.4.3 Quadtree Approach

In the quadtree approach, Figure 71.27a rectangular workspace is partitioned into four equal quadrants labeled A, B, C, D. Suppose the initial point is in quadrant A with the goal in quadrant B. If there are obstacles in quadrant A, it must be further partitioned into four quadrants AA, AB, AC, AD. Suppose the initial point is in AA, which also contains obstacles; then AA is further partitioned into quadrants AAA, AAB, AAC, AAD. This procedure terminates when there are no obstacles in the quadrant containing the initial point. A similar procedure is effected for the goal position. Based on this space decomposition, the quadtree shown in Figure 71.27b may be drawn. Now, tree search methods such as A^* may be used to determine the optimal obstacle-free path.

The quadtree approach has the advantage of partitioning the space only as finely as necessary. If any quadrant contains neither goal, initial point, nor obstacles, it is not further partitioned. If any quadrant containing the initial position or goal contains no obstacles, it is not further partitioned. In 3-D, this approach is called *octree*.

71.8.4.4   Maneuvering Board Solution for Collision Avoidance of Moving Obstacles

The techniques just discussed generate a set of via points between the initial and final positions. If there are moving obstacles within the free-space regions, one may often modify the paths between the via points online in real-time to avoid collision. If obstacles are moving with constant known velocities in the free space, a technique used by the U.S. Navy based on the *maneuvering board* can be used for on-line obstacle avoidance. Within a convex free-space region, generated for instance by the **cell decomposition** approach, one makes a *relative polar plot* with the moving robot at the center and other moving objects plotted as straight lines depending on their relative courses and speeds. A steady bearing and decreasing range (SBDR) indicates impending collision. Standard graphical techniques using a parallel ruler allow one to alter the robot's course and/or speed to achieve a prescribed CPA; these can be converted to explicit formulas for required course/speed changes. An advantage of this technique for mobile robots is that the coordinates of obstacles in the relative polar plot can be directly measured using onboard sonar and/or laser range finders. This technique can can be modified into a *navigational technique* when some of the stationary obstacles have fixed absolute positions, such obstacles are known as *reference landmarks*.

71.8.4.5   Potential Field Approach

The potential field approach [Arkin 1989] is especially popular in mobile robotics as it seems to emulate the reflex action of a living organism. A fictitious attractive potential field is considered to be centered at the goal position (Figure 71.28a). Repulsive fields are selected to surround the obstacles (Figure 71.28b). The sum of the potential fields (Figure 71.28c) produces the robot motion as follows. Using $\mathbf{F}(\mathbf{x}) = ma$, with m the vehicle mass and $\mathbf{F}(\mathbf{x})$ equal to the sum of the forces from the various potential fields computed at the current vehicle position $\mathbf{x}$, the required vehicle acceleration $a(\mathbf{x})$ is computed. The resulting motion avoids obstacles and converges to the goal position. This approach does not produce a global path planned a priori. Instead, it is a real-time on-line motion control technique that can deal with moving obstacles, particularly if combined with maneuvering board techniques. Various methods have been proposed for selecting the potential fields; they should be limited to finite influence distances, or else the computation of the total force $\mathbf{F}(\mathbf{x})$ requires knowledge of all obstacle relative positions.

The potential field approach is particularly convenient as the force $\mathbf{F}$ may be computed knowing only the *relative positions* of the goal and obstacles from the vehicle; this information is directly provided by onboard sonar and laser readings. The complete potential field does not need to be computed, only the force vector of each field acting on the vehicle. A problem with the potential field approach is that the vehicle may become trapped in *local minima* (e.g., an obstacle is directly between the vehicle and the goal); this can be corrected using various techniques, including adding a dither force to get the vehicle out of these false minima. The potential field approach can be combined with Lyapunov analysis techniques to integrate the path planning and trajectory following servocontrol functions of a mobile robot [Jagannathan et al. 1994].

In fact, Lyapunov functions and potential fields may simply be added in an overall controls design technique.

71.8.4.5.1   *Emergent Behaviors*

The responses to individual potential fields can be interpreted as *behaviors* such as seek goal, avoid obstacle, etc. Potential fields can be selected to achieve specialized behaviors such as *docking* (i.e., attaining a goal position with a prescribed angle of approach) and remaining in the center of a corridor (simply define repulsive fields from each wall). The sum of all of the potential fields yields an *emergent behavior* that has not been preprogrammed (e.g., seek goal while avoiding obstacle and remaining in the center of the hallway). This makes the robot exhibit behaviors that could be called intelligent or self-determined.

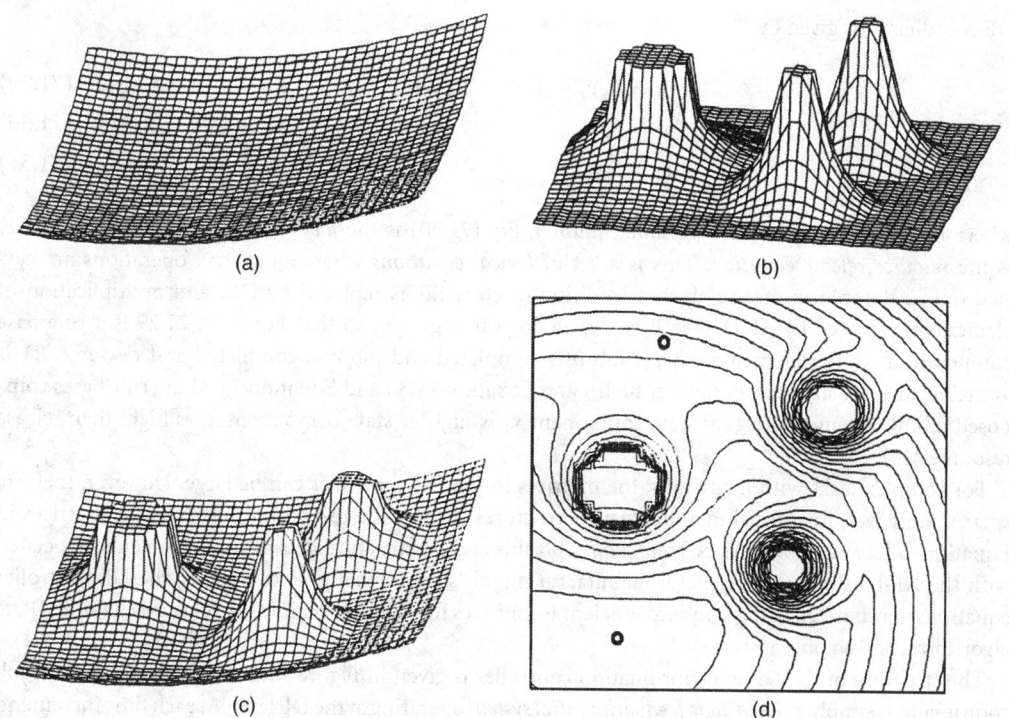

FIGURE 71.28 Potential field approach to navigation: (a) attractive field for goal at lower left corner, (b) repulsive fields for obstacles, (c) sum of potential fields, and (d) contour plot showing motion trajectory. (Courtesy of Zhou, C. 1996. Planning and intelligent control. In *CRC Handbook of Mechanical Engineering*. F. Kreith, Ed. CRC Press, Boca Raton, FL.)

71.9   Job and Activity Coordination

Coordination of workcell activities occurs on two distinct planes. On the *discrete event* (DE) or discrete activity plane, *job coordination* and sequencing, along with resource handling, is required. Digital input/output signals, or sequencing *interlocks*, are used between the workcell agents to signal job completion, resource availability, errors and exceptions, and so on. On a lower plane, servo-level motion/force coordination between multiple interacting robots is sometimes needed in special-purpose applications; this specialized topic is relegated to the end of this section.

71.9.1   Matrix Rule-Based Job Coordination Controller

The workcell DE coordinators must sequence the jobs according to the task plan, coordinating the agents and activities of the workcell. Discrete event workcell coordination occurs at two levels in Figure 71.7: coordination of *interagent* activities occurs within the job coordinator and coordination of *intra-agent* activities occurs within the virtual agent at the machine coordinator level. The latter might be considered as *reflex actions* of the virtual agent.

71.9.1.1   Rule-Based Matrix DE Coordinator

The workcell DE coordinators are easily produced using the four task plan matrices constructed by the task planner designed in the section on Petri net from task plan matrices. Given the job sequencing matrix F_v, the job start matrix S_v, the resource requirements matrix F_r, and the resource release matrix S_r, the

DE coordinator is given by

$$\bar{\mathbf{x}} = F_v \bar{v}_c + F_r \bar{r}_c + F_u \bar{u} + F_D \bar{u}_D \tag{71.29}$$

$$\mathbf{v}_s = S_v x \tag{71.30}$$

$$\mathbf{r}_s = S_r x \tag{71.31}$$

where Eq. (71.29) is the *controller state equation*, Eq. (71.30) is the *job start equation*, and Eq. (71.31) is the *resource release equation*. This is a set of *logical equations* where all matrix operations are carried out in the matrix or/and algebra; addition of elements is replaced by OR, and multiplication of elements is replaced by AND. Overbars denote logical negation, so that Equation 71.29 is a rule base composed of AND statements (e.g., if job b is completed and job c is completed and resource $R1$ is available, then set state component x_4 high), and Equation 71.30 and Equation 71.31 are rule bases composed of OR statements (e.g., if state component x_4 is high or state component x_7 is high, then release resource $M1$).

For complex tasks with many jobs, the matrices in the DE controller can be large. However, they are sparse. Moreover, for special manufacturing structures such as the re-entrant flow line, the matrices in Equation (71.29) are *lower block triangular*, and this special structure gets around problems associated with the NP-hard nature of general manufacturing job shops. Finally, as rule bases, the DE controller equations may be fired using standard efficient techniques for forward chaining, backward chaining (Rete algorithm), and so on.

The structure of the DE job coordination controller is given in Figure 71.29. This shows that the job coordinator is simply a *closed-loop feedback control system* operating at the DE level. At each time increment, workcell status signals are measured including the job complete status vector v_c (where entries of 1 denote jobs complete), the resource availability vector r_c (where entries of 1 denote resources available), and the part input vector u (where entries of 1 denote parts coming into the workcell). These signals are determined using workcell digital interlocks and digital input/output between the agents. Based on this workcell status information, the DE controller computes which jobs to start next and which resources to release. These commands are passed to the workcell in the job start vector v_s (where entries of 1 denote jobs to be started) and the resource release vector r_s (where entries of 1 denote resources to be released).

71.9.1.2 Deadlocks and Resource Dispatching Commands

Outer feedback loops are required to compute the *dispatching input* u_D. A separate dispatching input is required whenever a column of F_r contains multiple ones, indicating that the corresponding resource is a *shared resource* required for more than one job. In a manufacturing workcell, if care is not taken to assign shared resources correctly, *system deadlock* may occur. In deadlock, operation of a subsystem ceases as a *circular blocking* of resources has developed [Wysk et al. 1991]. In a circular blocking, each resource is waiting for each other resource, but none will ever again become available. In the example of Figure 71.24, a circular blocking occurs if machine $M1$ is waiting at b to be unloaded by $R1$, but $R1$ already has a part at d and the buffer $B2$ is full at e. On the other hand, buffer $B2$ cannot be unloaded at e since $M1$ already has a part at b.

There are well-established algorithms in industrial engineering for job dispatching [Panwalker and Iskander 1977], including first-in–first-out, earliest due date, last buffer first serve, etc. *Kanban systems* are pull systems where no job can be started unless a kanban card is received from a downstream job (possibly indicating buffer space or resource availability, or part requirements). A generalized kanban system that guarantees deadlock avoidance can be detailed in terms of the DE controller matrices, for it can be shown that all the *circular waits* of the workcell for a particular task are given in terms of the graph defined by $S_r F_r$, with S_r the resource release matrix and F_r the resource requirements matrix. Only circular waits can develop into circular blockings. Based on this, in Lewis and Huang [1995] a procedure known as maximum work-in-process (MAXWIP) is given that guarantees dispatching with no deadlock.

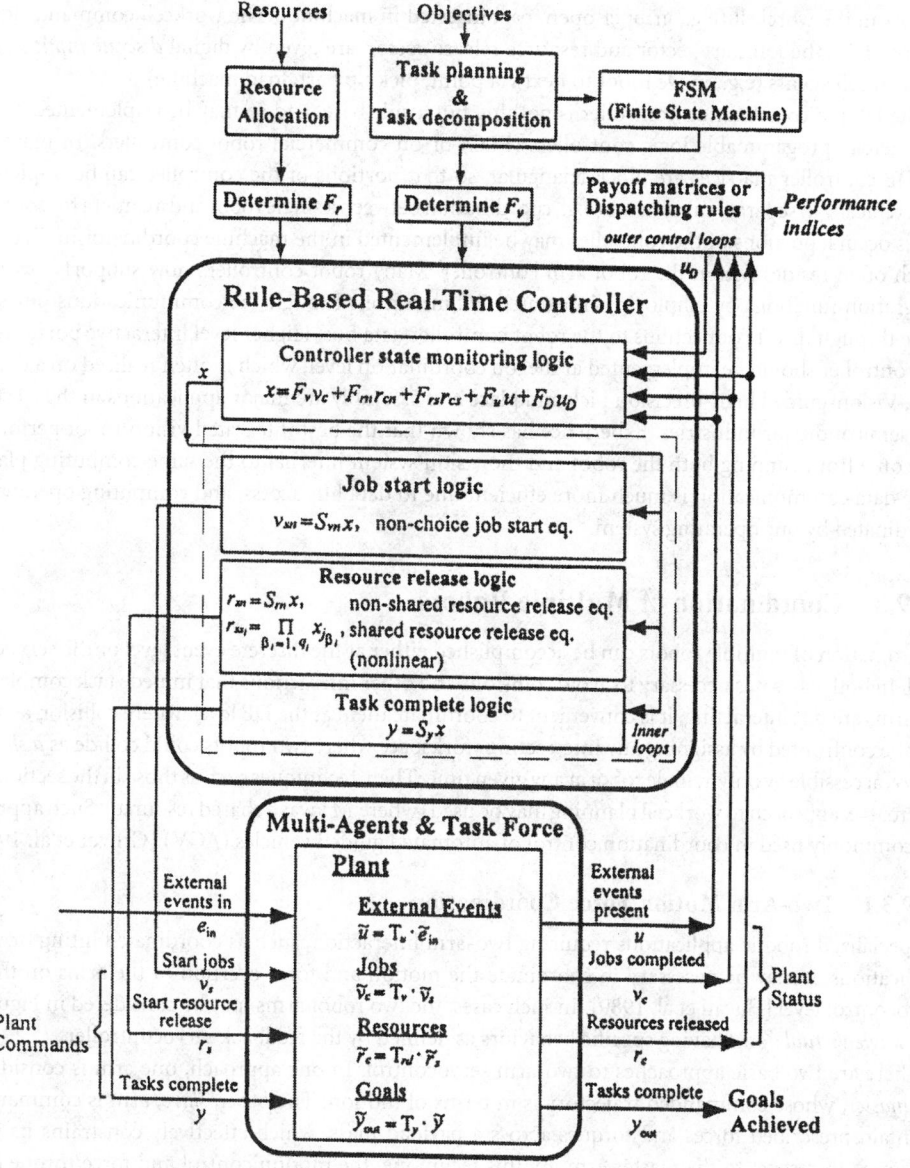

FIGURE 71.29 Matrix-based DE feedback controller for job sequencing and resource allocation.

71.9.2 Process Integration, Digital I/O, and Job Coordination Controller Implementation

Information integration is the process by which the activities and status of the various workcell agents interact. Subcategories of information integration include *sensor integration* and *sensor/actuator integration*. Motion/process integration involves coordinating manipulator motion with process sensor or process controller devices. The most primitive process integration is through discrete digital I/O, or sequencing interlocks. For example a machine controller external to the robot controller might send a 1-b signal indicating that it is ready to be loaded by the robot. The DE matrix-based job coordination controller provides an ideal technique for information integration. The workcell status signals required in Figure 71.29, the job completion vector and resource availability vector, are given by digital output signals provided by sensors in the worckell (e.g., gripper open, part inserted in machine). The workcell command signals in Figure 71.29, the job start vector and resource release vector, are given by digital *discrete input* signals to the workcell agents (e.g., move robot to next via point, pick up part, load machine).

The DE job coordination controller is nothing but a rule base, and so may be implemented either on commercial progammable logic controllers (PLC) or on commercial robot controllers. In many cases, the DE controller matrices are block triangular, so that portions of the controller can be implemented hierarchically in separate subsystems (e.g., coordination between a single robot and a camera for some jobs). If this occurs, portions of the controller may be implemented in the machine coordinator in Figure 71.7, which often resides within the robot arm controller. Many robot controllers now support information integration functions by employing integrated PC interfaces through the communications ports, or in some through direct connections to the robot controller data bus. Higher level interactive portions of the DE controller should be implemented at the job coordination level, which is often realized on a dedicated PLC. Vision-guided high-precision pick and place and assembly are major applications in the electronics and semiconductor industries. Experience has shown that the best integrated vision/robot performance has come from running both the robot and the vision system internal to the same computing platform, since data communication is much more efficient due to data bus access, and computing operations are coordinated by one operating system.

71.9.3   Coordination of Multiple Robots

Coordination of multiple robots can be accomplished either at the discrete event level or the servocontrol level. In both cases it is necessary to avoid collisions and other interactions that impede task completion. If the arms are not interacting, it is convenient to coordinate them at the DE level, where collision avoidance may be confronted by assigning any intersecting workspace where two robots could collide as *a shared resource*, accessible by only a single robot at any given time. Then, techniques such as those in the section on the task matrix approach to workcell planning may be used (where $M1$ was a shared resource). Such approaches are commonly used in coordination control of automated guided vehicles (AGV) [Gruver et al. 1984].

71.9.3.1   Two-Arm Motion/Force Coordination

In specialized robotic applications requiring two-arm interaction, such as coordinated lifting or process applications, it may be necessary to coordinate the motion and force exertion of the arms on the joint servocontrol level [Hayati et al. 1989]. In such cases, the two robot arms may be considered in Figure 71.7 as a *single virtual agent* having specific behaviors as defined by the feedback servocontroller.

There are two basic approaches to two-arm servocontrol. In one approach, one arm is considered as the *master*, whose commanded trajectory is in terms of motion. The other, *slave*, arm is commanded to maintain prescribed forces and torques across a payload mass, which effectively constrains its relative motion with respect to the master arm. By this technique, the motion control and force/torque control problems are relegated to different arms, so that the control objectives are easily accomplished by servo-level feedback controller design. Another approach to two-arm coordination is to treat both arms as equals, coordinating to maintain prescribed linear and angular motions of the center-of-gravity (c.g.) of a payload mass, as well as prescribed internal forces and torques across the payload. This approach involves complex analyses to decompose the payload c.g. motion and internal forces into the required motion of each arm; kinematic transformations and Jacobians are needed.

71.10   Error Detection and Recovery

The material in this section is modifed from Zhou [1996]. In the execution of a task, errors can occur. The errors can be classified into several categories: hardware error, software error, and operational error. The hardware errors include errors in mechanical and electrical mechanisms of the robot, such as failure in the drive system or sensing system. Software errors can be bugs in the application program or control software. Timing with cooperative devices can also be called software error. Operational errors are errors in the robot environment that are external to the robot system such as jamming of parts or collision with obstacles.

During this discussion one should keep in mind the PC&C structure in Figure 71.7. Error handling can be classified into two activities, *error detection* and *error recovery*. Error detection is composed of error sensing, interpretation, and classification. Error recovery is composed of decision making and corrective job assignment. While corrective jobs are being performed, the assigned task may be interrupted, or may continue to run at a reduced capability (e.g., one of two drilling machines may be down).

71.10.1 Error Detection

The sensors used in error detection can include all those discussed in Section 71.7 including tactile sensors for sensing contact errors, proximity sensors for sensing location or possible collision, force/torque sensors for sensing collision and jamming, and vision for sensing location, orientation and error existence. Once an error is sensed, it must be interpreted and classified. This may be accomplished by servo-level state observers, logical rule-based means, or using advanced techniques such as neural networks.

71.10.2 Error Recovery

The occurrence of an error usually causes interruption of the normal task execution. Error recovery can be done at three levels, where errors can be called *exceptions, faults, and failures*. At the lowest level the exception will be corrected automatically, generally in the real-time servocontrol loops, and the task execution continued. An example is jamming in the pin insertion problem, where a force/torque wrist sensor can indicate jamming as well as provide the information needed to resolve the problem. At the second level, the error is a fault that has been foreseen by the task planner and included in the task plan passed to the job coordinator.

The vector **x** in Equation 71.29 contains *fault states*, and logic is built into the task plan matrices to allow corrective job assignment. Upon detection of an error, jobs can be assigned to correct the fault, with the task subsequently continued from the point where the error occurred. At this level, the error detection/recovery logic can reside either in the machine coordinator or in the job coordinator.

At the highest level of recovery, the error was not foreseen by the task planner and there is no error state in the task plan. This results in a failure, where the task is interupted. Signals are sent to the planner, who must correct the failure, sometimes with external resources, and replan the task, passing another plan to the coordinator. In the worst case, manual operator intervention is needed. It can be seen that the flow of error signals proceeds upwards and of commands proceeds downwards, exactly as in the NASREM architecture in Figure 71.4.

At the lowest servocontrol level, additional sensory information is generally required for error recovery, as in the requirement for a wrist force/torque sensor in pin insertion. At the mid-level, additional logic is needed for error recovery. At the highest level, task replanning capabilities are needed.

71.11 Human Operator Interfaces

Human operator integration is critical to the expeditious setup, programming, maintenance, and sometimes operation of the robotic workcell. Especially important for effective human integration are the available human I/O devices, including the information available to the operator in graphical form and the modes of real-time control operation available for human interaction. Teaching, programming, and operational efforts are dramatically influenced by the type of user interface I/O devices available.

71.11.1 Levels of User Interface

Discounting workcell design and layout, operator interfaces occur at several levels in Figure 71.7 and may be classified into off-line and on-line activities. Off-line interfaces occur in task definition and setup, often consisting of teaching activities. In workcell management, user inputs include assignment of tasks, due dates, and so on. At the workcell planning level, user functions might be required in task planning, both in task decomposition/job sequencing and in resource assignment. Off-line CAD programs are often useful at this level. In path planning, the user might be required to teach a robot specific path via points for job accomplishment. Finally, if failures occur, a human might be required to clear the failure, reset the workcell, and restart the job sequence.

On-line user interfaces may occur at the discrete event level and the servocontrol level. In the former case, a human might perform some of the jobs requested by the job coordinator, or may be required to perform corrective jobs in handling foreseen faults. At the servocontrol level, a human might perform teleoperator functions, or may be placed in the inner feedback control loop with a machine or robotic device.

71.11.2 Mechanisms for User Interface

71.11.2.1 Interactive 3-D CAD

Computer integrated manufacturing operations require off-line programming and simulation in order to layout production facilities, model and evaluate design concepts, optimize motion of devices, avoid interference and collisions, minimize process cycle times, maximize productivity, and ensure maximum return on investment. Graphical interfaces, available on some industrial robots, are very effective for conveying information to the operator quickly and efficiently. A graphical interface is most important for design and simulation functions in applications which require frequent reprogramming and setup changes. Several very useful off-line programming software systems are available from third party suppliers (CimStation [SILMA 1992], ROBCAD, IGRIP). These systems use CAD and/or dynamics computer models of commercially available robots to simulate job execution, path motion, and process activities, providing rapid programming and virtual prototyping functions. Interactive off-line CAD is useful for assigning tasks at the management level and for task decomposition, job sequencing, and resource assignment at the task planning level.

71.11.2.2 Off-Line Robot Teaching and Workcell Programming

Commercial robot or machine tool controllers may have several operator interface mechanisms. These are generally useful at the level of off-line definition or teaching of jobs, which can then be sequenced by the job coordinator or machine coordinator to accomplish assigned tasks. At the lowest level one may program the robot in its operating language, specifying path via points, gripper open/close commands, and so on. Machine tools may require programming in CNC code. These are very tedious functions, which can be avoided by object-oriented and open architecture approaches in well-designed workcells, where such functions should be performed automatically, leaving the user free to deal with other higher level supervisory issues. In such approaches, macros or subroutines are written in machine code which *encapsulate the machine behaviors* (e.g., set speed, open gripper, go to prescribed point). Then, higher level software passes specific parameters to these routines to execute behaviors with specific location and motion details as directed by the job coordinator.

Many robots have a teach pendant, which is a low-level teleoperation device with push buttons for moving individual axes and other buttons to press commanding that certain positions should be memorized. On job execution, a playback mode is switched in, wherein the robot passes through the taught positions to sweep out a desired path. This approach is often useful for teaching multiple complex poses and Cartesian paths.

The job coordinator may be implemented on a programmable logic controller (PLC). PLC programming can be a tedious and time-consuming affair, and in well-designed flexible reconfigurable workcells an object-oriented approach is used to avoid reprogramming of PLCs. This might involve a programming scheme that takes the task plan matrices in Section 71.9 as inputs and automatically implements the coordinator using rule-based techniques (e.g., forward chaining, Rete algorithm).

71.11.2.3   Teleoperation and Man-in-the-Loop Control

Operator interaction at the servocontrol level can basically consist of two modes. In man-in-the-loop control, a human provides or modifies the feedback signals that control a device, actually operating a machine tool or robotic device. In teleoperation, an inner feedback loop is closed around the robot, and a human provides motion trajectory and force commands to the robot in a master/slave relationship. In such applications, there may be problems if extended communications distances are involved, since delays in the communications channel can destabilize a teleoperation system having force feedback unless careful attention is paid to designing the feedback loops to maintain *passivity*. See Lewis [1996] for more details.

71.12   Robot Workcell Programming

The robotic workcell requires programming at several levels [Leu 1985]. At the lower levels one generally uses commercial programing languages peculiar to device manufacturers of robots and CNC machine tools. At the machine coordination level, robot controllers are also often used with discrete I/O signals and decision making commands. At the job coordination level prorammable logic controllers (PLCs) are often used in medium complexity workcells, so that a knowledge of PLC programming techniques is required. In modern manufacturing and process workcells, coordination may be accomplished using general purpose computers with programs written, for instance, in C.

71.12.1   Robot Programming Languages

Subsequent material in this section is modified from Bailey [1996]. Each robot manufacturer has its own proprietary programming language. The variety of motion and position command types in a programming language is usually a good indication of the robot's motion generation capability. Program commands which produce complex motion should be available to support the manipulation needs of the application. If palletizing is the application, then simple methods of creating position commands for arrays of positions are essential. If continuous path motion is needed, an associated set of continuous motion commands should be available. The range of motion generation capabilities of commercial industrial robots is wide. Suitability for a particular application can be determined by writing test code.

The earliest industrial robots were simple sequential machines controlled by a combination of servomotors, adjustable mechanical stops, limit switches, and PLCs. These machines were generally programmed by a record and play-back method with the operator using a teach pendant to move the robot through the desired path. MHI, the first robot programming language, was developed at Massachusetts Institute of Technology (MIT) during the early 1960s. MINI, developed at MIT during the mid-1970s was an expandable language based on LISP. It allowed programming in Cartesian coordinates with independent control of multiple joints. VAL and VAL II [Shimano et al. 1984], developed by Unimation, Inc., were interpreted languages designed to support the PUMA series of industrial robots. **A manufacturing language** (AML) was a completely new programming language developed by IBM to support the R/S 1 assembly robot. It was a subroutine-oriented, interpreted language which ran on the Series/1 minicomputer. Later versions were compiled to run on IBM compatible personal computers to support the 7535 series of SCARA robots. Several additional languages [Gruver et al. 1984, Lozano-Perez 1983] were introduced during the late 1980s to support a wide range of new robot applications which were developed during this period.

71.12.2   V+, A Representative Robot Language

V+, developed by Adept Technologies, Inc., is a representative modern robot programming language with several hundred program instructions and reserved keywords. V+ will be used to demonstrate important features of robot programming. Robot program commands fall into several categories, as detailed in the following subsections.

71.12.2.1   Robot Control

Program instructions required to control robot motion specify location, trajectory, speed, acceleration, and obstacle avoidance. Examples of V+ robot control commands are as follows:

MOVE:	Move the robot to a new location.
DELAY:	Stop the motion for a specified period of time.
SPEED:	Set the speed for subsequent motions.
ACCEL:	Set the acceleration and deceleration for subsequent motions.
OPEN:	Open the hand.
CLOSE:	Close the hand.

71.12.2.2   System Control

In addition to controlling robot motion, the system must support program editing and debugging, program and data manipulation, program and data storage, program control, system definitions and control, system status, and control/monitoring of external sensors. Examples of V+ control instructions are as follows:

EDIT:	Initiate line-oriented editing.
STORE:	Store information from memory onto a disk file.
COPY:	Copy an existing disk file into a new program.
EXECUTE:	Initiate execution of a program.
ABORT:	Stop program execution.
DO:	Execute a single program instruction.
WATCH:	Set and clear breakpoints for diagnostic execution.
TEACH:	Define a series of robot location variables.
CALIBRATE:	Initiate the robot positioning system.
STATUS:	Display the status of the system.
ENABLE:	Turn on one or more system switches.
DISABLE:	Turn off one or more system switches.

71.12.2.3   Structures and Logic

Program instructions are needed to organize and control execution of the robot program and interaction with the user. Examples include familiar commands such as FOR, WHILE, IF as well as commands like the following:

WRITE:	Output a message to the manual control pendant.
PENDANT:	Receive input from the manual control pendant.
PARAMETER:	Set the value of a system parameter.

71.12.2.4   Special Functions

Various special functions are required to facilitate robot programming. These include mathematical expressions such as COS, ABS, and SQRT, as well as instructions for data conversion and manipulation, and kinematic transformations such as the following:

BCD:	Convert from real to binary coded decimal.
FRAME:	Compute the reference frame based on given locations.
TRANS:	Compose a transformation from individual components.
INVERSE:	Return the inverse of the specified transformation.

71.12.2.5 Program Execution

Organization of a program into a sequence of executable instructions requires scheduling of tasks, control of subroutines, and error trapping/recovery. Examples include the following:

PCEXECUTE:	Initiate the execution of a process control program.
PCABORT:	Stop execution of a process control program.
PCPROCEED:	Resume execution of a process control program.
PCRETRY:	After an error, resume execution at the last step tried.
PCEND:	Stop execution of the program at the end of the current execution cycle.

71.12.2.6 Example Program

This program demonstrates a simple pick and place operation. The values of position variables *pick* and *place* are specified by a higher level executive that then initiates this subroutine:

```
 1  .PROGRAM move.parts()
 2  ;                          Pick up parts at location "pick" and put them down at "place"
 3  parts = 100                                  ; Number of parts to be processed
 4  height1 = 25                                 ; Approach/depart height at "pick"
 5  height2 = 50                                 ; Approach/depart height at "place"
 6  PARAMETER.HAND.TIME = 16                      ; Setup for slow hand
 7  OPEN                                          ; Make sure hand is open
 8  MOVE start                                    ; Move to safe starting location
 9  For i = 1 TO parts                            ; Process the parts
10  APPRO pick, height1                           ; Go toward the pick-up
11  MOVES pick                                    ; Move to the part
12  CLOSEI                                        ; Close the hand
13  DEPARTS height1                               ; Back away
14  APPRO place, height2                          ; Go toward the put-down
15  MOVES place                                   ; Move to the destination
16  OPENI                                         ; Release the part
17  DEPARTS height2                               ; Back away
18  END                                           ; Loop for the next part
19  TYPE "ALL done.", /I3, parts, "parts processed"
20  STOP                                          ; End of the program
21  .END
```

71.12.2.7 Off-Line Programming and Simulation

Commercially available software packages (discussed in Section 71.11) provide support for off-line design and simulation of 3-D worccell layouts including robots, end effectors, fixtures, conveyors, part positioners, and automatic guided vehicles. Dynamic simulation allows off-line creation, animation, and verification of robot motion programs. However, these techniques are limited to verification of overall system layout and preliminary robot program development. With support for data exchange standards [e.g., **International Graphics Exchange Specification (IGES)**, **Virtual Data Acquisition and File Specification (VDAFS)**, **Specification for Exchange of Text (SET)**], these software tools can pass location and trajectory data to a robot control program, which in turn can provide the additional functions required for full operation (operator guidance, logic, error recovery, sensor monitoring/control, system management, etc.).

71.13 Mobile Robots and Automated Guided Vehicles

A topic which has always intrigued computer scientists is that of **mobile robots** [Zheng 1993]. These machines move in generally unstructured environments and so require enhanced decision making and sensors; they seem to exhibit various anthropomorphic aspects since vision is often the sensor, decision making mimics brain functions, and mobility is similar to humans, particularly if there is an onboard robot arm attached. Here are discussed mobile robot research and factory automated guided vehicle (AGV) systems, two widely disparate topics.

71.13.1 Mobile Robots

Unfortunately, in order to focus on higher functions such as decision making and high-level vision processing, many researchers treat the mobile robot as a dynamical system obeying Newton's laws $F = ma$ (e.g., in the potential field approach to motion control, discussed earlier). This simplified dynamical representation does not correspond to the reality of moving machinery which has nonholonomic constraints, unknown masses, frictions, Coriolis forces, drive train **compliance**, wheel slippage, and backlash effects. In this subsection we provide a framework that brings together three camps: computer science results based on the $F = ma$ assumption, nonholonomic control results that deal with a kinematic steering system, and full servo-level feedback control that takes into account all of the vehicle dynamics and uncertainties.

71.13.1.1 Mobile Robot Dynamics

The full dynamical model of a rigid mobile robot (e.g., no flexible modes) is given by

$$M(\mathbf{q})\ddot{\mathbf{q}} + V_m(\mathbf{q}, \dot{\mathbf{q}})\dot{\mathbf{q}} + F(\mathbf{q}, \dot{\mathbf{q}}) + \mathbf{G}(\mathbf{q}) + \tau_d = B(\mathbf{q})\tau - A^T(\mathbf{q})\lambda \qquad (71.32)$$

which should be compared to Equation 71.7 and Equation 71.14. In this equation, M is an inertia matrix, V_m is a matrix of Coriolis and centripetal terms, $\mathbf{F}$ is a friction vector, $\mathbf{G}$ is a gravity vector, and τ_d is a vector of disturbances. The n-vector $\tau(t)$ is the control input. The dynamics of the driving and steering motors should be included in the robot dynamics, along with any gearing. Then, τ might be, for example, a vector of voltage inputs to the drive actuator motors.

The vehicle variable $\mathbf{q}(t)$ is composed of Cartesian position (x, y) in the plane plus orientation θ. If a robot arm is attached, it can also contain the vector of robot arm joint variables. A typical mobile robot with no onboard arm has $\mathbf{q} = [x\ y\ \theta]^T$, where there are three variables to control, but only two inputs, namely, the voltages into the left and right driving wheels (or, equivalently, vehicle speed and heading angle).

The major problems in control of mobile robots are the fact that there are more degrees of freedom than control inputs, and the existence of nonholonomic constraints.

71.13.1.2 Nonholonomic Constraints and the Steering System

In Equation 71.12 the vector of constraint forces is λ and matrix $A(\mathbf{q})$ is associated with the constraints. These may include nonslippage of wheels and other *holonomic* effects, as well as the *nonholonomic constraints*, which pose one of the major problems in mobile robot control. Nonholonomic constraints are those which are nonintegrable, and include effects such as the impossibility of sideways motion (think of an automobile). In research laboratories, it is common to deal with omnidirectional robots that have no nonholonomic constraints, but can rotate and translate with full degrees of freedom; such devices do not correspond to the reality of existing shop floor or cross-terrain vehicles which have nonzero turn radius.

A general case is where all kinematic equality constraints are independent of time and can be expressed as

$$A(\mathbf{q})\dot{\mathbf{q}} = 0 \qquad (71.33)$$

Let $S(\mathbf{q})$ be a full-rank basis for the nullspace of $A(\mathbf{q})$ so that $AS = 0$. Then one sees that the linear and angular velocities are given by

$$\dot{\mathbf{q}} = S(\mathbf{q})\mathbf{v(t)} \qquad (71.34)$$

where $\mathbf{v(t)}$ is an auxiliary vector. In fact, $\mathbf{v(t)}$ often has physical meaning, consisting of two components: the commanded vehicle speed and the heading angle. Matrix $S(\mathbf{q})$ is easily determined independently of the dynamics 71.32 from the wheel configuration of the mobile robot. Thus, Equation 71.34 is a *kinematic* equation that expresses some simplified relations between motion $\mathbf{q}(t)$ and a fictitious *ideal speed and heading* vector $\mathbf{v}$. It does not include dynamical effects, and is known in the nonholonomic literature as the *steering system*. In the case of omnidirectional vehicles $S(\mathbf{q})$ is 3×3 and Equation 71.34 corresponds to the Newton's law model $F = ma$ used in, e.g., potential field approaches.

There is a large literature on selecting the command $\mathbf{v}(t)$ to produce desired motion $\mathbf{q}(t)$ in nonholonomic systems; the problem is that $\mathbf{v}$ has two components and $\mathbf{q}$ has three. Illustrative references include the chapters by Yamamato and Yun and by Canudas de Wit et al. in Zheng [1993], as well as Samson and Ait-Abderrahim [1991]. There are basically three problems considered in this work: following a prescribed path, tracking a prescribed trajectory (e.g., a path with prescribed transit times), and stabilization at a prescribed final docking position (x, y) and orientation θ. Single vehicle systems as well as multibody systems (truck with multiple trailers) are treated. The results obtained are truly remarkable and are in the vein of a path including the forward/backward motions necessary to park a vehicle at a given docking position and orientation. All of the speed reversals and steering commands are automatically obtained by solving certain coupled nonlinear equations. This is truly the meaning of intelligence and autonomy.

71.13.1.3 Conversion of Steering System Commands to Actual Vehicle Motor Commands

The steering system command vector obtained from the nonholonomic literature may be called $\mathbf{v}_c(\mathbf{t})$, the ideal desired value of the speed/heading vector $\mathbf{v}(\mathbf{t})$. Under the so-called perfect velocity assumption the actual vehicle velocity $\mathbf{v}(\mathbf{t})$ follows the command vector $\mathbf{v}_c(\mathbf{t})$, and can be directly given as control input to the vehicle. Unfortunately, in real life this assumption does not hold. One is therefore faced with the problem of obtaining drive wheel and steering commands for an actual vehicle from the steering system command $\mathbf{v}_c(\mathbf{t})$.

To accomplish this, premultiply Equation 71.32 by $S^T(\mathbf{q})$ and use Equation 71.34 to obtain

$$\overline{M}(\mathbf{q})\dot{\mathbf{v}} + \overline{V}_m(\mathbf{q}, \dot{\mathbf{q}})\mathbf{v} + \overline{\mathbf{F}}(\mathbf{v}) + \overline{\tau}_d = \overline{B}(\mathbf{q})\tau \tag{71.35}$$

where gravity plays no role and so has been ignored, the constraint term drops out due to the fact that $AS = 0$, and the overbar terms are easily computed in terms of original quantities. The true model of the vehicle is thus given by combining both Equation 71.34 and Equation 71.35. However, in the latter equation it turns out that $(\overline{B})(\mathbf{q})$ is square and invertible, so that standard computed torque techniques (see section on robot servo-level motion control) can be used to compute the required vehicle control τ from the steering system command $\mathbf{v}_c(\mathbf{t})$. In practice, correction terms are needed due to the fact that $\mathbf{v} \neq \mathbf{v}_c$; they are computed using a technique known as *integrator backstepping* [Fierro and Lewis 1995].

The overall controller for the mobile robot is similar in structure to the multiloop controller in Figure 71.13, with an inner nonlinear **feedback linearization** loop (e.g., computed torque) and an outer tracking loop that computes the steering system command. The robustifying term is computed using backstepping. Adaptive control and neural net control inner loops can be used instead of computed torque to reject uncertainties and provide additional dynamics learning capabilities. Using this multiloop control scheme, the idealized control inputs provided, e.g., by potential field approaches, can also be converted to actual control inputs for any given vehicle. A major criticism of potential field approaches has been that they do not take into account the vehicle nonholonomic constraints.

71.13.2 Automated Guided Vehicle Systems

Though research in mobile robots is intriguing, with remarkable results exhibiting intelligence at the potential field planning level, the nonholonomic control level, and elsewhere, few of these results make their way into the factory or other unstructured environments. There, reliability and repeatability are the main issue of concern.

71.13.2.1 Navigation and Job Coordination

If the environment is unstructured one may either provide sophisticated planning, decision making, and control schemes or one may force structure onto the environment. Thus, in most AGV systems the vehicles are guided by wires buried in the floor or stripes painted on the floor. Antennas buried periodically in the floor provide check points for the vehicle as well as transmitted updates to its commanded job sequence.

A single computer may perform scheduling and routing of multiple vehicles. Design of this coordinating controller is often contorted and complex in actual installed systems, which may be the product of several engineers working in an ad hoc fashion over several years of evolution of the system. To simplify and unify design, the discrete event techniques in the task matrix approach section may be used for planning. Track intersections should be treated as shared resources only accessible by a single vehicle at a time, so that on-line dispatching decisions are needed. The sequencing controller is then implemented using the approach in Section 71.9.

71.13.2.2 Sensors, Machine Coordination, and Servo-level Control

Autonomous vehicles often require extensive sensor suites. There is usually a desire to avoid vision systems and use more reliable sensors including contact switches, proximity detectors, laser rangefinders, sonar, etc. Optical bar codes are sometimes placed on the walls; these are scanned by the robot so it can update its absolute position. Integrating this multitude of sensors and performing coordinated activities based on their readings may be accomplished using simple decision logic on low-level microprocessor boards. Servo-level control consists of simple PD loops that cause the vehicle to follow commanded speeds and turn commands. Distance sensors may provide information needed to maintain minimum safe intervehicular spacing.

Defining Terms

Accuracy: The degree to which the actual and commanded position (of, e.g., a robot manipulator) correspond.

Adaptive control: A large class of control algorithms where the controller has its own internal dynamics and so is capable of learning the unknown dynamics of the robot arm, thus improving performance over time.

A manufacturing language (AML): A robot programming language.

Automatic programming of tools (APT): A robot programming language.

Cell decomposition: An approach to path planning where the obstacles are modeled as polygons and the free space is decomposed into cells such that a straight line path can be generated between any two points in a cell.

Compliance: The inverse of stiffness, useful in end effectors and tooling whenever a robot must interact with rigid constraints in the environment.

Computed torque control: An important and large class of robot arm controller algorithms that relies on subtracting out some or most of the dynamical nonlinearities using feedforward compensation terms including, e.g., gravity, friction, Coriolis, and desired acceleration feedforward.

End effector: Portion of robot (typically at end of chain of links) designed to contact the external world.

Feedback linearization: A modern approach to robot arm control that formalizes computed torque control mathematically, allowing formal proofs of stability and design of advanced algorithms using Lyapunov and other techniques.

Force control: A class of algorithms allowing control over the force applied by a robot arm, often in a direction normal to a prescribed surface while the position trajectory is controlled in the plane of the surface.

Forward kinematics: Identification of Cartesian task coordinates given robot joint configuration.

International Graphics Exchange Specification (IGES): A data exchange standard.

Inverse kinematics: Identification of possible robot joint configurations given desired Cartesian task coordinates.

Joint variables: Scalars specifying position of each joint, one for each degree of freedom. The joint variable for a revolute joint is an angle in degrees; the joint variable for a prismatic joint is an extension in units of length.

Learning control: A class of control algorithms for repetitive motion applications (e.g., spray painting) where information on the errors during one run is used to improve performance during the next run.

Linearity in the parameters: A property of the robot arm dynamics, important in adaptive controller design, where the nonlinearities are linear in the unknown parameters such as unknown masses and friction coefficients.

Manipulator Jacobian: A configuration-dependent matrix relating joint velocities to Cartesian coordinate velocities.

Mechanical part feeders: Mechanical devices for feeding parts to a robot with a specified frequency and orientation. They are classified as vibratory bowl feeders, vibratory belt feeders, and programmable belt feeders.

Mobile robot: A special type of manipulator which is not bolted to the floor but can move. Based on different driving mechanisms, mobile robots can be further classified as wheeled mobile robots, legged mobile robots, treaded mobile robots, underwater mobile robots, and aerial vehicles.

Path planning: The process of finding a continuous path from an initial robot configuration to a goal configuration without collision.

Prismatic joint: Sliding or telescoping robot joint that produces relative translation of the connected links.

Proportional-integral-derivative (PID) control: A classical servocontrol feedback algorithm where the actual system output is subtracted from the desired output to obtain a tracking error. Then, a weighted linear combination of the tracking error, its derivative, and its integral are used as the control input to the system.

Remote-center-of-compliance (RCC): A compliant wrist or end effector designed so that task-related forces and moments produce deflections with a one-to-one correspondence (i.e., without side effects). This property simplifies programming of assembly and related tasks.

Repeatability: The degree to which the actual positions resulting from two repeated commands to the same position (of, e.g., a robot manipulator) correspond.

Revolute joint: Rotary robot joint producing relative rotation of the connected links.

Robot axis: A direction of travel or rotation usually associated with a degree of freedom of motion.

Robot joint: A mechanism that connects the structural links of a robot manipulator together while allowing relative motion.

Robot link: The rigid structural elements of a robot manipulator that are joined to form an arm.

Robust control: A large class of control algorithms where the controller is generally nondynamic, but contains information on the maximum possible modeling uncertainties so that the tracking errors are kept small, often at the expense of large control effort. The tracking performance does not improve over time so the errors never go to zero.

SCARA: Selectively compliant assembly robot arm.

Singularity: Configuration for which the manipulator Jacobian has less than full rank.

Skew symmetry: A property of the dynamics of rigid-link robot arms, important in controller design, stating that $\dot{M} - \frac{1}{2}V_m$ is skew symmetric, with M the inertia matrix and V_m the Coriolis/centripetal matrix. This is equivalent to stating that the internal forces do no work.

Specification for Exchange of Text (SET): A data exchange standard.

Task coordinates: Variables in a frame most suited to describing the task to be performed by manipulator. They are generally taken as Cartesian coordinates relative to a base frame.

Virtual Data Acquisition and File Specification (VDAFS): A data exchange standard.

Visibility graph: A road map approach to path planning where the obstacles are modeled as polygons. The visibility graph has nodes given by the vertices of the polygons, the initial point, and the goal point. The links are straight line segments connecting the nodes without intersecting any obstacles.

Voronoi diagram: A road map approach to path planning where the obstacles are modeled as polygons. The Voronoi diagram consists of line as having an equal distance from adjacent obstacles; it is composed of straight lines and parabolas.

References

Albus, J. S. 1992. A reference model architecture for intelligent systems design. In *An Introduction to Intelligent and Autonomous Control.* P. J. Antsaklis and K. M. Passino, Eds., pp. 27–56. Kluwer, Boston, MA.

Antsaklis, P. J. and Passino, K. M. 1992. *An Introduction to Intelligent and Autonomous Control.* Kluwer, Boston, MA.

Arkin, R. C. 1989. Motor schema-based mobile robot navigation. *Int. J. Robotic Res.* 8(4):92–112.

Bailey, R. 1996. Robot programming languages. In *CRC Handbook of Mechanical Engineering.* F. Kreith, Ed. CRC Press, Boca Raton, FL.

Brooks, R. A. 1986. A robust layered control system for a mobile robot. *IEEE. J. Robotics Automation.* RA-2(1):14–23.

Craig, J. 1989. *Introduction to Robotics: Mechanics and Control.* Addison-Wesley, New York.

Decelle, L. S. 1988. Design of a robotic workstation for component insertions. *AT&T Tech. J.* 67(2): 15–22.

Edkins, M. 1983. Linking industrial robots and machine tools. *Robotic Technology.* A. Pugh, Ed. IEE control engineering ser. 23, Pergrinus, London.

Fierro, R. and Lewis, F. L. 1995. Control of a nonholonomic mobile robot: backstepping kinematics into dynamics, pp. 3805–3810. *Proc. IEEE Conf. Decision Control.* New Orleans, LA., Dec.

Fraden, J. 1993. *AIP Handbook Of Modern Sensors, Physics, Design, and Applications.* American Institute of Physics, College Park, MD.

Fu, K. S., Gonzalez, R. C., and Lee, C. S. G. 1987. *Robotics.* McGraw–Hill, New York.

Ghosh, B., Jankovic, M., and Wu, Y. 1994. Perspective problems in systems theory and its application in machine vision. *J. Math. Sys. Estim. Control.*

Groover, M. P., Weiss, M., Nagel, R. N., and Odrey, N. G. 1986. *Industrial Robotics.* McGraw–Hill, New York.

Gruver, W. A., Soroka, B. I., and Craig, J. J. 1984. Industrial robot programming languages: a comparative evaluation. *IEEE Trans. Syst., Man, Cybernetics* SMC-14(4).

Jagannathan, S., Lewis, F., and Liu, K. 1994. Motion control and obstacle avoidance of a mobile robot with an onboard manipulator. *J. Intell. Manuf.* 5:287–302.

Jamshidi, M., Lumia, R., Mullins, J., and Shahinpoor, M. 1992. *Robotics and Manufacturing: Recent Trends in Research, Education, and Applications,* Vol. 4. ASME Press, New York.

Hayati, S., Tso, K., and Lee, T. 1989. Dual arm coordination and control. *Robotics* 5(4):333–344.

Latombe, J. C. 1991. *Robot Motion Planning,* Kluwer Academic, Boston, MA.

Lee, K.-M. and Li, D. 1991. Retroreflective vision sensing for generic part presentation. *J. Robotic Syst.* 8(1):55–73.

Lee, K.-M. and Blenis, R. 1994. Design concept and prototype development of a flexible integrated vision system. *J. Robotic Syst.* 11(5):387–398.

Leu, M. C. 1985. Robotics software systems. *Rob. Comput. Integr. Manuf.* 2(1):1–12.

Lewis, F. 1996. Robotics. In *CRC Handbook of Mechanical Engineering,* Ed. F. Kreith. CRC Press, Boca Raton, FL.

Lewis, F. L., Abdallah, C. T., and Dawson, D. M. 1993. *Control of Robot Manipulators.* Macmillan, New York.

Lewis, F. L. and Huang, H.-H. 1995. Manufacturing dispatching controller design and deadlock avoidance using a matrix equation formulation, pp. 63–77. *Proc. Workshop Modeling, Simulation, Control Tech. Manuf.* SPIE Vol. 2596. R. Lumia, organizer. Philadelphia, PA. Oct.

Lewis, F. L., Liu, K., and Yeşildirek, A. 1995. Neural net robot controller with guaranteed tracking performance. *IEEE Trans. Neural Networks* 6(3):703–715.

Lozano-Perez, T. 1983. Robot programming. *Proc. IEEE* 71(7):821–841.

Nichols, H. R. and Lee, M. H. 1989. A survey of robot tactile sensing technology. *Int. J. Robotics Res.* 8(3):3–30.

Panwalker, S. S. and Iskander, W. 1977. A survey of scheduling rules. *Operations Res.* 26(1):45–61.

Pertin-Trocaz, J. 1989. Grasping: a state of the art. In *The Robotics Review 1*. O. Khatib, J. Craig and T. Lozano-Perez, Eds., pp. 71–98. MIT Press, Cambridge, MA.

Pugh, A., ed. 1983. *Robotic Technology*. IEE control engineering ser. 23, Pergrinus, London.

Samson, C. and Ait-Abderrahim, K. 1991. Feedback control of a nonholonomic wheeled cart in Cartesian space, pp. 1136–1141. *Proc. IEEE Int. Conf. Robotics Automation*. April.

Saridis, G. N. 1996. Architectures for intelligent control. In *Intelligent Control Systems*. M. M. Gupta and R. Sinha, Eds. IEEE Press, New York.

Shimano, B. E., Geschke, C. C., and Spalding, C. H., III. 1984. Val-II: a new robot control system for automatic manufacturing, pp. 278–292. *Proc. Int. Conf. Robotics*, March 13–15.

SILMA. 1992. SILMA CimStation Robotics Technical Overview, SILMA Inc., Cupertino, CA.

Snyder, W. E. 1985. *Industrial Robots: Computer Interfacing and Control*. Prentice–Hall, Englewood Cliffs, NJ.

Spong, M. W. and Vidyasagar, M. 1989. *Robot Dynamics and Control*. Wiley, New York.

Tzou, H. S. and Fukuda, T. 1992. *Precision Sensors, Actuators, and Systems*. Kluwer Academic, Boston, MA.

Warfield, J. N. 1973. Binary matrices in system modeling. *IEEE Trans. Syst. Man, Cybernetics*. SMC-3(5):441–449.

Wolter, J., Chakrabarty, S., and Tsao, J. 1992. Methods of knowledge representation for assembly planning, pp. 463–468. *Proc. NSF Design and Manuf. Sys. Conf.* Jan.

Wright, P. K. and Cutkosky, M. R. 1985. Design of grippers. In *The Handbook of Industrial Robotics*, S. Nof, Ed. Chap. 21. Wiley, New York.

Wysk, R. A., Yang, N. S., and Joshi, S. 1991. Detection of deadlocks in flexible manufacturing cells. *IEEE Trans. Robotics Automation* 7(6):853–859.

Zheng, Y. F., ed. 1993. *Recent Trends in Mobile Robots*. World Scientific, Singapore.

Zhou, C. 1996. Planning and intelligent control. In *CRC Handbook of Mechanical Engineering*. F. Kreith, Ed. CRC Press, Boca Raton, FL.

Zhou, M.-C., DiCesare, F., and Desrochers, A. D. 1992. A hybrid methodology for synthesis of Petri net models for manufacturing systems. *IEEE Trans. Robotics Automation* 8(3):350–361.

Further Information

For further information one is referred to the chapter on "Robotics" by F. L. Lewis in the *CRC Handbook of Mechanical Engineering*, edited by F. Kreith, CRC Press, 1996. Also useful are robotics books by Craig (1989), Lewis, Abdallah, and Dawson (1993), and Spong and Vidyasagar (1989).

VIII

Net-Centric Computing

The rapid evolution of the World Wide Web in the last decade has had enormous impact on the priorities for computer science research and application development. NSF's recent initiatives in this area are labeled "cyberinfrastructure," which has provided major support for research on the design and performance of the Web and its various uses. The chapters in this section encapsulate fundamental aspects of network organization, routing, security, and privacy concerns. They also cover contemporary issues and applications such as data mining, data compression, and malicious software (viruses and worms) and its detection.

72 **Network Organization and Topologies**   *William Stallings* **72**-1
Transmission Control Protocol/Internet Protocol and Open Systems
Interconnection • Network Organization

73 **Routing Protocols**   *Radia Perlman* **73**-1
Introduction • Bridges/Switches • Routers

74 **Network and Internet Security**   *Steven Bellovin* **74**-1
Introduction • General Threats • Routing • The Transmission Control Protocol/Internet
Protocol (TCP/IP) Protocol Suite • The World Wide Web • Using Cryptography
• Firewalls • Denial of Service Attacks • Conclusions

75 **Information Retrieval and Data Mining**   *Katherine G. Herbert,
Jason T.L. Wang, and Jianghui Liu* **75**-1
Introduction • Information Retrieval • Data Mining • Integrating IR and DM
Techniques into Modern Search Engines • Conclusion and Further Resources

76 **Data Compression**   *Z. Rahman* **76**-1
Introduction • Lossless Compression • Lossy Compression • Conclusion

77 **Security and Privacy**   *Peter G. Neumann* **77**-1
Introduction • Conclusions • Recommendations

78 **Malicious Software and Hacking**   *David Ferbrache and Stuart Mort* **78**-1
Background • Culture of the Underground • Techniques and Countermeasures
• The Future

79 **Authentication, Access Control, and Intrusion Detection**   *Ravi S. Sandhu
and Pierangela Samarati* **79**-1
Introduction • Authentication • Access Control • Auditing and Intrusion
Detection • Conclusion

72

Network Organization and Topologies

72.1   Transmission Control Protocol/Internet Protocol
and Open Systems Interconnection **72**-1
The Transmission Control Protocol/Internet Protocol
Architecture • The Open Systems Interconnection Model

72.2   Network Organization **72**-8
Traditional Wide-Area Networks • High-Speed Wide-Area
Networks • Traditional Local-Area Networks • High-Speed
Local-Area Networks

William Stallings
Consultant and Writer

72.1   Transmission Control Protocol/Internet Protocol and Open Systems Interconnection

In this chapter, we examine the communications software needed to interconnect computers, workstations, servers, and other devices across networks. Then we look at some of the networks in contemporary use. When communication is desired among computers from different vendors, the software development effort can be a nightmare. Different vendors use different data formats and data exchange protocols. Even within one vendor's product line, different model computers may communicate in unique ways.

As the use of computer communications and computer networking proliferates, a one at a time special-purpose approach to communications software development is too costly to be acceptable. The only alternative is for computer vendors to adopt and implement a common set of conventions. For this to happen, standards are needed. Such standards would have two benefits:

- Vendors feel encouraged to implement the standards because of an expectation that, because of wide usage of the standards, their products will be more marketable.
- Customers are in a position to require that the standards be implemented by any vendor wishing to propose equipment to them.

It should become clear from the ensuing discussion that no single standard will suffice. Any distributed application, such as electronic mail or client/server interaction, requires a complex set of communications functions for proper operation. Many of these functions, such as reliability mechanisms, are common across many or even all applications. Thus, the communications task is best viewed as consisting of a modular architecture, in which the various elements of the architecture perform the various required functions. Hence, before one can develop standards, there should be a structure, or *protocol architecture*, that defines the communications tasks.

Two protocol architectures have served as the basis for the development of interoperable communications standards: the transmission control protocol/Internet protocol (TCP/IP) protocol suite and the **open**

systems interconnection (OSI) reference model. TCP/IP is the most widely used interoperable architecture, especially in the context of **local-area networks (LANs)**. In this section, we provide a brief overview of the two architectures.

72.1.1 The Transmission Control Protocol/Internet Protocol Architecture

This architecture is a result of protocol research and development conducted on the experimental packet-switched network, ARPANET, funded by the Defense Advanced Research Projects Agency (DARPA), and is generally referred to as the TCP/IP protocol suite.

72.1.1.1 The Transmission Control Protocol/Internet Protocol Layers

In general terms, communications can be said to involve three agents: applications, computers, and networks. Examples of applications include file transfer and electronic mail. The applications that we are concerned with here are distributed applications that involve the exchange of data between two computer systems. These applications, and others, execute on computers that can often support multiple simultaneous applications. Computers are connected to networks, and the data to be exchanged are transferred by the network from one computer to another. Thus, the transfer of data from one application to another involves first getting the data to the computer in which the application resides and then getting it to the intended application within the computer.

With these concepts in mind, it appears natural to organize the communication task into four relatively independent layers:

- Network access layer
- Internet layer
- Host-to-host layer
- Process layer

The network access layer is concerned with the exchange of data between an end system (server, workstation, etc.) and the network to which it is attached. The sending computer must provide the network with the address of the destination computer, so that the network may route the data to the appropriate destination. The sending computer may wish to invoke certain services, such as priority, that might be provided by the network. The specific software used at this layer depends on the type of network to be used; different standards have been developed for **circuit switching, packet switching** (e.g., X.25), local-area networks (e.g., Ethernet), and others. Thus, it makes sense to separate those functions having to do with network access into a separate layer. By doing this, the remainder of the communications software, above the network access layer, need not be concerned about the specifics of the network to be used. The same higher layer software should function properly regardless of the particular network to which the computer is attached.

The network access layer is concerned with access to and routing data across a network for two end systems attached to the same network. In those cases where two devices are attached to different networks, procedures are needed to allow data to traverse multiple interconnected networks. This is the function of the Internet layer. The Internet protocol is used at this layer to provide the routing function across multiple networks. This protocol is implemented not only in the end systems but also in routers. A router is a processor that connects two networks and whose primary function is to relay data from one network to the other on its route from the source to the destination end system.

Regardless of the nature of the applications that are exchanging data, there is usually a requirement that data be exchanged reliably. That is, we would like to be assured that all of the data arrive at the destination application and that the data arrive in the order in which they were sent. As we shall see, the mechanisms for providing reliability are essentially independent of the nature of the applications. Thus, it makes sense to collect those mechanisms in a common layer shared by all applications; this is referred to as the host-to-host layer. The transmission control protocol provides this functionality.

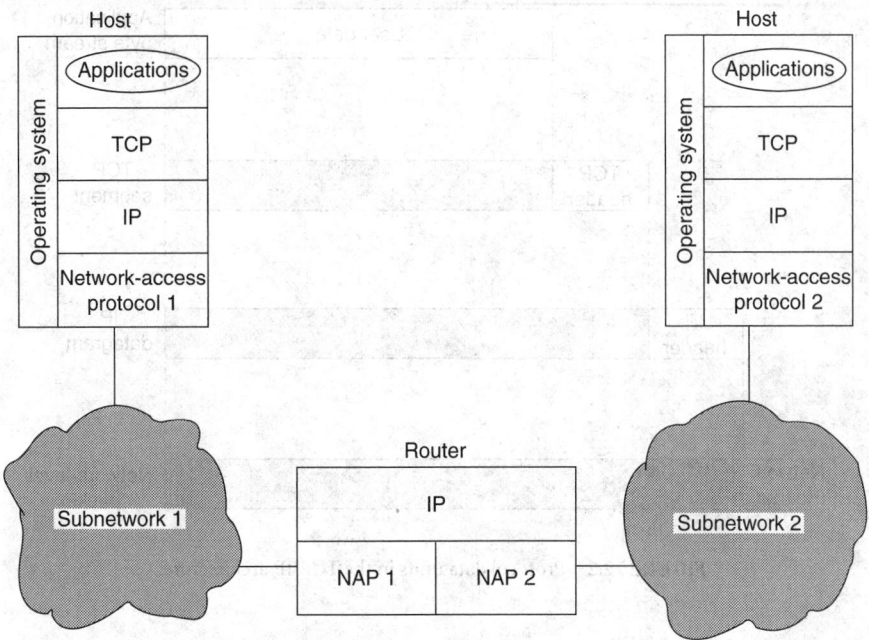

FIGURE 72.1 Communications using the TCP/IP protocol architecture.

Finally, the process layer contains the logic needed to support the various user applications. For each different type of application, such as file transfer, a separate module is needed that is peculiar to that application.

72.1.1.2 Operation of Transmission Control Protocol and Internet Protocol

Figure 72.1 indicates how these protocols are configured for communications. To make clear that the total communications facility may consist of multiple networks, the constituent networks are usually referred to as *subnetworks*. Some sort of network access protocol, such as the Ethernet logic, is used to connect a computer to a subnetwork. This protocol enables the host to send data across the subnetwork to another host or, in the case of a host on another subnetwork, to a router. IP is implemented in all of the end systems and the routers. It acts as a relay to move a block of data from one host, through one or more routers, to another host. TCP is implemented only in the end systems; it keeps track of the blocks of data to ensure that all are delivered reliably to the appropriate application.

For successful communication, every entity in the overall system must have a unique address. Actually, two levels of addressing are needed. Each host on a subnetwork must have a unique global Internet address; this allows the data to be delivered to the proper host. Each process with a host must have an address that is unique within the host; this allows the host-to-host protocol (TCP) to deliver data to the proper process. These latter addresses are known as ports.

Let us trace a simple operation. Suppose that a process, associated with port 1 at host A, wishes to send a message to another process, associated with port 2 at host B. The process at A hands the message down to TCP with instructions to send it to host B, port 2. TCP hands the message down to IP with instructions to send it to host B. Note that IP need not be told the identity of the destination port. All that it needs to know is that the data are intended for host B. Next, IP hands the message down to the network access layer (e.g., Ethernet logic) with instructions to send it to router X (the first hop on the way to B).

To control this operation, control information as well as user data must be transmitted, as suggested in Figure 72.2. Let us say that the sending process generates a block of data and passes this to TCP. TCP may break this block into smaller pieces to make it more manageable. To each of these pieces, TCP appends control information known as the TCP header, forming a *TCP segment*. The control information is to be

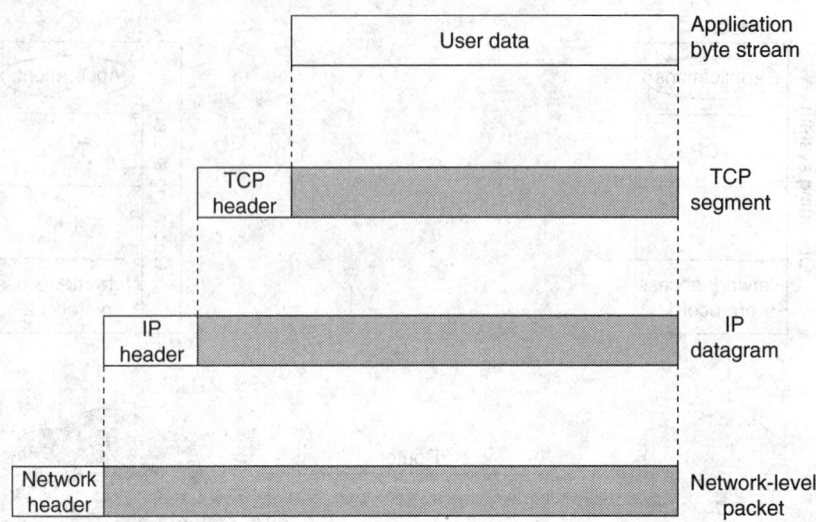

FIGURE 72.2   Protocol data units in the TCP/IP architecture.

used by the peer TCP protocol entity at host B. Examples of items that are included in this header include the following:

- *Destination port:* When the TCP entity at B receives the segment, it must know to whom the data are to be delivered.
- *Sequence number:* TCP numbers the segments that it sends to a particular destination port sequentially, so that if they arrive out of order, the TCP entity at B can reorder them.
- *Checksum:* The sending TCP includes a code that is a function of the contents of the remainder of the segment. The receiving TCP performs the same calculation and compares the result with the incoming code. A discrepancy results if there has been some error in transmission.

Next, TCP hands each segment over to IP, with instructions to transmit it to B. These segments must be transmitted across one or more subnetworks and relayed through one or more intermediate routers. This operation, too, requires the use of control information. Thus, IP appends a header of control information to each segment to form an *IP datagram*. An example of an item stored in the IP header is the destination host address (in this example, B).

Finally, each IP datagram is presented to the network access layer for transmission across the first subnetwork in its journey to the destination. The network access layer appends its own header, creating a packet, or frame. The packet is transmitted across the subnetwork to router X. The packet header contains the information that the subnetwork needs to transfer the data across the subnetwork. Examples of items that may be contained in this header include the following:

- *Destination subnetwork address:* The subnetwork must know to which attached device the packet is to be delivered.
- *Facilities requests:* The network access protocol might request the use of certain subnetwork facilities, such as priority.

At router X, the packet header is stripped off and the IP header is examined. On the basis of the destination address information in the IP header, the IP module in the router directs the datagram out across subnetwork 2 to B. To do this, the datagram is again augmented with a network access header.

When the data are received at B, the reverse process occurs. At each layer, the corresponding header is removed, and the remainder is passed on the next higher layer, until the original user data are delivered to the destination process.

72.1.1.3   Transmission Control Protocol/Internet Protocol Applications

A number of applications have been standardized to operate on top of TCP. We mention three of the most common here.

The simple mail transfer protocol (SMTP) provides a basic electronic mail facility. It provides a mechanism for transferring messages among separate hosts. Features of SMTP include mailing lists, return receipts, and forwarding. The SMTP protocol does not specify the way in which messages are to be created; some local editing or native electronic mail facility is required. Once a message is created, SMTP accepts the message, and makes use of TCP to send it to an SMTP module on another host. The target SMTP module will make use of a local electronic mail package to store the incoming message in a user's mailbox.

The file transfer protocol (FTP) is used to send files from one system to another under user command. Both text and binary files are accommodated, and the protocol provides features for controlling user access. When a user wishes to engage in file transfer, FTP sets up a TCP connection to the target system for the exchange of control messages. These allow user identifier (ID) and password to be transmitted and allow the user to specify the file and file actions desired. Once a file transfer is approved, a second TCP connection is set up for the data transfer. The file is transferred over the data connection, without the overhead of any headers or control information at the application level. When the transfer is complete, the control connection is used to signal the completion and to accept new file transfer commands.

TELNET provides a remote log-on capability, which enables a user at a terminal or personal computer to log on to a remote computer and function as if directly connected to that computer. The protocol was designed to work with simple scroll-mode terminals. TELNET is actually implemented in two modules: User TELNET interacts with the terminal input/output (I/O) module to communicate with a local terminal. It converts the characteristics of real terminals to the network standard and vice versa. Server TELNET interacts with an application, acting as a surrogate terminal handler so that remote terminals appear as local to the application. Terminal traffic between user and server TELNET is carried on a TCP connection.

72.1.2   The Open Systems Interconnection Model

The open systems interconnection (OSI) reference model was developed by the International Organization for Standardization (ISO) to serve as a framework for the development of communications protocol standards.

72.1.2.1   Overall Architecture

A widely accepted structuring technique, and the one chosen by ISO, is layering. The communications functions are partitioned into a hierarchical set of layers. Each layer performs a related subset of the functions required to communicate with another system. It relies on the next lower layer to perform more primitive functions and to conceal the details of those functions. It provides services to the next higher layer. Ideally, the layers should be defined so that changes in one layer do not require changes in the other layers. Thus, we have decomposed one problem into a number of more manageable subproblems.

The task of ISO was to define a set of layers and the services performed by each layer. The partitioning should group functions logically and should have enough layers to make each layer manageably small, but it should not have so many layers that the processing overhead imposed by the collection of layers is burdensome.

The resulting OSI architecture has seven layers, which are illustrated in Figure 72.3. Each computer contains the seven layers. Communication is between applications in the two computers, labeled application X and application Y in the figure. If application X wishes to send a message to application Y, it invokes the application layer (layer 7). Layer 7 establishes a peer relationship with layer 7 of the target computer, using a layer-7 protocol (application protocol). This protocol requires services from layer 6, so the two layer-6 entities use a protocol of their own, and so on down to the physical layer, which actually transmits bits over a transmission medium.

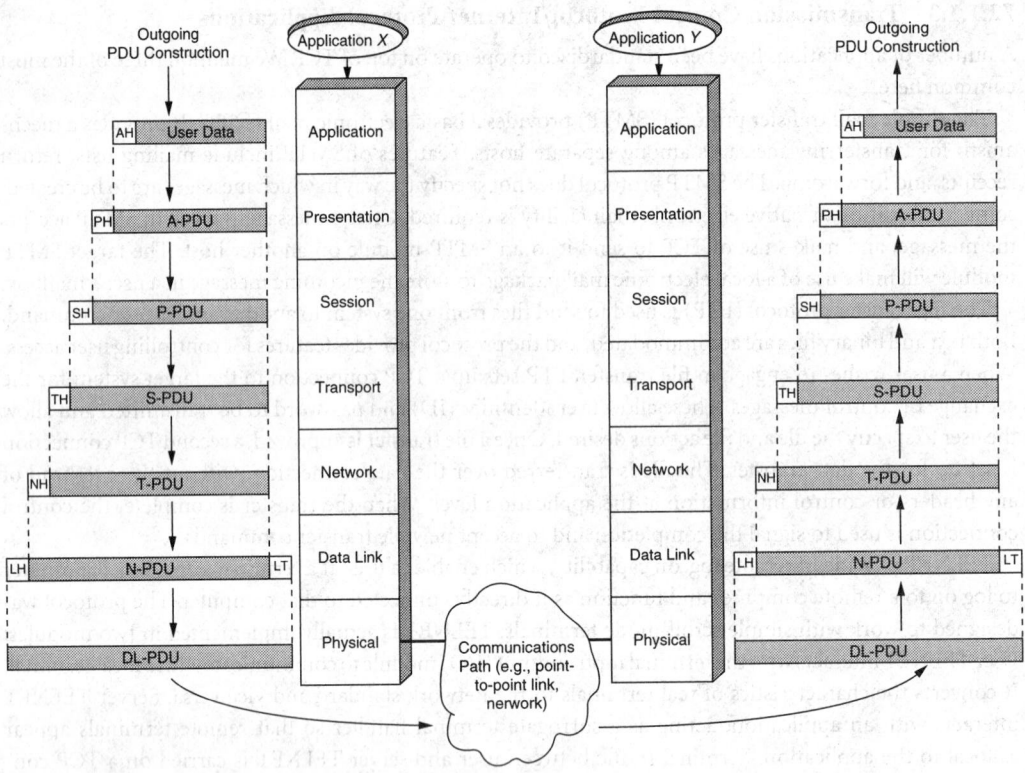

FIGURE 72.3   The OSI environment.

The figure also illustrates the way in which the protocols at each layer are realized. When application X has a message to send to application Y, it transfers those data to an application layer module. That module appends an application header to the data; the header contains the control information needed by the peer layer on the other side. The original data plus the header, referred to as an application protocol data unit (PDU), are passed as a unit to layer 6. The presentation module treats the whole unit as data and appends its own header. This process continues down through layer 2, which generally adds both a header and a trailer. This layer-2 protocol data unit, usually called a *frame*, is then transmitted by the physical layer onto the transmission medium. When the frame is received by the target computer, the reverse process occurs. As we ascend the layers, each layer strips off the outermost header, acts on the protocol information contained therein, and passes the remainder up to the next layer.

The principal motivation for development of the OSI model was to provide a framework for standardization. Within the model, one or more protocol standards can be developed at each layer. The model defines in general terms the functions to be performed at that layer and facilitates the standards-making process in two ways:

- Because the functions of each layer are well defined, standards can be developed independently and simultaneously for each layer. This speeds up the standards-making process.
- Because the boundaries between layers are well defined, changes in standards in one layer need not affect already existing software in another layer. This makes it easier to introduce new standards.

We now turn to a brief description of each layer and discuss some of the standards that have been developed for each layer.

72.1.2.2 Physical Layer

The physical layer covers the physical interface between a data transmission device and a transmission medium and the rules by which bits are passed from one to another. A well-known physical layer standard is RS-232-C.

72.1.2.3 Data Link Layer

The physical layer provides only a raw bit stream service. The data link layer attempts to make the physical link reliable and provides the means to activate, maintain, and deactivate the link. The principal service provided by the data link layer to higher layers is that of error detection and control. Thus, with a fully functional data link layer protocol, the next higher layer may assume error-free transmission over the link.

A well-known data link layer standard is high-level data link control (HDLC). For local area networks, the functionality of the data link layer is generally split into two sublayers: logical link control (LLC) and medium access control (MAC).

72.1.2.4 Network Layer

The network layer provides for the transfer of information between computers across some sort of communications network. It relieves higher layers of the need to know anything about the underlying data transmission and switching technologies used to connect systems. The network service is responsible for establishing, maintaining, and terminating connections across the intervening network. At this layer, the computer system engages in a dialogue with the network to specify the destination address and to request certain network facilities, such as priority.

There is a spectrum of possibilities for intervening communications facilities to be managed by the network layer. At one extreme, there is a direct point-to-point link between stations. In this case, there may be no need for a network layer because the data link layer can perform the necessary function of managing the link.

Next, the systems could be connected across a single network, such as a circuit-switching or packet-switching network. The lower three layers are concerned with attaching to and communicating with the network; a well-known example is the X.25 standard. The packets that are created by the end system pass through one or more network nodes that act as relays between the two end systems. The network nodes implement layers 1–3 of the architecture. The upper four layers are end-to-end protocols between the attached computers.

At the other extreme, two stations might wish to communicate but are not even connected to the same network. Rather, they are connected to networks that, directly or indirectly, are connected to each other. This case requires the use of some sort of internetworking technique, such as the use of IP.

72.1.2.5 Transport Layer

The transport layer provides a reliable mechanism for the exchange of data between computers. It ensures that data are delivered error free, in sequence, and with no losses or duplications. The transport layer also may be concerned with optimizing the use of network services and providing a requested quality of service. For example, the session layer may specify acceptable error rates, maximum delay, priority, and security features.

The mechanisms used by the transport protocol to provide reliability are very similar to those used by data link control protocols such as HDLC: the use of sequence numbers, error detecting codes, and retransmission after timeout. The reason for this apparent duplication of effort is that the data link layer deals only with a single, direct link, whereas the transport layer deals with a chain of network nodes and links. Although each link in that chain is reliable because of the use of HDLC, a node along that chain may fail at a critical time. Such a failure will affect data delivery, and it is the transport protocol that addresses this problem.

The size and complexity of a transport protocol depend on how reliable or unreliable the underlying network and network layer services are. Accordingly, ISO had developed a family of five transport protocol standards, each oriented toward a different underlying service.

72.1.2.6 Session Layer

The session layer provides the mechanism for controlling the dialogue between the two end systems. In many cases, there will be little or no need for session-layer services, but for some applications, such services are used. The key services provided by the session layer include the following:

- *Dialogue discipline:* This can be two-way simultaneous (full duplex) or two-way alternate (half-duplex).
- *Grouping:* The flow of data can be marked to define groups of data. For example, if a retail store is transmitting sales data to a regional office, the data can be marked to indicate the end of the sales data for each department. This would signal the host computer to finalize running totals for that department and start new running counts for the next department.
- *Recovery:* The session layer can provide a checkpointing mechanism, so that if a failure of some sort occurs between checkpoints, the session entity can retransmit all data since the last checkpoint.

ISO has issued a standard for the session layer that includes as options services such as those just described.

72.1.2.7 Presentation Layer

The presentation layer defines the format of the data to be exchanged between applications and offers application programs a set of data transformation services. For example, data compression or data encryption could occur at this level.

72.1.2.8 Application Layer

The application layer provides a means for application programs to access the OSI environment. This layer contains management functions and generally useful mechanisms to support distributed applications. In addition, general-purpose applications such as file transfer, electronic mail, and terminal access to remote computers are considered to reside at this layer.

72.2 Network Organization

Traditionally, data networks have been classified as either **wide-area network** (WAN) or local-area network. Although there has been some blurring of this distinction, it is still a useful one. We look first at traditional WANs and then at the more recently introduced higher speed WANs. The discussion then turns to traditional and high-speed LANs.

WANs are used to connect stations over a large area: anything from a metropolitan area to worldwide. LANs are used within a single building or a cluster of buildings. Usually, LANs are owned by the organization that uses them. A WAN may be owned by the organization that uses it (private network) or provided by a third party (public network); in the latter case, the network is shared by a number of organizations.

72.2.1 Traditional Wide-Area Networks

Traditional WANs are switched communications networks, consisting of an interconnected collection of nodes, in which information is transmitted from source station to destination station by being routed through the network of nodes. Figure 72.4 is a simplified illustration of the concept. The nodes are connected by transmission paths. Signals entering the network from a station are routed to the destination by being switched from node to node. Two quite different technologies are used in wide-area switched networks: circuit switching and packet switching. These two technologies differ in the way the nodes switch information from one link to another on the way from source to destination.

72.2.1.1 Circuit Switching

Circuit switching is the dominant technology for both voice and data communications today and will remain so for the foreseeable future. Communication via circuit switching implies that there is a dedicated

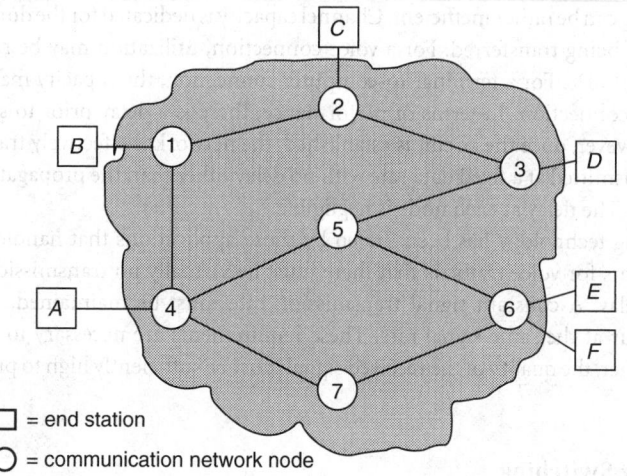

FIGURE 72.4 Simple switching network.

communication path between two stations. That path is a connected sequence of links between network nodes. On each physical link, a channel is dedicated to the connection. The most common example of circuit switching is the telephone network.

Communication via circuit switching involves three phases, which can be explained with reference to Figure 72.4. The three phases are as follows:

1. *Circuit establishment:* Before any signals can be transmitted, an end-to-end (station-to-station) circuit must be established. For example, station *A* sends a request to node 4 requesting a connection to station *E*. Typically, the link from *A* to 4 is a dedicated line, so that part of the connection already exists. Node 4 must find the next leg in a route leading to node 6. Based on routing information and measures of availability and perhaps cost, node 4 selects the link to node 5, allocates a free channel [using frequency-division multiplexing (FDM) or time-division multiplexing (TDM)] on that link and sends a message requesting connection to *E*. So far, a dedicated path has been established from *A* through 4 to 5. Because a number of stations may attach to 4, it must be able to establish internal paths from multiple stations to multiple nodes. The remainder of the process proceeds similarly. Node 5 dedicates a channel to node 6 and internally ties that channel to the channel from node 4. Node 6 completes the connection to *E*. In completing the connection, a test is made to determine if *E* is busy or is prepared to accept the connection.
2. *Information transfer:* Information can now be transmitted from *A* through the network to *E*. The transmission may be analog voice, digitized voice, or binary data, depending on the nature of the network. As the carriers evolve to fully integrated digital networks, the use of digital (binary) transmission for both voice and data is becoming the dominant method. The path is: *A*-4 link, internal switching through 4, 4-5 channel, internal switching through 5, 5-6 channel, internal switching through 6, 6-*E* link. Generally, the connection is full duplex, and signals may be transmitted in both directions simultaneously.
3. *Circuit disconnect:* After some period of information transfer, the connection is terminated, usually by the action of one of the two stations. Signals must be propagated to nodes, 4, 5, and 6 to deallocate the dedicated resources.

Note that the connection path is established before data transmission begins. Thus, channel capacity must be reserved between each pair of nodes in the path and each node must have available internal switching capacity to handle the requested connection. The switches must have the intelligence to make these allocations and to devise a route through the network.

Circuit switching can be rather inefficient. Channel capacity is dedicated for the duration of a connection, even if no data are being transferred. For a voice connection, utilization may be rather high, but it still does not approach 100%. For a terminal-to-computer connection, the capacity may be idle during most of the time of the connection. In terms of performance, there is a delay prior to signal transfer for call establishment. However, once the circuit is established, the network is effectively transparent to the users. Information is transmitted at a fixed data rate with no delay other than the propagation delay through the transmission links. The delay at each node is negligible.

Circuit-switching technology has been driven by those applications that handle voice traffic. One of the key requirements for voice traffic is that there must be virtually no transmission delay and certainly no variation in delay. A constant signal transmission rate must be maintained, because transmission and reception occur at the same signal rate. These requirements are necessary to allow normal human conversation. Further, the quality of the received signal must be sufficiently high to provide, at a minimum, intelligibility.

72.2.1.2 Packet Switching

A packet-switching network is a switched communications network that transmits data in short blocks called packets. The network consists of a set of interconnected packet-switching nodes. A device attaches to the network at one of these nodes and presents data for transmission in the form of a stream of packets. Each packet is routed through the network. As each node along the route is encountered, the packet is received, stored briefly, and then transmitted along a link to the next node in the route. Two approaches are used to manage the transfer and routing of these streams of packets: datagram and virtual circuit.

In the datagram approach, each packet is treated independently, with no reference to packets that have gone before. This approach is illustrated in Figure 72.5a. Each node chooses the next node on a packet's path, taking into account information received from neighboring nodes on traffic, line failures, and so on. So the packets, each with the same destination address, do not all follow the same route, and they may arrive out of sequence at the exit point. In some networks, the exit node restores the packets to their original order before delivering them to the destination. In other datagram networks, it is up to the destination rather than the exit node to do the reordering. Also, it is possible for a packet to be destroyed in the network. For example, if a packet-switching node crashes momentarily, all of its queued packets may be lost. Again, it is up to either the exit node or the destination to detect the loss of a packet and decide how to recover it. In this technique, each packet, treated independently, is referred to as a datagram.

In the virtual circuit approach, a preplanned route is established before any packets are sent. Once the route is established, all of the packets between a pair of communicating parties follow this same route through the network. This is illustrated in Figure 72.5b. Because the route is fixed for the duration of the logical connection, it is somewhat similar to a circuit in a circuit-switching network and is referred to as a virtual circuit. Each packet now contains a virtual circuit identifier as well as data. Each node on the pre-established route knows where to direct such packets; no routing decisions are required. At any time, each station can have more than one virtual circuit to any other station and can have virtual circuits to more than one station.

If two stations wish to exchange data over an extended period of time, there are certain advantages to virtual circuits. First, the network may provide services related to the virtual circuit, including sequencing and error control. *Sequencing* refers to the fact that, because all packets follow the same route, they arrive in the original order. Error control is a service that ensures not only that packets arrive in proper sequence but also that all packets arrive correctly. For example, if a packet in a sequence from node 4 to node 6 fails to arrive at node 6, or arrives with an error, node 6 can request a retransmission of that packet from node 4. Another advantage is that packets should transit the network more rapidly with a virtual circuit; it is not necessary to make a routing decision for each packet at each node.

One advantage of the datagram approach is that the call setup phase is avoided. Thus, if a station wishes to send only one or a few packets, datagram delivery will be quicker. Another advantage of the datagram service is that, because it is more primitive, it is more flexible. For example, if congestion develops in

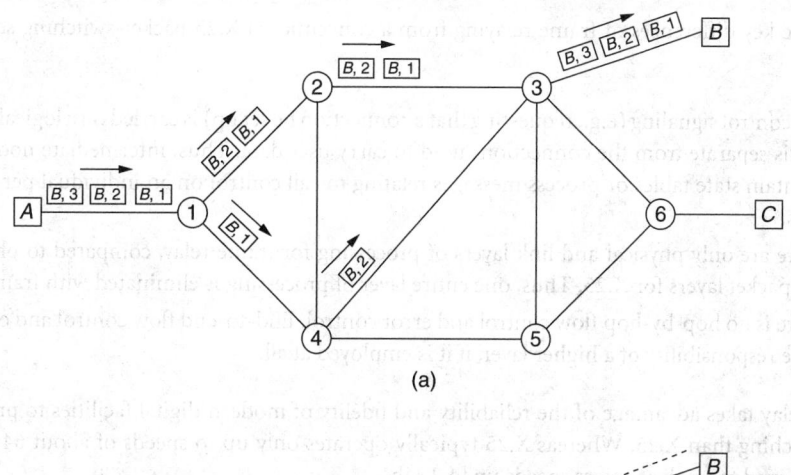

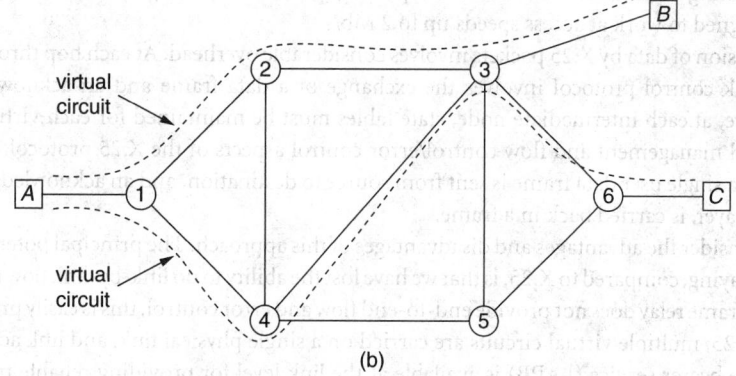

FIGURE 72.5 Virtual circuit and datagram operation: (a) datagram approach and (b) virtual circuit approach.

one part of the network, incoming datagrams can be routed away from the congestion. With the use of virtual circuits, packets follow a predefined route, and thus it is more difficult for the network to adapt to congestion. A third advantage is that datagram delivery is inherently more reliable. With the use of virtual circuits, if a node fails, all virtual circuits that pass through that node are lost. With datagram delivery, if a node fails, subsequent packets may find an alternative route that bypasses that node.

72.2.2 High-Speed Wide-Area Networks

As the speed and number of local-area networks continue their relentless growth, increasing demand is placed on wide-area packet-switching networks to support the tremendous throughput generated by these LANs. In the early days of wide-area networking, X.25 was designed to support direct connection of terminals and computers over long distances. At speeds up to 64 kb/s or so, X.25 copes well with these demands. As LANs have come to play an increasing role in the local environment, X.25, with its substantial overhead, is being recognized as an inadequate tool for wide-area networking. Fortunately, several new generations of high-speed switched services for wide-area networking are moving rapidly from the research laboratory and the draft standard stage to the commercially available, standardized-product stage. The two most important such technologies are **frame relay** and **asynchronous transfer mode (ATM)**.

72.2.2.1 Frame Relay

Frame relay provides a streamlined technique for wide-area packet switching, compared to X.25 [Black 1994b, Smith 1993]. It provides superior performance by eliminating as much as possible of the overhead of X.25. The key differences of frame relaying from a conventional X.25 packet-switching service are as follows:

- Call control signaling (e.g., requesting that a connection be set up) is carried on a logical connection that is separate from the connections used to carry user data. Thus, intermediate nodes need not maintain state tables or process messages relating to call control on an individual per-connection basis.
- There are only physical and link layers of processing for frame relay, compared to physical, link, and packet layers for X.25. Thus, one entire layer of processing is eliminated with frame relay.
- There is no hop-by-hop flow control and error control. End-to-end flow control and error control is the responsibility of a higher layer, if it is employed at all.

Frame relay takes advantage of the reliability and fidelity of modern digital facilities to provide faster packet switching than X.25. Whereas X.25 typically operates only up to speeds of about 64 kb/s, frame relay is designed to work at access speeds up to 2 Mb/s.

Transmission of data by X.25 packets involves considerable overhead. At each hop through the network, the data link control protocol involves the exchange of a data frame and an acknowledgment frame. Furthermore, at each intermediate node, state tables must be maintained for each virtual circuit to deal with the call management and flow control/error control aspects of the X.25 protocol. In contrast, with frame relay a single user data frame is sent from source to destination, and an acknowledgment, generated at a higher layer, is carried back in a frame.

Let us consider the advantages and disadvantages of this approach. The principal potential disadvantage of frame relaying, compared to X.25, is that we have lost the ability to do link-by-link flow and error control. (Although frame relay does not provide end-to-end flow and error control, this is easily provided at a higher layer.) In X.25, multiple virtual circuits are carried on a single physical link, and link access procedure to frame mode bearer service (LAPB) is available at the link level for providing reliable transmission from the source to the packet-switching network and from the packet-switching network to the destination. In addition, at each hop through the network, the link control protocol can be used for reliability. With the use of frame relaying, this hop-by-hop link control is lost. However, with the increasing reliability of transmission and switching facilities, this is not a major disadvantage.

The advantage of frame relaying is that we have streamlined the communications process. The protocol functionality required at the user–network interface is reduced, as is the internal network processing. As a result, lower delay and higher throughput can be expected. Preliminary results indicate a reduction in frame processing time of an order of magnitude.

The frame relay data transfer protocol consists of the following functions:

- Frame delimiting, alignment, and transparency
- Frame multiplexing/demultiplexing using the address field
- Inspection of the frame to ensure that it consists of an integer number of octets (8-b bytes) prior to zero-bit insertion or following zero-bit extraction
- Inspection of the frame to ensure that it is neither too long nor too short
- Detection of transmission errors
- Congestion control functions

This architecture reduces to the bare minimum the amount of work accomplished by the network. User data are transmitted in frames with virtually no processing by the intermediate network nodes other than to check for errors and to route based on connection number. A frame in error is simply discarded, leaving error recovery to higher layers.

The operation of frame relay for user data transfer is best explained by beginning with the frame format, illustrated in Figure 72.6. The format is similar to that of other data link control protocols, such as HDLC

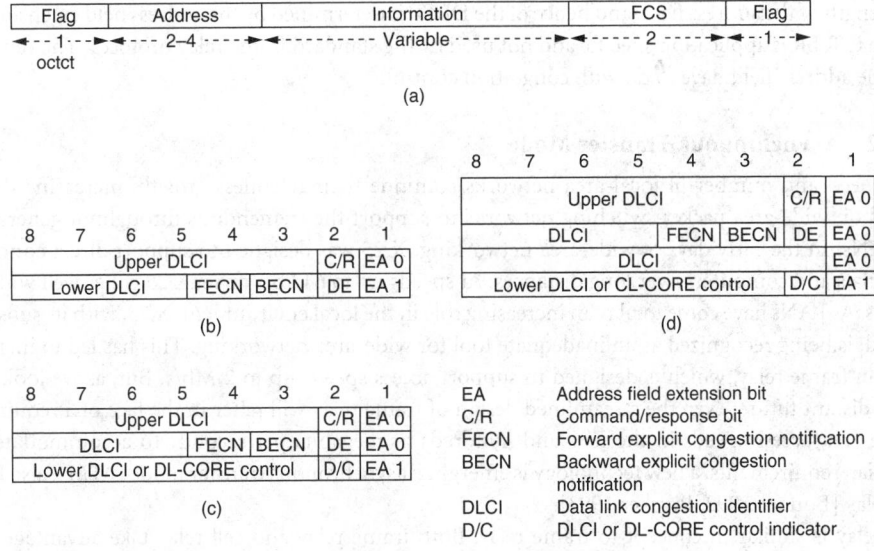

FIGURE 72.6   Frame relay formats: (a) frame format; (b) address field, 2 octets (default); (c) address field, 3 octets; and (d) address field, 4 octets.

and LAPB, with one omission: there is no control field. In traditional data link control protocols, the control field is used for the following functions:

- Part of the control field identifies the frame type. In addition to a frame for carrying user data, there are various control frames. These carry no user data but are used for various protocol control functions, such as setting up and tearing down logical connections.
- The control field for user data frames includes send and receive sequence numbers. The send sequence number is used to sequentially number each transmitted frame. The receive sequence number is used to provide a positive or negative acknowledgment to incoming frames. The use of sequence numbers allows the receiver to control the rate of incoming frames (flow control) and to report missing or damaged frames, which can then be retransmitted (error control).

The lack of a control field in the frame relay format means that the process of setting up and tearing down connections must be carried out on a separate channel at a higher layer of software. It also means that it is not possible to perform flow control and error control.

The flag and frame check sequence (FCS) fields function as in HDLC and other traditional data link control protocols. The flag field is a unique pattern that delimits the start and end of the frame. The FCS field is used for error detection. On transmission, the FCS checksum is calculated and stored in the FCS field. On reception, the checksum is again calculated and compared to the value stored in the incoming FCS field. If there is a mismatch, then the frame is assumed to be in error and is discarded.

The information field carries higher layer data. The higher layer data may be either user data or call control messages, as explained subsequently.

The address field has a default length of 2 octets and may be extended to 3 or 4 octets. It carries a data link connection identifier (DLCI) of 10, 17, or 24 b. The DLCI serves the same function as the virtual circuit number in X.25: it allows multiple logical frame relay connections to be multiplexed over a single channel. As in X.25, the connection identifier has only local significance: each end of the logical connection assigns its own DLCI from the pool of locally unused numbers, and the network must map from one to the other. The alternative, using the same DLCI on both ends, would require some sort of global management of DLCI values.

The length of the address field, and hence of the DLCI, is determined by the address field extension (EA) bits. The C/R bit is application specific and not used by the standard frame relay protocol. The remaining bits in the address field have to do with congestion control.

72.2.2.2 Asynchronous Transfer Mode

As the speed and number of local-area networks continue their relentless growth, increasing demand is placed on wide-area packet-switching networks to support the tremendous throughput generated by these LANs. In the early days of wide-area networking, X.25 was designed to support direct connection of terminals and computers over long distances. At speeds up to 64 kb/s or so, X.25 copes well with these demands. As LANs have come to play an increasing role in the local environment, X.25, with its substantial overhead, is being recognized as an inadequate tool for wide-area networking. This has led to increasing interest in frame relay, which is designed to support access speeds up to 2 Mb/s. But, as we look to the not-too-distant future, even the streamlined design of frame relay will falter in the face of a requirement for wide-area access speeds in the tens and hundreds of megabits per second. To accommodate these gargantuan requirements, a new technology is emerging: asynchronous transfer mode (ATM), also known as cell relay [Boudec 1992, Prycker 1993].

Cell relay is similar in concept to frame relay. Both frame relay and cell relay take advantage of the reliability and fidelity of modern digital facilities to provide faster packet switching than X.25. Cell relay is even more streamlined than frame relay in its functionality and can support data rates several orders of magnitude greater than frame relay.

ATM is a packet-oriented transfer mode. Like frame relay and X.25, it allows multiple logical connections to be multiplexed over a single physical interface. The information flow on each logical connection is organized into fixed-size packets, called cells. As with frame relay, there is no link-by-link error control or flow control.

Logical connections in ATM are referred to as virtual channels. A virtual channel is analogous to a virtual circuit in X.25 or a frame-relay logical connection. A virtual channel is set up between two end users through the network and a variable-rate, full-duplex flow of fixed-size cells is exchanged over the connection. Virtual channels are also used for user–network exchange (control signaling) and network–network exchange (network management and routing).

For ATM, a second sublayer of processing has been introduced that deals with the concept of virtual path. A virtual path is a bundle of virtual channels that have the same endpoints. Thus, all of the cells flowing over all of the virtual channels in a single virtual path are switched together.

Several advantages can be listed for the use of virtual paths:

- *Simplified network architecture:* Network transport functions can be separated into those related to an individual logical connection (virtual channel) and those related to a group of logical connections (virtual path).
- *Increased network performance and reliability:* The network deals with fewer, aggregated entities.
- *Reduced processing and short connection setup time:* Much of the work is done when the virtual path is set up. The addition of new virtual channels to an existing virtual path involves minimal processing.
- *Enhanced network services:* The virtual path is internal to the network but is also visible to the end user. Thus, the user may define closed user groups or closed networks of virtual channel bundles.

International Telecommunications Union–Telecommunications Standardization Sector (ITU-T) Recommendation I.150 lists the following as characteristics of virtual channel connections:

- *Quality of service:* A user of a virtual channel is provided with a quality of service specified by parameters such as cell loss ratio (ratio of cells lost to cells transmitted) and cell delay variation.
- *Switched and semipermanent virtual channel connections:* Both switched connections, which require call-control signaling, and dedicated channels can be provided.

- *Cell sequence integrity:* The sequence of transmitted cells within a virtual channel is preserved.
- *Traffic parameter negotiation and usage monitoring:* Traffic parameters can be negotiated between a user and the network for each virtual channel. The input of cells to the virtual channel is monitored by the network to ensure that the negotiated parameters are not violated.

The types of traffic parameters that can be negotiated would include average rate, peak rate, burstiness, and peak duration. The network may need a number of strategies to deal with congestion and to manage existing and requested virtual channels. At the crudest level, the network may simply deny new requests for virtual channels to prevent congestion. Additionally, cells may be discarded if negotiated parameters are violated or if congestion becomes severe. In an extreme situation, existing connections might be terminated.

Recommendation I.150 also lists characteristics of virtual paths. The first four characteristics listed are identical to those for virtual channels. That is, quality of service, switched and semipermanent virtual paths, cell sequence integrity, and traffic parameter negotiation and usage monitoring are all also characteristics of a virtual path. There are a number reasons for this duplication. First, this provides some flexibility in how the network manages the requirements placed on it. Second, the network must be concerned with the overall requirements for a virtual path and, within a virtual path, may negotiate the establishment of virtual circuits with given characteristics. Finally, once a virtual path is set up, it is possible for the end users to negotiate the creation of new virtual channels. The virtual path characteristics impose a discipline on the choices that the end users may make.

In addition, a fifth characteristic is listed for virtual paths:

- *Virtual channel identifier restriction within a virtual path:* One or more virtual channel identifiers, or numbers, may not be available to the user of the virtual path but may be reserved for network use. Examples would be virtual channels used for network management.

The asynchronous transfer mode makes use of fixed-size cells, consisting of a 5-octet header and a 48-octet information field (Figure 72.7). There are several advantages to the use of small, fixed-size cells. First, the use of small cells may reduce queuing delay for high-priority cells, because it waits less if it arrives slightly behind a lower priority cell that has gained access to a resource (e.g., the transmitter). Second, it appears that fixed-size cells can be switched more efficiently, which is important for the very high data rates of ATM.

Figure 72.7a shows the header format at the user–network interface. Multiple terminals may share a single access link to the network. The generic flow control field is to be used for end-to-end flow control. The details of its application are for further study. The field could be used to assist the customer in controlling the flow of traffic for different qualities of service. One candidate for the use of this field is a multiple-priority level indicator to control the flow of information in a service-dependent manner.

The virtual path identifier and virtual channel identifier fields constitute a routing field for the network. The virtual path identifier indicates a user-to-user or user-to-network virtual path. The virtual channel identifier indicates a user-to-user or user-to-network virtual channel. These identifiers have local significance (as with X.25 and frame relay) and may change as the cell traverses the network.

The payload type field indicates the type of information in the information field. A value of 00 indicates user information; that is, information from the next higher layer. Other values are for further study. Presumably, network management and maintenance values will be assigned. This field allows the insertion of network-management cells onto a user's virtual channel without impacting user's data. Thus, it could provide in-band control information.

The cell loss priority is used to provide guidance to the network in the event of congestion. A value of 0 indicates a cell of relatively higher priority, which should not be discarded unless no other alternative is available. A value of 1 indicates that this cell is subject to discard within the network. The user might employ this field so that extra information may be inserted into the network, with a CLP of 1, and delivered to the destination if the network is not congested. The network sets this field to 1 for any data cell that is in violation of a traffic agreement. In this case, the switch that does the setting realizes that the cell exceeds

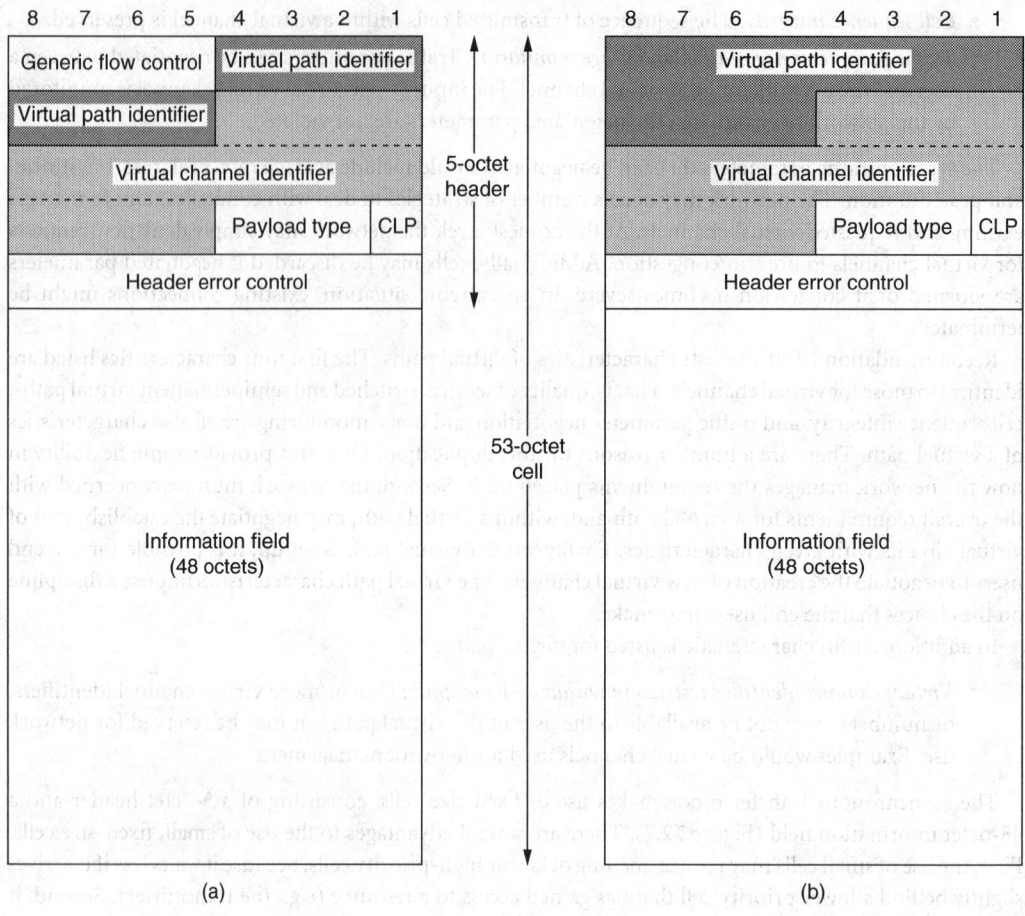

FIGURE 72.7   ATM cell format: (a) user–network interface and (b) network–network interface.

the agreed traffic parameters but that the switch is capable of handling the cell. At a later point in the network, if congestion is encountered, this cell has been marked for discard in preference to cells that fall within agreed traffic limits.

The header error control (HEC) field is an 8-b error code that can be used to correct single-bit errors in the header and to detect double-bit errors.

Figure 72.7b shows the cell header format internal to the network. The generic flow control field, which performs end-to-end functions, is not retained. Instead, the virtual path identifier field is expanded from 8 to 12 b. This allows support for an expanded number of virtual paths internal to the network to include those supporting subscribers and those required for network management.

72.2.3 Traditional Local-Area Networks

The two most widely used traditional LANs are carrier-sense multiple access/collision detection (CSMA/CD) (Ethernet) and token ring.

72.2.3.1 Carrier-Sense Multiple Access/Collision Detection (Ethernet)

The Ethernet LAN standards was originally designed to work over a bus LAN topology. With the bus topology, all stations attach, through appropriate interfacing hardware, directly to a linear transmission medium, or bus. A transmission from any station propagates the length of the medium in both directions and can be received by all other stations.

Transmission is in the form of frames containing addresses and user data. Each station monitors the medium and copies frames addressed to itself. Because all stations share a common transmission link, only one station can successfully transmit at a time, and some form of medium access control technique is needed to regulate access.

More recently, a star topology has been used. In the star LAN topology, each station attaches to a central node, referred to as the star coupler, via two point-to-point links, one for transmission in each direction. A transmission from any one station enters the central node and is retransmitted on all of the outgoing links. Thus, although the arrangement is physically a star, it is logically a bus: a transmission from any station is received by all other stations, and only one station at a time may successfully transmit. Thus, the medium access control techniques used for the star topology are the same as for bus and tree.

With CSMA/CD, a station wishing to transmit first listens to the medium to determine if another transmission is in progress (carrier sense). If the medium is idle, the station may transmit. It may happen that two or more stations attempt to transmit at about the same time. If this happens, there will be a collision; the data from both transmissions will be garbled and not received successfully. Thus, a procedure is needed that specifies what a station should do if the medium is found busy and what it should do if a collision occurs:

1. If the medium is idle, transmit
2. If the medium is busy, continue to listen until the channel is idle, then transmit immediately
3. If a collision is detected during transmission, immediately cease transmitting
4. After a collision, wait a random amount of time and then attempt to transmit again (repeat from step 1)

Figure 72.8 illustrates the technique. At time t_0, station A begins transmitting a packet addressed to D. At t_1, both B and C are ready to transmit. B senses a transmission and so defers. C, however, is still unaware of A's transmission and begins its own transmission. When A's transmission reaches C, at t_2, C detects the collision and ceases transmission. The effect of the collision propagates back to A, where it is detected some time later, t_3, at which time A ceases transmission.

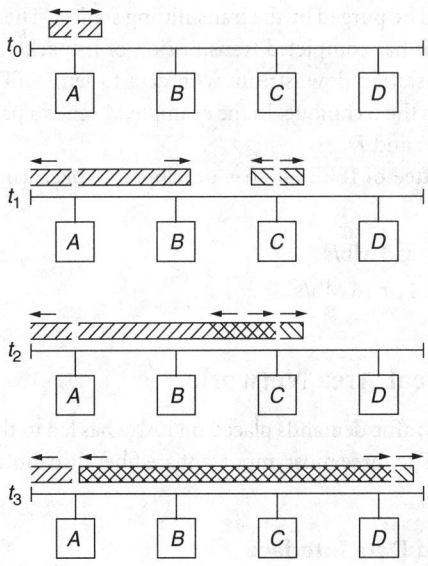

FIGURE 72.8 CSMA/CD operation.

The Institute of Electrical and Electronics Engineers (IEEE) LAN standards committee has developed a number of versions of the CSMA/CD standard, all under the designation IEEE 802.3. The following options are defined:

- 10-Mb/s bus topology using coaxial cable
- 10-Mb/s star topology using unshielded twisted pair
- 100-Mb/s star topology using unshielded twisted pair
- 100-Mb/s star topology using optical fiber

The last two elements in the list, both known as fast Ethernet, are the newest addition to the IEEE 802.3 standard. Both provide a higher data rate over shorter distances than traditional Ethernet. The token ring LAN standards operates over a ring topology LAN. In the ring topology, the LAN or metropolitan-area network (MAN) consists of a set of *repeaters* joined by point-to-point links in a closed loop. The repeater is a comparatively simple device, capable of receiving data on one link and transmitting it, bit by bit, on the other link as fast as it is received, with no buffering at the repeater. The links are unidirectional; that is, data are transmitted in one direction only and all oriented in the same way. Thus, data circulate around the ring in one direction (clockwise or counterclockwise).

72.2.3.2   Token Ring

Each station attaches to the network at a repeater and can transmit data onto the network through the repeater. As with the bus topology, data are transmitted in frames. As a frame circulates past all of the other stations, the destination station recognizes its address and copies the frame into a local buffer as it goes by. The frame continues to circulate until it returns to the source station, where it is removed.

Because multiple stations share the ring, medium access control is needed to determine at what time each station may insert frames.

The token ring technique is based on the use of a token packet that circulates when all stations are idle. A station wishing to transmit must wait until it detects a token passing by. It then seizes the token by changing 1 b in the token, which transforms it from a token to a start-of-packet sequence for a data packet. The station then appends and transmits the remainder of the fields (e.g., destination address) needed to construct a data packet.

There is now no token on the ring, so other stations wishing to transmit must wait. The packet on the ring will make a round trip and be purged by the transmitting station. The transmitting station will insert a new token on the ring after it has completed transmission of its packet. Once the new token has been inserted on the ring, the next station downstream with data to send will be able to seize the token and transmit. Figure 72.9 illustrates the technique. In the example, A sends a packet to C, which receives it and then sends its own packets to A and D.

The IEEE 802.5 subcommittee of IEEE 802 has developed a token ring standard with the following alternative configurations:

- Unshielded twisted pair at 4 Mb/s
- Shielded twisted pair at 4 or 16 Mb/s

72.2.4   High-Speed Local-Area Networks

In recent years, the increasing traffic demands placed on LANs has led to the development of a number of high-speed LAN alternatives. The three most important are fiber distributed data interface (FDDI), Fibre Channel, and ATM LANs.

72.2.4.1   Fiber Distributed Data Interface

One of the newest LAN standards is the fiber distributed data interface [Mills 1995]. The topology of FDDI is ring. The medium access control technique employed is token ring, with only minor differences from the

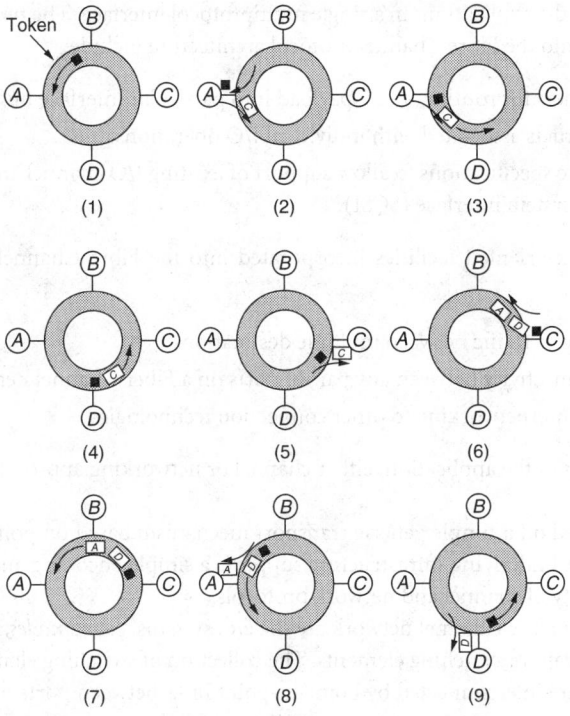

FIGURE 72.9 Token ring operation.

IEEE token ring specification. The medium specified is 100-Mb/s optical fiber. The medium specification specifically incorporates measures designed to ensure high availability.

72.2.4.2 Fibre Channel

As the speed and memory capacity of personal computers, workstations, and servers have grown, and as applications have become ever more complex with greater reliance on graphics and video, the requirement for greater speed in delivering data to the processor has grown. This requirement affects two methods of data communications with the processor: I/O channel and network communications.

An I/O channel is a direct point-to-point or multipoint communications link, predominantly hardware based and designed for high speed over very short distances. The I/O channel transfers data between a buffer at the source device and a buffer at the destination device, moving only the user contents from one device to another, without regard to the format or meaning of the data. The logic associated with the channel typically provides the minimum control necessary to manage the transfer plus simple error detection. I/O channels typically manage transfers between processors and peripheral devices, such as disks, graphics equipment, compact disc–read-only memories (CD-ROMs), and video I/O devices.

A network is a collection of interconnected access points with a software protocol structure that enables communication. The network typically allows many different types of data transfer, using software to implement the networking protocols and to provide flow control, error detection, and error recovery.

Fibre Channel is designed to combine the best features of both technologies: the simplicity and speed of channel communications with the flexibility and interconnectivity that characterize protocol-based network communications [Stephens and Dedek 1995]. This fusion of approaches allows system designers to combine traditional peripheral connection, host-to-host internetworking, loosely coupled processor clustering, and multimedia applications in a single multiprotocol interface. The types of channel-oriented facilities incorporated into the Fibre Channel protocol architecture include:

- Data-type qualifiers for routing frame payload into particular interface buffers
- Link-level constructs associated with individual I/O operations
- Protocol interface specifications to allow support of existing I/O channel architectures, such as the small computer system interface (SCSI)

The types of network-oriented facilities incorporated into the Fibre Channel protocol architecture include:

- Full multiplexing of traffic between multiple destinations
- Peer-to-peer connectivity between any pair of ports on a Fiber Channel network
- Capabilities for internetworking to other connection technologies

Depending on the needs of the application, either channel or networking approaches can be used for any data transfer.

Fibre Channel is based on a simple generic transport mechanism based on point-to-point links and a switching network. This underlying infrastructure supports a simple encoding and framing scheme that in turn supports a variety of channel and network protocols.

The key elements of a Fibre Channel network are the end systems, called *nodes,* and the network itself, which consists of one or more switching elements. The collection of switching elements is referred to as a *fabric.* These elements are interconnected by point-to-point links between ports on the individual nodes and switches. Communication consists of the transmission of frames across the point-to-point links.

Figure 72.10 illustrates these basic elements. Each node includes one or more ports, called N_Ports, for interconnection. Similarly, each fabric switching element includes one or more ports, called F_ports. Interconnection is by means of bidirectional links between ports. Any node can communicate with any other node connected to the same fabric using the services of the fabric. All routing of frames between N_Ports is done by the fabric. Frames are buffered within the fabric, making it possible for different nodes to connect to the fabric at different data rates.

A fabric can be implemented as a single fabric element, as depicted in Figure 72.10, or as a more general network of fabric elements. In either case, the fabric is responsible for buffering and routing frames between source and destination nodes.

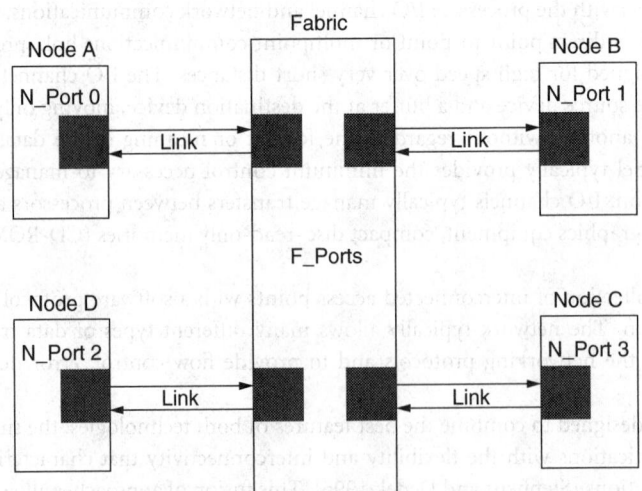

FIGURE 72.10   Fibre Channel port types.

The Fibre Channel network is quite different from the other LANs that we have examined so far. Fibre Channel is more like a tradional circuit-switching or packet-switching network in contrast to the typical shared-medium LAN. Thus, Fibre Channel need not be concerned with medium access control (MAC) issues. Because it is based on a switching network, the Fibre Channel scales easily in terms of both data rate and distance covered. This approach provides great flexibility. Fibre Channel can readily accommodate new transmission media and data rates by adding new switches and nodes to an existing fabric. Thus, an existing investment is not lost with an upgrade to new technologies and equipment. Further, as we shall see, the layered protocol architecture accommodates existing I/O interface and networking protocols, preserving the pre-existing investment.

72.2.4.3 Asynchronous Transfer Mode Local-Area Networks

High-speed LANs such as FDDI and Fiber Channel, provide a means for implementing a backbone LAN to tie together numerous small LANs in an office environment. However, there is another solution, known as the ATM LAN, that seems likely to become a major factor in local-area networking [Biagioni et al. 1993, Newman 1994]. The ATM LAN is based on the asynchronous ATM technology used in wide-area networks. The ATM LAN approach has several important strengths, two of which are as follows:

1. The ATM technology provides an open-ended growth path for supporting attached devices. ATM is not constrained to a particular physical medium or data rate. A dedicated data rate between workstations of 155 Mb/s is practical today. As demand increases and prices continue to drop, ATM LANs will be able to support devices at dedicated speeds, which are standardized for ATM, of 622 Mb/s, 2.5 Gb/s, and above.
2. ATM is becoming the technology of choice for wide-area networking. ATM can therefore be used effectively to integrate LAN and WAN configurations.

To understand the role of the ATM LAN, consider the following classification of LANs into three generations:

- *First generation:* Typified by the CSMA/CD and token ring LANs, the first generation provided terminal-to-host connectivity and supported client/server architectures at moderate data rates.
- *Second generation:* Typified by FDDI, the second generation responds to the need for backbone LANs and for support of high-performance workstations.
- *Third generation:* Typified by ATM LANs, the third generation is designed to provide the aggregate throughputs and real-time transport guarantees that are needed for multimedia applications.

Typical requirements for a third-generation LAN include the following:

1. They must support multiple, guaranteed classes of service. A live video application, for example, may require a guaranteed 2-Mb/s connection for acceptable performance, whereas a file transfer program can utilize a *background* class of service.
2. They must provide scalable throughput that is capable of growing both per-host capacity (to enable applications that require large volumes of data in and out of a single host) and aggregate capacity (to enable installations to grow from a few to several hundred high-performance hosts).
3. They must facilitate the interworking between LAN and WAN technology.

ATM is ideally suited to these requirements. Using virtual paths and virtual channels, multiple classes of service are easily accommodated, either in a preconfigured fashion (permanent connections) or on demand (switched connections). ATM is easily scalable by adding more ATM switching nodes and using higher data rates for attached devices. Finally, with the increasing acceptance of cell-based transport for wide-area networking, the use of ATM for a premises network enables seamless integration of LANs and WANs.

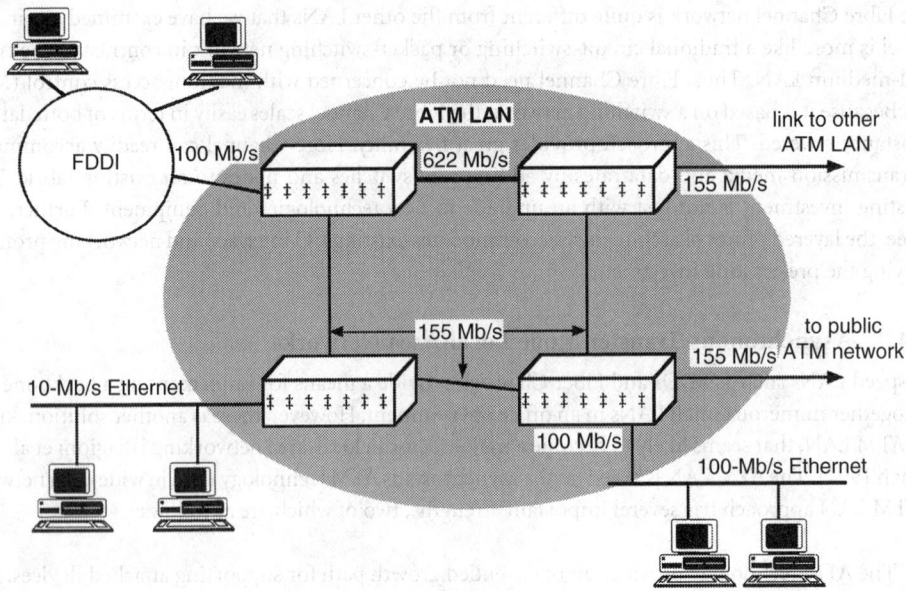

FIGURE 72.11   Example of ATM LAN configuration.

The term ATM LAN has been used by vendors and researchers to apply to a variety of configurations. At the very least, ATM LAN implies the use of ATM as a data transport protocol somewhere within the local premises. Among the possible types of ATM LANs are the following:

- *Gateway to ATM WAN:* An ATM switch acts as a router and traffic concentrator for linking a premises network complex to an ATM WAN.
- *Backbone ATM switch:* Either a single ATM switch or a local network of ATM switches interconnect other LANs.
- *Workgroup ATM:* High-performance multimedia workstations and other end systems connect directly to an ATM switch.

These are all pure configurations. In practice, a mixture of two or all three of these types of networks is used to create ATM LAN.

Figure 72.11 shows an example of a backbone ATM LAN that includes links to the outside world. In this example, the local ATM network consists of four switches interconnected with high-speed point-to-point links running at the standardized ATM rates of 155 and 622 Mb/s. On the premises, there are three other LANs, each of which has a direct connection to one of the ATM switches. The data rate from an ATM switch to an attached LAN conforms to the native data rate of that LAN. For example, the connection to the FDDI network is at 100 Mb/s. Thus, the switch must include some buffering and speed conversion capability to map the data rate from the attached LAN to an ATM data rate. The ATM switch must also perform some sort of protocol conversion from the MAC protocol used on the attached LAN to the ATM cell stream used on the ATM network. A simple approach is for each ATM switch that attaches to a LAN to function as a bridge or router.

An ATM LAN configuration such as that shown in Figure 72.11 provides a relatively painless method for inserting a high-speed backbone into a local environment. As the on-site demand rises, it is a simple matter to increase the capacity of the backbone by adding more switches, increasing the throughput of each switch, and increasing the data rate of the trunks between switches. With this strategy, the load on individual LANs within the premises can be increased and the number of LANs can grow.

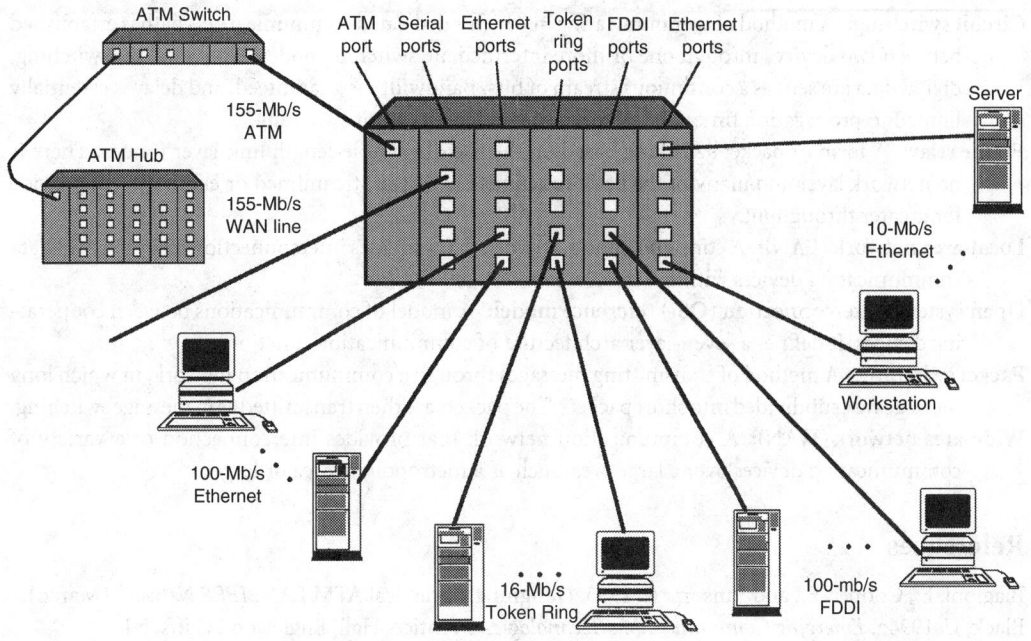

FIGURE 72.12   ATM LAN hub configuration.

However, this simple backbone ATM LAN does not address all of the needs for local communications. In particular, in the simple backbone configuration, the end systems (workstations, servers, etc.) remain attached to shared-media LANs with the limitations on data rate imposed by the shared medium.

A more advanced, and more powerful, approach is to use ATM technology in a hub. Figure 72.12 suggests the capabilities that can be provided with this approach. Each ATM hub includes a number of ports that operate at different data rates and use different protocols. Typically, such a hub consists of a number of rack-mounted modules, with each module containing ports of a given data rate and protocol.

The key difference between the ATM hub shown in Figure 72.12 and the ATM nodes depicted in Figure 72.11 is the way in which individual end systems are handled. Notice that in the ATM hub, each end system has a dedicated point-to-point link to the hub. Each end system includes the communications hardware and software to interface to a particular type of LAN, but in each case the LAN contains only two devices: the end system and the hub! For example, each device attached to a 10-Mb/s Ethernet port operates using the CSMA/CD protocol at 10 Mb/s. However, because each end system has its own dedicated line, the effect is that each system has its own dedicated 10-Mb/s Ethernet. Therefore, each end system can operate at close to the maximum 10-Mb/s data rate.

The use of a configuration such as that of either Figure 72.11 or Figure 72.12 has the advantage that existing LAN installations and LAN hardware, so-called legacy LANs, can continue to be used while ATM technology is introduced. The disadvantage is that the use of such a mixed-protocol environment requires the implementation of some sort of protocol conversion capability. A simpler approach, but one that requires that end systems be equipped with ATM capability, is to implement a pure ATM LAN.

Defining Terms

Asynchronous transfer mode (ATM):   A form of packet transmission using fixed-size packets, called cells. ATM is the data transfer interface for broadband-integrated services digital network (B-ISDN). Unlike X.25, ATM does not provide error control and flow control mechanisms.

Circuit switching: A method of communicating in which a dedicated communications path is established between two devices through one or more intermediate switching nodes. Unlike packet switching, digital data are sent as a continuous stream of bits. Bandwidth is guaranteed, and delay is essentially limited to propagation time. The telephone system uses circuit switching.

Frame relay: A form of packet switching based on the use of variable-length link-layer frames. There is no network layer and many of the basic functions have been streamlined or eliminated to provide for greater throughput.

Local-area network (LAN): A communication network that provides interconnection of a variety of data communicating devices within a small area.

Open systems interconnection (OSI) reference model: A model of communications between cooperating devices. It defines a seven-layer architecture of communication functions.

Packet switching: A method of transmitting messages through a communication network, in which long messages are subdivided into short packets. The packets are then transmitted as in message switching.

Wide-area network (WAN): A communication network that provides interconnection of a variety of communicating devices over a large area, such as a metropolitan area or larger.

References

Biagioni, E., Cooper, E., and Sansom, R. 1993. Designing a practical ATM LAN. *IEEE Network* (March).

Black, U. 1994a. *Emerging Communications Technologies.* Prentice–Hall, Englewood Cliffs, NJ.

Black, U. 1994b. *Frame Relay Networks: Specifications and Implementations.* McGraw–Hill, New York.

Boudec, J. 1992. The asynchronous transfer mode: a tutorial. *Comput. Networks ISDN Syst.* (May).

Comer, D. 1995. *Internetworking with TCP/IP, Volume I: Principles, Protocols, and Architecture.* Prentice–Hall, Englewood Cliffs, NJ.

Halsall, F. 1996. *Data Communications, Computer Networks, and Open Systems.* Addison–Wesley, Reading, MA.

Jain, B. and Agrawala, A. 1993. *Open Systems Interconnection.* McGraw–Hill, New York.

Mills, A. 1995. *Understanding FDDI.* Prentice–Hall, Englewood Cliffs, NJ.

Newman, P. 1994. ATM local area networks. *IEEE Commun. Mag.* (March).

Prycker, M. 1993. *Asynchronous Transfer Mode: Solution for Broadband ISDN.* Ellis Horwood, New York.

Smith, P. 1993. *Frame Relay: Principles and Applications.* Addison–Wesley, Reading, MA.

Stallings, W. 1997a. *Data and Computer Communications,* 5th ed. Prentice–Hall, Englewood Cliffs, NJ.

Stallings, W. 1997b. *Local and Metropolitan Area Networks,* 5th ed. Prentice–Hall, Englewood Cliffs, NJ.

Stephens, G. and Dedek, J. 1995. *Fiber Channel.* Ancot, Menlo Park, CA.

Further Information

For a more detailed discussion of the topics in this chapter, see the following: for TCP/IP, Stallings [1997a] and Comer [1995]; for OSI, Halsall [1996] and Jain and Agrawala [1993]; for traditional and high-speed WANs, Stallings [1997a] and Black [1994a]; for traditional and high-speed LANs, Stallings [1997b].

73

Routing Protocols

73.1 Introduction **73**-1
73.2 Bridges/Switches **73**-2
    Transparent Bridging • Spanning Tree Algorithm • Dealing
    with Failures • Source Route Bridging • Properties of Source
    Route Bridging
73.3 Routers **73**-7
    Types of Routing Protocols • Calculating Routes
    • Interdomain Routing

Radia Perlman
Sun Microsystems Laboratories

73.1 Introduction

Computer networking can be a very confusing field. Terms such as network, subnetwork, domain, **local area network** (**LAN**), internetwork, **bridge**, **router**, and **switch** are often ill-defined. Taking the simplest view, we know that the **data-link layer** of a network delivers a packet of information to a neighboring machine, the **network layer** routes through a series of packet switches to deliver a packet from source to destination, and the transport layer recovers from lost, duplicated, and out-of-order packets. But the bridge standards choose to place routing in the data-link layer, and the X.25 network layer puts the onus on the network layer to prevent packet loss, duplication, or misordering.

In this chapter we discuss routing protocols, attempting to avoid philosophical questions such as which layer something is, or whether something is an internetwork or a network. For a more complete treatment of these and other questions, see Perlman [1999] and other chapters in this section of this Handbook.

A network consists of several computers interconnected with various types of links. One type is a *point-to-point link*, which connects exactly two machines. It can either be a dedicated link (e.g., a wire connecting the two machines) or a dial-on-demand link, which can be connected when needed (e.g., when there is traffic to send over the link). Another category of link is a **multiaccess link**. Examples are LANs, asynchronous transfer mode (ATM), and X.25. Indeed, any network can be considered a link. When one protocol uses a network as a link, it is referred to as a **tunnel**. A multiaccess link presents special challenges to the routing protocol because it has two headers with addressing information. One header (which for simplicity we call the network layer header) gives the addresses of the ultimate source and destination. The other header (which for simplicity we call the data-link header) gives the addresses of the transmitter and receiver on that particular link. (See Figure 73.1, which shows a path incorporating a multiaccess link and a packet with two headers.) The terminology cannot be taken too seriously. It is not uncommon to tunnel IP over IP, for example, for an employee to connect to the corporate firewall across the Internet, using an encrypted connection. Although the "data link" in that case is an entire network, from the point of view of the protocols using it as a link, it can be considered just a data link.

We start by describing the routing protocols used by devices called bridges or switches, and then describe the routing protocols used by routers. Although the purpose of this chapter is to discuss the algorithms

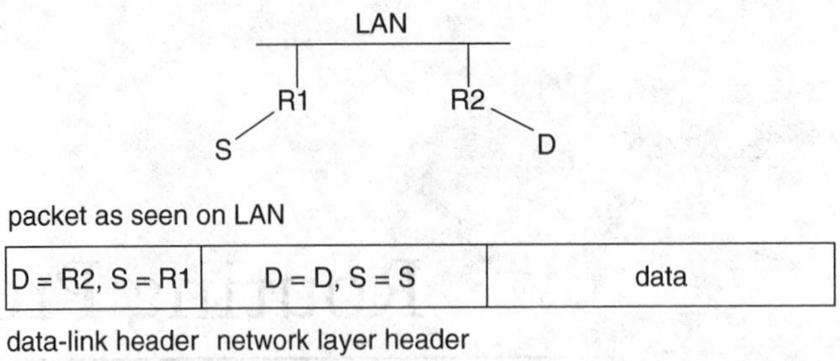

FIGURE 73.1   A multiaccess link and a packet with two headers.

generically, the ones we have chosen are ones that are in widespread use. In many cases we state timer values and field lengths chosen by the implementation, but it should be understood that these are general-purpose algorithms. The purpose of this chapter is not as a reference on the details of particular implementations, but to understand the variety of routing algorithms and the trade-offs between them.

73.2 Bridges/Switches

The term **switch** is generally a synonym for a bridge, although the term "switch" is sometimes used to refer to any forwarding box, including routers. So we use the term "bridge" to avoid ambiguity. The characteristic of a bridge that differentiates it from a router is that a bridge does its routing in the data-link layer, whereas a router does it in the network layer. But that becomes a matter of philosophy and history in which a standards body defined a particular algorithm rather than any property of the protocol itself. If the protocol was defined by a data-link layer standards body, the box implementing it becomes a bridge. If the same protocol were defined by a network layer standards body, the box implementing it would become a router. The algorithm itself could, in theory, be implemented in either layer.

There were routers before there were bridges. What happened was the invention of the so-called local area network, which is a multiaccess link. Unfortunately, the world did not think of a LAN as a multiaccess link, which would be a component in some larger network. Instead, perhaps because the inclusion of the word "network" into the name local area network, many systems were designed with the assumption that the LAN itself was the entire network, and the systems were designed without a network layer, making these protocols unroutable, at least through the existing network layer protocols.

There are two types of bridging technologies specified in the standards. One is known as the transparent bridge. This technology had as a goal complete backward compatibility with existing LAN-only systems. The other technique is known as source route bridging, which can only be considered different from a network layer protocol because the standards committee that adopted it was empowered to define data-link protocols, and because the fields necessary for running this protocol were stuck into a header that was defined as a data-link header. Source route bridges are becoming rare in practice, although they are still defined by the standards.

73.2.1 Transparent Bridging

The goal of transparent bridging was to invent a box that would interconnect LANs even though the stations were designed with protocols that only worked on a LAN; that is, they lacked a network layer able to cooperate with routers, devices that were designed for forwarding packets.

The basic idea of a transparent bridge is something that is attached to two or more LANs. On each LAN, the bridge listens promiscuously to all packets, stores them, and forwards each packet onto each other LAN when given permission by the LAN protocol on that LAN. An enhancement is to have the bridge learn, based on addresses in the LAN header, of packets received by the bridge, where the stations reside, so that the bridge does not unnecessarily forward packets. (See Figure 73.2, where a bridge has learned some of

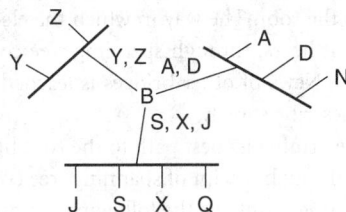

FIGURE 73.2   A bridge learning station addresses.

the station addresses.) The bridge learns from the source field in the LAN header, and forwards based on the destination address. For example, in Figure 73.2, when S transmits a packet with destination address D, the bridge learns which interface S resides on, and then looks to see if it has already learned where D resides. If the bridge does not know where D is, then the bridge forwards the packet onto all interfaces (except the one from which the packet was received). If the bridge does know where D is, then the bridge forwards it only onto the interface where D resides, or if the packet arrived from that interface, the bridge discards the packet.

This simple idea only works in a loop-free topology. Loops create problems:

- Packets that will not die: there is no **hop** count in the header, as there would be in a reasonable network layer protocol, to eliminate a packet that is traversing a loop.
- Packets that proliferate uncontrollably: network layer forwarding does not, in general, create duplicate packets, because each router forwards the packet in exactly one direction, and specifies the next recipient. With bridges, a bridge might forward a packet in multiple directions, and many bridges on the LAN might forward a packet. Each time a bridge forwards in multiple directions, or more than one bridge picks up a packet for forwarding, the number of copies of the packet grows.
- If there is more than one path from a bridge to a given station, that bridge cannot learn the location of the station, because packets from that source will arrive on multiple interfaces.

One possibility was to simply declare that bridged topologies must be physically loop-free. But that was considered unacceptable for the following reasons:

- The consequences of accidental misconfiguration could be disastrous; any loop in some remote section of the bridged network might spawn such an enormous number of copies of packets that it would bring down the entire bridged network. It would also be very difficult to diagnose and to fix.
- Loops are good because a loop indicates an alternate path in case of failure.

The solution was the spanning tree algorithm, which is constantly run by bridges to determine a loop-free subset of the current topology. Data packets are only transmitted along the tree found by the spanning tree algorithm. If a bridge or link fails, or if bridges or links start working, the spanning tree algorithm will compute a new tree. The spanning tree algorithm is described in the next section.

73.2.2   Spanning Tree Algorithm

The basic idea behind the spanning tree algorithm is that the bridges agree upon one bridge to be the root of the spanning tree. The tree of shortest paths from that root bridge to each LAN is the calculated spanning tree.

How do the bridges decide on the root? Each bridge comes, at manufacture time, with a globally unique 48-bit **IEEE 802 address**, usually one per interface. A bridge chooses one of the 48-bit addresses that it owns as its identifier (ID). Because each bridge has a unique number, it is a simple matter to choose the one with the smallest number. However, because some network managers like to easily influence which bridge will be chosen, there is a configurable priority value that acts as a more significant field tacked onto the ID. The concatenated number consisting of priority and ID is used in the election, and the bridge with the smallest value is chosen as the root. The way in which the election proceeds is that each bridge assumes itself to be the root unless it hears, through spanning tree configuration messages, of a bridge with a smaller value for priority–ID. News of other bridges is learned through receipt of spanning tree configuration messages, which we describe shortly.

The next step is for a bridge to determine its best path to the root bridge and its own cost to the root. This information is also discovered through receipt of spanning tree configuration messages.

A spanning tree configuration message contains the following, among other information:

- Priority–ID of best known root
- Cost from transmitting bridge to root
- Priority–ID of transmitting bridge

A bridge keeps the best configuration message received on each of its interfaces. The fields in the message are concatenated together, from most significant to least significant, as root's priority–ID to cost to root to priority–ID of transmitting bridge. This concatenated quantity is used to compare messages. The one with the smaller quantity is considered better. In other words, only information about the best-known root is relevant. Then, information from the bridge closest to that root is considered, and then the priority–ID of the transmitting bridge is used to break ties.

Given a best received message on each interface, B chooses the root as follows:

- Itself, if its own priority–ID beats any of the received value, else
- The smallest received priority–ID value

B chooses its path to the root as follows:

- Itself, if it considers itself to be the root, else
- The minimum cost through each of its interfaces to the best-known root

Each interface has a cost associated with it, either as a default or configured. The bridge adds the interface cost to the cost in the received configuration message to determine its cost through that interface.

B chooses its own cost to the root as follows:

- 0, if it considers itself to be the root, else
- The cost of the minimum cost path chosen in the previous step

B now knows what it would transmit as a configuration message, because it knows the root's priority–ID, its own cost to that root, and its own priority–ID. If B's configuration message is better than any of the received configuration messages on an interface, then B considers itself the *designated bridge* on that interface, and transmits configuration messages on that interface. If B is not the designated bridge on an interface, then B will not transmit configuration messages on that interface.

Each bridge determines which of its interfaces are in the spanning tree. The interfaces in the spanning tree are as follows:

- The bridge's path to the root: if more than one interface gives the same minimal cost, then exactly one is chosen. Also, if this bridge is the root, then there is no such interface.
- Any interfaces for which the bridge is designated bridge are in the spanning tree.

If an interface is not in the spanning tree, the bridge continues running the spanning tree algorithm but does not transmit any data messages (messages other than spanning tree protocol messages) to that interface, and ignores any data messages received on that interface.

If the topology is considered a graph with two types of nodes, bridges and LANs, the following is the reasoning behind why this yields a tree:

- The root bridge is the root of the tree.
- The unique parent of a LAN is the designated bridge.
- The unique parent of a bridge is the interface that is the best path from that bridge to the root.

73.2.3 Dealing with Failures

The root bridge periodically transmits configuration messages (with a configurable timer on the order of 1 s). Each bridge transmits a configuration message on each interface for which it is designated, after receiving one on the interface which is that bridge's path to the root. If some time elapses (a configurable value with default on the order of 15 s) in which a bridge does not receive a configuration message on an interface, the configuration message learned on that interface is discarded.

In this way, roughly 15 s after the root or the path to the root has failed, a bridge will discard all information about that root, assume itself to be the root, and the spanning algorithm will compute a new tree.

73.2.3.1 Eliminating Temporary Loops

In a routing algorithm, the nodes learn information at different times. During the time after a topology change and before all nodes have adapted to the new topology, there are temporary loops or temporary partitions (no way to get from some place to some other place). Because temporary loops are so disastrous with bridges (because of the packet proliferation problem), bridges are conservative about bringing an interface into the spanning tree. There is a timer (on the order of 30 s, but configurable). If an interface was not in the spanning tree, but new events convince the bridge that the interface should be in the spanning tree, the bridge waits for this timer to expire before forwarding data messages to and from the interface.

73.2.3.2 Properties of Transparent Bridges

Transparent bridges have some good properties:

- They are plug-and-play; that is, no configuration is required.
- They fulfill the goal of making no demands on end stations to interact with the bridges in any way.

They have some disadvantages:

- The topology is confined to a spanning tree, which means that some paths are not optimal.
- The spanning tree algorithm is purposely slow about starting to forward on an interface (to prevent temporary loops).

The overhead of the spanning tree algorithm is insignificant. The memory required for a bridge that has k interfaces is about $k * 50$ bytes, regardless of how large the actual network is. The bandwidth consumed per LAN (once the algorithm settles down) is a constant, regardless of the size of the network (because only the designated bridge periodically issues a spanning tree message, on the order of once a second). At worst, for the few seconds while the algorithm is settling down after a topology change, the bandwidth on a LAN is at most multiplied by the number of bridges on that LAN (because for a while more than one bridge on that LAN will think it is the designated bridge). The central processing unit (CPU) consumed by a bridge to run the spanning tree algorithm is also a constant, regardless of the size of the network.

73.2.4 Source Route Bridging

Source route bridging was a competing proposal in the IEEE 802 committee. Initially, 802.1, the committee standardizing bridges, chose transparent bridges, but the source routing proposal resurfaced in the 802.5 (Token Ring) committee as a method of interconnecting token rings.

Source route bridging did not have as a goal the ability to work with existing stations. As such, there is really no technical property of source route bridging that makes it natural for it to appear in the data-link layer. It places as much burden on a station as a network layer protocol (e.g., IP, IPX, Appletalk, DECnet). The only reason it is considered a bridging protocol rather than a routing protocol is that it was done within IEEE 802, a committee whose charter was LANs, rather than a network layer committee. The fields to support source route bridging also appear in the LAN header (because it was defined by a data-link layer committee), but the actual algorithm could, in theory, have been done in the network layer.

The idea behind source route bridging is that the data-link header is expanded to include a route. The stations are responsible for discovering routes and maintaining route caches. Discovery of a route to station D is done by source S launching a special type of packet, an *all-paths explorer* packet, which spawns copies every time there is a choice of path (multiple bridges on a LAN or a bridge with more than two ports). Each copy of the explorer packet keeps a history of the route it has taken. This process, although it might be alarmingly prolific in richly connected topologies, does not spawn infinite copies of the explorer packet for two reasons:

- The maximum length route is 14 hops; and so after 14 hops, the packet is no longer forwarded.
- A bridge examines the route before forwarding it onto a LAN, and will not forward onto that LAN if that LAN already appears in the route.

When D receives the (many) copies of the packet, it can choose a path based on criteria such as when it arrived (perhaps indicating the path is faster), or on length of path, or on maximum packet size along the route, which is calculated along with the route.

A route consists of an alternating list of LAN numbers and bridge numbers: 12 bits are allocated for the LAN number, and 4 bits for the bridge number. The bridge number at 4 bits will obviously not distinguish between all the bridges. Instead, the bridge number only distinguishes bridges that interconnect the same pair of LANs. For example, if the route is LAN A, bridge 3, LAN B, bridge 7, LAN C, and it is received by a bridge on the port that bridge considers to be LAN A, then the bridge looks forward in the route, finds the next LAN number (B), and then looks up the bridge number it has been assigned with respect to the LAN pair (A, B). If it has been assigned 3 for that pair, then it will forward the packet onto the port it has configured to be B.

There are three types of source route bridge packets:

1. Specifically routed: the route is in the header and the packet follows the specified route.
2. All-paths explorer: the packet spawns copies of itself at each route choice, and each copy keeps track of the route it has traversed so far.
3. Single copy broadcast: this acts as an all-paths explorer except that this type of packet is only accepted from ports in the spanning tree and only forwarded to ports in the spanning tree. A single copy will be delivered to the destination, and the accumulated route at the destination will be the path through the spanning tree.

To support the third type of packet, source route bridges run the same spanning tree algorithm as described in the transparent bridge section.

A source route bridge is configured with a 12-bit LAN number for each of its ports, along with a 4-bit bridge number for each possible pair of ports. In cases where a bridge has too many ports to make it feasible to configure a bridge number for each port pair, many implementations pretend there is an additional LAN inside the bridge, which must be configured with a LAN number, say, n. Paths through the bridge from LAN j to LAN k (where j and k are real LANs) look like they go from j to n to k. Each time a packet goes through such a bridge, it uses up another available hop in the route, but the advantage of this scheme is that because no other bridge connects to LAN n, the bridge does not need to be configured with any bridge numbers; it can always use 1.

The algorithm the source route bridge follows for each type of packet is as follows:

- A specifically routed packet is received on the port that the bridge considers LAN j: do a scan through the route to find j. If j is not found, drop the packet. If j is found, scan to the next LAN in the route. If the LAN is k, and this bridge has a port configured as k, then find the bridge number specified between j and k, say, b. If this bridge is configured to be b with respect to the pair (j, k), then forward the packet onto LAN k. Otherwise, drop the packet.
- An all-paths explorer packet is received on the port that the bridge considers LAN j: if j is not the last hop in the route, drop the packet. Otherwise, for each other port, forward the packet onto that port unless the destination port's LAN number is already in the accumulated route. If the destination LAN number, say, k, is not in the route, then append (b, k) to the route, where k is the

destination LAN number and b is the bridge's number with respect to (j, k). If the route is already full, then drop the packet.

- A single copy broadcast is received on the port that the bridge considers LAN j: if the port from which it was received is not in the spanning tree, drop the packet; otherwise, treat it as an all-paths explorer except do not forward onto ports in the spanning tree.

The standard was written from the point of view of the bridge and did not specify end-station operation. For example, there are several strategies end stations might use to maintain their route cache. If S wants to talk to D, and does not have D in its cache, S might send an all-paths explorer. Then D might at that point choose a route from the received explorers, or it might return each one to the source so that the source could make the choice. Or it might choose a route but send an explorer back to the source so that the source could independently make a route choice. Or maybe S, instead of sending an all-paths explorer, might send a single copy explorer, and D might respond with an all-paths explorer.

73.2.5 Properties of Source Route Bridging

Relative to transparent bridges, source route bridges have the following advantages:

- It is possible to get an optimal route from source to destination.
- It is possible to spread traffic load around the network rather than concentrating it into the spanning tree.
- It computes a maximum packet size on the path.
- A bridge that goes down will not disrupt conversations that have computed paths that do not go through that bridge.

Relative to transparent bridges, source route bridges have the following disadvantages:

- In a topology that is not physically a tree, the exponential proliferation of explorer packets is a serious bandwidth drain.
- It requires a lot of configuration.
- It makes end stations more complicated because they have to maintain a route cache.

Because source route bridging is a routing protocol that requires end-station cooperation, it must in fairness be compared as well against network layer protocols. Against a network layer protocol such as IP, IPX, DECnet, Appletalk, CLNP, etc., source route bridging has the following advantages:

- It computes the maximum packet size on the path.
- Although it requires significant configuration of bridges, it does not require configuration of endnodes (as in IP, although IPX, DECnet Phase V, and Appletalk also avoid configuration of endnodes).

However, relative to network layer protocols, source route bridging has the following disadvantages:

- The exponential overhead of the all-paths explorer packets.
- The delay before routes are established and data can be exchanged, unless data are carried as an all-paths explorer or single copy broadcast.

73.3 Routers

In this section we discuss network layer protocols and routing algorithms generically. Network layer protocols can be connection oriented or connectionless. A connection-oriented protocol sets up a path, and the routers along the path of a conversation keep state about the information. A connectionless protocol just puts a source and destination address on the packet and launches it. Each packet is self-contained and is routed independently of other packets from the same conversation. Different packets from the same source to the same destination might take different paths.

Another dimension in which network layer protocols can differ is whether they provide reliable or datagram service. Datagram is best-effort service. With a reliable service, the network layer makes sure that every packet is delivered and refuses to deliver packet n until it manages to deliver $n - 1$.

Examples of datagram connectionless network layers are IPv4, IPv6, IPX, DECnet, CLNP, and Appletalk. An example of a datagram connection-oriented network layer is ATM. An example of a reliable connection-oriented network layer is X.25. The last possibility, a reliable connectionless network layer, is not possible, and fortunately there are no examples of standards attempting to accomplish this.

The distinction between connection-oriented and connectionless network layers is blurring. In a connectionless network, routers often keep caches of recently seen addresses, and forward much more efficiently when the destination is in the cache. Usually all but the first packet of a conversation are routed very quickly because the destination is in the router's cache. As such, the first packet of the conversation acts as a route setup, and the routers along the path are keeping state, in some sense. Also, there is talk of adding lightweight connections to connectionless network layer protocols for bandwidth reservation or other reasons. In the header would be a field called something like *flow identifier*, which identifies the conversation, and the routers along the path would keep state about the conversation. Another connection-like feature sometimes implemented in routers is header compression, whereby neighbor routers agree on a shorthand for the header of recently seen packets. The first packet of a conversation alerts neighbors to negotiate a shorthand for packets for that conversation.

Whether the network layer is connection oriented or not (even if it is possible to categorize network layers definitively as one or the other) has no relevance to the fact that the network layer needs a routing protocol. Sometimes people think of a connection-oriented network as one in which all of the connections are already established, with a table of input port/connection ID mapping to output port/connection ID, and the only thing the router needs to do is a table lookup of input port/connection ID. If this were the case, then a router in a connection-oriented network would not need a routing protocol, but it is not the case. For the mapping table to be created, a route setup packet traverses the path to the destination, and a router has to make the same sort of routing decision as to how to reach the destination as it would on a per-packet basis in a connectionless network. Thus, whether the network is connectionless or not does not affect the type of routing protocol needed. Connectionless network layer protocols differ only in packet format and type of addressing. The type of routing protocol is not affected by the format of data packets and so for the purpose of discussing routing algorithms, it is not necessary for us to pick a specific network layer protocol.

73.3.1 Types of Routing Protocols

One categorization of routing protocols is distance vector vs. link state. Another categorization is intradomain vs. interdomain. We discuss these issues, but first we discuss the basic concepts of addressing and hierarchy in routing and addressing.

73.3.1.1 Hierarchy

A routing protocol can handle up to some sized network. Beyond that, many factors might make the routing protocol overburdened, including:

- Memory to hold the routing database
- CPU to compute the routing database
- Bandwidth to transmit the routing information
- The volatility of the information

It takes a routing protocol some amount of time to stabilize to new routes after a topology change. If the topology changes more frequently than the time it takes for the algorithm to settle, things will not work very well.

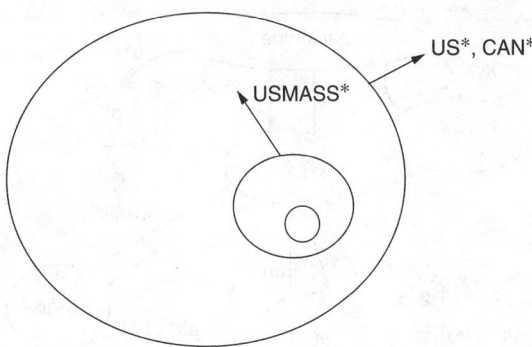

FIGURE 73.3 Network topology with addresses.

To support a larger network, portions of the network can be summarized to the outside world. This is similar to breaking the world up into countries, and within a country into states, and within a state into cities. It would be too difficult for the post office to know how to reach every street in the world (assuming that all of the streets had globally unique names); and so within a city, the post office knows all of the streets. But if it does not belong in your city, your post office just forwards it to the appropriate city, unless the letter belongs in a different state (or country). Then the post office just forwards it to the appropriate state (or country). It is routed to the appropriate state, then city, then street, and then finally to the destination.

With a network, assuming addresses are handed out sensibly, a logical circle can be drawn around a portion of the network, and all of the contents of the circle can be summarized with a small amount of information. For example, assuming postal addresses again, all of the U.S. could be summarized as: any addresses that start with the string US. All of Massachusetts, U.S., could be summarized as: any addresses that start with the string USMass'. Assuming North America was the logical place to draw a logical circle, then what would be advertised is: any addresses that start with the string US' or Canada' (see Figure 73.3).

A packet is routed according to the longest matching prefix that has been advertised. For instance, if the packet is addressed to: USMassLittletownMainStRadiaPerlman, and there are prefixes that have been advertised for US*, USMass*, then the packet will be routed toward USMass*. If the packet originated outside the U.S., then most likely the longest prefix seen by the routing protocol outside the U.S. would be US*'. Once it reached the U.S., the advertisement USMass' would be visible.

73.3.1.2 Hierarchical Addressing

To support hierarchical routing, addresses must be handed out so there is some mechanism for conveniently summarizing addresses in a region. The typical method is to give an organization a block of addresses that all start with the same prefix. The organization might then hand out blocks of the addresses it owns to suborganizations. For example, suppose there are three major backbone Internet providers, and each is given a block of addresses. Say the blocks are xyz*, a*, and b* (there is no reason why the prefix has to be the same length for each provider). The provider that has the block a* has some subscribing regional providers, and gives each of them a block of addresses to give to their customers. Say there are five regional providers, and the blocks given out are axyc*, an*, ak*, and adkfjlk*. The regional provider with the block axye* might give out blocks to each subscribing customer network that look like axyc1*, axyc2*, axyc3*, etc. One of those customers with a large network might be careful to assign addresses within the network so that the network can be broken again into pieces that are summarizable (see Figure 73.4).

With this assignment of addresses, provider 1 merely has to advertise: I can reach all addresses of the form xyz*. Typically, the network closer to the backbone advertises * outward, and a network advertises the block summarizing its own addresses toward the backbone. Thus, in Figure 73.4, R1 would most likely advertise * to R2, and R2 would advertise axye* to R1.

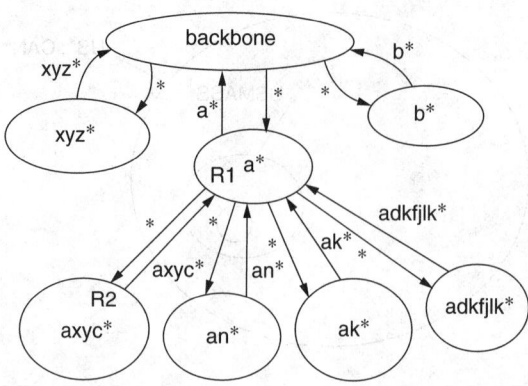

FIGURE 73.4 Block address assignment example.

73.3.1.3 Domains

What is a domain? It is a portion of a network in which the routing protocol that is running is called an intradomain routing protocol. Between domains one runs an interdomain routing protocol. Well, what is an intradomain protocol? It is something run within a domain. This probably does not help our intuition any.

Originally, the concept of a domain arose around the superstition that routing protocols were so complex that it would be impossible to get the routers from two different organizations to cooperate in a routing protocol, because the routers might have been bought from different vendors, and the metrics assigned to the links might have been assigned according to different strategies. It was thought that a routing protocol could not work under these circumstances. Yet it was important to have reachability between domains.

One possibility was to statically configure reachable addresses from other domains. But a protocol of sorts was devised, known as EGP, which was like a distance vector protocol but without exchanging metrics. It only specified what addresses were reachable, and only worked if the topology of domains was a tree (loop-free). EGP placed such severe restraints on the topology, and was itself such an inefficient protocol, that it was clear it needed to be replaced.

In the meantime, intradomain routing protocols were being specified well enough that multivendor operation was considered not only possible but mandatory. There was no particular reason why a different type of protocol needed to run between domains, except for the possible issue of policy-based routing.

The notion of policy-based routing is that it no longer suffices to find a path that physically works, or to find the minimum cost path, but that paths had to obey fairly arbitrary, complex, and eternally changing rules such as a particular country would not want its packets routed via some other particular country. The notion that there should be different types of routing protocols within a domain and between domains makes sense if we agree with all of the following assumptions:

- Within a domain, policy-based routing is not an issue; all paths are legal.
- Between domains, policy-based routing is mandatory, and the world would not be able to live without it.
- Providing for complex policies is such a burden on the routing protocol that a protocol that did policy-based routing would be too cumbersome to be deployed within a domain.

I happen not to agree with any of these assumptions, but these are beliefs, and not something subject to proof either way. Because enough of the world believes these assumptions, different protocols are being devised for interdomain vs. intradomain. At the end of this chapter we discuss some of the interdomain protocols.

73.3.1.4 Forwarding

In a connectionless protocol (such as IP, etc.), data is sent in chunks known as packets, together with information that specifies the source and destination. Each packet is individually addressed and two packets between the same pair of nodes might take different paths. Each router makes an independent decision as to where to forward the packet. The forwarding table consists of (destination, forwarding decision) pairs.

In contrast, in a connection-oriented protocol such as ATM, X.25, or MPLS, an initial packet sets up the path and assigns a connection identifier (also known as a label). Data packets only contain the connection identifier, rather than a source and destination address. Typically the connection identifier is shorter, and easier to parse (because it is looked up based on exact match rather than on longest prefix match). It would be very difficult to assign a connection identifier that was guaranteed not to be in use on any link in the path, so instead the connection identifier is only link-local, and is replaced at each hop.

In a connection-oriented protocol, the forwarding decision is based on the (input port, connection identifier) and the forwarding table will tell the router not only the outgoing port, but also what value the connection identifier on the outgoing packet should be.

A connection-oriented protocol still needs a routing protocol, and the initial path setup packet is routed similarly to packets in connectionless protocols.

Originally, connection-oriented forwarding was assumed preferable because it allowed for faster forwarding decisions. But today, connection-oriented forwarding is gaining popularity because it allows different traffic to be assigned to different paths. This is known as **traffic engineering**.

73.3.1.5 Routing Protocols

The purpose of a routing protocol is to compute a *forwarding database*, which consists of a table listing (destination, neighbor) pairs. When a packet needs to be forwarded, the destination address is found in the forwarding table, and the packet is forwarded to the indicated neighbor.

In the case of hierarchical addressing and routing, destinations are not exact addresses, but are rather address prefixes. The longest prefix matching the destination address is selected and routed forward.

The result of the routing computation, the forwarding database, should be the same whether the protocol used is distance vector or link state.

73.3.1.6 Distance Vector Routing Protocols

One class of routing protocol is known as distance vector. The idea behind this class of algorithm is that each router is responsible for keeping a table (known as a distance vector) of distances from itself to each destination. It computes this table based on receipt of distance vectors from its neighbors. For each destination D, router R computes its distance to D as follows:

- 0, if R = D
- The configured cost, if D is directly connected to R
- The minimum cost through each of the reported paths through the neighbors

For example, suppose R has four ports, a, b, c, and d. Suppose also that the cost of each of the links is, respectively, 2, 4, 3, and 5. On port a, R has received the report that D is reachable at a cost of 7. The other (port, cost) pairs R has heard are (b, 6), (c, 10), and (d, 2). Then the cost to D through port a will be 2 (cost to traverse that link) $+7$, or 9. The cost through b will be $4 + 6$, or 10. The cost through c will be $3 + 10$, or 13. The cost through d will be $5 + 2$, or 7. So the best path to D is through port d, and R will report that it can reach D at a cost of 7 (see Figure 73.5).

The spanning tree algorithm is similar to a distance vector protocol in which each bridge is only computing its cost and path to a single destination, the root. But the spanning tree algorithm does not suffer from the count-to-infinity behavior that distance vector protocols are prone to (see next section).

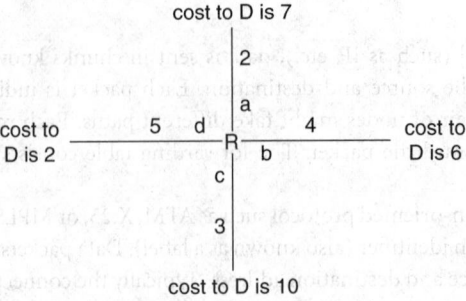

FIGURE 73.5 Example: distance vector protocol.

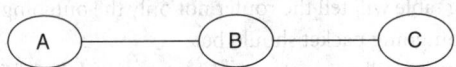

FIGURE 73.6 Example: the count-to-infinity problem.

73.3.1.7 Count-to-Infinity

One of the problems with distance vector protocols is known as the count-to-infinity problem. Imagine a network with three nodes, A, B, and C (see Figure 73.6).

Let us discuss everyone's distance to C. C will be 0 from itself, B will be 1 from C, and A will be 2 from C. When C crashes, B unfortunately does not conclude that C is unreachable, but instead goes to its next best path, which is via neighbor A, who claims to be able to reach C at a cost of 2. So now B concludes it is 3 from C, and that it should forward packets for C through A. When B tells A its new distance vector, A does not get too upset. It merely concludes its path (still through B) has gotten a little worse, and now A is 4 from C. A will report this to B, which will update its cost to C as 5, and A and B will continue this until they count to infinity. Infinity in this case is mercifully not the mathematical definition of infinity, but is instead a parameter (with a definite finite value such as 16). Routers conclude if the cost to something is greater than this parameter, that something must be unreachable.

A common enhancement that makes the behavior a little better is known as *split horizon*. The split horizon rule as usually implemented says that if router R uses neighbor N as its best path to destination D, R should not tell N that R can reach D. This eliminates loops of two routers. For instance, in Figure 73.6, A would not have told B that A could reach C. Thus, when C crashed, B would conclude B could not reach C at all; and when B reported infinity to A, A would conclude that A could not reach C either, and everything would work as we would hope it would.

Unfortunately, split horizon does not fix loops of three or more routers. Referring to Figure 73.7, and looking at distances to D, when D crashes, C will conclude C cannot reach D. (Because of the split horizon rule, A and B are not reporting to C that they can reach D.)

C will inform A and B that C can no longer reach D. Unfortunately, each of them thinks they have a next-best path through the other. Say A acts first, decides its best path is through B, and that A is now 3 from D. A will report infinity to B (because of split horizon), and report 3 to C. B will now (for a moment) think it cannot reach D. It will report infinity to A, but it is too late; A has already reported a finite cost to C. C will now report 4 to B, which will now conclude it is 5 from D.

Although split horizon does not solve the problem, it is a simple enhancement, never does any harm, does not add overhead, and helps in many cases.

Most of the distance vector protocols in use [router information protocol (RIP) for IP and IPX, and RTMP for Appletalk] are remarkably similar, and we call them the RIP-family of distance vector protocols. These protocols are simple to implement, but are very slow to converge after a topology change. The idea is

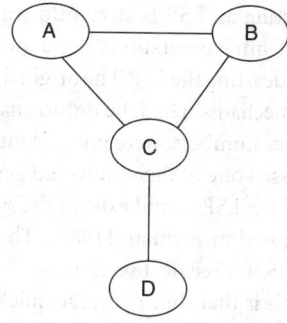

FIGURE 73.7　Example: loops with three or more routers.

that routing information is periodically transmitted, quite frequently (on the order of 30 s). Information is discarded if it has not been heard recently (on the order of 2 min). Most implementations only store the best path, and when that path fails they need to wait for their neighbors' periodic transmissions in order to hear about the second best path. Some implementations query their neighbors ("do you know how to reach D?") when the path to D is discarded. In some implementations when R discards its route to D (for instance, it times out), R lets its neighbors know that R has discarded the route, by telling the neighbors R's cost to D is now infinity. In other implementations, after R times out the route, R will merely stop advertising the route, so R's neighbors will need to time out the route (starting from when R timed out the route).

Distance vector protocols need not be periodic. The distance vector protocol in use for DECnet Phases 3 and 4 transmits routing information reliably, and only sends information that has changed. Information on a LAN is sent periodically (rather than collecting acknowledgments from all router neighbors), but the purpose of sending it periodically is solely as an alternative to sending acknowledgments. Distance vector information is not timed out in DECnet, as it is in a RIP-like protocol. Instead, there is a separate protocol in which Hello messages are broadcast on the LAN to detect a dead router. If a Hello is not received in time, the neighbor router is assumed dead and its distance vector is discarded.

Another variation from the RIP-family of distance vectors is to store the entire received distance vector from each neighbor, rather than only keeping the best report for each destination. Then, when information must be discarded (e.g., due to having that neighbor report infinity for some destination, or due to that neighbor being declared dead) information for finding an alternative path is available immediately.

There are variations proposed to solve the count-to-infinity behavior. One variation has been implemented in Border Gateway Protocol (BGP). Instead of just reporting a cost to destination D, a router reports the entire path from itself to D. This eliminates loops but has high overhead. Another variation proposed by Garcia-Luna [1989] and implemented in the proprietary protocol EIGRP involves sending a message in the opposite direction of D, when the path to D gets worse, and not switching over to a next-best path until acknowledgments are received indicating that the information has been received by the downstream subtree. These variations may improve convergence to be comparable to link state protocols, but they also erode the chief advantage of distance vector protocols, which is their simplicity.

73.3.1.8 Link State Protocols

The idea behind a link state protocol is that each router R is responsible for the following:

- Identifying its neighbors
- Constructing a special packet known as a *link state packet* (LSP) that identifies R and lists R's neighbors (and the cost to each neighbor)
- Cooperating with all of the routers to reliably broadcast LSPs to all the routers
- Keeping a database of the most recently generated LSP from each other router
- Using the LSP database to calculate routes

Identifying neighbors and constructing an LSP is straightforward. Calculating routes using the LSP database is also straightforward. Most implementations use a variation of an algorithm attributed to Dijkstra. The tricky part is reliably broadcasting the LSP. The original link state algorithm was implemented in the ARPANET. Its LSP distribution mechanism had the unfortunate property that if LSPs from the same source, but with three different sequence numbers, were injected into the network, these LSPs would turn into a virus. Every time a router processed one of them, it would generate more copies, and so the harder the routers worked, the more copies of the LSP would exist in the system. The problem was analyzed and a stable distribution scheme was proposed in Perlman [1983]. The protocol was further refined for the IS-IS routing protocol and copied in OSPF. (See next section.)

One advantage of link state protocols is that they converge quickly. As soon as a router notices one of its links has changed (going up or down), it broadcasts an updated LSP which propagates in a straight line outwards (in contrast to a distance vector protocol where information might sometimes be ping-ponged back and forth before proceeding further, or where propagation of the information is delayed waiting for news from downstream nodes that the current path's demise has been received by all nodes.)

Link state protocols have other advantages as well. The LSP database gives complete information, which is useful for managing the network, mapping the network, or constructing custom routes for complex policy reasons [Clark, 1989] or for sabotage-proof routing [Perlman, 1988].

73.3.1.9 Reliable Distribution of Link State Packets

Each LSP contains:

- Identity of the node that generated the LSP
- A sequence number, large enough to never wrap around except if errors occur (for example, 64 bits)
- An age field, estimating time since source generated the LSP
- Other information

Each router keeps a database of the LSP with the largest sequence number seen thus far from each source. The purpose of the age field is to eventually eliminate an LSP from a source that does not exist any more, or that has been down for a very long time. It also serves to get rid of an LSP that is corrupted, or for which the sequence number has reached the largest value.

For each LSP, a router R has a table, for each of R's neighbors, as to whether R and the neighbor N are *in sync* with respect to that LSP. The possibilities are as follows:

- R and N are in sync. R does not need to send anything to N about this LSP.
- R thinks N has not yet seen this LSP. R needs to periodically retransmit this LSP to N until N acknowledges it.
- R thinks N does not know R has the LSP. R needs to send N an acknowledgment (ack) for this LSP.

R goes through the list of LSPs round-robin, for each link, and transmits LSPs or acks as indicated. If R sends an ack to N, R changes the state of that LSP for N to be in sync.

The state of an LSP gets set as follows:

- If R receives a new LSP from neighbor N, R overwrites the one in memory (if any) with smaller sequence number, sets send ack for N, and sets send LSP for each of R's other neighbors.
- If R receives an ack for an LSP from neighbor N, R sets the flag for that LSP to be in sync.
- If R receives a duplicate LSP or older LSP from neighbor N, R sets the flag for the LSP in memory (the one with higher sequence number) to send LSP.
- After R transmits an ack for an LSP to N, R changes the state of that LSP to in sync.

If an LSP's age expires, it is important that all of the routers purge the LSP at about the same time. The age is a field that is set to some value by the source, and is counted down. In this way, the source can control how long it will last. If R decides that an LSP's age has expired, R refloods it to R's neighbors (by setting the state to send LSP). If R receives an LSP with the same sequence number as one stored, but the received one has zero age, R sets the LSP's age to 0 and floods it to its neighbors. If R does not have an LSP in memory, and receives one with zero age, R acks it but does not store it or reflood it.

73.3.2 Calculating Routes

Given an LSP database, the most popular method of computing routes is to use some variant of an algorithm attributed to Dijkstra. The algorithm involves having each router compute a tree of shortest paths from itself to each destination. Each node on the tree has a value associated with it which is the cost from the root to that node. The algorithm is as follows:

- Step 0: put yourself, with cost 0, on the tree as root.
- Step 1: examine the LSP of the node X just put on the tree. For each neighbor N listed in X's LSP, add X's cost in the LSP to the cost to X to get some number c. If c is smaller than any path to N found so far, place N tentatively in the tree, with cost c.
- Step 2: find the node tentatively in the tree with smallest associated cost c. Place that node permanently in the tree. Go to step 1.

73.3.3 Interdomain Routing

As stated previously, the only plausible technical difference between a routing protocol designed for inter- vs. intra-domain is support of policy-based routing. Policy-based routing is the ability to route according to exotic constraints, such as that one country does not want to see its traffic routed through some other country, some network might be willing to serve as a route-through carrier for only certain classes of traffic, or even that one backbone carrier has special rates at certain times of day and during those times someone would like to see all possible traffic routed via that carrier.

One method of accommodating all these policies is to have the source compute its own route, perhaps by being given the LSP database. Once the source computes the desired route that meets its constraints, it sets up the path by launching a special path setup packet that specifies the path, travels along the path, and has the router along the path keep track of the route. This is the basic approach taken by IDPR when the route that would be computed by the network would not be appropriate for whatever reason [Steenstrup, 1993].

The approach taken by BGP/IDRP [Lougheed, 1991] is to use a distance vector protocol in which the entire path is specified (the sequence of domains, not the sequence of routers because the path through a domain is assumed not to be of interest in terms of policy issues). Each router is configured with information that helps it evaluate choices of path to determine how it would like to route to the destination. For instance, if one neighbor says it can reach destination D through path XYZ, and another through path XQV, the router will use its configured information to decide which path it prefers. It chooses one, and that is the path it tells its neighbors. There are also policies that can be configured about which destinations can be reported to which neighbors. The assumption is that if you do not tell a neighbor you can get to a particular destination, that neighbor will not choose to route packets to that destination through you.

A link state approach makes support for many more different types of policies possible than a distance vector approach. With a distance vector approach, either a router chooses one path to the destination and only advertises that (as in BGP/IDRP), or the algorithm has exponential overhead (if every router advertises every possible path to the destination). If the router chooses one path, then it greatly limits the possible paths. For instance, suppose router R can reach destination D through A, B, X, or through A, Q, V. Suppose two sources that must get to D through R have different policies. One does not want to route through domain B. The other wants to avoid domain Q. Because R makes the choice, it makes the same choice for all sources. The ability for each source to have a custom route is supposedly one of the primary motivations behind policy-based routing.

BGP has a problem with convergence as described by Griffin and Wilfong [1999]. BGP makes forwarding decisions based on configured "policy," such as "do not use domain B" or "do not use domain B when going to destination D" or "only use domain B if there is no other choice." In contrast, other protocols using hop-by-hop forwarding base decisions on minimizing a metric. Minimizing a number is well-defined, and all routers will be making compatible decisions, once the routing information has propagated. However, with BGP, the routing decisions might be incompatible. A router R1, seeing the choices its neighbors have made for reaching D, bases its decision on those choices and advertises its chosen path. Once R1 does so, it is possible for that to affect the decision of another router R2, which will change what it advertises, which can cause R1 to change its mind, and so forth.

Defining Terms

Bridge: A box that forwards information from one link to another but only looks at information in the data link header.

Cloud: An informal representation of a multiaccess link. The purpose of representing it as a cloud is that what goes on inside is irrelevant to what is being discussed. When a system is connected to the cloud, it can communicate with any other system attached to the cloud.

Data-link layer: The layer that gets information from one machine to a neighbor machine (a machine on the same link).

IEEE 802 address: The 48-bit address defined by the IEEE 802 committee as the standard address on 802 LANs.

Hops: The number of times a packet is forwarded by a router.

Local area network (LAN): A multiaccess link with multicast capability.

MAC address: Synonym for IEEE 802 address.

Medium access control (MAC): The layer defined by the IEEE 802 committee that deals with the specifics of each type of LAN (for instance, token passing protocols on token passing LANs).

Multiaccess link: A link on which more than two nodes can reside.

Multicast: The ability to transmit a single packet that is received by multiple recipients.

Network layer: The layer that forms a path by concatenation of several links.

References

Clark, D. 1989. Policy routing in internet protocols. RFC 1102, May.

Garcia-Luna-Aceves, J. J. 1989. A unified approach to loop-free routing using distance vectors or link states. *ACM Sigcomm #89 Symp.*, Sept.

Griffin, Timothy and Wilfong, Gordon. 1999. An Analysis of BGP Convergence Properties. *ACM Sigcomm #99 Symp.*

Lougheed, K. and Rehkter, Y. 1991. A border gateway protocol 3 (BGP-3). RFC 1267, Oct.

Perlman, R. 1983. Fault-tolerant broadcast of routing information. *Comput. Networks*, Dec.

Perlman, R. 1988. Network layer protocols with byzantine robustness. *MIT Lab. Computer Science Tech. Rep. #429*, Oct.

Perlman, R. 1999. *Interconnections: Bridges, Routers, Switches, and Internetworking Protocols.* Addison-Wesley, Reading, MA.

Steenstrup, M. 1993. Inter-domain policy routing protocol specification: Version 1. RFC 1479, July.

74

Network and Internet Security

74.1	Introduction	74-1
74.2	General Threats	74-2
	Authentication Failures • User Authentication • Buggy Code	
74.3	Routing	74-3
74.4	The Transmission Control Protocol/Internet Protocol (TCP/IP) Protocol Suite	74-4
	Sequence Number Attacks • Connection Hijacking • The r-Commands • The X Window System • User Datagram Protocol (UDP) • Remote Procedure Call (RPC), Network Information Service (NIS), and Network File System (NFS)	
74.5	The World Wide Web	74-6
	Client Issues • Server Issues	
74.6	Using Cryptography	74-7
	Key Distribution Centers	
74.7	Firewalls	74-9
	Types of Firewalls • Limitations of Firewalls	
74.8	Denial of Service Attacks	74-11
74.9	Conclusions	74-11

Steven Bellovin
AT&T Research Labs

74.1 Introduction

Why is network security so hard, whereas stand-alone computers remain relatively secure? The problem of network security is hard because of the complex and open nature of the networks themselves.

There are a number of reasons for this. First and foremost, a network is designed to accept requests from outside. It is easier for an isolated computer to protect itself from outsiders because it can demand authentication — a successful log-in — first. By contrast, a networked computer expects to receive unauthenticated requests, if for no other reason than to receive electronic mail. This lack of authentication introduces some additional risk, simply because the receiving machine needs to talk to potentially hostile parties.

Even services that should, in principle, be authenticated often are not. The reasons range from technical difficulty (see the subsequent discussion of routing) to cost to design choices: the architects of that service were either unaware of, or chose to discount, the threats that can arise when a system intended for use in a friendly environment is suddenly exposed to a wide-open network such as the Internet.

More generally, a networked computer offers many different services; a stand-alone computer offers just one: log-in. Whatever the inherent difficulty of implementing any single service, it is obvious that adding more services will increase the threat at least linearly. In reality, the problem is compounded by the fact that different services can interact. For example, an attacker may use a file transfer protocol to upload some malicious software and then trick some other network service into executing it.

Additional problems arise because of the unbounded nature of a network. A typical local area network may be viewed as an implementation of a loosely coupled, distributed operating system. But in single-computer operating systems, the kernel can **trust** its own data. That is, one component can create a control block for another to act on. Similarly, the path to the disk is trustable, in that a read request will retrieve the proper data, and a write request will have been vetted by the operating system.

Those assumptions do not hold on a network. A request to a file server may carry fraudulent user credentials, resulting in access violations. The data returned may have been inserted by an intruder or by an authorized user who is nevertheless trying to gain more privileges. In short, the distributed operating system can not believe anything, even transmissions from the kernel talking to itself.

In principle, many of these problems can be overcome. In practice, the problem seems to be intractable. Networked computers are far more vulnerable than standalone computers.

74.2   General Threats

Network security flaws fall into two main categories. Some services do inadequate authentication of incoming requests. Others try to do the right thing; however, buggy code lets the intruder in. Strong authentication and **cryptography** can do nothing against this second threat; it allows the target computer to establish a well-authenticated, absolutely private connection to a hacker who is capable of doing harm.

74.2.1   Authentication Failures

Some machines grant access based on the network address of the caller. This is acceptable if and only if two conditions are met. First, the trusted network and its attached machines must both be adequately secure, both physically and logically. On a typical local area network (LAN), anyone who controls a machine attached to the LAN can reconfigure it to impersonate any other machine on that cable. Depending on the exact situation, this may or may not be easily detectable. Additionally, it is often possible to turn such machines into eavesdropping stations, capable of listening to all other traffic on the LAN. This specifically includes passwords or even encrypted data if the encryption key is derived from a user-specified password [Gong et al. 1993].

Network-based authentication is also suspect if the network cannot be trusted to tell the truth. However, such a level of trust is not tautological; on typical packet networks, such as the Internet, each transmitting host is responsible for putting its own reply address in each and every packet. Obviously, an attacker's machine can lie — and this often happens.

In many instances, a **topological defense** will suffice. For example, a router at a network border can reject incoming packets that purport to be from the inside network. In the general case, though, this is inadequate; the interconnections of the networks can be too complex to permit delineation of a simple border, or a site may wish to grant privileges — that is, trust — to some machine that really is outside the physical boundaries of the network.

Although **address spoofing** is commonly associated with packet networks, it can happen with circuit networks as well. The difference is in who can lie about addresses; in a circuit net, a misconfigured or malconfigured switch can announce incorrect source addresses. Although not often a threat in simple topologies, in networks where different switches are run by different parties address errors present a real danger. The best-known example is probably the phone system, where many different companies and organizations around the world run different pieces of it. Again, topological defenses sometimes work, but you are still limited by the actual interconnection patterns.

Even if the network address itself can be trusted, there still may be vulnerabilities. Many systems rely not on the network address but on the network *name* of the calling party. Depending on how addresses are mapped to names, an enemy can attack the translation process and thereby spoof the target. See Bellovin [1995] for one such example.

74.2.2 User Authentication

User authentication is generally based on any of three categories of information: something you know, something you have, and something you are. All three have their disadvantages.

The something you know is generally a password or personal identification number (PIN). In today's threat environment, passwords are an obsolete form of authentication. They can be guessed [Klein 1990, Morris and Thompson 1979, Spafford 1989a], picked up by network wiretappers, or simply *social engineered* from users. If possible, avoid using passwords for authentication over a network.

Something you have is a token of some sort, generally cryptographic. These tokens can be used to implement cryptographically strong challenge/response schemes. But users do not like token devices; they are expensive and inconvenient to carry and use. Nevertheless, for many environments they represent the best compromise between security and usability.

Biometrics, or something you are, are useful in high-threat environments. But the necessary hardware is scarce and expensive. Furthermore, biometric authentication systems can be disrupted by biological factors; a user with laryngitis may have trouble with a voice recognition system. Finally, cryptography must be used in conjunction with biometrics across computer networks; otherwise, a recording of an old fingerprint scan may be used to trick the authentication system.

74.2.3 Buggy Code

The Internet has been plagued by buggy network servers. In and of itself, this is not surprising; most large computer programs are buggy. But to the extent that outsiders should be denied access to a system, every network server is a privileged program.

The two most common problems are buffer overflows and shell escapes. In the former case, the attacker sends an input string that overwrites a buffer. In the worst case, the stack can be overwritten as well, letting the attacker inject code. Despite the publicity this failure mode has attracted — the Internet Worm used this technique [Spafford 1989a, 1989b, Eichin and Rochlis 1989, Rochlis and Eichin 1989] — new instances of it are legion. Too many programmers are careless or lazy.

More generally, network programs should check *all* inputs for validity. The second failure mode is simply another example of this: input arguments can contain shell metacharacters, but the strings are passed, unchecked, to the shell in the course of executing some other command. The result is that two commands will be run, the one desired and the one requested by the attacker.

Just as no general solution to the program correctness problem seems feasible, there is no cure for buggy network servers. Nor will the best cryptography in the world help; you end up with a secure, protected communication between a hacker and a program that holds the **back door** wide open.

74.3 Routing

In most modern networks of any significant size, host computers cannot talk directly to all other machines they may wish to contact. Instead, intermediate nodes — switches or routers of some sort — are used to route the data to their ultimate destination. The security and integrity of the network depends very heavily on the security and integrity of this process.

The switches in turn need to know the next hop for any given network address; whereas this can be configured manually on small networks, in general the switches talk to each other by means of **routing protocols**. Collectively, these routing protocols allow the switches to learn the topology of the network. Furthermore, they are dynamic, in the sense that they rapidly and automatically learn of new network nodes, failures of nodes, and the existence of alternative paths to a destination.

Most routing protocols work by having switches talk to their neighbors. Each tells the other of the hosts it can reach, along with associated cost metrics. Furthermore, the information is transitive; a switch will not only announce its directly connected hosts but also destinations of which it has learned by talking to other routers. These latter announcements have their costs adjusted to account for the extra hop.

An enemy who controls the routing protocols is in an ideal position to monitor, intercept, and modify most of the traffic on a network. Suppose, for example, that some enemy node X is announcing a very low-cost route to hosts A and B. Traffic from A to B will flow through X, as will traffic from B to A. Although the diversion will be obvious to anyone who checks the path, such checks are rarely done unless there is some suspicion of trouble.

A more subtle routing issue concerns the return data flow. Such a flow almost always exists, if for no other reason than to provide flow control and error correction feedback. On packet-switched networks, the return path is independent of the forward path and is controlled by the same routing protocols. Machines that rely on network addresses for authentication and authorization are implicitly trusting the integrity of the return path; if this has been subverted, the network addresses cannot be trusted either. For example, in the previous situation, X could easily impersonate B when talking to A or vice versa.

That is somewhat less of a threat on circuit-switched networks, where the call is typically set up in both directions at once. But often, the trust point is simply moved to the switch; a subverted or corrupt switch can still issue false routing advertisements.

Securing routing protocols is hard because of the transitive nature of the announcements. That is, a switch cannot simply secure its link to its neighbors, because it can be deceived by messages really sent by its legitimate and uncorrupted peer. That switch, in turn, might have been deceived by its peers, ad infinitum. It is necessary to have an authenticated chain of responses back to the source to protect routing protocols from this sort of attack.

Another class of defense is topological. If a switch has a priori knowledge that a certain destination is reachable only via a certain wire, routing advertisements that indicate otherwise are patently false. Although not necessarily indicative of malice — link or node failures can cause temporary confusion of the network-wide routing tables — such announcements can and should be dismissed out of hand. The problem, of course, is that adequate topological information is rarely available. On the Internet, most sites are *out there* somewhere; the false hop, if any, is likely located far beyond an individual site's borders. Additionally, the prevalance of redundant links, whether for performance or reliability, means that more than one path may be valid. In general, then, topological defenses are best used at **choke points: firewalls** (Section 74.7) and the other end of the link from a site to a network service provider. The latter allows the service provider to be a good network citizen and prevent its customers from claiming routes to other networks.

Some networks permit hosts to override the routing protocols. This process, sometimes called source routing, is often used by network management systems to bypass network outages and as such is seen as very necessary by some network operators.

The danger, though, arises because source-routed packets bypass the implicit authentication provided by use of the return path, as previously outlined. A host that does network address-based authentication can easily be spoofed by such messages. Accordingly, if source routing is to be used, address-based authentication must not be used.

74.4 The Transmission Control Protocol/Internet Protocol (TCP/IP) Protocol Suite

The **transmission control protocol** (TCP) suite is the basis for the Internet. Although the general features of the protocols are beyond the scope of this chapter (see Stevens [1995] and Wright and Stevens [1994] for more detail), the security problems of it are less well known.

The most important thing to realize about TCP/IP security is that since IP is a datagram protocol, one cannot trust the source addresses in packets. This threat is not just hypothetical. One of the most famous security incidents — the penetration of Tsutomu Shimomura's machines [Shimomura 1996, Littman 1996] — involved IP address spoofing in conjunction with a TCP sequence number guessing attack.

74.4.1 Sequence Number Attacks

TCP **sequence number attacks** were described in the literature many years before they were actually employed [Morris 1985, Bellovin 1989]. They exploit the predictability of the sequence number field in TCP in such a way that it is not necessary to see the return data path. To be sure, the intruder cannot get any output from the session, but if you can execute a few commands, it does not matter much if you see their output.

Every byte transmitted in a TCP session has a sequence number; the number for the first byte in a segment is carried in the header. Furthermore, the control bits for opening and closing a connection are included in the sequence number space. All transmitted bytes must be acknowledged explicitly by the recipient; this is done by sending back the sequence number of the next byte expected.

Connection establishment requires three messages. The first, from the client to the server, announces the client's initial sequence number. The second, from the server to the client, acknowledges the first message's sequence number and announces the server's initial sequence number. The third message acknowledges the second.

In theory, it is not possible to send the third message without having seen the second, since it must contain an explicit acknowledgment for a random-seeming number. But if two connections are opened in a short time, many TCP stacks pick the initial sequence number for the second connection by adding some constant to the sequence number used for the first.

The mechanism for the attack is now clear. The attacker first opens a legitimate connection to the target machine and notes its initial sequence number. Next, a spoofed connection is opened by the attacker, using the IP address of some machine trusted by the target. The sequence number learned in the first step is used to send the third message of the TCP open sequence, without ever having seen the second. The attacker can now send arbitrary data to the target; generally, this is a set of commands designed to open up the machine even further.

74.4.2 Connection Hijacking

Although a defense against classic sequence number attacks has now been found [Bellovin 1996], a more serious threat looms on the horizon: **connection hijacking** [Joncheray 1995]. An attacker who observes the current sequence number state of a connection can inject phony packets.

Again, the network in general will believe the source address claimed in the packet. If the sequence number is correct, it will be accepted by the destination machine as coming from the real source. Thus, an eavesdropper can do far worse than simply steal passwords; he or she can take over a session after log in. Even the use of a high-security log-in mechanism, such as a one-time password system [Haller 1994], will not protect against this attack. The only defense is full-scale encryption.

Session hijacking is detectable, since the acknowledgment packet sent by the target cites data the sender never sent. Arguably, this should cause the connection to be reset; instead, the system assumes that sequence numbers have wrapped around and resends its current sequence number and acknowledgment number state.

74.4.3 The r-Commands

The so-called **r-commands** — **r**sh and **r**login — use address-based authentication. As such, they are not secure. But too often, the alternative is sending a password in the clear over an insecure net. Neither alternative is attractive; the right choice is cryptography. But that is used all too infrequently.

In many situations, where insiders are considered reasonably trustworthy, use of these commands without cryptography is an acceptable risk. If so, a low-grade fire wall such as a simple **packet filter** *must* be used.

74.4.4   The X Window System

The paradigm for the X window system [Stubblebine and Gligor 1992] is simple: a server runs the physical screen, keyboard, and mouse; applications connect to it and are allocated use of those resources. Put another way, when an application connects to the server, it gains control of the screen, keyboard, and mouse. Whereas this is good when the application is legitimate, it poses a serious security risk if uncontrolled applications can connect. For example, a rogue application can monitor all keystrokes, even those destined for other applications, dump the screen, inject synthetic events, and so on.

There are several modes of access control available. A common default is no restriction; the dangers of this are obvious. A more common option is control by IP address; apart from the usual dangers of this strategy, it allows anyone to gain access on the trusted machine. The so-called **magic cookie** mechanism uses (in effect) a clear-text password; this is vulnerable to anyone monitoring the wire, anyone with privileged access to the client machines, and — often — anyone with network file system access to that machine. Finally, there are some cryptographic options; these, although far better than the other options, are more vulnerable than they might appear at first glance, as any privileged user on the application's machine can steal the secret cryptographic key.

There have been some attempts to improve the security of the X window system [Epstein et al. 1992, Kahn 1995]. The principal risk is the complexity of the protocol: are you sure that all of the holes have been closed? The analysis in Kahn [1995] provides a case in point; the author had to rely on various heuristics to permit operations that seemed dangerous but were sometimes used safely by common applications.

74.4.5   User Datagram Protocol (UDP)

The **user datagram protocol (UDP)** [Postel 1980] poses its own set of risks. Unlike TCP, it is not connection oriented; thus, there is no implied authentication from use of the return path. Source addresses cannot be trusted at all. If an application wishes to rely on address-based authentication, it must do its own checking, and if it is going to go to that much trouble, it may as well use a more secure mechanism.

74.4.6   Remote Procedure Call (RPC), Network Information Service (NIS), and Network File System (NFS)

The most important UDP-based protocol is **remote procedure call (RPC)** [Sun 1988, 1990]. Many other services, such as network information service (NIS) and **network file system (NFS)** [Sun 1989, 1990] are built on top of RPC. Unfortunately, these services inherit all of the weaknesses of UDP and add some of their own. For example, although RPC has an authentication field, in the normal case it simply contains the calling machine's assertion of the user's identity. Worse yet, given the ease of forging UDP packets, the server does not even have any strong knowledge of the actual source machine. Accordingly, no serious action should be taken based on such a packet.

There is a cryptographic authentication option for RPC. Unfortunately, it is poorly integrated and rarely used. In fact, on most systems only NFS can use it. Furthermore, the key exchange mechanism used is cryptographically weak [LaMacchia and Odlyzko 1991].

NIS has its own set of problems; these, however, relate more to the information it serves up. In particular, NIS is often used to distribute password files, which are very sensitive. Password guessing is very easy [Klein 1990, Morris and Thompson 1979, Spafford 1992]; letting a hacker have a password file is tantamount to omitting password protection entirely. Misconfigured or buggy NIS servers will happily distribute such files; consequently, the protocol is very dangerous.

74.5   The World Wide Web

The World Wide Web (WWW) is the fastest-growing protocol on the Internet. Indeed, in the popular press it *is* the Internet. There is no denying the utility of the Web. At the same time, it is a source of great danger. Indeed, the Web is almost unique in that the danger is nearly as great to clients as to servers.

74.5.1 Client Issues

The danger to clients comes from the nature of the information received. In essence, the server tells the client "here is a file, and here is how to display it." The problem is that the instructions may not be benign. For example, some sites supply troff input files; the user is expected to make the appropriate control entries to link that file type to the processor. But troff has shell escapes; formatting an arbitrary file is about as safe as letting unknown persons execute any commands they wish.

The problem of buggy client software should not be ignored either. Several major browsers have had well-publicized bugs, ranging from improper use of cryptography to string buffer overflows. Any of these could result in security violations.

A third major area for concern is **active agents:** pieces of code that are explicitly downloaded to a user's machine and executed. Java [Arnold and Gosling 1996] is the best known, but there are others.

Active agents, by design, are supposed to execute in a restricted environment. Still, they need access to certain resources to do anything useful. It is this conflict, between the restrictions and the resources, that leads to the problems; sometimes the restrictions are not tight enough. And even if they are in terms of the architecture, implementation bugs, inevitable in such complex code, can lead to security holes [Dean and Wallach 1996].

74.5.2 Server Issues

Naturally, servers are vulnerable to security problems as well. Apart from bugs, which are always present, Web servers have a challenging job. Serving up files is the easy part, though this, too, can be tricky; not all files should be given to outsiders.

A bigger problem is the so-called **common gateway interface (CGI) scripts**. CGI scripts are, in essence, programs that process the user's request. Like all programs, CGI scripts can be buggy. In the context of the Web, this can lead to security holes.

A common example is a script to send mail to some destination. The user is given a form to fill out, with boxes for the recipient name and the body of the letter. When the user clicks on a button, the script goes to work, parsing the input and, eventually, executing the system's mailer. But what happens if the user — someone on another site — specifies an odd-ball string for the recipient name? Specifically, what if the recipient string contains assorted special characters, and the shell is used to invoke the mailer?

Administering a WWW site can be a challenge. Modern servers contain all sorts of security-related configuration files. Certain pages are restricted to certain users or users from certain IP addresses. Others must be accessed using particular user ids. Some are even protected by their own password files.

Not surprisingly, getting all of that right is tricky. But mistakes here do not always lead to the sort of problem that generates user complaints; hackers rarely object when you let them into your machine.

A final problem concerns the uniform resource locators (URLs) themselves. Web servers are stateless; accordingly, many encode transient state information in URLs that are passed back to the user. But parsing this state can be hard, especially if the user is creating malicious counterfeits.

74.6 Using Cryptography

Cryptography, though not a panacea, is a potent solution to many network security issues. The most obvious use of cryptography is to protect network traffic from eavesdroppers. If two parties share the same secret key, no outsiders can intercept any messages. This can be used to protect passwords, sensitive files being transferred over a network, etc.

Often, though, secrecy is less important than authenticity. Cryptography can help here, too, in two different ways. First, there are cryptographic primitives designed to authenticate messages. Message authentication codes (MACs) are commonly used in electronic funds transfer applications to validate their point of origin.

More subtly, decryption with an invalid key will generally yield garbage. If the message is intended to have any sort of semantic or syntactic content, ordinary input processing will likely reject such messages. Still, care must be taken; noncryptographic **checksums** can easily be confused with a reasonable probability. For example, TCP's checksum is only 16 bits; if that is the sole guarantor of packet sanity, it will fail about once in 2^{16} packets.

74.6.1   Key Distribution Centers

The requirement that every pair of parties share a secret key is in general impractical for all but the smallest network. Instead, most practical systems rely on trusted third parties known as **key distribution centers** (**KDC**). Each party shares a long-term secret key with the KDC; to make a secure call, the KDC is asked to (in effect) introduce the two parties, using its knowledge of the shared keys to vouch for the authenticity of the call.

The Kerberos authentication system [Bryant 1988, Kohl and Neuman 1993, Miller et al. 1987, Steiner et al. 1988], designed at Massachusetts Institute of Technology (MIT) as part of Project Athena, is a good example. Although Kerberos is intended for user-to-host authentication, most of the techniques apply to other situations as well.

Each party, known as a *principal*, shares a secret key with the KDC. User keys are derived from a pass phrase; service keys are randomly generated. Before contacting any service, the user requests a **Kerberos ticket-granting ticket** (**TGT**) from the KDC,

$$K_c[K_{c,\mathrm{tgs}}K_{\mathrm{tgs}}[T_{c,\mathrm{tgs}}]]$$

where $K[X]$ denotes the encryption of X by key K. K_c is the client's key; it is used to encrypt the body of the message. In turn, the body is a ticket-granting ticket, encrypted by a key known only to the server, and an associated session key $K_{c,\mathrm{tgs}}$ to be used along with the TGT. TGTs and their associated session keys normally expire after about 8 hours, and are cached by the client during this time; this avoids the need for constant retyping of the user's password.

The TGT is used to request credentials — tickets — for a service s,

$$s, K_{\mathrm{tgs}}[T_{c,\mathrm{tgs}}], K_{c,\mathrm{tgs}}[A_c]$$

That is, the TGT is sent to the KDC along with an encrypted *authenticator* A_c. The authenticator contains the time of day and the client's IP address; this is used to prevent an enemy from replaying the message.

The KDC replies with

$$K_{c,\mathrm{tgs}}[K_s[T_{c,s}], K_{c,s}]$$

The session key $K_{c,s}$ is a newly chosen random key; $K_s[T_{c,s}]$ is the ticket for user c to access service s. It is encrypted in the key shared by the KDC and s; this ensures s of its validity. It contains a lifetime, a session key $K_{c,s}$ that is shared with c, and c's name. A separate copy of $K_{c,s}$ is included in the reply for use by the client. When transmitted by c to s, an authenticator is sent with it, encrypted by $K_{c,s}$; again, this ensures freshness.

Finally, c can ask s to send it a message encrypted in the same session key; this protects the client against someone impersonating the server.

There are several important points to note about the design. First, cryptography is used to create *sealed* packages. Tickets and the like are encrypted along with a checksum; this protects them from tampering. Second, care is taken to avoid repetitive password entry requests; human factors are quite important, as users tend to bypass security measures they find unpleasant. Third, messages must be protected against replay; an attacker who can send the proper message may not need to know what it says. Cut-and-paste attacks are a danger as well, though they are beyond the scope of this chapter.

It is worth noting that the design of cryptographic protocols is a subtle business. The literature is full of attacks that were not discovered until several years after publication of the initial protocol. See, for example, Bellovin and Merritt [1991] and Stubblebine and Gligor [1992] for examples of problems with Kerberos itself.

74.7   Firewalls

Firewalls [Cheswick and Bellovin 1994] are an increasingly popular defense mechanism. Briefly, a firewall is an electronic analog of the security guard at the entrance to a large office or factory. Credentials are checked, outsiders are turned away, and incoming packages — electronic mail — is handed over for delivery by internal mechanisms.

The purpose of a firewall is to protect more vulnerable machines. Just as most people have stronger locks on their front doors than on their bedrooms, there are numerous advantages to putting stronger security on the perimeter. If nothing else, a firewall can be run by personnel whose job it is to ensure security.

For many sites, though, the real issue is that internal networks *cannot* be run securely. Too many systems rely on insecure network protocols for their normal operation. This is bad, and everyone understands this; too often, though, the choice is between accepting some insecurity or not being able to use the network productively. A firewall is often a useful compromise; it blocks attacks from a high-threat environment, while letting people use today's technology.

Seen that way, a firewall works because of what it is not. It is not a general purpose host; consequently, it does not need to run a lot of risky software. Ordinary machines rely on networked file systems, remote log-in commands that rely on address-based authentication, users who surf the Web, etc. A firewall does none of these things; accordingly, it is not affected by potential security problems with them.

74.7.1   Types of Firewalls

There are four primary types of firewalls: packet filters, dynamic packet filters, **application gateways**, and **circuit relays**. Each has its advantages and disadvantages.

74.7.1.1   Packet Filters

The cheapest and fastest type of firewall is the packet filter. Packet filters work by looking at each individual packet, and, based on source address and destination addresses and port numbers, making a pass/drop decision. They are cheap because virtually all modern routers already have the necessary functionality; in effect, you have already paid the price, so you may as well use it. Additionally, given the comparatively slow lines most sites use for external access, packet filtering is fast; a router can filter at speeds higher than, say, a DSL line (1,500,000 bits/second).

The problem is that decisions made by packet filters are completely context free. Each packet is examined, and its fate decided, without looking at the previous input history. This makes it difficult or impossible to handle certain protocols. For example, file transfer protocol (FTP) [Mills 1985] uses a secondary TCP connection to transfer files; by default, this is an incoming call through the firewall [Bellovin 1994]. In this situation, the call should be permitted; the client has even sent a message specifying which port to call. But ordinary packet filters cannot cope.

Packet filters must permit not only outgoing packets but also the replies. For TCP, this is not a big problem; the presence of one header bit [the acknowledgment (ACK) bit] denotes a reply packet. In general, packets with this bit set can safely be allowed in, as they represent part of an ongoing conversation. Datagram protocols such as UDP do not have the concept of conversation and hence do not have such a bit, which causes difficulties: when should a UDP packet be allowed in? It is easy to permit incoming queries to known safe servers; it is much harder to identify replies to queries sent from the inside. Ordinary packet filters are not capable of making this distinction. At best, sites can assume that higher numbered ports are used by clients and hence are safe; in general, this is a bad assumption.

Services built on top of Sun's remote procedure call [Sun 1988, 1990] pose a different problem: the port numbers they use are not predictable. Rather, they pick more or less random port numbers and register with a directory server known as the portmapper. Would-be clients first ask portmapper which port number is in use at the moment, and then do the actual call. But since the port numbers are not fixed, it is not possible to configure a packet filter to let in calls to the proper services only.

74.7.1.2 Dynamic Packet Filters

Dynamic packet filters are designed to answer the shortcomings of ordinary packet filters. They are inherently stateful and retain the context necessary to make intelligent decisions. Most also contain application-specific modules; these do things like parse the FTP command stream so that the data channel can be opened, look inside portmapper messages to decide if a permitted service is being requested, etc. UDP queries are handled by looking for the outbound call and watching for the responses to that port number. Since there is no end-of-conversation flag in UDP, a timeout is needed. This heuristic does not always work well, but, without a lot of application-specific knowledge, it is the only possibility.

Dynamic packet filters promise everything: safety and full transparency. The risk is their complexity; one never knows exactly which packets will be allowed in at a given time.

74.7.1.3 Application Gateways

Application gateways live at the opposite end of the protocol stack. Each application being relayed requires a specialized program at the firewall. This program understands the peculiarities of the application, such as data channels for FTP, and does the proper translations as needed.

It is generally acknowledged that application gateways are the safest form of firewall. Unlike packet filters, they do not pass raw data; rather, individual applications, invoked from the inside, make the necessary calls. The risk of passing an inappropriate packet is thus eliminated.

This safety comes at a price, though. Apart from the need to build new gateway programs, for many protocols a change in user behavior is needed. For example, a user wishing to telnet to the outside generally needs to contact the firewall explicitly and then redial to the actual destination. For some protocols, though, there is no user visible change; these protocols have their own built-in redirection or proxy mechanisms. Mail and the World Wide Web are two good examples.

74.7.1.4 Circuit Relays

Circuit relays represent a middle ground between packet filters and application gateways. Because no data are passed directly, they are safer than packet filters. But because they use generic circuit-passing programs, operating at the level of the individual TCP connection, specialized gateway programs are not needed for each new protocol supported.

The best-known circuit relay system is socks [Koblas and Koblas 1992]. In general, applications need minor changes or even just a simple relinking to use the socks package. Unfortunately, that often means it is impossible to deploy it unless a suitable source or object code is available. On some systems, though, dynamically linked run-time libraries can be used to deploy socks.

Circuit relays are also weak if the aim is to regulate outgoing traffic. Since more or less any calls are permissible, users can set up connections to unsafe services. It is even possible to tunnel IP over such circuits, bypassing the firewall entirely. If these sorts of activities are in the threat model, an application gateway is probably preferable.

74.7.2 Limitations of Firewalls

As important as they are, firewalls are not a panacea to network security problems. There are some threats that firewalls cannot defend against.

The most obvious of these, of course, is attacks that do not come through the firewall. There are always other entry points for threats. There might be an unprotected modem pool; there are always insiders, and a substantial portion of computer crime is due to insider activity. At best, internal firewalls can reduce this latter threat.

On a purely technical level, no firewall can cope with an attack at a higher level of the protocol stack than it operates. Circuit gateways, for example, cannot cope with problems at the simple mail transfer protocol (SMTP) layer [Postel 1982]. Similarly, even an application-level gateway is unlikely to be able to deal with the myriad security threats posed by multimedia mail [Borenstein and Freed 1993]. At best, once such problems are identified a firewall may provide a place to deploy a fix.

A common question is whether or not firewalls can prevent virus infestations. Although, in principle, a mail or FTP gateway could scan incoming files, in practice it does not work well. There are too many ways to encode files, and too many ways to spread viruses, such as self-extracting executables.

Finally, firewalls cannot protect applications that must be exposed to the outside. Web servers are a canonical example; as previously described, they are inherently insecure, so many people try to protect them with firewalls. That does not work; the biggest security risk is in the service that of necessity must be exposed to the outside world. At best, a firewall can protect other services on the Web server machine. Often, though, that is like locking up only the bobcats in a zoo full of wild tigers.

74.8 Denial of Service Attacks

Denial of service attacks are generally the moral equivalent of vandalism. Rather than benefitting the perpetrator, the goal is generally to cause pain to the target, often for no better reason than to cause pain.

The simplest form is to flood the target with packets. If the attacker has a faster link, the attacker wins. If this attack is combined with source address spoofing, it is virtually untraceable as well.

Sometimes, denial of service attacks are aimed more specifically. A modest number of TCP open request packets, from a forged IP address, will effectively shut down the port to which they are sent. This technique can be used to close down mail servers, Web servers, etc.

The ability to interrupt communications can also be used for direct security breaches. Some authentication systems rely on primary and backup servers; the two communicate to guard against replay attacks. An enemy who can disrupt this path may be able to replay stolen credentials.

Philosphically, denial of service attacks are possible any time the cost to the enemy to mount the attack is less, relatively speaking, than the cost to the victim to process the input. In general, prevention consists of lowering your costs for processing unauthenticated inputs.

74.9 Conclusions

We have discussed a number of serious threats to networked computers. However, except in unusual circumstances — and they do exist — we do not advocate disconnection. Whereas disconnecting buys you some extra security, it also denies you the advantages of a network connection.

It is also worth noting that complete disconnection is much harder than it would appear. Dial-up access to the Internet is both easy and cheap; a managed connection can be more secure than a total ban that might incite people to evade it. Moreover, from a technical perspective an external network connection is just one threat among many. As with any technology, the challenge is to control the risks while still reaping the benefits.

Defining Terms

Active agents: Programs sent to another computer for execution on behalf of the sending computer.
Address spoofing: Any enemy computer's impersonation of a trusted host's network address.
Application gateway: A relay and filtering program that operates at layer seven of the network stack.
Back door: An unofficial (and generally unwanted) entry point to a service or system.
Checksums: A short function of an input message, designed to detect transmission errors.
Choke point: A single point through which all traffic must pass.

Circuit relay: A relay and filtering program that operates at the transport layer (level four) of the network protocol stack.

Common gateway interface (CGI) scripts: The interface to permit programs to generate output in response to World Wide Web requests.

Connection hijacking: The injection of packets into a legitimate connection that has already been set up and authenticated.

Cryptography: The art and science of secret writing.

Denial of service: An attack whose primary purpose is to prevent legitimate use of the computer or network.

Firewall: An electronic barrier restricting communications between two parts of a network.

Kerberos ticket-granting ticket (TGT): The cryptographic credential used to obtain credentials for other services.

Key distribution center (KDC): A trusted third party in cryptographic protocols that has knowledge of the keys of other parties.

Magic cookie: An opaque quantity, transmitted in the clear and used to authenticate access.

Network file system (NFS) protocol: Originally developed by Sun Microsystems.

Packet filter: A network security device that permits or drops packets based on the network layer addresses and (often) on the port numbers used by the transport layer.

r-Commands: A set of commands (**sh**, **rlogin**, **rcp**, **rdist**, etc.) that rely on address-based authentication.

Remote procedure call (RPC) protocol: Originally developed by Sun Microsystems.

Routing protocols: The mechanisms by which network switches discover the current topology of the network.

Sequence number attacks: An attack based on predicting and acknowledging the byte sequence numbers used by the target computer without ever having seen them.

Topological defense: A defense based on the physical interconnections of two networks. Security policies can be based on the notions of inside and outside.

Transmission control protocol (TCP): The basic transport-level protocol of the Internet. It provides for reliable, flow-controlled, error-corrected virtual circuits.

Trust: The willingness to believe messages, especially access control messages, without further authentication.

User datagram protocol (UDP): A datagram-level transport protocol for the Internet. There are no guarantees concerning order of delivery, dropped or duplicated packets, etc.

References

Arnold, K. and Gosling, J. 1996. *The Java Programming Language*. Addison–Wesley, Reading, MA.

Bellovin, S. M. 1989. Security problems in the TCP/IP protocol suite. *Comput. Commun. Rev.*, 19(2):32–48.

Bellovin, S. M. 1994. Firewall-Friendly FTP. Request for comments (informational) RFC 1579. Internet Engineering Task Force, Feb.

Bellovin, S. M. 1995. Using the domain name system for system break-ins, pp. 199–208. In *Proc. 5th USENIX Unix Security Symp*. Salt Lake City, UT, June.

Bellovin, S. M. 1996. Defending against sequence number attacks. RFC 1948. May.

Bellovin, S. M. and Merritt, M. 1991. Limitations of the Kerberos authentication system, pp. 253–267. In *USENIX Conf. Proc*. Dallas, TX, Winter.

Borenstein, N. and Freed, N. 1993. MIME (Multipurpose Internet Mail Extensions) Part One: Mechanisms for Specifying and Describing the Format of Internet Message Bodies. Request for comments (draft standard) RFC 1521, Internet Engineering Task Force, Sept. (obsoletes RFC 1341; updated by RFC 1590).

Bryant, B. 1988. Designing an authentication system: a dialogue in four scenes. Draft. Feb. 8.

Cheswick, W. R. and Bellovin, S. M. 1994. *Firewalls and Internet Security: Repelling the Wily Hacker*. Addison–Wesley, Reading, MA.

Dean, D. and Wallach, D. 1996. Security flaws in the HotJava web browser. In *Proc. IEEE Symp. Res. Security Privacy.* Oakland, CA, May.

Eichin, M. W. and Rochlis, J. A. 1989. With microscope and tweezers: an analysis of the Internet virus of November 1988, pp. 326–345. In *Proc. IEEE Symp. Res. Security Privacy.* Oakland, CA, May.

Epstein, J., McHugh, J., and Pascale, R. 1992. Evolution of a trusted B3 window system prototype. In *Proc. IEEE Comput. Soc. Symp. Res. Security Privacy.* Oakland, CA, May.

Gong, L., Lomas, M. A., Needham, R. M., and Saltzer, J. H. 1993. Protecting poorly chosen secrets from guessing attacks. *IEEE J. Select. Areas Commun.*, 11(5):648–656.

Haller, N. M. 1994. The S/Key one-time password system. In *Proc. Internet Soc. Symp. Network Distributed Syst. Security.* San Diego, CA, Feb. 3.

Joncheray, L. 1995. A simple active attack against TCP. In *Proc. 5th USENIX Unix Security Symp.* Salt Lake City, UT, June.

Kahn, B. L. 1995. Safe use of X window system protocol across a firewall, pp. 105–116. In *Proc. 5th USENIX Unix Security Symp.* Salt Lake City, UT, June.

Klein, D. V. 1990. Foiling the cracker: a survey of, and improvements to, password security, pp. 5–14. In *Proc. USENIX Unix Security Workshop.* Portland, OR, Aug.

Koblas, D. and Koblas, M. R. 1992. Socks, pp. 77–83. In *Unix Security III Symp.* Baltimore, MD, Sept. 14–17, USENIX.

Kohl, J. and Neuman, B. 1993. The Kerberos Network Authentication Service (V5). Request for comments (proposed standard) RFC 1510, Internet Engineering Task Force, Sept. 1993.

LaMacchia, B. A. and Odlyzko, A. M. 1991. Computation of discrete logarithms in prime fields. *Designs, Codes, Cryptography,* 1:46–62.

Littman, J. 1996. *Fugitive Game.* Little, Brown.

Miller, S. P., Neuman, B. C., Schiller, J. I., and Saltzer, J. H. 1987. Kerberos authentication and authorization system. In *Project Athena Technical Plan.* Sec. E.2.1, Massachusetts Institute of Technology, Cambridge, MA, Dec.

Mills, D. 1985. Network Time Protocol NTP. RFC 958, Internet Engineering Task Force, Sept. (obsoleted by RFC1059).

Morris, R. T. 1985. A Weakness in the 4.2BSD Unix TCP/IP Software. AT& T Bell Lab. Computing Science Tech. Rep. 117, Murray Hill, NJ, Feb.

Morris, R. H. and Thompson, K. 1979. Unix password security. *Commun. ACM,* 22(11):594.

Postel, J. 1980. User Datagram Protocol. Request for comments (standard) STD 6, RFC 768, Internet Engineering Task Force, Aug.

Postel, J. 1982. Simple Mail Transfer Protocol. Request for comments (standard) STD 10, RFC 821, Internet Engineering Task Force, Aug. (obsoletes RFC 0788).

Rochlis, J. A. and Eichin, M. W. 1989. With microscope and tweezers: the worm from MIT's perspective. *Commun. ACM,* 32(6):689–703.

Scheifler, R. W. and Gettys, J. 1992. *X Window System,* 3rd ed. Digital Press, Burlington, MA.

Shimomura, T. 1996. *Takedown.* Hyperion.

Spafford, E. H. 1989a. An analysis of the Internet worm. In *Proc. European Software Eng. Conf.* C. Ghezzi and J. A. McDermid, Eds. Lecture notes in computer science, 387, pp. 446–468. Warwick, England, Sept. Springer–Verlag.

Spafford, E. H. 1989b. The Internet worm program: an analysis. *Comput. Commun. Rev.,* 19(1): 17–57.

Spafford, E. H. 1992. Observations on reusable password choices, pp. 299–312. In *Proc. 3rd USENIX Unix Security Symp.* Baltimore, MD, Sept.

Steiner, J., Neuman, B. C., and Schiller, J. I. 1988. Kerberos: an authentication service for open network systems, pp. 191–202. In *Proc. Winter USENIX Conf.* Dallas, TX.

Stevens, W. R. 1995. *TCP/IP Illustrated,* Vol. 1. Addison–Wesley, Reading, MA.

Stubblebine, S. G. and Gligor, V. D. 1992. On message integrity in cryptographic protocols, pp. 85–104. In *Proc. IEEE Comput. Soc. Symp. Res. Security Privacy.* Oakland, CA, May.

Sun. 1988. RPC: Remote Procedure Call Protocol Specification Version 2. Request for comments (informational) RFC 1057, Internet Engineering Task Force, Sun Microsystems, June (obsoletes RFC 1050).

Sun. 1989. NFS: Network File System Protocol Specification. Request for comments (historical) RFC 1094, Internet Engineering Task Force, Sun Microsystems, March.

Sun. 1990. Network Interfaces Programmer's Guide. SunOS 4.1. Sun Microsystems. Mountain View, CA, March.

Wright, G. R. and Stevens, W. R. 1994. *TCP/IP Illustrated: The Implementation*, Vol. 2. Addison–Wesley, Reading, MA.

75

Information Retrieval
and Data Mining

75.1   Introduction 75-1
75.2   Information Retrieval 75-2
     Text Retrieval Issues • Text Retrieval Methods • Text Retrieval
     Systems and Models • Web and Multimedia Information
     Retrieval • Evaluating IR Systems
75.3   Data Mining 75-7
     Concept Description • Association Rule Mining
     • Classification and Prediction • Clustering
75.4   Integrating IR and DM Techniques into Modern
     Search Engines 75-11
     Web Mining and Retrieval • Vivisimo • KartOO
     • SYSTERS Protein Family Database • E-Commerce Systems
75.5   Conclusion and Further Resources 75-14

Katherine G. Herbert
New Jersey Institute of Technology

Jason T.L. Wang
New Jersey Institute of Technology

Jianghui Liu
New Jersey Institute of Technology

75.1 Introduction

With both commercial and scientific data sets growing at an extremely rapid rate, methods for retrieving knowledge from this data in an efficient and reliable manner are constantly needed. To do this, many knowledge discovery techniques are employed to analyze these large data sets. Generally, knowledge discovery is the process by which data is cleaned and organized, then transformed for use for pattern detection and evaluation tools and then visualized in the most meaningful manner for the user [13].

Two areas of research — information retrieval (IR) and data mining (DM) — are used to try to manage these data sets as well as gain knowledge from them. Data mining concentrates on finding and exploiting patterns found within a given data set to gain knowledge about that data set. As databases developed and became larger and more complex, the need to extract knowledge from these databases became a pressing concern. Data mining uses various algorithms that extract patterns from the data to gain knowledge about the data set. It borrows techniques from statistics, pattern recognition, machine learning, data management, and visualization to accomplish the pattern discovery task. Information retrieval is the study of techniques for organizing and retrieving information from databases [30]. Modern information retrieval concerns itself with many different types of databases. It studies returning information matching a user's query that is relevant in a reasonable amount of time. It also focuses on other complex problems associated with a static query that will be needed time and time again. In this chapter we explore both the topics of data mining and information retrieval. We discuss how these two approaches of obtaining knowledge from data can work in a complementary manner to create more effective knowledge discovery tools. We look at a common application of knowledge discovery tools where these approaches work together, namely search engines. Finally, we address future work in data mining and information retrieval.

75.2 Information Retrieval

As mentioned above, information retrieval investigates problems that are concerned with organizing and accessing information effectively. This is a broad area of research that currently encompasses many disciplines. Here, we primarily focus on text information retrieval, and then briefly mention emerging areas such as web and multimedia information retrieval.

75.2.1 Text Retrieval Issues

Text retrieval systems usually need to perform efficient and effective searches on large text databases, often with data that is not well organized. Text retrieval is generally divided into two categories: problems that concentrate on returning relevant and reliable information to the user and problems that concentrate on organizing data for long-term retrieval needs. Concerning the first problem, methods here usually investigate techniques for searching databases based on a user query. The user can enter a query and the text retrieval system searches the database, returning results based on the user's query. These results can be ranked or ordered according to how close the text retrieval system feels the results satisfy the query. Another type of text retrieval system is one that is used for long-term information needs. These employ text categorization, text routing, and text filtering techniques to enhance the user's ability to query the database effectively. These techniques essentially preprocess a portion of the querying process, whether it is classifying the text or creating a user profile to better semantically query the database or use filters on the database before beginning the search [30].

Text retrieval systems have many issues that they must address in order to effectively perform searches on a database, specifically text databases. Many of these issues result from the vernacular usage of words and phrases within a given language as well as the nature of the language. Two prominent issues that must be addressed by text retrieval systems related to this problem are synonymy and polysemy.

- *Synonymy* refers to the problem of when words or phrases mean similar things. This problem sometimes is solved and usually results in a text retrieval system needing to expand upon a query, incorporating a thesaurus to know which words or terms are similar to the words or terms in the user's query. This allows the system to return results that might be of interest to the user but would normally be returned for another word that has a similar meaning to the word or phrase used within the query.
- *Polysemy* refers to when one word or phrase has multiple meanings [30]. Work to address this problem has included creating user profiles so that the text retrieval system can learn what type of information the user is generally interested in as well as semantic analysis of phrases within queries [17].

Other common problems that text retrieval systems must be concerned with are phrases, object recognition, and semantics. Phrases within languages tend to have a separate meaning from what each individual word in the phrase means. Many text retrieval systems use phrase-based indexing techniques to manage phrases properly [10]. Object recognition usually concerns itself with a word or phrase. These word phrases usually have a specific meaning separate to itself from the meaning of the individual words. For example, the word "labor" means to work and the word "day" refers to a period of time. However, when these two words are placed next to each other to form "Labor Day," this refers to a holiday in September in the United States. Common parts of sentences that are considered objects are proper nouns, especially proper names, noun phrases, and dates.

A text retrieval system that can manage objects sometimes uses pattern recognition tools to identify objects [30]. All these problems can generally be thought of by considering how the word or phrase is used semantically by the user.

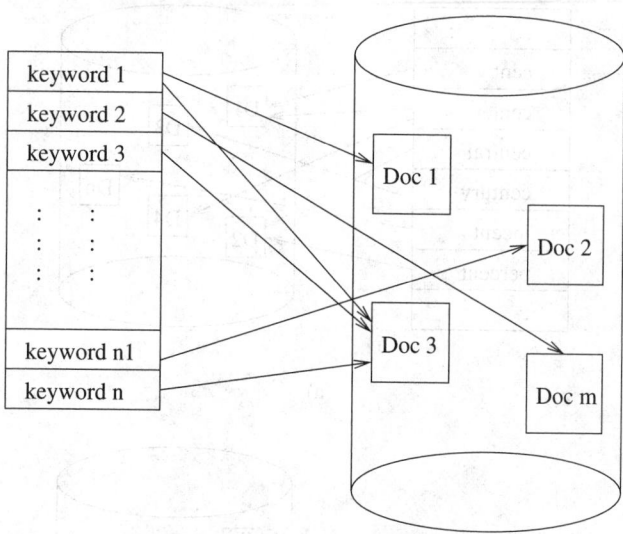

FIGURE 75.1   A general diagram for an inverted file.

75.2.2   Text Retrieval Methods

To address these problems, there are some common practices for processing and filtering data that help text retrieval tools be more effective. One very common practice is to use inverted files as an indexing structure for the database the tools search upon. An inverted file is a data structure that can index keywords within a text. These keywords are organized so that quick search techniques can be used. Once a keyword is found within the indexing structure, information is retained about the documents that contain this keyword and those documents are returned to fulfill the query.

Figure 75.1 illustrates the general concept behind inverted files. In this figure, there is an organized structuring of the keywords. This structuring can be formed through using various data structures, such as the B-tree or a hash table. These keywords have references or pointers to the documents where they occur frequently enough to be considered a content-bearing word for that document. Deciding whether or not a word is content bearing is a design issue for the information retrieval system. Usually, a word is considered to be content bearing if it occurs frequently within the document. Algorithms such as Zipf's law [39] or taking the term frequency with respect to the inverse document frequency (tf.idf) [31] can be used to determine whether or not a word is a content-bearing word for a document.

Other effective methods include stopword lists, stemming, and phrase indexing. Stopword lists, commonly seen when searching the World Wide Web using Google, exploit the idea that there are great occurrences of common words, such as "the," "a," and "with," within documents in a given text database. Because these words rarely add to the meaning of the query, they are disregarded and filtered out of the query so that the text retrieval tool can concentrate on the more important words within the query. Stemming utilizes the concept that many words within a query can be a variation on tense or case of another word. For example, the word "jumping" has the root word "jump" in it. The concepts related to "jumping" and "jump" are very similar. Therefore, if a query requests information about "jumping," it is highly likely that any information indexed for "jump" would also interest the user. Stemming takes advantage of this idea to make searching more efficient. Stemming can help improve space efficiency as well as help generalize queries. Generalized queries help to ensure documents that the user may want but might not have been included within the search results because of the wording of the query will be included. However, this can also lead to false positives if the stemming algorithm does not process a word properly [12,29].

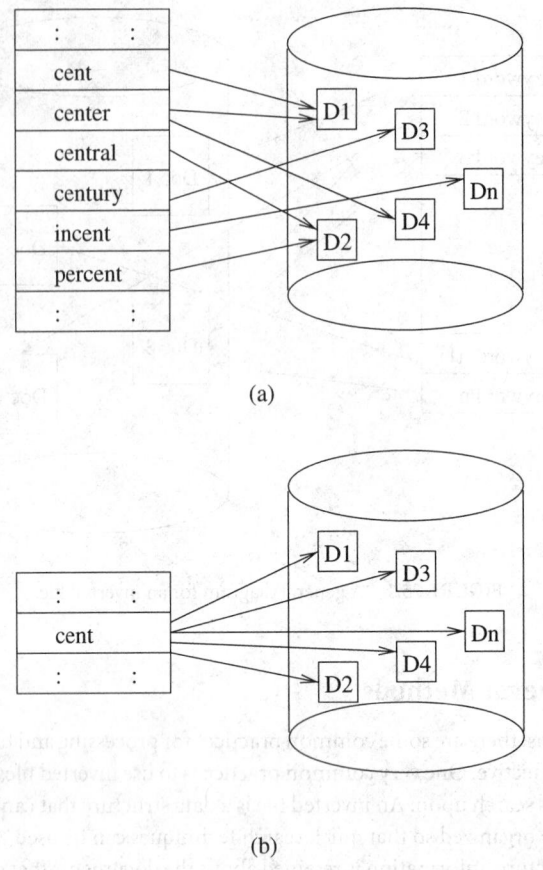

(a)

(b)

FIGURE 75.2   The effect of stemming on the inverted file: (a) represents an inverted file that does not use stemming; (b) represents an inverted file that uses stemming.

For example, Figure 75.2 demonstrates some of the issues stemming faces. Figure 75.2a represents how an index containing the words "cent," "center," "central," "century," "incent," and "percent" might be organized. Figure 75.2b demonstrates how each of these words can be reduced to the root "cent." Notice in Figure 75.2a that each word indexes different documents. However, in Figure 75.2b, all the documents that were reduced to the root "cent" are now indexed by cent. Stemming, while it might return related documents to the query, can also return many unrelated documents to the user. Also, while all of the terms in this figure can be reduced to the root "cent," it is not appropriate in some cases to do so. In the case of "percentage" and "incent," the issue of whether or not a prefix should be stripped arises. In this example, the prefixes are stripped off to demonstrate the problems of stemming. However, in general, prefixes are not stripped to reduce a word to its root because many prefixes change the meaning of the root.

75.2.3   Text Retrieval Systems and Models

Applying the above concepts to help filter and retrieve meaningful data, there are many methods for creating text retrieval systems. The most popular one is the Boolean keyword text retrieval system. In this system, the user enters a series of keywords joined by Boolean operators such as "and" or "or." These systems can be extended to employ ranking of search results as well as the ability to handle wildcard or "don't care" characters [33,40,41].

One popular model for text retrieval systems is the vector-space model [31,32]. In vector-space models, documents are viewed as vectors in N-dimensional space, where N is the number of keywords within the database. The values within the vector represent whether a particular keyword is present in the given document. These values can be as simple as 0 if the keyword is not present within the document or 1 if it is present. Also, the text retrieval system can use functions where the resulting value represents the importance of that keyword within the document. When querying the system, the user's query is transformed into a vector and then that vector is compared with the vectors within the database. Documents with similar vectors are returned to the user, usually ranked with respect to how similar the document's vector is to the original query. While this technique does allow for effective similarity searches, the dimensionality of the vectors can increase greatly, depending on the number of keywords.

Other popular text systems include probabilistic models as well as the employment of machine learning and artificial intelligence techniques. Probabilistic models test how well a document satisfies a user query. These techniques can employ Bayesian networks to represent both the document and the query. Machine learning and artificial intelligence techniques use natural language processing, rule-based systems, and case-based reasoning for information retrieval in text-based documents.

75.2.4 Web and Multimedia Information Retrieval

While there is great interest in the text-based methods previously mentioned, there is also currently a lot of research within the areas of Web-based information retrieval and multimedia information retrieval. Web-based information retrieval can best be seen in the techniques used to index Web sites for search engines as well as the techniques used to track usage [8,17]. Issues concerning this type of information retrieval will be reviewed later in this chapter. Multimedia information retrieval addresses issues involved with multimedia documents such as images, music, and motion pictures. Unlike text, multimedia documents use various formats to display information. Because this information is in different formats, before any retrieval algorithms are used on it, the documents must be mapped to a common format. Otherwise, there will be no standard representation of the document, making retrieval harder. Moreover, because most information retrieval algorithms apply to text documents, these algorithms either need to be modified or new algorithms must be developed to perform retrieval. Also, most tools can only process information about the format of the document, not its content [23].

In addition to issues concerning the data that is comprised in a multimedia database, other issues concerning the query also cause great difficulties. Because the query would be for multimedia information, naturally the user might want to use a multimedia format to author the query. This creates problems with user authoring tools. For example, if someone is searching for a particular piece of music and only knows a couple of notes from the song (and knows nothing about the title or artist), he or she must enter those notes. This creates issues concerning how the user enters these notes as well as how are these notes matched against a database of songs. Most multimedia databases allow only keyword search, thereby eliminating the authoring problem. Databases that allow only keyword search usually use whatever metadata about the images within the database (such as captions and file names) to index the data. The query is then matched against this data. A good example of this type of search tool is Google's Image Search (http://www.google.com). However, there are many research projects underway to allow users to author multimedia documents (such as images or segments of music) as queries for databases [20].

Currently there are many efforts for developing techniques to both process and retrieve multimedia data. These efforts combine numerous fields outside of information retrieval, including image processing and pattern recognition. Current popular techniques use relevance feedback as well as similarity measures such as Euclidean distance for multimedia information retrieval. Relevance feedback is a technique where the user continually interacts with the retrieval system, refining the query until he or she is satisfied with the results of the search. Similarity measures are used to match queries against database documents. However, current similarity measures, such as Euclidean distance and cosine distance for vector models, are based on the properties of text retrieval. Therefore, relevance feedback practices have performed better than the similarity measures [20]. One project by the Motion Picture Expert Group (MPEG) is called MPEG-7. MPEG-7 tries to create models for various types of multimedia documents so that all the information contained within the documents can be specified through metadata. This data can then be searched through the usual text retrieval methods [26,27].

75.2.5 Evaluating IR Systems

Information retrieval systems are evaluated based on many different metrics. The two most common metrics are recall and precision:

1. *Recall* measures the percentage of relevant documents retrieved by the system with respect to all relevant documents within the database. If the recall percentage is low, then the system is retrieving very few relevant documents.
2. *Precision* describes how many false hits the system generates. Precision equals the number of relevant documents retrieved divided by the total number of documents retrieved. If the precision percentage is low, then most of the documents retrieved were false hits.

Most IR systems experience a dilemma concerning precision and recall. To improve a system's precision, the system needs strong measures for deciding whether a document is relevant to a query. This will help minimize the false hits, but it will also affect the number of relevant documents that are retrieved. These strong measures can prevent some important relevant documents from being included within the set of documents that satisfy the query, thereby lowering the recall.

In addition to precision and recall, there are other measures that can be used to evaluate the effectiveness of an information retrieval system. One very important evaluation measure is ranking. *Ranking* refers to the evaluation techniques by which the search results are ordered and then returned to the user, presented in that order. In ranking, a rating is given to the documents that the information retrieval system considers a match for the query. This rating reflects how similar the matched document is to the user's query. One of the most popular algorithms for ranking documents on the World Wide Web is PageRank [5,28]. PageRank, developed by Page et al., is for understanding the importance for documents retrieved from the Web. It is similar to the citation method of determining the importance of a document. Basically, in this algorithm, the relevance of a Web site to a particular topic is determined by how many well-recognized Web pages (Web pages that are known to be a reliable reference to other pages) link to that page. In addition to precision, recall, and ranking, which are based on system performance, there are also many measures such as coverage ratio and novelty ratio that indicate the effectiveness of the information retrieval system with respect to the user's expectations [18]. Table 75.1 summarizes some of the measures that can be used to determine the effectiveness of an information retrieval system.

TABLE 75.1 Measures for Evaluating Information Retrieval Systems

Measure	Purpose
Precision	Describes the number of false hits.
Recall	Measures percentage of relevant documents retrieved with respect to all relevant documents within the database.
Citation	Measures importance of a document through the number of other documents referencing it.
Coverage	Measures the number of relevant documents retrieved the user was previously aware of.
Novelty	Measures the number of relevant documents retrieved the user was not previously aware of.
PageRank	This is similar to the citation measure, except for Web documents.

75.3 Data Mining

Data mining refers to the extraction or discovery of knowledge from large amounts of data [13,37]. Other terms with similar meaning include knowledge mining, knowledge extraction, data analysis, and pattern analysis. The main difference between information retrieval and data mining is their goals. Information retrieval helps users search for documents or data that satisfy their information needs [6]. Data mining goes beyond searching; it discovers useful knowledge by analyzing data correlations using sophisticated data mining techniques. Knowledge may refer to some particular patterns shared by a subset of the data set, some specific relationship among a group of data items, or other interesting information that is implicit or not directly inferable.

Data mining is an interdisciplinary field contributed to by a set of disciplines including database systems, statistics, machine learning, pattern recognition, visualization, and information theory. As a result, taxonomies of data mining techniques are not unique. This is due to the various criteria and viewpoints of each discipline involved with the development of the techniques. One generally accepted taxonomy is based on the data mining functionalities such as association rule mining [1], classification [7], clustering [4], and concept description. To be a comprehensive and effective data mining system, the above functionalities must be implemented within the system. These functionalities also give a portal to the understanding of general data mining system construction.

75.3.1 Concept Description

The explosive increase of data volume, especially large amounts of data stored in great detail, requires a succinct representation for the data. Most users prefer an overall picture of a class of data so as to distinguish it from other comparative classes. On the other hand, the huge volume of data makes it impossible for a person to give, intuitively, such a concise while accurate summarization for a given class of data. However, there exist some computerized techniques to summarize a given class of data in concise, descriptive terms, called concept description [8,9]. These techniques are essential and form an important component of data mining.

Concept description is not simply enumeration of information extracted from the database. Instead, some derivative techniques are used to generate descriptions for characterization and discrimination of the data. According to the techniques used to derive the summary, concept description can be divided into characterization analysis and discrimination analysis. *Characterization analysis* derives the summary information from a set of data. To do characterization, the data generalization and summarization-based method aims to summarize a large set of data and represent it at a higher, conceptual level. Usually, attribute-oriented induction is adopted to guide the summarization process from a lower conceptual level to a higher one by checking the number of distinct values of each attribute in the relevant set of data. For example, Table 75.2 shows the original data tuples in a transactional database for a chain company. If some generalization operation regarding the geographical locations of stores has already been established, then the store ID in the location field can be replaced by a higher-level description, namely geographical areas. In addition, generalization can be done on the time field by replacing it with a higher-level concept, say month. Table 75.3 shows generalized sales for the database in Table 75.2, where the generalizations are performed with respect to the attributes "time" and "location."

TABLE 75.2 Original Data in a Transactional Database

Item	Unit Price	Time	Payment	Location	Quantity
Printer	$45.00	14:02 7/5/2002	Visa	0089	1
Scanner	$34.56	11:09 8/1/2002	Cash	0084	1
Camcorder	$489.95	13:00 7/14/2002	Master	0100	1
⋮	⋮	⋮	⋮	⋮	⋮

TABLE 75.3   Generalized Sales for the Same Transactional Database in Table 75.2

Item	Unit Price	Time	Payment	Location	Quantity
Printer	$45.00	July, 2002	Visa	Essex County, NJ	1
Scanner	$34.56	August, 2002	Cash	Hudson County, NJ	1
Camcorder	$489.95	July, 2002	Master	Essex County, NJ	1
⋮	⋮	⋮	⋮	⋮	⋮

Discrimination analysis puts emphasis on the distinguishing features among sets of data. Discrimination analysis can be accomplished by extending the techniques proposed for characterization analysis. For example, by performing the generalization process among all data classes simultaneously and synchronously, the same level of generalization for all of the classes can be reached, thus making the comparison feasible. In previous examples, we assume the attributes selected for characterization or discrimination are always relevant. However, in many cases, not all the attributes are relevant for data characterization or comparison. Analytical characterization techniques are one kind of attribute relevance analysis. They are incorporated into data description or comparison to identify and exclude those irrelevant or weakly relevant attributes.

Concept description tries to capture the overall picture of a class of data by inducing the important features of it through conceptual generalization or comparison with a class of comparative data. By grasping the common features presented by the data class as a whole, it looks at the class of data as an entirety while ignoring the relationship among its component items. However, in many cases, exploring the relationship within component items is valuable. This forms another important data mining process: association rule mining.

75.3.2 Association Rule Mining

Association rule mining [2,3,22] is the process of finding interesting correlations among a large set of data items. For example, the discovery of interesting association relationships in large volumes of business transactions can facilitate decision making in marketing strategies. The general way of interpreting an association rule is that the appearance of the item(s) on the left-hand side of the rule implies the appearance of those item(s) on the right-hand side of the rule.

There are two parameters to measure the interestingness for a given association rule: support and confidence. For example, consider the following association rule discovered from a transaction database:

$$B \rightarrow C[support = 30\%, confidence = 66\%]$$

The usefulness of an association rule is measured by its support value. Given the above rule, it means that within the whole transactions of the database, 30% transactions contain both items B and C. The confidence value measures the certainty of the rule. Again, for the above rule, it means for all those transactions containing B, 66% of them also contain C. Figure 75.3 shows an example of finding association rules from a set of transactions. For rule $A \rightarrow C$, the number of transactions containing both A and C is 2, so the support for this rule is 2 divided by the total number of transactions (5), which is equivalent to 40%. To calculate confidence, we find the number of transactions containing A is 3, so we get the confidence as 66.7%.

An acceptable or interesting rule will have its two parameter values greater than a user-specified threshold. These two parameters are intuitively reasonable for measuring the interestingness of an association rule. The support parameter guarantees that there are statistically enough transactions containing the items appearing in the rule. The confidence parameter implies the validity of the right-hand side given the left-hand side of the rule, with certainty.

Given the two parameters, support and confidence, finding association rules requires two steps. First, find all frequent itemsets that contain all the itemsets so that, for each of them, its number of appearances

Transaction list: Association Rules:

1. (A, B, C)   A --> C [support=40%, confidence=66.7%]

2. (A, C)
                                     A --> B [support=40%, confidence=66.7%]
3. (D, E)

4. (B, C)   B --> C [support=40%, confidence=66.7%]

5. (A, B)

FIGURE 75.3   A simple example of finding association rules.

as a whole in the transactions must be greater than the support value. Next, generate association rules that satisfy the minimum support and minimum confidence, from the above frequent itemsets.

The well-known *a priori* [2,3,22] data-mining algorithm can demonstrate the principles underlying association rule mining. *A priori* is a classic algorithm to generate all frequent itemsets for discovering association rules given a set of transactions. It iteratively scans the transaction set to find frequent itemsets at one particular size a time. During each iteration process, new frequent candidate itemsets with size one larger than the itemsets produced at the previous iteration are generated; and the acceptable itemsets are produced and stored through scanning the set and calculating the support value for each of the candidate itemsets. If no new frequent itemsets can be produced, *a priori* stops by returning all itemsets produced from every iteration stage.

Given the frequent itemsets, finding association rules is straightforward. For each itemset, divide the items in it into two subsets with one acting as the left-hand side of the association rule and the other as the right-hand side. Different divisions will produce different rules. In this way, we can find all of the candidate association rules. It is obvious that each association rule satisfies the requirement of minimum support. By further verifying their confidence values, we can generate all the association rules.

Concept description and association rule discovery provide powerful underlying characteristics and correlation relationships from known data. They put emphasis on the analysis and representation of the data at hand while paying little attention in regard to constructing some kind of model for those data coming but still not available. This kind of model pays more attention to "future" data cases. In the data mining domain, classification and prediction accomplish the establishment of this kind of model. Many applications, such as decision making, marketing prediction, and investment assessment all benefit from these two techniques.

75.3.3 Classification and Prediction

In many cases, making a decision is related to constructing a model, such as a decision tree [25], against which unknown or unlabeled data could be categorized or classified into some known data class. For example, through the analysis of customer purchase behavior associated with age, income level, living area, and other factors, a model can be established to categorize customers into several classes. With this model, new customers can be classified properly so that an appropriate advertising strategy and effective promotion method could be set up for maximizing profit.

Classification is usually associated with finding a known data class for the given unknown data, which is analogous to labeling the unlabeled data. Therefore, the data values under consideration are always discrete and nominal. On the other hand, prediction aims to manage continuous data values by constructing a statistical regression model. Intuitively, a regression model tries to find a polynomial equation in the multidimensional space based on the given data. The trends presented by the equation give some possible predictions. Typical applications include investment risk analysis and economic growth prediction.

In the past, several classification approaches have been developed. The major models include decision tree induction, Bayesian classification, Bayesian belief networks, and neural network classification [35]. Although each model has its particular trait, all of them share a common two-step processing feature: a training stage and a classification stage. During the training stage, a model describing a predetermined set of data classes is established by analyzing database tuples comprised of attribute values. These tuples constitute the training data set. The acceptability of the model is measured in the classification stage where another data set, called the testing data set, is used to estimate the accuracy of the classification. If the model passes the classification stage, it means that its classification accuracy is acceptable and is ready to be used for classifying future data tuples or objects whose class labels are unknown.

In regard to prediction, the available regression techniques include linear regression, nonlinear regression, logistic regression, and Poisson regression [14]. Linear regression attempts to find a linear equation to represent the trend shown in the given database. Nonlinear regression uses a polynomial equation to represent the trend, instead of a linear equation, showing higher accuracy in those cases of complex trend prediction. Logistic regression and Poisson regression, also called generalized regression models, can be used to model both contiguous and discrete data.

As described above, classification starts with a set of known labeled data and its training stage is guided by the labeled data. We call this kind of training or learning "supervised learning," where both the label of each training datum and the number of data classes to be learned are known. On the other hand, there exist many cases in which the knowledge about the given set of data is very limited. Neither is the label for each datum known nor has the number of data classes been given. Clustering, known as "unsupervised learning," is aimed at handling those cases.

75.3.4 Clustering

Clustering is the process of grouping data objects into clusters without prior knowledge of the data objects [16,17,36]. It divides a given set of data into groups so that objects residing in the same group are "close" to each other while being far away from objects in other groups. Figure 75.4 illustrates the general concept underlying clustering. It has shown that object-dense regions, represented as point sets, are found and objects are clustered into groups according to the regions.

The objective of clustering is to enable one to discover distribution patterns and correlations among data objects by identifying dense vs. sparse regions in the data distribution. Unlike classification, which requires a training stage to feed predetermined knowledge into the system, clustering tries to deduce knowledge based on knowledge from which the clustering can proceed. Clustering analysis has a wide range of applications, including image processing, business transaction analysis, and pattern recognition.

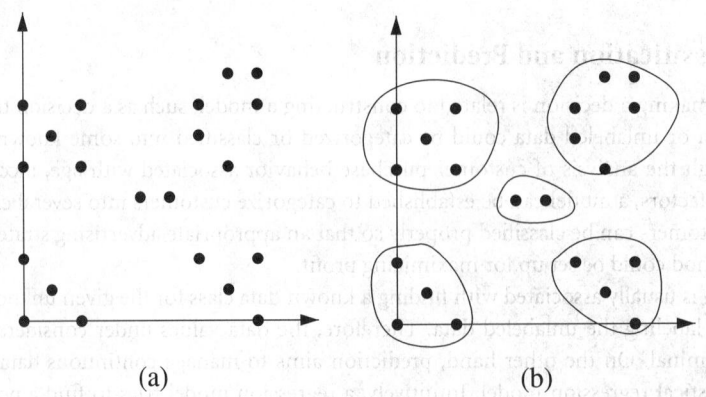

(a)   (b)

FIGURE 75.4 (a) A set of spatial points, and (b) a possible clustering for the spatial points.

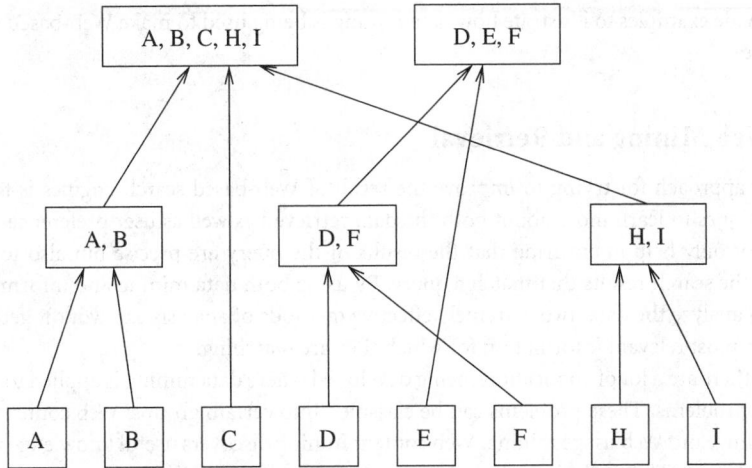

FIGURE 75.5 Agglomerative clustering.

The "learning from nothing" feature poses a set of typical requirements for an effective and efficient clustering analysis. These requirements, as discussed in [13], include scalability, the capability of dealing with different types of data, the ability to cope with noisy and high-dimensional data, the ability to be guided by clustering constraints, and the capability to cluster arbitrary shapes. To meet these requirements, researchers have proposed many clustering algorithms by taking advantage of the data under analysis and the characteristics of the application. The major categorization of clustering methods could be partition methods [21], hierarchical methods [16], and grid-based methods [38].

The well-known k-means algorithm [21] and its variation k-medoid [16] are two partition methods that accept *n* data objects and an integer *k*, and then divide the *n* objects into k groups satisfying the following two conditions. First, each group must contain at least one object. Second, each object must belong to exactly one group. During clustering, partition methods adopt iterative relocation techniques to try to find a different, more "reasonable" group for each data object and move data objects between groups until no group change occurs.

Hierarchical methods, such as agglomerative clustering, adopt a bottom-up strategy for tree construction. As shown in Figure 75.5, the leaf nodes are original objects. The clustering process goes from the bottom up along the tree, with each internal node representing one cluster. On the other hand, divisive clustering [16] uses top-down tactics to accomplish the same goal.

Both density-based methods and grid-based methods can handle arbitrary shape clustering. Density-based methods [11] accomplish this by distinguishing object-dense from object-sparse regions. On the other hand, grid-based methods use a multidimensional grid data structure to accommodate data objects. Through manipulation on the quantized grid cells, data objects are clustered. Model-based methods assume that the data is generated by a mixture of underlying probability distributions; thus, the goal of clustering becomes finding some mathematical model to fit the given data.

75.4 Integrating IR and DM Techniques into Modern Search Engines

With the development of the World Wide Web as well as developments in information retrieval and data mining, there are many applications in which one can exploit IR and DM techniques to help people discover knowledge they need. The most common instances of these applications tend to be in search tools. In this section we review some popular uses of information retrieval and data mining concerning the World Wide

Web. We provide examples to illustrate how data mining is being used to make Web-based search engines more effective.

75.4.1   Web Mining and Retrieval

One popular approach for trying to improve the recall of Web-based search engines is to employ data mining techniques to learn more about both the data retrieved as well as user preferences. Data mining techniques not only help in ensuring that the results of the query are precise but also to help the user sort through the search results that match a query. By using both data mining and information retrieval techniques to analyze the data, two extremely effective methods of analysis can work together to provide users with the most relevant information for which they are searching.

Currently, there are a lot of applications being developed where data mining is applied to Web information retrieval problems. These problems can be classified into certain groups: Web content mining, Web structure mining, and Web usage mining. Web content mining discovers useful knowledge within data on the World Wide Web. This analysis studies the content of Web sites as well as procedures for extracting and analyzing that content. Web structure mining looks at how various Web sites are related to one another. This analysis usually tries to discover the underlying connections of Web sites on the Internet (usually through the analysis of hyperlinks) so as to discover relationships and information about the Web sites. Finally, Web usage mining studies the behavior of a group of users with respect to the Web sites they view. From these studies, it can be observed what Web sites various groups of people with similar interests consider important. In this section, we concentrate solely on Web content mining because this type of mining is what most users have direct experience with [24].

One of the most popular applications of Web content data mining is clustering. Clustering algorithms are ideal for analyzing data on the Web. The premise behind clustering is that given a data set, find all groupings of data based on some data dimension. As discussed, clustering, unlike some other popular data mining techniques such as classification, does not require any mechanism for the tool to learn about the data.

Below we survey a couple of search engines that employ clustering to help users have meaningful and effective search experiences.

75.4.2   Vivisimo

Vivisimo (http://www.vivisimo.com) [34] is a meta-search engine that uses clustering and data mining techniques to help users have a more effective search experience. The search engine developed by the Vivisimo company offers users both on the Web and through enterprise solutions the ability to cluster information extracted by a search tool immediately or "on-the-fly" [34]. Concentrating on Vivisimo's Web-based search engine, this tool creates an extremely useful searching environment. In this Web search tool, the user can enter a query similarly to any popular search engine. When this query is entered, the Vivisimo search tool sends the query to its partner Web search tools. Some of Vivisimo's partners include Yahoo! (http://www.yahoo.com), GigaBlast (http://www.gigablast.com), DogPile (http://www.dogpile.com), MSN (http://www.msn.com), and Netscape (http://www.netscape.com). Once the results of the searches for the query on these search tools are complete, the results are returned to the Vivisimo search tool. Vivisimo then employs proprietary clustering techniques to the resulting data set to cluster the results of the search. The user can then search through the results either by browsing the entire list, as with most popular search tools, or by browsing the clusters created.

In addition to the traditional Web-based meta-search, Vivisimo allows users to search by specifying certain Web sites, especially news Web sites. For those users who want to search for very current information, this tool can search a specific Web site for that information. It also organizes the information categorically for the user. For example, a user can use the Vivisimo cluster tool to search a news Web site for current information in which he or she is interested. However, the user can only specify Web sites for this type of search from a list Vivisimo provides for the user.

75.4.3 KartOO

Another search tool that searches the Web using clustering techniques is KartOO (http://www. kartoo.com) [15]. KartOO is also a meta-search engine, similar to Vivisimo's search tool. However, KartOO's visualization methods add a new dimension for users to a data set. Similar to Vivisimo, KartOO uses a number of popular search engines for the initial search. The tool allows the user to select which search engines are included within the search. Once the results are returned, KartOO evaluates the result, organizing them according to relevance to the query. The links most relevant to the query are then returned as results. When the results are presented, the results are represented in an interactive graph for the user. Each node, or "ball" of the graph, represents a Web site that was returned to KartOO as fulfilling the query. Each node is connected to other nodes through edges that represent semantic links between the Web sites modeled within the node. The user can then browse the graph looking for the information in which he or she is interested. While browsing, if the user rolls the mouse over a node, information about the link it connects is displayed. When the user rolls the mouse over one of the semantic links, he or she can elect to refine the search to purposely include that semantic information within the query by clicking on the plus "+" sign or purposely exclude that semantic information within the query by clicking on the minus "−" sign. If the user does not want to include or exclude that information, he or she can take no action. Through this interaction with the semantic links, the user can refine his or her query in a very intuitive way. Moreover, with the graphical representation of the results, a user can see how various results are related to one another, thus identifying different clusters of results.

75.4.4 SYSTERS Protein Family Database

Looking at more domain-specific search tools, many research Web sites are also employing data mining techniques to make searching on their Web sites more effect. SYSTERS Protein Family Database (http://systers.molgen.mpg.de/) [19] is another interesting search tool that uses both clustering and classification to improve upon searching a data set. The SYSTERS Protein Family Database looks at clustering information about biological taxonomies based on genetic information and then classifies these clusters of proteins into a hierarchical structure. This database can then be searched using a variety of methods [19].

At the core of this tool are the clustering and classification algorithms. To place a protein sequence into the database, the database first uses a gapped BLAST search, which is a sequence alignment tool, to find what sequences the protein is similar to. However, because the alignment is asymmetric, this step is only used to narrow down the possible sequences the original might be similar to. Next, a pairwise local alignment is performed, upon which the clustering of the protein sequences will be based. Because these are biological sequences, all the sequences will have some measure of similarity. Sequences that are extremely similar are clustered together, creating superfamilies. These superfamilies are then organized hierarchically to classify their relationships. Users can then search this database on a number of key terms, including taxon (organism) names and cluster identification terms such as cluster number and cluster size. From this search, information about the specific query term is returned, as well as links to related information in the cluster. The user can then browse this information by traversing the links.

75.4.5 E-Commerce Systems

In addition to search tools, Web content data mining techniques can be used for a host of applications on the World Wide Web. Many E-commerce sites use association rule mining to recommend to users other items they might possibly like based on their previous selections. Association rule mining allows sites to also track the various types of usage on their site. For example, on an E-commerce site, information about the user's interactions with the site can help the E-commerce site customize the experience for the user, improve customer service for the user, and discover customer shopping trends [13]. Also, concerning financial issues, classification and clustering can be used to target new customers for products based on previous purchases. Data mining can also give companies insight into how well a marketing strategy is enticing customers into buying certain products [13].

75.5   Conclusion and Further Resources

Information retrieval and data mining are two very rich fields of computer science. Both have many practical applications while also having a rich problem set that allows researchers to continually improve upon current theories and techniques. In this chapter, we have looked at some of these theories and applications. Information retrieval has evolved from a field that was initially created by the need to index and access documents into a robust research area that studies techniques for not only retrieving data but also discovering knowledge in that data. Data mining, while a much younger field, has evolved to explore intriguing relationships in very complex data. The future promises to be very exciting with the developments in multimedia information retrieval and data mining as well as the movement toward trying to understand semantic meanings with the data.

While this chapter introduces these topics, there are many other resources available to readers who wish to study specific problems in-depth. In the field of information retrieval, there are a number of introductory texts that discuss information retrieval very comprehensively. Some excellent introductory texts include Salton's *Automatic Text Processing: The Transformation, Analysis, and Retrieval of Information by Computer* [31], Korfhage's *Information Storage and Retrieval* [18], and Baeza-Yates' and Ribeiro-Neto's *Modern Information Retrieval* [6]. Also, there are several conferences in which state-of-the-art research results are published. These include the Text Retrieval Conference (TREC, http://trec.nist.gov/) and ACM's Special Interest Group on Information Retrieval (SIGIR) Conference. Concerning data mining, there are also a number of introductory texts to this subject; see, for example, Han's and Kamber's *Data Mining: Concepts and Techniques* [13]. In addition, there are numerous data mining resources available as well as technical committees and conferences. A major conference is ACM's Special Interest Group on Knowledge Discovery in Data (SIGKDD) Conference, among others.

References

[1] Agrawal, R., Imielinski, T., and Swami, A., Mining association rules between sets of items in large databases, in *Proc. 1993 ACM-SIGMOD Int. Conf. on Management of Data*, Buneman, P. and Jajodia, S., Eds., ACM Press, Washington, D.C., 1993, 207.

[2] Agrawal, R. and Srikant, R., Fast algorithm for mining association rules, in *Research Report RJ 9839*, IBM Almaden Research Center, San Jose, CA, June 1994.

[3] Agrawal, R. and Srikant, R., Fast algorithm for mining association rules, in *Proc. of the 20th Int. Conf. on Very Large Databases*, Bocca, J.B., Jarke, M., and Zaniolo, C., Eds., Morgan Kaufmann, Santiago, Chile, 1994, 24.

[4] Agrawal, R. et al., Automatic subspace clustering of high dimensional data for data mining, in *Proc. of the ACM-SIGMOD Conf. on Management of Data*, Tiwary, A. and Franklin, M., Eds., ACM Press, Seattle, WA, 1998, 94.

[5] Arasu, A. et al., Searching the Web, *ACM Transactions on Internet Technology*, Kim, W., Ed., ACM Press, 2001, 2.

[6] Baeza-Yates, R. and Ribeiro-Neto, B., *Modern Information Retrieval*, ACM Press, Addison-Wesley, New York, 1999.

[7] Breiman, L. et al., *Classification and Regression Trees*, Wadsworth Publishing Company, Statistics/Probability Series, 1984.

[8] Chang, G., Healey, M.J., McHugh, J.A.M., and Wang, J.T.L., *Mining the World Wide Web: An Information Search Approach*, Kluwer Academic, Boston, 2001.

[9] Cleveland, W., *Visualizing Data*, Hobart Press, Summit, NJ, 1993.

[10] Croft, W.B., Turtle, H.R., and Lewis, D.D., The use of phrases and structured queries in information retrieval, in *Proc. 14th Annual Int. ACM SIGIR Conf. on Research and Development in Information Retrieval*, Bookstein, A. et. al., Eds., ACM Press, Chicago, IL, 1991, 32.

[11] Ester, M. et al., A density-based algorithm for discovering clusters in large spatial databases, in *Proc. 1996 Int. Conf. Knowledge Discovery and Data Mining(KDD'96)*, Simoudis, E., Han, J., and Fayyad, U.M., Eds., ACM Press, Portland, OR, 1996, 226.

[12] Frakes, W.B. and Baeza-Yates, R., Eds., *Information Retrieval: Data Structures and Algorithms*, Prentice Hall, Englewood Cliffs, NJ, 1992.

[13] Han, J. and Kamber, M., *Data Mining: Concepts and Techniques*. Morgan Kaufmann, San Francisco, CA, 2000.

[14] Johnson, R.A. and Wichern, D.A., *Applied Multivariate Statistical Analysis*, Prentice Hall, Upper Saddle River, NJ, 1992.

[15] KartOO Web site: http://www.kartoo.com.

[16] Kaufman, L. and Rousseeuw, P.J., *Finding Groups in Data: An Introduction to Cluster Analysis*, John Willey & Sons, New York, 1990.

[17] Kobayashi, M. and Takeda, K., Information retrieval on the web, *ACM Computing Surveys*, Wegner, P. and Israel, M., Eds., ACM Press, 2000, 144.

[18] Korfhage, R., *Information Storage and Retreival*, John Wiley & Son, New York, 1997.

[19] Krause, A. et al., SYSTERS, GeneNest, SpliceNest: exploring sequence space from genome to protein, *Nucleic Acids Research*, Oxford University Press, 2002, 299.

[20] Lay, J.A., Muneesawang, P., and Guan, L., Multimedia information retrieval, in *Proc. of Canadian Conference on Electrical and Computer Engineering*, Dunne, S., Ed., IEEE Press, Toronto, Canada, 2001, 619.

[21] MacQueen, J., Some methods for classification and analysis of multivariate observation, in *Proc. 5th Berkeley Symp. Math. Statist, Prob.*, Le Cam, L.M and Neyman, J., Eds., University of California Press, Berkeley, 1967, 281.

[22] Mannila, H., Toivonen, H., and Verkamo, A.I., Efficient algorithms for discovering association rules, in *Proc. of the AAAI Workshop on Knowledge Discovery in Databases*, Fayyad, U.M. and Uthurusamy, R., Eds., AAAI Press, Seattle, WA, 1994, 181.

[23] Meghini, C., Sebastiani, F., and Straccia, U., A model of multimedia information retrieval, *Journal of the ACM*, 48, 909, 2001.

[24] Mobasher, B. et al., Web data mining: effective personalization based on association rule discovery from web usage data, in *Proc of the 3rd Int. Work. on Web Information and Data Management*, ACM Press, Atlanta, GA, 2001, 243.

[25] Murthy, S.K., Automatic construction of decision trees from data: a multi-disciplinary survey, *Data Mining and Knowledge Discovery*, Kluwer Academic Publishing, 2, 345, 1998.

[26] Nack, F. and Lindsay, A., Everything you wanted to know about MPEG-7. Part 1, *IEEE Multimedia*, IEEE Press, July–September 1999, 65.

[27] Nack, F. and Lindsay A., Everything you wanted to know about MPEG-7. Part 2, *IEEE Multimedia*, IEEE Press, October–December 1999, 64.

[28] Page, L. et al., The pagerank citation ranking: bringing order to the Web, *Tech. Rep. Computer Systems Laboratory*, Stanford University, Stanford, CA., 1998.

[29] Riloff, E., Little words can make a big difference for text classification, in *Proc. 18th Annual Int. ACM SIGIR Conf. on Research and Development in Information Retrieval*, Fox, E.A., Ingwersen, P., and Fidel, R., Eds., ACM Press, Seattle, WA, 1995, 130.

[30] Riloff, E. and Hollaar, L.A., Text databases and information retrieval, *The Computer Science and Engineering Handbook*, Tucker, A.B., Ed., CRC Press, 1997, 1125–1141.

[31] Salton, G., *Automatic Text Processing: The Transformation, Analysis, and Retrieval of Information by Computer*, Addison-Wesley, Reading, MA, 1989.

[32] Salton, G., Wong, A., and Yang, C.S., A vector space model for automatic indexing, in *Communications of the ACM*, Ashenhurst, R., Ed., ACM Press, 18, 613, 1975.

[33] Shasha, D., Wang, J.T.L., and Giugno, R., Algorithmic and applications of tree and graph searching, in *Proc. of the 21st ACM SIGMOD-SIGACT-SIGART Symposium on Principles of Database Systems*, Abiteboul, S., Kolaitis, P.G., and Popa, L., Eds., ACM Press, Madison, WI, 2002, 39.

[34] Vivisimo Web site: http://www.vivisimo.com.

[35] Weiss, S.M. and Kulikowski, C.A., *Computer Systems that Learn: Classification and Prediction Methods for Statistics, Neural Nets, Machine Learning, and Experts Systems*, Morgan Kaufmann, San Francisco, CA, 1991.

[36] Wen, J.R, Nie, J.Y., and Zhang, H.J., Clustering user queries of a search engine, in *Proc. of the 10th Annual Int. Conf. on World Wide Web*, ACM Press, Hong Kong, China, 2001, 162.

[37] Wang, J.T.L, Shapiro, B.A., and Shasha, D., Eds., *Pattern Discovery in Biomolecular Data: Tools, Techniques, Applications*, Oxford University Press, New York, 1999.

[38] Wang, W., Yang, J., and Muntz, R., STING: a statistical information grid approach to spatial data mining, in *Proc. 1997 Int. Conf. Very Large Databases (VLDB'97)*, Jarke, M. et. al., Morgan Kaufmann, Athens, Greece, 1997, 186.

[39] Zipf, G.K., *Human Behavior and the Principle of Least Effort*, Addison-Wesley, Cambridge, MA, 1949.

[40] Zhang, K., Shasha, D., and Wang, J.T.L., Approximate tree matching in the presence of variable length don't cares, *J. of Algorithms*, Academic Press, 16, 33, 1994.

[41] Zhang, S., Herbert, K.G., and Wang, J.T.L., XML query by example, *Int. J. of Computational Intelligence and Applicantions*, Braga, A.D.P. and Wang, J.T.L, Eds., World Scientific Publishing, 2, 329, 2002.

76

Data Compression

76.1   Introduction 76-1
76.2   Lossless Compression 76-2
        Metric: Compression Ratio • Metric: Information • Methods
        • Dictionary-Based Techniques • Other Methods
76.3   Lossy Compression 76-18
        Data Domain Compression • Transform Domain
        Compression
76.4   Conclusion 76-39

Z. Rahman
College of William and Mary

76.1 Introduction

Over the past few years, data compression has become intimately integrated with information available in digital form; text, documents, images, sound, and music are all compressed before storage on a digital medium. However, depending on whether one is storing text, a document, or an image, there are different requirements on the type of compression algorithm that can be used. This is directly related to issues about the amount of compression that can be achieved and the quality with which the compressed data can be uncompressed. In this chapter we introduce the techniques behind data compression, especially those that apply to text and image compression.

Definition 76.1   Data compression is the method of representing a data object that takes B bits of storage, by a data object that takes B' bits of storage, where $B' < B$, often significantly.

Although there is no definitive taxonomy of data compression methods, they can be divided into two disjoint categories: informationally *lossless* methods and informationally *lossy* methods. Informationally lossless methods *exactly* reproduce the input data stream on decompression; that is, there is no loss of information between the application of the compression and decompression operations. Lossy methods produce a parameter-dependent approximation to the original data; that is, the compression and decompression operations cause information loss. For this reason, lossless compression is almost always used to compress text because text needs to be reproduced exactly, and lossy compression is useful in applications such as facsimile transmission (fax) where an approximation to the original data is acceptable. Under lossless compression are two major subcategories: *entropy coding* (Section 76.2.3) and *dictionary-based coding* (Section 76.2.4). There are other methods such as *run-length coding* (Section 76.2.5.2) that fall into neither of these sub-categories, and still others, such as *prediction with partial matching* (Section 76.2.5.1) that are a hybrid of these subcategories. The lossy compression category can also be subdivided into two major subcategories: *data domain* techniques (Section 76.3.1) and *transform domain* methods (Section 76.3.2). Because text compression algorithms invariably belong to the lossless compression category, and most image, document, audio, and video compression algorithms belong to the lossy compression subcategory, another way to categorize algorithms can be based on their use rather than on their structure.

76.2 Lossless Compression

One of the issues key in a discussion of lossless image compression methods is the evaluation of their performance. Some *metrics* need to be defined that can be used to judge the performance of the different algorithms. We use two metrics to judge the performance of the algorithms described in this chapter:

1. Compression ratio (CR)
2. Information content

In the next sections, we discuss these metrics in some detail.

76.2.1 Metric: Compression Ratio

Definition 76.2   Compression ratio C is defined as the ratio of the total number of bits used to represent the data before encoding to the total number of bits used to represent the data after encoding:

$$C = \frac{B'}{B}$$

where B is the size of the original data, and B' is the size of the compressed representation of the data.

The total number of bits $B' = B'_d + B'_o$, where B'_d is the number of bits used to represent the actual data and B'_o is the number of bits used to represent any additional information that is needed to decode* the data. B'_o is known as the *overhead*, and can be significant in computing the performance of compression algorithms. For example, if two compression schemes produce the same size compressed data representation, B'_d, then the one with larger overhead B'_o will produce the poorer compression ratio. Also, if B'_o is constant (i.e., all data regardless of its characteristics has the same associated overhead), then larger data sets will tend to have better data-to-overhead ratios, resulting in more efficient representation of encoded data.

This metric is universally applicable to both lossless and lossy compression techniques. However, one needs to be careful when applying this metric to lossy compression. Just because a compression method achieves better data compression does *not* automatically make it better overall because the quality of the decompressed data is significant. This observation is not relevant to lossless compression schemes because the decompressed data and the original data are identical.

76.2.2 Metric: Information

Central to the idea of compression is the concept of *information*. "Information" is a word that is part of most everybody's everyday lexicon. However, when one speaks of information in the context of data compression, one assigns it a very particular meaning. Consider a random experiment, and let $\mathcal{B}$ be a possible event in this random experiment. Let $p = \mathrm{Pr}\{\mathcal{B}\}$ be the probability that event $\mathcal{B}$ occurs. In this experiment, information depends only on the *probability* of occurrence of $\mathcal{B}$, and not on the *content* of $\mathcal{B}$. In other words, because we already know *what* $\mathcal{B}$ is, this does not provide any information. However, because we do not know *when* $\mathcal{B}$ occurs, the frequency with which $\mathcal{B}$ occurs (i.e., the probability with which it occurs) does give us insight, or information, about $\mathcal{B}$. So, we want to define a quantity that will measure the amount of this "information" associated with the probability of occurrence of $\mathcal{B}$.

Let $I(p)$ denote the information associated with $p(\mathcal{B})$, the probability that "$\mathcal{B}$ has occurred." We want to determine the functional form of $I(\cdot)$ by first listing a set of requirements that $I(\cdot)$ should satisfy.**

---

*We use *decode* and *decompress* interchangeably throughout this chapter.

**The material in this section is based, in part, on Chapter 9.3 in *A First Course in Probability*, third edition, by Sheldon Ross. Interested readers should consult this source for the proof of Theorem 76.1.

Definition 76.3 $I(p)$ is a non-negative real-valued function defined for all $0 < p \leq 1$, that satisfies the following requirements:

1. $I(1) = 0$ (i.e., if $\mathcal{B}$ is *certain* to occur), then "$\mathcal{B}$ has occurred" conveys no information.
2. $I(p)$ is a strictly monotonically *decreasing* function of p (i.e., the *more* likely $\mathcal{B}$ is to occur), the *less* is the information conveyed by "$\mathcal{B}$ has occurred." Formally, if $0 < p_1 < p_2 \leq 1$ then $I(p_1) > I(p_2)$.
3. $I(p)$ is a *continuous* function of p (i.e., small changes in p will not produce a large change in $I(p)$).
4. If $p = p_1 p_2$ with $0 < p_1 \leq 1$ and $0 < p_2 \leq 1$ then $I(p) = I(p_1 p_2) = I(p_1) + I(p_2)$.

The last requirement can be justified using the following argument. Suppose event $\mathcal{B}$ is the result of the *joint* occurrence of two independent, elementary events $\mathcal{B}_1$ and $\mathcal{B}_2$ with respective probabilities p_1 and p_2. Then, $\mathcal{B} = \mathcal{B}_1 \cap \mathcal{B}_2$, and

$$p = \Pr\{\mathcal{B}\} = \Pr\{\mathcal{B}_1 \cap \mathcal{B}_2\} = \Pr\{\mathcal{B}_1\} \Pr\{\mathcal{B}_2\} = p_1 p_2.$$

It is intuitive that the independence of events $\mathcal{B}_1$ and $\mathcal{B}_2$ should cause their associated information to *add* when they occur jointly.

Theorem 76.1   *The only function that satisfies the four requirements in Definition 76.3 is*

$$I(p) = -c \, \log_a(p)$$

where the constant c is positive but otherwise arbitrary, and a > 1.

The convention is to let the $c = 1$. The units of $I(p)$ are called *bits* when $a = 2$, *Hartleys* when $a = 10$, and *nats* when $a = e$.

Definition 76.4   Let X be a discrete random variable and let $\mathcal{X}$ be the associated set of possible values of X. For each $x \in \mathcal{X}$, the associated probability is $p(x) = \Pr\{X = x\}$ and the corresponding information is $I(p(x)) = -\log_2(p(x))$. The *expected* (average) information associated with X is then

$$\mathcal{H}(X) = -\sum_{x \in \mathcal{X}} p(x) \, \log_2(p(x)).$$

This expected value is known as the *entropy* of the random variable X.

To paraphrase Definition 76.4, if values of the random variable X are generated repeatedly and, for each observation x, the associated information $-\log_2(p(x))$ is computed, then the average over (infinitely) many observations would be $\mathcal{H}(X)$. If $|\mathcal{X}| = n$ then it can be shown that the largest possible value of $\mathcal{H}(X)$ is $\log_2(n)$ and this value is attained if, and only if, all n possible values are equally likely. In this case,

$$p(x) = \frac{1}{n} \quad \text{for all } x \in \mathcal{X}$$

and each possible value of X conveys exactly the same amount of information, namely:

$$\mathcal{H}(X) = -\sum_x p(x) \, \log_2(p(x)) = -\sum_x (1/n) \, \log_2(1/n) = \log_2(n).$$

The converse that all n values are equally likely when the entropy is maximum is also true. We will use this measure to define the efficiency of lossless compression algorithms in the next section.

76.2.3 Methods

Before delving into a description of what it means to losslessly compress a data stream, and the techniques used to achieve this goal, we need to define what we mean by a *data stream* and how it is represented.

Definition 76.5   A *data stream*, $d = s_1 s_2 s_3 \cdots s_n s_{n+1} \cdots$, is a sequence of *symbols* s_i drawn from an *alphabet* $\mathcal{A}$.

The index i does not represent the order in which the symbols occur in the alphabet; rather, it represents the order in which the symbols occur in the data stream. Input and output data streams draw symbols from different alphabets. For instance, when compressing English text, the input data stream comprises of all the letters of the English alphabet, the numbers, and punctuation marks. The output data stream can be a set of symbols derived from recurring patterns in the input data stream, or from the frequency of occurrence of symbols in the data stream. In either case, it depends on the characteristics of the input data rather than the raw data itself.

Definition 76.6   A *symbol s* can represent a single character, c, or a sequence of characters $c_1 c_2 \cdots c_n$ concatenated together.

Again, the differentiation is more in terms of symbols drawn from the alphabet for the input data stream vs. symbols drawn from the alphabet for the output data stream. Typically, the input alphabet $\mathcal{A}_i$ has symbols that represent a single character, whereas the output alphabet $\mathcal{A}_o$ can have symbols that represent concatenated strings of recurring characters. (See Section 76.2.4 for details).

Definition 76.7   An *alphabet* $\mathcal{A} = \{s_1, s_2, \ldots, s_S\}$ is the set of S possible symbols that can be present in a data stream.

Typically, alphabets for the input data stream are known *a priori*; that is, they are derived from a known source such as the English alphabet. The alphabet for the output data stream is generally generated on-the-fly from the patterns in the input data stream (LZW compression,[1] Section 76.2.4.1), or from the frequency of occurrence of symbols (Huffman coding[2] Section 76.2.3.2). However, output alphabets that have been determined previously are also used when *canonical* Huffman codes are used for entropy coding (see Section 76.2.3 and Section 76.3.2).

With these concepts in mind, lossless compression can be categorized into two major groups:

1. **Entropy coding.** In entropy-coding schemes, the number of bits used to represent a symbol (i.e., the length of the symbol) is proportional to its probability of its occurrence in the data stream. However, each symbol is considered independent of all previously occurring symbols.
2. **Dictionary-based coding.** In dictionary-based coding schemes, recurring patterns are assigned fewer bits.

76.2.3.1 Entropy Coding

Suppose we are attempting to compress an electronic version of this chapter. The symbol alphabet in this case comprises of the lower and upper case English characters, numbers, punctuation marks, and spaces. If we assume that every symbol has equal significance — it occurs with equal frequency in the text — then an equal number of bits, B, should be assigned to represent every symbol. This is known as uniform, fixed length coding.

Definition 76.8   If $b_{s_i}, i = 0, \ldots, S-1$ represents the number of bits that are used to represent (encode) the symbol s_i, and S is the total number of symbols in the alphabet, then for uniform, fixed-length coding, $b_{s_i} = B, i = 0, \ldots, S - 1$.

We can rewrite Definition 76.4 in terms of the above definitions:

$$\mathcal{H}(d_i) = -\sum_{s \in \mathcal{A}} p[s] \, \log_2(p[s]).$$

where, because of the (removable) singularity associated with $p[s] = 0$, the summation is only over those symbols s_i for which $p[s] > 0$. The entropy is, in a sense, a measure of how well data can be compressed. Shannon* showed that the best compression ratio that a lossless compression scheme can achieve is bounded above by the entropy of the original signal.[3] In other words, the best compression ratio is achieved when the average bits per symbol is equal to the entropy of the signal:

$$\bar{b} = \frac{1}{S} \sum_{i=0}^{S-1} b_{s_i} = \sum_{i=0}^{S-1} w[s_i] p[s_i] = \mathcal{H}(s_i),$$

where $w[s_i]$ is the length of the codeword representing symbol s_i.

Intuitively, then, to achieve the best possible lossless compression, the symbol distribution of the data needs to be examined and the number of bits assigned to represent each symbol set as a function of the probability of occurrence of that particular symbol; that is:

$$b_{s_i} = f(p_{s_i}) \quad i = 0, \ldots, S - 1,$$

where $p_{s_i} = p[s_i], i = 0, \ldots, S - 1$. The codes generated using this type of compression are called variable-length codes.

The above method outlines the basic idea of how to achieve maximum lossless compression but, aside from the vague *number of bits assigned should be inversely proportional to the frequency of occurrence*, it does not specify how such an assignment should be made. There are several ways in which this can be done:

1. The probability distribution of the data stream can be generated and then used to *manually* assign a unique code for each symbol. This technique would be efficient only for data streams with very small input alphabets.
2. A model-based approach can be used where the input data is assumed to have some standard probability distribution. The same set of encoded representations can then be used for all data. While this technique is automatic *once the initial encoding has been assigned*, it is inefficient because, in general, the symbols are encoded suboptimally.
3. An automatic technique that assigns minimum redundancy unique codes based upon the probability distribution of the input data stream, such as *Huffman coding*,[2] can be used.

76.2.3.2 Huffman Coding

Huffman codes belong to the class of *optimum* prefix codes.

Definition 76.9 An optimum code is a code whose average length, $\bar{b}$, does not exceed the average length of any other code, $\bar{b}_k$:

$$\bar{b} \leq \bar{b}_k \quad \forall k$$

and which has the following properties:

1. Symbols that occur more frequently have shorter associated codes.
2. The two symbols that occur least frequently have the same length code.
3. The two least frequently occurring symbols have a Hamming distance of 1; that is, they differ only in one bit location.

Huffman codes can be generated using the following algorithm:

Algorithm 76.1

1. Sort the S-element probability distribution array p in descending order; that is,

$$p'[0] = max(p[l]) \quad \text{and} \quad p'[S-1] = min(p[l]), \quad l = 0, \ldots, S - 1.$$

---

*Claude E. Shannon: father of modern communication theory.

2. Combine the last two elements of p' into a new element, and store it in the second to last location in p':

$$p'[S - 2] = p'[S - 1] + p'[S - 2]$$

reduce the number of elements in the array by one: $S = S - 1$. This operation of combining the last two elements into a new element and reducing the size of the array is called *Huffman contraction*.[4]

3. Assign the code $x[l]$ to each combined symbol by prefixing a 0 to the symbol(s) in the $p'[S - 1]$ location and a 1 to the symbol(s) in the $p'[S - 2]$ location.

4. Go to Step 1 and repeat until all the original symbols have been combined into a single symbol.

Example 76.1

Suppose we are given the probability distribution array:

l	0	1	2	3	4	5	6	7	8
$p[l]$	0.22	0.19	0.15	0.12	0.08	0.07	0.07	0.06	0.04

The entropy of this sequence is: $\mathcal{H} = -\sum_{p[l]>0} p[l] \log_2 \left(p[l]\right) = 2.703$. Let l_S represent the set of indices for the Huffman contracted arrays. Table 76.1 shows the process as the Huffman codes are generated one

TABLE 76.1   Huffman Coding

l_9	0	1	2	3	4	5	6	7	8
$p[l]$	0.22	0.19	0.15	0.12	0.08	0.07	0.07	0.06	0.04
l_8	0	1	2	3	(7 8)	4	5	6	
$x[l]$					(0 1)				
$p'[l]$	0.22	0.19	0.15	0.12	0.10	0.08	0.07	0.07	
l_7	0	1	2	(5 6)	3	(7 8)	4		
$x[l]$				(0 1)		(0 1)			
$p'[l]$	0.22	0.19	0.15	0.14	0.12	0.10	0.08		
l_6	0	1	(7 8 4)	2	(5 6)	3			
$x[l]$			(00 01 1)		(0 1)				
$p'[l]$	0.22	0.19	0.18	0.15	0.14	0.12			
l_5	(5 6 3)	0	1	(7 8 4)	2				
$x[l]$	(00 01 1)			(00 01 1)					
$p'[l]$	0.26	0.22	0.19	0.18	0.15				
l_4	(7 8 4 2)		(5 6 3)	0	1				
$x[l]$	(000 001 01 1)		(00 01 1)						
$p'[l]$	0.33		0.26	0.22	0.19				
l_3	(0 1)		(7 8 4 2)		(5 6 3)				
$x[l]$	(0 1)		(000 001 01 1)		(00 01 1)				
$p'[l]$	0.41		0.33		0.26				
l_2	(7 8 4 2 5 6 3)							(0 1)	
$x[l]$	(0000 0001 001 01 100 101 11)							(0 1)	
$p'[l]$	0.59							0.41	
l_1	7	8	4	2	5	6	3	0	1
$x[l]$	00000	00001	0001	001	0100	0101	011	10	11
l_1	7	8	4	2	5	6	3	0	1
l_0	0	1	2	3	4	5	6	7	8
$x[l]$	10	11	001	011	0001	0100	0101	00000	00001
$w[l]$	2	2	3	3	4	4	4	5	5

At each iteration the two symbols with the smallest probabilities are combined into a new symbol and the list resorted. Text in `teletype font` shows the symbols that have been combined so far, their combined probabilities, and the codeword assigned thus far.

symbol at a time. The average codeword length with this encoding is:

$$\bar{b} = \sum_{l=0}^{8} p[l]w[l] = 2(0.22 + 0.19) + 3(0.15 + 0.12) + 4(0.08 + 0.07 + 0.07) + 5(0.06 + 0.04) = 3.01$$

$$= 0.86\mathcal{H}$$

The compression ratio that the Huffman code achieves is $C_a = 4/3.01 = 1.33$, whereas the predicted compression ratio is $C_p = 4/2.70 = 1.48$. So, $C_a = 0.86C_p$, which says that Huffman coding is about 90% effective in compressing the input data stream. Because there are $S = 9$ symbols, the input alphabet is uniformly encoded at 4 bits per symbol, where the codewords are simply the binary representation symbols. The output (encoding) alphabet and associated codeword lengths are given in the last block of Table 76.1.

The code generated in Example 76.1 satisfies the criteria for an optimum code:

- The symbols with the largest frequency have the fewest bits assigned to them.
- The two lowest frequency symbols are represented with the same number of bits.
- The two longest codewords have a Hamming distance of 1.
- None of the codewords is a prefix of any other codeword, so the generated code is uniquely decodeable.

An alternative way of generating the Huffman code is to use a binary tree representation. This can be done because the Huffman code is a prefix code, so the insertion of a 0 or 1 at the beginning of the code is equivalent to going down another level in the tree. To generate the Huffman codes from the binary tree representation, use the following algorithm:

Algorithm 76.2

1. Traverse from the symbol to be encoded to the end (root) of the binary tree.
2. The codeword is formed by prefixing 0 or 1 to the codeword generated so far along the path, depending upon whether the left (0) or right (1) branch is taken.

Example 76.2

l :	0	1	2	3	4
$p[l]$:	0.4	0.2	0.2	0.1	0.1

To compute the code for $p[3]$ in Figure 76.1, traverse from $p'_3(0.1)$ to (1.0), passing through points a, b, c, d. Reading backward from d to a, $x[3] = 0010$. The rest of the codewords can be similarly found:

l :	0	1	2	3	4
$p[l]$:	0.4	0.2	0.2	0.1	0.1
$x[l]$:	1	01	000	0010	0011
$w[l]$:	1	2	3	4	4

This Huffman code is optimal based upon the criteria given in Definition 76.9. The entropy for this code is $\mathcal{H} = 2.12$ and the average code length is $\bar{b} = 2.20$. If fixed-length representation is used, $\bar{b} = 3.0$.

Definition 76.10 The difference between the entropy and the average code length is called the redundancy: $\mathcal{R} = \mathcal{H} - \bar{b}$.

Clearly, Huffman codes reduce the overall redundancy in the data when compared with a fixed-length encoding; see Examples 76.1 and 76.2. However, the redundancy is completely eliminated (i.e., $\mathcal{R} = 0$)

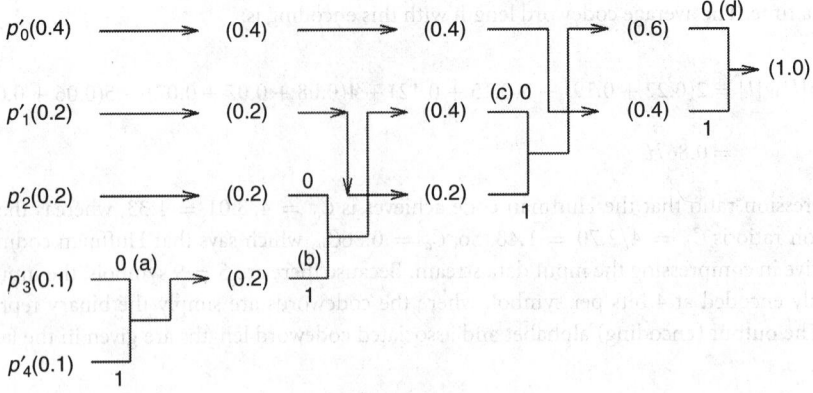

FIGURE 76.1   Generating Huffman codes using a binary tree representation.

if, and only if, the probabilities with which symbols are distributed are given by $p[l] = 2^{-k}, k \geq 0, l = 0, \ldots, L - 1$. In this case, $\mathcal{H} = \bar{b}$.

Once the Huffman code has been generated, encoding the data stream is simply the task of replacing each input symbol by its Huffman coded representation. In practice, because the codewords are of variable length and do not necessarily end on *word* boundaries, the encoded data needs to be buffered before transmission or storage, so that data-word boundaries are preserved.

Huffman decoding is very straightforward. It is best understood using the binary tree representation of the encoding process. This is described in the following algorithm.

Algorithm 76.3

1. Generate the binary tree representation of the Huffman code from the encoding dictionary.
2. Upon receipt of each bit, d_n, take the left or the right branch of the tree, depending on whether d_n was 0 or 1, respectively.
3. If the node is a terminal node (i.e., it has no children), then the value it contains is the decoded value.
4. If there is more data, go back to Step 1.
5. Terminate decoding.

Implicit in this algorithm is the assumption that the dictionary used to generate the Huffman code is available to the decoder.

Example 76.3

Following Example 76.2, if the sequence 0010 was received, then the path taken from the root node to $p[3]$ would pass through d, c, b, a, (Figure 76.1) which is the decoded sequence.

A major drawback of Huffman codes is that in order to be optimal, they must be data dependent. Because the generation of the codewords depends on the probability distribution of the data, a dictionary, or look-up table, must be built for each data stream that is being encoded. Since this dictionary is required for decoding, this means that the dictionary must be transmitted along with the data. This overhead significantly affects the overall compression ratio. An adaptive version of Huffman coding that builds the Huffman tree as data is received eliminates the need for transmitting the dictionary. However, its performance is always suboptimal compared to that of non-adaptive Huffman coding for a known probability distribution, and it is significantly slower in encoding and decoding the data.

Because of the substantial overhead that can occur with Huffman coding, it is typically not used for image compression directly. However, it is an integral part of image coding schemes such as JPEG compression,[5] where a fixed dictionary is used for all cases. This type of Huffman coding is called *canonical* Huffman coding. The idea is that given enough exposure to typical data streams, a model can be developed that is near-optimal for most data streams. The Huffman code for this model can then be used to encode all the data streams without incurring the overhead.

Because Huffman codes are built one symbol at a time, the smallest codeword length is 1 bit. So Huffman codes have a (loose) lower bound of 1 bit. In other words, for most cases, the average length of Huffman encoded data is bounded by $\overline{b} = \mathcal{H} + 1$. It can be shown that a tighter bound is $\overline{b} = \mathcal{H} + p_{max} + 0.086$, where p_{max} is the largest probability in the probability distribution.[6] For most general cases, the input alphabet $\mathcal{A}_i$ is large and, thus, p_{max} is fairly small. This implies that the difference between $\overline{b}$ and $\mathcal{H}$ is usually not very significant. This can be of significance, however, when encoding data with a *skewed* probability density distribution in the sense that one symbol occurs much more frequently than others. Because p_{max} is larger in such a case, Huffman codes would tend to be relatively inefficient. Better redundancy reduction can be achieved by encoding "blocks" of symbols: instead of using one symbol from the input data stream at a time, we use a pair of symbols, or three, or more. Encoding this requires the probability distribution of all possible combinations of the symbols in the original alphabet taken two at a time, or more, depending on the block size. For example, if the original input alphabet is $\mathcal{A}_i = \{a_1, a_2, a_3\}$, then the modified input alphabet would be $\mathcal{A}_i' = \{a_1 a_1, a_1 a_2, a_1 a_3, a_2 a_1, a_2 a_2, a_2 a_3, a_3 a_1, a_3 a_2, a_3 a_3\}$. Clearly, although the average word length of a Huffman code is much closer to the entropy, the size of the alphabet, and thus the overhead, increase exponentially.

76.2.3.3 Arithmetic Coding

Because of the exponentially increasing dictionary size, block Huffman coding is not used for large alphabets. What is needed is a method that can be used to encode blocks of symbols without incurring the exponential overhead of block Huffman. Arithmetic coding[6–9] is the name given given to a set of algorithms that generate unique output "tags" — codewords — for blocks of symbols from the input data stream. The key idea to comprehend in arithmetic coding is tag generation. In practice, the tag is a *binary* fraction representation of the input data sequence.

Suppose we have an input data sequence composed of symbols from an alphabet $\mathcal{A}_i = \{a_1, a_2, \ldots, a_S\}$. A typical input data stream can be $d_i = \cdots s_n s_{n+1} s_{n+2} s_{n+3} \cdots$, where $s_n \in \mathcal{A}_i$. The idea, then, is to generate a tag that uniquely identifies this, and only this, data stream. Because there is an infinite number of possible data streams, an infinite number of unique tags is needed. Any interval $\mathcal{I} = (\mathcal{I}_l, \mathcal{I}_h]$, where $\mathcal{I}_l$ is the lower limit of the interval and I_h is the upper limit of the interval, on the real number line provides a domain that can support this requirement. Without loss of generality, $(I_l, \mathcal{I}_h] = (0, 1]$. Let $p[a_i], i = 1, \ldots, S$ be the probabilities associated with the symbols $a_i, i = 1, \ldots, S$, and $\mathcal{P}_i, i = 0, \ldots, S$ be the cumulative probability density (CPD) defined as:

$$\mathcal{P}_0 = 0$$
$$\mathcal{P}_l = \mathcal{P}_{l-1} + p[a_l], \quad l = 1, \ldots, S - 1$$
$$\mathcal{P}_S = 1$$

where S is the total number of symbols in the alphabet. The CPD naturally partitions the unit interval $\mathcal{I} = (0, 1]$. The tag $\mathcal{T}$ that encodes the input data stream is generated using Algorithm 76.4.

Algorithm 76.4

1. Compute the CPD for the alphabet $\mathcal{A}_i$ and partition the $(0, 1]$ interval.
2. Read the next symbol $s_i = a_k, k \in \{1, \ldots, S\}$ from the data stream and move to the interval $(\mathcal{P}_{k-1}, \mathcal{P}_k]$.

3. Partition the $(\mathcal{P}_{k-1}, \mathcal{P}_k]$ interval into S partitions using:

$$\mathcal{P}_0 = \mathcal{P}_{k-1}$$

$$\mathcal{P}_l = \mathcal{P}_{l-1} + p[a_l]/(\mathcal{P}_k - \mathcal{P}_{k-1}), \quad l = 1, \ldots, S-1$$

$$\mathcal{P}_S = \mathcal{P}_k$$

4. Go back to Step 2 and repeat until the data stream is exhausted.
5. The tag $\mathcal{T}$ for the sequence read thus far is any point in the interval $(\mathcal{P}_{k-1}, \mathcal{P}_k]$. Typically, $\mathcal{T}$ is the binary representation of $\mathcal{T}_v = (\mathcal{P}_k + \mathcal{P}_{k-1})/2$, where $\mathcal{T}_v$ is the midpoint of the interval $\mathcal{I}$.

Example 76.4

Suppose we are encoding a sequence that contains symbols from the alphabet $\mathcal{A}_i = \{a_1, a_2, a_3\}$. Let $p_1 = 0.1$, $p_2 = 0.7$, and $p_3 = 0.2$, and $p_i = p[i]$. Then, $\mathcal{P}_0 = 0$, $\mathcal{P}_1 = 0.1$, $\mathcal{P}_2 = 0.8$, and $\mathcal{P}_3 = 1$. The data stream to be encoded is $a_2 a_3 a_2 a_1 a_1$. The progression of the encoding algorithm is shown in Table 76.2, which shows the current interval, the limits of the partitions contained in the interval, the $\mathcal{T}_v$ associated with each partition, and the selected partition (i.e., the one associated with the received symbol). The tag $\mathcal{T}$ is just the binary representation of the midpoint with the leading "0." dropped.

As the length of the sequence gets longer, one needs greater precision to generate the tag. When the last symbol of the sequence a_1 was received, the upper and lower limits of the interval started to differ only in the fourth decimal place. This suggests that the next symbol to be read would cause the interval to shrink even further. This, of course, is of considerable significance for a computer implementation of the arithmetic coding algorithm. Their are several modifications of the above algorithm needed for an efficient and finite precision implementation of the algorithm outlined above. The interested reader is referred to Sayood[6] and Bell[8] for a detailed discussion of such implementations.

If we look at Example 76.4 in terms of compression that has been achieved, then we see that a five-symbol sequence is now being represented by 11 bits. The choice of using 11 bits to represent the tag is dictated

TABLE 76.2 Arithmetic Coding

	$\mathcal{I}$			Partition $\mathcal{P}$			Selected
Symbol	$\mathcal{I}_l$	$\mathcal{I}_h$	k	$\mathcal{P}_{k-1}$	$\mathcal{P}_k$	$\mathcal{T}_v$	Partition
a_2	0.000000	1.000000	1	0.000000	0.100000	0.050000	
			2	0.100000	0.800000	0.450000	←
			3	0.800000	1.000000	0.900000	
a_3	0.100000	0.800000	1	0.100000	0.170000	0.135000	
			2	0.170000	0.660000	0.415000	
			3	0.660000	0.800000	0.730000	←
a_2	0.660000	0.800000	1	0.660000	0.674000	0.667000	
			2	0.674000	0.772000	0.723000	←
			3	0.772000	0.800000	0.786000	
a_1	0.674000	0.772000	1	0.674000	0.683800	0.678900	←
			2	0.683800	0.752400	0.718100	
			3	0.752400	0.772000	0.762200	
a_1	0.674000	0.683800	1	0.674000	0.674980	0.674490	←
			2	0.674980	0.681840	0.678410	
			3	0.681840	0.683800	0.682820	

$$\mathcal{T}_v = 0.67449; \quad \mathcal{T} = 10101100101$$

Each interval $\mathcal{I}$ is divided into $S = 3$ partitions, $(\mathcal{P}_{k-1}, \mathcal{P}_k]$, $k = 1, 2, 3$. The midpoint of each partition provides the tag value $\mathcal{T}_v$ at that point in the encoding process. The partition in which the CDF associated with the input symbol falls is marked with the ← symbol. This partition becomes the interval for the next iteration.

by the probability distribution of the symbols. Because the tag value had to fall *within* the final interval, a precision of (at least) 11 bits was required to represent the tag (i.e., at least 11 bit precision was needed to represent T_v). The entropy of the original data is $\mathcal{H} = 1.157$ bits per symbol. For the five symbols encoded thus far, the average length is $11/5 = 2.2$. This is considerably different from the entropy, but the discrepancy can be easily explained. Of the five symbols that were encoded, three (a_3, a_1, a_1) belonged to the group that had a combined probability of occurrence of 0.3; so they occurred twice as frequently as expected. However, the redundancy would decrease as the length of the sequence increases. For a sequence of infinite length, redundancy would be arbitrarily close to 0.

Decoding arithmetic encoded data is considerably more complicated than decoding Huffman encoded data. The following algorithm describes the decoding operation:

Algorithm 76.5

1. Initialize the current interval to the unit interval $(0, 1]$.
2. Obtain the decimal fraction representation T_v of the transmitted tag T.
3. Determine in which partition of the current interval T_v falls by picking that partition, $(P_{i-1}, P_i]$ which contains T_v.
4. Partition the current interval using the procedure outlined in the Step 4 of the encoding algorithm.
5. Using the value T_v from Step 2, go to Step 3 and repeat until the sequence has been decoded.

Example 76.5

We will decode the encoded sequence 10101100101 generated in Example 76.4, using the input alphabet and associated probabilities given in the example. The decoding sequence is shown in Table 76.3. The decoded sequence is, of course, identical to the sequence that was encoded in Example 76.4.

As is evident from Examples 76.1 and 76.2 for Huffman coding and decoding, and Examples 76.4 and 76.5 for arithmetic coding and decoding, the latter requires arithmetic computations at each step, whereas the former primarily requires comparisons only. For this reason, Huffman coding tends to be

TABLE 76.3 T_v Determines which Partition Contains the Tag T

$T = 10101100101;$ $T_v = 0.674805$					
Interval			Partition		
Lower Limit	Upper Limit	i	P_{i-1}	P_i	Symbol
0.000000	1.000000	1	0.0000000	0.100000	
		2	0.1000000	0.800000	a_2
		3	0.8000000	1.000000	
0.100000	0.800000	1	0.1000000	0.170000	
		2	0.1700000	0.660000	
		3	0.6600000	0.800000	a_3
0.660000	0.800000	1	0.6600000	0.674000	
		2	0.6740000	0.772000	a_2
		3	0.7720000	0.800000	
0.674000	0.772000	1	0.6740000	0.683800	a_1
		2	0.6838000	0.752400	
		3	0.7524000	0.800000	
0.674000	0.683800	1	0.6740000	0.674980	a_1
		2	0.6749800	0.681840	
		3	0.6818400	0.683800	

The associated symbol is decoded. The new limits of the interval $\mathcal{I}$ are the limits of the relevant partition $(P_{i-1}, P_i]$. The decoded sequence $d_o = a_2 a_3 a_2 a_1 a_1 = d_i$.

faster than arithmetic coding, both in the encoding and the decoding process. However, the superiority of the arithmetic code in terms of the overall redundancy reduction and the compression ratio make it a better encoder. The eventual choice of which entropy coder to use is, of course, application dependent. Applications where speed is more important than the compression ratio may rely on Huffman coding in preference to arithmetic coding, and vice versa. Both methods perform adequately when used for text and image compression. In most cases, images have fairly large alphabets and unskewed probability density distributions — univariate histograms — so both methods provide similar compression ratios.

76.2.4 Dictionary-Based Techniques

Entropy coding techniques exploit the frequency of distribution of symbols in a data stream but do not make use of structures or repeating patterns that the same data stream contains. There are other coding techniques that utilize the occurence of recurring patterns in the data to achieve better compression. If we can build a dictionary that allows us to map a block, or sequence, of symbols into a single codeword, then we can achieve considerable compression. This is the basic idea behind what is generally called *dictionary methods*. The following sections describe the *Lempel-Ziv-Welch* (LZW) compression[1] method, which is based on the seminal papers by Ziv and Lempel.[10,11] The methods described by Ziv and Lempel are popularly known as LZ77 and LZ78, where the digits refer to the year of publication. They also form the cornerstone of several other lossless image compression methods. LZW is part of Compuserve's Graphic Interchange Format (GIF), and is supported under the Tagged Image File Format (TIFF). LZ77, LZ78, LZW, and several variants are used for the UNIX COMPRESS, GNU's GZIP, BZIP, and the PKZIP and other lossless compression utilities commonly used for data archiving.

The idea for dictionary based techniques is quite straightforward and is best explained with an illustration for text compression.

Example 76.6

Suppose we are encoding a page of text from your favorite book. The alphabet, which is the set of all the symbols that can occur in the text, has a certain probability distribution that can be exploited by a coding technique such as Huffman coding to generate efficient codewords for the symbols. However, it is also obvious that there are a number of symbol pairs, digrams, that occur together with a high probability; for example, "th," "qu," and "in." If we could encode these digrams efficiently by representing them with a single codeword, then the encoding can become more efficient than those techniques such as Huffman coding that encode the data one symbol at a time. The same procedure could be performed for trigrams — combination of three letters from the alphabet at a time (e.g., "ing" and "the") — and for larger and larger sequences of symbols from the alphabet.

So, how can we construct such a dictionary? Suppose we consider the text that we are compressing to be comprised of symbols that are independent and identically distributed (*iid*).* Of course, this is not a realistic model of text, but it serves to make the point. Consider an alphabet** $\mathcal{A}_i$ that consists of just the lower-case English letters and the characters {⊔;,.}. Because there is a total of 30 symbols, a binary representation would require 5 bits per symbol, and the fact that the source is *iid* means that an entropy coding algorithms would generate equal-length codewords that are 5 bits long. Hence, $\mathcal{A}_i$ would contain 2^5 such symbols if we assume that the size of $\mathcal{A}_i$ is a power-of-two. Now also suppose that before encoding, we form a new alphabet $\mathcal{A}'_i$ where the symbols in $\mathcal{A}'_i$ are formed by by grouping symbols in $\mathcal{A}_i$ in blocks of four. Thus, each symbol in $\mathcal{A}'_i$ is 20 bits long. Again, with the *iid* assumption in mind, there are a

---

*This simply means that they have an equally likely probability of occurrence and the occurrence of one does not affect the probability of occurrence of another symbol.

**This material is derived substantially from Reference 6.

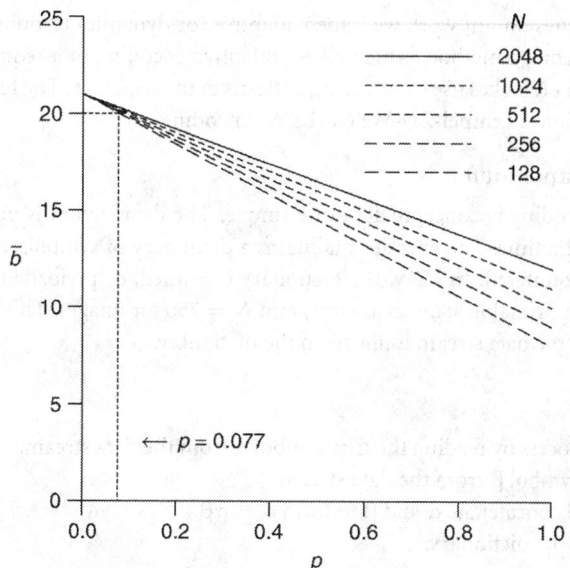

FIGURE 76.2 The average codeword length $\bar{b}$ as a function of the probability of the symbol being encoded in the dictionary p.

total of 2^{20} such symbols. If we build a dictionary of all these entries, then the dictionary would require $2^{20} = 1024^2 \approx 1,000,000$ entries! Suppose we perform the following exercise:

1. Put the N most frequently occurring patterns in a look-up table (i.e., a dictionary). The in-table entries can each be represented by $\log_2 N$ bits.
2. When the pattern to be encoded is found in the dictionary, transmit a 0 followed by the look-up table index.
3. When it is not in the dictionary, send a 1 followed by the 20-bit representation.

If the probability of finding a symbol in the dictionary is p, then the average number of bits needed to encode a sequence of symbols drawn from this alphabet is:

$$\bar{b} = p(\log_2 N + 1) + (1 - p)(20 + 1)$$
$$= 21 - (20 - \log_2 N)p$$

The average codeword length $\bar{b}$ as a function of p is shown in Figure 76.2. $\bar{b}$ is a linear, monotonically decreasing function of p. Setting $\bar{b} = 20$ and solving for p,

$$p_1 = p = \frac{1}{(20 - \log_2 N)}$$

As can be seen from Figure 76.2, for this experiment to be successful, $\bar{b} < 20$ and $p_1 > 0.062$. In addition, for *iid* symbols, the probability of finding a sequence in the look-up table is given by

$$p_2 = p = N/2^{20} = 2^{(\log_2 N - 20)} = \frac{1}{2^{(20 - \log_2 N)}}$$

This means that $p_2 \leq 0.00195$. So, whereas the first requirement dictates that $p \geq 0.062$, the reality is that $p \leq 0.00195$, about 32 times smaller! This situation is quite unlikely to occur because of the unrealistic *iid* assumption. In most practical applications, better context-dependent dictionaries can be (manually) constructed to exploit the structure of the text being encoded — novel, C program, etc. However, the problem of poor compression performance remains to some extent because of the static nature of the dictionary. Code words are defined that may never be used in the actual encoding of the symbol sequence.

When static techniques do not work well, then adaptive (or dynamic) techniques may do better. The problem, then, is to define a method, which allows adaptive encoding of a sequence of symbols drawn from an alphabet so that it makes use of recurring patterns in the sequence. The technique that allows one to perform this operation is Lempel-Ziv-Welch (LZW) encoding.

76.2.4.1 LZW Compression

The idea for LZW encoding is conceptually quite simple. The data stream is passed through the LZW encoder one symbol at a time. The encoder maintains a dictionary of symbols, or sequence of symbols, that it has already encountered. In LZW, the dictionary is primed, or preloaded, with N entries, where $N = 128$ for text (7-bit printable ASCII characters) and $N = 256$ for images. This priming means that the first symbol read from the data stream is *always* in the dictionary.

Algorithm 76.6

1. Initialize the process by reading the first symbol α from the data stream.
2. Read the next symbol β from the data stream.
3. Set $\alpha' = \alpha$, and concatenate α and β to form $\alpha = \alpha\beta$.
4. Check if α is in the dictionary.
5. If YES, go to Step 2.
6. If NO, then add α to the dictionary, output the code for α', set $\alpha = \beta$, and go to Step 2.
7. Repeat until the data stream is exhausted.

Thus, the dictionary for the LZW encoding is built *on-the-fly*, and exploits all the recurring patterns in the input data stream. The longer the recurring patterns, the more compact the final representation of the data stream. Example 76.7 illustrates the encoding process.

Example 76.7

Suppose we are encoding a text sequence that contains symbols derived from an alphabet $\mathcal{A}_i = \{c, e, i, h, m, n, o, r, y, ',', ⊔, '.', '-'\}$. The initial dictionary would then be primed with all the symbols in $\mathcal{A}_i$. Then the initial dictionary looks like this:

Index	1	2	3	4	5	6	7	8	9	10	11	12	13
Symbol	c	e	h	i	m	n	o	r	y	,	⊔	.	—

Suppose that the sequence we are encoding is:

$$chim\text{-}chimney, ⊔ \ chim\text{-}chimney, ⊔ \ chim\text{-}chim, ⊔ \ cheroo$$

Stepping through the sequence then, when the first symbol, '*c*' is received, the encoder checks to see if it is in the dictionary; because it is, the encoder reads in the next symbol forming the sequence '*ch*'. '*ch*' is not in the dictionary so it is added to the dictionary and assigned the index value 14. The code (index) for '*c*', 1, is sent to the output. The next symbol '*i*' is read and concatenated with '*h*' to form '*hi*'. Since '*hi*' is not in the dictionary, it is added in and assigned the next index which is 15. The code for '*h*' is then sent. This process is repeated until the sequence is completely encoded. The dictionary at that point looks like this:

1.	c	11.	⊔	21.	ne	31.	⊔ch
2.	e	12.	.	22.	ey	32.	him
3.	h	13.	-	23.	y,	33.	m-ch
4.	i	14.	ch	24.	,⊔	34.	him,
5.	m	15.	hi	25.	⊔c	35.	,⊔c
6.	n	16.	im	26.	chim	36.	che
7.	o	17.	m-	27.	m-c	37.	er
8.	r	18.	-c	28.	chimn	38.	ro
9.	y	19.	chi	29.	ney	39.	oo
10.	,	20.	imn	30.	y,⊔		

The sequence that is transmitted is:

1 3 4 5 13 14 16 6 2 9 10 11 19 17 26 21 23 25 15 27 32 24 14 2 8 7 7

The size of the dictionary, and hence the length of the codewords, are a function of the particular implementation. Typically, fixed-length codes are used to represent the symbols: LZW gets its compression from encoding *groups of symbols* (i.e., recurring patterns) efficiently and not from representing each symbol efficiently. The size of the dictionary is usually adaptive, and adjustments are made depending upon the number of symbols that are added to the dictionary as the encoding process continues. For example, the initial length of the codewords used for GIF and TIFF versions of LZW is 9 bits; that is, the dictionary has space for 512 entries. The first 128 elements in the dictionary are set to the ASCII code values. When the number of entries in the dictionary reaches 512, its size is doubled; that is, the codewords are represented as 10-bit numbers. This happens until the codewords are 16 bits wide. Beyond that, no new entries are added to the dictionary and it becomes static. Other implementations use different adaptation schemes for dictionary size.

The LZW decoder mimics the operations of the encoder. Because the encoder adds a symbol to the directory before it uses it for encoding, the decoder only sees symbols that already have an entry in the dictionary that it is building. The decoder dictionary is primed the same way as the encoder dictionary. Thus, the index of the first symbol to be decoded is always in the dictionary when the decoding process starts. Let the first symbol to be decoded be α. Then to decode the rest of the coded data stream, use Algorithm 76.7.

Algorithm 76.7

1. Read in the next codeword from the data stream, and decode the symbol β.
2. Concatenate β to α and add to the dictionary.
3. Set $\alpha = \beta$.
4. Go to Step 1 until the encoded data stream is exhausted.

Because the dictionaries at the encoder and decoder are built identically, they are in lock-step with each other during the encoding and decoding processes. Thus, any change on the encoder side is reflected immediately on the decoder side. This makes this method valid for any alphabet, including images.

Example 76.8 demonstrates the decoding operation.

Example 76.8

For the initial dictionary given in Example 76.7, and the transmitted sequence, we need to build the dictionary at the decoder so we can decode the transmitted sequence. We will ignore, without loss of generality, the buffering required to preserve the word boundaries in an actual software implementation. The first codeword received by the encoder is 1. The first codeword is always in the dictionary, so it is decoded as '*c*'. When the second codeword 3 is received, it is decoded as '*h*', which is then concatenated with '*c*' to form '*ch*' which is then added to the dictionary, and assigned index 14. In a similar manner, the arrival of codewords 4, 5, and 13 cause the sequences *hi*, *im*, and *m*- to be added to the dictionary with indices 15, 16, and 17, respectively. At this point, the cache consists of the symbol -. When symbol 14 is received, it is decoded as '*ch*'. The first character of the decoded sequence. '*c*', is concatenated with the '-' forming '-*c*'. The dictionary is searched to see if it already contains '-*c*'. Since it does not, '-*c*' is added to the dictionary and assigned the index 18, and the symbol string *ch* is cached. The next received index is 6 which is decoded as *i*. This is concatenated with the *ch*, forming *chi*, which is then added to the dictionary. When the final encoded symbol has been received, the dictionaries at the encoder and decoder are perfectly in sync with each other, so the sequence can be encoded at one end and decoded at the other completely losslessly.

It is difficult to predict the amount of compression an LZW encoder can achieve. LZW encodes increasingly longer sequences with a single codeword if the pattern being represented by the sequence occurs frequently in the data stream. So it can achieve "better"-than-entropy performance. Conversely, if the sequence does not contain many recurring patterns, then the performance of the LZW encoding is much poorer than entropy coding schemes. In fact, for data streams that contain few recurring patterns, the fixed-length dictionary indices can cause the data size of the compressed data stream to be larger than the original data stream! Compression ratios of between 2:1 and 4:1 are not uncommon with LZW for text: images, however, can be more difficult to compress because of the significantly larger alphabet — 8 bits per symbol as compared to 7 bits per symbol for text.

Although conceptually simple, LZW presents a number of programming challenges. The chief challenge is the search procedure used for checking if a symbol (or a sequence of symbols) already has an entry in the dictionary. Suppose after several presentations of symbols that the dictionary has grown to 8192 entries. At this point, for any incoming symbol, the entire dictionary must be searched to see if the symbol was previously encountered and thus already has an entry in the code book or dictionary. On average, this search will take $\sim$4096 comparisons before finding whether there is an entry for the symbol. This represents a considerable overhead for processing. Hash tables and other search space reduction techniques are often used to speed up the process.

76.2.5 Other Methods

76.2.5.1 Prediction with Partial Matching (PPM)

A technique that has been gaining popularity recently, and which has several variations, is *prediction with partial matching* (PPM). Conceptually, PPM is relatively straightforward and provides better compression than LZW. LZW achieves its compression by representing strings of symbols by fixed-length codewords. As the length of the string increases, so does the compression efficiency. Suppose we are encoding a data stream that consists of just two alternating symbols, $d_i = \alpha\beta\alpha\beta\ldots\alpha\beta\ldots$. While the compression efficiency increases as each new symbol is encoded, it turns out that for this particular d_i, each dictionary entry is used only once in the output data stream. The size of each entry also increases rapidly, and, of course, each entry is represented using (semi-)fixed encoding. PPM is a symbol-wise coding scheme and gets away from these problems inherent to the LZW encoding process.

The basic idea behind PPM is this: instead of allowing the strings of symbols to grow without a limit as in LZW, there is a maximum predefined length to which the strings can grow. Several tables are kept in which *contexts* for different symbols are stored for a given context length. For instance, a table for context-2 would have all the pairs of symbols that have occurred thus far that provide a context to the current symbol. For example, a context-2 table might contain an entry 'th' that provides a context to symbols 'e', 'i', 'o', and so on. For each of these symbols, a frequency count is maintained; and based upon the number of symbols and the frequency count for each context, an arithmetic code is generated. So, to encode a symbol, use Algorithm 76.8.

Algorithm 76.8

1. Let $\mathcal{N}$ be the maximum context length. The context tables store contexts and symbols encountered thus far that have occurred in those contexts.
2. Search in the context-$\mathcal{N}$ table for the occurrence of the string made up of contexts in that table and the current symbol.
3. If a match is found, then that context is used to arithmetic encode the symbol.
4. Otherwise, reduce context size by 1, $\mathcal{N} = \mathcal{N} - 1$, and go to Step 2.
5. If the symbol is not found in any previously encountered context, then a predetermined default representation is used.

The maximum size of the context is algorithm definable but typically does not exceed 5.

76.2.5.2 Run-Length Coding

LZW and other dictionary-based techniques achieve compression by exploiting the structure of recurring patterns in the input sequence. However, they do not exploit the spatial structure of symbol distribution to achieve redundancy reduction. Many images, for instance, contain regions where the intensity values are either unchanging or slowly changing. The structure of these slowly changing values is exploited by several schemes for redundancy reduction. One of the simplest of these schemes is run-length coding (RLC).

Definition 76.11 An image can be thought of as a two-dimensional matrix I where the elements of the matrix $I[i_1, i_2], i_1 = 0, \ldots, N_1 - 1; i_2 = 0, \ldots, N_2 - 1$ are the intensity values at each pixel location, and N_1, N_2 are the number of rows and the number of columns, respectively.

For simplicity we will assume that the images under consideration are grayscale, and $I[i_1, i_2] \in \{0, \ldots, 2^\kappa - 1\}$, where κ is the number of bits per pixel. For typical images, $\kappa = 8$.

Many images have large areas of constant intensity representing an area in a scene where there are few details. An example would be the part of the image that depicts the sky in an outdoor scene. RLC exploits this spatial structure by representing *runs* of a constant value by a pair of numbers (γ_k, λ_k), where $\gamma_k = I[i_1, i_2]$ at some $i_1 < N_1, i_2 < N_2$, and λ_k is the number of consecutive pixels that have the value γ_k.

Example 76.9

Suppose we have the following sequence that we need to encode:

$$3\,3\,3\,3\,4\,5\,5\,5\,5\,6\,6\,7\,7\,7\,7\,7\,7\,2\,2\,2.$$

Then the RLC representation would look like:

$$(3, 4)\,(4, 1)\,(5, 4)\,(6, 2)\,(7, 6)\,(2, 3).$$

The initial sequence is 20 symbols long, and each symbol is represented by 3 bits, for a total of 60 bits. Suppose we let the γ_k be represented by 3 bits, and the λ_k by 5 bits, allowing for a maximum run length of 32. With this representation, each (γ_k, λ_k) pair is represented by 8 bits. Because there are a total of six pairs needed to represent the original sequence, the total number of bits required to represent the RLC'd sequence is 48 bits long. So we have succeeded in compressing the data to 80% of its original size by this simple technique.

It is difficult to predict the compression ratio that can be achieved by RLC because it is highly image dependent. Images that are amenable to RLC generally have low mean spatial detail; such images have large areas of constant, or slowly changing, intensity values. Images with high mean spatial detail can actually cause the size of the encoded image to grow larger than the size of the unencoded image. This will happen when there are very short runs of gray level values; RLC doubles the amount of data if $\lambda_k = 1, \forall k$.

RLC is an integral part of a number of different encoding techniques, although it is seldom used as a stand-alone compression engine. It is used in JPEG before the data is entropy coded. It is also used in JBIG[6] (Joint Bi-level Imaging Group), which, among other things, provides the standard for facsimile (fax) transmission. We will not discuss the JBIG standard; however, RLC as it applies to binary $(b = 1)$ images, does deserve a comment.

When the image being encoded is binary (i.e., the pixels can only take the values 0 and 1), the RLC scheme can be further simplified. In such a situation, the representation (γ_k, λ_k) can be reduced to simply λ_k if we make the assumption that the first value in the image is a 0. The RLC'd image is then an alternating sequence of runs of 0 and runs of 1.

Example 76.10

Suppose we are encoding the following segment of a binary image:

$$1\,0\,0\,0\,0\,0\,0\,1\,1\,1\,0\,0\,0\,1\,0\,0\,1\,1\,1.$$

The encoded sequence would then be:

$$0\,1\,6\,3\,3\,1\,2\,3.$$

Note that the first element of the encoded sequence is a 0. This is because of our assumption that the first element of the image is a 0. Because the first element of the sequence we are encoding is a 1, this means there are 0 zeroes. For Example 76.10, the encoded sequence uses more bits than the original sequence. However, the RLC'd sequence can be entropy encoded, as is done in JPEG (Section 76.3.2) to gain further compression. Additionally, in actual situations where the binary version of RLC is used, the runs of numbers are substantially longer, thereby reducing the amount of data considerably. This is especially true for fax transmissions.

76.3 Lossy Compression

Thus far we have discussed lossless compression methods as they apply to text and image compression. We now move solely to the domain of image compression. Images are much more forgiving than text when it comes to compression. The human visual system (HVS) is capable of tolerating considerable loss in information before noticing a deterioration in image quality. Compression methods based on this observation adhere to the principle of *perceptual losslessness*.

Definition 76.12 An image is said to be encoded in a perceptually lossless manner if an observer does not *notice* any difference between the original image and the image decompressed from the encoded data.

There is, however, a loss in the total amount of information that is conveyed by the image. Most of this loss occurs in areas of high spatial details.

76.3.1 Data Domain Compression

As mentioned in the beginning of this chapter, lossy compression techniques can be partitioned into two distinct categories: data domain and transform domain. In this section we describe several data domain techniques.

Definition 76.13 Data domain techniques are those techniques that only operate on the unmodified input data stream, d_i.

In other words, the data is not preprocessed to make it more amenable to compression in any way.

76.3.1.1 Quantization: Pulse Code Modulation (PCM)

Pulse code modulation (PCM) is not an image compression technique *per se*; it is a way of representing the continuous, infinite precision domain of *visual scenes* by the discrete, finite precision domain of *digital images*. This process is also known as *scalar quantization* and *analog-to-digital (A/D)* conversion. The reason PCM is worth mentioning in the context of image compression is because it lies at the heart of image creation. The number of quantization levels $\mathcal{L}$ determine the quality of the formed image: the greater the number of levels, the finer the quantization and the fewer the errors — artifacts — that are introduced into the digital image. Images with fewer artifacts are typically more amenable to compression. However, images that have been formed with coarser quantization have less need to be compressed. So the number of quantization levels used in image formation is application dependent: applications where a high degree of fidelity between the scene and image is needed dictate higher quantization levels; applications where the emphasis is on having as little data as possible without a real regard to image fidelity dictate coarser quantization.

The quantization operation is defined using the following pair of equations:

$$I'[i_1, i_2] = \mathcal{Q}_\kappa \left(I[i_1, i_2] \right)$$
$$\mathcal{Q}_\kappa(x) = \left\lfloor \frac{\mathcal{L}x}{x_R} \right\rfloor$$

where $\kappa = \log_2(\mathcal{L})$ is the number of bits per pixel, x_R is the range of the variable x over which quantization is performed and is the floor function that returns the largest integer less than or equal to its argument.

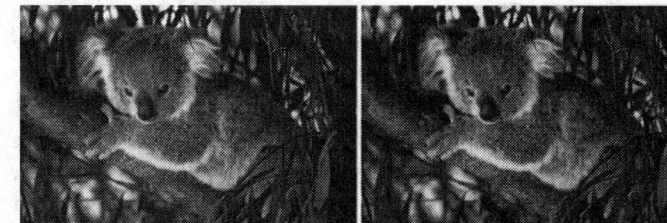

FIGURE 76.3 The original image is shown on the left, the quantized version with κ = 3 bits per pixel in the middle, and the difference image on the right. One can just begin to see artifacts in the lower left corner and in the details on the leaves.

This kind of quantization is called Uniform Quantization $\mathcal{U}$ because all the quantization intervals are of equal length. There are several other methods available for quantization, such as nonlinear quantization based on Gaussian, Laplacian, or Gamma distributions, and the Lloyd-Max optimal quantizer. However, the result of using these other techniques affects the type and amount of quantization error rather than the compression operation, which remains the same.

Definition 76.14 Quantization error $\mathcal{Q}_e$ is defined as the difference between the original image I and its quantized representation $\mathcal{Q}_\kappa(I)$:

$$\mathcal{Q}_e[i_1, i_2] = I[i_1, i_2] - \mathcal{Q}_\kappa(I[i_1, i_2])$$

Typically, the quantization error can be modeled as having a Gaussian distribution with mean $\mu_{\mathcal{Q}_e} = 0$ and a standard deviation $\sigma_{\mathcal{Q}_e}$ that is dependent on the quantization method.

The quantization operation itself, then, is the compression operation. For instance, if we start with an image at κ = 8 bits per pixel and quantize it to κ = 6 bits per pixel, then a compression ratio of $\mathcal{C} = 1.3$ has been achieved. The modified image may also be more amenable to entropy coding techniques.

Example 76.11

Figure 76.3 shows an original image at κ = 8 bits per pixel, a quantized version at κ = 3 bits per pixel, and the difference between by the two obtained by subtracting the quantized version from the original. The entropy is $\mathcal{H} = 7.435$ for the original image, and $\mathcal{H} = 2.476$ for the quantized version, giving a compression ratio $\mathcal{C} = 3$.

The difference between the original (left) and the quantized version (middle) is difficult to see with the naked eye. However, looking at the difference image (right), it is easy to see that the quantized version is actually substantially different from the original. The differences are primarily in areas of high spatial detail.

76.3.1.2 Differential Pulse Code Modulation (DPCM)

As a compression technique, PCM exploits neither the spatial structure in an image nor the frequency distribution of its symbols. If, however, one examines the spatial structure of the image (i.e., the regional distribution of intensity values), then one can exploit this structure to achieve better compression. For instance, in regions of slowly changing, or unchanging, intensity values, redundancy can be reduced if the gray levels are represented *relative* to a constant gray scale rather than absolutely. An example elucidates this concept.

Example 76.12

Suppose we have an initial data sequence $d_i = 5\ 5\ 4\ 4\ 6\ 6\ 4\ 3$. We would need 3 bits per symbol to represent these (absolute) values for fixed-length encoding. However, suppose we subtract 5 from each element, giving a new sequence $d_i' = 0\ 0\ -1\ -1\ 1\ 1\ -1\ 2$. The range of this new sequence is 3; thus, this new sequence can be represented by 2 bits per symbol. Given the subtrahend, 5 in this case, we can completely recover the original sequence by adding the subtrahend back into the new sequence.

Coding schemes that make use of this type of strategy are called differential pulse code modulation (DPCM) schemes. In general, however, instead of using a constant subtrahend for the entire sequence, differences between neighboring elements are generated. Using this method on the sequence shown in Example 76.12 results in $d_i' = 5\ 0\ 1\ 0\ -2\ 0\ 2\ 1$, where the new values $x_d[i], i = 1, \ldots, N$, N is the length of the sequence, are obtained from the elements $x[i]$ of the original sequence by:

$$x_d[i] = x[i] - x[i+1]; \quad i \geq 0; \quad x[0] = 0.$$

It might seem that this has not resulted in a gain in redundancy reduction because the dynamic range has increased to 7. However, if the first symbol is sent at the fixed-length representation of the original sequence, then the range of the remaining sequence is 4, which can be represented by 2 bits per symbol. For the sequence in Example 76.12, the second method does not seem to provide much of an improvement over the first method. In general, however, the second method is more effective, especially near edges in an image, where the intensity values change rapidly.

76.3.1.3 Predictive Differential Pulse Code Modulation (DPCM)

The method outlined in the previous section can also be described as a *predictive* DPCM (PDPCM) method. When we transmit the differences between neighboring samples, we are implicitly predicting the value of the current sample in terms of its predecessor. This is *nearest neighbor* prediction. If the original value of the current symbol is $x[i]$ and the predicted value is p, then the value being encoded (x_d) is the difference between the actual value of the symbol and the predicted value, which in this case is simply equal to the previous symbol $x[i-1]$. Thus,

$$x[0] = x[0]$$
$$x_d[i] = x[i-1] - x[i]; \quad 0 < i < N$$

which is essentially the same expression shown above for DPCM. PDPCM makes use of the spatial structure of the gray-scale distribution because it is most effective when neighboring pixel values are either close to each other or are the same.

There are several different ways in which the value of the current sample can be predicted, based upon previous samples. We have listed the simplest method above, where the predicted value of the current sample is equal to the previous sample value. More accurate estimates of the current sample value can be obtained using:

$$p = \rho x[i-1]$$

where ρ is the correlation coefficient of the image and measures the similarity between neighboring samples. The higher the correlation, the more similar the neighboring samples. Other prediction schemes use more than just the previous sample for predicting the current sample value to form even finer estimates of the current sample. In imaging applications, the current sample value is typically predicted based upon its three nearest past neighbors:

$$I_d[i_1, i_2] = I[i_1, i_2] - p[i_1, i_2]$$
$$p[i_1, i_2] = \rho_{xx} I[i_1, i_2 - 1] + \rho_{yy} I[i_1 - 1, i_2] + \rho_{xy} I[i_1 - 1, i_2 - 1]$$

where ρ_{xx}, ρ_{yy}, and ρ_{xy} are the horizontal, vertical, and diagonal correlation coefficients, respectively. Instead of computing these coefficients, several implementations assume $\rho_{xx} = \rho_{yy} = 1$ and $\rho_{xy} = -1$. The lossless compression mode in JPEG uses PDPCM for encoding. Algorithm 76.9 outlines the procedure for PDPCM encoding an $N_1 \times N_2$ image I.

Algorithm 76.9

1. Compute the prediction image, p using

$$p[0, i_2] = p[i_1, 0] = 0$$
$$p[i_1, i_2] = I[i_1, i_2] - (I[i_1 - 1, i_2] + I[i_1, i_2 - 1] - I[i_1 - 1, i_2 - 1])$$
$$\text{where } i_1 = 1, \ldots, N_1 - 1 \quad \text{and} \quad i_2 = 1, \ldots, N_2 - 1$$

   The first equation ensures that first row and first column of I_d is the same as the first row and column of I. These need to be transmitted with their original values to initialize the decoding process.
2. Form the difference image I_d:

$$I_d[i_1, i_2] = I[i_1, i_2] - p[i_1, i_2] \quad i_1 = 0, \ldots, N_1 - 1, i_2 = 0, \ldots, N_2 - 1$$

   The choice of unit magnitude weighting factors ensures that I_d has integer values provided that I has integer values.
3. Quantize I_d and transmit the result, $Q_\kappa[I_d]$, where $Q_\kappa[\ ]$ is the quantization process described previously.

Algorithm 76.10 outlines the procedure for decoding a PDPCM encoded image.

Algorithm 76.10

1. The first row and column of the decoded image I' are the first row and column of the received image $Q_\kappa[I_d]$. Dequantize the rest of the received image to form the difference image I'_d.
2. Predict $I'[i_1, i_2]$ based upon

$$I'[i_1, i_2] = I_d[i_1, i_2] + (I_d[i_1 - 1, i_2] + I_d[i_1, i_2 - 1] - I_d[i_1 - 1, i_2 - 1])$$
$$i_1 = 1, \ldots, N_1 - 1, i_2 = 1, \ldots, N_2 - 1,$$

   where the expression in parentheses is the predicted value of $I'[i_1, i_2]$.

We have glossed over two significant issues in the above algorithm description for PDPCM encoding and decoding: how localized is the error propagation, and how is the quantization performed? We address these issues in the next two comments.

Error propagation in PDPCM is cumulative. Only the first row and the first column of a (P)DPCM encoded image are transmitted without processing. Recall that the difference image is quantized before transmission. Depending on the severity of the quantization (i.e., the number of levels used to represent the differenced image), there will be an error when the difference image is dequantized. In other words:

$$Q[I_d[i_1, i_2]] - I_d[i_1, i_2] = e_Q$$

where e_Q is the quantization error. As can be seen from Algorithm 76.10, each predicted value after $p[1, 1]$ depends on values that have been reconstructed from previously reconstructed values. If there is a significant error in any of these reconstructed values, then this error will propagate to all the reconstruction values henceforth. This error accumulates with each step and the $p[i_1, i_2]$ are going to be poorer and poorer predictions of the original data. Certainly, we can perform (near)-perfect reconstruction if we have sufficient quantization resolution. However, the reason for performing DPCM and PDPCM is to make further gains in redundancy reduction. Thus, the number of quantization levels used to represent the encoded image *must* be less than the number of quantization levels used to represent the original image. If the probability density function of the error space is examined, then it is seen to closely follow the

Laplacian distribution $e^{-\alpha|x|}$. This means that the errors are clustered fairly tightly around a zero mean, and there are only a few outliers. This, of course, is exactly the behavior we hope to exploit. If the error function indeed clusters around the center, then the dynamic range within a few standard deviations from the mean can be well represented by a few quantization levels.

The magnitude of the outliers is really the major issue that needs to be resolved. For an $\mathcal{L}$-level quantization, the range of the error image is $2\mathcal{L}$. Thus, by using the differencing technique, we can potentially double the range we need to represent, at the same time that we need to reduce the number of quantization levels used to represent the data. The outliers in the error image typically occur near the edge locations in an image: the more *strong* edges there are in an image, the greater the number of these outliers. So how does one deal with these outliers? There is no simple answer! Several approaches could be adopted, each with its associated trade-offs:

1. Use $\mathcal{Q}_\kappa$ with $x_R = 2(\mathcal{L}/8)$. This would provide good results for all the values that fall within the specified range. However, this approach would work only if there were *very* few outliers, and hence only very few points where the residual error can increase.

2. Use the method outlined in Step 1, except use the following method to deal with the outliers. Instead of using a single codeword to represent the magnitude of the outliers, use a sequence of codewords. For example, if $\mathcal{L} = 256$ for the original image and the difference image is being encoded with $\mathcal{L}' = 64$ gray levels, then quantize all values of the difference image between ± 31 at 6 bits per symbol. If, however, $d[i_1, i_2] > 31$, say 120, then send 111111 as an escape and follow it by the 9-bit representation of the difference. The values between $\pm \mathcal{L}/8$ are mapped to $[0, \mathcal{L}/4]$, and conversely. A similar operation is performed for the 9-bit representation. This will result in (nearly) lossless reconstruction. However, the redundancy reduction is going to be reduced if there are a number of outliers.

76.3.1.4 Vector Quantization

The final technique that we will describe, very briefly, is vector quantization (VQ). In PCM, or scalar quantization, the data stream is quantized one symbol at a time, hence the name scalar quantization. The VQ process is conceptually very similar to PCM except that in VQ each data element is made of several elements from the input data stream. The idea, then, is to construct a dictionary of *prototype vectors* and encode the data using this dictionary. The encoding process is in Algorithm 76.11.

Algorithm 76.11

1. Construct a dictionary $\mathcal{D}$ of size N, where each entry $\mathcal{D}_i$ is a prototype vector of length L symbols.
2. Partition the input data stream into input vectors $v_k, k \geq 1$.
3. For the kth input vector v_k, find the $\mathcal{D}_i$ such that some metric between v_k and $\mathcal{D}_i$ is minimized. A typical metric is the normalized distance d between the vectors:

$$d = \frac{||v_k - \mathcal{D}_i||^2}{||\mathcal{D}_i||^2}$$

   where $||v||^2 = \sum_{j=0}^{L-1} v_j^2$, for some vector v of length L.
4. Encode v_k by the index i of $\mathcal{D}_i$ that minimizes d.

Hence, each vector of length L symbol is represented by a single index i. The compression ratio is $\mathcal{C} = Ll_i/l$, where l_i is the representation length of each symbol in $\mathcal{D}_i$ and $l = \log_2 N$. So, the smaller the size of the dictionary, the larger the compression ratio $\mathcal{C}$. However, the smaller the size of the dictionary, the larger the distortion error between the original and the decompressed data. Thus, there are two competing requirements — high compression ratio and low distortion error — both of which cannot be satisfied simultaneously.

Decoding VQ is very straightforward. The transmitted index i is used to retrieve the entry $\mathcal{D}_i$ from the dictionary $\mathcal{D}$. The decompressed vector is $v'_k = \mathcal{D}_i$. Thus, the larger the d, the less similar is the decompressed vector v'_k to v_k. This, of course, implies that the dictionary used to encode the data needs to be available to the decoder as well. One can transmit this dictionary as overhead, agree upon a pre-defined dictionary, or come up with adaptive methods to generate the dictionary on-the-fly. Different implementations use different methods to make the dictionary available to the encoder and the decoder.

Whereas it is easy to see that fewer dictionary entries lead to a better compression ratio, it may be not as obvious why smaller dictionary sizes lead to more distortion. The reason is, however, quite straightforward: With fewer dictionary entries, the distance d between the input vector v_k and the dictionary entry $\mathcal{D}_i$ is, on average, larger. Because $\mathcal{D}_i$ is the same as v'_k, the reconstructed data vector, the distortion error between v_k and v'_k is the same as d. Hence, smaller d leads to less distortion and better image quality.

We have completely skipped describing one of the aspects of VQ that is essential to the process: the construction of the dictionary. The basic idea is to derive N representative features from the data space. This can be done by taking all length L vectors v_k in the data space and then computing the mean vector μ_k. The vector μ_k can then be perturbed in a variety of ways to form two other vectors $\mu'_k = \mu_k + e$ and $\mu''_k = \mu_k - e$, where e is a perturbation vector. These two vectors can be similarly perturbed to form additional vectors, until the desired number of representative vectors has been generated. These representative vectors are then the entries of the dictionary $\mathcal{D}$. There are several other techniques used to generate the dictionary entries but we do not discuss them here. The interested reader is referred to References 6 and 12 for additional information.

VQ is a nonsymmetrical process in that whereas it takes computation of metrics and comparisons to find the encoding index, the decoding only requires an entry to be retrieved from the dictionary $\mathcal{D}$. Hence, VQ is slow on encoding and fast on decoding, a fact that is useful in practical implementations where more processor power is typically available for preparing the data for transmission than is available for decoding the compressed data. An example could be transmission of television signals where the transmitters (e.g., studios) can use high-powered computers to encode the data fast, but the receivers (e.g., televisions) do not have similar processor capacity and thus need to have relatively simple decoding techniques in order to process the data in real-time. There are several flavors of VQ that are now being used in practical systems. The interested reader is referred to Gersho and Gray[12] for a complete treatment of VQ.

76.3.2 Transform Domain Compression

In many cases, the original data d_i can be modified by the application of a transform $\mathcal{T}$ resulting in a new data stream $d_{\mathcal{T}} = \mathcal{T}[d_i]$ that is more amenable to compression. The transform $\mathcal{T}$ is usually reversible, that is, there exists an inverse transform $\mathcal{T}^{-1}$ such that $\mathcal{T}^{-1}[\mathcal{T}[d_i]] = d_i$. The transformed data contains all the information that was in the original data, although it is a completely different representation. One can exploit this representation to produce compressed data that, when uncompressed, is perceptually lossless. The degree of difference between the original and decompressed data depends on the degree of compression. Typically, the higher the compression, the higher the difference between the original and the decompressed image (Figure 76.4). In fact, the assumption about the compression being perceptually lossless quickly loses its validity as the compression ratios increase.

Example 76.13: JPEG

JPEG — Joint Photographic Experts Group — is the *de facto* image compression standard for still images.[5] It is based on the discrete cosine transform (described later) and attempts to incorporate the contrast sensitivity of the human visual response into the quantization maps used to quantize the luminance and the chrominance channels. It uses the YC_bC_r color space and quantizes the chrominance channels (C_b and C_r) at twice as coarse a resolution as it does the luminance component Y.

Figure 76.5 shows the contrast sensitivity of the HVS to different wavelengths as a function of the spatial frequency. Luminance sensitivity peaks between 2 and 5 cycles/degree. Sensitivity is greatly diminished for frequencies greater than 100 cycles/degree. Thus, intensity changes in objects of the size of about 0.2 cm,

FIGURE 76.4 The figure shows an original image (left image) that has been compressed in a perceptually lossless manner using JPEG (center image $C \approx 4$) and a compressed image also compressed using JPEG that exhibits obvious artifacts (right image $C \approx 20$).

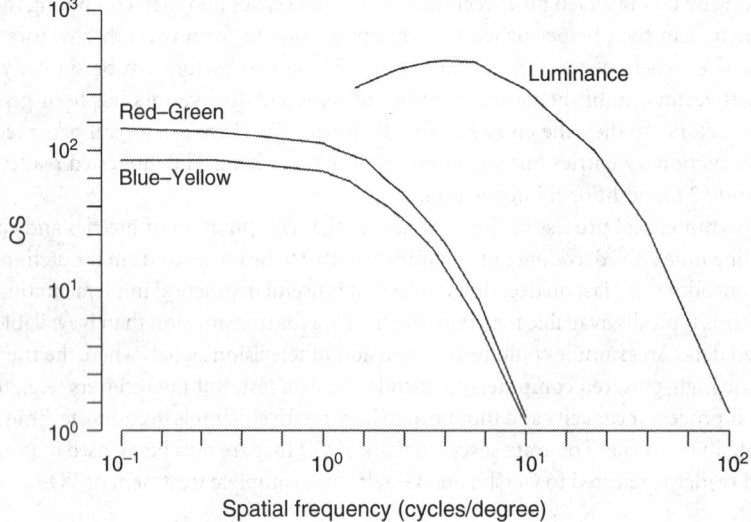

FIGURE 76.5 The contrast sensitivity (CS) of the human visual system as a function of the spatial frequency. The HVS is less sensitive to changes in color than in changes in luminance.

viewed from a distance of 1 m, can just be resolved; intensity changes in smaller objects cannot be resolved. Similarly, details finer than 0.01 cm cannot be resolved.

Chrominance (red-green; blue-yellow) response is coarser and peaks at about 1 cycle/degree. It becomes negligible at about 12 cycles/degree, which translates into an object of 0.07 cm viewed from a distance of 1 m. Hence, chrominance information can be more coarsely quantized than luminance information because the HVS is less sensitive to it. This is the justification for coarsely quantizing the chrominance maps in the JPEG compression algorithm.

At the heart of JPEG is the Discrete Cosine Transform (DCT).[13–15] DCT is one of many transforms used to convert the data domain representation into a form that is more amenable to compression. The best of such transforms is the Principal Component Transform (PCT),[15–17] which is also known as the Karhunin-Loeve Transform (KLT) or Hotelling Transform. The PCT is an optimal transform in the sense that it completely decorrelates the input data. In other words, all the information in the data domain representation gets transformed to a diagonal matrix representation in the transform domain, where the diagonal elements represent the standard deviation for the coefficient in that location. However, although the PCT is the optimal decorrelator, it is computationally intensive and image dependent. That is, the PCT must be computed afresh for each new data stream; it does not have a canonical, closed form that allows pre-computation and thus efficient implementation. The DCT decorrelates data almost as efficiently as the PCT for those data sets that have a high inter-coefficient correlation. And because the DCT has a fast implementation and the actual transform is not data dependent, it is preferred over the PCT for practical implementations.

The fast implementation of the DCT typically relies on the Fast Fourier Transform (FFT)[18–21] implementation of the Discrete Fourier Transform (DFT).[22–24] The DFT is a way of representing discrete data in terms of its spectral components.

Definition 76.15 The Discrete Fourier Transform (DFT) $\hat{X}[v]$ of the discrete data sequence $x[n]$ is given by:

$$\hat{X}[v] = \frac{1}{N} \sum_{n=0}^{N-1} x[n] \exp\left[-i2\pi vn/N\right], \quad v = 0, \ldots, N-1,$$

$$x[n] = \sum_{v=0}^{N-1} \hat{X}[v] \exp\left[i2\pi vn/N\right], \quad n = 0, \ldots, N-1.$$

where the second equation gives the Inverse DFT (IDFT).

The DFT transforms the data stream into its frequency domain representation given in terms of sums of sinusoids. High-frequency terms correspond with fine detail in the scene and the low-frequency terms with smoothness. The $\hat{X}[0]$ terms gives the mean value μ of the data stream. The DFT is an inherently complex transform, which means that each coefficient comprises a real and imaginary part. This is primarily the reason why the DFT itself is not used for compression; the conversion from data to transform domain doubles the size of the data. The DFT is also very sensitive to small changes in phase: small phase changes translate into large quality changes.

The two-dimensional (2-D) DFT is given by:

$$\hat{X}[v_1, v_2] = \frac{1}{N_1 N_2} \sum_{n_1=0}^{N_1-1} \sum_{n_2=0}^{N_2-1} x[n_1, n_2] \exp[-i2\pi(v_1 n_1/N_1 + v_2 n_2/N_2)],$$

$$v_1 = 0, \ldots, N_1 - 1, v_2 = 0, \ldots, N_2 - 1.$$

2-D DFT is a separable transform, which means that it can be implemented by first applying the 1-D DFT to the columns of the 2-D matrix x and then applying the 1-D transform to the rows of the result.[25,26]

Definition 76.16 The 2-D forward and reverse DCT is defined as:

$$\hat{C}[v_1, v_2] = \alpha[v_1, v_2] \sum_{i_1=0}^{N_1-1} \sum_{i_2=0}^{N_2-1} I[i_1, i_2] \cos\left[\frac{(2i_1 + 1)v_1\pi}{2N_1}\right] \cos\left[\frac{(2i_2 + 1)v_2\pi}{2N_2}\right]$$

$$I[i_1, i_2] = \sum_{v_1=0}^{N_1-1} \sum_{v_2=0}^{N_2-1} \alpha(v_1, v) C[v_1, v_2] \cos\left[\frac{(2i_1 + 1)v_1\pi}{2N_1}\right] \cos\left[\frac{(2i_2 + 1)v_2\pi}{2N_2}\right]$$

where

$$\alpha[v_1, v_2] = \begin{cases} \sqrt{\dfrac{1}{N_1 N_2}} & \text{for } v_1 = v_2 = 0 \\[3mm] \dfrac{2}{\sqrt{N_1 N_2}} & \text{for } v_1 \neq 0, v_2 = 0 \text{ or } v_2 \neq 0, v_1 = 0 \\[3mm] \sqrt{\dfrac{2}{N_1 N_2}} & \text{else} \end{cases}$$

$\hat{C}$ is the $N_1 \times N_2$ DCT of the $N_1 \times N_2$ image I, and $[v_1, v_2]$ are the (discrete) spatial frequency variables.

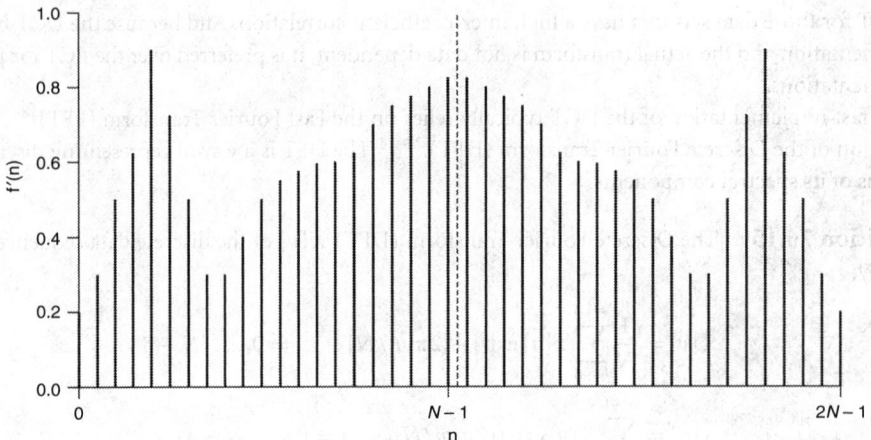

FIGURE 76.6   Generating the length-$2N$ sequence from the length-N sequence.

It is a common misconception that the DCT is the real part of the DFT. It is not! However, there is a very nice (mathematical) relationship between the two transforms. Consider a 1-D sequence $f[n], n = 0 \ldots, N - 1$, where f can be real or integer valued. Construct a new sequence f' such that

$$f'[n] = \begin{cases} f[n] & n = 0, \ldots, N - 1 \\ f[2N - n - 1] & n = N, \ldots, 2N - 1 \end{cases}$$

So, f' is formed by reflecting f about the imaginary line halfway between $k = N - 1$ and $k = N$. Figure 76.6 shows an instance of this process, with $N = 21$. Taking the Fourier transform of f'

$$\hat{F}[v] = \frac{1}{2N} \sum_{n=0}^{2N-1} f'[n] \exp[-(i2\pi vn)/2N]$$

Using the definition of f' from above and substituting,

$$\hat{F}[v] = \frac{1}{2N} \left[\sum_{n=0}^{N-1} f'[n] \exp[-(i2\pi vn)/2N] + \sum_{n=N}^{2N-1} f'[n] \exp[-(i2\pi vn)/2N] \right]$$

$$= \frac{1}{2N} \left[\sum_{n=0}^{N-1} f[n] \exp[-(i2\pi vn)/2N] + \sum_{n'=0}^{N-1} f[n'] \exp[-(i2\pi v(2N - n' - 1))/2N] \right]$$

$$= \frac{1}{2N} \left[\sum_{n=0}^{N-1} f[n] \exp[-(i2\pi vn)/2N] + \sum_{n'=0}^{N-1} f[n'] \exp[(i2\pi v(n' + 1))/2N] \right]$$

$$= \frac{1}{2N} \sum_{n=0}^{N-1} f[n] \left(\exp[-(i2\pi vn)/2N] + \exp[(i2\pi v(n + 1))/2N] \right)$$

Multiplying both sides of the equation by $\exp[-(i2\pi v)/2N]$,

$$\hat{F}[v] \exp[-(i2\pi v)/2N] = \frac{1}{2N} \sum_{n=0}^{N-1} f[n] \left(\exp\left[-\frac{i\pi v(2n + 1)}{2N} \right] + \exp\left[\frac{i\pi v(2n + 1)}{2N} \right] \right)$$

$$= \frac{2}{2N} \sum_{n=0}^{N-1} f[n] \cos\left[\frac{(2n + 1)v\pi}{2N} \right].$$

The expression on the right-hand side of the above equation should be recognizable as 1-D DCT, without the normalization factor α. The inverse kernel for the DCT is the same as the forward kernel. In other words, a DCT'd sequence passed back through the DCT returns the original sequence.

DCT is typically not applied to the image as a whole. Instead, the image is first broken up into non-overlapping square blocks*, and the DCT is performed on each block. At this point, the image has not been compressed; it has just been transformed to the DCT space. To achieve perceptually lossless compression, JPEG performs a series of operations on each block of the image.

Algorithm 76.12

1. Level shift the image by subtracting 2^{b-1} from each of its pixels, where $b = \log_2 L$ and L is the number of gray levels that a pixel in the image can take. This minimizes the variation in the DC component from block to block. The only impact this has on the DCT coefficients is to change the value of $C[0,0]$.
2. Partition the image into $\lambda = N_1 N_2 / B^2$ non-overlapping $B \times B$ blocks p_l. Then, $I = \bigcup p_l$, $l = 0, \ldots, \lambda - 1$.
3. Perform the 2-D DCT given in Definition 76.16 on each block, $p_l, l = 0, \ldots, \lambda - 1$.
4. Quantize the DCT coefficients $\hat{C}[\nu_1, \nu_2]$ using threshold coding:

$$Q[\nu_1, \nu_2] = \left\lfloor \frac{\hat{C}[\nu_1, \nu_2]}{Z[\nu_1, \nu_2]} + 0.5 \right\rfloor$$

where $Q[\nu_1, \nu_2]$ are the quantized DCT coefficients and the $Z[\nu_1, \nu_2]$ values are obtained from a quantization map.** The quantization maps are a function of the quality factor, and are derived from a pre-defined base representation. A typical quantization map is shown below.

$$Z[\nu_1, \nu_2] = \begin{bmatrix} 16 & 11 & 10 & 16 & 24 & 40 & 51 & 61 \\ 12 & 12 & 14 & 19 & 26 & 58 & 60 & 55 \\ 14 & 13 & 16 & 24 & 40 & 57 & 69 & 56 \\ 14 & 17 & 22 & 29 & 51 & 87 & 80 & 62 \\ 18 & 22 & 37 & 56 & 68 & 109 & 113 & 92 \\ 24 & 35 & 55 & 64 & 81 & 104 & 113 & 92 \\ 49 & 64 & 78 & 87 & 103 & 121 & 120 & 101 \\ 72 & 92 & 95 & 98 & 112 & 100 & 103 & 99 \end{bmatrix}$$

5. Zigzag scan each block as shown in Figure 76.7. Because the zigzag scan starts at the $Q[0,1]$ coefficient, it includes just the AC coefficients of the transformed block. The higher threshold values in the quantization map, Z, coincide with higher frequencies. This gives rise to regions of zero values. Raster scanning these regions from right-to-left results in sequence that contain runs of zeros interspersed with other values. Zigzag scanning the same block tends to clump the zeroes together without the interspersed values. These long(er) runs of zeroes can be run-length encoded more efficiently. Other scanning schemes such as Peano scanning[27,28] also provide regional clustering but have not been adopted by the JPEG standard.
6. The $\hat{C}[0,0]$ coefficient is the scaled DC component (mean) of each block and typically has the largest magnitude in the block. JPEG encodes the AC and DC coefficients differently.

   The DCT'd image has λ non-overlapping blocks, so there are a total of λ DC coefficients that need to be encoded. If a sequence $d[k], k = 0, \ldots, \lambda - 1$ of these DC components is formed, then

---

*A variation of this scheme for overlapping blocks is called the Lapped Orthogonal Transform.

**Mathematically, $\lfloor -|x| \rfloor = -(\lfloor |x| \rfloor + 1)$ and not $-\lfloor |x| \rfloor$. For example, $\lfloor -2.3 \rfloor = -3$ and not -2.

TABLE 76.4   Difference Categories for Canonical Huffman Encoding

Range	DC Difference Category	AC Category
0	0	N/A
$-1, 1$	1	1
$-3, -2, 2, 3$	2	2
$-7, \ldots, -4, 4, \ldots, 7$	3	3
$-15, \ldots, -8, 8, \ldots, 15$	4	4
$-31, \ldots, -16, 16, \ldots, 31$	5	5
$-63, \ldots, -32, 32, \ldots, 63$	6	6
$-127, \ldots, -64, 64, \ldots, 127$	7	7
$-255, \ldots, -128, 128, \ldots, 255$	8	8
$-511, \ldots, -256, 256, \ldots, 511$	9	9
$-1023, \ldots, -512, 512, \ldots, 1023$	A	A
$-2047, \ldots, -1024, 1024, \ldots, 2047$	B	B
$-4095, \ldots, -2048, 2048, \ldots, 4095$	C	C
$-8191, \ldots, -4096, 4096, \ldots, 8191$	D	D
$-16383, \ldots, -8192, 8192, \ldots, 16383$	E	E
$-32767, \ldots, -16384, 16384, \ldots, 32767$	F	N/A

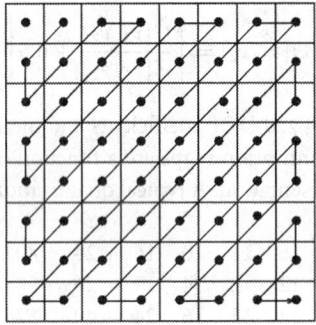

FIGURE 76.7   Zigzag scan for JPEG coefficient quantization.

we can use nearest neighbor DPCM to form a sequence with reduced redundancy, that encodes the difference between adjacent DC coefficients. The difference value Δ is encoded based upon the category in which it falls, and its position within that category. The categories are disjoint (i.e., each category represents differences that are not contained in any other category). The union of the categories, however, completely covers the dynamic range between $\mp 2^{15} - 1$. The categories and the corresponding difference values are tabulated in Table 76.4. For a difference value Δ, the category c is given by $c = \lfloor \log_2 |\Delta| \rfloor + 1$ in hexadecimal notation. The location within c is encoded as b_Δ, the binary representation of $|\Delta|$, if Δ is positive, or by $\overline{b_\Delta}$,* if Δ is negative. The canonical Huffman codes for encoding the categories are specified in Table 76.5. As an example, for a difference of 5, $c = \lfloor 2.32 \rfloor + 1 = 3$, which is encoded as **00** from Table 76.5. The location of $d = 5$ is given by **101**, the binary representation of 5. Hence, the codeword representing $d = 5$ is **00 101**. For a difference of -5, the codeword would be **00 010**.

7. The AC coefficients are encoded using Table 76.6. The value of the coefficient is used to obtain the category C to which the coefficient belongs, and the number Z of zero-valued coefficients

---

*$\overline{b}$ is the binary complement of b. For example, if $b = 0101$, then $\overline{b} = 1010$.

TABLE 76.5 Canonical Huffman Codes for Difference Categories

Category	Base Code	Length of Codeword	Category	Base Code	Length of Codeword
0	010	3	6	1110	10
1	011	4	7	11110	12
2	100	5	8	111110	14
3	00	5	9	1111110	16
4	101	7	A	11111110	18
5	110	8	B	111111110	20

TABLE 76.6 Canonical Huffman Codes for the AC Coefficients

Z/C	Codeword	Z/C	Codeword	$\cdots$	Z/C	Codeword
0/0 (EOB)	1010			$\cdots$	F/0 (ZRL)	11111111001
0/1	00	1/1	1100	$\cdots$	F/1	1111111111110101
0/2	01	1/2	11011	$\cdots$	F/2	1111111111110110
0/3	100	1/3	1111001	$\cdots$	F/3	1111111111110111
0/4	1011	1/4	111110110	$\cdots$	F/4	1111111111111000
$\vdots$	$\vdots$	$\vdots$	$\vdots$		$\vdots$	$\vdots$

preceding the non-zero coefficient to be coded form a pointer Z/C to a specific Huffman code as shown (partially) in Table 76.6.* This is done because threshold coding followed by zigzag scanning usually produces few non-zero values separated by runs of zeros. If we encode *only* these non-zero values and indicate how many zeros there were between the current and the previous non-zero value, we can code the scan very efficiently.

The symbols **EOB** and **ZRL** have special meaning: EOB (end-of-block) indicates that the rest of the coefficients in the block are 0, and ZRL indicates that a run of 15 zeros was encountered.

8. Once all the coefficients have been encoded, they can be transmitted or archived.

Example 76.14

As an example of the JPEG encoding process, consider the fragment of an image,

$$I[i_1, i_2] = \begin{bmatrix} 36 & 54 & 41 & 64 & 59 & 47 & 44 & 48 \\ 48 & 41 & 66 & 33 & 44 & 56 & 65 & 89 \\ 50 & 70 & 45 & 52 & 78 & 84 & 108 & 120 \\ 46 & 41 & 56 & 59 & 53 & 105 & 91 & 85 \\ 44 & 59 & 49 & 47 & 48 & 53 & 78 & 44 \\ 52 & 44 & 47 & 64 & 28 & 49 & 54 & 48 \\ 51 & 51 & 61 & 84 & 73 & 77 & 62 & 37 \\ 40 & 37 & 46 & 69 & 115 & 93 & 55 & 67 \end{bmatrix}.$$

According to step 0 of the algorithm, we need to subtract 2^{b-1} from each pixel. Because $b = 8$, this means subtracting 128 from each pixel. The only impact this operation has is to change the value of $\hat{C}[0,0]$ from

---

*For a complete table of these coefficients see, for instance, *Digital Image Processing*, R.C. Gonzalez and R.E. Woods, Addison-Wesley Publishers, 1996, pp. 398–399.

$\bar{\mu}_l \sqrt{B^2}$ to $(\bar{\mu}_l - 128)\sqrt{B^2}$, where $\bar{\mu}_l$ is the mean of the square block of side B. The output of this operation is not shown. The DCT of the mean adjusted fragment is:

$$\hat{C}[\nu_1, \nu_2] = \begin{bmatrix} -548.50 & -69.09 & -12.04 & 9.46 & -6.75 & 1.85 & -0.97 & -8.16 \\ -4.67 & -16.13 & 46.03 & -24.11 & -8.22 & -8.56 & -1.77 & 3.88 \\ -10.74 & 16.01 & -39.52 & 3.26 & 29.80 & -21.94 & 3.91 & 6.35 \\ -66.61 & 61.05 & -5.74 & -15.56 & -5.18 & 24.97 & -11.88 & -10.70 \\ -7.25 & -3.90 & -26.13 & 16.74 & 0.50 & -8.36 & -2.98 & -0.77 \\ 21.07 & -1.28 & -19.94 & 8.90 & -1.73 & 9.62 & -2.52 & -19.38 \\ 6.07 & -11.38 & 7.41 & 1.05 & 14.45 & -11.10 & -15.48 & -8.00 \\ 13.57 & 5.81 & -4.26 & 4.49 & -2.33 & -10.64 & -30.98 & 8.07 \end{bmatrix}.$$

Performing the threshold coding results in the quantized matrix:

$$Q[\nu_1, \nu_2] = \begin{bmatrix} -34 & -6 & -1 & 1 & 0 & 0 & 0 & 0 \\ 0 & -1 & 3 & -1 & 0 & 0 & 0 & 0 \\ -1 & 1 & -2 & 0 & 1 & 0 & 0 & 0 \\ -5 & 4 & 0 & -1 & 0 & 0 & 0 & 0 \\ 0 & 0 & -1 & 0 & 0 & 0 & 0 & 0 \\ 1 & 0 & 0 & 0 & 0 & 0 & 0 & 0 \\ 0 & 0 & 0 & 0 & 0 & 0 & 0 & 0 \\ 0 & 0 & 0 & 0 & 0 & 0 & 0 & 0 \end{bmatrix}.$$

Because we are dealing with a single block here, the DC coefficient vector consists of a single element, -34, which belongs to category 6. So it is represented by (see Step 5 of algorithm 76.1) **1110 011101** (category-code position-in-category).

Zigzag scanning the AC coefficients produces:

$$[-6\ 0\ -1\ -1\ -1\ 1\ 3\ 1\ -5\ 0\ 4\ -2\ -1\ 0\ 0\ 0\ 0\ 0\ 0\ 1\ 0\ 0\ -1\ -1\ 1\ 0\ \cdots\ 0]$$

The first element of the scan is -6, which belongs to category 3. The run of zeros preceding it is of length 0. So the code we use to represent -6 is the base code for $Z/C = 0/3$ to which is appended the code representing the position of -6 in category 3. This results in -6 being represented by **100 001**. The second non-zero coefficient in the scan is -1, which is preceded by 1 zero. Hence, it is encoded as **1100 0**—$Z/C = 1/1$. Encoding the rest of the non-zero elements in the sequence produces:

-1	-1	1	3	1	-5	4	-2	-1
00 0	00 0	00 1	01 11	00 1	100 010	1111001 100	01 01	00 0

1	-1	-1	1	EOB
1111011 1	11011 0	00 0	00 1	1010

So, the total number of bits used to represent the 64 coefficients is:

$$6 + 5 + 3 + 3 + 3 + 4 + 3 + 6 + 10 + 4 + 3 + 8 + 6 + 3 + 3 + 4 = 74$$

which gives a compression ratio of $C = (8)(64)/(74) = 7.92$ or 1.16 bits per pixel as opposed to 8 bits per pixel.

JPEG decompression is relatively straightforward when compared with JPEG compression. In this sense, the JPEG compression algorithm is asymmetric because it takes longer to compress the data than to decompress it.

Algorithm 76.13 To decompress JPEG compressed data we essentially reverse Algorithm 76.12.

1. The first step in the decompression process is to recreate the thresholded coefficients $Q[v_1, v_2]$. This is easily done via a lookup table because the JPEG compressed input sequence is Huffman coded.
2. The thresholded coefficients then need to be turned into the DCT coefficients $\hat{C}'[v_1, v_2]$ by applying

$$\hat{C}'[v_1, v_2] = Q[v_1, v_2] Z[v_1, v_2].$$

3. Performing the inverse DCT on $\hat{C}'$ gives I', which is the reconstructed representation of I after lossy JPEG compression.

Example 76.15

Continuing from Example 76.14, we first use the lookup table to generate the threshold coded image Q, which is exactly the same Q shown in Example 76.14. It is reproduced here for convenience:

$$Q[v_1, v_2] = \begin{bmatrix} -34 & -6 & -1 & 1 & 0 & 0 & 0 & 0 \\ 0 & -1 & 3 & -1 & 0 & 0 & 0 & 0 \\ -1 & 1 & -2 & 0 & 1 & 0 & 0 & 0 \\ -5 & 4 & 0 & -1 & 0 & 0 & 0 & 0 \\ 0 & 0 & -1 & 0 & 0 & 0 & 0 & 0 \\ 1 & 0 & 0 & 0 & 0 & 0 & 0 & 0 \\ 0 & 0 & 0 & 0 & 0 & 0 & 0 & 0 \\ 0 & 0 & 0 & 0 & 0 & 0 & 0 & 0 \end{bmatrix}$$

The coefficients are denormalized by the process described in Step 2 of Algorithm 76.12. For example, the DC coefficient is denormalized as

$$\hat{C}'[0,0] = Q[0,0]Z[0,0] = (-34)(16) = -544$$

Recall that $C[0,0] = -548.5$, so this is a relatively close approximation of the DC coefficient. Building the rest of the image,

$$\hat{C}'[v_1, v_2] = \begin{bmatrix} -544 & -66 & -10 & 16 & 0 & 0 & 0 & 0 \\ 0 & -12 & 42 & -19 & 0 & 0 & 0 & 0 \\ -14 & 13 & -32 & 0 & 40 & 0 & 0 & 0 \\ -70 & 68 & 0 & -29 & 0 & 0 & 0 & 0 \\ 0 & 0 & -37 & 0 & 0 & 0 & 0 & 0 \\ 24 & 0 & 0 & 0 & 0 & 0 & 0 & 0 \\ 0 & 0 & 0 & 0 & 0 & 0 & 0 & 0 \\ 0 & 0 & 0 & 0 & 0 & 0 & 0 & 0 \end{bmatrix}$$

Performing the inverse DCT on this data and adding 128 back to each pixel results in:

$$I'[i_1, i_2] = \begin{bmatrix} 50 & 46 & 53 & 66 & 58 & 39 & 40 & 57 \\ 54 & 45 & 41 & 48 & 54 & 60 & 71 & 83 \\ 58 & 54 & 47 & 48 & 67 & 92 & 105 & 104 \\ 49 & 58 & 59 & 55 & 69 & 94 & 98 & 84 \\ 39 & 54 & 56 & 45 & 47 & 61 & 60 & 47 \\ 51 & 56 & 54 & 45 & 42 & 47 & 47 & 43 \\ 56 & 50 & 53 & 66 & 74 & 69 & 61 & 59 \\ 39 & 30 & 45 & 84 & 105 & 91 & 69 & 60 \end{bmatrix}.$$

Subtracting I' from I shows the location and magnitude of reconstruction errors due to the lossy compression:

$$I[i_1, i_2] - I'[i_1, i_2] = \begin{bmatrix} -14 & 8 & -12 & -2 & 1 & 8 & 4 & -9 \\ -6 & -4 & 25 & -15 & -10 & -4 & -6 & 6 \\ -8 & 16 & -2 & 4 & 11 & -8 & 3 & 16 \\ -3 & -17 & -3 & 4 & -16 & 11 & -7 & 1 \\ 5 & 5 & -7 & 2 & 1 & -8 & 18 & -3 \\ 1 & -12 & -7 & 19 & -14 & 2 & 7 & 5 \\ -5 & 1 & 8 & 18 & -1 & 8 & 1 & -22 \\ 1 & 7 & 1 & -15 & 10 & 2 & -14 & 7 \end{bmatrix}.$$

One measure of the *quality* of reconstruction is the mean square error (MSE) between the original image I and the reconstructed image I'. The smaller the MSE, the *closer* the images are to each other. However, even large MSEs may be tolerable to the viewer, depending on the content of the image.

Example 76.16

The mean square error between the original image I in Example 76.14 and the reconstructed image I' in Example 76.15 is 95.6. The fidelity, which is a measure of how closely I and I' resemble each other, is 0.76.

The JPEG algorithm described here is the *sequential baseline system*. Other JPEG compression modes such as *progressive* compression also exist but will not be discussed in this introduction to JPEG. JPEG is a relatively simple compression scheme that results in significant compression, albeit lossy, of the original data stream. It is the *de facto* image compression standard. However, everything about JPEG is not sunshine and light, especially at high compression ratios. Because JPEG makes use of block DCT, at high compression ratios, the edges of these blocks become evident. This phenomenon is often referred to as *block artifacts* or, more affectionately(!), as *jpeggies*. The only way to eliminate these artifacts is to reduce the compression ratios (i.e., use more data).

JPEG is an evolving standard. In its (proposed) next version, the JPEG committee has recommended moving away from block DCT because of the jpeggies, and use wavelet compression. Wavelet encoding provides compression ratios equivalent to the block DCT based approach outlined in Algorithm 76.12, but does not produce the very annoying jpeggies at the higher compression ratios. However, it has its attendant problems, chief among which is the speed of compression. We introduce the basic ideas of wavelet compression in the remainder of this chapter.

76.3.2.1 Wavelets and Subband Coding

Wavelet analysis[29–36] is a general mathematical formulation that allows a signal to be analyzed at increasingly finer (or coarser) resolutions using basis functions generated by translating and scaling a *mother* wavelet. In most image compression applications, wavelet compression refers to a particular method known as subband coding.[37,38]

Definition 76.17 Subband coding, very simplistically, decomposes a signal into (approximately) disjoint frequency components that can be represented by fewer samples than the original, and which contain all the information required to reconstruct the original signal.

Suppose we have a discrete sequence $f[k], k = 0, \ldots, N - 1$. This sequence can be divided into two N-element sequences y and z generated in the following manner:

$$y[k] = \frac{f[k] + f[k+1]}{2}; \quad z[k] = \frac{f[k] - f[k+1]}{2}, \quad k = 0, \ldots, N - 1$$

In some sense, y is a smoothed version of f, and z is a sharpened version of f. In other words, y and z approximate the low and high passed components of f. Low passed versions are amenable to better

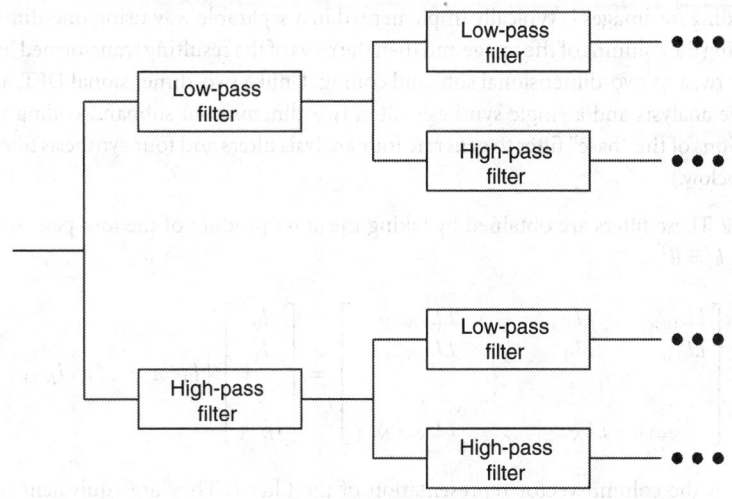

FIGURE 76.8   Subband decomposition of a signal.

entropy coding, and high passed version are similar to the DPCM scheme discussed in Section 76.3.1. The original sequence can be completely recovered by adding y and z; however, we have doubled the amount of data because both y and z are the same size as f. Suppose, now, that we send every other element in y and z, so that

$$y[2k] = \frac{f[2k] + f[2k+1]}{2}; \quad z[2k] = \frac{f[2k] - f[2k+1]}{2}, \quad k = 0, \ldots, N-1$$

Certainly, the total amount of data in y and z now is equal to the amount of data in f. But can f be recovered from this version of y and z? The answer is obviously "Yes!" Otherwise, we would not be talking about this compression method! To recover the original sequence, we first add the new y and z and then subtract them:

$$y[2k] + z[2k] = x[2k]; \quad y[2k] - z[2k] = x[2k+1], \quad k = 0, \ldots, N-1$$

So what this method allows us to do is to represent the original sequence by two sequences, each of which has characteristics that make it more amenable to compression than the original sequence, while keeping the total number of samples the same. In subband coding, each signal is decomposed into its low-frequency and high-frequency components. Each of these components can be further decomposed into low- and high-frequency components, leading to a cascade of stages where the spatial resolution is getting coarser and coarser. The scheme is shown in Figure 76.8.

The most popular pair of filters is known as the *quadrature mirror filter* (QMF) or the *conjugate mirror filter*.[37,39,40] These filters have a very nice property that relates the impulse response $h[n]$ of the low-pass filter l to the impulse response of the high-pass filter z; namely:

$$l[n] = h[n] \qquad\qquad n = 0, \ldots, N-1$$
$$z[n] = (-1)^n h[N-n-1]$$

QMFs can be symmetric, in which case

$$h[N-n-1] = h[n] \quad n = 0, \ldots, N/2-1$$

and only half the coefficients need to be defined; or they can be asymmetric, in which case all the coefficients need to be defined.

Subband coding for images is typically implemented in a separable way using one-dimensional filters, first transforming the columns of the image and then the rows of the resulting transformed image. However, there is a slight twist to two-dimensional subband coding. Unlike two-dimensional DFT, and DCT where there is a single analysis and a single synthesis filter, two-dimensional subband coding uses scaled and translated versions of the "base" filter to generate four analysis filters and four synthesis filters. These filters are described below.

1. *Low-low.* These filters are obtained by taking the *outer* product of the low-pass, one-dimensional filters: $LL = ll^T$

$$
\begin{bmatrix}
LL_{0,0} & LL_{0,1} & \cdots & LL_{0,N-1} \\
LL_{1,0} & LL_{1,1} & \cdots & LL_{1,N-1} \\
\vdots & \vdots & \ddots & \vdots \\
LL_{N-1,0} & LL_{N-1,1} & \cdots & LL_{N-1,N-1}
\end{bmatrix}
=
\begin{bmatrix}
l_0 \\
l_1 \\
\vdots \\
l_{N-1}
\end{bmatrix}
\begin{bmatrix} l_0 & l_1 & \cdots & l_{N-1} \end{bmatrix}
$$

   where l is the column vector representation of the filter l. They are equivalent to the low-pass, one-dimensional filter.
2. *High-low.* These filters are obtained by taking the outer product of the one-dimensional, high-pass filter h and low-pass filter l: $HL = hl^T$. These are *directional* filters and pick out the horizontal edges in an image.
3. *Low-high.* These filters are obtained by taking the outer product of the one-dimensional, low-pass filter l and high-pass filter h: $LH = lh^T$, where h is the column vector representation of the filter h. These are *directional* filters and pick out the vertical edges in an image.
4. *High-high.* These filters are obtained by taking the outer product of the one-dimensional, high-pass filter h: $HH = hh^T$. These are *directional* filters and pick out the diagonal edges in an image.

The subband coding algorithm can be described using the steps in Algorithm 76.14.

Algorithm 76.14

1. *Analysis.* Convolve the $N_1 \times N_2$ input image with the four filters described above. At this juncture, there are four $N_1 \times N_2$ filtered representations of the original image, which implies that there is a lot of redundancy that can be exploited for data compression.
2. *Subsampling or decimation.* To reduce the amount of total data generated by the convolution operations, subsample the filtered representations of the image by the factor $[S_1, S_2]$, where S_1 is the sampling rate in the N_1 dimension and S_2 is the sampling rate in the N_2 dimension. Typically, if there are K analysis filters, then $S_1 = S_2 = K^{1/d}$, where d is the number of dimensions of the signal. For example, for an image $d = 2$, and $S_1 = S_2 = \sqrt{K}$.
3. *Encoding or quantization.* The subband coding process is completely reversible if no encoding of the subsampled signals occurs. Several different encoding schemes can be employed at this point, with different bands being amenable to different types of encoding.
4. The encoded data can then be transmitted.

To reconstruct the data, the procedure outlined in Algorithm 76.14 is essentially applied in the reverse order.

Algorithm 76.15

1. *Decoding or dequantization.* The encoded data stream is decoded to form the $S_1 \times S_2$ subbands at each decomposition level (see Figure 76.9).
2. *Interpolation.* The subbands are interpolated (see Definition 76.19) to form signals that are the correct size for the level (i.e., the size they were before they were subsampled).
3. *Synthesis.* The synthesis filters are applied to combine the $S_1 \times S_2$ subbands into the single image I'.

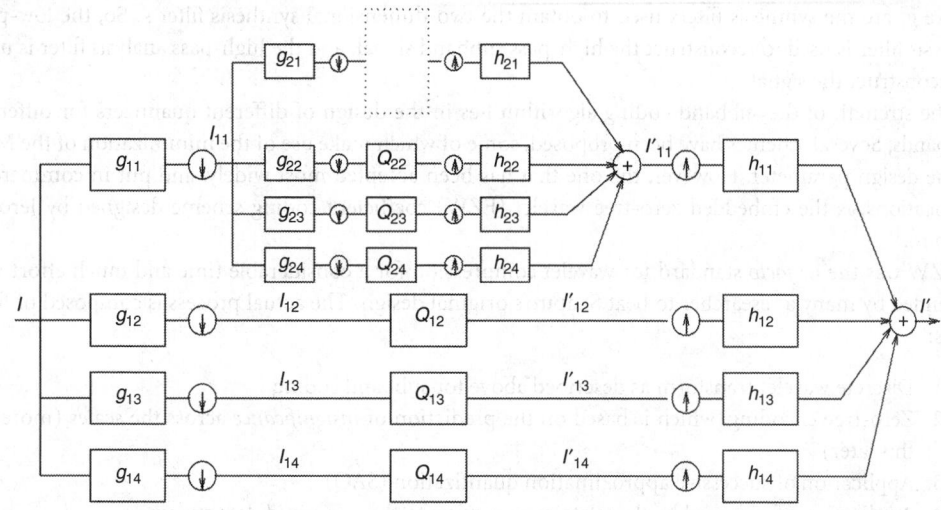

FIGURE 76.9 I is operated on by analysis filters g and down-sampled to produced critically sampled representations $I_{i_1 i_2}$: I_{11} is the LL version of I. The analysis filters are repeatedly applied to I_{kk} until the coarsest resolution has been achieved. The synthesis filters $h_{i_1 i_2}$ combined with the up-samplers generate the reconstructed image components $I'_{i_1 i_2}$, which are then combined to form the final image I'. The Q are the quantizers/encoders at different resolutions.

The general subband encoding and decoding processes are shown in Figure 76.9.

The $g_{i_1 i_2}$ are the analysis filters, where the index i_1 corresponds to the *level* of decomposition, and the index i_2 represents which of the analysis filters is being used: $i_2 = 1, 2, 3, 4$ implies LL, LH, HL, HH filters, respectively. The $Q_{i_1 i_2}$ are the quantization maps designed to maximize data compression and quality of transmission for each subband coded band, and the $h_{i_1 i_2}$ are the synthesis filters. I is the input image and I' is the image reconstructed from the lossy compressed data. The down-arrows in a circle represent subsampling and the up-arrows in a circle represent interpolation.

Definition 76.18   Subsampling by (S_1, S_2) means that sampled image I_s consists of the samples of I located at $(0, 0), (0, S_2), (0, 2S_2), \ldots, (0, \lfloor N_2/S_2 \rfloor S_2), \ldots, (\lfloor N_1/S_1 \rfloor S_1, \lfloor N_2/S_2 \rfloor S_2)$.

If an $N_1 \times N_2$ image is reconstructed from the subsampled data, it will contain *aliasing* artifacts. Aliasing is the phenomenon that results from not sampling the data at a high enough temporal or spatial resolution, that is, at a rate less than the Nyquist rate. The Nyquist rate is the minimum rate at which data can be sampled for artifact-free reconstruction and is given as twice the maximum frequency of the signal being sampled.

Definition 76.19   Interpolation is the phenomenon of providing sample values at locations where sample values are undefined by using the values at known locations. It is achieved by inserting in an image Q zeros between each sample in a row image, and P rows of zeros between each row, and then convolving by an interpolation kernel. Typical kernels are Gaussian, bilinear, and bi-cubic.

The combination of the application of analysis filters, down-sampling, up-sampling, and then the application of synthesis filters defines the subband coding operation. For this process to result in perfect reconstruction, certain constraints must be placed on the analysis and synthesis filters. As shown in comment 17, the impulse response of the high- and low-pass analysis filters have a certain relationship. Similarly, the synthesis filters and the analysis filters are related to each other: the impulse responses of the analysis filters are the time-reversed version of the impulse response of the synthesis filters:

$$g_1[n] = l[-n] = h[-n] \qquad = h[N - 1 - n] = (-1)^n z[n]$$
$$g_2[n] = z[-n] = h[n + 1 - N] = h[n] \qquad = l[n]$$

where g_i are the synthesis filters used to obtain the two-dimensional synthesis filter s. So, the low-pass analysis filter is used to reconstruct the high-pass subband signal, and the high-pass analysis filter is used to reconstruct the signal.

The strength of the subband coding algorithm lies in the design of different quantizers for different subbands. Several schemes have been proposed, some of which make use of the minimization of the MSE as the design parameter. However, the one that has been accepted most widely, and put in commercial applications, is the embedded zero-tree wavelet (EZW) coefficient coding scheme designed by Jerome Shapiro.[41]

EZW was the *de facto* standard for wavelet compression for a considerable time and much effort was expended by many a researcher to beat Shapiro's original design. The actual process is composed of four parts:

1. Discrete wavelet transform as described above for subband coding
2. Zero-tree encoding, which is based on the prediction of *insignificance* across the scales (more on this later)
3. Application of successive approximation quantization (SAQ)
4. Application of universal lossless data compression to the quantized data stream

We discuss parts 2 and 3 in more detail below.

Definition 76.20   Given an amplitude threshold T, a wavelet coefficient is said to be insignificant with respect to T if $|x| < T$. The SAQ sequentially applies a sequence of thresholds $T_0, \ldots, T_{N-1}$ to determine significance, where $T_i = T_{i-1}/2$. The initial threshold T_0 is chosen so that $|x_j| < 2T_0$ for all transform coefficients x_j.

Two lists are maintained during the encoding and decoding process:

1. The *dominant list* contains the coordinates of those coefficients that have not yet been found to be significant. The order of the list follows the order of the scan, which is carried out in raster order on each subband at a particular level. The scan moves from LL to HL to LH to HH.
2. The subordinate list contains the magnitudes of coefficients found to be significant.

Each list is scanned once for each threshold.

Carrying the idea of insignificance a little further, it is reasonable to hypothesize that if a coefficient at a coarser scale is insignificant with respect to a threshold T_i, then *all* the coefficients of similar orientation at the same spatial location at finer scales will also be insignificant with respect to T_i. Even more generally, the magnitude of a coefficient tends to be larger than that of its children. Each coefficient at a given scale can thus be related to a set of coefficients at the next finer scale (see Figure 76.10). The exception is, of course, the highest frequency subbands. This defines a parent–child relationship between the coefficients: the coefficient at the coarser resolution is called the parent, and its descendants are as illustrated in Figure 76.10. Thus, each coefficient in the subband QMF scheme we have looked at has four children with the exception of the LL band which has 3 children at the locations in HL, LH, and HH bands.

Definition 76.21   A coefficient is said to be an element of a zero-tree for threshold T if it, and all of its descendants, are insignificant with respect to T. A coefficient is a zero-tree root if it is not the descendant of a previously found zero-tree root (i.e., it is the coefficient at the coarsest scale for which it and its descendents are all insignificant with respect to T).

Definition 76.22   A coefficient is classified as an isolated zero if the coefficient under consideration is insignificant but its descendants are not all insignificant.

Definition 76.23   A coefficient is positive (negative) significant if it is not insignificant with respect to a threshold. If the value of the coefficient is greater (less) than zero, then it is positive (negative).

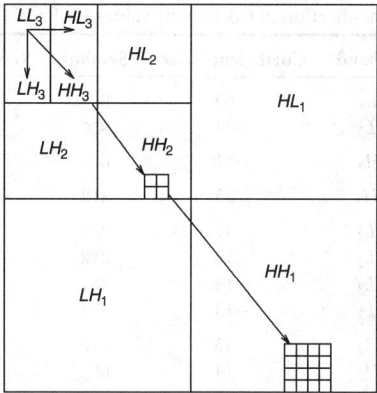

FIGURE 76.10 The relationship between coefficients at different levels of hierarchy in EZW.

The image is subband decomposed and then the coefficients are defined according to the above criteria. After the first pass, which is carried out with the highest threshold value, a list of significant coefficients is generated as well as the locations of zero-tree roots and isolated zeros. This allows the reconstruction of the signal at a coarse resolution. Further passes reduce the thresholds as defined in Definition 76.20. Only the insignificant coefficients are reevaluated, and the previously defined significant coefficients are treated as zeros. This allows us to build an increasingly finer resolution reconstruction. Adaptive arithmetic coding is used to actually encode the symbols that are generated in the subordinate and dominant passes. An example makes this process more clear.

Example 76.17*

Suppose the result of applying a three-level, two-dimensional subband decomposition to an 8×8 image produces the following decomposition:

63	−34	49	10	7	13	−12	7
−31	23	14	−13	3	4	6	−1
15	14	3	−12	3	−7	3	9
−9	−7	−14	8	4	−2	3	2
−5	9	−1	47	4	6	−2	2
3	0	−3	2	3	−2	0	4
2	−3	6	−4	3	6	3	6
5	11	5	6	0	3	−4	4

Because the maximum value is 63, then by Definition 76.20 the initial threshold is between 31.5 and 63. Let us choose, without loss of generality, $T_0 = 32$. Table 76.7 lists the coefficients and their classification based upon the selected threshold value. It has been reproduced from Shapiro.[41]

Comments:

1. The coefficient in the LL_3 has a value of $|63| > T_0$, so it is significant with respect to the threshold and positive. Hence, it is coded as POS. Similarly, HL_3, which has a value of −34, is encoded as NEG.
2. The coefficient in LH_3 has a value of $|-31| < T_0$, so it is insignificant. However, there is a coefficient that is significant among its children — 47 in LH_1; thus, this value is encoded as an isolated zero (IZ).

---

*Reproduced substantially from Shapiro.[41]

TABLE 76.7   Classification of Coefficient Values for EZW Compression

Comment	Subband	Coefficient Value	Symbol	Reconstruction Value
1	LL_3	63	POS	48
	HL_3	−34	NEG	−48
2	LH_3	−31	IZ	0
3	HH_3	23	ZTR	0
4	HL_2	49	POS	48
	HL_2	10	ZTR	0
	HL_2	14	ZTR	0
	HL_2	−13	ZTR	0
5	LH_2	15	ZTR	0
	LH_2	14	IZ	0
	LH_2	−9	ZTR	0
	LH_2	−7	ZTR	0
6	HL_1	7	Z	0
	HL_1	13	Z	0
	HL_1	3	Z	0
	HL_1	4	Z	0
7	LH_1	−1	Z	0
	LH_1	47	POS	48
	LH_1	−3	Z	0
	LH_1	2	Z	0

The encoding symbols are POS for coefficients that are positive and significant with respect to the threshold, NEG for significant and negative, IZ for isolated zero, ZTR for zero-tree root, and Z for zero.

3. The coefficient in the HH_3 subband has a value of $|-23| < T_0$, and all its descendants HH_2 and HH_1 are also insignificant with respect to T_0. Thus, it is encoded as a zero-tree root (ZTR). Because all of the coefficients in HH_2 and HH_3 are part of the zero-tree, they are not examined any further.

4. This coefficient in the HL_2 subband and all its descendants are insignificant with respect to the threshold, so it is labeled ZTR. Note, however, that one of its children, −12 in HL_3 has a larger magnitude, so it violates the assumption that coefficients at finer resolutions have smaller magnitudes than their parents at coarser resolutions.

5. The coefficient 14 in LH_2 is insignificant with respect to T_0 but one of its children, $-1, 47, -3, 2$ in LH_3 is significant, so it is encoded as an isolated zero.

6. Because HL_1 has no descendants, the ZTR and IZ classes are merged into a single class Z. This means that the coefficient has no descendants and it is insignificant with respect to the threshold.

7. This coefficient is significant with respect to the threshold, so it is encoded as POS. For future passes, its value is set to zero.

The first subordinate pass refines the reconstruction values of the four significant coefficients that were found in the first dominant pass. Because the range of the first dominant pass was between $[32, 64)$, the reconstruction value of 48 was used to represent the outputs. In the first subordinate pass, the range is divided into two parts, $[32, 48)$ and $[48, 64)$. The midpoints of these ranges are then used to reconstruct the coefficients that fall with the range. So, for instance, 63 gets reconstructed to 56, the midpoint of $[48, 64)$, while 47 gets reconstructed to 40, the midpoint of $[32, 48)$. Subsequent passes refine the ranges.

The second dominant pass is made with $T_1 = 16$. Recall that only those coefficients that were *not* significant in the first pass are considered in the second pass. Also, the significant coefficients from the previous pass(es) are set to zero for the current pass. Hence, each pass refines the reconstruction values from the previous pass and adds the less significant values to the data stream. Thus, decoding the first pass results in a very coarse decoded image that is successively refined as more coefficients are received.

EZW offers high data compression and precise control over the bit rate. Because of the use of SAQ, the process can stop exactly when the bit bucket is exhausted, and the image can be reconstructed from the coefficients transmitted up to that point.

The new JPEG standard, JPEG 2000,[42] has moved away from DCT-based compression to wavelet-based compression. JPEG performs very well for compression ratios of about 20:1. For lower bit rates, the artifacts are very evident. JPEG 2000 uses subband coding followed by Embedded Block Coding with Optimized Truncation (EBCOT). The basic idea of EBCOT is similar to that of EZW, in that each coefficient is encoded at increasingly better resolution using multiple passes over the coefficient data set. However, the notion of zero-trees is abandoned in EBCOT because of the criterion for optimal truncation. It has been shown that optimal truncation and embedded coefficient coding do not work well together.

76.4 Conclusion

We have introduced a number of different lossless and lossy compression techniques in this chapter. However, we have barely scratched the surface of a rich and complex field ripe for research and innovation. Each of the techniques mentioned in this chapter has several variations that have not been described. The bibliographic references provided give the interested reader a good starting point into the world of compression.

References

1. T.A. Welch, "A technique for high-performance data compression," *IEEE Computer*, pp. 8–19, June 1984.
2. D.A. Huffman, "A method for the construction of minimum redundancy codes," *Proceedings of the IRE*, 40(9), 1098–1101, 1952.
3. C. Shannon and W. Weaver, *The Mathematical Theory of Communication*. Urbana, IL: University of Illinois Press, 1964. Originally published in the *Bell System Technical Journal*, 27:379–423 and 28:623–656, 1948.
4. A. Papoulis, *Probability, Random Variables, and Stochastic Processes*. New York, NY: McGraw-Hill, 1984.
5. W.B. Pennebaker and J.L. Mitchell, *JPEG Still Image Data Compression Standard*. New York, NY: Van Nostrand Reinhold, 1992.
6. K. Sayood, *Introduction to Data Compression*. San Francisco, CA: Morgan Kaufmann, second ed., 2000.
7. G.G. Langdon, Jr., "An introduction to arithmetic coding," *IBM Jornal of Research and Development*, 28, 135–149, March 1984.
8. T.C. Bell, C.J.G., and I.H. Witten, *Text Compression*. Englewood Cliffs, NJ: Prentice Hall, 1990.
9. I.H. Witten, A. Moffat, and T.C. Bell, *Managing Gigabytes*. San Francisco, CA: Morgan Kaufmann, second ed., 1999.
10. J. Ziv and A. Lempel, "A universal algorithm for data compression," *IEEE Transactions on Information Theory*, IT-23, 337–343, May 1977.
11. J. Ziv and A. Lempel, "Compression of individual sequences via variable-rate coding," *IEEE Transactions on Information Theory*, IT-24, 530–536, September 1978.
12. A. Gersho and R.M. Gray, *Vector Quantization and Signal Compression*. Norwell, MA: Kluwer Academic Publishers, 1991.
13. N. Ahmed, T. Natarajan, and K.R. Rao, "Discrete cosine transform," *IEEE Transactions on Computers*, C-23, 90–93, January 1974.
14. N. Ahmed and K.R. Rao, *Orthogonal Transforms for Digital Signal Processing*. New York, NY: Springer-Verlag, 1975.
15. R.J. Clarke, *Transform Coding of Images*. Orlando, FL: Academic Press, 1985.
16. R.C. Gonzalez and R.E. Woods, *Digital Image Processing*. Reading, MA: Addison-Wesley, 1993.
17. J.T. Tou and P. Gonzalez, Rafael C., *Pattern Recognition Principles*. Reading, MA: Addison-Wesley, 1974.

18. W.W. Smith and J.M. Smith, *Handbook of Real-Time Fast Fourier Transforms*. New York, NY: IEEE Press, 1995.

19. D.F. Elliott and K.R. Rao, *Fast Transforms: Algorithms, Analyses, Applications*. Orlando, FL: Academic Press, 1982.

20. R.W. Ramirez, *The FFT Fundamentals and Concepts*. Englewood Cliffs, NJ: Prentice Hall, 1985.

21. E.O. Brigham, *The Fast Fourier Transform*. Englewood Cliffs, NJ: Prentice Hall, 1974.

22. A.V. Oppenheim and R.W. Schafer, *Digital Signal Processing*. Englewood Cliffs, NJ: Prentice Hall, 1975.

23. R.A. Haddad and T.W. Paaarsons, *Digital Signal Processing: Theory Applications and Hardware*. New York, NY: Computer Science Press, 1991.

24. J.G. Proakis and D.G. Manolakis, *Digital Signal Processing*. Upper Saddle River, NJ: Prentice Hall, 1996.

25. D.E. Dudgeon and R.M. Mersereau, *Multidimensional Digital Signal Processing*. Englewood Cliffs, NJ: Prentice Hall, 1984.

26. R.C. Gonzalez and P. Wintz, *Digital Image Processing*. Reading, MA: Addison-Wesley, second ed., 1987.

27. J.A. Provine and R.M. Rangayyan, "Effect of peanoscanning on image compression," in *Visual Information Processing II* (F. O. Huck and R. D. Juday, Eds.), pp. 152–159, Proc. SPIE 1961, 1993.

28. J. Quinqueton and M. Berthod, "A locally adaptive peano scanning algorithm," *IEEE Transactions on Pattern Analysis and Machine Intelligence*, PAMI-3, July 1981.

29. C.K. Chui, *An Introduction to Wavelets*. Orlando, FL: Academic Press, 1992.

30. C.K. Chui, Ed., *Wavelets: A Tutorial in Theory and Applications*. Orlando, FL: Academic Press, 1992.

31. S. Mallat, "Multifrequency channel decompositions of images and wavelet models," *IEEE Transactions on Acoustics, Speech and Signal Processing*, 37, 2091–2110, Decemeber 1993.

32. S. Mallat, "A theory for multiresolution signal decomposition: The wavelet representation," *IEEE Transactions on Pattern Analysis and Machine Intelligence*, 11, 674–693, July 1989.

33. J.J. Benedetto and M.W. Frazier, Eds., *Wavelets: Mathematics and Applications*. CRC Press, 1994.

34. G. Beylkin, "Wavelets and fast numerical algorithms," in *Different Perspectives on Wavelets* (I. Daubechies, Ed.), Providence, RI: American Mathematical Society, 1993.

35. I. Daubechies, *Ten Lectures on Wavelets*. Philadelphia, PA: Society for Industrial and Applied Mathematics, 1992.

36. G. Kaiser, *A Friendly Guide to Wavelets*. Boston, MA: Birkhäuser, 1994.

37. A.N. Akansu and R.A. Haddad, *Multiresolution Signal Decomposition*. Orlando, FL: Academic Press, 1992.

38. S. Mallat, "Multiresolution approximations and wavelets," *Transactions of the American Mathematical Society*, 315, 69–88, 1989.

39. P.P. Vaidyanathan, "Quadrature mirror filter banks, m-band extensions and perfect reconstruction techniques," *IEEE Signal Processing Magazine*, no. 7, 1987.

40. E.P. Simoncelli and E.H. Adelson, "Non-separable extensions of quadrature mirror filters to multiple dimensions," IEEE special issue on Multidimensional Signal Processing, 1990.

41. J.M. Shapiro, "Embedded image coding using zero trees of wavelet coefficients," *IEEE Transactions on Signal Processing*, 41, 3445–3462, 1993.

42. D.S. Taubman and M.W. Marcellin, *JPEG2000: Image Compression Fundamentals, Standards and Practice*. Norwell, MA: Kluwer Academic Publishers, 2002.

77

Security and Privacy

Peter G. Neumann
SRI International

77.1 Introduction 77-1
77.2 Conclusions 77-4
77.3 Recommendations 77-4

77.1 Introduction

This chapter provides an introduction to the concepts of security and privacy in computer-communication systems. Definitions tend to vary widely from one book to another, as well as from one system to another and from one application to another. The definitions used here are intuitively motivated and generally consistent with common usage, without trying to be overly precise.

Security is loosely considered as the avoidance of bad things. With respect to computers and communications, it encompasses many different attributes, and often connotes three primary attributes: confidentiality, integrity, and availability, with respect to various information entities such as data, programs, access control parameters, cryptographic keys, and computational resources (including processing and memory). Confidentiality implies that information cannot be read or otherwise acquired except by those to whom such access is authorized. Integrity implies that content (including software and data — including that of users, applications, and systems) cannot be altered except under properly authorized circumstances. Availability implies that resources are available when desired (for example, despite accidents and intentional denial-of-service attacks). All three attributes must typically be maintained in the presence of malicious users and accidental misuse, and ideally also under certain types of system failures. Which of these is most important depends on the specific application environments, as do the particular characteristics of each attribute.

Secondary attributes include authenticity, non-repudiability, and accountability, among others. Authenticity of a user, file, or other computational entity implies that the apparent identity of that entity is genuine. Non-repudiability implies that the authenticity is sufficiently trustworthy that later claims to its falsehood cannot be substantiated. Accountability implies that, whenever necessary, it is possible to determine what has transpired, in terms of who did what operations on what resources at what time, to whatever requisite level of detail, with respect to activities of users, systems, networks, etc., particularly in times of perceived or actual crises resulting from accidents or intentional misuse.

Trustworthiness is the notion that a given system, application, network, or other computational entity is likely to satisfy its desired requirements. The concept is essentially meaningless in the absence of well-defined requirements because there is no basis for evaluation. Furthermore, a distinction needs to be made between trustworthiness and trust. Something may be said to be trusted for any of several reasons; for example, it may justifiably be trusted because it is trustworthy; alternatively, it may have to be trusted because you have no alternative (you must depend on it), even if it is not trustworthy.

Authorization is the act of granting permission, typically based on authenticated identities and requested needs, with respect to a security policy that determines how authorizations may be granted. A fundamental distinction is made between policy and mechanism: policy implies what must occur or what must not occur, while mechanism determines how it is done. One of the fundamental challenges of computer-communication security is establishing the policy unambiguously before carrying out design and implementation, and then ensuring that the use of the mechanisms satisfies the required policies. A system cannot meaningfully be said to be secure unless it satisfies a well-defined security policy, both statically in terms of configured systems and dynamically in operation.

Security mechanisms can exist at many different hierarchical layers — such as hardware, operating system kernels, network software, database management, and application software. Each such layer may have its own security policy. In this way, it becomes clear (for example) that operating-system security depends on hardware and software mechanisms such as authentication and access control. Furthermore, cryptographic confidentiality and integrity of networked communications typically depend on the security of the underlying storage, processing, and network mechanisms, particularly with respect to protecting keys and exposed unencrypted information. By making careful distinctions among different hierarchical layers of abstraction and different entities in distributed environments, it is possible to associate the proper policies with the appropriate mechanisms, and to avoid circular dependencies.

Granularity is an important concept in security. It is possible to specify an access-control policy according to arbitrary granularities, with respect to individual bits, bytes, words, packets, pages, files, hyperfiles, subsystems, systems, collections of related system, and indeed entire networks of systems, as desired. Access permissions may themselves be subdivided into individually protecting operations such as read, write, append-only, execute, and perhaps the ability to execute without reading or to do a blind append without reading or overwriting. Each functional layer or subsystem may have protections according to its own operations and resources. For example, operations on files, database entities, electronic mail, and network resources such as Web pages all tend to be different. Digital commerce and network-mobile objects add further security requirements as well. Ideally, each type of object has its own set of protections defined according to the permissible operations on those objects.

Vulnerabilities are rampant in most computer-communication systems, whether stand-alone, accessible by dial-up maintenance paths, or connected directly to the Internet or private networks. Exploitations of those vulnerabilities are also rampant, including penetrations by outsiders, misuse by insiders, insertion of malicious Trojan horses and personal-computer viruses, financial fraud, and other forms of computer crime, as well as many accidental problems. When exploited, vulnerabilities can lead to a wide range of consequences — for example, resulting from a lack of confidentiality, integrity, availability, authentication, accountability, or other aspect of security, or a lack of reliability. Merely installing firewalls, virus checkers, and misuse-detection systems is nowhere near enough, especially if those systems are poorly configured — as often seems to be the case. (For numerous examples of vulnerabilities, threats, exploitations, and risks, see Neumann [1995]. Avoiding vulnerabilities is a very difficult matter because it depends on having sufficiently precise requirements, sufficiently flawless designs, sufficiently correct implementations, sensible system operations and administration, and aware users. All of those require strict adherence to high principles.

Various principles have evolved over the years, and are commonly observed — at least in principle. These include (1) separation of concerns according to types of system functions, user roles, and usage modes; (2) minimization of granted privileges; (3) abstraction; (4) modularity with encapsulation, as in hiding of implementation detail and minimization of adverse interactions; and (5) avoiding dependence on security by obscurity — that is, in essence, a belief that hiding your head in the sand makes you more secure. An example that encompasses (1) and (2) is to avoid using superuser privileges for nonprivileged operations. Overall, good software engineering practice (which includes (3) and (4)) can contribute considerably to the extent to which a system avoids certain characteristic flaws. Security by obscurity (5) is the generally invalid assumption that your antagonists know less than you do and always will; relying exclusively on that assumption is very dangerous, although security ultimately depends to some extent on staying ahead of your attackers. However, security is greatly increased when subsystem and network operations are well encapsulated; for example, if distributed implementations and remote operations can be hidden from the invoker.

TABLE 77.1   Generally Accepted System Security Principles (GASSP): Pervasive Principles

1. *Accountability:* information security accountability and responsibility should be explicit.
2. *Awareness:* principles, standards, conventions, mechanisms (PSCM) and threats should be known to those legitimately needing to know.
3. *Ethics:* information distribution and information security administration should respect rights and legitimate interests of others.
4. *Multidisciplinary:* PSCM should pervasively address technical, administrative, organizational, operational, commercial, educational, and legal concerns.
5. *Proportionality:* controls and costs should be commensurate with value and criticality of information, and with probability, frequency, and severity of direct and indirect harm or loss.
6. *Integration:* PSCM should be coordinated and integrated with each other and with organizational implementation of policies and procedures, creating coherent security.
7. *Timeliness:* actions should be timely and coordinated to prevent or respond to security breaches.
8. *Reassessment:* security should periodically be reassessed and upgraded accordingly.
9. *Democracy:* security should be weighed against relevant rights of users and other affected individuals.
10. *Competency:* information security professionals should be competent to fulfill their respective tasks.

Source: From GASSP, June 1997.

Some pervasive principles have been collected together as the emerging Generally Accepted System Security Principles [GASSP 1997], inspired by the National Research Council study, *Computers at Risk* [Clark et al. 1990], as summarized in Table 77.1.

One of the most important problems in ensuring security at every externally visible layer in a hierarchy and in all systems throughout a highly distributed environment is that of ensuring adequate authentication. Identities of all users, subsystems, servers, network nodes, and any other entities that might otherwise be spoofed or subverted should be authenticated with a level of certainty commensurate with the nature of the application, the potential untrustworthiness of the entity, and the risks of compromise — unless it can be demonstrated that no significant compromises can result. Fixed (reusable) passwords are extremely dangerous, particularly when they routinely traverse unencrypted local or global networks and can be intercepted. Similarly, although it is appealing to enhance ease of use, the single-sign-on concept is dangerous, except possibly within areas of common trust and trustworthiness. This is just one more example of the risks involved in trade-offs between security and ease of use. One-time tokens of some sort (e.g., cryptographically generated) are becoming highly desirable for authenticating users, systems, and in some cases even subsystems — particularly in distributed systems in which some of the entities are of unknown trustworthiness. Cryptographically based authentication is discussed in Chapter 79.

Privacy, and particularly electronic privacy, is an enormous problem area that is widely ignored. It is a socially motivated expectation that computer-communication systems will adequately enforce confidentiality against unauthorized people, and that authorized people will behave well enough. As is the case with trustworthiness, privacy is a meaningless concept in the absence of a well-defined policy defining the expectations that must be satisfied. Even then, many of the violations of typical privacy policies occur outside the confines of computers and communications, as results of actions of trusted insiders or untrustworthy outsiders.

Enforcement of privacy depends on system security, on adequate laws to discourage misuses that cannot otherwise be prevented, and on a society that is sufficiently orderly to follow the laws. Thus, privacy can be aided by computer-based controls, but has a substantial extrinsic component — it can be compromised externally by trusted people who are not trustworthy and by untrusted people who have been able to use whatever privileges they might have acquired or whatever security vulnerabilities they might have been able to exploit. Losses of privacy can have very serious consequences, although those consequences are not the subject of this chapter (see Chapter 2).

There are many techniques for enhancing security and reducing the risks associated with compromises of computer security and privacy. These techniques necessarily span a wide range, encompassing technological, administrative, and operational measures. Good system design and good software engineering practice can help considerably to increase security, but are by themselves inadequate. Considerable burden must also be placed in system administration. In addition, laws and implied threats of legal actions are necessary to discourage misuse and improper user behavior.

Inference and aggregation are problems arising particularly in distributed computer systems and database systems. Aggregation of diverse information items that individually are not sensitive can often lead to highly harmful results. Inferences can sometimes be drawn from just two pieces of information, even if they are seemingly unrelated. The absence of certain information can also provide information that cannot be gleaned directly from stored data, as can the unusual presence of an encrypted message. Such gleanings are referred to as exploitations of out-of-band information channels. Some channels are called covert channels, because they can leak information that cannot be derived explicitly — for example, as a result of the behavior of exception conditions (covert storage channels) or execution time (covert timing channels). Inferences can often be drawn from bits of gleaned information relating to improperly encapsulated implementations, such as exposed cryptographic keys. For example, various approaches such as differential power analysis, noise injection, and exposing specific spots on a chip to a flashbulb have all resulted in the extraction of secret keys from hardware devices.

Identity theft is increasingly becoming a serious problem, seriously exacerbated by the widespread availability of personal information and extremely bad policies of using that information not merely for identification, but also for authentication.

77.2 Conclusions

Attaining adequate security is a challenge in identifying and avoiding potential vulnerabilities and threats (see Neumann [1995]), and understanding the real risks that those vulnerabilities and threats entail. Security measures should be adopted whenever they protect against significant risks and their overall cost is commensurate with the expected losses. The field of risk management attempts to quantify risks. However, a word of warning is in order: if the techniques used to model risks are themselves flawed, serious danger can result — the risks of risk management may themselves be devastating. See Section 7.10 of Neumann [1995].

77.3 Recommendations

Security is a weak-link phenomenon. Weak links can often be exploited by insiders, if not by outsiders, and with very bad consequences. The challenge of designing and implementing meaningfully secure systems and networks is to minimize the presence of weak links whose accidental and intentional compromise can cause unpreventable risks. Considerable experience with past flaws and their exploitations, observance of principles, use of good software engineering methodologies, and extensive peer review are all highly desirable, but never by themselves sufficient to increase the security of the resulting systems and networks. Ultimately, security and privacy both depend ubiquitously on the competence, experience, and knowledge of many people — including those who establish requirements, design, implement, administer, and use computer-communication systems. Unfortunately, given the weak security that exists today, many risks arise from the same attributes of those who seek to subvert those systems and networks.

References

Amoroso, E. 1994. *Fundamentals of Computer Security Technology.* Prentice Hall, Englewood Cliffs, NJ.

Cheswick, W.R., Bellovin, S.M., and Rubin, A.D. 2003. *Firewalls and Internet Security.* Addison-Wesley, Reading, MA.

Clark, D.D., Boebert, W.E., Gerhart, S., et al. 1990. *Computers at Risk: Safe Computing in the Information Age.* Computer Science and Technology Board, National Research Council, National Academy Press, Washington, D.C.

Dam, K.W., Smith, W.Y., Bollinger, L., et al. 1996. *Cryptography's Role in Securing the Information Society.* Final Report of the National Research Council Crypto Study. National Academy Press, Washington, D.C.

Denning, P.J., Ed. 1990. *Computers Under Attack: Intruders, Worms, and Viruses*. ACM Press, New York and Addison-Wesley, Reading, MA.

Garfinkel, S. and Spafford, E. 1996. *Practical Unix Security*, 2nd ed. O'Reilly and Associates, Sebastopol, CA.

Gasser, M. 1988. *Building a Secure Computer System*. Van Nostrand Reinhold, New York.

Gasser, M., Goldstein, A., Kaufman, C., and Lampson, B. 1990. The digital distributed system security architecture. *Proc. 12th Nat. Comput. Security Conf.*

GASSP 1997. GASSP: Generally-Accepted System Security Principles, International Information Security Foundation, June 1997 (http://web.mit.edu/security/www/gassp1.html).

Hafner, K. and Markoff, J. 1991. *Cyberpunks*. Simon and Schuster, New York.

Hoffman, L.J., Ed. 1990. *Rogue Programs: Viruses, Worms, and Trojan Horses*. Van Nostrand Reinhold, New York.

Icove, D., Seger, K., and VonStorch, W. 1995. *Computer Crime*. O'Reilly.

Kocher, P. 1995. Cryptanalysis of Diffie-Hellman, RSA, DSS, and Other Systems Using Timing Attacks. Extended abstract, Dec. 7. 1995.

Landau, S., Kent, S., Brooks, C., Charney, S., Denning, D., Diffie, W., Lauck, A., Miller, D., Neumann, P., and Sobel, D. 1994. *Codes, Keys, and Conflicts: Issues in U.S. Crypto Policy*. ACM Press, New York. Summary available as Crypto Policy Perspectives. *Commun. ACM*, 37(8):115–121.

Morris, R. and Thompson, K. 1979. Password security: a case history. *Commun. ACM* 22(11):594–597.

Neumann, P.G. 1995. *Computer-Related Risks*. ACM Press, New York, and Addison–Wesley, Reading, MA.

Neumann, P.G. and Parker, D.B. 1990. A summary of computer misuse techniques. *Proc. 12th Nat. Comput. Security Conf.* Gaithersburg, MD. Oct. 10–13, 1989. National Institute of Standards and Technology.

Russell, D. and Gangemi, G.T. 1991. n.d. *Computer Security Basics*. O'Reilly and Associates, Sebastopol, CA.

Stoll, C. 1989. *The Cuckoo's Egg: Tracking a Spy through the Maze of Computer Espionage*. Doubleday, New York.

Thompson, K. 1984. Reflections on trusting trust. (1983 Turing Award Lecture) *Commun. ACM*, 27(8):761–763.

Further Information

In addition to the references, many useful papers on security and privacy issues in computing can be found in the following annual conference proceedings:

Proceedings of the IEEE Security and Privacy Symposia, Oakland, CA, each spring.

Proceedings of the SEI Conferences on Software Risk, Software Engineering Institute, Carnegie-Mellon University, Pittsburgh, PA.

[Peter G. Neumann is Principal Scientist in the Computer Science Laboratory at SRI International in Menlo Park, California, where he has been since 1971, after ten years at Bell Labs. His book, *Computer-Related Risks*, discusses security risks and other risks from a broad perspective, giving many examples.]

78

Malicious Software
and Hacking

78.1 Background..78-1
78.2 Culture of the Underground......................78-1
       Stereotypes • Hacker
78.3 Techniques and Countermeasures 78-3
       Malicious Software • Boot Sector Viruses • File Infector
       Viruses • Triggers and Payloads • Virus Techniques
       • Is the Threat of Viruses Real? • Protection Measures
       • Virus Construction Kits
78.4 The Future...78-17
       Computer Security • National Information Infrastructure

David Ferbrache
U.K. Ministry of Defence

Stuart Mort
*U.K. Defence and Evaluation
Research Agency*

78.1 Background

Since the advent of one of the first computer viruses on the IBM personal computer (PC) platform in 1986 the variety and complexity of malicious software has grown to encompass over 5000 viruses on IBM PC, Apple Macintosh, Commodore Amiga, Atari ST, and many other platforms. In addition to viruses a wide range of other disruptions such as Trojan horses, logic bombs, and e-mail bombs have been detected. In each case the software has been crafted with malicious intent ranging from system disruption to demonstration of the intelligence and creativity of the author.

The wide variety of malicious software is complemented by an extensive range of tools and methods designed to support unauthorized access to computer systems, misuse of telecommunications facilities and computer-based fraud. Behind this range of utilities lies a stratified and complex underculture: the computer underground. The underground embraces all age groups, motivations and nationalities, and its activities include software piracy, elite system hacking, pornographic bulletin boards, and virus exchange bulletin boards.

78.2 Culture of the Underground

An attempt to define the computer underground can produce a variety of descriptions from a number of sources. Many consider it a collection of friendless teenagers, who spend their time destroying people's data. To others, it is an elite society of computer gurus, whose expertise is an embarrassment to the legitimate bodies that continually try to extinguish their existence. However, the computer underground is really a collection of computer enthusiasts with as varied a collection of personalities as you would experience in any walk of life.

Not all members of the underground are computer anarchists; many use it as an environment in which to gather information and share ideas. However, many are in the following categories:

- *Hackers*, who try and break into computer systems for reasons such as gaining information or destroying data.
- *Malicious software writers*, who create software with a malicious intention. Viruses and Trojan horses are examples.
- *Phreakers*, who hack phones. This is done mainly to gain free phone calls, in order to support other activities such as hacking.

Some have described the inhabitants of the underground as information warriors; this is a too glamorous and inaccurate a term. It is true that many individuals' main cause is the freedom of information. These individuals may gain this information by breaking into a computer system, and extracting the stored information for distribution to any person who wants it. Many try and sell the information; these could be termed information brokers. Virus writers are certainly not information warriors, but may be information destroyers.

Thus, we have the person with the computer, surfing the net. An interesting site is stumbled across, with the electronic equivalent of a barbed-wire fence. Behind this fence there must be something interesting, otherwise, why so much security? The site is probed, in an attempt to challenge the security. Is this just a person's keen interest in the unknown, or is there a deeper malicious intent?

When security is breached, an assessment of the damage must be made. Was the availability of the system damaged? A virus could have destroyed vital files, crucial to the operation of the system. Has the integrity of data been compromised? An employee's salary could have been changed. Confidentiality lost? A company's new idea stolen. The cost of recovering from a security breach can be major: the time spent by an antivirus expert cleaning up machines after an infection, the time lost when employees could not work because their machines were inoperable. The cost mounts up. It is possible for a company dependent on its computer systems to go bankrupt after a security breach. It could also put peoples' lives at risk. The computer underground poses a significant threat to computer systems, of all descriptions, all over the world.

78.2.1 Stereotypes

The underground is a random collection of individuals, communicating over the Internet, bulletin boards, or occasionally face to face. Some individuals amalgamate to form a group. Groups sometimes compete with other groups, to prove they are the best. These competitions usually take the form of who can hack into the most computer systems. T-shirts even get printed to celebrate achievements.

A computer hacking group that did gain considerable recognition was the Legion of Doom (LoD). This group participated in a number of activities, including: obtaining money and property fraudulently from companies by altering computerized information, stealing computer source code from companies and individuals, altering routing information in computerized telephone switches to disrupt telecommunications, and theft or modification of information stored on individuals by credit bureaus. A member of LoD claims that curiosity was the biggest crime they ever committed!

Hacker groups do cause damage and disruption, wasting the resources of system administrators and law enforcement agencies worldwide. It has also been argued that hackers are responsible for closing security loopholes. Many hacker groups such as the Chaos Computer Club state that their members abide by a moral code.

78.2.2 Hacker

Simon Evans is a hypothetical example of a hacker with a broad level of computer expertise. Evans' history is one of persistent attempts to break into networks and systems; attempts which were often successful.

Evans' name first hit the press, when in 1980 a magazine wrote a cover story on an underground group. Evans had met the leader of this group a few years previously. Later in 1980, Evans and this hacker group broke into a computer system at U.S. Leasing. Not content with simple computer system breakins, Evans illegally entered an office of a telecom company in 1981 and stole documents and manuals. Following a tip off, Evans's home was searched and he was arrested, along with his accomplice and the leader of the hacker group. Evans was placed on one year's probation.

During his probation period, Evans managed to gain physical access to some university computers and started using them for hacking purposes. A computer crime unit pursued Evans and he was sentenced to six months in a juvenile prison for breaking probation. In 1984 Evans got a job, working for Great American Merchandising. From this company, he started making unauthorized credit checks, he was reported, and went into hiding.

In 1985, Evans came out of hiding and enrolled at a computer learning center. He fell for a girl, from whose address he hacked a system at Santa Cruz Operation. The call was traced, and Evans and his girlfriend were arrested. The girlfriend was released and Evans received 3 years probation, during which he married his girlfriend. In 1988, a friend of Evans started talking to the FBI, who subsequently arrested Evans for breaking into Digital Equipment Corporation's systems and stealing software. Evans got a year at a Californian prison, he and his wife then separated.

During Evans's probation in 1992 the FBI started probing again, and Evans went into hiding. In 1994 the Californian Department of Motor Vehicles issued a warrant for Evans' arrest. During the same year, Evans was accused of breaking into a security expert's system in San Diego, and stealing a large amount of information. He left a voice message made by a computer generated voice. Throughout the message he bragged about his expertise, and threatened to kill the security expert with the aid of his friends.

Evans made a mistake, he stored the data he stole from the computer expert on the Well, an on-line conferencing system. This information was spotted, and the security expert was alerted, who then subsequently monitored Evans' activities. Evans was tracked down, arrested and charged with 23 offenses, with a possibility of up to 20 years in prison for each offense.

Although the character described is fictional, the events are based on a real hacker's exploits.

78.3   Techniques and Countermeasures

78.3.1   Malicious Software

Malicious software is specifically written to perform actions that are not desired by the user of a computer. These actions could be passive, displaying a harmless message on the screen, or aggressive, reformatting a hard disk.

The programming abilities required to produce malicious software need not be at genius level. Little experience is required to use the toolkits that are currently available. A number of malicious software authors have taught themselves how to program. Some produce complex programs, which take time to analyze and demonstrate original programming concepts. Much malicious software, however, shows signs of bad programming, and does not execute correctly. Despite the varying quality, malicious software has found its way onto computers worldwide. Malicious software falls into a number of categories.

78.3.1.1   Trojan Horse

This software pretends to be something it is not. For example, on a disk operating system (DOS) machine, when we type DIR the contents of the current working directory is displayed. If, however, the contents of the directory were deleted, we would be witnessing a Trojan horse in action. It is a program that has the same name as a legitimate piece of software, but when executed, may perform an unexpected malicious act. This malicious act may not occur immediately, but on certain external conditions, for example, the user pressing ctrl-alt-delete (*logic bomb*) or the time being two minutes past midnight (*time bomb*).

78.3.1.2   Trojan Mule

When a computer is waiting to be logged into, a log-in screen is displayed. A user identification and a password usually needs to be entered in order to gain access to the system. If a piece of software is run that simulates the log-in screen, this would be a Trojan mule. A user would approach the computer, assume the screen was the genuine log-in screen, and enter their user identifier and password. The Trojan mule would record the data entered and terminate, usually informing the user that the log-in was incorrect. The effect of a Trojan mule is that users' passwords are captured by the person executing the Trojan mule.

78.3.1.3 Worm

A worm attacks computers that are connected by a network. A worm spreads by attacking a computer, then sending a copy of itself down the network looking for another machine to attack. An important difference exists between a worm and a virus (explained subsequently). A worm makes a copy of itself to spread, which is a standalone entity. A virus makes a copy of itself, but differs in that it needs to attach itself to a program, similar to a parasite attaching to a host. The most infamous example is the Internet worm which attacked computers connected to the Internet on November 2, 1988. It infected over 30% of Internet-connected computers and caused damage estimated at $10–$98 million.

78.3.1.4 E-Mail Bomb

The e-mail bomb is the electronic equivalent of a letter bomb. When the e-mail is read, an electronic bomb explodes. The result of the explosion may be degredation of system performance due to key system resources being used in the processing of the e-mail message; denial of service because the e-mail program does not filter out certain terminal control codes from e-mail messages, causing the terminal to hang; or something more serious due to the e-mail message containing embedded object code, which in turn contains malicious code (Trojan horse).

78.3.1.5 Malicious Scripts

These are constructed by the underground to aid an attack on a computer system. The script could take the form of a C program that takes advantage of a known vulnerability in an operating system. It could also be a simplification of a complex command sequence.

78.3.1.6 Viruses

Viruses have existed for some time and can cause a variety of annoyances to the user. They can produce amusing messages on a user's screen, delete files, and even corrupt the hard disk so that it needs refor-matting. Whatever its actions, the virus interferes with the correct operation of the computer without the authorization of the owner.

Many have compared computer viruses to human viruses. Thus the virus writer becomes the equivalent of an enemy waging germ warfare. The most vulnerable computer to virus infection at the moment is the PC running MS-DOS. Viruses do exist that can infect Macintosh, and other types of machines using different operating systems, such as OS/2. Viruses that infect Unix machines are in existence; most are laboratory viruses but there are new reports of one being in the wild, i.e., existing on innocent users machines that have not deliberately installed the virus.

In order to distinguish one virus from another, they are given names by the antivirus industry. Naming conventions vary considerably between antivirus software vendors. A virus author may include a text string in the virus which gives an obvious name, however unprintable. The classic definition of a virus is as follows.

> A virus is a self replicating program that can *infect* other programs, either by modifying them directly or by modifying the environment in which they operate. When an infected file is executed, this will cause virus code within the program to be run.

78.3.2 Boot Sector Viruses

A common form of PC virus is the boot sector virus. When a PC is booted a number of steps are followed. First, the power on self-test (POST) is executed, which tests the integrity of system memory and then initializes the hardware. Information stored in nonvolatile memory is collected, and finally POST sets up the basic input output system (BIOS) address in the interrupt table.

The A: drive is then checked, to see if a disk is present in the drive. This can be seen and heard when the A: drive's motor is started and the light flashes. If a disk is present in the drive, the first sector is read into memory and executed. If no disk is found, then the first sector of the hard disk is read. This sector is known as the master boot sector (MBS). The MBS searches for a pointer to the DOS boot sector (DBS), which is loaded into memory, and control is passed to it.

At this point an opportunity exists for virus infection. A boot sector virus can infect the MBS or the DBS of a hard disk, or the boot sector of the floppy disk. Consider a virus on a floppy first. A floppy with a virus resident on its boot sector is inserted into the A: drive (the original boot sector of the floppy is usually stored elsewhere on the floppy). The machine is booted, and the virus in the boot sector is loaded into memory and executed. The virus searches out the MBS or DBS, depending on the virus' plan, and copies itself to that sector. As with a floppy, the virus usually stores the original MBS or DBS elsewhere on the disk. When the virus has completed execution, it can load the original boot sector and pass control to it, making the actions of the virus invisible to the user. It is important to note that all DOS formatted floppies have a boot sector, even if the floppy is not a system disk.

If the virus infected the MBS of the hard disk (similarly, when the DBS is infected), how does the virus work? The computer is booted from the hard disk, i.e., there's no floppy in the A: drive. The virus code in the MBS is loaded into memory and executed. The virus loads any other sectors that it needs to execute, then loads the original boot sector into memory. The virus is active in memory and can now monitor any floppy disk read/write activity. When an uninfected floppy is detected, it can infect its boot sector. This allows the virus to spread from disk to disk and thus computer to computer.

78.3.3 File Infector Viruses

A file infector virus is basically a program that when executed seeks out another program to infect. When the virus finds a suitable program (the host) it attaches a copy of itself and may alter the host in some way. These alterations ensure that when the host is executed, the attached virus will also be executed. The virus can then seek out another host to infect, and so the process continues. The virus may attach itself to a host program in a number of ways, the most common types are the following:

Overwriting: the virus places its code over the host, thus destroying the host (Figure 78.1). When the virus has finished executing, control is returned to the operating system.

Appending: the virus places its code at the end of the host (Figure 78.2). When the host is executed, a jump instruction is usually executed which passes control to the virus. This jump instruction is placed at the start of the host by the virus, the original instructions that were at the start are stored in the body of the virus. During the virus's execution, it replaces the host's original start instructions, and on completion it passes control to these instructions. This process makes the virus invisible to the user until it triggers.

Prepending: the virus places its code at the start of the host (Figure 78.3). When the host is executed, the virus is executed first, followed by the host.

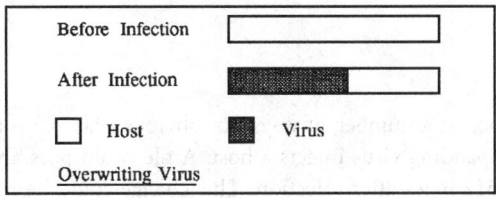

FIGURE 78.1   Overwriting virus.

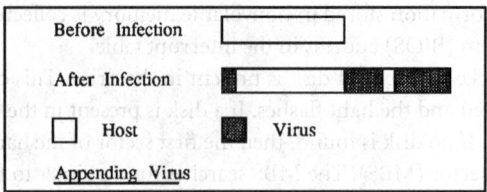

FIGURE 78.2   Appending virus.

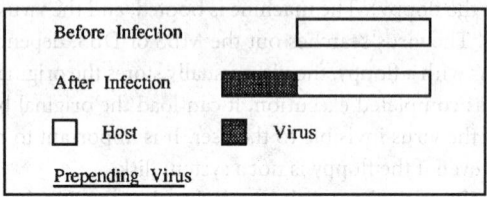

FIGURE 78.3   Prepending virus.

78.3.4   Triggers and Payloads

A trigger is the condition that must be met in order for a virus to release its payload, which is the malicious part of the virus. Some viruses simply display a message on the screen, others slow the operation of the computer, the nastier ones delete or corrupt files or reformat the hard disk. The trigger conditions are also only limited by the writer's imagination. It may be that a certain date causes the virus to trigger, a popular day is Friday 13th, or it may be a certain key sequence, such as control-alt-delete.

78.3.5   Virus Techniques

Viruses writers go to great lengths to hide the existence of their viruses. The longer a virus remains hidden, the further its potential spread. Once it is discovered, the virus' trail of infection comes to an end. Common concealment techniques include:

78.3.5.1   Polymorphism

Polymorphism is a progression from encryption (Figure 78.4). Virus writers started encrypting their viruses, so that when they were analyzed they appeared to be a collection of random bytes, rather than program instructions. Antivirus software was written that could decrypt and analyze these encrypted viruses. To combat this the writers developed polymorphic viruses.

   Polymorphism is the virus' attempt at making itself unrecognizable. It does this by encrypting itself differently every time it infects a new host. The virus can use a different encryption algorithm, as well as a different encryption key when it infects a new host. The virus can now encrypt itself in thousands of different ways.

78.3.5.2   Stealth

Viruses reveal their existence in a number of ways. An obvious example is an increase in the file size, when an appending or prepending virus infects a host. A file could possibly increase from 1024 bytes long before infection to 1512 bytes after infection. This change could be revealed during a DOS DIR command.

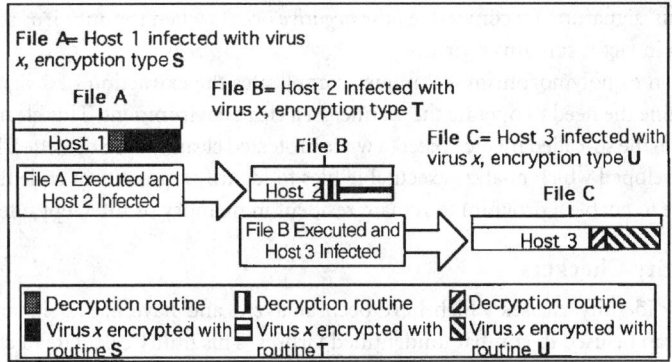

FIGURE 78.4 Polymorphic infection.

To combat this symptom of the virus' existence, the idea of stealth was created. As was mentioned earlier, the longer a virus remains hidden, the farther it spreads. Stealth can be described as a virus' attempt to hide its existence and activities from system services and/or virus detection software.

A virus, for example, to avoid advertising the increase in file size, would intercept the relevant system call and replace it with its own code. This code would take the file size of an infected file, subtract from it the size of the virus, and return the result, the original file size.

78.3.6 Is the Threat of Viruses Real?

Viruses are being written and released every day, in ever increasing numbers. Anyone with access to the Internet can download a virus, even the source code of the virus. These viruses can be run and can spread rapidly between machines. There are widely available electronic magazines such as *40-Hex* that deal with virus writing. They cover new techniques being developed, virus source code, and countermeasures to commercial antivirus software. The existence of magazines, books, and compact disk read-only memory (CD-ROM) information on viruses makes the task of virus construction considerably easier.

If someone has a knowledge of DOS and an understanding of assembly language then that person can write a virus. If someone can boot a PC, and run a file, then that person can create a virus using a toolkit. The costs to recover from a virus incident have been estimated as being as low as $17 and as high as $30,000.

78.3.7 Protection Measures

How can we stop a virus infecting a computer, and if infected, how can we get rid of it before it does any damage? Since prevention is better than cure, a wide range of antivirus software of varying effectiveness is available, commercially and as shareware. When the software has been purchased, follow the instructions. This usually involves checking the machine for viruses first, before installing the software. Antivirus software normally consists of one or more of the following utilities.

78.3.7.1 Scanner

Every virus (or file for that matter) is constructed from a number of bytes. A unique sequence of these bytes can be selected, which can be used to identify the virus. This sequence is known as the virus' signature. Therefore, any file containing these bytes may be infected with that virus. A scanner simply searches through files looking for this signature.

A scanner is the most common type of antivirus software in use, and is very effective. Unfortunately scanners occasionally produce *false positives*. That is, the antivirus product identifies a file as containing a virus, whereas in reality it is clean. This can occur by a legitimate file containing an identical sequence of bytes to the virus' signature. By contrast, a *false negative* occurs when the antivirus software identifies a file as clean, when in fact it contains a virus.

The introduction of polymorphism techniques complicates the extraction of a signature, and stealth techniques underline the need to operate the scanner in a clean environment. This clean environment is a system booted from the so-called magic object (a write protected clean system diskette). Heuristic scanners have also been developed which analyze executable files to identify segments of code that are typical of a virus, such as code to enable a program to remain resident in memory or intercept interrupt vectors.

78.3.7.2 Integrity Checkers

Scanners can only identify viruses which have been analyzed and have had a signature extracted. An integrity checker can be used to combat unidentified viruses. This utility calculates a checksum for every file that the user chooses, and stores these checksums in a file. At frequent intervals, the integrity checker is run again on the selected files, and checksums are recalculated. These recalculated values can be compared with the values stored in the file. If any checksums differ then it may be a sign that a virus has infected that file. This may not be the case of course, because some programs legitimately alter files during the course of their execution, and this would result in a different checksum being calculated.

78.3.7.3 Behavior Blocker

This utility remains in memory while the computer is active. Its task is to alert the user to any suspicious activity. An example would be a program writing to a file. The drawback of this is that user intervention is required to confirm an action to be taken, which can be an annoyance that many prefer to live without.

Fortunately, as viruses increase, so do the number of people taking precautions. With antivirus precautions in place the chance of virus infection can be kept to a minimum.

78.3.8 Virus Construction Kits

These kits allow anyone to create a computer virus. There are a number of types available, offering different functionality. Some use a pull down menu interface (such as the Virus Creation Laboratory), others (such as PS-MPC) use a text configuration file to contain a description of the required virus. Using these tools, anyone can create a variety of viruses in a minimal amount of time.

78.3.8.1 Hacking

Hacking is the unauthorized access to a computer system. Computer is defined in the broadest sense, and a fine line exists between hacking and telephone phreaking [unauthorized access to telephone switch, private automated branch exchange (PABX) or voice mail]. Routers, bridges, and other network support systems also increasingly use sophisticated computer bases, and are thus open to deliberate attack.

This section provides a hacker's eye view of a target system, indicating the types of probes and data gathering typically undertaken, the forms of penetration attack mounted, and the means of concealing such attacks. An understanding of these techniques is key to the placement of effective countermeasures and auditing mechanisms.

78.3.8.2 Anatomy of a Hack

An attack can be divided into five broad stages:

1. *Intelligence:* initial information gathering on the target system from bulletin board information swaps, technical journals, and social engineering aimed at extracting key information from current or previous employees. Information collection also includes searching through discarded information (dumpster diving) or physical access to premises.
2. *Reconnaissance:* using a variety of initial probes and tests to check for target accessibility, security measures, and state of maintenance and upgrade.

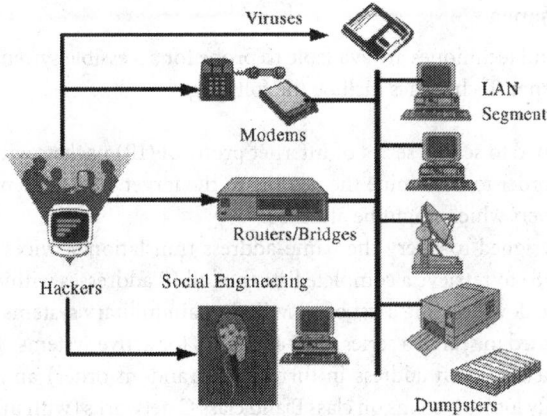

FIGURE 78.5 Possible attacks against IT systems.

3. *Penetration:* attacks to exploit known weaknesses or bugs in trusted utilities, the misconfiguration of systems, or the complete absence of security functionality.
4. *Camouflage:* modification of key system audit and accounting information to conceal access to the system, replacement of key system monitoring utilities.
5. *Advance:* subsequent penetration of interconnected systems or networks from the compromised system.

A typical hacking incident will contain all of these key elements. The view seen by a hacker attacking a target computer system is illustrated in Figure 78.5. There are many access routes which could be used.

78.3.8.3 Intelligence Gathering

A considerable amount of information is available on most commercial systems from a mix of public and semiopen sources. Examples range from monitoring posts on Usenet news for names, addresses, product information and technical jargon; probing databases held by centers such as the Internet Network Information Center (NIC); to the review of technical journals and professional papers. Information can be exchanged via hacker bulletin boards, shared by drop areas in anonymous FTP servers, or discussed on line in forums such as Internet Relay Chat (IRC).

Probably the most effective information gathering technique is known as social engineering. This basically consists of masquerade and impersonation to gain information or to trick the target into showing a chink in its security armor. Social engineering ranges from the shared drink in the local bar, to a phone call pretending to be the maintenance engineer, the boss, the security officer, or the baffled secretary who can't operate the system.

Techniques even include a brief spell as a casual employee in the target company. It is often surprising how much temporary staff, even cleaners, tend to be trusted. Physical access attacks include masquerading as legitimate employees (from another branch, perhaps) or maintenance engineers, to covert access using lockpicking techniques and tools (also available from bulletin boards).

Even if physical access to the interior of the building is impossible, access to discarded rubbish may be possible. So-called dumpster diving is a key part of a deliberate attack. Companies often discard key information including old system manuals, printouts with passwords/user codes, organization charts and telephone directories, company newsletters, etc. All this material lends credence to a social engineering attack, and may provide key information on system configuration which helps to identify exploitable vulnerabilities.

78.3.8.4   Reconnaissance

A wide variety of tools and techniques are available to probe for accessible systems on wide area networks (WAN) such as the Internet. Techniques include the follwing:

1. *Traceroute:* designed to send a series of Internet protocol (IP) packets with increasing time-to-live (TTL) values in order to determine the routing to the target, and to identify intermediate routers and Internet carriers which might be attacked.
2. *DNS dig tools:* designed to query the name-address translation services on the Internet domain name server (DNS) to retrieve a complete listing of all IP addresses within a specified domain. An example might be downloading a list of all MIL domain military systems.
3. *IP scanners:* designed to search a series of IP addresses for active systems. These operate by sending ICMP echo packets to each address in turn (or in random order) and awaiting a reply. These utilities can rapidly locate systems on class B and class C networks (with up to 65,534 and 254 hosts, respectively).
4. *Port scanners:* designed to search a specific system for transmission control protocol (TCP) and user datagram protocol (UDP) ports offering services. The port scanner will attempt to connect to each port in turn, verify whether the connection is accepted, and note which service is being offered. This can also include noting the version string which is sent by utilities such as Telnet (remote log in) and Sendmail [simple mail transfer protocol (SMTP) electronic mail] to check on possible vulnerabilities.
5. *RPCINFO:* designed to probe the portmapper on the remote system, which handles registration of remote procedure call (RPC) services. This allows the hacker to identify which RPC based services are being offered [such as the network filesystem (NFS) or network information system (NIS)].
6. *MOUNT:* designed to display the list of exported filesystems and associated security attributes, allowing the hacker to decide on the best target for NFS attacks.
7. *FINGER:* to check which users are active on the system (ruser is possible substitute), and to decide on busiest time and periods when no system administrator is logged on.

These probes allow an attacker to locate systems, identify which services are offered, gain some idea of the system usage patterns, and decide on an attack strategy.

78.3.8.5   Penetration

Once the hacker has gathered this key information, then exploitation of security weaknesses begins. If obvious configuration errors show up (such as a world exported NFS filesystem) or access via the anonymous file transfer protocol (FTP) or trivial file transfer protocol (TFTP) to the full filesystem, then this penetration is rapid. Otherwise the hacker has four courses of action:

1. Try to guess user code and passwords. Common default accounts such as ENGINEER, BIN, SYS, MAINT, GUEST, DIAG, ROOT, FIELD may have weak or default passwords. The hacker may make use of services such as FTP or rexec which do not log failed log-in attempts. Utilities such as *fbomb* use seed lists of common passwords to mount the attack.
2. Try to exploit vulnerabilities in network services such as Sendmail to gain access to the system. Key to the attack is a wide range of attack scripts available on the boards for swapping and trading, together with active participation in full disclosure security discussion lists such as INFOHAX and BUGTRAQ, which openly reveal the details of security holes.
3. Try to exploit weaknesses in the network protocols themselves such as the IP spoofing attack discussed subsequently.
4. Try to break into the network provider's system or network in order to capture user codes and passwords passed in clear across the provider's network. Once captured they can be used to break into the target system; failing which someone on the boards may be able to trade an account for other information.

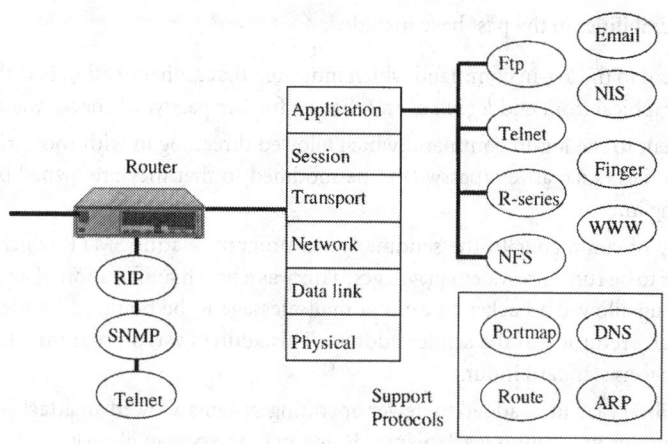

FIGURE 78.6   Network view of system services.

The proliferation of network services offered by systems, and the increasing intelligence of routers can assist the attacker. Figure 78.6 illustrates the range of services offered to the network by a typical Unix system. While many protocols are supported, each must be adequately secured to prevent system penetration.

78.3.8.6   Vulnerabilities and Exploitation

The increasing complexity and dynamicity of modern software is one of the key sources of software vulnerabilities. As an example, the Unix operating system now consists of over 2.8 million lines of C code, an estimated 67% of which execute with full privilege. With this size of code base there is a high likelihood of code errors which can open a window for remote exploitation via a network, or local exploitation by a user with an unprivileged account on the local system.

The main source of vulnerabilities are the assumptions made by system programmers about the operating environment of the software, these include:

- *Race conditions:* in which system software competes for access to a shared object. Unix-based operating systems do not support atomic transactions, and as such operations can be interrupted allowing malicious tampering with system resources. Race conditions are responsible for vulnerabilities in utilities such as expreserve, passwd, and mail. They have been widely exploited by the group known as the 8-legged groove machine (8lgm).

- *Buffer overruns:* allowing data inputs to overflow storage buffers and overwrite key memory areas. This form of attack was used by the Internet worm of November 1988 to overrun a buffer in the fingerd daemon causing the return stack frame to be corrupted, leading to a root privileged shell being run. Similar forms of attack were used against World Wide Web (WWW) servers.

- *Security responsibilities:* in which one component of a security system assumes the other is responsible for implementing access control or authentication, an example being the Berkeley Unix line printer daemon, which assumed that the front end utility (lpr) carried out security checks. This allowed an attacker to connect directly to the line printer and bypass security checks.

- *Argument checking:* utilities which do not fully validate the arguments they are invoked with, allowing illegal access to trusted files. An example was the failure of the TFTP daemon in AIX to check for inclusion of "…" components in a filename. This allowed access outside a secure directory area.

- *Privilege bracketing:* privileged utilities which fail to correctly contain their privilege, and in particular allow users to run commands in a privileged subshell via shell escapes.

Examples of vulnerabilities in the past have included:

- An argument to the log-in command which indicated that authentication had already been carried out by a graphical front end *logintool* and that no further password checks were needed.
- An argument to the log-in command which allowed direct log in with root privileges, or allowed key system files (such as /etc/passwd) to be modified so that they are owned by the unprivileged user logging in.
- A sequence of commands to the sendmail mail program via the SMTP which allowed arbitrary commands to be run with system privileges. This was a new manifestation of an old bug from 1988. The new bug allowed a hacker to cause a mail message to be bounced by the target system and automatically returned to the sender address. This address was a program which would take the mail message as standard input.
- Dynamic library facilities added to newer operating systems allowed an attacker to trick privileged setuid programs into running a Trojanized version of the system library.
- Bugs in the FTP server which allowed a user to begin anonymous (unprivileged) log in, overwrite the buffer containing the user information with the information for a privileged account, and then complete the log-in process. Since the server still believed this was an anonymous log in, no password was requested.

Vulnerabilities often manifest in other forms and on other operating systems. An example is the expreserve bug in the editor recovery software, which was first fixed by Berkeley, then fixed by Sun Microsystems, and finally in a slightly modified form by AT&T.

Penetration scripts circulate widely in the security and hacker community. These scripts effectively deskill hacking, allowing even novices to attack and compromise operating system security. The hacker threat is dynamic and rapidly evolving, new vulnerabilities are discovered, vendors promulgate patches, system administrators upgrade, hackers try again. A single operating system release had 1200 fielded patches, 35 of which were security critical. It is difficult, if not impossible, for system administrators to track patch releases and maintain a secure up-to-date operating system configuration. On a network as large and diverse as the Internet (6.5 million systems, 30 million users in 1995), hackers will always find a vulnerable target which is not correctly configured or upgraded.

78.3.8.7   Automated Penetration Tools

The task of verifying the security of a system configuration is complex. Security problems originate from insecure manufacturer default configurations; configurations drift as administrators upgrade and maintain the system; vulnerabilities and bugs in software. Tools have been developed to assist in this task, allowing the following:

- Checking of filesystem security settings and configuration files for obvious errors: these include the Computer Oracle & Password System (COPS), Tiger, and Security Profile Inspector (SPI).
- Verifying that operating system utilities are up to date and correctly patched: this is a function included in Tiger and in SPI which use checksum checks on utility files to verify patch installation, and in SATAN and the Internet Security Scanner (ISS) which directly probe for vulnerabilities in network servers.
- Monitoring system configuration to check for malicious alteration of key utilities: this is based on the use of integrity checkers generating checksums of software such as Tripwire.

These tools are extremely powerful, but represent a two-edged sword. Remote probe utilities such as SATAN can be used directly by hackers to scan remote systems for known vulnerabilities. The use of automated attack tools is allowing hackers to screen hundreds of systems within seconds. Initial probes were based on TFTP attack, but now cover a much wider range of protocols including NFS. The NFSBUG security tool allows rapid checking and exploitation of vulnerabilities including mounting of world accessible partitions, forgery of credentials, and exploitation of implementation errors.

78.3.8.8 Camouflage

After system penetration, hackers will attempt to conceal their presence on the system by altering system log files, audit trails, and key commands. Initial tools such as *cloak* work by altering log files such as *utmp* and *wtmp*, which record system log ins, and altering the system accounting trail.

Second-generation concealment techniques also modified key system utilities such as *ps*, *ls*, *netstat*, and *who* which report information on system state to the administrator. An example is the *rootkit* set of utilities which provides C source code replacements for many key utilities. Once rootkit is installed, the hacker becomes effectively invisible.

Hidden file techniques can also be used to conceal information (using the hidden attribute in DOS, the invisible attribute on the Macintosh, or the " . " prefix on Unix), together with more sophisticated concealment techniques based on steganographic methods such as concealing data in bit-mapped images.

78.3.8.9 Advance

Once hackers are active on a system there are a number of additional attacks open to them, including the following:

- Exploiting system configuration errors and vulnerabilities to gain additional privilege, such as breaking into the bin, sys, daemon, and root accounts
- Using Trojanized utilities such as telnet to record passwords used to access remote systems
- Implanting trapdoors, which allow easy access to hackers with knowledge of a special password
- Monitoring traffic on the local area network to gather passwords and access credentials passed in clear
- Exhaustive password file dictionary attacks

These attacks aim at breaking into systems on the same network or on networks accessed by users on the local system. Typically, a small network sniffer (such as sunsniff) would be run on the compromised system; this would monitor all TCP-based connections collecting the first 128 bytes of traffic on each virtual circuit in the hope of collecting passwords and user codes. These sniffers log information in hidden files for later recovery.

Vulnerabilities and system configuration errors may also allow the compromise of a privileged account, and subsequent alteration of key utilities such as telnet or FTP may demand passwords from users. These modified utilities can store information for later retrieval. The system log-in utility can be modified to add a simple trapdoor allowing privileged access even if the system administrator has revoked the compromised account originally used for access.

Systems within the same organization often trust each other through the use of explicit trust relationships. A typical example is the r-series protocols from Berkeley which allow a system administrator or user to specify a list of systems which are trusted to log in to corresponding user accounts on the local system without specifying a password. This is very powerful for simplifying administration but very dangerous once a system in the organization is penetrated. The growth of these trust relations over time has been likened to the growth of ivy on old oak trees.

Finally, older systems are susceptible to password cracking style attacks. In these attacks the hacker will take advantage of the one-way encryption algorithm used in Unix to hash plain text passwords to cipher text. This is done by exhaustively encrypting each word in a standard dictionary and then comparing the encrypted version against the stored password cipher text. If the two versions match, then the word can be used as the password for that account. Sophisticated tools are available for both the hacker (to attack accounts) and the system administrator (to proactively check for weak passwords). An example is the crack tool which allows for rule-based transformation of dictionary words to mimic common substitutions, such as replacing I by 1 or O by 0.

78.3.8.10   Countermeasures

Four major categories of countermeasure are available to counter the hacker threat, they are as follows:

- Firewalls: designed to provide security barriers to prevent unauthorized access to internal systems within an organization from an untrusted network.
- Audit and intrusion detection: designed to provide effective on-line monitoring of systems for unauthorized access.
- Configuration management: to ensure that systems are correctly configured and maintained.
- Community action: to jointly monitor hacker activities and take appropriate action.

78.3.8.10.1   *Firewall/Gateway Systems*

Firewall systems aim to defend internal systems against unauthorized access from an untrusted network. The firewall consists of two main components:

1. Screening or choke router is designed to prevent external access to an internal system other than the bastion host, in addition to restricting incoming protocols to a safer subset such as telnet, FTP and SMTP.
2. Bastion host or firewall authenticates and validates each external access attempt based on a predefined access control list. The bastion host also acts as a proxy by connecting legitimate external users to services on internal systems. Finally, the bastion host will also conceal the structure of the internal network by rewriting outgoing mail addresses and force all outgoing internal connections to be routed through the firewall proxy.

Firewalls are now being evaluated and certified under both the U.S. Trusted Computer Security Evaluation Criteria (TCSEC) and U.K. IT security evaluation criteria (ITSEC). A firewall is an effective barrier defense against external penetration, and if deployed internally within organizations it can also provide protection against insider threats. To be effective a firewall has to be supplemented by strong cryptographic authentication mechanisms based on challenge-response style mechanisms or one-time pads. Without such mechanisms user accounts and passwords will still be carrier unencrypted over Internet provider networks.

78.3.8.10.2   *Audit and Intrusion Detection Mechanisms*

Initial barrier defenses can be supplemented by effective monitoring of networks and systems. Traffic flow analysis can provide an indication of which systems are accessing the organization. Firewall products offer extensive logging of connection attempts, both failed and successful. Intrusion detection systems are being developed which permit user activity profiles to be constructed and allow deviation from established profiles of behavior to be flagged. Examples include the NIDES tool developed at SRI, and the Haystack tool from the U.S. Department of Energy. Key problems with intrusion detection are identifying effective behavior metrics and the processing of heterogeneous audit trails in a wide variety of vendor formats.

78.3.8.10.3   *Configuration Management*

The security of end systems can be improved through strict configuration management, and the use of a variety of proactive configuration checking tools such as COPS, Tiger, and SPI. These tools provide a means for rapidly checking filesystem security settings, verifying patch states, and correcting for the more obvious security blunders. Vendor specific tools such as Asset on the Sun platform are also available. The use of these tools should complement a policy that unnecessary functionality should be disabled on system installation.

78.3.8.10.4   *Community Action*

One of the most effective counters to hacker activity is community action. This takes two main forms: first, collectively reacting to security incidents; second, collectively investigating computer crime. Since the Internet worm of 1988, a series of incident response teams have been set up to deal with network intrusions. An example (and the first of its kind) was the Computer Emergency Response Team (CERT) set up by Carnegie-Mellon University. The CERT provides support to the Internet constituency, providing the following:

- Point of contact for system administrators believing that their system has been compromised
- Means of disseminating security advice and alerts
- Lobbying body for pressuring vendors to fix security problems rapidly
- Central repository of knowledge on the nature of the hacker threat

CERT produces an extensive range of security alerts and advisories giving information on security problems, available fixes, and vendor contact points. The incident response teams work together in a forum known as the Forum for Incident Response and Security Teams (FIRST) to exchange information and techniques. The FIRST group provides a worldwide community of concerned security professionals. FIRST maintains a WWW archive site at www.first.org, which carries advisories and security advice.

The second form of community action is cooperation among law enforcement agencies to investigate computer crime. Most countries in the world now have some form of computer abuse legislation, an example being the Computer Misuse Act in the U.K. The form of legislation varies considerably in its coverage, definition of abuse, extraterritorial extent, and its available penalties. For example, the U.K. act defines three categories of offense:

1. Unauthorized access to computer systems, carrying up to 6 months in prison or a fine
2. Unauthorized access to computer systems with intent to facilitate other criminal acts, such as collecting information from a computer to perpetrate a fraud, carrying up to 5 years in prison
3. Unauthorized modification of data held on a computer system, again with a maximum 5-year penalty

The U.K. act has extraterritorial scope in that a crime is committed if a U.K. system is penetrated or if the penetration attempt originates from the U.K. Clifford Stoll's experiences of attempting to investigate and track a hacker (related in *The Cuckoo's Egg*) clearly indicated the problem of persuading law enforcement agencies in the U.S., Canada, and Europe to work together to track an intruder. In the modern computer world, an attacker can go indirectly through Columbia, China, Nicaragua, and Brazil, making tracking and enforcement a world problem.

Interpol has established a computer crime working group which is attempting to build links between European police forces (including the former Soviet Union), in order to facilitate investigation, share experiences, and unify computer crime legislation.

78.3.8.11 Phreaking

In contrast to hacking, phreaking is the penetration and misuse of telephone systems. Since the replacement of older crossbar and Strowger switches by digital exchanges, this has become a very nebulous distinction.

Phreaking was born out of the early days of analog switches in which exchanges carried signaling information (such as the number dialed by a subscriber) in-band as part of the voice channel. Signaling information was passed by a series of dual-tone multifrequency (DTMF) codes. The U.S. telephone carriers making use of an international standard Consutative Committee of International Telephony and Telegraphy-5 (CCITT-5), which uses combinations of 700-, 900-, 1100-, 1300-, and 1500-Hz tones to signal line state and number dialed. There was little to prevent a phreaker from generating comparable tones via a PC sound card or a simple oscillator (or box).

In particular, U.S. carriers used a single 2600-Hz tone to signal to trunk equipment that a local call had finished, and that the trunk equipment should be placed in idle mode awaiting the next call. A hacker could inject a 2600-Hz tone, reset the trunk line, and then dial the number required in CCITT-5 format. This technique also allowed the phreaker to evade call billing, and thus proved quite popular with hackers needing to access remote computer systems or bulletin boards.

78.3.8.11.1 Colored Boxing

The history of phone phreaking revolved around 2600 Hz. Early stories include a famous phreaker named Captain Crunch, who discovered that a whistle in a breakfast cereal packet generated exactly the correct frequency, to a blind phone phreaker whose perfect pitch allowed him to whistle up the 2600-Hz carrier. A whole spectrum of boxes were built (with plans swapped on the boards) to generate tone sets such as CCITT-5 (a blue box) or the 2200-Hz pulses generated by U.S. coin boxes to signal insertion of coins (a red box) or a combination of both (a silver box). Instructions range from building lineman's handsets (to tap local loops), to ways of avoiding billing by simulating an on-hook condition while making calls, and to ways of disrupting telephone service by current pulses injected on telephone lines.

The move from in-band signaling to common carrier signaling is reducing the risk of boxing attacks. Common carrier signaling carries signaling information for a cluster of calls on a separate digital circuit rather than in-band where it may be susceptible to attack. Older switching equipment in developing countries, and in specialist networks (such as 1-800) are still being targeted.

78.3.8.11.2 War Dialers

The worlds of hacking and phone phreaking come together in the war dialer. A war dialer is a device designed to exhaustively scan a range of numbers on a specific telephone exchange looking for modems, fax machines, and interesting line characteristics. War dialers were also capable of randomizing the list of numbers to be searched (to avoid automated detection mechanisms), to automatically log line states, and to automatically capture the log-in screen presented by a remote computer system. While U.S. telecomm charging policies often meant that local calls were free (encouraging a bulletin board culture), to use long distance meant the use of phone phreaking to avoid call billing.

War dialers such as Toneloc and Phonetag offered an effective way of screening over 2000 lines per night. Boards regularly carried the detailed results of these scans for each area code in the U.S.

78.3.8.11.3 Modems

A war dial scan was likely to detect many modems with various levels of security, ranging from unprotected to challenge–response authentication. A popular defensive technique was the use of a dial-back facility, in which the user was identified to the modem, which would then ring the user back on a predefined number. If the modem used the same incoming line to initiate dial-back, a hacker could generate a simulated dial tone to trick the modem into believing that the line had been dropped and that dial-back could begin. The identification of modems also led to a range of other problems:

- Publicly accessible network gateways, which allowed an authorized user to access WAN functionality by dialing in: This led to system intrusion over the Internet which could not be traced back beyond the public dial in.
- Diagnostic modems for computer systems, for PABXs, and for PTT trunk switches: This opened the door to direct attacks on the digital switches with weak password security.

The details of such switches are widely exchanged in the underground (particularly the Unix-based 5-ESS switch from AT&T) with considerable knowledge of the methods for reconfiguration of switches to change quality of service on lines, or to compromise subscriber information. Subscriber information such as the reverse mapping between subscriber number and name/address has been openly sold by phreakers.

78.3.8.11.4 Phone Loops

Phone loops refer to a linked pair of subscriber circuits used for testing purposes. Callers dialing both circuits will be automatically interconnected. Numbers for phone loops were widely swapped among phreakers to provide a convenient forum for phreaker/hacker conferences.

78.3.8.11.5 PABX Penetration

Private automatic branch exchanges are open to a range of attacks including weak security on diagnostic modem lines and misconfiguration of switches. These systems are now a common target providing a convenient springboard for long-distance attacks. An example might be an attacker who calls into the PABX and then uses private wire circuits belonging to the firm to call out to countries overseas. Direct inward system access (DISA) facilities provide a rich facility set for legitimate company workers outside the office, including call diversion, conferencing, message pickup, etc. If misconfigured these facilities can compromise the security of the company and permit call fraud.

78.3.8.11.6 Cellular Phones

Cellular phone technology is still in its infancy in many countries with many analog cellular systems in common usage. Analog cellular phones are vulnerable in a number of areas:

- Call interception: no encryption or scrambling of the call is carried out, calls can therefore be directly monitored by an attacker with a VHF/UHF scanner. U.S. scanners are modified to exclude the cellular phone band, but the techniques for reversing the modification are openly exchanged.

- Signaling interception: signaling information for calls is also carried in clear including the telephone number/electronic serial number (ESN) pair used to authenticate the subscriber. This raises the risk of this information being intercepted and replayed for fraudulent use.

- Reprogramming: commercial phones are controlled by firmware in ROM (or flash ROM) and can be reprogrammed with appropriate interface hardware or access to the manufacturers security code.

Three forms of attack have been described: the simple reprogramming of a cellular phone to an ESN captured on-air or by interrogating a cellular phone; the tumbler, in which a telephone number/ESN pair is randomly generated; and the vampire, in which a modified phone rekeys itself with an ESN intercepted directly over the air.

The use of modified cellular phones provides a linkage between the phreaker community and organized crime. In particular, lucrative businesses have been set up in allowing immigrants to call home at minimal cost on stolen cellular phones or telephone credit cards. Newer digital networks such as GSM are encrypted and not open to the same form of direct interception attack.

78.3.8.11.7 Carding

The final category of malicious attack is aimed at the forgery of credit and telephone card information. A key desire is to make free phone calls to support hacking activities. Four techniques have been used:

1. Reading the magnetic stripe on the back of a credit card using a commercial stripe reader. This stripe can then be duplicated and affixed to a blank card or legitimate card.
2. Generating random credit card numbers for a chosen bank with a valid checksum digit. These numbers will pass the simple off-line authentication checks used by vendors for low-value purchases.
3. Using telephone card services to validate a series of randomly generated telephone card numbers generated by modem.
4. Compromising a credit card number in transit over an untrusted network, such as the Internet.

The last category is a growing problem with the increasing range of commercial agencies now attempting to carry out business on the Internet. The introduction of secure electronic funds transfer systems is key to supporting the growth of electronic commerce on the Internet.

78.4 The Future

78.4.1 Computer Security

The level of technical sophistication of computer systems, telecomm switches, router and network infrastructure continues to grow. With increasing complexity comes increasing vulnerability. The focus of hacker attacks has moved with improving security measures, as the attackers seek to find a weak point in system defenses. Common carrier signaling is more secure than in-band, but the digital switches are vulnerable.

Firewalls protect end systems but the network infrastructure can be attacked. Security is improving over time, but the level of technical attack sophistication continues to rise.

78.4.2 National Information Infrastructure

The U.S. vision of the information superhighways is leading to growing internetworking and a move toward ubiquitous computing. This move is increasing our use of and dependence on networks. Security will become a key issue on these networks, not just protection against casual penetration but also against deliberate motivated attack by organized crime, terrorists, or anarchists: the beginning of information warfare. The organization of effective coordinated defenses against threats against our infrastructure will be one of the key challenges.

Further Information

Bellovin, S. and Chiswick, B. 1994. *Firewalls and Internet Security.* Addison–Wesley, Reading, MA.

Brunner, J. Shockwave Rider. New York.

Chapman, B. and Zwicky, E. 1995. *Building Internet Firewalls.* O'Reilly & Associates, Sebastopol, CA.

Denning, P. 1990. *Computers Under Attack: Intruders, Worms and Viruses.* ACM Press, Addison–Wesley, Reading, MA.

Ferbrache, D. 1992. *A Pathology of Computer Viruses.* Springer–Verlag, London, England.

Garfinkel, S. and Spafford, G. 1996. Practical UNIX & Internet Security. O'Reilly & Associates, Sebastopol, CA.

Gibson, W. 1991. *Neuromancer.* Simon & Schuster, New York.

Hoffman, L. 1990. *Rogue Programs: Viruses, Worms and Trojan Horses.* Van Nostrand Reinhold, New York.

Littman, J. 1996. *The Fugitive Game: Online with Kevin Mitnick.* Little, Brown, Boston, MA.

Neumann, P. 1995. *Computer Related Risks.* Addison–Wesley, Reading, MA.

Shimomura, T. and Markoff, J. 1995. *Takedown: The pursuit and capture of Kevin Mitnick, America's most wanted computer outlaw.* Hyperion, New York.

Sterling, B. The Hacker Crackdown: Law and disorder on the electronic frontier. Available via WWW at http://www-swiss.ai.mit.edu/~bal/sterling/.

Stoll, C. 1989. *The Cuckoo's Egg.* Doubleday, Garden City, NY.

Virus Bulletin. Various issues, Virus Bulletin Ltd, England.

Weiner, L. 1995. *Digital Woes: Why we should not depend on software.* Addison–Wesley, Reading, MA.

79

Authentication, Access Control, and Intrusion Detection

Ravi S. Sandhu
George Mason University

Pierangela Samarati
Università degli Studi di Milano

79.1 Introduction **79**-1
79.2 Authentication **79**-3
   Authentication by Passwords • Token-Based Authentication • Biometric Authentication • Authentication in Distributed Systems
79.3 Access Control **79**-5
   The Access Control Matrix • Implementation Approaches • Access Control Policies • Administration of Authorization
79.4 Auditing and Intrusion Detection...................**79**-15
   Intrusion Detection Systems • Audit Control Issues
79.5 Conclusion **79**-21

79.1 Introduction

An important requirement of any information management system is to protect information against improper disclosure or modification (known as confidentiality and integrity, respectively). Three mutually supportive technologies are used to achieve this goal. Authentication, access control, and audit together provide the foundation for information and system security as follows. *Authentication* establishes the identity of one party to another. Most commonly, authentication establishes the identity of a user to some part of the system typically by means of a password. More generally, authentication can be computer-to-computer or process-to-process and mutual in both directions. *Access control* determines what one party will allow another to do with respect to resources and objects mediated by the former. Access control usually requires authentication as a prerequisite. The *audit* process gathers data about activity in the system and analyzes it to discover security violations or diagnose their cause. Analysis can occur off line after the fact or it can occur on line more or less in real time. In the latter case, the process is usually called *intrusion detection*. This chapter discusses the scope and characteristics of these security controls.

Figure 79.1 is a logical picture of these security services and their interactions. Access control constrains what a user can do directly as well as what programs executing on behalf of the user are allowed to do. Access control is concerned with limiting the activity of legitimate users who have been successfully authenticated. It is enforced by a reference monitor, which mediates every attempted access by a user (or program executing on behalf of that user) to objects in the system. The reference monitor consults an authorization database to determine if the user attempting to do an operation is actually authorized to perform that operation. Authorizations in this database are administered and maintained by a security administrator.

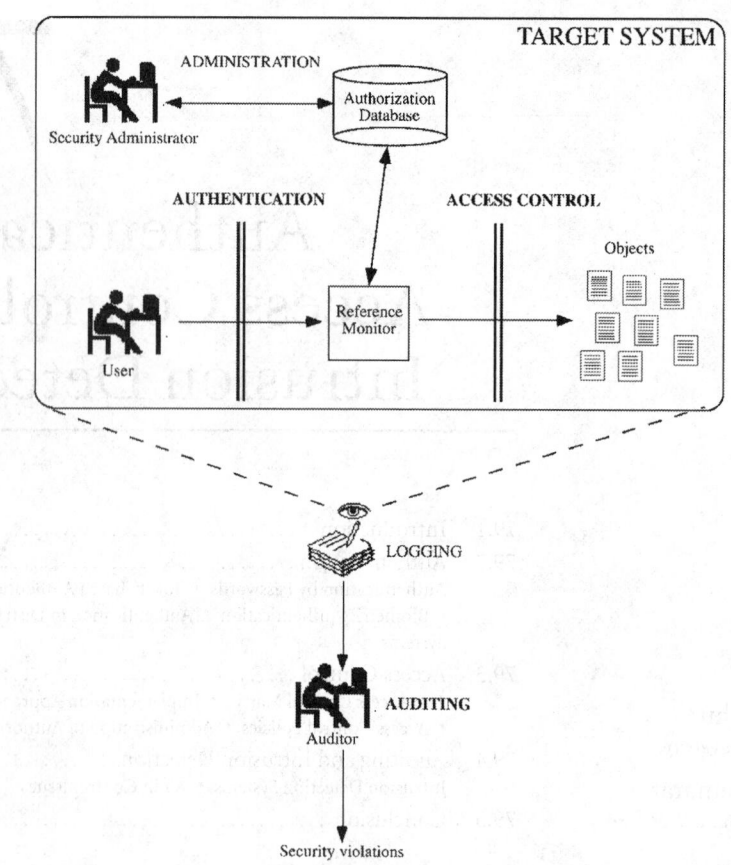

FIGURE 79.1   Access control and other security services.

The administrator sets these authorizations on the basis of the security policy of the organization. Users may also be able to modify some portion of the authorization database, for instance, to set permissions for their personal files. Auditing monitors and keeps a record of relevant activity in the system.

Figure 79.1 is a logical picture and should not be interpreted literally. For instance, as we will see later, the authorization database is often stored with the objects being protected by the reference monitor rather than in a physically separate area. The picture is also somewhat idealized in that the separation between authentication, access control, auditing, and administration services may not always be so clear cut. This separation is considered highly desirable but is not always faithfully implemented in every system.

It is important to make a clear distinction between authentication and access control. Correctly establishing the identity of the user is the responsibility of the authentication service. Access control assumes that identity of the user has been successfully verified prior to enforcement of access control via a reference monitor. The effectiveness of the access control rests on a proper user identification and on the correctness of the authorizations governing the reference monitor.

It is also important to understand that access control is not a complete solution for securing a system. It must be coupled with auditing. Audit controls concern a posteriori analysis of all of the requests and activities of users in the system. Auditing requires the recording of all user requests and activities for their later analysis. Audit controls are useful both as a deterrent against misbehavior and as a means to analyze the users' behavior in using the system to find out about possible attempted or actual violations. Auditing can also be useful for determining possible flaws in the security system. Finally, auditing is essential to ensure that authorized users do not misuse their privileges: in other words, to hold users accountable for their actions. Note that effective auditing requires that good authentication be in place; otherwise it is not possible to reliably attribute activity to individual users. Effective auditing also requires good access control; otherwise the audit records can themselves be modified by an attacker.

These three technologies are interrelated and mutually supportive. In the following sections, we discuss, respectively, authentication, access control, and auditing and intrusion detection.

79.2 Authentication

Authentication is in many ways the most primary security service on which other security services depend. Without good authentication, there is little point in focusing attention on strong access control or strong intrusion detection. The reader is surely familiar with the process of signing on to a computer system by providing an identifier and a password. In this most familiar form, authentication establishes the identity of a human user to a computer. In a networked environment, authentication becomes more difficult. An attacker who observes network traffic can replay authentication protocols to masquerade as a legitimate user.

More generally, authentication establishes the identity of one computer to another. Often, authentication is required to be performed in both directions. This is certainly true when two computers are engaged in communication as peers. Even in a client–server situation mutual authentication is useful. Similarly, authentication of a computer to a user is also useful to prevent against spoofing attacks in which one computer masquerades as another (perhaps to capture user identifiers and passwords).

Often we need a combination of user-to-computer and computer-to-computer authentication. Roughly speaking, user-to-computer authentication is required to establish identity of the user to a workstation and computer-to-computer authentication is required for establishing the identity of the workstation acting on behalf of the user to a server on the system (and vice versa). In distributed systems, authentication must be maintained through the life of a conversation between two computers. Authentication needs to be integrated into each packet of data that is communicated. Integrity of the contents of each packet, and perhaps confidentiality of contents, also must be ensured.

Our focus in this chapter is on user-to-computer authentication. User-to-computer authentication can be based on one or more of the following:

- Something the user knows, such as a password
- Something the user possesses, such as a credit-card sized cryptographic token or smart card
- Something the user is, exhibited in a biometric signature such as a fingerprint or voice print

We now discuss these in turn.

79.2.1 Authentication by Passwords

Password-based authentication is the most common technique, but it has significant problems. A well-known vulnerability of passwords is that they can be guessed, especially because users are prone to selecting weak passwords. A password can be snooped by simply observing a user keying it in. Users often need to provide their password when someone else is in a position to observe it as it is keyed in. Such compromise can occur without the user even being aware of it. It is also hard for users to remember too many pass-words, especially for services that are rarely used. Nevertheless, because of low cost and low technology requirements, passwords are likely to be around for some time to come.

An intrinsic problem with passwords is that they can be shared, which breaks down accountability in the system. It is all too easy for a user to give their password to another user. Sometimes poor system design actually encourages password sharing because there may be no other convenient means of delegating permissions of one user to another (even though the security policy allows the delegation).

Password management is required to prod users to regularly change their passwords, to select good ones, and to protect them with care. Excessive password management makes adversaries of users and security administrators, which can be counterproductive. Many systems can configure a maximum lifetime for a password. Interestingly, many systems also have a minimum lifetime for a password. This has come about to prevent users from reusing a previous password when prompted to change their password after its maximum life has expired. The system keeps a history of, say, eight most recently used passwords for each user. When asked to change the current password the user can change it eight times to flush the history and then resume reuse of the same password. The response is to disallow frequent changes to a user's password!

Passwords are often used to generate cryptographic keys, which are further used for encryption or other cryptographic transformations. Encrypting data with keys derived from passwords is vulnerable to so-called dictionary attacks. Suppose the attacker has access to known plaintext, that is, the attacker knows the encrypted and plaintext versions of data that were encrypted using a key derived from a user's password. Instead of trying all possible keys to find the right one, the attacker instead tries keys generated from a list of, say, 20,000 likely passwords (known as a dictionary). The former search is usually computationally infeasible, whereas the latter can be accomplished in a matter of hours using commonplace workstations. These attacks have been frequently demonstrated and are a very real threat.

Operating systems typically store a user's password by using it as a key to some cryptographic transformation. Access to the so-called encrypted passwords provides the attacker the necessary known plaintext for a dictionary attack. The Unix system actually makes these encrypted passwords available in a publicly readable file. Recent versions of Unix are increasingly using shadow passwords by which these data are stored in files private to the authentication system. In networked systems, known plaintext is often visible in the network authentications protocols.

Poor passwords can be detected by off-line dictionary attacks conducted by the security administrators. Proactive password checking can be applied when a user changes his or her password. This can be achieved by looking up a large dictionary. Such dictionaries can be very big (tens of megabytes) and may need to be replicated at multiple locations. They can themselves pose a security hazard. Statistical techniques for proactive password checking have been proposed as an alternative [Davies and Ganesan 1993].

Selecting random passwords for users is not user friendly and also poses a password distribution problem. Some systems generate pronounceable passwords for users because these are easier to remember. In principle this is a sound idea but some of the earlier recommended methods for generating pronounceable passwords have been shown to be insecure [Ganesan and Davies 1994]. It is also possible to generate a sequence of one-time passwords that are used one-by-one in sequence without ever being reused. Human beings are not expected to remember these and must instead write them down or store them on laptop hard disks or removable media.

79.2.2 Token-Based Authentication

A token is a credit-card size device that the user carries around. Each token has a unique private cryptographic key stored within it, used to establish the token's identity via a challenge-response handshake. The party establishing the authentication issues a challenge to which a response is computed using the token's private key. The challenge is keyed into the token by the user and the response displayed by the token is again keyed by the user into the workstation to be communicated to the authenticating party. Alternately, the workstation can be equipped with a reader that can directly interact with the token, eliminating the need for the user to key in the challenge and response. Sometimes the challenge is implicitly taken to be the current time, so only the response needs to be returned (this assumes appropriately accurate synchronization of clocks).

The private key should never leave the token. Attempts to break the token open to recover the key should cause the key to be destroyed. Achieving this in the face of a determined adversary is a difficult task. Use of the token itself requires authentication; otherwise the token can be surreptitiously used by an intruder or stolen and used prior to discovery of the theft. User-to-token authentication is usually based on passwords in the form of a personal identification number (PIN).

Token-based authentication is much stronger than password-based authentication and is often called strong as opposed to weak authentication. However, it is the token that is authenticated rather than the user. The token can be shared with other users by providing the PIN, and so it is vulnerable to loss of accountability. Of course, only one user at a time can physically possess the token.

Tokens can use secret key or public key cryptosystems. With secret key systems the computer authenticating the token needs to know the secret key that is embedded in the token. This presents the usual key distribution problem for secret key cryptography. With public key cryptography, a token can be authenticated by a computer that has had no prior contact with the user's token. The public key used to verify the response to a challenge can be obtained with public key certificates. Public key-based tokens have scalability advantages that in the long run should make them the dominant technique for authentication in large systems. However, the computational and bandwidth requirements are generally greater for public vs. secret key systems. Token-based authentication is a technical reality today, but it still lacks major market penetration and does cost money.

79.2.3 Biometric Authentication

Biometric authentication has been used for some time for high-end applications. The biometric signature should be different every time, for example, voice-print check of a different challenge phrase on each occasion. Alternately, the biometric signature should require an active input, for example, dynamics of handwritten signatures. Simply repeating the same phrase every time or using a fixed signature such as a fingerprint is vulnerable to replay attacks. Biometric authentication often requires cumbersome equipment, which is best suited for fixed installations such as entry into a building or room.

Technically the best combination would be user-to-token biometric authentication, followed by mutual cryptographic authentication between the token and system services. This combination may emerge sooner than one might imagine. Deployment of such technology on a large scale is certain to raise social and political debate. Unforgeable biometric authentication could result in significant loss of privacy for individuals. Some of the privacy issues may have technical solutions, whereas others may be inherently impossible.

79.2.4 Authentication in Distributed Systems

In distributed systems, authentication is required repeatedly as the user uses multiple services. Each service needs authentication, and we might want mutual authentication in each case. In practice, this process starts with a user supplying a password to the workstation, which can then act on the user's behalf. This password should never be disclosed in plaintext on the network. Typically, the password is converted to a cryptographic key, which is then used to perform challenge-response authentication with servers in the system. To minimize exposure of the user password, and the long-term key derived from it, the password is converted into a short-term key, which is retained on the workstation, while the long-term user secrets are discarded. In effect these systems use the desktop workstation as a token for authentication with the rest of the network. Trojan horse software in the workstation can, of course, compromise the user's long-term secrets.

The basic principles just outlined have been implemented in actual systems in an amazing variety of ways. Many of the early implementations are susceptible to dictionary attacks. Now that the general nature and ease of a dictionary attack are understood we are seeing systems that avoid these attacks or at least attempt to make them more difficult. For details on actual systems, we refer the reader to Kaufman et al. [1995], Neuman [1994], and Woo and Lam [1992].

79.3 Access Control

In this section we describe access control. We introduce the concept of an access matrix and discuss implementation alternatives. Then we explain discretionary, mandatory, and role-based access control policies. Finally, we discuss issues in administration of authorizations.

79.3.1   The Access Control Matrix

Security practitioners have developed a number of abstractions over the years in dealing with access control. Perhaps the most fundamental of these is the realization that all resources controlled by a computer system can be represented by data stored in objects (e.g., files). Therefore, protection of objects is the crucial requirement, which in turn facilitates protection of other resources controlled via the computer system. (Of course, these resources must also be physically protected so they cannot be manipulated, directly bypassing the access controls of the computer system.)

Activity in the system is initiated by entities known as subjects. Subjects are typically users or programs executing on behalf of users. A user may sign on to the system as different subjects on different occasions, depending on the privileges the user wishes to exercise in a given session. For example, a user working on two different projects may sign on for the purpose of working on one project or the other. We then have two subjects corresponding to this user, depending on the project the user is currently working on.

A subtle point that is often overlooked is that subjects can themselves be objects. A subject can create additional subjects in order to accomplish its task. The children subjects may be executing on various computers in a network. The parent subject will usually be able to suspend or terminate its children as appropriate. The fact that subjects can be objects corresponds to the observation that the initiator of one operation can be the target of another. (In network parlance, subjects are often called initiators, and objects are called targets.)

The subject–object distinction is basic to access control. Subjects initiate actions or operations on objects. These actions are permitted or denied in accord with the authorizations established in the system. Authorization is expressed in terms of access rights or access modes. The meaning of access rights depends on the object in question. For files, the typical access rights are read, write, execute, and own. The meaning of the first three of these is self-evident. Ownership is concerned with controlling who can change the access permissions for the file. An object such as a bank account may have access rights inquiry, credit and debit corresponding to the basic operations that can be performed on an account. These operations would be implemented by application programs, whereas for a file the operations would typically be provided by the operating system.

The access matrix is a conceptual model that specifies the rights that each subject possesses for each object. There is a row in this matrix for each subject and a column for each object. Each cell of the matrix specifies the access authorized for the subject in the row to the object in the column. The task of access control is to ensure that only those operations authorized by the access matrix actually get executed. This is achieved by means of a reference monitor, which is responsible for mediating all attempted operations by subjects on objects. Note that the access matrix model clearly separates the problem of authentication from that of authorization.

An example of an access matrix is shown in Figure 79.2, where the rights R and W denote read and write, respectively, and the other rights are as previously discussed. The subjects shown here are John, Alice, and Bob. There are four files and two accounts. This matrix specifies that, for example, John is the owner of file 3 and can read and write that file, but John has no access to file 2 or file 4. The precise meaning of ownership varies from one system to another. Usually the owner of a file is authorized to grant other users access to the file as well as revoke access. Because John owns file 1, he can give Alice the R right and Bob the R and W rights, as shown in Figure 79.2. John can later revoke one or more of these rights at his discretion.

	File 1	File 2	File 3	File 4	Account 1	Account 2
John	Own R W		Own R W		Inquiry Credit	
Alice	R	Own R W	W	R	Inquiry Debit	Inquiry Credit
Bob	R W	R		Own R W		Inquiry Debit

FIGURE 79.2   An access matrix.

The access rights for the accounts illustrate how access can be controlled in terms of abstract operations implemented by application programs. The inquiry operation is similar to read in that it retrieves information but does not change it. Both the credit and debit operations will involve reading the previous account balance, adjusting it as appropriate, and writing it back. The programs that implement these operations require read and write access to the account data. Users, however, are not allowed to read and write the account object directly. They can manipulate account objects only indirectly via application programs, which implement the debit and credit operations.

Also note that there is no own right for accounts. Objects such as bank accounts do not really have an owner who can determine the access of other subjects to the account. Clearly the user who establishes the account at the bank should not be the one to decide who can access the account. Within the bank different officials can access the account on the basis of their job functions in the organization.

79.3.2 Implementation Approaches

In a large system, the access matrix will be enormous in size, and most of its cells are likely to be empty. Accordingly, the access matrix is very rarely implemented as a matrix. We now discuss some common approaches to implementing the access matrix in practical systems.

79.3.2.1 Access Control Lists

A popular approach to implementing the access matrix is by means of access control lists (ACLs). Each object is associated with an ACL, indicating for each subject in the system the accesses the subject is authorized to execute on the object. This approach corresponds to storing the matrix by columns. ACLs corresponding to the access matrix of Figure 79.2 are shown in Figure 79.3. Essentially, the access matrix column for file 1 is stored in association with File 1, and so on.

By looking at an object's ACL, it is easy to determine which modes of access subjects are currently authorized for that object. In other words, ACLs provide for convenient access review with respect to an object. It is also easy to revoke all access to an object by replacing the existing ACL with an empty one. On the other hand, determining all of the accesses that a subject has is difficult in an ACL-based system. It is necessary to examine the ACL of every object in the system to do access review with respect to a subject. Similarly, if all accesses of a subject need to be revoked all ACLs must be visited one by one. (In practice,

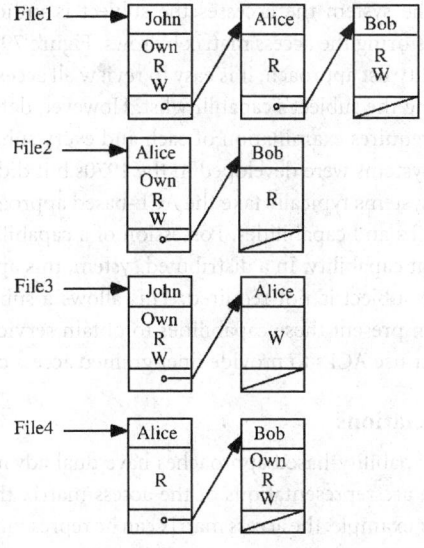

FIGURE 79.3 Access control lists for files in Figure 79.2.

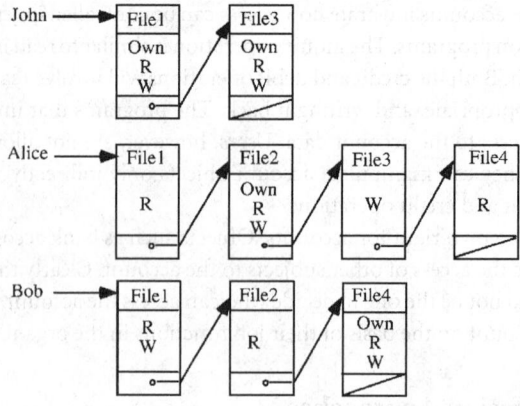

FIGURE 79.4　Capability lists for files in Figure 79.2.

revocation of all accesses of a subject is often done by deleting the user account corresponding to that subject. This is acceptable if a user is leaving an organization. However, if a user is reassigned within the organization it would be more convenient to retain the account and change its privileges to reflect the changed assignment of the user.)

Many systems allow group names to occur in ACLs. For example, an entry such as (ISSE, R) can authorize all members of the ISSE group to read a file. Several popular operating systems, such as Unix and VMS, implement an abbreviated form of ACLs in which a small number, often only one or two, of group names can occur in the ACL. Individual subject names are not allowed. With this approach, the ACL has a small fixed size so it can be stored using a few bits associated with the file. At the other extreme, there are a number of access control packages that allow complicated rules in ACLs to limit when and how the access can be invoked. These rules can be applied to individual users or to all users who match a pattern defined in terms of user names or other user attributes.

79.3.2.2 Capabilities

Capabilities are a dual approach to ACLs. Each subject is associated with a list, called the capability list, indicating for each object in the system the accesses the subject is authorized to execute on the object. This approach corresponds to storing the access matrix by rows. Figure 79.4 shows capability lists for the files in Figure 79.2. In a capability list approach, it is easy to review all accesses that a subject is authorized to perform by simply examining the subject's capability list. However, determination of all subjects who can access a particular object requires examination of each and every subject's capability list. A number of capability-based computer systems were developed in the 1970s but did not prove to be commercially successful. Modern operating systems typically take the ACL-based approach.

It is possible to combine ACLs and capabilities. Possession of a capability is sufficient for a subject to obtain access authorized by that capability. In a distributed system, this approach has the advantage that repeated authentication of the subject is not required. This allows a subject to be authenticated once, obtain its capabilities, and then present these capabilities to obtain services from various servers in the system. Each server may further use ACLs to provide finer grained access control.

79.3.2.3 Authorization Relations

We have seen that ACL- and capability-based approaches have dual advantages and disadvantages with respect to access review. There are representations of the access matrix that do not favor one aspect of access review over the other. For example, the access matrix can be represented by an authorization relation (or table), as shown in Figure 79.5. Each row, or tuple, of this table specifies one access right of a subject to an object. Thus, John's accesses to file 1 require three rows. If this table is sorted by subject, we get the

Subject	Access mode	Object
John	Own	File 1
John	R	File 1
John	W	File 1
John	Own	File 3
John	R	File 3
John	W	File 3
Alice	R	File 1
Alice	Own	File 2
Alice	R	File 2
Alice	W	File 2
Alice	W	File 3
Alice	R	File 4
Bob	R	File 1
Bob	W	File 1
Bob	R	File 2
Bob	Own	File 4
Bob	R	File 4
Bob	W	File 4

FIGURE 79.5   Authorization relation for files in Figure 79.2.

effect of capability lists. If it is sorted by object, we get the effect of ACLs. Relational database management systems typically use such a representation.

79.3.3 Access Control Policies

In access control systems, a distinction is generally made between policies and mechanisms. Policies are high-level guidelines that determine how accesses are controlled and access decisions are determined. Mechanisms are low-level software and hardware functions that can be configured to implement a policy. Security researchers have sought to develop access control mechanisms that are largely independent of the policy for which they could be used. This is a desirable goal to allow reuse of mechanisms to serve a variety of security purposes. Often, the same mechanisms can be used in support of secrecy, integrity, or availability objectives. On the other hand, sometimes the policy alternatives are so many and diverse that system implementors feel compelled to choose one in preference to the others.

In general, there do not exist policies that are better than others. Rather there exist policies that ensure more protection than others. However, not all systems have the same protection requirements. Policies suitable for a given system may not be suitable for another. For instance, very strict access control policies, which are crucial to some systems, may be inappropriate for environments where users require greater flexibility. The choice of access control policy depends on the particular characteristics of the environment to be protected.

We will now discuss three different policies that commonly occur in computer systems as follows:

- Classic discretionary policies
- Classic mandatory policies
- The emerging role-based policies

We have added the qualifier classic to the first two of these to reflect the fact that these have been recognized by security researchers and practitioners for a long time. However, in recent years there is increasing consensus that there are legitimate policies that have aspects of both of these. Role-based policies are an example of this fact.

It should be noted that access control policies are not necessarily exclusive. Different policies can be combined to provide a more suitable protection system. This is indicated in Figure 79.6. Each of the

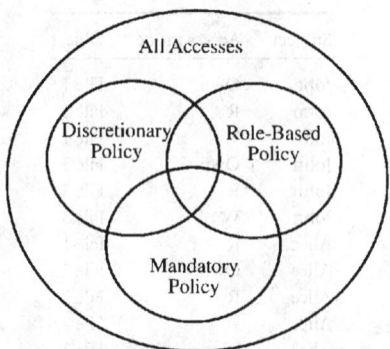

FIGURE 79.6   Multiple access control policies.

three inner circles represents a policy that allows a subset of all possible accesses. When the policies are combined, only the intersection of their accesses is allowed. Such a combination of policies is relatively straightforward so long as there are no conflicts where one policy asserts that a particular access *must* be allowed while another one prohibits it. Such conflicts between policies need to be reconciled by negotiations at an appropriate level of management.

79.3.3.1   Classic Discretionary Policies

Discretionary protection policies govern the access of users to the information on the basis of the user's identity and authorizations (or rules) that specify, for each user (or group of users) and each object in the system, the access modes (e.g., read, write, or execute) the user is allowed on the object. Each request of a user to access an object is checked against the specified authorizations. If there exists an authorization stating that the user can access the object in the specific mode, the access is granted, otherwise it is denied.

The flexibility of discretionary policies makes them suitable for a variety of systems and applications. For these reasons, they have been widely used in a variety of implementations, especially in the commercial and industrial environments.

However, discretionary access control policies have the drawback that they do not provide real assurance on the flow of information in a system. It is easy to bypass the access restrictions stated through the authorizations. For example, a user who is able to read data can pass it to other users not authorized to read it without the cognizance of the owner. The reason is that discretionary policies do not impose any restriction on the usage of information by a user once the user has read it, i.e., dissemination of information is not controlled. By contrast, dissemination of information is controlled in mandatory systems by preventing flow of information from high-level objects to low-level objects.

Discretionary access control policies based on explicitly specified authorization are said to be closed in that the default decision of the reference monitor is denial. Similar policies, called open policies, could also be applied by specifying denials instead of permissions. In this case, for each user and each object of the system, the access modes the user is forbidden on the object are specified. Each access request by a user is checked against the specified (negative) authorizations and granted only if no authorizations denying the access exist. The use of positive and negative authorizations can be combined, allowing the specification of both the accesses to be authorized as well as the accesses to be denied to the users. The interaction of positive and negative authorizations can become extremely complicated [Bertino et al. 1993].

79.3.3.2   Classic Mandatory Policies

Mandatory policies govern access on the basis of classification of subjects and objects in the system. Each user and each object in the system is assigned a security level. The security level associated with an object reflects the sensitivity of the information contained in the object, i.e., the potential damage that could result from unauthorized disclosure of the information. The security level associated with a user,

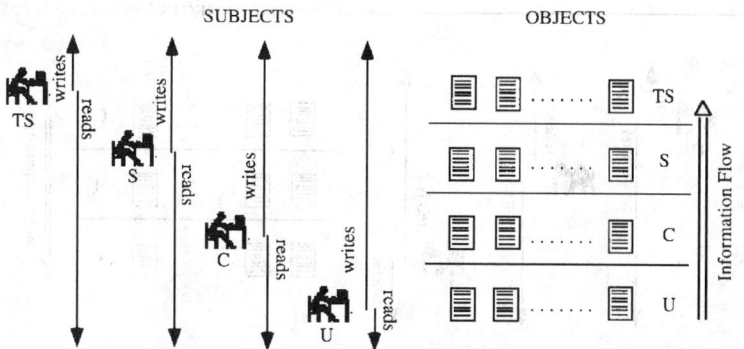

FIGURE 79.7 Controlling information flow for secrecy.

also called clearance, reflects the user's trustworthiness not to disclose sensitive information to users not cleared to see it. In the simplest case, the security level is an element of a hierarchical ordered set. In the military and civilian government arenas, the hierarchical set generally consists of top secret (TS), secret (S), confidential (C), and unclassified (U), where TS > S > C > U. Each security level is said to dominate itself and all others below it in this hierarchy.

Access to an object by a subject is granted only if some relationship (depending on the type of access) is satisfied between the security levels associated with the two. In particular, the following two principles are required to hold:

Read down: A subject's clearance must dominate the security level of the object being read.

Write up: A subject's clearance must be dominated by the security level of the object being written.

Satisfaction of these principles prevents information in high-level objects (i.e., more sensitive) to flow to objects at lower levels. The effect of these rules is illustrated in Figure 79.7. In such a system, information can flow only upward or within the same security class.

It is important to understand the relationship between users and subjects in this context. Let us say that the human user Jane is cleared to S and assume she always signs on to the system as an S subject (i.e., a subject with clearance S). Jane's subjects are prevented from reading TS objects by the read-down rule. The write-up rule, however, has two aspects that seem at first sight contrary to expectation:

- First, Jane's S subjects can write a TS object (even though they cannot read it). In particular, they can overwrite existing TS data and therefore destroy it. Because of this integrity concern, many systems for mandatory access control do not allow write up but limit writing to the same level as the subject. At the same time, write up does allow Jane's S subjects to send electronic mail to TS subjects and can have its benefits.

- Second, Jane's S subjects cannot write C or U data. This means, for example, that Jane can never send electronic mail to C or U users. This is contrary to what happens in the paper world, where S users can write memos to C and U users. This seeming contradiction is easily eliminated by allowing Jane to sign to the system as a C or U subject as appropriate. During these sessions, she can send electronic mail to C or U and C subjects.

In other words, a user can sign on to the system as a subject at any level dominated by the user's clearance. Why then bother to impose the write-up rule? The main reason is to prevent malicious software from leaking secrets downward from S to U. Users are trusted not to leak such information, but the programs they execute do not merit the same degree of trust. For example, when Jane signs on to the system at U level, her subjects cannot read S objects and thereby cannot leak data from S to U. The write-up rule also prevents users from inadvertently leaking information from high to low.

In addition to hierarchical security levels, categories (e.g., Crypto, NATO, Nuclear) can also be associated with objects and subjects. In this case, the classification labels associated with each subject and each object

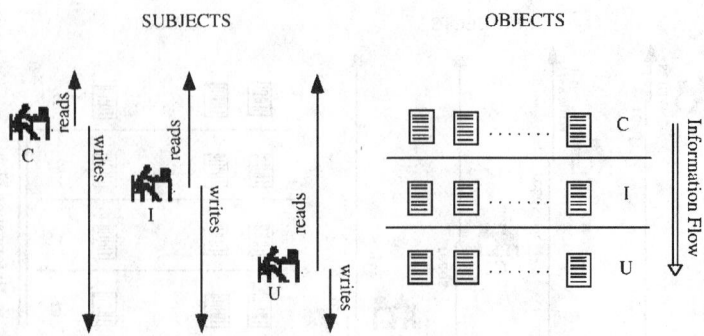

FIGURE 79.8   Controlling information flow for integrity.

consist of a pair composed of a security level and a set of categories. The set of categories associated with a user reflect the specific areas in which the user operates. The set of categories associated with an object reflect the area to which information contained in objects are referred. The consideration of categories provides a finer grained security classification. In military parlance, categories enforce restriction on the basis of the need-to-know principle, i.e., a subject should be given only those accesses that are required to carry out the subject's responsibilities.

Mandatory access control can as well be applied for the protection of information integrity. For example, the integrity levels could be crucial (C), important (I), and unknown (U). The integrity level associated with an object reflects the degree of trust that can be placed in the information stored in the object and the potential damage that could result from unauthorized modification of the information. The integrity level associated with a user reflects the user's trustworthiness for inserting, modifying, or deleting data and programs at that level. Principles similar to those stated for secrecy are required to hold, as follows:

> *Read up:* A subject's integrity level must be dominated by the integrity level of the object being read.
>
> *Write down:* A subject's integrity level must dominate the integrity level of the object being written.

Satisfaction of these principles safeguard integrity by preventing information stored in low objects (and therefore less reliable) to flow to high objects. This is illustrated in Figure 79.8. Controlling information flow in this manner is but one aspect of achieving integrity. Integrity in general requires additional mechanisms, as discussed in Castano et al. [1994] and Sandhu [1994].

Note that the only difference between Figure 79.7 and Figure 79.8 is the direction of information flow: bottom to top in the former case and top to bottom in the latter. In other words, both cases are concerned with one-directional information flow. The essence of classical mandatory controls is one-directional information flow in a lattice of security labels. For further discussion on this topic, see Sandhu [1993].

79.3.3.3   Role-Based Policies

The discretionary and mandatory policies previously discussed have been recognized in official standards, notably, the Orange Book of the U.S. Department of Defense. A good introduction to the Orange Book and its evaluation procedures is given in Chokhani [1992].

There has been a strong feeling among security researchers and practitioners that many practical requirements are not covered by these classic discretionary and mandatory policies. Mandatory policies rise from rigid environments, such as those of the military. Discretionary policies rise from cooperative yet autonomous requirements, such as those of academic researchers. Neither requirement satisfies the needs of most commercial enterprises. Orange Book discretionary policy is too weak for effective control of information assets, whereas Orange Book mandatory policy is focused on the U.S. Government policy for confidentiality of classified information. (In practice the military often finds Orange Book mandatory policies to be too rigid and subverts them.)

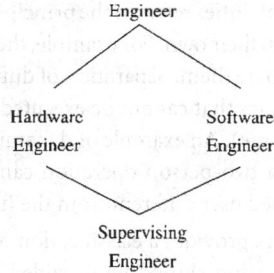

FIGURE 79.9 A role inheritance hierarchy.

Several alternatives to classic discretionary and mandatory policies have been proposed. These policies allow the specification of authorizations to be granted to users (or groups) on objects as in the discretionary approach, together with the possibility of specifying restrictions (as in the mandatory approach) on the assignment or on the use of such authorizations. One of the promising avenues that is receiving growing attention is that of role-based access control [Ferraiolo and Kuhn 1992, Sandhu et al. 1996].

Role-based policies regulate the access of users to the information on the basis of the activities the users execute in the system. Role-based policies require the identification of roles in the system. A role can be defined as a set of actions and responsibilities associated with a particular working activity. Then, instead of specifying all of the accesses each user is allowed to execute, access authorizations on objects are specified for roles. Users are given authorizations to adopt roles. A recent study by the National Institute of Standards and Technology (NIST) confirms that roles are a useful approach for many commercial and government organizations [Ferraiolo and Kuhn 1992].

The user playing a role is allowed to execute all accesses for which the role is authorized. In general, a user can take on different roles on different occasions. Also, the same role can be played by several users, perhaps simultaneously. Some proposals for role-based access control allow a user to exercise multiple roles at the same time. Other proposals limit the user to only one role at a time or recognize that some roles can be jointly exercised, whereas others must be adopted in exclusion to one another. As yet there are no standards in this arena, and so it is likely that different approaches will be pursued in different systems.

The role-based approach has several advantages. Some of these are discussed in the following:

- *Authorization management:* Role-based policies benefit from a logical independence in specifying user authorizations by breaking this task into two parts, one that assigns users to roles and one that assigns access rights for objects to roles. This greatly simplifies security management. For instance, suppose a user's responsibilities change, say, due to a promotion. The user's current roles can be taken away and new roles assigned as appropriate for the new responsibilities. If all authorization is directly between users and objects, it becomes necessary to revoke all existing access rights of the user and assign new ones. This is a cumbersome and time-consuming task.

- *Hierarchical roles:* In many applications, there is a natural hierarchy of roles based on the familiar principles of generalization and specialization. An example is shown in Figure 79.9. Here the roles of hardware and software engineer are specializations of the engineer role. A user assigned to the role of software engineer (or hardware engineer) will also inherit privileges and permissions assigned to the more general role of engineer. The role of supervising engineer similarly inherits privileges and permissions from both software-engineer and hardware-engineer roles. Hierarchical roles further simplify authorization management.

- *Least privilege:* Roles allow a user to sign on with the least privilege required for the particular task at hand. Users authorized to powerful roles do not need to exercise them until those privileges are actually needed. This minimizes the danger of damage due to inadvertent errors or by intruders masquerading as legitimate users.

- *Separation of duties:* Separation of duties refers to the principle that no users should be given enough privileges to misuse the system on their own. For example, the person authorizing a paycheck should not also be the one who can prepare them. Separation of duties can be enforced either statically (by defining conflicting roles, i.e., roles that cannot be executed by the same user) or dynamically (by enforcing the control at access time). An example of dynamic separation of duty is the two-person rule. The first user to execute a two-person operation can be any authorized user, whereas the second user can be any authorized user different from the first.

- *Object classes:* Role-based policies provides a classification of users according to the activities they execute. Analogously, a classification should be provided for objects. For example, generally a clerk will need to have access to the bank accounts, and a secretary will have access to the letters and memos (or some subset of them). Objects could be classified according to their type (e.g., letters, manuals) or to their application area (e.g., commercial letters, advertising letters). Access authorizations of roles should then be on the basis of object classes, not specific objects. For example, a secretary can be given the authorization to read and write the entire class of letters instead of being given explicit authorization for each single letter. This approach has the advantage of making authorization administration much easier and better controlled. Moreover, the accesses authorized on each object are automatically determined according to the type of the object without need of specifying authorizations on each object creation.

79.3.4 Administration of Authorization

Administrative policies determine who is authorized to modify the allowed accesses. This is one of the most important, and least understood, aspects of access controls.

In mandatory access control, the allowed accesses are determined entirely on the basis of the security classification of subjects and objects. Security levels are assigned to users by the security administrator. Security levels of objects are determined by the system on the basis of the levels of the users creating them. The security administrator is typically the only one who can change security levels of subjects or objects. The administrative policy is therefore very simple.

Discretionary access control permits a wide range of administrative policies. Some of these are described as follows:

- *Centralized:* A single authorizer (or group) is allowed to grant and revoke authorizations to the users.

- *Hierarchical:* A central authorizer is responsible for assigning administrative responsibilities to other administrators. The administrators can then grant and revoke access authorizations to the users of the system. Hierarchical administration can be applied, for example, according to the organization chart.

- *Cooperative:* Special authorizations on given resources cannot be granted by a single authorizer but needs cooperation of several authorizers.

- *Ownership:* Users are considered owners of the objects they create. The owner can grant and revoke access rights for other users to that object.

- *Decentralized:* In decentralized administration, the owner of an object can also grant other users the privilege of administering authorizations on the object.

Within each of these there are many possible variations.

Role-based access control has a similar wide range of possible administrative policies. In this case, roles can also be used to manage and control the administrative mechanisms.

Delegation of administrative authority is an important area in which existing access control systems are deficient. In large distributed systems, centralized administration of access rights is infeasible. Some existing systems allow administrative authority for a specified subset of the objects to be delegated by the central security administrator to other security administrators. For example, authority to administer objects in a particular region can be granted to the regional security administrator. This allows delegation of administrative authority in a selective piecemeal manner. However, there is a dimension of selectivity that is largely ignored in existing systems. For instance, it may be desirable that the regional security administrator be limited to granting access to these objects only to employees who work in that region. Control over the regional administrators can be centrally administered, but they can have considerable autonomy within their regions. This process of delegation can be repeated within each region to set up subregions, and so on.

79.4 Auditing and Intrusion Detection

Auditing consists of examination of the history of events in a system to determine whether and how security violations have occurred or been attempted. Auditing requires registration or logging of users' requests and activities for later examination. Audit data are recorded in an *audit trail* or *audit log*. The nature and format of these data vary from system to system.

Information that should be recorded for each event includes the subject requesting the access, the object to be accessed, the operation requested, the time of the request, perhaps the location from which the requested originated, the response of the access control system, the amount of resources [central processing unit (CPU) time, input/output (I/O), memory, etc.] used, and whether the operation succeeded or, if not, the reason for the failure, and so on.

In particular, actions requested by privileged users, such as the system and the security administrators, should be logged. First, this serves as a deterrent against misuse of powerful privileges by the administrators as well as a means for detecting operations that must be controlled (the old problem of guarding the guardian). Second, it allows control of penetrations in which the attacker gains a privileged status.

Audit data can become voluminous very quickly and searching for security violations in such a mass of data is a difficult task. Of course, audit data cannot reveal all violations because some may not be apparent in even a very careful analysis of audit records. Sophisticated penetrators can spread out their activities over a relatively long period of time, thus making detection more difficult. In some cases, audit analysis is executed only if violations are suspected or their effects are visible because the system shows an anomalous or erroneous behavior, such as continuous insufficient memory, slow processing, or nonaccessibility of certain files. Even in this case, often only a limited amount of audit data, namely, those that may be connected with the suspected violation, are examined. Sometimes the first clue to a security violation is some real-world event which indicates that information has been compromised. That may happen long after the computer penetration occurred. Similarly, security violations may result in Trojan horses or viruses being implanted whose activity may not be triggered until long after the original event.

79.4.1 Intrusion Detection Systems

Recent research has focused on the development of automated tools to help or even to carry out auditing controls. Automated tools can be used to screen and reduce audit data that need to be reviewed by humans. These tools can also organize audit data to produce summaries and measures needed in the analysis. This data reduction process can, for instance, produce short summaries of user behaviors, anomalous events, or security incidents. The auditors can then go over summaries instead of examining each single event recorded. Another class of automated tools is represented by the so-called *intrusion detection systems*. The purpose of these tools is not only to automate audit data acquisition and reduction but also its analysis. Some of the more ambitious efforts attempt to perform intrusion detection in real time.

Intrusion detection systems can be classified as *passive* or *active*. Passive systems, generally operating off line, analyze the audit data and bring possible intrusions or violations to the attention of the auditor, who then takes appropriate actions (see Figure 79.10). Active systems analyze audit data in real time. Besides bringing violations to the attention of the auditor, these systems may take an immediate protective response on the system (see Figure 79.11). The protective response can be executed ex post facto, after the violation has occurred, or preemptively, to avoid the violation being perpetrated to completion.

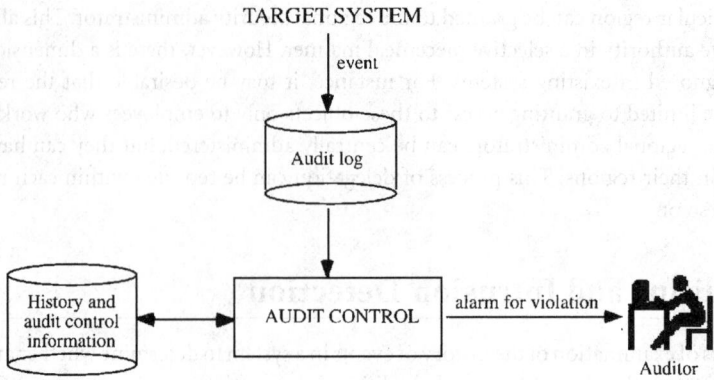

FIGURE 79.10   Passive intrusion detection.

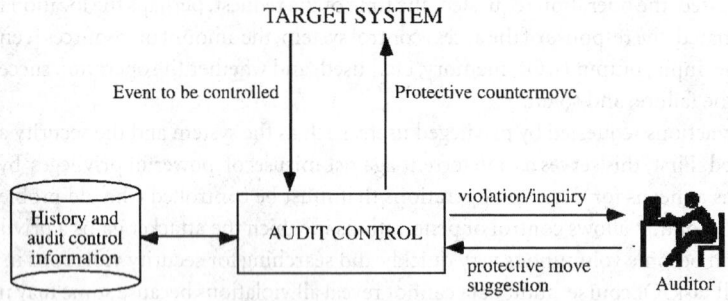

FIGURE 79.11   Active intrusion detection.

This latter possibility depends on the ability of the system to foresee violations. Protective responses include killing the suspected process, disconnecting the user, disabling privileges, or disabling user accounts. The response may be determined in total autonomy by the intrusion detection system or through interactions with the auditors.

Different approaches have been proposed for building intrusion detection systems. No single approach can be considered satisfactory with respect to different kinds of penetrations and violations that can occur. Each approach is appropriate for detecting a specific subset of violations. Moreover, each approach presents some pros and cons determined by the violations that can or cannot be controlled and by the amount and complexity of information necessary for its application. We now discuss the main intrusion detection approaches that have been attempted.

79.4.1.1   Threshold-Based Approach

The threshold-based approach is based on the assumption that the exploitation of system vulnerabilities involves abnormal use of the system itself. For instance, an attempt to break into a system can require trying several user accounts and passwords. An attempt to discover protected information can imply several, often denied, browsing operations through protected directories. A process infected by a virus can require an abnormal amount of memory or CPU resources.

Threshold-based systems typically control occurrences of specific events over a given period of time with respect to predefined allowable thresholds established by the security officer. For instance, more than three unsuccessful attempts to log in to a given account with the wrong password may indicate an attempt to penetrate that account. Multiple unsuccessful attempts to log in the system, using different accounts, concentrated in a short period of time, may suggest an attempt to break in.

Thresholds can also be established with respect to authorized operations to detect improper use of resources. For instance, a threshold can specify that print requests totaling more than a certain number of pages a day coming from the administrative office is to be considered suspicious. This misuse can be symptomatic of different kinds of violations such as the relatively benign misuse of the resource for personal use or a more serious attempt to print out working data for disclosure to the competition.

The threshold-based approach is limited by the fact that many violations occur without implying overuse of system resources. A further drawback of this approach is that it requires prior knowledge of how violations are reflected in terms of abnormal system use. Determining such connections and establishing appropriate thresholds are not always possible.

79.4.1.2 Anomaly-Based Approach

Like threshold-based controls, anomaly-based controls are based on the assumption that violations involve abnormal use of the system. However, whereas threshold-based systems define abnormal use with respect to prespecified fixed acceptable thresholds, anomaly-based systems define abnormal use as a use that is significantly different from that normally observed. In this approach, the intrusion detection system observes the behavior of the users in the target system and define profiles, i.e., statistical measures, reflecting the normal behavior of the users. Profiles can be defined with respect to different aspects to be controlled such as the number of events in a user session, the time elapsed between events in a user session, and the amount of resources consumed over a certain period of time or during execution of certain programs.

Construction of profiles from raw audit data is guided by rules that can be specified with respect to single users, objects, or actions as well as to classes of these. For instance, rules can state that profiles should be defined with respect to the number of pages printed every day by each user in the administration office, the number of resources per session and per day consumed by each user, the time elapsed between two log-in sessions for each single user, and some habit measures such as the time and the location from which a user generally logs in and the time the connections last. As users operate in the system, the intrusion detection system learns their behaviors with respect to the different profiles, thus defining what is normal and adapting the profiles to changes. Whenever a significant deviation occurs for a profile, an alarm is raised.

Statistical models that can be used include the *operational model*, the *mean and standard deviation model*, and *time series model*. With the operational model, an anomaly is raised when an observation exceeds a given acceptable threshold. This is similar to the threshold-based approach. With the mean and standard deviation model, an anomaly occurs when the observation falls outside an allowed confidence interval around the mean. For instance, an alarm can be raised if the CPU time consumed during a session for a user falls much below or above the CPU time generally consumed by the same user. With the time series model, an anomaly is raised when an event occurs at a given time at which the probability of its occurring is too low. For instance, a remote night-hour log-in request by a user who has never connected off hours or from outside the building may be considered suspicious.

The main advantage of the anomaly detection approach is that it does not require any a priori knowledge of the target system or of possible flaws from which the system may suffer. However, like the threshold-based approach, it can detect only violations that involve anomalous use. Moreover, some legitimate users may have a very erratic behavior (e.g., logging on and off at different hours or from different locations, varying their activity daily). For such users, no normal behavior can be actually established and misuse by them as well as by masqueraders exploiting their accounts would go undetected. The approach is also vulnerable from insiders who, knowing that behavior profiles are being defined, may either behave in a *bad* way from the beginning or slowly vary their behavior, going from *good* to *bad*, thus convincing the system that the bad behavior is normal.

79.4.1.3 Rule-Based Approach

In the rule-based approach, rules are used to analyze audit data for suspicious behavior independently from users' behavioral patterns. Rules describe what is suspicious on the basis of known past intrusions or known system vulnerabilities. This approach is generally enforced by means of expert systems encoding knowledge of the security experts about past intrusions in terms of sequences of events or observable properties characterizing violations.

For instance, a rule can specify that a sequence of browsing operations (e.g., **cd, ls,** and **more** commands in a Unix environment) coming off hours from a remote location may be symptomatic of an intrusion. Rules can also identify suspicious sequences of actions. For example, withdrawal of a large amount of money from an account and its deposit back a few days later may be considered suspicious.

The rule-based approach can detect violations that do not necessarily imply abnormal use of resources. Its main limitation is that the expert knowledge encoded in the rules encompasses only known system vulnerabilities and attack scenarios or suspicious events. The system can therefore be penetrated by attackers employing new techniques.

79.4.1.4   Model-Based Reasoning Approach

The model-based reasoning approach is based on the definition, by the security officers, of models of proscribed intrusion activities [Garvey and Lunt 1991]. Proscribed activities are expressed by means of sequences of user behaviors (single events or observable measures), called scenarios.

Each component of a scenario is therefore a high-level observation on the system and does not necessarily correspond to an audit record (which contains information at a lower level of specification). From these high-level specifications, the intrusion detection system generates, on the basis of specified rules, the corresponding sequences of actions at the level of the audit records. Each audit record produced on the observation of the system is controlled against the specified scenarios to determine if a violation is being carried out. Audit data reduction and analysis can be modeled in such a way that only events relevant to specific scenarios corresponding to intrusions probably being carried out are examined. When the probability of a given scenario being followed passes a specified threshold, an alarm is raised informing the auditor of the suspected violation.

The basis of this approach is essentially the same as the rule-based approach, the main difference being the way in which controls are specified. Whereas in the rule-based approach the security officer must explicitly specify the control rules in terms of the audit data, in the model-based approach the security officer specifies the scenario only in terms of high-level observable properties. This constitutes the main advantage of this approach, which allows the security officer to reason in terms of high-level abstractions rather than audit records. It is the task of the system to translate the scenarios into corresponding rules governing data reduction and analysis.

Like the rule-based approach, this approach can control only violations whose perpetration scenario (i.e., actions necessary to fulfill them) are known. By contrast, violations exploiting unknown vulnerabilities or not yet tried violations cannot be detected.

79.4.1.5   State Transition-Based Approach

In the state transition-based approach, a violation is modeled as a sequence of actions starting from an initial state to a final compromised state [Ilgun et al. 1995]. A state is a snapshot of the target system representing the values of all volatile, semipermanent, and permanent memory locations. Between the initial and the final states there are a number of intermediate states and corresponding transitions. State transitions correspond to key actions necessary to carry out the violation. Actions do not necessarily correspond to commands issued by users but, instead, refer to how state changes within the system are achieved. A single command may produce multiple actions. Each state is characterized as a set of assertions evaluating whether certain conditions are verified in the system. For instance, assertions can check whether a user is the owner of an object or has some privileges on it, whether the user who caused the last two transitions is the same user, or whether the file to which an action is referred is a particular file. Actions corresponding to state transitions are accesses to files (e.g., read, write, or execute operations), or actions modifying the permissions associated with files (e.g., changes of owners or authorizations), or some of the files' characteristics (e.g., rename operations).

As the users operate in the system, state transitions caused by them are determined. Whenever a state transition causes a final compromised state to be reached, an alarm is raised. The state transition-based approach can also be applied in a real-time active system to prevent users from executing operations that would cause a transition to a compromised state.

The state transition-based approach is based on the same concepts as the rule-based approach and therefore suffers from the same limitations, i.e., only violations whose scenarios are known can be detected. Moreover, it can be used to control only those violations that produce visible changes to the system state. Like the model-based approach, the state transition-based approach provides the advantage of requiring only high-level specifications, leaving the system the task of mapping state transitions into audit records and producing the corresponding control rules. Moreover, because a state transition can be matched by different operations at the audit record level, a single state transition specification can be used to represent different variations of a violation scenario (i.e., involving different operations but causing the same effects on the system).

79.4.1.6 Other Approaches

Other approaches have been proposed to complement authentication and access control to prevent violations from happening or to detect their occurrence.

One approach consists of preventing, rather than detecting, intrusions. In this class are tester programs that evaluate the system for common weaknesses often exploited by intruders and password checker programs that prevent users from choosing weak or obvious passwords (which may represent an easy target for intruders).

Another approach consists of substituting known bugged commands, generally used as trap doors by intruders, with programs that simulate the commands' execution while sending an alarm to the attention of the auditor. Other trap programs for intruders are represented by fake user accounts with *magic* passwords that raise an alarm when they are used.

Other approaches aim at detecting or preventing execution of Trojan horses and viruses. Solutions adopted for this include integrity checking tools that search for unauthorized changes to files and mechanisms controlling program executions against specifications of allowable program behavior in terms of operations and data flows.

Yet another intrusion detection approach is represented by the so-called keystroke latencies control. The idea behind the approach is that the elapsed time between keystrokes for regularly typed strings is quite consistent for each user. Keystroke latencies control can be used to cope against masqueraders. Moreover, they can also be used for authentication by controlling the time elapsed between the keystrokes when typing the password.

More recent research has interested intrusion detection at the network level [Mukherjee et al. 1994]. Analysis is performed on network traffic instead of on commands (or their corresponding low-level operations) issued on a system. Anomalies can then be determined, for example, on the basis of the probability of the occurrence of the monitored connections being too low or on the basis of the behavior of the connections. In particular, traffic is controlled against profiles of expected traffic specified in terms of expected paths (i.e., connections between systems) and service profiles.

79.4.2 Audit Control Issues

There are several issues that must be considered when employing intrusion detection techniques to identify security violations. These issues arise independently of the specific intrusion detection approach being utilized.

The task of generating audit records can be left to either the target system being monitored or the intrusion detection system. In the former case, the audit information generated by the system may need to be converted to a form understandable by the intrusion detection system. Many operating systems and database systems provide some audit information. However, this information often is not appropriate for security controls because it may contain data not relevant for detecting intrusions and omits details needed for identifying violations. Moreover, the audit mechanism of the target system may itself be vulnerable to a penetrator who might be able to bypass auditing or modify the audit log. Thus, a stronger and more appropriate audit trail might be required for effective intrusion detection.

Another important issue that must be addressed is the retention of audit data. Because the quantity of audit data generated every day can be enormous, policies must be specified that determine when historical data can be discarded.

Audit events can be recorded at different granularity. Events can be recorded at the system command level, at the level of each system call, at the application level, at the network level, or at the level of each keystroke. Auditing at the application and command levels has the advantage of producing high-level traces, which can be more easily correlated, especially by humans (who would get lost in low-level details). However, the actual effect of the execution of a command or application on the system may not be reflected in the audit records and therefore cannot be analyzed. Moreover, auditing at such a high level can be circumvented by users exploiting alias mechanisms or by directly issuing lower level commands. Recording at lower levels overcomes this drawback at the price of maintaining a greater number of audit records (a single user command may correspond to several low-level operations) whose examination by humans (or automated tools) therefore becomes more complicated.

Different approaches can be taken with respect to the time at which the audit data are recorded and, in the case of real-time analysis, evaluated. For instance, the information that a user has requested execution of a process can be passed to the intrusion detection system at the time the execution is required or at the time it is completed. The former approach has the advantage of allowing timely detection and, therefore, a prompt response to stop the violation. The latter approach has the advantage of providing more complete information about the event being monitored (information on resources used or time elapsed can be provided only after the process has completed) and therefore allows more complete analysis.

Audit data recording or analysis can be carried out indiscriminately or selectively, namely, on specific events, such as events concerning specific subjects, objects, or operations, or occurring at a particular time or in a particular situation. For instance, audit analysis can be performed only on operations on objects containing sensitive information, on actions executed off hours (nights and weekends) or from remote locations, on actions denied by the access control mechanisms, or on actions required by mistrusted users.

Different approaches can be taken with respect to the time at which audit control should be performed. Real-time intrusion detection systems enforce control in real time, i.e., analyze each event at the time of its occurrence. Real-time analysis of data brings the great advantage of timely detection of violations. However, because of the great amount of data to analyze and the analysis to be performs, real-time controls are generally performed only on selected data, leaving a more thorough analysis to be performed off line. Approaches that can be taken include the following:

- *Period driven:* Audit control is executed periodically. For example, every night the audit data produced during the working day are examined.
- *Session driven:* Audit control on a user's session is performed when a close session command is issued.
- *Event driven:* Audit control is executed upon occurrence of certain events. For instance, if a user attempts to enter a protected directory, audit over the user's previous and/or subsequent actions is initiated.
- *Request driven:* Audit control is executed upon the explicit request of the security officer.

The intrusion detection system may reside either on the target computer system or on a separate machine. This latter solution is generally preferable because it does not impact the target systems performance and protects audit information and control from attacks perpetrated on the target system. On the other hand, audit data must be communicated to the intrusion detection machine, which itself could be a source of vulnerability.

A major issue in employing an intrusion detection system is privacy. Monitoring user behavior, even if intended for defensive purposes, introduces a sort of Big Brother situation where a centralized monitor is watching everybody's behavior. This may be considered an invasion of individual privacy. It also raises concerns that audited information may be used improperly, for example, as a means for controlling employee performance.

79.5 Conclusion

Authentication, access control, and audit and intrusion detection together provide the foundations for building systems that can store and process information with confidentiality and integrity. Authentication is the primary security service. Access control builds directly on it. By and large, access control assumes authentication has been successfully accomplished. Strong authentication supports good auditing because operations can then be traced to the user who caused them to occur. There is a mutual interdependence between these three technologies, which can be often ignored by security practitioners and researchers. We need a coordinated approach that combines the strong points of each of these technologies rather than treating these as separate independent disciplines.

Acknowledgment

The work of Ravi Sandhu is partly supported by Grant CCR-9503560 from the National Science Foundation and Contract MDA904-94-C-6119 from the National Security Agency at George Mason University.

Portions of this paper appeared as Sandhu, R. S. and Samarati, P. 1994. Access control: principles and practice. *IEEE Commun.* 32(9):40–48. © 1994 IEEE. Used with permission.

References

Bertino, E., Samarati, P., and Jajodia, J. 1993. Authorizations in relational database management systems, pp. 130–139. In *1st ACM Conf. Comput. Commun. Security*. Fairfax, VA, Nov.

Castano, S., Fugini, M. G., Martella, G., and Samarati, P. 1994. *Database Security*. Addison–Wesley, Reading, MA.

Chokhani, S. 1992. Trusted products evaluation. *Commun. ACM* 35(7):64–76.

Davies, C. and Ganesan, R. 1993. Bapasswd: a new proactive password checker, pp. 1–15. In *16th NIST-NCSC Nat. Comput. Security Conf.*

Ferraiolo, D. F., Gilbert, D. M., and Lynch, N. 1993. An examination of federal and commercial access control policy needs, pp. 107–116. In *NIST-NCSC Nat. Comput. Security Conf.* Baltimore, MD, Sept.

Ferraiolo, D. and Kuhn, R. 1992. Role-based access controls, pp. 554–563. In *15th NIST-NCSC Nat. Comput. Security Conf.* Baltimore, MD, Oct. 13–16.

Ganesan, R. and Davies, C. 1994. A new attack on random pronouncable password generators, pp. 184–197. In *17th NIST-NCSC Nat. Comput. Security Conf.*

Garvey, T. D. and Lunt, T. 1991. Model-based intrusion detection, pp. 372–385. In *Proc. 14th Nat. Comput. Security Conf.* Washington, DC, Oct.

Ilgun, K., Kemmerer, R. A., and Porras, P. A. 1995. State transition analysis: a rule-based intrusion detection approach. *IEEE Trans. Software Eng.* 21(3):222–232.

Kaufman, C., Perlman, R., and Speciner, M. 1995. *Network Security*. Prentice–Hall, Englewood Cliffs, NJ.

Mukherjee, B., Heberlein, L. T., and Levitt, K. N. 1994. Network intrusion detection. *IEEE Network*, (May/June):26–41.

Neuman, B. C. 1994. Using Kerberos for authentication on computer networks. *IEEE Commun.* 32(9).

Sandhu, R. S. 1993. Lattice-based access control models. *IEEE Comput.* 26(11):9–19.

Sandhu, R. S. 1994. On five definitions of data integrity. In *Database Security VII: Status and Prospects*, T. Keefe and C. E. Landwehr, Eds., pp. 257–267. North-Holland.

Sandhu, R. S., Coyne, E. J., Feinstein, H. L., and Youman, C. E. 1996. Role-based access control models. *IEEE Comput.* 29(2):38–47.

Sandhu, R. S. and Samarati, P. 1994. Access control: principles and practice. *IEEE Commun.* 32(9):40–48.

Woo, T. Y. C. and Lam, S. S. 1992. Authentication for distributed systems. *IEEE Comput.* 25(1):39–52.

compromised information may be used inappropriately, for example, as a means for controlling employee performance.

79.6 Conclusion

Authentication, access control, and intrusion detection together provide a foundation for building systems that are secure and that restrict information to authorized users. Authentication, although the primary security service, access control builds directly on it. Once a larger base control is set, authentication has been successfully applied, then authorization can be implemented, and because operations run that have made the user who caused both to overcome this. Useful principles that address these three technologies, which can be often ignored by security requirements and consumer work. Much of consumer technology that combines with strong principles of which there remains a unit of this overarching security-independent deployment.

Acknowledgment

The work of the authors was supported by the Grant OCR 9202118 from the National Science Foundation and Contract MDA904-94-C-6118 from the National Security Agency at George Mason University. Parts of this chapter appeared as Sandhu, R. S. and Samarati, P. 1994, Access control principles and practice, IEEE Communications 32(9): 40-48 © 1991 IEEE. Used with permission.

References

Bertino, E., Samarati, P., and Jajodia, S. 1993. Authorization in relational database management systems, pp. 130-139 in 1st ACM Conf. on Computer and Communications Security, VA, Nov.

Castano, S., Fugini, M., Martella, G., and Samarati, P. 1994. Database Security, Addison-Wesley, Reading, MA.

Holberg, S. 1992. User authentication comparison, Commun. ACM 35, 2:164-65.

Davies, G. and Price, R. 1989. Basso sec. in new crypt, Wiley group word chapter, pp. 145-161, VSI WCIC World output. John's Cryst.

Ferriol, R., Gilbert, T.M., and Lynch, N. 1992. Administration of federal, federal, commercial, and control policy. April 10. Draft in WS/NCSC/98 to known 3, 4, with Camp. Rittman, MD 5 Op.

Ferrajolo, D. and Kuhn, R. 1992. Role-based access controls, pp. 554-563. 15th National Computer Security Conf., Baltimore, MD, Oct. 13-16.

Consortio, R. at intervening. 1994. Use, valuation and prob/inoculability system principles-coverage p 35. in 1994 WSA/NCSC Nat. Comp. Conf. Baltimore.

Denning, D. and Lunt, P. 1987. Multilevel secure information model, pp. 118-245, in Proc. Nat'l Int'l Computer Secure Conf., Washington DC, Oct.

Gati, S., Kemmerer, R. A., and Porras, P. 1995. State transition analysis: a rule-based intrusion detection approach. IEEE Trans. Software Eng. 21, 3:181-99.

Landwehr, C., Heitmeyer, C., and special ed. McLean. M. 1994. Computer railway Hall, Englewood Cliffs, NJ.

Matulevich, J., Hoffstadt, T., et al. and Revol, W. N. 1994. Pathways to practical realization Kurzfassung 39, 4.

Nyborg, S. et al. 1993. State-temp notes for authentication, security, computer networks, 12 Wed. Comm. and 520 Aug.

Sandhu, R. S. 1993. Lattice-based access control models. IEEE Comput. 26(11): 9-19.

Saltzer, J. 1991. On the definitions of confidentiality, In On Access, security, chapter 3, Oxford Univ. Press, pp. 7-30. ACM and Addison-Wesley Inc, pp. 21-270, Mass, England.

Sandhu, R., Coyne, E. J., Feinstein, H. L., and Youman, C. E. 1996. Role-based access control model, IEEE Computer 29(2):38-47.

Sandhu, R. S. and Samarati, P. 1994. Access control principles and practice, IEEE Communications 32(9):40-48.

Vijay, M., Coyne, E., and Jung, S. 1995. An approach to distributed system design. IEEE Comput. 13(1):37-47.

IX

Operating
Systems

Operating systems provide the software interface between the computer and its applications. This section covers the analysis, design, performance, and special challenges for operating systems in distributed and highly parallel computing environments. Much recent attention in operating system design is given to systems that control embedded computers, such as those found in vehicles. Also persistent are the particular challenges for synchronizing communication among simultaneously executing processes, managing scarce memory resources efficiently, and designing file systems that can handle massively large data sets.

80 What Is an Operating System?   *Raphael Finkel* **80**-1
Introduction • Historical Perspective • Goals of an Operating System • Implementing an Operating System • Research Issues and Summary

81 Thread Management for Shared-Memory Multiprocessors   *Thomas E. Anderson, Brian N. Bershad, Edward D. Lazowska, and Henry M. Levy* **81**-1
Introduction • Thread Management Concepts • Issues in Thread Management • Three Modern Thread Systems • Summary

82 Process and Device Scheduling   *Robert D. Cupper* **82**-1
Introduction • Background • Resources and Scheduling • Memory Scheduling • Device Scheduling • Scheduling Policies • High-Level Scheduling • Recent Work and Current Research

83 Real-Time and Embedded Systems   *John A. Stankovic* **83**-1
Introduction • Underlying Principles • Best Practices • Research Issues and Summary

84 Process Synchronization and Interprocess Communication   *Craig E. Wills* **84**-1
Introduction • Underlying Principles • Best Practices • Research Issues and Summary

85 Virtual Memory   *Peter J. Denning* **85**-1
Introduction • History • Structure of Virtual Memory • Distributed Shared Memory • The World Wide Web: Virtualizing the Internet • Conclusion

86 Secondary Storage and Filesystems   *Marshall Kirk McKusick* **86**-1
Introduction • Secondary Storage Devices • Filesystems

87 Overview of Distributed Operating Systems   *Sape J. Mullender* **87**-1
Introduction • Survey of Distributed Systems Research • Best Practice

88 Distributed and Multiprocessor Scheduling   *Steve J. Chapin and Jon B. Weissman* **88**-1
Introduction • Issues in Multiprocessor Scheduling • Best Practices • Research Issues and Summary

89 Distributed File Systems and Distributed Memory   *T. W. Doeppner Jr.* **89**-1
Introduction • Underlying Principles • Best Practices • Research Issues and Summary

XI

Operating Systems

Operating systems provide the essential interface between the computer and its applications. This section covers core analysis, design, performance, and special challenges for operating system, distributed and highly parallel computing environments, from touched central topics in operating systems design. It serves to explore the central emphasized central topics such as those found in modern uses. Also prominent are the essential challenges of synchronization, communication among multiple concurrently executing processes, managing expensive memory resources efficiently, and designed file systems that can handle massively large data sets.

80 What Is an Operating System?   Raymond Pike 80-1
    Introduction • Historical Perspective • Goals of an Operating System • Implementing an
    Operating System • Research Issues and Summary

81 Thread Management for Shared-Memory Multiprocessors   Thomas
    E. Anderson, Brian N. Bershad, Edward D. Lazowska, and Henry M. Levy 81-1
    Introduction • Shared-Memory Multiprocessors • Issues in Thread Management •
    Three Multithreaded Systems • Summary

82 Process and Device Scheduling   Robert D. Cupper 82-1
    Introduction • Background • Processes and Scheduling • Memory Scheduling •
    Device Scheduling • Shedding Policies • Highly Evolved Scheduling •
    Recent Work and Current Research

83 Real-Time and Embedded Systems   John A. Stankovic 83-1
    Introduction • Underlying Principles • Best Practices • Research Issues and Summary

84 Process Synchronization and Interprocess Communication   Craig E. Wills ... 84-1
    Introduction • Underlying Principles • Best Practices • Research Issues and Summary

85 Virtual Memory   Peter J. Denning 85-1
    Introduction • History • Structure of Virtual Memory • Distributed Shared
    Memory • The Whole World on Virtualization • Research • Conclusion

86 Secondary Storage and Filesystems   Marshall Kirk McKusick 86-1
    Introduction • Secondary Storage Devices • Filesystems

87 Overview of Distributed Operating Systems   Sape J. Mullender 87-1
    Introduction • Survey of Distributed Systems • Conclusion • Best Practices

88 Distributed and Multiprocessor Scheduling   Steve J. Chapin
    and Jon B. Weissman 88-1
    Introduction • Issues in Multiprocessor Scheduling • Best Practices •
    Research Issues and Summary

89 Distributed File Systems and Distributed Memory   T. W. Doeppner Jr. 89-1
    Introduction • Underlying Principles • Best Practices • Research Issues and Summary

80

What Is an Operating System?

80.1 Introduction 80-1
80.2 Historical Perspective 80-1
      Open Shop Organization • Operator-Driven Shop
      Organization • Offline Loading • Spooling Systems • Batch
      Multiprogramming • Interactive Multiprogramming
      • Graphical User Interfaces (GUIs) • Distributed Computing
80.3 Goals of an Operating System 80-7
      Abstracting Reality • Managing Resources • User Interface
80.4 Implementing an Operating System 80-11
      Processes • Virtual Machines • Components of the Kernel
80.5 Research Issues and Summary 80-14

Raphael Finkel
University of Kentucky

80.1 Introduction

In brief, an **operating system** is the set of programs that control a computer. Some operating systems you may have heard of are Unix (including SCO UNIX, Linux, Solaris, Irix, and FreeBSD); the Microsoft family (MS-DOS, MS-Windows, Windows/NT, Windows 2000, and Windows XP); IBM operating systems (MVS, VM, CP, OS/2); MacOS; Mach; and VMS. Some of these (Mach and Unix) have been implemented on a wide variety of computers, but most are specific to a particular architecture, such as the Digital Vax (VMS), the Intel 8086 and successors (the Microsoft family, OS/2), the Motorola 68000 and successors (MacOS), and the IBM 360 and successors (MVS, VM, CP).

Controlling the computer involves software at several levels. We distinguish kernel services, library services, and application-level services, all of which are part of the operating system. These services can be pictured as in Figure 80.1. Applications are run by processes, which are linked together with libraries that perform standard services such as formatting output or presenting information on a display. The kernel supports the processes by providing a path to the peripheral devices. It responds to service calls from the processes and interrupts from the devices.

This chapter discusses how operating systems have evolved, often in response to architectural advances. It then examines the goals and organizing principles of current operating systems. Many books describe operating systems concepts [4–6,17–19] and specific operating systems [1,2,9–11].

80.2 Historical Perspective

Operating systems have undergone enormous changes over the years. The changes have been driven primarily by hardware facilities and their cost, and secondarily by the applications that users have wanted to run on the computers.

1-58488-360-X/$0.00+$1.50
© 2004 by CRC Press, LLC

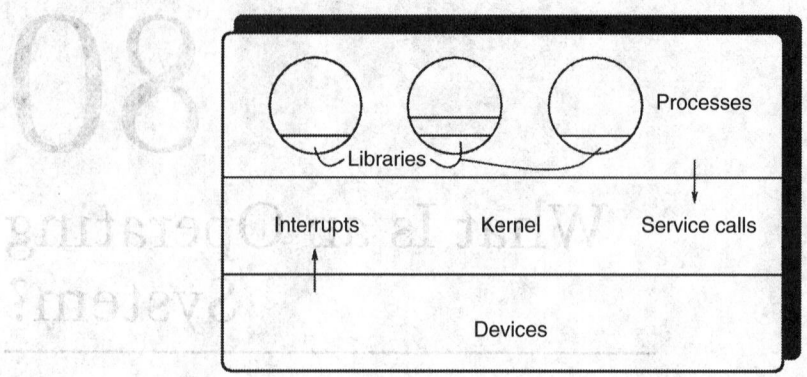

FIGURE 80.1 Operating system services.

80.2.1   Open Shop Organization

The earliest computers were massive, extremely expensive, and difficult to use. Users would sign up for blocks of time during which they were allowed "hands-on" exclusive use of the computer. The user would repeatedly load a program into the computer through a device such as a card reader, watch the results, and then decide what to do next.

A typical session on the IBM 1620, a computer in use around 1960, involved several steps in order to compile and execute a program. First, the user would load the first pass of the Fortran compiler.

This operation involved clearing main store by typing a cryptic instruction on the console typewriter; putting the compiler, a 10-inch stack of punched cards, in the card reader; placing the program to be compiled after the compiler in the card reader; and then pressing the "load" button on the reader. The output would be a set of punched cards called "intermediate output." If there were any compilation errors, a light would flash on the console, and error messages would appear on the console typewriter. If everything had gone well so far, the next step would be to load the second pass of the Fortran compiler just like the first pass, putting the intermediate output in the card reader as well. If the second pass succeeded, the output was a second set of punched cards called the "executable deck." The third step was to shuffle the executable deck slightly, load it along with a massive subroutine library (another 10 inches of cards), and observe the program as it ran.

The facilities for observing the results were limited: console lights, output on a typewriter, punched cards, and line-printer output. Frequently, the output was wrong. Debugging often took the form of peeking directly into main store and even patching the executable program using console switches. If there was not enough time to finish, a frustrated user might get a line-printer dump of main store to puzzle over at leisure. If the user finished before the end of the allotted time, the machine might sit idle until the next reserved block of time.

The IBM 1620 was quite small, slow, and expensive by our standards. It came in three models, ranging from 20K to 60K digits of memory (each digit was represented by 4 bits). Memory was built from magnetic cores, which required approximately 10 microseconds for a read or a write. The machine cost hundreds of thousands of dollars and was physically fairly large, covering about 20 square feet.

80.2.2   Operator-Driven Shop Organization

The economics of massive mainframe computers made idle time very expensive. In an effort to avoid such idleness, installation managers instituted several modifications to the open shop mechanism just outlined. An **operator** was hired to perform the repetitive tasks of loading jobs, starting the computer, and collecting the output. The operator was often much faster than ordinary users at such chores as mounting cards and magnetic tapes, so the setup time between job steps was reduced. If the program failed, the operator could have the computer produce a dump. It was no longer feasible for users to inspect main store or patch programs directly. Instead, users would submit their runs, and the operator would run them as soon as possible. Each user was charged only for the amount of time the job required.

The operator often reduced setup time by batching similar job steps. For example, the operator could run the first pass of the Fortran compiler for several jobs, save all the intermediate output, then load the second pass and run it across all the intermediate output that had been collected. In addition, the operator could run jobs out of order, perhaps charging more for giving some jobs priority over others. Jobs that were known to require a long time could be delayed until night. The operator could always stop a job that was taking too long.

The operator-driven shop organization prevented users from fiddling with console switches to debug and patch their programs. This stage of operating system development introduced the long-lived tradition of the users' room, which had long tables often overflowing with oversized fan-fold paper and a quietly desperate group of users debugging their programs until late at night.

80.2.3   Offline Loading

The next stage of development was to automate the mechanical aspects of the operator's job. First, input to jobs was collected **offline** by a separate computer (sometimes called a "satellite") whose only task was the transfer from cards to tape. Once the tape was full, the operator mounted it on the main computer. Reading jobs from tape is much faster than reading cards, so less time was occupied with input/output. When the computer finished the jobs on one tape, the operator would mount the next one. Similarly, output was generated onto tape, an activity that is much faster than punching cards. This output tape was converted to line-printer listings offline.

A small **resident monitor** program, which remained in main store while jobs were executing, reset the machine after each job was completed and loaded the next one. Conventions were established for control cards to separate jobs and specify their requirements. These conventions were the beginnings of command languages. For example, one convention was to place an asterisk in the first column of control cards, to distinguish them from data cards. The compilation job we just described could be specified in cards that looked like this:

```
*JOB SMITH            The user's name is Smith.
*    PASS CHESTNUT    Password so others can't use Smith's account
*    OPTION TIME=60   Limit of 60 seconds
*    OPTION DUMP=YES  Produce a dump if any step fails.
*STEP FORT1           Run the first pass of the Fortran compiler.
*    OUTPUT TAPE1     Put the intermediate code on tape 1.
*    INPUT FOLLOWS    Input to the compiler comes on the next cards.
     ...              Fortran program
*STEP FORT2           Run the second pass of the Fortran compiler.
*    OUTPUT TAPE2     Put the executable deck on scratch tape 2.
*    INPUT TAPE1      Input comes from scratch tape 1.
*STEP LINK            Link the executable with the Fortran library.
*    INPUT TAPE2      First input is the executable.
*    INPUT TAPELIB    Second input is a tape with the library.
*    OUTPUT TAPE1     Put load image on scratch tape 1.
*STEP TAPE1           Run whatever is on scratch tape 1.
*    OUTPUT TAPEOUT   Put output on the standard output tape.
*    INPUT FOLLOWS    Input to the program comes on the next cards.
     ...              Data
```

The resident monitor had several duties, including:

- Interpret the command language
- Perform rudimentary accounting
- Provide device-independent input and output by substituting tapes for cards and line printers

This last duty is an early example of information hiding and abstraction: programmers would direct output to cards or line printers but, in fact, the output would go elsewhere. Programs called subroutines provided by the resident monitor for input/output to both logical devices (cards, printers) and physical devices (actual tape drives).

The early operating systems for the IBM 360 series of computer used this style of control. Large IBM 360 installations could cost millions of dollars, so it was important not to let the computer sit idle.

80.2.4 Spooling Systems

Computer architecture advanced throughout the 1960s. (We survey computer architecture in Section II.) Input/output units were designed to run at the same time the computer was computing. They generated an interrupt when they finished reading or writing a record instead of requiring the resident monitor to track their progress. As mentioned, an interrupt causes the computer to save some critical information (such as the current program counter) and to branch to a location specific to the kind of interrupt. Device-service routines, known as **device drivers**, were added to the resident monitor to deal with these interrupts.

Drums and, later, disks were introduced as a secondary storage medium. Now the computer could be computing one job while reading another onto the drum and printing the results of a third from the drum. Unlike a tape, a drum allows programs to be stored anywhere, so there was no need for the computer to execute jobs in the same order in which they were entered. A primitive **scheduler** was added to the resident monitor to sort jobs based on priority and amount of time needed, both specified on control cards. The operator was retained to perform several tasks:

- Mount data tapes needed by jobs (specified on control cards, which caused request messages to appear on the console typewriter).
- Decide which priority jobs to run and which to hold.
- Restart the resident monitor when it failed or was inadvertently destroyed by the running job.

This mode of running a computer was known as a **spooling system**, and its resident monitor was the start of modern operating systems. (The word "spool" originally stood for "simultaneous peripheral operations on line," but it is easier to picture a spool of thread, where new jobs are wound on the outside, and old ones are extracted from the inside.) One of the first spooling systems was HASP (the Houston Automatic Spooling Program), an add-on to OS/360 for the IBM 360 computer family.

80.2.5 Batch Multiprogramming

Spooling systems did not make efficient use of all the hardware resources. The currently running job might not need the entire main store. A job performing input/output causes the computer to wait until the input/output finishes. The next software improvement, which occurred in the early 1960s, was the introduction of **multiprogramming**, a scheme in which more than one job is active simultaneously.

Under multiprogramming, while one job waits for an input/output operation to complete, another can compute. With luck, no time at all is wasted waiting for input/output. The more simultaneous jobs, the better. However, a **compute-bound** job (one that performs little input/output but much computation) can easily prevent **input/output-bound** jobs (those that perform mostly input/output) from making progress. Competition for the time resource and policies for allocating it are the main theme of Chapter 82.

Multiprogramming also introduces competition for memory. The number of jobs that can be accommodated at one time depends on the size of main store and the hardware available for subdividing that space. In addition, jobs must be secured against inadvertent or malicious interference or inspection by other jobs. It is more critical now that the resident monitor not be destroyed by errant programs, because not one but many jobs suffer if it breaks. In Chapter 85, we examine policies for memory allocation and how each of them provides security.

The form of multiprogramming we have been describing is often called **batch multiprogramming** because jobs are grouped into batches: those that need small memory, those that need customized tape mounts, those that need long execution, etc. Each batch might have different priorities and fee structures. Some batches (such as large-memory, long-execution jobs) can be scheduled for particular times (such as weekends or late at night). Generally, only one job from any batch can run at any one time.

Each job is divided into discrete steps. Because job steps are independent, the resident monitor can separate them and apply policy decisions to each step independently. Each step might have its own time, memory, and input/output requirements. In fact, two separate steps of the same job can be performed at the same time if they do not depend on each other. The term **process** was introduced in the late 1960s to mean the entity that performs a single job step. The operating system (as the resident monitor may now be called) represents each process by a data structure sometimes called a **process descriptor**, **process control block**, or **context block**. The process control block includes billing information (owner, time used), scheduling information, and the resources the job step needs. While it is running, a process may request assistance from the kernel by submitting a service call across the **process interface**. Executing programs are no longer allowed to control devices directly; otherwise, they could make conflicting use of devices and prevent the kernel from doing its work. Instead, processes must use service calls to access devices, and the kernel has complete control of the **device interface**.

Allocating resources to processes is not a trivial task. A process might require resources (such as tape drives) at various stages in its execution. If a resource is not available, the scheduler might block the process from continuing until later. The scheduler must take care not to block any process forever.

Along with batch multiprogramming came new ideas for structuring the operating system. The kernel of the operating system is composed of routines that manage central store, CPU time, devices, and other resources. It responds both to requests from processes and to interrupts from devices. In fact, the kernel runs only when it is invoked either from above, by a process, or below, by a device. If no process is ready to run and no device needs attention, the computer sits idle.

Various activities within the kernel share data, but they must not be interrupted when the data is in an inconsistent state. Mechanisms for **concurrency control** were developed to ensure that these activities do not interfere with each other. Chapter 84 introduces the mutual-exclusion and synchronization problems associated with concurrency control and surveys the solutions that have been found for these problems. The MVS operating system for the IBM 360 family was one of the first to use batch multiprogramming.

80.2.6 Interactive Multiprogramming

The next step in the development of operating systems was the introduction of **interactive multiprogramming**, also called **timesharing**. The principal user-oriented input/output device changed in the late 1960s from cards or tape to an interactive terminal. Instead of packaging all the data that a program might need before it starts running, the interactive user is able to supply input as the program wants it. The data can depend on what the program has produced thus far. Among the first terminals were teletypes, which produced output on paper at perhaps ten characters per second. Later terminals were called "glass teletypes" because they displayed characters on a television screen, substituting electronics for mechanical components. Like a regular teletype, they could not back up to modify data sitting earlier on the screen. Shortly thereafter, terminals gained cursor addressibility, which meant that programs could show entire "pages" of information and change any character anywhere on a page.

Interactive computing caused a revolution in the way computers were used. Instead of being treated as number crunchers, computers became information manipulators. Interactive text editors allowed users to construct data files online. These files could represent programs, documents, or data. As terminals improved, so did the text editors, changing from line- or character-oriented interfaces to full-screen interfaces.

Instead of representing a job as a series of steps, interactive multiprogramming identifies a **session** that lasts from initial connection ("login") to the point at which that connection is broken ("logout"). During login, the user typically gives two forms of identification: a user name and a password. The password is not echoed at the terminal, or is at least blackened by overstriking garbage, to avoid disclosing it to onlookers. This data is converted into a **user identifier** that is associated with all the processes that run on behalf of this user and all the files he or she creates. This identifier helps the kernel decide whom to bill for services and whether to permit various actions such as modifying files. (We discuss files in Chapter 86 and protection in Chapter 89.)

During a session, the user imagines that the resources of the entire computer are devoted to this terminal, although many sessions may be active simultaneously for many users. Typically, one process is created at login time to serve the user. That first process, which is usually a command interpreter, may start others as needed to accomplish individual steps.

Users need to save information from session to session. Magnetic tape is too unwieldy for this purpose. Disk storage became the medium of choice for data storage, both short term (temporary files used to connect steps in a computation), medium term (from session to session), and long-term (from year to year). Issues of disk space allocation and backup strategies needed to be addressed to provide this facility.

Interactive computing was sometimes added into an existing batch multiprogramming environment. For example, TSO ("timesharing option") was an add-on to the OS/360 operating system. The EXEC-8 operating system for Univac computers also included an interactive component.

Later operating systems were designed from the outset to support interactive use, with batch facilities added when necessary. TOPS-10 and Tenex (for the Digital PDP-10), and almost all operating systems developed since 1975, including Unix (first on the Digital PDP-11), MS-DOS (Intel 8086), OS/2 (Intel 286 family [10]), VMS (Digital VAX [9]), and all their descendents, were primarily designed for interactive use.

80.2.7   Graphical User Interfaces (GUIs)

As computers became less expensive, the time cost of switching from one process to another (which happens frequently in interactive computing) became insignificant. Idle time also became unimportant. Instead, the goal became helping users get their work done efficiently. This goal led to new software developments, enabled by improved hardware.

Graphics terminals, first introduced in the mid-1970s, have led to the video monitors that are now ubiquitous and inexpensive. These monitors allow individual control of multicolored pixels; a high-quality monitor (along with its video controller) can display millions of pixels in an enormous range of colors. Pointing devices, particularly the mouse, were developed in the late 1970s. Software links them to the display so that a visible cursor reacts to physical movements of the pointing device. These hardware advances have led to **graphical user interfaces** (GUIs), discussed in Chapter 48.

The earliest GUIs were just rectangular regions of the display that contained, effectively, a cursor-addressable glass teletype. These regions are called "windows." The best-known windowing packages were those pioneered by MacOS [15] and the later ones introduced by MS-Windows, OS/2 [10] and X Windows (for UNIX, VMS, and other operating systems [12]). Each has developed from simple rectangular models of a terminal to significantly more complex displays.

Programs interact with the hardware by invoking routines in libraries that know how to communicate with the display manager, which itself knows how to place bits on the screen. The early libraries were fairly low-level and difficult to use; toolkits (in the X Windows environment), especially ones with a fairly small interpreted language (such as Tcl/Tk [13] or Visual Basic), have eased the task of building good GUIs. Early operating systems that supported graphical interfaces, such as MacOS and MS-Windows, provided interactive computing but not multiprogramming. Modern operating systems all provide multiprogramming as well as interaction, allowing the user to start several activities and to switch attention to whichever one is currently most interesting.

80.2.8   Distributed Computing

At the same time that displays were improving, networks of computers were being developed. A network requires not only hardware to physically connect machines, but also protocols to use that hardware effectively, operating system support to make those protocols available to processes, and applications that make use of these protocols. Chapter 87 through Chapter 89 are devoted to the issues raised by networks.

Computers can be connected together by a variety of devices. The spectrum ranges from tight coupling, where several computers share main storage, to very loose coupling, where a number of computers belong to the same international network and can send one another messages.

The ability to send messages between computers opened new opportunities for operating systems. Individual machines become part of a larger whole and, in some ways, the operating system begins to span networks of machines. Cooperation between machines takes many forms.

- Each machine may offer **network services** to others, such as accepting mail, providing information on who is currently logged in, telling what time it is (quite important in keeping clocks synchronized), allowing users to access machines remotely, and transferring files.

- Machines within the same **site** (typically, those under a single administrative control) may **share file systems** in order to reduce the amount of disk space needed and to allow users to have accounts on multiple machines. Novell nets (MS-DOS), the Sun and Andrew network file systems (UNIX), and the Microsoft File-Sharing Protocol (Windows XP) are examples of such arrangements. Shared file systems are an essential component of a **networked operating system**.

- Once users have accounts on several machines, they want to associate graphical windows with sessions on different machines. The machine on which the display is located is called a **thin client** of the machine on which the processes are running. Thin clients have been available from the outset for X Windows; they are also available under Windows 2000 and successors.

- Users want to execute computationally intensive algorithms on many machines in parallel. **Middleware**, usually implemented as a library to be linked into distributed applications, makes it easier to build such applications. PVM [7] and MPI [14] are examples of such middleware.

- Standardized ways of presenting data across site boundaries developed rapidly. The File-Transfer Protocol (**ftp**) service was developed in the early 1970s as a way of transferring files between machines connected on a network. In the early 1990s, the **gopher** service was developed to create a uniform interface for accessing information across the Internet. Information is more general than just files; it can be a request to run a program or to access a database. Each machine that wishes to can provide a server that responds to connections from any site and communicate a menu of available information. This service was superseded in 1995 by the **World Wide Web**, which supports a GUI to gopher, ftp, and hypertext (documents with links internally and to other documents, often at other sites, and including text, pictures, video, audio, and remote execution of packaged commands).

Of course, all these forms of cooperation introduce security concerns. Each site has a responsibility to maintain security if for no other reason than to prevent malicious users across the network from using the site as a breeding ground for nasty activity attacking other sites. Security issues are discussed in Chapter 77 through Chapter 79.

80.3   Goals of an Operating System

During the evolution of operating systems, their purposes have also evolved. At present, operating systems have three major goals:

1. Hide details of hardware by creating abstractions.
2. Manage resources.
3. Provide a pleasant and effective user interface.

We address each of these goals in turn.

80.3.1   Abstracting Reality

We distinguish between the **physical** world of devices, instructions, memory, and time, and the **virtual** world that is the result of abstractions built by the operating system. An **abstraction** is software (often implemented as a subroutine or as a library of subroutines) that hides lower-level details and provides a set of higher-level functions. Programs that use abstraction can safely ignore the lower-level (physical) details; they need only deal with the higher-level (virtual) structures.

Why is abstraction important in operating systems? First, the code needed to control peripheral devices is often not standardized; it can vary from brand to brand, and it certainly varies between, say, disks and tape drives and keyboards. Input/output devices are extremely difficult to program efficiently and correctly. Abstracting devices with a uniform interface makes programs easier to write and to modify (e.g., to use a different device). Operating systems provide subroutines called **device drivers** that perform input/output operations on behalf of programs. The operations are provided at a much higher level than the device itself provides. For example, a program may wish to write a particular block on a disk. Low-level methods involve sending commands directly to the disk to seek to the right block and then undertake memory-to-disk data transfer. When the transfer is complete, the disk interrupts the running program. A low-level program needs to know the format of disk commands, which vary from manufacturer to manufacturer and must deal with interrupts. In contrast, a program using a high-level routine in the operating system might only need to specify the memory location of the data block and where it belongs on the disk; all the rest of the machinery is hidden.

Second, the operating system introduces new functions as it abstracts the hardware. In particular, operating systems introduce the "file" abstraction. Programs do not need to deal with disks at all; they can use high-level routines to read and write disk files (instead of disk blocks) without needing to design storage layouts, worry about disk geometry, or allocate free disk blocks.

Third, the operating system transforms the computer hardware into multiple virtual computers, each belonging to a different process. Each process views the hardware through the lens of abstraction; memory, time, and other resources are all tailored to the needs of the process. Processes see only as much memory as they need, and that memory does not contain the other processes (or the operating system) at all. They think that they have all the CPU cycles on the machine, although other processes and the operating system itself are competing for those cycles. Service calls allow processes to start other processes and to communicate with other processes, either by sending messages or by sharing memory.

Fourth, the operating system can enforce security through abstraction. The operating system must secure both itself and its processes against accidental or malicious interference. Certain instructions of the machine, notably those that halt the machine and those that perform input and output, are moved out of the reach of processes. Memory is partitioned so that processes cannot access each other's memory. Time is partitioned so that even a run-away process will not prevent others from making progress.

For security and reliability, it is wise to structure an operating system so that processes must use the operating system's abstractions instead of dealing with the physical hardware. This restriction can be enforced by the hardware, which provides several **processor states**. Most architectures provide at least two states: the **privileged state** and the **non-privileged state**.

Processes always run in non-privileged state. Instructions such as those that perform input/output and those that change processor state cause traps when executed in non-privileged state. Traps save the current execution context (perhaps on a stack), force the processor to jump to the operating system, and enter the privileged state. Once the operating system has finished servicing the trap or interrupt, it returns control to the same process or perhaps to a different one, resetting the computer into non-privileged state.

The core of the operating system runs in privileged state. All instructions have their usual, physical meanings in this state. The part of the operating system that always runs in privileged state is the **kernel** of the operating system. It only runs when a process has caused a trap or when a peripheral device has generated an interrupt. Traps do not necessarily represent errors; usually, they are service calls. Interrupts often indicate that a device has finished servicing a request and is ready for more work. The clock interrupts at a regular rate in order to let the kernel make scheduling decisions.

If the operating system makes use of this dichotomy of states, the abstractions that the operating system provides are presented to processes as **service calls**, which are like new CPU instructions. A program can perform high-level operations (such as reading a file) with a single service call. Executing the service call generates a trap, which causes a switch to the privileged state of the kernel. The advantage of the service-call design over a procedure-call design is that it allows access to kernel operations and data only through well-defined entry points.

Not all operating systems make use of non-privileged state. MS-DOS, for example, runs all applications in privileged state. Service calls are essentially subroutine calls. Although the operating system provides device and file abstractions, processes may interact directly with disks and other devices. One advantage of this choice is that device drivers can be loaded after the operating system starts; they do not need special privilege. One disadvantage is that viruses can thrive because nothing prevents a program from placing data anywhere it wishes.

80.3.2 Managing Resources

An operating system is not only an abstractor of information, but also an allocator that controls how processes (the active agents) can access resources (passive entities).

A **resource** is a commodity necessary to get work done. The computer's hardware provides a number of low-level resources. Working programs need to reside somewhere in main store (the computer's memory), must execute instructions, and need some way to accept data and present results. These needs are related to the fundamental resources of **memory**, **CPU time**, and **input/output**. The operating system abstracts these resources to allow them to be shared.

In addition to these physical resources, the operating system creates virtual, abstract resources. For example, **files** are able to store data. They abstract the details of disk storage. **Pseudo-files** (i.e., objects that appear to be data files on disk but are in fact stored elsewhere) can also represent devices, processes, communication ports, and even data on other computers. **Sockets** are process-to-process communication channels that can cross machine boundaries, allowing communication through networks such as the Internet. Sockets abstract the details of transmission media and network protocols.

Still higher-level resources can be built on top of abstractions. A **database** is a collection of information, stored in one or more files with structure intended for easy access. A **mailbox** is a file with particular semantics. A **remote file**, located on another machine but accessed as if it were on this machine, is built on both file and network abstractions.

The resource needs of processes often interfere with each other. Resource managers in the operating system include policies that try to be fair in giving resources to the processes and allow as much computation to proceed as possible. These goals often conflict.

Each resource has its own manager, typically in the kernel. The memory manager allocates regions of main memory for processes. Modern operating systems use address translation hardware that maps between a process's **virtual addresses** and the underlying **physical addresses**. Only the currently active part of a process's virtual space needs to be physically resident; the rest is kept on backing store (usually a disk) and brought in on demand. The virtual spaces of processes do not usually overlap, although some operating systems also provide **light-weight processes** that share a single virtual space. The memory manager includes policies that determine how much physical memory to grant to each process and which region of physical memory to swap out to make room for other memory that must be swapped in. For more information on virtual memory, see Chapter 85.

The CPU-time manager is called the **scheduler**. Schedulers usually implement a preemptive policy that forces the processes to take turns running. Schedulers categorize processes according to whether they are currently runnable (they may not be if they are waiting for other resources) and their priority.

The file manager mediates process requests such as creating, reading, and writing files. It validates access based on the identity of the user running the process and the permissions associated with the file. The file manager also prevents conflicting accesses to the same file by multiple processes. It translates input/output requests into device accesses, usually to a disk, but often to networks (for remote files) or other devices (for pseudo-files).

The device managers convert standard-format requests into the particular commands appropriate for individual devices, which vary widely among device types and manufacturers. Device managers may also maintain caches of data in memory to reduce the frequency of access to physical devices.

Although we usually treat processes as autonomous agents, it is often helpful to remember that they act on behalf of a higher authority: the human **users** who are physically interacting with the computer. Each process is usually "owned" by a particular user. Many users may be competing for resources on the same machine. Even a single user can often make effective use of multiple processes.

Each user application is performed by a process. When a user wants to compose a letter, a process runs the program that converts keystrokes into changes in the document. When the user mails that letter electronically, a process runs a program that knows how to send documents to mailboxes.

To service requests effectively, the operating system must satisfy two conflicting goals:

1. To let each process have whatever resources it wants
2. To be fair in distributing resources among the processes

If the active processes cannot all fit in memory, for example, it is impossible to satisfy the first goal without violating the second. If there is more than one process, it is impossible on a single CPU to give all processes as much time as they want; CPU time must be shared.

To satisfy the computer's owner, the operating system must also satisfy a different set of goals:

1. To make sure the resources are used as much as possible
2. To complete as much work as possible

These latter goals were once more important than they are now. When computers were all expensive mainframes, it seemed wasteful to let any time pass without a process using it, or to let any memory sit unoccupied by a process, or to let a tape drive sit idle. The measure of success of an operating system was how much work (measured in "jobs") could be finished and how heavily resources were used. Computers are now far less inexpensive; we no longer worry if computers sit idle, although we still prefer efficient use of resources.

80.3.3   User Interface

We have seen how operating systems are creators of abstractions and allocators of resources. Both of these aspects center on the needs of programmers and the processes that execute programs. But many users are not programmers and are uninterested in the process abstraction and in the interplay between processes and the operating system. They do not care about service calls, interrupts, and devices. Instead, they are interested in what might be termed the "look and feel" of the operating system.

The user interacts with the operating system through the **user interface**. Human–computer interaction is covered in detail in Section V of this Handbook. Here we will only point out some highlights.

The hardware for user interfaces has seen rapid change over the past 50 years, ranging over plugging wires into a plugboard (e.g., IBM 610, 1957), punching cards and reading printouts (IBM 1620, 1959), remote teletype (DEC PDP-10, 1967), monochrome glass teletypes (around 1973), monochrome graphics terminals with pointing devices (Xerox PARC's Alto computer, around 1974), color video CRTs (around 1980), and LCD displays (late 1980s).

User-interface software has steadily changed as well. Interactive text editors (WYLBUR and TECO, around 1975) replaced punched paper cards. Interactive command languages replaced job-control languages. Programming environments integrating editing, compiling, and debugging were introduced as early as 1980 (Smalltalk) and are still in heavy use (MetroWerks Code Warrior; Microsoft Visual Studio). Data entry moved from line-oriented to forms-based (by 1980) to Web-based (1995). Many user interfaces are now navigated without needing a keyboard at all; the user clicks a mouse to move to the next step in a process. Voice-activated commands are also gaining in popularity.

The "look and feel" of an operating system is affected by many components of the user interface. Some of the most important are the process launcher (a command interpreter, a menu-driven GUI, or clickable of icons); the file system (including remote files); online help; and application integration (such as ability to insert pictures in text files).

80.4 Implementing an Operating System

As mentioned, the core of the operating system is the **kernel**, a control program that functions in privileged state, reacting to interrupts from external devices and to service requests and traps from processes. In general, the kernel is a permanent resident of the computer. It creates and terminates processes and responds to their requests for service.

80.4.1 Processes

Each process is represented in the kernel by a collection of data called the **process descriptor**. A process descriptor includes such information as:

- Processor state: stored values of the program counter and registers, needed to resume execution of the process.
- Scheduling statistics, needed to determine when to resume the process and how much time to let it run.
- Memory allocation, both in main memory and backing store (disk), needed to accomplish memory management.
- Other resources held, such as locks or semaphores, needed to manage contention for such resources.
- Open files and pseudo-files (devices, communication ports), needed to interpret service requests for input and output.
- Accounting statistics, needed to bill users and determine hardware usage levels.
- Privileges, needed to determine if activities such as opening files and executing potentially dangerous service calls should be allowed.
- Scheduling state: running, ready, waiting for input/output or some other resource, such as memory.

The process descriptors can be saved in an array, in which case each process can be identified by the index of its descriptor in that array. Other structures are possible, of course, but the concept of **process number** is across operating systems. Some of the information in the process descriptor can be bulky, such as the page tables. Page tables for idle processes can be stored on disk to save space in main memory.

Resuming a process, that is, switching control from the kernel back to the process, is a form of **context switching**. It requires that the processor move from privileged to unprivileged state, that the registers and program counter of the process be restored, and that the address-translation hardware be set up to accomplish the correct mappings for this process. Switching back to the kernel is also a context switch; it can happen when the process tries to execute a privileged instruction (including the service call instruction) or when a device generates an interrupt.

Hardware is designed to switch context rapidly. For example, the hardware may maintain two sets of registers and address translation data, one for each privilege level. Context switches into the kernel just require moving to the kernel's set of registers. Resuming the most recently running process is also fast. Resuming a different process requires that the kernel load all the information for the new process into the second set of registers; this activity takes longer. For that reason, a **process switch** is often more expensive than two context switches.

80.4.2 Virtual Machines

Although most operating systems try to present to processes an enhanced and simplified view of the hardware, some take a different tack. They make the process interface look just like the hardware interface, except that the size of memory and the types, numbers, and sizes of input/output devices may be more or less than the physical resources. However, a process is allowed to use all the machine instructions, even the privileged ones.

Under this organization, the process interface is called a **virtual machine** because it looks just like the underlying machine. The kernel of such an operating system is called a **virtualizing kernel**. Each virtual machine runs its own ordinary operating system.

We examine virtual operating systems in some detail because they elucidate the interplay of traps, context switches, processor states, and the fact that a process at one level is just a data structure at a lower level. Virtualizing kernels were first developed (IBM VM, early 1970s) to allow operating system designers to experiment with new versions of an operating system on machines that were too expensive to dedicate to such experimentation. More importantly, virtualizing kernels allow multiple operating systems to run simultaneously on the same machine to satisfy a wide variety of users.

This idea is still valuable. The Wine program emulates the Win32 environment (used by Windows XP) as it runs as a process under UNIX, allowing a Unix user who has Windows programs to run them at the same time as other applications. Mach emulates Unix and can accept modules that emulate other operating systems as well. This emulation is at the library-routine level; service calls are converted to messages directed to a UNIX-emulator process that provides all the services. The NT [3] and OS/2 [10] operating systems for Intel computers also provide for virtual machines running other operating systems.

In a true virtualizing kernel, the hardware executes most instructions (such as arithmetic and data motion) directly. However, privileged instructions, such as the halt instruction, are just too dangerous to let processes use directly. Instead, the virtualizing kernel must run all processes in non-privileged state to prevent them from accidentally or maliciously interfering with each other and with the kernel itself.

To let each process P imagine it has control of processor states, the kernel keeps track of the virtual processor state of each P, that is, the processor state of the virtual machine that the kernel emulates on behalf of P. This information is stored in P's context block inside the kernel. All privileged instructions executed by P cause traps to the kernel, which then emulates the behavior of the hardware on behalf of P.

- If P is in virtual non-privileged state, the kernel emulates a trap for P. This emulation puts P in virtual privileged state, although it is still running in physical non-privileged state. The program counter for P is reset to the proper trap address within P's virtual space.

- If P is in virtual privileged state, the kernel emulates the action of the instruction itself. For example, it terminates P on a halt instruction, and it executes input/output instructions interpretively.

Some dangerous instructions are particularly difficult to emulate. Input/output can be very tricky. Address translation also becomes quite complex. A good test of a virtualizing kernel is to let one of its processes be another virtualizing kernel. For example, consider Figure 80.2, in which there are two levels

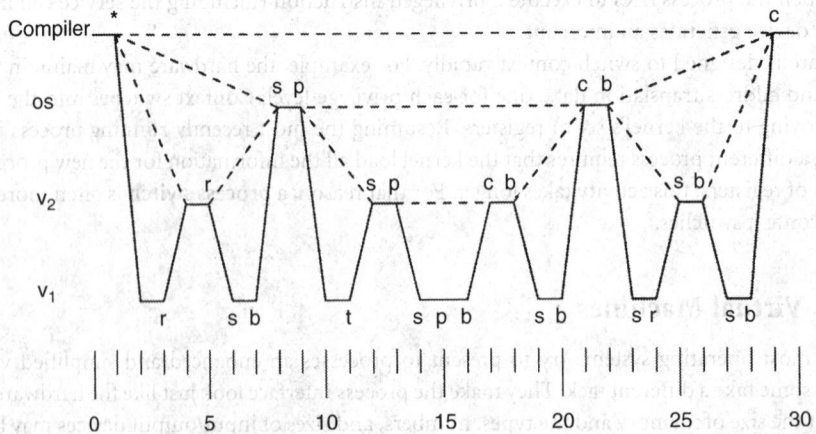

FIGURE 80.2   Emulating a service call.

of virtualizing kernel, V_1 and V_2, above which sits an ordinary operating system kernel, **OS**, above which a compiler is running.

The compiler executes a single service call (marked "*") at time 1. As far as the compiler is concerned, **OS** performs the service and lets the compiler continue (marked "c") at time 29. The dashed line at the level of the compiler indicates the compiler's perception that no activity below its level takes place during the interval.

From the point of view of **OS**, a trap occurs at time 8 (marked by a dot on the control-flow line). This trap appears to come directly from the compiler, as shown by the dashed line connecting the compiler at time 1 and the **OS** at time 8. **OS** services the trap (marked "s"). For simplicity, we assume that it needs to perform only one privileged instruction (marked "p") to service the trap, which it executes at time 9. Lower levels of software (which **OS** cannot distinguish from hardware) emulate this instruction, allowing **OS** to continue at time 21. It then switches context back to the compiler (marked "b") at time 22. The dashed line from **OS** at time 22 to the compiler at time 29 shows the effect of this context switch.

The situation is more complicated from the point of view of V_2. At time 4, it receives a trap that tells it that its client has executed a privileged instruction while in virtual non-privileged state. V_2 therefore reflects this trap at time 5 (marked "r") back to **OS**. Later, at time 12, V_2 receives a second trap, this time because its client has executed a privileged instruction in virtual privileged state. V_2 services this trap by emulating the instruction itself at time 13. By time 17, the underlying levels allow it to continue, and at time 18 it switches context back to **OS**. The last trap occurs at time 25, when its client has attempted to perform a context switch (which is privileged) when in virtual privileged state. V_2 services this trap by changing its client to virtual non-privileged state and switching back to the client at time 26.

V_1 has the busiest schedule of all. It reflects traps that arrive at time 2, 10, and 23. (The trap at time 23 comes from the context-switch instruction executed by **OS**.) It also emulates instructions for its client when traps occur at times 5, 14, 19, and 27.

This example demonstrates the principle that each software level is just a data structure as far as its supporting level is concerned. It also shows how a single privileged instruction in the compiler becomes two privileged instructions in **OS**, which becomes four in V_2 and eight in V_1. In general, a single privileged instruction at one level might require many instructions at its supporting level to emulate it.

80.4.3 Components of the Kernel

Originally, operating systems were written as a single large program encompassing hundreds of thousands of lines of assembly-language instructions. Two trends have made the job of implementing operating systems less difficult. First, high-level languages have made programming much easier. For example, more than 99% of the Linux variant of Unix is written in C. Complex algorithms can be expressed in a structured, readable fashion; code can be partitioned into modules that interact with each other in a well-defined manner, and compile-time typechecking catches most programming errors. Only a few parts of the kernel, such as those that switch context or modify execution priority, need to be written in assembly language.

Second, the discipline of structured programming has suggested a layered approach to designing the kernel. Each layer provides abstractions needed by the layers above it. For example, the kernel can be organized as follows:

- Context- and process-switch services (lowest layer)
- Device drivers
- Resource managers for memory and time
- File system support
- Service call interpreter (highest layer)

For example, the MS-DOS operating system provides three levels: (1) device drivers (the BIOS section of the kernel), (2) a file manager, and (3) an interactive command interpreter. It supports only one process and provides no security, so there is no need for context-switch services. Because service calls do not need to cross protection boundaries, they are implemented as subroutine calls.

The concept of layering allows the kernel to be small, because much of the work of the operating system need not operate in a protected and hardware-privileged environment. When all the layers listed above are privileged, the organization is called a **macrokernel**. UNIX is often implemented as a macrokernel.

If the kernel only contains code for process creation, inter-process communication, the mechanisms for memory management and scheduling, and the lowest level of device control, the result is a **microkernel**, also called a "communication kernel." Mechanisms are distinct from policies, which can be outside the kernel. Policies decide which resources should be allocated in cases of conflict, whereas mechanisms carry out those decisions. Mach [16] and QNX [8] follow the microkernel approach. In this organization, services such as the file system and policy modules for scheduling and memory are relegated to processes. These processes are often referred to as **servers**; the ordinary processes that need those services are called their **clients**. The microkernel itself acts as a client of the policy servers. Servers need to be trusted by their clients, and sometimes they need to execute with some degree of hardware privilege (for example, if they access devices).

The microkernel approach has some distinct advantages:

- It imposes uniformity on the requests that a process might make. Processes need not distinguish between kernel-level and process-level services because all are provided via messages to servers.
- It allows easier addition of new services, even while the operating system is running, as well as multiple services that cover the same set of needs, so that individual users (and their agent processes) can choose whichever seems best. For example, different file organizations for diskettes are possible; instead of having many file-level modules in the kernel, there can be many file-level servers accessible to processes.
- It allows an operating system to span many machines in a natural way. As long as inter-process communication works across machines, it is generally immaterial to a client where its server is located.
- Services can be provided by teams of servers, any one of which can help any client. This organization relieves the load on popular servers, although it often requires a degree of coordination among the servers on the same team.

A microkernel also has some disadvantages. It is generally slower to build and send a message, accept and decode the reply (taking about 100 µs), than to make a single service call (taking about 1 µs). However, other aspects of service tend to dominate the cost, allowing microkernels to be similar in speed to macrokernels. Keeping track of which server resides on which machine can be complex. This complexity may be reflected in the user interface. The perceived complexity of an operating system has a large effect on its acceptance by the user community.

Recently, people have begun to speak of **nanokernels**, which support only devices and communication ports. They sit at the bottom level of the microkernel, providing services for the other parts of the microkernel, such as memory management. All the competing executions supported by the nanokernel are called **threads**, to distinguish them from **processes**. Threads all share kernel memory, and they explicitly yield control in order to let other threads continue. They synchronize with each other by means of primitive locks or more complex semaphores. For more information on processes and threads, see chapters 93 and 97.

Although the trend toward microkernels is unmistakable, macrokernels are likely to remain popular for the forseeable future.

80.5 Research Issues and Summary

Operating systems have developed enormously in the past 45 years. Modern operating systems generally have three goals: (1) to hide details of hardware by creating abstractions, (2) to allocate resources to processes, and (3) to provide an effective user interface. Operating systems generally accomplish these goals by running processes in low privilege and providing service calls that invoke the operating system kernel in high privilege state. The recent trend has been toward increasingly integrated graphical user interfaces that encompass the activities of multiple processes on networks of computers. These increasingly sophisticated application programs are supported by increasingly small operating system kernels.

Current research issues revolve mostly around networked operating systems, including network protocols, distributed shared memory, distributed file systems, mobile computing, and distributed application support. There is also active research in kernel structuring, file systems, and virtual memory.

Defining Terms

The following terms may have more general definitions than shown here, and often have other narrow technical definitions. This list indicates how the terms have been used in this chapter.

Abstraction: An interface that hides lower-level details and provides a set of higher-level functions.

Batch multiprogramming: Grouping jobs into batches based on characteristics such as memory requirements.

Client: A process that requests services by sending messages to server processes.

Command interpreter: A program (usually not in the kernel) that interprets user requests and starts computations to fulfill those requests.

Commands: Instructions in a job-control language.

Compute-bound: A process that performs little input/output but needs significant execution time.

Concurrency control: Means to mediate conflicting needs of simultaneously executing threads.

Context block: Process descriptor.

Context switching: The action of directing the hardware to execute in a different context (kernel or process) from the current context.

Database: A collection of files for storing related information.

Device driver: An operating-system module (usually in the kernel) that deals directly with a device.

Device interface: The means by which devices are controlled.

File: A named, long-term repository for data.

ftp: The file-transfer protocol service.

Gopher: A network service that connects information providers to their users.

Graphical user interface: Interactive program that makes use of a graphic display and a mouse.

Input/output: A resource: ability to interact with peripheral devices.

Input/output-bound: A process that spends most of its time waiting for input/output.

Integrated application: An application that agrees on data formats with other applications so they can use each other's outputs.

Interactive multiprogramming: Multiprogramming in which each user deals interactively with the computer.

Job: A set of computational steps packaged to be run as a unit.

Job-control language: A way of specifying the resource requirements of various steps in a job.

Kernel: The privileged core of an operating system, responding to service calls from processes and interrupts from devices.

Lightweight process: A thread.

Macrokernel: A large operating-system core that provides a wide range of services.

Mailbox: A file for saving messages between users.

Memory: A resource: ability to store programs and data.

Microkernel: A small privileged operating-system core that provides process scheduling, memory management, and communication services.

Middleware: Program that provides high-level communication facilities to allow distributed computation.

Multiprogramming: Scheduling several competing processes to run at essentially the same time.

Nanokernel: A very small privileged operating-system core that provides simple process scheduling and communication services.

Network services: Services available through the network, such as mail and file transfer.

Network service: A facility offered by one computer to other computers connected to it by a network.

Networked operating system: An operating system that uses a network for sharing files and other resources.

Non-privileged state: An execution context that does not allow sensitive hardware instructions to be executed, such as the halt instruction and input/output instructions.

Offline: Handled on a different computer.

Operating system: A set of programs that controls a computer.

Operator: An employee who performs the repetitive tasks of loading and unloading jobs.

Physical: The material upon which abstractions are built.

Physical address: A location in physical memory.

Pipeline: A facility that allows one process to send a stream of information to another process.

Privileged state: An execution context that allows all hardware instructions to be executed.

Process: A program being executed; an execution context that is allocated resources such as memory, time, and files.

Process control block: Process descriptor.

Process descriptor: A data structure in the kernel that represents a process.

Process interface: The set of service calls available to processes.

Process number: An identifier that represents a process by acting as an index into the array of process descriptors.

Process switch: The action of directing the hardware to run a different process from the one that was previously running.

Processor state: Privileged or non-privileged state.

Pseudo-file: An object that appears to be a file on the disk but is actually some other form of data.

Remote file: A file on another computer that appears to be on the user's computer.

Resident monitor: A precursor to kernels; a program that remains in main store during the execution of a job to handle simple requests and to start the next job.

Resource: A commodity necessary to get work done.

Scheduler: An operating system module that manages the time resource.

Server: A process that responds to requests from clients via messages.

Service call: The means by which a process requests service from the kernel, usually implemented by a trap instruction.

Session: The period during which a user interacts with a computer.

Shared file system: Files residing on one computer that can be accessed from other computers.

Site: The set of computers, usually networked, under a single administrative control.

Socket: An abstraction for communication between two processes, not necessarily on the same machine.

Spooling system: Storing newly arrived jobs on disk until they can be run, and storing the output of old jobs on disk until it can be printed.

Thin client: A program that runs on one computer that allows the user to interact with a session on a second computer.

Thread: An execution context that is independently scheduled, but shares a single address space with other threads.

Time: A resource: ability to execute instructions.

Timesharing: Interactive multiprogramming.

User: A human being physically interacting with a computer.

User identifier: A number or string that is associated with a particular user.

User interface: The facilities provided to let the user interact with the computer.

Virtual: The result of abstraction; the opposite of physical.

Virtual address: An address in memory as seen by a process, mapped by hardware to some physical address.

Virtual machine: An abstraction produced by a virtualizing kernel, similar in every respect but performance to the underlying hardware.

Virtualizing kernel: A kernel that abstracts the hardware to multiple copies that have the same behavior (except for performance) of the underlying hardware.

World Wide Web: A network service that allows users to share multimedia information.

References

[1] Ed Bott, Carl Siechert, and Craig Stinson. *Microsoft Windows XP Inside Out.* Microsoft Press, deluxe edition, 2002.

[2] Daniel Pierre Bovet and Marco Cesati. *Understanding the LINUX Kernel: From I/O Ports to Process Management.* O'Reilly & Associates, 2000.

[3] Helen Custer. *Inside Windows NT.* Microsoft Press, 1993.

[4] William S. Davis and T. M. Rajkumar. *Operating Systems: A Systematic View.* Addison-Wesley, fifth ed. 2000.

[5] Raphael A. Finkel. *An Operating Systems Vade Mecum.* Prentice Hall, second edition, 1988.

[6] Ida M. Flynn and Ann McIver McHoes. *Understanding Operating Systems.* Brooks/Cole, 2000.

[7] Al Geist, Adam Beguelin, and Jack Dongarra, Eds. *PVM: Parallel Virtual Machine: A Users' Guide and Tutorial for Network Parallel Computing (Scientific and Engineering Computation).* MIT Press, 1994.

[8] D. Hildebrand. An architectural overview of QNX. *Proc. Usenix Workshop on Micro-Kernels and Other Kernel Architectures*, pages 113–126, 1992.

[9] Lawrence J. Kenah and Simon F. Bate. *VAX/VMS Internals and Data Structures.* Digital Equipment Corporation, 1984.

[10] Michael S. Kogan and Freeman L. Rawson. The design of operating system/2. *IBM Journal of Research and Development*, 27(2):90–104, June 1988.

[11] Samuel J. Leffler, Marshall Kirk McKusick, Michael J. Karels, and John S. Quarterman. *4.3BSD UNIX Operating System.* Addison-Wesley, 1989.

[12] Adrian Nye. *Xlib Programming Manual.* O'Reilly & Associates, third edition, 1992.

[13] John K. Ousterhout. *Tcl and the Tk Toolkit.* Addison-Wesley, 1994.

[14] Peter Pacheco. *Parallel Programming with MPI.* Morgan Kaufmann, 1997.

[15] David Pogue and Joseph Schorr. *Macworld Macintosh SECRETS.* IDG Books Worldwide, 1993.

[16] Richard Rashid. Threads of a new system. *UNIX Review*, pages 37–49, August 1986.

[17] Abraham Silberschatz, Peter B. Galvin, and Greg Gagne. *Operating Systems Concepts.* John Wiley & Sons, sixth ed. 2001.

[18] William Stallings. *Operating Systems: Internals and Design Principles.* Prentice Hall, fourth edition, 2000.

[19] Andrew S. Tanenbaum. *Modern Operating Systems.* Prentice Hall, second ed., 2001.

81

Thread Management for Shared-Memory Multiprocessors

Thomas E. Anderson
University of Washington

Brian N. Bershad
University of Washington

Edward D. Lazowska
University of Washington

Henry M. Levy
University of Washington

81.1 Introduction **81**-1
81.2 Thread Management Concepts **81**-2
    Address Spaces, Threads, and Multiprocessing
    • Basic Thread Functionality
81.3 Issues in Thread Management **81**-4
    Programmer Issues • Operating System Issues
    • Performance
81.4 Three Modern Thread Systems **81**-10
81.5 Summary **81**-11

81.1 Introduction

Disciplined concurrent programming can improve the structure and performance of computer programs on both uniprocessor and multiprocessor systems. As a result, support for *threads*, or lightweight processes, has become a common element of new operating systems and programming languages.

A thread is a sequential stream of instruction execution. A thread differs from the more traditional notion of a heavyweight process in that it separates the notion of execution from the other state needed to run a program (e.g., an address space). A single thread executes a portion of a program, while cooperating with other threads that are concurrently executing the same program. Much of what is normally kept on a per-heavyweight-process basis can be maintained in common for all threads in a single program, yielding dramatic reductions in the overhead and complexity of a concurrent program.

Concurrent programming has a long history. The operation of programs that must handle real-world concurrency (e.g., operating systems, database systems, and network file servers) can be complex and difficult to understand. Dijkstra [1968] and Hoare [1974, 1978] showed that these programs can be simplified when structured as cooperating sequential threads that communicate at discrete points within the program. The basic idea is to represent a single task, such as fetching a particular file block, within a single thread of control, and to rely on the thread management system to multiplex concurrent activities onto the available processor. In this way, the programmer can consider each function being performed by the system separately, and simply rely on automatic scheduling mechanisms to best assign available processing power.

In the uniprocessor world, the principal motivations for concurrent programming have been improved program structure and performance. Multiprocessors offer an opportunity to use concurrency in parallel programs to improve performance, as well as structure. Moderately increasing a uniprocessor's power can require substantial additional design effort, as well as faster and more expensive hardware components. But, once a mechanism for interprocessor communication has been added to a uniprocessor design, the system's peak processing power can be increased by simply adding more processors. A shared-memory multiprocessor is one such design in which processors are connected by a bus to a common memory.

Multiprocessors lose their advantage if this processing power is not effectively utilized. If there are enough independent sequential jobs to keep all of the processors busy, then the potential of a multiprocessor can be easily realized: each job can be placed on a separate processor. However, if there are fewer jobs than processors, or if the goal is to execute single applications more quickly, then the machine's potential can only be achieved if individual programs can be parallelized in a cost-effective manner. Three factors contribute to the cost of using parallelism in a program:

- **Thread overhead:** The work, in terms of processor cycles, required to create and control a thread must be appreciably less than the work performed by that thread on behalf of the program. Otherwise, it is more efficient to do the work sequentially, rather than use a separate thread on another processor.

- **Communication overhead:** Again in terms of processor cycles, the cost of sharing information between threads must be less than the cost of simply computing the information in the context of each thread.

- **Programming overhead:** A less tangible metric than the previous two, programming overhead reflects the amount of human effort required to construct an efficient parallel program.

High overhead in any of these areas makes it hard to build efficient parallel programs. Costly threads can only be used infrequently. Similarly, if arranging communication between threads is slow, then the application must be structured so that little interthread communication is required. Finally, if managing parallelism is tedious or difficult, then the programmer may find it wise to sacrifice some speedup for a simpler implementation. Few algorithms parallelize well when constrained by high thread, communication, and programming costs, although many can flourish when these costs are low.

Low overhead in these three areas is the responsibility of the thread management system, which bridges the gap between the physical processors (the suppliers of parallelism) and an application (its consumer). In this chapter, we discuss the issues that arise in designing a thread management system to support low-overhead parallel programming for shared-memory multiprocessors. In the next section, we describe the functionality found in thread management systems. Section 81.3 discusses a number of thread design issues. In Section 81.4, we survey three systems for shared-memory multiprocessors, Windows NT [Custer 1993], Presto [Bershad et al. 1988], and Multilisp [Halstead 1985], focusing our attention on how they have addressed the issues raised in this chapter.

81.2 Thread Management Concepts

81.2.1 Address Spaces, Threads, and Multiprocessing

An address space is the set of memory locations that can be generated and accessed directly by a program. Address space limitations are enforced in hardware to prevent incorrect or malicious programs in one address space from corrupting data structures in others. Threads provide concurrency within a program, while address spaces provide failure isolation between programs. These are orthogonal concepts, but the interaction between thread management and address space management defines the extent to which data sharing and multiprocessing are supported.

The simplest operating systems, generally those for older style personal computers, support only a single thread and a single-address space per machine. A single-address space is simpler and faster since it allows all data in memory to be accessed uniformly. Separate address spaces are not needed on dedicated systems to protect against malicious users; software errors can crash the system but at least are localized to one user, one machine.

Even single-user systems can have concurrency, however. More sophisticated systems, such as Xerox's Pilot [Redell et al. 1980], provide only one address space per machine, but support multiple threads within that single-address space. Because any thread can access any memory location, Pilot provides a compiler with strong type-checking to decrease the likelihood that one thread will corrupt the data structures of another.

Other operating systems, such as Unix, provide support for multiple-address spaces per machine, but only one thread per address space. The combination of a Unix address space with one thread is called a Unix *process*; a process is used to execute a program. Since each process is restricted from accessing data that belongs to other processes, many different programs can run at the same time on one machine, with errors confined to the address space in which they occur. Processes are able to cooperate by sending messages back and forth via the operating system. Passing data through the operating system is slow, however; only parallel programs that require infrequent communication can be written using threads in disjoint address spaces.

Instead of using messages to share data, processes running on a shared-memory multiprocessor can communicate directly through the shared memory. Some Unix systems allow memory regions to be set up as shared between processes; any data in the shared region can be accessed by more than one process without having to send a message by way of the operating system. The Sequent Symmetry's DYNIX [Sequent 1988] and Encore's UMAX [Encore 1986] are operating systems that provide support for multiprocessing based on shared memory between Unix processes.

More sophisticated operating systems for shared-memory multiprocessors, such as Microsoft's Windows NT and Carnegie Mellon University's Mach operating system [Tevanian et al. 1987] support multiple-address spaces *and* multiple threads within each address space. Threads in the same address space communicate directly with one another using shared memory; threads communicate across address space boundaries using messages. The cost of creating new threads is significantly less than that of creating whole address spaces, since threads in the same address space can share per-program resources. Figure 81.1 illustrates the various ways in which threads and address spaces can be organized by an operating system.

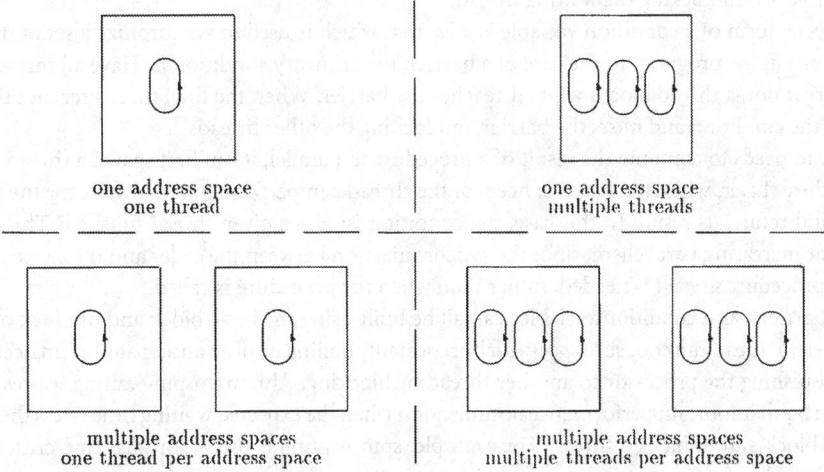

FIGURE 81.1   Threads and address spaces. MS-DOS is an example of a one address space, one thread system. A Java run-time engine is an example of one address space with multiple threads. The Unix operating system is an example of multiple address spaces, with one thread per address space. Windows NT is an example of a system that has multiple address spaces and multiple threads per address space.

81.2.2   Basic Thread Functionality

At its most basic level, a thread consists of a program counter (PC), a set of registers, and a stack of procedure activation records containing variables local to each procedure. A thread also needs a control block to hold state information used by the thread management system: a thread can be *running* on a processor, *ready-to-run* but waiting for a processor to become available, *blocked* waiting for some other thread to communicate with it, or *finished*. Threads that are ready-to-run are kept on a *ready-list* until they are picked up by an idle processor for execution. There are four basic thread operations:

- **Spawn:** A thread can create or spawn another thread, providing a procedure and arguments to be run in the context of a new thread. The spawning thread allocates and initializes the new thread's control block and places the thread on the ready-list.
- **Block:** When a thread needs to wait for an event, it may block (saving its PC and registers) and relinquish its processor to run another thread.
- **Unblock:** Eventually, the event for which a blocked thread is waiting occurs. The blocked thread is marked as ready-to-run and placed back on the ready-list.
- **Finish:** When a thread completes (usually by returning from its initial procedure), its control block and stack are deallocated, and its processor becomes available to run another thread.

When threads can communicate with one another through shared memory, *synchronization* is necessary to ensure that threads do not interfere with each other and corrupt common data structures. For example, if two threads each try to add an element to a doubly linked list at the same time, one or the other element may be lost, or the list could be left in an inconsistent state. *Locks* can solve this problem by providing mutually exclusive access to a data structure or region of code. A lock is acquired by a thread before it accesses a shared data structure; if the lock is held by another thread, the requesting thread blocks until the lock is released. (The code that a thread executes while holding a lock is called a *critical section*.) By serializing accesses, the programmer can ensure that threads only see and modify a data structure when it is in a consistent state.

When a program's work is split among multiple threads, one thread may store a result read by another thread. For correctness, the reading thread must block until the result has been written. This data dependency is an example of a more general synchronization object, the *condition variable*, which allows a thread to block until an arbitrary condition has been satisfied. The thread that makes the condition true is responsible for unblocking the waiting thread.

One special form of a condition variable is a *barrier*, which is used to synchronize a set of threads at a specific point in the program. In the case of a barrier, the arbitrary condition is: Have all threads reached the barrier? If not, a thread blocks when it reaches the barrier. When the final thread reaches the barrier, it satisfies the condition and *raises* the barrier, unblocking the other threads.

If a thread needs to compute the result of a procedure in parallel, it can first spawn a thread to execute the procedure. Later, when the result is needed, the thread can perform a *join* to wait for the procedure to finish and return its result. In this case, the condition is: Has a given thread finished? This technique is useful for increasing parallelism, since the synchronization between the caller and the callee takes place when the procedure's result is needed, rather than when the procedure is called.

Locks, barriers, and condition variables can all be built using the basic block and unblock operations. Alternatively, a thread can choose to *spin-wait* by repeatedly polling until an anticipated event occurs, rather than relinquishing the processor to another thread by blocking. Although spin-waiting wastes processor time, it can be an important performance optimization when the expected waiting time is less then the time it takes to block and unblock a thread. For example, spin-waiting is useful for guarding critical sections that contain only a few instructions.

81.3   Issues in Thread Management

This section considers the issues that arise in designing and implementing a thread management system as they relate to the programmer, the operating system, and the performance of parallel programs.

81.3.1 Programmer Issues

81.3.1.1 Programming Models

The flexibility to adapt to different programming models is an important attribute of thread systems. Parallelism can be expressed in many ways, each requiring a different interface to the thread system and making different demands on the performance of the underlying implementation. At the same time, a thread system that strives for generality in handling multiple models is likely to be well suited to none.

One general principle is that the programmer should choose the most restrictive form of synchronization that provides acceptable performance for the problem at hand. For coordinating access to shared data, messages are a more restrictive, and for many kinds of parallel programs, are a more appropriate form of synchronization than locks and condition variables. Threads share information by explicitly sending and receiving messages to one another, as if they were in separate address spaces, except that the thread system uses shared memory to efficiently implement message passing.

There are some cases where explicit control of concurrency may not be necessary for good parallel performance. For instance, some programs can be structured around a single instruction multiple data (SIMD) model of parallelism. With SIMD, each processor executes the same instruction in lockstep, but on different data locations. Because there is only one program counter, the programmer need not explicitly synchronize the activity of different processors on shared data, thus eliminating a major source of confusion and errors.

Perhaps the simplest programmer interface to the thread system is none at all: the compiler is completely responsible for detecting and exploiting parallelism in the application. The programmer can then write in a sequential language; the compiler will make the transformation into a parallel program. Nevertheless, the compiled program must still use some kind of underlying thread system, even if the programmer does not. Of course, there are many kinds of parallelism that are difficult for a compiler to detect, so automatic transformation has a limited range of use.

81.3.1.2 Language Support

Threads can be integrated into a programming language; they can exist outside the language as a set of subroutines that explicitly manage parallelism; or they can exist both within and outside the language, with the compiler and programmer managing threads together.

Language support for threads is like language support for object-oriented programming or garbage collection: it can be a mixed blessing. On one hand, the compiler can be made responsible for common bookkeeping operations, reducing programming errors. For example, locks can automatically be acquired and released when passing through critical sections. Further, the types of the arguments passed to a spawned procedure can be checked against the expected types for that procedure. This is difficult to do without compiler support.

On the other hand, language support for threads increases the complexity of the compiler, an important factor if a multiprocessor is to support more than one programming language. Further, the concurrency abstractions provided by a single parallel programming language may not do quite what the programmer wants or needs, making it necessary to express solutions in ways that are unnatural or inefficient.

A reasonable way of getting most of the benefits of language support without many of the disadvantages is to define both a language and a procedural interface to the thread management system. Common operations can be handled transparently by the compiler, but the programmer can directly call the basic thread management routines when the standard language support proves insufficient.

81.3.1.3 Granularity of Concurrency

The frequency with which a parallel program invokes thread management operations determines its *granularity*. A *fine-grained* parallel program creates a large number of threads, or uses threads that frequently block and unblock, or both. Thread management cost is the major obstacle to fine-grained parallelism. For a parallel program to be efficient, the ratio of thread management overhead to useful computation must be small. If thread management is expensive, then only *coarse-grained* parallelism can be exploited.

More efficient threads allow programs to be finer grained, which benefits both structure and performance. First, a program can be written to match the structure of the problem at hand, rather than the performance characteristics of the hardware on which the problem is being solved. Just as a singlethreaded environment on a uniprocessor can prevent the programmer from composing a program to reflect the problem's logical concurrency, a coarse-grained environment can be similarly restrictive. For example, in a parallel discrete-event simulation, physical objects in the simulated system are most naturally represented by threads that simulate physical interactions by sending messages back and forth to one another; this representation is not feasible if thread operations are too expensive.

Performance is the other advantage of fine-grained parallelism. In general, the greater the length of the ready-list, the more likely it is that a parallel program will be able to keep all of the available processors busy. When a thread blocks, its processor can immediately run another thread provided one is on the ready-list. With few threads though, as in a coarse-grained program, processors idle while threads do I/O or synchronize with one another.

The performance of a fine-grained parallel program is less sensitive to changes in the number of processors available to an application. For example, consider one phase of a coarse-grained parallel program that does 50 CPU-min worth of work. If the program creates five threads on a five processor machine, the phase finishes in just 10 min. But, if the program runs with only four processors, then the execution time of the phase *doubles* to 20 min: 10 min with four processors active followed by 10 min with one processor active. (Preemptive scheduling, which could be used to address this problem, has a number of serious drawbacks, which are discussed subsequently.) If the program had originally been written to use 50 threads, rather than 5, then the phase could have finished in only 13 min, a reasonable degradation in performance.

Of course, one could argue that the programmer erred in writing a program that was dependent on having exactly five processors. The program should have been parameterized by the number of processors available when it starts. But, even so, good performance cannot be ensured if that number can vary, as it can on a multiprogrammed multiprocessor. We consider further the issues of multiprogramming in the next section.

81.3.2 Operating System Issues

81.3.2.1 Multiprogramming

Multiprogramming on a uniprocessor improves system performance by taking advantage of the natural concurrency between computation and I/O. While one program waits for an I/O request, the processor can be running some other program. Because the processor and I/O devices are kept busy simultaneously, more jobs can be completed per unit time than if the system ran only one program at a time.

A multiprogrammed multiprocessor has an analogous advantage. Ideally, periods of low parallelism in one job can be overlapped with periods of high parallelism in another job. Further, multiprogramming allows the power of a multiprocessor to be used by a collection of simultaneously running jobs, none of which by itself has enough parallelism to fully utilize the multiprocessor.

81.3.2.2 Processor Scheduling

Processor scheduling can be characterized by whether physical processors are assigned directly to threads or are first assigned to jobs and then to threads within those jobs. The first approach, called *one-level* scheduling, makes no distinction between threads in the same job and threads in different jobs. Processors are shared across all runnable threads on the system so that all threads make progress at relatively the same rate. In this case, threads from all jobs are placed on one ready-list that supplies all processors, as shown in Figure 81.2. Although this scheme makes sense for a uniprocessor operating system, it has some unpleasant performance implications on a multiprocessor.

The most serious problem with one-level scheduling occurs when the number of runnable threads exceeds the number of physical processors, because preemptive scheduling is necessary to allocate processor time to threads in a fair manner. With preemption, a processor can be taken away from one thread and given to another at any time. In a sequential program, preemption has a well-defined effect: the program

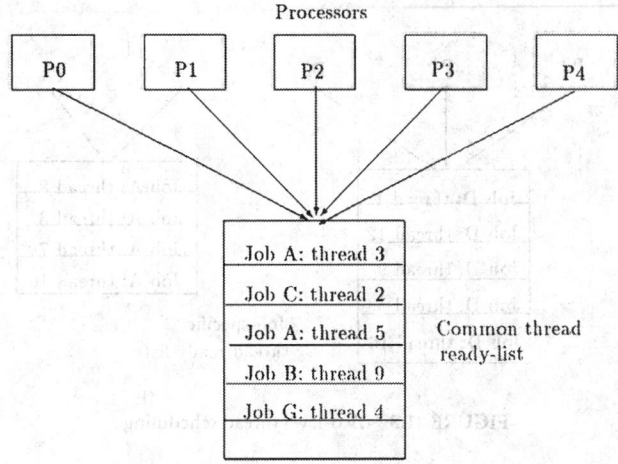

FIGURE 81.2 One-level thread scheduling.

goes from the running state to the not running state as its one thread is preempted. The effect of preemption on the performance of a sequential program is also well defined: if *n* CPU-intensive jobs are sharing one processor in a preemptive, round-robin fashion, then each job receives $1/n$th the processor and is slowed down by a factor of *n* (modulo the preemption and scheduling overhead).

For a parallel program, though, the effects of untimely processor preemption on performance can be more dramatic. In the previous section, we saw how a coarse-grained program can be slowed down by a factor of two when the number of processors is decreased from five to four. That program exemplified a problem that occurs more generally with preemption and barrier-based synchronization. The program had an implicit barrier, which was the final instruction in the phase. Until all threads reached that instruction, the program could not continue. When one processor was removed, it took twice as long to reach the barrier because not all threads within the job could make progress at an equal rate.

Preemptive multiprocessor scheduling also affects program performance when locks are used, but for a different reason than with barriers. Suppose a thread holding a lock while in a critical section is unexpectedly preempted by the operating system. The lock will remain held until the thread is rescheduled. As threads on other processors try to acquire the lock, they will find it held and be forced to block. It is even possible that, as more threads block waiting for the lock to be freed, the number of that job's runnable threads drops to zero and the application can make no progress until the preempted thread is rescheduled. The overhead of this unnecessary blocking and unblocking slows down the program's execution.

In the previous section, we saw how fine-grained parallelism can improve a program's performance by increasing the chance that a processor will find another runnable thread when its current thread blocks. Unfortunately, a fine-grained parallel program that packs the ready-list interacts badly with the behavior of a one-level scheduler. In particular, when a program's thread blocks in the kernel on an I/O request, the parallelism of the program can only be maintained if the kernel can schedule another of the program's threads in place of the one that blocked. This benefit, though, comes at the cost of increased preemption activity and diminished overall performance.

The problems of one-level scheduling are addressed by two-level schedulers. With a two-level scheduler, processors are first assigned to a job, and then threads within that job are executed only on the assigned processors. Each job has its own ready-list, which is used only by the job's processors, as shown in Figure 81.3. Thread preemption may no longer be necessary with a two-level scheduler since a preempted thread will only be replaced by another thread from the same job. Further, for long intervals, a processor runs only threads from the same application, and so the cost of switching between threads is kept low.

In a two-level scheduling system, processors can be allocated to jobs either statically or dynamically. A static two-level scheduler never changes the number of processors given to a job from its initial allocation;

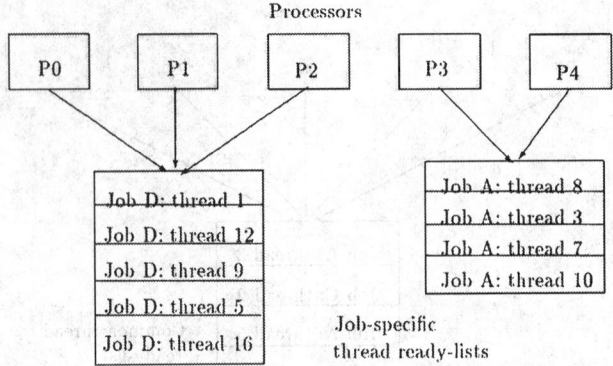

FIGURE 81.3 Two-level thread scheduling.

if some of those processors are needed by another job, the operating system must preempt all of the job's processors. A dynamic scheduler can adapt the number of processors assigned to each job according to changing conditions.

Dynamic two-level scheduling can give better performance, because it overlaps periods of poor parallelism in one job with periods of high parallelism in another. One difficulty with a dynamic scheduler is that it requires more information from an application describing the current processor requirements. As a result, though, dynamic scheduling can also more easily handle changes in the number of running jobs. For example, when a job finishes, its processors can be reallocated to a running job whose parallelism is increasing. To avoid the problems of one-level scheduling, though, it is crucial that the operating system coordinate with each application when it needs to preempt processors (e.g., to avoid preempting a processor when it would seriously affect performance). A dynamic scheduler always has the option, when it needs processors and no application has any available, of reverting to a static policy.

81.3.2.3 Kernel- vs User-Level Thread Management

Processor scheduling controls the allocation of processors to jobs. The operating system must be responsible for processor scheduling because processors are a hardware resource and shifting a processor from one job to another involves updating per-processor address space hardware registers. Spawning a thread so that it runs on an already allocated processor, however, does not require modifying privileged state. Thus, thread management and scheduling within a job can be done entirely by the application instead of by the operating system. In this case, thread management operations can be implemented in an application-level library. The library creates virtual processors using the operating system's processor scheduling interface, and schedules the application's threads on top of these virtual processors.

Unlike processor allocation, where a single systemwide scheduling policy can be used, thread scheduling policies benefit from being application specific. Some applications perform well if their threads are scheduled according to some fixed policy, such as first-in–first-out or last-in–first-out, but others need to schedule threads according to fixed, or even dynamically changing priorities. For example, consider a parallel simulation where each simulation object is represented by its own thread. Different objects become sequential bottlenecks at different times in the simulation; the amount of parallelism can be increased by preferentially scheduling these objects' threads.

It is difficult to provide sufficient thread scheduling flexibility with kernel-level threads. While the kernel could define an interface that allows each application to select its thread scheduling policy, it is unlikely that the system designer could foresee all possible application needs.

Thread management involves more than scheduling. A tradeoff exists between user- and kernel-level thread management. A user-level implementation provides more flexibility and better performance;

implementing threads in the kernel guarantees a uniformity that eases the integration of threads with system tools.

The downside of having many custom-built thread management systems is that there is no standard thread. By implication, a kernel-level thread management system defines a single, systemwide thread model that is used by all applications. Operating systems that support only one thread model, like those that support only one programming language, can more easily provide sophisticated utilities, such as debuggers and performance monitors. These utilities must rely on the abstraction and often the implementation of the thread model, and a single model makes it easier to provide complete versions of these tools since their cost can be amortized over a large number of applications. Peripheral support for multiple models is possible, but expensive.

A standard thread model also makes it possible for applications to use libraries, or canned software utilities. In the same sense that a standard procedure calling sequence sacrifices speed for the ability to call into separately compiled modules, a standard thread model allows one utility to call into another since they both share the same synchronization and concurrency semantics.

It is important to point out that two-level scheduling does not imply that threads are implemented at the application level; the job-specific ready queues shown in Figure 81.3 could be maintained either within the operating system or within the application. Also, a user-level thread implementation does not imply two-level scheduling, even though threads *are* being scheduled by the application. This implication only holds in the absence of multiprogramming, or in cases where processors are explicitly allocated to jobs. For example, a user-level thread implementation built on top of Unix processes that share memory suffers from the same problems relating to preemption and I/O as do one-level kernel threads because both are scheduled in a job-independent fashion.

81.3.3 Performance

The performance of thread operations determines the granularity of parallelism that an application can effectively use. If thread operations are expensive, then applications that have inherently fine-grained parallelism must be restructured (if that is even possible) to reduce the frequency of those operations. As the cost of thread operations begins to approach that of a few procedure calls, several issues become performance critical that, for slower operations, would merely be second-order effects.

Simplicity in the thread system's implementation is crucial to performance [Anderson et al. 1989]. There is a performance advantage to building multiple thread systems, each tuned for a single type of application. Even simple features that are needed by only some applications, such as saving and restoring all floating point registers on a context switch, will markedly affect the performance of applications that do not need the functionality. Each context switch takes only tens of instructions; a feature that adds even a few more instructions must have a large compensating advantage to be worthwhile. For example, the ability to preemptively schedule threads within each job makes the thread management system more sluggish at several levels, because preemption must be disabled (and then re-enabled) whenever scheduling decisions are being made. These scheduling decisions are on the critical path of all thread management operations.

Although kernel-level thread management simplifies the generation and maintenance of system tools, it increases the baseline cost of all thread management operations. Just trapping to the operating system can cost as much as the thread operation itself, making a kernel implementation unattractive for high-performance applications. Further, the generality that must be provided by a kernel-level thread scheduler hurts the performance of those applications needing only basic service. Kernel-level threads are less able to cut corners by exploiting application-specific knowledge. With a user-level thread system, the thread management system can be stripped down to provide exactly the functions needed by an application and no more. User-level thread operations also avoid the cost of trapping to the kernel.

Other performance issues have less to do with what a thread system does, than with how it goes about doing it. For example, using a centralized ready-list can limit performance for applications that have extremely fine-grained parallelism. The ready-list is a shared data structure that must be locked to prevent it from being modified by multiple processors simultaneously. Even if the ready-list critical sections consist only of simple enqueue and dequeue operations, they can become a sequential bottleneck, since there is little other work involved in spawning/finishing or blocking/unblocking a thread. An application for which thread overhead is 20% of the total execution time, and half of that overhead is spent accessing the ready-list, then its maximum speedup (the time of the parallel program on P processors divided by the time of the program on one processor) is limited to 10.

The bottleneck at the ready-list can be relieved by giving each processor its own ready-list. In this way, enqueueing and dequeueing of work can occur in parallel, with each processor using a different data structure. When a processor becomes idle, it checks its own list for work, and if that list is empty, it scans other processors' lists so that the workload remains balanced.

Per-processor ready-lists have another nice attribute: threads can be preferentially scheduled on the processor on which they last ran, thereby preserving cache state. Computer systems use caches to take advantage of the principle of *locality*, which says that a thread's memory references are directed to or near locations that have been recently referenced. By keeping references close to the processor in fast cache memory, the average time to access a memory location can be kept low. On a multiprocessor, a thread that has been rescheduled on a different processor will initially find fewer of its references in that processor's cache. For some applications, the cost of fetching these references can exceed the processing time of the thread operation that caused the thread to migrate.

The role of spin-waiting as an optimization technique changes in the presence of high-performance thread operations. If a thread needs to wait for an event, it can block, relinquishing its processor, or spin-wait. A thread must spin-wait for low-level scheduler locks, but in application code a thread should block instead of spin if the event is likely to take longer than the cost of the context switch. Even though context switches can be implemented efficiently, reducing the need to spin-wait, a hidden cost is that context switches also reduce cache locality.

81.4 Three Modern Thread Systems

We now outline three modern thread management systems for multiprocessors: Windows NT, Presto, and Multilisp. The choices made in each system illustrate many of the thread management issues raised in the previous section.

The thread management primitives for each of these systems are shown in Table 81.1. The table is organized to indicate how the primitives in one system relate to those in the others, as well as those provided by the basic thread interface outlined in the Basic Thread Functionality section.

Windows NT is an operating system designed to support Microsoft Windows applications on uniprocessors, shared memory multiprocessors, and distributed systems. Windows NT supports multiple threads within an address space. Its thread management functions are implemented in the Windows NT kernel. Since NT's underlying thread implementation is shared by all parallel programs, system services such as debuggers and performance monitors can be economically provided.

Windows NT's scheduler uses a priority-based one-level scheduling discipline. Because Windows NT allocates processors to threads in a job-independent fashion, a parallel program running on top of the Windows NT thread primitives (or even a user-level thread management system based on those primitives) can suffer from anomalous performance profiles due to ill-timed preemptive decisions made by the one-level scheduling system.

TABLE 81.1 The Basic Operations of Thread Management Systems

Basic	Windows NT	Presto	Multilisp
Spawn	thread_create;thread_resume	Thread::new; Thread::start	(future...)
Block	thread_suspend	Thread::sleep	*Touch unresolved future.*
Unblock	thread_resume	Thread::wakeup	*When future is resolved.*
Finish	thread_terminate	Thread::terminate	*Resolve this future.*

Presto is a user-level thread management system originally implemented on top of Sequent's DYNIX operating system, but later ported to DEC workstations. DYNIX provides a Presto program with a fixed number of Unix processes that share memory. The Presto run-time system treats these processes as virtual processors and schedules the user's threads among them. Presto's thread interface is nearly identical to Windows NT's.

Presto is distinguished from most other thread systems in that it is structured for flexibility. Presto is easy to adapt to application-specific needs because it presents a uniform object-oriented interface to threads, synchronization, and scheduling. The object-oriented design of Presto encourages multiple implementations of the thread management functions and so offers the flexibility to efficiently accommodate differing parallel programming needs.

Presto has been tuned to perform well on a multiprocessor; it tries to avoid bottlenecks in the thread management functions through the use of per-processor data structures. Presto does not provide true two-level scheduling, even though the thread management functions (e.g., thread scheduling) are implemented in an application library accessible to the user; DYNIX, the base operating system, schedules the underlying virtual processors (Unix processes) any way that it chooses. Although a Presto program can request that its virtual processors not be preempted, the operating system offers no solid guarantee. As a result, kernel preemption threatens the performance of Presto programs in the same was as it does Windows NT programs.

Although Windows NT and Presto are implemented differently, the interfaces to each represent a similar style of parallel programming in which the programmer is responsible for explicitly spawning new threads of execution *and* for synchronizing their access to shared data. This style is not accidental, but reflects the basic function of the underlying hardware: processors communicating through shared memory. One criticism often made of this style is that it forces the programmer to think about coordinating many concurrent activities, which can be a conceptually difficult task.

Multilisp demonstrates how thread support can be integrated into a programming language in order to simplify writing parallel programs. In Multilisp, a multiprocessor extension to LISP, the basic concurrency mechanism is the **future,** which is a reference to a data value that has not yet been computed. The **future** operator can be included in any Multilisp expression to spawn a new thread which computes the value of the expression in parallel. Once the value has been computed, the future *resolves* to that value. In the meantime, any thread that tries to use the future's value in an expression automatically blocks until the future is resolved. The language support provided by Multilisp can be implemented on top of a system like Windows NT or Presto using locks and condition variables.

With Multilisp, the programmer does not need to include any synchronization code beyond the future operator; the Multilisp interpreter keeps track of which futures remain unresolved. By contrast, using the Windows NT or Presto thread primitives, the programmer must add calls to the appropriate synchronization primitives wherever the data is needed. Multilisp, like Presto, uses per-processor ready-lists to reduce contention in scheduling operations.

81.5 Summary

This chapter has examined some of the key issues in thread management for shared-memory multiprocessors.

Shared-memory multiprocessors are now commonplace in both commercial and research computing. These systems can easily be used to increase throughput for multiprogrammed sequential jobs. However, their greatest potential — as yet not fully realized — is for accelerating the execution of single, parallelized programs.

As programmers make use of finer grained parallelism, the design and implementation of the thread management system becomes increasingly crucial. Modern thread management systems must address the programmer interface, the operating system interface, and performance optimizations; language support and scheduling techniques for multiprogrammed multiprocessors are two areas that require further research.

References

Anderson, T. E., Lazowska, E. D., and Levy, H. M. 1989. The performance implications of thread management alternatives for shared memory multiprocessors, pp. 49–60. In *ACM SIGMETRICS Perform. '89 Conf. Meas. Modeling Comput. Syst.* May.

Bershad, B., Lazowska, E., and Levy, H. 1988. PRESTO: a system for object-oriented parallel programming. *Software Prac. Exp.* 18(8):713–732.

Custer, H. 1993. *Inside Windows NT.* Microsoft Press.

Dijkstra, E. W. 1968. Cooperating sequential processes. In *Programming Languages*, pp. 43–112. Academic Press.

Encore. 1986. UMAX 4.2 Programmer's Reference Manual. Encore Computer Corp.

Halstead, R. 1985. Multilisp: A language for concurrent symbolic computation. *ACM Trans. Programming Lang. Syst.* 7(4):501–538.

Hoare, C. A. R. 1974. Monitors: an operating system structuring concept. *Commun. ACM* 17(10):549–557.

Hoare, C. A. R. 1978. Communicating sequential processes. *Commun. ACM* 21(8):666–677.

Redell, D. D., Dalal, Y. K., Horsley, T. R., Lauer, H. C., Lynch, W. C., McJones, P. R., Murray, H. G., and Purcell, S. C. 1980. Pilot: an operating system for a personal computer. *Commun. ACM* 23(2):81–92.

Sequent. 1988. Symmetry Technical Summary. Sequent Computer Systems, Inc.

Tevanian, A., Rashid, R. F., Golub, D. B., Black, D. L., Cooper, E., and Young, M. W. 1987. Mach threads and the Unix kernel: the battle for control, pp. 185–197. In *Proc. USENIX Summer Conf.*

82

Process and Device Scheduling

82.1   Introduction 82-1
82.2   Background 82-2
      Processes
82.3   Resources and Scheduling 82-4
      Processor Scheduling • Priority Dispatching Algorithms
      • Rotation Algorithms • Multilevel Dispatching
      • Dispatching Algorithms for Real-Time Systems
82.4   Memory Scheduling 82-11
82.5   Device Scheduling 82-12
      Scheduling Shareable Devices • Evaluation and Selection of a
      Disk Scheduling Algorithm • Scheduling Nonshareable
      Devices
82.6   Scheduling Policies 82-19
      Deadlock
82.7   High-Level Scheduling 82-24
82.8   Recent Work and Current Research 82-25
      Processor Scheduling • Disk Scheduling • Deadlock

Robert D. Cupper
Allegheny College

82.1 Introduction

High-level language programmers and computer users deal with what is really a virtual computer. That virtual computer they see is facilitated by a software bridge that plays the role of interlocutor between the actual computer hardware and the computer user's environment. This software, described in general in Chapter 80, is the operating system. The computer's operating system (OS) is made up of a group of systems programs that serve two basic ends:

- To control the allocation and use of the computing system's resources among the various users and tasks
- To provide an interface between the computer hardware and the programmer or user that simplifies and makes feasible the creation, coding, debugging, maintenance, and use of applications programs

Thus, the OS creates and maintains an environment in which users can have programs executed. That is, it provides a structure in which the user can request and monitor execution of his or her programs and can receive the resulting output. To this end, the OS must make available to the user's program the system resources needed for its execution. These system resources are the processor, primary memory, secondary memory (including the file system), and the various devices. Because most modern computing systems are powerful enough to allow multiple user programs or at least multiple tasks to execute in the same time

1-58488-360-X/$0.00+$1.50
© 2004 by CRC Press, LLC

frame, the OS must allocate these resources among the potentially competing needs of the multiple tasks in such a way as to ensure that all tasks can execute to completion. Furthermore, these resources must be allocated so that no one task is unnecessarily or unfairly delayed. This requires that the OS *schedule* its resources among the various and competing tasks. The detailed characterization of the problem of scheduling computer system resources in a number of settings; the techniques, algorithms, and policies that have been set forth for its solution; and the criteria and method of assessment of the efficacy of these solutions form the subject of this chapter.

The next section establishes the landscape for the discussion, with a brief review of delivery methods of computing services and a look at the essential concept of a process — a program in execution — the most basic unit of account in an OS. Then, we take a brief look at the components of the OS responsible for the execution of a process. Although this chapter is primarily concerned with the first of the two functions of an OS — that is, control of the allocation and use of computing system resources — it will become clear that the methods brought to bear on the simultaneous achievement of these two functions cannot treat them as wholly independent.

82.2   Background

Computer service delivery systems may be classified into three groups, which are distinguished by the nature of interaction that takes place between the computer user and his or her program during its processing. These classifications are batch, time-shared, and real-time.

In a *batch processing* OS environment, users submit jobs, which are collected into a batch and placed in an input queue at the computer where they will be run. In this case, the user has no interaction with the job during its processing, and the computer's response time is the turnaround time — the time from submission of the job until execution is complete and the results are ready for return to the person who submitted the job.

A second mode for delivering computing services is provided by a *time-sharing* OS. In this environment, a computer provides computing services to several users concurrently online. The various users share the central processor, the memory, and other resources of the computer system in a manner facilitated, controlled, and monitored by the operating system. The user, in this environment, has full interaction with the program during its execution, and the computer's response time may be expected to be no more than a few seconds.

The third class, the *real-time* OS, is designed to service those applications where response time is of the essence in order to prevent error, misrepresentation, or even disaster. Real-time operating systems are subdivided into what are termed *hard* real-time systems and *soft* real-time systems. The former provide for applications that cannot be compromised, such as airline reservations, machine tool control, and monitoring of a nuclear power station. The latter accommodate less critical applications, such as audio and video streaming. In either case, the systems are designed to be interrupted by external signals that require the immediate attention of the computer system.

In fact, many computer operating systems are *hybrids*, providing for more than one of these types of computing service simultaneously. It is especially common to have a background batch system running in conjunction with one of the other two on the same computer system.

Discussion of resource scheduling in this chapter is limited to uniprocessor and multiprocessor systems *sans* network connections. Resource scheduling in networking and distributed computing environments is considered in Chapter 87 and Chapter 88.

Programs proceed through the computer as *processes*. Therefore, the various computer system resources are to be allocated to processes. A thorough understanding of that concept is essential in all that follows here.

82.2.1   Processes

Most operating systems today are **multiprogramming** systems. Systems such as these, where multiple, independent programs are executing, must manage two difficult problems: concurrency and nondeterminacy. The *concurrency* problem arises from the coexistence of several active processes in the system during any given interval of time. *Nondeterminacy* arises from the fact that each process can be interrupted between any two of its steps. The unpredictability of these interruptions, coupled with the randomness that results from processes entering and leaving the system, makes it impossible to predict the relative speed of execution of interrelated processes in the system. A mechanism is needed to facilitate thinking about, and ultimately dealing with, the problems associated with concurrency and nondeterminacy. An important part of that mechanism is the conceptual and operational isolation of the fundamental unit of computation that the operating system must manage. This unit is called the *task* or **process**. Informally, a process is a program in execution.

This concept of process facilitates an understanding of the twin problems of concurrency and indeterminacy. Concurrency, as we have seen, occurs whenever there are two or more processes active within the system. Concurrency may be *real*, in the case where there is more than one processor and hence more than one process can execute simultaneously, or *apparent*, whenever there are more processes than processors. In the latter case, it is necessary for the OS to provide for the switching of processors from one process to another sufficiently rapidly to present the illusion of concurrency to system users. But this is difficult, for whenever a processor is assigned to a new process (called *context switching*), it is necessary to recall where the first process was stopped in order to allow that process, when it gets the processor back, to continue where it left off.

The idea of context switching implies that a particular process can be interrupted. Indeed, a process may be interrupted, as necessary, between individual steps (machine instructions). Such interruptions occur most often when a particular process has used up its quota of processor time or when it has requested and must wait for completion of an I/O operation. Nondeterminacy arises from the unpredictable order in which such interruptions can occur.

Because active processes in the system can be interrupted, each process can be in one of three states:

- **Running** — The process is currently executing on a processor.
- **Ready** — The process could use a processor if one were available.
- **Blocked** — The process is waiting for some event, such as I/O completion, to occur.

The relationship between these three states for a particular process is portrayed in Figure 82.1.

Here, we see that if a process is currently *running* and requests I/O, for example, it relinquishes its processor and goes to the *blocked* state. In order to maintain the illusion of concurrency, each process is assigned a fixed *quantum* of time, or *time-slice*, which is the maximum time a running process can control the processor. If a process is in the running state and does not complete or block before expiration of its time-slice, that process is placed in the *ready* state, and some other process is granted use of the processor for its quantum of time. A blocked process can move back to the ready state upon completion of the event that blocked it. A process in the ready state becomes running when it is assigned a processor by the system dispatcher.

All of these state changes are interrupt-driven. A request for I/O is effected by issuing a supervisor call via an I/O procedure, which causes a system interrupt. I/O completion is signaled by an I/O interrupt

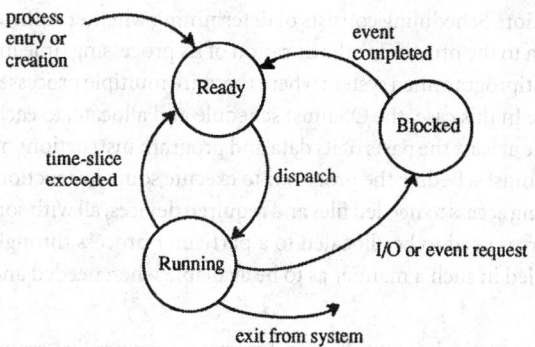

FIGURE 82.1 Process state transitions.

from a data channel* or device controller. Time-slice exceeded results in an external interrupt from the system's interval timer. And, of course, movement from ready state to running results from the dispatcher giving control of the processor to the most eligible ready process. In each case, when a process gives up the processor, it is necessary to save the particulars of where the process was in its execution when it was interrupted, so that it may properly resume later.

Each process within the system is represented by an associated *process control block* (PCB). The PCB is a data structure containing the essential information about an active process, including the following:

- Process ID
- Current state of the process
- Register save area
- A pointer to the process's allocated memory area
- Pointers to other allocated resources (disk, printer, etc.)

The last three contain the information necessary to restart an interrupted process. There is only one set of registers in the system, shared by all of the active processes. Therefore, the contents of these registers must be saved just before the context switch. Because memory is both space- and time-shared, it is necessary only to save pointers to the locations of the process' memory areas before its interruption. Devices vary. Some are shareable (e.g., disk devices) and so are treated like memory; others (e.g., the printer) are nonshareable and tied up by a process for as long as it is using them. In either case, it is necessary here to keep track only of the device ID and, perhaps, the current position in a file.

Thus, programs solve problems by being executed; they execute as processes. To this end, the OS must allocate, or schedule, to the process sufficient memory to hold its data and at least the part of the program immediately due for execution, the various devices needed, and a processor. Because there certainly will be multiple processes and possibly even multiple jobs, each made up of processes, it is necessary to have the OS schedule these resources in such a way as to enable all the jobs to run to completion. The next section deals with scheduling the processor among the processes competing for it to effect the execution of their parent programs.

82.3 Resources and Scheduling

Programs execute as processes using computer system resources, including a processor, primary memory, and most likely secondary memory, including files, and some devices. Thus, in order for the process to execute, it must enter the system and these resources must be allocated to the process. The OS must schedule the allocation of these resources to a given process so that this process, and any others in the system, may execute in a timely fashion.

The simplest case is a monoprogrammed system, where there is just one program executing in the system at a time. In this case, a process can be scheduled to the system whenever it becomes available following the previous process's execution. Scheduling consists of determining whether sufficient resources are available and, if so, allocating them to the process for the duration of its processing time in the system. The situation is more complex in a multiprogrammed system where there are multiple processes in the system competing for the various resources. In this case, the OS must schedule and allocate, to each active process, sufficient memory to accommodate at least the parts of its data and program instructions needed for execution in the near term. Then, the OS must schedule the processor to execute some instructions. In addition, there must be provision for scheduling access to needed files and required devices, all with some sort of time constraint.

Not all of these resources need to be allocated to a particular process throughout its life in the system, but they must be scheduled in such a manner as to be available when needed and in concert. Otherwise, a

---

*A data channel may be conceived as a small, special-purpose computer that executes programs to actually do I/O concurrently with the main processor's program execution. In today's desktop computers, this function is largely subsumed in the device controllers.

process may be stalled for lack of one or more resources, tying up other processes waiting for unavailable resources in the interim.

Resource scheduling and allocation is, from the performance point of view, perhaps the most important part of the OS. Good scheduling must consider the following objectives:

- Resource allocation that facilitates minimal average turnaround time
- Resource allocation that facilitates minimal response time
- Mutual exclusion of processes from nonshareable resources
- A high level of resource utilization
- **Deadlock** prevention, avoidance, or detection

It is clear that these objectives cannot necessarily be mutually satisfied. For example, a high level of resource utilization probably will mean a longer average wait for resources, thus lengthening both response and turnaround times. The choice may be, in part, a function of the particular service delivery system that the process has entered. A batch system scheduler would favor resource utilization, whereas a time-sharing system would need to be sensitive to response time, and at the extreme, a real-time system would minimize response time at the expense of resource utilization.

An *allocation mechanism* refers to the implementation of allocations. This includes the data structures used to represent the state of the various resources (shareable or nonshareable, available, busy, or broken), the methods used to assure mutual exclusion in use of nonshareable resources, and the technique for queuing waiting resource requests. The *allocation policy* refers to the rationale and ramifications of applying the mechanisms. Successful scheduling requires consideration of both.

The practices and policies regarding scheduling each resource class, the processor, primary memory, secondary memory and files, and devices, differ significantly and are next considered in turn. Because the processor is arguably the most important resource — certainly a process could not proceed without it — the discussion turns first to processor scheduling.

82.3.1 Processor Scheduling

In this section, it is assumed that adequate memory has been allocated to each process and the needed devices are available, to allow focus on the problems surrounding the allocation of the processor to the various processes. To distinguish scheduling programs from the input queue for entry into the computing system from the problem of allocating the processor among the active processes already in the system, the term *scheduler* is reserved for the former and *dispatcher* for the latter. The term *active process* refers to a process that has been scheduled, in this sense, into the system from the input queue, that is, has been allocated space in memory and has at least some of its needed devices allocated.

Development of methods for processor dispatching is motivated by a number of system performance goals, including the following:

- Reasonable turnaround and/or response time — here, as indicated previously, the tolerance is governed by the service delivery system (e.g., batch processing vs. time-sharing)
- Predictable performance
- Good absolute or relative throughput
- Efficient resource utilization (i.e., low CPU idle time)
- Proportional resource allocation
- Reasonable-length waiting queues
- Insurance that no process must wait forever
- Satisfaction of real-time constraints

Clearly, these goals sometimes conflict. For example, average response time can be improved, but at the expense of the very longest programs, which are likely to have to wait. Or minimization of response time could result in poor resource utilization if a long program that has many resources allocated to it is forced to wait for a long while, thus idling the resources allocated to it.

A processor scheduler, CPU scheduler, or dispatcher consists of two parts. One is a ready queue, consisting of the active processes that could immediately use a processor were one available. This queue is made up of all of the processes in the ready state of Figure 82.1.* The other part of the dispatcher is the algorithm used to select the process, from those on the ready queue, to get the processor next. A number of dispatching algorithms have been proposed and tried. These algorithms are classified here into three groups: priority algorithms, rotation algorithms, and multilevel algorithms.

82.3.2   Priority Dispatching Algorithms

These dispatching algorithms may be classified by queue organization, whether they are preemptive or nonpreemptive, and the basis for the priority.

The ready queue may be first-in, first-out (FIFO), priority, or unordered. The queue can be maintained in sorted form, which facilitates rapid location of the highest-priority process. However, in this case, inserting a new arrival is expensive because, on average, half of the queue must be searched to find the correct place for the insertion. Alternatively, new entries can simply and quickly be added to an unsorted ready queue. But the entire queue must be searched each time the processor is to be allocated to a new process. In fact, a compromise plan might call for a periodic sort, maintaining a short list of new arrivals on the top of the previously sorted queue. In this case, when a new process is to be selected, the priority of the process at the front of the sorted part of the queue is compared with each of the recently arrived unsorted additions, and the processor is assigned to the process of highest priority.

In a nonpreemptive algorithm, the dispatcher schedules the processor to the process at the front of the ready queue, and that process executes until it blocks or completes. A preemptive algorithm is the same, except that a process, once assigned a processor, will execute until it completes or is blocked, unless a process of higher priority enters the ready queue. In that case, the executing process is interrupted and placed on the ready queue and the now-higher priority process is allocated the processor.

82.3.2.1   First-Come, First-Served (FCFS) Dispatching

When the criterion for priority is arrival time, the dispatching algorithm becomes FCFS. In this case, the ready queue is a FIFO queue. Processor assignment is made to the process with its PCB at the front of the queue, and new arrivals are simply added to the rear of the queue. The algorithm is easy to understand and implement. FCFS is nonpreemptive, so a process, once assigned a processor, keeps it until it blocks (say, for I/O) or completes. Therefore, the performance of the system in this case is left largely in the hands of fate, that is, how jobs happen to arrive.

82.3.2.2   Shortest Job First (SJF) Dispatching

The conventional form of the SJF algorithm is a priority algorithm where the priority is inversely proportional to (user) estimated execution time. The relative accuracy of user estimates is enforced by what is, in effect, a penalty–reward system: too long an estimated execution time puts a job at lower priority than need be, and too short an estimate is controlled by aborting the job when the estimated time is exceeded. This effects a delay with penalty by forcing the user to rerun the job. There are preemptive and nonpreemptive forms of the SJF algorithm. In the nonpreemptive form, once a process is allocated a processor, the process runs until completion or block. The preemptive form of the algorithm allows a new arrival to the ready queue with lower estimated running time to preempt a currently executing process with a longer estimated running time.

---

*The ready queue contains representations of the ready processes, that is, the corresponding PCBs (or pointers to them).

SJF is clearly advantageous for short jobs. Because a typical execution time distribution is usually weighted toward shorter jobs — especially in an installation providing general computer services — one could argue that a SJF policy would benefit most users. But, as always, there is a trade-off. In this case, it is that long jobs get relatively poor service. This is especially true in the preemptive version, where preemption clearly favors short jobs, with the result that long jobs can be effectively starved out. Although the SJF rule provides the minimum average waiting time, it, like FCFS, appears to apply only to the batch processing system of service delivery.

82.3.2.3 Priority Dispatching

In this algorithm, process priorities are set based on criteria external to the system, reflecting the relative importance of the processes. These include such factors as memory size requirements, estimated job execution time, estimated amount of I/O activity, and/or some measure of the importance of the computation as set by the user or the institutional structure. For example, a class of jobs characterized by low execution time estimates combined with minimal resource requirements may be deemed high priority. Similarly, one might argue that particular systems programs, say device handlers, ought to have high priority. This type of dispatching algorithm can, like the previous ones, be either nonpreemptive or preemptive. In the nonpreemptive form, the highest-priority job is assigned to the CPU and runs until completion or block. In the preemptive form, the arrival of a higher-priority job at the ready queue results in the preemption of the currently running process for the higher-priority process.

It is clear that priority dispatching serves the highest-priority jobs optimally. Thus, to the extent that the externally set priorities reflect actual institutional priorities, priority dispatching is arguably best. However, there are two problems. First, low-priority jobs generally receive poor service. Especially in the preemptive form, low-priority jobs can be blocked indefinitely or starved out from the processor, violating the fairness criterion. Second, to the extent that priorities are set by users, based on the real or perceived importance of their programs without knowledge of the current system workload mix, the system can lose control of its performance parameters. For example, a high-priority process might get good service, but at the cost of other goals, such as maximum system throughput or resource utilization.

82.3.2.4 Dynamic Dispatching Priority Adjustment

Some of the shortcomings of the priority algorithms can be ameliorated by dynamic priority adjustment. In this case, decisions can be based on information about a process that is obtained while the process is in the system. Priorities determined from one of the previously described algorithms can be adjusted dynamically during the process' life in the system, according to a number of criteria, such as the number and type of resources currently allocated, accumulated waiting time since the job entered the system, amount of recent processing time, amount of recent I/O activity, total time in the system, etc.

One such plan advanced by Kleinrock [44] is to allow a process dispatching priority to increase at rate x while the process is in the ready queue, and rate y while it has a processor assigned to it. Priority then depends on the values of x and y. These could be set externally and differ for different jobs. They could also change dynamically over the time a given process is in the system. For example, the starvation problem can be eliminated if x of a high-priority process decreases over time and/or y of a low-priority job increases over time. Use of the SJF rule accompanied by nonlinear functions, where x and y decrease over time for a while and then jump to high values, will ensure that no jobs wait very long for service, yet still favors short jobs.

Another dynamic scheme would increase the priority of a process dynamically during periods of high I/O activity. That is, it would make priority inversely related to the time interval since the last I/O call. This favoring of I/O-bound jobs is desirable because it compensates for the speed disparity between the more mechanical I/O devices and the higher-speed electronic processor. It does this by keeping the devices running, thus minimizing the possibility that the CPU must wait for I/O completion. Moreover, the effect on other, more compute-bound jobs, is minimal because of the limited CPU time required to start I/O operations.

In general, priority algorithms are easy to understand and simple to implement. The main disadvantage is that low-priority jobs tend to get poor service. Moreover, the performance associated with this type of algorithm is not appropriate to some situations. The response times available to processes, especially lower-priority programs, would not be acceptable in a time-sharing or real-time environment. Apparently, there is a need to consider another approach to dispatching.

82.3.3 Rotation Algorithms

The essence of the rotation algorithms, as the name implies, is that the CPU is scheduled in rotation, so that each job in the ready queue is given some service in order to maintain a reasonable response interval. These algorithms are designed to apply to time-sharing systems.

82.3.3.1 Simple Round Robin (RR)

In the round-robin rotation algorithm, processor time is divided into time-slices, or quanta. The ready queue is treated like a circular queue, and each process in the ready queue is given one time-slice each rotation. If the process does not complete or block during its quantum, at the end of the quantum, the process is preempted and returned to the end of the ready queue. The idea is to provide response that is reasonably independent of job size and/or priority. In simple RR, all time-slices are the same size, say q. A typical q is 50 ms; the range is 10 ms to 100 ms. There is no static or dynamic priority information. Therefore, if there are k processes in the ready queue, each process is scheduled for q out of every kq milliseconds. Thus, users perceive processes running on a processor with $1/k$ of its actual processor speed. Response time is $(k - 1) \times q$ or less.

Clearly, performance is affected by two key parameters: the size of the ready queue, k and the quantum size, q. Response is inversely related to k, so as the system becomes loaded, response time deteriorates. While k is essentially determined externally, quantum size is a system parameter. If q is infinite, then round robin is FCFS. A large q tends to favor some jobs. If a number of processes in the ready queue are in blocked state or will block during their quanta, the remaining processes will cycle frequently, and the corresponding jobs will run to completion quickly and with excellent response times. On the other hand, arrival of new processes, each taking one of the longer quanta, causes the average response to deteriorate substantially. If q is small, these effects are decreased. For example, arrival of a new process to the ready queue will have a much smaller effect on average response time. A small q, in addition to providing a more consistent response, leads to a total waiting time for a job more proportional to the length of the job. At the extreme, a very small q will cause useful execution time of the quantum to be overwhelmed by the context switching delay. This would imply that the size of the quantum ought to be large relative to the time required for a context switch. Turnaround time is also affected by the size of the time-slice. Although one might expect that turnaround time would fall as the quantum size increases, this is not always true. Context switches add to turnaround, and, as a very long quantum moves RR toward FCFS, average turnaround can increase substantially. Silberschatz et al. [77] claim that 80% of the CPU executions should be shorter than q.

82.3.3.2 Round-Robin with Priorities

Simple round robin is designed to provide reasonable and fair response time for all active processes in the system. As such, it has no way of recognizing a "more important" process. Variants of simple round robin have been proposed to address this problem. *Biased round robin* allows for different length time-slices for different processes, based on some external priority. A process with a higher priority is assigned a longer q, allowing processes to proceed through the system in proportion to their priorities. An alternative, *selfish round robin*, is based on Kleinrock's linearly increasing priority scheme [44], discussed previously. Here, x and y are such that $0 \leq y < x$. The effect is that processes already in execution make entering processes wait in the ready queue until the priority of the new process increases to the value of the older processes' priority. This will eventually occur because y, the rate of priority increase of an executing process, is less than x, the rate of priority increase of a process in the ready queue.

82.3.3.3 Cycle-Oriented Round Robin

Simple round robin is subject to considerable performance degradation in terms of response time as more processes enter the system and enter the ready queue. This can lead to an increase in response time beyond that which is reasonable and expected for an interactive time-sharing environment. *Cycle-oriented round robin* was developed to obviate this problem. In this case, the longest tolerable response time, c, is set as the cycle time and becomes the basis for calculation of the quantum length, q. The time-slice, $q = c/k$, will guarantee that response time never exceeds the maximum tolerable, c. There are, however, two problems. Process arrivals during the cycle could cause the response time to exceed the acceptable limit. But this is easily resolved by denying new arrivals entry into the ready queue except at the end of a cycle, when the time-slice size is recalculated. The problem of system overload remains, however. As k becomes large, q becomes small, and k too large implies that q will be too small, leading to unacceptable overhead from context switching. The solution is to enforce a minimum time-slice size.

82.3.4 Multilevel Dispatching

Section 82.2 points out that a single computing system might be designed to provide for two (or more) of the computer service delivery systems, say time-sharing in the foreground and batch in the background. The two delivery systems have different service requirements. Consequently, taken individually, the systems most appropriately would employ different dispatching algorithms. When the two services are available on the same system, it is necessary to account for different scheduling needs and the priority of interactive jobs over background batch jobs.

This is accomplished by a *multilevel dispatching* system, in which the ready queue is partitioned into two (or more) separate queues: one for the interactive jobs, one for batch jobs, etc. Each process entering the system is assigned to one of the queues based on process type. Then, the processes on each of the several queues are scheduled by different algorithms as appropriate (e.g., RR for the interactive job queue and FCFS for the batch job queue). Moreover, a system of preset priorities with preemption is used for scheduling among the multiple queues. For example, there might be three separate queues, one for system processes, one for interactive processes, and one for batch processes. All processes in the system process queue must be completed before any processes from the interactive queue are dispatched to the processor. Similarly, the interactive queue must be empty before any processes from the batch queue are dispatched. A new system process entering the first queue would cause any interactive process or batch process to be preempted, and a new interactive process would preempt a batch process.

Alternately, dispatching could occur among the queues on a time-slice basis (i.e., so much time for the systems queue, then a quantum of time for the interactive process queue, etc.). Of course, the time-slice given to each queue could vary, reflecting the priority of processes assigned to that particular queue. For example, after processes in the system queue complete, 75% of the time could be allocated to interactive jobs and 25% to batch jobs, until another system process comes along.

82.3.4.1 Multilevel Feedback Queue Dispatching

The multilevel queue dispatching described in Section 82.3.4 relies on processes being assigned to one of the queues depending on some external factor, such as process type. Processes, once assigned to a particular queue, remain there for the duration of their time in the system. Multilevel feedback queue dispatching is similar, except that particular processes can move among the separate queues dynamically, based on some aspect or aspects of their progress through the system. For example, three queues could be established as before. Here, the queues are designated queue 0, queue 1, and queue 2, with priority highest for the lowest numbered queue, etc. Accordingly, the quantum q_0 would be shortest for the lowest numbered queue, which would be dispatched according to some form of RR algorithm. Queue 1 would also be RR, but with a longer quantum, q_1, and queue 2 FCFS with an even longer quantum, q_2. All new processes enter queue 0. Processes that complete or block within the quantum q_0 remain in this highest-priority queue 0. At q_0 time exceeded, those that do not complete or block are moved to queue 1. Similarly, those in queue 1 that do not complete or block in time q_1 are moved to queue 2. The queues, as before, are scheduled according to a system of preset priorities with preemption. All the processes in queue 0 must be completed or blocked before any processes from queue 1 are dispatched, and so forth. A new process arriving at queue 1 will preempt a currently executing process from queue 2.

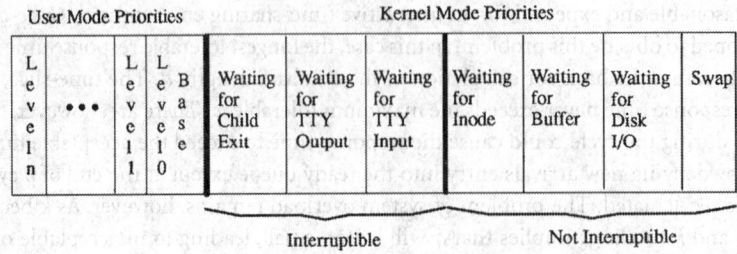

User Mode Priorities					Kernel Mode Priorities						
Level n	•••	Level 1	Level 0	Base	Waiting for Child Exit	Waiting for TTY Output	Waiting for TTY Input	Waiting for Inode	Waiting for Buffer	Waiting for Disk I/O	Swap

Interruptible   Not Interruptible

FIGURE 82.2   UNIX process priorities (adapted from Bach [1]).

It is evident that the multilevel feedback queue system of dispatching is the most complex. In addition to the three queues and the dispatching algorithms appropriate for each queue, the system requires a description of some algorithm to determine when to move a process to the next lower or next higher priority queue, and a method to determine which queue an entering process should initially join.

82.3.4.2 Process Scheduling in UNIX

The UNIX system uses a system of *decay usage scheduling* that might be characterized in terms of the previous classes as round robin with multilevel feedback [1]. Basically, each user process begins with a base level priority that puts it into a particular ready queue. Periodically, the priority is recalculated on the basis of recent CPU usage, increasing the priorities of processes inversely with recent CPU usage.

In UNIX, the possible priority levels are divided into two classes [1]: kernel priorities and user priorities. Within each class are a number of priority levels, as shown in Figure 82.2. Note that higher priorities are associated with lower priority values.

The basic operation is for the dispatcher to execute a context switch whenever the currently executing process blocks or exits, or when returning to user mode from kernel mode if a process with a priority higher than the process currently allocated the CPU is ready to run.

A process entering the system is assigned the highest user-level priority, called the *base-level priority*. If there are several processes in the highest user process level, the dispatcher selects the one that has been there longest. The CPU is dispatched to that process. The clock interrupts the executing process perhaps several times during its quantum. At each clock interrupt, the clock interrupt handler increases an execution time field in the PCB. Approximately once a second (in UNIX system V), the clock interrupt handler applies a decay function to the execution time field in the PCB of each active process in the system. Thus

$$C(\text{execution time field}) = \text{decay}(C(\text{execution time field}) = C(\text{execution time field})/2$$

At this same time, the process priority is recalculated based on recent CPU usage [1]:

$$\text{priority} = C(\text{execution time field})/2 + \text{base level priority}$$

A process that does not block or exit, and therefore uses its entire quantum, would be assigned a higher number, corresponding to a lower priority. If the process blocks, it is assigned a priority by the system call routine. The priority is not dependent on whether the process is I/O-bound or processor-bound, as in the multilevel feedback queue dispatching described previously. Rather, it is a function of the particular reason the process blocked. The kernel adjusts the priority of processes returning from kernel mode back to a user-level mode, reflecting the fact that it has just had access to kernel resources. Kernel-level priorities can only be obtained via a system call. When the event that caused the block has been completed, the process can resume execution, unless preempted by the dispatcher due to the arrival of a process with a higher priority in the ready queue. This could occur upon completion of some event that previously led to block state for some higher-priority process or if the periodic priority changes by the clock interrupt handler have made one of the other process's priority higher. In either case, the dispatcher initiates a context switch.

The periodic priority recalculation effectively redistributes the processes among the user-level priority levels. This, along with the policy of dispatching first the process longest in the highest-priority queue, assures round-robin scheduling for processes in user mode. It should be clear that this scheduler will provide preferential service to highly interactive tasks, such as editing, because their typically low ratio of CPU time to idle time causes their priority level to increase quickly, on account of their low recent CPU usage.

The UNIX process scheduler includes a feature allowing users to exercise some control over process priority. There is a system call

```
nice (priority_level)
```

that permits an additional element in the formula for recalculating priority [1]:

```
priority = C(execution time field)/2 + base level priority + nice
priority_level
```

This allows the user with a nonurgent process to increment the priority calculation "nicely," resulting in the process' moving to a lower level in the user process priority queue. A user cannot use `nice` to raise the priority level of a process or to lower the priority of any other process. Only the superuser can use `nice` to increase a process priority level, and even the superuser cannot use `nice` to give another process a lower priority level. But superuser can, of course, kill any process.

82.3.5 Dispatching Algorithms for Real-Time Systems

Priority dispatching algorithms are well suited for the batch method of service delivery, and rotation algorithms provide for the performance requirements imposed by multiprogramming time-sharing systems. But what about real-time systems, where the constraints on response time are very strict?

Real-time systems for critical systems appear as stand-alone, dedicated systems. In this case, when a requesting process arrives, the system must consider the parameters of the request in conjunction with its then current resource availabilities and accept the request only if it can service the request within the strict time constraint. However, other real-time processes, such as those associated with multimedia, interactive graphics, etc., can be handled by a real-time component of a system also providing time-sharing or batch services. However, combining real-time applications with others will lead inevitably to a degradation in response and/or turnaround time for the others. For the combination to be workable, real-time applications must have the highest priority, that priority must not be allowed to deteriorate over time (no aging), and the time required for the dispatcher to interrupt a process and start (or restart) the real-time application must be acceptably small.

The dispatch time problem is complicated by the fact that an effectively higher priority process, such as a system call, may be running when the real-time application arrives. One solution to this problem is to interrupt the system call process as soon as possible, that is, when it is not modifying some kernel data structure. This means that even systems applications would contain at least some "interrupt points," where they can be swapped out to make way for a real-time process. Another approach is to use IPC primitives (see Chapter 84) to guarantee mutual exclusion for critical kernel data structures, thus allowing systems programs to be interruptible. This latter is the technique used in Solaris 2. According to Silberschatz et al. [77], the Solaris 2 dispatch time with no preemption is around 100 ms, but with preemption and mutual exclusion protection, the dispatch time is closer to 2 ms.

82.4 Memory Scheduling

Memory allocation and scheduling are the functions of the OS's memory management subsystem. The options include real and virtual memory systems. The mechanisms for scheduling memory include the following:

- Data structures used to implement free block lists and page and/or segment tables, depending on the memory management system structure

- Cooperation with the processor scheduler to place a process waiting for a memory block, page, or segment in blocked state
- Cooperation with the I/O subsystem to queue processes waiting for page or segment transfers to wait for the concomitant I/O service from a disk drive

The details for the memory management subsystem are in Chapter 85 for standalone and networked systems and Chapter 87 for distributed systems.

82.5   Device Scheduling

The devices — sometimes called *peripheral* devices — that can be attached to a computer system are many and varied. They range from terminals, to tape drives and printers, plotters, etc., to disk drives. These devices differ significantly in terms of both type of operations and speed of operation. In particular, the various devices use different media, encodings, and formats for the data read, written, and stored. On this count, devices can be divided into two groups:

- **Block devices** — Devices that store and transfer information in fixed-sized blocks
- **Character devices** — Devices that transfer sequences of characters

Disks are block devices: they read, write, and store blocks ranging in size from 128 to 1024 bytes, depending on the system [80]. In a block device, each block can be read or written independently of the others. Terminals, tape drives, printers, and network interfaces are character devices. In these cases, data is transferred as a string of characters; there are no block addresses, and it is not possible to find a specific character by address on the device.

Devices also differ with respect to the speed of data transmission. A terminal may be able to send perhaps 1000 characters per second, as compared to a disk, which has a transfer rate of closer to a million characters per second. Finally, devices differ with respect to the operations they support. A disk drive facilitates arm movement of a head for a seek operation, and a tape drive allows rewind; neither operation is appropriate for the other device.

Devices in general consist of two parts: the physical device itself and a controller. The actual device is the mechanical part: the turntables and heads for a tape, the disk spindle, platters, heads, and arms for a disk. The controller is the electronic part — a small computer of sorts that contains the circuitry necessary to effect the operation of the one or more similar devices connected to it. Thus, it is the controller that initiates the operation of a device and translates a stream of bits to (input) or from (output) a block of information in the controller's local buffer. In addition, the controller has registers that are used to receive device operation commands and parameters from the operating system and hold status information regarding the device's most recent operation.

Although the buffer in the device's controller effectively separates the slower, mechanical operations of a device from the much faster, electronic CPU, data transfer can still exact a delay on system performance. For input, eventually the block of data must be transferred from the controller's buffer to main memory for use by the program requesting it. The transfer requires a loop involving the CPU in a character-by-character transfer. For this reason, many block device controllers contain a direct memory access (DMA) capability. In this instance, the CPU need only send to the controller the main memory start address for the block and the number of characters to send. The controller will effect the transfer asynchronously of CPU operation, until the transfer of the entire block is complete, at which time the controller sends an interrupt to the CPU.

This overview of device operation makes it clear that I/O is one of the more detailed and difficult parts of the programming process. This, plus the obvious need for control over device allocation and utilization, led to the early incorporation of low-level I/O programs as one of the principal functions of the operating system. Thus, it is the responsibility of the I/O subsystem to provide a straightforward interface to user programs, to translate user I/O requests into instructions to the device controller, to handle errors, and indicate completion to the user program.

In this context, several goals apply to the design of the I/O subsystem. Of course, the I/O subsystem should be efficient because all programs perform at least some I/O, and I/O, with the inherently slower devices, can often become a bottleneck to system performance. I/O software should provide for device independence in two ways. First, it should be possible for user programs to be written and translated so as to be independent of a particular device of a given type. That is, it should not be necessary to rewrite or retranslate a program to direct output to a printer used instead of another, available one. Moreover, it should be possible to have a user program independent even from device type. The program should not have to be rewritten or retranslated to obtain input from a file rather than the keyboard.

In the UNIX system, this is effected by treating devices as special files, allowing a uniform naming scheme, a *directory path*, which allows device independence to include files as well as devices. Similarly, user programs should be free from character code dependence. The user should not need to know or care about the particular codes associated with any one device. These requirements are equivalent to the goal of uniform treatment of devices. A good user interface should provide for simplicity and therefore minimization of error. The most obvious implication of these goals, especially the last, is that all device-specific information — that is, instructions for operation of the device, device character encodings, error handling, etc. — should be as closely associated with the specific device or device class as possible. This suggests a layered structure for design of the I/O subsystem.

The structure of the I/O subsystem can be seen as four layers [50] [80], proceeding from I/O functions employed in user programs to the device drivers and interrupt handlers that provide low-level program control of the various devices:

- User program calls to library I/O functions
- System I/O call interface (device-independent)
- Device drivers (device-dependent)
- Interrupt routines

User programs invoke I/O by means of calls to library functions. These functions, which reside in system libraries, are linked to user programs at execution time, and run outside the kernel, provide for two basic services: formatting I/O and setting parameters for the system I/O call interface. An example of a library function that facilitates this is `printf` in UNIX, which processes an input format string to format an ASCII string and then invokes the library function `write` to assemble the parameters for the system I/O call interface [80]. These parameters include the name of the logical device or file where I/O is to be done, the type of operation (e.g., read, write, seek, backspace), the amount of data to be transferred (number of characters or blocks), and the source or destination storage location into which (input) or from which (output) a data transfer is to occur.

The system I/O call interface, which is an integral part of the operating system, has three basic functions:

- Link the logical device or file specified in the user-level I/O function to an appropriate physical device
- Perform error checks on the parameters supplied by the user in the I/O function
- Set up and initiate the request for the physical I/O

Thus, after ensuring that the I/O parameters in the user request are consistent with the operation requested (e.g., the amount of data in the transfer request is equal to 1 for a character device, or some multiple of the block size for a block device), the I/O call interface assembles the needed parameters into what Lister [50] terms an *I/O request block* (IORB). In addition to the previous parameters, the IORB will contain pointers to the PCB of the requesting process and an error location for any error codes, should the requested operation turn out to be unsuccessful.

In order to maintain device independence and to accommodate the unique features and operations associated with the various types of devices, the I/O requests, I/O in progress, and the device characteristics are manifest in a number of data structures. The exact nature of these is, of course, dependent on the system and the computer. These data structures are associated with each device type in so far as is possible. A device control block, I/O block, I/O control block (IOCB), unit control block, or channel control block

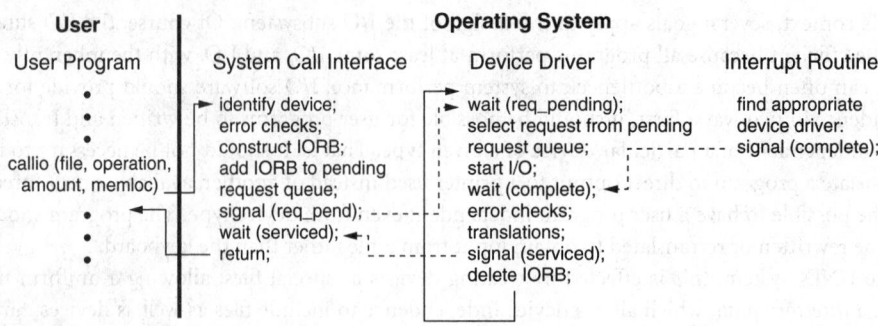

FIGURE 82.3   A sketch of the I/O subsystem (adapted from Lister [50]).

is used to parameterize the characteristics for each device (and/or control unit). In addition to the device identification, such characteristics include specific instructions to operate the device, the device status (available, busy, offline), pointers to character translation tables, and a pointer to the PCB of the process that has the device allocated to it. The IOCB also has a pointer to a queue of requests. This request-pending queue is a linked list of the IORBs prepared and linked by the system I/O call interface in response to user program I/O requests. There are two possibilities for the structure of the queue of pending requests. It could be organized as a single queue, holding IORBs for all devices, or as multiple queues: one for each specific device and/or controller type in the system.

The device drivers contain the code for operating the various controllers and devices, utilizing the parameters found in the IOCB. There is a separate device driver for each device type. The device drivers operate in a continuous loop, servicing requests from the pending-request queue and, in turn, notifying the user process that requested the I/O when the operation is complete. After selecting a request from the pending-request queue, the driver initiates the I/O operation. For some computers, this is done by issuing a particular machine instruction, such as start I/O (SIO) or, for computers that use memory mapped I/O, writing to registers in the device controller. The device driver then waits for I/O completion.

Upon completion of the physical I/O operation, the device controller generates an interrupt. The corresponding interrupt routine signals the device driver, using semaphores or some other IPC construct. The device driver translates and transmits the data to the destination indicated in the IORB and then signals the requesting user process of I/O completion. The whole process is summarized in Figure 82.3.

For purposes of scheduling, the various devices can be divided into two groups: shareable and nonshareable. Devices are *shareable* in the sense that primary memory is shared — more than one process can share space on the device and the several process' data transmissions to and from the device may be interleaved. In this sense, the disk, like primary memory, is a shareable device. Most other devices are *nonshareable*. A device is nonshareable because its physical characteristics make it impossible to share. Tape devices are nonshareable because it is impractical to switch tapes between characters. Similarly, line-printers are nonshareable because it is not practical to switch the paper between lines. In both cases, attempts to share the devices for output would result in a random intermixing of outputs from the processes writing the output.

Accordingly, techniques for scheduling the two classes of devices differ. By their nature, nonshareable devices cannot be dynamically allocated, that is, allocated by requested operation. These devices must be scheduled at a higher level (i.e., when processes enter the system, or at least on a longer term "capture and release" basis). Hence, scheduling in this case is analogous to the high-level scheduling that admits processes to the system from the input queue. Shareable devices, on the other hand, can be scheduled on an operation-by-operation basis. Since requests for these operations appear on an unpredictable and random basis, in exactly the same sense that I/O operations block and unblock process execution, the scheduling of shareable devices is a low-level function, analogous to processor scheduling as described previously.

Device scheduling is largely a question of queue organization and management. The problem varies in complexity. The simplest case is a monoprogrammed system: a job is admitted to the system, and needed resources, if available (connected and working), are scheduled to the process until its completion. Of course, even in this case, it should be possible to overlap I/O operations and execution of the main program — requiring interprocess communication so that the I/O subsystem can notify the main program of completion (or error). Multiprogramming requires that pending request queues be supported both for nonshareable devices that may be otherwise allocated or not ready (e.g., printer out of paper or tape not mounted on tape device) and shareable devices that may be busy serving another process's request. Thus, device scheduling, like process scheduling is essentially the question of selecting a request from the queue to be serviced next. The policy decision, as before, becomes choosing the logical and simple FCFS or some system based on a priority given to the requesting processes. But there are some additional considerations in the case of shareable devices, such as the disk, to which we now turn.

A shareable device not only has operations from various processes interleaved with one another, but as a consequence of this interleaving, must have subareas of the medium allocated to these various processes. Therefore, a file name alone, unlike a logical device name in the case of a nonshareable device, is insufficient to specify the particular location of the information on a shareable device. Such data areas are commonly called files. An I/O operation to a device that can accommodate files must first be able to find the specific location of the file on the medium. The file maintenance subsystem of the OS (see Chapter 86) keeps a directory of file names and their corresponding locations for this purpose. Therefore, on the occurrence of the first reference to a file operation, the system must reference the directory to obtain the device and the location of the needed file on that device before the actual I/O can be accomplished. This is termed *opening* the file. Because this requires a look-up and usually an additional disk read (to get the directory), it is a time-consuming operation. Doing this multiple referencing for every file operation (read or write) would be inefficient. Therefore, when a file is opened, a file descriptor is created. The file descriptor contains information to facilitate future references to the same file, including the device identification, the location of the file on that device, a description of the file organization, whether the file is open for read or write, and the position of the last operation. This file descriptor is referenced from the file or device descriptor created by the I/O subsystem when a device or file is allocated to a process, obviating the need to re-reference the directory to find the location of the file.

82.5.1 Scheduling Shareable Devices

The most important shareable device in the modern computer system is the disk. Disks provide the majority of the secondary storage used in support of virtual memory and the file system. It is appropriate, therefore, to single out the disk and consider disk scheduling as an example of shareable device scheduling.

A disk is organized as a set of concentric circles, called tracks, on a magnetic surface. Each track is divided into *sectors*, the physical unit in which data is transferred to or from the disk. A track normally has from 8 to 32 sectors per track [80]. The tracks are read from, or written to, by means of a head attached to a moveable arm allowing placement of the head over any track on the disk. A floppy disk, or diskette, consists of one such platter, which is removable from the drive unit itself. A so-called hard disk consists of a set of such platters stacked vertically around a common spindle. In this case, there are a number of surfaces on both sides of all but the top and bottom platters (which only record information on the inner surfaces). Corresponding to each usable surface, there is a read/write head. These read/write heads are fixed to a comblike set of arms that move the heads together, to position all of the heads over corresponding tracks on all of the disk surfaces. Taken together, these corresponding tracks are called a *cylinder*. Thus, all of the data stored on a particular cylinder can be referenced without moving the head assembly.

In this environment, information on a disk is referenced by an address consisting of several parts: disk drive number, disk surface, track, and sector. A sector is the smallest physical amount of information that can be read or written in a single disk operation. Sector size is a function of the particular disk drive; sizes vary from 32 to 4096 bytes, but the most common size is 512 bytes [77].

To read or write from or to the disk, the heads must be moved to the proper cylinder (track), then wait until the needed sector appears under the head. Then, the information can be transferred to the control unit buffer. The time for a disk transfer is made up of three components:

- **Seek time** — The time it takes to move the head assembly from its current location to the required cylinder
- **Latency** — The rotational delay needed for the selected sector to come under the read/write heads
- **Transfer time** — The time for the actual transfer of the data in the sector

In order to improve efficiency, data is usually read or written in blocks of one or more sectors. Still, for most disks, the seek time dominates, so one aim of disk scheduling algorithms is to reduce or minimize average disk seek time for each disk operation.

As with other system resource requests, if the needed disk and its controller are available, the disk request is attended to immediately. However, if the disk drive or its controller is busy servicing another request, new requests will be queued. Especially in a modern multiprogramming system, because most jobs depend on the disk for program loading and data files, it is likely that the disk request queue will be populated with new requests arriving regularly. When the current request is serviced, the system must choose the next to service from among those in the queue. Selection of the next disk request to service is controlled by a disk scheduling algorithm. Several possibilities exist:

First come, first served (FCFS) — The most obvious approach is to take disk requests from the disk request queue in the order of their arrival, that is, first come, first served. This algorithm is the easiest to understand and implement. Moreover, as in the case of any waiting line, FCFS appears to be naturally fair. Its performance, however, is a function of the likely random cylinder locations of the various requests on the queue. Particularly when successive requests are for disk blocks whose locations are on cylinders far removed from one another, seek time, which is roughly proportional to the distance that the read/write heads must move, will be significant. This leads to a high average seek time and consequently poor performance. Thus, performance depends more on the nature and requirements of currently active processes than on OS design. It is possible to improve upon the FCFS algorithm.

Shortest seek time first (SSTF) — The quest for shortest seek time would appear to favor selecting the request requiring the minimum head movement, the request at the closest cylinder (i.e., the request requiring the minimum seek time). SSTF will always provide better performance than FCFS. However, like its analog in processor scheduling SJF, SSTF takes requests found on nearby tracks, including newly arriving requests. This tends to provide poor service to requests for disk blocks that happen to start far away from the current read/write head position. Assuming that disk block allocation is evenly distributed across the disk, this is particularly true for requests at the extreme outside and inside cylinders, once the heads get positioned near the middle cylinders. Again, following the analogy to SJF, arrival of a string of requests for blocks in nearby cylinders may cause some requests to be starved under the SSTF algorithm. SSTF gains response time at the expense of fairness — the other extreme from FCFS.

Bidirectional scan scheduling (SCAN) — The search for requests at relatively nearby tracks without completely abandoning fairness can be accommodated by the bidirectional scan algorithm. SCAN works analogously to the scan/seek button on a modern digital audio tuner. That is, the read/write head mechanism begins a scan at one end of the disk (outside or inside) and scans, servicing requests from the queue as the heads come to the corresponding cylinder. The scan continues to the other end of the disk, assuring that no requesting process must wait too long for service. At the other end of the disk, the head assembly reverses and continues its scan in the opposite direction. This algorithm is sometimes called the *elevator algorithm* because its operation is nearly analogous to the operation of a building elevator. The analogy is complete, except for the fact that for elevator passengers, the elevator's direction at pickup makes a difference, depending on whether they wish to go up or down; disk requests are serviced equally well regardless of the direction from which

TABLE 82.1 Total Head Movements for Various Scheduling Algorithms

	FCFS	SSTF	SCAN	CSS	BBDS	BCSS
Pattern 1	540	195	186	223	186	208
Pattern 2	101	91	175	218	91	92
Pattern 3	381	162	186	242	180	216

Pattern 1   73, 125, 32, 127, 10, 120, 62

Pattern 2   81, 82, 83, 84, 85, 21, 22

Pattern 3   10, 46, 91, 124, 32, 85, 11

FIGURE 82.4 Sample patterns of disk requests.

the disk heads arrived. This latter point, however, is significant and a disadvantage of the SCAN scheme. If the requests are to disk blocks uniformly placed throughout the disk cylinders, the heads, once they reach one end, are unlikely to find many requests because the first cylinders scanned on the reverse trip are the ones just scanned. The requests now most in need of service, according to the fairness criteria, are at the other end of the disk head travel. This problem is addressed in a variant of SCAN.

Circular-scan scheduling (CSS) — This algorithm is similar to SCAN, except that when the scanning heads reach the end of their travel, rather than simply reversing direction, they return to the beginning end of the disk. The effect is a circular scan, where the first disk cylinder, in effect, immediately follows the last, and disk requests are provided with a more uniform wait time.

Bounded-scan scheduling (BSS) — Both SCAN and CSS were characterized as scanning from one end of the disk to the other, that is, from the lowest numbered cylinder to the highest. In fact, both SCAN and CSS do too much. They need only scan to the most extreme cylinder represented by the requests in the queue, that is, scan in one direction as long as there are requests beyond the current location, a bound represented by the block location of the highest or lowest numbered cylinder of a request in the queue. In actuality, both SCAN and CSS are implemented in these bounded (in this sense) versions.

Optimal scheduling — Optimal scheduling requires selecting the next request so that the total seek time is minimal. The problem with optimal scheduling is that the continuing arrival of new disk requests to the queue requires reordering the queue as each request arrives. To the extent that the disk is a heavily used resource with requests arriving continually and often, the computation needed to obtain optimal scheduling is not likely to be worth it. The concept, however, is useful as a reference for comparing the performance of the other algorithms.

82.5.2 Evaluation and Selection of a Disk Scheduling Algorithm

As with the processor scheduling algorithms, performance is related to the request pattern. It is possible to evaluate this performance by estimating or tracing a disk request sequence. Consider, as an example, a disk with 128 cylinders. For the three patterns of disk requests shown in Figure 82.4, the total head movements for each of the scheduling algorithms described above are shown in Table 82.1. In each case, assume that the disk head is initially at cylinder 58 and scanning in the direction of increasing cylinder numbers.

Though the FCFS scheme is easy to implement and conveys a sense of fairness, it leads to the poorest performance in each of the three example cases. SSTF seems, on cursory examination, to yield the best performance, but as the examples show, this is not always the case. Bounded bidirectional scheduling (BBDS) is as good or better in two of the three examples. From the table, it is clear that performance is a function of the intensity and order of the disk requests, as well as the disk scheduling algorithm. If disk usage is light and the request queue nearly always empty, then choice of a scheduling algorithm is insignificant, because they are all effectively the same. Heavy disk usage and short interarrival times of entries on the disk request queue effectively rule out FCFS and make the choice more critical. Because one of the primary uses of the disk is for the file system, disk performance is also influenced by the file space allocation technique. The examples of Table 82.1 suggest that contiguous allocation would result in less head movement and, consequently, in significantly better performance. Pattern 2, with its requests to sequentially numbered cylinders, is associated with better performance with all of the scheduling algorithms considered here. Directories and index blocks, if used, will cause extra disk traffic and, depending on their locations, could affect performance. Because references to files must pass through the directories and index blocks, placement of these near the center tracks of the disk would limit head movement to at most one-half of the total number of cylinders to find the file contents and, consequently, would lead to better performance, regardless of the scheduling algorithm.

Although seek times have been improved in modern disk drives, the combination of seek and latency times still dominates transfer time. This suggests an augmented SSTF to consider the sum of seek and latency times, called shortest access time first (SATF) or shortest positioning time first (SPTF).

82.5.3 Scheduling Nonshareable Devices

Because they cannot be scheduled dynamically on an operation-by-operation basis, nonshareable devices require a different approach to scheduling than shareable devices. This level of scheduling is particularly important for two reasons: most device types are nonshareable and scheduling of nonshareable resources can lead to deadlock — a situation in which the system is effectively locked up because none of the active processes can continue for want of some resource held exclusively by some other process.

By their very nature, nonshareable devices must be allocated to a single process for the duration of a use session. That is, a tape drive must be allocated to a particular process for the duration of a sequence of input and/or output operations, a printer for the duration of the print output of the process, etc. The duration can be explicitly indicated under program control by open and close operations or by default at the entry and exit of a process from the system. Regardless of the origin of the use session, the mechanism is described above. The particular device is described by an IOCB, and processes with requests for allocation of the device are queued in the pending requests queue attached to the IOCB. Nonshareability is enforced by an initial value of 1 for the queuing semaphore. Scheduling, then, amounts to waiting for the resource to be freed, then selecting a device request from the pending request queue FCFS, or by some measure of priority. In the latter case, the rationale or criteria for assignment of priorities clearly parallels that for processes in process scheduling. But the mutual exclusion requirement of nonshareability raises the possibility of deadlock. Therefore, dealing explicitly with this potential must be an integral part of the policy consideration in scheduling of nonshareable devices.

82.5.3.1 Files

The taxonomy used in this chapter for the classification of the computing system resources, processor: primary memory, and devices, with the subclassification of shareable and nonshareable devices, is not perfect. A single file may be shared among a number of processes and, in this sense, it also becomes a resource. The problem here is that though a particular file opened in read mode is a shareable resource, that same file, when opened by one process for writing, becomes a nonshareable resource. Thus, allocation of a file to a process's request may proceed only after appropriate read/write mode checks. A file in read mode may be scheduled (i.e., queued) as any other shareable resource — forcing the process to wait (in order) only for availability of the corresponding control unit and device. On the other hand, a process requesting a file already in use in write mode must wait until the process currently writing the file closes the file or exits the system — as in the case of any other nonshareable resource.

82.5.3.2 Virtual Devices

Process requests for nonshareable devices can often lead to unavoidable delays, the extent of which is a function of the activity of the process currently holding the needed nonshareable device. In the aggregate, such delays can have a significant adverse effect on overall system performance. There are two possible ways out of this difficulty: increase the number of the offending nonshareable devices (e.g., add more printers) or introduce virtual devices. In this latter case, a process's request for data transfer to a nonshareable device, such as a printer which is allocated elsewhere, is directed instead to some anonymous file on the disk, thus freeing the requesting process from the wait for the otherwise allocated nonshareable device. In this case, the file acts as a virtual printer. Then, a special process called a *spooling daemon*, or *spooler*, becomes responsible for scheduling and moving the data from the intermediate file to the printer when it becomes available. Of course, operation of the spooler admits yet another opportunity for scheduling — again, usually FCFS or priority, depending on the policy adopted.

82.6 Scheduling Policies

In the preceding sections, the various resources of the system were classified (i.e., processors, primary memory, devices, files, and virtual devices) and subclassified into shareable and nonshareable resources. Moreover, the subsystem in which allocation is performed was identified (i.e., process manager, memory manager, I/O subsystem, and file system). In each instance, resource allocation was characterized in terms of some form of a queue and an algorithm for managing the queue.

However, algorithm and mechanisms are not enough to make optimal scheduling unambiguous. What remains is to establish a policy framework that governs selection of the particular algorithms from the choices, setting and adjusting the priorities where appropriate, and other considerations necessary to keep the overall system running smoothly and serving its customers fairly and in a timely manner. Thus, it is not unlikely that use of the queuing algorithms described previously, to allocate resources as they and/or their associated software (e.g., process or memory manager) or hardware (memory, I/O device, or file) become available, can lead to a situation where the system is overcommitted to a particular resource (e.g., printer) or resource area (e.g., printers or network connections). This leads to diminished throughput for the system or to deadlock, where the entire system is halted due to a resource allocation state, where each of the processes has allocated to it some resource critically needed by another. Perhaps these policy decisions are most difficult, for they are not considered at the algorithmic/queuing level and are less amenable to quantification and algorithmic solution. In this regard, the problem of deadlock appears to be potentially the most debilitating.

82.6.1 Deadlock

Deadlock may occur when system resources are allocated solely on the basis of availability. The simplest example is when process 1 has been allocated nonshareable resource A, say a tape drive, and process 2 has been allocated nonshareable resource B, say a printer. Now, if it turns out that process 1 needs resource B (the printer) to proceed and process 2 needs resource A (the tape drive) to proceed and these are the only two processes in the system, each is blocking the other and all useful work in the system stops. This situation is termed deadlock. To be sure, a modern system is likely to have more than two active processes, and therefore the circumstances leading to deadlock are generally more complex; nonetheless, the possibility exists.

Deadlock is a possibility that must be considered in resource scheduling and allocation. For this chapter, the concern is with hardware resources, such as CPU cycles, primary memory space, printers, tape drives, and communications ports, but deadlock can also occur in the allocation of logical resources, such as files, semaphores, and monitors. Coffman et al. [14] identified four conditions necessary and sufficient for the occurrence of system deadlock:

- The resources involved are nonshareable.
- Requesting processes hold already allocated resources while waiting for requested resources.
- Resources already allocated to a process cannot be preempted.
- The processes in the system form a circular list when each process in the list is waiting for a resource held by the next process in the list.

There are basically four ways of dealing with the deadlock problem:

- Ignore deadlock
- Detect deadlock and, when it occurs, take steps to recover
- Avoid deadlock by cautious resource scheduling
- Prevent deadlock by resource scheduling so as to obviate at least one of the four necessary conditions

Each approach is considered, in order of decreasing severity, in terms of adverse effects on system performance.

82.6.1.1 Deadlock Prevention

Because all four conditions are necessary for deadlock to occur, it follows that deadlock may be prevented by obviating any one of the conditions. The first condition, that all resources involved be nonshareable, is difficult to eliminate because some resources, such as the tape drive and printer, are inherently nonshareable. However, in some cases this situation can be alleviated by spooling requests to a nonshareable device, such as the printer, to a temporary file on the disk for later transfer to the nonshareable device, as described previously.

There are two possibilities for elimination of the second condition. One is to require that a process request be granted all the resources it needs at once, before execution. An alternative is to disallow a process from requesting resources whenever it has previously allocated resources. That is, it must finish with those resources previously allocated and relinquish all of them before an additional resource request. Both methods assure that a process cannot be in possession of some resources while waiting for others. The first has the disadvantage of causing poor resource utilization by forcing allocation for a period likely to exceed, perhaps considerably, that needed for actual use of the resource. Moreover, depending upon the scheduling algorithm, some processes could face starvation from having to wait, perhaps indefinitely, for some resource or resources.

The third condition, nonpreemption, can be alleviated by forcing a process waiting for a resource that cannot immediately be allocated to relinquish all of its currently held resources, so other processes may use them to finish. Alternatively, the requesting process' request can be satisfied by preempting the requested resource from some currently blocked process, thus effectively eliminating the nonpreemption condition. The problem with these approaches is that some devices are simply not amenable to preemption — for instance, a printer, if preempted, will generate useless pages of output interleaved from several processes.

The fourth condition, the circular list, can be obviated by imposing an ordering on all of the resource types (presumably reflecting the order in which the resource types are likely to be used), and then forcing all processes to request the resources in order. Thus, if a process has resources of type m, it can only request resources of type $n > m$. If it needs more than one unit of a particular resource type, it must request all of them together. Although this will ensure that the circular list condition is denied, it can hurt resource utilization by requiring that a particular resource be allocated in advance of its logical need.

Deadlock prevention works by forcing rather severe constraints on resource allocation, leading to poor resource utilization and throughput. A less severe approach is to consider, individually, the implications of each resource request with respect to deadlock.

82.6.1.2 Deadlock Avoidance

This approach to the deadlock problem employs an algorithm to assess the possibility that deadlock could occur as a result of granting a particular resource request and acting accordingly. This method differs from deadlock prevention, which guarantees that deadlock cannot occur by obviating one of the necessary conditions, and from deadlock detection, in that it anticipates deadlock before it actually occurs.

The basic idea is to maintain a status indicator reflecting whether the current situation, with respect to resource availability and allocation, is safe from deadlock. When a resource request occurs, the system invokes the avoidance algorithm to determine whether granting the request for a set of resources would lead to an unsafe state and, if so, denies the request. The most common algorithm, due to Dijkstra [23], is called the *banker's algorithm*, because the process is analogous to that used by a banker in deciding whether a loan can be safely made. The algorithm is sketched in Figure 82.5. Here, **i** and **j** are the process and resource indices, and **N** and **R** are the number of processes and resources, respectively. Other variables and what they represent include the following:

maxneed[i, j] — The maximum number of resources of type **j** needed by process **i**

totalunits[j] — The total number of units of resource **j** available in the system

allocated[i, j] — The number of units of resource **j** currently allocated to process **i**

availableunits[i, j] — The number of units of resource **j** available after **allocated[i, j]** units are assigned to process **i**

needed[i, j] — The remaining need for resource **j** by process **i**

finish[i] — Represents the status of process **i**

> 0 if it is not clear that process **i** can finish
> 1 if process **i** can finish

What makes the deadlock avoidance strategy difficult is that granting a resource request that will lead to deadlock may not result in deadlock immediately. Thus, a successful strategy requires some knowledge about possible patterns of future resource needs. In the case of the banker's algorithm, that knowledge is the maximum quantity of each resource type that a particular process will need during its execution. As shown in Figure 82.5, the algorithm permits requests only when the current request added to the number of units already allocated is less than that maximum — and then only if granting the request still leaves some path for all the process in the system to complete, even if every one needs its maximum request.

But this last requirement — that each process know its maximum resource needs in advance, an unlikely supposition particularly for interactive jobs — severely limits the applicability of the banker's algorithm. Also, the interactive environment is characterized by a changing number of processes (i.e., **N** is not set) and a varying set of resources, **R**, as units occasionally malfunction and must be taken off-line. Further, even if the algorithm were to be applicable, Haberman [29] has shown that its execution has complexity proportional to N^2. Because the algorithm is executed each time a resource request occurs, the overhead is significant.

82.6.1.3 Deadlock Detection

An alternative to the costly prevention and avoidance strategies outlined previously is deadlock detection. This approach has two parts:

- An algorithm that tests the system status for deadlock
- A technique to recover from the deadlock

The detection algorithm, which could be invoked in response to each resource request (or if that is too expensive, at periodic time intervals), is similar in many ways to that used in avoidance. The basic idea is to check allocations against resource availability for all possible allocation sequences to determine whether the system is in a deadlocked state. There is no requirement that the maximum requests that a process will need must be stated here. The details of the algorithm are shown in Figure 82.6.

Of course, the deadlock detection algorithm is only half of this strategy. Once a deadlock is detected, there must be a way to recover. Several alternatives exist:

- Temporarily preempt resources from deadlocked processes. In this case, there must be some criteria for selecting the process and the resource affected. The criteria may include minimization of some cost function, based on parameters such as the number of resources a particular process holds and resource preemptability (e.g., a printer is difficult to preempt temporarily).
- Back off a process to some checkpoint, allowing preemption of a needed resource, and restart the process at the checkpoint later. In this case, the simplest way to find a safe checkpoint is to stop the process, return all of its allocated resources, and restart the process from the beginning at a later time.
- Successively kill processes until the system is deadlock-free.

```
const int safe = 1;
for (j = 1; j <= R; j++)                /* R resource types */
    available_units[j] = total_units[j];

for (j = 1; j <= R; j++)
    {
    for (i = 1; i <= N; i++)            /* N processes */
        {
        available_units[j] = available_units[j] — allocated[i, j];
          /*allocated [i, j] = number of units of resource j currently
            allocated to process i */
        finish[i] = 0;
          /* initialize — process i may not be able to finish */
        needed[i, j] = max_need[i, j] — allocated[i, j];
          /* needed is remaining need of process i for resource j;
        }
    not_done = 1;
    while (not_done)
        {
        not_done = 0;
        for (i = 1; i <= N; i++)
            if (! finish[i] && needed [i, j] <= available_units[j])
                {
                /* process i can finish */
                finish[i] = 1;
                available_units[j] = available_units[j] + allocated[i, j];
                /* give back process i's resources as done */
                not_done = 1;
                }
        }
    /* Continue loop until a process needed request cannot be met */
    /* Determine if all N processes could be completed. */
    if (allocated_units[j] = total_units[j])
        status = safe;
    else
        status = !safe;
    }
if (status == safe)
    /* allocate the requested resource */
```

FIGURE 82.5   The banker's algorithm for deadlock avoidance.*

These methods are expensive in the sense that the detection algorithm is rerun with each iteration until the system proves to be deadlock-free. The detection algorithm, like the avoidance algorithm of Figure 82.5, has time complexity N^2. Another potential problem is starvation — care must be taken to ensure that resources are not continually preempted from the same process or that the same process is not repeatedly backed off or killed.

---

*Based on statements of the algorithm in Dijkstra [23] and Tsichritzis and Bernstein [84].

```
for (j = 1; j <= R; j++)              /* R resource types */
    available_units[j] = total_units[j];

for (j = 1; j <= R; j++)
    {
    for (i = 1; i <= N; i++)          /* N processes */
        {
        available_units[j] = available_units[j] — allocated[i, j];
            /*allocated[i, j] = number of units of resource j currently
             allocated to process i */
        finish[i] = 0;
            /* initialize — process i may not be able to finish */
        }
    not_done = 1;
    while (not_done)
        {
        not_done = 0;
        for (i = 1; i <= N; i++)
            if (! finish[i] && request[i, j] <= available_units[j])
                {
                /* process i can finish */
                finish[i] = 1;
                available_units[j] = available_units[j] + allocated[i, j];
                /* give back process i's resources */
                not_done = 1;
                }
        }
deadlock = 0;
for (i = 1; i <= N && deadlock == 0; i++)
    if (finish[i] != 1)
        deadlock = 1;
if (deadlock == 1)
    /* system deadlocked */
```

FIGURE 82.6   Deadlock detection algorithm.*

82.6.1.4 Ignore Deadlock

The fourth approach — do nothing and hope — reflects the observation that deadlocks and their con-comitant effects including possible system crashes, do not occur with sufficient frequency to justify the expense in terms of system overhead required to handle deadlock by any of the other three approaches. This is the approach taken in the UNIX operating system.

82.6.1.5 The Eclectic Approach

It is clear that none of the conventional choices for dealing with the potential for deadlock is entirely satisfactory. An alternative strategy is to divide the system resources into classes and apply the most

---

*Based on statements of the algorithm in Dijkstra [23] and Tsichritzis and Bernstein [84].

appropriate technique for dealing with potential deadlocks in each class. Silberschatz et al. [77] suggest four classes:

1. In the case of resources, such as the process control blocks used by the system itself, deadlock prevention can be used by forcing resource ordering because there is no contention among pending requests.
2. Primary memory can be shared in the sense that an active process (or part of it) can be swapped out without destroying the process. Therefore, deadlock can be prevented by preemption.
3. For nonshareable resources, including devices such as the printer and writeable files, deadlock avoidance can be used as resource requirements become known.
4. Virtual memory space on the disk can be protected from deadlock by avoidance because maximum virtual storage requirements are generally known before execution.

In spite of this, most systems today take the optimistic approach and do nothing to prevent, avoid, or even detect deadlock.

82.7   High-Level Scheduling

Process and device scheduling operate in the context of a higher-level system scheduling. This higher-level scheduling deals with admitting new processes to the system, setting and changing priorities on the basis of system operation, and generally implementing the policy controls of system operation. In fact, because scheduling system resources is prerequisite to resource allocation, criteria and policy regarding scheduling are intimately related to system resource allocation. To this end, criteria for scheduling decisions match the objectives for the operating system as a whole, that is, to provide good service to processes within a context of operation at a high level of resource utilization. Other factors that cannot be overlooked are the need to assure mutual exclusion for nonshareable resources and prevention or mitigation of starvation and deadlock. A portion of the responsibility for this lies in a system scheduler process. Lister [50] suggests that the tasks of the scheduler include admission of processes into the system, determination of process priorities, and enforcement of system policies for resource allocation. In the first instance, the scheduler selects batch jobs from an input queue, using criteria such as resources required and priority. In a time-sharing environment, this task is more difficult because processes enter the system at user login. However, the scheduler can monitor resource utilization and deny access in order to maintain performance criteria for the currently active processes. Secondly, the scheduler determines the process priorities that govern placement in system queues. Because a primary system scheduling mechanism is manifest in these queues, the scheduler plays a key role here. Finally, the scheduler is responsible for system policies that deal with balance (ensuring that no resource class is either over- or underutilized) and deadlock.

The importance of these activities in determining overall system performance suggests that the scheduler itself should be a high-priority process. In particular, the scheduler executes whenever a resource request or release occurs or a process arrives at the system or terminates [50]. The lower-level schedulers discussed previously are analogously activated by interrupts — both are unpredictable, though interrupts may be expected more frequently than process entry and exit or resource request and release.

The criteria for high-level scheduling decisions are many and varied. System response time is enhanced by minimizing context switches and the associated overhead. Scheduling fewer new processes into the system is an obvious way to achieve this. But such a scheduling policy is likely to result in less-than-optimal resource utilization because fewer processes (a lower degree of multiprogramming) are together unlikely to use all the resources declared needed throughout these processes' time in the system. As more processes enter the system, the scheduler has more policy decisions to make when resources are requested and released. The criteria are numerous, and their effectiveness is generally affirmed or denied by empirical observation. Several possibilities are summarized in [50]:

1. Give priority in resource requests to processes already in possession of a significant quantity of resources, to advance their progress through the system in the hope of reclaiming the resources allocated at the earliest possible time.

2. Give resource-rich processes high dispatching priority, for the same reason as in **1**.
3. Use the working set principle [22] as a criterion for memory allocation.
4. Priorities should reflect process importance. Most system processes have higher priorities than user processes. Among system processes, the scheduler itself is high priority.
5. Device drivers should have high priority, and drivers for higher-speed devices should have relatively higher priority. This is designed to keep the devices running, mitigating the possibility of I/O delays.

In summary, process and device scheduling is a tricky business with goals that are not always mutually consistent. Different possibilities serve different needs. Experimentation and empirical observation are necessary to judge effectiveness in a particular situation.

82.8 Recent Work and Current Research

Although much of the work on process and device scheduling took place during the late 1960s and the 1970s, with the advent of multiprogramming and time-sharing systems and as the subject of operating systems came into its own, technical advances continue to spur new developments. In this section, some of the more significant recent discoveries are summarized, and pointers are given for further study.

82.8.1 Processor Scheduling

The goals listed previously, which underlie the development of techniques for processor scheduling, are often mutually conflicting. This inevitability has led to the priority and dynamic dispatching priority adjustment algorithms described earlier. Recent work in processor scheduling has continued to focus on refinement of dynamic scheduling schemes.

The problem with priority methods that set priority once at the process's entry into the system is inflexibility. Resource needs do not always match the priorities, and separate resources may require resource allocation mechanisms isolated from one another. The dynamic priority adjustment mechanisms described previously begin to address these problems. Recent alternatives include fair-share schedulers [33] [41], microeconomic schedulers [86], lottery schedulers [87], time-function scheduling [25], and observation-based improvements [39] [79].

A *fair-share scheduler* is designed to allocate resources to provide each user or group of users with an equitable share of the CPU over some period of time. This is accomplished by adding facilities to track the actual amount of processor usage allocated to each process and to adjust priorities dynamically to assure allocation of processor time to each process remains close to established shares.

Henry [33] describes a fair-share scheduler utilized at AT&T Bell Laboratories. This scheme divides users into groups and schedules, so that each group is allocated a fair share of CPU time for execution. Within a particular group, processes are scheduled according to the regular dispatcher relative to other processes in the group.

The implementation is a straightforward modification of the standard decay usage scheduling algorithm. In addition to the process's "execution time field," which is incremented at each clock interrupt, there is another "fair share group execution field" that is updated at clock interrupt when any process from the group is executing. The fair-share group execution field is decayed exactly as the individual process' execution time fields, once a second. Then, process priorities are recalculated using the modified formula:

$$\text{priority} = C(\text{execution time field})/2 + \text{base priority} + C(\text{fair share group execution field})/2$$

The result is that the more CPU time the processes from a particular group have used recently, the higher the priority number and the lower the priority for processes in that group. Overall, however, each group ultimately gets an equitable share of the CPU time.

Kay and Lauder [41] describe Share, a fair-share scheduler that is designed to work at both the individual user level and the group level. Unlike the AT&T fair-share scheduler, the Share scheduler explicitly considers use of resources other than the CPU. The goal is to have a scheduler that is fair to users, rather than just to processes. Therefore, the approach is designed to schedule resources so that users get a fair share of the machine over time, balancing an entitlement expressed as a number of shares and recent resource use history. Allocation of shares is, of course, an administrative matter and depends on the organization's entitlement and the user's entitlement within the organization.

The implementation is considerably more complex than that of the AT&T algorithm. There are basically three levels of operation. At the first (user) level, a figure representing usage is adjusted by both a decay factor and a measure of total resource consumption. Thus, every $t1$ seconds (several seconds, Kay and Lauder used 4)

$$\text{usage} = \text{usage} \times K1 + \text{charges}$$

where $K1$ is a suitable decay factor, and charges represents the sum of price times quantity for resources used since the last calculation.

Subsequently, every $t2$ seconds (e.g., $t2 = 1$ second), process priorities are decayed according to

$$\text{priority} = \text{priority} \times K2 \times (\text{nice} + K2')$$

where $K2$ and $K2'$ are factors reflecting priority decay rate and the effect of the UNIX `nice` command parameter (0 is default).

Finally, every $t3$ seconds (1/60 to 1/100 of a second), a priority recalculation occurs, taking into account recent usage and remaining shares:

$$\text{priority} = \text{priority} + (\text{usage} \times \#\text{active processes})/(\#\text{shares})^2$$

A user's process priority is not only a function of recent usage and entitlement, but also of the number of active processes. Thus, a single user's share is effectively spread among the number of currently active processes.

The Share system is also applicable to scheduling in a context where a computer is shared among a number of groups. In this case, shares are allocated to each group in proportion to the division of the machine. This requires more calculations to allocate and reallocate shares within a particular group when a new user logs on, subject to keeping the total number of shares allocated to the group constant. The system allows for different decay rates for usage in the various groups. The Share system does provide users with a sense of fairness, but at a cost of additional overhead.

The *lottery scheduling* scheme of Waldspurger and Weihl [87] is similar to the Kay and Lauder [41] Share scheduler in the sense that it provides for proportional share resource allocation. In this case, lottery "tickets," rather than shares, represent the resource entitlements. The scheme is designed to provide active and flexible control of resource allocation in proportion to share allocations.

The scheduler works in the context of a scheduling quantum of 10 ms. Resources are allocated by lottery in the sense that a resource is allocated to the process that holds the "winning ticket." The system is fair in that the probability that a particular process will be allocated a particular resource is proportional to the number of tickets held by that process. It follows that allocation of resources in general will be proportional to the number of tickets held, that is, to entitlement. This is so because the tickets are fungible in that they represent resource rights independent of the type or mix of resources needed.

Lottery scheduling has the advantage of being relatively straightforward to implement and efficient, in that its overhead is comparable to that of a standard time-sharing policy. By definition, the scheduling algorithm is randomized. Hence, actual allocation at any given moment may not match entitlement shares exactly. But allocations occur every 10 ms and fairness, in the sense of matching entitlements, increases quickly over time as the number of allocations grows. Moreover, the problem of starvation is obviated, because any client with tickets will eventually win a lottery and thus be allocated resources.

A more general approach to resource scheduling, called *time-function scheduling*, has been proposed by Fong and Squillante [25]. The method uses dynamically changing priorities based on general functions of such factors as the time a given process has waited for a particular resource. The idea is to support a wide and varying range of scheduling objectives by implementing time-functions reflecting these objectives.

Further, these time-functions are effectively variable in response to changes in the system in order to preserve the desired scheduling objectives. Fong and Squillante have found time-function scheduling to provide the same fair-share objectives as lottery scheduling, with less waiting time variance (resulting from the probabilistic nature of the lottery scheduling).

A number of empirical studies [69] [74] [78] have concluded that maximum CPU utilization and system throughput results from a CPU scheduling policy that gives preemptive priority to I/O bound jobs over compute bound jobs. Kamada [39] has used a Markovian model of job processing (identical to the finite source queuing model) to verify that scheduling policies which assure that I/O bound jobs are given preemptive priority over processor bound jobs provide maximum processor utilization and optimal throughput. Within the confines of the theoretical model and its concomitant assumptions, it is not possible to show that any additional sophistication of the sort described previously can improve upon this result. Kamada concludes that general applicability of the theoretical results is limited by the model and its underlying assumptions.

Suranauwarat and Taniguchi [79] describe an augmented process scheduler that facilitates both reduced execution time and wall clock time for programs that are run multiple times. The idea is to observe the context-switch behavior of a program and use this information to identify normal end of time-slice context switches that are closely followed by a process block (e.g., to wait for I/O completion). In these cases, they conjecture that it makes sense to extend the time-slice until the process blocks, thus improving run time for programs that are run repeatedly.

The method is to have the process scheduler log context switches for all the processes in the system and, at the end of a particular program's execution, create an experiential knowledge record called a program flow sequence (PFS) from the context switch log for the given program's processes. During subsequent executions of the program, when a process from the program is running and at the end of its current time-slice, the scheduler consults the PFS to determine whether a small (less than a proscribed factor) extension of the time-slice will, by allowing the process to proceed until it blocks, obviate a context switch. With each run of the program, its PFS is modified to develop an accurate portrayal of its context-switching behavior.

Experimentation with a parameterized test program and three existing programs showed that the knowledge-based scheduler provided for reduced run times for the test program and two out of three of the existing programs (from .05% to 13%). Moreover, the performance enhancement was a direct function of the maximum dispatch delay time permitted, and the information in the PFS for the program improved with the number of times the programs were run. However, it remains to assess more fully the impact of reducing the execution time of repeatedly run programs on the fairness of the overall system.

Other work in processor scheduling includes predictive deadline scheduling and dynamic mixed priority schemes for real-time systems. See Miller [59] [60].

82.8.2 Disk Scheduling

Recent work in disk scheduling continues the search for an optimal algorithm through experimental studies and refinement of the well known algorithms described previously, anticipatory algorithms causing movement of the disk heads in advance of the next disk I/O request, and use of application-specific requirements. Approaches to the former included extensive simulation studies [34] [82] [81] and analytical modeling [15]; anticipation algorithms have been suggested by King [43] and by Iyer and Druschel [37]; and algorithms based on use of application-specific requirements were presented by Cao et al. [11] and Shenoy and Vin [73].

Geist and Daniel [26] have proposed a disk scheduling scheme that is a varying mixture of SSTF and SCAN, designed to obviate both the considerable seek time variances and potential for starvation of SSTF and the potential delays for opposite end requests of SCAN. The continuum algorithm, $V(R)$, $R \in [0, 1]$, takes as its next request the one closest, in a recast sense of closest, to the current position of the head. As with SCAN, $V(R)$ keeps track of the current direction of head movement — but, in this case, the direction of the just-previous seek. The distance to the nearest request in that same direction is the number of cylinders away as before. But the distance to the nearest request in the direction opposite to that of the last seek is R times the total number of cylinders plus the number of cylinders to the request. Note that R is any real number in the interval 0 to 1. If $R = 0$, V(R) is SSTF, and if $R = 1$, V(R) is SCAN. For values of $0 < R < 1$, V(R) is somewhere between the two.

The implementation requires maintaining two request queues: one each for requests in either direction from the current head position. The requests in both queues are kept in order of increasing distance from the current head position. Selection of the next request to process is made by calculating the distance to the first request on each queue, adjusting the distance requiring a change in direction. The request associated with the least adjusted distance is selected, and the current head position and direction are updated. A simulation was used to test different values for R. With respect to improvements in mean waiting time and throughput, the simulation showed V(0.2) to be superior to FCFS and to either pure SSTF or SCAN.

In an effort to enhance the effectiveness of SSTF, Seltzer et al. [72] and Jacobson and Wilkes [38] have refined the idea to include rotation time. Thus, shortest access time first (SATF) serves first the request that has the least access time, compiled as the sum of seek and rotation times, from the current head position.

The efficacy of prerequest disk arm movement has been explored by King [43]. Anticipatory disk arm scheduling policies have been largely overlooked due to the supposition of a continuous substantial request queue length [21] [82]. This constant backlog would mean there was insufficient time to position the disk heads optimally for successive requests. However, Geist and Daniel [26] and Lynch [51] reported disk drive utilization rates on the order of 20% to 35%, suggesting that there is time, after all, that might profitably be used for positioning the disk heads in anticipation of the next request. This possibility was investigated by Coffman and Hofri [15], Hofri [34], and more recently by King [43]. King shows that the potential for performance improvements due to anticipatory scheduling is a function not only of disk idle time, but also of the pattern of the request locations over the physical disk space. If the data requests are spaced uniformly over the disk and a FCFS scheduling algorithm is used, prepositioning the disk heads at the middle cylinder results in 25% less head movement on average, which translates to a 13% reduction in mean seek time. In the case of a nonuniform distribution of request data locations on the disk, the optimal anticipatory location is between the most frequently accessed locations, so-called *hot spots*, and the midway point, the particular location a function of the probability of requests at the hot spots. Inasmuch as the locations and probabilities of these hot spots are known (perhaps from system performance data) and the request rate is sufficiently low as rarely to interfere with an anticipatory move, the anticipatory algorithm saves time.

As the frequency of request arrivals increases, so does the possibility that the arriving request will have to wait for completion of an anticipatory seek operation. Ultimately, the demand for disk service could become so high that, in the worst case, disk requests must wait the entire time needed for anticipatory repositioning. Possible ways out of this dilemma are suggested in King [43]. The first is to move the heads in short spurts toward the optimal anticipatory location, thus allowing for timely attention to currently arriving disk data requests. The second is to develop technology to facilitate an interruptible disk head seek. The former leads to seek time improvements of 8% to 10% for an anticipatory version of FCFS.

Simulation was used to assess the effect of anticipatory head motion in the case of a workload representing more realistic data from McNutt [58] — a workload that reflects the high probability that a given job will generate repeated requests to the same disk location. In this case, especially if the disk request rate is low, few requests will benefit from an anticipatory seek until that particular job completes. This suggests modifying the anticipatory algorithm by including a delay, representing job completion time, between a request and an anticipatory move. This modification also yielded a small performance improvement.

Other, more complex disk systems with two arms and duplexed disks with separate control and data lines to two disk arms were also considered. Modifications to the anticipatory algorithm yielded reductions of mean seek distances of 10% to 29% over more conventional disk scheduling algorithms.

Another approach to anticipatory scheduling, by Iyer and Druschel [37], is based on the observation that processes frequently issue multiple requests for disk service, separated by short spurts of computation time, and that a given process' disk requests exhibit spatial locality. They argue that, in an environment of queued disk service requests, a disk scheduler that immediately selects from its queue a request that arose from a process different from the one issuing the last request may incur an expensive disk seek and lose out on the opportunity to service a forthcoming request from the former process to a nearby location on the disk. In fact, they argue that waiting a short period to determine whether the original process will, in fact, issue another request is "deceptive idleness," that is, it may well pay off in overall throughput to do so. In particular, a seek reducing scheduler, such as SSTF, CSS, or the variant SPTF, is frustrated in its attempts to minimize the cost of head movement by immediately servicing a disk request from a new process. Similarly, a proportional-share scheduler, such as lottery scheduling [87] or Yet Another Fair Queuing [6], may find it impossible to provide disk service to multiple processes (applications) according to the preassigned ratios when it must select requests immediately from the disk request queue.

The solution proposed by Iyer and Druschel [37] is to wrap any given disk scheduling strategy in an anticipatory scheduling framework. This framework consists of an anticipatory scheduling core that envelopes an adaptive scheduler type–specific anticipation heuristic processor coupled with the original disk scheduler. The anticipatory core provides the generic waiting apparatus to be informed by the heuristic. A separate heuristic processor is implemented for each of the two classes of scheduler. For the seek-reducing schedulers, it determines the wait or no-wait decision based on the benefit of waiting (expected reduction in seek time) vs. the cost (potential waste of time). For the proportional schedulers, a decision is made to wait for the last process if it has received the smallest amount of disk service so far, perhaps with a relaxed time scale to enable some beneficial seek reduction. The waiting period is that where there is a 95% probability that a disk request will arrive from the last process. Experimental evaluation using FreeBSD confirms the effectiveness of this anticipatory scheduling framework. Throughput increases ranged from 8% for the Andrew file system benchmark, to up to 60% for variants of the TPC-B database benchmark, to between 29% and 71% for the Apache Web server. Moreover, proportional-share schedulers augmented by the anticipatory scheduling scheme were able to meet their objectives with reasonable overhead.

An alternative to anticipatory scheduling is to look explicitly ahead at two or three requests and service the most desirable set of requests by modifying existing disk scheduling algorithms. Thomasian and Liu [83] have done this for the CSS and the SATF algorithms to address the difficult case of random requests for small blocks of data, as in transaction systems, such as an airline reservation system. This lookahead scheme for CSS algorithms considers the next several pending requests and orders them so as to minimize the total of their service times. Subsequently, scanning continues in the original direction, possibly picking up newly arrived requests between those just processed. The SATF algorithm is augmented to look at several requests and process at least two at a time. Again, the two are chosen to minimize access time. SATF with lookahead is designed to consider several requests — those beyond the first two are evaluated with a discount factor. The third algorithm is an SATF algorithm with higher priority given to read requests. A higher-priority write request is processed first only if the ratio of its service time to that of the highest priority read is less than a certain experimentally determined factor. Performance of these disk scheduling algorithms was compared to each other and to their established (non-augmented) counterparts, using experiments based on a random number driven simulation model. This was done to assess performance, especially in cases of random requests. Response times for the modified CSS algorithm were better than the unmodified version, although service times were nearly equal. Response times for the modified SATF algorithm were at most 10% improved. Finally, the SATF algorithm modified to give read requests higher priority showed minimal improvement over a more conventional, nonpreemptive read priority system, in which write requests are serviced only when there are no read requests waiting.

Cao et al. [11] suggest a technique that combines application-controlled file caching, prefetching, and disk scheduling. Modern operating systems typically deal with the wide and increasing disparity in performance between disks and processors by establishing a file cache in memory to hold demand and prefetched file information. Cao et al. showed that a system which utilizes application-based caching and prefetching policies can outperform a more typical system utilizing predefined cache replacement (usually LRU) and prefetching (sequential or one-block lookahead) policies. Previous work [10], however, indicated that performance improvements can be obtained by utilizing application-specific replacement policies. They propose integration of a controlled-aggressive policy that they showed theoretically closely approximates optimal cache block replacement [10]. In this case, even with imperfect and limited lookahead knowledge and in spite of the complicating interplay between applications and between applications and the kernel in the multiprocess case, substantial improvements result when disk prefetches are batched and the disk scheduler assembles the requests according to increasing logical block numbers.

This is accomplished by using a two-level cache management strategy: a kernel-implemented global policy, which allocates cache blocks among the various processes, and a local policy, which allows each user process to control use of its own cache blocks. The kernel's global policy is critical in ensuring that no individual process's policy can negatively affect another. To this end, the kernel utilizes an LRU policy with swapping and place holders that allows the process with the least recently used block, using its application-specific knowledge, to replace a block other than that least recently used with a block less likely to be used by swapping the two and marking the position of the removed block with a place holder. This protects the process from continually having the least recently used block (if it were not swapped) and provides a place for cache replacement in case the displaced block is recalled. This provides for a more optimal choice while preserving fairness. Further, each process contributes to more efficient disk utilization by grouping prefetches into batches. Experimentation with an implemented prototype system showed an average 26% reduction in running times when just a single process is using the file system at any given time and an average 32% reduction when multiple processes are concurrently accessing the file system.

In a similar vein, Shenoy and Vin [73] have proposed a disk scheduling framework, Cello, designed to accommodate the varying service requirements of today's diverse set of applications while preserving efficient disk utilization. The Cello framework consists of a two-level disk scheduling framework: a low-level, class-independent scheduler, which maintains the queue of disk service requests processed by the disk device driver, and a higher-level set of application class specific schedulers, that maintain individual queues of disk requests organized according to needs specific to applications in that class. This two-level organization recognizes the disparate needs of categories of applications, such as hard real-time applications (e.g., machine tool control) and soft real-time applications (e.g., audio or video playback), typically utilizing algorithms such as earliest deadline first (EDF), priority SCAN (PSCAN), or feasible deadline SCAN (FD-SCAN), for interactive best-effort applications (e.g., word processors), and high throughput best-effort applications (e.g., file servers), which utilize FCFS, SSTF, SCAN, LOOK, V(R), SATF, or aged SAFT (ASATF).

In the Cello framework, the class-independent scheduler sets the portion of the available disk bandwidth for each of the class-specific schedulers. It does this either by proportionate time allocation or by proportionate space allocation. It tracks the various requests in a series of time intervals, ensuring that no class-specific scheduler uses more than its share. The class-independent scheduler interfaces with the class-specific scheduler by providing the current state of the disk device request queue it maintains, along with a measure of the constraints of those requests already queued — a factor it terms *slack*.

The application-specific schedulers maintain an ordering of disk requests according to the needs of applications in that particular class, while minimizing seek and rotational delay. For example, the best-effort class schedulers might use FCFS, SCAN, or SATF; the real-time schedulers might use EDF or SCAN-EDF. When requested by the class-independent scheduler, the application-specific scheduler calculates a position, perhaps using a first-fit, best-fit or other policy, for inserting the request at the head of its queue into the disk request queue of the class-independent scheduler. The latter checks that the insertion will not cause delays that would cause deadlines to be missed or the proportionate share of bandwidth to be exceeded (including changes in seek and rotational delay resulting from the proposed insertion).

Experiments with a trace-driven simulator and a prototype file system running on an Intel 233 MHz, Pentium machine running Solaris 2.5 allowed analysis of various parameters (e.g., time vs. space proportions, different weights and workloads), and established feasibility by showing a minimal increase in overhead running Cello.

The search for improvements in disk and file system performance has been marked by the development of redundant arrays of inexpensive disks (RAIDs) [63] [64]. The idea, as the name implies, is to connect a number of small disks to improve both reliability and performance. Initially, a RAID system was made up of a central controller connected to a host computer on one side and, on the other, to a number of disk controllers, each connected to a series of small disks. Performance is improved by distributing data across a number of disk drives, affording the possibility of parallel operation. Reliability is enhanced by *striping*, that is, storing a "stripe" of data blocks across a number of drives, thus minimizing the possibility of data loss from drive failure; *mirroring*, where a complete copy of the data is kept on separate drives; or *parity techniques*, with parity blocks on a dedicated drive or distributed throughout the disk array. Cao et al. [9] have proposed an architecture in which the central controller is replaced by a number of "array controller nodes," some of which are dubbed "worker" nodes and connect to local disks, and others "origination" nodes that provide for communication with the host computer. This latter arrangement enhances reliability by obviating the central controller — and thus the potential that its failure could disable the entire array. The "TickerTAIP" architecture also has the potential for improved performance by allowing for greater parallelism (e.g., for parity calculation).

The design of the RAID system preserves reliability and minimizes disruption from a failure of any single component: a disk failure, a worker failure, or an originator failure. The latter is assured by an *atomic write* policy: not allowing a write operation to make any changes until there is sufficient redundancy to guarantee successful completion of the write. Ensuring reliability becomes more difficult in the event of simultaneous failure of more than one component, a disaster according to Lampson and Sturgis [46]. However, following UNIX 4.2 BSD-type file systems [57], Cao et al. [9] propose controlling the effects of simultaneous component failure by using request sequencing, in essence assuming that no request can begin until all requests on which it depends (e.g., directory or inode requests) are complete.

Atomic requests and sequencing enable the TickerTAIP RAID to handle multiple requests requiring queues of requests at worker nodes. This leads to consideration again of request scheduling algorithms — this time, inside the disk array system. Thus, the worker system could use the previously described algorithms, such as FCFS, SSTF, or SATF. An additional algorithm [9], batched nearest neighbor (BNN), is a variant of SATF in that when a worker is available, it takes the entire batch of requests in its queue as a group and applies SATF until the entire batch is served, before picking up a new request. BBN provides much of the performance of SATF without the concomitant starvation problem.

Simulations showed that the distributed controller system of Cao et al. [9] provided greater throughput and lower response times with less CPU power. In addition, the tests showed that when the SATF and BNN algorithms were used, throughput improved substantially over FCFS (more so with SATF), especially when run with real world type workloads. Mean response time for the same workload tests were minimal with BNN, probably because of its built-in tendency to mitigate starvation.

Disk scheduling and allocation problems are, of course, closely related to both disk architecture and to the design and implementation of virtual memory and file systems. In particular, see the work of Patterson et al. [63] [64], Rosenblum and Ousterhout [67], and Hartman and Ousterhout [31]. See also Chapter 85 and Chapter 100 of this Handbook.

82.8.3 Deadlock

In Section 82.6.1.2 and Section 82.6.1.3, classical monoprocessor algorithms were shown to have time complexity $O(N^2)$, where N is the number of processes which, given that the algorithm must be executed for each resource request, presents unacceptable overhead. In fact, the best software algorithms, even for multiprocessor systems, have run-time complexities of $O(n \times r)$, where n is the number of processors and r is the number of resources [7] [42] [76]. Significant improvement apparently requires hardware implementation. Shiu et al. have developed a hardware implementation of a parallel deadlock detection algorithm [75]. Their simulated implementation experiments showed that this hardware detection algorithm could lower detection time by more than 99% and overall execution time by up to 68%. The algorithm uses a resource allocation graph [70], represented as an augmented adjacency matrix, to locate deadlocks by identifying cycles of resource requests and grants. The existence of such a cycle indicates a deadlock and vice versa. This enhanced adjacency matrix of elements representing requests and grants is further characterized as having [75]

Dead rows or columns — Those with no request or grant entries
Source rows or columns — Rows with at least one request, but no grants or columns with at least one grant but no requests
Sink rows or columns — Rows with at least one grant, but no requests or columns with at least one request but no grants

The implementation is a hardware parallel algorithm that identifies dead, source, and sink rows and columns and reduces the augmented adjacency matrix by iteratively eliminating these rows and columns as far as possible. Any remaining entries indicate existence of a deadlock; an empty matrix indicates freedom from deadlock. The algorithm has a run-time complexity of $O(\min(n, r))$.

Other recent work on deadlock is primarily focused on related areas:

Threads — [5] and Chapter 84 and Chapter 96 of this Handbook
Integrating independently designed software components — [3] [36]
Interconnection networks — [27] [65] and Chapter 72 and Chapter 73 of this Handbook
Programming languages — [28] [49] [24], Chapter 91, and Chapter 96 of this Handbook
Multiprocessor and parallel systems — [54] [61] [66] [71] [89], and Chapter 23 of this Handbook
Distributed systems — [2] [13] [40], Chapter 84, and Chapter 87 of this Handbook
Real-time systems — [55] [75] and Chapter 83 of this Handbook

Acknowledgments

Parts of Section 82.1 and Section 82.2 of this chapter are reprinted, with permission, from this author's contribution to [85].

Defining Terms

Deadlock: "A set of processes is deadlocked if each process in the set is waiting for an event [release] that only another process in the set can cause" [80, p 242].

Distributed system: "A *distributed computing system* consists of a number of computers that are connected and managed so that they *automatically* share the job processing load among the constituent computers or separate the job load as appropriate among particularly configured processors. Such a system requires an operating system which, in addition to the typical stand-alone functionality, provides coordination of the operations and information flow among the component computers" [85, p 403].

Multiprocessing: "A *multiprocessing* system is a computer hardware configuration that includes more than one independent processing unit" [85, p 403].

Multiprogramming: "A *multiprogramming* operating system is an OS which allows more than one active user program (or part of user program) to be stored in main memory simultaneously" [85, p 403].

Network: "A *networked computing system* is a collection of physically interconnected computers. The operating system of each of the interconnected computers must contain, in addition to its own stand-alone functionality, provisions for handling communication and transfer of programs and data among the other computers with which it is connected" [85, p 403].

Process: "A *process* is a series of operations associated with the execution of a sequence of instructions which effect a particular system or user action" [85, p 425].

References

[1] Bach, M.J., *The Design of the UNIX Operating System*, Prentice Hall, Englewood Cliffs, NJ, 1986.

[2] Badel, D.Z., "The Distributed Deadlock Detection Algorithm," *ACM Transactions on Computing Systems*, 4, 4 (November 1986), 320–337.

[3] Bernardo, M., P. Ciancarini, and L. Donatiello, "Architecting Families of Software Systems with Process Algebras," *ACM Transactions of Software Engineering and Methodology*, 11, 4 (October 2002), 386–426.

[4] Bic, L., and A.C. Shaw, *The Logical Design of Operating Systems*, 2nd edition, Prentice Hall, Englewood Cliffs, NJ, 1988.

[5] Boyapati, C., R. Lee, and M. Rinard, "Ownership Types for Safe Programming: Preventing Data Races and Deadlocks," *ACM SIGPLAN Notices, Proceedings of the 17th ACM Conference on Object-Oriented Programming, Systems, Languages, and Applications*, 37, 11 (November 2002), 211–230.

[6] Bruno, J., E. Gabber, B. Ozden, and A. Silberschatz, "Disk Scheduling with Quality of Service Guarantees," *IEEE ICMCS*, June 1999.

[7] Cahit, I., "Deadlock Detection Using (0, 1)-Labeling of Resource Allocation Graphs," *IEEE Proceedings: Computers and Digital Techniques*, 145, 1 (January 1998), 68–72.

[8] Calingaert, P., *Operating System Elements: A User Perspective*, Prentice Hall, Englewood Cliffs, NJ, 1982.

[9] Cao, P., S.B. Lim, S. Venkataraman, and J. Wilkes, "The TickerTAIP Parallel RAID Architecture," *ACM Transactions on Computing Systems*, 12, 3 (August 1994), 236–269.

[10] Cao, P., E.W. Felton, A.R. Karlin, and K. Liu, "A Study of Integrated Prefetching and Caching Strategies," *Proceedings of 1995 ACM SIGMETRICS*, ACM, New York, 188–197.

[11] Cao, P., E.W. Felton, and A.R. Karlin, "Implementation and Performance of Integrated Application-Controlled File Caching, Prefetching, and Disk Scheduling," *ACM Transactions on Computer Systems*, 14, 4 (November 1996), 311–343.

[12] Chandra, R., S. Devine, B. Verghese, A. Gupta, and M. Rosenblum, "Scheduling and Page Migration for Multiprocessor Compute Servers," Sixth International Conference on Architectural Support for Programming Languages and Operating Systems, proceedings published in *Operating Systems Review*, 28, 5 (December 1994), 12–24.

[13] Chandy, K.M., L.M. Haas, and J. Misra, "Distributed Deadlock Detection," *ACM Transactions on Computing Systems*, 1, 2 (May 1983), 144–156.

[14] Coffman, E.G., Jr., M. Elphick, and A. Shoshani, "System Deadlocks," *ACM Computing Surveys*, 3, 2 (June 1971), 67–78.

[15] Coffman, E.G., Jr., and M. Hofri, "A Class of FIFO Queues Arising in Computer Systems," *Operations Research*, 26, 5 (September–October 1978), 864–880.

[16] Coffman, E., and M. Hofri, "On the Expected Performance of Scanning Disks," *SIAM Journal of Computation*, 11 (1982), 60–70.

[17] Coffman, E.G., Jr., and L. Kleinrock, "Computer Scheduling Measures and Their Countermeasures," *Proceedings of AFIPS 32*, SJCC, 11–21.

[18] Coffman, E.G., Jr., and P.J. Denning, *Operating Systems Theory*, Prentice Hall, Englewood Cliffs, NJ, 1973.

[19] Daniel, S., and R. Geist, "V-SCAN: An Adaptive Disk Scheduling Algorithm," *Proceedings of the IEEE International Symposium on Computing Systems Organization*, New Orleans, LA, March 1983, 96–108.

[20] deJonge, W., M.F. Kaashoek, and W.C. Hsieh, "The Logical Disk: A New Approach to Improving File Systems," 14th ACM Symposium on Operating Systems Principles, proceedings published in *Operating Systems Review*, 27, 5 (December 1993), 15–28.

[21] Denning, P.J., "Effects of Scheduling on File Memory Operation," *Proceedings of the AFIPS Spring Joint Computer Conference 30* (1967), 9–21.

[22] Denning, P.J., "The Working Set Model for Program Behavior," *Communications of the ACM*, 11, 5 (May 1968), 323–333.

[23] Dijkstra, E.W., "Cooperating Sequential Processes," *Programming Languages*, F. Gehuys, Ed., Academic Press, New York, 43–112.

[24] Etalle, S., M. Gabbrielli, and M.C. Meo, "Transformations of CCP Programs," *ACM Transactions on Programming Languages and Systems*, 23, 3 (May 2001), 304–395.

[25] Fong, L.L., and M.S. Squillante, "Time-Function Scheduling: A General Approach To Controllable Resource Management," 15th ACM Symposium on Operating Systems Principles, proceedings published in *Operating Systems Review*, 29, 5 (December 1995), 230.

[26] Geist, R., and S. Daniel, "A Continuum of Disk Scheduling Algorithms," *ACM Transactions on Computing Systems*, 5, 1 (February 1987), 77–92.

[27] Gerla, M., P. Palnati, and S. Walton, "Multicasting Protocols for High-Speed, Wormhole-Routing Local Area Networks," *ACM SIGCOMM Computer Communication Review, Conference Proceedings on Applications, Technologies, Architectures, and Protocols for Computer Communications*, 26, 4 (August 1996), 10p.

[28] Groce, A., and W. Visser, "Model Checking Java Programs Using Structural Heuristics," *ACM SIGSOFT Software Engineering Notes, Proceedings of the International Symposium on Software Testing and Analysis*, 27, 4 (May 2002), 12–21.

[29] Habermann, A.N., "Prevention of System Deadlocks," *Communications of the ACM*, 12, 7 (July 1969), 373–377.

[30] Hartman, J.H., and J.K. Ousterhout, "The Zebra Striped Network File System," 14th ACM Symposium on Operating Systems Principles, proceedings published in *Operating Systems Review*, 27, 5 (December 1993), 29–43.

[31] Hartman, J.H., and J.K. Ousterhout, "The Zebra Striped Network File System," *ACM Transactions on Computing Systems*, 13, 3 (August 1995), 274–310.

[32] Hellerstein, J.L., "Achieving Service Rate Objectives with Decay Usage Scheduling," *IEEE Transactions on Software Engineering*, 19, 8 (August 1993), 813–825.

[33] Henry, G.J., "The Fair Share Scheduler," *AT&T Bell Laboratories Technical Journal*, 63, 8, Part 2 (October 1984), 1845–1857.

[34] Hofri, M., "Disk Scheduling: FCFS vs. SSTF Revisited," *Communications of the ACM*, 23, 11 (November 1980), 645–653.

[35] Hofri, M., "Should the Two-Headed Disk Be Greedy? Yes, It Should," *Information Processing Letters*, 16 (February 1983), 83–85.

[36] Inverardi, P., and M. Tivoli, "Automatic Synthesis of Deadlock Free Connectors for COM/DCOM Applications," ACM SIGSOFT *Software Engineering Notes, Proceedings of the 8th European Software Engineering Conference Held Jointly with 9th ACM SIGSOFT International Symposium on Foundations of Software Engineering*, 26, 5 (September 2001), 121–131.

[37] Iyer, S., and P. Druschel, "Anticipatory Scheduling: A Disk Scheduling Framework to Overcome Deceptive Idleness in Synchronous I/O," 18th ACM Symposium on Operating Systems Principles, proceedings published in *Operating Systems Review*, 35, 5 (December 2001), 117–130.

[38] Jacobson, D.M., and J. Wilkes, "Disk Scheduling Algorithms Based on Rotational Position," Technical Report HPL-CSP-91-7, Hewlett-Packard Laboratories, Palo Alto, California, 1991.

[39] Kameda, H., "Optimality of a Central Processor Scheduling Policy," *ACM Transactions on Computing Systems*, 2, 1 (February 1984), 78–90.

[40] Kaveth, N., and W. Emmerich, "Deadlock Detection in Distribution Object Systems," *ACM SIGSOFT Software Engineering Notes, Proceedings of the 8th European Software Engineering Conference Held Jointly with 9th ACM SIGSOFT International Symposium on Foundations of Software Engineering*, 26, 5 (September 2001), 44–51.

[41] Kay, J., and P. Lauder, "A Fair Share Scheduler," *Communications of the ACM*, 31, 1 (January 1988), 44–55.

[42] Kim, J.G., "Algorithmic Approach on Deadlock Detection for Enhanced Parallelism in Multiprocessing Systems," *Aizu International Symposium on Parallel Algorithms Architecture Synthesis*, IEEE, Piscataway, NJ, 1997, 233–238.

[43] King, R., "Disk Arm Movement in Anticipation of Future Requests," *ACM Transactions on Computing Systems*, 8, 3 (August 1990), 214–229.

[44] Kleinrock, L., "A Continuum of Time-Sharing Scheduling Algorithms," *Proceedings of the AFIPS Spring Joint Computer Conference*, AFIPS Press, Reston, VA, 36 (1970), 453–458.

[45] Koch, P.D.L., "Disk File Allocation Based on the Buddy System," *ACM Transactions on Computing Systems*, 5, 4 (November 1987), 352–370.

[46] Lampson, B.W., and H.E. Sturgis, "Atomic Transactions," in *Distributed Systems — Architecture and Implementation: An Advanced Course*, Lecture Notes in Computer Science, vol. 105, Springer-Verlag, New York, 1981, 246–265.

[47] Lane, M.G., and J.D. Mooney, *A Practical Approach to Operating Systems*, PWS-Kent, Boston, MA, 1989.

[48] Lazowska, E.D., J. Zahorjan, D.R. Cheriton, and W. Zwaenepoel, "File Access Performance of Diskless Workstations," *ACM Transactions on Computing Systems*, 4, 3 (August 1986), 238–268.

[49] Levine, T., "Deadlock Control with Ada95," ACM *SIGAda Letters*, 28, 2 (March 1998), 67–80.

[50] Lister, A.M., *Fundamentals of Operating Systems*, 2nd edition, Springer-Verlag, New York, 1979.

[51] Lynch, W.C., "Do Disk Arms Move?" *Performance Evaluation Review*, 1 (1972), 3–16.

[52] Mahlke, S.A., W.Y. Chen, R.A. Bringmann, R.E. Hank, W.W. Hwu, B.R. Rau, and M.S. Schlansker, "Sentinel Scheduling: A Model for Compiler-Controlled Speculative Execution," *ACM Transactions on Computing Systems*, 11, 4 (November 1993), 376–408.

[53] Mahoney, B., "An 'Open' Oriented File System," *Operating Systems Review*, 28, 1 (January 1994), 48–54.

[54] Mahmood, A., D.J. Lynch, and R.B. Shaffer, "Optimally Adaptive, Minimum-Distance, Circuit-Switched Routing in Hypercubes," *ACM Transactions on Computer Systems*, 15, 2 (May 1997), 166–193.

[55] Mays, D., and R.J. LeBlanc, "The Cyclefree Methodology: A Simple Approach to Building Reliable, Robust, Real Time Systems," *Proceedings of the 24th International Conference on Software Engineering*, May 2002, 567–575.

[56] McCann, C., R. Viswani, and J. Zahorjan, "A Dynamic Processor Allocation Policy for Multiprogrammed Shared-Memory Multiprocessors," *ACM Transactions on Computing Systems*, 11, 2 (May 1993), 146–178.

[57] McKusick, M.K., W.N. Joy, S.J. Leffler, and R.S. Fabry, "A Fast File System for UNIX," *ACM Transactions on Computing Systems*, 2, 3 (August 1984), 181–197.

[58] McNutt, B., "A Case Study of Access to VM Disk Volumes," in *CMG Proceedings* (December 1984), 175–180.

[59] Miller, F.W., "Predictive Deadline Multi-Processing," *Operating Systems Review*, 24, 4 (October 1990), 52–62.

[60] Miller, F.W., "The Performance of a Mixed Priority Real-Time Scheduling Algorithm," *Operating Systems Review*, 26, 4 (October 1992), 5–13.

[61] Mohapatra, P., "Wormhole Routing Techniques for Directly Connected Multicomputer Systems," *ACM Computing Surveys*, 30, 3 (September 1998), 374–410.

[62] Nieh, J., and M.S. Lam, "SMART: A Processor Scheduler for Multimedia Applications," 15th ACM Symposium on Operating Systems Principles, proceedings published in *Operating Systems Review*, 29, 5 (December 1995), 233.

[63] Patterson, D.A., G. Gibson, and R.H. Katz, "A Case for Redundant Arrays of Inexpensive Disks (RAID)," in *Proceedings of 1988 SIGMOD International Conference on Management of Data*, ACM, New York, 1988.

[64] Patterson, D.A., P. Chen, G. Gibson, and R.H. Katz, "Introduction to Redundant Arrays of Inexpensive Disks (RAID)," in *Spring COMPCON '89*, IEEE, New York, 1989, 112–117.

[65] Pinkston, T.M., and S. Warnakulasuriya, "On Deadlocks in Interconnection Networks," ACM SIGARCH *Computer Architecture News, Proceedings of the 24th International Symposium on Computer Architecture*, 25, 2 (May 1997), 38–49.

[66] Puente, V., C. Izu, J.A. Gregorio, R. Beivide, J.M. Prellezo, and F. Vallejo, "Improving Parallel System Performance by Changing the Arrangement of the Network Links," *Proceedings of the 14th International Conference on Supercomputing*, May 2000, 44–53.

[67] Rosenblum, M., and J.K. Ousterhout, "The Design and Implementation of a Log-Structured File System," *ACM Transactions on Computing Systems*, 10, 1 (February 1992), 26–52.

[68] Ruemmler, C., and J. Wilkes, "UNIX Disk Access Patterns," in *Proceedings of Winter 1993 USENIX*, USENIX Association, Berkeley, CA, 1993, 405–420.

[69] Ryder, K.D., "A Heuristic Approach to Task Dispatching," *IBM Systems Journal*, 9, 3 (1970), 189–198.

[70] Samadzadeh, M.H., and B.S. Koshy, "A Display and Analysis Tool for Process-Resource Graphs," *Operating Systems Review*, 30, 1 (January 1996), 39–62.

[71] Scott, M.L., "Non-Blocking Timeout in Scalable Queue-Based Spin Locks," *Proceedings of the 21st Annual Symposium on Principles of Distributed Computing*, July 2002, 31–40.

[72] Seltzer, M., P. Chen, and J. Ousterhout, "Disk Scheduling Revisited," *Proceedings of the Winter 1990 USENIX Conference*, USENIX Association, Berkeley, CA, 1990, 313–323.

[73] Shenoy, P.J., and H.M. Vin, "Cello: A Disk Scheduling Framework for Next Generation Operating Systems," ACM SIGMETRICS *Performance Evaluation Review*, proceedings of the 1998 ACM SIG-METRICS Joint International Conference on Measurement and Modeling of Computer Systems, 26, 1 (June 1998), 44–55.

[74] Sherman, S., F. Baskett, and J.C. Browne, "Trace Driven Modeling and Analysis of CPU Scheduling in a Multiprogramming System," *Communications of the ACM*, 15, 12 (December 1972), 1063–1069.

[75] Shiu, P.H., Y. Tan, and V.J. Mooney III, "A Novel Parallel Deadlock Detection Algorithm and Architecture," *Proceedings of the 9th International Symposium on Hardware/Software Codesign*, April 2001, 73–78.

[76] Shoshani, A., and E.G. Coffman, Jr., "Detection, Prevention and Recovery from Deadlocks in Multiprocess, Multiple Resource Systems," Princeton University, Technical Report Number 80, October 1969.

[77] Silberschatz, A., P.B. Galvin, and G. Gagne, *Operating System Concepts*, 6th edition, John Wiley & Sons, New York, 2003.

[78] Stevens, D.F., "On Overcoming High Priority Paralysis in Multiprogramming Systems: A Case History," *Communications of the ACM*, 11, 8 (August 1968), 539–541.

[79] Suranauwarat, S., and H. Taniguchi, "The Design, Implementation and Initial Evaluation of an Advanced Knowledge-Based Process Scheduler," *Operating Systems Review*, 35, 4 (October 2001), 61–81.

[80] Tanenbaum, A.S., *Modern Operating Systems*, 2nd edition, Prentice Hall, Englewood Cliffs, NJ, 2001.

[81] Teorey, T.J., "Properties of Disk Scheduling Policies in Multiprogrammed Computer Systems," in *Proceedings of the AFIPS Fall Joint Computer Conference*, AFIPS Press, Reston, VA, 1972.

[82] Teorey, T.J., and T.B. Pinkerton, "A Comparative Analysis of Disk Scheduling Policies," *Communications of the ACM*, 15, 3 (March 1972), 177–184.

[83] Thomasian, A., and C. Liu, "Disk Scheduling Policies with Lookahead," ACM SIGMETRICS *Performance Evaluation Review*, 30, 2 (September 2002), 31–40.

[84] Tsichritzis, D.C., and P.A. Bernstein, *Operating Systems*, Academic Press, New York, 1974.

[85] Tucker, A.B., R.D. Cupper, W.J. Bradley, R.G. Epstein, and C.F. Kelemen, *Fundamentals of Computing II: Abstraction, Data Structures, and Large Software Systems*, McGraw-Hill, New York, 1995.

[86] Waldspurger, C.A., T. Hogg, B.A. Huberman, J.O. Kephart, and W.S. Stornetta, "Spawn: A Distributed Computational Economy," *IEEE Transactions on Software Engineering*, 18, 2 (February 1992), 103–117.

[87] Waldspurger, C.A., and W.E. Weihl, "Lottery Scheduling: Flexible Proportional-Share Resource Management," First USENIX Symposium on Operating Systems Design and Implementation, proceedings published in *Operating Systems Review*, November 1994, 1–11.

[88] Worthington, B.L., G.R. Ganger, and Y.N. Patt, "Scheduling for Modern Disk Drive and Nonrandom Workloads," Technical Report #CSE-TR-194-94, Department of Electrical Engineering and Computer Science, University of Michigan, Ann Arbor, MI, March 1994.

[89] Wu, J., "A Deterministic Fault-Tolerant and Deadlock-Free Routing Protocol in 2-D Meshes Based on Odd-Even Turn Model," *Proceedings of the 16th International Conference on Supercomputing*, June 2002, 67–76.

Further Information

A good introduction to the practical problems in processor and device scheduling is presented in Tanenbaum [80], Silberschatz et al. [77], and Lister [50].

Current work is presented at the annual ACM Symposium on Operating System Principles, the USENIX Symposium on Operating System Design and Implementation, and the International Conference on Architectural Support for Programming Languages and Operating Systems. Copies of the proceedings are available from the ACM Special Interest Group on Operating Systems, ACM Headquarters, 1515 Broadway, New York, NY 10036.

The ACM quarterly *Transactions on Computer Systems* reports new developments in computer system scheduling. Also, the *Operating Systems Review*, a publication of the ACM Special Interest Group on Operating Systems, and ACM SIGMETRICS *Performance Evaluation Review* document current research in the area.

83

Real-Time and Embedded Systems

83.1 Introduction 83-1
83.2 Underlying Principles 83-2
83.3 Best Practices 83-5
    Real-Time Scheduling • Real-Time Kernels • Real-Time
    Architecture and Fault Tolerance • Real-Time
    Communications • Distributed Multimedia • Real-Time
    Databases • Real-Time Formal Verification, Design, and
    Languages • Real-Time Artificial Intelligence
83.4 Research Issues and Summary 83-12

John A. Stankovic
University of Virginia

83.1 Introduction

Real-time systems are defined as those systems in which the correctness of the system depends not only on the logical result of computation, but also on the time in which the results are produced [Stankovic 1988]. Real-time systems span a broad spectrum of complexity from very simple microcontrollers in **embedded systems** (such as a microprocessor controlling an automobile engine) to highly sophisticated, complex, and distributed systems (such as air traffic control for the continental U.S.). Other examples of real-time systems include command and control systems, process control systems, flight control systems, the Space Shuttle avionics system, flexible manufacturing applications, the space station, space-based defense systems, intensive care monitoring, collections of humans/robots coordinating to achieve common objectives (usually in hazardous environments such as undersea exploration or chemical plants), intelligent highway systems, mobile and wireless computing, and multimedia and high-speed communication systems. We are also beginning to see some of these real-time systems adding expert systems [Wright et al. 1986] and other artificial intelligence (AI) technology creating additional requirements and complexities. From this extensive list of applications we can see that real-time and embedded systems technology is a key *enabling* technology for the future in an ever growing domain of important applications.

At least three major trends in the real-time and embedded systems field have had major impacts on its technology. The first is the increased growth and sophistication of embedded systems; the second is the development of more scientific and technological results for **hard real-time systems**; and the third is the advent of distributed multimedia, a **soft real-time system**. In a hard real-time system there is no value to executing tasks after their deadlines have passed. A soft real-time system has tasks that retain some diminished value after their deadlines so these tasks should still be executed, even if they miss their deadlines.

Most embedded systems consist of a small microcontroller and limited software situated within some *product* such as a microwave oven or automobile. Often, the design of embedded systems is severely constrained by power, size, and cost constraints. However, to support increasing sophistication of embedded systems, we now see the common use of powerful microcontrollers and digital signal processor (DSP) chips, as well as the use of off-the-shelf real-time operating systems and design and debugging tools. Many people involved with embedded systems deal on a daily basis with sensors and data acquisition technology and systems; others construct architectures based on single board computers [many are still 68000 based, but reduced instruction set computer (RISC) processors are beginning to be used more and more] and busses such as the VME bus. Many people are involved with the programming and debugging of embedded systems, largely using the C programming language and cross development and debugging platforms. Embedded systems may or may not have real-time constraints.

In the hard real-time area, many fundamental results have been developed in real-time scheduling, operating systems, architecture and fault tolerance, communication protocols, specification and design tools, formal verification, databases and object-oriented systems. Increased emphasis on all of these areas is expected to continue for the foreseeable future. Many hard real-time systems are embedded systems.

Distributed multimedia has produced a new set of soft real-time requirements and when its potential is fully realized, it will fundamentally change how the world operates. Real-time principles lie at the heart of distributed multimedia, but without the concomitant high-reliability requirements found in safety-critical, hard real-time systems.

83.2 Underlying Principles

Typically, a real-time system consists of a *controlling system* and a *controlled system*. For example, in an automated factory, the controlled system is the factory floor with its robots, assembling stations, and the assembled parts; whereas the controlling system is the computer and human interfaces that manage and coordinate the activities on the factory floor. Thus, the controlled system can be viewed as the *environment* with which the computer interacts.

The controlling system interacts with its environment based on the information available about the environment from various **sensors**. It is imperative that the state of the environment, as perceived by the controlling system, be consistent with the actual state of the environment. Otherwise, the effects of the controlling systems' activities may be disastrous. Hence, periodic monitoring of the environment as well as timely processing of the sensed information is necessary.

Timing correctness requirements in a real-time system arise because of the *physical impact* of the controlling systems' activities upon its environment. For example, if the computer controlling a robot does not command it to stop or turn on time, the robot might collide with another object on the factory floor possibly causing serious damage. In many real-time systems even more severe consequences will result if timing as well as logical correctness properties of the system are not satisfied. For example, consider the effects of nuclear power plants or air traffic control systems failing.

Timing constraints for tasks can be arbitrarily complicated but the most common timing constraints for tasks are either *periodic*, *aperiodic*, or *sporadic*. A periodic task is one that is activated once every T units of time. The deadline for each activated instance may be less than, equal to, or greater than the period T. An aperiodic task is activated at unpredictable times. A sporadic task is an aperiodic task with an additional constraint that there is a minimum interarrival time between task activations.

Low-level application tasks such as those that process information obtained from sensors, or those that activate elements in the environment (through actuators), typically have stringent timing constraints dictated by the physical characteristics of the environment. A majority of sensory processing is periodic in nature. For example, a radar that tracks flights produces data at a fixed rate. A temperature monitor of a nuclear reactor core should be read periodically to detect any changes promptly. Some of these periodic tasks may exist from the point of system initialization whereas others may come into existence dynamically. The temperature monitor is an instance of a permanent task. An example of a dynamically created task is a (periodic) task that monitors a particular flight; this comes into existence when the aircraft enters an air traffic control region and will cease to exist when the aircraft leaves the region.

More complex types of timing constraints also occur. For example, spray painting a car on a moving conveyor must be started after time t_1 and completed before time t_2. Aperiodic requirements can arise from dynamic events, such as an object falling in front of a moving robot or a human operator pushing a button on a console.

Time related requirements may also be specified in indirect terms. For example, a value may be attached to the completion of each task where the value may increase or decrease with time; or a value may be placed on the *quality* of an answer whereby an inexact but fast answer might be considered more valuable than a slow but accurate answer. In other situations, missing X deadlines might be tolerated, but missing $X + 1$ deadlines cannot be tolerated.

What happens when timing constraints are not met? The answer depends, for the most part, on the type of application. A real-time system that controls a nuclear power plant or one that controls a missile, cannot afford to miss timing constraints of the **critical tasks**. Resources needed for critical tasks in such systems have to be preallocated so that the tasks can execute without delay. In many situations, however, some leeway does exist. For example, even on an automated factory floor, if it is estimated that the correct command to a robot cannot be generated on time, it may be appropriate to command the robot to stop (provided it will not cause other moving objects to collide with it and result in a different type of disaster), or to slow down (thereby dynamically generating more time to produce a correct command). Another example is a periodic task monitoring the position of an aircraft; depending on the aircraft's location and trajectory, missing the processing of one or two radar readings may not cause any problems.

In a real-time system, the characteristics of the various application tasks are usually known a priori and might be scheduled statically or dynamically. Whereas static specification of schedules is typically the case for periodic tasks, the opposite is true for aperiodic tasks. When the periodic temperature monitor of a nuclear reactor senses a problem in the core, it can invoke another (aperiodic) task to activate the appropriate elements of the reactor to correct the problem, for example, to force more coolant into the reactor core. In this case, the deadline for the aperiodic task can be statically determined as a function of the physical characteristics of the reactions within the core. On the other hand, the deadline of a task that controls a robot on a factory floor can be determined dynamically depending on the speed, direction, and weight of the robot. The command to the robot forcing it to turn right, left, or stop should be generated before this deadline.

In a real-time system that is designed in a static manner, the characteristics of the controlled system are assumed to be known a priori and, hence, the nature of activities and the sequence in which these activities take place can be determined off line before the system begins operation. Such systems are quite inflexible even though they may incur lower run-time overheads. In practice, most applications involve a number of components that can be statically specified along with many dynamic components. If handled appropriately, a system with high-resource utilization and low overheads can be produced for such applications.

Although a large proportion of currently implemented real-time systems are static in nature, many next generation systems will have to adopt solutions that are more dynamic and flexible. This is because such systems will be large and complex and they will function in environments that are both uncertain and physically distributed. More importantly, they will have to be maintainable and extensible due to their evolving nature and projected long lifetimes. Because of these characteristics, real-time systems, in general, and systems with the previously described characteristics, in particular, need to be *fast*, *predictable*, *reliable*, and *adaptive*.

One long-held misconception about real-time systems is that they only need to be fast to be effective. Basically, being fast is usually a necessary condition, but it is not sufficient. A real-time system has to meet explicit deadlines and being fast on average does not guarantee that a deadline will be met. If a real-time system can be shown to meet its deadlines (using a worst-case rather than an average-case behavior analysis), then we say that it is predictable. Predictability, itself, has many meanings and an entire journal article has been devoted to its meaning [Stankovic and Ramamritham 1990]. For purposes of this chapter it is sufficient to take a simplistic view of predictability. Consider that predictability means that when a task or set of tasks is activated it should be possible to determine their completion time subject to failure assumptions.

This must be done taking into account the state of the system (including the state of the operating system and the state of the resources controlled by the operating system) and the tasks' resource needs.

The task of building a real-time system can be very simple or it can be extremely complex. The difficulty depends on the characteristics of the real-time system along five dimensions, which we now discuss.

1. *Granularity of the deadline and laxity of the tasks:* In a real-time system some of the tasks have deadlines and/or periodic timing constraints. If the time between when a task is activated (required to be executed) and when it must complete execution is short, then the deadline is tight (i.e., the granularity of the deadline is small or the deadline is close). This implies that the operating system reaction time has to be short, and the scheduling algorithm to be executed must be fast and very simple. Tight time constraints may also arise when the deadline granularity is large (i.e., from the time of activation), but the amount of computation required is also great. In other words even large granularity deadlines can be tight when the laxity (deadline minus computation time) is small. In many real-time systems tight timing constraints predominate. Consequently, designers focus on developing very fast and simple techniques to react to this type of task activation. In general, the tighter the deadline the more difficult the design task.

2. *Strictness of deadline:* The strictness of the deadline refers to the value of executing a task after its deadline. For a hard real-time task there is no value to executing the task after the deadline has passed. A soft real-time task retains some diminished value after its deadline and so it should still be executed. Very different techniques are usually used for hard and soft real-time tasks. In many cases hard real-time tasks are preallocated and prescheduled resulting in 100% of them making their deadlines. Soft real-time tasks are often scheduled either with nonreal-time scheduling algorithms, with algorithms that explicitly address the timing constraints but aim only at good average-case performance, or with algorithms that combine importance and timing requirements. Hard real-time tasks are more difficult to deal with than soft real-time tasks, and systems which must deal with both types simultaneously are yet even more difficult. Multimedia in a timesharing environment is a soft real-time application, but multimedia in a real-time control environment such as an automated factory must deal with both hard and soft real-time constraints.

3. *Reliability:* Many real-time systems operate under severe reliability requirements. That is, if certain tasks, called critical tasks, miss their deadline then a catastrophe may occur. These tasks are usually guaranteed to make their deadlines by an off-line analysis and by schemes that reserve resources for these tasks even if it means that those resources are idle most of the time. In other words, the requirement for critical tasks should be that all of them always make their deadline (a 100% guarantee), subject to certain failure and workload assumptions. However, too many systems treat all of the tasks that have hard timing constraints as critical tasks (when, in fact, only some of those tasks are truly critical). This can result in erroneous requirements and an overdesigned and inflexible system. It is also common to see hard real-time tasks defined as those with both strict deadlines and critical importance. We prefer to keep a clear separation between these notions because they are not always related. Of course, many other reliability issues must also be resolved, but here we only mention the key issue that deals with timing constraints and reliability.

4. *Size of system and degree of coordination:* Real-time systems vary considerably in size and complexity. In most current real-time systems the entire system is loaded into memory; if there are well-defined phases, each phase is loaded just prior to the beginning of the phase. In many applications, subsystems are highly independent of each other and there is limited cooperation among tasks. The ability to load entire systems into memory and to limit task interactions simplifies many aspects of building and analyzing real-time systems. However, for future large, complex, real-time systems, having completely resident code and highly independent tasks will not always be practical. Moreover, solutions based on virtual memory are not acceptable because of the large degree of unpredictability associated with this technique. Consequently, increased size and coordination raise many new problems that must be addressed and further complicate the notion of predictability. Embedded systems may also have power, physical size, memory, and severe cost constraints adding to the difficulty of their design.

5. *Environment:* The environment in which a real-time system is to operate plays an important role in the design of the system. Many environments are very well defined (a laboratory experiment, an automobile engine, or an assembly line). Designers think of these as deterministic environments (even though they may not be intrinsically deterministic, they are well controlled and assumed to be deterministic). These environments give rise to small, static real-time systems where all deadlines are guaranteed a priori. Even in these simple environments we need to place restrictions on the inputs. For example, a particular assembly line may only be able to cope with five items per minute; given more than that, the system fails. Taking this approach enables an off-line quantitative analysis of the timing properties to be made. Since we know exactly what to expect given the assumptions about the well-defined environment we can usually design and build these systems to be predictable. However, the approaches taken in relatively small, static systems do not scale to other environments which are larger, much more complicated, and less controllable. Consider a next generation real-time system such as a team of cooperating mobile robots on the planet Mars. This system will be large, complex, distributed, adaptive, contain many types of timing constraints, need to operate in a highly nondeterministic environment, and evolve over a long system lifetime. It is not possible to assume that this environment is deterministic or to control it sufficiently well to make it look deterministic. If that were done, the system would be too inflexible and would not be able to react to unexpected events or combinations of events.

83.3 Best Practices

Now that we have presented some of the basic principles of real-time and embedded systems, we can discuss some of the applications in various areas of **real-time computing**, including: real-time scheduling, real-time kernels, real-time architectures and fault tolerance, real-time communications, distributed multimedia, **real-time databases**, real-time formal verification, design and languages, and real-time AI.

83.3.1 Real-Time Scheduling

Real-time scheduling results in recent years have been extensive. Theoretical results have identified worst-case bounds for dynamic on-line algorithms, and complexity results have been produced for various types of assumed task set characteristics. Queueing theoretic analysis has been applied to soft real-time systems covering algorithms based on real-time variations of **first-come–first-serve (FCFS)**, earliest deadline, and least laxity. We have seen the development of scheduling results for imprecise computation (a situation where tasks obtain a greater value the longer they execute, up to some maximum value) [Liu et al. 1991].

More applied scheduling results have also been produced with an extensive set of improvements to the rate monotonic algorithm (this includes the deferrable server and sporadic server algorithms [Sprunt et al. 1989], techniques to address the problem of priority inversion [Sha et al. 1990], and a set of algorithms that perform dynamic on-line planning [Ramamritham et al. 1990]). We have also seen practical application of a priori calculation of static schedules to provide what is called 100% guarantees for critical tasks. Although these a priori analyses are very valuable, system designers should not be lulled into thinking that 100% guarantees mean that no scheduling error can occur. It is important to know that these 100% guarantees are based on many (often unrealistic) assumptions. If the assumptions are a poor match for what can be expected from the environment (more and more likely in a distributed environment), then even with 100% guarantees the system can indeed miss deadlines. Hence, a key issue is to choose an algorithm whose assumptions provide the greatest coverage over what *really* happens in the environment. For all of these scheduling results outlined, the trend has been to deal with increasingly more complicated task set and environment characteristics (e.g., multiprocessing and distributed computing and tasks with precedence constraints).

An exciting trend is the extensive use of schedulability analysis for both static and dynamic real-time systems. For example, the Software Engineering Institute has developed a handbook [Klein et al. 1993] on rate monotonic analysis and gives seminars and tutorials regarding its use in real-time systems.

83.3.2   Real-Time Kernels

One focal point for next generation complex real-time systems is the operating system. The operating system must provide basic support for predictably satisfying real-time constraints, for fault tolerance and distribution, and for integrating time-constrained resource allocations and scheduling across a spectrum of resource types including sensor processing, communications, CPU, memory, and other forms of I/O. Toward this end, at least three major scientific issues need to be addressed:

- The *time dimension* must be elevated to a central principle of the system and should not be simply an afterthought. An especially perplexing aspect of this problem is that most system specification, design, and verification techniques are based on abstraction, which ignores implementation details. This is obviously a good idea; however, in real-time systems, timing constraints are derived from the environment and the implementation. This dilemma is a key scientific issue.

- The basic paradigms found in today's general purpose distributed operating systems must change. Currently, they are based on the notion that application tasks request resources as if they were random processes; operating systems are designed to expect random inputs and to display good average-case behavior. The new paradigm must be based on the delicate balance of *flexibility* and *predictability*: the system must remain flexible enough to allow a highly dynamic and adaptive environment, but at the same time be able to predict and possibly avoid resource conflicts so that timing constraints can be met. This is especially difficult in distributed environments where layers of operating system code and communication protocols interfere with predictability.

- A highly *integrated and time-constrained resource allocation approach* is necessary to adequately address timing constraints, predictability, adaptability, correctness, safety, and fault tolerance. For a task to meet its deadline, resources must be available *in time*, and events must be ordered to meet precedence constraints. Many coordinated actions are necessary for this type of processing to be accomplished on time. The state of the art lacks completely effective solutions to this problem.

For relatively small, less complex, real-time systems, it is often the case that real-time systems are supported by stripped down and optimized versions of timesharing operating systems. To reduce the run-time overheads incurred by the kernel and to make the system *fast*, the kernel underlying the real-time system:

- Has a fast context switch
- Has a small size (with its associated minimal functionality)
- Responds to external interrupts quickly
- Minimizes intervals during which interrupts are disabled
- Provides fixed or variable sized partitions for memory management (i.e., no virtual memory) as well as the ability to lock code and data in memory
- Provides special sequential files that can accumulate data at a fast rate

To deal with timing requirements, the kernel:

- Maintains a real-time clock
- Provides a priority scheduling mechanism
- Provides for special alarms and timeouts
- Permits tasks to invoke primitives to delay by a fixed amount of time and to pause/resume execution

In general, the kernels perform multitasking; intertask communication and synchronization are achieved via standard, well-known primitives such as mailboxes, events, signals, and semaphores. Examples of existing real-time kernels include: QNX, LynxOS, OS-9, VxWorks, and VRTXsa (over 70 commercial real-time kernels exist). Specialized kernels for DSP chips and homegrown kernels for microcontrollers are also still widely found.

Real-time kernels are also being extended to operate in highly cooperative multiprocessor and distributed system environments [Tokuda et al. 1990]. This means that there is an **end-to-end timing requirement** (in the sense that a set of communication tasks must complete before a deadline), i.e., a collection of activities must occur (possibly with complicated precedence constraints) before some deadline. Much research is being done on developing time constrained communication protocols to serve as a platform for supporting this user-level end-to-end timing requirement. However, while the communication protocols are being developed to support host-to-host bounded delivery time, using the current operating system (OS) paradigm of allowing arbitrary waits for resources or events, or treating the operation of a task as a *random process* causes great uncertainty in accomplishing the application level end-to-end requirements. As an example, the Mars project [Kopetz et al. 1989], the Spring project [Stankovic and Ramamritham 1991], and a project at the University of Michigan [Shin 1991] are all attempting to solve this problem. The Mars project uses an a priori analysis and then statically schedules and reserves resources so that distributed execution can be guaranteed to make its deadline. The Spring approach supports dynamic requests for real-time virtual circuits (guaranteed delivery time) and real-time datagrams (best effort delivery) integrated with CPU scheduling so as to guarantee the application level end-to-end timing requirements. The Spring project uses a distributed reflective memory based on a fiber optic ring to achieve lower level predictable communication properties. The Michigan work also supports dynamic real-time virtual circuits and datagrams, but it is based on a general multihop communication subnet.

Research is also being done on developing real-time object-oriented kernels to support the structuring of distributed real-time applications. As far as we know, no commercial products of this type are available. However, due to the major advantages of object orientation, it is likely that many such products will become available in the near future.

The diversity of the applications requiring predictable distributed systems technology will be significant. To handle this diversity, we expect the distributed real-time operating systems must use an *open system* approach, and the applications need to be portable. Regarding the open systems approach, it is important to avoid having to rewrite the operating system for each application area which may have differing timing and fault tolerance requirements. A library of real-time operating system objects might provide the level of functionality, performance, predictability, and portability required. We envision a Smalltalk-like system for hard real time, so that a designer can tailor the OS to the application without having to write everything from scratch. In particular, a library of real-time scheduling algorithms should be available that can be plugged in depending on the run-time task model being used and the load, timing, and fault tolerance requirements of the system.

Regarding the portability of applications, many real-time Unix operating systems are appearing [Furht et al. 1991], and a standard for real-time operating systems, called RT POSIX, is being developed [Gallmeister 1995]. Although such a standard facilitates porting the code, it is still an open issue on how to assess the timing properties of the ported application.

83.3.3 Real-Time Architecture and Fault Tolerance

Real-time systems are usually special purpose. In the past, architectures to support such applications tended to be special purpose too. The current trend is one in which more off-the-shelf components are being used to produce more generic architectures. Although considerable discussion could be given to real-time architectures, we shall consider only briefly how architecture impacts the computation of worst-case execution time and how it supports fault tolerance.

One aspect of architecture for real-time computing is the facility with which the worst-case execution time can be calculated. Worst-case execution times of programs are dependent on the system hardware, the operating system, the compiler used, and the programming language used. Many hardware features that have been introduced to speed up the average-case behavior of programs pose problems when information about worst-case behavior is sought (especially true on many state-of-the-art RISC CPUs). For instance, the ubiquitous caches, pipelining, dynamic random access memories (RAMs), and virtual (secondary) memory, lead to highly nondeterministic hardware behavior. Similarly, compiler optimizations tailored to make better use of these architectural enhancements, as well as more standard techniques such as

constant folding, which is the replacement of run-time computation by compile-time computation, contribute to poor predictability of code execution times. System interferences due to interrupt handling, shared memory references, and preemptions are additional complications. In summary, any approach to the determination of execution times of real-time programs has many complexities which must be solved.

Many real-time system architectures consist of multiprocessors, networks of uniprocessors, or networks of uni- and multiprocessors. Such architectures have potential for high fault tolerance, but they are also much more difficult to manage in a way such that deadlines are predictably met. Fault tolerance must be designed in at the start, must encompass both hardware and software, and must be integrated with timing constraints. In many situations, the fault tolerant design must be static due to extremely high data rates and severe timing constraints. Ultrareliable systems need to employ proof of correctness techniques to ensure fault tolerance properties [Vytopil 1993]. Primary and backup schedules computed off line are often found in hard real-time systems. We also see new approaches where on-line schedulers predict that timing constraints will be missed, enabling early action on such faults. Dynamic reconfigurability is needed but little progress has been reported in this area. Also, whereas considerable advance has been made in the area of software fault-tolerance, techniques that explicitly take timing into account are lacking.

Since fault tolerance is difficult, the trend is to let experts build the proper underlying support for it. For example, implementing checkpointing, reliable atomic broadcasts, logging, lightweight network protocols, synchronization support for replicas, and recovery techniques, and having these primitives available to applications, then simplifies creating fault tolerant applications. However, many of these techniques have not carefully addressed timing considerations nor the need to be predictable in the presence of failures. Many real-time systems, which require a high degree of fault tolerance, have been designed with significant architectural support, but the design and scheduling to meet deadlines is done statically, with all replicas in lock step. This may be too restrictive for many future applications. What is required is the integration of fault tolerance and real-time scheduling to produce a much more flexible system. For example, the use of the **imprecise computation** model [Liu et al. 1991], or a planning scheduler [Ramamritham et al. 1990] gives rise to a more flexible approach to fault tolerance than static schedules and fixed backup schemes. Adaptive fault tolerance with an explicit interaction with real-time constraints can be found in Bondavali et al. [1993].

83.3.4   Real-Time Communications

Distributed real-time systems require time-constrained message delivery. In many applications the communication protocols and network provide deterministic behavior. An alternative, applicable in other situations, is a best effort approach. Hybrid approaches also exist [Arvind et al. 1991]. Those systems requiring hard guarantees often use time-domain multiple access (TDMA), fiber distributed data interface (FDDI), Institute of Electrical and Electronics Engineers (IEEE) 802.4 token bus or 802.5 token ring [Strosnider and Marchok 1989]. Careful assumptions and analysis accompany the use of these networks to produce a deterministic guarantee. For best effort approaches, variations of carrier sense multiple access/collision detection (CSMA/CD) or window-based schemes can be used [Malcolm and Zhao 1995]. For distributed multimedia [Govindan and Anderson 1991, Clark et al. 1992], timing constraints on the network include end-to-end delays, minimum jitter, and interpacket maximum delays. Other requirements for transmitting audio, video, text, and data traffic include extremely high volume and high speeds. To support these requirements, asynchronous transfer mode (ATM) switches [Newman 1994] have been developed. With the advent of ATM technology we are seeing the projected use of ATMs as the local area network of real-time systems.

Specialized busses are still widely utilized today. The controller area network (CAN) bus for automobiles and the SAFEbus for commercial aircraft [Hoyme et al. 1991] are examples.

83.3.5   Distributed Multimedia

Many real-time control applications such as agile manufacturing and process control operate in highly nondeterministic environments under timing constraints of many types. Significant improvements in these applications can be created by embedding continuous and multimedia support in these applications. For example, in agile manufacturing, remote factories, each consisting of many automated workcells, must coordinate to handle new strategies for incoming product orders, to develop the design of new products, to schedule just-in-time deliveries of manufactured components, to monitor the plant operations, and to collaboratively solve difficult manufacturing floor problems.

To implement solutions cost effectively and to allow multimedia applications direct access to plant operational data, it is envisioned that the same computers would control the time-constrained operations in and across the workcells and support multimedia [Guha et al. 1995]. The backbone network would likely be ATM. Distributed multimedia over ATM networks has enormous potential to provide these applications with teleconferencing for real-time coordination, collaborative design, and a wealth of real-time information such as visual access to remote and local plant operations via cameras, as well as supporting real-time control. These applications would also benefit from an integrated design of a distributed database that contains such information as sensor information and control variables constrained by temporal validity intervals, plant operational data, information on availability of raw materials, inventory of products, customer orders, etc. The confluence of an integrated database, multimedia, and real-time control has great potential for moving application areas such as manufacturing and process control into the next generation.

It is important to point out that although the commercial world is developing multimedia support, most of this work is done in the context of general purpose timesharing systems and **not** integrated with real-time control applications. This difference is significant and will likely require solutions not likely to be developed for commercial multimedia. One reason for this is that the nonreal-time control applications can accept a probability that other applications that are executing can fail (or be late), and so it is likely that simpler solutions and less expensive systems can be built for these applications. This assumption is reasonable for many commercial multimedia-based systems, but not for real-time control applications. On the other hand, real-time control applications have focused on solutions for hard deadline systems, which would be too inefficient if they were used for multimedia. This new, distributed, multimedia, information technology environment will require greater support for soft real-time systems, especially in their continuous and multimedia aspects. Here, new models of guarantees and resource reservations are required.

Distributed multimedia applications require both high-performance operating system features (in order to respond to the high data and processing rates) as well as a **quality-of-service (QOS)** guarantee. A quality-of-service guarantee is a promise to provide a certain level of performance to a multimedia application. Quality of service may be defined in a number of ways even within the same system. Hence, it is important that the OS be flexible enough to accept different types of quality of service requests. Because QOS guarantees are end-to-end, this necessitates an integrated and synergistic approach to the problem, encompassing the scheduling algorithm, the underlying I/O system (e.g., mass storage), the database system, associated operating system components, and the network. The scheduling algorithms must be able to *cosupport* the hard deadline control tasks and the distributed multimedia aspects of the system. Solving the main issues for the I/O system includes interfacing to real-time databases, achieving adequate sustained transfer rates, I/O scheduling, data layout on the disk array, and I/O buffer management. Whereas some aspects of the operating system infrastructure required to meet these needs can be found in today's real-time kernels or as specially developed parts of commercial kernels, OS features typically available today are not sufficient especially in the context of real-time control applications with hard deadlines. Further, much of the current work has concentrated on either single site [Anderson et al. 1992], or network level end-to-end delays; what is required is support for distributed application level to application level processing.

Another issue is call admission in which the system dynamically decides whether to admit a new multimedia session or not. Note that this concept was previously used in hard real-time systems since 1984 [Ramamritham and Stankovic 1984]. In the context of multimedia, call admission requires that we determine that either the session being requested is possible (so we can admit it), or it is not possible even to an acceptable degraded level (so we reject the request, or identify what the degraded service will be). In control applications we may need to override previously admitted sessions or compute tradeoffs to maximize the likelihood of success of the overall mission.

83.3.6   Real-Time Databases

A real-time database is a database system where (at least some) transactions have explicit timing constraints such as deadlines and where data may become invalid with the passage of time. In such a system, transaction processing must satisfy not only the database consistency constraints, but also the timing constraints. Real-time database systems can be found, for instance, in program trading in the stock market, radar tracking systems, battle management systems, and computer integrated manufacturing systems. Some of these systems (such as program trading in the stock market) are soft real-time systems, because missing a deadline is not catastrophic. Usually, research into algorithms and protocols for such systems explicitly addresses deadlines and makes a best effort at meeting deadlines. In soft real-time systems there are no guarantees that specific tasks will make their deadlines.

In real-time databases there is a need for an integrated approach that includes time constrained protocols for concurrency control, conflict resolution, CPU and I/O scheduling, transaction restart and wakeup, deadlock resolution, buffer management, and commit processing. Many protocols based on locking, optimistic, and time-stamped concurrency control have been developed and evaluated in testbed or simulation environments [Abbott and Garcia-Molina 1988]. In most cases the optimistic approaches seem to work best [Huang et al. 1991].

In a typical database system a transaction is a sequence of operations performed on a database. Normally, consistency (serializability), atomicity, and permanence are properties supported by the transaction mechanism. Transaction throughput and response time are the usual metrics. In a soft real-time database, transactions have similar properties, but, in addition, have soft real-time constraints. Metrics include response time and throughput, but also include the percentage of transactions which meet their deadlines, or a weighted value function which reflects the value imparted by a transaction completing on time. On the other hand, in a hard real-time database, not all transactions have serializability, atomicity, and permanence properties. These requirements need to be supported only in certain situations. For example, hard real-time systems are characterized by their close interactions with the environment that they control. This is especially true for subsystems that receive sensory information or that control actuators. Processing involved in these subsystems is such that it is typically not possible to *rollback* a previous interaction with the environment.

Whereas the notion of consistency is relevant here (for example, the interactions of a real-time task with the environment should be consistent with each other), traditional approaches to achieving consistency, involving waits, rollbacks, and aborts are not directly applicable. Instead, compensating transactions may have to be invoked to nullify the effects of previously committed transactions. Also, another transaction property, namely, *permanence*, is of limited applicability in this context. This is because real-time data, such as those arriving from sensors, have limited *lifetimes*: they become obsolete after a certain point in time. Data received from the environment by the lower levels of a real-time system undergo a series of processing steps (e.g., filtering, integration, and correlation). Traditional transaction properties are less relevant at the lowest levels and become more relevant at higher levels in the system. Most hard real-time database systems are main memory databases of small size, with predefined transactions, and handcrafted for efficient performance.

A new trend is the use of active database technology for real-time databases. A **real-time active database (RTADB)** is a system where transactions have timing constraints such as deadlines, where data may become invalid with the passage of time, and where transactions may trigger other transactions. This type of database follows an event–condition–action paradigm subject to timing constraints. RTADBs are in their infancy and no commercial products yet exist, although many products contain *triggers*.

83.3.7   Real-Time Formal Verification, Design, and Languages

Today, when constructing a complex real-time system it is becoming more common to use a formal verification technique [Vytopil 1993, Jahanian and Mok 1986] for certain aspects of the design and to use a commercially available design tool [Kavi 1992]. Many formal verification approaches are available, e.g., those based on Petri nets, temporal logic, timed communicating sequential processes (CSP), probabilistic durational calculus, real-time logic (RTL), or prototype verification system (PVS). Although limitations still exist on the use of these techniques, the value of formal techniques early in the design process has been amply demonstrated. The trend is to develop formalisms that can directly address timing constraints.

Regarding design methods, commercial tools such as STATEMATE [Harel et al. 1990], CARDtools, or Control Shell provide graphical interfaces and many nice database features. Many design methodologies have been extended to deal with real-time systems and recently have included an object-oriented approach [Ellis 1994]. Since tools such as these are so important, continual improvements occur. The future should bring more understandable tools that better and better reflect and support the reliability and timing constraints of real-time systems.

A discussion on real-time languages (those specifically designed for real-time programming and specification) and languages used for real-time systems (assembler, C, ADA, etc.) can be found in Burns and Wellings [1989]. ADA and C are now commonly used for programming many complex real-time systems. Synchronous languages [Halbwachs 1993] are also widely used, mostly in Europe.

83.3.8 Real-Time Artificial Intelligence

Many complex real-time applications now require or will require knowledge-based on-line assistance operating in real time [Paul et al. 1991]. This necessitates a major change to some of the paradigms and implementations previously used by AI researchers. For example, AI systems must be made to run much faster (a necessary but not sufficient condition), allow preemption to reduce latency for responding to new stimuli, attain predictable memory management via incremental garbage collection or by explicit management of memory, include deadlines and other timing constraints in search techniques, develop anytime algorithms (algorithms where a nonoptimal solution is available at any point in time), develop time-driven inferencing, and develop time-driven planning and scheduling. Rules and constraints may also have to be imposed on the design, models, and languages used in order to facilitate predictability, e.g., limit recursion and backtracking to some fixed bound. Coming to grips with what predictability means in such applications is also very important.

In addition to these changes within AI, real-time AI (RTAI) techniques must be interfaced with lower level real-time systems technology to produce a functioning, reliable, and carefully analyzable system. Should the higher level RTAI techniques ignore the system level, treat it as a black box with *general* characteristics, or be developed in an integrated fashion with it so as to best build complex systems? What is the correct interface between these two traditionally separate systems? Integrating RTAI and low-level real-time systems software is quite a challenge because RTAI applications are operating in nondeterministic environments, there is missing or noisy information, some of the control laws are heuristic at best, objectives may change dynamically, partial solutions are sometimes acceptable (so that a tradeoff between the quality of the solution and the time needed to derive it can be made), the amount of processing is significant and highly data dependent, and the execution time of tasks may be difficult to determine. These demanding requirements will drive real-time research for many years to come.

Competing software architectures for real-time AI include production rule architectures, blackboard architectures, and a process trellis architecture. Some real-time AI systems have been built by carefully and severely restricting how production rules and blackboard systems are built and used. Research is ongoing to relax the restrictions so that the power of these architectures can be utilized, but at the same time providing a high degree of predictability. The process trellis architecture, used in the medical domain, is a highly static approach whereas the other two are much more dynamic. The trellis architecture (because it is static) has potential to provide static real-time guarantees for those applications characterized by enough time to completely compute results from a set of inputs before the next set of inputs arrive. This approach is suitable for certain types of real-time AI monitoring systems, but its generality for complex real-time AI systems has not been demonstrated.

In a distributed setting, high-level decision support requires organizing computations with networks of cooperative, semiautonomous agents, each capable of sophisticated problem solving. Theories of communication and organizational structure for groups of cooperative problem solving agents must be developed. These theories must include problem solving under uncertainty and under timing constraints.

83.4   Research Issues and Summary

Many research issues exist in all areas of real-time computing. Many of the problems are of a fundamental nature and others are more applied. It is impossible to list all of the key research issues; instead we identify representative examples of open research problems.

Although many interesting scheduling results have been produced, the state of the art still provides piecemeal solutions. Many realistic issues have not yet been addressed in an integrated and comprehensive manner. The real-time scheduling area still requires analyzable scheduling approaches (it may be a collection of algorithms) that are comprehensive and integrated. For example, the overall approach must be comprehensive enough to handle:

- Preemptable and nonpreemptable tasks
- Periodic and nonperiodic tasks
- Tasks with multiple levels of importance (or a value function)
- Groups of tasks with a single deadline
- End-to-end timing constraints
- Precedence constraints
- Communication requirements
- Resource requirements
- Placement constraints
- Fault tolerance needs
- Tight and loose deadlines
- Normal and overload conditions

The solution must be integrated enough to handle the interfaces between:

- CPU scheduling and resource allocation
- I/O scheduling and CPU scheduling
- CPU scheduling and real-time communication scheduling
- Local and distributed scheduling
- Static scheduling of critical tasks and dynamic scheduling of essential and nonessential tasks

One key issue is the need to provide predictability. Predictability requires bounded operating system primitives, some knowledge of the application, proper scheduling algorithms, and a viewpoint based on a *team* attitude between the operating system and the application. For example, simply having a very primitive kernel that is itself predictable is only the first step. More direct support is needed for developing predictable and fault tolerant real-time applications. One aspect of this support comes in the form of scheduling algorithms. For example, if the operating system is able to perform integrated CPU scheduling and resource allocation in a planning mode so that collections of cooperating tasks can obtain the resources they need at the right time, in order to meet timing constraints, this facilitates the design and analysis of real-time applications. Further, if the operating systems retains information about the importance of a task and what actions to take if the task is assessed as not being able to make its deadline, then a more intelligent decision can be made as to alternative actions, and graceful degradation of the performance of the system can be better supported (rather than a possible catastrophic collapse of the system if no such information is available). Kernels which support retaining and using semantic information about the application are sometimes referred to as *reflective* kernels [Stankovic and Ramamritham 1995].

Basic research is also required in many areas of distributed multimedia, including:

- Specification of quality of service
- Algorithmic and kernel support to actually achieve this guaranteed service and to dynamically negotiate other levels of service, if necessary

- How to perform reservations of sets of resources
- Integrated scheduling across a set of resources (e.g., so that CPU, I/O buffer, disk controller, network bandwidth, and resources at the receiver are reserved together)
- End-to end scheduling
- Atomic guarantees for sets of tasks (This supports the call admission policies needed in multimedia.)

Obviously, achieving complex, real-time systems is nontrivial and will require research breakthroughs in many aspects of system design and implementation. For example, good design methodologies and tools which include programming rules and constraints must be used to guide real-time system developers so that subsequent implementation and *analysis* can be facilitated. This includes proper application decomposition into subsystems and allocation of those subsystems onto distributed architectures. The programming language must provide features tailored to these rules and constraints, must limit its features to enhance predictability, and must provide the ability to specify timing, fault tolerance, and other information for subsequent use at run time. Many language features are continuously being proposed, although few of them are currently used in practice. Execution time of each primitive of the kernel must be bounded and predictable, and the operating system should provide explicit support for all of the requirements including the real-time requirements. New trends in the OS area include the use of microkernels, support for multiprocessors and distributed systems, and real-time thread packages. The architecture and hardware must also be designed to support predictability and facilitate analysis. For example, hardware should be simple enough so that predictable timing information can be obtained. This has implications for how to deal with caching, memory refresh and wait states, pipelining, and some complex instructions, which all contribute to timing analysis difficulties. The resulting system must be scalable to account for the significant computing needs that occur both initially and as the system evolves. An insidious aspect of critical real-time systems, especially with respect to their real-time requirements, is that the weakest link in the entire system can undermine careful design and analysis at other levels. Research is required to address all of these issues in an integrated fashion.

Finally, a number of new trends involve the use of formal verification for real-time systems and the development of real-time databases, real-time object-oriented systems, and real-time artificial intelligence. Since these areas are very new, many open problems exist.

Defining Terms

Constant folding: A compiler optimization technique where run-time computations are replaced by compile-time computations.

Critical real-time task: One in which missing its deadline may cause a catastrophe or total failure of the system.

Embedded system: A system that is a component of a larger system, e.g., an automobile cruise control system or the navigation system of the space shuttle.

End-to-end timing requirement: A single overall timing requirement for a set of tasks that operate with some precedence constraints. This is typical of a distributed set of tasks communicating over a network.

First-come–first-served (FCFS): Scheduling policy that chooses which task to execute based on the earliest time of arrival.

Hard real-time task: One in which there is no value in continuing to execute the task after the deadline has passed.

Imprecise computation: One where if the computation terminates before completion, the intermediate result produced is usable.

Quality-of-service (QOS): Guarantee that is a promise to provide a certain level of performance to a multimedia application.

Real-time active database: One where transactions have timing constraints such as deadlines, where data may become invalid with the passage of time, and where transactions or events may trigger other transactions.

Real-time computing: That computing where the results must be logically correct and produced on time.

Real-time database: One where transactions have timing constraints such as deadlines and where data may become invalid with the passage of time.

Sensor: A device that outputs a signal for the purpose of detecting or measuring a physical property.

Soft real-time task: One that retains some diminished value after its deadline so it should still be executed even if its deadline has passed.

References

Abbott, R. and Garcia-Molina, H. 1988. Scheduling real-time transactions: a performance evaluation, pp. 1–12. *Proc. Very Large Databases Conf.*

Anderson, D., Osawa, O., and Govindan, R. 1992. A file system for continuous media. *ACM Trans. Comput. Syst.* 10(4):311–377.

Arvind, K., Ramamritham, K., and Stankovic, J. 1991. A local area network architecture for communication in distributed real-time systems. *Real-Time Syst.* 3(2).

Bondavali, A., Stankovic, J., and Strigini, L. 1993. Adaptable fault tolerance for real-time systems. *IEEE 3rd Int. Workshop Responsive Comput. Syst.*

Burns, A. and Wellings, A. 1989. *Real-Time Systems and Their Programming Languages.* Addison–Wesley, Reading, MA.

Clark, D., Shenkar, S., and Zhang, L. 1992. Supporting real-time applications in an integrated services packet network: architecture and mechanism. *Proc. ACM SIGCOMM.*

Ellis, J. 1994. *Objectifying Real-Time Systems.* SIGS Books, New York.

Furht, B., Grostick, D., Gluch, D., Rabbat, G., Parker, J., and McRoberts, M. 1991. *Real-Time Unix Systems, Design and Application Guide.* Kluwer Academic, Boston, MA.

Gallmeister, B. 1995. *POSIX.4: Programming for the Real World.* O'Reilly and Associates.

Govindan, R. and Anderson, D. 1991. Scheduling and IPC mechanisms for continuous media. *Proc. Symp. Operating Syst. Principles.* ACM.

Guha, A., Pavan, A., Liu, J., Rastogi, A., and Steeves, T. 1995. Supporting real-time and multimedia applications on the Mercuri testbed. *IEEE J. Selec. Areas Commun.* 13(4):749–763.

Halbwachs, N. 1993. *Synchronous Programming of Reactive Systems.* Kluwer Academic, Boston, MA.

Harel, D., Lachover, H., Naamad, A., Pnueli, A., Politi, M., Sherman, R., Shtull-Trauring, A., and Trakhten-brot, M. 1990. Statemate: a working environment for the development of complex reactive systems. *IEEE Trans. Software Eng.* 16(4):413–414.

Hoyme, K., Driscoll, K., Herrlin, J., and Radke, K. 1991. ARINC 629 and SAFEbus: data buses for commercial aircraft. *Sci. Honeyweller* 57–70.

Huang, J., Stankovic, J., Ramamritham, K., and Towsley, D. 1991. Experimental evaluation of optimistic concurrency control. *Proc. Very Large Databases Conf.*

Jahanian, F. and Mok, A. 1986. Safety analysis of timing properties in real-time systems. *IEEE Trans. Software Eng.* 12(9):890–904.

Kavi, K. 1992. *Real-Time Systems: Abstractions, Languages, and Design Methodologies.* IEEE Computer Society Press, Los Alamitos, CA.

Klein, M., Ralya, T., Pollak, B., Obenza, R., and Gonzales Harbour, M. 1993. *A Practitioner's Handbook for Real-Time Analysis.* Kluwer Academic, Norwell, MA.

Kopetz, H., Damm, A., Koza, C., and Mulozzani, D. 1989. Distributed fault tolerant real-time systems: the Mars approach. *IEEE Micro.* 9(1):25–40.

Liu, J., Lin, K., Shih, W., Yu, A., Chung, J., and Zhao, W. 1991. Algorithms for scheduling imprecise computations. *IEEE Comput.* 24(5):58–68.

Malcolm, N. and Zhao, W. 1995. Hard real-time communication in multiple access networks. *Real-Time Syst.* 8(1):35–78.

Newman, P. 1994. Traffic management for ATM local area networks. *IEEE Commun. Mag.* 32(8):44–51.

Paul, C., Acharya, A., Black, B., and Strosnider, J. 1991. Reducing problem-solving variance to improve predictability. *Commun. ACM* 34(8).

Ramamritham, K. and Stankovic, J. 1984. Dynamic task scheduling in distributed hard real-time systems. *IEEE Software* 1(3):65–75.

Ramamritham, K., Stankovic, J., and Shiah, P. 1990. Efficient scheduling algorithms for real-time multiprocessor systems. *IEEE Trans. Parallel Distributed Comput.* 1(2):184–194.

Sha, L., Rajkumar, R., and Lehoczky, J. 1990. Priority inheritance protocols: an approach to real-time synchronization. *IEEE Trans. Comput.* 39(9):1175–1185.

Shin, K. 1991. HARTS: a distributed real-time architecture. *IEEE Comput.* 24(5).

Sprunt, B., Sha, L., and Lehoczky, J. 1989. Aperiodic task scheduling for hard real-time systems. *Real-Time Syst.* 1:27–60.

Stankovic, J. 1988. Misconceptions about real-time computing: a serious problem for next generation systems. *IEEE Comput.* 21(10).

Stankovic, J. and Ramamritham, K. 1990. What is predictability for real-time systems. *Real-Time Syst. J.* 2:247–254.

Stankovic, J. and Ramamritham, K. 1991. The spring kernel: a new paradigm for real-time systems. *IEEE Software* 8(3):62–72.

Stankovic, J. and Ramamritham, K. 1995. A reflective architecture for real-time operating systems. In *Advances in Real-Time Systems*, pp. 487–507. Prentice–Hall, Englewood Cliffs, NJ.

Strosnider, J. and Marchok, T. 1989. Responsive, deterministic IEEE 802.5 token ring scheduling. *Real-Time Syst.* 1(2):133–158.

Tokuda, H., Nakajima, T., and Rao, P. 1990. Real-time MACH: towards a predictable real-time system. *Proc. USENIX MACH Workshop.*

Vytopil, J. 1993. *Formal Techniques in Real-Time and Fault Tolerant Systems.* Kluwer Academic, Boston, MA.

Wright, M., Green, M., Fiegl, G., and Cross, P. 1986. An expert system for real-time control. *IEEE Software* 3(2):16–24.

Further Information

Tutorial Texts

A good introduction to real-time systems can be found in *Hard Real-Time Systems* by John A. Stankovic and Krithi Ramamritham. More recent research results in the field can be found in a follow-on text entitled *Advances in Real-Time Systems* by the same authors. Yann Hang Lee and C. M. Krishna have edited a tutorial text entitled *Readings in Real-Time Systems* where papers on key real-time concepts are presented. Krishna Kavi has also produced a tutorial text on real-time systems entitled *Real-Time Systems, Abstractions, Languages, and Design Methodologies.* This latter book approaches real-time systems from the design point of view, whereas the others mentioned take a systems implementation perspective. All of these texts are published by the IEEE Computer Society Press.

Other Information

Proceedings of the Real-Time Systems Symposium are published annually by the IEEE Computer Society. The conference is held each December and research papers are presented.

A new annual IEEE Computer Society Symposium was started in 1995 titled the Real-Time Technology and Applications Symposium. This symposium focuses on the interaction between industry and academia; a proceedings is published.

An archival journal entitled *Real-Time Systems* is published six times a year by Kluwer Academic Publishers. Kluwer also publishes an international book series on real-time computing. There are approximately 20 volumes in this series. For subscription information for the journal or for information on the book series contact: Kluwer Academic Publishers, 101 Philip Drive, Assinippi Park, Norwell, MA 02061.

Embedded Systems Programming is published monthly by Miller Freeman Inc., 600 Harrison St., San Francisco, CA 94107.

Embedded Systems Conference East and Embedded Systems Conference West are each held once per year. These conferences include product exhibitions and technical instruction.

84

Process Synchronization and Interprocess Communication

84.1 Introduction 84-1
84.2 Underlying Principles 84-2
Synchronization Problems • Synchronization Issues
• Interprocess Communication Problems • Interprocess
Communication Issues
84.3 Best Practices 84-7
Synchronization Mechanisms • Interprocess Communication
(IPC) Mechanisms • Classic Problems • Deadlock and
Starvation
84.4 Research Issues and Summary 84-20

Craig E. Wills
Worcester Polytechnic Institute

84.1 Introduction

Process **synchronization** (also referred to as process coordination) is a fundamental problem in operating system design and implementation. It is a situation that occurs when two or more processes coordinate their activities based on a condition. An example is when one process must wait for another process to place a value in a buffer before the first process can proceed. A specific problem of synchronization is **mutual exclusion**, which requires that two or more concurrent activities do not simultaneously access a shared resource. This resource may be shared data among a set of processes where the instructions that access these shared data form a **critical region** (also referred to as a critical section). A solution to the mutual exclusion problem guarantees that among the set of processes, only one process is executing in the critical region at a time.

Processes involved in synchronization are indirectly aware of each other by waiting on a condition that is set by another process. Processes can also communicate directly with each other through **interprocess communication (IPC)**. IPC causes communication to be sent between two (or more) processes. A common form of IPC is message passing.

The origins of process synchronization and IPC are work in concurrent program control by people such as Dijkstra, Hoare, and Brinch Hansen. Dijkstra described and presented a solution to the mutual exclusion problem [Dijkstra 1965] and proposed other fundamental synchronization problems and solutions such as the dining philosophers problem [Dijkstra 1971] and semaphores [Dijkstra 1968]. Brinch Hansen [1972] and Hoare [1972] suggested the concept of a critical region. Brinch Hansen published a classic text with many examples of concurrency in operating systems [Brinch Hansen 1973]. Hoare [1974] provided a complete description of monitors following work by Brinch Hansen [1973].

Modern work in the area includes development of multithreaded, message-based operating systems executing on multiprocessors. Many of the primitives used for synchronizing processes also work for synchronizing threads, which are discussed in Chapter 76. Investigation is being done on mechanisms that work in a multiprocessor environment and extend to a distributed one. Concurrent programming using synchronization and IPC in a distributed environment, discussed in Chapters 85 and 98, is more complex than the mechanisms described in this chapter because of additional complications. In a distributed environment, there may be a failure by some processes participating in synchronization while others continue, or it is possible for messages to be lost or delayed in an IPC mechanism.

The remainder of this chapter discusses the underlying principles and practices commonly used for synchronization and IPC. Section 84.2 identifies fundamental problems needing solutions and issues that arise in considering various solutions. Section 84.3 discusses specific solutions to synchronization and IPC problems and discusses their relative merits in terms of these issues. The chapter concludes with a summary, a glossary of terms that have been defined, references, and sources of further information.

84.2 Underlying Principles

Process synchronization and IPC arose from the need to coordinating concurrent activities in a multiprogrammed operating system. This section defines and illustrates the fundamental synchronization and IPC problems and characterizes the issues on which to compare the solutions.

84.2.1 Synchronization Problems

A fundamental synchronization problem is mutual exclusion, as described by Dijkstra [1965]. Lamport [1986] provides a formal treatment of the problem, with Anderson [2001] surveying Lamport's contributions on this topic. In this problem, multiple processes wish to coordinate so that only one process is in its critical region of code at any one time. During this critical region, each process accesses a shared resource such as a variable or table in memory. The use of a mutual exclusion primitive for access to a shared resource is illustrated in Figure 84.1, where two processes have been created that each access a shared global variable through the routine **Deposit()**.

The routines **BeginRegion()** and **EndRegion()** define a critical region ensuring that **Deposit()** is not executed simultaneously by both processes. With these routines, the final value of **balance** is always 20, although the execution order of the two processes is not defined.

To illustrate the need for mutual exclusion, consider the same example without the **BeginRegion()** and **EndRegion()** routines. In this case, the execution order of the statements for each process is time dependent. The final value of **balance** may be 20 if **Deposit()** is executed to completion for each process, or the value of **balance** may be 10 if the execution of **Deposit()** is interleaved for the two processes. This example illustrates a **race condition**, where multiple processes access and manipulate the same data with the outcome dependent on the relative timing of these processes. The use of a critical region avoids a race condition.

Many solutions have been proposed for the implementation of these two primitives; the most well known of which are given in the following section. Solutions to the mutual exclusion synchronization problem must meet a number of requirements, which were first set forth in Dijkstra [1965] and are summarized in Stallings [2000]. These requirements are as follows:

- Mutual exclusion must be enforced so that at most one process is in its critical region at any point in time.
- A process must spend a finite amount of time in its critical region.

```
int balance = 0;   /* global shared variable */

ProcessA()
{
    Deposit(10);
    cout << "Balance is " << balance << '\n';
}

ProcessB()
{
    Deposit(10);
    cout << "Balance is " << balance << '\n';
}

Deposit(int deposit)
{
    int newbalance; /* local variable */

    BeginRegion();   /* enter critical region */
    newbalance = balance + deposit;
    balance = newbalance;
    EndRegion();     /* exit critical region */
}
```

FIGURE 84.1   Shared variable access handled as a critical region.

- The solution must make no assumptions about the relative speeds of the processes or the number of processes.
- A process stopped outside of its critical region must not lead to blocking of other processes.
- A process requesting to enter a critical region held by no other process must be permitted to enter without delay.
- A process requesting to enter a critical region must be granted access within a finite amount of time.

Another fundamental synchronization problem is the producer/consumer problem. In this problem, one process produces data to be consumed by another process. Figure 84.2 shows one form of this problem where a *producer* process continually increments a shared global variable and a *consumer* process continually prints out the shared variable. This variable is a fixed-size buffer between the two processes, and hence this specific problem is called the bounded-buffer producer/consumer problem.

The ideal of this example is for the consumer to print each value produced. However, the processes are not synchronized and the output generated is timing dependent. The number 0 is printed 2000 times if the consumer process executes before the producer begins. At the other extreme, the number 2000 is printed 2000 times if the producer process executes before the consumer begins. In general, increasing values of **n** are printed with some values printed many times and others not at all. This example illustrates the need for the producer and consumer to synchronize with each other.

The producer/consumer problem is a specific type of synchronization that is needed between two processes. In general, many types of synchronization between processes can be expressed with a *synchronization graph*, such as shown in Figure 84.3. A synchronization graph is a directed graph showing the relative execution order for a set of actions (code segments). In the example, actions B and C execute after action

```
int n = 0; /* shared by all processes */

main()
{
    int producer(), consumer();
    CreateProcess(producer);
    CreateProcess(consumer);
    /* wait until done */
}
producer() // produce values of n
{
    int i;
    for (i=0; i<2000; i++)
        n++;   // increment n
}
consumer() // consume and print values of n
{
    int i;
    for (i=0; i<2000; i++)
        printf("n is %d\n", n); /* print value of n */
}
```

FIGURE 84.2 Example of producer/consumer synchronization problem.

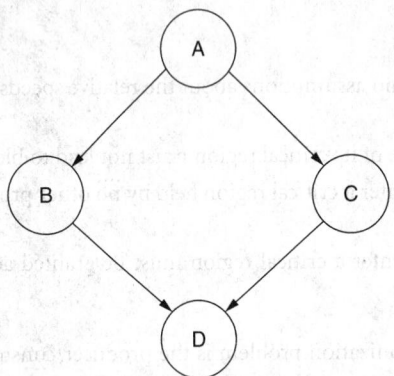

FIGURE 84.3 Synchronization graph for four actions.

A completes, with action D executing after both B and C are complete. Solutions to the synchronization problem need to allow for implementation of this type of problem.

84.2.2 Synchronization Issues

There are many approaches available for solving mutual exclusion and synchronization problems. There are a number of issues concerning the implementation of solutions to these problems, with the primary ones described in the following:

- *Processor and store synchronicity.* Solutions to synchronization problems that are processor synchronous require that processors can be made uninterruptable. These solutions work for uniprocessor machines but not for multiprocessors where disabling the interrupts on one processor does not affect another. Solutions to synchronization problems that are store synchronous, which assumes individual references to main memory are atomic, are applicable to both uni- and multiprocessor machines.

- *Busy waiting.* Another issue in synchronization solutions is the consumption of CPU resources while a process is waiting for a condition to occur. Some solutions require **busy waiting**, the continued polling of a condition variable. These solutions are less efficient, particularly on a uniprocessor where busy waiting continues until the time slice of the waiting process is complete.

- *Programmer errors.* Another issue is the potential programming errors inherent in using a particular synchronization approach. Ordering of synchronization primitives is necessary for correct solutions with some primitives. Other primitives for synchronization have been specifically designed to minimize the possibility of such a programmer error.

- *Starvation.* It is possible for synchronization solutions to lead to the condition of **starvation**. Starvation occurs when a process is indefinitely denied access to a resource while other processes are granted access to the resource. Starvation is an issue in mutual exclusion if it is possible for one process to be indefinitely denied access to its critical region while access is granted to other processes.

- *Deadlock.* Another fundamental problem confronted by solutions is **deadlock**. Deadlock is the condition when a set of processes using shared resources or communicating with each other are permanently blocked. Coffman et al. [1971] describe three necessary, but not sufficient, conditions that must exist for deadlock to occur among a set of processes sharing resources:

1. *Mutual exclusion.* A resource may be used by only one process at a time.

2. *Hold and wait.* Processes holding resources can request new resources.

3. *No preemption.* A resource given to a process cannot be taken back. For deadlock to actually take place, a fourth condition must also exist:

4. *Circular wait.* A set of processes are in a circular wait; that is, there is a set of processes $\{p_0, p_1, \ldots, p_n\}$ where p_i is waiting for a resource held by p_{i+1}, and p_n is waiting for a resource held by p_0.

When combined with the three necessary conditions, the presence of the circular wait condition indicates that a deadlock has occurred and that no means exists for breaking the deadlock. When designing solutions to synchronization problems such as mutual exclusion, preventing deadlock involves disallowing one or more of the preceding conditions. Specific solutions to the deadlock problem are given in Section 84.3.

84.2.3 Interprocess Communication Problems

Interprocess communication problems generally involve direct communication between two or more processes, in contrast to synchronization where processes communicate indirectly by waiting on or setting a condition. Thus, communication is indirect between the processes when using synchronization, whereas IPC mechanisms generally communicate by passing messages directly between processes. The primitives for handling messages are

- **send(pid, &message)**
- **pid = receive(&message)**

where **send()** sends the given message to a specific destination process and **receive()** receives the last message sent to it returning the process identifier (pid) of the sender.

Message passing is commonly used in operating systems and user applications for communicating between processes. In a message-based operating system where server processes perform many of the

```
int serverpid;          /* well-known pid of the server */

Server()                /* server process code */
{
    int pid;            /* process id of the requesting process */
    Message request_msg, reply_msg;

    while (TRUE) {
        pid = receive(&request_msg);          /* receive message */
        /* handle request and build reply message */
        send(pid, &reply_msg);                      /* send the reply */
    }
}

int ServiceRequest(args)    /* service request function with arguments */
{
    Message request_msg, reply_msg;

    /* load args into request_msg */
    send(serverpid, &request_msg);
    (void)receive(&reply_msg);
    return(reply_msg.status);
}
```

FIGURE 84.4   Service request for a message-based operating system.

functions associated with an operating system, message passing is used to pass requests to the appropriate server and to receive replies, as is shown in the example of Figure 84.4.

84.2.4   Interprocess Communication Issues

As described in the previous example, IPC mechanisms typically involve the exchange of messages from one process to another. There are a number of issues concerning the implementation of a message passing mechanism, with the primary ones described in the following. Implementations of IPC mechanisms that address these issues are described in Section 84.3.

- *Direct vs. indirect communication.* A key issue in a message passing system is how messages are addressed to their recipient. In a direct message passing scheme, the delivery address is a process id. This approach was illustrated in Figure 84.4. In an indirect scheme, the address is an intermediate repository, such as a mailbox or port, that is a drop-off point for the message for later pick-up by the recipient.
- *Buffering of data.* Another issue is whether the message passing mechanism allows messages to be buffered if the receiving process is currently not ready to receive a message. If buffering is allowed, then another issue is the size of this buffer.
- *Blocking vs. nonblocking operations.* Related to the issue of buffering is the semantics of the **send()** operation when the message cannot be delivered. The operation can either block until the message can be delivered or the operation can immediately return with an error if blocking would occur. Similarly, the **receive()** operation can be defined to either block and wait for message delivery if no message is available or not block and return an error.
- *Fixed- or variable-size messages.* Fixed-size messages allow for easier implementation but may require more work for the programmer. In contrast, variable-size messages are easier to program but require more work to implement.

- *Synchronous vs. asynchronous reception.* Message passing mechanisms allow more process control if messages are received only when the **receive()** operation is invoked. However, some mechanisms allow for communication to be received when it arrives through an asynchronous approach using message handlers.

84.3   Best Practices

This section discusses specific solutions for synchronization and IPC. Each solution contains an example of its use, its relative merits, and problems for which it is useful. More examples of synchronization mechanisms can be found in operating system texts [Silberschatz et al. 2002, Tanenbaum 2001, Stallings 2000, Flynn and McHoes 1997, Finkel 1988]. Comer [1984] and Tanenbaum [1987] present code for actual operating systems to show how synchronization and IPC mechanisms can be implemented. More concurrent programming examples can be found in Ben-Ari [1982, 1990], Raynal [1986], and Brinch Hansen [1973]. Andrews and Schneider [1983] provide a survey of synchronization and IPC techniques. The chapter concludes with more classic synchronization problems and specific solutions to the deadlock problem.

84.3.1   Synchronization Mechanisms

The mechanisms for synchronization are divided into four types based on their level of implementation and support: software only, hardware support, operating system support, and language support. In addition, hybrid solutions exist that combine more than one of these approaches. Each approach shows a solution to the mutual exclusion problem along with other applicable synchronization problems and discusses the relevant synchronization issues.

84.3.1.1   Software Solutions

Software-based synchronization solutions require only that multiple processes can access shared global variables. The solutions use these variables to control access to the critical region. Dekker was the first to devise a software solution that correctly handles the mutual exclusion problem among a set of processes. A discussion of this solution is given in Dijkstra [1965]. Peterson [1981] provided a simpler solution of the same problem, which is given in Figure 84.5 for two processes in terms of the **BeginRegion()** and **EndRegion()** routines.

This solution works correctly on any hardware (uni- or multiprocessor) in which references to main memory are atomic. The main disadvantage is that it requires busy waiting, continued polling of a status variable, before gaining access to the critical region if another process is already in the critical region. The solution can generalize to more processes, but better solutions are available.

84.3.1.2   Solutions Using Hardware Support

Not all problems of synchronization, particularly ones in the operating system itself, can be handled solely with software-based solutions. As in other aspects of computing, hardware support can not only make the task easier but is also necessary for some levels of synchronization within the operating system.

One of the simplest ways to enforce mutual exclusion is to disable hardware interrupts at the start of the critical region, thus ensuring that the process does not give up the CPU (through a context switch) before completing the critical region. When the critical region is done, the process re-enables interrupts. The **BeginRegion()** and **EndRegion()** routines with this approach are shown in Figure 84.6.

This approach is fast and is used for manipulation of shared operating system data structures on a uniprocessor but in general has many disadvantages for general-purpose use:

- Programmers must be careful not to disable interrupts for too long; devices that raise interrupts need to be serviced.
- The programmer must be careful about nesting. Activities that disable interrupts must restore them to their previous settings. In particular, if interrupts are already disabled before entering a critical

```
int turn;                           /* whose turn is it? */
int flag[2];                        /* want the mutex? Initially FALSE */

BeginRegion(int pid)                /* pid is 0 or 1 */
{
    int other;                      /* pid of other process */

    other = 1 - pid;               /* the opposite of pid */
    flag[pid] = TRUE;              /* express interest in mutex */
    turn = pid;                    /* set flag */
    while ((turn == pid) && (flag[other] == TRUE))
        ;                          /* busy wait */
}

EndRegion(int pid)
{
    flag[pid] = FALSE;            /* drop interest in mutex */
}
```

FIGURE 84.5 Peterson's solution of mutual exclusion for two processes. Based on statements of the algorithm in Tanenbaum [2001].

```
BeginRegion()
{
    DisableInterrupts();
}

EndRegion()
{
    EnableInterrupts();
}
```

FIGURE 84.6 Mutual exclusion by disabling/enabling hardware interrupts.

region, they must remain disabled after leaving the critical region. Code in one critical region may call a routine that executes a different critical region.

- Disabling interrupts prevents all other activities, even though many may never execute the same critical region. Disabling interrupts is like using a sledge hammer; it is a powerful tool but bigger than is needed for most jobs.
- The technique is ineffective on multiprocessor architectures, where disabling interrupts on one processor still allows other processes to run on other processors.

Rather than perform mutual exclusion by controlling interrupts, a common approach is to use special instructions provided by the hardware to implement mutual exclusion. One such instruction is **Test_and_Set,** which is functionally defined by the procedure of Figure 84.7. It returns the previous value of a target variable and sets the target to the given value. Most importantly, this instruction is performed in an atomic manner so a context switch cannot occur in the middle of it. In addition, operations performed on two separate processors of a multiprocessor are guaranteed to occur in sequential order because of store synchronicity. Figure 84.7 shows how the **BeginRegion()** and **EndRegion()** primitives are implemented using this instruction. The variable **mutex** is also referred to as a lock variable, and this approach to mutual exclusion is called a **spin lock** because a process spins in an infinite loop waiting for the lock. When the **mutex** variable is set to **FALSE** by a process exiting from its critical region, another process is allowed to gain access to its critical region.

```
int mutex = FALSE;                          /* global variable for mutex */

int Test_and_Set(int *pVar, int value) /* atomic machine instruction */
{
    int temp;

    temp = *pVar;
    *pVar = value;
    return(temp);
}

BeginRegion()        /* Loop until safe to enter */
{
    while (Test_and_Set(&mutex, TRUE)) ;
        ; /* Loop until return value is FALSE */
}

EndRegion()
{
    mutex = FALSE;
}
```

FIGURE 84.7 Mutual exclusion using the test-and-set machine instruction.

Solutions using other machine instructions are also available. Another common instruction is **EXCH**, which swaps the contents of two memory locations in an atomic fashion. This instruction also can be used to implement mutual exclusion. The advantages of these machine-instruction approaches are their simplicity and the fact they work for any number of processes in either a uni- or a multiprocessor environment. Through the use of more than one mutex variable, multiple critical regions can be easily created.

The primary disadvantage of this approach is its use of busy waiting, thus wasting CPU resources. The use of busy waiting also allows for process starvation if multiple processes are contending for a critical region. Finally, deadlock is possible on a uniprocessor machine if a lower priority process gets interrupted in the middle of its critical region and then a higher priority process tries to gain access to the same critical region. The higher priority process busy waits forever because the lower priority process never runs.

84.3.1.3 Operating System Support

All of the synchronization approaches shown thus far can be implemented with the bare features of the hardware. The approaches also cause busy waiting if another process already is in its critical region. **Semaphores**, an important synchronization primitive, can be constructed by adding process coordination support to the operating system. Semaphores are data structures consisting of an identifier, a counter, and a queue; where processes waiting on a semaphore are blocked and placed on the queue, processes signaling a semaphore may unblock and remove a process from the queue, and the counter maintains a count of waiting processes.

The concept of a semaphore was first introduced by Dijkstra [1968]. Dijkstra defined two atomic semaphore operations: wait and signal, which he termed the P-operation (for wait; from the Dutch word *proberen*, to test) and the V-operation (for signal; from the Dutch word *verhogen*, to increment).

A restricted version of a semaphore, called a *binary semaphore*, limits the value of the counter to 0 and 1. However, the more general case is to use a *counting semaphore*, which has the following properties concerning the counter:

- A nonnegative count always means that the queue is empty.
- A count of negative **n** indicates that the queue contains **n** waiting processes.
- A count of positive **n** indicates that **n** resources are available and **n** requests can be granted without delay.

```
int semid;

Initialization()
{
semid = semcreate(1);    /* initialize the semaphore count to 1 */
}

BeginRegion()
{
    semwait(semid);
}

EndRegion()
{
    semsignal(semid);
}
```

FIGURE 84.8   Mutual exclusion using semaphores.

There are four basic operations defined for creating, deleting, waiting on, and signaling a semaphore:

1. **semid = semcreate(val):** create a semaphore with the given initial value for the counter.
2. **semdelete(semid) :** Delete a semaphore.
3. **semwait(semid) :** wait on a semaphore. Decrement the semaphore counter. If the counter is negative, then suspend execution of the process and place it in the semaphore queue.
4. **semsignal(semid) :** signal a semaphore. Increment the semaphore counter. Make the first process in the semaphore queue ready for execution.

Given these operations, a simple solution to the mutual exclusion problem is shown in Figure 84.8. Unlike previous solutions, a process waiting for its critical region does not busy wait. While it is waiting, it is in a suspended state, allowing the CPU to perform other activities. Because semaphores are provided by the operating system, they work correctly on either uni- or multiprocessor machines.

Semaphores also provide a mechanism to solve the bounded-buffer producer/consumer problem, which was introduced in the example of Figure 84.2. A solution for this problem using semaphores is shown in Figure 84.9. This solution ensures that the consumer process prints all integer values from 1 to 2000 by alternating between the producer and consumer.

Although semaphores provide straightforward solutions to both of these problems, semaphores themselves must be implemented in the operating system with lower level primitives such as the **Test_and_Set** machine instruction. Semaphores work well for processes, that are allowed to block, but must not be used in code that cannot block, such as interrupt service routines.

84.3.1.4   Language Constructs

The solutions presented thus far to the mutual exclusion problem each require the programmer to take explicit action to ensure mutual exclusion. To guarantee mutual exclusion, some programming languages provide constructs to implicitly guarantee mutual exclusion. One such construct is a **monitor** [Hoare 1974], which permits only one process to be executing in a monitor at a time.

Monitors are a programming language construct, similar to abstract data types in that the programmer defines a set of data types and procedures that can manipulate the data, procedures can be exported to other modules, and the system invokes an initialization routine before execution begins. Monitors differ in that they support guard procedures. When a process invokes a guard procedure, its execution is delayed until no other processes are executing a guard procedure within the monitor. The Java programming language

```
int psem, csem;  /* semaphores */
int n = 0;       /* shared by all processes */
main()
{
    int producer(), consumer();
    csem = semcreate(0);  /* consumer initially blocks */
    psem = semcreate(1);  /* producer initially allowed to run */
    CreateProcess(producer);
    CreateProcess(consumer);
    /* wait until done */
}
producer()
{
    int i;
    for (i=0; i<2000; i++) {
        wait(psem);
        n++;  /* increment n by 1 */
        signal(csem);
    }
}
consumer()
{
    int i;
    for (i=0; i<2000; i++) {
        wait(csem);
        printf("n is %d\n", n);  /* print value of n */
        signal(psem);
    }
}
```

FIGURE 84.9   Bounded-buffer producer/consumer problem with semaphores.

supports monitors using the **synchronized** primitive on methods within a class to prevent threads from simultaneously access of these methods [Arnold and Gosling 1998]. Figure 84.10 revisits the mutual exclusion problem of Figure 84.1 using a solution written in Java. The definition of a monitor guarantees that the variable **balance** is not updated and read simultaneously.

Monitors are a programming language construct that must be implemented using a lower level facility provided by the hardware or operating system, such as semaphores. Although monitors provide mutual exclusion, they need additional primitives to provide synchronization. To do so, monitors are defined to have condition variables, which are waited on and signaled similar to semaphores. Java provides a single, unnamed condition for a class with the **wait()** method used by a thread to wait on the condition and the **notify()** or **notifyAll()** methods to wake up one or all waiting threads. These primitives allow other synchronization problems such as the producer/consumer problem to be implemented with monitors.

84.3.1.5 Hybrid Solutions

Modern operating systems have migrated from monolithic systems written for a uniprocessor in which critical regions were used to access shared data structures. These critical regions were guarded by setting the interrupt level appropriately. Current operating systems must not only support multiprocessors but also provide real-time capabilities, thus leading to multithreaded designs. In these designs, the use of interrupts to control access to shared data does not work. Rather, these systems have moved to solutions using complex locks, which combine the use of spin locks with the semantics of semaphores.

```
public class Account {
    private int balance;

    public Account() {
        balance = 0;          // initialize balance to zero
    }

    // use synchronized to prohibit concurrent access of balance
    public synchronized void Deposit(int deposit) {
        int newbalance; // local variable

        newbalance = balance + deposit;
        balance = newbalance;
    }

    public synchronized int GetBalance() {
        return balance;           // return current balance
    }
}
```

FIGURE 84.10 Mutual exclusion using monitors.

As one example, the Solaris 2 operating system [Eykholt et al. 1992] uses adaptive mutex locks, a type of complex lock, to protect access to shared data among a set of threads. The adaptive lock starts out executing like a standard spin lock. If the lock is currently free, then the issuing thread immediately obtains the lock. However, if the lock is in use, then the operating system checks the status of the thread holding the lock. If this thread is currently in the run state (as could be the case on a multiprocessor), then the issuing thread continues in a spin lock waiting for what is expected to be a short time for the lock to be released. If the holding thread is not in the run state (as would always be the case on a uniprocessor), then the issuing thread is suspended until the lock is released. The rationale is to use a spin lock if the wait for the lock is expected to be short and to actually suspend the thread if the wait is expected to be longer.

This hybrid approach tries to minimize overhead and maximize performance. If the size of a critical region is large (hundreds of instructions), then the adaptive mutex lock is less desirable compared to a lock that simply causes a thread to suspend when the lock is not available.

Another type of complex lock used in multithreaded operating systems is a read/write lock. These locks allow either a single writer or multiple readers to simultaneously hold the lock, thus increasing the parallelism when reading of a shared data structure predominates. Writers must wait until all readers have released the lock before obtaining the lock, whereas readers are granted immediate access to the lock in the absence of a writer. To prevent starvation of a writer, all read requests after a write request has been issued are queued until the write request has been satisfied. Starvation of readers in the face of multiple writers is similarly avoided.

84.3.1.6 Other Solutions

Many other synchronization primitives have been proposed but, in general, can be expressed in terms of the solutions already given. A sampling of these primitives are critical regions [Hoare 1972, Brinch Hansen 1972], serializers [Atkinson and Hewitt 1979], path expressions [Campbell and Habermann 1974], and event counts and sequencers [Reed and Kanodia 1979].

84.3.2 Interprocess Communication (IPC) Mechanisms

As with synchronization, a variety of IPC mechanisms are available. The following discusses a number of these mechanisms and how they handle the IPC issues raised in Section 84.2.

84.3.2.1 Direct Message Passing

The simplest form of message passing is to send messages directly from one process to another. An example of this approach is the low-level message passing facility used in the Xinu operating system [Comer 1984]. The primitive operations used are:

- **send(pid, msg)**
- **msg = receive()**

where **send()** sends a fixed, integer-size message to a specific process and **receive()** returns the last message sent to it. A process can buffer only one message. If **send()** detects a message already buffered at the process, then it returns immediately with an error, not delivering the message. If no message is buffered, then **send()** buffers the message and readies the receiving process if it is waiting for a message. The **receive()** operation blocks if a message is not available.

Another direct message passing mechanism is implemented in Minix [Tanenbaum 1987]. The primitives for handling messages are:

- **send(destpid, &message)**
- **receive(srcpid, &message)**

where **send()** sends the given message to a specific destination process and **receive()** receives a message from a particular process. The source process for **receive()** can contain a wildcard value of **ANY,** indicating that messages from any process are accepted. Messages are a fixed size, but there is no buffering. Rather, the **send()** and **receive()** operations *rendezvous* so that both operations block until the receiving process has actually copied the message from the sender. The use of rendezvous explicitly synchronizes the execution of the sending and receiving process.

84.3.2.2 Mailboxes/Ports

Rather than send directly to process, a more common approach is to define another operating system abstraction called a **mailbox** (also referred to as a port). Mailboxes are buffers that hold messages sent by one process to be received by another process. Thus, there is indirect communication between the two processes. The primitives for handling messages are:

- **send(mailbox, &message)**
- **receive(mailbox, &message)**

where **send()** buffers the message in the given mailbox and **receive()** removes a message from the mailbox.

As an example, the Unix operating system provides ports to allow for intra- and inter-machine communication between processes. Ports often represent well-known services where a server process binds to a port and client processes of the service send requests to the port. The port buffers communication sent to the buffer until it is read by the receiving process. The messages sent to the port can be of variable size.

Message passing can also be implemented using shared memory and semaphores, illustrating the equivalence of synchronization and IPC primitives. Figure 84.11 shows message passing with shared memory and semaphores to implement a set of mailboxes. The mechanism uses fixed-size messages with each of four mailboxes containing eight message buffers. The **send()** operation blocks if there is no buffer space in the mailbox; similarly, the **receive()** operation blocks if there is no message available in the mailbox. The mutex semaphore is needed to guarantee that no more than one process tries to send to or receive from a mailbox at the same time. This semaphore would not be needed if only one process could send to and receive from a mailbox. The mutex semaphore would also not be needed if a mailbox contained only one buffer

```
#define N 8                    /* number of msgs buffered in a mailbox */
#define M 4                    /* number of mailboxes */

Message mailboxes[M][N];       /* shared memory for mailboxes of
                                  messages */

int semidMsg[M];               /* message available */
int semidSlot[M];              /* slot available */
int semidMutex[M];             /* controls access to critical region */


Initialization()
{
    int i;

    for (i = 0; i < M; i++) {
        semidMsg[i] = semcreate(0);   /* no messages are available */
        semidSlot[i] = semcreate(N);  /* N slots are available
                                         for messages */
        semidMutex[i] = semcreate(1); /* one process can enter
                                         region */
        /* initialize indices for inserting/deleting in
 mailboxes[i] */
    }
}

send(int m, Message *pmsg)     /* send message to mailbox m */
{
    semwait(semidSlot[m]);         /* is a slot available */
    semwait(semidMutex[m]);        /* enter critical region */
    addmessage(m, pmsg);           /* add msg to circular
                                      queue for mailbox m */
    semsignal(semidMutex[m]);      /* exit critical region */
    semsignal(semidMsg[m]);        /* signal message available */
}


receive(int m, Message *pmsg)  /* retrieve message from
                                  mailbox m */
{
    semwait(semidMsg[m]);          /* is a message available */
    semwait(semidMutex[m]);        /* enter critical region */
    removemessage(m, pmsg);        /* remove next msg from queue for
                                      mailbox m */
    semsignal(semidMutex[m]);      /* exit critical region */
    semsignal(semidSlot[m]);       /* signal slot available */
}
```

FIGURE 84.11 Message passing with semaphores and shared memory.

```
#define DATA "hello world"
#define BUFFSIZE 1024

int rgfd[2];              /* file descriptors of pipe ends */

main()
{
    char sbBuf[BUFFSIZE];

    pipe(rgfd);           /* create a pipe returning two file desciptors */
    if (fork()) {         /* parent, read data from pipe */
        close(rgfd[1]); /* close write end */
        read(rgfd[0], sbBuf, BUFFSIZE);
        printf("Pipe contents: %snn", sbBuf);
        close(rgfd[0]);
    }
    else {                /* child, write data to pipe */
        close(rgfd[0]); /* close read end */
        write(rgfd[1], DATA, sizeof(DATA));  /* write data to pipe */
        close(rgfd[1]);
        exit(0);
    }
}
```

FIGURE 84.12   Pipe example in the Unix operating system.

slot. This is also an example of a bounded-buffer producer/consumer problem where sending processes produce messages and receiving processes consume them.

84.3.2.3 Pipes

A special case of IPC is the pipe abstraction available in the Unix operating system. A *pipe* is a unidirectional, stream communication abstraction. One process writes data to the write end of the pipe, and a second process reads data from the read end of the pipe. The pipe itself is a buffer between the two processes that causes the reader to block if no data is available and the writer to block if the buffer is full. As it implements a stream abstraction, there is no notion of fixed-size messages. A pipe is another example of a solution to the bounded-buffer producer/consumer problem where the writing process is a producer and the reading process is a consumer.

Figure 84.12 shows a simple example of the use of pipes in the UNIX operating system. Pipes are typically requested and set up by a UNIX command interpreter, but the example shows one process creating another process with **fork()** with a pipe between them. A string of characters is then written to and read from the pipe.

84.3.2.4 Software Interrupts

Software interrupts are a primitive form of IPC. They are similar to hardware interrupts in that when an interrupt of a process occurs, an interrupt handler routine corresponding to the type of interrupt is invoked. Interrupts are asynchronous; so when an interrupt is received, execution of the process stops and is restarted after the interrupt handling routine has been executed. Software interrupts are sent to a process using the process id of the process. Many interrupts are used for well-known functions, such as when the user types the interrupt key, a child process completes, or an alarm scheduled by the process has expired.

```
#include <signal.h>

int n;

main(int argc, char **argv)
{
    void InterruptHandler(), InitHandler();

    n = 0;

    signal(SIGINT, InterruptHandler);  /* signal 2 */
    signal(SIGHUP, InitHandler);       /* signal 1 */
    while (1) {
        n++;
        sleep(1);                      /* sleep for one second */
    }
}

void InterruptHandler()
{
    printf("The current value of n is %dnn", n);
    exit(0);
}

void InitHandler()
{
    printf("Resetting the value of n to zeronn");
    n = 0;
}
% cc -o signalex signalex.c
% signalex
^C              (interrupt character)
The current value of n is 3
% signalex &
[1] 20822
% kill -1 20822
Resetting the value of n to zero
% kill -2 20822
The current value of n is 19
[1]   Done                    signalex
```

FIGURE 84.13 Software interrupt program and script in the Unix operating system.

Two routines are used to send and handle software interrupts:

1. **SendInterrupt(pid, num)** : an interrupt of type **num** is sent to process **pid.** In the Unix operating system this routine is **kill().**
2. **HandleInterrupt(num, handler)** : this specifies that user supplied **handler** routine should be invoked when interrupt of type **num** occurs. Typical handlers are to ignore the interrupt, terminate the process, or execute a user-supplied interrupt handler. In the Unix operating system, this routine is **signal().**

Figure 84.13 shows a sample program for the UNIX operating system with software interrupts. It sets up two interrupt handlers for signals 1 and 2 and then goes into an infinite loop where it updates a counter and sleeps for 1 s. Figure 84.13 also shows a command line script with this program. Invoking the interrupt character from the command interpreter causes interrupt 2 to be sent to the process. The second invocation of the program causes it to be run in the background with a process id of 20822. The **kill** program is then used to send interrupts to the background process.

84.3.3 Classic Problems

Two classic synchronization problems — the critical region and bounded-buffer producer/consumer — have already been discussed. A slight variation of the producer/consumer problem is to use an unbounded buffer, in which case the producer never blocks because the buffer never fills. Many other classic synchronization problems have been proposed and solved. The following describes two such problems.

84.3.3.1 Readers/Writers Problem

The readers/writers problem occurs when multiple readers and writers want access to a shared object such as a database. The problem was introduced in Courtois et al. [1971]. In the problem, multiple readers are allowed to access the database simultaneously, but a writer must have exclusive access to the database before performing any updates for consistency. A practical example of this problem is an airline reservation system with many readers and an occasional update of the information.

Figure 84.14 shows a solution to this problem for multiple reader and writer processes with semaphores. The solution allows multiple readers access to the database at a time. A writer process can gain access only after all reader processes have relinquished the database. The solution gives priority to reader processes, who can gain access to the database even if a writer process is already requesting access to the database. Solutions giving more balanced priority to each type of process can also be constructed, as discussed in Flynn and McHoes [1997].

84.3.3.2 Dining Philosophers Problem

The dining philosophers problem was proposed and solved by Dijkstra [1971]. It consists of five philosophers sitting at a round table. The philosophers each have a bowl of rice in front of them and there is a chopstick in between each bowl (alternately, the problem is described using plates of spaghetti and forks). The problem is illustrated in Figure 84.15 with the philosophers' bowls labeled A through E and the chopsticks 1 through 5. These philosophers have two functions in life: (1) *think*, requiring no interaction with colleagues, and (2) *eat*, requiring the philosopher to pick up the chopstick on the left and right.

This classic synchronization problem has potential for both deadlock and starvation (literally!). The straightforward solution for a philosopher to eat is to first pick up the left chopstick and then the right chopstick. However, if all philosophers pick up their left chopstick at the same time, they will all deadlock when they go to pick up their right chopstick. A simple modification to this approach is for the philosophers to put down the left chopstick if the right chopstick is not available, wait for some time, and try again. However, there is still a chance that the philosophers will operate in lock step and no philosophers will acquire both chopsticks. This condition is called **livelock** and occurs when attempts by two or more processes (philosophers) to acquire a resource (the left and right chopsticks) run indefinitely without any process succeeding. The dining philosophers problem will be used as a guide in the following discussion on deadlock and starvation.

84.3.4 Deadlock and Starvation

Deadlock occurs when a set of processes using shared resources are permanently blocked trying to gain access to those resources. Classic papers on this topic are Coffman et al. [1971] and Holt [1972]. Zobel and Koch [1988] contains a more up-to-date annotated bibliography on the subject. Deadlock can occur with synchronization such as when two processes each need to gain access to two separate critical regions for execution. If one process gains access to the first critical region and the other process gains access to

```
int readercount = 0;      /* number of readers currently reading */
int readermutex;          /* semaphore mutex for reader count */
int dbaccess;             /* semaphore to control access to database */

main()
{
    readermutex = semcreate(1); /* mutex for reader count */
    dbaccess = semcreate(1);    /* mutex for database */
    CreateProcess(reader);      /* create a reader process */
    CreateProcess(writer);      /* create a writer process */
}

reader()
{
    while (TRUE) {
        semwait(readermutex);        /* get access to readercount */
        readercount++;               /* increment count */
        if (readercount == 1)        /* if first reader ... */
            semwait(dbaccess);       /* gain access to database */
        semsignal(readermutex);      /* done with count */
        /* read database */
        semwait(readermutex);        /* get access to readercount */
        readercount--;               /* decrement count */
        if (readercount == 0)        /* if last reader ... */
            semsignal(dbaccess);     /* relinquish access to database */
        semsignal(readermutex);      /* done with count */
        /* use data read */
    }
}

writer()
{
    while (TRUE) {
        /* generate new data */
        semwait(dbaccess);              /* gain access to database */
        /* update database */
        semsignal(dbaccess);            /* relinquish access to database */
    }
}
```

FIGURE 84.14 Solution to the readers/writers problem using semaphores.

the second critical region, then these two processes will be in deadlock when they attempt to acquire the other needed critical region.

There are four principles used for dealing with the issue of deadlock in operating systems. These principles are prevention, detection, avoidance, and recovery and are summarized in Isloor and Marsland [1980]. Solutions to the deadlock problem exemplifying these principles are described in the following. Each of these solutions prevents deadlock by precluding one or more of the four conditions for deadlock given in Section 84.2. The solutions are characterized as being conservative or liberal, depending on the degree of concurrency they allow.

The most liberal solution is to allocate requested resources to processes if the resources are available. This approach was given as an initial solution when introducing the dining philosophers problem. As was shown,

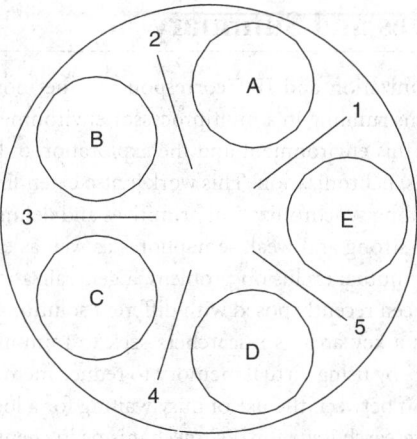

FIGURE 84.15 Illustration of the dining philosophers problem.

it can lead to deadlock and requires that a deadlock detection process be periodically invoked. Deadlock detection involves detecting the circular wait condition through any algorithm for detecting cycles in directed graphs. Once deadlock occurs, a deadlock recovery method must be invoked, either to take away resources from a victim process or to terminate the process altogether. In the dining philosophers problem, a recovery method would be to take away a chopstick from one of the philosophers.

At the other extreme, the most conservative deadlock prevention solution is *serialization*. This deadlock prevention approach guarantees no deadlock by allowing only one process to acquire resources at a time. In the dining philosophers problem, this approach means that only one philosopher is allowed to eat at any time. The approach prevents any concurrency and can lead to starvation if a philosopher is constantly passed over in obtaining chopsticks.

Another solution that allows more concurrency but still prevents deadlock is *one-shot allocation*. This approach requires that a process obtain all of its resources at once. Using this approach, a dining philosopher must obtain both chopsticks at the same time. It avoids deadlock by preventing the hold and wait condition. This solution prevents deadlock but not starvation. It also requires that a process obtain all of its resources at the same time, even if it does not currently need all of them.

A still more liberal solution that prevents deadlock is *hierarchical allocation*. This solution requires that all resources have a number associated with them as in Figure 84.15. Hierarchical allocation prevents deadlock by requiring that processes can acquire only resources with a higher number than any resource it currently holds. Thus, in Figure 84.15, each philosopher must first acquire the left chopstick and then the right except for Philosopher E, who must first acquire the right chopstick. This solution is still conservative and requires processes to acquire resources not necessarily in the order they are needed but in the order resources are numbered. It avoids deadlock by preventing the circular wait condition.

As opposed to deadlock prevention, deadlock avoidance policies do not prevent deadlock *a priori*, but monitor the allocation requests as they are made so as not to allow deadlock to occur. A solution using this approach is the *bankers algorithm* (also called the advanced claim algorithm) introduced in Dijkstra [1968]. This solution is the most liberal that avoids deadlock but requires that each process know the maximum number from a class of resources that it may request at any time during execution. The algorithm performs as a banker, giving out loans of money in that any resource requests are granted only if they:

1. Do not exceed the total number of resources in a class
2. Do not exceed the maximum number of resources for that process
3. Lead to a *safe state* where a sequence of resource deallocations and allocations can allow all processes to complete without deadlock

84.4   Research Issues and Summary

The research issues in synchronization and IPC correspond to the movement toward multithreaded, message-based operating systems running in a multiprocessor environment. The adoption of traditional synchronization primitives for this environment and the exploration of better primitives are leading to work on hybrid approaches for synchronization. This work is also extending to distributed environments.

Research continues on extending synchronization primitives and defining new problems. Recent work has examined the semantics of strong and weak semaphores as well as extending semaphores to define tagged semaphores. The group mutual exclusion problem, a generalization of the mutual exclusion and readers/writers problems, has been recently posed with different solutions under investigation.

The issue of performance is a key area as researchers seek to minimize the cost of busy waiting in shared memory multiprocessors by using virtual memory to reduce memory access costs. Other research is investigating optimal trade-off between the use of busy waiting for a lock vs. suspending the thread or process. Performance is also a research issue for IPC mechanisms in client/server systems as they execute in intra- and inter-machine environments.

An ongoing research issue is correctness, particularly as the synchronization problems of modern operating systems become more complex. Using language constructs to better encapsulate the synchronization details is being explored, but this approach can lead to trade-offs with performance. Tools for detecting synchronization errors are also an ongoing area of research.

In summary, synchronization and IPC are fundamental to multiprogrammed operating system design. Many primitives to solve fundamental problems such as mutual exclusion and producer/consumer exist, ranging from software-only approaches to special hardware instructions, to primitives constructed by the operating system and programming languages. Many of the primitives are equivalent in terms of their semantics, and one can be implemented in terms of another. Examples include implementing monitors with semaphores or message passing with shared memory and semaphores. Modern operating systems are using hybrid approaches, which adaptively switch between techniques according to the runtime operating environment.

Defining Terms

Busy waiting: Situation in synchronization when a process continuously polls the status of a condition variable.

Critical region: Set of instructions for a process that access data shared with other processes. Only one process may execute its critical region at a time.

Deadlock: Condition when a set of processes using shared resources or communicating with each other are permanently blocked.

Interprocess communication (IPC): Communication between two (or more) processes directly aware of each other.

Livelock: Condition when attempts by two or more processes to acquire a resource run indefinitely without any process succeeding.

Mailbox: An operating system abstraction containing buffers to hold messages. Messages are sent to and received from the mailbox by processes.

Monitor: Programming language construct providing abstract data types and mutually exclusive access to a set of guard procedures.

Mutual exclusion: A synchronization problem requiring that two or more concurrent activities do not simultaneously access a shared resource.

Race condition: Situation where multiple processes access and manipulate shared data with the outcome dependent on the relative timing of these processes.

Semaphore: Synchronization primitive consisting of an identifier, a counter, and a queue where processes waiting on a semaphore are blocked and placed on the queue; processes signaling a semaphore may unblock and remove a process from the queue, and the counter maintains a count of waiting processes.

Spin lock: Mutual exclusion mechanism where a process spins in an infinite loop waiting for the value of a lock variable to indicate availability.

Starvation: Condition when a process is indefinitely denied access to a resource while other processes are granted access to the resource.

Synchronization: Situation when two or more processes coordinate their activities based upon a condition.

References

Anderson, J. H. 2001. Lamport on mutual exclusion: 27 years of planting seeds. In *Proc. ACM Symposium on Principles of Distributed Computing*, pp. 3–12.

Andrews, G. R. and Schneider, F. B. 1983. Concepts and notations for concurrent programming. *Comput. Surv.*, 15(1):3–43.

Arnold, K. and Gosling, J. 1998. *The Java Programming Language*, 2nd ed. Addison–Wesley, Reading, MA.

Atkinson, R. and Hewitt, C. 1979. Synchronization and proof techniques for serializers. *IEEE Trans. Software Eng.*, 5(1):10–23.

Ben-Ari, M. 1982. *Principles of Concurrent Programming*. Prentice Hall, Englewood Cliffs, NJ.

Ben-Ari, M. 1990. *Principles of Concurrent and Distributed Programming*. Prentice Hall, Englewood Cliffs, NJ.

Brinch Hansen, P. 1972. Structured multiprogramming. *Commun. ACM*, 15(7):574–578.

Brinch Hansen, P. 1973. *Operating Systems Principles*. Prentice Hall, Englewood Cliffs, NJ.

Campbell, R. H. and Habermann, A. N. 1974. The specification of process synchronization by path expressions. In *Operating Systems*, E. Gelenbe and C. Kaiser, Eds., pp. 89–102. Springer-Verlag, Berlin.

Coffman, E., Elphick, M., and Shoshani, A. 1971. System deadlocks. *ACM Comput. Surv.*, 3(2):67–78.

Comer, D. 1984. *Operating System Design, the Xinu Approach*. Prentice Hall, Englewood Cliffs, NJ.

Courtois, P., Heymans, F., and Parnas, D. L. 1971. Concurrent control with "readers" and "writers." *Commun. ACM*, 14(10):667–668.

Dijkstra, E. W. 1965. Solution of a problem in concurrent programming control. *Commun. ACM*, 8(9):569.

Dijkstra, E. W. 1968. Co-operating sequential processes. In *Programming Languages*, F. Genuys, Ed., pp. 43–112. Academic Press, New York. Reprint of Technical Report EWD-123, Technological University, Eindhoven, The Netherlands (1965).

Dijkstra, E. W. 1971. Hierarchical ordering of sequential processes. *Acta Informatica*, 2(1):115–138.

Eykholt, J. R., Kleiman, S. R., Barton, S., Faulkner, S., Shivalingiah, A., Smith, M., Stein, D., Voll, J., Weeks, M., and Williams, D. 1992. Beyond multiprocessing: multithreading the SunOS kernel. In *Proc. Summer USENIX Conf.* USENIX Association, pp. 11–18.

Finkel, R. A. 1988. *An Operating Systems VADE MECUM*. Prentice Hall, Englewood Cliffs, NJ.

Flynn, I. M. and McHoes, A. M. 1997. *Understanding Operating Systems*, 2nd ed. PWS Publishing Company, Boston, MA.

Hoare, C. A. R. 1972. Towards a theory of parallel programming. In *Operating Systems Techniques*. C. A. R. Hoare and R. H. Perrott, Eds., pp. 61–71. Academic Press, New York.

Hoare, C. A. R. 1974. Monitors: an operating system structuring concept. *Commun. ACM*, 17(10):549–557; Erratum 1975. *Commun. ACM*, 18(2):95.

Holt, R. 1972. Some deadlock properties of computer systems. *ACM Comput. Surv.*, 4(3):179–196.

Isloor, S. S. and Marsland, T. A. 1980. The deadlock problem: an overview. *IEEE Comput.*, 13(9):58–78.

Lamport, L. 1986. The mutual exclusion problem. *J. ACM*, 33(2):313–348.

Peterson, G. L. 1981. Myths about the mutual exclusion problem. *Inf. Process. Lett.*, 12(3):115–116.

Raynal, M. 1986. *Algorithms for Mutual Exclusion*. Wiley, New York.

Reed, D. P. and Kanodia, R. K. 1979. Synchronization with eventcounts and sequencers. *Commun. ACM*, 22(2):81–92.

Silberschatz, A., Galvin, P. B., and Gagne, G. 2002. *Operating System Concepts*, 6th ed. Addison-Wesley, Reading, MA.

Stallings, W. 2000. *Operating Systems: Internals and Design Principles*, 4th ed. Prentice Hall, Upper Saddle River, NJ.

Tanenbaum, A. 1987. *Operating Systems: Design and Implementation*. Prentice Hall, Englewood Cliffs, NJ.

Tanenbaum, A. 2001. *Modern Operating Systems*, 2nd ed. Prentice Hall, Upper Saddle River, NJ.

Zobel, D. and Koch, C. 1988. Resolution techniques and complexity results with deadlocks: a classifying and annotated bibliography. *Operating Syst. Rev.*, 22(1):52–72.

Further Information

Many good textbooks on operating systems, such as those by Silberschatz et al. [2002], Tanenbaum [2001], and Stallings [2000] exist, which describe problems, issues, and solutions for synchronization and IPC. The book *Principles of Concurrent and Distributed Programming* by M. Ben-Ari [1990] contains a number of problems and worked-out solutions for both concurrent and distributed programming. *Algorithms for Mutual Exclusion* by M. Raynal [1986] presents a comprehensive treatment of solutions for the mutual exclusion problem. "Concepts and Notations for Concurrent Programming" by Andrews and Schneider [1983] provides a survey of processes, synchronization, and interprocess communication.

The Association for Computing Machinery (ACM) Special Interest Group on Operating Systems (SIGOPS) publishes *Operating Systems Review* four times a year. This publication contains work on a variety of operating system topics, including synchronization. This group also sponsors the biennial ACM Symposium on Operating Systems Principles that covers the latest developments in the field of operating systems. Its proceedings are published in an issue of *Operating Systems Review*. Another ACM publication, the *ACM Transactions on Computer Systems* is good source for relevant work.

The USENIX Association sponsors a number of conferences on operating system-related topics. A general technical conference is sponsored each year. Two conferences, the Symposium on Operating Systems Design and Implementation (OSDI) and Workshop on Hot Topics in Operating Systems (HOTOS), are specific to operating system issues.

85
Virtual Memory

85.1 Introduction 85-1
85.2 History 85-1
85.3 Structure of Virtual Memory 85-4
      Paging • Translation Lookaside Buffer • Cache Memories
      • Object-Oriented Virtual Memory • Protection
      • Multiprogramming • Performance
85.4 Distributed Shared Memory 85-11
85.5 The World Wide Web: Virtualizing the Internet 85-12
85.6 Conclusion 85-12

Peter J. Denning
Naval Postgraduate School

85.1 Introduction

Virtual memory, since the 1960s a standard feature of nearly every operating system and computer chip, is now invading the Internet through the **World Wide Web (WWW)**. Once the subject of intense controversy, it is now so ordinary that few people think much about it. That this has happened is one of the engineering triumphs of the computer age.

Virtual memory is the simulation of a storage space so large that software programmers and document authors do not need to rewrite their works when the internal structure of a program module, the capacity of a local memory, or the configuration of a network changes. The name, borrowed from optics, recalls the virtual images formed in mirrors and lenses: objects that are not there but behave as if they are. The story of virtual memory, from the Atlas Computer at the University of Manchester in the 1950s to the multicomputers and World Wide Web of the 2000s, is not simply a story of progress in automatic storage allocation; it is a fascinating story of machines helping programmers to protect information, reuse and share objects, and link software components.

Virtual memory is the first instance of caching, which is one of the great principles of information technology. Caching says that a computer will perform better if the most-used data is held in fast storage close to the processor and the least-used data is held in slow storage more distant from the processor. Caching works because all computations exhibit **locality**, a strong tendency to cluster references to subsets of address space within extended time intervals. Locality is a manifestation of how the human mind organizes to solve problems.

85.2 History

From their beginnings in the 1940s, electronic computers had two-level storage systems. The **main memory** was then magnetic cores and is now random access memories (**RAMs**); the secondary memory was then magnetic drums and is now disks and an array of other media including tapes, CDs, and remote servers in the Internet. The processor [central processing unit (**CPU**)] could address only the main memory.

A major part of a programmer's job was to devise a good way to divide a program into blocks and to schedule their moves between the levels. The blocks were called *segments* or *pages* and the movement operations *overlays* or *swaps*. This was a complex task even for small programs. The designers of the Atlas Computer at the University of Manchester invented virtual memory in the 1950s to eliminate two looming programming problems: (1) planning and scheduling data transfers between main and **secondary memory** and (2) recompiling programs for each change of size of main memory. They dreamt of automating it all.

In 1959, the Atlas computer system was the first working prototype of a virtual memory [Fotheringham 1961, Kilburn et al. 1962]. Its designers called it a *one-level storage system*. At the heart of their idea was a radical innovation: a distinction between *address* and *memory location*. It led them to three inventions: (1) they built hardware that automatically translated each address generated by the processor to its current memory location; (2) they devised **demand paging**, an interrupt mechanism triggered by the address translator that moved a missing page of data from secondary to main memory; and (3) they built the first replacement algorithm, a procedure to detect and move the least useful pages back to secondary memory.

Despite the success of the Atlas memory system, the literature of 1961 records a spirited debate about the feasibility of automatic storage allocation in general-purpose computers. By that time, Cobol, Algol, Fortran, and Lisp had become the first widely used higher level programming languages. These languages made storage allocation harder because programs were larger, more portable, more modular, and their dynamics more dependent on their input data. Through the 1960s there were dozens of experimental studies that sought to either affirm or deny the hypothesis that operating systems could do a better job at storage allocation than any compiler or programmer [Denning 1970]. The matter was finally laid to rest — in favor of automatic storage allocation — by an extensive study of system performance by an IBM research team led by David Sayre [Sayre 1969].

Convinced that virtual memory was the right way to go, the makers of major commercial computers adopted it in the 1960s; these included the IBM 360/67, CDC 7600, Burroughs 5500/6500, RCA Spectra/70, and Multics for the GE 645. By the mid-1970s the IBM 370, DEC VMS, DEC TENEX, and Unix had joined the crowd. These systems all used multiprogramming; their designers turned to virtual memory to solve not only the storage allocation problem, but also the more critical memory protection problem. To their designers' dismay, and to the delight of virtual memory's critics, these systems all exhibited **thrashing**, a condition of near-total performance collapse triggered when the multiprogrammed load became too high [Denning 1968]. Thrashing motivated a long line of experiments and models seeking to understand it and to design effective load control systems. This was finally accomplished by the late 1970s; near-optimal **throughput** will result when the virtual memory guarantees each active process just enough space to hold its working set [Denning 1980].

Virtual memory attracted hardware designers as well as software designers. In 1965, Maurice Wilkes proposed the *slave memory*, a small high-speed store included in the processor to hold, close by, a small number of most recently used blocks of program code and data. Slave memory used address translation, demand loading, and usage-based replacement. Wilkes said that, by eliminating many data transfers between processor and the main memory, slave memory would allow the system to run within a few percent of the full processor speed at a cost within a few percent of the main memory [Wilkes 1965]. The term *cache memory* replaced *slave memory* in 1968 when IBM introduced cache memory in its 360/85 machine. Cache memory is now a standard principle of computer architecture [Hennessey and Patterson 1990].

If it ended here, this story would already have guaranteed virtual memory a place in history. But the designers of the 1960s were no less inventive than those of the 1950s. Just as the designers of the 1950s sought a solution to the problem of storage allocation, the designers of the 1960s sought solutions to two new kinds of programming problems: (1) shareable, reusable, and recompilable program modules, and (2) packages of procedures hiding the internal structure of classes of objects (abstract data types). The first of these led to the **segmented address space**, the second to the architecture that was first called **capability-based addressing** and later **object-oriented programming**.

In 1965 the designers of Multics at the Massachusetts Institute of Technology (MIT) sought systems to support large programs built from separately compiled, shareable modules linked together on demand

[Dennis 1965, Organick 1972]. To them, virtual memory as a pure computational storage system was too restrictive; they held that **modular programming** would not become a reality as long as programmers had to manually bind the component files of an address space (by a linking loader or makefile program). Their innovation was to add a second dimension of addressing to the virtual address space, enabling it to span segment names as well as within-segment addresses. A program could refer to a variable X within a module S by the two-part name (S, X); the symbols S and X were retained by the compiler and converted to the hardware addresses for S and X by the virtual memory on first reference (a *linkage fault*). The Multics virtual memory demonstrated sophisticated forms of sharing, reuse, access control, and protection. This innovation, however, did not find its way beyond Multics; programmers were content with one private, linear address space and a handful of open files. As will be discussed, the World Wide Web [Berners-Lee 1996] is changing this: programs and documents contain *hypertext links* — symbolic pointers to other objects that are not linked until the program references them for the first time.

In 1966 Jack Dennis and Earl Van Horn published a prescient paper that initiated a new line of computer architectures: machines that help programmers create managers of classes of objects. They anticipated what is now called object-oriented programming. They were especially concerned that objects be freely reusable and shareable and, at the same time, be protected from internal access by anyone except their authorized managers. They proposed an extension of virtual memory that would map a process' local name for an object into an internal bit pattern called a *capability*; a capability contained a type indicator, an access code, and a unique name. Capability addressing offered an elegant solution to the problem of sharing and reusing modules. Several commercial and academic capability systems were built in the 1970s: notably the Plessey 250, IBM System 38, Cambridge CAP, Intel 432, SWARD, and Hydra. In these systems, capabilities were implemented as long addresses (e.g., 64 bits), which the hardware protected from alteration. (See Fabry [1974], Myers [1982], and Wilkes and Needham [1979].) The reduced instruction set computer (**RISC**) microprocessor, with its simplified instruction set, coupled with programming languages with type checking, rendered capability-managing hardware obsolete by the mid-1980s. But software-managed capabilities, now called **handles**, are indispensable in modern object-oriented programming systems, databases, and distributed operating systems [Chase et al. 1994]. The same conceptual structure has recently reappeared in a proposal to manage objects and intellectual property in the Internet [Kahn and Wilensky 1995]. It is a powerful structure indeed.

You may have wondered why virtual memory, so popular in the operating systems of the 1960s and 1970s, was not present in the personal computer (PC) operating systems of the 1980s. The pundits of the microcomputer revolution proclaimed bravely that personal computers would not succumb to the diseases of the large commercial operating systems; the personal computer would be simple, fast, and cheap. Bill Gates, who said that no user of a personal computer would ever need more than 640K of main memory, brought out the Microsoft Disk Operating System (DOS) in 1982 without most of the common operating system functions, including virtual memory. Over time, however, programmers of personal computers encountered exactly the same programming problems as their predecessors in the 1950s, 1960s, and 1970s. That put pressure on the major PC operating system makers (Apple, Microsoft, and IBM) to add multiprogramming and virtual memory to their operating systems. These makers were able to respond positively because the major chip makers had not lost faith; Intel offered virtual memory and cache in its 80386 chip in 1985; Motorola did likewise in its 68020 chip. Apple offered multiprogramming in its MultiFinder and virtual memory in its System 6 operating system. Microsoft offered multiprogramming in Windows 3.1 and virtual memory in Windows 95. IBM offered multiprogramming and virtual memory in OS/2.

A similar pattern appeared in the early development of distributed-memory multicomputers beginning in the mid-1980s. These machines allowed for a large number of computers, sharing a high-speed interconnection network, to work concurrently on a single problem. Around 1985, Intel and N-Cube introduced the first hypercube machines consisting of 128 component microcomputers. Shortly thereafter, Thinking Machines produced the first commercial supercomputer of this genre, the Connection Machine, with as many as 65,536 component computer chips. These machines soon challenged the traditional supercomputer by offering the same aggregate processing speed at a lower cost [Denning and Tichy 1990]. Their designers initially eschewed virtual memory, believing that address translation and page swapping would seriously detract from the machine's performance. But they quickly encountered new programming problems having to do with synchronizing the **processes** on different computers and exchanging data among them. Without a common address space, their programmers had to pass data in messages. Message operations copy the same data three times: first from the sender's local memory to a local buffer, then across the network to a buffer in the receiver, and then to the receiver's local memory. The designers of these machines began to realize that virtual memory can reduce communication costs by as much as two thirds because it copies the data once at the time of reference. Tanenbaum [1995] describes a variety implementations under the topic of distributed shared memory.

The WWW, started in 1991 by Tim Berners-Lee, extends virtual memory to the world. The Web allows an author to embed, anywhere in a document, a **uniform resource locator** (URL), which is an Internet address of a file. The WWW appeals to many people because it replaces the traditional **processor-centered view** of computing with a **data-centered view** that sees computational processes as navigators in an immense space of shared objects. To avoid the problem of URLs becoming invalid when the object's owner moves it to a new machine, Kahn and Wilensky proposed that objects be named by globally unique handles; handles are translated with a two-level mapping scheme first into a URL, and then into the machine hosting the object [Kahn and Wilensky 1995]. This scheme recalls the Dennis-Van-Horn object-oriented virtual memory of the 1960s but now with worldwide, decentralized mapping systems. With its Java language, Sun Microsystems has extended WWW links to address programs as well as documents; when a Java interpreter encounters the URL of another Java program, it brings a copy of that program to the local machine and executes it [Gilder 1995]. These technologies, now seen as essential for the Internet, vindicate the view of the Multics designers in 1965 — that many large-scale computations will consist of many processes roaming a large space of shared objects.

From time to time over the past 50 years, various people have argued that virtual memory is not really necessary because advancing memory technology would soon permit us to have all the random-access main memory we could possibly want. Each new generation of users has discovered that its ambitions for processing, memory, and sharing led it to virtual memory. It is unlikely that today's predictions of the passing of virtual memory will prove to be any more reliable than similar predictions made in 1960, 1965, 1970, 1975, 1980, 1985, 1990, 1995, and 2000. Virtual memory accommodates essential patterns in the way people use computers to communicate and share information. It will still be used when we are all gone.

85.3   Structure of Virtual Memory

Figure 85.1 shows a system consisting of a processor, main memory, and secondary memory. Main memory is typically RAM and secondary memory disk. The access time of the RAM is on the order of 0.1 to 0.01 μs and of the disk 10 to 100 ms, giving speed ratios from 10^5 to 10^7. In the early computers, the speed ratio was on the order of 10^4. The penalty for referencing an item in secondary memory is even more severe than in the computers of the 1960s. This does not make virtual memory very attractive to those who believe that the primary purpose of virtual memory is to swap pages between main and secondary memory.

85.3.1   Paging

The computer hardware addresses bytes (8 bits of data). Data is usually stored and moved as blocks of contiguous bytes. In the simplest case, called paging, all the blocks are of the same size, say, 256 bytes. The main memory is divided similarly into blocks of **locations**, called **page frames**. Any page can be loaded in any page frame. We will consider first how a paged virtual memory works and then later examine variations that accommodate segments and other objects of variable sizes.

The set of addresses generatable by the processor is called the *address space*; all programs must be compiled to generate addresses within this space. Similarly, the set of main memory locations is called the *memory space*. The virtual memory system defines and maintains a dynamic map f from address to memory space, so that the hardware can quickly convert an address x to a memory location $y = f(x)$. The *page table* in which the map is stored has one entry for each page of the address space.

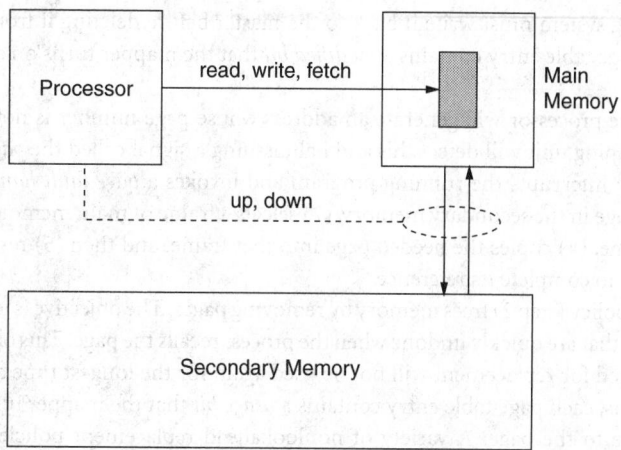

FIGURE 85.1 A processor executes a program from main memory. If the main memory is too small to hold the whole program, portions will be in secondary memory. Without virtual memory, the programmer would have to encode the commands to move blocks of data up and down the memory hierarchy.

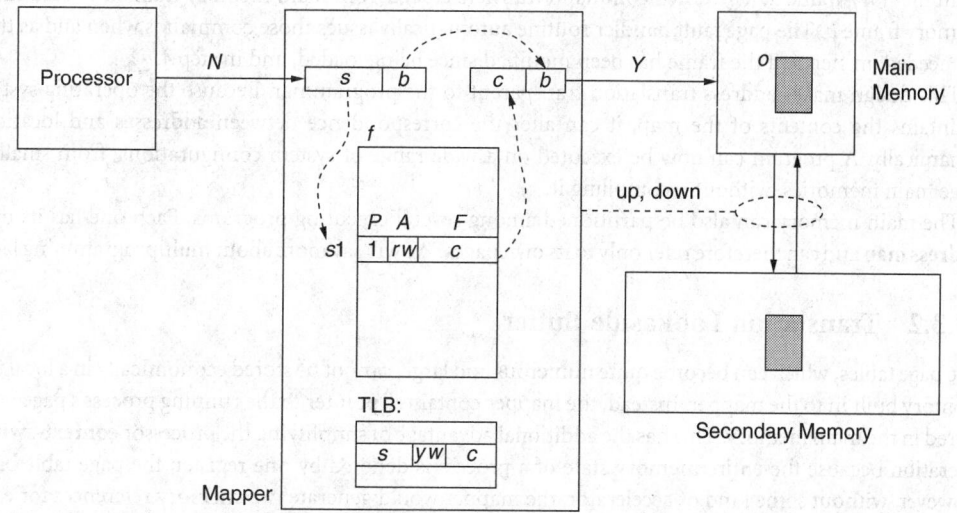

FIGURE 85.2 To convert a virtual address from the process into a real address for the main memory, the mapper refers to a page table f. A presence bit P indicates that the page is present in main memory; if so, the bits in the field F indicate which frame. An access code field A tells whether the page can be read (r) or written (w). A translation lookaside buffer (TLB) accelerates mapping by bypassing the page table on repeat access to pages.

An example will clarify the relationships among these elements. Suppose that the virtual addresses are 32 bits, main memory addresses are 24 bits, and the page size is $256 = 2^8$ bytes. In any 32-bit virtual address, the 8 low-order bits select a byte within a page and the 24 high-order bits select a page. Thus, the address space contains 2^{24} (about 16.8 million) pages and the memory space contains 2^{16} (about 65,000) page frames. The page table f contains 2^{24} entries, each of which is a 16-bit word. During address translation, the 24 high-order bits of the virtual address are replaced with the corresponding 16 bits from the mapping table. All of this is illustrated in Figure 85.2.

A master copy of all the pages of the address space resides in the secondary memory; a subset of those pages will be in main memory. Any page modified by the processor will cease to be an exact copy of the master; the operating system must write it back to the master before deleting it from main memory. To support this, each page table entry contains a *modified bit* that the mapper turns on automatically during any write to the page.

Sooner or later the processor will generate an address whose page number is not mapped (indicated by $P = 0$). The mapping unit will detect this and halt, issuing a signal called the *page fault*. In response, the operating system interrupts the running program and invokes a *page fault handler* routine that (1) locates the needed page in the secondary memory, (2) selects a frame of main memory to put that page in, (3) empties that frame, (4) copies the needed page into that frame, and then (5) restarts the interrupted program, allowing it to complete its reference.

The replacement policy (step 2) frees memory by removing pages. The objective is to minimize mistakes, that is, replacements that are quickly undone when the process recalls the page. This objective is met ideally when the page selected for replacement will not be used again for the longest time among all the loaded pages. To support this, each page table entry contains a *usage bit* that the mapper turns on automatically during any reference to the page. A variety of nonlookahead replacement policies have been studied extensively to see how close they come to this ideal in practice. When the memory space allocated to a process is fixed in size, this usually is least recently used (LRU); when space can vary, it is **working set** (WS) [Denning 1980].

The controller of the channel between main and secondary memory can accept commands of the form (*up, a, b*) and (*down, a, b*). The up command transfers a page from secondary memory frame b into main memory frame a. The down command transfers a page from main memory frame a to secondary memory frame b. The page fault handler routine automatically issues those commands when and as they are needed: in step 3, if the frame has been modified since being loaded, and in step 4.

This design makes address translation transparent to the programmer. Because the operating system maintains the contents of the map, it can alter the correspondence between addresses and locations dynamically. A program can now be executed on a wide range of system configurations, from small to large main memories, without recompiling it.

The main memory can also be partitioned among several executing programs. Each one has its own **address map** and can therefore refer only to its own pages. We will say more about multiprogramming later.

85.3.2 Translation Lookaside Buffer

The page tables, which can become quite numerous and large, cannot be stored economically in a local fast memory built in to the mapper. Instead, the mapper contains a pointer to the running process's page table stored in the main memory. This has the additional advantage of simplifying the processor **context-switch** operation because the entire memory state of a process is denoted by one register, the page table base. However, without some kind of accelerator, the mapper would generate two memory references for each virtual address, running the program at half-speed.

Virtual memory mappers use a small cache, called a **translation lookaside buffer** (TLB), as this accelerator. The TLB is a high-speed hardware associative memory that holds a small number of most recently mapped *paths*. A path consists of a page number (more generally, an object number) and the corresponding memory location: $(a, f(a))$. If the TLB already contains the path being attempted, the mapper bypasses the table lookups. In practice, small TLBs (e.g., 64 or 128 cells) give high enough hit ratios that average mapping speeds are within 1% to 3% of main memory speeds [Hennessey and Patterson 1990]. The TLB is a powerful and cost-effective performance accelerator. Without it, address mapping would be intolerably slow.

85.3.3 Cache Memories

The caching principle used in the translation lookaside buffer was first proposed by Wilkes [1965] as a direct hardware method for speeding up memory accesses (Figure 85.3). The main memory and cache are divided into equal-size blocks. The cache memory, which is attached directly to the processor, holds a subset of the blocks of the main memory; the block stored in a cache slot is indicated in a tag register associated with the slot. When the processor generates address (a, b) — meaning byte b of block a — the

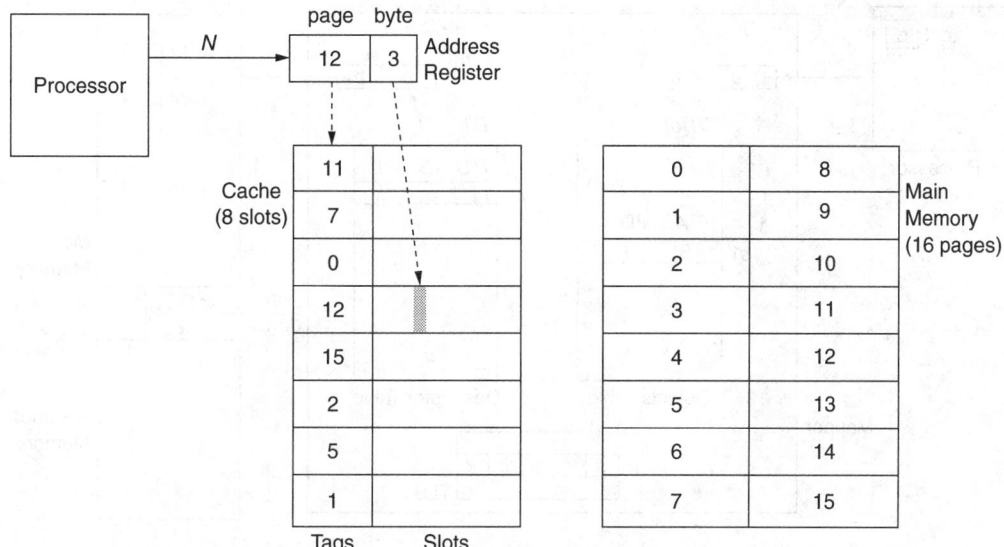

FIGURE 85.3 In a cache memory, the indices of pages (blocks) stored in cache slots are held in tag registers. The address hardware searches the tag registers in parallel for a match on the addressed page, and then uses the remaining address bits to select a byte of that page. The search can be made faster by dividing the tags into 2^m sets, using the m low-order bits of the block number to select the set, and restricting the parallel search to that set. (The figure is drawn for $m = 0$.) This partitions the blocks equally among the sets and thereby limits the number of slots into which a given block may be placed. In the worst case, when 2^m equals the number of cache slots, the set size is 1 and each block can be loaded into one slot only.

addressing hardware searches all of the tag registers in parallel for a match on a. If there is a match, it addresses byte b of that slot. If not, the hardware copies block a into a slot, sets the slot's tag to a, and then addresses byte b of that slot. This is just like paging except that the page table is inverted — that is, the page numbers are the results obtained by looking up frame numbers. (Hennessey and Patterson [1990].)

Note how the caching principle appears twice in a virtual memory: once when a subset of address space is loaded into main memory, and again when a subset of address-paths are loaded into the TLB. Caching paths in the TLB enables fast address translation; caching pages in main memory enables fast program execution. The storage system will run at nearly full speed even when these caches are a fraction of their maximum potential size. Because the caching store is small compared to the total size of the data, its cost is fixed and controllable.

85.3.4 Object-Oriented Virtual Memory

Many computing environments offer abstractions and functions that require virtual addressing but are not easily accommodated by paging. These include program objects such as arrays, procedures, structures, processes, message buffers, files, and directories; concurrent processes (threads) with varying **permissions** sharing the same address space; modular programs; and very large address spaces containing many objects shared among many users. The designers of early virtual memories anticipated these uses with segmented and capability-based virtual memories [Dennis 1965, Dennis and Van Horn 1966, Fabry 1974]. The resulting storage systems are called object-oriented virtual memories.

The earliest form of object-oriented virtual memory was the segmented address space. It appeared as a collection of named blocks (segments) of various sizes. Each segment was a container for a program object. In the Burroughs B5000 and later series, for example, the Algol compiler created program segments containing procedures and data segments containing array rows [Organick 1973]. The compiler generated virtual addresses of the form (s, b), meaning byte b of segment s. The size of each segment was explicitly

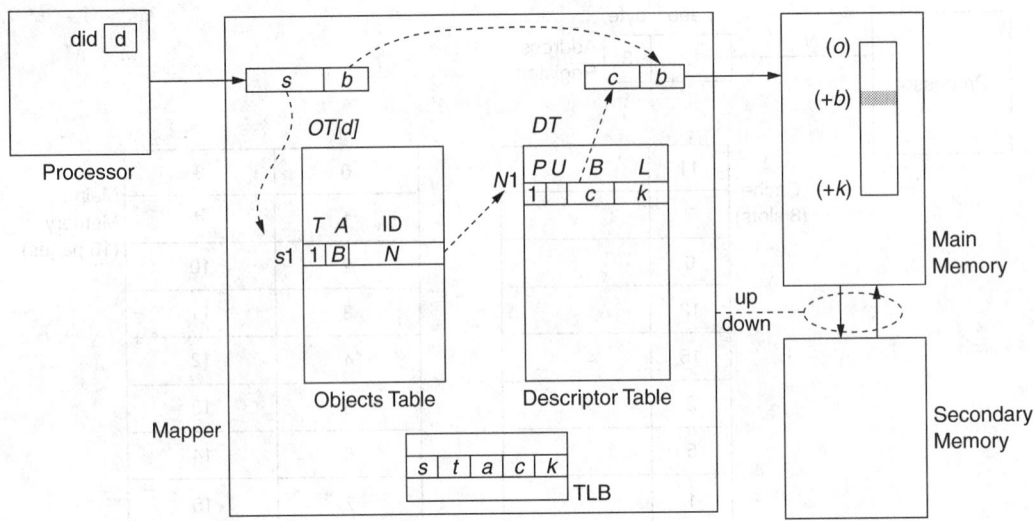

FIGURE 85.4   The mapping for object-oriented virtual memory operates in two levels. The first maps a local object number *s* of type *t* to an object unique identifier *x*, which in turn locates the object's descriptor in a descriptor table. An object's descriptor contains a presence bit *P* and a base-limit pair that designates a memory region of *k* bytes starting at address *c*. The byte number *b* must be less than *k*. A TLB that holds paths (*s, t, a, c, k*) accelerates the mapping. Different processes can have different segment numbers for the same segment: sharing of objects is possible without prior arrangements about the names each process will use internally. Shared objects can be relocated simply by updating the base address *c* in the descriptor.

recorded in the mapping table so that the mapper could reject out-of-bounds addresses *b*. Multics let the programmer, rather than the compiler, define the segments. In Multics PL/I, operands had symbolic two-part names (see earlier discussion); the operating system used a *linkage fault* to invoke a routine that mapped a symbolic (*S, X*) name-pair to its corresponding virtual address (*s, b*) as previously described [Organick 1972].

Multics also combined segmentation with paging by allowing each segment to be divided into pages. The offset into a segment was subdivided into a page and line number, and the page number mapped to a frame via the page table associated with the segment. This **two-level mapping** scheme made it easy to share segments: the first level of mapping would direct two users with different segment numbers to the same page table. The page table in turn mapped page numbers to page frames. Any changes made by the operating system to the page table would be seen immediately by all users sharing the segment. There would be no need for the operating system to locate all the users and update their segment tables.

Today's **object-oriented addressing** is the descendent of the Multics two-level segment address scheme. An object is stored in an address space segment. Each address space has an object table mapping object names to *descriptors*. A descriptor is a system object that contains all the information about the physical location of the object in the memory system. Two address spaces can assign different names to a shared object, which their object tables map to the object's single descriptor. Figure 85.4 shows the scheme. The mapper follows this routine:

```
processor places (s,b) in address register
  if ((t,a,c,k) = LOOKUP(s) undefined)
  then
    (t,a,x) := OT[d,s]
    (P,c,k) := DT[x]
    if (P = 0) then ADDRESS FAULT
    DT[x].U := 1
```

```
      LOAD(s,t,a,c,k)
   endif
if (b≥k) then BOUNDS FAULT
if (request not allowed by (t,a)) then PROTECTION FAULT
place c+b in memory address register
```

The operation **LOOKUP(s)** scans all the TLB cells in parallel and returns the contents of the cell whose key matches **s.** The operation **LOAD** replaces the least recently used cell of TLB with **(s,t,a,c,k).** The mapper sets the usage bit **U** to 1 whenever the entry is accessed so that the replacement algorithm can detect unused objects.

Object addressing creates a new problem of storage allocation in main memory: finding unused *holes* in which to place object-containing segments loaded from secondary memory. This is only a problem if the average size of an object is larger than about 10% of the memory size. In that case, the problem can be alleviated by paging each segment. In Figure 85.4, this would be implemented by adding a third level of mapping: the base address *c* points to a page table, and offset *b* is mapped to a frame in the same way as with pure paging. The combination of segmentation and paging is not often used.

85.3.5 Protection

One of the fundamental requirements of an operating system is that users cannot interfere with each other. By default, they cannot see each other's address spaces. The virtual memory system plays an integral role in meeting this requirement. The images of address spaces are always in disjoint regions of main memory. This feature of virtual memory is called **logical partitioning**.

With virtual memory, a processor can address only the objects listed in its object table, and only then in accord with the **access codes** of the objects. In effect, the operating system walls off each process, giving it no chance to read or write the private objects of any other process. This has important benefits for system reliability. Should a process run amok, it can damage only its own objects: a program crash does not imply a system crash. This benefit is so important that many systems use virtual memory even if they allocate enough main memory to hold a process' entire address space.

85.3.6 Multiprogramming

Multiprogramming is a mode of operation in which the main memory is partitioned among the address spaces of different processes. Each user can start multiple processes. Multiprogramming allows users to switch among active programs such as word processor, spreadsheet, and print spooler. It also provides a supply of programs ready to be resumed next by the operating system, thus maintaining high processor unit efficiency. As noted in Section 85.3.5, virtual memory confines each process to its assigned address space. Virtual memory provides an elegant and flexible way of partitioning a multiprogrammed memory.

Multiprogramming can be done with fixed or variable regions. Fixed regions are easier to implement but variable regions offer much better performance. With variable regions, the operating system can adjust the size of the region so that the rate of **address faults** stays within acceptable limits. The operating system can transfer space from processes with small memory needs to processes with large memory needs. Variable partitions often improve over fixed partitions even when the variation is random [Denning 1980]. System throughput will be near-optimal when the virtual memory guarantees each active process just enough space to hold its working set [Denning 1980].

85.3.7 Performance

Because most virtual memory systems represent objects with pages (one or more to an object), we can discuss performance of these systems in terms of decisions about loading, mapping, and replacing pages. Loading refers to the costs of bringing a page into main memory. Most systems load pages only on demand because no method of predicting future page use has proved reliable enough. However, because the time to load a page is usually 4 to 6 orders of magnitude higher than the time to access a byte on a loaded page, a few bad guesses are more expensive than simply waiting for page faults.

Mapping refers to the cost of tracing address paths through the mapping tables. As noted, the translation lookaside buffer reduces these costs to a small percentage of the main memory access time.

Replacement refers to the costs of removing pages from main memory. Each removal has a cost because a modified page must be copied back to the master copy in secondary storage and because most pages will be recalled into main memory by future page references. Because this is an ongoing process — removal followed by recall — replacement is the major determinant of virtual memory performance.

The main metrics of system performance are **throughput** and **response time**, and the main metric of memory usage is **space-time**. Throughput is measured as jobs or transactions completed per second, response time as the average number of seconds to complete a job or transaction, and space-time as the total number of byte (or page) seconds accumulated by a job while it holds main memory. Space-time is a like "rent" paid by a job for memory usage. If n jobs complete in T seconds, the throughput is $X = n/T$. When the total memory is M bytes, the total space-time available in the system is MT, and therefore the space-time per job is $Y = MT/n$. These definitions imply:

$$M = XY$$

This invariant relation is fundamental: it says that minimizing space-time is the same as maximizing throughput for a given amount of main memory.

When a program has a fixed memory allocation, the minimum space-time occurs when the addressing faults are minimum. The ideal policy — let us call it MIN — replaces the page that will not be used again for the longest time. Unfortunately, such a policy cannot be implemented because the operating system cannot know the future reference pattern of a program. Among the implementable policies, LRU attempts to implement the MIN rule by predicting that time until next reference to an object is same as time since last reference. Although LRU is not as good as MIN, it has been found to be quite robust over a range of programs, typically doing as well or better than other nonlookahead policies, such as first-in-first-out (FIFO). LRU is often used in caches.

If we remove the constraints that memory size is fixed and that replacements occur only at page-fault times, we can do better than MIN. The ideal policy — let us call it variable-space MIN (VMIN) — operates as follows [Prieve and Fabry 1974]. After each object reference, VMIN looks ahead to the moment of the next reference to that page; if the time to that reference exceeds a threshold T, VMIN immediately replaces the page; otherwise, it retains the page in memory until that next reference. As T gets larger, VMIN retains more pages and generates fewer page faults. Thus, the parameter T gives a way to trade off the amount of memory used against the amount of paging generated. The reason no other policy can do as well is that VMIN minimizes space-time at every reference: VMIN retains a page only if the cost of recovering it by page-fault at the next reference exceeds the cost of retaining it.

Although we cannot implement VMIN because we cannot predict future page references, we can implement a good approximation. Simply define a process's **working set** as the set of pages referenced in a window of size T looking *backwards* from the current time. The **working-set (WS) replacement policy** removes a page as soon as it leaves the working set. Under WS replcement, a page will cause a page fault at its next reference only if the time since previous reference exceeds T. In other words, WS replacement generates exactly the same faults as VMIN. The difference between WS and VMIN is due solely to WS's retaining pages for T seconds after last use when VMIN would replace those pages. A long line of experimental studies in the 1970s showed that there is not much one can do about clipping off the memory "overshoot" of WS relative to VMIN. WS is about as good as one can do in a replacement policy [Denning 1980]. The working-set policy was first defined in 1967 [Denning 1968, 1970, 1980].

The WS replacement policy works especially well in a multiprogrammed system. Using a common window size T, the system measures every program's working set dynamically. The scheduler admits waiting processes to the active state, one at a time, until the available memory space is filled with working sets. The window size can be adjusted empirically until it maximizes system throughput. System throughput

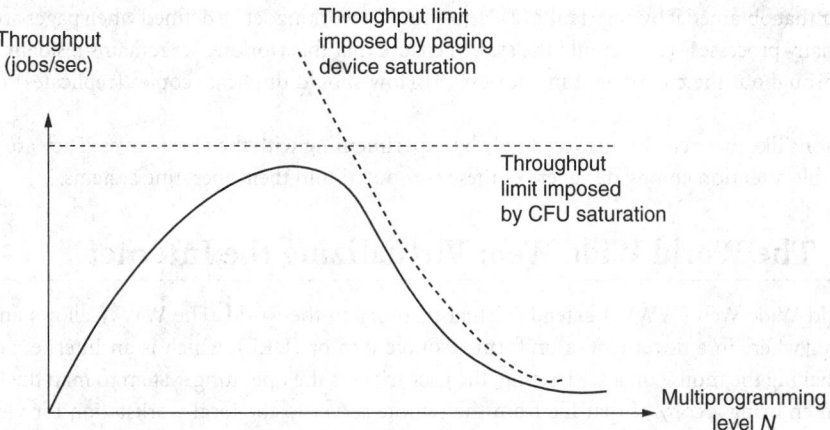

FIGURE 85.5 The system throughput is depicted as a function of the multiprogramming level N, which is the number of active programs among which main memory is partitioned. When N is too large, each program has so little space that it is forced toward a high paging rate. This makes the paging device the bottleneck, which slows down the system, producing thrashing. The ideal load control dynamically adjusts N to be constantly near the peak throughput.

may be improved further, but only by 5 to 10% at most, by measuring each running process with its own private window size [Denning 1980].

Many early systems using multiprogrammed virtual memory attempted to extend the LRU policy, which works very well in fixed partitions, by lumping all pages in main memory into a single global program managed by LRU. This strategy does not have the built-in load control of the working set policy: the scheduler can keep on activating more programs without limit. Each activation reduces the average space available to each program and increases the paging rate. These policies were therefore subject to thrashing (Figure 85.5). Thrashing can be avoided by limiting the multiprogramming level either by a fixed limit or by a working-set policy. The working-set policy generally leads to more stable performance with higher throughput.

85.4 Distributed Shared Memory

Starting in the mid-1980s, Sequent, Intel, Thinking Machines, N-Cube, and then later IBM, Cray, Kendall Square, and a few others introduced commercial multicomputers. These machines allowed for a large number of computers, sharing a high-speed interconnection network, to work concurrently on a single problem. They soon began to challenge the traditional supercomputer by offering the same aggregate processing speed at a lower cost [Denning and Tichy 1990]. But they also introduced a host of new programming problems having to do with synchronizing the processes on the different computers and exchanging data among them. Because these machines offered no common address space among all of the component computers, their programmers could share data only by copying it between machines. Not only does this complicate the job of programming the machine, but it also causes high message overhead.

It is possible to merge the address spaces of the component machines into a single global address space. The mapping mechanism uses a standard memory reference for a local object, or a message requesting data from the machine containing a nonlocal object. The resulting virtual memory is often referred to as **distributed shared memory** [Tanenbaum 1995].

The performance problems of distributed share memory are different because the other computers do not form a memory hierarchy, but are instead peers at the same level of memory hierarchy. Typical questions include: (1) Should a page be moved to the computer that most recently referenced it? Should it be moved only to the computer with the highest density of recent references to it? Should it be left in the computer that obtained it by page fault? (2) How should a working set be defined when pages are shareable among many processes? How should the system ensure that this working set remains resident even while spread throughout the component memories? (3) How should duplicate copies (replicates) of pages be treated?

Questions like these can be answered only by experimenting with the alternatives. They are subjects of considerable attention among designers of these computers and their operating systems.

85.5 The World Wide Web: Virtualizing the Internet

The World Wide Web (WWW) extends virtual memory to the world. The WWW allows an author to embed, anywhere in a document, a uniform resource locator (URL), which is an Internet address of a file. By clicking the mouse on a URL string, the user triggers the operating system to map the URL to the file and then bring a copy of that file from the remote server to the local workstation for viewing. The WWW appeals to many people because it replaces the traditional processor-centered view of computing with a data-centered view that sees computational processes as navigators in an enormous space of shared objects.

A URL is invalidated when the object's owner moves or renames the object. To overcome this problem, Kahn and Wilensky [1995] have proposed a scheme that refers to mobile objects by location-independent handles and, with special servers, tracks the correspondence between handles and object locations. Their method is functionally similar to that described in Figure 85.4: first, it maps a URL to a handle and then it maps the handle to the Internet location of the object. Unlike Figure 85.4, however, their method does not rely on central databases to store the mapping information.

The WWW is being extended to programs as well as documents. Sun Microsystems has taken the lead with its Java language. The URL of a Java program can be embedded in another program; exercising the link brings the Java program to a local interpreter, which executes it. The Java interpreter is encapsulated so that imported programs cannot access local objects other than those given it as parameters. Java programs organized in this way are called applets.

85.6 Conclusion

Virtual memory systems are used to meet one or more of the following needs:

1. *Automatic storage allocation.* Solving the overlay problem that arises when a program exceeds the size of the computational store available to it. Also includes the problems of relocation and partitioning arising with multiprogramming.

2. *Protection.* Each process is given access to a limited set of objects, its protection domain. The operating system enforces the rights granted in a protection domain by restricting references to the memory regions in which objects are stored and by permitting only the types of reference stated for each object (e.g., read, write, or apply a function). These constraints are easily checked by the hardware in parallel with the main computation. These same principles are being used for efficient implementations of object-oriented programs.

3. *Modular programs.* Programmers should be able to combine separately compiled, reusable, and shareable components into programs without prior arrangements about anything other than interfaces, and without having to link the components manually into an address space.

4. *Object-oriented programs.* Programmers should be able to define managers of classes of objects and be assured that only the manager can access and modify the internal structures of objects [Myers 1982]. Objects should be freely shareable and reusable throughout a distributed system (Chase et al. 1994, Tanenbaum 1995). (This is an extension of the modular programming objective.)

5. *Data-centered programming.* Computations in the World Wide Web tend to consist of many processes navigating through a space of shared, mobile objects. Objects can be bound to a computation only on demand.

6. *Parallel computations on multicomputers.* Scalable algorithms that can be configured at runtime for any number of processors are essential to mastery of highly parallel computations on multicomputers. Virtual memory joins the memories of the component machines into a single address space and reduces communication costs by eliminating some of the copying inherent in message passing.

Virtual memory, once the subject of intense controversy, is now so ordinary that few people think much about it. That this has happened is one of the engineering triumphs of the computer age. Virtual memory accommodates essential patterns in the way people use computers.

Defining Terms

Access control: A means of allowing access to an object based on the type of access sought, the accessor's privileges, and the owner's wishes.

Address fault: An error that halts the mapper when it cannot locate a referenced object in main memory; it invokes an interrupt, whose handler corrects the condition by loading the missing object.

Address map: A table that associates an object (or page) number with the main memory locations containing the object.

Address space: The set of all addresses that a processor can issue while processing a program.

Bounds fault: An error that halts the mapper when it detects that the offset requested into an object exceeds the object's size; it invokes an interrupt that terminates the program.

Capability: A systemwide unique identifier for an object; the bits of a capability are protected from alteration.

Context-switch: An operation that switches the CPU from one process to another, by saving all of the CPU registers for the first and replacing them with the CPU registers for the second.

CPU: Central processing unit, or processor.

Data-centered view: A view of computing that emphasizes navigation of many concurrent processes within a large space of objects.

Handle: A systemwide unique identifier for an object, like a capability without the system guarantee of integrity.

Location: A memory register with its own address.

Logical partitioning: A property of virtual memory whereby the address spaces of different jobs are mapped into disjoint regions of memory.

Main memory: The highest level of the memory hierarchy; all CPU memory references are directed to main memory; CPU can access objects only when they are loaded in main memory.

Memory hierarchy: A system of memory devices of different speeds and capacities; allows for trading off between capacity and speed, and between volatility and persistence.

Memory space: The set of all hardware addresses of memory locations in RAM available to a given address space.

Modular programming: Programs are divided into parts that can be shared, reused, and recompiled without affecting other parts of the system as long as the interfaces to modules are unchanged.

Object-oriented addressing: A form of virtual addressing in which object numbers are mapped to memory regions and internal object references are mapped to offsets within an object's memory region.

Object-oriented programming: A form of programming in which data is organized into classes of objects, each with a specific set of functions that can be applied to the objects.

Page: a fixed size unit of storage and transfer in a memory hierarchy.

Page frame: A contiguous block of memory locations used to hold a page.

Paging: A method of virtual memory in which address space and memory space are paged.

Partition: A division of memory space into disjoint subsets of pages for each address space.

PC: Personal computer.

Permissions: Access rights granted by an object's owner and represented as bits in the object's access code.

Process: An abstraction of the execution of a program, usually represented as the sequence of values of its CPU state as the program traces through its intruction sequence.

Processor-centered view: A view of computing that emphasizes the work of a processor.

Protection fault: An error condition detected by the address mapper when the type of request is not permitted by the object's access code.

RAM: Random access memory.

Response time: The time from when a command is submitted to a computer until the computer responds with the result.

RISC: Reduced instruction set computer (e.g., PowerPC, Sun SPARC, DEC Alpha, MIPS).

Secondary memory: Lower, large-capacity level of a memory hierarchy, usually a set of disks.

Segmentation: An approach to virtual memory when the mapped objects were variable-size memory regions rather than fixed-size pages; superseded by object-oriented addressing.

Slave memory: A hardware cache attached to a CPU, enabling fast access to recently used pages and lowering traffic on the CPU-to-main-memory bus.

Space-time: The accumulated product of the amount of memory and the amount of time used by a process.

Thrashing: A condition of performance collapse in a multiprogramming system when the number of active programs gets too large.

Throughput: The number of jobs (or transactions) per second completed by a computer system.

TLB: Translation lookaside buffer, a cache that holds the most recently followed address paths in the mapper.

Two-level map: A two-tiered mapping mapping scheme; the upper tier converts local object numbers into system unique handles, and the second tier converts handles to the memory regions containing the objects. Essential for sharing.

URL: Uniform resource locator (in the WWW).

Working set: The smallest subset of a program's pages that must be loaded into main memory to ensure acceptable processing efficiency; changes dynamically.

Working-set (WS) policy: A memory allocation strategy that regulates the amount of main memory allocated to a process, so that the process is guaranteed a minimum level of processing efficiency.

World Wide Web (WWW): A set of servers in the Internet and an access protocol that permits fetching documents by following hypertext links on demand.

References

Berners-Lee, T. 1996. The Web Maestro. *Technology Review*, July.

Chase, J. S., Levy, H. M., Feeley, M. J., and Lazowska, E. D. 1994. Sharing and protection in a single-address-space operating system. *ACM TOCS*, 12(4):271–307.

Denning, P. J. 1968. Thrashing: its causes and prevention, pp. 915–922. *Proc. AFIPS FJCC 33*.

Denning, P. J. 1970. Virtual memory. *Comput. Surv.*, 2(3):153–189.

Denning, P. J. 1976. Fault tolerant operating systems. *Comput. Surv.*, 8(3).

Denning, P. J. 1980. Working sets past and present. *IEEE Trans. on Software Eng.*, SE-6(1):64–84.

Denning, P. J. and Tichy, W. F. 1990. Highly parallel computation. *Science*, 250:1217–1222.

Dennis, J. B. 1965. Segmentation and the design of multiprogrammed computer systems. *J. ACM*, 12(4):589–602.

Dennis, J. B. and Van Horn, E. 1966. Programming semantics for multiprogrammed computations. *ACM Commun.*, 9(3):143–155.

Fabry, R. S. 1974. Capability-based addressing. *ACM Commun.*, 17(7):403–412.

Fotheringham, J. 1961. Dynamic storage allocation in the Atlas computer, including an automatic use of a backing store. *ACM Commun.*, 4(10):435–436.

Gilder, G. 1995. The coming software shift. *Forbes ASAP*, Aug. 5.

Hennessey, J. and Patterson, D. 1990. *Computer Architecture: A Quantitative Approach*. Morgan-Kaufmann.

Kahn, R. and Wilensky, R. 1995. A framework for distributed object services. Technical Note 95-01, Corporation for National Research Initiutives, Reston. VA. See also www.handle.net.

Kilburn, T., Edwards, D. B. G., Lanigan, M. J., and Sumner, F. H. 1962. One-level storage system. *IRE Trans.*, EC-11(2):223–235.

Myers, G. J. 1982. *Advances in Computer Architecture*, 2nd ed. Wiley, New York.

Organick, E. I. 1972. *The Multics System: An Examination of Its Structure*. MIT Press, Cambridge, MA.

Organick, E. I. 1973. *Computer System Organization: The B5700/B6700 System*. Academic Press, New York.

Prieve, B. and Fabry, R. 1974. VMIN: an optimal variable space page replacement algorithm. *ACM Commun.*, 19(5):295–297.

Sayre, D. 1969. Is automatic folding of programs efficient enough to displace manual? *ACM Commun.*, 12(12):656–660.

Tannenbaum, A. S. 1995. *Distributed Operating Systems*. Prentice–Hall, Englewood Cliffs, NJ.

Wilkes, M. V. 1965. Slave memories and dynamic storage allocation. *IEEE Trans.*, EC-14:270–271.

Wilkes, M. V. 1975. *Time Sharing Computer Systems*, 3rd ed. Elsevier/North-Holland.

Wilkes, M. V. and Needham, R. 1979. *The Cambridge CAP Computer and Its Operating System*. North-Holland.

86

Secondary Storage and Filesystems

86.1 Introduction 86-1
86.2 Secondary Storage Devices 86-2
    Magnetic Disks • Redundant Array of Inexpensive Disks (RAID) • CD-ROM Disks • Tapes
86.3 Filesystems 86-6
    Directory Structure • Describing a File on Disk • Filesystem Input/Output • Disk Space Management • Log-Based Systems • Versioning Systems

Marshall Kirk McKusick
Consultant

86.1 Introduction

The memory on a computer is organized into a hierarchy of storage [Smith, 1981]. This storage ranges from small and fast to large and slow. Figure 86.1 shows a typical hierarchy. It is composed of two main parts: the primary store and the secondary store.

The main components of this hierarchy include:

1. The first level of the primary store is the *cache memory*. It is often contained on the same chip as the central processing unit (CPU), or on other nearby chips that can be connected to the CPU with a minimum of delay. Because it must be able to run at close to the speed of the CPU, with access times of as little as a few nanoseconds, cache memory is typically small, rarely exceeding a few megabytes (Mbytes). The cache is never used for permanent storage; it holds values that are actively being processed by the CPU.

2. The second level of the primary store is the *main memory* on the computer. It currently runs with access times of 6 to 7 microseconds that may delay a CPU by 5 to 100 instruction cycles. The size of main memory ranges from a few hundred Mbytes up to several hundred gigabytes (Gbytes). Like the cache, main memory is not used for permanent storage; it holds the active part of running programs. Inactive parts of running programs are swapped out of the main memory to disk when the main memory becomes full. Thus, the size of a program is not constrained to the size of the main memory.

3. The first level of the secondary store is usually built from one or more *disk drives*. These disk drives are usually connected directly to the computer, although they may be located across a fast network on a central storage server. Disks are used for intermediate to long-term data storage. Access time for a fast disk is currently about 1 millisecond; thus, a CPU that needs to access data that is on disk will have to wait thousands of instruction cycles. Modern multitasking operating systems will suspend a program that awaits a disk access and run another program. Some time after the disk access has completed, the program that requested the data will begin to run again.

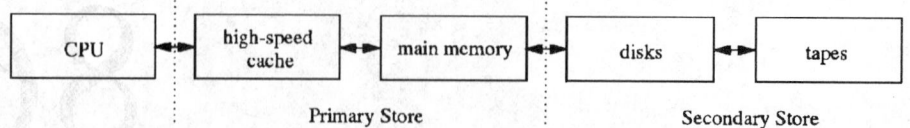

FIGURE 86.1   Computer memory hierarchy.

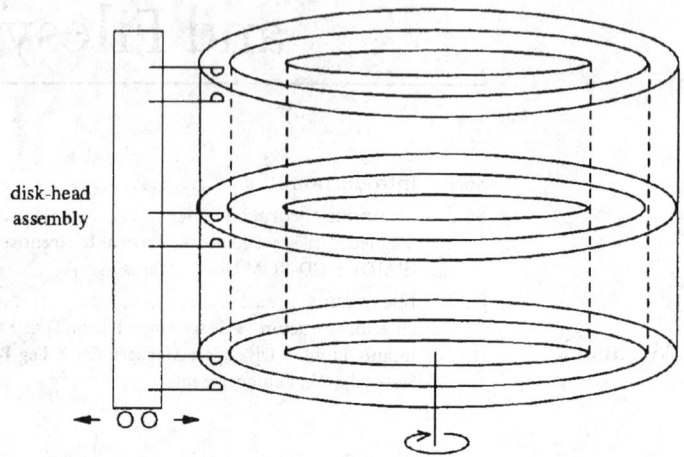

FIGURE 86.2   Construction of a magnetic disk.

4. The second level of the secondary store consists of *tape drives*. The tape drives are used for archival and backup storage. The access time to get to the start of and begin reading a file stored on a robotically managed tape system is several minutes. If a human operator must get involved, the access time takes longer. Historically, actively running programs directly manipulated data on tapes. Today, most applications arrange to have data read from tapes onto disk before beginning to access that data. Tapes are used primarily to archive data that is not currently accessed.

This chapter is concerned with the secondary part of the storage hierarchy; Chapter 19 and Chapter 108 discuss primary storage and its management. In particular, disks can be used as a temporary store when the main memory becomes full. This chapter considers disks solely from the perspective of their use as long-term storage media. The first half of this chapter discusses the hardware used to support secondary storage. The second half of the chapter discusses filesystem software used to access and manage secondary storage.

86.2   Secondary Storage Devices

Many types of hardware are used to support secondary storage. This section describes the most commonly used devices — magnetic disks, compact disk–read-only memory (CD-ROM) disks, and various types of tape devices.

86.2.1   Magnetic Disks

Disks are the most frequently used form of secondary storage. The most common type of disk is the magnetic disk [White, 1980]. Figure 86.2 shows the construction of a typical magnetic disk.

A disk is built from one or more metal platters coated on both sides with an iron oxide compound. The platters are mounted on a spindle connected to a motor that spins them. Data is read and written on the platters by magnetic heads mounted in a head assembly that runs on a track that enables the heads to be moved between the inner and outer edges of the platter. The operation of moving the head assembly is referred to as a *seek*. Separate heads scan the top and bottom of each platter. Each platter is divided into a set of concentric circles referred to as *tracks*. Each track is subdivided into fixed-size blocks referred to as *sectors*. The set of all tracks that are accessible from a single seek position are referred to as a *cylinder*.

Moving a head to any particular location on the disk to read or write data is a two step process. First, the head assembly must be moved to the desired cylinder. Once the head assembly is located at the correct cylinder, the head for the requested track is enabled. The second step, called *rotational delay*, is to wait for the desired sector on the disk to rotate under the head. Reading is done by the head by measuring the polarization of the iron oxide. Writing is done by sending a current through the head to generate a magnetic field that polarizes the iron oxide.

The access time for a disk is a function of the seek time plus the rotational delay. If the entire contents of a chunk of data can be stored in the same cylinder as the current head position, then it can be accessed with no seek delay. If the data cannot be placed entirely within the same cylinder, it should be placed on nearby cylinders because seek time is a function of distance. Rotational delay can be minimized by placing data on contiguous sectors within a track. A disk can only read or write data from one head at a time; however, switching between heads is almost instantaneous. If the contents of a chunk of data can be stored contiguously on several tracks within a cylinder, it can be accessed without any rotational delay because the disk can switch between heads at the end of each track.

Early disk technology used platters up to a meter in diameter spinning at a few hundred revolutions per minute. Modern technology has been steadily shrinking the size of the platters so that today they are only a few centimeters in diameter. The smaller platters have allowed the head assemblies to shrink in size and mass. Because the lighter head assemblies can be moved more quickly and have shorter distances to travel, the seek times of modern disks are a hundredth that of early disks. The heads do not touch the disk surface; instead, they float on a cushion of air a few microns above the surface. The air cushion pushed ahead of the heads creates friction that must be dissipated as heat. Because the amount of friction rises by the cube of the speed of the platter, even a small increase in rotational speed causes the generation of a lot more heat. Over time, the speed of rotation has increased much more slowly than the seek speed. Access times on early disks were dominated by seek times while in modern disks, the rotational delay is the dominating factor.

Data is transferred between the main memory and the disk by a disk controller. Most disk controllers transfer data between the disk and main memory using direct memory access. The CPU issues a command specifying the range of main memory addresses to be written or read and the disk sectors to or from which they should be transferred. The controller is responsible for issuing a seek command to the disk if necessary, waiting for the desired sectors to rotate under the head, transferring the data between the specified main memory locations, and interrupting the CPU to signal that the transfer has been completed.

To improve performance, disk controllers often provide a *track cache*, which holds the contents of the track containing the currently requested sector. For example, consider a file that fits on one track of a disk. Suppose the disk controller has been requested to read the first part of the file, but when the seek completes, the head is sitting over the middle of the file half a rotation away from its beginning. Instead of waiting for the disk to rotate the beginning of the file into position, the controller immediately begins reading the track into its cache. As the beginning of the file passes under the head, it is transferred into the main memory as requested and the CPU is notified of its completion. Following the end of the requested data transfer, if there are no further requests awaiting the controller, it reads the remaining quarter of the track into its cache. Because files are often read sequentially, the controller will probably be asked to read the next quarter of the file. Instead of having to wait for the disk to rotate back into position, it can satisfy this request from its cache. Thus, after one disk revolution, the controller can produce any part of the file with no rotational delay. Because the controller can transfer data from its cache to main memory much faster than data is delivered by the disk, the disk cache produces an even greater performance gain than just that measured by eliminating rotational delay.

Using a cache to speed up disk writes is a bit more difficult. If the controller completes a seek half a rotation ahead of the spot that requests the write, there is nothing useful that it can do; it must wait for the requested position to rotate into place. Normally, the controller waits until the write completes before issuing a completion interrupt to the CPU. Typically, a millisecond or so of CPU processing occurs before it issues the next write request. If the next write request contiguously follows the previous write, then the disk head will have just missed the start of the block and will incur nearly an entire rotation's delay reaching the correct starting point. When writing a large contiguous file, the controller will consistently be delayed by a full rotation on each block, which leads to poor throughput. One approach to correcting this problem is to transfer the requested block into the controller cache and then issue the completion interrupt before the block has been placed entirely on the disk. Here, the CPU can prepare and issue the next request for the controller while the previous block is being written. With the new request in hand, the controller can start its transfer to the disk without loss of a revolution.

This approach has a serious failing if the controller does not have a non-volatile cache. Many applications such as databases depend on the completion interrupt to let them know that critical data such as a transaction log has been stored in a location that will survive a system failure. If the data is only in the volatile controller cache and the system fails before the cache is written to disk, then the database will be unable to recover. Thus, early completion interrupts must be used only if the controller cache uses non-volatile memory and has software to restart it and write out any incomplete blocks after a system failure.

A much better alternative is to use a technique called tag queuing. Here, several requests are given to the disk controller, each of which is identified with a unique tag. When a request has been written to the disk, an interrupt is generated which presents the tag of the completed transfer to the system. Tag queuing allows the disk to write contiguously without the need to depend on premature and possibly incomplete I/O. Thus, applications can reliably know that their data is on stable store. Unfortunately, tag queuing is available only on high-end SCSI disks and is not generally available on the cheaper and more ubiquitous IDA disks usually found on personal computers.

The capacity of disks has been rising steadily; single disks today hold hundreds of Gbytes of data. Unfortunately, disks are still able to transfer data from only one location at a time. As the amount of data on the disk grows, this serialized access has become more and more of a bottleneck. To compensate, some systems deliberately use smaller capacity disks to increase the total number of disks on the system, which allows more parallel data access.

86.2.2   Redundant Array of Inexpensive Disks (RAID)

The biggest problem with magnetic disks is dealing with their inevitable failures. Because the heads fly only microns above the disk surface and because of the need to dissipate the heat that this small gap produces, even the best-designed disks have a lifetime of only 5 to 7 years. On a large file server containing 50 to 100 disks, a disk failure is expected every few weeks. Traditionally, recovery from such failures were handled by use of a tape backup system. Every few weeks a complete copy of each disk would be made on tape. Each day, a backup would be made of everything that had changed since the last complete copy had been made. When a disk failed, it would be replaced with a new disk; then the contents of the complete copy, followed by all the daily changes, would be made to complete the recovery process.

There are three problems with this approach:

1. *Capacity limits.* As the capacity of disks has increased, the amount of data that must be backed up on tape has exploded. The large server above would have several terabytes (Tbytes) of disk space connected to it. Even with the high-density tapes available today, it would take hundreds of tapes to back it up. The server would require several tape drives running continuously just to keep up with the backups.
2. *Data loss.* At best, most systems can only schedule backups once per day. If the disk crashes shortly before it is scheduled for its daily backup, everything that was done that day will be lost. For many businesses, losing even a day's work is unacceptable.

3. *Recovery delay*. Replacing a disk and recovering its contents from backup tapes takes several hours. During the recovery period, the disk is completely unavailable. This recovery delay is often unacceptable for time-critical applications.

Two approaches have been taken to avoid these problems. The first of these approaches is a brute-force solution called *mirroring*. The number of disks on the system is doubled. Disks are paired off and each pair keeps a copy of its partner. Thus, each time an application does a write, the changed data is written to both disks in the pair. Reads can be done from either disk because they both contain the same data. If one disk in the pair fails, the other disk can continue servicing requests without interruption. When the failed disk is replaced, its initial contents are copied in from its operating partner. Although it may take an hour or two to do the replacement and contents copy, users are unaware of the delay because they are running from the remaining good drive. Additionally, no data is lost because there is no dependence on backup tapes for recovery. Mirroring has traditionally been used for time- or business-critical data such as that handled by banking and airline reservation systems where the extra cost can be justified.

The second approach to avoiding the tape backup problem is to collect several disks together and use one of them to store a parity of the others. Such an organization is referred to as a Redundant Array of Inexpensive Disks (RAID) [Patterson et al., 1988; Chen et al., 1994]. A typical RAID cluster will contain five disks. Four of the disks contain data and the fifth contains a parity of the data on the other four. Each time data is written to any of the other four disks, a new parity must be computed and written to the fifth disk. In practice, the parity is not stored entirely on one disk as it would become the bottleneck when trying to write to one of the other four disks. Instead, each disk in a five-disk RAID cluster would be divided so that 20% stores parity and 80% stores data that would be covered by parity on the other four disks.

Recovery from disk failure in a RAID cluster is not as transparent as it is with mirroring. Access to the RAID cluster can continue, but at about half of its regular access rate while the broken disk is replaced and rebuilt. The replacement disk is initialized by reading the other four disks in the cluster and computing what value should be put onto the new disk. In data communications, parity can be used to detect errors, but not to correct them. That is because in data communications, the receiver does not know which bit is in error. Parity can be used for error correction for a RAID cluster because the cluster knows which disk failed. Thus, it can recompute the correct value for the failed disk using the data on the other drives. Failure recovery on a RAID cluster typically takes one to two hours.

Once the recovery is complete, the cluster returns to the state it was in just before the disk failed. Thus, RAID clusters solve two of the three tape backups problems. They avoid the need for daily backups and they avoid losing data when they fail. While access to the data is slow during the recovery period, that period is typically about half the time required for a tape recovery. A RAID cluster is considerably cheaper than a mirroring strategy because there is only a 25% redundancy of hardware rather than a 100% redundancy. Thus, RAID cluster usage is increasing in less time-critical environments.

Even with RAID clusters, the need for tape backups is not completely eliminated. Full backups need to be taken and stored off-site for recovery if a major catastrophe such as a fire destroys the disks in a machine room. Tapes are also needed to recover from user errors where an important file is accidentally deleted. Neither of these problems can be handled by RAID clusters.

Mirroring can handle the catastrophic failure if the mirror disks are kept in physically separate locations. For genuine safety, the mirrors should be in different buildings that are several miles apart. The communication costs of the high-bandwidth network connection that is required usually makes a long-distance mirroring solution unacceptably expensive. And, distance mirroring does not provide recovery from user errors where an important file is accidentally deleted.

86.2.3 CD-ROM Disks

The ubiquitous CD-ROM plays a small but important role in computer systems today [Asthana, 1995]. Its primary use is as a software distribution medium where it has replaced tapes and floppy disks. It has the benefits of being cheap, reliable, and easily mass-produced. It has the random access features of a magnetic disk that makes it much more convenient than a tape. Because the software is often used directly from the CD-ROM rather than being loaded onto the system disk, large software packages can be used on systems that are otherwise short of disk space.

The CD-ROM disks are more compact to store than tapes. They are expected to hold data reliably for 50 to 100 years, compared to tapes, which can only hold data reliably for 5 to 8 years.

86.2.4 Tapes

Tapes remain the most commonly used form of fourth level storage. Early tape technology used 12-inch reels of half-inch tape that stored a little over 100 Mbytes of data. Current tape technology uses DLT cartridges that store hundreds of Gbytes of data. The rule of thumb has been that the largest tapes hold about the same amount of data as the largest disks.

Data transfer to and from tapes tends to be slower than data transfer to and from disks. Random access to tapes is much slower than disks. Even modern DLT drives take about a minute to seek from one end of a tape to the other. The big benefit of tapes is that they are a tenth the cost of disks per Mbyte of storage. Also, by installing a robotic tape system with a capacity of several hundred tapes, it is possible to create a file store capable of storing 100 Tbytes of data. While the access time to the data may be a minute or two, it is far cheaper than storing a similar amount of data on disk.

In practice, tape store is generally used as the final repository for data. Recently accessed data is stored on disk where it is more readily available. When the disks become full, the least recently accessed data is copied to tape and deleted from the disk. If it is later needed, it is reloaded onto the disk, displacing other less recently accessed data. To maintain reasonable access times, most systems arrange to have enough disk space to keep data disk resident for at least a month.

86.3   Filesystems

Most applications that users run on their computer do not write data directly onto the disk. The operating system provides a filesystem that organizes the data into files. The filesystem is responsible for deciding where the file contents should be placed on the disk. The filesystem provides several important services:

- *Protection*. Most filesystems allow users to control access to their files. At a minimum, they can restrict access to themselves, a defined group of other users, or all other users of the system.

- *Organization*. Data in each file can be manipulated independently of data in other files. Data can be added or deleted from one file without affecting the data contained in other files. In particular, users need not be concerned that one file might run into another file on the disk as they would if they were directly placing the data on the disk themselves.

- *Multiple access*. Modern filesystems allow multiple files to be accessed at the same time and even allow separate programs to consistently access the same file. Consistent access means that if one application writes a file while another is reading the same part of the file, the reader will see the file either before the write has been started or after it has finished, but not in a partially written state.

- *Space management*. The filesystem tracks the used and free space on the disk. When any file in the filesystem needs to grow, the filesystem finds an appropriate-sized piece of free space and allocates it to the file. As long as there is free space left on the disk, the filesystem is prepared to allocate it to any file within the filesystem that wants to grow or to create a new file and allocate the space to that new file.

In addition, a filesystem is expected to optimize the use of disk bandwidth. Thus, it must not only find space to allocate to files, but it must also try to find space that is contiguous or at least rotationally close together to minimize the time that it takes to read and write the file. Issues of space management are discussed below.

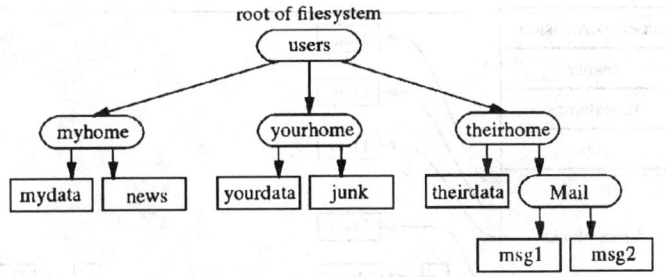

FIGURE 86.3   A set of files and directories.

86.3.1   Directory Structure

Most filesystems allow files to be grouped together into directories. These directories can then be further grouped together into other directories. The files and directories are usually grouped together in a tree hierarchy; Figure 86.3 shows a typical filesystem hierarchy. The rounded boxes represent directories; the square boxes represent files. The top of the tree is referred to as the *filesystem root*. Files are accessed by giving the set of names of directories from the root of the tree down to the desired file, separated by forward slashes; this name is called the *pathname*. For example, access to the file *mydata* in Figure 86.3 would use the pathname */users/myhome/mydata* [IEEE, 1994].

Many filesystems maintain a current working directory for the user. Instead of having to always specify a file by its complete pathname, the system keeps track of which directory the user is currently referencing, and does all filename translation relative to that directory. The user initially specifies a current directory using a complete pathname. Using the filesystem shown in Figure 86.3, the user might request that the current directory be set to */users/myhome*. Thereafter, the reference to the file *mydata* can be used without specifying its entire path because it is resident in the current directory.

86.3.2   Describing a File on Disk

To allow both multiple file allocation and random access, most systems uses a data structure similar to that shown in Figure 86.4 to describe the contents of a file. This structure includes:

- Access permission for the file
- The file's owner
- The time the file was last read and written
- The size of the file in bytes

Notably missing in this structure is the filename. Filenames are usually maintained in directories rather than in this structure because a file may have many names, or links, and the name of a file may be large (often up to 255 bytes in length).

The file structure also contains an array of pointers to the blocks in the file. The system can convert from a *logical block* number to a *physical block* number by indexing into this array using the logical block number. A null array entry shows that no block has been allocated, and will cause a block of zeros to be returned on a read. On a write of such an entry, a new block is allocated, the array entry is updated with the new block number, and the data is written on that physical block of the disk.

Because the filesystem must support many files and most files that it stores are small, the file structure has a small array of pointers for efficient use of space. The first few array entries are allocated in the file structure itself. For typical filesystems, these array entries allow the first 100 Kbytes of data to be located directly using a simple indexed lookup.

For somewhat larger files, Figure 86.4 shows how the file structure contains a pointer to a single *indirect block* of pointers to data blocks. To find the hundredth logical block of a file, the system first fetches the

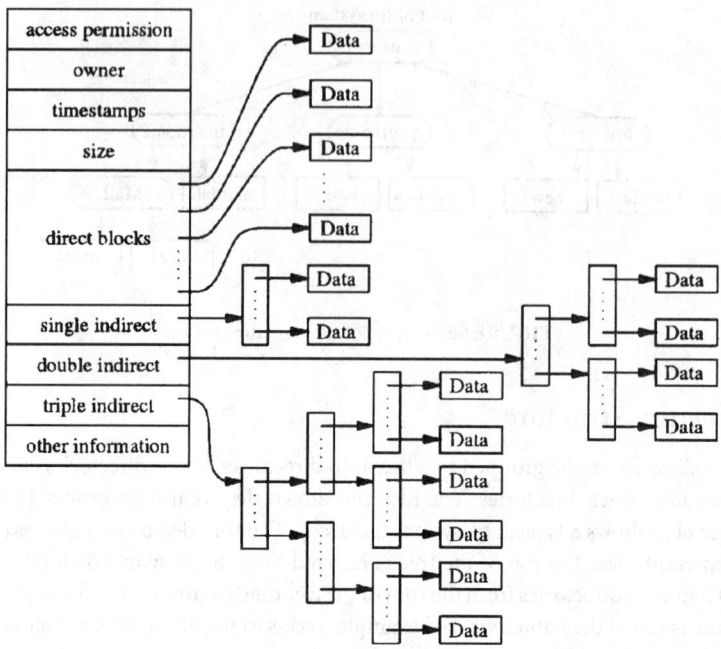

FIGURE 86.4　Extensible data structure used to describe a file.

block identified by the indirect pointer, then indexes into it by 100 minus the number of direct pointers, and fetches that data block.

For files that are larger than a few Mbytes, this single indirect block is eventually exhausted. These files must resort to using a double indirect block, which is a pointer to a block of pointers to pointers to data blocks. For files of multiple Gbytes, the system uses a triple indirect block, which contains three levels of pointers leading to the data block.

Although indirect blocks appear to increase the number of disk accesses required to reach a block of data, the overhead for this transfer is typically much lower. Most filesystems maintain a memory-based cache of recently read disk blocks. The first time that a block of indirect pointers is needed, it is brought into the filesystem cache. Further accesses to the indirect pointers find the block already resident in memory; thus, they require only a single disk access to reach the data.

The filesystem handles the allocation of new blocks to files as they grow. Simple filesystem implementations, such as those used by early microcomputer systems, allocate files contiguously, one after the next, until the files reach the end of the disk. As files are removed, gaps occur. To reuse this freed space, the system must compact the disk to move all the free space to the end. Files can be created only one at a time; to increase the size of a file (other than the last one on the disk), it must be copied to the end and then expanded. For the more complex file structure just described, the locations of the data blocks in each file are given by its block pointers. Although the filesystem may cluster the blocks of a file to improve I/O performance, the file structure can reference blocks scattered anywhere throughout the disk. Thus, multiple files can be written simultaneously, and all the disk space can be used without the need for compaction.

86.3.3　Filesystem Input/Output

The filesystem implementation converts the file from the user abstraction as an array of bytes to the structure imposed by the underlying physical medium. Consider a typical medium of a magnetic disk with fixed-size sectoring. Although the user may wish to write a single byte to a file, the disk supports reading

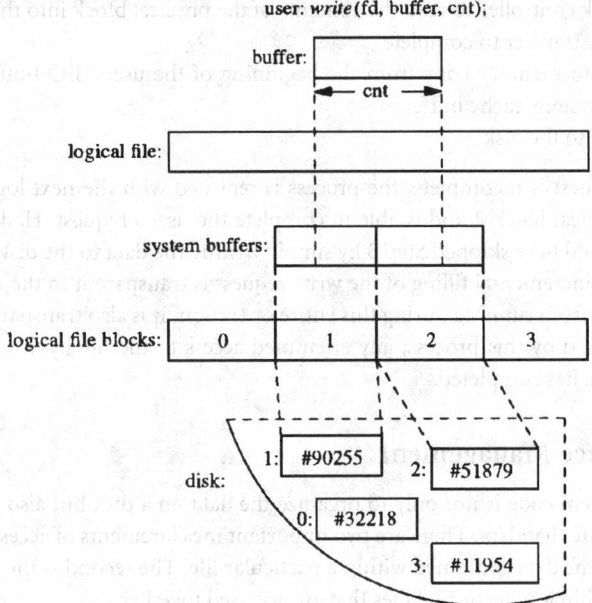

FIGURE 86.5 The block I/O system.

and writing only in multiples of sectors. Here, the system must read the sector containing the byte to be modified, replace the affected byte, and write the sector back to the disk. This operation of converting a random access within an array of bytes to reads and writes of disk sectors is called *block I/O*.

First, the system breaks the user's request into a set of operations to be done on logical blocks of the file. Logical blocks describe block-sized pieces of a file. The system calculates the logical blocks by dividing the array of bytes into filesystem-sized pieces. Thus, if a filesystem's block size is 8192 bytes, logical block 0 would contain bytes 0 to 8191, logical block 1 would contain bytes 8192 to 16383, etc.

Figure 86.5 shows the flow of information and work required to access the file on the disk. The abstraction shown to the user is an array of bytes. These bytes are collectively described by a file descriptor that refers to some location in the array. The user can request a write operation on the file by presenting the system with a pointer to a buffer, with a request for some number of bytes to be written. Figure 86.5 shows that the requested data need not be aligned with the beginning or end of a disk sector. Further, the size of the request is not constrained to a single disk sector. In the example shown, the user has requested data to be written to parts of logical blocks 1 and 2.

The data in each logical block is stored in a physical block on the disk. A physical block is the location on the disk to which the system maps a logical block. A physical disk block is constructed from one or more contiguous sectors. For a disk with 512-byte sectors, an 8192-byte filesystem block would be built from 16 contiguous sectors. Although the contents of a logical block are contiguous on disk, the logical blocks of the file need not be laid out contiguously.

Returning to our example in Figure 86.5, we now know which logical blocks are to be updated. Because the disk can transfer data only in multiples of sectors, the filesystem must first arrange to read in the data for any part of the block that is to be left unchanged. The system must arrange an intermediate staging area for the transfer. This staging is done through one or more *system-cache buffers*.

In our example, the user wishes to modify data in logical blocks 1 and 2. The operation iterates over five steps:

1. Allocate a system-cache buffer
2. Determine the location of the corresponding physical block on the disk

3. Request the disk controller to read the contents of the physical block into the system-cache buffer and wait for the transfer to complete
4. Do a memory-to-memory copy from the beginning of the user's I/O buffer to the appropriate portion of the system-cache buffer
5. Write the block to the disk

Because the user's request is incomplete, the process is repeated with the next logical block of the file. The system fetches logical block 2 and is able to complete the user's request. Had an entire block been written, the system could have skipped Step 3 by simply writing the data to the disk without first reading the old contents. This incremental filling of the write request is transparent to the user's process because that process is blocked from running during this entire operation. It is also transparent to other processes because the file is locked by this process; any attempted access to the file by any other process will be blocked until this write has completed.

86.3.4   Disk Space Management

The role of the filesystem code is not only to organize the data on a disk but also to minimize the time it takes to read and write that data. There are two important measurements of access time. The first is the time it takes to access the data contained within a particular file. The second is the time it takes to access the data contained within a collection of files that are accessed together.

Examples of collections of files accessed together include files that are collected together in a spool area such as those destined to be sent to a printer or those being batched to be sent as a collection of electronic mail messages. Another example might be a collection of files that make up the components of a spreadsheet, a document, or a program. Directories provide a strong clue that a set of files will be accessed together. Thus, many filesystems will try to allocate all the files contained within a directory in close physical proximity to each other on the disk. If they are accessed together, the disk will not need to make long or multiple seeks to get from one to the next.

Most filesystems put their greatest effort into optimizing the layout of individual files. The default assumption is that the files will be accessed sequentially, starting at their beginning. Certain files, such as those that contain a large database, may have randomly scattered accesses. The optimal layout for such files depends on their access patterns, which may be known only to the database or application program, or may not be known at all. For such files, the filesystem may allow the database or application to direct the layout of the file. Alternatively, it may simply layout the file sequentially, and assume that the database or application will attempt to cluster related information within the file to minimize the time it takes to seek between file locations.

For a sequentially accessed file, the usual strategy is to allocate a contiguous piece of space on the disk on which it is stored. When a disk is first put into service, doing sequential allocation for files is easy. The filesystem has a large area of contiguous physical free space, and allocates pieces out of that space for each new file. Unfortunately, this allocation approach quickly uses up all the physically contiguous space. As old files are deleted, they return the space that they were using. However, this unused space will be randomly scattered throughout the disk. Eventually, the disk will become completely fragmented, with no large contiguous pieces of space remaining, thus making it impossible for the filesystem to allocate new files contiguously.

Filesystems often use defensive algorithms to reduce the rate of fragmentation of the free space on the disk. A key to reducing fragmentation is to observe that most of the files within a filesystem are small. When a new file is created, the filesystem can assume that it will be small. Instead of allocating its initial disk space in a large contiguous area of the disk, the file will be allocated in a smaller fragment that may have been freed when another small file was deallocated. If the file turns out to be small as expected, then it will nicely fit in the small space that it was initially allocated. If it continues to grow, then the filesystem can move it to a fragment left by a somewhat larger file. Only when the file grows large is it finally relocated to the large contiguous space that the filesystem has now been able to hoard. By only allowing large files to be allocated in the large contiguous space, the filesystem is better able to ensure that some large contiguous space will be available when it is needed.

Moving files around on the disk can be a potentially slow and expensive operation if the file must be read and rewritten every time the filesystem wants to move it. To reduce the relocation cost, the filesystem will attempt to defer writing the file until it has determined its final location on the disk. The deferral is done by holding blocks of the file in system-cache buffers until its size can be determined (see Figure 86.5). The steps involved in allocating space to a file are as follows:

- When the first block is written, a system-cache buffer is allocated, A single block-sized piece of free disk space is found, the address of that piece of disk space is assigned to the system-cache buffer, but the buffer is not written to the disk.

- If the file continues to grow, a second system-cache buffer is allocated. The filesystem finds a two block-sized piece of free space, frees the single block-sized piece of space originally assigned to the file, and assigns the addresses of the new, larger block to the two buffers. As before, neither buffer is written to the disk.

- This process continues until the file has grown to the maximum-sized block allowed (typically the size of a disk track) or the file has ceased to grow, at which point the buffers are written to their final destination.

If the system-cache buffers are needed for another purpose, or the application explicitly requests that the file be stored on disk (for example, it is a transaction log that must be in stable storage before the application can proceed), the system-cache buffers always have a location to which they can be written. The only implication of doing the write early is the loss of performance that comes from writing the data to disk more than once.

This algorithm has the additional benefit that it allows even slowly growing files to be written contiguously. If a file such as one holding accounting records grows at the rate of a few Kbytes per hour, it will be relocated to larger contiguous blocks periodically. Here, the relocation usually involves reading and rewriting the data, because the turnover in system-cache buffers causes them to be flushed before the new disk addresses for the data are assigned. Because the reallocation occurs only a few times per hour, the added I/O overhead does not adversely affect system performance.

To ensure successful layouts, a filesystem must have a quick method for finding contiguous and nearby free blocks on the disk and then allocating them. The most common data structure used for describing the free disk blocks is an array of bits. The blocks on the disk are sequentially numbered; the corresponding bit in the array is set to 1 if the block is being used and 0 if it is free. Block allocation involves setting the bits corresponding to the blocks being allocated; block deallocation involves clearing the bits corresponding to the blocks being freed. Finding other nearby blocks can be done by looking for 0 bits in the array near the location of the most recently allocated block in the file. Clusters of blocks can be identified by looking for strings of 0 bits in the array.

The free disk block array is large for a big disk. Exhaustive searches of the array (for example, when the disk is nearly full and there are few free blocks remaining) would slow filesystem performance unacceptably. Consequently, most filesystems maintain auxiliary data structures that summarize the contents of subranges of this bit array. Such summaries include the number of free blocks and the maximum-sized contiguous piece within each subrange of the bitmap. When looking for a block or a cluster of blocks, the filesystem first scans the summary information to find a subrange of the bitmap that has the necessary free space. Once it finds the needed space, the filesystem can narrow its search to that subrange of the bitmap rather than searching the whole space.

86.3.5 Log-Based Systems

Logging has long been used in database systems to provide recovery after a system failure [Date, 1995]. The database periodically does a *checkpoint* of its on-disk data structures to ensure that they are in a consistent state. Following the checkpoint, the database keeps a log of every change that it commits. The log is usually written serially onto a disk that is dedicated to the log. Thus, the log can be written quickly because the disk head is always within no more than a rotational delay from the next location to be written.

If the system fails, the log is used to do a *roll forward* operation on the database to bring it back to consistency. The roll forward works by going through the log and ensuring that all the changes that it lists are reflected in the on-disk database state. The time required to recover the database after a system failure is bounded by the frequency of checkpoints. By checkpointing the database and resetting the log every few minutes, the recovery time for a disk failure can be kept to a few minutes or less.

The idea of using a log to recover a database has also been applied to filesystems. Like a database, the filesystem periodically checkpoints its state. It then writes all changes after that point, both to its log and to the filesystem itself. When the log becomes full or a time limit is reached, the filesystem checkpoints itself again and resets its log. As with the database, the log is rolled forward after a system failure to ensure that all the changes made since the checkpoint are reflected in the filesystem. The log only needs to record modifications to the filesystem. While it has no effect on the speed with which data can be read from the filesystem, logging does introduce additional overhead when files are written. However, if the log is stored on a separate disk from the rest of the filesystem, writing to the log is seldom the limiting function on the speed with which data are placed into the filesystem.

Although a log provides fast recovery after a system failure, it does not provide the data recovery of mirroring or RAID that can recover from catastrophic disk failure. The reason that logging does not help with catastrophic disk failure is that data may be lost from parts of the disk that the filesystem thinks are stable and hence did not enter in the log.

A more recent filesystem design has taken the use of a log to its logical conclusion. Instead of using a log as an adjunct to an existing filesystem implementation, the entire filesystem is implemented as a log [Rosenblum and Ousterhout, 1992]. The log-structured filesystem is being used in commercial products [Wilkes et al., 1995]. The fundamental idea of a log-structured filesystem is to improve filesystem performance by storing all filesystem data in a single, contiguous log. A log-structured filesystem is optimized for writing, and no seek is required between writes, regardless of the file to which the writes belong. It is also optimized for reading files written in their entirety over a brief period of time (as is the norm in workstation environments), because the files are placed contiguously on disk. A log-structured filesystem places files created at the same time together on disk. If an application reads a set of files written at different times, the filesystem potentially will have to seek around on the disk to read them. Log-structured filesystems expect to avoid the need for many seeks when reading files because they assume that all files that are actively being read will reside in the system cache.

The underlying structure of a log-structured filesystem is that of a sequential, append-only log. The disk is statically partitioned into fixed-size contiguous segments that are generally 0.5 to 1 Mbyte. In ideal operation, a log-structured filesystem accumulates *dirty blocks* in memory. When enough blocks have been accumulated to fill a segment, they are written to the disk in a single, contiguous I/O operation. All writes to the disk are appended to the logical end of the log.

Although the log logically grows forever, portions of the log that have already been written must be made available periodically for reuse because the disk is not infinite in length. This process is called *cleaning*, and the utility that performs this reclamation is called the *cleaner*. The need for cleaning is the reason that the disk is logically divided into segments. Because the disk has reasonably large static areas, it is easy to segregate the portions of the disk that are currently being written from those that are currently being cleaned. The logical order of the log is not fixed, and the log can be viewed as a linked-list of segments, with segments being periodically cleaned, detached from their current position in the log, and reattached after the end of the log.

As a log-structured filesystem writes dirty data blocks to the logical end of the log (that is, into the next available segment), modified blocks will be written to the disk in locations different from those of the original blocks. This behavior is called a *no-overwrite policy*, and it is the responsibility of the cleaner to reclaim space resulting from deleted or rewritten blocks. Generally, the cleaner reclaims space in the filesystem by reading a segment, discarding dead blocks (blocks that belong to deleted files or that have been superseded by rewritten blocks), and rewriting any live blocks to the end of the log.

Cleaning must be done often enough that the filesystem does not fill up; however, the cleaner can have a devastating effect on performance. One study shows that cleaning segments while a log-structured filesystem is active (i.e., writing other segments) can result in a performance degradation of about 35 to 40% for some transaction-processing-oriented applications. This degradation is largely unaffected by how full the filesystem is; it occurs even when the filesystem is half empty [Seltzer et al., 1995]. Another study shows that typical workstation workloads can permit cleaning during disk idle periods without introducing any user-noticeable latency [Blackwell et al., 1995]. The effect of filesystem cleaning on performance is still hotly debated.

Like a conventional filesystem that has a log associated with it, a log-structured filesystem must periodically do a checkpoint that synchronizes the information on disk so that all disk data structures are completely consistent. The frequency with which checkpoints are done affects the time needed to recover the filesystem after system failure. The more frequently they are done, the shorter the time it takes to recover. Checkpoints must also be taken whenever an application requests that one of its files be moved to stable storage. For example, an editor will usually request that a new version of a file be moved to stable storage before it deletes the old copy of the file. A conventional filesystem must be checkpointed whenever its logging disk becomes full. In the absence of application-requested checkpoints, a log-structured filesystem is only required to checkpoint a segment between the time that it is last written and the time that it is cleaned.

Recovery after a system failure is handled by rolling the filesystem log forward from the last checkpoint. In a conventional filesystem, the changes listed in the log are applied to the filesystem data structures. In a log-structured filesystem, the filesystem is the log, so rolling it forward simply means discarding any incomplete operations.

86.3.6 Versioning Systems

Users often want to retain previous versions of files, such as released versions of programs or documents. Usually they use a revision control utility that maintains a database which stores and retrieves the selected versions. Most version control systems only keep a complete copy of the original file. Each successive version is then stored as a set of differences from the previous version. Some version control systems operate by storing each version in a separate file, allowing fast and easy access to older versions (because it is just a matter of finding and reading the desired version). However, this method of version control is wasteful of space, especially if the file being versioned is large and the changes between versions are minimal.

The whole file copy version of revision control can be done by the filesystem itself. Each time a file is opened for writing, the contents of the old file are saved instead of being overwritten. The old file is kept around until the filesystem runs out of space. When additional space is needed for new files, the user must either explicitly delete unneeded earlier versions of files or request the filesystem to reclaim the space from the oldest of the earlier file versions. Log-structured filesystems are particularly good at providing versioning because they never overwrite data in files. Older versions of files can be retrieved by going backward through the log. Space reclamation in a log-structure filesystem is done only by the cleaner process. Thus, an application can control when old versions are reclaimed by controlling when the cleaner process runs. The cleaner process can also be customized to skip over older versions of selected files that the user wishes to retain for a longer period of time.

The user interface for accessing older files may simply provide a list of different versions that are available and allow the user to specify the one that they want. Or it may be more sophisticated, allowing the user travel in time. For example, the user might be able to request all the files making up a particular program as they existed at the release date for the program.

Another variant of versioning involves taking a *snapshot* of the filesystem. Logically, the snapshot represents a frozen copy of the filesystem as it existed at the time that the snapshot was taken. Snapshots are typically compact, as they reference the filesystem on a copy-on-write basis. So, the snapshot only needs to make copies of the disk blocks that are modified. Thus, it is possible to take a snapshot of a filesystem every few hours during the day. Often, the snapshots are put in a location accessible to the users so that they can go retrieve older versions of their files without the intervention of a system administrator. Thus, if a user wrote a file in the morning and accidentally overwrote it in the afternoon, he could retrieve the original copy from the late-morning snapshot.

Defining Terms

Block I/O: The conversion of application reads and writes of records with arbitrary numbers of bytes into reads and writes that can be done based on the block size and alignment required by the underlying hardware.

Checkpoint: The writing of all modified data associated with a filesystem to stable storage (either non-volatile memory or the disk). A checkpoint ensures that all operations completed before the checkpoint will be recovered following a system failure.

Dirty blocks: In computer systems, modified. A system usually tracks whether or not an object has been modified — is dirty — because it needs to save the object's contents before reusing the space held by the object. For example, in the filesystem, a system-cache buffer is dirty if its contents have been modified. Dirty buffers must be written back to the disk before they are reused.

Filesystem root: The starting point for all absolute pathnames.

Indirect block: A filesystem data structure composed of an array of pointers to disk blocks used to locate the data blocks associated with a file.

Logging: Writing data to a file where existing data are never overwritten; the system thus modifies the file only by appending new data.

Logical block: The sequential fixed-size pieces of a file. The logical block associated with a given byte offset in a file is calculated by dividing the offset by the filesystem block size. For example, byte 20000 is located in the third logical block of a file residing on a filesystem with 8-Kbyte blocks.

No-overwrite policy: The filesystem never rewrites existing data in a file. New data is always written into a new location on the disk.

Physical block: The disk sector addresses associated with a logical block of a file. The filesystem finds the contents of a logical block in a file using the logical block number as an index into an indirect block to find the disk sector address holding the requested data.

Roll forward: Used to recover after a system failure. The operation of rerunning the update operations stored in a log file against a filesystem or database to bring it to a consistent state as of the last update completed in the log.

System-cache buffers: System memory used to hold recently used data. For example, in the filesystem, system-cache buffers are used to hold recently accessed disk blocks.

References

P. Asthana, "Superdense Optical Storage," *IEEE Spectrum*, 32(8), 25–31 (August 1995).

T. Blackwell, J. Harris, and M. Seltzer, "Heuristic Cleaning Algorithms in Log-Structured File Systems," *USENIX Association Conference Proceedings*, p. 277–288 (January 1995).

P. Chen, E. Lee, G. Garth, R. Katz, and D. Patterson, "RAID: High-Performance, Reliable Secondary Storage," *ACM Computing Surveys*, 26(2), 145–185 (June 1994).

C. J. Date, *An Introduction to Database Systems*, sixth ed., Addison-Wesley, Reading, MA (1995).

IEEE, *POSIX: Part 1: System Application Program Interface*, Institute of Electrical and Electronics Engineers, 324 East 47th Street, New York, (1994).

M. K. McKusick, M. J. Karels, K. Bostic, and J. S. Quarterman, *The Design and Implementation of the 4.4BSD Operating System*, Addison-Wesley, Reading, MA (1996).

D. Patterson, G. Garth, and R. Katz, "A Case for Redundant Arrays of Inexpensive Disks (RAID)," *SIGMOD Record*, 17(3), 109–116 (June 1988).

M. Rosenblum and J. Ousterhout, "The Design and Implementation of a Log-Structured File System," *ACM Transactions on Computer Systems*, 10(1), 26–52, Association for Computing Machinery (February 1992).

M. Seltzer, K. Smith, H. Balakrishnan, J. Chang, S. McMains, and V. Padmanabhan, "File System Logging Versus Clustering: A Performance Comparison," *USENIX Association Conference Proceedings*, p. 249–264 (January 1995).

A. Silberschatz and P. Galvin, *Operating System Concepts*, fourth ed., Addison-Wesley, Reading, MA (1994).

A. J. Smith, "Bibliography on File and I/O System Optimizations and Related Topics," *Operating Systems Review*, 14(4), 39–54 (October 1981).

R. M. White, "Disk Storage Technology," *Scientific American*, 243(2), 138–148 (August 1980).

J. Wilkes, R. Golding, C. Staelin, and T. Sullivan, "The HP AutoRAID Hierarchical Storage System," *ACM Operating System Review*, 29(5), 96–108 (December 1995).

Further Information

A good overview of filesystems can be found in Chapter 3 of Silberschatz and Galvin [1994]. Most operating systems today use filesystem designs similar to those found in McKusick et al. [1996]. This chapter summarizes information on file layout on disk described in Section 2 of Chapter 7; filesystem naming described in Section 3 of Chapter 7; the traditional disk space management described in Section 2 of Chapter 8; and a log-structured filesystem described in Section 3 of Chapter 8.

87

Overview of Distributed Operating Systems

87.1 Introduction 87-1
87.2 Survey of Distributed Systems Research 87-2
 Naming • Communication • Transactions • Group
 Management
87.3 Best Practice 87-13
 Failure Models • Naming • Communication • Binding
 • Transactions • Group Communication

Sape J. Mullender
Lucent Technologies

87.1 Introduction

The advent of distributed systems came hand in hand with that of workstations and personal computers. The presence of many computers interconnected by a network opened up several new possibilities:

1. Every user could have personal dedicated computing cycles, while the network still allowed sharing data or devices through centralized file servers, printers, etc.
2. Reliability could be increased by arranging for computers to take over from each other in the case of crashes.
3. Performance could be increased by allowing software to make use of many processors in parallel.
4. Systems could grow incrementally by adding computers one at a time.

These possibilities triggered research in the new field of *distributed and network operating systems*. The first projects started in the middle 1970s, but the bulk of activity in the area took place in the 1980s. Now, halfway through the 1990s, the activity in distributed systems research seems to be declining.

A distinction between distributed operating systems on the one hand and network operating systems on the other has sometimes been made. A network operating system is essentially a centralized operating system whose components have been distributed over multiple nodes, whereas a distributed system is one in which this distribution, combined with replication, plays a role in achieving fault tolerance as well.

The distribution of components of an operating system over multiple nodes requires splitting up the traditional operating system into its constituent components, leaving only a small amount of machine-dependent and resource-protecting code in a *microkernel* [Accetta et al. 1986, Mullender et al. 1990, Rozier et al. 1988]. Microkernels can thus, to some extent, be viewed as a consequence of introducing network operating system or distributing operating system functionality.

Another distinction is necessary between distributed systems and *parallel* systems. In parallel systems research, the focus is very much on completing computations in minimal time, by exploiting the presence of multiple processing nodes. Distributed systems also exploit parallelism, but there is at least as much concentration on fault tolerance.

Early research in distributed systems focused very much on functionality: better mechanisms for sharing, fault tolerance, communication, and parallelism. Later, attention shifted to keeping the same functionality while improving performance. The declining interest in distributed-systems research today may have a lot to do with the rapid increase in reliability of hardware components during the past 10 years.

87.2 Survey of Distributed Systems Research

In this section we survey the contributions that important projects have made for the distributed systems research community. Only a few of these projects have resulted in complete systems that are in use, but most of them have contributed from their key features to commercial operating systems. It is because of this that we do not survey the important projects one by one, but that, instead, we survey them area by area.

In the following sections we look at some of the important projects in the areas of naming, communication, transactions, and group communication. We do not discuss security in this chapter, even though fault tolerance properly implies security as well. The subject of security is in a separate chapter in this handbook. Research on distributed shared memory has also been left out: distributed shared memory (DSM) makes parallel programs run better, but does not contribute to fault tolerance — it does the opposite, if anything. DSM has been delegated to the parallel processing chapter.

87.2.1 Naming

87.2.1.1 Amoeba

Amoeba was a research project, initially only of the Vrije Universiteit, later also of CWI, the Centre for Mathematics and Computer Science, both in Amsterdam [Mullender 1985, Mullender et al. 1990, Mullender and Tanenbaum 1986, Tanenbaum et al. 1990]. Together with the V-system [Cheriton 1988], it was one of the first standalone distributed systems.

Amoeba has two levels of naming, which were both innovative. At the system's level, all objects are named using *capabilities*, which are managed in user space. A capability consists of four fields, as illustrated in Figure 87.1. The *service* field, also known as the service's *port*, identifies the service that manages the object. This field is used by the remote-operations protocol to deliver messages to a server process (see the communication subsection). The *object* field identifies the object to the service, and the *rights* (Rts) field indicates what operations the holder of the capability may carry out on the object. The *check* field prevents forging capabilities; it is calculated by the server, using a *secure hash* of object and rights fields, plus possibly a per-object secret *random number* maintained by the service.

Service ports are 48-b random numbers and, if you know a service's port, you can send messages to it. Services can be made private by keeping their ports secret.

Amoeba uses *request/reply* communication between *clients* and *servers*. A request contains a capability whose port names a service, while its remainder names an object maintained by that service. The Amoeba system finds a server for the service by broadcasting the port and waiting for location information from the servers (clients maintain a server-location cache to save broadcasts). When a server has been found, the request is sent to it; the server processes the request and returns a reply. Replies are addressed to the client's port.

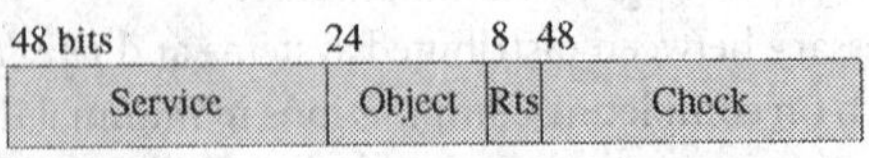

48 bits	24	8	48
Service	Object	Rts	Check

FIGURE 87.1   Layout of an Amoeba capability.

In Amoeba, ports and capabilities are the names for services and objects, respectively. At the operating-system's application programmer's interface (API), they are the only names supported. For operating system and application software development, these fixed-length names are quite convenient.

For human beings, a directory service is available which maps hierarchical path names onto capabilities. A directory entry consists of a name and a *list* of capabilities. Normally, these capabilities all refer to the same object, but carry different rights; they are not all equally powerful. Depending on the rights in the capability of a directory, a client will be allowed to retrieve subsets of the capabilities in its entries. A powerful capability to a directory allows a client to see most or all of the capabilities, a weak capability allows it to see only small or empty subsets of, presumably, weak capabilities.

87.2.1.2 DEC Global Name Server (GNS)

The DEC global name server (GNS) [Lampson 1986] is an example of a design that was intended to be scalable to worldwide size and millions of nodes.

The members of the design team (Andrew Birrell, Butler Lampson, Roger Needham, and Michael Schroeder) had been involved with Grapevine, Xerox's name server [Birrell et al. 1982]. One could say that Grapevine succumbed under its own success: its popularity caused it to grow to a size for which it had not been designed, revealing several deficiencies in its scalability [Birrell et al. 1984].

GNS was designed to scale both in size and, as it were, in time. Scaling in size means that the naming database must be able to grow to billions of entries stored at millions of nodes. Scaling in time means that the name space can cope with large structural changes at any level in the naming tree (as, for instance, the unification of East and West Germany and the split up of Czechoslovakia).

The name space of GNS is necessarily hierarchical; no other structure could scale to the desired size. The hierarchy is not necessarily geographically determined; GNS has no problems coping with multinational organizations. Each directory entry is essentially a list of attribute {*name, value*} pairs. An entry **/nl/utwente/cs/sape** could describe a user and have attributes such as **mailbox** and **certificate**. A user's public-key certificate could then be retrieved under the name **/nl/utwente/cs/sape/certificate**.

Each directory has a unique *directory identifier* (DI). A directory is referred to by its parent via a *directory reference* (DR), which contains a DI. A *full name* in GNS consists of a DI and a *pathname*; the pathname is resolved starting at the directory named by the DI.

The naming database is organized such that every system can retrieve the directories it controls and all parent directories up to the global root by their DIs. However, the flat — and *pure** — name space of DIs alone cannot be used to find directories elsewhere in the naming tree. For this purpose, a DR not only contains the DI of the named directory, but also a list of servers that store copies of the directory. Server names are stored as full names also, and, if one is not careful, looking them up can result in endless lookup loops.

The designers of GNS recognized that availability of the name server is of crucial importance. Directories that are essential for the operation of a system must be available locally. As a consequence, directories near the root of the naming tree will be very highly replicated indeed; the root will be replicated everywhere. With such high degrees of replication, consistent update is not possible.

GNS, therefore, defines a form of loose consistency that may be formulated as follows: "If the supply of updates stops, there will eventually be glorious consistency" [Needham 1993]. This is achieved by making sure that updates have the following properties:

1. Every update eventually reaches every replica.
2. Two updates can be applied in any order and yield identical results (the *set* of updates matters, not the order in which they are made).
3. Updates are *idempotent*: applying an update twice has the same effect as applying it only once.

To achieve property 1, updates are distributed among the copies of a directory by a *sweep algorithm*: A sweep operation visits every directory copy, collects a complete set of updates, and then writes this set

---

*Pure names are explained in a subsequent section on naming (Section 87.3).

back to every copy. Property 3 implies that directory replicas need not keep track of the identity of every update received.

Achieving property 2 is subtle: Normally, you cannot update a directory before it has been created, and so the order of directory-create and directory-update operations does seem to matter. This is solved by the following ruse: If you update an item and it, or one of the directories on its path, does not exist, then that directory is created and updated, but it is marked *absent*. When a create operation is performed, the item is created if it does not exist, and it is marked *present*. Queries will only find and traverse paths with *present* items.

Furthermore, updates are *time stamped* so that conflicting updates yield the result of the one with the highest time stamp. This also causes the desired idempotency.

DEC GNS was used initially within DEC only, but since it has been adopted by the Open Software Foundation as part of its Distributed Computing Environment, it is becoming more widespread.

87.2.1.3 Domain Name Server (DNS)

The Domain Name Server [RFC-1035, RFC-1034] is the world's most widely distributed name service. It is used to name and locate hosts, services, and mailboxes on the Internet. It uses a hierarchical name space and pathnames with components separated by dots. Like postal addresses, but unlike most computer-related naming hierarchies, DNS puts the highest level domain at the end to give names, such as **amstel.cs.utwente.nl.** for a machine called Amstel in the Computer-Science Department of the University of Twente in The Netherlands.

Pathnames (at most 255 characters) are composed of a sequence of labels (of at most 63 characters) separated by dots (the character " . "). The root of the name space has an empty name, and so full names end in a dot. Relative names are resolved starting at the current domain. Names that cannot be resolved relative to the current domain are resolved as full names (by appending an imaginary dot).

DNS was designed to scale with the number of hosts and users in the global Internet. Most of the data in the system is expected to change very slowly, but small subsets of the name space may change rapidly, e.g., on the order of seconds to minutes, and DNS should be capable of keeping up with those changes. In doing this, however, DNS depends on the rapidly changing subset being small; if it were large, the update traffic would saturate the Internet.

DNS has a worldwide distributed implementation. Obviously, not all servers can be trusted. Clients can indicate which name servers they trust and name servers in this trusted set are always consulted before any others. Just like GNS, DNS is also based on the assumption that old data are better than no data; consistency guarantees are sacrificed to obtain maximum availability.

Queries are first sent to one of the local name servers. Servers only contain subsets of the name space, and so most names can only be resolved by consulting several servers. Normally, servers do not query other servers on behalf of clients, but they pass the partially resolved query back to the client and let the client resolve the query by iterating over multiple servers. Thus, clients can build up a cache that can tell them which servers serve a particular subdomain of the name space.

Each domain has an administrator who maintains a master naming database. The master database also identifies the master databases of each of the subdomains. An *authoritative* answer to a query is obtained by consulting master databases. Servers not only maintain master databases, but they also cache information from other servers. The system allows administrators to configure their servers to cache certain information from other servers permanently. This mechanism allows names to be resolved even though the domain cannot be reached or the authoritative server is down.

The naming database is maintained by a set of servers. A program which is called a *resolver* queries one or more servers in order to resolve a name. Clients give their queries to a resolver; thus, they do not have to iterate over multiple servers themselves.

Each internal node or leaf node in the name space corresponds to a *resource set*, which is stored by one or more name servers. The name servers make no distinction between internal nodes and leaf nodes (but resolvers do). A resource set contains zero or more *resource records*. A resource record consists of a *type*, a *class*, a *time to live* (TTL), and *type*-dependent (and sometimes also *class*-dependent) *data*.

Types and classes are represented by 16 b. Types and classes are defined per domain (including all subdomains). A new global type or class can only be introduced by the administrator of the root domain. This is the Network Information Center (NIC).

Important types are **A** for host addresses, **CNAME** for naming aliases, **HINFO** for host information [central processing unit (CPU) and operating system], **MX** for mail-handling agents, and **NS** for the authoritative name server for a domain. There are more types. Classes can be used to distinguish between different (sub)networks.

The time to live entry in a resource record tells name servers and resolvers how long it is safe to cache the resource record. When a resource record is cached, the authoritative server is not consulted, and so updates are not seen until the TTL has expired. Administrators must choose the TTL to balance between the two evils of increased name server traffic caused by rapid expiry of caches and increased inconsistency caused by slow expiry.

Queries are for {*name, type*} pairs. To find a mail handler for **sape@cs.utwente.nl**, one posts a query for {**cs.utwente.nl, MX**}. This query will yield a list of mail handlers, for instance:

```
cs.utwente.nl               preference = 0,
                            mail exchanger = utrhcs.cs.utwente.nl
cs.utwente.nl               preference = 10,
                            mail exchanger = driene.student.utwente.nl
utrhcs.cs.utwente.nl        inet address = 130.89.10.247
driene.student.utwente.nl   inet address = 130.89.220.2
```

In this case, two possible mail handlers are produced, the first one being preferred over the second, and for each one, the **A**-type record is also produced — a useful optimization, since the Internet address will be needed to send the mail message.

The Domain Name Service must be one of the most heavily used distributed applications in the world (along with e-mail, the World Wide Web, and net news) and it works remarkably well. This is quite surprising given the sheer size of the worldwide DNS database today, and the unavoidable variations in professionality of the administrators. The reliability and scalability of DNS show off the skill of its designers.

87.2.1.4 Plan 9

The group that designed Unix in the late 1960s and early 1970s has built a new operating system named Plan 9 from Bell Laboratories. It is an elegant little system available on CD for academic and noncommercial use [Harcourt 1995]. The Unix philosophy of using the filename space for naming everything* has been preserved and developed further which resulted in a very elegant design.

In Plan 9, each server can export a name space. Clients can refer to objects maintained by such a server by name. These name spaces are hierarchical and singly rooted. A process can access the name spaces of several servers by *grafting* them onto its own name space. This grafting is called *mounting*. A *mount table* maintains which server name spaces are mounted where in the naming tree. This resembles the way in which Sun Network File System (NFS) servers can be mounted in Unix.

But where Unix maintains a single mount table per machine, Plan 9 can have a mount table per process. When a process is created, it normally inherits its parent's mount table and then shares it with its parent. However, processes can also inherit a *copy* of the mount table so that mount and unmount operations of the parent are not visible by the child and vice versa. They do this by starting a new *process group*; Plan 9 maintains a mount table per process group. Here, the analogy with Unix file descriptors is enlightening: normally, children inherit the open files from the parent, but, if the parent so chooses, it can modify the set of open files between *fork* and *exec* (using *close*, *open*, and *dup*, usually). This can be done with mount tables in an analogous way (using *newpgrp*, followed by *mount* and *unmount* operations).

---

*The addition of networking and environment variables to Unix has diluted this philosophy to some extent.

As a result, each process group can create its own private name space. A parent can, for instance, *encapsulate* a child process by mounting an encapsulation server in its root directory before starting it. The encapsulation server can monitor all of the child's file input/output (I/O) requests and name space operations and process (some of) them in the parent's name space. Such encapsulation servers can be very useful for debugging, collecting statistics on application file usage, or for checking out imported software for anomalies such as Trojan horses.

The handle that specifies the server to be mounted is a *connection* in Plan 9 (identified by a file descriptor). Connections to local servers are essentially *pipes*; connections to remote ones are network connections.

There is a standard protocol for accessing the objects maintained by a server. This protocol contains operations normally associated with files: *open*, *close*, *read*, *write*, *seek*, etc. However, they need not be files. The Domain Name Server in Plan 9, for instance, can present itself as a file system that allows users to open and read a file such as **com/bell-labs/plan 9**. The mouse server implements a file **mouse** which is conventionally mounted as **/dev/mouse** and can be read to give the position of the mouse.

The Plan 9 window system, known as 8 $1/2$,* illustrates the use of the Plan 9 name space very elegantly. The device drivers for screen, keyboard, and mouse are represented by files mounted in **/dev**: /dev/bitblt (for bit-blit operations to write to the screen), **/dev/keyboard** and **/dev /mouse**. The window manager, $8^1/2$, uses these devices. When $8^1/2$ creates a window, it forks the child process that runs in that window, and gives the process a new mount table (by creating a new process group for it); $8^1/2$ then mounts itself as a service onto the new process's **/dev** where it implements new versions of /dev/bitblt, **/dev/keyboard** and **/dev/mouse**. A process running in a window, thus, does not read from the hardware mouse directly, but from one synthesized by the window manager. The window manager only gives it the mouse clicks and movements related to a particular window. A happy consequence of this organization is that the window manager can run as a window in itself. X-window applications run on Plan 9, through an X server running in an $8^1/2$ window.**

Plan 9 does not support a global name space in the sense that all objects have the same name everywhere. In fact, it makes explicit use of the fact that different objects have the same name in different places (e.g., **/dev/mouse** in different windows). Rob Pike, a principal designer of Plan 9, claims that, even in global naming schemes, sharing only becomes practical if everybody adheres to certain naming conventions. Naming conventions allow you to find things in the name space by guessing. Plan 9 uses naming conventions explicitly; the user who mounts **/dev/mouse** as **/dev/keyboard** should not be surprised if certain things fail to work properly.

87.2.2 Communication

It is likely that the research into efficient communication for distributed systems in the late 1970s came as a reaction to the cumbersomeness of the standards being developed by the International Organization for Standardization and the Consultative Committee on International Telephony and Telegraphy (CCITT). In any case, the early 1980s saw a race by a number of research groups to develop the fastest protocols for supporting remote procedure call. Early participants in the race were the V systems group at Stanford, led by Cheriton [1988]; and the Amoeba group at the Vrije Universiteit Amsterdam, where van Renesse et al. [1988] made their record attempts for fastest remote procedure call (RPC).

In the mid-1980s, Schroeder and Burrows [1989] thoroughly analyzed the performance of the RPC implementation of DEC Systems Research Center's (SRC's) Firefly multiprocessor. This resulted in a significantly better understanding of the design issues for interprocess communication. Hutchinson and Peterson [1988] at the University of Arizona then designed an extremely flexible framework for building efficient protocol stacks, the *x*-kernel. This is now widely used by researchers in a large number of research systems.

---

*Also after a movie.

**It was noticed that X servers were not designed to deal gracefully with resize operations of what they perceive to be their screen.

87.2.2.1 The V System

The V system, during its heyday, ran on SUN workstations connected by an Ethernet. It is believed that the multicast capability of Ethernets inspired a communications infrastructure that made heavy use of it for delivering requests to replicated services.

V uses remote operations for all communication. The common form is a remote operation between two processes, but there is also a form where a request is *multicast* to a set of processes with multiple replies as a result. Requests can thus be addressed to individual processes or to *process groups*.

Processes and process groups in V share a name space; clients sending requests to a process group multicast the request to all its members. The reliability of the multicast is that of the underlying hardware multicast mechanism (i.e., the reliability of Ethernet multicast). A few of the servers, therefore, may not receive the request, but this is viewed as normal and replicated services must take this into account. One, several, or all of the servers can return a response; this is up to the service designer.

The request and reply messages in V consist of a 32-byte *fixed-size message* with an optional *data segment* that can be as large as 16 kilobytes. A minimal message consists of just the fixed-size message. This separation of data into two kinds is useful because it can prevent quite a lot of copying and allows *scatter-gather* operations to collect the data to be sent or to deliver the data to be received. Amoeba does this too.

In the V system, a request is sent and a reply is received with one operation, called *send*. A single system call thus suffices to complete a remote operation, which, in V, is called a *message transaction*. All other interactions with the operation system take place using message transactions.

87.2.2.2 Amoeba

Amoeba [Mullender 1985, Mullender et al. 1990, Tanenbaum et al. 1990] was designed and implemented as a general-purpose distributed operating system combining fault tolerance to high performance. Its interprocess-communication mechanisms were based on the exclusive use of *remote operations* for all communication: In a remote operation, one process, the client, sends a request to a *service*; exactly one of the service's server processes receives and processes the request and returns a reply.

Requests and replies in Amoeba consist of two parts, a parameter area and a body part, very similar to V. Messages can be of arbitrary length so that a wide range of service types can easily be realized. Also, Amoeba implemented *at-most-once* semantics for its remote operations. Under this regime, requests will normally be executed exactly once. When a failure occurs, however, no reply reaches the client, while the request may not have been carried out at all, or it may have been carried out partially, or even completely. A failure can be caused by a crash of the server process, a crash of the server host or operating system, or a network failure. Sometimes, for instance, when a server cannot be reached, it is possible to provide exact feedback to the client about whether or not its request was acted on, but in most cases a client must find out using indirect means.

In this respect, Amoeba differs from both SUN RPC and V: SUN RPC implements an *at-least-once* strategy in that it retransmits a request until it gets a reply; in the process, it is possible that the request will be executed partially or wholly multiple times. V multicasts a request to multiple servers, with the possibility of multiple, parallel executions.

SUN RPC and V message transactions have the advantage of achieving a higher probability of success, but applications have to use requests with *idempotent* semantics; that is, semantics where executing a request partially or wholly several times, followed by a final, complete execution has the same result as exactly one complete execution. Reading from a file is an example of an idempotent operation: as long as you get the data in the end, semantically, it does not matter if you tried in vain a couple of times.

Amoeba's at-most-once semantics also allows nonidempotent requests to be used (e.g., transfer $10 from my account to that of the Red Cross). The reason for putting so much emphasis on these subtle differences in semantics for different communication mechanisms is that one cannot move an application from a system that implements at-most-once to one that implements at-least-once with impunity. Program correctness can crucially depend on the difference.

Guaranteeing at-most-once behavior during communication with a replicated service requires some care: Suppose a request is sent to one of the servers, but no reply is received. The client may not retransmit to a different server, because the absence of a reply does not indicate that the request was not acted on (the server may have crashed after executing it). At the same time, Amoeba does have to deal with *migrating* server (and client) processes, and so, although retransmissions may not be sent to a different server, it is possible that they are sent to the same server on a different *host*.

Making this work requires individual servers to be named. Therefore, in addition to the *service port* (see preceding section), server processes also have a *unique port*. When a client initiates a remote operation, the system locates a server by broadcasting a *locate message* that contains the server port. Servers respond with a message containing their location and their unique port. After selecting a server (typically by using the first response to arrive), the system commits to that particular server for the duration of the remote operation. If the server migrates, the new location can be found by broadcasting a locate message containing the server's unique port.

Mappings from server port to unique port to current location are, of course, cached. When a host receives a message for a port no longer serviced by it, it returns a not-here reply which invalidates the cache. When a not-here response arrives for the initial transmission of a request, the client may try to locate another server; when it arrives for a retransmission or a control message, the client must try to relocate the server at another address.

87.2.2.3 Firefly Remote Procedure Call

The Firefly was a shared-memory multiprocessor workstation built at the DEC Systems Research Center in the early 1980s. Its operating system, Taos, implemented a remote-procedure-call mechanism that was used for all communication. When one takes the speed of its processors into account, Firefly RPC may have had the lowest latency implementation of its day. Its performance was the result of carefully crafting the interactions among application threads, kernel threads, and interrupt routines. Schroeder and Burrows [1989] measured and analyzed the performance and their report has provided useful insights into high-performance communication architectures for later generations.

Firefly RPC allows request and reply messages of at most one Ethernet packet in size. The packets must also contain user datagram protocol (UDP) and Internet protocol (IP) headers, so the payload could not exceed 1440 bytes. Remote procedure calls with more data were split up into a sequence of RPCs. To take advantage of the multiprocessor capability of the Firefly, it was common practice to carry out large-data transfers with multiple parallel threads making multiple single-packet RPCs in parallel. The resulting throughput is a significant fraction of the capacity of the Ethernet, but is achieved by loading five processors at each end almost to capacity. Amoeba, Sprite, and V achieved similar throughput on much less loaded uniprocessors (but with processors that ran three to five times faster).

In their paper, Schroeder and Burrows [1989] measured the performance of a *null RPC* (minimum-size request and reply) and of a *maxresult RPC* (minimum-size request, maximum-size reply). The roundtrip latencies for these were 2.66 ms and 6.35 ms, respectively. With four threads doing maxresult RPCs in parallel, the throughput is 4.65 Mb/s.

A remote procedure call can be separated into four activities, *marshalling*, *protocol processing*, *hardware processing*, and *synchronization*.

The code for marshalling parameters into and out of network packets is produced by a *stub compiler* and runs in the application address space. The time needed for marshalling depends on the complexity and size of the parameters to be marshalled. The latency of 6.35 ms for a maxresult RPC was achieved with a (Modula-2) **var array[0..1439] of char** parameter that took 550 μs to marshal; that is, some 10 percent of the RPC latency. For complex parameters of the same size, the marshalling time can be a multiple of this. As more and more 10-Mb/s local area networks (LANs) are now being replaced by much faster networks, the significance of marshalling times will increase. It is, thus, worthwhile to invest in well-designed RPC type systems that use carefully tuned stub compilers.

The protocol processing consists of filling in IP and UDP headers and calculating the UDP checksum. This costs between 50 and 450 μs, depending on packet size. UDP-checksum verification costs the same amount of time.

The hardware processing time can be divided in two parts: the time the driver takes to enqueue the packets and process the interrupt, and the time the hardware itself needs to transmit the packets. The driver time was some 240 μs, and the hardware time 210 μs for a minimum packet and 2880 μs for a maximum packet of 1500 bytes.

Finally, time is needed for synchronization: A user thread must be woken up when its data have arrived and, on the Firefly, an interprocessor interrupt is needed to activate the processor that operates the Ethernet device. The time for this is on the order of 350 μs, where the bulk of the time is used to wake up the receiving thread.

An important thing to notice is that the time the hardware uses to transmit the packets in an RPC call makes up only half of the RPC latency; the other half of the time is spent in software. With faster networks, the software overhead will increase even more. Protocols that spend a large amount of effort to optimize the use of the network hardware are, therefore, in many cases self-defeating. In local-area networks, it pays to use protocols that are as lightweight and simple as possible. The next section describes an excellent project on streamlining protocol stacks.

87.2.2.4 The *x*-Kernel

The *x*-kernel is a configurable operating system kernel designed specifically to simplify the process of implementing network protocols [Hutchinson et al. 1989]. Its structure allows flexible configuration of protocol stacks, if necessary even at run time, and combines this with excellent performance. This has made it popular in the operating systems research community and, since it became available to researchers, it has been incorporated into several distributed systems.

The *x*-kernel derives its flexibility and performance from several features. The first, and most important, is that there is a uniform interface to all protocol layers. This allows layers to be stacked arbitrarily (although there are, of course, many protocol combinations that make no semantic sense) and it allows one layer in a stack to be replaced by another.

Protocol layers can be *bound late*; that is, a protocol stack can be constructed at run time, when a connection is established. Late binding is exploited through the use of *virtual protocols*. Virtual IP (VIP), for instance, is a protocol layer that provides an IP interface, but uses dynamic binding to other protocol layers to achieve the actual transport. For destinations on the Internet, VIP would use IP itself, but for destinations on the local Ethernet, or dial-up telephone lines, other protocols can be used which provide the best possible performance for the media used.

Another technique exploiting late binding is decomposing protocols into sublayers. A single protocol often combines several functions, for example, *(de)multiplexing*, *fragmentation*, and *(re)transmission*. Sometimes, higher layers only require a subset of these functions. By decomposing a protocol in separate (dynamically bindable) sublayers, protocol stacks can be composed that have no unnecessary functions or header fields.

Using late binding, a transport protocol can use different lower level protocols, depending on which network is used to reach the destination. An RPC transport protocol, for instance, can use UDP/IP for its data transport when the destination has to be reached over the Internet, but use Ethernet packets directly when the destination is on the same Ethernet. Late binding allows network-dependent optimizations without any loss in flexibility.

Protocol layers in the *x*-kernel have a simple procedural interface. One thread of control can traverse several protocol layers in order to send or deliver packets. This reduces the number of context switches and enhances performance.

87.2.3 Transactions

87.2.3.1 Locus

The Locus operating system [Walker et al. 1983], developed at the University of California, Los Angeles, (UCLA) in the early 1980s, represents an important advance in distributed systems. It demonstrated that a standard Unix interface can be implemented on a fault-tolerant distributed platform, so that unmodified Unix applications can exploit many of the advantages of an operating system that masks failures.

Two basic mechanisms in Locus are used to assist in failure recovery. One is a replicated storage facility, the other is a nested-transaction mechanism. Replicated storage allows data to survive system and media failures. Transactions allow applications to recover from system crashes.

The system does not replicate computations, so a crash will stop all application processes on a machine. Crash recovery is, therefore, necessary, but the transaction mechanisms can be used to leave the system in a consistent state when the crash occurs.

Replicated files are normally updated consistently, but, when a failure partitions the network, separating replicas, then updates are allowed on the accessible subset of the replicas. This can cause inconsistencies among the replicas. These are detected when the network becomes whole again and then reconciliated.

The automatic reconciliation mechanism tries to repair inconsistencies. When it cannot do so for lack of relevant information, it refers the reconciliation up to a mechanism at a higher level. At higher levels, the reconciliation mechanisms become more specialized and distinguish between directories, mailboxes, database files, and other files. Semantic knowledge of a file type allows more reconciliations to happen. Remaining inconsistencies are referred to the human owner of the file, who is considered to be the ultimate reconciliator.

Locus provides nested transactions [Mueller et al. 1983] for failure atomicity. Nested transactions are transactions within other transactions. The outermost transaction is the *top-level transaction*. When a transaction commits, the resulting state is only visible in the enveloping transaction. When a top-level transaction commits, its effects and the effects of all committed subtransactions become globally visible.

Transactions modify files. Files opened as part of a transaction are locked by that transaction and unlocked when the transaction aborts, or the enveloping top-level transaction either commits or aborts. When a transaction commits, its locks are thus inherited by its supertransaction.

Partitions form a complicating factor in the realization of transactions, because the transaction coordinator may become separated from some of the files that form part of the transaction. Subtransactions, when they are separated from their callers, are simply aborted, and the supertransaction is informed. Transactions are also aborted when they are separated away from a file for which they hold a lock and no other replica of the file is accessible.

If the owner of a transaction is separated from the transaction coordinator, then the owner cannot find out the fate of the transaction until the partition is repaired. Thus, under certain circumstances, applications will be blocked waiting for communication to be restored. This blocking problem occurs in all transaction systems, but in some it is worse than in others.

Locus has been in use as a general-purpose fault-tolerant Unix system for a number of years and it has had a significant influence on many other systems, including Quicksilver.

87.2.3.2 Quicksilver

The Quicksilver project of the IBM Almaden Research Center [Cabrera and Wyllie 1987, Haskin et al. 1988, Schmuck and Wyllie 1991] has demonstrated that it is possible to build general-purpose support in distributed systems for fault tolerance. The design of Quicksilver was guided by the following principles [Haskin et al. 1988]:

1. Servers and other applications should be resilient to external failures and be able to recover the resources associated with failed components.
2. The operating system should not contain any code to aid the error recovery for particular servers: each server should contain its own recovery code.
3. The point in item 2 notwithstanding, the system should offer a systemwide uniform error recovery architecture to prevent ad hoc proliferation of error recovery mechanisms.
4. There is a mechanism that allows a client to perform a group of logically related activities in interaction with a set of different servers as a single *atomic* operation (all of the activities should succeed, or none).

With respect to fault tolerance, four categories of applications are recognized in Quicksilver: (1) Those that manage volatile internal state that does not have to be recovered in a crash; after a crash, the server is simply started afresh (example: window servers). (2) Servers that manage replicated volatile state; when a single server crashes, it can recover from one of its replicas; when all servers crash, e.g., in a systemwide power failure, they are started afresh (example: the Quicksilver binding agent where servers register themselves so that clients can find them; after a systemwide crash, all servers must reregister). (3) Servers that manage a recoverable state; that is, a state that may not be lost as the result of a crash (example: the file system). (4) Long-running applications that need periodic checkpointing to make their state recoverable (example: simulations).

Quicksilver offers mechanisms for *atomic transactions* to these applications, very similar to the atomic transactions of database systems. Application classes 1 and 2 only use a subset of the mechanisms described subsequently; the others can use the full set.

There are servers that make transaction-based recovery possible:

1. The *transaction manager* is replicated over all nodes and coordinates transaction commit by communicating with other transaction managers.
2. The *log manager* implements a recovery log for the transaction manager's commit log and for servers' recovery data.
3. The *deadlock detector* detects global deadlocks and resolves them by judiciously aborting transactions.

The messages used by clients to communicate with servers carry a *transaction identifier* (*tid*). Servers thus know to which transaction a client request belongs; they can tag the state information they keep with the associated tids. The interprocess communication (IPC) protocols keep track of the servers addressed as part of a particular transaction so that the appropriate transaction managers can be invoked at commit.

The commit protocol messages are used as a mechanism both for transaction synchronization and for failure notification. Before commit, servers maintaining recoverable state make use of the log manager to store the recoverable data.

These three services make up the *recovery manager*: with this recovery manager, Quicksilver concentrates the recovery functions in one place; servers can use them or not use them, according to their needs; applications can choose between transaction-protocol variants, such as one-phase or two-phase, as appropriate to their function.

Servers communicate with their local recovery manager. The recovery managers at different nodes communicate among themselves to achieve atomicity or recovery.

Processes using transactions use the primitives *begin*, *commit*, and *abort* to manage them. Begin allocates a new tid and makes the invoked transaction manager the coordinator for the transaction just begun.

Transactions in Quicksilver typically have an overhead of between 5 and 100 ms above the time required for the operations that were done as part of the transaction.* This overhead is a very acceptable price to pay for an excellent, well-structured, fault-tolerant mechanism.

87.2.4 Group Management

The technique of replicating computations over multiple, independently failing processing nodes is not new. It has been in use for a long time in safety-critical real-time applications, such as fly-by-wire aircraft control. In real-time environments, the processors are dedicated to running the application and the replicated computation runs in lock step.

Important techniques for managing fault-tolerance for *non*-real-time applications by replicating computations in more relaxed synchrony were first explored by Birman [1985] in the ISIS system [Birman and Joseph 1987, Joseph and Birman 1986]. The ISIS project has inspired research on the theoretical

---

*On an RT-PC.

foundations of replicated computations, causality and virtual synchrony, and models of fault tolerance. This has made it one of the most important projects in distributed systems research.

87.2.4.1 ISIS

The goal of the ISIS project is to provide a system that automates the "transformation of fault-*intolerant* program specifications into fault-tolerant implementations" [Birman 1985]. This is done by taking a sequential program and replicating its code and data over a number of nodes.

The failure model underlying ISIS is *fail-silent*, that is, processors fail by stopping, not by giving wrong results. The surviving processes find out about the crash using a *failure detector* (which uses *time out* to detect processors that are no longer responding). Network failures are transformed into processor failures by declaring unreachable processors crashed; when the network is repaired, such processors learn about their crash and execute a crash-recovery protocol to synchronize themselves to the rest of the replicated computation again.

Computations manipulate *objects* which are made *resilient* by replicating them over multiple sites. *K*-resiliency means that the replicated object behaves like its nondistributed, sequential counterpart running to completion; that, when *k* or fewer replicas fail, the object continues to accept and process requests and does not block, and that recovering replicas can rejoin the group of replicas; and that, when there are more than *k* failures, the replicas restart when all failures are repaired.

Applications can group operations on objects into (nested) atomic transactions. For this purpose, the system provides operations for starting, committing, and aborting transactions and for locking objects.

Replicated objects coordinate their actions by *broadcasting* the relevant information. The broadcast operations are all *reliable*; that is, if one working replica receives the broadcast, all of them will (see subsection "Group Communication," or Hadzilacos and Toueg [1993]). There are three types of reliable broadcast; they are called *Bcast*, *OBcast*, and *GBcast* and they differ in the *ordering* semantics; that is, in the way delivery of broadcast messages is ordered relative to the delivery of other broadcast messages.

The Bcast primitive achieves a total ordering of broadcast deliveries: if a broadcast message is delivered at one site before another, then it is also delivered before the other at all of the other sites. Such a broadcast operation is known as *atomic reliable broadcast* (see subsection "Group Communication," or Hadzilacos and Toueg [1993]) for details).

The ordering semantics of the Bcast primitive are quite strong: two totally unrelated broadcasts are still forced to be processed in the same order everywhere. Relaxed ordering semantics can be implemented more efficiently. The OBcast primitive is one that does not induce a total order, but instead induces order only on broadcasts that could be related in a cause-and-effect manner. Such broadcast primitives, known as *causal broadcast*, use a logical-time stamp on each message and deliver them in increasing time-stamp order. The logical clock, from which the logical-time stamps are derived, is maintained independently by each process; it is incremented on every broadcast operation and always set to a value that is higher than that in received broadcast messages.*

Finally, there is a GBcast primitive which is used to inform the members of a broadcast group (that is, the collection of processes receiving the broadcast messages) of changes in the composition of the group. When a replica joins, it tells the rest of the group with the GBcast operation; when a replica crashes, one of the remaining processes will notice and send a GBcast message on behalf of the crashed process. The GBcast broadcasts are ordered with respect to other broadcast messages: a GBcast message informing of a crash will be delivered after all extant broadcasts (of any kind) from the crashed process have been delivered and a GBcast announcing the joining of the group will be delivered before any messages from the new member. Thus, GBcast messages are totally ordered with respect to group = membership changes; they are also totally ordered with respect to other GBcasts.

---

*Logical clocks are but one way of enforcing causality. Since clock values are rarely the same, most messages will still have a delivery order forced on them even if they are not causally related. A better way to maintain time stamps is the maintenance of *vector clocks* (see Hadzilacos and Toueg [1993] for details).

ISIS uses OBcast wherever it can, because the extra asynchrony allowed by it causes less waiting of processes for each other. It thus provides for more concurrency. Crashes are rare, so the GBcast operation will only rarely be invoked.

ISIS applications are built using an object-oriented style of programming. Each object can receive requests from other objects, which are processed and responded to. Each replica of a replicated object will receive all requests (via one of the broadcast primitives) and will also coordinate with the other replicas using broadcast messages.

Knowing which primitive to use in a particular situation is not trivial and ISIS has been criticized for this [Cheriton and Skeen 1993]. There are claims that transactions can be used to manage replicated objects just as well. This may be the case, but the fact remains that ISIS has been more influential in the development of distributed-systems theory and in increasing the understanding of concurrency, fault tolerance, and causality than any other system. The commercial success of ISIS in stock-market applications proves that ISIS certainly is not a toy.

87.3 Best Practice

Most computers are connected to networks now so that all systems are becoming, to a greater or lesser extent, distributed. Most system builders, therefore, need some knowledge of distributed systems in their baggage and distributed-systems research is becoming mixed with other research areas.

A major — probably the major — motivation for distributed systems research used to be the quest for dependable systems, systems that would tolerate failures in order to become more reliable than their parts. This quest has largely succeeded in that we now have a wide range of techniques and algorithms that work.

However, the subsequent integration of such techniques and algorithms in everyday systems has largely failed. We find two important causes for this. One is the reliability of current computer hardware, the other is the difficulty to change systems that have become accepted as standards.

Computer hardware is now very reliable. Disk manufacturers claim mean times between failure of 200,000 h and more so that very few disks ever fail during their operational lifetime. Because of this, in most situations, there is little need for replicated data storage. Highly distributed services, such as electronic mail and the domain name service, have their specialized fault-tolerance mechanisms. In the World Wide Web no fault tolerance exists at the moment, but some replication will likely occur in the next few years. It appears that only a small set of specialized applications and application domains need mechanisms that provide reliability beyond what networked, but nondistributed, systems can give today.

The other reason is that the world is currently burdened with a few operating system standards that cannot easily be extended with fault-tolerance mechanisms without major change. There is such an investment in existing software that any short-term changes are unlikely. In any case, the world's most widely used operating systems have many, more urgent problems to solve before increased fault tolerance will be noticeable.

87.3.1 Failure Models

Distributed systems can grow to a very large size. But large systems tend to be more complicated than small ones and the consequence of a fault is more difficult to control. One of the most desirable properties for a distributed system is that the failure of any component does not cause other components to fail as well.

Fault-tolerant applications must be distributed over multiple processors so that the crash of one or a few processors does not bring down the application as a whole. The state that the application maintains must be distributed as well, with enough added redundancy to recover from failures.

A *failure model* describes what failures are expected to occur. If it is assumed, for instance, that nodes fail by crashing, then recovery is usually simpler than when they can fail by producing erroneous data. In safety-critical applications, such as a fly-by-wire system or a system controlling railway signalling, the failure model typically assumes that a limited number of *arbitrary* failures may occur. Arbitrary failures are also known as *Byzantine* failures — after the problem of the Byzantine generals [Lamport et al.

1982] — or *malicious* failures. For other applications, however, a *fail-stop* model is common: processors fail by stopping. For a more detailed discussion of failure models we refer to Schneider [1993].

87.3.2 Naming

Without a naming mechanism to allow information sharing, a distributed system cannot exist. Names provide a level of indirection in referring to objects, processes, services, and data that is crucial when entities can be relocated or replicated.

Needham [1993] distinguishes between *pure* names and names of the other sort, *impure* names. A pure name only identifies, an impure name also guides: File names (/usr/bin/sort), URLs (http://www.pegasus. esprit.ec.org/sape), and IP addresses (130.89.181.118) are all examples of impure names, because they lead a system toward the location of the named thing. Impure names have the disadvantage that moving an object often requires renaming it.

Pure names do not have this problem. If a name only identifies, there is no relationship between an object's location and its name. Unfortunately, the scale of many systems makes the use of pure names impractical; in the best case, finding the location of an object requires an $O(\sqrt{n})$ search, where n is the number of possible locations [Mullender and Vitányi 1988]. This makes using a pure name from Europe to locate a file in Australia very expensive.

Distributed systems are, for this reason, forced to use impure names. However, some names are impurer than others. While one name space, because of its structure, fixes the location of objects in exactly one location, another name space might allow objects to move within an organization, a local-area network, or a group of nodes.

Name services should be among the most available services in a distributed system. Availability is an even more important requirement for a name service than correctness: When name resolution does not work, not much else can work either, but if a name service occasionally gives wrong or outdated answers, an application might be guided to the wrong place where the error will usually be discovered.

To be very available, a name service usually is highly replicated. Particularly, name services that are accessible worldwide tend to be enormously replicated. The list of top-level domains of the Internet, for example, is replicated at practically every site; that is, hundreds of thousands of times. Under such massive replication, maintaining consistency is impossible.

Most of the data in large name spaces is fairly static, so inconsistencies will be rare. Smaller name spaces, such as a name space for a distributed file system, will have much higher rates of change. Consistency is also more of an issue in such name spaces. Fortunately, the limited size and replication of such name spaces make it easier to maintain consistency.

Schroeder [1993] claims that "an object should have the same name everywhere" so that sharing becomes possible. Pike et al. [1993] have countered that sharing is facilitated primarily by *naming conventions:* Usually, sharing is achieved by knowing where to look for the shared object; that is, you guess most of its name; it is only sometimes that somebody tells you.

The hierarchical name spaces of Digital's GNS, X.500, and DNS are singly rooted; one name has the same meaning everywhere. Plan 9 [Presotto et al. 1991] uses a two-level naming scheme. One name space names *servers*, each server maintains a local name space, and applications construct a name space by *mounting* some of these servers in a private name space. The servers are named in a global name space (which Plan 9 defined somewhat ad hoc), and conventions play an important role in deciding where users are expected to mount certain servers. The result is a name space in which, for the objects that matter, objects have the same name everywhere.

Amoeba [Mullender et al. 1990] was one of the few systems to use pure names. In this system, services were named using 48-b *ports*. Ports were located using broadcast. As a result, Amoeba did not scale to a size beyond a local network composed of bridged Ethernet segments. Within an Amoeba system, however, objects could be arbitrarily relocated, and processes could be migrated without naming inconsistencies.

An excellent example of a very large-scale name service is the design of Digital's Global Name Service [Lampson 1986]. The update operations in GNS were carefully chosen to be commutative and idempotent.

Commutativity lets only the set of updates determine the state of the naming database, not the order in which they are applied. The idempotency allows updates to be carried out more than once without effect in the state, so that the distribution of updates does not have to happen too carefully.

87.3.3 Communication

The primary function of the interprocess communication mechanisms in distributed systems is to transport information between processes on the nodes of a distributed system. Processes on the same node can also interact in other ways, such as through shared memory. But even within a node, it is useful to use network communication mechanisms between processes. Doing this achieves three things:

1. The system can be reconfigured in such a way that processes, previously running on one node can run on different nodes.
2. Crashes of one process do not have to bring down another.
3. The message-passing interface between processes provides a convenient place for carrying out sanity checks on the data transmitted.

In effect, the interprocess communication mechanism also functions as a *fire wall* that protects one process from bad influences from another. This fire-wall function is an important one in distributed systems, because it helps to get the property of independent failure, it helps to hide the location of a process or function and, thus, it helps the realization of fault tolerance.

It is almost always assumed that networks can fail. They lose and corrupt packets, and sometimes network links may fail altogether. Corrupted packets are detected by making use of checksums; and they are then treated as lost.

Recovery from lost packets can simply be done by numbering the packets, detecting gaps in the number sequence of received packets, and requesting retransmission. When a connection is set up, packet numbers are usually initialized at zero. It is important that packets from one connection cannot mistakenly be received as part of another. There is this risk especially when a connection is set up to replace one that broke in a host or network crash.

Host and network crashes bring about fundamental uncertainty about the state of the system in the surviving nodes. When receipt of messages is acknowledged by a receiving host which crashes, can the sending host deduce from the receipt of an acknowledgment that its message was acted on? The answer is no, because the recipient may have crashed the microsecond after sending the acknowledgment. Can a host deduce anything from *not* receiving an acknowledgment then? Again, the answer is no; the acknowledgment may have been lost.

When an operation is executed remotely, acknowledgments for the arrival of packets or messages are not useful. The only truly useful information is a message that the operation was completed. This is the *end-to-end argument* of Saltzer et al. [1984] who stated that, in a protocol stack, lower layers can never be used to recover from all of the faults and crashes of higher layers. In other words: the highest level of protocol, which necessarily is in the application domain, must have some error recovery if the application is to be fault tolerant.

The opposite is not true; on the contrary: higher layers of protocol are always used to recover from some of the faults in lower layers: when an IP packet is corrupted, TCP will do the recovery.

The end-to-end argument suggests treating the messages belonging to the execution of a remote operation as belonging together in a single protocol unit, directly used by the application, and this is exactly what most distributed systems do. The protocols that handle such groups are known as *message transaction* [Cheriton 1988, Mullender and Tanenbaum 1986], *remote operations*, or *remote procedure call* [Birrell and Nelson 1984].

The basic idea is that activities in a distributed system are structured around the notion of sending a *request* to carry out an operation to a server and receiving a *response* from that server when the operation is completed. The response serves as the fundamental acknowledgment that the request was carried out. Acknowledgments for message arrival may also be used, but they should be viewed as optimizations only.

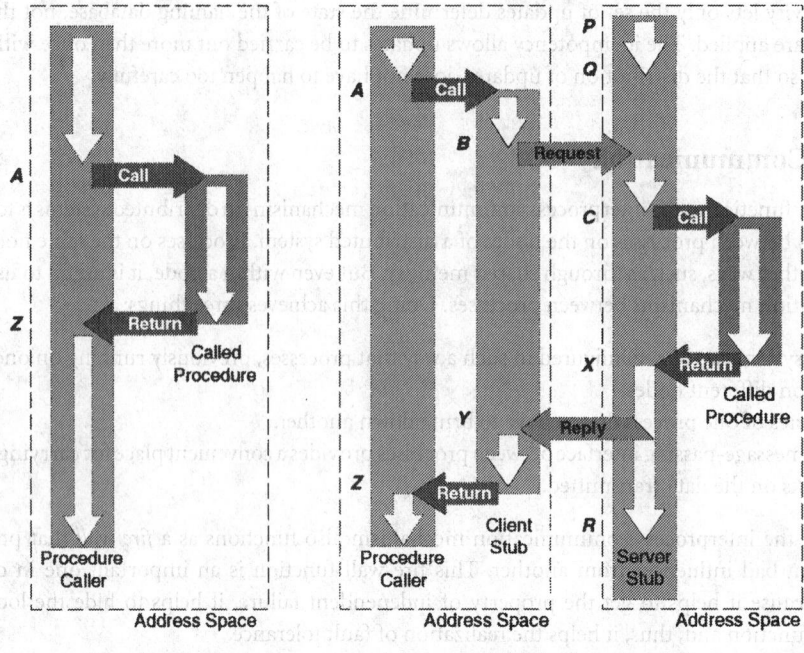

FIGURE 87.2 Structure of remote procedure call.

Amoeba [Mullender et al. 1990] and the V system [Cheriton 1986] were among the first systems to enter the race for implementing remote operations with very low-response times [Nordmark and Cheriton 1989, van Renesse et al. 1988]. The race was joined by many others, but the most interesting work was published by Schroeder and Burrows [1989], who gave an excellent account for where an implementation spends its time and Hutchinson et al. [1989] who built a framework for high-performance protocol stacks that is still being incorporated in many research systems and some commercial ones.

When remote operations are adorned with a mechanism to pass parameters between caller and callee, one has the ingredients for remote procedure call [Birrell and Nelson 1984]. The mechanism is illustrated in Figure 87.2. On the left is a normal procedure call: The procedure is called at point A by the processor's *call-subroutine* instruction sequence. At point Z, the called procedure returns control to the caller, using the processor's *return-from-subroutine* instruction sequence.

On the right of Figure 87.2, one can see how remote procedure call works. In the address space of the caller, a *client stub* is linked in, and in the address space of the called procedure, a *server stub* is linked in. On the server side, the server stub begins executing at P; it announces its presence to its clients (presumably via a name server, we shall discuss the mechanisms for this presently) and, at point Q, it asks the system for incoming client messages and blocks.

On the client side, the caller calls the stub at point A, just as it would have called the procedure in the normal procedure-call case. The stub retrieves the parameters from the stack, puts them in a request message, and sends that to the server at point B. This is called *parameter marshalling*. After marshalling, the client stub blocks itself awaiting the reply message.

This unblocks the server stub. The server stub *unmarshals* the parameters — it takes the parameters from the request message, and pushes them onto the stack — and calls the remote procedure at C, using the standard calling sequence. The procedure returns at point X, the server stub retrieves the return parameters, puts them in a reply message, and sends it back to the client at point Y. The server stub does not block at this point. It carries out some clean-up operations, if necessary, and, at point R, it could jump back to point P or Q to await the next client request.

The arrival of the reply message unblocks the client stub, which then retrieves the return parameters for the reply message and it uses them to make a normal return from procedure at point *Z*.

From the point of view of the calling program and the called procedure, calling a remote procedure appears to be exactly the same as calling a conventional one. This is not the case, however. When remote procedures are used across address space or even machine boundaries, crashes need only affect the caller or the callee. Thus, when remote procedures are used callers must anticipate the possibility that the procedure does not return a value as expected, but that a crash, or communication failure is reported instead.

Another difference between calls within an address space and calls between address spaces is that in the latter case, it is pointless for caller and callee to exchange a pointer to an object in a call. A pointer refers to a different thing in a different address space. Remote procedure call thus imposes restrictions on the kinds of parameters that can be passed that normal procedure-calling sequences do not have.

When remote procedures are used between hosts with different architectures, or between processes written in different programming languages, stubs can be used to convert between the different representations that the parameters may have. This conversion can only be carried out when the parameter types are known.

Given a procedure *signature*,* client and server stubs can be built automatically from an interface definition specified in an *interface definition language* (IDL). Interface definition languages are an almost vital tool in building large distributed applications. They specify the interfaces at module boundaries, the fire walls where type checking is so very important.

Examples of IDLs are HP/Apollo's NCS which is now part of the Distributed Computing Environment (DCE) of the Open Software Foundation, SUN RPC [Sun 1985], Mercury [Liskov et al. 1987], Flume [Birrell et al. 1984], Courier [Xerox 1981], and Middl [Roscoe 1994].

In client/server settings, it can be useful to put some of the server functionality in the client stubs. The client stubs then no longer provide a direct mapping of client calls to the stub and remote calls to the server's procedure. When this is the case, we use the term *clerk* or *agent* rather than the word stub.

Clerks can help with the implementation of automatic rebinding, should a server crash [Schroeder 1993]. The clerk provides a good point for this, because semantic knowledge of the service's behavior can be built into it. Alternatively, clerks can try to provide a (degraded) service while a server is down. A name-server stub, for instance, can provide information for its cache which may be obsolete; clients are better off with old data than no data in this case.

Clerks can also help with performance improvements through caching. A client-side cache for a file server can be viewed as a clerk for the file server. Clients make calls to the clerk and the clerk passes some of them on the the server.

87.3.4 Binding

In a distributed system, services do not always have to reside in one location. Reconfiguration can cause services to be moved, or a service can be restarted on a different machine when the original crashes.

Binding and naming are closely related. Binding is the process of mapping the name of a *service* onto a connection with a provider of that service, a *server*.** The service can be anything: a mail-delivery service for a particular user, a service that can get a file printed nearby, or another part of a distributed application. A name can vary from something as specific as an IP address plus port number to something as vague as the nearest printer that can render PostScript.

Thus, when a binding is created, a specification of a service is converted into a connection to a particular server. A service can be characterized by two things: one is what it can do for its clients, the service's *function*, the other is how its clients can make it do those things, the service's *interface*.

---

*The signature of a procedure is the procedure's name and type: the number and type of parameters and the type of the return value.

**We exercise some care in distinguishing *service* and *server*. Service represents the abstract notion of a set of operations that can be carried out on a set of objects. Server represents a processing entity (a process or processor) that can carry these operations out.

A service has a *state* and, through the service interface, clients can query or modify that state. The interface describes the syntax and semantics of the interactions between client and service. The semantics describe how the operations that the service can carry out modify the state and and what values will be returned.

In the binding process, a service is usually sought that has a particular interface and a distinguished state: When delivering mail, we look for a mail service whose state indicates that it works for a particular user; when connecting to a file service, we want one whose state contains the file we aim to read.

A mechanism is needed to *name* the things one binds to, explicitly or implicitly. An example of explicit naming occurs in binding to an NFS file server, the name of the service corresponds directly to the server.* Slightly more implicit is the way one names a mail box, sape@cs.utwente.nl, for instance. This name does not directly name a particular server, but a *set* of them. During the binding process, the servers in the set are tried until a binding is established. Bindings can be named even more implicitly: The ANSA Trader [API 1989] allows associative names; for instance, one can name a printer with certain properties: it should be on the fourth floor and it should be capable of printing dvi files.

The binding process consists of three stages. First, a set of servers must be found that implement the service. Then, a connection must be created between client and one of these servers, and finally client and server must initialize their mutual binding state. We shall discuss these in turn.

Finding a server for a service can be done by clients and servers themselves, or with the help of a separate *binding agent*. The V system [Cheriton 1988] and Amoeba [Mullender 1985] are examples of systems where clients locate servers without a separate binding agent. Both systems were designed for use in a local network, and clients found servers by *broadcasting* for the service on the local network. In the V system, service requests were broadcast and, if there were multiple servers for the service, it was up to these servers to decide which one would respond; in some cases all servers would respond. In Amoeba, before a request was sent a *locate* message was broadcast and the first server to respond would be chosen to send the request to.

In far-flung systems, usage of broadcast for the location of a server is not doable. A service with the function of broker is needed to bind clients to servers. The idea is that servers, when they become active, notify the broker service of the service they perform and their location. Naturally, most bindings between clients and servers take place within a node or a local network, but bindings that span the globe occur too. For efficiency, it is common practice to make use of a hierarchy of brokers in such a way that the brokers needed to bring about a binding are as near to client and server as they are to each other.

Binding in large systems cannot be done by making use of a pure** name space for the identification of services. Brokerage is best done with a hierarchically organized set of broker services. The Domain Name Service [RFC-1035, RFC-1034] is the almost universally used broker service at the moment.

Binding to the broker presents a bootstrapping problem that is solved by putting the broker at a well-known address, or providing the addresses of a set of brokers in a (well-known) file.

After a server has been found, the second step in the binding process is setting up a connection. This is straightforward and does not need further discussion.

When a connection exists between client and server, negotiation can take place concerning protocol parameters, such as packet sizes, window sizes, network data representations, etc. A step that is becoming increasingly important now that most hosts are connected to the Internet is carrying out an authentication handshake and, if necessary, establishing encryption keys. When all of this is done, a binding exists, and clients can start sending requests to the server.

Client or server crashes result in broken bindings. The role of a server is usually such that, when a client crashes, no attempt is made to create a new binding. The server may, of course, use the report of a broken binding to clean up the connection state. When a server crashes, a client may attempt to bind to another server for the same service in order to be able to continue getting service.

---

*Provided the server is up; if it is down, binding will fail.

**Cf. Section 87.3 subsection on naming.

87.3.5 Transactions

Process and node crashes can leave the state of a distributed computation in an inconsistent, unknown state. When a process in a distributed application crashes, the others must find out where that process got to when it crashed in order to do recovery. Consider, as a trivial example, a request made to the bank's computer to transfer a sum of money from one account to another. If that computer fails, it can leave the database in at least four states: (1) nothing was done yet, (2) the money was removed from one account but not yet added to the other, (3) the money was added to one account but not yet removed from the other, and (4) the transaction was completed.

Without maintaining extra administration, it is not possible to find out how much of a mess a crashed process leaves behind in a system. This administration could be maintained in an application-dependent manner, but, as it turns out, there are excellent general-purpose mechanisms as well.

In this section and in the next, we discuss these general-purpose mechanisms. In "Group Communication," we show how computations can be replicated and how communication can be structured so that all replicas are guaranteed to receive all relevant information in the correct order.

The mechanisms discussed in this section are based on the notion that distributed applications query and modify a distributed database of some sort and that they structure the update operations in such a manner that, after a crash, a consistent state of the system can always be restored.

The database is organized in such way that updates on it succeed completely or fail completely; that is, if an update fails, it leaves the database in the state it had before the update started. We can call such updates *atomic updates*, because they appear to happen all at once.

Atomicity is also an important structuring mechanism for the management of multiple simultaneous updates. Suppose that, in our bank-account example, two updates on a single bank account happen simultaneously, one depositing and one withdrawing. The updates both proceed by reading the balance, computing the new balance, and writing it back. When the two updates both read before the first writes back, the balance of the account becomes inconsistent, a euphemism for wrong.

The update consists of a group of operations that belong together (a read operation and a write operation, in the example). We call such a group a *transaction*. Database systems usually have operations *transaction-begin* and *transaction-end* to allow applications to indicate the grouping.

By making transactions be — or appear to be — atomic, the effect of simultaneous transactions is to *serialize* the updates; that is, the result would be exactly the same if one transaction finished before the other started. Thus, applications do not have to be aware of concurrent transactions, which makes programming them much simpler.

Transactions have, what is often called, the ACID property: they are *atomic, consistent, isolated,* and *durable*. By consistency, we mean that, provided each transaction by itself maintains consistency, the combination of multiple, concurrent transactions also maintains consistency. Isolation means that transactions do not interfere with each other, they are serialized so that the effect is that of one transaction finishing before the next one starts. Finally, durability means that the updates made by a transaction last: when a transaction finishes successfully, all updates are safely stored on stable media.

So far, all of this is just as relevant to centralized systems as it is to distributed ones. In distributed systems, however, the additional problem is to realize atomic transactions also in the face of failures: host crashes, communication failures, or media failures.

To deal with media failures, data can be replicated. Full replication can be done by *disk mirroring*: storing all data on two, identical, disks. Another popular replication technique is redundant array of inexpensive disks (RAID) [Chen et al. 1988]. Here, a parity disk is added to a small number of disks (say, four) and the blocks on the parity disk consist of the exclusive-or of the corresponding blocks on the data disks. When (a block on) a single disk fails, its contents can be calculated by computing the XOR of the corresponding blocks of the other disks.

Node crashes or communication failures can make it impossible to finish a transaction successfully: a node may store information that is needed to complete the transaction, for instance. Transactions that cannot be completed are *aborted*. Those that can are *committed*.

Atomicity of transactions, like so many other things in computer science, is made possible by introducing a level of indirection: a pointer refers to the data. Updates are made by copying a portion of the databases and modifying it. The updates are then committed by changing the pointer to point to the modified data. The rewrite of a single pointer is an atomic operation (Lampson and Sproull [1979] describe how atomic operations can be implemented on replicated disks); since all data are accessed via the pointer, the whole database undergoes atomic modification as well.

In distributed systems the technique for atomic update is quite similar to the one previously described. The pointers, of course, must be implemented as {server name, data pointer} pairs so that they can refer to remote objects as well as local ones.

A technique known as *two-phase commit* (2PC) works as follows. The database is distributed over a number of servers. A client, wishing to make an update, contacts one of the servers and issues a transaction-begin operation. The client receives a *transaction identifier* and makes read and write requests, labeled with the transaction identifier, to the various database servers. One of the database servers becomes the transaction coordinator; it forms the point where the decision is made to commit or abort the transaction.

All the database servers make their updates as described earlier: both the old state and the new state are stored and the flipping of a pointer will switch between the old and the new contents of the data involved in the transaction. In the distributed case, a *lock* is added. This lock is used to synchronize the commit operations of all of the participants: first all servers lock the data, denying access from outside the transaction, then the pointers are flipped, committing each participant to the transaction's updates, and finally, the participants unlock the data again.

Two-phase locking uses this mechanism as follows. If the client aborts the transaction, the coordinator only has to tell the other servers that the transaction has been aborted. They can then do garbage collection. If the client asks for a commit, the coordinator sends a prepare-to-commit request to all participants. The participants write all buffered data to disk to get ready to flip their pointer, they lock the database, and they send an *okay* reply to the coordinator. Naturally, if a server or a network connection fails, no okay message will be forthcoming.

When the coordinator has received the okay from all participants, it commits the transaction locally (by flipping the pointer) and sends a commit message to all participants. The participants also commit and they can unlock.

This is how transactions are committed using 2PC in normal failure-free circumstances. Let us now consider what havoc failures wreak. Suppose a participant crashes before sending its okay message. In this case, the coordinator will receive one okay too few and refuse to complete the commit. The coordinator will broadcast an abort message to the participants, which leaves them in the state before the transaction started. Now, suppose that the participant crashes after sending the okay. Sending an okay implies that both old and new data are on stable storage, so the transaction can (and probably will) still be committed. The coordinator commits and broadcasts the commit message. When the crashed participant comes back up, it finds its database still locked (the locks are also on stable storage) and it can find out from the coordinator, or another participant, what the outcome of the commit operation was.

Now, suppose the coordinator fails. If it fails before sending the prepare message, the transaction can be aborted without further ado. If it crashes after sending out (some) of the prepare messages, the remaining participants can compare notes. If there is at least one that has not responded okay, then it is safe to abort the transaction. But if all remaining participants have sent their okay, then it is not known whether or not the coordinator had already committed internally. The other participants can only wait until the coordinator comes back up. If the coordinator crashes after sending at least one commit message, then, of course, the other participants can finish the transaction, knowing the coordinator must have already committed internally.

Two-phase commit, thus, can *block* on the failure of the coordinator. The probability that this happens is reasonably small, because the time between receiving the last okay and sending the first commit can be kept quite short.

A technique that further reduces the probability of blocking requires an extra round of communication and is therefore known as *three-phase commit* [Skeen 1982].

Two-phase and three-phase commits are important techniques for making updates in distributed databases atomic. In addition to atomicity mechanisms, however, we also need mechanisms to *serialize* the updates; that is, make sure that the effect of two concurrent updates will be the same as that of either first doing one and then the other, or vice versa.

There are two important approaches to causing serializability, one is based on locking out all access that can cause nonserializable access, the other is to check for serializability at commit time and aborting in case of nonserializability. These approaches are widely know as *pessimistic* and *optimistic*, respectively. Locking out access is pessimistic because it also locks out some accesses which, in retrospect, could be serializable after all; checking at commit time is optimistic because it is used under the assumption that serializability conflicts are rare.

A serializability conflict between two transactions can only occur when they access the same data; otherwise, they are independent and it is obvious that, carrying out the two transactions one after the other or concurrently makes no difference. One transaction only influences another if the former writes data that the latter reads.

This observation suggests a simple test for making the decision whether to allow a transaction to commit in an optimistic setting [Kung and Robinson 1981, Schlageter 1981, Strom and Yemeni 1985]: When a transaction is about to commit, a test is made to see if the intersection of the data read by the current transaction (its *read set*) with the data written by transactions that committed after the start of the current transaction is empty. If it is empty, the current transaction has not read any data that was modified afterwards; therefore, the current transaction would have proceeded no differently if the other transactions had already committed before the current one started and, thus, the current transaction is serializable after the other ones. Transactions under optimistic concurrency control are serialized in commit order.

In pessimistic concurrency control, locking is used to make sure that no transaction proceeds beyond the point where serializability is endangered.

A perfectly safe way of locking is to acquire all of the necessary locks when the transaction begins and to release them when it commits or aborts. This is not possible, however, when it is only during the course of the transaction that the dataset accessed by the transaction becomes known. Transactions thus need to acquire locks dynamically. Locks cannot, however, also be released dynamically, at least not carelessly: Suppose one transaction first modifies datum a and then datum b, while another transaction does the same in the opposite order. If both transactions acquire and release locks on datum a or b before acquiring the locks for the other datum, they deadlock.

A technique for dynamic locking that guarantees serializability is *two-phase locking* (2PL), not to be confused with two-phase commit: every transaction consists of two phases, one in which locks are acquired and one in which they are released. No lock may be acquired if even a single one has already been released. As long as it is not known whether more locks are needed, none should be released, so that many transaction systems do not release locks until commit (or abort).

The alternatives of total success or total failure that transactions provide are not always desirable. When a participant in a transaction fails, it is sometimes possible to use an alternative participant without having the whole transaction abort. Mechanisms that support splitting up large transactions into smaller ones are *nested transactions* [Reed 1978]. Subtransactions can commit or abort inside the main transaction. When the main transaction aborts, however, all of its subtransactions are aborted as well. When it successfully commits, however, only the committed subtransactions stay committed.

The ACID properties of transactions make structuring fault-tolerant systems much easier. Transactions have not only been found useful in distributed and centralized databases, but also in distributed file systems and even in operating systems. Examples of systems with atomic properties are Argus from MIT [Liskov and Scheifler 1983], Clouds from Georgia Institute of Technology [McKendry 1984], the Amoeba file system [Mullender and Tanenbaum 1985], Camelot and Avalon from Carnegie-Mellon University (CMU) [Eppinger et al. 1991], and Quicksilver from IBM Almaden [Haskin et al. 1988].

For more information on distributed databases and transactions we refer to books by Bernstein et al. [1987] and by Gray and Reuter [1992]. A compact introduction to distributed transactions can be found in Weihl [1993] and a rigorously formal treatment in Lynch et al. [1993].

87.3.6 Group Communication

In the previous section, we showed how distributed databases can be kept consistent by using transactions that atomically transform the database from one consistent state to another. Many applications can benefit from this way of structuring updates, but not all.

Transactions keep stably stored data consistent, but when applications must manage dynamic data structures consistently and reliably, or deliver exactly the same information to multiple locations, other mechanisms are required.

Group communication forms a class of mechanisms that allows delivering messages to groups of processes or machines. We shall see that group communication allows many more semantic variations than point-to-point communication and that different forms of group communication can be used to solve problems in very different settings.

As a first example, consider the design of a safety-critical control application, such as a fly-by-wire control system. Safety-critical systems must continue to function in the face of processor and communication failures of all kinds, not merely crash failures, but Byzantine failures.

The way in which this is typically done is to run the identical control program on a number of processors. Each of the processors starts out from exactly the same state and is fed with exactly the same information. Consequently, each of the processors will normally produce exactly the same results. These are compared and, if one result differs from the others, the processor with the dissenting result must have failed.

The minimum number of processors for this approach is three — with two processors, when the results differ, one cannot tell which one is wrong — and such an arrangement of three processors is known as *triple-modular redundancy* (TMR). When more processors are used, the configuration is usually labeled *n-modular redundancy* (NMR).

An NMR-based control system will get its inputs from a number of sources (*sensors*) and deliver its outputs to a number of destinations (*actuators*). Each of the sensors will deliver its data to all of the processors, but this, by itself, is not enough to guarantee that the processors will run in lock step. Additionally, all processors must receive the data from the sensors in exactly the same order.

The broadcast system that delivers sensor readings to all processors is known as an atomic broadcast system, because it happens indivisibly: no other broadcast or message delivery can break in and be delivered between the reception of the broadcast by one processor and another.

Another example of an application that uses broadcast is the system that broadcasts the Internet news worldwide. Anyone, anywhere in the world, can send messages and everyone, anywhere, can read them. A news message is labeled with a broad subject classification, the *news group* and with a *subject line* that is supposed to give some clue to its contents. When somebody reacts to a message, they send a *follow-up* message which contains a reference to the original message. Discussions on news net can create long chains of follow-up messages.

When reading the news, it makes sense to read messages in the order in which they follow one another up. It can be confusing to read somebody's reaction to something you have not seen yet. For news delivery, a broadcast system that maintains *causal order* makes much sense. Two messages sent independently can then arrive at different sites in a different order, but a message that causes a follow up must be delivered before the follow up everywhere.

Replicated systems that must withstand crash failures can use broadcast protocols, such as causal broadcast, to organize the communication between replicas and with clients. The participants are then organized as a group of processes, and both communication and membership changes are ordered according to the semantics chosen.

The earliest system that experimented with group communication was ISIS, developed at Cornell under the supervision of Birman [1985]. The idea in ISIS and in follow-up projects worldwide is to create, through reliable, ordered communication an illusion of synchrony: *virtual synchrony*. Examples of other well-known projects that make use of group communication and virtual synchrony are Paralex [Babaoglu et al. 1991], Relacs [Babaoglu et al. 1995], Delta-4 [Veríssimo et al. 1991], and Horus [van Renesse et al. 1995].

Broadcast protocols can be classified according to their ordering semantics, their reliability semantics, and their timing semantics. Ordering is established by making message *reception* and message *delivery* separate operations, so that after a message has been received it is possible to postpone delivery until other messages can be delivered.

Basically, a broadcast protocol is reliable when the following properties are satisfied: (1) If a correct process broadcasts message *m*, then all correct processes eventually deliver *m* (*validity*); (2) if a correct process delivers *m*, then all correct processes eventually deliver *m* (*agreement*); (3) for any message *m*, every correct process delivers *m* at most once, and only if some process broadcasts *m* (*integrity*) [Hadzilacos and Toueg 1993].

Reliability, as defined here, is only concerned with *correct* processes; these are the processes that correctly and completely execute the reliable-broadcast protocol at hand. It is thus possible that a process delivers a message *m* and crashes immediately afterwards so that no other process delivers *m*. This process may even act on the information in *m* before crashing. In some cases this is undesirable and, in such cases, reliability can be extended with *uniformity*, where the agreement rule changes into: if a process (correct or not) delivers a message, then all correct processes do so too; and the integrity rule changes into: any message is delivered to a process (correct or faulty) at most once, and only if it was broadcast by a process (correct or faulty).

The ordering semantics can be classified as (1) *no order*; (2) *first-in–first-out* (*FIFO*) *order*: if a process broadcasts *m* before *m'*, then no correct process delivers *m'* before *m*; (3) *causal order*: if the broadcast of *m* causally precedes* the broadcast of *m'*, then no correct process delivers *m'* before *m*; (4) *atomic broadcast*: if correct processes *p* and *q* both deliver *m* and *m'*, then *p* delivers *m* before *m'* if and only if *q* delivers *m* before *m'*.

Atomic broadcast does not relate broadcast ordering to delivery order; but it does say that the delivery order must be the same everywhere. Thus, atomicity can be combined with FIFO order, FIFO atomic broadcast, or with causal order, causal atomic broadcast, (causality does imply FIFO, by the way).

Real-time applications require messages being delivered within a bounded time after the broadcast. A protocol that does this is a **timed broadcast protocol**.

Reliability, ordering, and timing requirements can be combined to make, for example, *uniform timed causal atomic broadcast*.

A set of processes in a distributed application can use an appropriate broadcast protocol to maintain a replicated state. When a process crashes, the others must be informed reliably. It is often particularly important that the surviving processes agree on the moment of the crash with respect to the broadcasts made.

The combination of a set of protocols for broadcast (multicast) and a set of protocols to maintain membership state of a group of communicating processes is referred to as group communication. ISIS was the earliest group-communication system and it is still the best known.

Other groups have taken this work further. The Relacs system of Babaoglu et al. [1995] was designed to overcome the problems of scale that were present in ISIS. The Delta-4 project [Powell 1991] has explored group communication in the context of dependable computing.

References

Accetta, M., Baron, R., Bolosky, W., Golub, D., Rashid, R., Tevanian, A., and Young, M. 1986. Mach: a new kernel foundation for UNIX development. *Proc. Summer Usenix Conf*. Atlanta, GA, July.

API. 1989. The *ANSA Reference Manual*. Vol. Release 1.1, Architecture projects management, Poseidon House, Castle Park, Cambridge, UK.

---

*An event *e* *causally precedes* an event *f* if and only if [Hadzilacos and Toueg 1993]: (1) a process executed both *e* and *f*, and in that order; (2) *e* is the broadcast of some message *m*, and *f* is the delivery of *m* at some process; or (3) there is an event *h*, such that *e* precedes *h* and *h* precedes *f*.

Babaoglu, Ö., Alvisi, L., Amoroso, A., and Davoli, R. 1991. Mapping parallel computations onto distributed systems in paralex. Invited paper, *Proc. IEEE CompEuro '91*. Bologna, Italy, May.

Babaoglu, Ö., Davoli, R., Giachini, L. A., and Baker, M. G. 1995. Relacs: a communication infrastructure for constructing reliable applications in large-scale distributed systems, pp. 612–621. *Proc. 28th Hawaii Int. Conf. Syst. Sci.* II.

Bernstein, P. A., Hadzilacos, V., and Goodman, N. 1987. *Concurrency Control and Recovery in Database Systems*. Addison–Wesley, Reading, MA.

Birman, K. P. 1985. Replication and fault tolerance in the ISIS system. *ACM Operating Syst. Rev.* 19(5):79–86. *Proc. 10 Symp. Operating Syst. Principles*. Orcas Island, WA.

Birman, K. P. and Joseph, T. A. 1987. Exploiting virtual synchrony in distributed systems. *ACM Operating Syst. Rev.* 21(5):123–138. *Proc. 11th Symp. Operating Syst. Principles*. Austin, TX.

Birrell, A. D., Lazowska, E. D., and Wobber, E. 1984. Flume — remote procedure call stub generator for Modula-2+. Topaz manpage.

Birrell, A. D., Levin, R., Needham, R. M., and Schroeder, M. D. 1982. Grapevine: an exercise in distributed computing. *Commun. ACM* 25:260–274. [Presented at the 8th ACM Symp. Operating Syst. Principles (1981).]

Birrell, A. D., Levin, R., Needham, R. M., and Schroeder, M. D. 1984. Experience with Grapevine: the growth of a distributed system. *ACM Trans. Comput. Syst.* 2(1):3–23.

Birrell, A. D. and Nelson, B. J. 1984. Implementing remote procedure calls. *ACM Trans. Comput. Syst.* 2:39–59.

Cabrera, L. F. and Wyllie, J. 1987. *QuickSilver Distributed File Services: An Architecture for Horizontal Growth*. Computer Science Department, IBM Almaden Research Center, RJ5578.

Chen, P., Gibson, G., Katz, R. H., Patterson, D. A., and Schulze, M. 1988. Two papers on RAIDs. *Comput. Sci. Div.* EECS, UCB, UCB/CSD 88/479, CA.

Cheriton, D. R. 1986. VMTP: a transport protocol for the next generation of communication systems. *Proc. SIGCOMM '86*. Aug 5–7. ACM.

Cheriton, D. R. 1988. The V distributed system. *Commun. ACM* 31:314–333.

Cheriton, D. R. and Skeen, D. 1993. Understanding the limitations of causally and totally ordered communication. *ACM Operating Syst. Rev.* 27(5): 44–57. *Proc. 14th Symp. Operating Syst. Principles*. Asheville, NC.

Eppinger, J. L., Mummert, L. B., and Spector, A. Z. 1991. *Camelot and Avalon: a Distributed Transaction Facility*. Morgan Kaufmann.

Gray, J. and Reuter, A. 1992. *Transaction Processing: Techniques and Concepts*. Morgan Kaufmann.

Hadzilacos, V. and Toueg, S. 1993. Fault-tolerant broadcasts and related problems. In *Distributed Systems*. S. J. Mullender, Ed., 2nd ed., pp. 97–145. ACM Press, New York.

Harcourt. 1995. *Plan 9, Manuals, Documents and CD-ROM*. Harcourt Brace.

Haskin, R., Malachi, Y., Sawdon, W., and Chan, G. 1988. Recovery management in Quicksilver. *ACM Trans. Comput. Syst.* 6(1):82–108.

Hutchinson, N. and Peterson, L. 1988. Design of the x-kernel, pp. 65–75. In *Proc. SIGCOMM '88, Symp. Commun. Architectures and Protocols*. Stanford, CA, Aug.

Hutchinson, N. C., Peterson, L. L., Abbott, M. B., and O'Malley, S. 1989. RPC in the x-kernel: evaluating new design techniques. *ACM Operating Syst. Rev.* 23(5):91–101. *Proc. 12th Symp. Operating Syst. Principles*.

Joseph, T. A. and Birman, K. P. 1986. Low cost management of replicated data in fault-tolerant distributed systems. *ACM Trans. Comput. Syst.* 4(1):54–70.

Kung, H. T. and Robinson, J. T. 1981. On optimistic methods for concurrency control. *ACM Trans. Database Syst.* 6(2):213–226.

Lamport, L., Shostak, R., and Pease, M. 1982. The Byzantine generals problem. *ACM Trans. Programming Lang. Syst.* 4(3):382–401.

Lampson, B. W. 1986. Designing a global name service, pp. 1–10. In *Proc. 5th ACM Annu. Symp. Principles Distributed Comput.* Calgary, Canada, Aug.

Lampson, B. W. and Sproull, R. F. 1979. An open operating system for a single user machine. *ACM Operating Syst. Rev.* 13(5):98–105. *Proc. 7th Symp. Operating Syst. Principles.*

Liskov, B., Bloom, T., Gifford, D., Scheifler, R., and Weihl, W. E. 1987. Communication in the Mercury System. *Programming Methodology Group Memo* 59-1 MIT LCS, Cambridge, MA.

Liskov, B. H., and Scheifler, R. W. 1983. Guardians and actions: linguistic support for robust, distributed programs. *ACM Trans. Programming Lang. Syst.* 5(3):381–404.

Lynch, N. A., Merritt, M., Weihl, W. E., and Fekete, A. 1993. *Atomic Transactions.* Morgan Kaufmann.

McKendry, M. S. 1984. Clouds: a fault-tolerant distributed operating systems. *IEEE Tech. Commun. Distributed Process. Newsletter* 2(6).

Mueller, E. T., Moore, J. D., and Popek, G. J. 1983. A nested transaction mechanism for LOCUS. *ACM Operating Syst. Rev.* 17(5):71–90. *Proc. 9th Symp. Operating Syst. Principles.* Bretton Woods, NH.

Mullender, S. J. 1985. *Principles of Distributed Operating System Design,* Ph.D. Thesis, Vrije Universiteit, Amsterdam, Oct.

Mullender, S. J. and Tanenbaum, A. S. 1985. A distributed file service based on optimistic concurrency control. *ACM Operating Syst. Rev.* 19(5):51–62. *Proc. 10th Symp. Operating Syst. Principles.* Orcas Island, WA.

Mullender, S. J. and Tanenbaum, A. S. 1986. The design of a capability-based distributed operating system. *Comput. J.* 29(4):289–300.

Mullender, S. J., van Rossum, G., Tanenbaum, A. S., van Renesse, R., and van Staveren, J. M. 1990. Amoeba — a distributed operating system for the 1990s. *IEEE Comput.* 23(5).

Mullender, S. J. and Vitányi, P. M. B. 1988. Distributed match-making. *Algorithmica.* 3:367–391.

Needham, R. M. 1993. Names. In *Distributed Systems,* S. J. Mullender, ed., 2nd ed., pp. 315–327. ACM Press, New York.

Nordmark, E. and Cheriton, D. R. 1989. Experiences from VMTP: how to achieve low response time. *Proc. IFIP WG6.1/WG6.4 Int. Workshop Protocols for High-Speed Networks.* H. Rudin and R. Williamson, eds., Zürich, Switzerland, May.

Pike, R., Presotto, D., Thompson, K., Trickey, H., and Winterbottom., P. 1993. The use of name spaces in Plan 9. *ACM Operating Syst. Rev.* 27(2):72–76. *Proc. 5th ACM SIGOPS European Workshop.* Mont Saint-Michel.

Powell, D., ed. 1991. *Delta-4 — A Generic Architecture for Dependable Distributed Computing,* ESPRIT Research Rep. Springer Verlag.

Presotto, D., Pike, R., Thompson, K., and Trickey, H. 1991. Plan 9, a distributed system, pp. 43–50. In *Proc. Spring 1991 EurOpen Conf.* Tromsø, Norway.

Reed, D. P. 1978. *Naming and Synchronization in a Decentralized Computer System.* Ph.D. dissertation, MIT. Available as *Tech. Rep.* MIT/LCS/TR-205. Cambridge, MA.

RFC-1034. Domain Names — Concepts and Facilities.

RFC-1035. Domain Names — Implementation and Specification.

Roscoe, T. 1994. Linkage in the Nemesis single address space operating system. *ACM Operating Syst. Rev.* 28(4):48–55.

Rozier, M., Abrossimov, V., Armand, F., Boule, I., Gien, M., Guillemont, M., Hermann, F., Kaiser, C., Langlois, S., Léonard, P., and Neuhauser, W. 1988. CHORUS Distributed Operating Systems. *Chorus Systemes Rep.* CS/TR-88-7.6, Paris.

Saltzer, J. H., Reed, D. P., and Clark, D. D. 1984. End-to-end arguments in system design. *ACM Trans. Comput. Syst.* 2:277–278.

Schlageter, G. 1981. Optimistic methods for concurrency control in distributed database systems. *Proc. VLDB Conf.*

Schmuck, F. and Wyllie, J. 1991. Experience with transactions in QuickSilver. *ACM Operating Syst. Rev.* 25(5). *Proc. 13th Symp. Operating Syst. Principles.* Pacific Grove, CA.

Schneider, F. B. 1993. What good are models and what models are good? In *Distributed Systems.* S. J. Mullender, ed., 2nd ed., pp. 7–26. ACM Press, New York.

Schroeder, M. D. 1993. A state-of-the-art distributed system: computing with BOB. In *Distributed Systems*. S. J. Mullender, ed., 2nd ed., pp. 1–16. ACM Press, New York.

Schroeder, M. D. and Burrows, M. 1989. Performance of Firefly RPC. *ACM Operating Syst. Rev.* 23(5):83–90. *Proc. 12th Symp. Operating Syst. Principles*.

Skeen, D. 1982. *Crash Recovery in a Distributed Database System*. Ph.D. dissertation. University of California, Berkeley.

Strom, R. and Yemeni, S. 1985. Optimistic recovery in distributed systems. *ACM Trans. Comput. Syst.* 3(3):204–226.

Sun. 1985. Remote Procedure Call Protocol Specification. Sun Microsystems, Inc.

Tanenbaum, A. S., van Renesse, R., van Staveren, J. M., Sharp, G. J., Mullender, S. J., Jansen, A. J., and van Rossum, G. 1990. Experiences with the Amoeba distributed operating system. *Commun. ACM* 33(12):46–63.

van Renesse, R., Birman, K. P., Glade, B. B., Guo, K., Hayden, M., Hickey, T., Malki, D., Vaysburd, A., and Vogels, W. 1995. Horus: A Flexible Group Communications System. *Tech. Rep.* TR 95-1500. Cornell University. March.

van Renesse, R., van Staveren, H., and Tanenbaum, A. S. 1988. Performance of the world's fastest distributed operating system. *ACM Operating Sys. Rev.* 22(4):25–34.

Veríssimo, P., Rodrigues, L., and Rufino, J. 1991. The Atomic Multicast protocol (AMp). In *Delta-4 — A Generic Architecture for Dependable Distributed Computing*. D. Powell, ed. Springer–Verlag.

Walker, B., Popek, G., English, R., Kline, C., and Thiel, G. 1983. The LOCUS distributed operating system. *ACM Operating Syst. Rev.* 17(5):49–70. *Proc. 9th Symp. Operating Syst. Principles*. Bretton Woods, NH.

Weihl, W. E. 1993. Transaction-processing techniques. In *Distributed Systems*, S. J. Mullender, ed., 2nd ed., pp. 329–352. ACM Press, New York.

Xerox. 1981. Courier: The Remote Procedure Call Protocol, *Xerox Syst. Integration Std.* XSIS-038112, Xerox Corp. Stamford, CT.

88

Distributed and Multiprocessor Scheduling

Steve J. Chapin
Syracuse University

Jon B. Weissman
University of Minnesota–Twin Cities

88.1 Introduction 88-1
88.2 Issues in Multiprocessor Scheduling 88-2
   Distributed Scheduling • Scheduling for Shared-Memory
   Parallel Systems
88.3 Best Practices 88-5
   Parallel Scheduling • Distributed Scheduling
   Algorithms • Wide-Area Distributed Scheduling:
   The Grid
88.4 Research Issues and Summary 88-15
   Algorithms • Distributed Scheduling Mechanisms

88.1 Introduction

This chapter discusses CPU scheduling in parallel and distributed systems. CPU scheduling is part of a broader class of resource allocation problems, and is probably the most carefully studied such problem. The main motivation for multiprocessor scheduling is the desire for increased speed in the execution of a workload. Parts of the workload, called *tasks*, can be spread across several processors and thus be executed more quickly than on a single processor. In this chapter we examine techniques for providing this facility.

The scheduling problem for multiprocessor systems can be generally stated as: "How can we execute a set of tasks T on a set of processors P subject to some set of optimizing criteria C?" The most common goal of scheduling is to minimize the expected runtime of a task set. Examples of other scheduling criteria include minimizing the cost, minimizing communication delay, giving priority to certain users' processes, or needs for specialized hardware devices. The scheduling policy for a multiprocessor system usually embodies a mixture of several of these criteria.

Section 88.2 outlines general issues in multiprocessor scheduling and gives background material, including issues specific to either parallel or distributed scheduling. Section 88.3 describes the best practices from prior work in the area, including a broad survey of existing scheduling algorithms and mechanisms. Section 88.4 outlines research issues and gives a summary. Section 88.5 lists the terms defined in this chapter, and is followed by references to important research publications in the area.

88.2 Issues in Multiprocessor Scheduling

There are several issues that arise when considering scheduling for multiprocessor systems. First, we must distinguish between **policy** and **mechanism**. Mechanism gives us the ability to perform an action, while policy decides what we do with the mechanism. Most automobiles have the power to travel at speeds over 150 kilometers per hour or 90 miles per hour (the mechanism), but legal speed limits are usually set well below that (the policy). We will see examples of both scheduling mechanisms and scheduling policies.

Next, we distinguish between **distributed** and **parallel** systems. Past distinctions have been based on whether an interrupt is required to access some portion of memory; in other words, whether communication between processors is via shared memory (also known as **tight coupling**) or via message passing (also known as **loose coupling**). Unfortunately, while this categorization applies well to systems such as shared-memory symmetric multiprocessors (obviously parallel) and networks of workstations (obviously distributed), it breaks down for message-passing multiprocessors such as hypercubes. By common understanding, the hypercube is a parallel machine; but by the memory test, it is a distributed system.

The true test of whether a system is parallel or distributed is the support for **autonomy** of the individual nodes. Distributed systems support autonomy, while parallel systems do not. A node is autonomous if it is free to behave differently than other nodes within the system.* By this test, a hypercube is classified as a parallel machine. There are four components to the autonomy of a multiprocessor system: **design autonomy, communication autonomy, execution autonomy**, and **administrative autonomy**.

Design autonomy frees the designers of individual systems from being bound by other architectures; they can design their hardware and software to their own specifications and needs. Design autonomy gives rise to heterogeneous systems, both at the level of the operating system software and at the underlying hardware level. Communication autonomy allows each node to choose what information to send, and when to send it. Execution autonomy permits each processor to decide whether it will honor a request to execute a task. Furthermore, the processor has the right to stop executing a task it had previously accepted. With administrative autonomy, each system sets its own resource allocation policies, independent of the policies of other systems. The local policy decides what resources are to be shared. In effect, execution autonomy allows each processor to have a local scheduling policy; administrative autonomy allows that policy to be different from other processors within the system.

A **task** is the unit of computation in our computing systems, and several tasks working toward a common goal are called a **job**. There are two levels of scheduling in a multiprocessor system: **global** scheduling and **local** scheduling [Casavant and Kuhl, 1988]. Global scheduling involves assigning a task to a particular processor within the system. This is also known as **mapping, task placement**, and **matching**. Local scheduling determines which of the set of available tasks at a processor runs next on that processor.

Global scheduling takes places before local scheduling, although **task migration**, or dynamic reassignment, can change the global mapping by moving a task to a new processor. To migrate a task, the system freezes the task, saves its state, transfers the saved state to a new processor, and restarts the task. There is substantial overhead involved in migrating a running task.

Given that we have several jobs, each composed of many tasks, competing for CPU service on a fixed set of processors, we have two choices as to how we allocate the tasks to the processors. We can assign several processors to a single job, or we can assign several tasks to a single processor. The former is known as **space sharing** and the latter is called **time sharing**.

Under space sharing, we usually arrange things so that the job has as many processors as it has tasks. This allows all the tasks to run to completion, without any tasks from competing jobs being run on the processors assigned to this job. In many ways, space sharing is similar to old-fashioned batch processing, applied to multiprocessor systems. Under time sharing, tasks may be periodically preempted to allow other tasks to run. The tasks may be from a single job or from multiple jobs. Generally speaking, space sharing is a function of the global scheduling policy, while timesharing is a function of local scheduling.

---

*We speak of behavior at the operating system level, not at the application level.

One of the main uses of global scheduling is to perform **load sharing** between processors. Load sharing allows busy processors to offload some of their work to less busy, or even idle, processors. **Load balancing** is a special case of load sharing, in which the goal of the global scheduling algorithm is to keep the load even (or balanced) across all processors. **Sender-initiated** load sharing occurs when busy processors try to find idle processors to offload some work. **Receiver-initiated** load sharing occurs when idle processors seek busy processors. It is now accepted wisdom that, while load *sharing* is worthwhile, load *balancing* is generally not worth the extra effort, as the small gain in execution time of the tasks is more than offset by the effort expended in maintaining the balanced load.

A global scheduling policy may be thought of as having four distinct parts: the **transfer policy**, the **selection policy**, the **location policy**, and the **information policy**. The transfer policy decides when a node should migrate a task, and the selection policy decides which task to migrate. The location policy determines a partner node for the task migration, and the information policy determines how node state information is disseminated among the processors in the system. For a complete discussion of these components, see [Singhal and Shivaratri, 1994, Chap. 11].

An important feature of the selection policy is whether it restricts the candidate set of tasks to new tasks that have not yet run, or allows the transfer of tasks that have begun execution. **Nonpreemptive** policies only transfer new jobs, while **preemptive** policies will transfer running jobs as well. Preemptive policies have a larger set of candidates for transfer, but the overhead of migrating a job that has begun execution is higher than for a new job because of the accumulated state of the running job (such as open file descriptors, allocated memory, etc.).

As the system runs, new tasks arrive while old tasks complete execution (or, equivalently, are served). If the arrival rate is greater than the service rate, then the process waiting queues within the system will grow without bound and the system is said to be **unstable**. If, however, tasks are serviced at least as fast as they arrive, the queues in the system will have bounded length and the system is said to be **stable**. If the arrival rate is just slightly less than the service rate for a system, it is possible for the additional overhead of load sharing to push the system into instability. A stable scheduling policy does not have this property, and will never make a stable system unstable.

88.2.1 Distributed Scheduling

In most cases, work in distributed scheduling concentrates on global scheduling because of the architecture of the underlying system. Casavant and Kuhl [1988] define a taxonomy of task placement algorithms for distributed systems, which we have partially reproduced in Figure 88.1. The two major categories of global algorithms are static and dynamic.

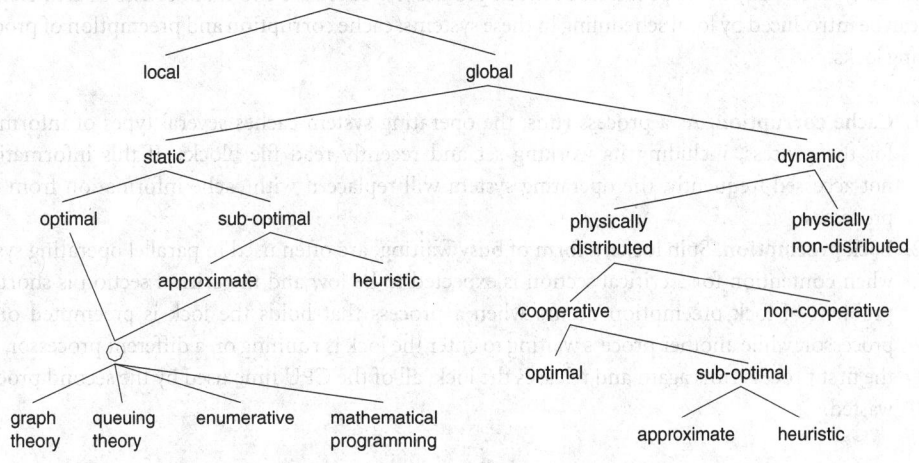

FIGURE 88.1 A taxonomy of distributed scheduling algorithms.

Static algorithms make scheduling decisions based purely on information available at compilation time. For example, the typical input to a static algorithms would include the machine configuration and the number of tasks and estimates of their running time. Dynamic algorithms, on the other hand, take factors into account such as the current load on each processor. Adaptive algorithms are a special subclass of dynamic algorithms, and are important enough that they are often discussed separately. Adaptive algorithms go one step further than dynamic algorithms, in that they may change the policy based on dynamic information. A dynamic load-sharing algorithm might use the current system state information to seek out a lightly loaded host, while an adaptive algorithm might switch from sender-initiated to receiver-initiated load sharing if the system load rises above a threshold.

In **physically non-distributed**, or centralized, scheduling policies, a single processor makes all decisions regarding task placement — this has obvious implications for the autonomy of the participating systems. Under **physically distributed** algorithms, the logical authority for the decision-making process is distributed among the processors that constitute the system.

Under **non-cooperative** distributed scheduling policies, individual processors make scheduling choices independent of the choices made by other processors. With **cooperative** scheduling, the processors subordinate local autonomy to the achievement of a common goal.

Both static and cooperative distributed scheduling have **optimal** and **suboptimal** branches. Optimal assignments can be reached if complete information describing the system and the task force is available. Suboptimal algorithms are either **approximate** or **heuristic**. Heuristic algorithms use guiding principles, such as assigning tasks with heavy inter-task communication to the same processor, or placing large jobs first. Approximate solutions use the same computational methods as optimal solutions, but use solutions that are within an acceptable range, according to an algorithm-dependent metric.

Approximate and optimal algorithms employ techniques based on one of four computational approaches: (1) enumeration of all possible solutions, (2) graph theory, (3) mathematical programming, or (4) queuing theory. In the taxonomy, the subtree appearing below optimal and approximate in the static branch is also present under the optimal and approximate nodes on the dynamic branch; it is elided in Figure 88.1 to save space.

In subsequent sections, we examine several scheduling algorithms from the literature in light of this taxonomy.

88.2.2   Scheduling for Shared-Memory Parallel Systems

Researchers working on shared-memory parallel systems have concentrated on local scheduling because of the ability to trivially move processes between processors. There are two main causes of artificial delay that can be introduced by local scheduling in these systems: cache corruption and preemption of processes holding locks.

1. **Cache corruption.** As a process runs, the operating system caches several types of information for the process, including its working set and recently read file blocks. If this information is not accessed frequently, the operating system will replace it with cache information from other processes.

2. **Lock preemption.** Spin locks, a form of busy waiting, are often used in parallel operating systems when contention for a critical section is expected to be low and the critical section is short. The problem of lock preemption occurs when a process that holds the lock is preempted on one processor, while another process waiting to enter the lock is running on a different processor. Until the first process runs again and releases the lock, all of the CPU time used by the second process is wasted.

In an upcoming section, we examine methods to alleviate or avoid these delays.

88.3 Best Practices

In this section we examine the current state-of-the-art in multiprocessor scheduling. We first consider the techniques used in parallel systems and then examine scheduling algorithms for message-passing systems. Finally, we study scheduling support mechanisms and algorithms for distributed systems, including computational grids.

88.3.1 Parallel Scheduling

We examine three aspects of scheduling for parallel systems: local scheduling for shared-memory systems such as the SGI Altix family; static analysis tools that are beneficial for producing global schedules for parallel systems; and dynamic scheduling for distributed-memory systems.

88.3.1.1 Local Scheduling for Parallel Systems

For most shared-memory timesharing systems, there is no explicit global placement: all processors share the same ready queue, and so any task can be run on any processor. In contrast, local scheduling is crucial for these systems, while it is nonexistent in space-sharing systems. We examine several local scheduling techniques for parallel systems. In general, these techniques are attempting to eliminate one of the causes of delay mentioned previously. All of these techniques are discussed in Chapter 17 of Singhal and Shivaratri [1994].

Co-scheduling, or gang scheduling, schedules the entire pool of subtasks for a single task simultaneously. This can work well with fine-grained applications where communication dominates computation, so that substantial work can be accomplished in a single time slice. Without co-scheduling, it is easy to fall into a pattern where subtasks are run on a processor, only to immediately block waiting for communication. In this way, co-scheduling combines aspects of both space sharing and time sharing.

Smart scheduling tries to avoid the preemption of a task that holds a lock on a critical section. Under smart scheduling, a process sets a flag when it acquires a lock. If a process has its flag set, it will not be preempted by the operating system. When a process leaves a critical section, it resets its flag.

The Mach operating system uses scheduler hints to inform the system of the expected behavior of a process. Discouragement hints inform the system that the current thread should not run for awhile, and hand-off hints are similar to co-routines in that they "hand off" the processor to a specific thread.

In affinity-based scheduling, a task is said to have an affinity for the processor on which it last ran. If possible, a task is rescheduled to run on the processor for which it has affinity. This can ameliorate the effects of cache corruption. The disadvantage of this scheme is that it diminishes the chances of successfully doing load sharing because of the desire to retain a job on its current processor. In effect, affinity-based scheduling injects a measure of global scheduling into the local scheduling policy.

88.3.1.2 Static Analysis

There are several systems that perform static analysis on a task set and generate a static task mapping for a particular architecture. Examples of such systems include Parallax, Hypertool, Prep-P, Oregami, and Pyrros (see Shirazi et al. [1995] for individual articles on these systems). Each of these tools represents the task set as a directed acyclic graph, with the nodes in the graph representing computation steps. Edges in the graph represent data dependencies or communication, where the result of one node is made ready as input for another node. These tools attempt to map the static task graph onto a given machine according to an optimizing criterion (usually, minimal execution time, although other constraints such as minimizing the number of processors used can also be included). The scheduling algorithm then uses some heuristic to generate a near-optimal mapping.

Parallax (Lewis and El-Rewini) is a partitioning and scheduling system that implements seven different heuristic policies. The input to the system is a graph representing the structure of the tasks and a user-selectable representation of the machine architecture. Parallax will then generate schedules based on each of the available heuristics, and present the expected results to the user. This system is unique in its ability to permit the user to explore different combinations of scheduling heuristics and machine architecture for a given task set.

Hypertool (Wu and Gajski) takes as input C source code and generates the task graph representing the program. This is distinct from Parallax, where the user must supply the task graph (and may therefore study the behavior of algorithms that have not been explicitly expressed in any particular programming language). Hypertool then schedules the derived task graph on a hypercube.

Sarkar and Hennessy built one of the first tools to extract parallelism from a functional program, partition the individual tasks into jobs, and then place the jobs on a multiprocessor. They developed a new representation for parallel computation called Macro-Dataflow, and applied their work to programs written in SISAL for the VAX.

Prep-P (Berman and Stramm) is a mapping tool that runs in conjunction with the Poker programming environment for the Pringle machine. Prep-P was one of the earliest program mapping tools, and uses a Graph Description Language as input to describe the program structure. Prep-P uses an iterative partitioning algorithm, wherein an initial partitioning is derived and the system repeatedly attempts to improve upon the partition by moving a task from one partition into another. Whenever the proposed move results in a lower-cost schedule, the move is kept.

Oregami (Lo et al.) is similar to Prep-P in functionality, with the addition of a new model for representing the computation called the Temporal Communications Graph (TCG). The TCG represents each event (computation, message send, or message receipt) as a node within a directed acyclic graph. Thus, it represents a combination of the static task graph with Lamport's process-time graphs.

In many ways, Pyrros (Yang and Gersoulis) represents a merger between several ideas from earlier static scheduling systems. Pyrros uses Sarkar and Hennessy's Macro-Dataflow model, and is targeted for a hypercube architecture. The system takes a Macro-Dataflow graph as input, and then performs partitioning and scheduling. It can produce optimal schedules for several restricted classes of algorithms.

88.3.1.3 Distributed-Memory Systems

In distributed-memory systems, such as hypercube systems, global scheduling is done. Most hypercube systems use space sharing, in that they reserve subcubes of the larger hypercube for use by a single application. Several algorithms have been proposed for this, including that found in Huang et al. [1989]. A typical algorithm maintains a binary tree listing the various sizes of hypercubes available in the system. When a request for an m-dimensional cube is made, the scheduling system searches the tree to see if a hypercube of that exact size is available, and if so, allocates it. If no such hypercube is available, the system splits the smallest hypercube of dimension greater than m into multiple hypercubes, allocates one, and updates the binary tree to reflect the new set of available hypercubes.

For example, consider a request for four processors from a 16-processor hypercube, with all nodes currently free. Four processors comprise a two-dimensional hypercube (a square). The scheduling system would split the 16 processors into two 8-processor cubes, and then split one of the 8-processor cubes into two 4-processor squares. One of the squares would be allocated to the job, leaving one two-dimensional and one three-dimensional hypercube for other jobs.

It is interesting to examine the coexistence of space sharing and time sharing on a single machine. The ASCI Red machine at Sandia National Laboratories is a distributed-memory system that divides its nodes into two partitions: a service partition that runs a general-purpose, full-featured operating system (Unix), and a compute partition that runs a special-purpose, highly efficient operating system (Cougar). The service partition runs in time-sharing mode, while Cougar provides low-latency communication under a mixed space-sharing and time-sharing paradigm (nodes are dedicated to a single job, but that job can run multiple tasks on a single node). Users launch their jobs from the service partition, and the scheduling system reserves a portion of the compute partition to run the tasks for those jobs.

TABLE 88.1 Summary of Distributed Scheduling Survey

Method	Distributed	Heterogeneous	Overhead	Scalable
Blake [1992] (NS, RS)	Y	N	Y	Y
(ABS, EBS)	Y	N	N	Y
(CBS)	N	N	N	N
Casavant and Kuhl [1984]	Y	N	x	P
Ghafoor and Ahmad [1990]	Y	N	Y	P
Wave Scheduling [Van Tilborg and Wittie, 1984]	Y	N	x	P
Ni and Abani [1981] (LED)	Y	N	x	N
(SQ)	Y	N	Y	Y
Stankovic and Sidhu [1984]	Y	N	x	P
Stankovic [1985]	Y	N	x	N
Andrews et al. [1982]	Y	x	x	Y
Greedy Load-Sharing [Chowdhury, 1990]	Y	N	X	Y
Gao et al. [1984] (BAR)	Y	N	x	N
(BUW)	Y	N	x	N
Stankovic [1984]	Y	N	x	P
Chou and Abraham [1983]	Y	N	x	Y
Bryant and Finkel [1981]	Y	N	x	Y
Casey [1981] (dipstick, bidding)	Y	N	x	N
(adaptive learning)	Y	N	Y	Y
Klappholz and Park [1984]	Y	N	x	Y
Reif and Spirakis [1982]	Y	N	x	N
Ousterhout et al., see Singhal and Shivaratri [1994]	N	N	x	N
Hochbaum and Shmoys [1988]	N	Y	x	x
Hsu et al. [1989]	N	Y	x	x
Stone [1977]	N	Y	x	x
Lo [1988]	N	Y	x	x
Price and Salama [1990]	N	Y	x	x
Ramakrishnan et al. [1991]	N	Y	x	x
Sarkar and Hennessy, in Shirazi et al. [1995]	N	Y	x	x

88.3.2 Distributed Scheduling Algorithms

Many researchers have devised algorithms for task placement in distributed systems. This section categorizes several of these techniques in terms of the taxonomy presented earlier.

Table 88.1 displays information garnered from a survey of existing scheduling algorithms. For each algorithm, an entry indicates whether the method is distributed or centralized, supports heterogeneity, minimizes overhead, or supports scalability. Entries are either *Y*, *N*, *P*, or *x*, indicating that the answer is yes, no, partially, or not applicable. The remainder of this section contains a brief description of each method, with a discussion of its place in the taxonomy and its salient properties. Interested readers are referred to the cited publications to obtain full details about the algorithms.

88.3.2.1 Dynamic, Distributed, Cooperative, Suboptimal Algorithms

Blake [1992] describe four suboptimal, heuristic algorithms. Under the first algorithm, Non-Scheduling (NS), a task is run where it is submitted. The second algorithm is Random Scheduling (RS), wherein a processor is selected at random and is forced to run a task. The third algorithm is Arrival Balanced Scheduling (ABS), in which the task is assigned to the processor that will complete it first, as estimated by the scheduling host. The fourth method uses receiver-initiated load balancing, and is called End Balanced Scheduling (EBS). NS, RS, and ABS use one-time assignment; EBS uses dynamic reassignment.

Casavant and Kuhl [1984] describe a distributed task execution environment for Unix System 7, with the primary goal of load balancing without altering the user interface to the operating system. As such, the system combines mechanism and policy. This system supports execution autonomy, but not communication autonomy or administrative autonomy.

Ghafoor and Ahmad [1990] describe a bidding system that combines mechanism and policy. A module called an Information Collector/Dispatcher runs on each node and monitors the local load and that of the node's neighbors. The system passes a task between nodes until either a node accepts the task or the task reaches its transfer limit, in which case the current node accepts the task. This algorithm assumes homogeneous processors and has limited support for execution autonomy.

Van Tilborg and Wittie [1984] present Wave Scheduling for hierarchical virtual machines. The task force is recursively subdivided and the processing flows through the virtual machine like a wave, hence the name. Wave Scheduling combines a non-extensible mechanism with policy, and assumes the processors are homogeneous.

Ni and Abani [1981] present two dynamic methods for load balancing on systems connected by local area networks: Least Expected Delay and Shortest Queue. Least Expected Delay assigns the task to the host with the smallest expected completion time, as estimated from data describing the task and the processors. Shortest Queue assigns the task to the host with the fewest number of waiting jobs. These two methods are not scalable because they use information broadcasting to ensure complete information at all nodes. Ni and Abani [1981] also present an optimal stochastic strategy using mathematical programming.

The method described in Stankovic and Sidhu [1984] uses task clusters and distributed groups. Task clusters are sets of tasks with heavy inter-task communication that should be on the same host. Distributed groups also have inter-task communication but execute faster when spread across separate hosts. This method is a bidding strategy and uses non-extensible system and task description messages.

Stankovic [1985] lists two scheduling methods. The first is adaptive with dynamic reassignment, and is based on broadcast messages and stochastic learning automata. This method uses a system of rewards and penalties as a feedback mechanism to tune the policy. The second method uses bidding and one-time assignment in a real-time environment.

Andrews et al. [1982] describes a bidding method with dynamic reassignment based on three types of servers: free, preferred, and retentive. Free server allocation will choose any available server from an identical pool. Preferred server allocation asks for a server with a particular characteristic, but will take any server if none is available with the characteristic. Retentive server allocation asks for particular characteristics, and if no matching server is found, a server, busy or free, must fulfill the request.

Chowdhury [1990] describes the Greedy load-sharing algorithm. The Greedy algorithm uses system load to decide where a job should be placed. This algorithm is non-cooperative in the sense that decisions are made for the local good, but it is cooperative because scheduling assignments are always accepted and all systems are working toward a global load balancing policy.

Gao et al. [1984] describe two load-balancing algorithms using broadcast information. The first algorithm balances arrival rates, with the assumption that all jobs take the same time. The second algorithm balances unfinished work. Stankovic [1984] gives three variants of load-balancing algorithms based on point-to-point communication that compare the local load to the load on remote processors. Chou and Abraham [1983] describe a class of load-redistribution algorithms for processor-failure recovery in distributed systems.

The work presented in Bryant and Finkel [1981] combines load balancing, dynamic reassignment, and probabilistic scheduling to ensure stability under task migration. This method uses neighbor-to-neighbor communication and forced acceptance to load balance between pairs of machines.

Casey [1981] gives an earlier and less complete version of the Casavant and Kuhl taxonomy, with the term *centralised* replacing *non-distributed* and *decentralised* substituting for *distributed*. This article also lists three methods for load balancing — Dipstick, Bidding, and Adaptive Learning — and then describes a load-balancing system wherein each processor includes a 2-byte status update with each message sent. The Dipstick method is the same as the traditional watermark processing found in many operating systems. The Adaptive Learning algorithm uses a feedback mechanism based on the run queue length at each processor.

88.3.2.2 Dynamic Non-cooperative Algorithms

Klappholz and Park [1984] describe Deliberate Random Scheduling (DRS) as a probabilistic, one-time assignment method to accomplish load balancing in heavily loaded systems. Under DRS, when a task is spawned, a processor is randomly selected from the set of ready processors, and the task is assigned to the selected processor. DRS dictates a priority scheme for time-slicing, and is thus a mixture of local and global scheduling. There is no administrative autonomy or execution autonomy with this system because DRS is intended for parallel machines.

Reif and Spirakis [1982] present a Resource Granting System (RGS) based on probabilities and using broadcast communication. This work assumes the existence of either an underlying handshaking mechanism or of shared variables to negotiate task placement. The use of broadcast communication to keep all resource providers updated with the status of computations in progress limits the scalability of this algorithm.

88.3.2.3 Dynamic Non-distributed Algorithms

Ousterhout et al. (see [Singhal and Shivaratri, 1994]) describe Medusa, a distributed operating system for the Cm* multiprocessor. Medusa uses static assignment and centralized decision making, making it a combined policy and mechanism. It neither supports autonomy, nor is the mechanism scalable.

In addition to the four distributed algorithms already mentioned, Blake [1992] describes a fifth method called Continual Balanced Scheduling (CBS) that uses a centralized scheduler. Each time a task arrives, CBS generates a mapping within two time quanta of the optimum, and causes tasks to be migrated accordingly. The centralized scheduler limits the scalability of this approach.

88.3.2.4 Static Algorithms

All the algorithms in this section are static and, as such, are centralized and without support for autonomy. They are generally intended for distributed-memory parallel machines, in which a single user can obtain control of multiple nodes through space sharing. However, they can be implemented on fully distributed systems.

Hochbaum and Shmoys [1988] describes a polynomial-time, approximate, enumerative scheduling technique for processors with different processing speeds, called the dual-approximation algorithm. This algorithm solves a relaxed form of the bin packing problem to produce a schedule within a parameterized factor, ϵ, of optimal. That is, the total runtime is bounded by $(1 + \epsilon)$ times the optimal run time.

Hsu et al. [1989] describe an approximation technique called the critical sink underestimate method. The task force is represented as a directed acyclic graph, with vertices representing tasks and edges representing execution dependencies. If an edge (α, β) appears in the graph, then α must execute before β. A node with no incoming edges is called a *source*, and a node with no outgoing edges is a *sink*. When the last task represented by a sink finishes, the computation is complete; this last task is called the critical sink. The mapping is derived through an enumerative state space search with pruning, which results in an underestimate of the running time for a partially mapped computation and, hence, the name critical sink underestimate.

Stone [1977] describes a method for optimal assignment on a two-processor system based on a Max Flow/Min Cut algorithm for sources and sinks in a weighted directed graph. A maximum flow is one that moves the maximum quantity of goods along the edges from sources to sinks. A minimum cutset for a network is the set of edges with the smallest combined weighting, which, when removed from the graph, disconnects all sources from all sinks. The algorithm relates task assignment to commodity flows in networks, and shows that deriving a Max Flow/Min Cut provides for optimal mapping.

Lo [1988] describes a method based on Stone's Max Flow/Min Cut algorithm for scheduling in heterogeneous systems. This method utilizes a set of heuristics to map from a general system representation to a two-processor system so that Stone's work applies.

Price and Salama [1990] describe three heuristics for assigning precedence-constrained tasks to a network of identical processors. With the first heuristic, the tasks are sorted in increasing order of communication, and then are iteratively assigned so as to minimize total communication time. The second heuristic creates pairs of tasks that communicate, sorts the pairs in decreasing order of communication, then groups the pairs into clusters. The third method, simulated annealing, starts with a mapping and uses probability-based functions to move toward an optimal mapping.

Ramakrishnan et al. [1991] present a refinement of the A* algorithm that can be used either to find optimal mappings or to find approximate mappings. The algorithm uses several heuristics based on the sum of communication costs for a task, the task's estimated mean processing cost, a combination of communication costs and mean processing cost, and the difference between the minimum and maximum processing costs for a task. The algorithm also uses ϵ-relaxation similar to the dual-approximation algorithm of Hochbaum and Shmoys [1988].

Sarkar and Hennessy (in Shirazi et al. [1995]) describe the GR graph representation and static partitioning and scheduling algorithms for single-assignment programs based on the SISAL language. In GR, nodes represent tasks and edges represent communication. The algorithm consists of four steps: cost assignment, graph expansion, internalization, and processor assignment. The cost assignment step estimates the execution cost of nodes within the graph, and communication costs of edges. The graph expansion step expands complex nodes (e.g. loops) to ensure that sufficient parallelism exists in the graph to keep all processors busy. The internalization step performs clustering on the tasks, and the processor assignment phase assigns clusters to processors so as to minimize the parallel execution time.

88.3.3 Wide-Area Distributed Scheduling: The Grid

Recent advances in distributed scheduling algorithms have occurred in a distributed computing environment known as the Grid, an ensemble of shared, geographically dispersed computers that may not be fully under the control of a single scheduler.

Scheduling in a Computational Grid environment introduces several challenges that go beyond that of traditional distributed system schedulers: heterogeneity, scale, and ownership. In a Grid, resources are highly heterogeneous, ranging from supercomputers to PCs, from high-end mass storage devices to low-end disks. Grid resources may number in hundreds or thousands making efficient run-time scheduling more difficult. Finally, in a Grid, resources may be under the control of resource owners, putting constraints on how schedulers can use them. The scheduling approaches described below deal with these issues in different ways and to differing degrees.

88.3.3.1 Messiahs

The first scheduling system for grids, then called metasystems, was the MESSIAHS (**M**echanism **E**ffecting **S**cheduling **S**upport **I**n **A**utonomous, **H**eterogeneous **S**ystems) system [Chapin and Spafford, 1994]. MESSIAHS provided extensive support for autonomy, using hierarchically structured systems based on administrative domains. This structuring was based on an observation of a social aspect of computing: people are willing to allow outside utilization of their unused resources, as long as they maintain control of the system. This means that the local systems are autonomous and the local administrators can set their own access policies. An example distributed autonomous system is illustrated in Figure 88.2.

Each node within the MESSIAHS system is a **virtual system**, which represents a subset of the resources of one or more real systems, and has a hierarchical structure modeling the administrative hierarchies of computer systems and institutional organization. Virtual systems can be combined into encapsulating virtual systems. For example, in Figure 88.2, the University, National Laboratory, and Industry are each virtual systems, and are collected into a single large distributed (virtual) system. Within the University, National Lab, and Industry virtual systems are other virtual systems, giving a hierarchical structure. These intermediate groupings may correspond to divisions that contain departments, and the departments may contain research groups, etc. At the lowest level of grouping, each virtual system typically consists of a subset of the capabilities of a single machine. Together, these virtual systems form a directed acyclic graph topology.

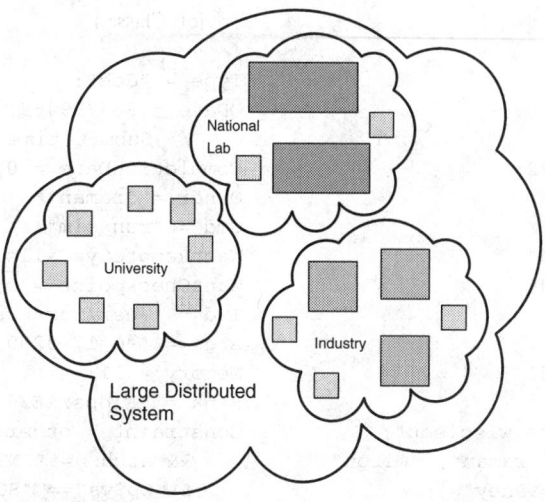

FIGURE 88.2   A sample virtual system.

There is no centralized resource management in MESSIAHS. Instead, each virtual system runs a **scheduling module**, which maintains the system description information for that node. The scheduling module also exchanges service requests with neighboring virtual systems within the hierarchy, and is responsible for starting and stopping jobs.

MESSIAHS provides two interfaces with which administrators can define the scheduling policy for a system: an interpreted language called the MESSIAHS Interface Language (MIL) dedicated to implementing schedulers, and a library of functions for the C programming language. Using these interfaces, administrators write small event handlers that are linked into the scheduling system. Thus, MESSIAHS has extensive support for autonomy.

88.3.3.2   Matchmaking

Matchmaking is a highly distributed scheduling process that accounts for the heterogeneity of resources and the distribution of ownership. This approach consists of a semi-structured data model that combines schema, data, and query in a specification language, and a separation of matching and claiming phases of resource allocation. Both resources and jobs use a mechanism known as classads to specify attributes and constraints (Figure 88.3). A classad is a mapping from attribute names to expressions. This mechanism allows the heterogeneity and ownership issues inherent to the Grid to be addressed. Issues of scale are not addressed by this mechanism, and the locality of resources is not captured by classads.

Scheduling is accomplished by a matchmaker that matches resource classads to job classads, satisfying the constraints of both. The implementation of the decision is negotiated between the resource and the job using matchmaking and resource claiming protocols. The matching (aka scheduling) algorithm first determines feasible matches (schedules) by determining compatible resource and job classads via the Constraint expression. The next phase of the algorithm is to choose among compatible matches by maximizing the Rank attribute of the job. This scheme has been designed for localized resources within a single Grid site.

88.3.3.3   AppLeS

The AppLeS scheduling approach is a conceptual framework for the design of Grid schedulers tuned specifically for the application. The goal is to select a potentially efficient configuration of resources based on load and availability, evaluate the potential performance of such configurations based on application-specific performance criteria, and interact with the relevant resource management systems to carry out

```
         Machine Classad                       Job Classad
[                                       [
Type = "Machine";                       Type = "Job";
Activity = "Idle";                      QDate = 886799469;
DayTime = 36107;                            // Submit time secs. past 1/1/1970
KeyboardIdle = 1432;                    CompletionDate = 0;
Disk = 323496;                          Owner = "raman";
Memory = 64;                            Cmd = "run_sim";
State = "Unclaimed";                    WantRemoteSysCalls = 1;
LoadAvg = 0.042969;                     WantCheckpoint = 1;
Mips = 104;                             Iwd = "/usr/raman/sim2";
Arch = "INTEL";                         Args = "-Q 17 3200 10";
OpSys = "SOLARIS251";                   Memory = 31;
KFlops = 21893;                         Rank = KFlops/1E3 + other.Memory/32;
Name = "leonardo.cs.wisc.edu";          Constraint = other.Type == "Machine"
ResearchGroup = { "raman", "miron",         && Arch == "INTEL"
    "solomon", "jbasney" };                 && OpSys == "SOLARIS251"
Friends = { "tannenba", "wright" };         && Disk >= 10000
Untrusted = { "rival", "riffraff" };        && other.Memory >= self.Memory;
Rank = member(other.Owner,              ]
ResearchGroup)
    *10+member(other.Owner, Friends);
Constraint = !member(other.Owner,
Untrusted)
    && Rank >= 10 ? true
    : Rank > 0 ? LoadAvg<0.3
    && KeyboardIdle>15*60
    : DayTime < 8*60*60
    DayTime > 18*60*60;
]
```

FIGURE 88.3 Classads describing a machine and a job.

the decision. AppLeS agents perform the scheduling task for the user and exploit several generic components such as the Network Weather Service [NWS, 1999] which provides resource load predictions. Specific AppLeS scheduling agents have been developed for tightly-coupled parallel scientific computations
[Berman et al., 1996], data replica selection [Su et al., 1999], parameter sweeps [Casanova et al., 2000],
and gene sequence comparison [Spring and Wolski, 1998]. These AppLeS agents address the issue of Grid
scale by considering a subset of Grid resources and the issue of heterogeneity by dynamic cost modeling via
static and dynamic cost functions. An example of an AppLeS scheduling algorithm is shown in Figure 88.4.
The issue of ownership is not addressed. The more recent thrust in this project is to generalize the AppLeS
approach to build template schedulers applicable to a wider class of applications. Templates have been
constructed for master-slave applications and parameter sweep applications [Casanova et al., 2000].

88.3.3.4 Prophet and Gallop

Data parallel applications are a large and important class of parallel scientific applications. Efficiently
scheduling these applications in a Grid is difficult due to the potential sharing and heterogeneity of resources. In addition, selecting the appropriate number of processors and data domain decomposition are
all difficult problems that depend on the available resources, their load and heterogeneity, network load,
and problem characteristics. Most systems leave this complex problem up to the end user. Even if the user
figures out the best resource combination by running an application multiple times, this same resource
combination may not be available all the time due to resource sharing; and even if it was, it is unlikely to
be best for a different problem instance of the same application. Finally, such optimizations are tailored to

Resource Selection

Let *locus* = machine having the maximum criterion value
Let *list* = a sort of the remaining machines according to
    their logical distance
begin **for** $k = 0$ **to** $I - 1$
    let S = *locus* + the first k elements of *list*
    parameterize C_i and P_i for $1 \leq i \leq |S|$
    with Weather Service forecasts
    solve linear system of equations using this
        parameterization
    if (not all $A_i > 0$)
        then reject partitioning as infeasible **fi**
        else if (there exists an A_i that does not fit
            in free memory of processor i)
            then reject partitioning as infeasible **fi**
        else record expected execution time for subset S
end
Implement the partitioning corresponding to the minimum
execution time using the S for which it was computed

Performance Model

T_i = time for processor i to compute region i
A_i = the area of region i
P_i = time for processor i to compute a point
C_i = time for processor i to send/recv its borders

$T_i = A_i \times P_i + C_i$
Require $T_1 = T_2 = \cdots = T_I$ s.t. $\sum_{i=1}^{I} A_i = N \times M$
$C_i = f\{\text{Recv}\,(i \pm 1, i), \text{Send}\,(i, i \pm 1)\}$
Send/Recv $(i, j) = N \times$ sizeof(element) $\times$ Bandwidth(i, j)

FIGURE 88.4 Jacobi resource selection and performance model.

Local SM Component

1. Local SM receives WA_sched (Job_class) request
2. Local SM chooses k candidate $SM_1 \ldots SM_k$
3. For each SM_i
    call $best_i = SM_i.$get_best(Job_class)
4. The_best = min($best_i$.elapsed_time) over all i in $1 \ldots k$
5. For each SM_i
    if (SM_i == the_best_site)
        result = SM_i.go_SM(Job_class, the_best)
    else
        SM_i.no_go_SM(Job_class, the_best)
6. Return result and scheduling info to front-end

Remote SM Component

1. Remote SM receives get_best(Job_class) from local SM
2. Call scheds = LS.get_scheds(Job_class)
3. Call best = LS.eval_scheds(scheds, Job_class)
4. Lock best, return best to local SM
5. If remote SM receives go_SM(Job_class, the_best)
    result = LS.go_ls(Job_class, the_best)
    return result to local SM
    Else if remote SM receives no_go_SM(Job_class, the_best)
    store the_best in table
6. Release lock on best

FIGURE 88.5 Sample wide-area scheduling algorithm.

the specific application and are not available to another user's application. Prophet [Weissman, 1999] is an automated scheduler for data parallel applications that utilizes a performance model for predicting application performance on different resource combinations that automates processor selection, task placement, and data domain decomposition for data parallel applications. Prophet uses a callback mechanism to allow the runtime scheduler to obtain application-specific information to construct cost functions.

Gallop [Weissman, 1998] is a wide-area Grid scheduling system that provides a federated scheduling model across different sites in a Grid. In the Gallop model, each site can run its own local scheduler, and decide which resources are to be made available to Gallop. Gallop utilizes a bid-based distributed scheduling algorithm to decide which site to select from among a network of sites (see Figure 88.5.) A multi-site Gallop testbed was constructed in which the local sites each ran a version of Prophet to provide performance prediction estimates for using their resources. Gallop demonstrated that the overhead of remote execution can be tolerated for applications of sufficiently large granularity in the Grid. It achieved performance gains up to 25% over local site execution for a range of applications: a Poisson solver, an image processing pipeline, and a genomic sequence comparison.

88.3.3.5 Stochastic Scheduling

Stochastic scheduling [Schopf and Berman, 1999] harnesses the variability inherent in Grid computing to produce performance-efficient schedules. Stochastic scheduling models the performance variance of a resource using a stochastic value (i.e., distribution) and then proposes scheduling heuristics that use this value. By parameterizing models with stochastic information, the resulting performance prediction is also a stochastic value. Such information can be more useful to a scheduler than point predictions with an unknown range of accuracy. The authors introduce a tuning factor that represents the variability of the system as a whole as some constant number of standard deviations away from a mean value. Resource platforms with a smaller variability are given scheduling priority in this scheme. Extensions of well-known scheduling methods such as time-balancing are shown to achieve better performance under stochastic time-balancing.

88.3.3.6 Co-allocation

Resource co-allocation of multiple resources introduces a new scheduling problem for Grid applications. In the co-allocation model, the Grid application requires access to a specific set of resources concurrently. Several approaches for addressing this problem have been proposed [Chapin et al., 1999, Foster et al., 1999]. Scheduling techniques include atomic all-or-nothing semantics in which either all resources are acquired and the application starts, or it must wait. If the application starts and one or more resources are taken away or fail, then the application must abort. Another approach is advanced reservations in which resources can be locked at some future time so that they are available together. Co-allocation-based reservation schemes usually try to provide the soonest time a reservation of the desired length can be accomodated across all desired resources (see Algorithm 88.1).

> **begin**
>   rh-a ← CreateReservation(*contact-a*, "*&(reservation_type=compute)*
>   *(start_time="10:30pm") (duration="1 hour") (nodes=32)*");
>   **if** rh-a *is null* **then exit**;
>   **repeat**
>   (contact-b, id-b, contact-net) ← FindNextCandidate();
>   rh-b ← CreateReservations (*contact-b*, "*&(reservation_type=compute)*
>   *(start_time="10:30pm") (duration="1 hour") (percent_cpu=75)*");
>   **if** rh-b *is null* **then** *continue*;
>   rh-net ← CreateReservation(*contact-net*, "*&(reservation_type=network)*
>   *(start_time="10:30pm") (duration="1 hour") (bandwidth=200)*
>   *(endpoint-a=id-a)(endpoint-b=id-b)*");
>   **if** rh-net *is null* **then** CancelReservation(rh-b);
>   **until** rh-b *and* rh-net *are defined*;
>   **if** rh-b *is null* **then** signal that search failed;
> **end**

Algorithm 88.1:   Pseudo-code for a ReserveResources subroutine.

Legion [Chapin et al., 1999] provided the first co-allocation system for the Grid, and supports both co-allocation and advanced reservations. Legion provides full autonomy support, allowing a scheduling request to return a scheduling token, which is later redeemed when the job is actually started. Legion supports multiple simultaneous schedulers, including per-application, per-user, and default schedulers; these can in fact implement the techniques described here, such as Prophet or stochastic scheduling. Under Legion, schedulers produce a list of proposed target schedules ranked by desirability, and an Enactor component is responsible for verifying that the resources are in fact available and implementing the schedule. This allows maximum flexibility and support for autonomy.

88.4 Research Issues and Summary

The central problem in distributed scheduling is assigning a set of tasks to a set of processors in accordance with one or more optimizing criteria. We have reviewed many of the algorithms and mechanisms developed thus far to solve this problem. However, much work remains to be done.

88.4.1 Algorithms

Until now, scheduling algorithms have concentrated mainly on systems of homogeneous processors. This has worked well for parallel machines, but has proved to be unrealistically simple for distributed systems. Some of the simplifying assumptions made include constant-time or even free inter-task communication, processors with the same instruction set, uniformity of available files and devices, and the existence of plentiful primary and secondary memory. In fact, the vast majority of algorithms listed in the survey model the underlying system only in terms of CPU speed and a simplified estimate of interprocessor communication time.

These simple algorithms work moderately well to perform load balancing on networks of workstations that are, in most senses, homogeneous. However, as we look to the future and attempt to build wide-area distributed systems composed of thousands of heterogeneous nodes, we will need policies capable of making good scheduling decisions in such complex environments.

88.4.2 Distributed Scheduling Mechanisms

As we have seen, current scheduling systems do a good job of meeting the technical challenges of supporting the relatively simplistic scheduling policies that have been developed to date. Future work will expand in new directions, especially in the areas of heterogeneity, security, and the social aspects of distributed computing.

Just as scheduling algorithms have not considered heterogeneity, distributed scheduling mechanisms have only just begun to support scheduling in heterogeneous systems. Some of the major obstacles to be overcome include differences in the file spaces, speeds of the processors, processor architectures (and possible task migration between them), operating systems and installed software, devices, and memory. Future support mechanisms will have to make this information available to scheduling algorithms to fully utilize a large, heterogeneous distributed system.

The challenges in the preceding list are purely technical. Another set of problems arises from the social aspects of distributed computing. Large-scale systems will be composed of machines from different administrative domains; the social challenge will be to ensure that computations that cross administrative boundaries do not compromise the security or comfort of users inside each domain. To be successful, a distributed scheduling system must provide security, both for the foreign task and for the local system; neither should be able to inflict harm upon the other. In addition, the scheduling system will have to assure users that their local rules for use of their machines will be followed. Otherwise, the computing paradigm will break down, and the large system will disintegrate into several smaller systems under single administrative domains.

Defining Terms

Autonomy: The freedom to be different or behave differently than other nodes within the system.

Centralized mechanisms: Mechanisms in which data is stored at a single node, or which pass all data to a single node for a decision.

Distributed mechanisms: Mechanisms in which decisions are made on the local system, based on data located on that system.

Distributed memory: A system in which processors have different views of memory. Often, each processor has its own memory and cannot directly access another processor's local memory.

Distributed systems: Systems with a high degree of autonomy.

Global scheduling: The assignment of tasks to processors (also called task placement and matching).

Heterogeneous systems: The property of having different underlying machine architectures or systems software.

Job: A group of tasks cooperating to solve a single problem.

Load balancing: A special form of load sharing in which the system attempts to keep all nodes equally busy.

Load sharing: The practice of moving some of the work from busy processors to idle processors. The system does not necessarily attempt to keep the load equal at all processors; instead, it tries to avoid the case where some processors are heavily loaded while others sit idle.

Local scheduling: The decision as to which task, of those assigned to a particular processor, will run next on that processor.

Loosely coupled hardware: A message-passing multiprocessor.

Mechanism: The ability to perform an action.

Parallel systems: Systems with a low degree of autonomy.

Policy: A set of rules that decides what action will be performed.

Shared memory: A system in which all processors have the same view of memory. If processors have local memories, then other processors may still access them directly.

Space sharing: A system in which several jobs are each assigned exclusive use of portions of a common resource. For example, if some of the processors in a parallel machine are dedicated to one job, while another set of processors is dedicated to a second job, the jobs are space sharing the CPUs.

Stability: The property of a system that the service rate is greater than or equal to the arrival rate. A stable scheduling algorithm will not make a stable system unstable.

Task: The unit of computation in a distributed system; an instance of a program under execution.

Task migration: The act of moving a task from one node to another within the system.

Tightly coupled hardware: A shared-memory multiprocessor.

Time sharing: A system in which jobs have the illusion of exclusive access to a resource, but in which the resource is actually switched among them.

References

[Andrews et al., 1982] G. R. Andrews, D. P. Dobkin, and P. J. Downey. Distributed allocation with pools of servers. In *Proceedings of the Symposium on Principles of Distributed Computing*, pp. 73–83. ACM, August 1982.

[Berman et al., 1996] F. Berman, R. Wolski, S. Figueira, J. Schopf, and G. Shao. Application-Level Scheduling on Distributed Heterogeneous Networks. *Proceedings of Supercomputing 1996*.

[Blake, 1992] B. A. Blake. Assignment of Independent Tasks to Minimize Completion Time. *Software — Practice and Experience*, 22(9):723–734, September 1992.

[Bryant and Finkel, 1981] R. M. Bryant and R. A. Finkel. A stable distributed scheduling algorithm. In *Proceedings of the International Conference on Distributed Computing Systems*, pp. 314–323. IEEE, April 1981.

[Casanova et al., 2000] H. Casanova, G. Obertelli, F. Berman, and R. Wolski, The AppLeS Parameter Sweep Template: User-Level Middleware for the Grid. *Proceedings of SC 2000*, 2000.

[Casavant and Kuhl, 1984] T. L. Casavant and J. G. Kuhl. Design of a loosely-coupled distributed multi-processing network. In *Proceedings of the International Conference on Parallel Processing*, pp. 42–45. IEEE, August 1984.

[Casavant and Kuhl, 1988] T. L. Casavant and J. G. Kuhl. A Taxonomy of Scheduling in General-Purpose Distributed Computing Systems. *IEEE Transactions on Software Engineering*, 14(2):141–154, February 1988.

[Casey, 1981] L. M. Casey. Decentralised scheduling. *Australian Computer Journal*, 13(2):58–63, May 1981.

[Chapin et al., 1999] S. J. Chapin, D. Katramatos, and J. Karpovich. The Legion Resource Management System, *Proceedings of the IPPS Workshop on Job Scheduling Strategies for Parallel Processing*, pp. 105–114, San Juan, Puerto Rico, 1999.

[Chapin and Spafford, 1994] S. J. Chapin and E. H. Spafford. Support for Implementing Scheduling Algorithms Using MESSIAHS. *Scientific Programming*, 3:325–340, 1994.

[Chou and Abraham, 1983] T. C. K. Chou and J. A. Abraham. Load redistribution under failure in distributed systems. *IEEE Transactions on Computers*, C-32(9):799–808, September 1983.

[Chowdhury, 1990] S. Chowdhury. The Greedy Load Sharing Algorithm. *Journal of Parallel and Distributed Computing*, 9:93–99, 1990.

[Foster et al., 1999] I. Foster et al. A Distributed Resource Management Architecture that Supports Advanced Reservations and Co-Allocation. *Proceedings of the International Workshop on Quality of Service*, 1999.

[Gantz et al., 1989] C. A. Gantz, R. D. Silverman, and S. J. Stuart. A Distributed Batching System for Parallel Processing. *Software–Practice and Experience*, 19, 1989.

[Gao et al., 1984] C. Gao, J. W. S. Liu, and M. Railey. Load Balancing Algorithms in Homogeneous Distributed Systems. In *Proceedings of the International Conference on Parallel Processing*, pp. 302–306. IEEE, August 1984.

[Ghafoor and Ahmad, 1990] A. Ghafoor and I. Ahmad. An Efficient Model of Dynamic Task Scheduling for Distributed Systems. In *Computer Software and Applications Conference*, pp. 442–447. IEEE, 1990.

[Hochbaum and Shmoys, 1988] D. Hochbaum and D. Shmoys. A Polynomial Approximation Scheme for Scheduling on Uniform Processors: Using the Dual Approximation Approach. *SIAM Journal of Computing*, 17(3):539–551, June 1988.

[Hsu et al., 1989] C. C. Hsu, S. D. Wang, and T. S. Kuo. Minimization of Task Turnaround Time for Distributed Systems. In *Proceedings of the 13th Annual International Computer Software and Applications Conference*, 1989.

[Huang et al., 1989] C. H. Huang, T. L. Huang, and J. Y. Juang. On Processor Allocation in Hypercube Systems. In *Proceedings of the 13th annual international computer software and Applications Conference*, 1989.

[Klappholz and Park, 1984] D. Klappholz and H. C. Park. Parallelized Process Scheduling for a Tightly-Coupled MIMD Machine. In *Proceedings of the International Conference on Parallel Processing*, pp. 315–321. IEEE, August 1984.

[Litzkow, 1987] M. J. Litzkow. Remote UNIX: Turning Idle Workstations Into Cycle Servers. In *USENIX Summer Conference*, pp. 381–384, 2560 Ninth Street, Suite 215, Berkeley, CA 94710, 1987. USENIX Association.

[Lo, 1988] V. M. Lo. Heuristic Algorithms for Task Assignment in Distributed Systems. *IEEE Transactions on Computers*, 37(11):1384–1397, November 1988.

[Matchmaking, 1998] Matchmaking: Distributed Resource Management for High Throughput Computing. *Proceedings of the 7th IEEE International Symposium on High Performance Distributed Computing*, 1998.

[NWS, 1999] R. Wolski, N. T. Spring, and J. Hayes. The Network Weather Service: A Distributed Resource Performance Forecasting Service for Metacomputing. *The Journal of Future Generation Computing Systems*, 1999.

[Ni and Abani, 1981] L. M. Ni and K. Abani. Nonpreemptive load balancing in a class of local area networks. In *Proceedings of the Computer Networking Symposium*, pp. 113–118. IEEE, December 1981.

[Nichols, 1987] D. A. Nichols. Using Idle Workstations in a Shared Computing Environment. In *Proceedings of the Eleventh ACM Symposium on Operating Systems Principles*, pp. 5–12. ACM, 1987.

[Price and Salama, 1990] C. C. Price and M. A. Salama. Scheduling of Precedence-Constrained Tasks on Multiprocessors. *Computer Journal*, 33(3):219–229, June 1990.

[Ramakrishnan et al., 1991] S. Ramakrishnan, I. H. Cho, and L. Dunning. A Close Look at Task Assignment in Distributed Systems. In *INFOCOM '91*, pp. 806–812, Miami, FL, April 1991. IEEE.

[Reif and Spirakis, 1982] J. Reif and P. Spirakis. Real Time Resrouce Allocation in Distributed Systems. In *Proceedings of the Symposium on Principles of Distributed Computing*, pp. 84–94. ACM, August 1982.

[Schopf and Berman, 1999] J. Schopf and F. Berman. Stochastic Scheduling. *Proceedings of SC99*, November 1999.

[Shirazi et al., 1995] B. A. Shirazi, A. R. Hurson, and K. M. Kavi, Editors. *Scheduling and Load Balancing in Parallel and Distributed Systems*. IEEE Computer Society Press, 1995.

[Singhal and Shivaratri, 1994] M. Singhal and N. G. Shivaratri. *Advanced Concepts in Operating Systems*. McGraw-Hill, 1994.

[Spring and Wolski, 1998] N. Spring and R. Wolski. Application Level Scheduling of Gene Sequence Comparison on Metacomputers. *Proceedings of the 12th ACM Conference on Supercomputiing*, July 1998.

[Squillante et al., 2001] M. S. Squillante, Y. Zhang, A. Sivasubramaniam, N. Gautam, H. Franke, and J. Moreira. Modeling and Analysis of Dynamic Coscheduling in Parallel and Distributed Environments. In *Proceedings of the 2002 ACM SIGMETRICS International Conference on Measurement and Modeling of Computer Systems*, 2002.

[Stankovic, 1984] J. A. Stankovic. Simulations of Three Adaptive, Decentralized Controlled, Job Scheduling Algorithms. *Computer Networks*, 8(3):199–217, June 1984.

[Stankovic and Sidhu, 1984] J. A. Stankovic and I. S. Sidhu. An Adaptive Bidding Algorithm for Processes, Clusters and Distributed Groups. In *Proceedings of the International Conference on Distributed Computing Systems*, pp. 49–59. IEEE, May 1984.

[Stankovic, 1985] J. A. Stankovic. Stability and Distributed Scheduling Algorithms. In *Proceedings of the 1985 ACM Computer Science Conference*, pp. 47–57. ACM, March 1985.

[Stone, 1977] H. S. Stone. Multiprocessor Scheduling with the Aid of Network Flow Algorithms. *IEEE Transactions on Software Engineering*, SE-3(1):85–93, January 1977.

[Stumm, 1988] M. Stumm. The Design and Implementation of a Decentralized Scheduling Facility for a Workstation Cluster. In *Proceedings of the 2nd IEEE Conference on Computer Workstations*, pp. 12–22. IEEE, March 1988.

[Su et al., 1999] A. Su, F. Berman, R. Wolski, and M. Mills Strout. Using AppLeS to Schedule Simple SARA on the Computational Grid. *International Journal of High Performance Computing Applications*, 13(3):253–262, 1999.

[Swanson et al., 1993] M. Swanson, L. Stoller, T. Critchlow, and R. Kessler. The Design of the Schizophrenic Workstation System. In *Proceedings of the Mach III Symposium*, pp. 291–306. USENIX Association, 1993.

[Theimer and Lantz, 1989] M. M. Theimer and K. A. Lantz. Finding Idle Machines in a Workstation-Based Distributed System. *IEEE Transactions on Software Engineering*, 15(11):1444–1458, November 1989.

[Van Tilborg and Wittie, 1984] A. M. Van Tilborg and L. D. Wittie. Wave scheduling — decentralized scheduling of task forces in multicomputers. *IEEE Transactions on Computers*, C-33(9):835–844, September 1984.

[Weissman, 1999] J.B. Weissman. Prophet: Automated Scheduling of SPMD Programs in Workstation Networks. *Concurrency: Practice and Experience*, 11(6):301–321, 1999.

[Weissman, 1998] J.B. Weissman. Gallop: The Benefits of Wide-Area Computing for Parallel Processing. *Journal of Parallel and Distributed Computing*, Vol. 54, No. 2, pp. 183–205, November 1998.

[Zhou et al., 1993] S. Zhou, X. Zheng, J. Wang, and P. Delisle. Utopia: a Load Sharing Facility for Large, Heterogeneous Distributed Computer Systems. *Software — Practice and Experience*, 23(12):1305–1336, 1993.

For Further Information

Many of the seminal theoretical papers in the area are contained in *Scheduling and Load Balancing in Parallel and Distributed Systems*, edited by Shirazi, Hurson, and Kavi [Shirazi et al., 1995]. This volume contains many of the papers cited in this chapter, and is an excellent starting point for those interested in further reading in the area.

Advanced Concepts in Operating Systems by Singhal and Shivaratri [Singhal and Shivaratri, 1994] contains two chapters discussing scheduling for parallel and distributed systems. These two references contain pointers to much more information than could be presented here.

Descriptions of other distributed scheduling systems may be found in papers describing Stealth [Singhal and Shivaratri, 1994] [Ch. 11], Utopia [Zhou et al., 1993], as well as in [Theimer and Lantz, 1989, Litzkow, 1987, Stumm, 1988]. The Global Grid Forum (http://www.gridforum.org) is codifying best practices for all aspects of Grid systems. More information about scheduling in distributed operating systems such as Sprite, the V System, Locus, and MOSIX can also be found in [Singhal and Shivaratri, 1994].

89

Distributed File Systems and Distributed Memory

89.1   Introduction **89**-1
89.2   Underlying Principles **89**-2
     Coherency of Data • Coherency of File Attributes
     • Performance • Resilience • Naming • Replication
     • Disconnected Operation • Security
89.3   Best Practices **89**-10
     NFS • Distributed File System • IVY • Munin
89.4   Research Issues and Summary **89**-16

T. W. Doeppner Jr.
Brown University

89.1   Introduction

The model of a single file system shared by all users of a computer is not only convenient but expected by most computer users. It seems natural to extend this model across multiple computers so that all users on a collection of computers share the same file system, thus forming a *distributed file system*. Similarly, the model of a collection of threads of control sharing the same address space as they cooperate in a computation is attractive for exploiting concurrency. This single-address-space abstraction is certainly the natural model for use on a shared-memory multiprocessor. Its convenience for programming is so compelling that it is used increasingly to take advantage of parallelism on distributed systems, where it is called *distributed memory*.

Primarily because of the ubiquity of Sun's Network File System (NFS), programmers have become accustomed to distributed file systems; they realize how much more convenient such a system is than explicitly copying files across machines. Whereas distributed memory is not so commonplace (most existing implementations are research projects), it shows great promise for taking advantage of the collective power of networked computers for computationally intensive problems.

The traditional means for implementing parallel applications on distributed systems is to use explicit message passing or remote procedure calls. These certainly give programmers full control over the locations of data and processing but also force them to be concerned about such details. The promise of distributed memory is that some of these details can be handled by the underlying implementation. In particular, programmers need not be concerned about the transfer of data among computers. Instead, data appear as needed merely because the program has referenced it. As with traditional virtual memory, exceptionally poor performance can arise, but most programs exhibit reasonable locality of reference and thus work quite well.

In both distributed file and distributed memory systems, the usual implementation model is one of clients obtaining data from servers. Clients typically maintain a cache of data recently obtained from servers: if data have been fetched previously, reads (in file systems) and loads (in memory systems) can often be satisfied directly from the cache without contacting the server. Writes (in file systems) and stores (in memory systems) can be applied to data in the cache and only later made visible to others by updating the server. For file systems, servers are typically distinct from clients and files are permanently assigned to servers, so that clients always contact the same server for the same file. However, an approach pioneered with distributed memory systems [Li and Hudak 1989] and recently adopted for distributed file systems [Anderson et al. 1995] distributes the server role among all of the clients: the home of a shared-memory segment is the client who last modified it.

Our primary concerns are performance, how (and whether) the clients' views of data are kept coherent, and how machine crashes and network outages are handled. Among the performance concerns are minimizing both network traffic and latency of responses to user requests. If the data necessary to handle a user request (such as a read operation on a file or a load from memory) are available locally, latency is minimal, but if not, it can be considerable. However, data can now be transferred over high-speed networks faster than from disks. Thus, it may be quicker to obtain data over the network from the primary storage of a server machine than from a local disk. In either case, since the impact of network traffic on overall performance is more dependent on the number of messages being transmitted than on their size, it is advantageous to transmit data in batches.

In the remainder of this chapter, we first discuss the underlying principles of distributed file and memory systems, including issues of coherency, performance, resilience, naming, replication, disconnected operation, and security. Next is a section on best practices in which we discuss two commercially distributed file systems — Network File System (NFS) and Distributed File System (DFS) — and two research distributed memory systems — IVY and Munin. Finally, we present a summary of the research issues in the field.

89.2 Underlying Principles

What distributed file systems and distributed memory systems have in common is an architectural model in which some number of computers have common access to data. The difference between the two lies primarily in their intended use. Distributed file systems provide data in the form of *files* whose lifetimes are usually far longer than the lifetimes of the processes accessing them. The usual emphasis in a distributed file system is on providing files for the private use of clients and providing shared files for the read-only use of clients. There might be some degree of support for read–write access to shared files, but such use is typically rare. In distributed memory systems the emphasis is on the read–write sharing of data organized into *segments*, and little attention is typically paid to data permanence; the lifetime of data is roughly equal to that of the processes sharing it. In distributed file systems a major concern is resiliency: coping with server and client crashes and network outages. This typically has not been a major concern of distributed memory systems, though there is no reason why it could not be. One usually thinks of the data provided by distributed memory systems as being mapped into the address spaces of the client processes, whereas the files provided by distributed file systems are thought of as being accessed by explicit input/output (I/O) calls (e.g., read and write), but a client operating system could provide a filelike interface to shared memory (though this would be unusual) and files could be mapped into a client's address space (a not uncommon technique).

What is usually desired of distributed file systems and distributed memory systems is that they be **access transparent**: that programs access data on remote computers in the same way as they would locally. Thus, programs use standard read- and write-type file-system calls or use standard load- and store-type operations on memory. This rules out approaches based on explicit file transfer, such as the Internet's file transfer protocol (FTP) and Unix's remote copy (rcp). The underlying implementation is responsible for whatever data movement may be required; the remote data appear to be local to the application program.

Another important property of files is *permanence*: they must be able to continue to exist even when they have no active users. Thus, files are stored on some form of nonvolatile storage, and this ties them to a particular site; files rarely move. Files require administration and maintenance: limitations on the use of available space may have to be enforced. In some circumstances groups of files may have to be moved to different storage devices, perhaps attached to different computers. Files must be backed up, i.e., copied, say to tape, to guard against loss of data because of loss of media or other problems.

It may also be convenient to replicate files and make them available from multiple sources to prevent bottlenecks and protect against loss of access due to server failure. Thus, the location of a file might change over time. An important property of a distributed file system for making such changes of location tolerable is **location transparency:** how one refers to a file, i.e., its name, should not depend on its location.

In the typical distributed memory system, permanence is not an issue. In a number of systems there is no fixed distinction made between clients and servers; instead, a segment has a moveable home, typically on the last computer to have modified it.

In the next few pages we examine the issues that arise in the design of distributed file and memory systems. For both, the idealization against which our designs are compared is the **single-system model**: the behavior observed by parties executing on a distributed system should be identical to the behavior they would observe if all were on a single computer.* In practice, some aspects of this ideal are not achievable or not even desirable, but it is our basis for examining the various approaches.

In the next few pages we discuss the major issues in the design of both distributed file systems and distributed memory systems. We start by discussing coherency, first of data, then of file attributes, and then we look at performance. The two concerns are somewhat at odds, and so we examine the interplay. Next we look at resilience, which is also interrelated with the first two concerns. We then look at naming issues, replication, disconnected operation, and finally security issues.

89.2.1   Coherency of Data

A major concern in both distributed file systems and distributed memory systems is coping with concurrent access to files or memory and still providing adequate performance and resilience. Strict adherence to the single-system model can be expensive. However, we can often weaken this model to provide improved performance without making sacrifices in other areas.

One way of achieving the ideal of the single-system model is for a system to be **strictly coherent:** whenever a thread running on a node reads from a file or loads from memory, the value it retrieves is the value produced by the most recent write or store to that location. This, of course, is exactly what happens in the single-system model. What makes strict coherency nontrivial to achieve is, of course, the distributed nature of the underlying architecture: it takes time to make modifications to data visible to all nodes. When a write or store is executed, we say that the value produced becomes *visible* when it can be retrieved by reads or loads from this location by other processors. We distinguish between when an instruction that modifies memory or files is *issued* and when its effect becomes visible to others. Thus, with strict coherency, something must be done to ensure that the effect of a write is visible to the next read to the same location.

Strict coherency, however, turns out to be stronger than necessary. A somewhat weaker requirement still equivalent to the single-system model is **sequential coherency** [Lamport 1979]: the effect of any execution of threads on a collection of nodes is one that could have happened had all been executed on the same processor. The idea here is that a read does not need to return the results of the most recent write if it was possible for that write to have occurred after the read; if it was just an accident of time that the write preceded the read, then there is no reason to require the read to return the write's results. However, if it was no accident that the write preceded the read — if due to synchronization or other mechanisms the write was required to precede the read — then the read must return the write's results.

To see the distinction between strict coherency and sequential coherency, consider the time lines for three nodes in Figure 89.1. Each makes the sequence of accesses (reads and writes) indicated by the subscripts;

---

*The word *computer* can be a bit ambiguous: it can mean, among other things, a standalone system or a single processor from a parallel computer. We often use the word *node* instead, which for our purposes is a system on which the single-system model is easily implemented (for example, a node might be a personal computer or a workstation; a multiprocessor is a node if it is a shared-memory multiprocessor). Thus, a distributed system consists of a number of nodes interconnected by some means for reasonably high-speed communication.

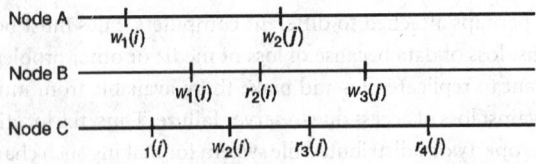

FIGURE 89.1 Concurrent access by three nodes.

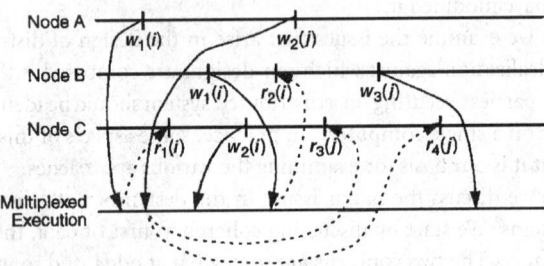

FIGURE 89.2 A possible execution in the sequentially coherent model.

the location accessed is given in parentheses. In a strictly coherent model, $r_1(i)$ of node C must retrieve what was written by $w_1(i)$ of node A, $r_2(i)$ of node B must retrieve what was written by $w_2(i)$ of node C, $r_3(j)$ of node C must retrieve what was written by $w_2(j)$ of node A, and $r_4(j)$ of node C must retrieve what was written by $w_3(j)$ of node B.

In a sequentially coherent model, however, any interleaving of the accesses of nodes A, B, and C is possible, as long as the relative order of the accesses by each node is preserved. The possible interleaving shown in Figure 89.2 demonstrates how the accesses might be multiplexed if all were run on a single node. The solid arrows indicate where write accesses would take place, and the dashed arrows indicate which values are retrieved by read accesses. The use of this model could certainly result in a number of possible outcomes of the accesses; if determinism is desired, then explicit synchronization is necessary.

Sequential coherency is certainly a weaker condition than strict coherency: under what circumstances can this weakness be exploited to yield more efficient systems? The fact that loads need not necessarily retrieve the values produced by the most recent stores suggests a technique for improving performance: stores of each processor may be collected into batches and made visible all at once; loads simply retrieve the values produced by the most recent visible stores. We discuss subsequently how we can guarantee that this technique results in sequentially coherent execution.

Unfortunately, there are further complications. Files are typically organized in terms of *blocks*: space may be allocated on disk in blocks and, more importantly for this discussion, files are transferred in block-sized pieces. Similarly, memory is usually organized in terms of *pages*: data are transferred in page-size units. Assume that a file server blindly obeys the requests of its clients to update blocks of a file with blocks supplied by the client (this is, in fact, how most file servers operate). One process might modify one data structure within a file with the knowledge that no other process is accessing that data structure. But some other process might be modifying a different data structure in the same file that happens to share the file-system block occupied by the first data structure. If these processes reside on different nodes, and if both modify their data structures at roughly the same time, a *race* results when each node sends its updated file-system block to the server. The server copies both versions of the file-system block to its (permanent) copy of the file, one at a time, probably in the order received. But since each copy of the block contains only a portion of the total update, the permanent copy of the file on the server will contain the result of one of the two updates, but not both, since the copy of the block it writes second overwrites the changes that came with the first copy. This problem, which occurs in distributed memory systems as well, is known as **false sharing**.

Some sort of synchronization is necessary to deal with the false-sharing problem. This synchronization can be provided automatically by the underlying distributed memory or file system. If a write access to any location within a unit is taking place, no other write or read access is allowed to take place at the same time to a location within the same unit. This ensures not only sequential coherency but also strict coherency. Furthermore, it allows writes to be made visible in batches.

However, even if the correctness issues of false sharing are adequately dealt with, there remains a performance problem: no concurrency of reads and writes to the same unit by different nodes is permitted. In many cases, particularly for file systems, such loss of concurrency is a minor problem since concurrent read/write access is rare. If concurrent access is frequent, however, then this loss of concurrency is serious.

Performance improvements are possible, even with false sharing, if we do not require the distributed file or memory system to take sole responsibility for sequential coherency but require user programs to assist. Thus, we define the notion of **weak coherency**, meaning that sequential coherency can be attained if additional instructions, not needed on a single-processor system, are executed by the program. In sequentially coherent systems we must make certain that, whenever a load takes place, it retrieves a value it could have retrieved on a single-node system. The underlying system has no means of determining ahead of time when loads will be taking place and thus must be ready to cope with them at all times. However, a programmer or compiler has knowledge of a program that can be used to advantage. For example, if certain locations are *private*, i.e., are used by only one thread, there is no need to ensure that their values are up to date in all nodes' views. Locations that are shared by multiple threads might be accessed only when they are protected by some sort of synchronization primitive, e.g., a mutex or a semaphore. Since they are not accessed when not so protected and are accessed by only the thread that arranged for the protection when they are protected, we can incorporate into the synchronization primitives code to ensure coherency. In particular, when a thread performs a lock operation to gain mutually exclusive access to a shared data structure, it might issue a *flushr* instruction (defined subsequently) that ensures that subsequent loads can retrieve data recently made visible. As part of an unlock operation, it might issue a *flushw* instruction (also defined subsequently), which ensures that its changes become visible.

How can a distributed file or memory system not be sequentially coherent? Consider the example in Figure 89.3, which is in terms of a memory system, though it applies equally well to file systems. Here we have two processors, each executing two instructions. Assume that the initial values in locations a and b are both 0. In the sequential coherency model, there are exactly three possible outcomes of the execution of the processors' instructions:

1. Processor 1's load returns 0, processor 2's load returns 1. This happens when processor 1's load occurs before processor 2's store.
2. Processor 1's load returns 1, processor 2's returns 0. This happens when processor 2's load occurs before processor 1's store.
3. Both loads return 1. This happens when neither processor 1's load occurs before processor 2's store nor processor 2's load occurs before processor 1's store.

However, if caching has delayed the effects of stores, has caused loads to retrieve old data, or both, there is a fourth possibility:

4. Both loads return 0. This happens when processor 1's load occurs after processor 2's store and processor 2's load occurs after processor 1's store. Since the effect of the stores is delayed, neither processor's load retrieves the value being set by the other processor's store.

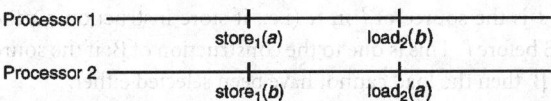

FIGURE 89.3 Potential violation of sequential coherency.

Caching is beneficial for performance, but if it can cause unanticipated results, it must be done with care. In our example we can eliminate the unanticipated results, yet still allow caching, by flushing the cache before each load. We now show that such flushing will always produce sequential coherency.

We first define exactly what we mean by *flush*. We assume that when writes and stores become visible, they do so for all nodes at once; this is true in most, if not all, distributed file and memory systems today and greatly simplifies our exposition. The value produced by a store instruction is not necessarily visible immediately after the instruction is issued. However, once a *flushw* is executed by a processor, the values produced by all stores issued by that processor prior to the issuance of the flushw become visible (if they are not already). A flushw is said to be *ordered* if it causes the values produced by stores to become visible in the same order in which the corresponding stores were issued. When processor A issues a load instruction to obtain a value from some specific location, the value obtained is the most recent visible value stored there before t time units in the past by some other processor or, if it occurred more recently, the most recent value stored there by processor A. The parameter t is unspecified; it may have an upper bound, but there is a fair amount of uncertainty about its value. However, if a *flushr* is executed immediately prior to the load, then t is zero. A *flush* is the combination of a flushw and a flushr. An *ordered flush* is the combination of an ordered flushw and a flushr.

Consider a distributed system X in which the order in which stored values are made visible is the order in which the stores were issued, and in which each load is preceded by an ordered flush. Is this system sequentially coherent? To show that it is, we must show, for each execution in this system, that there is an equivalent one in a system Y in which all processors execute the same instructions in the same order as they do in X, but in which stores become visible when they are issued.

Our first concern is to define what we mean by an execution in X. Since X is a distributed system, its executions consist of the executions of its component nodes. To simplify the presentation, we assume that each of these nodes consists of a single processor. An execution of a single processor is represented as the sequence of the instructions that were executed. Thus, if α_i is the execution sequence of processor i, then α, an execution of the distributed system, is the collection of all α_i for each processor i.

There is clearly a total order on the instructions in any execution sequence α_i. But because the processors of X execute in parallel, there is no total order on all of the instructions of the components of α; however, a number of partial orders can be defined. In particular, a store instruction supplies the value for a load instruction if the value retrieved from a location by the load was placed there by the store. We say that the store was the *source* of the load.

Our goal is to show that, for any execution α on distributed system X, there is a valid, equivalent execution sequence β on the single-processor system Y consisting of some interleaving of the execution sequences α_i of the component processors of X. Note that the store instructions in α become visible in some particular order. (If the store instructions of two different processors become visible simultaneously, we assume that their effect is as if one became visible before the other; we use this effective order to define our order of becoming visible.) We construct a proposed sequence β of instructions for system Y as follows (again to simplify the presentation, we assume that each of the α_i starts with a store): The first instruction of β is the first visible store of α. Following this in β are all of the initial instructions of each α_i up to, but not including, either the first store or the first load to retrieve a value not produced by the first visible store, whichever comes first. Next in β comes the next visible store from α, followed by all of the subsequent instructions of each α_i starting with the first not previously selected and continuing up to, but not including, either the first store or the first load to retrieve a value not produced by an already selected visible store, whichever comes first. The remaining instructions of β are sequenced accordingly.

We claim that β is not only a valid instruction sequence for the single-processor system Y but also is equivalent to α. To show this, we show that each load in β retrieves the same value as the corresponding load in α.

We first show that if s is the source of l in α (i.e., if store instruction s produced the value fetched by l), then s appears in β before l. This is due to the construction of β: if the source of a load has not been selected for inclusion in β, then the load cannot have been selected either.

Next we show that if a load appears in β, then any conflicting store (i.e., a store that places a value into the location being accessed by the load) that becomes visible after the source of the load became visible must appear after the load in β. To see this, again consider the construction of β. Suppose there is a load l in β, from processor i, for which a conflicting store appears between it and its source store. For this to happen, l could not have been selected for β as part of the instructions selected along with the source store. Thus, during the construction of β, there must have existed some positive number of stores and unsatisfied loads (i.e., loads whose source has not yet been selected for β) in the initial unselected portion of α_i appearing before l.

All of the stores in this portion must have been nonconflicting, because if they had been conflicting, since they occur in α_i and became visible after the source store, they would have overridden the value provided by the source store in α and prevented it from being the source store. If a conflicting store of some other processor had become visible before any of these nonconflicting stores of α_i and had thus been selected in β before l, then it too would have overridden the source store in α. This is because the flush executed before each load guarantees that any store instruction w in α_i appearing before l becomes visible before l is executed, and thus that if a conflicting store of another processor becomes visible before w, it will override l's source before l is executed.

Consider now the loads in the initial unselected portion of α_i. If they accessed the same location as l, they must have the same source store as l, since a different source store would have become visible after the source store of l and thus would have overridden it in α. If any of these load instructions was accessing a different location, both they and l could not have been selected for β until after the sources of these loads became visible. If a conflicting store of l comes before any of these stores and after l's source, then it over-rides this source in α. Thus, there cannot appear between a load in β and its source store a conflicting store.

Since we have shown that the source of a load appears before it in β and that no conflicting stores appear between the source and the load in β, we have what we were after: that loads in β produce the same values as loads in α. Because of this and the fact that the instructions of each α_i appear in the same relative order in β as they do in α_i, each execution in X has an equivalent execution in Y. Thus, X is sequentially coherent.

89.2.2 Coherency of File Attributes

An issue that affects file systems is maintaining file *attributes:* information about files, such as their sizes and the times of the most recent read and write accesses. This information is used often by clients and thus is often cached. Even if the underlying model is strictly coherent, the clients' views of the file system could be at odds with the single-system model if the attributes are not properly maintained.

For example, the following scenario has occurred (and has caused problems) in NFS: file X contains a sequence of records. A process P makes a private copy Y of X and records X's attributes at the time of the copy. P then edits Y, deleting some of the records contained in it (there are no consistency issues, since Y is a private file). P then uses some sort of mechanism to gain mutually exclusive access to X and replaces the contents of X with the contents of Y if X has not been modified (as reflected in the time of last modification stored in its attributes) since the copy Y was produced. If X has been modified (e.g., a new record has been added), then, rather than replacing X, P reproduces the changes it made to Y in the current version of X. Now, suppose a new record is added to X just before P gains mutually exclusive access to X. If P's copy of X's attributes is not appropriately updated, then, regardless of the memory model, P might replace X's contents with Y's rather than merge its edits to Y with the modified X. The result is that the record added to X is lost.

89.2.3 Performance

The primary performance concerns for distributed file and memory systems are minimizing both network traffic and latency of responses to user requests. If the data necessary to handle a user request (such as a read operation on a file or a load from memory) are available locally, latency is minimal, but if not, latency can be considerable. Since the impact of network traffic on overall performance is more dependent on the number of messages being transmitted than on their size, it is advantageous to transmit data in batches.

To reduce latency and network loads, most systems use *caching*: portions of files or segments are automatically maintained on client nodes, either on disk or in primary memory. Assuming that local access is quicker than remote access, this clearly helps reduce average latency, since many reads and writes (and loads and stores) can be handled directly from the cache. This also helps to reduce server and network loads since files and segments tend to be used repeatedly: once a file or segment has been transferred to a node, it is quite likely to be used many times, so that the number of transfers required over the network can be greatly reduced. In fact, many files, though they reside on server nodes, are accessed only on a single client node; thus, there is rarely a need for data transfer between server and client. Furthermore, network bandwidth can often be better utilized with caching, since large amounts of data can be transferred at once. However, data can now be transferred over high-speed networks faster than from disks. Thus, it may be quicker to obtain data over the network from the primary storage of a server machine than from a local disk.

To further improve latency, many systems, particularly file systems, use *prefetching*: a file or portion of a file is fetched before it is needed. Some operating systems provide asynchronous I/O facilities with which one can explicitly request data to be transferred without having to wait for it. One can use multithreaded programming techniques to get the same effect. These approaches require a very knowledgeable programmer and are often difficult to take advantage of. Some operating systems, such as Unix, attempt to predict what data are needed next and fetch it automatically before the program needs it. These techniques are effective only if files are being accessed sequentially (which is the only case for which the operating system can make predictions of data needs), but this happens frequently enough to be quite useful. Prefetching from the disk to primary storage is common in local file systems, and this notion is extended to prefetch from server to client in most distributed file systems.

One uses distributed memory systems to take advantage of parallelism. Thus, the most important measure of the performance of such systems is the speedup obtained when running a program on multiple nodes. This, of course, depends on the program being run, but we can compare actual performance with ideal performance: the time required if the nodes actually shared memory and all memory access times were those of local access.

89.2.4 Resilience

The next major concern is resilience. Failure is easy to cope with in the single-system model: either all components are running or all components are down — a crash crashes everything. Recovering from a crash entails restarting everything; merely resuming operations where things had left off is generally not possible. But this is one aspect of the model that we do not want to mimic in a distributed system: crashes of individual nodes should be handled gracefully by the other nodes. Functionality and data that are available only on the down nodes are lost for the duration of the downtime, but all other functionality and data should continue to be available. Moreover, operations emanating from client nodes that were in progress on a server node when it went down should be able to continue when the server comes back up, if they were not already completed on some other node.

Crash recovery involves first recognizing that there was a crash and then coping with it. Recognizing that there has been a crash is not simple, since a crashed node normally does not broadcast that it is about to crash before doing so. One node might suspect that another is down when nothing is received from the other node, despite expectations, after a reasonable period of time. But this unresponsiveness could be due to a communication failure having nothing to do with the node, or it could arise because the node is merely extremely busy, not down. Thus, the only sure way of determining that a node is down is for it to announce that it has been down as it restarts itself.

Coping with a server crash on a client might be as simple as waiting until the server restarts, then continuing normal operations, as if nothing had happened. Alternatively, the client operation might timeout, passing to the caller the decision on what to do. Yet another approach is to transfer requests from a down server to another server that provides equivalent files (i.e., replicas of the files provided by the down server).

A server's recovery from its own crash depends on how much state information was lost with the crash and hence must be recovered. In the simplest case, servers are *stateless*: there is no state information on the server, thus there is no state information that must be restored, and thus crash recovery involves simply restarting the server.

For servers that maintain state, the state information must be recovered. Such state information could be maintained on nonvolatile storage where it could survive a crash. This is generally not done: doing so would be expensive, and, moreover, losing state information in a crash can be advantageous. Suppose a client crashes while the server is down, thus invalidating its contribution to the server's state. If the server's state were maintained in nonvolatile storage, then, when it comes back to life, it must first determine that the client had crashed and then reset the client's contribution to the state. Since the server's recovery of state information depends on information supplied by clients when the server comes back to life, if the down client provides no information, there is no information that the server must reset.

For a client crash, the general assumption is that the operations in progress on the client cannot be resumed, and thus the client is simply restarted. The onus falls on the server to recognize a client crash and to react by restoring any client-specific state information to some initial value and recovering any resources that may have been allocated to the client. For stateless servers, nothing need be done, but for other types of servers a fair amount of work may be required.

89.2.5 Naming

Our next concern is the naming of files and how it relates to locating files. In the single-system model, we assume that a single tree-structured directory hierarchy (extended to a directed acyclic graph if links are allowed) is used for file naming. Files are identified by their pathnames in the directory hierarchy; this identification also serves as the means for locating files in secondary storage.

The entire name space can be viewed as a disjoint collection of subtrees joined together to form a tree: the root of one subtree is somehow connected to or superimposed on a directory of another. Each server provides some number of these subtrees to its clients. At issue are the mechanism for joining the subtrees into a single tree and the appearance of the tree to the clients. One approach is for each client to piece together the subtrees itself, independent of the other clients. This approach, which is used by NFS, allows each client node to tailor the combination of subtrees. Thus, though all client nodes might share access to all files, the pathnames of these files might differ across nodes (though, in practice, nodes are typically configured so this is not the case, i.e., any shared file has the same pathname on all nodes).

Another approach is that the connections between subtrees are built into the subtrees themselves. For example, in DFS, subtrees contain links to other subtrees; the subtrees are connected into a single tree whose appearance is identical to all clients.

It is also important that client nodes have private files, containing information to be used only on the node. This could be done by giving each client a subtree and position within the global name space, e.g., */NodeA/. . . , /NodeB/. . . .* But what is usually done is to provide a private name space so that the same name can refer to different files on different nodes. For example, on Unix systems, whether using NFS or DFS (or both), */etc/passwd* refers to the password file on the node on which the pathname is used. This provides what we call **inverse location transparency:** programs referring to node-private files (e.g., */etc/passwd*) can use the same pathname regardless of the node on which they are running. The disadvantage is that one cannot easily refer to node-private files on other nodes, but this is rarely important.

89.2.6 Replication

One of the advantages of distribution is that multiple copies of files can be maintained on separate servers, i.e., files can be *replicated*. This can be taken advantage of both for performance and for reliability: the task of providing files to clients can be spread over several servers, thereby reducing the load on each of them. If one server goes down, the others can take over its load.

With location transparency we have the notion that the name of a file or group of files does not tie it to a particular site. Thus, if for administrative or other reasons the files must be moved, they need not be renamed. If the naming technique permits a group of equivalent files to be given a single name, then any of the group can be used to satisfy a read of the files associated with the name. This allows easy **failover** to an alternative server if one holding a copy of the file fails.

89.2.7   Disconnected Operation

Another concern is support for *disconnected operation:* Can a client continue to operate when disconnected from a file server for extended periods of time? For this to take place, those files needed by the client must be somehow cached on the client. In connected operation clients and servers cooperate to maintain the consistency of the various cached copies of files with the copies maintained on servers, but in disconnected operation, such consistency management is not possible. The same sort of consistency management performed in connected operation could be performed for the disconnected case during the possibly brief periods for which a node is connected to the server. The difference, of course, is the time scale involved. In a typical scenario a client node might load its cache from the server, disconnect from the network, and then operate on the cached files for hours or days before reconnecting with the server. Any attempt to provide single-system semantics will fail in this sort of environment.

89.2.8   Security

Our final concern is security. In the single-system model all files and accessors appear to be on the same node and thus access is controlled by a single operating system. There is no opportunity to circumvent security measures other than by successfully masquerading as another user (perhaps by guessing a password). For distributed file systems things are more difficult. File-system data and other information are transmitted across communication networks and are thus subject to being read and modified by malicious third parties. Thus, providing a single-node-system level of security requires measures far beyond those required on such a system.

Providing security across nodes can be expensive. Such expense is justifiable in many, but certainly not all, situations. One approach, as suggested in Hartman and Ousterhout [1995], is to have fairly relaxed security within local clusters of nodes (whose users presumably trust one another) and to apply more stringent security measures to accesses between such clusters.

89.3   Best Practices

In this section we look at the application of the principles discussed in the previous section by examining two commercially available distributed file systems — Sun's Network File System (NFS) and OSF DCE's Distributed File System (DFS) — and two distributed memory systems that are products of university research — IVY from Yale University and Munin from Rice University. NFS, which has been the standard for use in distributed Unix systems for the past decade, is a relatively simple system that has undergone much scrutiny as it has evolved and improved. DFS, which was developed by Transarc and is an outgrowth of their Andrew file system (AFS), is considerably more complex than is NFS and comes closer to achieving our single-system-model ideal in some respects, though not in others. IVY, the seminal research implementation of distributed memory, shows how a virtual memory implementation could be extended to support a strictly coherent distributed memory. Munin takes a somewhat different approach: its model is weakly coherent and sequential coherency is achieved by a variety of techniques, chosen depending on the types of shared data structures and programmer-supplied information on how such data structures are used.

89.3.1   NFS

NFS has four component protocols: a *mount protocol* for making collections of files stored on servers available to client nodes, a *file protocol* for accessing and manipulating files and directories, a *lock* and *status protocol* for locking files over the network and recovering lock state after failures, and an *automount protocol* to support replication and name-space management.

Servers divide their files into disjoint collections called *file systems*, each of which contains a rooted directory tree naming all its members. Servers, as specified in their local /etc/exports files, specify which file systems or subtrees within a file system are available to which clients. By cooperating with a server via the mount protocol, a client can then *mount* a remote file system. This entails superimposing the root of the remote file system on top of a directory in the client's current naming tree. The root of the remote file system (the mounted file system) effectively replaces the mounted-on directory. Thus, the remote file system is attached to the client's naming tree at the mounted-on directory (and the previous contents of that directory are invisible as long as the remote file system remains mounted). The mount protocol additionally provides some minimal security by allowing administrators to specify for each file system on a server which client nodes are allowed to mount it.

Once a client has mounted a remote file system, its processes may access files in this system. This is where the file protocol is used: it provides a remote-procedure-call interface on the server for the client to use. A client process initiates activity on a file by opening it, an action that involves presenting a pathname to its local operating system, which follows the path, into remote file systems as necessary. If the path terminates in a remote file system, a *file handle* is returned to the client operating system identifying the remote file. The server, being stateless, keeps no record of the fact that the file is being used by the client. The file handle is used to identify the file to the server for all operations, including reading and writing.

The original NFS protocol took advantage of a cache of recently accessed file blocks maintained in the client's primary storage, known as the *buffer cache*, which allowed the prefetching of file-system blocks before they are needed (if the file is being accessed sequentially) and the write-behind of modified blocks. The buffer cache has been augmented in the most recent version of the protocol (version 3) with an on-disk cache, used primarily for files that rarely change. The buffer cache suffers from cache-coherency problems. NFS keeps such problems to a normally unnoticeable minimum, but occasionally false sharing and other problems can give the programmer unpleasant surprises.

The NFS approach to cache coherency is for the client to check with the server periodically to determine if items in its cache are stale and, if so, remove them. This is done by maintaining an *attributes cache* on each client that contains, for each file for which the client is caching blocks, the file's attributes, consisting of the file's size and access and modification times. When a file's attributes are fetched, they are tagged with the current time. Blocks of the file in the cache may be used on the client until the attributes expire, usually in a few seconds (the expiration time is a function of how recently the file was last modified; a file modified in the recent past is likely to be modified again soon). When an attempt is made to access blocks in the cache after the attributes have expired, a call is made to the server for the current attributes. If the file has been modified since the previous attributes were fetched, all blocks of the file are removed from the cache (modified blocks are sent back to the server to update the file). Thus, the cache is always close to a consistent state, but its consistency is never guaranteed.

Note that NFS is weakly coherent only if used via its mapped-file interface. On Unix systems, one can map an NFS-provided file into a client's address space. If one is careful to lay out data structures to prevent false-sharing problems, one can use flushr- and flushw-type instructions to achieve program-assisted sequential coherency. Though a flushw-type instruction exists for the file interface (i.e., accessing files via reads and writes), there is no flushr instruction.

An important aspect of performance is the minimization of messages transferred between client and server. NFS is normally layered on top of Sun RPC, which in turn is layered on the user datagram protocol (UDP) of the Internet suite. Sun RPC, particularly when layered on top of UDP, is quite simple: for each request, a response is expected. If the response is not received in a reasonable amount of time, the request is retransmitted, and again and again until a response is received. The usual case, of course, is that the request and response are each transmitted only once, and thus a minimum number of messages is transmitted.

There is a potential problem here. If a response is not received, then the request is retransmitted. If the reason that the response was not received is that the request was lost, then, assuming the retransmitted request arrives at the server, only one instance of the request is executed on the server. But suppose not the first request, but the first response was lost. When the retransmitted request arrives at the server, it is executed, with the result that two instances of the same request are executed on the server. If the request has a cumulative effect, such as transferring $100 from one bank account to another, this double execution would be a serious bug. However, in NFS most requests are designed to be **idempotent** — the effect of executing such a request twice is exactly the same as executing it once. For example, a request to write to a file specifies which file, the location in the file, and the data to be written. Executing the request twice causes the same data to be written to the same place twice — inefficient, but not harmful.

Recent improvements in the performance of transmission control protocol (TCP) implementations have caused some vendors to run NFS on TCP. Although this protocol guarantees reliability, idempotency is still an issue. If a server crashes while performing an operation, the client is uncertain whether the operation took place. Idempotency allows the client simply to repeat the operation once the server is running again, without fear of causing harm.

NFS is the prime example of the advantages of statelessness. Servers running NFS's file protocol need not retain any information about their clients or what their clients are doing: any information about clients maintained on a server is an optimization; it can be regenerated if necessary. Thus, no recovery actions are required of the file protocol on the server after the server has crashed.

As previously mentioned, an issue for the client side is determining *if* the server has crashed. With stateless servers, the client application (as opposed to client operating system) can be oblivious to whether a server has crashed; the operating system simply repeatedly retries file-protocol RPCs to the server until a response is received; the application perceives a slowdown but no loss of function or data. However, via the mount protocol, a client node can specify how a server's crash (or, more accurately, server unresponsiveness) is to be dealt with, by specifying how the remote file system is mounted. One option, known as the *hard mount*, is for the client to retry operations repeatedly, as just described, until they succeed. This has the advantage that, other than a delay in execution, the client process's semantics are unaffected. However, it has the disadvantage that it can cause a client process to block indefinitely, waiting for the result of the operation. Thus, another option, known as the *interruptible hard mount*, allows the remote operation to be aborted while in progress. This has the benefit of releasing a process (and perhaps an entire user session) from the clutches of a server that will be down for an indefinite period but certainly has a dramatic effect on the process's semantics. The final option, the *soft mount*, is that the operation returns an error code indicating that no response was obtained after a reasonable number of retries (note that the operation may or may not have successfully taken place on the server — all the client knows is that there has been no response). This is a useful alternative if the client application takes the trouble to notice such error codes and is prepared to deal with them, but otherwise could cause serious problems. The great advantage of the first two options is that from the client application's point of view, no special treatment is required for server failures — the system is **failure-transparent**.

Coping with a client crash is equally simple since servers are stateless. When a client crashes, whatever was running on it at the time of the crash is lost; when the client restarts, no information is available on what it was doing at the time of the crash. Thus, if the client had state on the server that must be cleaned up after the crash, it cannot direct the cleanup, since it has no knowledge of what this state is. Since servers are stateless, this is not a problem: clients simply restart and servers are oblivious to the fact that anything happened.

However, some sort of state information is required to support file locking: we must keep track of whether files are locked or not. NFS deals with this with a pair of protocols: one for maintaining locks (the *lock protocol*) and one for keeping track of whether clients have restarted (the *status protocol*). A file's server maintains the state information associated with locking. If a client crashes while holding a lock on a file, this lock must be removed. As noted earlier, the server cannot use the lack of communication from the client as an indication that the client has crashed; it requires definite notification, which is supplied when the client restarts (thus, files remain locked while the client is down). Manual intervention may be necessary if the client is down for an extended period of time.

A server crash is more difficult to deal with: the state information concerning locks is kept in volatile storage, which is lost after a crash. If it were kept in nonvolatile storage, then if both a client and the server crashed, there would be difficulties in determining that the client's locks must be removed; the server would realize that certain items are locked but would not realize that not only should they no longer be locked but also that the client is unaware that the items are locked. Thus, when a server crashes, all locks are removed by default. It is then up to the clients, if they remain up, to restore the lock information on the server. When a server restarts after a crash, it informs all its clients via the status protocol that it is now back in operation. It also establishes a grace period during which clients must notify it of locks they had held before the crash, so as to reclaim them. During this period no new lock requests are honored. After the grace period, all state information is recovered (that not recovered, presumably due to down clients, is lost) and the server goes back to normal operation.

The final aspect of NFS has to do with the name space. There are two issues here. It is the responsibility of each node to set up its own name space by mounting into its hierarchy file systems provided by servers. The number of such file systems could be huge, even though, for any one particular node, many are rarely, if ever, accessed. Thus, rather than attempting to use the mount protocol to mount all conceivable file systems when a node is booted, the *automount protocol* is used to mount file systems when needed.

The other issue, also dealt with by the automount protocol, is to support the replication of read-only file systems. If a server providing an important file system, such as the binary images of system commands, crashes, it is important that there be some sort of failover facility so that an alternative server can provide the file system. To accomplish this transparently to the clients, both instances of the file system should appear to have the same name, i.e., the same pathname in a client's directory tree. When a client node mounts a remote file system using the automount protocol, it broadcasts a request asking for providers of the desired file system. There could be many potential providers; the first one whose response is received is chosen. Thus, if one provider is down, another can be used.

The only drawback of this approach occurs when the provider of the important file system crashes while it is mounted on a client's naming tree. There is no support for automatically unmounting the file system and looking for a new provider. Thus, though a new provider is available, it cannot be used. What is done is to unmount file systems automatically if they have not been used after a period of time, typically five minutes. An automount broadcast is then performed on the next access to the file system. This approach works well for file systems that are used sporadically but does not solve the problem for file systems in constant use.

89.3.2 Distributed File System

OSF DCE's DFS goes much farther in some areas than NFS toward achieving the single-system ideal, though not as far in others. Assuming all components (clients, servers, and network) stay up and running, its variance from the ideal is small. DFS achieves this even though it employs large client caches on local disks to hold all files (split into chunks) being used on each client. It maintains consistency via a token-passing algorithm, which produces much state information on servers that must be restored after a server crash and cleared after a client crash.

A DFS installation is an optional part of a DCE *cell*, which is a potentially large (thousand-node) collection of computers sharing a common security database. Each cell supporting DFS has a single DFS name space used by all nodes. As in NFS, the name space consists of a collection of *file sets* (a better term than the ambiguous file system used in NFS and other Unix file systems). The file sets are connected together into a single tree structure by storing mounting information not at each node but in the file sets themselves. This mount information is represented via a form of *symbolic link* that provides the name of the file set mounted at this directory. Unlike NFS, in which the root of the mounted file set replaces the mounted-on directory, here the root of the mounted file set becomes the child of the mounted-on directory. In fact, one mounted-on directory can contain links to any number of file sets.

DFS exploits these links to help provide for the replication of read-only file sets. One can create any number of read-only copies of a file set. Each file set has a short, descriptive name; a read-only copy has the suffix "readonly" at the end of its name. Encoded in the mounting information stored in a file-set link is an indication of whether it is a read–write mount point, so that the read–write version is required, or that it is a read-only mount point, so that any of the read-only replicas may be used.*

---

*There is also a regular mount point that is equivalent to a read–write mount point if it (the mount point) resides in a read–write file set, and to a read-only mount point if it resides in a read-only replica.

A special replicated database, the file-set location database (FLDB), provides a mapping from file-set names to the names of the servers that hold the various replicas of the file set. This database is used by client-side DFS code when following a path: when a file-set link is encountered, the client code looks up the file-set name stored in the link in FLDB (these lookups are cached so that name-to-file-set translation is not terribly expensive). Depending on the type of mount point, it selects either the server containing the read–write replica of the file set or any of the servers containing read-only replicas. If the client is using a read-only replica and the server becomes unresponsive (perhaps because it has crashed), the client operating system simply finds from FLDB another server containing a read-only replica and (quietly) switches to it; the client application is oblivious. Thus, when using read-only file sets, failover is automatic. No attempt is made to support read–write replication.

DFS maintains strict coherency. Its clients maintain caches on their local disks that store files in chunks of typically 64 kilobytes each. As in NFS, prefetching and write-behind are used to overlap I/O and computation. Servers maintain the consistency of the chunks from their files that appear on client caches by controlling which clients may read and modify the chunks. This is accomplished via a token-passing algorithm: for a client to perform an operation on some portion of a file in its cache, it must have, from the server, a token that grants it the necessary permission (note that this is not access permission in the sense of whether or not the client is authorized to access the file but merely indicates whether an access can now be done in a strictly coherent fashion).

Various forms of permissions are represented by tokens. There are tokens that allow a client node to read data from a portion of a file and tokens that allow a client node to modify data in a portion of a file. There are tokens for locking a file (or portions of a file), both for shared (read) locks and for exclusive (write) locks. Tokens are required for reading and setting file attributes (such as modification time, access time, and file size). To maintain strict coherency, if any client node is modifying a portion of a file, no other node may be reading or modifying that portion of the file. Of course, any number of nodes may be reading a portion of a file at once, as long as no node is modifying it. This is controlled through the distribution of tokens: to modify a portion of a file in its cache, a client node must obtain a write token from the file's server. To read the data in its cache, the node must have a read token. The server is responsible for making certain that if a write token for a particular portion of a file has been granted, then no read tokens for that portion are outstanding, and so forth. If, for example, a write token is outstanding and some node wishes to read the file, the server contacts the holder of the write token to revoke it and then gives the other node a read token.

Similarly, to modify a file's attributes, a client must have an appropriate token; the server will grant the token only if no tokens are outstanding for reading the attributes. A small problem here is that included with file attributes is the file access time, which should be updated every time the file is read. But doing so requires a *status write* token, which cannot be granted if there are *status read* or status write tokens outstanding. Thus, for single-system semantics, multiple nodes cannot be reading the same portion of a file at once, since doing so would require that they all have status write tokens. Because of this, DFS must deviate from single-system semantics and not maintain the access time of a file exactly as done on a single system. Instead, the update of the access time may be delayed, so that the file may have been accessed some time ago, but the access time does not yet reflect it.

Because DFS must maintain a lot of state information (e.g., what tokens are out), its crash recovery is much more complicated than NFS's. Three independent things can go wrong:

- *A client can crash:* Thus, the server will need to reclaim all of the tokens that were held by the client.
- *A server can crash:* Token information is not held in nonvolatile storage. It must somehow be re-created when the server comes back up.
- *The network can fail:* Though both client and server remain up, neither can communicate with the other.

Suppose that a server is unable to contact a client. If the client has possession of tokens and another client wishes to perform an operation that conflicts with the first client's tokens, then the server would like to revoke the tokens. Since the client is unresponsive, the server must do this unilaterally.

Conversely, suppose that a client is unable to contact a server. As long as the client has what it needs in its cache, then it really has no need to contact the server — it can get along quite well on its own. When the server comes back to life, it recovers its tokens using an approach similar to that used in NFS's lock protocol: the clients notify the servers of the tokens they possess.

Thus, we have two potentially conflicting points of view:

- The client should be able to use its cache even if the server is down or not accessible.
- The server should be able to revoke tokens from a client even if the client is down or not accessible.

If either the server or the client has crashed, then providing the other's point of view is not difficult. But if a network outage occurs and the server and client become separated from each other but both continue to run, then the two points of view conflict with each other.

DFS uses a compromise approach. If the client cannot contact the server, it continues to use its cache until its tokens expire; they are typically good for two hours, though they are normally refreshed every minute or so. However, say the server is actually up and running but is somehow disconnected from the client. If the server has no need to revoke tokens, then it does nothing. But if some other client that is communicating with the server attempts an operation that conflicts with the unresponsive client's tokens, then the server is forced to take action.

If the server has not heard from the client for a few minutes, it can revoke the client's tokens unilaterally. This means that when the client does resume communication with the server, it may discover that not only are some of its tokens no longer good, but some of its modifications to files may be rejected.

To protect client applications from such unexpected bad news, the client-side DFS code causes attempts to modify a file to fail if it has discovered that the server is not responding. A client program can take measures to deal with this problem by repeatedly retrying operations until the server comes back to life. This does not provide the transparency of the NFS hard-mount approach, but it does allow the client to use its cache to satisfy reads while the server is down.

89.3.3 IVY

IVY [Li and Hudak 1989] is one of the earliest distributed memory systems. A research project that included a study of various techniques for handling the visibility of writes while maintaining strict coherency, it helped to establish the viability of the concept of distributed memory by demonstrating impressive performance.

A primary concern was how to maintain strict coherency yet propagate the effects of writes efficiently. All of their solutions involve the notion of a node owning a shared page. Ownership is dynamic: to modify a page, a node must first own it. Other nodes can have copies of a page in their cache; when the owner modifies a page, it *invalidates* the copies in the other nodes' caches by sending them invalidation messages.

To modify a page, a node must first become its owner. This involves first determining the present owner, so that an ownership transfer can be arranged. Thus, some sort of mapping must be established associating page number with owning node. One technique is to have a central server handling the mapping chores: a node attempting to modify a page of which it is not the owner contacts the single mapping node to find the location of the owner, establishes a new mapping giving itself as the owner, and finally contacts the current owner to take over the page. The technique has the advantage of being simple but the disadvantage of creating a performance bottleneck and a single point of failure.

An alternative technique is to distribute the mapping functionality. The simplest implementation of this idea is to spread the mapping chores across all nodes but to have a fixed assignment of page number to node mapping the page to its owner. The developers' experiments showed that this technique was superior to the central mapping technique, but they found it difficult to obtain a fixed assignment that fit all applications well.

The final technique involves a dynamic approach to mapping page number to owner node: each node maintains a table mapping each page to its *probable* owner. Requests for change of ownership are sent to the probable owner who, if not the actual owner, forwards the request to the node it believes to be the owner, and so forth until the actual owner is reached. In the worst case, $N - 1$ messages are required to locate a page's owner, assuming N nodes. The ownership information is updated whenever a node receives an invalidation request, relinquishes ownership, or suffers a write fault, so that the worst-case number of messages for locating a page's owner rarely occurs. Li and Hudak [1989] show that the maximum number of messages required to locate the owner of a single page k times is $O(N + k \log N)$; in practice, a page's owner can be found fairly quickly.

89.3.4 Munin

Munin [Carter et al. 1995], developed at Rice University and the University of Utah, is an example of the use of weak coherency and other techniques to improve the performance of a distributed memory system. Rather than use a single software approach to ensure sequential consistency, multiple approaches are used, depending on the access patterns to the shared data. The loss of concurrency due to false sharing is minimized by using a *write-shared protocol* that merges together modifications made to the same page by separate nodes. In addition, an *update-with-timeout* mechanism removes copies of shared data from caches of nodes that have not used them for a while — this reduces the number of nodes whose caches must be invalidated when there is an update.

Research by the developers suggests that support for five access patterns can significantly improve performance in a distributed memory system:

- *Conventional shared variables* are treated just like shared data in IVY: to modify such a variable, a node must be the sole owner of the page containing it. When modifications occur, invalidation messages are sent to nodes containing copies of the page.
- *Read-only data*, once initialized, are never modified. Thus, no overhead is required to maintain coherency.
- *Migratory data* are used by one node at a time. Thus, once a node that has not been accessing such a data item starts accessing it (via a read or a write), the entire item is transferred to the new node and the original copy in the old node is invalidated.
- *Write-shared data* is a collection of items, each being modified by a single node, that share a page. Without special treatment there would be a false-sharing problem, but, by using the *write-shared protocol*, full concurrency can be achieved: when a node propagates its changes to write-shared data, it transmits merely its changes to the page rather than the entire page. Receivers of these updates can then merge these changes with their own. Conflicting changes can be detected and flagged as run-time errors.
- *Synchronization variables* are specialized implementations of three common synchronization constructs: *locks* (otherwise known as *mutexes*), *barriers*, and *condition variables*. They allow the programmer to take advantage of weak coherency by modifying shared data only when they are protected by the synchronization variables. Sequential coherency is obtained by making modifications visible at appropriate moments (e.g., during an unlock operation).

The net effect of these features of Munin is to decrease significantly the number of messages from those required by strictly coherent distributed memory systems. The developers' measurements show their system to be within 5% of implementations of a number of applications done with explicit message passing and within roughly 30% for others. They argue that further enhancements will improve the latter results substantially.

89.4 Research Issues and Summary

The notion of distributed file systems is a straightforward extension of the notion of local file systems: functionality available on nonnetworked individual computers has been made available to networked computers collectively. But computing has changed a great deal since the days when the only file systems were local file systems — it is no longer enough for data, organized into files, to be made available strictly to a collection of machines statically interconnected with cable. For the notion of a distributed file system to remain relevant, it must be useful in a world in which mobile computing, multimedia, and enterprise computing play key roles.

In mobile computing, client machines are often disconnected from servers. When they are connected, it might be via relatively low-speed links, such as direct and cellular telephone connections. Some commercial products, such as Lotus Notes®, provide some support for data access by mobile users, but they operate in specialized domains (electronic mail, simple databases) and do not provide the general-purpose foundation for constructing applications that is the promise of distributed file systems. Research is under way in adapting distributed file systems to mobile use (e.g., Satyanarayanan et al. [1993] and Huston and Honeyman [1995]). One goal of such research projects is to extend to mobile computing the functionality currently available to clients statically connected to their servers via high-speed networks. The issues being grappled with include providing some degree of coherency (sequential coherency may prove completely infeasible in such environments), loading a client's cache with appropriate files for a lengthy disconnected session, and scheduling communications traffic so that information needed urgently is given top priority. Dealing with coherency may involve resolving the effects of conflicting changes to files by multiple clients. This may be automatic in some cases, but many currently require assistance from the user. Preloading the cache often requires user input. Scheduling communications traffic allows currently needed data to be fetched prior to write-backs of modified data (though this, of course, has an effect on coherency).

Distributed file systems supporting multimedia have stringent performance demands. Not only must data be transferred quickly, but the transmission rate must be reliable so that data arrive before they are needed. A system specialized as a video server developed at Microsoft [Rashid 1994] transmits movies chosen from a library to a large number of clients at an appropriate rate and allows VCR-like functions such as fast forward and reverse. The demands on this system could well prevent its functionality from being subsumed by a general-purpose distributed file system, but some of the techniques used could be of use in general-purpose systems for handling more moderate numbers of clients.

Many of the issues arising in large enterprise systems have to do with scaling, taken here as the number of clients that can use a distributed file system. One point of view is that the collected file systems of a large company (or the entire world, for that matter) should be transparently accessible by anyone from anywhere subject, of course, to security constraints. A relatively small-scale implementation of this view involves users of Transarc's AFS system (the latest incarnation of the forerunner of DFS). Users around the world see one another's files as fitting in a single-directory hierarchy. A program running on a workstation in Providence, RI, can access files in Sydney, Australia, using exactly the same steps as to access files on a local disk. This works because, though the participating sites are widespread, they are relatively few; each site maintains a list of all other sites. If the entire world is to be made available in this fashion, other means will be necessary for sites to locate one another.

With large numbers of clients potentially accessing individual files, improvements in caching and load balancing become essential. Whereas both NFS and DFS support the replication of read-only collections of files, the processes of determining how many replicas to make and where to place them are entirely manual. At the very least, monitoring tools must be available to let an administrator react quickly to changing access patterns, creating and placing replicas, and deleting unneeded ones. Ideally, this sort of load balancing would be automatic.

The promise of distributed memory systems is to provide an easy-to-use programming paradigm for highly parallel systems. For the promise to become reality, the amount of parallelism must approach levels of interest to those with significant computational demands. As the numbers of processors increase, it may become important to integrate some degree of fault tolerance into the model so that individual processor failures do not take down the entire system.

In architecture there are two separate trends, one toward the interconnection of a collection of workstations into a distributed memory system and the other toward a system consisting of a large number of processors (perhaps all in one box) designed from the beginning to be a distributed memory system. In the former, the distributed memory is implemented entirely with software (other than hardware support for virtual memory); the latter typically uses hardware support.

Utilizing a collection of workstations as a parallel computer has been a goal of a number of researchers and many reasonably successful implementations exist (e.g., IVY and Munin). The attraction is that an organization might have a large number of workstations, many of which are unused over long stretches of time (overnight, for example). Among the longstanding issues only partly resolved in current systems are identifying available workstations and coping when workstations become unavailable (i.e., forcing off distributed memory users when the workstation is needed for other work). Adequate resolution of these issues could make this model of computation much more common.

In dedicated distributed memory systems in which processors are physically close to one another, interprocessor communication speeds can be quite high, rivaling processor-memory communication speeds. Such systems are beginning to appear on the market. Among the research issues still requiring attention are means for adequately distributing the parallel components of a computation, balancing the loads of the various processors, and devising and popularizing programming techniques for exploiting large-scale parallelism.

Defining Terms

Access transparency: A system property by which the complications involved in providing access to something (e.g., data) are not apparent to the accessor.

Failover: A system property by which, in the event of a failure of one component, the function provided by that component is taken over by another.

Failure transparency: A system property by which no special actions are required of clients to cope with failures of servers.

False sharing: What happens when two data structures share the same unit of storage, e.g., blocks in file systems in pages in memory systems.

Idempotency: The effect of performing an operation many times in succession (with no conflicting operation appearing between the repetitions) is the same as the effect of performing the operation once.

Inverse location transparency: A system property by which a given name refers to a different item on each node. Thus, a name can be used transparently by a program to refer to information specific to the node on which the program is running.

Location transparency: A system property by which how one refers to something depends on the location of neither the subject nor the object.

Sequential coherency: The property of a distributed file or memory system in which the effect of any execution is an effect that could have happened had all computation taken place on a single processor.

Single-system model: The computational model in which all computation takes place on a single processor.

Strict coherency: The property of a distributed file or memory system in which each read or load retrieves the value produced by the most recent write or store to that location.

Weak coherency: The property of a distributed file or memory system that does not necessarily provide sequential coherency by itself but that can provide sequential coherency if certain additional instructions are executed by any program running on the system.

References

Anderson, T. E., Dahlin, M. D., Neefe, J. M., Patterson, D. A., Roselli, D. S., and Wang, R. Y. 1995. Serverless network file systems, pp. 109–126. In *Proc. 15th ACM Symp. Operating Syst. Principles*. ACM, New York, Dec.

Carter, J. B., Bennett, J. K., and Zwaenepoel, W. 1995. Techniques for reducing consistency-related communication in distributed shared-memory systems. *ACM Trans. Comput. Syst.* 13(3):205–243.

Hartman, J. H. and Ousterhout, J. K. 1995. The zebra striped network file system. *ACM Trans. Comput. Syst.* 13(3):274–310.

Huston, L. B. and Honeyman, P. 1995. Partially connected operation. *USENIX Comput. Syst.* 8(4):365–380.

Lamport, L. 1979. How to make a multiprocessor computer that correctly executes multiprocess programs. *IEEE Trans. Comput.* C-28(9):690–691.

Li, K. and Hudak, P. 1989. Memory coherence in shared memory virtual memory systems. *ACM Trans. Comput. Syst.* 27(4):321–359.

Rashid, R. 1994. Microsoft's tiger media server. *1st Networks Workstations Workshop Rec.* Oct.

Satyanarayanan, M., Kistler, J. J., Mummert, L. B., Ebling, M. R., Kumar, P., and Lu, Q. 1993. Experience with disconnected operation in a mobile environment, pp. 11–28. In *Proc. ACM Symp. Mobile Location-Independent Comput.* USENIX, Berkeley, CA, Aug.

Further Information

Good coverage of a number of the issues covered in this chapter can be found in *Distributed Systems,* edited by Sape Mullender, Addison–Wesley, 1993.

Both ACM's bi-yearly *Symposium on Operating Systems Principles* and the ACM journal *Transactions on Computer Systems* often contain a number of excellent papers on issues related to distributed file and memory systems. Copies of the proceedings and subscription information for the journal can be obtained from ACM, 1515 Broadway, 17th Floor, New York 10036, (212) 869-7440.

The USENIX journal *Computing Systems* also often contains excellent papers on issues related to distributed file and memory systems. Subscription information can be obtained from The MIT Press Journals, 55 Hayward Street, Cambridge, MA 02142, (617) 253-2866, journals-info@mit.edu.

Programming Languages

The area of Programming Languages includes programming paradigms, language implementation, and the underlying theory of language design. Today's prominent paradigms include imperative (with languages like COBOL, FORTRAN, and C), object-oriented (C++ and Java), functional (Lisp, Scheme, ML, and Haskell), logic (Prolog), and event-driven (Java and Tcl/Tk). Scripting languages (Perl and Javascript) are a variant of imperative programming for Web applications. Event-driven programming is useful in Web-based and embedded applications, and concurrent programming serves applications in parallel computing environments. This section also provides a balanced treatment of the underlying theories of language design and implementation such as type systems, semantics, memory management, and compilers.

90 **Imperative Language Paradigm**   *Michael J. Jipping and Kim Bruce* 90-1
Introduction · Data Bindings: Variables, Type, Scope, and Lifetime
· Control Structures · Best Practices · Research Issues and Summary

91 **The Object-Oriented Language Paradigm**   *Raimund Ege* 91-1
Introduction · Underlying Principles · Best Practices · Language Implementation
Issues · Research Issues

92 **Functional Programming Languages**   *Benjamin Goldberg* 92-1
Introduction · History of Functional Languages · The Lambda Calculus: Foundation
of All Functional Languages · Pure Versus Impure Functional Languages · SCHEME: A
Functional Dialect of LISP · Standard ML: A Strict Polymorphic Functional
Language · Nonstrict Functional Languages · HASKELL: A Nonstrict Functional
Language · Research Issues in Functional Programming

93 **Logic Programming and Constraint Logic Programming**   *Jacques Cohen* 93-1
Introduction · An Introductory Example · Features of Logic Programming
Languages · Historical Remarks · Resolution and Unification · Procedural
Interpretation: Examples · Impure Features · Constraint Logic Programming · Recent
Developments in CLP (2002) · Applications · Theoretical Foundations · Metalevel
Interpretation · Implementation · Research Issues · Conclusion

94 **Scripting Languages**   *Robert E. Noonan and William L. Bynum* 94-1
Introduction · Perl · Tcl/Tk · PHP · Summary

95 **Event-Driven Programming**   *Allen B. Tucker and Robert E. Noonan* 95-1
Foundations: The Event Model · The Event-Driven Programming
Paradigm · Applets · Event Handling · Example: A Simple GUI
Interface · Event-Driven Applications

96 **Concurrent/Distributed Computing Paradigm** *Andrew P. Bernat and Patricia Teller* **96**-1
Introduction • Hardware Architectures • Software Architectures • Distributed Systems • Formal Approaches • Existing Languages with Concurrency Features • Research Issues • Summary

97 **Type Systems** *Luca Cardelli* **97**-1
Introduction • The Language of Type Systems • First-Order Type Systems • First-Order Type Systems for Imperative Languages • Second-Order Type Systems • Subtyping • Equivalence • Type Inference • Summary and Research Issues

98 **Programming Language Semantics** *David A. Schmidt* **98**-1
Introduction • A Survey of Semantics Methods • Semantics of Programming Languages • Applications of Semantics • Research Issues in Semantics

99 **Compilers and Interpreters** *Kenneth C. Louden* **99**-1
Introduction • Underlying Principles • Best Practices • Incremental Compilation • Research Issues and Summary

100 **Runtime Environments and Memory Management** *Robert E. Noonan and William L. Bynum*.. **100**-1
Introduction • Runtime Stack Management • Pointers and Heap Management • Garbage Collection • Summary

90

Imperative Language Paradigm

90.1 Introduction 90-1
90.2 Data Bindings: Variables, Type, Scope,
and Lifetime 90-2
Binding Time • Variables • Types • Scope • Execution
Units: Expressions, Statements, Blocks, and Programs
90.3 Control Structures 90-6
Conditional Structures • Iterative Structures • Unconstrained
Control Structures: Goto and Exceptions • Procedural
Abstraction • Data Abstraction
90.4 Best Practices 90-15
Data Bindings: Variables, Types, Scope, and Lifetime
• Execution Units • Control Structures • Procedural
Abstraction • Data Abstraction and Separate Compilation
90.5 Research Issues and Summary 90-20

Michael J. Jipping
Hope College

Kim Bruce
Williams College

90.1 Introduction

In the 1940s, John von Neumann pioneered the design of basic computer architecture by structuring computers into two major units: a central processing unit (CPU), responsible for computations, and a data storage unit, or memory. This architecture is demand driven, based on a command and instruction-oriented computing model. The basic unit cycle of execution, typically composed of a single instruction, consists of four steps:

1. Obtain the addresses of the result and operands.
2. Obtain the operand data from the operand location(s).
3. Compute the result data from the operand data.
4. Store the result data in the result location.

Note in this sequence how separation of the execution unit from the memory unit has structured the sequence. Data must be located and piped from memory, operated on, and transferred back to memory to be available for the next operation. All operations in a von Neumann machine operate this way, in a stepwise, structured manner. The von Neumann model has been the basis of nearly every computer built since the 1940s.

Imperative programming languages are modeled after the von Neumann model of machine execution and were invented to provide the abstractions of machine components and actions in order to make it easier to program computers. Abstractions such as variables (which model memory cells), assignment statements (which model data transfer), and other language statements are all abstractions of the basic von Neumann approach.

In this chapter we address the fundamental principles underlying imperative programming languages and examine the way the constructs of imperative languages are represented in several languages. We devote special attention to features of more modern imperative programming languages, among them support for abstract data types and newer control constructs such as iterators and exception handling. Examples in this chapter are given in a variety of imperative programming languages, including FORTRAN, Pascal, C, C++, MODULA-2, and Ada 83. In the Best Practices section we explore in more detail the languages FORTRAN IV (chosen for historical reasons), C and C++ (its imperative parts), and Ada 83.

90.2 Data Bindings: Variables, Type, Scope, and Lifetime

In this section we discuss some of the fundamental properties of imperative programming languages. In particular, we address issues related to binding time, the properties of variables, **types**, scope, and lifetime.

90.2.1 Binding Time

We will find it useful to classify many of the differences in programming languages based on the notion of binding time. A **binding** is the association of an attribute to a name. The time at which a binding takes place is an important consideration. There are many times when a binding can occur. Some of these follow:

- Language definition: when the language is designed. An example is the binding of the constant name true to the corresponding Boolean value.
- Language implementation: when a compiler or interpreter is written. An example is the binding of the representation of values of various types.
- Compile time: when a program is being translated into machine language. For example, the type of a variable in a statically typed language is bound at compile time. In statically typed languages, overloaded functions are bound at compile time.
- Load time: when the executable machine language image of the program is loaded into the memory for execution by the execution unit. The location of global variables is bound at load time.
- Procedure or function invocation time: the time a program is being executed. Actual parameters are bound to formal parameters and local variables are bound to locations at procedure invocation time.
- Run time: any time during the execution of a program. A new value can be bound to a variable at run time. In dynamically typed languages, overloaded functions are bound at run time.

As we examine fundamental issues in the definition of imperative programming languages, we will keep in mind the distinctions between languages based on differences in binding time.

90.2.2 Variables

Imperative languages support computation by executing commands whose purpose is to change the underlying state of the computer on which they are executed. The *state* of a computer encompasses the contents of memory and also includes both data which are about to be read from outside of the computer and data which have been output.

Variables are central to the definition of imperative languages as they are objects whose values are dependent on the contents of memory. A variable is characterized by its attributes, which generally include its name, location in memory, value, type, scope, and lifetime.

Depending on context, the meaning of a variable may be considered to be either its value or its location. For instance, in the assignment statement, $x := x + 1$, the meaning of the occurrence of the variable x to the left of the assignment symbol is its location (sometimes called the l-value of x), whereas the meaning of the occurrence on the right side is its value, that is, the value stored at the location corresponding to x (sometimes called the r-value). The location of global variables is bound at load time, whereas the location of local variables and reference parameters is typically bound at procedure entry. The value of the variable can be changed at any point during execution of the program.

90.2.3 Types

Types in programming languages are abstractions which represent sets of values and the operations and relations which are applicable to them. Types can be used to hide the representation of the primitive values of a language, allow type checking at either compile time or run time, help disambiguate overloaded operators, and allow the specification of constraints on the accuracy of computations. Types also can play an important role in compiler optimization.

Types in a programming language include both simple and composite types. The use of *simple types* such as integer, real, Boolean, and character types allows the user to abstract away from the actual computer representation of these values, which may differ from computer to computer. The operations on simple types may or may not be supported directly by the underlying hardware. For instance, many early microprocessors supported only real or floating-point operations in software.

Some languages (e.g., those derived from Pascal) allow the programmer to define their own simple *enumerated* types by simply listing the values of the type. The ordering of elements in this enumeration is significant as these types typically support successor and predecessor functions as well as ordering relations. Later we will discuss mechanisms for supporting **abstract data types**, another way of constructing types which can be used as though they were primitive to a language.

Many languages support the creation of *subrange* types, which allows a programmer to define a new type as a copy of a type with a subset of its values. The new type comes equipped with the same operators as its parent type and is usually compatible with the original type.

Composite or structured data types can be created from simple types using *type constructors*. Typical composite types include arrays, records (or structures), variant records (or unions), sets, subranges, pointer types, and, in a few languages, function or procedure types. For instance, arrays are typically constructed from two types: a subrange type which provides the set of indices of the array, and another type representing the values stored in the array. Not all languages support all these type constructors. For instance, function and procedure types are provided by MODULA-2 but are not available in Ada 83. Many languages support strings as special types of composite types, for instance, as arrays of characters, but they may also be provided as builtin types.

Most imperative languages bind types to variables statically. These bindings are usually specified in declarations, but some languages, such as FORTRAN, allow implicit declaration of variables, with the type binding determined by the name of the identifier (e.g., in FORTRAN if the name starts with I through N then the variable is an integer, otherwise real).

An important issue in type-checking programming languages is type equivalence. When do two terms have equivalent types? The two extremes in the definitions of **type equivalence** are structural and name equivalence:

- Structural equivalence: Two types are said to be *structurally (or domain) equivalent* if they have the same structure. That is, they are built from the same type constructors and builtin types in the same way.
- Name equivalence: Two types are *name equivalent* if they have the same name.

The language C uses structural equivalence, whereas Ada 83 uses name equivalence. There are also a range of possibilities between these two extremes. For instance, Pascal and MODULA-2 use *declaration equivalence*: two types are declaration equivalent if they are name equivalent or they lead back to the same structure declaration by a series of redeclarations.

Inequivalent types may be compatible in certain situations. For instance, two types are assignment compatible if an expression of one type may be assigned to a variable of another. For instance, in Pascal a subrange of integer is assignment compatible with integer, even though the types are not equivalent.

An application of these ideas can be found in the rules for determining whether a particular actual parameter may be used in a procedure call for a particular formal parameter. In Pascal, if the formal parameter is a reference parameter then the actual parameter must be a variable of equivalent type. If the formal parameter is a value parameter then the actual parameter must be assignment compatible.

As mentioned earlier, some languages support the creation of subrange types. The new subrange type is usually assignment compatible with the original type. Because of this compatibility, the new type is called a **subtype** of the parent in Ada. Another mechanism available in Ada, called **derived typing**, defines a new type by constructing an exact copy of a type that already exists. However, the resulting new type is distinct and is not type equivalent or even assignment compatible with the existing type.

The type equivalence rules are the cause of one of the greatest limitations in the use of Pascal. If a formal parameter has an array type, then the actual parameter must have an equivalent type. In particular, the subscript ranges of the two arrays must be identical. Thus, it is impossible to write a procedure in Pascal which can be used to sort different-sized arrays of real numbers. (Actually, the current ANSI standard Pascal provides a special mechanism to allow exceptions to this rule.)

Ada escapes from this problem by designating some properties of types to be static, while others are dynamic. For example, in a type defined to be a subrange of integers, the underlying static type is integer while the subrange bounds are a dynamic property. Only the static properties of types are considered at compile time by the type checker, whereas restrictions due to dynamic properties are checked at run time.

Consider the following Ada declarations as an example of type bindings:

```
type COINS is (PENNY, NICKEL, DIME, QUARTER);
subtype SILVER is COINS range (NICKEL..QUARTER);
type CHANGE is new COINS;

C1, C2; COINS;
S: SILVER;
CH: CHANGE;
```

COINS is an enumerated type, defined by the programmer to allow assignments such as

```
C1 := DIME;
```

SILVER is a subrange of **COINS**, which includes only the values **NICKEL**, **DIME**, and **QUARTER**. **CHANGE** is a derived type taken from **COINS**.

Because Ada employs name equivalence, only **C1** and **C2** are equivalent, but **S** is assignment compatible with them. If Ada used structural equivalence, then variables **C1**, **C2**, and **CH** would be equivalent.

90.2.4 Scope

The scope of a binding is the area or section of a program in which that particular binding is effective. The method and extent of **scope rules** that define a binding scope will, to a large degree, affect the usefulness and applicability of a language. If, for instance, the rules allow the scope of a binding to be determined by the execution path of a program, the language might be more flexible, yet the code becomes harder to understand.

Scope rules are tied tightly to concepts of binding time. *Static scope rules* determine the scope of a binding at compile time and are based on the lexical structure of the program. *Dynamic scope rules* determine the scope of a binding at run time. Thus, an occurrence of a variable name in a procedure may refer to one variable the first time it is evaluated yet refer to an entirely different variable the next time, depending on the execution path at run time. Most imperative languages use static scope rules.

```
with TEXT_IO; use TEXT_IO;
procedure SCOPED is
   package INT_IO is new INTEGER_IO (integer); use INT_IO;
   I,J: integer;

   procedure P is begin put (J); new_line; end P;
   begin
      J := 0;
      I := 10;
      declare   -- Block 1
         J: integer;
      begin
         j := I; -- reference point A
         P;
      end;
      put (J); new_line;
      declare   -- Block 2
         I: Integer
      begin
         I := 5
         J := I + 1; -- reference point B
         P;
      end;
      put (J); new_line;
   end;
```

FIGURE 90.1   Scoping rules in Ada.

As an example of scope rules in Ada, consider the code in Figure 90.1. Static scope rules are determined by the program block structure, which does not change while the program runs. Therefore, the call to procedure **P** prints the variable **J** defined in the outer, main program, no matter where it is called from. Likewise, the assignment in block 1 at reference point A changes **J** from the block and not from the main program. Dynamic scope rules, on the other hand, typically follow dynamic call paths to determine variable bindings. If Ada used dynamic scope rules, the first call to **P** from block 1 would print the value 10 corresponding to the **J** from block 1, whereas the second call to **P** would print the value 3 corresponding to the **J** from the main program.

90.2.5   Execution Units: Expressions, Statements, Blocks, and Programs

An *expression* is a program phrase which returns a value. Expressions are built up from constants and variables using operators. As described earlier, variables may represent two values, depending on context: their location and the value stored at that location. Operators may be builtin, like the arithmetic and comparison operators, or may be user-defined functions.

Reflecting the sequential order of von Neumann computation, an imperative language specifies the order in which operations are evaluated. Typically, evaluation order is determined by precedence rules. A typical precedence rule set for arithmetic expressions might be the following:

1. Subexpressions inside parentheses are evaluated first (according to the precedence rules).
2. Instances of unary negation are evaluated next.
3. Then, multiplication ($*$) and division ($/$) operators are evaluated in left to right order.
4. Finally, addition ($+$) and subtraction ($-$) are evaluated left to right.

Although procedure rules are commonly used by imperative languages, some languages use other conventions to avoid precedence rules. For example, PostScript uses postfix notation for expressions, while LISP uses prefix notation. APL evaluates all expressions from right to left without regard to precedence, using only parentheses to change the evaluation order.

The fundamental unit of execution in an imperative programming language is the *statement*. A statement is an abstraction of machine language instructions, grouped together to form a single logical activity.

The simplest and most fundamental statement in imperative programming languages is the assignment statement. This statement, typically written in the form **x** **:=** **e** or **x** **=** **e** with **x** a variable (or other expression representing a location) and **e** an expression, is usually interpreted by evaluating **e** and copying its value into the location represented by **x**. This is known as the *copy semantics* for assignment.

Less common are languages which use the sharing interpretation of assignment. In these languages, variables generally represent references to objects which contain the actual values. The assignment $x := y$ would then be interpreted as binding the object referred to by y to x rather than its value. Since both variables refer to the same object, they share the same value. If the value of one is changed, the value of the other will also change. This is the *sharing semantics* for assignment.

Declarations and statements may be grouped together to form a *block*. Procedure and function bodies are represented as blocks, whereas **control structures** (discussed subsequently) can also be understood as acting on blocks of statements (generally without declarations). The most general form of a block contains a *declarative* section, which contains the declarations that define the bindings that are effective in the block, and an *executable* section, which contains the statements over which the binding is to hold, i.e., the scope of the declarations.

In so-called block-structured languages (including most languages descended from ALGOL 60, e.g., Pascal, Ada, and C), blocks may be nested. Within any block, therefore, there can be two kinds of bindings in force: *local bindings*, which are specified by the declarative sections associated with the block, and *nonlocal bindings* (also known as *global* bindings), which are bindings defined by declarative sections of blocks within which the specific block is nested.

Consider again the code from Figure 90.1. The first two assignments of the main program assign **J** from the main program the value 0 and **I** from the main program the value 10. The next assignment assigns the value 10, derived from the global **I**, to the variable **J** from the first inner block. When the definition of the second inner block is encountered, the variable **I** is found in the local scope, while **J** is found in the *outer* scope, that of the main program. The value 6 will be printed for **J** at the end of the main program.

90.3 Control Structures

By adopting the semantics of the basic execution cycle of a von Neumann architecture, an imperative language adopts a strict sequential ordering for its statements. By default, the next statement to execute is the next physical statement in the program. Control structures in imperative languages provide ways to alter this strict sequential ordering. The most common control structures are *conditional structures* and *iterative structures*. *Unconstrained control structures* are also allowed in most languages through the use of goto statements.

90.3.1 Conditional Structures

Conditional control structures (also known as *selection statement*) determine whether or not a block of statements is executed based on the result of one or several tests. These structures fall into one of two classes:

90.3.1.1 If Statements

All imperative languages include some form of if statement. This control structure provides a text and a single statement or statement block to be executed if the test evaluates to a true value. Optionally, the programmer may provide another block of statements which can be executed only if the test evaluates to false. The following is a simple example from Ada:

```
if (x = 2) then
   y := 3;
else
   y := 6;
end if;
```

The variable **y** is set to either 3 or 6 depending on the value of **x**.

In most languages, if statements can be nested within other control structures, including other if statements. However, nested if statements can result in awkward, deeply nested code. Thus, many languages provide a special construct (e.g., **elsif** in Ada) to represent *else if* constructs without requiring further nesting. The two Ada examples given next are equivalent semantically, though the first, which uses **elsif**, is easier to read than the second, which uses nested conditionals:

```
if (x = 2) then           if (x = 2) then
   y := 3;                    y := 3;
elsif (x = 3) then        else
   y := 15;                  if (x = 3) then
elsif (x = 5) then            y := 15;
   y :=18;                  else
else                          if (x = 5) then
   y := 6;                       y := 18;
end if;                       else
                                y := 6;
                            end if;
                          end if;
                        end if;
```

90.3.1.2 Case Statements

This conditional combines case-by-case expression examination with a restricted multiway conditional. This conditional may be seen to be simply a syntactic convenience, but in many cases its implementation results in a much faster determination at run time of the actual block of code to be executed. Consider the following case statement from Ada:

```
case y is
   when 2 => y := 3;
   when 3 => y := 15;
   when 15 => y := 18;
   when others => y := 6;
end case;
```

An expression (**y** in this case) of an ordinal type occurs after the keyword **case**. Each **when** clause contains a guard, which is a list of one or more constants of the same type as the expression. Most languages require that there be no overlap between these guards. The expression after the keyword **case** is evaluated, and the resulting value is compared to the guards. The block of statements connected with the first matched alternative is executed. If the value does not correspond to any of the guards, the statements in the **others** clause is executed. Note that the semantics of this example is identical to that of the previous example.

The case statement may be implemented in the same way as a multiway if statement, but in most languages it will be implemented via table lookup, resulting in a constant time determination of which block of code is to be executed.

C's switch statement differs from the case previously described in that if the programmer does not explicitly exit at the end of a particular clause of the switch, program execution will continue with the code in the next clause.

90.3.2 Iterative Structures

One of the most powerful features of an imperative language is the specification of *iteration* or statement repetition. *Iterative structures* can be classified as either definite or indefinite, depending on whether the number of iterations to be executed is known before the execution of the iterative command begins:

- Indefinite iteration: The different forms of indefinite iteration control structures differ by where the test for termination is placed and whether the success of the test indicates the continuation or termination of the loop. For instance, in Pascal the **while-do** control structure places the test before the beginning of the loop body (a pretest), and a successful test determines that the execution of the loop shall continue (a continuation test). Pascal's **repeat-until** control structure, on the other hand, supports a posttest, which is a termination test. That is, the test is evaluated at the end of the loop and a success results in termination of the loop.

  Some languages also provide control structures which allow termination anywhere in the loop. The following example is from Ada:

  ```
  loop
     ...
     exit when test;
     ...
  end loop
  ```

  The **exit when test** statement is equivalent to **if test then exit.**

  A few languages also provide a construct to allow the programmer to terminate the execution of the body of the loop and proceed to the next iteration (e.g., C's **continue** statement), whereas some provide a construct to allow the user to exit from many levels of nested loop statements (e.g., Ada's named **exit** statements).

- Definite iteration: The oldest form of iteration construct is the definite or fixed-count iteration form, whose origins date back to FORTRAN. This type of iteration is appropriate for situations where the number of iterations called for is known in advance. A variable, called the *iteration control variable* (ICV), is initialized with a value and then incremented or decremented by regular intervals for each iteration of the loop. A test is performed before each loop body execution to determine if the ICV has gone over a final, boundary value. Ada provides fixed-count iteration as a for loop; an example is shown next.

  ```
  for i in 1..10 loop
     y := y + i;
     z := z * i;
  end loop;
  ```

  Here, **i** is initialized to 1, and incremented by 1 for each iteration of the loop, until it exceeds 10. Note that this type of loop is a pretest iterative structure and is essentially syntactic sugar for an equivalent **while** loop.

  An ambiguity that may arise with a for loop is what value the iteration control variable has after termination of the loop. Most languages specify that the value is formally undetermined after termination of the loop, though in practice it usually contains either the upper limit of the ICV or the value assigned which first passes the boundary value. Ada eliminates this ambiguity by treating the introduction of the control variable as a variable declaration for a block containing only the for loop.

Some modern programming languages have introduced a more general form of for loop called an *iterator* construct. Iterators allow the programmer to control the scheme for providing the iteration control variable with successive values. The following example is from CLU [Liskov et al. 1977]. We first define the iterator:

```
string_chars = iter (s : string) yields (char);
  index: Int  := 1;
  limit: Int  := string$size (s);
  while index <= limit do
    yield (string$fetch(s, index));
    index := index + 1;
  end;
end string_chars;
```

which can be used in a **for** loop as follows:

```
for c: char in string_chars(s) do LoopBody end;
```

When the for loop controlled by an iterator is encountered, control is passed to the iterator, which runs until a **yield** statement is executed. The value associated with the **yield** statement is used as the initial value of the iterator control variable **c**, and the body of the loop is executed. Control is then passed back to the iterator, which resumes execution with the statement following the **yield**. Control is passed to the loop body each time a yield statement is executed and back to the iterator each time the loop body finishes execution. Thus, iterators behave as a restricted form of coroutine, passing control back and forth between the two blocks of code. The loop is terminated when the iterator runs to completion. In the preceding examples this will occur when **index > limit**.

90.3.3 Unconstrained Control Structures: Goto and Exceptions

Unconstrained control structures, generally known as goto constructs, cause control to be passed to the statement labeled by the **identifier** or line number given in the goto statement. Dijkstra [1968] first questioned the use of goto statements in his famous letter, "Goto statement considered harmful," to the editor of the *Communications of ACM*. The controversy over the goto mostly centers on readability of code and handling of the arbitrary transfer of control into and out of otherwise structured sections of program code.

For example, if a goto statement passes control into the middle of a loop block, how is the loop to be initialized, especially if it is a fixed-count loop? Even worse, what happens when a goto statement causes control to enter or exit in the middle of a procedure or function? The problems with readability arise because a program with many goto statements can be very hard to understand if the dynamic (run time) flow of control of the program differs significantly from the static (textual) layout of the program. Programs with undisciplined use of gotos have earned the name of *spaghetti code* for their similarity in structure to a plate of spaghetti.

Although some argue for the continued importance of goto statements, most languages either greatly restrict their use (e.g., do not allow gotos into other blocks) or eliminate them altogether. In order to handle situations where gotos might be called for, other, more restrictive language constructs have been introduced to make the resulting code more easily readable. These include the **continue** and **exit** statements (particularly labeled **exit** statements) referred to earlier.

Another construct which has been introduced in some languages in order to replace some uses of the goto statement is the *exception*. An exception is a condition or event that requires immediate action on the part of the program. An exception is *raised* or *signaled* implicitly by an event such as arithmetic overflow or an index out of range error, or it can be explicitly raised by the programmer.

The raising of an exception results in a search for an exception *handler*, a block of code defined to handle the exceptional condition and (hopefully) allow normal processing to resume. The search for an appropriate handler generally starts with the routine which is executing when the exception is raised. If no appropriate handler is found there, the search continues with the routine which called the one which contained the exception. The search continues through the chain of routine calls until an appropriate handler is found, or the end of call chain is passed without finding a handler.

If no handler is found the program terminates, but if a handler is found the code associated with the handler is executed. Different languages support different models for resuming execution of the program. The termination model of exception handling results in termination of the routine containing the handler, with execution resuming with the caller of that routine. The continuation model typically resumes execution at the point in the routine containing the handler which occurs immediately after the statement whose execution caused the exception.

The following is an example of the use of exceptions in Ada (which uses the termination model):

```
procedure pop(s: stack) is
  begin
    if empty(s) then raise emptyStack
                else ...
  end;
procedure balance (parens: string) return boolean is
    pStack: stack
  begin
    ...
    if ... then pop(s) ...
  exception
    when emptyStack => return false
  end
```

Many variations on exceptions are found in existing languages. However, the main characteristics of exception mechanisms are the same. When an exception is raised, execution of a statement is abandoned and control is passed to the nearest handler. (Here "nearest" refers to the dynamic execution path of the program, not the static structure.) After the code associated with the handler is executed, normal execution of the program resumes.

The use of exceptions has been criticized by some as introducing the same problems as goto statements. However, it appears that disciplined use of exceptions for truly exceptional conditions (e.g., error handling) can result in much clearer code than other ways of handling these problems.

We complete our discussion of control structures by noting that, although many control structures exist, only a very few are actually necessary. At the one extreme, simple conditionals and a goto statement are sufficient to replace any control structure. On the other hand, it has been shown [Boehm and Jacopini 1966] that a two-way conditional and a while loop are sufficient to replace any control structure. This result has led some to point out that a language has no need for a goto statement; indeed, there are languages that do not have one.

90.3.4 Procedural Abstraction

Support for abstraction is very useful in programming languages, allowing the programmer to hide details and definitions of objects while focusing on functionality and ease of use. **Procedural abstraction** [Liskov and Guttag 1986] involves separating out the details of an execution unit into a procedure and referencing this abstraction in a program statement or expression. The result is a program that is easier to understand, write, and maintain.

The role of procedural abstraction is best understood by considering the relationships between the four levels of execution units described earlier: expressions, statements, blocks, and programs. A statement can contain several expressions; a block contains several statements; a program may contain several blocks. Following this model, a procedural abstraction replaces one execution unit with another one that is simpler. In practice, it typically replaces a block of statements with a single statement or expression.

The *definition* of a procedure binds the abstraction to a name and to an executable block of statements called the *body*. These bindings are compile-time, declarative bindings. In Ada, such a binding is made by specifying code such as the following:

```
procedure area (height, width: real; result: out real) is
   begin
      result := height * width;
   end;
```

The *invocation* of a procedure creates an activation of that procedure at run time. The *activation record* for a procedure contains data bound to a particular invocation of a procedure. It includes slots for parameters, local variables, other information necessary to access nonlocal variables, and data to enable the return of control to the caller. In languages supporting recursive procedures, more than one activation record can exist at the same time for a given procedure. In those languages, the lifetime of the activation record is the duration of the procedure activation.

Although scoping rules provide access to nonlocal variables, it is generally preferable to access nonlocal information via **parameter** passing. Parameter-passing mechanisms can be classified by the direction in which the information flows: *in parameters*, where the caller passes data to the procedure, but the procedure does not pass data back; *out parameters*, where the procedure returns data values to the caller, but no data are passed in; and *in out parameters*, where data flow in both directions.

Formal parameters are specified in the declaration of a procedure. The *actual parameters* to be used in the procedure activation are specified in the procedural invocation. The procedure passing mechanism creates an association between corresponding formal and actual parameters. The precise information flow which occurs during procedure invocation depends on the parameter passing mechanism.

The association or mapping of formal to actual parameters can be done in one of three ways. The most common method is *positional parameter association*, where the actual parameters in the invocation are matched, one by one in a left-to-right fashion, to the formal parameters in the procedural definition. *Named parameter association* also can be used, where a name accompanies each actual parameter and determines to which formal parameter it is associated. Using this method, any ordering can be used to specify parameter values. Finally, *default parameter association* can be used, where some actual parameter values are given and some are not. In this case, the unmatched formal parameters are simply given a default value, which is generally specified in the formal parameter declaration.

Note that in a procedural invocation, the actual parameter for an in parameter may be any expression of the appropriate type, since data do not flow back, but the actual parameter for either an out or an in out parameter must be a variable, because the data that are returned from a procedural invocation must have somewhere to go.

Parameter passing is usually implemented as being one of copy, reference, and name. There are two copy parameter passing mechanisms. The first, labeled *call-by-value*, copies a value from the actual to the formal parameter before the execution of the procedure's code. This is appropriate for in parameters. A second mode, called *call-by-result*, copies a value from the formal parameter to the actual parameter after the termination of the procedure. This is appropriate for out parameters. It is also possible to combine these two mechanisms, obtaining *call-by-value-result*, providing a mechanism which is appropriate for in out parameters.

The *call-by-reference* passes the address of the actual parameter in place of its value. In this way, the transfer of values is not by copying but occurs by virtue of the formal parameter and the actual parameter referencing the same location in memory. Call-by-reference makes the sharing of values between the formal and actual a two-way, immediate transfer, because the formal parameter becomes an alias for the actual parameter.

Call-by-name was introduced in ALGOL 60 and is the most complex of the parameter passing mechanisms described here. Although it has some theoretical advantages, it is both harder to implement and generally more difficult for programmers to understand. In call-by-name, the actual parameter is re-evaluated every time the formal parameter is referenced. If any of the constituents of the actual parameter expression has changed in value since the last reference to the formal parameter, a different value may be returned at successive accesses of the formal parameter. This mechanism also allows information to flow back to the main program with an assignment to a formal parameter. Although call-by-name is no longer used in most imperative languages, a variant is used in functional languages which employ lazy evaluation (see Chapter 92).

Several issues crop up when we consider parameters and their use. The first is a problem called *aliasing*, where the same memory location is referenced with two or more names. Consider the following Ada code:

```
procedure MAIN is
   a: integer;
   procedure p(x, y: in out integer) is
   begin
       a := 2;
       x := y + a;
   end;
begin
   a := 10;
   p(a,a);
   ...
end;
```

During the call of `p(a, a)` the actual parameter **a** is bound to both of the formal parameters **x** and **y**. Because **x** and **y** are in out parameters, the value for **a** will change after the procedure returns. It is not clear, however, which value **a** will have after the procedure call. If the parameter passing mechanism is call-by-value-result then the semantics of this program depend on the order in which values are copied back to the caller. If they are copied into the parameters from left to right, the value of **a** will be 10 after the call. The results with call-by-reference will be unambiguous (though perhaps surprising to the programmer), with the value of **a** being 4 after the call. In Ada, a parameter specified to be passed as in out may be passed using either call by value-result or call by reference. The preceding code provides an example where, because of aliasing, these parameter passing mechanisms give different answers. Ada terms such programs to be erroneous and considers them not to be legal, even though the compiler may not be able to detect such programs.

Most imperative programming languages support the use of procedures as parameters (Ada is one of the few exceptions). In this case the parameter declaration must include a specification of the number and types of parameters of the procedure parameter. MODULA-2, for example, supports procedure types which may be used to specify procedural parameters. There are few implementation problems in supporting procedure parameters, though the implementation must ensure that nonlocal variables are accessed properly in the procedure passed as a parameter.

There are two kinds of procedural abstractions. One kind, usually known simply as a *procedure*, is an abstraction of a program statement. Its invocation is like a statement, and control passes to the next statement after the invocation. The other type is called a *value returning procedure* or *function*. Functions are abstractions for an operand in an expression. They return a value when invoked, and, upon return, evaluation of the expression containing the call continues.

Many programming languages restrict the values which may be returned by functions. In most cases, this is simply a convenience for the compiler implementor. However, although common in functional languages, most imperative languages which support nested procedures or functions do not allow functions to return other functions or procedures. The reason has to do with the stack-based implementation of block-structured languages. If allowed, it might be possible to return a nested procedure or function which depends on a nonlocal variable which is no longer available when the procedure is actually invoked.

To avoid confusion, most languages allow a name to be bound to only one procedural abstraction within a particular scope. Some languages, however, permit the *overloading* of names. Overloading permits several procedures to have the same name as long as they can be distinguished in some manner. Distinguishing characteristics may include the number and types of parameters or the data type of the return value for a function. In some circumstances, overloading can increase program readability, whereas in others it can make it difficult to understand which operation is actually being invoked.

Program mechanisms to support concurrent execution of program units are discussed in Chapter 98. However, we mention briefly *coroutines* [Marlin 1980], which can be used to support pseudoparallel execution on a single processor. The normal behavior for procedural invocation is to create the procedural instance and its activation record (runtime environment) upon the call and to destroy the instance and the activation record when the procedure returns. With coroutines, procedural instances are first created and then invoked. Return from a coroutine to the calling unit only suspends its execution; it does not destroy the instance. A resume command from the caller results in the coroutine resuming execution at the statement after the last return.

Coroutines provide an environment much like that of parallel programming; each coroutine unit can be viewed as a process running on a single processor machine, with control passing between processes. Despite their interesting nature (and clear advantages in writing operating systems), most programming languages do not support coroutines. MODULA-2 is an example of a language which supports coroutines. As mentioned earlier, iterators can be seen as a restricted case of coroutines.

90.3.5 Data Abstraction

Earlier in this chapter, we introduced the idea of data types as specifying a set of values and operations on them. Here we extend that notion of values and operations to abstract data types and their definitional structures in imperative languages.

The primitive data types of a language are specified by both a set of values and a collection of operations which may be applied to them. Clearly, the set of integers would be useless without the simultaneous provision of operation on those integers. It is characteristic of primitive data types that the programmer is not allowed access to their representations.

Many modern programming languages provide a mechanism for a programmer to specify a new type which behaves as though it were a primitive type. An abstract data type (ADT) is a collection of data objects and operations on those data objects whose representation is hidden in such a way that the new data objects may be manipulated only using the operations provided in the ADT. ADTs abstract away the implementation of a complex data structure in much the same way primitive data types abstract away the details of the underlying hardware implementation.

The *specification* of an ADT presents interface details relevant to the users of the ADT, whereas the *implementation* contains the remaining implementation details that should not be exported to users of the ADT. *Encapsulation* involves the bundling that together of all definitions in the specification of the ADT in one place. Because the specification does not depend on any implementation details, the implementation of the ADT may be included in the same program unit with the specification or it may be contained in a separately compiled unit. This encapsulation of the ADT typically supports information hiding so that the user of the ADT (1) need not know the hidden information in order to use the ADT, and (2) is forbidden from using the hidden information so that the implementation can be changed without impact on correctness to users of the ADT (at least if the specifications of the operations are still satisfied in the new implementation). Of course, one would expect a change in implementation to affect the efficiency of programs using the ADT. A further advantage of information hiding is that, by forbidding direct access to the implementation, it is also possible to protect the integrity of the data structure.

CLU was among the earliest languages providing explicit language support for ADTs through its clusters. Ada also provides facilities to support ADTs via packages. Let us consider an example from MODULA-2, a successor language to Pascal designed by Niklaus Wirth. The stack ADT provides a stack type as well as operations init, push, pop, top, and empty. A MODULA-2 specification for a stack of integers resembles the following:

```
DEFINITION MODULE StackADT;
    TYPE stack;
    PROCEDURE init (VAR s: stack);
    PROCEDURE push (VAR s: stack; elt: INTEGER);
    PROCEDURE pop (VAR s: stack);
    PROCEDURE top (s: stack): INTEGER;
    PROCEDURE empty (s: stack): BOOLEAN
END StackADT.
```

Note the declaration includes the type name and procedural *headers* only. The type **stack** included in the preceding specification is called an *opaque* type in MODULA-2, because users cannot determine the actual implementation of the type from the specification. This ADT specification uses information hiding to get rid of irrelevant detail. Now, this specification can be placed in a separate file and made available to programmers. By including the following declaration:

```
FROM StackADT IMPORT stack, init, push, pop, top, empty;
```

at the beginning of a module, a programmer could use each of these names as though the complete specification was included in the module. Thus, the user can write

```
var s1, s2: Stack;
begin
  push(s1, 15);
  push(s2, 20);
  if not empty(s1) then pop(s1) ...
```

is such a module.

In MODULA-2 the complete definitions of the type **stack** and its associated operations are provided in an implementation module, which typically is stored in a separate file.

```
    IMPLEMENTATION MODULE StackADT;
    TYPE stack = POINTER TO stackRecord;
        stackRecord = RECORD
                    top: 0..100;
                    values: ARRAY[1..100] of INTEGER;
                    END;
    PROCEDURE init(VAR s: stack);
        BEGIN
            ALLOCATE(s, SIZE(stackRecord));
            S^.top :=0
        END
    PROCEDURE push (VAR s: stack; elt: INTEGER);
        BEGIN
            S^.top := S^.top + 1;
            S^.value[S^.top] := elt
        END;
    PROCEDURE pop(VAR s: stack); ...
    PROCEDURE top(s: stack): INTEGER; ...
    PROCEDURE empty(s: stack) BOOLEAN; ...
    END StackADT.
```

Notice that the type name **stackRecord** is not exported.

The specification module must be compiled before any module that imports the ADT and before its implementation module, but importing modules and the implementation module of the ADT can be compiled in any order. As previously suggested, the implementation is irrelevant to writing and compiling a program using the ADT, though, of course, the implementation must be compiled and present when the final program is linked and loaded in preparation for execution.

There is one important implementation issue which arises with the use the language mechanisms supporting ADTs. When compiling a module which includes variables of an opaque type imported from an ADT (e.g., **stack**), the compiler must determine how much space to reserve for these variables. Either the language must provide a linguistic mechanism to provide the importing module with enough information to compute the size required for values of each type or there must be a default size which is appropriate for every type defined in an ADT. CLU and MODULA-2 use the latter strategy. Types declared as CLU clusters are represented implicitly as pointers, whereas in MODULA-2 opaque types must be represented explicitly using pointer types as in the **Stack ADT** example just given. In either case, the compiler need reserve for a variable of these types only an amount of space sufficient to hold a pointer. The memory needed to hold the actual data pointed to is allocated from the heap at run time. As discussed later, Ada uses a language mechanism to provide size information for each type to importing units.

The definition of ADTs can be *parameterized* in several languages, including CLU, Ada, and C++. Consider the definition of the **stack** ADT. Although the preceding example was specifically given for an integer data type, the implementations of the data type and its operations do not depend essentially on the fact that the stack holds integers. It would be more desirable to provide a parameterized definition of **stack** ADT which can be instantiated to create a stack of any type T.

Allocating space for these parameterized data types raises the same problems as previously discussed for regular ADTs. C++ and Ada resolve these difficulties by requiring parameterized ADTs to be instantiated at compile time, whereas CLU again resolves the difficulty by implementing types as implicit references.

90.4 Best Practices

In this section, we will examine three quite different imperative languages to evaluate how the features of imperative languages have been implemented in each. The example languages are FORTRAN (FORTRAN IV for illustrative purposes), Ada 83, and C++. We chose FORTRAN to give a historical perspective on early imperative languages. Ada 83 is chosen as one of the most important modern imperative languages which supports ADTs. C++ might be considered a controversial choice for the third example language, as it is a hybrid language that supports both ADT-style and object-oriented features. Nevertheless, the more modern feature contained in the C++ language design makes it a better choice than its predecessor, C (though many of the points that will be made about C++ also apply to C). In this discussion we ignore most of the object-oriented features of C++, as they are covered in more detail in Chapter 96.

90.4.1 Data Bindings: Variables, Types, Scope, and Lifetime

Like most imperative languages, all three of our languages use static binding and static scope rules. FORTRAN is unique, however, because it supports the *implicit declaration* of variables. An identifier whose name begins with any of the letters I through L is implicitly declared to be of type integer, whereas any other identifier is implicitly declared to be of type real. These implicit declarations can be overridden by explicit declaration. Therefore, in the fragment:

```
INTEGER A
I = 0
A = I
B = C + 2.3
```

A and **I** are integer variables, while **B** and **C** are of type real. Most other statically typed languages (including Ada and C++) require *explicit declaration* of each identifier before use. Aside from providing better documentation, these declarations lessen the danger of errors due to misspellings of variable names.

FORTRAN has a relatively rich collection of numerical types, including integer, *double precision* (real), and *complex*. *Logical* (Boolean) is another builtin type, but FORTRAN IV provided no direct support for characters. However, characters could be stored in integer variables. (The *character* data type was added by FORTRAN 77). FORTRAN IV supported arrays of up to three dimensions but did not support records. Strings were represented as arrays of integers. FORTRAN IV did not provide any facilities to define new named types.

Later languages provided much richer facilities for defining data types. Pascal, C, MODULA-2, Ada, and C++ all provided a full range of primitive types as well as constructors for arrays, records (structures in C and C++), variant records (unions in C and C++), and pointers (access types in Ada). All provided facilities for naming new types and for constructing types hierarchically (constructing nested types). Variant records or unions opened up holes in the static type systems of most of these languages, but Ada (and CLU before it) provided restrictions on the access to variants and builtin run time checks in order to prevent type insecurities.

The scope rules for each language, though static in nature, differ significantly. In FORTRAN, the rules are the simplest. The unit of scope for an identifier is either the main program or the procedural unit in which it is declared. Declarations of procedures (subroutines) and functions are straightforward as well in FORTRAN, with no nesting and all parameters passed by reference. Whereas FORTRAN does not support access to nonlocal variables through scoping rules, it allows the programmer to explicitly declare that certain variables are to be more globally available. When two subprograms need to share certain variables, they are listed in a *common* statement which is included in each subprogram. If different combinations of subprograms need to share different collections of variables, several distinct common blocks can be set up, with each subprogram specifying which blocks it wishes access to.

Ada (like Pascal) supports *nested declarations* of procedures and functions. As a result, block structure becomes extremely important to scope determination.

Whereas C and C++ do not provide for nested procedures and functions, they do share with Ada the ability to include declaration statements in local blocks of code. In C++, blocks are syntactically enclosed in bracket {...} symbols, and any declarations that occur between the symbols hold for the duration of the block. Consider the following code:

```
for (i = 0; i<20; i++) {
  int i = 1, j;
  j = 0;
  while (i < 25) {
    j += i*2;
    i ++;
  }
}
```

One might think that when the inner loop is done, the outer loop also will be done, because **i** has the value 26. But since the inner block of statements redeclared **i**, the scope rules state that the new, inner **i** was manipulated, leaving the outer **i** untouched and free to correctly manipulate the for loop.

In FORTRAN IV the lifetime of all variables is the lifetime of the program. As a result, all memory in a program could be statically allocated, including activation records. Because each subprogram has only one activation record, FORTRAN could not support recursion.

Pascal, Ada, C, and C++ all support recursive functions and procedures. In order to support these, implementations generally rely on stack-allocated activation records. Thus, the lifetime of a local variable in a procedure *p* extends from the call of *p* to the return to its caller. Each of these languages also supports pointer or access types, generally representing data accessed from the heap (though C and C++ also allow pointers to stack-allocated memory). The lifetime of these variables is generally from the time that the programmer executes a creation instruction until a corresponding destruction statement is executed.

90.4.2 Execution Units

FORTRAN and Ada make a strong distinction between expressions and statements, with expressions simply returning a value, but with statements forming the basic unit for program execution. In C and C++, however, these two units of execution are merged, with statements treated as expressions. The statement **x = 5** assigns the value 5 to the variable x. But, in C++, the = sign is also an operator, and this assignment statement is actually an expression that returns the value being assigned. Thus, the statement **y = x = 5** assigns the value 5 to *both* x and y, because the value 5 is assigned to x and the expression $x = 5$ returns 5, which is assigned to y. Although interesting, it can also be very confusing. Because many expressions will have side effects, the order of evaluation will affect the value returned from an expression. Consider the code

```
if ( (y = ++x) == (x + 6)) { ... }
```

This code actually has two statements embedded in it; first, **++x** increments **x**, then this value is assigned to **y**, then the value assigned is tested against the value of **x + 6**. If the compiler decides to change the order in which the subexpressions are evaluated (a not unheard of occurrence in C++ compilers), it may change whether the guard on the if statement is true or false.

Allowing statements to be part of expressions also means that typographical errors are more likely to give rise to syntactically correct (but logically incorrect) statements. For instance, if one of the = signs in

```
if (x == 6) { ... }
```

is omitted, then it will assign of value 6 to x and the conditional will always evaluate to true as all non-0 integers in C and C++ are treated as representing true.

90.4.3 Control Structures

Because FORTRAN was one of the earliest high-level languages, it is not surprising that its control structures are much closer to the underlying machine language instructions. Aside from the do loop (which was similar to the for loop in Pascal and Ada), most other control structures were based on the use of goto statements. Thus, the if statement of FORTRAN IV evaluated an integer expression and, depending on whether the result was negative, zero, or positive, resulted in a jump to one of three statement labels included with the statement. Aside from the usual goto statement, FORTRAN IV also included assigned and computed gotos, which provided some of the flexibility of case statements. FORTRAN 77 and the more recent FORTRAN 90 provide more modern control structures such as the if and while statements of other languages.

The control constructs in Ada are similar to those of Pascal, including if, case, while, and for loops, as well as indefinite loops, which are terminated with exit statements. Several of these were described earlier in the general discussion of control structures.

C and C++ include if statements and a switch construct which is similar to the case statement. The while loop is similar to that in Pascal and Ada, but for the loops in C and C++ are more general than those in most other languages. For loops have the form:

```
for (E₁; E₂; E₃) S;
```

where E_1 is initialization code, E_2 is a test for termination, E_3 contains code to update variables for the next iteration of the loop, and S represents the code to be executed each time through the loop. The test for termination is executed before the update code. Thus, a statement of the form

```
for (i = 1; i < 10, i++) S;
```

will result in *S* being executed once for each value of **i** from 1 to 10. (The expression $i++$ is an expression which increments the value of **i**.) However, much more flexible statements are also possible

```
for (i = 1; not done and i < 1024; i = 2 * i) S;
```

This statement repeatedly executes **S** while **i** ranges through the powers of 2 from 1 to 1024. If done is ever true, it will terminate early.

90.4.4 Procedural Abstraction

Each of Ada, FORTRAN, C, and C++ provides procedural abstraction. Ada and FORTRAN distinguish between functions and procedures, whereas C and C++ do not since procedures are just functions which return an element of type void. FORTRAN IV also supported single-line statement functions, which could be defined local to a program or subprogram. As noted earlier, Ada, C, and C++ all support recursive functions and procedures, whereas FORTRAN does not.

The languages differ in minor ways in how they return values from functions. FORTRAN, like Pascal, treats the name of the function as a pseudovariable which can be assigned to. An explicit return statement returns control to the calling program unit. When the function returns, the last value stored in the function name is returned as the value of the function. Ada, C, and C++ use return statements of the form return exp to return control to the calling program unit. The value of the expression associated with the return statement is the value returned from the function.

Most programming languages provide system-defined overloaded functions, such as arithmetic operators ($+, -, *$, etc.) and comparison functions (e.g., $=, <$, etc.). Ada and C++ are relatively unusual, though, in allowing user-defined overloading. In both, the compiler must be able to disambiguate at compile time whichever of the versions of the overloaded operator are called for at each of its occurrences. C++ determines which version is called for by looking at the number and types of the actual parameters. Ada goes further and can also use the return type to determine which version works in the particular context in which it is found. Thus, in Ada one may overload the $+$ operator to take two integer parameters and return a user-defined rational type, even though there already exists a built-in version of $+$ which takes two integer parameters and returns an integer. If $+$ occurs in a context in which only an integer result would make sense, the builtin version would be selected. If $+$ occurs in a context in which only a rational value would make sense the user-defined version would be selected. If the system cannot tell which should be used, then an error will occur at compile time.

Unlike FORTRAN, Pascal, and C, both Ada and C++ provide language support for exceptions. Ada and C++ both use the termination model for program resumption after handling the exception.

90.4.5 Data Abstraction and Separate Compilation

FORTRAN provides support for separate compilation of subroutines and functions but provides no type checking across compilation unit boundaries. Thus, the main program may call a function F with two real parameters, but the definition of F in a separately compiled unit may have only one formal parameter, and it might be an integer. Because FORTRAN does not support the definition of new named types, it provides no support for abstract data types.

C and C++ provide slightly better support for separate compilation by allowing the programmer to put external function and procedure declarations in a header file, which may be included into compilation units which use them. The header files are treated as though they were textually part of the compilation unit into which they are included. C provides no support for abstract data types, though C++ does provide strong support through its class facilities. C++ classes give the programmer control over which aspects of a data type the user will be allowed to see and use. Because C++ is described in some depth in Chapter 96, we omit a detailed description here.

Standard Pascal provides no support for separate compilation, though virtually all Pascal implementations provide support for separately compiled units. These units typically include both interface and implementation sections. The interface of a unit may be explicitly imported into another compilation unit with a *uses* statement. This provides for separate, but not independent, compilation without requiring the programmer to create individual header files by hand. These units generally do not provide support for information hiding.

Ada provides both separate compilation and strong support for abstract data types. Like MODULA-2's modules described earlier, Ada packages come in two separately compiled units, the specification and body. Only items listed in the nonprivate part of the package specification are accessible at compile time to units which import the package. The following is an Ada package specification for stacks:

```
package StackADT is
    type stack is private;
    procedure push(s: in out stack; elt: in integer);
    procedure pop(s: in out stack);
    procedure top(s: in stack) return integer;
    procedure empty(s: in stack) return boolean;
  private
    type stack is record
       top: integer := 0;
       values: array (1..100) of integer
    end record
  end StackADT;
```

The private section of a package specification is necessary to provide a description of private types. This is necessary so that importing programs know how much space to provide for a variable of that type. This is not as clean as the MODULA-2 solution, since any change of representation of the type will require the recompilation of the specification and hence of any program which imports the package. For this particular representation of stack, no initialization routine is necessary because top is initialized to 0 in the declaration of the type.

If the implementation of the private variable is a pointer, then only partial type information need be provided. That is, if we replace the private part of the preceding example by

```
private
   type stackRecord;
   type stack is access StackRecord;
end StackADT;
```

then this would provide sufficient information for importing programs to determine the memory needs for a variable of this type (i.e., the amount of space necessary to hold a pointer).

An implementation of the original package specification is given next:

```
package body StackADT is
  procedure push (s: in out stack; elt: in integer) is
    begin
        s.top := s.top + 1;
        s.value[s.top] := elt
    end push;
  procedure pop(s: in out stack); ...
  function top(s: in stack): integer; ...
  function empty(s: in stack) boolean; ...
end StackADT;
```

The **StackADT** package specification can be imported into an Ada program unit by including with **StackADT** at the beginning of the unit. Components can be referred to as record components, e.g., **StackADT.stack** and **StackADT.push**. The package name prefix can be omitted if use **Stack-ADT** is also included at the beginning of the unit.

Both Ada and C++ provide mechanisms for supporting parameterized packages (or classes in the case of C++). The C++ template mechanism is quite primitive, with template instantiations being treated as being similar to compile-time macroexpansions. The template is never type checked, only its instantiations. Ada also requires its generic packages to be instantiated at compile time, but the generics are type checked before, rather than after, instantiation. Thus, a generic package can be compiled and later used in another until which does not have access to the implementation.

The following is an example of the header of a generic **BinarySearchTree** package:

```
generic
   type Element is private;
   with function LessThan (x, y: Element) return boolean;
package BinarySearchTree is
   type BSTree is private;
   ...
end BinarySearchTree;
```

This can be used in another unit by instantiating it with a type and appropriate function, for example,

```
package PeopleDict is new BinarySearchTree(People, PeopleComp)
```

where **PeopleComp** is a function taking pairs of type **People** and returning a Boolean. **PeopleDict** can then be used like any other package. The ability to require generic package instantiations to include necessary functions and values as well as types ensures that they will not be instantiated with types which do not support the appropriate operations.

90.5   Research Issues and Summary

Research issues in imperative languages in recent years have tended to focus on many of the new constructs presented in this chapter. These include support for exceptions, iterators, abstract data types, and parameterized or generic types. It is fair to say that most current research in programming languages is devoted to implementation and environment issues or to other programming paradigms. There are not many new concepts currently being introduced into imperative programming languages. Many languages which formerly were purely imperative have recently been extended to include object-oriented concepts (e.g., Object Pascal, Objective C, C++, Ada 95). Another series of extensions has provided features for concurrent and distributed programming. Discussions of these two different kinds of extensions can be found in Chapters 96 and 98 of this Handbook.

From our earlier discussion, it is clear that support for abstraction plays an important role in imperative language design and use. Variables abstract away details of memory usage; data types (and in particular abstract data types) abstract from the representation of values to provide support for operations that are independent from the actual implementation; execution units abstract away details of machine instruction execution and expression computation while providing clean interfaces for sharing information between caller and callee.

A second major focus in the development of modern imperative programming languages has been the enrichment of type systems, especially static type systems. Abstract data types can be understood as the enrichment of type systems with so-called existential types, in which the existence of a type is revealed be instantiated with any type (or in Ada and CLU's case any type which comes supplied with the appropriate operations). These more flexible type systems allow for the construction of safe statically typed programming languages which are more expressive than their predecessors. There is hope that we are moving forward to a time when most programmers will see such secure languages as assisting them in their goal of creating correct and efficient software, rather than getting in the way. (See Chapter 104 for a further discussion of type systems.)

We have surveyed the class of programming languages modeled after the sequential organization of the von Neumann architecture. The imperative programming language paradigm is characterized by its sequential, stepwise statement execution.

As discussed in the Section 90.4 the imperative programming constructs are implemented in a variety of ways in different languages. There are many languages to choose from; choosing the right language for the applications at hand is an important first step to software implementation.

It could be argued that the object-oriented paradigm is simply a minor variation on the imperative paradigm in which remote procedure and function calls replace the more familiar imperative calls. However, the object-oriented paradigm requires an entirely different way of thinking about the organization of a program, with the traditional conception of a program as a series of operations being applied to values replaced in the object-oriented view by an organization of more distributed responsibility. In this view, values (typically referred to as objects) are responsible for knowing how to perform their own operations, and the programmer is responsible for bringing together a group of objects with appropriate capabilities and organizing a program which relies on these distributed capabilities to accomplish a task. Subtyping and inheritance provide important organizing tools and promote code reuse in ways unavailable in traditional imperative languages.

Most programmers today are initially taught to program in imperative languages. Thus, these languages reflect the way that most programmers currently think about algorithm construction and program execution. Whether this will continue in the face of the challenge of the object-oriented paradigm will be interesting to see.

Defining Terms

Abstract data type: A collection of data type and value definitions and operations on those definitions which behaves as a primitive data type. The specifications of these types, values, and operations are generally collected in one place, with the implementations hidden from the user.

Binding: A connection between an abstraction used in the language and a data object as it exists in the computer hardware. The usage, establishment, and number of these bindings characterize the various imperative languages and affect their ease of use and performance.

Control structures: Structures or statements that alter the strict sequential ordering in an imperative program, presenting alternatives to sequential control. Control structures can be conditional, iterative, or unconstrained.

Derived type: A new data type constructed by copying a type that already exists. The resulting new type is distinct and not identified as being copied from the existing type, though operations on the old type are automatically inherited in the new type.

Identifier: The name bound to an abstraction.

Parameters: Data objects passed between the caller and the called procedural abstraction.

Procedural abstraction: Separating out the details of an execution unit in such a way that it may be invoked in a program statement or expression.

Scope rules: Rules in a language that define the area or section of a program in which a particular binding is effective.

Subtype: A new data type defined as a copy of another defined type, typically with a restricted subset of its values. It may generally be used in the same contexts as its parent type.

Type: A collection of values with an associated collection of primitive operations on those values.

Type equivalence: Rules that govern when variables or values from two different data types may be used together.

Variable: An abstraction used in imperative languages for a memory location or cell.

References

Boehm, C. and Jacopini, G. 1966. Flow diagrams, Turing machines, and languages with only two formation rules. *Commun. ACM* 9(5):366–371.

Dijkstra, E. W. 1968. Goto statement considered harmful. *Commun. ACM* 11(3):147–148.

Liskov, B. H. and Guttag, J. V. 1986. *Abstraction and Specification in Program Development*. MIT Press, Cambridge, MA.

Liskov, B., Snyder, A., Atkinson, R., and Schaffert, C. 1977. Abstraction mechanisms in CLU. *IEEE Trans. Software Eng.* SE-5(6):546–558.

Marlin, C. D. 1980. *Coroutines*. Lecture notes in computer science 95. Springer–Verlag, New York.

Further Information

A good examination of imperative languages, as well as other paradigms, can be found in the following texts:

Dershem, H. L. and Jipping, M. J. 1995. *Programming Languages: Structures and Models*, 2nd ed. PWS, Boston, MA.

Louden, K. C. 2003. *Programming Languages: Principles and Practice*, 2nd ed. PWS-Kent, Boston, MA.

Pratt, T. W. and Zelkowitz, M. V. 2001. *Programming Languages: Design and Implementation*, 4th ed. Prentice Hall, Englewood Cliffs, NJ.

Sebesta, R. 2003. *Concepts of Programming Languages*, 2nd ed. Benjamin-Cummings.

Sethi, R. 1996. *Programming Languages: Concepts and Constructs*, 6th ed. Addison–Wesley, Reading, MA.

Several journals are devoted to programming languages and language design. *ACM Transactions on Programming Languages and Systems* and *Computer Languages* both feature referred papers on programming languages. *ACM SIGPLAN Notices* is a collection of unreferenced papers from the ACM Special Interest Group on Programming Languages. Proceedings of the ACM conferences, *Principles of Programming Languages (POPL)*, and *Programming Language Design and Implementation*, provide a good presentation of current research in programming languages.

91

The Object-Oriented Language Paradigm

	91.1	Introduction **91**-1
	91.2	Underlying Principles **91**-3
	91.3	Best Practices **91**-6
		Smalltalk • C++ • Java • C#
Raimund Ege	91.4	Language Implementation Issues **91**-24
Florida International University	91.5	Research Issues.................................... **91**-25

91.1 Introduction

During the 1990s, **object-oriented programming** (OOP) established itself as the dominant programming paradigm. Although it was first viewed as a revolutionary new programming paradigm, such a characterization is only partly accurate.

OOP is, to be sure, a paradigm in the current sense of that term. It embodies a way of organizing and representing knowledge, "a way of viewing the world" [Budd 2001], that encompasses a wide range of programming activities, including program analysis, design, and implementation. The paradigm derives its power from its view of computation as the simulation of real-world entities. That is, according to Dan Ingalls, "Instead of a bit grinding processor ... plundering data structures, we have a universe of well-behaved objects that courteously ask each other to carry out their various desires" [Ingalls 1981].

Central to this view of computation is the notion of self-contained little systems that work together. OOP tools and languages facilitate the description of **objects** as self-contained systems that maintain their own internal state (data), perform actions (**methods**) in their own interest, and interact with other objects (by sending **messages** to one another). Objects can be low-level programming tools, such as lists, stacks, and trees, akin to traditional abstract data types (ADTs). They can also be higher-level abstractions that reflect what a program is intended to model: an automated teller machine (ATM), a deck of playing cards, an elevator, or a collection of graphical objects on a screen.

The primary power of OOP derives from the fact that, once defined, objects enjoy a type-like status. (As we will explain later, objects are defined via **classes**, which are very much like types). That is:

Objects can be used without the user's knowing the details of their implementation and can be properly protected from their consumers.

Objects can be used according to a standard notation, using names, symbols, and operators in conventional ways.

Objects can be combined with other objects and types in expressive and efficient ways (composition and hierarchy) to define new, more complex types.

Whereas structured programming languages such as C and PASCAL allowed a programmer to define new types, these were primarily a notational convenience. In such languages, user-defined type names served as shorthand to improve readability and as aids to compilers for recognizing type equivalences. These types, though, did not enjoy the status or flexibility of the built-in types. That is, one could not overload operators to apply to these new types and, more importantly, one could not easily hide the implementation details of a new type from its consumers.

The seeds for the object-oriented paradigm can be found in the languages that developed in the 1970s and 1980s to provide more direct support for building user-defined ADTs. ADA-83, for example, allows one to build clearly specified, modular software components that effectively hide their implementation details. ADA-83 uses the package construct to describe type specifications and subprograms that can belong to a user-defined ADT. Languages that provide this level of support for ADTs are now called object-based.

In many ways, the evolution of the OOP paradigm extends that of the structured paradigm which preceded it. We saw the structured paradigm, as manifested in a succession of increasingly high-level programming languages, progress from straight-line code laced with unconditional transfers of control, to block-oriented code that exploited control structures, to finer-grained code built from simple subprograms, to top-down, structured code that relied on parameterized and separately developed libraries of code which distinguished a subprogram's interface from its implementation. The object-based paradigm followed, to enable the capture of ADTs. Finally, OOP combines procedural and data abstraction into the notion of a class that can be part of a class hierarchy.

Hindsight being what it is, we can now see how the seeds of OOP have been nurtured by developments in many subfields of computer science over the past three decades: the basic idea of computation as simulation was first popularized in the programming language SIMULA 67, which also introduced the concepts of class, object, and message. Hardware architectures were developed in the 1970s using an object-based approach to describe the components of the machine and their interactions at a higher, more natural level. This development was further motivated by operating systems and applications that depended increasingly on graphical interfaces, which lend themselves directly to object-oriented descriptions. In the entity-relationship database model, in which a collection of information is described in terms of entities, attributes, and relations among entities, we see yet another manifestation of objects (at least the information content aspect of them). Even in knowledge representation schemes that supported work in artificial intelligence (frames, scripts, and semantic networks), we can see clearly this object orientation. Most clearly, the first purely object-oriented language, Smalltalk, has been around since the early 1970s.

Still, all of the premonitions of OOP did not coalesce into a recognizable influence on programming language and practice until they were motivated to do so by our collective practical experience in software engineering. As we became increasingly adept at exploiting the procedural paradigm and its languages, we recognized more directly their shortcomings. The more we pushed that paradigm, the more difficult it was to support data abstraction to that same degree that algorithm abstraction could be supported. Further, methods of encapsulation were relatively primitive, thus making if difficult to develop safe, reusable code. Finally, software analysis and design techniques that relied on procedural abstraction as the basis for performing decomposition of complex systems were proving increasingly awkward to apply to domains that were increasingly dependent on data, as opposed to algorithms.

In summary, structured methods flourished in the era of application domains that were either algorithm-centric or information-centric. These same methods prove less useful in an increasingly complex application world, where functionality and information structure must be considered in context. The concept of an object is a simple and elegant combination of all informational and algorithmic aspects of an application entity.

Today, all of these technologies have matured and all of the motivations have been established to the extent that OOP is a viable, recognizable, and popular approach to software development. Programming languages and commercial development environments abound. Smalltalk, as an example of a pure object-oriented programming language, is still commercially available; however, Java and C# have gained significant popularity. A variety of hybrid OOP languages (like C++, ANSI COBOL, and Visual Basic, which extended existing languages to incorporate essential OOP features) dominates the programming language marketplace.

Because OOP is rightly referred to as a paradigm, it has spawned the development of many software analysis and design techniques that support the identification and description of objects in a problem specification. As with all paradigms, OOP languages and techniques are best suited to problems that match the paradigm's view of the world. The use and popularity of OOP rises with the increased demand for complex, interface-intensive systems, those that can be modeled in real-world terms.

Finally, if OOP raises the level of abstraction to bridge the gap between programmer and machine, it may be that novice programmers would stand to benefit the most from its use. Indeed, Smalltalk was developed based on research detailing how young children describe and interact with the world in solving problems. OOP is now influencing significantly how we teach and learn programming. There is a growing recognition that object orientation makes learning to program easier for the novice. Universities are now teaching object orientation as part of the computer science (CS) introduction. Java is now widely used in CS curricula.

However, it is also recognized that experienced programmers and software engineers must be retrained significantly: not because object orientation is hard, but because of their experience in traditional, function-centered (or information-centered) problem solving. Experienced programmers must, as Bertrand Meyer puts it, reacquire an object-oriented frame of mind.

Proponents of OOP claim that the paradigm represents the state of the art in terms of bridging the language gap between programmer and machine. It offers the prospect for achieving many of the software quality goals to which all programmers aspire: easily designed, safe, efficient, uniform software.

91.2   Underlying Principles

OOP takes a quite natural, but also quite different, view of programming. In a process-state view of computing (which reflects both the machine's fetch-execute cycle and its data processing nature), one thinks about programs as lists of instructions executed sequentially, and serving to manipulate and change the state of memory is the fundamental strategy of imperative languages. In contrast, OOP sees a software system as set of collaboration objects. Old-style programming involves thinking in terms of algorithms and data structures separately. OOP languages, on the other hand, permit the programmer to take a broader view and think of a program as a collection of cooperating objects, each of which encapsulates both structure and function.

Each object has three aspects: what it is, what it does, and what it is called. We illustrate these aspects by considering a geometric object, a circle. Most programming languages do not have circles as a built-in data type, so if we want to write a program that manipulates circles, we must write our own description of circle objects into a program. The descriptive information associated with an object, called its state, includes its properties and the values of these properties. For most objects, the state properties do not change, although the state values may very well be modified during the object's lifetime. For instance, a circle has a radius and a center, and at a given time, these might have the values 4.67 and (0, 0). A circle object will always have a radius and a center, although their values will change if the circle's size or location changes.

The collection of actions an object may perform on itself is called its behavior. In our example, we may want to use a circle as part of a program that draws on a computer screen, and so we would need to be able to set the center and radius of the object and instruct the circle to draw itself on the screen. Notice that we said a circle would draw itself on the screen. This is an important feature of the object-oriented approach. In a language with an algorithmic approach, one could describe circles by radius and center, as we have, but the description of the actions to perform on a circle would not be as tightly associated with the state data as it is here. That is, the drawing command would not be part of the circle object itself. Encapsulating data and actions together in an object helps us design and understand a program because, for instance, all circle-related data and actions are collected in one place. If we had to modify a circle object's definition, we would know exactly where the modifications would go. Furthermore, we would know that we would not have to modify any other part of the program.

Finally, each object has an identity, which serves to distinguish it from all others. One can refer to an object uniquely via its identity. In most programming languages, we identify an object by giving it a name, so we might call our circle theCircle — theCircle is our reference to the object.

It is entirely reasonable to assume that a program that would use one circle object might use several. All of these circle objects would be similar, in that they would have the same collection of member data and functions. It would clearly be a wasteful duplication of effort to describe each separately, since they would differ only in their data values and names. Object-oriented languages allow one to describe a template, if you will, for an entire set of objects. Such a template is called a class.

A class is used to generate a set of objects sharing common properties. Continuing our example, we could describe a class, Circles, by describing the state properties (but generally not the state values) and behaviors of all objects that belong to that class. Then, we could describe as many circle objects as we needed merely by declaring that they are instances of the Circles class, that is, they are Circle objects. An object can never exist in isolation; every object must be an instance of some class. This means that if we are going to use an object in a program, we must first provide the class to which the object belongs.

One can think of a class as a means for implementing an abstract data type. That is, to describe a class we identify the properties that any instance of the class (a particular object) must have. These properties consist of state information (referred to as member, or state, data) and a collection of behaviors (member functions, or methods) that such objects are capable of performing. Methods typically involve accessing, setting, or otherwise manipulating the object's member data.

Methods are invoked by an object in response to a message (i.e., a request of the object to perform one of its methods). Message passing as a means for invoking methods differs subtly from subprogram invocation in most other programming paradigms. In OOP, there is a formal distinction between the subprogram call (the sending of a message) and the method of invocation (the receiving of the message). It is the responsibility of the receiver of a message to interpret the message, that is, to determine which of its methods to perform in response. Think of OOP as programming by sending messages to objects.

Taken together, these two mechanisms support the description of full-fledged ADTs in a more general and extensible way than was provided by the first object-based languages. In particular, classes provide an abstraction mechanism that can be used to model a wide range of information structures and information processing agents. Using classes as the primary descriptive vehicle also characterizes an OOP approach to program design that has proved extremely effective for many real-world applications.

Not only are classes a rich descriptive tool; they are safe, in the software engineering sense of that term. That is, classes effectively encapsulate the abstractions they model by controlling (either implicitly or explicitly, depending on the language being used) access to an object's members. Some of an object's members will be public in nature (visible and accessible to other objects), and some will be private to the object itself (visible and accessible only to other members). Many OOP languages also support a formal distinction between a class's interface and its implementation.

Classes are also an efficient means of representation by virtue of the fact that they can be related to one another both compositionally (an object of one class can serve as a member of an object of another class) and hierarchically (to express *is-a* relationships). This latter feature is unique to OOP, and it affords a means for describing directly any information that is hierarchical in nature.

In cases where classes are arranged in a hierarchy, a **subclass** is said to **inherit** from its superclasses. That is, instances of the subclass contain (either directly as copies or indirectly via the superclass) both the data and methods described by the superclass. These same general properties are associated with all subclasses of the particular superclass. Each subclass extends the descriptions of its superclasses by adding specialized member data and methods that differentiate it from the superclass and from other subclasses.

Going back to our geometric example, we do not have to restrict our objects to circles; we could equally well have objects that are rectangles, triangles, and so on. We could augment our program by including declarations and definitions for the classes Rectangles and Triangles, just as we did when we defined the Circles class. In doing so, we notice that all of these are what we might call geometric objects.

Object-oriented languages allow a class like Rectangles to inherit the methods and data from a parent class, such as GeometricObjects. For example, because every object in our hierarchy has an extent (a bounding rectangle inside which the object fits as closely as possible), we could declare the data bounds_top, bounds_left, bounds_bottom, bounds_right, and the method ReportBounds within a superclass named GeometricObjects. These would apply to all the derived classes Rectangles, Triangles, and Circles. Doing this, we would not need to re-declare these data items in any of these subclasses. They would be inherited from the base class GeometricObjects. Indeed, it is sometimes useful to define a superclass, such as GeometricObjects, knowing full well that we will never be interested in generating direct instances of it. That is, the only instances of GeometricObjects that we will use in our programming are those that are also instances of its subclasses: Circles, Triangles, or Rectangles. In such cases, superclasses are referred to as **abstract classes**.

Inheritance is not only distinctive of OOP languages; it is also the means by which the paradigm addresses many software engineering concerns. For example, inheritance is an efficient, non-redundant notation for representing information that is hierarchical in nature. It is natural in the sense that it reflects how we humans tend to describe such information (which explains in part why OOP has spawned so many related program design techniques). The efficiency of the notation derives from the ability to reuse code that was used in defining a superclass when defining a subclass. This, in turn, allows the inherited members of classes that are derived from one another to project consistent interfaces to their consumers. That is, member data and methods can be referred to by the same names in both the super- and subclasses.

There are times, though, when a base class action or state might not be appropriate for use by a derived class. All of the classes in our hierarchy would have Draw methods, but drawing a rectangle might involve different strategies than drawing a circle. OOP languages allow a derived class to redefine, or **override**, a method inherited from a base class. If, for example, we have an object called theShape and send it a message to draw itself, how is the computer to know which Draw action to use? Whereas procedural languages and object-based ones would resolve this ambiguity statically (i.e., based on the declared type of theShape), an OOP language solves this problem at run time, by looking up the class to which theShape belongs and finding a reference to an appropriate Draw method.

This ability to use the same name for actions on objects of different classes is an example of a kind of **polymorphism** (called inclusion polymorphism) that is common to all OOP languages. In the OOP sense of the term, polymorphism describes the fact that a superclass includes all **instances** of its subclasses. Thus, the notion of a geometric object contains all circle, rectangle, and triangle objects. As a result, an entity can take on different types of information during the execution of a program. For example, a geometric object can be a circle, a rectangle, or a triangle at any point during our program.

The concept of **interface** is related to the polymorphic use of the class concept. An interface is a listing of methods where, for each method, one only lists its name and any potential parameters. OO languages allow classes to **implement** an interface; that is, the classes provide explicit bodies for all methods listed in the interface.

Other forms of polymorphism often supported in OO languages include

Overloading — Two or more functions or operators can share the same name, so the + operator is allowed to apply to both integers and real numbers.

Parametric polymorphism — A parameter is used to establish choices: for example, generic packages — or C++ class templates — use a parameter to allow another level of abstraction.

Coercion polymorphism — C type casting is an example.

The following characteristics are the essence of OOP:

Objects — As a means for encapsulation of state and behavior

Classes — As templates for generating objects

Message passing — As the mechanism by which computation takes place

Dynamic method invocation — As the means by which methods are bound and interpreted

Inheritance — As a technique for modifying classes and creating subclasses

Inclusion polymorphism — As a means for allowing expressions of one type to be used in place of expressions of a supertype

Each contributes significantly to the overall utility of the paradigm, and each allows the paradigm to address one of the many software engineering concerns that motivated it. Whereas different programming languages implement them in various combinations and to varying degrees, any language that implements them all is considered object oriented.

91.3 Best Practices

To illustrate these characteristics of object-oriented programming, let us construct a very simple application to deal with queues of packets as they might appear in a network simulation. Packets in our program maintain their names and priorities, and allow their observation and comparison. Different kinds of packets are modeled as subclasses: one for packets that carry protocol information (Ack) and one for packets that carry data (Data). The subclasses specialize how packets are observed. The second set of classes models the queue concept. Class FifoQueue represents the algorithmic and data abstraction of a standard first-in-first-out queue. Internally, it employs a doubly linked list — anchored by a *head* and *tail* **member field** — to maintain the packets currently queued. The member functions *enter* and *leave* implement the standard protocol of such a queue. The FifoQueue class ignores the packet's priority information but also serves as a superclass for two additional subclasses, PriQueue and QueuePri, which use the packet's comparison abilities to handle packets of different degrees of importance.

We will develop the example in terms of Smalltalk, C++, Java, and C#, four of the major object-oriented programming languages in use today. Our intention here is to provide quick overviews of these languages and to illustrate the different notations and styles for implementing the object-oriented paradigm.

91.3.1 Smalltalk

Smalltalk is an integrated programming environment and language [Goldberg and Robson 1981]. It led the OOP evolution by incorporating the ideas of personal computer, interactive computing, graphical user interface (GUI), and OOP. The underlying design principle was to raise the level of abstraction substantially, so that program elements could communicate at a level closer to that of human problem solvers.

The language is purely object oriented. Every entity in a program is an object — everything from windows and projects to integers. Smalltalk defines only a limited syntax for declaring variables, assigning objects to variables, and sending messages to objects. Control structures are accomplished by message passing, where methods in standard classes define their behavior.

Smalltalk was originally developed as a research tool, and it is thoroughly integrated with its complex and powerful graphical programming environment. It spurred the subsequent development of all current commercial OOP languages and has seen some commercial acceptance, especially in the financial services industry. A variety of implementations are available.

Let us develop a Smalltalk program for our network/packet/queue example. We start by describing the class Packet:

```
Object subclass: #Packet
  instanceVariableNames: 'name priority '
  classVariableNames: ''
  poolDictionaries: ''
  category: 'MyFifoQueue-Examples'
```

A packet is defined as a subclass of Object. Smalltalk enforces a single class hierarchy, that is, all classes need a superclass. Object is the anchor of the existing Smalltalk class hierarchy. The # character before Packet is needed as part of Smalltalk syntax (i.e., it defines a symbol). The Packet class defines two instance variables (member fields): *name* and *priority*. Smalltalk does not type its member fields; in effect, they can refer to an instance of any object within Smalltalk.

Class variables are fields that do not carry unique values per object, such as **instance variables**, but rather one value per class. For our Packet class, we do not need to define any class variables; *pool dictionaries* are another way to refer to variables, and *category* defines that this new class belongs to a set of classes called MyFifoQueue-Examples. Categories are used in the Smalltalk programming environment to find classes quickly.

The packet class needs member functions — *instance methods* in Smalltalk terminology. The first instance method, *list*, is defined as follows:

```
!Packet methodsFor: 'printing'!
list
       Transcript show: name, ' packet: '.
```

The first line constructs the relationship of this method to the packet class (the exclamation marks differentiate it from regular Smalltalk syntax). The method is part of the *printing* protocol (Smalltalk categorizes related methods into protocols). The method has the name *list*; no arguments are listed. The body of the method consists of a single statement: it instructs to send message *show:* to object Transcript. Transcript is a preexisting Smalltalk object that records text and displays it to the user.

In Smalltalk syntax, a method name that ends with a colon (:) signifies that a parameter is expected. Here, a string constructed from the *name* of the packet and the text *packet* is sent along with the message as a parameter. The comma (,) is used to concatenate the two parts of the string. And finally, a Smalltalk statement is delimited by a period (.).

Next, we define two methods that allow us to compare packets: *less than or equal to,* written as <=, and *greater than,* written as >.

```
!Packet methodsFor: 'access'!
<= aPacket
  ^   (priority <= aPacket priority).
> aPacket
  ^   (priority > aPacket priority).
priority
  ^   priority.
```

Here, <= is defined with a single parameter, that is, *aPacket*. Again, Smalltalk does not type method parameters: any object can be sent along as parameter. The body of the first method lists a single *return* statement (^ is used to denote return). Returned is the expression that is listed in parentheses: a comparison of the instance variable *priority* to the priority of the *aPacket* parameter. Smalltalk hides all instance fields inside an object. Therefore, one packet object has no access to the *priority* of another packet. It is therefore necessary to define an instance method *priority*, which is used by both the <= and > methods. The *priority* method just returns the *priority* instance variable.

Finally, the *initialize* method allows one to set a packet's instance variables. It uses two parameters: one for *name* and one for *priority*. Smalltalk uses a special style for methods with multiple parameters: they are embedded into the method name. The complete name of this method is *initialize:priority:*, where after each colon, Smalltalk expects a parameter:

```
!Packet methodsFor: 'initializing'!
initialize: aName priority: pri
 name := aName.
 priority := pri.
```

Methods can be defined at the instance and at the class level. *Instance methods* are those that are invoked when an object receives a message. Class methods are typically used to create new instances of a class. For example, a new packet object would be created with

```
Packet new.
```

The *new* method is not defined for class Packet; rather, it is inherited from its superclass Object. Here, *new* creates and returns a new instance of class Packet. It can be initialized with our *initialize:priority:* method, as in

```
Packet new initialize: 'one' priority: 10.
```

Because all packets should be initialized when created, we can define a new class method for Packet that enables the creation and initialization of Packet objects in a single method:

```
!Packet class methodsFor: 'initializing'!
new: aName priority: pri
 ^ (Packet new initialize: name priority: pri).
```

The name of this method is *new:priority:*. It does not override the *new* method that Packet inherits from its superclass; rather, it uses it internally. The body of the method is a single line that proceeds in three steps:

1. It sends the message *new* to the Packet class, which creates a new instance and returns it.
2. The new instance then receives an *initialize:priority:* message to initialize the new packet.
3. The method then returns the new and initialized object.

Subclass Data is based on the Packet class. It adds two more instance variables, *body* and *length*:

```
Packet subclass: #Data
  instanceVariableNames: 'body length '
  classVariableNames: ''
  poolDictionaries: ''
  category: 'MyFifoQueue-Examples'
```

Because data packets have additional instance variables, it makes sense to define an *initialize* method different from the one that is inherited from the Packet class. It takes three parameters: one each for the instance variables name, priority, and body (the full name of this method is now *initialize:body: priority:*):

```
!Data methodsFor: 'initializing'!
initialize: aName body: aBody priority: pri
  body := aBody.
  length := body size.
  self initialize: aName priority: pri.
```

The *body* instance variable is set from the parameter. The length is calculated by sending a *size* message to the *body* parameter. Note that although parameters are not typed, there is an assumption that it belongs to a class that defines the *size* method.

The third line in the body shows how one method delegates to another. Because *name* and *priority* still need to be set, which we already spelled out in the *initialize:priority:* method in the Packet superclass, we send message *initialize: aName priority: pri* to object self. Self refers to the Data object itself that is currently executing the method. It will now execute the method inherited from its superclass. And finally, the *list* method

```
!Data methodsFor: 'printing'!
list
  super list.
  Transcript show: body.
```

uses the keyword *super*, which is similar to *self*; it allows one to send a message to the current object. However, the corresponding method that is executed is not found in the class of the current object, but rather in its superclass, that is, the Packet class. In effect, the *list* method defined here extends the list behavior of the superclass.

The class methods for Data allow for the creation of Data objects with or without an explicit priority:

```
!Data class methodsFor: 'initializing'!
new: aBody
    ^  super new initialize: 'Data' body: aBody priority:5.
new: aBody priority: pri
    ^  super new initialize: 'Data' body: aBody priority: pri.
```

The *super* keyword is needed here. Otherwise, if we sent a new message to the Data class, we would have an infinite loop because we are currently defining the *new* class method for Data. *Super* will ensure that the *new* method defined in the superclass is executed.

Subclass Ack is also based on Packet. It defines no further instance variables, but it redefines the *list* method:

```
Packet subclass: #Ack
  instanceVariableNames: ''
  classVariableNames: ''
  poolDictionaries: ''
  category: 'MyFifoQueue-Examples'
!Ack methodsFor: 'printing'!
list
  Transcript show: 'acknowledged'.
```

The class method for Ack ensures that Ack objects are created with priority 10:

```
!Ack class methodsFor: 'initializing'!
new
    ^ (super new initialize: 'Ack' priority: 10).
```

Here, the *new* method overrides the *new* method inherited from its superclass. It also uses the *super* keyword to invoke the *new* method defined by its superclass.

The Packet class hierarchy is now complete. Before we construct our Queue class, we need one more helper class: Node, which will be used to implement a doubly linked list. The Node class defines three instance variables: *value* to hold a packet object, *next* to refer to the next node in the list, and *previous* to refer to the previous node in the doubly linked list. The instance methods allow for the manipulation of all instance variables.

```
Object subclass: #Node
  instanceVariableNames: 'value next previous'
  classVariableNames: ''
  poolDictionaries: ''
  category: 'MyFifoQueue-Examples'
!Node methodsFor: 'access'!
next
    ^    next.
next: aNode
  next := aNode.
previous
    ^    previous.
previous: aNode
  previous := aNode.
value
```

```
      ^ value.
  value: aValue
    value := aValue.
  !Node methodsFor: 'initializing'!
  initialize: aValue
    value := aValue.
  !Node class methodsFor: 'initializing'!
  new: aValue
    ^ super new initialize: aValue.
```

Now, we can construct the class FifoQueue, which defines two instance variables: *head* to refer to the beginning of the list of nodes, and *tail* to hold on to the end of the list.

```
    Object subclass: #FifoQueue
      instanceVariableNames: 'head tail'
      classVariableNames: ''
      poolDictionaries: ''
      category: 'MyFifoQueue-Examples'
```

The first instance method enters a packet object into the queue. It uses a local variable *tmp* for a newly created node object, which is then inserted at the *tail* of the linked list.

```
    enter: aPacket
      | tmp |
    tmp := Node new: aPacket.
    tmp previous: tail.
    tail := tmp.
    (head notNil)
      ifTrue: [ tmp previous next: tail ]
      ifFalse: [ head := tmp ].
```

Local variables in methods are listed between vertical bars (|). Here, *tmp* is created as a new instance of class Node by sending message *new:* to the Node class. The *tmp* object receives the message *previous:* to attach it to the rest of the linked list content (*tail*).

The last statement in the method body is an example of an *if-then-else* construct. First, *head* receives message *notNil*, which returns Boolean objects *true* or *false*, depending on whether *head* refers to a *Node* object or is undefined. The returned Boolean object then receives the *ifTrue:ifFalse:* message. The parameters to *ifTrue:ifFalse:* are delimited by brackets, which denote blocks of Smalltalk statements. The execution logic is as follows: if (*head notNil*) returns Boolean object *true*, then the parameter after *ifTrue* is executed. If the returned object is *false*, then the parameter after *ifFalse* is executed. In effect, we have a regular *if-then-else* construct modeled in a completely object-orientated fashion as a set of communicating objects.

The second instance method, *leave*, removes a packet object from the queue. It uses a local variable *it* to store that packet object, which is maintained as the value of the *head* node of the list. The instance method finally returns packet *it*, after it has reconnected the list.

```
    leave
      | it |
      it := head value.
      (head = tail) ifTrue: [ head := tail := nil ]
      ifFalse: [ head next notNil ifTrue: [
              head := head next.
              head previous: nil.
              ]
      ].
      ^ it.
```

The third instance method, *list,* allows us to observe the current status of the queue and its contents. It assumes that each object in the list understands a *list* message. In our example, we will store packet objects in the queue, and all packet classes define a *list* method. This illustrates one of the major features of object-oriented programming languages in general and Smalltalk in particular: flexibility. At this point, we need not worry about which types of packets will actually be stored in a queue: Packet objects, Data objects, or Ack objects are welcome. Moreover, we might even have more packet subclasses in future versions of our software. The *list* method also illustrates a major feature of Smalltalk, that is, uncertainty. If we enter objects into the queue that do not understand the list message, then this code will fail when the queue is asked to *list*. Smalltalk represents a very flexible approach to software modeling. As we will see later, other object-oriented programming languages, notably Java, add considerably more safety and predictability.

```
list
  | tmp |
  tmp := head.
  [tmp notNil] whileTrue: [
    tmp value list.
    tmp := tmp next.
  ].
```

The body of the list method uses a *while* loop: [*tmp notNil*] is a block that is executed for each iteration of the loop. If it results in *true,* then the body of the loop (listed after the *whileTrue:* marker) is executed. If it is *false,* then the loop terminates.

Now that we have described four classes, it is possible to exercise them. Let us create a queue object and some packets (two Data objects and two Ack objects), enter them into the queue, and observe the current queue content by sending the appropriate messages:

```
| q w1 c1 w2 c2 |
q := FifoQueue new.
w1 := Data new: 'first packet'.
c1 := Ack new.
w2 := Data new: 'second packet' priority: 6.
c2 := Ack new.
q enter: w1.
q enter: c1.
q enter: w2.
q enter: c2.
q list.
```

The output from *q list* is

```
Data packet: first packet
acknowledged
Data packet: second packet
acknowledged
```

Then, we ask the queue to remove the packets and list them as we go,

```
q leave list.
q leave list.
q leave list.
q leave list.
```

Not surprisingly, the packet objects are stored by the queue object in the order in which they were entered, and then they are released in exactly the same order. This is what we would expect of a first-in-first-out queue abstraction.

To extend our example, we can now extend our queue abstraction by taking the importance (i.e., priority) of packets into account. Class PriQueue is defined as a subclass of Queue. It inherits all instance variables and methods, and specializes the *enter* method:

```
FifoQueue subclass: #PriQueue
    instanceVariableNames: ''
    classVariableNames: ''
    poolDictionaries: ''
    category: 'MyFifoQueue-Examples'.

!PriQueue methodsFor: 'access'!
enter: aPacket
    | tmp p |
    tmp := Node new: aPacket.
    tail isNil ifTrue: [
        tail := tmp.
        head := tmp.
    ] ifFalse: [
        p := tail.
        [(p notNil) and: (aPacket > (p value))] whileTrue: [
            p := p previous
        ].
        p isNil ifTrue: [
            tmp next: head.
            head previous: tmp.
            head := tmp.
        ] ifFalse: [
            tmp previous: p.
            tmp next: p next.
            (p = tail) ifTrue: [
                tail := tmp.
            ] ifFalse: [
                p next previous: tmp.
            ].
            p next: tmp.
        ]
    ].
```

The implementation of *enter* for priority queues is significantly more complex. It traverses the existing list for each packet that is to be entered to determine its relative importance. The *enter* method uses the > instance method defined for packet objects. Again, it is important that all objects stored in a PriQueue understand such a message. The PriQueue class specializes the *enter* instance method of class Queue. Another way of providing a queue with priority handling would be to specialize the *leave* instance method rather than *enter*. Class QueuePri, described next, does that.

```
FifoQueue subclass: #QueuePri
    instanceVariableNames: ''
    classVariableNames: ''
    poolDictionaries: ''
    category: 'MyFifoQueue-Examples'

!QueuePri methodsFor: 'access'!
leave
```

```
      | it p |
      it := head.
      head = tail ifTrue: [
            head := tail := nil.
      ] ifFalse: [
            p := head.
            [p notNil] whileTrue: [
                  (p value) > (it value) ifTrue: [ it := p].
                  p := p next.
            ].
            it = tail ifTrue: [
                  tail := it previous.
                  tail next: nil.
            ] ifFalse: [
                  it = head ifTrue: [
                        head := it next.
                        it next previous: nil.
                  ] ifFalse: [
                        it previous next: it next.
                        it next previous: it previous.
                  ]
            ]
      ].
      ^  it value.
```

The strategy for implementing a priority queue is straightforward. Whenever a packet is to be removed from the queue, we traverse the linked list and determine which of the packets has the highest priority. The good news is that Queue, PriQueue, and QueuePri objects can now be used interchangeably, depending on what kind of queuing strategy is desired. All three classes provide the same protocol; that is, their objects understand the same set of messages.

In summary, Smalltalk is much more than just a programming language. It is also a very elaborate programming environment that includes a large library of ready-to-use classes and allows for the interactive and incremental development of Smalltalk programs. The Smalltalk language is a truly object-oriented programming language. It hides all informational detail of objects and makes all instance methods freely available. Smalltalk allows only one form of inheritance, as seen in the example, where a class is defined as a subclass of a single superclass. Other programming languages allow multiple inheritance, where a subclass may have more than one superclass.

Smalltalk is very flexible. All message requests are resolved when an object receives a message. Other programming languages enforce this OOP principle to varying degrees, thus allowing for trade-offs between flexibility and safety. The creation of objects is defined by programmers. The deletion of objects, however, is left unspecified. Smalltalk automatically detects if objects are obsolete and reclaims them. This capability of object-oriented run-time support systems is called *garbage collection*.

91.3.2   C++

Our next OOP language is C++. C++ is the evolutionary enhancement to the C language, originally developed at Bell Laboratories, to support OOP features. It is most accurately described as a hybrid language (i.e., one that extends a traditional procedural language and remains compatible with that language, to include a variety of OOP features). Basically, this is C with classes, inheritance, and polymorphism. As Bjarne Stroustrup [Stroustrup 1994], the designer of the language, likes to point out, C++ is not an object-oriented programming language, but a language with which one can write an object-oriented program.

In C++, a class specification is separated into two files: a header file that contains a class declaration (listing the class name, fields, and methods) and a body file that contains all method bodies. C++ uses the C preprocessor to unite the two files upon compilation. It also enables that client code need only to include the header file to be able to use a class.

As our first example, we list the header for class Packet. The keyword *public* indicates that the instance methods, now called member functions in C++ terminology, are publicly visible. The instance variables (or, as referred to in C++, member field*s*) are *private*. That is, they are not visible outside of this class.

```
class Packet {
  int priority;
  char *name;
public:
  Packet(char *n, int p) {
  name=n;
  priority=p;
      }
  virtual void list();
  int operator>(Packet &);
  int operator<=(Packet &);
};
```

The class header declares the member fields and member functions, along with their accessibility. *Priority* is defined as an integer number (*int*); name is a string (*char* *). The comparisons are defined as operator functions — the ampersand (&) denotes call by reference. The header shows one other important member function: it has the same name as the class and is referred to as the **constructor**. The constructor serves to initialize a new instance of the class when it is created. Here, a new packet needs a name and a priority.

The keyword *virtual* in the declaration of member function *list* enables **dynamic binding** in C++. It is needed to allow inclusion polymorphism to work. Without it, C++ will use static binding, as in conventional programming languages.

The class body elaborates the bodies of the member functions:

```
int Packet::operator>(Packet & other) {
  return priority > other.priority;
}
int Packet::operator<=(Packet & other) {
  return priority <= other.priority;
}
void Packet::list() {
  cout << name << " packet: ";
}
```

where *cout* is the predefined output object in C++. It receives message << and prints its parameters.

C++ differs from Smalltalk in its support of encapsulation. Here, *private* does not mean that only a single object has access to its private data, but rather that it is private to the class. All other instances of class Packet have access to each others' *priority* and *name* fields. The member functions that implement the comparison capabilities take advantage of this fact.

Subclass *Ack* is very simple:

```
class Ack: public Packet {
public:
  Ack():Packet("Ack,''10){}
  void list() {
      cout << "acknowledged\n";
```

```
   }
};
```

Its constructor uses a special syntax to invoke the constructor of its superclass Packet. This class header also shows that a member function can (if it is short enough) be fully defined right away.

Similarly, for class Data

```
class Data: public Packet {
   char *body;
   int length;
public:
   Data(char *b, int p = 5):Packet("Data,'' p) {
      length = strlen(b) + 1;
      body = new char [length];
      strcpy(body, b);
   }
   void list();
};
void Data::list() {
   Packet::list();
    cout << body << endl;
   }
```

The constructor for class Data is defined with two parameters: one to initialize the *body* field and one for the *priority*, which will default to 5 if no value is provided in the call. The member function *list* refers to *Packet::list()*, which is the *list* function defined in the Packet class. Again, Smalltalk uses a simpler and cleaner approach by using the keyword *super* to refer to a method defined in a superclass. The word *endl* ensures that the output ends with a new line.

The next class, FifoQueue, contains and hides the definition of class Node (C++ would also support parameterized generic classes). The member fields *head* and *tail* are specified as *protected*. C++ allows one to control explicitly how subclasses have access to inherited members. Protected members are accessible in subclasses, whereas private members are not.

```
class FifoQueue {
protected:
   struct Node {
      Packet & value;
      Node *next, *previous;
      Node(Packet &p):value(p), next(nil), previous(nil){};
   } *head, *tail;
public:
   FifoQueue():head(nil),tail(nil){};
   virtual void enter(Packet &);
   virtual Packet & leave();
   void list();
};
```

Here, the constructor *Node* uses a shortcut to initialize the *value*, *next*, and *previous* member fields. Therefore, the body of the constructor can remain empty.

The body of class FifoQueue details the member functions *enter*, *leave*, and *list*:

```
void FifoQueue::enter(Packet &it) {
   Node *tmp = new Node(it);
   tmp->previous = tail;
```

```
    tail = tmp;
    if (head)
      tmp->previous->next = tail;
    else
      head = tmp;
  }
  Packet & FifoQueue::leave() {
    Packet &it = head->value;
    if (head == tail)
      head = tail = nil;
    else if (head->next) {
      head = head->next;
      head->previous = nil;
    }
    return it;
  }
  void FifoQueue::list() {
    for (Node* tmp = head; tmp; tmp=tmp->next)
      tmp->value.list();
  }
```

We can now exercise our objects. This program produces the same output as our previous example written in Smalltalk:

```
main(){
  FifoQueue q;
  Data w1("first packet");
  Ack c1, c2;
  Data w2("second packet,''6);
  q.enter(w1);
  q.enter(c1);
  q.enter(w2);
  q.enter(c2);
  q.list();
  q.leave().list();
  q.leave().list();
  q.leave().list();
  q.leave().list();
}
```

Again, we can use FifoQueue as the base class for subclass PriQueue to refine the *enter* member function:

```
class PriQueue: public FifoQueue {
  public:
    void enter(Packet &) {
    //logic as before
    }
};
```

And, of course, a similar class QueuePri is also possible, as follows:

```
class QueuePri: public FifoQueue {
public:
  Packet & leave() {
```

```
      //logic as before
   }
};
```

In summary, C++ provides detailed support for specifying the degree of access to its members. C++ goes beyond what we have illustrated here. It allows one to specify the type of inheritance that is used: public, protected, or private. All our examples use public inheritance, which propagates the accessibility of members to the subclass. Protected and private inheritance allow one to hide the fact that a class is based on a superclass. C++ supports both single and multiple inheritance. It requires that dynamic binding (i.e., the object-oriented behavior of an object to search for a suitable method for a message at run-time) be explicitly requested per member function. C++ uses the keyword *virtual* to request dynamic binding; otherwise, it defaults to static binding. C++ leaves memory management to the programmer, as garbage collection is not supported.

91.3.3   Java

Our next "best practice" programming language, Java, is perhaps the most compelling entry into the landscape of object-oriented programming languages. Java was developed at Sun Microsystems [Sun Microsystems]. Although its popularity stems from its ability to create highly interactive World Wide Web content, we discuss it here because it is a truly object-oriented programming language.

From an object-oriented perspective, Java can be seen as a distillation of many of the good features from Smalltalk and C++. From C++ it inherits its style of syntax, but with great simplifications: Java does not support pointers, and it has only single inheritance, no arbitrary typecasts, no class templates, no implicit type conversions defined by constructors, and no destructors. Method overloading is supported, but not operator overloading. Emphasis is placed on the readability and understandability of the source code.

From Smalltalk it inherits its execution model: all objects carry a unique identifier (reference), and all methods are dynamically bound (or *virtual*, in C++ terms). It is compiled into byte codes that are interpreted within the target environment, and all class information is available at run time, which provides additional type-checking capability and robustness. Java also provides automatic garbage collection, which simplifies a programmer's task significantly and tends to reduce many errors related to memory management. And of course, Java comes with a large class library that contains support for GUIs, networking, Web software development, and more.

Consider this Java version of our Packet class:

```java
public class Packet {
    int priority;
    String name;
    public Packet(String n, int p) {
        name = n;
        priority = p;
    }
    public void list() {
        System.out.print(name + " packet: ");
    }
    boolean more(Packet other) {
        return priority > other.priority;
    }
    boolean less(Packet other) {
        return priority <= other.priority;
    }
}
```

The resemblance to C++ is clear: most of the basic syntax, including declarations and control structures, use the C++ style. Missing are pointers and arrays based on pointers. Java supports actual arrays, but for character strings it has a built-in String class. All object handling is done by reference. Access specifiers (like private or public) are listed per field or function. Java comes with a significant set of predefined classes to allow input and output and, of course, to allow one to build *applets,* Java programs that can run within a Web browser that supports a Java interpreter.

Java insists on a closed-class hierarchy: all classes must have a superclass. If a superclass is not specified in the class declaration, then it defaults to class Object. In effect, all Java objects can be thought of as instances of that class (much as in Smalltalk). This enables broad run-time support, such as automatic garbage collection.

As in C++, Java member fields are explicitly typed. For example, *priority* is of atomic type *int*. Class Ack is defined as a subclass to Packet. The *extends* clause defines the inheritance relationship among classes. Java supports only single inheritance among classes.

```
public class Ack extends Packet {
    public Ack() {
        super("Ack,"10);
    }
    public void list() {
        System.out.println(" acknowledged");
    }
}
```

Class Data is just slightly more complicated. It inherits from class Packet using the explicit *extends* keyword. Like Smalltalk, Java uses the keyword *super* to refer to an instance method defined in the superclass.

```
public class Data extends Packet {
    String body;
    int length;
    public Data(String b) {
        super("Data," 5);
        length = b.length();
        body = new String(b);
    }
    public Data(String b, int p) {
        super("Data," p);
        length = b.length();
        body = new String(b);
    }
    public void list() {
        super.list();
        System.out.println(body);
    }
}
```

Again, before we can define the FifoQueue class, we need the Node class. Class Node is defined here as a simple class with an instance variable *value* of class Packet. This limits our queues to contain only Packets. Notable here are the types associated with fields *next* and *previous*. Both are defined as being of type *Node*. This does not mean that a Node object will contain other Node objects. They will, however, contain the object identifiers of other Node objects. Thinking of object identifiers as pointers to objects yields the conventional linked list metaphor. Moreover, a true object-oriented programming language does not need pointers at all. Since all objects carry their unique identity, that can be used instead. That is why Java and Smalltalk need not support pointers.

```
class Node {
      Packet value;
      Node next, previous;
      Node(Packet p) {
          value = p;
          next = null;
          previous = null;
      }
}
```

Class FifoQueue makes use of these classes to implement our first-in-first-out queue abstraction. The data features *head* and *tail* are defined as *protected*, that is, these fields will be accessible in subclasses. The *list*, *enter*, and *leave* methods are defined as *public*.

```
public class FifoQueue {
    protected Node head, tail;
    public FifoQueue() {
        head = tail = null;
    }
    public void enter(Packet it) {

    Node tmp = new Node(it);
    tmp.previous = tail;
    tail = tmp;
    if (head != null)
        tmp.previous.next = tail;
    else
        head = tmp;
    }
public Packet remove() {
    Packet it = head.value;
    if (head == tail)
        head = tail = null;
    else if (head.next != null) {
        head = head.next;
        head.previous = null;
    }
    return it;
    }
public void list() {
    for (Node tmp = head; tmp != null; tmp=tmp.next)
        tmp.value.list();
    }
}
```

As in the Smalltalk version of the example, we can assemble a few objects and send messages. In method *enter*, local object *tmp* is created as an instance of class Node using the C++ style *new* operator. Method *list* uses a *for* loop construct: before the loop starts, *tmp* is initialized to the value of *head*; the loop will continue to execute its body while *tmp* is not undefined (i.e., is not equal to *null*).

The next class doubles as the *main* program. Java does not allow the definition of stand-alone functions. The class has a single static method called *main*, which creates a few objects and starts execution. All objects must be explicitly created using the *new* operator. No object is allocated by default. The rest of the program reflects the same logic as illustrated in the Smalltalk and C++ examples. We enter four objects into the queue, remove them, and observe the queue and its contents.

```java
class Main {
      public static void main(String args[]) {
      System.out.println("Starting Main ... ");
      FifoQueue q = new FifoQueue();
      Data w1 = new Data("first packet");
      Ack c1 = new Ack(), c2 = new Ack();
      Data w2 = new Data("second packet,"6);
      q.enter(w1);
      q.enter(c1);
      q.enter(w2);
      q.enter(c2);
      System.out.println("The queue:");
      q.list();
      System.out.println("Order of leaving:");
      q.remove().list();
      q.remove().list();
      q.remove().list();
      q.remove().list();
}
```

Classes PriQueue and QueuePri can be defined as follows. Both classes inherit from FifoQueue, one redefining method *enter* and the other *leave*.

```java
public class PriQueue extends FifoQueue {
      public void enter(Packet it) {
            //logic as in Smalltalk example
      }
}
public class QueuePri extends FifoQueue {
      public Packet remove() {
            //logic as in Smalltalk example
      }
}
```

In summary, Java is a complete and truly object-oriented programming language. It supports a very open and flexible style of encapsulation. In addition to the three access rights of C++ — public, protected and private — Java defines a fourth: package. All fields and methods are of access right *package* unless explicitly stated otherwise. All classes in Java belong to packages, and all package-defined fields and methods are accessible from within any class within the package. Java also uses packages to organize its source and compiled code.

Java does not support multiple inheritance, where a class can have more than one superclass. Java supports the notion of interface, a specification of the public methods that an object can respond to. The interface does not include any member fields or method bodies. Multiple inheritance is supported for interfaces. Classes can be declared to implement interfaces. Interfaces allow the programmer to establish declared relationships between modules, which in turn can change their underlying implementation as the class changes.

Java also provides garbage collection as a means to reclaim obsolete objects. Java does not have a delete operator. Objects cannot be deleted explicitly.

91.3.4   C#

C# (pronounced *C sharp*) is the latest entry into the landscape of object-oriented programming languages. C# was developed at Microsoft [Microsoft 2003]. From an object-oriented perspective, C# can be seen as a distillation of many of the good features of Java and C, plus influences from Delphi [Cantu 2001], the object-oriented programming environment for ObjectPascal.

Like Java, C# is compiled into byte codes from a common language run-time (CLR) specification. All class information is available at run time, which provides additional type-checking capability and robustness. C# also provides automatic garbage collection, which simplifies a programmer's task significantly and tends to reduce many errors related to memory management. And, of course, C# comes with a large class library, called *common language infrastructure* (CLI), which contains support for common data structures, GUIs, database access, networking, etc.

Consider this C# version of our Packet class:

```
public class Packet {
    int priority;
    protected string name;
    public Packet(string n, int p){
        name = n;
        priority = p;
    }
    virtual public void print(){
        Console.Write("{0} packet:  ," name);
    }
    public static bool operator>(Packet p1, Packet p2) {
        return p1.priority > p2.priority;
    }
    public static bool operator<(Packet p1, Packet p2) {
        return p1.priority < p2.priority;
    }
}
```

The resemblance to Java is clear. The basic syntax, including declarations and control structures, use the C++ style. All object handling is done by reference. Access specifiers (like protected or public) are listed per field or function. As in C++, dynamic binding must be requested using the *virtual* keyword. C# comes with a significant set of predefined classes to allow input and output. The example lists the Console class, which defines the *Write* method. C# insists on a closed-class hierarchy: all classes must have a superclass. If a superclass is not specified in the class declaration, then it defaults to class Object.

A C# subclass specification resembles C++ more closely than Java. In the following example, class Ack is defined as a subclass of Packet. The colon is used to designate the superclass. The constructor uses the *base* reference to denote the invocation of the superclass constructor. The *list* method is explicit about redefining the superclass's virtual *list* method by using the keyword *override*. If the superclass did not have a *list* method, the compiler would flag an error.

```
public class Ack: Packet {
    public Ack():base("Ack,"10){}
    override public void list(){
        Console.WriteLine(" acknowledged");
    }
}
```

Class Data is just slightly more complicated. It inherits from class Packet and redefines the *list* method. The keyword *base* is used to refer to methods defined in the superclass.

```
public class Data: Packet {
    String body;
    int length;
    public Data(String b):base("Data,"5){
    length = b.Length;
    body = String.Copy(b);
    }
    public Data(String b, int p):base("Data,"p){
    length = b.Length;
    body = String.Copy(b);
    }
    override public void list(){
    base.list();
    Console.WriteLine(body);
    }
}
```

Again, before we can define the FifoQueue class, we need the Node class. Class Node is defined here as a simple class with an instance variable *value* of class Packet. This limits our queues to contain only Packet objects. Like Java and Smalltalk, C# does not need pointers.

```
class Node {
    public Packet value;
    public Node next, previous;
    public Node(Packet p) {
        value = p;
        next = null;
        previous = null;
    }
}
```

Class FifoQueue makes use of these classes to implement our first-in-first-out queue abstraction. The code closely resembles Java. Note, however, that the methods *enter* and *leave* are not defined as *virtual*. Therefore, dynamic binding will not be available: subclasses will not be able to override the methods. C# uses the keyword *new* for methods in subclasses that just want to specialize methods of superclasses. More detail is forthcoming.

```
public class FifoQueue {
    protected Node head, tail;
    public FifoQueue() {
        head = tail = null;
    }
    public void enter(Packet it) {
        Node tmp = new Node(it);
        tmp.previous = tail;
        tail = tmp;
        if (head != null)
        tmp.previous.next = tail;
        else
        head = tmp;
    }
```

```
        public Packet remove() {
            Packet it = head.value;
            if (head == tail)
                head = tail = null;
            else if (head.next != null) {
                head = head.next;
                head.previous = null;
            }
            return it;
        }
    public void list() {
            for (Node tmp = head; tmp != null; tmp=tmp.next)
                tmp.value.list();
        }
    }
}
```

As in the Java version of the example, our main program is a class with a static method *main*, where we assemble a few objects and send messages to test our Packet and FifoQueue classes:

```
class main {
    public static void Main(String[] args) {
        Console.WriteLine("Starting Main... ");
        FifoQueue q = new FifoQueue();
        Data w1 = new Data("first packet");
        Ack c1 = new Ack(), c2 = new Ack();
        Data w2 = new Data("second packet","6");
        q.enter(w1);
        q.enter(c1);
        q.enter(w2);
        q.enter(c2);
        Console.WriteLine("The queue:");
        q.list();
        Console.WriteLine("Order of leaving:");
        q.remove().list();
        q.remove().list();
        q.remove().list();
        q.remove().list();
    }
}
```

Classes PriQueue and QueuePri can be defined as follows. Both classes inherit from FifoQueue, one redefining method *enter* and the other *leave*. Because the superclass did not define either method as virtual, C# does not allow the keyword *override* here. Instead, we use the keyword *new*. The difference is the fact that dynamic binding is not enabled for the *enter* and *leave* method. That is, given a variable of type FifoQueue that holds a reference to a PriQueue, it would respond to an incoming message *enter* with the method defined in the FifoQueue class.

```
public class PriQueue: FifoQueue {
    new public void enter(Packet it) {
        //logic as before
    }
}
```

```
public class QueuePri: FifoQueue {
    new public Packet remove() {
        //logic as before
    }
}
```

In summary, C# is a complete and truly object-oriented programming language. C# also provides garbage collection as a means to reclaim obsolete objects. C# also supports the notion of interface, a specification of the public methods that an object can respond to. The interface does not include any member fields or method bodies. Multiple inheritance is supported for interfaces. Classes can be declared to implement interfaces. Interfaces allow the programmer to establish declared relationships between modules, which in turn can change their underlying implementation as the class changes.

C# supports some additional novel features: it allows programs to treat values of atomic types, such as *int* or *char*, as objects, with a process called boxing. Boxing automatically converts an atomic value that is being stored in the execution stack to wrapper objects that are allocated on the heap and referenced from the stack. The reverse process, unboxing, is also done automatically. The advantage of this boxing feature is that values of atomic types can be passed by reference to methods. C# also supports C-like *structs*, which allow one to build object-like value sets that are allocated on the stack rather than the heap. Other performance-enhancing features of C# include the capability to declare unsafe code, where C-like syntax, with direct pointer addressing and arithmetic, is allowed.

91.4   Language Implementation Issues

In evolutionary terms, object-oriented programming languages can be described as either pure or hybrid. The hybrid approach adds object-oriented features on top of a procedural core: C++ and ObjectPascal are representatives of this approach. The pure approach, Smalltalk and Java, limits the language features to those that are strictly object oriented. C# presents a compromise in that it enforces its object-oriented principles but allows unsafe code, as long as it is clearly labeled.

In theoretical terms, the primary distinction between these kinds of languages lies in their interpretation of the concept of type. Because of their origins, hybrid languages rely on a traditional approach to typing. That is, the type of something resides with its container. Thus, when we declare an integer variable *i* in C++, for example, as *int i;*, variable *i* is of basic type *integer*. In Smalltalk, variables do not carry the type of what they contain. That is, variables are untyped: they can contain any object. Rather, the command *Integer new* returns an integer object. The type in this case is seen to reside with the object, and if we store a reference to it in a variable, the variable could be said to be of type *Integer*.

In implementation terms, representing values of basic types as objects carries penalties in amount of storage and performance: if an integer value is stored as an object, it will occupy about double the amount of storage, let us say 4 bytes for its value and 4 bytes for the object reference. To access the value as an object, we first must dereference the object reference, then retrieve the value from the object. Java stores all values of basic types as simple values and does not treat them as objects. The programmer must explicitly create a wrapper object to treat the value as an object. C# does this implicitly through its boxing and unboxing mechanisms.

Another aspect of whether type resides with a variable or with an object is that, in Smalltalk for example, the implementation must simply ensure that every object knows its class (and, of course, somehow its superclasses). This can be implemented either by encoding into classes knowledge of (pointers to) their superclasses, or by allowing classes to contain directly all information that they inherit.

Hybrid languages must provide support for both notions of type: variables and objects can have type. In C++, object type information is added either via a *vtable* per object (which contains pointers to all functions that apply to the object) or, in newer implementations, by using run-time type information (RTTI) extensions.

A second critical distinction between OOP languages is in how they handle the binding of messages to methods. In languages such as Smalltalk that perform this binding dynamically (at run time), an object receives a message and the search for an appropriate method begins at the class of the object. The search continues through superclasses until the message can be processed. Although this approach affords the programmer tremendous flexibility, it has clear practical downsides. A more efficient approach is to leave the binding choice to the system (i.e., the compiler and linker). That is, the system tries to determine at compile time which method should be invoked for each message sent. In cases where inclusion polymorphism is used (and it may be unclear which class to refer to), binding can be performed at run time using techniques varying from simple case statements to a more complex system of virtual method tables. In any case, it is still up to the compiler to detect and indicate the need for run-time binding.

Another important issue to consider in the context of language implementation is the approach one adopts to memory management. OO languages are relatively uniform in their approaches to memory allocation (object creation). Creating objects and all that that entails (determining how much memory to allocate, the types of member fields, etc.) is performed by the system. Initialization, on the other hand, is left to the programmer. Many languages provide direct support for initializing objects: we have constructors in C++, Java, and C# and initialize methods in Smalltalk. There are two common approaches to deleting objects from memory (deallocation). In the programmer-controlled approach, one uses a *delete* operator. The system-enabled approach to deleting objects uses the concept of garbage collection [Jones and Lins 1996]. All objects that are unknown to other objects are garbage. The system needs a way to detect that fact: reference counters, memory mirroring, mark and sweep, etc., are examples of algorithms that enable it. In principle, the system sweeps all objects constantly to determine which are reclaimable. Practically speaking, this is quite compute-intensive. To lighten the impact on system performance, garbage collection is typically done either during times of idling or when some threshold of memory usage is reached. Just before objects are removed from memory, the destructor is invoked: it describes what cleanup steps must be done. The problem here is that the programmer now needs to know all possible circumstances in which objects of the class will (ever) be used. In practice, this has been shown to be a problematic approach. Memory leaks — memory that is occupied by deleted objects — can occur easily in large bodies of C++ code.

91.5 Research Issues

Object-oriented programming languages have matured dramatically and have gained a significant degree of acceptance over their almost 40 years of history. Perhaps the most dramatic test of the paradigm will be in how it responds to demands imposed on it by the coming advances in computing hardware, architectures, and operating systems. Currently, the execution model for the OOP paradigm is strictly sequential and synchronous. Computation starts at one object, which sends a message to the next, and so on. Language constructs, such as threads in Java, and architectures, such as the common object request broker architecture (CORBA), attempt to go beyond these constraints, pushing OOP into the worlds of distributed and parallel computing.

Although viewed mostly as a programming language with a sequential execution model, Java actually allows the construction of parallel programs. Java supports threads. In addition to the conventional single thread of execution, it allows the creation, starting, and management of multiple threads. The Java run-time system then schedules the threads with the underlying operating system. If the underlying hardware has multiple processors, then it truly will execute in parallel. CORBA [Object Management Group 2003] enables communication among distributed objects, where these objects do not need to be from a single programming language. C++ objects can communicate with Smalltalk objects. The basic idea is to set up an object request broker that relays messages from a sender to a receiver object. Implementations of CORBA and similar architectures are widely available.

In summary, many of the original motivations for the OOP paradigm have been justified by our practical experience with it. OOP has been seen to

Embody useful analysis and design techniques
Revise the traditional software life cycle toward analysis and design, instead of coding, testing, and
    debugging

Encourage software reuse through the development of useful code libraries of related classes

Improve the workability of a system so that it is easier to debug, modify, and extend

Appeal to human instincts in problem solving and description — in particular, to problems that model real-world phenomena

Defining Terms

Abstract class: A class that has no direct instances but is used as a base class from which subclasses are derived. These subclasses will add to its structure and behavior, typically by providing implementations for the methods described in the abstract class.

Class: A description of the data and behavior common to a collection of objects. Objects are instances of classes.

Constructor: An operation associated with a class that creates and/or initializes new instances of the class.

Dynamic binding: Binding performed at run time. In OOP, this typically refers to the resolution of a particular name within the scope of a class, so that the method to be invoked in response to a message can be determined by the class to which it belongs at run time.

Inheritance: A relationship among classes, wherein one class shares the structure or behavior defined in an *is-a* hierarchy. Subclasses are said to inherit both the data and methods from one or more generalized superclasses. The subclass typically specializes its superclasses by adding to its state data and by redefining its behavior.

Instance: A specific example that conforms to a description of a class. An instance of a class is an object.

Interface: A named listing of method headers to be implemented by a class.

Member field: The data items that are associated with (and are local to) each instance of a class.

Message: A means for invoking a subprogram or behavior associated with an object.

Method: A procedure or function that is defined as part of a class and is invoked in a message-passing style. Every instance of a class exhibits the behavior described by the methods of the class.

Object: An object is an instance of a class described by its state, behavior, and identity.

Object-oriented programming (OOP): A method of implementation in which a program is described as a sequence of messages to cooperating collections of objects, each of which represents an instance of some class. Classes can be related through inheritance, and objects can exhibit polymorphic behavior.

Override: The action that occurs when a method in a subclass with the same name as a method in a superclass takes precedence over the method in the superclass.

Polymorphism (or *many shapes*): That feature of a variable that can take on values of several different types or a feature of a function that can be executed using arguments of a variety of types.

Subclass: A class that inherits variables and methods from another class (called the superclass).

Virtual function: Most generally, a method of a class that may be overridden by a subclass to the class. In languages in which dynamic binding is not the default, this may also mean that a function is subject to dynamic binding.

References

Budd, T. 2001. *The Introduction to Object-Oriented Programming (3rd edition)*. Addison-Wesley, Reading, MA.

Cantu, M. 2001. *Mastering Delphi*. Sybex, Alameda, CA.

Cardelli, L., and Wegner, P. 1985. On understanding types, data abstraction, and polymorphism. *ACM Comput. Surv.*, 17(4).

Goldberg, A., and Robson, D. 1983. *Smalltalk-80: The Language and Its Implementation*. Addison-Wesley, Reading, MA.

Ingalls, D. 1981. Design principles behind Smalltalk. *Byte*, 6(8).

Jones, R and Lins D. 1996. *Garbage Collection: Algorithms for Automatic Dynamic Memory Management*. John Wiley & Sons, New York.

Meyer, B. 2000. *Object-Oriented Software Construction (2nd edition)*. Prentice Hall, Englewood Cliffs, NJ.

Microsoft. 2003. C# introduction and overview. http://msdn.microsoft.com/vstudio/techinfo/articles/ upgrade/Csharpintro.asp.

Nygaard, K., and Dahl, O.J. 1981. The development of the Simula languages. In *History of Programming Languages*. R. Wexelblat, Ed. Academic Press, New York.

Object Management Group. 2003. CORBA basics. http://www.omg.org/gettingstarted/corbafaq.htm.

Stroustrup, B. 1994. *The Design and Evolution of C++*. Addison-Wesley, Reading, MA.

Stroustrup, B. 2000. *The C++ Programming Language (special 3rd edition)*. Addison-Wesley, Reading, MA.

Sun Microsystems. The Java language: an overview. Sun Microsystems, http://java.sun.com/docs/ overviews/java/java-overview-1.html.

Further Information

That object-oriented programming has become the predominant software development paradigm is evidenced by the multitude of conferences, journals, texts, and Web sites devoted to both general and language-specific topics.

The two most prominent conferences, both of which address a wide range of OOP issues, are the conference on Object-Oriented Programming Systems, Languages, and Applications (OOPSLA, www.oopsla.org) and the European Conference on Object-Oriented Programming (ECOOP, www.ecoop.org).

The *Journal of Object Technology*, published by Bertrand Meyer, provides contemporary coverage of OOP languages, applications, and research (www.jot.fm).

Perhaps the most general of the references listed are Budd [2001] and Meyer [2002].

92
Functional Programming Languages

92.1   Introduction **92**-1
92.2   History of Functional Languages **92**-3
92.3   The Lambda Calculus: Foundation of All
       Functional Languages **92**-4
92.4   Pure Versus Impure Functional Languages **92**-5
92.5   SCHEME: A Functional Dialect of LISP **92**-5
       SCHEME Data Types • SCHEME Syntax • Predefined
       Functions • Impure Features in SCHEME: Assignment
       and I/O
92.6   Standard ML: A Strict Polymorphic Functional
       Language **92**-10
       Predefined Types in ML • Expressions in ML
       • Declarations in ML • Pattern Matching
       • Type Definitions • Type Variables and Parametric
       Polymorphism • Type Constructors • The ML Module
       System • Impurities in ML: References and I/O
92.7   Nonstrict Functional Languages **92**-20
92.8   HASKELL: A Nonstrict Functional Language **92**-22
       The HASKELL Class System • User-Defined Types
       • Instance Declarations • List Comprehensions in HASKELL
       • Functional I/O in HASKELL
92.9   Research Issues in Functional Programming **92**-25
       Program Analysis and Optimization • Parallel Functional
       Programming • Partial Evaluation • State in Functional
       Programming

Benjamin Goldberg
New York University

92.1   Introduction

Functional languages are a class of languages based on the **lambda calculus**, a very simple but powerful model of computation. Proponents claim that the use of a functional language supports faster production of software, shorter programs, and more readable and verifiable code than the use of conventional so-called imperative programming languages. Furthermore, in the research community functional languages have been used as the basis of study on advanced type systems, parallel computing, program optimization, and programming language semantics.

Within the class of functional languages there is substantial variety. In this chapter, we describe three popular languages that are representative of the class: SCHEME, a dialect of LISP; Standard ML; and HASKELL. Although these languages differ in significant ways, they all exhibit the necessary properties in order to be considered functional.

A program written in a functional language consists of function definitions and function applications. As in mathematics, a function is an entity that maps each input to a single output. This is in stark contrast to imperative languages such as C and FORTRAN in which a function is simply a collection of statements which may modify variables, allowing the same input to be mapped to different outputs over the course of the computation.

Consider the factorial function. It is described formally by

$$n! = \begin{cases} 1 & \text{if } n = 0 \\ n(n-1)! & \text{otherwise} \end{cases}$$

In a functional language, in this case Standard ML, the executable definition of factorial is usually written as

```
fun factorial(0) = 1
  | factorial(n) = n * factorial(n-1)
```

and follows directly from the formal definition. In contrast, the factorial function is typically written in an imperative language, in this case American National Standards Institute (ANSI) C, as

```
int factorial(int n)
{ int prod = 1;
  for (int i = 1; i <=n; i++)
    prod = prod * i;
  return(prod);
}
```

where the programmer specifies that a variable holds the running product and is modified in each iteration of the loop. Although a version similar to the first program could be written in most imperative languages, the syntax and traditional programming style of these languages encourage the writing of the second.

Because functional languages, and the programs written in them, are intended to have nice mathematical properties — such as the properties that functions written in functional languages are functions in the true mathematical sense — there is no explicit notion of memory and the modification of memory. As in mathematics, a variable in a functional program simply represents a value, it is not a cell in memory that can be modified. There is no need for an assignment operator, and thus functional languages do not provide one (at least **pure functional languages** do not, see Section 92.4). In Standard ML, for example, the operator = does not express assignment but rather an equation. Consider the following statement in C:

```
x = x + 1;
```

It is obvious to a C programmer that this statement specifies the modification of an existing variable **x**. Notice, though, that in mathematics this is an equation that has no solution. There is no **x** that satisfies the equation. The functional programming community has long argued that by departing from a mathematical interpretation of programs, the use of imperative languages leads to complex, poorly understood programs.

By preventing the modification of existing variables, functional languages exhibit what is known as **referential transparency**, which essentially means that in a functional program, equal expressions can be interchanged. Consider, for example, the following expression (in Standard ML):

```
let
  val x = f(a)
in
  ... x + x ...
end
```

This introduces a new variable **x** whose value is the result of the function call **f(a)**, and then evaluates an expression containing the sum **x + x**. Because **x** cannot be modified, a reader of the program can be sure that each occurrence of **x** has the value of **f(a)**, and thus the sum would have the same result as if the programmer had written **f(a) + f(a)**.

In an imperative program, it is difficult for the reader to be sure that the value of **x** had not changed, either by a direct assignment to **x** or indirectly via call to some procedure that modifies the value of **x**. Expressions containing these modifications, either direct or indirect, are known as **side effects** because, aside from returning a value, these expressions have the side effect of changing a variable's value. Functional programmers argue that it is side effects (and the corresponding loss of referential transparency) that lead to incomprehensible large programs.

An additional property that functional languages exhibit is the ability to treat functions as data. That is, functions can be passed as parameters to other functions, returned as the result of function calls, and stored in data structures. Thus, functions are said to be **first-class objects** since their use is no more restricted than other kinds of data. This is attractive for philosophical reasons, since functions are mathematical entities just like integers and Booleans, and for practical reasons, it increases the flexibility of code.

For instance, all functional languages provide a construct for specifying a function value without having to declare the function's name, equivalent to lambda abstractions in the lambda calculus. In Standard ML, such an expression is of the form

```
fn(x) => e
```

and denotes a function whose formal parameter is x and whose body is the expression e. This function value can be used in larger expressions, function calls, etc.

Consider again the factorial function. It might be argued that the formal definition of factorial just given was tailored to suit the recursive nature of the definition of factorial in the functional language, and that a more reasonable and common definition of factorial is

$$n! = \prod_{i=1}^{n} i$$

The product operator $\prod$ is a very useful operator for defining a wide range of functions and has the general form

$$\prod_{i=m}^{n} f(i)$$

for some initial value m, some final value n, and some function f. In a functional language, $\prod$ would be written as a **higher order function**, namely, a function that takes a function as a parameter or returns a function as its result. In particular, $\prod$ takes three parameters, **m**, **n**, and **f** and could be written in Standard ML as

```
fun prod(m,n,f) = if m = n then f(m) else f(m) * prod(m+1,n,f)
```

Thus, factorial can simply be defined as

```
fun fac(n) = prod(1,n, fn i => i)
```

and the exponentiation function computing x^n can be defined as

```
fun power(x,n) = prod(1,n,fn(i) => x)
```

92.2 History of Functional Languages

All functional languages trace their roots to the lambda calculus, developed by the logician Alonzo Church [1941], in the 1930s. This simple model, which describes computation as a series of syntactic conversions between expressions, was developed in order to gain a deeper understanding into computation and what it means for functions to be computable, rather than as a programming language (since it obviously predates computers).

The first programming language that at least resembled the lambda calculus was LISP, developed by John McCarthy in the late 1950s [McCarthy et al. 1962]. It differs from the lambda calculus in several important ways: It was dynamically scoped (although McCarthy attributes this to a bug in the initial implementation), and provided an assignment operator. McCarthy states that although the lambda calculus served as an influence on the syntax of LISP, it was not the primary factor in the design of LISP's semantic features [McCarthy 1978]. LISP, however, has had a tremendous influence on modern functional languages. In 1975, Steele and Sussman designed SCHEME [Sussman and Steele 1975], a dialect of LISP that fixed some of the problems of earlier LISPs, such as dynamic scoping, and now its pure subset serves as the most LISP-like of all functional languages.

Another early language to have a great impact on the design of modern functional languages, especially ML and HASKELL, was ISWIM, developed by Landin [1966]. It was an explicit attempt to create a language whose semantics mirrored those of the lambda calculus, provided more convenient syntax and programming features, and was able to be implemented efficiently. Prior to the development of ISWIM, Landin [1964] had developed an abstract machine model, called the SECD machine, which specified how the conversion rules of the lambda calculus could be efficiently executed. Thus, the behavior of ISWIM operators could be described by their effect on the SECD machine.

The visibility of functional languages received a large boost in 1978, when John Backus, the designer of FORTRAN and the recipient of the 1978 A.M. Turing Award (computer science's highest award), chose to describe a new functional language, FP [Backus 1978], in his invited talk upon receiving the award. FP was a language of less expressive power than other functional languages of its time, since it did not provide user-defined higher order functions but rather supplied a fixed number of higher order *combining forms* used to create complex functions out of simple ones, and was heavily influenced by the APL programming language. Despite its limitations, and despite being of little interest today, FP was very influential in attracting researchers to the field of functional programming due Backus's stature, background, and convincing arguments in its favor.

During the 1970s and 1980s, functional languages, both strict and nonstrict, proliferated. Receiving a fair amount of attention and popularity were languages such as ML, SASL, HOPE, Lazy ML, and MIRANDA. Because of this proliferation, there was a movement to create standardized functional languages. The results of these standardization movements were a standardized definition of SCHEME [Rees et al. 1992]; Standard ML [Milner et al. 1990, Milner and Tofte 1991], now the standard strict functional language; and HASKELL [Hudak et al. 1992], now the standard nonstrict functional language. It is these languages that we have chosen to describe in this chapter.

92.3   The Lambda Calculus: Foundation of All Functional Languages

No description of functional languages is complete without the introduction of the lambda calculus, a simple but powerful model of computation. The reader is referred elsewhere in this handbook and to Barendregt [1984] for a detailed description of the lambda calculus. The important points to note about the lambda calculus are the following:

1. All functional languages are simply syntactically sugared versions of the lambda calculus, in some cases typed versions of the lambda calculus. Thus, any property that holds for the lambda calculus, such as its computational power, also holds for functional languages.

2. The lambda calculus is Turing complete: Every computable function can be expressed in the lambda calculus, and thus it is as least as powerful as any other computational model (such as Universal Turing Machines, for example).

3. One of two common evaluation orders in the lambda calculus, applicative order and normal order, have been adopted by almost all functional languages. Those functional languages which use

applicative-order evaluation, in which the arguments in a function call are evaluated before the body of the function (as is the case with all imperative languages), are called **strict functional languages**. Those functional languages which use **normal-order evaluation**, in which the arguments in a function call are only evaluated if and when needed in the body of the function, are called **nonstrict functional languages**.

4. The first Church–Rosser theorem about the lambda calculus states that no matter which evaluation order is chosen, the result of functional program will be the same as long as the program terminates. Not all evaluation orders are equally likely to terminate, however, and the second Church–Rosser theorem states that the evaluation order that is most likely to lead to termination is normal-order evaluation.

The three languages described here, SCHEME, Standard ML, and HASKELL, are all based on the lambda calculus. They differ primarily in three ways: their syntax, their type systems, and whether they are strict or nonstrict.

92.4 Pure Versus Impure Functional Languages

Of the three functional languages described in this chapter, only one, HASKELL, is purely functional. That is, only HASKELL does not provide any mechanism for performing side effects. Both SCHEME and ML provide mechanisms for performing assignment to variables, although ML's mechanism is far more limited. However, SCHEME and ML deserve to be included in this chapter because good practice dictates that programs written in these languages are generally purely functional and side effects are used only where the programmer considers them absolutely necessary. At the end of the sections on SCHEME and ML, some of their impure features will be described.

A side-effect mechanism that is quite difficult to omit from a language is input/output (I/O). From an external viewpoint, such as the view of the operating system handling I/O requests from a functional program, input and output operations change the state of the input and output buffers (for the terminal, printer, etc.). However, to see why conventional I/O routines, such as read and print, do not support referential transparency within the program, consider

```
let x = read()
in x + x
end
```

where **read()** reads data from the standard input and returns the value read. If referential transparency were preserved, this code could be replaced by

```
read() + read()
```

which is clearly not the case.

SCHEME and ML adopt relatively conventional I/O routines, sacrificing referential transparency in expressions involving I/O. HASKELL, however, uses a more novel approach to support I/O in a referentially transparent manner.

92.5 SCHEME: A Functional Dialect of LISP

SCHEME is a dialect of LISP which differs from the more traditional LISPs (including Common LISP) primarily in that it is *statically scoped*. It is also a smaller, simpler language than most other LISPs. SCHEME, as defined in the IEEE Standard, is not a purely functional language. It supports the **set!** operator, similar to **SETQ** in traditional LISPs, which performs assignment on variables. However, SCHEME programmers tend to write in a functional style, and at least one SCHEME compiler (see Kranz et al. [1986]) performs optimizations specifically targeted at programs written in a functional style.

In this chapter, we will focus on *pure* SCHEME, a subset of SCHEME that is purely functional. Pure SCHEME differs from SCHEME only in that it omits the few side-effect operators that SCHEME provides. By doing so, the mathematical properties of pure SCHEME mirror those of the lambda calculus.

Like all LISPs, SCHEME adopts a prefix notion for all syntactic entities, thus looks strikingly different from conventional languages and other functional languages. The beauty of LISP and SCHEME syntax is that there are very few syntactic rules, thus learning the syntax of the language is trivial. Furthermore, the appearance of SCHEME data structures and SCHEME programs is quite similar, leading to the ability to manipulate programs as data, as is the case with interpreters, compilers, program verifiers, and program transformers.

Also like all LISPs, but unlike the other functional languages described in this chapter, SCHEME has **latent types**, which means that types are associated with values, not variables. Type checking occurs at run time, not compile time (which is why SCHEME is often called a *dynamically typed* language) and a type error is signaled only when a primitive operator (such as **+**, **−**, etc.) has been applied to a value of an inappropriate type. There are no type declarations, and the types of user-defined functions and variables are not specified. Variables can be bound to values of different types over the course of the computation.

92.5.1　SCHEME Data Types

There are two kinds of types in SCHEME (as in LISP), atomic types known as *atoms*, and *pairs*. The atomic types include numbers (floating point numbers and arbitrarily large integers), Booleans (written **#t** and **#f**), character strings, and a type that is peculiar to LISP dialects, namely, symbols. Symbols are objects that have only one property, their name. Two symbols are equivalent if and only if they have the same name. SCHEME symbols are different from those of traditional LISPs, since LISP symbols often have many properties associated with them.

The other kind of type, a pair, is a two-element record. This record is generally referred to as a *cons cell*. Each element can be of any type and is generally implemented as seen in Figure 92.1a. The first element is known as the *car* and the second is known as the *cdr*. There is a constant **( )**, called the *empty list*. Any collection of pairs of the form pictured in Figure 92.1b where the cdr of each cons cell is either **( )** or points to another cons cell, is called a *list*. The list is the primary aggregate data structure in SCHEME (and all

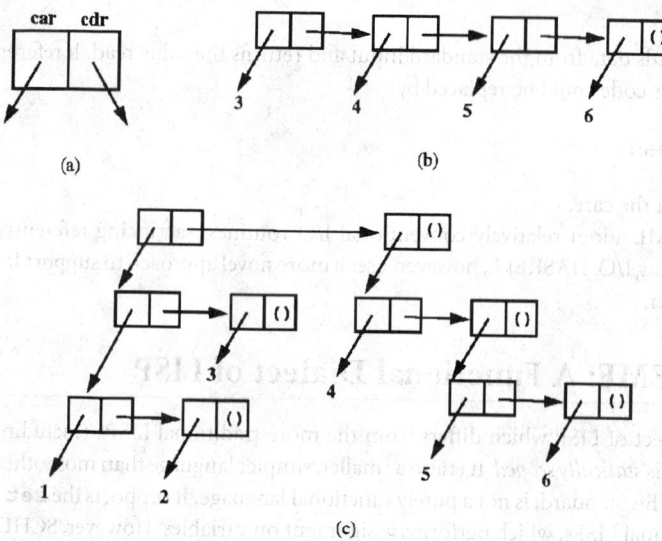

FIGURE 92.1　List structures.

functional languages). It is a very flexible data structure, since each element of a list can itself be a list. The list pictured in Figure 92.1b would be printed as

```
(3 4 5 6)
```

and the list in Figure 92.1c would be printed as

```
(((1 2) 3) (4 (5 6)))
```

92.5.2   SCHEME Syntax

In SCHEME, every language construct is an expression that returns a value. An expression is either an *atomic expression* or a *combination*. An atomic expression can either be a numeric literal, a string literal, or an identifier (representing both symbols and variables).

A combination is an expression consisting of an open parenthesis followed by some subexpressions and then a close parenthesis. These represent a range of expressions, including function calls, definitions, conditionals, etc.

A function call is of the form

$$(e_0 \ e_1 \ \ldots \ e_n)$$

where each e_i is an expression. The expression e_0 should evaluate to a function, which is then applied to the values of the expressions $e_1 \ldots e_n$. A call to a function with no arguments is simply written as (e_0). The function call syntax is used for all user-defined functions as well as all predefined functions, including the arithmetic operators.

There are two forms of conditional expressions. The simplest is the **if**, of the form

```
(if e0 e1 e2)
```

If e_0 evaluates to any value other than **#f**, then the value of e_1 is computed and returned as the result of the entire expression. Otherwise, the value of e_2 is computed and returned.

A more general conditional, the **cond** expression, has the form

```
(cond (c0 e0)
      (c1 e1)
      ...
      (cn en))
```

The expressions $c_0, c_1, \ldots, c_n$ are evaluated in order until the first c_i, for some i, evaluates to a value other than **#f**. The value of e_i is then computed and returned. The expression c_n may be replaced by the keyword **else**, in which case the value of e_n is returned if none of $c_0, \ldots, c_{n-1}$ evaluates to a value other than **#f**.

In the construct

```
(quote exp)
```

exp is treated as data — either a symbol, number, or list — instead of an expression to be evaluated. Thus,

```
(quote a)
```

returns the symbol **a**, not the value of the variable **a**. The result of the expression

```
(quote (f a b c))
```

is the list containing the symbols **f**, **a**, **b**, and **c**. It is not a call to function **f** with arguments **a**, **b**, and **c**. Because *quote* is used so extensively in SCHEME and LISP, a syntactic shorthand is provided. The construct *'exp* is simply shorthand for **(quote** *exp***)**. Thus, **'a** is equivalent to **(quote a)** and **'(a b c)** is equivalent to **(quote (a b c))**. A nested list is easily specified, for example, **'(a b (c d))**.

SCHEME's **lambda** expression is an expression whose value is a function. It is of the form

```
(lambda (x₁...xₙ) e)
```

and evaluates to a function whose formal parameters are $x_1 \ldots x_n$ and whose body is the expression e. A lambda expression without parameters would be of the form **(lambda () e)**.

A definition of the form

```
(define x e)
```

introduces a new variable x and binds it to the result of evaluating the expression e. The variable x is visible during the evaluation of e, thus allowing for recursive function definitions such as

```
(define fac (lambda (x) (if (= x 0) 1 (* x (fac (- x 1))))))
```

In most implementations, **define** is only allowed at the top level, i.e., not nested inside any other expression. In these cases, the variable introduced is global.

As a syntactic convenience, functions can also be defined using the form

```
(define (f x₁ ... xₙ) e)
```

which is equivalent to **(define f (lambda (x₁ ... xₙ) e))**. Thus, the factorial function given previously is generally written

```
(define (fac x) (if (= x 0) 1 (* x (fac (- x 1)))))
```

The **let** construct, of the form

```
(let ((x₁ e₁)
      (x₂ e₂)
      ...
      (xₙ eₙ))
  e)
```

is used to introduce the variables $x_1 \ldots x_n$ and bind them to the values of the expressions $e_1 \ldots e_n$, respectively. The value of e is then computed and returned as the value of the entire **let** expression. The scope of the new variables $x_1 \ldots x_n$ is just the body of e. Thus, these variables cannot be referenced in expressions $e_1 \ldots e_n$. This means that none of $x_1 \ldots x_n$ can be defined recursively.

The **letrec** construct can be used to introduce recursively defined local variables. It has the same form as the **let** construct, except that the keyword **letrec** is used instead of **let**. In this case, the expressions $e_1 \ldots e_n$ are defined in an environment in which each of $x_1 \ldots x_n$ are visible and thus can be referenced. Here is an example of a use of **letrec**,

```
(letrec ((f (lambda (x) (if (= x 0) 1 (g (- x 1)))))
         (g (lambda (y) (if (= y 0) 1 (f (- y 1))))))
  (+ (f 3) (g 5)))
```

where **f** and **g** are mutually recursive functions.

92.5.3 Predefined Functions

SCHEME provides a large number of predefined functions. The usual collection of arithmetic and logical operators, +, -, =, !=, >, <, etc., are provided and can be applied to any numeric values. Examples of their use include **(+ 3 4)**, **(> 6.2 5.1)**, and **(!= 4 5)**.

A function commonly used to create lists is **list**. It takes an arbitrary number of arguments and creates a list containing their values. Thus, for example,

```
(list 'a (+ 2 5) b (list 6 2))
```

would return the list **(a 7 v 6 2)**, where v is the value of the variable **b**.

The most heavily used list construction function is **cons**. It takes two arguments and, as its name implies, creates a cons cell whose car is the value of the first parameter and whose cdr is the value of the second. For example, here is a function that takes parameters **N** and **M** and constructs the list of integers between **N** and **M**, inclusive.

```
(define (listof N M)
  (cond ((> N M) '())
        (else (cons N (listof (+ N 1) M)))
        ))
```

To access the car and cdr fields of a cons cell, SCHEME provides the functions **car** and **cdr**, respectively. For example, **(car '(3 4 5 6))** returns **3** and **(cdr '(3 4 5 6))** returns the list **(4 5)**. If the first element of a list l_1 is itself a list l_2, then **car** applied to l_1 returns l_2, as one would expect. For example,

```
(car '((1 2) (3 4) 5))
```

returns the list (1 2) and

```
(cdr '((1 2) (3 4) 5))
```

returns ((3 4) 5).

The predicate **null?** is used to test for an empty list. Here **(null? x)** returns **#t** if the value of **x** is the empty list and returns **#f** otherwise. Here is an example of the use of **car**, **cdr**, and **null?**: Given a list of numbers, the function **sumof** returns the sum of the elements of the list.

```
(define (sumof l)
    (cond ((null? l) 0)
          (else (+ (car l) (sumof (cdr l))))
          ))
```

Also, **cons** is useful for constructing lists one element at a time. Another useful predefined function is **append**. It takes as parameters two lists l_1 and l_2 and returns a list containing the elements of l_1 followed by the elements of l_2. For example,

```
(append '(1 2 3 4) '((a b) c d))
```

returns the list **(1 2 3 4 (a b) c d)**. Although **append** is always provided by SCHEME implementations, it is not *primitive* in the sense that it can easily be written in SCHEME.

```
(define (append x y)
  (cond ((null? x) y)
        (else (cons (car x) (append (cdr x) y)))))
```

Another predefined function that can easily be written in SCHEME is **reverse**. This function takes a list l and returns a new list with the same elements as l, but in reverse order. For example,

```
(reverse '(1 2 (3 4) 5))
```

returns the list **(5 (3 4) 2 1)**. Notice that nested lists, such as the third element of the previous input list, are not recursively reversed. The reverse function can be defined in SCHEME as

```
(define (reverse 1)
  (cond ((null? 1) '())
        (else (append (reverse (cdr 1)) (list (car 1))))))
```

Unfortunately, the cost of this function is proportional to the square of the length of the input list. This can be seen by noting that **append** is linear in the size of its argument and is called each time that **reverse** is called recursively. The depth of the recursion in **reverse** is proportional to the length of its argument. A more efficient **reverse**, whose cost is linear in the length of its argument, is

```
(define (reverse 1)
  (rev 1 '()))

(define (rev 1 accum)
  (cond ((null? 1) accum)
        (else (rev (cdr 1) (cons (car 1) accum)))))
```

One can think of **rev** as successively taking the elements of **1** and putting them at the front of the list **accum**. Thus, when **1** is empty **accum** will contain the elements of **1** in reverse order.

The function **map** is a commonly used predefined function. It takes two parameters, a function f and a list l, and returns a list resulting from applying f to each element of l. For example,

```
(map (lambda (x) (* x 2)) '(3 4 5 6))
```

returns the list **(6 8 10 12)**. It can be written in SCHEME as

```
(define (map f 1)
  (cond ((null? 1) '())
        (else (cons (f (car 1)) (map f (cdr 1))))
        ))
```

92.5.4 Impure Features in SCHEME: Assignment and I/O

The most heavily used impure SCHEME construct is **set!**. It is SCHEME's variant of **SETQ** in LISP and is used to modify the value of an existing variable. That is, **(set!** x *exp***)** evaluates *exp* and assigns the result to the variable x. Other side-effect operators include **(set-car!** l *exp***)** and **(set-cdr!** l *exp***)**, which assign the value of *exp* to the car and cdr fields of the list l, respectively.

There are a number of I/O routines provided in SCHEME, including those for opening and reading or writing to files. The simplest routines, however, are **(read)** which reads a scheme object (either an atom or a list) from the standard input and returns the object as the result of the call and **(write** *exp***)** which writes the value of *exp* to the standard output. Here **(newline)** starts a new line on the standard output.

92.6 Standard ML: A Strict Polymorphic Functional Language

Standard ML is a popular functional language that uses applicative-order evaluation and has a flexible but static type system. It has a more conventional syntax than SCHEME and provides a pattern-matching facility for programming in an equational style. It also has an exception facility and a sophisticated module system supporting the development of large programs. Several robust implementations of Standard ML implementations exist and are used primarily at universities and research laboratories around the world.

Since the dynamic behavior specified by expressions in ML is based on the lambda calculus, and is thus similar to SCHEME, we will concentrate on ML's syntax and its type system.

92.6.1   Predefined Types in ML

ML provides the usual primitive types, **int**, **real**, **bool**, and **string**. Its aggregate types include lists, tuples, and records. A list is homogeneous, meaning that, unlike SCHEME, all elements of the list must be of the same type. The type for a list of integers is written **int list**, the type for a list of Booleans is written **bool list**, and so on. Literals for lists start and end with square brackets and the elements are separated by commas. Examples of list literals include **[1,2,3]**, **[true,false,true]**, and **[[1,2,3],[4,5,6]]**. The types of these lists are **int list**, **bool list**, and **int list list**, respectively. The literal **[]** denotes the empty list.

A tuple is an ordered collection of elements. Tuples are heterogeneous, their elements can be of different types. A tuple type is written as the element types separated by *****. Thus, **(int * bool * real)** is a tuple type whose first element is an integer, second element is a Boolean, and third element is a real. Tuple literals are written in the same way as list literals, except that parentheses are used instead of square brackets. For example, **(true, 3, [4.2])** denotes a tuple whose type is **bool * int * real list**. The elements of a tuple are accessed either by position or, more commonly, using patterns as described later in this section.

Records are similar to tuples except that, like in most languages, their elements are named. The type written {**a: int, b: real, c: string**} is a record type with field names **a**, **b**, and **c**, whose types are **int**, **real**, and **string**, respectively.

Being a functional language, ML provides higher order functions. These functions have types like any other object. The type of a function that takes a parameter of type a and returns a parameter of type b is written a **->** b. Examples of function types are **int -> bool**, **real -> int -> bool**, and **int * real -> bool list**. The **->** is right associative, so the second example is equivalent to **real -> (int -> bool)**. This is a type describing functions that take a **real** as a parameter and return a function taking an **int** as a parameter and returning a **bool**.

Here, **->**, *****, and **list** are known as *type constructors* because they are not types themselves, but rather construct new types (such as **int list** or **bool -> real**) when combined with existing types (such as **int**, **bool**, and **real**).

92.6.2   Expressions in ML

Arithmetic and logical expressions are written using the familiar infix notation. Examples include **a+b**, **4>b**, and **c andalso (d = 5)**. The conditional expression is written

> if *condition* then *exp* else *exp*

and function application is written simply as the juxtaposition of the function and the argument. For example,

> f x

is the application of the function **f** to **x**. Often, ML programmers will put the argument in parentheses as was done in the factorial example in Section 92.1. Simply placing parentheses around an expression has no effect on the value of the expression. Function application is left associative, thus

> g 4 5

is equivalent to

> (g 4) 5

List construction and selection are similar to that of SCHEME. The **::** operator is identical to SCHEME's **cons**, so that **x :: xs** returns a list whose first element is **x** and whose subsequent elements are those of the list **xs**. For example, the value of the expression

> 3 :: [4,5,6]

is the list **[3,4,5,6]**.

The **@** operator is identical to SCHEME's **append** function. For example, the value of

```
[3,4,5] @ [6,7,8]
```

is the list **[3,4,5,6,7,8]**.

The ML functions **hd** and **tl** are identical to SCHEME's **car** and **cdr**, respectively. For example, the value of **hd [3,4,5,6]** is **3** and the value of **tl [3,4,5,6]** is **[4,5,6]**.

Function expressions, corresponding to **lambda** expressions in SCHEME, are written in the form

```
fn arg => body
```

Examples are

```
fn x => x + 1
fn a => fn b => a + (b * 2)
```

where **=>** is right associative, and so the second example is equivalent to

```
fn a => (fn b => a + (b * 2)).
```

92.6.3 Declarations in ML

Variables and functions are declared using the **let** construct, much like SCHEME's **let**. It has the form

```
let declaration₁
    declaration₂
    ...
    declarationₙ
in
    exp
end
```

where each *declaration$_i$* defines a new variable or function, and *exp* is the body of the **let**.

A variable declaration has the form

```
val x = e
```

in which case the expression *e* is evaluated and the variable *x* is given the resulting value. A function declaration has the form

$$\text{fun } f x_1 \ldots x_n = e$$

where $x_1 \ldots x_n$ are the formal parameters and *e* is the body of the function.

Here is an example of a let expression:

```
let val x = 6
    val g = fn z => z + 2
    fun fac n = if n = 0 then 1 else n * fac (n-1)
in
    fac (g x)
end
```

Notice that the variable **g** is bound to a function of type **int -> int**. The use of the keyword **fun** (as in the succeeding line) provides two conveniences: first, the formal parameters appear to the left of the **=**, and second, it supports the definition of recursive functions. In the declaration of **g** using the keyword **val**, **g** cannot appear on the right-hand side of the definition. The keyword **fun** was necessary in the recursive definition of **fac**.

In ML all functions take a single parameter. Thus, the declaration of the function

```
fun f x y = x + y + 2
```

is just shorthand for

```
fun f x = fn y => x + y + 2
```

This function has type **int -> int -> int** and when it is applied to a single argument, it returns a function of type **int -> int**. A function, such as **f**, that can be applied to fewer parameters than appear in the declaration is called a **curried function**, after the logician HASKELL Curry.

92.6.4   Pattern Matching

One of the nicest features of ML is its pattern-matching facility. In function definitions, the formal parameter name can be replaced by a pattern. In the introduction to this chapter, the factorial function was written as

```
fun fac 0 = 1
  | fac n = n * fac(n-1)
```

in which factorial is defined by two clauses separated by a |. In the first clause, the formal parameter is replaced by the literal **0**. When **fac** is called, if the argument has the value **0**, then the right-hand side of the first clause is evaluated. Otherwise, the formal parameter **n** in the second clause is bound to the value of the argument and the right-hand side of the second clause is evaluated.

Consider a function that computes the sum of the elements of a list.

```
fun sum [] = 0
  | sum l = hd l + sum (tl l)
```

The literal pattern **[]** in the first clause is used to determine if the argument is the empty list. Instead of using **hd** and **tl** to select the components of **l** in the second clause, **l** could be replaced by a pattern that accomplishes the same thing:

```
fun sum [] = 0
  | sum (x::xs) = x + sum xs
```

In this case, the pattern **(x::xs)** matches any nonempty list and binds **x** to the head of the list and **xs** to the tail.

A tuple can also be used as a pattern. It was previously mentioned that

```
fun f x y = x + y + 2
```

is a curried function of type **int -> int -> int**, and that it is legal to apply **f** to just one argument. If the programmer knows that **f** will always be called with both arguments, then it is generally more efficient to define **f** as taking a single argument which is a tuple:

```
fun f (x,y) = x + y + 2
```

In this case, **f** has type **int*int->int** and a call to **f** would look like **f(3,4)**. This example also demonstrates how a pattern is used to access the individual elements of a tuple, in this case as **x** and **y**.

92.6.5 Type Definitions

There are several ways to introduce new type names in ML. The simplest way is to create a type synonym, i.e., to define a new name for an existing type. This is accomplished by a declaration of the form

```
type name = type_exp
```

which introduces the new name *name* for the type described by *type_exp*. Some examples are

```
type foo = int * bool * real
type bar = string
type personnel_record = { name: string, salary: int, ss_num:
                          string }
```

No new type is created. Thus, **foo** and **int*bool*real** describe the same type and can be used interchangeably in the program.

New types are created using the **datatype** construct. In its simplest form, a data type declaration specifies all of the elements of the type, much like an enumerated type in PASCAL or ADA.

```
datatype stoplight = Red | Green | Yellow
```

defines a new type **stoplight** whose values are **Red**, **Green**, and **Yellow**.

In the more general form of a data type declaration, the components on the right-hand side can be *value constructors*. Instead of being values themselves, such as **Red** or **Green**, value constructors take parameters and construct values of the new type. Consider,

```
datatype tree = Empty | Leaf of int | Node of tree * tree
```

Here, **Leaf** is a value constructor taking an integer parameter and **Node** is a value constructor taking a tuple of two trees. **Empty**, like **Red**, **Green**, and **Yellow** previously is simply a value constructor that takes no parameters. The declaration of type **tree** says that a value of that type can be the empty tree, a leaf with an integer label, or an interior node with two subtrees.

The expression **(Leaf 5)** constructs a value of type tree which is a leaf node with the label 5. The expression

```
Node (Node (Leaf 5, Node (Leaf 6, Empty)), Leaf 7)
```

constructs the tree shown in Figure 92.2.

Value constructors can be used in patterns, as in

```
fun drive Red = "stop"
  | drive Green = "go"
  | drive Yellow = "go faster"
```

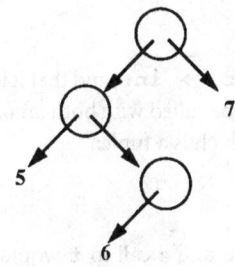

FIGURE 92.2 Tree created by **Node (Node (Leaf 5, Node (Leaf 6, Empty)), Leaf 7)**.

The type of the function **drive** is **stoplight -> string**. Pattern matching can also be used to select out the parameters associated with value constructors. The fringe function, defined by

```
fun fringe Empty = []
  | fringe (Leaf x) = [x]
  | fringe (Node (left,right)) = fringe(left) @ fringe(right)
```

returns a list of the labels associated with the leaves of a tree. If the tree is empty, then the empty list is returned. If the tree consists of just a leaf, then the variable **x** would be bound to the value of the leaf's label and the list containing **x** would be returned. Otherwise, if the tree consists of a node with left and right subtrees, the variables **left** and **right** are bound to those subtrees and their fringes are computed. The two resulting lists are then appended to form the result. The call

```
fringe (Node (Node (Leaf 5, Node (Leaf 6, Empty)), Leaf 7))
```

would return the list **[5,6,7]**.

92.6.6 Type Variables and Parametric Polymorphism

Consider the **length** function, which computes the length of a list:

```
fun length [] = 0
  | length (x::xs) = 1 + length xs
```

What is it's type? Clearly, it can take a list of any type, since the type of the elements of the list has no effect on its length. Thus, we say that the type of **length** is

$$\forall \alpha. \alpha \text{ list} \to \text{int}$$

which means that for all types α, **length** can take an α list and returns an integer. In ML, this type is written

```
'a -> int
```

where the **'** signifies that **a** is a universally quantified type variable rather than some type **a** previously defined.

Because **length** can be applied to many different types of arguments, as in

```
length [1,2,3] + length [[4,5],[6]] + length [true,false,true]
```

we say that length is **polymorphic** (meaning many shaped). In ML, any object whose type contains a type variable, such as **'a**, is polymorphic. All others are said to be *monomorphic*. This kind of polymorphism is called *parametric polymorphism* because in some theoretical type models, type variables occur as extra formal parameters in a function definition.

ML supplies a **map** function much likes SCHEME's **map**. It can be written as

```
fun map f [] = []
  | map f (x::xs) = f x :: map f xs
```

The type of **map** is **('a -> 'b) -> 'a list -> 'b list**, because for any types **'a** and **'b**, **map** takes a function **f** of type **'a -> 'b** and a list of type **'a list** and applies **f** to each element of the list. The result of each application is of type **b**, and since **map** returns the list of the results, the return type of **map** is **'b list**. The result of

```
map (fn n => n+1) [1,2,3] @ map (fn l => length l)
                               [[2.2,3.3],[4.4]]
```

would be **[2,3,4,2,1]**.

In all the ML examples so far, the programmer never specified the types of the functions, variables, or expressions. If desired, one could do so explicitly, as in

```
val a: int list = [1,2,3]
```

and

```
fun f (x:int) (y:real) = (if x = 1 then y + 1.2 else y - 1.7): real
```

In general, however, the ML compiler can infer the types of the functions and variables by the way they are defined and used. This process is called **type inference** and the ML type system, based on work by Hindley and by Milner, ensures that type inference can be safely performed. Furthermore, the type that is inferred for an object is the most general type possible, allowing that object to be used as polymorphically as possible. For example, if type inference had inferred the type of the **length** function to be **int list->int**, then the length function would have been restricted to lists of integers. Instead, type inference infers the more general type **'a list -> int**, allowing **length** to be used on all types of lists.

92.6.7   Type Constructors

Type variables can also be used to parameterize data type declarations. Earlier, we defined a tree type whose leaves were labeled with integers:

```
datatype tree = Empty | Leaf of int | Node of tree * tree
```

Instead, we can write

```
datatype 'a tree = Empty | Leaf of 'a | Node of 'a tree * 'a tree
```

which says that for all types **'a**, an **'a tree** is either empty, a leaf labeled with a value of type **'a**, or a node with two subtrees of type **'a tree**. In this case, **tree** is a type constructor because many different tree types can be constructed by instantiating **'a** with different types. For example, if the programmer writes

```
Node (Leaf 3.2, Empty)
```

the compiler can infer the type of this expression to be **real tree**. Similarly,

```
Node (Leaf[3,4,5], Leaf[4,5,6])
```

describes an **int list tree**. The type variable **'a** can only be instantiated one way within a single type, so that

```
Node (Leaf 4, Leaf true)
```

is illegal.

Interestingly, the type of the expression

```
Empty
```

is **'a tree**, thus **Empty** is polymorphic but is not a function (another example of this is **[]: 'a list**).

Polymorphic functions work well in the presence of type constructors. The **fringe** function seen earlier,

```
fun fringe Empty = []
  |  fringe (Leaf x) = [x]
  |  fringe (Node (left,right)) = fringe(left) @ fringe(right)
```

now has type **'a tree -> 'a list** and can work on any kind of tree.

92.6.8 The ML Module System

In order to support large-scale programs and separate compilation, ML provides a sophisticated module system. As in other languages, a module consists of a *body*, a collection of definitions of types, variables, etc., and an *interface*, specifying which components of the body are visible outside the module. In ML, a module body and a module interface are separate entities. Thus, many different module bodies can share the same interface, and different modules might share a body but have different interfaces.

A module interface, called a *signature* in ML, is described by a signature expression of the form

```
sig decl₁
    decl₂
    . . .
    declₙ
end
```

where each *decl*ᵢ is usually a declaration of the name and type of an object or the name of a type. To give a name to a signature, a declaration of the form

```
signature name = sig_exp
```

is used, where *sig_exp* is a signature expression. For example, the interface for a module implementing a (functional!) stack might be

```
signature STACK = sig
                type 'a stack
                val empty: 'a stack
                val push: ('a * 'a stack) -> 'a stack
                val pop: 'a stack -> ('a * 'a stack)
                val isempty: 'a stack -> bool
                exception stack_underflow
            end
```

A module body, known as a *structure* in ML, is described by a structure expression of the form

```
struct def₁
      def₂
      . . .
      defₙ
end
```

where each *def*ᵢ is a definition. To give a structure a name, a declaration of the form

```
structure name = struct_exp
```

is used, where *struct_exp* is a structure expression or the name of a previously defined structure.

For example, a structure implementing a stack might look like

```
structure StackImp = struct
                exception stack_underflow
                type 'a stack = 'a list
                val empty = []
                fun push(x,s) = x::s
                fun pop [] = raise stack_underflow
                  | pop (x::rest) = (x,rest)
                fun isempty [] = true
                  | isempty l = false
            end
```

At this point, there is no connection between the signature **STACK** and the structure **StackImp**. Thus, all the components of **StackImp** are visible. To create a stack implementation with the signature **STACK**, we can write:

```
structure Stack:STACK = StackImp
```

The signature **STACK** can be reused in a different implementation of a stack, as in

```
structure NewStack : STACK =
  struct
    exception stack_underflow
    datatype 'a stack = empty | non_empty of int * 'a list
    fun push (x,empty) = non_empty (1,[x])
      | push (x,non_empty(n, s)) = non_empty (n+1, x::s)
    fun pop empty = raise stack_underflow
      | pop (non_empty(n,x::rest)) = (x, if n=1 then empty else
                                            non_empty(n-1,rest))
    fun isempty empty = true
      | isempty l = false
  end
```

Once defined, a component *x* of a module *m* is referenced by the expression *m*. *x*. For convenience, the components of a module may be referenced without the module name, if the module is first opened via the command

```
open name
```

where *name* is the name of the module.

Modules are not values in ML. That is, they cannot be passed to functions, stored in lists, etc. ML does provide something similar to a function from modules to modules. It is called a *functor* and supports code reuse by allowing the definition of a module in terms of other modules. All modules, even those resulting from functor applications, are instantiated at compile time and therefore cannot depend on values computed during execution.

A functor definition has the form

$$\text{functor} \quad f(s_1 : sig_1, \ldots, s_n : sig_n) : sig\_exp = struct\_exp$$

where $s_1 \ldots s_n$ are the formal parameter names that will be bound to structures in a functor application and $sig_1 \ldots sig_n$ are the signatures that $s_1 \ldots s_n$ must conform to, respectively. Here, *sig_exp* is the signature of the result of the functor and *struct_exp* is the definition of the resulting structure. For example, suppose one wanted to create an implementation of a queue based on an existing stack implementation, such that the signature of a queue is

```
signature QUEUE =
  sig
    exception queue_underflow
    type 'a queue
    val empty: 'a queue
    val enqueue: ('a * 'a queue) -> 'a queue
    val dequeue: 'a queue -> 'a * 'a queue
    val isempty: 'a queue -> bool
  end
```

Next is a functor definition that takes any structure that conforms to the previous **STACK** signature and creates an implementation of a queue, using the data structures and routines supplied by the stack argument.

```
functor MakeQueue(Stack: STACK): QUEUE =
  struct
    exception queue_underflow
    type 'a queue = 'a Stack.stack * 'a Stack.stack
    val empty = (Stack.empty, Stack.empty)

    fun reverse_stack(from, to) =
      if Stack.isempty from then to
        else let val (x, new_from) = Stack.pop from
          in reverse_stack(new_from, Stack.push(x,to))
          end

    fun enqueue(x,(s1,s2)) = (s1, Stack.push(x,s2))
    fun dequeue (s1,s2) =
      if Stack.isempty s1 then
        if Stack.isempty s2 then raise queue_underflow
        else dequeue (reverse_stack (s2, Stack.empty),
                  Stack.empty)
      else
        let val (x,new_s1) = Stack.pop s1
        in (x, (new_s1,s2))
        end
    fun isempty(s1,s2) = Stack.isempty s1 andalso
                    Stack.isempty s2
  end
```

To create an actual queue module the functor must be invoked, as in

```
structure Queue1 = MakeQueue(Stack)
```

Another implementation of a queue, based on a different stack implementation, is created by

```
structure Queue2 = MakeQueue(NewStack)
```

Functors are commonly used in ML to support separate compilation. They allow a module to be written and compiled despite referring to components of modules that are not yet implemented. Note that, for example, the code for the previous functor **Queue** could have been written, type checked, and compiled before any structure with the signature **STACK** was implemented. Only the signature **STACK** had to exist before compiling **Queue**.

92.6.9 Impurities in ML: References and I/O

As previously mentioned, ML supports conventional I/O, which violates referential transparency. For example,

```
val x = (print "hello"; print "world"; true)
```

where **hello world** would be printed and the value of **x** would be **true**. The semicolon is used to separate statements that are executed sequentially.

The other impure feature of ML is *references*. In conventional languages, these would be considered constant pointers to assignable locations. The expression

```
ref exp
```

allocates a new cell *c* in memory and places the value of *exp* in *c*. The address of *c* is returned as the value of the entire expression. The type of the expression is *t* **ref**, where *t* is the type of *exp*. Given, for example, the declaration

```
val x = ref 6
```

the value of **x** is a new cell containing **6**, and the type of **x** is **int ref**.

The value stored in this location may be changed, using the expression

$$exp_1 := exp_2$$

where the value of exp_1 must be a reference of type **ref** *t* and *t* is the type of exp_2. Evaluating this expression causes the value of exp_2 to be stored in the location denoted by exp_1. For example,

```
x := 7
```

changes the value referenced by **x** to **7**. The dereference operator is **!**. In the expression **!***exp*, *exp* must evaluate to a reference and the value of the entire expression is the value contained in the referenced cell. Thus, after the assignment to **x**, the value of **!x** is **7**. Since the value of an expression of type *t* **ref** is essentially a pointer, references can be used for aliasing. For example, given the code

```
val x = ref 10
val y = x
```

the variable **y**, of type **int ref**, would point to the same location (containing **10**) that **x** does. Thus, the result of the expression

```
(x := !x+1; !y)
```

would be **11**.

92.7   Nonstrict Functional Languages

Before describing HASKELL in detail, it is worthwhile examining the costs and benefits of a nonstrict language, i.e., a language based on normal-order evaluation. Normal-order evaluation specifies that an argument in a function call is evaluated only when the value of the corresponding formal parameter is needed. In most implementations of nonstrict functional languages, any subsequent reference to the formal parameter uses the already computed value of the argument rather than re-evaluating it. This more efficient mechanism for supporting normal-order evaluation is called **lazy evaluation** (it is also sometimes referred to as *call-by-need*). Nonstrict functional languages are often informally referred to as **lazy functional languages**, even though laziness is a property of the implementation rather than the language.

In a nonstrict language, even using lazy evaluation, there is a significant overhead cost to delaying the evaluation of an actual parameter until the corresponding formal parameter is needed. This cost arises due to the fact that an object representing the delayed argument must be constructed when the function is called. This object might be a closure (i.e., a parameterless procedure, generally known as a *thunk*) that will be invoked when the value of the argument is needed, or it might be a graph representation of the delayed expression (this is found in systems that use a technique called graph reduction). In each case, the overhead cost can be substantial.

The benefit of a nonstrict language is that its programs are more likely to terminate: for example, if the evaluation of an argument might never terminate but the argument is not needed by the function. However, since the vast majority of popular programming languages are strict, it might appear that this particular termination issue is unimportant. However, when used properly, nonstrictness frees the programmer from worrying about some control issues, such as interleaving the execution of producer and consumer procedures. The other benefit of using a nonstrict language is that it allows the programmer to create infinite data structures. To illustrate this, consider the following definition:

```
fun numsfrom n = n :: numsfrom (n+1)
```

This function, given an integer argument n, creates the list $[n, n + 1, n + 2, \ldots]$. In ML, the call

```
numsfrom 1
```

would not terminate until memory was exhausted because all of the (infinite number of) elements of the list would have to be created before the call returned.

In a nonstrict language, the cons function **: :** does not evaluate its arguments. Thus, the expression

```
n :: numsfrom (n+1)
```

would create a list whose head is **n** and whose tail is specified by **numsfrom (n+1)** but is left unevaluated. Only when the value of the tail of this list is demanded using the **tl** function is the call **numsfrom (n+1)** actually evaluated. The result of that call, then, is a list whose head is **(n+1)** and whose tail is described by the unevaluated expression **numsfrom (n+2)**. In a nonstrict language, the call

```
numsfrom 1
```

would return almost immediately with a delayed list representing **[1,2,3,...]**. These infinite, but delayed, lists are generally known as **streams**.

The function

```
fun sumstream s 0 = 0
  | sumstream s n = hd s + sumstream (tl s) (n - 1)
```

takes a stream s and an integer n and computes the sum of the first n elements of s. The result of

```
sumstream (numsfrom 1) 10
```

would compute the sum of the first 10 elements of **(numsfrom 1)**, namely, 55.

From a programmers point of view, the use of infinite data structures provides a nice separation between the production of data (by **numsfrom**, in this case) and the consumption of the data (by **sumstream**). The producer does not need to know how much data the consumer will need, nor does it have to worry about buffering data that is already produced but not consumed. The data is produced only when demanded by the consumer.

A more substantial example is the program that computes the infinite list of primes using the Sieve of Erostosthenes.

```
let fun numsfrom n = n :: numsfrom (n+1)
    fun filter f (x::xs) = if f x then x :: filter f xs
                           else filter f xs
    fun remove_multiples (x :: xs) =
    let fun is_multiple n = (n mod x) <> 0
    in x :: remove_multiples (filter is_mult xs)
    end
in
    remove-multiples (numsfrom 2)
end
```

92.8 HASKELL: A Nonstrict Functional Language

Aside from being a nonstrict functional language, HASKELL features a sophisticated type system that extends the ML-style (i.e., Hindley–Milner) type system to incorporate dynamic overloading.

HASKELL's syntax differs somewhat from that of ML, although the programs have a similar look. A few of the more important syntactic differences are as follows:

- Identifiers representing types and value constructors are capitalized. Identifiers representing type variables and values are not capitalized.
- Function and variable definitions do not begin with a keyword (whereas ML uses **fun** and **val**).
- Type constructors precede their arguments, as in **List Int** (in contrast to **int list** in ML).
- HASKELL uses **:** and **::** in precisely the opposite way from ML. The **:** is the *cons* operator and the **::** is used to associate a type with an expression, as in

  ```
  (4:[5,6]) :: List Int
  ```

- Indentation can be used to begin and end new blocks. For example, in

  ```
  let f x = let z = x + 3
                in x * y
            y = a * b
  in
      f y
  ```

the indentation specifies that the names **f** and **y** are defined at the same level.

92.8.1 The HASKELL Class System

HASKELL's most interesting feature, other than its nonstrict semantics, is its *class* system for supporting *dynamic overloading* in a systematic way. The term overloading refers to the ability to give the same name to two distinct entities in a program. Overload resolution is the process by which the use of that name is disambiguated. For example, in ADA the programmer can define two or more different functions with the same name. When encountering that name in a function call, the compiler performs overload resolution to determine which function is being called by examining the number and types of the actual parameters. If the compiler is unable to resolve the overloading, the program is rejected. Because the overload resolution occurs at compile time, this kind of overloading is called *static overloading*.

In almost all languages, there is some form of overloading. For example, most languages use **+** as the name of the addition operators for both integers and reals, even though they are different operators. This is the case in ML, for example. However, the mixing of static overloading and type inference causes a problem in ML. Consider the following definition in ML:

```
fun f x y = x + y
```

The ML compiler rejects this definition because it cannot determine which addition operator, integer or real, is specified by **+**. The parameters **x** and **y** are either integers or reals, but it cannot be determined which. The ML programmer has to provide explicit type information, such as:

```
fun f (x:int) y = x + y
```

In the first (erroneous) definition, the type of **f** is clearly not

$$\forall \alpha. \alpha \rightarrow \alpha \rightarrow \alpha$$

since **+** cannot be applied to every type α. We would like to be able to say that **f** is of type

$$\forall \alpha \text{ for which } + \text{ is defined.} \ \alpha \rightarrow \alpha \rightarrow \alpha$$

and to allow the programmer to define the meaning of **+** for any type (complex numbers, sets, etc.) desired. Then, when **f** is called, the choice of which **+** to use in the body of **f** depends on the types of the arguments to **f**. Since **f** is polymorphic, albeit in a restricted way, it can be applied to many different types of arguments and thus the choice of **+** has to be made at run time. This kind of overloading is called dynamic overloading and is seen, in a different framework, in object-oriented languages.

HASKELL uses type classes to support dynamic overloading. A type class is a way to specify what operations must be supported by a particular collection of types. For example, the equality class, **Eq**, defined in HASKELL by

```
class Eq a where
   (==) :: a->a->Bool
```

specifies that every type **a** in class **Eq** must provide a definition for the infix equality operator **==** of type **a -> a -> Bool**. One can then write a polymorphic function that uses **==**, for example

```
f :: (Eq a) => a->a->int
f x y = if x == y then 1 else 2
```

The first line gives the type of **f**, which is **a -> a -> int** for any type **a** in class **Eq**.

The notation **(Eq a)** is called a *context* and indicates that **a** is in class **Eq**. Like ML, HASKELL is designed to support type inference. In fact, the first line declaring the type of **f** can be omitted. The HASKELL compiler will infer that the type of the parameters must be in class **Eq**.

92.8.2   User-Defined Types

New types in HASKELL are defined using the **data** construct which is analogous to the **datatype** construct in ML. For example,

```
data IntTree = Empty | Leaf Int | Node IntTree IntTree
```

defines the same integer-labeled tree type seen earlier. HASKELL also provides type constructors, so

```
data Tree a = Empty | Leaf a | Node (Tree a) (Tree a)
```

defines the **Tree** type constructor parameterized by the label type **a**.

92.8.3   Instance Declarations

Once a new type has been created it can be declared to be an *instance* of a type class, in which case the definition of the required operators must be provided. For example, declaring type **IntTree** to be in class **Eq** might look like

```
instance Eq IntTree where
   Empty == Empty               = True
   Leaf x == Leaf y             = x == y
   (Node l1 r1) == (Node l2 r2) = l1 == l2 && r1 = r2
   t1 == t2                     = false
```

The code following the **where** keyword is simply the definition of the **==** operator using pattern matching on the value constructors of **IntTree**. Since **==** is infix, the definition is in infix form. Once **IntTree** has been declared to be in class **Eq**, an **IntTree** value can be passed to any function expecting a type in class **Eq**.

We would also like to declare that any type constructed from the **Tree** type constructor is in class **Eq**. The declaration

```
instance Eq (Tree a) where
    Empty == Empty              = True
    Leaf x == Leaf y            = x == y
    (Node l1 r1) == (Node l2 r2)= l1 == l2 && r1 = r2
    t1 == t2                    = false
```

is incorrect because the definition of **==** requires, in the second clause, that the labels **x** and **y** be compared using **==**. Thus, not only must **==** be defined on **(Tree a)** for any type **a**, **==** must also be defined on **a**. Thus, **a** must already be an instance of class **Eq**. The correct instance declaration requires a context as follows:

```
instance (Eq a) => Eq (Tree a) where
    Empty == Empty              = True
    Leaf x == Leaf y            = x == y
    (Node l1 r1) == (Node l2 r2)= l1 == l2 && r1 = r2
    t1 == t2                    = false
```

This should be read as "For all types **a**, if **a** is in class **Eq** then **(Tree a)** is in class **Eq** with **==** defined as follows...."

HASKELL also provides a form of inheritance, in which one class can be used to define another class. For example, the class definition

```
class (Eq a) => Ord a where
    (<), (<=), (>=), (>)    :: a->a->Bool
    max, min                :: a->a->a
```

defines the class **Ord** of ordered types in terms of the class **Eq**. In this case, a type **a** can be in class **Ord** if it is in class **Eq** and supports the additional operators previously mentioned. We say that **Eq** is the *superclass* of **Ord** and **Ord** is the *subclass* of **Eq**.

92.8.4 List Comprehensions in HASKELL

Another nice feature of HASKELL is its *list comprehensions*. These are concise expressions for constructing entire lists, resembling set notation in mathematics. For example, the expression

```
[f x | x <- xs]
```

computes the list of all values returned by **(f x)**, for each **x** taken from the list **xs**. Thus

```
[ x * 2 | x <- numsfrom 1]
```

returns the infinite list **[2, 4, 6,...]**. Naturally, the elements of the list are not computed until needed, just as in the preceding case of the infinite list resulting from **numsfrom 1**.

List comprehensions can also include guards. For example,

```
psort []     = []
psort (x:xs) = psort [y | y <- xs, y < x] ++ [x] ++
               psort [y | y <- xs, y >= x]
```

defines the partition sort (often mislabeled quicksort), where **++** is HASKELL's append operator. The first list comprehension in **psort**

```
[y | y <- xs, y < x]
```

contains the guard **y < x**, so that the only elements **y** taken from **xs** are those that are less than **x**.

92.8.5 Functional I/O in HASKELL

In order for HASKELL's I/O facility to maintain referential transparency, the input to a program is considered a stream, a possibly infinite list. Like other infinite lists, the entire input list is not immediately available at the start of execution, but the elements are supplied over the course of the computation: for example, as the user enters data from the keyboard. Similarly, the output of a program is also a stream. A program, then, can be viewed as a mapping from input streams to output streams. In actuality, HASKELL's I/O system is substantially more complicated in order to support error handling, files, and channels. It makes heavy use of *continuations*, and the reader is referred to a nice introduction to the language in Hudak and Fasel [1992].

92.9 Research Issues in Functional Programming

The functional language research community is very active in a number of areas. Of particular interest is improving the speed of functional language implementations. There are two primary approaches to this: through compiler-based program analysis and optimization techniques and through the parallelization of functional programs. Another area of research is to increase the expressiveness of functional languages, particularly in applications in which side effects are seen as necessary in conventional programs. In this section, we provide a brief description of these research areas and refer the reader to the literature in order to gain a deeper understanding of the issues.

92.9.1 Program Analysis and Optimization

Because of the solid mathematical foundation and well-defined semantics of functional languages, functional programs are particularly good candidates for compile-time analysis and optimization. An analysis technique that has been used in a large number of optimizations is *abstract interpretation* (see Abramsky and Hankin [1987]). It is a technique in which the program is executed at compile time, but rather than operating on the usual kinds of values, it operates on, and returns, only particular pieces of information desired by the compiler. If the desired information is sufficiently restricted, or an approximate answer is acceptable, then the program execution can be shown to terminate and is thus useful at compile time. The example generally used to illustrate abstract interpretation is overly trivial but useful nonetheless: the rule of signs in multiplication. This rule determines the sign of the result of a multiplication, using only the signs of the operands:

```
0 × x = 0 for any x
x × 0 = 0 for any x

+ × + = +
− × + = −
+ × − = −
```

If only the sign of the result of a multiplication is desired, then the rule of signs provides a form of execution that is much less expensive then performing the actual multiplication and then taking the sign of the result.

An early, but still important, application of abstract interpretation arises in *strictness analysis*. Strictness analysis is an analysis used for a nonstrict language in order to transform, where possible, normal-order evaluation into applicative-order evaluation. Essentially, we want to know if a function f will always require the value of its argument, in which case we say that f is *strict*. If we know a function is strict, then we can go ahead and evaluate its argument before calling the function, without changing the result computed by the program.

More formally, we say that f is strict if, even using normal-order evaluation,

$$f(\perp) = \perp$$

where $\perp$ represents a nonterminating computation. Intuitively, this says that if f is applied to an argument whose evaluation would not terminate and yet f does terminate, then it is clear that f does not require the value of that argument and is thus not strict. Abstract interpretation is used to find the strictness property of a function by executing an abstract version f' of f function over the domain of values $\{0, 1\}$ where 0 represents nontermination and 1 represents possible termination. If

$$f'(0) = 0$$

reflects the behavior of f in the previous equation, then we know that f is strict.

92.9.2 Parallel Functional Programming

The attractiveness of functional languages for writing programs for parallel machines arises from the first Church–Rosser theorem. The theorem states that given a function call, the order in which the arguments and the body of the function are evaluated will not effect the final answer, assuming the program terminates. Thus, given the expression

```
(f x) + (g y)
```

the expressions **(f x)** and **(g y)** can be evaluated *in parallel*. In this case, it is clear that both operands to + are needed, so that there will be no wasted effort.

There are two major approaches to parallel programming using functional languages (see Kelly [1989] for extended reading in this area). The first involves programming in a standard functional language, such as ML or HASKELL, and using a compiler and run time system that will partition the program into parallel threads and execute them. Because of the Church–Rosser property, it is not difficult for the compiler to determine which expressions can be executed in parallel.

The difficult part is determining the appropriate granularity of the parallel version of the program. Granularity is the measure of the size of the tasks into which the program is decomposed; the finer the granularity, the smaller and more numerous the tasks, and the greater the degree of parallelism. However, there is a cost associated with creating each task, whether due to communication over a network, increased contention for a shared memory, or context switching in the operating system.

The second approach to parallel computing using functional languages involves adding constructs to a functional language for expressing parallelism. These constructs might specify which expressions should be evaluated in parallel (and, thus, which are not worth spawning as their own tasks), which processor an expression should be evaluated on, and, in the case of languages which contain impure features, the creation and use of channels for communication.

92.9.3 Partial Evaluation

Partial evaluation [Bjorner et al. 1988] is the technique where, if part of the input to a program is known at compile time, the program is evaluated as much as possible using the input available. The result is a new version of the program, called the specialized program, that is ready to accept the rest of the input and return the same result as the original program would have on the entire input. The process of specialization creates a more efficient program than the original because the known data have already been integrated into the program, reducing the amount of interpretation that the program has to perform on its input.

92.9.4 State in Functional Programming

An active area of research in functional programming is the development of new techniques for reasoning about state and changes in state in a manner that preserves referential transparency. For example, these techniques, most based on the concept of a monad [Wadler 1992], allow programmers to manipulate arrays in a pure functional language in a style that looks familiar to imperative language programmers [Peyton Jones and Wadler 1993]. Furthermore, this work boosts the efficiency of functional arrays by allowing the arrays to be updated in place, rather than copied each time an update operation is performed (which is normally necessary to preserve referential transparency). This is accomplished by using the type system to encapsulate the array such that it cannot be shared. Thus, since a previous version of the array can never be referenced (because the array is not shared), the new version of the array can be created simply by modifying the previous version.

Defining Terms

Applicative-order evaluation: An execution order in which the arguments in a function call are evaluated before the body of the function.

First-class object: An object that can be stored in data structures, passed as arguments, and returned as the result of function calls. In functional languages, functions are first-class objects.

Higher order function: A function that takes another function as a parameter or returns a function as its result.

Lambda calculus: A simple syntactic model of computation equal in power to the Turing Machine.

Latent type system: A type system where types are associated with values, not variables. This usually requires run time type checking which is why latently typed languages such as SCHEME are often referred to as *dynamically typed*.

Lazy evaluation: An evaluation technique for nonstrict functional languages.

Lazy functional language: Informal but common name for a nonstrict functional language.

Nonstrict functional languages: A function language adopting normal-order evaluation.

Normal-order evaluation: An execution order in which the arguments in a functional call are only evaluated if and when needed in the body of the function.

Polymorphism: A property of a languages type system in that an objects type may include type variables which can range over an infinite number of types. Most such polymorphic objects are functions, which can be applied to arguments of many different types.

Pure functional languages: Functional languages that provide absolutely no mechanism for performing side effects, and thus exhibit referential transparency.

Referential transparency: The property of a language that states that equal expressions can be interchanged with each other.

Side effect: A change in the value of a variable as the result of evaluating an expression, e.g., if the expression contains an assignment operation.

Strict functional language: A function language adopting applicative-order evaluation.

Type inference: A process in which the compiler determines the types of objects in a program without the programmer having to declare them explicitly.

References

Abramsky, S. and Hankin, C., Eds. 1987. *Abstract Interpretation of Declarative Languages*. Ellis Horwood.

Backus, J. 1978. Can programming be liberated from the von Neumann style? A functional style and its algebra of programs. *Commun. ACM* 21(8):613–641.

Barendregt, H. P. 1984. *The Lambda Calculus: Its Syntax and Semantics*. North-Holland.

Bird, R. and Wadler, P. 1988. *Introduction to Functional Programming*. Prentice–Hall, Englewood Cliffs, NJ.

Bjorner, D., Ershov, A., and Jones, N. 1988. *Partial Evaluation and Mixed Computation*. North-Holland.

Church, A. 1941. *The Calculi of Lambda-Conversion*. Princeton University Press, Princeton, NJ.

Hudak, P. 1989. The conception, evolution, and application of functional programming languages. *ACM Comput. Surv.* 21(3):359–411.

Hudak, P. et al. 1992. Report on the programming language Haskell. *SIGPLAN Notices* 27(5):Section R.

Hudak, P. and Fasel, J. 1992. A gentle introduction to Haskell. *ACM SIGPLAN Notices* 27(5):1–53.

Kelly, P. 1989. *Functional Programming for Loosely-Coupled Multiprocessors*. Pitman.

Kranz, D., Kelsey, R., Rees, J., Hudak, P., Philbin, J., and Adams, N. 1986. Orbit: an optimizing compiler for Scheme, pp. 219–233. In *SIGPLAN'86 Symp. Compiler Construction*. June. *ACM SIGPLAN Notices* 21(7).

Landin, P. 1964. The mechanical evaluation of languages. *Comput. J.* 6(4):308–320.

Landin, P. 1966. The next 700 programming languages. *Commun. ACM* 9(3):157–166.

McCarthy, J. 1978. The history of LISP. In *Proc. ACM SIGPLAN Symp. History Programming Lang.*

McCarthy, J. et al. 1962. *LISP 1.5 Programmers Manual.* MIT Press.

Milner, R. and Tofte, M. 1991. *Commentary on Standard ML.* MIT Press.

Milner, R., Tofte, M., and Harper, R. 1990. *The Definition of Standard ML.* MIT Press.

Paulson, L. 1991. *ML for the Working Programmer.* Cambridge University Press.

Peyton Jones, S. 1987. *The Implementation of Functional Programming Languages.* Prentice–Hall, Englewood Cliffs, NJ.

Peyton Jones, S. and Wadler, P. 1993. Imperative functional programming. In *Proc. 20th ACM Symp. Principles Programming Lang.*

Rees, J., Clinger, W. et al. 1992. Revised Report on the Algorithmic Language Scheme. *Artificial Intelligence Lab. Tech. Rep.* Massachusetts Institute of Technology, Cambridge, Nov.

Sussman, G. and Abelson, H. 1985. *Structure and Interpretation of Computer Programs.* MIT Press.

Sussman, G. and Steele, G. Jr., 1975. Scheme: An Interpreter for Extended Lambda Calculus. *Artificial Intelligence Lab. Tech. Rep. Memo* 349. Massachusetts Institute of Technology, Cambridge.

Ullman, J. 1994. *Elements of ML Programming.* Prentice–Hall, Englewood Cliffs, NJ.

Wadler, P. 1992. The essence of functional programming, pp. 1–14. In *Proc. 19th ACM Symp. Principles Programming Lang.*

Further Information

There are several textbooks that provide an overview of the development and use of functional programming languages. Among these are Bird and Wadler [1988], Paulson [1991], Sussman and Abelson [1985], and Ullman [1994]. The reader is also referred to [Hudak 1989], an excellent survey paper. Peyton Jones [1987] provides a description of how functional programming languages are implemented.

There are a number of professional journals that include papers on functional languages. The more eminent of these are *The Journal of Functional Programming*, *ACM Transactions on Programming Languages and Systems*, and *The Journal of LISP and Symbolic Computation*. Furthermore, recent results in functional programming research can be found in the proceedings of several important annual symposia, including The ACM Symposium on Principles of Programming Languages, The International Conference on Functional Programming, and The ACM Symposium on Programming Language Design and Implementation.

93

Logic Programming and Constraint Logic Programming

93.1   Introduction **93**-1
93.2   An Introductory Example **93**-2
93.3   Features of Logic Programming Languages **93**-4
93.4   Historical Remarks **93**-4
93.5   Resolution and Unification **93**-6
       Resolution • Unification • Combining Resolution and Unification
93.6   Procedural Interpretation: Examples **93**-12
93.7   Impure Features **93**-15
93.8   Constraint Logic Programming **93**-16
93.9   Recent Developments in CLP (2002) **93**-17
       CSP and CLP • Special Constraints • Control Issues • Optimization • Soft Constraints • SATisfiability Problems • Future Developments
93.10 Applications **93**-21
93.11 Theoretical Foundations **93**-21
93.12 Metalevel Interpretation **93**-23
93.13 Implementation **93**-25
       Warren Abstract Machine (WAM) • Parallelism • Design and Implementation Issues in Constraint Logic Programming • Optimization Using Abstract Interpretation
93.14 Research Issues **93**-27
       Resolution Beyond Horn Clauses • Concurrent Logic Programming and Constraint Logic Programming • Interval Constraints • Constraint Logic Programming Language Design
93.15 Conclusion **93**-29

Jacques Cohen
Brandeis University

93.1   Introduction

Logic programming (LP) is a language paradigm based on logic. Its constructs are Boolean *implications* (e.g., *q implies p* meaning that *p* is true if *q* is true), compositions using the Boolean operators *and* (called conjunctions) and *or* (called disjunctions). LP can also be viewed as a procedural language in which the procedures are actually Boolean functions, the result of a program always being either true or false.

1-58488-360-X/$0.00+$1.50
© 2004 by CRC Press, LLC

In the case of implications, a major restriction applies: when q implies p, written $p :- q$, then q can consist of conjunctions but p has to be a singleton, representing the (sole) Boolean function being defined. The Boolean operator *not* is disallowed but there is a similar construct that may be used in certain cases.

The Boolean functions in LP may contain parameters, and the parameter matching mechanism is called **unification**. This type of general pattern matching implies, for example, that a variable representing a formal parameter may be bound to another variable, or even to a complex data structure, representing an actual parameter (and vice versa). When an LP program yields a *yes* answer, the bindings of the variables are displayed, indicating that those bindings make the program logically correct and provide a solution to the problem expressed as a logic program.

An important recent extension of LP is **constraint LP** (CLP). In this extension, unification can be replaced or complemented by other forms of constraints, depending on the domains of the variables involved. For example, in CLP a relationship such as $X > Y$ can be expressed even in the case where X and Y are unbound real variables. As in LP, a CLP program yields answers expressing that the resulting constraints (e.g., $Z < Y + 4$) must be satisfied for the program to be logically correct.

This chapter includes sections describing the main aspects of LP and CLP. It includes examples, historical remarks, theoretical foundations, implementation techniques, metalevel interpretation, and concludes with the most recent proposed extensions to this language paradigm.

93.2 An Introductory Example

Let us consider a simple program written in PROLOG, the main representative among logic programming languages. The program's objective is to check if a date given by the three parameters: *Month* (a string), *Day*, and *Year* (integers) is valid. For example, *date* (*oct*, 15, 1996) is valid whereas *date* (*june*, 31, 1921) is not. The program can be expressed as:

> *date* (*Month, Day, Year*) :- *member* (*Month*, [*jan, march, may, july, aug, oct, dec*]),
>   *comprised* (*Day*, 1,31).
> *date* (*Month, Day, Year*) :- *member* (*Month*, [*april, june, oct, sept, nov*]),
>   *comprised* (*Day*, 1,30).
> *date* (*feb, Day, Year*) :- *leap* (*Year*), *comprised* (*Day*, 1,29).
> *date* (*feb, Day, Year*) :- *comprised* (*Day*, 1,28).
> *comprised* (*Day, Start, End*) :- *Day* >=*Start*,   *Day* =<*End*.
> *leap* (*Year*) :- (*Year*/4) × 4 = *Year*.

The variables in the preceding program start with a capital letter, e.g., *Month*; in this simple example the constants either start with small case letters (strings) or are integers. The symbol :- can be read as is defined by, and it separates the procedure heading from its body, which consists of calls to other procedures. More precisely, it is convenient to view every procedure as a Boolean function yielding *true* or *false*. The body of such a function is made up of conjunctions of calls to functions. Therefore, a comma separating the calls on the right-hand side correspond to the Boolean operator *and*.

The following remarks also apply to the program. There are multiple definitions of *date*, each of which tests for the corresponding month and the compatible number of days. Therefore, the program is non-deterministic in the sense that several possibilities have to be considered either one at a time, or in parallel. In many cases there might be several solutions to a given program. Also note that the right-hand side of a rule could be empty. In that case, the rule indicates that the left-hand side is always true.

Let us for the time being assume that *member* (*Element, List*) is a built-in function testing if an *Element* is a member of a *List*. The Boolean function *leap* tests if its parameter is a multiple of 4. One could also add a function *year* (*Year*), which would test if an integer defined by *Year* is positive. Notice that the ordering of the statements in this particular version of the program is important, since a leap year is tested before a nonleap year. Ideally, pure PROLOG programs should be declarative; from a syntactic point of view, that means that the order of the rules and the order of the calls on the right-hand sides should not matter. A new version of the program satisfying this criteria is one in which the Boolean function *nonleap (Year)* :- *(Year/4)* × 4 = *Year* is called in the definition applicable to the month of February, in the case of a nonleap year.

The preceding program is used with queries (equivalent to main programs) to test the validity of given dates. The answers are either *yes* or *no*. Note that in PROLOG one is allowed to have queries containing variables. In the present example, the query *date (X, 31, 1996)* yields as results the successive strings that the variable *X* is bound to for the months that have 31 days. More specifically, the results are:

yes $X = jan$, More?
yes $X = march$, More?
. . .
Yes $X = dec$, More?
No.

Similarly, the query *date (X, 29, 1996)* yields all the months in a year. Note that the built-in functions such as =< and >= require that their arguments be ground; that is, they cannot contain unbound variables. This is satisfied by the query *date (X, 31, 1996)*. However, a call *date (feb, Y, 1996)* entails a problem that provides insight into the desirability of extending PROLOG to handle constraints such as it is done in PROLOG IV. The last query fails in standard PROLOG because the variable *Y* remains unbound. The right answer as provided by a constraint LP processor consists of the two constraints:

Yes $Y >= 1, Y =< 29$, More?
No.

Similarly, the correct answer to the query *date (feb, Y, Z)* submitted to a CLP processor should yield two pairs of constraints corresponding to leap and nonleap years represented by *Z*. These results are obtained by keeping lists of satisfiable constraints and triggering a failure when the constraints become unsatisfiable. This failure may entail exploring the remaining nondeterministic possibilities.

It is now appropriate to present the recursive Boolean function *member*. It has two parameters: an *Element* and a *List*; the function succeeds if the *Element* appears in the *List*. Lists are represented using the constructor *cons* as in functional languages. This means that a list [a, b] can be represented by:

$$cons\,(a, cons\,(b, nil))$$

Note that the *cons* in the preceding representation is simply any user-selected identifier describing a data structure (or record) consisting of two fields: (1) a string and (2) a pointer to another such record or to *nil*. In the case of lists there are special builtin features such as [a, b] simplifying their description. The program for *member* consists of two rules:

member (X, cons (X, T)).
member (X, cons (Y, T)) :- X /= Y, *member (X, T)*.

The first rule states that if the head of a list contains the element *X*, then *member* succeeds. The second rule uses recursion to inspect the remaining elements of the list. Note that the query *member (X, cons (a, cons(b, nil))*, equivalent to *member (X, [a, b])*, provides two solutions: namely, $X = a$ or $X = b$. It should be obvious that *member (Y, [a, a])* also yields two (identical) solutions.

Finally, notice that *member* can search for more complex data structures. For example, *member (member (a, Y), Z)* succeeds if the list *Z* contains an element such as *member (a, U)*; if so, the variable *Y* is bound to the variable *U*. The reader should consider the embedded term *member* as the identifier of a record or data structure having two fields. In contrast, the first identifier *member* represents a Boolean function having two parameters. This example illustrates that, in PROLOG, program and data have the same form.

An inquisitive reader will remark that *member* can also be used to place elements in a list if they are not already present in that list. A clue in understanding this property is that the query *member (a, Z)* will bind

Z to a record *cons*(*a*, *W*) in which *W* is a new unbound variable created by the PROLOG processor when applying the first rule defining the Boolean function *member*.

93.3 Features of Logic Programming Languages

Summarized next are some of the features whose combination renders PROLOG unique among languages:

1. Procedures may contain parameters that are both input and output.
2. Procedures may return results containing unbound variables.
3. **Backtracking** is built in, therefore allowing the determination of multiple solutions to a given problem.
4. General pattern-matching capabilities operate in conjunction with a goal-seeking search mechanism.
5. Program and data are presented in similar forms.

The preceding listing of the features of PROLOG does not fully convey the subjective advantages of the language. There are at least three such advantages:

1. Having its foundations in logic, PROLOG encourages the programmer to describe problems in a logical manner that facilitates checking for correctness and, consequently, reduces the debugging effort.
2. The algorithms needed to interpret PROLOG programs are particularly amenable to parallel processing.
3. The conciseness of PROLOG programs, with the resulting decrease in development time, makes it an ideal language for prototyping.

Another important characteristic of PROLOG that deserves extension, and is now being extended, is the ability to postpone variable bindings as much as is deemed necessary (lazy evaluation). Failure and backtracking are triggered only when the interpreter is confronted with a logically unsatisfiable set of constraints. In this respect, PROLOG's notion of variables approaches that used in mathematics.

The price to be paid for the advantages offered by the language amounts to the increasing demands for larger memories and faster central processing units (CPUs). The history of programming language evolution has demonstrated that, with the consistent trend toward less expensive and faster computers with larger memories, this price becomes not only acceptable but also advantageous because the savings achieved by program conciseness and by a reduced programming effort largely compensate for the space and execution time overheads. Furthermore, the quest for increased efficiency of PROLOG programs encourages new and important research in the areas of optimization and parallelism.

93.4 Historical Remarks

The birth of logic programming (LP) can be viewed as the confluence of two different research endeavors: one in artificial or natural language processing, and the other in automatic theorem proving. These endeavors contributed to the genesis of the PROLOG language, the principal representative of LP.

Alain Colmerauer, assisted by Philippe Roussel, is credited as the originator of PROLOG, a language that was first developed in the early 1970s and continues to be substantially extended. Colmerauer's contributions stemmed from his interest in language processing using theorem proving techniques. Robert Kowalski was also a major contributor to the development of LP. Kowalski had an interest in logic and theorem proving [Cohen 1988, Bergin 1996]. In their collaboration, Kowalski and Colmerauer became interested in problem solving and automated reasoning using resolution theorem proving.

Kowalski's main research was based on the work of Alan Robinson [1965]. Robinson had the foresight to distinguish the importance of two components in automatic theorem proving: a single inference rule called *resolution* and the testing for equality of trees called *unification*.

Theorems to be proved using Robinson's approach are placed in a special form consisting of conjunctions of *clauses*. Clauses are disjunctions of positive or negated literals; in the case of Boolean algebra (propositional calculus), the literals correspond to variables. In the more general case of the predicate calculus, literals correspond to a potentially infinite number of Boolean variables, one for each combination of values that the literal has as parameters. A *Horn clause* is one containing (at most) one positive literal; all of the others (if any) are negated.

According to the informal description in the introductory example, a Horn clause corresponds to the definition and body of a Boolean function. The positive literal is the left-hand side of a rule (called *Head*); the negative literals appear on the right-hand side of the rule (called *Body*). A non-Horn clause is one in which (logical) negations can appear qualifying a call in the *Body*. An example of a Horn clause is:

$$date\,(feb,\,Day,\,Year)\,:\text{-} leap\,(Year),\,comprised\,(Day,\,1,\,29).$$

because it is equivalent to:

$$leap\,(Year)\ \text{and}\ comprised\,(Day,\,1,\,29)\ \text{implies}\ date\,(feb,\,Day,\,Year)$$

or

$$date\,(feb,\,Day,\,Year)\ \text{or}\ \text{not}\,(leap\,(Year))\ \text{or}\ \text{not}\,(comprised\,(Day,\,1,\,29))$$

If one wished to express the clause

$$date\,(feb,\,Day,\,Year)\,:\text{-} \text{not}\ leap\,(Year),\,comprised\,(Day,\,1,\,28).$$

in which the *not* is the logical Boolean operation, the preceding example would be logically equivalent to:

$$date\,(feb,\,Day,\,Year)\ \text{and}\ leap\,(Year)\ \text{or}\ comprised\,(Day,\,1,\,28)$$

which is not a Horn clause because it has two (positive) elements in the head (and there is no procedural equivalent to it).

Kowalski concentrated his research on reducing the search space in resolution-based theorem proving. With this purpose, he developed with Kuehner a variant of the linear resolution algorithm called SL resolution (for linear resolution with selection function [Kowalski and Kuehner 1970]). Kowalski's view is that, from the automatic theorem-proving perspective, this work paved the way for the development of PROLOG. Having this more efficient (but still general) predicate calculus theorem prover available to them, the Marseilles and Edinburgh groups started using it to experiment with problem-solving tasks. To further increase the efficiency of their prover, the Marseilles group resorted to daring simplifications that would be inadmissible to logicians. These audacious attempts turned out to open new vistas for the future of LP.

Several formulations for solving a given problem were attempted. Almost invariably, the formulations that happened to be written in Horn clause form turned out to be much more natural than those that used non-Horn clauses. A case in which the Horn clause formulation was particularly effective occurred in parsing strings defined by grammar rules. Recall that context-free grammars have a single nonterminal being defined on the left-hand side of each of its rules, therefore establishing a strong similarity with Horn clauses.

The SL inference mechanism applicable to a slight variant of Horn clauses led to the present PROLOG inference mechanism: *selective linear definite* (*SLD*) clause resolution. The word *definite* refers to Horn clauses with exactly one positive literal, whereas general Horn clauses may contain entirely negative clauses. (Because the term "Horn clause" is more widely used than definite clause, the former is often used to denote the latter.)

As a final historical note one should mention that LP gained renewed impetus by its adoption as the language paradigm for the Japanese Fifth Generation Program. PROLOG and, in particular, its CLP extensions now count on a significant number of loyal and enthusiastic users.

93.5   Resolution and Unification

Resolution and unification appear in different guises in various algorithms used in computer science. This section first describes these two components separately and then their combination as it is used in LP. In doing so it is useful to consider first the case of the propositional calculus (Boolean algebra) in which unification is immaterial. It is well known that there exist algorithms that can always decide if a system of Boolean formulas is satisfiable or not, albeit with exponential complexity.

In terms of the informal example considered in the introduction, one can view resolution as a (nondeterministic) call of a user-defined Boolean function. Unification is the general operation of matching the formal and actual parameters of a call. Consequently, unification does not occur in the case of parameterless Boolean functions.

The *predicate calculus* includes the quantifiers $\forall$ and $\exists$; it can be viewed as a general case of the propositional calculus for which each predicate variable (a literal) can represent a potentially infinite number of Boolean variables. Unification is only used in this latter context. Theorem-proving algorithms for the predicate calculus are not guaranteed to provide a yes-or-no answer because they may not terminate.

93.5.1   Resolution

In the propositional calculus, a simple form of resolution is expressed by the inference rule:

$$\textit{if } a \rightarrow b \textit{ and } b \rightarrow c \textit{ then } a \rightarrow c, \textit{ or}$$

$$(\neg a \vee b) \wedge (\neg b \vee c) \rightarrow (\neg a \vee c)$$

Recall that *a implies b* is equivalent to *not a or b*. The final disjunction $\neg a \vee c$ is called a *resolvent*. In particular, resolving $a \wedge \neg a$ implies the empty clause (i.e., falsity).

To better understand the meaning of the empty clause, consider the implication $a \rightarrow a$, which is equivalent to *(not a) or a*. This expression is always true; therefore, its negation *not ((not a) or a)* equivalent to *(a and (not a))* is always false. If a Boolean expression is always *true*, its negation is always *false*. Resolution theorem proving consists of showing that if the expression is always true, its negation results in contradictions of the type *(a and (not a))*, which is always false. The empty clause is simply the resolvent of *(a and (not a))*.

Observe the similarity between resolution and the elimination of variables in algebra, for example:

$$a + b = 3 \textit{ and } -b + c = 5 \textit{ imply } a + c = 8$$

Another intriguing example occurs in matching a procedure definition with its call. Consider, for example,

$$\textit{procedure } b; a$$

$$\cdots$$

$$\textit{call } b; c$$

in which a is the body of b, and c is the code to be executed after the call of b. If one views the definition of b and its call as complementary, a (pseudo)resolution yields: $a; c$ in which concatenation is noncommutative and the resolution corresponds to replacing a procedure call by its body. Actually, the last example provides an intuitive procedural view of resolution as used in PROLOG.

In the case of pure PROLOG programs, only Horn clauses are allowed. For example, if a, b, c, d, and f are Boolean variables (literals), then

$$b \wedge c \wedge d \rightarrow a \quad \textit{and} \quad f$$

are readily transformed in Horn clauses because they correspond, respectively, to

$$a \vee \neg b \vee \neg c \vee \neg d \quad \textit{and} \quad f$$

where the clause f is called unit clause or a fact. The preceding example is written in PROLOG as

$$a :\text{-} b, c, d. \quad \text{and} \quad f.$$

where the symbols :- and " , " correspond to the logical connectors *only if* and *and*. They are read as: a is true only if b and c and d are true. The above conjunction also requires that f be true; equivalently $b \land c \land d \to a$ and f are true.

The resolution mechanism applicable to Horn clauses takes as input a conjunction of Horn clauses $H = h_1 \land h_2 \land \cdots \land h_n$, and a query Q, which is the negation of a theorem to be proved. Q consists of the negation of a conjunction of positive literals or, equivalently, a disjunction of negated literals. Therefore, a query is itself in a Horn clause form in which the head is empty.

A theorem is proved by contradiction, namely, the goal is to prove that $H \land Q$ is inconsistent, implying that the result of successive resolutions involving the negated literals of Q inevitably — in the case of the propositional calculus — leads to falsity (i.e., the empty clause). In other words, if H implies the nonnegated Q is always true, then H *and* the negated Q is always false.

Consider for example the query *date* (*oct*, 15, 1996) in our introductory example. Its negation is *not date* (*oct*, 15, 1996). This resolves with the first rule yielding the bindings *Month* = *oct*, *Day* = 15, *Year* = 1996. The resolvant is the disjunction *not member* (*oct*, [*jan*, *march*, *may*, *july*, *aug*, *oct.*, *dec*]) or *not comprised* (15, 1, 31).

Although not elaborated here, the reader can easily find out that the successive resolutions using the definition of *member* will fail because *oct* is a member of the list of months containing 31 days. Similarly, the day 15 is comprised between 1 and 31. Therefore, the empty (falsity) clause will be reached for both disjuncts of the resolvant.

In what follows, the resolution inference mechanism is applied to Horn clauses representing a PROLOG program. One concrete syntax for PROLOG rules is given by:

$$\text{<}rule\text{>} ::= \text{<}clause\text{>} \, . \, | \, \text{<}unit\ clause\text{>} \, .$$

$$\text{<}clause\text{>} ::= \text{<}head\text{>} :\text{-} \, \text{<}tail\text{>}$$

$$\text{<}head\text{>} ::= \text{<}literal\text{>}$$

$$\text{<}tail\text{>} ::= \text{<}literal\text{>} \, \{, \text{<}literal\text{>}\}$$

$$\text{<}unit\ clause\text{>} ::= \text{<}literal\text{>}$$

where the braces { } denote any number of repetitions (including none) of the sequence enclosed by the brackets <>. First consider the simplest case, where a *literal* is a single letter. For example, consider the following PROLOG program in which rules are numbered for future reference:

1. $a :\text{-} b, c, d.$
2. $a :\text{-} e, f.$
3. $b :\text{-} f.$
4. $e.$
5. $f.$
6. $a :\text{-} f.$

In the first rule, a is the <*head*>, and b, c, d is the <*tail*>, also called the *body*. The fourth and fifth rules are unit clauses; that is, the body is empty. A query Q is syntactically equivalent to a <*tail*>. For example, $a, e.$ is a query and it corresponds to the Horn clause $\neg a \lor \neg e$. The result of querying the program is one (or multiple) *yes* or a single *no* answer indicating the success or failure of the query. In this particular example, the query $a, e.$ yields two *yes* answers. Note that a is defined by three rules; the second and the third yield the two solutions; the first fails because c is undefined. The successful sequence of the list of goals is

Solution 1: $a, e \Rightarrow e, f, e \Rightarrow f, e \Rightarrow e \Rightarrow nil;$

Solution 2: $a, e \Rightarrow f, e \Rightarrow e \Rightarrow nil.$

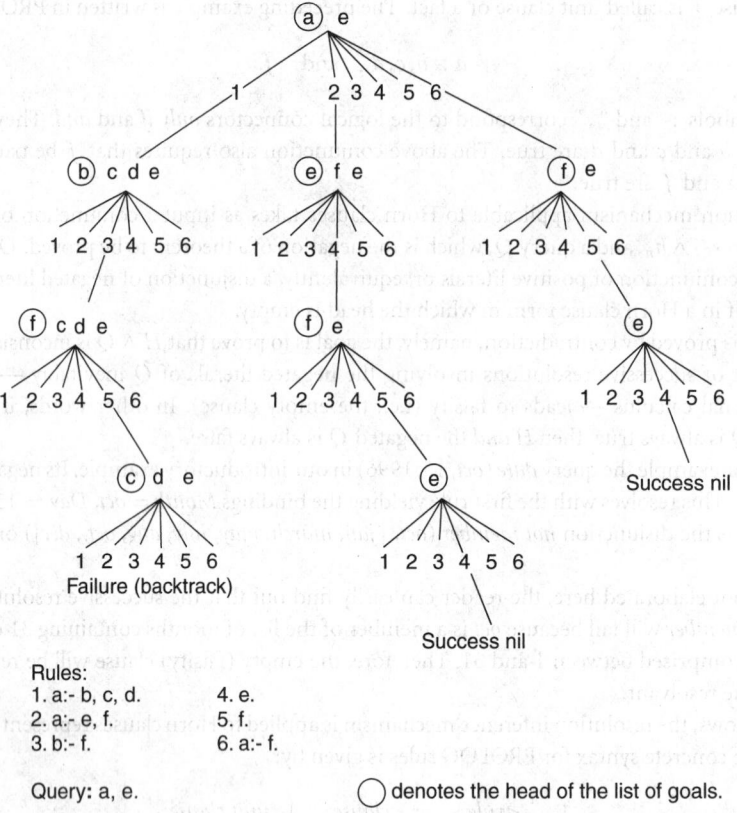

Rules:
1. a:- b, c, d.   4. e.
2. a:- e, f.   5. f.
3. b:- f.   6. a:- f.

Query: a, e.   ◯ denotes the head of the list of goals.

FIGURE 93.1   Tree of choices.

Let us follow in detail the development of the second solution. The negated query *not a or not e* resolves with the sixth rule *a or not f*, yielding the resolvant *not f or not e*. Now the last expression is resolved with the *f* in the fifth rule, yielding *not e* as the resolvant. Finally, *not e* is resolved with the *e* in the fourth rule, yielding the empty clause, which implies the falsity of the negation of the query, using the program rules.

The entire search space is shown in Figure 93.1 in the form of a tree. In nondeterministic algorithms, that tree is called the tree of choices. Its leaves are nodes representing failures or successes. The internal nodes are labeled with the list of goals that remain to be satisfied. Note that if the tree of choices is finite, the order of the goals in the list of goals is irrelevant insofar as the presence and number of solutions are concerned. Figure 93.2 shows a proof tree for the first solution of the example. The proof tree is an *and* tree depicting how the proof has been achieved.

There are three ways of interpreting the semantics of PROLOG rules and queries. The first is based on logic, in this particular case on Boolean algebra. The literals are Boolean variables, and the rules express formulas. The PROLOG program is viewed as the conjunction of formulas it defines. The query Q succeeds if it can be implied from the program. In a second (called procedural) interpretation of a PROLOG rule, it is assumed that a *<literal>* is a goal to be satisfied. For example, the first rule states that:

goal a can be satisfied if goals b, c, and d can be satisfied.

The unit clause states that the defined goal can be satisfied. The program defines a conjunction of goals to be satisfied. The query succeeds if the goals can be satisfied using the rules of the program. Finally, a third interpretation is based on the similarity between PROLOG rules and context-free grammar rules. A PROLOG program is associated with a context-free grammar in which a *<literal>* is a nonterminal and a *<rule>* corresponds to a grammar rule in which the *<head>* rewrites into the *<tail>*; a unit

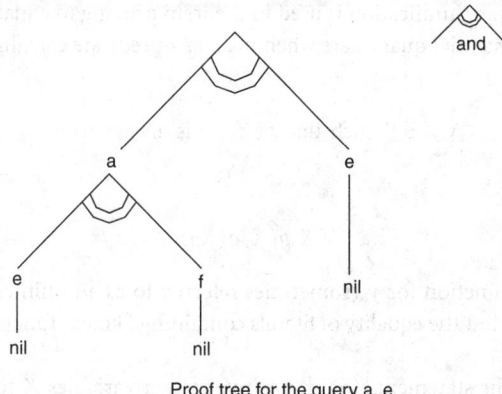

Proof tree for the query a, e

FIGURE 93.2 Proof tree.

clause is viewed as a grammar rule in which a nonterminal rewrites into the empty symbol ε. Under this interpretation, a query succeeds if it can be rewritten into the empty string.

Although the preceding three interpretations are all helpful in explaining the semantics of this simplified version of PROLOG, the logic interpretation is the most widely used among theoreticians, and the procedural by language designers and implementors.

The algorithms that test if a query Q can be derived from a Horn clause program P can be classified in various manners. An analogy with parsing algorithms is relevant: P corresponds to a grammar G, and Q corresponds to the string (of nonterminals) to be parsed (i.e., the sequence of nonterminals that may be rewritten into the empty string using G). *Backward-chaining* theorem provers correspond to top-down parsers and are by far the preferred approach presently used in logic programming. *Forward-chaining* provers correspond to bottom-up parsers. Hybrid algorithms have also been proposed.

In a top-down algorithm, the list of goals in a query is examined from left to right and the corresponding (recursive) procedures are successively called, the equivalent of resolution, until the list of goals becomes empty. Note that the algorithm is essentially nondeterministic because there are usually several choices for the goals (see rules 1 and 2). Another nondeterministic choice occurs when selecting the (next) element of the list of goals to be considered after processing a goal.

Notice that if one had a program consisting of the sole rule a :- a. and the query a, the top-down approach would not terminate. The program states that either a or $\neg a$ is true. Because the query does not specify a, it could be either true or false. A semantically correct interpreter should provide the following constraints as answers $a = true$ or $a = false$. PROLOG programmers have learned to live with these unpalatable characteristics of top-down provers.

Contrary to what usually happens in parsing, bottom-up provers can be very inefficient unless the algorithm contains selectivity features that prevent the inspection of dead-end paths. In the case of database applications, the bottom-up approach using magic sets has yielded interesting results [Minker 1987].

Also note the correspondence between nondeterministic grammars and nondeterministic PROLOG programs. In many applications the programs can be made deterministic and therefore more efficient. However, ambiguous grammars that correspond to programs having multiple solutions are also useful and therefore nondeterminism has its advantages.

93.5.2 Unification

The operation of unification is akin to that of pattern matching. Robinson's unification algorithm brings rigor to the notion of pattern matching and it has a deep meaning: it replaces a set of fairly intricate predicate calculus axioms specifying the equality of trees by an efficient algorithm, which is easily implementable in a computer. As mentioned earlier, unification can be viewed as a very general matching of actual and formal parameters.

From the logic point of view, unification is used in theorem proving to equate terms that usually result from the elimination of existential quantifiers when placing a predicate calculus formula in clausal form. For example:

$$\forall X \exists Y \text{ such that } p(X, Y) \text{ is always true}$$

is replaced by

$$\forall X \; p(X, g(X))$$

where $g(x)$ is the Skolem function for y (sometimes referred to as an uninterpreted function symbol). The role of unification is to test the equality of literals containing Skolem functions and, in so doing, bind values to variables.

Consider, for example, the statement "for all positive integer variables X there is always a variable Y representing the successor of X." The predicate p expressing this statement is $p(X, s(X)) :\text{-} integer(X)$, where $s(X)$ is the Skolem function representing Y, the successor of X. This representation is commonly used to specify positive integers from a theoretical perspective. [It is called a Peano representation of integers (e.g., $s(s(0))$ represents the integer 2).]

To show the effect of unification, the definition of a literal in the previous subsection on resolution is now generalized to encompass labeled tree structures.

$$<literal> ::= <composite>$$

$$<composite> ::= <functor>(<term>\{, <term>\}) \mid <functor>$$

$$<functor> ::= <lower\ case\ identifiers>$$

$$<term> ::= <constant> \mid <variable> \mid <composite>$$

$$<constant> ::= <integers\ and\ lower\ case\ identifiers>$$

$$<variable> ::= <identifiers\ starting\ with\ an\ upper\ case\ letter\ or\_>$$

Examples of terms are $constant$, Var, 319, $line(point(X, 3), point(4, 3))$. It is usual to refer to a single rule in a PROLOG program as a clause. A PROLOG procedure (or predicate) is defined as the set of rules whose head has the same $<functor>$ and arity. Unification tests whether two terms $T1$ and $T2$ can be matched by binding some of the variables in $T1$ and $T2$. The simplest algorithm uses the rules summarized in Table 93.1 to match the terms. If both terms are composite and have the same $<functor>$, it recurses on their components. This algorithm only binds variables if it is *absolutely necessary*, so there may be variables that remain unbound. This property is referred to as the most general unifier (mgu).

One can write a recursive function *unify* which, given two terms, tests for the result of unification using the contents of Table 93.1. The unification of the two terms $f(X, g(Y), T)$ *and* $f(f(a, b), g(g(a, c), Z))$

TABLE 93.1 Summary of Tests for the Unification of Two Terms

Terms 1↓, 2→	$<constant>$ C2	$<variable>$ X2	$<composite>$ T2
$<constant>$ C1	succeed if $C1 = C2$	succeed with $X2 := C1$	fail
$<variable>$ X1	succeed with $X1 := C2$	succeed with $X1 := X2$	succeed with $X1 := T2$
$<composite>$ T1	fail	succeed with $X2 := T1$	succeed if (1) $T1$ and $T2$ have the same functor and arity (2) the matching of corresponding children succeeds

succeeds with the following bindings $X = f(a, b), Y = g(a, c)$, and $T = Z$. Note that if one had $g(Y, c)$ instead of $g(a, c)$ in the second term, the binding would have been $Y = g(Y, c)$.

PROLOG interpreters would usually carry out this circular binding, but would soon get into difficulties because most implementations of the described unification algorithm cannot handle circular structures (called **infinite trees**). Manipulating these structures (e.g., printing, copying) would result in an infinite loop unless the so-called **occur check** were incorporated to test for circularity. This is an expensive test: unification is linear with the size of the terms unified, and incorporation of the occur check renders unification quadratic.

There are versions of the unification algorithm that use a representation called *solved form*. Essentially, new variables are created whenever necessary to define terms; for example, the system:

$$X1 = f(X2, g(X3, X2)) \qquad X5 = g(X4, X2)$$

A solved form is immediately known to be satisfiable because $X2$, $X3$, and $X4$ can be bound to any terms formed by using the function symbols f and g and again replacing (*ad infinitum*) their variable arguments by the functions f and g. This is called an element of the **Herbrand universe** for the given set of terms. The solved form version of the unification algorithm is presented by Lassez in Minker [1987].

To further clarify the notion of solved forms, consider the equation in the domain of reals specified by the single constraint $X + Y = 5$. This is equivalent to $X = 5 - Y$, which is in solved form and satisfiable for any real value of Y. Basically a constraint in solved form contains definitions in terms of free variables, i.e., those that are not constrained. This form is very useful when unification is extended to be applicable to domains other than trees. For example, linear equations may be expressed in solved form.

93.5.3 Combining Resolution and Unification

Consider now general clauses in which the literals contain arguments which are represented by terms. The result of resolving

$$p(\ldots) \quad \vee \quad r(\ldots)$$

with

$$\neg p(\ldots) \quad \vee \quad s(\ldots)$$

is

$$r(\ldots) \quad \vee \quad s(\ldots)$$

if and only if the arguments of p and $\neg p$ are unifiable and the resulting variables are substituted in the corresponding arguments of r and s.

Using programming language terminology, the combination of resolution and unification corresponds to a Boolean function call in which a variable in the head of a rule is matched (i.e., unified) with its corresponding actual parameter; once this is done, the value to which the variable is bound replaces the instances of that variable appearing in the body. Note that a variable can be bound to another variable or to a complex term. Remark also that when unification fails, another head of a rule has to be tried and so on. For example, consider the query $p(a)$ and the program:

$$p(b).$$
$$p(X) :- q(X).$$
$$q(a).$$

The first unification of the query $p(a)$ with $p(b)$ fails. The second attempt binds X to a starting a search for $q(a)$ which succeeds. Had the query been $p(Y)$, the two solutions would be $Y = b$ or $Y = a$. The first solution is a consequence of the unit clause $p(b)$, the second follows from binding Y to a (new) variable

X in the second clause. A search is then made for *q*(*X*), which succeeds with the last clause, therefore binding both *Y* and *X* to *a*.

From a logical point of view, when the result of a resolution is the empty clause (corresponding to the head of a rule with an empty body) and no more literals remain to be examined, the list of bindings is presented as the answer, that is, the bindings (**constraints**) that result from proving that the query is deductible from the program.

93.6   Procedural Interpretation: Examples

It is worthwhile to first understand the importance of the procedure *append* as used in symbolic functional languages. In PROLOG the word procedure is a commonly used misnomer because it always corresponds to a Boolean function, as mentioned in the introductory example; *append* is often used to concatenate two lists. Because one has to avoid destructive assignments, which have very complex semantic meaning, *append* resorts to copying the first list to be appended and making its last element point to the second list. This preserves the original first list. It should be kept in mind that the avoidance of a destructive assignment forces one to produce results that are *cons*tructed from a given data using a LISP-like *cons*. Notice that LP also avoids destructive assignment by having variables bound only once during unification.

This section first demonstrates the transformation of a functional specification of *append* into its PROLOG counterpart corresponding to the previous clauses. Examples of inventive usage of *append* are then presented to further illustrate the procedural interpretation of logic programs. Consider the LISP-like function *append* that concatenates two lists, *L*1 and *L*2:

> function *append*(*L*1, *L*2: *pLIST*): *pLIST*;
> > if *L*1 = *nil* then *append*: = *L*2
> > > else *append*: = *cons* (*head* (*Ll*), *append* (*tail* (*Ll*), *L*2));

In the preceding display, *L*1 and *L*2 are pointers to lists; a list is a record containing the two fields *head* and *tail*, which themselves contain pointers to other lists or to atoms. The tail must point to a list or to the special atom *nil*. The constructor *cons* (*H*, *T*) creates the list whose *head* and *tail* are, respectively, *H* and *T*.

The function *append* can be rewritten into a procedure having an explicit third parameter *L*3 that will contain the desired result. The local variable *T* is used to store intermediate results.

> procedure *append* (*Ll*, *L*2: *pLIST*: var *L*3: *pLIST*);
> > begin local *T*:   *pLIST*;
> > if *L*1 = *nil* then *L*3: = *L*2
> > > else begin *append*(*tail* (*Ll*), *L*2, *T*); *L*3 := *cons*(*head*(*Ll*), *T*) end
> > end

The former procedure can be transformed into a Boolean function that, in addition to building *L*3, checks if *append* produces the correct result.

> function *append*(*Ll*, *L*2: *pLIST*: var *L*3: *pLIST*): Boolean;
> > begin local *H*1, *T*1, *T*: *pLIST*;
> > > if *L*1 = *nil*   then begin *L*3 := *L*2; *append* := true end
> > > > else if {There exists an *H*1 and a *T*1 such that
> > > > > *H*1 = *head*(*L*1) and *T*1 = *tail*(*L*1)}
> > > > > then begin *append* := *append*(*T*1, *L*2, *T*); *L*3 := *cons*(*H*1, *T*) end
> > > > > else *append* := *false*
> > end

The Boolean in the conditional surrounded by braces has been presented informally, but it could actually have been programmed in detail. Note that the assignments in the preceding program are executed at most once for each recursive call. The function returns *false* if $L1$ is not a list (e.g., if $L1 = cons(a, b)$ for some atom $b \neq nil$).

One can now transform the last function into a PROLOG-like counterpart in which rules of assignments and conditionals are subsumed by unification and specified by the equality sign. The statement $E1 = E2$ succeeds if $E1$ and $E2$ can be matched. In addition, some of the variables in $E1$ and $E2$ may be bound if necessary.

$append(L1, L2, L3)$ is true if $L1 = nil$ and $L3 = L2$

otherwise

$append(L1, L2, L3)$ is true

if $L1 = cons(H1, T1)$ and $append(T1, L2, T)$ and $L3 = cons(H1, T)$

otherwise *append* is false.

The reader can now compare the preceding results with the previous description of the subset of PROLOG. This comparison yields the PROLOG program

$append(L1, L2, L3) :- L1 = nil,\ L3 = L2.$

$append(L1, L2, L3) :- L1 = cons(H1, T1), append(T1, L2, T), L3 = cons(H1, T).$

This version is particularly significative because it clearly separates resolution from unification. In this case, the equality sign is the operator that commands a unification between its left and right operands. This could also be done using the unit clause *unify* (X, X) and substituting $L1 = nil$ by *unify* (Ll, nil), and so on. Replacing $L1$ and $L3$ with their respective values on the right-hand side of a clause, one obtains:

$append(nil, L2, L2).$

$append(cons\ (H1, T1), L2, cons\ (H1, T)) :- append\ (T1, L2, T).$

The explicit calls to *unify* have now been replaced by implicit calls that will be triggered by the PROLOG interpreter when it tries to match a goal with the head of a rule. Notice that a *<literal>* now becomes a function name followed by a list of parameters, each of which is syntactically similar to a *<literal>*. In the predicate calculus, the preceding program corresponds to:

$\forall\ L2\ append\ (nil, L2, L2).$

$\forall\ H1, T1, L2, T\ append\ (cons\ (H1, T1), L2, cons\ (H1, T)) \vee \neg\ append\ (T1, L2, T)$

A word of caution about types is in order. The preceding version of *append*, would produce a perhaps unwanted result if the list $L2$ is a term, say, $g(a)$. For example, *append* $(nil, g(a), Z)$ would yield as result $Z = g(a)$. To ensure that this result would not be considered valid, one would have to make explicit the calls to functions that test if the two lists to be appended are indeed lists, namely,

$is\_a\_list\ (nil).$

$is\_a\_list\ (cons\ (H, T)) :- is\_a\_list\ (T).$

The Edinburgh PROLOG representation of $cons(H, T)$ is $[H \mid T]$, and *nil* is $[\ ]$. The Marseilles counterparts are $H.T$ and *nil*. In the Edinburgh dialect, *append* is presented by

$append\ ([\ ], L2, L2).$

$append\ ([H1 \mid T1], L2, [H1 \mid T]) :- append\ (T1, L2, T).$

The query applicable to the program that uses the term *cons* is *append* (*cons* (*a*, *cons* (*b*, *nil*)), *cons* (*c*, *nil*), *Z*) and it yields *Z* = *cons* (*a*, *cons* (*b*, *cons* (*c*, *nil*))). In Edinburgh PROLOG, the preceding query is stated as *append* ([*a*, *b*], [*c*], *Z*), and the result becomes *Z* = [*a*, *b*, *c*].

A remarkable difference between the original PASCAL-like version and the PROLOG version of *append* is the ability of the latter to determine (unknown) lists that, when appended, yield a given list as a result. For example, the query *append*(*X*, *Y*, [*a*]) yields two solutions: *X* = [ ] *Y* = [*a*] and *X* = [*a*] *Y* = [ ].

The preceding capability is due to the generality of the search and pattern matching mechanism of PROLOG. An often-asked question is: Is the generality useful? The answer is definitely yes! A few examples will provide supporting evidence. The first is a procedure for determining a list *LLL*, which is the concatenation of *L* with *L* the result itself being again concatenated with *L*:

$$triple\ (L, LLL) :- append\ (L, LL, LLL),\ append\ (L, L, LL).$$

Note that the first *append* is executed even though *LL* has not yet been bound. This amounts to copying the list *L* and having the variable *LL* as its last element. After the second append is finished, *LL* is bound, and the list *LLL* becomes fully known. This property of postponing binding times can be very useful. For example, a dictionary may contain entries whose values are unknown. Identical entries will have values that are bound among themselves. When a value is actually determined, all communally unbound variables are bound to that value.

Another interesting example is *sublist* (*X*, *Y*), which is true when *X* is a sublist of *Y*. Let *U* and *W* be the lists at the left and right of the sublist *X*. Then the program becomes:

$$sublist\ (X, Y) :- append\ (Z, W, Y),\ append\ (U, X, Z).$$

where the variables represent the sublists indicated as follows:

An additional role for *append* is to simulate the behavior of *member* as used in the introductory example. One could state that *member* (*X*, *L*) :- *append* (*Start*, [*X*|*End*], *L*). in which *Start* and *End* are variables representing lists which, when surrounding *X*, produce *L*. This approach is expensive because it may use a quadratic number of *conses*.

A final example is a bubblesort program. The specification of two adjacent elements *A* and *B* in a list *L* is done by a call:

$$append\ (\_, [A, B\ |\ \_], L)$$

The underscores stand for different variables whose names are irrelevant to the computation, and the notation [*A*, *B* | *C*] stands for *cons*(*A*, *cons*(*B*, *C*)). The rules to bubblesort then become:

$$bsort(L, S) :- append\ (U, [A, B\ |\ X], L),\ B<A,\ append\ (U, [B, A\ |\ X], M),\ bsort(M, S).$$
$$bsort(L, L).$$

The first *append* generates all pairs of adjacent elements in *L*. The literal *B* < *A* is a built-in predicate that tests whether *B* is lexicographically smaller than *A*. (Note that both *A* and *B* must be bound, otherwise a failure occurs! There will be more discussion of this limitation later.) The second *append* reconstructs the modified list, which becomes the argument in a recursive call to *bsort*. If the first clause is no longer applicable, then all pairs of adjacent elements are in order, and the second clause then provides the desired result. This version of bubblesort is space and time inefficient because U, the initial segment of the list, is copied twice at each level of recursion. However, the brevity of the program is indicative of the savings that can be accrued in programming and debugging. It should be now clear to the reader that the translation of a functional program into its PROLOG counterpart is easily done but the reverse translation becomes more difficult because the functional program has to simulate nondeterminism and backtracking.

A unique property of some PROLOG programs is their ability to perform inverse computations. For example, if $p(X, Y)$ defines a procedure p taking an input X and producing an output Y, it can (in certain cases) determine which input X produces Y. Therefore, if p is a differentiation program, it can also perform integration. This is easier said than done, because the use of impure features and simplifications may result in operations that are not correctly backtrackable.

The inverse computation of parsing is string generation, and parsers are now available to perform both operations. A compiler carefully written in PROLOG can also be useful in decompiling. The difficulties encountered in doing the inverse operations are frequently due to the use of impure features.

93.7 Impure Features

In the previous sections, the so-called pure features of PROLOG were described, namely, those that conform with the logic interpretation. A few impure features have been added to the language to make its use more practical. (This situation parallels the introduction of *setq*, *rplaca*, and other impure LISP functions.) However, a word of warning is in order. Some of these impure features vary from one implementation to another (O'Keefe [1990] is an excellent reference on extralogical features).

The most prominent of these impure features is the *cut*, represented by an exclamation point. Its purpose is to let the programmer change the search control mechanism embodied by the procedure solve in Figure 93.3. Reconsider the example in Section 93.5. Assuming that a :- $b, !, c, d$., then in the forward mode, $b, c,$ and d would, as before, be placed in the list of goals and matched with heads of clauses in the database. If, however, a goal following the cut fails (e.g., c, or d), then no further matching of clauses defining a or b would take place. The cut is often used to increase the efficiency of programs and to prevent the consideration of alternate solutions. PROLOG purists avoid the use of cuts.

Another useful predicate is *fail*, which automatically triggers a failure. To implement it, one simply forbids its presence in the database. Other built-in predicates that need to be introduced are the input–output commands *read* and *write*.

Once *cut* and *fail* are available, negation by failure is accomplished by the clauses:

$$not\,(X) :\text{-}\, X, !, fail.$$

$$not\,(X).$$

which fails if X succeeds and vice versa. It is important to note that this artifice does not always follow the rules of true negation in logic. This version of negation illustrates that a term in the head of a rule

```
procedure solve. L: pLIST/;
  begin local i: integer;
    if L ≠ nil
    then
      for i := 1 to n do
        if match(head(Rule[i]), head(L)) then
          solve(append(tail(Rule[i]), tail(L)));
      else write('yes')
  end;
```

FIGURE 93.3 An initial version of the interpreter.

can appear in the body as a call to a function; therefore, in PROLOG, program and data have the same form.

The built-in predicates *assert* and *retract* are used to add or remove a clause from the database; they are often used to simulate assignments. For example, the unit clause *value* (*Variable, Value*). can be asserted using bound parameters [e.g., *value*(Z, 37)]; it can be subsequently retracted to change the actual *Value*, and then reasserted with a new value. Another use of *assert* and *retract* is associated with the built-in function *setof* that collects the multiple answers of a program and places them in a list.

The assignment is introduced using a built-in binary infix operator such as *is*. For example, Y is $X + 1$ is only valid if X has been bound to a number in which case the right-hand side is evaluated and unified with Y; otherwise the *is* fails. To have a fully backtrackable addition, one would have to use CLP. Note that $I = I + 1$ is invalid in CLP, as it should be. The equality $Z = X + Y$ in CLP is, of course, valid with some or all variables unbound.

93.8 Constraint Logic Programming

Major extensions of the unification component of PROLOG became very significant and resulted in a new area of LP called constraint logic programming or CLP. It may well have had its roots in Colmerauer's approach in generalizing the unification algorithm by making it capable of determining the satisfiability of equalities and disequalities of infinite trees (these are actually graphs containing special loops.) However, the notion of backpropagation (Sussman and Steele) dates back to the 1970s. In the case of PROLOG, the backtracking mode is triggered only in the case of unsatisfiability of a given unification. A logical extension of this approach is to make similar backtracking decisions for other (built-in) predicates, say, inequalities (i.e., $\leq, \geq, \ldots$) in the domain of rationals.

The first CLP example presented here is the classic program for computing Fibonacci series. Before presenting the program, it is helpful to consider the program's PROLOG counterpart (the annotation *is* corresponds to an assignment and it has been discussed in Section 93.7):

$$fib\,(0,1).$$
$$fib\,(1,1).$$
$$fib\,(N, R) :\text{-} N1 \text{ is } N - 1.$$
$$fib\,(N1, R1),$$
$$N2 \text{ is } N - 2,$$
$$fib\,(N2, R2),$$
$$R \text{ is } R1 + R2.$$

The *is* predicate prevents the program from being invertible: the query *fib* (10, X) succeeds in producing $X = 89$ as a result, but the query *fib* (Y, 89) yields an error because N is unbound and the assignment to $N1$ is not performed. Note that if we had placed the predicate $N1 \geq 2$ prior to the first recursive call, the query *fib* (Y, 89) would also lead to an error, because the value of N is unbound and the test of inequality cannot be accomplished by the PROLOG interpreter.

The modified CLP version of the program illustrates the invertibility capabilities of CLP interpreters ($N \geq 2$ is a constraint):

$$fib\,(0,1).$$
$$fib\,(1,1).$$
$$fib\,(N, R1 + R2) :\text{-} N \geq 2,$$
$$fib\,(N - 1, R1),$$
$$fib\,(N - 2, R2).$$

The query ?– *fib* (10, *Fib*) yields *Fib* = 89, and the query ?– *fib* (N, 89) yields N = 10. The latter result is accomplished by solving systems of linear equations and inequations that are generated when an explicit or implicit constraint must be satisfied. In this example, the matching of actual and formal parameters results in equations. Let us perform an initial determination of those equations. The query ?– *fib* (N, 89) matches only the third clause. New variables $R1$ and $R2$ are created as well as the constraint $R1 + R2 = 89$. The further constraint $N \geq 2$ is added to the list of satisfiable constraints, which are now $R1 + R2 = 89$ and $N \geq 2$.

The recursive calls of *fib* generate further constraints that are added to the previous ones. These are $N1 = N - 1, N2 = N - 2$, and so forth. Recall that each recursive call is equivalent to a call by value in which new variables are created and new constraints are added (see the section on Implementation).

Therefore, unification is replaced by testing the satisfiability of systems of equations and inequations. A nontrivial implementation problem is how to determine if the constraints are satisfiable, only resorting to expensive general methods such as Gaussian elimination and the simplex method as a last resource.

The second example presented is a sorting program. For the purposes of this presentation, it is unnecessary to provide the code of this procedure [Sterling and Shapiro 1994]; *qsort* (L, S) sorts an input list L by constructing the sorted list S. When L is a list of variables, a PROLOG interpreter would fail because unbound variables cannot be compared using the relational operator $\leq$. In CLP, the query ?– *qsort* ([X1, X2, X3], S). yields as result $S = [X1, X2, X3], X1 \leq X2, X2 \leq X3$. When requested to provide all solutions, the interpreter will generate all the permutations of L as well as the applicable constraints.

Jaffar and Lassez [1987] proved that the theoretical foundations of LP languages (see Section 93.11) remain valid for CLP languages. Several CLP languages are now widely used among LP practioners. Among them one should mention PROLOG III and IV designed by the Marseilles group, CLP(R) designed in Australia and at IBM, CHIP created by members of the European Research Community, and CLP (BNR) designed in Canada at Bell Northern Research.

The subsequent summary describes the main CLP languages and their domains:

PROLOG IV: trees, reals, intervals, linear constraints, rationals, finite domains (including Booleans), strings

CLP(R): trees, linear constraints, floating point arithmetic

CHIP: trees, linear constraints, floating point arithmetic, finite domains

CLP(BNR): trees, intervals

The languages considering intervals (defined by their lower and upper bounds) deal with numeric nonlinear constraints; symbolic linear constraints are handled by the first three languages. (Additional information about interval constraints is provided in Section 93.14.)

93.9 Recent Developments in CLP (2002)

The single most important development in CLP is its amalgamation with an area of artificial intelligence (AI) that has been consistently explored since the late 1970s, namely, Constraint Satisfaction Problems or CSP. This merging has shown to be very fruitful because the two areas CLP and CSP share many common goals, such as concise declarative statements of combinatorial problems and efficient search strategies. In addition, the new combined areas benefit from research that has been done for decades in the field of operations research (OR).

The following subsections summarize the recent developments under the umbrella of the CLP-CSP paradigms. Highly recommended material has appeared in the textbook by Marriot [1998] and in the survey by Rossi [2000]. Research articles have been published in the proceedings of the conferences on Principles and Practice of Constraint Programming, the most recent one held in 2002 [van Hentenryck 2002].

93.9.1 CSP and CLP

Consider a set of constraints C containing a set of variables x_i; the x_i's are defined in their respective domains D_i. Usually, D_i is a finite set of integers. A CSP involves assigning values to the variables so that C is satisfied. CSP is known to be an NP–complete problem (see Chapter 5 of this *Handbook*).

Consider now the cases where the constraints C are (1) unary, that is, they contain a single variable (e.g., $x > 0$), and (2) binary, that is, they contain two variables (e.g., $x^2 + y = 92$) and so forth. It is possible to specify a set of constraints by an undirected graph whose nodes designate each variable and an edge specifies a constraint that is applicable between any two variables. This is often called a constraint network.

The notions of node consistency, arc consistency, and path consistency follow from the above concepts. Furthermore, one can also illustrate the concept of constraint propagation, also called narrowing: it amounts to deleting from a given domain the values that do not satisfy the constraints being specified.

More specifically node consistency consists of considering each node of the constraint graph and its associated unary constraints and performing a narrowing: let $c(x_i)$ be the unary constraint applicable to node x_i; then the values satisfying $not(c(x_i))$ are deleted from D_i.

Arc consistency consists of considering each edge, the associated binary constraint, and performing a narrowing. In the binary case, the constraints are $c(x_i, x_j)$ and the values satisfying $not(c(x_i, x_j))$ are deleted from D_i and D_j.

The above can be generalized to define k-consistency by considering constraints involving k variables (the case $k = 3$ is called path consistency, and $k > 3$ hyper-arc consistency).

One should view node and arc consistency as preprocessing the domains of a set of variables in an attempt to eliminate unfeasible values. How much narrowing can be accomplished depends on the constraint problem being considered. However, if after narrowing there remain no feasible values specified for one or more variables, then there is obviously no solution to the given set of constraints.

It is straightforward to write a CLP program that, given the unary and binary constraints pertaining to a problem, would attempt to reduce the domains of its variables. Then one could check if any of the domains becomes empty thus determining that the given constraints are unfeasible. Note that if there is a single value remaining in a domain, then one could replace the corresponding variable by its value and re-attempt narrowing using node and arc consistency in the reduced constraint network. This is a typical situation that can occur in many constraint domains, including the case of systems of linear equations and inequations.

To further narrow the values of the variables, one could resort to k-consistency. However, it can be shown that for k greater or equal to 3, the consistency check is itself exponential. Therefore, one has to ask the question: is it worth to attempt further narrowing, known to be computational expensive and possibly fruitless, or should one proceed directly to an exhaustive search and determine if a given CSP is satisfiable or not? The answer is, of course, problem dependent.

Another type of narrowing occurs when dealing with interval constraints, those whose variables are integers or floating-point numbers defined within lower and upper bounds. Whenever an arithmetic expression is evaluated, the variables' bounds may narrow according to the operations being performed. This is called *bound consistency*, and as in the case of other type of consistencies, it is incomplete: failures may be detected, (e.g., if the upper bound becomes lower than the lower bound), but bounds may remain unchanged. In those cases, splitting the intervals is the only resource in attempting to solve an interval constraint problem.

Much of the effort in CSP has been done in pruning the search tree by avoiding backtracking as much as possible. The reader is referred to Chapter 63 and Chapter 67 for details on how pruning can be accomplished.

93.9.2 Special Constraints

One can distinguish two kinds of special constraints: the first, called global constraints, are very useful in solving operations research problems; the second, quadratic constraints, can be viewed as an effort to widen the scope of linear constraints without resorting to methods for handling general nonlinear constraints (e.g., interval constraints in Section 93.14.3)

Global constraints are essentially consistency constraints involving many variables and are implemented using specific and efficient algorithms. They perform the narrowing to each variable given as a parameter. A typical example is the predicate *all_different*, having n variables as parameters; *all_different* only succeeds if all its variables are assigned different values within a given domain.

Another useful global constraint is *cumulative*. Its parameters specify (1) the possible starting times for n tasks, (2) their durations, (3) the resources, machines, or personnel needed to perform each task, and (4) the total resources available. This constraint has proven its usefulness in large scheduling problems.

Algorithms for solving linear equations and inequations have been thoroughly studied in algebra and in operations research. The latter field has also evidenced significant interest in quadratic programming, whereby one extends the notion of optimality to quadratic objective functions.

A way of handling quadratic constraints is through the use of interval constraints and linearization, followed by multiple applications of the simplex method. More specifically, one can attempt to linearize quadratic terms by introducing new variables and finding their minimal and maximal values using the classic simplex algorithm. This process is repeated until convergence is reached. (See paper by Lebbah et al. in van Hentenryck [2002]). Therefore, quadratic constraints can be viewed as a variant of global constraints. An earlier interesting algebraic approach to quadratic constraints is described in Pesant [1994].

93.9.3 Control Issues

In LP or CLP, one has limited control of control viewed as a component of the pseudo-equations $LP = logic + control$ or $CLP = constraints + control$. In the LP–CLP paradigm, control is achieved through (1) ordering of rules, (2) ordering of predicates on the right-hand side of a rule, (3) usage of the *cut* or special predicates like *freeze*, and (4) metalevel interpretation. Most processors for LP–CLP utilize a rather rigid, depth-first, top-down search. (A notable exception is the XSB (Stony Brook) PROLOG processor that uses a tabling approach.)

The issues of backtracking and control have been more thoroughly investigated in CSP than in LP–CLP. This is not surprising because general LP-CLP processors usually utilize a fixed set of strategies. Control in CSP is achieved through (1) ordering of variables, (2) ordering of the values to be assigned to each variable, (3) deciding how much backtracking is performed before resuming a forward execution, and (4) utilizing information gathered at failure states to avoid them in future searches.

In CSP processors, efforts have been made to have language constructs that specify various control strategies available to a user. These have been called *indexicals* and dictate the order in which various alternative narrowing methods are applied.

A recent trend in control strategies is to use stochastic searches: random numbers and probability criteria are used to select starting nodes of the search tree; information gathered at failure states is also used to redirect searches. Stochastic searches are not complete (i.e., they do not ensure a solution if one exists). Therefore, one must specify a limited time for performing random searches; once that time is reached, the search can restart using new random numbers and information gathered in previous runs. Stochastic searches have proven highly successful in dealing with very large CSP.

Frequently, problems expressed as CSP have multiple successful and unsuccessful situations that are symmetric. The goal of *symmetry breaking* is to bypass searches that, due to the symmetrical nature of a problem, are uninteresting to pursue. A typical example is the n-queens problem in which a multitude of equivalent symmetrical solutions are found but have to be discarded; the same occurs with unsuccessful configurations that have already been proven to yield no solutions.

93.9.4 Optimization

CSP may have a large number of valid solutions. This parallels the situation where LP–CLP processors are used to solve combinatorial problems. In those cases, it is important to select one or a few among multiple solutions by specifying an objective function that has to be maximized or minimized. In the case of finite domains or interval constraints, one may have to resort to exhaustive searches.

For interval constraints this amounts to splitting a given interval into two or more subintervals. When dealing with small finite domains, one may have to explore each possible value within the domain. The reader is referred to Chapter 15 and Chapter 63 of this Handbook where searches and combinatorial optimization are covered in greater detail. In particular, the *branch-and-bound* strategies are often used in CSP with objective functions. In the CLP paradigm, it is often up to the user to perform optimization searches by developing specific programs.

93.9.5 Soft Constraints

It is not uncommon that, in trying to establish a set of constraints, one is confronted with an over-constrained configuration that is unsatisfiable. It is then relevant to attempt to modify certain constraints so that the system becomes satisfiable.

There are several criteria for "relaxing" an over-constrained system. In general, those criteria involve some sort of optimization; for example, one may choose to violate the least number of individual constraints. Another possibility is to assign a weight to each constraint, specifying the tolerated "degree of unsatisfaction." In that case, one minimizes some function combining those specified weights.

Alternatively, one can assign probabilities to the desirability that a given constraint should be satisfied and then maximize a function expressing the combined probability for satisfying all the constraints (this is referred to as fuzzy constraints).

The usage of "preferences" is another way of loosening over-constrained systems. Consider, for example, the case of the n-queens problem: one may wish to allow solutions that accept attacking queens provided they are far apart. Whenever there are solutions with non-attacking queens, those are preferred over the ones having attacking but distant queens. Preferences are specified by an ordering stipulating that constraint C_1 is preferable to constraint C_2 (e.g., $C_1 > C_2$).

Recent work in the area of soft constraints provides a general framework for relaxing over-constrained systems (see Bistarelli et al. in van Hentenryck [2002]). That framework ensures that if certain algebraic (semi-ring) properties are respected, one can define soft constraints expressing most variants of the above-illustrated constraint relaxation approaches.

93.9.6 SATisfiability Problems

The satisfiability problem can be described as: given a Boolean formula B with n variables, are there *true* or *false* assignments to the variables that render B true? (See Chapters 6 and 66 of this *Handbook* for further details.) It is usual to have B expressed in Conjunctive Normal Form (CNF), consisting of conjunctions of disjunctions of variables or their negations.

Boolean formulas in CNF can be further transformed into equivalent ones whose disjunctions contain three variables negated or not, or the constants *true* and *false*. That variant of the satisfiability problem is known as 3SAT and enjoys remarkable properties. The class NP-complete (see Chapter 6) congregates combinatorial problems that can be reduced to SAT problems in polynomial time. Therefore, 3SAT can be viewed as a valid standard for studying the practical complexity of hard problems. Furthermore, it is possible to show that 3SAT problems can be transformed into CSP.

Consider a random 3SAT problem with n variables and m clauses. It is intuitive to check that, when n is large and m is small, the likelihood of satisfiability is high. On the other hand, when m is large and n is small, that likelihood is low. The curve expressing the ratio m/n versus the probabilities of satisfaction consists of two relatively flat horizontal components indicating high or low probabilities of satisfaction. It has been empirically determined that in the region around $m/n = 4.5$, the curve decreases sharply; problems in that region have around 50% probability of being satisfiable. These are the computationally intensive problems for 3SAT. The problems where m/n is small or large are in general easily solvable.

The implication of this result is that there are "islands of tractability" within very hard problems. From a theoretical viewpoint, 2SAT is polynomial and the determination of other tractability islands remains of great interest. Unfortunately, the transformation of an NP-complete problem into 3SAT distorts the corresponding values of the m/n ratio. It has become important to find the boundaries of individual NP problems that allow their solutions using inexpensive, moderate, or expensive computational means.

The term *phasetransition* establishes an analogy between algorithmic complexity and the physical properties of matter where temperatures and pressures are used to distinguish gaseous, liquid, and solid forms of states. Finding the boundary values delimiting regions of computational easiness and difficulty has become one of the important aims of CSP.

93.9.7 Future Developments

The novel and valuable future developments in CLP-CSP appear to be oriented in melding machine learning and data-mining approaches to the existing models. These developments parallel those that have occurred in the area of Inductive Logic Programming (ILP) vis-a-vis LP. In ILP, Prolog programs are generated from positive and negative examples ([Muggleton 1991] and [Bratko 2001]).

In the case of CLP-CSP the major question becomes: *Can constraints be learned or inferred from data?* If that is the case then probabilistic or soft constraints are likely to play important roles in providing answers to that quest.

93.10   Applications

The main areas in which LP and CLP have proved successful are summed up in the following:

Symbolic manipulation. Although LISP and PROLOG are currently the main languages in this area, it is probable that a CLP language may replace PROLOG in the next few years. There is a close relationship between the aims of CLP and symbolic languages such as MAPLE, MATHEMATICA, and MACSYMA.

Numerical analysis and operations research. The proposed CLP languages allow their users to generate and refine hundreds of equations and inequations having special characteristics (e.g., the generation of linear equations approximating Laplace's differential equations). The possibility of expressing inequations in a computer language has attracted the interest of specialists in operations research. Difficult problems in scheduling have been solved using CLP in finite domains.

Combinatorics. Nondeterministic languages such as PROLOG have been successful in the solution of combinatorial problems. The availability of constraints extends the scope of problems that can be expressed by CLP programs.

Artificial intelligence applications. Boolean constraints have been utilized in the design of expert systems. Constraints have also been used in natural language processing and in parsing. The increased potential for invertibility makes CLP languages unique in programming certain applications. For example, the inverse operation of parsing is string generation.

Deductive databases. These applications have attracted a considerable number of researchers and developers who are now extending the database (DB) domains to include constraints. The language DATALOG is the main representative, and its programs contain only variables or constants (no composite terms are allowed).

Engineering applications. The ease with which CLP can be used for generating and refining large numbers of equations and inequations makes it useful in the solution of engineering problems. Ohm's and Kirchhoff's laws can readily be used to generate equations describing the behavior of electrical circuits.

93.11   Theoretical Foundations

This section provides a summary of the fundamental results applicable to logic programs [Apt 1990, Lloyd 1987]. It will be shown later that these results remain applicable to a wide class of constraint logic programs.

The semantics of LP and CLP are usually defined using either logic or sets. In the logic approach, one establishes that given both (1) a Horn clause program P and (2) a query Q, Q can be shown to be a consequence of P. In other words, $\neg P \vee Q$ is always true, or equivalently, using contradiction, $P \wedge \neg Q$ is always false. This relates the logical and operational meaning of programs; that is, that Q is a consequence of P can be proved by a resolution-based **breadth-first** theorem prover. (This is because a **depth-first** prover could loop in trying to determine a first solution, being therefore incapable of finding other solutions that may well exist.)

Logic-based semantics are accomplished in two steps. The first considers that a program yields a *yes* answer. In that case, the results (i.e., the bindings of variables to terms as the result of successive unifications) are the constraints that P and Q must satisfy so that Q becomes deducible from P. This is accomplished by Horn clause resolutions that render $P \wedge \neg Q$ unsatisfiable.

The second step of the proof is concerned with logic programs that yield a *no* answer and therefore do not specify bindings or constraints. In that case, it becomes important to make a stronger statement about the meaning of a clause.

Recall that in the case of *yes* answers, a Horn clause specifies that the *Head* is a consequence of the *Body*, or equivalently that the *Head* is true if the *Body* is true. In the case of a *no* answer, the so-called Clark completion becomes applicable. That means that the *Head* is true *if and only if* the *Body* is true. Then the semantics of programs yielding a *no* answer amount to proving that *not Q* is a consequence of the completion of P. This amounts to considering the implication in *Body implies Head* in every clause of P as being replaced by *Body equivalent to Head*.

It should be noted that the preceding results are only applicable to queries that do not contain logic negation. For example, the query $\neg\, date\,(X, Y, Z)$ in the introductory example is invalid because a negative query is not in Horn clause form. Therefore, only positive queries are allowed and the behavior of the prover satisfies the so-called closed word assumption: only positive queries deducible from the program provide *yes* answers. Recent developments in the so-called nonmonotonic logic extend programs to handle negative queries.

The second approach in defining the semantics of LP and CLP uses sets. Consider the set $S0$ of all unit clauses in a program P. This set involves assigning any variables in these unit clauses to elements of the Herbrand universe. Consider then the clauses whose bodies contain elements of that initial set $S0$. Obviously the *Head* of those clauses is now deducible from the program and the new set $S1$ is constructed by taking the union of $S0$ with the heads of clauses that have been found to be true.

This process continues by computing the sets $S2, S3$, etc. Notice that $S(i)$ always contains $S(i-1)$. Eventually these sets will not change because all the logical information about a finite program P is contained in them. This is called a least fixed point. Then Q is a consequence of P if and only if each conjunct in Q is in the least fixed point of P.

Consider, for example, the PROLOG program for adding two positive numbers specified by a successor function $s(X)$ denoting the successor of X:

$$add\,(0, X, X).$$
$$add\,(s(X), Y, s(Z)) :\!\text{-}\ add\,(X, Y, Z).$$

First notice the similarity of the preceding example with *append*. Adding 0 to a number X yields X (first rule). Adding the successor of X to a number Y amounts to adding one to the result Z obtained by adding X to Y.

In this simple example, the Herbrand universe consists of 0, $s(0)$, $s(s(0))$, and so on, namely, the positive natural numbers including the constant 0. The so-called Herbrand base considers all of the literals (in this case *add*) for which a binding of a variable to elements of the Herbrand universe satisfy the program rules. Using the set approach $S0$ consists of all natural numbers because any number can be added to zero. The fixed point corresponds to the infinite set of bindings of X, Y, and Z to elements of the Herbrand universe, which satisfies both rules.

The meaning of a program P and query Q yielding a *no* answer can also be specified using set theory. In that case, one starts with the set corresponding to the Herbrand universe $H0$. Then this set is reduced to a smaller set by using once the rules in P. Call this new set $H1$. By applying again the rules in P, one obtains $H2$, and so on. In this case there is not necessarily a fixed point Hn. The property pertaining to programs yielding *no* answers then consists of the statement: *not Q* is a logical consequence of the completion of Q if and only if some conjunct of Q is not a member of Hn.

In addition to programs yielding *yes* or *no* answers, there are those which loop. The halting problem tells us that we cannot hope to detect all of the programs which will eventually loop. Let us consider, as an example, the program P_1:

$$p(a).$$
$$p(b) :- p(b).$$

As expected, the queries $Q_1 : p(a)$ and $Q_2 : p(c)$ yield, respectively, *yes* and *no* because $p(a)$ is a consequence of P_1, and $\neg p$ (c) is a consequence of the completion of p_1; that is,

$$p(X) \equiv (X = a) \vee (X = b \wedge p(b))$$

However, the interpreter will loop for the query $Q_3 : p(b)$, or when all solutions of $Q_4 : p(X)$ are requested.

As mentioned earlier, the preceding theoretical results can be extended to CLP languages. Jaffar and Lassez [1987] established two conditions that a CLP extension to PROLOG must satisfy so that the semantic meaning (using logic or sets) is still applicable. The first is that the replacement of unification by an algorithm that tests the satisfiability of constraints should always yield a *yes* or *no* answer. This property is called *satisfaction-completeness* and it is obviously satisfied by the unification algorithm in the domain of trees. Similarly, the property applies to systems of linear equations in the domain of rationals and even to polynomial equations but with a significantly larger computational cost.

The second of Jaffar and Lassez's conditions is called *solution-compactness*. It basically states that elements of a domain (say, irrational numbers) can be defined by a potentially infinite number of more stringent constraints that bound their actual values by increasingly finer approximations. For example, the real numbers satisfy this requirement. For more detail on Jaffar and Lassez's theory, see Jaffar and Maher [1994] and Cohen [1990]. Existing CLP languages satisfy the two conditions established by Jaffar and Lassez.

The beauty of the Jaffar and Lassez metatheory is that they have established conditions under which the basic theorems of logic programming remain valid, provided that the set of proposed axioms specifying constraints satisfy the described properties.

A convenient (although incomplete) taxonomy for CLP languages is to classify them according to their domains or combinations thereof. One can have CLP (rationals), or CLP (Booleans, reals). PROLOG can be described as CLP (trees) and CLP (R) as CLP (reals, trees). A complete specification of CLP language would also have to include the predicates and operations allowed in establishing valid constraints.

From the language-design perspective, the designer would have to demonstrate the correctness of an efficient algorithm implementing the test for constraint satisfiability. This is equivalent to proving the satisfiability of the constraints specified by the axioms.

93.12 Metalevel Interpretation

Metalevel interpretation allows the description of interpreters for the languages (such as LISP or PROLOG), using the languages themselves. In PROLOG, the metalevel interpreter for pure programs consists of a few lines of code. The procedure solve has as a parameter a list of PROLOG goals to be processed. The interpreter assumes that the program rules are stored as unit clauses:

clause (Head, Body).

each corresponding to a rule: *Head :- Body.*, where *Head* is a literal and *Body* is a list of literals. Unit clauses are stored as *clause (Head, [ ]).* The interpreter is:

solve ([ ]).

solve ([Goal | Restgoal]) :- solve (Goal), solve (Restgoal).

solve (Goal) :- clause (Goal, Body), solve (Body).

The first rule states that an empty list of goals is logically correct. (In that case, the interpreter should print the latest bindings of the variables.) The second rule states that when processing (i.e., *solving*) a list of goals, the *head* and then the *tail* of the list should be processed. The third rule specifies that when a single *Goal* is to be processed, one has to look-up the database containing the clauses and process the *Body* of the applicable clause. In the preceding interpreter, metainterpreter unification is implicit. One could write metainterpreters in which the built-in unification is replaced by an explicit sequence of PROLOG constructs using the impure features.

A very useful extension often incorporated into interpreters is the notion of co-routining, or lazy evaluation. The built-in procedure *freeze*(X, P) tests whether the variable X has been bound. If so, P is executed; otherwise, the pair (X, P) is placed in a freezer. As soon as X becomes bound, P is placed at the head of the list of goals for immediate execution.

The procedure *freeze* can be easily implemented by expressing it as a variant of *solve* also written in PROLOG. Although this metalevel programming will of course considerably slow down the execution, this capability can and has been used for fast prototyping extensions to the language [Sterling and Shapiro 1994, Cohen 1990].

Another important application of metalevel programming is partial evaluation. Its objective is to transform a given program (a set of procedures) into an optimized version in which one of the procedures has one or more parameters that are bound to a known value. An example of partial evaluation is the automatic translation of a simple (inefficient) pattern matching algorithm which tests if a given pattern appears in a text. When the pattern is known, a partial evaluator applied to the simple matching algorithm produces the equivalent of the more sophisticated Knuth–Morris–Pratt pattern matching algorithm.

In a metalevel interpreter for a CLP language, a rule is represented by: *clause* (*Head, Body, Constraints*). Corresponding to a rule: *Head :- Body* {*Constraints*}.

The modified procedure solve contains three parameters: (1) the list of goals to be processed, (2) the current set of constraints, and (3) the new set of constraints obtained by updating the previous set. The metalevel interpreter for CLP, written in PROLOG becomes:

> *solve* ([ ], C, C).
>
> *solve* ([*Goal* | *Restgoal*], *Previous_C, New_C*) :-
>> *solve* (*Goal, Previous_C, Temp_C*),
>> *solve* (*Restgoal, Temp_C, New_C*).
>
> *solve* (*Goal, Previous_C, New_C*) :-
>> *clause* (*Goal, Body, Current_C*),
>> *merge_constraints* (*Previous_C, Current_C, Temp_C*),
>> *solve* (*Body, Temp_C, New_C*).

The heart of the interpreter is the procedure *merge_constraints*, which merges two sets of constraints: (1) the previous constraints, *Previous_C*, and (2) the constraints introduced by the current clause, *Current_C*. If there is no solution to this new set of constraints, the procedure fails; otherwise, it simplifies the resulting constraints, and it binds any variables that have been constrained to take a unique value. For example, the constraints $X \leq 0 \wedge X \geq 0$ simplify to the constraint $X = 0$, which implies that X can now be bound to 0.

The design considerations that influence the implementation of this procedure will be discussed in Section 93.13. Note that the controversial unit logical inference steps per second (LIPS), often used to estimate the speed of PROLOG processors, loses its significance in the case of a constraint language. The number of LIPS is established by counting how many times per second the procedure *clause* is activated; in the case of CLP, this time to process *clause* and *merge_constraints* may vary significantly, depending on the constraints being processed.

93.13   Implementation

It is worthwhile to present the basic LP implementation features by describing an interpreter for the simplified PROLOG of Section 93.5 written in a Pascal or C-like language. The reader should note the similarities between the metalevel interpreter *solve* of the previous section and the one about to be described.

The rules will be stored sequentially in a database implemented as a one-dimensional array *Rule*[*l* . .*n*] and containing pointers to a special type of linear list. Such a list is a record with two fields, the first storing a letter and the second being either *nil* or a pointer to a linear list. Let the (pointer) function *cons* be the constructor of a list element, and assume that its fields are accessible via the (pointer) functions *head* and *tail*. The first rule is stored in the database by

$$Rule[1] := cons('a', cons('b', cons('c', cons('d', nil))))).$$

The fifth rule defining a unit clause is stored as $Rule[5] := cons('e', nil)$. Similar assignments are used to store the remaining rules.

The procedure *solve* that has as a parameter a pointer to a linear list is capable of determining whether or not a query is successful. The query itself is the list with which *solve* is first called. The procedure uses two auxiliary procedures *match* and *append*; *match* (A, B) simply tests if the alphanumeric A equals the alphanumeric B; *append* (Ll, $L2$) produces the list representing the concatenation of Ll with $L2$ (this is equivalent to the familiar *append* function in LISP: it basically copies Ll and makes its last element point to $L2$).

The procedure *solve*, written in a Pascal-like language, appears in Figure 93.3. Recall that the variable n represents the number of rules stored in the array *Rule*. The procedure performs a depth-first search of the problem space where the local variable is used for continuing the search in case of a failure. The head of the list of goals L is matched with the head of each rule. If a match is found, the procedure is called recursively with a new list of goals formed by adding (through a call of *append*) the elements of the tail of the matching rule to the goals that remain to be satisfied. When the list of goals is *nil*, all goals have been satisfied and a success message is issued. If the attempts to match fail, the search is continued in the previous recursion level until the zeroeth level is reached, in which case no more solutions are possible. For example, the query $a, e.$ is expressed by *solve* (*cons* ('a', *cons* ('e', *nil*))) and yields the two solutions presented in Section 93.5 on resolution of Horn clauses.

Note that if the tree of choices is finite, the order of the goals in the list of goals is irrelevant insofar as the presence and number of solutions are concerned. Thus, the order of the parameters of *append* in Figure 93.3 could be switched, and the two existing solutions would still be found. Note that if the last rule were replaced by $a :- f, a.$, the tree of choices would be infinite and solutions similar to the first solution would be found repeatedly. The procedure *solve* in Figure 93.3 can handle these situations by generating an infinite sequence of solutions. However, had the preceding rule appeared as the first one, the procedure *solve* would also loop, but without yielding any solutions. This last example shows how important the ordering of the rules is to the outcome of a query. This explains Kowalski's dictum program = logic + control, in which control stands for the ordering and (impure) control features such as the cut [Kowalski 1979].

It is not difficult to write a recursive function *unify*, which, given two terms, tests for the result of unification using the contents of Table 93.1 (Section 93.5). For this purpose, one has to select a suitable data structure. In a sophisticated version, terms are represented by variable-sized records containing pointers to other records, to constants, or to variables. Remark that the extensive updating of linked data structures inevitably leads to unreferenced structures that can be recovered by a garbage collection. It is frequently used in most PROLOG and CLP processors.

A simpler data structure uses linked lists and the so-called Cambridge Polish notation. For example, the term $f(X, g(Y, c))$ is represented by ($f(var\ x)(g(var\ y)(const\ c))$), which can be constructed with *cons*es.

As mentioned in Section 93.5, if the result of unification results in the binding $Y := g(Y, c)$, then (most) PROLOG interpreters would soon get into difficulties because most implementations of the described unification algorithm cannot handle circular structures. Manipulating these structures (e.g., printing, copying) would result in an infinite loop unless the so-called occur check is incorporated to test for circularity.

The additional machinery needed to incorporate unification into the procedure *solve* of Figure 93.3 is described in Cohen [1985]. An important remark is in order: when introducing unification, it is necessary to *copy* the clauses in the program and introduce new variables (which correspond to parameters that should be called by value). The frequent copying and updating of lists makes it almost mandatory to use garbage collection, which is often incorporated to LP processors.

93.13.1 Warren Abstract Machine (WAM)

D.H.D. Warren, a pioneer in the compilation of PROLOG programs, proposed in 1983 a set of primitive instructions that can be generated by a PROLOG compiler, usually written using PROLOG. (Warren's approach parallels that of P-code used in early Pascal compilers.) The **Warren abstract machine** (WAM) primitives can be efficiently interpreted using specific machines.

The main data structures used by the WAM are (1) the recursion stack, (2) the heap, and (3) the trail. The heap is used for storing terms and the trail for backtracking purposes. The WAM uses the copying approach mentioned in the beginning of this section. A local garbage collector takes advantage of the cut by freeing space in the trail.

The WAM has been used extensively by various groups developing PROLOG compilers. Its primitives are of great efficiency in implementating features such as tail-recursion elimination, indexing of the head of the clause to be considered when processing a goal, the cut, and other extralogical features of PROLOG. A useful reference in describing the WAM is the one by Ait-Kaci [1991]. A recent reference on implementation is by Van Roy [1994].

93.13.2 Parallelism

Whereas for most languages it is fairly difficult to write programs that automatically take advantage of operations and instructions that can be executed in parallel, PROLOG offers an abundance of opportunities for parallelization. There are at least three possibilities for performing PROLOG operations in parallel:

1. *Unification.* Because this is one of the most frequent operations in running PROLOG programs, it would seem worthwhile to search for efficient parallel unification algorithms. Some work has already been done in this area [Jaffar et al. 1992]. However, the results have not been encouraging.
2. *And-parallelism.* This consists of simultaneously executing each procedure in the tail of a clause. For example, in $a(X, Y, U) :- b(X, Z), c(X, Y), d(T, U)$, an attempt is made to continue the execution in parallel for the clauses defining b, c, and d. The first two share the common variable X; therefore, if unification fails in one but not in the other, or if the unification yields different bindings, then some of the labor done in parallel is lost. However, the last clause in the tail can be executed independently because it does not share variables with the other two.
3. *Or-parallelism.* When a given predicate is defined by several rules, it is possible to attempt to apply the rules simultaneously. This is the most common type of parallelism used in PROLOG processors.

Kergommeaux and Codognet [1994] is a recommended survey of parallelism in PROLOG.

93.13.3 Design and Implementation Issues in Constraint Logic Programming

There is an important implementation consideration that appears to be fulfilled in both CLP(R) and PRO-LOG IV: the efficiency of processing PROLOG programs (without constraints) should approach that of current PROLOG interpreters; that is, the overhead for *recognizing* more general constraints should be small.

There are three factors that should be considered when selecting algorithms for testing the satisfiability of systems of constraints used in conjunction with CLP processors. They are (1) incrementality, (2) simplification, and (3) canonical forms. The first is a desirable property that allows an increase in efficiency of multiple tests of satisfiability (by avoiding recomputations). This can be explained in terms of the metalevel interpreter for CLP languages described in Section 93.11: if the current system of constraints S is known to be satisfiable, the test of satisfiability should be incremental, minimizing the computational effort required to check if the formula remains satisfiable or not. Classical PROLOG interpreters have this property because previously performed unifications are not recomputed at each inference step. There are modifications of Gaussian methods for solving linear equations that also satisfy this property. This is accomplished by introducing temporary variables and replacing the original system of equations by an equivalent solved form (see Section 93.5): *variable = linear terms involving only the temporary variables*.

The simplex method can also be modified to satisfy incrementality. Similarly, the SL resolution method for testing the satisfiability of Boolean equations and the Gröbner method for testing the satisfiability of polynomial equations have this property.

In nearly all the domains considered in CLP, it may be possible to replace a set of constraints by a simpler set. This simplification can be time-consuming, but is sometimes necessary. The implementor of CLP languages may have to make a difficult choice as to what level of simplification should occur at each step verifying constraint satisfaction. It may turn out that a system of constraints eventually becomes unsatisfiable, and all of the work done in simplification is lost. When a final result has to be output, it becomes essential to simplify it and present it to the reader in the clearest, most readable form.

An important function of simplification is to detect the assignment of a variable to a single value (e.g., from $X \geq 1$ *and* $X \leq 1$ *one infers* $X = 1$). This property is essential when implementing a modified simplex method that detects when a variable is assigned to a single value. Note that this detection is necessary when using lazy evaluation.

The incremental algorithms for testing the satisfiability of linear equations and inequations, as well as that used in the Gröbner method for polynomial equations, are capable of discarding redundant equations; therefore, they perform some simplifications [Sato and Aiba 1993].

The canonical (solved) forms referred to earlier in this section can be viewed as (internal) representations of the constraints which facilitate both the tests of satisfiability and the ensuing simplifications. For example, in the case of the Gröbner method for solving polynomial equations, the input polynomials are internally represented in a normal form, such that variables are lexicographically ordered and the terms of the polynomials are ordered according to their degrees. This ordering is essential in performing the required computations. Also note that if two seemingly different constraints have the same canonical form, only one of them needs to be considered. Therefore, the choice of appropriate canonical forms deserves an important consideration in the implementation of CLP languages [Jaffar and Maher 1994].

93.13.4 Optimization Using Abstract Interpretation

Abstract interpretation is an enticing area of computer science initially developed by Cousot and Cousot [1992]; it consists of considering a subdomain of the variables of a program (usually a Boolean variable, e.g., one representing the evenness or oddness of the final result). Program operations and constructs are performed using only the desired subdomain. Cousot proved that if certain conditions are applicable to the subdomains and the operations acting on their variables, the execution is guaranteed to terminate. Dataflow analyses, partial evaluation, detection of safe parallelism, etc. can be viewed as instances of abstract interpretation. The research group at the University of Louvain, Belgium, has been active in exploring the capabilities of abstract interpretation in LP and CLP.

93.14 Research Issues

It is worthwhile to classify the numerous extensions of PROLOG into three main categories, namely, those related to (1) resolution beyond Horn clauses, (2) unification, and (3) others (e.g., concurrency). Major extensions of the unification became very significant and resulted in a new area of LP called constraint logic programming or CLP that was dealt with in Section 93.8; nevertheless, the more recent addition to CLP dealing with the domain of intervals is discussed in this section.

93.14.1   Resolution Beyond Horn Clauses

Several researchers have suggested extensions for dealing with more general clauses and for developing semantics for negation that are more general than that of negation by failure (see Section 93.7). Experience has shown that the most general extension, that is, to the general predicate calculus, poses difficult combinatorial search problems. Nevertheless, substantial progress has been made in extending LP beyond pure Horn clauses. Two such extensions deserve mention: stratified programs and generalized predicate calculus formulas in the *body* part of a clause.

Stratified programs are variants of Horn clause programs that are particularly applicable in deductive databases; true negation may appear in the body of clauses, provided that it satisfies certain conditions. These stratified programs have clean semantics based on logic and avoid the undesirable features of negation by failure. (See Minker [1987].)

To briefly describe the second extension, it is worthwhile to recall that the procedural interpretation of resolution applied to Horn clauses is based on the substitution model: a procedure call consists of replacing the call by the body of the procedure in which the formal parameters are substituted by the actual parameters via unification. The generalization proposed by Ueda and others can use the substitution model to deal with the clauses of the type *head* :-, (*a general formula in the predicate calculus containing quantifiers and negation*).

93.14.2   Concurrent Logic Programming and Constraint Logic Programming

A significant extension of LP has been pursued by several groups. A premise of their effort can be stated as: a programming language worth its salt should be expressive enough to allow its users to write complex but efficient operating systems software (as is the case of the C language). With that goal in mind, they incorporated into LP the concepts of **don't care nondeterminism** as advocated by Dijkstra. The resulting languages are called concurrent LP languages. The variants proposed by these groups were implemented and refined; they have now converged to a common model that is a specialized version of the original designs.

Most of these concurrent languages use special punctuation marks "?" and "|". The question mark is a shorthand notation for *freezes*. For example, the literal $p(X?, Y)$ can be viewed as a form of *freeze* $(X, p(X, Y))$. The vertical bar is called *commit* and usually appears once in the tail of clauses defining a given procedure. Consider, for example,

$$a :- b, c \mid d, e.$$

$$a :- p \mid q.$$

The literals b, c, and d, e are executed using *and* parallelism. However, the computation using *or* parallelism for the two clauses defining a continues only with the clause that first reaches the *commit* sign. For example, if the computation of b, c proceeds faster than p, then the second clause is abandoned, and execution continues with d, e only (see Saraswat 1993).

93.14.3   Interval Constraints

The domain of interval arithmetic has become a very fruitful area of research in CLP. Older has been a pioneer in this area [Older and Vellino 1993]. This domain specifies reals as being defined between lower and upper bounds that can be large integers or rational numbers. The theory of solving most nonlinear and trigonometric equations using intervals guarantees that *if* there is a solution, that solution must lie within the computed intervals. Furthermore, it is also guaranteed that *no* solution exists outside the computed interval or unions of intervals.

The computations involve the operation of *narrowing* that consists of finding new bounds for a quantity denoting the result of an operation (say, +, *, sin, etc.) involving operands, which are also defined by their lower and upper bounds. The narrowing operation also involves intersecting intervals obtained by various computations defining the same variable. The intersection may well fail (e.g., the equality operation applying to operands whose intervals are disjoint). The narrowing is guaranteed to either converge or fail. This, however, may not be sufficient to find possible solutions of interest. One can nevertheless split a given interval into two or more unions of intervals and proceed to find a more precise solution, if one exists. This is akin to enumeration of results in CLP.

The process of splitting is a **don't know nondeterministic** choice, an existing component of LP. The failure of the narrowing operation is analogous to that encountered in CLP when a constraint is unsatisfiable and backtracking occurs. Therefore, there is a natural interaction between LP and the domain of intervals.

Interval arithmetic is known to yield valuable results in computing the satisfiability of nonlinear constraints or in the case of finite domains. Its use in linear constraints is an active area of research because results indicate a poor convergence of narrowing. In the case of polynomials constraints interval arithmetic may well be a strong competitor to Gröebner base techniques.

93.14.4 Constraint Logic Programming Language Design

A current challenge in the design and implementation of CLP is to blend computations in different domains in a harmonious and sound manner. For example, the reals can be represented by intervals whose bounds are floating-point numbers (these have to be carefully implemented to retain soundness due to rounding operations). Actually, floating-point numbers are nothing more than (approximate) very large integers or fractions. This set is, of course, a superset of finite domains, which in turn is a superset of Booleans. Problems in CLP language design that still remain to be solved include how to handle the interaction of these different domains and subdomains. This situation is further complicated by efficiency considerations. Linear inequations, equations, and disequations can be efficiently solved using rational arithmetic but research remains to be done in adapting simplex-like methods to deal with interval arithmetic.

93.15 Conclusion

As in most sciences, there has always been a valuable symbiosis among the theoretical and experimental practitioners of computer science, including, of course, those working in logic programming. Three examples come to mind: the elimination of the occur test in unifications, the cut, and the *not* operator as defined in PROLOG. These features were created by practical programmers and are here to stay. They provide a vast amount of food for thought for theoreticians. As mentioned earlier, the elimination of the occur test was instrumental in the development of algorithms for unification of **infinite trees**. Although the concept of the cut has resisted repeated attempts for a clean semantic definition, its use is unavoidable in increasing the efficiency of programs. Finally, PROLOG's *not* operator has played a key role in extending logic programs beyond Horn clauses.

CLP is one of the most promising and stimulating new areas in computer science. It amalgamates the knowledge and experience gained in areas as varied as numerical analysis, operations research, artificial languages, symbolic processing, artificial intelligence, logic, and mathematics.

During the past 20 years, LP has followed a creative and productive course. It is not unusual for a fundamental scientific endeavor to branch out into many interesting subfields. An interesting aspect of these developments is that LP's original body of knowledge actually branched into subareas, which joined previously existing research areas. For example, CLP is being merged with the area of constraint satisfaction problems (CSP); LP researchers are interested in modal, temporal, intuitionistic, and linear logic; relational database research now includes constraints; operations research and CLP have found previously unexplored similarities, and so on.

The several subfields of LP now include research on CLP in various domains, typing, non-monotonic reasoning, inductive LP, semantics, concurrency, nonstandard logic, abstract interpretation, partial evaluation, blending with functional and with object-oriented language paradigms. It will not be surprising if each of these subfields will become fairly independent from their LP roots and the various specialized groups will organize autonomous journals and conferences. The available literature on LP is abundant and it is likely to be followed by a plentiful number of publications in its autonomous subfields.

Defining Terms

Backtracking: A manner to handle (*don't know*) nondeterministic situations by considering *one* choice at a time and storing information which is necessary to restore a given state of the computation. PROLOG interpreters often use backtracking to implement nondeterministic situations.

Breadth first: A method for traversing trees in which all of the children of a node are considered simultaneously. OR-Parallel PROLOG interpreters use breadth-first traversal.

Clause: A general normal form for expressing predicate calculus formulas. It is a disjunction of literals $(P_1 \vee P_2 \vee \cdots)$ whose arguments are terms. The terms are usually introduced by eliminating existential quantifiers.

Constraint logic programming languages: PROLOG-like languages in which unification is replaced or complemented by constraint solving in various domains.

Constraints: Special predicates whose satisfiability can be established for various domains. Unification can be viewed as equality constraints in the domain of trees.

Cut: An annotation used in PROLOG programs to bypass certain nondeterministic computations.

Depth first: A method for traversing trees in which the leftmost branches are considered first. Most sequential PROLOG interpreters use depth-first traversal.

Don't care nondeterminism: The arbitrary choice of one among multiple possible continuations for a computation.

Don't know nondeterminism: Situations in which there are equally valid choices in pursuing a computation.

Herbrand universe: The set of all terms that can be constructed by combining the terms and constants which appear in a logic formula.

Horn clause: A clause containing (at most) one positive literal. The term *definite clause* is used to denote a clause with exactly one positive literal. PROLOG programs can be viewed as a set of definite clauses in which the positive literal is the head of the rule and the negative literals constitute the body or tail of the rule.

Infinite trees: Trees that can be unified by special unification algorithms which bypass the occur-check. These trees constitute a new domain, different from that of usual PROLOG trees.

Metalevel interpreter: An interpreter written in L for the language L.

Occur-check: A test performed during unification to ensure that a given variable is not defined in terms of itself [e.g., $X = f(X)$ is detected by an occur-check, and unification fails].

Predicate calculus: A calculus for expressing logic statements. Its formulas involve:

- *atoms:* $P(T_1, T_2, \ldots)$ where P is a predicate symbol and the T_i are terms.
- *Boolean connectives:* conjunction ($\wedge$), disjunction ($\vee$), implication ($\rightarrow$), and negation ($\neg$).
- *literals:* atoms or their negations.
- *quantifiers:* for all ($\forall$), there exists ($\exists$).
- *terms* (also called *trees*): constructed from constants, variables, and function symbols.

Resolution: A single inference step used to prove the validity or predicate calculus formulas expressed as clauses. In its simplest version: $P \vee Q$ and $\neg P \vee R$ imply $Q \vee R$, which is called the resolvant.

SLD resolution: Selective linear resolution for definite clauses inference step used in proving the validity of Horn clauses.

Unification: Matching of terms used in a resolution step. It basically consists of testing the satisfiability or the equality of trees whose leaves may contain variables. Unification can also be viewed as a general parameter matching mechanism.

Warren abstract machine (WAM): An intermediate (low-level) language that is often used as an object language for compiling PROLOG programs. Its objective is to allow the compilation of efficient PROLOG code.

References

Ait-Kaci, H. 1991. *The WAM: A (Real)Tutorial*. MIT Press, Cambridge, MA.

Apt, K. R. 1990. Logic programming. In *Handbook of Theoretical Computer Science*, J. van Leewun, Ed., pp. 493–574. North Holland, Amsterdam.

Bergin, T. J. 1996. History of programming languages HOPL 2. Addison-Wesley, Reading, MA.

Borning, A. 1981. The programming language aspects of Thing-Lab, a constraint-oriented simulation laboratory. *ACM TOPLAS*, 3(4):252–387.

Bratko, I., 2001. *Prolog Programming for Artificial Intelligence*, 3rd ed., Addison-Wesley, Reading, MA.

Clocksin, W. F. and Mellish, 1984. *Programming in PROLOG*, 2nd ed. Springer-Verlag, New York.

Cohen, J. 1985. Describing PROLOG by its interpretation and compilation. *Commun. ACM*, 28(12):1311–1324.

Cohen, J. 1988. A view of the origins and development of PROLOG. *Commun. ACM*, 31(1):26–36.

Cohen, J. 1990. Constraint logic programming languages. *Commun. ACM* (July):52–68.

Cohen, J. and Hickey, T. J. 1987. Parsing and compiling using PROLOG. *ACM Trans. Programming Lang. Syst.*, 9(2):125–163.

Colmerauer, A. 1990. An introduction to PROLOG III. *Commun. ACM*, 33(7).

Cousot, P. and Cousot, R. 1992. Abstract interpretation and applications to logic programs. *J. Logic Programming*, 13(2/3):103–179.

Dincbas, M., Van Hentenryck, P., Simonis, H., Aggoun, A., Graf, T., and Berthier, F. 1988. The constraint logic programming language CHIP, pp. 693–702. In FGCS'88, *Proc. Int. Conf. Fifth Generation Comput. Syst.*, Vol. 1. Tokyo, Japan, December.

Jaffar, J. and Lassez, J.-L. 1987. Constraint logic programming, pp. 111–119. In *Proc. 14th ACM Symp. Principles Programming Lang.*, Munich.

Jaffar, J. and Maher, M. 1994. Constraint logic programming, a survey. *J. Logic Programming*, 503–581.

Jaffar, J., Michaylov, S., and Yap, R. H. C. 1992. The CLP language and system. *ACM Trans. Programming Lang. Syst.*, 14(3):339–395.

Kergommeaux, J. C. and Codognet, P. 1994. Parallel LP systems. *Comput. Surv.*, 26(3).

Kowalski, R. A. 1979. Algorithm = logic + control. *Commun. ACM*, 22(7):424–436.

Kowalski, R. and Kuehner, D. 1970. Resolution with selection function. *Artif. Intell.*, 3(3):227–260.

Lloyd, J. W. 1987. *Foundations of Logic Programming*. Springer-Verlag.

Marriot, K. and Stuckey, P. J. 1998. *Programming with Constraints: An Introduction*. MIT Press, Cambridge, MA.

Minker, J., Ed. 1987. *Foundations of Deductive Databases and Logic Programming*. Morgan Kaufmann.

Muggleton, S. 1991. Inductive Logic Programming. *New Generation Computing*, 8(4):295–318.

O'Keefe, R. A. 1990. *The Craft of PROLOG*. MIT Press, Cambridge, MA.

Older, W. and Vellino, A. 1993. Constraint arithmetic on real intervals. In *Constraint Logic Programming: Selected Research*. F. Benhamou and A. Colmerauer, Eds., MIT Press, Cambridge, MA.

Pesant, G. and Boyer, M. 1994. QUAD-CLP(R): Adding the Power of Quadratic Constraints. In *Principles and Practice of Constraint Programming, Second International Workshop*. A. Borning, Ed., Lecture notes in Computer Science, 1865:40–74. Springer.

Robinson, J. A. 1965. A machine-oriented logic based on the resolution principle. *J. ACM*, 12(1):23–41.

Rossi, F. 2000. Constraint (Logic) Programming: A Survey on Research and Applications. In *New Trends in Constraints: Joint ERCIM/Compulog Net Workshop*. K. Apt et al., Eds., Springer.

Saraswat, V. A. 1993. *Concurrent Constraint Programming Languages*. MIT Press.

Sato, S. and Aiba, A. 1993. An application of CAL to robotics. In *Constraint Logic Programming: Selected Research*. F. Benhamou and A. Colmerauer, Eds., pp. 161–174. MIT Press, Cambridge, MA.

Shapiro, E. 1989. The family of concurrent LP languages. *Comput. Surv.*, 21(3):413–510.

Sterling, L. and Shapiro, E. 1994. *The Art of PROLOG*. MIT Press.

Van Hentenryck, P. 1989. *Constraint Satisfaction in Logic Programming*. Logic programming series, MIT Press, Cambridge, MA.

Van Hentenryck, Ed. 2002. *Principles and Practice of Constraint Programming: Proc. 8th International Conference*, Lecture Notes in Computer Science. Springer.

Van Roy, P. 1994. The wonder years of sequential PROLOG implementation, 1983–1993. *J. Logic Programming*, 19(20):385–441.

Warren, D. H. D. 1983. An Abstract PROLOG Instruction Set. *Tech. Note* 309, SRI International, Menlo Park, CA.

Further Information

There are several journals specializing in LP and CLP. Among them we mention the *Journal of Logic Programming* (North-Holland), *New Generation Computing* (Springer-Verlag), and *Constraint* (Kluwer).

Most of the conference proceedings have been published by MIT Press. Recent proceedings on constraints have been published in the Lecture Notes in Computer Science (LNCS) series published by Springer-Verlag. A newsletter is also available (Logic Programming Newsletter, alp@doc.ic.ac.uk)

Among the references provided, the following relate to CLP languages: PROLOG III [Colmerauer 1990], CLP(R) [Jaffar et al. 1992], CHIP [Dincbas et al. 1988], CAL [Sato and Aiba 1993], finite domains [Van Hentenryck 1989], and Intervals [Older and Vellino 1993]. The recommended textbooks include Clocksin and Mellish [1984] and Sterling and Shapiro [1994]. The theoretical aspects of LP are well covered in Apt [1990] and Lloyd [1987] and implementation in Ait-Kaci [1991], Kergommeaux and Codognet [1994], Van Roy [1994], and Warren [1983].

94

Scripting Languages

Robert E. Noonan
College of William and Mary

William L. Bynum
College of William and Mary

94.1 Introduction **94**-1
94.2 Perl **94**-3
94.3 Tcl/Tk **94**-7
    The Main Program • The `lookup_uids` Procedure
    • The `lookup_by_name` Procedure
    • The `find_uid_from_name` Procedure • Summary
94.4 PHP **94**-14
94.5 Summary **94**-17

94.1 Introduction

According to WebMonkey [10]:

> "A scripting language is a simple programming language used to write an executable list of commands, called a script. A scripting language is a high-level command language that is interpreted rather than compiled, and is translated on the fly rather than first translated entirely. JavaScript, Perl, VBscript, and AppleScript are scripting languages rather than general-purpose programming languages."

The major characteristic of scripting languages is that they often serve as *glue* for connecting existing components or applications together. Scripting languages usually have powerful string processing operations, because text strings are a fairly universal communication medium.

Scripting languages, in the form of job command languages, have existed from the time of the earliest operating systems. However, these early scripting languages lacked variables, conditional statements, and loops. With the advent of Unix [5] in the 1970s, job command languages began to emerge as true scripting languages. Both the early Bourne shell and later C shell had variables and control flow constructs. Later Unix scripting languages included sed and AWK [3]. While early Unix scripting languages had support for variables, conditional statements, and loops, later versions added support for functions, procedures, and parameters.

One characteristic of scripting languages is that they are usually interpreted rather than requiring compilation. For example, Perl is dynamically compiled to byte code and then interpreted; however, there are also compilers for Perl that produce an executable. Most conventional programming languages are compiled to an executable. However, the conventional programming languages Java and C# are compiled to byte code and then interpreted. While Perl does not require compilation, Java and C# do. A more extensive discussion of compilation versus interpretation can be found in Chapter 99.

Another characteristic of scripting languages is that variables need not be declared and are *typeless* (or *dynamically typed*). By this we mean that the programmer does not declare a variable to be of a fixed type (typeless), but rather the type of the variable is allowed to vary according to the type of the value currently assigned (dynamically typed). Thus, values have associated with them a runtime type identification. Typed values also appear in non-scripting languages; for example, Java associates type identifiers with objects and Lisp (a language used in artificial intelligence) does so for all values.

However, most conventional programming languages (including Java) require that each variable be explicitly declared and that a type be associated with the variable. Such languages are said to be *statically typed*, in that the association of a type with a variable is done in the source code. This allows the compiler to check the usage of a variable at compile time to ensure that the operations performed on a variable are consistent with its type. In the development of large programs there is an enormous economic benefit to the early detection and correction of errors.

In contrast, the necessity to declare each variable and associate with it a static type adds statements to a program. In a large program, the cost to write these extra statements is negligible. However, most scripts are relatively small, intended primarily to connect or *glue* applications or components together. Many scripts are only a printed page or two of text. The hallmark of a scripting language is to facilitate directness of expression in the process of getting disparate applications or components to communicate. Brevity of expression and a simple development cycle is preferred for these simple tasks. Hence, scripting languages prefer dynamically typed variables and interpretation.

There is clearly a runtime performance penalty for this flexibility. However, for most scripted applications, most of the time is spent in the underlying components, which are commonly written in conventional programming languages and thus compiled. Offsetting the performance penalty is the dramatic increase in computer power and the subsequent decline in cost per computer instruction executed. A script that might have taken minutes or hours to run a decade ago is often instantaneous on a modern processor. Thus, faster processors are a major factor in enabling the effectiveness of scripting technology.

At the current time, scripting languages and conventional programming languages are complementary in nature. Conventional programming languages are used to develop large components or applications in which efficiency is a major consideration. In contrast, scripting languages are used to develop applications in which the primary task is to provide the ability for existing applications to communicate with each other. In the latter case, either computer efficiency is not a major issue or most of the time is spent in the preexisting applications. Scripting languages promote rapid development because of their use of interpretation and dynamic typing.

Many argue that scripting is a technology whose time has come. Certainly, the use of scripting languages has increased dramatically in the last decade. While conventional programming languages will continue to be used for creating components for some time, the preferred method of integrating components is to use scripting.

A number of developments appear to be driving this trend toward the use of scripting languages. First, the systems task of installing and maintaining applications has increased enormously. Two decades ago, a company would typically have had a single, central computer on which all applications would be run. Commands for installing and updating an application would have been entered manually. Today, such a company has hundreds or thousands of personal computers. The cost of having a systems staff person enter a sequence of commands on each of thousands of computers is unworkable. Instead, scripts are used. Scripting languages that are prototypical of this area include Perl, Python, and Rexx. In Section 94.2, we examine Perl.

Before the widespread adoption of personal computer software, a company deployed applications on its single, central computer that were customized, either by developing the application internally or by acquiring an industry-specific application. With the widespread deployment of personal computers, cost considerations have led to an increased use of commercial, off-the-shelf software. If the company needs to customize the application, an event-driven scripting language such as Tcl/Tk or Visual Basic is often used. Section 94.3 examines a typical usage of Tcl/Tk.

Another major driving force in the increasing use of scripting languages has been the Web. Companies have found great benefit in making their own computer applications more accessible by exposing them to the Web via a scripting language. Also, there has been a demand for simple Web applications, such as putting the company phone book online. More complex applications would include retail shopping via the

Web; for efficiency reasons, these larger, complex Web applications are often programmed in conventional programming languages such as Java.

Originally, many simple Web applications were programmed using system scripting languages such as Perl via the Common Gateway Interface (CGI). More recently, scripting languages specifically designed for Web applications have been developed, including PHP and ColdFusion. In Section 94.4 we examine the language PHP via a simple Web application.

Until fairly recently, scripting languages have often been thought of as niche languages: Perl, Python, or Rexx for systems tasks, Tcl/Tk or Visual Basic for simple GUIs, PHP or ColdFusion for Web pages. However, there is an increasing preference for scripting languages over conventional programming languages even when creating components. One cause of this trend is that many scripting languages now support object-oriented programming. An example is Perl, whose version 5 now supports object-oriented programming. One recent scripting language, namely Ruby, is a pure object-oriented language, much like Smallltalk.

The reason for this trend is clear: economics. Computers are constantly getting faster (and thus, cheaper), while people are getting relatively more expensive. Also, the skill level required to develop small scripts is considerably less than that required to develop large programs in conventional programming languages. Thus, more and more applications will be developed using scripting languages.

94.2 Perl

"Larry Wall ... created Perl when he was trying to produce some reports from a Usenet-news-like hierarchy of files for a bug-reporting system, and *awk* ran out of steam. Larry, being the lazy programmer that he is, decided to over-kill the problem with a general purpose tool that he could use in at least one other place. The result was the first version of Perl." [8]

Although Perl has its roots as a Unix scripting language, it is now widely available for most major computing systems, including Linux, Macintosh, and Windows. In this section we focus on the use of Perl as a typical scripting language for gluing applications together. Such applications include systems administration tasks, string processing, etc. Other scripting languages such as Python, Rexx, and Tcl can be used as well.

The authors themselves have used such scripting languages for:

- Systems administration tasks on a network of Unix computers, including:
  - Setting up new user accounts, including e-mail aliases and creating home directories
  - Running backups
  - Reconfiguring servers
  - Installing software
- Class management, including:
  - Converting electronic class rolls (off a mainframe computer) to a more convenient form
  - Managing an e-mail alias for each class
  - Grading support
- Replacing programs previously written in conventional programming languages such as C and Pascal. Such programs fall into a variety of subject domains, including simple utility programs, computer science research, etc.
- Web software (using the CGI interface), including:
  - Running the program portion of a national computer conference
  - Helping to manage the authors' home Web site
  - Running an annual survey

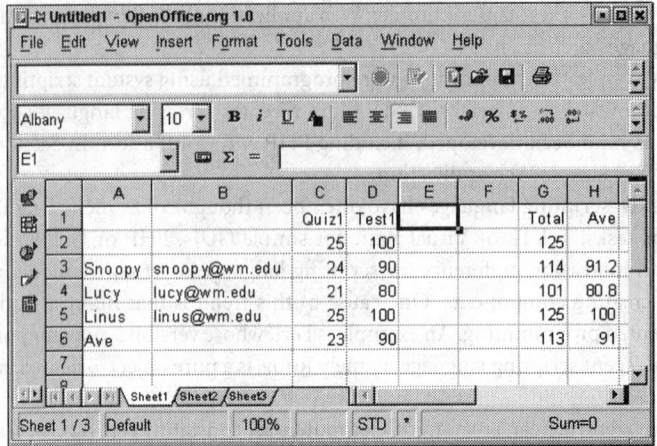

FIGURE 94.1   Spreadsheet Containing Grade Information.

As a typical systems (or utility) usage, we present one of our classroom management scripts. The problem is to maintain a grade book consisting of numeric grades earned by each student on each assignment, quiz, test, and exam. Each grade may have different weights; for example, an exam may be more heavily weighted than a quiz. We would also like to provide a secure way for students to electronically access their grade records.

Specialized Web servers built for schools and universities, such as Blackboard and WebCt, provide such facilities through Web site login. However, the system described here evolved before the use of the Web became widespread. It relies instead on the use of e-mail to send grades to each student after each assignment or test is recorded.

We use a spreadsheet to do the grade management itself, one spreadsheet per class. An example spreadsheet is given in Figure 94.1. In this spreadsheet, column one is used for student names and column two is used for e-mail addresses. The rightmost two columns are weighted totals and averages. The intervening columns are used for grades on assignments and tests; at any point in time, any number of grade columns may be blank (empty).

Similarly, a perusal of the rows reveals that the first row is used for a descriptive title for each grade, and the second row for the maximum points for each grade. The last row is used to compute class averages for each assignment or test. The intervening rows contain the students records, one row per student.

The problem is to move the grade values from the spreadsheet to the e-mail system. The process consists of exporting the data in textual form from the spreadsheet and then using a Perl script to generate the e-mail. The spreadsheet output form used is comma-separated values, but using a colon as a field separator (because student names often have commas in them). We have found over the years that each spreadsheet application has its own, often unique definition of the format of comma-separated values. For our current spreadsheet the third row would appear as:

```
"Snoopy":"snoopy@wm.edu":24:90:::114:91.2
```

In the remainder of this section, we describe a Perl script to solve this problem as a means of examining some of the features of the Perl language. Familiarity with programming, particularly the C language, is assumed. Finally, we summarize some of the features of Perl.

In Unix/Linux, a script may be marked as being executable, in which case the first line of the Perl script is a comment that gives the full path name of the Perl interpreter:

```
#! /usr/bin/perl
```

The sharp sign is the comment symbol, with comments terminating at the end of line. The exclamation mark is required by Unix/Linux to indicate that this comment is giving the path of the Perl interpreter.

In Windows or Mac OS (prior to Mac OS X), the Perl interpreter would be explicitly invoked and passed the script as an argument.

Next we set any global constants that are likely to change. For this script, there is only one, namely, the field separator character; currently, we use a colon for the field separator because student names often contain commas:

```
$sep = ":";
```

Perl variables that take on a scalar value such as a number or string must be prefixed with a dollar sign. Perl statements end with a semicolon. Note that the scalar variable $sep is not declared and has the type of its current value.

Next we read the header row and the max values row, do some processing of each, split each row into fields (original columns), and store each field into an array. Because the processing of each line is identical, we use a subroutine:

```
@title = &readSplit();
@max = &readSplit();
push(@max, "100");
```

In Perl, array variables are prefixed with an at symbol and subroutine calls with an ampersand. The third line above adds the string value denoting 100% to the end of the maximum grade value array.

Although it physically appears at the end of the script, we next examine the readSplit subroutine:

```
sub readSplit {
    $_ = <>;
    chomp;
    tr /"//d;
    return split(/$sep/);
}
```

The first line of the subroutine reads a line from the file STDIN (file handles are enclosed in the less than–greater than symbols and STDIN is the default file handle for input) and assigns it to the default scalar named $_. By default; the first command line argument to the script is opened using the file handle STDIN. The second line deletes the end of line character or characters from the line; because no subject or target is specified, the default scalar is used. The third line deletes all double quotes from the line. Finally, the split operator is used to split the fields of the default scalar into an array, which is then returned.

This subroutine highlights Perl's cryptic style, including the implicit use of the default scalar in the second through last lines of the subroutine. Similarly, the parentheses are unnecessary in the invocation of split; we tend to use parentheses even when unnecessary, as we believe it improves the readability of the script. While some programmers love Perl's brevity of expression, others feel the need for extra syntactic sugar to improve readability. One of the hallmarks of Perl as a language is its many alternative ways of doing the same thing.

Next, we get to the portion of the script that reads and processes each student. At first glance, it would appear that the e-mail message could be generated as each student record is read. Unfortunately, the class average for each grade is in the last line of input. So the line for each student is read, the end of line deleted, double quotes deleted, and the line stored in the student array, which is neither declared nor initialized. The loop terminates when the end of file is reached on STDIN:

```
while (<>) {
    chomp;
    tr /"//d;
    push(@student, $_);
}
@ave = split(/$sep/, pop(@student));
```

The last line removes the class averages from the last line of the student array and splits the fields (columns) into an average array, one field per array position.

Finally, we traverse the student array generating an e-mail for each student using Perl's foreach loop:

```
foreach (@student) {
    &sendMail(split(/$sep/));
}
exit;
```

Because no iteration variable is explicitly stated for the loop, the default scalar variable is used. The exit is unnecessary, but it serves to visually separate Perl's main program from its subroutines.

All that remains is to examine the subroutine that actually sends the e-mail message to each student:

```
sub sendMail {
    $name = shift;
    $email = shift;
    open(MAIL, "| mail -s Grades $email")
        || die "Cannot fork mail: $!\n";
    print MAIL "Re: Grades for $name\n\n";
    print MAIL "GRADE\t\tYOUR\tMAX\tCLASS\n";
    print MAIL "NAME\t\tSCORE\tSCORE\tAVE\n\n";

    $i = 1;
    foreach (@_) {
        $i++;
        next unless $title[$i];
        print MAIL "$title[$i]\t";
        print MAIL "\t" if length($title[$i]) < 8;
        if (/^\d/) { print MAIL int($_ + 0.5); }
        else { print MAIL $_; }
        print MAIL "\t$max[$i]\t";
        print MAIL int($ave[$i] + 0.5), "\n";
    }
    close(MAIL);
}
```

Perl does not declare formal parameters for its subroutines. Instead, actual arguments are passed through the default array @_. So the first two lines shift off the array the first two values passed, namely, the student's name and e-mail address.

The third line opens the file handle MAIL for output and pipes it to the Unix/Linux mail command. If the open fails, then the program dies with a descriptive message. The next three lines write some preliminary information that is part of the e-mail message, including column headers. The final line of the subroutine closes the file MAIL.

The foreach loop writes the grade information to the e-mail message, one grade per iteration. The first actual grade corresponds to the third column of the spreadsheet; like C, Perl uses zero-based arrays, so the first index used for the title, max, and ave arrays is two. Note that array references (e.g., line 2 of the loop) are made using C's bracket notation, but the subscripted array reference is treated as a scalar, and, hence starts with a dollar sign rather than an at sign. This is a common point of confusion for Perl novices.

The second line of the foreach skips the remainder for the loop if there is no grade title, implying that the grade column is empty. The fifth and the last lines of the loop convert numbers with decimal places to rounded whole numbers; the test used in the if in the fifth line is a pattern that tests whether the value starts with a digit.

We have presented a typical glue program in order to discuss some of the salient features of Perl. In roughly 36 lines of Perl (not counting blank lines, comment lines, and closing brace lines), we have presented a script that takes a spreadsheet containing student grade information and e-mails each student his or her grades, together with individual, assignment, and class averages. Roughly half the lines of code are devoted to generating the information in the e-mail message itself.

This typical glue script linking two disparate applications exposes many commonly used features of Perl:

- Perl supports a wide variety of alternative ways of coding the same basic idea. This makes Perl a more difficult language to learn. It also makes it more difficult to read Perl code, because the code can use unfamiliar features.

- Perl does not require declaration of variables and supports dynamic typing. The same value can be treated either as a string on as a number (provided it can be interpreted as a valid number).

- Perl supports a wide variety of string operations. It also supports pattern matching as found in the Unix utilities grep and sed.

- Many Unix utilities such as tr are included as Perl operators.

- Convenient access is provided for executing system utilities, either providing its input or capturing its output.

- Perl provides both dynamically sized arrays and associative arrays (hash tables).

Despite some of the criticisms above, Perl is one of the most widely used scripting languages for non-GUI applications. In the next section, we explore scripting GUI applications.

94.3 Tcl/Tk

In the late 1980s, John Ousterhout and his students found themselves repeatedly developing interactive tools for integrated circuit design. This activity led Ousterhout to perceive the need for a re-usable scripting language built as an extensible C library package, and he developed the Tcl language as a result. The introduction of HyperCard by Apple Computer in 1987 provided the impetus for Ousterhout to develop a component-based graphical toolkit, Tk, based on Tcl and the X11 windowing system.

Tcl/Tk [6] has grown into a scripting language framework within which graphical user interfaces can be easily developed. Perhaps the most widely used Tcl/Tk application is the exmh mail user interface that is included in most Linux distributions, written by Brent Welsh (a Ph.D. student of Ousterhout's). The authors of this article have used Tcl/Tk to produce X11 GUIs in class management tools for configuring course submission software [2] and in the debugging tool for the BACI mutual exclusion toolbox [1].

In the remainder of this section we present a GUI program derived from a more extensive one developed for configuring a student project submission system [2]. A class roll supplied by the Registrar contains the student's first and last names, but does not contain the student's login name for the Computer Science LAN. The program discussed here searches the system login name file for a first and/or last name fragment and lists the login name and full name fields of all entries that match the name fragments. For example, Figure 94.2 shows the result of a (case-insensitive) search for the first name string bill. Figure 94.3 shows the result of a subsequent search for the last name noonan.

On many UNIX systems, user information is kept in a system text file named /etc/passwd. A typical line of this file has the form:

```
bynum:abcdefghij:34:20:Bill Bynum:/home/f85/bynum:/bin/tcsh
```

The fields of each line of /etc/passwd are colon separated, as in the spreadsheet example discussed in Section 94.2. The contents of these fields are login name (user id), encrypted password, user id number, group id number, user's full name, login directory, and default shell.

In this application a search is initiated by clicking the Find button. The results pane is cleared by clicking the Clear button. The program is terminated by clicking the Quit button.

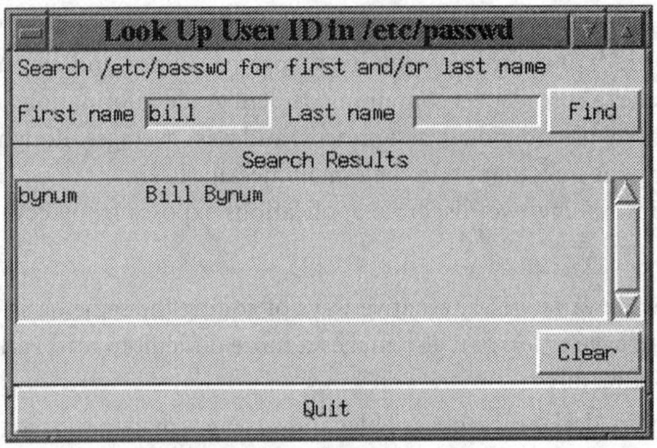

FIGURE 94.2   Searching for first name bill in /etc/passwd.

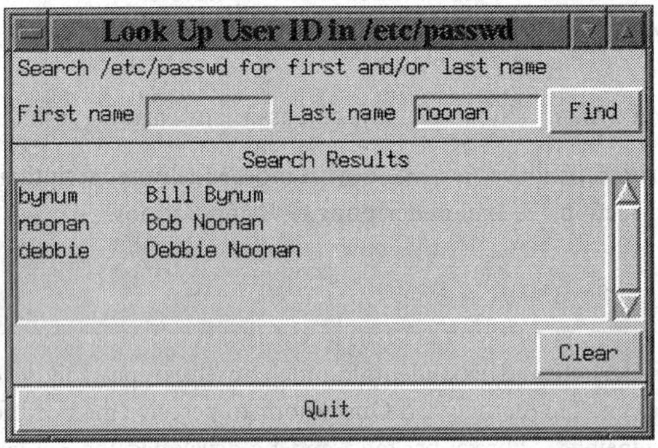

FIGURE 94.3   Searching for last name noonan in /etc/passwd.

To illustrate the capabilities offered by Tcl/Tk for handling keyboard events, this program binds the Tab keypress event so that the keyboard-oriented user can use the Tab key and the space bar to interact with the program, rather than having to alternate between the keyboard and mouse. When the program begins, the first name entry field is the focus of the window. When the user presses the Tab key, the focus shifts to the last name entry field. A subsequent press of the Tab key shifts the focus to the Find button. Another Tab press shifts the focus to the Clear button. Subsequent Tab keypresses shift the focus to the Quit button and then back to the first name entry field. The user can cause a button press from the keyboard by pressing the space bar when the window focus is on the button.

In the subsections that follow, we present each of the parts of the program, starting with the main program. We conclude by comparing Tcl/Tk with Perl.

94.3.1   The Main Program

The main program is quite modest, consisting of only five lines:

```
#!/usr/bin/wish
option add *Font "-misc-fixed-medium-r-normal-*-14-*-*-*-*-*-*-*"
wm withdraw .
lookup_uids
exit
```

Like the Perl script in Section 94.2, the first line gives the path to the interpreter to be used, in this case the Tk WIndowing SHell. The second line establishes the font to be used in the window created by the application. In the absence of any action by the user, the windowing shell puts up a default window for the application to use. The third line instructs the user's window manager not to show this default window, because the parameterless lookup_uids procedure called in the fourth line will create its own window. The exit statement in the fifth line terminates the program and the windowing shell. As with Perl, the sharp sign is the comment symbol, with comments terminating at the end of line.

94.3.2 The lookup_uids Procedure

The lookup_uids procedure is mainly responsible for laying out the widgets seen in Figures 94.2 and 94.3. Unlike Perl, procedures in Tcl/Tk use formal parameter lists much as in C. The lookup_uids procedure has no parameters, so its header has an empty list of formal parameters. The rightmost left brace begins the source block for the procedure and its matching right brace ends the procedure:

```
proc lookup_uids { } {
    toplevel .luu
    wm title .luu "Look Up User ID in /etc/passwd"
    wm geometry .luu +250+50
    # insert code discussed below
}   ;# lookup_uids
```

First, the procedure describes the position of the window in the window hierarchy ("toplevel") and gives the title of the window and its geometry.

Next, the procedure builds the window used by the application from top to bottom in horizontal strips, using the default Tcl/Tk geometry manager called the "packer." Tcl/Tk frames are used for the more complicated horizontal strips that contain several widgets.

The first widget placed in the window is an explanatory label:

```
label .luu.l1 -text "Search /etc/passwd for first and/or last name"
pack .luu.l1 -side top -anchor w -padx 1
```

Notice that the label is constructed with the label statement, and then the label is packed into the .luu window on the "top" side of the window and is anchored on the left (or "west") side (that is, is left-justified in the window). The padx option of the pack statement adds a 1-pixel horizontal (x) pad on each side of the label in the window.

The first frame of the window .luu.f1 holds the properly labeled entry widgets for the first and last names to be searched for, plus the Find button. First, the frame must be declared; then widgets are added to the frame from left to right, starting with the label for the first name entry widget and the entry widget.

```
# frame .luu.f1 holds the first & last name and the Find button
frame .luu.f1
label .luu.f1.l1 -text "First name"
pack .luu.f1.l1 -side left -in .luu.f1 -anchor w -padx 1 -pady 3
entry .luu.f1.e1 -relief sunken -width 10 -textvariable lfname
pack .luu.f1.e1 -in .luu.f1 -side left -padx 1 -pady 3
```

The label command labels the entry field for the first name. In the following pack command, the -in option tells the packer to place the label in the .luu.f1 frame. The entry command creates a user-modifiable text field where the user can enter the first name of the person whose login name is to be searched for. The -textvariable lfname option of this command creates a global variable lfname, which holds the text the user enters. The pack command is required to pack the first name entry field in the frame to the right of the label.

The label and entry for the last name field are created and packed into the frame in a similar manner:

```
label .luu.f1.12 -text "Last name"
pack .luu.f1.12 -side left -in .luu.f1 -anchor w -padx 5 -pady 3
entry .luu.f1.e2 -relief sunken -width 10 -textvariable llname
pack .luu.f1.e2 -side left -in .luu.f1 -padx 1 -pady 3
```

The .luu.f1 frame is completed with the addition of the Find button:

```
button .luu.f1.b -text Find -command {lookup_by_name}
pack .luu.f1.b -side left -in .luu.f1 -padx 1 -pady 3
pack .luu.f1 -side top -fill x -expand true
```

Construction of the button widget is accomplished with the button command. The -text Find option specifies the text that should appear on the button. The -command option specifies the procedure that should be called when the button is clicked, lookup_by_name in this case, which will be discussed below.

The scrollable pane where the search results are displayed is created in the .luu.f2 frame. Construction is started by declaring the frame and then packing a "Search Results" label to identify the frame to the user.

```
# frame .luu.f2 holds the scrollable results window
frame .luu.f2
label .luu.f2.1 -text "Search Results"
pack .luu.f2.1 -in .luu.f2 -side top
```

The label is centered in the window, because no other widgets are packed with it and no -anchor is used.

The scrollable area is created with a combination of the listbox and scrollbar commands to create and interlink the listbox and scrollbar widgets.

```
listbox .luu.f2.lb -width 40 -height 5 -yscrollcommand .luu.f2.sy set
pack .luu.f2.lb -side left -fill both -expand true -in .luu.f2
scrollbar .luu.f2.sy -orient vert -command .luu.f2.lb yview
pack .luu.f2.sy -side right -fill y -in .luu.f2
pack .luu.f2 -side top -fill both -expand true
```

A listbox widget displays a collection of strings, allowing the user to scroll the list. The -yscrollcommand option identifies the scrollbar widget that will scroll the listbox widget. Whenever the view in the listbox changes, the listbox calls the .luu.f2.sy set command with the four numbers: the total number of entries currently in the listbox, the number of entries that fit in the window at any one time, and the indices of the entries currently visible in the window.

The next frame holds the Clear button:

```
# frame .luu.f3 holds the Clear button
frame .luu.f3
button .luu.f3.b -text Clear -command { .luu.f2.lb delete 0 end; \
    if [info exists puid] { unset puid }; \
    if [info exists pname] { unset pname } }
pack .luu.f3.b -in .luu.f3 -side right -anchor e -padx 1 -pady 3
pack .luu.f3 -side top  -expand true -fill x
```

The -command option of the button command illustrates that the command activated by clicking a button can be given directly by code snippet, instead of a procedure call, as we saw in the creation of the Find button above. The .luu.f2.lb delete 0 end; command deletes all text from the beginning of the .luu.f2.lb listbox to the end. The backslash at the end of the line is the Tk line continuation character.

The if [info exists puid] { unset puid } command is an if statement where the variable puid is checked for existence with the built-in info function. The set command is the assignment operator, and the unset command is its inverse. These two if statements ensure that the puid and pname global variables are cleared.

The last widget to be added to the window is the Quit button:

```
button .luu.qb -text "Quit"  -command { destroy .luu }
pack .luu.qb -side top -fill x -expand true
```

The Quit button spans the window because of the -fill x option in the pack statement.

Finally, we add a series of bindings of the window widgets to Tab keypress events so that the user can cycle through the widgets of the window by pressing the Tab key.

```
bind .luu.f1.e1 <Tab> \
    { focus .luu.f1.e2; .luu.f1.e2 select range 0 end; break }
    # select what's in the .luu.f1.e2 field so user can change it
bind .luu.f1.e2 <Tab> { focus .luu.f1.b; break }
bind .luu.f1.b <Tab> { focus .luu.f3.b; break }
bind .luu.f3.b <Tab> { focus .luu.qb; break }
bind .luu.qb <Tab> \
    { focus .luu.f1.e1; .luu.f1.e1 select range 0 end; break }
```

The first bind command specifies that when the cursor is in the .luu.f1.e1 first name entry widget and a Tab keypress event occurs, then the window focus should shift to the .luu.f1.e2 last name entry widget and the text in that widget from first to last character should be selected, allowing the user to edit the existing text in the widget. The break command terminates the current binding and suppresses bindings from any remaining widgets in the binding list. The bind .luu.f1.b is the only statement that really needs a break statement. If this break statement were missing, the focus would shift to the next element of the window, the first line of the .luu.lb listbox instead of jumping over the intermediate widgets to the Clear button.

The lookup_uids procedure ends with the statements:

```
    # start with the first name entry field
    focus .luu.f1.e1
    # wait here for the window .luu to be destroyed
    tkwait window .luu
```

The focus command sets the window focus in the first name entry field, where the user can enter the student's first name. The tkwait statement causes the procedure's thread of execution to wait at this point until the window is destroyed. If the statement were not present, the procedure would probably terminate before (or soon after) the window that it creates is displayed. Recall that the command action of the Quit button was also to destroy the window. When the user clicks the Quit button, the .luu window disappears, the lookup_uids procedure returns to the exit statement in the main program, and the program terminates.

94.3.3 The lookup_by_name Procedure

The purpose of lookup_by_name procedure is to display any matches found in the scrollable text widget. The procedure is parameterless, using the global variables lfname, llname, pname, and puid for its input parameters. It is unnecessary to declare local variables that are used in a procedure; by default, all variables used in a procedure are local, unless they are explicitly declared as global.

```
proc lookup_by_name { } {
    global lfname
    global llname
    global pname
    global puid
```

```
        set ret [find_uid_from_name $lfname $llname]
        # set and if statements -- see below
    }   ;# lookup_by_name
```

The set command is the assignment statement. In this case, the actual work of looking up the first name and last name strings is performed by the find_uid_from_name procedure with the two parameters $lfname and $llname. Like Perl, the $ symbol indicates the current values of the variables. The square brackets signify a function call. The result of the call is stored in the ret variable.

The remainder of the procedure consists of an if statement to deal with the different possible values of the ret value returned by the call:

```
if {$ret == 0} {
    puts "firstname: \"$lfname\" lastname: \"$llname\" was not found"
} elseif {$ret >= 1} {
    for {set i 0 } { $i < $ret} {incr i} {
        .luu.f2.lb insert end [format "%-10s %-20s" $puid($i) $pname($i)]
        .luu.f2.lb see end
    }
} else {      ;# should never happen
    puts "Weird return: $ret"
}
```

In case the value returned is zero, the message that the first name/last name combination was not found in is written with a puts statement to standard output; in the original application, this message is displayed in a pop-up window. The use of puts considerably simplifies this example.

The results of a successful search in find_uid_from_name call are passed back to this procedure through the puid and pname variables. These variables are actually Tcl/Tk associative arrays, indexed by the strings 0, 1, and so on. The syntax of the Tcl/Tk for loop is quite similar to the syntax of the C for loop. The {set i 0} sets the loop variable to zero at the beginning of the loop. The {$i < $ret} is the continuation test of the loop, and the {incr i} instruction is the action performed each time the body of the loop is executed; in this case, the action consists of incrementing the loop variable by one. The .luu.f2.lb see end makes sure that the listbox keeps the last line of the listbox visible.

The find_uid_from_name procedure never returns a negative value, so the final else should never be executed. Its inclusion here is strictly a good defensive strategy.

94.3.4 The **find_uid_from_name** Procedure

The find_uid_from_name procedure performs all the non-GUI real work of searching the login name file for matches:

```
proc find_uid_from_name { fname lname } {
    global puid
    global pname
    if [info exists puid] { unset puid }    ;# clear global arrays
    if [info exists pname] { unset pname }
    if [catch [list exec grep -i "$fname $lname" /etc/passwd] pwinfo ] {
        return 0
    } else { ;# got a hit
        # split result of grep into lines
        set lnamelist [split $pwinfo "\n"]
        # first look through to see if we get an exact first/last name
        # match.  If so, then return the uid and first/last name
```

```
        # in puid(0) and pname(0)
        for {set i 0} { $i < [llength $lnamelist] } { incr i} {
            set pwentry [lindex $lnamelist $i]
            set pwline [split $pwentry : ]
            set tuid($i) [ lindex $pwline 0 ]
            set tname($i) [lindex $pwline 4]
            if { "$fname" == "[lindex $tname($i) 0]" } {
                if { "$lname" == "[lindex $tname($i) 1]" } {
                    set puid(0) $tuid($i)
                    set pname(0) $tname($i)
                    return 1
                }
            } ;# if exact match
        } ;# for
        # return all hits in the puid and pname arrays
        set lastuid $i
        for {set i 0} { $i < $lastuid } { incr i } {
            set puid($i) $tuid($i)
            set pname($i) $tname($i)
        }
        return  $lastuid
    } ;# end else got a hit
} ;# find_uid_from_name
```

The procedure first globalizes the puid and pname arrays that will be used to return the search results to its caller. The first two if statements clear the two arrays with the unset command.

The catch statement is the exception-handling statement. It has two arguments. The first argument is a **list** (a blank-separated collection of words) for the Tcl/Tk interpreter to execute and the second argument is a variable into which the result of the execution is placed. If there is an error in executing the first argument, then the catch statement returns **true** (or 1) and the second argument contains the error message from the execution; otherwise, catch returns 0 and leaves the result of the execution in the second argument. In this function, the catch clause simply returns 0 to the procedure's caller if there is an error.

On the other hand, if the execution of the case-insensitive grep of /etc/passwd is successful, then the thread of execution moves to the else clause. The lnamelist list variable is created by splitting the string returned by the grep on the newline character (\n). The first for loop looks through the lnamelist list of lines for a first name/last name match. The llength function returns the length of a list. The set pwentry statement stores the i-th element of the lnamelist list in the pwentry variable. The set pwline creates the pwline line from /etc/passwd on the colon symbol. The i-th entry of the tuid array is obtained from the 0-th entry of the pwline list, and the first/last name list is stored into the i-th entry of the tname array from the 4-th entry of the pwline list.

On the other hand, if the first for loop terminates without finding such a match, then the second for loop transfers the matches that were found to the puid and pname arrays and returns the number of matches found.

94.3.5 Summary

In this section we have presented a typical GUI-based event-driven application in the scripting language Tcl/TK in order to demonstrate features of the language. Of the approximately 80 lines of code, approximately 70% are devoted to various aspects of the GUI. In the actual application, the percentage is higher due to the use of pop-up windows and the omission of some features.

We note both similarities and differences when comparing Tcl/Tk to Perl:

- Unlike Perl, Tcl/Tk does not offer the programmer a wide variety of different ways of coding the same construct. Although the programming functionality offered by Tcl/Tk is comparable to that of Perl, Tcl/Tk usually offers only one way to code each construct.
- Like Perl, Tcl/Tk does not require declaration of variables and supports dynamic typing.
- Tcl/Tk, like Perl, supports a wide variety of string operations, although some programmers feel that the Tcl/Tk string operations are more awkward to use.
- Like most scripting languages, a convenient method is provided to execute system utilities, supplying input and capturing their output.
- Tcl/Tk, like Perl, provides dynamically sized arrays and associative arrays.
- Unlike Perl, Tcl/Tk subroutines use formal parameters to declare the arguments in the subroutine source. Also the actual parameters are passed explicitly in the call, rather than passing the parameters through a default array.
- Although not explicitly event-driven, both Perl and Python provide an interface to the Tk toolkit in order to support the development of GUI-based applications.

In the next section, we explore a language explicitly developed for supporting server-side Web applications.

94.4 PHP

According to its developer, Rasmus Lerdorf, the motivation for developing PHP [4] was the following:

"As the Web caught on, the number of non-coders creating Web content grew exponentially. . . . But soon they were asked to add dynamic content to their sites. . . . This is where PHP found its niche. . . . I had written all sorts of CGI [Common Gateway Interface] programs in C, and found that I was writing the same code over and over. What I needed was a simple wrapper that would enable me to separate the HTML portion of my CGI scripts from my C code . . . This concept became PHP."

PHP was initially developed by in 1994, but within a few years usage grew beyond the abilities of a single developer, so it became an open-source product. PHP is a server-side scripting language intended as an alternative to using CGI programming. As such, PHP is intended for Web pages with dynamic content, including forms processing and database access.

At our site PHP is installed on the university's main Web server. Typical PHP usage includes:

- Dynamic content such as including an image or news item of the day.
- Forms processing, including forms validation.
- Database access, using several distinct databases.

The PHP processor takes a document file as input and produces HTML as output. Like JavaScript, the input file consists of a mixture of HTML and PHP script code, which is marked by special HTML-like tags. The major difference from JavaScript is that in the case of PHP the script is executed on the server, not on the client. This provides a level of security that is unachievable with JavaScript, since the PHP script is never downloaded to the Web browser. This is particularly important in the case of database access and updating.

The PHP processor operate in two modes of operation. It begins in copy mode in which HTML tags and text directly to the output. When it encounters the special tags:

```
<?
    # one or more lines of PHP script
?>
```

the PHP processor switches to script mode, which interprets the script. The output from the script, whatever is written to STDOUT, replaces the script in the resulting HTML page. As with Perl, the hash mark is used to denote a comment, which continues until the end of the line.

As in the two previous sections, we will use a single example as a vehicle for exploring the features of PHP. Familiarity with both HTML tags and C programming is presumed. In this example, we examine the use of a PHP script in conjunction with a database to produce department directories, one each for the faculty, staff, and graduate teaching assistants. The directory desired is specified as a parameter in the URL; for example,

```
http://www.cs.wm.edu/people/index.php?id=Faculty
```

The script then accesses the appropriate database table, in this case the faculty table, and generates the appropriate HTML output.

Because these directories are fairly static, the question arises: why not maintain the information as static HTML pages. One answer is that in the current setup the staff people who maintain this information only deal with a form that interfaces to the database; they need not know or care about HTML. Second, using PHP allows the Webmaster to more easily maintain a consistent look and feel to these Web pages. Third, the information is used in other portions of the Web site.

A PHP page begins in HTML mode; this would be used to set up the page in a site-specific standard format, including the title, background color, navigation buttons, etc. Because HTML lacks an include facility, a common use of PHP is to set up the page via parameterized header and trailer files. In this case, we begin by including a header:

```
<?   $title = $id;
     include "header.inc";
?>
```

The header expects a `title` variable to be set. In this case, the value comes from the variable whose name is identical to the parameter in the URL.

Before we begin setting up the body of the page, we want to check the `id` parameter for a valid value. Because this value is used to access a database table, a malicious user could attempt to attack the database by providing an unexpected value:

```
<?   if (!eregi(":Faculty:Staff:GradTA:", ":$id:")
          errorPage($id);
?>
```

In this case, we test the value supplied for the `id` parameter using a pattern match against the three legal values. If the pattern match fails, an error routine is called, which generates an error page and exits with no further processing.

Otherwise, database access code is included, which sets the name of the database, security information including userid and password, etc.:

```
<?   include "dbaccess.inc";
     $result = mysql_query("select * from $id");
?>
```

The `select` SQL statement is shown, which in this example accesses the faculty directory in the database.

At the current time, each of these three database tables contains the following information for each person:

- The person's name
- A URL, if they have a home page
- Their office address
- Their phone number
- Their e-mail address

The information is to be generated as an HTML table with one row per person. Each column should have an appropriate heading. So the next part of the script is pure HTML code:

```
<center>
<table cellspacing=5 cellpadding=5 border=2>
<tr align=left><th>Name</th><th>Office</th>
    <th>Phone</th><th>Email</th></tr>
```

Next comes the main logic of the script. A while loop is used to fetch one row of the result of the SQL query representing one person. That result is returned as an array, so the list function is used to assign the array values to conveniently named variables. The body of the loop consists mostly of print statements to write the appropriate columns:

```
<? while (list($name,$url,$office,$phone,$email)
         =mysql_fetch_array($result)) {
    print "<tr align=left>\n";
    $lname = $name;
    if ($url != "")
        $lname = "<a href=$url>$name</a>";
    print "<td>$lname</td>\n";
    print "<td>$office</td>\n";
    print "<td>$phone</td>\n";
    print "<td><a href=mailto:$email>$email</a></td>\n";
    print "</tr>\n";
}
?>
```

As with Perl, variable references may be freely embedded in double-quoted strings. The name field is made into a link if the person has a non-empty URL field.

All that remains is to close the table and center HTML tags and invoke the standard Web site trailer:

```
</table>
</center>
<? include "trailer.inc"; ?>
```

This script is a typical example of server-side scripting and exposes commonly used features:

- PHP scripts freely alternate between pure HTML and PHP.
- PHP does not require the declaration of variables and supports dynamic typing. The same value can be treated both as a string and as a number (provided it can be interpreted as a valid number).
- PHP supports a wide variety of string operations. It also supports pattern matching as found in the Unix utilities grep and sed.
- PHP provides both dynamically sized arrays and associative arrays (hash tables).
- PHP directly supports accessing information from a database.
- Security is a major concern in Web scripting.

PHP and Macromedia's ColdFusion are widely used for producing dynamic Web content. Their ability to freely switch between HTML coding and scripting makes them very useful for Web scripting.

94.5   Summary

In the past 15 years, scripting languages have emerged as more than merely enhanced job command languages. In this chapter we have explored the usage of scripting for gluing components together, for building simple GUI interfaces, and for developing Web applications. The phenomenal increase in computer power has enabled the use of scripting for developing relatively small applications. We expect the trend toward using scripting languages over conventional programming languages for application development to continue.

Defining Terms

Common Gateway Interface (CGI): CGI is essentially what a Web server must provide in order to allow an external script or program to create Web pages:

- An environment containing server information, including the HTTP request
- An input file containing form data if the request used the post method
- An output file for the CGI program to write its response; this output is returned as the resulting Web page

Dynamically typed: In a dynamically typed language, each value is typed; a variable has the type of its current value. *See also* typeless.

Script: A program written in a scripting language.

Scripting: The act of writing a program in a scripting language.

Statically typed: In a statically typed language each variable be explicitly declared and that a type be associated with the variable. This allows the compiler to check the usage of a variable at compile time to ensure that the operations performed on a variable are consistent with its type.

Typeless: From a programmer's perspective, the variables in a dynamically typed language appear to be typeless in that there types are not declared.

References

[1]   Bynum, Bill and Tracy Camp. After you, Alfonse: a mutual exclusion toolkit. *Proceedings of the 28th SIGCSE Technical Symposium on Computer Science Education*, 1996, pp. 170–174.

[2]   Bynum, William L., Robert E. Noonan, and Richard H. Prosl. Using a project submission tool across the curriculum. *The Journal of Computing in Small Colleges*, Vol. 15, No. 5, 2000, pp. 96–104.

[3]   Dougherty, Dale and Arnold Robbins. *sed & awk*. O'Reilly, 1990.

[4]   Hughes, Sterling. *PHP Developer's Cookbook*. SAMS, 2001.

[5]   Kernighan, Brian W. and Rob Pike. *The UNIX Programming Environment*. Prentice Hall, 1984.

[6]   Osterhout, John K. *Tcl and the Tk Toolkit*. Addison-Wesley, 1994.

[7]   Osterhout, John K. Scripting: higher level programming for the 21st century. *IEEE Computer*, 31, 3 (March 1998), pp. 23–30.

[8]   Schwartz, Randal L. *Learning Perl*. O'Reilly & Associates, 1993.

[9]   Wall, Larry, Tom Christiansen, and Randall L. Schwartz. *Programming Perl*, 2nd edition. O'Reilly, 1996.

[10]   Web Monkey, http://www.webmonkey.com.

Further Information

The computer section of your favorite bookstore is a good place to find books on the more popular scripting languages. Online bookstores such as amazon.com or buy.com contain more extensive collections but lack the ability to peruse books of interest.

There is also a great deal of information available on the Web. For perusing information in general, we recommend Yahoo. For general searching about a specific scripting language, we recommend Google.

The standard reference book on Perl is [9], although some programmers find it a bit overwhelming. Many more gentle introductions to Perl are available, including [8]. To obtain more information on Perl or to obtain a Perl distribution, see the Web site www.perl.org. The online comprehensive Perl archive for contributed programs and packages is at www.cpan.org.

The standard reference book on Tcl/Tk is [6], although some programmers find it a bit overwhelming. The online Tcl developer's exchange is at www.tcl.tk.

There are many good books on PHP. The official PHP Web site is at www.php.net. There is an excellent online tutorial at www.phpbuilder.com.

Finally, an excellent online resource for open source scripting languages is www.devshed.com.

95

Event-Driven Programming

95.1 Foundations: The Event Model 95-2
95.2 The Event-Driven Programming Paradigm 95-2
95.3 Applets 95-4
95.4 Event Handling 95-6
　　Mouse Clicks • Mouse Motion • Buttons • Labels, TextAreas, and TextFields • Choices
95.5 Example: A Simple GUI Interface 95-11
95.6 Event-Driven Applications 95-18
　　Interactive Games: Tic-Tac-Toe • Automated Teller Machine • Home Security Alarm System

Allen B. Tucker
Bowdoin College

Robert E. Noonan
College of William and Mary

The *event-driven* programming paradigm turns the fundamental model of computation inside out, in that event-driven programs do not predict the control sequence that will occur. Instead, they are written in a way that the program reacts reasonably to any particular sequence of events that may occur once execution begins. In this way, the input data govern the particular sequence of control that is actually carried out by the program. Moreover, execution of an event-driven program does not typically terminate; such a program is designed to run for an arbitrary period of time, often indefinitely.

The most widespread example of an event-driven program is the mouse- and windows-driven graphical user interface (GUI) found on most desktop and laptop computers in use today. Event-driven programs also drive Web-based applications. For example, an on-line student registration system must be prepared to interact with a student no matter what her next action is: adding a course, dropping a course, determining the classroom where a course meets, and so forth. An on-line airline reservation system, similarly, must be prepared to respond to various sequences of user events, like changing the date of travel, the destination city, or the seating preference.

Although the event-driven programming paradigm has been in use much longer than the Web, it has only recently become prominent in the eyes of programmers because of the Web. Before the Web (if we can imagine such a time!), event-driven programs were found embedded in a variety of vehicles and devices, such as airplanes and home security systems. In these environments, the events that trigger programmed responses include a change in direction, wind speed, or temperature; by their nature, these events also do not occur in any predictable order.

To provide effective support for event-driven programming, some languages have developed some basic terminology and principles of design. Most recently, these principles have appeared in Java, although other languages, like Visual Basic and Tcl/Tk, also support event-driven programming. In this chapter, we use Java as the primary vehicle for illustrating the principles and practice of event-driven programming.

1-58488-360-X/$0.00+$1.50
© 2004 by CRC Press, LLC

95.1   Foundations: The Event Model

The traditional programming paradigms have more clearly defined lineage than the event-driven paradigm. For example, functional programming has clear and traceable roots in the lambda calculus, as logic programming has in Horn clause logic. However, event-driven programming is in a more infantile stage of development, so its theoretical foundations are less clear and not as universally understood or accepted at this time.

One model of event-driven programming, offered by Stein [1999], explains event-driven programming by contrasting it with the traditional view of computation (which embodies the imperative, functional, and object-oriented paradigms) [Stein 1999, p. 1]:

"Computation is a function from its input to its output. It is made up of a sequence of functional steps that produce — at its end — some result as its goal.... These steps are combined by temporal sequencing."

Stein argues that modern computations are embedded in physical environments where the temporal sequencing of events is unpredictable and (potentially) without an end. In order to cope with this unpredictability, Stein claims that computation needs to be modeled as *interaction* [Stein 1999, p. 8]:

"Computation is a community of 'persistent entities coupled together by their ongoing interactive behavior....' Beginning and end, when present, are special cases that can often be ignored."

This view is well supported by the wide range of applications for which computer programs are now being designed, including robotics, video games, global positioning systems, and home security alarm systems.

A more extreme view is offered by Wegner [1997], who claims that interaction is a fundamentally more powerful metaphor than the traditional notion of algorithm (i.e., anything that can be modeled by a Turing machine). This view, which claims that there are interactive programs representative of a more powerful genre that cannot be systematically reduced to Turing machines, has received considerable recent discussion in the literature. Wegner has made efforts to further develop the underlying theory that would support it [Wegner 1999]. If successful, this work may have significant and long-lasting impact on our fundamental understanding of computing theory.

These concerns notwithstanding, the Wegner–Stein approach to describing computation as interaction provides a fairly rigorous foundation for modeling event-driven programming as it is practiced today.

95.2   The Event-Driven Programming Paradigm

The event-driven paradigm is different from the imperative paradigm in a way that is summarized in Figure 95.1.

Here, we see that the imperative paradigm models a computation as a series of steps that have a discrete beginning and ending in time. Input is generally gathered near the beginning of the time period, and results

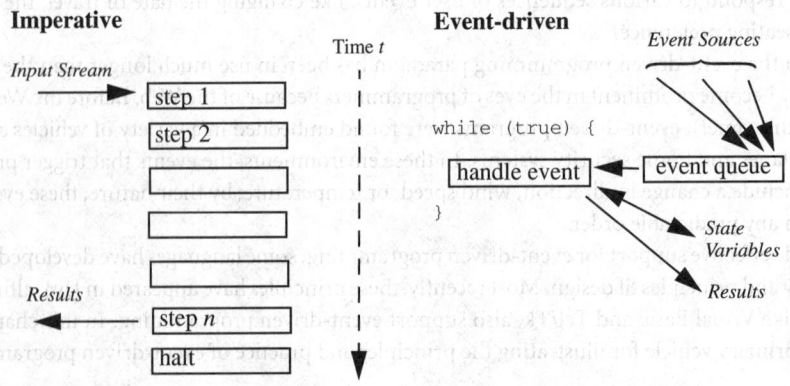

FIGURE 95.1   The imperative and event-driven paradigms contrasted.

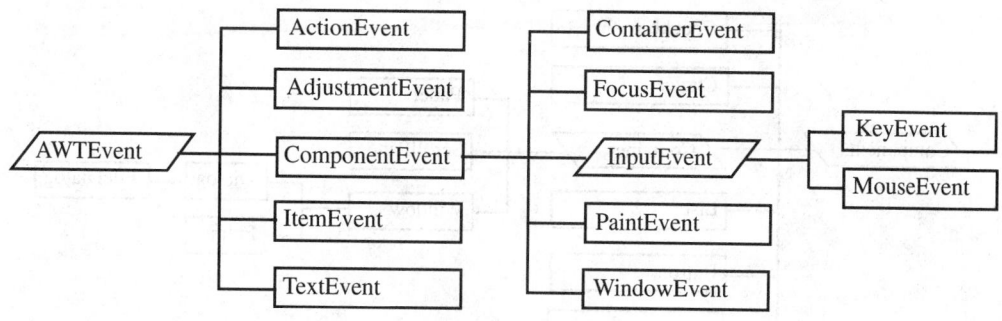

FIGURE 95.2 Java class AWTEvent and its subclasses.*

are generally emitted near the end. Some variations, of course, would have input and results continuously occurring, but nevertheless, the process has a distinct ending time.

In contrast, the input to an event-driven program comes from different autonomous *event sources*, which may be sensors on a robot or buttons in an interactive frame in a Web browser. These events occur asynchronously, so each one enters an event queue whenever it occurs. As time passes, a simple control loop receives the next event by removing it from this queue and handling it. In the process of handling the event, the program may consult and/or change the value of a *state variable* or even produce intermediate *results*. Importantly, we see that the event-driven program is designed to run forever, with no predefined stopping point, as appears in an imperative program.

Java provides direct support for event-driven programming by providing certain classes and methods that can be used to design an interaction. When we design an interaction, our program must classify the events that can occur, associate those event occurrences with specific objects in the frame, and then handle each event effectively when it does occur.

The types of events that can occur in Java are defined by the subclasses of the predefined abstract class AWTEvent. These subclasses are summarized in Figure 95.2.

Every event source in an interaction can generate an event that is a member of one of these classes. For instance, if a button is an event source, it generates events that are members of the `ActionEvent` class. We shall discuss the details of this relationship below.

The objects themselves that can be event sources are members of subclasses of the abstract class `Component`. A summary of these classes is given in Figure 95.3.

Here we see, for example, that any button be selected by the user in an interaction is declared as a variable in the `Button` class.

In order for a program to handle an event, it must be equipped with appropriate listeners that will recognize when a particular event, such as a click, has occurred on an object that is an *event source*. The `EventListener` class contains subclasses that play this role for each of the event classes identified previously. These are summarized in Figure 95.4.

For example, to equip a button so that the program can "hear" an occurrence of that button's selection, the program needs to send it the message *addActionListener*. If this is not done, button events will not be heard by the program. This is more fully discussed in Section 95.3.

Finally, in order to respond to events that are initiated by objects in these classes, we need to implement special methods called *handlers*. Each class of events predefines the name(s) of the handler(s) that can be written for it. A summary of the handlers that are preidentified for button selections, choice (menu) selections, text typing, and mouse events is given in Figure 95.5.

In the next section, we illustrate how these classes come together to support the event-driven design process in Java.

---

*In these class diagrams, abstract classes are enclosed in parallelograms, while nonabstract classes are enclosed in rectangles.

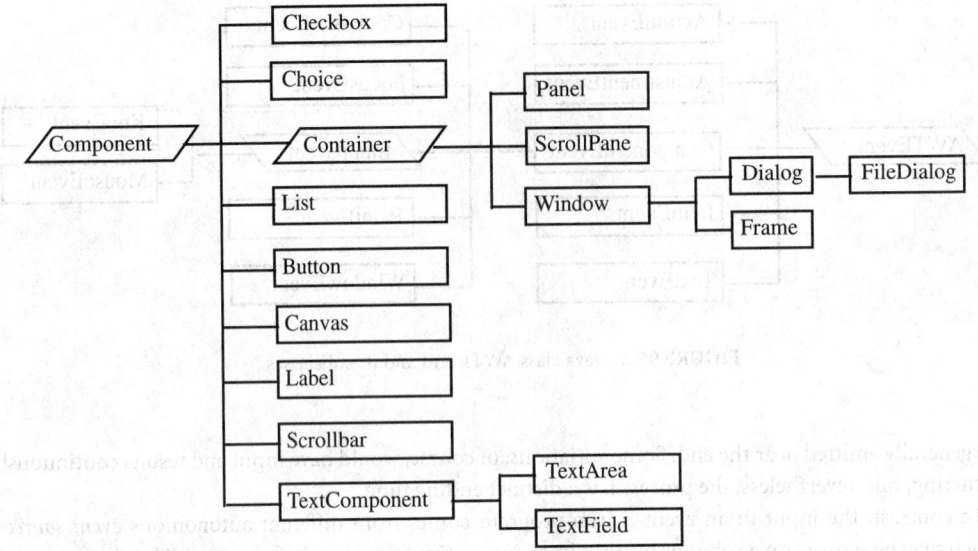

FIGURE 95.3 Subclasses of Component that can be sources of events.

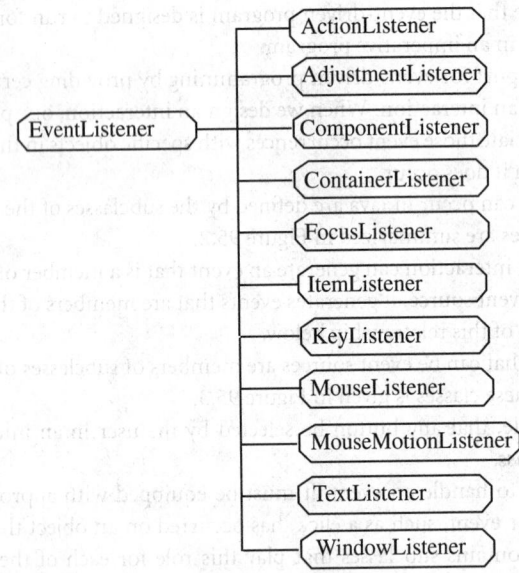

FIGURE 95.4 Java EventListener class interface and its subclasses.*

95.3 Applets

An *applet* is a Java program that runs inside a Web browser. It provides a framework for programmers to design event-driven programs where the source of the events is the user interface.

Because they are designed to react to events rather than initiate them, applets have a slightly different structure than Java applications; this structure is sketched in Figure 95.6. The first three lines shown

---

*Enclosing a class in a hexagon distinguishes it as a class interface, rather than a regular class.

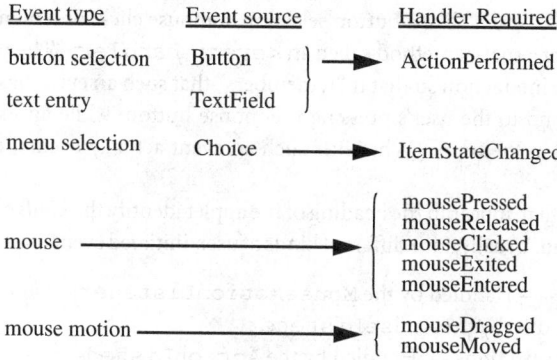

Event type	Event source	Handler Required
button selection	Button	ActionPerformed
text entry	TextField	
menu selection	Choice	ItemStateChanged
mouse		mousePressed mouseReleased mouseClicked mouseExited mouseEntered
mouse motion		mouseDragged mouseMoved

FIGURE 95.5 Handlers required for button, menu, text typing, and mouse events.

```
import java.applet.*;
import java.awt.*;
import java.awt.event.*;
public class <classname> extends Applet
                      implements <listeners>{
  <state variable declarations>
  public void init (){
    <code to initialize the applet>
  }
  <event handlers>
}
```

FIGURE 95.6 Overall structure of a Java applet.

in this figure are necessary to provide access to the Java class libraries, where the various event-driven programming classes are located.

The `<state variable declarations>` define various objects that can be placed in the applet, as well as other values that help the program remember what has occurred so far. As with any other class, these variables collectively define the *state* of the interaction, just as in Java application programs. Moreover, variables may also represent buttons, menus, text fields, and other objects that have special functions as the program and the user interact.

An applet is different from other classes in that it has no constructor. Instead, when the applet begins executing, the `init` method is executed first. It has the responsibility of placing objects in the frame, such as buttons, choices, labels, and text fields. In this sense, the `init` method takes the place of a constructor. The `init` method also attaches specific event listeners to different objects in the frame, so that these objects can be equipped with "ears" that recognize when they have been selected by the user. Event listeners are nothing more than objects that implement a particular interface; they can be the applet itself, as indicated above, or they can be separate classes. As we will see, such classes often need access to various applet methods, so they are commonly implemented as inner classes.

As a GUI, an applet often has a `paint` method, which is invoked whenever the applet needs to repaint itself. This can occur for a number of reasons: a window partially obscured the applet for a time, the applet was first iconified and then deiconified, etc. Components such as buttons, text fields, etc., will repaint themselves. However, anything done directly to the frame using `Graphics` methods will be lost during a repaint operation if not done in the paint method. Other major applet methods, `start`, `stop`, and `destroy`, are not discussed here.

For each specific kind of event (like a button selection, a mouse click, or a menu choice selection), the program must implement a special method called an `<event handler>`. The purpose of the handler is to change the state of the interaction so that it "remembers" that such an event has occurred. One handler is programmed to respond to the user's pressing the mouse button, while another may respond to the user's selecting a button in the frame. Whenever such an event actually occurs, its associated handler is executed one time.

The `<listeners>` that appear in the heading of the applet identify the *kinds* of events to which the applet is prepared to respond. Usually, four different kinds of user-initiated events can be handled by an applet:

Mouse motion events — Handled by the `MouseMotionListener`
Mouse events — Handled by the `MouseListener`
Button and text field selections — Handled by the `ActionListener`
Selections from a menu of choices — Handled by the `ItemListener`

Importantly, the program cannot know, or predict, the order in which these different events will occur or the number of times each one will be repeated; it must be prepared for all possibilities. That is the essence of event-driven programming.

95.4   Event Handling

In this section, we describe the basic Java programming considerations for responding to various types of user-initiated events that occur while an applet is running — mouse events, button selections, text areas, and choice (menu) selections.

95.4.1   Mouse Clicks

For the program to handle mouse clicks, the `MouseListener` interface must be implemented, either by the applet itself or by a mouse handler class. If the listener is being handled directly by the applet, then the applet must specify the listener in its `implements` clause and activate the listener, usually inside the init method:

```
public class MyApplet extends Applet implements MouseListener {
  public void init() {
...
    addMouseListener(this);
...
  }
```

The alternative is to use a separate class to handle mouse clicks. Commonly, this class is an inner class to the applet, so that the mouse handler has access to all of the applet's methods, particularly the graphics context:

```
public class MyApplet extends Applet {
  public void init() {
...
    addMouseListener(new MouseHandler());
...
  }
  private class MouseHandler implements MouseListener {
...
  }
```

If an external class is used, it is common for the applet to pass itself in the call to the mouse handler constructor; the applet object can then be saved to an instance variable of the mouse handler class.

Whichever alternative is used, all of the following methods must be added to the class that implements the `MouseListener`, and at least one must have some statements that respond to the event that it represents:

```
public void mousePressed(MouseEvent e) { }
 public void mouseReleased(MouseEvent e) { }
 public void mouseClicked(MouseEvent e) { }
 public void mouseExited(MouseEvent e) { }
 public void mouseEntered(MouseEvent e) { }
```

An advantage to using a separate class is that Java provides a `MouseAdapter` class, which is precisely the trivial implementation of `MouseListener` given previously. This means that the separate class can extend the `MouseAdapter` class (which an applet cannot do), overriding exactly the methods for which actions are to be provided. In most instances, this is usually only the `mouseClicked` method.

```
public class MyApplet extends Applet {
 public void init() {
...
 addMouseListener(new MouseHandler());
...
 }
 private class MouseHandler extends MouseAdapter {
 public void mouseClicked(MouseEvent e){
 <action>
 }
 }
```

For instance, the typical response to a mouse event is to capture the $x - y$ pixel coordinates where that event occurred on the frame. To do this, we use the `getX` and `getY` methods of the `MouseEvent` class. For example, the following handler responds to a mouse click by storing the $x - y$ coordinates of the click in the applet's instance variables x and y.

```
public void mouseClicked(MouseEvent e) {
 x = e.getX();
 y = e.getY();
 }
```

95.4.2 Mouse Motion

Similar to mouse clicks, for the program to handle mouse motion, the `MouseMotionListener` interface must be implemented either by the applet itself or by a mouse motion handler class. To activate the listener, the following calls must be placed inside the `init` method:

```
addMouseMotionListener(<listener>);
```

The class that implements the listener must implement all of the following methods; at least one must have some statements that respond to the event that it represents:

```
public void mouseDragged(MouseEvent e) { }
 public void mouseMoved(MouseEvent e) { }
```

An advantage to using a separate class is that Java provides a `MouseMotionAdapter` class, which is precisely the trivial implementation of `MouseMotionListener` given previously. This means that the separate class can extend the `MouseMotionAdapter` class (which an applet cannot do), overriding exactly the methods for which actions are to be provided. In most instances, this is usually only the `mouseDragged` method.

95.4.3   Buttons

A button is an object on the screen which is named and can be selected by a mouse click. Because any number of variables can be declared with class `Button` and placed in the applet, button handlers are usually implemented via a separate class, rather than by the applet itself. A button is declared and initialized as follows:

```
Button <variable> = new Button(<string>);
```

Here is an example:

```
Button clearButton = new Button("Clear");
```

A button is placed in the applet by including the following inside the `init` method:

```
add(<variable>);
```

For example:

```
add(clearButton);
```

To be useful, a button must have a listener attached so that the button responds to mouse clicks. This is normally done by including the following inside the applet's `init` method:

```
<variable>.addActionListener(<listener>);
```

For example:

```
clearButton.addActionListener(new ClearButtonHandler());
```

The class that handles the user's selection of the button must implement the `ActionListener` interface. The listener class must implement an `actionPerfomed` method to handle the button selection event:

```
public void actionPerformed (ActionEvent e) {
 if (e.getSource() == <variable>) {
 <action>
 }
}
```

Here, the *<variable>* refers to the name of the `Button` variable as declared and initialized, and *<action>* defines what to do whenever that event occurs. If unique handlers are created for each button, the *if* test to determine which button was selected can be omitted; this is normally preferred. Following, for example, is a handler written as an inner class to the applet class that clears the screen whenever the `clearButton` is clicked by the user:

```
private class ClearButtonHandler implements ActionListener {
 public void actionPerformed(ActionEvent e) {
 repaint();
 }
}
```

Note that the repaint method is an applet method; thus, this class works as written only if it is an inner class to applet. Using an external class requires writing a constructor that takes the applet as an argument. Using our Clear button as an example, the `addActionListener` code in applet's `init` method would appear as follows:

```
clearButton.addActionListener(new ClearButtonHandler(this));
```

The button handler class would then appear as:

```
public class ClearButtonHandler implements ActionListener {
  Applet applet;
  public ClearButtonHandler(Applet a) { applet = a; }
  public void actionPerformed(ActionEvent e) {
  applet.repaint();
  }
}
```

Note that using an external class makes it more difficult for a listener class to determine which button was selected, because the buttons themselves are declared within the applet class.

95.4.4 Labels, TextAreas, and TextFields

A label is an object whose string value can be placed inside a frame to label another object, such as a `TextField`. It can be added to the frame from within the `init` method. For example, the statement

```
add(new Label("Fahrenheit"));
```

would place the message *Fahrenheit* in the frame. Labels cannot have a listener attached to them.

A `TextArea` is a rectangular object on the screen which is named and can accept or display text messages. It is a scrollable object, so users have a complete record of the text. A `TextField` is an object into which the user can type a single line of text from the keyboard; it raises an `ActionEvent` when the user presses the Enter key at the end of the line. `TextAreas` and `TextFields` are declared as follows:

```
TextArea <variable> ;
TextField <variable> ;
```

For example:

```
TextArea echoArea;
TextField typing;
```

`TextArea` and `TextField` objects are normally placed in the applet as part of the `init` method in your program. Initialization requires the number of lines of text (for `TextAreas`) and the number of characters per line to be specified.

```
<variable1> = new TextArea(<lines>, <chars>);
add(<variable1>);
<variable2> = new TextField(<chars>);
add(<variable2>);
```

For example:

```
echoArea = new TextArea(5, 40);
add(echoArea);
typing = new TextField(40);
add(typing);
```

In this example, we declare and place a 5-line by 40-character `TextArea` and a 40-character `TextField` in the current frame.

When the user types in the `TextField` and hits the Enter key, the applet can handle that event by writing additional code in its `actionPerformed` event handler:

```
public void actionPerformed (actionEvent e) {
if (e.getSource() == <variable>) {
String s = <variable>.getText();
<action>
}
}
```

Here, *<variable>* refers to the name of the `TextArea` variable that was declared and placed in the frame by the `init` method, like `typing` in the example. When the event occurs, the string value typed in the text area is assigned to the variable *s* and the *<action>* is then executed.

A better solution is to use an internal class that listens specifically to the `TextField typing`:

```
private class TextHandler implements ActionListener {
public void actionPerformed (actionEvent e) {
String s = <variable>.getText();
<action>
}
}
```

In this case, the handler need not check for the source of the triggering event; it must be the user pressing the Enter key in the `TextField typing`.

As an example of an *<action>*, the user's typing can be immediately echoed in the `TextArea` by concatenating it with all the text that is already there. The `append` method is useful for this purpose:

```
echoArea.append(s + "\n");
```

If this line is added as the *<action>* in the previous code, the user's typing will be echoed on a new line inside the `TextArea` object named `echoArea`.

95.4.5 Choices

A choice is an object on the screen that offers several options to be selected by the user. It is like a pull-down menu, but it can be placed anywhere within the applet, not just at the top. A choice is declared as follows:

```
Choice <variable> ;
```

For example:

```
Choice choice;
```

The choice is named and placed in the frame as part of the `init` method in your program. The different selections are assigned to the `Choice` variable using the `addItem` method, and interaction is triggered by adding a listener to the frame for the choice.

```
<variable> = new Choice();
<variable>.addItem(<string1>);
...
add(<variable>);
<variable>.addItemListener(<listener>);
```

For example:

```
choice = new Button();
choice.addItem("North");
choice.addItem("East");
choice.addItem("South");
choice.additem("West");
add(choice);
choice.addItemListener(new Choice Handler());
```

In this case, an inner class to the applet itself is assumed to be handling the choice event. The applet or item listener event handler must implement the `ItemListener` interface. When the user selects one of the choices, the event is handled by an `itemStateChanged` method:

```
private class ChoiceHandler implements ItemListener {
public void itemStateChanged (ItemEvent e) {
String s = (String)e.getItem();
if (s.equals(<string1>) {
<action1> in response to a selection of <string1>
}
else if (s.equals(<string2>) {
<action2> in response to a selection of <string2>
}
...
}
}
```

When the event of selecting a choice occurs, this handler is executed. The string s gets the value of the choice the user selected, which is passed to the handler by way of the method call `e.getItem()`. This choice is used in a series of if statements to select the appropriate *<action>*.

95.5   Example: A Simple GUI Interface

The process of event-driven program design involves anticipating the various states and state transitions that can occur as the program runs.

Consider the design of a simple drawing tool, in which the user can draw rectangles and type texts in arbitrary locations of the frame. The user should be able to accomplish this as simply as possible, so providing buttons, menus, and text typing areas and handling mouse click actions on the screen are essential. An initial frame design to support this activity is shown in Figure 95.7.

This frame has four objects: a Clear button, a Choice menu, a `TextArea` for communicating with the user as events are initiated, and a `TextField` in which the user can enter messages. Thus, we

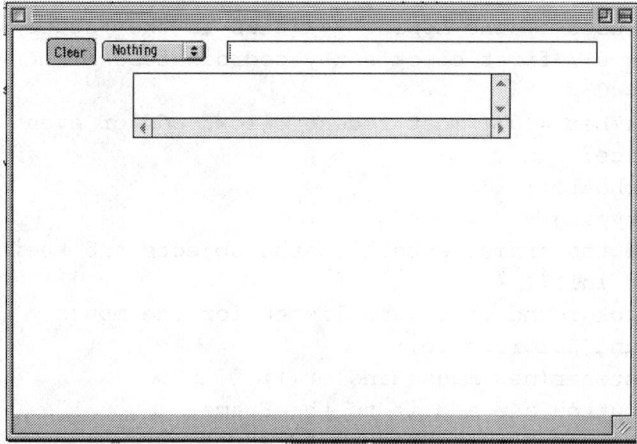

FIGURE 95.7   Initial frame design for a graphical drawing tool.

```
// Variables used by the Applet -- this is the "state"
    int lastX = 0;                    //first click's x-y coordinates
    int lastY = 0;
    int clickBumber = 0;        // most recent click; odd or even
    Choice choice;
    TextArea echoArea;
    TextField typing;
```

FIGURE 95.8 Code to define the state for the interaction.

can begin our design of defining the state of the computation with the following objects and interpretations:

> **choice** — The user can select Nothing, Rectangle, or Message from this menu.
> **echoArea** — This is a `TextArea` for reporting the most recent event that has occurred.
> **typing** — This is a `TextField` for entering user input.

The button need not be part of the global state, because its function is merely to clear the screen, and this can be done completely within the event handler for the button. No other object or handler need be aware of the Clear button.

The state of the computation for this problem must also keep track of all information that is relevant to the user's accomplishing the next task, be it drawing a rectangle or locating a text somewhere in the frame. To draw a rectangle, the system must have two pieces of information: the $x - y$ coordinates of the upper left-hand corner of the rectangle and the $x - y$ coordinates of the lower right. These two points can be retrieved via a mouse event and the `mouseClicked` handler, but the first $x - y$ coordinate pair must be stored globally within the applet's state. Furthermore, a count of the mouse clicks must also be stored globally, so that the first click can be distinguished from the second in determining the corners of the rectangle. Thus, additional global state information includes

```
lastX, lastY X-y coordinates of last mouse click
clickNumber The number of mouse clicks (odd or even)
```

The x and y coordinates of a mouse click are reported in the `TextArea` by the program whenever a mouse click event occurs. Informative messages to the user are also displayed there.

The code that defines the state is shown in Figure 95.8, and the code that initializes the interaction is shown in Figure 95.9.

```
//Variables used by the Applet --- this is the "state"
  int lastX = 0;//first click's x-y coordinates
  int lastY = 0;
  int clickBumber = 0;//most recent click; odd or even
  Choice choice;
  TextArea echoArea;
  TextField typing;
//Initialize the frame: establish the objects and their listeners
  public void init() {
//Set the background color and listen for the mouse
  setBackground(Color.white);
  addMouseListener(new MouseHandler());
//Create a button and add it to the Frame.
  Button clearButton = new Button("Clear");
  clearButton.setForeground(Color.black);
  clearButton.setBackground(Color.lightGray);
```

```
// Initialize the frame: establish the objects and their listeners
    public void init () {
        // Set the background color and listen for the mouse
        setBackground (Color.White);
        addMouseListener (new MouseHandler());

        // Create a button and add it to the Frame.
        Button clearButton = new Button ("Clear");
        clearButton.setForeground (Color.black) ;
        clearButton.setBackground (Color.lightGray) ;
        add (clearButton) ;
        clearButton.addActionListener (new ClearButtonHandler()) ;

        // Create a menu of user choices and add it to the Frame.
        choice = new Choice () ;
        choice.addItem ("Nothing") ;
        choice.addItem ("Rectangle");
        choice.addItem ("Message");
        add(choice);
        choice.AddItemListener (new ChoiceHandler());

        // Create a TextField and a TextArea and add them to the Frame.
        typing = new TextField (40);
        add(typing);
        typing.addActionListener (new TestHandler());
        echoArea = new TextArea (2, 40);
        echoArea.setEditable(false);
        add(echoArea);
    }
```

FIGURE 95.9 Code to initialize the interaction.

```
    add(clearButton);
    clearButton.addActionListener(new ClearButtonHandler());
  //Create a menu of user choices and add it to the Frame.
    choice = new Choice();
    choice.addItem("Nothing");
    choice.addItem("Rectangle");
    choice.addItem("Message");
    add(choice);
    choice.addItemListener(new ChoiceHandler());
  //Create a TextField and a TextArea and add them to the Frame.
    typing = new TextField(40);
    add(typing);
    typing.addActionListener(new TextHandler());
    echoArea = new TextArea(2, 40);
    echoArea.setEditable(false);
    add(echoArea);
    }
```

Designing the event handlers is a more intricate process. For each event that occurs, an appropriate handler must contain code to distinguish that event from the rest and then change the state and/or generate output into the frame appropriately. Consider the following scenario.

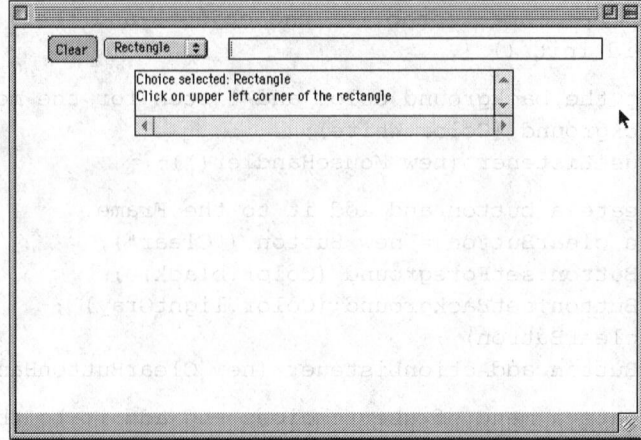

FIGURE 95.10   First step in an interaction: the user selects Rectangle from the menu.

```
public void itemStateChanged (ItemEvent e) {
    String currentChoice = (String) (e.getItem());
    echoArea.setText ("Choice selected: " + currentChoice);
    clickNumber = 0;
        // prepare to handle first mouse click for this choice
    if (currentChoice.equals("Rectangle"))
        echoArea.append(
            "\nClick to set upper left corner of the rectangle");
    else if (currentChoice.equals ("Message"))
        echoArea.append(
            "\nEnter a massage in the text area");
    }
}
```

FIGURE 95.11   ItemStateChanged handler for this interaction.

The user first selects Rectangle from the Choice menu. To properly handle that event, the system must store that selection and then prompt the user to click the mouse at the point where the upper left-hand corner of the rectangle should be drawn, as shown in Figure 95.10.

What code do we write to respond to this selection? We write code inside an `itemStateChanged` event handler of the `ChoiceHandler` class, because the object selected, choice, is a `Choice`. The complete `itemStateChanged` handler for this problem is shown in Figure 95.11.

```
//Called when the user makes a Choice selection
private class ChoiceHandler implements ItemListener {
 public void itemStateChanged (ItemEvent e) {
 String currentChoice = (String) (e.getItem());
 echoArea.setText("Choice selected: " + currentChoice);
 clickNumber = 0;
            //prepare to handle first mouse click for this choice
 if (currentChoice.equals("Rectangle"))
 echoArea.append(
 "\nClick to set upper left corner of the rectangle");
 else if (currentChoice.equals("Message"))
```

```
private class MouseHandler extends MouseAdapter {
    public void mouseClicked(MouseEvent e) {
        int x = e.getX();
        int y = e.getY();
        echoArea.setText ("Mouse Clicked at " +
                           e.getX() + " , " + e.getY() + "\n");
        Graphics g = getGraphics();
        if (choice.getSelectedItem().equals("Rectangle")) {
            clickNumber = clickNumber + 1;
            // is it the first click?
            if (clickNumber % 2 == 1) {
              echoArea.setText
                  ("Click to set lower right corner of the rectangle");
              lastX = x:
              lastY = y;
            }
            // or the second?
            else g.drawRect(Math.min(lastX,x), Math.min(lastY,y),
                    Math.abs(x-lastX), Math.abs(y-lastY));
        }
    // for a message, display it
        else if (choice.getSelectedItem().equals("Message"))
            g.drawString(currentMessage, x, y);
    }
}
```

FIGURE 95.12   Details of the mouseClicked handler for this interaction.

```
echoArea.append(
"\nEnter a message in the text area");
}
}
```

While this code is a bit obscure, it does reveal that the itemStateChanged handler must prepare for *any* menu event that can occur. It distinguishes among the possibilities by checking the parameter e for the item selected, which is assigned to the state variable currentChoice. If the choice is a Rectangle, the handler dutifully prompts the user to click to set the location where the upper left-hand corner of the rectangle should be located in the frame. If the choice is a Message, a different prompt goes to the user. In either case, this handler echoes the nature of this event in the echoArea.

Assuming that the user clicks the mouse in the frame, the system must respond by storing the $x - y$ coordinates of that click in the applet's state variables lastX, lastY and prompt the user to click to set the lower right-hand corner of the desired rectangle. But that should be done *only* if the current choice is a Rectangle. By the time the mouse click event happens, the only source of information about what event immediately preceded it comes from the applet's state. That is, the click could have been preceded by a different event, which should provoke a different response from the interaction. Where is this all sorted out? In the mouseClicked handler, as shown in Figure 95.12.

```
private class MouseHandler extends MouseAdapter {
  public void mouseClicked(MouseEvent e) {
  int x = e.getX();
  int y = e.getY();
  echoArea.setText("Mouse Clicked at " +
```

```
        e.getX() + ," " + e.getY() + "\n");
   Graphics g = getGraphics();
     if (choice.getSelectedItem().equals("Rectangle")) {
     clickNumber = clickNumber + 1;
//is it the first click?
     if (clickNumber% 2 == 1) {
     echoArea.setText
     ("Click to set lower right corner of the rectangle");
     lastX = x;
     lastY = y;
     }
//or the second?
     else g.drawRect(Math.min(lastX,x), Math.min(lastY,y),
     Math.abs(x-lastX), Math.abs(y-lastY));
     }
//for a message, display it
     else if (choice.getSelectedItem().equals("Message"))
     g.drawString(currentMessage, x, y);
     }
   }
```

This handler must also be prepared for anything, because it doesn't implicitly know what events occurred immediately before this particular mouse click. The state variable clickNumber helps to sort things out, because its update value will have an odd number for the first click of a pair and an even number for the second. Thus, the upper left-hand corner of a rectangle is indicated for odd values, and the drawing of a complete rectangle, using the *x* and *y* coordinates of the current click together with the *x* and *y* coordinates of the previous click (stored in lastX and lastY), is indicated for even clicks.

The remainder of this event handler should be fairly readable. The effect of drawing a rectangle after the user has clicked twice is shown in Figure 95.13. Here, the arrow in the figure shows the location of the second click, whose *x* and *y* coordinates are 215 and 204, respectively.

The next task in designing this interaction is to implement the event handler that responds to the user's selecting the Clear button or typing text in the typing area. The actionPerformed method for this event is shown in Figure 95.14. Note here the simplicity of attaching a unique handler to the Clear button;

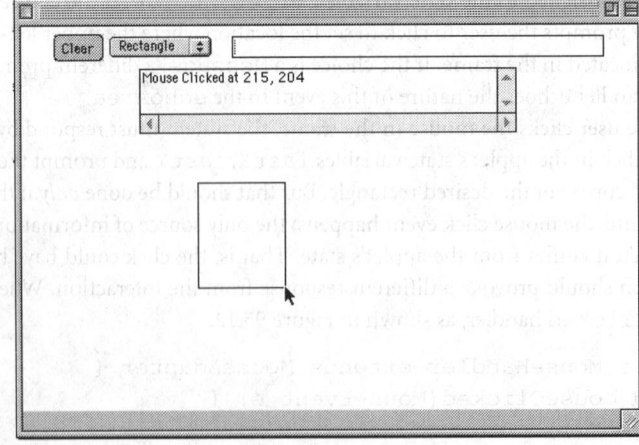

FIGURE 95.13   Effect of selecting Rectangle choice and clicking the mouse twice.

```
        private class ClearButtonHandler implements ActionListener {
            public void actionPerformed (ActionEvent e) {
                echoArea.setText ("Clear button selected ");
                repaint();
            }
        }
```

FIGURE 95.14 ActionPerformed handler for the Clear button.

```
    private class TextHandler implements ActionListener {
        public void actionPerformed (ActionEvent e) {
            echoArea.setText ("Text entered: " + typing.getText());
            if (choice.getSelectedItem().equals("Message"))
                echoArea.append("\nNow click to place this message");
        }
    }
```

FIGURE 95.15 ActionPerformed handler for the Enter key in the typing area.

it does not have to test what triggered the event (the button or the Enter key in the typing area), merely to clear the screen via a repaint.

```
    private class ClearButtonHandler implements ActionListener {
    public void actionPerformed (ActionEvent e) {
    echoArea.setText("Clear button selected ");
    repaint();
    }
    }
```

Responding to text typed by the user is straightforward, needing only to store the text for later use and prompting the user to click the mouse to locate the text in the frame, as shown in Figure 95.15.

```
    private class TextHandler implements ActionListener {
    public void actionPerformed (ActionEvent e) {
    echoArea.setText("Text entered: " + typing.getText());
    if (choice.getSelectedItem().equals("Message"))
    echoArea.append("\nNow click to place this message");
    }
    }
```

Note that both the TextHandler and ClearButtonHandler classes implement the ActionListener interface and both have actionPerformed methods. Neither handler has to be aware of the other, as each is listening for separate events.

The net effect of typing text and clicking to locate it below the rectangle that had been placed in the frame is shown in Figure 95.16. There, the location of the mouse click is indicated by the arrow in the figure, which is at $x - y$ coordinates 162 and 221.

This sketch provides a concrete example of the process of designing an event-driven program. The program runs indefinitely — the user can spend hours drawing rectangles, placing messages, and clearing the frame, and only the user decides what event sequence will occur at any particular point in time.

The observant reader will have noticed that this applet has no paint method. Thus, iconifying the window or overlaying another window on top of this window will eventually necessitate a repaint, causing

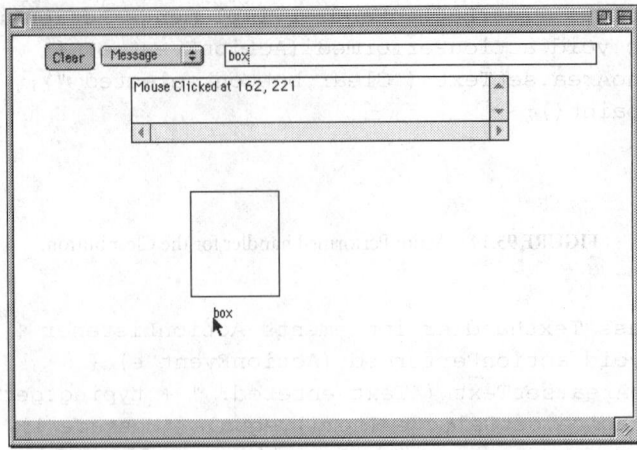

FIGURE 95.16   Net effect of user's placing a text in the frame.

all the drawing of rectangles and placing of messages to disappear. If the applet must be able to repaint what has been drawn to the screen, then the applet must remember what it wrote to the screen. We will see examples of this in later programs.

95.6   Event-Driven Applications

Event-driven programs occur in many domains besides those which are Web-based. Here are three examples (and an example of a Web-based program):

Web-based interactive games
Automated teller machines (ATMs)
Home security alarm systems
A supermarket checkout station

In this section, we briefly describe the first three of these interactions, focusing on their state variables and the kinds of events that should be handled in the event loop to maintain integrity among the state variables.

95.6.1   Interactive Games: Tic-Tac-Toe

Consider designing an event-driven program that monitors an interactive game between two people. For example, the game tic-tac-toe has a 3 × 3 board and two players, who alternate turns. Each turn forces the board into a new state, which differs from the old state by the contents of one square. A game is over either when the board is full or when one of the players has placed three markers (X or O) in a row, either horizontally, vertically, or diagonally. The interaction can run indefinitely, since the players can clear the board and start a new game at any time. Thus, the state of a tic-tac-toe game naturally includes the following:

The state of the board
Whose turn it is
Whether the game is over
A New Game button that clears the board and starts a new game

An event in this setting is either a player move (clicking the mouse in an unoccupied square to place the next X or O) or a player selecting the New Game button. The sequencing of these events is entirely unpredictable by the program — either one is equally likely while a game is going on.

The results displayed by the program are the board itself and a message area which is used to report whose turn it is, the winner, and other information as the game proceeds. The major visual design element for this game is a 3 × 3 grid:

Each player takes a turn by clicking the mouse to place an X or an O on one of the unoccupied squares; player X goes first. The winner is the player who first places three Xs or 3 Os in a row, either horizontally, vertically, or diagonally. A tie game occurs when the board is full and no one has three Xs or Os in a row.

The program can use a `TextArea` for displaying appropriate messages (for instance, when a player attempts to make an illegal move). It should allow each player to click on an empty square and replace that square with an X or an O. The program thus keeps track of whose turn it is and reports that information in the `TextArea` at the beginning of each turn. Thus, the state variables for this game include the following:

A `Grid` variable for the 3 × 3 board
A `TextArea` variable for sending messages to the players
A variable that determines whose turn it is, which flips back and forth whenever a player has completed
    a legal move

The central event-driven code for this game appears in the handler for mouse clicks. Some of these clicks reflect legal moves (i.e., a mouse click from a player on an unoccupied square of the board), and others reflect illegal moves (e.g., a mouse click outside the board). Another distinct event is the selection of the Clear button to signal ending the game. These events can occur in any order, so the program must be equipped to handle them sensibly, change the state appropriately, and detect when the game is over.

95.6.2 Automated Teller Machine

An ATM is driven by a program that runs 7 days a week and 24 hours a day. The program must be able to interact with each user who has a proper ATM bank card, and it helps the machine conduct transactions with the user. The essential elements of a typical ATM display are shown in Figure 95.17. (In practice, the details are different, but the elements shown here are adequate to characterize what happens during an ATM transaction).

Welcome to AnyBank ATM

Account number [＿＿＿＿＿＿]

⊙ Deposit　　　　Amount [＿＿＿＿＿＿]
⊙ Withdrawal　　Message [＿＿＿＿＿＿]
⊙ Balance Inquiry
⊙ No more transactions

FIGURE 95.17 Elements of a typical ATM transaction user interface.

The state of this interaction is captured partially by the objects in this display and partially by other information that relates to the particular user who is at the machine. A basic collection of state variables required for an ATM transaction includes the following:

Account — The user's account number
Type — The type of transaction (deposit, withdrawal, etc.)
Amount — The amount of the transaction
Message — A message from the bank to the user
Balance — The user account's available balance

The last variable in this list, the available balance, brings into play a new dimension for event-driven programming. That is, the program must interact not only with the user but also with the bank's database of all its accounts and current balances. A program that interacts with such a database, which may reside on an entirely different computer, or *server*, on the bank's network, is called a *client–server* application. Client–server applications exist in a wide variety of systems, including airline reservation systems, on-line textbook ordering systems, and inventory systems. They are discussed more directly in other chapters of this Handbook.

In this example, we can characterize the different events that can occur, together with how they should be handled (that is, the effect they should have on the state of the interaction).

Event: User enters an account number (swipes her card).
   Handled by: Program checks that account is a valid number, sets balance, and issues the message
    "Choose a transaction."
Event: User selects a button (deposit, withdrawal, etc.).
   Handled by: Program checks to see that a valid account has been entered.
    If so: Save the type of button selected.
      If deposit or withdrawal, issue the message "Enter an amount."
      If Balance Inquiry, display the balance.
      If No more transactions, clear the account.
    If not, issue the Message "Enter an account number."
Event: User enters an amount.
   Handled by: Program checks that user has selected a deposit or withdrawal type.
    If deposit, add the amount to the balance.
    If withdrawal:
      If balance is greater than amount, subtract amount from balance.
      Otherwise, issue the message "Insufficient funds."
    Otherwise, issue the message "Select a transaction type (deposit or withdrawal)."

The key insight with this design is that the system does not anticipate the type of transaction or the order in which the events will occur. It responds to every different possibility and updates the state of the interaction appropriately.

95.6.3 Home Security Alarm System

Home security systems provide homeowners with a modest warning that an unwanted event, such as a break-in, fire, or flood, has occurred. These systems are programmed to react to sensors that can detect smoke, water, or motion, and are placed strategically around the house. They can be programmed by homeowners so that their own motion about the house is not taken as an unwanted break-in. They also can be connected to the local fire and police stations so that notification of an unwanted event can receive immediate response.

The overall design for such a system is sketched in Figure 95.18.

Here, the sensors and the user interface supply events to the program; alarms and the user interface receive responses from the program. Responses at the user interface are displayed on an LCD display.

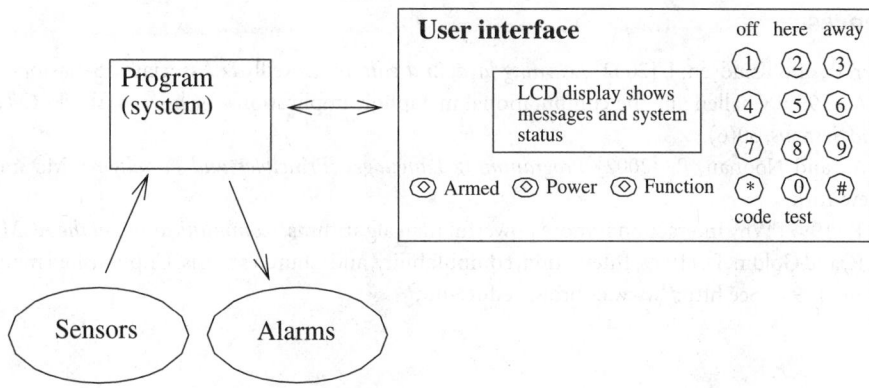

FIGURE 95.18 Overall design of a home security system.

Because this system must be able to receive handle events in parallel (e.g., signals from two different sensors may occur simultaneously), the program must embody both the event-driven paradigm discussed in this chapter and the parallel programming paradigm (discussed in Chapter 96). We ignore the parallel programming dimensions of this problem in the current discussion.

The state variables for this program are several:

Password — The user's password
User — The state of the user (here or away)
Armed — The state of the system (armed or unarmed)
Sensors — The state of each sensor (active or inactive)
Alarms — The state of each alarm (active or inactive)
Message — A message from the system to the control panel

Here are some of the events that can reasonably occur, with a sketch of what should happen to the state in response to each event.

Event: User enters password.
   Handled by: Program checks that password is valid and displays message.
Event: User enters function "away."
   Handled by: Program checks that password has been entered and changes the state of all sensors to "active," the state of the system to "armed," and displays message.
Event: Sensor receives signal.
   Handled by: If system is armed, program sends appropriate alarm and displays a message.
Event: User enters function "test."
   Handled by: Program disarms system.
Event: User enters function "here."
   Handled by: Program disables motion-detection sensors, enables all others, and changes state of the system to "armed."

In these examples, we have greatly generalized the details in order to simplify the discussion and focus on the main ideas about event-driven programming that these problems evoke. Interested readers should consult the references for more detailed discussions of event-driven programming applications.

Acknowlegdments

This chapter is an adaptation of a chapter in the authors' book *Programming Languaages*, and is published with permission from McGraw-Hill.

References

Niemeyer, P., and Knudsen, J. [2002] *Learning Java, 2nd edition*. O'Reilly & Associates, Sebastopol, CA.

Stein, L.A. [1999] Challenging the computational metaphor: implications for how we think. *Cybernetics and Systems*, 30(6).

Tucker, A., and Noonan, R. [2002] *Programming Languages: Principles and Paradigms*. McGraw-Hill, New York.

Wegner, P. [1997] Why interaction is more powerful than algorithms. *Communications of the ACM*, 40(5).

Wegner, P., and Goldin, D. [1999] Interaction, computability, and Church's thesis. Unpublished manuscript (June 1999). See http://www.cs.brown.edu/~pw.

96

Concurrent/ Distributed Computing Paradigm

96.1 Introduction 96-1
96.2 Hardware Architectures 96-4
96.3 Software Architectures 96-4
        Busy-Wait: Concurrency without Abstractions • Semaphores
        • Monitors • Message Passing
96.4 Distributed Systems 96-19
96.5 Formal Approaches 96-19
96.6 Existing Languages with Concurrency Features 96-20
96.7 Research Issues 96-21
96.8 Summary 96-22

Andrew P. Bernat
Computer Research Association

Patricia J. Teller
University of Texas at El Paso

96.1 Introduction

Concurrent computing is the use of multiple, simultaneously executing processes or tasks to compute an answer or solve a problem. The original motivation for the development of concurrent computing techniques was for timesharing multiple users or jobs on a single computer. Modern workstations use this approach in a substantial manner. Another advantage of concurrent computing, and the reason for much of the current attention to the subject, is that it seems obvious that solving a problem using multiple computers is faster than using just one. Similarly, there is a powerful economic argument for using multiple inexpensive computers to solve a problem that normally requires an expensive supercomputer. Additionally, the use of multiple computers can provide fault tolerance.

Moreover, there is an additional powerful argument for concurrent computing — the world is inherently concurrent. Just as each of us engages in a large number of concurrent tasks (hearing while seeing while reading, etc.), operating systems need to handle multiple, simultaneously executing tasks; robots need to engage in a multiplicity of actions; database systems must simultaneously handle large numbers of users accessing and updating information; etc. Often, breaking a problem into concurrent tasks provides a simpler, more straightforward solution.

As an example, consider Conway's problem: input is in the form of 80-character records (card images in the original problem, which gives an idea of how long it has been around); output is to be in the form of 120-character records; each pair of dollar signs, '$$', is to be replaced by a single dollar sign, '$'; and a space, ' ', is to be added at the end of each input record. In principle, a sequential solution may be developed, but the complications introduced require complex and non-obvious buffer manipulations. Moreover, a concurrent solution consisting of three processes is both simpler and more elegant. The three processes execute within infinite loops performing the following actions:

1. Process1 reads 80-character records into an 81-character buffer, places a space character in location 81, and then outputs single characters from the buffer sequentially.
2. Process2 reads single characters and copies them to output, but uses a simple state machine to substitute a single '$' for two consecutive '$$'.
3. Process3 reads single characters, saves them in a buffer, and outputs 120-character records.

To develop an implementable solution, we need to decide how the independently executing processes communicate. A simple, widely used approach is to add two buffers: Buffer1 stores output characters from Process1 to be input to Process2; Buffer2 stores output characters from Process2 to be input to Process3. For simplicity, assume that Buffer1 and Buffer2 each hold a single character. Thus:

1. Process1 reads 80-character records into an 81-character internal buffer, places a space character in location 81, and sequentially places in Buffer1 single characters from the internal buffer.
2. Process2 reads single characters from Buffer1 and places them into Buffer2, but uses a simple state machine to substitute a single '$' for two consecutive '$$'.
3. Process3 reads single characters from Buffer2, saves them in an internal 120-character buffer, and outputs 120-character records.

This solution demonstrates the essence of the concurrent paradigm: individual sequential processes that cooperate to solve a problem. The exemplified concurrency is pipelined concurrency, where the input of all processes but the first is provided by another process. Cooperation, in this and all other cases, requires that the processes:

1. Share information and resources
2. Not interfere during access to shared information or resources

In the Conway solution, information is readily shared via the buffers. The chief problem is to ensure that concurrent accesses to the two buffers do not conflict; for example, Process2 does not attempt to retrieve a character from Buffer1 before it has been placed there by Process1 (which would lead to garbage characters), and Process1 does not attempt to place a character into Buffer1 before the previous character has been retrieved by Process2 (which would lead to lost characters).

A simpler example of interference is provided by the following simple program (where the statements within the cobegin–coend pair are to be executed simultaneously):

```
x := 0
cobegin
  x := x + 1
  x := x + 2
coend
```

Consider the value of x at the end of execution. Because each assignment statement is actually a sequence of machine-level instructions, various interleavings of the execution of these instructions result in different final values for x (i.e., 1, 2, or 3). Clearly, this is unacceptable!

In each of these examples, it is clear that there are *critical regions* in which two (or more) processes have sections of code that may not be executed concurrently; we must have *mutual exclusion* between the critical regions. In the Conway example, critical regions include:

- Process1 placing a value into Buffer1
- Process2 retrieving a value from Buffer1
- Process2 placing a value into Buffer2
- Process3 retrieving a value from Buffer2

In the simple example above, each of the two assignment statements are critical regions. The essence of avoiding interference is to discover the critical regions and isolate them. This isolation takes the form of an "entry protocol" to announce entry into a critical region and an "exit protocol" to announce that the execution of the critical region has completed (below the '#' introduces a comment and the '...' represents the appropriate program code):

```
# entry protocol
...
# critical region code
...
# exit protocol
...
```

This is the basic model used by the *busy-wait* and *semaphore* approaches (discussed below). It is a low-level model in the sense that careful attention must be paid to the placement of the entry and exit protocols to ensure that critical regions are properly protected.

There are other implementation approaches to concurrency that solve the critical region problem by prohibiting any direct interference between concurrent processes. This is done by not allowing any sharing of variables. The *monitor* approach places all shared variables and other resources under the control of a single monitor module, which is accessed by only a single process at a time. The *message-passing* approach is to share information only through messages passed from process to process. Both of these approaches are discussed in this chapter.

As well as avoiding interference in data access, we must avoid interference in the sharing of resources (e.g., keyboard input for multiple processes). Also, we must ensure that any physical actions of concurrent processes, such as movement of robotic arms, are appropriately synchronized.

Thus, to develop concurrent solutions, we require notations to:

1. Specify which portions of our processes can run concurrently
2. Specify which information and resources are to be shared
3. Prevent interference by concurrent processes by ensuring mutual exclusion
4. Synchronize concurrent processes at appropriate points

Further, any proposed solution to a concurrent problem must have certain properties (see, for example, [Ben-Ari, 1990]):

1. Safety: this property must always be true; examples include:
   a. Non-interference
   b. No *deadlock*, which occurs when no process can continue because all processes are waiting upon conditions that can never occur
   c. Partial correctness: whenever the program terminates, it has the correct answer
2. Liveness: this property must be true eventually; examples include:
   a. Program terminates (if it also has the correct answer, this is total correctness)
   b. No *race*: nondeterministic behavior caused by concurrently executing processes
   c. *Fairness*: each process has an opportunity to execute (this is affected by implementation and process/thread scheduling)

The verification or proof that solutions satisfy these properties is vastly complicated by the concurrent execution of code: particular orderings of code execution may exhibit interference or deadlock while others proceed nicely to termination. Returning to Conway's problem, suppose the execution of Process1 and Process2 are matched evenly so that each character placed by Process1 into Buffer1 is retrieved by Process2 before Process1 is ready to output another character. In this case, when tested, the program exhibits the desired correctness properties, lack of deadlock, etc. But if, due to a variation in processor workload or type, Process1 runs faster, then characters will be overwritten and lost; on the other hand, if Process2 runs faster, characters will be repeated. The fact that we tested the program under one particular set of circumstances (even for all possible inputs) is irrelevant to this issue. Sufficient testing is impossible because of the exponential explosion in the number of possible interleavings of instruction execution that can occur. The only fully satisfactory approach is to use formal methods (techniques that are still predominantly under development), which are touched on later in this chapter.

This chapter focuses on the software architectures used for concurrency, using a set of archetypical problems and their solutions for illustration. These problems are chosen because of the frequency with which they arise in computing; careful study of actual problems frequently leads to the realization that a seemingly complicated problem is, at heart, one of these archetypes. First, we briefly explore hardware architectures and their impact on software.

96.2   Hardware Architectures

Hardware can influence synchronization and communication primarily with respect to efficiency. *Multiprogramming* is the interleaving of the execution of multiple programs on a processor; on a uniprocessor, a time-sharing operating system implements multiprogramming. Although such an approach on a uniprocessor does not provide the execution speedup discussed in the introduction, it does provide the possibility of elegance and simplicity in problem solution, which is the second argument for the concurrent paradigm.

By employing multiple computers, we have *multiprocessing*, or parallel processing. Multiprocessing can involve multiple computers working on the same program or on different programs at the same time. If a multiprocessor system is built so that processors share memory, then processes can communicate via global variables stored in shared memory; otherwise, they communicate via messages passed from process to process. In contrast to a multiprocessor system, a distributed system is comprised of multiple computers that are remote from each other. This chapter focuses on multiprogramming and multiprocessing systems with a short introduction to the additional problems associated with distributed systems. In addition (but outside the scope of this chapter), a wide variety of hybrid hardware/software approaches exist.

96.3   Software Architectures

To specify a software architecture for implementing concurrency, we must provide the syntax and semantics to:

1. Specify which information and resources are to be shared
2. Specify which portions of processes can run concurrently
3. Prevent interference by concurrent processes by ensuring mutual exclusion
4. Synchronize concurrent processes at appropriate points

The first feature requires no special notation (shared variables are simply global), and the third and fourth are usually merged into one. A large number of software mechanisms have been proposed to support these features; in this chapter we explore the most widely used among them:

1. *Busy-wait*: implementable on virtually any processor without operating system support; this is concurrency without abstractions
2. *Semaphores*: historically the oldest satisfactory mechanism
3. *Monitors*: modules that encapsulate concurrent access to shared data
4. *Message passing*: a higher-level abstraction widely used in distributed systems

The references at the end of the chapter provide pointers to a number of other mechanisms, such as Unix fork/join, conditional critical regions, etc.

96.3.1 Busy-Wait: Concurrency without Abstractions

To illustrate the busy-wait mechanism, we use (following [Ben-Ari, 1982], a very simple example consisting of two concurrent processes, each with a single critical region. The only assumption made is that each memory access is atomic; that is, it proceeds without interruption. Our task is to ensure mutual exclusion; the purpose of the exercise is to demonstrate the care with which a solution must be crafted to ensure the safety and liveness properties discussed above.

Our first approach, which follows, is to ensure that the processes, p1 and p2, simply take turns in their critical regions.

```
global var turn := 1

process p1
  while true do ->
    # non-critical region
    ...
    # entry protocol
      while turn = 2 do ->
        <nothing>                        # wait for turn
    # critical region
    ...
    # exit protocol
      turn := 2
    # rest of computation
    ...
  end p1

process p2
  while true do ->
    # non-critical region
    ...
    # entry protocol
      while turn = 1 do ->
        <nothing>                        # wait for turn
    # critical region
    ...
    # exit protocol
      turn := 1
    # rest of computation
    ...
  end p2
```

This approach meets the desired properties but has a fundamental flaw: processes must take turns entering their critical regions. If p1 is ready and needs to execute its critical region at a higher frequency than p2, it cannot. The processes are an example of *co-routines*, historically one of the first approaches to concurrency.

If we modify the solution to allow each process to proceed into its critical region if the other process is not in its critical region, and to then notify the other process, we obtain the following (where c_i is used to signify that p_i is in its critical region):

```
global var c1 := false, c2 := false

process p1
  while true do ->
```

```
      # non-critical region
      ...
    # entry protocol
      while c2 do ->
         <nothing>                     # wait for turn
         c1 := true                    # p1 in critical region
      # critical region
      ...
    # exit protocol
      c1 := false                      # p1 out of critical region
    # non-critical region
  end p1

process p2
  while true do ->
    # non-critical region
    ...
    # entry protocol
      while c1 do ->
         <nothing>                     # wait for turn
         c2 := true                    # p2 in critical region
      # critical region
      ...
    # exit protocol
      c2 := false                      # p2 out of critical region
    # non-critical region
  end p2
```

Now, however, we have the possibility that the mutual exclusion requirement of the critical region can be violated; that is, both processes can be in their critical regions at the same time. (For example, suppose both $c1$ and $c2$ are false; $p1$ checks $c2$ via the loop and decides that it may enter its critical region; before it sets $c1$ to true, $p2$ checks $c1$ via its loop and decides that it may enter its critical region).

As shown next, this disastrous possibility can be eliminated by having a process announce its intent to enter into its critical region before checking whether it can enter:

```
global var c1 := false, c2 := false

process p1
  while true do ->
    # non-critical region
    ...
    # entry protocol
      c1 := true                       # signal intent to enter
      while c2 do ->
         <nothing>                     # wait for turn
      # critical region
      ...
    # exit protocol
      c1 := false                      # p1 out of critical region
    # non-critical region
  end p1
process p2
```

```
    while true do ->
      # non-critical region
      ...
      # entry protocol
      c2 := true                  # signal intent to enter
        while c1 do ->
          <nothing>               # wait for turn
      # critical region
      ...
      # exit protocol
      c2 := false                 # p1 out of critical region
    # non-critical region
  end p2
```

But now we have raised the possibility of a race (when p1 sets c1 to true *and* p2 sets c2 to true).

A possible solution to this difficulty, which appears below, moves the announcement statement into the loop, together with a random delay:

```
global var c1 := false, c2 := false

process p1
  while true do ->
    # non-critical region
    ...
    # entry protocol
    c1 := true                  # signal intent to enter
      while c2 do ->
        c1 := false             # give up intent if p2 already
                                # in critical region
        <delay>
        c1 := true              # try again
    # critical region
    ...
    # exit protocol
    c1 := false                 # p1 out of critical region
    # non-critical region
    ...
end p1
process p2
  while true do ->
    # non-critical region
    ...
    # entry protocol
    c2 := true                  # signal intent to enter
      while c1 do ->
        c2 := false             # give up intent if p1 already
                                # in critical region
        <delay>
        c2 := true              # try again
    # critical region
    ...
    # exit protocol
```

```
            c2 := false                    # p2 out of critical region
       # non-critical region
          ...
     end p2
```

But this is not a satisfactory solution because it exhibits a race in the (unlikely) situation that the two loops proceed in perfect synchronization.

A valid solution, such as that which appears below, can be developed by returning to the concept of taking turns when applicable, which ensures mutual exclusion while not requiring alternating turns (thus allowing true concurrency):

```
global var c1 := false, c2 := false, turn := 1

process p1
  while true do ->
    # non-critical region
      ...
    # entry protocol
      c1 := true                    # signal intent to enter
      turn := 2                     # give p2 priority
      while c2 and turn = 2 do ->
        <nothing>                   # wait if p2 in critical region
    # critical region
      ...
    # exit protocol
      c1 := false                   # p1 out of critical region
    # non-critical region
      ...
  end p1
process p2
  while true do ->
    # non-critical region
      ...
    # entry protocol
      c2 := true                    # signal intent to enter
      turn := 1                     # give p1 priority
      while c1 and turn = 1 do ->
        <nothing>                   # wait if p2 in critical region
    # critical region
      ...
    # exit protocol
      c2 := false                   # p1 out of critical region
    # non-critical region
      ...
  end p2
```

This solution is due to Peterson [1983]; the first valid solution was presented by Dekker.

The importance of the busy-wait approach is threefold:

1. It provides a nice introduction to the problems inherent in designing concurrent solutions.
2. It is executable on virtually every machine architecture without additional further software support and is, thus, suitable for micro-controllers, etc.
3. Variants are frequently used in hardware implementations.

However, this approach also suffers from two difficulties:

1. It is very inefficient: machine cycles are expended when executing busy-wait loops.
2. Programming at such a low level is highly prone to error.

96.3.2 Semaphores

Dijkstra [1968] presented the first abstract mechanism for synchronization in concurrent programs. The *semaphore*, so named in direct relation to the semaphores used on railroad lines to control traffic over a single track, is a non-negative integer-valued abstract data type with two operations:

```
P(s) : delay until s > 0, then s := s - 1
V(s) : s := s + 1
```

When a process delays on a semaphore, it is awakened only when another process executes a V operation on that semaphore. Thus, it uses no machine cycles to check if it can proceed. If more than one process is delaying on a semaphore, only one (which one is implementation dependent) can be awakened by a V operation.

Additionally, the value of s can be set at instance creation via the semaphore declaration; if set to 0, then some process must execute the V(s) operation before any processes first executing the P(s) operation can continue. With this abstract data type, we have a mechanism that handles both interference and synchronization.

Additional notes:

1. These are the *only* two synchronization operations defined; in particular, the value of s is not determinable.
2. Implementation of these operations must be either in the hardware or in the (non-interruptible) system kernel.
3. By sleeping while waiting for a semaphore (the delay in P(s)), a process does not waste machine cycles by repeated checking.
4. The operation names (P and V) come from the Dutch words *passeren* (to pass) and *vrygeven* (to release); sometimes, `signal` and `wait` are used in place of P and V, respectively.
5. Each of the P and V operations proceeds atomically; that is, it may not be interrupted by another process.

The use of the semaphore in concurrent programming relates directly to the railroad analogy. Each critical section looks like the following:

```
global var s : semaphore := 1

# entry protocol
  P(s)
# critical region
  ...
# exit protocol
  V(s)
```

The initialization of s to 1 ensures that the first process executing P(s) will continue. (Deadlock arises if s were initialized to 0.) Only the first process to reach its P(s) statement is allowed to proceed, as subsequent processes find s = 0 and delay. When the first process finishes its critical region, it executes V(s), which sets s to 1. One of the waiting processes is awakened, find s > 0, decrements s, and proceeds. Note the importance that these operations are atomic; this ensures that two processes cannot wake up and each find s > 0.

96.3.2.1 Semaphores and Producer-Consumer

The Producer-Consumer problem arises whenever one process is creating values to be used by another process. Examples are Conway's problem and buffers of various kinds, etc. Here we first look at the multi-element buffer version of this problem and then add multiple producers and consumers as a refinement.

```
# define the buffer
  const N := ...                    # size
  var buf[N] : int                  # buffer
      front := 1                    # pointers
      rear := 1
  semaphore empty := N              # counts the number of empty slots
                                        in the buffer
              full := 0             # counts the number of items
                                        in the buffer
process producer
var x : int
while true do ->
  # produce x
    ...
  P(empty)                          # delay until there is space in the buffer
  buf[rear] := x                    # place value in the buffer
  V(full)                           # signal that the buffer is non-empty
  rear := rear mod N + 1            # update buffer pointer
end producer

process consumer
var x : int
while true do ->
  P(full)                          # delay until a value is in the buffer
  x := buf[front]                  # obtain value
  V(empty)                         # signal that the buffer is not full
  front := front mod N + 1         # update buffer pointer
  # consume x
    ...
end consumer
```

The buffer processing is conventional; only the actual buffer access must be placed into a critical region because there is no possibility of interference between the assignments to `rear` and `front`. Note also the use of two semaphores: `empty` to signal that the producer can proceed because there is at least one empty slot in the buffer and `full` to signal that the consumer can proceed because there is at least one item in the buffer. Although it is possible to solve this problem with one semaphore, less concurrency results. Note that the `empty` semaphore is initialized to N, the size of the buffer. The producer process can run up to N steps ahead of the consumer process.

To allow multiple producers and/or consumers, we must protect the actual buffer operations with additional semaphores to prevent, for example, two producers from accessing `rear` simultaneously with read and assignment operations. These semaphores, `mutexR` and `mutexF`, guarantee mutual exclusion of access to the `rear` and `front` pointers, respectively. It is not sufficient to use `empty` here because up to N producers will be able to continue through the P(empty) statement.

```
# define the buffer as previously
semaphore empty := N, full := 0
```

```
semaphore mutexR := 1        # mutual exclusion on rear pointer
          mutexF := 1        # mutual exclusion on front pointer

process pi                   # one for each producer
var x : int
while true do ->
  # produce x
  ...
  P(empty)                   # delay until there is space in the buffer
  P(mutexR)                  # delay until rear pointer is not in use
  # place value in the buffer and modify pointer
    buf[rear] := x; rear := rear mod N + 1
  V(mutexR)                  # release rear pointer
  V(full)                    # signal that the buffer is non-empty
end pi

process ci                   # one for each consumer
var x : int
while true do ->
  P(full)                    # delay until a value is in the buffer
  P(mutexF)                  # delay until front pointer is not in use
  # access the value in the buffer and modify pointer
  x := buf[front]; front := front mod N + 1
  P(mutexF)                  # release front pointer
  V(empty)                   # signal that there is space in the buffer
  # consume x
  ...
end ci
```

96.3.2.2 Semaphores and Readers-Writers

The Readers-Writers model captures the fundamental actions of a database; i.e.,

- No exclusion between readers
- Exclusion between readers and a writer
- Exclusion between writers

In other words, the software must guarantee only one update of a database record at a time, and no reading of that record while it is being updated.

The simplest semaphore solution is to wait only for the first reader; subsequent readers need not check because no writer can be writing if there is already a reader reading (here, nr and nw are the numbers of active readers and writers, respectively):

```
...
nr := nr + 1
if nr = 1 -> P(rw)    # if no one is presently reading,
                      # then ensure no one is writing
                      # before proceeding
# access database
...
nr := nr - 1
if nr = 0 -> V(rw)    # if no more are reading, possibly wake up
                      # writer, or prepare for next reader
```

```
   ...
   P(rw)                      # delay until no readers or writers
   # access database
   ...
   V(rw)                      # wake up delayed reader or writer, or prepare
                              # for next reader or writer
```

This solution gives readers preference over writers: new readers continually freeze out waiting writers. Extending this solution to other kinds of preferences, such as writer preference or first-come-first-served preference is cumbersome.

A more general approach is known as "passing the baton"; it is easily extended to other kinds of preferences because control is explicitly handed from process to process. Although a careful explanation of the approach is not given here, the concept is easily summarized. A process must check to ensure that it can legally proceed before doing so; if it cannot proceed, the process waits upon a semaphore assigned to it. For example, a writer process checks to see if no readers or writers are executing on the database before it proceeds; if they are executing on the database, then the writer process sleeps, waiting upon the semaphore assigned to it. When a process is finished accessing the database, it checks the conditions and wakes up (via signaling on the appropriate semaphore) one of the processes waiting upon the condition. This last operation essentially "passes the baton" from one process to another. The key is that first a check is made to ensure that it is legal for the other process to wake up. The strength of the "passing the baton" approach emerges when its flexibility is used to develop more general solutions. Details may be found in Andrews [1991].

96.3.2.3 Difficulties with Semaphores in Software Design

While the use of semaphores does provide a complete solution to the interference problem, the correctness of the solution directly depends on the correct usage of the semaphore operations, which are fairly low-level and unstructured. Semaphores and shared variables are global to all processes and, like any global data structure, their correct usage requires considerable discipline by the programmer. Additionally, if a large system is to be built, any one implementor is likely responsible for only a portion of the semaphore usage so that correct pairing of Ps and Vs may be difficult. Despite this difficulty, semaphores are a widely used construct for concurrency.

96.3.3 Monitors

A more structured approach is to encapsulate the shared data/resources and their operations into a single module called a *monitor*. A monitor can contain non-externally accessible data and procedures that handle the state of resources. External access is strictly controlled through procedure calls to the monitor; mutual exclusion is ensured because procedure execution within the monitor is not concurrent.

Monitors have the traditional advantages of abstract data types, but they must also deal with two issues rising from their use by concurrently executing processes: avoiding interference and providing synchronization. This section illustrates some sample applications of monitors and how they internally handle concurrency.

Returning to the Producer-Consumer problem, we implement a monitor for handling shared access to the buffer. The monitor requires a synchronization mechanism to ensure that the Producer cannot overfill the buffer and that the Consumer cannot retrieve from an empty buffer. Monitors implement *condition variables*, the values of which are queues of processes delayed upon the corresponding conditions. Two standard operations defined on conditional variable cv are:

1. wait(cv): causes the executing process to delay and to be placed at the end of cv's queue; in order to allow eventual awakening of the process, the process must relinquish exclusive access to the monitor when it executes a wait.
2. signal(cv): causes the process at the head of cv's queue to be awakened; if the queue is empty, there is no effect.

Although these operations mirror those of semaphores, there is a key difference: the `signal` operation has no memory.

96.3.3.1 Monitors and Producer-Consumer

The buffer monitor can be defined as follows:

```
monitor Buffer
  # define the buffer
    const N := ..              # size of the buffer
    var buf[N] : int           # buffer
        front := 1             # buffer pointers
        rear := 1
  # define the condition variables
    var not_full,              # signaled when count < N
        not_empty : cv         # signaled when count > 0
  procedure deposit(data : int)
    if count = N               # check for space
       then wait(not_full)     # delay if no space
    buf[rear] := data
    rear := (rear mod N) + 1
    N := N + 1
    signal(not_empty)          # signal non-empty
  end
  procedure fetch(var data : int)
    if count = 0               # check for not empty
       then wait(not_empty)    # delay if empty
    data := buf[front]
    front := (front mod N) + 1
    N := N - 1
    signal(not_full)           # signal not full
  end Buffer
```

Using this monitor, the producer and consumer tasks can be redone as follows:

```
process Producer
var x : int
while true do ->
  # produce x
  ...
  deposit(x)
end Producer

process Consumer
var x : int
while true do ->
  fetch(x)
  # consume x
  ...
end Consumer
```

Now it is clear that programming (outside of the monitor) can now be done at a more abstract level, which will lead to more reliable software.

96.3.3.2 Difficulties with Monitors

There are difficulties with monitors as well. Consider the case where we have two consumers, C1 and C2. If the buffer is empty when C1 executes fetch, then C1 will delay on not_empty. If the producer then executes deposit (note that deposit and fetch cannot be executed concurrently), it will eventually signal(not_empty), which will awaken C1. But if C2 executes fetch before C1 continues execution and its call to fetch proceeds, then C1 will access an empty buffer. Hence, the signal operation must be considered to be a *hint* that proceeding with execution is possible, but not that it is correct. The following two approaches are used to solve this problem:

1. Replace the check on the condition variable with a check inside a loop to ensure that the condition is true before execution proceeds. For example:

```
procedure deposit(data : int)
   while count = N do ->      # check for space
      then wait(not_full)     # delay if no space
   buf[rear] := data
   rear := (rear mod N) + 1
   N := N + 1
   signal(not_empty)         # signal non-empty
end
```

2. Give the highest priority to awakening processes so that intervening access to the monitor is not possible; this also requires that the signal operation be the last operation executed in any procedure in which it occurs (to ensure that two processes will not be executing within the monitor).

Monitors form the basis for concurrent programming in a number of systems and provide an efficient, high-level synchronization mechanism. They have the further advantage, as do other abstract data types or objects, of allowing for local modification and tuning without affecting the remainder of the system.

96.3.4 Message Passing

Consider a hardware architecture with multiple independent computers. Creating a semaphore to be efficiently accessed by processes running on separate computers is a difficult problem. We need a new abstraction for this case: message passing in which a sending process outputs a message to a *channel* and a receiving process inputs the message from this same channel. There are a large number of variations of this basic concept, depending on the semantics of the operations and the channels.

The basic primitives are:

1. Channel declaration
2. send <channel> <message>
3. receive <channel> <variable>

If both sending and receiving processes block upon reaching their corresponding message-passing operation, we have *synchronous* communication; if the sending process can send a message and continue without waiting for receipt, the system is *asynchronous*. Analogies are telephone communication and the postal system. The synchronous approach allows for ready synchronization of processes (at the instant of message passing we know where both are in their execution). This was the approach chosen by Hoare [1985] for his communicating sequential processes model and its subsequent implementation in the *occam* language [Jones and Goldsmith, 1988]. If we desire asynchronicity, we can add intermediate buffer processes to the synchronous approach. An advantage of synchronous message passing is that it often simplifies analysis of an algorithm because it is known where the sending and receiving processes are in their execution at the moment the message is passed.

Further variations arise, depending on whether channels are one process-to-one process or one-to-many, statically instantiated at load time or dynamically created during execution, bi-directional or uni-directional, whether the receiving process must be named by the sending process, etc. However, the basic concept is the same in all cases; ease of use and efficiency of implementation vary.

Further variations include *remote procedure call* (RPC), which is the core of many distributed systems, and *rendezvous*, the approach used in Ada. We further explore these approaches after looking more closely at simple message passing.

Note that, in the message-passing approach, there are no shared variables so interference is not an issue. The critical section issue does not arise because there is no way for concurrent processes to interfere with each other. This is one of the major motivating factors for the use of message-passing software architectures.

96.3.4.1 Message Passing and Producer-Consumer

If the message-passing system is asynchronous, as demonstrated below, we can rely on the system itself to buffer values:

```
channel P2C

process Producer
int x
while true do ->
  # produce x
  send P2C x
end Producer

process Consumer
int x
while true do ->
  receive P2C x
  # consume x
end Consumer
```

Using this approach, the `Producer` sends a message over channel P2C and continues producing and sending (up to channel capacity at which point the system blocks), while the `Consumer` blocks at the `receive` statement if no messages are available.

If our system is synchronous, then as shown below, we create a separate buffer process:

```
channel P2B, B2C

process Buffer
# create the buffer
  const N := ..
  var buffer[N] : int
      front := 1
      rear := 1
      count := 0                 # number of items in the buffer
while true do ->
  if
    # there is room and the producer is sending
      count < n and receive P2B buffer[rear] ->
        count++; rear := rear mod n + 1
  else
    # there are items and the consumer is receiving
      count > 0 and send B2C buffer[front] ->
        count--; front := front mod n + 1
end Buffer
```

```
process Producer
var x : int
while true do ->
  # produce x
    ...
  send P2B x
end Producer

process Consumer
var x : int
while true do ->
  receive B2C x
  # consume x
end Consumer
```

Above the `if` statement is nondeterministic; that is, any `true` clause can be selected. The Boolean conditions in the clauses are called *guards*. The clauses are:

- If there is room and the producer wishes to send a character
- If there are items to retrieve and the consumer wishes to receive a character

For implementation efficiency reasons, actual programming languages do not allow guards for both input and output statements, so we must modify our solution; for example, as shown below, we can modify the buffer and consumer processes to eliminate the output guard:

```
channel P2B, B2C, C2B

process Buffer
# define the buffer
  var buffer[n] : int
  var front := 1
      rear := 1
      count := 0
while true do - >
  if
    # there is room and the producer is sending
      count < n and receive P2B buffer[rear] ->
        count++; rear := rear mod n + 1
  else
    # there are items and the consumer is requesting
      count > 0 and receive C2b buffer[front] ->
        send B2C buffer[front]
        count--; front := front mod n + 1
end Buffer

process Producer
var x : int
while true do ->
  # produce x
```

```
      ...
   send P2B x
end Producer

process Consumer
var int : x
while true do ->
   send C2B NIL          # announce ready for input
   receive B2C x
   # consume x
   ...
end Consumer
```

Above, the Consumer process first announces its intention to receive a value from the Buffer process (`send C2B NIL`; the NIL signifying that no message need be actually exchanged) and then actually receives the value (`receive B2C x`).

This program is an example of *client/server* programming. The Consumer process is a client of the Buffer process; that is, it requests service from the buffer, which provides it. Client/server programming is widely used to provide services across a network and is based on the message-passing paradigm.

96.3.4.2 Message Passing and Readers-Writers

The message-passing approach to Readers-Writers is straightforward: do not accept a message from a reader or writer if a writer is writing; do not accept a message from a writer if a reader is reading. The solution, shown below, is simple if we adopt synchronous message passing and the notion of the database as a server:

```
channel Rrequests, Rreceives, Wsends

Reader
   send Rrequests <request message>
   receive Rreceives <data>

Writer
   send Wsends <write message>

Server
   if
       # there are no writers, accept reader requests
       nw = 0 ->
           receive Rrequests <request message>
           # access the database
           ...
           send Rreceives <data>
       # there are no readers or writers, accept writer
         requests
         nr = 0 and nw = 0 ->
           receive Wsends <write message>
           # modify the database
           ...
```

96.3.4.3 Message Passing and Semaphore Simulation

Of course, as we show next, message passing can simulate a semaphore (and vice versa if need be):

```
channels P, V, initSemaphore

process Semaphore
var s : int
receive initSemaphore i
s := i
while true do ->
  if
      # semaphore is non-zero accept P operation
        s > 0 and receive P NIL->
          s--
      # always accept V operation
        receive V NIL ->
          s++
end Semaphore
```

96.3.4.4 The Remote Procedure Call and Rendezvous Abstractions

The remote procedure call, or RPC, abstraction is widely used to provide client/server services in a distributed system. Revisiting the client/server examples above, it is clear that the client executes a `send-receive` pair while the server executes a `receive-send` pair. Using the standard procedure model to capture the server's actions, a call statement to capture the client's actions and parameters to capture the messages being sent, we have:

```
Client
  ...
  call Server(args)
  ...

Server(formal args)
  ...
  return
```

which mirrors traditional procedure calls. The difference is that the `Server` procedure can be on a machine remote to the `Client` process. Indeed, the `Server` is implemented as a process that is always delayed until a `Client` executes a call. If multiple `Client`s concurrently execute calls to a `Server`, the `Server` must be re-entrant or must provide protection for shared information. The RPC approach forms the basis for distributed systems programs on a wide variety of platforms; its relationship to monitors should be clear.

The calling process and procedure are not truly concurrent in the sense used throughout this chapter, in that the calling process delays once the call is made, the procedure does not execute until called, the procedure delays when the return is executed, and the calling process resumes execution only upon the return from the procedure. The model is similar to that of synchronous message passing if the execution of the procedure is viewed as a component of the message-passing process (essentially, the procedure creates the return message).

We can increase the power of this approach if we modify the procedure into a process and have both processes executing concurrently. When a call is made, execution of the calling process delays while execution of the called process continues until it is ready to accept the call (via a special statement). The called process continues execution, performing actions or calculating values for the return message. The return message is sent back to the caller, the called process continues executing, and the calling process resumes execution once the message is received. Because there is an extended time period during which the two processes are synchronized (from called accept through called return), this model of concurrency is termed *rendezvous*. It is the basis for the model of concurrency used in the Ada language. The Ada model is not symmetric: the calling process must know the name of the process it is calling, but the called process need not know its caller. Accept statements may have guards, as discussed above for message passing, in order to control acceptance of calls. The complexity of these guards, and their priority, must be carefully followed during program implementation.

There are several advantages to this approach, all based on the possibility of the called routine using multiple accept statements:

1. The called routine can provide different responses to the calling process at different stages of its execution.
2. The called routine can respond differently to different calling processes.
3. The called routine chooses when it will receive a call.
4. Different accept statements can be used to provide different services in a clear fashion (rather than through parameter values).

96.3.4.5 Difficulties with Message Passing

Message-passing systems are frequently inefficient during execution unless the algorithm is developed carefully. This is because messages take time to propagate, and this time is essentially overhead. For example, a single element buffer version of Conway's problem spends significantly more time exchanging messages than any other operation.

96.4 Distributed Systems

In addition to the difficulties inherent in developing and understanding concurrent solutions, distributed systems contain the fundamental problem of identifying global state. For example, how do we determine if a program has terminated? In the sequential case, this is obvious, we execute the exit or end statement. In the concurrent case, we must ensure that all processes are ready to terminate. In the multiprogramming case, we can do this by checking the ready queue; if it is empty, then there are no processes waiting to run, which ensures that no process will ever be added to the ready queues (if no process can run, then there can be no changes to create another ready process). But if we are in a distributed system, there is no single ready queue to examine. If a process is in the suspended queue on its processor, it may be made ready by a message from a process on a different processor.

Similarly, we may still require mutual exclusion on a system resource — how do we ensure access across processors? The solution is to develop a method of determining global state; see, for example, Ben-Ari [1990].

While a true "distributed" paradigm has not yet emerged in the programming paradigms domain, it will most likely evolve in the area of operating systems; for more information on distributed computing, readers are encouraged to look at Chapter 108 in this *Handbook*.

96.5 Formal Approaches

We argued above that software verification in concurrent programming must take into account the enormous number of possible interactions between concurrent processes. Obviously, traditional testing only demonstrates the presence of "good" execution histories and is not a mechanism to verify any solution — sequential or concurrent. The use of a trace routine to generate execution histories is a standard sequential technique that becomes infeasible in the concurrent domain. Consider, for example, that n processes each executing m atomic actions generates $(n*m)!/(m!)^n$ histories. For three processes, each executing only two actions, this is a total of 90 possible histories!

The alternative is to use a formal, mathematically rigorous method to develop a solution and/or to verify a complete solution. Two approaches have been applied to verifying concurrent software:

1. Axiomatic or assertional
2. Process algebraic

The axiomatic approach develops assertions in the predicate logic that characterize the possible states of a computation. The actions of a program are viewed as predicate transformers that move the computation from one state to another. The beginning state is specified by the pre-condition of the computation, and the final state is characterized by the post-condition. This approach has been exploited for some time in the sequential paradigm; see Schneider [1997] for a comprehensive introduction to the field in the context of concurrency.

The process algebraic approach was pioneered by Hoare [1985], who also pioneered the coarse-grained model of concurrency. The concept is that the interactions between a system and its environment (which are all that is ultimately observable) can be modeled via a mathematical abstraction called a process (this is the abstraction of the computing process as used above). Processes can be combined via algebraic laws to form systems. Communication between processes is an example of this interaction. By building up a system through these mathematical laws and then transforming the abstract mathematics into an implementable language, one arrives at a correct solution. The *occam* language was designed to match the algebraic laws devised by Hoare; transformations exist between these laws and *occam* programming constructs (but the transformations are not perfect due to practicalities of implementation) [Hinchey and Jarvis, 1995]. A number of subsequent efforts developed process algebras with varying properties [Milner, 1989]; see Magee and Kramer [1999] for the use of a process algebra in the development of Java programs.

Although both approaches are in active use, they are not typically applied in the concurrent paradigm with any greater frequency than they are in the sequential paradigm, and they remain primarily research tools. The fundamental difficulty is that theoreticians search for the "fundamental particles" of computing to develop mathematical laws enabling formal reasoning. Practical languages are (inherently) extremely complex mixtures of these fundamental particles and laws in order to have sufficient power to solve real-world problems. Theoretical tools do not yet scale to these large, complex problems.

96.6 Existing Languages with Concurrency Features

A large number of languages have been developed to use the concurrency paradigm; most have remained in the laboratory environment. If the underlying operating system provides the requisite support, then semaphores can be implemented in any language via system calls. Higher-level concurrency control structures require modification of the underlying sequential language; for example, Concurrent Pascal [Brinch Hansen, 1975] uses monitors while Concurrent C [Gehani and Roome, 1986] is based on the rendezvous. By beginning with a widely used sequential programming language, a designer has a large community from which to draw users to the new language. The Ada (concurrency based upon the rendezvous) and SR (which includes structures for all of the approaches discussed in this chapter and is therefore particularly useful for exploring concurrent programming) [Andrews and Olsson, 1993; see Hartley, 1995, for extensive examples] languages are examples of sequential languages with concurrent structures included from the initial stages of development.

Object-oriented languages have similarly had concurrency features added. For example, Smalltalk has the Process and Semaphore classes to provide for the dynamic creation of independent processes and their interaction using the semaphore approach [Goldberg and Robson, 1989].

Languages based on an inherently concurrent model include Linda (more a language-independent philosophy than a language) [Ahuja et al., 1986] and *occam* (synchronous message passing) [Jones and Goldsmith, 1988].

A different approach is to provide a standardized interface (an application program interface or API) that is language independent. A language implementation then provides a set of library routines to implement this API. Thus, programmers can use a language of their choice while being assured that their program will function correctly. Currently, the two main paradigms that are the basis for writing parallel programs are message passing and shared memory. A hybrid paradigm is used in systems comprised of shared-memory multiprocessor nodes that communicate via message passing. For writing message-passing programs, MPI (Message Passing Interface) [http://www-unix.mcs.anl.gov/mpi/index.html] is a widely used standard; many variants of MPI exist, including MPICH, CH for Chameleon, which is a complete, freely-available implementation of the MPI specification, targeted at high performance [http://www-unix.mcs.anl.gov/mpi/mpich/].

MPI's interface includes features of a number of message-passing systems and attempts to provide portability and ease-of-use. The MPI programming model is an MPMD (multiple program multiple data) model, in which every MPI process can execute a different program. A computation is envisioned as one or more processes that communicate by calling library routines to send and receive messages to other processes. In general, a fixed set of processes, one for each processor, is created at program initialization (versions of MPI that will support dynamic creation and termination of processes are anticipated). Local and global communication (e.g., broadcast and summation) is provided by point-to-point and collective communication operations, respectively. The former is used to send messages from one named process to another, while the latter is used to provide message passing among a group of processes. Most parallel algorithms are readily implemented using MPI. If an algorithm creates just one task per processor, it can be implemented directly with point-to-point or collective communication routines that meet its communication requirements. In contrast, if tasks are created dynamically or if several tasks are executed concurrently on a processor, the algorithm must be refined to permit an MPI implementation.

The OpenMP API is becoming a standard that supports multi-platform shared-memory parallel programming in C/C++ and Fortran on all architectures, including Unix and Windows NT platforms. OpenMP is a portable, scalable model that gives shared-memory parallel programmers a simple and flexible interface for developing parallel applications for platforms ranging from the desktop to the supercomputer [http://www.openmp.org/]. This API is jointly defined by a group of major computer hardware and software vendors. OpenMP can be used to explicitly direct multi-threaded, shared memory parallelism. It is comprised of three primary API components: compiler directives, runtime library routines, and environment variables. Using the fork/join model of parallel execution, an OpenMP program begins as a single master thread. The master thread creates or forks a set of parallel threads, which concurrently execute a parallel region construct. On completion, the threads parallel threads join (i.e., synchronize and terminate), leaving only the master thread. The API supports nested parallelism and dynamic threads, that is, dynamic alternation of the number of active threads. Variable scoping, for example, declaration of private and shared data, parallelism, and synchronization are specified through the use of compiler directives. By itself, OpenMP is not meant for distributed memory parallel systems. For example, for high-performance cluster architectures such as the IBM SP, where intranode communication is accomplished via shared memory and internode communication is performed via message passing, OpenMP is used within a node while MPI is used between nodes.

There are many parallel programming tools available that help the user parallelize her/his application and then easily port it to a parallel machine. These machines can be shared-memory machines or a network of workstations.

96.7 Research Issues

While it is clear that concurrency is a necessary technique for the solution of many problems, it also is clear that progress must be made in order to ensure its effective application. That this is still a research issue is clear whenever an operating system crashes due to system processes that interfere with each other or we discover someone in our airplane seat due to concurrent access to the airline's database. This required progress falls into three categories:

1. Theoretical advances must be made to develop formal techniques that scale to real-world applications. For example, process interference checkers exist, but operate essentially by checking all

possible interactions between processes to check for deadlock, etc. This approach rapidly develops combinatorial explosion.

2. Design tools that provide development support for concurrent solutions. For example, debuggers that capture the concurrent computation without overwhelming the user with information.

3. Languages with powerful structures to support the correct application of concurrency. For example, the development of concurrent object-oriented languages appears straight-forward: simply allow each object to run concurrently because each object is logically autonomous. However, there are a number of issues that need resolution, including:

   a. Not all objects need to run concurrently because the majority of computation will still be sequential (thereby incurring no scheduler overhead).

   b. If we consider multiple concurrent objects attempting to communicate with the same object:

      i. Acceptance of a message must delay all other messages in order to correctly preserve the internal state of the object.

      ii. Ordering of message acceptance must be synchronized to ensure computations are correct.

      iii. Acceptance of messages must occur only at appropriate points in the object's execution.

   c. Inheritance through the class hierarchy creates problems because it will mix this synchronization with object behavior.

96.8   Summary

The single outstanding problem with concurrency is the development of correct solutions (as it is in all software systems): the state of development of both formal methods and software engineering tools for concurrent solutions lags behind the sequential world in this regard and well behind hardware advances.

Defining Terms

Asynchronous message passing: The message-sending process allows messages to be buffered and the sending process may continue after the send is initiated; the receiving process blocks if the message queue is empty.

Channel: The data structure, which may be realized in hardware, over which processes send messages.

Client/server: The software architecture in which clients are able to request services of processes executing on remote machines.

Condition variables: A variable used within a monitor to delay an executing process.

Critical regions: A section of code that must appear to be executed indivisibly.

Deadlock: The state in which processes are waiting for events that can never occur; that is, the processes cannot progress.

Distributed processing: The use of multiple processors that are remote from each other.

Fairness: Processes will eventually be able to progress, that is, enter their critical regions.

Message passing: A technique for providing mutual exclusion, communication, and synchronization among concurrent processes via sending messages between processes.

Monitor: An encapsulation of a resource and the operations on that resource that serve to ensure mutual exclusion.

Multiprocessing: The use of multiple processors.

Multiprogramming: Simulating concurrency by interleaving instruction execution from multiple programs; time sharing or time slicing.

Mutual exclusion: The property ensuring that a critical region is executed indivisibly by one process or thread at a time.

Race: Nondeterministic behavior caused by incorrectly synchronized concurrent processes.

Remote procedure call: The message-passing architecture in which processes request services of processes executing procedures on remote machines.

Rendezvous: The message-passing construct used in the Ada language.

Semaphore: A nonnegative integer-valued variable on which two operations are defined: P and V to signal intent to enter and exit, respectively, a critical region.

Synchronous message passing: The message-sending process requires both sender and receiver to synchronize at the moment of message transmission.

References

Journals

Ahuja, S., Carriero, N., and Gelernter, D. 1986. Linda and Friends. *Computer*, 19(8):26–34.

Andrews, G. R. and Schneider, F. B. 1983. Concepts and notations for concurrent programming. *Comp. Surv.*, 15(1):3–43; reprinted in Gehani, N. and McGettrick, A. D. 1988. *Concurrent Programming*. Addison-Wesley, New York.

Brinch Hansen, P. 1975. The Programming Language Concurrent Pascal. *IEEE Trans. on Software Engineering*, 1(2):199–207; reprinted in Gehani, N. and McGettrick, A. D. 1988. *Concurrent Programming*. Addison-Wesley, New York.

Dijkstra, E. W. 1968. The structure of the T. H. E. multiprogramming system. *CACM*, 11:341–346.

Gehani, N. H. and Roome, W. D. 1986. Concurrent C. *Software: Practice and Experience*, 16(9):821–844; reprinted in Gehani, N. and McGettrick, A. D. 1988. *Concurrent Programming*. Addison-Wesley, New York.

Peterson, G. L. 1983. A new solution to Lamport's concurrent programming problem using small shared variables. *ACM Trans. Prog. Lang. and Syst.*, 5(1):56–55.

Books

Andrews, G. R. 2000. *Foundations of Multithreaded, Parallel, and Distributed Programming*. Benjamin-Cummings, New York.

Andrews, G. R. and Olsson, R. A. 1993. *The SR Programming Language*. Benjamin-Cummings, New York.

Ben-Ari, M. 1982. *Principles of Concurrent Programming*. Prentice Hall, London.

Ben-Ari, M. 1990. *Principles of Concurrent and Distributed Programming*. Prentice Hall, London.

Bernstein, A. J. and Lewis, P. M. 1993. *Concurrency in Programming and Database Systems*. Jones and Bartlett, Boston.

Filman, R. E. and Friedman, D. P. 1984. *Coordinated Computing*. McGraw-Hill, New York.

Gehani, N. and McGettrick, A. D. 1988. *Concurrent Programming*. Addison-Wesley, New York.

Goldberg, A. and Robson, D. 1989. *Smalltalk–80 The Language*. Addison-Wesley, New York.

Hartley, S. J. 1995. *Operating Systems Programming*. Oxford, New York.

Hinchey, M. G. and Jarvis, S. A. 1995. *The CSP Reference Book*. McGraw-Hill, New York.

Hoare, C. A. R. 1985. *Communicating Sequential Processes*. Prentice Hall, London.

Jones, G. and Goldsmith, M. 1988. *Programming occam 2*. Prentice Hall, New York.

Lester, B. P. 1993. *The Art of Parallel Programming*. Prentice Hall, New Jersey.

Magee, J. and Kramer, J. 1999. *Concurrency: State Models and Java Programs*. Wiley, West Sussex.

Milner, R. 1989. *Communication and Concurrency*. Addison-Wesley, New York.

Schneider, F. 1997. *On Current Programming*. Springer-Verlag, New York.

Wilkinson, B. and Allen, M. 1999. *Parallel Programming: Techniques and Applications Using Networked Workstations and Parallel Computers*. Prentice Hall, New Jersey.

Further Information

Further information can be gleaned from a number of sources; particularly recommended are Andrews [2000] for a comprehensive view of the field with an axiomatic flair, including a fascinating bibliography with historical notes and extensive problem sets; Schneider [1997] for a graduate level treatise on axiomatic semantics in the context of concurrency; Ben-Ari [1982] for a nice introduction including problem sets; Ben-Ari [1990], which adds Ada code examples, correctness arguments, and distributed computing; the process algebra approach is developed in Hoare [1985] and Milner [1989] and demonstrated in Magee and Kramer [1999]; Filman and Friedman [1984] emphasize the various models of concurrent computation; Lester [1993] provides a comprehensive introduction including efficiency considerations, but without correctness arguments; Bernstein and Lewis [1993] use the axiomatic approach to develop concurrent solutions to a variety of problems with an emphasis on databases; Gehani and McGettrick [1988] reprint a number of the classic papers in the field. Wilkinson and Allen [1999] demonstrate parallel programming for a wide range of problems.

The journal *Concurrency: Practice and Experience* focuses on practical experience with concurrent machines and concurrent solutions to problems; concurrency is also frequently dealt with in a large number of society journals.

In addition, there are a large number of resources available via the Web that may be discovered through the use of the various search techniques.

97

Type Systems

97.1 Introduction 97-1
    Execution Errors • Lack of Safety • Should Languages
    Be Safe? • Should Languages Be Typed? • Expected
    Properties of Type Systems • How Type Systems
    Are Formalized • Type Equivalence

97.2 The Language of Type Systems 97-8
    Judgments • Well Typing and Type Inference
    • Type Soundness

97.3 First-Order Type Systems 97-11

97.4 First-Order Type Systems for Imperative
    Languages 97-18

97.5 Second-Order Type Systems 97-19

97.6 Subtyping 97-22

97.7 Equivalence 97-25

97.8 Type Inference 97-25
    The Type Inference Problem

97.9 Summary and Research Issues 97-28
    What We Learned • Future Directions

Luca Cardelli
Microsoft Research

97.1 Introduction

The fundamental purpose of a **type system** is to prevent the occurrence of *execution errors* during the running of a program. This informal statement motivates the study of type systems, but requires clarification. Its accuracy depends, first of all, on the rather subtle issue of what constitutes an execution error, which we will discuss in detail. Even when that is settled, the absence of execution errors is a nontrivial property. When such a property holds for all the program runs that can be expressed within a programming language, we say that the language is **type sound**. It turns out that a fair amount of careful analysis is required to avoid false and embarrassing claims of type soundness for programming languages. As a consequence, the classification, description, and study of type systems has emerged as a formal discipline.

The formalization of type systems requires the development of precise notations and definitions, and the detailed proof of formal properties that give confidence in the appropriateness of the definitions. Sometimes the discipline becomes rather abstract. One should always remember, however, that the basic motivation is pragmatic: the abstractions have arisen out of necessity and can usually be related directly to concrete intuitions. Moreover, formal techniques need not be applied in full in order to be useful and influential. A knowledge of the main principles of type systems can help in avoiding obvious and not-so-obvious pitfalls, and can inspire regularity and orthogonality in language design.

When properly developed, type systems provide conceptual tools with which to judge the adequacy of important aspects of language definitions. Informal language descriptions often fail to specify the type structure of a language in sufficient detail to allow unambiguous implementation. It often happens that different compilers for the same language implement slightly different type systems. Moreover, many language definitions have been found to be type unsound, allowing a program to crash even though it is judged acceptable by a **typechecker**. Ideally, formal type systems should be part of the definition of all typed programming languages. This way, typechecking algorithms could be measured unambiguously against precise specifications and, if at all possible and feasible, whole languages could be shown to be type sound.

In this introductory section we present an informal nomenclature for typing, execution errors, and related concepts. We discuss the expected properties and benefits of type systems, and we review how type systems can be formalized. The terminology used in the introduction is not completely standard; this is due to the inherent inconsistency of standard terminology arising from various sources. In general, we avoid the words *type* and *typing* when referring to runtime concepts; for example, we replace dynamic typing with dynamic checking and avoid common but ambiguous terms such as strong typing. The terminology is summarized in the "Defining Terms" section.

In Section 97.2 we explain the notation commonly used for describing type systems. We review **judgments**, which are formal assertions about the typing of programs; **type rules**, which are implications between judgments; and **derivations**, which are deductions based on type rules. In Section 97.3 we review a broad spectrum of simple types, the analog of which can be found in common languages, and we detail their type rules. In Section 97.4 we present the type rules for a simple but complete imperative language. In Section 97.5 we discuss the type rules for some advanced type constructions: *polymorphism* and *data abstraction*. In Section 97.6 we explain how type systems can be extended with a notion of *subtyping*. Section 97.7 is a brief commentary on some important topics that we have glossed over. In Section 97.8 we discuss the *type inference* problem and present type inference algorithms for the main type systems that we have considered. Finally, Section 97.9 presents a summary of achievements and future directions.

97.1.1 Execution Errors

The most obvious symptom of an execution error is the occurrence of an unexpected software fault, such as an illegal instruction fault or an illegal memory reference fault.

There are, however, more subtle kinds of execution errors that result in data corruption without any immediate symptoms. Moreover, there are software faults, such as divide by zero and dereferencing *nil*, that are not normally prevented by type systems. Finally, there are languages lacking type systems where, nonetheless, software faults do not occur. Therefore, we need to define our terminology carefully, beginning with what is a type.

97.1.1.1 Typed and Untyped Languages

A program variable can assume a range of values during the execution of a program. An upper bound of such a range is called a **type** of the variable. For example, a variable x of type *Boolean* is supposed to assume only Boolean values during every run of a program. If x has type *Boolean*, then the Boolean expression *not* (x) has a sensible meaning in every run of the program. Languages for which variables can be given (nontrivial) types are called **typed languages**.

Languages that do not restrict the range of variables are called **untyped languages**; they do not have types or, equivalently, have a single universal type that contains all values. In these languages, operations may be applied to inappropriate arguments; the result may be a fixed arbitrary value, a fault, an exception, or an unspecified effect. The pure λ-calculus is an extreme case of an untyped language where no fault ever occurs; the only operation is function application and, because all values are functions, that operation never fails.

A type system is that component of a typed language that keeps track of the types of variables and, in general, of the types of all expressions in a program. Type systems are used to determine whether programs are **well behaved** (as discussed subsequently). Only program sources that comply with a type system should be considered real programs of a typed language; the other sources should be discarded before they are run.

A language is typed by virtue of the existence of a type system for it, whether or not types actually appear in the syntax of programs. Typed languages are **explicitly typed** if types are part of the syntax, and

implicitly typed otherwise. No mainstream language is purely implicitly typed, but languages such as ML and Haskell support writing large program fragments where type information is omitted; the type systems of those languages automatically assign types to such program fragments.

97.1.1.2 Execution Errors and Safety

It is useful to distinguish between two kinds of execution errors: the ones that cause the computation to stop immediately, and the ones that go unnoticed (for awhile) and later cause arbitrary behavior. The former are called **trapped errors**, whereas the latter are **untrapped errors**.

An example of an untrapped error is improperly accessing a legal address, for example, accessing data past the end of an array in absence of runtime bounds checks. Another untrapped error that may go unnoticed for an arbitrary length of time is jumping to the wrong address; memory there may or may not represent an instruction stream. Examples of trapped errors are division by zero and accessing an illegal address: the computation stops immediately (on many computer architectures).

A program fragment is *safe* if it does not cause untrapped errors to occur. Languages for which all program fragments are safe are called **safe languages**. Therefore, safe languages rule out the most insidious form of execution errors: the ones that may go unnoticed. Untyped languages may enforce **safety** by performing runtime checks. Typed languages may enforce safety by statically rejecting all programs that are potentially unsafe. Typed languages may also use a mixture of runtime and **static checks**.

Although safety is a crucial property of programs, it is rare for a typed language to be concerned exclusively with the elimination of untrapped errors. Typed languages usually aim to rule out also large classes of trapped errors, along with the untrapped ones. We discuss these issues next.

97.1.1.3 Execution Errors and Well-Behaved Programs

For any given language, we can designate a subset of the possible execution errors as **forbidden errors**. The forbidden errors should include all the untrapped errors, plus a subset of the trapped errors. A program fragment is said to have **good behavior**, or equivalently to be well behaved, if it does not cause any forbidden error to occur. (The contrary is to have *bad behavior*, or equivalently to be *ill behaved*.) In particular, a well-behaved fragment is safe. A language in which all the (legal) programs have good behavior is called **strongly checked**.

Thus, with respect to a given type system, the following holds for a strongly checked language:

- No untrapped errors occur (safety guarantee).
- None of the trapped errors designated as forbidden errors occur.
- Other trapped errors may occur; it is the programmer's responsibility to avoid them.

Typed languages can enforce good behavior (including safety) by performing static (i.e., compile time) checks to prevent unsafe and ill-behaved programs from ever running. These languages are **statically checked**; the checking process is called **typechecking**, and the algorithm that performs this checking is called the **typechecker**. A program that passes the typechecker is said to be **well typed**; otherwise, it is **ill typed**, which may mean that it is actually ill behaved, or simply that it could not be guaranteed to be well behaved. Examples of statically checked languages are ML, Java, and Pascal (with the caveat that Pascal has some unsafe features).

Untyped languages can enforce good behavior (including safety) in a different way, by performing sufficiently detailed runtime checks to rule out all forbidden errors. (For example, they may check all array bounds, and all division operations, generating recoverable exceptions when forbidden errors would happen.) The checking process in these languages is called **dynamic checking**; LISP is an example of such a **dynamically checked language**. These languages are strongly checked although they have neither static checking nor a type system.

Even statically checked languages usually need to perform tests at runtime to achieve safety. For example, array bounds must in general be tested dynamically. The fact that a language is statically checked does not necessarily mean that execution can proceed entirely blindly.

TABLE 97.1   Safety

	Typed	Untyped
Safe	ML, Java	LISP
Unsafe	C	Assembler

Several languages take advantage of their static type structures to perform sophisticated dynamic tests. For example, Simula67's INSPECT, Modula-3's TYPECASE, and Java's instanceof constructs discriminate on the runtime type of an object. These languages are still (slightly improperly) considered statically checked, partially because the dynamic type tests are defined on the basis of the static type system. That is, the dynamic tests for type equality are compatible with the algorithm that the typechecker uses to determine type equality at compile time.

97.1.2   Lack of Safety

By our definitions, a well-behaved program is safe. Safety is a more primitive and perhaps more important property than good behavior. The primary goal of a type system is to ensure language safety by ruling out *all* untrapped errors in all program runs. However, most type systems are designed to ensure the more general good-behavior property and, implicitly, safety. Thus, the declared goal of a type system is usually to ensure good behavior of all programs, by distinguishing between well-typed and ill-typed programs.

In reality, certain statically checked languages do not ensure safety. That is, their set of forbidden errors does not include all untrapped errors. These languages can be euphemistically called weakly checked (or *weakly typed*, in the literature), meaning that some unsafe operations are detected statically and some are not detected. Languages in this class vary widely in the extent of their weakness. For example, Pascal is unsafe only when untagged variant types and function parameters are used, whereas C has many unsafe and widely used features, such as pointer arithmetic and casting. It is interesting to notice that the first five of the ten commandments for C programmers [Spencer] are directed at compensating for the weak-checking aspects of C. Some of the problems caused by weak checking in C have been alleviated in C++, and even more have been addressed in Java, confirming a trend away from weak checking. Modula-3 supports unsafe features, but only in modules that are explicitly marked as unsafe, and prevents safe modules from importing unsafe interfaces.

Most untyped languages are, by necessity, completely safe (e.g., LISP). Otherwise, programming would be too frustrating in the absence of both compile time and runtime checks to protect against corruption. Assembly languages belong to the unpleasant category of untyped unsafe languages (see Table 97.1).

97.1.3   Should Languages Be Safe?

Some languages, like C, are deliberately unsafe because of performance considerations: the runtime checks needed to achieve safety are sometimes considered too expensive. Safety has a cost even in languages that do extensive static analysis: tests such as array bounds checks cannot be, in general, completely eliminated at compile time.

Still, there have been many efforts to design safe subsets of C, and to produce development tools that try to execute C programs safely by introducing a variety of (relatively expensive) runtime checks. These efforts are due to two main reasons: (1) the widespread use of C in applications that are not largely performance critical, and (2) the security problems introduced by unsafe C programs. The security problems include buffer overflows and underflows caused by pointer arithmetic, and lack of array bounds checks that can lead to overwriting areas of memory and that can be exploited for attacks.

Safety is cost-effective according to different measures than just pure performance. Safety produces fail-stop behavior in case of execution errors, reducing debugging time. Safety guarantees the integrity of runtime structures, and therefore enables garbage collection. In turn, garbage collection considerably reduces code size and code development time, at the price of some performance. Finally, safety has emerged as a necessary foundation for system security, particularly for systems (such as operating system kernels and Web browsers) that load and run foreign code. System security is becoming one of the most expensive aspects of program development and maintenance, and safety can reduce these costs.

Thus, the choice between a safe and unsafe language may be ultimately related to a trade-off between development and maintenance time, and execution time. Although safe languages have been around for many decades, it is only recently that they have become mainstream, uniquely because of security concerns.

97.1.4 Should Languages Be Typed?

The issue of whether programming languages should have types is still subject to some debate. There is little doubt that production code written in untyped languages can be maintained only with great difficulty. From the point of view of maintainability, even weakly checked unsafe languages are superior to safe but untyped languages (e.g., C vs. LISP). Following are the arguments that have been put forward in favor of typed languages, from an engineering point of view:

- *Economy of execution.* Type information was first introduced in programming to improve code generation and runtime efficiency for numerical computations, for example, in FORTRAN. In ML, accurate type information eliminates the need for *nil*-checking on pointer dereferencing. In general, accurate type information at compile time leads to the application of the appropriate operations at runtime without the need for expensive tests.

- *Economy of small-scale development.* When a type system is well designed, typechecking can capture a large fraction of routine programming errors, thus eliminating lengthy debugging sessions. The errors that do occur are easier to debug, simply because large classes of other errors have been ruled out. Moreover, experienced programmers adopt a coding style that causes some logical errors to show up as typechecking errors: they use the typechecker as a development tool. (For example, by changing the name of a field when its invariants change even though its type remains the same, so as to get error reports on all its old uses.)

- *Economy of compilation.* Type information can be organized into *interfaces* for program modules, for example, as in Modula-2 and Ada. Modules can then be compiled independently of each other, with each module depending only on the interfaces of the others. Compilation of large systems is made more efficient because, at least when interfaces are stable, changes to a module do not cause other modules to be recompiled.

- *Economy of large-scale development.* Interfaces and modules have methodological advantages for code development. Large teams of programmers can negotiate the interfaces to be implemented, and then proceed separately to implement the corresponding pieces of code. Dependencies between pieces of code are minimized, and code can be locally rearranged without fear of global effects. (These benefits can also be achieved by informal interface specifications; but in practice, typechecking helps enormously in verifying adherence to the specifications.)

- *Economy of development and maintenance in security areas.* Although safety is necessary to eliminate security breaches such as buffer overflows, typing is necessary to eliminate other catastrophic security breaches, such as the following. If there is any way at all, no matter how convoluted, to cast an integer into a value of pointer type (or object type), the entire system is compromised. If that is possible, attackers can access any data anywhere in the system, even within the confines of an otherwise typed language, using any type they choose to view the data. Another helpful technique is to convert a given typed pointer into an integer, and then into a pointer of different type as above. The most cost-effective way to eliminate these security problems, in terms of maintenance and execution efficiency, is to employ typed languages. Still, security is a problem at all levels of a system; typed languages provide an excellent foundation, but not a complete solution.

- *Economy of language features*. Type constructions are naturally composed in orthogonal ways. For example, in Pascal, an array of arrays models two-dimensional arrays; in ML, a procedure with a single argument that is a tuple of n parameters models a procedure of n arguments. Thus, type systems promote orthogonality of language features, question the utility of artificial restrictions, and thus tend to reduce the complexity of programming languages.

97.1.5   Expected Properties of Type Systems

In the remainder of this chapter we proceed under the assumption that languages should be both safe and typed, and therefore that type systems should be employed. In the study of type systems, we neither distinguish between trapped and untrapped errors, nor between safety and good behavior; we concentrate on good behavior, and we take safety as an implied property.

Types, as normally intended in programming languages, have pragmatic characteristics that distinguish them from other kinds of program annotations. In general, annotations about the behavior of programs can range from informal comments to formal specifications subject to theorem proving. Types sit in the middle of this spectrum: they are more precise than program comments, and more easily mechanizable than formal specifications. Here are the basic properties expected of any type system:

- Type systems should be *decidably verifiable*: there should be an algorithm (called a typechecking algorithm) that can ensure that a program is well behaved. The purpose of a type system is not simply to state programmer intentions, but to actively capture execution errors before they happen. (Arbitrary formal specifications do not have these properties.)
- Type systems should be *transparent*: a programmer should be able to predict easily whether a program will typecheck. If it fails to typecheck, the reason for the failure should be self-evident. (Automatic theorem proving does not have these properties.)
- Type systems should be *enforceable*: type declarations should be statically checked as much as possible, and otherwise dynamically checked. The consistency between type declarations and their associated programs should be routinely verified. (Program comments and conventions do not have these properties.)

97.1.6   How Type Systems Are Formalized

As we have discussed, type systems are used to define the notion of well typing, which is itself a static approximation of good behavior (including safety). Safety facilitates debugging because of fail-stop behavior, and enables garbage collection by protecting runtime structures. Well typing further facilitates program development by trapping execution errors before runtime.

But how can we guarantee that well-typed programs are really well behaved? That is, how can we be sure that the type rules of a language do not accidentally allow ill-behaved programs to slip through?

Formal type systems are the mathematical characterizations of the informal type systems that are described in programming language manuals. Once a type system is formalized, we can attempt to prove a *type soundness* theorem stating that *well-typed programs are well behaved*. If such a soundness theorem holds, we say that the type system is sound. (Good behavior of all programs of a typed language and soundness of its type system mean the same thing.)

To formalize a type system and prove a soundness theorem, we must in essence formalize the whole language in question, as we now sketch.

The first step in formalizing a programming language is to describe its syntax. For most languages of interest, this reduces to describing the syntax of types and *terms*. Types express static knowledge about programs, whereas terms (statements, expressions, and other program fragments) express the algorithmic behavior.

The next step is to define the *scoping* rules of the language, which unambiguously associate occurrences of identifiers to their binding locations (the locations where the identifiers are declared). The scoping needed for typed languages is invariably *static*, in the sense that the binding locations of identifiers must be determined before runtime. Binding locations can often be determined purely from the syntax of a language, without any further analysis; static scoping is then called *lexical scoping*. The lack of static scoping is called *dynamic scoping*.

Scoping can be formally specified by defining the set of *free variables* of a program fragment (which involves specifying how variables are bound by declarations). The associated notion of *substitution* of types or terms for free variables can then be defined.

When this much is settled, one can proceed to define the type rules of the language. These describe a relation *has-type* of the form $M : A$ between terms M and types A. Some languages also require a relation *subtype-of* of the form $A <: B$ between types, and often a relation *equal-type* of the form $A = B$ of type equivalence. The collection of type rules of a language forms its type system. A language that has a type system is called a typed language.

The type rules cannot be formalized without first introducing another fundamental ingredient that is not reflected in the syntax of the language: *static typing environments*. These are used to record the types of free variables during the processing of program fragments; they correspond closely to the symbol table of a compiler during the typechecking phase. The type rules are always formulated with respect to a static environment for the fragment being typechecked. For example, the has-type relation $M : A$ is associated with a static typing environment Γ that contains information about the free variables of M and A. The relation is written in full as $\Gamma \vdash M : A$, meaning that M has type A in environment Γ.

The final step in formalizing a language is to define its semantics as a relation *has-value* between terms and a collection of *results*. The form of this relation depends strongly on the style of semantics that is adopted. In any case, the semantics and the type system of a language are interconnected: the types of a term and of its result should be the same (or appropriately related); this is the essence of the soundness theorem.

The fundamental notions of type system are applicable to virtually all computing paradigms (functional, imperative, concurrent, etc.). Individual type rules can often be adopted unchanged for different paradigms. For example, the basic type rules for functions are the same whether the semantics are call-by-name or call-by-value or, orthogonally, functional or imperative.

In this chapter we discuss type systems independently of semantics. It should be understood, however, that ultimately a type system must be related to a semantics, and that soundness should hold for those semantics. Suffice it to say that the techniques of structural operational semantics deal uniformly with a large collection of programming paradigms, and fit very well with the treatment found in this chapter.

97.1.7 Type Equivalence

As mentioned above, most nontrivial type systems require the definition of a relation *equal type* of type equivalence. This is an important issue when defining a programming language: when are separately written type expressions equivalent? Consider, for example, two distinct type names that have been associated with similar types:

$$type\ X = Bool$$
$$type\ Y = Bool$$

If the type names X and Y match by virtue of being associated with similar types, we have *structural equivalence*. If they fail to match by virtue of being distinct type names (without looking at the associated types), we have *by-name equivalence*.

In practice, a mixture of structural and by-name equivalence is used in most languages. Pure structural equivalence can be easily and precisely defined by means of type rules, while by-name equivalence is more difficult to pin down, and often has an algorithmic flavor. Structural equivalence has unique advantages when typed data must be stored or transmitted over a network; in contrast, by-name equivalence cannot deal easily with interacting programs that have been developed and compiled separately in time or space.

We assume structural equivalence in what follows (although this issue does not arise often). Satisfactory emulation of by-name equivalence can be obtained within structural equivalence, as demonstrated by the Modula-3 *branding* mechanism.

97.2   The Language of Type Systems

A type system specifies the type rules of a programming language independently of particular typechecking algorithms. This is analogous to describing the syntax of a programming language by a formal grammar, independently of particular parsing algorithms.

It is both convenient and useful to decouple type systems from typechecking algorithms: type systems belong to language definitions, while algorithms belong to compilers. It is easier to explain the typing aspects of a language by a type system, rather than by the algorithm used by a given compiler. Moreover, different compilers may use different typechecking algorithms for the same type system.

As a minor problem, it is technically possible to define type systems that admit only unfeasible type-checking algorithms, or no algorithms at all. The usual intent, however, is to allow for efficient typechecking algorithms.

97.2.1   Judgments

Type systems are described by a particular formalism, which we now introduce. The description of a type system starts with the description of a collection of formal utterances called *judgments*. A typical judgment has the form:

$$\Gamma \vdash \mathcal{J} \quad \text{where } \mathcal{J} \text{ is an assertion; the free variables of } \mathcal{J} \text{ are declared in } \Gamma.$$

We say that Γ *entails* $\mathcal{J}$. Here, Γ is a static typing environment; for example, an ordered list of distinct variables and their types, of the form $\emptyset, x_1 : A_1, \ldots, x_n : A_n$. The empty environment is denoted by $\emptyset$, and the collection of variables $x_1 \cdots x_n$ declared in Γ is indicated by $dom(\Gamma)$, the domain of Γ. The form of the *assertion* $\mathcal{J}$ varies from judgment to judgment, but all the free variables of $\mathcal{J}$ must be declared in Γ.

The most important judgment, for our present purposes, is the *typing judgment*, which asserts that a term M has a type A with respect to a static typing environment for the free variables of M. It has the form:

$$\Gamma \vdash M : A \qquad M \text{ has type } A \text{ in } \Gamma$$

Examples.

$$\emptyset \vdash true : Bool \qquad\qquad true \text{ has type } Bool$$
$$\emptyset, x : Nat \vdash x + 1 : Nat \qquad x + 1 \text{ has type } Nat, \text{ provided that } x \text{ has type } Nat$$

Other judgment forms are often necessary; a common one asserts simply that an environment is **well formed:**

$$\Gamma \vdash \diamond \qquad \Gamma \text{ is well-formed (i.e., it has been properly constructed)}$$

Any given judgment can be regarded as *valid* (e.g., $\Gamma \vdash true : Bool$) or *invalid* (e.g., $\Gamma \vdash true : Nat$). Validity formalizes the notion of well-typed programs. The distinction between valid and invalid judgments could be expressed in a number of ways; however, a highly stylized way of presenting the set of valid judgments has emerged. This presentation style, based on type rules, facilitates stating and proving technical lemmas and theorems about type systems. Moreover, type rules are highly modular: rules for different constructs can be written separately (in contrast to a monolithic typechecking algorithm). Therefore, type rules are comparatively easy to read and understand.

97.2.1.1   Type Rules

Type rules assert the validity of certain judgments on the basis of other judgments that are already known to be valid. The process gets off the ground by some intrinsically valid judgment (usually: $\emptyset \vdash \diamond$, stating that the empty environment is well formed).

The general form of a type rule is:

$$\text{(Rule name) (Annotations)}$$
$$\frac{\Gamma_1 \vdash \mathfrak{J}_1 \; \ldots \; \Gamma_n \vdash \mathfrak{J}_n \; \text{(Annotations)}}{\Gamma \vdash \mathfrak{J}}$$

Each type rule is written as a number of *premise* judgments $\Gamma_i \vdash \mathfrak{J}_i$ above a horizontal line, with a single *conclusion* judgment $\Gamma \vdash \mathfrak{J}$ below the line. When all the premises are satisfied, the conclusion must hold; the number of premises may be zero. Each rule has a name. (By convention, the first word of the name is determined by the conclusion judgment; for example, rule names of the form "(Val ...)" are for rules whose conclusion is a value typing judgment.) When needed, conditions restricting the applicability of a rule, as well as abbreviations used within the rule, are annotated next to the rule name or the premises.

For example, the first of the following two rules states that any numeral is an expression of type *Nat*, in any well-formed environment Γ. The second rule states that two expressions M and N denoting natural numbers can be combined into a larger expression $M + N$, which also denotes a natural number. Moreover, the environment Γ for M and N, which declares the types of any free variable of M and N, carries over to $M + N$.

$$\text{(Val } n) \; (n = 0, 1, \ldots) \qquad \text{(Val +)}$$
$$\frac{\Gamma \vdash \diamond}{\Gamma \vdash n : Nat} \qquad\qquad \frac{\Gamma \vdash M : Nat \quad \Gamma \vdash N : Nat}{\Gamma \vdash M + N : Nat}$$

A fundamental rule states that the empty environment is well formed, with no assumptions:

$$\text{(Env } \emptyset)$$
$$\frac{}{\emptyset \vdash \diamond}$$

A collection of type rules is called a (*formal*) *type system*. Technically, type systems fit into the general framework of *formal proof systems*: collections of rules used to carry out step-by-step deductions. The deductions carried out in type systems concern the typing of programs.

97.2.1.2 Type Derivations

A derivation in a given type system is a tree of judgments with leaves at the top and a root at the bottom, where each judgment is obtained from the ones immediately above it by some rule of the system. A fundamental requirement on type systems is that it must be possible to check whether or not a derivation is properly constructed.

A **valid judgment** is one that can be obtained as the root of a derivation in a given type system. That is, a valid judgment is one that can be obtained by correctly applying the type rules. For example, using the three rules given previously, we can build the following derivation, which establishes that $\emptyset \vdash 1 + 2 : Nat$ is a valid judgment. The rule applied at each step is displayed to the right of each conclusion:

$$\frac{\dfrac{\emptyset \vdash \diamond \qquad \text{by (Env } \emptyset)}{\emptyset \vdash 1 : Nat \qquad \text{by (Val } n)} \qquad \dfrac{\emptyset \vdash \diamond \qquad \text{by (Env } \emptyset)}{\emptyset \vdash 2 : Nat \qquad \text{by (Val } n)}}{\emptyset \vdash 1 + 2 : Nat \qquad \text{by (Val +)}}$$

97.2.2 Well Typing and Type Inference

In a given type system, a term M is well typed for an environment Γ if there is a type A such that $\Gamma \vdash M : A$ is a valid judgment; that is, if the term M can be given some type.

The discovery of a derivation (and hence of a type) for a term is called the **type inference** problem. In the simple type system consisting of the rules (Env $\emptyset$), (Val n), and (Val $+$), a type can be *inferred* for the term $1 + 2$ in the empty environment. This type is *Nat*, by the preceding derivation.

Suppose we now add a type rule with premise $\Gamma \vdash \diamond$ and conclusion $\Gamma \vdash true : Bool$. In the resulting type system, we cannot infer any type for the term $1 + true$ because there is no rule for summing a natural number with a Boolean. Because of the absence of any derivations for $1 + true$, we say that $1 + true$ is *not typeable*, or that it is ill typed, or that it has a **typing error**.

We could further add a type rule with premises $\Gamma \vdash M : Nat$ and $\Gamma \vdash N : Bool$, and with conclusion $\Gamma \vdash M + N : Nat$ (e.g., with the intent of interpreting *true* as 1). In such a type system, a type could be inferred for the term $1 + true$, which would now be well typed.

Thus, the type inference problem for a given term is very sensitive to the type system in question. An algorithm for type inference may be very easy, very difficult, or impossible to find, depending on the type system. If found, the best algorithm may be very efficient, or hopelessly slow. Although type systems are expressed and often designed in the abstract, their practical utility depends on the availability of good type inference algorithms.

The type inference problem for explicitly typed procedural languages such as Pascal is fairly easily solved; we treat it in Section 97.8. The type inference problem for implicitly typed languages such as ML is much more subtle, and we do not treat it here. The basic algorithm is well understood (several descriptions of it appear in the literature) and is widely used. However, the versions of the algorithm that are used in practice are complex and are still being investigated.

The type inference problem becomes particularly difficult in the presence of **polymorphism** (discussed in Section 97.5). The type inference problems for the explicitly typed polymorphic features of Ada, CLU, and Standard ML are treatable in practice. However, these problems are typically solved by algorithms, without first describing the associated type systems. The purest and most general type system for polymorphism is embodied by a λ-calculus discussed in Section 97.5. The type inference algorithm for this polymorphic λ-calculus is fairly easy, and we present it in Section 97.8. The simplicity of the solution, however, depends on impractically verbose typing annotations. To make this general polymorphism practical, some type information must be omitted. Such type inference problems are still an area of active research.

97.2.3 Type Soundness

We have now established all of the general notions concerning type systems, and we can begin examining particular type systems. Starting in Section 97.3, we review some very powerful but rather theoretical type systems. The idea is that by first understanding these few systems, it becomes easier to write the type rules for the varied and complex features that one may encounter in programming languages.

When immersing ourselves in type rules, we should keep in mind that a sensible type system is more than just an arbitrary collection of rules. Well typing is meant to correspond to a semantic notion of good program behavior. It is customary to check the internal consistency of a type system by proving a type soundness theorem. This is where type systems meet semantics. For denotational semantics we expect that if $\emptyset \vdash M : A$ is valid, then $[[M]] \in [[A]]$ holds (the value of M belongs to the set of values denoted by the type A); and for operational semantics, we expect that if $\emptyset \vdash M : A$ and M reduces to M', then $\emptyset \vdash M' : A$. In both cases, the type soundness theorem asserts that well-typed programs compute without execution errors. See Gunter [1992] and Wright and Felleisen [1994] for surveys of techniques, as well as state-of-the-art soundness proofs.

97.3 First-Order Type Systems

The type systems found in most common procedural languages are called **first order**. In type-theoretical jargon, this means that they lack type parameterization and type abstraction, which are **second-order** features. First-order type systems include (rather confusingly) higher-order functions. Pascal and Algol68 have rich first-order type systems, whereas FORTRAN and Algol60 have very poor ones.

A minimal first-order type system can be given for the untyped λ-calculus, where the untyped λ-abstraction $\lambda x.M$ represents a function of parameter x and result M. Typing for this calculus requires only function types and some base types; we will see later how to add other common type structures.

The first-order typed λ-calculus is called system F_1. The main change from the untyped λ-calculus is the addition of type annotations for λ-abstractions, using the syntax $\lambda x : A.M$, where x is the function parameter, A is its type, and M is the body of the function. (In a typed programming language we would likely include the type of the result, but this is not necessary here.) The step from $\lambda x.M$ to $\lambda x : A.M$ is typical of any progression from an untyped to a typed language: bound variables acquire type annotations.

Because F_1 is based mainly on function values, the most interesting types are function types: $A \rightarrow B$ is the type of functions with arguments of type A and results of type B. To get started, however, we also need some basic types over which to build function types. We indicate by *Basic* a collection of such types, and by $K \in Basic$ any such type. At this point, basic types are purely a technical necessity, but shortly we consider interesting basic types such as *Bool* and *Nat*.

The syntax of F_1 is given in Table 97.2. It is important to comment briefly on the role of syntax in typed languages. In the case of the untyped λ-calculus, the context-free syntax describes exactly the legal programs. This is not the case in typed calculi, because good behavior is not (usually) a context-free property. The task of describing the legal programs is taken over by the type system. For example, $\lambda x : K.x(y)$ respects the syntax of F_1 given in Table 97.2, but is not a program of F_1 because it is not well typed, since K is not a function type. The context-free syntax is still needed, but only in order to define the notions of free and bound variables; that is, to define the scoping rules of the language. Based on the scoping rules, terms that differ only in their bound variables, such as $\lambda x : K.x$ and $\lambda y : K.y$, are considered *syntactically identical*. This convenient identification is implicitly assumed in the type rules (one may have to rename bound variables in order to apply certain type rules).

The definition of free variables for F_1 is the same as for the untyped λ-calculus, simply ignoring the typing annotations.

We need only three simple judgments for F_1; they are shown in Table 97.3. The judgment $\Gamma \vdash A$ is in a sense redundant, since all syntactically correct types A are automatically well formed in any environment Γ. In second-order systems, however, the well formedness of types is not captured by grammar alone, and the

---

TABLE 97.2 Syntax of F_1

$A, B ::=$		Types
K	$K \in Basic$	basic types
$A \rightarrow B$		function types
$M, N ::=$		Terms
x		variable
$\lambda x : A.M$		function
$M\,N$		application

TABLE 97.3 Judgments for F_1

$\Gamma \vdash \diamond$	Γ is a well-formed environment
$\Gamma \vdash A$	A is a well-formed type in Γ
$\Gamma \vdash M : A$	M is a well-formed term of type A in Γ

TABLE 97.4   Type Rules for F_1

$$\frac{}{\emptyset \vdash \diamond}\ (\text{Env } \emptyset)$$

$$\frac{\Gamma \vdash A \qquad x \notin dom(\Gamma)}{\Gamma, x : A \vdash \diamond}\ (\text{Env } x)$$

$$\frac{\Gamma \vdash \diamond \qquad K \in Basic}{\Gamma \vdash K}\ (\text{Type Const})$$

$$\frac{\Gamma \vdash A \qquad \Gamma \vdash B}{\Gamma \vdash A \to B}\ (\text{Type Arrow})$$

$$\frac{\Gamma', x : A, \Gamma'' \vdash \diamond}{\Gamma', x : A, \Gamma'' \vdash x : A}\ (\text{Val } x)$$

$$\frac{\Gamma, x : A \vdash M : B}{\Gamma \vdash \lambda x : A.M : A \to B}\ (\text{Val Fun})$$

$$\frac{\Gamma \vdash M : A \to B \qquad \Gamma \vdash N : A}{\Gamma \vdash M N : B}\ (\text{Val Appl})$$

TABLE 97.5   A Derivation in F_1

$\emptyset \vdash \diamond$	by (Env $\emptyset$)	$\emptyset \vdash \diamond$ by (Env $\emptyset$)	$\emptyset \vdash \diamond$ by (Env $\emptyset$)		$\emptyset \vdash \diamond$	by (Env $\emptyset$)
$\emptyset \vdash K$	by (Type Const)	$\emptyset \vdash K$ by (Type Const)	$\emptyset \vdash K$ by (Type Const)		$\emptyset \vdash K$	by (Type Const)
$\emptyset \vdash K \to K$			$\emptyset \vdash K \to K$			by (Type Arrow)
$\emptyset, y : K \to K \vdash \diamond$		by (Type Arrow)	$\emptyset, y : K \to K \vdash \diamond$			by (Env x)
$\emptyset, y : K \to K \vdash K$		by (Env x)	$\emptyset, y : K \to K \vdash K$			by (Type Const)
$\emptyset, y : K \to K, z : K \vdash \diamond$		by (Type Const)	$\emptyset, y : K \to K, z : K \vdash \diamond$			by (Env x)
$\emptyset, y : K \to K, z : K \vdash y : K \to K$		by (Env x)	$\emptyset, y : K \to K, z : K \vdash z : K$			by (Val x)
		by (Val)				
$\emptyset, y : K \to K, z : K \vdash y(z) : K$						by (Val Appl)
$\emptyset, y : K \to K \vdash \lambda z : K.y(z) : K \to K$						by (Val Fun)

judgment $\Gamma \vdash A$ becomes essential. It is convenient to adopt this judgment now, so that later extensions are easier.

Validity for these judgments is defined by the rules in Table 97.4. The rule (Env $\emptyset$) is the only one that does not require assumptions (i.e., it is the only *axiom*). It states that the empty environment is a valid environment. The rule (Env x) is used to extend an environment Γ to a longer environment $\Gamma, x : A$, provided that A is a valid type in Γ. Note that the assumption $\Gamma \vdash A$ implies, inductively, that Γ is valid. That is, in the process of deriving $\Gamma \vdash A$, we must have derived $\Gamma \vdash \diamond$. Another requirement of this rule is that the variable x must not be defined in Γ. We are careful to keep variables distinct in environments, so that when $\Gamma, x : A \vdash M : B$ has been derived, as in the assumption of (Val Fun), we know that x cannot occur in $dom(\Gamma)$.

The rules (Type Const) and (Type Arrow) construct types. The rule (Val x) extracts an assumption from an environment: we use the notation $\Gamma', x : A, \Gamma''$, rather informally, to indicate that $x : A$ occurs somewhere in the environment. The rule (Val Fun) gives the type $A \to B$ to a function, provided that the function body receives the type B under the assumption that the formal parameter has type A. Note how the environment changes length in this rule. The rule (Val Appl) applies a function to an argument: the same type A must appear twice when verifying the premises.

Table 97.5 shows a rather large derivation where all of the rules of F_1 are used.

Now that we have examined the basic structure of a simple first-order type system, we can begin enriching it to bring it closer to the type structure of actual programming languages. We are going to add a set of rules for each new type construction, following a fairly regular pattern. We begin with some basic data types: the type *Unit*, whose only value is the constant *unit*; the type *Bool*, whose values are *true* and *false*; and the type *Nat*, whose values are the natural numbers.

The *Unit* type is often used as a filler for uninteresting arguments and results; it is called *Void* or *Null* in some languages. There are no operations on *Unit*, so we need only a rule stating that *Unit* is a legal type, and one stating that *unit* is a legal value of type *Unit* (Table 97.6).

We have a similar pattern of rules for *Bool*, but Booleans also have a useful operation, the conditional, that has its own typing rule (Table 97.7). In the rule (Val Cond), the two branches of the conditional must have the same type A, because either may produce the result.

TABLE 97.6 *Unit* Type

(Type Unit)	(Val Unit)
$\Gamma \vdash \diamond$	$\Gamma \vdash \diamond$
$\overline{\Gamma \vdash Unit}$	$\overline{\Gamma \vdash unit : Unit}$

TABLE 97.7 *Bool* Type

(Type Bool)	(Val True)	(Val False)
$\Gamma \vdash \diamond$	$\Gamma \vdash \diamond$	$\Gamma \vdash \diamond$
$\overline{\Gamma \vdash Bool}$	$\overline{\Gamma \vdash true : Bool}$	$\overline{\Gamma \vdash false : Bool}$

(Val Cond)

$$\frac{\Gamma \vdash M : Bool \qquad \Gamma \vdash N_1 : A \qquad \Gamma \vdash N_2 : A}{\Gamma \vdash (if_A\ M\ then\ N_1\ else\ N_2) : A}$$

TABLE 97.8 *Nat* Type

(Type Nat)	(Val Zero)	(Val Succ)
$\Gamma \vdash \diamond$	$\Gamma \vdash \diamond$	$\Gamma \vdash M : Nat$
$\overline{\Gamma \vdash Nat}$	$\overline{\Gamma \vdash 0 : Nat}$	$\overline{\Gamma \vdash succ\ M : Nat}$

(Val Pred)	(Val IsZero)
$\Gamma \vdash M : Nat$	$\Gamma \vdash M : Nat$
$\overline{\Gamma \vdash pred\ M : Nat}$	$\overline{\Gamma \vdash isZero\ M : Bool}$

The rule (Val Cond) illustrates a subtle issue about the amount of type information needed for type-checking. When encountering a conditional expression, a typechecker has to infer separately the types of N_1 and N_2, and then find a single type A that is compatible with both. In some type systems it might not be easy or possible to determine this single type from the types of N_1 and N_2. To account for this potential typechecking difficulty, we use a subscripted type to express additional type information: if_A is a hint to the typechecker that the result type should be A, and that types inferred for N_1 and N_2 should be separately compared with the given A. In general, we use subscripted types to indicate information that may be useful or necessary for typechecking, depending on the whole type system under consideration. It is often the task of a typechecker to synthesize this additional information. When it is possible to do so, subscripts may be omitted. (Most common languages do not require the annotation if_A.)

The type of natural numbers, *Nat* (Table 97.8), has 0 and *succ* (successor) as generators. Alternatively, as we did earlier, a single rule could state that all numeric constants have type *Nat*. Computations on *Nat* are made possible by the *pred* (predecessor) and *isZero* (test for zero) primitives; other sets of primitives can be chosen.

Now that we have a collection of basic types, we can begin looking at structured types, starting with *product types* (Table 97.9). A product type $A_1 \times A_2$ is the type of pairs of values with first component of type A_1 and second component of type A_2. These components can be extracted with the projections *first* and *second*, respectively. Instead of (or in addition to) the projections, one can use a *with* statement that decomposes a pair M and binds its components to two separate variables x_1 and x_2 in the scope N. The *with* notation is related to pattern matching in ML, but also to Pascal's *with*; the connection with the latter will become clearer when we consider record types.

Product types can be easily generalized to *tuple types* $A_1 \times \cdots \times A_n$, with corresponding generalized projections and generalized *with*.

Union types (Table 97.10) are often overlooked, but are just as important as product types for expressiveness. An element of a union type $A_1 + A_2$ can be thought of as an element of A_1 tagged with a *left*

TABLE 97.9   Product Types

(Type Product)	(Val Pair)
$$\dfrac{\Gamma \vdash A_1 \qquad \Gamma \vdash A_2}{\Gamma \vdash A_1 \times A_2}$$	$$\dfrac{\Gamma \vdash M_1 : A_1 \qquad \Gamma \vdash M_2 : A_2}{\Gamma \vdash \langle M_1, M_2 \rangle : A_1 \times A_2}$$
(Val First)	(Val Second)
$$\dfrac{\Gamma \vdash M : A_1 \times A_2}{\Gamma \vdash \mathit{first}\ M : A_1}$$	$$\dfrac{\Gamma \vdash M : A_1 \times A_2}{\Gamma \vdash \mathit{second}\ M : A_2}$$

(Val With)
$$\frac{\Gamma \vdash M : A_1 \times A_2 \qquad \Gamma, x_1 : A_1, x_2 : A_2 \vdash N : B}{\Gamma \vdash (\mathit{with}\ (x_1 : A_1, x_2 : A_2) := M\ \mathit{do}\ N) : B}$$

TABLE 97.10   Union Types

(Type Union)	(Val inLeft)	(Val inRight)
$$\dfrac{\Gamma \vdash A_1 \qquad \Gamma \vdash A_2}{\Gamma \vdash A_1 + A_2}$$	$$\dfrac{\Gamma \vdash M_1 : A_1 \qquad \Gamma \vdash A_2}{\Gamma \vdash \mathit{inLeft}_{A_2}\ M_1 : A_1 + A_2}$$	$$\dfrac{\Gamma \vdash A_1 \qquad \Gamma \vdash M_2 : A_2}{\Gamma \vdash \mathit{inRight}_{A_1}\ M_2 : A_1 + A_2}$$
(Val isLeft)	(Val isRight)	
$$\dfrac{\Gamma \vdash M : A_1 + A_2}{\Gamma \vdash \mathit{isLeft}\ M : Bool}$$	$$\dfrac{\Gamma \vdash M : A_1 + A_2}{\Gamma \vdash \mathit{isRight}\ M : Bool}$$	
(Val asLeft)	(Val asRight)	
$$\dfrac{\Gamma \vdash M : A_1 + A_2}{\Gamma \vdash \mathit{asLeft}\ M : A_1}$$	$$\dfrac{\Gamma \vdash M : A_1 + A_2}{\Gamma \vdash \mathit{asRight}\ M : A_2}$$	

(Val Case)
$$\frac{\Gamma \vdash M : A_1 + A_2 \qquad \Gamma, x_1 : A_1 \vdash N_1 : B \qquad \Gamma, x_2 : A_2 \vdash N_2 : B}{\Gamma \vdash (\mathit{case}_B\ M\ \mathit{of}\ x_1 : A_1\ \mathit{then}\ N_1 \mid x_2 : A_2\ \mathit{then}\ N_2) : B}$$

token (created by *inLeft*), or an element of A_2 tagged with a *right* token (created by *inRight*). The tags can be tested by *isLeft* and *isRight*, and the corresponding value extracted with *asLeft* and *asRight*. If *asLeft* is mistakenly applied to a right-tagged value, a trapped error or exception is produced; this trapped error is not considered a forbidden error. Note that it is safe to assume that any result of *asLeft* has type A_1, because either the argument is left tagged, in which case the result is indeed of type A_1, or it is right tagged, in which case there is no result. Subscripts are used to disambiguate some of the rules, as we discussed in the case of the conditional.

The rule (Val Case) describes an elegant construct that can replace *isLeft*, *isRight*, *asLeft*, *asRight*, and the related trapped errors. (It also eliminates any dependence of union operations on the *Bool* type). The *case* construct executes one of two branches, depending on the tag of M, with the untagged contents of M bound to x_1 or x_2 in the scope of N_1 or N_2, respectively. A vertical bar separates the branches.

In terms of expressiveness (if not of implementation), note that the type *Bool* can be defined as *Unit* + *Unit*, in which case the *case* construct reduces to the conditional. The type *Int* can be defined as *Nat* + *Nat*, with one copy of *Nat* for the nonnegative integers and the other for the negative ones. We can define a prototypical trapped error as $\mathit{error}_A = \mathit{asRight}\,(\mathit{inLeft}_A(\mathit{unit}\,)) : A$. Thus, we can build an error expression for each type.

Product types and union types can be iterated to produce tuple types and multiple unions. However, these derived types are rather inconvenient, and are rarely seen in languages. Instead, *labeled* products and unions are used: they go under the names of *record types* and *variant types*, respectively.

A record type is the familiar named collection of types, with a value-level operation for extracting components by name. The rules in Table 97.11 assume the syntactic identification of record types and records up to reordering of their labeled components; this is analogous to the syntactic identification of functions up to renaming of bound variables.

TABLE 97.11 Record Types

(Type Record) (l_i distinct)	(Val Record) (l_i distinct)
$$\frac{\Gamma \vdash A_1 \cdots \Gamma \vdash A_n}{\Gamma \vdash Record(l_1 : A_1, \ldots, l_n : A_n)}$$	$$\frac{\Gamma \vdash M_1 : A_1 \cdots \Gamma \vdash M_n : A_n}{\Gamma \vdash record(l_1 = M_1, \ldots, l_n = M_n) : Record(l_1 : A_1, \ldots, l_n : A_n)}$$

(Val Record Select)
$$\frac{\Gamma \vdash M : Record(l_1 : A_1, \ldots, l_n : A_n) \qquad j \in 1 .. n}{\Gamma \vdash M.l_j : A_j}$$

(Val Record With)
$$\frac{\Gamma \vdash M : Record(l_1 : A_1, \ldots, l_n : A_n) \qquad \Gamma, x_1 : A_1, \ldots, x_n : A_n \vdash N : B}{\Gamma \vdash (with(l_1 = x_1 : A_1, \ldots, l_n = x_n : A_n) := M \, do \, N) : B}$$

TABLE 97.12 Variant Types

(Type Variant) (l_i distinct)	(Val Variant) (l_i distinct)
$$\frac{\Gamma \vdash A_1 \cdots \Gamma \vdash A_n}{\Gamma \vdash Variant(l_1 : A_1, \ldots, l_n : A_n)}$$	$$\frac{\Gamma \vdash A_1 \cdots \Gamma \vdash A_n \quad \Gamma \vdash M_j : A_j \quad j \in 1 .. n}{\Gamma \vdash variant_{(l_1:A_1, \ldots, l_n:A_n)}(l_j = M_j) : Variant(l_1 : A_1, \ldots, l_n : A_n)}$$

(Val Variant Is)
$$\frac{\Gamma \vdash M : Variant(l_1 : A_1, \ldots, l_n : A_n) \qquad j \in 1 .. n}{\Gamma \vdash M \, is \, l_j : Bool}$$

(Val Variant As)
$$\frac{\Gamma \vdash M : Variant(l_1 : A_1, \ldots, l_n : A_n) \qquad j \in 1 .. n}{\Gamma \vdash M \, as \, l_j : A_j}$$

(Val Variant Case)
$$\frac{\Gamma \vdash M : Variant(l_1 : A_1, \ldots, l_n : A_n) \quad \Gamma, x_1 : A_1 \vdash N_1 : B \cdots \Gamma, x_n : A_n \vdash N_n : B}{\Gamma \vdash (case_B \, M \, of \, l_1 = x_1 : A_1 then \, N_1 \mid \cdots \mid l_n = x_n : A_n \, then \, N_n) : B}$$

TABLE 97.13 Reference Types

(Type Ref)	(Val Ref)
$$\frac{\Gamma \vdash A}{\Gamma \vdash Ref \, A}$$	$$\frac{\Gamma \vdash M : A}{\Gamma \vdash ref \, M : Ref \, A}$$
(Val Deref)	(Val Assign)
$$\frac{\Gamma \vdash M : Ref \, A}{\Gamma \vdash deref \, M : A}$$	$$\frac{\Gamma \vdash M : Ref \, A \quad \Gamma \vdash N : A}{\Gamma \vdash M := N : Unit}$$

The *with* statement of product types is generalized to record types in (Val Record With). The components of the record M labeled $l_1, \ldots, l_n$ are bound to the variables $x_1, \ldots, x_n$ in the scope of N. Pascal has a similar construct, also called *with*, but where the binding variables are left implicit. (This has the rather unfortunate consequence of making scoping depend on typechecking, and of causing hard-to-trace bugs due to hidden variable clashes.)

Product types $A_1 \times A_2$ can be defined as *Record* (*first* : A_1, *second* : A_2).

Variant types (Table 97.12) are named disjoint unions of types; they are syntactically identified up to reordering of components. The *is l* construct generalizes *isLeft* and *isRight*, and the *as l* construct generalizes *asLeft* and *asRight*. As with unions, these constructs may be replaced by a *case* statement, which now has multiple branches.

Union types $A_1 + A_2$ can be defined as *Variant*(*left* : A_1, *right* : A_2). Enumeration types, such as {*red*, *green*, *blue*}, can be defined as *Variant*(*red* : *Unit*, *green* : *Unit*, *blue* : *Unit*).

Reference types (Table 97.13) can be used as the fundamental type of mutable locations in imperative languages. An element of *Ref*(*A*) is a mutable cell containing an element of type *A*. A new cell can be

TABLE 97.14   An Implementation of Arrays

$Array(A) \triangleq$	Array type
$\quad Nat \times (Nat \to Ref(A))$	a bound plus a map from indices less than
	the bound to refs
$array_A(N, M) \triangleq$	Array constructor (for N refs initialized to M)
$\quad let\ cell_0 : Ref(A) = ref(M)\ and \ldots$	
$\quad and\ cell_{N-1} : Ref(A) = ref(M)$	
$\quad in \langle N, \lambda x : Nat.if\ x = 0\ then\ cell_0\ else\ if \ldots$	
$\qquad else\ if\ x = N - 1\ then\ cell_{N-1}\ else\ error_{Ref(A)} \rangle$	
$bound(M) \triangleq$	Array bound
$\quad first\ M$	
$M[N]_A \triangleq$	Array indexing
$\quad if\ N < first\ M$	
$\quad then\ deref((second\ M)(N))$	
$\quad else\ error_A$	
$M[N] := P \triangleq$	Array update
$\quad if\ N < first\ M$	
$\quad then\ ((second\ M)(N)) := P$	
$\quad else\ error_{Unit}$	

TABLE 97.15   Array Types (Derived Rule)

(Type Array)

$$\frac{\Gamma \vdash A}{\Gamma \vdash Array(A)}$$

(Val Array)	(Val Array Bound)
$\dfrac{\Gamma \vdash N : Nat \quad \Gamma \vdash M : A}{\Gamma \vdash array(N, M) : Array(A)}$	$\dfrac{\Gamma \vdash M : Array(A)}{\Gamma \vdash bound\ M : Nat}$
(Val Array Index)	(Val Array Update)
$\dfrac{\Gamma \vdash N : Nat \quad \Gamma \vdash M : Array(A)}{\Gamma \vdash M[N] : A}$	$\dfrac{\Gamma \vdash N : Nat \quad \Gamma \vdash M : Array(A) \quad \Gamma \vdash P : A}{\Gamma \vdash M[N] := P : Unit}$

allocated by (Val Ref), updated by (Val Assign), and explicitly dereferenced by (Val Deref). Because the main purpose of an assignment is to perform a side effect, its resulting value is chosen to be *unit*. Common mutable types can be derived from *Ref*. Mutable record types, for example, can be modeled as record types containing *Ref* types.

More interestingly, arrays and array operations can be modeled as in Table 97.14, where *Array(A)* is the type of arrays of elements of type *A* of some length. (The code uses some arithmetic primitives and local *let* declarations.) The code in Table 97.14 is, of course, an inefficient implementation of arrays, but it illustrates a point: the type rules for more complex constructions can be derived from the type rule for simpler constructions. The typing rules for array operations shown in Table 97.15 can be easily derived from Table 97.14, according to the rules for products, functions, and refs.

In most programming language, types can be defined recursively. Recursive types are important because they make all of the other type constructions more useful. They are often introduced implicitly, or without precise explanation, and their characteristics are rather subtle. Hence, their formalization deserves particular care.

The treatment of recursive types requires a rather fundamental addition to F_1: environments are extended to include type variables X. These type variables are used in recursive types of the form $\mu X.A$ (Table 97.16), which intuitively denote solutions to recursive equations of the form $X = A$ where X may occur in A. The operations *unfold* and *fold* are explicit coercions that map between a recursive type $\mu X.A$ and its

TABLE 97.16   Recursive Types

(Env X)	(Type Rec)
$\dfrac{\Gamma \vdash \diamond \quad X \notin dom(\Gamma)}{\Gamma, X \vdash \diamond}$	$\dfrac{\Gamma, X \vdash A}{\Gamma \vdash \mu X.A}$
(Val Fold)	(Val Unfold)
$\dfrac{\Gamma \vdash M : [\mu X.A/X]A}{\Gamma \vdash fold_{\mu X.A} M : \mu X.A}$	$\dfrac{\Gamma \vdash M : \mu X.A}{\Gamma \vdash unfold_{\mu X.A} M : [\mu X.A/X]A}$

TABLE 97.17   List Types

$List_A \triangleq \mu X.Unit + (A \times X)$
$nil_A : List_A \triangleq fold(inLeft\ unit)$
$cons_A : A \to List_A \to List_A \triangleq \lambda hd : A.\lambda tl : List_A.fold(inRight\langle hd, tl\rangle)$
$listCase_{A,B} : List_A \to B \to (A \times List_A \to B) \to B \triangleq$
$\quad \lambda l : List_A.\lambda n : B.\lambda c : A \times List_A \to B.$
$\quad\quad case\ (unfold\ l)\ of\ unit\ :\ Unit\ then\ n \mid p\ :\ A \times List_A\ then\ c\ p$

TABLE 97.18   Encoding of Divergence and Recursion via Recursive Types

$\bot_A : A \triangleq (\lambda x : B.(unfold_B x)\ x)\ (fold_B(\lambda x : B.(unfold_B\ x)\ x))$
$\mathbf{Y}_A : (A \to A) \to A \triangleq \lambda f : A \to A.(\lambda x : B.f((unfold_B\ x)\ x))\ (fold_B(\lambda x : B.f((unfold_B\ x)\ x)))$
where $B \equiv \mu X.X \to A$,   for an arbitrary A

TABLE 97.19   Encoding the Untyped λ-Calculus via Recursive Types

V	$\mu X.X \to X$	the type of untyped λ-terms
$\langle\!\langle x \rangle\!\rangle$	x	translation $\langle\!\langle - \rangle\!\rangle$ from untyped λ-terms to V elements
$\langle\!\langle \lambda x.M \rangle\!\rangle$	$fold_V(\lambda x : V.\langle\!\langle M \rangle\!\rangle)$	
$\langle\!\langle MN \rangle\!\rangle$	$(unfold_V\ \langle\!\langle M \rangle\!\rangle)\ \langle\!\langle N \rangle\!\rangle$	

unfolding $[\mu X.A/X]A$ (where $[B/X]A$ is the substitution of B for all free occurrences of X in A), and vice versa. These coercions do not have any run time effect (in the sense that $unfold(fold(M)) = M$ and $fold(unfold(M')) = M'$). They are usually omitted from the syntax of practical programming languages, but their existence makes formal treatment easier.

A standard application of recursive types is in defining types of lists and trees, in conjunction with products and union types. The type $List_A$ of lists of elements of type A is defined in Table 97.17, together with the list constructors *nil* and *cons*, and the list analyzer *listCase*.

Recursive types can be used together with record and variant types to define complex tree structures such as abstract syntax trees. The *case* and *with* statements can then be used to analyze these trees conveniently.

When used in conjunction with function types, recursive types are surprisingly expressive. Via clever encodings, one can show that recursion at the value level is already implicit in recursive types: there is no need to introduce recursion as a separate construct. Moreover, in the presence of recursive types, untyped programming can be carried out within typed languages. More precisely, Table 97.18 shows how to define, for any type A, a divergent element $\bot_A$ of that type, and a fixpoint operator $\mathbf{Y}_A$ for that type. Table 97.19 shows how to encode the untyped λ-calculus within typed calculi. (These encodings are for call-by-name; they take slightly different forms in call-by-value.)

Type equivalence becomes particularly interesting in the presence of recursive types. We have sidestepped several problems here by not dealing with type definitions, by requiring explicit *fold–unfold* coercions between a recursive type and its unfolding, and by not assuming any identifications between recursive types except for renaming of bound variables. In the current formulation we do not need to define a formal judgment for type equivalence: two recursive types are equivalent simply if they are structurally identical (up to renaming of bound variables). This simplified approach can be extended to include type definitions and type equivalence up to unfolding of recursive types [Amadio and Cardelli 1993].

97.4 First-Order Type Systems for Imperative Languages

Imperative languages have a slightly different style of type systems, mostly because they distinguish commands, which do not produce values, from expressions, which do produce values. (It is quite possible to reduce commands to expressions by giving them type *Unit*, but we prefer to remain faithful to the natural distinction.)

As an example of a type system for an imperative language, we consider the untyped imperative language summarized in Table 97.20. This language permits us to study type rules for declarations, which we have not considered so far. The treatment of procedures and data types is very rudimentary in this language, but the rules for functions and data described in Section 97.3 can be easily adapted. The meaning of the features of the imperative language should be self-evident.

The judgments for our imperative language are listed in Table 97.21. The judgments $\Gamma \vdash C$ and $\Gamma \vdash E : A$ correspond to the single judgment $\Gamma \vdash M : A$ of F_1, since we now have a distinction between commands C and expressions E. The judgment $\Gamma \vdash D \therefore S$ assigns a *signature S* to a *declaration D*; a signature is essentially the type of a declaration. In this simple language a signature consists of a single component, for example, $x : Nat$, and a matching declaration could be var $x : Nat = 3$. In general, signatures would consist of lists of such components, and would look very similar or identical to environments Γ.

TABLE 97.20   Syntax of the Imperative Language

$A ::=$	Types
Bool	Boolean type
Nat	natural numbers type
Proc	procedure type (no arguments, no result)
$D ::=$	Declarations
proc $I = C$	procedure declaration
var $I : A = E$	variable declaration
$C ::=$	Commands
$I := E$	assignment
$C_1; C_2$	sequential composition
begin D in C end	block
call I	procedure call
while E do C end	while loop
$E ::=$	Expressions
I	identifier
N	numeral
$E_1 + E_2$	sum of two numbers
E_1 not $= E_2$	inequality of two numbers

TABLE 97.21   Judgments for the Imperative Language

$\Gamma \vdash \diamond$	Γ is a well-formed environment
$\Gamma \vdash A$	A is a well-formed type in Γ
$\Gamma \vdash C$	C is a well-formed command in Γ
$\Gamma \vdash E : A$	E is a well-formed expression of type A in Γ
$\Gamma \vdash D \therefore S$	D is a well-formed declaration of signature S in Γ

TABLE 97.22   Type Rules for Imperative Language

(Env $\emptyset$)	(Env I)	
	$\dfrac{\Gamma \vdash A \quad I \notin dom(\Gamma)}{\Gamma, I : A \vdash \diamond}$	
$\dfrac{}{\emptyset \vdash \diamond}$		
(Type Bool)	(Type Nat)	(Type Proc)
$\dfrac{\Gamma \vdash \diamond}{\Gamma \vdash Bool}$	$\dfrac{\Gamma \vdash \diamond}{\Gamma \vdash Nat}$	$\dfrac{\Gamma \vdash \diamond}{\Gamma \vdash Proc}$
(Decl Proc)	(Decl Var)	
$\dfrac{\Gamma \vdash C}{\Gamma \vdash (proc\ I = C) \therefore (I : Proc)}$	$\dfrac{\Gamma \vdash E : A \quad A \in \{Bool, Nat\}}{\Gamma \vdash (var\ I : A = E) \therefore (I : A)}$	
(Comm Assign)	(Comm Sequence)	
$\dfrac{\Gamma \vdash I : A \quad \Gamma \vdash E : A}{\Gamma \vdash I := E}$	$\dfrac{\Gamma \vdash C_1 \quad \Gamma \vdash C_2}{\Gamma \vdash C_1; C_2}$	
(Comm Block)	(Comm Call)	(Comm While)
$\dfrac{\Gamma \vdash D \therefore (I : A) \quad \Gamma, I : A \vdash C}{\Gamma \vdash begin\ D\ in\ C\ end}$	$\dfrac{\Gamma \vdash I : Proc}{\Gamma \vdash call\ I}$	$\dfrac{\Gamma \vdash E : Bool \quad \Gamma \vdash C}{\Gamma \vdash while\ E\ do\ C\ end}$
(Expr Identifier)	(Expr Numeral)	
$\dfrac{\Gamma_1, I : A, \Gamma_2 \vdash \diamond}{\Gamma_1, I : A, \Gamma_2 \vdash I : A}$	$\dfrac{\Gamma \vdash \diamond}{\Gamma \vdash N : Nat}$	
(Expr Plus)	(Expr NotEq)	
$\dfrac{\Gamma \vdash E_1 : Nat \quad \Gamma \vdash E_2 : Nat}{\Gamma \vdash E_1 + E_2 : Nat}$	$\dfrac{\Gamma \vdash E_1 : Nat \quad \Gamma \vdash E_2 : Nat}{\Gamma \vdash E_1\ not{=}\ E_2 : Bool}$	

Table 97.22 lists the type rules for the imperative language.

The rules (Env ...), (Type ...), and (Expr ...) are straightforward variations on the rules we have seen for F_1. The rules (Decl ...) handle the typing of declarations. The rules (Comm ...) handle commands; notice how (Comm Block) converts a signature to a piece of an environment when checking the body of a block.

97.5   Second-Order Type Systems

Many modern languages include constructs for type parameters, type abstraction, or both. Type parameters can be found in the module system of several languages, where a generic module, class, or interface is parameterized by a type to be supplied later. Planned extensions of Java and C# use type parameters at the class and interface level. (C++ templates are similar to type parameters, but are actually a form of macro-expansion, with very different properties.) Polymorphic languages such as ML and Haskell use type parameters more pervasively, at the function level. Type abstraction can be found in conjunction with modules, in the form of opaque types in interfaces, as in Modula-2 and Modula-3. Languages such as CLU use type abstraction at the data level to obtain abstract data types. These advanced features can be modeled by so-called second-order type systems.

Second-order type systems extend first-order type systems with the notion of *type parameters*. A new kind of term, written $\lambda X.M$, indicates a program M that is parameterized with respect to a type variable X that stands for an arbitrary type. For example, the identity function for a fixed type A, written $\lambda x : A.x$, can be turned into a parametric identity function by abstracting over A and writing $id \triangleq \lambda X.\lambda x : X.x$. One can then instantiate such a parametric function to any given type A by a *type instantiation*, written $id\ A$, which produces back $\lambda x : A.x$.

Corresponding to the new terms $\lambda X.M$ we need new *universally quantified* types. The type of a term such as $\lambda X.M$ is written $\forall X.A$, meaning that *forall* X, the body M has type A (here, M and A may contain occurrences of X). For example, the type of the parametric identity is $id : \forall X.X \to X$.

TABLE 97.23   Syntax of F_2

A, B ::=		Types
	X	type variable
	$A \to B$	function type
	$\forall X.A$	universally quantified type
M, N ::=		Terms
	x	variable
	$\lambda x : A.M$	function
	$M\ N$	application
	$\lambda X.M$	polymorphic abstraction
	$M\ A$	type instantiation

TABLE 97.24   Judgments for F_2

$\Gamma \vdash \diamond$	Γ is a well-formed environment
$\Gamma \vdash A$	A is a well-formed type in Γ
$\Gamma \vdash M : A$	M is a well-formed term of type A in Γ

TABLE 97.25   Type Rules for F_2

(Env $\emptyset$)

$$\emptyset \vdash \diamond$$

(Env x)

$$\frac{\Gamma \vdash A \qquad x \notin dom(\Gamma)}{\Gamma, x : A \vdash \diamond}$$

(Env X)

$$\frac{\Gamma \vdash \diamond \qquad X \notin dom(\Gamma)}{\Gamma, X \vdash \diamond}$$

(Type X)

$$\frac{\Gamma', X, \Gamma'' \vdash \diamond}{\Gamma', X, \Gamma'' \vdash X}$$

(Type Arrow)

$$\frac{\Gamma \vdash A \qquad \Gamma \vdash B}{\Gamma \vdash A \to B}$$

(Type Forall)

$$\frac{\Gamma, X \vdash A}{\Gamma \vdash \forall X.A}$$

(Val x)

$$\frac{\Gamma', x : A, \Gamma'' \vdash \diamond}{\Gamma', x : A, \Gamma'' \vdash x : A}$$

(Val Fun)

$$\frac{\Gamma, x : A \vdash M : B}{\Gamma \vdash \lambda x : A.M : A \to B}$$

(Val Appl)

$$\frac{\Gamma \vdash M : A \to B \qquad \Gamma \vdash N : A}{\Gamma \vdash M\ N : B}$$

(Val Fun2)

$$\frac{\Gamma, X \vdash M : A}{\Gamma \vdash \lambda X.M : \forall X.A}$$

(Val Appl2)

$$\frac{\Gamma \vdash M : \forall X.A \qquad \Gamma \vdash B}{\Gamma \vdash M\ B : [B/X]A}$$

The pure second-order system F_2 (Table 97.23) is based exclusively on type variables, function types, and quantified types. Note that we are dropping the basic types K, because we can now use type variables as the basic case. It turns out that virtually any basic type of interest can be encoded within F_2 [Böhm and Berarducci 1985]. Similarly, product types, sum types, existential types, and some recursive types can be encoded within F_2: polymorphism has an amazing expressive power. Thus, there is little need, technically, to deal with these type constructions directly.

Free variables for F_2 types and terms can be defined in the usual fashion; suffice it to say that $\forall X.A$ binds X in A and $\lambda X.M$ binds X in M. An interesting aspect of F_2 is the substitution of a type for a type variable that is carried out in the type rule for type instantiation, (Val Appl2).

The judgments for F_2 (Table 97.24) are the same ones as for F_1, but the environments are richer. With respect to F_1, the new rules (Table 97.25) are: (Env X), which adds a type variable to the environment; (Type Forall), which constructs a quantified type $\forall X.A$ from a type variable X and a type A where X may occur; (Val Fun2), which builds a polymorphic abstraction; and (Val Appl2), which instantiates a polymorphic abstraction to a given type, where $[B/X]A$ is the substitution of B for all the free occurrences of X in A. For example, if *id* has type $\forall X.X \to X$ and A is a type, then by (Val Appl2) we have that *id A* has type $[A/X](X \to X) \equiv A \to A$. As a simple but instructive exercise, the reader may want to build the derivation for $id(\forall X.X \to X)(id)$.

TABLE 97.26   Existential Types

(Type Exists)	(Val Pack)

$$\frac{\Gamma, X \vdash A}{\Gamma \vdash \exists X.A} \qquad\qquad \frac{\Gamma \vdash [B/X]M : [B/X]A}{\Gamma \vdash (pack_{\exists X.A}\, X = B\ with\ M) : \exists X.A}$$

(Val Open)

$$\frac{\Gamma \vdash M : \exists X.A \quad \Gamma, X, x : A \vdash N : B \quad \Gamma \vdash B}{\Gamma \vdash (open_B M\ as\ X, x : A\ in\ N) : B}$$

As extensions of F_2 we could adopt all the first-order constructions that we already discussed for F_1. A more interesting extension to consider is *existentially quantified* types, also known as type abstractions; see Table 97.26.

To illustrate the use of existentials, we consider an **abstract type** for Booleans. As we said earlier, Booleans can be represented as the type *Unit + Unit*. We can now show how to hide this representation detail from a client who does not care how Booleans are implemented, but who wants to make use of *true, false,* and *cond* (conditional). We first define an interface for such a client to use,

$$BoolInterface \triangleq \exists Bool.\ Record(true: Bool, false: Bool, cond: \forall Y.\ Bool \to Y \to Y \to Y)$$

This interface declares that there exists a type *Bool* (without revealing its identity) that supports the operations *true, false,* and *cond* of appropriate types. The conditional is parameterized with respect to its result type Y, which may vary depending on the context of usage.

Next we define a particular implementation of this interface; one that represents *Bool* as *Unit + Unit*, and that implements the conditional via a case statement. The Boolean representation type and the related Boolean operations are packaged together by the *pack* construct:

$$boolModule : BoolInterface \triangleq$$
$$pack_{BoolInterface}\ Bool = Unit + Unit$$
$$with\ record($$
$$\qquad true = inLeft(unit),$$
$$\qquad false = inRight(unit),$$
$$\qquad cond = \lambda Y.\lambda x : Bool.\lambda y_1 : Y.\lambda y_2 : Y.$$
$$\qquad\qquad case_Y\ x\ of\ x_1 : Unit\ then\ y_1 \mid x_2 : Unit\ then\ y_2)$$

Finally, a client could make use of this module by opening it, and thus getting access to an abstract name *Bool* for the Boolean type, and a name *boolOp* for the record of Boolean operations. These names are used in the next example for a simple computation that returns a natural number. (The computation following *in* is, essentially, *if boolOp.true then 1 else 0*.)

$$open_{Nat}\ boolModule\ as\ Bool, boolOp : Record(true : Bool, false : Bool, cond :$$
$$\forall Y.Bool \to Y \to Y \to Y)in\ boolOp.cond(Nat)(boolOp.true)(1)(0)$$

The reader should verify that these examples typecheck according to the rules previously given. Note the critical third assumption of (Val Open), which implies that the result type B cannot contain the variable X. That assumption forbids writing, for example, *boolOp.true* as the body of *open* in the preceding example, since the result type would be the variable *Bool*. Because of that third assumption, the abstract name of the representation type (*Bool*) cannot escape the scope of *open*, and therefore values having the representation type cannot escape either. A restriction of this kind is necessary; otherwise, the representation type might become known to clients.

97.6 Subtyping

Typed object-oriented languages have particularly interesting and complex type systems. There is little consensus about what characterizes these languages, but at least one feature is almost universally present: **subtyping**. Subtyping captures the intuitive notion of inclusion between types, where types are seen as collections of values. An element of a type can also be considered an element of any of its supertypes, thus allowing a value (object) to be used flexibly in many different typed contexts.

When considering a subtyping relation, such as the one found in object-oriented programming languages, it is customary to add a new judgment $\Gamma \vdash A <: B$ stating that A is a subtype of B (Table 97.27). The intuition is that any element of A is an element of B or, more appropriately, any program of type A is also a program of type B.

One of the simplest type systems with subtyping is an extension of F_1 called $F_{1<:}$. The syntax of F_1 is unchanged, except for the addition of a type *Top* that is a supertype of all types. The existing type rules are also unchanged. The subtyping judgment is independently axiomatized, and a single type rule, called **subsumption**, is added to connect the typing judgment to the subtyping judgment.

The subsumption rule states that if a term has type A, and A is a subtype of B, then the term also has type B. That is, subtyping behaves very much like set inclusion when type membership is seen as set membership.

The subtyping relation in Table 97.28 is defined as a reflexive and transitive relation with a maximal element called *Top*, which is therefore interpreted as the type of all well-typed terms.

The subtype relation for function types says that $A \to B$ is a subtype of $A' \to B'$ if A' is a subtype of A, and B is a subtype of B'. Note that the inclusion is inverted (**contravariant**) for function arguments, while it goes in the same direction (**covariant**) for function results. Simple-minded reasoning reveals that this is the only sensible rule. A function M of type $A \to B$ accepts elements of type A; obviously, it also accepts elements of any subtype A' of A. The same function M returns elements of type B; obviously, it returns elements that belong to any supertype B' of B. Therefore, any function M of type $A \to B$, by virtue of accepting arguments of type A' and returning results of type B', also has type $A' \to B'$. The latter is compatible with saying that $A \to B$ is a subtype of $A' \to B'$.

In general, we say that a type variable occurs contravariantly within another type of F_1, if it always occurs on the left of an odd number of arrows (double contravariance equals covariance). For example, $X \to Unit$ and $(Unit \to X) \to Unit$ are contravariant in X, whereas $Unit \to X$ and $(X \to Unit) \to X$ are covariant in X.

Ad hoc subtyping rules can be added on basic types, such as $Nat <: Int$ [Mitchell 1984].

All of the structured types we considered as extensions of F_1 admit simple subtyping rules; therefore, these structured types can be added to $F_{1<:}$ as well (Table 97.29). Typically, we need to add a single

TABLE 97.27   Judgments for Type Systems with Subtyping

$\Gamma \vdash \diamond$	Γ is a well-formed environment
$\Gamma \vdash A$	A is a well-formed type in Γ
$\Gamma \vdash A <: B$	A is a subtype of B in Γ
$\Gamma \vdash M : A$	M is a well-formed term of type A in Γ

TABLE 97.28   Additional Rules for $F_{1<:}$

(Sub Refl)	(Sub Trans)	(Val Subsumption)
$\dfrac{\Gamma \vdash A}{\Gamma \vdash A <: A}$	$\dfrac{\Gamma \vdash A <: B \quad \Gamma \vdash B <: C}{\Gamma \vdash A <: C}$	$\dfrac{\Gamma \vdash a : A \quad \Gamma \vdash A <: B}{\Gamma \vdash a : B}$
(Type Top)	(Sub Top)	(Sub Arrow)
$\dfrac{\Gamma \vdash \diamond}{\Gamma \vdash Top}$	$\dfrac{\Gamma \vdash A}{\Gamma \vdash A <: Top}$	$\dfrac{\Gamma \vdash A' <: A \quad \Gamma \vdash B <: B'}{\Gamma \vdash A \to B <: A' \to B'}$

TABLE 97.29 Additional Rules for Extensions of $F_{1<:}$

(Sub Product)	(Sub Union)
$\dfrac{\Gamma \vdash A_1 <: B_1 \quad \Gamma \vdash A_2 <: B_2}{\Gamma \vdash A_1 \times A_2 <: B_1 \times B_2}$	$\dfrac{\Gamma \vdash A_1 <: B_1 \quad \Gamma \vdash A_2 <: B_2}{\Gamma \vdash A_1 + A_2 <: B_1 + B_2}$

(Sub Record) (l_i distinct)
$$\frac{\Gamma \vdash A_1 <: B_1 \cdots \Gamma \vdash A_n <: B_n \quad \Gamma \vdash A_{n+1} \cdots \Gamma \vdash A_{n+m}}{\Gamma \vdash Record(l_1 : A_1, \ldots, l_{n+m} : A_{n+m}) <: Record(l_1 : B_1, \ldots, l_n : B_n)}$$

(Sub Variant) (l_i distinct)
$$\frac{\Gamma \vdash A_1 <: B_1 \cdots \Gamma \vdash A_n <: B_n \quad \Gamma \vdash B_{n+1} \cdots \Gamma \vdash B_{n+m}}{\Gamma \vdash Variant(l_1 : A_1, \ldots, l_n : A_n) <: Variant(l_1 : B_1, \ldots, l_{n+m} : B_{n+m})}$$

TABLE 97.30 Environments with Bounded Variables

(Env $X <:$)	(Type $X <:$)	(Sub $X <:$)
$\dfrac{\Gamma \vdash A \quad X \notin dom(\Gamma)}{\Gamma, X <: A \vdash \diamond}$	$\dfrac{\Gamma', X <: A, \Gamma'' \vdash \diamond}{\Gamma', X <: A, \Gamma'' \vdash X}$	$\dfrac{\Gamma', X <: A, \Gamma'' \vdash \diamond}{\Gamma', X <: A, \Gamma'' \vdash X <: A}$

TABLE 97.31 Subtyping Recursive Types

(Type Rec)	(Sub Rec)
$\dfrac{\Gamma, X <: Top \vdash A}{\Gamma \vdash \mu X.A}$	$\dfrac{\Gamma \vdash \mu X.A \quad \Gamma \vdash \mu Y.B \quad \Gamma, Y <: Top, X <: Y \vdash A <: B}{\Gamma \vdash \mu X.A <: \mu Y.B}$

subtyping rule for each type constructor, taking care that the subtyping rule is sound in conjunction with subsumption. The subtyping rules for products and unions work componentwise. The subtyping rules for records and variants also operate lengthwise: a longer record type is a subtype of a shorter record type (additional fields can be forgotten by subtyping), whereas a shorter variant type is a subtype of a longer variant type (additional cases can be introduced by subtyping). For example,

$$WorkingAge \triangleq Variant(student: Unit, adult: Unit)$$
$$Age \triangleq Variant(child: Unit, student: Unit, adult: Unit, senior: Unit)$$
$$Worker \triangleq Record(name: String, age: WorkingAge, profession: String)$$
$$Person \triangleq Record(name: String, age: Age)$$

Then,

$$WorkingAge <: Age$$
$$Worker <: Person$$

Reference types do not have any subtyping rule: $Ref(A) <: Ref(B)$ holds only if $A = B$ (in which case $Ref(A) <: Ref(B)$ follows from reflexivity). This strict rule is necessary because references can be both read and written, and hence behave both covariantly and contravariantly. For the same reason, array types have no additional subtyping rules.

As was the case for F_1, a change to the structure of environments is necessary when considering recursive types. This time, we must add *bounded variables* to environments (Table 97.30). Variables bound by *Top* correspond to our old unconstrained variables. The soundness of the subtyping rule (Sub Rec) for recursive types (Table 97.31) is not obvious, but the intuition is fairly straightforward. To check whether $\mu X.A <: \mu Y.B$, we assume $X <: Y$ and we check $A <: B$; the assumption helps us when finding matching occurrences of X and Y in A and B, as long as they are in covariant contexts. A simpler rule asserts that

TABLE 97.32 Syntax of $F_{2<:}$

$A, B ::=$	Types
X	type variable
Top	the biggest type
$A \to B$	function type
$\forall X <: A.B$	bounded universally quantified type
$M, N ::=$	Terms
x	variable
$\lambda x : A.M$	function
$M\,N$	application
$\lambda X <: A.M$	bounded polymorphic abstraction
$M\,A$	type instantiation

TABLE 97.33 Rules for Bounded Universal Quantifiers

(Type Forall<:)

$$\frac{\Gamma, X <: A \vdash B}{\Gamma \vdash \forall X <: A.B}$$

(Sub Forall<:)

$$\frac{\Gamma \vdash A' <: A \quad \Gamma, X <: A' \vdash B <: B'}{\Gamma \vdash (\forall X <: A.B) <: (\forall X <: A'.B')}$$

(Val Fun2<:)

$$\frac{\Gamma, X <: A \vdash M : B}{\Gamma \vdash \lambda X <: A.M : \forall X <: A.B}$$

(Val Appl2<:)

$$\frac{\Gamma \vdash M : \forall X <: A.B \quad \Gamma \vdash A' <: A}{\Gamma \vdash M\,A' : [A'/X]B}$$

TABLE 97.34 Rules for Bounded Existential Quantifiers (Derivable)

(Type Exists<:)

$$\frac{\Gamma, X <: A \vdash B}{\Gamma \vdash \exists X <: A.B}$$

(Sub Exists<:)

$$\frac{\Gamma \vdash A <: A' \quad \Gamma, X <: A \vdash B <: B'}{\Gamma \vdash (\exists X <: A.B) <: (\exists X <: A'.B')}$$

(Val Pack<:)

$$\frac{\Gamma \vdash C <: A \quad \Gamma \vdash [C/X]M : [C/X]B}{\Gamma \vdash (pack_{\exists X <: A.B} X <: A = C \; with \; M) : \exists X <: A.B}$$

(Val Open<:)

$$\frac{\Gamma \vdash M : \exists X <: A.B \quad \Gamma \vdash D \quad \Gamma, X <: A, x : B \vdash N : D}{\Gamma \vdash (open_D M \; as \; X <: A, x : B \; in \; N) : D}$$

$\mu X.A <: \mu X.B$ whenever $A <: B$ for any X, but this rule is unsound when X occurs in contravariant contexts (e.g., immediately on the left of an arrow).

The bounded variables in environments are also the basis for the extension of F_2 with subtyping, which gives a system called $F_{2<:}$ (Table 97.32). In this system, the term $\lambda X <: A.M$ indicates a program M parameterized with respect to a type variable X that stands for an arbitrary subtype of A. This is a generalization of F_2, since the F_2 term $\lambda X.M$ can be represented as $\lambda X <: Top.M$. Corresponding to the terms $\lambda X <: A.M$, we have bounded type quantifiers of the form $\forall X <: A.B$.

Scoping for $F_{2<:}$ types and terms is defined similarly to F_2, except that $\forall X <: A.B$ binds X in B but not in A, and $\lambda X <: A.M$ binds X in M but not in A.

The type rules for $F_{2<:}$ consist of most of the type rules for $F_{1<:}$ (namely, (Env $\emptyset$), (Env x), (Type Top), (Type Arrow), (Sub Refl), (Sub Trans), (Sub Top), (Sub Arrow), (Val Subsumption), (Val x), (Val Fun), and (Val Appl)), plus the rules for bounded variables (namely, (Env $X<:$), (Type $X<:$), and (Sub $X<:$)), and the ones listed in Table 97.33 for bounded polymorphism.

As for F_2, we do not need to add other type constructions to $F_{2<:}$, since all of the common ones can be expressed within it (except for recursion). Moreover, it turns out that the encodings used for F_2 satisfy the expected subtyping rules. For example, it is possible to encode bounded existential types so that the rules described in Table 97.34 are satisfied. The type $\exists X <: A.B$ represents a *partially abstract type*, whose representation type X is not completely known, but is known to be a subtype of A. This kind of partial abstraction occurs in some languages based on subtyping (e.g., in Modula-3).

Some nontrivial work is needed to obtain encodings of record and variant types in $F_{2<:}$ that satisfy the expected subtyping rules, but even those can be found [Cardelli and Wegner 1985].

97.7 Equivalence

For simplicity, we have avoided describing certain judgments that are necessary when type systems become complex and when one wishes to capture the semantics of programs in addition to their typing. We briefly discuss some of these judgments.

A *type equivalence* judgment, of the form $\Gamma \vdash A = B$, can be used when type equivalence is nontrivial and requires precise description. For example, some type systems identify a recursive type and its unfolding, in which case we would have $\Gamma \vdash \mu X.A = [\mu X.A/X]A$ whenever $\Gamma \vdash \mu X.A$. As another example, type systems with type operators $\lambda X.A$ (functions from types to types) have a reduction rule for operator application of the form $\Gamma \vdash (\lambda X.A) B = [A/X] B$. The type equivalence judgment is usually employed in a *retyping rule* stating that if $\Gamma \vdash M : A$ and $\Gamma \vdash A = B$, then $\Gamma \vdash M : B$.

A *term equivalence* judgment determines which programs are equivalent with respect to a common type. It has the form $\Gamma \vdash M = N : A$. For example, with appropriate rules we could determine that $\Gamma \vdash 2 + 1 = 3 : Int$. The term equivalence judgment can be used to give typed semantics to programs: if N is an irreducible expression, then we can consider N as the resulting value of the program M.

97.8 Type Inference

Type inference is the problem of finding a type for a term within a given type system, if any type exists. In the type systems we have considered earlier, programs have abundant type annotations. Thus, the type inference problem often amounts to little more than checking the mutual consistency of the annotations. The problem is not always trivial but, as in the case of F_1, simple typechecking algorithms may exist.

A harder problem, called *typability* or **type reconstruction**, consists in starting with an untyped program M, and finding an environment Γ, a type-annotated version M' of M, and a type A such that A is a type for M' with respect to Γ. (A type-annotated program M' is simply one that stripped of all type annotations reduces back to M.) The type reconstruction problem for the untyped λ-calculus is solvable within F_1 by the Hindley–Milner algorithm used in ML [Milner 1978]; in addition, that algorithm has the property of producing a unique representation of all possible F_1 typings of a λ-term. The type reconstruction problem for the untyped λ-calculus, however, is not solvable within F_2 [Wells 1994]. Type reconstruction within systems with subtyping is still largely an open problem, although special solutions are beginning to emerge [Aiken and Wimmers 1993, Eifrig et al. 1995, Gunter and Mitchell 1994, Palsberg 1994].

We concentrate here on the type inference algorithms for some representative systems: F_1, F_2, and $F_{2<:}$. The first two systems have the unique type property: if a term has a type, it has only one type. In $F_{2<:}$ there are no unique types, simply because the subsumption rule assigns all of the supertypes of a type to any term that has that type. However, a minimum type property holds: if a term has a collection of types, that collection has a least element in the subtype order [Curien and Ghelli 1992]. The minimum type property holds for many common extensions of $F_{2<:}$ and of $F_{1<:}$ but may fail in the presence of ad-hoc subtypings on basic types.

97.8.1 The Type Inference Problem

In a given type system, given an environment Γ and a term M, is there a type A such that $\Gamma \vdash M : A$ is valid? The following are examples:

- In F_1, given $M \equiv \lambda x : K.x$ and any well-formed Γ we have that $\Gamma \vdash M : K \to K$.
- In F_1, given $M \equiv \lambda x : K.y(x)$ and $\Gamma \equiv \Gamma', y : K \to K$ we have that $\Gamma \vdash M : K \to K$.
- In F_1, there is no typing for $\lambda x : B.x(x)$, for any type B.

TABLE 97.35　　Type Inference Algorithm for F_1

$Type(\Gamma, x) \triangleq$
　　$if\ x : A \in \Gamma\ for\ some\ A\ then\ A\ else\ fail$

$Type(\Gamma, \lambda x : A.M) \triangleq$
　　$A \rightarrow Type((\Gamma, x : A), M)$

$Type(\Gamma, M\ N) \triangleq$
　　$if\ Type(\Gamma, M) \equiv Type(\Gamma, N) \rightarrow B\ for\ some\ B\ then\ B\ else\ fail$

TABLE 97.36　　Type Inference Algorithm for F_2

$Good(\Gamma, X) \triangleq X \in dom(\Gamma)$

$Good(\Gamma, A \rightarrow B) \triangleq Good(\Gamma, A)\ and\ Good(\Gamma, B)$

$Good(\Gamma, \forall X.A) \triangleq Good((\Gamma, X), A)$

$Type(\Gamma, x) \triangleq$
　　$if\ x : A \in \Gamma\ for\ some\ A\ then\ A\ else\ fail$

$Type(\Gamma, \lambda x : A.M) \triangleq$
　　$if\ Good(\Gamma, A)\ then\ A \rightarrow Type((\Gamma, x : A), M)\ else\ fail$

$Type(\Gamma, M\ N) \triangleq$
　　$if\ Type(\Gamma, M) \equiv Type(\Gamma, N) \rightarrow B\ for\ some\ B\ then\ B\ else\ fail$

$Type(\Gamma, \lambda X.M) \triangleq$
　　$\forall X.Type((\Gamma, X), M)$

$Type(\Gamma, M\ A) \triangleq$
　　$if\ Type(\Gamma, M) \equiv \forall X.B\ for\ some\ X, B\ and\ Good(\Gamma, A)\ then\ [A/X]B\ else\ fail$

- However, in $F_{1<:}$ there is the typing $\Gamma \vdash \lambda x : Top \rightarrow B.x(x) : (Top \rightarrow B) \rightarrow B$, for any type B, since x can also be given type Top.
- Moreover, in F_1 with recursive types, there is the typing $\Gamma \vdash \lambda x : B.(unfold_B x)(x) : B \rightarrow B$, for $B \equiv \mu X.X \rightarrow X$, since $unfold_B\ x$ has type $B \rightarrow B$.
- Finally, in F_2 there is the typing $\Gamma \vdash \lambda x : B.x(B)(x) : B \rightarrow B$, for $B \equiv \forall X.X \rightarrow X$, since $x(B)$ has type $B \rightarrow B$.

(An alternative formulation of the type inference problem requires Γ to be found, instead of given. However, in programming practice, one is interested only in type inference for programs embedded in a complete programming context, where Γ is therefore given.)

We begin with the type inference algorithm for pure F_1, given in Table 97.35. The algorithm can be extended in straightforward ways to all of the first-order type structures studied earlier. This is the basis of the typechecking algorithms used in Pascal and all similar procedural languages.

The main routine $Type(\Gamma, M)$, takes an environment Γ and a term M and produces the unique type of M, if any. The instruction *fail* causes a global failure of the algorithm: it indicates a typing error. In this algorithm, as in the ones that follow, we assume that the initial environment parameter Γ is well formed so as to rule out the possibility of feeding invalid environments to internal calls. (For example, we may start with the empty environment when checking a full program.) In any case, it is easy to write a subroutine that checks the well-formedness of an environment, from the code we provide. The case for $\lambda x : A.M$ should have a restriction requiring that $x \notin dom(\Gamma)$, since x is used to extend Γ. However, this restriction can be easily sidestepped by renaming, for example, by making all binders unique before running the algorithm. We omit this kind of restriction from Tables 97.35 through 97.37.

TABLE 97.37 Type Inference Algorithm for $F_{2<:}$

$Good(\Gamma, X) \triangleq X \in dom(\Gamma)$

$Good(\Gamma, Top) \triangleq true$

$Good(\Gamma, A \to B) \triangleq Good(\Gamma, A)$ and $Good(\Gamma, B)$

$Good(\Gamma, \forall X <: A.B) \triangleq Good(\Gamma, A)$ and $Good((\Gamma, X <: A), B)$

$Subtype(\Gamma, A, Top) \triangleq true$

$Subtype(\Gamma, X, X) \triangleq true$

$Subtype(\Gamma, X, A) \triangleq$ *for $A \neq X, Top$*
    if $X <: B \in \Gamma$ for some B then $Subtype(\Gamma, B, A)$ else false

$Subtype(\Gamma, A \to B, A' \to B') \triangleq$
    $Subtype(\Gamma, A', A)$ and $Subtype(\Gamma, B, B')$

$Subtype(\Gamma, \forall X <: A.B, \forall X' <: A'.B') \triangleq$
    $Subtype(\Gamma, A', A)$ and $Subtype((\Gamma, X' <: A'), [X'/X]B, B')$

$Subtype(\Gamma, A, B) \triangleq false$ *otherwise*

$Expose(\Gamma, X) \triangleq$ if $X <: A \in \Gamma$ for some A then $Expose(\Gamma, A)$ else fail

$Expose(\Gamma, A) \triangleq A$ *otherwise*

$Type(\Gamma, x) \triangleq$
    if $x : A \in \Gamma$ for some A then A else fail

$Type(\Gamma, \lambda x : A.M) \triangleq$
    if $Good(\Gamma, A)$ then $A \to Type((\Gamma, x : A), M)$ else fail

$Type(\Gamma, M\,N) \triangleq$
    if $Expose(\Gamma, Type(\Gamma, M)) \equiv A \to B$ for some A, B
    and $Subtype(\Gamma, Type(\Gamma, N), A)$ then B else fail

$Type(\Gamma, \lambda X <: A.M) \triangleq$
    if $Good(\Gamma, A)$ then $\forall X <: A.Type((\Gamma, X <: A), M)$ else fail

$Type(\Gamma, M\,A) \triangleq$
    if $Expose(\Gamma, Type(\Gamma, M)) \equiv \forall X <: A'.B$ for some X, A', B
    and $Good(\Gamma, A)$ and $Subtype(\Gamma, A, A')$ then $[A/X]B$ else fail

---

As an example, let us consider the type inference problem for term $\lambda z : K.y(z)$ in the environment $\emptyset, y : K \to K$, for which we gave a full F_1 derivation in Section 97.3. The algorithm proceeds as follows:

$Type((\emptyset, y : K \to K), \lambda z : K.y(z))$

$= K \to Type((\emptyset, y : K \to K, z : K), y(z))$

$= K \to (if\ Type((\emptyset, y : K \to K, z : K), y) \equiv Type((\emptyset, y : K \to K, z : K), z)$
     $\to B$ for some B then B else fail)

$= K \to (if\ K \to K \equiv K \to B$ for some B then B else fail)   (taking $B \equiv K$)

$= K \to K$

The type inference algorithm for F_2 (Table 97.36) is not much harder than the one for F_1, but it requires a subroutine $Good(\Gamma, A)$ to verify that the types encountered in the source program are well formed. This check is necessary because types in F_2 contain type variables that might be unbound. A substitution subroutine must also be used in the type instantiation case, MA.

The type inference algorithm for $F_{2<:}$, given in Table 97.37, is more subtle. The subroutine $Subtype$ (Γ, A, B) attempts to decide whether A is a subtype of B in Γ, and is at first sight straightforward. It has been shown, however, that $Subtype$ is only a semialgorithm: it may diverge on certain pairs A, B that are not in subtype relation. That is, the typechecker for $F_{2<:}$ may diverge on ill-typed programs, although it will still converge and produce a minimum type for well-typed programs. More generally, there is no decision procedure for subtyping: the type system for $F_{2<:}$ is undecidable [Pierce 1992]. Several attempts have been made to cut $F_{2<:}$ down to a decidable subset; the simplest solution at the moment consists of requiring equal quantifiers bounds in (Sub Forall<:). In any case, the bad pairs A, B are extremely unlikely to arise in practice. The algorithm is still sound in the usual sense: if it finds a type, the program will not go wrong. The only troublesome case is in the subtyping of quantifiers; the restriction of the algorithm to $F_{1<:}$ is decidable and produces minimum types.

$F_{2<:}$ provides an interesting example of the anomalies one may encounter in type inference. The type inference algorithm given in Table 97.37 is theoretically undecidable but is practically applicable. It is convergent and efficient on virtually all programs one may encounter; it diverges only on some ill-typed programs, which should be rejected anyway. Therefore, $F_{2<:}$ sits close to the boundary between acceptable and unacceptable type systems, according to the criteria enunciated in the introduction.

97.9   Summary and Research Issues

97.9.1   What We Learned

Natural questions for a beginner programmer include the following. What is an error? What is type safety? What is type soundness? (perhaps phrased, respectively, as: Which errors will the computer tell me about? Why did my program crash? Why does the computer refuse to run my program?). The answers, even informal ones, are surprisingly intricate. We have paid particular attention to the distinction between type safety and type soundness, and we have reviewed the varieties of static checking, dynamic checking, and absence of checking for program errors in various kinds of languages.

The most important lesson to remember from this chapter is the general framework for formalizing type systems. Understanding type systems, in general terms, is as fundamental as understanding BNF (Backus–Naur Form): it is hard to discuss the typing of programs without the precise language of type systems, just as it is hard to discuss the syntax of programs without the precise language of BNF. In both cases, the existence of a formalism has clear benefits for language design, compiler construction, language learning, and program understanding. We described the formalism of type systems and how it captures the notions of type soundness and type errors.

Armed with formal type systems, we embarked on the description of an extensive list of program constructions and of their type rules. Many of these constructions are slightly abstracted versions of familiar features, whereas others apply only to obscure corners of common languages. In both cases, our collection of typing constructions is meant as a key for interpreting the typing features of programming languages. Such an interpretation may be nontrivial, particularly because most language definitions do not come with a type system, but we hope to have provided sufficient background for independent study. Some of the advanced type constructions will appear, we expect, more fully, cleanly, and explicitly in future languages.

In the latter part of the chapter, we reviewed some fundamental type inference algorithms: for simple languages, for polymorphic languages, and for languages with subtyping. These algorithms are very simple and general, but are mostly of an illustrative nature. For a host of pragmatic reasons, type inference for real languages becomes much more complex. It is interesting, however, to be able to describe concisely the core of the type inference problem and some of its solutions.

97.9.2   Future Directions

The formalization of type systems for programming languages, as described in this chapter, evolved as an application of type theory. Type theory is a branch of formal logic. It aims to replace predicate logics and set theory (which are untyped) with typed logics, as a foundation for mathematics.

One of the motivations for these logical type theories, and one of their more exciting applications, lies in the mechanization of mathematics via proof checkers and theorem provers. Typing is useful in theorem provers for exactly the same reasons it is useful in programming. The mechanization of proofs reveals striking similarities between proofs and programs: the structuring problems found in proof construction are analogous to the ones found in program construction. Many of the arguments that demonstrate the need for typed programming languages also demonstrate the need for typed logics.

Comparisons between the type structures developed in type theory and in programming are, thus, very instructive. Function types, product types, (disjoint) union types, and quantified types occur in both disciplines, with similar intents. This is in contrast, for example, to structures used in set theory, such as unions and intersections of sets, and the encoding of functions as sets of pairs, that have no correspondence in the type systems of common programming languages.

Beyond the simplest correspondences between type theory and programming, it turns out that the structures developed in type theory are far more expressive than the ones commonly used in programming. Therefore, type theory provides a rich environment for future progress in programming languages.

Conversely, the size of systems that programmers build is vastly greater than the size of proofs that mathematicians usually handle. The management of large programs, and in particular the type structures needed to manage large programs, is relevant to the management of mechanical proofs. Certain type theories developed in programming, for example, for object-orientation and for modularization, go beyond the normal practices found in mathematics, and should have something to contribute to the mechanization of proofs.

Therefore, the cross-fertilization between logic and programming will continue, within the common area of type theory. At the moment, some advanced constructions used in programming escape proper type-theoretical formalization. This could be happening either because the programming constructions are ill conceived, or because our type theories are not yet sufficiently expressive: only the future will tell. Examples of active research areas are the typing of advanced object-orientation and modularization constructs and the typing of concurrency and distribution.

Defining Terms

Abstract type: A data type whose nature is kept hidden in such a way that only a predetermined collection of operations can operate on it.

Contravariant: A type that varies in the inverse direction from one of its parts with respect to subtyping. The main example is the contravariance of function types in their domain. For example, assume $A <: B$ and vary X from A to B in $X \rightarrow C$; we obtain $A \rightarrow C :> B \rightarrow C$. Thus, $X \rightarrow C$ varies in the inverse direction of X.

Covariant: A type that varies in the same direction as one of its parts with respect to subtyping. For example, assume $A <: B$ and vary X from A to B in $D \rightarrow X$; we obtain $D \rightarrow A <: D \rightarrow B$. Thus, $D \rightarrow X$ varies in the same direction as X.

Derivation: A tree of judgments obtained by applying the rules of a type system.

Dynamic checking: A collection of runtime tests aimed at detecting and preventing forbidden errors.

Dynamically checked language: A language where good behavior is enforced during execution.

Explicitly typed language: A typed language where types are part of the syntax.

First-order type system: One that does not include quantification over type variables.

Forbidden error: The occurrence of one of a predetermined class of execution errors; typically the improper application of an operation to a value, such as *not* (3).

Good behavior: Same as being well-behaved.

Ill typed: A program fragment that does not comply with the rules of a given type system.

Implicitly typed language: A typed language where types are not part of the syntax.

Judgment: A formal assertion relating entities such as terms, types, and environments. Type systems prescribe how to produce valid judgments from other valid judgments.

Polymorphism: The ability of a program fragment to have multiple types (opposite of monomorphism).

Safe language: A language where no untrapped errors can occur.

Second-order type system: One that includes quantification over type variables, either universal or existential.

Static checking: A collection of compile-time tests, mostly consisting of typechecking.

Statically checked language: A language where good behavior is determined before execution.

Strongly checked language: A language where no forbidden errors can occur at runtime (depending on the definition of forbidden error).

Subsumption: A fundamental rule of subtyping, asserting that if a term has a type A, which is a subtype of a type B, then the term also has type B.

Subtyping: A reflexive and transitive binary relation over types that satisfies subsumption; it asserts the inclusion of collections of values.

Trapped error: An execution error that immediately results in a fault.

Type: A collection of values. An estimate of the collection of values that a program fragment can assume during program execution.

Type inference: The process of finding a type for a program within a given type system.

Type reconstruction: The process of finding a type for a program where type information has been omitted, within a given type system.

Type rule: A component of a type system. A rule stating the conditions under which a particular program construct will not cause forbidden errors.

Type safety: The property stating that programs do not cause untrapped errors.

Type soundness: The property stating that programs do not cause forbidden errors.

Type system: A collection of type rules for a typed programming language. Same as static type system.

Typechecker: The part of a compiler or interpreter that performs typechecking.

Typechecking: The process of checking a program before execution to establish its compliance with a given type system and therefore to prevent the occurrence of forbidden errors.

Typed language: A language with an associated (static) type system, whether or not types are part of the syntax.

Typing error: An error reported by a typechecker to warn against possible execution errors.

Untrapped error: An execution error that does not immediately result in a fault.

Untyped language: A language that does not have a (static) type system, or whose type system has a single type that contains all values.

Valid judgment: A judgment obtained from a derivation in a given type system.

Weakly checked language: A language that is statically checked but provides no clear guarantee of absence of execution errors.

Well behaved: A program fragment that will not produce forbidden errors at runtime.

Well formed: Properly constructed according to formal rules.

Well-typed program: A program (fragment) that complies with the rules of a given type system.

References

Aiken, A. and Wimmers, E. L. 1993. Type inclusion constraints and type inference, pp. 31–41. In *Proc. ACM Conf. Functional Programming Comput. Architecture.*

Amadio, R. M. and Cardelli, L. 1993. Subtyping recursive types. *ACM Trans. Programming Lang. Syst.,* 15(4):575–631.

Birtwistle, G. M., Dahl, O.-J., Myhrhaug, B., and Nygaard, K. 1979. Simula Begin. Studentlitteratur.

Böhm, C. and Berarducci, A. 1985. Automatic synthesis of typed λ-programs on term algebras. *Theor. Comput. Sci.,* 39:135–154.

Cardelli, L. 1987. Basic polymorphic typechecking. *Sci. Comput. Programming,* 8(2):147–172.

Cardelli, L. 1994. Extensible records in a pure calculus of subtyping. In *Theoretical Aspects of Object-Oriented Programming,* C. A. Gunter and J. C. Mitchell, Eds., pp. 373–425. MIT Press, Cambridge, MA.

Cardelli, L. and Wegner, P. 1985. On understanding types, data abstraction and polymorphism. *ACM Comput. Surv.,* 17(4):471–522.

Curien, P.-L. and Ghelli, G. 1992. Coherence of subsumption, minimum typing and typechecking in $F_{\leq}$. *Math. Struct. Comput. Sci.,* 2(1):55–91.

Dahl, O.-J., Dijkstra, E. W., and Hoare, C. A. R. 1972. *Structured Programming*. Academic Press.

Eifrig, J., Smith, S., and Trifonov, V. 1995. Sound polymorphic type inference for objects, pp. 169–184. In *Proc. OOPSLA '95*.

Gunter, C. A. 1992. *Semantics of Programming Languages: Structures and Techniques*. MIT Press, Cambridge, MA.

Girard, J.-Y., Lafont, Y., and Taylor, P. 1989. *Proofs and Types*. Cambridge University Press, Cambridge, England.

Gunter, C. A. and Mitchell, J. C., Eds. 1994. *Theoretical Aspects of Object-Oriented Programming*. MIT Press, Cambridge, MA.

Huet, G., Ed. 1990. *Logical Foundations of Functional Programming*. Addison-Wesley, Reading, MA.

Jensen, K. 1978. *Pascal User Manual and Report*, 2nd ed. Springer-Verlag, New York.

Liskov, B. H. 1981. *CLU Reference Manual*. Lecture Notes in Computer Science 114. Springer-Verlag, New York.

Milner, R. 1978. A theory of type polymorphism in programming. *J. Comput. Syst. Sci.*, 17:348–375.

Milner, R., Tofte, M., and Harper, R. 1989. *The Definition of Standard ML*. MIT Press, Cambridge, MA.

Mitchell, J. C. 1984. Coercion and type inference, pp. 175–185. In *Proc. 11th Annu. ACM Symp. Principles Programming Lang.*

Mitchell, J. C. 1990. Type systems for programming languages. In *Handbook of Theoretical Computer Science*, J. van Leeuwen, Ed., pp. 365–458. North Holland, Amsterdam.

Mitchell, J. C. 1996. *Foundations for Programming Languages*. MIT Press, Cambridge, MA.

Mitchell, J. C. and Plotkin, G. D. 1985. Abstract types have existential type. In *Proc. 12th Annu. ACM Symp. Principles Programming Lang.*, pp. 37–51.

Nordström, B., Petersson, K., and Smith, J. M. 1990. *Programming in Martin-Löf's Type Theory*. Oxford Science.

Palsberg, J. 1995. Efficient inference for object types. *Inf. Comput.* 123(2):198–209.

Pierce, B. C. 1992. Bounded quantification is undecidable. In *Proc. 19th Annu. ACM Symp. Principles Programming Lang.*, pp. 305–315.

Pierce, B. C. 2002. *Types and Programming Languages*. MIT Press, Cambridge, MA.

Reynolds, J. C. 1974. Towards a theory of type structure. In *Proc. Colloquium sur la programmation*. Lecture Notes in Computer Science 19, pp. 408–423. Springer-Verlag, New York.

Reynolds, J. C. 1983. Types, abstraction, and parametric polymorphism. In *Information Processing*, R. E. A. Mason, Ed., pp. 513–523. North Holland, Amsterdam.

Schmidt, D. A. 1994. *The Structure of Typed Programming Languages*. MIT Press, Cambridge, MA.

Spencer, H. The ten commandments for C programmers. annotated ed. (available on the World Wide Web).

Tofte, M. 1990. Type inference for polymorphic references. *Inf. Comput.*, 89:1–34.

Wells, J. B. 1994. Typability and type checking in the second-order λ-calculus are equivalent and undecidable, pp. 176–185. In *Proc. 9th Annu. IEEE Symp. Logic Comput. Sci.*

Wijngaarden, V., Ed. 1976. *Revised Report on the Algorithmic Language Algol68*.

Wright, A. K. and Felleisen, M. 1994. A syntactic approach to type soundness. *Inf. Comput.* 115(1):38–94.

Further Information

For a complete background on type systems, one should read (1) some material on type theory, which is usually rather difficult; (2) some material connecting type theory to computing; and (3) some material about programming languages with advanced type systems.

The book edited by Huet [1990] covers a variety of topics in type theory, including several tutorial articles. The book edited by Gunter and Mitchell [1994] contains a collection of papers on object-oriented type theory. The book by Nordström et al. [1990] provides summary of Martin-Löf's work. Martin-Löf proposed type theory as a general logic that is firmly grounded in computation. He introduced the systematic notation for judgments and type rules used in this chapter. Girard et al. [1989] and Reynolds

[1974] developed the polymorphic λ-calculus (F_2), which inspired much of the work covered in this chapter.

A modern exposition of technical issues that arise from the study of type systems can be found in Pierce's book [2002], in Gunter's book [1992], in Mitchell's [1990] article in the *Handbook of Theoretical Computer Science*, and in Mitchell's book [1996].

Closer to programming languages, rich type systems were pioneered in the period between the development of Algol and the establishment of structured programming [Dahl et al. 1972], and were developed into a new generation of richly typed languages, including Pascal [Jansen 1978], Algol68 [Wijngaarden 1976], Simula [Birtwistle et al. 1979], CLU [Liskov 1981], and ML [Milner et al. 1989]. Reynolds gave type theoretical explanations for polymorphism and data abstraction [Reynolds 1974, Reynolds 1983]. (On that topic, see also Cardelli and Wegner [1985] and Mitchell and Plotkin [1985].) The book by Schmidt [1994] covers several issues discussed in this chapter, and provides more details on common language constructions.

Milner's article on type inference for ML [Milner 1978] brought the study of type systems and type inference to a new level. It includes an algorithm for polymorphic type inference, and the first proof of type soundness for a (simplified) programming language, based on a denotational technique. A more accessible exposition of the algorithm described in that article can be found in Cardelli [1987]. Proofs of type soundness are now often based on operational techniques [Tofte 1990, Wright and Felleisen 1994]. Currently, Standard ML is the only widely used programming language with a formally specified type system [Milner et al. 1989], although similar work has now been carried out for large fragments of Java.

98

Programming Language Semantics

98.1   Introduction 98-1
98.2   A Survey of Semantics Methods 98-2
     Operational Semantics • Denotational Semantics
     • Natural Semantics • Axiomatic Semantics
98.3   Semantics of Programming Languages 98-6
     Language Syntax and Informal Semantics • Domains for
     Denotational Semantics • Denotational Semantics of
     Programs • Semantics of the While-Loop
     • Action Semantics • The Natural Semantics of the Language
     • The Operational Semantics of the Language
     • An Axiomatic Semantics of the Language
98.4   Applications of Semantics 98-14
98.5   Research Issues in Semantics........................ 98-15

David A. Schmidt
Kansas State University

98.1   Introduction

A programming language possesses two fundamental features: syntax and semantics. Syntax refers to the appearance of the well-formed programs of the language, and semantics refers to the meanings of these programs. A language's syntax can be formalized by a grammar or syntax chart; such a formalization is found in the back of almost every language manual. A language's semantics should be formalized as well, so that it can appear in the language manual, too. This is the topic of this chapter.

It is traditional for computer scientists to calculate the semantics of a program by using a test-case input and tracing the program's execution with a state table and flowchart. This is one form of semantics, called *operational semantics*, but there are other forms of semantics that are not tied to test cases and traces; we will study several such approaches.

Before we begin, we might ask: What do we gain by formalizing the semantics of a programming language? Before we answer, we might consider the related question: What was gained when language syntax was formalized? The formalization of syntax, via Backus–Naur Form (BNF) rules, produced the following benefits:

- The syntax definition standardizes the official syntax of the language. This is crucial to users, who require a guide to writing syntactically correct programs, and to implementors, who must write a correct parser for the language's compiler.

- The syntax definition permits a formal analysis of its properties, such as whether the definition is $LL(k)$, $LR(k)$, or ambiguous.

- The syntax definition can be used as input to a compiler front-end generating tool, such as YACC or Bison. In this way, the syntax definition is also the implementation of the front end of the language's compiler.

There are similar benefits to providing a formal semantics definition of a programming language:

- The semantics definition standardizes the official semantics of the language. This is crucial to users, who require a guide to understanding the programs that they write, and to implementors, who must write a correct code generator for the language's compiler.
- The semantics definition permits a formal analysis of the language's properties, such as whether the language is strongly typed, is stack- or heap-allocated, or is single- or multi-threaded.
- The semantics definition can be used as input to a compiler back-end generating tool, such as the Semantics Implementation System (SIS) [Mosses 1976]. In this way, the semantics definition is also the implementation of the back end of the language's compiler.

Programming language syntax was studied intensively in the 1960s and 1970s, and presently programming language semantics is undergoing similar intensive study. Unlike the acceptance of BNF as a standard definition method for syntax, it is unlikely that a single definition method will take hold for semantics — semantics is more difficult to formalize than syntax, and it has a wider variety of applications.

Semantics definition methods fall roughly into three groups:

1. **Operational**. The meaning of a well-formed program is the trace of computation steps that results from processing the program's input. Operational semantics is also called *intensional* semantics because the sequence of internal computation steps (the intension) is most important. For example, two differently coded programs that both compute the factorial function have different operational semantics.
2. **Denotational**. The meaning of a well-formed program is a mathematical function from input data to output data. The steps taken to calculate the output are unimportant; it is the relation of input to output that matters. Denotational semantics is also called *extensional* semantics because only the extension — the visible relation between input and output — matters. Thus, two differently coded versions of factorial have nonetheless the same denotational semantics.
3. **Axiomatic**. A meaning of a well-formed program is a logical proposition (a specification) that states some property about the input and output. For example, the proposition $\forall x.\, x \geq 0 \supset \exists y.\, y = x!$ is an axiomatic semantics of a factorial program.

98.2 A Survey of Semantics Methods

We survey the three semantics methods by applying each of them in turn to the world's oldest and simplest programming language, arithmetic. The syntax of our arithmetic language is:

$$E ::= N \mid E_1 + E_2$$

where N stands for the set of numerals $\{0, 1, 2, \ldots\}$. Although this language has no notion of input data and output data, it does require computation, so it is useful for initial case studies.

98.2.1 Operational Semantics

There are several versions of operational semantics for arithmetic. The one that you learned as a child is called a *term rewriting system*. A term rewriting system uses rewriting rule schemes to generate computation steps. There is just one rewriting rule scheme for arithmetic:

$$N_1 + N_2 \Rightarrow N'$$

where N' is the sum of the numerals N_1 and N_2. This rule scheme states that the addition of two numerals is a computation step. One use of the scheme would be to rewrite $1 + 2$ to 3; that is, $1 + 2 \Rightarrow 3$. An operational semantics of a program is the sequence of computation steps generated by the rewriting rule schemes. For example, the operational semantics of the program $(1 + 2) + (4 + 5)$ goes as follows:

$$(1 + 2) + (4 + 5) \Rightarrow 3 + (4 + 5) \Rightarrow 3 + 9 \Rightarrow 12$$

The semantics shows the three computation steps that led to the answer 12. An intermediate expression such as $3 + (4 + 5)$ is a *state*, and so this operational semantics is a trace of the states of the computation.

Perhaps you noticed that another legal semantics for the example is $(1+2)+(4+5) \Rightarrow (1+2)+9 \Rightarrow 3+9 \Rightarrow 12$. The outcome is the same in both cases, but sometimes operational semantics must be forced to be *deterministic*; that is, a program has exactly one operational semantics — one trace.

A **structural operational semantics** is a term rewriting system plus a set of inference rules that state precisely the context in which a computation step can be undertaken.* Say that we desire left-to-right computation of arithmetic expressions. This is encoded as follows:

$$N_1 + N_2 \Rightarrow N'$$

where N' is the sum of N_1 and N_2.

$$\frac{E_1 \Rightarrow E_1'}{E_1 + E_2 \Rightarrow E_1' + E_2} \qquad \frac{E_2 \Rightarrow E_2'}{N + E_2 \Rightarrow N + E_2'}$$

The first rule is as before; the second rule states that if the left operand of an addition expression can be rewritten, then the addition expression should be revised to show this. The third rule is the crucial one: if the right operand of an addition expression can be rewritten *and* the left operand is a numeral (that is, it is completely evaluated), then the addition expression should be revised to show this. Working together, the three rules force left-to-right evaluation of expressions.

Now, each computation step must be deduced by these rules. For our example, $(1 + 2) + (4 + 5)$, we must deduce this initial computation step:

$$\frac{1 + 2 \Rightarrow 3}{(1 + 2) + (4 + 5) \Rightarrow 3 + (4 + 5)}$$

Thus, the first step is $(1+2)+(4+5) \Rightarrow 3+(4+5)$; we *cannot* deduce that $(1 + 2)+(4+5) \Rightarrow (1+2)+9$. The next computation step is justified by this deduction:

$$\frac{4 + 5 \Rightarrow 9}{3 + (4 + 5) \Rightarrow 3 + 9}$$

The last deduction is simply $3 + 9 \Rightarrow 12$, and we are finished. The example shows why the semantics is structural: a computation step, such as an addition, which affects a small part of the overall program, is explicitly embedded into the structure of the overall program.

Operational semantics can also be used to represent internal data structures, such as instruction counters, storage vectors, and stacks. For example, say that the semantics of arithmetic must show that a stack is used to hold intermediate results. Thus, we use a state of the form $\langle s, c \rangle$, where s is the stack and c is the arithmetic expression to be executed. A stack containing n items is written $v_1 :: v_2 :: \ldots :: v_n :: nil$, where v_1 is the topmost item and *nil* marks the bottom of the stack. The c component will be written as a stack as well. The initial state for an arithmetic expression p is written $\langle nil, p :: nil \rangle$, and computation proceeds until the state appears as $\langle v :: nil, nil \rangle$; we say that the result is v.

---

*A structural operational semantics is sometimes called a *small-step semantics* because each computation step is a small step toward the final answer.

The semantics uses three rewriting rules:

$$\langle s, \text{N} :: c \rangle \Rightarrow \langle \text{N} :: s, c \rangle$$

$$\langle s, E_1 + E_2 :: c \rangle \Rightarrow \langle s, E_1 :: E_2 :: add :: c \rangle$$

$$\langle N_2 :: N_1 :: s, add :: c \rangle \Rightarrow \langle N' :: s, c \rangle$$

where N' is the sum of N_1 and N_2. The first rule says that a numeral is evaluated by pushing it on the top of the stack. The second rule states that the addition of two expressions is decomposed into first evaluating the two expressions and then adding them. The third rule removes the top two items from the stack and adds them. Here is the previous example, repeated:

$$\langle nil, (1 + 2) + (4 + 5) :: nil \rangle$$

$$\Rightarrow \langle nil, 1 + 2 :: 4 + 5 :: add :: nil \rangle$$

$$\Rightarrow \langle nil, 1 :: 2 :: add :: 4 + 5 :: add :: nil \rangle$$

$$\Rightarrow \langle 1 :: nil, 2 :: add :: 4 + 5 :: add :: nil \rangle$$

$$\Rightarrow \langle 2 :: 1 :: nil, add :: 4 + 5 :: add :: nil \rangle$$

$$\Rightarrow \langle 3 :: nil, 4 + 5 :: add :: nil \rangle \quad \Rightarrow \ldots \Rightarrow \langle 12 :: nil, nil \rangle$$

This form of operational semantics is sometimes called a *state transition semantics* because each rewriting rule operates upon the entire state. With a state transition semantics, there is no need for structural operational semantics rules.

The three example semantics just shown are typical of operational semantics. When one wishes to prove properties of an operational semantics definition, the standard proof technique is *induction on the length of the computation*. That is, to prove that a property P holds for an operational semantics, one must show that P holds for all possible computation sequences that can be generated from the rewriting rules. For an arbitrary computation sequence, it suffices to show that P holds no matter how long the computation runs. Therefore, one shows (1) P holds after zero computation steps, that is, at the outset; and (2) if P holds after n computation steps, it holds after $n + 1$ steps. See Nielson and Nielson [1992] for examples.

98.2.2   Denotational Semantics

A drawback of operational semantics is the emphasis it places on state sequences. For the arithmetic language, we were distracted by questions regarding order of evaluation of subphrases, although this issue is not central to arithmetic. Further, a key aspect of arithmetic, the property that the meaning of an expression is built from the meanings of its subexpressions, was obscured by the operational semantics.

Denotational semantics handles these issues by emphasizing that a program has an underlying mathematical meaning that is independent of whatever computation strategy is taken to uncover it. In the case of arithmetic, an expression such as $(1 + 2) + (4 + 5)$ has the meaning 12, and we need not worry about the internal computation steps that were taken to discover this.

The assignment of meaning to programs is performed in a *compositional* manner: the meaning of a phrase is built from the meanings of its subphrases. We can see this in the denotational semantics of the arithmetic language: first, we note that meanings of arithmetic expressions are natural numbers, $Nat = \{0, 1, 2, \ldots\}$, and we note that there is a binary function, $plus : Nat \times Nat \to Nat$, which maps a pair of natural numbers to their sum.

The denotational semantics definition of arithmetic is simple and elegant:

$$\mathcal{E} : Expression \to Nat$$

$$\mathcal{E}[\![\text{N}]\!] = \text{N}$$

$$\mathcal{E}[\![E_1 + E_2]\!] = plus\,(\mathcal{E}[\![E_1]\!], \mathcal{E}[\![E_2]\!])$$

The first line states merely that $\mathcal{E}$ is the name of the function that maps arithmetic expressions to their meanings. Because there are just two BNF constructions for expressions, $\mathcal{E}$ is completely defined by the two equational clauses. The interesting clause is the one for $E_1 + E_2$; it says that the meanings of E_1 and E_2 are combined compositionally by *plus*. Here is the denotational semantics of the running example:

$$
\begin{aligned}
\mathcal{E}[\![(1+2)+(4+5)]\!] &= plus\,(\mathcal{E}[\![1+2]\!], \mathcal{E}[\![4+5]\!]) \\
&= plus\,(plus\,(\mathcal{E}[\![1]\!], \mathcal{E}[\![2]\!]), plus\,(\mathcal{E}[\![4]\!], \mathcal{E}[\![5]\!])) \\
&= plus\,(3,9) = 12
\end{aligned}
$$

One might read the preceding example as follows: the meaning of $(1+2)+(4+5)$ equals the meanings of $1+2$ and $4+5$ added together. Because the meaning of $1+2$ is 3, and the meaning of $4+5$ is 9, the meaning of the overall expression is 12. This reading says nothing about order of evaluation or run time data structures — it emphasizes underlying mathematical meaning.

Here is an alternative way of understanding the semantics; write a set of simultaneous equations based on the denotational definition

$$
\begin{aligned}
\mathcal{E}[\![(1+2)+(4+5)]\!] &= plus\,(\mathcal{E}[\![1+2]\!], \mathcal{E}[\![4+5]\!]) \\
\mathcal{E}[\![1+2]\!] &= plus\,(\mathcal{E}[\![1]\!], \mathcal{E}[\![2]\!]) \\
\mathcal{E}[\![4+5]\!] &= plus\,(\mathcal{E}[\![4]\!], \mathcal{E}[\![5]\!]) \\
\mathcal{E}[\![1]\!] = 1 \qquad & \mathcal{E}[\![2]\!] = 2 \\
\mathcal{E}[\![4]\!] = 4 \qquad & \mathcal{E}[\![5]\!] = 5
\end{aligned}
$$

Now solve the equation set to discover that $\mathcal{E}[\![(1+2)+(4+5)]\!]$ is 12.

Because denotational semantics states the meaning of a phrase in terms of the meanings of its subphrases, its associated proof technique is structural induction. That is, to prove that a property P holds for all programs in the language, one must show that the meaning of each construction in the language has property P. Therefore, one must show that each equational clause in the semantic definition produces a meaning with property P. In the case that a clause refers to subphrases (e.g., $\mathcal{E}[\![E_1 + E_2]\!]$), one can assume that the meanings of the subphrases have property P. Again, see Nielson and Nielson [1992] for examples.

98.2.3 Natural Semantics

A "hybrid" semantics method, **natural semantics**, attempts to combine the advantages of both operational semantics and denotational semantics. Like structural operational semantics, natural semantics shows the context in which a computation step occurs, and, like denotational semantics, natural semantics emphasizes that the computation of a phrase is built from the computations of its subphrases.

A natural semantics is a set of inference rules, and a complete computation in natural semantics is a single derivation. The natural semantics rules for the arithmetic language are

$$
N \Downarrow N
$$

$$
\frac{E_1 \Downarrow n_1 \qquad E_2 \Downarrow n_2}{E_1 + E_2 \Downarrow m}
$$

where m is the sum of n_1 and n_2. Read a configuration of the form $E \Downarrow n$ as "E evaluates to n." The rules resemble a denotational semantics written in inference rule form; this is no accident: natural semantics can be viewed as a denotational semantics variant where the internal calculations of meaning are made explicit. These internal calculations are seen in the natural semantics derivation of the example

$$
\frac{\dfrac{1 \Downarrow 1 \qquad 2 \Downarrow 2}{(1+2) \Downarrow 3} \qquad \dfrac{4 \Downarrow 4 \qquad 5 \Downarrow 5}{(4+5) \Downarrow 9}}{(1+2)+(4+5) \Downarrow 12}
$$

Unlike denotational semantics, natural semantics does not claim that the meaning of a program is necessarily mathematical. And unlike structural operational semantics, where a configuration $e \Downarrow e'$ says that e transits to an intermediate state e', in natural semantics $e \Downarrow v$ asserts that the final answer for e is v. For this reason, a natural semantics is sometimes called a *big-step semantics*. An interesting limitation of natural semantics is that semantics derivations can be drawn only for terminating programs.

The usual proof technique for proving properties of a natural semantics definition is induction on the height of the derivation trees that are generated from the semantics. Once again, see Nielson and Nielson [1992].

98.2.4 Axiomatic Semantics

An axiomatic semantics deduces *properties* of programs rather than meanings. Derivation of these properties is done with an inference rule set that portrays a *logic* for the programming language.

As an example, say that we wish to deduce even–odd properties of programs in arithmetic. The set of properties is simply $\{is\_even, is\_odd\}$. We define an axiomatic semantics to do this:

$$N : is\_even \text{ if } N \bmod 2 = 0 \qquad N : is\_odd \text{ if } N \bmod 2 = 1$$

$$\frac{E_1 : p_1 \quad E_2 : p_2}{E_1 + E_2 : p_3} \quad \text{where } p_3 = \begin{cases} is\_even & \text{if } p_1 = p_2 \\ is\_odd & \text{otherwise} \end{cases}$$

The derivation of the even–odd property of the example is:

$$\frac{\dfrac{1 : is\_odd \quad 2 : is\_even}{1 + 2 : is\_odd} \quad \dfrac{4 : is\_even \quad 5 : is\_odd}{4 + 5 : is\_odd}}{(1 + 2) + (4 + 5) : is\_even}$$

In the usual case, the properties proved of programs are expressed in predicate logic. (See the subsection on axiomatic semantics later in this chapter.) Axiomatic semantics has strong ties to the *abstract interpretation* of denotational and natural semantics definitions [Cousot and Cousot 1977; Nielson, Nielson, and Hanki 1998].

98.3 Semantics of Programming Languages

The semantics methods shine when they are applied to a realistic programming language: the primary features of the programming language are proclaimed loudly and subtle features receive proper mention. Ambiguities and anomalies stand out like the proverbial sore thumb. In this section we give the semantics of a block-structured imperative language. Emphasis is placed on the denotational semantics method, but excerpts from the other semantics formalisms are provided for comparison.

98.3.1 Language Syntax and Informal Semantics

The syntax of the programming language is presented in Figure 98.1. As stated in the figure, there are four levels of syntax constructions in the language, and the topmost level, Program, is the primary one.* The language is a while-loop language with local, nonrecursive procedure definitions. For simplicity, variables are predeclared and there are just three of them: **X, Y,** and **Z.** A program, C., operates as follows: an input number is read and assigned to **X**'s location. Then the body, C, of the program is evaluated, and, on completion, the storage vector holds the results. For example, this program computes n^2 for a positive

---

*The Identifier and Numeral sets are collections of words — terminal symbols — and not phrase-level *syntax constructions* in the sense of this chapter.

P ∈ Program
D ∈ Declaration
C ∈ Command
E ∈ Expression
I ∈ Identifier = upper-case alphabetic strings
N ∈ Numeral = {0, 1, 2, ...}

P ::= C.
D ::= **proc** I = C
C ::= I := E | C$_1$; C$_2$ | **begin** D **in** C **end** | **call** I | **while** E **do** C **od**
E ::= N | E$_1$ + E$_2$ | E$_1$ **not** = E$_2$ | I

FIGURE 98.1   Language syntax rules.

input *n*; the result is found in **Z**'s location:

```
begin proc INCR = Z:= Z+X; Y:= Y+1
   in Y:= 0; Z:= 0; while Y not=X do call INCR od end.
```

It is possible to write nonsense programs in the language; an example is **A:=0; call B**. Such programs have no meaning, and we will not attempt to give semantics to them. Nonsense programs are trapped by a type checker, and an elegant way of defining a type checker is by a set of typing rules for the programming language; see Chapter 97 for details.

98.3.2   Domains for Denotational Semantics

To give a denotational semantics to the sample language, we must state the sets of meanings, called *domains*, that we use. Our imperative, block-structured language has two primary domains: (1) the domain of storage vectors, called *Store*; and (2) the domain of symbol tables, called *Environment*. There are also secondary domains of Booleans and natural numbers. The primary domains and their operations are displayed in Figure 98.2.

The domains and operations deserve study. First, the *Store* domain states that a storage vector is a triple, for example, $\langle 1, 3, 5 \rangle$. (Recall that programs have exactly three variables.) The operation *lookup* extracts a value from the store, e.g., $lookup(2, \langle 1, 3, 5 \rangle) = 3$, and *update* updates the store, for example, $update(2, 6, \langle 1, 3, 5 \rangle) = \langle 1, 6, 5 \rangle$. Operation *init_store* creates a starting store. We examine *check* momentarily.

The environment domain states that a symbol table is a list of identifier-value pairs. For example, if variable **X** is the name of location 1, and **P** is the name of a procedure that is a no-op, then the environment that holds this information would appear (**X**, 1) :: (**P**, *id*) :: *nil*, where *id*(*s*) = *s*. (Procedures will be discussed momentarily.) Operation *find* locates the binding for an identifier in the environment, for example, *find* (**X**, (**X**, 1) :: (**P**, *id*) :: *nil*) = 1, and *bind* adds a new binding, for example, *bind*(**Y**, 2, (**X**, 1) :: (**P**, *id*) :: *nil*) = (**Y**, 2) :: (**X**, 1) :: (**P**, *id*) :: *nil*. Operation *init_env* creates an environment to start the program.

In the next section, we see that the job of a command, for example, an assignment, is to update the store. That is, the meaning of a command is a function that maps the current store to the updated one. (That is why a no-op command is the identity function, $id(s) = s$, where $s \in Store$.) But sometimes commands loop and no updated store appears. We use the symbol ⊥, read bottom, to stand for a looping store, and we use $Store_\perp$ to stand for the set of possible outputs of commands. Therefore, the meaning of a command is a function of the form $Store \rightarrow Store_\perp$.

It is impossible to recover from looping, so that if there is a command sequence C$_1$; C$_2$ and C$_1$ is looping, then C$_2$ cannot proceed. The *check* operation is used in the next subsection to watch for this situation.

$$Store = \{\langle n_1, n_2, n_3 \rangle \mid n_i \in Nat, i \in 1..3\}$$

$$lookup : \{1, 2, 3\} \times Store \rightarrow Nat$$
$$lookup(i, \langle n_1, n_2, n_3 \rangle) = n_i$$
$$update : \{1, 2, 3\} \times Nat \times Store \rightarrow Store$$
$$update(1, n, \langle n_1, n_2, n_3 \rangle) = \langle n, n_2, n_3 \rangle$$
$$update(2, n, \langle n_1, n_2, n_3 \rangle) = \langle n_1, n, n_3 \rangle$$
$$update(3, n, \langle n_1, n_2, n_3 \rangle) = \langle n_1, n_2, n \rangle$$
$$init\_store : Nat \rightarrow Store$$
$$init\_store(n) = \langle n, 0, 0 \rangle$$
$$check : (Store \rightarrow Store_\perp) \times Store_\perp \rightarrow Store_\perp \quad \text{where} \quad Store_\perp = Store \cup \{\perp\}$$
$$check(c, a) = \text{if } (a = \perp) \text{ then } \perp \text{ else } c(a)$$

$$Environment = (Identifier \times Denotable)^*$$
$$\text{where } A^* \text{ is a list of } A\text{-elements,} \quad a_1 :: a_2 :: \ldots :: a_n :: nil, n \geq 0$$
$$\text{and } Denotable = \{1, 2, 3\} \cup (Store \rightarrow Store_\perp)$$

$$find : Identifier \times Environment \rightarrow Denotable$$
$$find(i, nil) = 0$$
$$find(i, (i', d) :: rest) = \text{ if } (i = i') \text{ then } d \text{ else } find(i, rest)$$
$$bind : Identifier \times Denotable \times Environment \rightarrow Environment$$
$$bind(i, d, e) = (i, d) :: e$$
$$init\_env : Environment$$
$$init\_env = (\mathbf{X}, \ 1) :: (\mathbf{Y}, \ 2) :: (\mathbf{Z}, \ 3) :: nil$$

FIGURE 98.2   Semantic domains.

Finally, here are two commonly used notations. First, functions such as $id(s) = s$ are often reformatted to read $id = \lambda s. s$; in general, for $f(a) = e$, we write $f = \lambda a. e$, that is, we write the argument to the function to the right of the equals sign. This is called *lambda notation* and stems from the *lambda calculus*, an elegant formal system for functions. (See Chapter 96 of this Handbook.) The notation $f = \lambda a. e$ emphasizes that (1) the function $\lambda a. e$ is a value in its own right, and (2) the function's name is f.

Second, it is common to revise a function that takes multiple arguments, for example, $f(a, b) = e$, so that it takes the arguments one at a time: $f = \lambda a. \lambda b. e$. Therefore, if the arity of f was $A \times B \rightarrow C$, its new arity is $A \rightarrow (B \rightarrow C)$. This reformatting trick is called *Currying*, after Haskell Curry, one of the developers of lambda calculus.

98.3.3   Denotational Semantics of Programs

Figure 98.3 gives the denotational semantics of the programming language. Because the syntax of the language has four levels, the semantics is organized into four levels of meaning. For each level, we define a *valuation function* that produces the meanings of constructions at that level. For example, at the expression level, the constructions are mapped to their meanings by $\mathcal{E}$.

What is the meaning of the expression, say, **X+5**? This would be $\mathcal{E}[\![\mathbf{X+5}]\!]$, and the meaning depends on which location is named by **X** and what number is stored in that location. Therefore, the meaning depends on the current value of the environment and the current value of the store. Thus, if the current environment is $e_0 = (\mathbf{P}, \lambda s.s) :: (\mathbf{X}, 1) :: (\mathbf{Y}, 2) :: (\mathbf{Z}, 3) :: nil$ and the current store is $s_0 = \langle 2, 0, 0 \rangle$, then the meaning of **X+5** is 7,

$$\mathcal{E}[\![\mathbf{X+5}]\!]e_0\, s_0 = plus(\mathcal{E}[\![\mathbf{X}]\!]e_0\, s_0, \mathcal{E}[\![\mathbf{5}]\!]e_0\, s_0)$$
$$= plus(lookup(find(\mathbf{X}, \ e_0), s_0), 5)$$
$$= plus(lookup(1, s_0), 5) = plus(2, 5) = 7$$

$$\mathcal{P} : Program \rightarrow Nat \rightarrow Nat_{\perp}$$
$$\mathcal{P}[\![C.]\!] = \lambda n. \mathcal{C}[\![C]\!]init\_env \ (init\_store \ n)$$
$$\mathcal{D} : Declaration \rightarrow Environment \rightarrow Environment$$
$$\mathcal{D}[\![\textbf{proc } I = C]\!] = \lambda e. \ bind(I, \mathcal{C}[\![C]\!]e, e)$$
$$\mathcal{C} : Command \rightarrow Environment \rightarrow Store \rightarrow Store_{\perp}$$
$$\mathcal{C}[\![I := E]\!] = \lambda e. \lambda s. \ update \ (find \ (I, e), \mathcal{E}[\![E]\!]e \ s, s)$$
$$\mathcal{C}[\![C_1; C_2]\!] = \lambda e. \lambda s. \ check \ (\mathcal{C}[\![C_2]\!] \ e, \ \mathcal{C}[\![C_1]\!]e \ s)$$
$$\mathcal{C}[\![\textbf{begin } D \textbf{ in } C \textbf{ end}]\!] = \lambda e. \lambda s. \mathcal{C}[\![C]\!](\mathcal{D}[\![D]\!]e)s$$
$$\mathcal{C}[\![\textbf{call } I]\!] = \lambda e. \ find(I, e)$$
$$\mathcal{C}[\![\textbf{while } E \textbf{ do } C \textbf{ od}]\!] = \lambda e. \ \bigcup_{i \geq 0} w_i$$
$$w_0 = \lambda s. \perp$$
$$\text{where}$$
$$w_{i+1} = \lambda s. \ \textbf{if} \ \mathcal{E}[\![E]\!]e \ s \ \textbf{then} \ check(w_i, \mathcal{C}[\![C]\!]e \ s) \ \textbf{else} \ s$$
$$\mathcal{E} : Expression \rightarrow Environment \rightarrow Store \rightarrow (Nat \ \cup Bool)$$
$$\mathcal{E}[\![N]\!] = \lambda e. \lambda s. \ N$$
$$\mathcal{E}[\![E_1 + E_2]\!] = \lambda e. \lambda s. \ plus(\mathcal{E}[\![E_1]\!]e \ s, \mathcal{E}[\![E_2]\!]e \ s)$$
$$\mathcal{E}[\![E_1 \ \textbf{not=} \ E_2]\!] = \lambda e. \lambda s. \ notequals(\mathcal{E}[\![E_1]\!]e \ s, \mathcal{E}[\![E_2]\!]e \ s)$$
$$\mathcal{E}[\![I]\!] = \lambda e. \lambda s. \ lookup(find(I, e), s)$$

FIGURE 98.3 Denotational semantics.

As this simple derivation shows, data structures such as the symbol table and storage vector are modeled by the environment and store arguments. This pattern is used throughout the semantics definition.

As noted in the previous section, a command updates the store. Precisely stated, the valuation function for commands is $\mathcal{C} : Command \rightarrow Environment \rightarrow Store \rightarrow Store_{\perp}$. For example, for e_0 and s_0 given previously, we see that:

$$\mathcal{C}[\![\textbf{Z:=X+5}]\!]e_0 \ s_0 = update(\ find(\textbf{Z}, \ e_0), \mathcal{E}[\![\textbf{X+5}]\!]e_0 \ s_0, s_0) = update(3, 7, s_0) = \langle 2, 0, 7 \rangle$$

But a crucial point about the meaning of the assignment is that it is a function upon stores. That is, if we are uncertain of the current value of store, but we know that the environment for the assignment is e_0, then we can conclude that:

$$\mathcal{C}[\![\textbf{Z:=X+5}]\!]e_0 = \lambda s. \ update(3, plus(lookup(1, s), 5), s)$$

That is, the assignment with environment e_0 is a function that updates a store at location 3.

Next consider this example of a command sequence:

$$\mathcal{C}[\![\textbf{Z:=X+5; call P}]\!]e_0 \ s_0 = check(\mathcal{C}[\![\textbf{call P}]\!]e_0, \mathcal{C}[\![\textbf{Z:=X+5}]\!]e_0 \ s_0)$$
$$= check(find(\textbf{P}, e_0), \langle 2, 0, 7 \rangle) = check(\lambda s. s, \langle 2, 0, 7 \rangle)$$
$$= (\lambda s. s)\langle 2, 0, 7 \rangle = \langle 2, 0, 7 \rangle$$

As noted in the earlier section, the *check* operation verifies that the first command in the sequence produces a proper output store; if so, the store is handed to the second command in the sequence. Also, we see that the meaning of **call P** is the store updating function bound to **P** in the environment.

Procedures are placed in the environment by declarations, as we see in this example: let e_1 denote $(\mathbf{X},\ 1) :: (\mathbf{Y},\ 2) :: (\mathbf{Z},\ 3) :: nil$,

$$\mathcal{C}[\![\texttt{begin proc P = Y:=Y in Z:=X+5; call P end}]\!]e_1\,s_0$$
$$= \mathcal{C}[\![\texttt{Z:=X+5; call P}]\!](\mathcal{D}[\![\texttt{proc P = Y:=Y}]\!]e_1)s_0$$
$$= \mathcal{C}[\![\texttt{Z:=X+5; call P}]\!](bind(\mathbf{P},\ \mathcal{C}[\![\texttt{Y:=Y}]\!]e_1, e_1))s_0$$
$$= \mathcal{C}[\![\texttt{Z:=X+5; call P}]\!](bind(\mathbf{P},\ \lambda s.\,update(2, lookup(2,s), s), e_1))s_0$$
$$= \mathcal{C}[\![\texttt{Z:=X+5; call P}]\!]((\mathbf{P},\ id) :: e_1)s_0$$
$$\qquad \text{where } id = \lambda s.\,update(2, lookup(2,s), s) = \lambda s.\,s \qquad\qquad (*)$$
$$= \mathcal{C}[\![\texttt{Z:=X+5; call P}]\!]e_0\,s_0 = \langle 2, 0, 7\rangle$$

The equality marked by $(*)$ is significant; we can assert that the function $\lambda s.\,update(2, lookup(2,s), s)$ is identical to $\lambda s.\,s$ by appealing to the *extensionality* law of mathematics: if two functions map identical arguments to identical answers, then the functions are themselves identical. The extensionality law can be used here because in denotational semantics the meanings of program phrases are mathematical — functions. In contrast, the extensionality law cannot be used in operational semantics calculations.

Finally, we can combine our series of little examples into the semantics of a complete program,

$$\mathcal{P}[\![\texttt{begin proc P = Y:=Y in Z:=X+5; call P end.}]\!]2$$
$$= \mathcal{C}[\![\texttt{begin proc P = Y:=Y in Z:=X+5; call P end}]\!]init\_env\,(init\_store\,2)$$
$$= \mathcal{C}[\![\texttt{begin proc P = Y:=Y in Z:=X+5; call P end}]\!]e_1\,s_0$$
$$= \langle 2, 0, 7\rangle$$

98.3.4 Semantics of the While-Loop

The most difficult clause in the semantics definition is the one for the while-loop. Here is some intuition: to produce an output store, the loop **while** E **do** C **od** must terminate after some finite number of iterations. To measure this behavior, let **while**$_i$ E **do** C **od** be a loop that can iterate at most i times; if the loop runs more than i iterations, it becomes exhausted and its output is $\perp$. For example, for input store $\langle 4, 0, 0\rangle$, the loop **while**$_k$ **Y not = X do Y: = Y+1 od** can produce the output store $\langle 4, 4, 0\rangle$ only when k is greater than 4. (Otherwise, the output is $\perp$.)

It is easy to conclude that the family, **while**$_i$ E **do** C **od,** for $i \geq 0$, can be written equivalently as:

$$\textbf{while}_0\ \text{E}\,\textbf{do}\,\text{C}\,\textbf{od} = \textit{"exhausted"} \quad \text{(that is, its meaning is } \lambda s.\ \perp)$$
$$\textbf{while}_{i+1}\ \text{E}\,\textbf{do}\,\text{C}\,\textbf{od} = \textbf{if}\,\text{E}\,\textbf{then}\,\text{C};\ \textbf{while}_i\ \text{E}\,\textbf{do}\,\text{C}\,\textbf{od}\,\textbf{else}\,\textbf{skip}\,\textbf{fi}$$

When we refer back to Figure 98.3, we draw these conclusions:

$$\mathcal{C}[\![\textbf{while}_0\ \text{E}\,\textbf{do}\,\text{C}\,\textbf{od}]\!]\,e = w_0$$
$$\mathcal{C}[\![\textbf{while}_{i+1}\ \text{E}\,\textbf{do}\,\text{C}\,\textbf{od}]\!]e = w_{i+1}$$

Because the behavior of a while-loop must be the union of the behaviors of the **while**$_i$**-loops,** we conclude that $\mathcal{C}[\![\textbf{while}\ \text{E}\,\textbf{do}\,\text{C}\,\textbf{od}]\!]e = \bigcup_{i \geq 0} w_i$. The semantic union operation is well defined because each w_i is a function from the set $Store \to Store_\perp$, and a function can be represented as a set of argument-answer pairs. (This is called the *graph of the function*.) Thus, $\bigcup_{i \geq 0} w_i$ is the union of the graphs of the w_i functions.[*]

---

[*]Several important technical details have been glossed over. First, pairs of the form $(s, \perp)$ are ignored when the union of the graphs is performed. Second, for all $i \geq 0$, the graph of w_i is a subset of the graph of w_{i+1}; this ensures the union of the graphs is a function.

The definition of $C[\![\mathbf{while}\ E\ \mathbf{do}\ C\ \mathbf{od}]\!]$ is succinct, but it is awkward to use in practice. An intuitive way of defining the semantics is:

$$C[\![\mathbf{while}\ E\ \mathbf{do}\ C\ \mathbf{od}]\!]e = w$$

where $w = \lambda s.\ \text{if}\ \mathcal{E}[\![E]\!]e\ s\ \text{then}\ check(w, C[\![C]\!]e\ s)\ \text{else}\ s$

The problem here is that the definition of w is circular, and circular definitions can be malformed. Fortunately, this definition of w can be claimed to denote the function $\bigcup_{i\geq 0} w_i$ because the following equality holds:

$$\bigcup_{i\geq 0} w_i = \lambda s.\ \text{if}\ \mathcal{E}[\![E]\!]e\ s\ \text{then}\ check\left(\bigcup_{i\geq 0} w_i, C[\![C]\!]e\ s\right)\ \text{else}\ s$$

Thus, $\bigcup_{i\geq 0} w_i$ is a solution — a *fixed point* — of the circular definition, and in fact it is the smallest function that makes the equality hold. Therefore, it is the *least fixed point*.

Typically, the denotational semantics of the while-loop is presented by the circular definition, and the claim is then made that the circular definition stands for the least fixed point. This is called **fixed-point semantics**. We have omitted many technical details regarding fixed-point semantics; these are available in several texts [Gunter 1992, Mitchell 1996, Schmidt 1986, Stoy 1977, Winskel 1993].

98.3.5   Action Semantics

One disadvantage of denotational semantics is its dependence on functions to describe all forms of computation. As a result, the denotational semantics of a large language is often too dense to read and too low level to modify. **Action semantics** is an easy-to-read denotational semantics variant that rectifies these problems by using a family of standard operators to describe standard forms of computation in standard languages [Mosses 1992].

In action semantics, the standard domains are called *facets* and are predefined for expressions (the *functional facet*), for declarations (the *declarative facet*), and for commands (the *imperative facet*). Each facet includes a set of standard operators for consuming values of the facet and producing new ones. The operators are connected together by combinators (*pipes*), and the resulting action semantics definition resembles a data-flow program. For example, the semantics of assignment reads as follows:

$$execute[\![I := E]\!] = (find\ I\ \text{and}\ evaluate[\![E]\!])\ \text{then}\ update$$

One can naively read the semantics as an English sentence, but each word is an operator or a combinator: execute is C, evaluate is $\mathcal{E}$, find is a declarative facet operator, update is an imperative facet operator, and and and then are combinators. The equation accepts as its inputs a declarative facet argument (that is, an environment) and an imperative facet argument (that is, a store) and pipes them to the operators. Thus, find consumes its declarative argument and produces a functional-facet answer, and, independently, evaluate$[\![E]\!]$ consumes declarative and imperative arguments and produces a functional answer. The and combinator pairs these, and the then combinator transmits the pair to the update operator, which uses the pair and the imperative-facet argument to generate a new imperative result.

The important aspects of an action semantics definition are (1) standard arguments, such as environments and stores, are implicit; (2) standard operators are used for standard computation steps (e.g., find and update); and (3) combinators connect operators together seamlessly and pass values implicitly. Lack of space prevents a closer examination of action semantics, but see Watt [1991] and Mosses [1992] for details.

98.3.6   The Natural Semantics of the Language

We can compare the denotational semantics of the imperative language with a natural semantics formulation. The semantics of several constructions appear in Figure 98.4.

$$e \vdash \mathbf{proc}\ \mathrm{I} = \mathrm{C} \Downarrow \textit{bind}(\mathrm{I}, (e, \mathrm{C}), e) \qquad \frac{e \vdash \mathrm{D} \Downarrow e' \quad e', s \vdash \mathrm{C} \Downarrow s'}{e, s \vdash \mathbf{begin}\ \mathrm{D}\ \mathbf{in}\ \mathrm{C}\ \mathbf{end} \Downarrow s'}$$

$$\frac{l\ =\ \textit{find}(\mathrm{I}, e) \quad e, s \vdash \mathrm{E} \Downarrow n}{e, s \vdash \mathrm{I} := \mathrm{E} \Downarrow \textit{update}(l,\ n,\ s)} \qquad \frac{e, s \vdash \mathrm{C}_1 \Downarrow s' \quad e, s' \vdash \mathrm{C}_2 \Downarrow s''}{e, s \vdash \mathrm{C}_1\ ;\ \mathrm{C}_2 \Downarrow s''}$$

$$\frac{(e', \mathrm{C}')\ =\ \textit{find}(\mathrm{I}, e) \quad e', s \vdash \mathrm{C}' \Downarrow s'}{e, s \vdash \mathbf{call}\ \mathrm{I} \Downarrow s'} \qquad \frac{e, s \vdash \mathrm{E} \Downarrow \textit{false}}{e, s \vdash \mathbf{while}\ \mathrm{E}\ \mathbf{do}\ \mathrm{C}\ \mathbf{od} \Downarrow s}$$

$$\frac{e, s \vdash \mathrm{E} \Downarrow \textit{true} \quad e, s \vdash \mathrm{C} \Downarrow s' \quad e, s' \vdash \mathbf{while}\ \mathrm{E}\ \mathbf{do}\ \mathrm{C}\ \mathbf{od} \Downarrow s''}{e, s \vdash \mathbf{while}\ \mathrm{E}\ \mathbf{do}\ \mathrm{C}\ \mathbf{od} \Downarrow s''}$$

FIGURE 98.4   Natural semantics.

$$\begin{aligned}
\mathbf{let}\ e_0 &= (\mathbf{X},\ 1) :: (\mathbf{Y},\ 2) :: (\mathbf{Z},\ 3) :: \textit{nil}\\
s_0 &= \langle 2, 0, 0\rangle, \quad s_1 = \langle 2, 1, 0\rangle\\
\mathrm{E}_0 &= \mathbf{Y}\ \mathbf{not=1}, \quad \mathrm{C}_0 = \mathbf{Y}\mathbf{:=}\mathbf{Y}\mathbf{+1}\\
\mathrm{C}_{00} &= \mathbf{while}\ \mathrm{E}_0\ \mathbf{do}\ \mathrm{C}_0\ \mathbf{od}
\end{aligned}$$

$$e_0, s_0 \vdash \mathrm{E}_0 \Downarrow \textit{true} \qquad \frac{2 = \textit{find}(\mathbf{Y},\ e_0) \quad e_0, s_0 \vdash \mathbf{Y}\mathbf{+1} \Downarrow 1}{e_0, s_0 \vdash \mathrm{C}_0 \Downarrow s_1} \qquad \frac{e_0, s_1 \vdash \mathrm{E}_0 \Downarrow \textit{false}}{e_0, s_1 \vdash \mathrm{C}_{00} \Downarrow s_1}$$
$$e_0, s_0 \vdash \mathrm{C}_{00} \Downarrow s_1$$

FIGURE 98.5   Natural semantics derivation.

A command configuration has the form $e, s \vdash \mathrm{C} \Downarrow s'$, where e and s are the inputs to command C and s' is the output. To understand the inference rules, read them bottom up. For example, the rule for $\mathrm{I} := \mathrm{E}$ says, given the inputs e and s, one must first find the location l, bound to I, and then calculate the output n, for E. Finally, l and n are used to update s, producing the output.

The rules are denotational-like, but differences arise in several key constructions. First, the semantics of a procedure declaration binds I not to a function but to an environment–command pair called a *closure*. When procedure I is called, the closure is disassembled, and its text and environment are executed. Because a natural semantics does not use function arguments, it is called a *first-order semantics*. (Denotational semantics is sometimes called a *higher order semantics*.)

Second, the while-loop rules are circular. The second rule states that, in order to derive a while-loop computation that terminates in s'', one must derive (1) the test, E is true, (2) the body C, outputs s', and (3) using e and s', one can derive a terminating while-loop computation that outputs s''. The rule makes one feel that the while-loop is running backward from its termination to its starting point, but a complete derivation, such as the one shown in Figure 98.5, shows that the iterations of the loop can be read from the root to the leaves of the derivation tree.

One important aspect of the natural semantics definition is that derivations can be drawn only for terminating computations. A nonterminating computation is equated with no computation at all.

98.3.7   The Operational Semantics of the Language

A fragment of the structural operational semantics of the imperative language is presented in Figure 98.6.

For expressions, a computation step takes the form $e \vdash \langle \mathrm{E}, s\rangle \Rightarrow \mathrm{E}'$, where e is the environment, E is the expression that is evaluated, s is the current store, and E' is E rewritten. In the case of a command C, a step appears $e \vdash \langle \mathrm{C}, s\rangle \Rightarrow \langle \mathrm{C}', s'\rangle$, because computation on C might also update the store. If the computation step on C uses up the command, the step appears $e \vdash \langle \mathrm{C}, s\rangle \Rightarrow s'$.

$e \vdash \langle n_1 + n_2, s \rangle \Rightarrow n_3$ where n_3 is the sum of n_1 and n_2

$$\frac{e \vdash \langle E, s \rangle \Rightarrow E'}{e \vdash \langle I := E, s \rangle \Rightarrow \langle I := E', s \rangle}$$

$e \vdash \langle I := n, s \rangle \Rightarrow update(l, n, s)$ where $find(I, e) = l$

$$\frac{e \vdash \langle C_1, s \rangle \Rightarrow \langle C_1', s' \rangle}{e \vdash \langle C_1; C_2, s \rangle \Rightarrow \langle C_1'; C_2, s' \rangle} \qquad \frac{e \vdash \langle C_1, s \rangle \Rightarrow s'}{e \vdash \langle C_1; C_2, s \rangle \Rightarrow \langle C_2, s' \rangle}$$

$e \vdash \langle \textbf{while } E \textbf{ do } C \textbf{ od, } s \rangle \Rightarrow \langle \textbf{if } E \textbf{ then } C; \textbf{while } E \textbf{ do } C \textbf{ od else skip fi, } s \rangle$

$e \vdash \langle \textbf{call } I, s \rangle \Rightarrow \langle \textbf{use } e' \textbf{in } C', s \rangle$ where $find(I, e) = (e', C')$

$$\frac{e \vdash \langle C, s \rangle \Rightarrow \langle C', s' \rangle}{e \vdash \langle \textbf{use } e' \textbf{ in } C, s \rangle \Rightarrow \langle \textbf{use } e' \textbf{ in } C', s' \rangle} \qquad \frac{e' \vdash \langle C, s \rangle \Rightarrow s'}{e \vdash \langle \textbf{use } e' \textbf{ in } C, s \rangle \Rightarrow s'}$$

$e \vdash \textbf{proc } I = C \Rightarrow bind(I, (e, C), e)$

$$\frac{e \vdash D \Rightarrow e'}{e \vdash \langle \textbf{begin } D \textbf{ in } C \textbf{ end, } s \rangle \Rightarrow \langle \textbf{use } e' \textbf{ in } C, s \rangle}$$

FIGURE 98.6 Structural operational semantics.

The rules in the figure are more tedious than those for a natural semantics because the individual computation steps must be defined, and the order in which the steps are undertaken must also be defined. This complicates the rules for command composition, for example. On the other hand, the rewriting rule for the while-loop merely decodes the loop as a conditional command.

The rules for procedure call are awkward; as with the natural semantics, a procedure I is represented as a closure of the form (e', C'). Because C' must execute with environment e', which is different from the environment that exists where procedure I is called, the rewriting step for **call** I must retain *two* environments; a new construct, **use** e' **in** C', remembers that C' must use e' (and not e). A similar trick is used in **begin** D **in** C **end.**

Unlike a natural semantics definition, a computation can be written for a nonterminating program; the computation is a state sequence of countably infinite length.

98.3.8 An Axiomatic Semantics of the Language

An axiomatic semantics uses properties of stores, rather than stores themselves. For example, we might write the predicate $\textbf{X} = 3 \wedge \textbf{Y} > 0$ to assert that the current value of the store contains 3 in $\textbf{X}$'s location and a positive number in $\textbf{Y}$'s location. We write a configuration $\{P\}C\{Q\}$ to assert that if predicate P holds true prior to evaluation of command C, then predicate Q holds upon termination of C (if C does indeed terminate). For example, we can write $\{\textbf{X} = 3 \wedge \textbf{Y} > 0\}\textbf{Y} := \textbf{X} + \textbf{Y}\{\textbf{X} = 3 \wedge \textbf{Y} > 3\}$, and indeed this holds true.

There are three ways of stating the semantics of a command in an axiomatic semantics:

1. *Relational semantics.* The meaning of C is the set of P, Q pairs for which $\{P\}C\{Q\}$ holds.
2. *Postcondition semantics.* The meaning of C is a function from an input predicate to an output predicate. We write $slp(P, C) = Q$; this means that $\{P\}C\{Q\}$ holds, and for all Q' such that $\{P\}C\{Q'\}$ holds, it is the case that Q implies Q'. This is also called *strongest liberal postcondition semantics.* When termination is demanded also of C, the name becomes **strongest postcondition semantics**.
3. *Precondition semantics.* The meaning of C is a function from an output predicate to an input predicate. We write $wlp(C, Q) = P$; this means that $\{P\}C\{Q\}$ holds, and for all P' such that $\{P'\}C\{Q\}$ holds, it is the case that P' implies P. This is also called *weakest liberal precondition semantics.* When termination is demanded also of C, the name becomes **weakest precondition semantics**.

It is traditional to study relational semantics first, so we focus on it here.

$$\{[E/I]P\}I \; := \; E\{P\}$$

$$\frac{P \supset P' \quad \{P'\}C\{Q'\} \quad Q' \supset Q}{\{P\}C\{Q\}} \qquad \frac{\{P\}C_1\{Q\} \quad \{Q\}C_2\{R\}}{\{P\}C_1;C_2\{R\}}$$

$$\frac{\{P \wedge E\}C_1\{Q\} \quad \{P \wedge \neg E\}C_2\{Q\}}{\{P\}\textbf{if } E \textbf{ then } C_1 \textbf{ else } C_2 \textbf{ fi } \{Q\}} \qquad \frac{\{P \wedge E\}C\{P\}}{\{P\}\textbf{ while } E \textbf{ do } C \textbf{ od } \{P \wedge \neg E\}}$$

FIGURE 98.7   Axiomatic semantics.

If the intended behavior of a program C is written as a pair of predicates P, Q, a relational semantics can be used to verify that $\{P\}C\{Q\}$ holds. For example, we might wish to show that an integer division subroutine **DIV** that takes inputs **NUM** and **DEN** and produces outputs **QUO** and **REM** has this behavior:

$$\{\neg(\textbf{DEN} = 0)\}\textbf{DIV}\{\textbf{QUO} \times \textbf{DEN} + \textbf{REM} = \textbf{NUM}\}$$

A proof of this claim is a derivation built with the rules in Figure 98.7.

Figure 98.7 displays the rules for the primary command constructions. The rule for I := E states that a property P about I will hold upon completion of the assignment if $[E/I]P$ (that is, P restated in terms of E) holds beforehand. $[E/I]P$ stands for the substitution of phrase E for all free occurrences of I in P. For example, $\{\textbf{X} = 3 \wedge \textbf{X+Y} > 3\}\textbf{Y:=X+Y}\{\textbf{X} = 3 \wedge \textbf{Y} > 3\}$ holds because $[\textbf{X+Y/Y}](\textbf{X} = 3 \wedge \textbf{Y} > 3)$ is $\textbf{X} = 3 \wedge \textbf{X+Y} > 3$.

The second rule lets us weaken a result. For example, because $(\textbf{X} = 3 \wedge \textbf{Y} > 0) \supset (\textbf{X} = 3 \wedge \textbf{X+Y} > 3)$ holds, we deduce that $\{\textbf{X} = 3 \wedge \textbf{Y} > 0\}\textbf{Y:=X+Y}\{\textbf{X} = 3 \wedge \textbf{Y} > 3\}$ holds.

The properties of command composition are defined in the expected way, by the third rule. The fourth rule, for the if-command, makes a property Q hold upon termination if Q holds regardless of which arm of the conditional is evaluated. Note that each arm of the conditional uses information about the result of the conditional's test.

The most fascinating rule is the last one, for the while-loop. If we can show that a property P is preserved by the body of the loop, then we can assert that no matter how long the loop iterates, P must hold upon termination. P is called the **loop invariant**. The rule is an encoding of a mathematical induction proof: to show that P holds upon completion of the loop, we must prove (1) the basis case: P holds upon loop entry (that is, after zero iterations), and (2) the induction case: if P holds after i iterations, then P holds after $i + 1$ iterations as well. Therefore, if the loop terminates after some number k of iterations, the induction proof ensures that P holds.

Here is an example that shows the rules in action. We wish to verify that

$$\{\textbf{X} = \textbf{Y} \wedge \textbf{Z} = 0\}\textbf{while Y not=0 do Y:=Y-1; Z:=Z+1 od } \{\textbf{X} = \textbf{Z}\}$$

holds true. The key to the proof is determining a loop invariant; here, a useful invariant is $\textbf{X} = \textbf{Y+Z}$, because $\textbf{X} = \textbf{Y+Z} \wedge \neg(\textbf{Y not=0})$ implies $\textbf{X} = \textbf{Z}$. This leaves us $\{\textbf{X} = \textbf{Y+Z} \wedge \textbf{Y not=0}\}\textbf{Y:=Y-1; Z:=Z+1}$ $\{\textbf{X} = \textbf{Y+Z}\}$ to prove. We work backward: the rule for assignment gives us: $\{\textbf{X} = \textbf{Y+(Z+1)}\}\textbf{Z:=Z+1}\{\textbf{X} = \textbf{Y + Z}\}$, and we can also deduce that $\{\textbf{X} = (\textbf{Y} - 1) + (\textbf{Z} + 1)\}\textbf{Y:=Y-1}\{\textbf{X} = \textbf{Y} + (\textbf{Z} + 1)\}$ holds. Because $\textbf{X} = \textbf{Y+Z} \wedge \textbf{Y not=0}$ implies $\textbf{X} = (\textbf{Y} - 1) + (\textbf{Z} + 1)$, we can assemble a complete derivation; it is given in Figure 98.8.

98.4   Applications of Semantics

Language designers have used semantics definitions to formalize their creations. An early example was the formalization of a large subset of Ada in denotational semantics [Donzeau-Gouge 1980]. The semantics definition was then prototyped using Mosses' SIS compiler generating system [Mosses 1976]. Scheme is another widely used language that has been given a standardized denotational semantics [Rees and

$$\text{let } P_0 \text{ be } X = Y + Z$$
$$P_1 \text{ be } X = Y + (Z + 1), \quad P_2 \text{ be } X = (Y - 1) + (Z + 1)$$
$$E_0 = \textbf{Y not=0}, \quad C_0 = \textbf{Y:=Y-1;Z:=Z+1}$$

$$(P_0 \wedge E_0) \supset P_2 \quad \dfrac{\dfrac{\{P_2\}\textbf{Y:=Y-1}\{P_1\} \quad \{P_1\}\textbf{Z:=Z+1}\{P_0\}}{\{P_2\}C_0\{P_0\}}}{\dfrac{\{P_0 \wedge E_0\}C_0\{P_0\}}{\{P_0\} \textbf{ while } E_0 \textbf{ do } C_0 \textbf{ od } \{P_0 \wedge \neg E_0\}}} \quad P_0 \supset P_0$$

FIGURE 98.8   Axiomatic semantics derivation.

Clinger 1986]. Another notable example is the formalization of the complete Standard ML language in structural operational semantics [Milner et al. 1990]. The plethora of object-oriented languages are being untangled with the assistance of formal semantics definitions [Abadi and Cardelli 1996, Bruce 2002, Pierce 2002].

Another significant application of semantics definitions has been to rapid prototyping — the synthesis of an implementation for a newly defined language. Notable prototyping systems are SIS [Mosses 1976], PSI [Nielson and Nielson 1988], MESS [Lee 1989], Actress [Brown et al. 1992], and Typol [Despeyroux 1984]. The first two process denotational semantics, the second two process action semantics, and the last one handles natural semantics. SIS and Typol are interpreter generators; that is, they interpret a source program with the semantics definition; and PSI, MESS, and Actress are compiler generators, that is, compilers for the source language are synthesized.

A major success of formal semantics is the analysis and synthesis of data-flow analysis and type-inference algorithms from semantics definitions. This subject area, called *abstract interpretation* [Abramsky and Hankin 1987, Cousot and Cousot 1977, Muchnick and Jones 1981, Nielson, Nielson, and Hankin 1998], supplies precise techniques for analyzing semantics definitions, extracting properties from the definitions, applying the properties to data-flow and type inference, and proving the soundness of the code-improvement transformations that result. Abstract interpretation provides the theory that allows a compiler writer to prove the correctness of compilers as well as validate correctness properties of specific programs.

Finally, axiomatic semantics is a long-standing fundamental technique for validating the correctness of programs. Recent emphasis on large-scale, distributed, and safety-critical systems has again spotlighted this technique, and as Chapter 97 of this Handbook indicates, there is now a marriage between data-type checking techniques and axiomatic-semantic deduction.

98.5   Research Issues in Semantics

The techniques in this chapter have proved highly successful for defining, analyzing, and implementing procedural programming languages. But new language paradigms present new challenges to the semantics methods.

In the functional programming paradigm, a higher-order functional language can use functions as arguments to other functions. This makes the language's domains more complex than those in Figure 98.2. Denotational semantics can be used to understand these complexities; an applied mathematics called *domain theory* [Gunter 1992, Mitchell 1996] is used to formalize the domains with algebraic equations. For example, the *Value* domain for a higher-order, Scheme-like language takes the form:

$$Value = Nat + (Value \rightarrow Value)$$

That is, legal values are numbers or functions on values. Of course, Cantor's theorem makes it impossible to find a set that satisfies this equation, but domain theory uses the concept of continuity from topology to restrict the size of *Value* so that a solution can be found as that in the subsection on the semantics of the while-loop; namely,

$$Value = \lim_{i \geq 0} V_i, \quad \text{where} \quad \begin{aligned} V_0 &= \{\bot\} \\ V_{i+1} &= Nat \uplus (V_i \to {}^{ctn} V_i) \end{aligned}$$

where $V_i \to {}^{ctn} V_i$ denotes the topologically continuous functions on V_i.

Challenging issues also arise in the object-oriented programming paradigm: objects are constructed from templates called *classes*, and classes can be declared incrementally and even rewritten (*overridden*) by means of *subclassing*. This arrangement makes understanding procedure (method) invocation fiendishly difficult, and denotational and natural semantics have been applied to stating precisely what it means to construct an object from incrementally defined classes and to invoke its methods [Bruce 2002, Gunter and Mitchell 1994, Mitchell 1996]. Also, the data-type checking techniques proposed for existing object-oriented languages contain flaws, and semantic methods have been applied to developing correct data-typing systems based on *parametric* and *inclusion* polymorphism [Abadi and Cardelli 1996, Pierce 2002].

Another challenging topic is parallelism and communication as it arises in the distributed and reactive programming paradigm, where multiple processes may run in parallel, react to each other's outputs, and synchronize. Structural operational semantics is used to formalize systems of processes and study their interaction, and new semantical systems have been developed specifically for this subject. (See Chapter 96 of this Handbook.)

Finally, a long-standing research topic is the relationship between the different forms of semantic definitions. If one has, say, both a denotational semantics and an axiomatic semantics for a programming language, in what sense do the semantics agree? Agreement is crucial, because a programmer might use axiomatic semantics to reason about the properties of programs, whereas a compiler writer might use denotational semantics to implement the language. In mathematical logic, one uses the concepts of *soundness* and *completeness* to relate a logic's proof system to its interpretation, and in semantics there are similar notions of *soundness* and *adequacy* to relate one semantics to another [Gunter 1992, Ong 1995].

A standard example is proving the soundness of a structural operational semantics to a denotational semantics: for program P and input v, $(P, v) \Rightarrow v'$ in the operational semantics implies $\mathcal{P}[\![P]\!](v) = v'$ in the denotational semantics. Adequacy is a form of inverse: if $\mathcal{P}[\![P]\!](v) = v'$, and v' is a primitive value (e.g., an integer or Boolean), then $(P, v) \Rightarrow v'$. There is a stronger form of adequacy, called *full abstraction* [Stoughton 1988], which has proved difficult to achieve for realistic languages, but recent research on viewing computation as a form of "game playing," where a program interacts with its external environment to decide which computation step to make next, has produced some solutions to the full abstraction problem and has suggested yet another format for expressing programming language semantics [Abramsky et al. 1994, Abramsky and McCusker 1998].

Acknowledgments

Brian Howard and Anindya Banerjee provided helpful criticism.

Defining Terms

Action semantics: A variation of denotational semantics where low-level details are hidden by use of modularized sets of operators and combinators.

Axiomatic semantics: The meaning of a program as a property or specification in logic.

Denotational semantics: The meaning of a program as a compositional definition of a mathematical function from the program's input data to its output data.

Fixed-point semantics: A denotational semantics where the meaning of a repetitive structure, such as a loop or recursive procedure, is expressed as the smallest mathematical function that satisfies a recursively defined equation.

Loop invariant: In axiomatic semantics, a logical property of a while-loop that holds true no matter how many iterations the loop executes.

Natural semantics: A hybrid of operational and denotational semantics that shows computation steps performed in a compositional manner. Also known as a big-step semantics.

Operational semantics: The meaning of a program as calculation of a trace of its computation steps on input data.

Strongest postcondition semantics: A variant of axiomatic semantics where a program and an input property are mapped to the strongest proposition that holds true of the program's output.

Structural operational semantics: A variant of operational semantics where computation steps are performed only within prespecified contexts. Also known as a small-step semantics.

Weakest precondition semantics: A variant of axiomatic semantics where a program and an output property are mapped to the weakest proposition that is necessary of the program's input to make the output property hold true.

References

Abadi, M. and Cardelli, L., 1996. *A Theory of Objects*. Springer-Verlag, New York.

Abramsky, S. and McCusker, G., 1987. Game Semantics. In H. Schwichtenberg and U. Berger, Editors, *Logic and Computation: Proceedings of the 1997 Marktoberdorf Summer School*. Springer-Verlag, New York.

Abramsky, S. and Hankin, C., Eds. 1987. *Abstract Interpretation of Declarative Languages*. Ellis Horwood, Chichester, England.

Abramsky, S., Jagadeesan, R., and Malacaria, P. 1994. Full abstraction for PCF. In *TACS94, Proc. Theor. Aspects Comput. Software*. Lecture Notes in Computer Science 789, pp. 1–15. Springer.

Apt, K. 1981. Ten years of Hoare's logic: a survey — Part 1. *ACM Trans. Programming Lang. Syst.*, 3:431–484.

Brown, D. F., Moura, H., and Watt, D. A. 1992. ACTRESS: an action semantics directed compiler generator. In *CC'92, Proc. 4th Int. Conf. Compiler Construction*, Paderborn. Lecture Notes in Computer Science 641, pp. 95–109. Springer-Verlag, New York.

Bruce, K., 2002. *Foundations of Object-Oriented Programming Languages: Types and Semantics*. MIT Press, Cambridge, MA.

Cousot, P. and Cousot, R. 1977. Abstract interpretation: a unified lattice model for static analysis of programs, pp. 238–252. In *Proc. 4th ACM Symp. Principles Programming Lang.* ACM Press.

Despeyroux, T. 1984. Executable specification of static semantics. In *Semantics of Data Types*, G. Kahn, D. B. MacQueen, and G. Plotkin, Eds. Lecture Notes in Computer Science 173, pp. 215-234. Springer-Verlag, New York.

Dijkstra, E. W. 1976. *A Discipline of Programming*. Prentice Hall, Englewood Cliffs, NJ.

Donzeau-Gouge, V. 1980. On the formal description of Ada. In *Semantics-Directed Compiler Generation*. N. D. Jones, Ed. Lecture Notes in Computer Science 94, Springer-Verlag, New York.

Dromey, G. 1989. *Program Derivation*. Addison-Wesley, Sydney.

Gries, D. 1981. *The Science of Programming*. Springer.

Gunter, C. 1992. *Foundations of Programming Languages*. MIT Press, Cambridge, MA.

Gunter, C. and Mitchell, J. 1994. *Theoretical Aspects of Object-Oriented Programming*. MIT Press, Cambridge, MA.

Hennessy, M. 1991. *The Semantics of Programming Languages: An Elementary Introduction Using Structured Operational Semantics*. Wiley, New York.

Hoare, C. A. R. 1969. An axiomatic basis for computer programming. *Commun. ACM*, 12:576–580.

Hoare, C. A. R. and Wirth, N. 1973. An axiomatic definition of the programming language Pascal. *Acta Informatica*, 2:335–355.

Jones, C. B. 1980. *Software Development: A Rigorous Approach*. Prentice Hall, Englewood Cliffs, NJ.

Kahn, G. 1987. Natural semantics. In *Proc. STACS '87*. Lecture Notes in Computer Science 247, pp. 22–39. Springer, Berlin.

Lee, P. 1989. *Realistic Compiler Generation*. MIT Press, Cambridge, MA.

Milner, R., Tofte, M., and Harper, R. 1990. *The Definition of Standard ML*. MIT Press, Cambridge, MA.

Mitchell, J., 1996. *Foundations for Programming Languages*. MIT Press, Cambridge, MA.

Morgan, C. 1994. *Programming from Specifications*, 2nd ed. Prentice Hall, Englewood Cliffs, NJ.

Mosses, P. D. 1976. Compiler generation using denotational semantics. In *Mathematical Foundations of Computer Science*. A. Mazurkiewicz, Ed. Lecture Notes in Computer Science 45, pp. 436–441. Springer, Berlin.

Mosses, P. D. 1990. Denotational semantics. In *Handbook of Theoretical Computer Science*, J. van Leeuwen, Ed. Vol. B, chap. 11, pp. 575–632. Elsevier.

Mosses, P. D. 1992. *Action Semantics*. Cambridge University Press, Cambridge, England.

Muchnick, S. and Jones, N. D., Eds. 1981. *Program Flow Analysis: Theory and Applications*. Prentice Hall, Englewood Cliffs, NJ.

Nielson, F. and Nielson, H. R. 1988. Two-level semantics and code generation. *Theor. Comput. Sci.*, 56(1): 59–133.

Nielson, H. R. and Nielson, F. 1992. *Semantics with Applications, a Formal Introduction*. Wiley Professional Computing. Wiley, New York.

Nielson, H. R., Nielson, F., and Hankin, C., 1998. *Principles of Program Analysis*. Springer-Verlag, New York.

Ong, C. H.-L. 1995. Correspondence between operational and denotational semantics. In *Handbook of Computer Science*. S. Abramsky, D. Gabbay, and T. Maibaum, Eds. Vol. 4. Oxford University Press, Rio de Janeiro, Brazil.

Pierce, B., 2002. *Types and Programming Languages*. The MIT Press, Cambridge, MA.

Plotkin, G. D. 1981. A Structural Approach to Operational Semantics. *Tech. Rep.* FN-19, DAIMI, Aarhus, Denmark, Sept.

Rees, J. and Clinger, W. 1986. Revised 3 report on the algorithmic language Scheme. *SIGPLAN Notices*, 21:37–79.

Schmidt, D. A. 1986. *Denotational Semantics: A Methodology for Language Development*. Allyn and Bacon.

Schmidt, D. A. 1994. *The Structure of Typed Programming Languages*. MIT Press, Cambridge, MA.

Stoughton, A. 1988. *Fully Abstract Models of Programming Languages*. Research Notes in Theoretical Computer Science. Pitman/Wiley.

Stoy, J. E. 1977. *Denotational Semantics*. MIT Press, Cambridge, MA.

Tennent, R. D. 1991. *Semantics of Programming Languages*. Prentice Hall International, Englewood Cliffs, NJ.

Watt, D. 1991. *Programming Language Syntax and Semantics*. Prentice Hall International, Englewood Cliffs, NJ.

Winskel, G. 1993. *Formal Semantics of Programming Languages*. MIT Press, Cambridge, MA.

Further Information

A good starting point for further reading is the comparative semantics text of H. R. Nielson and F. Nielson [1992], which thoroughly develops the topics in this chapter. Mitchell's [1996] and Reynolds's [1998] texts provide in-depth presentations.

Operational semantics has a long history, and good introductions are Hennessey's [1991] text and Plotkin's report on structural operational semantics [1981]. The principles of natural semantics are documented by Kahn [1987].

Mosses' [1990] paper is a useful introduction to denotational semantics; textbook-length treatments include those by Schmidt [1986], Stoy [1977], Tennent [1991], and Winskel [1993]. Gunter's [1992] text uses denotational-semantics-based mathematics to compare several of the semantics approaches, and Schmidt's [1994] and Mitchell's [1996] texts show the influences of data-type theory on denotational semantics, which is developed in detail by Bruce [2002] and Pierce [2002]. Action semantics is surveyed by Watt [1991] and defined by Mosses [1992].

Of the many textbooks on axiomatic semantics, one might start with the books by Dromey [1989] or Gries [1981]; both emphasize precondition semantics, which is most effective at deriving correct code. Apt's [1981] paper is an excellent description of the formal properties of relational semantics, and Dijkstra's [1976] text is the standard reference on precondition semantics. Hoare's [1969, 1973] landmark papers on relational semantics are worth reading as well. Many texts have been written on the application of axiomatic semantics to systems development; two samples are by Jones [1980] and Morgan [1994].

99

Compilers and Interpreters

Kenneth C. Louden

San Jose State University

99.1 Introduction 99-1
99.2 Underlying Principles 99-4
    Finite Automata • Context-Free Grammars
    • Parsing Algorithms • Attribute Grammars
99.3 Best Practices 99-10
    Specifying Syntax Using Regular Expressions and Grammars
    • Lex/Flex and Yacc/Bison • Purdue Compiler Construction
    Tool Set (PCCTS) • Dealing with Ambiguity • Attribute
    Analysis • Intermediate Representations and Code
    Generation • Code Optimization • Error Recovery
99.4 Incremental Compilation 99-25
99.5 Research Issues and Summary 99-26

99.1 Introduction

Compilers and interpreters are language translators that have many functions in common, in that both must read and analyze source code. A compiler, however, produces a program equivalent to the source program in a target language, usually object or assembly code but also sometimes C, whereas an interpreter directly executes the source program. Any programming language may be either compiled or interpreted, but languages with significant static properties (e.g., FORTRAN, Ada, C++) are almost always compiled, whereas languages that are more dynamic in nature (e.g., LISP, Smalltalk) are more likely to be interpreted. Languages that differ substantially from the standard von Neumann model of most architectures (e.g., PROLOG) may also be interpreted rather than compiled. A performance penalty is incurred by interpretation over compilation, so in cases where speed is critical, compilation is to be preferred. By mixing compilation and interpretation, this performance penalty can be reduced, usually to well within an order of magnitude. The advantage to interpretation is that the compilation step is avoided (useful during program development), and an interpreter offers greater control over the execution environment (useful for complex run-time environments) and greater flexibility in adapting to different architectures.

The first translators were developed in the 1950s. Prior to the development of high-level languages, a compiler was essentially what is known as a linker today: it compiled a collection of machine-language routines from a library to form a single program. A team at IBM under the direction of John Backus is generally credited with developing the first commercial compiler for a high-level language, during the period 1954 to 1957 [Backus et al., 1957]. The language translated by this first compiler was FORTRAN, which was designed simultaneously with the compiler (and is also credited with being the first high-level language). Modern translation techniques were first used in Algol60 compilers a few years later

1-58488-360-X/$0.00+$1.50
© 2004 by CRC Press, LLC

(see e.g., [Randell and Russell, 1964]), when the relationship of language translation to the theory of finite automata and context-free grammars became better understood. The study of these subjects was further stimulated by this relationship, and by the early 1970s most of the standard techniques in use today were known.

Since then, general improvements in translators have come in the following areas:

The automation of a significant portion of the construction of a translator

Improvements in the speed of the target code, due to the increased application of code-improving (or *optimizing*) algorithms

A greater ability for compilers to be relatively easily **retargeted**, or rewritten for a new target machine language, due in part to automation and in part to a better understanding of the required compiler structure

Improvements have also come in the implementation of special language features, such as exception handling, generics (parametric polymorphism), object-oriented features (such as dynamic binding), and parallelization, due to increasing understanding of these mechanisms. One area in which significant theoretical advances have taken place in the last 20 years is in the translation of functional languages, with new algorithms for type checking, type classes, and interpretation by tree reduction (see e.g., [Peyton Jones, 1987]). So far, however, these techniques have remained outside of the mainstream of languages and translators.

An important aspect of the technology of translators is its strong interaction with language design. For example, the introduction of block structure in the Algol language family gave impetus to the development of stack-based translation algorithms. In turn, the stack-based algorithms influenced the development of language semantics specifically designed to take advantage of the algorithms, such as the lexical scope rule and the principle of syntax-directed translation, in which the semantics of a program (i.e., its execution behavior) is directly reflected in its syntax, or structure, so that translation can be guided by this structure. (This rule could equally well have been formulated as *semantics-based syntax*.) Because of its early appearance, FORTRAN was not designed with these principles in mind, and it remains a somewhat more difficult language to translate with the standard techniques than later languages of the Algol family.

Another aspect to the development of language translators has been the tendency of the computing community to constantly rediscover or recreate translation techniques (other than the basic, well known ones outlined in many texts). Primarily, this is due to the lack of detailed documentation of specific translators in the computer science literature. This, in turn, is largely due to the commercial and/or proprietary nature of most translators. Two translators that historically were reasonably well documented and exerted a corresponding influence on subsequent translator construction, were the portable C compiler [Johnson, 1978] and the PASCAL P-compiler [Wirth, 1971] [Nori et al., 1981]. Lately, the availability of public-domain software from the Internet has greatly improved the opportunity to study existing translators. Particularly well-documented and of high quality are the compilers distributed as part of the GNU software project of the Free Software Foundation [Stallman, 1994].

Common to almost all modern compilers is a conceptual structure in which the tasks performed are divided into **phases**, or logically complete processing steps. The standard phase structure is shown in outline in Figure 99.1.

The first phase, called the scanner, or lexical analyzer, is the only phase directly involved in reading the source program. It converts the characters of the source program into tokens, sequences of characters that represent the basic units of program structure. Typical tokens are keywords such as *while*, numeric literals such as 3.14159, and identifiers.

The second phase in a compiler is the parser, or syntactic analyzer, which collects sequences of tokens into complete units, such as expressions, statements, and declarations. The output of the parser (either explicitly or implicitly) is a tree or other equivalent data structure representing the structure of the programming unit just recognized. This tree is called the *syntax tree*.

The third phase in a compiler is the semantic analyzer, which computes attributes, or properties of each programming construct, as well as its effect on the attributes of other constructs. Typical attributes include

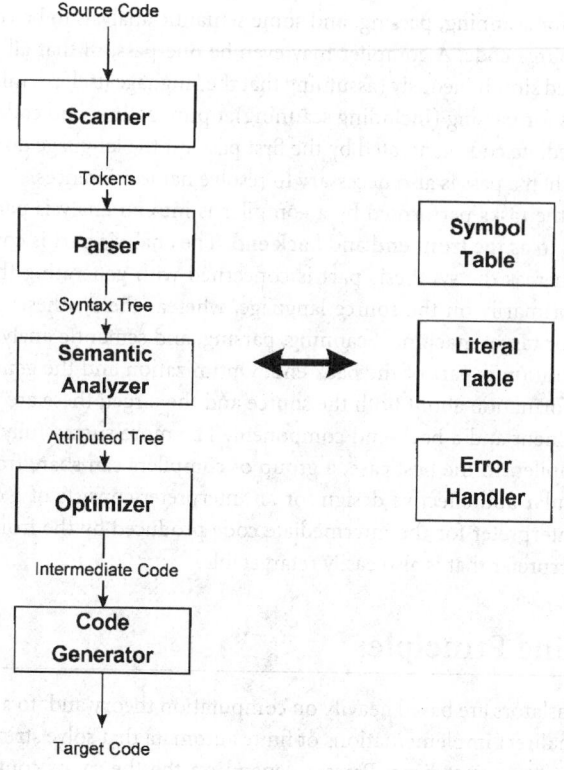

FIGURE 99.1   The phases of a compiler.

data types of expressions and identifiers, memory sizes, and actual or potential memory locations. The semantic analyzer also determines if the construct makes sense according to these attributes. That is, it performs consistency checks, such as type checking or range checking. The output of the semantic analyzer can be represented by an attributed tree representing the original tree modified by the computed attributes.

The fourth phase in a compiler is called the optimizer in Figure 99.1. This phase usually generates some form of linear code, called **intermediate code**, from the representation passed to it by the semantic analyzer. It also applies some forms of code-improving algorithms, either before or after generating intermediate code (or both). This phase of compilation is the most variable in different compilers; it may even be absent altogether.

The fifth and final compiler phase shown in Figure 99.1 is the code generator. This phase generates the final target code and may also perform some additional improvements to the code.

All of the phases of a compiler interact with various tables and handlers within the compiler. Figure 99.1 shows three important examples of these other components: the symbol table, the literal table, and the error handler. The figure indicates their interaction with the phases by a large double-headed arrow. The symbol table maintains the names defined by the program under translation, as well as possible predefined names in the language, and associates the names with their attributes, which may include scope, memory location, data type, and memory size. The symbol table may be monolithic, or it may be separated into a tree or graph of smaller tables, representing different scopes in the program. A similar table is needed for literal values that appear in the program, such as strings and numeric literals. The third component shown in Figure 99.1 is the error handler, whose job it is to generate different kinds of error messages but, more importantly, to provide error recovery, so that translation may continue (at least as far as is practical) in the presence of errors.

It must be emphasized that the compiler phases are logical units only and may not correspond to any actual grouping of operations within the compiler itself, or to any temporal sequencing of these operations.

Indeed, it is customary for scanning, parsing, and some semantic analysis to be completely integrated in a single pass over the source code. A compiler may even be one-pass, in that all phases, including code generation, are performed simultaneously (assuming that the language itself permits it). More likely is that there are separate passes for parsing (including scanning), optimization, and code generation (with later passes using the intermediate code generated by the first pass). If the language does not require names to be declared before use, then a pass is also necessary to resolve name references.

A useful division of the tasks performed by a compiler is into an analysis part and a synthesis part, sometimes also referred to as the **front end** and **back end**. The analysis part is concerned with analyzing the source program, whereas the synthesis part is concerned with generating the target program. The analysis part depends primarily on the source language, whereas the synthesis part depends primarily on the target language or target machine. Scanning, parsing, and semantic analysis are part of the front end, whereas code generation is part of the back end. Optimization and the generation of intermediate code usually require information about both the source and the target; these are more difficult to divide into a front-end component and a back-end component. The more successfully this is done, the easier it is to retarget the compiler. In the best case, a group of compilers can share front ends and back ends interchangeably. A popular and effective design for an interpreter consists of a compiler front end and a back end that is an interpreter for the intermediate code produced by the front end. This results in a reasonably efficient interpreter that is also easily retargetable.

99.2   Underlying Principles

Algorithms used in translators are based heavily on computation theory and, to a lesser extent, on formal semantics. Scanners are direct implementations of finite automata that solve string recognition problems through nonrecursive pattern matching. Parsers depend on the theory of context-free grammars and pushdown automata, which solve recursive recognition problems through stack-based pattern matching. Semantic analysis depends on solving sets of tree equations called **attribute grammars**. Code generation and interpretation can also be seen as applications of attribute grammars. It is possible to use even more formal semantic specifications, particularly denotational specifications, of the source and target languages to construct semantic analyzers and code generators (see, e.g., [Polak, 1981] [Lee, 1989]). The advantage to doing so is that the compiler can be proved correct (i.e., the semantics of the source and target programs are guaranteed to be the same). However, these techniques have not become popular, and we do not discuss them further. In the remainder of this section, we will discuss each of the areas mentioned in a little greater detail.

99.2.1   Finite Automata

A finite automaton is an abstract computational machine consisting of a finite number of states and transitions between states based on input symbols. The machine runs by beginning in the starting state and consuming input symbols while entering new states via corresponding transitions, until either an error state or an accepting state is reached. At that point, it may declare success or failure, or possibly continue executing. Each state represents stored knowledge about the computation up to that point. A finite automaton can handle arbitrary repetition by a fixed, finite set of input symbols, but it cannot handle recursive processes, because that would involve an unpredictable number of states.

Finite automata are the basis for the recognition of tokens within a scanner. Based on the well known correspondence from computation theory between finite automata and regular expressions, tokens are usually given initially by regular expressions, which are specifications for the string patterns, or **lexemes**, that a token represents. The mathematical theory of regular expressions limits itself to the consideration of three matching operations: the choice between two alternatives, indicated by the vertical bar | (similar to the logical OR operation); the concatenation or sequencing of two strings (with no operator symbol); and the repetition of a pattern, indicated by a postfix asterisk * (sometimes called the closure or Kleene closure operation). Parentheses are also used to group subexpressions together. As an example of the use of regular expressions to represent tokens, the following regular expression for a token represents simple, unsigned numbers consisting of a sequence of one or more decimal digits:

(0|1|2|3|4|5|6|7|8|9) (0|1|2|3|4|5|6|7|8|9)*

Such a regular expression can be converted into a finite automaton by one of several standard algorithms. The basic method is to use Thompson's construction [Aho et al., 1986, p. 122] to derive a nondeterministic automaton (i.e., one with an unpredictable next state) from the regular expression, and then to use the subset construction [Aho et al., 1986, p. 118] to derive an equivalent deterministic automaton from the nondeterministic one. Other algorithms exist that perform this conversion in one step and also construct an automaton with a minimal number of states. Whereas these algorithms can sometimes be useful for the design of scanners, their primary use is in the construction of scanner generators such as Lex (discussed in Section 99.3.2).

99.2.2 Context-Free Grammars

The theory of context-free grammars extends the ideas of finite automata and regular expressions to recursive situations. A context-free grammar is a collection of named recursive rules of the form $A \rightarrow \alpha_1 | \ldots | \alpha_n$, where A is the name of the rule and the α are strings of tokens and names (including possibly A itself) representing the different possible choices for the structure of A. The names are called **nonterminals** and the tokens are called **terminals**, for technical reasons explained shortly. Grammar rules are also sometimes referred to in the theory as **productions**. Each such rule represents the fact that a structure represented by A may become any one of the structures represented by an α. The absence of context for these potential replacements of A is what makes these grammars context free. It is possible to define grammars that are more general than context-free grammars; these can be arranged according to increasing generality in what is known as the Chomsky hierarchy. However, context-free grammars are the most useful within the computational constraints of language translation.

Every nonterminal in a context-free grammar defines a set of token strings: the strings of tokens that can be legally parsed by the grammar rule for that nonterminal (and associated rules). The most general structure that is defined by the grammar is usually singled out as the structure representing the entire language, and its associated nonterminal is called the *start symbol*. A legal parse is represented by a sequence of replacements of nonterminals by choices of right-hand sides of their associated grammar rules. This sequence of replacements is called a *derivation*. In a derivation, every nonterminal must eventually be eliminated by replacing it with another string, whereas a terminal, once it appears, is never replaced; this may be seen as a justification for the names *terminal* and *nonterminal*. In general, there are many possible derivations for the same string, which can be constructed by varying the order in which the replacements are made. Two kinds of derivations that are important for parsing use fixed orders for the replacements. In the first, called a leftmost derivation, the leftmost nonterminal in the current string is always the one to be replaced at the next step. Correspondingly, a rightmost derivation is one in which the rightmost nonterminal is always the one to be replaced at the next step.

Although derivations are useful for expressing exactly which steps are taken in a parse, they do not express the structure of the parsed string very well. A more useful representation of this structure is the parse tree, which represents each terminal and nonterminal by labeled nodes, and each replacement step in a derivation by the construction of a set of children of the node label being replaced. In a parse tree, each leaf node is labeled by a terminal, and each interior node is labeled by a nonterminal. Even parse trees have more detail than is actually needed to determine the meaning of a structure, and a condensed form of a parse tree, called a syntax tree, is the most useful data structure for translation. In a syntax tree, nodes may have a more diverse structure than in a parse tree: they may have more than one label, and both terminals and nonterminals may be used as labels of interior nodes.

As a brief example of the grammar concepts we have summarized, consider the grammar

exp → *exp* **+** **num** | **num**

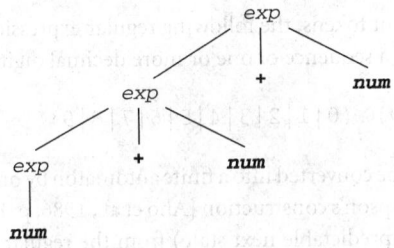

FIGURE 99.2   A parse tree for the string 3 + 4 + 5.

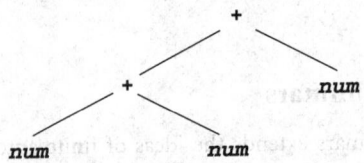

FIGURE 99.3   A syntax tree for the string 3 + 4 + 5.

which has one nonterminal *exp*, two terminals + and *num*, and a single production with two choices. (Here *num* represents a number token with lexemes such as 42 or 7.) An example of a legal string for this grammar is 3 + 4 + 5. A leftmost (and rightmost) derivation is

$$exp \Rightarrow exp + num \Rightarrow exp + num + num \Rightarrow num + num + num$$

A parse tree for the same string is given in Figure 99.2. A syntax tree is given in Figure 99.3.

99.2.3   Parsing Algorithms

Algorithms designed to match an input string of tokens based on a grammar and, either implicitly or explicitly, to construct a parse or syntax tree, are called parsing algorithms. Parsing algorithms come in two general varieties: **top-down** and **bottom-up**. Top-down algorithms construct a parse tree from the root to the leaves by guessing which structures are about to appear, based on the next part of the input string and the structure expected to be seen. Bottom-up algorithms construct a parse tree from the leaves to the root by consuming the input and forming a set of subtrees until the next structural element can be guessed from the structures seen so far and the next part of the input string. Because of the recursive nature of context-free grammars, both kinds of algorithms must use a stack, either explicitly or implicitly, to hold partial results. When it is constructed explicitly, this stack is referred to as the *parsing stack*, and it will possibly contain terminals, nonterminals, or other symbols representing the state of the parser. Because of the nature of their operation, top-down parsers trace out the steps of a leftmost derivation, and bottom-up parsers trace out in reverse the steps of a rightmost derivation.

There are algorithms of both varieties that will parse any context-free grammar. Early's algorithm [Early, 1970] is the most well-known bottom-up algorithm. A general top-down algorithm may be found in [Graham et al., 1980]. General algorithms may run in significantly slower than linear time (Early's algorithm requires cubic time in general), so they are rarely used in practice. Standard algorithms for top-down parsing are the **LL(k)** algorithms (parsing the input from left to right, giving a leftmost derivation, using *k* symbols of lookahead), and the **LR(k)** algorithms for bottom-up parsing (parsing the input from left to right, giving a rightmost derivation, using *k* symbols of lookahead). Both kinds of algorithms require that a grammar satisfy extra conditions to be parsable. The LL(*k*) algorithms in particular are quite restrictive, although easy to use. The LR(*k*) algorithms, although less restrictive, are more complex.

One top-down parsing method that is more flexible than the LL(k) methods is called **recursive-descent**. A bottom-up algorithm that is simpler than the LR(k) algorithms is called LookAhead LR(1) (**LALR(1)**) [DeRemer, 1971] [DeRemer and Pennello, 1982] and is normally restricted to one symbol of lookahead. Because these algorithms have proved themselves to be the most effective and easiest to use in practice, we discuss them in a little more detail.

In recursive-descent parsing, the grammar rules are viewed as prescriptions for the code of a set of mutually recursive procedures, one for each nonterminal. Recursive-descent, although suffering from some of the same problems as LL(k) parsing, is more flexible and can use simple *ad hoc* techniques to solve many of the problems of LL(k) parsing [Wirth, 1976]. For instance, simple left recursion, which cannot be handled directly by an LL(k) parser, can be handled in recursive-descent by noting that a left-recursive rule $A \rightarrow A\alpha \mid \beta$ is equivalent to a parsing procedure that first recognizes β and then a sequence of zero or more α's (because the grammar rule generates strings of the form $\beta\alpha\alpha...$). Thus, a recursive-descent procedure for the grammar *exp* → *exp* + *num* | *num* can be written using a *while* loop, as follows:

```
void exp(void)
{ match(NUM);
  while (nextToken == PLUS)
  { match(PLUS);
    match(NUM);
  }
}
```

An LALR(1) parser uses an explicit parsing stack instead of recursion. The state of a parse can be expressed by a finite automaton whose states consist of sets of so-called **items**, each item consisting of a production choice, a distinguished position indicated by a period (representing the point of progress in recognizing the rule), and an associated set of lookahead tokens legal at that point in the parse. (In the following discussion, we will use the so-called LR(0) items that lack a lookahead component; although LALR(1) items are more complex, the basics of the LALR(1) algorithm can be understood using these simpler items.)

Consider, for instance, the grammar *exp* → *exp* + *num* | *num*, which we will write for convenience in the form $E \rightarrow E + n \mid n$. There are six LR(0) items: $E \rightarrow .E + n$, $E \rightarrow E. + n$, $E \rightarrow E +. n$, $E \rightarrow E + n.$, $E \rightarrow .n$, and $E \rightarrow n.$. The start of a parse is indicated by beginning in a state represented by a new rule representing a start symbol that cannot appear elsewhere, and which we will write as $E' \rightarrow E$ in the example. The corresponding initial item is $E' \rightarrow .E$. Because this rule represents the fact that we may be about to recognize an E, we must also include the items $E \rightarrow .E + n$ and $E \rightarrow .n$ in this state. Transitions to new states are then given by moving the period past the symbols that follow it. For instance, there is a transition on the symbol E from the start state to the state containing the item $E \rightarrow E. + n$, and a transition on the symbol n from the start state to the state containing the item $E \rightarrow n.$. The complete finite automaton of items is given in Figure 99.4.

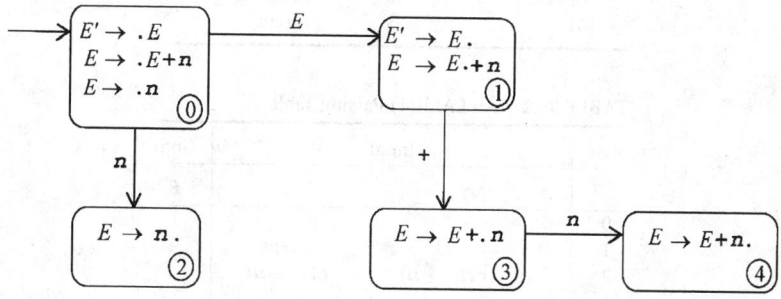

FIGURE 99.4 A DFA of sets of LR(0) items.

This finite automaton has no accepting states and is only used to keep track of the state of the parser. It is used in conjunction with a parser stack that holds the state numbers that the parser has passed through while parsing the input. It is used as follows. First, the initial state is pushed onto the stack. Then, the next input token is consulted. If there is a transition on this token from the current state (on top of the stack), then the token is removed from the input and the new state is pushed onto the stack; this is called a *shift operation*. If, however, there is an item in the current state of the form $A \rightarrow \alpha$. (a so-called final item), then this indicates that the string α has already been recognized, and it can be replaced by A; this is called a *reduce operation*.

A reduce operation is performed as follows. The states on the stack corresponding to α are popped from the stack (one state for each symbol in α). The state remaining at the top of the stack must have a transition on A, which is then taken, and the new state is pushed onto the stack.

As an example of this process, consider the automaton of Figure 99.4, and suppose that the input string is $n + n$. We depict the initial state of the parse as follows:

Parsing Stack	Input
$0	n + n$

The $ in this representation is used to indicate both the bottom of the stack and the end of the input. The first step in the parse is a shift on n from state 0 to state 2. Then, a reduction by $E \rightarrow n$ takes place, and the parser moves to state 1. At that point, the + and n are shifted, and the parser is in state 4. Then, a reduction by $E \rightarrow E + n$ is made, popping states 4, 3, and 1, revealing again state 0. Again, the E transition is followed into state 1. At this point the end of the input is encountered, and a reduction by $E' \rightarrow E$ is made, which corresponds to accepting the input.

The complete set of actions of the parser is given in Table 99.1, in which shift actions also include the new state number. Table 99.2 shows the LALR(1) parsing table for this simple grammar, which is used by the parser to select the actions indicated in Table 99.1. This table is two dimensional and is indexed by state and lookahead token. Each table entry contains an action; shift entries are indicated by an s and the new state number; reduce entries are indicated by an r and the rule to be reduced; empty entries are errors. Whereas this table is closely related to the automaton of Figure 99.4, the exact entries can only be inferred from a lookahead component, which we did not compute here.

An additional area in Table 99.2 is called the *goto area*. In this area are the transitions on nonterminals that are performed during reductions. These are essentially the same as shift operations, except that no

TABLE 99.1   The Actions of an LALR(1) Parser

Parsing Stack	Input	Action
$0	n + n$	shift 2
$02	+ n$	reduce $E \rightarrow n$
$01	+ n$	shift 3
$013	n$	shift 4
$0134	$	reduce $E \rightarrow E + n$
$01	$	accept

TABLE 99.2   An LALR(1) Parsing Table

State	Input			Goto
	n	+	$	E
0	s2			1
1		s3	accept	
2		r $(E \rightarrow n)$	r $(E \rightarrow n)$	
3	s4			
4		r $(E \rightarrow E + n)$	r $(E \rightarrow E + n)$	

input is consumed. A parser may also choose to condense this table by the use of default entries. For example, because state 2 only has reduce entries by the same rule, this could be made the default, in which case even on input token **n** the parser will perform the given reduction. This has the effect of postponing the declaration of error, but it cannot result in an incorrect parse. A similar default can be used for state 4.

One final bottom-up parsing method that deserves mention is operator-precedence parsing. This is a method that predates the LALR(1) method, but it can be used effectively on expression grammars involving infix operators; see [Aho et al., 1986, pp. 203–215] for a description.

99.2.4   Attribute Grammars

Whereas context-free grammars are generally accepted as the standard way to describe the syntax of a programming language, there is no equivalently accepted method for describing semantics. Formal methods for describing semantics, such as operational semantics or denotational semantics, have not met with universal acceptance, and although translators can be derived from such specifications, this is rarely done. Instead, various *ad hoc* mechanisms are used in translators to perform semantic analysis and code generation or interpretation.

One method that has proved useful for the translator writer is to use a so-called attribute grammar to describe the semantics of a programming language [Knuth, 1968]. An attribute grammar associates to each grammar rule a set of equations describing the computational relationships among a set of attributes attached to the symbols in the rule. These attributes can be anything from the data type of a variable to the value of an expression; even the target code generated by a compiler can be represented as a string attribute in an attribute grammar. Most often, attributes are used to represent the static, rather than the dynamic, properties of programs, and they are viewed as fixed values attached to the nodes of a syntax tree. Indeed, attribute values are usually written using a dot notation similar to that of record fields, so that $X.a$ means the value of attribute a of symbol X. Attributes may be implemented as fields in syntax tree nodes, or they may be stored in the symbol table or other data structures elsewhere in the translator.

Given a set of attributes $a_1, \ldots, a_k$ and a grammar rule choice $X_0 \to X_1 X_2 \ldots X_n$, the j-th attribute at the i-th symbol is given in an attribute grammar by an equation of the general form

$$X_i.a_j = f_{ij}(X_0.a_1, \ldots, X_0.a_k, X_1.a_1, \ldots, X_1.a_k, \ldots, X_n.a_1, \ldots, X_n.a_k) \qquad (99.1)$$

where f_{ij} is a mathematical function. An attribute grammar is thus written in purely functional style without side effects.

As an example, consider the grammar $exp \to exp + \textbf{\textit{num}} \mid \textbf{\textit{num}}$, which we may assume expresses the summation of a series of numbers. An attribute grammar for the numeric value of an expression defined by this grammar is given in Table 99.3. Note that the two instances of the nonterminal *exp* in the first grammar rule must be distinguished by subscripting, and that the terminal **num** is assumed to have its numeric value (called *lexval* in Table 99.3) precomputed, possibly by the scanner.

Of particular importance to the translator writer are the kinds of dependencies that the attribute equations create among the attributes of different symbols in a parse tree, because these dependencies determine when and how — or even if — the attributes can be computed during translation. A primary requirement is that the attribute grammar not have any circular dependencies. In practical situations, this requirement is virtually guaranteed unless an error has been made. Attributes whose dependencies flow

TABLE 99.3   An Attribute Grammar for a Simple Expression Grammar

Grammar Rule	Attribute Equations
$exp_1 \to exp_2 + \textbf{\textit{num}}$	$exp_1.val = exp_2.val + \textbf{\textit{num}}.lexval$
$exp \to \textbf{\textit{num}}$	$exp.val = \textbf{\textit{num}}.lexval$

from right to left in the grammar rules (i.e., those whose Equations 99.1 all have only the symbol X_0 on the left) are called **synthesized** attributes; any other attributes are called **inherited**. Synthesized attributes can be computed bottom-up during a parse, or by postorder traversal of the syntax tree, whereas inherited attributes require a more complex computation scheme. Indeed, as the name implies, inherited attributes are often passed down the syntax tree from parent to child, or from sibling to sibling, and so can be computed by some form of modified preorder traversal of the syntax tree.

A great deal of effort can be expended to ensure that all attribute values are computable during the parsing phase, to avoid having to construct the entire syntax tree and to avoid having to make additional passes over the input. The requirements that this places on the attribute grammar vary, depending on the parsing method employed. First, because virtually all parsers read the input from left to right, the attributes must be computable from left to right. This means that all inherited attributes at a symbol must depend only on the attribute values of its left siblings and the inherited attributes of its parent. In terms of Equation 99.1, this means that each equation for an inherited attribute ($X_i.a_j$ is on the left and $i > 0$) must have the form

$$X_i.a_j = f_{ij}(X_0.a_1, \ldots, X_0.a_k, X_1.a_1, \ldots, X_1.a_k, \ldots, X_{i-1}.a_1, \ldots, X_{i-1}.a_k) \qquad (99.2)$$

and that only inherited attributes of X_0 may appear in the f_{ij}. An attribute grammar in which all equations for inherited attributes are of this form is called **L-attributed**.

A further requirement for attribute evaluation during parsing is a form of strong noncircularity, in which an order for attribute evaluation can be fixed in advance without incurring any cycles (naturally occurring attribute grammars satisfy this noncircularity requirement, too).

The particulars of the parsing algorithm can also have a significant effect on which attributes are computable during parsing. Recursive-descent parsers are the most flexible; in the recursive routines, inherited attributes can be implemented as passed parameters, whereas synthesized attributes become returned values [Katayama, 1984]. Bottom-up parsers most naturally compute synthesized attributes. They do this by maintaining a stack of attribute values in parallel with the parsing stack. New synthesized attributes are computed on this stack at each reduction step. It is also possible to evaluate certain inherited attributes during a bottom-up parse, but this often requires that the grammar be rewritten, so that the attribute equations can be converted to a manageable form. Indeed, it is theoretically possible to rewrite a grammar so that all attributes become synthesized [Knuth, 1968]. However, the grammar thus produced bears little resemblance to the original. Thus, in difficult situations, it may be preferable to delay an attribute computation until after the parse and avoid rewriting the grammar into an unrecognizable form.

99.3 Best Practices

The theory of scanning, parsing, and attribute analysis implies that, in principle, a compiler or interpreter can be generated automatically from descriptions of the source language, attributes, and target machine. Whereas a number of compilers have been generated in this way [Farrow, 1984], it is not common. This is partially because the necessary tools (particularly for optimization and code generation) are complex and have not become standard, and partially because of the need for efficiency, both in translation and in terms of the generated code, which is more difficult to achieve using general tools. At this writing, it is common to automate only the construction of the scanner and the parser, although at least partial automation of code generation has become more common with the increasing importance of retargetability.

Whereas a large number of scanner and parser generators have been written, only a few have gained broad acceptance. We describe here two sets of tools that have been used in many compilers: the Unix tools Lex and Yacc (and their public domain versions Flex and Bison), and the Purdue Compiler Construction Tool Set (PCCTS). Yacc produces an LALR(1) parser, whereas PCCTS produces a recursive-descent parser.

Both sets of tools were designed initially to generate C code, but versions of both exist that can generate C++ code. PCCTS has also been extended to generate Java code. Another tool, not discussed here, is JavaCC: it is written in Java and generates a recursive-descent parser in Java. It should be noted that many commercial compilers and interpreters have been written by hand without the use of any tools at all; in such cases, the parsers are usually written using recursive-descent.

99.3.1   Specifying Syntax Using Regular Expressions and Grammars

In order to automate the task of generating a scanner and a parser, the first step is to specify the tokens using regular expressions and the syntax using context-free grammar rules. Although it is essential to understand the mathematical theory of both of these mechanisms, the theory ignores extensions and features of practical importance. We mention a few of these here.

The theory of regular expression relies on only three operations: concatenation, choice, and repetition. Whereas these are enough to match any string recognizable by a finite automaton, most pattern-matching systems extend this set of operators in many ways. As a typical example, consider the regular expression

```
[0-9]+(\.[0-9]+)?
```

which is written using standard conventions for the Unix tools Lex and Grep. This expression specifies a pattern for a simple floating point constant without exponential part: the expression [0-9] refers to a choice from the range of characters 0 to 9 (i.e., a digit), the + indicates a repetition of one or more ($a+$ is equivalent to aa^*), the backslash in front of the period *escapes* the metacharacter meaning of the period (otherwise, it would match any character), and the question mark indicates an optional component.

Even with these extensions, it is sometimes difficult to write regular expressions for certain patterns, even when such patterns do exist, and one may choose to apply an *ad hoc* recognition process instead of using a regular expression. A notorious case in point is that of C comments, which can be loosely described as /* (not */) */. The trouble is expressing (not */) as a regular expression. More generally, the nonexistence of sequences of more than one character in a string is a difficult property to express as a regular expression. Fortunately, these situations do not occur very often in real languages.

It is also necessary to be aware that the theory of regular expressions can easily be extended to cover simple nonregular situations. For instance, nested comments require recursion and so cannot be directly expressed as a regular expression. Nevertheless, adding a simple counter variable to a scanner permits it to recognize potentially nested comments.

Similar considerations arise in defining syntax using context-free grammars. Consider the following grammar, which specifies a simple four-function floating-point calculator program with a single memory:

```
session → asgn  nl  session  | nl
asgn → = expr | expr
expr → expr addop term | term
addop → + | -
term → term mulop factor | factor
mulop → * | /
factor → - factor | ( expr ) | NUMBER | m
```

A session consists of a sequence of assignments (asgn) followed by newlines (indicated in the grammar by **nl**), or just a newline (this is used to end the session). The arithmetic operations have their usual meanings. The optional **=** sign at the beginning of an asgn indicates assignment of the value of the following expression to the memory. The single letter **m** in a factor fetches the value of this memory. The token **NUMBER** is the only token with more than one possible lexeme, and it is assumed to be given by the regular expression given earlier in this section.

A sample session with an interpreter for this grammar might look as follows:

```
= 3.14 - 2
  1.14
= m*3.14 + 3
  6.5796
m*3.14 - 4
  16.6599
```

This represents a computation of the value of the polynomial $x^3 - 2x^2 + 3x - 4$ at the point $x = 3.14$ using Horner's rule $(x^3 - 2x^2 + 3x - 4 = ((x - 2)x + 3)x - 4)$.

This grammar is written in a style called Backus normal form or Backus-Naur form (**BNF**). In it, the only metasymbols are the arrow and the vertical bar (sometimes ::= or : or = is used instead of the arrow, and sometimes nonterminals are distinguished more directly from terminals by surrounding them with angle brackets $<\ldots>$). Repetition is expressed by recursion, and optional features are expressed by writing separate choices. This grammar also expresses the associativity and precedences of the operators: a left-recursive grammar rule such as expr → expr addop term indicates left associativity, and the different precedence levels expr, term, and factor cause the operators at each level to be given precedences in ascending order (lowest precedence first). When recursion is used only to express repetition, it can be written on the right or the left: the rule for a session is written right-recursively for more or less arbitrary reasons.

An alternative to this grammar would be to condense it to the following grammar (still in BNF):

```
session → asgn nl session | nl
asgn → = expr | expr
expr → expr op expr
     | - expr
     | ( expr )
     | NUMBER
     | m
op → + | - | * | /
```

This grammar is ambiguous, however, in that the order in which the operations are applied is not specified, and a legal string may have many different parse trees. This grammar can still be used as the basis for the calculator, as long as separate **disambiguating rules** are stated, giving the associativities and precedences of the operators. The advantage to this is that the grammar itself becomes shorter and easier to understand.

A different variety of context-free grammar is obtained by adding metasymbols representing optional and repeated constructs. One standard version of this is called **extended BNF** (**EBNF**), which uses square brackets [...] to surround optional constructs, and braces {...} to surround repeated constructs (one could just as well use parentheses, ?, and * as in regular expressions). The calculator grammar in EBNF becomes

```
session → { asgn nl } nl
asgn → [ = ] expr
expr → term { addop term }
addop → + | -
term → factor { mulop factor }
mulop → * | /
factor → - factor | ( expr ) | NUMBER | m
```

In such a grammar, the associativity of the operators is now suppressed in favor of showing the repetition directly. A standard convention is to assume left associativity of operators in any rule using the braces (since a session has no operators, the associativity of its rule is of no consequence).

99.3.2   Lex/Flex and Yacc/Bison

Lex [Lesk, 1975] and Yacc [Johnson, 1975] are scanner and parser generators that are a part of most Unix distributions. Both have public domain versions, Flex (Fast Lex [Paxson, 1990], based on ideas of [Jacobson, 1987]) and Bison, which run under a variety of operating systems. Each of these programs reads a definition file and produces as output a C source code file containing a scanning/parsing procedure. The definition file for each has the same basic format:

```
{definitions}
%%
{rules}
%%
{auxiliary routines}
```

We discuss the contents of the definition files first for Lex and then for Yacc, using our running example of a simple calculator program whose tokens and grammar were described previously.

The Lex definition file for the calculator scanner is given in Table 99.4. In this example, the definitions section contains a `#include` directive inside the brackets `%{` and `%}` and the definitions of the three tokens `digit`, `number`, and `whitespace`, using regular expressions written with previously described metasymbols (`whitespace`, for example, is defined to be a sequence of one or more blanks or tabs). All code inside the special brackets is inserted directly at the beginning of the C output file, thus allowing the user to provide declarations/definitions that may be used by the rest of the C code. In this case, the only insertion is to indicate the inclusion of the file `y.tab.h`. This file can be generated by Yacc, and it contains the definitions of tokens and other globals that permit communication between the Lex-generated scanner and the Yacc-generated parser. In our example (and for one particular version of Yacc), this file is as follows:

```
/* file y.tab.h */
extern double yylval;
extern int yylineno;
#define NUMBER 258
#define UNARY 259
```

The subsequent Lex code uses only the definition of `NUMBER` and `yylval` from this file.

The rules section specifies the actions that the Lex scanner is to take when each token is recognized. These actions are placed inside a C block and are inserted directly at the appropriate places in the scanner. In Table 99.4, the specified actions are as follows. All `whitespace` is skipped (empty action). A `number`

TABLE 99.4   A Lex Definition File

```
/*****************************************************************
CALC.L
*****************************************************************/
%{
#include "y.tab.h"
%}
digit        [0-9]
number       {digit}+(\.{digit}+)?
whitespace   [ \t]+
%%
{whitespace}    {/* skip */}
{number}        { sscanf(yytext,"%lf", &yylval);
                  return NUMBER; }
\n              { return '\n'; }
.               { return yytext[0]; }
```

causes the scanner to compute the floating point value from the yytext string and place it in yylval, and then return the NUMBER token. (The yytext string contains the lexeme matched from the input.) The newline character \n is singled out for special handling, because it is not placed in the yytext string; it is returned directly. Finally, the period indicates a default action (it matches any character), and this causes the character value itself to be returned (yytext[0]).

This concludes the description of the calculator Lex file in Table 99.4. Note that this file contains no auxiliary routines section, and the %% symbol separating this section from the previous rules section is also omitted.

We turn now to a description of the Yacc definition file, which is in Table 99.5. The definitions section contains two lines of C code to be inserted in the output file. The first defines YYSTYPE to be double; this is the type used to define the Yacc value stack, which needs to be double, because expressions compute floating point values. The second line defines a static variable mem, which is to be used as the actual memory location for the single calculator memory. The definitions section also contains the token definitions, indicated by the %token directive. In this example, only the NUMBER token need be defined; other tokens are single-character and may be referred to directly. Finally, the definitions section contains a description of the associativity and precedence of the arithmetic operators; these are necessary disambiguating rules, because the rules section uses the ambiguous form of the calculator grammar. The order in which the operators are listed determines their precedence (with lowest precedence listed first). The %left directive indicates that an operator is left associative. Finally, the UNARY token is implicitly

TABLE 99.5 A Yacc Definition File

```
/*******************************************************************
CALC.Y
*******************************************************************/
%{
#define YYSTYPE double
static double mem;
%}
%token NUMBER
%left '+','-'
%left '*','/'
%left UNARY
%%
session :/* empty */
        | session '\n' { exit(0); }
        | session asgn '\n' {printf("\t%g\n", $2);}
        | session error '\n' {yyerrok;}
        ;
asgn    : '=' expr { mem = $2; $$ = $2;}
        | expr { $$ = $1; }
        ;
expr    : expr '+' expr { $$ = $1 + $3; }
        | expr '-' expr { $$ = $1 - $3; }
        | expr '*' expr { $$ = $1 * $3; }
        | expr '/' expr { $$ = $1/$3; }
        | '-' expr %prec UNARY { $$ = - $2; }
        | '(' expr ')' { $$ = $2; }
        | NUMBER { $$ = $1; }
        | 'm' { $$ = mem; }
        ;
%%
int main()
{ yyparse(); return 0; }

int yyerror(char* t)
{ fprintf(stderr,"%s\n", t); return 0;}
```

defined by the last definition as having the highest precedence. It will be used to give unary minus a higher precedence than any of the binary operators.

The rules section of the Yacc specification contains the grammar in a modified BNF format, with actions contained in braces. Since Yacc is a bottom-up parser generator, it is easiest to use to compute synthesized attributes. In the calculator grammar, the value attribute is synthesized, and it is this attribute that we use Yacc's value stack to compute. The stored memory value, which has an inherited component, is handled directly by using the defined mem variable. The action code refers to the attribute values on the value stack by using symbols beginning with $. The symbol $$ refers to the (synthesized) value to be computed for the nonterminal defined by the rule. Each of the symbols $1, $2, etc., refers to the attribute value computed for each symbol on the right-hand side of the grammar rule. Thus, $$ = $1 + $3 in the rule exp : exp + exp indicates that the value of the first and third symbols (the right-hand expressions) are to be added to get the value of the result expression. This convention allows Yacc rules to be written in a style very close to the synthesized rules of an attribute grammar.

A few changes have been made to the grammar to make it more usable with Yacc. One is that the operators are written directly into the expressions, instead of being listed separately; this eliminates the need for a separate character attribute for an operator rule. Two additional changes have been made to the grammar rule for a session. First, the rule is written in left-recursive instead of right-recursive form. A right-recursive rule causes the parsing stack to grow without limit; this means that a very long session might cause a stack overflow (a similar situation is caused by tail recursive procedure calls). Thus, all right-recursive rules that do not reflect an associativity requirement should be rewritten in left-recursive form. The remaining change is the addition of the rule

```
session : session error '\n' {yyerrok;}
```

to the choices for a session. This is an error-handling production. It matches any line that is not a legal assignment to the internally defined Yacc nonterminal error, and the action yyerrok resets the Yacc parser to accept more input. Yacc also automatically calls a yyerror procedure that can print an error message or perform some other action.

Finally, the auxiliary routines section of Table 99.5 contains a main procedure which just calls the parsing procedure yyparse which is generated by Yacc. (The scanning procedure generated by Lex is called yylex and is automatically called at the appropriate times by yyparse.) The yyerror procedure is also defined in this section; it simply prints an error message (supplied by Yacc).

Assuming that the Lex definition for the calculator is in the file calc.l, and the Yacc definition is in the file calc.y, then a running calculator program can be built in Unix with the commands

```
lex -I calc.l
yacc -d calc.y
cc y.tab.c lex.yy.c -ll -ly
```

The file y.tab.c is the output file produced by Yacc, and the file lex.yy.c is the output file produced by Lex. The option -I causes Lex to produce an interactive scanner (i.e., one with lazy lookahead), the option -d causes Yacc to produce the file y.tab.h automatically, and the options -ll and -ly cause the C compiler to consult the Lex and Yacc libraries when linking (if necessary).

One additional Yacc feature is the verbose option -v. If we give the command

```
yacc -v calc.y
```

then a file y.output is produced that contains a description of the parsing table used by the Yacc-generated parser, similar to Table 99.2. This can be useful in tracking down exactly the behavior of the parser.

99.3.3 Purdue Compiler Construction Tool Set (PCCTS)

PCCTS [Parr et al., 1992] is a set of tools for automatically generating scanners and recursive-descent parsers. Although these parsers suffer from some of the same restrictions as other top-down parsers,

TABLE 99.6 A PCCTS Definition File

```
/*************************************************************
CALC.G
*************************************************************/
#header <<
typedef double Attrib;
#define zzcr_attr(a,tok,t)        sscanf(t,"%lf",''a);
>>

<<
static double mem;
int main()
{ ANTLR(session(),stdin);
  return 0;
}
>>
#token WhiteSpace "[\t\ ]*" << zzskip(); >>
#token NUMBER "[0-9]+{.[0-9]+}"
session  : (asgn "\n" << printf("\t%g\n", $1); >>)*
           "\n" << exit(0); >>
         ;
asgn     : "\=" expr << mem = $2; $asgn = $2; >>
         | expr              << $asgn = $1; >>
         ;
expr     : term << $expr = $1; >>
           ("\+" term << $expr += $2; >>
           | "\-" term << $expr -= $2; >>
           )*
         ;
term     : factor << $term = $1; >>
           ("\*" factor << $term *= $2; >>
           | "\/" factor << $term /= $2; >>
           )*
         ;
factor   : "\-" factor << $factor = - $2; >>
         | NUMBER << $factor = $1; >>
         | "\(" expr "\)" << $factor = $2; >>
         | "\m" << $factor = mem; >>
         ;
```

PCCTS offers some advantages over Lex and Yacc. First, the parser and scanner generators are more fully integrated, so that only one definition file must be created; this file contains both the token and the grammar definitions. Second, there are tools for applying automatic disambiguating rules using syntactic predicates (checked automatically by the parser) or semantic predicates (supplied by the user), or by increasing the number of lookahead symbols (PCCTS parsers are not restricted to single symbol lookaheads). Finally, there are additional tools within PCCTS for the automatic generation of both syntax trees and target code.

As with Yacc and Lex, we give a description of a PCCTS definition file using the calculator grammar as an example. The only file needed for PCCTS is shown in Table 99.6. The general form for this file (somewhat simplified) is

```
{header}
{actions or token definitions}
{rules or token definitions}
{actions or token definitions}
```

The header contains C definitions that will go both in the scanner code and the parser code. Actions are C code for auxiliary procedures and data. Token definitions contain regular expressions for tokens, together with actions to be taken on recognition. Rules are grammar rules in a modified EBNF form that also contain actions for execution during the recognition process. Actions must be contained inside the bracketing metasymbols <<...>>. Table 99.6 contains a header, an action with the definitions of the mem variable and the `main` procedure, two token definitions, and five rules. There are no actions or token definitions after the rules.

The header section contains a `typedef` of `Attrib`, which is the internal name for the returned value of the recursive-descent procedures (this corresponds directly to the Yacc value stack type `YYSTYPE`). The header also contains a definition of the internal macro `zzcr_attr`, which is used whenever a token string is to be converted to a value of type `Attrib`. There is only one such token — NUMBER — and `zzcr_attr` is identified with a call to the C string scan function `sscanf` that converts a string to a `double`. The `main` program is defined in the first action section using the macro

```
ANTLR(session(),stdin);
```

ANother Tool for Language Recognition (ANTLR) refers to the parser generation utility of PCCTS. The parameter `session()` indicates that the start symbol of the grammar is `session`, and so a call to the corresponding procedure begins the parse. The second parameter indicates the input file from which the program is to be taken, in this case `stdin`.

The token definitions consist of the symbol `#token`, followed by the name of the token, a regular expression in quotes defining the token (using conventions similar to Lex and other regular expression processors), and an optional action section. In Table 99.6 two tokens are defined, `WhiteSpace` and `NUMBER`; the action `zzskip()` for `WhiteSpace` causes this input to be discarded, and the token `NUMBER` has no action (recall that `zzcr_attr` already tells the scanner how to convert a NUMBER to a `double`).

Finally, the rules are given in a modified EBNF form, where (...) * indicates a repetition of 0 or more times and {...} indicates an optional part. Again, as in the previous Yacc solution, we have rewritten the `session` rule so that it is the EBNF equivalent of a left-recursive rule, which saves space on the call stack. We have also included the operator tokens directly in the rules for expressions, saving some steps. Note that PCCTS also allows tokens to be written directly into the rules using double quotes (the \ is used to avoid any metasymbol interpretation). Note how the actions for expressions and terms are embedded within the rules to achieve the desired results.

Assuming that the PCCTS definition file is called `calc.g`, a running calculator program can be built with the Unix commands:

```
antlr calc.g
dlg -i parser.dlg scan.c
cc calc.c scan.c err.c
```

The first line calls the main PCCTS parser generator tool ANTLR. ANTLR generates several files, the most important of which are `calc.c`, containing the parser coded in C, and `parser.dlg`, which is a scanner description specifically intended for input into the PCCTS scanner generator DLG (DFA-based lexical analyzer generator). DLG is then called in the next line, producing the scanner in the output file `scan.c` (the `-i` option indicates that the input will be interactive). Finally, the C compiler is called on `calc.c`, `scan.c`, and a third file `err.c`, which was also produced by ANTLR.

99.3.4 Dealing with Ambiguity

Although the expectation is that a language grammar should be unambiguous, in that each legal input string will correspond to exactly one syntax tree, there are many practical situations in which it is difficult or impossible to write a grammar unambiguously. In such cases, disambiguating rules must be stated and built into the parser in some way. We have already seen how Yacc can accommodate precedence and associativity disambiguating rules for operators. We give three additional examples of the use of disambiguating rules.

The dangling-else ambiguity is typical in languages such as C and PASCAL, which allow both simple and compound statements in control structures. In C the dangling-else ambiguity appears in statements like

```
if(e) if(f) {/*then-part*/} else {/*else-part*/}
```

The question is: should the else-part be executed when f is false or when e is false? The standard answer is that it should be executed when f is false (and when e is true). This is called the most closely nested disambiguating rule for the *if* statement, because it implies that the else-part structure is to be associated in the parse tree with the closest previous *if* statement that does not have an else-part. This also corresponds to always preferring the second choice in the BNF rule

```
if_statement → if (expression) statement
             | if (expression) statement else statement
```

This disambiguating rule could actually be incorporated directly into the BNF, but it is slightly complicated (and not very illuminating), and the rule as stated is easily implemented in a parser. It is worth noting that languages in which the *if* statement has a closing keyword (such as **end** or **endif**) do not have the dangling-else ambiguity; Ada is an example of such a language.

A second kind of ambiguity is represented by the availability in C++ of function-style casts, where one can write int(x) to mean the same thing as the C cast (int)x. Now consider the code fragment

```
typedef int (*F)();
void **x;
void p(void)
{ F (*x)();
  ...
}
```

Is the first code line inside p a cast of global *x to type F, after which this function value is called, or is it a declaration of a local variable x of type "pointer to function returning a value of type F?" The C++ standard says it is the latter: any statement that may be interpreted as a declaration *is* a declaration. This means that all grammar rules for declarations are to be preferred to all other grammar rules. Note that this is an ambiguity *between* nonterminals, rather than an ambiguity within a single nonterminal, as with the dangling-else ambiguity. It is also impossible to remove this ambiguity directly within the grammar, because the order of rules in a BNF description is immaterial to the language defined.

The final example of an ambiguity that we present here is also one that arises with casts, but this time in C. Consider the C statement

```
(t)-x
```

If t is a type name declared in a typedef, then this is a cast of the value -x to type t. On the other hand, if t is a variable, then this is the value obtained by subtracting x from t. In order to parse this correctly, it is necessary to consult the symbol table to see whether t is defined as a type. Thus, the symbol table (or at least the typedef part of it) must be built as parsing proceeds. Unlike the previous examples of ambiguity, this is a context ambiguity that cannot be resolved by purely syntactic means.

In an LR parser, these ambiguities, or parsing conflicts, can be of two different kinds. The first, more common, situation is when both a shift and a reduction by a grammar rule choice are called for on the same input token; this is a **shift-reduce conflict**. The other case is when reductions by two different rules are called for on the same input token; this is a **reduce-reduce conflict**.

Shift-reduce conflicts are usually resolved by selecting the shift over the reduce. This ensures that the longest possible string will be matched by each rule. The typical example is the dangling-else ambiguity, where preferring the shift corresponds exactly to the most closely nested disambiguating rule. Reduce-reduce conflicts, on the other hand, have no natural disambiguating rule. Parsers generally adopt an *ad hoc* rule, in which the reduction by the lowest-numbered rule is preferred (where the rules are numbered in the order in which they are considered by the parser). The C++ ambiguity between cast expressions and declarations (also described previously) can be resolved in this way, in favor of the declarations (as the C++ definition requires) by listing the declaration rules before the expression rules. Yacc adopts both of these disambiguating rules as stated. Unfortunately, the third ambiguity mentioned previously, a cast vs. an arithmetic expression in C, cannot be resolved by either of these means. Typically, this is handled by having the scanner consult the symbol table when recognizing an identifier and returning a different token for a type name than for a variable.

An alternative to these disambiguating rules is to build predicate testing directly into the parser generator in order to give the programmer control over the disambiguating mechanism. This is the case for PCCTS [Parr and Quong, 1994], where the ANTLR parser generator allows both syntactic and semantic predicates as disambiguating rules. For instance, the C++ ambiguity between declarations can be resolved by adding a syntactic predicate in the ANTLR definition file as follows:

```
stat:  (declaration)? declaration <<...>>
    |  expression <<...>>
    ;
```

Here, the parentheses and question mark indicate that the parser should try to match a declaration; if that fails, it should go on to try to match an expression.

The third ambiguity can be solved similarly, but with a semantic predicate supplied by the user:

```
var :  <<isvar(LATEXT(1))>>? ID <<...>>
    ;
typename
    :  <<istype(LATEXT(1))>>? ID <<...>>
    ;
```

Again, the question mark indicates a predicate, but this time the predicate is user-supplied, as indicated by enclosing it in brackets (`LATEXT(1)` is the lexeme of the next token in the input, made available by the scanner).

99.3.5   Attribute Analysis

Yacc and ANTLR restrict the kinds of attributes that can be reasonably computed during a parse, because of the requirements of their parsing algorithms. In cases of difficult attribute computations, these limitations are overcome either by providing *ad hoc* solutions using external data structures or by constructing an intermediate representation such as a syntax tree during the parse, and then writing specialized procedures that perform semantic analysis by traversing the intermediate form in one or more passes. At the time of this writing, tools for automating the computation of general attributes have not been widely used, partially due to the many different varieties of intermediate representations, and partially due to the variety and complexity of the semantic attributes of different languages. Some notable systems that do permit the automation of this step include LINGUIST [Farrow, 1984], GAG [Kastens et al., 1982], and Eli [Gray et al., 1992]. Code generation is a special case of this problem, and special methods for automating the code-generation step have been developed. We describe these next.

99.3.6   Intermediate Representations and Code Generation

The abstract syntax tree is a convenient way to represent the source program within a translator, particularly for semantic attribute analysis. However, it is less suited to code generation. Although target code can be generated as a form of attribute analysis, either during parsing or by traversal of the syntax tree, the quality of this code is usually poor, and further processing must be done before an acceptable level of target code is produced. Typically, this is achieved by generating some form of intermediate code that is closer to the code of the target machine but still abstract enough that the compiler can perform code-improving transformations on it. Many choices have been used for this intermediate code. Some of the more common are sequences of expression trees, sequences of postfix expressions, an abstract linearized form of syntax tree called *three-address code*, and actual code for a hypothetical target machine, such as P-code for the stack-based P-machine of many PASCAL-related compilers [Nori et al., 1981].

An example of three-address code corresponding to the source code line = 3.1 + m/2 for the calculator language described previously is

```
t1 := m/2
t2 := 3.1 + t1
m := t2
```

Here, the identifiers t1 and t2 are temporaries introduced by the compiler that can be thought of as pseudoregisters, and later assigned either to actual registers or to temporary locations in memory.

The equivalent (annotated) P-code for this same source code is as follows:

```
ldo r,m ; load real value onto stack from static location m
ldc r,2.0 ; load constant real value 2.0 onto stack
dvr ; divide two reals on top of stack, push result
ldc r,3.1 ; load constant real value 3.1 onto stack
adr ; add two reals on top of stack, push result
sro r,m ; store real from stack to static location m, pop stack
```

The use of the code of a precisely defined, simple machine such as the P-machine as the intermediate target language has several benefits. First, the compiler can be run in interpreter mode, in which the P-code functions as the target code, and a P-machine simulator is used to execute the P-code on the actual target machine. The target code is then kept as P-code, which is generally much more compact than actual executable machine code. Also, in order to retarget the compiler to a new machine, it is only necessary to rewrite the P-machine simulator. Second, in order to obtain a native-code compiler from the P-code compiler, one need only write a translator from P-code to the native code of the target machine. This is usually done by writing individual procedures for each P-code instruction that perform something like a macro expansion of each P-code instruction into target code. By itself, this process will produce very poor code, but it can be combined with a static simulation of the P-machine itself, whose stack can be used to discover opportunities for optimization. An improvement on P-code that adds significant attributes for tracking address calculations and optimizing transformations is U-code [Perkins and Sites, 1979].

An intermediate language that is compact, capable of efficient interpretation, and easily retargeted to different architectures is of special interest for heterogeneous networks, where the platform on which the intermediate code is to be executed or compiled to native code may not be known in advance. If the intermediate code is given a numeric machine code–like encoding (sometimes called *bytecodes*), and techniques to improve interpretation performance are used (such as on-the-fly compilation or threaded code interpretation [Bell, 1973]), then good performance on a variety of machines across a network can be achieved. This is, for example, the basis for Java implementations [Gosling, 1995]. Although this approach incurs some performance penalty, the latest research is showing that this penalty can be reduced to very acceptable levels [Bacon et al., 2002].

A similar process of macro expansion works for three-address code and its variants, although in this case the underlying abstract machine and its state are not made explicit (as they are for P-code), and the accumulated information during code generation must be stored as attributes in data structures associated with the intermediate code. Typically, these include address descriptors, which record the locations at which named quantities (such as variables) can be found, and register descriptors, which record information about the values that can be statically predicted to be in registers at particular points in the target code. Address descriptors can be kept in the symbol table; register descriptors can be maintained in an array indexed by the register numbers.

Unfortunately, both of these code-generation schemes spread the information about the target machine throughout the code generator. This means that retargeting the compiler is more costly than if the target machine information were collected compactly in one place. Several approaches have been taken toward achieving this.

The first method for improving the retargetability of the code generator involves writing a syntactic description of the macro expansion process for the intermediate code together with code-emitting actions, much as a Yacc or PCCTS description associates semantic actions with syntactic rules [Glanville and Graham, 1978]. This method permits the automatic generation of the code generator from this description, using a tool similar to a parser generator. This method has been expanded in [Ganapathi and Fischer, 1985] to include semantic predicates.

As an example, consider generating code for the three-address instruction

```
t1 := m/2
```

A rule that would handle this instruction for the VAX might appear as

```
reg(n) → adr(a)/const(i) dead?(n) float?(a) #emit("divd3 #i.,a,rn")
```

This rule establishes that emitting the VAX instruction `divd3 #i.,a,rn` allows the expression consisting of an address a divided by a constant i to be reduced to a register n, provided n is dead (i.e., contains a value that is no longer needed) and provided that a is the address of a real-valued quantity. Matching this rule to the previous three-address instruction causes `t1` to be identified with `reg(n)`, m to be identified with `adr(a)` (thus computing a as the address of m), and identifying 2 with `const(i)` (thus setting i to 2). The resulting line of VAX code generated is

```
divd3 #2.,m,r1
```

In the absence of predicates, such a set of rules describing the target machine could be written as a Yacc input. The difference between this kind of grammar and a grammar describing the syntax of the source language is that the target machine grammar usually comprises thousands rather than hundreds of rules, and the grammar is highly ambiguous, because there are usually many ways to generate target code to achieve a specific effect. Parser generators such as Yacc are inefficient and cumbersome when dealing with such grammars, and a code generator generator specifically tuned to this situation is more appropriately used [Henry, 1984].

The second variety of retargeting mechanism is that used in the portable C compiler [Johnson, 1978]. In this mechanism, a special template language is used to describe the process of matching intermediate code to target code. The code generator then consults a table of these templates during code generation in an attempt to find the optimal match.

The third variety of retargeting mechanism is similar to the second, except that both the intermediate code and the target code templates are written in the template language, and the matching routine runs statically, usually during the installation of the compiler, and selects a match between certain sequences of intermediate code and target code which is then fixed once and for all. This method was used in the Production-Quality Compiler Compiler (PQCC) Project [Cattell, 1980] [Leverett et al., 1980]. A related method has also been used in the GNU C++ compiler [Stallman, 1994], based on ideas of Davidson and Fraser [Davidson and Fraser, 1984]. In the GNU compiler, the template language is called the register transfer language (RTL). This language is written in LISP-like prefix form. An example of an instruction template is as follows:

```
(define_insn "divdf3"
  [(set (match_operand:DF 0 "register_operand" "=f")
   (over:DF (match_operand:DF 1 "register_operand" "f")
            (match_operand:DF 2 "register_operand" "f")))]
  "! TARGET_SOFT_FLOAT"
  "divd3%1,%2,%0"
  [(set_attr "type" "arith")]
)
```

Instruction templates are introduced using the `define_insn` operator. This operator is followed by a symbolic name for the instruction template, a pattern with predicates (such as `register_operand`), constraints (such as f, indicating the register may be used for floating point values), and extra conditions (such as `! TARGET_SOFT_FLOAT`, indicating that there must be a hardware floating point arithmetic unit available in the processor). Following this, there is a pattern for the instruction to be generated, with numbers referring to the operands of the pattern (in this example, it is the VAX instruction divd3 %1, %2, %0). Finally, there is a specification of attributes for the template.

Such instruction templates are used to generate target code in RTL format directly from C code during a parse. Optimizing steps are then applied directly to the RTL intermediate code, and then templates are again used to generate assembly output. Specialized RTL attribute descriptions are also used to guide the code-generation process.

99.3.7   Code Optimization

A code generator that naively expands intermediate code into target code will produce code that is extremely inefficient, both in terms of execution speed and target code size. A production-quality compiler must include processing steps that improve its ability to generate good target code, so that it more nearly resembles the code that would be produced by an assembly language programmer. Such processing steps are usually referred to as *optimization*, though they almost never produce truly optimal code. Indeed, the production of mathematically optimal code, except in the very simplest cases, is known to be computationally intractable (NP-complete), so to attempt this would result in unacceptably slow compilation speed.

Optimization steps can be built into almost all the phases of a compiler, from parsing to final target code generation. If an optimization pass is performed separately, it usually occurs after intermediate code generation but before target code generation. Such optimizations are generally *source-level* in that they do not depend heavily on the details of the target machine. Some optimizations, however, are *target-level* and require that the details of the target code be known. Such optimizations are referred to as *peephole optimizations*, because they were originally (and are sometimes still) performed by examining small sequences of target code and replacing them with more efficient code. This term can be misleading, however, because modern compilers sometimes perform considerably more sophisticated analysis of the target code than the name might imply (an example is the GNU C++ compiler). One complex aspect of the scheduling of optimizations during compilation is that some optimizations may uncover opportunities for other optimizations and vice versa (this is sometimes called the *phase problem*). This leads some compilers to repeat certain optimizations in an attempt to catch such cascaded situations.

Optimizations are classified according to the region of the program about which information is gathered in order to perform the code improvement. Local optimizations consider only straight-line segments of code, that is, those not involving jumps or calls. Global optimizations consider the code of a single procedure. Interprocedural optimizations consider the entire program or compilation unit. Most compilers perform some kinds of local optimizations. A more heavily optimizing compiler will perform global optimizations, typically using an information-collection method called *dataflow analysis*, which passes information around a flow graph representing the flow of control through a procedure. It is a rare compiler that performs interprocedural optimization, and for a very good reason: such optimizations are largely ineffective unless they are postponed to link time, because many procedures will not be available in a single compilation unit. This means that the linker must be closely coupled with the compiler; in particular, the system linker cannot be used. This raises the level of complexity of the compilation environment considerably.

In the remainder of this brief discussion of optimizations, we list the principal sources of code improvement over a naive code-generation strategy.

Register allocation. This is the most important and pervasive issue in the generation of quality code. Keeping temporaries in registers is indispensable, especially for reduced instruction set computer (RISC) architectures. In order to extend register allocation to include local variables, parameters, and global variables, a compiler may permanently allocate certain registers to heavily used quantities. An alternative is to build an interference graph and assign registers by graph coloring, permitting noninterfering variables to share the same register. Good global register allocation also requires some form of dataflow analysis in order to identify values that have no further uses in subsequent code and that can be safely overwritten if they are in registers.

Common subexpression elimination. This refers to the identification of expression values that are recomputed one or more times in a program and the avoidance of such recomputation by suitable storing and reuse of the computed value. Although one might naively think that a good programmer should not write code that contains common subexpressions, in fact, most common subexpressions are due to address computations of array and record references that cannot be expressed adequately in source code and that therefore require elimination by an optimizer. Common subexpressions are relatively easy to identify in straight line code; the principal difficulty in the general case is to determine what subexpressions may have changed between two computations of the same expression. Thus, a compiler may choose to perform only local common subexpression elimination.

Copy propagation. This optimization tracks code regions in which two or more variables have the same value. After an assignment x = y (in C syntax), x and y have the same value in subsequent expressions until one of them acquires a new value; in this region, any use of x can be replaced by a use of y, which can lead to better code if, for example, y is already in a register. As with common subexpressions, copy statements may exist in the intermediate code even when they are not written by the programmer. A special case of copy propagation is constant propagation: after an assignment such as x = 2, uses of x can be replaced by the constant 2. As with common subexpressions, copy propagation may be done as either a local or a global optimization.

Reduction in strength. This refers to the replacement of arithmetic expressions by equivalent expressions that execute faster. A typical example is the replacement of multiplication or division by a power of 2 by a shift operation. Another example occurs in loop optimization, where a linear combination calculation is replaced by a simple addition (see the subsequent paragraph on loop optimization).

Jump optimization. Opportunities for improving jump code occur when a sequence of jumps can be replaced by a single jump, or when a jump can be eliminated by rearranging the code. For instance, the code sequence

```
goto lab1
...
lab1: if x = y goto lab2
lab3: ...
```

can be replaced by the more efficient

```
if x = y goto lab2
goto lab3
...
lab3: ...
```

Algebraic Laws. An optimizer can look for special cases in expressions such as $x * 1$ and $x + 0$ (replacing these with x). Sometimes it is also worthwhile to replace a computation $x + y$ or $x * y$ with its commutative equivalent $y + x$ or $y * x$. More difficult is to discover opportunities to use distributive laws, such as replacing $x * y + z * y$ with the more efficient $(x + z) * y$.

Loop optimization. Loops are traditionally an area where a great deal of attention has been paid to making code as efficient as possible, because programs tend to spend a lot of time in loops. In programs with goto statements, it is a nontrivial operation even to discover loops, and this usually must be done by building the flow graph. Fortunately, in modern languages it is reasonable to depend on syntax (such as keywords *while* and *do*) to locate loops. Typical loop optimizations include attempting to discover invariant computations inside loops (i.e., those that always yield the same value, regardless of the loop iteration), and move their computation to just before the loop entry. Another optimization seeks to discover so-called induction variables, which are linear combinations $a * i + b$, where i is the loop control variable. Because such an expression is incremented by a fixed amount with each iteration, its computation can be reduced in strength to a simple addition.

Constant folding. This optimization seeks to replace constant expressions (e.g., $3 + 5$) with their resulting values (e.g., 8). Such an optimization could, in theory, replace all computations whose results are predictable at compile time by their results alone. In practice, this would involve repeated applications of constant propagation and folding, so this optimal result is rarely achieved.

Dead code elimination. This optimization seeks to skip code generation for those statements that are either never reached during execution or whose actions have no effect on the results of the program. The first case happens when compile-time constants are set to select certain actions over others (e.g., to suppress the collection of run-time statistics). The second occurs if common subexpression elimination or copy propagation makes a computation or assignment unnecessary.

99.3.8 Error Recovery

An important practical problem in the design of a translator is its response to errors in the input program. Although many translators have been constructed so that they stop at the first error encountered, it is generally considered better to attempt to discover as many errors as possible in the input before halting. Thus, the response to errors includes error recovery, so that translation may continue, as well as the generation of informative error messages. A further step in an error handler may also be error repair, where the translator attempts to correct at least some of the errors. Because it is difficult to infer from an erroneous program what the actual intended program was, error repair is usually not attempted, except in special cases, such as missing punctuation.

Errors may be classified into lexical, syntactic, semantic, and run-time errors. An interpreter must have a reasonable response to run-time errors that involves at least generating an error message and exiting gracefully. A compiler need only recover from static errors, although it may need to generate code to report run-time errors. Semantic errors are relatively easy to recover from, because the syntax tree can still be used to guide the remainder of translation. Lexical errors can simply be passed to the parser by the scanner as error tokens. Thus, the principal difficulty in building an error handler occurs in dealing with parsing errors, where the structure of the input is disturbed, and there may be no obvious way to restart the parser.

Several criteria apply to any method for recovery from parse errors. First, it is useful to try to detect errors as early as possible, because continuing to process the input can make it more difficult to determine where the error occurred and to generate an appropriate error message. Second, the translator should attempt to discard as little input as possible when recovering from an error, because discarding input can lead to other errors and mislead the programmer. Third, the translator should avoid generating large numbers of error messages caused by a single error. In particular, a translator must never get into an infinite loop when recovering from an error. This usually requires that at least some input be discarded during error recovery.

A standard technique for error recovery is called *panic mode*, in which tokens are simply discarded, and the parsing stack accordingly adjusted until the parse can resume. This method can be made sophisticated enough so that it becomes better than its name implies. It can be used either as the standard error recovery technique or as a fall-back technique for a more complex method. The important feature of panic mode is the proper computation of a synchronizing set of tokens, which are used to determine when the parser should stop discarding the input and attempt to resume the parse. Panic mode in recursive-descent parsing is discussed in [Wirth, 1976, Section 5.9].

Yacc uses a slightly different method, in which an error pseudotoken is made available, with which important error recovery locations can be marked in so-called error productions in the input grammar. For example, in the Yacc definition file of Table 99.5, the error production

```
session : session error '\n' {yyerrok;}
```

was included in the specification.

Yacc's behavior on encountering an error is as follows. First, it sets a flag to enter an error phase, and then it begins popping the parsing stack until a state is found where the error pseudotoken can be successfully shifted. Then, input tokens are discarded until three consecutive tokens can be shifted successfully, whence the error phase is canceled and normal parsing is resumed. The error phase can also be canceled manually by calling the Yacc procedure `yyerrok`. For example, the previous error production for a calculator session causes Yacc to discard all tokens until a newline is reached, and then to resume the parse. The result is that any incorrect line of input is deleted. (Yacc also automatically calls `yyerror` whenever the error phase is entered, which can perform other actions and print an error message.)

Other error recovery and repair mechanisms are discussed in [Dion, 1982] [Graham et al., 1979] [Roehrich, 1980].

99.4 Incremental Compilation

Not only must compilers generate high-quality code; they must do it in a reasonable amount of time. Historically, the speed of compilation was often inversely proportional to the quality of the generated code, with the compilation of a large program under full optimization sometimes taking hours or even days. With modern processors, compilation speed has become less of an issue, but strong optimization still requires significant amounts of processor time. For this reason, compilers usually offer an unoptimized mode that not only avoids code improvements but is also tuned to the speediest possible generation of runnable code; such an option can be very useful for development or instructional work. Even so, large programs can require significant amounts of time to recompile if all of the code must be retranslated during each compilation step. A significant improvement in compiler technology is therefore the ability of a compiler to operate incrementally, that is, to avoid recompilation of code that has not changed since the previous compilation step. Alternatively, compilers that are linked to an editor as part of an interactive development environment (IDE) can compile program text in the background as it is entered by the programmer. Either of these mechanisms is called **incremental compilation**.

One of the original motivations for separate compilation and linking was to allow a form of incremental compilation, where only files that had changed needed to be recompiled, and the results of this recompilation would be combined with previously compiled files by a linker. Of course, this approach is subject to error, because the linker usually has no knowledge of the source language and cannot find incompatibilities between the recompiled code and the unchanged code. Additionally, the programmer has to keep track manually of which files have changed and which have not.

The first major advance over this situation was the Unix *make* utility [Feldman, 1979], varieties of which are part of virtually every IDE today (usually called *build* or *make project*). Make itself is still heavily in use in command-line development environments. Make keeps track automatically of file changes by using file system time stamps: a later time stamp on a source code file over its corresponding object code file indicates the need for recompilation. Additionally, make can keep track of dependencies in a set of source code files (made necessary by the convention of using file inclusion to check incompatibilities in C), although such dependencies must still be maintained manually by a programmer (in a so-called makefile). The makelike features that are built into IDEs usually automate the process of maintaining dependencies even further, particularly in languages with strong interface requirements like Ada and Java (but unlike C or C++).

It is possible, and often even desirable, to go beyond the capabilities of an automated make utility in either or both of the two different approaches to incremental compilation mentioned previously. First, a compiler could implement a finer-grained incremental approach than the file-level increments of make, such as statement-level or declaration-level increments (i.e., recompiling only those statements or function definitions within a file that have changed, rather than the entire file). Such a mechanism requires much more retained information about code structure than just object files and time stamps, however, and can lead to a significant increase in the amount of storage space required for a program or project. Second, the compiler could run in the background behind an editor, recompiling code as it is edited, rather than waiting for a compile command from the user. (Obviously, this approach makes the most sense if finer-grained incremental compilation is also implemented.) However, this approach can also mean more compiler interference in the process of editing or creating code, where an editor may refuse to accept program text that the compiler cannot understand. This can easily create programmer resistance to the use of such systems.

Examples of compilers that offer incremental compilation are the Visual Age Java compiler from IBM and the Visual Studio C# Compiler from Microsoft. These compilers also provide a contrast in how they mix the two approaches to incremental compilation described in the previous paragraph. The Visual Age compiler is invoked automatically by the IDE whenever a file save operation is performed (or when a file is added or imported into a project). It performs only file-level incremental compilation. However, it immediately reports any compilation problems to the user and will even prevent the completion of a file save operation if syntax errors occur in significant places. In contrast, the C# compiler will perform method-level (function-level) incremental compilation, but only at the direction of the user during a requested build.

99.5   Research Issues and Summary

Language translators can be decomposed into the phases of scanning (lexical analysis), parsing (syntactic analysis), semantic analysis, optimization, and code generation. A scanner breaks the input program into tokens using the theory of regular expressions and finite automata. A parser constructs, implicitly or explicitly, a representation for the syntactic structure of the program, using the theory of context-free grammars. The construction of both a scanner and a parser can be easily automated, using tools such as Lex and Yacc or PCCTS. Yacc constructs an LALR(1) bottom-up parser; PCCTS constructs a recursive-descent top-down parser. Bottom-up parsers are generally too complex to construct by hand, but recursive-descent can be used to hand-construct a parser. Implementations of scanners as finite automata are also relatively easy to construct by hand.

The semantic analysis and code generation steps of a compiler can be modeled theoretically by an attribute grammar, which expresses in equational form the relationships among the various attributes of language entities. In fact, attribute grammars can be used as a basis for automating the construction of an entire compiler, but this has not become common, possibly because of the complexity of representing the entire semantics of a language as an attribute grammar, and possibly because of the difficulty of producing optimized target code. It may be that other semantic definition mechanisms, such as denotational semantics, will result in better automation techniques, but this remains for future study. Current methods typically construct hand-generated semantic analyzers which operate during parsing using auxiliary data structures, such as the symbol table, or analyzers that perform recursive traversals of a syntax tree.

Some success has been achieved in automating the code-generation step, with easy retargeting as the primary goal. These methods include the syntax-based approach of [Glanville and Graham, 1978] and the semantic approach of [Ganapathi and Fischer, 1985]. An alternative to use a symbolic machine description language to describe the target machine, which is then used by the code generator to produce target code. Effective use of this method has been made in the widely retargeted GNU C++ compiler.

An important aspect of the automation and retargetability of a compiler is the choice of an appropriate intermediate code representation for the source code. The best choice appears to be a symbolic code for a hypothetical abstract machine. One may the choose either to interpret the intermediate code using a simulator or to perform code generation based on a static simulation of this machine. The primary requirements of such an intermediate code are flexibility, security, and the availability of enough information to provide good optimization over a wide variety of target architectures. A significant challenge for future translator technology is to develop a standard intermediate code that can be generated by many different language front ends, and which can also be efficiently and safely interpreted and compiled on many different architectures under many different operating systems. Optimized versions of the Java **Virtual Machine**, such as the Jikes Research Virtual Machine [Bacon et al., 2002], hold the promise that this may happen in the not-too-distant future.

Defining Terms

Attribute grammar: A set of equations associated to the grammar rules of a context-free grammar that define a collection of attributes associated to the terminals and nonterminals of the grammar. Attribute equations are written in purely functional form (i.e., without side effects) and may be "solved" for the actual attribute values by different kinds of traversals of the parse or syntax tree.

Alternatively, attribute values may be computed by replacing the attribute equations with equivalent side effect–generating code using separate data structures such as the symbol table.

Back end: The part of a compiler that depends only on the target language and is independent of the source language. The back end receives the intermediate code produced by the front end and translates it into the target language.

BNF: Backus normal form or Backus-Naur form. A notation for context-free grammar rules first used in the Algol60 report to describe syntax. It comprises only two metasymbols, usually written $\rightarrow$ or ::= and |.

Bottom-up: A parsing algorithm that constructs the parse tree from the leaves to the root. Bottom-up algorithms include LR and LALR parsers, such as those produced by Yacc.

Disambiguating rule: A rule stated separately from the rules of a context-free grammar which specifies the correct choice of syntax tree structure when more than one structure is possible.

EBNF: Extended BNF. Adds bracketing metasymbols [...] and {...} to BNF to indicate optional and repeated structures, respectively. (These can also be written as (...)? and (...)* to remain consistent with standard regular expression notation.)

Front end: The part of a compiler that depends only on the source language and is independent of the target language. The front end translates and analyzes the source program.

Incremental compilation: Compilation or recompilation that occurs in small steps based on incremental changes made during the editing of a program and which encompasses only those parts of the program affected by the changes.

Inherited attribute: An attribute whose value depends on attribute values at syntax tree nodes which are not descendants. A nonsynthesized attribute.

Intermediate code or representation: An internal data structure or sequence of abstract instructions representing a program as produced by the parser or other phase inside a compiler.

Item: A grammar rule choice with a distinguished position (usually indicated by a period), indicating that a parse has reached that position in attempting to recognize the rule. Sets of items are used by a bottom-up parser to record a state reached during a parse. Items may have 0 or more tokens of lookahead attached to them. LR(0) items contain no lookahead, whereas LR(1) items contain one token of lookahead.

L-attributed grammar: An attribute grammar whose attributes may be computed by a left-to-right traversal of the source program. An attribute grammar must be L-attributed for the attributes to be computable during a parse that processes the input from left to right (as most parsers do). Synthesized attributes are always L-attributed.

LALR(1): The lookahead LR(1) algorithm invented by DeRemer [DeRemer, 1971]. The algorithm used in many bottom-up parser generators, including Yacc. A language is also called LALR(1) if it can be parsed unambiguously by the LALR(1) algorithm.

Lexeme: The actual character string read from the input when recognizing a token. The lexeme of an identifier token is the identifier name.

LL(k): A top-down parsing algorithm that processes the input from left to right, producing a leftmost derivation using k tokens of lookahead. The term can also be applied to a language that can be unambiguously parsed using this algorithm.

LR(k): A bottom-up parsing algorithm that processes the input from left to right, producing a rightmost derivation (in reverse) using k tokens of lookahead. The term can also be applied to a language that can be unambiguously parsed using this algorithm.

Nonterminal: A name for a structure defined by a context-free grammar rule. Interior nodes of parse trees are labeled by nonterminals.

Phase: A logical unit of a compiler. Typical phases include scanning, parsing, semantic analysis, and code generation. Phases are distinct from passes, which comprise a complete sequential processing of the input program. Phases may or may not correspond to physical code units within the compiler.

Production: Another term for a context-free grammar rule or grammar rule choice.

Recursive-descent: A top-down parsing algorithm that translates context-free grammar rules into a set of mutually recursive procedures, with each procedure corresponding to a nonterminal. Recursive-descent parsing is usually the method of choice when writing a parser by hand.

Reduce-reduce conflict: In bottom-up parsers, a property of a state in which a parser has a choice of two productions that can be used to reduce the parsing stack, and both are legal for the amount of lookahead allowed. Reduce-reduce conflicts have no natural disambiguating rule.

Retargeting: The process of changing a compiler to produce target code (assembly or machine code) for a different machine. This may involve rewriting the compiler back end or creating a machine definition file for the new machine.

Shift-reduce conflict: In bottom-up parsers, a property of a state in which a parser has a choice of either reducing the parsing stack using a production or shifting a token from the input, and both are legal for the amount of lookahead allowed. A natural disambiguating rule is to prefer the shift, thus allowing the parser to match the longest possible input string at each point.

Synthesized attribute: An attribute whose value depends only on the attribute values of descendants in the parse or syntax tree. Synthesized attributes are the easiest to compute during a parse, requiring no special data structures or techniques. The syntax tree itself is the most important example of a synthesized attribute.

Terminal: Another term for a token in a context-free grammar. Leaf nodes of parse trees are labeled by terminals.

Top-down: A parsing algorithm that constructs the parse tree from the root to the leaves. Top-down algorithms include LL parsers and recursive-descent parsers, such as those produced by PCCTS.

Virtual machine: An interpreter for intermediate code, such as the Java Virtual Machine which executes Java bytecode. A virtual machine provides transparent cross-platform retargetability for compilers and other software.

References

Aho, A.V., Sethi, R., and Ullman, J.D. 1986. *Compilers: Principles, Techniques, and Tools*. Addison-Wesley, Menlo Park.

Backus, J., et al. 1957. The FORTRAN automatic coding system. *Western Joint Computer Conf.*: 188–198. Reprinted in Rosen, S., Ed., *Programming Systems and Languages*, McGraw-Hill, New York, 1967, p. 29–47.

Bacon, D., Fink S., and Grove, D. 2002. Space- and time-efficient implementation of the Java object model. *European Conference on Object-Oriented Programming (ECOOP 2002)*, pp. 111–132. Lecture Notes in Computer Science #2374, Springer-Verlag, New York.

Bell, J.R. 1973. Threaded code. *Commun. of the ACM*, 16(6): 370–372.

Cattell, R.G.G. 1980. Automatic derivation of code generators from machine descriptions. *ACM Trans. on Prog. Langs. and Systems*, 2(2): 173–190.

Davidson, J.W., and Fraser, C.W. 1984. Code selection through object code optimization. *ACM Trans. on Prog. Langs. and Systems*, 6(4): 505–526.

DeRemer, F.L. 1971. Simple LR(k) grammars. *Comm. of the ACM*, 14(7): 453–460.

DeRemer, F.L., and Pennello, T. 1982. Efficient computation of LALR(1) lookahead sets. *ACM Trans. on Prog. Lang. and Systems*, 4(4): 615–645.

Dion, B.A. 1982. *Locally Least-Cost Error Correctors for Context-Free and Context-sensitive Parser*. Univ. of Michigan Research Press.

Early, J. 1970. An efficient context-free parsing algorithm. *Comm. ACM*, 13(2): 94–102.

Farrow, R. 1984. Generating a production compiler from an attribute grammar. *IEEE Software*, 1(10): 77–93.

Feldman, S. 1979. Make — a program for maintaining computer programs. *Software: Practice & Experience*, 9(4): 255–265.

Ganapathi, M.J., and Fischer, C.N. 1985. Affix grammar driven code generation. *ACM Trans. on Prog. Lang. and Sys.*, 7(4): 560–599.

Glanville, R.S., and Graham, S.L. 1978. A new method for compiler code generation. In *Fifth Annual ACM Symposium on Principles of Programming Languages.*

Gosling, J. 1995. Java intermediate bytecodes. *ACM SIGPLAN Notices*, 30(3): 111–118.

Graham, S.L., Haley, C.B., and Joy, W.N. 1979. Practical LR error recover. *SIGPLAN Notices*, 14(8): 168–175.

Graham, S.L., Harrison, M.A., and Ruzzo, W.L. 1980. An improved context-free recognizer. *ACM Trans. on Prog. Lang. and Sys.*, 2(3): 415–462.

Gray et al. 1992. Eli: a complete, flexible compiler construction system. *Commun. of the ACM*, 35(2): 121–131.

Henry, R.R. 1984. *Graham-Glanville Code Generators*. Ph.D. thesis, Univ. of Calif., Berkeley.

Jacobson, V. 1987. Tuning Unix Lex, or it's not true what they say about Lex. *Proceedings of the Winter USENIX Conf.*

Johnson, S.C. 1975. Yacc — yet another compiler-compiler. *CS Technical Report #32*, Bell Laboratories, Murray Hill, NJ.

Johnson, S.C. 1978. A portable compiler: theory and practice. In *Fifth Annual ACM Symposium on Principles of Programming Languages*, pp. 97–104. ACM Press, New York.

Kastens, U., Hutt, B., and Zimmermann, E. 1982. *GAG: A Practical Compiler Generator*. Lecture Notes in Computer Science #141, Springer-Verlag, New York.

Katayama, T. 1984. Translation of attribute grammars into procedures. *ACM Trans. on Prog. Lang. and Systems*, 6(3): 345–369.

Knuth, D.E. 1968. Semantics of context-free languages. *Math. Systems Theory*, 2(2): 127–145. Errata 5(1) (1971): 95–96.

Lee, Peter 1989. *Realistic Compiler Generation*. MIT Press, Cambridge, MA.

Lesk, M. 1975. Lex — a lexical analyzer generator. *CS Technical Report #39*, Bell Laboratories, Murray Hill, NJ.

Leverett, B.W., et al. 1980. An overview of the production-quality compiler-compiler project. *IEEE Computer*, 13(8): 38–49.

Nori, K.V., et al. 1981. Pascal P implementation notes. In *Pascal — The Language and Its Implementation*, Barron, D.W. Ed., Wiley, Chichester, UK.

Parr, T.J., Dietz, H.G., and Cohen, W.E. 1992. PCCTS reference manual. *ACM SIGPLAN Notices*, 27(2): 88–165.

Parr, T.J., and Quong, R.W. 1994. Adding semantic and syntactic predicates to LL(k): pred-LL(k). *Int'l. Conf. on Compiler Construction*. April, 1994.

Paxson, V. 1998. Flex users manual. Available at http://www.gnu.org/software/flex/manual.

Perkins, D.R., and Sites, R.L. 1979. Machine independent Pascal code optimization. *ACM SIGPLAN Notices*, 14(8): 201–207.

Peyton Jones, S.L. 1987. *The Implementation of Functional Programming Languages*. Prentice Hall, Englewood Cliffs, NJ.

Polak, W. 1981. *Compiler Specification and Verification*. Lecture Notes in Computer Science #124, Springer-Verlag, New York.

Randell, B., and Russell, L.J. 1964. *Algol60 Implementation*. Academic Press, New York.

Roehrich, J. 1980. Methods for the automatic construction of error-correcting parsers. *Acta Informatica*, 13(2): 115–139.

Stallman, R. 1999. *Using and Porting GNU CC*. GNU Press, Boston, MA..

Wirth, N. 1971. The design of a Pascal compiler. *Software — Practice and Experience*, 1(4): 309–333.

Wirth, N. 1976. *Algorithms + Data Structures = Programs*. Prentice Hall, Englewood Cliffs, NJ.

Further Information

The standard text in compiler design has for many years been *Compilers: Principles, Techniques, and Tools* by Alfred V. Aho, Ravi Sethi, and Jeffrey D. Ullman (Addison-Wesley, 1986). This book is now showing its age but is still useful as a reference, because it was fairly comprehensive at the time it was written. A modern introductory text that is based on the C language is *Compiler Construction: Principles and Practice* by

Kenneth C. Louden (International Thomson, 1997). A more comprehensive treatment is *Modern Compiler Implementation in Java, 2nd edition* by Andrew W. Appel and Jens Palsberg (Cambridge University Press, 2002). This book also treats object-oriented issues in some detail, for both the implementation language and the target language. For a more detailed study of optimization, see *Advanced Compiler Design and Implementation* by Steven S. Muchnick (Morgan Kaufmann, 1997). For those interested in functional languages, *Modern Compiler Implementation in ML* by Andrew W. Appel (Cambridge University Press, 1997) may be of interest. A book that presents a full C compiler in complete detail is *A Retargetable C Compiler: Design and Implementation* by Christopher W. Fraser and David Hanson (Addison-Wesley, 1995).

Aside from these and many other texts, the best place to locate the latest information on compiler design is the comp.compilers newsgroup on the Internet.

Research papers on language translation can be found in publications by the IEEE and the ACM, particularly the conference proceedings published as part of the ACM SIGPLAN Notices (especially the annual Programming Languages Design and Implementation [PLDI] conference), the ACM Annual POPL Conference proceedings, and the ACM TOPLAS journal. For information, contact the ACM via its Web site, http://www.acm.org (or by e-mail at acmhelp@acm.org) and the IEEE via its Web site http://www.ieee.org.

General information on interpreters, insofar as they differ from compilers, is more difficult to find in one place. A Scheme-based introduction can be found in *Structure and Interpretation of Computer Programs, 2nd edition* by H. Abelson and G. J. Sussman with J. Sussman (MIT Press, 1996). Advanced techniques both for the compilation and interpretation of functional languages can be found in *The Implementation of Functional Programming Languages* by Simon L. Peyton Jones (Prentice Hall, 1987) and *Implementing Functional Languages* by Simon L. Peyton Jones and David Lester (Prentice Hall, 1992). The latter text concentrates more on implementation issues; the former gives a more theoretical description.

A brief but useful introduction to the use of Lex and Yacc can be found in *The Unix Programming Environment* by Brian W. Kernighan and Rob Pike (Prentice Hall, 1984). A more detailed study is contained in *Lex & Yacc* by John R. Levine, Tony Mason, and Doug Brown. (O'Reilly and Associates, 1992). Manuals for the GNU versions Flex and Bison can be found at http://www.gnu.org/manual. Information about PCCTS can be found in the comp.compilers.tools.pccts newsgroup, at http://www.polhode.com/pccts.html, and at http://www.antlr.org. The Java parser generator JavaCC can be found at http://javacc.dev.java.net.

100

Runtime Environments and Memory Management

Robert E. Noonan
College of William and Mary

William L. Bynum
College of William and Mary

100.1 Introduction **100**-1
100.2 Runtime Stack Management **100**-2
  Procedures and Functions • Design Issues • Parameter
  Passing • Activation Records • Activation Stack
  • The Sun SPARC Architecture
100.3 Pointers and Heap Management **100**-11
100.4 Garbage Collection **100**-12
  Reference Counting • Mark-Sweep • Copy Collection
100.5 Summary **100**-14

100.1 Introduction

Objects in programming languages can be assigned to any of three different types of memory: static memory, the runtime stack, and the heap. It is the responsibility of the programmer to decide which objects to assign to each memory type and to manage memory usage effectively.

Objects are usually assigned to static memory when the object must be accessed globally throughout the program. A restriction associated with using static memory is that such an object must have a constant size.

In contrast, the runtime stack is used for dispatching active functions and methods, including space for their local variables, and their parameter-argument linkages, as they gain and relinquish control to the caller. A *stack frame,* is pushed onto the top of the stack whenever a method is called, and popped from the stack when the method returns control to the caller. Thus, memory usage in the runtime stack is predictable and regular.

The heap is the least structured of the three memory types. The heap is used for all other objects that are dynamically allocated during the runtime life of the program. The heap becomes fragmented as it is used for the dynamic allocation and deallocation of storage blocks. At any point in time, the active memory blocks in the heap may not be contiguous. For that reason, *garbage collection* of unused heap blocks, whether manual or automatic, is a major issue in modern programming.

The principal goal of this chapter is to give the reader an overview of runtime memory management. First, we discuss management of the runtime stack with respect to the calling of procedures and functions and the passing of parameters and return values. After a review of the general principles involved, an example

1-58488-360-X/$0.00+$1.50
© 2004 by CRC Press, LLC

is presented for the Sun SPARC architecture. Next, we discuss the use of pointer variables and manual heap allocation and deallocation. Finally, we review some widely used garbage collection algorithms.

100.2 Runtime Stack Management

The purpose of this section is to provide a general introduction to the principles used in designing a runtime stack for programming languages, with a particular emphasis on procedures and parameters. The sections that follow introduce the basic terminology and major ideas in the design and implementation of such stacks. Texts that cover this material in more detail include Sebesta [2002], Louden [1993], Pratt and Zelkowitz [2001], and MacLennan [1987]. Examples are taken primarily from the language C [Kernighan and Ritchie 1988].

100.2.1 Procedures and Functions

In this section we are concerned with the characteristics of a procedure, subprogram, or function. For our purposes, such an entity has the following characteristics: (1) it has a single entry point, (2) execution of the calling unit is suspended, and (3) in the absence of a nonlocal goto, control is returned to the caller upon normal completion of the procedure. In the case of a function, a value is returned to the caller.

Figure 100.1 shows a C program that calls a local **bcopy** function to copy a string. Routine **bcopy** is a subroutine or procedure.

Line 16 is said to contain a *call* of the function. When this line is executed, execution of the calling routine, **main** in this case, is explicitly suspended. Program control is transferred to the *body* of routine **bcopy**, in this case everything between the opening brace on line 6 and the closing brace on line 9. Line 2 is called the *header* of the routine; it gives the name of the function, the return type (in this case, the **void** type indicates that the function returns no value), and the names and types of the arguments of the routine.

The variables **s**, **d**, and **n** in line 2 are said to be the *formal arguments* or *formal parameters* of the routine. The variable **hello** in the call on line 16 is said to be the *actual parameter* of the call corresponding to

```
1 /* bcopy -- copy source block to destination block */
2 void bcopy (char *s, char *d, int n)
3     /* s -- pointer to source block */
4     /* d -- pointer to destination block */
5     /* n -- number of bytes to copy */
6 {
7     for (; n> 0; n--)
8         *(d++) = *(s++);
9 }
10
11 char hello[] = { "Hello, world!"};
12 /* 1234567890123 */
13 main()
14 {
15     char newhellow[50];
16     bcopy(hello, newhellow, 14);
17     printf("%s\n", newhellow);
18 }
```

FIGURE 100.1   The **bcopy** function in ANSI C.

the formal parameter **s**. Similarly, the variable **newhello** on line 16 is the actual parameter of the call corresponding to the formal parameter **d**, and the integer constant **14** is the actual parameter of the call corresponding to the formal parameter **n**. In this example, although there are only 13 characters in the string to be copied, the constant **14** must be used in the call to be sure to copy the null termination byte with which the C language indicates the end of a string.

In the case of this example when line 16 is executed, control is passed from the calling routine, **main** in this case, to the called routine, **bcopy**, starting at the opening brace on line 6 and continuing until the closing brace is executed (an implicit return). Although this example does not contain one, the called routine can also return to its caller through an explicit **return** statement. After program control has entered line 6 and before the implicit return on line 9, **bcopy** is said to be *active*.

100.2.2   Design Issues

In designing a runtime stack for a given programming language, a number of design issues come in to play. The first of these is whether allocation of variables local to a procedure is static or dynamic. If the language permits recursive calls of a procedure, then each activation of the procedure will have its own set of local variables. In this case, the local variables are allocated dynamically on the runtime stack. Referencing of the local environment is done relative to the stack pointer. On the other hand, if variables local to a procedure are required to remember their values from one activation to the next, then the local variables must be allocated statically.

Static allocation generally permits faster referencing of the local environment. However, dynamic allocation makes better use of memory. Languages such as FORTRAN 77 [Zirkel 1994] and COBOL [Welburn 1995] use static allocation by default. Languages such as C, C++, and Ada use dynamic allocation by default but permit the programmer to mark any specific local variable as static. In the example shown in Figure 100.1, the variable **hello** is allocated statically, whereas the variable **newhello** is allocated dynamically.

A second significant design issue is the method or methods by which actual parameters or arguments are passed from the caller to the called routine. These methods are treated in detail in the next section.

A third significant design issue concerns the degree to which it is possible within a procedure to reference nonlocal variables, that is, variables that are neither passed as parameters nor declared within the procedure. In the C program given in Figure 100.1, the variable **hello** is global to the procedure **bcopy**. Because **hello** is global to the entire compilation unit, the variable **hello** could have been directly referenced within **bcopy**. Such nonlocal access is usually avoided, because it complicates the referencing environment, both for the compiler and for those who must read the source code.

A fourth significant design issue concerns *scoping*, the method used to determine how a reference to a nonlocal name within a procedure is resolved. Programming languages use two types of scoping: *static* and *dynamic*.

Most compiled languages that allow nonlocal references, including Ada, C, and C++, use static scoping. With static scoping, reference to a nonlocal name within a procedure is resolved at compile time by referring to the static parent of the procedure: the statically enclosing block or procedure. This process is repeated as needed until the nonlocal name is found.

Within static scoping, there is the further issue of whether static nesting of procedures is allowed. C and C++ do not allow static nesting, but Ada [Ada 1991, Barnes 1995], Pascal [Jensen et al. 1991], and Modula [Carmony 1990] do. To implement static nesting of procedures, as in Ada and Pascal, it is necessary to add another pointer to the activation record, the *static link*. The runtime system sets the static link of a procedure to point to the activation record of the statically enclosing procedure. In this way, references to nonlocal variables in the procedure will be resolved in a way that is consistent with the static structure of the program. The dynamic link is not adequate to find the activation record of the statically enclosing block when a procedure is called by a program unit that is not its static parent. In C and C++, where static nesting of procedures is not allowed, no static link is needed, because a nonlocal name referenced within a procedure must be global and can be easily located with no additional information.

The LISP [Steele 1990] and APL [Gilman 1992] programming languages use dynamic scoping, in which reference to a nonlocal name within a procedure is resolved at runtime by referring to the activation record of the *dynamic parent* of the procedure, the procedure or block that called the procedure with the nonlocal reference. With dynamic scoping, the program unit whose activation record is consulted to resolve a nonlocal reference cannot be determined statically at compile time, but must determined at runtime through the configuration of the stack of *activation records*.

For a more thorough discussion of static and dynamic scoping and the static link, the interested reader should consult Sebesta [2002], Pratt and Zelkowitz [2001], or Tucker and Noonan [2002].

100.2.3   Parameter Passing

Most programming languages have based their parameter (or argument) passing convention on how parameter passing is actually implemented. In contrast, Ada bases its parameter passing convention on the semantics of the use of the parameter and leaves it to the compiler implementor to pick an appropriate implementation model.

Ada distinguishes three distinct semantic models for parameter passing: (1) *in*, (2) *out*, and (3) *in–out*. A formal parameter specified as *in* means that a value will be supplied to the called routine by the caller. No attempt is made by the called routine to change the value of the parameter. A formal parameter tagged as *out* means that the called routine will set the value of the parameter before any attempt to reference the value of the parameter. Thus, the calling routine must supply a location in which to store the *out* parameter, but need not initialize that location. A formal parameter tagged as *in–out* means that the called routine expects the parameter to be initialized and expects to return a value to the calling routine. Thus, the calling routine must supply an initialized location.

Most older languages defined their parameter passing conventions in terms of how they are implemented. The major ones in use today include (1) *call-by-value*, (2) *call-by-result*, (3) *call-by-value-result*, and (4) *call-by-address* (or *call-by-reference* or *call-by-location*).

In *call-by-value*, the calling routine supplies a value to the called routine. The formal parameter is treated as a local variable that is initialized at the time of the call to the value supplied by the caller. Thereafter, there is no further link between the actual and formal parameter. The called routine is free to change the value of the formal parameter; this has no effect whatever on the actual parameter either during the activation of the called routine or at return time. *Call-by-value* can be thought of as a copy-in process.

In contrast, *call-by-result* can be thought of as a copy-out process. Except for this respect, it is very similar to *call-by-value*. Again, the formal parameter is treated as an uninitialized local variable. It is the responsibility of the called routine to initialize a result parameter before referencing it. When control is about to be returned to the caller, the final value of a result parameter is copied to the corresponding actual parameter. Except for this final copy, there is no other correspondence between the actual and formal parameters during the duration of the call.

A *value-result* parameter is one that combines the two previous methods. The formal parameter is treated during the activation of the called routine as a purely local variable, which is initialized at the time of the call to the corresponding actual parameter, and the final value of the formal parameter is copied back to the actual parameter at the time of the return. Thus, *call-by-value-result* is a copy-in, copy-out process.

In *call-by-address* (or *reference*), the calling routine provides the address of each actual argument. In effect, the formal parameter is effectively a constant pointer to the actual argument. References to the formal parameter are compiled to do a dereferencing so that the location of the actual argument is used in all cases. It is impossible to change the value of the pointer, only what the pointer points to. *Call-by-address* is used by C++ [Stroustrup 1995] for reference parameters, Pascal for **var** parameters, and FORTRAN 77 [Zirkel 1994] for all parameters.

100.2.4   Activation Records

Each time a routine is called, an activation record is created. Although this record may not physically all be stored in the same place, it is still useful to think of it as a single, logical record. Minimally, this record must

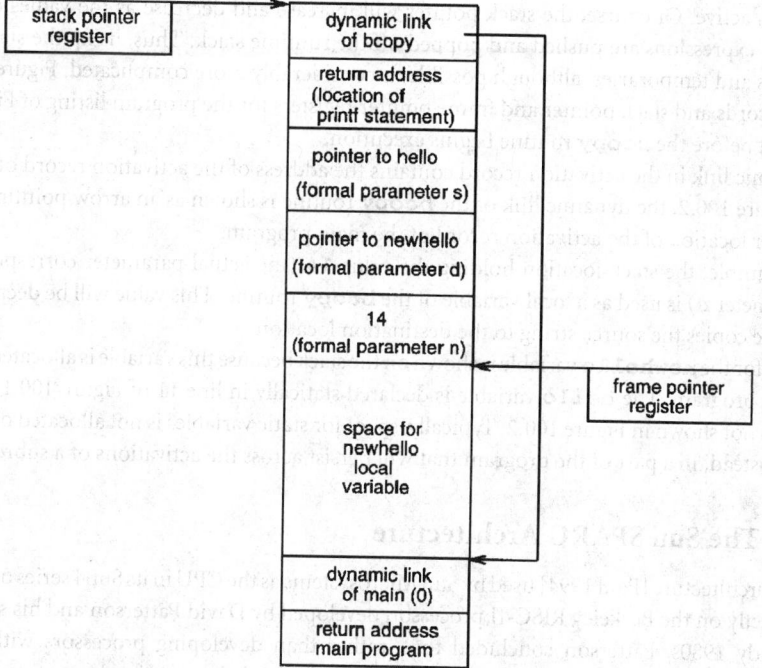

FIGURE 100.2 Activation stack for calling the **bcopy** function.

contain (1) the return address of the calling routine, (2) space for the formal parameters, and (3) space for the local variables of the called routine. An example of an activation record for the function **bcopy** of Figure 100.1 is given in Figure 100.2, which is discussed in detail in the next section.

If the activation record is allocated statically, then variables local to the called routine can remember their values from one activation to the next. This was a feature of early versions of FORTRAN. The disadvantage of static allocation is that such routines cannot be called recursively.

Most modern languages allocate the activation record on the runtime stack, thus allowing recursive routines. C/C++ allow the programmer to specify that some local variables should be allocated statically; by default local variables are allocated dynamically.

At the time of the call, the calling routine must (1) save its own execution status, including any registers it wants saved; (2) carry out the parameter passing process; (3) pass the return address; and (4) finally transfer control to the called routine. In many modern architectures, steps 3 and 4 are combined into a single machine instruction.

At the time of return, first, the called routine must make available the final values of any result or value-result arguments. Second, if the routine is a function, then the computed value of the function must be made available. Next, the execution status of the caller must be restored. And finally, control must be transferred back to the caller.

100.2.5 Activation Stack

Most modern architectures provide some direct support for procedure calls and parameter passing. A typical *activation stack* using the program of Figure 100.1 is given in Figure 100.2.

The figure assumes that execution is on line 7 of the activation of the **bcopy** routine (see Figure 100.1). In many implementations, a frame pointer register points to the base of the activation record on the runtime stack. The stack pointer, of course, points to the logical top of the activation record. Addressing of locals and formal arguments would normally be done relative to the frame pointer, which is constant while the routine is active. Of course, the stack pointer will increase and decrease as the values of temporary variables and expressions are pushed and popped off the runtime stack. Thus, using the stack pointer to address locals and temporaries, although possible, is considerably more complicated. Figure 100.2 shows activation records and stack pointer and frame pointer registers for the program listing of Figure 100.1 at the point just before the **bcopy** routine begins execution.

The dynamic link in the activation record contains the address of the activation record of the routine's caller. In Figure 100.2, the dynamic link of the **bcopy** routine is shown as an arrow pointing back to the frame pointer location of the activation record of the main program.

In this example, the stack location holding the value **14** (the actual parameter corresponding to the formal parameter **n**) is used as a local variable of the **bcopy** routine. This value will be decremented to **0** as the routine copies the source string to the destination location.

The space for the **newhello** variable is shown on the stack because this variable is allocated dynamically in the **main** program. The **hello** variable is declared statically in line 11 of Figure 100.1, so that stack space for it is not shown in Figure 100.2. Typically, space for static variables is not allocated on the runtime stack, but, instead, in a part of the program that will persist across the activations of a subroutine.

100.2.6 The Sun SPARC Architecture

The SPARC architecture [Paul 1994] used by Sun Microsystems as the CPU in its Sun4 series of workstations is based directly on the Berkeley RISC-II processor, developed by David Patterson and his students.

In the early 1980s, Patterson concluded that, rather than developing processors with increasingly complex instruction sets, better performance could be obtained through implementing a reduced instruction set computer, with simpler, more regular machine instructions executing at a higher clock rate and a larger number of registers.

The SPARC CPU uses the reduced instruction set, the LOAD/STORE architecture, pipelined instruction execution, instruction and data caching, the delayed branch, and a large register set with register windows, all present in Berkeley RISC-II design. Sun Microsystems has added several features to the SPARC architecture not present in the Berkeley RISC-II processor.

100.2.6.1 SPARC Register Set

The SPARC register set contains from 48 to 528 32-bit registers. Each register window consists of 24 registers (8 input registers, 8 local registers, and 8 output registers), plus 8 global registers common to all procedures.

One of the innovations in the Berkeley RISC processors was the use of overlapping *register windows* to use in passing parameters in a subroutine call. A sketch of the SPARC register set is shown in Figure 100.3. Each procedure is allocated a group of 32 registers, separated into four groups: *input* registers, *output* registers, *local* registers, and *global* registers. The CPU manipulates the register windows so that the input registers of the called procedure are identical to the output registers of the calling procedure. In this way, the runtime stack is not needed to transmit actual parameters from the caller to the called procedure. This reduces the amount of time required for a procedure call.

The register windows are numbered from the number of windows (**NWINDOWS**) minus 1 down to 0. The SPARC CPU has a processor state register (**PSR**) that contains 4 bits of arithmetic logic unit flags, called the *integer condition codes field*, and a 5-bit field called the *Current Window Pointer* (**CWP**) that holds the number of the current register window being used. Because the output registers of one procedure overlap the input registers of the procedures that it calls, the number of 48 registers allows for two nonoverlapping sets of 16 registers, plus one overlapping set of 8 registers, plus the 8 global registers; that is, a SPARC CPU with 48 registers can implement 2 register windows. A SPARC CPU with 528 registers can implement 32 register windows.

The current window pointer is used by the SPARC CPU to point to the currently active register window. Its value is modified by the programmer through the **save** and **restore** instructions that are discussed subsequently.

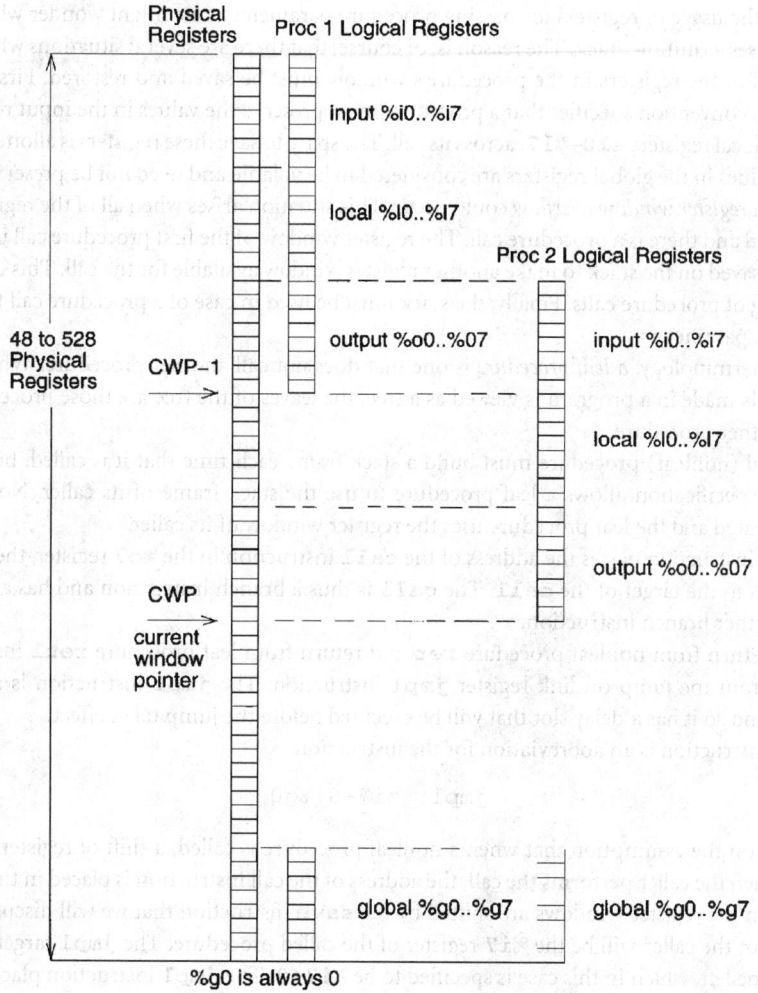

FIGURE 100.3 SPARC register windows.

The SPARC input registers for a given procedure are denoted by **%i0–%i7**, the local registers of the procedure are denoted by **%l0–%l7**, the output registers are denoted by **%o0–%o7**, and the global registers that all procedures can access are denoted by **%g0–%g7**. Several registers are dedicated to special uses. Register **%o6** is the stack pointer register and is usually denoted by **%sp**. The **%i6** register is the procedure's frame pointer register and is denoted by **%fp**. The **%fp** register is actually the caller's stack pointer register because of the register window overlap between the caller's and callee's register windows (see Figure 100.3).

Register **%o7** is the *link register*, from which the return address of a procedure call can be calculated. The **call** instruction places the address of the **call** instruction in **%o7** before jumping to the target of the call. A procedure would place its first six parameters of a procedure call in registers **%o0–%o5** (in left-to-right order, Pascal or C). Parameters after the sixth must be passed on the stack.

100.2.6.2 SPARC Procedure Call Protocol

The SPARC procedure call protocol depends on several factors: the semantics of the call/return instructions, the manipulation of the register windows through the **save** and **restore** instructions, and the layout of the SPARC stack frame.

In view of the usage of registers for passing procedure parameters, one might wonder why the SPARC architecture uses a runtime stack. The reason is, of course, that there are several situations when the values of some or all of the registers in the procedure's window must be saved and restored. First, the SPARC procedure call convention specifies that a procedure must preserve the values in the input registers **%i0–%i7**, and the local registers **%l0–%l7**, across its call. The space to save these registers is allotted in the stack frame. The values in the global registers are considered to be volatile and need not be preserved across the call. Second, a *register window overflow* could occur. This situation arises when all of the register windows have been used and there is a procedure call. The register window of the first procedure call in the chain of calls must be saved on the stack to make another register window available for the call. This can happen in a deep nesting of procedure calls. Finally, the stack must be used in case of a procedure call that has more than six input parameters.

In SPARC terminology, a *leaf procedure* is one that does not call another procedure. When a chain of procedure calls made in a program is viewed as a tree, the leaves of the tree are those procedures that do not call any other procedure.

The normal (nonleaf) procedure must build a stack frame each time that it is called, but the SPARC architectural specification allows a leaf procedure to use the stack frame of its caller. No new register window is created and the leaf procedure uses the register window of its caller.

The **call** instruction places the address of the **call** instruction in the **%o7** register, the link register, then branches to the target of the **call**. The **call** is thus a branch instruction and has a branch delay slot like any other branch instruction.

Both the return from nonleaf procedure **ret** and return from leaf procedure **retl** instructions are synthesized from the jump on link register **jmpl** instruction. The **jmpl** instruction is also a branch instruction, and so it has a delay slot that will be executed before the jump takes effect.

The **ret** instruction is an abbreviation for the instruction

```
jmpl   %i7+8,%g0
```

This is based on the assumption that when a nonleaf procedure is called, a shift of register windows has occurred. When the caller performs the call, the address of the call instruction is placed in the caller's **%o7** register. When the register windows are shifted by the **save** instruction that we will discuss shortly, the **%o7** register of the caller will be the **%i7** register of the called procedure. The **jmpl** target is the source operand on the left, which in this case is specified to be **%i7+8**. The **jmpl** instruction places the address of the **jmpl** instruction in the destination register, which in this case is the **%g0** register. Because the **%g0** is permanently zero, its use here indicates that the address of the **jmpl** instruction should not be saved.

The **retl** is for returning from a leaf procedure. It is an abbreviation for the instruction

```
jmpl   %o7+8,%g0
```

This is consistent with the view that the no shift of register windows has occurred with a leaf procedure. The leaf procedure is using the register window of its caller, and so the address of the **call** instruction is expected to be in the **%o7** register where the **call** instruction placed it.

When a nonleaf procedure performs a **call**, the nonleaf procedure does not need to know whether it is calling a leaf procedure or a nonleaf procedure. The problem is resolved by the called procedure's choice between the **ret** and **retl** instructions.

100.2.6.3   SPARC save and restore Instructions

The **save** instruction is used as the first instruction of a nonleaf procedure to build its stack frame. As mentioned previously, the stack frame will be used to save registers that must be preserved across the procedure call, in the case of register window overflow or in case the procedure calls a procedure having more than six input parameters.

The SPARC stack frame is shown in Figure 100.4. The smallest allowable stack frame that a nonleaf procedure can allocate is 92 bytes. This includes 64 bytes to store the procedure's input and local registers, plus a 6-word (24-byte) storage area into which any callee of the procedure can store arguments, and a

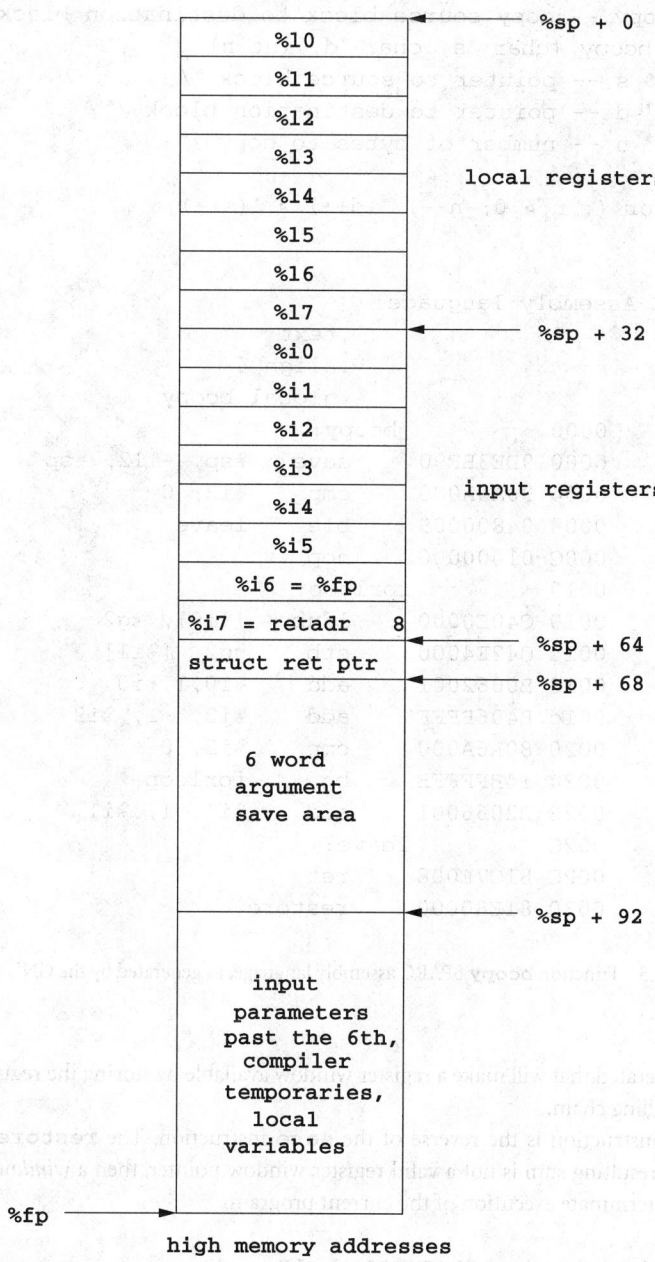

low memory addresses

%l0	← %sp + 0
%l1	
%l2	
%l3	
%l4	local registers
%l5	
%l6	
%l7	
%i0	← %sp + 32
%i1	
%i2	
%i3	input registers
%i4	
%i5	
%i6 = %fp	
%i7 = retadr 8	← %sp + 64
struct ret ptr	← %sp + 68
6 word argument save area	
	← %sp + 92
input parameters past the 6th, compiler temporaries, local variables	

high memory addresses

FIGURE 100.4 SPARC stack frame.

1-word (4-byte) location to contain a pointer to an *aggregate return value*. This pointer would be used by a function whose return value was a **struct** or **RECORD** that did not fit in 32 bits.

The **save** instruction for a nonleaf procedure using a minimal stack frame would be

```
save    %sp,-92,%sp
```

The **save** instruction subtracts one from **CWP**, the current register window pointer. If the result of this subtraction is a valid register window number, then the new value is stored into **CWP**; otherwise, a *window*

```
C source:
/* bcopy -- copy source block to destination block */
void bcopy (char *s, char *d, int n)
    /* s -- pointer to source block */
    /* d -- pointer to destination block /*
    /* n -- number of bytes to copy /*
{
    for (; n > 0; n--) *(d++) = *(s++);
}

SPARC Assembly language:
                            .text
                            .align 4
                            .global_bcopy
        0000        _bcopy:
        0000 9DE3BF90    save    %sp, -112, %sp
        0004 90A6A000    cmp     %i2, 0
        0008 04800009    ble     leave
        000C 01000000    nop
        0010        forloop:
        0010 C40E0000    ldub    [%i0], %g2
        0014 C42E4000    stb     %g2, [%i1]
        0018 B0062001    add     %i0,1,%i0
        001c B406BFFF    add     %i2, -1, %i2
        0020 80A6A000    cmp     %i2, 0
        0024 14BFFFFB    bg      forloop
        0028 B2066001    add     %i1, 1, %i1
        002C        leave:
        002C 81C7E008    ret
        0030 81E80000    restore
```

FIGURE 100.5 Function **bcopy** SPARC assembly language, as generated by the GNU C compiler.

overflow trap is generated that will make a register window available by storing the registers of the earliest procedure in the calling chain.

The **restore** instruction is the reverse of the **save** instruction. The **restore** instruction adds one to **CWP**. If the resulting sum is not a valid register window pointer, then a *window underflow* trap is generated that will terminate execution of the current program.

100.2.6.4 Function bcopy as a SPARC Nonleaf Procedure

The source code and the assembly language code for **bcopy** generated by the GNU C compiler is shown in Figure 100.5. Notice that because of the **save**, **restore**, and **ret** instructions, the procedure is cast as a nonleaf procedure, even though it does not call any other procedure. This is a conservative decision, apparently based on the belief that every procedure should have a stack frame. This sort of conservative design is perhaps one reason why the code that the GNU C generates is so robust.

The parameters of the call are in the input registers now: **%i0** is the **s** parameter, **%i1** is the **d** parameter, and **%i2** is the **n** parameter of the call. The GNU compiler was not able to fill the branch delay slot of the **ble** branch at location **000C** with a useful instruction, so it inserted a no-operation **nop** instruction there. Notice that the **add** instruction at location **0028** that accomplishes the **d++** of the C program is executed in the delay slot of the **bg** instruction at location **0024**. This ensures that the **d** formal parameter has the correct value if the loop is executed again. Notice also that the **restore** instruction at location **0030** is located in the delay slot of the **ret** instruction, so that the **restore** is executed to restore the register window of the caller before the **ret** returns control to the caller.

100.3   Pointers and Heap Management

In addition to space on the runtime stack, modern languages such as Ada, C, C++, and Java also provide for allocating space dynamically from heap memory. Such space is commonly used for arrays whose size is determined dynamically at runtime and for various linked structures such as linked lists, trees, and graphs. A variable that contains a heap memory reference literally contains a memory address and is commonly referred to as a pointer*.

A simple linked list would consist of nodes, each of which would contain a value field and a pointer to the next node in the list. Using an integer value field as an example, such a node would be defined in C++:

```
struct Node{
    int value;
    Node* next;
};
```

Heap allocation is provided by the new operator, which in this case allocates heap space sufficient to hold an integer and a pointer and returns the address of the space allocated. Given the following auxiliary function:

```
Node* mkNode(int val, Node* nxt) {
    Node* p = new node;
    p->value = val;
    p->next = nxt;
    return p;
}
```

a linked list containing a sequence of integers read from standard input can be constructed as follows:

```
Node* list = NULL;
int x;
while (cin >> x) {
    list = mkNode(x, list);
}
```

This code snippet stores the integers in the reverse order read. A common error in programming linked data structures is to forget to set the last pointer in the list to the NULL value. In this case, the error would occur if the variable list had not been initialized to NULL.

Suppose the program snippet above read the sequence of integers {5, 3, 2} and the program logic called for removing the node containing the 3 from the list. One way to accomplish this task is:

```
list->next = list->next->next;
```

This results in the situation shown in Figure 100.6, in which the second node of the list is no longer accessible to the program. Any allocated memory block that is inaccessible to the program is denoted as *garbage.*

---

*Note that in C/C++ and Ada, a pointer may also contain a non-heap memory address, for example, a runtime stack address.

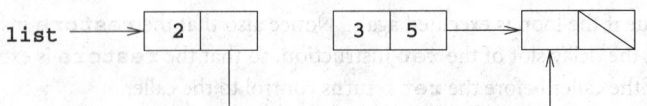

FIGURE 100.6   A linked list with a deleted node.

Unlike the program stack, heap space does not usually become unusable in the reverse order space is allocated. C++ provides an explicit operator named `delete` for explicitly returning unused space to the heap. The above code snippet for removing the second node of the list and returning it to the heap can be rewritten:

```
Node* p = list->next;
list->next = p->next;
delete p;
```

Failure to return unused memory to the heap is termed a *memory leak*. Such memory leaks can be a problem in long-lived programs such as Web browsers and Web servers.

Even more problematic in a large, complicated program is determining whether or not a node is truly garbage or is still accessible via some other reference. Returning memory space to the heap via a `delete` that is still accessible via another reference makes the latter a *dangling reference*. In the code snippet above, after the `delete`, any pointer containing the same value as p is a dangling reference; in C++ the `delete` operator sets the pointer to `NULL`, but other references to the node may still exist. For a more formal treatment of heap management and the `new` and `delete` operators, see Chapter 5 of Tucker and Noonan [2002].

Pointers and heap memory management are especially difficult and have made programming in C/C++ and Pascal unreliable. Newer languages such as Java have attempted to partially remedy this situation by removing the burden of reclaiming unused heap space via automatic garbage collection, which is the subject of the next section.

100.4   Garbage Collection

Automatic garbage collection has been a hallmark of functional languages since the introduction of Lisp in the early 1960s. Automatic garbage collection has traditionally been viewed as too slow for use in conventional programming languages. However, newer scripting languages such as Perl and Python and object-oriented languages such as Smalltalk and Java have all included automatic garbage collection as one of their features.

In Smalltalk and Java, all objects are allocated out of the heap. As a program executes, many objects become inactive and are no longer accessible to the program. Reclaiming the space used by these inaccessible objects, or garbage, is the task of the garbage collector.

The garbage collection algorithm is automatically applied as needed to reclaim inaccessible heap space. Modern algorithms are derived from similar work for functional languages. We examine the three major strategies known as: *reference counting*, *mark-sweep*, and *copy collection*.

100.4.1   Reference Counting

In reference counting, each node allocated out of the heap has an extra, hidden field that contains a count of the number of direct references to the node. When a node is initially allocated via a `new` operation, the count is initialized to zero.

The semantics of a heap pointer assignment are modified as follows. Consider the assignment:

```
p = q;
```

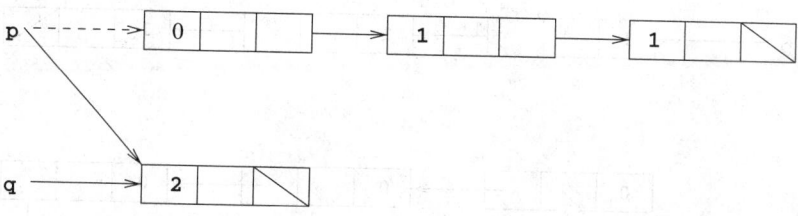

FIGURE 100.7   Garbage collection using reference counting.

where p and q are both pointers to the heap. Normally, the address stored in q is simply copied to p. However, to keep the reference counts updated, the node referenced by p before the assignment must have its count decremented by one and the node referenced by q must have its count incremented by one. A complication is that on exiting a program function, all local pointers are effectively set to NULL, thus having the counts in the referenced nodes decremented by one.

When a node has its counts decremented to zero, it is returned as free space to the heap manager. This is the situation depicted in Figure 100.7, in which the dotted line represents the value of p before the above assignment is executed. The first node in the list has a reference count of zero, and is thus freed. This causes the reference count in the second node to also be set to zero, and is thus also freed. The process repeats itself for each node in the list originally pointed at by p until the entire list is freed.

However, imagine the situation where p's list was circular, that is, the last node in the list pointed to the first node. Then, after executing the above assignment, the reference count in the first node would be one (not zero), because it still has a direct reference, namely, the last node in the list. The list, however, would still be garbage but the reference counting method would not detect this fact.

This simple example serves to illustrate both the strengths and weaknesses of the method of garbage collection using reference counts. A major strength is that the time spent garbage collecting is distributed across the program execution, occurring naturally whenever a pointer assignment is made or a program scope is closed. A disadvantage is the storage cost of the reference count associated with each node. A more serious weakness is the inability to free nodes that occur in circular lists and graphs.

In contrast to reference counting, the remaining two algorithms are invoked only when a new operation would fail because there is insufficient free space remaining in the heap. In the next section we examine the *mark-sweep* algorithm.

100.4.2 Mark-Sweep

The mark-sweep algorithm is basically a two-pass algorithm. In the first pass, the references from each program pointer are traced, marking each node reached as accessible. In the second pass, each heap node or block is examined; if its mode bit is marked inaccessible, then it is returned to the free space list. The mark-sweep algorithm is invoked when a heap space request via a new operation cannot be satisfied.

As with reference counting, each heap node has an extra hidden field, which is used to keep track of whether the node is accessible to the program. Thus, the extra space required is logically only a single bit, denoted the mark bit. Each memory block in the free space and each allocated node has its mark bit set to inaccessible or zero.

When the mark pass is begun, it follows each chain from every active program pointer variable, changing the mark bit of all heap nodes visited from inaccessible (zero) to accessible (one). The end of a mark pass is depicted in Figure 100.8; the nodes in the list referenced by p are marked accessible, while the others remain inaccessible.

The sweep pass references every node or memory block in the heap. Any such node whose mark bit is set to inaccessible is returned to the free space list. The accessible nodes have their mark bit reset to inaccessible. As can be seen from Figure 100.8, circular chains or cycles present no problem.

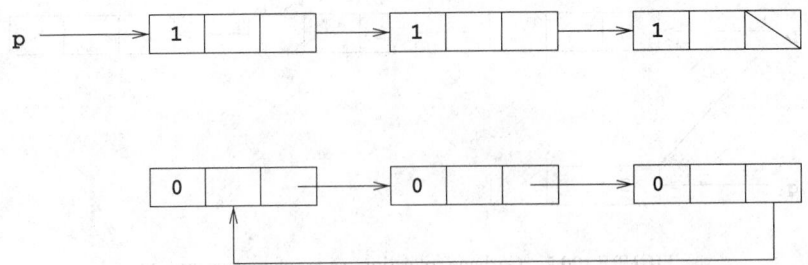

FIGURE 100.8   End of mark pass of mark-sweep algorithm.

Once the sweep pass is complete and garbage has been returned to the free space list, the attempt to complete the new operation is again attempted. If the free space is still insufficient, the new operation terminates in error, the details of which are language dependent.

The strength and weaknesses of mark-sweep are the exact opposite of reference counting. The major weakness of mark-sweep is the fact that garbage collection is postponed until the free space in the heap is exhausted. This can result in a noticeable interruption in the program execution while the garbage collection algorithm is being run. On the other hand, all garbage is identified and returned to the free space list. Another advantage of mark-sweep is that the storage overhead needed is only a single bit per node.

100.4.3   Copy Collection

An alternative to the mark-sweep algorithm is the copy collection algorithm. The latter represents a time/space compromise relative to mark-sweep.

For copy collection, the heap is divided in half, termed the *active space* and the *inactive space*. A free space list is maintained for the active space.

As with mark-sweep, requests for space are made out of the active space until a space request cannot be satisfied. At that point, the copy collection algorithm is invoked to garbage collect inaccessible memory blocks.

As with mark-sweep, pointers are traced from every active program variable. Instead of being marked, the accessible nodes are copied from the active space to the inactive space; the associated pointers are updated to reflect this copying. At the end of the copy pass, all the active space is freed and the roles of the active and inactive spaces are reversed. One of the complications of the algorithm is that it must keep a list of all nodes that have been moved from the active space to the inactive space. Then when a reference is found to a moved node, that reference is updated, rather than (incorrectly) moving the node again.

The copy collection algorithm involves a classic time/space trade-off versus the mark-sweep algorithm. By dividing the heap in half, garbage collection will be invoked more frequently. On the other hand, copy collection is considerably faster because it makes only a single pass over the heap nodes. An extensive discussion of the efficiency of these garbage collection strategies can be found in Jones [1996].

100.5   Summary

In this chapter the memory management principles of the runtime stack have been discussed; the application of these principles to the Sun SPARC architecture has been shown. An understanding of these principles and their application to a specific architecture is important in developing debuggers, working with multiple threads, and developing other tools that must interact with the runtime environment. We have further discussed memory management of the heap and examined common garbage collection algorithms.

One of the important areas of research is better garbage collection algorithms. Another area is *points-to analysis*, which determines where various pointers point. This technique is used both for better memory usage and for identifying memory leaks.

Another continuing area of research in runtime environments is the development of compilers and related tools for both massively parallel and distributed computer systems. The goal is to develop languages and compilers so that programming systems can be developed and debugged on traditional single-CPU workstations and then recompiled and run on either a massively parallel computer system or a distributed computer system. This approach is particularly attractive for the so-called grand challenge problems of science. Wolfe [1996] is one text in this area.

Meek [1995] discusses the work of the International Standard Organization (ISO) in attempting to standardize the notion of a procedure call and parameter passing even in a distributed environment.

Defining Terms

Activation record: A record containing all the information associated with an activation or call of a procedure or function. This information includes the return address of the caller, the procedure's parameters and local variables, and the frame pointer of the caller.

Activation stack: A stack of activation records, one for each active procedure call.

Actual parameter or argument: A parameter that appears in a call of the procedure or function.

Call-by-address parameter: A method of parameter transmission whereby the address of the actual parameter is copied to the formal parameter at the time of the call. The formal parameter is effectively a constant pointer variable. Any reference to the formal parameter is treated as a reference to the actual parameter. Also known as call-by-reference or call-by-location.

Call-by-result parameter: A method of parameter transmission whereby the value of the formal parameter is copied back to the actual parameter at the time of the return. Prior to executing the return statement, there is no correspondence between the actual and formal parameters.

Call-by-value parameter: A method of parameter transmission whereby the value of the corresponding actual parameter is copied to the formal parameter at the time of the call. Thereafter, there is no correspondence between the actual and formal parameters. In particular, changes to the formal parameter have no effect on the actual parameter.

Call-by-value-result parameter: A method of parameter transmission that combines call-by-value and call-by-result.

Dangling reference: A pointer that still references a memory block that has been been returned to the heap as free space.

Formal parameter or argument: A parameter name that appears in the declaration or header of a procedure or function.

Frame pointer: A register that normally points to the base or beginning of the activation record on the top of the activation stack.

Garbage collecton: The process of collecting portions of the heap that are no longer referenced by the program.

Heap: The portion of memory assigned to an executing program to use for dynamic memory allocation.

Leaf procedure: A procedure that does not call another procedure.

Memory leak: Failure to return unused space to the heap.

Register window: A collection of registers assigned to an executing process in the SPARC architecture.

Stack frame: The activation record on the runtime stack for a method or function.

Stack pointer: A register that points to the top of the activation stack.

References

Ada 1991. *The Annotated Ada Reference Manual*, 2nd ed. Grebyn, Vienna, VA.

Barnes, J. G. P. 1995. *Programming in Ada 95*. Addison-Wesley, Reading, MA.

Gilman, L. 1992. *APL, an Interactive Approach*. Krieger, Malabar, FL.

Goldberg, A. 1985. *Smalltalk-80: the Language and its Implementation*. Addison-Wesley, Reading, MA.

Jones, R. R. and Lins, R. 1996. *Garbage Collection*. John Wiley, New York.

Kernighan, B. W. and Ritchie, D. M. 1988. *The C Programming Language*, 2nd ed. Prentice Hall, Englewood Cliffs, NJ.

Louden, K. C. 1993. *Programming Languages: Principles and Practice*. PWS-Kent, Boston, MA.

MacLennan, B. J. 1987. *Principles of Programming Languages*, 2nd ed. Holt, Rinehart and Winston, New York.

Meek, B. L. 1995. What is a procedure call? *SIGPLAN Notices*, 30(9):33–40.

Motorola. 1993. *PowerPC 601: RISC Microprocessor's User's Manual*. Motorola, Phoenix, AZ.

Paul, R. P. 1994. *SPARC Architecture Assembly Language Programming and C*. Prentice Hall, Englewood Cliffs, NJ.

Pratt, T. W. and Zelkowitz, M. V. 2001. *Programming Languages Design and Implementation*, 4th ed. Prentice Hall, Englewood Cliffs, NJ.

Sankaran, N. 1994. Bibliography on garbage collection and related topics. *SIGPLAN Notices*, 29(9):149–158.

Sebesta, R. W. 2002. *Concepts of Programming Languages*, 5th ed. Addison-Wesley, Reading, MA.

Steele, G. L. 1990. *Common LISP: the Language*. Digital Press, Bedford, MA.

Stroustrup, B. 1995. *The C++ Programming Language*, 2nd ed., reprinted with corrections. Addison-Wesley, Reading, MA.

Tucker, A. and Noonan, R. 2002. *Programming Languages: Principles and Paradigms*. McGraw-Hill, New York.

Welburn, T. 1995. *Structured COBOL: Fundamentals and Style*. McGraw-Hill, New York.

Wolfe, M. 1996. *High Performance Compilers for Parallel Computing*. Addison-Wesley, Reading, MA.

Zirkel, G. 1994. *Understanding FORTRAN 77 and 90*. PWS, Boston, MA.

Further Information

The *ACM Transactions on Programming Languages and Systems* (TOPLAS) is probably the leading theoretical journal in the area of run time environments and memory management. Many important papers in the areas of compiler optimization and points-to analysis are published here, as well as various ACM conferences.

A more applied journal is *Software Practice & Experience*, which is aimed at the practitioner rather than the theoretician. The scope of this journal (as stated on the inner cover) is "practical experience with new and established software for both systems and applications." Thus, although the scope is broader, techniques of various architectural features are illustrated. One example might be the simulation of the improvement gained in execution speed by adding register windows to a given architecture.

The journal *Computer Languages* sits squarely between the previous two, in that it contains both theoretical and applied papers, but its scope is more narrowly focused.

Finally, the *Journal of Systems and Software* includes both research papers as well as reports on the state of the art and practical experience. Topic areas include programming methodology and related hardware-software issues, including programming environments.

XI

Software Engineering

The subject area of Software Engineering is concerned with the application of a systematic approach to the analysis, design, implementation, deployment, and evolution of software. There are various models of the software process and many methodologies for the different phases of the process. Also, the process often involves the use of tools and techniques, such as software architecture and formal methods, to build computing systems that are correct, reliable, extensible, reusable, secure, efficient, and easy to use. This section provides a modern treatment of software engineering, both from the technical point of view and from the project management point of view.

101 Software Qualities and Principles   *Carlo Ghezzi, Mehdi Jazayeri, and Dino Mandrioli* **101**-1
Classification of Software Qualities • Representative Software Qualities • Quality Assurance • Software Engineering Principles in Support of Software Quality • A Case Study in Compiler Construction • Summarizing Remarks

102 Software Process Models   *Ian Sommerville* **102**-1
Introduction • Specification-Driven Models • Evolutionary Development Models • Iterative Models • Formal Transformation • The Cleanroom Process • Process Model Applicability • Research Issues and Summary

103 Traditional Software Design   *Steven A. Demurjian Sr.* **103**-1
Introduction • The High-Tech Supermarket System • Traditional Approaches to Design • Design by Encapsulation and Hiding • Mathematical and Analytical Design

104 Object-Oriented Software Design   *Steven A. Demurjian Sr. and Patricia J. Pia* ... **104**-1
Introduction • Object-Oriented Concepts and Terms • Choosing Classes • Inheritance: Motivation, Usage, and Costs • Categories of Classes and Design Flaws • The Unified Modeling Language • Design Patterns

105 Software Testing   *Gregory M. Kapfhammer* **105**-1
Introduction • Underlying Principles • Best Practices • Conclusion

106 Formal Methods   *Jonathan P. Bowen and Michael G. Hinchey* **106**-1
Introduction • Underlying Principles • Best Practices • A Case Study • Technology Transfer and Research Issues • Glossary of Z Notation

107 Verification and Validation   *John D. Gannon* **107**-1
Introduction • Approaches to Verification • Verifying Specifications of Systems • Verifying Programs • Current Status

108 Development Strategies and Project Management   *Roger S. Pressman* **108**-1
Development Strategies • The Management Spectrum • Software Project Management • Software Quality Assurance • Software Configuration Management • Summary

109 Software Architecture *Stephen B. Seidman* **109**-1
    Introduction • Underlying Principles • Describing Individual
    Software Architectures • Architecture Description Languages
    • Describing Architectural Styles

110 Specialized System Development *Osama Eljabiri and Fadi P. Deek* **110**-1
    Introduction • Principles of Specialized System Development • Application-Based
    Specialized Development • Research Issues and Summary

101

Software Qualities and Principles

101.1 Classification of Software Qualities **101**-2
Product and Process Qualities • External vs. Internal
Qualities

101.2 Representative Software Qualities **101**-3
Correctness, Reliability, and Robustness • Performance
• Usability • Verifiability • Security • Maintainability
• Reusability • Portability • Understandability
• Interoperability • Productivity • Timeliness
• Visibility

101.3 Quality Assurance............................... **101**-13

101.4 Software Engineering Principles in Support
of Software Quality **101**-14
Rigor and Formality • Separation of Concerns
• Modularity • Abstraction • Anticipation of Change
• Generality • Incrementality

101.5 A Case Study in Compiler Construction **101**-21
Rigor and Formality • Separation of Concerns
• Modularity • Abstraction • Anticipation of Change
• Generality • Incrementality

101.6 Summarizing Remarks **101**-24

Carlo Ghezzi
Politecnico di Milano

Mehdi Jazayeri
Technical University of Vienna

Dino Mandrioli
Politecnico di Milano

The goal of any engineering activity is to build something — an artifact or a product. The civil engineer builds a bridge, the aerospace engineer builds an airplane, and the electrical engineer builds a circuit. The product of software engineering is a software application or software system. It is not as tangible as the other products, but it is a product nonetheless. It serves a function.

In some ways, software products are similar to other engineering products, and in other ways they are very different. The characteristic that perhaps sets software apart from other engineering products the most is that software is *malleable*. We can modify the product itself — as opposed to its design — rather easily. This makes software quite different from other products, such as cars or ovens.

The malleability of software is often misused. Even though it is possible to modify a bridge or an airplane to satisfy some new need — for example, to make the bridge support more traffic or the airplane carry more cargo — such a modification is never taken lightly and certainly is not attempted without first making a design change and verifying the impact of the change extensively. Software engineers, on the other hand, are often asked to perform such modifications on software. Software's malleability sometimes leads people to think that it can be changed easily. In practice, it cannot.

We may be able to change code easily with a text editor, but meeting the need for which the change was intended is not necessarily done so easily. Indeed, we must treat software like other engineering products in this regard. A change in software must be viewed as a change in the design rather than in the code, which is just an instance of the product. We can exploit the malleability property, but we must do so with discipline.

Another characteristic of software is that its creation is *human intensive*. It requires mostly engineering, rather than manufacturing, resources. In most other engineering disciplines, the manufacturing process determines the final cost of the product. Also, the process must be managed closely to ensure that defects are not introduced into the product, for example, through the use of faulty parts. The same considerations apply to computer hardware products. For software, on the other hand, "manufacturing" is a trivial process of duplication. The software production process deals with design and implementation, rather than with manufacturing. This process must meet certain criteria to ensure the production of high-quality software.

Any product is expected to fulfill some need and meet some acceptance standards that dictate the qualities it must have. A bridge performs the function of making it easier to travel from one point to another; one of the qualities it is expected to have is that it will not collapse when the first strong wind blows or when a convoy of trucks travels across it. In traditional engineering disciplines, the engineer has tools for describing the qualities of the product distinctly from those of the design of the product. In software engineering, the distinction is not so clear. The functional requirements of the software product are often intermixed in specifications with the qualities of the design.

To achieve the desired qualities, the construction of any nontrivial product must follow sound *design principles*. These principles apply to any engineering discipline, including software engineering. In applying such principles to software engineering, they must be customized to deal with the peculiar characteristics of software.

Software engineering principles deal with both the **software process** of software engineering and the final **software product**. The right process helps to produce the right product, but the desired product also affects the choice of which process to use.

In this chapter, we first examine the qualities that are relevant to software products and software production processes (Section 101.1 and Section 101.2) and the means to assess them (Section 101.3). In Section 101.4, we present general principles that may be applied throughout the process of software construction and management in order to achieve the desired qualities of software products. Finally, Section 101.5 presents a case study of the use of software engineering principles for compiler development.

Throughout the chapter, we assume that the software product to be developed is large and complex. The application of engineering principles is indispensable in the development of such products. In general, the choice of principles and techniques is determined by the software quality goals. Software for critical applications, where the effects of errors are serious, even disastrous, imposes stricter reliability requirements than noncritical applications.

Principles, however, are not sufficient. In fact, they are general and abstract statements describing desirable properties of software processes and products. To apply principles, the software engineer uses appropriate methods and specific techniques that help to incorporate the desired properties into processes and products. Sometimes, methods and techniques are packaged together to form a **methodology**. The purpose of a methodology is to promote a certain approach to solving a problem by preselecting the methods and techniques to be used. Some important software design methods will be the issue of specific chapters of this Handbook. Many methodologies are supported by software tools.

101.1 Classification of Software Qualities

There are many desirable software qualities. Some apply both to the product and to the process used to produce the product. The user wants the software product to be reliable, efficient, and easy to use. The producer of the software wants it to be verifiable, maintainable, portable, and extensible. The manager of the software project wants the process of software development to be productive, predictable, and easy to control. In this section, we consider two different classifications of software-related qualities: internal vs. external and product vs. process.

101.1.1 Product and Process Qualities

We use a process to produce the software product. We distinguish between qualities that apply to products and those that apply to processes. For example, we may want the product to be user-friendly and we may want the process to be efficient. Often, however, process qualities are closely related to product qualities. For example, if the process requires careful planning of test data before any design and development of the system starts, product reliability will increase. Moreover, there are qualities, such as efficiency, that can refer both to the product and the process.

It is useful to examine the word *product* here. It usually refers to what is delivered to the customer. Even though this is an acceptable definition from the customer's perspective, it is not adequate for the developer who produces a number of intermediate products in the course of the software process. The customer-visible product consists perhaps of the executable code and the user manual, but the developers produce a number of other artifacts, such as requirements and design documents, test data, etc. We refer to these intermediate products as *work products* or artifacts to distinguish them from the end product delivered to the customer. Work products are often subject to similar quality requirements as the end product. Given the existence of many work products, it is possible to deliver different subsets of them to different customers.

For example, a computer manufacturer might sell to a process control company the object code to be installed in the specialized hardware for an embedded application. It might sell the object code and the user's manual to software dealers. It might even sell the design and the source code to software vendors, who modify them to build other products. In this case, the developers of the original system see one product, the salespersons in the same company see a set of related products, and the end user and the software vendor see still other, different products.

Process quality has received increasing attention over time, as its impact on product quality has been recognized and observed.

101.1.2 External vs. Internal Qualities

We can divide software qualities into external and internal qualities. *External qualities* are visible to the users of the system; *internal qualities* concern the developers of the system. In general, users of software care only about external qualities, but it is the internal qualities — which deals largely with the structure of the software — that help developers achieve the external qualities. For example, the internal quality of verifiability is necessary for achieving the external quality of reliability. In many cases, however, the qualities are closely related, and the distinction between internal and external may depend on the user and the delivered product. For instance, a well documented design is usually an internal quality; in some cases, however, users want the delivery of design documentation as an essential part of the product (e.g., for military products). In this case, this quality becomes external.

101.2 Representative Software Qualities

In this section, we present the most important qualities of software products and processes. Where appropriate, we analyze a quality with respect to the classifications discussed in Section 101.1. Several software quality models are proposed in the literature. Most of them share the essential aspects but differ in emphasis and organization. ISO 9126 states, "The maturity of the models, terms, and definitions does not yet allow them to be included in a standard" [ISO 9126]. Furthermore, some application areas emphasize certain qualities and may require more specialized qualities. We give some examples of such qualities when we describe specific qualities.

101.2.1   Correctness, Reliability, and Robustness

The terms **correctness**, **reliability**, and **robustness** are related and collectively characterize a quality of software which implies that the application performs its functions as expected. In this section, we define these three terms and discuss their relationships to one another.

101.2.1.1   Correctness

A program is written to provide functions specified in its functional requirements specifications. Often, there are other requirements — such as performance and scalablity — that do not pertain to the functions of the system. We call these kinds of requirements *nonfunctional requirements*. A program is *functionally correct* if it behaves according to its stated functional specifications. It is common simply to use the term *correct* rather than *functionally correct*; similarly, in this context, the term *specifications* implies *functional requirements specifications*. We shall follow this convention when the context is clear.

The definition of correctness assumes that specifications for the system are available and that it is possible to determine unambiguously whether a program meets the specifications. Such specifications rarely exist for most current software systems. If a specification does exist, it is usually written in an informal style using natural language. Therefore, it is likely to contain many ambiguities. Regardless of these difficulties with current specifications, however, the definition of correctness is useful because it captures a desirable goal for software systems.

Correctness establishes the equivalence between the software and its specification. Obviously, we can be more systematic and precise in assessing correctness, depending on how rigorous we are in specifying functional requirements. As we shall see in Chapter 113, we may assess the correctness of a program through a variety of methods, some based on an experimental approach (e.g., testing), others based on an analytic approach (e.g., inspections or formal verification of correctness). Correctness can be enhanced by using appropriate tools, such as high-level languages — particularly those supporting extensive static analysis. Likewise, correctness can be improved by using standard proven algorithms or libraries of standard modules, rather than inventing new ones. Finally, correctness can be enhanced by using proven methodologies and processes.

101.2.1.2   Reliability

Informally, software is reliable if the user can depend on it; in fact, the attribute *dependable* is also used in such a case. The specialized literature on software reliability, however, defines reliability in terms of statistical behavior — the probability that the software will operate as expected over a specified time interval. According to this definition, reliability is a more specialized — and measurable — quality than the generic term dependability. For the purpose of this chapter, however, the informal definition is sufficient, and we consider the two terms as synonymous.

Correctness is an absolute quality: any deviation from its requirements makes a system incorrect, regardless of how minor or serious the consequence of the deviation is. The notion of reliability is, on the other hand, relative. If the consequence of a software error is not serious, the incorrect software may still be reliable.

Engineering products are *expected* to be reliable. Unreliable products, in general, disappear quickly from the marketplace. Unfortunately, software products have not yet achieved this enviable status of reliability. Software products are commonly released along with a list of known bugs. Users of software take it for granted that Release 1 of a product is "buggy." This is one of the most striking symptoms of the immaturity of the software engineering field as an engineering discipline.

In classic engineering disciplines, a product is not released if it has bugs. You do not expect to take delivery of an automobile along with a list of shortcomings or a bridge with a warning not to lean over the railing. In the classic disciplines, design errors are extremely rare and generate news headlines. A collapsed bridge may even cause the designers to be prosecuted in court.

In contrast, software design errors are generally treated as unavoidable. Far from being surprised when we find software errors, we *expect* them. Instead of a guarantee of reliability, with software we get a disclaimer

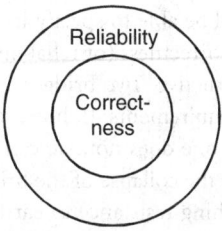

FIGURE 101.1   Idealized relationship between correctness and reliability.

that the manufacturer is not responsible for any damages due to product errors. Software engineering is still trying to achieve software reliability comparable to the reliability of other products.

Figure 101.1 illustrates the relationship between reliability and correctness, under the assumption that the functional requirements specification indeed captures all the desirable properties of the application and that no undesirable properties are erroneously specified in it. The figure shows that the set of all reliable programs includes the set of correct programs, but not vice versa. Unfortunately, things are different in practice. In fact, the specification is a model of what the user wants, but the model may or may not be an accurate statement of the user's actual requirements. All the software can do is meet the specified requirements of the model; it cannot assure the accuracy of the model.

Thus, Figure 101.1 represents an idealized situation wherein the requirements are themselves assumed to be correct; that is, they are a faithful representation of what the implementation must ensure in order to satisfy the needs of the expected users. Often, however, there are insurmountable obstacles to achieving this goal. The upshot is that we sometimes have correct applications designed for "incorrect" requirements, so that correctness of the software may not be sufficient to guarantee the user that the software behaves as expected. This situation is discussed in Section 101.2.1.3.

Reliability has several subcomponents that assume importance depending on the application domain. Three qualities that contribute to reliability are fault-tolerance, availability, and safety. *Fault-tolerance* refers to the ability of the software to detect faults in the hardware environment, such as transient power failures, and continue operation in the face of them. *Availability* refers to the property of the system that ensures that the system's services are available to users close to 100% of the time. Thus, a highly available system must not only be able to deal with faults and errors; it also must allow system diagnostic and maintenance procedures to be carried out while the system is in normal operations. Systems such as telephone systems and banking systems must be both fault-tolerant and highly available. High availability and fault tolerance are usually accomplished by distributed architectures that use redundant resources.

In the case of highly critical systems, the term *safety* is often used to denote the absence of undesirable behaviors that can cause system hazards. Safety deals with requirements other than the primary mission of a system and requires that the system execute without causing unacceptable risk. Unlike functional requirements, which describe the intended correct behavior in terms of input–output relationships, safety requirements describe what should *never* happen while the system is executing. In some sense, they are negative requirements: they specify the states the system must never enter. For example, an X-ray medical system must observe the safety property that the radiation it applies never exceeds a certain threshold.

101.2.1.3   Robustness

A program is robust if it behaves "reasonably," even in circumstances that were not anticipated in the requirements specification — for example, when it encounters incorrect input data or some hardware malfunction (say, a disk crash). A program that assumes perfect input and generates an unrecoverable run-time error as soon as the user inadvertently types an incorrect command is not robust. It might be correct, though, if the requirements specification does not state what the program should do in the face of incorrect input. Obviously, robustness is difficult to define; after all, if we could state precisely what we should do to make an application robust, we would be able to specify its "reasonable" behavior completely. Thus, robustness would become equivalent to correctness (or reliability, in the sense of Figure 101.1).

Again, an analogy with bridges is instructive. Two bridges connecting two sides of the same river are both correct if each satisfies the stated requirements. If, however, during an unexpected, unprecedented earthquake, one collapses and the other one does not, we can call the latter more robust than the former. Notice that the lesson learned from the collapse of the bridge will probably lead to more complete requirements for future bridges, establishing resistance to earthquakes as a correctness requirement. In other words, as the phenomenon under study becomes better known, we approach the ideal case shown in Figure 101.1, where specifications capture the expected requirements exactly.

The means to achieve robustness depend on the application area. For example, a system written for novice computer users must be more prepared to deal with ill formatted input than an embedded system that receives its input from a sensor. This, of course, does not imply that embedded systems do not need to be robust. On the contrary, embedded systems often control critical devices and require extra robustness.

In conclusion, we can see that robustness and correctness are strongly related, without a sharp dividing line between them. If we put a requirement in the specification, its accomplishment becomes an issue of correctness; if we leave the requirement out of the specification, it may become an issue of robustness. The border between the two qualities is thus determined by the specification of the system. Finally, reliability comes in because not all incorrect behaviors signify equally serious problems; that is, some incorrect behaviors may be tolerable.

We may also use the terms correctness, robustness, and reliability in relation to the software production process. A process is robust, for example, if it can accommodate unanticipated changes in the environment, such as a new release of the operating system or the sudden transfer of half the employees to another location. A process is reliable if it consistently leads to the production of high-quality products.

101.2.2 Performance

Any engineering product is expected to operate at a certain level of **performance**. Unlike other disciplines, software engineering often equates performance with efficiency, but they are not the same. Efficiency is an internal quality and refers to how economically the software utilizes the resources of the computer. Performance, on the other hand, is an external quality based on user requirements. For example, a customer may specify a performance requirement for a telephone switch to be able to process 10,000 calls per hour. An efficiency requirement for the same switch may state that processing a call must use a minimum amount of memory and take a minimum amount of time. Efficient use of resources often affects the performance of a system. For example, conserving memory may increase the response time of a switch and its ability to process a high number of calls per second.

Performance may affect the usability of the system. If a software system is too slow, it reduces the users' productivity, possibly to the point of not meeting their needs. Performance also affects the scalability of a software system. An algorithm that is quadratic may work on small inputs but not work at all on larger inputs.

There are three basic approaches to evaluating the performance of a system: measurement, analysis, and simulation. We can measure the actual performance of a system by means of hardware and software monitors that collect data while the system is running and thereby allow us to discover bottlenecks in the system. The second approach is to build a model of the product and mathematically analyze it. The third approach is to use the model to simulate the product.

In some application areas, performance criteria are standardized and may be used as the basis for comparing different products. For example, in telephone switching systems, number of calls processed per second and the maximum number of simultaneous calls processed are two common measures. For information systems, number of transactions per second is a common performance measure and is used as the basis for product selection.

In many software development projects, performance is addressed only after the initial version of the product is implemented. At that point, it is very difficult — sometimes even impossible — to achieve significant improvements in performance without redesigning the software. Instead, even a simple model is useful for predicting a system's performance and guiding design choices so as to minimize the need for redesign.

In some complex projects, in which the feasibility of the performance requirements is not clear, much effort is devoted to building performance models. Such projects start with a performance model and use it initially to answer feasibility questions and later in making design decisions. These models can help to resolve such issues as whether a function should be provided by software or by a special-purpose hardware device.

The notion of performance also applies to a development process, in which case we call it *productivity*. Productivity is important enough to be treated as an independent quality and is discussed in Section 101.2.11.

101.2.3 Usability

A software system is usable — or user-friendly— if its human users find it easy to use. This definition reflects the subjective nature of **usability**.

The *user interface* is an important component of user-friendliness. Properties that make an application user-friendly to novices are different from those desired by expert users. For example, a software system that presents the novice user with a graphical interface is friendlier than one that requires the user to enter a set of one-letter commands. On the other hand, experienced users might prefer a set of commands that minimize the number of keystrokes, rather than a fancy graphical interface through which they must navigate to get to the command that they knew all along they wanted to execute.

There is more to user-friendliness, however, than the user interface. For example, an embedded software system does not have a human user interface. Instead, it interacts with hardware and perhaps other software systems. In this case, usability is reflected in the ease with which the system can be configured and adapted to the hardware environment.

In general, the user-friendliness of a system depends on the consistency and predictability of its user and operator interfaces. Clearly, however, the other qualities mentioned — such as correctness and performance — also affect user-friendliness. A software system that produces wrong answers is not friendly, regardless of how fancy its user interface is. Also, a software system that produces answers more slowly than the user requires is not friendly, even if the answers are displayed in a beautiful color.

Usability is also discussed under the subject of *human factors*. Human factors and usability engineering play a major role in many engineering disciplines. For example, automobile manufacturers devote significant effort to deciding the positions of the various control knobs on the dashboard. Television manufacturers and microwave oven makers also try to make their products easy to use. User interface decisions in these classical engineering fields are made after extensive study of user needs and attitudes by specialists in fields such as industrial design or psychology.

Interestingly, ease of use in many engineering disciplines is achieved through standardization of the human interface. Once a user knows how to use one television set, that user can operate almost any other television set. There is a clear trend in software applications toward more uniform and standard user interfaces, as seen, for example, in Web browsers. There are also usability labs that attempt to measure the usability of software products.

101.2.4 Verifiability

A software system is verifiable if its properties can be verified easily. For example, it is important to be able to verify the correctness or the performance of a software system. Verification can be performed by formal and informal analysis methods or through testing.

Verifiability is usually an internal quality, although it sometimes becomes an external quality also. For example, in many security-critical applications, the customer requires the verifiability of certain properties. The highest level of the security standard for a trusted computer system requires the verifiability of the operating system kernel.

101.2.5 Security

A system is secure if it provides its services only to its authorized users and protects the rights and information of those users. Although **security** is important in any system, it has gained importance as applications have become increasingly distributed and offered over public networks. Security is an important quality of information systems, in which users trust the safeguarding of their data to the system. For example, a banking system that maintains customer account data and provides access to that data through the Internet is required to be secure. Two qualities related to security are data integrity and privacy. *Data integrity* ensures that once the user data is committed to the system, it will not be modified or destroyed through system malfunction or unintended or malicious acts of other users. *Privacy* ensures that user transactions and user data are protected from unauthorized users and will be not be used for unauthorized purposes. For example, credit card numbers entered into an electronic commerce system must be used only for the purpose of the transaction for which they were entered.

The combination of security and verifiability enables security properties to be verified. Such a combination is relevant in financial and military systems.

101.2.6 Maintainability

The term *software maintenance* is commonly used to refer to the modifications that are made to a software system after its initial release. Maintenance used to be viewed as merely "bug fixing," and it was distressing to discover that so much effort was being spent on fixing defects. Studies have shown, however, that the majority of time spent on maintenance is, in fact, spent on enhancing the product with features that were not in the original specifications or that were stated incorrectly there.

Maintenance is indeed not the proper word to use with software. First, as it is used today, the term covers a wide range of activities, all having to do with modifying an existing piece of software in order to make an improvement. A term that captures the essence of this process better is *software evolution*. Second, in other engineering products, such as computer hardware, automobiles, or washing machines, maintenance refers to the upkeep of the product in response to the gradual deterioration of parts due to extended use of the product. For example, transmissions are oiled and air filters are dusted and periodically changed. To use the word *maintenance* with software gives the wrong connotation, because software does not wear out. Unfortunately, the term is used widely, and we will continue using it here.

There is evidence that maintenance costs exceed 60 percent of the total costs of software. To analyze the factors that affect such costs, it is customary to divide software maintenance into three categories: corrective, adaptive, and perfective.

Corrective maintenance has to do with the removal of residual errors that are present in the product when it is delivered, as well as errors introduced into the software during its maintenance. Corrective maintenance accounts for about 20% of maintenance costs.

Adaptive maintenance, which accounts for nearly another 20% of maintenance costs, involves adjusting the application to changes in the environment (e.g., a new release of the hardware or the operating system or a new database system).

Finally, *perfective maintenance*, which absorbs over 50% of maintenance costs, involves changing the software to improve some of its qualities. Here, changes are due to the need to modify the functions offered by the application, add new functions, improve the performance of the application, make it easier to use, etc. The requests to perform perfective maintenance may come directly from the software engineer, in order to improve the status of the product on the market, or they may come from the customer, to meet some new requirements.

The term *legacy software* refers to software that already exists in an organization and usually embodies much of the organization's processes and knowledge. Such software holds considerable value for the organization, represents past investments, and may not be replaced easily. On the other hand, because of its age, it is usually written in older languages and uses older software engineering technology. Legacy software is, therefore, difficult to maintain. For example, an old personnel system may embody an organization's operational procedures and personnel policies. Such legacy systems represent a challenge to software evolution. *Reverse engineering* and *reengineering* techniques and technologies aim at uncovering the structure of legacy software and restructuring or in some way improving it.

Maintainability can be seen as two separate qualities: **reparability** and **evolvability**. Software is reparable if it allows the fixing of defects; it is evolvable if it allows changes that enable it to satisfy new or modified requirements.

The distinction between reparability and evolvability is not always clear. For example, if the requirements specifications are vague, it may not be clear whether a change is made to fix a defect or to satisfy a new requirement. In general, however, the distinction between the two qualities is useful.

Both reparability and evolvability are improved by suitable modularization in the software structure. As we shall see later, the right modularization may help to locate errors more easily. It may also help to encapsulate the changeable parts in a separate module, making it easier to apply changes.

101.2.7 Reusability

Reusability is akin to evolvability. In product evolution, we modify a product to build a new version of that same product; in product reuse, we use the product — perhaps with minor changes — to build another product. Reusability may be applied at different levels of granularity — from whole applications to individual routines — but it appears to be more applicable to software components than to whole products.

A good example of a reusable product is the UNIX shell, which is a command language interpreter; that is, it accepts user commands and executes them. But it is designed to be used both interactively and in batch. The ability to start a new shell with a file containing a list of shell commands allows us to write programs (scripts) in the shell command language. We can view the program as a new product that uses the shell as a component. By encouraging standard interfaces, the UNIX environment in fact supports the reuse of any of its commands, as well as the shell, in building powerful utilities.

Numeric libraries were the first examples of reusable components. Several large FORTRAN libraries, now rewritten in other languages, have existed for many years. Users buy these libraries and use them to build their own products, without having to reinvent or recode well known algorithms. Several companies are devoted to producing just such libraries. Nowadays, reusable libraries exist for different areas, such as graphical user interfaces, simulation, etc. One of the goals of reusability researchers is to increase the granularity of components that may be reused. One of the goals of object-oriented programming (see Chapter 91) is to achieve both reusability and evolvability.

Reusability is difficult to achieve *a posteriori*. Reusable components must be designed with reusability as a primary design goal. Reusable components are abstractions of useful concepts, are general, and have clear and usable interfaces.

Languages that support generic components, such as C++ and Ada, enable higher levels of reusability in software components. Most C++ libraries commonly consist of template modules. The field of component-based software engineering (CBSE) has the goal of building applications by assembling a set of ready-made, reusable, off-the-shelf components. (See Chapter 117 on component-based computing.) Object-oriented languages, such as C++ and Java, help in unifying the qualities of evolvability and reusability by extensive use of interfaces, inheritance, and polymorphism.

Reusability has broader applicability than just code. It may occur at different levels and may affect both product and process. In general, any of the artifacts of the software process, such as the requirements specification, may be reused. For example, at the requirements level, when a new application is conceived, we may try to identify parts that are similar to parts used in a previous application. Thus, we may reuse parts of the previous requirements specification instead of developing an entirely new one. Clearly, the more modularly designed the work products are, the more likely it is that they, or parts of them, may be reused in the future.

Reusability applies to the software process, as well. Indeed, the various software methodologies can be viewed as attempts to reuse the same process for building different products. Life-cycle models are also attempts at reusing higher-level processes.

Reusability of standard parts characterizes the maturity of an industrial field. We see high degrees of reuse in such mature areas as the automobile industry and consumer electronics. For example, a car is constructed by assembling many components that are highly standardized and used across many models produced by the same industry. Certainly, the designs are routinely reused from model to model. Finally, the manufacturing process is often reused. The level of reuse is increasing in software, but it still is short of that of other established engineering disciplines.

101.2.8 Portability

Software is portable if it can run in different environments. The term *environment* may refer to a hardware platform or a software environment, such as a particular operating system. **Portability** is economically important because it helps amortize the investment in the software system across different environments and different generations of the same environment. Many applications are independent of the actual hardware platform, because the operating system provides portability across hardware platforms. These days, the applications' dependencies are on operating systems and other software systems, such as databases and user interface systems. Portability may be achieved by modularizing the software so that dependencies on the environment are isolated in only a few, well designated modules. To port the software to a new environment, only these environment-dependent modules need to be modified. With the proliferation of networked systems, portability has taken on new importance because the execution environment is naturally heterogeneous, consisting of many different kinds of computers and operating systems. In addition, the delivery devices have become diverse. For example, Internet browsers must be able to run not only on workstations and personal computers, but also on palmtops and even mobile phones.

Some software systems are inherently machine-specific. For example, an operating system is written to control a specific computer, and a compiler produces code for a particular machine. Even in these cases, however, it is possible to achieve some level of portability. UNIX and its variant, Linux, are examples of an operating system that has been ported to many different hardware systems. Of course, the porting effort may require months of work. Still, we can call the software portable because writing the system from scratch for the new environment would require much more effort than porting it.

101.2.9 Understandability

Some software systems are easier to understand than others. Of course, some tasks are inherently more complex than others. For example, a system that does weather forecasting, no matter how well it is written, will be harder to understand than one that prints a mailing list. We can follow certain guidelines to produce more understandable designs and to write more understandable programs. Systematic documentation of both the design and the program is clearly very important. Furthermore, abstraction and modularity enhance a system's understandability.

The activity of software maintenance is dominated by the subactivity of *program understanding*. Maintenance engineers spend most of their time trying to uncover the logic of the application and a smaller portion of their time applying changes to the application.

Understandability is an internal product quality, and it helps in achieving many of the other qualities, such as evolvability and verifiability. From an external point of view, the user considers a system understandable if it has predictable behavior. External understandability is a factor in a product's usability.

101.2.10 Interoperability

Interoperability refers to the ability of a system to coexist and cooperate with other systems — for example, a word processor's ability to incorporate a chart produced by a graphics package, the graphics package's ability to graph the data produced by a spreadsheet, or the spreadsheet's ability to process an image scanned by a scanner. Interoperability can be seen as reusability at the application level.

Interoperability abounds in other engineering products. For example, stereo systems from various manufacturers work together and can be connected to television sets and video recorders. In fact, stereo systems produced decades ago accommodate new technologies such as compact discs! In contrast, early operating systems had to be modified — sometimes significantly — before they could work with new devices. The generation of plug-and-play operating systems attempts to solve this problem by automatically detecting and working with new devices.

The UNIX environment, with its standard interfaces, offers a limited example of interoperability within a single environment. UNIX encourages software engineers to design applications so that they have a simple, standard interface, which allows the output of one application to be used as the input to another. The UNIX standard interface is a primitive, character-oriented one. It falls short when one application needs to use structured data — say, a spreadsheet or an image — produced by another application.

With interoperability, a vendor can produce different products and allow the user to combine them if necessary. This makes it easier for the vendor to produce the products, and it gives the user more freedom in exactly what functions to pay for and to combine. Interoperability can be achieved through standardization of interfaces. An example of such interoperability is the Web browser application that provides plug-in interfaces for different applications. For example, a new audio player provided by one vendor may be added to the browser provided by another vendor.

A concept related to interoperability is that of an *open system* — an extensible collection of independently written applications that function as an integrated system. An open system allows the addition of new functionality by independent organizations, after the system is delivered. This can be achieved, for example, by releasing the system together with a specification of its open interfaces. Any application developer can then take advantage of these interfaces, some of which may be used for communication between different applications or systems. Open systems allow different applications, written by different organizations, to interoperate.

An interesting requirement of open systems is that new functionality may be added without taking the system down. An open system is analogous to a growing (social) organization that evolves over time, adapting to changes in the environment. The importance of interoperability has sparked a growing interest in open systems, producing some recent efforts at standardization in this area. For example, the CORBA standard defines interfaces that support the development of components that may be used in open distributed systems.

101.2.11 Productivity

Productivity is a quality of the software production process, referring to its efficiency and performance. An efficient process results in faster delivery of the product.

Individual engineers produce software at a certain rate, although there are great variations among individuals of different ability. When individuals are part of a team, the productivity of the team is some function of the productivity of the individuals. Very often, the combined productivity is much less than the sum of the parts. Management tries to organize team members and adopt processes in such manner as to capitalize on the individual productivity of the members.

Productivity offers many trade-offs in the choice of a process. For example, a process that requires specialization of individual team members may lead to high productivity in producing a certain product, but not in producing a variety of products. Software reuse is a technique that increases the overall productivity of an organization in producing a collection of products, but the cost of developing reusable modules can be amortized only over many products.

While software productivity is of great interest due to the increasing cost of software, it is difficult to measure. Clearly, we need a metric for measuring productivity — or any other quality, for that matter — if we are to have any hope of comparing different processes in terms of their productivity. Early metrics, such as the number of lines of code produced, have many shortcomings.

As with other engineering disciplines, the efficiency of the process is affected strongly by automation. Many modern software engineering tools and environments help in achieving increases in productivity.

101.2.12 Timeliness

Timeliness is a process-related quality that refers to the ability to deliver a product on time. Historically, timeliness has been lacking in software production processes, leading to the "software crisis," which in turn led to the need for — and birth of — software engineering itself. Today, due to increased competitive market pressures, software projects face even more stringent time-to-market challenges. Being late may sometimes preclude market opportunities. Although on-time delivery of a product that is lacking in other qualities, such as reliability or performance, may be pointless, some argue that the early delivery of a preliminary and still unstable version of a product may favor the later acceptance of the final product. The Internet has facilitated this approach. Vendors can place early versions of products on the Internet, enabling potential users to try the product, providing feedback to the vendor.

Timeliness requires careful scheduling, accurate estimation of work, and clearly specified and verifiable milestones. All engineering disciplines use standard project management techniques to achieve timeliness. They are sometimes difficult to apply in software engineering because of its human intensive nature. There are no objective or standard ways of defining, predicting, and measuring the amount of work required to produce a given piece of software, the productivity of software engineers, and the milestones in the development process.

One technique for achieving timeliness is through the *incremental delivery* of the product. Progressively larger subsets of the product are developed and delivered as new increments at each stage. Each increment provides additional functionality and becomes closer to the final product. Obviously, incremental delivery depends on the ability to break down the set of required system functions into subsets that can be delivered in increments. Incremental delivery allows (parts of) the product to become available earlier; and the use of the early increments helps in refining the requirements incrementally.

The biggest challenge to achieving timeliness is to ensure that other qualities, those of both product and process, are not jeopardized by focusing only on timeliness.

101.2.13 Visibility

A software development process has **visibility** if all of its steps and its current status are documented clearly. Another term used to characterize this property, also used in business processes and organizations, is *transparency*. The idea is that the steps and the status of the project are available and easily accessible for external examination.

In many software projects, most engineers and even managers are unaware of the exact status of the project. Some may be designing, others coding, and still others testing, all at the same time. This, in itself, is not bad. Yet, if an engineer starts to redesign a major part of the code just before the software is supposed to be delivered for integration testing, the risk of serious problems and delays will be high.

Visibility allows engineers to weigh the impact of their actions and thus guides them in making decisions. It allows the members of the team to work in the same direction, rather than in opposing directions. The most common example of the latter situation is when the integration group has been testing a version of the software, assuming that the next version will involve fixing defects, while the engineering group decides to do a major redesign to add some functionality. This tension, created when one group is trying to stabilize the software while another group is destabilizing it, is common. The process must encourage a consistent view of the status and current goals among all participants.

One of the benefits of the *synch-and-stabilize* software build process, popularized by Microsoft, is that it adds some visibility to the process. In this approach, the software product is rebuilt (integrated) each day, exposing problems in components and their interactions as soon as they occur. These problems may be symptoms of deeper problems, but the process helps to uncover them early.

Visibility is not only an internal quality; it is also external. During the course of a long project, many requests arise about the status of the project. Sometimes the requests come from the organization's management for future planning, and at other times they come from the outside, perhaps from the customer. If the software development process has low visibility, either these status reports will not be accurate, or they will require a lot of effort to prepare each time.

A difficulty in managing large projects is dealing with personnel turnover. With many software projects, critical information about the software requirements and design has the form of folklore, known only to people who have been with the project either from the beginning or for a sufficiently long time. In such situations, recovering from the loss of a key engineer or adding new engineers to the project is very difficult. In fact, adding new engineers often reduces the productivity of the whole project, as the folklore is being transferred slowly from the existing crew of engineers to the new engineers.

The preceding discussion points out that visibility of the process requires not only that all of its steps be documented, but also that the current status of the intermediate products, such as requirements specifications and design specifications, be maintained accurately; that is, visibility of the product is required, as well. In other words, it must be understandable (see Section 101.2.9).

101.3 Quality Assurance

Once we have decided on the qualities that are the goals of software engineering, we need principles and techniques to help us achieve them. We also need to be able to measure a given quality. In software organizations, this activity is called **quality assurance**.

If we identify a quality as important, we must be ready to measure it to determine how well we are achieving it. This, in turn, requires that we define each quality precisely, so that it is clear what we should be measuring. Without measurements, any claims of improvement are without basis. But without defining a quality precisely, there is no hope that we can measure it precisely — let alone quantitatively.

The established engineering disciplines have standard techniques for measuring quality. For example, the reliability of many artifacts is commonly characterized by mean time to failure (MTTF), that is, the average time between two consecutive failures of the product. Although some software qualities, such as performance, are measured relatively easily, most software qualities unfortunately have no universally accepted metrics. For example, whether a given system will evolve more easily than another is usually determined subjectively. Nevertheless, metrics are needed, and indeed, much research work is currently under way for defining objective metrics.

The current state of practice of quality assurance is a collection of verification and monitoring methods. Some of them are based on objective evaluations, such as testing, and others are based on informal procedures, such as walkthroughs; a few, more ambitious methods aim at exploiting mathematical analysis during software verification. Chapters 111, 113, and 114 relate, to different extents and from different points of view, to the quality assurance problem.

Quality assurance for software processes has taken the form of process assessment procedures whose aim is to measure the ability of software organizations and their processes. For example, the *capability maturity model* (CMM) classifies software development organizations on the basis of detailed evaluation of their processes. The model defines five levels of maturity which may be used to characterize the software process quality of an organization, referred to as its *maturity level*. The maturity level of an organization is intended to reflect the ability of the organization to predict its costs and schedules. The levels are as follows:

1. **Initial** — The organization has no statistical control over its processes and no predictability.
2. **Repeatable** — The organization has a stable, repeatable process, supported by rigorous project management controls.
3. **Defined** — The organization has defined a base process that is applied consistently for managing different projects.
4. **Managed** — The process is systematically monitored and its performance measured with appropriate metrics.
5. **Optimizing** — The process measurements are used for continuous improvement of the process.

It turns out that most organizations fall into maturity levels 2 and 3. The CMM is being used by organizations to document their competence and by customers to qualify their vendors. Although the particular way of measuring process maturity and the appropriate metrics to be used are sometimes controversial, the impact of process quality on product quality is generally accepted to be significant.

An indispensable technology that supports disciplined processes and is a primary factor that differentiates mature from immature organizations is *configuration management,* which is concerned with controlled change management. In the software production process, configuration management is concerned with maintaining and controlling the relationship between all the work products of the various versions of a product. Configuration management tools allow the maintenance of families of products and their components. They help in controlling and managing changes to work products.

101.4   Software Engineering Principles in Support of Software Quality

So far, we have discussed a number of important software qualities. How can we achieve these qualities? In this section, we discuss seven general principles that help in achieving software quality. These principles may be applied throughout the software development process and are not limited to a particular phase of the process. The principles deal with rigor and formality, separation of concerns, modularity, abstraction, anticipation of change, generality, and incrementality. By its very nature, the list cannot be exhaustive, but it does cover the important areas of software engineering. The principles are, of course, strongly related and together form a set of guidelines for the engineer to follow.

101.4.1   Rigor and Formality

Software development is a creative activity. In any creative process, there is an inherent tendency to be neither precise nor accurate, but to follow the inspiration of the moment in an unstructured manner. *Rigor* — defined as precision and exactness — on the other hand, is a necessary complement to creativity in every engineering activity. It is only through a rigorous approach that we can repeatedly produce reliable products, control their costs, and increase our confidence in their reliability. Rigor need not constrain creativity. Rather, it can be used as a tool to enhance creativity. The engineer can be more confident of the results of a creative process after performing a rigorous assessment of those results.

Paradoxically, rigor is an intuitive principle that cannot be defined in a rigorous way. Also, various degrees of rigor can be achieved. The highest degree is what we call *formality*. Thus, formality is a stronger requirement than rigor. It requires the software process to be driven and evaluated by mathematical laws. Of course, formality implies rigor, but the converse is not true. One can be rigorous and precise even in an informal setting.

In every engineering field, the design process proceeds as a sequence of well defined, precisely stated, and supposedly sound steps. In each step, the engineer follows some method or applies some technique. The methods and techniques applied may be based on some combination of theoretical results derived by some formal modeling of reality, empirical adjustments that take care of phenomena not dealt with by the model, and rules of thumb that depend on past experience. The blend of these factors results in a rigorous and systematic approach — the methodology— that can be easily explained and applied time and again.

There is no need to be always formal during design, but the engineer must know how and when to be formal, should the need arise. For example, the engineer can rely on past experience and rules of thumb to design a small bridge, to be used temporarily to connect the two sides of a creek. If the bridge were a large and permanent one, on the other hand, the engineer would instead use a mathematical model to verify whether the design was safe. An exceptionally long bridge or one built in an area of much seismic activity would require a more sophisticated mathematical model. In that case, the mathematical model would take into account factors that could be ignored in the previous case.

Another — perhaps striking — example of the interplay between rigor and formality may be observed in mathematics. Textbooks on functional calculus are rigorous but seldom formal. Proofs of theorems are done in a very careful way, as sequences of intermediate deductions that lead to the final statement; each deductive step relies on an intuitive justification that should convince the reader of its validity. Almost never, however, is the derivation of a proof stated in a formal way, in terms of mathematical logic. This means that very often the mathematician is satisfied with a rigorous description of the derivation of a proof, without formalizing it completely. In critical cases, however, in which the validity of some intermediate deduction is unclear, the mathematician may try to formalize the informal reasoning to assess its validity.

These examples show that the engineer (and the mathematician) must be able to identify and understand the level of rigor and formality that should be achieved, depending on the conceptual difficulty and criticality of the task. The level may even vary for different parts of the same system. For example, critical parts — such as the scheduler of a real-time operating systems kernel or the security component of an electronic commerce system — may merit a formal description of their intended functions and a formal approach to their assessment. Well understood and standard parts would require simpler approaches.

If we examine this issue in the context of software specifications, we see, for example, that the description of what a program does may be given in a rigorous way by using natural language; it can also be given formally by providing a formal description in a language of logical statements. The advantage of formality over rigor is that formality may be the basis of mechanization of the process. For instance, one may hope to use the formal description of the program to derive the program (if the program does not yet exist) or to show that the program corresponds to the formal description (if the program and its formal specification exist).

Traditionally, there is only one phase of software development in which a formal approach is used: programming. In fact, programs are formal objects. They are written in a language whose syntax and semantics are fully defined. Programs are formal descriptions that may be automatically manipulated by compilers. They are checked for formal correctness, transformed into an equivalent form in another language (assembly or machine language), "pretty-printed" so as to improve their appearance, etc. These mechanical operations, which are made possible by the use of formality in programming, can improve the reliability and verifiability of software products.

Rigor and formality are not restricted to programming: They should be applied throughout the software process. Chapter 111 shows these concepts in action in the case of software specifications. Chapter 107 does the same for software verification. Chapter 106 is about formal methods.

Rigor and formality also apply to software processes. Rigorous documentation of a software process helps in reusing the process in other, similar projects. On the basis of such documentation, managers may foresee the steps through which a new project will evolve, assign appropriate resources as needed, etc. Similarly, rigorous documentation of the software process may help to maintain an existing product. If the various steps through which a project evolved are documented, one can modify an existing product, starting from the appropriate intermediate level of its derivation, not the final code. Finally, if the software process is specified rigorously, managers may monitor it accurately, in order to assess its timeliness and improve productivity.

101.4.2 Separation of Concerns

Separation of concerns allows us to deal with different aspects of a problem to dominate its complexity, so that we can concentrate on each aspect individually. Separation of concerns is a commonsense practice that we try to follow in our everyday lives to overcome the difficulties we encounter. The principle should also be applied to software development, to master its inherent complexity.

More specifically, there are many decisions that must be made in the development of a software product. Some of them concern features of the product: functions to offer, expected reliability, efficiency with respect to space and time, the product's relationship with the environment (i.e., the special hardware or software resources required), user interfaces, etc. Others concern the development process: the development environment, the organization and structure of teams, scheduling, control procedures, design strategies, error recovery mechanisms, etc. Still others concern economic and financial matters. These different decisions may be unrelated to one another. In such a case, it is obvious that they should be treated separately.

Very often, however, many decisions are strongly related and interdependent. For instance, a design decision (e.g., swapping some data from main memory to disk) may depend on the size of the memory of the selected target machine (and hence, the cost of the machine). This, in turn, may affect the policy for error recovery. When different design decisions are strongly interconnected, it would be useful to take all the issues into account at the same time and by the same people, but this is not usually possible in practice.

There are various ways in which concerns may be separated. First, one can separate them in *time*. Temporal separation of concerns allows for the precise planning of activities and eliminates overhead that would arise through switching from one activity to another in an unconstrained way. In fact, it is the underlying motivation of the software life-cycle models, each of which defines a sequence of activities that should be followed in software production (see Chapter 110).

Another type of separation of concerns is in terms of *qualities* that should be treated separately. For example, in the case of software, we might wish to deal separately with the efficiency and the correctness of a given program. One might decide first to design software in such a careful and structured way that its correctness is expected to be guaranteed *a priori* and then to restructure the program partially to improve its efficiency. Similarly, in the verification phase, one might first check the functional correctness of the program and then its performance. Both activities can be done rigorously, applying some systematic procedures, or even formally (i.e., using formal correctness proofs and complexity analysis). Verification of program qualities is the subject of Chapter 113 and Chapter 114.

Another important type of separation of concerns allows different views of the software to be analyzed separately. For example, when we analyze the requirements of an application, it may be helpful to concentrate separately on the flow of data from one activity to another in the system and the flow of control that governs how different activities are synchronized. Both views help us understand the system we are working on better, although neither one gives a complete view of it.

Still another type of separation of concerns allows us to deal with parts of the same system separately; here, separation is in terms of size. This is a fundamental concept that we must master to dominate the complexity of software production. Indeed, it is so important that we prefer to detail it shortly as a separate point under modularity (see Section 101.4.3).

There is an inherent disadvantage in separation of concerns. By separating two or more issues, we might miss some global optimization that would be possible by tackling them together. While this is true in principle, our ability to make optimized decisions in the face of complexity is rather limited. If we consider too many concerns simultaneously, we are likely to be overwhelmed by the amount of detail and complexity we face. Some of the most important decisions in design concern which aspects to consider together and which separately.

Perhaps the most important application of separation of concerns is to separate problem-domain concerns from implementation-domain concerns. Problem-domain properties hold in general, regardless of the implementation environment. For example, in designing a personnel-management system, we must separate issues that are true about employees in general from those which are a consequence of our implementation of the employee as a data structure or object. In the problem domain, we may speak of the relationship between employees, such as "employee A reports to employee B"; in the implementation domain we may speak of one object pointing to another. These concerns, unfortunately, are often intermingled in many projects.

As a final remark, we note that separation of concerns may result in separation of responsibilities in dealing with separate issues. Thus, the principle is the basis for dividing the work on a complex problem into specific assignments, possibly for different people with different skills. For example, by separating managerial and technical issues in the software process, we allow two types of people to cooperate in a software project. Or, having separated requirements analysis and specification from other activities in a software life cycle, we may hire specialized analysts with expertise in the application domain, instead of relying on internal resources. The analyst, in turn, may concentrate separately on functional and nonfunctional system requirements.

101.4.3 Modularity

A complex system may be divided into simpler pieces called *modules*. A system that is composed of modules is called *modular*. The main benefit of modularity is that it allows the principle of separation of concerns to be applied in two phases:

- When dealing with the details of each module in isolation (and ignoring details of other modules)
- When dealing with the overall characteristics of all modules and their relationships in order to integrate them into a coherent system

If the two phases are executed in sequence, first concentrating on modules and then on their composition, then we say that the system is designed from the *bottom up*. The converse, when we decompose the system into modules first and then concentrate on individual module design, is *top-down* design.

Modularity is an important property of most engineering processes and products. For example, in the automobile industry, the construction of cars proceeds by assembling building blocks that are designed and built separately. Furthermore, parts are often reused from model to model, perhaps after minor changes. Most industrial processes are essentially modular, made out of work packages that are combined in simple ways (sequentially or overlapping) to achieve the desired result. Although modularity is often discussed in the context of software design, it is not only a desirable design principle; it permeates the whole of software production. In particular, modularity provides four main benefits in practice:

- The capability of decomposing a complex system into simpler pieces
- The capability of composing a complex system from existing modules
- The capability of understanding the system in terms of its pieces
- The capability of modifying a system by modifying only a small number of its pieces

The *decomposability* of a system is based on dividing the original problem, top-down, into subproblems and then applying the decomposition to each subproblem recursively. This procedure reflects the well known Latin motto *divide et impera* (divide and conquer), which describes the philosophy followed by the ancient Romans to dominate other nations: divide and isolate them first, and then conquer them individually.

The *composability* of a system is based on starting from the bottom up with elementary components and combining them in steps toward finally producing a finished system. As an example, a system for office automation may be designed by assembling existing hardware components, such as personal workstations, a network, and peripherals; system software, such as the operating system; and productivity tools, such as document processors, databases, and spreadsheets.

Ideally, in software production we would like to be able to assemble new applications by taking modules from a library and combining them to form the required product. Such modules should be designed with the express goal of being reusable. By using reusable components, we may speed up both the initial system construction and its fine-tuning. For example, it would be possible to replace a component with another that performs the same function, but differs in computational resource requirements.

Modularity supports the capability of understanding and modifying a system. If the entire system can be understood only in its entirety, modifications are likely to be difficult to apply, and the result will probably be unreliable. When it is necessary to repair a defect or enhance a feature, proper modularity helps to confine the search for the fault or enhancement to single components. Modularity thus forms the basis for software evolution.

To achieve modular composability, decomposability, understandability, and modifiability, the software engineer must design the modules with the goal of high cohesion and low coupling.

A module has *high cohesion* if all of its elements are related strongly. Elements of a module (e.g., statements, procedures, and declarations) are grouped together in the same module for a logical reason, not just by chance. They cooperate to achieve a common goal, which is the function of the module.

Whereas **cohesion** is a property about the internal structure of a module, **coupling** characterizes a module's relationship to other modules. Coupling measures the interdependence of two modules (e.g., module A calls a routine provided by module B or accesses a variable declared by module B). If two

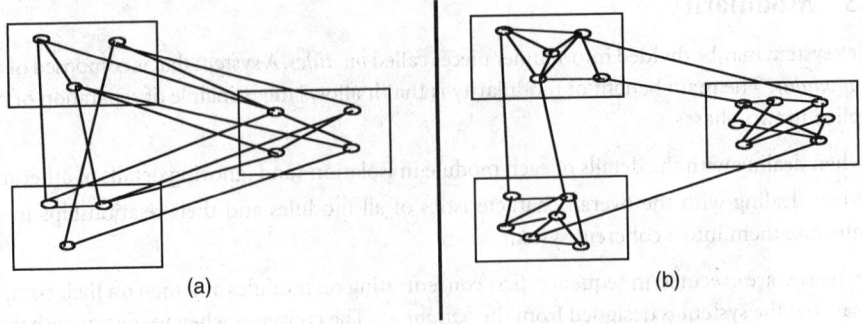

FIGURE 101.2   Graphical description of cohesion and coupling. (a) A highly coupled structure. (b) A structure with high cohesion and low coupling.

modules depend on each other heavily, they have high coupling. Ideally, we would like modules in a system to exhibit *low coupling*, because it will then be possible to analyze, understand, modify, test, and reuse them separately. Figure 101.2 provides a graphical view of cohesion and coupling. The practical design guideline derived from the high-cohesion/low-coupling rule is that two units should be put in the same module if they are tightly dependent on each other; otherwise they should be placed in different modules. Indeed, this is achieved by object-oriented programming, which groups both the data and the routines that manipulate them in a single module.

A good example of a system that has high cohesion and low coupling is the electric subsystem of a house. Because it is made out of a set of appliances with clearly definable functions and interconnected by simple wires, the system has low coupling. Because each appliance's internal components are there exactly to provide the service the appliance is supposed to provide, the system has high cohesion.

Modular structures with high cohesion and low coupling allow us to see modules as black boxes when the overall structure of a system is described and then deal with each module separately when the module's functionality is described or analyzed. In other words, modularity supports the application of the principle of separation of concerns.

101.4.4   Abstraction

Abstraction is a fundamental principle for understanding, designing, and analyzing complex problems. In applying abstraction, we identify the important aspects of a phenomenon and ignore its details. Thus, abstraction is a special case of separation of concerns wherein we separate the concern of the important aspects from the concern of the less important details.

What we abstract away and consider as a detail that may be ignored depends on the purpose of the abstraction. For example, consider a digital watch. A useful abstraction for the owner is a description of the effects of pushing its various buttons, which allow the watch to enter various modes of functioning and react differently to sequences of commands. A useful abstraction for the person in charge of maintaining the watch is a box that can be opened in order to replace the battery. Still other abstractions of the device are useful for understanding the watch and performing the activities that are needed to repair it (let alone design it). Thus, there may be many different abstractions of the same reality, each providing a view of the reality and serving some specific purpose.

Abstraction is a powerful technique practiced by engineers of all fields for mastering complexity. For example, the representation of an electrical circuit in terms of resistors, capacitors, etc., each characterized by a set of equations, is an idealized abstraction of a device. The equations are a simplified model that approximates the behavior of the real components. The equations often ignore details, such as the fact that there are no "pure" connectors between components and that connectors should also be modeled in terms of resistors, capacitors, etc. The designer ignores both of these facts, because the effects they describe are negligible in terms of the observed results.

This example illustrates an important general idea. The models we build of phenomena — such as the equations for describing devices — are an abstraction from reality, ignoring certain facts and concentrating on others that we believe are relevant. The same holds true for the models built and analyzed by software engineers. For example, when the requirements for a new application are analyzed and specified, software engineers build a model of the proposed application. As shown in Chapter 111, this model may be expressed in various forms, depending on the required degree of rigor and formality. No matter what language we use for expressing requirements — be it natural language or the formal language of mathematical formulas — what we provide is a model that abstracts away from a number of details that we decide can be ignored safely.

Abstraction permeates the whole of programming. The programming languages that we use are abstractions built on top of the hardware. They provide us with useful and powerful constructs so that we can write (most) programs ignoring such details as the number of bits that are used to represent numbers or the specific computer's addressing mechanism. This helps us to concentrate on the solution to the problem we are trying to solve, rather than on the way to instruct the machine on how to solve it. The programs we write are themselves abstractions. For example, a computerized payroll procedure is an abstraction of the manual procedure it replaces. It provides the *essence* of the manual procedure, not its exact details. When applied judiciously in design and programming, abstraction affects all software qualities in the product. For example, proper abstraction leads to modularity, which aids in achieving maintainability and reusability.

Abstraction is an important principle that applies to both software products and software processes. For example, the comments that we often use in the header of a procedure are an abstraction that describes the effect of the procedure. When the documentation of the program is analyzed, such comments are supposed to provide all the information that is needed to understand the use of the procedure by the other parts of the program.

As an example of the use of abstraction in software processes, consider the case of cost estimation for a new application. One possible way of doing cost estimation is to identify some key factors of the new system — for example, the number of engineers on the project and the expected size of the final system — and to extrapolate from the cost profiles of previous similar systems. The key factors used to perform the analysis are an abstraction of the system for the purpose of cost estimation.

101.4.5 Anticipation of Change

Anticipation of change is perhaps the one principle that distinguishes software the most from other types of industrial productions. In fact, software undergoes changes constantly, and anticipation of change is a principle that we can use to achieve evolvability.

The ability of software to evolve does not happen by accident or sheer luck — it requires a special effort to anticipate how and where changes are likely to occur. Designers should try to identify likely future changes and take special care to make these changes easy to apply. Software should be designed such that likely changes that we anticipate in the requirements, or modifications that are planned as part of the design strategy, may be incorporated into the application smoothly and safely. Basically, likely changes should be isolated in specific portions of the software in such a way that changes will be restricted to such small portions. In other words, anticipation of change should be the basis for our modularization strategy. (See Chapter 103 on software design).

In many cases, a software application is developed before its requirements are completely understood. Then, after being released, on the basis of feedback from the users, the application must evolve as new requirements are discovered or old requirements are updated. In addition, applications are often embedded in an environment, such as an organizational structure. The environment is affected by the introduction of the application, and this generates new requirements that were not known initially.

Anticipation of change also affects the management of the software process. For example, managers should anticipate the effects of personnel turnover. Also, when the life cycle of an application is designed, it is important to take maintenance into account. Depending on the anticipated changes, managers must estimate costs and design the organizational structure that will support the evolution of the software. Finally, managers should decide whether it is worthwhile to invest time and effort in the production of reusable components, either as a by-product of a given software development project or as a parallel development effort.

101.4.6 Generality

The principle of generality may be stated as follows:

> Every time you are asked to solve a problem, try to focus on the discovery of a more general problem that may be hidden behind the problem at hand. It may happen that the generalized problem is not more complex — indeed, it may even be simpler — than the original problem. Moreover, it is likely that the solution to the generalized problem has more potential for being reused. It may even happen that the solution is already provided by some off-the-shelf package. Also, it may happen that, by generalizing a problem, you end up designing a module that is invoked at more than one point of the application, rather than having several specialized solutions.

A generalized solution may of course be more costly, in terms of speed of execution, memory requirements, or development time, than the specialized solution that is tailored to the original problem. Thus, it is necessary to evaluate the trade-offs of generality with respect to cost and efficiency, in order to decide whether it is worthwhile to solve the generalized problem instead of the original problem.

Generality is a fundamental principle that allows us to develop software components for the market. Such general-purpose, off-the-shelf products represent a rather general trend in software. For every specific application area, general packages that provide standard solutions to common problems are increasingly available. If the problem at hand may be restated as an instance of a problem solved by a general package, it may be convenient to adopt the package instead of implementing a specialized solution. For example, we may use macros to specialize a spreadsheet application to be used as an expense-report application.

101.4.7 Incrementality

Incrementality characterizes a process that proceeds in a stepwise fashion, in *increments*. We try to achieve the desired goal by successively closer approximations to it. Each approximation is an increment over the previous one.

Incrementality applies to many engineering activities. When applied to software, it means that the desired application is produced as a result of an evolutionary process.

One way of applying the incrementality principle is to identify useful *early subsets* of an application that may be developed and delivered to customers, in order to get *early feedback*. This allows the application to evolve in a controlled manner in cases where the initial requirements are not stable or fully understood. The motivation for incrementality is that in most practical cases there is no way of getting all the requirements right before an application is developed. Rather, requirements emerge as the application — or some part of it — is available for practical experimentation. Consequently, the sooner we can receive feedback from the customer concerning the usefulness of the application, the easier it is to incorporate the required changes into the product. Thus, incrementality is intertwined with anticipation of change and is one of the cornerstones upon which evolvability may be based.

Incrementality applies to many of the software qualities discussed in the early part of this chapter. We may progressively add functions to the application being developed, starting from a kernel of functions that would still make the system useful, though incomplete. For example, in a business automation system, some functions would still be done manually, whereas others would be done automatically by the application. We can also add performance in an incremental fashion. That is, the initial version of the application might emphasize user interfaces and reliability more than performance, and successive releases would then improve space and time efficiency.

When an application is developed incrementally, intermediate stages may constitute prototypes of the end product; that is, they are just an approximation of it. The idea of rapid **prototyping** is often advocated as a way of progressively developing an application hand in hand with an understanding of its requirements. Obviously, a software life cycle based on prototyping is rather different from the typical waterfall model, wherein we first do a complete requirements analysis and specification and then start developing the application. Instead, prototyping is based on a more flexible and iterative development model. This difference affects not only the technical aspects of projects, but also the organizational and managerial issues. The unified process, presented in Chapter 110, is based on **incremental development**.

Evolutionary software development requires special care in the management of documents, programs, test data, etc., developed for the various versions of software. Each meaningful incremental step must be recorded, documentation must be easily retrieved, changes must be applied in a controlled way, and so on. If these are not done carefully, an intended evolutionary development may quickly turn into undisciplined software development, and all the potential advantages of evolvability will be lost.

101.5 A Case Study in Compiler Construction

In this section, we will show the application of the principles we have just presented to the practical case of compiler construction.

101.5.1 Rigor and Formality

There are many reasons that compiler designers should be rigorous and, possibly, formal. First, a compiler is a critical product. A compiler that generates incorrect code is as serious a problem as a processor that executes an instruction incorrectly. An incorrect compiler can generate incorrect applications, regardless of the quality of the application itself. Second, when a compiler is used to generate code for mass-produced software, such as databases or word processors, the effect of an error in the compiler is multiplied on a mass scale. Thus, in general, it is important to approach the development of a compiler rigorously, with the aim of producing a high-quality compiler.

Compiler construction is one of the fields in computer science where formality has been exploited well for a long time. In fact, formal languages and automata theory were largely motivated by the need for making compiler construction more effective and reliable. Nowadays, the syntax of programming languages is formally defined through Backus–Naur form (BNF) or an equivalent formalism. It is not by chance that, most often, problems associated with compiler correctness are related to the semantic aspects of the language, which are usually defined informally, rather than the syntactic aspects, which are well defined by BNF.

101.5.2 Separation of Concerns

As with most nontrivial engineering artifacts, the construction of a compiler involves several concerns. Correctness (i.e., producing an object code consistent with the source code and producing appropriate error messages in the case of erroneous source programs) is, as usual, a primary concern. Other important issues are efficiency and user-friendliness. Efficiency could be related to compile time (in which case it amounts to performing source code analysis and translation quickly or using little memory) or to run time (in which case it involves producing an object code that is itself efficient). User-friendliness also has several aspects, ranging from the precision, thoroughness, and helpfulness of the diagnostics to the ease of interacting with the human–computer interface (e.g., through well designed windows and other graphical aids).

These and other aspects of the compiler should be analyzed separately, as far as possible. For instance, there is no reason to worry about diagnostic messages while one is designing a sophisticated algorithm to optimize register allocation. This is not to say, as we already noted in general, that different concerns do not affect each other. Typically, in an attempt to make object code as efficient as possible, we might incorrectly overload some register. Also, attempts to produce good run-time diagnostics (e.g., checking that array indexes are within their bounds) may produce run-time inefficiencies.

Run-time diagnostics and efficiency are a typical example of when separation of concerns can and should be applied while keeping in mind the mutual dependencies between the different aspects. In this case, in fact, the two concerns are often well separated by offering the user the option of enabling or disabling run-time checks. During the development and verification phases, when correctness is still being established and is a major concern, the user turns on run-time checks, making diagnostics the prevailing concern for the compiler. Once the program has been thoroughly checked, efficiency becomes the major concern for its user and, therefore, for the compiler, too; thus, the user could turn off the generation of run-time checks by the compiler.

101.5.3 Modularity

A compiler can be modularized in several ways. Here, we propose a fairly simplistic and traditional modularization based on the several "passes" performed by the compiler on the source code. Such a modular structure should be good enough for our initial purposes. According to this scheme, each pass performs a partial translation from an intermediate representation to another one, the last pass transforming the final intermediate representation to the object code.

The following are the usual compiler phases:

Lexical analysis — Analyzes program identifiers, replaces them with an internal representation, and builds a symbol table with their description. It also produces a first set of diagnostic messages if the source code contains lexical errors (e.g., ill formed identifiers).

Syntax analysis or parsing — Takes the output of the lexical analysis and builds a syntax tree, describing the syntactic structure of the original code. It also produces a second set of diagnostic messages related to the syntactic structure of the program (e.g., missing parentheses).

Code generation — Produces the object code. This last phase is usually done in several steps. For example, a machine-independent intermediate code is produced as a first step, followed by a translation into machine-oriented object code. Each of these partial translations may include an optimizing phase that rearranges the code to make it more efficient.

The foregoing description suggests a corresponding modular description of the structure of the compiler, depicted graphically in Figure 101.3a.

Despite the oversimplification present in the figure, we can already derive a few distinguishing features of modular design:

- System *modules* can be drawn naturally as boxes.
- Module *interfaces* can be drawn as directed lines connecting the boxes representing the modules.

An interface is an item that somehow connects different modules; it represents anything that is communicated or shared by them. Notice that the graphical metaphor suggests that everything that is inside a box is hidden from the outside; modules interact with each other exclusively through interfaces. In the figure, it is convenient to represent interfaces with arrows to emphasize the fact that the item they describe is the output of some module and the input to another one. In other cases, the notion of an interface may be more symmetric (e.g., a shared data structure); in such cases, it is more convenient to represent the item with an *undirected* line.

Notice also that the lines representing the source code, the diagnostic messages, and the object code are the input or output of the whole "system." They are, therefore, drawn without source and target, respectively.

The modular structure of Figure 101.3a lends itself to a natural iteration of the decomposition process. For instance, according to the description of the code-generation phase, the box representing this pass can be refined as suggested in Figure 101.3b.

101.5.4 Abstraction

Abstraction can be applied in compiler design along several directions. From a syntactic point of view, it is fairly typical to distinguish between concrete and abstract syntax. Abstract syntax aims at focusing on the essential features of language constructs, neglecting details that do not affect a program's structure.

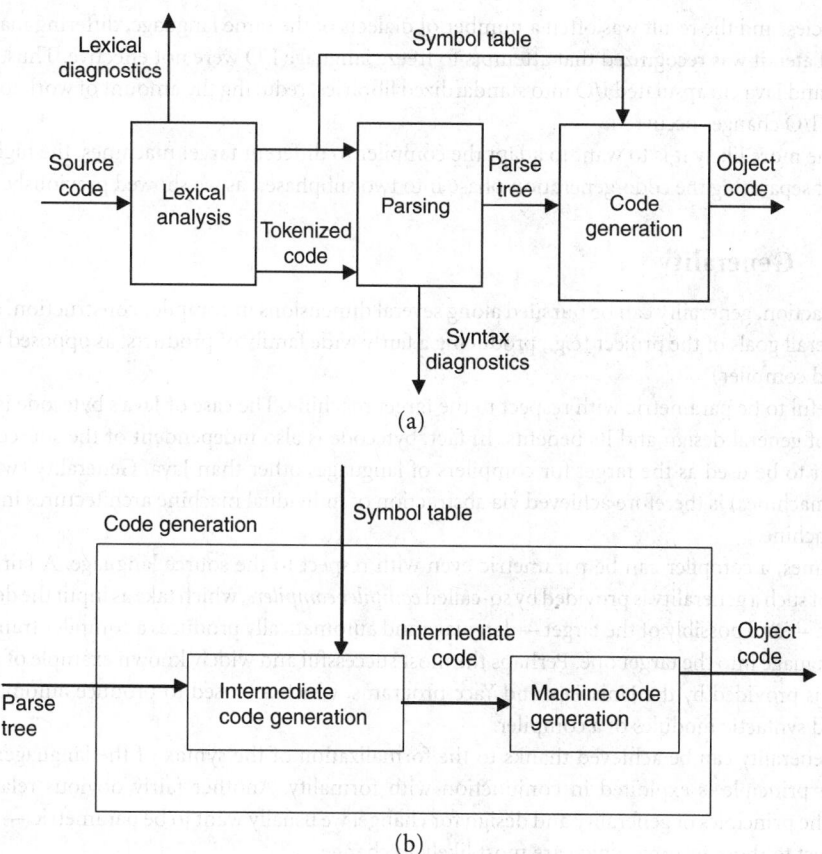

FIGURE 101.3 (a) The modular structure of a compiler. (b) A further modularization of the code-generation module.

For instance, a conditional statement consists of a condition, a statement to be executed if the condition holds and, possibly, a statement to be executed if the condition does not hold. This description remains valid both if we include the keyword *then* before the positive statement, as it happens in Pascal, and if we do not, as it happens in C. Similar remarks apply to the use of the C-like pair {,} and of the Algol-like pair *begin-end* to bracket sequences of statements.

Another typical abstraction is often applied with respect to the target code: as we saw in the previous section, the first phase of code generation produces an intermediate code, which can be viewed as the code for an *abstract machine*. The second phase then translates the code of this abstract machine into code for the concrete target machine. In this way, a major part of the compiler construction abstracts away from the peculiarities of the particular processor that must run the object code. The Java language, indeed, defines a Java Virtual Machine, whose code (Java bytecode) can be executed by interpreting it on different concrete machines.

101.5.5 Anticipation of Change

Several changes may occur during the lifetime of a compiler:

- New releases of the target processors may become available with new, more powerful, instructions.
- New input–output (I/O) devices may be introduced, requiring new types of I/O statements.
- Standardization committees may define changes and extensions to the source language.

The design of the compiler should anticipate such changes. For instance, the Pascal language tried to "freeze" I/O statements within a rigid language definition. This decision conflicted with typical machine dependencies, and the result was often a number of dialects of the same language, differing mainly in the I/O part. Later, it was recognized that attempts to freeze language I/O were not effective. Thus, languages such as C and Java encapsulated I/O into standardized libraries, reducing the amount of work to be redone whenever I/O changes occurred.

Also, the more likely it is to want to adapt the compiler to different target machines, the higher are the benefits of separating the code-generation phase into two subphases, as we showed previously.

101.5.6 Generality

Like abstraction, generality can be pursued along several dimensions in compiler construction, depending on the overall goals of the project (e.g., producing a fairly wide family of products, as opposed to a highly specialized compiler).

It is useful to be parametric with respect to the target machine. The case of Java's bytecode is a striking example of general design and its benefits. In fact, bytecode is also independent of the source language, allowing it to be used as the target for compilers of languages other than Java. Generality (with respect to target machines) is therefore achieved via abstraction of individual machine architectures into just one virtual machine.

Sometimes, a compiler can be parametric even with respect to the source language. A fairly extreme example of such a generality is provided by so-called *compiler compilers*, which take as input the definition of the source — and possibly of the target — language and automatically produces a compiler translating the source language into the target one. Perhaps the most successful and widely known example of a compiler compiler is provided by the Unix Lex and Yacc programs, which are used to produce automatically the lexical and syntactic modules of a compiler.

Such generality can be achieved thanks to the formalization of the syntax of the language. Thus, the generality principle is exploited in conjunction with formality. Another fairly obvious relation exists between the principles of generality and design for change: we usually want to be parametric — general — with respect to those features which are most likely to change.

101.5.7 Incrementality

Incrementality, too, can be pursued in several ways. For instance, we can first deliver a kernel version of a compiler that recognizes only a subset of the source language and then follow that with subsequent releases that recognize increasingly larger subsets of the language. Alternatively, the initial release could offer just the very essentials: translation into a correct object code and a minimum of diagnostics. Then, we can add more diagnostics and better optimizations in further releases. The systematic use of libraries offers another natural way to exploit incrementality. It is quite common that the first release of a new compiler includes a very minimum of such libraries (e.g., for I/O and memory management); later, new or more powerful libraries are released (e.g., graphical and mathematical libraries).

101.6 Summarizing Remarks

In this chapter, we have discussed important software engineering qualities and principles. Qualities are the ultimate goal we want to achieve; principles provide the basis for concrete means to reach the goal.

Because of their general applicability, we have presented the principles separately, as the cornerstones of software engineering, rather than in the context of any specific phase of the software life cycle. We used a case study of compiler construction to show the application of the general principles. We emphasized the role of general principles without presenting specific methods, techniques, or tools. The reason is that software engineering — like any other branch of engineering — must be based on a sound set of principles. In turn, principles are the basis for the set of methods used in the discipline and for the specific techniques and tools used in everyday life.

As technology evolves, software engineering tools will evolve. As our knowledge about software engineering increases, methods and techniques will evolve, too — though less rapidly than tools. Principles, on the other hand, will remain more stable; they constitute the foundation upon which all the rest may be built. They form the basis for the concepts discussed in the remainder of this Handbook. For instance, Chapter 104 presents object-oriented design, a popular methodology that stresses the qualities of reusability and evolvability, as well as the principles of modularity, anticipation of change, generality, and incrementality.

Defining Terms

Cohesion: Property of a modular software system that measures the logical coherence of a module.

Correctness: Software is (functionally) correct if it behaves according to the specification of the functions it should provide.

Coupling: Property of a modular software system that measures the amount of mutual dependence among modules.

Evolvability: Ease of software evolution.

Incremental development: A software process that proceeds by producing progressively larger subsets of the desired product by delivering new increments at each stage. Each increment provides additional functionality and brings the currently available subset closer to the desired one.

Interoperability: Ability of a software system to coexist and cooperate with other systems.

Maintainability: Ease of maintaining software. It can be further decomposed into evolvability and reparability.

Methodology: A combination of methods and techniques promoting a disciplined approach to software development.

Performance: In software engineering, *performance* is a synonym for *efficiency*. It refers to how economically the software utilizes the resources of the computer.

Portability: Software is portable if it can run on different machines.

Productivity: Efficiency of the software process.

Prototyping: A development process in which early executable versions of the end product are delivered as prototypes, with the main purpose of verifying the adequacy of specifications and driving further development.

Quality assurance: The process of verifying whether a software product meets the required qualities.

Reliability: Software is reliable if the user can depend on it.

Reparability: A software system is reparable if it allows the correction of its defects with a limited amount of work.

Reusability: Ease of reusing software components in more than one product. Reusability can also refer to other artifacts (such as requirements, design, etc.).

Robustness: Software is robust if it behaves "reasonably," even in circumstances that were not anticipated in the requirements specification.

Security: A system is secure if it protects its data and services from unauthorized access and modification.

Software process: Activities through which a software product is developed and maintained.

Software product: All of the artifacts produced by a software process. This definition encompasses not only the executable code and user manuals that are delivered to the customer, but also requirements and design documents, source code, test data, etc.

Timeliness: A process-related quality meaning the ability to deliver a product on time.

Usability: A software system is usable — or user-friendly — if its human users find it easy to use. This definition reflects the subjective nature of usability.

Verifiability: A software system is verifiable if its properties can be verified easily.

Visibility: A process-related quality meaning that all steps and the current process status are documented clearly.

References

Boehm et al. [1978]. B.W. Boehm, J.R. Brown, H. Kaspar, M. Lipow, G. MacLeod, and M.J. Merritt, *Characteristics of Software Quality*, volume 1 of TRW Series on Software Technology, North-Holland, Amsterdam.

Fenton and Pfleeger [1998]. N.E. Fenton and S.L. Pfleeger, *Software Metrics: A Rigorous and Practical Approach*, 2nd ed., PWS Publishing, Boston, MA.

Garg and Jazayeri [1996]. P. Garg and M. Jazayeri, *Process-Centered Software Engineering Environments*, IEEE Computer Society Press.

Ghezzi et al. [2002]. C. Ghezzi, M. Jazayeri, and D. Mandrioli, *Fundamentals of Software Engineering*, 2nd edition. Prentice Hall.

Hoffman and Weiss [2001]. D.M. Hoffman and D.M. Weiss, Eds., *Software Fundamentals — Collected Papers by David L. Parnas*, Addison-Wesley, Reading, MA.

Humphrey [1989]. W.S. Humphrey, *Managing the Software Process*, Addison-Wesley, Reading, MA.

ISO 9126 [1991]. ISO/IEC 9126, *Information Technology — Software Product Evaluation — Quality Characteristics and Guidelines for Their Use*, 1991-12-15.

Jazayeri [1995]. M. Jazayeri, "Component programming — a fresh look at software components," in *Proceedings of the 5th European Software Engineering Conference*, Lecture Notes in Computer Science 989, Springer-Verlag, pages 457–478.

Neumann [1995]. P.G. Neumann, *Computer-Related Risks*. Addison-Wesley, Reading, MA.

Parnas [1978]. D.L. Parnas, "Some software engineering principles," in *Structured Analysis and Design, State of the Art Report*, INFOTECH International, pages 237–247.

Further Information

This chapter is adapted from Chapters 2 and 3 of Ghezzi et al. [2002], which is a general textbook on software engineering.

A classification of software qualities is presented and discussed in detail by Boehm et al. [1978]. The international standard [ISO 9126] also provides a list and a discussion of major software qualities. Fenton and Pfleeger [1998] give a comprehensive study of software metrics for quality. Neumann [1995] illustrates the consequences of lack of quality in software. These real-life situations should concern all software professionals.

Humphrey [1989] defined the software process maturity model. The book is devoted to improving software quality through better processes and the assessment of the process. Garg and Jazayeri [1996] is a collection of articles on software environments that integrate and automate the software process.

Parnas's work on design methods is the major source of insight into the concepts of separation of concerns, modularity, abstraction, and anticipation of change. In particular, Parnas [1978] illustrates important software engineering principles. Thirty of Parnas's important papers have been collected in Hoffman and Weiss [2001], along with some updates and commentaries.

Jazayeri [1995] discusses and gives concrete examples of the power of the principle of generality in design and programming.

102

Software Process Models

102.1 Introduction **102**-1
102.2 Specification-Driven Models **102**-2
102.3 Evolutionary Development Models **102**-7
102.4 Iterative Models **102**-10
    Incremental Development • The Spiral Model
102.5 Formal Transformation **102**-14
102.6 The Cleanroom Process **102**-14
102.7 Process Model Applicability **102**-15
102.8 Research Issues and Summary **102**-16

Ian Sommerville
Lancaster University

102.1 Introduction

The development of anything but trivial software systems is a structured activity. Various steps are involved where the software is designed, programmed, and validated. This sequence of activities and their inputs and outputs make up the **software process**. Every organization has its own specific software process, but these individual approaches usually follow some, more abstract, generic process model. These generic **software process models** are the subject of this chapter.

A generic software process model is an abstract representation of the activities and deliverables in the software process. Depending on the level of detail, the model may also show the roles responsible for these activities, the tools used to carry out these activities, communications of different types between activities, and roles and exceptions which must be handled as part of the process. However, in the examples here, the software process is considered at a fairly abstract level and only process activities and their inputs and outputs are discussed.

Software processes are immensely complex. The activities involved in these processes are intellectually demanding and may require significant creativity on the part of the process participants. These processes have a number of attributes as shown in Table 102.1.

If we wish to compare or reason about software processes using these attributes, we must be able to make some assessment of them. Because processes are so complex, a software process model is essential for this purpose. If we wish to improve the software process in some way in order to deliver software more quickly, produce software at lower cost, or deliver software with fewer defects, a defined process model is necessary as a starting point for the improvement process.

Similarly, if we need to communicate and exchange information about software processes, we need a process model. A detailed software process model is an important source of organizational knowledge

TABLE 102.1   Software Process Attributes

Process Attribute	Description
Understandability	Is the process clearly and explicitly defined and is the process definition understandable?
Visibility	Does each activity in the process have a well-defined endpoint and results so that progress is clearly visible?
Supportability	To what extent can the process activities be supported by CASE tools?
Acceptability	Do the engineers who are responsible for the software development find the process acceptable and a realistic match to their everyday activities?
Reliability	Is the process designed so that process errors are avoided or trapped before they lead to errors in the software being developed?
Robustness	Do unexpected problems cause process delays or can the process cope with these?
Maintainability	Can the process evolve to meet changing organizational requirements or process improvements for lower costs, higher quality, or faster delivery?
Rapidity	Are there inherent delays in the process which affect the overall development time from system specification to product delivery?

and is used to communicate organizational standards, procedures, and practices to new engineers and managers.

One of the most important functions of a software process model is to facilitate process management. Process management involves scheduling the process activities, estimating the resources required for each activity, assigning people to carry out the activities, and ensuring that appropriate quality procedures are applied to both the process and the developed product. The process should be designed so that managers can check progress against the plan and accurately judge how resources have been deployed in the process.

Consequently, managers express their plan in terms of their model of the development process. If there is no explicit model, the manager must make assumptions about the process and make estimates on the basis of these assumptions. However, the software developers may actually use a completely different process, so the plan is therefore a misleading guide for project management. An explicit process model, agreed upon by managers and software developers, is an invaluable tool for project communication.

There are, currently, no accepted standards for describing process models. The vast majority of models are expressed informally, using diagrams and descriptive text. Notations used in software design such as data-flow diagrams and entity-relation diagrams may be used to show the flow of work and the relationships between activities, and deliverables. Petri nets have been used to show the timing dependencies between activities, and various more-specialized process description notations have been proposed, although they are not widely used. The best known of these is that suggested by Christie [Christie 1994]. Armenise et al. [Armenise et al. 1992] summarize other process modelling notations.

In the remainder of this chapter, several types of generic process model are discussed. These include the "classical" waterfall model and its derivative, the V-model, models centered on software prototyping, and models which have been developed to accommodate change and uncertainty in a structured way. In the final sections, an assessment is made of the domains where each process model is most applicable, and issues such as process improvement models are briefly discussed.

102.2   Specification-Driven Models

In the 1950s and 1960s, software development was largely an informal process which was undertaken as part of other engineering activities. Rather than being developed according to some formal process standard, software was simply programmed using an informal description of what was required. However, as software systems grew in size and complexity, this informal approach became increasingly inadequate. There were a number of large software project failures where software failed to deliver the required functionality and performance, where the software was unreliable and expensive to maintain, and where the project was completed (or more often, abandoned) years behind schedule.

These failures led a number of organizations to adopt a more structured software development process which was based on the process used for the development of other engineered systems. Thus, there were

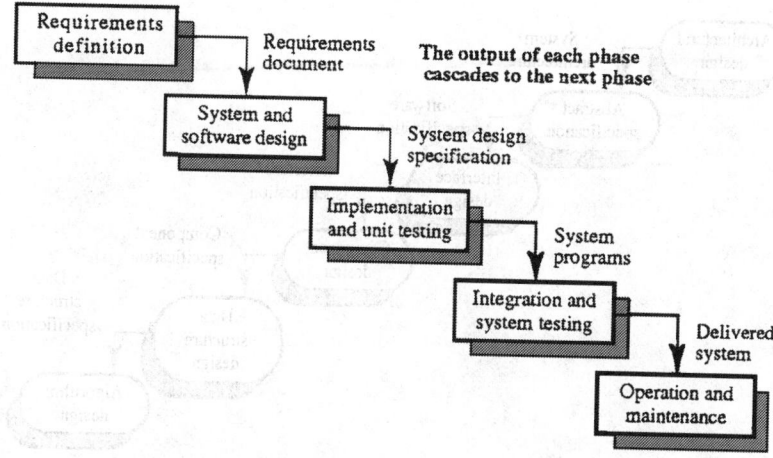

FIGURE 102.1 Waterfall model of the software process.

clearly defined specification, design, and development activities with a "signing-off" process between these activities. When documents were "signed-off," they were deemed to be complete and the next stage of the process could then begin.

The first public discussion of this software "life-cycle" was due to Royce [Royce 1970], who described experience with this approach in defense projects. The initial life-cycle model is now termed the **waterfall model**. This model consists of a set of stages, starting with system specification, with results cascading from one stage to another. Figure 102.1 illustrates this waterfall model.

The waterfall model includes the following phases:

1. *Requirements definition.* The services that the system must provide and its operational constraints are defined.
2. *System and software design.* The overall structure of the system is designed and software subsystems are identified. Depending on the organization, the design may be fairly abstract or developed in detail. Structured design methods may be used to develop the software design.
3. *Implementation and unit testing.* The modules making up the system are individually developed in some programming language and tested.
4. *Integration and system testing.* The system modules are integrated into a complete system and this is tested as a whole.
5. *Operation and maintenance.* The software is delivered to the customer and put into use. During its lifetime, it is modified to meet changing requirements and to repair errors discovered in use.

Of course, this is a very abstract view of the software process. Each of these phases must be decomposed into a set of more detailed activities which, in themselves, may be major tasks involving several person-years of work. Further decomposition is necessary to establish the activities undertaken by each engineer. The essence of "waterfall" decomposition is that further cascades of activities should be identified. For example, Figure 102.2 illustrates the decomposition of the system and software design phase shown in Figure 102.1. Here, design is considered to involve six separate activities, each of which adds detail to the design. The input to each activity is the document produced by the previous design phase and its output is a document for the next phase. Naturally, each of these design activities might be further decomposed into another cascade, but this would usually only be necessary for large software systems.

The waterfall model of the software process has been widely accepted and, in one form or another, has been incorporated in many process standards such as the U.S. Defense standard, MIL-STD-2167A. In these standards, of course, the model is expressed in detail with the deliverables from each phase carefully defined.

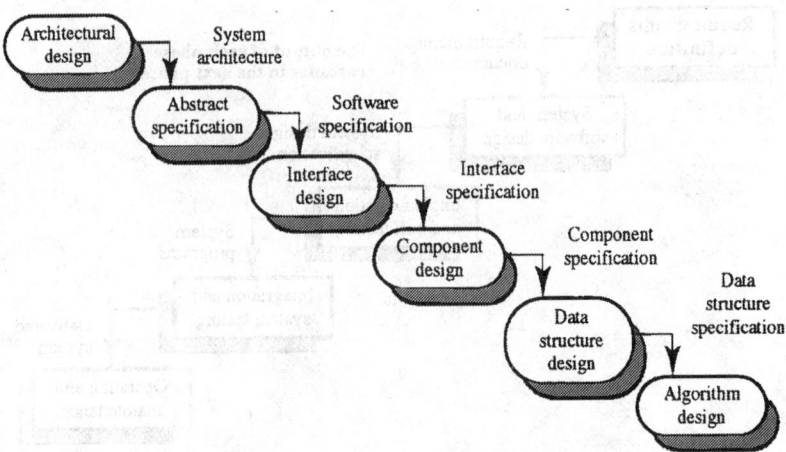

FIGURE 102.2   Waterfall model of the design process.

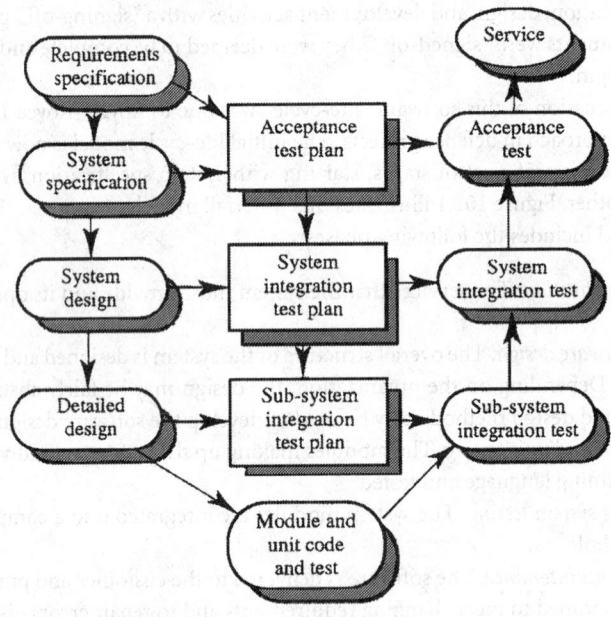

FIGURE 102.3   V-model of development.

There are numerous variants of the waterfall model which propose different breakdowns of the basic activities of specification, design, implementation, and testing. Perhaps the best known of these is the so-called **V-model** of development (the basis of a German government standard) which explicitly links stages in the development process with validation activities. This is illustrated in Figure 102.3.

The V-model maintains the development phases of the waterfall model but links specific validation activities and validation plans with stages in the specification and design process. Therefore, from Figure 102.3, we can see that the acceptance test plan is generated from the system requirements document and an abstract system specification. This is then used to drive the acceptance testing activity after the software has been delivered to the customer. Similarly, the system integration test plan is associated with the system design activity and the subsystem integration test plan with detailed design.

The V-model illustrates one of the omissions of the simple waterfall model. This model suggests that there is a single output from a phase which is the single input to the next phase. Rather, activities have multiple inputs and outputs which include both product and process information. In the V-model, some of this process information is specified. As part of each activity, a plan for the validation of the results of that activity should be produced. This plan should define the validation process and should set out tests which may be applied to the system to check that the implementation conforms to the system specification.

The stages shown in the simple waterfall model closely correspond to the stages in the engineering process of developing some product to be manufactured. There are, however, critical differences between software and manufactured products which mean that the waterfall approach is an over-simplified process representation for software development. These differences are:

1. Manufactured products usually deal either with tangible materials (such as machined components, say) or with things which obey physical laws (such as electricity). It is possible to reason about and visualize their operation. By contrast, software deals with information, which is intangible and almost infinitely flexible. People find it very difficult to anticipate what information they need to help them with their work. They cannot specify their requirements with a reasonable level of confidence.

2. There are generic designs for a most manufactured products (e.g., pumps, amplifiers, wings, etc.). Both the specification and the system design can often be expressed as differences from these basic designs. Those involved in specifying and designing the system have a solid and agreed basis for the design. By contrast, there are few "standard" software system designs which are understood by all software engineers. This means that it is usually much more difficult for those involved in the process to communicate about the software which is being designed.

3. There is no real equivalent in software development to the manufacturing stage of product development. Manufacturing involves expensive outlays in raw materials and tooling. Once a product design has been released for manufacture, making changes to the design is extremely expensive and change is usually contemplated only if there are serious design errors which make the product unfit for its original purpose. By contrast, programming is really a more detailed design phase and changes to the software can be accommodated at any stage in the process.

As a consequence of these differences, there tends to be much more design and specification uncertainty in software systems and changes to the system may be required at any stage in the process.

The difficulties in specifying software with any degree of confidence and the inherent flexibility of software led to a modified form of the waterfall model where there is iteration between the stages in the model. As problems are discovered, they are fed back to earlier process stages for correction and revised documents are issued. New customer requirements also result in change and the production of a revised requirements document. This waterfall model with iteration is illustrated in Figure 102.4.

In principle, the introduction of iteration ought to address the feedback problems of the waterfall model. As problems are discovered in later phases, earlier phases are re-entered and the documents which define the system are modified. Later phases then continue with these modified documents.

The introduction of iteration into the waterfall model is intended to address some of the problems of requirements change but brings with it its own problems:

1. *When should iteration stop?* In principle, the introduction of iteration means that the development process can continue indefinitely. In practice, this is obviously impossible, so there have to be a planned number of iterations in the process. It is very difficult to assess the rate of change of the system, so managers simply plan for some arbitrary number of iterations (normally one or two) during the lifetime of the project. Any further iterations are likely to cause the project to go over budget or to fall behind schedule.

2. *How can change costs be controlled?* There is a high cost involved in analyzing the impact of system changes and reworking the system to incorporate these changes. As changes are, by their very nature, unpredictable, it is very difficult for managers to budget for the costs of iteration.

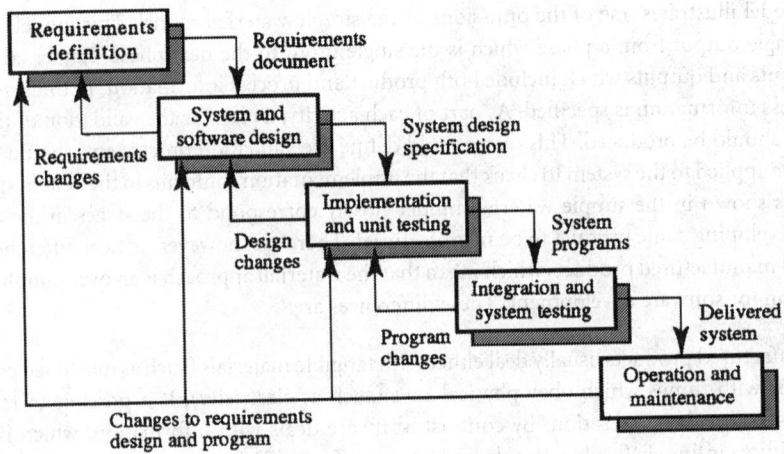

FIGURE 102.4   Iteration in the waterfall model.

3. *How can parallel development of the system be supported?* The sequential nature of the waterfall model suggests that, during iteration, activity within a phase should stop until problems have been resolved and revised documents from earlier phases have been produced. In practice, this is completely unrealistic and, normally, development continues while the problems are being resolved. This often results in a great deal of rework, as changes to the system mean that much of this work has to be thrown away.

In small projects, these problems are often addressed informally. Rather than a formal process iteration, those working on the project work together to resolve the difficulties and to make changes to the software. The real problems arise in large projects, particularly those where different parts of the software are developed by different organizations. In these cases, a formal iteration and change mechanism must be introduced. There is much less scope for informal communication between specifiers, designers, and programmers. Iteration in these projects is therefore very expensive and often results in cost and schedule overruns. Customers, therefore, try to avoid process iterations wherever possible.

While this is understandable, there are some circumstances which can arise where this approach can lead to dramatic project failure:

1. Where the system requirements are poorly understood. This may be because the software system is an innovative one and end users have no intuition of how it will be incorporated into their work; it may be because the software is required to integrate hardware which is incompletely specified; it may be because the software is a product in a market which is changing rapidly. Whatever the reason, if requirements are poorly understood, this means that software change is inevitable. The waterfall model (in any form) does not cope well with frequent change.
2. Where the original system requirements are unrealistic and cannot be met, given the available development budget. Problems of this type are not discovered until the programming of the system is under way and initial versions are available for experiment. Either the system development has to be abandoned or the requirements have to be drastically modified to reflect what may be implemented.

It can be argued that one or both of the above are true for most software projects, so the waterfall model is probably inappropriate for almost all large-scale software development and the deficiencies of the waterfall model are now widely accepted. More and more systems are interactive systems, with complex graphical user interfaces. Experience has shown that the waterfall model is particularly inappropriate for this type of system development.

Nevertheless, in spite of its disadvantages, the waterfall model or a variant of it is still very widely used, particularly for large projects. There are several reasons for this:

1. The use of an engineering model means that the development of the software can be integrated with other engineering activities. Although there are problems in committing to a set of requirements or a design, this is sometimes necessary to allow parallel development of the subsystems in a large system.
2. The model is document-based with one or more documents being produced at each stage in the model. This makes the project visible to management and makes it possible to assess progress against budget and schedule estimates.
3. The model supports contractor/subcontractor relationships, which are normal in large projects. As the output of a stage is documented, the contract for developing the next stage may be let to some subcontractor.

The waterfall model of software development is part of many software development standards and is compatible with the process used to develop hardware systems. It allows for process management and it is familiar to engineers from all disciplines. In spite of its deficiencies, it is therefore likely to remain in use for large systems engineering projects for the foreseeable future.

A further reason for the continued use of the waterfall model is the development of "offshore" software engineering [Dedene and De Vreese 1995]. A system is specified in one country but is designed and implemented in some other country with lower labor costs. The document-based nature of the waterfall model and the separation of the development phases means that it may be applied to support this form of development.

The weakness of the waterfall model is its inability to cope with change, so it is most appropriate for systems whose requirements are well understood and which can be specified in detail with some degree of confidence. Generally, it may be successfully applied to small and medium-sized systems which automate well-understood business processes and which are re-implementations of existing systems or prototypes.

102.3 Evolutionary Development Models

The fundamental difficulties with an approach to development based on clearly defined phases such as specification, design, etc., are that stakeholders in a system find it difficult to articulate their requirements in advance and that these requirements change during the development process. The waterfall model forces premature commitment to a set of system requirements. This means that requirements change and rework is almost inevitable. Furthermore, the system which is finally developed may not meet the real needs of the customers buying that system.

To address this problem, an evolutionary approach to development may be adopted. A rudimentary system is initially produced and evolves, according to the customer's needs, to the final required system. An evolutionary approach to development does not require detailed prespecification of the system requirements, nor does it usually involve the (expensive) production of detailed documentation. These stages in this evolutionary approach are:

1. Formulate an outline of the system requirements. This need be neither complete nor consistent but should give developers some guidance as to what the system should do.
2. Develop a system, as rapidly as possible, based on this outline specification.
3. Evaluate this system with users and modify the system until the system functionality meets the users' needs. This involves modifying the initial functionality of the system and adding new functionality as required.

There are several advantages to this evolutionary approach to system development. Users do not have to develop a detailed requirements specification but need only have some general ideas on the software support which they need. A version of the software is delivered quickly so customers can gain business value from it even when it is incomplete. They also find it much easier to refine their requirements when they have a system for experiment. Ideas can be tried out and refined and the interactions between the

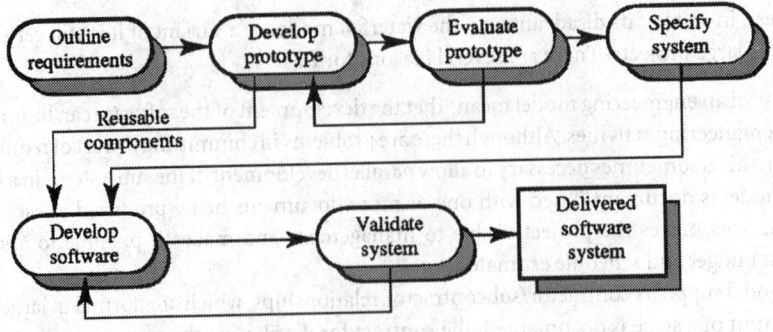

FIGURE 102.5   Throw-away prototyping.

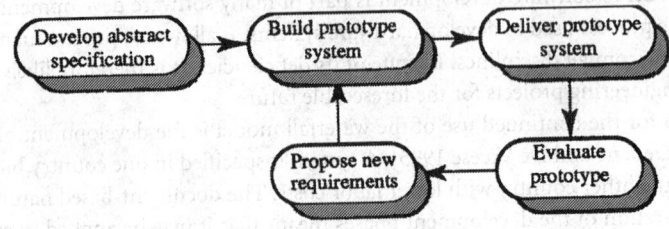

FIGURE 102.6   Evolutionary prototyping.

different parts of the software can be visualized. If the software proves to be unsuitable, it can be discarded at a relatively early stage in the process.

For interactive systems, evolutionary development is particularly important. It is impossible to develop graphical user interfaces according to a waterfall model because of the very high specification uncertainty. People cannot predict, in advance, whether or not the look and feel of the interface will be acceptable and whether users will be able to access system functionality in an effective way. User interface specification and design is so dependent on the application domain and the particular system users that it must be carried out using some experimental prototype system.

Within this general evolutionary framework, there are two generic models which are commonly used. These are:

1. *Throw-away prototyping*. The objective of this evolutionary development process is to understand the customer's real software requirements. A prototype system is developed, then discarded once this understanding has been achieved.
2. *Evolutionary prototyping*. The objective of this process is to develop a software system for delivery to the customer. The prototype system which is developed becomes the production system used by the customer.

These two approaches to development are illustrated in Figure 102.5 and Figure 102.6. Although they appear to be similar, the fact that these approaches have different ultimate goals means that they must be approached in quite different ways.

The throw-away prototyping process has the goal of developing a better understanding of the customer's requirements so that a better system specification can be produced. This specification may then be implemented using some other development approach where the software is designed, implemented, and validated according to a waterfall-like process model. As the prototype is not intended for regular use, its development cost may be reduced by leaving out facilities such as error messages and on-line help. Nonfunctional requirements such as performance and reliability may be relaxed.

By contrast, the evolutionary prototyping approach has the goal of producing a final system for delivery to the customer. This system must include all required functionality and must meet the customer's nonfunctional requirements for reliability, availability, usability, etc. These must always be borne in mind when prototyping the system's functionality. As a consequence, the development of this type of prototype is approached quite differently from the development of a throw-away system:

1. In a throw-away system, the primary objective is to understand the customer's real requirements. Therefore, prototyping should start with those requirements which are *poorly understood* so that customers have as much time as possible for requirements refinement. Well-understood requirements need not be incorporated into the prototype, so it never becomes a fully functional system.
2. In evolutionary prototyping, the primary objective is to deliver useful functionality to the customer as soon as possible. Therefore, prototyping should start with those customer requirements which are *best understood*. Requirements which are less well understood should be added incrementally as customers become familiar with the system.

An evolutionary prototyping approach is now widely used for small and medium size system development, particularly for business systems. In this application domain, prototyping systems such as rapid application development toolkits [Guerrieri 1994] and 4GLs [Wojtkowski and Wojtkowski 1994] have been developed to support the process. These systems make use of the fact that many operations are similar in that they involve retrieving information from a database and formatting that on-screen or in reports. These development systems are therefore centered on a database and include a high-level database query language, tools for screen design, and report generators.

As I have already suggested, evolutionary development is also the normal development approach for interactive systems where the user interface design is a critical part of the system. In this approach, the user interface is designed using an evolutionary approach and the functionality of the system is initially simulated. As the user interface design is refined, more and more functionality is added to the system until a final system is produced. To support this process of development, very high-level languages and environments such as Smalltalk, Lisp or, increasingly, Visual Basic may be used. Alternatively, graphical user interface builders [Colebourne et al. 1993] which allow users to layout interface components and associate functionality with them may be used to support the process.

While an evolutionary development process is more likely to lead to software which is better suited to the *end user's* requirements, it has a number of problems in its own right:

1. It has an end-user focus so critical organizational requirements (such as the need for interoperability, say) may not be given sufficient priority.
2. The constant change to software degrades its structure so that the end result is often difficult and expensive to change. Consequently, the software is expensive to maintain and may have to be completely rewritten after a relatively short lifetime. Organizations are now encountering significant maintenance problems with systems written a few years ago in 4GLs which are no longer supported or which are not available on modern PCs.
3. The process does not have a high visibility and it is difficult for managers to assess how well development is proceeding. As a result, many organizations are reluctant to use an evolutionary approach for large systems where management is the principal problem.
4. It is very difficult to apply this approach to large systems which require a significant infrastructure (e.g., a database and communications) before any system functionality can be provided. If any hardware interfacing is involved, this may have to be specified early to allow the hardware to be designed and manufactured.

Prototyping with user involvement is essential for the development of interactive user interfaces so, for interactive systems in general, evolutionary prototyping is the best generic process to adopt. For small and medium-sized business systems, evolutionary development using 4GLs means that systems can be delivered more quickly and development costs are significantly reduced compared with development in a conventional programming language using the waterfall model. It must be accepted, however, that the lifetime of systems developed using this approach is, inevitably, relatively short.

In applications which have a long lifetime or which have very high performance or reliability requirements, throw-away prototyping may be used for part or all of the system development process. The prototype may be developed using some rapid development method and the final system re-implemented using a programming language which allows better-structured, more-maintainable, and more-efficient systems to be built. For large systems, it is unusual to develop a complete system prototype. Rather, specific parts of the system with high specification uncertainty (such as the user interface) are prototyped before the final system specification is produced.

102.4   Iterative Models

The need for model iteration was discussed earlier in this chapter, where the principal problem with waterfall-type models was the lack of support which they provide for process iteration. Models based on evolutionary development do support iteration but lack visibility and may be difficult to manage. Iterative models are designed to address the need to plan for and accommodate change yet still provide a structured and manageable approach to development.

In this section, I discuss two iterative process models. These are:

1. The incremental development model, where the software is broken down into a set of separately developed increments.
2. The spiral model, where different parts of the system are built in different ways depending on an identification of the risks involved.

These are complementary rather than opposing models, so that a spiral model could be used to develop system increments. I am not aware of any published work which describes such experience, but there is clearly scope for some integration of these approaches.

102.4.1   Incremental Development

The waterfall model of development requires commitment to be made early in the development process. The customer for the system must commit to a set of requirements before design begins, and the designer must commit to particular design strategies before implementation. During the development process, changes in the requirements are normal and require rework of the requirements, design, and implementation.

The incremental approach to development (Figure 102.7) was suggested by Mills [Mills et al. 1980] as a means of reducing rework in the development process and giving customers some opportunities to delay decisions on their detailed requirements until they had some experience with the system. It is an intrinsic part of the Cleanroom process discussed later in this chapter.

In an incremental development process, customers identify, in outline, the services to be provided by the system and they prioritize these services. That is, they identify which of the services are most important and which are least important to them. A number of delivery increments are then identified where each increment provides some subset of the total system functionality. Naturally, the allocation of services to increments depends on the service priority. The highest-priority services are delivered first to the customer.

After the required set of services has been identified, an overall system architectural design is produced where the set of services is partitioned and assigned to different parts of the system. The subsystems making up the system and their relationships are identified. System services may make use of common facilities such as database facilities, network support, etc., so these are also identified at this stage.

Once the system increments have been identified, the requirements for the services to be delivered in the first increment are defined in detail and that increment is developed using the most appropriate development process. During that development, further requirements analysis for later increments can take place, but no requirements changes for the increment which is currently being developed are accepted. Once an increment is completed and delivered, customers can put it into service. This means that they

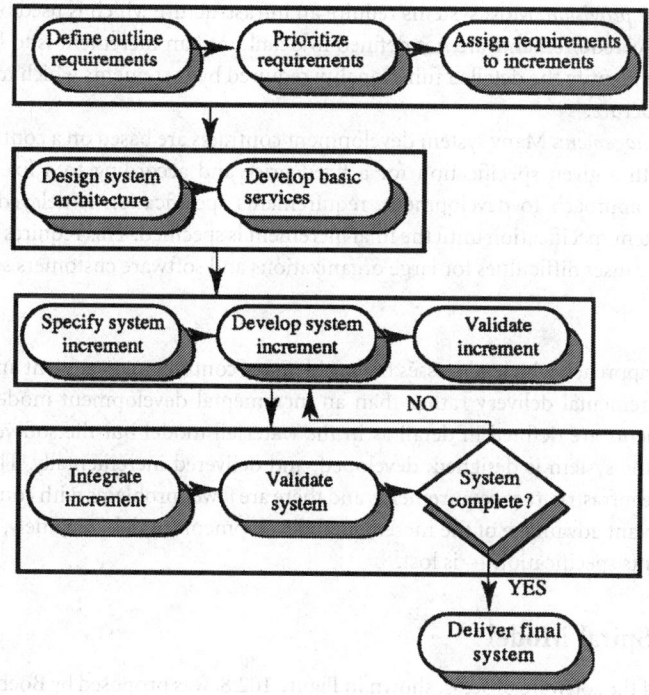

FIGURE 102.7 An incremental development process.

take early delivery of part of the system functionality. They can experiment with the system, which helps them clarify their requirements for later increments and for later versions of the current increment. As new increments are completed, they are integrated with existing increments so that the system functionality improves with each delivered increment. The common services may be implemented early in the process or may be implemented incrementally as functionality is required by an increment.

There is no need to use the same process for the development of each increment. Where the services in an increment have a well-defined specification, a waterfall model of development may be used for that increment. Where the specification is unclear, an evolutionary development model may be used.

This incremental development process has a number of advantages:

1. Customers do not have to wait until the entire system is delivered until they can gain value from it. The first increment satisfies their most critical requirements, so the software can be immediately put into use and contribute to the work of the customer buying that software.
2. Customers can use the early increments as a form of prototype and thus gain experience which informs the requirements for later system increments.
3. There is a lower risk of overall project failure. Although problems may be encountered in some increments, it is likely that some will be successfully delivered to the customer.
4. As the highest priority services are delivered first and later increments are integrated with them, it is inevitable that the most important system services receive the most testing. This means that customers are less likely to encounter software failures in the most important parts of the system.

However, this approach to software development does have some problems. These include:

1. *Increment identification.* Increments should be relatively small (no more than 20,000 lines of code) and each increment should deliver some system functionality. It is sometimes difficult to map the customer's requirements neatly onto increments. Requirements are not independent and, in order to satisfy one requirement, several others may also have to be implemented.

2. *Infrastructure provision.* Most systems require an infrastructure which is used by different parts of the system. As requirements are not defined in detail until an increment is to be implemented, it is difficult to identify the detailed functionality required by increments which must be provided in this infrastructure.

3. *Contract management.* Many system development contracts are based on a contractor's developing a system with a given specification for a fixed price and according to a fixed schedule. In the incremental approach to development, requirements specification is delayed so there is not a compete system specification until the final increment is specified. This requires a new form of contract, which causes difficulties for large organizations and software customers such as government agencies.

A variant of this approach which addresses the problems of contract management and common service provision is an incremental delivery rather than an incremental development model. In this case, the customer requirements are defined in detail as in the waterfall model but the software development is structured so that the system is designed, developed, and delivered incrementally. This means that it is easier to identify the infrastructure requirements and there are fewer problems with contract management. However, an important advantage of the incremental development model — namely, the ability to delay detailed requirements specification — is lost.

102.4.2   The Spiral Model

The **spiral model** of the software process, shown in Figure 102.8, was proposed by Boehm in 1988 [Boehm 1988]. The model views the software development process as a spiral where development spirals from initial conception to final system deployment. The spiral model was developed for use in large defense contractors where some form of waterfall model was the normal software process used and where process standards

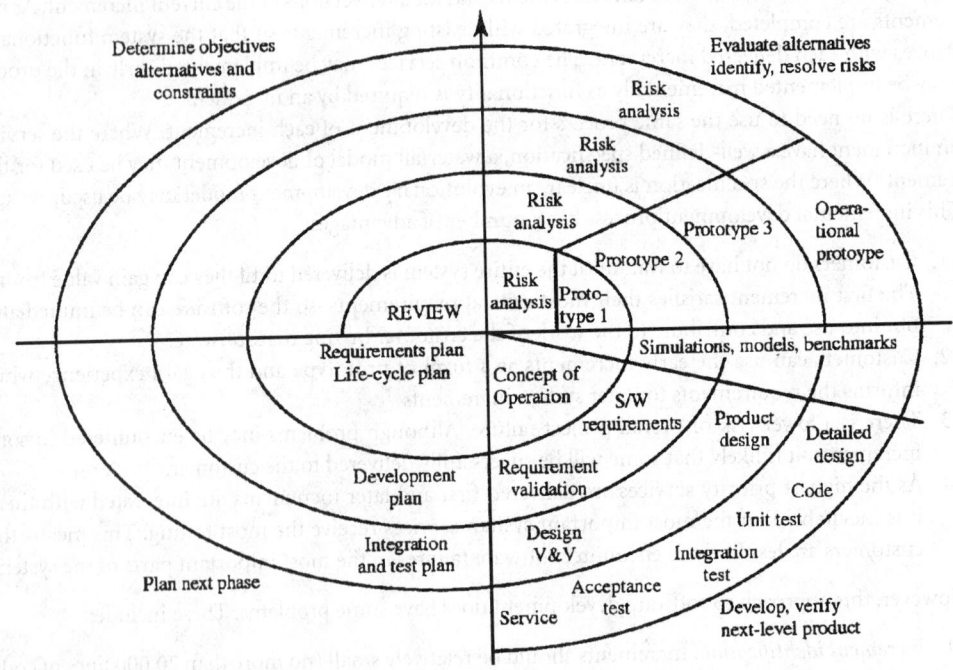

FIGURE 102.8   Boehm's spiral model of the software process. (*From:* Boehm, B. W. 1988. A spool model of software development and enhancement. *IEEE Computer* 21(5). © IEEE, 1988. With permission.)

such as MIL-STD-2167A may have to be applied. It therefore identifies the broad process activities of requirements specification, design, implementation, etc.

The developers of the spiral model recognized the inadequacies of the waterfall approach and designed the model to include activities to resolve design and specification uncertainties. It also allows for regular reviews of progress against well-defined objectives and hence does not suffer from some of the management problems of prototyping approaches.

The spiral model is a risk-driven model where project risks which are applicable to that phase are identified and resolved before progress is made to the next stage. Each phase of the process corresponds to a round of a spiral. Within that round, there are four stages, shown in separate quadrants in Figure 102.8:

1. An objective-setting phase, where the objectives of the round and the development constraints are established.
2. A risk-analysis phase, where the project risks are assessed against these objectives and where, if necessary, risk-resolution activities such as prototyping are carried out.
3. A development phase, which may encompass design, programming, and validation activities. Within this phase, a waterfall development model may be adopted.
4. A planning phase, where plans for the next iteration of the spiral are drawn up.

The most important contribution of the spiral model is its explicit recognition of risk in the software process. In the risk-analysis phase, the critical risks which affect that phase of development are identified and work is done to resolve these risks. Risks are considered to be a lack of information, so risk resolution involves information gathering and analysis.

For example, consider a situation where a system is intended to be a "walk up and use" system, where members of the general public use the system to retrieve information of some kind. The obvious risk here is that the user interface is not appropriate for casual use, so that people without computer experience cannot actually retrieve the information that they require. The risk-resolution activities here would involve the creation of mock-up systems and the development of prototypes which are extensively tested before the final development of the system takes place.

The spiral model is an adaptable model which can encompass other process models. For example, for systems where there is a low specification risk, the model can be considered as a waterfall model with no need for prototyping and risk-analysis activities. Where there is a high specification risk, for example, in the development of user interfaces for interactive systems, the model can be considered as an evolutionary development model with the prototyping activities dominating and relatively little development in the third quadrant.

The spiral model was designed in a large defense contracting organization so, as you would expect, it explicitly addresses the needs of project management. The first phase, which sets objectives, identifies alternatives, etc., identifies a baseline for the management process. Progress can be assessed against this. The second phase of risk analysis identifies risks and provides information to managers about these risks. The development phase may produce project documents, and the final phase is explicitly concerned with the management planning of future phases.

There are two main problems with the spiral model:

1. *The difficulties of risk analysis.* This is a very difficult and specialized process and there are relatively few people with this type of experience.
2. *Contractual problems.* This model is explicitly designed to avoid premature decision making, but the nature of system engineering contracts may make this essential. The model can really be used only where there is trust between client and contractor.

The spiral model has been enormously influential in making the notion of risk explicit to the software engineering community. There is now a widespread recognition of the need for software project risk identification and analysis. While this explicit risk analysis is important for all projects, the spiral model itself, with its emphasis on management, is most suited as an alternative to the waterfall model in large system engineering projects.

102.5   Formal Transformation

There are some classes of system, particularly safety-critical systems, where it is very important that the system conforms to its specification. Classically, this conformance is demonstrated by testing the software using test data which is comparable to that which the software must process when it is in use. However, testing can only demonstrate the presence of software errors and cannot prove their absence. Therefore, when it is essential that the software must conform to its specification, it has been argued that the conventional waterfall model, with its emphasis on system validation, is inappropriate.

Rather, an alternative model may be used which does not include an explicit testing phase to discover errors in the implemented system. In this model, a formal mathematical specification is produced and this is systematically transformed into a system implementation. The specification may be mathematically analyzed to discover inconsistencies, and these are removed at this stage. The specification transformations are correctness-preserving, so that the developed software is an exact implementation of the specification.

As current methods of formal specification and analysis do not allow the compilation of a mathematical specification into an efficient implementation, the transformation of a specification into an implementation is a multistage process. Detail is added to the specification at each stage, and the transformations which are carried out may be either automated or manual with some automated assistance.

This formal transformation process is not widely used, although some organizations which develop safety-critical systems (such as railway signalling systems) are now starting to introduce it for part of their software development. However, a variant of this approach is an important part of the Cleanroom process discussed in the next section.

102.6   The Cleanroom Process

The **Cleanroom process** was developed at IBM's Federal Systems Division with the objective of dramatically reducing the number of faults in the software delivered to customers. It combined aspects of the incremental development and the formal transformation model which have already been covered in this chapter. The Cleanroom process takes its name from the cleanroom used in semiconductor fabrication, where the objective is to provide an environment where defects are not introduced into the semiconductor wafers which are being fabricated.

The Cleanroom process was developed from work in the 1970s on structured programming [Linger et al. 1979] and is described in a number of papers by Mills and Linger [Mills et al. 1987, Mills 1988, Cobb and Mills 1990, Linger 1994]. The model is illustrated in Figure 102.9.

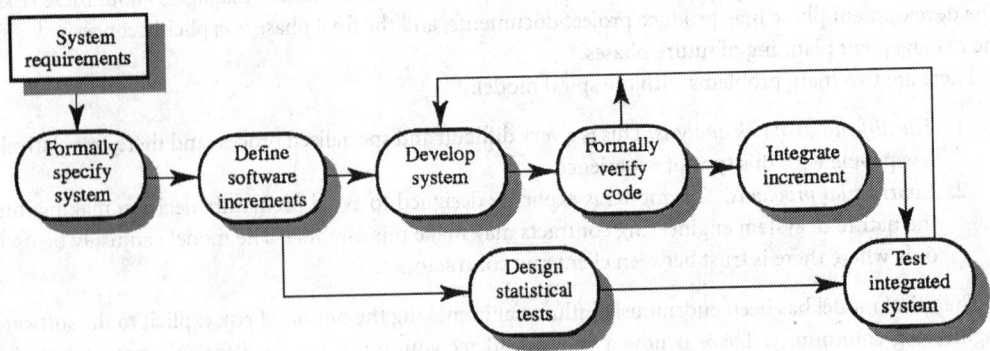

FIGURE 102.9   The Cleanroom process.

The essential characteristics of the Cleanroom approach to software development are:

1. *Formal software development.* The software is mathematically specified and a development process is used where mathematical arguments are used to demonstrate that the developed software conforms to its specification. This is a weaker approach to the formal transformation model discussed above in that there is no systematic, correctness-preserving specification transformation process. However, the approaches are conceptually similar in that neither model includes a defect-testing activity but both rely on mathematical arguments to demonstrate that a program meets its specification.

2. *Incremental development.* An essential characteristic of the Cleanroom process is the incremental development and integration of software. If the software was not structured into increments, it is unlikely that a manageable formal specification and associated mathematical correctness arguments could be produced.

3. *Statistical testing.* The testing process has the goal of validating the reliability of the software rather than defect discovery. Test data are based on an **operational profile**, which is a set of test data which reflects the frequency of actual inputs that the system must process. The number of failures detected processing these inputs reflects the software's reliability. Reliability is predicted using **reliability growth models** [Littlewood 1990].

Reports of the Cleanroom process by its developers suggest that it is very successful in producing software which has a low number of defects. Independent assessment [Selby et al. 1987] confirmed that the Cleanroom process resulted in software which had fewer defects than an approach based on defect testing.

However, the process relies on staff who have the training and ability to work with mathematical specifications and mathematical correctness arguments. This restricts its applicability to organizations which are willing to accept the relatively high training costs involved in introducing this process. It is also unclear how the process may be applied to the development of user interface software, which is an increasingly significant component of most software systems. Formal specification techniques have not been developed for specifying system interaction.

102.7 Process Model Applicability

The most appropriate software process model depends on the organization developing the software, the type of software to be developed, and the capabilities of the staff involved. There is no "ideal" model, and it makes little sense to try and fit all development in an organization to a single approach. The most appropriate process model should be chosen depending on the type of project, the application domain, and the skills and experience of the staff available.

Table 102.2 summarizes the applicability of the different models discussed here.

Large systems normally include subsystems of different types. Rather than impose a single process model for the whole system, each subsystem should be developed according to the most appropriate model. Well-understood parts of the system may be developed using some form of the waterfall model, and those subsystems whose requirements are difficult to predict may be developed using an evolutionary approach.

An issue which has not really been addressed is the relationship between these models and object-oriented development, where there is a blurred boundary between analysis, design, and implementation activities and where there is a significant potential for object reuse. Clearly, incremental development models are suited to this approach, but it is less clear how other process models relate to object-oriented development. This issue is most significant for waterfall models, which are embedded in many standards (such as the US MIL-STD-2167A) and which propose a development process with clear separations between phases. These cannot simply be discarded, and work is required to investigate how to integrate them with object-oriented development.

TABLE 102.2   Process Model Applicability

Process Model	Applicability
Waterfall model	Development of systems whose specification is well understood. This approach may also be used in systems where development is subcontracted, although it is not always technically appropriate in these cases. The waterfall approach may be required by some government organizations.
Throw-away prototyping	Parts of large systems, such as the user interface or expert system components, whose specification cannot be drawn up in advance. Software products where a prototype is developed for test marketing.
Evolutionary development	Interactive systems with a relatively short lifetime. Small to medium-sized business systems based around a database.
Incremental development	The development of large systems whose functionality can be readily partitioned and systems which have well-understood (e.g., a standard DBMS) infrastructure requirements. This approach is particularly appropriate for internal use in an organization; contractual problems are not then an issue.
Spiral model	Software which is part of a large systems engineering project and so involves development by a number of interdisciplinary teams. Again, contractual problems are avoided if the model is used within an organization.
Formal transformations	The development of relatively small safety-critical software systems or systems with very high reliability requirements.
Cleanroom process	Large systems whose functionality can be partitioned and which have very high reliability requirements. Unsuitable for the development of interactive components of these systems.

102.8 Research Issues and Summary

This chapter has discussed a number of generic models of the software process and has suggested where these models are most likely to be appropriate. These have included specification-driven models such as the waterfall model and the V-model, evolutionary and throw-away prototyping, the incremental development and spiral models, and models based on formal software specification.

The three most important factors affecting the choice of process model are:

1. *The project size*. Large projects are dominated by management concerns, so a process model which takes these into account must be used.
2. *The specification uncertainty*. If there is a lot of specification uncertainty, a process based on some form of prototyping must be used for development. In general, the interactive parts of systems always fall into this category.
3. *The nature of the development contract*. The procurer of the software may demand that a particular approach to development is used.

It is now generally recognized that there is no "ideal" process model, so research in this area is not much concerned with the development of new generic models. Rather, current research in the software process is mostly oriented around two related themes:

1. *Process support technology*. What software tools and techniques may be deployed to support the software process and to reduce overall process costs and development schedules?
2. *Process improvement*. How can existing software processes be improved to achieve the same objectives and to develop software with fewer defects?

These themes are closely related as, obviously, one approach to process improvement is process automation where technology is used to take over potentially slow and error-prone human activities.

The notion of **process support technology** is a development of the work in the 1980s on software engineering environments [Taylor et al. 1988, Bott 1989, Thomas 1989, Brown et al. 1992]. These environments were collections of CASE tools to support specific process activities with some degree of integration between these tools. However, while individual CASE tools have been successful in providing support for specific process activities, integrated environments are not widely used. They are not seen as cost-effective by most software development organizations.

There are a number of reasons for this, not least the very rapid change in hardware technology, which has meant that more and more systems are interactive systems for personal computers. These make use of built-in libraries and are often concerned with integrating a number of existing software packages rather than in developing a complete application from scratch. Integrated environments are, in essence, designed to support a waterfall model of development which is not really applicable to this class of system.

It has also been argued that another reason for the lack of use of these large-scale software engineering environments is the fact that they do not provide facilities for process definition and support. Users of these environments should be able to define a detailed model of the software process to be used in a project, the development standards and tools to be used, and the people responsible for each task. The environment should automatically schedule tasks and distribute information, as required, to the engineers involved.

Over the past few years, there has been a great deal of research into this notion of process modelling and associated support technology [Curtis et al. 1992, Krasner et al. 1992, Huff 1995]. There have been a number of experimental environments developed [Finkelstein et al. 1994], and tools such as Process Weaver [Fernstrom 1993] may be used to provide some measure of process automation. However, at the time of writing (1995), this technology is immature and is not widely used. It is debatable whether it will ever become mainstream software technology, as it does not appear to be particularly suited to the development of small and medium-sized application systems. However, for large software and systems engineering projects which require complex configuration management and where tens and perhaps hundreds of developers must be coordinated, process automation technology may have a role to play.

As discussed above, the notion of process improvement, particularly process improvement for defect reduction, is one which has been widely accepted in a number of industries. As failures in software systems are a result of human design errors rather than material failure (say), there is particular scope for improving products by modifying the software process so that product defects are avoided.

In this area, the most influential work has been done by the Software Engineering Institute at Carnegie Mellon University, which published the **capability maturity model** (CMM) for software process improvement [Humphrey 1988, Paulk et al. 1993, Paulk et al. 1995]. This model identifies and classifies key process activities for large-scale projects and suggests that the capability of an organization is a reflection of the number of these processes which are incorporated in the organizational software development process. Other approaches to maturity assessment, such as the Bootstrap approach [Haase et al. 1994] have also been developed.

The capability maturity model is applicable to the improvement of large-scale processes but less appropriate for small organizations concerned with smaller project development. To address this, Humphrey [Humphrey 1995] has proposed an approach to developing a personal software process and process improvement strategies.

The importance of the software process and software process models is now generally recognized. Evolving software processes to meet new demands for rapid delivery of high-quality interactive software is perhaps the major challenge which we face in the future.

Defining Terms

Capability maturity model: A process improvement model which defines a number of levels of process maturity in terms of the key processes undertaken at these levels.

Cleanroom process: An approach to software development based on fault avoidance. The software is split into increments and each increment is formally specified, verified, and statistically tested.

Operational profile: A set of test data whose frequency reflects the actual usage of the system.

Process support technology: Any CASE tools or methods used to support software process activities.

Reliability growth model: A mathematical model which is used to predict when a given level of software reliability is likely to be reached.

Software process: The activities and their inputs and outputs which are involved in developing a software system from initial conception through to final delivery to a customer.

Software process model: An abstract model of the software process which identifies the principal activities and their deliverables. It may also include information about the tools and development environment used and about the roles of the people responsible for particular activities.

Spiral model: An incremental process model based on cycles where each cycle includes objective setting, risk analysis, development, and planning of the next cycle.

V-model: A derivative of the waterfall model where specification, design, and development activities are explicitly linked to validation activities through deliverables.

Waterfall model: A software process model based on engineering models where the system is specified, designed, implemented, and tested in separate phases.

References

Armenise, P., Bandinelli, S., Ghezzi, C., and Mortenzi, A. 1992. Software process representation languages: survey and assessment. In *Proc. 4th Int. Conf. Software Engineering Knowledge Engineering*. Capri, Italy.

Boehm, B. W. 1988. A spiral model of software development and enhancement. *IEEE Comput.* 21(5):61–72.

Bott, M. F. 1989. *The ECLIPSE Integrated Project Support Environment*. Peter Perigrinus, Stevenage, UK.

Brown, A. W., Earl, A. N., and McDermid, J. A. 1992. *Software Engineering Environments*. McGraw–Hill, London.

Christie, A. 1994. *A Practical Guide to the Technology and Adoption of Software Process Automation*. Software Engineering Institute. Carnegie–Mellon University, Pittsburgh, PA.

Cobb, R. H. and Mills, H. D. 1990. Engineering software under statistical quality control. *IEEE Software* 7(6):44–54.

Colebourne, A., Sawyer, P., and Sommerville, I. 1993. MOG user interface builder: a mechanism for integrating application and user interface. *Interacting Computers* 5(3):315–332.

Curtis, B., Kellner, M. I., and Over, J. 1992. Process modeling. Commun. ACM 35(9):75–90.

Dedene, G. and De Vreese, J.-P. 1995. Realities of off-shore engineering. *IEEE Software* 12(1):35–45.

Fernstrom, C. 1993. Process Weaver: adding process support to Unix. In *Proc. 2nd Int. Conf. Software Process*, Berlin.

Finkelstein, A., Kramer, J., and Nuseibeh, B., Eds. 1994. *Software Process Modelling and Technology*. Wiley, New York.

Guerrieri, E. 1994. Case study: Digital's application generator. *IEEE Software* 11(5):95–96.

Haase, V., Messnarz, R., Koch, G., Kugler, H. J., and Decrinis, P. 1994. Bootstrap: fine tuning process assessment. *IEEE Software* 11(4):25–35.

Huff, K. E. 1995. Software process modeling. In *Trends in Software Process*, A. Fuggetta and A. Wolf, Eds., pp. 1–24. Wiley, New York.

Humphrey, W. S. 1988. Characterizing the software process. *IEEE Software* 5(2):73–79.

Humphrey, W. S. 1995. *A Discipline for Software Engineering*. Addison–Wesley, Reading, MA.

Krasner, H., Terrel, J., Linehan, A., Arnold, P., and Ett, W. 1992. Lessons from a learned software process modeling system. *Commun. ACM* 35(9):91–100.

Linger, R. C. 1994. Cleanroom process model. *IEEE Software* 11(2):50–58.

Linger, R. C., Mills, H. D., and Witt, B. I. 1979. *Structured Programming — Theory and Practice*. Addison–Wesley, Reading, MA.

Littlewood, B. 1990. Software reliability growth models. In *Software Reliability Handbook*. P. Rook, Ed., pp. 401–412. Elsevier, Amsterdam.

Mills, H. D. 1988. Stepwise refinement and verification in box-structured systems. *IEEE Comput.* 21 (6):23–37.

Mills, H. D., Dyer, M., and Linger, R. 1987. Cleanroom software engineering. *IEEE Software* 4(5):19–25.

Mills, H. D., O'Neill, D., Linger, R. C., Dyer, M., and Quinnan, R. E. 1980. The management of software engineering. *IBM Sys. J.* 24(2):414–477.

Paulk, M. C., Curtis, B., Chrissis, M. B., and Weber, C. V. 1993. Capability maturity model, version 1.1. *IEEE Software* 10(4):18–27.

Paulk, M. C., Weber, C. V., Curtis, B., and Chrissis, M. B. 1995. *The Capability Maturity Model: Guidelines for Improving Software Process.* Addison–Wesley, Reading, MA.

Royce, W. W. 1970. Managing the development of large software systems: concepts and techniques, pp. 1–9. In *Proc. IEEE WESTCON*, Los Angeles, CA.

Selby, R. W., Basili, V. R., and Baker, F. T. 1987. Cleanroom software development: an empirical evaluation. *IEEE Trans. Software Eng.* SE-13(9):1027–1037.

Taylor, R. N., Selby, R. W., Young, M., Belz, F. C., Clarke, L. A., Wileden, J. C., Osterweil, L., and Wolf, A. L. 1988. Foundations for the Arcadia environment architecture. *SIGSOFT Software Engineering Notes* 13(5):1–13.

Thomas, I. 1989. PCTE interfaces: supporting tools in software engineering environments. *IEEE Software* 6(6):15–23.

Wojtkowski, W. G. and Wojtkowski, W. 1994. *4GL Tools and Methods.* Boyd and Fraser, Boston, MA.

Further Information

Process models and the software process in general are covered in most software engineering textbooks [Pressman 2003, Sommerville 2000]. Research in software process issues is covered in recent books such as those by Finkelstein et al. [Finkelstein et al. 1994] and by Fuggetta and Wolf [Fuggetta and Wolf 1996]. There is a series of international and European workshops on the software process and software process technology and, more recently, an international software process conference has been established. Proceedings of the European workshops have been published by Springer; proceedings of the international workshops and process conference are available from the IEEE Computer Society.

The SEI approach to process improvement is well documented [Paulk et al. 1993], and reports of practical experience of the applicability of these models are available directly from the SEI (accessible through the World Wide Web at http://www.sei.cmu.edu/FrontDoor.html). Smaller-scale process improvement is covered in Humphrey's book on personal software processes [Humphrey 1995]. Process measurement and process improvement are discussed in the ami handbook (Addison–Wesley 1995) and in a series of papers by Basili [Basili and Rombach 1988, Basili and Green 1993].

Basili, V. and Green, S. 1993. Software process improvement at the SEL. *IEEE Software* 11(4):58–66.

Basili, V. R. and Rombach, H. D. 1988. The TAME project: towards improvement-oriented software environments. *IEEE Trans. Software Eng.* 14(6):758–773.

Fuggetta, A. and Wolf, A., Eds. 1996. *Trends in Software: The Software Process.* Wiley, Chichester, UK.

Pressman, R. S. 2003. *Software Engineering — A Practitioners Approach.* 5th ed. McGraw–Hill, New York.

Sommerville, I. 2000. *Software Engineering.* 6th ed. Addison–Wesley, Wokingham, UK.

103

Traditional Software Design

103.1	Introduction	**103**-1
103.2	The High-Tech Supermarket System	**103**-2
103.3	Traditional Approaches to Design..................	**103**-3
	Top-Down and Bottom-Up Design • Data-Flow Diagrams • Entity-Relationship Diagrams • Finite-State Machines	
103.4	Design by Encapsulation and Hiding..............	**103**-9
	Modules • Abstract Data Types	
103.5	Mathematical and Analytical Design	**103**-12
	Queueing Network Models • Time-Complexity Analysis • Simulation Models	

Steven A. Demurjian Sr.
University of Connecticut

103.1 Introduction

Software design techniques span a wide spectrum, and they have incrementally evolved as the discipline has matured over the years. In the early 1960s, flowcharts were the most heavily used design technique for programming, and they subsequently evolved through the 1960s and into the mid-1970s into approaches such as data-flow and entity-relationship diagrams. At this same time, parallel efforts began on approaches for design using **modules** [Parnas 1972] and **abstract data types** (**ADTs**) [Liskov and Zilles 1975, Liskov et al. 1977]. Module concepts were further explored in the late 1970s [Wirth 1977], taking us into the early 1980s, where these design concepts were supported in programming languages such as Smalltalk-80 [Goldberg 1989], Ada [Barnes 1991], and Modula-2 [Wirth 1985]. While it would be impossible to review this entire history of traditional software design in a single chapter, we will introduce and trace the important concepts and techniques.

Software design is not an isolated activity, and some believe it is one of the most important aspects of the overall design, development, and maintenance process. In an oft-cited article, F. Brooks presents the notion that there is no *silver bullet* to solve all of the problems related to software design and development [Brooks 1987]. In the article, Brooks establishes a fundamental challenge across application domains, as follows:

> The hardest single part of building a software system is deciding precisely what to build....
> Therefore, the most important function that the software builder performs for the client is the iterative extraction and refinement of the product requirements.... I would go a step further and assert that it is really impossible for a client, even working with a software engineer, to specify completely, precisely, and correctly the exact requirements of a modern software product before trying some versions of the product. [Brooks 1987, p. 17].

Brooks believes that the focus must be on the design process. Specifically, there is a shortage of what Brooks calls "great designers," the one or two individuals or software engineers who are head-and-shoulders above the other team members and consequently drive the successful completion of a software system. These great designers must be identified, recognized, and rewarded for their expertise. The key is that we cannot separate the individual (software engineer) from the techniques and processes. Software design approaches are irrelevant without knowledgeable individuals who can utilize and exploit the techniques to their fullest extent. Our discussion of the various approaches and techniques will also include, where appropriate, indications of their strengths and weaknesses.

This chapter contains four sections. To serve as a basis for discussion, Section 103.2 introduces the High-Tech Supermarket System, **HTSS**, which is used as an explanation vehicle for the different design approaches presented in this chapter, and in the next chapter on object-oriented software design. In Section 103.3, we review traditional approaches for software design, including **top-down**, **bottom-up**, **data-flow diagrams (DFDs)**, **entity-relationship (ER) diagrams**, and **finite-state machines (FSMs)**, highlighting their strengths and weaknesses. The techniques reviewed in Section 103.3 are often intended for conceptual software design used as software engineers first attempt to understand the system components, structure, and interactions. Section 103.4 examines techniques for encapsulation and hiding via modules and ADTs. These techniques often follow the use of DFDs, ERs, and FSMs, because they allow detailed system structure and interactions to be defined, and serve as a basis for object-oriented software engineering (see Chapter 104). Section 103.5 reviews mathematical and analytical design techniques, specifically, **queueing network models**, **time-complexity analysis**, and **simulation models**. These techniques have been a part of computer science and engineering since its earliest days. They are relevant and important for software design because they offer the ability to predict and estimate performance, a key concern for software engineers.

103.2   The High-Tech Supermarket System

The High-Tech Supermarket System, **HTSS**, uses the newest and most up-to-date computing technology to support inventory control and to assist customers in their shopping. **HTSS** utilizes computing technology in a positive way to enhance and facilitate the shopping experience for customers by integrating inventory control with:

1. The cashier's functions for checking out customers to automatically update inventory when an item is sold
2. A user-friendly grocery item locator that indicates textually and graphically where items are in the store and if the item is out of stock
3. A fast-track deli orderer (deli orders are entered electronically), with the shoppers allowed to pick up the order, weighed and packaged, without waiting

The inventory control aspect of the proposed system would maintain all inventory for the store and alert the appropriate store personnel whenever the amount of an item drops to its reorder limit. The system should also have extensive query capabilities that allow store personnel to investigate the status of the inventory and to track sales for the store over various time periods and other restrictions. Finally, note that **HTSS** and its functional components are based on an actual store that opened in Connecticut in the spring of 1993. Thus, the concepts that are presented have their basis in an actual "real-world" application.

To support the functional and operational requirements of **HTSS**, from an end-user perspective, there must be a set of user–system interfaces, including:

- Cash register/universal product code (UPC) scanner: used to process an order, which includes recording individual items, totaling them, deducting coupons, and taking payment. As each item is scanned, it must be deducted from the inventory, so that values are always consistent and up-to-date.
- Displays for inventory querying: to access and manage the inventory, a separate display is needed. Through this display, orders and updates can be made. Only authorized individuals will be allowed to enter new orders or update the inventory when a shipment arrives.

- Shopper interface for locator: used by customers to locate where (aisle, shelf) a particular item is displayed in a store.
- Shopper interface for orderer: through this interface, customers can place orders for the deli (e.g., meats, cheeses, salads, etc.). These orders are then filled and the customer picks up the order at some later time.
- Deli interface for orderer: this interface is needed by store employees who work in the deli department to scan and fill customer orders.

We have chosen this set on the basis of both their differences (they all have unique requirements for their operation) and similarities (they all share common requirements regarding response time, throughput, and user-friendliness). Response time and throughput are important for the first two interfaces because there are likely to be multiple cash registers that must work in parallel with many inventory displays. User friendliness is also important, for new employees using cash registers, and especially for customers using the different shopper interfaces. Clearly, **HTSS** contains multiple types of data that must interact, performance constraints on throughput and the number of concurrent users, persistence for multiple databases, and a wide variety of users with different capabilities and access requirements.

103.3 Traditional Approaches to Design

Traditional design approaches (e.g., top-down, bottom-up, DFDs, etc.) focus on developing a functional characterization of an application. Historically, there are close ties between these approaches and imperative or procedural programming languages such as FORTRAN, Pascal, and C. The reason is that there is a direct correspondence between the design for an application using one of the approaches and its realization as a working piece of software or program, at both a conceptual level and from the perspective of the coding and organizational techniques that are utilized to develop software using an imperative language. This section is a case study of traditional design approaches — namely, top-down, bottom-up, data-flow diagrams (DFDs), entity-relationship (ER) diagrams, and finite-state machines (FSMs). Each approach can be used in many different ways, and is well suited to developing the solution to a problem from a specific perspective. Each approach can also be used to conceptualize a design at various levels of granularity.

103.3.1 Top-Down and Bottom-Up Design

Regardless of which traditional approach is chosen, the common theme of a functional characterization of the system or application persists. In the case of **HTSS**, a functional list of a subset of the basic system components would include:

1. Check-out customers: the actions that must be taken by a cashier to process a customer's order.
2. Locate items: the actions that are taken to locate specific items in the store. This could be initiated as a result of a customer using the item locator user interface or a bagger needing to check a price or get an item for a customer.
3. Order deli meats, cheeses, and salads: these actions support the deli orderer, and allow the order to be transferred to deli employees for processing.
4. Update and query inventory: these actions are needed by management and stock-room personnel to track and maintain the status of the inventory.

These four major components are determinable based on a *top-down* design examination of the problem at hand. Once these components have been identified, they can each be expressed as a set of detailed tasks via the process of *stepwise refinement*. For **HTSS**, the first component can be refined as the tasks:

1a. Scan UPCs for all items
1b. Modify inventory
1c. Maintain running total

    1d. Subtotal/coupon adjustment
    1e. Final total and take payment
    1f. Etc. etc. etc.

Tasks 1a, 1b, and 1c are repeated to process all items for an order, followed by tasks 1d and 1e. Each of these tasks can, in turn, be refined and expanded by an iterative and incremental process that can evolve the design toward an implementation. For example, as part of task 1e, if a noncash option is chosen, it may be necessary to verify the credit card, automated teller machine (ATM) card, or checking account status. This top-down process proceeds from the general to the specific to arrive at a solution.

    The complement of top-down is the *bottom-up* approach, which, while still functionally oriented, is driven strongly by information and its usage. For example, given the four components previously reviewed (i.e., check-out customers; locate items; order deli meats, cheeses, and salads; and update and query inventory), and the general description of **HTSS** in Section 103.2, it is likely that a data structure can be defined that maintains grocery items. In **HTSS**, each **Item** should have a UPC for unique identification, a **Name**, various **Costs** (e.g., **Wholesale** and **Retail**), a **Size** or **Weight**, the **Amount** on shelves or in the stockroom, and so on. Given this information, the major components of **HTSS** can be examined to identify their access requirements. For example:

- UPC scanner: must scan the UPC on an **Item**, verify it against the database for the inventory, and then return all appropriate information on an **Item** to be used in checking-out a customer's order.
- Locator: once an **Item** has been selected by a customer, the shelf **Amounts** can be accessed to display quantity and location.
- Inventory control: all of the responsibilities associated with managing the inventory, which include creating new entries for **Items**, updating existing entries, deleting **Items**, querying for both scanner and locator, and so on.

From these requirements, the commonalities can be identified and synthesized in a bottom-up process to arrive at a set of functions that can support access to **Items**. For example, **Get_UPC_Code()** and **Get_Shelf_Amount()** are two such functions. These low-level functions are used as building blocks to develop higher-level procedures and functions, which can then support the components of **HTSS**.

    Whether top-down or bottom-up is utilized for design and implementation, there are still a number of important considerations that are not addressed by either approach. First, as new refinements are made (top-down) or higher-level tasks are determined (bottom-up), there is no way to identify when we are done or whether the design matches the specification. Second, both approaches seem counterproductive with respect to user-interface design, because they are prone to separate system functions from user interface needs. This often leads to user interfaces that are evolved rather than formally planned. Top-down and bottom-up design are both suited to smaller, well-defined problems. Top-down and bottom-up design as principles are very important in many other design approaches. For example, they are both critical when developing solutions using an ADT or module approach, as we will discuss in Section 103.4. In addition, in object-oriented approaches (see Chapter 104), top-down design for specialization and bottom-up design for generalization are two critical design concepts used in the construction of inheritance hierarchies.

103.3.2 Data-Flow Diagrams

Another approach to design user *data-flow diagrams (DFDs)*, which describe system operations by means of a high-level characterization of information input/output and the identification of major functional actions and informational flow. In the former, the emphasis is on what information must be input, stored, and displayed, so that it can be effectively used. In the latter, the focus is on how the information is used, by displays, individuals, other systems functions, and so on. DFDs as a design technique are very versatile. To represent high-level system behavior, a macroscopic view of an application, DFDs can characterize major

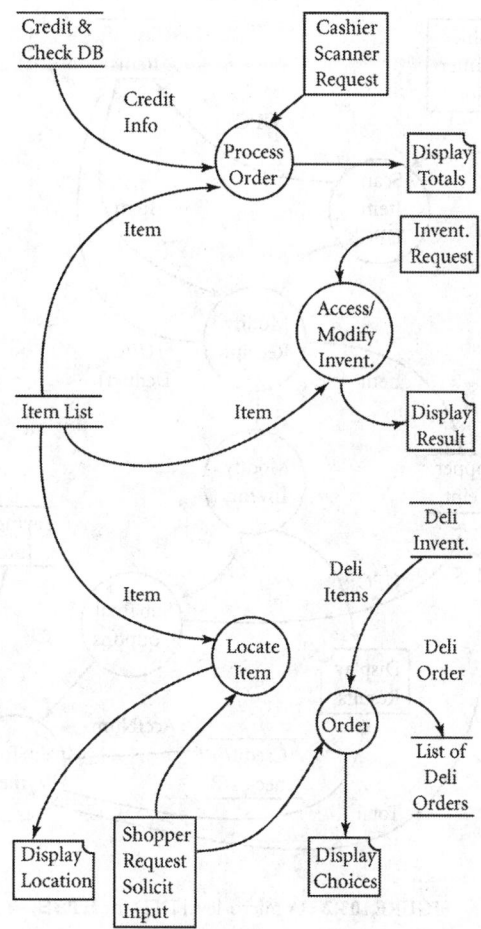

FIGURE 103.1   A macro-level DFD for **HTSS**.

system components, as shown for **HTSS** in Figure 103.1.* The major components or *functions* of **HTSS** are represented in a DFD using circles. Actions for *input* by a user or system are found in the rectangular boxes. *Databases*, or repositories of information, are indicated by the parallel lines (open boxes) that enclose a phrase. *Displayed* or output information is identified by the rectangular box with the upper right corner squiggled. The arrows indicate data or information flow, with labels provided to indicate what is flowing between the various portions of a DFD.

The four functions represented in Figure 103.1 correspond to four of the five user–system interfaces presented for **HTSS** in Section 103.2. The **Process Order** function represents all of the actions taken by the **Cashier** to total a customer's order. This includes getting the **Items** from a database that the customer is buying and verifying payment information by means of a **Credit & Check** database. Other functions on the diagram represent inventory controller actions, and requests by shoppers to either locate **Items** in the store or to **Order Deli Items**. Clearly, the DFD as presented is a gross-level description of the major actions for **HTSS**.

---

*We have utilized concepts and notation for DFDs from Ghezzi et al. [2003].

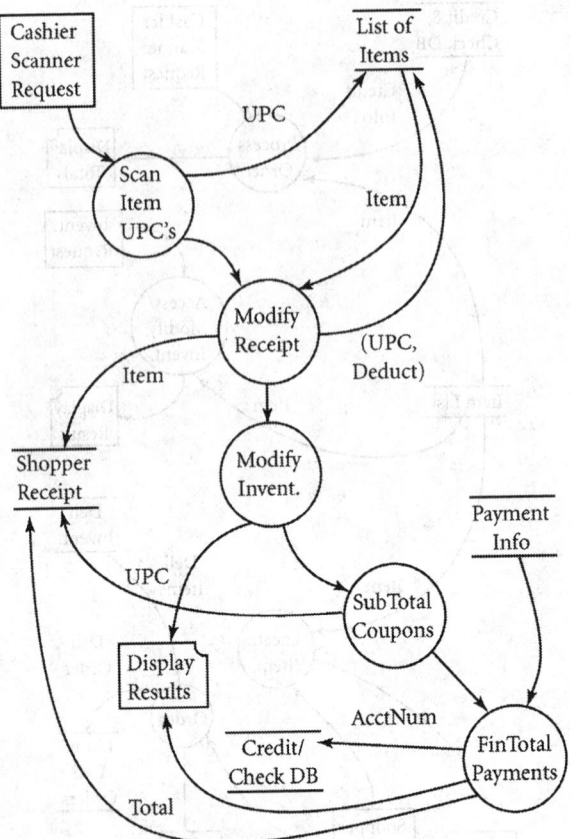

FIGURE 103.2   A micro-level DFD for **HTSS**.

DFDs can also be utilized to expand a certain function of a system in greater detail. For **HTSS**, Figure 103.2 contains a DFD that might represent the **Process Order** function from Figure 103.1. There are three tasks to process an order, represented by five separate functions. First, each item must be scanned (one function), recorded on the receipt (second function), and updated in the inventory (third function). Once all items have been processed, the second task is to subtotal and subtract all appropriate coupons (fourth function). The third and final task completes the order with payment by the customer and verification of valid credit by the cashier (fifth function). To support the five functions, databases are accessed, output is displayed, and flow occurs between them. Overall, the actions in a high-level DFD as given in Figure 103.1 can be decomposed into greater detail as shown in Figure 103.2 as the software engineer incrementally works toward the problem solution.

DFDs are still very popular today because they are very easy to use, learn, and understand, even for individuals who do not have a computer science background. Thus, DFDs are a critical communication medium between the technical (designers and engineers) and nontechnical (customers and end users) individuals who participate in the software design process. However, one important omission in DFDs is the inability to easily specify sequencing and iteration among the various tasks. Consequently, for the DFDs in Figures 103.1 and 103.2, the flow of control across the diagram is neither obvious nor always inferable. FSMs can also be utilized for a software design to capture flow between various system components, supplementing DFDs. Ghezzi also notes that DFDs can only be considered as "... semi-formal notation," and must be used as part of a bigger picture where system structure not represented by DFDs can be captured by other techniques [Ghezzi et al. 2003], which argues for DFDs, ER diagrams, and so on to be used in conjunction with other software design techniques to describe the different facets of an application. When collected, these representations are an overall characterization of an application from complementary and supplementary perspectives.

103.3.3 Entity-Relationship Diagrams

Since DFDs only support the representation of information at a coarse granularity level (i.e., large categories of information with little regard to its makeup and content), they can be complemented with *entity-relationship* (ER) *diagrams* for supporting a detailed conceptual modeling of the database requirements for an application. The ER approach was originally proposed by Chen [1976]. The basic building blocks of the ER approach are entities and relationships. *Entities* are used to model static information aggregations that represent meaningful information components of an application from a database perspective. Entities can have one or more *attributes* associated with them to characterize their structural content. ER diagrams utilize *relationships* to model static information associations between entities. Relationships have cardinalities (e.g., one-to-one, one-to-many, many-to-many) that define the associations. While the original ER approach did not contain the ability to specify inheritance among entities, later extensions have provided that modeling choice. **Inheritance** between entities is available in modern versions of ER diagrams to capture the commonalities that exist from a data/attribute perspective among different entities.

Figure 103.3 is an ER diagram for **HTSS**. In the figure, there are entities for **Item**, **DeliItem**, **CustomerOrder**, **DailySales**, **CreditInfo**, **CreditCard**, **CheckInfo**, and **DebitCard**,

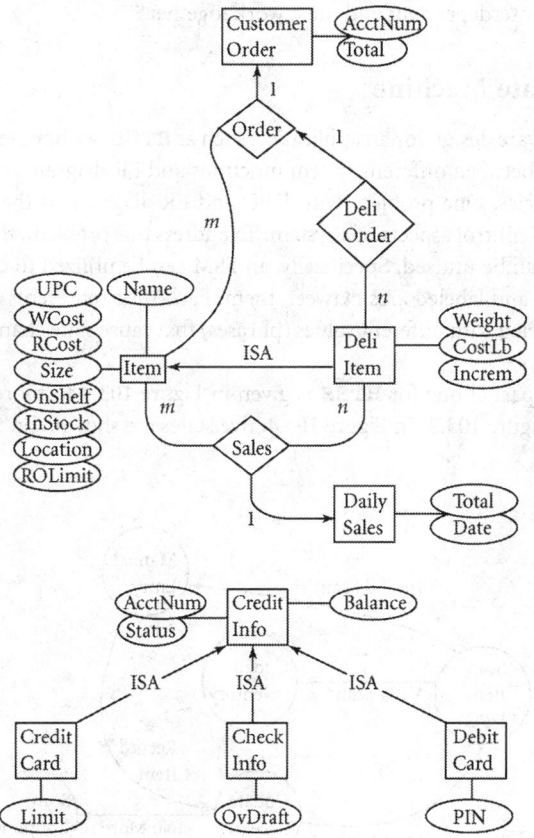

FIGURE 103.3 An ER diagram for **HTSS**.

which are shown in rectangular boxes. The attributes for each entity are enclosed in ovals and connected to each entity via lines. For example, the **Item** entity has attributes for **UPC**, **Name**, **W(holesale)Cost**, **R(etail)Cost**, and so on. Relationships are enclosed using diamonds, and include **Order**, **DeliOrder**, and **Sales** in the figure. **Order** is a one-to-(one-to-many) relationship between **CustomerOrder** and **Item** (one-to-many), and **CustomerOrder** and **DeliOrder** (one-to-one) signifying that one customer order has many **Items** and one **DeliOrder**. A **DeliOrder** is, in turn, composed of many **DeliItems**. Numbers (1, n, m) are used on the diagram to indicate these cardinalities. Finally, inheritance is used to abstract out commonalities across multiple entities. For example, when paying for an order by a noncash method, the account number, status, and account balance are all common and placed in one entity, **CreditInfo**. The information in this entity can then be inherited by other entities, in this case, **CreditCard**, **CheckInfo**, and **DebitCard**. These other entities in turn have their own unique attributes. Inheritance is represented by lines labeled with **ISA** in Figure 103.3.

As a design technique, ER diagrams have many advantages. First, they are an excellent technique for conceptual database design that are easily utilized to represent information, as shown for **HTSS** in Figure 103.3. As with DFDs, both technical and nontechnical individuals utilize ER diagrams as a means to communicate and exchange ideas on the software design. Second, both the information and its interdependencies can be identified and modeled. Third, by supporting inheritance, generalization is promoted to reduce information redundancies. Despite these advantages, there are also many drawbacks. First, by focusing on information and ignoring functional requirements and usage, it is very possible that one can arrive at an ER diagram that does not meet the operational needs of the application. Second, ER diagrams lack the ability of DFDs to represent interactions with other system components (e.g., user interfaces, systems functions, etc.). This is critical for applications such as **HTSS**, where all of the diverse system components are interdependent and must work together.

103.3.4 Finite-State Machines

When developing a software design for an application such as **HTSS**, we have seen that DFDs can capture the flow of information between different system functions and ER diagrams can represent the database structure and dependencies. One problem with DFD and ER diagrams is that neither is well suited to the representation of the control aspects of a system. To address this problem, the design technique *finite-state machines* (FSMs) can be utilized. Specifically, an FSM can be utilized to capture control, by means of a diagram with states and labeled arcs between them. Each *state* represents a functional aspect of the design, with each *arc* labeled with different values (phrases) that cause state changes or *transitions* between states.

To illustrate FSMs, a partial one for **HTSS** is given in Figure 103.4 to more accurately represent the flow from the DFD in Figure 103.2. In Figure 103.4, five states are shown. On the basis of input, control

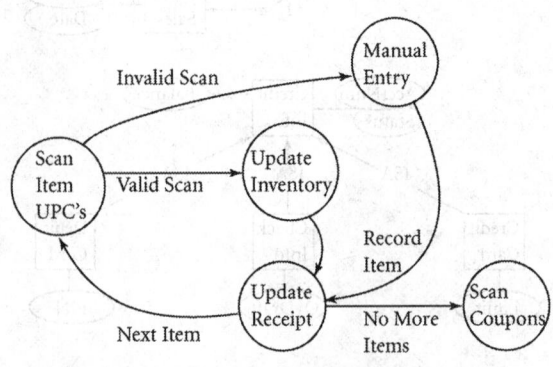

FIGURE 103.4   FSM for totaling a customer's order.

transfers from one state to another. When scanning an **Item**'s **UPC**, alternative actions are taken on the basis of whether the **Item** was found, for example, new items or on-sale items are sometimes omitted from inventory control databases. As the **Items** for a customer's order are scanned and processed, the inventory is updated and a receipt is created and generated. As long as there are **Items** to be processed, control will keep looping through the left portion of the FSM. As soon as all items are processed, control will change and go on to the next step to scan and deduct all coupons. The strength of FSMs is in their ability to capture detailed flow to supplement DFDs and ERs. The weaknesses of FSMs are complementary to the advantages of earlier techniques. First, FSMs lack the data-representational capabilities of ER diagrams. Second, they tend to be more detailed and geared toward software engineers and other technical individuals, unlike DFDs and ERs, which are an excellent discussion medium between software engineers and customers.

103.4   Design by Encapsulation and Hiding

Software design using modules and ADTs is guided by a number of classic software engineering concepts, which are briefly reviewed for completeness.

- **Separation of concerns** and **modularity.** Any domain or application can be divided and decomposed into major building blocks and components (separation of concerns). This decomposition allows the application requirements to be further defined and refined, while partitioning these requirements into a set of interacting components (modularity). Changes to the application are (it is hoped) localized. In addition, team-oriented design and development can proceed with different team members concentrating on particular components.

- **Abstraction** and **representation independence.** Through abstraction, the details of an application's components can be hidden, providing a broad perspective on the design. This in turn allows changes to be made to the internal structure and function of each component, achieving representation independence, because the external view of a module/ADT is not impacted.

- **Incrementality/anticipation of change.** The design process at all times is iterative or incremental. This is true whether a given set of modules/ADTs represents an initial or final design for an application. There is an expectation that components will be changed, added, refined, etc., as needed to support evolving requirements.

- **Cohesion/coupling.** An application is cohesive if each component does a single well-defined task. Cohesion has a long history in computing. In the "early days," the rule of thumb was that each procedure or function should be limited to one output page (approximately 60 lines). If so, then the resulting system was deemed cohesive. *Coupling* is used to signify the interdependencies of components. Coupling is often considered the complement of cohesion, and an application that minimizes coupling has components that require little or no interaction. When the application also exhibits high cohesion, the end result is a well-defined system with well-understood interactions between its components.

These terms and concepts occur repeatedly throughout the remainder of this section as modules and ADTs are presented and discussed.

103.4.1   Modules

Module concepts for design were first proposed by Parnas [1972] in the early 1970s, and were realized in languages such as Modula-2 and Ada. As a concept, the motivation of modules can be tracked to the examination of existing programs that seemed to share similar solution approaches despite their different domains. Specifically, a given data type (e.g., record, structure, etc.) in a program always seemed to have a set of dedicated procedures and functions that represented its capabilities. Informally, this situation was often organized in separate files during implementation in a programming language such

```
MODULE Cashier;                          MODULE Scanner;
  IMPORT Get_Item(UPC),                  ...
         Get_Credit_Info(AcctNum),       END Scanner;
         Get_Deli_Item(UPC), ...;
  EXPORT Display_SubTot(),               MODULE Locator;
         Display_Total(),                ...
         Generate_Receipt(), ...;        END Locator;

  {Code for internal data and for        MODULE Orderer;
   procedures and functions}             ...
  END Cashier;                           END Orderer;

MODULE Items;                            MODULE DeliItems;
  IMPORT {various I/O routines};         ...
  EXPORT Get_Item(UPC),                  END DeliItems;
         New_Item(UPC, Name, ...),
         Modify_Item(UPC, Integer),      MODULE CreditCheck;
         Delete_Item(UPC), ...;          ...
  anItem = RECORD                        END CreditCheck;
     UPC : INTEGER;
     NAME: STRING;
     SIZE: REAL;
     ...
  END;

  {Code for Procedures/Functions}
  END Items;
```

FIGURE 103.5   Sample modules for **HTSS**.

as C. Modules formalized this ad hoc process by recognizing that most programs are partitionable into discrete program units, with well-defined interactions. Individually, each program unit **encapsulated** the required functionality for a given task, and provided an *interface* for a user, while **hiding** its *implementation*. Representation independence is achieved because changes can be made to the implementation that have no impact on the interface and its users. Collectively, the ability to encapsulate functionality and to support future changes was the driving force behind introducing modules as an improved design approach.

In many ways, design using modules can mirror the top-down approach of stepwise refinement. The process begins with the identification of major system tasks. These system tasks might correspond to information or operation. For **HTSS**, as shown in Figure 103.5, modules for system tasks such as **Cashier**, **Scanner** (used by **Cashier**, **Locator**, **Orderer**), and so on, are needed, as well as modules for information such as **Items** and **DeliItems** (which are used by many other modules). Modules can also be utilized in a bottom-up direction, by first identifying the lower-level shared modules, and then incrementally combining modules to represent higher-level system functions. A top-down, bottom-up, or even mixed approach to software design using modules is often dictated by the preferences and experiences of software engineers.

In some ways, modules appear to encompass both DFD and ER approaches. Each task in a system will be realized by a module that contains a dedicated set of information and a set of procedures and/or functions that operate on the information to accomplish its task. A subset of a module's information, procedures, and/or functions is identified or tagged for *export* to other modules. Exported portions of a module represent its interface to the "world." Finally, because a module might interact with other modules, a module may *import* information and functionality for standard actions (e.g., say, for input/output [I/O] of data, printing, strings, etc.) or application-specific functions. Note that portions imported by one module must have been exported by other modules. Concepts for defining a module and importing (exporting) from (to) other modules are shown in Figure 103.5.

When using modules, designers strive for low coupling and high cohesion. Low coupling implies that the interdependencies of modules with respect to exchanging information are minimal. High cohesion refers to the ability of a module to characterize a single well-defined task. Thus, through modules, controlled sharing is promoted; portions that are not exported from a module are hidden from other modules, and representation independence is facilitated. Through either a top-down or bottom-up approach, modules provide a technique that stresses the breakdown of a software design into logical discrete components, and are intended for software designers and engineers rather than customers and end users.

103.4.2   Abstract Data Types

ADTs were first proposed by Liskov for the CLU programming language [Liskov and Zilles 1975, Liskov et al. 1977]. An ADT is characterized by a set of operations (procedures and functions) referred to as the **public interface**, which represents the behavior of a data type. The **private implementation** of the data type is hidden from the programmer or software engineer who wishes to use the ADT. System-available ADTs have been extensively utilized in programming languages for many years. For example, in Pascal, when using the integer data type, the software engineer is able to utilize all of the appropriate operations against integers (e.g., **+, -,** ∗, **div, mod**, etc.), without needing to know the implementation of integers in the underlying machine-dependent architecture. Designer-defined ADTs are readily available today in languages such as Ada, C++, and Java.

ADTs are a design technique that allows software engineers to define their own data types. For example, the classic ADT is a **stack** that contains operations for **push, pop, top, initialize, isempty**, and so on. These operations serve as the *public interface* for the software engineer, and they typically include the type of the stack (e.g., integer, real, etc.) and the parameters and return types of the public operations. However, the *private implementation* of **stack**, which includes the implementation of all operations and the data representation (e.g., array or list, etc.), is hidden from the user. ADTs achieve representation independence by means of a hidden private implementation, and abstraction by means of a visible public interface. Moreover, this allows implementation changes to be made (e.g., say, from an array to a list) as long as these changes are transparent to the public interface (i.e., the operations and their names, parameters, and return types cannot change).

ADTs promote the design and development of applications from the perspective of information and its usage. From an information perspective, there are ties between ADTs and the ER approach. From a usage perspective, the functional characteristics can be explored in either a top-down or a bottom-up direction. However, ADTs take a combined view that focuses on information and its manipulation, which can yield a different design solution than an approach that considers each facet individually.

In the ADT design process, there are a number of considerations that must be addressed:

- *Identify the major information units*: determines the ADTs that are needed for an application or system.
- *Describe the purpose of each unit*: indicates the overall responsibility for each ADT in the application.
- *Define manipulation techniques for each unit*: for ADTs, this corresponds to the operations or methods that must be characterized, including the parameters, return type, etc.
- *Encapsulate and hide*: representation of each unit and its manipulation are both encapsulated and hidden within the ADT.

In addition to describing the design for individual ADTs, we also note the iterative or cyclical process that can be utilized to arrive at a design solution. We have taken a bottom-up approach to ADT design for **HTSS**, as shown in Figure 103.6 and Figure 103.7. In Figure 103.6, the lowest level of ADTs is shown, with an emphasis on the information components for an application. Thus, we have ADTs for **Item, DeliItem, Receipt**, and so on. Given this lowest level, other ADTs can be designed that combine and utilize multiple low-level ADTs, as shown in Figure 103.7. In this case, the **Process_Order** ADT uses multiple ADTs,

```
ADT Item;
    PRIVATE DATA: SET OF Item(s), Each Item Contains:
                  UPC, Name, WCost, RCost, OnShelf, InStock,
                  Location, ROLimit;
             PTR TO Current_Item;
    PUBLIC OPS: Create_New_Item(UPC, ...) : RETURN Status;
                Get_Item_NameCost(UPC) : RETURN (STRING, REAL);
                Modify_Inventory(UPC, Delta) : RETURN Status ;
                Get_InStock_Amt(UPC) : RETURN INTEGER;
                Get_OnShelf_Amt(UPC) : RETURN INTEGER;
                Check_If_On_Shelf(UPC): RETURN BOOLEAN;
                Time_To_Reorder(UPC): RETURN BOOLEAN;
                Get_Item_Profit(UPC): RETURN REAL;
                Get_Item_Location(UPC): RETURN Location;
                ...
    END Item;

    ADT DeliItem;                              ADT CustomerInfo;
        PRIVATE DATA: SET OF (Item, Weight,    ...
                      CostLb, Increm);         END CustomerInfo;
        ...
    END DeliItem;                              ADT Shelf_Info;
                                               ...
    ADT Receipt;                               END Shelf_Info;
        PRIVATE DATA: SET OF Items;
                      SET OF Coupons; {An ADT} ADT Sales_Info;
                      SubTotal, Total, PayType; ...
        ...                                    END Sales_Info;
    END Receipt;
```

FIGURE 103.6   Low-level ADTs for **HTSS**.

while the **Sales_Info** ADT only uses **Receipt**. Current and lower levels are incrementally combined to increase ADT functionality. Eventually, an ADT that describes the uppermost level of system behavior will be specified. Note that a top-down approach to ADTs is also reasonable and feasible.

Clearly, the advantages of ADTs are similar to those of modules. However, because the techniques are somewhat ad hoc, the decisions made at higher levels (for the bottom-up approach) are impacted by lower levels, i.e., if ADTs at the lowest level are wrong, those errors are carried through all subsequent levels. In addition, the lack of inheritance for ADTs will likely result in design redundancies, even though there is **software reuse**.

103.5   Mathematical and Analytical Design

From a more quantitative perspective, there is also a wide range of techniques available to mathematically analyze the different components of a software design. These design analysis techniques play a critical role in ensuring that the requirements identified in the specification are attainable in the application design. For example, queueing network models are utilized to estimate performance of systems on the basis of various load and resource access patterns. Also, individual algorithms can be analyzed in detail to arrive at an understanding of their algorithmic complexity with respect to time (absolute) or as compared to other algorithms (relative). Evaluating alternative design solutions to problems is an important activity

```
ADT Process_Order; {Middle-Level ADT}
    PRIVATE DATA: {Local variables to process an order.}
    PUBLIC OPS : {What do you think are appropriate?}

    {This ADT uses the ADT/PUBLIC OPS from Item, Deli_Item, Receipt,
     Coupons, and Customer_Info to process and total an Order. each
     Receipt must be cataloged and stored when an Order has been
     completed.}
    ...
END Process_Order;

ADT Sales_Info; {Middle-Level ADT}
    PRIVATE DATA: {Local variables to collate sales information.}
    PUBLIC OPS : {What do you think are appropriate?}

    {This ADT uses the ADT/PUBLIC OPS from Receipt so that
     the sales information for the store can be maintained.}
    ...
END Sales_Info;

ADT Cashier; {High-Level ADT}
    PRIVATE DATA: {Local variables used by a cashier.}
    PUBLIC OPS : {What do you think are appropriate?}

    {This ADT uses the ADT/PUBLIC OPS from the middle-level ADTs
     (Process_Order, Sales_Info, etc.), and from the low-level ADTs.}
    ...
END Cashier;
```

FIGURE 103.7   Middle- and high-level ADTs for **HTSS**.

that occurs during the early stages of software design. In the remainder of this section, queueing network models, time-complexity analysis, and simulation models are reviewed.

103.5.1   Queueing Network Models

Throughout the history of computer science and engineering, queueing networks have been extensively utilized to model computer systems [Kleinrock 1975, Kleinrock 1976]. Often, after a specification has been written and either prior or concurrent with the software design process, queueing network models can be utilized as an important tool to estimate and predict performance, including potential bottlenecks. In a basic *queueing network model*, jobs arrive and are queued for processing. Processing commences when all required resources are available. During the processing, it may be necessary to wait for a resource (e.g., I/O device), or a job may be returned to the queue if it has exceeded its allowable processing time slice. The former characterizes the *I/O-bound jobs*, the latter, the *computation-bound jobs*. To represent these alternative possibilities, probabilities can be assigned to jobs that correspond to their waiting times, say, *probability r* for the jobs' *I/O waiting time* and *probability* $(1 - r)$ for the jobs' *CPU waiting time*. For design analysis and performance estimation during software design, both open and closed queueing network models can be utilized. In a *closed* model, jobs neither enter nor depart from the network. On the other hand, an *open* queueing network is characterized by one or more jobs coming into the network and by one or more jobs departing from the same network.

The behavior of jobs within the network is characterized by the *probability of transitions* of jobs between servers. When modeling systems, two of the key variables that must be specified are the *scheduling discipline* and the *size of the queue*. Typically, we can use first-come-first-serve scheduling and unlimited-capacity queues. For an open queueing network model, a characterization of *job-arrival processes* must be defined. For a closed queueing network model, the *number of jobs* in the network must be given.

Queueing network models are especially useful during the software design of **HTSS** because there are so many obvious places where jobs queue up for service. For example, there are multiple cash registers, with multiple customers queued to have their orders processed and totaled. As each order is processed by all cashiers in a concurrent fashion, there are concerns about the performance of simultaneous database access when **Items** are scanned. The overall throughput of the system is critical, to ensure that customers are processed in a timely fashion when the system is under maximum loading. Another aspect of **HTSS** where performance is critical is when deli orders are queued by customers to be filled by deli workers. The number of deli workers (servers), the average number of orders in the queue, and the time needed to fill an average deli order can be used in conjunction to estimate the time delay needed before a customer can proceed to the deli counter to pick up the filled order. The results of queueing network models are an important input to the software design process and can definitely influence and guide software design decisions.

103.5.2   Time-Complexity Analysis

The analysis of the complexity of an algorithm can be used to evaluate the performance of the algorithm [Horowitz et al. 1997], and more importantly, from a software design perspective, to compare performance of multiple algorithms. Algorithm analysis can occur during software design even before code has been written. In one approach, equations can be developed that represent the time spent in carrying out an algorithm. In this case, we may be given a description of the algorithm in pseudocode, in software architecture, or in hardware organization, and from this description, we develop *time-complexity equations* that represent the time spent by the algorithm. There are three different types of time-complexity equations that can be defined: the *best-case time* represents the time to execute an algorithm under ideal conditions, the *average-case time* represents the time to execute an algorithm under typical or average conditions, and the *worst-case time* represents the time to execute an algorithm under the exhaustive or worst conditions. During software design, the type of application can dictate the degree of freedom in an algorithm's performance. For example, in a medical, life-critical application, the worst-case time for an algorithm might be the guiding factor, to ensure that lives are not lost under any conditions. In an application such as **HTSS**, the average-case time may be sufficient.

Time-complexity equations can be developed for two different purposes during software design. On the one hand, time-complexity equations can be used for the *case study* of an algorithm, for example, the best-case, average-case, and worst-case times for the algorithm. On the other hand, time-complexity equations can be used to provide a *relative analysis* of the time spent in different designs of the same algorithm. For example, suppose we wish to compare two different storage/search designs for the database of **Items** in **HTSS**. In one approach, suppose that a sequential data structure (array) is utilized, with **Items** stored in sorted order by **UPC**. Further suppose that for locating an **Item**, the best available search method requires an $O(\log n)$ algorithm, where n is the number of **Items**. On the other hand, suppose a heavily indexed data structure is utilized, which keeps indices on data and uses the indices to optimize the storage for fast retrieval. Suppose that the best retrieval method in this case requires an $O(\log_i n)$ algorithm, where i represents the efficiency of the indexing technique. For values of i that exceed 2, the second approach is superior. The trade-offs between algorithm complexity (both time and space) can be evaluated by a software engineer to determine the algorithm that best matches the problem constraints.

103.5.3   Simulation Models

There are two major types of simulation languages: *continuous simulation languages* and *discrete simulation languages* [Hoover and Perry 1989, Sammet 1969]. In continuous simulation languages, the variables of the simulation change in a uniform fashion for a given uniform increment of time. That is, the variables of the simulation are continuous functions over time. In discrete simulation languages, there is a nonuniform change for the simulation variables over uniform increments of time. That is, the variables of the simulation can be specified as step-functions, where the distance between the steps is discrete. Continuous simulation languages are the oldest, and they have mostly been used for the simulation of analog computers [Sammet 1969].

The GPSS simulation language is characterized as a *block-diagram* or *flowchart-oriented* simulation language [Gordon 1975]. In block-diagram simulation languages, the system to be simulated is decomposed into a fixed number of blocks. Then, to define the simulation structure, the transition or flow between the blocks is specified. The simulation execution involves moving objects between blocks as governed by the transition requirements and restrictions until a termination condition is satisfied. The SIMULA simulation language is characterized as a *process-oriented* simulation language [Hoover and Perry 1989]. In process-oriented simulation languages, the system to be simulated is represented by a fixed number of processes that operate in parallel. The modeling of each process is characterized as a sequence of events. The simulation execution involves moving objects through the process organization of the system. As before, execution terminates when a criterion is met.

Like queueing network models, simulation models can then be utilized to predict and estimate performance under varying system load conditions. Simulation techniques offer a software engineer a more fine-grained estimate of load, because a software design can be decomposed into a number of components that can be analyzed both individually and collectively. Results of simulation models are used in a similar way to queueing network models, allowing a software engineer to understand the software design in greater detail, and leading to informed and justified decisions during the software design process.

Defining Terms

Abstract data type (ADT): A software design approach that decomposes a problem into components by identifying the public interface and private implementation. An ADT allows software engineers to define their own data types to support the informational and behavioral needs of their applications.

Abstraction: See opening paragraph of Section 103.4 for a precise definition.

Anticipation of change: See opening paragraph of Section 103.4 for a precise definition.

Bottom-up design: A software design approach that decomposes a problem into components in a process that proceeds from the low-level specific components to general high-level components.

Cohesion: Please see opening paragraph of Section 103.4 for a precise definition.

Coupling: Please see opening paragraph of Section 103.4 for a precise definition.

Data-flow diagram (DFD): A software design approach that emphasizes the decomposition of a problem on the basis of input information, functional actions, information flow, and output results.

Encapsulation: The software engineering concept that is utilized by ADTs and modules to group together information and methods/operations within a single component. Once these are encapsulated, access to the component can be controlled.

Entity-relationship (ER) diagram: A software design approach that identifies the database components of an application using entities to model information aggregations, attributes on entities to model information content, and relationships between entities to model information associations.

Finite-state machine (FSM): A software design approach that is used to precisely capture control flow using a combination of states, which are the functional components of a design, and arcs, which are labeled with the actions that cause state changes or transitions.

Hiding: Related to encapsulation, hiding involves the private implementation of an ADT or module, which is inaccessible to all other components of an application.

Incrementality: See opening paragraph of Section 103.4 for a precise definition.

Inheritance: A modeling technique used in ER diagrams to represent commonalities that exist between various components in an application. In ER diagrams, these components focus on shared information.

Method: Contains the definition of the actions required for a particular operation against the private data of an ADT. Methods can be in either the public interface or the private implementation of an ADT.

Modularity: Please see opening paragraph of Section 103.4 for a precise definition.

Module: A software design approach that functionally partitions the components of an application into design/program units, which, like ADTs, have a public interface and private implementation. All services that are available from a module are exported to other modules, while all services needed by a module must be imported from other modules.

Queueing network models: A mathematically based software analysis/design technique for estimating and predicting performance for computing systems. Queueing models allow software engineers to identify bottlenecks and determine components that are I/O or computation bound early in the software design process.

Private implementation: That portion of an ADT or module that is hidden from the other portions of an application. Critical for representation independence.

Public interface: That portion of an ADT or module that contains the permissible operations (methods, functions) that are available for use by other portions of an application. Critical for abstraction.

Representation independence: See opening paragraph of Section 103.4 for a precise definition.

Separation of concerns: See opening paragraph of Section 103.4 for a precise definition.

Simulation models: A software analysis/design technique for predicting and estimating performance under varying system load conditions.

Software reuse: The process that describes the ability to reuse existing software in new applications. When software is reused in its entirety without changes, a gain in productivity is attained. Critical for ADT/module-based design.

Time-complexity analysis: A software analysis/design technique where the performance of individual algorithms can be precisely determined from a timing perspective. The best-case, average-case, and worst-case times of different algorithms can be compared against one another to assist a software engineer in making the correct choice of an algorithm for an application.

Top-down design: A software design approach that decomposes a problem into components in a process that proceeds from high-level general components to specific low-level components.

References

Barnes, J. G. P. 1991. *Programming in Ada plus Language Reference Manual*, 3rd ed. Addison-Wesley, Reading, MA.

Brooks, F. 1987. No silver bullet — essence and accidents of software engineering. *IEEE Comput.* 20(4): 10–19.

Chen, P. 1976. The entity-relationship model — toward a unified view of data. *ACM Trans. Database Syst.*, 1(1):9–36.

Ghezzi, C., Jazayeri, M., and Mandrioli, D. 2003. *Fundamentals of Software Engineering*, 2nd ed. Prentice Hall, Englewood Cliffs, NJ.

Goldberg, A. 1989. *Smalltalk-80: The Language*. Addison-Wesley, Reading, MA.

Gordon, G. 1975. *The Application of GPSS V to Discrete System Simulation*. Prentice Hall, Englewood Cliffs, NJ.

Hoover, S. and Perry, R. 1989. *Simulation: A Problem Solving Approach*. Addison-Wesley, Reading, MA.

Horowitz, E., Sahni, S., and Rajasekaran, S. 1997. *Computer Algorithms*. Computer Science Press, Rockville, MD.

Kleinrock, L. 1975. *Queueing Systems I*. Wiley, New York.

Kleinrock, L. 1976. *Queueing Systems II*. Wiley, New York.

Liskov, B. and Zilles, S. 1975. Specification techniques for data abstraction. *IEEE Trans. Software Eng.* 1(1):7–19.

Liskov, B. et al. 1977. Abstraction mechanisms in CLU. *Commun. ACM* 20(8):564–576.

Parnas, D. 1972. A technique for software module specification with examples. *Comm. ACM*, 15(5): 330–336.

Pressman, R. 2001. *Software Engineering: A Practitioner's Approach*, 5th ed. McGraw-Hill, New York.

Sammett, J. R. 1969. *Programming Languages: History and Fundamentals*. Prentice Hall, Englewood Cliffs, NJ.

Schach, S. 2002. *Object-Oriented and Classical Software Engineering*, 5th ed. McGraw-Hill, New York.

Sethi, R. 1996. *Programming Languages: Concepts and Constructs*, 2nd ed. Addison-Wesley, Reading, MA.

Sommerville, I. 2001. *Software Engineering*, 6th ed. Addison-Wesley, Reading, MA.

Tucker, A. and Noonan, R. 2002. *Programming Languages: Principles and Paradigms*, McGraw-Hill, New York.

Wirth, N. 1977. Modula: a language for modular multiprogramming. *Software — Practice and Experience* 7:3–35.

Wirth, N. 1985. *Programming in Modula-2*, 3rd ed. Springer-Verlag.

Further Information

The interested reader is referred to software engineering and programming language textbooks for a more in-depth coverage of these and other design techniques. A sampling of representative textbooks includes [Ghezzi et al. 2003, Pressman 2001, Schach 2002, Sethi 1996, Sommerville 2001, Tucker and Noonan 2002]. In addition, the two main computing organizations, the Association for Computing Machinery (ACM) and the Institute of Electrical and Electronics Engineers (IEEE) Computer Society, both have publications that are targeted to software engineering and design, discussed along with URLs given below. Electronically, there are a variety of resources available on the World Wide Web. R. S. Pressman & Associates, Inc., maintains the site:

http://www.rspa.com/spi/index.html

Topics of interest for this chapter at this site are wide ranging, comprehensive, and relate to all aspects of software engineering, including software design. Upon choosing these topics, the reader is directed to other Web sites with presentations, papers, and discussions.

The ACM maintains the sites:

http://www.acm.org
http://www.acm.org/pubs/journals.html
http://www.acm.org/sigs/guide98.html

The first site listed is the home page for the ACM. At the pubs/journals site, topics for the magazine *Communications of ACM* and the journal *Transactions on Software Engineering and Methodology* are relevant. At the special interest groups (SIG) site sigs/guide, the *SIGSOFT* topic at http://www.acm.org/sigsoft/ is for software engineering.

The IEEE Computer Society also maintains three sites of interest for this chapter:

http://www.computer.org/
http://www.computer.org/publications/
http://www.computer.org/tab/tclist/index.htm

The first site listed is the home page for the IEEE Computer Society. At the publications site, topics for the magazines *Computer* and *IEEE Software* are relevant. This site also has topics for the journals *Transactions on Software Engineering* and *Transactions on Knowledge & Data Engineering*. The site tab/tclist maintains the technical committees supported by the IEEE Computer Society, including the *Software Engineering* topic. Finally, note that all ACM and IEEE Computer Society publications are also available in digital form at most major college and university libraries.

104

Object-Oriented Software Design

104.1 Introduction **104**-1
104.2 Object-Oriented Concepts and Terms **104**-2
104.3 Choosing Classes **104**-3
104.4 Inheritance: Motivation, Usage, and Costs **104**-4
104.5 Categories of Classes and Design Flaws **104**-6
104.6 The Unified Modeling Language **104**-7
Use-Case Diagrams • The Class Diagram • Sequence
Diagrams • Other UML Diagrams
104.7 Design Patterns **104**-11
Defining a Design Pattern • Pattern Catalogs • Design
Patterns to Solve Design Problems • Using a Design Pattern

Steven A. Demurjian Sr.
University of Connecticut

Patricia J. Pia
University of Connecticut

104.1 Introduction

Object-oriented design techniques evolved from abstract data types (ADTs) [Liskov and Zilles 1975, Liskov et al. 1977], and embody the concept of a **class** (replacing the ADT) as the unit of abstraction, partitionable into a **public interface**, which represents the behavior of a data type, and a **private implementation**, which is hidden from the software engineer using the class. Both the interface and implementation can be composed of data members (attributes) and operations (methods). Object-oriented and ADT approaches promote: *representation independence* for changes to the internal structure and function of each component without impacting the external view of a class; *incrementality/anticipation of change* for the addition of functionality, with the expectation that components will be changed, added, refined, etc., as needed to support evolving requirements; high *cohesion* (each class performs a single well-defined task) and low *coupling* (controlled interactions among classes). The distinguishing factor between ADTs and the object-oriented approach is *inheritance*, which allows controlled sharing between classes, permitting the passing of data and/or operations from the parent (superclass) to the child (subclass).

Proponents of object-oriented design agree that it provides a clearer and easier conceptualization of the intended application. Further, the increased emphasis on design (in both time and effort) is supposed to lead to a reduction in implementation effort. Thus, object-oriented design is advocated for a number of important reasons, including:

- Stresses **modularity**: achieved by the class concept and encapsulation.
- Increases **productivity**: while difficult to prove, this is a long-standing claim of object-oriented design.

- Controls **information** consistency: this is attained, since hiding allows the access to the private implementation of a class to be managed.
- Promotes **software reuse**: software engineers can reuse existing classes for solving other problems. In addition, through inheritance, software engineers can define new classes that acquire the characteristics of existing classes without violating the hidden implementation.
- Facilitates **software evolution**: abstraction, encapsulation, hiding, and inheritance allow minor changes to private implementations to be transparent, while major increases in functionality can be realized by extending the existing **class library** through inheritance.

Testing cuts across all these claims: done on a class-by-class basis (modularity); once tested, a class can be used and reused (productivity); testing of changes is limited to the private implementation as long as the public interface has not been changed (evolution).

From a historical perspective, the object-oriented approach for software design and development emerged in the mid-1980s and has become dominant. There are a wide range of object-oriented programming languages, for example, Java [Deitel and Deitel 1997, Campione et al. 2001], Ada 95 [Barnes 1996, Department of Defense 1995], Modula-3 [Harbison 1992], C++ [Stroustrup 1986], Eiffel [Meyer 1992], Smalltalk [Goldberg 1989], and Object Pascal [Tesler 1985], etc., formalized in Wegner [1990]. There was work done in object-oriented database systems (e.g., Ontos [Ontologic 1991], Gemstone [Bretl et al. 1989], O2 [Deux et al. 1991], Orion [Kim 1990], ObjectStore [Lamb et al. 1991], etc.) with formal underpinnings in Kim [1990] and Zdonik and Maier [1990]. These techniques continue to evolve today, as evidenced by the emergence of components and extensions to relational database systems that offer object-oriented capabilities.

To investigate, explain, and discuss object-oriented software design, this chapter contains six sections. Section 104.2 reviews key object-oriented design concepts, to establish the terminology. Section 104.3 examines the issues and factors involved in choosing classes. Section 104.4 focuses on the motivation, usage, and costs of inheritance. Section 104.5 examines design considerations and flaws related to determining classes and utilizing inheritance. Section 104.6 explores the **unified modeling language** (UML), which came to prominence in the mid-to-late 1990s to unify the approaches of Rumbaugh et al. [1991], Jacobson et al. [1992], Booch [1991], and others [Coleman 1994, Meyer 1988, Lieberherr 1996, Wirfs-Brock et al. 1990] into a standard for object-oriented design [Booch et al. 1999]. Finally, Section 104.7 examines the technique of **design patterns**, based on the idea that recurring patterns in object-oriented design and development can be generalized and categorized to leverage reuse [Gamma et al. 1995]. To facilitate the discussion throughout this chapter, the High-Tech Supermarket System, **HTSS**, from Section 103.2 is employed.

104.2 Object-Oriented Concepts and Terms

This section summarizes the concepts and terminology for object-oriented design:

- **Class:** used to model the features (information) and behavior (methods) for an application, and partitioned into a public interface and private implementation.
- **Private implementation:** the portion of a class that is hidden from all other parts (classes) of an application, containing information and/or methods.
- **Public interface:** the portion of a class that contains the permissible operations (methods) that are available for use by other parts (classes) of an application, which may also contain information.
- **Information:** typically, the private data of a class. Information represents the different internal data components that define the class and characterize all of its instances. When public, the information content and consistency cannot be guaranteed.
- **Method:** contains the definition of the actions required for a particular operation against the private and/or public data of a class.
- **Encapsulation:** the coupling of information and methods within a class.

- **Hiding:** controlling access to the information and methods of a class.
- **Inheritance:** the controlled sharing of information/methods between related classes of an application. In inheritance relationships, the parent is referred to as the superclass, while the child is referred to as the subclass.
- **Inheritance hierarchy:** all inheritance relationships between classes that share a common parent (or grandparent) form a hierarchy with an identifiable root (ancestor). Inheritance hierarchies are simply the trees that organize the sharing between all related classes.
- **Instance/object:** an occurrence of a class, or the actual information/data.
- **Message:** an action (method call) that is initiated by an instance on itself or by other instances.
- **Class library:** all classes and inheritance hierarchies for an application form a common library for use by tools and end users.

Advanced concepts that are important to fully appreciate the potential and power of object-oriented design include **generics** and **dispatching.** A *generic* is a type-parameterizable class. For example, instead of having a stack that is bound to a specific data type (say, **integer**), the stack can require that the data type be provided as part of its initialization. Thus, the creation of a stack [e.g., **Stack(Real)**, **Stack(Char)**, etc.] binds the stack's methods to the appropriate types. *Dispatching* is the runtime or dynamic choice of the method to be called on the basis of type of the calling instance. As a concept, the effective use of dispatching is tightly bound with inheritance, and it offers many benefits: versatility in the design and use of inheritance hierarchies; promotion of reuse and evolution of code, allowing hierarchies to be defined and evolved over time as needs and requirements change; and development of code that is highly generic and easier to debug (and hence reuse/evolve).

104.3 Choosing Classes

The first and most frequent question asked by newcomers to object-oriented design is a variant of *How are classes chosen?*. Typical (lazy) answers to this question echo software engineering mantra (e.g., strive for encapsulation with high cohesion and low coupling) or rely on that old-time favorite, "As you gain more experience with objects, your ability to identify them will also improve." However, neither answer is really satisfactory. A "better" answer should relate the following:

> Choosing classes is not a first step in the design and development process, but rather must follow in a logical fashion from earlier efforts. In practice, the choice must be guided by a specification for an application that contains the intent and requirements. The specification will make use of other software design techniques (e.g., data-flow diagrams, ER diagrams, etc., see Chapter 103) to define the scope and breadth of functions, user interfaces, required user/system interactions, and so on. As the specification gains in content and complexity, the relevant classes begin to define themselves as a natural side effect. One hopes that this leads to an object-oriented design. This in turn will be explored, refined, and evolved into a detailed design, which can then be transitioned to an implementation.

The moral is that it is unrealistic to "jump" to an object-oriented design from only a basic understanding of an application. It would be just as unreasonable to make such a jump using any software design technique.

Instead, one must acquire an understanding of what is appropriate to put into a class. Three possibilities are illustrated below:

```
                    Private Data    Public Interface

      Employee      Name            Create_Employee()
      Class         Address         Give_Raise(Amount)
                    SSN             Change_Address(New_Addr)
                    . . .           . . .
```

```
ATM_Log      Acct_Name        Check_Database(Name)
Class        PIN_Number       Verify_PIN(PIN)
             ...              Log_on_Actions(Name, PIN)
                               ...

ATM_User     Action           Log_on_Steps()
Class        Balance          Acct_WithDrawal(Amt)
             WD_Amt           Check_Balance(Number)
                               ...
```

The first class was designed from an information perspective, and to track **Employees**, standard data and operations are needed. The second class, **ATM Log**, embodies the functions for an individual to initiate an ATM session, which also requires information to capture user input for verifying status. The third class, **ATM User**, represents a user interface by capturing the different interactions that are supported.*

During this design process, there are a number of possible design flaws. First, a software engineer places too much functionality in one class. In this situation, the class can be split into two or more classes, or the functionality can be absorbed into other, existing, classes. The latter leads to a second design flaw: a class lacks functionality. In this situation, two or more classes are often merged to result in a more cohesive class.

104.4   Inheritance: Motivation, Usage, and Costs

Inheritance distinguishes object-oriented design from its ADT ancestor. To successfully utilize inheritance, an iterative process to identify commonalties (generalization) and distinguish differences (specialization) is undertaken. For example, in **HTSS**, a first attempt at needed classes, concentrating only on information in each class could be:

```
SnackItem     LiquorItem        MeatItem        ... Other Items ...
  Name          Name              Name
  UPC           UPC               UPC
  ShelfLife     SpecialTaxes      ExpireDate
```

From the above example, two data components, **Name** and **UPC** are common to all **Item**-related classes and can be (generalized) into the **Item** class:

```
Item          SnackItem:Item     LiquorItem:Item      MeatItem:Item ...
  Name          ShelfLife          SpecialTaxes          ExpireDate
  UPC
```

The original classes (e.g., **MeatItem:Item**) now inherit private data (e.g., **Name** and **UPC**) from their ancestor (e.g., **Item**).

In general, inheritance-related decisions are often based on the overlaps that exist between information and/or operations across multiple classes. The end result of this modeling activity is one or more inheritance hierarchies that are extensible, evolvable, and reusable. To complement generalization, specialization distinguishes classes by pushing down differences to lower levels of the hierarchy. Thus, the superclass is refined and focused to represent the shared characteristics required by all descendants. For example, in the **HTSS** hierarchy,

---

*Any discussion on object-oriented concepts would not be complete without an ATM example. Mercifully, this is the only one in this chapter.

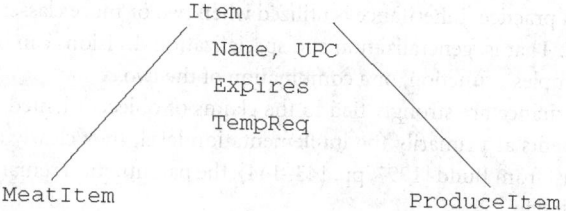

the **Expires** private data of **Item** should be moved to **MeatItem**, which requires explicit expiration dates, while **TempReq** should be moved to **ProduceItem** to track produce that requires refrigerated versus room-temperature storage. Conversely, generalization operates bottom-up by examining a set of classes in an attempt to identify commonalties, which are then pushed into a superclass. For example, in **HTSS**, the classes

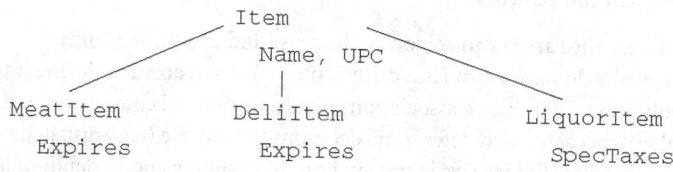

could be revised into the hierarchy

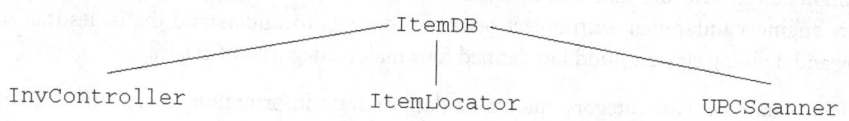

which defines a new **Item** superclass. Note that it might also be necessary to define a new level that is a parent of **MeatItem** and **DeliItem**, and a child of **Item** (e.g., a **PerishItem** for perishable foods). Note also that in these and other examples, the names of variables have been conveniently the same, which, in practice, may not be the case.

Another type of inheritance is based on the concept that classes with significantly different and unrelated abstract views might require access to the same underlying implementation. For example, in **HTSS**, an **ItemDB** is required to track all items that are sold. There are multiple abstract views for **ItemDB** that can be represented in an inheritance hierarchy:

```
                          ItemDB
           _____/   |   _____
          /                  |                  \
   InvController         ItemLocator          UPCScanner
```

In this example, the user interfaces for managing the inventory, locating items, and scanning UPCs all need access to the same information, but they clearly have different intent and usage within **HTSS**. Another type of inheritance combines two or more classes, allowing software engineers a unified view, which can be achieved via *multiple inheritance*; for example, **DeliItem** would inherit from both **ProduceItem** and **MeatItem**.

In summary, inheritance can be defined as a relationship where the subclass acquires information and/or operations from a superclass. Inheritance is a transitive operation, allowing behavior to be passed from parent to child to grandchild, etc. Because the subclass is more refined than the superclass, it has been *specialized*. Alternatively, because the superclass contains common characteristics of all of its subclasses, it has been *generalized*. In practice, inheritance is utilized when two or more classes are functionally and/or informationally related. That is, generalization and specialization decisions can be based on information (as in the previous examples), function, or a combination of the two.

The benefits of inheritance are strongly tied to the claims of object-oriented design; and while Budd characterizes these benefits at primarily the implementation level, they clearly also apply to the design level. In the following list from Budd [1997, pp. 143–144], the parenthetical remarks represent the benefits of inheritance during design:

- *Software (design) reusability.* When a set of one or more classes is reused from an earlier effort, there is a high degree of assurance that compiled and tested code/behavior has been provided. (Similar reuse during design accrues corresponding benefits regarding the completeness of a design component in addition to downstream benefits for the implementation.)

- *Code (design) sharing.* When two or more subclasses of the same superclass share code as the result of a specialization or generalization, redundancy can be reduced; that is, there is only one copy of code to be implemented/tested. (It can be strongly argued that the sharing of code must be identified during the object-oriented design process. Otherwise, the resulting design had many flaws that were not found until the implementation process began.)

- *Software components.* These promote the concept of reusable components and the software factory. (Software factory ideas should not be limited to code, since a "component" may represent a portion of a design or an implementation. In fact, the design patterns discussed in Section 104.7 are an example of design components.)

These benefits for inheritance are key to achieving the reuse and evolution claims.

However, these benefits do have a cost [Budd 1997, pp. 145–146], accrued during implementation and runtime for the application. Inheritance associations require additional compile time (e.g., for overloaded names, multiple inheritance, etc.) and defer some decisions to runtime (e.g., dispatching). As application complexity increases, the class library size increases, both in numbers and in depth of inheritance hierarchies, which impacts on the compile and runtime environments; that is, there is a heavy cost for dynamic linking. At a more practical level, new and experienced software engineers often define too many operations with too little functionality in a class. This affects the activation records in the runtime environment, especially when the operations call other operations, resulting in nested activation records. In fact, it is often the case that the activation record requires significantly more memory during runtime than the code of a poorly defined "small" operation.

104.5 Categories of Classes and Design Flaws

In Chapter 3 of Budd [1997], other important considerations regarding the categories of classes [pp. 48–49] and common design flaws are examined. Both of these considerations play an important role in assisting a software engineer unfamiliar with object-oriented concepts to understand the issues that impact on designing and defining classes. Budd has defined four major categories of classes:

1. *Data managers.* This category maintains data or state information to capture and contain the functionality for an application.
2. *Data sinks/data sources.* This category either produces or processes data, and it may be generated on demand to meet a short-term need.
3. *View/observer.* This category can serve as an interface for an end user or might be used to collect the public interfaces of multiple classes into a single view.
4. *Facilitator/helper.* This category has no functionality when viewed independently, but exists solely to support other classes. For example, typical facilitators include a string class library, I/O handler classes, etc.

Note that classes spanning two or more categories should be decomposed, because they likely contain excessive functionality; that is, they are noncohesive. While it is important to specify each class and to understand to which category a given class belongs, it is more critical to know the context of the class within the overall application. The reason that the class has been specified, the role it will play in the application, and the other classes that it interacts with should all be clearly understood when examining the class.

Like any other approach, object-oriented design is not intended to allow a software engineer to arrive at the "completed" design in one step from the specification. Rather, object-oriented software engineering encourages and promotes an incremental and iterative process, allowing an application to evolve from its specification into an object-oriented design. Thus, when defining classes in the aforementioned categories, there are a number of common design flaws that might occur after the first few iterations of the design process, including:

1. *Classes that directly modify the private data of other classes.* This is a **major** error in the use of object-oriented design techniques. To rectify this situation, the public interface must be upgraded to include operations that encapsulate the modification of private data.
2. *Too much functionality in one class.* In this situation, the obvious solution is to split the class into two or more classes that exhibit higher cohesion. The result should also be lowly coupled.
3. *An class that lacks functionality.* The reverse of the previous case requires the merging of two or more classes. Any time that classes are merged, the impact on existing classes and inheritance hierarchies must be carefully examined.
4. *Classes that have unused functionality.* In this situation, the key is to understand the reason for the unused functions. Have they been duplicated elsewhere? Were they needed in an earlier prototype? The answers to these questions will dictate the choices made in this case.
5. *Classes that duplicate functionality.* As in the previous case, it must be understood why the duplication has occurred. Was it due to a specification problem? Did two software engineers unintentionally define the same class?

It is expected that the initial attempts at an object-oriented design for an application will not be perfect. The key issue is to learn to recognize imperfections so that they can be corrected in subsequent iterations. The list of common design flaws is not comprehensive; rather, it defines the most commonly occurring errors so that they can be easily eliminated by a novice and avoided as a software engineer gains experience.

104.6   The Unified Modeling Language

The *unified modeling language* (UML) is a language for specifying, visualizing, constructing, and documenting software artifacts [Booch et al. 1999]. UML unified the approaches of Rumbaugh et al. [1991], Jacobson et al. [1992], Booch [1991], and others [Coleman 1994, Meyer 1988, Lieberherr 1996, Wirfs-Brock et al. 1990] into a standard. This was motivated by the fact that the individual methods of object-oriented design were evolving toward one another; and a unified semantics and notation would bring stability to an object-oriented design market leading to the "best" features of the constituent models included in the UML. The end result is the availability of a wide variety of UML modeling tools; namely, Together Control Center, Rational Rose, Paradigm Plus, Softmodeler, GDPro, etc. To accompany the unification, there were some important goals for UML, including:

1. A ready-to-use, expressive visual modeling language that promotes design and development via an exchange of ideas, concepts, solutions, etc.
2. Independent of programming languages and development processes, and suitable for multiple application domains.
3. Encourage the growth of object-oriented tools market, which has been facilitated of late with the emergence of XML as an data exchange standard, allowing models created in one tool (e.g., Rational Rose) to be imported into another tool (e.g., Together Control Center).

In the UML, diagrams are utilized to create *views* into a model, to provide different perspectives for different **stakeholders** (e.g., users, designers, software engineers, developers, etc.), and to offer alternate representations of the different parts of the system (e.g., data-flow diagrams, finite-state machines, ER

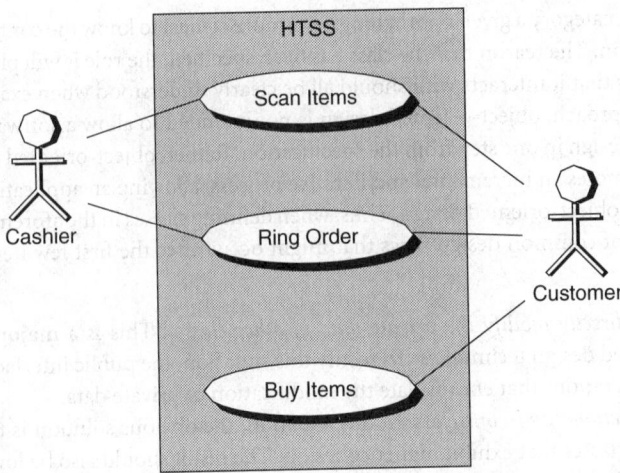

FIGURE 104.1   High-Level Use Cases for **HTSS**.

diagrams, etc.). There are nine standard diagrams in UML: static views of use case, class, object, component, and deployment diagrams; and dynamic views of sequence, collaboration, statechart, and activity
diagrams. In the remainder of this section, selected diagrams and associated modeling techniques for
UML are explored. First, the *use-case diagram* for modeling the interactions of users with system components, is explained. Next, the *class diagram* for the static structure of the conceptual model is examined.
Then, one of the dynamic views, the *sequence diagram*, is detailed, which provides a characterization of
object interactions over time. Finally, we conclude with a brief discussion of the other remaining UML
diagrams and their usage. Note also that as a brief review of UML, many details and capabilities have been
omitted; the reader is referred to Booch et al. [1999] for a comprehensive discussion of UML.

104.6.1   Use-Case Diagrams

Use-case diagrams track the interaction of users with system components, and are comprised of three
different types of elements: actors, systems, and use cases; as shown in Figure 104.1. An **actor** is an
external entity that interacts with software at some level, to represent the simulation of possible events
(business processes) in the system. Actors can be people, classes, software tools, etc. The events that actors
interact with are referred to as **use cases**, which represent discrete (and possibly related) functions for
an application that is being modeled. Use cases can be collected into a **system** when they are related to
one another. Collectively, a **use-case diagram** is a graph of actors and use cases (enclosed by optional
system boundaries), which represents a black-box view of system components. The focus in a use-case
diagram is on the actions, methods, functions, etc., that are utilized by different actors, typically derived
via user/customer interviews.

  The granularity of use cases is variable, depending on the situation being modeled. To illustrate use
cases and the granularity differences, consider Figure 104.1 and Figure 104.2, which contain use cases
for **HTSS**. Note that these figures and all other UML figures in this chapter have been constructed
using the Together (registered) Control Center™ (version 6.2) UML tool (www.togethersoft.com). In
Figure 104.1, a high-level use case for the system **HTSS** is shown, with **Cashier** and **Customer** actors, and use cases to **Scan Items**, **Ring Order**, and **Buy Items**. Interactions between actors
and use cases are shown with lines from the actor to each use case. Conversely, in Figure 104.2 a lower
level use case of the system **Process Order** is shown, with **Supervisor**, **Sales**, and **Customer**
actors, and use cases to **Establish Credit**, **Order**, **Place Order**, **Fill Order**, and
Check Status. In addition to actor/use case interactions, Figure 104.2 also illustrates the three

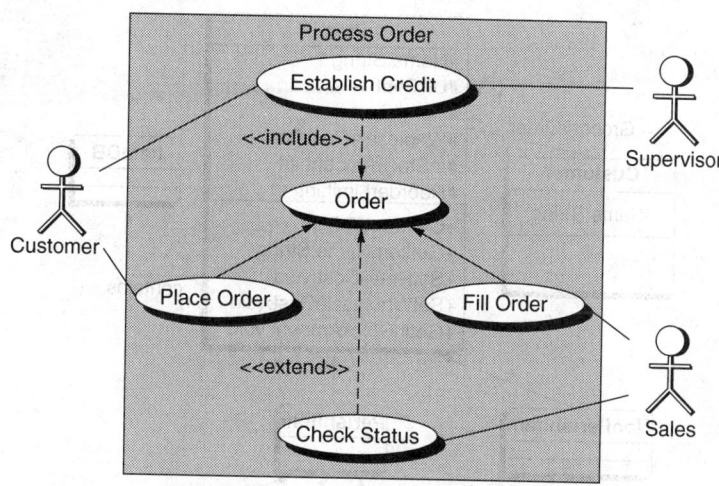

FIGURE 104.2   Use Cases for `Process Order` of `HTSS`.

dependencies among use cases: **Check Status** *extends* **Order** is a relationship where the **Check Status** use case adds behavior to the **Order** use case; **Establish Credit** *includes* **Order** is a relationship that denotes the inclusion of the behavior sequence of the **Order** use case into the interaction sequence of the **Establish Credit use** case; and **Place Order(Fill Order)** *generalizes* **Order** and relates the specialized use case **Place Order(Fill Order)** to a more general **Order** use case.

104.6.2   The Class Diagram

A **class diagram** in UML is utilized to capture the static structure of the conceptual model for an application, describing its classes and the static relationships among them. A class diagram in UML contains classes that have attributes and operations, where each attribute/operation can be distinguished as public (available for all to use), private (hidden from use), and *protected* (available to descendants via inheritance). Classes can be logically grouped into packages and related to one another via associations, generalizations, and other dependencies. In UML, classes are graphically represented as boxes with compartments for class name, attributes, and operations, and the ability to track properties, responsibilities, rules, modification history, etc. Over time, a software designer develops classes incrementally, adding features to existing classes, creating new classes, defining new relationships, etc.

To illustrate a class diagram, Figure 104.3 contains a UML diagram for **HTSS**. In the diagram, we have a class inheritance hierarchy containing **Item** (parent), **NonPerishItem** and **PerishItem** (children of **Item**), and **DeliItem**, **MeatItem**, **ProduceItem** (children of **PerishItem**). In addition, there are classes for a **Customer**; **ItemDB**, which is the collection of all **Items** for a supermarket; and **DeliOrder**, which represents the **Items** to be filled by a deli worker. There is a **GroceryOrder** association between a **Customer** and the Items being purchased, and contains aggregations (between **ItemDB** and **Item**, and **DeliOrder** and **DeliItem**). In all classes, attributes are listed under the class name, followed by operations (all separated by horizontal lines). Private members (attributes or operations) are prefaced by a minus sign, protected members by a sharp, and public members by a plus sign. In the **Item** class, the attributes are all protected (inheritable by descendants), with data types as given, while the operations are all public with return types listed. The **Item** class is also **abstract** (not shown), which indicates that it cannot be instantiated, while the **DeliItem**, **MeatItem**, and **ProduceItem** classes are **final** (not shown), which means that they cannot have children.

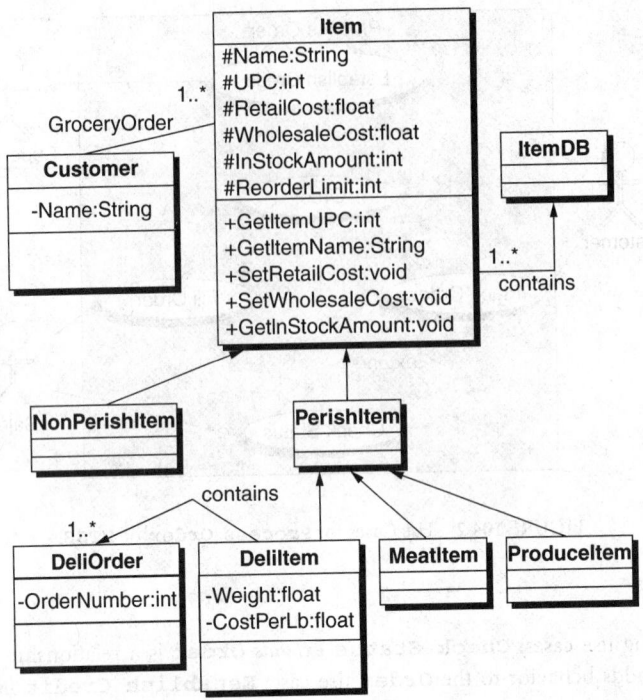

FIGURE 104.3   Classes for **HTSS**.

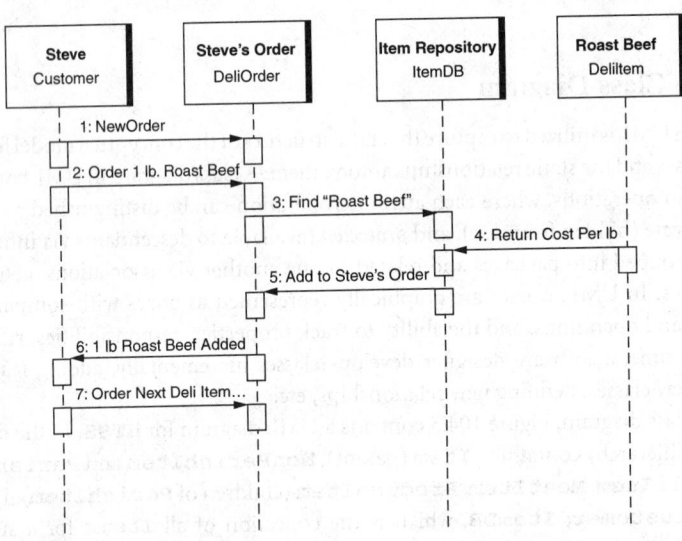

FIGURE 104.4   A sequence diagram for **HTSS**.

104.6.3   Sequence Diagrams

A **sequence diagram** in UML is intended to capture and represent the dynamic behavior of instances (objects) of the class diagram (see Figure 104.3). For a given task, a sequence diagram indicates the object interactions over time to accomplish the task. The purpose of a sequence diagram is to model flow of control and, in doing so, to illustrate a typical scenario or processing, thereby providing perspective on usage and flow across the various objects that comprise an application. Figure 104.4 contains a sequence diagram for

HTSS to represent the actions taken when a **Customer** is performing the actions associated with ordering deli meats. In the figure, there are four objects, **Steve, Steve's Order, Item Repository**, and **Roast Beef**, which are, respectively, instances of **Customer, DeliOrder, ItemDB**, and **Deli-Item**. (Note that Figure 104.4 also illustrates the objects of an object diagram.) Connecting the objects, in a numbered sequence, are messages that indicate the actions (messages 1 to 7) that are taken for **Steve** to enter his **DeliOrder** over time (from top to bottom in the figure). At this early stage of the design process, the messages are strings of text. At later stages, they can be replaced with public method calls where appropriate. In reading the seven messages, it is clear on the flow, the actions by each object, and the dependencies among the objects, which must all eventually be captured in the software (code) as it is developed.

104.6.4 Other UML Diagrams

In this section, UML diagrams that have not been discussed so far are reviewed. In addition to use-case, class, and object diagrams, other static diagrams include:

- A **component diagram** captures the high-level interaction and dependencies among software components that represent the physical structure of the implementation which is built as part of the architectural specification of an application. The purposes of the component diagram are to organize source code, construct an executable release, and specify a physical database.
- A **deployment diagram** allows software architects and network/performance engineers to focus on the placement and configuration of components at runtime, including the topology of an application's hardware. The purpose of the deployment diagram is to specify the distribution of components and, as a result, to strive to identify potential performance bottlenecks.

The sequence diagram is one of four total dynamic diagrams that track behavior from different perspectives, including:

- A **collaboration diagram** is structured from the perspective of interactions among objects and captures message-oriented behavior. The purposes of a collaboration diagram are to model flow of control, to illustrate the coordination of object structure and control, and to track the objects that interact with other objects. While sequence diagrams track object interactions versus time, collaboration diagrams focus on messages that pass between objects.
- A **statechart diagram** tracks the states that an object goes through and in the process captures event-oriented behavior. The purposes of a statechart diagram are to model the object life cycle and reactive objects such as user interfaces, external devices, etc. Statecharts are similar to finite-state machines, and contrast with the other dynamic diagrams by focusing on the events that occur.
- An **activity diagram** represents the performance of operations and transitions that are triggered in order to capture activity-oriented behavior, in a similar fashion to a Petri-Net. The purposes of an activity diagram are to model business workflows and individual operations, from an action perspective.

Overall, the nine different UML diagrams are an excellent vehicle for conceptual modeling in support of object-oriented design.

104.7 Design Patterns

The architect Christopher Alexander is credited with laying the foundations for patterns in software development via a series of books [Alexander et al. 1977, Alexander 1979] in that he describes problems that occurred over and over again within the context of designing and constructing buildings and towns. The ideas were adapted to software development with the introduction of five patterns that dealt with the design of user interfaces [Beck 1994]. Other prominent researchers also discovered and documented design patterns [Coad 1992, Coad et al. 1995, Coplien 1992, Gamma et al. 1995, Schmidt 1995]. However, the work that paved the way for the wide acceptance and usage of patterns in the software engineering research and development communities was published by E. Gamma, R. Helm, R. Johnson, and J. Vlissides, a group of experts who became known as the "Gang-of-Four" (GOF) [Gamma et al. 1995]. This resulted in subsequent work in architecture patterns [Buschmann et al. 1996, Schmidt et al. 2000], the emergence of a community of pattern developers who work together to document their experiences in patterns [Buschmann et al. 1996], and the support of design patterns in most UML tools. In the remainder of this section, the capabilities and usage of patterns for software design are explored. First, the concept of a design pattern is defined. Next, pattern catalogs for organizing and categorizing patterns are examined. Then, the way that design patterns solve design problems is presented. Finally, the usage of design patterns to solve a design problem, including an example, is detailed.

104.7.1 Defining a Design Pattern

Almost since a programmer wrote the first program, there has been the observation that similar code appears across multiple programs. If a programmer had a set of C functions for manipulating a stack of integers, that code was typically used over and over again by the programmer, changing and adapting the code to different requirements. Now, fast forward to the 1990s when object-oriented software design and development emerged as the preferred approach. In an object-oriented setting, software engineers have noticed that similar types of classes and objects, and their interactions, occur over and over again when solving similar problems in different applications, which has been formalized in a *design pattern* by the GOF as:

> ... descriptions of communicating objects and classes that are customized to solve a general design problem in a particular context. [Gamma et al. 1995]

For software engineers, design patterns offer simple, elegant solutions to specific problems in an object-oriented software design for an application [Gamma et al. 1995]. The unique nature of a design pattern is that it describes both the problem and its solution. Design patterns provide solutions that have been developed and refined over time by experienced software engineers and developers, and that have been found to be useful in several application contexts. Design patterns provide design alternatives for software engineers that can be used to create software that is elegant, reusable, and flexible.

For a software engineer to successfully utilize design patterns, they must be described in a consistent format at a level of abstraction that transcends single classes, instances, or components [Gamma et al. 1995]. Design patterns must provide a common vocabulary for software engineers to communicate about designs and discuss design alternatives at a raised level of abstraction, allowing information about design trade-offs and decision alternatives to be easily explored. To address this issue, *design pattern descriptions* [Gamma et al. 1995] have been offered and contain four essential elements: pattern name, problem element, solution element, and consequences element. The *pattern name* is an intuitive name for the pattern being captured. The *problem element* describes the context of the problem and when the pattern should be applied. Remember, to be a design pattern, it must be abstract enough so that it can be applied to many situations. The *solution element* describes the way that classes and objects can be used to solve the problem, with UML class and the sequence diagrams utilized to indicate class structure, class hierarchy, and the communication behavior between the objects within the pattern. Finally, the *consequences element* describes both the pros and the cons of using the pattern. Collectively, these four elements document a design pattern, providing stakeholders with different information on what a pattern can and cannot do, based on their specific needs.

104.7.2 Pattern Catalogs

Once a software engineer understands what design patterns are and the way that they can be utilized effectively to solve problems, the next step is to provide a means to categorize and make patterns available in an easy-to-use fashion. This is typically accomplished via a **pattern catalog**, which is the collection of all design patterns organized into categories that are easily accessible for use by software engineers during object-oriented design and development. The pattern catalog must be browsable, to allow a software engineer to easily view and select patterns as needed. From a practical perspective, pattern catalogs are supported by most UML tools such as Together CC and Rational Rose.

There are many different ways to categorize patterns within a catalog. In Gamma et al. [1995], GOF has defined three categories for its patterns: creational, structural, and behavioral. A *creational design pattern* focuses on describing the way that objects in the pattern are instantiated, with the idea that a software engineer may browse based on instantiation behavior to choose a pattern. A *structural design pattern* organizes patterns based on the manner in which classes and objects are composed with one another, again providing a different perspective for browsing. *Behavioral design patterns* offer a comprehensive breakdown on what a pattern actually does, via the algorithms, the assignment of responsibilities to objects, and the communications between objects.

In Buschmann et al. [1996], architectural design patterns, which specify the system-wide structural properties of an application, are partitioned into categories: structural composition, organization of work, access control, management, communication, from mud to structure, distributed systems, interactive systems, and adaptable systems. Again, these categorizations are critical for allowing software engineers to easily find the patterns that are suitable, which then allows them to make informed decisions within a category to select the pattern that best fits an application's needs and requirements.

104.7.3 Design Patterns to Solve Design Problems

To effectively utilize design patterns to solve design problems, software engineers must rely heavily on (and understand) the critical object-oriented concepts discussed in earlier portions of this chapter, including polymorphism, inheritance, interfaces, abstract classes, and runtime vs. compile-time object composition. Many, if not all, of these concepts are integrally tied to two important goals of the object-oriented approach, namely, reuse and flexibility. The purpose of this section is to highlight a select set of the specific techniques embodied within design pattern solutions that achieve the goals of reuse and flexibility. To do so, we rely on four critical points related to design and its evolutionary process that have been identified in Gamma et al. [1995].

1. *Find objects that do not occur in nature.* During design, software engineers often start by identifying objects that correspond to the real world. However, abstractions such as processes (e.g., business or engineering) and algorithms, two critical aspects of any problem solution, do not occur in nature. To address this issue, select design patterns provide solutions for designing processes and algorithms in a flexible manner. From this perspective, the utilization of a design pattern is a *reuse* of proven ideas and code.

2. *Reduce implementation dependencies between subsystems by programming to an interface instead of a particular implementation.* Software engineers can employ inheritance to define families of classes/objects with the same interfaces. Then, polymorphism can be used to provide a client (user object) with an interface containing a collection of methods. As a result, the client does not know any specifics about the class it is using, other than the interface that it provides to the client. Again, select design patterns can provide a solution template that supports the use of inheritance and interfaces, promoting both reuse (inheritance) and flexibility (interfaces).

3. *Maximize reuse by favoring object composition over class inheritance.* In object-oriented software engineering, both inheritance and object composition are methods for reuse. Recall that inheritance allows for the implementation of the parent class to be reused by the subclass, is implemented at compile time, and is part of a instance's physical representation. A change in the parent class can cause the subclass to change, leading to a best case of recompilation and to the worst case of major redesign. On the other hand, object composition dynamically composes classes at runtime by using existing objects. Only the interfaces of objects are known to the composing objects. Any object can be replaced by an object of the same type (having the same interface) without redesign of classes.

In GOF design patterns, object composition is often favored, because it provides for maximum flexibility as a design evolves over time.

4. *Avoid redesign when client, classes, or other factors change.* Software engineers, once a design pattern has been chosen, must take great care when changes occur that have the potential to affect the pattern; that is, once a suitable pattern has been chosen, replacing the pattern (redesign) can have a significant impact. To avoid this situation, it is often very useful to decouple behavior. Some examples that occur in practice include: decouple the sender of a request from its receiver using a layer of abstraction, which can be easily found and replaced; and, user interfaces, external operating system interfaces, and interfaces to algorithms can be decoupled from their use by inserting a layer of abstraction between the request for the use and the implementation of the service. Decoupling supports flexibility, allowing the design to adapt with a minimum of change and impact.

Each of the four points discussed above encourages software engineers to make informed decisions that maximize flexibility and/or reuse, always keeping in mind an application's needs and requirements.

104.7.4   Using a Design Pattern

There are many different factors that a software engineer must consider when choosing and then using a design pattern [Gamma et al. 1995], including: understanding the way that patterns solve problems; targeting the consequences element that contains both the pros and cons of using the pattern; familiarity with the pattern catalog to know what is available; and, identifying the variables that may change (and may not change) in the design, because the ones that do change have the potential to impact the chosen pattern in the future, thereby requiring redesign. Once a software engineer chooses a design pattern, it must be applied to the design, and this is often facilitated via the expression of a pattern's solution using UML or other modeling constructs. These constructs should be used as the basis for defining classes, class attributes, and behaviors, including interactions between objects, which customizes the pattern to a specific context. Several resources exist that have implementation code for design patterns in various languages such as C++ [Gamma et al. 1995], Smalltalk [Alpert et al. 1998], and Java [Grand et al. 1998].

To illustrate a design pattern, consider the **Observer** pattern from GOF [Gamma et al. 1995], which is used to define a one-to-many relationship between objects. When one object changes state, all of its dependents are notified and updated automatically. The **Observer** pattern allows a design to loosely couple objects so that when one object changes its state (subject), many objects can be notified (observer). The observer is a passive object and is notified whenever the subject object changes. The subject is an active object that does not have to know how many objects want to be notified or make any assumptions about the classes of objects that must be notified. The observer object simply implements an interface that specifies methods for the way that it will be notified of a change in the subject. The subject allows any observer to be registered (or unregistered with it) and will send notifications to each observer automatically each time it is updated. One advantage of **Observer** is that it allows for an abstract decoupling between subject and observer that facilitates system layering. One disadvantage is that it is possible for the subject to change very frequently, which can cause a cascade of updates to the observers, which has the potential to impact performance. The UML class diagram for the **Observer** pattern is shown in Figure 104.5.

In Figure 104.5, the **Concrete Subject** stores the state that is of interest to **ConcreteObserver** objects and sends a notification to its observers when its state changes. The **Concrete Observer** maintains a reference to a **ConcreteSubject** object and stores the state that should stay consistent with the subjects. It also implements the **Observer** updating interface to keep its state consistent with the subjects. An **Observer** pattern would be useful in **HTSS** in a number of different ways. First, whenever the retail price of an **Item** (in **ItemDB**) is changed, there can be notification to the store manager and stock clerks (which are objects in **HTSS**) so that shelf and item prices can be updated. Likewise, whenever an **Item** is recalled (e.g., due to health concerns), there can be notification to inventory control personnel

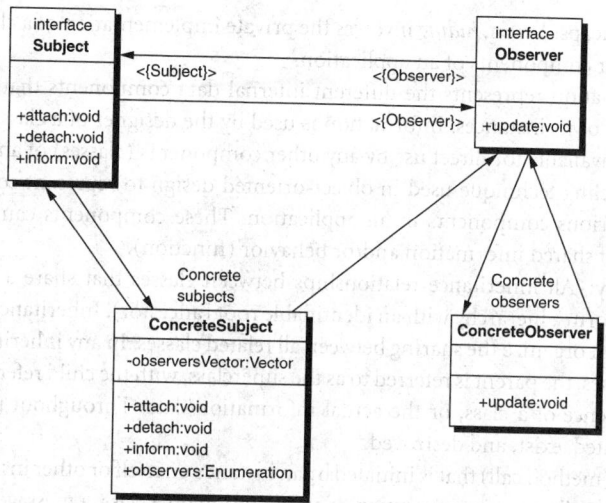

FIGURE 104.5 Classes for the **Observer** Pattern.

and stock clerks (again, **HTSS** objects) to pull the item off the shelf. In fact, if **HTSS** is truly computerized, then it may also be possible to use the **Observer** pattern to notify customers (again objects) that have purchased the recalled item.

Defining Terms

Abstract data type (ADT): A software design approach that decomposes a problem into components by identifying the public interface and private implementation. An ADT allows software engineers to define their own data types to support the informational and behavioral needs of their applications.

Class: The major unit of abstraction in the object-oriented design approach, used to model the features (information) and behavior (methods) for an application. Each class is partitionable into a public interface and private implementation.

Class diagram: A UML diagram that represents the classes and the relationships among classes (generalizations and other associations) in an object-oriented design for an application.

Class library: The collection of all classes and their inheritance hierarchies form an application library for use by software engineers, classes, tools, and end users.

Cohesion: The ability of a class for an application to perform a highly cohesive well-defined task.

Coupling: The dependencies that exist among the different classes of an application.

Design pattern: A template of objects and classes that represents a predefined set of behavior which has been observed and generalized from existing systems and can be applied to solve new problems with similar characteristics.

Dispatching: The runtime determination of the method to be called, based on the type of the invoking instance. Dispatching allows code to behave differently, based on who is calling particular methods. Dispatching is tightly bound with inheritance and promotes reuse and evolution of code, allowing inheritance hierarchies to be defined and evolved over time as needs and requirements change.

Encapsulation: The concept utilized by classes to group together information and methods/operations within a single component. Once these are encapsulated, access to the component can be controlled.

Generic: A type-parameterizable piece of a software or design that will be utilized by multiple components. For example, a linked-list data structure can be developed generically, independent of the type of element in the list. When the list is declared and then used, it is customized on the basis of type to meet the application's specific needs.

Hiding: Related to encapsulation, *hiding* involves the private implementation of a class, which is inaccessible to all other components of an application.

Information: Information represents the different internal data components that define the class and characterize all of its instances. Information is used by the designer of a class to maintain internal state and is unavailable for direct use by any other components (classes) of an application.

Inheritance: A modeling technique used in object-oriented design to represent commonalties that exist between various components in an application. These components can be determined by a combination of shared information and/or behavior (function).

Inheritance hierarchy: All inheritance relationships between classes that share a common parent (or grandparent) form a hierarchy with an identifiable root (ancestor). Inheritance hierarchies are simply the trees that organize the sharing between all related classes. In any inheritance relationship between two classes, the parent is referred to as the superclass, with the child referred to as the subclass.

Instance: An occurrence of a class, or the actual information/data. Throughout runtime, instances of classes are created, exist, and destroyed.

Message: An action (method call) that is initiated by an instance on itself or other instances. While method refers to the compile-time interpretation of an operation on a class, a message is the corresponding runtime concept.

Method: Contains the definition of the actions required for a particular operation against the private data of a class. Methods can be in either the public interface or the private implementation of a class.

Pattern catalog: The collection of design patterns organized into a coherent structure that is easily accessible for use by software engineers to select patterns for during object-oriented design and development. Supported by most UML tools.

Private implementation: That portion of an object-oriented class that is hidden from the other portions of an application. Critical for achieving representation independence.

Public interface: That portion of an object-oriented class that contains the permissible operations (methods, functions) that are available for use by other portions of an application. Critical for achieving abstraction.

Representation independence: The ability to change the hidden private implementation without impacting the visible public interface.

Sequence diagram: A UML diagram that represents the sequence of actions between objects and classes for a particular method invocation (message).

Software evolution: The process that describes the ability to change an application over time as new requirements are identified, when major upgrades occur, or when significant flaws are corrected. As a concept, software evolution is promoted heavily for object-oriented design.

Software reuse: The process that describes the ability to reuse existing software in new applications. When software is reused in its entirety without changes, a gain in productivity is attained. Critical for object-oriented design.

Unified modeling language (UML): A language for specifying, visualizing, constructing, and documenting software artifacts. Emerged as the *de facto* standard.

Use-case diagram: A UML diagram that represents the interaction of users (actors) with system components, defining use-cases of behavior.

UML diagram: Represents the static application structure via a use case, class, object, component, and deployment diagrams, and the dynamic application content via sequence, collaboration, statechart, and activity diagrams.

References

Alexander, C. 1979. *The Timeless Way of Building.* Oxford University Press.

Alexander, C. et al. 1977. *A Pattern Language Towns-Buildings-Construction.* Oxford University Press.

Alpert, S. et al. 1998. *The Design Patterns Smalltalk Companion.* Addison-Wesley, Reading, MA.

Barnes, J. G. P. 1991. *Programming in Ada plus Language Reference Manual*, 3rd ed. Addison-Wesley, Reading, MA.

Barnes, J. G. P. 1996. *Programming in Ada 95*. Addison-Wesley, Reading, MA.

Beck, K. 1994. Patterns generate architectures. *Proc. of the European Conf. for Object-Oriented Programming*. Oct., pp. 139–149.

Beck, K. and Cunningham, W. 1989. A laboratory for teaching object-oriented thinking. *Proc. 1989 OOPSLA Conf.*, Oct., pp. 1–6.

Booch, G. 1991. *Object-Oriented Design with Applications*. Benjamin/Cummings, Redwood City, CA.

Booch, G., Jacobson, I., and Rumbaugh, J. 1999. *Unified Modeling Language Reference Manual*. Addison-Wesley, Reading, MA.

Booch, G. and Rumbaugh, J. 1995. *Unified Method for Object-Oriented Development*. Rational Software Corporation, Santa Clara, CA, Tech. Rep.

Bretl, R. et al. 1989. The GemStone Data Management System. In *Object-Oriented Concepts, Databases and Applications*, W. Kim and F. Lochovsky, Eds., pp. 283–308. ACM Press, Addison-Wesley, Reading, MA.

Budd, T. 1997. *An Introduction to Object-Oriented Programming*, 2nd ed. Addison-Wesley, Reading, MA.

Buschmann, F. et al. 1996. *Pattern-Oriented Software Architecture — A System of Patterns*. John Wiley & Sons.

Campione, M. et al. 2001. *The Java Tutorial: A Short Course on the Basics*, 3rd ed. Addison-Wesley, Reading, MA.

Coad, P. 1992. Object-oriented patterns. *Commun. ACM* 35(9):152–159.

Coad, P. et al. 1995. *Object Models Strategies, Patterns, and Applications*. Prentice Hall, Englewood Cliffs, NJ.

Coleman, D. 1994. *Object-Oriented Development — The Fusion Method*. Prentice Hall, Englewood Cliffs, NJ.

Coplien, J. 1992. *Advanced C++ — Programming Styles and Idioms*. Addison-Wesley, Reading, MA.

Deitel, H. and Deitel, P. 1997. *Java: How to Program, 1/e*. Prentice Hall, Englewood Cliffs, NJ.

Department of Defense, 1995. *Ada 95 Reference Manual*, International Standard, ANSI/ISO/IEC-8652:1995, Jan.

Deux, O. et al. 1991. The O2 system. *Commun. ACM* 34(10):34–48.

Gamma, E. et al. 1995. *Design Patterns: Elements of Reusable Object-Oriented Software*. Addison-Wesley, Reading, MA.

Goldberg, A. 1989. *Smalltalk-80: The Language*. Addison-Wesley, Reading, MA.

Gordon, G. 1975. *The Application of GPSS V to Discrete System Simulation*. Prentice Hall, Englewood Cliffs, NJ.

Grand, M. et al. 1998. *Patterns in Java. Volume 1, A Catalog of Reusable Design Patterns Illustrated with UML*. John Wiley & Sons.

Harbison, S. 1992. *Modula-3*. Prentice Hall, Englewood Cliffs, NJ.

Hoover, S. and Perry, R. 1989. *Simulation: A Problem Solving Approach*. Addison-Wesley, Reading, MA.

Jacobson, I. et al. 1992. *Object-Oriented Software Engineering: A Use Case Driven Approach*. Addison-Wesley, Reading, MA.

Kim, W. 1990. Object-oriented databases: definition and research directions. *IEEE Trans. Knowledge Data Eng.* 2(3):327–341.

Lamb, C. et al. 1991. The ObjectStore database system. *Commun. ACM* 34(10):50–63.

Lieberherr, K. 1996. *Adaptive Object-Oriented Software: The Demeter Method with Propagation Patterns*. PWS Publishing, Boston, MA.

Liskov, B. and Zilles, S. 1975. Specification techniques for data abstraction. *IEEE Trans. Software Eng.* 1(1):7–19.

Liskov, B. et al. 1977. Abstraction mechanisms in CLU. *Commun. ACM* 20(8):564–576.

Meyer, B. 1988. *Object-Oriented Software Construction*. Prentice Hall, Englewood Cliffs, NJ.

Meyer, B. 1992. *Eiffel: The Language*. Prentice Hall, Englewood Cliffs, NJ.

Ontologic. 1991. ONTOS object database documentation, Release 2.1. Ontologic, Burlington, MA.

Pressman, R. 2001. *Software Engineering: A Practitioner's Approach*, 5th ed. McGraw-Hill, New York.

Rumbaugh, J. et al. 1991. *Object-Oriented Modeling and Design*. Prentice Hall, Englewood Cliffs, NJ.

Sammett, J. R. 1969. *Programming Languages: History and Fundamentals*. Prentice Hall, Englewood Cliffs, NJ.

Schach, S. 2002. *Object-Oriented and Classical Software Engineering*, 5th ed. McGraw-Hill, New York.

Schmidt, D. 1995. Using design patterns to develop reusable object-oriented communication software. *Commun. ACM* 38(10):65–74.

Schmidt, D. et al. 2000. *Pattern-Oriented Software Architecture: Patterns for Concurrent and Networked Objects*. John Wiley & Sons.

Sethi, R. 1996. *Programming Languages: Concepts and Constructs*, 2nd ed. Addison-Wesley, Reading, MA.

Sommerville, I. 2001. *Software Engineering*, 6th ed. Addison-Wesley, Reading, MA.

Stroustrup, B. 1986. *The C++ Programming Language*. Addison-Wesley, Reading, MA.

Tesler, L. 1985. *Object Pascal Report*. Apple Computer, Santa Clara, CA.

Tucker, A. 2002. *Programming Languages: Principles and Paradigms*, McGraw-Hill, New York.

Wegner, P. 1990. Concepts and paradigms of object-oriented programming. *OOPS Messenger*. 1(1):7–87.

Wirfs-Brock, R., Wilkerson, B., and Weiener, R. 1990. *Designing Object-Oriented Software*. Prentice Hall, Englewood Cliffs, NJ.

Zdonik S. and Maier, D. 1990. Fundamentals of object-oriented databases. In *Readings in Object-Oriented Database Systems*, S. Zdonik and D. Maier, Eds. pp. 1–34. Morgan Kaufmann.

Further Information

The interested reader is referred to object-oriented software engineering textbooks — both past and present [Booch 1991, Booch et al. 1999, Coleman 1994, Jacobson et al. 1992, Lieberherr 1996, Meyer 1988, Rumbaugh et al. 1991] and programming language textbooks on Java [Deitel and Deitel 1997, Campione et al. 2001], Ada 95 [Barnes 1996, Department of Defense 1995], Modula-3 [Harbison 1992], C++ [Stroustrup 1986], Eiffel [Meyer 1992], Smalltalk [Goldberg 1989], and Object Pascal [Tesler 1985] for a more in-depth coverage of various approaches. Further, the reader is referred to the many different electronic sources for software engineering given at the end of Chapter 103 which all track object-oriented concepts and techniques.

In addition to these resources, there are also dedicated resources for the Unified Modeling Language (UML) and design patterns. For UML, the main Web site is http://www.uml.org, which is under the auspices of the Object Management Group (OMG) and contains the UML standard along with documents, white papers, tutorials, links to other organizations and resources, etc. In addition, there is a yearly UML conference for research and practice advances (http://www.umlconference.org). For design patterns, there are numerous sites, including http://hillside.net/patterns/, which is very comprehensive with numerous links to conferences, discussion groups, journals, research projects, etc.; http://www.cs.wustl.edu/schmidt/patterns.html, and http://c2.com/ppr/index.html, which are examples of Web sites maintained by faculty researchers; and http://jerry.cs.uiuc.edu/plop/, which is the Web page of the Pattern Languages of Programs Conference.

105

Software Testing

105.1 Introduction **105**-1
105.2 Underlying Principles **105**-2
  Terminology • Model of Execution-Based
  Software Testing
105.3 Best Practices **105**-6
  An Example Program • Fault/Failure Model • Program
  Building Blocks • Test Adequacy Metrics • Test Case
  Generation • Test Execution • Test Adequacy Evaluation
  • Regression Testing • Recent Software Testing Innovations
105.4 Conclusion **105**-35

Gregory M. Kapfhammer
Allegheny College

I shall not deny that the construction of these testing programs has been a major intellectual effort: to convince oneself that one has not overlooked "a relevant state" and to convince oneself that the testing programs generate them all is no simple matter. The encouraging thing is that (as far as we know!) it could be done.

— Edsger W. Dijkstra, 1968

105.1 Introduction

When a program is implemented to provide a concrete representation of an algorithm, the developers of this program are naturally concerned with the correctness and performance of the implementation. Software engineers must ensure that their software systems achieve an appropriate level of quality. **Software verification** is the process of ensuring that a program meets its intended specification [Kaner et al., 1993]. One technique that can assist during the specification, design, and implementation of a software system is software verification through **correctness proof**. **Software testing**, or the process of assessing the functionality and correctness of a program through execution or analysis, is another alternative for verifying a software system. As noted by Bowen and Hinchley [1995] and Geller [1978], software testing can be appropriately used in conjunction with correctness proofs and other types of formal approaches to develop high-quality software systems. Yet it is also possible to use software testing techniques in isolation from program correctness proofs or other formal methods.

Software testing is not a "silver bullet" that can guarantee the production of high-quality software systems. While a "correct" correctness proof demonstrates that a software system (which exactly meets its specification) will always operate in a given manner, software testing that is not fully exhaustive can only suggest the presence of flaws and cannot prove their absence. Moreover, Kaner et al. [1993] have noted that it is impossible to completely test an application because (1) the domain of program inputs is too large, (2) there are too many possible input paths, and (3) design and specification issues are difficult to

1-58488-360-X/$0.00+$1.50
© 2004 by CRC Press, LLC

test. The first and second points present obvious complications and the final point highlights the difficulty in determining if the specification of a problem solution and the design of its implementation are also correct.

Using a thought experiment developed by Beizer, we can explore the first assertion by assuming that we have a method that takes a `String` of ten characters as input and performs some arbitrary operation on the `String`. To test this function exhaustively, we would have to input 2^{80} `Strings` and determine if they produce the appropriate output.* Thus, exhaustive testing is an intractable problem because it is impossible to solve with a polynomial-time algorithm [Binder, 1999; Neapolitan and Naimipour, 1998]. The difficulties alluded to in the second assertion are exacerbated by the fact that certain execution paths in a program could be infeasible. Finally, software testing is an algorithmically unsolvable problem because there may be input values for which the program does not halt [Beizer, 1990; Binder, 1999].

Thus far, we have provided an intuitive understanding of the limitations of software testing. However, Morell [1990] has proposed a theoretical model of the testing process that facilitates the proof of pessimistic theorems that clearly state the limitations of testing. Furthermore, Hamlet [1994] and Morell [1990] have formally stated the goals of a software testing methodology and implicitly provided an understanding of the limitations of testing. Young and Taylor [1989] have also observed that every software testing technique must involve some trade-off between accuracy and computational cost because the presence (or lack thereof) of defects within a program is an undecidable property. The theoretical limitations of testing clearly indicate that it is impossible to propose and implement a software testing methodology that is completely accurate and applicable to arbitrary programs [Young and Taylor, 1989].

While software testing is certainly faced with inherent limitations, there are also a number of practical considerations that can hinder the application of a testing technique. For example, some programming languages might not readily support a selected testing approach, a test automation framework might not easily facilitate the automatic execution of certain types of test suites, or there could be a lack of tool support to test with respect to a specific test adequacy criterion. Although any testing effort will be faced with significant essential and accidental limitations, the rigorous, consistent, and intelligent application of appropriate software testing techniques can improve the quality of the application under development.

105.2 Underlying Principles

105.2.1 Terminology

The IEEE standard defines a **failure** as the external, incorrect behavior of a program [IEEE, 1996]. Traditionally, the anomalous behavior of a program is observed when incorrect output is produced or a runtime failure occurs. Furthermore, the IEEE standard defines a **fault** as a collection of program source code statements that causes a failure. Finally, an **error** is a mistake made by a programmer during the implementation of a software system [IEEE, 1996]. The purpose of software testing is to reveal software faults in order to ensure that they do not manifest themselves as runtime failures during program usage. Throughout this chapter, we use P to denote the program under test and F to represent the specification that describes the behavior of P. Furthermore, we use $T = \langle t_1, \ldots, t_n \rangle$ to denote the test suite created to test program P. Also, we use C for the adequacy criterion that formalizes an understanding of what attributes a "good" test suite should have. Finally, we use $S = \langle s_0, s_1, \ldots, s_n \rangle$ to denote the set of publicly visible states during the execution of T and we require that $S_i = T_i(s_{i-1})$.

Building on the definitions of a test case used in Kapfhammer and Soffa [2003] and Memon [2001], we formalize a test for an arbitrary software system in Definition 105.1. Also, Definition 105.2 describes

---

*Suppose that our `String` is encoded in ASCII. There are $256 = 2^8$ different ASCII characters characters. Because we are dealing with a `String` of ten characters, there are 2^{80} possible unique `Strings`.

a restricted type of test suite where each test case returns the application under test back to the initial state, S_0, before it terminates [Pettichord, 1999]. If a test suite T is not **independent**, we do not place any restrictions upon the $\langle S_1, \ldots, S_n \rangle$ produced by the test cases and we simply refer to it as a **non-restricted** test suite. Our discussion of test execution in Section 105.3.6 will reveal that the JUnit test automation framework [Hightower, 2001; Jeffries, 1999] facilitates the creation of test suites that adhere to Definition 105.1 and are normally either independent or non-restricted in nature (although, JUnit encourages the creation of independent test suites).

Definition 105.1 A test case, $t_i \in T$, is a triple $\langle s_0, \langle o_1, o_2, \ldots, o_m \rangle, \langle s_1, s_2, \ldots, s_m \rangle \rangle$, consisting of an initial state, s_0, a test operation sequence $\langle o_1, o_2 \ldots, o_m \rangle$ for state s_0, and expected internal states $\langle s_1, s_2, \ldots, s_m \rangle$ where $s_j = o_j(s_{j-1})$ for $j = 1, \ldots, m$.

Definition 105.2 A test suite T is independent if and only if for all $\gamma = 1, \ldots n$, $S_\gamma = S_0$.

Figure 105.1 provides a useful hierarchical decomposition of different testing techniques and their relationship to different classes of test adequacy criteria. While our hierarchy generally follows the definitions provided by Binder [1999] and Zhu et al. [1997], it is important to note that other decompositions of the testing process are possible. This chapter focuses on **execution-based software testing** techniques. However, it is also possible to perform **non-execution-based software testing** through the use of **software inspections** [Fagan, 1976]. During a software inspection, software engineers examine the source code of a system and any documentation that accompanies the system. A software inspector can be guided by a **software inspection checklist** that highlights some of the important questions that should be asked about the artifact under examination [Brykczynski, 1999]. While an inspection checklist is more sophisticated than an ad-hoc software inspection technique, it does not dictate how an inspector should locate the required information in the artifacts of a software system. **Scenario-based reading** techniques, such as Perspective-Based Reading (PBR), enable a more focused review of software artifacts by requiring inspectors to assume the perspective of different classes of program users [Laitenberger and Atkinson, 1999; Shull et al., 2001].

Because the selected understanding of adequacy is central to any testing effort, the types of tests within T will naturally vary based upon the chosen adequacy criterion C. As shown in Figure 105.1, all

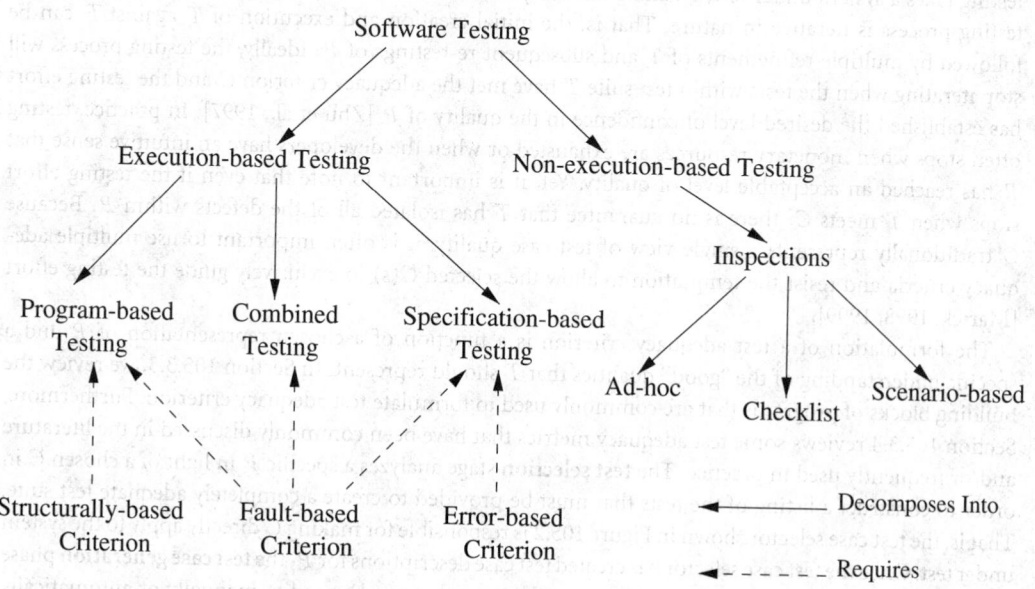

FIGURE 105.1 Hierarchy of software testing techniques.

execution-based testing techniques are either **program-based**, **specification-based**, or **combined** [Zhu et al., 1997]. A program-based testing approach relies on the structure and attributes of P's source code to create T. A specification-based testing technique simply uses F's statements about the functional and/or nonfunctional requirements for P to create the desired test suite. A combined testing technique creates a test suite T that is influenced by both program-based and specification-based testing approaches [Zhu et al., 1997]. Moreover, the tests in T can be classified based on whether they are **white-box**, **black-box**, or **grey-box** test cases. Specification-based test cases are black-box tests that were created without knowledge of P's source code. White-box (or, alternatively, **glass-box**) test cases consider the entire source code of P, while so-called grey-box tests only consider a portion of P's source code. Both white-box and grey-box approaches to testing would be considered program-based or combined techniques.

A complementary decomposition of the notion of software testing is useful to highlight the centrality of the chosen test adequacy criterion. The tests within T can be viewed based on whether they are "good" with respect to a **structurally-based**, **fault-based**, or **error-based** adequacy criterion [Zhu et al., 1997]. A structurally based criterion requires the creation of a test suite T that solely requires the exercising of certain control structures and variables within P. Thus, it is clear that a structurally based test adequacy criterion requires program-based testing. Fault-based test adequacy criterion attempt to ensure that P does not contain the types of faults that are commonly introduced into software systems by programmers [DeMillo et al., 1978; Morell, 1990; Zhu et al., 1997]. Traditionally, a fault-based criterion is associated with program-based testing approaches. However, Richardson et al. [1989] have also described fault-based testing techniques that attempt to reveal faults in F or faults in P that are associated with misunderstandings of F. Therefore, a fault-based adequacy criterion C can require either program-based, specification-based, or combined testing techniques. Finally, error-based testing approaches rely on a C that requires T to demonstrate that P does not deviate from F in any typical fashion. Thus, error-based adequacy criteria necessitate specification-based testing approaches.

105.2.2 Model of Execution-Based Software Testing

Figure 105.2 provides a model of execution-based software testing. Because there are different understandings of the process of testing software, it is important to note that our model is only one valid and useful view of software testing. Using the notation established in Section 105.2.1, this model of software testing takes a system under test, P, and a test adequacy criterion, C, as input. This view of the software testing process is iterative in nature. That is, the initial creation and execution of T against P can be followed by multiple refinements of T and subsequent re-testings of P. Ideally, the testing process will stop iterating when the tests within test suite T have met the adequacy criterion C and the testing effort has established the desired level of confidence in the quality of P [Zhu et al., 1997]. In practice, testing often stops when monetary resources are exhausted or when the developers have an intuitive sense that P has reached an acceptable level of quality. Yet, it is important to note that even if the testing effort stops when T meets C, there is no guarantee that T has isolated all of the defects within P. Because C traditionally represents a single view of test case quality, it is often important to use multiple adequacy criteria and resist the temptation to allow the selected $C(s)$ to exclusively guide the testing effort [Marick, 1998; 1999].

The formulation of a test adequacy criterion is a function of a chosen representation of P and a specific understanding of the "good" qualities that T should represent. In Section 105.3.3, we review the building blocks of programs that are commonly used to formulate test adequacy criterion. Furthermore, Section 105.3.4 reviews some test adequacy metrics that have been commonly discussed in the literature and/or frequently used in practice. The **test selection** stage analyzes a specific P in light of a chosen C in order to construct a listing of the tests that must be provided to create a completely adequate test suite. That is, the test case selector shown in Figure 105.2 is responsible for making C directly apply to the system under test. Once the test case selector has created test case descriptions for P, the **test case generation** phase can begin. Section 105.3.5 examines different techniques that can be used to manually or automatically generate test cases.

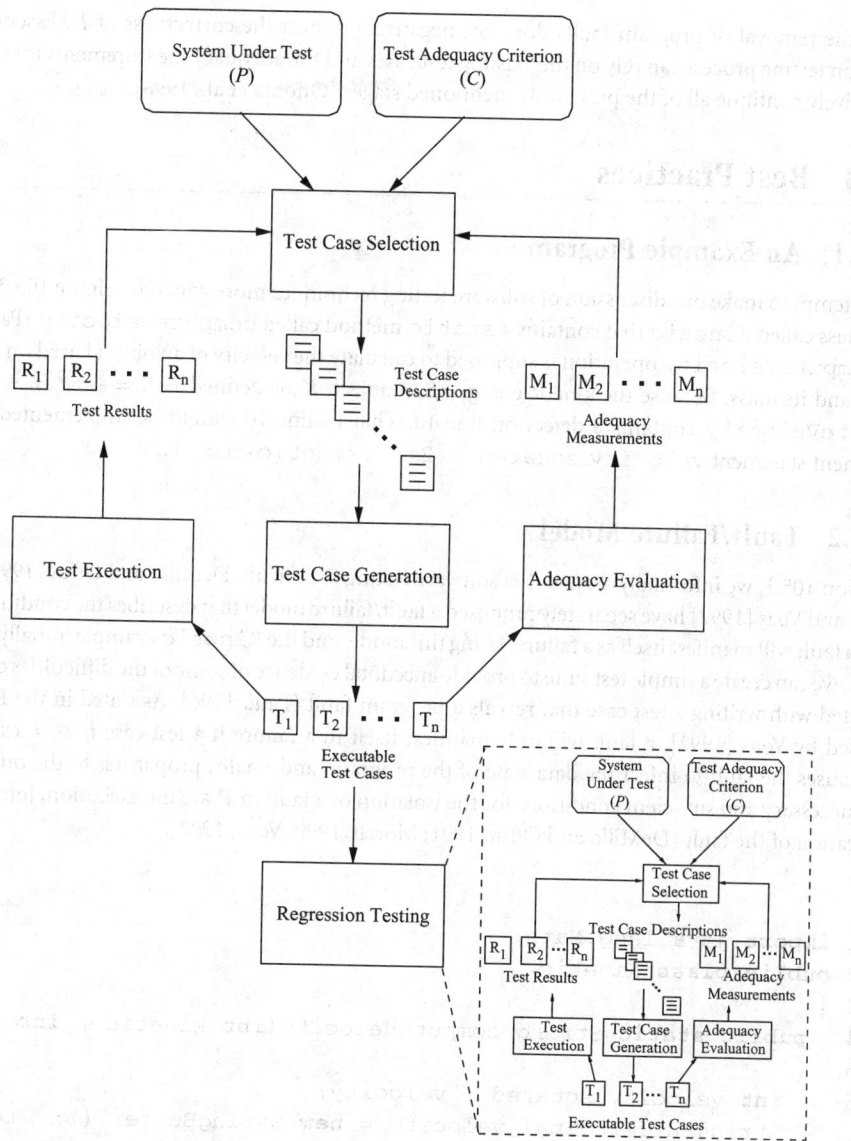

FIGURE 105.2   A model of the software testing process.

After the test cases have been generated, it is possible to perform **test execution**. Once again, the execution of the tests within T can be performed in a manual or automated fashion. Also, the executable test cases that were constructed during the generation phase can be analyzed by a **test adequacy evaluator** that measures the quality of T with respect to the test case descriptions produced by the test selector. The process of test execution is detailed in Section 105.3.6 and the test adequacy evaluation phase is described in Section 105.3.7. Of course, the test results from the execution phase and the adequacy measurements produced by the evaluator can be used to change the chosen adequacy criteria and/or augment the listing of test case descriptions that will be used during subsequent testing.

The iterative process of testing can continue throughout the initial development of P. However, it is also important to continue the testing of P after the software application has been released and it enters the **maintenance phase** of the **software lifecycle** [Sommerville, 2000]. **Regression testing** is an important software maintenance activity that attempts to ensure that the addition of new functionality and/or the removal of program faults does not negatively impact the correctness of P. Essentially, the regression testing process can rely on the existing test cases and the adequacy measurements for these tests to iteratively continue all of the previously mentioned stages [Onoma et al., 1998].

105.3 Best Practices

105.3.1 An Example Program

In an attempt to make our discussion of software testing techniques more concrete, Figure 105.3 provides a Java class called Kinetic that contains a static method called computeVelocity [Paul, 1996]. The computeVelocity operation is supposed to calculate the velocity of an object based on its kinetic energy and its mass. Because the kinetic energy of an object, K, is defined as $K = \frac{1}{2}mv^2$, it is clear that computeVelocity contains a defect on line 10. That is, line 10 should be implemented with the assignment statement velocity_squared = 2 * (kinetic/mass).

105.3.2 Fault/Failure Model

In Section 105.1, we informally argued that software testing is difficult. DeMillo and Offut [1991], Morell [1990], and Voas [1992] have separately proposed a fault/failure model that describes the conditions under which a fault will manifest itself as a failure. Using this model and the Kinetic example initially proposed by Paul, we can create a simple test suite to provide anecdotal evidence of some of the difficulties commonly associated with writing a test case that reveals a program fault [Paul, 1996]. As stated in the **PIE model** proposed by Voas [1992], a fault will only manifest itself in a failure if a test case $t_i \in T$ executes the fault, causes the fault to infect the data state of the program, and finally, propagates to the output. That is, the necessary and sufficient conditions for the isolation of a fault in P are the execution, infection, and propagation of the fault [DeMillo and Offut, 1991; Morell, 1990; Voas, 1992].

```
1  import java.lang.Math;
2  public class Kinetic
3  {
4    public static String computeVelocity(int kinetic , int mass)
5    {
6      int velocity_squared , velocity;
7      StringBuffer final_velocity = new StringBuffer ();
8      if( mass != 0 )
9      {
10       velocity_squared = 3 * ( kinetic / mass );
11       velocity = (int)Math.sqrt( velocity_squared );
12       final_velocity.append(velocity);
13     }
14     else
15     {
16       final_velocity.append("Undefined");
17     }
18     return final_velocity.toString();
19   }
20 }
```

FIGURE 105.3 The Kinetic class that contains a fault in computeVelocity.

```
1  import junit.framework.*;
2  public class KineticTest extends TestCase
3  {
4    public KineticTest(String name)
5    {
6      super(name);
7    }
8    public static Test suite()
9    {
10     return new TestSuite(KineticTest.class);
11   }
12   public void testOne()
13   {
14     String expected = new String("Undefined");
15     String actual = Kinetic.computeVelocity(5,0);
16     assertEquals(expected, actual);
17   }
18   public void testTwo()
19   {
20     String expected = new String("0");
21     String actual = Kinetic.computeVelocity(0,5);
22     assertEquals(expected, actual);
23   }
24   public void testThree()
25   {
26     String expected = new String("4");
27     String actual = Kinetic.computeVelocity(8,1);
28     assertEquals(expected, actual);
29   }
30   public void testFour()
31   {
32     String expected = new String("20");
33     String actual = Kinetic.computeVelocity(1000,5);
34     assertEquals(expected, actual);
35   }
36 }
```

FIGURE 105.4 A JUnit test case for the faulty Kinetic class.

Figure 105.4 provides the source code for KineticTest, a Java class that adheres to the JUnit test automation framework [Hightower, 2001; Jeffries, 1999]. Using our established notation, we have $T = \langle t_1, t_2, t_3, t_4 \rangle$ with each $t_i \in T$ containing a single testing operation o_1. For example, t_1 contains the testing operation String actual = Kinetic.computeVelocity (5, 0). Thus, each t_i contains a set $\langle s_0, s_1 \rangle$ of publicly visible states, where $s_1 = o_1(s_0)$. It is important to distinguish between the data states that arise during the execution of a test case and the internal data states that result after the execution of a single line within the method under test [Voas, 1992]. To this end, we use $\Delta = \{\delta_1, \delta_2, \ldots, \delta_b^e\}$ to denote the set of internal data states associated with a specific method under test within P. We require $\delta_b \in \Delta$ to correspond to the internal data state after the execution of line b in the method under test. Finally, we use δ_c^e to denote the expected data state that would normally result from the execution of a non-faulty version of line b.

Using an adaptation of the notation proposed by Voas [1992], Equation 105.1 describes the publicly observable state before the execution of o_1 in t_1. Equation 105.2 provides the state of the Kinetic class after the faulty computeVelocity method has been executed.* It is important to note that this test case causes the computeVelocity method to produce the data state s_2 that correctly corresponds to the expected data state. In this example, we are also interested in the internal states $\delta_{10} \in \Delta$ and δ_{10}^e, which correspond to the actual and expected data states after the execution of line 10. However, since t_1 does not execute the defect on line 10, the test does not produce the internal data states that could result in the isolation of the fault.

$$s_0 = \{(\text{actual}, \; null), \; (K, 5), (m, 0)\} \tag{105.1}$$

$$s_1 = \{(\text{actual}, \; Undefined), (K, 5), (m, 0)\} \tag{105.2}$$

The execution of t_2 corresponds to the initial and final date states as described by Equation 105.3 and Equation 105.4, respectively. In this situation, it is clear that the test case produces a final state s_1 that correctly matches the expected date state. Equation 105.5 and Equation 105.6 state the actual and expected data states that result after the execution of the faulty line in the method under test. While the execution of this test case does cause the fault to be executed, the faulty statement does not infect the data state (i.e., δ_{10} and δ_{10}^e are equivalent). Due to the lack of infection, it is impossible for t_2 to detect the fault in the computeVelocity method.

$$s_0 = \{(\text{actual}, \; null), \; (K, 0), (m, 5)\} \tag{105.3}$$

$$s_1 = \{(\text{actual}, \; 0), (K, 0), (m, 5)\} \tag{105.4}$$

$$\delta_{10} = \{(K, 0), (m, 5), (v^2, 0), (v, 0)(v_f, 0)\} \tag{105.5}$$

$$\delta_{10}^e = \{(K, 0), (m, 5), (v^2, 0), (v, 0)(v_f, 0)\} \tag{105.6}$$

Equations 105.7 and 105.8 correspond to the initial and final data states when t_3 is executed. However, state s_1 still correctly corresponds to the expected output. In this situation, the test case does execute the fault on line 10 of computeVelocity. Because Equation 105.9 and Equation 105.10 make it clear that the data states δ_{10} and δ_{10}^e are different, we know that the fault has infected the method data state. However, the cast to an int on line 11 creates a **coincidental correctness** that prohibits the fault from manifesting itself as a failure [Voas, 1992]. Due to the lack of propagation, this test case has not isolated the fault within the computeVelocity method.

$$s_0 = \{(\text{actual}, null), (K, 8), (m, 1)\} \tag{105.7}$$

$$s_1 = \{(\text{actual}, 4), (K, 8), (m, 1)\} \tag{105.8}$$

$$\delta_{10} = \{(K, 8), (m, 1), (v^2, 24), (v, 0), (v_f, 0)\} \tag{105.9}$$

$$\delta_{10}^e = \{(K, 8), (m, 1), (v^2, 16), (v, 0), (v_f, 0)\} \tag{105.10}$$

Test case t_4 produces the initial and final states that are described in Equation 105.11 and Equation 105.12. Because s_1 is different from the expected data state, the test is able to reveal the fault in the computeVelocity method. This test case executes the fault and causes the fault to infect the data state because the

---

*For the sake of brevity, our descriptions of the publicly visible and internal data states use the variables K, m, v^2, v, and v_f to mean the program variables kinetic, mass, velocity_squared, velocity, and final_velocity, respectively.

δ_{10} and δ_{10}^e provided by Equation 105.13 and Equation 105.14 are different. Finally, the internally visible data state δ_{10} results in the creation of the publicly visible state s_1. Due to the execution of line 10, the infection of the data state δ_{10}, and the propagation to the output, this test case is able to reveal the defect in `computeVelocity`. The execution of the `KineticTest` class in the JUnit test automation framework described in Section 105.3.6 will confirm that t_4 will reveal the defect in `Kinetic`'s `computeVelocity` method.

$$s_0 = \{(\texttt{actual}, null), (K, 1000), (m, 5)\} \tag{105.11}$$

$$s_1 = \{(\texttt{actual}, 24), (K, 1000), (m, 5)\} \tag{105.12}$$

$$\delta_{10} = \{(K, 1000), (m, 5), (v^2, 600), (v, 0), (v_f, 0)\} \tag{105.13}$$

$$\delta_{10}^e = \{(K, 1000), (m, 5), (v^2, 400), (v, 0), (v_f, 0)\} \tag{105.14}$$

105.3.3 Program Building Blocks

As noted in Section 105.2.2, a test adequacy criterion depends on the chosen representation of the system under test. We represent the program P as an **interprocedural control flow graph** (ICFG). An ICFG is a collection of control flow graphs (CFGs) $G_1, G_2, \ldots, G_u$ that correspond to the CFGs for methods $m_1, m_2, \ldots, m_u$, respectively. We define control flow graph G_v so that $G_v = (N_v, E_v)$ and we use N_v to denote a set of CFG nodes and E_v to denote a set of CFG edges. Furthermore, we assume that each $n \in N_v$ represents a statement in method m_v and each $e \in E_v$ represents a transfer of control in method m_v. Also, we require each CFG G_v to contain unique nodes $entry_v$ and $exit_v$ that demarcate the entrance and exit points of method m_v, respectively. We use the sets $pred(n) = \{m \mid (m, n) \in E_v\}$ and $succ(n) = \{m \mid (n, m) \in E_v\}$ to denote the set of predecessors and successors of node n. Finally, we require $N = \cup\{N_v \mid v \in [1, u]\}$ and $E = \cup\{E_v \mid v \in [1, u]\}$ to contain all of the nodes and edges in the interprocedural control flow graph for program P.

Figure 105.5 provides the control flow graphs G_{cv} and G_{t_1} for the `computeVelocity` method and the `testOne` method that can be used to test `computeVelocity`. Each of the nodes in these CFGs are labeled with line numbers that correspond to the numbers used in the code segments in Figure 105.3 and Figure 105.4. Each of these control flow graphs contain unique entry and exit nodes, and G_{t_1} contains a node n_{15} labeled "Call `computeVelocity`" to indicate that there is a transfer of control from G_{t_1} to G_{cv}. Control flow graphs for the other methods in the `KineticTest` class would have the same structure as the CFG for `testOne`. Although these control flow graphs do not contain iteration constructs, it is also possible to produce CFGs for programs that use `for`, `while`, and `do while` statements. Control flow graphs for programs that contain significantly more complicated conditional logic blocks with multiple, potentially nested, `if`, `else if`, `else`, or `switch` statements can also be created.

When a certain test adequacy criterion and the testing techniques associated with that criterion require an inordinate time and space overhead to compute the necessary test information, an **intraprocedural control flow graph** for a single method can be used. Of course, there are many different graph-based representations for programs. Harrold and Rothermel [1996] survey a number of graph-based representations and the algorithms and tool support used to construct these representations [Harrold and Rothermel, 1995]. For example, the class control flow graph (CCFG) represents the static control flow between the methods within a specific class [Harrold and Rothermel, 1996; 1994]. This graph-based representation supports the creation of class-centric test adequacy metrics that only require a limited interprocedural analysis. The chosen representation for the program under test influences the measurement of the quality of existing test suites and the generation of new tests. While definitions in Section 105.3.4 are written in the context of a specific graph-based representation of a program, these definitions are still applicable when different program representations are chosen. Finally, these graph-based representations can be created with a program analysis framework like Aristotle [Harrold and Rothermel, 1995] or Soot [Vallée-Rai et al., 1999].

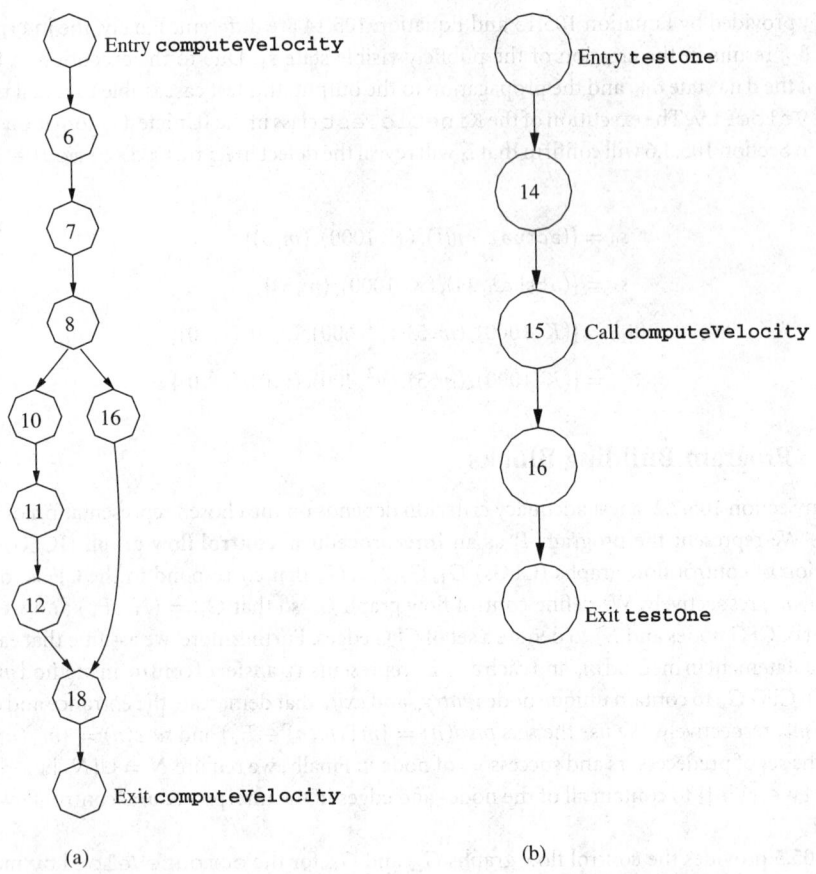

(a)   (b)

FIGURE 105.5   The control flow graphs for the computeVelocity and testOne methods.

105.3.4   Test Adequacy Metrics

As noted in Section 105.2.1, test adequacy metrics embody certain characteristics of test case "quality" or "goodness." Test adequacy metrics can be viewed in light of a program's control flow graph and the program paths and variable values that they require to be exercised. Intuitively, if a test adequacy criterion C_α requires the exercising of more path and variable value combinations than criterion C_β, it is "stronger" than C_β. More formally, a test adequacy criterion C_α **subsumes** a test adequacy criterion C_β if every test suite that satisfies C_α also satisfies C_β [Clarke et al., 1985; Rapps and Weyuker, 1985]. Two adequacy criteria C_α and C_β are equivalent if C_α subsumes C_β, and vice versa. Finally, a test adequacy criterion C_α **strictly subsumes** criterion C_β if and only if C_α subsumes C_β and C_β does not subsume C_α [Clarke et al., 1985; Rapps and Weyuker, 1985].

105.3.4.1   Structurally-Based Criterion

Some software test adequacy criteria are based on the control flow graph of a program under test. Control flow-based criteria solely attempt to ensure that test suite T covers certain source code locations and values of variables. While several control flow-based adequacy criteria are relatively easy to satisfy, others are so strong that it is generally not possible for a T to test P and satisfy the criterion. Some control flow-based adequacy criteria focus on the control structure of a program and the value of the variables that are used in conditional logic predicates. Alternatively, data flow-based test adequacy criteria require coverage of the control flow graph by forcing the selection of program paths that involved the definition and/or usage of program variables.

105.3.4.2 Control Flow-Based Criterion

Our discussion of control flow-based adequacy criterion will use the notion of an arbitrary path through P's interprocedural control flow graph $G_1, \ldots, G_u$. We distinguish π as a **complete path** in an interprocedural control flow graph or other graph-based representation of a program module. A complete path is a path in a control flow graph that starts at the program graph's entry node and ends at its exit node [Frankl and Weyuker, 1988]. Unless otherwise stated, we will assume that all of the paths required by the test adequacy criterion are complete. For example, the interprocedural control flow graph described in Figure 105.5 contains the complete interprocedural path $\pi = \langle entry_{t_1}, n_{14}, n_{15}, entry_{cv}, n_6, n_7, n_8, n_{16}, n_{18}, exit_{cv}, n_{16}, exit_{t_1} \rangle$. Note that the first n_{16} in π corresponds to a node in the control flow graph for computeVelocity and the second n_{16} corresponds to a node in testOne's control flow graph. Because the fault/failure model described in Section 105.3.2 indicates that it is impossible to reveal a fault in P unless the faulty node from P's CFG is included within a path that T produces, there is a clear need for a test adequacy criterion that requires the execution of all statements in a program. Definition 105.3 explains the **all-nodes** (or, alternatively, **statement coverage**) criterion for a test suite T and CFG in a program under test P.

Definition 105.3 A test suite T for P's control flow graph $G_v = (N_v, E_v)$ satisfies the all-nodes test adequacy criterion if and only if the tests in T create a set of complete paths $\Pi_{N_v} = \{\pi_1, \ldots, \pi_q\}$ that include all $n \in N_v$ at least once.

Intuitively, the all-nodes criterion is weak because it is possible for a test suite T to satisfy this criterion and still not exercise all the transfers of control (i.e., the edges) within the control flow graph [Zhu et al., 1997]. For example, if a test suite T tests a program P that contains a single while loop, it can satisfy statement coverage by only executing the iteration construct once. However, a T that simply satisfies statement coverage will not execute the edge in the control flow graph that returns execution to the node that marks the beginning of the while loop. Thus, the **all-edges** (or, alternatively, **branch coverage**) criterion described in Definition 105.4 requires a test suite to exercise every edge within a control flow graph.

Definition 105.4 A test suite T for program P's control flow graph $G_v = (N_v, E_v)$ satisfies the all-edges test adequacy criterion if and only if the tests in T create a set of complete paths $\Pi_{E_v} = \{\pi_1, \ldots, \pi_q\}$ that include all $e \in E_v$ at least once.

Because the inclusion of every edge in an interprocedural control flow graph implies the inclusion of every node within the same CFG, it is clear that the branch coverage criterion subsumes the statement coverage criterion [Clarke et al., 1985; Zhu et al., 1997]. However, it is still possible to cover all the edges within a control flow graph and not cover all the unique paths from the entry point to the exit point of a CFG. For example, if a test suite T is testing a program P that contains a single while loop, it can cover all the edges in the control flow graph by executing the iteration construct twice. Yet, a simple program with one while loop contains an infinite number of unique paths because each iteration of the looping construct creates a new path. Definition 105.5 explores the **all-paths** test adequacy criterion that requires the execution of every path within a CFG.

Definition 105.5 A test suite T for program P's control flow graph $G_v = (N_v, E_v)$ satisfies the all-paths test adequacy criterion if and only if the tests in T create a set of complete paths $\Pi_v = \{\pi_1, \ldots, \pi_q\}$ that include all of the execution paths beginning at the unique entry node $entry_v$ and ending at the unique exit node $exit_v$.

In Definition 105.5, we use $entry_v$ and $exit_v$ to denote the unique entry and exit points in CFG G_v. Clearly, the all-paths criterion subsumes both all-edges and all-nodes. However, it is important to note that it is possible for a test suite T to be unable to satisfy even all-nodes or all-edges if P contains infeasible paths. Yet, it is often significantly more difficult (or impossible) for a test suite to satisfy all-paths while still being able to satisfy both all-edges and all-nodes. The distinct difference in the "strength" of all-paths and the "weaker" all-edges and all-nodes presents a need for alternative adequacy criteria to "stand in the gap" between these two criteria.

As noted in Michael et al. [2001], there is also a hierarchy of test adequacy criteria that strengthen the all-edges criterion described in Definition 105.4. The **multiple condition coverage** criterion requires a test suite to account for every permutation of the Boolean variables in every branch of the program's control flow graph, at least one time. For example, if a and b are boolean variables, then the conditional logic statement if (a && b) requires a test suite T that covers the $2^2 = 4$ different assignments to these variables. The addition of a single Boolean variable to a conditional logic statement doubles the number of assignments that a test suite must produce in order to fulfill multiple condition coverage.

Another test adequacy criterion related to the all-edges criterion is **condition-decision coverage**. We use the term **condition** to refer to an expression that evaluates to true or false while not having any other Boolean-valued expressions within [Michael et al., 2001]. Intuitively, condition-decision coverage requires a test suite to cover each of the edges within the program's control flow graph and ensure that each condition in the program evaluates to true and false at least one time. For the example conditional logic statement if (a && b), a test suite T could satisfy the condition-decision adequacy criterion with the assignments $(a = 0, b = 0)$ and $(a = 1, b = 1)$. It is interesting to note that these assignments also fulfill the all-edges adequacy criterion. Yet, $(a = 1, b = 1)$ and $(a = 1, b = 0)$ fulfill all-edges without meeting the condition-decision criterion. Moreover, for the conditional logic statement if (alwaysFalse(a,b)) (where alwaysFalse(boolean a, boolean b) simply returns false), it is possible to fulfill condition-decision coverage with the assignments $(a = 0, b = 0)$ and $(a = 1, b = 1)$ and still not meet the all-edges criterion. These examples indicate that there is no clear subsumption relationship between all-edges and condition-decision. While multiple condition subsumes both of these criteria, it is clearly subsumed by the all-paths test adequacy criterion.

105.3.4.3 Data Flow-Based Criterion

Throughout our discussion of data flow-based test adequacy criteria, we will adhere to the notation initially proposed in Frankl and Weyuker [1988] and Rapps and Weyuker [1982, 1985]. For a standard program, the occurrence of a variable on the left-hand side of an assignment statement is called a **definition** of this variable. Also, the occurrence of a variable on the right-hand side of an assignment statement is called a **computation-use** (or **c-use**) of this variable. Finally, when a variable appears in the predicate of a conditional logic statement or an iteration construct, we call this a **predicate-use** (or **p-use**) of the variable.

As noted in Section 105.3.3, we will view a method in an application as a control flow graph $G_v = (N_v, E_v)$, where N_v is the set of CFG nodes and E_v is the set of CFG edges. For simplicity, our explanation of data flow-based test adequacy criteria will focus on data flow information within a single control flow graph of the entire ICFG for program P. However, our definitions can be extended to the interprocedural control flow graph. In practice, interprocedural data flow analysis often incurs significant time and space overhead, and many testing techniques limit their adequacy metrics to only require intraprocedural or limited interprocedural analysis. Yet, there are interprocedural data flow analysis techniques, such as the exhaustive algorithms proposed by Harrold and Soffa [1994] and the demand-driven approach discussed by Duesterwald et al. [1996], that can be used to compute data flow-based test adequacy criteria for an ICFG.

We define a **definition clear path** for variable x as a path $\langle entry_v, n_1, n_2, \ldots, n_m, exit_v \rangle$ in a CFG G_v, such that every node from entry point $entry_v$ to exit point $exit_v$ does not contain a definition of program variable x. Furthermore, we define the **def-c-use** association as a triple (n_d, n_{c-use}, x) where a definition of variable x occurs in node n_d and a c-use of x occurs in node n_{c-use}. Also, we define the **def-p-use** association as the two triples $(n_d, (n_{p-use}, t), x)$ and $(n_d, (n_{p-use}, f), x)$ where a definition of variable x occurs in node n_d and a p-use of x occurs during the true and false evaluations of a predicate at node n_{p-use} [Frankl and Weyuker, 1988; Rapps and Weyuker, 1982, 1985]. A complete path π^x **covers** a def-c-use association if it has a definition clear sub-path, with respect to x and the method's CFG, that begins with node n_d and ends with node n_{c-use}. Similarly, π^x covers a def-p-use association if it has definition clear sub-paths, with respect to x and the program's CFG, that begin with node n_d and end with the true and false evaluations of the logical predicate contained at node n_{p-use} [Frankl and Weyuker, 1988].

In Rapps and Weyuker [1982, 1985], the authors propose a family of test adequacy measures based on data flow information in a program. Among their test adequacy measures, the **all-uses** data flow adequacy criteria require a test suite to cover all the def-c-use and def-p-use associations in a program. The all-uses criterion is commonly used as the basis for **definition-use testing**. Alternatively, the **all-c-uses** criterion requires the coverage of all the c-use associations for a given method under test and the **all-p-uses** adequacy criterion requires the coverage of all the p-use associations. Furthermore, the **all-du-paths** coverage criterion requires the coverage of all the paths from the definition to a usage of a program variable [Rapps and Weyuker, 1982; 1985]. Definition 105.6 through Definition 105.9 define several important test adequacy criteria that rely on data flow information. However, we omit a formal definition of certain data flow-based test adequacy criterion, such as **all-c-uses/some-p-uses** and **all-p-uses/some-c-uses** [Rapps and Weyuker, 1982; 1985]. In each definition, we use U to refer to the universe of live program variables for program under test P.

Definition 105.6 A test suite T for control flow graph $G_v = (N_v, E_v)$ satisfies the all-c-uses test adequacy criterion if and only if for each association $\langle n_d, n_{c-use}, x \rangle$, where $x \in U$ and $n_d, n_{c-use} \in N_v$, there exists a test $t_i \in T$ to create a complete path π^x in G_v that covers the association.

Definition 105.7 A test suite T for control flow graph $G_v = (N_v, E_v)$ satisfies the all-p-uses test adequacy criterion if and only if for each association $(n_d, (n_{p-use}, t), x)$ and $(n_d, (n_{p-use}, f), x)$ where $x \in U$ and $n_d, n_{p-use} \in N_v$, there exists a test $t_i \in T$ to create a complete path π^x in G_v that covers the association.

Definition 105.8 A test suite T for control flow graph $G_v = (N_v, E_v)$ satisfies the all-uses test adequacy criterion if and only if for each association (n_d, n_{c-use}, x), $(n_d, (n_{p-use}, t), x)$ and $(n_d, (n_{p-use}, f), x)$ where $x \in U$ and $n_d, n_{c-use}, n_{p-use} \in N_v$, there exists a test $t_i \in T$ to create a complete path π_x in G that covers the association.

Definition 105.9 A test suite T for control flow graph $G_v = (N_v, E_v)$ satisfies the all-du-paths test adequacy criterion if and only if for each association (n_d, n_{c-use}, x), $(n_d, (n_{p-use}, t), x)$ and $(n_d, (n_{p-use}, f), x)$ where $x \in U$ and $n_d, n_{c-use}, n_{p-use} \in N_v$, the tests in T create a set of complete paths $\Pi_v^x = \{\pi_1^x, \ldots, \pi_q^x\}$ that include all of the execution paths that cover the associations.

Our discussion of subsumption for these test adequacy criteria is limited to the traditional understanding of data flow-based testing. A review of the **feasible data flow testing criteria**, a family of test adequacy criteria that only require the coverage of associations that are actually executable, and the subsumption hierarchy associated with these feasible criteria is provided in Frankl and Weyuker [1988]. Intuitively, the all-paths criterion subsumes all-du-paths because there might be complete paths within P that do not involve the definition and usage of program variables. Furthermore, all-du-paths subsumes all-uses because all-uses only requires one complete path π^x to cover the required associations and all-du-paths requires all of the $\Pi_v^x = \{\pi_1^x, \ldots, \pi_q^x\}$ that cover the association. It is clear that all-uses subsumes both all-p-uses and all-c-uses and that there is no subsumption relationship between the all-p-uses and all-c-uses. Because a p-use requires the evaluation of the true and false branches of a conditional logic statement, it can be shown that all-p-uses subsumes both the all-edges and the all-nodes criterion [Frankl and Weyuker, 1988]. Because structurally based test adequacy criteria that use control flow and data flow information are theoretically well-founded [Fenton, 1994; Nielson et al., 1999; Parrish and Zweben, 1991; Weyuker, 1986] and the criteria themselves can be organized into often meaningful subsumption hierarchies, it is clear that they will continue to be an important component of future research and practice.

105.3.4.4 Fault-Based Criterion

Mutation adequacy is the main fault-based test adequacy criterion. The conception of a mutation adequate test suite is based on two assumptions called the **competent programmer hypothesis** and the **coupling effect** [DeMillo et al., 1988]. The competent programmer hypothesis assumes that competent programmers create programs that compile and very nearly meet their specification. The coupling effect assumption indicates that test suites that can reveal simple defects in a program under test can also reveal more complicated combinations of simple defects [DeMillo et al., 1988]. Therefore, fault-based test adequacy criterion attempt to ensure that a test suite can reveal all of the defects that are normally introduced into software systems by competent programmers.

Definition 105.10 describes a test suite T that is **adequate** and Definition 105.11 defines the notion of **relative adequacy** (or, alternatively, **mutation adequacy**) for test suites [DeMillo and Offut, 1991]. In these definitions, we use the notation $Q(D)$ and $F(D)$ to mean the "output" of a program Q and the specification F on the entire domain of possible program inputs. Furthermore, we use the notation $Q(t)$ and $F(t)$ to denote the "output" of a program Q and a specification F on a single test $t \in T$ that provides a single input to Q and F. If a test suite T is adequate for a program under test P, then T is able to distinguish between all of the incorrect implementations of specification F. If a test suite T is relative-adequate (or, mutation-adequate) for a program under test P, then T is able to distinguish between a finite set $\Phi_P = \{\phi_1, \ldots, \phi_s\}$ of incorrect implementations of specification F.

Definition 105.10 If P is a program to implement specification F on domain D, then a test set $T \subset D$ is adequate for P and F if $(\forall \text{ programs } Q), [Q(D) \neq F(D)] \Rightarrow [(\exists t \in T)(Q(t) \neq F(t))]$.

Definition 105.11 If P is a program to implement specification F on domain D and Φ is a finite collection of programs, then a test set $T \subset D$ is adequate for P relative to Φ if $(\forall \text{ programs } Q \in \Phi)$, $[Q(D) \neq F(D)] \Rightarrow [(\exists t \in T)(Q(t) \neq F(t))]$.

Clearly, an adequate test suite is "stronger" than a relative-adequate one. However, through the judicious proposal and application of **mutation operators** that create Φ_P, the finite set of syntactically incorrect programs, we can determine if a test suite can demonstrate that there is a difference between each $\phi_r \in \Phi_P$ and the actual program under test. Under a **strong mutation** test adequacy criterion, a test suite T is strong mutation adequate if it can **kill** each mutant in Φ_P by showing that the output of the mutant and the program under test differ. If a specific mutant $\phi_r \in \Phi_P$ remains alive after the execution of T, this could indicate that T does not have the ability to isolate the specific defect that ϕ_r represents [DeMillo and Offut, 1991; Hamlet and Maybee, 2001]. Alternatively, it is possible that ϕ_r represents an **equivalent mutant**, or a program that is syntactically different from P while still having the ability to produce the same output as P.

Because strong mutation adequacy is often difficult to fulfill, the **weak mutation** test adequacy criterion only requires that the data states that occur after the execution of the initial source code location and the mutated code location differ [Hamlet and Maybee, 2001; Zhu et al., 1997]. Using the terminology established in our discussion of the fault/failure model in Section 105.3.2, weak mutation adequacy simply requires the execution of the source code mutation and the infection of the data state of each mutant program $\phi_r \in \Phi_P$. On the other hand, strong mutation adequacy requires the execution of the source code mutation, the infection of the data state of each mutant program, and the propagation of the mutant to the output. Both the strong and weak variants of mutation adequacy require the ability to automatically construct the mutants within Φ_P. While strong mutation adequacy only requires the ability to inspect the output of the program under test, weak mutation adequacy also requires additional instrumentation to reveal the data states that occur after a source code mutation is executed.

Ideally, mutation operators should produce the kinds of mutant programs that software engineers are most likely to create. A **fault model** can be used to describe the types of programming pitfalls that are normally encountered when a software system is implemented in certain types of programming languages. In a procedural programming language such as C, it might be useful to include mutation operators that manipulate the relational operators found within a program. Therefore, the conditional logic statement `if (a < b)` in the program under test could require the creation of mutants that replace `<` with every other relational operator in the set $\Re = \{<=, ==, >=, >, ! =\}$ [Hamlet and Maybee, 2001]. Other mutation operators might involve the manipulation of scalar and Boolean variables and constants [Jezequel et al., 2001]. An arithmetic mutation operator designed to change arithmetic operators could mutate the assignment statement `a = b + 1` with the inverse of the `+` operator and create the mutant `a = b - 1` [Jezequel et al., 2001].

Of course, the inheritance, polymorphism, and encapsulation provided by object-oriented languages makes it more challenging to propose, design, and implement mutation operators for programming languages such as Java. Alexander et al. [2000] and Kim et al. [2000] discuss some traditional "pitfalls" found in software systems that are implemented with object-oriented programming languages. However, these mutation operators lack generality because they are not supported by a fault model that clearly describes the class of potential faults related to the unique features of object-oriented programming languages. Kim et al. [1999] propose certain mutation operators for the Java programming language after conducting a Hazard and Operability (HAZOP) study [Leveson, 1995] of the Java programming language. However, the authors omit the description of operators for some fundamental aspects of the object-oriented programming paradigm, such as certain forms of method overriding, and they do not include operators for select facets of the Java programming language [Kim et al., 1999]. Yet, Offutt et al. [2001] have proposed a fault model for object-oriented programming languages that describes common programmer faults related to inheritance and polymorphism.

Using the fault model described in Offutt et al. [2001], Ma et al. [2002] include a comprehensive listing of object-oriented mutation operators for the Java programming language. The proposed mutation operators fall into the following six types of common object-oriented programming mistakes: (1) information hiding (access control), (2) inheritance, (3) polymorphism, (4), overloading, (5) Java-specific features, and (6) common programming mistakes [Ma et al., 2002]. The Access Modifier Change (AMC) mutation operator is an example of an operator that focuses on information hiding mistakes within object-oriented programs. For example, if a `BankAccount` class contained the declaration of the field `private double balance`, the AMC operator would produce the following mutants: `public double balance`, `protected double balance`, and `double balance`. The Java `this` keyword deletion (JTD) mutation operator removes the usage of `this` inside of the constructor(s) and method(s) provided by a Java class. For example, if the `BankAccount` class contained a `public BankAccount(double balance)` constructor that used the `this` keyword to disambiguate between the parameter `balance` and the instance variable `balance`, the JTD operator would remove the keyword [Ma et al., 2002].

Figure 105.6 describes an algorithm that can be used to calculate the mutation adequacy score, $MS(P, T, M_o)$, for program P, test suite T, and a set of mutation operators M_o [DeMillo and Offut, 1991]. Our description of the *CalculateMutationAdequacy* algorithm measures strong mutation adequacy and thus requires the outputs of P and the automatically generated mutants to differ. However, the algorithm could be revised to compute a weak mutation adequacy score for the test suite. The algorithm in Figure 105.6 uses a $n \times s$ matrix, $\mathcal{D}$, to store information about the dead mutants and the specific tests that killed the mutants. The s column vectors in the test information matrix $\mathcal{D}$ indicate whether each of the tests within T were able to kill one of the mutants in the set $\Phi_P = \{\phi_1, \ldots, \phi_s\}$. We use the notation $\mathcal{D}[i][r]$ to denote access to the ith row and the rth column within $\mathcal{D}$ and we use a 1 to indicate that a mutant was killed and a 0 to indicate that test t_i left mutant ϕ_r alive. Finally, the algorithm to calculate $MS(P, T, M_o)$ uses $\mathcal{Z}_{n \times s}$ to denote the $n \times s$ zero matrix and $\mathcal{Z}_s$ to denote a single column vector composed of s zeros.

If a specific test $t_i \in T$ is not able to kill the current mutant ϕ_r, it is possible that ϕ_r and P are equivalent. Because the determination of whether ϕ_r is an **equivalent mutant** is generally undecidable [Zhu et al., 1997], it is likely that the execution of the *IsEquivalentMutant* algorithm on line 11 of

Algorithm *CalculateMutationAdequacy*(T, P, M_o)
(∗ Calculation of Strong Mutation Adequacy ∗)
Input: Test Suite T;
      Program Under Test P;
      Set of Mutation Operators M_o
Output: Mutation Adequacy Score; $MS(P, T, M_o)$
1. $\mathcal{D} \leftarrow \mathcal{Z}_{n \times s}$
2. $\mathcal{E} \leftarrow \mathcal{Z}_s$
3.   **for** $l \in ComputeMutationLocations(P)$
4.   **do** $\Phi_P \leftarrow GenerateMutants(l, P, M_o)$
5.   **for** $\phi_r \in \Phi_P$
6.   **do for** $t_i \in T$
7.   **do** $R_i^P \leftarrow ExecuteTest(t_i, P)$
8. $R_i^{\phi_r} \leftarrow ExecuteTest(t_i, \phi_r)$
9.   **if** $R_i^P \neq R_i^{\phi_r}$
10.   **do** $\mathcal{D}[i][r] \leftarrow 1$
11.   **else if** $IsEquivalentMutant(P, \phi_r)$
12.   **do** $\mathcal{E}[r] \leftarrow 1$
13. $D \leftarrow \sum_{r=1}^{s} pos\left(\sum_{f=1}^{n} \mathcal{D}[f][r]\right)$
14. $E \leftarrow \sum_{r=1}^{s} \mathcal{E}[r]$
15. $MS(P, T, M_o) \leftarrow \frac{D}{(|\Phi_P| - E)}$
16.   **return** $MS(P, T, M_o)$

FIGURE 105.6 Algorithm for the computation of mutation adequacy.

CalculateMutationAdequacy will require human intervention. When a mutant is not killed and it is determined that ϕ_r is an equivalent mutant, we place a 1 in $\mathcal{E}[r]$ to indicate that the current mutant is equivalent
to P. The mutation testing information collected in $\mathcal{D}$ and $\mathcal{E}$ is complete once the algorithm has executed
every test case against every mutant for every mutation location in program P. Line 13 computes the
number of dead mutants, D, by using the *pos* function to determine if the sum of one of the s column
vectors is positive. We define the *pos* function to return 1 if $\sum_{i=1}^{n} \mathcal{D}[i][r] > 0$ and 0 otherwise. Finally,
line 14 computes the number of equivalent mutants, E, and line 15 uses this information to calculate the
final mutation adequacy score for program P and test suite T [DeMillo and Offut, 1991].

    While the calculation of mutation test adequacy is conceptually simple, it is computationally expensive.
Choi et al. [1989] attempted to improve the cost-effectiveness and practicality of measuring mutation adequacy by parallelizing the steps in algorithm *ComputeMutationAdequacy* and scheduling the computation
on the hypercube parallel computer architecture. Krauser et al. [1991] have also investigated a number of
techniques that can improve the performance of mutation analysis on a single instruction multiple data
(SIMD) parallel computer architecture and evaluated these techniques with a detailed system model that
supported a simulation-based empirical analysis. As noted in Zhu et al. [1997], early attempts to improve
the cost-effectiveness and practicality of mutation testing through parallelization had limited impact because they required specialized hardware and highly portable software. Yet it is likely that the availability of
general-purpose distributed computing middleware like Jini and JavaSpaces [Arnold et al., 1999; Edwards,
1999; Freemen et al., 1999] and high-throughput computing frameworks like Condor [Epema et al., 1996]
will facilitate performance improvements in the algorithms that measure mutation adequacy.

    In another attempt to make mutation adequacy analysis more cost-effective and practical, Offutt et al.
have investigated the **N-selective mutation testing** technique that removes the N mutation operators that
produce the most mutants of the program under test [Offutt et al., 1996; 1993; Zhu et al., 1997]. Although
all mutation operators make small syntactic changes to the source code of the program under test, some
operators are more likely to produce a greater number of mutants. Furthermore, the **semantic impact**,

or the change in meaning of the program under test, associated with small syntactic changes can vary from one mutation operator to another [Ma et al., 2002]. N-selective mutation test adequacy attempts to compute a high-fidelity mutation adequacy score without executing the mutation operators that create a high number of mutants that do not truly shed light of the defect-revealing potential of the test suite. Ma et al. [2002] observe that their AMC operator creates the most mutants out of their set of object-oriented mutation operators. They also note that the Java `static` modified change (JSC) operator and the AMC operator have a tendency to produce a significant number of equivalent mutants. Because little is known about the subsumption relationship between different mutation-based test adequacy criteria and between mutation adequacy and other notions of test adequacy, it is clear that mutation analysis is a promising practical technique that requires further implementation, empirical analysis, and theoretical study.

105.3.4.5 Error-Based Criterion

Error-based test adequacy criteria require a test suite T to demonstrate that the program under test P does not deviate from F, the program's specification, in a certain number of predefined ways. Certain elements of the **category-partition method** proposed by Balcer et al. [1989] and Ostrand and Balcer [1988] are indicative of error-based test adequacy criteria. The category-partition method requires the analysis of F to create a partition of the input domain of the program under test. By relying on the guidance of the tester, the category-partition method identifies the parameters and environment conditions, known as **categories**, that impact the behavior of the program under test. Next, the tester is responsible for the decomposition of each category into mutually exclusive **choices** that will be used to describe the partitions of the input within the category [Balcer et al., 1989].

The **test specification language** (TSL) provides a mechanism that allows a tester to write succinct descriptions of the categories and choices that state the input and output of the program under test. It is also possible for the tester to provide a description of the constraints that control the requirement of specific values within the choices and categories. In Ostrand and Balcer [1988], the authors describe the categories and choices that might be used for the `find <pattern> <file>` program that is often distributed with the Unix and GNU/Linux operating systems. For example, the specification for `find` might require that `<file>` is a valid file name. Thus, it would be important for the tests within T to ensure that `find` can handle the situations when (1) there is a valid file associated with the provided name, (2) there is no file with the stated name, and (3) the file name is omitted when the usage of `find` occurs [Ostrand and Balcer, 1988].

Error-based test adequacy criteria judge a test suite to be "stronger" if it covers more of the identified categories and choices. Because there is no general technique to automatically create an error-based test adequacy criterion from F, most error-based adequacy metrics, such as those used in the category-partition method, require human intervention [Zhu et al., 1997]. However, with the complete description of F and the specification of the test adequacy criterion in TSL or some other test case specification language, it is possible to automatically generate a test suite that will fulfill the adequacy criterion. For example, the AETG system of Cohen et al. and the PairTest system of Lei et al. enable the generation of combinatorially balanced test suites from test specifications [Cohen et al., 1996, 1997; Lei and Tai, 1998]. While combinatorial test case generation is not discussed in more detail in this chapter, more information about test data generation algorithms and their relation to test adequacy criteria is provided in Section 105.3.5.

105.3.4.6 Comparing Test Adequacy Criteria

As noted in Section 105.3.4, many test adequacy criterion can be related in subsumption hierarchies. However, Ntafos [1988] has argued that some test adequacy criteria are incomparable under the subsumption relation. Ntafos has also observed that even when it is possible to order test adequacy criteria through subsumption, it is likely that this ordering will not provide any direct indication of either the effectiveness of test suites that fulfill the adequacy criteria or the costs associated with testing to the selected criteria. Weyuker et al. [1991] categorized all comparisons of test adequacy criteria as being either **uniform** or **pointwise**. A uniform comparison of test adequacy criteria C_α and C_β attempts to relate the requirements of the criteria in light of all possible programs P and all test suites T. The usage of the subsumption relation can be seen as a type of uniform comparison of test adequacy criteria. However, a pointwise comparison of test adequacy criteria C_α and C_β compares the behavior of the two criteria with respect to a single (or, limited number of) P and T. While pointwise comparisons often provide interesting insights into the effectiveness of selected test adequacy criteria for certain programs, these types of comparisons often do not provide results that can be generalized. However, as noted by Weyuker et al. [1991], there are also severe limitations associated with uniform comparisons:

> [...] We can conclude that a uniform comparison that guarantees one criterion to be more
> effective at detecting program defects than another for all programs is no comparison at all. This
> is a convincing argument against the use of comparisons that attempt to guarantee the relative
> fault-exposing ability of criteria for all programs.

In light of these concerns about comparing test adequacy criteria, Weyuker et al. [1991] conclude their thoughts with the following observation: "We see effectiveness and cost as the two most meaningful bases by which test criteria can be compared; effectiveness is our ultimate concern." To this end, Frankl, Weyuker, Weiss, and Hutchins et al. have conducted both analytical and empirical investigations of the effectiveness of test adequacy criteria [Frankl and Weiss, 1993; Frankl and Weyuker, 1993; Hutchins et al., 1994]. Hutchins et al. [1994] compared the effectiveness of the all-edges and **all-DU** test adequacy criteria. The all-DU test adequacy criterion is a modified version of all-uses that simply requires the coverage of def-use associations without distinguishing between a p-use and a c-use.

The experimental design of Hutchins et al. [1994] used the TSL system described in Section 105.3.4.3 to automatically generate an **initial test pool** (ITP) that was analyzed to determine the level of achieved adequacy. Next, an **additional test pool** (ATP) was created to ensure that each of the exercisable coverage units within their subject programs was touched by at least 30 test cases. After the construction of a large test universe from the union of the ITP and the ATP, test sets of specific sizes were randomly selected from the test universe. Furthermore, the eight base programs selected by Hutchins et al. were seeded with defects that the researchers deemed to be representative in terms of their similarity to real-world defects and the "difficulty" associated with the isolation of the faults. An experimental design of this nature enabled Hutchins et al. to examine the relationship between the **fault detection ratio** for a testing technique and the adequacy and size of the resulting test suites. The fault detection ratio is the ratio between the number of test suites that contain a fault-revealing test case and the total number of test suites whose adequacy or size is in a specific interval.

The empirical study conducted by Hutchins et al. [1994] reveals some interesting trends. For example, the fault detection ratios for their candidate programs rose sharply as the test adequacy increased above 80 or 90%. Furthermore, as the size of a test suite increased, the fault detection ratio also increased. However, the fault detection ratio for test suites that were completely adequate with respect to the all-edges and all-DU criteria varied significantly. Indeed, Hutchins et al. [1994] observe that "the fault detection ratio of test sets with 100% DU coverage varied from 0.19 to 1.0, with an average of 0.67 for the 31 faults in the DU class." Perhaps more interesting is the following observation from Hutchins et al.:

> [...] Rather, code coverage seems to be a good indicator of test inadequacy. If apparently thorough
> tests yield only a low coverage level, then there is good reason to continue testing and try to raise
> the coverage level. The value of doing this can be seen by examining the detection ratios of test
> sets as their coverage levels approach 100%.

In a comparison of their operation difference (OD) test case selection technique, Harder et al. [2003] propose a new **area** and **stacking** approach for comparing test adequacy criteria. Indeed, they contend that the fault detection ratio is an inappropriate measure of the "efficiency" of test suites generated with respect to a specific test adequacy criterion for two reasons. First, because the fault detection ratio vs. test suite size curve relationship is not necessarily linear in nature, there is no guarantee that a doubling in the size of a test suite will always double the ability of the tests to detect faults. Second, and more importantly, the fault detection ratio does not directly reveal which test adequacy criterion is most likely to engender the production of test suites that reveal the most defects.

Suppose that we were interested in comparing test adequacy criteria C_α and C_β using the area and stacking technique. To do so, we could use C_α and C_β to guide the manual and/or automatic generation of test suites T_α and T_β that obtain a specific level of adequacy with respect to the selected criteria. It is likely that the size of the generated test suites will vary for the two criteria and we refer to $|T_\alpha|$ and $|T_\beta|$ as the **natural size** of the tests for the chosen criteria [Harder et al., 2003]. In an attempt to fairly compare C_α and C_β, Harder et al. advocate the construction of two new test suites T_α^β and T_β^α, where T_α^β denotes a test suite derived from T_α that has been stacked (or, reduced) to size $|T_\beta|$, and T_β^α is a test suite derived from T_β that has been stacked (or, reduced) to the size $|T_\alpha|$. Harder et al. [2003] propose stacking as a simple technique that increases or decreases the size of a base test suite by randomly removing tests or adding tests using the generation technique that created the base test suite.

In a discussion about the size of a test suite, Harder et al. [2003] observed that "comparing at any particular size might disadvantage one strategy or the other, and different projects have different testing budgets, so it is necessary to compare the techniques at multiple sizes." Using the base and faulty versions of the candidate programs produced by Hutchins et al., the authors measured the number of faults that were detected for the various natural sizes of the tests produced with respect to certain adequacy criterion. In our example, we could plot the number of revealed defects for the four test suites T_α, T_β, T_α^β, and T_β^α at the two sizes of $|T_\alpha|$ and $|T_\beta|$. The calculation of the area underneath the two fault-detection vs. test suite size curves can yield a new view of the effectiveness of the test adequacy criteria C_α and C_β. However, the area and stacking technique for comparing test adequacy criteria has not been applied by other researchers, and there is a clear need for the comparison of this design with past experimental designs. While the comparison of test adequacy criteria is clearly important, it is also an area of software testing that is fraught with essential and accidental difficulties.

105.3.5 Test Case Generation

The generation of test cases can be performed in a manual or automated fashion. Frequently, **manual test generation** involves the construction of test cases in a general purpose programming language or a test case specification language. Although the `KineticTest` class in Figure 105.4 adheres to the JUnit testing framework, it could have also been specified in a programming language-independent fashion by simply providing the class under test, the method under test, the method input, and the expected output. This specification could then be transformed into a language-dependent form and executed in a specific test execution infrastructure. Alternatively, test cases can be "recorded" or "captured" by simply using the program under test and monitoring the actions that were taken during usage [Steven et al., 2000].

An automated solution to the test data generation problem attempts to automatically create a T that will fulfill selected adequacy criterion C when it is used to test program P. While it is possible for C to be an error-based criterion, automated test data generation is more frequently performed with fault-based and structurally based test adequacy criteria. There are several different techniques that can be used to automatically generate test data. **Random**, **symbolic**, and **dynamic** test data generation approaches are all alternatives that can be used to construct a T that adequately tests P. A random test data generation approach relies on a random number generator to simply generate test input values.* For complex (and, sometimes quite simple) programs, it is often difficult for random test data generation techniques to produce adequate test suites [Korel, 1996].

Symbolic test data generation attempts to express the program under test in an abstract and mathematical fashion. Intuitively, if all the important aspects of a program can be represented as a system of one or more linear equations, it is possible to use algebraic techniques to determine the solution to these equations

---

*Traditionally, random test data generation has been applied to programs that accept numerical inputs. However, these random number generators could be used to produce `Strings` if we treat the numbers as ASCII or Unicode values. Furthermore, the random number generator could create complex abstract data types if we assign a semantic meaning to specific numerical values.

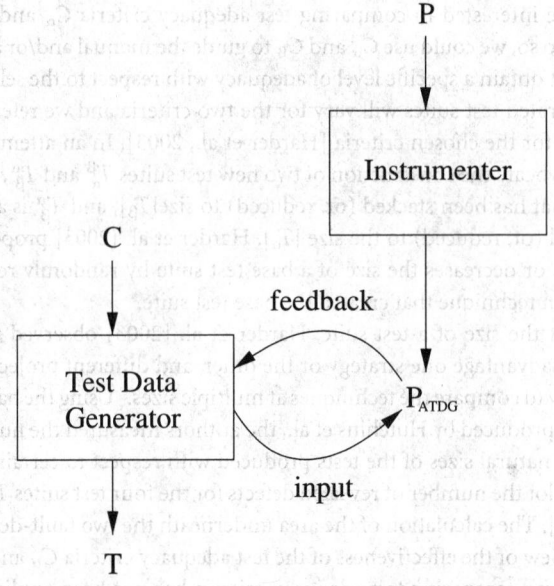

FIGURE 105.7   Dynamic software test data generation.

[Clarke, 1976; Ramamoorty et al., 1976]. Symbolic test data generation is appealing because it does not require the execution of the program under test. However, this type of symbolic generation is of limited practical value due to the problems associated with iteration constructs and arrays that depends on other program variables and pointers [Michael et al., 2001]. Alternatively, Howe et al. [1997] and Memon et al. [2001b] have chosen to represent aspects of the program under test with a specification of preconditions and postconditions and then express the desired test cases in terms of initial program states and goal program states. This type of abstract program representation facilitates the usage of an **artificial intelligence** (AI) **planner** that can automatically produce a test case that causes the program to progress from the start state to the goal state using the formally specified preconditions and postconditions.

Dynamic test data generation is an approach that actually relies on the execution of P (or some instrumented version of P) to generate an adequate T [Gupta et al., 1998, 1999; Korel, 1996; Michael et al., 2001]. While any dynamic test data generation approach must incur the overhead associated with actually executing the program under test, this execution can reveal additional insights about which test requirements have not been satisfied. Furthermore, dynamic test data generation techniques can use information gathered from program executions to determine how "close" the generator is to actually creating a test suite that satisfies C. Furthermore, dynamic test generation methods can handle arrays and pointer references because the values of array indices and pointers are know throughout the generation process. Figure 105.7 provides a view of the dynamic test data generation process that relies on a program instrumenter to produce P_{ATDG}, a version of P that contains statements that allow P to interact with the test generation subsystem. The test data generator that takes C as input can interact with P_{ATDG} to facilitate the automatic generation of test suite T.

Frequently, dynamic test data generation is viewed as a function minimization problem [Ferguson and Korel, 1996; Korel, 1996; Michael et al., 2001]. However, it is important to note that some test data generation techniques that use functions to represent a program do not focus on minimizing the selected functions [Fisher et al., 2002b]. For the purposes of our discussion, we will examine automated test data generation approaches that attempt to generate T by minimizing functions that describe P. For example, suppose that the selected adequacy criterion requires the execution of the `true` branch of the conditional logic statement `if( mass != 0 )` provided on line 8 of the `computeVelocity` method listed in Figure 105.3. We can ensure the execution of the `true` branch of this conditional logic statement by

Decision Example	Objective Function
if(c <= d)	$F(x) = \begin{cases} c_i(x) - d_i(x), & \text{if } c_i(x) > d_i(x); \\ 0, & \text{otherwise.} \end{cases}$
if(c >= d)	$F(x) = \begin{cases} d_i(x) - c_i(x), & \text{if } c_i(x) < d_i(x); \\ 0, & \text{otherwise.} \end{cases}$
if(c == d)	$F(x) = \|c_i(x) - d_i(x)\|$
if(c != d)	$F(x) = -\|c_i(x) - d_i(x)\|$
if(b)	$F(x) = \begin{cases} 1000, & \text{if } b_i(x) = \textit{false}; \\ 0, & \text{otherwise.} \end{cases}$

FIGURE 105.8   General form of selected objective functions.

minimizing the objective function $F(x) = -\|m_8(x)\|$ where $m_8(x)$ denotes the value of the variable mass on line 8 that was induced by test input value x. Because there is a wide range of program inputs that can lead to the satisfaction of this conditional logic statement, we can actually select any inputs that cause $F(x)$ to evaluate to a negative output and not directly focus on finding the minimum of the function. Thus, automated test data generation often relies on constrained function minimization algorithms that search for minimums within certain bounds [Ferguson and Korel, 1996].

Figure 105.8 provides the general form of the objective function for several different conditional logic forms. Our formulation of an objective function is based on the "fitness functions" described by Michael et al. [2001] and the "branch functions" used by Ferguson and Korel [1996]. In this figure, we use the notation $c_i(x)$ to denote the value of variable c on line i that was induced by test input x. Because dynamic test data generation attempts to minimize objective functions while executing the program under test, the actual value of either $c_i(x)$ or $d_i(x)$ is known during the attempt to generate an adequate T. Yet, the $c_i(x)$ and $d_i(x)$ used in objective functions might be complex functions of the input to the program under test. In this situation, an objective function cannot be minimized directly, and it can only be used in a heuristic fashion to guide the test data generator [Michael et al., 2001]. Yet, even when $c_i(x)$ and $d_i(x)$ are complicated functions of test input x, an objective function can still be used to provide "hints" to the test data generator in an attempt to show whether the modification of the test input is improving test suite adequacy.

It is important to note that the formulations of the objective functions for the conditional logic statements if(c >= d), if(c <= d), if(c == d), and if(e) are constructed in a fashion that causes the function to reach a minimum and then maintain this value. However, the conditional logic statement if (c != d) can create functions that have no specific minimum value and therefore must be minimized in a constrained fashion. Each of the functions in Figure 105.8 describe an $F(x)$ that must be minimized in order to take the true branch of the condition; the objective functions for the false branches could be developed in an analogous fashion. Our discussion of objective functions for conditional logic predicates omits the details associated with dynamic test data generation for conjunctive and disjunctive predicates. However, Fisher et al. [2002b] propose a type of objective function that can handle more complicated predicates. While we omit a discussion of the objective functions for conditional logic predicates that include

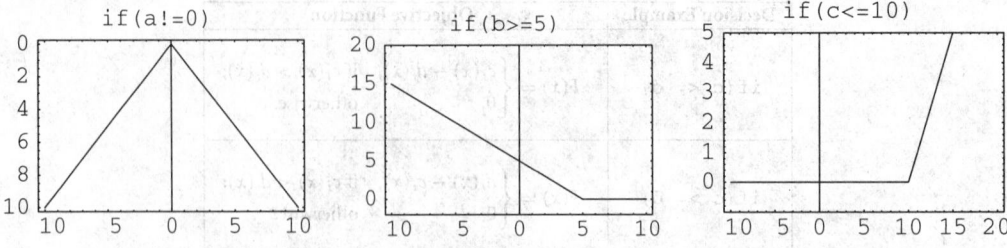

FIGURE 105.9   Objective functions for selected conditional logic statements.

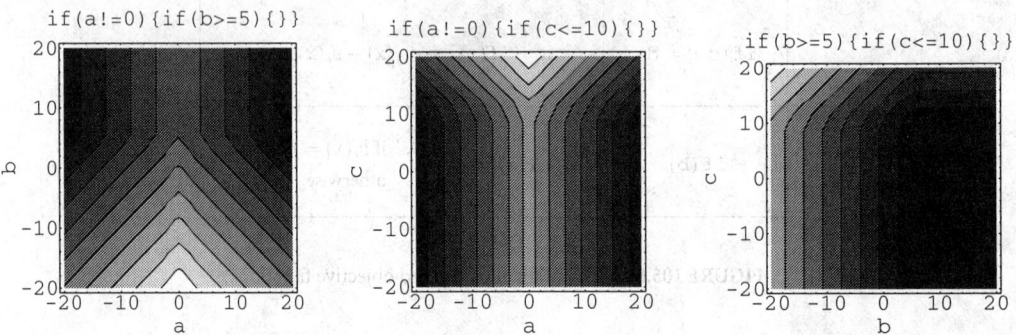

FIGURE 105.10   Objective functions for nested conditional logic statements.

the $>$ and $<$ relational operators, Ferguson and Korel [1996] describe the form of the functions for conditional logic predicates that use these operators.

Figure 105.9 depicts the graphs of the objective functions for several different conditional logic statements. These graphs can be viewed as specific instantiations of the general objective functions described in Figure 105.8. The first graph represents the objective function for the conditional logic statement if (a != 0) where variable a is a component of test input x. The second and third graphs offer the plots of the objective functions associated with the conditional logic statements if (b >= 5) and if (c <= 10), respectively. Of course, P might have multiple conditional logic statements that are nested in an arbitrary fashion. For example, the conditional logic block associated with the statement if (a != 0) might have a conditional logic block for if (b >= 5) located at an arbitrary position within its own body. If $F_a(x)$ corresponds to the objective function for if (a != 0) and $F_b(x)$ represents the objective function for if (b >= 5), then the objective function $F_a(x) + F_b(x)$ must be minimized in an attempt to exercise the true branches of the nested conditional logic statement if (a != 0) {<...>if (b >= 5) {<...>}} [Michael et al., 2001].

Figure 105.10 includes the contour plots for the three of the unique combinations of the objective functions described in Figure 105.9 (there are three alternative conditional logic nestings that we do not consider). For simplicity, we assume that the objective functions are direct functions of the test inputs so that $a_i(x) = a$, $b_i(x) = b$, and $c_i(x) = c$ and that the test input x is composed of three values (a, b, c). Because the addition of two objective functions for different program variables creates a three-dimensional function, these contour plots represent "high" areas in the combined objective function with light colors and "low" areas in the combined function with dark colors. Furthermore, each contour plot places the outer conditional logic statement on the x-axis and the inner conditional logic statement on the y-axis.

As expected, the contour plot for the nested conditional logic statement if (a != 0) {<...>if (b >= 5) {<...>}} is at minimum levels in the upper left and right corners of the plot. Because our automated test data generation algorithms attempt to seek minimums in objective functions, either of these dark regions could lead to the production of test data that would cause the execution of the `true` branches of both conditional logic statements. While the contour plot associated with the nested conditional logic statement `if(a != 0){<...>if(c <= 5){<...>}}` also contains two dark regions on the left and right sections of the graph, the statement `if(b >= 0){<...>if(c <= 10){<...>}}` only has a single dark region on the right side of the plot.

105.3.6 Test Execution

The execution of a test suite can occur in a manual or automated fashion. For example, the test case descriptions that are the result of the test selection process could be manually executed against the program under test. However, we will focus on the automated execution of test cases and specifically examine the automated testing issues associated with the JUnit test automation framework [Hightower, 2001; Jeffries, 1999]. JUnit provides a number of `TestRunners` that can automate the execution of any Java class that `extends` `junit.framework.TestCase`. For example, it is possible to execute the `KineticTest` provided in Figure 105.4 inside of either the `junit.textui.TestRunner`, `junit.awtui.TestRunner`, or the `junit.swingui.TestRunner`. While each `TestRunner` provides a slightly different interface, they adhere to the same execution and reporting principles. For example, JUnit will simply report "okay" if a test case passes and report a failure (or error) with a message and a stack trace if the test case does not pass. Figure 105.11 shows the output resulting from the execution of the `KineticTest` provided in Figure 105.4 of Section 105.3.2.

The JUnit test automation framework is composed of a number of Java classes. Version 3.7 of the JUnit testing framework organizes its classes into nine separate Java packages. Because JUnit is currently released under the open source Common Public License 1.0, it is possible to download the source code from `http://www.junit.org` to learn more about the design and implementation choices made by Kent Beck and Erich Gamma, the creators of the framework. In this chapter, we highlight some of the interesting design and usage issues associated with JUnit. More details about the intricacies of JUnit can be found in Gamma and Beck [2003] and Jackson [2003].

The `junit.framework.TestCase` class adheres to the Command design pattern and thus provides the `run` method that describes the default manner in which tests can be executed. As shown in Figure 105.4, a programmer can write a new collection of tests by creating a subclass of `TestCase`. Unless a `TestCase` subclass provides a new implementation of the `run` method, the JUnit framework is designed to call the default `setUp`, `runTest`, and `tearDown` methods. The `setUp` and `tearDown` methods are simply

```
  .... F
 Time: 0.026
 There was 1 failure:

1) testFour(KineticTest)junit.framework.AssertionFailedError : expected:<20>
     but was:<24>

             at KineticTest.testFour(KineticTest.java:48)
             at sun. reflect.NativeMethodAccessorImpl.invoke0(Native Method)
             at sun. reflect.NativeMethodAccessorImpl.invoke
                (NativeMethodAccessorImpl.java:39)
             at sun. reflect.DelegatingMethodAccessorImpl.invoke
                (DelegatingMethodAccessorImpl.java:25)

 FAILURES!!!
 Tests run : 4,   Failures : 1,   Errors : 0
```

FIGURE 105.11 The results from executing `KineticTest` in the `junit.textui.TestRunner`.

responsible for creating the state of the class(es) under test and then "cleaning up" after a single test has executed. That is, JUnit provides a mechanism to facilitate the creation of independent test suites, as defined in Definition 105.2 of Section 105.2.1.

JUnit uses the `Composite` design pattern to enable the collection of `TestCases` into a single `Test-Suite`, as described by Definition 105.1 in Section 105.2.1. The `Test` interface in JUnit has two subclasses: `TestCase` and `TestSuite`. Like `TestCase`, a `TestSuite` also has a `run` method. However, `Test-Suite` is designed to contain 1 to n `TestCases` and its `run` method calls the `run` method of each of the instances of `TestCase` it contains [Jackson, 2003]. A `TestCase` can describe the tests that it contains by providing a `suite` method. JUnit provides an interesting "shorthand" that enables a subclass of `Test-Case` to indicate that it would simply like to execute all of the tests that it defines. The statement `return new TestSuite(KineticTest.class)` on line 10 of the `suite` method in Figure 105.4 requires the JUnit framework to use the Java reflection facilities to determine, at runtime, the methods within `KineticTest` that start with the required "`test`" prefix.

The `run` method in the `Test` superclass has the following signature: `public void run(Test Result result)` [Gamma and Beck, 2003]. Using the terminology established in Beck [1997] and Gamma and Beck [2003], `result` is known as a "collecting parameter" because it enables the collection of information about whether the tests in the test suite passed or caused a **failure** or an **error** to occur. Indeed, JUnit distinguishes between a test that fails and a test that raises an error. JUnit test cases include assertions about the expected output of a certain method under test or the state of the class under test and a failure occurs when these assertions are not satisfied. On the other hand, the JUnit framework automatically records that an error occurred when an unanticipated subclass of `java.lang.Exception` is thrown by the class under test. In the context of the terminology established in Section 105.2.1, JUnit's errors and failures both reveal faults in the application under test.

JUnit also facilitates the testing of the expected "exceptional behavior" of a class under test. For example, suppose that a `BankAccount` class provides a `withdraw(double amount)` method that raises an `OverdraftException` whenever the provided `amount` is greater than the `balance` encapsulated by the `BankAccount` instance. Figure 105.12 provides the `BankAccountTest` class that tests the normal and exceptional behavior of a `BankAccount` class. In this subclass of `TestCase`, the `testInvalidWithdraw` method is designed to fail when the `withdraw` method does not throw the `OverdraftException`. However, the `testValidWithdraw` method will only throw an exception if the `assertEquals(expected, actual, NO_DELTA)` is violated. In this test, the third parameter in the call to `assertEquals` indicates that the test will not tolerate any small difference in the double parameters `expected` and `actual`.

105.3.7 Test Adequacy Evaluation

It is often useful to determine the adequacy of an existing test suite. For example, an automated test data generation algorithm might be configured to terminate when the generated test suite reaches a certain level of adequacy. If test suites are developed in a manual fashion, it is important to measure the adequacy of these tests to determine if the program under test is being tested "thoroughly." In our discussion of test adequacy evaluation, we use $R(C, P)$ to denote the set of test requirements for a given test adequacy criterion C and a program under test P. If the all-nodes criterion was selected to measure the adequacy of a T used to test the `computeVelocity` method in Figure 105.3, then we would have $R(C, P) = \{enter_{cv}, n_6, n_7, n_8, n_{10}, n_{11}, n_{12}, n_{16}, n_{18}, exit_{cv}\}$. Alternatively, if C was the all-uses test adequacy criterion, then $R(C, P)$ would contain all of the def-c-use and def-p-use associations within `computeVelocity`. For example, line 11 of `computeVelocity` contains a definition of the variable `velocity` and line 12 contains a computation-use of `velocity` and this def-use association would be included in $R(C, P)$.

Normally, the adequacy of test suite T is evaluated by instrumenting the program under test to produce $P_{R(C,P)}$, a version of P that can report which test requirements are covered during the execution of T. Pavlopoulou and Young [1999] have proposed, designed, and implemented a **residual test adequacy**

```
1  import junit.framework.*;
2  public class BankAccountTest extends TestCase
3  {
4    private BankAccount account = new BankAccount(1000);
5    private static double NO_DELTA = 0.0;
6    public BankAccountTest(String name)
7    {
8      super(name);
9    }
10   public static Test suite()
11   {
12     return new TestSuite(BankAccountTest.class);
13   }
14
15   public void testValidWithdraw()
16   {
17     double expected = 500.00;
18     account.withdraw(500);
19     double actual = account.getBalance();
20     assertEquals(expected, actual, NO_DELTA);
21   }
22   public void testInvalidWithdraw()
23   {
24     try
25     {
26         account.withdraw(1500);
27         fail("Should_have_thrown_OverdraftException");
28     }
29     catch(OverdraftException e)
30     {
31         // test is considered to be successful
32     }
33   }
34 }
```

FIGURE 105.12 The BankAccountTest with valid and invalid invocations of withdraw.

evaluator that can instrument the program under test and calculate the adequacy of the test suites used during development. Figure 105.13 provides a high-level depiction of this test adequacy evaluation system for Java programs. The residual coverage tool described by these authors can also measure the coverage of test requirements after a software system has been deployed and is being used in the field. Finally, this test coverage monitoring tool provides the ability to incrementally remove the test coverage probes placed in the program under test after the associated test requirements have been exercised [Pavlopoulou and Young, 1999]. Pavlopoulou and Young [1999] report that the removal of the probes used to monitor covered test requirements often dramatically reduces the overhead associated with test adequacy evaluation.

105.3.8 Regression Testing

After a software system experiences changes in the form of bug fixes or additional functionality, a software maintenance activity known as *regression testing* can be used to determine if these changes introduced

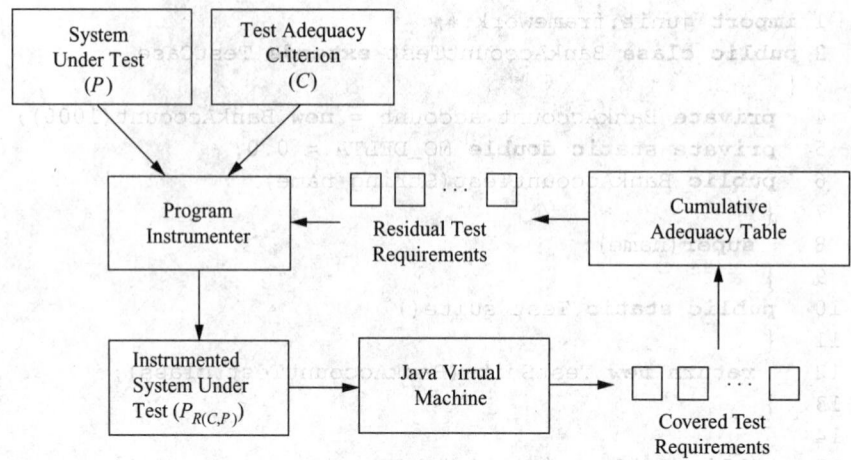

FIGURE 105.13   The iterative process of residual test coverage monitoring. (From Christina Pavlopoulou and Michal Young. Residual test coverage monitoring. In *Proceedings of the 21st International Conference on Software Engineering*, pages 277–284. IEEE Computer Society Press, 1999. With permission.)

defects. As described in Section 105.2.2 and depicted in Figure 105.2, the regression testing process applies all of the other software testing stages whenever the program under test changes. The creation, maintenance, and execution of a regression test suite helps to ensure that the evolution of an application does not result in lower quality software. The industry experiences noted by Onoma et al. [1998] indicate that regression testing often has a strong positive influence on software quality. Indeed, the importance of regression testing is well understood. However, as noted by Beizer [1990] and Leung and White [1989], many software development teams might choose to omit some or all of the regression testing tasks because they often account for as much as one half the cost of software maintenance. Moreover, the high costs of regression testing are often directly associated with the execution of the test suite. Other industry reports from Rothermel et al. [1999] show that the complete regression testing of a 20,000-line software system required seven weeks of continuous execution. Because some of the most well-studied software failures, such as the Ariane-5 rocket and the 1990 AT&T outage, can be blamed on the failure to test changes in a software system [Hamlet and Maybee, 2001], many techniques have been developed to support efficient regression testing.

Several different methods have been developed in an attempt to reduce the cost of regression testing. Regression **test selection** approaches attempt to reduce the cost of regression testing by selecting some appropriate subset of the existing test suite [Ball, 1998; Rothermel and Harrold, 1997; Vokolos and Frankl, 1997]. Test selection techniques normally use the source code of a program to determine which tests should be executed during the regression testing stage [Rothermel and Harrold, 1996]. Regression **test prioritization** techniques attempt to order a regression test suite so that those tests with the highest priority, according to some established criterion, are executed earlier in the regression testing process than those with lower priority [Elbaum et al., 2000; Rothermel et al., 1999]. By prioritizing the execution of a regression test suite, these methods hope to reveal important defects in a software system earlier in the regression testing process. Regression **test distribution** is another alternative that can make regression testing more practical by more fully utilizing the computing resources that are normally available to a testing team [Kapfhammer, 2001].

Throughout our discussion of regression testing, we continue to use the notation described in Section 105.2.1 and extend it with additional notation used in Rothermel and Harrold [1996]. Therefore, we will use P' to denote a modified version of program under test P. Problem 105.1 characterizes the regression testing problem in a fashion initially proposed in Rothermel and Harrold [1994]. It is important to note that any attempt to solve the regression testing problem while attempting to make regression testing more cost-effective could use regression test selection, prioritization, and distribution techniques in conjunction with or in isolation from one another. In Section 105.3.8.1 through Section 105.3.8.3, we iteratively construct a regression testing solution that can use test selection, prioritization, and distribution techniques.

Problem 105.1

Given a program P, its modified version P', and a test set T that was used to previously test P, find a way to utilize T to gain sufficient confidence in the correctness of P'.

105.3.8.1 Selective Regression Testing

Selective retest techniques attempt to reduce the cost of regression testing by identifying the portions of P' that must be exercised by the regression test suite. Intuitively, it might not be necessary to re-execute test cases that test source code locations in P' that are the same as the source locations in P. Any selective regression testing approach must ensure that it selects all the test cases that are defect-revealing for P', if there are any defects within P'. Selective retesting is distinctly different from a retest-all approach that conservatively executes every test in an existing regression test suite. Figure 105.14 uses the *RTS* algorithm to express the steps that are commonly associated with a regression testing solution that uses a test selection approach [Rothermel and Harrold, 1996, 1998].

Each step in the *RTS* algorithm addresses a separate facet of the regression testing problem. Line 1 attempts to select a subset of T that can still be used to effectively test P'. Line 3 tries to identify portions of P' that have not been sufficiently tested and then seeks to create these new regression tests. Line 2 and Line 4 focus on the efficient execution of the regression test suite and the examination of the testing results, denoted R_1 and R_2, for incorrect results. Finally, Line 5 highlights the need to analyze the results of all previous test executions, the test suites themselves, and the modified program in order to produce the final test suite. When the *RTS* algorithm terminates, modified program P' becomes the new program under test and T_L is now treated as the test suite that will be used during any further regression testing that occurs after the new P changes. Traditionally, regression test selection mechanisms limit themselves to the problem described in Line 1. Furthermore, algorithms designed to identify T' must conform to the **controlled regression testing assumption**. This assumption states that the only valid changes that can be made to P in order to produce P' are those changes that impact the source code of P [Rothermel and Harrold, 1997].

105.3.8.2 Regression Test Prioritization

Regression test prioritization approaches assist with regression testing in a fashion that is distinctly different from test selection methods. Test case prioritization techniques allow testers to order the execution of a

Algorithm $RTS(T, P, P')$
($*$ Regression Testing with Selection $*$)
Input: Initial Regression Test Suite T;
      Initial Program Under Test P;
      Modified Program Under Test P'
Output: Final Regression Test Suite T_L
1. $T' \leftarrow SelectTests(T, P)$
2. $R_1 \leftarrow ExecuteTests(T', P)$
3. $T'' \leftarrow CreateAdditionalTests(T', P)$
4. $R_2 \leftarrow ExecuteTests(T'', P)$
5. $T_L \leftarrow CreateFinalTests(T, T', R_1, T'', R_2, P')$
6.   **return** T_L

FIGURE 105.14 Regression testing with selection.

Algorithm *RTSP(T, P, P')*
(∗ Regression Testing with Selection and Prioritization ∗)
Input: Initial Regression Test Suite *T*;
          Initial Program Under Test *P*;
          Modified Program Under Test *P'*
Output: Final Regression Test Suite T_L
1. $T' \leftarrow SelectTests(T, P)$
2. $T'_r \leftarrow PermuteTests(T', P)$
3. $R_1 \leftarrow ExecuteTests(T'_r, P)$
4. $T'' \leftarrow CreateAdditionalTests(T'_r, P)$
5. $R_2 \leftarrow ExecuteTests(T'', P)$
6. $T_L \leftarrow CreateFinalTests(T, T'_r, R_1, T'', R_2, P')$
7.   **return** T_L

FIGURE 105.15   Regression testing with selection and prioritization.

regression test suite in an attempt to increase the probability that the suite might detect a fault at earlier testing stages [Elbaum et al., 2000; Rothermel et al., 1999, 2001a; Wong et al., 1997]. Figure 105.15 uses the formalization of Elbaum et al. [2000] and Rothermel and Harrold [1996] to characterize the typical steps taken by a technique that employs both selection and prioritization to solve the regression testing problem.

The expression of the *RTSP* algorithm in Figure 105.15 is intended to indicate that a tester can use regression test selection and prioritization techniques in either a collaborative or independent fashion. Line 1 allows the tester to select *T'* such that it is a proper subset of *T* or such that it actually contains every test that *T* contains. Furthermore, Line 2 could produce T'_r such that it contains an execution ordering for the regression tests that is the same or different than the ordering provided by *T'*. The other steps in the *RTSP* algorithm are similar to those outlined in Figure 105.14, the regression testing solution that relied only on test selection.

The test case prioritization approaches developed by Elbaum et al. restrict themselves to the problem described in Line 2. Testers might desire to prioritize the execution of a regression test suite such that code coverage initially increases at a faster rate. Or, testers might choose to increase the rate at which high-risk faults are first detected by a test suite. Alternatively, the execution of a test suite might be prioritized in an attempt to ensure that the defects normally introduced by competent programmers are discovered earlier in the testing process. Current techniques that prioritize a regression test suite by the fault-exposing-potential of a test case also rely on the usage of mutation analysis [Elbaum et al., 2000; Rothermel et al., 2001a], as previously described in Section 105.3.4.2. When a regression test suite for a given program *P* is subjected to a mutation analysis, a finite set of syntactically different versions of *P* are produced. Then, the entire regression test suite is executed for each one of the mutated versions of *P* in order to determine if the tests can detect the fault that each mutant represents. Although mutation analysis has proven to be a viable regression test prioritization technique, its practicality is limited because it is so computationally intensive [Elbaum et al., 2000; Rothermel et al., 2001a].

Regression test suite prioritization algorithms are motivated by the empirical investigations of test adequacy criteria, as discussed in Section 105.3.4.4, which indicate that tests that are highly adequate are often more likely to reveal program defects. In an empirical evaluation of regression test suite prioritization, techniques that create a prioritized test suite can be evaluated based upon the weighted average of the percentage of faults detected over the life of the test suite, or the **APFD** [Elbaum et al., 2003]. For example, suppose that a regression test suite *T* was prioritized by its code coverage ability in order to produce (T'^c_r and was also prioritized by its fault exposing potential, thereby creating T'^{fep}_r. If the faults in the program under test are known and the APFD of T'^{fep}_r is greater than the APFD of T'^c_r, then it is likely that we would prefer the usage of mutation adequacy over code coverage to prioritize the execution of *T*. Yet, Elbaum et al. [2003] caution against this interpretation when they observe that "to assume that a higher APFD

Test Case	Faults				
	f_1	f_2	f_3	f_4	f_5
1			×	×	
2	×	×			
3	×	×	×		
4			×	×	×
5		×	×		

FIGURE 105.16 The faults detected by a regression test suite $T = \langle t_1, \ldots, t_5 \rangle$.

implies a better technique, independent of cost factors, is an oversimplification that may lead to inaccurate choices among techniques."

However, because the APFD metric was used in early studies of regression test suite prioritization techniques and because it can still be used as a basis for more comprehensive prioritization approaches that use cost-benefit thresholds [Elbaum et al., 2003], it is important to investigate it in more detail. If we use the notation established in Section 105.2.1 and we have $T = \langle t_1, \ldots, t_n \rangle$ and a total of g faults within program under test P, then Equation 105.15 defines the $APFD(T, P)$ [Elbaum et al., 2003]. We use $reveal(h, T)$ to denote the position within T of the first test that reveals fault i.

$$APFD(T, P) = 1 - \frac{\sum_{h=1}^{g} reveal(h, T)}{ng} + \frac{1}{2n} \qquad (105.15)$$

For example, suppose that we have the test suite $T = \langle t_1, \ldots, t_5 \rangle$ and we know that the tests detect faults $f_1, \ldots, f_5$ in P according to Figure 105.16. Next, assume that $PermuteTests(T', P)$ creates a $T'_{r_1} = \langle t_1, t_2, t_3, t_4, t_5 \rangle$, thus preserving the ordering of T. In this situation, we now have $APFD(T'_{r_1}, P) = 1 - .4 + .1 = .7$. However, if $PermuteTests(T', P)$ does change the order of T to produce $T'_{r_2} = \langle t_3, t_4, t_1, t_2, t_5 \rangle$, then we have $APFD(T'_{r_2}, P) = 1 - .2 + .1 = .9$. In this example, T'_{r_2} has a greater weighted average of the percentage of faults detected over its life than T'_{r_1}. This is due to the fact that the tests which are able to detect all of the faults in P, t_4 and t_5, are executed first in T'_{r_2}. Therefore, if the prioritization technique used to produce T'_{r_2} is not significantly more expensive than the one used to create T'_{r_1}, then it is likely a wise choice to rely on the second prioritization algorithm for our chosen P and T.

105.3.8.3 Distributed Regression Testing

Any technique that attempts to distribute the execution of a regression test suite will rely on the available computational resources during Line 2, Line 3, and Line 5 of algorithm *RTSP* in Figure 105.15. That is, when tests are being selected, prioritized, or executed, distributed regression testing relies on all of the available testing machines to perform the selection, prioritization, and execution in a distributed fashion. If the changes that are applied to P to produce P' involve the program's environment and this violates the controlled regression testing assumption, a distribution mechanism can be used to increase regression testing cost-effectiveness. When the computational requirements of test case prioritization are particularly daunting, a distributed test execution approach can be used to make prioritizations based upon coverage-levels or fault-exposing-potential more practical [Kapfhammer, 2001]. In situations where test selection and/or prioritization are possible, the distributed execution of a regression test suite can be used to further enhance the cost-effectiveness of the regression testing process. When only a single test machine is available for regression testing, the distribution mechanism can be disabled and the other testing approaches can be used to solve the regression testing problem.

105.3.9 Recent Software Testing Innovations

The software testing and analysis research community is actively proposing, implementing, and analyzing new software testing techniques. Recent innovations have been both theoretical and practical in nature.

In this section, we summarize a selection of recent testing approaches that do not explicitly fit into the process model proposed in Section 105.2.2. Yet, it is important to note that many of these techniques have relationship(s) to the "traditional" phases described by our model.

105.3.9.1 Robustness Testing

A software system is considered to be **robust** if it can handle inappropriate inputs in a graceful fashion. Robustness testing is a type of software testing that attempts to ensure that a software system performs in an acceptable fashion when it is provided with anomalous input or placed in an inappropriate execution environment. Robustness testing is directly related to the process of hardware and software **fault injection**. For example, the FUZZ system randomly injects data into selected operating system kernel and utility programs in order to facilitate an empirical examination of operating system robustness [Miller et al., 1990]. Initial studies indicate that it was possible to crash between 25 and 33% of the utility programs that were associated with 4.3 BSD, SunOS 3.2, SunOS 4.0, SCO Unix, AOS Unix, and AIX 1.1 Unix [Miller et al., 1990]. Subsequent studies that incorporated additional operating systems, such as AIX 3.2, Solaris 2.3, IRIX 5.1.1.2, NEXTSTEP 3.2, and Slackware Linux 2.1.0, indicate that there was a noticeable improvement in operating system utility robustness during the intervening years between the first and second studies [Miller et al., 1998]. Interestingly, the usage of FUZZ on the utilities associated with a GNU/Linux operating system showed that these tools exhibited significantly higher levels of robustness than the commercial operating systems included in the study [Miller et al., 1998].

Other fault injection systems, such as FTAPE, rely on the usage of computer hardware to inject meaningful faults into a software system [Tsai and Iyer, 1995]. Yet, most recent fault injection systems, such as Ballista, do not require any special hardware in order to perform robustness testing. Ballista can perform robustness testing on the Portable Operating System Interface (POSIX) API to assess the robustness of an entire operating system interface [Koopman and DeVale, 2000]. Instead of randomly supplying interfaces with test data, Ballista builds test cases that are based on the data types that are associated with procedure parameters. Ballista associates each of the possible parameter data types with a finite set of test values that can be combined to generate test inputs for operations.

A Ballista test case can be automatically executed in an infrastructure that supports the testing of operating systems such as AIX 4.41, Free BSD 2.2.5, HP-UX 10.20, Linux 2.0.18, and SunOS 5.5 [Koopman and DeVale, 2000]. Finally, if a test causes the operating system to perform in an anomalous fashion, the test records the severity of the test results based on the **CRASH scale** that describes *Catastrophic*, *Restart*, *Abort*, *Silent*, and *Hindering* failures. According to Biyani and Santhanam [1997], Krop et al. [1998], and Koopman and DeVale [2000], a catastrophic failure occurs when a utility program (or an operating system API) causes the system to hang or crash while a restart failure happens when the procedure under test never returns control to the test operation in a specific robustness test. Furthermore, abort failures are traditionally associated with "core dumps," and silent failures occur when the operating system does not provide any indication of the fact that the robustness testing subject just performed in an anomalous fashion. Finally, hindering failures happen when the procedure under test returns an error code that does not properly describe the exceptional condition that arose due to robustness testing. An empirical analysis of Ballista's ability to detect robustness failures indicates that a relatively small number of POSIX functions in the candidate operating systems did not ever register on the CRASH scale [Krop et al., 1998].

Several robustness testing systems have also been developed for the Windows NT operating system [Ghosh et al., 1998, 1999; Tsai and Singh, 2000]. For example, Ghosh et al. [1999] discuss the fault injection simulation tool (FIST) and a wrapping technique that can improve the robustness of Windows NT applications. The authors describe the usage of FIST to isolate a situation in which Microsoft Office 97 performs in a non-robust fashion and then propose a wrapping technique that can enable the Office application to correctly handle anomalous input. Furthermore, an empirical analysis that relied on the application of a fault injection tool called the Random and Intelligent Data Design Library Environment (RIDDLE) revealed that GNU utilities ported to Windows NT appear to be less robust than the same utilities on the Unix operating system [Ghosh et al., 1998]. Finally, the dependability test suite (DTS) is another fault injection tool that can be used to determine the robustness of Windows NT applications such as Microsoft IIS Server and SQL Server [Tsai and Singh, 2000]. While robustness testing and software fault injection frequently focus their testing efforts on the operating system level, Haddox et al. [2001, 2002] have developed techniques that use fault injection to test and analyze software **commercial off-the-shelf (COTS)**, components of a finer level of granularity than a complete operating system API or an entire application.

105.3.9.2 Testing Spreadsheets

Our discussion of software testing has generally focused on the testing and analysis of programs written in either procedural or object-oriented programming languages. However, Rothermel et al. [1997, 2000, 2001b] and Fisher et al. [2000a, 2000b] have focused on the testing and analysis of programs written in spreadsheet languages. The form-based visual programming paradigm, of which spreadsheet programs are a noteworthy example, is an important mode of software development. Indeed, Brooks [1995] makes the following observation about spreadsheets and databases: "These powerful tools, so obvious in retrospect and yet so late in appearing, lend themselves to a myriad of uses, some quite unorthodox." However, until recently, the testing and analysis of spreadsheet programs has been an area of research that has seen little investigation. Rothermel et al. [2000] echo this sentiment in the following observation:

> Spreadsheet languages have rarely been studied in terms of their software engineering properties. This is a serious omission, because these languages are being used to create production software upon which real decisions are being made. Further, research shows that many spreadsheets created with these languages contain faults. For these reasons, it is important to provide support for mechanisms, such as testing, that can help spreadsheet programmers determine the reliability of values produced by their spreadsheets.

Rothermel et al. [1997] have described some of the differences between form-based programming languages and imperative programming language paradigms, noting that these "programs" are traditionally composed of cells that have formulas embedded within them. These authors have proposed the **cell relation graph** (CRG) as an appropriate model for a form-based program that is loosely related to the control-flow graph of an imperative program. The CRG includes information about the control flow within the formulas that are embedded within cells and the dependencies between the program's cells. Rothermel et al. [1997] have also examined the usefulness of different understandings of test adequacy for spreadsheet programs, such as traditional node and edge-based criteria defined for a program's CRG. However, they concluded that a data flow-based test adequacy criterion is most appropriate for form-based programs because it models the definition and usage of the cells within a program. Furthermore, because form-based programs do not contain constructs such as array indexing operations and pointer accesses, it is often much easier to perform data flow analysis on the CRG of a spreadsheet program than it would be to perform a data flow analysis on the ICFG of an imperative program [Rothermel et al., 1997].

Rothermel et al. [2001b] provide all of the algorithms associated with a testing methodology for the Forms/3 spreadsheet language. In this chapter, we focus on the issues associated with automated test data generation [Fisher et al., 2002b] and test reuse [Fisher et al., 2002a] for programs written in spreadsheet languages. The "What You See Is What You Test" (WYSIWYT) spreadsheet testing methodology proposed by Fisher et al. and Rothermel et al. includes a test data generation mechanism for the **output-influencing-all-du-pairs** (or, oi-all-du-pairs) test adequacy criterion that is based on the CRG representation of a spreadsheet [Fisher et al., 2002b; Rothermel et al., 2001b]. This test adequacy criterion is similar to the standard all-uses and all-DUs criteria, except that it requires the coverage of the def-use associations in a spreadsheet's CRG that influence the cells that contain user-visible output. The goal-oriented test data generation approach implemented by the authors is similar to Ferguson and Korel's [1996] chaining approach because it represents a spreadsheet as a collection of branch functions. The constrained linear search procedure described by Fisher et al. [2002b] attempts to exercise uncovered output influencing def-use-associations by causing the branch functions associated with the selected CRG path to take on positive values.

Test suite reuse is also an important facet of spreadsheet testing because end users often share spreadsheets, and production spreadsheets must be revalidated when new versions of commercial spreadsheet engines are released [Fisher et al., 2002a]. Current spreadsheet reuse algorithms are similar to the regression test suite selection techniques presented in Section 105.3.8.1. When a spreadsheet user modifies a spreadsheet application, the reuse algorithms must determine which portion of the test suite must be executed in order to validate the program in a manner that is cost-effective and practical. Because the WYSIWYT tools are designed to be used by a spreadsheet developer in a highly interactive fashion, the timely retesting of a spreadsheet often can only occur if the test suite is reused in an intelligent fashion. Fisher et al. [2002a] propose specific test reuse actions that should occur when the spreadsheet developer deletes a cell, changes the formula within a spreadsheet cell, or inserts a new cell into a spreadsheet. For example, if the programmer changes the formula within a specific spreadsheet cell, the test reuse algorithms must select all the existing test(s) that validate the output of the modified cell and any other impacted cells. Alternatively, if the formula changes for a certain cell change cause the formula to rely on new and/or different spreadsheet input cells, some test cases might be rendered obsolete [Fisher et al., 2002a].

105.3.9.3 Database-Driven Application Testing

Even simple software applications have complicated and ever-changing operating environments that increase the number of interfaces and the interface interactions that must be tested. Device drivers, operating systems, and databases are all aspects of a software system's environment that are often ignored during testing [Whittaker, 2000; Whittaker and Voas, 2000]. Yet, relatively little research has specifically focused on the testing and analysis of applications that interact with databases. Chan and Cheung [1999a, 1999b] have proposed a technique that can test database-driven applications written in a general-purpose programming language such as Java, C, or C++, while also include embedded Structured Query Language (SQL) statements that are designed to interact with a relational database. In their approach, Chan and Cheung transform the embedded SQL statements within a database-driven application into general purpose programming language constructs. Chan and Cheung [1999a] provide C code segments that describe the selection, projection, union, difference, and cartesian product operators that form the relational algebra and thus heavily influence the SQL. Once the embedded SQL statements within the program under test have been transformed into general-purpose programming language constructs, it is possible to apply traditional test adequacy criteria, as described in Section 105.3.4, to the problem of testing programs that interact with one or more relational databases.

Chays et al. [2000, 2002] and Chays and Deng [2003] have described the challenges associated with testing database-driven applications and proposed the AGENDA tool suite as a solution to a number of these challenges. In fact, Chays et al. [2000] observe that measuring the "correctness" of database-driven applications might involve the following activities: (1) determining whether the program behaves according to specification, (2) deciding whether the relational database schema is correctly designed, (3) showing that the database is secure, (4) measuring the accuracy of the data within the database, and (5) ensuring that the database management system correctly performs the SQL operations required by the application itself. Chays et al. [2000] also propose a partially automatable software testing methodology, inspired by the category-partition method described in Section 105.3.4.3, that addresses the first understanding of database-driven application correctness. When provided with the relational schema of the database(s) used by the program under test and a description of the categories and choices for the attributes required by the relational tables, the AGENDA tool can generate meaningful test databases [Chays et al., 2002; Chays and Deng, 2003]. AGENDA also provides a number of database testing heuristics, such as "determine the impact of using attribute boundary values" or "determine the impact of null attribute values" that can enable the tester to gain insight into the behavior of a program when it interacts with a database that contains "interesting" states [Chays et al., 2000, 2002; Chays and Deng, 2003].

While the AGENDA tool suite does provide innovative techniques for populating the relational database used by a database-driven application, it does not explicitly test the interactions between the program and the database. As noted by Daou et al. [2001] and Kapfhammer and Soffa [2003], a database interaction point in a program can be viewed as an interaction with different entities of a relational database, depending on the granularity with which we view the interaction. That is, we can view an SQL statement's interaction with a database at the level of databases, relations, records, attributes, or attribute values. Kapfhammer and

```
1   public boolean lockAccount(int card_number)
2     throws SQLException
3   {
4     boolean completed = false;
5     String qu_lock =
6       "UPDATE_UserInfo_SET_acct_lock=1_WHERE_card_number="+
7         card_number + ";";
8     Statement update_lock = m_connect.createStatement();
9     int result_lock = update_lock.executeUpdate(qu_lock);
10    if( result_lock == 1)
11    {
12      completed = true;
13    }
14    return completed;
15  }
```

FIGURE 105.17   The lockAccount in an ATM application.

Soffa [2003] propose an approach that can enumerate all the relational database entities that a database-driven program interacts with and then create a **database interaction control flow graph** (DICFG) that specifically models the actions of the program and its interactions with a database. As an example, the lockAccount method provided in Figure 105.17 could be a part of a database-driven ATM application. Line 9 of this program contain a database interaction point where the lockAccount method sends an SQL update statement to a relational database. Figure 105.18 offers a database interaction control flow graph for the lockAccount method that represents the operation's interaction with the relational database at the level of the database and attribute interactions.*

After proposing a new representation for database-driven applications, Kapfhammer and Soffa also describe a family of test adequacy criteria that can facilitate the measurement of the test suite quality for programs that interact with relational databases. The **all-database-DUs, all-relation-DUs, all-record-DUs, all-attribute-DUs**, and **all-attribute-value-DUs** test adequacy criteria that are extensions of the traditional all-DUs proposed by Hutchins et al. and discussed in Section 105.3.4.4. Definition 105.12 defines the all-relation-DUs test adequacy criterion. In this definition, we use R_l to denote the set of all the database relations that are interacted with by a method in a database-driven application application. While Kapfhammer and Soffa [2003] define the all-relation-DUs and other related test adequacy criteria in the context of the database interaction control flow graph for a single method, the criteria could be defined for a "database enhanced" version of the class control flow graph or the interprocedural control flow graph.

Definition 105.12   A test suite T for database interaction control flow graph $G_{DB} = (N_{DB}, E_{DB})$ satisfies the all-relation-DUs test adequacy criterion if and only if for each association (n_d, n_{use}, x), where $x \in R_l$ and $n_d, n_{use} \in N_{DB}$, there exists a test in $t_i \in T$ to create a complete path π^* in G_{DB} that covers the association.

105.3.9.4   Testing Graphical User Interfaces

The graphical user interface (GUI) is an important component of many software systems. While past estimates indicated that an average of 48% of an application's source code was devoted to the interface

---

*In this example, we assume that the program interacts with a single relational database called Bank. Furthermore, we assume that the Bank database contains two relations, Account and UserInfo. Finally, we require the Account relation to contain the attributes id, acct_name, balance, and card_number.

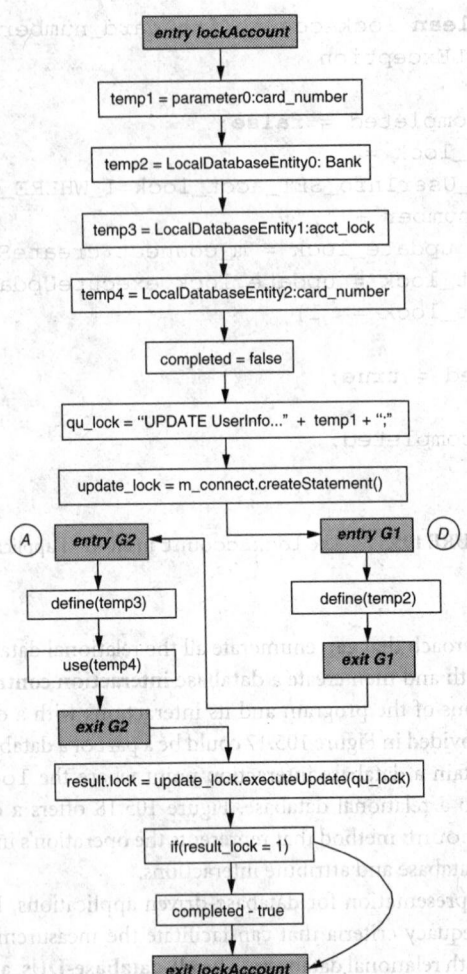

FIGURE 105.18 A DICFG for lockAccount. (From Gregory M. Kapfhammer and Mary Lou Soffa. A Family of Test Adequacy Criteria for Database-Driven Applications. In *Proceedings of the 9th European Software Engineering Conference and the 11th ACM SIGSOFT Symposium on the Foundations of Software Engineering.* ACM Press. With permission.)

[Myers and Rosson, 1992], current reports reveal that the GUI represents 60% of the overall source of a program [Memon, 2002]. While past research has examined user interface (UI) usability and widget layout [Sears, 1993], interactive system performance [Endo et al., 1996], and GUI creation framework performance [Howell et al., 2003], relatively little research has focused on the testing and analysis of GUIs. Memon et al. [2001a, 2001b] have conducted innovative research that proposes program representations, test adequacy criteria, and automated test data generation algorithms that are specifically tailored for programs with GUIs. As noted in the following observation from Memon et al. [1999], the testing of GUIs is quite challenging:

> In particular, the testing of GUIs is more complex than testing conventional software, for not only does the underlying software have to be tested but the GUI itself must be exercised and tested to check for bugs in the GUI implementation. Even when tools are used to generate GUIs automatically, they are not bug free, and these bugs may manifest themselves in the generated GUI, leading to software failures.

To complicate matters further, the space of possible GUI states is extremely large. Memon et al. [1999, 2001a] chose to represent a GUI as a series of operators that have preconditions and postconditions related to the state of the GUI. This representation classifies the GUI events into the categories of menu-open events, unrestricted-focus events, restricted-focus events, and system-interaction events [Memon et al., 2001a]. Menu-open events are normally associated with the usage of the pull-down menus in a GUI and are interesting because they do not involve interaction with the underlying application. While unrestricted-focus events simply expand the interaction options available to a GUI user, restricted-focus events require the attention of the user before additional interactions can occur. Finally, system interaction events require the GUI to interact with the actual application [Memon et al., 2001a]. To perform automated test data generation, Memon et al. rely on artificial intelligence planners that can use the provided GUI events and operations to automatically produce tests that cause the GUI to progress from a specified initial GUI state to a desired goal state.

In an attempt to formally describe test adequacy criteria for GUI applications, Memon et al. [2001b] propose the event-flow graph as an appropriate representation for the possible interactions that can occur within a GUI component. Furthermore, the integration tree shows the interactions between all the GUI components that comprise a complete graphical interface. Using this representation, Memon et al. [2001b] define intra-component and inter-component test adequacy criteria based upon GUI event sequences. The simplest intra-component test adequacy criterion, **event-interaction coverage**, requires a test suite to ensure that after a certain GUI event e has been performed, all events that directly interact with e are also performed [Memon et al., 2001b]. The **length-n event sequence** test adequacy criterion extends the simple event-interaction coverage by requiring a context of n events to occur before GUI event e actually occurs. Similarly, Memon et al. [2001b] propose inter-component test adequacy criteria that generalize the intra-component criteria and must be calculated using the GUI's integration tree.

105.4 Conclusion

Testing is an important technique for the improvement and measurement of a software system's quality. Any approach to testing software faces essential and accidental difficulties and, as noted by Edsger Dijkstra [1968], the construction of the needed test programs is a "major intellectual effort." While software testing is not a "silver bullet" that can guarantee the production of high-quality applications, theoretical and empirical investigations have shown that the rigorous, consistent, and intelligent application of testing techniques can improve software quality. Software testing normally involves the stages of test case selection, test case generation, test execution, test adequacy evaluation, and regression testing. Each of these stages in our model of the software testing process plays an important role in the production of programs that meet their intended specification. The body of theoretical and practical knowledge about software testing continues to grow as research expands the applicability of existing techniques and proposes new testing techniques for an ever-widening range of programming languages and application domains.

Defining Terms

All-edges/branch coverage: A test adequacy criterion that requires the execution of all the branches within the program under test.

All-nodes/statement coverage: A test adequacy criterion that requires the execution of all the statements within the program under test.

All-uses: A test adequacy criterion, used as the basis for definition-use testing, that requires the coverage of all the definition-c-use and definition-p-use associations within the program under test.

Category-partition method: A partially automatable software testing technique that enables the generation of test cases that attempt to ensure that a program meets its specification.

Coincidental correctness: A situation when a fault in a program does not manifest itself in a failure although the fault has been executed and it has infected the data state of the program.

Commercial off-the-shelf component: A software component that is purchased and integrated into a system. Commercial off-the-shelf components do not often provide source code access.

Competent programmer hypothesis: An assumption that competent programmers create programs that compile and very nearly meet their specification.

Complete path: A path in a control flow graph that starts at the program graph's entry node and ends at its exit node.

Condition: An expression in a conditional logic predicate that evaluates to `true` or `false` while not having any other Boolean valued expressions within it.

Condition-decision coverage: A test adequacy criterion that requires a test suite to cover all the edges within a program's control flow graph and to ensure that each condition evaluates to `true` and `false` at least one time.

Coupling effect: An assumption that test suites that can reveal simple defects in a program can also reveal more complicated combinations of simple defects.

Equivalent mutant: A mutant that is not distinguishable from the program under test. Determining whether a mutant is equivalent is generally undecidable. When mutation operators produce equivalent mutants, the calculation of mutation adequacy scores often requires human intervention.

Error: A mistake made by a programmer during the implementation of a software system.

Failure: The external, incorrect behavior of a program.

Fault: A collection of program source statements that cause a program failure.

Fault detection ratio: In the empirical evaluation of test adequacy criteria, the ratio between the number of test suites whose adequacy is in a specific interval and the number of test suites that contain a fault-revealing test case.

Interprocedural control flow graph: A graph-based representation of the static control flow for an entire program. In object-oriented programming languages, the interprocedural control flow graph is simply a collection of the intraprocedural control flow graphs for each of the methods within the program under test.

Multiple condition coverage: A test adequacy criterion that requires a test suite to account for every permutation of the Boolean variables in every branch of a program.

Mutation adequacy/relative adequacy: A test adequacy criterion, based on the competent programmer hypothesis and the coupling effect assumption, that requires a test suite to differentiate between the program under test and a set of programs that contains common programmer errors.

Mutation operator: A technique that modifies the program under test in order to produce a mutant that represents a faulty program that might be created by a competent programmer.

N-selective mutation testing: A mutation testing technique that attempts to compute a high-fidelity mutation adequacy score without executing the mutation operators that create the highest number of mutants and do not truly shed light on the defect-revealing potential of the test suite.

PIE model: A model proposed by Voas that states that a fault will only manifest itself in a failure when it is executed, it infects the program data state, and finally propagates to the output.

Regression testing: An important software maintenance activity that attempts to ensure that the addition of new functionality and/or the removal of program faults does not negatively impact the correctness of the program under test.

Regression test selection: A technique that attempts to reduce the cost of regression testing by selecting some appropriate subset of an existing test suite for execution.

Regression test suite distribution: A technique that attempts to make regression testing more cost-effective and practical by using all the available computational resources during test suite selection, prioritization, and execution.

Regression test suite prioritization: A technique that attempts to order a regression test suite so that the test cases that are most likely to reveal defects are executed earlier in the regression testing process.

Residual test adequacy evaluator: A test evaluation tool that can instrument the program under test to determine the adequacy of a provided test suite. A tool of this nature inserts probes into the program under test to measure adequacy and can remove these probes once certain test requirements have been covered.

Robustness testing/fault injection: A software testing technique that attempts to determine how a software system handles inappropriate inputs.

Software testing: The process of assessing the functionality and correctness of a program through execution or analysis.

Software verification: The process of ensuring that a program meets its intended specification.

Strong mutation adequacy: A test adequacy criterion that requires that the mutant program and the program under test produce different output. This adequacy criterion requires the execution, infection, and propagation of the mutated source locations within the mutant program.

Subsumption: A relationship between two test adequacy criterion. Informally, if test adequacy criterion C_α subsumes C_β, then C_α is considered "stronger" than C_β.

Test adequacy evaluation: The measurement of the quality of an existing test suite for a specific test adequacy criterion and a selected program under test.

Test case generation: The manual or automatic process of creating test cases for the program under test. Automatic test case generation can be viewed as an attempt to satisfy the constraints imposed by the selected test adequacy criteria.

Test case selection: The process of analyzing the program under test, in light of a chosen test adequacy criterion, in order to produce a list of tests that must be provided in order to create a completely adequate test suite.

Weak mutation adequacy: A test adequacy criterion that requires that the mutant program and the program under test produce different data states after the mutant is executed. This test adequacy criterion requires the execution and infection of the mutated source locations within the mutant program.

References

R.T. Alexander, J.M. Bieman, and J. Viega. Coping with Java programming stress. *IEEE Computer*, 33(4): 30–38, April 2000.

K. Arnold, B. O'Sullivan, R.W. Scheifler, J. Waldo, and A. Wollrath. *The Jini Specification*. Addison-Wesley, Reading, MA, 1999.

M. Balcer, W. Hasling, and T. Ostrand. Automatic generation of test scripts from formal test specifications. In *Proceedings of the ACM SIGSOFT Third Symposium on Software Testing, Analysis, and Verification*, pages 210–218. ACM Press, 1989.

T. Ball. The limit of control flow analysis for regression test selection. In *Proceedings of the International Symposium on Software Testing and Analysis*, pages 134–142. ACM Press, March 1998.

K. Beck. *Smalltalk Best Practice Patterns*. Prentice Hall, 1997.

B. Beizer. *Software Testing Techniques*. Van Nostrong Reinhold, New York, 1990.

R.V. Binder. *Testing Object-Oriented Systems: Models, Patterns, and Tools*. Addison-Wesley, Boston, MA, 1999.

R. Biyani and P. Santhanam. TOFU: Test optimizer for functional usage. *Software Engineering Technical Brief*, 2(1), 1997.

J.P. Bowen and M.G. Hinchley. Ten commandments of formal methods. *IEEE Computer*, 28(4):56–63, April 1995.

F.P. Brooks Jr. *The Mythical Man-Month*. Addison-Wesley, Reading, MA, 1995.

B. Brykczynski. A survey of software inspection checklists. *ACM SIGSOFT Software Engineering Notes*, 24 (1):82, 1999.

M. Chan and S. Cheung. Applying white box testing to database applications. Technical Report HKUST-CS9901, Hong Kong University of Science and Technology, Department of Computer Science, February 1999a.

M. Chan and S. Cheung. Testing database applications with SQL semantics. In *Proceedings of the 2nd International Symposium on Cooperative Database Systems for Advanced Applications*, pages 363–374, March 1999b.

D. Chays, S. Dan, P. G. Frankl, F. I. Vokolos, and E. J. Weyuker. A framework for testing database applications. In *Proceedings of the 7th International Symposium on Software Testing and Analysis*, pages 147–157, August 2000.

D. Chays and Y. Deng. Demonstration of AGENDA tool set for testing relational database applications. In *Proceedings of the International Conference on Software Engineering*, pages 802–803, May 2003.

D. Chays, Y. Deng, P.G. Frankl, S. Dan, F.I. Vokolos, and E.J. Weyuker. AGENDA: A test generator for relational database applications. Technical Report TR-CIS-2002-04, Department of Computer and Information Sciences, Polytechnic University, Brooklyn, NY, August 2002.

B. Choi, A. Mathur, and B. Pattison. PMothra: scheduling mutants for execution on a hypercube. In *Proceedings of the Third ACM SIGSOFT Symposium on Software Testing, Analysis, and Verficiation*, pages 58–65, December 1989.

L.A. Clarke. A system to generate test data symbolically. *IEEE Transactions on Software Engineering*, 2(3): 215–222, September 1976.

L.A. Clarke, A. Podgurski, D.J. Richardson, and S.J. Zeil. A comparison of data flow path selection criteria. In *Proceedings of the 8th International Conference on Software Engineering*, pages 244–251. IEEE Computer Society Press, 1985.

D.M. Cohen, S.R. Dalal, M.L. Fredman, and G.C. Patton. The combinatorial design approach to automatic test generation. *IEEE Software*, 13(5):83–87, September 1996.

D.M. Cohen, S.R. Dalal, M.L. Fredman, and G.C. Patton. The AETG system: an approach to testing based on combinatorial design. *IEEE Transactions on Software Engineering*, 23(7):437–443, July 1997.

B. Daou, R.A. Haraty, and N. Mansour. Regression testing of database applications. In *Proceedings of the 2001 ACM Symposium on Applied Computing*, pages 285–289. ACM Press, 2001.

R.A. DeMillo, D.S. Guindi, W.M. McCracken, A.J. Offutt, and K.N. King. An extended overview of the Mothra software testing environment. In *Proceedings of the ACM SIGSOFT Second Symposium on Software Testing, Analysis, and Verficiation*, pages 142–151, July 1988.

R.A. DeMillo, R. J. Lipton, and F.G. Sayward. Hints on test data selection: help for the practicing programmer. *IEEE Computer*, 11(4):34–41, April 1978.

R.A. DeMillo and A.J. Offutt. Constraint-based automatic test data generation. *IEEE Transactions on Software Engineering*, 17(9):900–910, September 1991.

E.W. Dijkstra. The structure of the THE multiprogramming system. *Communications of the ACM*, 11(5): 341–346, 1968.

E. Duesterwald, R. Gupta, and M.L. Soffa. A demand-driven analyzer for data flow testing at the integration level. In *Proceedings of the 18th International Conference on Software Engineering*, pages 575–584. IEEE Computer Society Press, 1996.

W.K. Edwards. *Core Jini*. Prentice Hall PTR, Upper Saddle River, NJ, 1999.

S. Elbaum, A.G. Malishevsky, and G. Rothermel. Prioritizing test cases for regression testing. In *Proceedings of the International Symposium on Software Testing and Analysis*, pages 102–112. ACM Press, August 2000.

S. Elbaum, G. Rothermel, S. Kanduri, and A.G. Malishevsky. Selecting a cost-effective test case prioritization technique. Technical Report 03-01-01, Department of Computer Science and Engineering, University of Nebraska–Lincoln, January 2003.

Y. Endo, Z. Wang, J.B. Chen, and M. Seltzer. Using latency to evaluate interactive system performance. In *Proceedings of the Second USENIX Symposium on Operating Systems Design and Implementation*, pages 185–199. ACM Press, 1996.

D. Epema, M. Livny, R.V. Dantzig, X. Evers, and J. Pruyne. A worldwide flock of Condors: load sharing among workstation clusters. *Journal on Future Generations of Computer Systems*, 12(1):53–65, December 1996.

M. Fagan. Design and code inspections to reduce errors in program development. *IBM Systems Journal*, 15(3):182–211, 1976.

N. Fenton. Software measurement: a necessary scientific basis. *IEEE Transactions on Software Engineering*, 20(3):199–206, March 1994.

R. Ferguson and B. Korel. The chaining approach for software test data generation. *ACM Transactions on Software Engineering and Methodology*, 5(1):63–86, 1996.

M. Fisher, D. Jin, G. Rothermel, and M. Burnett. Test reuse in the spreadsheet paradigm. In *Proceedings of the 13th International Symposium on Software Reliability Engineering*, Annapolis, MD, November 2002a.

M. Fisher, M. Cao, G. Rothermel, C.R. Cook, and M.M. Burnett. Automated test case generation for spreadsheets. In *Proceedings of the 24th International Conference on Software Engineering*, pages 141–153. ACM Press, 2002b.

P.G. Frankl and S. Weiss. An experimental comparison of the effectiveness of branch testing and data flow testing. *IEEE Transactions on Software Engineering*, 19(8):774–787, August 1993.

P.G. Frankl and E.J. Weyuker. An applicable family of data flow testing criteria. *IEEE Transactions on Software Engineering*, 14(10):1483–1498, October 1988.

P.G. Frankl and E.J. Weyuker. A formal analysis of the fault-detecting ability of testing methods. *IEEE Transactions on Software Engineering*, 19(3):202–213, March 1993.

E. Freemen, S. Hupfer, and K. Arnold. *JavaSpaces: Principles, Patterns, and Practice*. Addison-Wesley, Inc., Reading, MA, 1999.

E. Gamma and K. Beck. JUnit: a cook's tour. 2004. http://www.junit.org/.

M. Geller. Test data as an aid in proving program correctness. *Communications of the ACM*, 21(5):368–375, 1978.

A.K. Ghosh, M. Schmid, and F. Hill. Wrapping Windows NT software for robustness. In *Twenty-Ninth Annual International Symposium on Fault-Tolerant Computing*, June 1999.

A.K. Ghosh, M. Schmid, and V. Shah. Testing the robustness of Windows NT software. In *Proceedings of the Ninth International Symposium on Software Reliability Engineering*, pages 231–235. IEEE Computer Society, November 1998.

N. Gupta, A.A. Mathur, and M.L. Soffa. Automated test data generation using an iterative relaxation method. In *Proceedings of the 5th ACM SIGSOFT Symposium on the Foundations of Software Engineering*, November 1998.

N. Gupta, A.P. Mathur, and M.L. Soffa. UNA based iterative test data generation and its evaluation. In *Proceedings of the 14th International Conference on Automated Software Engineering*, pages 224–232, October 1999.

J. Haddox, G.M. Kapfhammer, and C.C. Michael. An approach for understanding and testing third-party software components. In *48th Reliability and Maintainability Symposium*, January 2002.

J. Haddox, G.M. Kapfhammer, C.C. Michael, and M. Schatz. Testing commercial-off-the-shelf components with software wrappers. In *Proceedings of the 18th International Conference on Testing Computer Software*, Washington, D.C., June 2001.

D. Hamlet. Foundations of software testing: dependability theory. In *Proceedings of the 2nd ACM SIGSOFT Symposium on Foundations of Software Engineering*, pages 128–139. ACM Press, 1994.

D. Hamlet and J. Maybee. *The Engineering of Software*. Addison-Wesley, Boston, MA, 2001.

M. Harder, J. Mellen, and M.D. Ernst. Improving test suites via operational abstraction. In *Proceedings of the 24th International Conference on Software Engineering*, pages 60–71. IEEE Computer Society Press, 2003.

M.J. Harrold and G. Rothermel. Performing data flow testing on classes. In *Proceedings of the 2nd ACM SIGSOFT Symposium on Foundations of Software Engineering*, pages 154–163. ACM Press, 1994.

M.J. Harrold and G. Rothermel. Aristotle: a system for research on and developement of program-analysis-based tools. Technical Report OSU-CISRC-3/97-TR17, The Ohio State University, Department of Computer and Information Science, March 1995.

M.J. Harrold and G. Rothermel. A coherent family of analyzable graphical representations for object-oriented software. Technical Report Technical Report OSU-CISRC-11/96-TR60, Department of Computer and Information Sciences, Ohio State University, November 1996.

M.J. Harrold and M.L. Soffa. Efficient computation of interprocedural definition-use chains. *ACM Transactions on Programming Languages and Systems*, 16(2):175–204, 1994.

R. Hightower. *Java Tools for Extreme Programming: Mastering Open Source Tools, Including Ant, JUnit, and Cactus.* John Wiley & Sons, New York, 2001.

A.E. Howe, A. von Mayrhauser, and R.T. Mraz. Test case generation as an AI planning problem. *Automated Software Engineering: An International Journal*, 4(1):77–106, January 1997.

C.J. Howell, G.M. Kapfhammer, and R.S. Roos. An examination of the run-time performance of GUI creation frameworks. In *Proceedings of the 2nd ACM SIGAPP International Conference on the Principles and Practice of Programming in Java*, Kilkenny City, Ireland, June 2003.

M. Hutchins, H. Foster, T. Goradia, and T. Ostrand. Experiments of the effectiveness of dataflow- and controlflow-based test adequacy criteria. In *Proceedings of the 16th International Conference on Software Engineering*, pages 191–200. IEEE Computer Society Press, 1994.

IEEE. *IEEE Standard Glossary of Software Engineering Terminology.* ANSI/IEEE Std 610.12-1990, 1996.

D. Jackson. Lecture 17: Case study: JUnit. 2003. http://ocw.mit.edu/6/6.170/f01/lecture-notes/index.html.

R.E. Jeffries. Extreme testing. *Software Testing and Quality Engineering*, March/April 1999.

J.-M. Jezequel, D. Deveaux, and Y. Le Traon. Reliable objects: lightweight testing for OO languages. *IEEE Software*, 18(4):76–83, July/August 2001.

C. Kaner, J. Falk, and H.Q. Hguyen. *Testing Computer Software.* International Thompson Computer Press, London, U.K., 1993.

G.M. Kapfhammer. Automatically and transparently distributing the execution of regression test suites. In *Proceedings of the 18th International Conference on Testing Computer Software*, Washinton, D.C., June 2001.

G.M. Kapfhammer and M.L. Soffa. A family of test adequacy criteria for database-driven applications. In *Proceedings of the 9th European Software Engineering Conference and the 11th ACM SIGSOFT Symposium on Foundations of Software Engineering.* ACM Press, 2003.

S. Kim, J.A. Clark, and J.A. McDermid. The rigorous generation of Java mutation operators using HAZOP. In *Proceedings of the 12th International Conference on Software and Systems Engineering and their Applications*, December 1999.

S. Kim, J.A. Clark, and J.A. McDermid. Class mutation: mutation testing for object-oriented programs. In *Proceedings of the Object-Oriented Software Systems, Net.ObjectDays Conference*, October 2000.

P. Koopman and J. DeVale. The exception handling effectiveness of POSIX operating systems. *IEEE Transactions on Software Engineering*, 26(9):837–848, September 2000.

B. Korel. Automated test data generation for programs with procedures. In *Proceedings of the International Symposium on Software Testing and Analysis*, pages 209–215. ACM Press, 1996.

E.W. Krauser, A.P. Mathur, and V.J. Rego. High performance software testing on SIMD machines. *IEEE Transactions on Software Engineering*, 17(5):403–423, May 1991.

N.P. Krop, P.J. Koopman, and D.P. Siewiorek. Automated robustness testing of off-the-shelf software components. In *Proceedings of the 28th Fault Tolerant Computing Symposium*, pages 230–239, June 1998.

O. Laitenberger and C. Atkinson. Generalizing perspective-based inspection to handle object-oriented development artifacts. In *Proceedings of the 21st International Conference on Software Engineering*, pages 494–503. IEEE Computer Society Press, 1999.

Y. Lei and K.C. Tai. In-parameter-order: a test generation strategy for pairwise testing. In *Proceedings of the High-Assurance Systems Engineering Symposium*, pages 254–261, November 1998.

H.K.N. Leung and L.J. White. Insights into regression testing. In *Proceedings of the International Conference on Software Maintenance*, pages 60–69. IEEE Computer Society Press, October 1989.

N.G. Leveson. *Safeware: System safety and computers.* Addison-Wesley, Reading, MA, September 1995.

Y.-S. Ma, Y.-R. Kwon, and J. Offutt. Inter-class mutation operators for Java. In *Proceedings of the Twelfth International Symposium on Software Reliability Engineering*, November 2002.

B. Marick. When should a test be automated? In *Proceedings of the 11th International Quality Week Conference*, San Francisco, CA, May 26–29, 1998.

B. Marick. How to misuse code coverage. In *Proceedings of the 16th Interational Conference on Testing Computer Software*, June 1999.

A.M. Memon. *A Comprehensive Framework for Testing Graphical User Interfaces*. Ph.D. thesis, University of Pittsburgh, Department of Computer Science, 2001.

A.M. Memon. GUI testing: pitfalls and process. *IEEE Computer*, 35(8):90–91, August 2002.

A.M. Memon, M.E. Pollack, and M.L. Soffa. Using a goal-driven approach to generate test cases for GUIs. In *Proceedings of the 21st International Conference on Software Engineering*, pages 257–266. IEEE Computer Society Press, 1999.

A.M. Memon, M.E. Pollock, and M.L. Soffa. Heirarchical GUI test case generation using automated planning. *IEEE Transactions on Software Engineering*, 27(2):144–155, 2001a.

A.M. Memon, M.L. Soffa, and M.E. Pollack. Coverage criteria for GUI testing. In *Proceedings of the 8th European Software Engineering Conference and the 9th ACM SIGSOFT International Symposium on Foundations of Software Engineering*, pages 256–267. ACM Press, 2001b.

C.C. Michael, G. McGraw, and M. Schatz. Generating software test data by evolution. *IEEE Transactions on Software Engineering*, 27(12):1085–1110, December 2001.

B. Miller, L. Fredriksen, and B. So. An empirical study of the reliability of operating system utilities. *Communications of the ACM*, 33:32–44, December 1990.

B. Miller, D. Koski, C. Lee, V. Maganty, A. Natarajan, and J. Steidl. Fuzz revisited: a re-examiniation of the reliability of UNIX utilities and services. Technical Report 1268, University of Wisconsin–Madison, May 1998.

L.J. Morell. A theory of fault-based testing. *IEEE Transactions on Software Engineering*, 16(8):844–857, 1990.

B.A. Myers and M.B. Rosson. Survey on user interface programming. In *Proceedings of the SIGCHI Conference on Human Factors in Computing Systems*, pages 195–202. ACM Press, 1992.

R. Neapolitan and K. Naimipour. *Foundations of Algorithms*. Jones and Bartlett Publishers, Boston, MA, 1998.

F. Nielson, H.R. Nielson, and C. Hankin. *Principles of Program Analysis*. Springer-Verlag, Berlin, Germany, 1999.

S.C. Ntafos. A comparison of some structural testing strategies. *IEEE Transactions on Software Engineering*, 14(6):868–874, June 1988.

A.J. Offutt, A. Lee, G. Rothermel, R.H. Untch, and C. Zapf. An experimental determination of sufficient mutant operators. *ACM Transactions on Software Engineering and Methodology*, 5(2):99–118, 1996.

A.J. Offutt, G. Rothermel, and C. Zapf. An experimental evaluation of selective mutation testing. In *Proceedings of the 15th International Conference on Software Engineering*, pages 100–107, May 1993.

A.J. Offutt, R. Alexander, Y. Wu, Q. Xiao, and C. Hutchinson. A fault model for subtype inheritance and polymorphism. In *Proceedings of the Twelfth International Symposium on Software Reliability Engineering*, pages 84–95, November 2001.

A.K. Onoma, W.-T. Tsai, M. Poonawala, and H. Suganuma. Regression testing in an industrial environment. *Communications of the ACM*, 41(5):81–86, 1998.

T.J. Ostrand and M.J. Balcer. The category-partition method for specifying and generating fuctional tests. *Communications of the ACM*, 31(6):676–686, 1988.

A. Parrish and S.H. Zweben. Software test data adequacy properties. *IEEE Transactions on Software Engineering*, 17(6):565–581, June 1991.

A.S. Paul. SAGE: A static metric for testability under the PIE model. Technical Report 96-5, Allegheny College, Department of Computer Science, 1996.

C. Pavlopoulou and M. Young. Residual test coverage monitoring. In *Proceedings of the 21st International Conference on Software Engineering*, pages 277–284. IEEE Computer Society Press, 1999.

B. Pettichord. Seven steps to test automation success. In *Proceedings of the International Conference on Software Testing, Analysis, and Review*, San Jose, CA, November 1999.

C.V. Ramamoorty, S.F. Ho, and W.T. Chen. On the automated generation of program test data. *IEEE Transactions on Software Engineering*, 2(4):293–300, December 1976.

S. Rapps and E.J. Weyuker. Data flow analysis techniques for test data selection. In *Proceedings of the 6th International Conference on Software Engineering*, pages 272–278. IEEE Computer Society Press, 1982.

S. Rapps and E.J. Weyuker. Selecting software test data using data flow information. *IEEE Transactions on Software Engineering*, 11(4), April 1985.

D. Richardson, O. O'Malley, and C. Tittle. Approaches to specification-based testing. In *Proceedings of the ACM SIGSOFT Third Symposium on Software Testing, Analysis, and Verification*, pages 86–96. ACM Press, 1989.

G. Rothermel and M.J. Harrold. A framework for evaluating regression test selection techniques. In *Proceedings of the Sixteenth International Conference on Software Engineering*, pages 201–210. IEEE Computer Society Press, May 1994.

G. Rothermel and M.J. Harrold. Analyzing regression test selection techniques. *IEEE Transactions on Software Engineering*, 22(8):529–551, August 1996.

G. Rothermel and M.J. Harrold. A safe, efficient regression test selection technique. *ACM Transactions on Software Engineering and Methodology*, 6(2):173–210, April 1997.

G. Rothermel and M.J. Harrold. Empirical studies of a safe regression test selection technique. *IEEE Transactions on Software Engineering*, 24(6):401–419, June 1998.

G. Rothermel, L. Li, and M. Burnett. Testing strategies for form-based visual programs. In *Proceedings of the 8th International Symposium on Software Reliability Engineering*, pages 96–107, Albuquerque, NM, November 1997.

G. Rothermel, R.H. Untch, C. Chu, and M.J. Harrold. Test case prioritization: an empirical study. In *Proceedings of the International Conference on Software Maintenance*, pages 179–188, August 1999.

G. Rothermel, R.H. Untch, C. Chu, and M.J. Harrold. Prioritizing test cases for regression testing. *IEEE Transactions on Software Engineering*, 27(10):929–948, October 2001a.

G. Rothermel, M. Burnett, L. Li, C. Dupuis, and A. Sheretov. A methodology for testing spreadsheets. *ACM Transactions on Software Engineering and Methodology*, 10(1):110–147, January 2001b.

K.J. Rothermel, C.R. Cook, M.M. Burnett, J. Schonfeld, T.R.G. Green, and G. Rothermel. WYSIWYT testing in the spreadsheet paradigm: an empirical evaluation. In *Proceedings of the 22nd International Conference on Software Engineering*, pages 230–239. ACM Press, 2000.

A. Sears. Layout appropriateness: a metric for evaluating user interface widget layout. *IEEE Transactions on Software Engineering*, 19(7):707–719, 1993.

F. Shull, I. Rus, and V. Basili. Improving software inspections by using reading techniques. In *Proceedings of the 23rd International Conference on Software Engineering*, pages 726–727. IEEE Computer Society, 2001.

I. Sommerville. *Software Engineering*. Addison-Wesley, 6th edition, August 2000.

J. Steven, P. Chandra, B. Fleck, and A. Podgurski. jRapture: a capture/replay tool for observation-based testing. In *Proceedings of the International Symposium on Software Testing and Analysis*, pages 158–167. ACM Press, 2000.

T. Tsai and R. Iyer. Measuring fault tolerance with the FTAPE fault injection tool. In *Proceedings of the 8th International Conference on Modeling Techniques and Tools for Computer Performance Evaluation*, pages 26–40, 1995.

T. K. Tsai and N. Singh. Reliability testing of applications on Windows NT. In *Proceedings of the International Conference on Dependable Systems and Networks*, New York City, June 2000.

R. Vallée-Rai, L. Hendren, V. Sundaresan, P. Lam, E. Gagnon, and P.Co. Soot — a Java optimization framework. In *Proceedings of CASCON 1999*, pages 125–135, 1999.

J.M. Voas. PIE: a dynamic failure-based technique. *IEEE Transactions on Software Engineering*, 18(8): 717–735, 1992.

F. Vokolos and P. Frankl. Pythia: a regression test selection tool based on textual differencing. In *Third International Conference of Reliability, Quality, and Safety of Software Intensive Systems*, May 1997.

E.J. Weyuker. Axiomatizing software test data adequacy. *IEEE Transactions on Software Engineering*, (12): 1128–1138, December 1986.

E.J. Weyuker, S.N. Weiss, and D. Hamlet. Comparison of program testing strategies. In *Proceedings of the Symposium on Testing, Analysis, and Verification*, pages 1–10. ACM Press, 1991.

J.A. Whittaker. What is software testing? and why is it so hard? *IEEE Software*, 17(1):70–76, January/ February 2000.

J.A. Whittaker and J.M. Voas. Toward a more reliable theory of software reliability. *IEEE Computer*, 32(12): 36–42, December 2000.

W.E. Wong, J.R. Horgan, S. London, and H. Agrawal. A study of effective regression testing in practice. In *Proceedings of the 8th International Symposium on Software Reliability Engineering*, pages 230–238, November 1997.

M. Young and R.N. Taylor. Rethinking the taxonomy of fault detection techniques. In *Proceedings of the 11th International Conference on Software Engineering*, pages 53–62. ACM Press, 1989.

H. Zhu, P.A.V. Hall, and J. H.R. May. Software unit test coverage and adequacy. *ACM Computing Surveys*, 29(4):366–427, 1997.

Further Information

Software testing and analysis is an active research area. The ACM/IEEE International Conference on Software Engineering, the ACM SIGSOFT Symposium on the Foundations of Software Engineering, the ACM SIGSOFT International Symposium on Software Testing and Analysis, and the ACM SIGAPP Symposium on Applied Computing's Software Engineering Track are all important forums for new research in the areas of software engineering and software testing and analysis. Other important conferences include IEEE Automated Software Engineering, IEEE International Conference on Software Maintenance, the IEEE International Symposium on Software Reliability Engineering, the IEEE/NASA Software Engineering Workshop, and the IEEE Computer Software and Applications Conference.

There are also several magazines and journals that provide archives for important software engineering and software testing research. The *IEEE Transactions on Software Engineering* and the *ACM Transactions on Software Engineering and Methodology* are two noteworthy journals that often publish software testing papers. Other journals include *Software Testing, Verification, and Reliability; Software: Practice and Experience; Software Quality Journal, Automated Software Engineering: An International Journal*, and *Empirical Software Engineering: An International Journal*. Magazines that publish software testing articles include *Communications of the ACM, IEEE Software, IEEE Computer*, and *Better Software* (formerly known as *Software Testing and Quality Engineering*). ACM SIGSOFT also sponsors the bi-monthly newsletter called *Software Engineering Notes*.

106

Formal Methods

106.1 Introduction 106-1
106.2 Underlying Principles 106-1
       Formal Methods
106.3 Best Practices 106-3
       Specification Languages • Modeling Systems • Conclusion
106.4 A Case Study 106-6
       Basic Types • Abbreviation Definitions • Generic
       Definitions • Abstract System State • Operations
       • Error Conditions • Status Operations • Conclusion
106.5 Technology Transfer and Research Issues 106-13
106.6 Glossary of Z Notation 106-14
       Names • Definitions • Logic • Sets and Expressions
       • Relations • Functions • Numbers • Sequences
       • Bags • Schema Notation • Conventions

Jonathan P. Bowen
London South Bank University

Michael G. Hinchey
NASA Goddard Space Flight Center

106.1 Introduction

Computers do not make mistakes, or so we are told. However, computer software is written by, and hardware systems are designed and assembled by, humans, who certainly *do* make mistakes.

Errors in a computer system can occur as a result of misunderstood or contradictory requirements, unfamiliarity with the problem, or simply human error during design or coding of the system. Alarmingly, the costs of maintaining software — the costs of rectifying errors and adapting the system to meet changing requirements or changes in the environment of the system — greatly exceed the original implementation costs.

As computer systems are being used increasingly in safety-critical applications — that is, systems where a failure could result in the loss of human life, mass destruction of property, or significant financial loss — both the media (e.g., Gibbs [1994]) and various regulatory bodies involved with standards, especially covering safety-critical and security applications (e.g., see Bowen and Stavridou [1993]), have considered formal methods and their role in the specification and design phases of system development.

106.2 Underlying Principles

There may be some confusion over what is meant by a "specification" and a "model." Parnas [1995] differentiates between specification and descriptions or models as follows:

- A **description** is a statement of some of the actual attributes of a product, or a set of products.
- A **specification** is a statement of properties required of a product, or a set of products.
- A **model** is a product, neither a description nor a specification. Often it is a product that has some, but not all, of the properties of some "real product."

1-58488-360-X/$0.00+$1.50
© 2004 by CRC Press, LLC

Others use the terms *specification* and *model* more loosely; a model can sometimes be used as a specification. The process of developing a specification into a final product is one in which a model can be used along the way or even as a starting point.

106.2.1   Formal Methods

Over the past 50 years, computer systems have increased rapidly in terms of both size and complexity. As a result, it is both naive and dangerous to expect a development team to undertake a project without stating clearly and precisely what is required of the system. This is done as part of the **requirements specification** phase of the software life cycle, the aim of which is to describe *what* the system is to do, rather than *how* it will do it.

The use of natural language for the specification of system requirements tends to result in ambiguity and requirements that may be mutually exclusive. Formal methods have evolved in an attempt to overcome such problems, by employing discrete mathematics to describe the function and architecture of a hardware or software system, and various forms of **logic** to reason about requirements, their interactions, and validity.

The term **formal methods** is itself misleading; it originates from formal logic but is now used in computing to refer to a plethora of mathematically-based activities. For our purposes, a formal method consists of notations and tools with a mathematical basis that are used to unambiguously specify the requirements of a computer system and that support the **proof** of properties of this specification and proofs of correctness of an eventual implementation with respect to the specification.

Indeed, it is true to say that so-called "formal methods" are not so much methods as formal systems. Although the popular formal methods provide a **formal notation**, or formal specification language, they do not adequately incorporate many of the methodological aspects of traditional development methods.

Even the term **formal specification** is open to misinterpretation by different groups of people. Two alternative definitions for "formal specification" are given in a glossary issued by the IEEE [1991]:

1. A specification written and approved in accordance with established standards.
2. A specification written in a formal notation, often for use in proof of correctness.

In this chapter we adopt the latter definition, which is the meaning assumed by most formal methods users.

The notation employed as part of a formal method is "formal," in that it has mathematical semantics so that it can be used to express specifications in a clear and unambiguous manner and allow us to abstract from actual implementations to consider only the salient issues. This is something that many programmers find difficult to do because they are used to thinking about implementation issues in a very concrete manner.

While programming languages are formal languages, they are generally not used in formal specification, as most languages do not have full formal semantics, and they force us to address implementation issues before we have a clear description of what we want the system to do. Instead, we use the language of mathematics, which is universally understood, well established in notation, and, most importantly, enables the generalization of a problem so that it can apply to an unlimited number of different cases [Dijkstra 1981]. Here we have the key to the success of formal specification — one must *abstract* away from the details of the implementation and consider only the essential relationships of the data, and we can model even the most complex systems using simple mathematical objects: for example, **sets**, **relations**, **functions**, etc.

At the specification phase, the emphasis is on clarity and precision, rather than efficiency. Eventually, however, one must consider how a system can be implemented in a programming language that, in general, will not support abstract mathematical objects (functional programming languages are an exception) and will be efficient enough to meet agreed requirements and concrete enough to run on the available hardware configuration.

As in structured design methods, the formal specification must be translated to a design — a clear plan for implementation of the system specification — and eventually into its equivalent in a programming language. This approach is known as **refinement** [Back and von Wright 1998, Sekerinski and Sere 1999].

The process of **data refinement** involves the transition from abstract data types such as sets, sequences, and mappings to more concrete data types such as arrays, pointers, and record structures, and the subsequent verification that the concrete representation can adequately capture all of the data in the formal specification. Then, in a process known as **operation refinement**, each **operation** must be translated so that it operates on the concrete data types. In addition, a number of **proof obligations** must be satisfied, demonstrating that each concrete operation is indeed a "refinement" of the abstract operation — that is, performing at least the same functions as the abstract equivalent, but more concretely, more efficiently, involving less nondeterminism, etc.

Many specification languages have relatively simple underlying mathematical concepts involved. For example, the Z (pronounced "zed") notation [Spivey 1992, ISO 2002] is based on (typed) set theory and first-order predicate logic, both of which *could* be taught at school level. The problem is that many software developers do not currently have the necessary education and training to understand these basic principles, although there have been attempts to integrate suitable courses into university curricula [Almstrum et al. 2001, Bjørner and Cuéllar 1999, Bowen 2001, Garlan 1995].

It is important for students who intend to become software developers to learn how to abstract away from implementation detail when producing a system specification. Many find this process of **abstraction** a difficult skill to master. It can be useful for reverse engineering as part of the software maintenance process, to produce a specification of an existing system that requires restructuring. Equally important is the skill of refining an abstract specification toward a concrete implementation, in the form of a program for development purposes [Morgan 1994].

The process of refinement is often carried out informally because of the potentially high cost of fully formal refinement. Given an implementation, it is theoretically possible, although often intractable, to *verify* that it is correct with respect to a specification, if both are mathematically defined. More usefully, it is possible to *validate* a formal specification by formulating required or expected properties and formally proving, or at least informally demonstrating, that these hold. This can reveal omissions or unexpected consequences of a specification. **Verification** and **validation** are complementary techniques, both of which can expose errors.

106.3 Best Practices

Technology transfer (e.g., see Craigen et al. [1995] and Caldwell [1998]) has always been an issue with formal methods, largely because of the significant training and expertise that is necessary for their use. Most engineering disciplines accept mathematics as the underpinning foundations, to allow calculation of design parameters before the implementation of a product [Hoare 1996b]. However, software engineers have been somewhat slow to accept such principles in practice, despite the very mathematical nature of all software. This is partly because it *is* possible to produce remarkably reliable systems without using formal methods [Hoare 1996a]. In any case, the use of mathematics for software engineering is a matter of continuing debate [Tremblay 2000].

However, there are now well-documented examples of cases in which a formal approach has been taken to develop significant systems in a beneficial manner, examples which are easily accessible by professionals (e.g., see Hinchey and Bowen [1995, 1999], Hall [1996], and Larsen et al. [1996]). Formal methods, including formal specification and modeling, should be considered as one of the possible techniques to improve software quality, where it can be shown to do this cost-effectively.

In fact, just using formal specification within the software development process has been shown to have benefits in reducing the overall development cost [Bowen and Stavridou 1993]. Costs tend to increase early in the life cycle, but are reduced later on at the programming, testing, and maintenance stages, where correction of errors is far more expensive. An early and widely publicized example was the IBM CICS (Customer Information Control System) project, where Z was used to specify a portion of this large transaction processing system with an estimated 9% reduction in development costs [Houston and King 1991]. There were approximately half the usual number of errors discovered in the software, leading to increased software quality.

Writing a good specification is something that comes only with practice, despite the existence of guidelines. However, there are some good reasons why a mathematical approach might be beneficial in producing a specification, including:

Precision. Natural language and diagrams can be very ambiguous. A mathematical notation allows the specifier to be very exact about what is specified. It also allows the reader of a specification to identify properties, problems, etc., that may not be obvious otherwise.

Conciseness. A formal specification, although precise, is also very concise compared with an equivalent high-level language program, which is often the first formalization of a system produced if formal methods are not used. Such a specification can be an order of magnitude smaller than the program that implements it, and hence is that much easier to comprehend.

Abstraction. It is all too easy to become bogged down in detail when producing a specification, making it very confusing and obscure to the reader. A formal notation allows the writer to concentrate on the essential features of a system, ignoring those that are implementation details. However, this is perhaps one of the most difficult skills in producing a specification.

Reasoning. Once a formal specification is available, mathematical reasoning is possible to aid in its validation. This is also useful for discussing implications of features, especially within a team of designers.

A design team that understands a particular formal specification notation can benefit from the above improvements in the specification process. It should be noted that much of the benefit of a formal specification derives from the process of producing the specification, as well as the existence of the formal specification after this [Hall 1990].

106.3.1 Specification Languages

The choice of specification language is likely influenced by many factors: previous experience, availability of tools, standards imposed by various regulatory bodies, and the particular aspects that must be addressed by the system in question. Another consideration is the degree to which a specification language is executable. This is the subject of some dispute, and the reader is directed elsewhere for a discussion of this topic [Hayes and Jones 1989, Fuchs 1992, Bowen and Hinchey 1999].

Indeed, the development of any complex system is likely to require the use of multiple notations at different stages in the process and to describe different aspects of a system at various levels of abstraction [Hoare 1987]. As a result, over the past 25 years, the vast majority of mainstream formal methods have been extended and re-interpreted to address issues of concurrency [Hoare 1985, Milner 1989], real-time behavior [Joseph 1996], and object orientation [Duke and Rose 2000, Smith 2000].

There is always, necessarily, a certain degree of trade-off between the expressiveness of a specification language and the levels of abstraction that it supports [Wing 1990]. While certain languages may have wider "vocabularies" and constructs to support the particular situations with which we wish to deal, they are likely to force us toward particular implementations; while they will shorten a specification, they will make it less abstract and more difficult for reasoning [Bowen and Hinchey, 1995a].

Formal specification languages can be divided into essentially three classes:

1. **Model-oriented approaches** as exemplified by ASM (Abstract State Machines) [Börger and Stärk 2003]; B-Method [Abrial 1996, Bicarregui et al. 1997, Schneider 2001]; RAISE (Rigorous Approach to Industrial Software Engineering) [RAISE Language Group 1992, Dang Van et al. 2002]; VDM (Vienna Development Method) [Jones 1991, ISO 1996]; and the Z notation [Spivey 1992, ISO 2002]. These approaches involve the derivation of an explicit model of the system's desired behavior in terms of abstract mathematical objects.

2. **Property-oriented approaches** using **axiomatic semantics** (such as Larch [Guttag 1993]), which use first-order predicate logic to express **preconditions** and **postconditions** of operations over abstract data types, and **algebraic semantics** (such as the OBJ family of languages including CafeOBJ

[Diaconescu and Futatsugi 1998, Futatsugi et al. 2000]), which are based on multisorted algebras and relate properties of the system in question to equations over the entities of the system.

3. **Process algebras** such as CSP (Communicating Sequential Processes) [Hoare 1985, Schneider 1999] and CCS (Calculus of Communicating Systems) [Milner 1989, Bruns 1996], which have evolved to meet the needs of concurrent, distributed, and real-time systems, and which describe the behavior of such systems by describing their algebras of communicating processes.

Unfortunately, it is not always possible to classify a formal specification language in just one of the categories above. LOTOS (Language Of Temporal Ordering Specifications) [ISO 1989, Turner 1993], for example, is a combination of ACT ONE and CCS. While it can be classified as an algebraic approach, it also exhibits many properties of a process algebra. Similarly, the RAISE development method is based on extending a model-based specification language (specifically, VDM-SL) with concurrent and temporal aspects.

As well as the basic mathematics, a specification language should also include facilities for structuring large specifications. Mathematics alone is all very well in the small, but if a specification is a thousand pages long (and formal specifications of this length exist), there must be aids to organize the inevitable complexity. Z provides the **schema notation** for this purpose, which packages the mathematics so that it can be subsequently reused in the specification. A number of schema operators, many matching logical connectives, allow recombination in a flexible manner.

A formal specification should also include an informal explanation to put the mathematical description into context and help the reader understand the mathematics. Ideally, the natural language description should be understandable on its own, although the formal text is the final arbiter as to the meaning of the specification. As a rough guide, the formal and informal descriptions should normally be of approximately the same length. The use of mathematical terms should be minimized, unless explanations are being included for didactic purposes.

Formal methods have proved useful in embedded systems and control systems (e.g., see Tiwari et al. [2003] and Tretmans et al. [2001]). Synchronous languages, such as Esterel, Lustre and Signal, have also been developed for **reactive systems** requiring continuous interaction with their environment [Benveniste et al. 2003]. Specialist and combined languages may be needed for some systems. More recently, **hybrid systems** [Davoren and Nerode 2000, Lee and Cha 2000] extend the concept of **real-time systems** [Fidge et al. 1997]. In the latter, time must be considered, possibly as a continuous variable. In hybrid systems, the number of continuous variables may be increased. This is useful in control systems where a digital computer is responding to real-world analog signals.

More visual formalisms, such as Statecharts [Harel 1987], are available and are appealing for industrial use, with associated STATEMENT tool support [Harel and Politi 1998] that is now part of the widely used Unified Modeling Language (UML). However, the reasoning aspects and the exact semantics are less well-defined. Some specification languages, such as SDL (Specification and Design Language) [Turner 1993], provide particularly good commercial tool support, which is very important for industrial use.

There have been many attempts to improve the formality of the various structural design notations in widespread use [Semmens et al. 1992, Bowen and Hinchey 1999]. UML includes the Object Constraint Language (OCL) developed by IBM [Warmer and Kleppe 1998], an expression language that allows constraints to be formalized, but this part of UML is underutilized with no tool support in most major commercial UML tools and is only a small part of UML in any case.

Object orientation is an important development in programming languages that has also be reflected in specification languages. For example, Object-Z is an object-oriented version of the Z notation that has gained some acceptance [Smith 2000, Duke and Rose 2000, Derrick and Boiten 2001]. More recently, the Perfect Developer tool has been developed by Escher Technologies Limited to refine formal specifications to object-oriented programming languages such as Java.

106.3.2 Modeling Systems

As previously discussed, the difference between specification and modeling is open to some debate. Different specification languages emphasize and allow modeling to different extents. Algebraic specification eschews the modeling approach, but other specification languages such as Z and VDM actively encourage it.

Some styles of modeling have been formulated for specific purposes. For example, Petri nets may be applied in the modeling of concurrent systems using a specific diagrammatic notation that is quite easily formalizable. The approach is appealing but the complexity can become overwhelming. Features such as deadlock are detectable but full analysis can be intractable in practice because the problem of scaling is not well addressed.

Mathematical modeling allows reasoning about (some parts of) a system of interest. Here, aspects of the system are defined mathematically, allowing the behavior of the system to be predicted. If the prediction is correct, this reinforces confidence in the model. This approach is familiar to many scientists and engineers.

Executable models allow rapid prototyping of systems [Fuchs 1992]. A very high-level programming language such as a functional program or a logic program (which have mathematical foundations) can be used to check the behavior of the system. Rapid prototyping can be useful in demonstrating a system to a customer before the expensive business of building the actual system is undertaken. Again, scientists and engineers are used in carrying out experiments using such models.

A branch of formal methods known as **model checking** allows systems to be tested exhaustively [Grumberg et al. 2000, Bérard et al. 2001]. Most computer-based systems are far too complicated to test completely because the number of ways the system could be used is far too large. However, a number of techniques, **Binary Decision Diagrams** (BDDs) for example, allow relatively efficient checking of significant systems, especially for hardware [Kropf 2000]. An extension of this technique, known as **symbolic model checking**, allows even more generality to be introduced.

Mechanical tools exist to handle BDDs and other model-checking approaches efficiently. SPIN is one of the leading general model-checking tools that is widely used [Holzmann 2003]. A more specialist tool based on CSP [Hoare 1985], known as FDR (Failure Divergence Refinement), from Formal Systems (Europe) Ltd., allows model checking to be applied to concurrent systems that can be specified in CSP.

106.3.3 Conclusion

Driving forces for best practice include standards, education, training, tools, available staff, certification, accreditation, legal issues, etc. [Bowen and Stavridou 1993]. A full discussion of these is outside the scope of this chapter. Aspects of best practice for specification and modeling depend significantly on the selected specification notation. One of the more popular formal specification notations used in industry is Z. To illustrate the way in which this notation is typically used, a case study using Z is presented in the next section. This demonstrates both some of the underlying principles and best practice when employing Z for specification and modeling.

106.4 A Case Study

The Z notation [Spivey 1992, ISO 2002] is one of the most widely used formal specification languages and is normally used in a modeling style. An abstract state is first formulated, and then operations on that state are specified. In this section we present a case study using Z to illustrate this style of specification. The example does not exhaustively present the features of Z, but gives a flavor of the style of presentation of a typical Z specification, with extra informal explanation on the notation and conventions where required. Z constructs are introduced as the example is presented and a glossary of Z notation is provided at the end of the chapter for the convenience of the reader. A basic understanding of set theory and logic will help in understanding the specification.

Window management systems are now used extensively for user interfaces to computer systems. The specification given here is of a (fictitious) window system. For more realistic examples of some implemented systems presented in Z, see Bowen [1996], especially section VI, which includes some actual windows systems, such as the X window system.

106.4.1 Basic Types

Z is a typed language that allows a certain amount of consistency checking by a mechanical type-checker. However, the only predefined type is the set of integers, denoted $\mathbb{Z}$. Further types must be defined for a particular specification. These **basic types** (also known as **given sets**) may be introduced as follows:

$$[Position, Value]$$

This provides a set of pixel (picture element) positions (e.g., coordinates on a screen), together with possible pixel values (e.g., colors). Note that we are no more specific than this in the specification presented here. It is important not to introduce irrelevant implementation details into a specification because this restricts the eventual implementer of the system and clutters the specification with information that is not required at a high level of abstraction.

106.4.2 Abbreviation Definitions

It is often useful to include definitions in a specification for commonly used concepts. This helps reduce the size of the specification and introduces important concepts to the reader in one place, allowing them to be used later within the specification. Pixel maps, relating pixels to their associated values, are an integral part of most window systems. In fact, each pixel has, at most, one value (assuming it is defined), so we can model a pixel map as a **partial function** from pixel positions to their values:

$$Pixmap == Position \nrightarrow Value$$

106.4.3 Generic Definitions

Z has its own library of "tool-kit" operators, formally defined in terms of more basic mathematical concepts, as presented in Spivey [1992]. Sometimes it is helpful to extend this library with further **generic definitions** that can be used to define a family of generic constants, applicable to a variety of basic types. Such definitions may be useful for other specifications as well as the one being constructed, allowing reuse of specification components.

For example, a **sequence** of pixel maps can be overlaid in the order given by the sequence to produce a new pixel map. An operator to do this could equally well apply to other partial functions as well as pixel maps, so we can define it generically, using a "distributed overriding" operator:

$$
\begin{array}{l}
\rule{6cm}{0.4pt}[P, V]\rule{6cm}{0.4pt} \\
\quad \oplus/ : \mathrm{seq}\,(P \nrightarrow V) \to (P \nrightarrow V) \\
\rule{10cm}{0.4pt} \\
\quad \oplus/\langle\rangle = \varnothing \\
\quad \forall p : P \nrightarrow V \bullet \oplus/\langle p \rangle = p \\
\quad \forall s, t : \mathrm{seq}\,(P \nrightarrow V) \bullet \oplus/(s \frown t) = (\oplus/s) \oplus (\oplus/t)
\end{array}
$$

Here, the base cases for the empty sequence $\langle\rangle$ and a singleton sequence $\langle p \rangle$ are considered, followed by the more general case of two arbitrary sequences concatenated together $s \frown t$. Distributed overriding is particularly useful for the specification presented here in specifying the view on a screen of a display, given a sequence of possibly overlapping pixel maps.

Z tool-kit operators normally have a number of laws associated with them that are helpful in reasoning about specifications. For example, the following law applies for the distributed overriding operator:

$$p_1, p_2 : Pixmap \vdash \oplus/\langle p_1, p_2 \rangle = p_1 \oplus p_2$$

Such laws must be proved from the original definition; for example, in this case:

$$\oplus/\langle p_1, p_2 \rangle$$
$$= \oplus/(\langle p_1 \rangle \frown \langle p_2 \rangle) \qquad\qquad\qquad\qquad\qquad \text{[property of } \frown\text{]}$$
$$= (\oplus/\langle p_1 \rangle) \oplus (\oplus/\langle p_2 \rangle) \qquad\qquad\qquad \text{[by the general case definition]}$$
$$= p_1 \oplus p_2 \qquad\qquad \text{[by the second base case definition, substituting twice]}$$

If the windows in a sequence overlap, it is useful to be able to move selected windows so that their contents can be viewed (or hidden). This is analogous to stuffing a pile of sheets of paper (windows) on a desk (screen). Note that the sheets of paper may be of different sizes and in different positions on the desk.

For example, the following function can be used to move a selected window number in the sequence (if it exists) to the top of the pile (i.e., the end of the sequence). This can also be defined generically:

$$\begin{array}{|l|}\hline
\underline{\quad[W]\quad} \\[4pt]
top : \mathbb{N} \rightarrow \text{seq } W \rightarrow \text{seq } W \\[6pt]
\hline
\forall n : \mathbb{N}, s : \text{seq } W \bullet \\[4pt]
\quad top\ ns = \textbf{if } n \in \text{dom } s \textbf{ then } squash\ (\{n\} \lhd s) \frown \langle s(n) \rangle \textbf{ else } s \\[4pt]
\hline
\end{array}$$

If the window number n is in the sequence of windows s, then it is removed from the sequence (by eliminating that element and squashing the resulting function back into a sequence). This element is then concatenated to the end of the sequence. If the window number is not valid, the sequence of windows is unaffected. The exact technical details require some knowledge of Z, but the above example illustrates the fact that important concepts can be captured formally using relatively short definitions.

In this simple example, we ignore the complication of window identifiers. We simply use the position of the window in the sequence to identify it, assuming that the user of the system keeps track of which window is which.

106.4.4 Abstract System State

The window display can be modeled as a sequence of windows against a background "window" that is the size of the display screen itself. The order of the sequence defines which windows are on top in the case of overlapping windows, in ascending order. Only parts of windows that are contained within the background area are displayed.

$$\begin{array}{|l|}\hline
\underline{\quad SYS\quad} \\[4pt]
windows : \text{seq } Pixmap \\[4pt]
screen,\ background : Pixmap \\[6pt]
\hline
screen = background \oplus (\text{dom } background \lhd \oplus /windows) \\[4pt]
\hline
\end{array}$$

In the specification of the abstract state above, the components *windows* (a sequence of pixel maps), *screen* (as displayed to the user), and *background* (the display if no windows are present) are packaged together in a **schema** box called *SYS*. The **declarations** with their associated type information are above the line and **predicates** defining constraints between these components are (optionally) included below the line. Here, what appears on the display screen is defined in terms of the background pixel map overridden by the sequence of windows in the system, constrained to the background area as defined by its domain of pixel positions.

The screen area is the same as the background area. This can be formalized as follows:

$$SYS \vdash \text{dom } screen = \text{dom } background$$

It is useful to prove such properties correct, either informally, even just mentally, or formally, in order to validate that the specification behaves as expected. Discovering that an expected property does not hold might expose an error in a specification, perhaps in the form of an extra constraint that is required but has been omitted.

Note that the user can only see the display screen. We can specify this view formally by hiding (existentially quantifying) other components in the *SYS* schema to produce a new *View* schema, defined horizontally:

$$View \cong SYS \setminus (windows, background)$$

Initially there are no windows in the system:

$$InitSYS \cong [SYS' \mid windows' = \langle\rangle]$$

It is important to define the **initial state** and also to ensure that it exists (otherwise, the system cannot start to operate). The state at initialization normally consists of the abstract state for the system with some extra constraints. By including *SYS'* in the definition above, all the components in *SYS* are defined, with the extra decoration ′ added to each component name. The prime (or dash) ′ is used by convention in Z to indicate the state *after* an operation. Here we are interested in the state after initialization. Again, we can formulate a property to check that our intuition about the specification is correct:

$$InitSYS \vdash screen' = background'$$

That is, the screen display at initialization should consist of just the background.

Next we can define general properties about the **change of state** for operations on the system. By convention in Z, Δ ("delta") is used to indicate a change of state where both an unprimed **before state** *SYS* (for example) and a primed **after state** (e.g., *SYS'*) are defined:

$$\Delta SYS \cong [SYS; SYS' \mid background' = background]$$

Here we have added the extra constraint that the background never changes for any operation. This means that we do not have to consider and define this for each individual operation that uses ΔSYS subsequently, because the predicate *background'* = *background* is automatically included (conjoined) with any other predicates that are defined.

Note that if the before and after states are not related in an operation scheme, then the after state can take on any value. This is the opposite of most programming languages, where unreferenced variables retain their values by default. However, this style is useful in specifications because it allows **nondeterminism** to be included easily, where more than one outcome of an operation is allowed. Eventually, of course, the implementer will have to choose a particular outcome; but if the choice is not important at the specification level, then leaving the options open gives the implementer a greater choice, possibly allowing different optimization strategies in different implementations of the same specification.

Some operations may leave the state of the system unchanged, for example during a status operation or if an error in the input is detected. Here, the Ξ ("xi") convention is used in Z:

$$\Xi SYS \cong [\Delta SYS \mid \theta SYS' = \theta SYS]$$

The predicate θSYS' = θSYS is a shorthand way of ensuring that the tuple formed from all the *SYS* components is the same as that formed from all the primed *SYS'* components.

106.4.5   Operations

To use the system, we must have the ability to create windows. These are created on top of all the existing windows. We specify that they must fit within the display background:

AddWindows0

ΔSYS

$window? : Pixmap$

$dom\ window? \subseteq dom\ background$

$windows' = windows \frown \langle window? \rangle$

In the definition above, the ΔSYS component automatically includes all the SYS and SYS' (unprimed before and primed after state) components, together with the constraint that the background remains unchanged. The *window?* component is an input (as indicated by the Z convention of the added "?"). A **precondition** for the operation is that the window area must be contained within (i.e., be a subset of) the background area. If this is so, then the sequence of windows after the operation has the required window concatenated to it. This means that the window appears on top of any other windows already in the system. Note that, by default, predicates on separate lines in a schema are conjoined using "∧."

The ability to update windows is very useful. This may involve changing the size of the window or its contents, or moving it about the screen. Again, the updated window must still fit within the display area.

Update Window0

ΔSYS

$which? : \mathbb{N}$

$window? : Pixmap$

$which? \in dom\ windows$

$dom\ window? \subseteq dom\ background$

$windows' = windows \oplus \{which? \mapsto window?\}$

Here, two inputs are provided — both which window number is to be updated and the new value for the window. The window to be updated must already be in the system and, as for the *AddWindow* 0 schema, the new window must be within the background for the update to be successful. The sequence of windows is overridden with a new entry for the selected window.

It is desirable to be able to uncover a window that may be partially or even totally obscured by other windows. This can be done by moving the window to the end of the sequence of displayed windows:

ExposeWindow0

ΔSYS

$which? : \mathbb{N}$

$which? \in dom\ windows$

$windows' = top\ which?\ windows$

Sometimes it is useful to simply rotate the order of the displayed windows, one at a time, moving the bottommost window to the top.

RotateWindows0

ΔSYS

$windows \neq \langle\rangle$

$windows' = top\ 1\ windows$

Note that sequences are numbered from one updates in Z.

We also wish to be able to delete windows. For instance, we could delete the topmost window (the last window in the sequence):

```
┌─RemoveTop0──────────────────────────────────────
│ ΔSYS
├──────────────────────────────────────────────
│ windows ≠ ⟨⟩
│ windows′ = front windows
└──────────────────────────────────────────────
```

Alternatively, we may wish to specify which window is to be removed:

```
┌─RemoveWindow0───────────────────────────────────
│ ΔSYS
│ which? : ℕ
├──────────────────────────────────────────────
│ which? ∈ dom windows
│ windows′ = squash({which?} ⩤ windows)
└──────────────────────────────────────────────
```

The above schema definitions give a flavor of the way operations are typically presented in a Z specification. They are intended to illustrate that a number of different operations on a system can be specified succinctly using Z, providing a suitable abstract state has been formulated.

106.4.6 Error Conditions

The operations covered so far detail what should happen in the event of no errors. Normally, operations can also handle error conditions in some controlled manner. It is useful to report the status of an operation. For example, the following reports could be issued:

$$Report ::= \text{“OK”}$$
$$| \text{“Not a window”}$$
$$| \text{“No windows”}$$
$$| \text{“Invalid window”}$$

Here, a **free type** definition defines *Report* to be a set with four possible unique values.

It is helpful to report the fact that the operation was successful if this is the case:

```
┌─Success─────────────────────────────────────────
│ rep! : Report
├──────────────────────────────────────────────
│ rep! = “OK”
└──────────────────────────────────────────────
```

By convention in Z, "!" indicates an output from an operation.

If errors do occur, then these need to be reported. For example, an invalid window can be specified:

```
┌─NotAWindow──────────────────────────────────────
│ ΞSYS
│ which? : ℕ
│ rep! : Report
├──────────────────────────────────────────────
│ which? ∉ dom windows
│ rep! = “Not a window”
└──────────────────────────────────────────────
```

In this case, no change of state occurs, as specified by ΞSYS above. As a precondition, a check is made on whether the window number supplied as an input is not a valid existing window in the system; and if this is so, an appropriate report is issued as an output.

It is possible that there are no windows displayed when one is required:

```
┌─NoWindow──────────────────────────────────────
│ ΞSYS
│ rep! : Report
├───────────────────────────────────────────────
│ windows = ⟨⟩
│ rep! = "No windows"
└───────────────────────────────────────────────
```

A specified window may not be within the background area:

```
┌─BadWindow─────────────────────────────────────
│ ΞSYS
│ window? : Pixmap
│ rep! : Report
├───────────────────────────────────────────────
│ ¬ (dom window? ⊆ dom background)
│ rep! = "Invalid window"
└───────────────────────────────────────────────
```

We can include these errors with the previously defined operations that ignored error conditions to produce **total** operations:

$$AddWindow\,1 \cong (AddWindow\,0 \land Success) \lor BadWindow$$
$$Update\,Window\,1 \cong (Update\,Window\,0 \land Success) \lor BadWindow \lor NotAWindow$$
$$Expose\,Window\,1 \cong (Expose\,Window\,0 \land Success) \lor NotAWindow$$
$$Rotate\,Windows\,1 \cong (Rotate\,Windows\,0 \land Success) \lor NoWindows$$
$$RemoveTop\,1 \cong (RemoveTop\,0 \land Success) \lor NoWindows$$
$$Remove\,Window\,1 \cong (Remove\,Window\,0 \land Success) \lor NotAWindow$$

Here, the schema operators of conjunction ($\land$) and disjunction ($\lor$) are used to combine schemas. For both operators, components are merged. If components have the same name, then they must be type-compatible or the specification becomes meaningless. Using schema conjunction, predicates in each schema are logically conjoined. Similarly, if schema disjunction is used, then the predicates in the two schemas are combined using logically disjunction.

The operations are total in that their preconditions are true. This can be checked by calculation, which is a useful way of ensuring that all error conditions have been handled. This is something that is very easily overlooked if only informal specification using natural language and/or diagrams is used.

106.4.7 Status Operations

The contents of an existing window may be of interest:

```
┌─GetWindow0────────────────────────────────────
│ ΞSYS
│ which? : ℕ
│ window! : Pixmap
├───────────────────────────────────────────────
│ which? ∈ dom windows
│ window! = windows which?
└───────────────────────────────────────────────
```

Using ΞSYS, the state of the system does not change during this operation. Status operations normally have one or more outputs returning some aspect of the state of the system. Here, a particular window is returned.

We can make this operation total as well:

$$GetWindow \mathrel{\widehat{=}} (GetWindow\,0 \land Success) \lor NotAWindow$$

106.4.8 Conclusion

Given the abstract state, initial state, and operation schemas defined in this section, the operation of the system consists of starting in the initial state, followed by an arbitrary sequence of the specified operations on the state, as allowed by the preconditions of the operations. If the preconditions of all the operations are true, then any order of operations is allowed.

This section has presented the use of the Z notation [Spivey 1992, ISO 2002] in a modeling style, as it is widely used for specifying systems. It should be remembered that Z is a general-purpose specification language and can be used in other styles if desired. However, the use of an abstract state and operations on that state has been found to be a style that is easy to understand (once the notation and conventions have been learned), and this is the approach that is often adopted in practice.

For those wishing to learn Z, there are many textbooks available (e.g., see Jacky [1997], Lightfoot [2001], and Woodcock and Davies [1996]). An international standard for Z is available [ISO 2002] and an earlier *de facto* standard for Z, with a matching type-checker called *f*UZZ by the same author, is also widely used [Spivey 1992].

106.5 Technology Transfer and Research Issues

Claims that formal methods can guarantee correct hardware and software, eliminate the need for testing, etc., have led some to believe that formal methods are something almost magical [Hall 1990]. More significantly, beliefs that formal methods are difficult to use, delay the development process, and raise development costs [Bowen and Hinchey 1995b] have led many to believe that formal methods offer few advantages over traditional development methods. Formal methods are not a panacea; they are just one of a range of techniques that, when correctly applied, have proven themselves to result in systems of the highest integrity [Bowen and Hinchey 1995a].

Hitherto, the uptake of formal methods has been hindered, at least in part by a lack of tools. Many of the successful projects discussed in Hinchey and Bowen [1995, 1999] required significant investment in tool support. Just as the advent of compiler technology was necessary for the uptake of high-level programming languages, and CASE (Computer Aided Software Engineering) technology provided the impetus for the emergence of structural design methodologies in the 1970s, a significant investment in formal methods tools is required for formal methods to be practical at the level of industrial application.

In the future, we envisage greater emphasis on IFDSEs (Integrated Formal Development Support Environments) that will support formal specification and development, based on an emerging range of high-quality, stand-alone tools. The framework for this could well be based around the Extended Markup Language (XML) that has emerged from the World Wide Web community. For example, ZML provides an XML-based markup for the Z notation that could be useful in the communication of specifications between tools [Utting et al. 2003].

Method integration is one approach that might aid in the acceptance of formal methods and might help in the technology transfer from academic theory to industrial practice. This has the advantage of providing multiple views of a system, for example, incorporating a graphical representation that is likely to be more acceptable to non-specialists, while retaining the ability to propose and prove system properties and to demonstrate that requirements are not contradictory before, rather than after, implementation.

The Unified Modeling Language (UML) provides a popular software development framework that could benefit from the formal methods approach. For example, amid concerns over the lack of formality (or even uniform interpretation) in UML, the OMG (Object Management Group) has issued a request for proposals on re-architecting UML version 2.0. Several groups are interested in making the notation more formal, understandable, and unambiguous, including the 2U Consortium (Unambiguous UML) and the pUML group (precise UML).

Cleanroom is a method that provides a middle road between correctness proofs and informal development by stipulating significant checking of programs before they are first run [Prowell et al. 1999]. The testing phase then becomes more like a certification phase because the number of errors should be much reduced. **Static analysis** involved rigorous checking of programs without actually executing them. SPARK Ada [Barnes 2003] is a restricted version of the Ada programming language that includes additional comments that facilitate formal tool-assisted analysis, especially worthwhile in high-integrity system development. Such approaches may be more cost-effective than full formal development using refinement techniques.

In any case, full formal development is typically not appropriate in most software systems. However, many systems could benefit from some use of formal methods at some level (perhaps just specification) in their most critical parts. This approach has been dubbed "lightweight" formal methods [Saiedian 1996, Feather 1998]. In particular, many errors are introduced at the requirements stage and some formality at this level could have very beneficial results because the system description is still relatively simple [Easterbrook et al. 1998].

Formal methods are complementary to testing in that they aim to avoid the introduction of errors whereas testing aims to remove errors that have been introduced during development. The best balance of effort between these two approaches is a matter for debate [King et al. 2000]. In any case, the existence of a formal specification can benefit the testing process by providing an objective and exact description of the system against which to perform subsequent program testing. It can also guide the engineer in deciding which tests are worthwhile (for example, by considering disjunctive preconditions in operations and ensuring that there is full test coverage of these).

In practical industrial use, formal methods have proved to have a niche in high-integrity systems such as safety-critical applications where standards may encourage or mandate their use in software at the highest levels of criticality. Formal methods are also being successfully used in security applications such as smart cards, where the technology is simple enough to allow fully formal development. They are also useful in discovering errors during cryptographic protocol analysis [Meadows 2003].

Formal methods have largely been used for software development but are arguably even more successful in hardware design where engineers may be more open to the use of rigorous approaches because of their background and training. Formal methods can be used for the design of microprocessors where errors can be costly because of the large numbers involved and also because of their possible use in critical applications [Jones et al. 2001]. Fully formal verification of significant hardware systems is possible even within the limits of existing proof technology (e.g., see Hunt and Sawada [1999]).

Full formal refinement is the ideal but is expensive and can sometimes be impossible to achieve in practice. Retrenchment [Banach and Poppleton 1999] is a suggested liberalization of refinement designed for formal description of applications too demanding for true refinement. Examples are the use of infinite or continuous types or models from classical physics and applications, including inconsistent requirements. In retrenchment, the abstraction relation between the models is weakened in the operation postcondition by a *concession* predicate. This weakened relationship allows approximating, inconsistent, or exceptional behaviors to be described, in which a *false* concession denotes a refinement.

There are many different formal methods for different purposes, including specification (e.g., the Z notation) and refinement (e.g., the B-Method). There are some moves to produce more general formal approaches (e.g., see B#, combining ideas from B with some concepts from Z [Abrial 2003]). There have also been moves toward related different semantic theories like algebraic, denotational, and operational approaches [Hoare and He 1998]. In any case, it is clear that there are still many research and technology transfer challenges ahead in the field of formal methods (e.g., see Hoare [2003]).

106.6 Glossary of Z Notation

A glossary of the Z mathematical and schema notation is included here for the reader's convenience. For more information on the Z notation, see the *Z Reference Manual* [Spivey 1992] and, more recently, the ISO international standard [ISO 2002].

Names

a, b	Identifiers
d, e	Declarations (e.g., $a : A$; $b, \ldots : B \ldots$)
f, g	Functions
m, n	Numbers
p, q	Predicates
s, t	Sequences
x, y	Expressions
A, B	Sets
C, D	Bags
Q, R	Relations
S, T	Schemas
X	Schema text (e.g., $d, d \mid p$ or S)

Definitions

$a == x$	Abbreviated definition
$a ::= b \mid \ldots$	Free type definition (or $a ::= b\langle\langle x\rangle\rangle \mid \ldots$)
$[a]$	Introduction of a given set (or $[a, \ldots]$)
a_-	Prefix operator
$_-a$	Postfix operator
$_-a_-$	Infix operator

Logic

true	Logical true constant
false	Logical false constant
$\neg p$	Logical negation
$p \wedge q$	Logical conjunction
$p \vee q$	Logical disjunction
$p \Rightarrow q$	Logical implication ($\neg p \vee q$)
$p \Leftrightarrow q$	Logical equivalence ($p \Rightarrow q \wedge q \Rightarrow p$)
$\forall X \bullet q$	Universal quantification
$\exists X \bullet q$	Existential quantification
$\exists_1 X \bullet q$	Unique existential quantification
let $a == x; \ldots \bullet p$	Local definition

Sets and Expressions

$x = y$	Equality of expressions
$x \neq y$	Inequality ($\neg(x = y)$)
$x \in A$	Set membership
$x \notin A$	Nonmembership ($\neg(x \in A)$)
$\emptyset$	Empty set
$A \subseteq B$	Set inclusion
$A \subset B$	Strict set inclusion ($A \subseteq B \wedge A \neq B$)
$\{x, y, \ldots\}$	Set of elements
$\{X \bullet x\}$	Set comprehension
$\lambda X \bullet x$	Lambda-expression — function
$\mu X \bullet x$	Mu-expression — unique value
let $a == x; \ldots \bullet y$	Local definition

if p then x else y	Conditional expression
$(x, y, \ldots)$	Ordered tuple
$A \times B \times \ldots$	Cartesian product
$\mathbb{P} A$	Power set (set of subsets)
$\mathbb{P}_1 A$	Nonempty power set
$\mathbb{F} A$	Set of finite subsets
$\mathbb{F}_1 A$	Nonempty set of finite subsets
$A \cap B$	Set intersection
$A \cup B$	Set union
$A \backslash B$	Set difference
$\bigcup A$	Generalized union of a set of sets
$\bigcap A$	Generalized intersection of a set of sets
first x	First element of an ordered pair
second x	Second element of an ordered pair
$\#A$	Size of a finite set

Relations

$A \leftrightarrow B$	Relation ($\mathbb{P}(A \times B)$)
$a \mapsto b$	Maplet ((a, b))
$\operatorname{dom} R$	Domain of a relation
$\operatorname{ran} R$	Range of a relation
$\operatorname{id} A$	Identity relation
$Q \,;\, R$	Forward relational composition
$Q \circ R$	Backward relational compositon ($R \,;\, Q$)
$A \lhd R$	Domain restriction
$A \ntriangleleft R$	Domain anti-restriction
$A \rhd R$	Range restriction
$A \ntriangleright R$	Range anti-restriction
$R(\!(A)\!)$	Relational image
iter n R	Relation composed n times
R^n	Same as *iter* n R
$R^\sim$	Inverse of relation (R^{-1})
R^*	Reflexive-transitive closure
R^+	Irreflexive-transitive closure
$Q \oplus R$	Relational overriding (($\operatorname{dom} R \ntriangleleft Q) \cup R$)
$a \underline{R} b$	Infix relation

Functions

$A \nrightarrow B$	Partial functions
$A \rightarrow B$	Total functions
$A \rightarrowtail\!\!\!\!\shortmid B$	Partial injections
$A \rightarrowtail B$	Total injections
$A \twoheadrightarrow\!\!\!\!\shortmid B$	Partial surjections
$A \twoheadrightarrow B$	Total surjections
$A \rightarrowtail\!\!\!\!\!\twoheadrightarrow B$	Bijective functions
$A \nrightarrow\!\!\!\!\shortmid B$	Finite partial functions
$A \rightarrowtail\!\!\!\!\!\!\twoheadrightarrow B$	Finite partial injections
$f x$	Function application (or $f(x)$)

Numbers

$\mathbb{Z}$	Set of integers
$\mathbb{N}$	Set of natural numbers $\{0, 1, 2, \ldots\}$
$\mathbb{N}_1$	Set of nonzero natural numbers ($\mathbb{N} \backslash \{0\}$)
$m + n$	Addition
$m - n$	Subtraction
$m * n$	Multiplication
m div n	Division
m mod n	Modulo arithmetic
$m \leq n$	Less than or equal
$m < n$	Less than
$m \geq n$	Greater than or equal
$m > n$	Greater than
succ n	Successor function $\{0 \mapsto 1,\ 1 \mapsto 2, \ldots\}$
$m .. n$	Number range
min A	Minimum of a set of numbers
max A	Maximum of a set of numbers

Sequences

seq A	Set of finite sequences
$\text{seq}_1\ A$	Set of nonempty finite sequences
iseq A	Set of finite injective sequences
$\langle \rangle$	Empty sequence
$\langle x, y, \ldots \rangle$	Sequence $\{1 \mapsto x,\ 2 \mapsto y, \ldots\}$
$s \frown t$	Sequence concatenation
$\frown / s$	Distributed sequence concatenation
head s	First element of sequence ($s(1)$)
tail s	All but the head element of a sequence
last s	Last element of sequence ($s(\#s)$)
front s	All but the last element of a sequence
rev s	Reverse a sequence
squash f	Compact a function to a sequence
$A \upharpoonright s$	Sequence extraction ($squash\ (A \lhd s)$)
$s \upharpoonright A$	Sequence filtering ($squash\ (s \rhd A)$)
s prefix t	Sequence prefix relation ($s \frown v = t$)
s suffix t	Sequence suffix relation ($u \frown s = t$)
s in t	Sequence segment relation ($u \frown s \frown v = t$)
disjoint A	Disjointness of an indexed family of sets
A partition B	Partition an indexed family of sets

Bags

bag A	Set of bags or multisets ($A \nrightarrow \mathbb{N}_1$)
$[[\,]]$	Empty bag
$[[\, x, y, \ldots]]$	Bag $\{x \mapsto 1,\ y \mapsto 1, \ldots\}$
count $C\ x$	Multiplicity of an element in a bag
$C \sharp x$	Same as *count* $C\ x$
$n \otimes C$	Bag scaling of multiplicity

$x \in C$	Bag membership
$C \sqsubseteq D$	Subbag relation
$C \uplus D$	Bag union
$C \cup D$	Bag difference
items s	Bag of elements in a sequence

Schema Notation

Vertical Schema

$$\begin{array}{|l} S \\\hline d \\\hline p \end{array}$$

New lines denote ";" and "∧." The schema name and predicate part are optional. The schema may subsequently be referenced by name in the document.

Axiomatic Definition

$$\begin{array}{|l} d \\\hline p \end{array}$$

The definitions may be nonunique. The predicate part is optional. The definitions apply globally in the document.

Generic Definition

$$\begin{array}{|l} [a, \ldots] \\\hline d \\\hline p \end{array}$$

The generic parameters are optional. The definitions must be unique. The definitions apply globally in the document.

$S \cong [X]$	Horizontal schema
$[T; \ldots | \ldots]$	Schema inclusion
$z.a$	Component selection (given $z : S$)
θS	Tuple of components
$\neg S$	Schema negation
pre S	Schema precondition
$S \wedge T$	Schema conjunction
$S \vee T$	Schema disjunction
$S \Rightarrow T$	Schema implication
$S \Leftrightarrow T$	Schema equivalence
$S \backslash (a, \ldots)$	Hiding of component(s)
$S \upharpoonright T$	Projection of components
$S; T$	Schema composition (S then T)
$S \gg T$	Schema piping (S outputs to T inputs)
$S[a/b, \ldots]$	Schema component renaming (b becomes a, etc.)
$\forall X \bullet S$	Schema universal quantification
$\exists X \bullet S$	Schema existential quantification
$\exists_1 X \bullet S$	Schema unique existential quantification

Conventions

$a?$	Input to an operation
$a!$	Output from an operation
a	State component before an operation
a'	State component after an operation
S	State schema before an operation
S'	State schema after an operation
ΔS	Change of state (normally $S \wedge S'$)
ΞS	No change of state (normally $[S \wedge S' \mid \theta S = \theta S']$)
$d \vdash p$	Theorem

Defining Terms

Formal methods: Techniques, notations, and tools with a mathematical basis, used for specification and reasoning in software or hardware system development.

Formal notation: A language with a mathematical semantics, used for formal specification, reasoning, and proof.

Logic: A scheme for reasoning, proof, inference, etc. Two common schemes are **propositional logic** and **predicate logic**, which is propositional logic generalized with quantifiers. Other logics, such as modal logics, including **temporal logics** that handle time, are also available. Examples include TLA (Temporal Logic of Actions), ITL (Interval Temporal Logic), and more recently Duration Calculus. Schemes may use **first-order logic** or **higher-order logic**. In the former, functions are not allowed on predicates, simplifying matters somewhat; but in the latter they are, thus providing greater power. Logics include a calculus that allows reasoning in the logic.

Operation: The performance of some desired action. This may involve the change of state of a system, together with inputs to the operation and outputs resulting from the operation. To specify such an operation, the **before state** (and inputs) and the **after state** (and outputs) must be related with constraining predicates.

Precondition: The predicate that must hold before an operation for it to be successful. Compare **postcondition**, which is the predicate that must hold after an operation.

Predicate: A constraint between a number of variables that produces a truth value (e.g., *true* of *false*).

Proof: A series of mathematical steps forming an argument of the correctness of a mathematical statement or theorem. For example, the **validation** of a desirable property for a formal specification could be undertaken by proving it correct. Proof can also be used to perform a formal **verification** that an implementation meets a specification. A less formal style of reasoning is **rigorous argument**, where a proof outline is sketched informally, which may be done if the effort of undertaking a fully formal proof is not considered cost-effective.

Refinement: The stepwise transformation of a specification toward an implementation (i.e., as a program). Compare **abstraction**, where unnecessary implementation detail is ignored in a specification.

Relation: A connection or mapping between elements in a number of sets. Often, two sets (a **domain** and a **range**) are related in a **binary relation**. A special case of a relation is a **function** where individual elements in the domain can only be mapped to, at most, one element in the range of the function. Functions can be further categorized. For example, a **partial function** may not map all possible elements that could be in the domain of the function, whereas a **total function** maps all such elements.

Set: A collection of distinct objects or **elements**, which are also known as **members** of the set. In a typed language, types may consist of maximal sets, as in the Z notation.

Specification: A description of *what* a system is intended to do, as opposed to *how* it does it. A specification can be *formal* (mathematical) or *informal* (natural language, diagrams, etc.). Compare an **implementation** of a specification, such as a program, which actually performs and executes the actions required by a specification.

State: A representation of the possible values that a system might have. In an abstract specification, this can be modeled as a number of sets. By contrast, in a concrete program implementation, the state typically consists of a number of data structures, such as arrays, files, etc. When modeling sequential systems, each operation can include a **before state** and an **after state**, which are related by some constraining predicates. The system will also have an **initial state**, normally with some additional constraints, from which the system starts at initialization.

References

Abrial, J.-R. 1996. *The B-Book*. Cambridge University Press, Cambridge, U.K.

Abrial, J.-R. 2003. B#: towards a synthesis between Z and B. In [Bert et al. 2003], pp. 168–177.

Abrial, J.-R., Börger, E., and Langmaack, H., Eds. 1996. *The Steam Boiler Control Specification Problem*. Lecture Notes in Computer Science 1165. Springer-Verlag.

Almstrum, V. L., Dean, C. N., Goelman, D., Hilburn, T. B., and Smith, J. 2001. Working group reports from ITiCSE on innovation and technology in computer science education. In *Annual Joint Conference Integrating Technology into Computer Science Education*, pp. 71–88. ACM Press, New York.

Arts, T. and Fokkink, W. 2003. *Eighth International Workshop on Formal Methods for Industrial Critical Systems*, Roros, Norway, June 5–7, 2003. European Research Consortium for Informatics and Mathematics (ERCIM).

Back, R.-J. and von Wright, J. 1998. *Refinement Calculus: A Systematic Introduction*, Graduate Texts in Computer Science, Springer-Verlag.

Banach, R. and Poppleton, M. 1999. Sharp Retrenchment, Modulated Refinement and Simulation, *Formal Aspects of Computing*, 11(5):498–540.

Barnes, J. 2003. *High Integrity Software: The SPARK Approach to Safety and Security*, Addison-Wesley.

Benveniste, A., Caspi, P., Edwards, S. A., Halbwachs, N., Le Guernic, P., and de Simone, R. 2003. The synchronous languages 12 years later. *Proc. of the IEEE*, 91(1):64–83.

Bérard, B., Bidoit M., Finkel, A., Laroussinie, F., Petit, A., Petrucci, L., Schnoebelen, Ph., and McKenzie, P. 2001. *Systems and Software Verification: Model-Checking Techniques and Tools*. Springer-Verlag.

Bert, D., Bowen, J. P., King, S., and Waldén, M., Eds. 2003. *ZB2003: Formal Specification and Development in Z and B*. Lecture Notes in Computer Science 2651. Springer-Verlag.

Bicarregui, J. C., Clutterbuck, D. L., Finnie, G., Haughton, H., Lano, K., Lesan, H., Marsh, D. W. R. M., Matthews, B. M., Moulding, M. R., Newton, A. R., Ritchie, B., Rushton, T. G. A., and Scharbach, P. N. 1997. Formal methods into practice: case studies in the application of the B method. *IEE Proceedings – Software* 144(2):119–133.

Bjørner, D. 2000. Pinnacles of software engineering: 25 years of formal methods. *Annals of Software Eng.*, 10(1–4):11–66.

Bjørner, D. and Cuéllar, J. R. 1999. Software engineering education: roles of formal specification and design calculi. *Annals of Software Eng.*, 6(1–4):365–409.

Börger, E. and Stärk, R. 2003. *Abstract State Machines: A Method for High-Level System Design and Analysis*. Springer-Verlag.

Bowen, J. P. 1996. *Formal Specification and Documentation Using Z: A Case Study Approach*. International Thomson Computer Press, London.

Bowen, J. P. 2001. Experience teaching Z with tool and web support. *ACM SIGSOFT Software Eng. Notes*, 26(2):69–75.

Bowen, J. P. and Hinchey, M. G. 1995a. Ten commandments of formal methods. *IEEE Comput.*, 28(4):56–63.

Bowen, J. P. and Hinchey, M. G. 1995b. Seven more myths of formal methods. *IEEE Software*, 12(4):34–41.

Bowen, J. P. and Hinchey, M. G. 1999. *High-Integrity System Specification and Design*. FACIT Series, Springer-Verlag, London.

Bowen, J. P. and Stavridou, V. 1993. Safety-critical systems, formal methods and standards. *IEE/BCS Software Eng. J.*, 8(4):189–209.

Bruns, G. 1996. *Distributed System Analysis with CCS*. Prentice Hall International Series in Computer Science. Hemel Hempstead, U.K.

Butler, M., Petre, L., and Sere, K., Eds. 2002. *Integrated Formal Methods*. Lecture Notes in Computer Science 2335. Springer-Verlag.

Caldwell, J. L. 1998. Formal methods technology transfer: a view from NASA. *Formal Methods in System Design*, 12(2):125–137.

Clarke, E. M., Wing, J. M. et al. 1996. Formal methods: state of the art and future directions. *ACM Computing Surveys*, 28(4):626–643.

Craigen, D., Gerhart, S. L., and Ralston, T. J. 1995. Formal methods reality check: industrial usage. *IEEE Trans. Software Eng.*, 21(2):90–98.

Crow, J. and Di Vito, B. 1998. Formalizing space shuttle software requirements: four case studies. *ACM Trans. Software Eng. and Methodology*, 7(3):296–332.

Dang Van, H., George, C., Janowski, T., and Moore, R., Eds. 2002. *Specification Case Studies in RAISE*. FACIT Series, Springer-Verlag, London.

Davoren, J. M. and Nerode, A. 2000. Logics for hybrid systems. *Proc. of the IEEE* 88(7):985–1010.

Derrick, J. and Boiten, E. A. 2001. *Refinement in Z and Object-Z*. FACIT Series, Springer-Verlag, London.

Derrick, J., Boiten, E. A., Woodcock, J., and von Wright, J. 2002. *REFINE 2002: The BCS FACS Refinement Workshop*. Electronic Notes in Theoretical Computer Science 70(3). Elsevier Science Publishers.

Diaconescu, R. and Futatsugi, K. 1998. *CafeOBJ Report: The Language, Proof Techniques, and Methodologies for Object-Oriented Algebraic Specification*. AMAST Series in Computing, Vol. 6, World Scientific Publishing Co.

Dijkstra, E. W. 1981. Why correctness must be a mathematical concern. In *The Correctness Problem in Computer Science*, R. S. Boyer and J. S. Moore, Eds., pp. 1–6. Academic Press, London.

Duke, R. and Rose, G. 2000. *Formal Object-Oriented Specification using Object-Z*. Cornerstones of Computing Series, MacMillan.

Easterbrook, S., Lutz, R., Covington, R., Kelly, J., Ampo, Y., and Hamilton, D. 1998. Experiences using lightweight formal methods for requirements modeling. *IEEE Trans. Software Eng.*, 24(1):4–14.

Eriksson, L.-H. and Lindsay, P. A., Eds. 2002. *FME 2002: Formal Methods – Getting IT Right*. Lecture Notes in Computer Science 2391. Springer-Verlag.

Feather, M. S. 1998. Rapid application of lightweight formal methods for consistency analyses. *IEEE Trans. Software Eng.*, 24(11):949–959.

Fidge, C., Kearney, P., and Utting, M. 1997. A formal method for building concurrent real-time software. *IEEE Software*, 14(2):99–106.

Frappier, M. and Habrias, H., Eds. 2001. *Software Specification Methods: An Overview Using a Case Study*. FACIT Series, Springer-Verlag, London.

Fuchs, N. E. 1992. Specifications are (preferably) executable. *IEE/BCS Software Eng. J.*, 7(5):323–334.

Futatsugi, K., Nakagawa, A. T., and Tamai, T., Eds. 2000. *CAFE: An Industrial-Strength Algebraic Formal Method*. Elsevier Health Sciences.

Garlan, D. 1995. Making formal methods effective for professional software engineers. *Inf. Software Tech.*, 37(5/6):261–268.

George, C. and Miao, H., Eds. 2002. *Formal Methods and Software Engineering*. Lecture Notes in Computer Science 2495. Springer-Verlag.

Gibbs, W. W. 1994. Software's chronic crisis. *Sci. Am.*, 271(3):86–95.

Goguen, J. and Winkler, T. 1988. *Introducing OBJ3*. SRI International, Menlo Park, CA. Tech. Rep. SRI-CSL-88-9.

Grumberg, O., Peled, D., and Clarke, E. M. 2000. *Model Checking*. MIT Press.

Guttag, J. V. 1993. *Larch: Languages and Tools for Formal Specification*. Springer–Verlag.

Hall, J. A. 1990. Seven myths of formal methods. *IEEE Software*, 7(5):11–19.

Hall, J. A. 1996. Using formal methods to develop an ATC information system. *IEEE Software*, 13(2): 66–76.

Hansen, K. M., Ravn, A. P., and Stavridou, V. 1998. From safety analysis to software requirements. *IEEE Trans. Software Eng.*, 24(7):573–584.

Harel, D. 1987. Statecharts: a visual formalism for complex systems. *Sci. Comput. Program.*, 8:231–274.

Harel, D. and Politi, M. 1998. *Modeling Reactive Systems with Statecharts: The Statemate Approach*. McGraw-Hill, New York.

Hayes, I. J. and Jones, C. B. 1989. Specifications are not (necessarily) executable. *IEE/BCS Software Eng. J.*, 4(6):330–338.

Heitmeyer, C., Kirby, J., Jr., Labaw, B., Archer, M., and Bharadwaj, R. 1998. Using abstraction and model checking to detect safety violations in requirements specifications. *IEEE Trans. Software Eng.*, 24(11):927–948.

Hinchey, M. G. and Bowen, J. P., Eds. 1995. *Applications of Formal Methods*. Prentice Hall International Series in Computer Science. Hemel Hempstead, U.K.

Hinchey, M. G. and Bowen, J. P., Eds. 1999. *Industrial-Strength Formal Methods in Practice*. FACIT Series, Springer-Verlag, London.

Hoare, C. A. R. 1985. *Communicating Sequential Processes*. Prentice Hall International Series in Computer Science. Hemel Hempstead, U.K.

Hoare, C. A. R. 1987. An overview of some formal methods for program design. *IEEE Comput.*, 20(9):85–91.

Hoare, C. A. R. 1996a. How did software get so reliable without proof? In *FME '96: Industrial Benefit and Advances in Formal Methods*. M.-C. Gaudel and J. Woodcock, Eds., pp. 1–17. Lecture Notes in Computer Science 1051. Springer-Verlag.

Hoare, C. A. R. 1996b. The logic of engineering design. *Microprocessing Microprogramming*, 41(8/9):525–539.

Hoare, C. A. R. 2003. The verifying compiler: a grand challenge for computing research. *Journal of the ACM*, 50(1):63–69.

Hoare, C. A. R. and He, J. 1998. *Unified Theories of Programming*. Prentice Hall International Series in Computer Science. Hemel Hempstead, U.K.

Holzmann, G. 2003. *The Spin Model Checker – Primer and Reference Manual*. Addison–Wesley.

Houston, I. S. C. and King, S. 1991. CICS project report: experiences and results from the use of Z in IBM. In *VDM '91: Formal Software Development Methods*, Vol. 1, S. Prehn and W. Toetenel, Eds. Lecture Notes in Computer Science 551, pp. 588–596. Springer-Verlag.

Hunt, W. A., Jr. and Sawada, J. 1999. Verifying the FM9801 microarchitecture. *IEEE Micro*, 19(3):47–55.

IEEE. 1991. IEEE standard glossary of software engineering terminology. In *IEEE Software Engineering Standards Collection*. Elsevier Applied Science, Amsterdam.

ISO. 1989. *Information Processing Systems – Open Systems Interconnection – LOTOS – A formal description technique based on the temporal ordering of observational behaviour*. International Standard ISO 8807:1989, International Organization for Standardization, Switzerland.

ISO. 1996. *Information Technology – Programming languages, their environments and system software interfaces – Vienna Development Method – Specification Language – Part 1: Base language*. International Standard ISO/IEC 13817-1:1996, International Organization for Standardization, Switzerland.

ISO. 2002. *Information Technology – Z Formal Specification Notation – Syntax, Type System and Semantics*. International Standard ISO/IEC 13568:2002. International Organization for Standardization, Switzerland.

Jacky, J. 1997. *The Way of Z: Practical Programming with Formal Methods*, Cambridge University Press, U.K.

Jones, C. B. 1991. *Software Development Using VDM*, 2nd ed. Prentice Hall International Series in Computer Science. Hemel Hempstead, U.K.

Jones, R. B., O'Leary, J. W., Seger, C.-J. H., Aagaard, M. D., and Melham, T. F. 2001. Practical formal verification in microprocessor design. *IEEE Design & Test of Computers*, 18(4):16–25.

Joseph, M., Ed. 1996. *Real-Time Systems: Specification, Verification and Analysis.* Prentice Hall International Series in Computer Science. Hemel Hempstead, U.K. URL: *http://www.tcs.com/techbytes/htdocs/book_mj.htm* (2001).

King, S., Hammond, J., Chapman, R., and Pryor, A. 2000. Is proof more cost-effective than testing? *IEEE Trans. Software Eng.*, 26(8):675–686.

Kropf, T. 2000. *Introduction to Formal Hardware Verification.* Springer-Verlag.

Larsen, P. G., Fitzgerald, J., and Brookes, T. 1996. Applying formal specification in industry. *IEEE Software,* 13(7):48–56.

Lee, J.-S. and Cha, S.-D. 2000. Qualitative formal method for requirements specification and validation of hybrid real-time safety systems. *IEE Proceedings – Software,* 147(1):1–11.

Lightfoot, D. 2001. *Formal Specification using Z.* 2nd ed. Grassroots Series, Palgrave.

Lugi and Goguen, J. A. 1997. Formal methods: promises and problems. *IEEE Software,* 14(1): 73–85.

Meadows, C. 2003. Formal methods for cryptographic protocol analysis: emerging issues and trends. *IEEE Journal on Selected Areas in Communications,* 21(1):44–54.

Milner, R. 1989. *Communication and Concurrency.* Prentice Hall International Series in Computer Science. Hemel Hempstead, U.K.

Morgan, C. 1994. *Programming from Specifications,* 2nd ed. Prentice Hall International Series in Computer Science. Hemel Hempstead, U.K. URL: *http://web.comlab.ox.ac.uk/oucl/publications/books/PfS/* (1998).

Palshikar, G. K. 2001. Applying formal specifications to real-world software development. *IEEE Software,* 18(6):89–97.

Parnas, D. 1995. Using mathematical models in the inspection of critical software. In *Applications of Formal Methods,* M. G. Hinchey and J. P. Bowen, Eds., pp. 17–31. Prentice Hall International Series in Computer Science. Hemel Hempstead, U.K.

Peled, D. and Vardi, M. Y., Eds. 2002. *Formal Techniques for Networked and Distributed Systems – FORTE 2002.* Lecture Notes in Computer Science 2529. Springer-Verlag.

Pfleeger, S. L. and Hatton, L. 1997. Investigating the influence of formal methods. *IEEE Comput.,* 30(2): 33–43.

Prowell, S. J., Trammell, C. J., Linger, R. C., and Poore, J. H. 1999. *Cleanroom Software Engineering: Technology and Process.* Addison-Wesley.

RAISE Language group. 1992. *The RAISE Specification Language.* BCS Practitioner Series. Prentice-Hall, Hemel Hempstead, U.K.

Saiedian, H., Ed. 1996. An invitation to formal methods. *IEEE Comput.,* 29(4):16–30.

Schneider, S. 1999. *Concurrent and Real-time Systems: The CSP Approach.* John Wiley & Sons.

Schneider, S. 2001. *The B-Method: An Introduction.* Cornerstones of Computing Series, MacMillan.

Sekerinski, E. and Sere, K., Eds. 1999. *Program Development by Refinement.* FACIT Series, Springer-Verlag, London.

Semmens, L. T., France, R. B., and Docker, T. W. G. 1992. Integrated structural analysis and formal specification techniques. *The Computer Journal,* 35(6):600–610.

Smith, G. 2000. *The Object-Z Specification Language.* Advances in Formal Methods Series. Kluwer Academic Publishers.

Spivey, J. M. 1992. *The Z Notation: A Reference Manual,* 2nd ed. Prentice Hall International Series in Computer Science, Hemel Hempstead, U.K.

Tiwari, A., Shankar, N., and Rushby, J. 2003. Invisible formal methods for embedded control systems. *Proc. of the IEEE,* 91(1):29–39.

Tremblay, G. 2000. Formal methods: Mathematics, computer science or software engineering? *IEEE Trans. on Education,* 43(4):377–382.

Tretmans, J., Wijbrans, K., and Chaudron, M. 2001. Software engineering with formal methods: the development of a storm surge barrier control system revisiting seven myths of formal methods. *Formal Methods in System Design,* 19(2):195–215.

Turner, K. J., Ed. 1993. *Using Formal Description Techniques: An Introduction to Estelle, LOTOS and SDL*. John Wiley & Sons, Chichester, U.K.

Utting, M., Toyn, I., Sun, J., Martin, A., Dong, J. S., Daley, D., and Currie, D. 2003. ZML: XML support for standard Z. In [Bert et al. 2003], pp. 437–456.

Warmer, J. and Kleppe, A. 1998. *The Object Constraint Language: Precise Modeling with UML*. Addison-Wesley.

Wing, J. M. 1990. A specifier's introduction to formal methods. *IEEE Comput.*, 23(9):8–24.

Wing, J. M. and Woodcock, J. 2000. Special issues for FM '99: The First World Congress on Formal Methods in the Development of Computing Systems. *IEEE Trans. Software Eng.*, 26(8):673–674.

Wing, J. M., Woodcock, J., and Davies, J., Eds. 1999. *FM'99 — Formal Methods*. Lecture Notes in Computer Science 1708 & 1709. Springer-Verlag.

Woodcock, J. and Davies, J. 1996. *Using Z: Specification, Refinement, and Proof*. Prentice Hall International Series in Computer Science. Hemel Hempstead, U.K.

Further Information

A number of organizations have been established to meet the needs of formal methods practitioners; for example:

- Formal Methods Europe (FME) organizes a regular conference (e.g., Eriksson and Lindsay [2002] and Wing et al. [1999]), formerly the VDM symposia, and other activities for users of various formal methods.

- The British Computer Society Specialist Group on Formal Aspects of Computing Science (BCS-FACS) organizes workshops and meetings on various aspects of formal methods, as well as a series of Refinement Workshops (e.g., see Derrick et al. [2002]).

- The Z User Group (ZUG) has organized a regular international conference, historically known as the Z User Meeting (ZUM), attracting users of the Z notation from all over the world. The International B Conference Steering Committee (Association de Pilotage des Conférences B, APCB) has organized a similar International B Conference series. Since 2000 these have been combined into a single conference (e.g., see Bert et al. [2003]).

There are now a number of journals devoted specifically to formal methods. These include *Formal Methods in System Design* and *Formal Aspects of Computing*. The *FAC* journal is published by Springer-Verlag in association with BCS-FACS. Other European-based journals, such as *The Computer Journal, IEE Proceedings–Software* (formerly the *Software Engineering Journal*), and *Information and Software Technology*, publish articles on, or closely related to, formal methods, and they have run special issues on the subject.

While there are no U.S.-based journals that deal specifically with formal methods, they regularly are featured in popular periodicals such as *IEEE Computer* (e.g., Bowen and Hinchey [1995a], Hoare [1987], Pfleeger and Hatton [1997], Saiedian [1996], and Wing [1990]) and *IEEE Software* (e.g., Bowen and Hinchey [1995b], Hall [1990, 1996], Larsen et al. [1996], Luqi and Goguen [1997], and Palshikar [2001]), as well as in journals such as the *Annals of Software Engineering* (e.g., Bjørner and Cuéllar [1999] and Bjørner [2000]) *IEEE Transactions on Software Engineering* (e.g., Craigen et al. [1995], Easterbrook et al. [1998], Feather [1998], Hansen et al. [1998], Heitmeyer et al. [1998], and Wing and Woodcock [2000]), *ACM Transactions on Software Engineering and Methodology* [Crow and Di Vito 1998], and the *Journal of the ACM*. A classic paper on the state of the art in formal methods has also appeared in the *ACM Computing Surveys* [Clarke et al. 1996].

In addition to the conferences mentioned earlier, the IFIP (International Federation of Information Processing) FORTE international conference concentrates on **Formal Description Techniques** (FDTs, e.g., see Peled and Vardi [2002]). The International Conference on Formal Engineering Methods series (ICFEM) has also been established more recently (e.g., see George and Miao [2002]). A number of more specialist conferences on formal methods have been established. For example, the Integrated Formal Methods (IFM) International Conference concentrates on the use of formal methods with other approaches (e.g., see Butler et al. [2002]). The International Workshop on Formal Methods for Industrial Critical Systems (FMICS) concentrates on industrial applications, especially using tools [Arts and Fokkink 2003].

Some more wide-ranging conferences give particular attention to formal methods; primary among these are the ICSE (International Conference on Software Engineering) and ICECCS (International Conference on Engineering of Complex Computer Systems) series of conferences. Other specialist conferences in the safety-critical sector, such as SAFECOMP, and SSS (the Safety-critical Systems Symposium) also regularly cover formal methods.

There have been some collections of case studies on formal methods with various aims and themes. For some industrial applications, see Bowen and Hinchey [1995, 1999]. Solutions to a control specification problem using a number of different formal approaches are presented in Abrial et al. [1996]. Frappier and Habrias [2001] collected together a number of formal specification methods applied to an invoicing case study where the presentations concentrate on the *process* of producing a formal description, including the questions raised along the way.

A number of electronic forums are available as online newsgroups:

`comp.specification.misc`	Formal specification
`comp.specification.larch`	Larch
`comp.specification.z`	Z notation

In addition, the following electronic mailing lists are available, among others:

`formal-methods-request@cs.uidaho.edu`	Formal methods
`procos-request@jiscmail.ac.uk`	Provably Correct Systems
`vdm-forum-request@jiscmail.ac.uk`	VDM
`zforum-request@comlab.ox.ac.uk`	Z (gatewayed to `comp.specification.z`)

For up-to-date online information on formal methods in general, readers are directed to the following World Wide Web URL (Uniform Resource Locator), which provides formal methods links as part of the WWW Virtual Library:

`http://vl.fmnet.info/`

107

Verification and Validation

107.1   Introduction **107**-1
107.2   Approaches to Verification **107**-1
107.3   Verifying Specifications of Systems **107**-2
       General Properties • Specific Properties
107.4   Verifying Programs **107**-7
       General Properties • Specific Properties
107.5   Current Status **107**-22

John D. Gannon
University of Maryland

107.1   Introduction

Verification and *validation* are terms that are sometimes used interchangeably. In Ghezzi et al. [1991], *verification* is used to describe "all activities that are undertaken to ascertain that the software meets its objectives," and *validation* is not used at all. In Rushby [1993], specification validation is a two-component process of seeking assurance that a specification means something (i.e., is consistent), and that it means what is intended. We use *verification* to describe the process of demonstrating that a description of a software system guarantees particular properties. General properties may be derived from the form of the description (e.g., that functions are total, axioms are consistent, or variables are initialized before they are referenced), and specific properties may be derived from the problem domain. The latter case involves the comparison of two objects, a detailed description of a software system, and a more abstract description of its intended properties.

In Section 107.2 we briefly describe validation and verification approaches. Section 107.3 and Section 107.4 deal with the verification of general and specific properties of specifications and programs, respectively. We pay particular attention to opportunities for automating verification activities. We conclude with a short discussion of the current verification practices.

107.2   Approaches to Verification

A variety of analysis activities may be used to verify software artifacts. In software inspections, teams of software developers manually examine artifacts for defects. If a requirements document or a design is written in a formal language, it may be possible to use it as a prototype for the system by simulating the description for some test cases. General properties of software artifacts may be verified automatically by static analysis of the artifact. State-exploration or theorem-proving techniques can be used to prove specific properties of system descriptions.

Inspections have proven to be an effective method for detecting software defects because they subject a software artifact to the scrutiny of several people, some of whom did not participate in the artifact's design. Early requirements inspections catch errors before they propagate into designs and implementations, making them less costly to repair. Fagan [1976] describes a six-stage inspection process:

- A determination is made that a software artifact is ready for an inspection, and an inspection team is assembled.
- The artifact's author provides reviewers an overview of the artifact.
- The team members individually study the artifact and record potential defects.
- A fixed-length inspection meeting is held. A moderator controls the discussion. The designer presents and explains his work. Participants identify errors (but not solutions), which are recorded by a secretary.
- The author fixes any errors.
- The moderator checks the new version of the artifact and determines if another inspection is necessary.

Successful **software inspections** depend on the experience levels of the participants and the quality of the artifacts. Requirements inspections should include participants who are future users of the system who will help the software developers judge if the system will function as intended. Requirements notations for embedded systems (e.g., the Software Cost Reduction (SCR) notation [Heitmeyer et al. 1995, Heniger 1980], the Requirements State Machine Language [Leveson et al. 1994], and Statecharts [Harel et al. 1990]) describe systems as sets of concurrently executing state machines responding to events in their environments. Finite state machines have precise meanings, but they are also easy to understand because they can be described in tabular or graphic formats.

In order to maximize the benefits of inspections, participants may be given lists of questions about the artifact that they must answer to ensure that they are sufficiently prepared for an inspection. For code inspections, participants may receive checklists of potential errors that they are to check are not present in the implementation.

Simulation of a software artifact helps software developers determine if the system behaves as expected by producing results like those which will be produced by the eventual implementation of the system. Such operational descriptions of systems give recipes for achieving desired results rather than just describing properties of final results. Simulations of state-machine descriptions of systems are easy to perform; however, simulating more detailed descriptions may require developers to sacrifice abstraction in favor of executability. Being able to reverse a simulation may permit analysts to determine how potentially **hazardous states** may be reached [Ratan et al. 1996].

If state machines manipulate few variables with simple data types (i.e., types with finite numbers of values), properties such as deadlock freedom or mutual exclusion can be verified by using state-space enumeration and exploration techniques. More detailed system descriptions with richer data types correspond to infinite-state machines. While more specific properties can be stated and verified for infinite-state machines, analysis techniques must either investigate approximations of the state space by folding states together [Young and Taylor 1989] or reason with compact descriptions of the entire state space (i.e., assertions).

107.3   Verifying Specifications of Systems

With appropriate abstraction, synchronous communicating processes in distributed systems can be described by sets of state-transition diagrams.

For example, Figure 107.1 describes a producer/consumer system with one producer and two consumer processes. *Consumer1* decides either to remain in an idle state (labeled *i*) or to move to a state in which it requests output from the producer (labeled *r1*). When the producer process grants the consumer's request (by entering its state labeled *p1*), *Consumer1* returns to its idle state. *Consumer2* behaves in a similar

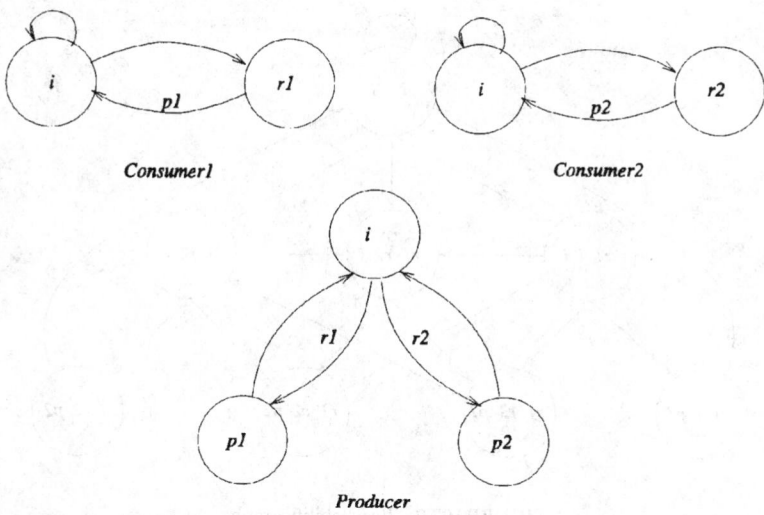

FIGURE 107.1 A simple system with two consumers and one producer.

manner. *Producer* starts in an idle state (also labeled *i*), and moves into a production state (labeled *p1* or *p2*) in response to one of the consumer processes moving into its request state. After satisfying the request, *Producer* returns to its idle state.

107.3.1 General Properties

We can derive general properties for the transitions in our model (e.g., that they are deterministic or total). In Figure 107.1, unlabeled arcs are considered to be labeled "true." That is, these transitions may always be taken. *Consumer1* has nondeterministic transitions because there are two arcs with the same transition conditions (i.e., true) leaving its idle state and ending in different states. We might always want to ensure that some transition is always enabled from each state. To check this property we compute the logical operation for all the conditions on transitions leaving a state, and we check that its result is identical to true. *Consumer1*'s idle state satisfies this property, but its request state does not, because the only transition leaving this state occurs when *p1* is true (i.e., when the *Producer* is in its state labeled *p1*). Although these are very simple properties, they are valuable checks, particularly for large systems because they do not require construction of the system's state space.

107.3.2 Specific Properties

107.3.2.1 Reachability Analysis

Reachability analysis is performed to determine if potentially hazardous states (e.g., those representing deadlock or mutual-exclusion failures) are reachable. To perform this analysis, a reachability graph representing the global behavior of the system is constructed and exhaustively examined to determine if a hazardous state is reachable.

Figure 107.2 represents the reachability graph for our producer/consumer system. Each state is labeled with three properties corresponding to properties of *Consumer1*, *Consumer2*, and *Producer*. In the reachability graph's initial state, labeled (i, i, i), each of the processes is in its idle state. Three transitions leave this state, corresponding to either or both *Consumer1* or *Consumer2* issuing requests for output. There are no transitions from the initial state to other states labeled either $(i, i, p1)$ or $(i, i, p2)$ because the producer process waits for either of the consumers to issue a request before moving to a state in which it produces output. Thus we could verify that objects are produced only in response to requests by determining that no reachable states are labeled either $(i, i, p1)$ or $(i, i, p2)$.

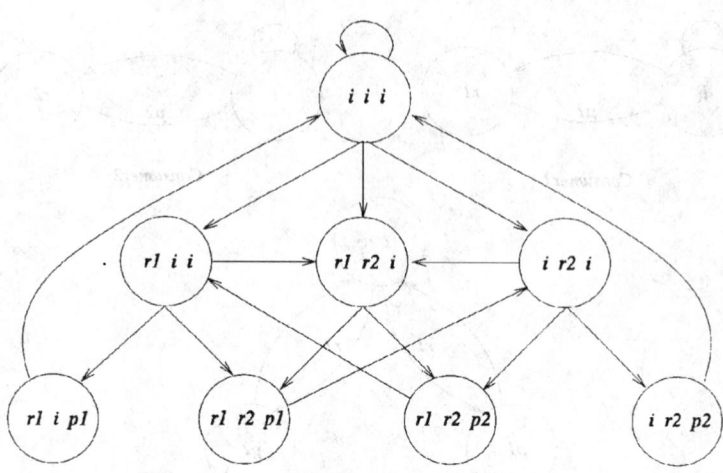

FIGURE 107.2 Reachability graph.

107.3.2.2 Model Checking

Although it is a useful verification technique, reachability analysis can be used only to verify properties specified as propositional logic formulas quantified over all the states in the graph. We might also like to assert properties about sequences of events, e.g., "if the consumer requests an output, the producer always supplies one." Pneuli [1981] showed how **temporal logic** can be used to state such properties and to reason about concurrent systems.

A temporal logic is a propositional logic with additional temporal operators to express concepts such as "always," "eventually," and "until" to assert that formulas are true in all or some future states. Two major types of temporal logic are used in specifications: linear time logic and branching time logic. In linear time logic, states have unique pasts and futures. To prove that a property is invariantly true, the property must be proved over all possible execution paths of the system. In branching time temporal logics, states have unique pasts but many possible futures. Thus, assertions may be made about properties holding on some future executions or on all future executions. The latter assertions are invariants.

Computational tree logic (CTL) is a propositional branching time logic, whose operators permit explicit quantification over all possible futures [Clarke et al. 1986]. The syntax for CTL formulas is summarized below:

1. Every atomic proposition is a CTL formula.
2. If f and g are CTL formulas, then so are: $\sim f$, $f \wedge g$, $f \vee g$, $f \rightarrow g$, AXf, EXf, $A[fUg]$, $E[fUg]$, AFf, EFf, AGf, and EGf.

Note that temporal operators occur only in pairs in which a quantifier A (always) or E (exists) is followed by F (future), G (global), U (until), or X (next). The logical operators have their usual meanings. The meanings of the temporal operators are described below.

Concept	Operator	Meaning
Next	AXf	Formula f holds in every next state.
	EXf	Formula f holds in some next state.
Until	$A[fUg]$	Along every path, there exists some future state s in which g is true, and f is true in every state on the path until s.
	$E[fUg]$	Along some path, there exists some future state s in which g is true, and f is true in every state on the path until s.
Eventually	AFf	Along every path, f is true in some state.
Possibly	EFf	Along some path, f is true in some state.
Invariance	AGf	Along every path, f is true in every state.
Possible invariance	EGf	Along some path, f is true in every state.

The specification "if the consumer requests an output, the producer always supplies one" can be written as the CTL formula: $AG((r1 \rightarrow AF(p1)) \wedge (r2 \rightarrow AF(p2)))$. That is, it is invariantly true that if *Consumer1* makes a request (represented by a state in which $r1$ is true), eventually (i.e., along every path starting at such states) the producer supplies an output for *Consumer1* (a state is encountered in which $p1$ is true), and similarly for *Consumer2*.

If formula f is true in state s of model M, we write $M, s \models f$. A formula f is true for the model, if it is true in the model's start state, i.e., $M, s_0 \models f$. When we are concerned with a single model, we abbreviate $M, s \models f$ as $s \models f$.

Introduced by Clarke and Emerson [Clarke et al. 1986] and by Quielle and Sifakis [1981], model checking determines the value of a formula f for a particular model by building a reachability graph and computing the set of states in which the formula is true, i.e., $\{s \mid s \models f\}$. For example, formula $AF(f)$ represents the set of states from which a state satisfying f can be reached in *some* number of state transitions along *all* paths from the state.

$$f \vee AXf \vee AX(AXf) \vee \dots$$

where f is the set of states in which f is true, i.e., the set of states from which an f-state can be reached in zero-state transitions. $AX(f)$ is the set of states all of whose transitions reach an f-state. $AX(AXf)$ is the set of states from which any two state transitions reach an f-state, etc.

This set of states can be computed using the following **least fixpoint** computation:

```
Y = { };
Y' = {s | s ⊨ f};
while ( Y ≠ Y' ) do {
   Y = Y';
   Y' = Y' ∪ {s | all successors of s are in Y};
}
```

By way of example, consider computing the set of states in which $AF(p1)$ for the model in Figure 107.2. The first iteration computes the set of states in which the formula $p1$ holds. This set, which is shaded in Figure 107.3, corresponds to computing $p1 \vee AX\ false$. During the second iteration, the predecessors of the states already in the set are examined. The state labeled $(r1, i, i)$ is added to the set because the formula is true in all of its successors. At this point, we have computed the set of states satisfying $p1 \vee AX(p1)$. The remaining iterations are summarized in the following table.

Iteration	State Set	Remarks
1	$\{(r1, i, p1), (r1, r2, p1)\}$	The only states in which $p1$ is true.
2	$\{(r1, i, p1), (r1, r2, p1),$ $(r1, i, i)\}$	$AX(p1)$ is true for the state labeled $(r1, i, i)$ since $p1$ is true in all its successors.
3	$\{(r1, i, p1), (r1, r2, p1),$ $(r1, i, i), (r1, r2, p2)\}$	$AX(AX(p1))$ is true for the state labeled $(r1, i, i)$ since $AX(p1)$ is true in all its successors.
4	$\{(r1, i, p1), (r1, r2, p1),$ $(r1, i, i), (r1, r2, p2),$ $(r1, r2, i)\}$	For the state labeled $(r1, r2, i)$, $AX(AX(AX(p1)))$ is true for one of its successors and $AX(p1)$ is true for the other.
5	$\{(r1, i, p1), (r1, r2, p1),$ $(r1, i, i), (r1, r2, p2),$ $(r1, r2, i)\}$	Each of the possible new predecessors, the states labeled state (i, i, i) and $(i, r2, i)$, have infinitely long paths before reaching a state in which $p1$ is true.

Since no new states are added during the fifth iteration, we reach a fixpoint and the algorithm terminates. Unfortunately, the set of states computed does not contain the start state, s_0 (the state labeled (i, i, i)), so $s_0 \not\models AF(p1)$. However, the model checker gives a specific counter-example (i.e., the loop $(i, i, i), (i, r2, i), (i, r2, p2), \dots$) showing why the formula is not satisified.

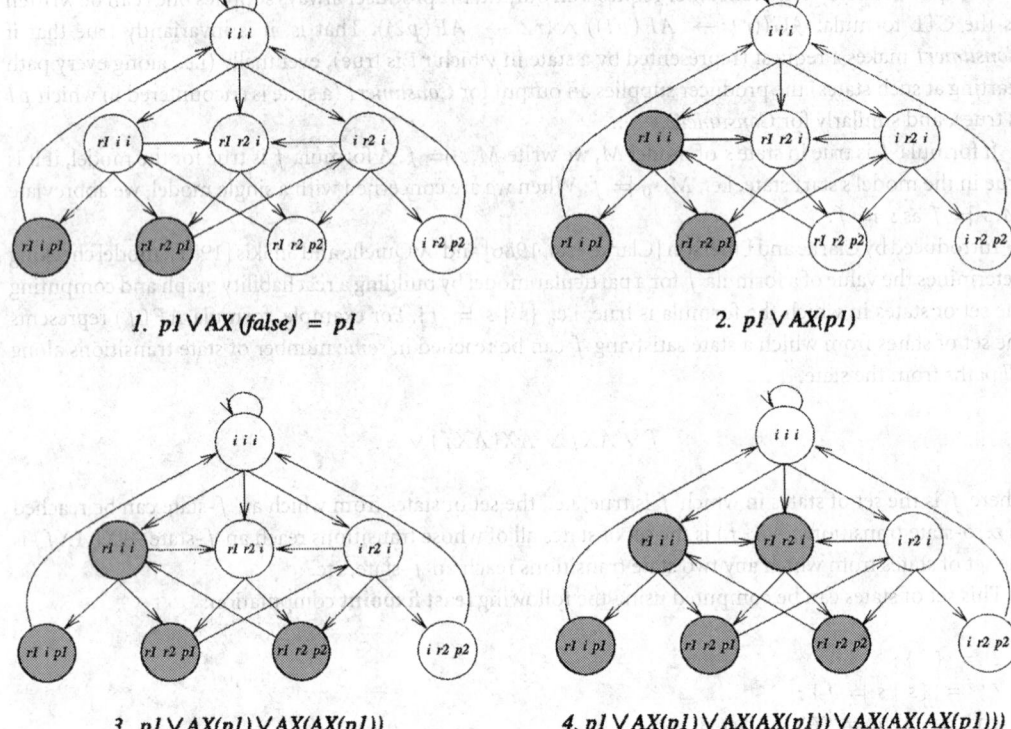

1. $p1 \lor AX$ (false) = $p1$ 2. $p1 \lor AX(p1)$

3. $p1 \lor AX(p1) \lor AX(AX(p1))$ 4. $p1 \lor AX(p1) \lor AX(AX(p1)) \lor AX(AX(AX(p1)))$

FIGURE 107.3   Model checking $s_0 \models AF(p1)$.

To check $AG((r1 \rightarrow AF(p1)) \land (r2 \rightarrow AF(p2)))$, we calculate the sets of states for which the innermost, simplest formulas hold and work our way outward, calculating sets of states for more complex formulas. Thus we might perform the following computations:

Step	Formula	Set of States
1	$p1$	$\{(r1,i,p1),(r1,r2,p1)\}$
2	$AF(p1)$	$\{(r1,i,p1),(r1,r2,p1),(r1,i,i),$ $(r1,r2,p2),(r1,r2,i)\}$
3	$r1$	$\{(r1,i,p1),(r1,r2,p1),(r1,i,i),$ $(r1,r2,p2),(r1,r2,i)\}$
4	$r1 \rightarrow AF(p1)$	All states
5	$p2$	$\{(i,r2,p2),(r1,r2,p2)\}$
6	$AF(p2)$	$\{(i,r2,p2),(r1,r2,p2),(i,r2,i),$ $(r1,r2,p1),(r1,r2,i)\}$
7	$r2$	$\{(i,r2,p2),(r1,r2,p2),(i,r2,i),$ $(r1,r2,p1),(r1,r2,i)\}$
8	$r2 \rightarrow AF(p2)$	All states
9	$(r1 \rightarrow AF(p1)) \land (r2 \rightarrow AF(p2))$	All states
10	$AG((r1 \rightarrow AF(p1)) \land (r2 \rightarrow AF(p2)))$	All states

The formula $r1 \rightarrow AF(p1)$ holds in all states because $r1$ and $AF(p1)$ are true in identical sets of states, and because $r1$ is false in all other states of the model. A similar analysis determines that $r2 \rightarrow AF(p2)$ also holds in all states, so the entire formula is true in all states satisfying the invariant operator AG.

Automating model checking is quite easy, except that the entire state space of the model is constructed before the fixpoint algorithms can be applied. However, model checking can also be done symbolically by manipulating quantified Boolean formulas without constructing a model's state space [McMillan 1993]. To perform symbolic model checking, sets of states and transition relations are represented by formulas, and set operations are defined in terms of formula manipulations. A CTL formula f is evaluated for a model by deriving a propositional logic expression that describes the set of states satisfying f for the model and verifying that the interpretation of the model's initial state satisfies the expression.

107.4 Verifying Programs

107.4.1 General Properties

Probably the best-known property of programs which is verified is that a program is type safe. In statically typed languages, a data type is associated with a variable in a declaration. During its lifetime, the variable may be assigned only values of the same type. The context of each appearance of a variable in a statement implies a type, which can be checked against its declared type. Violations are reported as syntax errors.

Other important properties depend on data and control flow, e.g., each variable is assigned a value before the value is used in an expression. Such properties are verified by static checkers, which analyze a program's syntax tree or control flow graph; no test data are used during these checks. Static checkers fold different states together to make analysis tractable. For example, to check uninitialized variables, we may care only if a variable has been assigned a value or not. Different integer values are all folded to a single "defined" value to reduce the size of the state space. In the following program fragment, x will always have a value when control reaches the final write statement, since either $i < j$ and x is assigned the value 1 or $i \geq j$ and x is assigned the value 2. However, a static checker keeping track of whether or not x had been assigned a value on every potential path through the program would conclude that the write statement might be executed with an undefined value.

Statement	Defined Values
`read(i);`	{i}
`read(j);`	{i, j}
`if (i < j)`	
`    x = 1;`	{i, j, x}
`fi;`	{i, j} ∩ {i, j, x} = {i, j}
`if (i >= j)`	
`    x = 2;`	{i, j, x}
`fi;`	{i, j} ∩ {i, j, x} = {i, j}
`write(x)`	{i, j}

In each if statement, x is defined on only one path through the statement. Thus, the static checker intersects the sets of variables defined on each of its paths to determine the set of variables which are certain to have values. This approximation of the state space is called conservative or pessimistically inaccurate because it preserves states which potentially contain errors. In this case, it preserved a state in which x is undefined which can be reached only on an infeasible path (i.e., when $i \geq j \wedge i < j$). Programmers using static checkers must examine error messages to determine if an anomaly exists before trying to repair their programs.

A syntax-directed definition can be used to describe the analysis needed to verify this property. A syntax-directed definition extends a context-free grammar by associating attributes with grammar symbols. The value of an attribute in a syntax tree is defined by rules associated with each production used at the particular node of the tree. Attributes can be values of any type. In this example, we use two sets of identifiers: In (representing the set of identifiers defined on all paths leading to the current statement) and Out (representing the set of identifiers defined on all paths after executing the current statement). The syntax-directed definition for the simple language is shown below.

Production	Attributes
P ::= SL	SL.In = ∅
SL₁ ::= S ';' SL₂	S.In = SL₁.In
	SL₂.In = S.Out
	SL₁.Out = SL₂.Out
SL ::= ε	SL.Out = SL.In
S ::= id '=' Exp	S.Out = {id} ∪ S.In
S ::= 'if' Exp 'then' SL 'fi'	S.In = S.In
	S.Out = S.In
S ::= 'if' Exp 'then' SL₁ 'else' SL₂ 'fi'	SL₁.In = S.In
	SL₂.In = S.In
	S.Out = SL₁.Out ∩ SL₂.Out
S ::= 'while' Exp 'do' SL 'od'	SL.In = S.In
	S.Out = S.In
S ::= 'read' '(''id'')'	S.Out = {id} ∪ S.In
S ::= 'write' '(''id'')'	S.Out = S.In

The production **P ::= SL** initializes **SL's In** attribute to the empty set. When identical nonterminals appear in a production, the instances are numbered so their attributes can be distinguished. For example, the production **SL ::= S ';' SL** is rewritten as **SL₁ ::= S ';' SL₂**. The set of defined variables available to **S** is the same as that available to **SL₁**. However, the defined variables for **SL₂** are those from **S.Out**. Each **S**-production copies its incoming set of definitions (**In**) to its **Out** attribute and adds any new definitions made in the statement. Thus, for example, the statement **read(i)** adds the variable **i** to the empty set of definitions that reaches it. The set of outgoing definitions from the if statement if i < j then x = 1 fi; is {**i, j**}, since the statement contains an execution path on which x is not defined.

Figure 107.4 shows the attributes evaluated on the syntax tree corresponding to our sample program. Since the write statement's **In** attribute contains only {**i, j**}, a static checker would say that definition-before-use property was violated for **x**.

Abstract interpretation [Cousot and Cousot 1976] is a method for computing approximate semantics of programs in order to provide safe answers to questions about their run-time behaviors. In an abstract interpretation of a program, "abstract" values are associated with program variables instead of the actual execution values, and a programming language's operators are redefined to manipulate the abstract values. An abstract interpretation of a program computes a fixpoint approximation of the abstract program state at different points in the program, and properties of the program are verified with respect to these states.

Consider the following example [Cousot and Cousot 1977], which searches a list for a particular value. One property we would like to ensure is that the value of **p** is never **NULL** when it is dereferenced on lines 3 or 4.

```
1. p = L; b = TRUE;
2. while (p ≠ NULL & b)
3.     if (p → v == n) then b = FALSE;
4.     else p = p → next;
```

The pointer variable **p** may have one of four possible values: undefined (⊥), **NULL**, **nonNULL**, or **NULL** or **nonNULL** (⊤). These values form the complete lattice in Figure 107.5. Initially, a pointer variable has the value ⊥. Pointer variables may be assigned on of the following values: **NULL**, **nonNULL** (i.e., the result produced by a new operation), or the value of another pointer variable. Dereferencing a pointer whose value is **nonNULL** yields the value ⊤, and dereferencing a pointer with any other value yields ⊥.

Each node in a program's control flow graph defines an output state in terms of its input state. Assignment nodes' output states are identical to their input states except for the value of the variable on the left side of the assignment operation. The output state for a join node is the union of the respective values in its input states. The output state corresponding to the true outcome of a decision node labeled **p! = NULL** is calculated by creating a new state which is identical to the input state except that **p** has value **nonNULL** and intersecting the new state with the input state. The output state corresponding to the false outcome of the

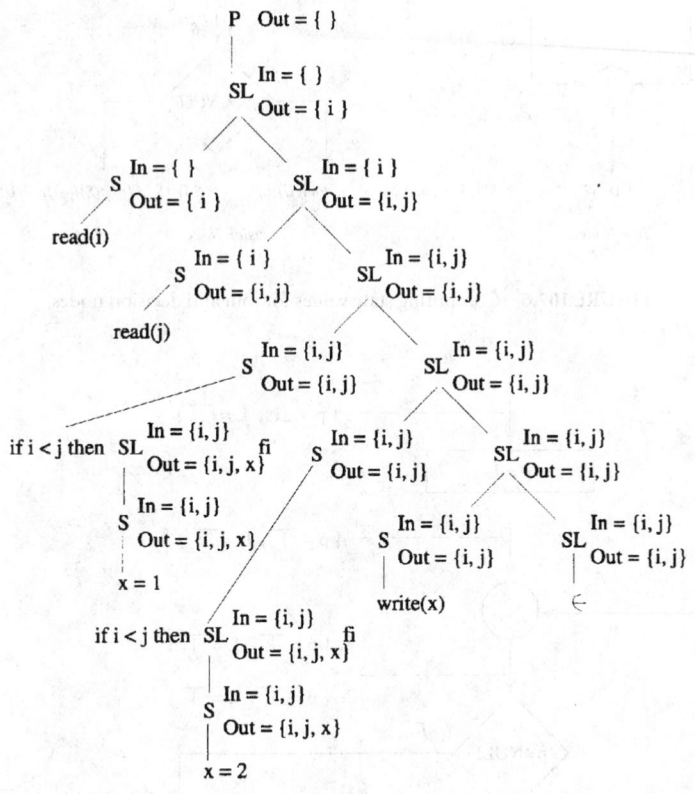

FIGURE 107.4 Attributes evaluated on a syntax tree.

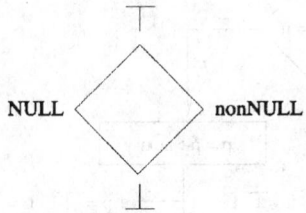

FIGURE 107.5 Lattice of pointer values.

decision node is calculated in a similar manner except that **p** has value **NULL** in the newly created state. These computations are depicted in Figure 107.6, and the definitions for ∪ and ∩ are given in the following tables. The **X**s in the table for ∩ represent error entries corresponding to infeasible paths.

x ∪ y	⊥	NULL	nonNULL	⊤
⊥	⊥	NULL	nonNULL	⊤
NULL	NULL	NULL	⊤	⊤
nonNULL	nonNULL	⊤	nonNULL	⊤
⊤	⊤	⊤	⊤	⊤

x ∩ y	⊥	NULL	nonNULL	⊤
⊥	⊥	⊥	⊥	⊥
NULL	⊥	NULL	X	NULL
nonNULL	⊥	X	nonNULL	nonNULL
⊤	⊥	NULL	nonNULL	⊤

Since pointer variables have a finite number of abstract values, system states do not have an infinite increasing chain of values.

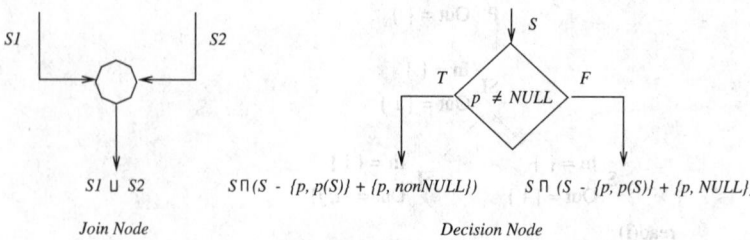

FIGURE 107.6   Computing state values for join and decision nodes.

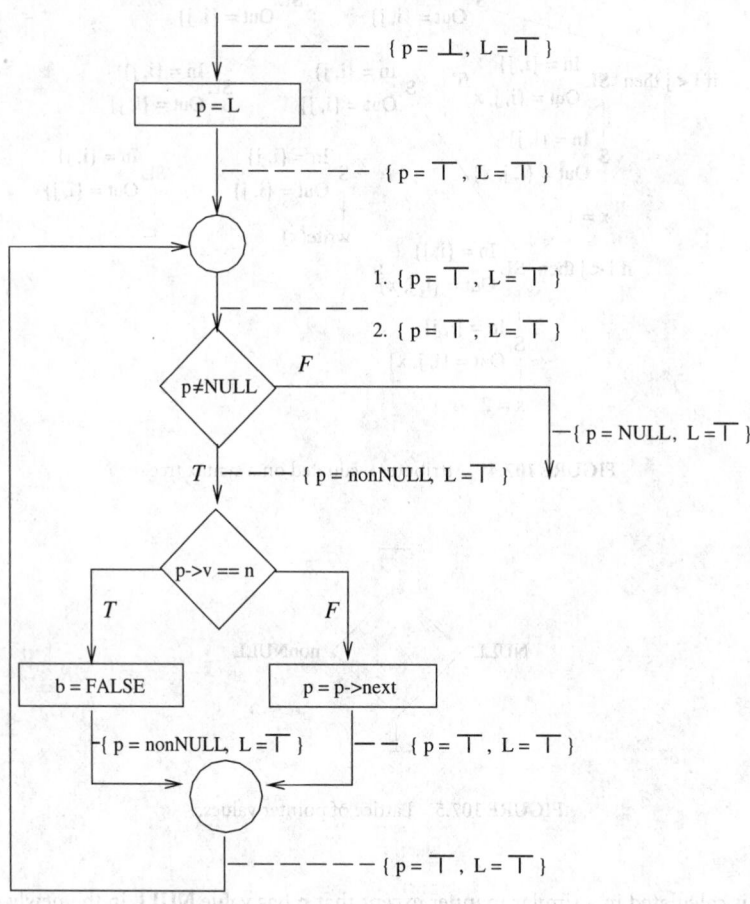

FIGURE 107.7   Calculated state values for sample program.

Figure 107.7 shows the control flow graph for the sample program with edges labeled with computed state values for the pointer variables **p** and **L**. All edges are initially labeled $\{p = \perp, L = \perp\}$ except for the first arc, where we assume **L** has value $\top$. The first time the arc leaving the while statement's join node is reached, the state value is the set labeled "1" in Figure 107.7. This set of values results from the union of the output state of the assignment statement $p = L$ and the default program state corresponding to the output state of the if statement's join node.

```
while join   {p = ⊤, L = ⊤} ∩ {p = ⊥, L = ⊥} = {p = ⊤, L = ⊤}
```

Interpreting the while statement's decision node creates two states:

```
Test succeeds {p = T, L = T} ∩ {p = nonNULL, L = T} = {p = nonNULL, L = T}
Test fails     {p = T, L = T} ∩ {p = NULL, L = T} = {p = NULL, L = T}
```

Since **p** has value **nonNULL** before being dereferenced in the expression **p → next**, the dereference operation yields the value T which is bound to **p** in the assignment statement. At the if statement's join node, the union of two input states creates the output value.

```
if join  {p = nonNULL, L = T} ∩ {p = T, L = T} = {p = T, L = T}
```

The output statement of the if statement's join node is unioned with the output state of the assignment **p = L**, yielding the state labeled "2" at the while statement's join node. Since this state is identical to the previous state at this point, we have reached a fixpoint and the computation ceases. We can now check that both pointer dereferences (i.e., **p → v** and **p → next**) occurs in input state where **p** has the value **nonNULL**, so we know no **NULL**-value pointers are dereferenced.

Static checkers have difficulty dealing with pointer variables. General properties involving pointers are more easily verified at run time. Two such properties are freedom from memory access errors or **memory leaks**. A memory access error occurs when a reference containing a valid address for a block of storage which has already been freed is used to read from or write to the address. A memory leak occurs when storage is allocated but not freed before the last reference to it is lost. The following program illustrates these problems.

```
void leak(){
  node* q = new node; // the storage referenced by q is a
                      memory leak
}
void remove (node* p){
  delete p; //p is a reference to freed memory
}
void main(){
  node* p = new node;
  p→data = 3;
  remove(p);
  p→data = 4;
  leak();
}
```

In procedure remove, the storage referenced by **p** is freed, but **p** retains its value. When remove returns, **p** is dereferenced to store a value in one of the freed memory locations. When procedure leak is called, **q** is assigned the address of newly allocated storage. When leak returns, **q**'s storage is reclaimed but the storage it referenced remains allocated. Over time, the build-up of state caused by memory leaks can lead to program crashes.

Tools (e.g., Purify [Hasting and Joyce 1992]) redefine memory management routines to record information needed to check these properities. A table corresponding with a single bit for each byte of memory can be used to determine if a memory location is allocated or not. Accesses to unallocated memory are reported. Standard mark and sweep algorithms for **garbage collection** are modified to detect memory leaks. The mark phase recursively follows pointers from the stack and marks referenced heap locations. This phase is conservative because pointers cannot be distinguished from other data, and an integer with a value that appears to be a valid address will cause freed data to be erroneously marked. The sweep phase steps through the heap and reports blocks that are no longer referenced. By labeling each block with the return address of the the functions on the call stack, useful diagnostics about offending statments can be printed.

107.4.2 Specific Properties

Floyd [Floyd 1967] introduced assertional reasoning for sequential programs represented as flowcharts. Hoare formulated this as a logic for program test. A Hoare-triple has the form $\{P\}S\{Q\}$, where P and Q are assertions about program states, and S is a statement. This expression is interpreted as "If P, called the *precondition*, is true before executing S and S terminates normally, then Q, called the *postcondition*, will be true." This concept is called "partial" correctness because S's **termination** is not guaranteed.

107.4.2.1 Axioms and Rules of Inference

The assignment axiom schema:

$$\text{Assignment:} \quad \{P_y^x\}x = y\{P\}$$

defines the effect of the assignment statement on postcondition P. That is, if we want P to be true after executing the assignment statement $x = y$, then P_y^x, P with all free (i.e., unquantified) occurrences of x replaced by y must be true before executing the assignment. The assignment axiom is a schema that must be instantiated for individual assignment statements. For example, $\{y - 1 \geq 0\}\ x = y - 1\ \{x \geq 0\}$. The assignment axiom allows us to calculate an assertion which describes the set of input states for which the assignment statement will produce the desired result if it terminates. Thus, if we want $x \geq 0$ to hold in all program states after $x = y - 1$ terminates, $y \geq 1$ must hold in all states before this statement executes.

Results for verifying two statements are composed using the following rule of inference:

$$\text{Composition:} \quad \frac{\{P\}S_1\{Q\},\ \{Q\}S_2\{R\}}{\{P\}S_1; S_2\{R\}}$$

If the formula above the line (the antecedent) is true, then we may conclude that the statement below the line (the consequent) is true. The rule of composition allows us to combine the results of executing two statements to conclude that if P is true and the execution of S_1 followed by S_2 terminates normally, then R will be true. To reach this conclusion, the antecedent requires us to show that the postcondition of S_1 is the same as the precondition of S_2.

The postcondition of one statement is rarely identical to the precondition of another state, so we have rules of consequence to weaken the postcondition or strengthen the precondition of a statement. The rules of consequence build on predicate logic rules of inference.

$$\text{Rules of consequence:} \quad \frac{\{P\}S\{R\},\ R \to Q}{\{P\}S\{Q\}} \quad \frac{P \to R,\ \{R\}S\{Q\}}{\{P\}S\{Q\}}$$

Each programming language statement has a separate rule of inference.

$$\text{If statement}_1: \quad \frac{\{P \wedge B\}\ S\ \{Q\},\ P \wedge \sim B \to Q}{\{P\}\text{ if } B \text{ then } S\{Q\}}$$

$$\text{If statement}_2: \quad \frac{\{P \wedge B\}\ S_1\ \{Q\},\ \{P \wedge \sim B\}\ S_2\ \{Q\}}{\{P\}\text{ if } B \text{ then } S_1 \text{ else } S_2\ \{Q\}}$$

$$\text{While statement:} \quad \frac{\{P \wedge B\}\ S\ \{P\}}{\{P\}\text{ while } B \text{ do } S\ \{P \wedge \sim B\}}$$

A rule of inference captures the statement's semantics. For example, to conclude the consequent of If statement$_2$, we have to show that for each execution path through the statement that if we start execution with assertion P being true and execution of the path terminates normally, then assertion Q will be true. On one path, we assume P and B are true before S_1 is executed, and on the other path we assume P and $\sim B$ are true before S_2 is executed. Each of these paths appears in the left half of Figure 107.8. Defining a rule for the while statement is more difficult because the number of paths to verify is potentially infinite.

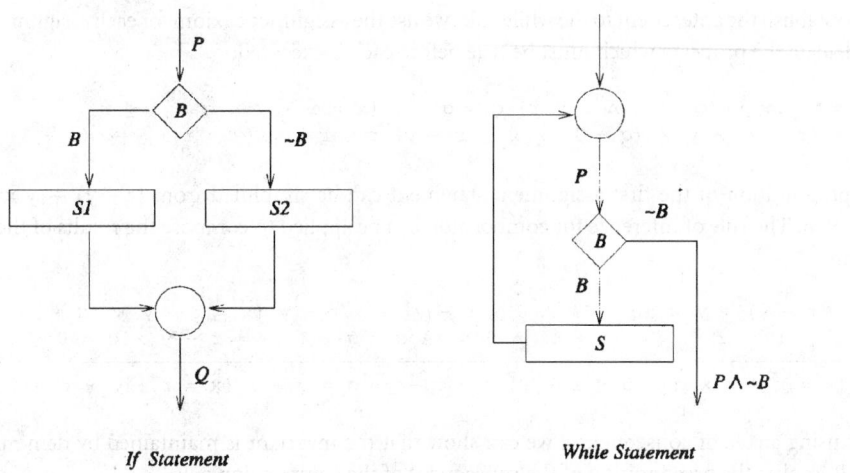

FIGURE 107.8   Flow of control for if and while statements.

The while statement rule of inference resorts to induction to solve this problem. We assume a property *P* (called the invariant) is true when the while statement begins execution and show that it is still true when *S* terminates normally. We conclude that *P* is true after zero or more executions of *S*, and that *B* must be false when the statement terminates.

107.4.2.2   Verifying a Small Program

Consider the following example of a Hoare-style proof of partial correctness of a program which uses repeated subtractions to compute the remainder (**r**) and quotient (**q**) obtained be dividing the integer **x** by the integer **y**:

```
{x ≥ 0 ∧ y > 0}
q = 0;
r = x;
while y ≤ r do {
   r = r - y;
   q = q + 1;
   }
{x = r + y* q ∧ 0 ≤ r ∧ r < y}
```

The postcondition characterizes the desired relationship between values in order for **r** and **q** to represent the remainder and quotient, respectively.

For the postcondition to be true, it must be so as a result of application of the while rule of inference. To apply the rule, we must identify the loop invariant (*P*) in the rule's antecedent. One way to do this is to "remove" ~ *B* from the postcondition and check if the remainder of the postcondition is invariant. Since *B* is **y** ≤ **r**, ~ *B* is **y** ¿ **r** or **r** ¡ **y**, and *P* is x = r + y × q ∧ 0 ≤ r. Hence the inference rule which could be applied would be:

$$\frac{\{x = r + y \times q \wedge 0 \le r \wedge r \ge y\} \quad r = r - y;\ q = q + 1 \quad \{x = r + y \times q \wedge 0 \le r\}}{\{x = r + y \times q \wedge 0 \le r\ \} \quad \text{while } r \ge y \text{ do } \{r = r - y\ ;\ q = q + 1\} \quad \{x = r + y \times q \wedge 0 \le r \wedge r < y\}}$$

To establish the antecedent to the while rule, we use the assignment axiom for each assignment statement to calculate the property which must be true before each is executed.

$$\{x = r + y \times (q + 1) \wedge 0 \leq r\} \; q = q + 1 \; \{x = r + y \times q \wedge 0 \leq r\}$$
$$\{x = (r - y) + y \times (q + 1) \wedge 0 \leq r - y\} \; r = r - y \; \{x = r + y \times (q + 1) \wedge 0 \leq r\}$$

The precondition of the first assignment statement can be simplified from $(r - y) + y \times (q + 1)$ to $r - y \times q$. The rule of inference for composition can be applied to compose the results of the assignment axioms:

$$\{x = (r - y) + y \times (q + 1) \wedge 0 \leq r - y\} \; r = r - y \; \{x = r + y \times (q + 1) \wedge 0 \leq r\},$$
$$\{x = r + y \times (q + 1) \wedge 0 \leq r\} \; q = q + 1 \; \{x = r + y \times q \wedge 0 \leq r\}$$

$$\overline{\{x = r - y \times q \wedge 0 \leq r - y\} \; r = r - y; \; q = q + 1 \; \{x = r + y \times q \wedge 0 \leq r\}}$$

Now, using a rule of consequence, we can show that the invariant is maintained by demonstrating that $P \wedge B$ implies the precondition of the consequent of the composition rule.

$$(x = r + y \times q \wedge 0 \leq r \wedge r \geq y) \rightarrow (x = r - y \times q \wedge 0 \leq r - y),$$
$$\{x = r - y \times q \wedge 0 \leq r - y\} \; r = r - y; \; q = q + 1 \; \{x = r + y \times q \wedge 0 \leq r\}$$

$$\overline{\{x = r + y \times q \wedge 0 \leq r \wedge r \geq y\} \; r = r - y; \; q = q + 1 \; \{x = r + y \times q \wedge 0 \leq r\}}$$

We observe that $0 \leq r - y$ is true because $r \geq y$. This establishes the antecedent of the while rule of inference.

Now we must determine whether or not the initialization steps in the program make the precondition of the while statement's consequent true. We use the assignment axiom twice to calculate the the property which must be true before each is executed.

$$\{x = x + y \times q \wedge 0 \leq x\} \; r = x \; \{x = r + y \times q \wedge 0 \leq r\}$$
$$\{x = x + y \times 0 \wedge 0 \leq x\} \; q = 0 \; \{x = x + y \times q \wedge 0 \leq x\}$$

After simplifying $x = x + y \times 0$ to true, we use the rule of inference for composition first to compose the results of the two assignment axioms:

$$\{0 \leq x\} \; q = 0 \; \{x = x + y \times q \wedge 0 \leq x\},$$
$$\{x = x + y \times q \wedge 0 \leq x\} \; r = x \; \{x = r + y \times q \wedge 0 \leq r\}$$

$$\overline{\{0 \leq x\} \; q = 0; \; r = x \; \{x = r + y \times q \wedge 0 \leq r\}}$$

and then to compose this result with that of the while rule of inference:

$$\{0 \leq x\} \; q = 0; \; r = x \; \{x = r + y \times q \wedge 0 \leq r\},$$
$$\{x = r + y \times q \wedge 0 \leq r\} \; \text{while } r \geq y \text{ do } \{ r = r - y; \; q = q + 1 \} \; \{x = r + y \times q \wedge 0 \leq r \wedge r < y\}$$

$$\overline{\{0 \leq x\} \; q = 0; \; r = x; \; \text{while } r \geq y \text{ do } \{ r = r - y; \; q = q + 1 \} \; \{x = r + y \times q \wedge 0 \leq r \wedge r < y\}}$$

We use a rule of consequence to show that the program's precondition is stronger than the property we have calculated and must be true before executing the program in order to make the program's postcondition true.

$$(x \geq 0 \wedge y > 0) \rightarrow 0 \leq x,$$
$$\{0 \leq x\}$$
$$q = 0; \; r = x; \; \text{while } r \geq y \text{ do } \{r = r - y; \; q = q + 1\}$$
$$\{x = r + y \times q \wedge 0 \leq r \wedge r < y\}$$

$$\overline{\{x \geq 0 \wedge y > 0\}}$$
$$q = 0; \; r = x; \; \text{while } r \geq y \text{ do } \{r = r - y; \; q = q + 1\}$$
$$\{x = r + y \times q \wedge 0 \leq r \wedge r < y\}$$

107.4.2.3 Program Termination

The stated precondition for the program $(x \geq 0 \wedge y \ \text{¿} \ 0)$ is more restrictive (i.e., describes a smaller set of program states) than the precondition $(x \geq 0)$ we calculated was necessary for the program to execute and produce a set of states satisfying its postcondition. The difference between these assertions highlights the difference between partial and total program correctness. For states satisfying the calculated precondition but not the original precondition (i.e., those in which $x \geq 0 \wedge y \leq 0$), the program would produce the desired result if it halted, but it does not. When values of y which are less than or equal to 0 are subtracted from r, the difference between r and y does not decrease, so the while statement fails to terminate.

To demonstrate that the while statement "while B do S" terminates, we show that B must eventually evaluate to false. To do this, we derive an expression from B whose value is bounded below by 0, and we show that on each path through S the value of the expression decreases. Since it has a lower bound, the expression cannot decrease infinitely, so the while statement must terminate. In our example program, we want to show that $r \geq y$ cannot remain true indefinitely. We can form a termination test expression by subtracting y from both sides of the while statement predicate to obtain $r - y \geq 0$. There is only a single path in S on which r is decremented by y. As long as y is positive, $r - y$ will decrease and the while statement will terminate. Thus, we add the assertion $y \ \text{¿} \ 0$ to the calculated assertion $x \geq 0$ to guarantee total correctness.

107.4.2.4 Advanced Language Features

107.4.2.4.1 Arrays

Proofs of programs manipulating scalar variables are relatively straightforward. However, to prove realistic programs, axioms and rules of inference must be devised for all language features. In this section, we discuss arrays and procedure calls, two features that complicate verifications.

Using the axiom of assignment to reason about assignments to variables which are array elements can lead to unsound reasoning. The following code fragment assigns the value 4 to a[i] and a[j] because the first assignment statement ensures that i and j have identical values. `i = j; a[i] = 3; a[j] = 4;` However, using the axioms and rules of inference introduced thus far, we can prove that no matter in what state the program begins execution (i.e., the precondition is true) this code fragment finishes execution with the postcondition a[i] ¿ a[j]. `{true} i = j; {3 < 4} a[i] = 3; {a[i] < 4} a[j] = 4; {a[i] < a[j]}`.

To avoid this problem, we need to consider an array as a function which maps its indices to values, and an assignment statement as an operation which assigns a new function to the array. For example, a[i] = 3 assigns a new function to the array a which is identical to the old function except that it maps i to 3. That is,

$$\alpha(a,i,x)[j] = \begin{cases} x & \text{when } i = j \\ a[j] & \text{when } i \neq j \end{cases}$$

Using this definition, we can work out the value of subscripted array expressions, e.g., $\alpha(\alpha(a, 3, x), 4, y)$ [3] $= \alpha(a, 3, x)$ [3] $= x$. Our new assignment axiom schema is

$$\text{Array assignment:} \quad \left\{ P^a_{\alpha(a,i,x)} \right\} a[i] = x \{P\}$$

With this new axiom and the previous rules of inference, we can reason safely about programs which alter arrays. `{α(a,j,4)[i] < α(a,j,4)[j]} a[j] = 4; {a[i] < a[j]}` We can simplify $\alpha(a, j, 4)[j]$ to **4** using the definition of α and continue our verification. `{α(α(a,i,3),j, 4)[i] < 4} a[i] = 3; {α(a,j,4)[i] < 4} {α(α(a,j,3),j,4)[j] < 4} i = j; {α(α(a,i,3),j,4)[i] < 4}` Simplifying $\alpha(\alpha(a, j, 3), \ j, 4)[j]$ yields the value 4. Thus there are no states in which the program begins execution (i.e., the precondition 4 < 4 is false) for which this code fragment finishes execution with the postcondition a[i] ¿ a[j].

107.4.2.4.2 *Procedure Invocations*

In verifications involving procedures, our goal is to verify a procedure's body once, and then use this result at each point at which the procedure is invoked. We have two new rules of inference for procedures: one rule handles the substitution of actual parameters for formal parameters, and the other rule relates the procedure's precondition and postcondition to the assertion which must be true after the procedure's invocation [Hoare 1971]. If all our parameters are passed by reference, we can use the following simplified rule of substitution:

$$\text{Substitution:} \quad \frac{\{R\}p(f)\{S\}}{\{R_{k'\ a}^{k\ x}\}p(a)\{S_{k'\ a}^{k\ x}\}}$$

where f and a are the lists of formal and actual parameters, respectively. The procedure's body may not reference nonlocal variables, and each variable in a must be unique. Symbols which are free in R and S but do not appear in the actual parameter list (i.e., k) are renamed. The rule's antecedent requires verification of the procedure's body once using the names of formal parameters.

A procedure's postcondition is rarely identical to the assertion which must be true after the call, since the procedure may be called from many different locations. Thus we need a rule similar to the rule of consequence to adapt the results of the procedure body to the different assertions needed to hold after invocations.

$$\text{Adaptation:} \quad \frac{\{R\}p(a)\{S\}}{\{\exists k\,(R \wedge \forall a(S \to T))\}p(a)\{T\}}$$

In this rule, the names of actual parameters have a different meaning in R than they do in S and T. The name of a parameter in R represents a value before the call, but the same name in S or T represents a value after the call. These values may be different because parameters are transmitted by reference and may be changed by the procedure's body. Names of actual parameters are free variables in R and universally quantified variables in S and T. Thus even if name appears in R and S or T, its meaning is different. Initial values of variables often appear in a procedure's precondition or postcondition, but not in a or T. These names are existentially quantified because some such value must exist.

By way of example, assume we have verified the body of a procedure **swap(x, y)** whose precondition is $\{x = x' \wedge y = y'\}$ and whose postcondition is $\{x = y' \wedge y = x'\}$, and we want to verify the following code fragment. {true} a = 1; b = 2; swap(a, b); {a = 2 ∧ b = 1} Having verified the body of swap, we can use the rule of substitution to replace swap's formal parameters with the actual parameters of the call.

$$\frac{\{x = x' \wedge y = y'\}\ \text{swap}(x, y)\ \{x = y' \wedge y = x'\}}{\{a = x' \wedge b = y'\}\ \text{swap}(a, b)\ \{a = y' \wedge b = x'\}}$$

Substitution's consequent is the antecedent of the rule of adaptation. When we apply adaptation we need to existentially quantify **x'** and **y'** (the initial values of **a** and **b** in the precondition), and universally quantify **a** and **b** (the values of the parameters after the call).

$$\frac{\{a = x' \wedge b = y'\}\ \text{swap}(a, b)\ \{a = y' \wedge b = x'\}}{\{\exists,\ x',\ y'\ (a = x' \wedge b = y' \wedge (\forall\ a,\ b,\ (a = y' \wedge b = x') \to (a = 2 \wedge b = 1)))\}}$$
$$\text{swap}(a, b)\ \{a = 2 \wedge b = 1\}$$

We can pick values for x' and y' (e.g., $x' = 1$ and $y' = 2$) to simplify the precondition of the adaptation rule's consequent.

$$(a = 1 \land b = 2 \land (\forall a,b, (a = 2 \land b = 1) \rightarrow (a = 2 \land b = 1)))$$

Clearly, this precondition is established by the sequence of assignment statements.

107.4.2.4.3   User-Defined Data Types

Modern programming languages provide special constructs such as classes to implement user-defined data types. These constructs are specifically designed to hide the representation of a value of the type from users who manipulate values of the type only through operations provided by the special constructs. Hoare [Hoare 1972] divided the verification of such programs into two parts.

1. Each operation's preconditions and postconditions are specified using values and operations from well-defined mathematical domains (e.g., sets or lists), and user-level code is verified with these assertions.
2. A representation mapping is defined to relate implementation-level values (e.g., arrays or linked representations) to user-level values. User-level variables in preconditions and postconditions are replaced by the corresponding mapped implementation-level variables, and the implementations of the operations are verified using the techniques described in the previous section.

Guttag et al. [1985] replaced model-oriented, user-level specifications with property-oriented specifications. Property-oriented specifications describe aspects of values in terms of properties they possess. In this approach, called algebraic specification, properties of operations of user-defined types are defined in terms of how they interact with each other.

Algebraic Specifications. Algebraic specifications have syntactic and semantic parts. The syntactic description, often referred to as the type's signature, describes the domains and ranges of the type's operations. For example, some operations on objects of type "stack of integer" are listed below.

```
estack    → Stack
push      Stack × integer → Stack
pop       Stack → Stack
top       Stack → natural
empty     Stack → Boolean
depth     Stack → natural
=         Stack × Stack → Boolean
```

Axioms describe the meanings of operators in terms of how they interact with one another. Axioms appear as equations; each left side contains a composition of operators manipulating implicitly universally quantified variables, and each right side contains a description of how the composition behaves in terms of the type's operators and simple "if-then-else" expressions. The axioms for the operations of type Stack appear below.

```
1.    pop(estack)          =    estack
2.    pop(push(S, X))      =    S
3.    top(estack)          =    0
4.    top(push(S, X))      =    X
5.    empty(estack)        =    true
6.    empty(push(S, X))    =    false
7.    depth(estack)        =    0
8.    depth(push(S, X))    =    depth(S) + 1
9.    T = estack           =    depth(T) = 0
10.   T = push(S, X)       =    top(T) = X ∧ pop(T) = S
```

where **S** and **T** are objects of types **Stack**, and **X** is an integer. Axiom 2 describes the value computed by pushing an arbitrary value on **Stack S** followed by popping the resulting **Stack** object as being equal to the original value of **S**. Pushing a value on a **Stack** object increases the depth of the object by one according to Axiom 8. Axiom 10 asserts that **Stacks T** and **push(S, X)** are equal if their respective top values (**top(T)** and **X**) and remaining values (**pop(T)** and **S**) are equal.

We can use equational reasoning, replacing a term with an equal term, to validate that the axioms behave as intended. For example, we could check that popping a nonempty **Stack** object decreases its depth by picking a particular **Stack** object (e.g., **push(estack, X)**) and reasoning equationally as follows:

Term	Axiom
`(~(push(estack, X) = estack)→(depth(pop(push(estack, X)))<depth(push(estack, X))))`	
`(~(depth(push(estack, X)) = 0)→(depth(pop(push(estack, X)))<depth(push(estack, X))))`	9
`(~(depth(estack) + 1 = 0)→(depth(pop(push(estack, X)))<depth(push(estack, X))))`	8
`(~(0 + 1 = 0)→(depth(pop(push(estack, X)))<depth(push(estack, X))))`	7
`true→(depth(pop(push(estack, X)))<depth(push(estack, X))))`	1≠0
`true→(depth(estack)<depth(push(estack, X))))`	2
`true→(0<depth(push(estack, X))))`	7
`true→(0<depth(estack)+1)`	8
`true→(0<0 + 1)`	0<1
`true→true`	

Axioms are inconsistent when an operation is overspecified. This occurs when two rules can be used to rewrite the same combination of arguments to different values. For example, if we added the following axiom:

`top(pop(push(S, X)))` = X to our previous axioms we would be able to rewrite the term **top(pop(push(estack, 5)))** to two different values.

`top(pop(push(estack, 5)))`	⇒	5	New axiom
`top(pop(push(estack, 5)))`	⇒ `top(estack)` ⇒	0	Axiom 2 then 3

Overspecification can be detected by a superposition algorithm [Knuth and Bendix 1970] which uses unification to detect overlapping axioms which produce different results.

Axioms are incomplete when an operation is underspecified (i.e., when no rule can be used to rewrite some combination of arguments). The specification of a type is *sufficiently complete*, if it assigns a value to each term of the type [Guttag and Horning 1978]. All **Stack** values can be built by a finite number of compositions of push operations on **estack** values, since any stack either is empty or is obtainable by pushing some element on some other stack. Operations **estack** and **push** are called *constructors*, and the remaining operations are called *defined* operations.

An algorithm exists for detecting underspecified operations [Huet and Hullot 1982]. The variables on the left side of each axiom must be unique, and a recursive test ensuring that all permutations of constructors may appear in the operation's argument positions must succeed. This test would succeed for Stack's pop operation for any of the following left sides of axioms:

Left sides	Reason for success
`{pop(estack), pop(push(S, X))}`	All constructors
`{pop(estack), pop(S)}`	S represents estack and push(S', X)
`{pop(S), pop(push(S, X))}`	S represents estack and push(S', X)
`{pop(S)}`	S represents estack and push(S', X)

Of course, it is much easier to write the right sides of axioms for some of these sets of left sides than for others.

The algorithm of Huet and Hullot works as follows. A set of *n*-tuples is formed from the *n* arguments which appear in each of the operation's axioms. The set of arguments in the tuple's first positions is constructed. The test fails if the set does contain either a variable or an instance of each constructor.

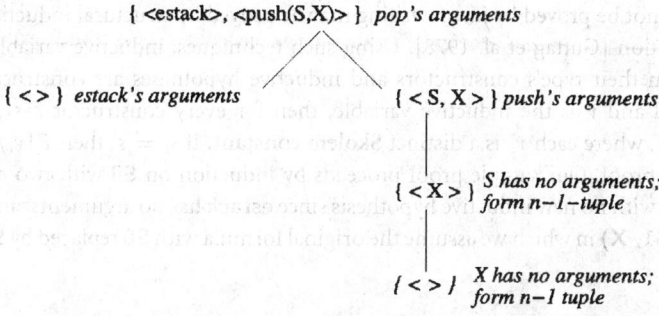

FIGURE 107.9   Verifying that **pop**'s axioms are not underspecified.

The *n*-tuple set is divided into subsets on the basis of which constructor appeared in the first position. Each of these sets is augmented by the set *n*-tuples with variables in the first position, since the variable could represent an instance of the particular constructor. Each tuple in the set for a constructor with *p* arguments is transformed as follows. If the first element of the tuple is the constructor, it is replaced by its *p* arguments forming a new $n - 1 + p$-tuple. If the first element of the tuple is a variable, it is replaced by *p* fresh variables forming a new $n - 1 + p$-tuple. The test is repeated on each new tuple subset, and succeeds for tuples of length zero.

Figure 107.9 shows the results of applying the test to the **Stack** axioms which define **pop**. The initial set of tuples is {¡estack¿, ¡push(S, X)¿}. The set of arguments in the first positions of these tuples contains an instance of each constructor, so we continue by dividing the set and constructing new tuples. The set of new tuples formed from the tuples with **estack** in the first position is {¡ ¿}, while the set formed from tuples with **push** in the first position is {¡S, X¿}. The test succeeds for the zero-length tuple. For {¡S, X¿}, the set of arguments in the first position contains a variable, so we form the new set of tuples {¡X¿}. Repeating the step once more leads to a set of zero-length tuples. If one of **push**'s arguments had not been a variable (e.g., **push(estack, X)**), the algorithm would fail because we could not determine a meaning for some terms (e.g., **pop(push(push(estack), 1), 2))**).

To validate sets of axioms, we change each equation of the form $t_1 = t_2$ into a rewrite rule of the form $t_1 \Rightarrow t_2$. A rewrite rule allows the replacement of an instance of t_1 with the corresponding instance of t_2, but it forbids replacement in the opposite direction. Orienting equations transforms an algebraic specification into a term rewriting system [Dershowitz and Jouannaud 1990] that supports automated verification and validation.

Two crucial properties must hold when equations are oriented. Two terms provably equal by equational reasoning should have a common third term to which both can be rewritten. This property is referred to as **confluence** or *Church–Rosser*. Also, there should be a finite number of rewriting steps that can be applied to a term. This property is referred to as *termination* or *Noetherianity*. To ensure the confluence of a constructor-based specification it is sufficient to avoid overspecification. To ensure sufficient completeness it is necessary, but not sufficient, to avoid underspecification. Although underspecification and overspecification can be checked, the termination of a rewrite system is undecidable [Dershowitz 1987].

Term rewriting allows us to verify that a property is true for all values of a type rather than just testing that the property holds for particular values. However, when we try to verify that popping a nonempty **Stack** object decreases its depth for an arbitrary **Stack** value **S0**, we quickly reach a point where no more rewriting can be performed.

```
(~(S0 = estack) → (depth(pop(S0)) < depth(S0)))      = true
(~(depth(S0) = 0)) → (depth(pop(S0)) < depth(S0))) = true
```

Equations that cannot be proved by just rewriting may be proved by structural induction [Burstall 1969] or data type induction [Guttag et al. 1978]. Using such techniques, inductive variables are replaced by terms derived from their type's constructors and inductive hypotheses are constructed. If F is a formula to be proved and v is the inductive variable, then for every constructor $c(s_1, \ldots, s_n)$ we prove $F[c(v_1, \ldots, v_n)/v]$, where each v_i is a distinct **Skolem constant**. If $s_i = s$, then $F[v_i/v]$ is an inductive hypothesis for the proof. Our sample proof proceeds by induction on **S0** with two cases: one with **S0** replaced by **estack** with no new inductive hypothesis since **estack** has no arguments, and another with **S0** replaced by **push(S1, X)** in which we assume the original formula with **S0** replaced by **S1** as the inductive hypothesis.

```
imply(negate(depth(S0) = 0))→(depth(pop(S0)) < depth(S0)))          =   true
```

```
Case S0 = estack
```

```
(~((depth(estack) = 0))→(depth(pop(estack)) < depth(estack)))       =   true
(~(0 = 0)  →  (depth(pop(estack)) < depth(estack)))                 =   true
(~(true)  →  (depth(pop(estack)) < depth(estack)))                  =   true
(false  →  (depth(pop(estack)) < depth(estack)))                    =   true
(false  →  (depth(estack) < depth(estack)))                         =   true
(false  →  0 < 0)                                                   =   true
(false  →  false)                                                   =   true
true                                                                =   true
```

```
Case S0 = push(S1, X)
Inductive hypothesis:
    (~(depth(S1) = 0)→(depth(pop(S1)) < depth(S1)))                 =   true
```

```
(~((depth(push(S1, X)) = 0))→(depth(pop(push(S1, X)))
                                   < depth(push(S1, X))))           =   true
(~((succ(depth(S1)) = 0))→(depth(pop(push(S1, X)))
                                   < depth(push(S1, X))))           =   true
(~(false)  →  (depth(pop(push(S1, X))) < depth(push(S1, X))))       =   true
(true  →  (depth(pop(push(S1, X))) < depth(push(S1, X))))           =   true
(true  →  (depth(S1) < succ(depth(S1))))                            =   true
(true  →  true)                                                     =   true
true                                                                =   true
```

107.4.2.4.4 *Verifying User-Level Programs*

Having defined type **Stack**, we can use the operations of type **Stack** in the preconditions and post-conditions of the procedures of the implementation. In the example below, identifier **s** represents each operation's implicit first formal parameter of type **Stack**. The specification of **Push** states that if the value of **s** before invocation is the term **s'** and **Push** terminates normally, then the value of **s** after **Push** will be equal to the term **push(s', x)**

```
class Stack {
private:
    int* v;
    int top;
public:
    void Push(int x) {
        /* pre: s = s'; post: s = push(s', x) */
        ...
    void Pop( );
        /* pre: ~empty(s') ∧ s = s'; post: s = pop(s') */
        ...
};
```

Using the rules of inference for procedure call, we can verify the following code fragment using the stated preconditions and postconditions.

```
{s = A} s.Push(x); s.Pop(); {s = A}
```

First we use the rule of adaptation, to relate **pop**'s precondition and postcondition to the program's postcondition.

$$\frac{\{\sim\text{empty}(s') \land s = s'\} \; \text{pop}(s) \; \{s = \text{pop}(s')\}}{\{\exists s' \; (\sim\text{empty}(s') \land s = s' \land (\forall s, s = \text{pop}(s') \rightarrow s = A))\} \; \text{pop}(s) \; \{s = A\}}$$

Picking **s' = push(A,x)** permits us to begin simplifying the precondition of the adaptation rule's consequent.

Term	Axiom
$(\sim\text{empty}(\text{push}(A, x)) \land s = \text{push}(A, x) \land (\forall s, s = \text{pop}(\text{push}(A, x)) \rightarrow s = A)) \Rightarrow$	2
$(\sim\text{empty}(\text{push}(A, x)) \land s = \text{push}(A, x) \land (\forall s, s = A \rightarrow s = A)) \Rightarrow$	6
$(\sim\text{false} \land s = \text{push}(A, x)) \Rightarrow s = \text{push}(A, x)$	

Since **s = push(A, x)** → **(∃ s' (∼empty(s') ∧ s = s' ∧ (∀s, s = pop(s') → s = A)))** we use a rule of consequence to conclude:

$$\frac{\{s = \text{push}(A, x) \rightarrow (\exists s' (\sim\text{empty}(s') \land s = s' \land (\forall s, s = \text{pop}(s'))), (\exists s' (\sim\text{empty}(s') \land s = s' \land (\forall s, s = \text{pop}(s') \rightarrow s = A)))\} \; \text{pop}(s) \; \{s = A\}}{\{s = \text{push}(A, x)\} \; \text{pop}(s) \; \{s = A\}}$$

Applying the rule of adaptation to the invocation of the push operation results in the following rule of inference.

$$\frac{\{s = s'\} \; \text{push}(s, x) \; \{s = \text{push}(s', x)\}}{\{\exists s' (s = s' \land (\forall s, s = \text{push}(s', x) \rightarrow (s = \text{push}(A, x))\} \; \text{push}(s, x) \; \{s = \text{push}(A, x)\}}$$

Picking **s' = A** permits us to simplify the previous precondition to **S = A**. Using rules of consequence and composition, we conclude:

$$\frac{\{s = A\} \; \text{push}(s, x) \; \{s = \text{push}(A, x)\}, \; \{s = \text{push}(A, x)\} \; \text{pop}(s) \; \{s = A\}}{\{s = A\} \; \text{push}(s, x); \; \text{pop}(s) \; \{s = A\}}$$

107.4.2.4.5 Verifying Implementation-Level Programs

In the second part of verifying implementations of user-defined data types, implementations manipulating concrete objects must satisfy preconditions and postconditions containing terms defined by the axioms. The implementation of the **Stack** operation **Push** is shown below:

```
void Push(int x) {
   /* pre: s = s'; post: s = push(s',x) */
   s.top = s.top + 1;
   s.v[s.top] = x;
}
```

Hoare [1972] introduced representation mappings to map implementation-level objects to their corresponding user-level objects. In the implementation of type **Stack**, an instance **s** of type **Stack** consisted of an array of integers **s.v** and an integer **s.top** indicating the topmost value. To verify the correctness of an implementation of type **Stack** we define a representation mapping $\mathcal{A}$ which maps an array and an integer to its corresponding user-level value.

```
1.   A(s.v, 0)              =   estack
2.   A(s.v, s.top + 1)  =   push(A(s.v, s.top), s.v[s.top + 1])
```

We replace instances of the user-level value **s** with corresponding instances of mapped implementation-level values. The proof obligation for **Push** is $\{\mathcal{A}(\text{s.v, s.top}) = \text{s'}\}$ s.top = s.top + 1; s.v[s.top] = x; $\{\mathcal{A}(\text{s.v, s.top}) = \text{push(s', x)}\}$ After using axioms of assignment (for both scalar and array values) and composition, the final step in the verification is an application of a rule of consequence.

$$\mathcal{A}(\text{s.v, s.top}) = \text{s'} \rightarrow (\mathcal{A}(\alpha(\text{s.v, s.top + 1, x}), \text{s.top + 1}) = \text{push(s', x)}),$$
$$\{\mathcal{A}(\alpha(\text{s.v, s.top + 1, x}), \text{s.top + 1}) = \text{push(s', x)}\}$$
$$\text{s.top = s.top + 1; s.v[s.top] = x; } \{\mathcal{A}(\text{s.v, s.top}) = \text{push(s', x)}\}$$

$\{\mathcal{A}(\text{s.v, s.top}) = \text{s'}\}$ s.top = s.top + 1; s.v[s.top] = x; $\{\mathcal{A}(\text{s.v, s.top}) = \text{push(s', x)}\}$

To show that the antecedent is true, we need to axiomatize the array assignment and subscript operations:

```
1. newarray[J] = 0
2. α(A,I,X)[J] = (if I = J then X else A[J])
```

We continue using term rewriting:

Term	Axiom
$\mathcal{A}(\alpha(\text{s.v, s.top + 1, x}), \text{s.top + 1}) \Rightarrow$	Map 2
push($\mathcal{A}(\alpha(\text{s.v, s.top + 1, x}), \text{s.top})$,	
$\quad$ α(s.v, s.top + 1,x)[s.top + 1]) $\Rightarrow$	Array 2
push($\mathcal{A}(\alpha(\text{s.v, s.top + 1, x}), \text{s.top})$,	
$\quad\quad$ (if s.top + 1 = s.top + 1 then x else s.v[s.top + 1])) $\Rightarrow$	x = x
push($\mathcal{A}(\alpha(\text{s.v, s.top + 1, x}), \text{s.top})$, x)	

At this point, we need to reduce $\mathcal{A}(\alpha(\textbf{s.v, s.top + 1, x}), \textbf{s.top})$ to $\mathcal{A}(\textbf{s.v, s.top})$ to achieve equality. Since the representation mapping only maps values **s.v[i]** for values of i in the range $1 \leq i \leq$ **s.top**, we can reach this conclusion by proving the following theorem.

Term
(i < s.top + 1 → (α(s.v, s.top + 1,x)[i] = s.v[i]))
(i < s.top + 1 → (if s.top + 1 = i then x else s.v[i]) = s.v[i])
(i < s.top + 1 → (if s.top + 1 = i then x = s.v[i] else s.v[i] = s.v[i]))
(i < s.top + 1 → (if s.top + 1 = i then x = s.v[i] else true))
(i < s.top + 1 ∧ s.top + 1 = i → x = s.v[i]) ∧
$\quad\quad\quad\quad\quad\quad\quad\quad$ (i < s.top + 1 ∧ ∼(s.top + 1 = i) → true)
(false → x = s.v[i]) ∧ (i < s.top + 1 ∧ ∼(s.top + 1 = i) → true)
true ∧ (i < s.top + 1 ∧ ∼(s.top + 1 = i) → true)
true ∧ true

107.5   Current Status

General properties of programs, particularly that variables are initialized before they are used and that no invalid memory references or memory leaks occur, are easy to check automatically. However, verification of the most problem-specific properties is still carried out manually by inspections. As a result, unqualified guarantees about properties cannot be made because the software artifacts inspected may contain deficiencies or the inspectors may not have been thorough enough to find obscure failures.

Verifying specific properties of programs has proved too difficult a task for the average software developer. Program proofs are often more detailed than the programs being verified. When proofs are done manually, they are subject to the same human fallibilities that plague inspections. Automated support for theorem proving includes verification condition generators which apply rules of inference to produce the set of theorems which need to be proved manually by rules of consequence, proof checkers which check that steps in a proof are justified by lemmas from an existing library, and deductive systems which search for proofs by means of simplifications (like term rewriting) and heuristics for generating inductive proofs of necessary lemmas. Theorem provers eliminate errors of omission, but they require proofs of many low-level, uninteresting lemmas. However, skilled users have used theorem provers to verify complex problems such as a Byzantine fault-tolerant algorithm for synchronizing clocks in replicated computers [Rushby and von Henke 1993]. Such proofs have generally been carried out only for safety-critical applications because even automated proofs require highly skilled experts who know how to use a theorem prover and understand the application domain.

Model checking has proven successful in the design of hardware; it has been used to find bugs in pipelined microprocessors [Burch and Dill 1994] and cache coherence protocols [Clarke et al. 1986]. More recently it has been used to analyze software artifacts, e.g., software architecture designs [Allen and Garlan 1994] and editors [Jackson and Damon 1996], and distributed file system cache coherence protocols [Wing and Vaziri-Farahani 1995]. Software developers may be more likely to understand a proof technique like model checking, which is based on search and which produces counter-examples when proofs fail, than a technique based inductive theorem proving. Model checking an abstraction of a system rather than the system itself raises the level at which we apply formal verification. The key to success in these endeavors is creating an appropriate abstraction of a system so that results obtained from analyzing the abstraction also apply to the system.

Defining Terms

Abstract interpretation: The abstract interpretation of a program is a symbolic interpretation of the program using abstract values rather than actual values. Abstract values represent sets or ranges of actual values, and operators are redefined to manipulate abstract values. At program decision points, separate copies of the program state are created, corresponding to the true and false outcomes of the predicate, and computation continues along each path with the different states. At program join points, values of the states computed along different computation paths are unioned to produce a single state. The union operation for joins must reduce the number of distinct values in the program state in order to compute a fixpoint approximation of the state.

Confluence: A term-rewriting system is confluent if a term t can be rewritten into different forms t_1 and t_2, then we can prove by rewriting that $t_1 = t_2$. For example, the term **top(push(pop(push (estack, 1)), 2)** can be rewritten as either **2** using stack axiom 4 or as **top(push(estack, 2))** using stack axiom 2. We can show these two terms are equal by applying stack axiom 2 to the later term.

Garbage collection: Programming languages that do not require users to explicitly free unneeded storage reclaim this storage through a process called garbage collection. Storage is identified as unneeded (i.e., garbage) when there are no pointers to it. Determining if pointers reference a storage location can be accomplished by reference counting (i.e., explicitly keeping track of the number of pointers to the location) or tracing (i.e., starting with variables on the stack, mark heap locations they reference, and continue marking other heap locations reachable from marked heap locations). The collector either copies reachable values to new storage locations to create a large unmarked region or consolidates adjacent unmarked regions and creates a list of free storage.

Hazardous state: A hazardous state is one in which the occurrence of certain events would lead to a mishap.

Least fixpoint: A fixpoint of a function $f : T \rightarrow T$ is an element $t \in T$ such that $f(t) = t$. A least fixpoint of f is the least element of the set of all fixpoints of f.

Memory leak: A memory leak occurs when storage is allocated but not freed before the last reference to it is lost. Memory leaks may cause long-running programs to crash when storage allocation requests fail because of insufficient memory. The storage allocation request triggering the program crash may not be part of a memory leak, so identifying the cause of a memory leak is difficult.

Partial correctness: A program which meets its specification for all specified input values for which it terminates is called partially correct. A program which is partially correct and which terminates for all its inputs is said to be totally correct.

Reachability analysis: In reachability analysis, a graph representing the states and state-transitions of a system is constructed and exhaustively searched to determine if states with particular properties are reachable.

Skolem constant: An automated theorem prover simplifies a formula by replacing a universally quantified variable with a symbolic constant (called a Skolem constant) which represents an arbitrary value of the same type as the variable. For example, if x is a natural number, to prove $(\forall x, x > 0)$, picking a particular natural number to substitute for x might make the formula either true or false. Instead, we pick an arbitrary constant c and try to prove the quantifier-free formula $c > 0$.

Software inspections: Software artifacts are usually verified by people using informal analysis techniques called *inspections*. In inspections, teams of software developers either follow sequences of state changes or possible execution paths resulting from a particular series of events or inputs, or, using a checklist of potential errors, determine if similar errors are present in the artifact.

Temporal logic: A temporal logic is a propositional logic with additional temporal operators to express concepts such as a formula will always be true in the future, or a formula will eventually be true in the future. The value of a temporal logic formula is defined with respect to a finite-state model. If formula f is true in state s of model M, we write $M, s \models f$. A formula f is true for the model if it is true in the model's start state. Temporal logic allows reasoning about state changes rather than just the function computed by the program.

Termination: To demonstrate that the while statement "while B do S" terminates, we show that B must eventually evaluate to false. To do this, we derive an expression from B whose value is bounded below by 0, and we show that on each path through S the value of the expression decreases.

References

Allen, R. and Garlan, D. 1994. Formalizing architectural connection, pp. 71–80. In *Proc. 16th Int. Conf. Software Eng.*

Burch, J. R. and Dill, D. L. 1994. Automatic verification of pipelined microprocessor control. In *Lecture Notes in Computer Science* 818, D. Dill, ed., pp. 68–80. Springer–Verlag.

Burstall, R. 1969. Proving properties of programs by structural induction. *Comput. J.* 12(1):41–48.

Clarke, E., Emerson, E., and Sistla, A. 1986. Automatic verification of finite state concurrent systems using temporal logic specifications. *ACM Trans. Program. Lang. Syst.* 8(2):244–263.

Clarke, E. M., Grumberg, O., Hiraishi, H., Jha, S., Long, D. E., McMillan, K. L., and Ness, L. A. 1990. Verification of the futurebus+ cache coherence protocol, pp. 15–30. In *Proc. 11th Int. Symp. Comput. Hardware Description Lang. Appl.* L. Claesen, ed., North-Holland, Amsterdam.

Cousot, P. and Cousot, R. 1976. Static determination of dynamic properties of programs. In *Proc. "Colloque sur la Programmation."*

Cousot, P. and Cousot, R. 1977. Static determination of dynamic properties of generalized type unions. *SIGPLAN Notices* 12(3):77–94.

Dershowitz, N. 1987. Termination of rewriting. *J. Symb. Comput.* 3:69–116.

Dershowitz, N. and Jouannaud, J. 1990. Rewrite systems. In *Handbook of Theoretical Computer Science B: Formal Methods and Semantics*, J. van Leeuwen, ed. Ch. 6, pp. 243–320. North Holland, Amsterdam.

Fagan, M. E. 1976. Design and code inspections to reduce errors in program development. *IBM Syst. J.* 15(3):182–211.

Floyd, R. W. 1967. Assigning meaning to programs. *Symp. Appl. Math.* 19:19–32.

Ghezzi, C., Jazayeri, M., and Mandrioli, D. 1991. *Fundamentals of Software Engineering.* Prentice–Hall, Englewood Cliffs, NJ.

Guttag, J. V. and Horning, J. J. 1978. The algebraic specification of abstract data types. *Acta Informatica* 10:27–52.

Guttag, J. V., Horning, J. J. and Wing, J. M. 1985. The Larch family of specification languages. *IEEE Software* 2(5):24–36.

Guttag, J. V., Horowitz, E., and Musser, D. 1978. Abstract data types and software validation. *Commun. ACM* 21:1048–1064.

Harel, D., Lachover, H., Naamad, A., Pnueli, A., Politi, M., Sherman, R., Shtull-Trauring, A., and Trakhtenbrot, M. 1990. Statemate: a working environment for the development of complex reactive systems. *IEEE Trans. Software Eng.* 16(4):403–414.

Hastings, R. and Joyce, R. 1992. Purify: fast detection of memory leaks and access errors. In *Proc. Winter 1992 USENIX Conf.*, pp. 125–136.

Heitmeyer, C., Labaw, B., and Kiskis, D. 1995. Consistency checking of scr-style requirements specifications. In *Proc. RE' 95 Int. Symp. Req. Eng.*

Heninger, K. 1980. Specifying software requirements for complex systems: new techniques and their applications. *IEEE Trans. Software Eng.* SE-6(1):2–12.

Hoare, C. A. R. 1971. Procedures and parameters, an axiomatic approach. In *Symposium on the Semantics of Algorithmic Languages*, E. Engler, ed., pp. 102–116. Springer–Verlag.

Hoare, C. A. R. 1972. Proof of correctness of data representations. *Acta Inf.* 1(4):271–281.

Huet, G. and Hullot, J.-M. 1982. Proofs by induction in equational theories with constructors. *JCSS* 25(1):239–266.

Jackson, D. and Damon, C. A. 1996. Elements of style: analyzing a software design feature with a counterexample detector, pp. 239–249. In *Proc. 1996 Int. Symp. Software Test. and Anal. (ISSTA).*

Knuth, D. E. and Bendix, P. B. 1970. *Simple Word Problems in Universal Algebras*, pp. 263–297. Pergamon, Oxford, U.K.

Leveson, N. G., Heimdahl, M. P. E., Hildreth, H., and Reese, J. D. 1994. Requirements specification for process-control systems. *IEEE Trans. Software Eng.* 20(9):684–706.

McMillan, K. L. 1993. *Symbolic Model Checking.* Kluwer Academic Publishers, Boston, MA.

Pneuli, A. 1981. A temporal logic of concurrent programs. *Theor. Comput. Sci.* 13:45–60.

Quielle, J. P. and Sifakis, J. 1981. Specification and verification of concurrent systems in cesar. In *Proc. 5th Int. Symp. Program.*

Ratan, V., Partridge, K., Reese, J., and Leveson, N. 1996. Safety analysis tools for requirements specifications. In *Proc. 11th Conf. Comput. Assurance.*

Rushby, J. 1993. *Formal Methods and the Certification of Critical Systems.* SRI International, Palo Alto, CA. Tech Rep.

Rushby, J. M. and von Henke, F. 1993. Formal verification of algorithms for critical systems. *IEEE Trans. Software Eng.* 19(1):13–23.

Wing, J. and Vaziri-Farahani, M. 1995. Model checking software systems: a case study, pp. 128–139. In *Proc. 3rd Symp. Found. Software Eng.*

Young, M. and Taylor, R. N. 1989. Rethinking the taxonomy of fault detection techniques, pp. 53–62. In *Proc. 11th Int. Conf. Software Eng.*

Further Information

The monthly journals *IEEE Transactions on Software Engineering* and *ACM Transactions on Software Engineering and Methodology* contain articles on software verification. Papers on this topic are frequently presented at the International Conference on Software Engineering, the ACM's Foundations of Software Engineering, the ACM's International Conference on Software Analysis and Testing, and the IEEE Conference on COMPuter ASSurance (COMPASS).

Two recent books on the analysis of software systems are:

Leveson, N. G. 1995. *Safeware: System Safety and Computers*. Addison–Wesley, Reading, MA.

Rushby, J. 1995. *Formal Methods and the Certification of Critical Systems*. Cambridge University Press, Cambridge, U.K.

Readers interested in automated analysis tools should investigate both automated theorem provers and model checkers. Representative theorem provers include the following:

PVS (Prototype Verification System) is a theorem prover based on classical typed higher-order logic developed at the SRI International Computer Science Laboratory.

EVES unites the Verdi specification language based on set theory and an automated deduction system, called NEVER. This system is available from Mark Saaltink and Dan Craigen of ORA, Ottawa, Ontario, Canada.

LP, the Larch Prover, is an interactive theorem proving system for multisorted first-order logic. It was developed by Stephen Garland and John Guttag at the MIT Laboratory for Computer Science, Cambridge, MA.

Interesting model checkers include:

HyTech (The Cornell HYbrid TECHnology Tool) computes the condition under which a linear hybrid system satisfies a temporal requirement. Hybrid systems are specified as collections of automata with discrete and continuous components.

The SMV (Symbolic Model Verifier) model checker verifies formulas written in a propositional branching-time temporal logic. It is available from Ed Clarke at Carnegie Mellon University, Pittsburgh, PA.

Murphi is a symbolic model checker developed by David Dill at Stanford University, Stanford, CA.

108

Development Strategies and Project Management

108.1	Development Strategies **108**-1	
	The Linear Sequential Model • The Prototyping Model • The RAD Model • Evolutionary Software Process Models	
108.2	The Management Spectrum **108**-11	
	People • The Problem • The Process	
108.3	Software Project Management **108**-14	
	Measurement and Metrics • Project Estimating • Risk Analysis • Scheduling • Tracking and Control	
108.4	Software Quality Assurance...................... **108**-17	
108.5	Software Configuration Management **108**-18	
108.6	Summary **108**-18	

Roger S. Pressman

R.S. Pressman & Associates, Inc.

Successful planning, control, and tracking of a software project is accomplished when a project manager defines an effective development strategy. Once the strategy has been established, software project management commences. The intent of this chapter is to: (1) describe the generic development strategies that are available to software project teams, and (2) present an overview of the tasks required to perform good software project management.

108.1 Development Strategies

A development strategy for **software engineering** integrates a **process model** and the technical methods and tools that populate the model. A process model for software engineering is chosen on the basis of the nature of the project and application, the methods and tools to be used, and the controls and deliverables that are required. Four classes of process models have been widely discussed (and debated). A brief overview of each is presented in the sections that follow.

All software development can be characterized as a problem-solving loop (Figure 108.1) in which four distinct stages are encountered: status quo, problem definition, technical development, and solution integration [Raccoon 1995]. Status quo represents state of the project; problem definition identifies the specific problem to be solved; technical development solves the problem through the application of some technology; and solution integration delivers the results (e.g., documents, programs, data, new business function, new product) to those who requested the solution in the first place.

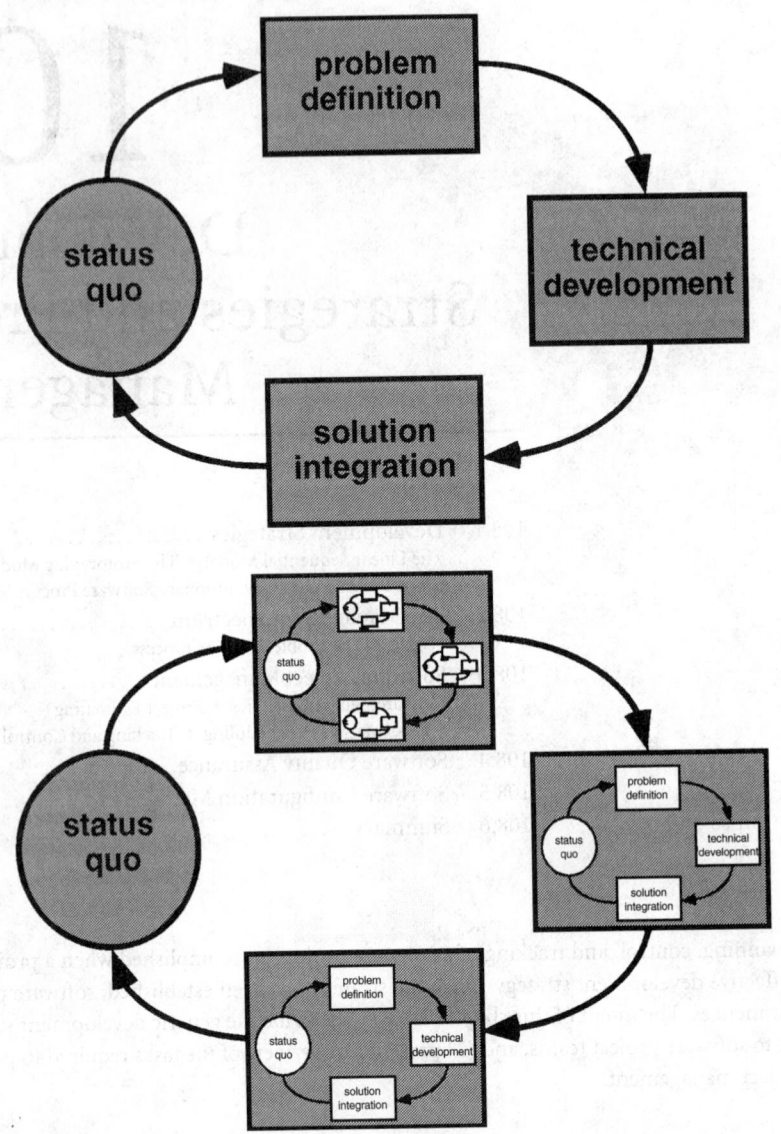

FIGURE 108.1   The phases of a problem-solving loop [Raccoon 1995].

The problem-solving loop described above applies to software engineering work at many different levels of resolution. It can be used at the macro level when the entire application is considered, at a middle level when program components are being engineered, and even at the line of code level.

In the sections that follow, a variety of different process models for software engineering are discussed. Each represents an attempt to bring order to an inherently chaotic activity. It is important to remember that each of the models has been characterized in a way that (one hopes) assists in the control and coordination of a real software project. Each represents a different development strategy, and yet, at their core, all of the models exhibit characteristics of the problem-solving loop described above.

108.1.1   The Linear Sequential Model

Figure 108.2 illustrates the **linear sequential model** for software engineering. Sometimes called the "**classic life cycle**" or the "waterfall model," the linear sequential model demands a systematic, sequential

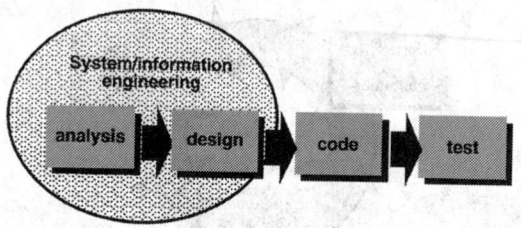

FIGURE 108.2 The linear sequential model.

approach to software development that begins at the system level and progresses through analysis, design, coding, **testing**, and **maintenance**. Modeled after the conventional engineering cycle, the linear sequential model encompasses the following activities: system/information engineering and modeling, software **requirements analysis**, design, code generation, and testing and maintenance/**reengineering**.

The linear sequential model is the oldest and the most widely used development strategy. However, criticism of the paradigm has caused even active supporters to question its efficacy [Hanna 1995]. Among the problems that are sometimes encountered when the linear sequential model is applied are the following:

1. Real projects rarely follow the sequential flow that the model proposes. Although the linear model can accommodate iteration, it does so indirectly. As a result, changes can cause confusion as the project team proceeds.
2. It is often difficult for the customer to state all requirements explicitly. The linear sequential model requires this and has difficulty accommodating the natural uncertainty that exists at the beginning of many projects.
3. The customer must have patience. A working version of the program(s) will not be available until late in the project time-span. A major blunder, if undetected until the working program is reviewed, can be disastrous.

In an interesting analysis of actual projects [Bradac et al. 1994], Bradac found that the linear nature of the classic life cycle leads to "blocking states" in which some project team members must wait for other members of the team to complete dependent tasks. In fact, the time spent waiting can exceed the time spent on productive work! The blocking state tends to be more prevalent at the beginning and end of a linear sequential process.

Each of these problems is real. However, the linear development strategy has a definite and important place in software engineering work. It provides a template into which methods for analysis, design, coding, testing, and maintenance can be placed.

108.1.2 The Prototyping Model

Often, a customer defines a set of general objectives for software but does not identify detailed input, processing, or output requirements. In other cases, the developer may be unsure of the efficiency of an algorithm, the adaptability of an operating system, or the form that human–machine interaction should take. In these, and many other situations, a **prototyping** paradigm may offer the best approach.

The prototyping strategy (Figure 108.3) begins with gathering requirements. Developer and customer meet and define the overall objectives for the software, identify whatever requirements are known, and outline areas where further definition is mandatory. A "quick design" then occurs. The quick design focuses on a representation of those aspects of the software that will be visible to the customer/user (e.g., input approaches and output formats). The quick design leads to construction of a prototype. The prototype is evaluated by the customer/user and is used to refine requirements for the software to be developed.

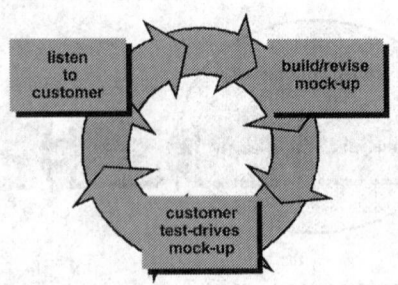

FIGURE 108.3 The prototyping paradigm.

Iteration occurs as the prototype is tuned to satisfy the needs of the customer while enabling the developer to better understand what needs to be done.

Ideally, the prototype serves as a mechanism for identifying software requirements. If a working prototype is built, the developer attempts to make use of existing program fragments or applies tools (e.g., report generators, window managers, etc.) that enable working programs to be generated quickly.

Both customers and developers like the prototyping paradigm. Users get a feel for the actual system and developers get to build something immediately. Yet, prototyping can also be problematic for the following reasons:

1. The customer sees what appears to be a working version of the software, unaware that the prototype is held together "with chewing gum and baling wire," unaware that in the rush to get it working we have not considered overall software quality or long-term maintainability. When informed that the product must be rebuilt so that high levels of quality can be maintained, the customer cries foul and demands that "a few fixes" be applied to make the prototype a working product. Too often, software development management relents.

2. The developer often makes implementation compromises in order to get a prototype working quickly. An inappropriate operating system or programming language may be used simply because it is available and known; an inefficient algorithm may be implemented simply to demonstrate capability. After a time, the developer may become familiar with these choices and forget all the reasons why they were inappropriate. The less-than-ideal choice has now become an integral part of the system.

Although problems can occur, prototyping can be an effective paradigm for software engineering. The key is to define the rules of the game at the beginning; that is, the customer and developer must both agree that the prototype is built to serve as a mechanism for defining requirements. It is then discarded (at least in part) and the actual software is engineered with an eye toward quality and maintainability.

108.1.3   The RAD Model

Rapid Application Development (RAD) is a linear sequential software development process model that emphasizes an extremely short development cycle. The RAD model is a "high speed" adaptation of the linear sequential model in which rapid development is achieved by using a component-based construction approach. If requirements are well understood and project scope is constrained,* the RAD process enables a development team to create a "fully functional system" within very short time periods (e.g., 60–90 days)

---

*These conditions are by no means guaranteed. In fact, many software projects have poorly defined requirements at the start. In such cases, prototyping or evolutionary approaches are much better process options.

[Martin 1991]. Used primarily for information systems applications, the RAD approach encompasses the following phases [Kerr and Hunter 1994]:

Business modeling. The information flow among business functions is modeled in a way that answers the following questions: What information drives the business process? What information is generated? Who generates it? Where does the information go? Who processes it?

Data modeling. The information flow defined as part of the business modeling phase is refined into a set of data objects that are needed to support the business. The characteristics (called attributes) of each object are identified and the relationships between these objects are defined.

Process modeling. The data objects defined in the data modeling phase are transformed to achieve the information flow necessary to implement a business function. Processing descriptions are created for adding, modifying, deleting, or retrieving a data object.

Application generation. RAD assumes the use of fourth-generation techniques (Fourth Generation Techniques subsection of the following section). Rather than creating software using conventional third-generation programming languages, the RAD process works to reuse existing program components (when possible) or create reusable components (when necessary). In all cases, automated tools are used to facilitate construction of the software.

Testing and turnover. Since the RAD process emphasizes reuse, many of the program components have already been tested. This reduces overall testing time. However, new components must be tested and all interfaces must be fully exercised.

The RAD process model is illustrated in Figure 108.4. Obviously, the time constraints imposed on a RAD project demand "scalable scope" [Kerr and Hunter 1994]. If a business application can be modularized in a way that enables each major function to be completed in less than three months (using the approach described above), it is a candidate for RAD. Each major function can be addressed by a separate RAD team and then integrated to form a whole.

Like all process models, the RAD approach has drawbacks [Butler 1994]:

- For large, but scalable, projects, RAD requires sufficient human resources to create the right number of RAD teams.

- RAD requires developers and customers who are committed to the rapid-fire activities necessary to get a system complete in a much-abbreviated time frame. If commitment is lacking from either constituency, RAD projects will fail.

- Not all types of applications are appropriate for RAD. If a system cannot be properly modularized, building the components necessary for RAD will be problematic. If high performance is an issue, and performance is to be achieved through tuning the interfaces to system components, the RAD approach may not work.

- RAD is not appropriate when development technical risks are high. This occurs when a new application makes heavy use of new technology or when the new software requires a high degree of interoperability with existing computer programs.

RAD emphasizes the development of reusable program components. Reusability is the cornerstone of object technologies and is encountered in the component assembly strategy discussed later in this chapter.

108.1.4 Evolutionary Software Process Models

There is growing recognition that software, like all complex systems, evolves over a period of time [Gilb 1988]. Business and product requirements often change as development proceeds, making a straight line path to an endproduct unrealistic; tight market deadlines make completion of a comprehensive software product impossible, but a limited version must be introduced to meet competitive or business pressure; a set of core product or system requirements is well understood, but the details of product or system extensions have yet to be defined. In these and similar situations, software engineers need a development strategy that has been explicitly designed to accommodate a product that evolves over time.

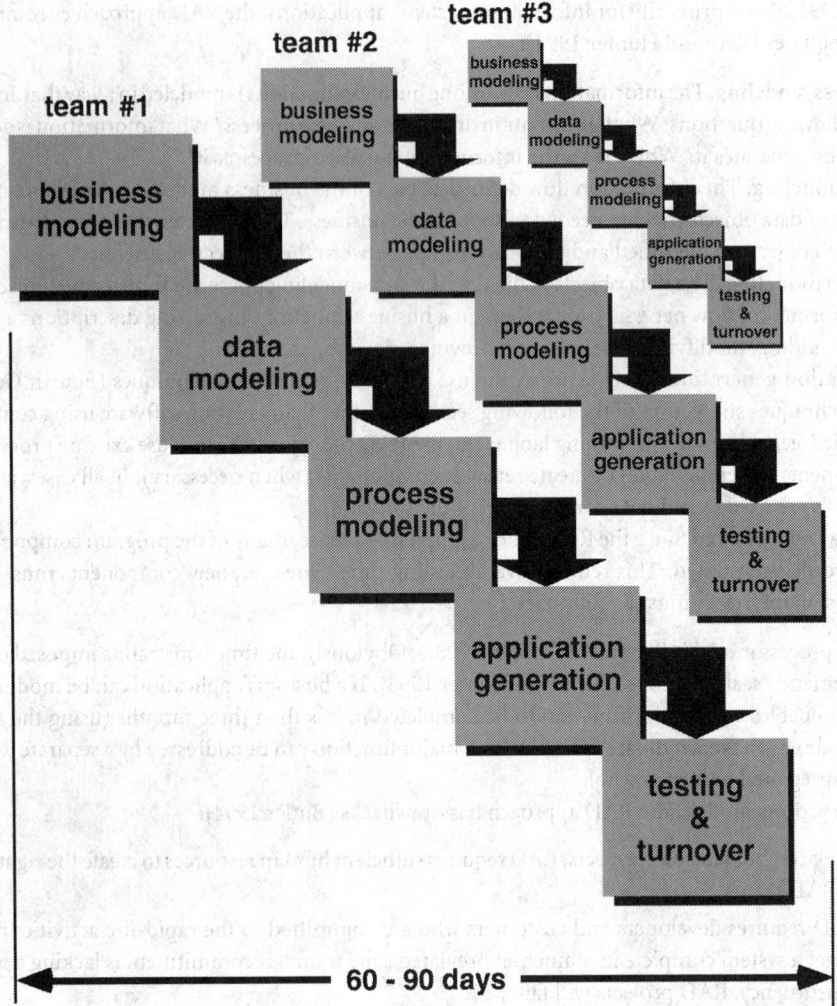

FIGURE 108.4 The RAD model.

Evolutionary models meet this need. They are characterized in a manner that enables software engineers to develop increasingly more complete versions of the software in an iterative manner.

108.1.4.1 The Incremental Model

The **incremental model** combines elements of the linear sequential model (applied repetitively) with the iterative philosophy of prototyping. Referring to Figure 108.5, the incremental model applies linear sequences in a staggered fashion as calendar time progresses. Each linear sequence produces a deliverable "increment" of the software [McDermid and Rook 1993]. For example, wordprocessing software developed using the incremental paradigm might deliver basic file management, editing, and document production functions in the first increment; more sophisticated editing and document production capabilities in the second increment; spelling and grammar checking in the third increment; and advanced page layout capability in the fourth increment. It should be noted that the process flow for any increment can incorporate the prototyping paradigm.

When an incremental model is used, the first increment is often a *core product*; that is, basic requirements are addressed, but many supplementary features (some known, others unknown) remain undelivered. The core product is used by the customer (or undergoes detailed review). As a result of use and/or evaluation,

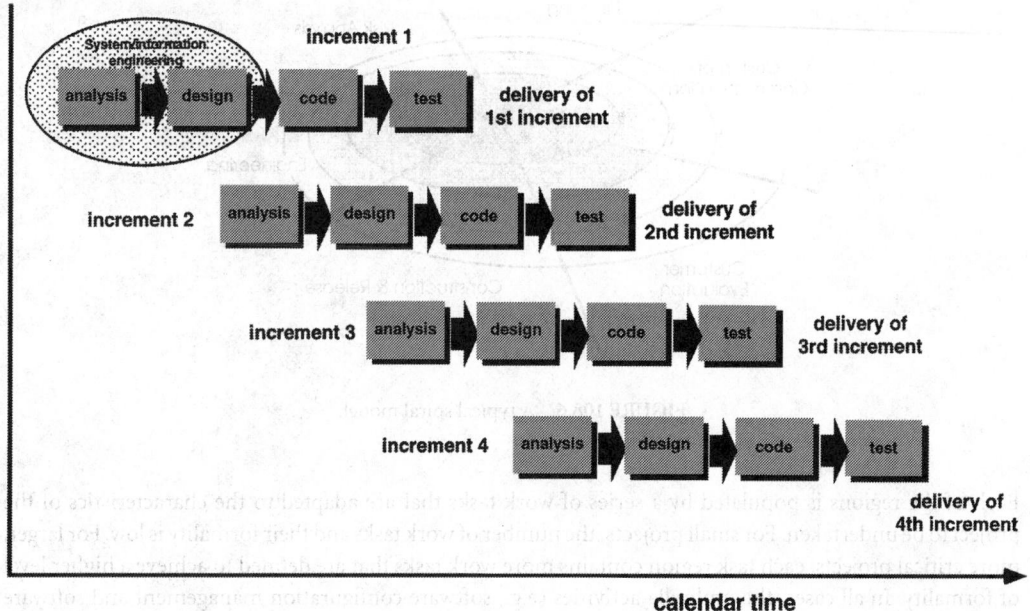

FIGURE 108.5 The incremental model.

a plan is developed for the next increment. The plan addresses the modification of the core product to better meet the needs of the customer and the delivery of additional features and functionality. This process is repeated following the delivery of each increment, until the complete product is produced.

Incremental development is particularly useful when staffing is unavailable for a complete implementation by the business deadline that has been established for the project. Early increments can be implemented with fewer people. If the core product is well received, then additional staff (if required) can be added to implement the next increment. In addition, increments can be planned to manage technical risks.

108.1.4.2 The Spiral Model

The **spiral model**, originally proposed by Boehm [Boehm 1988], is an evolutionary software process model that couples the iterative nature of prototyping with the controlled and systematic aspects of the linear sequential model. It provides the potential for rapid development of incremental versions of the software. Using the spiral model, software is developed in a series of incremental releases. During early iterations, the incremental release might be a paper model or prototype. During later iterations, increasingly more complete versions of the engineered system are produced.

The spiral model is divided into a number of *framework activities*, also called *task regions*. Typically, there are between three and six task regions. Figure 108.6 depicts a spiral model that contains six task regions:

- **Customer communication.** Tasks required to establish effective communication between developer and customer.
- **Planning.** Tasks required to define resources, timelines, and other project-related information.
- **Risk analysis.** Tasks required to assess both technical and management risks.
- **Engineering.** Tasks required to build one or more representations of the application.
- **Construction and release.** Tasks required to construct, test, install, and provide user support (e.g., documentation and training).
- **Customer evaluation.** Tasks required to obtain customer feedback based on evaluation of the software representations created during the engineering stage and implemented during the installation stage.

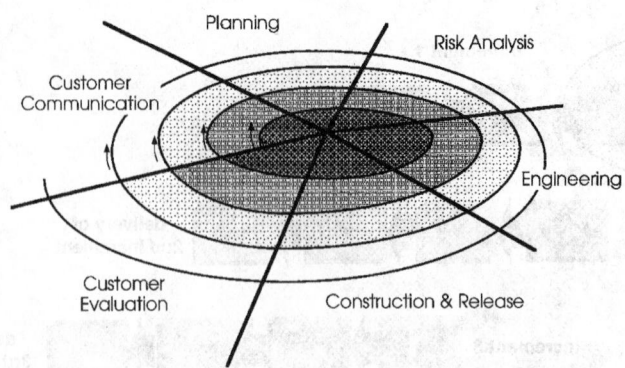

FIGURE 108.6   A typical spiral model.

Each of the regions is populated by a series of work tasks that are adapted to the characteristics of the project to be undertaken. For small projects, the number of work tasks and their formality is low. For larger, more critical projects, each task region contains more work tasks that are defined to achieve a higher level of formality. In all cases, the umbrella activities (e.g., software configuration management and software quality assurance) are performed.

As this evolutionary process begins, the software engineering team moves around the spiral in a clockwise direction, beginning at the center. The first circuit around the spiral might result in development of a product specification; subsequent passes around the spiral might be used to develop a prototype and then progressively more sophisticated versions of the software. Each pass through the planning region results in adjustments to the project plan. Cost and schedule are adjusted on the basis of feedback derived from customer evaluation. In addition, the project manager adjusts the planned number of iterations required to complete the software.

The spiral model is a realistic approach to the development of large-scale systems and software. Because software evolves as the process progresses, the developer and customer better understand and react to risks at each evolutionary level. The spiral model uses prototyping as a risk reduction mechanism but, more importantly, enables the developer to apply the prototyping approach at any stage in the evolution of the product. It maintains the systematic stepwise approach suggested by the classic life cycle but incorporates it into an iterative framework that more realistically reflects the real word. The spiral model demands a direct consideration of technical risks at all stages of the project and, if properly applied, should reduce risks before they become problematic.

But like other paradigms, the spiral model is not a panacea. It may be difficult to convince customers (particularly in contract situations) that the evolutionary approach is controllable. It demands considerable risk assessment expertise, and it relies on this expertise for success. If a major risk is not uncovered and managed, problems will undoubtedly occur. Finally, the model itself is relatively new and has not been used as widely as the linear sequential or prototyping paradigms. It will take a number of years before efficacy of this important new paradigm can be determined with absolute certainty.

108.1.4.3   The Component Assembly Model

Object technologies provide the technical framework for a component-based process model for software engineering. The object-oriented paradigm emphasizes the creation of classes that encapsulate both data and the algorithms that are used to manipulate the data. If properly designed and implemented, object-oriented classes are reusable across different applications and computer-based system architectures.

The **component assembly model** (Figure 108.7) incorporates many of the characteristics of the spiral model. It is evolutionary in nature [Nierstrasz 1992], demanding an iterative approach to the creation of software. However, the component assembly model composes applications from prepackaged software components (called "classes" in Figure 108.7).

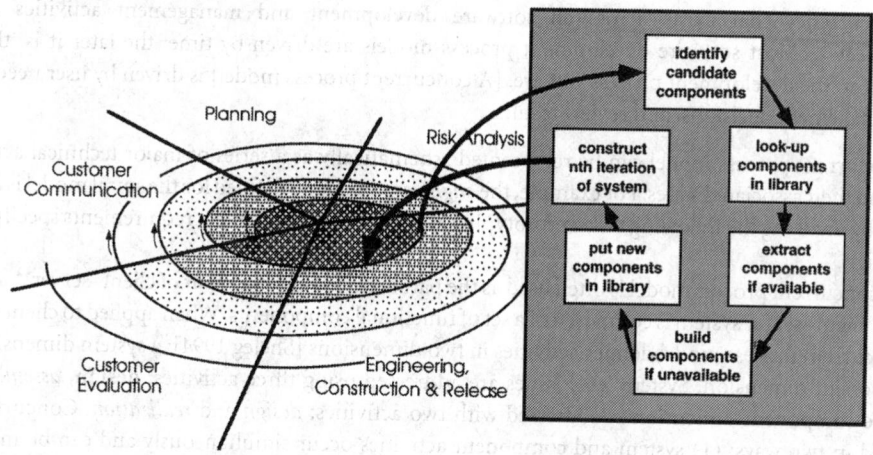

FIGURE 108.7   The component assembly model.

The engineering activity begins with the identification of candidate classes. This is accomplished by examining the data that are to be manipulated by the application and the algorithms that will be applied to accomplish the manipulation. Corresponding data and algorithms are packaged into a class.

Classes created in past software engineering projects are stored in a *class library* or repository. Once candidate classes are identified, the class library is searched to determine if these classes already exist. If they do, they are extracted from the library and reused. If a candidate class does not reside in the library, it is engineered using **object-oriented** methods. The first iteration of the application to be built is then composed, using classes extracted from the library and any new classes built to meet the unique needs of the application. Process flow then returns to the spiral and will ultimately reenter the component assembly iteration during subsequent passes through the engineering activity.

The component assembly model leads to software reuse, and reusability provides software engineers with a number of measurable benefits. Based on studies of reusability, QSM Associates reports [Yourdon 1994] component assembly leads to a 70% reduction in development cycle time, an 84% reduction in project cost, and a productivity index of 26.2, compared with an industry norm of 16.9. Although these results are a function of the robustness of the component library, there is little question that the component assembly model provides significant advantages for software engineers.

108.1.4.4   The Concurrent Development Model

The *concurrent development model*, sometimes called *concurrent engineering*, has been described in the following manner by Davis and Sitaram [Davis and Sitaram 1994]:

> Project managers who track project status in terms of the major phases [of the classic life cycle] have no idea of the status of their projects. These are examples of trying to track extremely complex sets of activities using overly simple models. Note that although … [a large] project is in the coding phase, there are personnel on the project involved in activities typically associated with many phases of development simultaneously. For example, … personnel are writing requirements, designing, coding, testing, and integration testing [all at the same time]. Software engineering process models by Humphrey and Kellner [Humphrey and Kellner 1989] have shown the concurrency that exists for activities occurring during any one phase. Kellner's more recent work [Kellner 1991] uses statecharts [a notation that represents the states of a process] to represent the concurrent relationship existent among activities associated with a specific event (e.g., a requirements change during late development), but fails to capture the richness of

concurrency that exists across all software development and management activities in project ... Most software development process models are driven by time; the later it is, the later in the development process you are. [A concurrent process model] is driven by user needs, management decisions, and review results.

The concurrent process model can be represented schematically as a series of major technical activities, tasks, and their associated states. For example, the *engineering* activity defined for the spiral model is accomplished by invoking the following tasks: prototyping and/or analysis modeling, requirements specification, and design.*

The concurrent process model is often used as the paradigm for development of client–server** applications. A client–server system is composed of a set of functional components. When applied to client–server, the concurrent process model defines activities in two dimensions [Sheleg 1994]: a system dimension and a component dimension. System level issues are addressed using three activities: **design**, *assembly,* and *use*. The component dimension is addressed with two activities: *design* and *realization*. Concurrency is achieved in two ways: (1) system and component activities occur simultaneously and can be modeling using the state-oriented approach described above; (2) a typical client–server application is implemented with many components, each of which can be designed and realized concurrently.

In reality, the concurrent process model is applicable to all types of software development and provides an accurate picture of the current state of a project. Rather than confining software engineering activities to a sequence of events, it defines a network of activities. Each activity on the network exists simultaneously with other activities. Events generated within a given activity or at some other place in the activity network trigger transitions among the states of an activity.

108.1.4.5 The Formal Methods Model

The **formal methods** model encompasses a set of activities that leads to formal mathematical specification of computer software. Formal methods enable a software engineer to specify, develop, and verify a computer-based system by applying a rigorous, mathematical notation. A variation on this approach, called *cleanroom engineering* [Mills et al. 1987, Dyer 1992], is currently applied by some software development organizations.

When formal methods (Chapter 107) are used during development, they provide a mechanism for eliminating many of the problems that are difficult to overcome by using other software engineering paradigms. Ambiguity, incompleteness, and inconsistency can be discovered and corrected more easily — not through *ad hoc* review, but through the application of mathematical analysis. When formal methods are used during design, they serve as a basis for program verification and therefore enable the software engineer to discover and correct errors that might otherwise go undetected.

108.1.4.6 Fourth Generation Techniques

The term **fourth generation techniques** (4GT) encompasses a broad array of software tools that have one thing in common: each enables the software engineer to specify some characteristic of software at a high level. The tool then automatically generates source code based on the developer's specification. There is little debate that the higher the level at which software can be specified to a machine, the faster a program can be built. The 4GT paradigm for software engineering focuses on the ability to specify software using specialized language forms or a graphic notation that describes the problem to be solved in terms that the customer can understand.

Currently, a software development environment that supports the 4GT paradigm includes some or all of the following tools: nonprocedural languages for database query, report generation, data manipulation,

---

*It should be noted that analysis and design are complex tasks that require substantial discussion.

**In client–server applications, software functionality is divided between clients (normally PCs) and a server (a more powerful computer) that typically maintains a centralized database.

screen interaction and definition, and code generation; high-level graphics capability; and spreadsheet capability. Initially, many of the tools noted above were available only for very specific application domains, but today 4GT environments have been extended to address most software application categories.

Like other paradigms, 4GT begins with a requirements gathering step. Ideally, the customer would describe requirements and these would be directly translated into an operational prototype. But this is unworkable. The customer may be unsure of what is required, may be ambiguous in specifying facts that are known, and may be unable or unwilling to specify information in a manner that a 4GT tool can consume. For this reason, the customer–developer dialog described for other process models remains an essential part of the 4GT approach.

For small applications, it may be possible to move directly from the requirements gathering step to implementation using a nonprocedural *fourth generation language* (4GL). However, for larger efforts, it is necessary to develop a design strategy for the system, even if a 4GL is to be used. The use of 4GT without design (for large projects) will cause the same difficulties (poor quality, poor maintainability, poor customer acceptance) that we have encountered when developing software by using conventional approaches.

Implementation using a 4GL enables the software developer to represent desired results in a manner that results in automatic generation of code to generate those results. Obviously, a data structure with relevant information must exist and be readily accessible by the 4GL.

To transform a 4GT implementation into a product, the developer must conduct thorough testing, develop meaningful documentation, and perform all other solution integration activities that are also required in other software engineering paradigms. In addition, the 4GT-developed software must be built in a manner that enables maintenance to be performed expeditiously.

Like all software engineering paradigms, the 4GT model has advantages and disadvantages. Proponents claim dramatic reduction in software development time and greatly improved productivity for people who build software. Opponents claim that current 4GT tools are not all that much easier to use than programming languages, that the resultant source code produced by such tools is "inefficient," and that the maintainability of large software systems developed using 4GT is open to question.

There is some merit in the claims of both sides, and it is possible to summarize the current state of 4GT approaches:

1. The use of 4GT has broadened considerably over the past decade and is now a viable approach for many different application areas. Coupled with *computer-aided software engineering* (CASE) tools and code generators, 4GT offers a credible solution to many software problems.

2. Data collected from companies that are using 4GT indicate that time required to produce software is greatly reduced for small and intermediate applications and that the amount of design and analysis for small applications is also reduced.

3. However, the use of 4GT for large software development efforts demands as much or more analysis, design, and testing (software engineering activities) to achieve substantial time saving that can be achieved through the elimination of coding.

To summarize, 4GT have already become an important part of software development. When coupled with component assembly approaches, the 4GT paradigm may become the dominant software development strategy in the 21st century.

108.2 The Management Spectrum

Effective software project management focuses on the three P's: *people, problem,* and *process*. The order is not arbitrary. The manager who forgets that software engineering work is an intensely human endeavor will never have success in project management. A manager who fails to encourage comprehensive customer communication early in the evolution of a project risks building an elegant solution for the wrong problem. Finally, the manager who pays little attention to the process runs the risk of inserting competent technical methods and tools into a vacuum.

108.2.1 People

The cultivation of motivated, highly skilled software people has been discussed since the 1960s (e.g., [Cougar and Zawacki 1980, DeMarco and Lister 1987, Weinberg 1988]). The Software Engineering Institute has sponsored a *people management maturity model* "to enhance the readiness of software organizations to undertake increasingly complex applications by helping to attract, grow, motivate, deploy, and retain the talent needed to improve their software development capability" [Curtis 1989].

The people management maturity model defines the following key practice areas for software people: recruiting, selection, performance management, training, compensation, career development, organization, and team and culture development. Organizations that achieve high levels of maturity in the people management area have a higher likelihood of implementing effective software engineering practices.

108.2.2 The Problem

Before a project can be planned, objectives and scope should be established, alternative solutions should be considered, and technical and management constraints should be identified. Without this information, it is impossible to develop reasonable estimates of the cost, a realistic breakdown of project tasks, or a manageable project schedule that provides a meaningful indication of progress.

The software developer and customer must meet to define project objectives and scope. In many cases, this activity occurs as part of a structured customer communication process such as *joint application design* (JAD) [Wood and Silver 1994]. JAD is an activity that occurs in five phases: project definition, research, preparation, the JAD meeting, and document preparation. The intent of each phase is to develop information that helps better define the problem to be solved or the product to be built.

108.2.3 The Process

A software process can be characterized as shown in Figure 108.8. A small number of *framework activities* are applicable to all software projects, regardless of their size or complexity. A number of *task sets* — tasks, milestones, deliverables, and quality assurance points — enable the framework activities to be adapted to the characteristics of the software project and the requirements of the project team. Finally, umbrella activities — such as software quality assurance, software configuration management, and measurement — overlay the process model. Umbrella activities are independent of any one framework activity and occur throughout the process.

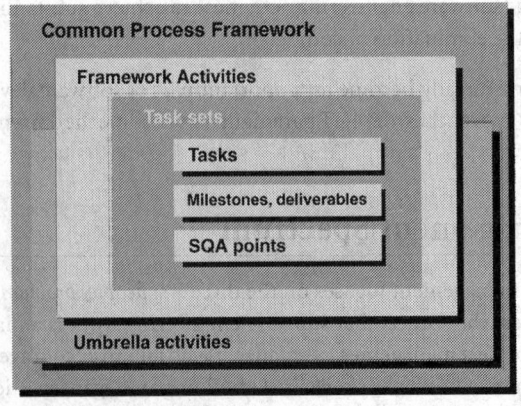

FIGURE 108.8   A common process framework.

In recent years, there has been a significant emphasis on process "maturity" [Paulk et al. 1993]. The Software Engineering Institute (SEI) has developed a comprehensive assessment model that is predicated on a set of software engineering capabilities that should be present as organizations reach different levels of process maturity. To determine an organization's current state of process maturity, the SEI uses an assessment questionnaire and a five-point grading scheme. The grading scheme determines compliance with a **capability maturity model** [Paulk et al. 1993] that defines key activities required at different levels of process maturity. The SEI approach provides a measure of the global effectiveness of a company's software engineering practices and establishes five process maturity levels that are defined in the following manner:

Level 1, Initial: The software process is characterized as *ad hoc* and occasionally even chaotic. Few processes are defined, and success depends on individual effort.

Level 2, Repeatable: Basic project management processes are established to track cost, schedule, and functionality. The necessary process discipline is in place to repeat earlier successes on projects with similar applications.

Level 3, Defined: The software process for both management and engineering activities is documented, standardized, and integrated into an organization-wide software process. All projects use a documented and approved version of the organization's process for developing and maintaining software. This level includes all characteristics defined for level 2.

Level 4, Managed: Detailed measures of the software process and product quality are collected. Both the software process and products are quantitatively understood and controlled by using detailed measures. This level includes all characteristics defined for level 3.

Level 5, Optimizing: Continuous process improvement is enabled by quantitative feedback from the process and from testing innovative ideas and technologies. This level includes all characteristics defined for level 4.

The five levels defined by the SEI are derived as a consequence of evaluating responses to the SEI assessment questionnaire that is based on the capability maturity model. The results of the questionnaire are distilled to a single numerical grade that provides an indication of an organization's process maturity.

The SEI has associated *key process areas* (KPAs) with each of the maturity levels. The KPAs describe those software engineering functions (e.g., software project planning, requirements management) that must be present to satisfy good practice at a particular level. Each KPA is described by identifying the following characteristics:

• *Goals:* The overall objectives that the KPA must achieve.

• *Commitments:* Requirements (imposed on the organization) that must be met to achieve the goals and provide proof of intent to comply with the goals.

• *Abilities:* Those things that must be in place (organizationally and technically) that will enable the organization to meet the commitments.

• *Activities:* The specific tasks that are required to achieve the KPA function.

• *Methods for monitoring implementation:* The manner in which the activities are monitored as they are put into place.

• *Methods for verifying implementation:* The manner in which proper practice for the KPA can be verified.

Eighteen KPAs (each defined using the structure noted above) are defined across the maturity model and are mapped into different levels of process maturity.

Each of the KPAs is defined by a set of *key practices* that contribute to satisfying its goals. The key practices are policies, procedures, and activities that must occur before a key process area has been fully instituted. The SEI defines *key indicators* as "those key practices or components of key practices that offer the greatest insight into whether the goals of a key process area have been achieved." Assessment questions are designed to probe for the existence (or lack thereof) of a key indicator.

108.3 Software Project Management

Software project management encompasses the following activities: measurement, project estimating, risk analysis, scheduling, tracking, and control. A comprehensive discussion of these topics is beyond the scope of this chapter, but a brief overview of each topic will enable the reader to understand the breadth of management activities required for a mature software engineering organization.

108.3.1 Measurement and Metrics

To be most effective, software metrics should be collected for both the process and the product. Process-oriented metrics [Hetzel 1993, Jones 1991] can be collected during the process and after it has been completed. Process metrics collected during the process focus on the efficacy of quality assurance activities, change management, and project management. Process metrics collected after a project has been completed examine the efficacy of various software engineering activities and productivity. Process measures are normalized using either lines of code or function points [Dreger 1989], so that data collected from many different projects can be compared and analyzed in a consistent manner. Product metrics measure technical characteristics of the software that provide an indication of software quality [Fenton 1991, Zuse 1990, Lorenz and Kidd 1994]. Measures can be applied to models created during analysis and design activities, during code generation, and during testing. The mechanics of measurement and the specific measures to be collected are beyond the scope of this chapter.

108.3.2 Project Estimating

Scheduling and budgets are often dictated by business issues. The role of estimating within the software process often serves as a "sanity check" on the predefined deadlines and budgets that have been established by management. (Ideally, the software engineering organization should be intimately involved in establishing deadlines and budgets, but this is not a perfect or fair world.)

All software project **estimation** techniques require that the project have a bounded scope, and all rely on a high-level functional decomposition of the project and an assessment of project difficulty and complexity. There are three broad classes of estimation techniques [Pressman 1993] for software projects:

Effort estimation techniques. The project manager creates a matrix in which the left-hand column contains a list of major system functions derived using functional decomposition applied to project scope. The top row contains a list of major software engineering tasks derived from the common process framework. The manager (with the assistance of technical staff) estimates the effort required to accomplish each task for each function.

Size-oriented estimation. A list of major system functions derived using functional decomposition applied to project scope. The "size" of each function is estimated by using either lines of code (LOC) or function points (FP). Average productivity data (e.g., function points per person month) for similar functions or projects are used to generate an estimate of effort required for each function.

Empirical models. Using the results of a large population of past projects, an empirical model that relates product size (in LOC or FP) to effort is developed, using a statistical technique such as regression analysis. The product size for the work to be done is estimated and the empirical model is used to generate projected effort.

In addition to the above techniques, a software project manager can develop estimates by analogy; that is, by examining similar past projects and projecting effort and duration recorded for these projects to the current situation.

108.3.3 Risk Analysis

Almost five centuries have passed since Machiavelli said: "I think it may be true that fortune is the ruler of half our actions, but that she allows the other half to be governed by us ... [fortune] is like an impetuous river ... but men can make provision against it by dykes and banks." Fortune (we call it risk) is in the back of every software project manager's mind, and that is often where it stays. And as a result, risk is never adequately addressed. When bad things happen, the manager and the project team are unprepared.

In order to "make provision against it," a software project team must conduct risk analysis explicitly. **Risk analysis** [Charette 1990, Jones 1994] is actually a series of steps that enable the software team to perform risk identification, risk assessment, risk prioritization, and risk management. The goals of these activities are: (1) to identify those risks that have a high likelihood of occurrence; (2) to assess the consequence (impact) of each risk should it occur; and (3) to develop a plan for mitigating the risks when possible, monitoring factors that may indicate their arrival, and developing a set of contingency plans should they occur.

Risk identification is a systematic attempt to specify threats to the project plan (estimates, schedule, resource loading, etc.). By identifying known and predictable risks, the project manager takes a first step toward avoiding them when possible and controlling them when necessary.

There are two distinct types of risks for each of the categories that have been presented: generic risks and product-specific risks. Generic risks are a potential threat to every software project. Product-specific risks can be identified only by those with a clear understanding of the technology, the people, and the environment that are specific to the project at hand. To identify product-specific risks, the project plan and the software statement of scope are examined and an answer to the following question is developed: "What special characteristics of this product may threaten our project plan?"

Both generic and product-specific risks should be identified systematically. Gilb [Gilb 1988] drives this point home when he states: "If you don't actively attack the risks, they will actively attack you."

Risk projection, also called *risk estimation,* attempts to rate each risk in two ways — the *likelihood* or probability that the risk is real and the *consequences* of the problems associated with the risk, should it occur. The project planner, along with other managers and technical staff, performs four risk projection activities [Babich 1986]: (1) establish a scale that reflects the perceived likelihood of a risk; (2) delineate the consequences of the risk; (3) estimate the impact of the risk on the project and the product; and (4) note the overall accuracy of the risk projection so that there will be no misunderstandings.

All of the risk analysis activities presented to this point have a single goal — to assist the project team in developing a strategy for dealing with risk. An effective strategy must consider three issues:

- Risk avoidance
- Risk monitoring, and
- Risk management and contingency planning.

The manner in which each of these issues is to be addressed is documented in a plan for *risk mitigation, monitoring, and management.*

108.3.4 Scheduling

Fred Brooks, the well-known author of The Mythical Man-Month [Brooks 1975], was once asked how software projects fall behind schedule. His response was as simple as it was profound: "One day at a time."

The reality of a technical project (whether it involves building a hydroelectric plant or developing an operating system) is that hundreds of small tasks must occur to accomplish a larger goal. Some of these tasks lie outside the mainstream and may be completed without worry about impact on project completion date. Other tasks lie on the "critical path."* If these "critical" tasks fall behind schedule, the completion date of the entire project is put into jeopardy.

---

*The critical path is the sequence of project tasks that nust be closely monitored by the project manager.

The objective of the project manager is to define all project tasks, identify the ones that are critical, and then track their progress to ensure that delay is recognized "one day at a time." To accomplish this, the manager must have a schedule that has been defined at a degree of resolution that enables the manager to monitor progress and control the project.

Software project **scheduling** is an activity that distributes estimated effort across the planned project duration by allocating the effort to specific software engineering tasks. It is important to note, however, that the schedule evolves over time. During early stages of project planning, a *macroscopic schedule* is developed. This type of schedule identifies all major software engineering activities and the product functions to which they are applied. As the project gets under way, each entry on the macroscopic schedule is refined into a *detailed schedule*. Here, specific software tasks (required to accomplish an activity) are identified and scheduled.

Scheduling for software development projects can be viewed from two rather different perspectives. In the first, an end-date for release of a computer-based system has already (and irrevocably) been established. The software organization is constrained to distribute effort within the prescribed time frame. The second view of software scheduling assumes that rough chronological bounds have been discussed but that the end-date is set by the software engineering organization. Effort is distributed to make best use of resources and an end-date is defined after careful analysis of the software. Unfortunately, the first situation is encountered far more frequently than the second.

As in all other areas of software engineering, a number of basic principles guide software project scheduling:

Compartmentalization: The project must be compartmentalized into a number of manageable activities and tasks. To accomplish compartmentalization, both the product and the process are decomposed (Chapter 3).

Interdependency: The interdependencies of each compartmentalized activity or task must be determined. Some tasks must occur in sequence, whereas others can occur in parallel. Some activities cannot commence until the work product produced by another is available. Other activities can occur independently.

Time allocation: Each task to be scheduled must be allocated some number of work units (e.g., person-days of effort). In addition, each task must be assigned a start date and a completion date that is a function of the interdependencies and whether work will be conducted on a full-time or part-time basis.

Effort validation. Every project has a defined number of staff members. As time allocation occurs, the project manager must ensure that no more than the allocated number of people have been allocated at any given time. For example, consider a project that has three assigned staff members (e.g., 3 person-days are available per day of assigned effort).* On a given day, seven concurrent tasks must be accomplished. Each task requires 0.50 person-day of effort. More effort has been allocated than there are people to do the work.

Defined responsibilities: Every task that is scheduled should be assigned to a specific team member.

Defined outcomes: Every task that is scheduled should have a defined outcome. For software projects, the outcome is normally a work product (e.g., the design of a module) or a part of a work product. Work products are often combined in *deliverables*.

Defined milestones: Every task or group of tasks should be associated with a project milestone. A milestone is accomplished when one or more work products have been reviewed for quality and have been approved.

Each of the above principles is applied as the project schedule evolves.

---

*In reality, less than 3 person-days are available because of unrelated meetings, sickness, vacation, and a variety of other reasons. For our purposes, however, we assume 100% availability.

108.3.5   Tracking and Control

Project tracking and control is most effective when it becomes an integral part of software engineering work. A well-defined development strategy should provide a set of milestones that can be used for project tracking. Control focuses on two major issues: quality and change.

To control quality, a software project team must establish effective techniques for software quality assurance, and, to control change, the team should establish a software configuration management framework.

108.4   Software Quality Assurance

In his landmark book on quality, Crosby [Crosby 1979] states:

> The problem of quality management is not what people don't know about it. The problem is what they think they do know . . .
>
> In this regard, quality has much in common with sex. Everybody is for it. (Under certain conditions, of course.) Everyone feels they understand it. (Even though they wouldn't want to explain it.) Everyone thinks execution is only a matter of following natural inclinations. (After all, we do get along somehow.) And, of course, most people feel that problems in these areas are caused by other people. (If only *they* would take the time to do things right.)

There have been many definitions of software **quality** proposed in the literature. For our purposes, software quality is defined as: *Conformance to explicitly stated functional and performance requirements, explicitly documented development standards, and implicit characteristics that are expected of all professionally developed software.*

There is little question that the above definition could be modified or extended. If fact, a definitive definition of software quality could be debated endlessly. But the definition stated above does serve to emphasize three important points:

1. Software requirements are the foundation from which *quality* is assessed. Lack of conformance to requirements is lack of quality.
2. A mature software process model defines a set of development criteria that guide the manner in which software is engineered. If the criteria are not followed, lack of quality will almost surely result.
3. There is a set of *implicit requirements* that often goes unmentioned (e.g., the desire for good maintainability). If software conforms to its explicit requirements, but fails to meet implicit requirements, software quality is suspect.

Software quality is designed into a product or system. It is not imposed after the fact. For this reason, **software quality assurance (SQA)** actually begins with the set of *technical methods and tools* that help the analyst to achieve a high quality specification and the designer to develop a high quality design.

Once a specification (or prototype) and design have been created, each must be assessed for quality. The central activity that accomplishes quality assessment is the **formal technical review** (FTR). The FTR — conducted as a *walkthrough* or an *inspection* [Freedman and Weinberg 1990] — is a stylized meeting conducted by technical staff with the sole purpose of uncovering quality problems. In many situations, formal technical reviews have been found to be as effective as testing in uncovering errors in software [Gilb and Graham 1993].

Software testing combines a multistep strategy with a series of test case design methods that help ensure effective error detection. Many software developers use software testing as a quality assurance "safety net." That is, developers assume that thorough testing will uncover most errors, thereby mitigating the need for other SQA activities. Unfortunately, testing, even when performed well, is not as effective as we might like for all classes of errors. A much better strategy is to find and correct errors (using FTRs) before getting to testing.

The degree to which formal *standards and procedures* are applied to the software engineering process varies from company to company. In many cases, standards are dictated by customers or regulatory mandate. In other situations standards are self-imposed. An assessment of compliance to standards may be conducted by software developers as part of a formal technical review, or, in situations where independent verification of compliance is required, the SQA group may conduct its own *audit*.

A major threat to software quality comes from a seemingly benign source: *changes*. Every change to software has the potential for introducing error or creating side effects that propagate errors. The **change control** process contributes directly to software quality by formalizing requests for change, evaluating the nature of change, and controlling the impact of change. Change control is applied during software development and, later, during the software maintenance phase.

Measurement is an activity that is integral to any engineering discipline. An important object of SQA is to track software quality and assess the impact of methodological and procedural changes on improved software quality. To accomplish this, **software metrics** must be collected.

Record keeping and recording for SQA provide procedures for the collection and dissemination of SQA information. The results of reviews, audits, change control, testing, and other SQA activities must become part of the historical record for a project and should be disseminated to development staff on a need-to-know basis. For example, the results of each formal technical review for a procedural design are recorded and can be placed in a "folder" that contains all technical and SQA information about a module.

108.5   Software Configuration Management

Change is inevitable when computer software is built. And change increases the level of confusion among software engineers who are working on a project. Confusion arises when changes are *not* analyzed before they are made, recorded before they are implemented, reported to those who should be aware that they have occurred, or controlled in a manner that will improve quality and reduce error. Babich [Babich 1986] discusses this when he states:

> The art of coordinating software development to minimize . . . confusion is called *configuration management*. Configuration management is the art of identifying, organizing, and controlling modifications to the software being built by a programming team. The goal is to maximize productivity by minimizing mistakes.

Software configuration management (SCM) is an umbrella activity that is applied throughout the software engineering process. Because change can occur at any time, SCM activities are developed to (1) identify change, (2) control change, (3) ensure that change is being properly implemented, and (4) report change to others who may have an interest.

A primary goal of software engineering is to improve the ease with which changes can be accommodated and reduce the amount of effort expended when changes must be made.

108.6   Summary

The role of a software project manager is to understand the scope of the problem to be solved and, knowing this, to select an appropriate development strategy for the problem. Once a strategy is selected, software project management activities are conducted. Project management encompasses the measurement of the process and the product, estimation, risk analysis, scheduling, and tracking. To control the project, software quality assurance and software configuration management also must be conducted.

Acknowledgment

This chapter has been adapted from selected excerpts of R. S. Pressman, *Software Engineering: A Practitioner's Approach*, 4th ed. McGraw–Hill, New York, 1997.

Defining Terms

Capability maturity model: Defines key activities required at different levels of software process maturity.

Change control: An umbrella process that enables a project team to accept, evaluate, and act on changes in a systematic manner.

Classic life cycle: A linear, sequential approach to process modeling.

Common process framework: A process model that encompasses a limited set of problem-solving activities populated by tasks, milestones, SQA points, and deliverables.

Component assembly model: A process model that encourages construction of software from reusable software components.

Design: An activity that translates the requirements model into a more detailed model that is the guide to implementation of the software.

Errors: A lack of conformance found before software is delivered to the customer.

Estimation: A project planning activity that attempts to project effort and cost for a software project.

Evolutionary model: A process model that is designed with the recognition that software evolves through a number of iterations.

Formal methods: A mathematical approach to the specification and validation of computer-based systems.

Formal technical review: A structured meeting conducted by software engineers and others with the intent of uncovering errors in some deliverable or work product.

Fourth generation techniques: Encompasses a broad array of software tools that enables the software engineer to specify some characteristic of software at a high level of abstraction.

Incremental model: A process model that results in delivery of versions of an application that provide increasingly greater functionality.

Linear sequential model: A process model that defines a set of linear activities for developing computer software.

Maintenance: The activities associated with changes to software after it has been delivered to end-users.

Measurement: Collecting quantitative data about the software or the software engineering process.

Object-oriented: An approach to software development that makes use of a classification approach and packages data and processing together.

Process model: A model that outlines the major activities and work flow for software development and acts as a framework for project management.

Prototyping: The creation of a mock-up of an application with the intent of helping a customer to better identify requirements.

Quality: The degree to which a product conforms to both explicit and implicit requirements.

Rapid Application Development (RAD): A linear sequential software development process model that emphasizes an extremely short development cycle.

Reengineering: A series of activities that transform old systems (with poor maintainability) into software that exhibits high quality.

Requirements analysis: A modeling activity whose objective is to understand what the customer really wants.

Risk analysis: The set of activities that identify and evaluate a potential problem or occurrence that may put a project in jeopardy.

Scheduling: The activity that lays out a timeline for work to be conducted on a project.

Software engineering: A discipline that encompasses process, methods, and tools.

Software metrics: Quantitative measures of the process or the product.

Software quality assurance (SQA): A series of activities that assist an organization in producing high-quality software.

Spiral model: An evolutionary software engineering paradigm.

Testing: A set of activities that attempt to find errors.

Work breakdown structure (WBS): The set of work tasks required to build the software; defined as part of the process model.

References

Babich, W. 1986. *Software Configuration Management*. Addison–Wesley, Reading, MA.

Boehm, B. 1988. A spiral model for software development and enhancement. *Computer* 21(5): 61–72.

Bradac, M., Perry, D., and Votta, L. 1994. Prototyping a process monitoring experiment. *IEEE Trans. Software Eng.* 20(10):774–784.

Brooks, M. 1975. *The Mythical Man-Month*. Addison–Wesley, Reading, MA.

Butler, J. 1994. Rapid application development in action. *Managing System Development, Applied Computer Research* 14(5):6–8.

Charette, R. 1990. *Application Strategies for Risk Analysis*. McGraw–Hill, New York.

Cougar, J. and Zawacki, R. 1980. *Managing and Motivating Computer Personnel*. Wiley, New York.

Crosby, P. 1979. *Quality is Free*. McGraw–Hill, New York.

Curtis, B. 1989. People management maturity model. *Proc Intl. Conf. Software Eng.*, Pittsburgh.

Davis, A. and Sitaram, P. 1994. A concurrent process model for software development. *Software Eng. Notes* 19(2):38–51.

DeMarco, T. and Lister, T. 1987. *Peopleware*. Dorset House.

Dreger, J. B. 1989. *Function Point Analysis*. Prentice–Hall, Englewood Cliffs, NJ.

Dyer, M. 1992. *The Cleanroom Approach to Quality Software Development*. Wiley, New York.

Fenton, N. E. 1991. *Software Metrics*. Chapman & Hall, New York.

Freedman, D. and Weinberg, G. 1990. *The Handbook of Walkthroughs, Inspections and Technical Reviews*. Dorset House.

Gilb, T. 1988. *Principles of Software Engineering Management*. Addison–Wesley, Reading, MA.

Gilb, T. and Graham, D. 1993. *Software Inspection*. Addison–Wesley, Reading, MA.

Hanna, M. 1995. Farewell to waterfalls, pp. 38–46. *Software Magazine*. May.

Hetzel, B. 1993. *Making Software Measurement Work*. QED Publishing.

Humphrey, W. and Kellner, M. 1989. Software process modeling: principles of entity process models, pp. 331–342. In *Proc. 11th Intl. Conf. Software Eng.* IEEE Computer Society Press.

Jones, C. 1991. *Applied Software Measurement*. McGraw–Hill, New York.

Jones, C. 1994. *Assessment and Control of Software Risks*. Yourdon Press.

Kellner, M. 1991. Software process modeling support for management planning and control, pp. 8–28. In *Proc. 1st Intl. Conf. Software Process*. IEEE Computer Society Press.

Kerr, J. and Hunter, R. 1994. *Inside RAD*. McGraw–Hill, New York.

Lorenz, M. and Kidd, J. 1994. *Object-Oriented Software Metrics*. Prentice–Hall, Englewood Cliffs, NJ.

Martin, J. 1991. *Rapid Application Development*. Prentice–Hall, Englewood Cliffs, NJ.

McDermid, J. and Rook, P. 1993. Software development process models, pp. 15/26–15/28. In *Software Engineer's Reference Book*. CRC Press, Boca Raton, FL.

Mills, H. D., Dyer, M., and Linger, R. 1987. Cleanroom software engineering. *IEEE Software* X(Y): 19–25.

Nierstrasz. 1992. Component-oriented software development. *Commun ACM* 35(9):160–165.

Paulk, M. et al. 1993. *Capability Maturity Model for Software*. Software Engineering Institute, Carnegie Mellon University, Pittsburgh, PA.

Pressman, R. S. 1993. *A Manager's Guide to Software Engineering*. McGraw–Hill, New York.

Raccoon, L. B. S. 1995. The chaos model and the chaos life cycle. *ACM Software Eng. Notes* 20(1): 55–66.

Sheleg, W. 1994. Concurrent engineering: a new paradigm for C/S development. *App. Dev. Trends* 1(6): 28–33.

Weinberg, G. 1988. *Understanding the Professional Programmer*. Dorset House.

Wood, J. and Silver, D. 1994. *Joint Application Design*, 2nd ed. Wiley, New York.

Yourdon, E. 1994. Software reuse. *App. Dev. Strategies* VI(12):1–16.

Zuse, H. 1990. *Software Complexity*. deGruyer, Berlin.

Further Information

The current state of the art in software engineering can best be determined from monthly publications such as *IEEE Software, Computer,* and the *IEEE Transactions on Software Engineering.* Industry periodicals such as *Application Development Trends* and *Software Development* often contain articles on software engineering topics. The discipline is "summarized" every year in the *Proceedings of the International Conference on Software Engineering,* sponsored by the IEEE and ACM, and is discussed in depth in journals such as *ACM Transactions on Software Engineering and Methodology, ACM Software Engineering Notes,* and *Annals of Software Engineering.*

Many software engineering books have been published in recent years. Some present an overview of the entire process, whereas others delve into a few important topics to the exclusion of others. Three anthologies that cover a wide range of software engineering topics are:

Keyes, J., ed. 1993. *Software Engineering Productivity Handbook.* McGraw–Hill, New York.

McDermid, J., ed. 1993. *Software Engineer's Reference Book.* CRC Press, Boca Raton, FL.

Marchiniak, J. J., ed. 1994. *Encyclopedia of Software Engineering.* Wiley, New York.

An excellent three-volume series written by Weinberg (1992, 1993, 1994. *Quality Software Management.* Dorset House) introduces basis systems thinking and management concepts, explains how to use measurements effectively, and addresses "congruent action," the ability to establish "fit" between the manager's needs, the needs of technical staff, and the needs of the business. It will provide both new and experienced managers with useful information. Fred Brooks (1995. *The Mythical Man-Month,* Anniversary Edition, Addison–Wesley, Reading, MA) has updated his classic book to provide new insight into software project and management issues. S. Purba (1995. *How to Manage a Successful Software Project.* Wiley, New York) presents a number of case studies that indicate why some projects succeed and others fail. E. Bennatan (1995. *Software Project Management in a Client/Server Environment.* Wiley, New York) discussed special management issues associated with the development of client/server systems.

R. House (1988. *The Human Side of Project Management.* Addison–Wesley, Reading, MA) and P. Crosby (1989. *Running Things: The Art of Making Things Happen.* McGraw–Hill, New York) provide practical advice for managers who must deal with human as well as technical problems. Books by T. DeMarco and T. Lister [1987] and G. Weinberg [1988] provide useful insight into software people and the way in which they should be managed.

Pragmatic guidance on project management is presented by F. O'Connell (1994. *How To Run Successful Projects.* Prentice–Hall, Englewood Cliffs, NJ). Still another take on project management in the software world is provided by L. Constantine (1995. *Constantine on Peopleware.* Prentice–Hall, Eaglewood Cliffs, NJ).

A wide variety of information sources on software engineering and the software process is available on the internet. An up-to-date list of World Wide Web references that are relevant to the software process can be found at http://www.rspa.com.

Further Information

The literature of the software engineering field can best be determined from monthly publications, such as *IEEE Software, Computer*, and the *IEEE Transactions on Software*. Presenting industry-oriented articles, *Application Development Trends* and *Software Development* often contain articles on software engineering topics. The distribution is summarized yearly. Most of the *Proceedings of the International Conference on Software Engineering*, sponsored by the IEEE and ACM, are discussed in depth in journals such as *ACM Transactions on Software Engineering and Methodology*. *IEEE Software* covers a range of topics in a more timely manner.

Many software engineering books have been published in recent years. Some present an overview of the entire process; others range from one or a few important topics to the status of one or others. Three authhors offer a broad range of software engineering topics are:

Keyes, J. ed. 1993. *Software Engineering Productivity Handbook*. New York: McGraw-Hill, New York.
McDermid, J., ed. 1991. *Software Engineer's Reference Book*. The CRC Press, Boca Raton, FL.
Marciniak, J.J., ed. 1994. *Encyclopedia of Software Engineering*. New York: Wiley & Sons.

An excellent three-volume series written by Weinberg (1992, 1993, 1994), *Quality Software Management*, Dorset House, introduces basic systems thinking and management concepts, explains how to use measurement effectively, and addresses the critical action of the ability to establish trust between the manager and staff. The book's freshness and treatment of the human side will provide a balanced and experienced approach, and useful information. Pfleeger, H., and Brooks, T. (1996), *A New Man, Month*. Anniversary Edition. Addison-Wesley, Reading, VA. Discusses and illustrates the various people and management project and management issues. Similarly, Brooks's famous book, *The Mythical Man-Month* (Wiley, New York), presents a number of case studies that indicate why some projects succeed and others fail. E. Bergman, *Better Software Project Management* (John Wiley & Sons, New York), presents and discusses management issues associated with the development of their software.

R. House. 1988. *The Human Side of Project Management*. Addison-Wesley, Reading, MA; and P. Crosby (1984). *Running Things: The Art of Making Things Happen*. McGraw-Hill, New York, provide practical advice for managers who must deal with human issues as well as technical problems. Books by DeMarco and T. Lister (1987), and T. Weinberg (1988) provide useful insight into software people and the way in which they should be managed.

Pragmatic guidelines on project management is presented by P.C. Connell (1994). *How to Run Successful Projects* (Prentice-Hall, Englewood Cliffs). This will ensure practice on project management in the software world is provided by T. Constantine (1995). *Constantine on Peopleware* (Prentice-Hall, Englewood Cliffs).

A wide variety of information on software engineering and the software process is available on the Internet. A good source is the World Wide Web reference with reference to the software process can be found at http://www.rspa.com.

109

Software Architecture

	109.1	Introduction	**109**-1
	109.2	Underlying Principles	**109**-2
	109.3	Describing Individual Software Architectures	**109**-3
Stephen B. Seidman	109.4	Architecture Description Languages	**109**-4
New Jersey Institute of Technology	109.5	Describing Architectural Styles	**109**-6

109.1 Introduction

A **software architecture** (henceforth *architecture*) is an abstraction that allows a designer to ignore low-level implementation issues, such as programming languages, hardware and device requirements, and communication protocols. Garlan and Perry [1995] state that architectures "simplify our ability to comprehend large systems by presenting them at a level of abstraction at which a system's high-level design can be understood." The idea of abstracting away detail to uncover the essential structure of a complex system is very old. The classical notion of architecture abstracts the structure of a building or other human construction away from the entity itself. By 1980, this idea had been adopted by computer engineers and network engineers. Common examples in these domains include RISC architectures, instruction set architectures, shared-memory architectures [Hennessy and Patterson, 1996], layered architectures, TCP/IP architectures, and IP forwarding architectures [Leon-Garcia and Widjaja, 2000].

Software architecture comprises two kinds of entities: **components** that perform computation and **connectors** that express relationships (typically communication) between the components. An architectural description must also include syntactic and semantic information about the components and connectors. This information constrains how an assemblage of components and connectors can be formed and when it can be regarded as an architecture. These constraints are embodied in an **architecture description language** (ADL).

Software architectures allow designers to codify their expertise. One way to do this is by recognizing and defining software architectural styles: collections of architectural features that tend to co-occur. The concept of *style* is familiar from building architecture; examples include Gothic, Tudor, and skyscraper. Software architectural styles can be abstractions of descriptors like *client–server model*, uses *remote procedure calls*, or *pipeline*. They can also be embodied in collections of rules encapsulated in a specific paradigm for software development.

Architecture description languages assist with the process of describing and developing software architectures and styles. They incorporate formal foundations that support **architectural analysis**: reasoning about descriptions of architectures and architectural styles. Many formal techniques have been used for this purpose, including specification languages, process algebras, graph grammars, and a variety of logics. These techniques are used to investigate properties of architectures and styles along several dimensions. **Functional properties** include the semantics of the components that make up an architecture. **Structural properties** deal with the types of interactions supported by the components. **Nonfunctional properties** are more difficult to address formally; they include reliability, robustness, ease of use, conformance to standards, hardware requirements, and security [Shaw and Garlan, 1996].

Architectural abstraction can potentially reduce testing costs. Because an architecture is often used to develop multiple systems, the cost of any architecture level testing effort is amortized across all of these systems [Richardson and Wolf, 1996]. It can also aid in software evolution and reuse. Finally, it is important to note that much software is domain-specific. By creating architectural abstractions that are specific to their domain, software organizations can combine the best aspects of standard platforms and standardized components to create and specialize their domain-related product families [Garlan and Perry, 1995].

109.2   Underlying Principles

By the late 1970s, software engineers realized that there was a critical distinction between **programming in the small** (using conventional programming languages) and **programming in the large** (building industrial-strength software systems). In order to deal with the issues raised by programming in the large, software systems were regarded as collections of modules, with communication between modules mediated by clearly defined interfaces. **Module interconnection languages** (MILs) were developed to specify the characteristics of these interfaces and particularly the resources required by and provided to the modules. A MIL can be used to describe the global structure of a software system. Examples of early MILs are MIL75 [deRemer and Kron, 1976] and INTERCOL [Tichy, 1979].

By 1985, several researchers had come to realize that module specifications needed to express aspects of interface semantics that could not be accommodated within the requires/provides model used by early MILs. This realization gave rise to much research. For example, the Conic configuration language developed at Imperial College [Magee et al., 1989] specifies how the components of a distributed system are assembled. The STILE system described in Stovsky and Weide [1990] uses a graphical language that embodies a similar approach.

Software systems specified in Conic are assemblages of *group modules*, which are hierarchical entities made up of smaller modules. Lower-level modules may also be group modules, or they may be *task modules*, which are specified using a programming language. The external interfaces of Conic modules consist of *exit ports* and *entry ports*. Ports in turn have *notify* and *request–reply* modes. Communication between Conic modules is specified by linking ports in their external interfaces that have the same type and mode. A single notify exit port may be linked to a set of notify entry ports (or the reverse). A request–reply exit port may be linked to only one request–reply entry port. Module specifications in Conic describe the syntax of permissible connections between modules.

In general, the purpose of MILs is to describe the syntax of module interconnections. Alternative attributes associated with module interfaces may be *provides* and *requires* (INTERCOL), *notify* and *request–reply* (Conic), or *entry* and *exit* (Conic). Rice and Seidman [1994] presented a general model of MILs that can express the specific syntactic constraints used by many MILs. The model is based on a number of generic Z schemas [Spivey, 1989] that can be customized for a particular MIL. It has been used to model Conic and Stile.

For similar reasons, Batory and O'Malley [1992] developed a model of hierarchical software systems. This model assumes that software systems have an explicitly layered organization. Components are selected at each layer from one or more domain-specific lists of choices. Communications between components may only take place by invoking operations that access components that are adjacent in the hierarchy. The Batory model was first applied to file structures and database management systems.

These models can be used in two distinct ways: to analyze the features of a particular MIL and to explore the design space for MILs. In either case, they provide formalism that describes the way in which a particular MIL permits modules to be composed to form a software system. By analogy with conventional programming languages, MILs can be used to describe the syntax that governs the construction of a system of modules.

Software architectures are abstract models that contain syntactic and semantic information about the components of software systems and the relationships between those components. Since MILs only give syntactic information, they cannot be used to represent software architectures. Modeling a software architecture requires a way to describe the semantics associated with the composition of modules in a software system. A similar observation was made in Abowd et al. [1995].

Software architectures can be described at several levels [Shaw and Garlan, 1996, p. 130]. First, one can describe the architecture of a particular software system. Second, one can describe a family of architectures as an architectural style. Third, one can develop an architecture description language (ADL) that is based on a formal theory of software architectures. Finally, one can define semantics for ADLs. Formalisms that can be used for all of these descriptions include specification languages (e.g., Z), process algebras (e.g., CSP, π-calculus), graph grammars and regular expressions, and partial orders.

109.3 Describing Individual Software Architectures

The software architecture modeling process will be illustrated by a relatively early example that is particularly well worked out. In the late 1980s, Garlan and Delisle developed a formal model for the architecture of a digital oscilloscope [Delisle and Garlan, 1990]; also see the discussions in Garlan and Shaw [1996, Sections 3.2, 6.2]. The goal of their project was to develop an oscilloscope system architecture that would increase reuse and make configuration easier. Several models were initially proposed but rejected. For example, an object-oriented model identified relevant data types but could not explain how they fit together. In the end, the architecture was represented by a pipe-and-filter model modified to allow an external entity to set filter parameters. The processing done by a digital oscilloscope is modeled as a pipeline of nodes that successively transform the signals. The processing carried out by the individual nodes can be configured by a user through parameter settings. For example, the waveform display is parameterized by user-defined factors that support zooming and panning. The description given below is taken from Section 6.2 of Shaw and Garlan [1996].

A portion of the model is shown in Figure 109.1 (taken from Figure 6.1 in Shaw and Garlan [1996]). The four nodes successively subtract a DC offset from a signal (*Couple*), extract a time-sliced waveform from a signal (*Acquire*), create a trace by converting (time, voltage) pairs to horizontal and vertical values (*WaveformToTrace*, or $W \rightarrow T$), and clip it to a display screen (*Clip*).

The oscilloscope's inputs (signals), internal representations (waveforms), and outputs (traces) are modeled as functions. The model uses Z formalism [Spivey, 1989], in which $\rightarrow$ represents a function and $\nrightarrow$ a partial function defined on a subset of its domain. A brief summary of Z notation is given in the appendix for this chapter. This distinguishes waveforms, which are only defined on a specific time interval, from signals. Signals, waveforms, and traces are specified to be *Signal == AbsTime $\rightarrow$ Volts*, *Waveform == AbsTime $\nrightarrow$ Volts*, and *Trace == Horiz $\nrightarrow$ Vert*.

For our present purposes, it will be sufficient to give a precise characterization of the first node of the pipeline. The Couple transformer subtracts a DC offset from a signal. The user has three parameter choices: DC, AC, Ground. Choosing DC leaves the signal unchanged; choosing AC subtracts a DC offset, and choosing Ground sets the signal to zero. The formal specification of this architectural element is given in Figure 109.2, where *Coupling* is specified to be either *DC, AC*, or *GND*.

Similar formal descriptions can be given for the *Acquire, WavefrontToTrace*, and *Clip* nodes, but lack of space precludes their inclusion here. The pieces are assembled into a system by the specification illustrated in Figure 109.3.

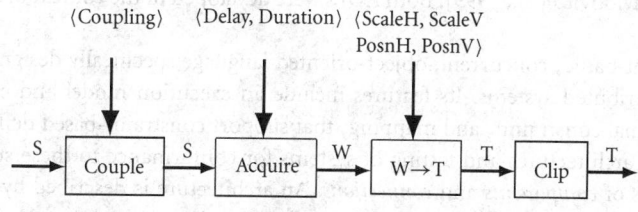

FIGURE 109.1   The architecture of a digital oscilloscope.

$$Couple : Coupling \rightarrow Signal \rightarrow Signal$$

$$Couple\ DC\ s = s$$
$$Couple\ AC\ s = (\lambda\ t:\ AbsTime \cdot s(t) - dc(s))$$
$$Couple\ GND\ s == (\lambda\ t:\ AbsTime \cdot 0)$$

FIGURE 109.2   Z specification for the Couple transformer.

ChannelParameters

> *c: Coupling*
> *delay, dur : RelTime*
> *scaleH : RelTime*
> *scaleV : Volts*
> *posnV : Vert*
> *posnH : Horiz*

$$ChannelConfigurations : ChannelParameters \rightarrow TriggerEvent \rightarrow Signal \rightarrow Time$$

$$ChannelConfiguration == (\lambda\ trig:\ TriggerEvent \cdot$$
$$Clip \circ WaveformToTrace\ (p.scaleH, p.scaleV, p.posnH, p.posnV)$$
$$\circ\ Acquire\ (p.delay, p.dur)\ trig\ Couple\ p.c)$$

FIGURE 109.3   Z specification for digital oscilloscope.

This description represents the high-level structure of the digital oscilloscope software. It describes what the software is to do and how it is to be organized. It can be used as the basis for low-level design and code development, while also serving as the basis for evolutionary change. This perspective has recently been elaborated into the idea of software *product families* (see Jayazeri et al. [2000]).

109.4 Architecture Description Languages

Since the early 1990s, many groups of investigators have proposed ADLs for describing software architectures. At first glance, these languages are quite different from each other. The goal of Medvidovic and Taylor [2000] is to provide some clarity by giving a framework for classifying and comparing ADLs. This framework looks at the way individual ADLs model architectural elements (components and connectors) and the way these elements are configured to form architectures. Model features considered by the framework include interfaces, types, semantics, constraints, evolution, and nonfunctional properties. Configuration features treated include understandability, refinement and traceability, heterogeneity, scalability, evolution, and dynamism. Medvidovic and Taylor also consider the tool support provided by various ADLs. In this section, we will give brief treatments of two rather different ADLs: Rapide [Luckham et al., 1995] and C2SADEL [Medvidovic et al., 1999]. Both ADLs were developed in the context of specific application domains.

Rapide is an event-based, concurrent, object-oriented language specifically designed for prototyping architectures of distributed systems. Its features include an execution model and executable architecture constructs, formal constraints, and mappings that support constraint-based definition of (industry standard) reference architectures and testing of systems for conformance to these standards. A Rapide architecture consists of *components* and *connections*. An architecture is described by specifying module (component) interfaces, connection rules that define communication between the interfaces, and formal constraints that define legal and illegal patterns of communication. These architectural aspects are

```
Type Application is interface
  extern action Request (p: params);
  public action Results (p: params);
behavior
  (?M in String) Receive (?M) => Results (?M);;
end Application

Type Resource is interface
  extern action Results (Msg: String);
  public action Receive (Msg: String);
end Resource
```

FIGURE 109.4 Examples of Rapide module interfaces.

```
architecture AP_RM_Only return X/Open is
    P: Application; Q:Resource; ...
    connect (?M in String)
        P.Request(?M) to Q.Receive (?M);
    end AP_RM_Only
```

FIGURE 109.5 Defining the flow of events among Rapide components.

expressed using distinct sublanguages. The interfaces of a Rapide architecture provide a template for a family of systems; specific systems are instantiated by assigning modules to interfaces. The architecture and each of its instances can be analyzed and executed.

An executing Rapide architecture generates a partially ordered set (poset) of events that interact with threads of control (called *processes*, which may be module processes, architecture connections, or interface behavior transitions). Rapide architectural models use an execution semantics built on three event dependence relations:

A and *B* are generated by the same process, and *A* is generated before *B*.

A process observes *A* and generates *B*.

A process generates *A* and writes to a variable *v*; another process within the same module reads *v* before any intervening write to *v*, and then generates *B*.

Module interfaces are specified by a *types language*. The example shown in Figure 109.4 [Luckham et al., 1995] defines Application and Resource components. In Application, the observation of a Receive event causes a Results event to be generated with the same parameter.

An *architecture language* is used to describe the flow of events between components. In the simple example shown in Figure 109.5 [Luckham et al., 1995], a connection is defined between Application module *P* and Resource module *Q*. Whenever *P* generates a Request event, *Q* will observe a Receive event with the same data.

More complex connection patterns are possible. Algebraic constraints and pattern constraints can be used to control how events may occur. Figure 109.6 shows a specification [Luckham et al., 1995] in which

The –> operator requires that the Receive and Result events occur in dependent pairs.

The * operator asserts that any number of pairs may occur.

The ~ operator asserts that the events in the pairs must be distinct.

The ?S parameter is used to require that the Receive and Results events have the same parameter.

Rapide has been applied to specify the X/Open distributed transaction processing (DTP) reference architecture. Interfaces in the published X/Open standard are formalized as Rapide interface types, and textual

```
type Resource is interface
  public action Receive (Msg: String);
  extern action Results (Msg: String);
  constraint
    match
      ((?S in String) (Receive (?S) -> Results (?S)))^(*~);
end Resource
```

FIGURE 109.6   Complex connections among Rapide components.

descriptions of calling sequence protocols are formalized as Rapide connection rules and pattern constraints. Rapide has been used to test the conformance of a combination of two X/Open DTP subsystems with the reference architecture. This is done by constructing a map between the two Rapide models that preserves the necessary structures and constraints.

C2ADEL [Medvidovic et al., 1999] was developed explicitly for modeling architectures within the C2 architectural style [Taylor et al., 1996], which was itself developed for use in the GUI application domain. A C2 architecture is specified as a topology that defines its components and connectors and their interconnections. The components and connectors are instantiated from type definitions. This makes it possible to use subtyping and type checking. The example illustrated in Figure 109.7, taken from Medvidovic et al. [1999], shows how an architecture is specified in C2SADEL. Note the explicit top and bottom connections, which enforce the requirement of the C2 style that components are linearly ordered into layers that may only communicate with components immediately above or below them.

Components may be virtual (not defined within C2SADEL) or external (defined elsewhere in a specification). Figure 109.8 gives a specification for the DeliveryPort component taken from Medvidovic et al. [1999]. Note that # denotes set cardinality and ~ indicates the value of a variable after an operation has been performed. The invariant requires that the current capacity of a port be between zero and the maximum capacity. The example illustrates how operations required and provided by a component can be specified.

Many ADLs use formal methods and notation that are likely to be unfamiliar to practitioners in industry. Examples include posets in Rapide and first-order logic in C2SADEL. This may present an obstacle to widespread use of software architecture modeling. An alternative approach is presented in Robbins et al. [1998]. The authors show how universal modeling language (UML) metamodels and stereotypes can be used to represent C2 architecture models in UML.

109.5   Describing Architectural Styles

Architectural styles are "sets of design rules that identify the kinds of components and connectors that may be used to compose a system or subsystem, together with local or global constraints on the way the composition is done" [Shaw and Clements, 1996]. The importance of identifying useful styles is widely recognized [Abowd et al., 1995]. This section presents a formalism and methodology for describing both software architectural styles and architectures built within specific styles. The approach is embodied in a language called ASDL [Rice and Seidman, 1996], which can be used to compare architectural styles and particular architectures, and also to gain an increased understanding of complex architectural styles. One important feature of ASDL is its powerful and flexible capabilities for abstraction and representation of hierarchy in descriptions of software architectures.

An ASDL description of a software architecture or architectural style is made up of three elements: templates, settings, and units. *Templates* represent interfaces of components that are available for inclusion into a software architecture. *Settings* represent architectures that have been built by instantiating templates. *Units* represent system hierarchy: they can encapsulate settings so that their interfaces can be used in turn as templates or represent interfaces designed in a top-down manner. Each element is associated with a

```
Architecture CargoRouteSystem is
  component_types {
     component DeliveryPort is extern (Port.cs;)
     component graphicsBinding is virtual ()
  }
  connector_types {
     connector FiltConn is (filter msg_filter;)
     connector RegConn is {filter no_filter)
  }
  architectural_topology {
     component_instances {
     Runway : DeliveryPort
     Binding : GraphicsBinding;
     }
  connector_instances {
     UtilityConn : FiltConn;
     BindingConn: RegConn;
     }
     connections {
        connector UtilityConn {
           top SimClock, DistanceCalc;
           bottom Runway, Truck;
        }
        connector BindingConn {
           top LayoutArtist, RouteArt;
           bottom Binding;
        }
     }
  }
}
```

FIGURE 109.7 C2SADEL architecture specification.

generic Z schema [Spivey, 1989], which is invariant across all styles. A brief introduction to Z syntax is given in the appendix to this chapter.

Structural aspects common to all styles are expressed by the generic schemas, whereas features that are characteristic of a particular style are expressed by style-specific versions of the schemas. This is done by specifying the parameters of the generic schemas and by adding style-specific declarations, constraints, and operations. The resulting schema versions constrain the configuration of the elements that make up an architecture. The ASDL schemas use a common formalism to describe both syntax and semantics of architectures and styles. The use of Z gives the language a flexible and modifiable foundation. ASDL also provides operations that support construction of a software architecture within a style.

The Z schemas of ASDL are similar to those used in Rice and Seidman [1994] to describe MILs. Figure 109.9 shows the generic schema that describes ASDL templates. It provides a collection (library) of templates that represent interfaces of components available for inclusion in a software architecture. The schema parameters (*Indices, Attributes, Parts*) are used to customize the schema to describe a specific style. The schema components define the template's interface, which consists of a finite set of ports that can be used for data communication or other relationships between components. The *port-attr* function determines the characteristics of a port by giving it attribute values (taken from the style-specific parameter *Attributes*) with respect to members of a set of indices (represented by the style-specific parameter *Indices*).

```
Component DeliveryPort is
  subtype CargoRouteEntity {int \and beh {
    state {
       cargo : \set Shipment;
       selected   : Integer;
    }
  invariant {
     (cap >= 0) \and (cap < = max_cap);
     connector RegConn is {filter no_filter)
  }
  interface {
     prov ip_selshp:     Select (sel: Integer);
     req      ir_clktck: ClockTick();
  }
  operations {
     prov op_selshp: ;
        let num : Integer ;
        pre num <= #cargo ;
        post ~selected = num ;
     }
     req    or_clktck {
        let time : STATE_VARIABLE;
        post ~time = time + 1;
     }
  }
  map {
        ip_selshp -> op_selshp (sel -> num);
        ir_clktck -> or_clktck () ;
     }
}
```

FIGURE 109.8   C2SADEL component specification.

ASDL_Library [Indices, Attributes, Parts]

---

interfaces : Templates $\rightarrowtail$ $\mathbb{F}_1$ Ports
port-attr : Ports $\nrightarrow$ Indices $\rightarrow$ Attributes
part   : Templates $\nrightarrow$ Parts
interp   : Templates $\nrightarrow$ Interpretations
Collection : $\mathbb{F}_1$ Templates

Primitives $\subseteq$ Collection
Collection $= dom$ interfaces $= dom$ interp $= dom$ part
disjoint ran interfaces
dom port-attr $= \cup$ *ran* interfaces
$\{dir, type\} \subseteq$ Indices $\land \{in, out\} \subseteq$ Attributes
$\forall p \in dom$ port-attr $\cdot$ port-attr$(p).dir \in \{in, out\}$

FIGURE 109.9   Z schema defining ASDL templates.

interfaces($filter_f$) = $\{p_f, q_f\}$
interfaces($split$) = $\{p_s, q_s, r_s\}$
interfaces($merge$) = $\{p_m, q_m, r_m\}$
port-attr(p_f)(dir) = in; port-attr(q_f)(dir) = out
port-attr(p_s)(dir) = in; port-attr(q_s)(dir) = port-attr(r_s)(dir) = out
port-attr(pm)(dir) = port-attr(q_m)(dir) = in; port-attr(r_m)(dir) = out
interp($filter_f$) = *(p_f ? x → q_f ! $f(x)$ → **SKIP**)
interp($split$) = *(p_s ? x → (q_s ! x → **SKIP**||r_s ! x → **SKIP**))
interp($merge$) = *((p_m ? x → r_m ! x → **SKIP**) $\square$ (q_m ? x → r_m ! x → **SKIP**))
part($filter_f$) = part($split$) = part($merge$) = $filter$

FIGURE 109.10   Pipe-and-filter interfaces.

The function *part* assigns each template to a style-specific category. The function *interp* defines a template's interface semantics by associating the template with a composition of guarded CSP processes [Hoare, 1985]. For example, if a template τ has ports p and q with direction attributes *in* and *out*, respectively, then the CSP process *interp* (τ) = *(p ? x → q ! $f(x)$ → **SKIP**) specifies that the template acts as a filter represented by the function f.*

Some templates are identified as primitive templates; these correspond to interfaces of software system components that have been preloaded into the library. *Collection* represents the set of templates that can be used to construct software systems. The members of *Collection\Primitives* are templates that correspond to interfaces of encapsulated composite modules. Members of *Templates\Collection* can serve as *reference templates*, which correspond to interfaces designed in a top-down fashion.

The schema constraints define the requirements needed for any style. For example, they require that each template has a nonempty interface, a category assignment, and interface semantics; that the interfaces of distinct templates are disjoint; that attribute values are defined for all interfaces; and that category and semantics are given for each template. Furthermore, they require that *dir* and *type*, which give a port's direction and data type, are indices supplied for all styles, and that the only acceptable attribute values for *dir* are *in* and *out*.

As an example, we will consider the pipe-and-filter architectural style (see also Allen [1997], Abowd et al. [1995], and Shaw and Garlan [1996]). Our version of this style uses a *filter* component that applies a function f to its input and produces an output, a *split* component that sends its input to each of its two outputs, and a *merge* component that performs a nondeterministic merge of its two inputs. To specify the interfaces of these components in ASDL, we define the schema parameters to be *Indices* = $\{dir, type\}$, *Attributes* = $\{in, out, float\}$, and *Parts* = $\{filter\}$. We then assume that ($filter_f$, *split*, *merge*) $\subseteq$ *Templates* (where f is a member of a set that includes all of the filter functions needed by an application), and that the schema components satisfy the requirements given in Figure 109.10.**

Settings (defined in the Z schema of Figure 109.11) represent architectures that have been built by instantiating templates as computational nodes. The nodes correspond to the components of a software architecture. A node has external interfaces called *slots* that correspond to (and inherit attributes from) the ports on the node's underlying template. Slots can be labeled; shared labels are used to represent relationships among nodes, such as data communication.

The schema components define the essential syntactic features of the nodes representing the components of an architecture: the templates from which the nodes were instantiated, the node interfaces (represented

---

*The process p ? x → q ! $f(x)$ → **SKIP** receives x on channel p, sends $f(x)$ on channel q, then terminates. The operator * indicates indefinite repetition of the filter process.

**In the *split* process, the || operator indicates concurrency, so that the input of x on the channel p_s is followed by the output of x on channels q_s and r_s. In the *merge* process, the operator $\square$ indicates deterministic choice, so that the first of the channels p_m or q_m to receive x outputs it on channel r_m.

ASDL_Setting [Indices, Attributes, Parts]

ASDL_Library [Indices, Attributes, Parts]
node-parent   :   Nodes $\longmapsto$ Templates
slots   : $\mathbb{F}$(Nodes $\times$ Ports)
slot-attr   :   Nodes $\times$ Ports $\longmapsto$ Indices $\rightarrow$ Attributes
label   :   Nodes $\times$ Ports $\longmapsto$ Labels
comp-expr   :   ProcessExpressions
semantic-descr :   Labels $\longmapsto$ SemanticDescriptions

---

slots $= dom$ slot-attr
dom label $\subseteq$ slots
dom semantic-descr $= ran$ label
$\forall_n \in dom$ node-parent $\bullet$
    node-parent$(n) \in$ Collection $\wedge p \in$ interfaces(node-parent(n))
        $\Rightarrow (n, p) \in dom$ slot-attr $\wedge$ slot-attr$(n, p) =$ port-attr(p)

FIGURE 109.11   Z schema defining ASDL schemas.

as ordered node–port pairs), and the interfaces' characteristics and labels. Other schema components contain information that can be used to determine the semantics of the architecture represented by the setting. The *semantic description* mapping assigns a semantic abbreviation to each label used in a module, and the *composition expression* specifies how the nodes in a setting are composed for execution purposes. A composition expression is a CSP process in which node names are viewed as processes. For example, it may specify that the nodes in a setting will be executed in parallel. The members of *ProcessExpressions* are described in Rice and Seidman [1996]. The node constraints require that the slots representing the interfaces of a node *n* consist of the pairs (n, p), where *p* is a port of node-parent(n), that the slot attributes are inherited from those of the corresponding ports, and that all slot labels are associated with semantic abbreviations.

Note that ASDL uses the Z and CSP formalisms orthogonally, so that there is no need to propose a common semantic domain for the two formalisms. The use of CSP is confined to providing a process algebra value for the *comp-expr* variable of the **ASDL_Setting** schema and for the interpretation of each template. The character strings assigned to these elements correspond to CSP process algebra expressions.

The semantic abbreviation associated with a label represents a communication protocol, as well as additional style-dependent information. The set *SemanticDescriptions* contains abbreviations that correspond to a variety of communication capabilities, and the mapping *semantic-descr* assigns an abbreviation to each label in a setting. For example, the abbreviations **uac** and **usc** represent unidirectional asynchronous and synchronous communication, respectively, and **brod** represents broadcast of input data. Each abbreviation *a* has a *meaning* [a] and a set of associated *properties*, including its text description. For example, the meaning of **usc** is described by the CSP expression $[\textbf{usc}] = {}^*(in\,?\,x \rightarrow (out\,!\,x \rightarrow \textbf{SKIP}))$.

The associated properties may include an alphabet like $\{in, out\}$ or an alternate specification of the meaning, such as $out \leq in$ (each trace on *out* is a prefix of a trace on *in*). Other properties might include timing information or a restriction on the buffer size for an asynchronous protocol. In some cases, the meaning of an abbreviation is parameterized by a potential set of connections. For example, the meaning of the broadcast abbreviation is defined by

$$[\textbf{brod}](S) = {}^*(in\,?\,x(out_s\,!x \rightarrow \textbf{SKIP} : s \in S))$$

The execution semantics of a module can be derived from the semantic interpretations of the templates underlying the nodes, the composition expression, and the semantic descriptions of the labels that specify

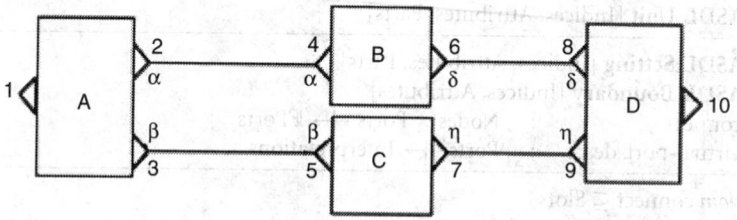

FIGURE 109.12 A pipe-and-filter architecture.

the connections between nodes. The **ASDL_Setting** schema therefore contains the basic components and the information needed to simulate the execution of the module.

To illustrate these concepts, we return to the pipe-and-filter style. Figure 109.12 shows a pipe- and-filter architecture with four components: A sends its input to B and C, which are filters (using the functions f, g, respectively) whose results are merged by D.

An ASDL description of this architecture corresponds to a setting that uses the nodes A, B, C, and D. The function *node-parent* is defined by the ordered pairs (A, split), (B, filter$_{1,1,f}$), (C, filter$_{1,1,g}$), (D, merge). The setting uses the following ten slots:

$1 = (A, p_s)$	$2 = (A, q_s)$	$3 = (A, r_s)$	$4 = (B, p_f)$	$5 = (B, q_f)$
$6 = (C, p_g)$	$7 = (C, q_g)$	$8 = (D, p_m)$	$9 = (D, q_m)$	$10 = (D, r_m)$

In the figure, Greek letters are used to represent slot labels, and shared labels indicate data communication between slots. The function *label* associates labels with slots; it is defined by the ordered pairs (2, α), (3, β), (4, α), (5, β), (6, δ), (7, η), (8, δ), (9, η). The *semantic-descr* function assigns the abbreviation **usc** (unidirectional synchronous communication) to all four labels. The composition expression associated with this setting represents the concurrent composition of its components.

The following four constraints are associated with this style:

- $b \in ran$ label $\Rightarrow$ | label $(b)| \leq 2$
- $\forall b \in ran$ label . semantic-descr$(b) = $ **usc**
- $n \in dom$ node-parent $\Rightarrow$ part (node-parent$(n)) = filter$
- $\forall s, t \in dom$ label . label$(s) = $ label$(t) \Rightarrow$ port-attr(second(s)).dir $\neq$ port-attr(second(t)).dir

The first three constraints require that no more than two slots can share the same label, that all labels represent unidirectional synchronous communication, and that all nodes are instantiated from filter templates. The final constraint states that if two slots share a label, the underlying ports must have the opposite direction.

The **ASDL_Setting** schema represents an architecture as a self-contained computational unit without any external connections. Figure 109.13 shows the **ASDL_Unit** and **ASDL_Boundary** schemas that describe these connections and the associated interface semantics. They include a set of *virtual ports* that represent the public interfaces of the unit and a mapping that specifies the attributes of these ports. The mapping *virtual-port-descr* assigns semantics to each virtual port in a unit. The *connect* mapping describes the links between slots and virtual ports.

ASDL_Unit imposes only a minimal restriction on the interface, which enforces consistency with respect to the direction of data movement. Further restrictions are based on style-dependent information about the desired behavior of units. For example, type-consistency requirements may be placed on the *connect* mapping, and the *virtual-port-descr* mapping may specify broadcasting or multiplexing behavior for a virtual port.

ASDL_Unit [Indices, Attributes, Parts]

ASDL_Setting [Indices, Attributes, Parts]
ASDL_Boundary [Indices, Attributes]
connect   :   Nodes × Ports $\rightarrowtail$ $\mathbb{F}$Ports
virtual-port-descr   :   Ports $\rightarrowtail$ Interpretations

---

dom connect ⊆ Slots
∪ *ran* connect ⊆ virtual-ports
dom virtual-port-descr = virtual-ports
$\forall_p$ ∈ virtual-ports . {interface-attr(p).dir} = {slot-attr(s).dir : p ∈ connect(s)}

ASDL_Boundary [Indices, Attributes]

---

interface-attr   : Ports $\rightarrowtail$ Indices → Attributes
virtual-ports   : $\mathbb{F}$Ports

---

virtual-ports = *dom* interface-attr

FIGURE 109.13   Unit and boundary schemas.

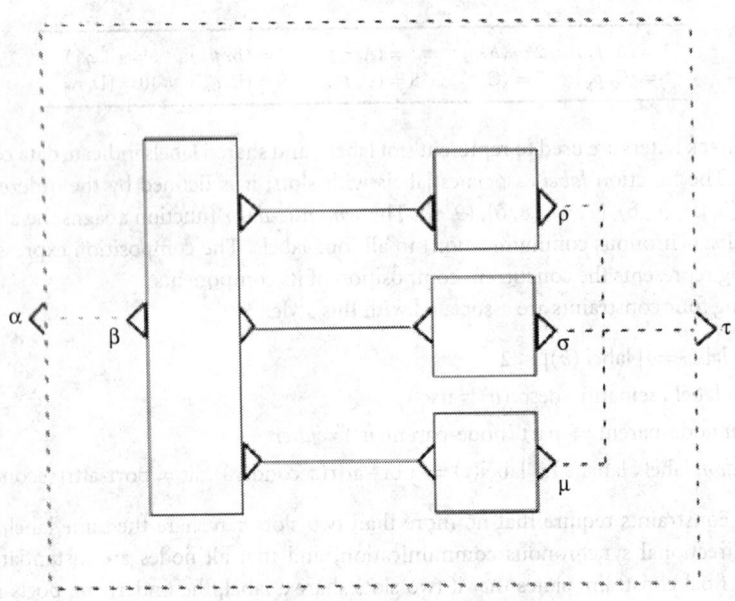

FIGURE 109.14   Multiplexing semantics in a unit.

Because ASDL operations can be used to add the boundary of a unit to the library as a template, units provide powerful and flexible support for hierarchical descriptions of software architectures. The mapping of slots to virtual ports specified by the *connect* mapping need not be one-to-one. A software architect can use the *virtual-port-descr* mapping to specify the attributes of virtual ports and derive the semantics of the virtual ports from information about the setting associated with the unit: the interpretation of the templates used to instantiate the setting's nodes, the semantic abbreviations assigned to slot labels, and the setting's composition expression.

Figure 109.14 shows an architecture that has been converted into a unit that uses a multiplexing semantics. In this example, *virtual-ports* = {α, τ} and the *connect* mapping is constructed from the ordered pairs (β, α), (ρ, τ), (σ, τ), and (μ, τ). Note that this mapping is not one-to-one. The semantics of the virtual ports are defined by

$$\text{virtual-port-descr}(\alpha) = {}^*(a\,?\,x \rightarrow b\,!\,x \rightarrow \textbf{SKIP})$$

$$\text{virtual-port-descr}(\tau) =$$

$${}^*(((r\,?\,x \rightarrow \textbf{SKIP})\,\square\,(s\,?\,x \rightarrow \textbf{SKIP})\,\square\,(m\,?\,x \rightarrow \textbf{SKIP}))\,;\,(t\,!\,x \rightarrow \textbf{SKIP})$$

where a, b, r, s, m, and t are CSP channels corresponding to the slots and ports in the ASDL description.

ASDL contains a number of operations that support the incremental specification of software architectures. These serve as guides for the design of style-dependent operations that are constructed by adding new signatures and constraints to existing operations or by incorporating existing operations into a new operation. The operations include *setting operations* to create and delete nodes and pseudonodes, assign labels to slots, specify a composition expression, and select semantic abbreviations; *interface operations* to specify virtual ports, attributes, links, and virtual port descriptions; an *encapsulation operation* to create a new library template based on a unit; and operations that define the units needed to support a top-down design methodology.

For example, the encapsulation operation **ASDL_External** creates a new library template from an existing unit type. The virtual ports of the unit type become the ports of the template, and the attributes of these ports are derived from the unit's interface. The template's interpretation is derived from the interpretations of the templates underlying the nodes, the composition expression, the abbreviations of labels, and the semantics of the virtual ports. This represents a complex synthesis of the semantics of the entities associated with the unit.

The new library template can in turn be used to create a node in another module. ASDL permits a style-dependent interpretation of the extent to which the internal structure of the node is visible in the new module. For example, if encapsulation requires that each virtual port is linked to a node in the underlying unit, then one interpretation is that only the resources of the nodes linked to the port can be accessed through the port. On the other hand, if encapsulation permits a virtual port with no links, then another interpretation may allow characteristics of the node to be modified by using the port.

Defining Terms

Architectural analysis: Reasoning about descriptions of architectures and architectural styles.

Architectural style: A set of design rules that identify the kinds of components and connectors that may be used to compose a system or subsystem, together with local or global constraints on the way the composition is done.

Architectural view: A partial description of a software architecture that features architectural aspects useful for specific purposes or for specific categories of users.

Architecture description language (ADL): Language designed for describing software architectures.

Component: Software architecture entity that abstracts a computational activity.

Connector: Software architecture entity that abstracts a relationship between components.

Functional properties: The semantics of a software architecture's components.

Module interconnection language (MIL): A language that describes the properties of the interfaces between the modules of a software system.

Nonfunctional properties: Deal with an architecture's reliability, robustness, ease of use, conformance to standards, hardware requirements, and security.

Programming in the large: Building industrial-strength software systems.

Programming in the small: Development of small software applications.

Software architecture: An abstraction of a software system that allows a designer to ignore low-level implementation issues, such as programming languages, hardware and device requirements, and communication protocols.

Structural properties: The nature of the interactions supported by the architecture's components.

References

Abowd, G., Allen, R.J., and Garlan, D., 1995. Formalizing Style to Understand Descriptions of Software Architecture, *IEEE Trans. Softw. Engr.* 4: 319–364.

Allen, R.J., 1997. A Formal Approach to Software Architecture, Carnegie Mellon University School of Computer Science Technical Report CMU-CS-97-144.

Bass, L., Clements, P., and Kazman, R., 1998. *Software Architecture in Practice*, Addison-Wesley, Reading, MA.

Batory, and O'Malley, S., 1992. The Design and Implementation of Hierarchical Software Systems with Reusable Components, *ACM Trans. Softw. Engr. Meth.* 1: 355–398.

Buschmann, F., Meunier, R., Rohnert, H., Sommerlad, P., and Stal, M., 1996. *Pattern-Oriented Software Architecture: A System of Patterns*, John Wiley & Sons, Chichester, U.K.

Clements, P., Bachmann, F., Bass, L., Garlan, D., Ivers, J., Little, R., Nord, R., and Stafford, J., 2003. *Documenting Software Architectures: Views and Beyond*. Addison-Wesley, Boston, MA.

Clements, P., Kazman, R., and Klein, M., 2002. *Evaluating Software Architectures: Methods and Case Studies*, Addison-Wesley, Boston, MA.

Delisle, N., and Garlan, D., 1990. A Formal Specification of an Oscilloscope, *IEEE Softw.* 7: 29–36.

DeRemer, F., and Kron, H., 1976. Programming-in-the-large versus programming-in-the-small, *IEEE Trans. Soft. Engr.* 2: 80–86.

Garlan, D., and Perry, D., 1995. Introduction to the Special Issue on Software Architecture, *IEEE Trans. Softw. Engr.* 21: 269–274.

Hennessy, J.L., and Patterson, D., 1996. *Computer Architecture: A Quantitative Approach, 2nd edition*, Morgan Kaufmann, San Francisco, CA.

Hoare, C.A.R., 1985. *Communicating Sequential Processes*, Prentice Hall, New York.

IEEE, 2000. *IEEE Standard 1471-2000: IEEE Recommended Practice for Architectural Description of Software-Intensive Systems*, IEEE, New York.

Jayazeri, M., Ran, A., van der Linden, F., and van der Linden, P., 2000. *Software Architectures for Product Families: Principles and Practice*, Addison-Wesley, Boston, MA.

Leon-Garcia, A., and Widjaja, I., 2000. *Communication Networks: Fundamental Concepts and Key Architectures*, McGraw-Hill, New York.

Luckham, D., Kenney, J., Augustin, L., Vera, J., Bryan, D., and Mann, W., 1995. Specification and Analysis of System Architecture Using Rapide, *IEEE. Trans. Softw. Engr.* 21: 336–355.

Magee, J., Kramer, J., and Sloman, M., 1989. Constructing Distributed Systems in Conic, *IEEE Trans. Softw. Engr.* 15(6): 663–675.

Medvidovic, N., Rosenblum, D., and Taylor, R., 1999. A Language and Environment for Architecture-Based Software Development and Evolution, *Proceedings of 1999 Intl. Conf. on Softw. Engr.*, pp. 44–53.

Medvidovic, N., and Taylor, R., 2000. A Classification and Comparison Framework for Software Architecture Description Languages, *IEEE. Trans. Softw. Engr.* 26: 70–93.

Rice, M., and Seidman, S., 1994. A Formal Model for Module Interconnection Languages, *IEEE Trans. Softw. Engr.* 20: 88–101.

Rice, M., and Seidman, S., 1996. Using Z as a Substrate for an Architectural Style Description Language, Technical Report CS-96-120, Department of Computer Science, Colorado State University.

Richardson, D., and Wolf, A., 1996. Software Testing at the Architectural Level, *Proc. of the 2nd Intl. Softw. Arch. Workshop (ISAW-2)*, San Francisco, CA, pp. 68–71.

Robbins, J.E., Medvidovic, N., Redmiles, D.F., and Rosenblum, D.S., 1998. Integrating Architecture Description Languages with a Standard Design Method, *Proc. of the 20th Intl. Conf. on Softw. Engr.*, pp. 209–218.

Shaw, M., and Clements, P., 1996. Toward Boxology: Preliminary Classification of Architectural Styles, *Proc. of the 2nd Intl. Softw. Arch. Workshop*, pp. 50–54.

Shaw, M., and Garlan, D., 1996. *Software Architecture: Perspectives on an Emerging Discipline*, Prentice Hall, Upper Saddle River, NJ.

Spivey, J.M., 1989. *The Z Notation: A Reference Manual*, Prentice Hall, New York.

Stovsky, M., and Weide, B., 1990. Building Interprocess Communication Models Using STILE, in *Visual Programming Environments: Paradigms and Systems*, Ed. E. Glinert, pp. 566–574. IEEE Computer Society Press, Los Alamitos, CA.

Taylor, R., Medvidovic, N., Anderson, K., Whitehead, E., Robbins, J., Nies, K., Oreizy, P., and Dubrow, D., 1996. A Component- and Message-Based Architectural Style for GUI Software. *IEEE Trans. Softw. Engr.* 22: 390–406.

Tichy, W., 1979. Software Development Control Based on Module Interconnection. *Proc. of the 4th Intl. Conf. on Softw. Engr.*, pp. 29–41.

Further Information

An excellent source for the foundations of software architectures and architectural styles is *Software Architecture: Perspectives on an Emerging Discipline* [Shaw and Garlan, 1996]. This book has a strong research flavor and contains a good summary of early software architecture research. Although the Shaw and Garlan book is an essential basic reference, its applicability to realistic industrial situations may be less clear. This deficiency is remedied by *Software Architecture in Practice* [Bass et al., 1998], which provides a useful bridge between software architecture research and industrial practice. It contains significant industrial case studies and also deals with issues of architectural analysis and reuse. The recent book *Documenting Software Architectures* [Clements et al., 2003] addresses the question of how software architectures should be described to different communities of potential users. They advocate the explicit adoption of disparate **architectural views** and discuss the use of styles within each view.

Software architectures play a critical role in the software life cycle, and the adoption of an architectural perspective is now a recommended practice. This has been recognized in the recently adopted *IEEE Recommended Practice for Architectural Description of Software-Intensive Systems* [IEEE, 2000].

A critical activity in the software life cycle is the evaluation of potential architectures. This topic is treated extensively in *Evaluating Software Architectures: Methods and Case Studies* [Clements et al., 2002]. The book presents several architecture evaluation methodologies and illustrates them with case studies.

The use of architectural ideas in the design of software product families has received much attention in recent years. Research on this topic has a strong industrial flavor. An overview of the area can be obtained from the papers in Jayazeri et al. [2000].

Patterns are another way of capturing design knowledge, and they are often regarded as architectural in nature. If code construction is regarded as a low-level endpoint of the design axis, architectural styles can be found near the opposite endpoint. Architectures are close to styles, and patterns closer to construction. A good survey of patterns and their relationship to architectures can be found in Buschmann et al. [1996].

The Software Engineering Institute maintains a useful bibliography on software architecture (http://www.sei.cmu.edu/architecture/bibliography.html). The best place to look for current research on software architectures is in the proceedings of special-purpose conferences. Of particular interest is the Working IEEE/IFIP Conference on Software Architecture, which tends to include both industrial and academic papers (http://wicsa3.cs.rug.nl). Other, more general software engineering conferences that usually include papers on software architecture are ICSE (International Conference on Software Engineering, http://www.icse-conferences.org) and FSE (Foundations of Software Engineering, e.g., http://www.cs.pitt.edu/FSE-10).

Appendix: A Quick Introduction to Z Notation

The following discussion will introduce the reader to the basic elements of Z notation that are used in this paper. A more complete treatment can be found in Spivey [1989]. Z is based on typed set theory and uses *schemas* to define functions and types. A schema associates declarations of typed variables with predicates that constrain their possible values. The simplest variable types name familiar sets such as the natural numbers $\mathbb{N}$. More complex types are built using type constructors, which are analogues of familiar set operations:

power set formation ($\mathbb{P}$), products ($\cdot$), and function space formation ($\rightarrow$). A *schema definition* assigns a name to a group of variable declarations and predicates relating these variables. It has the following form:

S
declarations
predicates

A schema S can be used as a type, and the notation $w : S$ declares a variable w whose components are declared in S. For example, if x is a variable declared in S, then $w.x$ denotes the x component of w. A schema definition may also use *generic parameters* $X_1, X_2, \ldots, X_n$ associated with the schema name: $S[X_1, X_2, \ldots, X_n]$. These parameters are set constants that can be used in the schema definition.

A schema S can be included in the declarations of another schema T, in which case the declarations of S are merged with the declarations of T and the predicates of S and T are conjoined. An inclusion of S in T has the following form:

T
S
declarations
predicates

The schemas formed by merging the declarations of S and T are denoted by

$S \wedge T$ if the respective predicates are conjoined
$S \vee T$ if the respective predicates are disjoined

Schema definitions may be used to specify *operations* on a state specified by another schema. In this case, the following conventions are used for variable names:

undashed	state before	ending in ?	inputs
dashed'	state after	ending in !	outputs

Given a schema S, S' is the schema obtained by renaming all variables declared in S with dashes'.

$\mathbb{N}$	Set of natural numbers $\{0, 1, \ldots\}$
$\mathbb{Z}$	Set of integers $\{\ldots, -1, 0, 1, \ldots\}$

Given a set S

#S	Cardinality of S
$\mathbb{P}S$ ($\mathbb{F}S$)	Set of all subsets (all finite subsets) of S
$\mathbb{P}_1 S$ ($\mathbb{F}_1 S$)	Set of all nonempty subsets (all nonempty finite subsets) of S

Given $G \in \mathbb{P}S$

$\cup G$	Union of all subsets in the family G
disjoint G	Predicate which is true if and only if G is a pairwise-disjoint family

Given sets S and T

$S \times T$	Cartesian product of S and T
$S \setminus T$	Difference of S and T
$\exists x : T.P$	There exists x of type T such that P holds (a unique x if $\exists_1$ is used)
$\forall x : T.P$	For all x of type T, P holds
$\{x : T \mid P\}$	Set of all x's of type T such that P holds

Given sets S and T

    $S \leftrightarrow T$ Set of all relations from S to T

    $S \rightarrow T$ Set of all *total* mappings from S to T

    $S \rightarrowtail T$ Set of all *partial* mappings from S to T

    $S \nrightarrow T$ Set of all partial mappings from S to T with finite domains

The additional symbol $>$ on the left end (right end) of an arrow denotes a one-to-one (onto) mapping. For example, $\leftarrowtail$ denotes a one-to-one partial mapping with a finite domain. All the function operators are right-associative. For example, $f: S \rightarrow T \rightarrow V$ means that for each $x \in S$, $f(x): T \rightarrow V$. In this case, we write $f(x).y$ instead of $f(x)(y)$.

    Given $f: S \rightarrowtail g$

dom $f = \{x \in g : f(x) \text{ is defined}\}$
ran $f = \{f(x) \in T : x \in dom\ f\}$

Given $f: S \rightarrowtail g$ and $g \rightarrowtail V$

        $g \circ f: S \rightarrowtail V$ composition of f and g with *dom* $(g \circ f) = \{x \in dom\ f : f(x) \in dom\ g\}$.

Functions can also be defined by *lambda abstraction*, as in

$$square == \lambda n : \mathbb{N} . n * n$$

The domain is the set of natural numbers, and the expression defines the function.

110

Specialized System Development

110.1   Introduction **110**-1
110.2   Principles of Specialized System Development **110**-2
     Roots of Specialized System Development • Generic
     vs. Specialized Development • The Context of Problem
     Solving in Specialized System Development
110.3   Application-Based Specialized Development **110**-5
     Pervasive Software Development • Real-Time Software
     Development • Web-Based Software Development
     • Security-Driven Software Development
110.4   Research Issues and Summary **110**-14

Osama Eljabiri
New Jersey Institute of Technology

Fadi P. Deek
New Jersey Institute of Technology

110.1   Introduction

Software development is a complex problem-solving process with various interdisciplinary variables driving its evolution. Such variables are either problem-related or solution-based. *Problem-related variables* set the criteria for solution characteristics and help designers tailor solutions to specific problems. *Solution-based variables* explain current options, assist in future forecasting, and facilitate scaling solutions to problems. The issue of whether to find generic prescriptions to common problems (i.e., bottom-up generalization) or derive domain-dependent solutions to specific problems (i.e., top-down specialization) is debatable.

One viewpoint considers modern software engineering to be a standardized response that uses generic methodologies and strategies, as opposed to the nonsystematic approaches that characterized earlier software development. Standardization implies the use of generic rules, procedures, theories, and notations that mark a milestone in the development of any discipline. When standardization came, software development witnessed a paradigm shift from trial-and-error experimentation to scientific maturity, from differing representation and implementation of concepts to unified modeling and cross-platform independency, and from vague economic considerations to well defined, software-driven business models. The competing viewpoint of "one size fits all" has not proved to be practical in real-world software development (Glass, 2000). There is no one methodology appropriate for every case, no strategy that works perfectly for every problem, no off-the-shelf-prescriptions that can be applied directly without scalability, tailorability, or customization. Even specific approaches that fit certain situations do not necessarily fit them all the time, because change is the only constant in contemporary business. Evolving needs accompany innovation and emerging technologies. It can be argued that a balanced approach between generalization and specialization can be adopted to achieve effectiveness in software development.

This chapter addresses the notion of specialized system development. The field of **system specialization** has been overlooked in the software engineering literature since the discipline was formally launched. Also, generic software development had only provided a **weak strategy** (Vessey and Glass, 1998) to solve problems because it only supplies guidance for solving problems and not actual solutions to problems at hand. Scalability, tailorability, and specialization have become relevant issues in the software industry and software engineering research. Even general applications are not actually generic. Many current applications support customization features. Additionally, these systems are released in various modes, which range from standard to professional to enterprise editions, suiting a diversity of needs and problem complexity. Such applications also evolve over time to reflect changes in business requirements and technological capabilities. Subsequent sections of this chapter define specialized system development, discuss its drivers, present its advantages and disadvantages, and explore the types of specialized system development and its categories. We also consider the need for specialized system development and how that can be mapped to team structures.

110.2   Principles of Specialized System Development

According to the Merriam-Webster dictionary, to *specialize* is to concentrate one's efforts in a special activity or field or to change in an adaptive manner. Concentration leads to more attention to detail and presumably enables more efficient problem solving. Specialization links theory to practice and makes it more meaningful. Generally speaking, specialized system development is about developing software systems with focus. The focus may be on the application domain, a certain phase of the development cycle, or a specific system development methodology. An example of application-domain focus is software development for **pervasive computing**, including wireless and portable systems. An example of development-phase focus is special emphasis on project management, requirements analysis, or architectural design, as opposed to generic knowledge in software engineering. An example of methodology focus is systems development using structured or object-oriented strategies. However, specialization in methodology spans a wider array of approaches and tools. This includes software development process models (i.e., problem-solving strategies), CASE tools, and implementation techniques. Application-focused software development is the most frequently used definition for specialized system development in the current software industry.

Application-focused development can be classified into two categories: application-oriented and infrastructure-oriented. Each of these two categories can have a problem focus or a solution focus. Problem focus can be based on the type of industry involved or the application domain. Solution focus can be based on custom development, package development, or development aid (Glass and Vessey, 1995).

110.2.1   Roots of Specialized System Development

The history of specialized system development is tightly coupled with the evolution of computer hardware and technology advancement.

110.2.1.1   Domain-Dependent Era (Pre–Software Development Methodology)

During the period of 1955 to 1965, computer hardware was application-dependent. It was virtually impossible to develop business and scientific applications on the same machine. Medical applications and unusual applications were two examples of the distinct focus of application development during this era. Problem-oriented languages, such as Fortran, COBOL, and ALGOL, were developed to translate old software to be compatible with the requirements of new computers. **Domain-specific focus** was the major driver in building successful software systems. New disciplines emerged to support these systems, such as numerical analysis and information retrieval (Vessey, 1997).

110.2.1.2   Domain-Independent Era (Early Software Development Methodology)

This era comprised the period between 1964 and 1980. In 1964, the IBM 360 was introduced, including the lower-midrange model 40 and the model 67, shipped with hardware to support virtual memory. The IBM

360 combined scientific and business applications in one machine. The sociology of software development was strongly influenced by the 360's ability to end the separation between scientific and business applications. Generic applications were possible when the software business became independent of hardware vendors. Competitive advantage in software development became directly proportional to the interdependency of standards, hardware, and platforms. This era witnessed many attempts to institutionalize application-independent software development strategies (Vessey, 1997) and led to the building of a solid foundation for the next era of methodology-intensive software development.

110.2.1.3 Generic Applications Era (Methodology-Intensive Software Development)

The period of 1980 to 1995 witnessed the birth and evolution of desktop PC computing and laptop computing. With the availability of computers and the high degree of usability, user involvement became more dominant, the availability of technology facilitated automation efforts in software implementation, and nontechnical users became active participants in the process (Glass, 1998). User-friendly GUIs took over job control language (JCL) taking human–computer interaction (HCI) to a new level. Some attempts at developing application-dependent software (such as fourth-generation languages, rule-based languages, and simulation languages) were also carried out (Vessey, 1997).

110.2.1.4 Return to Application-Focused Development (Software Development Post-Methodology)

From 1995 to the present, the evolution of networked hardware architecture has been dominant. Developing Web-based applications marked a milestone in this era, along with the emergence of Web-driven tools and programming languages (i.e., HTML, Java, JavaScript, XML, VML, etc.), the evolution of friendly Web interfaces through Internet browsers and e-mail agents, the emergence of Web-based software engineering as a software development methodology, the increasing demand for software that balances speed and quality, and the synchronization of business processes and software evolution.

110.2.2 Generic vs. Specialized Development

The shift from domain-specific computers to application-independent ones was an important event for software development. The subsequent advancement of application-independent computers into desktop and notebook computing was another milestone, which marked a shift toward generic infrastructure systems, applications, and components with notable advantages such as the following:

Portability — Software can be used virtually anytime and anywhere because of the development of generic Web-based downloading and installation protocols.

Compatibility — One operating system can host a vast number of applications, regardless of their vendors. Generic operating systems are a central repository for shared components across applications.

Reusability — One application or one module can be used across computer models, organizations, and user groups. It can be distributed over an organizational network or the World Wide Web. It can also be reused to develop new releases of software implementations. Furthermore, with few modifications through built-in preferences or options, the same application can be customized or tailored to a variety of specific needs.

Ease of training — Generic applications are easier to learn because of their availability, and training material is inexpensive (or even free), due to the use of mass production techniques.

Cost-effectiveness — Because operational costs are generally lower with mass production and sales volume is usually high, products can be sold at competitive prices to the end user.

Generic applications also have disadvantages worth noting. For example, such applications are based on the assumption that there are no significant differences among individuals and/or organizations that require special tailorability or scalability. This assumption applies also to generic methodologies and strategies in software development. These methodologies are rarely based on the type or size of the project or on technology environments and organizational settings. Such methodologies are considered

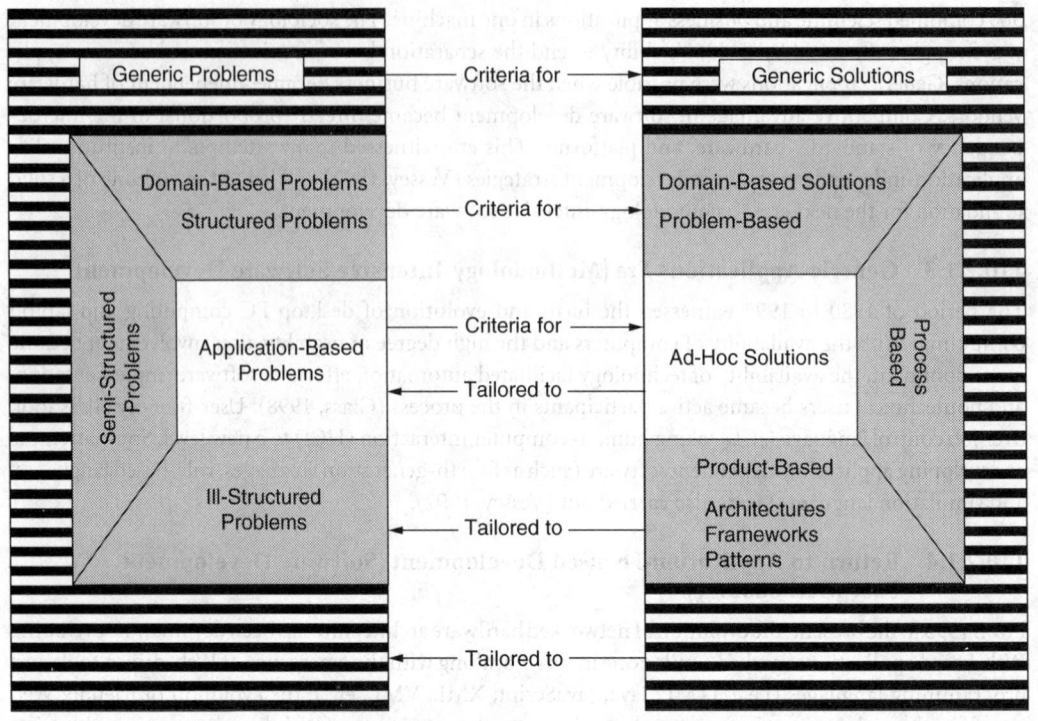

FIGURE 110.1   Generic and specialized software development in the problem-solving context.

one-dimensional approaches, because they often do not mirror a particular organization's underlying social, political, and organizational development dimensions (Avison and Fitzgerald, 2003). Generic applications also assume that businesses or individuals should be able to adapt to the infrastructure and functionalities of generic applications with limited room for changes. This assumption can be true within the same application domain, but it may be untrue for another, causing extreme ineffectiveness.

Additionally, the assumption that business processes can easily be changed to fit a generic software product is unrealistic and costly. Diversity of goals, market demands, stakeholder requirements, architectural specifications, nonfunctional requirements, and organizational cultures across business domains and specializations makes generic development strategies impractical. For some organizations, adopting a specific methodology may not lead to the desired result, and it can lead to rejecting methodologies altogether (Avison and Fitzgerald, 2003). Agile software development may be viewed as a response to this difficulty.

110.2.3   The Context of Problem Solving in Specialized System Development

Because software development essentially aims to solve problems, it is important to view specialized system development in the problem-solving context. Basically, solving problems involves two key elements: the ability to comprehend the problem and the capability to solve it. Hence, specialized system development is either problem-focused or solution-driven (see Figure 110.1). Because problem types and solution strategies in software engineering vary, effective understanding of their diversity is a precondition to successful specialized system development. In fact, this diversity is a major driver of specialized development, because differences are the catalyst for any specialization.

Relevant challenges for the developer include understanding how specialization, in identifying problem characteristics, can help to evaluate existing options, select the most proper options, and use domain analysis and requirements engineering to develop effective solutions. How can software products or solutions be adequately used, reused, customized, personalized, reengineered, or redeveloped based on application-driven or domain-specific specialization? How can specialization in problem, method, product, or domain analysis assist in proper selection or successful construction of computer-based solutions that utilize suitable methods, process models, techniques, and tools?

Careful examination of problem and solution diversity reveals three key drivers for specialized system development: characteristics of the system to be developed (as well as characteristics of the system's anticipated users); solution-driven capabilities, experience, and knowledge; and characteristics of system developers.

110.2.3.1 Characteristics of the System to Be Developed

This is a problem-focused category. Diversity of software systems in terms of size, complexity, time constraints, scope, underlying technology, business goals, and problem environment are its most critical drivers. Problems range from structured, at the operational levels of organizations, to semistructured at the tactical level, to ill structured at the top management or strategic level (**vertical specialization**). Problem specialization can be between organizations in the same industry, across industries (external **horizontal specialization**) or within the same organization across its various functional departments or key business processes (internal horizontal specialization).

110.2.3.2 Characteristics of System's Anticipated Users

This is also a problem-focused category. Some of the drivers in this category are age considerations, gender considerations, purpose in using the system (i.e., personal vs. business users), user background (i.e., technical vs. nontechnical users), and user environment. User environment includes, but is not limited to, cultures, languages, geographic locations, technical resources, financial resources, human resources, and legal and ethical issues. Each of these creates certain needs in systems development and therefore triggers specific specializations in responding to these requirements.

110.2.3.3 Solution-Driven Capabilities, Experience, and Knowledge

System specialization under this category is based on tools and resources, rather than on application domain. This includes capabilities and experience in project management tools, requirements analysis techniques, architectural models, user interface approaches, database management strategies, implementation languages, development tools, development methodologies, and process models. These capabilities affect numerous specializations in the solution area.

110.3 Application-Based Specialized Development

The convergence of three traditional computing specializations — personal, networking, and embedded — produced a new computing era referred to as *pervasive computing*. Mobile computing, wireless devices, PDAs, Pocket PCs, and Tablet PCs are all examples of pervasive computing products. Software applications are important components of these products, and the distinct nature of these applications brings a new set of challenges to software development.

110.3.1 Pervasive Software Development

Pervasive applications can be distinguished by the following characteristics: ubiquity, interconnectedness, and dynamism. These applications strive to be embedded, distributed, nonintrusive, and cost-effective (Ciarletta and Dima, 2000). This implies that software economics, system architecture, and security are significant issues in pervasive software engineering. A conceptual model is suggested to highlight the aspects of pervasive system development in which four layers have been identified: physical, resource,

TABLE 110.1   The Roles of Pervasive System Development Layers

Layer	Rationale	Software Development Ramifications
Physical	The flow of control in pervasive applications may depend on signals received from or by the user's physical body. Excellent software architecture is ineffective in pervasive devices, unless it is well supported by hardware design that mirrors physical characteristics of humans.	Designing effective hardware architectures is crucial to software design because software effectiveness is dependent on hardware usability and hardware is irreplaceable (in contrast with desktop computing).
Resource	Represents the infrastructure of pervasive software applications (operating systems, logical devices, system API, user interface, network protocol).	ROM-based operating systems should be reliable with early releases because it will be very costly to make any upgrades thereafter. System resources must be matched to user goals and needs. User interfaces must be intuitive and consistent. They must accommodate users' language and physical limitations. Networking features should be automatically available, self-configuring, and compatible with existing technology. System storage must enable users to access, retrieve, and organize information in a way that suits their requirements. Execution environment and volatile memory should be responsive and provide both speed and sense of control via multithreading and multitasking.
Abstract	Represents the direct software application that the user will use.	Maintaining compatibility between users' mental model expectations and application logic "state." Shorter time frames are available to pervasive system users for learning about the system, compared with desktop users. More difficult physical conditions are encountered by mobile users of pervasive systems. User involvement and participation is much more critical in pervasive applications than in traditional applications.
Intentional	Represents user goals and purposes in using the pervasive system.	Analyzing the system to determine user goals and designing the system to fulfill these goals.

abstract, and intentional (Ciarletta and Dima, 2000). Table 110.1 describes the roles of these layers in specialized pervasive system development.

A framework of four levels can provide a sound process for developing effective mobile commerce (m-commerce) applications (Varshney and Vetter, 2001). These four levels are as follows:

M-commerce applications — These modify e-commerce applications for a mobile environment.

Wireless user infrastructure — New m-commerce applications should support the capabilities of user infrastructures. For example, m-commerce applications must be effective for such mobile devices as PDAs and cell phones.

Mobile middleware — The new m-commerce applications must have better response time and reliability when deployed, because the middleware will be used to connect e-commerce applications to different wireless networks.

Wireless network infrastructure — Networking requirements must be fulfilled by the m-commerce applications being deployed. Such requirements include quality of service, network reliability, location management, roaming across multiple networks, and multicast support.

Effective m-commerce applications can be deployed if network reliability and redundancy are increased. Furthermore, creating m-commerce applications requires unique knowledge and needs-specific networking support to create effective applications (Kalakota et al., 2000), which includes wireless quality of service (QoS), efficient location management, and reliable and survivable networks.

110.3.2 Real-Time Software Development

Real-time software development originated in the 1970s and continues to evolve today. The development of real-time systems requires consideration of three basic issues (Felder, 2002): complex timing (at the higher requirements specification levels), resource constraints (at lower design levels), and scheduling constraints (at lower design levels).

Goulding and Lawson (1976) describe a disciplined engineering approach to real-time software development with a focus on a process design methodology. The basis for this approach is a *process performance requirement*, a document describing the software interfaces, the software functional and performance requirements, the operating rules, and the data processor hardware description. The goal of **process design engineering** is the development of an automated approach to the evolutionary design, implementation, and testing of real-time software. Goulding and Lawson define the crucial aspects of effective real-time software development to include four important components:

Transformational technology — Enables traceable transformation from functional requirements to a software structure for a given computer

Architectural approach — Requires top-down design, implementation, and testing techniques supported by a single process design language

Simulation technology — Provides a capability for evaluating trial designs for real-time software processes

Supporting tools — Automate such functions as requirements traceability, configuration management, library management, simulation control, and data collection and analysis.

An early software development life-cycle method for real-time systems was proposed by Gomaa (1986). This method attempts to tailor generic software development methodology to reflect the special needs of real-time software development. Table 110.2 describes this method, its phases, and its applications.

110.3.3 Web-Based Software Development

Web-based software development is growing at a faster rate than any other domain. Software systems with Web capabilities can maximize the business added value more effectively, with their ability to reach customers and partners and to enrich the business process with information (Evans and Wurster, 1999). Three criteria to assess business value in IT-based systems are productivity, business profitability, and consumer surplus (Hit and Brynjolfsson, 1996). Web applications extend traditional business goals to encompass measures of customer satisfaction, internal processes, and the organization's innovation activities. These operational measures affect organizational financial performance (Der Zee and de Jong, 1999). Efficiency, quality, market share, and penetration have emerged as important measures and goals of business (Singleton et al., 1988) that can be improved by Web-based systems. These influences have motivated industry to integrate Internet/intranet information systems in their businesses and adopt new management techniques to align new technology with the organizational structure.

110.3.3.1 E-business Software Systems

There are now more demands on quality and reliability for Web-based software development than ever before. Successful configuration of Web applications requires special attention to several interrelated strategies that leverage Web engineering to the level of competitive advantage. Development teams, legacy systems, value chain, and business integration and management structures drive these strategies.

TABLE 110.2   Life-Cycle Phases for Real-Time Software Systems

Phase	Phase Definition	Phase Application
Requirements analysis and specification	As in other approaches, user requirements are analyzed, and system specifications are formulated to elaborate on these requirements.	State transition diagrams are used to describe the different states of the system to the user. Object-oriented UML-based state transition diagrams carry out this technique more effectively. Any operator interaction with the system should also be explicitly specified. Throwaway repaid prototyping techniques have proved to be extremely effective in requirements analysis for real-time systems.
System design	While the system is structured into tasks as in other software systems, real-time systems are designed with a specific focus on concurrent processes and task interfaces.	The asynchronous nature of the functions within the system is a key characteristic that distinguishes decomposing real-time software systems into concurrent tasks. Data flow diagrams and event-trace diagrams are effective techniques in mapping this phase.
Task design	Each task is structured into modules, and module interfaces are defined.	Task-structure charts with intensive project and team management elements are essential to carrying out task design efficiently.
Module construction	Detailed design, coding, and unit testing of each module are carried out.	This is similar to module construction in other system development approaches.
Task and system integration	Modules are integrated and tested to form tasks, which in turn are gradually integrated and tested to form the total system.	Incremental system development is used to achieve task and system integration.
System testing	The whole system or major subsystems are tested to verify conformance with functional specifications. To achieve greater objectivity, system testing is best performed by independent test teams.	Automated testing is widely used for real-time systems.
Acceptance testing	This is performed by the user.	Extends user involvement to the validation and verification stages after system delivery.

110.3.3.1.1 *Skills, Structure, and Management of the Development Team*

In Web-driven software development projects, skillful staff can significantly boost performance. Training programs and availability of necessary resources have a strong influence on the quality of e-business applications, reducing the development time for tailoring solutions to application needs. Effective management can create the right team structure and the necessary synergy from diverse abilities.

110.3.3.1.2 *Legacy Applications*

The scope and domain of legacy systems shape the strategies needed to solve e-business software problems. The negative correlation between organizational complexity and the impact of technical change is a disputable one (Keen, 1981), because the more complex the organization, the more ill structured its business problems are (Mitroff and Turoff, 1973). Even though this influences the ability to tackle such problems smoothly, information technology enables a complex organization to redesign its business processes so that it can manage complexity more effectively (Davenport and Stoddard, 1994).

110.3.3.1.3 Value Chain and Logistics Management

Value chain is the set of activities business requires to achieve its objectives, by adding values as activities proceed from one phase to another. E-business applications utilize Internet technology to cover products and services that require integration of business processes, the logistics of end users, and original suppliers. Effective management of the entire process can add considerable value to consumers by organizing, coordinating, and controlling supply chain activities and logistics (Turban et al., 2000). This defines certain criteria for effective Web-based development, which encompass flexibility, quality, dependability, agility, and efficiency. Optimality can be assessed in terms of delivering the right product at the right time at each level of the supply chain (Vokurka et al., 2002). The value chain concept can be further utilized to build decision support systems that enhance the decision-making process at the tactical and strategic management levels (Haavengen et al., 1996). Also, electronic product development (EPD) is another aspect of e-business growth that relies on a holistic perspective of the entire product value chain (encompassing customers, designers, suppliers, manufacturers, and logistics providers) to develop more successful mass customization (Helander and Jiao, 2000).

110.3.3.1.4 Aligning E-business Applications with Organizational Goals

E-business solutions can effectively serve organizational goals and marketing requirements. Strategies that integrate the Internet and traditional advantages are expected to create potential advantages for existing corporations (Porter, 2001). E-business software systems rely on both the internal preparation of the company and the readiness of its customers and suppliers to engage in electronic interactions. By committing resources to the business problem, management can create a value driver that boosts business readiness for e-commerce challenges (Barua et al., 2001). These e-commerce solutions link customers, suppliers, partners, and interorganizational departments in one or more unified value chains. If these links are not well managed and efficiently aligned in synchronized frameworks, delays will occur and costs will exceed profits, resulting in financial loss and customer dissatisfaction.

Other issues may have an indirect effect on the success of e-business applications. These are supply chain management (SCM) and enterprise resource management (ERM), which can help explain the impact of legacy business applications on the success of e-business development. Better understanding of customer and supplier needs, and the effect of current business processes on the overall methods of supply chain and resource management, can lead to flexible and manageable utilization of information technology to help reengineer business processes (Daoud, 2000).

110.3.3.2 Object-Oriented Development for Web Applications

Gellersen and Gaedke (1999) propose a Web composition model that defines an object-oriented approach to Web development based on Web implementation models. This model was developed to provide developers with the capabilities of object-oriented concepts, such as reusability, inheritance, improved modifiability, and extensibility. Conallen (1999) addresses object-oriented Web application architecture through a UML-based approach. This approach aims to facilitate managing complexity for Web applications and to enable enhanced reusability. The approach, which works in conjunction with CASE tool support, integrates three models of Web application architecture: business model, navigation model, and implementation model.

110.3.3.3 Customizable Web Applications

Several approaches to modeling and implementing customizable Web applications have been proposed. These approaches all share those characteristics of Web development environments (Kappel et al., 2000) that explicitly consider user context for customization. This view reflects the need for personalization, both for individuals and for classes of users, and it includes network and device contexts. *Network context* is related to network settings; *device context* is based on multidelivery of different devices or classes of devices. However, they have different degrees of location context (related to mobile computing and portability) and temporal context (based on time constraints).

TABLE 110.3 The Four Generations of Information Systems Security Approaches

Generation	Drivers	Strategies	Techniques	Problems
First (early 1980s)	Generic thinking Common sense principles	Linking requirements (what to do) with existing capabilities (what can be done)	Risk analysis	Gaps between generic strategies and special needs
Second (late 1980s)	Some focus on organization requirements	Formal methods	Control points and checklists	Considering natural, functional, and technical requirements while ignoring the social nature of organizations
Third (early 1990s)	Business processes Focus on specific organization requirements	Information systems modeling	Responsibility modeling Security semantics Logical approach ERM, DFD, OO and business process modeling for security	Not enough focus on social requirements of organizations
Fourth (late 1990s to present)	Socio-technical design User participation Strong focus on specific organizational requirements	Domain-specific and application-driven design for information systems security	Responsibility modeling Viable information systems	Still in its first phases

110.3.4 Security-Driven Software Development

Software systems have evolved into global networked infrastructures, multidimensional databases, and enterprise data warehouses that interconnect individuals, businesses, organizations, competing supply chains, numerous mobile and wireless applications, and even countries. The software engineering literature typically includes security as one of the measures of quality and reliability in software products. Moreover, the software engineering field addresses the security issue as a part of the **risk analysis** process, in order to minimize the likelihood of intrusions, attacks, hacking, or fraud in information systems. The issue of security in contemporary software applications is a critical component of business survival. There is a need to protect organizational strategic assets, such as information. In e-commerce, for example, customers, who are more aware than ever of the ramifications of unsecured personal or private information, tend to trust businesses with sufficient security measures, policies, and standards.

The area of information systems security has evolved across paradigms and strategies (Siponen, 2002). These range from the generic, based on common sense, to the specific, based on organizational culture and needs, as summarized in Table 110.3. Security-driven systems are receiving greater attention in current software development strategies, and the reengineering of existing systems adds security features, builds security-based applications to ensure security in systems (such as antiviruses and firewalls), adds features that enhance individuals' privacy, and builds surveillance-based applications that can help detect or protect against crime and terrorism. Computer vision, image processing, and multimedia-based technologies play a significant role in these applications.

As with all forms of software development, the design of such systems is not without challenges. The trade-off between open communication channels and the potential for security threats through these same

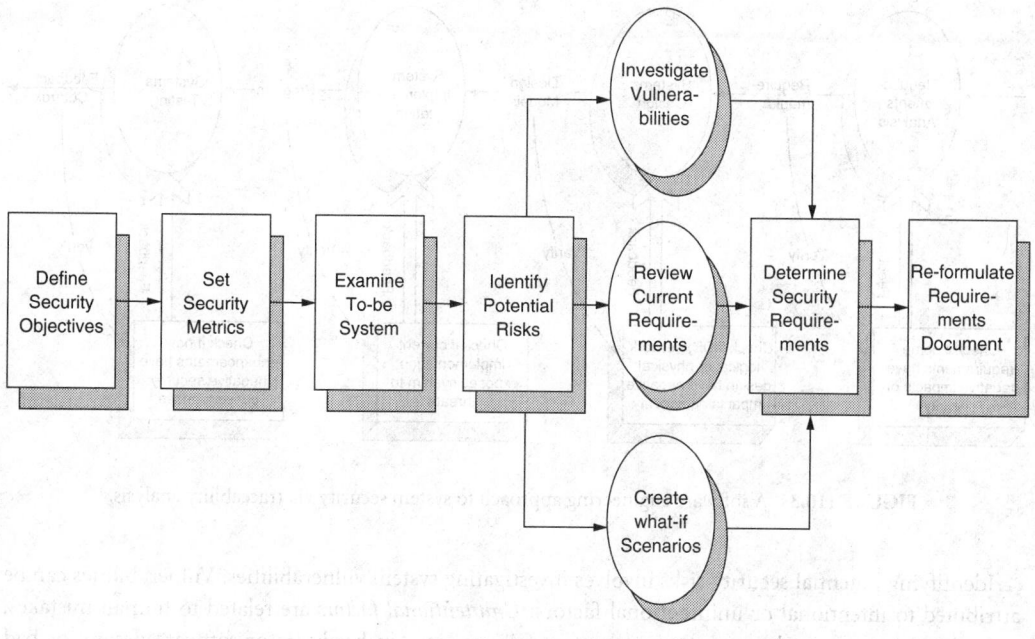

FIGURE 110.2 Security-driven requirements analysis process.

channels is one example. The remaining parts of this section present a framework for dealing with security considerations in the software development process, particularly in terms of the analysis and design of such systems.

110.3.4.1 Security-Driven Requirements Analysis

Because a large portion of software engineering literature was developed before the Web era, investigating vulnerabilities was rarely addressed adequately. Web-driven applications and infrastructures have necessitated a change. For example, in terms of security, while Web connectivity increased access to public information, it exposed the very same information and information systems to more risks and vulnerabilities (Deswarte, 1997). In some software engineering methodologies, security requirements are addressed in the analysis phase as nonfunctional requirements, because software systems must comply with internal and external security standards. Sommerville (1996) classifies security requirements as external, nonfunctional safety and privacy requirements. This view is true from a categorization perspective, but it needs to consider that even functional requirements should be guided by security metrics; otherwise, they may increase system vulnerabilities. Additional requirements or flexible requirements could expose the system to unexpected risks (Smith, 1991; Pfleeger, 1997).

Security-driven requirements analysis involves defining security objectives, setting their metrics, identifying potential risks, investigating vulnerabilities, creating *what-if* scenarios, reviewing current requirements, and reformulating requirements to reflect the input of the analysis phase. Analysis output becomes the guideline for designing security-driven solutions. Additional details are shown in Figure 110.2.

Security objectives are usually based on organizational standards, underlying technology, and the magnitude of the anticipated threat. Because security breaches are highly unpredictable and their nature and scope can change over time, organizations must adapt to new threats and be able to adjust their objectives to meet the demands of evolving challenges. Once objectives are determined, quantitative and qualitative measurements should be derived to establish evaluation metrics to verify quality of software products in terms of security requirements. The major task in security-driven analysis is to identify potential security risks. Risk assessment is essential, because an organization may be attacked from both inside and outside its network (Philips and Swiler, 1998).

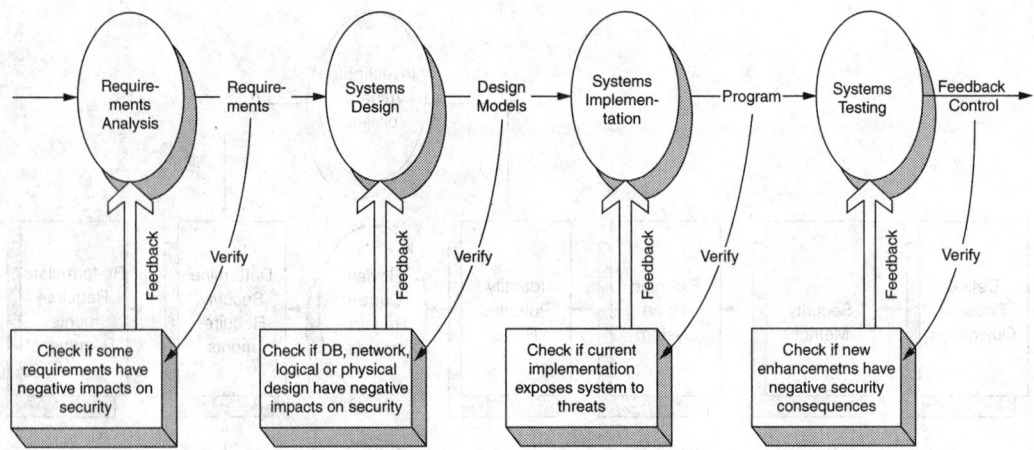

FIGURE 110.3 A software engineering approach to system security via traceability analysis.

Identifying potential security risks involves investigating system vulnerabilities. Vulnerabilities can be attributed to intentional or unintentional factors. *Unintentional factors* are related to human mistakes, exceptional hazards in the environment, system failures, gaps in hardware or software design, or bad requirements specifications. While external factors contribute to the existence of vulnerabilities, it is the analysis, design, implementation, and usability of the system that enable the vast majority of security threats in most organizations. For example, a problem in data collection, data entry, data distribution, referential integrity, or authorization can result in breaches that put data into risky situations. The growing concern about **infrastructure vulnerabilities**, where more damage can be done with a keyboard than with a bomb (Baskerville, 1993), is an important issue for organizational management. Tracing and tracking leakages, security gaps, and security-related problems across the software development process are ways to ensure security in software systems. The traceability process shown in Figure 110.3 offers a strategy for a software engineering approach to system security via traceability analysis.

Intentional factors that threaten system security include data theft, data abuse, source code theft, deliberate data manipulation, data tampering, malicious damage, virus and attack destruction, cyber crimes, terror attacks, and other miscellaneous computer crimes. Computer crimes range from using the computer or computer network as a target, to using the computer as a medium (i.e., giving misleading information), to using the computer as a planning or deception tool (Turban et al., 2002). One of the current and serious challenges of information systems is to discover how information and communication technologies can contribute to public safety (Shneiderman, 2002). Recent efforts focus on enhancing security at the technical level (i.e., network-based security), while paying some attention to security at the analysis and architectural levels. Antiterror system development relies not only on solution-focused capabilities but also on profound comprehension of the problem domain by studying the attacker's behavior (Erland and Olovsson, 1997). **System vulnerabilities** or security gaps in any information system provide opportunities to carry out attacks or steal critical information. Identifying and securing these gaps will minimize potential risks. Holmes (2001) points out the need to assess system security breaching motives in order to protect and then manage the system's infrastructure according to the vulnerabilities of those motives. Salenger (1997) relates the level of organizational Internet security to the relative "functional uses" of the Internet. Engineering secure systems requires managing infrastructure vulnerability (Demuth and Rieke, 2000).

Three models suggested in designing a secure environment (Salter et al., 1998) are the adversary model, the vulnerabilities model, and the methodology model. The adversary model includes an understanding of the motives for threat potentials: what they can do, what they are willing to do, and what they want to do. The vulnerabilities model suggests three steps for any successful attack: analyze the targeted system to find weaknesses, gain access quietly, and execute the attack. The methodology model categorizes attacks based on their characteristics and aims to find the best protective countermeasure. Although the adversary

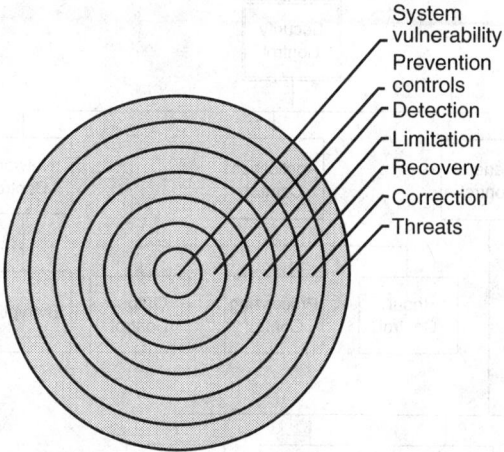

System
vulnerability
Prevention
controls
Detection
Limitation
Recovery
Correction
Threats

FIGURE 110.4   Seven-layer conceptual model for defense strategies in security-focused system design.

model is based on **information gathering**, the vulnerabilities model is driven by risk analysis, and the methodology model is related to **response procedure** and **recovery**.

110.3.4.2   Security-Driven Systems Design

Designing security-focused solutions for software systems can be done at two different levels: conceptual and technical. The conceptual level provides the architectural foundation for the technical level. The key concept for security-focused architectures is defense strategies. The ability of a software system to withstand threats is tightly coupled with its capability to reduce vulnerabilities and provide protection shields that prevent, eliminate, or deal effectively with breaches and attacks. Figure 110.4 depicts this concept as a seven-layer conceptual model for defense strategies in security-focused system design.

In this model, five key defense strategies — prevention control, detection, limitation, recovery, and correction — can be used separately or together to minimize system vulnerabilities or weaknesses (Turban et al., 2002). Prevention control is the most effective strategy, whether it prevents human error, external attack, or unauthorized use. Access control also plays a significant role in this defense strategy. Figure 110.5 provides a basic taxonomy of various types of security controls in software systems. An *intrusion detection system* (IDS) is a system that can distinguish authorized uses, misuses, and abuses of computers, either by authorized users or by external perpetrators. Intrusions can be classified into three categories: single intruder, single terminal (SIST); single intruder, multiple terminal (SIMT); and multiple intruder, multiple terminal (MIMT) (Puketza et al., 1996).

Object-oriented and component-based architectures have proved to be maintainable structures, because they allow easy replacing of defective components. Distributed object architecture and design standards provide an adequate level for generic distributed applications. But these are only the first step in building application-specific software architectures for achieving overall system development objectives. Although commercial enterprise application integration (EAI) tools and workflow management system (WFMS) products can help advance basic distributed standards to the commercial level, they are still far below the mission-critical needs of business and information security processes. System designers should employ security solutions that reinforce each other, define relationships based on trust, and use protective countermeasures to prevent attacks.

The effectiveness of database and network design plays a crucial role in reducing system vulnerabilities. For instance, cryptographic protocol design is frequently cited in the literature as a source of distributed systems vulnerabilities. Yet, analysis and design techniques have proved useful in detecting protocol vulnerabilities (Stubblebine and Wright, 2002).

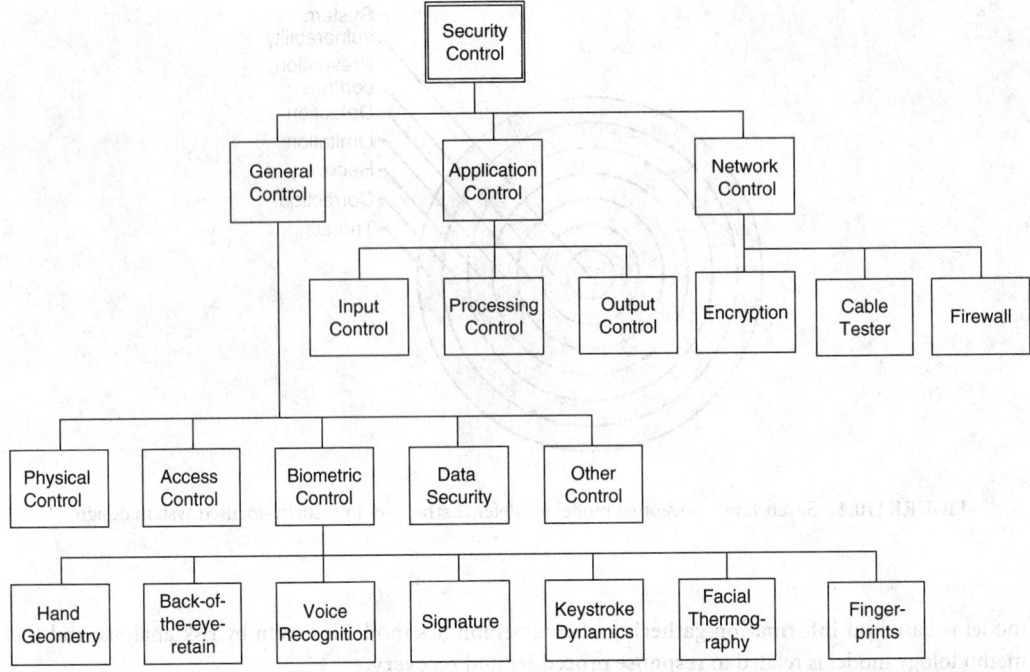

FIGURE 110.5 A basic taxonomy for security control techniques in software systems.

Cybenko and Jiang (2000) discussed the vulnerabilities of the Internet and proposed a six-stage protection process to counteract malicious uses. *Information-gathering techniques* are the first essential step, and they include intelligence reports, unusual-incident analysis, and automated information harvesting from the Web and news services. The second essential step is a thorough *risk assessment* of the current system to find vulnerable areas. This risk assessment includes modeling an attack, modeling failure of the main system, and modeling subsidiary failures due to main system failures. The third step is **interdiction**, which includes being able to use current prevention methods that are already available. The fourth step is *detection* of attacks through early warning systems and monitoring resources. Monitoring subsystems can take actions while an attack is underway, whereas a warning system can attempt to prevent an attack before it happens (Salter et al., 1998). The fifth step is implementing the proper *response procedure* once an attack has been acknowledged. Response procedures, which Cybenko and Jiang call *forensic challenges*, can only be implemented when an attack is already underway. Once an attack is detected, the system should be able to trace the attack. The final stage in Cybenko and Jiang's approach is *recovery*, which includes learning from the attack and documenting its characteristics for future reference in a knowledge base.

110.4 Research Issues and Summary

The area of specialized system development can be characterized as new, huge, and crucial. This field is evolving as the importance of scalability and tailorability in software development, as opposed to generic strategies and approaches, is realized. The theoretical foundations of specialized system development will continue to evolve and provide a roadmap for new research and development. This will provide new challenges and opportunities to the software engineering community, because specialized system development is not only new but also critical for many contemporary software applications.

An important consideration is that future research and development of specialized systems concerns the government, industry, and academia alike. The government's role in information decryption on the

Internet is crucial (Fox, 2001). For example, some of the intelligence issues and policies to be further addressed (Artz, 2001; Wilson, 2000; Zorpette, 2002) include the human role in information analysis, gaps in technical intelligence, and cooperation between organizations and services that collect intelligence. While there is some need to define the role of the government, other needs require a clearer definition of organizational roles. Salenger (1997) states that the level of security implemented by organizations is directly proportional to two factors: size and income. Larger companies have the people and the resources required to establish and run a secure Internet environment, whereas smaller companies may not. Better protocols for defining and enforcing standards are expected to continue to emerge.

Defining Terms

Attentive systems: Systems that can be used to understand user trends or log and track Internet use across multiple sources.

Cognitive fit: An approach in specialized system development where the goal is to match, as closely as possible, the representation to the task and the user. The key concept is that there should be harmony among three variables: the user's cognitive skills, the task, and the representation of the task (as presented to the user).

Horizontal specialization: Specialization across various functional departments or business needs within the organization, across various domains of an industry, or between industries.

Infrastructure vulnerabilities: Weak points and security gaps in the physical or logical architecture of information systems that may enhance opportunities to carry out attacks or steal critical information.

Interdiction: The ability to make use of prevention methods that are already available.

Pamela: Process abstraction method for embedded large applications.

Pervasive computing: The convergence of three traditional computing specializations (personal, networked, and embedded) to produce a new computing era marked by wireless and portable hardware and software.

Process design engineering: An automated engineering approach to the evolutionary design, implementation, and testing of real-time software.

SCR: Software cost reduction.

Steganography: Hiding data within data.

System specialization: Concentration on unique problems and the techniques for comprehending and solving them.

Vertical specialization: Specialization in the different levels of problem complexity across the interorganizational pyramid, from operations to top management.

Weak strategy: A generic approach to problem solving that is not tailored to specific problem domains.

References

Artz, D. 2001. Digital Steganography: Hiding Data within Data. *IEEE Internet Computing*. 5(3): 75–80.

Avison, D., and Fitzgerald, G. 2003. Where Now for Development Methodologies? *Communications of the ACM*. 46(1): 48–72.

Barua, A., Konana, P., Whinston, A., and Yin, F. 2001. Driving E-business Excellence. *MIT Sloan Management Review*. Cambridge, MA. 43(1): 36–44.

Baskerville, R. 1993. Information Systems Security Design Methods: Implications for Information Systems Development. *ACM Computing Surveys*. 25(4): 365–414.

Ciarletta, L., and Dima, A. 2000. A Conceptual Model for Pervasive Computing Parallel Processing. *Proceedings of the 2000 International Workshops on Parallel Processing*. 9–15.

Conallen, J. 1999. Modeling Web Application Architecture with UML. *Communications of the ACM*. 42: 63–70.

Cybenko, G., and Guofei, J. 2000. Developing a Distributed System for Infrastructure Protection. *IT Professional*. 2(4): 17–23.

Cybenko, G., and Jiang, G. 2000. Developing a Distributed System for Infrastructure Protection. *IEEE IT Professional.* 4:17–22.

Daoud, F. 2000. Electronic Commerce Infrastructure. *IEEE Potentials.* 19(1): 30–33.

Davenport, T., and Stoddard, D. 1994. Reengineering: Business Change of Mythic Proportions? *MIS Quarterly.* 18(2): 121–127.

Demuth, T., and Rieke, A. 2000. Bilateral Anonymity and Prevention of Abusing Logged Web Addresses. *21st Century Military Communications Conference Proceedings.* 1: 435–439.

Der Zee, J.T.M. van, and de Jong, B.M. 1999. Alignment is not enough-Integrating IT in the Balanced Business Scorecard. *Journal of Management Information Systems.*

Deswarte, Y. 1997. Internet Security despite Untrustworthy Agents and Components. *Proceedings of the 6th IEEE Computer Society Workshop on Future Trends of Distributed Computing Systems.* 53: 218–219.

Erland, J., and Olovsson, T. 1997. A Quantitative Model of the Security Intrusion Process Based on Attacker Behavior. *IEEE Transactions on Software Engineering.* 23(4): 235–245.

Evans, P., and Wurster, T. 1999. Getting Real about Virtual Commerce. *Harvard Business Review.* 77(6): 85–94.

Felder, M. 2002. A Formal Design Notation for Real-Time Systems. *ACM Transactions on Software Engineering and Methodology (TOSEM).* 11(2): 149–190.

Fox, R. 2001. Privacy Tradeoff Fighting Terrorism. *Communications of the ACM.* 44(12): 9–10.

Gaulding, S.N., and Lawson, J.D. 1976. Process design system: an integrated set of software development tools. *Proceedings of the 2nd International Conference on Software Engineering.* San Francisco, 86–90.

Gellersen, H., and Gaedke, M. 1999. Object-Oriented Web Application Development. *IEEE Internet Computing.* 3(1): 60–68.

Glass, R. 1998. *In the Beginning: Recollections of Software Pioneers.* IEEE Computer Society Press, Los Alamitos, CA.

Glass, R., and Vessey, I. 1995. Contemporary Application-Domain Taxonomies. *IEEE Software.* 12(4): 63–76.

Glass, R.L. 2000. Process Diversity and a Computing Old Wives'/Husbands' Tale. *IEEE Software.* 17(4): 128–129.

Gomaa, H. 1986. Software Development of Real-Time Systems. *Communications ACM.* 29(7): 657–668.

Haavengen, B., Olsen, D., and Sena, J. 1996. The Value Chain Component in a Decision Supports System: A Case Example. *IEEE Transactions on Engineering Management.* 43(4): 418–428.

Helander, M., and Jiao, J. 2000. E-Product Development (EPD) for Mass Customization. Management of Innovation and Technology, *Proceedings of the 2000 IEEE International Conference on ICMIT 2000.* 2 (2): 848–854.

Hitt, L., and Brynjolfsson, E. 1996. Productivity, Business Profitability, and Consumer Surplus: Three Different Measures of Information Technology Value. *MIS Quarterly.* 20(2): 121–142.

Holmes, N. 2001. Terrorism, Technology and the Profession. *Computer.* 34(11): 134–136.

Kalakota, R., Varshney, U., and Vetter, R. 2000. Mobile Commerce: A New Frontier. *IEEE Computer Society: Special Issue on E-commerce.* 33(10): 32–38.

Kappel, G., Retschitzegger, W., and Schwinger, W. 2000. Modeling Customizable Web Applications — A Requirements Perspective. *Proceedings of the International Conference on Digital Libraries: Research and Practice.* Kyoto.

Keen, P. 1981. Information Systems and Organizational Change. *Communications of the ACM.* 24(1): 24–33.

Kelly, J. 1987. A Comparison of Four Design Methods for Real-Time Systems. *IEEE Proceedings of the 9th International Conference on Software Engineering.* 238–252.

Kelsey, J., and Bruce, S. 1999. Secure Audit Logs to Support Computer Forensics. *ACM Transactions on Information and System Security (TISSEC).* 2(2): 159–176.

Mitroff, I., and Murray, T. 1973. Technological Forecasting and Assessment: Science and/or Mythology? *Journal of Technological Forecasting and Social Change.* 5(1): 113–134.

Phillips, C.A., and Swiler, L.P. 1998. A Graph-based System for Network Vulnerability Analysis. *Proceedings of the 1998 Workshop on New Security Paradigms.* 71–79.

Pfleeger, C.P. 1997. The Fundamentals of Information Security. *IEEE Software.* 14(1): 15–16.

Porter, M.E. 2001. Strategy and the Internet. *Harvard Business Review.* 79: 63–78.

Puketza, N., Zhang, K., Chung, M., Mukherjee, B., and Olsson, R. 1996. A Methodology for Testing Intrusion Detection Systems. *IEEE Transactions on Software Engineering.* 22(10): 719–729.

Salenger, D. 1997. Internet Environment and Outsourcing. *International Journal of Network Management.* 7(6): 300–304.

Salter, C.O.S., Schneier, B., and Wallner, J. 1998. Toward a Secure Engineering Methodology. *Proceedings of the 1998 Workshop on New Security Paradigms.* 2–10.

Shneiderman, B. 2002. ACM's Computing Professionals Face New Challenges. *Communications of the ACM.* 31–34.

Singleton, J., McLean, E. and Altman, E. 1988. Measuring Information Systems Performance: Experience with the Management by Results System at Security Pacific Bank. *MIS Quarterly.* 12(2): 325–337.

Siponen, M. 2002. Designing Secure Information Systems and Software: Critical Evaluation of the Existing Approaches and a New Paradigm. Unpublished Ph.D. dissertation. University of Oulu.

Sommerville, I. 1996. *Software Engineering.* 5th ed., Addison-Wesley, Workingham, U.K.

Stubblebine, S., and Wright, R. 2002. Authentication Logic with Formal Semantics Supporting Synchronization, Revocation, and Recency. *IEEE Transactions on Software Engineering.* 28(3): 265–285.

Smith, G.W. 1991. Modeling Security-Relevant Data Semantics. *IEEE Transactions on Software Engineering.* 17(11): 1195–1203.

Turban, E., Lee, J., King, D., and Chung, H. 2000. *Electronic Commerce: A Management Perspective.* Prentice Hall, Upper Saddle River, NJ.

Turban E., Rainer, K., and Potter, R. 2002. *Introduction to Information Technology, 2nd edition.* Wiley, New York.

Varshney U., and Vetter, R. 2001. A Framework for the Emerging Mobile Commerce Applications. *Proceedings of the 34th Hawaii International Conference on System Sciences (HICSS 34).* IEEE Computer Society.

Varshney U., and Vetter, R. 2000. Emerging Mobile and Wireless Networks. *Communications of the Association of Computing Machinery (ACM).* 43(6): 73–81.

Vessey, I. 1997. Problems versus Solutions: The Role of the Application Domain in Software. *Proceedings of the 7th Workshop on Empirical Studies of Programmers,* Virginia. 233–240.

Vessey, I., and Glass, R. 1998. Strong vs. Weak: Approaches to Systems Development. *Communications of the ACM.* 41(4): 99–102.

Vokurka, R., Gail, M., and Carl, M. 2002. Improving Competitiveness through Supply Chain Management: A Cumulative Approach. *Competitiveness Review.* 12(1): 14–24.

Wilson, C. 2000. Holding Management Accountable: A New Policy for Protection against Computer Crime. *National Aerospace and Electronics Conference. Proceedings of the IEEE 2000.* 272–281.

Zorpette, G. 2002. Making Intelligence Smarter. *IEEE Spectrum.* 39(1): 38–43.

Further Information

A good survey of industry frameworks is presented in the article Contemporary Application-Domain Taxonomies by Glass and Vessey, published in *IEEE Software* in 1995. The authors pay particular attention to representative taxonomies, the IBM industry's taxonomy, digital industry's taxonomy, digital application taxonomy, and Reifer's application taxonomy.

Here are some other good sources:

SIMS: A Secure Information Management System for Large-Scale Dynamic Coalitions by Jiang and Dasgupta published in the IEEE Proceedings of DARPA Information Survivability Conference and Exposition (DISCEX II), June 2001. The article discusses security of large-scale systems.

Attack Detection in Large Networks by Peterson and Bauman, published by the IEEE Proceedings of DARPA Information Survivability Conference and Exposition (DISCEX II), June 2001. This article addresses the impact of large systems' characteristics on security.

Security of Distributed Object-Oriented Systems by MacDonnell et al., published by the IEEE Proceedings of DARPA Information Survivability Conference and Exposition (DISCEX II), June 2001. This article addresses object-oriented security mechanisms that can provide scalable, fine-grained access control both in applications and at the boundary controller, using CORBA and Java.

Appendix A:
Professional Societies
in Computing

A.1   The Association for Computing Machinery
(ACM) **A-1**

A.2   The Computing Research Association (CRA) **A-1**

A.3   The Institute of Electrical and Electronics
Engineers (IEEE) Computer Society **A-2**

A.4   The British Computer Society (BCS) **A-2**

A.5   Computer Professionals for Social
Responsibility (CPSR) **A-2**

A.6   The American Association for Artificial
Intelligence (AAAI) **A-2**

A.7   Special Interest Group on Computer
Graphics (SIGGRAPH) **A-2**

A.8   The Society for Industrial and Applied
Mathematics (SIAM) **A-3**

A.1   The Association for Computing Machinery (ACM)

"Founded in 1947, ACM is a major force in advancing the skills of information technology professionals and students worldwide. Today, our 75,000 members and the public turn to ACM for the industry's leading portal to computing literature, authoritative publications and pioneering conferences, providing leadership for the 21st century." More complete information on ACM can be obtained by visiting its Web page, from which the preceding quotation was taken: http://www.acm.org.

A.2   The Computing Research Association (CRA)

"The Computing Research Association (CRA) is an association of more than 200 North American academic departments of computer science, computer engineering, and related fields; laboratories and centers in industry, government, and academia engaging in basic computing research; and affiliated professional societies. CRA's mission is to strengthen research and education in the computing fields, expand opportunities for women and minorities, and improve public and policymaker understanding of the importance of computing and computing research in our society." More information about the CRA can be obtained by visiting its Web page, from which the preceding quotation was taken: http://www.cra.org.

A.3 The Institute of Electrical and Electronics Engineers (IEEE) Computer Society

"With nearly 100,000 members, the IEEE Computer Society is the world's leading organization of computer professionals. Founded in 1946, it is the largest of the 37 societies of the Institute of Electrical and Electronics Engineers (IEEE). The Computer Society's vision is to be the leading provider of technical information and services to the world's computing professionals." More information about the IEEE Computer Society can be obtained by visiting its Web page, from which the preceding quotation was taken: http://www.computer.org.

A.4 The British Computer Society (BCS)

"The British Computer Society (BCS) is the only Chartered Engineering Institution for Information Technology (IT). With members in over 100 countries around the world, the BCS is the leading professional and learned Society in the field of computers and information systems." More information about the BCS can be found by visiting its Web page, from which the preceding quotation was taken: http://www1.bcs.org.uk.

A.5 Computer Professionals for Social Responsibility (CPSR)

"CPSR is a public-interest alliance of computer scientists and others concerned about the impact of computer technology on society . . . . As technical experts, CPSR members provide the public and policymakers with realistic assessments of the power, promise, and limitations of computer technology. As concerned citizens, we direct public attention to critical choices concerning the applications of computing and how those choices affect society." More information about CPSR can be found by visiting its Web page, from which the preceding quotation was taken: http://www.cpsr.org.

A.6 The American Association for Artificial Intelligence (AAAI)

"Founded in 1979, the American Association for Artificial Intelligence (AAAI) is a nonprofit scientific society devoted to advancing the scientific understanding of the mechanisms underlying thought and intelligent behavior and their embodiment in machines. AAAI also aims to increase public understanding of artificial intelligence, improve the teaching and training of AI practitioners, and provide guidance for research planners and funders concerning the importance and potential of current AI developments and future directions." More information about AAAI can be found by visiting its Web page, from which the preceding quotation was taken: http://www.aaai.org.

A.7 Special Interest Group on Computer Graphics (SIGGRAPH)

"ACM SIGGRAPH is dedicated to the generation and dissemination of information on computer graphics and interactive techniques. We are a membership organization that values passion, integrity, excellence, volunteerism, and cross-disciplinary interaction in all of our activities." More information about SIGGRAPH can be found by visiting its Web page, from which the preceding quotation was taken: http://www.siggraph.org.

A.8 The Society for Industrial and Applied Mathematics (SIAM)

"To ensure the strongest interactions between mathematics and other scientific and technological communities, it remains the policy of SIAM to: advance the application of mathematics and computational science to engineering, industry, science, and society; promote research that will lead to effective new mathematical and computational methods and techniques for science, engineering, industry, and society; provide media for the exchange of information and ideas among mathematicians, engineers, and scientists." More information about SIAM can be found by visiting its Web page, from which the preceding quotation was taken: http://www.siam.org.

Appendix B: The ACM Code of Ethics and Professional Conduct

B.1   Preamble B-1
B.2   General Moral Imperatives B-2
B.3   More Specific Professional Responsibilities B-4
B.4   Organizational Leadership Imperatives B-5
B.5   Compliance with the Code B-7
B.6   Acknowledgments B-7

B.1 Preamble

Commitment to ethical professional conduct is expected of every member (voting members, associate members, and student members) of the Association for Computing Machinery (ACM).

This Code, consisting of 24 imperatives formulated as statements of personal responsibility, identifies the elements of such a commitment. It contains many, but not all, issues professionals are likely to face. Section 1 outlines fundamental ethical considerations, whereas section 2 addresses additional, more specific considerations of professional conduct. Statements in section 3 pertain more specifically to individuals who have leadership roles, whether in the workplace or in a volunteer capacity such as with organizations like ACM. Principles involving compliance with this Code are given in section 4.

The Code shall be supplemented by a set of guidelines, which provide explanations to assist members in dealing with the various issues contained in the Code. It is expected that the guidelines will be changed more frequently than the Code.

The Code and its supplemented guidelines are primarily intended to serve as a basis for ethical decision making in the conduct of professional work. Secondarily, they may serve as a basis for judging the merit of a formal complaint pertaining to violation of professional ethical standards.

It should be noted that although computing is not mentioned in the imperatives of section 1.0, the Code is concerned with how these fundamental imperatives apply to one's conduct as a computing professional. These imperatives are expressed in a general form to emphasize that ethical principles that apply to computer ethics are derived from more general ethical principles.

It is understood that some words and phrases in a Code of ethics are subject to varying interpretations, and that any ethical principle may conflict with other ethical principles in specific situations. Questions related to ethical conflicts can best be answered by thoughtful consideration of fundamental principles, rather than reliance on detailed regulations.

B.2   General Moral Imperatives: *As an ACM member I will* . . .

B.2.1   Contribute to Society and Human Well-Being

This principle concerning the quality of life of all people affirms an obligation to protect fundamental human rights and to respect the diversity of all cultures. An essential aim of computing professionals is to minimize negative consequences of computing systems, including threats to health and safety. When designing or implementing systems, computing professionals must attempt to ensure that the products of their efforts will be used in socially responsible ways, will meet social needs, and will avoid harmful effects to health and welfare.

In addition to a safe social environment, human well-being includes a safe natural environment. Therefore, computing professionals who design and develop systems must be alert to, and make others aware of, any potential damage to the local or global environment.

B.2.2   Avoid Harm to Others

Harm means injury or negative consequences, such as undesirable loss of information, loss of property, property damage, or unwanted environmental impacts. This principle prohibits use of computing technology in ways that result in harm to any of the following: users, the general public, employees, employers. Harmful actions include intentional destruction or modification of files and programs leading to serious loss of resources or unnecessary expenditure of human resources such as the time and effort required to purge systems of computer viruses.

Well-intended actions, including those that accomplish assigned duties, may lead to harm unexpectedly. In such an event the responsible person or persons are obligated to undo or mitigate the negative consequences as much as possible. One way to avoid unintentional harm is to carefully consider potential impacts on all those affected by decisions made during design and implementation.

To minimize the possibility of indirectly harming others, computing professionals must minimize malfunctions by following generally accepted standards for system design and testing. Furthermore, it is often necessary to assess the social consequences of systems to project the likelihood of any serious harm to others. If system features are misrepresented to users, co-workers, or supervisors, the individual computing professional is responsible for any resulting injury.

In the work environment the computing professional has the additional obligation to report any signs of system dangers that might result in serious personal or social damage. If one's superiors do not act to curtail or mitigate such dangers, it may be necessary to blow the whistle to help correct the problem or reduce the risk. However, capricious or misguided reporting of violations can, itself, be harmful. Before reporting violations, all relevant aspects of the incident must be thoroughly assessed. In particular, the assessment of risk and responsibility must be credible. It is suggested that advice be sought from other computing professionals. See principle 2.5 regarding thorough evaluations.

B.2.3   Be Honest and Trustworthy

Honesty is an essential component of trust. Without trust an organization cannot function effectively. The honest computing professional will not make deliberately false or deceptive claims about a system or system design, but will instead provide full disclosure of all pertinent system limitations and problems.

A computer professional has a duty to be honest about his or her own qualifications, and about any circumstances that might lead to conflicts of interest.

Membership in volunteer organizations such as ACM may at times place individuals in situations where their statements or actions could be interpreted as carrying the weight of a larger group of professionals. An ACM member will exercise care to not misrepresent ACM or positions and policies of ACM or any ACM units.

B.2.4   Be Fair and Take Action Not to Discriminate

The values of equality, tolerance, respect for others, and the principles of equal justice govern this imperative. Discrimination on the basis of race, sex, religion, age, disability, national origin, or other such factors is an explicit violation of ACM policy and will not be tolerated.

Inequities between different groups of people may result from the use or misuse of information and technology. In a fair society, all individuals would have equal opportunity to participate in, or benefit from, the use of computer resources regardless of race, sex, religion, age, disability, national origin, or other such similar factors. However, these ideals do not justify unauthorized use of computer resources nor do they provide an adequate basis for violation of any other ethical imperatives of this Code.

B.2.5   Honor Property Rights Including Copyrights and Patents

Violation of copyrights, patents, trade secrets, and the terms of license agreements is prohibited by law in most circumstances. Even when software is not so protected, such violations are contrary to professional behavior. Copies of software should be made only with proper authorization. Unauthorized duplication of materials must not be condoned.

B.2.6   Give Proper Credit for Intellectual Property

Computing professionals are obligated to protect the integrity of intellectual property. Specifically, one must not take credit for other's ideas or work, even in cases where the work has not been explicitly protected by copyright, patent, etc.

B.2.7   Respect the Privacy of Others

Computing and communication technology enables the collection and exchange of personal information on a scale unprecedented in the history of civilization. Thus there is increased potential for violating the privacy of individuals and groups. It is the responsibility of professionals to maintain the privacy and integrity of data describing individuals. This includes taking precautions to ensure the accuracy of data, as well as protecting it from unauthorized access or accidental disclosure to inappropriate individuals. Furthermore, procedures must be established to allow individuals to review their records and correct inaccuracies.

This imperative implies that only the necessary amount of personal information be collected in a system, that retention and disposal periods for that information be clearly defined and enforced, and that personal information gathered for a specific purpose not be used for other purposes without consent of the individual(s). These principles apply to electronic communications, including electronic mail, and prohibit procedures that capture or monitor electronic user data, including messages, without the permission of users or bona fide authorization related to system operation and maintenance. User data observed during the normal duties of system operation and maintenance must be treated with strictest confidentiality, except in cases where it is evidence for the violation of law, organizational regulations, or this Code. In these cases, the nature or contents of that information must be disclosed only to proper authorities.

B.2.8   Honor Confidentiality

The principle of honesty extends to issues of confidentiality of information whenever one has made an explicit promise to honor confidentiality or, implicitly, when private information not directly related to the performance of one's duties becomes available. The ethical concern is to respect all obligations of confidentiality to employers, clients, and users unless discharged from such obligations by requirements of the law or other principles of this Code.

B.3   More Specific Professional Responsibilities: *As an ACM computing professional I will . . .*

B.3.1   Strive to Achieve the Highest Quality, Effectiveness, and Dignity in Both the Process and Products of Professional Work

Excellence is perhaps the most important obligation of a professional. The computing professional must strive to achieve quality and to be cognizant of the serious negative consequences that may result from poor quality in a system.

B.3.2   Acquire and Maintain Professional Competence

Excellence depends on individuals who take responsibility for acquiring and maintaining professional competence. A professional must participate in setting standards for appropriate levels of competence and strive to achieve those standards. Upgrading technical knowledge and competence can be achieved in several ways: doing independent study; attending seminars, conferences, or courses; and being involved in professional organizations.

B.3.3   Know and Respect Existing Laws Pertaining to Professional Work

ACM members must obey existing local, state, province, national, and international laws unless there is a compelling ethical basis not to do so. Policies and procedures of the organizations in which one participates must also be obeyed. But compliance must be balanced with the recognition that sometimes existing laws and rules may be immoral or inappropriate and, therefore, must be challenged. Violation of a law or regulation may be ethical when that law or rule has inadequate moral basis or when it conflicts with another law judged to be more important. If one decides to violate a law or rule because it is viewed as unethical, or for any other reason, one must fully accept responsibility for one's actions and for the consequences.

B.3.4   Accept and Provide Appropriate Professional Review

Quality professional work, especially in the computing profession, depends on professional reviewing and critiquing. Whenever appropriate, individual members should seek and utilize peer review as well as provide critical review of the work of others.

B.3.5   Give Comprehensive and Thorough Evaluations of Computer Systems and Their Impacts, Including Analysis of Possible Risks

Computer professionals must strive to be perceptive, thorough, and objective when evaluating, recommending, and presenting system descriptions and alternatives. Computer professionals are in a position of special trust and therefore have a special responsibility to provide objective, credible evaluations to employers, clients, users, and the public. When providing evaluations the professional must also identify any relevant conflicts of interest, as stated in imperative 1.3.

   As noted in the discussion of principle 1.2 on avoiding harm, any signs of danger from systems must be reported to those who have opportunity and/or responsibility to resolve them. See the guidelines for imperative 1.2 for more details concerning harm, including the reporting of professional violations.

B.3.6 Honor Contracts, Agreements, and Assigned Responsibilities

Honoring one's commitments is a matter of integrity and honesty. For the computer professional this includes ensuring that system elements perform as intended. Also, when one contracts for work with another party, one has an obligation to keep that party properly informed about progress toward completing that work.

A computing professional has a responsibility to request a change in any assignment that he or she feels cannot be completed as defined. Only after serious consideration and with full disclosure of risks and concerns to the employer or client should one accept the assignment. The major underlying principle here is the obligation to accept personal accountability for professional work. On some occasions other ethical principles may take greater priority.

A judgment that a specific assignment should not be performed may not be accepted. Having clearly identified one's concerns and reasons for that judgment, but failing to procure a change in that assignment, one may yet be obligated, by contract or by law, to proceed as directed. The computing professional's ethical judgment should be the final guide in deciding whether or not to proceed. Regardless of the decision, one must accept the responsibility for the consequences.

However, performing assignments against one's own judgment does not relieve the professional of responsibility for any negative consequences.

B.3.7 Improve Public Understanding of Computing and Its Consequences

Computing professionals have a responsibility to share technical knowledge with the public by encouraging understanding of computing, including the impacts of computer systems and their limitations. This imperative implies an obligation to counter any false views related to computing.

B.3.8 Access Computing and Communication Resources Only When Authorized to Do So

Theft or destruction of tangible and electronic property is prohibited by imperative 1.2: "Avoid harm to others." Trespassing and unauthorized use of a computer or communication system is addressed by this imperative. Trespassing includes accessing communication networks and computer systems, or accounts and/or files associated with those systems, without explicit authorization to do so. Individuals and organizations have the right to restrict access to their systems so long as they do not violate the discrimination principle (see 1.4). No one should enter or use another's computer system, software, or datafiles without permission. One must always have appropriate approval before using system resources, including .rm57 communication ports, filespace, other system peripherals, and computer time.

B.4 Organizational Leadership Imperatives: *As an ACM member and an organizational leader, I will . . .*

B.4.1 Background Note

This section draws extensively from the draft IFIP Code of Ethics, especially its sections on organizational ethics and international concerns. The ethical obligations of organizations tend to be neglected in most codes of professional conduct, perhaps because these codes are written from the perspective of the individual member. This dilemma is addressed by stating these imperatives from the perspective of the organizational leader. In this context *leader* is viewed as any organizational member who has leadership or educational responsibilities. These imperatives generally may apply to organizations as well as their leaders. In this context *organizations* are corporations, government agencies, and other employers, as well as volunteer professional organizations.

B.4.2 Articulate Social Responsibilities of Members of an Organizational Unit and Encourage Full Acceptance of Those Responsibilities

Because organizations of all kinds have impacts on the public, they must accept responsibilities to society. Organizational procedures and attitudes oriented toward quality and the welfare of society will reduce harm to members of the public, thereby serving public interest and fulfilling social responsibility. Therefore, organizational leaders must encourage full participation in meeting social responsibilities as well as quality performance.

B.4.3 Manage Personnel and Resources to Design and Build Information Systems That Enhance the Quality of Working Life

Organizational leaders are responsible for ensuring that computer systems enhance, not degrade, the quality of working life. When implementing a computer system, organizations must consider the personal and professional development, physical safety, and human dignity of all workers. Appropriate human–computer ergonomic standards should be considered in system design and in the workplace.

B.4.4 Acknowledge and Support Proper and Authorized Uses of an Organization's Computing and Communication Resources

Because computer systems can become tools to harm as well as to benefit an organization, the leadership has the responsibility to clearly define appropriate and inappropriate uses of organizational computing resources. Whereas the number and scope of such rules should be minimal, they should be fully enforced when established.

B.4.5 Ensure That Users and Those Who Will Be Affected by a System Have Their Needs Clearly Articulated during the Assessment and Design of Requirements; Later the System Must Be Validated to Meet Requirements

Current system users, potential users, and other persons whose lives may be affected by a system must have their needs assessed and incorporated in the statement of requirements. System validation should ensure compliance with those requirements.

B.4.6 Articulate and Support Policies That Protect the Dignity of Users and Others Affected by a Computing System

Designing or implementing systems that deliberately or inadvertently demean individuals or groups is ethically unacceptable. Computer professionals who are in decision-making positions should verify that systems are designed and implemented to protect personal privacy and enhance personal dignity.

B.4.7 Create Opportunities for Members of the Organization to Learn the Principles and Limitations of Computer Systems

This complements the imperative on public understanding (2.7). Educational opportunities are essential to facilitate optimal participation of all organizational members. Opportunities must be available to all members to help them improve their knowledge and skills in computing, including courses that familiarize them with the consequences and limitations of particular types of systems. In particular, professionals must be made aware of the dangers of building systems around oversimplified models, the improbability of anticipating and designing for every possible operating condition, and other issues related to the complexity of this profession.

B.5 Compliance with the Code: *As an ACM member I will...*

B.5.1 Uphold and Promote the Principles of This Code

The future of the computing profession depends on both technical and ethical excellence. Not only is it important for ACM computing professionals to adhere to the principles expressed in this Code, each member should encourage and support adherence by other members.

B.5.2 Treat Violations of This Code as Inconsistent with Membership in the ACM

Adherence of professionals to a Code of ethics is largely a voluntary matter. However, if a member does not follow this Code by engaging in gross misconduct, membership in ACM may be terminated.

B.6 Acknowledgments

Adopted by ACM Council Oct. 16, 1992; Copyright 1993 by ACM, all rights reserved; reprinted with permission from ACM; originally published in 1993 *Communications of the ACM* 36(2) and also available on the ACM Web site http://www.acm.org.

Appendix C: Standards-Making Bodies and Standards

C.1   The International Organization for
       Standardization (ISO) C-1
C.2   The American National Standards Institute (ANSI) ... C-2
C.3   The IEEE Standards Association C-2
C.4   The World Wide Web Consortium (W3C) C-2
C.5   The American Standard Code for Information
       Interchange (ASCII) C-2
C.6   The UNICODE Standard C-3
C.7   Floating-Point Arithmetic C-3

International and national standards play an important role in computer science and engineering. Standards help unify the definition and implementation of complex systems, especially in the areas of architecture, human–computer interaction, operating systems and networks, programming languages, and software engineering.

Principal roles in standardization for computer science and engineering are played by the International Standards Organization (ISO), the American National Standards Institute (ANSI), and the Institute of Electrical and Electronics Engineers (IEEE). These organizations are briefly described in the following sections, with pointers to their Web pages provided for further information.

C.1   The International Organization for Standardization (ISO)

"ISO is a network of the national standards institutes of 147 countries, on the basis of one member per country, with a Central Secretariat in Geneva, Switzerland, that coordinates the system.

ISO is a non-governmental organization: its members are not, as is the case in the United Nations system, delegations of national governments. Nevertheless, ISO occupies a special position between the public and private sectors. This is because, on the one hand, many of its member institutes are part of the governmental structure of their countries, or are mandated by their government. On the other hand, other members have their roots uniquely in the private sector, having been set up by national partnerships of industry associations."

Some of the countries represented in ISO and their respective member bodies (in parentheses) are listed below.

Country	Member Body	Country	Member Body
Australia	(SAI)	Ireland	(NSAI)
Brazil	(ABNT)	Israel	(SII)
Canada	(SCC)	Italy	(UNI)
China	(SAC)	Japan	(JISC)
Czech Republic	(COSMT)	Netherlands	(NEN)
Denmark	(DS)	Sweden	(SIS)
Egypt	(EOS)	Switzerland	(SNV)
Finland	(SFS)	USA	(ANSI)
France	(AFNOR)	Ukraine	(DSSU)
Germany	(DIN)	United Kingdom	(BSI)
India	(BIS)		

More information about the ISO can be can be obtained by visiting its Web page: http://www.iso.ch.

C.2 The American National Standards Institute (ANSI)

"The American National Standards Institute (ANSI) is a private, non-profit organization (501(c)3) that administers and coordinates the U.S. voluntary standardization and conformity assessment system. The Institute's mission is to enhance both the global competitiveness of U.S. business and the U.S. quality of life by promoting and facilitating voluntary consensus standards and conformity assessment systems, and safeguarding their integrity." (See http://www.ansi.org for more details.)

ANSI standards in computer science exist in the areas of architecture, graphics, and programming languages. The International Committee for Information Technology Standards (INCITS) is accredited by ANSI to create and maintain standards in information technology, including the various areas of computer science. More information about specific ANSI standards in these and other areas can be obtained by visiting the INCITS Web site: http://www.incits.org.

C.3 The IEEE Standards Association

The IEEE Standards Association also develops standards for certain areas of computer science and engineering, especially the areas of architecture, networks, and software engineering. For more information, visit the Web page: http://standards.ieee.org.

C.4 The World Wide Web Consortium (W3C)

The World Wide Web Consortium was created in October 1994 to lead the World Wide Web to its full potential by developing common protocols that promote its evolution and ensure its interoperability. W3C has around 400 Member organizations from all over the world and has earned international recognition for its contributions to the growth of the Web. For more information, see the Web site: http://www.w3.org.

C.5 The American Standard Code for Information Interchange (ASCII)

The American Standard Code for Information Interchange (ASCII) is a standard representation scheme for English language text based information storage and network transfer. The ASCII standard was established in 1968, and the current version of the standard is ANSI X3.110-1983.

C.6   The UNICODE Standard

Unicode is a standard representation scheme for every character, no matter what the language. The Unicode Standard has been adopted by the major technology vendors, and is required by modern standards such as XML, CORBA, and Java. It is supported in many operating systems, all modern browsers, and many other products. For more information, see the Web site: http://www.unicode.org.

C.7   Floating-Point Arithmetic

Computer implementations of floating-point numbers and arithmetic generally follow the IEEE floating-point standards ANSI/IEEE 754-1985 (R1991) and ANSI/IEEE 854-1988 (R1994). The 754 standard has been adopted by nearly every computer manufacturer since about 1980. It uses a 32- and 64-bit binary word as the basis for representing a floating-point number. The 854 standard restates this representation in a radix-independent style.

Appendix D:
Common Languages
and Conventions

D.1   ADA D-1
D.2   C D-2
D.3   C++ D-2
D.4   COBOL D-2
D.5   EIFFEL D-3
D.6   Extensible Markup Language (XML) D-3
D.7   FORTRAN D-3
D.8   Hypertext Markup Language (HTML) D-3
D.9   Java D-3
D.10 LaTeX D-4
D.11 LISP D-4
D.12 ML D-4
D.13 OpenGL D-5
D.14 PASCAL D-5
D.15 PERL D-5
D.16 PostScript and PDF D-5
D.17 PROLOG D-5
D.18 SCHEME D-6
D.19 Tcl/Tk D-6
D.20 X Windows D-6

This appendix contains brief descriptions of several computer languages, with pointers to their standard versions and current Web pages added for further information. Each of these languages is supported by texts and professional references as well as compilers and interpreters. Readers interested in learning about current texts or implementations for a programming language are encouraged to consult the Web pages and Usenet news groups listed below.

D.1   ADA

ADA was designed during the late 1970s in a collaborative effort sponsored by the U.S. Department of Defense. Its purpose was to provide a common high-level language in which systems programs could be designed and implemented, with special features that support concurrency, data abstraction, and software reuse. ADA was first implemented in the early 1980s and was first standardized in 1983 as a U.S. military standard. Since then, a variety of ADA implementations have emerged and many new features have been added. The current international ADA standard definition is found in ANSI/ISO/IEC 8652-1995.

ADA is an imperative programming language whose recently added features also support object-oriented programming. Its syntax is in the PASCAL tradition, and its semantics supports strong typing, data abstraction and encapsulation, and concurrency control for real-time systems. ADA applications now cover a wide range, including military, commercial, and other large software systems. ADA compilers are available for a wide variety of computing platforms. Compilers for the 1995 standard version of ADA are also available. For more information about ADA and its implementations, consult the Usenet news group comp.lang.ada or the following Web page: http://www1.acm.org/sigs/sigada.

D.2   C

The language C was designed in 1969 as a systems programming language to support programmers who were implementing the Unix operating system. Its usage grew rapidly alongside that of Unix itself and today is probably the most widely used systems programming language. C was standardized in 1990, and its current standard version is ANSI/ISO 9899-1990.

C is also used widely in the sciences and other programming application areas. It is a high-level imperative language with extensive function libraries and unusually efficient implementations. C compilers run on most modern computers and operating systems. For more information about C, consult the Usenet news group comp.lang.c or the following Web page: http://www.gnu.org/software/gcc/gcc.html.

D.3   C++

C++ was designed by Bjarne Stroustrup in the early 1980s. It is an extension of C that adds new features for data abstraction, object-oriented programming, and a number of other improvements over traditional C constructs.

C++ is a hybrid language, including facilities for both imperative and object-oriented programming. It is a very widely used language, especially in areas of software design that require object-oriented techniques. C++ implementations exist for nearly all modern platforms, including Unix and non-Unix operating systems. A standard definition of C++ was adopted by ANSI and ISO in 1998. For more information about C++, consult the Usenet news group comp.lang.c or the following Web page: http://www.gnu.org/software/gcc/gcc.html.

D.4   COBOL

Common business-oriented language (COBOL) was first designed in 1960 as a high-level programming language for data processing applications. Its use grew rapidly in the 1960s, and it has been the most widely used language in business applications throughout the past three decades. The first ANSI standard version of COBOL was developed and published in 1968, and subsequent extensions to the standard were published in 1974 and 1985. The current standard version of COBOL is ANSI X3.23-1985 (R1991).

COBOL programs are written in an abbreviated and stylized form of English. The text of a program has four main parts or divisions. The identification division serves to identify the program, its author, and other documentary information. The environment division characterizes the features of the computer on which the program is run. The data division describes the variables, data structures, and files that the program uses, and the procedure division contains the executable code for the program. COBOL is an imperative language, so that programs are the result of procedural decomposition as a design methodology. For more information about COBOL, consult the Usenet news group comp.lang.cobol or the following Web page: http://www.cobolportal.com.

D.5 EIFFEL

EIFFEL is an object-oriented programming language that enforces principles of software design, especially reliability and reuse. It was invented by Bertrand Meyer in the late 1980s, but at this time no EIFFEL standard either exists or is in development.

EIFFEL programs are written using the philosophy of "design by contract." This means that each object's state during execution conforms to a predetermined set of constraints that are defined by method preconditions and postconditions and a class invariant. Before a method can be applied to an object, the object must be in a state that satisfies the class invariant and the method's preconditions. Similarly, after a method has been applied to an object, assurance is guaranteed that the state satisfies the class invariant and the method's postconditions. EIFFEL's type system ensures that type errors are caught at compile time, and EIFFEL provides automatic garbage collection so that programs need not use a destructor to take an object out of use. EIFFEL is implemented on a wide variety of platforms. For more information about EIFFEL, consult the Usenet news group comp.lang.eiffel or the following Web page: http://www.eiffel.com.

D.6 Extensible Markup Language (XML)

XML is a flexible text formatting language derived from SGML (ISO 8879). Originally designed for large-scale electronic publishing applications, XML is now playing an important role in the definition and exchange of a wide variety of data on the Web. More information about XML can be found at the Web site: http://www.w3c.org/XML.

D.7 FORTRAN

Designed by John Backus in 1954, formula translating system (FORTRAN) has become the most widely used scientific and engineering programming language of the past three decades. Its early versions were standardized in 1966 and a more extended version was standardized in 1977. The current FORTRAN standards are defined in ISO/IEC 1539:1991 and ISO/IEC 1539-2:1994 (Part 2: varying length character strings). A new draft Fortram standard is due to be published in 2004.

FORTRAN is an imperative language, with extensive facilities and libraries to support scientific and engineering applications. Vast amounts of FORTRAN software exist in government and industrial computing laboratories. FORTRAN is implemented efficiently and widely, with compilers available on all contemporary platforms and operating systems. For more information about FORTRAN, consult the Usenet news group comp.lang.fortran or the following Web page: http://www.fortran.com.

D.8 Hypertext Markup Language (HTML)

HTML is the standard language for preparing documents to be published on the World Wide Web. It is nonproprietary, and it can be created and processed by a wide range of word and document processing tools. HTML uses tags such as <h1> and </h1> to structure text into headings, paragraphs, lists, hypertext links, graphics, sound, and video. HTML was standardized by ISO in the year 2000 as ISO-15445. More information can be found at the Web site: http://www.w3c.org/MarkUp.

D.9 Java

Java was designed in the early 1990s by a team at Sun Microsystems headed by James Gosling. Designed to facilitate interactive programming on the Internet and World Wide Web, Java was rapidly disseminated among systems programmers at major companies in the technology industry. Java programs, or *applets*, can be embedded in appliances and HTML documents, providing interactive executable programs for users on the Web.

According to the description in Sun's white paper, "Java is a simple, object-oriented, distributed, interpreted, robust, secure, architecture neutral, portable, high-performance, multithreaded, and dynamic language." Java is based on C++ but excludes much of the baggage that makes C++ so cumbersome to use. Absent from Java are pointers; all objects are dynamic, and automatic garbage collection eliminates the need for destructors. Because Java is designed for use in networked environments, its designers included facilities for security. For more information about Java, consult the Usenet news group comp.lang.Java or the following Web page: http://java.sun.com.

D.10 LaTeX

LaTeX is a markup language and system for document typesetting. It is implemented as a macro package that extends the TeX system, allowing a wide range of scientific documents to be easily prepared for typesetting. Tex was designed in 1970 by Donald Knuth. Many of the chapters in this *Handbook* were prepared using LaTeX.

The LaTeX language can be used to describe the typesetting characteristics (e.g., boldface words, numbered lists) of a document. It is particularly good for describing mathematical formulas, maintaining bibliographies, and managing number streams (such as section numbers, figure numbers, etc.). LaTeX supports the automatic insertion of PostScript figures, and has many other features. For more information about LaTeX, consult the following Web page: http://www.latex-project.org.

D.11 LISP

List processor (LISP) was designed by John McCarthy in the late 1950s. LISP has been used predominantly in the artificial intelligence area and developed rapidly throughout the 1960s and 1970s. Two dominant dialects of LISP evolved during that period: MACLISP and INTERLISP. An effort to unify these dialects and develop a single standard resulted in Common LISP, first implemented in the 1980s. Common LISP was finally standardized in 1994 as the standard ANSI X3.226-1994. More recently, object-oriented extensions to LISP have been developed under the rubric Common LISP Object System (CLOS). Thus, one can view CLOS as a hybrid functional/object-oriented programming language. Both Common LISP and CLOS are implemented on a wide range of platforms.

LISP is a functional programming language based on the application of functions written in the form of lambda expressions using prefix notation. It is particularly useful in areas of artificial intelligence programming that require the representation of symbolic expressions for mechanical reasoning and knowledge representation. Many illustrations of the functional programming paradigm appear in Chapter 92, and examples of LISP programs appear among the chapters of the Intelligent Systems section of this *Handbook*.

For more information about LISP and CLOS, consult the Usenet news groups comp.lang.lisp and comp.lang.clos as well as the following Web page: http://www.lisp.org.

D.12 ML

Meta language (ML) was developed by Robin Milner and others as a functional programming language with imperative features and an unusually advanced concept of type. Its current version was defined in 1983, modestly updated in 1997, and is called Standard ML. It is a compiled language whose major applications are in the computer science education and research communities.

ML has a simple syntax and yet supports data abstraction through its strong static typing system and type inference mechanism, polymorphism, exceptions, and rule-based specifications. ML is widely implemented on the major computing platforms, including Unix, Mac, and PC machines. Standard ML of New Jersey is a free implementation of ML, developed jointly by Princeton University and AT&T. For more information about ML, consult the Usenet news group comp.lang.ml or the following Web page: http://www.smlnj.org.

D.13 OpenGL

OpenGL is the most widely supported environment for developing portable, interactive graphics applications. Since its introduction in 1992, OpenGL has become the most widely used graphics application programming interface (API), incorporating a broad set of rendering, texture mapping, special effects, and other visualization functions. Developers can use OpenGL on all popular desktop and workstation platforms. For more information, consult the following Web page: http://www.opengl.org.

D.14 PASCAL

PASCAL was designed by Niklaus Wirth in the early 1970s as a language for teaching principles of computer science and imperative programming. It was the main language for expressing algorithms in computer science curricula throughout the 1970s and 1980s. However, wide PASCAL usage has given way to the rapid rise of the object-oriented programming paradigm and related languages such as C++ and Java. PASCAL's current standard version is defined in the document ANSI/ISO/IEC 7185–1990.

As a language designed for teaching, PASCAL is characterized by a strong type system support for modularity, simple syntax, and robust compile and runtime programming environments. Its features have evolved over the past two decades, and nonstandard extensions are available that support a wide range of library functions as well as object-oriented programming. PASCAL was also used as a basis for the design of the language ADA. For more information about PASCAL, consult the Usenet news group comp.lang.pascal.misc or the following Web page: http://www.pascal-central.com.

D.15 PERL

PERL is a special-purpose language designed for text processing applications, especially those that require text search, extraction, and text-based reporting. Its syntax is similar to that of C, and it is usually implemented in Unix environments. However, PERL is an interpreted language, designed for rapid prototyping, so that its programs will not run as fast as comparable C programs.

Optimized for text processing, PERL employs sophisticated pattern matching techniques to speed up text search. It also does not arbitrarily limit the size of a file or the depth of a recursive call, as long as memory is available. For more information about PERL, consult the Usenet news group comp.lang.perl.misc or the following Web page: http://www.perl.com/.

D.16 PostScript and PDF

PostScript is both a graphics standard and a programming language for page layout and typesetting text and graphics on laser printers. For example, most of the figures in this *Handbook* were separately created in PostScript and then embedded in a word processing document at the time it was typeset. Portable Document Format (PDF) is a universal file format that preserves the fonts, images, graphics, and layout of any source document, regardless of the application and platform used to create it. For more information about PostScript and PDF, consult the Web page: http://www.adobe.com.

D.17 PROLOG

Programming in logic (PROLOG) was developed in the early 1970s by Philippe Roussel. It is primarily an interpreted logic programming language, designed for use in such artificial intelligence applications as problem solving, expert systems, knowledge representation, and natural language processing. PROLOG is implemented on a wide variety of computers, and its general core is defined by the standard ISO/IEC 13211-1:1995.

The syntax of PROLOG is based on logic expressions, and its semantics is defined using the concepts of *resolution and unification*. Chapter 93 in this *Handbook* provides a tutorial introduction to the logic programming paradigm, with many PROLOG examples provided as illustrations. For more information about PROLOG, consult the Usenet news group comp.lang.prolog or the following Web page: http://www.afm.sbu.ac.uk/logic-prog/.

D.18   SCHEME

SCHEME is a dialect of LISP that developed in the 1970s, designed for educational use, widely implemented, and having a simple syntax and semantics. SCHEME was standardized by ANSI and IEEE in 1991 (ANSI/IEEE 1178-1991).

SCHEME is distinguished from LISP by its small size, static scoping, and more flexible treatment of functions (i.e., a SCHEME function can be a list element, the value of a variable, the value of an expression, or passed as a parameter). For more information about SCHEME, consult the following Web page: http://www.swiss.ai.mit.edu/projects/scheme.

D.19   Tcl/Tk

The Tcl/TK programming system was developed by John Ousterhout. It has two parts: the programming language Tcl and the toolkit of *widgets* called Tk, which supports the programming of interactive graphical user interfaces (GUIs). A main goal of Tcl/Tk is to support the rapid development and prototyping of such interfaces, so that Tcl programs are usually run in interpretive mode. Tcl/Tk can also be used in coordination with other languages, and it is implemented on a variety of platforms.

Tcl is an imperative language, with modest support for handling types and data abstractions. It may not be an ideal language for writing large, complex programs; its narrow focus is to facilitate the rapid development of user interfaces. Tk widgets include labels, messages, listboxes, texts, frames, scrollbars, buttons, and other elements that commonly appear in user interfaces. A wide range of applications for languages such as Tcl/Tk are discussed in the Human–Computer Interaction section of this *Handbook*. For more information about Tcl/Tk, consult the following Web page: http://www.tcl.tk.

D.20   X Windows

X Windows is a standard technology for windowing systems that was developed at MIT and is now maintained by the consortium X.Org. X Windows consists of a library of graphics function calls, called Xlib, written in C, that is freely available. Application programs that require graphics can call functions from this library. The functions in Xlib are simpler than those in GKS or PHIGS. They are also more stylized to the needs of interactive user interface programming, such as creating a window or sampling the mouse pointer. On the other hand, X Windows functions are not as extensive as PHIGS functions in the area of graphics applications. For more information about X Windows, consult the following Web page: http://www.x.org.

Index

A

A* algorithm, **55**-12, **63**-10–11
ab initio methods
  organic compound modeling, **32**-12–16
  spectroscopic analysis, **32**-16–19
Abstract classes, **91**-5
Abstract data types, **4**-2, **90**-3, **90**-13, **103**-1, **103**-11–12
Abstract interpretation, **93**-27, **98**-15, **107**-8
Abstraction, in software design, **101**-18–19
Abstractions
  concurrent computing without, **96**-5–9
  data, **90**-13–15, **90**-18–20
  definition of, **1**-7, **80**-7–8
  procedural, **90**-10–13, **90**-18
  rendezvous, **96**-18–19
Abstract system state, **106**-8–9
Abstract type, **97**-21
(a, b)-tree, dictionary realization with, **4**-17–20
Academic programs, **1**-4–5
Acceleration anomaly, **63**-15
Acceleration function, **20**-4
Acceleration rendering, **41**-13–15
Access codes, **85**-9
Access controls
  administration of authorization, **79**-14–15
  authorization relations, **79**-8–9
  capabilities, **79**-8
  description of, **60**-3, **79**-5
  discretionary, **60**-3–5, **60**-9–10, **79**-10, **79**-14
  implementation of, **79**-7–9
  lists, **79**-7–8
  mandatory, **60**-5–7, **60**-11, **79**-10–12, **79**-12
  matrix of, **79**-6–7
  overview of, **79**-1–3
  policies for, **79**-9–14
  role-based policies for, **79**-12–14
  schematic diagram of, **79**-2
Access transparent, **89**-2
Accountability, **2**-11–12
ACID properties, **56**-1–2
ACM Curriculum Committee on Computer Science, **1**-3
ACM/IEEE-CS Task Force, **1**-3–4

Acoustic trackers, **42**-8
Actinide, **32**-15
Action semantics, **98**-11
Activation records, **100**-4–5
Activation stack, **100**-5–6
Active agents, **74**-7
Active contours, **43**-13–14
Activity diagram, **104**-11
Actuators, **87**-22
Act-utilitarians, **2**-4
ADA, **90**-16–17, D%-1–2
Adaptive grid generation, **26**-11–12
Adaptive modeling, **28**-21
Adaptive optics, **33**-3
Adaptive supersampling, **39**-18
Adders, **16**-8–10, **16**-15–16
Adder/subtractor, **16**-15–16
Addition
  fixed point algorithms, **22**-4–11
  floating point algorithms, **22**-18–19
Additive-weighted Voronoi diagram, **11**-16
Address faults, **85**-9
Addressing
  definition of, **17**-7
  format, **17**-7–8
  object-oriented, **85**-8
Address locking, **19**-15
Address spaces
  definition of, **81**-2–3
  segmented, **85**-2
Address spoofing, **74**-2
Adjacency list, **7**-2, **10**-19
Adjacency matrix, **7**-2, **10**-19–20
Admissibility condition, **63**-11
Advanced database management systems, **60**-11–12
Advanced encryption standard, **57**-7
Advanced Research Projects Agency Spoken Language System Project, **64**-11–14
Advancing front procedure, **26**-12–13
Adversary games, **63**-18–21
Affine transformations, **35**-2
Aho–Corasick algorithm, **13**-10–13, **13**-46
Algebraic space, **55**-6–8

Algorithm(s)
  A*, 55-12, **63**-10–11
  Aho–Corasick, **13**-10–13, **13**-46
  alpha-beta, **63**-11–12
  analysis of
    description of, **3**-2–5
    examples, **3**-5–9
  approximate, **88**-4
  a priori data mining, 75-9
  bankers, **82**-20–21, **84**-19
  Bellman–Ford, **7**-8–9
  biased round robin, 82-8
  bipartite matching, **7**-13–14
  Boyer–Moore, **13**-5–8, **13**-45
  breadth-first search, 7-6, **10**-20–22
  brute-force, 13-2
  characteristic set, 8-10
  classification of, 3-1
  closest insertion, 3-20
  clustering, 75-122
  connected components, 10-22, **10**-26
  copy collection, 100-14
  costs assigned to, 10-8
  CSS, 82-29
  cycle-oriented round robin, **82**-8–9
  Delaunay, **26**-19–23
  depth-first search, **7**-3–4
  depth of, 10-8
  description of, 1-11
  deterministic, 1 2-6–7
  digital signature, 9-19
  Dijkstra's, **7**-7–8
  dispatching, 82-11
  divide-and-conquer, **3**-10–12
  domainwise, 28-17
  Durand–Kerner, 8-8
  dynamic non-cooperative, 88-9
  dynamic non-distributed, 88-9
  dynamic programming, **3**-12–17
  edge detection, 43-2
  elementwise parallel, 28-17
  ellipsoid, **15**-9–10
  fixed point addition, **22**-4–11
  floating point, **22**-18–20
  form factor, 38-22
  garbage collection
    copy collection, 100-14
    mark-sweep, **100**-13–14
    reference counting, **100**-12–13
    software program verification using, **107**-11
  Gaussian elimination, 8-4, **10**-36
  genetic
    applications of, 14-4
    automatic programming uses, **14**-7–8
    best practices, **14**-4–9
    description of, 14-1, 63-15
    function optimization, **14**-5–6
    mathematical analysis of, **14**-9–12
    modeling uses of, 14-9
    ordering problems using, **14**-6–7
    overview of, 14-2

  performance of, **14**-3–4
  principles of, **14**-1–4
  research issues regarding, 14-12
gradient descent, 66-11
greedy heuristics, **3**-17–21
growth rates for, 3-2
GSAT, **61**-11, **61**-13
hashing, **18**-4
hemicube, 38-21
heuristic, **88**-4
Hirschberg, **13**-28–29
history of, 6-1
Huffman, 3-19, **13**-36–40, **13**-46
inference, **70**-4–5
Jenkins–Traub, 8-8
Karmarkar's, **15**-12
Karp–Rabin, **13**-3–4
k-means, 75-11
k-medoid, 75-11
Knuth–Morris–Pratt, **13**-4–5
Kruskal's, **7**-10–11
Las Vegas, **11**-9, 1 2-6
Levenberg–Marquardt, **66**-11
linear programming
  description of, 15-5
  ellipsoid, **15**-9–10
  Fourier's scheme for linear inequalities, **15**-5–6
  semidefinite programming, **15**-10
  simplex method, **15**-7–8
Linear time, **11**-22
load-balancing, **88**-8
Malhotra–Kumar–Maheshwari blocking flow,
    7-19–20
Marching Cubes, **41**-3
Mark-sweep, **100**-13–14
matching, **7**-13–14
McCreight, 13-20
MergeHull, **10**-33–34
Min-Cut, 1 2-4–6
model formulation, **62**-9–10
Monte Carlo, **11**-9, 1 2-6, **32**-5
MTD(f), **63**-14, **63**-22
nodewise, 28-17
painter's, 37-16
p-approximate, 15-30
parallel, *see* Parallel algorithms
Prim's, **3**-19–20, **7**-9–10
priority dispatching, **82**-6–8
priority queues, **3**-7–9
progressive refinement, 38-25
projected tetrahedra, **41**-20–21
pruning, 66-12
PVSplit, **63**-18–20
QuickHull, **10**-32–33
quick search, **13**-8–10
quorum-based voting, 58-11
radix sort parallel, **10**-28–29
randomized, *see* Randomized algorithms
recurrences
  divide-and-conquer, **3**-4–5
  linear, **3**-2–4

reference counting, 100-12–13
remove-duplicate, 10-17–19
Robinson's unification, 61-3
rotation, 82-8–9
SATF, 82-29
shading, 38-3
shadow vertex simplex, 15-8
Shift-Or, 13-30–31
simple round robin, 82-8
sorting, 3-5–9
spanning tree, 73-3–4
SSS*, 63-12–14
static, 88-9–10
string-matching, *see* String-matching algorithms
unification, 61-3
universal, 6-4–7
Wu–Manber, 13-34–36
Z-buffer, 38-3
Zhu–Takaoka, 13-14–15
Algorithm-based fault tolerance, 25-11
Aliasing, 20-13, 39-13–14, 39-26, 41-5
Alignment
  definition of, 13-20, 13-22
  global, 13-23–25
  Hirschberg algorithm, 13-28–29
  local, 13-25–27
  longest common subsequence of two strings, 13-27–28
Allocation mechanism, 82-5
All-prefix-sums, 10-15
Alphabet, 5-2, 6-11
Alpha-beta algorithm, 63-11–12
Alpha shapes, 27-14–15
Alternate Turing machine, 5-5, 5-19
Altimeter, 31-10
Altruistic locking, 57-15
Ambient illumination, 37-16
Ambiguity, 99-17–19
Ambiguous context-free grammar, 6-19
American Association for Artificial Intelligence, A%-2
American National Standards Institute, C%-2
American Standard Code for Information Interchange, C%-2
Amoeba, 87-2–3, 87-6–8
Amorphous phenomena, 41-29
Amulet framework, 48-16
Analog-to-digital conversion, 76-18–19
Analytix tableaux, 61-7–11
AND/OR trees, 63-3
Angular motion grippers, 71-22
Angular springs, 40-6
Animation
  algorithms for, 40-12–14
  artificial life, 40-8–9
  balance control, 40-6
  behavioral methods of, 40-8–9, 40-14
  character deformations, 40-5–6
  classification of, 40-2
  cloth, 40-7–8
  collision detection and response, 40-7
  crowds, 40-9–10
  definition of, 49-8–9

description of, 27-18
dynamic simulation in, 40-6–7
facial, 40-10–12
geometric methods of, 40-2–5
groups, 40-9–10
inverse kinematics, 40-4–5
keyframe, 40-4
kinematic methods of, 40-2–5
motion capture, 40-2–3
motion control, 40-2
motion retargeting, 40-3–4
overview of, 40-1–2
physics-based methods of, 40-6–8
procedural animation, 40-5
virtual life, 40-8–9
Annihilators, 3-3
Annotation-based video database models, 59-16–17
Anomaly-based approach to intrusion detection, 79-17
Anonymity, 50-10–11
Antialiased voxelization, 41-25–26
Antialiasing
  adaptive supersampling, 39-18
  area sampling, 39-16–17
  definition of, 39-14
  description of, 38-3
  example of, 39-29
  in image scaling, 39-21–25
  point sampling, 39-15
  supersampling, 39-17
Anticipatory scheduling, 82-28–29
Antivirus software, 78-7–8
Appearance-based matching, 43-19–20
Appending, 78-5
AppLeS, 88-11–12
Applets, 95-4–6
Application gateways, 74-10
Application programming interface
  description of, 34-5, 96-20–21
  OpenMP, 96-21
Application semantics, 48-17
Application specific integrated circuits, 16-31
Applicative-order evaluation, 92-5
Apply-to-each construct, 10-10
Appropriate mapping, 20-4
Approximate algorithms, 88-4
Approximation
  Born–Oppenheimer, 32-6
  finite-difference, 29-14–15
  finite-volume, 29-15–16
  Hartree–Fock, 32-16
A priori data mining algorithm, 75-9
Arbitration, 67-12
Arborescences, 7-11
Architecture
  computer
    addressing
      definition of, 17-7
      format, 17-7–8
    description of, 17-1
    execution hazards
      control, 17-14–15

data, 17-12–13
description of, 17-12
structural, 17-15
instruction execution
description of, 17-9–11
execution unit, 17-11
instruction decode unit, 17-11
instruction fetch unit, 17-11
storeback unit, 17-11–12
instruction set
arithmetic logic unit, 17-3
complex, 17-2–3
control transfer, 17-4–5
definition of, 17-2
memory, 17-3–4
memory referencing, 17-3–4
reduced, 17-2
memory
amount of, 17-6–7
cache, 17-6
description of, 17-5
dynamic random-access memory, 17-6
main, 17-6
physical, 17-8–9
register file, 17-5–6
virtual, 17-8–9
processor-program interface, 17-1–2
secondary storage, 17-7
superscalar design, 17-15
very long instruction word computers, 17-15
description of, 1-11
parallel
description of, 23-1
MIMD, 23-8–11
MISD, 23-8
network interconnections, 23-11–12
SIMD, 23-5–8
SISD, 23-2–5
stream model, 23-1–2
software
analysis, 109-1
definition of, 109-1
description languages, 109-4–6
elements of, 109-1
principles of, 109-2–3
properties of, 109-1–2
styles of, 109-6–13
types of, 109-3–4
Area sampling, 39-16–17
ARIES, 56-14–15
Arithmetic and Logic Unit, 16-22–23, 16-29
Arithmetic coding, 76-9–12
Arithmetic logic unit, 17-2–3
Array(s)
bucket, 4-25
definition of, 4-3
field programmable gate
architecture of, 16-31–35
calculator mapping to, 16-35
configurable logic blocks, 16-31–33
definition of, 16-31

interconnect, 16-33–34
Xilinix input/output block, 16-34–35
sequence implementation with, 4-5
sorted, 4-14–15
verification of, 107-15
Array multipliers, 22-13
Array processors, 23-5–6
Artificial intelligence
constraint logic programming and, 93-17
definition of, 61-1
real-time, 83-11
Assignment problem, 7-14–16
Association for Computing Machinery, A%-1
Association rule mining, 57-22, 75-8–9
Astronomical databases, 33-1–4
Astrophysics, computational
astronomical databases, 33-1–4
data analysis
cosmic microwave background, 33-8
description of, 33-4–5
gamma-ray burst, 33-8–9
time series, 33-9–10
data mining, 33-6–7
description of, 33-1
multi-wavelength studies, 33-7–8
parallel computation in, 33-15–16
theoretical modeling
compact objects, 33-14–15
cosmology, 33-13
galactic dynamics, 33-13–14
galaxy clusters, 33-13–14
galaxy formation, 33-13–14
gravitational n-body, 33-10
hydrodynamics, 33-10–11
magnetohydrodynamics, 33-11
numerical relativity, 33-14
planetary dynamics, 33-11–12
radiative transfer, 33-11
simulations, 33-10
solar system dynamics, 33-11–12
star formation, 33-12–13
stellar astrophysics, 33-12
Asynchronous bus protocol, 19-7–8
Asynchronous inputs, 16-26
Asynchronous transfer mode
description of, 72-14–16
local-area networks, 72-21–23
Atmosphere models, 31-5–8
Atomic expression, 92-7
Atomicity, 58-9, 87-19–20
Attribute, 52-6
Attribute grammars, 99-4, 99-9–10
Attribute value independence assumption, 55-13
Attribute-value representation, 65-1–2
Audio content analysis, 59-8–9
Audio displays, 42-6
Audit events, 79-20
Auditing
control issues for, 79-19–21
data from, 79-15
data recording, 79-20

description of, **79**-2–3
elements of, **79**-15
Auditory displays, **20**-21–22
Audit records, **79**-19
Augmented reality, **20**-18
Augmenting paths
applications of, **7**-12
definition of, **7**-11
Authentication
biometric, **79**-5
definition of, **79**-3
in distributed systems, **79**-5
failure of, **74**-2–3
overview of, **79**-1–3
password, **79**-3–4
token-based, **79**-4–5
user, **74**-3
Authoring software for multimedia systems, **49**-23
Authorization
administration of, **79**-14–15
definition of, **77**-2
Authorization relations, **79**-8–9
Automated guided vehicle systems, **71**-51–52
Automated reasoning, **61**-1
Automatic processing, **49**-15
Automatic programming, **14**-7–8
AVL-tree
definition of, **4**-20
description of, **4**-3
dictionary realization with, **4**-20–25
example of, **4**-22
Axiomatic semantics, **98**-2, **98**-6, **98**-13–14,
98-16

B

Backchannel responses, **50**-6
Backplane bus, **19**-1–2
Backtracking, **63**-2, **93**-4
Backus–Naur form, **98**-1
Balance control, **40**-6
Balance property, **4**-20
Bandlimited, **39**-3
Bankers algorithm, **82**-20–21, **84**-19
Banking, **18**-8
Baroclinic model, **31**-4
Barotropic model, **31**-4. **31**-9–10
Barrier, **81**-4
Baseband signaling, **24**-5–6
Batch multiprogramming, **80**-4–5
Batch processing, **21**-3
Bayesian networks
action calculus, **70**-11–12
causal information carried by, **70**-6–12
counterfactuals, **70**-13–15
description of, **70**-1–2
historical background of, **70**-2–3
history of, **70**-12
illustration of, **70**-2
probabilistic information carried on, **70**-3–6
Bayesian viewpoint, **66**-14–15

Behavioral animation, **40**-2, **40**-14
Bell–LaPadula model, **60**-5, **60**-12
Bellman–Ford algorithm, **7**-8–9
Bender's representation, **15**-19–20
Berry's paradox, **6**-8
Best-first search, **63**-8–10
Bezier curves, **36**-2, **36**-8–10
Bezier surfaces, **36**-12–14
BGP/IDRP, **73**-15
Biased round robin algorithm, **82**-8
Bichromatic closest pair problem, **11**-13
Bichromatic line segment intersection problem, **11**-23
Biconnected graph, **7**-5
Bidirectional reflectance distribution function, **35**-7
Bidirectional scan scheduling, **82**-16–17
Bidirectional search, **63**-5–6
Big Bang Nucleosynthesis, **33**-13
Bilinear interpolation, **38**-13
Bimanual input, **20**-11
Binary decision diagrams, **106**-6
Binary numbers, negative, **16**-17–18
Binary property, **4**-20
Binary search, **3**-10–11
Binary search tree, **3**-13
Binary semaphore, **84**-9
Binary tree, **4**-3
Binary voxelization, **41**-24
Binding, **87**-17–18
Bindings
best practices, **90**-15–20
binding time, **90**-2
control structures in
best practices for, **90**-17–18
conditional structures
case statements, **90**-7–8
if statements, **90**-6–7
iterative structures, **90**-8–9
data abstraction, **90**-13–15, **90**-18–20
description of, **90**-6
exceptions, **90**-9–10
Goto statement, **90**-9–10
procedural abstraction, **90**-10–13, **90**-18
unconstrained, **90**-9–10
declarations, **90**-6
execution units, **90**-5–6, **90**-17
research issues, **90**-20–21
scope of, **90**-4–5
types, **90**-3–4
variables, **90**-2–3, **90**-15–16
Binding time, **90**-2
Biology, computational
databases
access to, **34**-5
data modeling, **34**-5–6
description of, **34**-3–5
representation, **34**-5–6
genomics
definition of, **34**-11
genetic mapping, **34**-12
sequence analysis, **34**-13–14
sequence assembly, **34**-12–13

imaging, **34**-6–7
microscopy, **34**-6–7
nuclear magnetic resonance, **34**-7–8
overview of, **34**-1–2
protein folding, **34**-10–11
tomography, **34**-6–7
x-ray crystallography, **34**-8–9
Biometric authentication, **79**-5
Bipartite graphs, **7**-1
Bipartite matching algorithm, **7**-13–14
Bits per inch, **21**-5
Blackman window, **39**-12
Block ciphers, **9**-10
Blocking, **18**-12
Blocking flows, **7**-19–20
Block operations, **41**-26–28
Block-structured languages, **90**-6
b-matching problem, **7**-16–17, **15**-26
Boltzmann machine, **66**-16
Bonding, **32**-14–15
Boolean circuit, **5**-20, **6**-31
Boolean logic
    axioms of, **16**-10–11
    realization of, **16**-8–10
    theorems of, **16**-10–11
Boolean operators, **16**-2–3, **36**-7
Boot sector viruses, **78**-4–5
Born–Oppenheimer approximation, **32**-6
Bottom-up software design, **103**-3–4
Boundary element method, **28**-6
Bounded bidirectional scheduling, **82**-17
Bounded-scan scheduling, **82**-17
Box filter, **39**-8–9
Boyer–Moore algorithm, **13**-5–8, **13**-45
Branch and bound method, **15**-27–29
Branch and cut method, **15**-29–30
Branchings, **7**-11
Branch predication, **17**-14–15
Breadth-first search, **7**-6, **10**-20–22, **63**-4
Breadth-first theorem, **93**-22
Breadth-first tree, **7**-6
Brent's scheduling principle, **8**-5–6
Brent's theorem, **10**-8–9
Bridges, **24**-11, **24**-14
Bridging
    source route, **73**-2, **73**-5–7
    transparent, **73**-2–3, **73**-5
British Computer Society, A%-2
Broyden–Fletcher–Goldfarb–Shanno method, **28**-11
Brute-force algorithm, **13**-2
Bryan–Cox–Semtner *z*-level model, **31**-10–13
B-spline curve, **36**-10–11
B-spline surfaces, **36**-14–15
B-tree, **53**-9
B+ trees, **21**-3
Bucket array, **4**-25
Buffer cache, **89**-11
Buffer management, **56**-6–7
Buffer tree, **4**-3
Buggy code, **74**-3
Building blocks hypothesis, **14**-12

Bump maps, **35**-7
Bus
    backplane, **19**-1–2
    CCL-XMP, **19**-15–17
    cross-coupling effects, **19**-5
    definition of, **10**-4, **19**-1
    designing of, **19**-2
    historical perspectives, **19**-17–19
    local, **19**-1
    overview of, **19**-1–2
    research issues, **19**-17–19
    signal reflections, **19**-4
    signal skew, **19**-5
    symmetric multiprocessor
        arbitration, **19**-12–13
        bandwidth, **19**-13–14
        cache coherence protocols, **19**-10–12
        description of, **19**-9–11
        design issues for, **19**-10
        locking, **19**-14–15
        memory access latency, **19**-14
        synchronization, **19**-14–15
    system, **19**-1
    transmission-line concepts of, **19**-2–3
    wire-OR glitches, **19**-4
Bus arbitration, **19**-2
    centralized, **19**-5–6
    decentralized, **19**-6–7
    definition of, **19**-5
    distributed, **19**-6
    symmetric multiprocessor, **19**-12–13
Bus bandwidth, **19**-13–14
Business modeling, **108**-5
Bus master, **19**-2
Bus parking, **19**-10
Bus protocol, **19**-2
    asynchronous, **19**-7–8
    definition of, **19**-7
    split-transaction, **19**-8–9
    synchronous, **19**-8
Bus slaves, **19**-2

C

C, D%-2, **90**-17
C#, **91**-21–24
C++, D%-2, **90**-17, **91**-13–18
Cable modems, **24**-10
Cache coherence protocols, **19**-10–12
Cache coherency, **23**-9
Cache corruption, **88**-4
Cache memory, **17**-6, **18**-3–7, **85**-6–7, **86**-1
Cache misses, **18**-13
Caching, **21**-16, **89**-7–8
Calculator
    arithmetic functions of, **16**-22–23
    4-bit, **16**-28–30
    mapping to field programmable gate array, **16**-35
Call admission, **83**-9
Call-back procedures, **48**-11
Call-by-address parameter passing, **100**-4

Call-by-name, **90**-11

Call-by-reference, **90** -11

Call-by-result parameter passing, **100**-4

Call-by-value parameter passing, **100**-4

Call-by-value-result parameter passing, **100**-4

Canny edge detector, **43**-3, **43**-6

Canonical complexity classes, **5**-7

Capability, **85**-3

Capability-based addressing, **85**-2

Capability maturity model, **47**-11, **101**-13, **102**-17, **108**-13

Capacity constraint, **7**-18

Carding, **78**-17

Carrier-sense multiple access/collision detection, **72**-16–18

Carry lookahead adder, **22**-6–7

Carry select adder, **22**-10–11

Carry skip adder, **22**-7, **22**-9–10

Cartesian coordinate robots, **71**-9, **71**-11

Cartesian grid, **26**-2–3

Carving, **41**-30

Case-based planning, **67**-18

Case statements, **90**-7–8

Cathode-ray tube, **35**-12–13

Causal broadcast, **87**-12

Causal reasoning, **62**-10–11

CCL-XMP bus, **19**-15–17

CD-ROM, **49**-7–8, **86**-5–6

Cell decomposition, **71**-40

Cell relation graph, **105**-31

Cellular phones, **78**-17

Center of interest, **35**-10

Center of projection, **35**-10, **38**-9

Central differencing, **41**-6

Centralized bus arbitration, **19**-5–6

Centralized serial priority arbitration, **19**-6

Central processing unit, **21**-2

Chained declustering, **21**-19

Characteristic polynomial, **8**-3

Characteristic set algorithm, **8**-10

Characterization analysis, **75**-7

Checkpoint, **86**-11

Checkpointing, **25**-12, **57**-2, **57**-15

Chemical reaction sequences, **32**-4

Chemistry, computational

chemical reaction sequences, **32**-4

in education, **32**-2–4

inorganic chemistry, **32**-13–16

JCE software, **32**-4

Journal of Chemical Education, **32**-2–3

molecular dynamics simulations

applications of, **32**-8

description of, **32**-5–6

methodology of, **32**-6–8

Monte Carlo methods, **32**-5

numerical solution of differential equations, **32**-4–5

organic compound modeling

ab initio methods, **32**-12–13

description of, **32**-9–10

empirical solutions, **32**-10–11

semiempirical methods, **32**-12

organometallic chemistry, **32**-13–16

overview of, **32**-1–2

principles of, **32**-2

Project SERAPHIM, **32**-3–4

Chimera grid, **26**-11

Chinese postman problem, **7**-20

Choices, **95**-10–11

Choke points, **74**-4

Chomsky hierarchy, **6**-16–17

Chomsky normal form, **6**-21

Chromatic adaptation, **20**-15

Chromosomes, **14**-2

Church–Rosser property, **107**-19

Church–Turing thesis, **6**-1

Ciphertext, **9**-2, **9**-12–13

Circuit(s)

Boolean, **5**-20, **6**-31

complexity of, **5**-20–21

Hamiltonian, **5**-7–8

sequential

concept of, **16**-23–24

data flip-flop, **16**-26–27

data register, **16**-27–28

definition of, **16**-8

D latch, **16**-25–26

shift register, **16**-27–28

SR latch, **16**-24–25

stack, **16**-27–28

Circuit family, **5**-20

Circuit model, **10**-7

Circuit relays, **74**-10

Circuit switching, **24**-3

definition of, **72**-2

wide-area networks, **72**-8–10

Circular-scan scheduling, **82**-17

Class

complexity

alternating, **5**-19–20

canonical, **5**-7

characteristics of, **5**-6–8

complementation of, **5**-10

co-NP, **5**-10

constructibility of, **5**-8–9

definition of, **5**-1

diagonalization, **5**-2, **5**-11

hierarchy theorems, **5**-10–11

NP, **5**-1–2, **5**-7, **5**-17

P, **5**-1, **5**-7, **5**-17

padding argument, **5**-2, **5**-11–12

probabilistic, **5**-22–23

relationships, **5**-8–12

research issues, **5**-26–27

space-constructible, **5**-8

time-constructible, **5**-8

definition of, **4**-2, **104**-3–4

Class diagram, **104**-9–10

Classical logic, **61**-1, **61**-4–5

Classifier, **68**-2

Class variables, **91**-7

Clauses, **5**-15

Cleanroom process, **102**-14–15

Client/server architecture, **58**-5
Clipping, **38**-2–6
  and culling, **38**-6–9
Closed loop, **67**-5
Closed queuing network model, **57**-18–19
Closed under complementation, **5**-10
Closest insertion algorithm, **3**-20
Closest pair problem, **10**-30–31, **11**-12–13
CLU, **90**-13
Clustered RAID, **21**-20
Clustering, **53**-9–10, **75**-10–12
Clustering algorithm, **75**-122
Clustering index, **53**-10
CMOS binary logic, **16**-4–5
CMOS switching model, **16**-5
Co-allocation, **88**-14
Coamplitude and cophase, **31**-10
COBOL, D%-2
Code of ethics, B%-1–7
Code optimization, **99**-22–24
Cognition theories, **49**-13–16
Cognitive band, **69**-5
Cognitive modeling
  best practices, **69**-8–10
  description of, **62**-16, **69**-1–2
  principles of
    computer science, **69**-4
    description of, **69**-2
    environmental psychology, **69**-4–5
    evolutionary psychology, **69**-4–5
    neuroscience, **69**-3–7
    psychology, **69**-2–3
  research issues, **69**-5–8
  Soar architecture, **69**-8–9
  symbol processing, **69**-5–7
  unified theories of cognition, **69**-7
Cognitive psychology, **46**-6–7
Cognitive task analysis, **46**-13
Coherence principle, **49**-16
Cohesion, **101**-17
Coincidental correctness, **105**-8
Collaboration diagram, **104**-11
Collaborative filtering, **63**-22
Collaborative work, computer-supported
  anonymity, **50**-10–11
  communication
    auditory cues, **50**-4–5
    facilitation of, **50**-3
    video vs. audio-only, **50**-5
    visual cues, **50**-4–5
  definition of, **50**-1
  description of, **50**-1–2
  environmental factors that affect
    asynchrony, **50**-8–9
    auditory cues, **50**-4–5
    description of, **50**-2–3
    dialog structure, **50**-6
    facilitation of, **50**-3
    information richness, **50**-8
    input/output rates, **50**-8–9
    managerial behavior, **50**-8

    social context cues, **50**-6–8
    visual cues, **50**-4–5
  free-riding, **50**-10–11
  groupware, **50**-14–15
  information sharing, **50**-12–14
  media factors in, **50**-2–9
  opinion formation, **50**-13–14
  process gains and losses, **50**-9–10
  process structure, **50**-11–12
  process support tools, **50**-12
  production blocking, **50**-10
  proxemic effects, **50**-5–6
  research issues, **50**-15–16
  task structure, **50**-12
Collision-resistant hash function, **9**-11
Collocated work, **50**-2
Color displays, **20**-14
Color gamut, **20**-14–15
Color information coding, **20**-16–17
Color quantization, **59**-3
Color space, **59**-3
Color specification, **20**-14
Column generation, **15**-13–14
Combinational circuits, **16**-7
Combinatorial logic, **16**-1
Combinatorial optimization
  algorithms for, **15**-35–36
  approximation in
    description of, **15**-30–31
    Lagrangian relaxation, **15**-34–35
    LP relaxation, **15**-31
    neighborhood search, **15**-33–34
    primal–dual, **15**-31–32
    randomized rounding, **15**-31
    semidefinite relaxation, **15**-32–33
  description of, **15**-1–2
  integer linear programs
    Bender's representation, **15**-19–20
    covering and packing problems, **15**-15–16
    description of, **15**-15
    inference problems, **15**-17–18
    Jeroslow's representability theorem, **15**-19
    multiprocessor scheduling, **15**-18
    plant location problems, **15**-17
    satisfiability problems, **15**-17–18
  integer programming, **15**-35–36
  large-scale linear programming in
    compact representations, **15**-15
    cutting stock problem, **15**-14–15
    decomposition, **15**-15
    description of, **15**-13–14
  mixed integer linear program, **15**-2
  paradigms in, **15**-2
  partial enumeration methods
    branch and bound, **15**-27–29
    branch and cut, **15**-29–30
    description of, **15**-27
  polyhedral combinatorics
    b-matching problem, **15**-26
    cutting plane method, **15**-25–26
    definition of, **15**-3

integral polyhedra, **15**-20–23
matroids, **15**-23–24
special structures, **15**-20–23
valid inequalities, **15**-24–27
Commercial off-the-shelf applications, **47**-5–6
Commercial robots
Cartesian coordinate, **71**-9–10
commercial configurations of, **71**-7–13
drive types of, **71**-11
dynamics of, **71**-15–16
error detection and recovery, **71**-44–45
fingers of, **71**-22–23
grippers, **71**-22–23
human operator interfaces with, **71**-45–47
kinematics of, **71**-8, **71**-13–14
manipulator performance of, **71**-8
mobile, **71**-49–51
motion trajectory generation, **71**-18–21
multiple, **71**-44
part fixtures, **71**-21
SCARA, **71**-8–11
sensors
acceleration, **71**-26
data processing, **71**-27–28
description of, **71**-25
distance, **71**-26
force, **71**-26–27
photoelectric, **71**-27
position, **71**-26
proximity, **71**-26
tactile, **71**-26
torque, **71**-26–27
types of, **71**-25–31
velocity, **71**-26
servo-level motion control, **71**-16–18
tooling, **71**-21
tooling process integration and coordination,
71-24–25
trajectory generation, **71**-18–21
types of, **71**-7–13
vision for, **71**-29–31
wrist mechanisms, **71**-23–24
Common gateway interface scripts, **74**-7
Common information, **50**-13
Common-mode failures, **25**-14
Communication protocol, **24**-3
Compact objects, **33**-14–15
Comparative analysis, **62**-12
Competitive analysis, in usability engineering, **45**-5–6
Compilers
ambiguity in, **99**-17–19
attribute analysis, **99**-19
attribute grammars, **99**-4, **99**-9–10
best practices, **99**-10–25
case study construction of, **101**-21–24
code generation, **99**-19–22
code optimization, **99**-22–24
context-free grammars, **99**-5–6
development of, **99**-1–2
error recovery, **99**-24–25
finite automata, **99**-4–5

incremental compilation, **99**-25–26
intermediate representations, **99**-19–22
Lex/Flex, **99**-13–15
overview of, **99**-1–4
parsing algorithms, **99**-6–9
phases of, **99**-2–3
principles of, **99**-4–10
Purdue Compiler Construction Tool Set, **99**-15–17
research issues, **99**-26
syntax specification using expressions and grammars,
99-11–12
Yacc/Bison, **99**-13–15
Complete languages, **5**-14
Complex instruction set computers, **17**-2
Complexity
circuit, **5**-20–21
description of, **1**-11
Kolmogorov, **5**-25–26
space, **5**-6, **6**-29
time, **5**-6, **6**-29
Complexity class
alternating, **5**-19–20
canonical, **5**-7
characteristics of, **5**-6–8
complementation of, **5**-10
co-NP, **5**-10
constructibility of, **5**-8–9
definition of, **5**-1
diagonalization, **5**-2, **5**-11
hierarchy theorems, **5**-10–11
NP, **5**-1–2, **5**-7, **5**-17
P, **5**-1, **5**-7, **5**-17
padding argument, **5**-2, **5**-11–12
probabilistic, **5**-22–23
relationships, **5**-8–12
research issues, **5**-26–27
space-constructible, **5**-8
time-constructible, **5**-8
Complexity theory, **61**-20
overview of, **5**-1–2
research issues, **5**-26–27
Component assembly model, **108**-8–9
Component diagram, **104**-11
Composite attributes, **52**-7
Composite classifiers, **65**-13–14
Compositionality, **62**-2
Compositional modeling, **62**-8–9
Compound tasks, **20**-3
Compression of data
definition of, **76**-1
description of, **76**-1–2
lossless
arithmetic coding, **76**-9–12
compression ratio, **76**-2
description of, **76**-2
dictionary-based techniques for, **76**-12–16
entropy coding, **76**-4–5
Huffman coding, **76**-5–9
information content, **76**-2–3
Lempel–Ziv–Welsh compression, **13**-40–43, **76**-12,
76-14–16

methods of, **76**-3–12
  prediction with partial matching, **76**-16
  run-length coding, **76**-16–18
lossy
  data domain compression, **76**-18–23
  differential pulse code modulation, **76**-19–20
  JPEG, **76**-23–32, **76**-39
  predictive differential pulse code modulation,
    76-20–22
  pulse code modulation, **76**-18–19
  subband coding, **76**-32–39
  transform domain compression, **76**-23–39
  vector quantization, **76**-22–23
  wavelets, **76**-32–39
Compression ratio, **76**-2
Computability, **6**-2–6
Computational astrophysics
  astronomical databases, **33**-1–4
  data analysis
    cosmic microwave background, **33**-8
    description of, **33**-4–5
    gamma-ray burst, **33**-8–9
    time series, **33**-9–10
  data mining, **33**-6–7
  description of, **33**-1
  multi-wavelength studies, **33**-7–8
  parallel computation in, **33**-15–16
  theoretical modeling
    compact objects, **33**-14–15
    cosmology, **33**-13
    galactic dynamics, **33**-13–14
    galaxy clusters, **33**-13–14
    galaxy formation, **33**-13–14
    gravitational *n*-body, **33**-10
    hydrodynamics, **33**-10–11
    magnetohydrodynamics, **33**-11
    numerical relativity, **33**-14
    planetary dynamics, **33**-11–12
    radiative transfer, **33**-11
    simulations, **33**-10
    solar system dynamics, **33**-11–12
    star formation, **33**-12–13
    stellar astrophysics, **33**-12
Computational biology
  databases
    access to, **34**-5
    data modeling, **34**-5–6
    description of, **34**-3–5
    representation, **34**-5–6
  genomics
    definition of, **34**-11
    genetic mapping, **34**-12
    sequence analysis, **34**-13–14
    sequence assembly, **34**-12–13
  imaging, **34**-6–7
  microscopy, **34**-6–7
  nuclear magnetic resonance, **34**-7–8
  overview of, **34**-1–2
  protein folding, **34**-10–11
  tomography, **34**-6–7
  x-ray crystallography, **34**-8–9

Computational chemistry
  chemical reaction sequences, **32**-4
  in education, **32**-2–4
  inorganic chemistry, **32**-13–16
  JCE software, **32**-4
  Journal of Chemical Education , **32**-2–3
  molecular dynamics simulations
    applications of, **32**-8
    description of, **32**-5–6
    methodology of, **32**-6–8
  Monte Carlo methods, **32**-5
  numerical solution of differential equations,
    32-4–5
  organic compound modeling
    ab initio methods, **32**-12–13
    description of, **32**-9–10
    empirical solutions, **32**-10–11
    semiempirical methods, **32**-12
  organometallic chemistry, **32**-13–16
  overview of, **32**-1–2
  principles of, **32**-2
  Project SERAPHIM, **32**-3–4
Computational electromagnetics
  characteristic-based formulation, **29**-5–7
  description of, **29**-1–2
  eigenvalues, **29**-10–11
  eigenvectors, **29**-10–11
  finite-difference approximation, **29**-14–15
  finite-volume approximation, **29**-15–16
  flux-vector splitting, **29**-11–14
  governing equations, **29**-3–4
  Maxwell equations
    in curvilinear frame, **29**-7–9
    description of, **29**-2–3
  numerical accuracy, **29**-2
  research issues, **29**-16–17
Computational field simulation, **26**-1
Computational fluid dynamics
  background of, **30**-2–6
  geometry, **30**-6–8
  nonlinear methods
    description of, **30**-9–10
    Euler-equation, **30**-11–13
    Navier–Stokes equation method, **30**-13–15
    nonlinear potential-equation method, **30**-10–11
  overview of, **30**-1
  panel methods, **30**-6, **30**-8–9
  principles of, **30**-2–8
Computational geometry problems
  convex hull, **11**-11–12
  decomposition, **11**-21–23
  description of, **11**-1–2
  geometric optimization, **11**-20–21
  geometric searching, **11**-24–26
  intersection, **11**-23–24
  motion planning, **11**-18–20
  parallel algorithms
    closest pair problem, **10**-30–31
    description of, **10**-29–30
    planar convex hull problem, **10**-31–35
  point location, **11**-17–18

problem-solving techniques for
  divide-and-conquer, 11-5–7
  dynamization, 11-8–9
  geometric duality, 11-4–5
  incremental construction, 11-2–3
  locus, 11-5
  plane sweep, 11-3–4
  prune-and-search, 11-7–8
  random sampling, 11-9–10
proximity
  bichromatic closest pair, 11-13
  closest pair, 11-12–13
  Voronoi diagrams, *see* Voronoi diagrams
range searching, 11-25–26
Computational languages, 5-2–3
Computational media aesthetics, 59-13–14
Computational models
  description of, 5-2
  finite automata, 6-22–26
  pointer machines, 6-31
  random access machines, 6-30
  Turing machines
    alternate, 5-5, 5-19
    BPP-, 5-22
    computation path of, 5-4
    configuration of, 5-4
    description of, 5-2
    deterministic, 5-4, 5-19
    elements of, 5-3–4
    k-alternating, 5-20
    k-worktape, 5-3–4
    nondeterministic, 5-4, 5-19
    oracle, 5-5
    PP-, 5-22
    probabilistic, 5-22
    RP-, 5-22
    universal, 5-4–5
Computational neurobiology, 34-2–3
Computational ocean modeling
  application-oriented, 31-4–5
  atmosphere models coupled with, 31-5–8
  baroclinic model, 31-4
  barotropic model, 31-4, 31-9–10
  best practices, 31-8–19
  comprehensive, 31-3
  computational issues, 31-17–19
  data assimilation, 31-15–17
  deep basins, 31-2
  free surface, 31-3
  global vs. regional, 31-2
  isopycnal models, 31-15
  layered models, 31-14–15
  long-term climate studies, 31-3–4
  Nowcast/Forecast in Gulf of Mexico, 31-19–21
  overview of, 31-1–2
  primitive-equation-based methods, 31-4
  process-studies-oriented, 31-4–5
  purely dynamical, 31-3
  quasigeostrophic models, 31-4
  rigid lid, 31-3
  shallow coastal, 31-2–3

  short-term simulations, 31-3–4
  sigma-coordinate models, 31-13–14
  z-level models, 31-10–13
Computational problems
  description of, 5-2–3, 6-3
  difficulty of, 5-6
  partially decidable decision problems, 6-3–4
  resource-bounded reducibilities, 5-12–14
  resources, 5-6
  space, 5-6
  time, 5-6
  undecidable
    definition of, 6-5
    diagonalization, 6-7–9
    self-reference, 6-7–9
Computational science
  definition of, 27-2–3
  description of, 1-9, 1-12
  visualization and, 27-3–4
Computational structural mechanics
  adaptive modeling, 28-21
  adaptive strategies, 28-8
  discretization techniques, 28-6–7
  dynamic analysis
    central difference explicit scheme, 28-13
    description of, 28-11–15
    direct integration techniques, 28-12–13
    energy balance in, 28-15
    model superposition method, 28-14–15
    Newmark's method, 28-13–14
    transient, 28-15
  eigenvalue problems, 28-16
  error estimation, 28-8
  future engineering systems for, 28-19–20
  hierarchical modeling, 28-21
  high-fidelity modeling, 28-21
  history of, 28-18–19
  integrated multiple methods, 28-21
  mathematical models, 28-6
  mesh generation, 28-7
  model generation, 28-7
  multidisciplinary analysis and design optimization, 28-22
  nondeterministic analysis, 28-22
  numerical algorithms, 28-17–18
  numerical simulations, 28-22
  overview of, 28-1–2
  primary pacing, 28-20–23
  problems
    classification of, 28-2–3
    formulation of, 28-3–5
  quality assessments, 28-8
  sensitivity analysis, 28-16–17
  stability analysis, 28-15–16
  static analysis, 28-10–11
  steps involved in, 28-5–6
  superconvergent recovery techniques, 28-8
Computation-use, 105-12
Compute-bound job, 80-4
Computer-aided design, 51-1
Computer architecture

addressing
    definition of, **17**-7
    format, **17**-7–8
description of, **17**-1
execution hazards
    control, **17**-14–15
    data, **17**-12–13
    description of, **17**-12
    structural, **17**-15
instruction execution
    description of, **17**-9–11
    execution unit, **17**-11
    instruction decode unit, **17**-11
    instruction fetch unit, **17**-11
    storeback unit, **17**-11–12
instruction set
    arithmetic logic unit, **17**-3
    complex, **17**-2–3
    control transfer, **17**-4–5
    definition of, **17**-2
    memory, **17**-3–4
    memory referencing, **17**-3–4
    reduced, **17**-2
memory
    amount of, **17**-6–7
    cache, **17**-6
    description of, **17**-5
    dynamic random-access memory, **17**-6
    main, **17**-6
    physical, **17**-8–9
    register file, **17**-5–6
    virtual, **17**-8–9
processor-program interface, **17**-1–2
secondary storage, **17**-7
superscalar design, **17**-15
very long instruction word computers, **17**-15
Computer Emergency Response Team, **78**-15
Computer ethics, *see also* Ethics
    description of, **2**-2
    paramedic method for, **2**-6
    privacy issues, **2**-10
    property rights, **2**-11
    software ownership, **2**-11
Computer-generated visualization, **27**-2–3
Computer industry, **1**-2, **1**-5–6
Computer Professionals for Social Responsibility, **A%**-2
Computer science
    abstraction in, **1**-7
    academic programs, **1**-4–5
    academic research and development, **1**-5–6
    challenges for, **1**-7–8
    curriculum development, **1**-3–4
    description of, **1**-1–2
    design in, **1**-7
    growth of, **1**-2–6
    perspectives in, **1**-6–7
    professional context of, **1**-7
    risks associated with, **2**-11
    social context of, **1**-7
    theory of, **1**-6
Computer-supported collaborative work

anonymity, **50**-10–11
communication
    auditory cues, **50**-4–5
    facilitation of, **50**-3
    video vs. audio-only, **50**-5
    visual cues, **50**-4–5
definition of, **50**-1
description of, **50**-1–2
environmental factors that affect
    asynchrony, **50**-8–9
    auditory cues, **50**-4–5
    description of, **50**-2–3
    dialog structure, **50**-6
    facilitation of, **50**-3
    information richness, **50**-8
    input/output rates, **50**-8–9
    managerial behavior, **50**-8
    social context cues, **50**-6–8
    visual cues, **50**-4–5
free-riding, **50**-10–11
groupware, **50**-14–15
information sharing, **50**-12–14
media factors in, **50**-2–9
opinion formation, **50**-13–14
process gains and losses, **50**-9–10
process structure, **50**-11–12
process support tools, **50**-12
production blocking, **50**-10
proxemic effects, **50**-5–6
research issues, **50**-15–16
task structure, **50**-12
Computer vision
    digital images used in, **43**-2
    goals of, **43**-1
    high-level
        appearance-based matching, **43**-19–20
        dense feature matching, **43**-18–19
        description of, **43**-14
        interpretation tree approach, **43**-15–16
        invariants, **43**-17–18
        k-tuple search, **43**-16–17
        object recognition, **43**-14–15
        transformation space search, **43**-16
    low-level
        Canny edge detectors, **43**-3, **43**-6
        description of, **43**-2–3
        image smoothing and filtering, **43**-5–6
        local edge detectors, **43**-3–5
        multiscale processing, **43**-6–7
        optical flow, **43**-7–9
        visual motion, **43**-7–9
    middle-level
        active contours, **43**-13–14
        description of, **43**-9
        snakes, **43**-13–14
        stereopsis, **43**-9–12
        structure from motion techniques, **43**-12–13
    overview of, **43**-1–2
Computing Research Association, **A%**-1, **1**-6
"Computing the Future," **1**-8
Concatenation, **6**-11–12

Concurrency control
  best practices, **56**-8–13
  description of, **56**-1–2
  distributed databases, **58**-7–8
  hierarchical locking, **56**-12–13
  isolation levels, **56**-10–12
  multiversion, **56**-13
  optimistic, **56**-13, **57**-3
  parallel databases, **58**-7–8
  principles of, **56**-4–6
  research issues, **56**-17
  semantics-based, **57**-15
  serializability of, **56**-4
  timestamp-ordering, **57**-3
  transaction processing, **57**-12–15
  transaction schedules, **56**-4–5
  tuning of, **53**-6–8
  two-phase locking, **56**-8–10
Concurrent computing
  description of, **96**-1
  distributed systems, **96**-19
  example of, **96**-1–4
  formal approaches to, **96**-19–20
  hardware architectures for, **96**-4
  in object-oriented programming languages, **96**-19–20
  research issues for, **96**-21–22
  software architectures for
    busy-wait, **96**-5–9
    description of, **96**-4
    message passing, **96**-14–19
    monitors, **96**-12–14
    semaphores, **96**-9–12
  without abstractions, **96**-5–9
Concurrent development model, **108**-9–10
Concurrent error detection, **25**-9
Concurrent execution, **56**-2
Concurrent logic programming, **93**-28
Concurrent programming, **81**-1–2
Concurrent-read concurrent-write parallel random access
        machines, **10**-6
Conditional control transfer, **17**-4
Conditional hook, **10**-25
Condition-decision coverage, **105**-12
Condition number, **8**-4
Condition variables, **81**-4
Confidentiality, **77**-1
Configurable logic blocks, **16**-31–33
Conflict of interest, **2**-9
Conflict ratio, **57**-22
Confluences, **62**-3, **107**-19
Conformance testing, **51**-15–16
Conjunctive normal form, **61**-2
Connected components algorithms, **10**-22, **10**-26
Connected graphs, **7**-1
Connection hijacking, **74**-5
Cons cell, **92**-6
Constant angular velocity, **21**-5
Constant coefficients, linear recurrence with, **3**-2
Constant folding, **83**-8
Constraint languages, **48**-15
Constraint logic programming, **15**-36

  applications of, **93**-21
  artificial intelligence and, **93**-17
  characteristics of, **93**-16–17
  constraints, **93**-18–20
  control issues for, **93**-19
  definition of, **93**-2
  design issues, **93**-26–27, **93**-29
  developments in, **93**-17–21
  examples of, **93**-16–17
  implementation issues, **93**-26–27
  interval constraints, **93**-28–29
  optimization, **93**-19–20
  satisfiability problems, **93**-20–21
  soft constraints, **93**-20
Constraint satisfaction problems, **93**-17–18
Constructive solid geometry objects, **36**-7–8
Constructive solid modeling, **41**-26–28
Constructive volume geometry, **41**-26
Constructor, **91**-14
Containers, **4**-1
Content addressable memory, **69**-5
Context-free grammars, **6**-18–21, **48**-14, **99**-5–6
Context-free languages, **6**-16
Context-sensitive language, **6**-16
Context switching, **80**-11
Continuous Dijkstra, **11**-18
Contraction
  definition of, **10**-17
  deterministic graph, **10**-24–26
  random mate graph, **10**-22–23
Contrario, **13**-36
Contravariant, **97**-22
Control hazards, **17**-14–15
Controlled regression testing assumption, **105**-27
Control transfer instructions, **17**-4–5
Convex hull, **11**-11–12
Cook-Levin theorem, **5**-14–15
Cook reducible, **5**-13
Copy collection algorithm, **100**-14
Copy propagation, **99**-23
Coroutines, **90**-13
Correctness of software, **101**-4
Correctness proof, **105**-1
Co-scheduling, **88**-5
Cosmic microwave background data analysis, **33**-8
Cosmology, **33**-13
Cost function
  definition of, **66**-8
  gradients of, **66**-10–11
  likelihood-based, **66**-9–10
Counter, **6**-28
Counterfactuals, **70**-13–15
Counting problems, **11**-23–24
Counting semaphore, **84**-9
Count-key-data, **21**-6
Count-mode, **11**-24
Count-to-infinity problem, **73**-12–13
Courant–Friedrichs–Lewy number, **29**-16, **30**-12
Covariant, **97**-22
Covert channels, **60**-6
CPU, **82**-12, **85**-1

CPU scheduling, **88**-1
Crash recovery, **89**-8–9
CRASH scale, **105**-30
Critical decision method, **46**-9
Critical points, **41**-17
Critical region, **84**-1
Critical tasks, **83**-3
Crossbar switch, **24**-11
Cross-coupling, **19**-5
Crossover, **14**-2
Crowd animation, **40**-9–10
Crucial rescan operation, **13**-21
Cryptography
 classical, **9**-2
 computational notion of, **9**-3–4
 digital signature scheme, **9**-2, **9**-17–20
 hash functions, **9**-10–11
 information-theoretic notions of, **9**-2–3
 key distribution centers, **74**-8–9
 message authentication, **9**-14–15
 message authentication codes, **9**-1
 network security using, **74**-7–9
 notation, **9**-4
 one-way functions, **9**-5–6
 overview of, **9**-1–2
 primitives
  block ciphers, **9**-10
  description of, **9**-7
  pseudorandom functions, **9**-9–10
  pseudorandom generators, **9**-8–9
 private-key encryption, **9**-1, **9**-11–13
 public-key encryption, **9**-2, **9**-15–17
 security uses of, **9**-2–4
 transaction processing, **57**-7–9
 trapdoor permutations, **9**-6–7
CSS algorithm, **82**-29
Cubic convolution, **39**-9
Culling and clipping, **38**-6–9
Curse of dimensionality, **66**-5
Cut-through switches, **24**-14
Cutting plane method, **15**-25–26
Cutting stock problem, **15**-14–15
Cyberinfrastructure, **1**-9
Cycle, **7**-2
Cycle-oriented round robin algorithm, **82**-8–9
Cyclical SCAN, **21**-8
Cylinder, **82**-15, **86**-3
Cylinder skew, **21**-5

D

Dadda fast multiplier, **22**-13–14
Daisy-chain arbitration, **19**-6
Data abstraction, **90**-13–15, **90**-18–20
Data access skew, **21**-8
Database
 definition of, **80**-9
 distributed
  architecture of, **58**-4–16
  characteristics of, **58**-2–3

concurrency control in, **58**-7–8
 data integration in, **58**-6–7
 data placement, **58**-12–13
 description of, **58**-1
 load balancing in, **58**-15–16
 principles of, **58**-1–4
 query optimization in, **55**-16
 query processing and optimization in,
  58-13–15
 reliability of, **58**-8–11
 replication of, **58**-11–12
 technology, **58**-4–16
 real-time, **83**-10
Database-driven application testing, **105**-32–33
Database security
 access controls
  description of, **60**-3
  discretionary, **60**-3–5, **60**-9–10
  mandatory, **60**-5–7, **60**-11
 advanced database management systems,
  60-11–12
 assurance, **60**-7–8
 information privacy principles, **60**-8–9
 overview of, **60**-1–2
 principles of, **60**-2
Database tuning
 applications of, **53**-2
 best practices, **53**-4–15
 description of, **53**-1
 performance inhibitors, **53**-2–3
 principles of, **53**-2–4
Data bindings
 best practices, **90**-15–20
 binding time, **90**-2
 control structures in
  best practices for, **90**-17–18
  conditional structures
   case statements, **90**-7–8
   if statements, **90**-6–7
  iterative structures, **90**-8–9
  data abstraction, **90**-13–15, **90**-18–20
  description of, **90**-6
  exceptions, **90**-9–10
  Goto statement, **90**-9–10
  procedural abstraction, **90**-10–13, **90**-18
  unconstrained, **90**-9–10
 declarations, **90**-6
 execution units, **90**-5–6, **90**-17
 research issues, **90**-20–21
 scope of, **90**-4–5
 types, **90**-3–4
 variables, **90**-2–3, **90**-15–16
Data channel, **82**-4
Data coherency, **89**-3–7
Data compression
 definition of, **76**-1
 description of, **76**-1–2
 lossless
  arithmetic coding, **76**-9–12
  compression ratio, **76**-2
  description of, **76**-2

dictionary-based techniques for, **76**-12–16
  entropy coding, **76**-4–5
  Huffman coding, **76**-5–9
  information content, **76**-2–3
  Lempel–Ziv–Welsh compression, **13**-40–43, **76**-12, **76**-14–16
  methods of, **76**-3–12
  prediction with partial matching, **76**-16
  run-length coding, **76**-16–18
lossy
  data domain compression, **76**-18–23
  differential pulse code modulation, **76**-19–20
  JPEG, **76**-23–32, **76**-39
  predictive differential pulse code modulation, **76**-20–22
  pulse code modulation, **76**-18–19
  subband coding, **76**-32–39
  transform domain compression, **76**-23–39
  vector quantization, **76**-22–23
  wavelets, **76**-32–39
Data contention, **57**-14–15
Data domain compression, **76**-18–23
Data-driven parallelism, **63**-1
Data encryption standard, **57**-7
Data filtering, **27**-6
Data flip-flop, **16**-26–27
Dataflow diagrams, **103**-4–7
Dataflow packages, **27**-21–22
Data hazards, **17**-12–13
Data integration, **58**-6–7
Data-link layers, **24**-13–14
Data mining
  a priori algorithm, **75**-9
  association rule mining, **75**-8–9
  classification, **75**-9–10
  clustering, **75**-10–12
  concept description, **75**-7–8
  definition of, **75**-7
  description of, **57**-22, **75**-1
  prediction, **75**-10
  search engine integration of, **75**-11–14
  World Wide Web, **75**-12
Data model
  definition of, **52**-1
  object-based
    database programming languages, **52**-12
    description of, **52**-5–6
    entity-relationship model, **52**-6–10
    object-oriented model, **52**-10–13
    object-relational data models, **52**-13–14
    relational, **52**-1–5
    XML, **52**-14–17
Data modeling, **108**-5
Data refinement, **106**-3
Data register, **16**-27–28
Data representation
  baseband methods, **24**-5–6
  modulation, **24**-5–8
Data server, **57**-4
Data structures
  abstract data types, **4**-2

containers, **4**-1
  dictionary, *see* Dictionary
  dynamic, **4**-2
  explicit, **4**-2
  external memory, **4**-2
  implicit, **4**-2
  internal memory, **4**-2
  issues regarding, **4**-2–3
  locators, **4**-1
  overview of, **4**-1
  positions, **4**-1
  priority queue, *see* Priority queue
  randomized algorithms, 1 **2**-9–13
  sequence, *see* Sequence
  space vs. time, **4**-2–3
  static, **4**-2
Data transfer rate, **21**-4
Data types
  abstract, **4**-2, **90**-3, **90**-13, **103**-1, **103**-11–12
  definition of, **4**-2, **90**-3
  user-defined, **107**-17–20
Data visualization tools, **48**-19
Davis-Putnam procedure, **61**-12–13
DCE, **89**-13–15
Deadlock
  avoidance of, **82**-20–21
  definition of, **57**-3, **57**-5, **58**-8, **71**-42, **84**-5, **84**-17–19
  description of, **82**-19–20
  detection of, **82**-21–23
  ignoring of, **82**-23
  prevention of, **82**-20
  recent work in, **82**-31–32
  synchronization, **84**-5
DEC, **16**-19–20
Decay usage scheduling, **82**-10
Decentralized bus arbitration, **19**-6–7
DEC global name server, **87**-3–4
Decisional Diffie-Hellman assumption, **9**-17
Decision making, ethical, **2**-7
Decision problems, **5**-2–3, **6**-3–4
Decision trees
  constructing of, **65**-2–3
  cost-complexity pruning, **65**-5–6
  definition of, **66**-7
  description of, **65**-2, **66**-7
  efficiency issues for, **65**-9–10
  extensions, **65**-7–10
  minimum description length pruning, **65**-6
  missing attribute values, **65**-6–7
  multiclass problems, **65**-8–9
  multiple, **65**-9
  overfitting, **65**-5–6
  peepholing, **65**-9–10
  pruning of, **65**-5–6
  tests
    linear multiattribute, **65**-7–8
    selection of, **65**-3–4
    subset, **65**-7
    symbolic multiattribute, **65**-8
Declaration equivalence, **90**-3

Declarations
    definition of, **90**-6
    in standard ML, **92**-12–13
Declarative languages, **48**-14–15
Decoder, **16**-19
Decomposable searching problems, **11**-8
Decomposition, **11**-21–23, **15**-15
Decrypting, **9**-2
Default parameter association, **90**-11
Definite iteration, **90**-8
Definition clear path, **105**-12
Degree, **7**-2
Degree order, **8**-12–13
Deictic reference, **50**-2
Deindividuation, **50**-7
Delaunay algorithm, **26**-19–23
Delaunay triangulation, **26**-12
Delay fault, **25**-3
Dell, **1**-2
Delta form factor, **38**-23
Demand paging, **17**-8, **85**-2
DEMUX/DECODER, **16**-21–22
Denial of service attacks, **74**-11
Denotational semantics
    axiomatic semantics and, **98**-16
    characteristics of, **98**-4–5
    definition of, **98**-2
    domains for, **98**-7–8
    of programs, **98**-8–10
Dense feature matching, **43**-18–19
Dense matrix, **8**-5
Density estimation, **66**-2–3
Density functional theory, **32**-14
Deontological theories of ethics, **2**-4–5
Depersonalization, **50**-7–8
Deployment diagram, **104**-11
Depth-first search
    algorithm, **7**-3–4
    analysis of, **7**-4–5
    definition of, **7**-3
    description of, **63**-4–5
    directed, **7**-5–6
    sample execution of, **7**-4
De-randomization, **5**-23
Derivation tree, **6**-19
Derived attribute, **52**-7
Derived typing, **90**-4
Description, **106**-1
Design, **1**-7
Design autonomy, **88**-2
Design diversity, **25**-14
Design–use cycle, **51**-13
Destination vertex, **10**-20
Determinant, **8**-3
Deterministic algorithm, **12**-6–7
Deterministic fractals, **37**-2–3
Deterministic graph contraction, **10**-24–26
Deterministic Turing machines, **5**-4, **5**-19, **6**-27
Device drivers, **80**-4
Device scheduling
    deadlock

    avoidance of, **82**-20–21
    description of, **82**-19–20
    detection of, **82**-21–23
    eclectic approach to, **82**-23–24
    ignoring of, **82**-23
    prevention of, **82**-20
    description of, **82**-12–15
    disk scheduling algorithm, **82**-17–18
    high-level, **82**-24–25
    nonshareable devices, **82**-14, **82**-18–19
    policies, **82**-19–24
    queue organization, **82**-15
    shareable devices, **82**-14–17
DFS forest, **7**-3
Diagonalization, **5**-2, **5**-11, **6**-7–9
Diagram(s)
    activity, **104**-11
    binary decision, **106**-6
    class, **104**-9–10
    collaboration, **104**-11
    component, **104**-11
    dataflow, **103**-4–7
    deployment, **104**-11
    entity-relationship, **103**-7–8
    function-flow, **46**-11
    influence, **70**-2
    sequence, **104**-10–11
    statechart, **104**-11
    use-case, **104**-8–9
    Voronoi
        additive-weighted, **11**-16
        construction of, **11**-14–15
        definition of, **71**-38
        farthest-neighbor, **11**-15
        multiplicative-weighted, **11**-16
        weighted, **11**-15–16
Diagrammatic reasoning, **62**-7
Dialogue modeling, **64**-8
Dictionary
    definition of, **4**-3, **4**-12
    operations, **4**-12–13
    priority queue realization with, **4**-12
    realization of
        with a hash table, **4**-25–27
        with an (a, b)-tree, **4**-17–20
        with an AVL-tree, **4**-20–25
        with a search tree, **4**-15–17
        with a sequence, **4**-13–15
Dictionary-based techniques, for lossless data compression,
        76-12–16
Diels–Alder reaction, **32**-12
Differential pulse code modulation
    description of, **76**-19–20
    predictive, **76**-19–20
Diffuse shading, **37**-16
Digital audio, **49**-9–10
Digital differential analyzer, **41**-13
Digital gate
    active logic, **16**-4
    CMOS binary logic, **16**-4–5
    CMOS switching model, **16**-5

concept of, **16**-2–3
  multiple inputs, **16**-6
  primitives for, **16**-6
Digital signature algorithm, **9**-19
Digital signature schemes, **9**-2, **9**-17–20
Digital-to-analog converters, **35**-14
Dijkstra's algorithm, **7**-7–8
Dimensionality curse, **59**-19
Dimensionality reduction, **59**-19–20
Dining philosophers problem, **84**-17
Directed acyclic graph, **7**-1
Directed depth-first search, **7**-5–6
Direct graphical specification, **48**-17–18
Direct manipulation, **51**-1
Direct message passing, **84**-13
Direct numerical simulations, **30**-14–15
Direct volume rendering, **41**-3
Disambiguating rules, **99**-12
Discourse, **64**-2
Discrete cosine transform, **76**-24
Discrete Fourier transform, **10**-36, **76**-25
Discrete logarithms, **9**-5–6
Discretionary access controls, **60**-3–5, **60**-9–10, **79**-10
Discretization techniques, **28**-6–7
Discriminant function, **66**-4
Discrimination analysis, **75**-8
Disfluencies, **64**-6
Disjunctive augmentation, **68**-13–14
Disk
  data transfer on, **86**-3
  definition of, **82**-12
  organization of, **82**-15
  storage uses of, **86**-2–4
Disk arm prepositioning, **21**-10–11
Disk arm scheduling, **21**-7–10
Disk defragmentation, **21**-11
Disk drives, **86**-1
Disk mirroring, **87**-19
Disk performance
  disk arm prepositioning, **21**-10–11
  disk reorganization and defragmentation, **21**-11
Disk positioning time, **21**-4
Disk scheduling
  algorithm for, **82**-17–18
  description of, **21**-10
  recent work in, **82**-27–31
Disk zoning, **21**-6
Dispatching
  first-come, first-served, **82**-6
  multilevel, **82**-9–11
  multilevel feedback queue dispatching, **82**-9–10
  processor, **82**-5
  shortest job first, **82**-6–7
Dispatching algorithms, **82**-11
Displacement maps, **35**-7
Display(s)
  audio, **42**-6
  auditory, **20**-21–22
  color, **20**-14
  force, **20**-20–21
  large-format, **20**-19–20

  multiple, **20**-19
  properties of, **20**-12–13
  scale in, **20**-18–20
  small, **20**-18–19
  spatial resolution of, **20**-13
  tactile, **20**-20–21
  temporal resolution of, **20**-13
  virtual reality, **42**-3–4
Display devices, **35**-12–13
Distance vector routing protocols, **73**-11
Distributed arbitration, **19**-6
Distributed computing, **80**-6–7
Distributed databases
  architecture of, **58**-4–16
  characteristics of, **58**-2–3
  concurrency control in, **58**-7–8
  data integration in, **58**-6–7
  data placement, **58**-12–13
  description of, **58**-1
  load balancing in, **58**-15–16
  principles of, **58**-1–4
  query optimization in, **55**-16
  query processing and optimization in, **58**-13–15
  reliability of, **58**-8–11
  replication of, **58**-11–12
  technology, **58**-4–16
Distributed directory, **19**-18
Distributed file systems
  data coherency, **89**-3–7
  DCE, **89**-13–15
  disconnected operation, **89**-10
  file attribute coherency, **89**-7
  multimedia support, **89**-17
  naming, **89**-9
  NFS, **89**-10–13
  overview of, **89**-1–2
  performance, **89**-7–8
  principles of, **89**-2–10
  replication, **89**-9
  research issues, **89**-16–18
  resilience, **89**-8–9
  security of, **89**-10
Distributed memory systems
  data coherency, **89**-3–7
  description of, **88**-6
  disconnected operation, **89**-10
  file attribute coherency, **89**-7
  IVY, **89**-15–16
  Munin, **89**-16
  naming, **89**-9
  NFS, **89**-10–13
  overview of, **89**-1–2
  performance, **89**-7–8
  principles of, **89**-2–10
  replication, **89**-9
  research issues, **89**-16–18
  resilience, **89**-8–9
  security of, **89**-10
Distributed multimedia, **83**-8–9, **83**-12–13
Distributed operating systems
  Amoeba, **87**-2–3, **87**-6–8, **87**-14

binding, **87**-17–18
communication, **87**-6–9, **87**-15–17
DEC global name server, **87**-3–4
description of, **87**-1–2
domain name server, **87**-4–5
failure models, **87**-13–14
Firefly remote procedure call, **87**-8–9
group communication, **87**-22–23
group management, **87**-11–13
ISIS project, **87**-12–13
Locus system, **87**-9–10
naming, **87**-2–6, **87**-14–15
parallel systems vs., **87**-1–2, **88**-2
Plan 9, **87**-5–6
Quicksilver, **87**-10–11
research, **87**-2–13
transactions, **87**-9–11, **87**-19–21
V system, **87**-7
x-kernel, **87**-9
Distributed ray tracing, **38**-19
Distributed scheduling
algorithms, **88**-7–10
description of, **88**-3–4
mechanisms, **88**-15
wide-area, **88**-10–14
Distributed shared memory, **85**-11–12
Distributed systems, **47**-3–4
Distributed transaction processing, **57**-4–7
Distributed virtual reality, **42**-16
Distributional complexity, 1 **2**-8
Diversified parallel design, **45**-8
Divide-and-conquer
algorithms, **3**-10–12, **10**-11–13
problem solving by, **11**-5–7
recurrences, **3**-4–5
Division
fixed point algorithms, **22**-15–18
floating point algorithms, **22**-19
D latch, **16**-25–26
Document type definition, **52**-16–17
Domain(s)
definition of, **73**-10
search, **6**-5
stochastic, **67**-18–19
volume rendering, **41**-13
Domain name server, **87**-4–5
Domain-specific focus, **110**-2
Domain theories, **62**-2, **62**-8–9
Domain theory, **68**-2, **98**-15
Domain volume rendering, **41**-13
Domainwise algorithms, **28**-17
Don't cares, **16**-15, **93**-28–29
Doubly linked list, sequence implementation using, **4**-5–6
DSL, **24**-10–11
DSPACE, **5**-7
DTIME, **5**-7, **5**-17
Dual coding theory, **49**-14
Duality, **16**-10
Duplication, **25**-9
Durand–Kerner algorithm, **8**-8
DVD, **49**-8

Dynamical system, **67**-3–4
Dynamic analysis
central difference explicit scheme, **28**-13
description of, **28**-11–15
direct integration techniques, **28**-12–13
energy balance in, **28**-15
model superposition method, **28**-14–15
Newmark's method, **28**-13–14
transient, **28**-15
Dynamic analysis and replanning tool, **64**-13
Dynamic binding, **91**-14
Dynamic checking, **97**-3
Dynamic data structure, **4**-2
Dynamic dispatching priority adjustment, **82**-7–8
Dynamic method invocation, **91**-5
Dynamic non-cooperative algorithms, **88**-9
Dynamic non-distributed algorithms, **88**-9
Dynamic packet filters, **74**-10
Dynamic programming
algorithms, **55**-8–11
description of, **3**-12–17, **6**-21
Dynamic random-access memory, **17**-6, **21**-2
Dynamic scope rules, **90**-4
Dynamic scoping, **100**-3
Dynamic search, **63**-20–21
Dynamic simulation, in animation, **40**-6–7
Dynamization, **11**-8–9

E

Ear decomposition, **10**-14
Early ray termination, **41**-8
E-business, **110**-7–9
E-carriers, **24**-8
Echo system, **14**-9
Eclectic approach, **82**-23–24
Ecology biology, **34**-3
E-commerce, **75**-13–14
Edge detection algorithm, **43**-2
Edge detectors
Canny, **43**-3, **43**-6
local, **43**-3–5
Edge list, **10**-19
Edge recombination, **14**-6–7
Edge swapping, **26**-22
Effective core potentials, **32**-14
Effective latency, **18**-3
Egoism, **2**-4
EIFFEL, D%-3
Eigenvalue problems, **28**-16
Eigenvalues, **29**-10–11
matrix, **8**-6–8
Eigenvectors, **8**-6, **29**-10–11
Electromagnetic spectral, **33**-3
Electromagnetic trackers, **42**-8
Electronic circuit design, **46**-16
Electronic communities, **2**-12
Electrostatic potentials, **32**-20
Elementwise parallel algorithms, **28**-17
Elimination ordering, **8**-4

Ellipsoid algorithm, **15**-9–10
Elliptical arcs, **36**-4
El Niño, **31**-5
E-mail bomb, **78**-4
Embedded systems, **83**-1
Employee monitoring, **2**-3
Employers, **2**-3
ENC, **16**-19
Encapsulation
    definition of, **53**-15, **90**-13, **104**-2
    software design by, **103**-9–12
Encoder, **16**-19
Encryption
    definition of, **9**-2
    mode of, **9**-13
    private-key, **9**-1, **9**-11–13
    public-key, **9**-2, **9**-15–17
End effectors, **71**-2
Endogenous fitness function, **14**-9
End-to-end argument, **87**-15
End-to-end timing requirement, **83**-7
Entity, **52**-6
Entity-relationship diagrams, **103**-7–8
Entity-relationship model, **52**-6–10
Entropy coding, **76**-4–5
Envisionment, **62**-6
Epipolar lines, **43**-11
Epistemic gestures, **20**-10
Equivalence
    term, **97**-25
    type, **90**-3, **97**-7, **97**-25
Equivalent, **5**-13, **6**-13
Equivocality, **50**-8
Ergonomics, **20**-9
Ergotic gestures, **20**-10
Error
    definition of, **25**-3, **105**-2
    masking of, **25**-7–8
    recovery from, **99**-24–25
Error correcting codes, **21**-6, **25**-8
Error on output, **25**-6–7
Ethernet NIC, **24**-13
Ethical issues
    computer-related
        description of, **2**-2, **2**-9–10
        privacy, **2**-10
        property rights, **2**-11
        software ownership, **2**-11
    overview of, **2**-1–3
    uniqueness of, **2**-2
Ethics
    code of, B%-1–7
    decision making based on, **2**-7
    deontological theories, **2**-4–5
    paramedic method for, **2**-6
    professional, **2**-7–9
    purpose of, **2**-3
    social contract theories, **2**-5
    utilitarianism, **2**-3–4
Ethnography, **46**-7
Euler angles, **42**-7

Euler equations, **30**-5, **30**-11–13
Eulerian graph, **7**-20
Euler tour, **10**-14
Euler tour problem, **7**-20
European Council Directive, **51**-10–11
Event-driven programming
    applets, **95**-4–6
    applications of, **95**-18–21
    description of, **95**-1–2
    event handling
        buttons, **95**-8–9
        choices, **95**-10–11
        labels, **95**-9
        mouse clicks, **95**-6–7
        mouse motion, **95**-7–8
        textareas, **95**-9–10
        textfields, **95**-9–10
    graphical user interface, **95**-11–18
    model of, **95**-2
    paradigm for, **95**-2–3
    Web applications of, **95**-1
Event languages, **48**-14
Event schedule, **11**-3
Event tree, **46**-15
Evolutionary biology, **34**-3
Evolutionary computation, **14**-2
Evolutionary development software process models,
        102-7–10, **108**-5–11
Evolutionary prototyping, **47**-8–9, **102**-8–9
Example space, **68**-3
Exception handler, **90**-9
Exceptions, **90**-9–10
Exclusive-read exclusive-write parallel random
        access machines, **10**-6
Execution-based software testing, **105**-4–6
Execution errors, **97**-2–4
Execution hazards
    control, **17**-14–15
    data, **17**-12–13
    description of, **17**-12
    structural, **17**-15
Execution units, **17**-11, **90**-5–6, **90**-17
Explanation-based learning
    background, **68**-3–5
    characteristics of, **68**-2
    empirical learning vs., **68**-2, **68**-5–7
    explanations
        construction of, **68**-8–9
        properties of, **68**-7–8
    generalizing
        description of, **68**-9–11
        disjunctive augmentation, **68**-13–14
        identity elimination, **68**-11–12
        irrelevant feature elimination, **68**-11
        operationality pruning, **68**-12–13
    overview of, **68**-1–2
    resources for, **68**-16
    steps involved in, **68**-7–8
    utility problem, **68**-14–16
Explicit data structure, **4**-2
Expression, **90**-5–6

Extended BNF, **99**-12
Extensible Markup Language, *see* XML
External memory, **4**-2
Extreme point, **15**-4

F

Facial animation, **40**-10–12
Factoring
    definition of, **9**-5
    trapdoor permutations based on, **9**-7
Failover, **89**-9
Failure
    common-mode, **25**-14
    definition of, **25**-2, **105**-2
    routing, **73**-5
Failure detector, **87**-12
Failure-transparent system, **89**-12
False sharing, **89**-4
Farthest-neighbor Voronoi diagram, **11**-15
Fast fourier transform, **8**-2, **8**-9, **10**-36
Fast polynomial arithmetic, **8**-9
Fathoming, **15**-28
Fat-tree network, **10**-4–5
Fault avoidance, **25**-6
Fault detection ratio, **105**-17
Fault injection, **25**-14–15
Fault model, **105**-15
Faults, **25**-4–5
Fault secure techniques, **25**-8–11
Fault tolerance
    algorithm-based, **25**-11
    definition of, **25**-1
    elements of, **25**-2
    errors, **25**-3
    evaluation, **25**-5–6
    failures, **25**-2
    faults, **25**-4–5
    metrics, **25**-4
    overview of, **25**-1–2
    real-time systems, **83**-7–8
    software, **101**-5
    system recovery, **25**-11–12
    system repair
        built-in self-test and diagnosis, **25**-12
        fail-soft techniques, **25**-12
        self-repair techniques, **25**-13–14
    system responses
        error masking, **25**-7–8
        error on output, **25**-6–7
        fail-safe techniques, **25**-11
        fault secure techniques, **25**-8–11
Fault tree, **46**-15
Feasible data flow testing criteria, **105**-13
Feasible vertex-labeling, **7**-15
Features, **66**-15
Feedback, **20**-7–8
Feed-forward-back-propagation network, **69**-2
Fetch-and-add operation, **10**-16
Fiber distributed data interface, **72**-18–19

Fibonacci heap, **3**-8–9
Fibonacci recurrence, **3**-3
Fibre Channel, **72**-19–21
Field programmable gate array
    architecture of, **16**-31–35
    calculator mapping to, **16**-35
    configurable logic blocks, **16**-31–33
    definition of, **16**-31
    interconnect, **16**-33–34
    Xilinix input/output block, **16**-34–35
File(s)
    naming of, **89**-9
    permanence of, **89**-2–3
    scheduling of, **82**-17
File attributes, **89**-7
File sector viruses, **78**-5
File systems
    defensive algorithms used by, **86**-10
    definition of, **86**-6
    directory structure of, **86**-7
    disk space management, **86**-10–11
    distributed
        data coherency, **89**-3–7
        DCE, **89**-13–15
        disconnected operation, **89**-10
        file attribute coherency, **89**-7
        multimedia support, **89**-17
        naming, **89**-9
        NFS, **89**-10–13
        overview of, **89**-1–2
        performance, **89**-7–8
        principles of, **89**-2–10
        replication, **89**-9
        research issues, **89**-16–18
        resilience, **89**-8–9
        security of, **89**-10
    file, **86**-7–8
    input/output, **86**-8–10
    log-based systems, **86**-11–13
    uses of, **86**-6
    versioning systems, **86**-13–14
Fingerprinting technique, 1 **2**-15–19
Finite automata, **6**-17, **6**-22–26, **99**-4–5
Finite-difference approximation, **29**-14–15
Finite-state machines, **6**-17, **6**-26, **16**-24, **103**-8–9
Finite-volume approximation, **29**-15–16
Firefly remote procedure call, **87**-8–9
Firewalls
    application gateways, **74**-10
    circuit relays, **74**-10
    definition of, **74**-9
    dynamic packet filters, **74**-10
    hack prevention using, **78**-14
    limitations of, **74**-10–11
    packet filters, **74**-9–10
    types of, **74**-9–10
First-class objects, **92**-3
First come, first served, **82**-16, **83**-5
First-come, first-served dispatching, **82**-6
First-order logic, **61**-3–4
First-order semantics, **98**-12

First-order type systems
  characteristics of, **97**-11–18
  for imperative languages, **97**-18–19
Fitness function, **14**-2
Fitness landscape, **14**-10
Fitts' law, **20**-5, **42**-4
Five-minute rule, **21**-4
Fixed block architecture, **21**-6
Fixed point addition algorithm, **22**-4–11
Fixed point arithmetic algorithms
  addition, **22**-4–11
  division, **22**-15–18
  multiplication, **22**-12–15
  subtraction, **22**-11
Fixed point number systems
  description of, **22**-1–2
  one's complement, **22**-3
  sign magnitude, **22**-2
  two's complement, **22**-2
Fixed-point semantics, **98**-11
Flat shading, **38**-16
Flat transactions, **57**-3
Flexible image transport system, **33**-4
Flicker fusion frequency, **20**-13
Floating point algorithms, **22**-18–20
Floating-point arithmetic, **22**-18–20, C%-3
Floating-point unit, **17**-2
Flow conservation, **7**-18
Fluid dynamics, computational
  background of, **30**-2–6
  geometry, **30**-6–8
  nonlinear methods
    description of, **30**-9–10
    Euler-equation, **30**-11–13
    Navier–Stokes equation method, **30**-13–15
    nonlinear potential-equation method,
        30-10–11
  overview of, **30**-1
  panel methods, **30**-6, **30**-8–9
  principles of, **30**-2–8
Flux-vector splitting, **29**-11–14
FMEA, **62**-1
Foiling an adversary, 1 **2**-6–7
Font cache, **36**-2
Forbidden errors, **97**-3
Force displays, **20**-20–21
Forest, **7**-3
Fork-Join synchronization, **21**-23–24
Formal languages
  definition of, **6**-12
  description of, **6**-1–2, **6**-11–12
  representation of, **6**-12–16
Formal methods
  best practices, **106**-3–6
  case study of, **106**-6–13
  description of, **106**-1
  modeling systems, **106**-5–6, **108**-10
  notation for, **106**-2
  principles of, **106**-1–3
  specification languages, **106**-4–5
Formal specification

description of, **106**-2
  languages, **106**-4–5
Formal technical review, **108**-17
Form factor algorithm, **38**-22
FORTRAN, **90**-15–16, D%-3
Forwarding, **73**-11
Fourier's scheme for linear inequalities, **15**-5–6
Fourier transforms, **39**-3
Fourth generation techniques, **108**-10–11
f-padded version, **5**-11
Fractals, **37**-2–4
Framebuffers, **35**-13–14
Frame relay, **72**-11–14
Freeblock scheduling, **21**-11–12
Free edges, **7**-11
Free-riding, **50**-10–11
Free vertex, **7**-11
Freivalds' technique, 1 **2**-15–16
Frequency leakage, **39**-7
Frontier molecular orbitals, **32**-12
Full adders, **16**-8–10, **16**-15–16, **22**-4–5
Full strip writes, **21**-15
Functional dependencies, **52**-5
Functionality
  design of, **46**-2–4
  misdefined, **46**-3
Function allocation, **46**-11
Functional programming languages
  HASKELL
    class system, **92**-22–23
    description of, **92**-22
    functional I/O in, **92**-25
    instance declarations, **92**-23–24
    list comprehensions in, **92**-24
    user-defined types, **92**-23
  history of, **92**-3–4
  impure, **92**-5
  lambda calculus, **92**-1, **92**-4–5
  lazy, **92**-20
  nonstrict, **92**-5, **92**-20–21
  overview of, **92**-1–2
  parallel, **92**-26
  partial evaluation, **92**-26
  program analysis and optimization, **92**-25–26
  pure, **92**-2, **92**-5
  SCHEME
    data types, **92**-6–7
    description of, **92**-5–6
    impure features in, **92**-10
    predefined functions, **92**-8–10
    syntax, **92**-7–8
  standard ML
    declarations in, **92**-12–13
    description of, **92**-10
    expressions in, **92**-11–12
    impurities in, **92**-19–20
    I/O, **92**-19–20
    module system, **92**-17–19
    parametric polymorphism, **92**-15–16
    pattern matching, **92**-13
    predefined types in, **92**-11

references, **92**-19–20
type constructors, **92**-16
type definitions, **92**-14–15
type variables, **92**-15–16
state in, **92**-26–27
Function-flow diagrams, **46**-11
Function optimization using genetic algorithms, **14**-5–6
Fundamental interpersonal relations orientation–behavior survey, **47**-14–15
Future engineering systems, **28**-19–20
Fuzzy logic, **61**-1, **61**-15–17

G

Gain, **20**-4
Gain ratio, **65**-4
Galactic dynamics, **33**-13–14
Galaxy clusters, **33**-13–14
Galaxy formation, **33**-13–14
Gallop, **88**-13
Game-tree search
  alpha-beta algorithm, **63**-11–12
  MTD(f) algorithm, **63**-14, **63**-22
  SSS* algorithm, **63**-12–14
Gamma correction, **20**-16
Gamma-ray burst data analysis, **33**-8–9
Garbage collection
  algorithms
    copy collection, **100**-14
    mark-sweep, **100**-13–14
    reference counting, **100**-12–13
    software program verification using, **107**-11
  definition of, **100**-12
  description of, **100**-1, **100**-12
Gaussian, **43**-5–6
Gaussian basis functional, **32**-17
Gaussian elimination algorithm, **8**-4, **10**-36
Generalizations, **52**-11
Generally accepted system security principles, **77**-3
General mixture models, **66**-7–8
General relativity, **33**-14
Generic definitions, **106**-7
Genetic algorithms
  applications of, **14**-4
  automatic programming uses, **14**-7–8
  best practices, **14**-4–9
  description of, **14**-1, **63**-15
  function optimization, **14**-5–6
  mathematical analysis of, **14**-9–12
  modeling uses of, **14**-9
  ordering problems using, **14**-6–7
  overview of, **14**-2
  performance of, **14**-3–4
  principles of, **14**-1–4
  research issues regarding, **14**-12
Genetic mapping, **34**-12
Genetic programs, **14**-8
Genomics
  definition of, **34**-11
  genetic mapping, **34**-12

sequence analysis, **34**-13–14
sequence assembly, **34**-12–13
Genotype, **14**-2
Geodesic path, **11**-18
Geometric duality, **11**-4–5
Geometric hashing, **43**-18
Geometric invariants, **43**-17–18
Geometric modeling
  description of, **37**-1
  fractals, **37**-2–4
  grammar-based models, **37**-4–6
  implicit surfaces, **37**-12–14
  particle systems, **37**-14–16
  procedural volumetric models, **37**-6–12
Geometric optimization problems, **11**-20–21
Geometric primitives
  Bezier curves, **36**-8–10
  B-spline curve, **36**-10–11
  B-spline surfaces, **36**-14–15
  description of, **36**-1–2
  elliptical arcs, **36**-4
  implicit, **36**-6–7
  lines, **36**-3–4
  marker, **36**-4
  parametric curves, **36**-8–11
  parametric surfaces, **36**-12–15
  polygons, **36**-4–5
  polylines, **36**-3–4
  rendering using, **41**-2–3
  screen specification, **36**-2
  standardization of, **36**-15–16
  text, **36**-2
  triangular facet, **36**-5–6
  wireframes, **36**-4
Geometric searching, **11**-8, **11**-24–26
Geometry management, **48**-11
Gestures, in virtual reality, **42**-10
Gibbs phenomenon, **39**-6
Gini index, **65**-4
Global alignment, **13**-23–25
Global coordinate systems, **35**-3
Globally networked telecommunications, **2**-12
Globally optimum decisions, **3**-17
Global positioning systems, **42**-9
Global query optimization, **55**-17
Global reflection models, **38**-2
Global scheduling, **88**-2
Glyph, **20**-17
GOMS models, **46**-7–8, **46**-14
GOMS task analysis, **46**-15–21
GOTO language, **6**-2–3
Goto statement, **90**-9–10
Gouraud shading, **38**-16–17
GPR, **17**-7–8
Graceful degradation, **25**-12, **69**-5
Gradient descent algorithm, **66**-11
Gradient magnitude, **43**-3
Grain assumption, **62**-9
Grammar(s)
  attribute, **99**-4, **99**-9–10
  context-free, **6**-18–21, **48**-14, **99**-5–6

definition of, **6**-11
description of, **6**-2
formal
definition of, **6**-12
description of, **6**-1–2, **6**-11–12
representation of, **6**-12–16
hierarchy of, **6**-16–18
Grammar-based models, **37**-4–6
Granularity, **77**-2, **81**-5
Graph(s)
biconnected, **7**-5
bipartite, **7**-1
cell relation, **105**-31
chromatic number of, **15**-22
connected, **7**-1
connected components, **10**-22, **10**-26
covering and packing problems in, **15**-16
definition of, **10**-19
description of, **10**-19
deterministic, **10**-24–26
directed, **7**-1
directed acyclic, **7**-1
Eulerian, **7**-20
interprocedural control flow, **105**-9
intraprocedural control flow, **105**-9
perfect matchings in, 1 **2**-19
representation of, **10**-19–20
sparse, **7**-1
strongly connected, **7**-1
synchronization, **84**-3–4
triangulated, **66**-18
undirected, **7**-1, **10**-19
uses for, **7**-1
visibility, **71**-38
weighted, **10**-19
Graph contraction, **10**-14
Graphical user interface
definition of, **48**-1
development of, **80**-6
direct graphical specification, **48**-17–18
event-driven, **95**-11–18
higher-level tools, **48**-12–19
research issues, **48**-20–22
software
description of, **48**-3–20
models of, **48**-20
technology transfer, **48**-20
testing of, **105**-33–35
toolkits, **48**-9–12
tools for, **48**-2–20
types of, **48**-1–2
windowing systems, **48**-3–9
for World Wide Web, **48**-19–20
Gravitational *n*-body, **33**-10
Greedy heuristics, **3**-17–21
Green–Lagrange strain, **28**-5
Grids
Cartesian, **26**-2–3
characteristics of, **26**-5–7
chimera, **26**-11
C-type, **26**-3

geometry preparation, **26**-7–8
H-type, **26**-3
hybrid, **26**-23–24
irregular, **41**-20–22
O-type, **26**-3
research issues, **26**-26–27
structured
adaptive grid generation, **26**-11–12
algebraic generation of, **26**-8–9
best practices, **26**-15–19
description of, **26**-2–4
elliptic generation of, **26**-9–10, **26**-18–19
fluid-mechanical calculations for constructing,
30-7–8
hyperbolic generation of, **26**-10
multiblock systems for, **26**-10–11
transfinite interpolation method, **26**-16
systems of, **26**-25–26
terminology of, **26**-5–7
unstructured
adaption methods, **26**-13–14
advancing front procedure for, **26**-12–13
Delaunay triangulation of, **26**-12
description of, **26**-4–5
fluid-mechanical calculations for constructing, **30**-8
generation of, **26**-12–14
Grippers, **71**-22–23
Gröbner bases, **8**-10, **8**-12–14
Group animation, **40**-9–10
Group communication, **87**-22–23
Grouped shortest time first, **21**-9
Group polarization, **50**-2
Group rotate declustering, **21**-18
Groupware, **50**-14–15
GSAT algorithm, **61**-11, **61**-13

H

Hack
attack
advance, **78**-13
stages of, **78**-8–9
camouflaging of, **78**-13
countermeasures against, **78**-14–15
intelligence gathering techniques, **78**-9
penetration
automated tools for, **78**-12
description of, **78**-10–11
reconnaissance techniques for, **78**-10
vulnerabilities to, **78**-11–12
Hackers
definition of, **78**-2, **78**-8
media example of, **78**-2–3
overview of, **78**-1–2
stereotypes of, **78**-2
underground world of, **78**-2–3
Half adders, **16**-8–10
Hamiltonian circuit, **5**-7–8
Hamming distance, **13**-29–30
Hamming windows, **39**-11–12

HAN31, **68**-11
Hand gestures, **20**-10–11
Handles, **85**-3
Handshaking protocol, **19**-7
Hankel matrix, **8**-5
Hann window, **39**-11–12
Haptic displays, **42**-7
Hard modems, **24**-9
Hardware tuning, **53**-4–5
Hartree–Fock approximation, **32**-16
Hashed files, **21**-3
Hash functions
    collision-resistant, **9**-11
    cryptographic, **9**-10–11
    definition of, **53**-9
Hashing, **4**-25–27, **10**-18–19
Hashing algorithm, **18**-4
Hash structures, **53**-9
Hash tables
    definition of, **4**-3
    dictionary realization with, **4**-25–27
    removing duplicates using, **10**-18–19
HASKELL
    class system, **92**-22–23
    description of, **92**-22
    functional I/O in, **92**-25
    instance declarations, **92**-23–24
    list comprehensions in, **92**-24
    user-defined types, **92**-23
Hausdorff distances, **43**-18–19
Hazardous states, **107**-2
Heap
    definition of, **3**-6–7, **4**-3, **4**-8, **53**-10
    Fibonacci, **3**-8–9
    garbage collection of, **100**-1, **100**-12
    management of, **100**-11–12
    priority queues, **3**-7–9, **4**-8–11
Heap file, **21**-3
Hemicube algorithm, **38**-21
Herbrand universe, **93**-11, **93**-22
Hermite matrix, **8**-7
Hessian matrices, **66**-11–12
Heuristic algorithms, **88**-4
Heuristic evaluations, **45**-9–10, **51**-15–16
Heuristic search methods
    A* algorithm, **55**-12, **63**-10–11
    best-first search, **63**-8–10
    description of, **63**-6–7
    hill climbing, **63**-7–8
Hidden Markov modeling, **64**-6
Hidden Markov models, **66**-11, **66**-18
Hidden nodes, **66**-18
Hidden-surface removal, **38**-3, **38**-17–18
Hidden units, **66**-6
Hierarchical addressing, **73**-9
Hierarchical allocation, **84**-19
Hierarchical locking, **56**-12–13
Hierarchical modeling, **28**-21
Hierarchical solution method, **57**-19
Hierarchical task analysis, **46**-14
Hierarchy

Chomsky, **6**-16–17
    of grammars, **6**-16–18
Hierarchy theorems, **5**-10–11
Higher order function, **92**-3
Higher order semantics, **98**-12
High-level computer vision
    appearance-based matching, **43**-19–20
    dense feature matching, **43**-18–19
    description of, **43**-14
    interpretation tree approach, **43**-15–16
    invariants, **43**-17–18
    k-tuple search, **43**-16–17
    object recognition, **43**-14–15
    transformation space search, **43**-16
High-level GOMS analysis, **46**-16–21
High-level image content analysis, **59**-5
High-level scheduling, **82**-24–25
High Performance Computing and Communications
        Program, **1**-7
High-speed computer arithmetic, **22**-1–4
High-speed local-area networks, **72**-18–23
High-speed wide-area networks, **72**-11–16
High-Tech Supermarket System, **103**-2–3
Hilbert irreducibility theorem, **8**-17
Hill climbing, **63**-7–8
Hindcast, **31**-4
Hirschberg algorithm, **13**-28–29
Histograms, **55**-13–14
History trajectory, **67**-5
Hit ratio, **53**-5
Hook edges, **10**-24
Horizontal prototyping, **45**-11
Horizontal specialization, **110**-5
Horn clauses, **93**-28
Horn theories, **68**-8
Hotelling transform, **76**-24
Hough transform, **43**-16
HSV color system, **35**-7–8
HTML, **48**-19–20, **57**-10, D%-3
HTTPR, **57**-11
Hubble Space telescope, **33**-2–3
Hubs, **24**-11–12
Huffman coding
    algorithm, **3**-19, **13**-36–40, **13**-46
    description of, **76**-5–9
Human–computer interaction
    acquisition phase of, **44**-12
    description of, **1**-12, **44**-1–2
    development
        custom, **44**-8
        organizational contexts of, **44**-6–9
    history of, **44**-6–9
    impacts of, **44**-13
    implementation and use phase of, **44**-12–13
    information system design methodologies, **44**-10
    initiation phase of, **44**-11
    organizational analysis in, **44**-2
    organizational contexts
        description of, **44**-3
        of development, **44**-6–9
        of use, **44**-9–11

organizational modeling, **44**-5–6
organizations
   characteristics of, **44**-2–3
   components of, **44**-3–5
   conflicts in, **44**-3
   context for, **44**-3
   definition of, **44**-2
   differences in, **44**-5
   groups in, **44**-4–5
   performance of, **44**-13
   task analysis, **46**-6–8
Human dynamics research, **47**-2–3
Human factors, **46**-4–6
Hybrid grid, **26**-23–24
Hybrid redundancy, **25**-13
Hydrodynamics, **33**-10–11
Hypercube, **10**-4–5
Hypermedia, **49**-4
Hyperplanes, **14**-10
Hypertext markup language, **48**-19–20, **57**-10, D%-3
Hypertextures, **41**-29
Hypertool, **88**-6
Hypothesis space, **68**-3

I

IBM 1620, **80**-2
Ideal low-pass filter, **39**-5
Idempotent, **89**-11
Identity elimination, **68**-11–12
IEEE 802 address, **73**-3
IEEE Computer Society, A%-2, C%-2, **1**-3
IFIP reference model, **51**-3–7
If statements, **90**-6–7
iKP protocol, **57**-8–9
Image content analysis
   high-level, **59**-5
   low-level, **59**-2–5
   mid-level, **59**-5
Image models
   object-oriented, **59**-15
   object-relational, **59**-14–15
Imagens, **49**-14
Image-order volume rendering, **41**-6–10
Image reduction and analysis facility, **33**-5
Image scaling, **39**-21–25
Immersion, **49**-4
Imperative programming languages
   data bindings, *see* Data bindings
   description of, **90**-1
   first-order type systems for, **97**-18–19
   research issues, **90**-20–21
Implicit data structure, **4**-2
Implicit object, **36**-6
Implicit primitives, **36**-6–7
Implicit surfaces, **35**-6, **37**-12–14
Imprecise computation, **83**-8
Impurity reduction, **65**-4
Inclusion polymorphism, **91**-5
Incremental compilation, **99**-25–26

Incremental construction, **11**-2–3
Incremental development, **102**-10–12
Incremental-iterative procedures, **28**-10–11
Incremental model, **108**-6–7
Indefinite iteration, **90**-8
Indexes
   clustering, **53**-10
   data structures, **53**-9
   decision support applications, **53**-11
   definition of, **53**-8
   nonclustering, **53**-10
Indexing
   metric spaces, **59**-20–21
   phrase, **75**-3
Indivisible symbols, **6**-11
Inductive bias, **68**-5
Inertial tracking, **42**-9
Inference algorithms, **70**-4–5
Inference rules, **61**-3
Infinite lights, **35**-9
Infinite trees, **93**-11, **93**-29
Influence diagrams, **70**-2
Information coding
   color, **20**-16–17
   issues in, **20**-16
Information integration, **71**-43
Information management, **1**-13
Information privacy, **60**-8–9
Information retrieval
   description of, **75**-1
   multimedia, **75**-5–6
   search engine integration of, **75**-11–14
   systems for, **75**-6
   text retrieval
      issues regarding, **75**-2
      methods of, **75**-3
      models for, **75**-4–5
      systems, **75**-4–5
   World Wide Web, **75**-5–6
Information richness, **50**-8
Information sharing, in computer-supported collaborative
      work, **50**-12–14
Information technology
   capability maturity model, **47**-11
   changes in, **47**-1–3
   commercial off-the-shelf applications, **47**-5–6
   description of, **1**-9, **47**-1
   development of, **47**-3–7
   distributed systems, **47**-3–4
   factors that affect, **47**-16
   growth of, **1**-9
   human dynamics research, **47**-2–3
   Internet, **47**-6–7
   management, **47**-7–12
   personal computer growth, **47**-4–5
   personal metrics, **47**-10–11
   process improvement initiatives, **47**-10–12
   research issues, **47**-23–24
   team dynamics, **47**-16–17
   team process metrics, **47**-10–11
   users, **47**-17–18

Information technology professionals
  developers, **47**-18–20
  educators, **47**-22–23
  managers, **47**-21–22
  personality types, **47**-12–15, **47**-22
  psychology of, **47**-15–16
Information visualization
  augmented reality, **20**-18
  definition of, **20**-16
  information coding, **20**-16–17
  integrated control/display objects, **20**-17–18
  three-dimensional graphics, **20**-18
  virtual environment, **20**-18
Inherently ambiguous grammar, **6**-21
Inheritance, **91**-5, **104**-4–6
Inheritance hierarchy, **104**-3
Inorganic chemistry, **32**-13–16
In parallel do, **10**-10
Input assignment, **5**-20
Input media, **49**-5–6
Input/output
  bound job, **80**-4
  filesystems, **86**-8–10
Input/output devices
  feedback, **20**-7–8
  joystick, **20**-7
  keyboards, **20**-8–9
  mouse, **20**-6
  overview of, **20**-1–2
  pen input, **20**-7
  perception, **20**-7–8
  pointing devices, **20**-6–7
  properties of, **20**-3–6
  states of, **20**-5
  tablets, **20**-6–7
  text entry, **20**-8–9
  trackball, **20**-6
Input switches, **10**-5
Instance-based approaches
  attributes
    continuous, **65**-10–11
    discrete, **65**-11–12
    irrelevant, **65**-13
  neighbors, **65**-12–13
  overview of, **65**-10
  similarity metric, **65**-10–12
Instance variables, **91**-7
Instruction decode unit, **17**-11
Instruction execution
  description of, **17**-9–11
  execution unit, **17**-11
  instruction decode unit, **17**-11
  instruction fetch unit, **17**-11
  storeback unit, **17**-11–12
Instruction fetch unit, **17**-11
Instruction-level parallelism, **23**-3
Instruction set
  arithmetic logic unit, **17**-3
  complex, **17**-2–3
  control transfer, **17**-4–5
  definition of, **17**-2

memory, **17**-3–4
  memory referencing, **17**-3–4
  reduced, **17**-2
Integer linear programs
  Bender's representation, **15**-19–20
  covering and packing problems, **15**-15–16
  description of, **15**-15
  inference problems, **15**-17–18
  Jeroslow's representability theorem, **15**-19
  multiprocessor scheduling, **15**-18
  plant location problems, **15**-17
  satisfiability problems, **15**-17–18
Integer programming, **15**-35–36
Integers, **5**-3
Integral polyhedra, **15**-20–23
Integrated circuits
  application specific, **16**-31
  small-scale, **16**-30
  very large-scale, **16**-30–31
Integrated control/display objects, **20**-17–18
Integrated multiple methods, **28**-21
Intelligent agents, **63**-22
Intelligent learning environments, **62**-15–16
Intelligent systems, **1**-13
Intelligent tutoring systems, **62**-15–16
Interaction modalities
  bimanual input, **20**-11
  gestures, **20**-10–11
  hand gestures, **20**-10–11
  passive measurement, **20**-11–12
  pen-based gestures, **20**-10–11
  voice and speech, **20**-10
Interaction tasks
  composition of, **20**-3
  description of, **20**-2–3
Interactive multimedia, **49**-4
Interactive multiprogramming, **80**-5–6
Interactive proofs, **5**-23–25
Interactive proof system, **5**-24
Interactive steering, **27**-19
Interconnect, **16**-33–34
Interdiction, **110**-14
Interdomain routing, **73**-15–16
Interface builder, **48**-11, **48**-18
Interface definition language, **87**-17
Interface surveys, **46**-10
Interior point methods, **15**-11–13
Interleaved declustering, **21**-18
Interleaved memory, **18**-7–10
Interleaving, **18**-8
Intermediate code, **99**-3
Internal memory, **4**-2
International Organization for Standardization, C%-1–2
International usability standards
  best practices for, **51**-11–16
  conformance testing, **51**-15–16
  content of, **51**-8–10
  context of use, **51**-12–13
  definition of, **51**-1
  description of, **51**-1–2
  design–use cycle, **51**-13

ergonomic, **51**-2
European Council Directive, **51**-10–11
guidelines use of, **51**-11
heuristic evaluations, **51**-15–16
IFIP reference model, **51**-3–7
principles of, **51**-3–11
research issues, **51**-16–17
structure of, **51**-8–10
technical, **51**-2
test criterion
conformity in, **51**-13–15
description of, **51**-7–8
uncertainty principle, **51**-13
user interface reference models, **51**-3
user participation, **51**-12
Internet, **2**-12, **47**-6–7, *see also* World Wide Web
Internet SCSI, **21**-25
Interpolation
bilinear, **38**-13
volumetric function, **41**-3–6
Interpretation, **61**-2
Interpretation tree, **43**-15–16
Interpreters, **99**-1–4
Interprocedural control flow graph, **105**-9
Interprocess communication
definition of, **84**-1
issues regarding, **84**-6–7
mechanisms, **84**-13–17
problems, **84**-5–6
Inter-query parallelism, **58**-2
Interrupts, **17**-4
Intersection computation, **11**-24
Intersection detection, **11**-23–24
Intersection reporting, **11**-23–24
Intersection testing, **38**-20–21
Interstellar medium, **33**-3, **33**-12–13
Intra-operation parallelism, **58**-2
Intraprocedural control flow graph, **105**-9
Intra-query parallelism, **58**-2
Intrusion detection
active systems for, **79**-15–16
anomaly-based approach to, **79**-17
definition of, **79**-1
model-based reasoning approach to, **79**-18
passive systems for, **79**-15–16
privacy issues, **79**-20–21
rule-based approach to, **79**-17–18
state transition-based approach to, **79**-18–19
systems for, **79**-15–19
threshold-based approach to, **79**-16–17
Invariances, **66**-15
Invariants, **43**-17–18
Inverse kinematics, **40**-4–5, **71**-14
Inverse location transparency, **89**-9
I/O call interface, **82**-13
I/O request block, **82**-13
Irregular grids, **41**-20–22
Irrelevant feature elimination, **68**-11
ISIS project, **87**-12–13
Isocontour, **41**-3
Isopycnal models, **31**-15

Isopycnals, **31**-3
ISO 9000 standards, **47**-10
Iso-surface rendering, **41**-6–7
Isosurfaces, **27**-13–14, **37**-12
Iteration control variable, **90**-8
Iterative deepening, **63**-5
Iterative method, **11**-2
Iterative software process models, **102**-10–14
Iterative structures, **90**-8–9
IVY, **89**-15–16

J

Jamming, **15**-11
Java
applets, **95**-4–6
description of, **91**-17–20
history of, D%-3–4
JCE software, **32**-4
Jenkins–Traub algorithm, **8**-8
Jeroslow's representability theorem, **15**-19
Joint application development, **47**-9
Journal of Chemical Education , **32**-2–3
Joystick, **20**-7
JPEG, **76**-23–32
Judgments, **97**-8–9
Jump optimization, **99**-23
Junction tree, **66**-18

K

Kaiser window, **39**-12–13
Kalman filtering, **34**-8
Karhunin–Loeve transform, **76**-24
Karmarkar's algorithm, **15**-12
Karnaugh map, **16**-11–12
Karp–Rabin algorithm, **13**-3–4
Karp reducible, **5**-13
KartOO, **75**-13
k differences, **13**-32–34
Kerberos ticket-granting ticket, **74**-8
Kernel(s)
components of, **80**-13–14
macrokernel, **80**-14
microkernel, **80**-14, **87**-1
nanokernel, **80**-14
real-time, **83**-6–7
reconstruction
Blackman window, **39**-12
box filter, **39**-8–9
cubic convolution, **39**-9
description of, **39**-8
Hamming windows, **39**-11–12
Hann window, **39**-11–12
Kaiser window, **39**-12–13
Lanczos window, **39**-13
triangle filter, **39**-9
windowed sinc function, **39**-9–11
virtualizing, **80**-12
x-, **87**-9

Kernel-level thread management, 81-8–9
Kernel splatting, 41-10
Keyboards, 20-8–9
Key distribution centers, 74-8–9
Keyframe
    description of, 40-4
    segment-based extraction, 59-11–12
Key process areas, 108-13
Keystroke level, 46-16
Kinesthetic correspondence, 20-8
Kleene closure, 6-12
K-maps, 16-16
k-means algorithm, 75-11
k-medoid algorithm, 75-11
k mismatches, string matching with, 13-31–32
Knockout switch, 24-14
Knuth–Morris–Pratt algorithm, 13-4–5
Kochanek–Bartels spline interpolation, 40-12–13
Kolmogorov complexity, 5-25–26
k-opt heuristic, 15-33
Kruskal's algorithm, 7-10–11
Krylov matrix, 8-3
Krylov sequence, 8-3
k-tuple search, 43-16–17
k-worktape Turing machine, 5-3–4

L

Lagrangian description, 28-5
Lagrangian relaxation, 15-34–35
Laguerre's method, 8-8
Lambda calculus, 92-1, 92-4–5
Lanczos window, 39-13
Landmark values, 62-3
Language model, 10-8
Languages, *see also specific language*
    complete, 5-14
    computational, 5-2–3
    context-free, 6-16
    context-sensitive, 6-16
    definition of, 6-11
    NP-complete, 5-14
    NP-hard, 5-14
    pattern, 6-14
    real-time, 83-10–11
    regular, 6-13
    representation of, 6-12–16
    right-linear, 6-13
Large-eddy simulations, 30-14
Large-format displays, 20-19–20
Large-scale linear programming, 15-13–15
Large-scale query processing, 58-17–18
Largest empty circle problem, 11-21
Las Vegas algorithms, 11-9, 1 2-6
Latent types, 92-6
LaTeX, D%-4
L-attributed grammars, 99-10
Layered models, of ocean modeling, 31-14–15
Lazy evaluation, 92-20
Lazy functional languages, 92-20

Leaf procedure, 100-8
Learning
    empirical, 68-2, 68-5–7
    explanation-based
        background, 68-3–5
        characteristics of, 68-2
        empirical learning vs., 68-2, 68-5–7
        explanations
            construction of, 68-8–9
            properties of, 68-7–8
        generalizing
            description of, 68-9–11
            disjunctive augmentation, 68-13–14
            identity elimination, 68-11–12
            irrelevant feature elimination, 68-11
            operationality pruning, 68-12–13
        overview of, 68-1–2
        resources for, 68-16
        steps involved in, 68-7–8
        utility problem, 68-14–16
Leaves, 7-2
Legacy software, 101-8–9
Lempel–Ziv–Welsh compression, 13-40–43,
        76-12, 76-14–16
Lennard–Jones potential, 32-7
Level property, 4-8
Levenberg–Marquardt algorithm, 66-11
Levenshtein distance, 13-29–30
Lexemes, 99-4
Lex/Flex, 99-13–15
Lexicographic order, 8-12–13
Lifting lemma, 61-3
Light extinction values, 41-7
Lights, in three-dimensional computer graphics
        system, 35-8–10
Limit analysis, 62-11
Limit points, 62-3
Linear combination of atomic orbitals, 32-16
Linear discriminants, 66-3–5
Linear inequalities, 15-5–6
Linear programming
    algorithms for
        description of, 15-5
        ellipsoid, 15-9–10
        Fourier's scheme for linear inequalities, 15-5–6
        semidefinite programming, 15-10
        simplex method, 15-7–8
    definition of, 15-2
    description of, 1 2-13–15
    interior point methods, 15-11–13
    large-scale, 15-13–15
    optimization problems of, 15-4
    primer on, 15-3–13
    relaxation, 15-3
    solvability problems of, 15-4
    submodular set functions, 15-11
Linear recurrences, 3-2–4
Linear regression, 66-3–5
Linear scaleup, 58-4
Linear sequential model of software development,
        108-2–3

Linear speedup, **58**-4
Linear systems of equations
  singular, **8**-4–5
  sparse, **8**-5
Linear time algorithm, **11**-22
Linkage mapping, **34**-3
Link analysis, **46**-14–15
Link register, **100**-7
Link state packets, **73**-14–15
Link state routing protocols, **73**-13–14
LISP, D%-4
List ranking, **10**-16–17
Little's results, **57**-17
Live insertion, **19**-17–18
Livelock, **84**-17
Load balancing
  algorithms for, **88**-8
  description of, **10**-13, **58**-15–16, **88**-3, **88**-8
Load sharing, **88**-3
Local alignment, **13**-25–27
Local-area networks
  asynchronous transfer mode, **72**-21–23
  carrier-sense multiple access/collision detection,
    72-16–18
  definition of, **24**-2–3
  fiber distributed data interface, **72**-18–19
  Fibre Channel, **72**-19–21
  high-speed, **72**-18–23
  switching devices, **24**-11–12
  traditional, **72**-16–18
Local bindings, **90**-6
Local coordinate systems, **35**-2
Local edge detectors, **43**-3–5
Locality, **85**-1
Locality of reference principle, **18**-2
Locally optimum decisions, **3**-17
Local memory machine, **10**-3
Local reflection models, **38**-2, **38**-13–16
Local scheduling, **88**-2
Location transparency, **89**-3
Locators, **4**-1
Lock preemption, **88**-4
Locks, **81**-4
Locus, **11**-5
Locus operating system, **87**-9–10
Log-based systems, **86**-11–13
Logging, **56**-7–8, **86**-11–12
Logic
  classical, **61**-1, **61**-4–5
  combinatorial, **16**-1
  default, **61**-2
  first-order, **61**-3–4
  fuzzy, **61**-1, **61**-15–17
  model finding in, **61**-11–13
  multiple-valued, **61**-14–15
  nonclassical, **61**-13–20
  nonmonotonic, **61**-18–20
  nonstandard, **61**-1
  overview of, **16**-2
  propositional, **61**-2, **61**-11–13
  research issues, **61**-20–21

  saturated, **16**-4
  temporal, **107**-4
Logical consequence, **61**-2
Logical logging, **56**-7–8
Logical partitioning, **85**-9
Logic gates
  active, **16**-4
  definition of, **16**-3
  saturated, **16**-4
Logic minimization, **16**-8–10
Logic programming
  applications of, **93**-21
  Boolean functions in, **93**-2
  concurrent, **93**-28
  constraint
    applications of, **93**-21
    artificial intelligence and, **93**-17
    characteristics of, **93**-16–17
    constraints, **93**-18–20
    control issues for, **93**-19
    definition of, **93**-2
    design issues, **93**-26–27, **93**-29
    developments in, **93**-17–21
    examples of, **93**-16–17
    implementation issues, **93**-26–27
    interval constraints, **93**-28–29
    optimization, **93**-19–20
    satisfiability problems, **93**-20–21
    soft constraints, **93**-20
  definition of, **93**-1
  example of, **93**-2–4
  features of, **93**-4
  historical descriptions of, **93**-4–5
  implementation of, **93**-25–27
  impure features, **93**-15–16
  metalevel interpretation, **93**-23–24
  overview of, **93**-1–2
  parallelism, **93**-26
  procedural interpretation, **93**-12–15
  PROLOG processor, **93**-4
  research issues, **93**-27–29
  resolution, **93**-6–8, **93**-11–12
  summary of, **93**-29
  theoretical foundations of, **93**-21–23
  unification, **93**-9–12
  Warren Abstract Machine, **93**-26
Logogens, **49**-14
Log record, **56**-7
Log sequence number, **56**-7
Log-space reducible, **5**-13
Log-space uniform, **5**-21
Log-structured file systems, **21**-11
Longest common subsequence of two strings,
  13-27–28
Long-term memory, **49**-13–14
Lookahead, **21**-9
Loop invariant, **98**-14
Lossless data compression
  arithmetic coding, **76**-9–12
  compression ratio, **76**-2
  description of, **76**-2

dictionary-based techniques for, **76**-12–16
entropy coding, **76**-4–5
Huffman coding, **76**-5–9
information content, **76**-2–3
Lempel–Ziv–Welsh compression, **13**-40–43, **76**-12, **76**-14–16
methods of, **76**-3–12
prediction with partial matching, **76**-16
run-length coding, **76**-16–18
Lossy data compression
data domain compression, **76**-18–23
differential pulse code modulation, **76**-19–20
JPEG, **76**-23–32, **76**-39
predictive differential pulse code modulation, **76**-20–22
pulse code modulation, **76**-18–19
subband coding, **76**-32–39
transform domain compression, **76**-23–39
vector quantization, **76**-22–23
wavelets, **76**-32–39
Lottery scheduling, **82**-26
Low-level computer vision
Canny edge detectors, **43**-3, **43**-6
description of, **43**-2–3
image smoothing and filtering, **43**-5–6
local edge detectors, **43**-3–5
multiscale processing, **43**-6–7
optical flow, **43**-7–9
visual motion, **43**-7–9
Low-level image content analysis, **59**-2–5
LP relaxation, **15**-31
L-systems, **40**-9
Luminance, **20**-14

M

Mach banding, **38**-16
Macrokernel, **80**-14
Macromolecules, x-ray crystallography of, **34**-8–9
Magnetic disks, **86**-2–4
Magnetic motion capture systems, **40**-3
Magnetic random-access memory, **21**-2
Magnetohydrodynamics, **33**-11
Mailboxes, **80**-9, **84**-13–14
Mainstream rendering, **38**-26
Maintainability of software, **101**-8–9
Malhotra–Kumar–Maheshwari blocking flow algorithm, **7**-19–20
Malicious scripts, **78**-4
Malicious software
definition of, **78**-3
e-mail bomb, **78**-4
malicious scripts, **78**-4
purpose of, **78**-3
Trojan horse, **78**-3
Trojan mule, **78**-3–4
viruses
antivirus software for, **78**-7–8
boot sector, **78**-4–5
construction kits for, **78**-8
definition of, **78**-4
file sector, **78**-5

integrity checkers for, **78**-8
naming of, **78**-4
payloads, **78**-6
polymorphism of, **78**-6
protection measures against, **78**-7–8
stealth nature of, **78**-6–7
threats caused by, **78**-7
trigger of, **78**-6
worm, **78**-4
writers of, **78**-2
Mandatory access controls, **60**-5–7, **60**-11, **79**-12
Mandatory work first scheme, **63**-18
Manipulator Jacobian, **71**-15
Many-body potentials, **32**-7
Many-one reducibility, **5**-12
Mapping cardinalities, **52**-8
Marching Cubes algorithm, **41**-3
Markov chain, **57**-19
Markov network, **70**-4
Mark-sweep algorithm, **100**-13–14
Masking, **25**-7
Master, **16**-26
Matched edges, **7**-11
Matched vertex, **7**-11
Matching algorithms, **7**-13–14
Matching problems
B-, **7**-16–17
definitions, **7**-11
description of, **7**-11
Matchings
applications of, **7**-12
assignment problem, **7**-14–16
definition of, **7**-12
Matchmaking, **88**-11
Mate, **7**-13
Matrix
adjacency, **10**-19
dense, **8**-5
Hankel, **8**-5
Hermite, **8**-7
Krylov, **8**-3
products of, **8**-2–3
Smith, **8**-7
sparse, **8**-3, **8**-5
structured, **8**-3, **8**-5
Sylvester, **8**-10
Toeplitz, **8**-5, **8**-14
Matrix computations
description of, **8**-1–2
multiplication, **8**-3
parallel, **8**-5–6
rational, **8**-6
Matrix DE controller, **71**-41–42
Matrix eigenvalues, **8**-6–8
Matroids, **15**-23–24
Max-flow min-cut theorem, **7**-18
Maximum a posteriori estimation, **66**-8
Maximum intensity projection, **41**-6–7
McCreight algorithm, **13**-20
M-commerce, **110**-6
Mean access time, **21**-4

Means–ends analysis, **63**-2
Mean time to data loss, **21**-17
Mean time to failure, **21**-16
Mean time to repair, **21**-16
Mechanical linkage trackers, **42**-8
Media, **24**-8–9
Media aesthetics, **59**-13–14
Member field, **91**-6
Memory
  amount of, **17**-6–7
  bandwidth of, **18**-1
  cache, **17**-6, **18**-3–7, **85**-6–7, **86**-1
  definition of, **16**-7
  description of, **17**-5, **18**-1–2
  distributed
    data coherency, **89**-3–7
    description of, **88**-6
    disconnected operation, **89**-10
    file attribute coherency, **89**-7
    IVY, **89**-15–16
    Munin, **89**-16
    naming, **89**-9
    NFS, **89**-10–13
    overview of, **89**-1–2
    performance, **89**-7–8
    principles of, **89**-2–10
    replication, **89**-9
    research issues, **89**-16–18
    resilience, **89**-8–9
    security of, **89**-10
  distributed shared, **85**-11–12
  dynamic random-access memory, **17**-6
  external, **4**-2
  hierarchies of, **18**-2–3
  interleaved, **18**-7–10
  internal, **4**-2
  latency of, **18**-1
  locality of reference principle, **18**-2
  main, **17**-6, **18**-7–10, **85**-1
  parallel, **18**-7–10
  physical, **17**-8–9
  random-access
    dynamic, **17**-6, **21**-2
    magnetic, **21**-2
  register file, **17**-5–6
  research issues regarding, **18**-12–13
  secondary, **85**-2
  spatial, **20**-18
  summary of, **18**-13–14
  virtual
    cache memory, **85**-6–7
    definition of, **85**-1
    description of, **17**-8–9, **18**-10–12
    multiprogramming, **85**-9
    object-oriented, **85**-7–9
    paging, **85**-4–6
    performance, **85**-9–11
    protection, **85**-9
    structure of, **85**-4–11
    translation lookaside buffer, **85**-6
    uses of, **85**-12–13

  World Wide Web, **85**-12
Memory access latency, **19**-14
Memory consistency, **23**-9
Memory controller, **18**-9
Memory deferred, **17**-7
Memory referencing instructions, **17**-3–4
Memory scheduling, **82**-11–12
Memory space, **85**-4
Memory word, **18**-1
Mental coding, **49**-20
MergeHull algorithm, **10**-33–34
Mesh generation, **28**-7
MESI coherence protocol, **19**-11
Message authentication codes, **9**-1, **57**-8
Message passing, **91**-5, **96**-14–19
Message passing interface, **96**-21
Message processing regions, **57**-4
Messiahs, **88**-10–11
Metaballs, **40**-5
Meta language, *see* ML
Metalevel interpretation, **93**-23–24
Meta problems, **6**-4
Methodology, **101**-2
Metric diagram/place vocabulary model, **62**-7
Metrics, **25**-4
Metric spaces, indexing of, **59**-20–21
Metropolitan area networks, **24**-2–3
Micro-electromechanical system-based storage, **21**-2
Microkernel, **80**-14, **87**-1
Microscopy, **34**-6–7
Microsoft, **1**-2
Middle-level computer vision
  active contours, **43**-13–14
  description of, **43**-9
  snakes, **43**-13–14
  stereopsis, **43**-9–12
  structure from motion techniques, **43**-12–13
MIDI, **49**-9–10
Mid-level image content analysis, **59**-5
MIMD processors, **23**-8–11
Min-Cut algorithm, **12**-4–6
Minimality criterion, **41**-23
Minimiax principle, **12**-7–9
Minimum annulus covering problem, **11**-21
Minimum cost spanning tree, **11**-20
Minimum description length pruning, **65**-6
Minimum diameter spanning tree, **11**-20
Minimum enclosing circle problem, **11**-20–21
Minimum link path problem, **11**-19
Minimum spanning tree
  description of, **10**-26–27
  Kruskal's algorithm for, **7**-10–11
  Prim's algorithm for, **7**-9–10
Mirrored disks, **21**-17–19
Mirroring, **86**-5
Misaligned benefit, **50**-15
MISD, **23**-8
Mixed integer linear program, **15**-2
ML
  declarations in, **92**-12–13
  description of, **92**-10, D%-4

expressions in, **92**-11–12
impurities in, **92**-19–20
I/O, **92**-19–20
module system, **92**-17–19
parametric polymorphism, **92**-15–16
pattern matching, **92**-13
predefined types in, **92**-11
references, **92**-19–20
type constructors, **92**-16
type definitions, **92**-14–15
type variables, **92**-15–16
Mobile databases, **58**-16–17
Mobile middleware, **110**-6
Mobile robots, **71**-49–51
Modality principle, **49**-16
Model-based reasoning approach to intrusion
    detection, **79**-18
Model checking, **107**-4–7
Model finding, **61**-11–13
Model formulation algorithm, **62**-9–10
Model fragments, **62**-8
Modeling assumptions, **62**-9
Model-view-controller concept, **48**-20
Modems, **78**-16
    cable, **24**-10
    voice, **24**-9–10
Modified value difference metrix, **65**-11
Modularity, **101**-17–18
Modular memory machine, **10**-3
Modular programming, **85**-3
Modulation, **24**-5–8
Molecular dynamics simulations
    applications of, **32**-8
    description of, **32**-5–6
    methodology of, **32**-6–8
Molecular mechanics, **32**-10
Monitors
    concurrent computing use of, **96**-12–14
    definition of, **84**-10; **96**-12
Monoprogrammed system, **82**-4
Monte Carlo algorithms, **11**-9, 1 2-6, **32**-5
Monte Carlo simulations, **41**-18
Morse potential, **32**-7
Motion, **27**-15
Motion blur, **20**-13
Motion capture, **40**-2–3
Motion path planning, **71**-37
Motion planning problems, **11**-18–20
Motion retargeting, **40**-3–4
Mouse, **20**-6
Move To Front, **13**-44
MPEG-2, **49**-8
MPEG-7, **59**-22
MTD(f) algorithm, **63**-14, **63**-22
Multiaccess link, **73**-1
Multiblock grid, **26**-10
Multidimensional indexing
    description of, **59**-17–18
    dimensionality curse, **59**-19
    dimensionality reduction, **59**-19–20
    tree-based index structures, **59**-18–19

Multilayer perceptron, **66**-6
Multilevel dispatching, **82**-9–11
Multilevel feedback queue dispatching, **82**-9–10
Multilevel transactions, **57**-3
Multimedia
    application of, **49**-12
    authoring software for, **49**-23
    case study of, **49**-16–22
    cognition theories, **49**-13–16
    distributed, **83**-8–9, **83**-12–13
    distributed file systems that support, **89**-17
    future of, **49**-24
    hardware
        animations, **49**-8–9
        audio technology, **49**-9–10
        CD-ROM, **49**-7–8
        scanner, **49**-11
        three-dimensional trackers, **49**-11–12
        video storage and manipulation, **49**-8
    information retrieval methods for, **75**-5–6
    input media, **49**-5–6
    interactive, **49**-4
    interfaces, **49**-1–3, **49**-12
    ISO design standard for, **49**-12–13
    media types, **49**-3–7
    operator performance and, **49**-21–22
    output media, **49**-4–5
    overview of, **49**-1–3
    query processing, **59**-21
    semantic gap in, **59**-12–14
    wearable computers, **49**-6
Multiple-client/multiple-server architecture, **58**-5
Multiple-valued logic, **61**-14–15
Multiplexer, **16**-20–21
Multiplexing, **24**-7–8
Multiplication
    fixed point algorithm, **22**-12–15
    floating point algorithms, **22**-19–20
Multiplicative-weighted Voronoi diagram, **11**-16
Multiprefix operations, **10**-7, **10**-16
Multiprocessor scheduling
    distributed scheduling
        algorithms, **88**-7–10
        description of, **88**-3–4
        mechanisms, **88**-15
        wide-area, **88**-10–14
    issues in, **88**-2–3
    parallel scheduling, **88**-5–7
Multiprogramming
    batch, **80**-4–5
    definition of, **81**-6, **82**-2–3
    interactive, **80**-5–6
    virtual memory, **85**-9
Multistage network, **10**-4–5
Multiversion concurrency control, **56**-13
Munin, **89**-16
Muscle memory, **20**-4
Mutation, **14**-3
Mutation adequacy, **105**-14
Mutual exclusion, **84**-1–2
MUX, **16**-20–21

M-way set-associative cache map, **18**-5
Myers-Briggs Type Indicator, **47**-12–15

N

NAND, **16**-6–7
Nanokernel, **80**-14
Natural join operation, **52**-3
Natural semantics, **98**-5–6, **98**-11–12
Navier–Stokes equations, **30**-2, **30**-13–14
N-best interface, **64**-7
n-body, **33**-10
Nearly random permutations, **21**-20
Necker cube, **69**-3
Negation normal form, **61**-2, **61**-7–8
Negative binary numbers, **16**-17–18
Neighborhood search, **15**-33–34
Nested declarations, **90**-16
Nested transactions, **57**-3, **87**-21
NetBill, **57**-9
Net-centric computing, **1**-10, **1**-13
Network
    definition of, **10**-3
    depth of, **10**-5
    fat-tree, **10**-4–5
    local-area
        asynchronous transfer mode, **72**-21–23
        carrier-sense multiple access/collision
            detection, **72**-16–18
        definition of, **24**-2–3
        fiber distributed data interface, **72**-18–19
        Fibre Channel, **72**-19–21
        high-speed, **72**-18–23
        switching devices, **24**-11–12
        traditional, **72**-16–18
    neural, *see* Neural networks
    topology of, **10**-3–6
    wide-area
        description of, **24**-2–3, **72**-8
        high-speed, **72**-11–16
        traditional, **72**-8–11
Network attached storage, **21**-25
Network file system, **74**-6
Network flows
    description of, **7**-17
    problems, **7**-11, **7**-18–20
Network interface cards, **24**-13
Network layer, **24**-14–15
Network security
    authentication failures, **74**-2–3
    buggy code, **74**-3
    cryptography for, **74**-7–9
    denial of service attacks, **74**-11
    firewalls, **74**-9–11
    overview of, **74**-1–2
    routing, **74**-3–4
    threats, **74**-2–3
    user authentication, **74**-3
    vulnerabilities, **77**-2
Neural networks

Bayesian viewpoint, **66**-14–15
complexity control, **66**-12–14
cost functions
    definition of, **66**-8
    gradients of, **66**-10–11
    likelihood-based, **66**-9–10
decision trees, **66**-7
definition of, **66**-1
density estimation, **66**-2–3
general mixture models, **66**-7–8
graphical models, **66**-16–19
Hessian matrices, **66**-11–12
invariances, **66**-15
linear discriminants, **66**-3–5
linear regression, **66**-3–5
nonlinear classification, **66**-5–7
nonlinear regression, **66**-5–7
optimization algorithms, **66**-11
overfitted, **66**-12
overview of, **66**-1–2
preprocessing, **66**-15
prior knowledge, **66**-15–16
Newmark's method, **28**-13–14
Newton–Raphson divider, **22**-16–18
Newton–Raphson technique, **28**-11
Newton's iteration, **8**-8
NFS, **89**-10–13
n-gram language model, **64**-6
N-modular redundancy, **87**-22
Nodewise algorithms, **28**-17
Noetherianity property, **107**-19
Nonclustering index, **53**-10
Nondeterminism
    definition of, **106**-9
    don't care, **93**-28–29
Nondeterministic finite automaton, **6**-22–23
Nondeterministic fractals, **37**-3
Nondeterministic Turing machines, **5**-4, **5**-19, **6**-27–28
Nonideal reconstruction, **39**-6–8
Nonlinear classification, **66**-5–7
Nonlinear equations
    description of, **8**-10
    Gr"bner bases for solving, **8**-10, **8**-12–14
    resultant methods for solving, **8**-10–12
Nonlinear regression, **66**-5–7
Nonlocal bindings, **90**-6
Non-maximum suppression, **43**-6
Nonmonotonic logic, **61**-18–20
Nonpreemption, **82**-20
Nonrestoring divider, **22**-15–16
Nonspeech auditory displays, **20**-21
Nonterminals, **99**-5
Nonvolatile storage, **21**-2, **56**-6
No-overwrite policy, **86**-12
Normalization, **52**-5
Normal-order evaluation, **92**-5
Normative influences, **50**-7
Nowcast, **31**-4
NP-completeness, **5**-14–16
N-selective mutation testing, **105**-16
NSPACE, **5**-7

NTIME, 5-7
Nuclear magnetic resonance, 34-2, 34-7–8
Numerical parallel algorithms
   description of, 10-35
   discrete Fourier transform, 10-36
   matrix operations, 10-36
Numerical relativity, 33-14
Numerical solution of differential equations, 32-4–5
Nusselt analog, 38-22
NVS cache, 21-18
Nyquist limit, 41-5
Nyquist rate, 39-4

O

Object(s)
   compact, 33-14–15
   implicit, 36-6
   integrated control/display, 20-17–18
Object-based data models
   database programming languages, 52-12
   description of, 52-5–6
   entity-relationship, 52-6–10
   object-oriented model, 52-10–13
   object-relational, 52-13–14
Object-order volume rendering, 41-10–12
Object-oriented addressing, 85-8
Object-oriented data model, 52-10–13, 59-15
Object-oriented encapsulation, 53-15
Object-oriented programming, 85-2
   abstract classes, 91-5
   best practices, 91-6–24
   C#, 91-21–24
   C++, 91-13–18
   characteristics of, 91-5–6
   classes, 91-1, 91-4
   concurrency features of, 96-20
   C vs., 91-2
   description of, 91-1
   development of, 91-2
   implementation issues for, 91-24–25
   Java, 91-17–20
   objects, 91-1, 91-3–4
   Pascal vs., 91-2
   polymorphism in, 91-5
   principles of, 91-3–6
   research issues, 91-25–26
   Smalltalk, 91-6–13
   subclass, 91-4
Object-oriented software design
   advantages of, 104-1–2
   classes, 104-3–4, 104-6–7
   concepts and terms associated with, 104-2–3
   description of, 104-1
   design flaws, 104-6–7
   design patterns, 104-11–15
   inheritance, 104-4–6
   unified modeling language
      activity diagram, 104-11
      characteristics of, 104-7–8

class diagram, 104-9–10
   collaboration diagram, 104-11
   component diagram, 104-11
   deployment diagram, 104-11
   description of, 104-2
   history of, 104-7
   sequence diagram, 104-10–11
   stakeholders, 104-7
   statechart diagram, 104-11
   use-case diagrams, 104-8–9
   Web applications, 110-9
Object-oriented virtual memory, 85-7–9
Object recognition, 43-14–15
Object-relational data models, 52-13–14, 59-14–15
Occur check, 93-11
Ocean modeling, computational
   application-oriented, 31-4–5
   atmosphere models coupled with, 31-5–8
   baroclinic model, 31-4
   barotropic model, 31-4, 31-9–10
   best practices, 31-8–19
   comprehensive, 31-3
   computational issues, 31-17–19
   data assimilation, 31-15–17
   deep basins, 31-2
   free surface, 31-3
   global vs. regional, 31-2
   isopycnal models, 31-15
   layered models, 31-14–15
   long-term climate studies, 31-3–4
   Nowcast/Forecast in Gulf of Mexico, 31-19–21
   overview of, 31-1–2
   primitive-equation-based methods, 31-4
   process-studies-oriented, 31-4–5
   purely dynamical, 31-3
   quasigeostrophic models, 31-4
   rigid lid, 31-3
   shallow coastal, 31-2–3
   short-term simulations, 31-3–4
   sigma-coordinate models, 31-13–14
   z-level models, 31-10–13
Oceanography, 31-1
Octree, 41-14–15
ODBC, 57-4
Off-chip cache memory, 18-7
Offline loading, 80-3–4
Off-line path planning, 71-37
OLAP, 57-22
Onboard disk cache, 21-6
One-level scheduling, 81-6–7
One's complement numbers, 22-3
One-time pad, 9-3
One-way functions, 9-5–6
On-line checking, 25-9
Online transaction processing, 21-3, 57-3–4
Ontological assumption, 62-9
Ontology, 62-5
OpenGL, D%-5, 38-26–30
Open loop, 67-5
OpenMP application programming interface, 96-21
Open queuing network models, 57-16–17

Open Systems Interconnection
  architecture of, 72-5–6
  description of, 72-1–2
  layers of, 72-7–8
Operating systems
  abstracting reality, 80-7–9
  definition of, 1-14, 80-1
  distributed
    Amoeba, 87-2–3, 87-6–8, 87-14
    binding, 87-17–18
    communication, 87-6–9, 87-15–17
    DEC global name server, 87-3–4
    description of, 87-1–2
    domain name server, 87-4–5
    failure models, 87-13–14
    Firefly remote procedure call, 87-8–9
    group communication, 87-22–23
    group management, 87-11–13
    ISIS project, 87-12–13
    Locus system, 87-9–10
    naming, 87-2–6, 87-14–15
    parallel systems vs., 87-1–2
    Plan 9, 87-5–6
    Quicksilver, 87-10–11
    research, 87-2–13
    transactions, 87-9–11, 87-19–21
    V system, 87-7
    x-kernel, 87-9
  goals of, 80-7–11
  historical development of
    batch multiprogramming, 80-4–5
    description of, 80-1
    distributed computing, 80-6–7
    graphical user interfaces, 80-6
    interactive multiprogramming, 80-5–6
    offline loading, 80-3–4
    open shop organization, 80-2
    operator-driven shop organization, 80-2–3
    spooling systems, 80-4
  implementing of, 80-11–14
  kernel
    components of, 80-13–14
    description of, 80-8
    virtualizing, 80-12
  managing resources, 80-9–10
  processes, 80-11, 82-2–3
  real-time, 82-2
  research issues, 80-14–15
  synchronization support, 84-9–10
  thread, 81-6–10
  tuning, 53-5–6
  user interface, 80-10
  virtual machines, 80-11–13
Operationality boundary, 68-13
Operationality pruning, 68-12–13
Operational profile, 102-15
Operational semantics, 98-1–4, 98-12–13
Operational sequence diagrams, 46-13
Operation FIND, 4-15
Operation INSERT, 4-9, 4-16
Operation refinement, 106-3

Operation REMOVE, 4-10, 4-17
Operation REMOVEMAX, 4-10
Operative diagnosis tasks, 62-14
Ops, 8-2–3
Optical fiber, 24-9
Optical flow, 43-7–9
Optical flow constraint equation, 43-7
Optical motion capture systems, 40-3
Optimality principle, 3-14, 3-17
Optimal scheduling, 82-17
Optimistic concurrency control, 56-13, 57-3
Optimization
  algorithms, 66-11
  jump, 99-23
  query
    algebraic space, 55-6–8
    distributed databases, 55-16, 58-13–15
    dynamic programming algorithms, 55-8–11
    global, 55-17
    in noncentralized environments, 55-15–16
    optimizer architecture for, 55-4–6
    overview of, 55-1–3
    parallel databases, 55-15–16, 58-13–15
    parametric/dynamic, 55-17–18
    Planner, 55-8–13
    randomized algorithms, 55-11–12
    search strategies, 55-12–13
    semantic, 55-16–17
    size-distribution estimator, 55-13–15
  using abstract interpretation, 93-27
Optimizing, 67-2
Optimum tree, 3-19
Oracle Turing machine, 5-5
Ordering problems using genetic algorithms, 14-6–7
Organic compound modeling
  ab initio methods, 32-12–13
  description of, 32-9–10
  empirical solutions, 32-10–11
  semiempirical methods, 32-12
Organometallic chemistry, 32-13–16
Output switches, 10-5
Output tape, 5-4
Overfitting, 68-2
Overwriting, 78-5

P

Packet, 91-6
Packet filters
  description of, 74-9–10
  dynamic, 74-10
Packet routing, 23-11
Packet switching, 24-3
  definition of, 72-2
  wide-area networks, 72-10–11
Padding argument, 5-2, 5-11–12
Page frames, 85-4
Pages, 18-11
Page table, 85-4
Paging, 85-4–6

Painter's algorithm, 37-16
Pair potentials, 32-7
Panel methods, 30-6, 30-8–9
Panic mode, 99-24
p-approximate algorithm, 15-30
Parallax, 88-5–6
Parallel algorithms
  computational geometry
    closest pair problem, 10-30–31
    description of, 10-29–30
    planar convex hull problem, 10-31–35
  divide-and-conquer, 10-11–13
  list ranking, 10-16–17
  minimum spanning tree, 10-26–27
  numerical
    description of, 10-35
    discrete Fourier transform, 10-36
    matrix operations, 10-36
  parallel pointer techniques, 10-13–14
  radix sort, 10-28–29
  randomization, 10-13
  remove-duplicate algorithm, 10-17–19
  scans, 10-15
  sorting of, 10-28–29
  spanning tree, 10-26
  sums, 10-15
Parallel architectures
  description of, 23-1
  MIMD, 23-8–11
  MISD, 23-8
  network interconnections, 23-11–12
  SIMD, 23-5–8
  SISD, 23-2–5
  stream model, 23-1–2
Parallel A* search, 63-16–17
Parallel aspiration window search, 63-19
Parallel complexity theory, 10-37
Parallel computations
  in astrophysics, 33-15–16
  matrix, 8-5–6
  modeling, 10-2–11
  multiprocessor models
    definition of, 10-2
    emulations among, 10-8–9
    local memory machine, 10-3
    modular memory machine, 10-3
    types of, 10-2–3
  network topology, 10-3–6
  primitive operations, 10-6–7
  work-depth models, 10-7–8
Parallel databases
  architecture of, 58-4–16
  characteristics of, 58-2–3
  concurrency control in, 58-7–8
  data integration in, 58-6–7
  data placement, 58-12–13
  description of, 58-1
  load balancing in, 58-15–16
  principles of, 58-1–4
  query optimization in, 55-15–16
  query processing and optimization in, 58-13–15

  reliability of, 58-8–11
  replication of, 58-11–12
  technology, 58-4–16
Parallel-do construct, 10-10
Parallel functional programming, 92-26
Parallel IDA* search, 63-17
Parallelism, 93-26
Parallel memory, 18-7–10
Parallel pointer techniques, 10-13–14
Parallel random access machines
  concurrent-read concurrent-write, 10-6
  definition of, 10-2–3
  exclusive-read exclusive-write, 10-6
  queue-read queue write, 10-6
Parallel scheduling, 88-5–7
Parallel search
  adversary games, 63-18–21
  description of, 63-15–16
  single-agent, 63-16–18
Parallel systems
  distributed systems vs., 87-1–2, 88-2
  shared-memory, 88-4
Parallel window search, 63-17–18
Paramedic method for computer ethics, 2-6
Parameter passing, 100-4
Parameters, 90-11
Parametric curves, 36-8–11
Parametric/dynamic query optimization, 55-17–18
Parametric polymorphism, 92-15–16
Parametric surfaces, 35-6
Parametri keyframe animation, 40-4
Paramodulation, 61-6
Parity prediction, 25-9–10
Parity striping, 21-14
Parse tree, 6-19
Parsing
  algorithms for, 99-6–9
  definition of, 6-18
Partial c-path, 61-8
Partial differential equations, 26-1
Partial enumeration
  branch and bound method of, 15-27–29
  branch and cut method of, 15-29–30
  description of, 15-27
Partial function, 106-7
Partially decidable decision problems, 6-3–4
Partial order bounding, 63-14
Partial order property, 4-8
Partial plans, 67-9
Particle systems, 37-14–16
Partitioning, 11-21–22
PASCAL, 90-18–19, D%-5
Passband, 33-3, 39-7
Passive measurement, 20-11–12
Password authentication, 79-3–4
Path, 7-2
Path dissolution, 61-3, 61-10–11
Path finding, 11-18–19
Pattern, 6-14
Pattern languages, 6-14
Pattern matching, 13-2

Pattern matching algorithms
  Bird/Baker algorithm, 13-15–17
  description of, 13-14
  two-dimensional, 13-14–17
  Zhu–Takaoka algorithm, 13-14–15
Pattern sensitive fault, 25-3
Pattern system, 6-14
Payloads, 78-6
PDF, D%-5
Peano representation, 93-10
Peepholing, 65-9–10
Peer-to-peer programming, 57-5
Pen-based gestures, 20-10–11
Penetration scripts, 78-12
Pen input device, 20-7
Perception, 20-7–8, 27-8
Perceptualization, 27-4
Perceptual losslessness, 76-18
Perfect matchings in graphs, 12-19
Performance of software, 101-6–7
PERL, 94-3–7, 94-14, D%-5
Permissions, 85-7
Persistent data structure, 4-2
Personal area networks, 24-2
Personal computers, 47-4
Perspective assumption, 62-9
Pervasive computing, 110-2
Petri nets, 46-11, 71-36–37
Phase problem, 34-9
Phone loops, 78-16
Phong shading, 38-17
Photo mapping, 41-28
PHP, 94-14–16
Phrase indexing, 75-3
Phreakers, 78-2
Phreaking, 78-15–17
Physical layers
  DSL, 24-10–11
  hubs, 24-11–12
  media, 24-8–9
  regenerators, 24-9
  repeaters, 24-9
  voice modems, 24-9–10
Physical mapping, 34-3
Physical memory, 17-8–9
Physical property sensed, 20-4
Physical space, 41-2
Physiological logging, 56-8
PIE model, 105-6
Pipelined divide-and-conquer, 10-13
Pipelining, 17-10, 23-2
Pipes, 84-15
Pixel maps, 106-7
Pixels, 43-2
Pixel spacing, 42-6
Pixmap, 35-12
Plaintext, 9-2
Plan 9, 87-5–6
Planar convex hull, 10-31–35
Plane sweep, 11-3–4
Planets, 33-11–12

Planning and scheduling
  definition of, 67-1
  deterministic dynamics
    planning with, 67-7–15
    scheduling with, 67-15–16
  disciplines, 67-2–3
  distinctions, 67-2–3
  efficiency improvements, 67-17–18
  practical, 67-20
  problems
    categories of, 67-6
    classifying of, 67-3–6
    complexity of, 67-7
    description of, 67-1–2
    dynamical system representation, 67-3–4
    hybrid refinements, 67-14
    incomplete information, 67-14–15
    performance measurements, 67-5–6
    refinement search, 67-10–11
    representing plans of action, 67-5
    task-reduction refinements, 67-13–14
  refinement method, 67-6
  research issues, 67-20–21
  stochastic domains, 67-18–19
Plan-space refinements, 67-11–12
Platters, 21-5
Plotting libraries, 27-20
Pointed polyhedron, 15-7
Pointer jumping, 10-14, 10-16
Pointer machines, 6-31
Pointers, 100-11–12
Pointing devices, 20-6–7
Point lights, 35-9
Point location problems, 11-17–18
Point sampling, 39-15
Polling, 42-11–13
Polygon mesh objects
  definition of, 38-1
  rendering of
    clipping, 38-3–6
    clipping and culling, 38-6–9
    description of, 38-2–3
    hidden-surface removal, 38-17–18
    local reflection models, 38-13–16
    projective transformation, 38-9–11
    shading algorithm, 38-11–17
    three-dimensional screen space, 38-9–11
    viewing, 38-3–6
Polygons, 36-4–5
Polyhedral combinatorics
  cutting plane method, 15-25–26
  definition of, 15-3
  integral polyhedra, 15-20–23
  matroids, 15-23–24
  special structures, 15-20–23
  valid inequalities, 15-24–27
Polyhedral cone, 15-3–4
Polyhedron
  definition of, 15-3
  pointed, 15-7
Polyinstantiation, 60-6–7

Polylines, 36-3–4
Polymorphisms
    definition of, **97**-10
    inclusion, **91**-5
    parametric, **92**-15–16
Polynomial factorization
    overview of, **8**-14–15
    polynomials in many variables, **8**-18
    polynomials in single variable
        over a finite field, **8**-15–16
        over fields of characteristic zero, **8**-16–17
    polynomials in two variables, **8**-17–18
Polynomial hierarchy, **5**-18
Polynomials, fingerprinting technique application to,
        1 2-16–18
Polynomial-time approximation scheme, **15**-30
Polynomial zeros, **8**-8–9
Polytope, **15**-3
Pop, **16**-29
Popout phenomenon, **20**-16
Portability
    of software, **101**-10
    of spoken language systems, **64**-17
Portable pixmap, **35**-14–15
Positional parameter association, **90**-11
Positions, **4**-1
Position tracking, in virtual reality, **42**-7–11
Postcondition semantics, **98**-13
PostScript, D%-5
Potential temperature, 6
Preattentive discrimination, **20**-16
Precondition semantics, **98**-13
Predication-use, **105**-12
Prediction with partial matching, **76**-16
Predictive differential pulse code modulation,
        76-20–22
Prefetching, **18**-12, **89**-8
Prefiltering
    definition of, **39**-14
    description of, **39**-18
    pyramids, **39**-18–20
    summed-area tables, **39**-20
Prefix code, **13**-36
Preimage, **39**-16
Prependnig, **78**-5
Presto, **81**-10–11
Primal–dual approximation, **15**-31–32
Primary copy locking, **57**-6
Primary pacing, **28**-20–23
Primitive operations, **10**-6–7
Primitives
    Bezier curves, **36**-8–10
    B-spline curve, **36**-10–11
    B-spline surfaces, **36**-14–15
    constructive solid geometry objects, **36**-7–8
    description of, **11**-21, **36**-1–2
    elliptical arcs, **36**-4
    implicit, **36**-6–7
    lines, **36**-3–4
    marker, **36**-4
    parametric curves, **36**-8–11

    parametric surfaces, **36**-12–15
    polygons, **36**-4–5
    polylines, **36**-3–4
    rendering using, **41**-2–3
    screen specification, **36**-2
    standardization of, **36**-15–16
    text, **36**-2
    triangular facet, **36**-5–6
    wireframes, **36**-4
Prim's algorithm, **3**-19–20, **7**-9–10
Principal component transform, **76**-24
Priority dispatching algorithms, **82**-6–8
Priority encoder, **16**-22
Priority queues
    definition of, **4**-3, **4**-6
    description of, **3**-7–9
    operations, **4**-6–7
    realization of
        with a dictionary, **4**-12
        with a heap, **4**-8–11
        with a sequence, **4**-7–8
        time complexity, **4**-11
Prior knowledge, **66**-15–16
Privacy, **60**-8–9
    description of, **77**-1–4
    enforcement of, **77**-3
    intrusion detection and, **79**-20–21
Privacy issues, **2**-10
Private automatic branch exchanges, **78**-16–17
Private implementation, **104**-2
Private-key encryption, **9**-1, **9**-11–13
Probabilistically checkable proof, **5**-25
Probabilistic Turing machines, **5**-22
Procedural abstraction, **90**-10–13, **90**-18
Procedural animation, **40**-5
Procedural volumetric models, **37**-6–12
Process, **82**-2–3
Process algebras, **106**-5
Process descriptor, **80**-11
Process design engineering, **110**-7
Process gains, **50**-2, **50**-9–10
Process loss, **50**-2, **50**-9–10
Process modeling, **108**-5
Processor dispatching, **82**-5
Processor-program interface, **17**-1–2
Processor scheduling, **81**-6–8, **82**-5–6, **82**-25–27
Production blocking, **50**-10
Productions, **99**-5
Professional codes, **2**-8
Professional ethics
    conflict of interest, **2**-9
    description of, **2**-7–8
    whistle blowing, **2**-9
Professional organizations, **2**-8
Professional societies, A%-1–3
Program, **6**-2, *see also specific program*
Program flow sequence, **82**-27
Programmable switch matrices, **16**-33
Programming languages, *see also specific language*
    description of, **1**-13
    logic, *see* Logic programming

object-oriented, *see* Object-oriented programming
    safety of, **97**-4–5
    typed, **97**-5–6
    types in, **90**-3–4
Progression, **67**-8
Progressive refinement algorithm, **38**-25
Projected tetrahedra algorithm, **41**-20–21
Projective transformation, **38**-3, **38**-9–11
Project SERAPHIM, **32**-3–4
PROLOG, **93**-4, D%-5–6
Property rights, **2**-11
Prophet, **88**-12–13
Proportional-integral-derivative controller, **71**-6
Propositional logic, **61**-2, **61**-11–13
Protein folding, **34**-10–11
Protocol stacks, **24**-3–4
Prototyping
    evolutionary, **102**-8–9
    rapid, **98**-15, **101**-20–21
    software development based on, **108**-3–4
    throw-away, **102**-8–9
    tools for, **48**-17–19
    in usability engineering, **45**-11–12
    user involvement in, **102**-9
Proxemic effects, **50**-5–6
Proximity clouds, **41**-14
Prune-and-search, **11**-7–8
Pruning algorithms, **66**-12
Pseudo-files, **80**-9
Pseudorandom functions, **9**-9–10
Pseudorandom generators, **9**-8–9
Public interface, **104**-2
Public-key encryption, **9**-2, **9**-15–17
Puck, **20**-6
Pulse code modulation
    description of, **76**-18–19
    differential, **76**-19–20
    predictive differential, **76**-19–20
Purdue Compiler Construction Tool Set, **99**-15–17
Purely cognitive task, **46**-12
Push, **16**-29
Pushdown automata, **6**-17, **6**-28–29
Pushdown store, **6**-28
PVSplit algorithm, **63**-18–20
Pyramids, **39**-18–20

Q

Quadtree approach, **71**-39–40
Qualitative process automation, **62**-14
Qualitative proportionalities, **62**-4
Qualitative reasoning
    applications of
        cognitive modeling, **62**-16
        design, **62**-15
        intelligent learning environments, **62**-15–16
        intelligent tutoring systems, **62**-15–16
        monitoring, control, and diagnosis, **62**-13–15
    definition of, **62**-1
    techniques

        causal reasoning, **62**-10–11
        comparative analysis, **62**-12
        data interpretation, **62**-12–13
        description of, **62**-9
        model formulation, **62**-9–10
        planning, **62**-13
        simulation, **62**-11–12
        spatial reasoning, **62**-13
        teleological reasoning, **62**-12
Qualitative representations
    behaviors, **62**-6
    definition of, **62**-2
    mathematical relationships, **62**-3–5
    ontology, **62**-5
    quantity, **62**-2–3
    shapes, **62**-7–8
    space, **62**-7
    state, **62**-5–6
    time, **62**-6
Qualitative state, **62**-5–6
Quality assurance, **101**-13–14, **108**-17–18
Quality of service, **83**-9
Quasigeostrophic models, **31**-4
Quaternions, **42**-8
Queries, **4**-1
Query optimization
    algebraic space, **55**-6–8
    distributed databases, **55**-16, **58**-13–15
    dynamic programming algorithms,
        55-8–11
    global, **55**-17
    in noncentralized environments, **55**-15–16
    optimizer architecture for, **55**-4–6
    overview of, **55**-1–3
    parallel databases, **55**-15–16, **58**-13–15
    parametric/dynamic, **55**-17–18
    Planner, **55**-8–13
    randomized algorithms, **55**-11–12
    search strategies, **55**-12–13
    semantic, **55**-16–17
    size-distribution estimator, **55**-13–15
Query optimizer architecture, **55**-4–6
Query processing
    distributed databases, **58**-13–15
    large-scale, **58**-17–18
    multimedia, **59**-21
    parallel databases, **58**-13–15
Queue
    definition of, **6**-28
    priority
        definition of, **4**-3, **4**-6
        description of, **3**-7–9
        operations, **4**-6–7
        realization of
            with a dictionary, **4**-12
            with a heap, **4**-8–11
            with a sequence, **4**-7–8
            time complexity, **4**-11
Queued transaction processing, **57**-5–6
Queue-read queue write parallel random access
        machines, **10**-6

Queuing network models, **103**-13–14
  closed, **57**-18–19
  open, **57**-16–17
QuickHull algorithm, **10**-32–33
Quick search algorithm, **13**-8–10
Quicksilver, **87**-10–11
Quicksort, **10**-28
Quorum-based voting algorithm, **58**-11
Quorum Consensus Protocol, **57**-6

R

Radial basis function network model, **66**-6
Radiative transfer, **33**-11
Radiosity method
  definition of, **38**-2
  rendering using, **38**-21–25
Radix sort, **10**-28–29
Radix sort parallel algorithm, **10**-28–29
RAID
  caching, **21**-16
  characteristics of, **86**-4–5
  classification, **21**-14–16
  definition of, **21**-2, **21**-12
  description of, **82**-30
  fault-tolerance, **21**-14–16
  motivation for, **21**-12–13
  performance of, **21**-22–23
  reliability modeling, **21**-16–17
  striping, **21**-13–14
RAID1, **21**-17–19
RAID5
  analysis of, **21**-23–24
  description of, **21**-19–21
Random access machines
  definition of, **10**-2
  description of, **6**-30
  parallel
    concurrent-read concurrent-write, **10**-6
    definition of, **10**-2–3
    exclusive-read exclusive-write, **10**-6
    queue-read queue write, **10**-6
Random-access memory
  dynamic, **17**-6, **21**-2
  magnetic, **21**-2
Random gradation, 1 **2**-11
Randoming sampling, 1 **2**-2–3
Randomization, **10**-13
Randomized algorithms, **8**-18
  algebraic methods, 1 **2**-15–19
  classification of, 1 **2**-6
  data structures, 1 **2**-9–13
  definition of, 1 **2**-1
  fingerprinting technique, 1 **2**-15–19
  foiling an adversary, 1 **2**-6–7
  Freivalds' technique, 1 **2**-15–16
  linear programming, 1 **2**-13–15
  min-cut algorithm, 1 **2**-4–6
  minimiax principle, 1 **2**-7–9
  query optimization, **55**-11–12

randoming sampling, 1 **2**-2–3
random reordering, 1 **2**-13–15
Randomized rounding, **15**-31
Randomized selection, 1 **2**-3–4
Random mate graph contraction, **10**-22–23
Random reordering, 1 **2**-13–15
Random sampling, **10**-13, **11**-9–10
Range searching, **11**-25–26
Rank subdeterminant computation, **8**-11
Rapid application development, **47**-9, **108**-4–5
Rapid prototyping, **98**-15, **101**-20–21
Raster, **35**-13
Rasterization, **38**-11–12
Rational matrix computations, **8**-6
Ray casting, **41**-21
Ray tracer, **37**-13
Ray tracing
  definition of, **38**-2
  distributed, **38**-19
  intersection testing, **38**-20–21
  rendering using, **38**-18–21
RBFM, **63**-22
r-commands, **74**-5
Reachability analysis, **107**-3
Read-after-write hazard, **17**-12
Readers-Writers, **96**-11–12, **96**-17
Readers/writers problem, **84**-17
Read-Once/Write-All protocol, **58**-11
Read/write heads, **21**-5
Read/write lock, **84**-12
Reality, virtual, *see* Virtual reality
Realization
  dictionary
    with a hash table, **4**-25–27
    with an (a, b)-tree, **4**-17–20
    with an AVL-tree, **4**-20–25
    with a search tree, **4**-15–17
    with a sequence, **4**-13–15
  priority queues
    with a dictionary, **4**-12
    with a heap, **4**-8–11
    with a sequence, **4**-7–8
    time complexity, **4**-11
Real-time active database, **83**-10
Real-time software development, **110**-7
Real-time systems
  architecture, **83**-7–8
  artificial intelligence, **83**-11
  building of, **83**-4
  communications, **83**-8
  databases, **83**-10
  definition of, **83**-1
  description of, **82**-2
  design, **83**-10–11
  dispatching algorithms for, **82**-11
  environment of, **83**-5
  fault tolerance, **83**-7–8
  formal verification, **83**-10–11
  granularity, **83**-4
  hard, **83**-1
  kernels, **83**-6–7

languages, **83**-10–11
misconceptions regarding, **83**-3
principles of, **83**-2–5
reliability of, **83**-4
research issues, **83**-12–13
scheduling, **83**-5
size of, **83**-4
soft, **83**-1
Rebalancing, **4**-23–24
Receiver-initiated load sharing, **88**-3
Receptive field, **66**-16
Reconstruction
conditions for, **39**-4
definition of, **39**-4
ideal low-pass filter, **39**-5
in image scaling, **39**-21–25
nonideal, **39**-6–8
sinc function, **39**-5–6
summary of, **39**-26–27
Reconstruction kernels
Blackman window, **39**-12
box filter, **39**-8–9
cubic convolution, **39**-9
description of, **39**-8
Hamming windows, **39**-11–12
Hann window, **39**-11–12
Kaiser window, **39**-12–13
Lanczos window, **39**-13
triangle filter, **39**-9
windowed sinc function, **39**-9–11
Recovery
ARIES, **56**-14–15
best practices, **56**-14–17
buffer management, **56**-6–7
principles of, **56**-6–8
REDO, **56**-15–16
research issues, **56**-17
restart, **56**-14–15
UNDO, **56**-16–17
Rectangle intersection reporting problem, **11**-24
Rect function, **39**-10
Recurrences
divide-and-conquer, **3**-4–5
linear, **3**-2–4
Recursive-descent methods, **99**-7
Recursive models, **70**-2
REDO, **56**-15–16
Reduced instruction set computers, **17**-2
Reduce-reduce conflict, **99**-18
Reducibility
closed under a, **5**-13
Cook, **5**-13
log-space, **5**-13
many-one, **5**-12
resource-bounded, **5**-12–14
Turing, **5**-12
Reduction
definition of, **6**-9
Turing, **6**-9
Redundant data, **52**-4
Reference counting algorithm, **100**-12–13

Referential transparency, **92**-2
Refinement, **106**-2–3
Refinement search, **67**-10–11
Regenerators, **24**-9
Register allocation, **99**-22
Register file, **17**-5–6
Regression, **67**-8
Regression testing
description of, **105**-5–6, **105**-25–27
distributed, **105**-29
selective, **105**-27
test prioritization, **105**-26–29
Regular expression, **6**-13
Regularization, **66**-13
Regular languages, **6**-13
Relation, **52**-2
Relational database, **52**-4–5
Relational model, **52**-1–5
Relational semantics, **98**-13
Relation schema, **52**-3
Relationship, **52**-6
Relationship set, **52**-6
Relative adequacy, **105**-14
Relaxation
definition of, **15**-28
Lagrangian, **15**-34–35
LP, **15**-31
semidefinite, **15**-32–33
Relevance feedback, **75**-5
Reliability growth models, **102**-15
Reliability of software, **101**-4–5
Remote-center-of-compliance wrist, **71**-21–22
Remote file, **80**-9
Remote procedural call, **74**-6
Remove-duplicate algorithm, **10**-17–19
Rendering
definition of, **38**-1
direct volume, **41**-3
geometric primitives for, **41**-2–3
hardware for, **41**-19–20
mainstream, **38**-26
OpenGL example of, **38**-26–30
polygon mesh objects
clipping, **38**-3–6
clipping and culling, **38**-6–9
description of, **38**-2–3
hidden-surface removal, **38**-17–18
local reflection models, **38**-13–16
projective transformation, **38**-9–11
shading algorithm, **38**-11–17
three-dimensional screen space, **38**-9–11
viewing, **38**-3–6
radiosity method for, **38**-21–25
ray tracing for, **38**-18–21
three-dimensional computer graphics, **35**-10–11
time-critical, **42**-12–13
volume, **27**-11–12
description of, **41**-6
domain, **41**-13
hybrid techniques for, **41**-12

image-order techniques, **41**-6–10
object-order techniques, **41**-10–12
Repeaters, **24**-9
Repeating history, **56**-15
Report-mode, **11**-24
Representations
attribute-value, **65**-1–2
qualitative
behaviors, **62**-6
definition of, **62**-2
mathematical relationships, **62**-3–5
ontology, **62**-5
quantity, **62**-2–3
shapes, **62**-7–8
space, **62**-7
state, **62**-5–6
time, **62**-6
scale space, **43**-6
Research and development, **1**-5–6
Resident monitor, **80**-3–4
Residual test adequacy evaluator, **105**-24–25
Residue codes, **25**-10
Resolution
definition of, **93**-6–8, **93**-11–12
logic, **61**-5–7
video, **42**-5
Resource, **80**-9
Resource-bounded reducibilities, **5**-12–14
Resource release matrix, **71**-36
Resource requirements matrix, **71**-35–36
Restart recovery, **56**-14–15
Reusability of software, **101**-9–10
Rewrite rules, **61**-3
Reynolds-averaged Navier–Stokes equations, **30**-4
Reynolds number, **30**-3
RGB system, **35**-7
Right-linear languages, **6**-13
Ripple carry adder, **22**-5
Risk analysis, **108**-14–15
Robinson's unification algorithm, **61**-3
Robot(s)
Cartesian coordinate, **71**-9–10
commercial configurations of, **71**-7–13
dynamics of, **71**-15–16
error detection and recovery, **71**-44–45
fingers of, **71**-22–23
grippers, **71**-22–23
human operator interfaces with, **71**-45–47
kinematics of, **71**-8, **71**-13–14
manipulator performance of, **71**-8
mobile, **71**-49–51
motion trajectory generation, **71**-18–21
multiple, **71**-44
part fixtures, **71**-21
SCARA, **71**-8–11
sensors
acceleration, **71**-26
data processing, **71**-27–28
description of, **71**-25
distance, **71**-26
force, **71**-26–27

photoelectric, **71**-27
position, **71**-26
proximity, **71**-26
tactile, **71**-26
torque, **71**-26–27
types of, **71**-25–31
velocity, **71**-26
servo-level motion control, **71**-16–18
tooling, **71**-21
tooling process integration and coordination,
71-24–25
trajectory generation, **71**-18–21
types of, **71**-7–13
vision for, **71**-29–31
wrist mechanisms, **71**-23–24
Robot controllers, **71**-11–13
Robot force, **71**-18
Robot Jacobians, **71**-14–15
Robot T matrix, **71**-14
Robot workcells
activity coordination, **71**-41–44
description of, **71**-2–4
digital I/O, **71**-43–44
hybrid system design, **71**-5–6
intelligent control architectures, **71**-4
job coordination, **71**-41–44
planning of, **71**-6–7, **71**-31–41
agents, **71**-32
behaviors, **71**-32
path, **71**-37–41
task decomposition and planning, **71**-32–33
task matrix approach, **71**-33–37
process integration, **71**-43–44
programming of, **71**-47–49
Robustness of software
description of, **101**-5–6
testing of, **105**-30–31
Role-based policies, for access control, **79**-12–14
Rollback, **25**-12
Roll forward, **86**-12
Roll-mode, **21**-6
Rotational delay, **86**-3
Rotation algorithms, **82**-8–9
Rotational latency, **21**-5
Rotational positing sensing, **21**-22
Routers, **73**-7–8
Routing protocols
calculating of, **73**-15
distance vector, **73**-11
domains, **73**-10
failures, **73**-5
forwarding, **73**-11
hierarchical addressing, **73**-9
hierarchy, **73**-8–9
interdomain, **73**-15–16
link state, **73**-13–14
network security, **74**-3–4
overview of, **73**-1–2
source route bridging, **73**-2, **73**-5–7
spanning tree algorithm, **73**-3–4
switch, **73**-2

transparent bridging, **73**-2–3, **73**-5
types of, **73**-8–15
RSA assumption
description of, **9**-6
trapdoor permutations based on, **9**-6–7
Rule-based approach to intrusion detection, **79**-17–18
Rules of inference, **107**-12–13
Rule-utilitarians, **2**-4
Runge–Kutta method, **30**-13
Run-length coding, **76**-16–18
Run Length Encoding, **13**-43–44
Runtime stack management
activation records, **100**-4–5
activation stack, **100**-5–6
description of, **100**-2
design issues, **100**-3–4
functions, **100**-2–3
parameter passing, **100**-4
procedures, **100**-2–3
SPARC architecture, **100**-6–11
Run-time type information, **91**-24

S

Safe languages, **97**-3
Safety of programming languages, **97**-4–5
Salient object-based video database models, **59**-17
Sampling theory, **39**-2–4, **39**-26–27
SATF algorithm, **82**-29
Satisficing, **67**-2
Saturated logic gate, **16**-4
Savitch's theorem, **5**-9
Sawing, **41**-30
Scalar processors, **23**-4
Scalar quantization, **76**-18–19
Scale space representation, **43**-6
Scanning confocal microscopy, **34**-7
Scans, **10**-15
Scheduler, **80**-9
Scheduling
algorithms, **88**-15
anticipatory, **82**-28–29
bidirectional scan, **82**-16–17
bounded-scan, **82**-17
circular-scan, **82**-17
CPU, **88**-1
decay usage, **82**-10
device
deadlock
avoidance of, **82**-20–21
description of, **82**-19–20
detection of, **82**-21–23
eclectic approach to, **82**-23–24
ignoring of, **82**-23
prevention of, **82**-20
description of, **82**-12–15
disk scheduling algorithm, **82**-17–18
high-level, **82**-24–25
nonshareable devices, **82**-14, **82**-18–19
policies, **82**-19–24

queue organization, **82**-15
shareable devices, **82**-14–17
disk
algorithm for, **82**-17–18
description of, **21**-10
recent work in, **82**-27–31
distributed
algorithms, **88**-7–10
description of, **88**-3–4
mechanisms, **88**-15
wide-area, **88**-10–14
high-level, **82**-24–25
memory, **82**-11–12
optimal, **82**-17
parallel, **88**-5–7
priority queue for, **4**-6
processor, **81**-6–8, **82**-5–6, **82**-25–27
real-time, **83**-5
stochastic, **88**-14
Scheduling module, **88**-11
Schema, **14**-10
Schemas, **49**-14
SCHEME, D%-6
data types, **92**-6–7
description of, **92**-5–6
impure features in, **92**-10
predefined functions, **92**-8–10
syntax, **92**-7–8
Scientific illustration, **27**-4
Scientific visualization
accuracy, **27**-7–8
animation, **27**-18
application of, **27**-4
computational science and, **27**-3–4
data filtering, **27**-6
definition of, **27**-4
evolution of, **27**-5
examples of, **27**-24–28
goal of, **27**-5–6
historic overview of, **27**-1–5
motivation for, **27**-2–3
principles of, **27**-5–8
representation
description of, **27**-6–7
form, **27**-10–16
techniques of, **27**-8–16
research issues, **27**-29–30
smog, **27**-28
still imagery, **27**-17
tools for
animation, **27**-22–23
dataflow packages, **27**-21–22
plotting libraries, **27**-20
turnkey visualization package, **27**-20–21
Scope rules, for data bindings, **90**-4–5
Scoping, **100**-3
Scripting languages
characteristics of, **94**-1–2
history of, **94**-1
increased use of, **94**-2
Perl, **94**-3–7, **94**-14

PHP, **94**-14–16
Tcl/Tk, **94**-7–14
Web applications of, **94**-2–3
Scrubbing, **25**-12
SDMA, **24**-8
Search
best-first, **63**-8–10
bidirectional, **63**-5–6
breadth-first, **63**-4
depth-first, **63**-4–5
description of, **63**-1–2
dynamic, **63**-20–21
game-tree
alpha-beta algorithm, **63**-11–12
MTD(f) algorithm, **63**-14, **63**-22
SSS* algorithm, **63**-12–14
heuristic methods
A* algorithm, **55**-12, **63**-10–11
best-first search, **63**-8–10
description of, **63**-6–7
hill climbing, **63**-7–8
parallel
adversary games, **63**-18–21
description of, **63**-15–16
single-agent, **63**-16–18
refinement, **67**-10–11
state-space, **63**-2–4
strategies for, **63**-2
Search domain, bounding of, **6**-5
Search engines
data mining integrated into, **75**-11–14
definition of, **63**-22
information retrieval integrated into, **75**-11–14
Search tree
definition of, **4**-3, **4**-15
dictionary realization with, **4**-15–17
Secondary ion mass spectrometry, **32**-6
Secondary memory, **85**-2
Secondary precondition, **67**-12
Secondary storage
CD-ROM disks, **86**-5–6
description of, **17**-7, **21**-1–2, **86**-1
disk drives, **86**-1
magnetic disks, **86**-2–4
nonvolatile, **21**-2
RAID, *see* RAID
tape drives, **86**-2
tapes, **86**-6
Second-order type systems, **97**-19–21
Secrecy, **9**-1
Secure electronic transaction, **57**-8
Secure sockets layer, **57**-8
Security
database
access controls
description of, **60**-3
discretionary, **60**-3–5, **60**-9–10
mandatory, **60**-5–7, **60**-11
advanced database management systems, **60**-11–12
assurance, **60**-7–8
information privacy principles, **60**-8–9

overview of, **60**-1–2
principles of, **60**-2
description of, **77**-1–4
distributed file systems, **89**-10
future of, **78**-17–18
generally accepted system security principles, **77**-3
granularity in, **77**-2
Internet
description of, **74**-6–7
overview of, **74**-1–2
mechanisms of, **77**-2
network
authentication failures, **74**-2–3
buggy code, **74**-3
cryptography for, **74**-7–9
denial of service attacks, **74**-11
firewalls, **74**-9–11
overview of, **74**-1–2
routing, **74**-3–4
threats, **74**-2–3
user authentication, **74**-3
vulnerabilities, **77**-2, **110**-12
software, **101**-8
software development motivated by, **110**-10–14
Transmission Control Protocol/Internet Protocol
connection hijacking, **74**-5
description of, **74**-4–5
network file system, **74**-6
network information service, **74**-6
r-commands, **74**-5
remote procedural call, **74**-6
sequence number attacks, **74**-5
user datagram protocol, **74**-6
X window system, **74**-6
vulnerabilities, **77**-2, **110**-12
Seeheim model, **48**-20
Seeking, **53**-5–6
Seek time, **21**-5
Segmentation-based video database models, **59**-16
Segment-based keyframe extraction, **59**-11–12
Segmented address space, **85**-2
Segment tree, **11**-3
Selection, **14**-3
Self-centering, **20**-4
Self-explanatory simulators, **62**-12
Self-purging redundancy, **25**-14
Self-reference, **6**-7–9
Self-shadowing, **37**-16
Semantic analyzer, **99**-2–3
Semantic query optimization, **55**-16–17
Semantics
action, **98**-11
applications of, **98**-14–15
axiomatic, **98**-2, **98**-6, **98**-13–14, **98**-16
definition of, **98**-1, **98**-2
denotational
axiomatic semantics and, **98**-16
characteristics of, **98**-4–5
definition of, **98**-2
domains for, **98**-7–8
of programs, **98**-8–10

first-order, **98**-12
fixed-point, **98**-11
higher order, **98**-12
informal, **98**-6–7
natural, **98**-5–6, **98**-11–12
operational, **98**-1–4, **98**-12–13
postcondition, **98**-13
precondition, **98**-13
relational, **98**-13
research issues in, **98**-15–16
of the While-Loop, **98**-10–11
Semaphores
for concurrency computing, **96**-9–12
definition of, **84**-9, **84**-14
Semidefinite programming, **15**-10
Semidefinite relaxation, **15**-32–33
Semiotic gestures, **20**-10
Semiquantitative simulators, **62**-12
Semi-structured data, **52**-17
Sender-initiated load sharing, **88**-3
Sensitivity analysis, **28**-16–17
Sensors
acceleration, **71**-26
data processing, **71**-27–28
description of, **71**-25
distance, **71**-26
force, **71**-26–27
photoelectric, **71**-27
position, **71**-26
proximity, **71**-26
tactile, **71**-26
torque, **71**-26–27
types of, **71**-25–31
velocity, **71**-26
Sentential form, **6**-15
Separability criterion, **41**-23
Separation/branching, **15**-28–29
Separation of concerns, **101**-15–16
Separation problem, **11**-13
Sequence
definition of, **4**-3–4
dictionary realization using, **4**-13–15
implementation of
with an array, **4**-5
with doubly linked list, **4**-5–6
with singly linked list, **4**-5
operations, **4**-4
Sequence diagrams, **104**-10–11
Sequence number attacks, **74**-5
Sequencing state vector, **71**-34
Sequential booth multiplier, **22**-12–13
Sequential circuits
concept of, **16**-23–24
data flip-flop, **16**-26–27
data register, **16**-27–28
definition of, **16**-8
D latch, **16**-25–26
shift register, **16**-27–28
SR latch, **16**-24–25
stack, **16**-27–28
Sequential classifiers, **65**-13

Sequential coherency, **89**-3–4
Sequential modified booth multiplier, **22**-13
Serial execution, **56**-2
Serializability
in concurrency control, **56**-4
definition of, **53**-6, **56**-3
strict two-phase locking and, **57**-2
testing for, **56**-5–6
Service calls, **80**-8
Shading algorithm, **38**-3, **38**-11–17
Shadow feeler, **38**-20
Shadow vertex simplex algorithm, **15**-8
Shareable devices scheduling, **82**-14–17
Shared-memory architecture, **58**-6
Shared memory multiprocessors, **21**-3
Shared-memory parallel systems, **88**-4
Shared-nothing architecture, **58**-6
Shift-Or algorithm, **13**-30–31
Shift-reduce conflict, **99**-18
Shift register, **16**-27–28
Shock capturing, **30**-5
Shock fitting, **30**-5
Shortest access time first, **21**-8
Shortest job first dispatching, **82**-6–7
Shortest positioning time first, **21**-8
Shortest seek time first, **82**-16
Sigma-coordinate models, of ocean modeling, **31**-13–14
Signaling channel, **60**-11
Signal reflections, **19**-4
Signal skew, **19**-5
Sign magnitude numbers, **22**-2
Signs, **61**-14
SIMD, **23**-5–8
Similarity transformation, **8**-7
Simple round robin algorithm, **82**-8
Simplex method, **15**-7–8
Simulated annealing, **34**-9–10, **63**-8
Simulation, **62**-11–12
Simulation models, for software design, **103**-14–15
Sinc function
definition of, **39**-5–6
description of, **39**-26
windowed, **39**-9–10
Single-address space, **81**-2–3
Single-agent parallel search, **63**-16–18
Single disk
organization of, **21**-4–12
performance of, **21**-21–22
Single instruction multiple data, **81**-5
Single-source shortest paths
Bellman–Ford algorithm, **7**-8–9
description of, **7**-7
Dijkstra's algorithm, **7**-7–8
Single stuck-at fault, **25**-3
Single value decomposition, **8**-7–8
Singly linked list, sequence implementation using, **4**-5
Singularity, **8**-4
Singular linear systems of equations, **8**-4–5
SISD, **23**-2–5
Size-distribution estimator, **55**-13–15
Skewed-associative caches, **18**-12

Skew symmetry constraint, **7**-18
Skip list, 1 **2**-9, 1 **2**-12
Skolem constant, **107**-20
Skolem standard form, **61**-4
Slave, **16**-26
Small-scale integrated circuits, **16**-30
Smalltalk, **91**-6–13
Smith matrix, **8**-7
Smoothness criterion, **41**-23
Snakes, **43**-13–14
Snooping bus, **19**-11
SOAP, **57**-10
Soar architecture, **69**-8–9
Social context cues, **50**-6–8
Social contract theories, **2**-5
Society for Industrial and Applied Mathematics, **A%**-3
Soft modems, **24**-9
Software architecture
　analysis, **109**-1
　definition of, **109**-1
　description languages, **109**-4–6
　elements of, **109**-1
　principles of, **109**-2–3
　properties of, **109**-1–2
　styles of, **109**-6–13
　types of, **109**-3–4
Software configuration management, **108**-18
Software design
　abstract data types for, **103**-11–12
　analytical, **103**-12–15
　bottom-up, **103**-3–4
　dataflow diagrams, **103**-4–7
　encapsulation, **103**-9–12
　entity-relationship diagrams, **103**-7–8
　finite-state machines, **103**-8–9
　High-Tech Supermarket System, **103**-2–3
　mathematical, **103**-12–15
　module concepts for, **103**-9–11
　object-oriented
　　advantages of, **104**-1–2
　　classes, **104**-3–4, **104**-6–7
　　concepts and terms associated with, **104**-2–3
　　description of, **104**-1
　　design flaws, **104**-6–7
　　design patterns, **104**-11–15
　　inheritance, **104**-4–6
　　unified modeling language,
　　　see Unified modeling language
　overview of, **103**-1–2
　queuing network models, **103**-13–14
　simulation models for, **103**-14–15
　time-complexity analysis, **103**-14
　top-down, **103**-3–4
Software development
　component assembly model, **108**-8–9
　concurrent development model, **108**-9–10
　evolutionary software process models, **108**-5–11
　formal methods model, **108**-10
　fourth generation techniques, **108**-10–11
　incremental model, **108**-6–7
　linear sequential model, **108**-2–3

pervasive, **110**-5–7
problem-solving loop conceptualization of, **108**-1–2
prototyping model of, **108**-3–4
rapid application development, **108**-4–5
real-time, **110**-7
security-driven, **110**-10–14
specialized
　advantages of, **110**-3
　application-based, **110**-5–14
　description of, **110**-1
　generic development vs., **110**-3–4
　history of, **110**-2–3
　pervasive software development, **110**-5–7
　principles of, **110**-2–5
　problem solving in, **110**-4–5
　real-time software development, **110**-7
　research issues, **110**-14–15
　security-driven software development, **110**-10–14
　Web-based software development, **110**-7–10
　strategies for, **108**-1–11
　Web-based, **110**-7–10
Software engineering, **1**-13–14, **108**-1
Software engineering license, **2**-8
Software inspections, **105**-3, **107**-2
Software interrupts, **84**-15–17
Software ownership, **2**-11
Software process
　attributes of, **102**-2
　Cleanroom process, **102**-14–15
　definition of, **101**-2, **102**-1
　formal transformation, **102**-14
　models
　　applicability of, **102**-15–16
　　description of, **102**-1
　　evolutionary development, **102**-7–10, **108**-5–11
　　iterative, **102**-10–14
　　selection criteria for, **102**-16
　　specification-driven, **102**-2–7
　　spiral, **102**-12–13, **108**-7–8
　research issues, **102**-16–17
　technology for, **102**-16–17
Software project management
　measurement and metrics, **108**-14
　people involved in, **108**-12
　problems associated with, **108**-12
　process of, **108**-12–13
　project estimating, **108**-14
　quality assurance, **108**-17–18
　risk analysis, **108**-14–15
　scheduling, **108**-15–16
　spectrum of, **108**-11–13
　tracking and control, **108**-17
Software qualities
　classification of, **101**-2–3
　correctness, **101**-4
　engineering principles in support of
　　abstraction, **101**-18–19, **101**-22–23
　　anticipation of change, **101**-19–20, **101**-23–24
　　formality, **101**-14–15, **101**-21
　　generality, **101**-20, **101**-24
　　incrementality, **101**-20–21, **101**-24

modularity, **101**-17–18, **101**-22
rigor, **101**-14–15, **101**-21
separation of concerns, **101**-15–16
fault-tolerance, **101**-5
interoperability, **101**-10–11
maintainability, **101**-8–9
overview of, **101**-1–2
performance, **101**-6–7
portability, **101**-10
productivity, **101**-11
reliability, **101**-4–5
reusability, **101**-9–10
robustness, **101**-5–6
security, **101**-8
timeliness, **101**-12
understandability, **101**-10
usability, **101**-7
verifiability, **101**-7
visibility, **101**-12–13
Software quality assurance, **101**-13–14, **108**-17–18
Software testing
definition of, **105**-1
execution-based, **105**-4–6
fault/failure model, **105**-6–9
hierarchy of, **105**-3
innovations in
database-driven application testing, **105**-32–33
graphical user interfaces, **105**-33–35
overview of, **105**-29–30
robustness testing, **105**-30–31
spreadsheet testing, **105**-31–32
principles of, **105**-2–6
program building blocks, **105**-9–10
regression testing
description of, **105**-25–27
distributed, **105**-29
selective, **105**-27
test prioritization, **105**-26–29
terminology associated with, **105**-2–4
test
adequacy evaluation, **105**-24–25
adequacy metrics for
comparisons of, **105**-17–19
control flow-based criterion, **105**-11–12
data flow-based criterion, **105**-12–14
description of, **105**-10
error-based criterion, **105**-17
fault-based criterion, **105**-14–17
structurally-based criterion, **105**-10–11
case generation, **105**-19–23
execution of, **105**-23–24
Software verification
approaches to, **107**-1–2
arrays, **107**-15
definition of, **105**-1, **107**-1
general properties, **107**-3
model checking, **107**-4–7
programs, **107**-7–22
reachability analysis, **107**-3
specifications, **107**-2–7
status, **107**-22–23

Solar system, **33**-11–12
Solid texturing, **41**-28–29
Sorted array, **4**-14–15
Sorted sequence
for dictionary realization, **4**-13–14
for priority queue realization, **4**-7
Sorting
description of, **3**-5–7
priority queue for, **4**-6–8
Source route bridging, **73**-2, **73**-5–7
Space complexity, **5**-6, **6**-29
Space-constructible complexity class, **5**-8
Space division switching, **24**-12
Space sharing, **88**-2
Spanning tree, **10**-26
Spanning tree algorithm, **73**-3–4
SPARC architecture, **100**-6–11
Sparse graph, **7**-1
Sparse indexes, **53**-9–10
Sparse matrix, **8**-3, **8**-5
Spatial access methods, **59**-17
Spatial constancy, **42**-2
Spatial contiguity principle, **49**-16
Spatial frequency, **20**-12–13
Spatialized auditory displays, **20**-22
Spatial memory, **20**-18
Spatial presence, **42**-1
Spatial reasoning, **62**-13
Spatial resolution, **20**-13
Spawn, **81**-4
Special Interest Group on Computer Graphics, A%-2
Specializations, **52**-11
Specialized system development
advantages of, **110**-3
application-based, **110**-5–14
description of, **110**-1
generic development vs., **110**-3–4
history of, **110**-2–3
pervasive software development, **110**-5–7
principles of, **110**-2–5
problem solving in, **110**-4–5
real-time software development, **110**-7
research issues, **110**-14–15
security-driven software development, **110**-10–14
Web-based software development, **110**-7–10
Specifications
definition of, **106**-1
formal, **106**-2
software process models driven by, **102**-2–7
Spectroscopic analysis, **32**-16–19
Specular highlights, **37**-16
Speech auditory displays, **20**-21–22
Spin lock, **84**-8
Spin-waiting, **81**-10
Spiral model, **102**-12–13, **108**-7–8
Splitting criterion, **65**-4
Split-transaction bus protocol, **19**-8–9
Spoken language systems
Advanced Research Projects Agency Spoken Language
System Project, **64**-11–14
architecture, **64**-3–4

best practices, **64**-11–14
data collection, **64**-5
definition of, **64**-1
development of, **64**-4–5
dialogue, **64**-8
discourse, **64**-8
evaluation of, **64**-8–11
language understanding, **64**-7
natural language integration with, **64**-7–8
overview of, **64**-2–3
portability of, **64**-17
problems associated with, **64**-1–2
real domains for, **64**-14–15
research issues, **64**-3–4, **64**-14–17
speech recognition, **64**-6–7
spoken language generation, **64**-16
SUNDIAL program, **64**-14
text-to-speech synthesis, **64**-3
unknown words, **64**-15
Spooling system, **80**-4
Spreadsheet testing, **105**-31–32
SQL, **57**-4
SQL language, **52**-4
SR latch, **16**-24–25
SSS* algorithm, **63**-12–14
Stability analysis, **28**-15–16
Stable database, **58**-9
Stack, **16**-27–28
Stack frame, **100**-1
Standard ML
    declarations in, **92**-12–13
    description of, **92**-10
    expressions in, **92**-11–12
    impurities in, **92**-19–20
    I/O, **92**-19–20
    module system, **92**-17–19
    parametric polymorphism, **92**-15–16
    pattern matching, **92**-13
    predefined types in, **92**-11
    references, **92**-19–20
    type constructors, **92**-16
    type definitions, **92**-14–15
    type variables, **92**-15–16
Standards-making bodies, C%-1–3
Standby spares, **25**-13
Star graph, **10**-25
Stars, **33**-12–13
Start symbol, **6**-14
Starvation, **84**-5
Statechart diagram, **104**-11
State difference heuristics, **67**-11
State-space operators, **67**-4
State-space refinements, **67**-11
State-space search, **63**-2–4
State-space trajectory, **67**-5
State transition-based approach to intrusion detection, **79**-18–19
State-transition function, **67**-3
State transition networks, **48**-13
Static algorithms, **88**-9–10
Static analysis, **28**-10–11, **88**-5–6, **106**-14

Static checks, **97**-3
Static data structure, **4**-2
Static scope rules, **90**-4
Static scoping, **100**-3
s–t cut, **7**-18
Stellar astrophysics, **33**-12
Stemming, **75**-3–4
Stereopsis, **43**-9–12
Still imagery, **27**-17
Stochastic domains, **67**-18–19
Stochastic scheduling, **88**-14
Stopband, **39**-7
Stopping criterion, **65**-2
Stopword lists, **75**-3
Storage management, **21**-26
Storage network
    definition of, **21**-25
    requirements of, **21**-25–26
Storeback unit, **17**-11–12
Store-through, **18**-6
Streakline, **27**-13
Stream buffers, **18**-12
Streamlines, **27**-13
Stream model, **23**-1–2
Strict coherency, **89**-3–4
Strict two-phase locking, **57**-2
String
    alignment of, *see* Alignment
    definition of, **5**-2
    longest common subsequence of two strings, **13**-27–28
String matching
    approximate, **13**-29–30
    with *k* differences, **13**-32–34
    with *k* mismatches, **13**-31–32
String-matching algorithms
    Aho–Corasick, **13**-10–13, **13**-46
    Boyer–Moore, **13**-5–8, **13**-45
    description of, **13**-2
    experimental results, **13**-10
    Karp–Rabin, **13**-3–4
    Knuth–Morris–Pratt, **13**-4–5
    quick search, **13**-8–10
    Shift-Or, **13**-30–31
    Wu–Manber, **13**-34–36
Striping, **21**-13–14
Strongly connected graphs, **7**-1
Structural equivalence, **90**-3
Structural hazards, **17**-15
Structural mechanics, computational
    adaptive modeling, **28**-21
    adaptive strategies, **28**-8
    discretization techniques, **28**-6–7
    dynamic analysis
        central difference explicit scheme, **28**-13
        description of, **28**-11–15
        direct integration techniques, **28**-12–13
        energy balance in, **28**-15
        model superposition method, **28**-14–15
        Newmark's method, **28**-13–14
        transient, **28**-15

eigenvalue problems, **28**-16
error estimation, **28**-8
future engineering systems for, **28**-19–20
hierarchical modeling, **28**-21
high-fidelity modeling, **28**-21
history of, **28**-18–19
integrated multiple methods, **28**-21
mathematical models, **28**-6
mesh generation, **28**-7
model generation, **28**-7
multidisciplinary analysis and design optimization, **28**-22
nondeterministic analysis, **28**-22
numerical algorithms, **28**-17–18
numerical simulations, **28**-22
overview of, **28**-1–2
primary pacing, **28**-20–23
problems
    classification of, **28**-2–3
    formulation of, **28**-3–5
quality assessments, **28**-8
sensitivity analysis, **28**-16–17
stability analysis, **28**-15–16
static analysis, **28**-10–11
steps involved in, **28**-5–6
superconvergent recovery techniques, **28**-8
Structural operational semantics, **98**-3
Structural risk minimization, **66**-14
Structured grids
    adaptive grid generation, **26**-11–12
    algebraic generation of, **26**-8–9
    best practices, **26**-15–19
    description of, **26**-2–4
    elliptic generation of, **26**-9–10, **26**-18–19
    fluid-mechanical calculations for constructing, **30**-7–8
    hyperbolic generation of, **26**-10
    multiblock systems for, **26**-10–11
    transfinite interpolation method, **26**-16
Structured matrix, **8**-3
Structured transaction definition language, **57**-4
Subband coding, **76**-32–39
Submodular set functions, **15**-11
Subresultant theory, **8**-11
Subroutines, **6**-5–6
Subspace techniques, for appearance-based matching, **43**-19–20
Substitution, **61**-4
Subsumption, **97**-22
Subtraction, fixed point algorithm for, **22**-11
Subtype, **90**-4
Subtyping, **97**-22–25
Suffix tree
    characteristics of, **13**-17–19
    construction of, **13**-21
    definition of, **13**-10
    McCreight algorithm, **13**-20
Summed-area tables, **39**-20
Summons-answer sequence, **50**-6
Sum of Products, **16**-14
Sums, **10**-15
Sum-squared difference, **43**-8

SUNDIAL program, **64**-14
Superclass, **52**-11, **104**-5
Superpipelining, **17**-15
Supersampling
    adaptive, **39**-18
    definition of, **39**-17
Superscalar design, **17**-15
Superscalar processors, **23**-5
Surface nets, **41**-25
Sutherland–Hodgman clipper, **38**-7–8
Switch, **73**-2
Switch points, **11**-9
Sylvester matrix, **8**-10
Sylvester resultant, **8**-10–11
Symbolic model checking, **106**-6
Symbols, **6**-11
Symmetric multiprocessor
    bus
        arbitration, **19**-12–13
        bandwidth, **19**-13–14
        cache coherence protocols, **19**-10–12
        description of, **19**-9–11
        design issues for, **19**-10
        locking, **19**-14–15
        memory access latency, **19**-14
        synchronization, **19**-14–15
    definition of, **19**-2
Symmetry breaking, **10**-13
Synch-and-stabilize software building process, **101**-12
Synchronization
    deadlock considerations, **84**-5
    definition of, **84**-1
    hardware support, **84**-7–8
    hybrid solutions, **84**-11
    issues regarding, **84**-4–5
    language constructs, **84**-10–11
    mechanisms, **84**-7–12
    operating system support, **84**-9–10
    problems, **84**-2–4, **84**-17
    software solutions, **84**-7
Synchronization graph, **84**-3–4
Synchronized multimedia integration language, **49**-23
Synchronous bus protocol, **19**-8
Syntax
    definition of, **98**-1
    expressions and grammars for specifying, **99**-11–12
    language, **98**-6–7
Syntax tree, **99**-2
System specialization, **110**-2
SYSTERS Protein Family Database, **75**-13

T

Tableau method, **61**-3, **61**-8–10
Tables, **52**-1
Tablets, **20**-6–7
Tactile displays, **20**-20–21

Tag queuing, **86**-4
Tape alphabet, **6**-2
Tape drives, **86**-2
Target concept, **68**-4
Task, **88**-2
Task analysis
  best practices for, **46**-8–15
  cognitive, **46**-13
  critical role of, **46**-2–4
  data collection, **46**-9–10
  definition of, **45**-3, **46**-1
  in development, **46**-3–4
  formalized methods for, **46**-8
  GOMS, **46**-15–21
  GOMS models for, **46**-7–8, **46**-14
  hierarchical, **46**-14
  human–computer interaction contributions to, **46**-6–8
  human factor contributions to, **46**-4–6
  research issues, **46**-22
  task decomposition, **46**-10
  usability engineering, **45**-3–4
  user's task, **46**-12–15
  at whole-system level, **46**-11–12
Task migration, **88**-2
Task performance, **51**-7
T-carriers, **24**-8
Tcl/Tk, **94**-7–14, D%-6
TCP/IP, *see* Transmission Control Protocol/Internet
          Protocol
TDMA, **83**-8
Technology transfer, **106**-3, **106**-13–14
Teleological reasoning, **62**-12
Templates, **50**-12
Temporal contiguity principle, **49**-16
Temporal locality, **18**-2
Temporal logic, **107**-4
Temporal resolution, **20**-13
Term equivalence, **97**-25
Terminals, **99**-5
Ternary relationship, **52**-7
Test adequacy evaluator, **105**-5
Text compression
  description of, **13**-36
  Huffman coding, **13**-36–40, **13**-46
  Lempel–Ziv–Welsh method, **13**-40–43
  Move To Front, **13**-44
  Run Length Encoding, **13**-43–44
  software for, **13**-36
Text entry, **20**-8–9
Text retrieval
  issues regarding, **75**-2
  methods of, **75**-3
  models for, **75**-4–5
  systems, **75**-4–5
Texts, processing of, **13**-1–2
Text-to-speech synthesis, **64**-3
Texture, **59**-3–4
Texture mapping, **41**-28–29
Texture maps, **35**-7
Thermodynamics, **32**-15
Thin client, **57**-4, **80**-7

Thrashing, **57**-20, **85**-2
Thread
  definition of, **80**-14, **81**-1
  functionality of, **81**-4
  language support for, **81**-5
Thread management
  address spaces, **81**-2–3
  functionality, **81**-4
  issues in, **81**-4–10
  kernel-level, **81**-8–9
  language support, **81**-5
  modern systems of, **81**-10–11
  operating system issues, **81**-6–10
  overview of, **81**-1–2
  performance, **81**-9–10
  programming models, **81**-5
  simplicity, **81**-9
  user-level, **81**-8–9
Thread overhead, **81**-2
Three-address code, **99**-19–20
Three-dimensional computer graphics
  description of, **35**-1
  display devices, **35**-12–13
  framebuffers, **35**-13–14
  image files, **35**-14–17
  rendering, **35**-10–11
  scene specification
    coordinate systems for, **35**-2–6
    description of, **35**-2
    geometric modeling, **35**-6
    lights, **35**-8–10
    materials, **35**-7–8
  storage and display, **35**-11–17
  system for, **35**-1–16
Three-dimensional graphics, **20**-18
Three-dimensional screen space, **38**-9–11
Three-dimensional trackers, **49**-11–12
Three-phase commit, **87**-20–21
Three-tiered architecture, **57**-4
Threshold-based approach to intrusion detection,
          79-16–17
Throughput, **85**-2, **85**-10
Throw-away prototyping, **102**-8–9
Time complexity, **5**-6, **6**-29
Time-complexity analysis, **103**-14
Time-constructible complexity class, **5**-8
Time-critical computing, **42**-13
Time-critical rendering, **42**-12–13
Timed broadcast protocol, **87**-23
Time-function scheduling, **82**-26
Timeline analysis, **46**-13
Timesharing, **80**-5
Time sharing, **88**-2
Toeplitz matrix, **8**-5, **8**-14
Token-based authentication, **79**-4–5
Token ring, **72**-18
Tomography, **34**-6–7
Toolkits, **48**-9–12
Top-down software design, **103**-3–4
Topological defense, **74**-2
Topological order, **7**-5

Torsion, 32-10
Torus, 10-5
Tour problems, 7-20
Tracers, 27-13
Trackball, 20-6
Track cache, 86-3
Track skew, 21-5
Tracks per inch, 21-5
Traffic engineering, 73-11
Training set, 68-1
Transaction(s)
    flat, 57-3
    multilevel, 57-3
    nested, 57-3
    properties of, 57-2
Transaction commit, 57-2
Transaction processing
    concurrency control, 57-12–15
    cryptography, 57-7–9
    data contention, 57-14–15
    data replication, 57-6
    distributed, 57-4–7
    iKP protocol, 57-8–9
    infrastructure for, 57-3–4
    NetBill, 57-9
    online, 57-3–4
    overview of, 57-1–2
    performance degradation, 57-20–22
    properties of, 57-2
    queued, 57-5–6
    secured, 57-7–9
    system performance analysis, 57-16–22
    web services, 57-9–12
Transaction processing monitors, 57-4
Transfer function, 20-4
Transfer functions, 41-8
Transfinite interpolation method, 26-16
Transformation function, 5-12
Transformation space search, 43-16
Transform domain compression, 76-23–39
Translation lookaside buffer, 17-9, 18-11, 85-6
Translators, 99-2, 99-26, *see also* Compilers
Translucent rendering, 41-6–7
Transmission Control Protocol/Internet Protocol
    applications, 72-5
    communications in, 72-3
    description of, 72-1–2
    layers of, 72-2–3
    operation of, 72-3–4
    security of
        connection hijacking, 74-5
        description of, 74-4–5
        network file system, 74-6
        network information service, 74-6
        r-commands, 74-5
        remote procedural call, 74-6
        sequence number attacks, 74-5
        user datagram protocol, 74-6
        X window system, 74-6
Transmission electron microscopy, 34-6–7
Transparency, 101-12

Transparent bridging, 73-2–3, 73-5
Transport level security, 57-8
Transposition table-driven scheduling, 63-21
Trapped execution errors, 97-3
Traveling salesman problem, 7-20, 14-6, 15-34–35
Travel planning problem, 67-1–2
Tree, *see specific tree*
Tree-based index structures, 59-18–19
Tree contraction, 10-14
Tree nodes, 3-9
Tree path, 61-9
Tree traversals, 7-2–3
Triangle filter, 39-9
Triangular facet, 36-5–6
Triangulated graph, 66-18
Triangulation
    Delaunay, 26-12
    description of, 11-21–22
Triggers, 78-6
Triple modular redundancy, 25-7
Triple-modular redundancy, 87-22
Trojan horse, 60-5, 78-3
Trojan mule, 78-3–4
Tropism, 37-6
Trusted computing base, 60-7
Trustworthiness, 77-1
Truth table, 16-11
Tuning
    application interface, 53-15–18
    concurrency control, 53-6–8
    database
        applications of, 53-2
        best practices, 53-4–15
        description of, 53-1
        performance inhibitors, 53-2–3
        principles of, 53-2–4
    hardware, 53-4–5
    indexes, 53-8–12
    monitoring tools, 53-18
    operating system, 53-5–6
    rules of thumb for, 53-18–19
Tuning table, 53-12–15
Tuples, 5-4
Turbulence model, 30-13–14
Turing machines
    alternate, 5-5, 5-19
    BPP-, 5-22
    computation path of, 5-4
    configuration of, 5-4
    definition of, 6-17
    description of, 5-2
    deterministic, 5-4, 5-19, 6-27
    elements of, 5-3–4, 6-27
    finite control, 6-27
    history of, 6-1
    k-alternating, 5-20
    k-worktape, 5-3–4
    nondeterministic, 5-4, 5-19, 6-27–28
    oracle, 5-5
    PP-, 5-22
    probabilistic, 5-22

RP-, **5**-22
universal, **5**-4–5
Turing reducibility, **5**-12
Turing reduction, **6**-9
Turnkey visualization package, **27**-20–21
Two-dimensional mesh network topology, **10**-4–5
Two-level mapping, **85**-8
Two-level multistage network, **10**-4
Two-level scheduling, **81**-7–8
Two-phase commit, **58**-9, **87**-20–21
Two-phase concurrency control, **57**-12–13
Two-phase locking
  description of, **56**-8–10
  strict, **57**-2
Two's complement, **22**-2
Two-tiered architecture, **57**-4
Two-way set-associative cache map, **18**-5
Typechecker, **97**-2
Type constructors, **90**-3
Type derivations, **97**-9
Typed languages, **97**-2–3
Type equivalence, **90**-3, **97**-7, **97**-25
Type inference, **97**-10, **97**-25–28
Type reconstruction, **97**-25
Type rules, **97**-8–9
Type sound, **97**-1
Type systems
  benefits of, **97**-5–6
  definition of, **97**-1
  description of, **97**-1–2
  execution errors, **97**-2–4
  first-order
    characteristics of, **97**-11–18
    for imperative languages, **97**-18–19
  formalizing of, **97**-6–7
  judgments, **97**-8–9
  lack of safety, **97**-4–5
  properties expected of, **97**-6
  research issues for, **97**-28–29
  second-order, **97**-19–21
  subtyping, **97**-22–25
  summary of, **97**-28
  type derivations, **97**-9
  type rules, **97**-8–9
  type soundness, **97**-10
Type theory, **97**-29
Typing error, **97**-10

U

Ubiquitous computing, **49**-6
UDDI, **57**-10
UML, *see* Unified Modeling Language
Unambiguous grammar, **6**-19
Uncertainty principle, **51**-13
Unconditional hook, **10**-26
Underground world, **78**-2–3
Undirected graph, **7**-1, **10**-19
UNDO, **56**-16–17
UNICODE standard, C%-3

Unification, **93**-2, **93**-9–12
Unification algorithm, **61**-3
Unified modeling language
  activity diagram, **104**-11
  characteristics of, **104**-7–8
  class diagram, **104**-9–10
  collaboration diagram, **104**-11
  component diagram, **104**-11
  deployment diagram, **104**-11
  description of, **52**-11–12, **104**-2, **106**-13
  history of, **104**-7
  sequence diagram, **104**-10–11
  stakeholders, **104**-7
  statechart diagram, **104**-11
  use-case diagrams, **104**-8–9
Unified theories of cognition, **69**-7
Unifier, **61**-4
Uniform resource locator, **85**-4
Uniform timed causal atomic broadcast, **87**-23
Unique information, **50**-13
Universal algorithm
  definition of, **6**-4–5
  example of, **6**-6–7
Universal Turing machine, **5**-4–5
UNIX
  description of, **80**-12, **101**-10–11
  device scheduling in, **82**-13
  process scheduling in, **82**-10–11
Unsorted sequence, **4**-7
Unstructured grids
  adaption methods, **26**-13–14
  advancing front procedure for, **26**-12–13
  Delaunay triangulation of, **26**-12
  description of, **26**-4–5
  fluid-mechanical calculations for constructing,
    30-8
  generation of, **26**-12–14
Untrapped execution errors, **97**-3
Untyped languages, **97**-2–3
Update locks, **57**-3
Updates, **4**-1
Usability engineering
  competitive analysis, **45**-5–6
  consistency in, **45**-8–9
  coordination of, **45**-8–9
  definition of, **45**-1
  follow-up studies, **45**-19
  functional analysis, **45**-4
  goal setting in, **45**-6–8
  heuristic evaluation, **45**-9–10
  iterative design, **45**-18–19
  lifecycle of, **45**-1–2
  parallel design process, **45**-7–8
  prototyping, **45**-11–12
  severity ratings, **45**-15–16
  task analysis, **45**-3–4
  testing by users, **45**-13–18
  total interface, **45**-8–9
  users
    characteristics of, **45**-2–4
    testing by, **45**-13–18

Usability laboratories, **45**-16–18
Usability of software, **101**-7
Usability quality model, **51**-7
Usability standards
  best practices for, **51**-11–16
  conformance testing, **51**-15–16
  content of, **51**-8–10
  context of use, **51**-12–13
  definition of, **51**-1
  description of, **51**-1–2
  design–use cycle, **51**-13
  ergonomic, **51**-2
  European Council Directive, **51**-10–11
  guidelines use of, **51**-11
  heuristic evaluations, **51**-15–16
  IFIP reference model, **51**-3–7
  principles of, **51**-3–11
  research issues, **51**-16–17
  structure of, **51**-8–10
  technical, **51**-2
  test criterion
    conformity in, **51**-13–15
    description of, **51**-7–8
  uncertainty principle, **51**-13
  user interface reference models, **51**-3
  user participation, **51**-12
Usability test criterion
  conformity in, **51**-13–15
  description of, **51**-7–8
Usability testing, **45**-13–18
Use-case diagrams, **104**-8–9
User authentication, **74**-3
User datagram protocol, **74**-6
User identifier, **80**-5–6
User interface
  definition of, **48**-1, **80**-10
  direct graphical specification, **48**-17–18
  graphical
    definition of, **48**-1
    development of, **80**-6
    direct graphical specification,
      48-17–18
    event-driven, **95**-11–18
    higher-level tools, **48**-12–19
    research issues, **48**-20–22
    software
      description of, **48**-3–20
      models of, **48**-20
    technology transfer, **48**-20
    testing of, **105**-33–35
    toolkits, **48**-9–12
    tools for, **48**-2–20
    types of, **48**-1–2
    windowing systems, **48**-3–9
    for World Wide Web, **48**-19–20
  higher-level tools, **48**-12–19
  IFIP reference model, **51**-3–7
  international usability standards, **51**-3
  research issues, **48**-20–22
  robots, **71**-45–46
  software

    description of, **48**-3–20
    models of, **48**-20
    toolkits, **48**-9–12
    tools for, **48**-2–20
    types of, **48**-1–2
    windowing systems, **48**-3–9
    for World Wide Web, **48**-19–20
User interface design environment, **48**-16–17
User interface management system, **48**-13
User participation, **51**-12
Utilitarianism
  criticisms of, **2**-4–5
  principle of, **2**-3–4
Utility problem, **68**-14–16

V

Vacationing server model, **21**-24
Valence bond theory, **32**-9
Validation, **107**-1
Valid judgment, **97**-9
Valuation function, **98**-8
Value returning procedure, **90**-12
Vandermonde matrix, 1 **2**-17
van der Waals forces, **32**-10
Variables, **90**-2–3, **90**-15–16
VC dimension, **66**-14
Vector approximation file, **59**-20
Vector model, **10**-8
Vector processor, **18**-9
Vector processors, **23**-6–8
Vector quantization, **76**-22–23
Vectors, **8**-2–3, **27**-11–13
Verifiability of software, **101**-7
Verification
  approaches to, **107**-1–2
  arrays, **107**-15
  definition of, **105**-1, **107**-1
  general properties, **107**-3
  model checking, **107**-4–7
  programs, **107**-7–22
  reachability analysis, **107**-3
  specifications, **107**-2–7
  status, **107**-22–23
Versioning, **86**-13–14
Vertex
  destination, **10**-20
  free, **7**-11
  matched, **7**-11
Vertical partitioning, **53**-15
Vertical prototyping, **45**-11
Vertical specialization, **110**-5
Very large-scale integrated circuits, **16**-30–31
Very long instruction word computers, **17**-15
Vibrational state, **32**-18–19
Video abstraction, **59**-9–12
Video content analysis, **59**-5–8
Video database models
  annotation-based, **59**-16–17
  description of, **59**-15–16

salient object-based, **59**-17
segmentation-based, **59**-16
Video scene and event duration, **59**-7–8
Video shot detection, **59**-6–7
Video skimming, **59**-9–10
Video storage, **49**-8
Video summarization, **59**-10–12
Video tracking, **42**-8–9
Viewing transformation, **38**-2
Virtual camera, **38**-3
Virtual circuits, **24**-3
Virtual device scheduling, **82**-17–18
Virtual environment, **20**-18, **42**-5–7
Virtualizing kernel, **80**-12
Virtual machines, **80**-11–13
Virtual memory
    cache memory, **85**-6–7
    definition of, **85**-1
    description of, **17**-8–9, **18**-10–12
    multiprogramming, **85**-9
    object-oriented, **85**-7–9
    paging, **85**-4–6
    performance, **85**-9–11
    protection, **85**-9
    structure of, **85**-4–11
    translation lookaside buffer, **85**-6
    uses of, **85**-12–13
    World Wide Web, **85**-12
Virtual reality
    application evaluation and design,
        42-16–18
    architectural walkthrough case study of,
        42-18
    audio displays, **42**-6–7
    best practices in, **42**-5–11
    case studies of, **42**-18
    definition of, **42**-1
    displays in, **42**-3–4
    distributed, **42**-16
    frame rate, **42**-3
    gesture recognition, **42**-10
    haptic displays, **42**-7
    head-mounted display, **42**-3
    human depth cues, **42**-4
    navigation in, **42**-13–14
    position tracking, **42**-7–11
    principles of, **42**-2–5
    research issues regarding, **42**-18–19
    software architectures for
        description of, **42**-11
        navigation, **42**-13–14
        polling, **42**-11–13
        time-critical computing, **42**-13
        time-critical rendering, **42**-12–13
        time-critical structures, **42**-12
    spatial constancy in, **42**-2
    spatial presence, **42**-1
    summary of, **42**-19
    virtual environment
        description of, **42**-5–7
        design concepts for, **42**-15

distributed, **42**-16
    objects in, **42**-14–15
    virtual wind tunnel case study of, **42**-18
Virtual sensors, **40**-9
Virtual synchrony, **87**-22
Virtual system, **88**-10
Virtual toolkits, **48**-12
Viruses
    antivirus software for, **78**-7–8
    boot sector, **78**-4–5
    construction kits for, **78**-8
    definition of, **78**-4
    file sector, **78**-5
    integrity checkers for, **78**-8
    naming of, **78**-4
    payloads, **78**-6
    polymorphism of, **78**-6
    protection measures against, **78**-7–8
    stealth nature of, **78**-6–7
    threats caused by, **78**-7
    trigger of, **78**-6
Visibility graph, **71**-38
Visible nodes, **66**-18
Vision, computer
    digital images used in, **43**-2
    goals of, **43**-1
    high-level
        appearance-based matching, **43**-19–20
        dense feature matching, **43**-18–19
        description of, **43**-14
        interpretation tree approach, **43**-15–16
        invariants, **43**-17–18
        k-tuple search, **43**-16–17
        object recognition, **43**-14–15
        transformation space search, **43**-16
    low-level
        Canny edge detectors, **43**-3, **43**-6
        description of, **43**-2–3
        image smoothing and filtering, **43**-5–6
        local edge detectors, **43**-3–5
        multiscale processing, **43**-6–7
        optical flow, **43**-7–9
        visual motion, **43**-7–9
    middle-level
        active contours, **43**-13–14
        description of, **43**-9
        snakes, **43**-13–14
        stereopsis, **43**-9–12
        structure from motion techniques, **43**-12–13
    overview of, **43**-1–2
Visual computing, **1**-12
Visual idioms, **27**-8
Visualization
    accuracy, **27**-7–8
    animation, **27**-18
    application of, **27**-4
    computational science and, **27**-3–4
    data filtering, **27**-6
    definition of, **27**-4
    evolution of, **27**-5
    examples of, **27**-24–28

goal of, 27-5–6
historic overview of, 27-1–5
motivation for, 27-2–3
principles of, 27-5–8
representation
  description of, 27-6–7
  form, 27-10–16
  techniques of, 27-8–16
research issues, 27-29–30
scientific
  accuracy, 27-7–8
  animation, 27-18
  application of, 27-4
  computational science and, 27-3–4
  data filtering, 27-6
  definition of, 27-4
  evolution of, 27-5
  examples of, 27-24–28
  goal of, 27-5–6
  historic overview of, 27-1–5
  motivation for, 27-2–3
  principles of, 27-5–8
  representation
    description of, 27-6–7
    form, 27-10–16
    techniques of, 27-8–16
  research issues, 27-29–30
  smog, 27-28
  still imagery, 27-17
  tools for
    animation, 27-22–23
    dataflow packages, 27-21–22
    plotting libraries, 27-20
    turnkey visualization package, 27-20–21
  smog, 27-28
  still imagery, 27-17
  tools for
    animation, 27-22–23
    dataflow packages, 27-21–22
    plotting libraries, 27-20
    turnkey visualization package, 27-20–21
volume, *see* Volume visualization
Visual perception, 20-12–13
Visual programming, 48-15
Vital force, 32-9
Viterbi search, 64-6
Vivisimo, 75-12
V-model, 102-4–5
Voice modems, 24-9–10
Volatile database, 58-9
Volatile storage, 56-6
Volume buffer, 41-2
Volume graphics
  amorphous phenomena, 41-29
  antialiased voxelization, 41-25–26
  binary voxelization, 41-24
  block operations, 41-26–28
  constructive solid modeling, 41-26–28
  description of, 41-22–23
  natural phenomena, 41-29–30
  solid texturing, 41-28–29

texture mapping, 41-28–29
3-D discrete topology, 41-23–24
volume sculpting, 41-30
voxelization, 41-23
Volume rendering, 27-11–12
  description of, 41-6
  domain, 41-13
  hybrid techniques for, 41-12
  image-order techniques, 41-6–10
  object-order techniques, 41-10–12
Volume sculpting, 41-30
Volumetric function interpolation, 41-3–6
Volumetric global illumination, 41-18–19
Volume visualization
  acceleration techniques, 41-13–15
  classification function, 41-16–18
  definition of, 41-1
  high-dimensional data, 41-22
  irregular grids, 41-20–22
  multivalued data, 41-22
  overview of, 41-1–2
  rendering
    direct volume, 41-3
    geometric primitives for, 41-2–3
  rendering hardware, 41-19–20
  transfer function, 41-16–18
  volume graphics
    amorphous phenomena, 41-29
    antialiased voxelization, 41-25–26
    binary voxelization, 41-24
    block operations, 41-26–28
    constructive solid modeling, 41-26–28
    description of, 41-22–23
    natural phenomena, 41-29–30
    solid texturing, 41-28–29
    texture mapping, 41-28–29
    3-D discrete topology, 41-23–24
    volume sculpting, 41-30
    voxelization, 41-23
  volume rendering
    description of, 41-6
    domain, 41-13
    hybrid techniques for, 41-12
    image-order techniques, 41-6–10
    object-order techniques, 41-10–12
  volumetric data, 41-2
  volumetric function interpolation, 41-3–6
  volumetric global illumination, 41-18–19
von Koch curve, 37-2
von Neumann, John, 90-1
Voronoi diagram
  additive-weighted, 11-16
  construction of, 11-14–15
  definition of, 71-38
  farthest-neighbor, 11-15
  multiplicative-weighted, 11-16
  weighted, 11-15–16
Voxelization
  antialiased, 41-25–26
  binary, 41-24
  definition of, 41-23

V system, **87**-7
Vulnerabilities, security, **77**-2, **110**-12

W

Wait-depth limited concurrency control, **57**-12
Waits-for-graphs, **57**-5
Walk, **7**-2
Wallace tree, **22**-13–14
War dialers, **78**-16
Warren Abstract Machine, **93**-26
Waterfall life cycle model, **47**-8
Waterfall model, **102**-3, **102**-6
Wavelet analysis, **76**-32–39
Wavelet theory, **43**-7
Wavelet transform, **41**-13
Weak coherency, **89**-5
Web services, **57**-9–10
Weighted graph, **10**-19
Weighted path length, **3**-14
Weighted shortest time first, **21**-9–10
Weyl's quadtree construction, **8**-9
While loop, **10**-18–19
Whistle blowing, **2**-8–9
Wide-area distributed scheduling, **88**-10–14
Wide-area networks
    description of, **24**-2–3, **72**-8
    high-speed, **72**-11–16
    traditional, **72**-8–11
Widget set, **48**-11
Windowed sinc function, **39**-9–11
Windows NT, **81**-10
Wireframes, **36**-4
Wireless network infrastructure, **110**-6
Wireless user infrastructure, **110**-6
Wire-OR glitches, **19**-4
Workcells
    activity coordination, **71**-41–44
    description of, **71**-2–4
    digital I/O, **71**-43–44
    hybrid system design, **71**-5–6
    intelligent control architectures, **71**-4
    job coordination, **71**-41–44
    planning of, **71**-6–7, **71**-31–41
        agents, **71**-32
        behaviors, **71**-32
        path, **71**-37–41
        task decomposition and planning, **71**-32–33
        task matrix approach, **71**-33–37
    process integration, **71**-43–44
    programming of, **71**-47–49
Work-depth models, **10**-7–8
Workflow management systems, **44**-6

Working set, **85**-6, **85**-10–11
Worktapes, **5**-3
World Wide Web
    data mining, **75**-12
    description of, **47**-6, **85**-1
    event-driven programming for, **95**-1
    information retrieval, **75**-5–6
    scripting language use, **94**-2–3
    security of, **74**-6–7
    software development, **110**-7–10
    transaction processing on, **57**-9–12
    virtual memory, **85**-12
World Wide Web Consortium, C%-2
Worm, **78**-4
Write-after-read hazard, **17**-13
Write-after-write hazard, **17**-13
Write-ahead logging, **56**-8
Write-through, **18**-6
WSDL, **57**-10
Wu–Manber algorithm, **13**-34–36

X

Xilinix input/output block, **16**-34–35
x-kernel, **87**-9
XML, **52**-14–17, D%-3
XPath, **52**-17
XQuery, **52**-17
X-ray crystallography, **34**-2, **34**-8–9
X-ray rendering, **41**-7
X Windows, D%-6
X window system, **74**-6

Y

Yacc/Bison, **99**-13–15
Yao's Minimax principle, 1 **2**-8
Young brothers wait concept, **63**-20

Z

Z-buffer, **38**-17–18
Z-buffer algorithm, **38**-3
Zero, **16**-2
Zero-latency, **21**-6
Zhu–Takaoka algorithm, **13**-14–15
z-level models, **31**-10–13
Z notation, **106**-6–7, **106**-14–19
Zoned bit recording, **21**-6
Zoned constant angular velocity, **21**-6
Zwicker, William, **6**-8–9